# Neurological Disorders
## Course and Treatment
### Second Edition

# Neurological Disorders
## *Course and Treatment*
**Second Edition**

EDITED BY

**Thomas Brandt**
*Department of Neurology*
*Ludwig-Maximilians University*
*Munich, Germany*

**Louis R. Caplan**
*Department of Cerebrovascular Disease*
*Beth Israel Deaconess Medical Center,*
*and Department of Neurology*
*Harvard Medical School*
*Boston, Massachusetts*

**Johannes Dichgans**
*Department of Neurology*
*University of Tübingen*
*Tübingen, Germany*

**H. Christoph Diener**
*Department of Neurology*
*University of Essen*
*Essen, Germany*

**Christopher Kennard**
*Unit of Neuroscience*
*Charing Cross Hospital*
*London, United Kingdom*

## ACADEMIC PRESS
An imprint of Elsevier Science

Amsterdam   Boston   London   New York   Oxford   Paris   San Diego   San Francisco   Singapore   Sydney   Tokyo

Academic Press
*An Elsevier Science Imprint*
525 B Street, Suite 1900, San Diego, California 92101-4495, USA
http://www.academicpress.com

Academic Press
84 Theobalds Road WC1X 8RR, UK
http://www.academicpress.com

Library of Congress Catalog Card Number: 2002100861

International Standard Book Number: 0-12-125831-9

PRINTED IN THE UNITED STATES OF AMERICA
03   04   05   06   07   9   8   7   6   5   4   3   2   1

# CONTENTS

CONTRIBUTORS xi

PREFACE xvii

PREFACE TO THE FIRST EDITION xix

## Acute and Chronic Pain

1. Migraine 1
   H. Christoph Diener and
   Peter J. Goadsby

2. Cluster Headache and
   Paroxysmal Hemicrania 17
   Peter J. Goadsby and Thomas Brandt

3. Tension-Type Headache 23
   Rigmor Jensen, Jes Olesen, and
   H. Christoph Diener

4. Medication Overuse Headache 31
   H. Christoph Diener and Ninan Mathew

5. Other Forms of Headache 37
   Stephen D. Silberstein and
   Volker Limmroth

6. Hormone-Sensitive Headache
   in Women 47
   Stephen D. Silberstein and
   Helene Massiou

7. Atypical Facial Pain 59
   Marianne Dieterich and
   Charlotte Feinmann

8. Trigeminal and Glossopharyngeal
   Neuralgia 67
   G. D. Perkin and Thomas Brandt

9. Postlumbar Puncture Headache
   Syndrome 77
   Marianne Dieterich and G. D. Perkin

10. Whiplash Injury 83
    Matthias Keidel

11. Acute and Chronic Pain 95
    Geoffrey K. Gourlay and
    H. Christoph Diener

## Cranial Nerves and Brain Stem

12. Optic Nerve Lesions 115
    Valérie Biousse and Nancy J. Newman

13. Ocular Motor Disorders 125
    Andreas Straube and R. John Leigh

14. Idiopathic (Bell's) Facial Palsy 135
    Amparo Gutierrez and
    Austin J. Sumner

15. Dizziness and Vertigo 139
    Thomas Brandt

16. Tinnitus 165
    Ulrich Büttner and Alan H. Lockwood

17. Anosmia, Ageusia, and Other
    Disorders of Chemosensation 171
    Richard L. Doty and
    Steven M. Bromley

18. Hiccup 185
    Michael Fetter and
    Christopher Kennard

## Cognitive and Behavioral Disorders

19. Sleep and Its Disorders 191
    Christian Guilleminault and
    Ilonka Eisensehr

20. The Epilepsies                                    207
    *Samden D. Lhatoo and*
    *Josemir W. A. S. Sander*

21. Surgical Therapy of Epilepsy                      235
    *Soheyl Noachtar, Peter A. Winkler, and*
    *Hans Otto Lüders*

22. Dysarthria and Dysphonia                          245
    *Hermann Ackermann and*
    *T. A. T. Hughes*

23. Dysphagia                                         249
    *T. A. T. Hughes and Hermann Ackermann*

24. Disorders of Higher Visual
    Function                                          255
    *Josef Zihl and Christopher Kennard*

25. Aphasia                                           265
    *David Glenn Clark and*
    *Jeffrey L. Cummings*

26. Disorders of Spatial Orientation                  277
    *Hans-Otto Karnath and Josef Zihl*

27. Memory: Disturbances and
    Therapy                                           287
    *Hans J. Markowitsch*

28. Delirium                                          303
    *Marshal Folstein and Louis R. Caplan*

29. Diagnosis and Treatment of
    Dementia                                          311
    *Margaret M. Swanberg and*
    *Jeffrey L. Cummings*

## Cerebrovascular Disorders

30. Acute Ischemic Stroke                             327
    *Louis R. Caplan and Werner Hacke*

31. Prevention of Stroke                              349
    *H. Christoph Diener and*
    *Gregory W. Albers*

32. Microangiopathic Diseases of
    the Brain                                         363
    *Martin Dichgans*

33. Brain Embolism                                    373
    *Louis R. Caplan and Werner Hacke*

34. Spinal Cord Ischemia                              393
    *Louis R. Caplan*

35. Therapy of Intracerebral
    Hemorrhage                                        403
    *Joao A. Gomes, Carlos S. Kase, and*
    *Louis R. Caplan*

36. Subarachnoid Hemorrhage                           421
    *J. van Gijn*

37. Cerebral Venous and Sinus
    Thrombosis                                        447
    *Michael Strupp, Karl M. Einhäupl, and*
    *Marie-Germaine Bousser*

38. The Vasculitides                                  461
    *Patricia M. Moore and Arno Villringer*

39. Giant Cell Arteritis and
    Polymyalgia Rheumatica                            475
    *James J. Corbett and A. Melms*

40. Interventional Neuroradiology
    in the Brain, Head, and
    Neck Region                                       481
    *T. Yousry and R. T. Higashida*

41. Vascular Malformations and
    Interventional Neuroradiology
    of the Spinal Cord                                517
    *Armin K. Thron and Louis R. Caplan*

## Infections and Inflammatory Diseases

42. Bacterial Infections                              529
    *Hans-Walter Pfister and T. P. Bleck*

43. Brain and Spinal Cord Abscess                     545
    *Stefan Kastenbauer, Hans-Walter Pfister,*
    *and W. Michael Scheld*

44. Tuberculous Meningitis                            559
    *Karen L. Roos and Hans-Walter Pfister*

45. Neurosarcoidosis                                  567
    *Barney J. Stern*

46. Neurosyphilis                                     575
    *H. Prange*

47. Lyme Neuroborreliosis,
    Leptospirosis and
    Relapsing Fever                                   583
    *Klaus Hansen and Hans-Walter Pfister*

48. Parasitic Infections    595
Raad A. Shakir and Martin Rösener

49. Acute Viral Infections of the
Central Nervous System    601
R. Malessa and Kenneth L. Tyler

50. Fungal Infections of the Central
Nervous System    615
Martin Rösener and Geraint Fuller

51. Human Immunodeficiency Virus
Infections—Neurological
Manifestations    623
Hadi Manji and Roberto Guiloff

52. Tropical Neurology    647
J. S. Chopra and I. M. S. Sawhney

53. Multiple Sclerosis    677
Roland Martin, Reinhard Hohlfeld, and
H. F. McFarland

54. Prion Diseases    707
Richard Knight and Bob Will


Intensive Care in Neurology

55. Critical Care Neurology    721
S. Schwab, R. Kollmar, D. F. Hanley,
and Werner Hacke

56. Increased Intracranial Pressure    749
Mark T. Keegan and
Eelco F. M. Wijdicks

57. Traumatic Brain Injury    765
Andres M. Salazar, Patrick Cooper, and
James Ecklund

58. Malignant Hyperthermia and
Neuroleptic Malignant
Syndrome    783
Michael Weller and
Frank Lehmann-Horn

59. Acute Intoxications    791
Friedrich von Rosen and
John C. M. Brust

60. Hydrocephalus    805
Joseph R. Madsen, Rodolfo Hakim,
and R. Michael Scott

Tumors and Developmental Disorders

61. Palliative Care in the Terminal
Stage of Neurological Diseases    813
Raymond Voltz,
Gian Domenico Borasio, and
Russell Portenoy

62. Primary Tumors of the Central
and Peripheral Nervous System    827
Michael Weller and
David G. T. Thomas

63. Primary and Secondary
Lymphoma of the Central
Nervous System    865
Martin Begemann, Martin Schabet, and
Lisa M. DeAngelis

64. Brain Metastases from Systemic
Solid Tumors    881
Michael Weller

65. Leptomeningeal Metastasis    897
Michael Weller

66. Paraneoplastic Syndromes    911
Jerome B. Posner

67. Pseudotumor Cerebri
(Idiopathic Intracranial
Hypertension)    921
Ulrich Wüllner and James J. Corbett

68. The Syndromes of Spinal Cord
Dysfunction    925
Volker Dietz and Robert R. Young

69. Syringomyelia and
Syringobulbia    939
Louis R. Caplan and
Henry J. M. Barnett

70. Malformations and
Neurocutaneous Disorders    947
M. Bähr and B. L. Schlaggar


Metabolic and Progressive Disorders

71. Acute and Chronic Alcohol-
Related Disorders    971
Peter Thier

72. Metabolic and Toxic
    Encephalopathies   991
    *Jörg B. Schulz and Allen I. Arieff*

73. Inherited and Noninherited
    Ataxias   1011
    *Thomas Klockgether and
    Johannes Dichgans*

## Movement Disorders

74. Parkinsonism   1021
    *Wolfgang H. Oertel and Stanley Fahn*

75. Atypical Parkinsonism   1081
    *Werner Poewe and
    Gregor K. Wenning*

76. Deep Brain Stimulation for
    Movement Disorders   1099
    *Kai Bötzel and Paul Krack*

77. Normal Pressure Hydrocephalus   1113
    *Joachim K. Krauss and
    Michael Strupp*

78. Dyskinesias   1123
    *Helge Topka, J. Jankovic, and
    Johannes Dichgans*

79. Huntington's Disease and
    Sydenham's Chorea   1149
    *Thomas Gasser and Karl Kieburtz*

80. Wilson's Disease   1157
    *Andreas Straube and
    Phillip D. Swanson*

81. Upper and Lower Motor
    Neuron Disorders   1165
    *Gian Domenico Borasio and
    Stanley H. Appel*

82. Restless Legs Syndrome   1177
    *Claudia Trenkwalder and
    Arthur S. Walters*

83. Rehabilitation of
    Motor Function   1185
    *A. J. Thompson and S. Hesse*

84. Neural Prostheses   1199
    *J. Quintern, R. J. Jaeger, and
    U. Baumann*

85. Myoclonus   1221
    *Mark Hallett and Helge Topka*

86. Tremor   1233
    *Peter Bain, Prithiva Navan, and
    Tipu Aziz*

87. The Syndrome of Spastic Paresis   1247
    *Volker Dietz and Robert R. Young*

## Muscle and Peripheral Nervous System

88. Inflammatory and Infectious
    Polyneuropathy   1259
    *Ian Sutton and John B. Winer*

89. Noninflammatory
    Polyneuropathy   1281
    *Michael Donaghy*

90. Disorders of Nerve Roots
    Caused by Bony and
    Disk Diseases   1289
    *Michael Swash*

91. Compression Neuropathies of
    Peripheral Nerves and
    Compartment Syndromes   1303
    *Manfred Stöhr, David M. Dawson, and
    Michael Swash*

92. Nerve Injury   1325
    *Rolfe Birch and Praveen Anand*

93. Delayed Radiotherapy Injury   1337
    *Manfred Stöhr*

94. Therapy of Myasthenia Gravis
    and Myasthenic Syndromes   1341
    *Reinhard Hohlfeld, A. Melms, C. Schneider,
    K. V. Toyka, and D. B. Drachman*

95. Inflammatory Myopathies   1363
    *Marinos C. Dalakas and
    Dieter Pongratz*

96. Myopathies   1369
    *Richard W. Orrell and
    Robert C. Griggs*

97. Metabolic Myopathies   1385
    *Thomas Klopstock and
    Salvatore DiMauro*

**98.** Myotonias    1407
*Kenneth Ricker*

**99.** Dyskalemic Periodic Paralyses    1413
*Frank Lehmann-Horn and
Karin Jurkat-Rott*

**100.** Cramps    1421
*Helge Topka and E. Logigian*

**101.** Muscle Stiffness    1429
*Philip D. Thompson and
Hans-Michael Meinck*

### Endocrine and Autonomic Function

**102.** Neuroendocrine Disorders    1435
*Oliver Kastrup and A. Bernard Pleet[†]*

**103.** Autonomic Dysfunction    1453
*C. J. Mathias and R. Freeman*

**104.** Neurogenic Disorders of
Micturition, Defecation, and
Sexual Function    1479
*M. Harper, D. E. Andrich, and
C. J. Fowler*

**105.** Syncope    1495
*N. Colman, W. Wieling, and
R. Freeman*

### Neurological Side Effects of Therapy

**106.** Neurological and General Side
Effects of Drug Therapy    1507
*H. Christoph Diener and Oliver Kastrup*

### Molecular Genetics of Neurological Disorders

**107.** Molecular Genetic Diagnosis of
Neurological Diseases    1525
*Thomas Gasser and N. W. Wood*

INDEX    1539

# CONTRIBUTORS

*Numbers in parentheses indicate the pages on which the authors' contributions begin.*

HERMANN ACKERMANN (245, 249), Department of Neurology, University of Tübingen, 72076 Tübingen, Germany

GREGORY W. ALBERS (349), Stanford Stroke Center, Palo Alto, California 94304

PRAVEEN ANAND (1325), ICSM at Hammersmith Hospital Campus, London W12 OHS, United Kingdom

D. E. ANDRICH (1479), Department of Neurology, The National Hospital for Neurology and Neurosurgery, London WC1N 3BG, United Kingdom

STANLEY H. APPEL (1165), Department of Neurology, Baylor College of Medicine, Houston, Texas 77030

ALLEN I. ARIEFF (991), Department of Medicine, University of California School of Medicine at San Francisco, San Francisco, California 94143

TIPU AZIZ (1233), Division of Neuroscience and Psychological Medicine, Imperial College School of Medicine, London W6 8RF, United Kingdom

M. BÄHR (947), Department of Neurology, University Hospital Göttingen, 37075 Göttingen, Germany

PETER BAIN (1233), Division of Neuroscience and Psychological Medicine, Imperial College School of Medicine, London W6 8RF, United Kingdom

HENRY J. M. BARNETT (939), Department of Clinical Neurology, The University of Western Ontario, London, Ontario, Canada N6A 4B8

U. BAUMANN (1199), ENT Department, Ludwig-Maximilians University, 81377 Munich, Germany

MARTIN BEGEMANN (865), Memorial Hospital, New York, New York 10021

VALÉRIE BIOUSSE (115), Department of Ophthalmology and Neurology, Emory University School of Medicine, Atlanta, Georgia 30322

ROLFE BIRCH (1325), The Royal National Orthopaedic Hospital Trust, Peripheral Nerve Injury and Children's Hand Unit, Stanmore, Middlesex HA7 4LP, United Kingdom

T. P. BLECK (529), University of Virginia, Health Sciences Center, School of Medicine, Charlottesville, Virginia 22908

GIAN DOMENICO BORASIO (813, 1165), Interdisciplinary Palliative Care Unit and Department of Neurology, Ludwig-Maximilians University, 81377 Munich, Germany

KAI BÖTZEL (1099), Department of Neurology, Ludwig-Maximilians University, 81377 Munich, Germany

MARIE-GERMAINE BOUSSER (447), Service de Neurologie, Hôpital Lariboisière, Paris Cedex 10, 75475, France

THOMAS BRANDT (17, 67, 139), Department of Neurology, Ludwig-Maximilians University, 81377 Munich, Germany

STEVEN M. BROMLEY (171), Neurological Institute, Department of Neurology, Columbia Presbyterian Medical Center, New York, New York 10040

JOHN C. M. BRUST (791), Department of Neurology, Harlem Hospital Center, New York, New York 10037

ULRICH BÜTTNER (165), Department of Neurology, Ludwig-Maximilians University, 81377 Munich, Germany

LOUIS R. CAPLAN (303, 327, 373, 393, 403, 517, 939), Department of Cerebrovascular Disease, Beth Israel Deaconess Medical Center and Department of Neurology, Harvard Medical School, Boston, Massachusetts 02215

J. S. CHOPRA (647), Department of Neurology, Postgraduate Institute of Medical Education and Research, Chandigarh 160 012, India

DAVID GLENN CLARK (265), Department of Neurology, UCLA School of Medicine, Los Angeles, California 90095

N. COLMAN (1495), Department of Internal Medicine, Academic Medical Centre, 1105 AZ Amsterdam, The Netherlands

PATRICK COOPER (765), Neurosurgery Service, National Capital Consortium (Walter Reed-Bethesda), Washington, DC 20307

JAMES J. CORBETT (475, 921), Department of Neurology, University of Mississipi Hospitals, Jackson, Mississippi 39216

JEFFREY L. CUMMINGS (265, 311), Reed Neurological Research Center, UCLA School of Medicine, Los Angeles, California 90095

MARINOS C. DALAKAS (1363), National Institute of Neurological Disorders and Stroke, National Institutes of Health, Bethesda, Maryland 20892

DAVID M. DAWSON (1303), VA Hospital, Boston, Massachusetts 02115

LISA M. DEANGELIS (865), Memorial Hospital, New York, New York 10021

JOHANNES DICHGANS (1011, 1123), Department of Neurology, University of Tübingen, 72076 Tübingen, Germany

MARTIN DICHGANS (363), Department of Neurology, Ludwig-Maximilians University, 81377 Munich, Germany

H. CHRISTOPH DIENER (1, 23, 31, 95, 349, 1507), Department of Neurology, University of Essen, 45122 Essen, Germany

MARIANNE DIETERICH (59, 77), Department of Neurology, Johannes Gutenberg University, 55131 Mainz, Germany

VOLKER DIETZ (925, 1247), ParaCare, Paraplegic Center, University Hospital Balgrist, Zürich CH-8008, Switzerland

SALVATORE DIMAURO (1385), Department of Neurology, Columbia University College of Physicians & Surgeons, New York, New York 10032

MICHAEL DONAGHY (1281), Department of Neurology, Radcliffe Infirmary, Oxford OX2 6HE, United Kingdom

RICHARD L. DOTY (171), Smell & Taste Center and Department of Otorhinolaryngology: Head & Neck Surgery, School of Medicine, University of Pennsylvania, Philadelphia, Pennsylvania 19104

D. B. DRACHMAN (1341), Neurology-Neuromuscular Unit, Johns Hopkins University School of Medicine, Baltimore, Maryland 21205

JAMES ECKLUND (765), Neurosurgery Service, National Capital Consortium, (Walter Reed-Bethesda), Walter Reed Army Medical Center, Washington, DC 20307

KARL M. EINHÄUPL (447), Department of Neurology, Charité Hospital, 10117 Berlin, Germany

ILONKA EISENSEHR (191), Department of Neurology, Ludwig-Maximilians University, 81377 Munich, Germany

STANLEY FAHN (1021), Neurological Institute, Columbia University, New York, New York 10032

CHARLOTTE FEINMANN (59), Joint Department of Maxillofacial Surgery, Eastman Dental Hospital, London WC1X 8LD, United Kingdom

MICHAEL FETTER (185), Department of Neurology, Reha-Klinik, 76307 Karlsbad-Langensteinbach, Germany

MARSHAL FOLSTEIN (303), Department of Psychiatry, New England Medical Center, Boston, Massachusetts 02111

C. J. FOWLER (1479), Department of Uro-Neurology, The National Hospital, London WC1N 3BG, United Kingdom

R. FREEMAN (1453, 1495), Harvard Medical School and Center for Autonomic and Peripheral Nervous System Disorders, Beth Israel Deaconess Medical Center, Boston, Massachusetts 02215

GERAINT FULLER (615), Department of Neurology, Gloucestershire Royal Hospital, Gloucester GL1 3NN, United Kingdom

THOMAS GASSER (1149, 1525), Department of Neurology, University of Tübingen, 72076 Tübingen, Germany

PETER J. GOADSBY (1, 17), Institute of Neurology, University College London, The National Hospital for Neurology and Neurosurgery, London WC1N 3BG, United Kingdom

JOAO A. GOMES (403), Department of Neurology, Boston University School of Medicine, Boston, Massachusetts 02215

GEOFFREY K. GOURLAY (95), Pain Management Unit, Flinders Medical Center, Bedford Park, South Australia 5042, Australia

ROBERT C. GRIGGS (1369), Department of Neurology, University of Rochester School of Medicine,

Rochester, and Strong Memorial Hospital, Rochester, New York 14642

CHRISTIAN GUILLEMINAULT (191), Sleep Disorders Clinic, Stanford University Medical Center, Palo Alto, California 94304

ROBERTO GUILOFF (623), Department of Neurology, Charing Cross Hospital, London W6 8RF, United Kingdom

AMPARO GUTIERREZ (135), Louisiana State University Health Science Center, New Orleans, Louisiana 70112

WERNER HACKE (327, 373, 721), Department of Neurology, University of Heidelberg, 69120 Heidelberg, Germany

RODOLFO HAKIM (805), Clinical Pediatric Neurosurgery, Children's Hospital Boston, Boston, Massachusetts 02115

MARK HALLETT (1221), Human Motor Control Section, National Institute of Neurological Disorders and Stroke, National Institutes of Health, Bethesda, Maryland 20892

D. F. HANLEY (721), Johns Hopkins, Division of Neurosciences Critical Care, Baltimore, Maryland 21287

KLAUS HANSEN (583), Department of Neurology, The Neuroscience Center, Copenhagen University Hospital Rigshospitalet, Copenhagen 2100, Denmark

M. HARPER (1479) Department of Uro-Neurology, The National Hospital for Neurology and Neurosurgery, London WC1N 3BG, United Kingdom

S. HESSE (1185), Klinik Berlin, Department of Neurological Rehabilitation, Free University, 14089 Berlin, Germany

R. T. HIGASHIDA (481), University of California San Francisco Medical Center, Neuro Interventional Radiology, San Francisco, California 94143

REINHARD HOHLFELD (677, 1341), Institute for Clinical Neuroimmunology, Ludwig-Maximilians University, 81377 Munich, Germany

T. A. T. HUGHES (245, 249), University Hospital of Wales, Heath Park, Cardiff CF14 4XN, United Kingdom

R. J. JAEGER (1199), Research Service, Edward Hines, Jr. Hospital; Department of Veterans Affairs, Hines, Illinois 60141

J. JANKOVIC (1123), Parkinson's Disease Center and Movement Disorder Clinic, Baylor College of Medicine, Houston, Texas 77030

RIGMOR JENSEN (23), Department of Neurology, Glostrup Hospital, Glostrup DK-2600, Denmark

KARIN JURKAT-ROTT (1413), Department of Applied Physiology, University of Ulm, 89069 Ulm, Germany

HANS-OTTO KARNATH (277), Department of Cognitive Neurology, University of Tübingen, 72076 Tübingen, Germany

CARLOS S. KASE (403), Department of Neurology, Boston University School of Medicine, Boston, Massachusetts 02118

STEFAN KASTENBAUER (545), Department of Neurology, Ludwig-Maximilians University, 81377 Munich, Germany

OLIVER KASTRUP (1435, 1507), Department of Neurology, University of Essen, 45122 Essen, Germany

MARK T. KEEGAN (749), Anesthesiology/Critical Care, Mayo Clinic, Rochester, Minnesota 55905

MATTHIAS KEIDEL (83), Departments of Neurology and Neurorehabilitation, District Hospital of Bayreuth, 35445 Bayreuth, Germany

CHRISTOPHER KENNARD (185, 255), Unit of Neuroscience, Charing Cross Hospital, London W6 8RF, United Kingdom

KARL KIEBURTZ (1149), Department of Neurology, University of Rochester, New York 14642

THOMAS KLOCKGETHER (1011), Department of Neurology, University of Bonn, 53105 Bonn, Germany

THOMAS KLOPSTOCK (1385), Department of Neurology, Ludwig-Maximilians University, 81377 Munich, Germany

RICHARD KNIGHT (707), National CJD Surveillance Unit, Western General Hospital, Edinburgh EH4 2XU, United Kingdom

R. KOLLMAR (721), Department of Neurology, University of Heidelberg, 69120 Heidelberg, Germany

PAUL KRACK (1099), Service de Neurologie, Centre Hospitalier Universitaire de Grenoble, 38043 Grenoble, France

JOACHIM K. KRAUSS (1113), Department of Neurosurgery, University Hospital, Klinikum Mannheim, 68167 Mannheim, Germany

FRANK LEHMANN-HORN (783, 1413), Institute of Applied Physiology, University of Ulm, 89069 Ulm, Germany

R. JOHN LEIGH (125), Department of Neurology, Case Western Reserve University and Veterans Affairs Medical Center, Cleveland, Ohio 44106

SAMDEN D. LHATOO (207), Department of Neurology, Institute of Clinical Neurosciences and University of Bristol, Frenchay Hospital, Bristol BS16 1LE, United Kingdom

VOLKER LIMMROTH (37), Department of Neurology, University of Essen, 45122 Essen, Germany

ALAN H. LOCKWOOD (165), Center for PET (II5P), VA Western NY Healthcare System, Buffalo, New York 14215

E. LOGIGIAN (1421), Department of Neurology, Neuromuscular Disease Unit, University of Rochester School of Medicine and Dentistry, Rochester, New York 14642

HANS OTTO LÜDERS (235), Department of Neurology, Cleveland Clinic Foundation, Cleveland, Ohio 44195

JOSEPH R. MADSEN (805), Clinical Pediatric Neurosurgery, Children's Hospital Boston, Boston, Massachusetts 02115

R. MALESSA (601), Department of Neurology, Sophen and Hufeland Hospital Weimar, Teaching Hospital of the Friedrich Schiller University Jena, 99245 Weimar, Germany

HADI MANJI (623), Department of Neurology, National Hospital for Neurology and Neurosurgery, London; and The Ipswich Hospital, Suffolk IP4 5PD, United Kingdom

HANS J. MARKOWITSCH (287), Physiological Psychology, University of Bielefeld, 33615 Bielefeld, Germany

ROLAND MARTIN (677), Cellular Immunology Section; Neuroimmunology Branch, NINDS, National Institutes of Health, Bethesda, Maryland 20892

HELENE MASSIOU (47), Department of Neurology, Hôpital Lariboisiere, Paris, France

NINAN MATHEW (31), Houston Headache Clinic, Houston, Texas 77004

C. J. MATHIAS (1453), Pickering Unit, St. Mary's Hospital, London W2 1NY, United Kingdom

H. F. MCFARLAND (677), Neuroimmunology Branch, National Institute of Neurological Disorders and Stroke, National Institutes of Health, Bethesda, Maryland 20892

HANS-MICHAEL MEINCK (1429), Department of Neurology, University of Heidelberg, 69120 Heidelberg, Germany

A. MELMS (475, 1341), Department of Neurology, University of Tübingen, 72076 Tübingen, Germany

PATRICIA M. MOORE (461), Department of Neurology, University of Pittsburgh, Pittsburgh, Pennyslvania 15260

PRITHIVA NAVAN (1233), Division of Neuroscience and Psychological Medicine, Imperial College, London W6 8RP, United Kingdom

NANCY J. NEWMAN (115), Neuro-Ophthalmology Unit, Emory University School of Medicine, Atlanta, Georgia 30322

SOHEYL NOACHTAR (235), Department of Neurology, University of Munich, 81377 Munich, Germany

WOLFGANG H. OERTEL (1021), Department of Neurology, Philipps-University Marburg, 35033 Marburg, Germany

JES OLESEN (23), Department of Neurology, Glostrup Hospital, Glostrup 2600, Denmark

RICHARD W. ORRELL (1369), University Department of Clinical Neurosciences, Royal Free and University College Medical School, University College London and Royal Free Hospital, London NW3 2PF United Kingdom

G. D. PERKIN (67, 77), Department of Neurology, Charing Cross Hospital, London W6 8RF, United Kingdom

HANS-WALTER PFISTER (529, 545, 559, 583), Department of Neurology, Ludwig-Maximilians University, 81377 Munich, Germany

A. BERNARD PLEET[‡](1435), Division of Neurology, Baystate Medical Center, Springfield, Massachusetts 01199

WERNER POEWE (1081), Department of Neurology, University Hospital of Innsbruck, A-6020 Innsbruck, Austria

DIETER PONGRATZ (1363), Friedrich-Baur Foundation and Department of Neurology, Ludwig-Maximilians University, 80366 Munich, Germany

RUSSELL PORTENOY (813), Department of Pain Medicine and Palliative Care, Beth Israel Medical Center, New York, New York 10003

JEROME B. POSNER (911), Cornell University Medical School, Memorial Sloan-Kettering Cancer Center, New York, New York 10021

H. PRANGE (575), Department of Neurology, University of Göttingen, 37075 Göttingen, Germany

J. QUINTERN (1199), Neurological Hospital Bad Aibling, 83043 Bad Aibling, Germany

[‡]Deceased.

KENNETH RICKER (1407), Department of Neurology, University of Würzburg, 97070 Würzburg, Germany

KAREN L. ROOS (559), Department of Neurology, Indiana University School of Medicine, Indianapolis, Indiana 46202

MARTIN RÖSENER (595, 615), Department of Neurology, University of Tübingen, 72076 Tübingen, Germany

ANDRES M. SALAZAR* (765), Defense and Veterans Head Injury Program, Washington, DC 20008

JOSEMIR W. A. S. SANDER (207), Epilepsy Research Group, University College London, Institute of Neurology, London WC1N 3BG, United Kingdom

I. M. S. SAWHNEY (647), Morriston Hospital, Swansea SA6 6NL, United Kingdom

MARTIN SCHABET (865), Department of Neurology, Klinikum Ludwigsburg, 71640 Ludwigsburg, Germany

W. MICHAEL SCHELD (545), Division of Infectious Diseases, University of Virginia Health Science Center, Charlottesville, Virginia 22908

B. L. SCHLAGGAR† (947), St. Louis Children's Hospital, St. Louis, Missouri 63110

C. SCHNEIDER (1341), Department of Neurology, Heinrich-Heine University, 40225 Düsseldorf, Germany

JÖRG B. SCHULZ (991), Department of Neurology, University of Tübingen, 72076 Tübingen, Germany

S. SCHWAB (721), Department of Neurology, University of Heidelberg, 69120 Heidelberg, Germany

R. MICHAEL SCOTT (805), Clinical Pediatric Neurosurgery, Children's Hospital Boston, Boston, Massachusetts 02115

RAAD A. SHAKIR (595), Charing Cross Hospital, London, London W6 8RF, United Kingdom

STEPHEN D. SILBERSTEIN (37, 47), Jefferson Headache Center, Philadelphia, Pennsylvania 19107

BARNEY J. STERN (567), Department of Neurology, The Emory Clinic, Atlanta, Georgia 30322

MANFRED STÖHR (1303, 1337), Department of Neurology and Clinical Neurophysiology, Klinikum Augsburg, 86156 Augsberg, Germany

ANDREAS STRAUBE (125, 1157), Department of Neurology, Ludwig-Maximilians University, 81377 Munich, Germany

MICHAEL STRUPP (447, 1113), Department of Neurology, Ludwig-Maximilians University, 81377 Munich, Germany

AUSTIN J. SUMNER (135), Neurology Department, LSU Medical University, New Orleans, Louisiana 70112

IAN SUTTON (1259), Department of Neurology, Queen Elizabeth Hospital, Birmingham B15 2TH, United Kingdom

MARGARET M. SWANBERG (311), Department of Neurology, UCLA School of Medicine, Los Angeles, California 90095

PHILLIP D. SWANSON (1157), Department of Neurology, University of Washington, Seattle, Washington 98195

MICHAEL SWASH (1289, 1303), Department of Neurology, Royal London Hospital, London E1 1BB, United Kingdom

PETER THIER (971), Department of Neurology, University of Tübingen, 72076 Tübingen, Germany

DAVID G. T. THOMAS (827), Department of Neurosurgery, Insititute of Neurosurgery, University College London, London WC1N 3BG, United Kingdom

A. J. THOMPSON (1185), Department of Neuro-Rehabilitation, Institute of Neurology, National Hospital for Neurology and Neurosurgery, London WC1N 3BG, United Kingdom

PHILIP D. THOMPSON (1429), The University Department of Medicine, The Royal Adelaide Hospital, Adelaide, South Australia 5000, Australia

ARMIN K. THRON (517), Department of Neuroradiology, University Hospital RWTH Aachen, 52074 Aachen, Germany

HELGE TOPKA (1123, 1221, 1421), Department of Neurology and Clinical Neurophysiology, Academic Hospital Munich Bogenhausen, 81925 Munich, Germany

K. V. TOYKA (1341), Department of Neurology, University of Würzburg, 97080 Würzburg, Germany

CLAUDIA TRENKWALDER (1177), Department of Clinical Neurophysiology, University of Göttingen, 37075 Göttingen, Germany

KENNETH L. TYLER (601), Department of Neurology, University of Colorado Health Science Center, Denver, Colorado 80262

*Current address: Ribopharm, Inc., Washington, DC 20008
†Current address: Departments of Neurology, Radiology, and Pediatrics, Washington University School of Medicine, St. Louis, Missouri

J. VAN GIJN (421), Department of Neurology, University of Utrecht, NL-3584 CX Utrecht, The Netherlands

ARNO VILLRINGER (461), Department of Neurology, Charité Hospital, Berlin, Germany

RAYMOND VOLTZ (813), Institute of Clinical Neuroimmunology and Department of Neurology, Ludwig-Maximilians University, 81366 Munich, Germany

FRIEDRICH VON ROSEN (791), Neurological Hospital, Bad Aibling, 83043 Bad Aibling, Germany

ARTHUR S. WALTERS (1177), John Fitzgerald Kennedy New Jersey Neuroscience Institute, Edison, New Jersey 08818

MICHAEL WELLER (783, 827, 881, 897), Department of Neurology, University of Tübingen, 72076 Tübingen, Germany

GREGOR K. WENNING (1081), Department of Neurology, University of Innsbruck, Innsbruck A-6020, Austria

W. WIELING (1495), Department of Internal Medicine, Academic Medical Centre, 1105 AZ Amsterdam, The Netherlands

EELCO F. M. WIJDICKS (749), Department of Neurology W8A, Mayo Clinic, Rochester, Minnesota 55905

BOB WILL (707), National CJD Surveillance Unit, Western General Hospital, Edinburgh EH4 2XU, United Kingdom

JOHN B. WINER (1259), Department of Neurology, Queen Elizabeth Hospital, Birmingham B15 2TH, United Kingdom

PETER A. WINKLER (235), Department of Neurosurgery, Ludwig-Maximilians University, 81377 Munich, Germany

N. W. WOOD (1525), Institute of Neurology, University College London, London WC1N 3BG, United Kingdom

ULRICH WÜLLNER (921), Department of Neurology, University of Tübingen, 72076 Tübingen, Germany

ROBERT R. YOUNG (925, 1247), Department of Neurology, University of California, Irvine, California 92717

T. YOUSRY (481), Departments of Radiology and Neurology, University College London, WC1N 3BG London, United Kingdom

JOSEF ZIHL (255, 277), Department of Psychology, Neuropsychology Unit, Ludwig-Maximilians University, 80802 Munich, Germany

# PREFACE

The first edition of *Neurological Disorders, Course and Treatment* was so well received by the medical community that the editors were encouraged to keep the same uniform structure of the chapters: Clinical Aspects, Natural Course, Principles of Therapy, Practical Management, and Treatments No Longer Recommended. However, due to the numerous advances made in neurology in the interim, almost all the previous chapters required extensive updating. Also, several new authors have participated in this edition, reflecting our efforts to win more internationally recognized experts. The number of chapters has increased from 95 to 107. The new chapters include *Palliative care in the terminal stage of neurologic disease; Prion diseases; Deep brain stimulation for movement disorders; Atypical parkinsonism; Rehabilitation of motor function; Metabolic myopathies; Muscle stiffness; Microvascular diseases of the brain; Spinal cord ischemia; CNS parasitic infestations*; and also more specific conditions such as *Hormone-sensitive headache in women; Anosmia, ageusia, and other disorders of chemosensation;* as well as *Disorders of spatial orientation.* Because of the way our colleagues use the book, we have continued to list the most important symptoms and diagnostic criteria under the section "Clinical Aspects," which we consider to be an integral part of the management of individual disorders. In daily practice, diagnosis and therapy cannot be separated. It seemed especially important to include all therapeutic alternatives and special exceptions under the heading of therapeutic possibilities, since they are usually missing from popular textbooks. Unfortunately, in striving for a high standard of accuracy, we have not achieved our goal of concise formulation, for the total pages in the book have increased from 1150 to over 1500. Nevertheless, we hope the uniform organization of the chapters, the numerous tables, and the expanded index will still allow rapid reference and effective use.

Please note that in the sections "Principles of Therapy" and "Practical Management," the authors have indicated whenever possible the strength of the evidence on which the recommended treatment is based. The following system has been used: * designates retrospective studies, non-randomized studies, or empirical recommendation without scientific proof; ** a randomized, prospective study; and *** more than one prospective, randomized, placebo-controlled study or a meta-analysis. These symbols appear without any further clarification in tables and the text next to the relevant study.

Once again we thank all the authors, the editorial staff at Academic Press, in particular Noelle Gracy, and our own assistant Judy Benson, who again mastered the awesome task of organizing and coordinating the second edition, and finally, after many reminders, succeeded in obtaining the very last manuscripts from the over-committed experts. We would sincerely appreciate hearing your suggestions or corrections to improve the book's usefulness in any future edition.

*Thomas Brandt*
*Louis R. Caplan*
*Johannes Dichgans*
*H. Christoph Diener*
*Christopher Kennard*

# PREFACE TO THE FIRST EDITION

Most neurological disorders pose an intellectual challenge to the physician to make the correct neurological diagnosis. The patient, on the other hand, is more concerned about the course and effective treatment of the condition. In the last analysis, it is only this effective treatment of diseases or symptoms that justifies our very expensive health care system and rewards the physician for his day-to-day efforts.

Twenty years ago, neurology was considered a science of intriguing, but frequently untreatable, disorders, with an "aesthetics of diagnosis" that exercised a unique fascination. Fortunately, this has undergone a fundamental change. Today the exciting prospect of new therapies is at the center of the neurologist's work. Indeed, the effectiveness of treatment in neurology corresponds today to that of internal medicine. Striking examples of this are the successful management of epilepsy, meningitis, Parkinson's disease, headache, vertigo, and myasthenia gravis, the best elucidated model of the human autoimmune disorders.

The remarkable diversity of treatable diseases involving the peripheral and central nervous systems requires a clear and concise presentation, especially if this wealth of information is packed into about 1000 pages. Our aim is to present a comprehensive source of neurological therapy, applicable not only to the average patient but also to exceptional cases and covering alternative therapies and complications as well. In addition, we have included specific data on the expected success of therapy, prognostic factors, and limitations, as well as a short comparison of natural course and course under therapy, if available. Organized for ease of use and quick reference, each chapter of the book presents a neurological disorder or major symptom under the following aspects:

*Clinical Aspects.* Since therapy is based on the appropriate diagnosis, a definition of the disease is given along with the relevant clinical and diagnostic features. Also, the differential diagnosis relevant to therapy is mentioned.

*Natural Course.* Here we include facts and figures on incidence and prevalence, age of onset, sex, genetics, epidemiology, and provoking or aggravating factors. This section also contains a short description of the natural course, the frequency of symptoms, and complications or causes of morbidity and death. Specific details are given on how the natural course is improved by therapy.

*Principles of Therapy.* This section describes the pathomechanism of the disorder on which the principles of therapy have been developed. At this stage the rationale and results from controlled studies are presented, along with hypothetical or experimental therapies not yet tried or tested.

*Practical Management.* In this "cookbook section" the reader can find immediate guidance concerning situations when a prompt decision is required. Here specific guidelines are given, detailing which drug is used at which dosage, how the dosage should be increased, the possible side effects and contraindications of the drug, if any, and alternative substances if the drug of first choice cannot be prescribed or is ineffective. This section also includes various approaches to management of different stages and variants of a disorder (e.g., epilepsy, Parkinson's disease). Tables and flow-charts are employed to illustrate the various stages of therapy as well as alternatives to the medical, physical, and operative management. Indications for surgery are made clear. Also included are those special situations in which routine therapy must be reconsidered, such as pregnancy, surgery, anesthesia, allergy, dependency, typical combinations of disorders, or simply differences of treatment in relation to age. It is this combination of precise guidelines for routine management, alternative therapies, and variations of therapy in exceptional situations that makes our book different.

*Treatments No Longer Recommended.* This might be the most problematic section. Clear judgments are made on forms of therapy that are still in use but have never been shown to be effective or, even worse, have been shown to be ineffective or harmful. The benefit-to-risk ratio of such therapies is included, if known.

The last chapter of the book summarizes most of the relevant side effects of drugs used in neurology. A final table provides the reader with a selected list of trade names for the generic drugs mentioned in the single chapters. Since these were our aims, we hope it is understandable and excusable that the book in its current version consists of 1122 pages.

Our special thanks to the authors, above all, for their patience in the face of the editors' provoking criticism during the review process of the manuscripts. In the ensuing discussions and confrontations they often showed that they had been right in the beginning! Our sincere thanks also to our assistant, Judy Benson, who has survived the difficult tasks of organizing and coordinating 95 chapters, helped resolve some difficult language-editing problems, and politely but relentlessly rounded up all delinquent manuscripts. Last but certainly not least, we also express our appreciation to Academic Press, above all Graham Lees, his assistant Karen Dempsey, and their coworkers, for their continuing enthusiastic support of this project since its inception in 1992.

We have tried our best to produce a therapy book for neurology that is as practical and specific as possible while also advising the neurologist in the clinic and private practice on the value of different methods. Of course, nothing worthwhile is accomplished simply, and many nerves were at times quite frayed. The following excerpt from an author's response to a query about the whereabouts of his manuscript perhaps best suggests the trials and tribulations of such a project:

> I received the chapter in question from Dr. X last March. I found it unacceptable. I asked Dr. Y what to do, and he suggested that I recommend extensive changes. I did that, and after a long time Dr. X wrote back and suggested that I rewrite the chapter. I rewrote the chapter and sent it to Dr. X in July. He found it unacceptable and wrote back asking for extensive changes. I made major changes, trying to be fair to both points of view, and sent Drs. X, Y, and Z a final version in September. Since then, Dr. Y has told me the chapter is wonderful, Dr. Z has rejected it, and you act as if you haven't even seen it! I suspect that it will not be possible to come to an agreement on this chapter.

We are thankful that the prediction of the exasperated author has not come true. Nevertheless, our respect for the difficulty of this project did not diminish over the past four years. If anything, we have become even more aware of the remaining uncertainties and inadequacies (for example, on the border areas of neurosurgery, child neurology, and psychiatry), but above all of the transience of concepts that are currently still valid. Thus, the editors would be thankful for any constructive suggestions from critical readers that would allow us to eliminate any remaining weaknesses, correct errors, and in general improve a later edition.

*Thomas Brandt*
*Louis R. Caplan*
*Johannes Dichgans*
*H. Christoph Diener*
*Christopher Kennard*

*Acute and Chronic Pain*

# CHAPTER 1
# Migraine

## H. Christoph Diener and Peter J. Goadsby

The International Headache Society defines migraine as a disease characterized by intermittent attacks of headache, often nausea, and associated features of sensory sensitivity, such as photophobia, phonophobia, and sensitivity to head movement (Headache Classification Committee of the International Headache Society, 1988). The most frequent types of migraine are migraine without aura and migraine with aura.

## CLINICAL ASPECTS

### Migraine without Aura

Migraine without aura (formerly common migraine) is characterized by repeated attacks of headache typically lasting from 4 to 72 h. In 60% of the attacks the headache is restricted to one side. It is pulsating, of moderate to severe intensity, and aggravated by physical activity. In its typical form it is accompanied by nausea, vomiting, photophobia, and phonophobia. Migraine without aura should be diagnosed only after five attacks.

### Migraine with Aura

Migraine with aura (formerly migraine accompanée or classic migraine) is characterized by neurological symptoms and deficits, which are preferentially localized in cortical areas (scotomas, hemianopsia, sensory deficits, aphasia) or in the brain stem (paraparesis, vertigo with nystagmus, ataxia, diplopia). These symptoms develop within a time period of 5 to 20 min and disappear completely within less than 60 min. Concomitant with or within 1 h after cessation of neurological symptoms the typical headache starts. It may be accompanied by autonomic disturbances. In some cases,

an isolated aura without headache may occur, especially in patients with the onset of migraine after the age of 20. Some of the subtypes of migraine currently recognized are: migraine with prolonged aura (>60 min), familial hemiplegic migraine (with an idetified gene defect on chromosome 19 in 50% of the cases) (Battistini *et al.*, 1999; Joutel *et al.*, 1993; Ophoff *et al.*, 1997; Terwindt *et al.*, 1996), basilar migraine, and retinal migraine (monocular scotoma, monocular amaurosis).

### Complications of Migraine

A complicated form of migraine attack is defined by neurological deficits persisting for 7 or more days after a migraine attack has ended or an ischemic lesion is found in a CT or MRI scan while other reasons for a cerebral infarction are excluded. Migraine is a risk factor for stroke. Women who suffer from migraine with aura, are overweight, smoke, and take the pill have an increased stroke risk (Bousser, 1999; Buring *et al.*, 1995; Chang *et al.*, 1999; Merikangas *et al.*, 1997). Focal dysrhythmia in the EEG may persist after the end of a migraine attack even without any ischemic lesion in the CT scan.

### Migraine in Children

Pediatric migraine is characterized by shorter attacks and generalized headache. Nausea, vomiting, abdominal pain, or general impairment of well-being may predominate (Gladstein *et al.*, 1993). After a short period of sleep the headache is often reduced. Some forms of migraine in children are characterized by an isolated aura without headache (e.g., benign paroxysmal vertigo in children) (Abu-Arafeh and Russell, 1995; Lindskog *et al.*, 1999).

## Diagnosis of Migraine

The diagnosis of migraine can be made only by taking a careful prior history and physical examination. Further investigations, such as CT, MRI, EEG, or evoked potentials are not helpful (Quality Standards Subcommittee of the American Academy of Neurology, 1994). The following clinical aspects would be unusual for migraine: duration of headache <2 h, no change in side of headache, first manifestation after the age of 50 years, or fever as an additional symptom.

## NATURAL COURSE

The prevalence of migraine in children is 2–4% (Abu-Arefeh and Russell, 1994). Boys and girls are equally affected (Bille, 1962). In about 50% of children migraine attacks cease completely or are reduced significantly after puberty. Migraine attacks persisting after puberty show the typical female predominance of migraine known in adults.

The one year prevalence of migraine after puberty is 12–14% for women and 7–8% for men (Rasmussen, 1993; Russel et al., 1995; Silberstein and Lipton, 1993; Stewart et al., 1992). Women are affected two to three times more often than men. Adult migraine in women usually begins during puberty. In men its onset is between 20 and 30. First manifestations of migraine after the age of 45 are rare and should prompt a carefull consideration of the diagnosis. During pregnancy, migraine is reduced in about one-third of the patients, remains unchanged in one-third, and worsens in one-third (Marcus et al., 1999). The natural history may be extremely variable from one patient to the other. An increased number of migraine attacks, e.g., >10/month, is most likely due to abuse of anti-migraine drugs (see Chapter 4).

## THE MIGRAINE ATTACK

Migraine therapy is best understood in two parts: preventative agents are likely to act in the brain on the region, or regions, that promote or generate the attacks, while acute attack therapies specifically arrest the attacks and thus probably act somewhere in the pain pathway. The pharmacology of these agents is different, and taken with the clinical picture of migraine when compared to other primary neurovascular headache (Goadsby, 2000), such as cluster headache (see Chapter 2), it seems that the treatment differences reflect some underlying pathophysiological principles. The essential elements to be considered in understanding migraine are (Goadsby et al., 2002):

- anatomy of head pain, particularly that of the trigeminovascular system;
- physiology and pharmacology of activation of the peripheral branches of ophthalmic branch of the trigeminal nerve as marked by dural plasma protein extravasation and cranial neuropeptide release;
- physiology and pharmacology of the trigeminal nucleus, in particular its caudalmost part, the trigeminocervical complex;
- brain-stem and diencephalic modulatory systems that modulate trigeminal pain processing.

### Headache—Anatomy

Surrounding the large cerebral vessels, pial vessels, large venous sinuses, and dura mater is a plexus of largely unmyelinated fibers that arise from the ophthalmic division of the trigeminal ganglion and in the posterior fossa from the upper cervical dorsal roots. Trigeminal fibers innervating cerebral vessels arise from neurons in the trigeminal ganglion that contain substance P and calcitonin gene-related peptide (CGRP), both of which can be released when the trigeminal ganglion is stimulated. Stimulation of the cranial vessels, such as the superior sagittal sinus (SSS), is painful in humans. Human dural nerves that innervate the cranial vessels largely consist of small-diameter myelinated and unmyelinated fibers that almost certainly subserve a nociceptive function.

What then is the source of pain in migraine? If the carotid artery is occluded ipsilateral to the side of headache in migraineurs then two-thirds will experience relief, although this does not account for the other one-third. It must then be borne in mind that the pain process is likely to be a combination of direct factors, i.e., activation of the nociceptors of pain-producing intracranial structures, in concert with a reduction in the normal functioning of the endogenous pain control pathways that normally modulates pain input. Although a simple question, the source of the pain is a difficult issue in primary headache (Table 1).

### Headache Physiology—Peripheral Connections

#### Plasma Protein Extravasation

Moskowitz (1990) has provided an elegant series of experiments to suggest that the pain of migraine may be a form of sterile neurogenic inflammation in the dura mater. Neurogenic plasma extravasation can be seen

**TABLE I** Neuroanatomical Processing of Vascular Head Pain

| Target innervation | Structure | Comments |
|---|---|---|
| • Cranial vessels<br>• Dura mater | Ophthalmic branch<br>of trigeminal nerve | |
| First | Trigeminal ganglion | Middle cranial fossa |
| Second | Trigeminal nucleus<br>(quintothalamic<br>tract) | Trigeminal n. caudalis<br>and $C_1/C_2$ dorsal<br>horns |
| Third | Thalamus | Ventrobasal complex<br>medial n. of<br>posterior group<br>intralaminar<br>complex |
| Final | Cortex | • Insulae<br>• Frontal cortex<br>• Anterior cingulate<br>cortex<br>• Basal ganglia |

during electrical stimulation of the trigeminal ganglion in the rat and can be blocked by ergot alkaloids, indomethacin, acetylsalicylic acid, and the triptans, serotonin-5-$HT_{1B/1D}$ agonists (see below in Acute Treatment of Migraine). In addition there are structural changes seen in the dura mater that are seen with trigeminal ganglion stimulation and these include mast cell degranulation and changes in postcapillary venules including platelet aggregation. While it is generally accepted that the initiation of a sterile inflammatory response would cause pain, it is not clear whether this is sufficient in itself or requires other stimulators or promoters. Clearly blockade of neurogenic plasma protein extravasation is not completely predictive of anti-migraine efficacy in humans as evidenced by the failure in clinical trials of a number of substances which can attenuate plasma protein extravasation in experimental settings.

*Neuropeptide Studies*

Electrical stimulation of the trigeminal ganglion in both humans and the cat leads to increases in extracerebral blood flow and local release of both CGRP and SP. Stimulation of the more specifically vascular pain-producing superior sagittal sinus increases cerebral blood flow and jugular vein CGRP levels. Human evidence that CGRP is elevated in the headache phase of migraine, cluster headache, and chronic paroxysmal hemicrania, but not in nonthrobbing tension-type headache, supports the view that the trigeminovascular system may be activated in neurovascular headache (Edvinsson and Goadsby, 1994; Goadsby and Edvinsson, 1993). It seems likely that central processing

of the peripheral pain signals is the crucial and differentiating factor for neurovascular headaches.

**Headache Physiology—Central Connections**

*The Trigeminocervical Complex*

The sites within the brain stem that are responsible for craniovascular pain have begun to be mapped using Fos immunohistochemistry. After stimulation of the superior sagittal sinus Fos-like immunoreactivity is seen in the trigeminal nucleus caudalis and in the dorsal horn at the $C_1$ and $C_2$ levels in the cat and monkey (Goadsby and Zagami, 1991). These latter findings are consistent with a view of the trigeminal nucleus extending beyond the traditional nucleus caudalis to the dorsal horn of the high cervical region in a functional continuum that includes a cervical extension, and could be collectively known as the trigeminocervical complex. Stimulation of the greater occipital nerve, a branch of $C_2$, increases metabolic activity in the same regions (Goadsby et al., 1997). Thus the same group of cells has input from both supratentorial structures and from branches of the nerve roots of the high cervical spinal cord, which accounts for the referral of pain to the neck so commonly in the clinic. Moreover, there is a substantial literature to demonstrate that these trigeminocervical neurons can be inhibited by specific acute attack anti-migraine compounds, such as triptans and ergotamine.

**Higher Order Processing**

Following transmission in the caudal brain stem and high cervical spinal cord, information is relayed in a group of fibers (the quintothalamic tract) to the thalamus. Processing of trigeminovascular pain in the thalamus probably includes areas in the ventroposteromedial thalamus, medial nucleus of the posterior complex, and the intralaminar thalamus. Human imaging studies have confirmed activation of thalamus contralateral to pain in acute migraine (Bahra et al., 2001).

**Central Modulation of Trigeminal Pain**

It has been shown in the experimental animal that stimulation of a discrete nucleus in the brain stem, nucleus locus coeruleus (the main central noradrenergic nucleus), reduces cerebral blood flow in a frequency-dependent manner through an $\alpha_2$-adrenoceptor-linked mechanism. This reduction is maximal in the occipital cortex. While a 25% overall reduction in cerebral blood flow is seen, extracerebral vasodilatation occurs in parallel. In addition the main serotonin-containing nucleus in the brain stem, the midbrain dorsal raphe nucleus, can increase cerebral blood flow when activated. Consistent with these animal studies is the description in humans of activation of a brain-stem region in PET studies during migraine without aura (Weiller *et al.*, 1995) that would include the periaqueductal gray in the region of the dorsal raphe nucleus and in the dorsal pons near the locus coeruleus (Weiller *et al.*, 1995). Moreover, these areas are active immediately after successful treatment of the headache but are not active interictally, suggesting a role that is more than simply a response to pain. These areas are not active in cluster headache (see Chapter 2) and may represent the site of reversible dysfunction that is key to migraine and may well be the target for preventative treatments.

## Principles of Therapy

### Analgesics

Analgesics, nonsteroidal antiinflammatory drugs (NSAIDs) and acetylsalicylic acid (ASA) are thought to act via the inhibition of prostaglandin synthesis and may affect peripheral receptors and the release of inflammatory mediators. Buzzi *et al.* (1989) showed, that ASA blocks neurogenic plasma protein extravasation in rat dura mater after trigeminal stimulation. Furthermore, there is evidence that ASA has central actions. Kaube *et al.* (1993) found in the cat that ASA exerts inhibitory effects on the central processing of trigeminal nociceptive input. There are fewer controlled trials with analgesics than with "triptans" for the treatment of acute migraine attacks. Reasonbly well conducted studies are available for paracetamol (acetaminophen), ASA, ibuprofen, naproxen, tolfenamic acid, and diclofenac (Dahlöf and Björkman, 1993; Hakkarainen *et al.*, 1979; Havanka-Kanniainen, 1989; Henry *et al.*, 1995; Karabetsos *et al.*, 1997; Karachalios *et al.*, 1992; Kellstein *et al.*, 2000; Kloster *et al.*, 1992; Larsen *et al.*, 1990; MacGregor *et al.*, 1993; Myllylä *et al.*, 1998; Tfelt-Hansen and Olesen, 1980; The Diclofenac-K/Sumatriptan Migraine Study Group, 1999; Tokola *et al.*, 1984). The combination of acetaminophen, ASA, and caffeine was tested against placebo but not against a triptan or each compound alone (Lipton *et al.*, 1998). The combination of ASA and metoclopramide is of comparable efficacy to the specific migraine drug sumatriptan (Tfelt-Hansen *et al.*, 1995; The Oral Sumatriptan and Aspirin plus Metoclopramide Comparative Study Group, 1992). Recently, a (oral) rapid release form of tolfenamic acid was shown to be comparable in efficacy to sumatriptan (Myllylä *et al.*, 1998). In some countries ASA is available through iv injection and has shown to be superior to subcutaneous ergotamine but less effective than subcutaneus sumatriptan (Diener for the ASASUMAMIG Study Group, 1999; Limmroth *et al.*, 1999b). Due to its severe adverse effects ketolorac has been withdrawn from the market in some countries.

### Ergotamine and Dihydroergotamine (DHE)

Ergotamine and DHE are vasoconstrictors and inhibit the aseptic perivascular inflammation in animal models (Moskowitz, 1992) and the release of CGRP in animals and humans (Edvinsson and Goadsby, 1995). Dahlöf (1993) performed an analysis of the available studies comparing ergotamine with placebo or control. On the evidence of these studies ergotamine is of limited efficacy in migraine. This, however, is due to the poor quality of older headache trials. A consensus group reached the conclusion that ergots are inferior to triptans for the treatment of migraine attacks (Tfelt-Hansen *et al.*, 2000). No placebo-controlled trials were performed with the suppository formulation of ergotamine. DHE is available in oral form (1 or 2 mg), in many countries in fixed combinations with caffeine or analgesics, as intranasal spray (2 mg) (DiSerio *et al.*, 1989; Krause and Bleicher, 1985; Paiva *et al.*, 1985; Rohr and Dufresne, 1985; Tulunay *et al.*, 1987), and for parenteral administration (0.5–2 mg) for intravenous, subcutaneous, or intramuscular application.

### Serotonin 5-HT$_{1B/D}$-Receptor Agonists (Triptans)

All triptans including sumatriptan act postjunctionally at 5-HT$_{1B}$-receptors in the vessel wall (Ferro *et al.*, 1995; Martin, 1997). In animal models they lead to a constriction of arteries, which is more pronounced at cerebral and dural arteries than at coronary or peripheral arteries. In addition the triptans inhibit the aseptic perivascular inflammation of the dura mater induced by stimulation of the trigeminal ganglion via presynaptic 5-HT$_{1D}$ (and 5-HT$_{1F}$) receptors (Buzzi *et al.*, 1991). An increase in the concentration of CGRP could be observed in the jugular vein of humans during migraine attacks which decreased after injection of sumatriptan (Goadsby and Edvinsson, 1993). Sumatriptan does not cross the intact blood–brain barrier. The newer 5-HT$_{1B/1D}$ agonists are able to penetrate the blood–brain barrier and bind to neurons in the trigeminal nucleus caudalis of the brain stem and upper cervical cord (Cumberbatch *et al.*, 1997; Goadsby and Hoskin, 1996). Which of these mechanisms of action is crucial is still under discussion.

Soon after its introduction sumatriptan became the "classical" triptan and was followed by several other triptans, (zolmitriptan, naratriptan, rizatriptan, almotriptan, frovatriptan, and eletriptan). After 1 h, most oral 5-HT$_{1B/1D}$ agonists lead to an improvement of headache (from severe or moderate to mild or none) in 30–40% of all patients and migraine attacks. After 2 h the success rates amount to 50–70% (for details see below and (Diener, 2000a). Due to its long half-life and slow absorption naratriptan needs 4 h to reach this degree of efficacy (Klassen et al., 1997). Nausea, vomiting, and photo- as well as phonophobia were also improved and the need for rescue medication was significantly less than with placebo. If the first application of a triptan was ineffective, a second dose did not improve the outcome. A particular problem of all triptans is headache recurrence. Headache recurrence (better named secondary treatment failure) is defined as recurring headache of moderate to severe intensity within 24 h after an initially good response to medication. This happens in about 20 to 40% of patients (Visser et al., 1996). Headache may reoccur, because migraine drugs ameliorate the symptoms of an attack, but are not able to turn off the "migraine generator" in the brain stem (Weiller et al., 1995). Perhaps a combination of a NSAID with a tritpan could decrease the frequency of recurrences (Krymchantowski et al., 1999). Another problem of all triptans—well known from ergotamine derivatives—seems to be the potential danger to lead to a drug-induced headache when taken frequently (Gaist et al., 1998; Kaube et al., 1994; Limmroth et al., 1999a).

*Sumatriptan* offers a wide range of dosages and formulations, enabling individually tailored therapy (for a review see Diener, 2000d; Tfelt-Hansen, 1998). DHE given subcutaneously is less effective than sumatriptan by the same route, but has a significantly lower rate of headache recurrence (Winner et al., 1996).

*Zolmitriptan* is available as tablet with 2.5 and 5 mg and soon as nasal spray. Its efficacy is not different from sumatriptan (Nappi and Johnson, 2000). It is, however, effective in some patients who do not respond to sumatriptan (and vice versa). Whether this is related to the central action of zolmitriptan is not known. *Naratriptan* (2.5 mg) is less effective than sumatriptan (Goadsby, 1998; Göbel et al., 1997) but causes significantly fewer side effects. Again, it may be effective in patients who do not respond to sumatriptan (Stark et al., 2000). *Rizatriptan* has both 5 and 10 mg doses. Its onset of action is sooner than sumatriptan (Goldstein et al., 1998b; Tfelt-Hansen et al., 1998). It has a high consistency across migraine attacks (Kramer et al., 1998). Rizatriptan is also available as a wafer which dissolves instantly on the tongue without liquid. Efficacy is the same as with the tablet (Adelman et al., 2000). Rizatriptan has an interaction with propranolol (not with metoprolol). Patients who are on propranolol prophylaxis should only take rizatriptan 5 mg. In two trials *eletriptan* was superior to sumatriptan in oral doses of

40 and 80 mg (Diener, 2000b; Goadsby et al., 2000). In the 80-mg group, however, more side effects were reported than in the sumatriptan (100 mg) group. Like rizatriptan, eletriptan has a fast absorption and a fast onset of action. Frovatritpan has a long half-life which theoretically should reduce headache recurrence (Goldstein et al., 1998a). Due to the low efficacy this has not been shown in prospective trials. Almotriptan has an efficacy which is not different from sumatriptan (Cabarrocas et al., 1998a,b).

Typical but rare side effects of 5-HT$_{1B/1D}$ agonists are: fatigue (5–9%), dizziness (3–6%), throat symptoms (3%), weakness (3%), neck pain (2–3%), sedation (2–3%), and chest symptoms (2–3%). After subcutaneous injection of sumatriptan the following side effects were observed: injection side reaction (40%), tingling (9%), warm or hot feeling (8%), dizziness or vertigo (7%), heaviness (7%), pressure (6%), flushing (6%), chest symptoms (5%), neck pain (3%), tightness (3%) (Brown et al., 1991). Most reported side effects could not be distinguished from symptoms of the migraine attack itself. Only 2 to 6% of patients were dropouts in long-term trials due to side effects. The few serious side effects like myocardial infarction, angina pectoris, or cardiac arrhythmia occurred in patients who had clear contraindications (O'Quinn et al., 1999) or a disease other than migraine (ischemic stroke, subarachnoidal hemorrhage). Ideally a migraine drug should have the efficacy of sumatriptan but no vasoconstrictive side effects. However, powerful inhibitors of neurogenic inflammation without vasoconstrictive properties like substance-P antagonists (Diener for the RPR100893-201 Migraine Study Group, 1995; Goldstein et al., 1997), endothelin antagonists (May et al., 1996), and CP122,288 (Roon et al., 2000) were not effective in migraine.

### Antiemetics

Most patients suffer from gastrointestinal symptoms like nausea, vomiting, and diarrhea. Of significant importance when treating migraine is gastric stasis, which may result in delayed or abolished absorption of migraine drugs. Fast absorption with an early peak in plasma concentration is very important for efficacy. Antiemetics like metoclopramide and domperidone attenuate the autonomic dysfunction and are able to speed up gastric emptying. In the USA where domperidone is not available, antidopaminergic drugs, such as chlorpromazine or prochlorperazine, are used.

### Practical Management

The management of migraine includes offering advice on lifestyle issues that can be helpful in avoiding triggers and treating the disorder using acute attack treatments and preventative agents (Lance and Goadsby, 1998).

### General Measures and Patient Education

Patients with migraine have a genetically determined susceptibility to headache. They are *headachy* in a broad sense and certain key things need to be understood by the patient, notably:

- migraine is an inherited tendency to headache and cannot be cured, *but*;
- migraine can be modified and controlled by lifestyle adjustment and the use of medicines;
- migraine is neither life-threatening nor associated with serious illness, with the exception of females who smoke and are on the oestrogenic oral contraceptives, who are at increased risk for stroke, but *migraine can and often does make life a misery*;
- migraine management takes time and cooperation when information, such as that from a headache diary, has to be collected.

### Nonpharmacological Management

Put very simply nonpharmacological management of migraine is to help the patient identify things that make the problem worse and encourage them to avoid or modify these. It is important to explain to the patient that the tendency to suffer an attack probably varies because of some changes in the brain that are not well understood and certainly vary with time. This is why avoiding some things on some days will prevent attacks and, perversely, enjoying the same things on other days produces no headache. The crucial piece of the puzzle is what the brain is doing and this is subject of intense study. Rather than make a long list of things to avoid, patients should at first be encouraged to have regular habits and not exceed their limits. Regular sleep, exercise, meals, work habits, and some time for relaxation will be very rewarding in terms of reducing headache frequency. Conversely, avoiding overtiredness, sleeping in, overexercise, skipping meals, excess use of stimulants, such as caffeine, and excess stress is a good start in reducing headache frequency.

All migraine patients should be warned about the overuse of acute-attack treatments, particularly those obtained over the counter, and compounded analgesics with caffeine, or opiates, or both. Regular, more than twice a week, or increasing consumption with time are alarm bells that require attention.

The evidence base for complementary medicines is poor and our experience with their use limited. With the exception of the use of riboflavin (vitamin $B_2$; 400 mg

daily) as a preventative agent in migraine (Schoenen *et al.*, 1997), placebo-controlled evidence is lacking. Conflicting study results showing an effect and no effect when compared to placebo are available for feverfew and magnesium (Peikert *et al.*, 1996; Pfaffenrath *et al.*, 1996) as preventatives (see Migraine Prophylaxis).

### Abortive Agents

Acute-attack treatments for migraine can be usefully divided into disease-nonspecific treatments, analgesics and NSAIDS, and disease-specific treatments, ergot-related compounds and triptans (Table II). The number of treatments that can now be offered is substantial and each has some particular benefit to patients.

Given the array of options to control an acute attack of migraine, how does one start? The simplest approach to treatment has been described as *"step care."* In this model all patients are treated, assuming no contraindications, with the simplest treatment, such as aspirin, 900–1000 mg, with an antiemetic, an effective strategy proven by double-blind controlled clinical trials. The alternative strategy is known as *"stratified care,"* by which the physician determines, or stratifies, treatment at the start based on likelihood of response to levels of care. Lipton and Stewart have proposed a migraine disability scale (MIDAS, Migraine Disability Assessment Scale) (Stewart *et al.*, 1999). The scale is easy to use and freely available. It can be used to predict outcome in that the acute attacks in more disabled patients are less likely to respond to simple treatment measures (Lipton *et al.*, 2000). Treatment selection in reality needs to be

#### TABLE II   Oral Acute Migraine Treatments

| Nonspecific treatments | Specific treatments |
| --- | --- |
| Often used with anti-emetic/prokinetics, such as domperidone (10 mg)[*] or metoclopramide (10 mg)[**] | |
| Aspirin (900–1000 mg)[**] | *Ergot derivatives* |
| Paracetamol (1000 mg)[**] | Ergotamine (1–2 mg)[**] |
| NSAIDS | |
| Naproxen (500–1000 mg)[**] | *Triptans*[***] |
| Ibuprofen (400–800 mg)[**] | Sumatriptan (50 or 100 mg) |
| Tolfenamic acid (200 mg)[**] | Naratriptan (2.5 mg) |
| | Rizatriptan (10 mg) |
| | Zolmitriptan (2.5 or 10 mg) |
| | Eletriptan (20, 40, or 80 mg) |
| | Almotriptan (12.5 mg) |
| | Frovatriptan (2.5 mg) |

[*]Treatment based on personal experience, expert opinion, retrospective, or nonrandomized studies.

[**]Treatment is based on at least one randomized prospective study of reasonable size.

[***]Treatment recommendations based on more than one well-designed randomized, placebo-controlled trial or a metaanalysis.

a hybrid of both approaches since individuals may have attacks of different severity over time and will need more than one strategy to cover all their attacks.

**Nonspecific Treatments.** Simple things, such as aspirin (900–1000 mg) and paracetamol (1000 mg), are cheap, can be very effective, and can be usefully employed in many patients. The addition of domperidone (10 mg po) or metoclopramide (10 mg po) can be very helpful. NSAIDS can be very useful when tolerated. Their success is often limited by inappropriate dosing, but naproxen (500–1000 mg po or pr with an antiemetic), ibuprofen (400–800 mg po), or tolfenamic acid (200 mg po) can be extremely effective. Tolfenamic acid has been shown in a double-blind placebo-controlled study to have comparable efficacy to sumatriptan 100 mg, a result that reinforces the general clinical view that NSAIDs can be very useful compounds in migraine. Typical adverse effects of ASA and NSAIDs are abdominal discomfort and gastroesophageal reflux. Contraindications are gastrointestinal disorders (ulcers), gastritis, coagulopathies, tinnitus, and asthma. ASA can be administered in situations where ergotamine, DHE, or triptans may be contraindicated, e.g., coronary heart disease, prior myocardial infarction, hypertension, Raynauds disease, intermittent claudication, and the presence of multiple vascular risk factors.

**Specific Treatments.** When simple measures fail or more aggressive treatment is required, the specific treatments are required (Table III). While ergotamine remains a useful antimigraine compound, its place as the first-choice has slipped in recent years and this trend is likely to continue (Tfelt-Hansen *et al.*, 2000). There are particular situations in which ergotamine is very useful but its use must be strictly controlled as ergotamine overuse produces dreadful headache in addition to a host of vascular problems (Andersson, 1988; Senter and Lieberman, 1976). Ergotamine, and to a lesser extent DHE, have many adverse effects, including nausea, vomiting, worsening of headache, numbness, dizziness, vertigo, gastric symptoms, dry mouth, and restlessness. In addition, regular intake of ergotamine makes prophylaxis of migraine ineffective. Contraindications for ergotamine and DHE include ischemic heart disease, myocardial infarction, intermittent claudication, Raynaud's disease, hypertension and pregnancy. Ergotamine should not be given within 12 h after any of the triptans or within 24 h of naratriptan.

The triptans have revolutionized the life of many patients with migraine and are clearly the most powerful option available to stop a migraine attack (Goadsby *et al.*, 2002). They can be rationally applied by considering their pharmacological, physicochemical, and pharmacokinetic features, as well as the formulations that are available.

**TABLE III** Clinical Stratification of Acute Specific Migraine Treatments

| Clinical situation | Treatment options |
|---|---|
| Failed analgesics/NSAIDS | *First tier*<br>Sumatriptan 50 mg po<br>Rizatriptan 10 mg po<br>Zolmitriptan 2.5 mg po<br>Almotriptan 12.5 mg po<br>Eletriptan 40 mg po<br>*Better tolerability/rapid onset*<br>Almotriptan 12.5 mg<br>*Slower effect/better tolerability*<br>Naratriptan 2.5 mg<br>Frovatriptan 2.5 mg<br>*Infrequent headache*<br>Ergotamine 1–2 mg po<br>Dihydroergotamine nasal spray 2 mg |
| Early nausea or difficulties taking tablets | *Rapidly dissolving tablets*<br>Rizatriptan 10 mg wafer<br>Zolmitriptan 2.5 mg wafer[a]<br>Nasal sprays<br>Sumatriptan 20 mg nasal spray<br>Zolmitriptan 5 mg nasal spray[a] |
| Headache recurrence | Ergotamine 1–2 mg<br>(perhaps most effective pr/usually with caffeine)<br>Almotriptan 12.5 mg[a]<br>Eletriptan 80 mg[a]<br>Naratriptan 2.5 mg po<br>Dihydroergotamine 1 mg imi<br>Triptan plus NSAID |
| Tolerating acute treatments poorly | Almotriptan 12.5 mg[a]<br>Naratriptan 2.5 mg<br>Frovatriptan 2.5 mg[a] |
| Early vomiting | Sumatriptan 6 mg sc<br>Dihydroergotamine 1 mg imi |
| Very rapidly developing symptoms | Sumatriptan 6 mg sc<br>Dihydroergotamine 1 mg imi |

[a]Based on study data as these medications were not widely clinically available at the time of writing.

It is relatively easier to pick a triptan if a particular clinical situation arises (Table IV) than to provide a blanket rule. It is our certain experience that should one triptan fail, often another oral triptan will work. Whether this reflects differences in the compounds or simply the varying natural history of migraine is a vexed question. We suggest, based unfortunately on experience, not evidence, that patients should be offered at least three oral triptans before switching formulations. It can be said, based on repeat-attack controlled studies, that patients have an approximate 10% chance of responding to an oral triptan if they have failed in three attacks consecutively, so any compound can be tried three times and then evaluated by the practitioner. An alternative view would be to change triptan formulation

TABLE IV  Special Populations

| Clinical situation | Treatment options |
|---|---|
| Menstrually-related headache | Triptans |
| | Dihydroergotamine nasal spray/im |
| Children (<12 years) | Sumatriptan nasal spray |
| | Triptan wafer formulations[a] |
| Elderly without contraindications (>65 years) | Sumatriptan 50 mg |

[a]Wafers have the advantage of being easily administered to children but have thus far not been formally studied in this age group.

after three attacks rather than move to a new triptan. Preference studies consistently report that migraine patients prefer oral medications, which provides some rationale for switching triptans rather than formulations.

### Treatment No Longer Recommended

Opioids and neuroleptics are still frequently used in emergency rooms. They offer no advantage over specific migraine drugs. Opioids have addictive properties and can lead to drug-induced headache. Neuroleptics have poor analgesic properties and the risk of early or late dyskinesias.

## MIGRAINE: PREVENTIVE TREATMENT

The prophylactic treatment of migraine is an appealing therapeutic approach. The ideal drug or procedure would completely abolish migraine attacks, resolving patient's symptoms. To date, this goal has been unattainable, with few drugs being more than 50% effective and with patients still requiring acute treatment.

### Clinical Aspects

The mode of action for most drugs used in migraine prophylaxis is not known and animal models to test the action of drugs used for migraine prophylaxis are not available. A powerful placebo effect exists (de Craen et al., 2000; McQuay et al., 1995), which can lead to a decrease in migraine frequency of up to 70% within 3 months (Migraine-Nimodipine European Study Group, 1989a,b). Couch (1987) analyzed some parallel and double-blind studies on the prevention of migraine. The effect of placebo in the reduction of frequency and severity of migraine attacks ranged from 11 to 36%.

This effect decreased after 3 months. It is still not known whether prophylaxis is more effective using a combination of drugs rather than monotherapy. However, it is preferable to avoid combinations of drugs because, in such instances, adverse effects cannot be attributed to a single substance.

The different spectrum of adverse effects is another problem in patients with migraine pain. Several studies demonstrated that adverse effects are more frequent in patients with migraine than in those taking the same drugs for other reasons. Compliance with migraine prophylaxis is low because patients first experience adverse effects and only later any beneficial effect of the drug. The following section focuses on the most important aspects of migraine prophylaxis and the different compounds are considered in the order of their efficacy and risk–benefit ratio.

### Natural Course

The highest migraine frequency is encountered between the ages of 35 and 45 years. Later on the frequency of migraine becomes lower. Therefore, migraine prophylaxis should be interrupted in regular intervals to see if it is still needed.

### Principles of Therapy

#### Beta-Adrenoceptor Antagonists

Prophylaxis of migraine headache by beta blockers was detected incidentally in patients who were treated for hypertension and who also suffered from migraine headaches. Propranolol (Holroyd et al., 1991; Kangasniemi and Hedman, 1984) and metoprolol (Ljung, 1980; Olsson et al., 1984; Steiner et al., 1988) have both been convincingly shown to have migraine prophylactic activity. This was shown in placebo-controlled trials and in comparative trials against flunarizine, pizotifen, and methysergide (Gawel et al., 1992; Holroyd et al., 1992; Lücking et al., 1988; Ludin, 1989; Shimell et al., 1990; Sorensen and the Danish Migraine Study Group, 1989; Sorensen et al., 1991; Vilming et al., 1985; Wörz et al., 1991). Administration, adverse effects and contraindications for propranolol and metoprolol are listed in Table V. Atenolol (Forssman et al., 1983; Stensrud and Sjaastad, 1980), timolol (Briggs and Millac, 1979; Stellar et al., 1984; Tfelt-Hansen et al., 1984), nadolol (Freitag and Diamond, 1984), and bisoprolol (van de Ven et al., 1997; Wörz et al., 1991) are beta blockers with a possible prophylactic action. No migraine prophylactic activity has been shown for acebutolol, alprenolol,

TABLE V  Migraine Prophylaxis Effective Substances in the Order of Therapeutic Choice, First Choice: Beta-Receptor Blockers

| Substance | Dose/day (mg) | Remarks/Mechanism of action |
|---|---|---|
| Metoprolol*** | Initially 50; later 150–200 | Beta-1-selective |
| Propranolol*** | Initially 40; later 160–200 | Nonselective |

***Treatment recommendations based on more than one well-designed, placebo-controlled trial or a meta-analysis.

TABLE VI  Migraine Prophylaxis, Second Choice: Calcium-Antagonists and Valproic Acid

| Substance | Dose/day (mg) | Remarks/Mechanism of action |
|---|---|---|
| Flunarizine*** | Initially 5; later 10 in man, 5 in women | Long half-life; calcium-channel antagonist and 5-HT antagonist |
| Valproic acid*** | 500–600 | GABAergic-acting drug/ channel modulation |

***Treatment recommendations based on more than one well-designed, placebo-controlled trial or a meta-analysis.

oxprenolol, and pindolol (Olesen *et al.*, 1998). Holroyd performed a metaanalysis for propranolol in the prophylaxis of migraine (Holroyd *et al.*, 1991). The 53 studies included in the metaanalysis reported 2403 patients who were treated with either the beta blocker (propranolol 160 mg) or a reference substance or placebo. On average, propranolol yielded a 44% reduction in migraine activity when daily headache recordings were used to assess treatment outcome, and a 65% reduction of migraine activity when clinical ratings of improvement and global patient reports were used. The dropout rate due to side effects was 5.3%.

### Calcium Channel Blockers

Flunarizine was introduced as a calcium-antagonist for the prophylaxis of migraine on the hypothetical basis of being protective against brain hypoxia. It has, however, a large spectrum of side effects: antidopaminergic (extrapyramidal motor system), antiserotonergic (sedation, weight gain), and antinoradrenergic (depression) properties. Flunarizine is not approved in all countries for migraine prophylaxis despite the fact that it has shown its efficacy in many controlled trials (Diener, 2000c). Verapamil has a very poor evidence for efficacy (Markley *et al.*, 1984; Solomon *et al.*, 1983), while nifedipine (Albers *et al.*, 1989; Scholz *et al.*, 1987), and nimodipine (Migraine-Nimodipine European Study Group, 1989a,b) are ineffective. Verapamil is used in countries were flunarizine is not available. Cyclandelate is another calcium channel blocker which was recently investigated for migraine prophylaxis with different results. In a large recently conducted multicenter trial the substance was not superior to placebo.

### Dihydroergotamine (DHE)

DHE is still widely used as a prophylactic medication in some European countries, but has been poorly studied using modern double-blind or placebo-controlled techniques (Grotemeyer *et al.*, 1989; Neumann *et al.*, 1986). DHE was significantly more effective than placebo in reducing the frequency of

TABLE VII  Migraine Prophylaxis, Third Choice: ASA, NSAIDs, and Serotonin Antagonists

| Substance | Dose/day (mg) | Remarks/Mechanism of action |
|---|---|---|
| Acetylsalicylic acid** | 300 | Inhibitor of prostaglandin synthesis |
| Naproxen** | 3 × 250 | Inhibitor of prostaglandin synthesis |
| Methysergide*** | 2–8 | Not >6 months (retroperitoneal fibrosis) Serotonin-antagonist (5-HT2) |
| Pizotifen** | 3 × 0.5 | Serotonin-antagonist (5-HT2) |
| Alpha-dihydro-ergocryptine* | 1 × 10 | Dopamine-agonist |
| Lisuride* | 3 × 0.025 | Dopamine-agonist |

*Treatment based on personal experience, expert opinion, retrospective, or nonrandomized studies.
**Treatment is based on at least one randomized prospective study of reasonable size.
***Treatment recommendations based on more than one well-designed randomized, placebo-controlled trial or a meta-analysis.

migraine attacks. It should, however, be kept in mind that the prolonged use of DHE can lead to chronic daily headache.

### Serotonin Receptor Antagonists

Methysergide and pizotifen (Table VII) are thought to act as 5-HT antagonists. Both substances are clearly effective, but have a high frequency of adverse effects (Silberstein, 1998). Methysergide can (rarely) lead to retroperitoneal fibrosis and therefore should not be given for longer than 6 months. It raises the mean blood pressure and constricts peripheral vessel (e.g., coronary arteries) and is therefore contraindicated in patients with cardiovascular risk factors. Further adverse events are sedation, dizziness, and weight gain. Its use should now be restricted to patients with cluster headache and to migraine patients who do not respond to other prophylactics. Pizotifen does not lead to retroperitoneal fibrosis, but shows the same profile of adverse events like methysergide and is to be a drug of third choice as well.

### Aspirin and NSAIDs

NSAIDs and ASA are well established in the acute treatment of migraine attacks. The first indications that the regular intake of ASA could act as a migraine prophylaxis came from the Physicians' Health study (Buring et al., 1990). This study investigated the use of 325 mg ASA (every second day) versus placebo in 22071 male physicians for the prevention of myocardial infarction and stroke. ASA resulted in a 20% reduction of migraine attacks in the 661 men with migraine. In a controlled study with crossover design, Grotemeyer et al. (1990) compared 200 mg metoprolol and ASA, 1500 mg per day. A significant reduction in migraine frequency could be observed in 67% of the patients in the metoprolol group compared with only 14% in the ASA group. A recent study compared 300 mg aspirin with 200 mg metoprolol in 270 migraine patients. Aspirin was less effective (responder rate 29.6% vs 45.2%) compared to metoprolol, but had fewer side effects (Diener et al., 2001). Naproxen sodium is established in migraine prophylaxis as well (Table VII) (Behan and Connelly, 1986; Bellavance and Meloche, 1990; Sargent et al., 1985; Ziegler and Ellis, 1985). Studies indicate that naproxen is as effective as pizotifen and can be used for the prophylaxis of menstrual migraine (Sances et al., 1990). Other NSAIDs like ketoprofen, mefenamic acid, tolfenamic acid, and lornoxicam are also effective but the number of controlled trials is much smaller than for naproxen (Diamond et al., 1990; Johnson et al., 1986; Mikkelsen et al., 1986). The clinical experience, however, shows that some patients with migraine do not tolerate NSAIDs for a longer time due to their gastrointestinal adverse effects.

### Other Prophylactic Treatments

Lisuride is approved in some countries for the prophylaxis of migraine (Table VI). Its action is probably mediated through dopamine and 5-HT receptors. Amitriptyline has been used in only a few controlled trials (Couch and Hassanein, 1979; Ziegler et al., 1987, 1993). Its efficacy in migraine prophylaxis is low, though it can be recommended in selected patients with a combination of tension-type headache and infrequent migraine attacks. It is more popular in the United States, where many prophylactic agents used in other countries are not available. Selective serotonin reuptake inhibitors are not effective in migraine prophylaxis. Dopamine agonists like α-DHEC are probably effective (Bussone et al., 1999).

### Antiepileptic Drugs

Recent studies indicate that valproic acid has a prophylactic value in migraine (Hering and Kurzitzky, 1992; Jensen et al., 1994; Kaniecki, 1997; Klapper on behalf of the Divalproex Sodium in Migraine Prophylaxis Study Group, 1997; Mathew et al., 1995). Valproic acid reduces the frequency of migraine attacks, but not their severity and duration. Doses as low as 500 mg/day seem to be as effective as higher doses.

### Other Drugs

High-dose magnesium has shown conflicting results. One placebo-controlled trial in patients from private practice found a positive effect on migraine frequency (Peikert et al., 1996), while another study in patients from headache centers was unable to replicate this result (Pfaffenrath et al., 1996).

### Practical Management

Prophylaxis of migraine should be considered when:

- three or more migraine attacks occur per month,
- attack duration is >48 h,
- there is extreme headache severity,
- inadequate relief is obtained from acute medication,
- migraine attacks occur following a prolonged aura,
- acute migraine treatment results are unsatisfactory or result in unacceptable adverse effects.

### Priority of Drugs Used for Migraine Prophylaxis

Prior to the start of migraine prophylaxis the patient should note the frequency, duration, and severity of migraine attacks in a diary. This diary may help to verify effects of therapy. The dose of any medication used for prophylaxis must be slowly increased. Patients must be informed about possible adverse effects. Migraine prophylaxis should be carried out for 9–12 months until the dose of beta blockers can be slowly decreased. The natural history of migraine should then be assessed for 2–3 months. If a substance is not effective after 3–5 months, another group of substances should be tested.

Migraine prophylaxis should be started with the compound with the smallest probability of adverse effects and highest probability of success. Beta blockers should be tried first. They are most successful if arterial hypertension or an anxiety disorder exist. Beta blockers should not be used, however, in the patient with preexisting orthostatic hypotension, sleep disturbances, or impotence. Contraindications are heart failure, AV block, insulin-dependent diabetes, and asthma. Flunarizine is best used in patients with anorexia or sleeping problems. It should not be used if obesity or depressive disorders are present. Contraindications for flunarizine are extrapyramidal disorders, tremor and depression. Serotonin antagonists are drugs of third choice. Adverse effects like sedation, dizziness, weight gain, and depression are frequent. Contraindications include pregnancy, coronary heart disease, peripheral vascular disease, hypertension, connective tissue disease, obesity, and impaired hepatic or renal function.

### Pitfalls in Migraine Prophylaxis

As with acute migraine treatment frequent mistakes in migraine prophylaxis include the following:

- Incorrect diagnosis: In some patients the diagnosis may be wrong. Tension-type headache does not respond to beta blockers or flunarizine. Migraine prophylaxis is not effective in drug-induced headache.
- Substances with unproven efficacy: Some patients are treated with ineffective substances, e.g., barbiturates or carbamazepine.
- Incorrect usage: Inappropriate priority of prophylactic medication can be another problem. Treatment should start with beta blockers or flunarizine, not with methysergide.
- Excessive dose: Many patients stop their treatment because of intolerable adverse effects at the beginning of therapy. Migraine prophylaxis should start with very low doses (e.g., 25 mg metoprolol, 20 mg propranolol). Migraine patients have more adverse effects than other patients taking the same drugs for other indications.
- Inadequate trial of treatment: sometimes, treatment is stopped either by the physician or the patient after only a few weeks. The minimal time interval for migraine prophylaxis to become effective is 3 months, and may be longer with flunarizine.
- Long-term use: In cases of effective treatment patients have continued migraine prophylaxis for years. Migraine prophylaxis should be stopped after 9–12 months and the patient reassessed.
- Expectations: Patients sometimes expect a "cure" from migraine prophylaxis, but this is impossible to achieve. Migraine prophylaxis only reduces the frequency and severity of migraine attacks.
- Adverse events: Patients should be informed about adverse events. Sometimes, adverse effects can be used to treat another disorder in the same patient: beta blockers are best used in patients with hypertension or an anxiety disorder. Flunarizine can be used in patients with insomnia or anorexia. Aspirin is useful if a patient has several vascular risk factors.

**TABLE VIII**  Special Populations in Migraine Prophylaxis

| Clinical situation | Treatment options |
|---|---|
| Menstrually related headache | NSAIDs with cycle |
| | Ergotamine po nocte |
| | Oestrogen patches |
| | Low-dose sumatriptan |
| Children (<12 years) | Beta blockers |
| | Low-dose DHE |
| Elderly (>65 years) | Beta-blockers in patients with hypertension |

### Prophylaxis of Menstrual Migraine

Prophylaxis is necessary whenever acute migraine attacks during menstruation cannot be treated with success (Table VIII). First, oral contraceptives should be stopped. As second choice beta blockers or flunarizine can be used. Prophylaxis may sometimes be achieved by naproxen, 250 mg twice daily, given 4 days before to 3 days after menstruation. An open pilot study described a prophylactic action of 25 mg sumatriptan tid taken 2–3 days prior to menstruation (Newman et al., 1998).

### Prophylaxis of Pediatric Headache

There are few controlled trials of migraine prophylaxis in children. Drugs include beta blockers, low-dose flunarizine (5 mg/day), low doses of amitriptyline (1 mg/kg/body weight)(Hershey et al., 2000), or Pizotifen DHE, 1–2 mg/day (Lanzi et al., 1996). Behavioral therapy is the treatment of choice in children.

### Prophylaxis in the Elderly

In elderly patients without heart block, diabetes, or asthma the drugs of first choice are beta blockers. Flunarizine may lead to brady-kinesia rigidity, and tremor and 5-HT-anatgonists can induce depression. Valproic acid is well tolerated in the elderly, but dose increases should be performed more slowly than in younger people.

### Behavioral Therapy

Drug treatment of migraine can be supported by non-medical procedures like coping with stress or biofeedback (Andrasik, 1996; Blanchard, 1992; Blanchard and Andrasik, 1985; Blanchard et al., 1983; McGrath et al., 1992; McGrath and Sorbi, 1993; Reid and McGrath, 1996). Biofeedback methods include electromyographic feedback of frontal or temporal muscles and cephalic artery feedback. Changes in frontal skin temperature are monitored and used as a basis for biofeedback.

The evaluation of these methods is difficult due to the marked placebo effect. Relaxation techniques seem to be more effective than other techniques of coping with stress (Sorbi and Tellegen, 1988). Behavioral therapies are of major importance in the treatment of pediatric migraine. (Duckro and Cantwell-Simmons, 1989). Most psychological therapies are extremely time consuming and only a few qualified therapists are available.

### Acupuncture

There are many open and uncontrolled trials as to the efficacy of acupuncture in migraine. The results are mostly presented in a positive to enthusiastic way. In most cases, the designs do not fulfill the criteria for good migraine studies. This holds true for the better studies by Borglum-Jensen et al. (1979) and Lenhard and Waite (1984), too. A recent review and metaanlysis found only a marginal efficacy of acupuncture versus sham acupuncture or control treatment in migraine (Melchart et al., 1999). Acupuncture should be restricted to patients who cannot be treated otherwise and should be performed by a physician.

### Treatment No Longer Recommended

Drug treatment with clonidine, diuretics, hormones, lamotrigine, lithium, neuroleptics, reserpine are no longer recommended. The following procedures are without value: autogenic feedback, chiropractic therapy, injections into the skin of the head or into the neck musculature, homeopathy (Ernst, 1999; Walach et al., 1997; Whitmarsch et al., 1997), magnetic currents, massage, neural therapy, ozone therapy, oxygen inhalation, hyperbaric oxygen.

## ACKNOWLEDGMENTS

The work of P.J.G. described herein has been supported by the Wellcome Trust and the Migraine Trust.

## REFERENCES

Abu-Arafeh, I., and Russell, G. (1995). Paroxysmal vertigo as a migraine equivalent in children: A population-based study. Cephalalgia 15, 22–25.

Abu-Arefeh, J., and Russel, G. (1994). Prevalence and migraine in schoolchildren. Br. Med. J. 309, 765–769.

Adelman, J. U., Mannix, L. K., and Seggern von, R. L. (2000). Rizatriptan tablet versus wafer: Patient preference. Headache 40, 371–372.

Albers, G. W., Simon, L. T., Hamik, A., and Peroutka, S. J. (1989). Nifedipine versus propranolol for the initial prophylaxis of migraine. Headache 29, 214–217.

Andersson, P. G. (1988). Ergotism—the clinical picture. In "Drug-Induced Headache" (H. C. Diener and M. Wilkinson, Eds.), pp. 16–19. Springer, Heidelberg, New-York.

Andrasik, F. (1996). Behavioral mangement of migraine. Biomed. Pharmacother. 50, 52–57.

Bahra, A., Matharu, M., Büchel, C., Frakowiak, R. S. J., and Goadsby, P. J. (2001). Brainstem activation specific to migraine headache. Lancet 357, 1016–1017.

Battistini, S., Stenirri, S., Piatti, M., Gelfi, C., Righetti, P. G., Rocchi, R., Giannini, F., Battistini, N., Guazzi, G. C., Ferrari, M., and Carrera, P. (1999). A new CACNA1A gene mutation in acetazolamide-responsive familial hemiplegic migraine and ataxia. Neurology 53, 38–43.

Behan, P. O., and Connelly, K. (1986). Prophylaxis of migraine: A comparison between naproxen sodium and pizotifen. Headache 26, 237–239.

Bellavance, A. J., and Meloche, J. P. (1990). Comparative study of naproxen sodium, pizotyline and placebo in migraine prophylaxis. Headache 30, 710–715.

Bille, B. (1962). Migraine in school children. Acta Pediatr. Scand. 51, 13640.

Blanchard, E. B. (1992). Psychological treatment of benign headache disorders. J. Consult. Clin. Psychol. 60, 537–551.

Blanchard, E. B., and Andrasik, F. (1985). "Management of Chronic Headaches. A Psychological Approach." Pergamon Press, Elmsford, NY.

Blanchard, E. B., Andrasik, F., Arena, J. G., Neff, D. F., Jurish, S. E., Teders, S. J., Barron, K. D., and Rodichok, L. D. (1983). Nonpharmacologic treatment of chronic headache: Prediction of outcome. Neurology 33, 1596–1603.

Borglum-Jensen, B. L., Mersen, B., and Jensen, S. B. (1979). Effect of acupuncture on headache measured by reduction in mumber of attacks and use of drugs. Scand. J. Dent. Res. 87, 373–380.

Bousser, M.-G. (1999). Migraine, female hormons, and stroke. Cephalalgia 19, 75–79.

Briggs, R. S., and Millac, P. A. (1979). Timolol in migraine prophylaxis. Headache 19, 379–381.

Brown, E. G., Endersby, C. A., Smith, R. N., and Talbot, J. C. C. (1991). The safety and tolerability of sumatriptan: An overview. Eur. Neurol. 31, 339–344.

Buring, J. E., Hebert, P., Romero, J., Kittors, A., Cook, N., Manson, J., Peto, R., and Hennekens, C. (1995). Migraine and subsequent risk of stroke in the physicians' health study. Arch. Neurol. 52, 129–134.

Buring, J. E., Peto, R., and Hennekens, C. H. (1990). Low-dose aspirin for migraine prophylaxis. JAMA 264, 1711–1713.

Bussone, G., Cerbo, R., Martucci, N., Micieli, G., Zanferrari, C., Grazzi, L., Fabbrini, G., Cavallini, A., Granella, F., Ambrosoli, L., Mailland, F., Poli, A., and Manzoni, G. (1999). Alphadihydroergocryptine in the prophylaxis of migraine: A multicenter double-blind study versus flunarizine. Headache 39, 426–431.

Buzzi, G., Sakas, D. E., and Moskowitz, M. A. (1989). Indomethacin and acetylsalicylic acid block neurogenic plasma protein extravasation in rat dura mater. Eur. J. Pharmacol. 165, 251–258.

Buzzi, M. G., Moskowitz, M. A., Peroutka, S. J., and Byun, B. (1991). Further characterization of the putative 5-HT receptor which mediates blockade of neurogenic extravasation in rat dura mater. Br. J. Pharmacol. 103, 1421–1428.

Cabarrocas, X., Zayas, J. M., on behalf of the Almotriptan Oral Study Group (1998a). Efficacy data on oral almotriptan, a novel 5-HT1B/D agonist. Headache 38, 377.

Cabarrocas, X., Zayas, J. M., Suris, M., on behalf of the Almotriptan Oral Study Group (1998b). Equivalent efficacy of oral almotriptan, a new 5-HT1B/D agonist, compared with sumatritpan 100 mg. Headache 38, 377–378.

Chang, C. L., Donaghy, M., Poulter, N., and World Health Organisation Collaborative Study of Cardiovascular Disease and Steroid Hormone Contraception (1999). Migraine and stroke in young women: Case–control study. *Br. Med. J.* **318,** 13–18.

Couch, J. R. (1987). Placebo effect and clinical trials in migraine therapy. *Neuroepidemiology* **6,** 178–185.

Couch, J. R., and Hassanein, R. S. (1979). Amitriptyline in migraine prophylaxis. *Arch. Neurol.* **36,** 695–699.

Cumberbatch, M. J., Hill, R. G., and Hargreaves, R. J. (1997). Rizatriptan has central antinociceptive effects against durally evoked responses. *Eur. J. Pharmacol.* **328,** 37–40.

Dahlöf, C. (1993). Placebo-controlled clinical trials with ergotamine in the acute treatment of migraine. *Cephalalgia* **13,** 166–171.

Dahlöf, C., and Björkman, R. (1993). Diclofenac-K (50 and 100 mg) and placebo in the acute treatment of migraine. *Cephalalgia* **13,** 117–123.

de Craen, A. J. M., Tijssen, J. G. P., de Gans, J., and Kleijnen, J. (2000). Placebo effect in the acute treatment of migraine: Subcutaneous placebos are better than oral placebos. *J. Neurol.* **247,** 183–188.

Diamond, S., Freitag, F. G., Gallagher, R. M., and Feinberg, D. T. (1990). Ketoprofen in the prophylaxis of migraine. *Headache Q.* **1,** 75–77.

Diener, H. C. (2000a). Drug treatment of migraine and other headaches. Karger, Basel.

Diener, H. C. (2000b). Eletriptan—Therapy. *In* "Drug Treatment of Migraine and Other Headaches" (H. C. Diener, Ed.), pp. 184–189. Karger, Basel.

Diener, H. C. (2000c). Flunarizine for migraine prophylaxis. *In* "Drug Treatment of Migraine and Other Headaches" (H. C. Diener, Ed.), pp. 269–278. Karger, Basel.

Diener, H. C. (2000d). Sumatriptan—Therapy. *In* "Drug Treatment of Migraine and Other Headaches" (H. C. Diener, Ed.), pp. 93–109. Karger, Basel.

Diener, H. C., for the ASASUMAMIG Study Group (1999). Efficacy and safety of intravenous acetylsalicylic acid lysinate compared to subcutaneous sumatriptan and parenteral placebo in the acute treatment of migraine. A double-blind, double-dummy, randomized, multicenter, parallel group study. *Cephalalgia* **19,** 581–588.

Diener, H. C., for the RPR100893-201 Migraine Study Group (1995). Substance P antagonist RPR100893-201 is not effective in human migraine attacks. *6th International Headache Research Seminar, Copenhagen,* 245–371.

Diener, H. C., Hartung, E., Chrubasik, J., Evers, S., Schoenen, J., Latta, G., Gendolla, A., for the Study Group (2001). A comparative study of acetysalicylic acid and metoprolol for the prophylactic treatment of migraine. A randomised, controlled, double-blind parallel phase III study. *Cephalalgia* **21,** 140–144.

DiSerio, F., Patin, J., and Friedman, A. (1989). USA trials of dihydroergotamine nasal spray in the acute treatment of migraine headache. *Cephalalgia* **9** (Suppl. 10), 344–345.

Duckro, P. N., and Cantwell-Simmons, E. (1989). A review of studies evaluating biofeedback and relaxation training in the management of pediatric headache. *Headache* **29,** 428–433.

Edvinsson, L., and Goadsby, P. J. (1994). Neuropeptides in migraine and cluster headache. *Cephalalgia* **14,** 320–327.

Edvinsson, L., and Goadsby, P. J. (1995). Neuropeptides in the cerebral circulation: Relevance to headache. *Cephalalgia* **15,** 272–276.

Ernst, E. (1999). Homeopathic prophylaxis of headache and migraine. A systematic review. *J. Pain Symptom Manage.* **18,** 353–357.

Ferro, A., Longmore, J., Hill, R. G., and Brown, M. J. (1995). A comparison of the contractile effects of 5-hydroxytryptamine, sumatriptan and MK-462 on human coronary artery in vitro. *Br. J. Clin. Pharmacol.* **40,** 245–251.

Forssman, B., Lindblad, C. J., and Zbornikova, V. (1983). Atenolol for migraine prophylaxis. *Headache* **23,** 188–190.

Freitag, F. G., and Diamond, S. (1984). Nadolol and placebo comparison study in the prophylactic treatment of migraine. *J. Am. Osteopath. Assoc.* **84,** 343–347.

Gaist, D., Tsiropoulus, I., Sindrup, S. H., Hallas, J., Rasmussen, B. K., and Kragstrup, J. (1998). Inappropriate use of sumatriptan: Population based register and interview study. *Br. J. Med.* **316,** 1352–1353.

Gawel, M. J., Kreeft, J., Nelson, R. F., Simard, D., and Arnott, W. S. (1992). Comparison of the efficacy and safety of flunarizine to propranolol in the prophylaxis of migraine. *Can. J. Neurol. Sci.* **19,** 340–345.

Gladstein, J., Holden, E. W., Peralta, L., and Raven, M. (1993). Diagnoses and symptom patterns in children presenting to a pediatric headache clinics. *Headache* **33,** 497–500.

Goadsby, P. J. (1998). A triptan too far? *J. Neurol. Neurosurg. Psychiatry* **64,** 143–147.

Goadsby, P. J. (2000). The pharmacology of headache. *Prog. Neurobiol.* **62,** 509–525.

Goadsby, P. J., and Edvinsson, L. (1993). The trigeminovascular system and migraine: Studies characterizing cerebravascular and neuropeptide changes seen in humans and cats. *Ann. Neurol.* **33,** 48–56.

Goadsby, P. J., Ferrari, M., and Lipton, R. B. (2002). Migraine: Current understanding and management. *N. Engl. J. Med.* **346,** 257–270.

Goadsby, P. J., Ferrari, M. D., Olesen, J., Stovner, L. J., Senard, J. M., Jackson, N. C., Poole, P. H., for the Eletriptan Steering Committee (2000). Eletriptan in acute migraine: A double-blind, placebo-controlled comparison to sumatritpan. *Neurology* **54,** 156–163.

Goadsby, P. J., and Hoskin, K. J. (1996). Inhibition of trigeminal neurons by intravenous administration of the serotonin (5-HT)1B/D receptor agonist zolmitriptan (311C90): Are brain stem sites therapeutic target in migraine? *Pain* **67,** 355–359.

Goadsby, P. J., Knight, Y. E., and Hoskin, K. L. (1997). Stimulation of the greater occipital nerve increases metabolic activity in the trigeminal nucleus caudalis and cervical dorsal horn of the cat. *Pain* **73,** 23–28.

Goadsby, P. J., and Zagami, A. S. (1991). Stimulation of the superior sagittal sinus increases metabolic activity and blood flow in certain regions of the brainstem and upper cervical spinal cord of the cat. *Brain* **114,** 1001–1011.

Göbel, H., Boswell, D., Winter, P. D. O., and Crisp, A. (1997). A comparison of the efficacy and tolerability of naratriptan and sumatriptan among migraineurs with a history of frequent (50% of attacks) headache recurrence. *Poster, VIIIth IHS Congress, Amsterdam,* 163–165.

Goldstein, D. J., Wang, O., Saper, J. R., Stoltz, R., Silberstein, S. T., and Mathew, N. T. (1997). Ineffectiveness of neurokinin-1 antagonist in acute migraine: Crossover study. *Cephalalgia* **17,** 785–790.

Goldstein, J., Elkind, A., Keywood, C., Klapper, J., and Ryan, R. (1998a). A low dose range-finding study of frovatriptan: A potent selective 5-HT1B/D agonist for the acute treatment of migraine. *Headache* **38,** 382–383.

Goldstein, J., Ryan, R., Jiang, K., Getson, A., Norman, B., Block, G., Lines, C., and the Rizatriptan Protocol 046 Study Group (1998b). Crossover comparison of rizatriptan 5 mg and 10 mg versus sumatriptan 25 and 50 mg in migraine. *Headache* **38,** 737–747.

Grotemeyer, K., Scharafinski, H., Schlake, H., and Husstedt, I. W. (1990). Acetylsalicylic acid vs metoprolol in migraine prophylaxis—A double blind cross-over study. *Headache* **30,** 639–641.

Grotemeyer, K.-H., Schlake, H.-P., and Husstedt, I. W. (1989). Etilefrine pivalate vs. dihydroergotamin and flunarizin in prophylactic treatment of migraine in patients with low blood pressure—A randomized double-blind-study. *Cephalalgia* **9,** 433–434.

Hakkarainen, H., Vapaatalo, H., Gothoni, G., and Paratainen, J. (1979). Tolfenamic acid is as effective as ergotamine during migraine attacks. *Lancet* ii, 326–327.

Havanka-Kanniainen, H. (1989). Treatment of acute migraine attack: Ibuprofen and placebo compared. *Headache* 29, 507–509.

Headache Classification Committee of the International Headache Society (1988). Classification and diagnostic criteria for headache disorders, cranial neuralgias and facial pain. *Cephalalgia* 8, 1–93.

Henry, P., Hiesseprovost, O., Dillenschneider, A., Ganry, H., and Insuaty, J. (1995). Efficacy and tolerance of effervescent aspirin metoclopramide association in the treatment of migraine attack. *Presse Med.* 24, 254–258.

Hering, R., and Kurzitzky, A. (1992). Sodium valproate in the prophylactic treatment of migraine: A double-blind study versus placebo. *Cephalalgia* 12, 81–84.

Hershey, A. D., Powers, S. W., Bentti, A. L., and Grauw de, T. J. (2000). Effectiveness of amitriptyline in the prophylactic management of childhood headaches. *Headache* 40, 539–549.

Holroyd, K. A., Penzien, D. B., and Cordingley, G. E. (1991). Propranolol in the management of recurrent migraine: A meta-analytic review. *Headache* 31, 333–340.

Holroyd, K. A., Penzien, D. B., Rokicki, L. A., and Cordingley, G. E. (1992). Flunarizine vs propranolol: a meta-analysis of clinical trials. *Headache* 32, 256.

Jensen, R., Brinck, T., and Olesen, J. (1994). Sodium valproate has a prophylactic effect in migraine without aura: A triple-blind, placebo-controlled crossover study. *Neurology* 44, 647–651.

Johnson, R. H., Hornabrook, R. W., and Lambie, D. G. (1986). Comparison of mefenamic acid and propranolol with placebo in migraine prophylaxis. *Acta Neurol. Scand.* 73, 490–492.

Joutel, A., Bousser, M., Biousese, V., Labauge, P., Chabriat, H., Nibbio, A., Maciazek, J., Meyer, B., Bach, M., Weissenbach, J., Lathrop, G. M., and Tournier-Lasserve, E. (1993). A gene for familial hemiplegic migraine maps to chromosome 19. *Nat. Genet.* 5, 40–45.

Kangasniemi, P., and Hedman, C. (1984). Metoprolol and propranolol in the prophylactic treatment of classical and common migraine. A double-blind study. *Cephalalgia* 4, 91–96.

Kaniecki, R. G. (1997). A comparision of divalproex with propranolol and placebo for the prophylaxis of migraine without aura. *Arch. Neurol.* 54, 1141–1145.

Karabetsos, A., Karachalios, G., Bourlinou, P., Reppa, A., Koutri, R., and Fotiadou, A. (1997). Ketoprofen versus paracetamol in the treatment of acute migraine. *Headache* 37, 12–14.

Karachalios, G. N., Fotiadou, A., Chrisikos, N., Karabetsos, A., and Kehagioglou, K. (1992). Treatment of acute migraine attack with diclofenac sodium: A double-blind study. *Headache* 32, 98–100.

Kaube, H., Hoskin, K. L., and Goadsby, P. J. (1993). Intravenous acetylsalicylic acid inhibits central trigeminal neurons in the dorsal horn of the upper cervical spinal cord in the cat. *Headache* 33, 541–544.

Kaube, H., May, A., Diener, H. C., and Pfaffenrath, V. (1994). Sumatriptan misuse in daily chronic headache. *Br. Med. J.* 308, 1573.

Kellstein, D. E., Lipton, R. B., Geetha, R., Koronkiewicz, K., Evans, F. T., Stewart, W. F., Wilkes, K., Furey, S. A., Subramanian, T., and Cooper, S. A. (2000). Evaluation of a novel solubilized formulation of ibuprofen in the treatment of migraine headache: A randomized, double-blind, placebo-controlled, dose-ranging study. *Cephalalgia* 20, 233–243.

Klapper, J., on behalf of the Divalproex Sodium in Migraine Prophylaxis Study Group (1997). Divalproex sodium in migraine prophylaxis: A dose-controlled study. *Cephalalgia* 17, 103–108.

Klassen, A., Elkind, A., Asharnejad, M., Webster, C., Laurenza, A., on behalf of the Naratriptan S2WA3001 Study Group (1997). Naratriptan is effective and well tolerated in the acute treatment of migraine. Results of a double-blind, placebo-controlled, parallel-group study. *Headache* 37, 630–645.

Kloster, R., Nestvold, K., and Vilming, S. T. (1992). A double-blind study of ibuprofen versus placebo in the tratment of acute migraine attacks. *Cephalalgia* 12, 169–171.

Kramer, M. S., Matzura-Wolfe, D., Polis, A., Getson, A., Amaraneni, P. G., Solbach, M. P., McHugh, W., Feighner, J., Silberstein, S., Reines, S. A., and the Rizatriptan Multiple Attack Study Group (1998). A placebo-controlled crossover study of rizatriptan in the treatment of multiple migraine attacks. *Neurology* 51, 773–781.

Krause, K., and Bleicher, M. A. (1985). Dihydroergotamine nasal spray in the treatment of migraine attacks. *Cephalalgia* 5, 138–139.

Krymchantowski, A. V., Adriano, M., and Fernandes, D. (1999). Tolfenamic acid decreases migraine recurrence when used with sumatriptan. *Cephalalgia* 19, 186–187.

Lance, J. W., and Goadsby, P. J. (1998). "Mechansims and Management of Headache." Butterworth-Heinemann, London.

Lanzi, G., Balottin, U., Zambrino, C. A., Cernibori, A., Del Bene, E., Gallai, V., Guidetti, V., and Sorge, F. (1996). Guidelines and recommendations for the treatment of migraine in pediatric and adolescent patients. *Funct. Neurol.* 11, 269–275.

Larsen, B. H., Christiansen, L. V., Andersen, B., and Olesen, J. (1990). Randomized double-blind comparison of tolfenamic acid and paracetamol in migraine. *Acta Neurol. Scand.* 81, 464–467.

Lenhard, L., and Waite, P. M. E. (1984). Acupuncture in the prophylactic treatment of migraine headache. *NZ. Med. J.* 96, 663–666.

Limmroth, V., Kazarawa, S., Fritsche, G., and Diener, H. C. (1999a). Headache after frequent use of new 5-HT agonists zolmitriptan and naratriptan. *Lancet* 353, 378.

Limmroth, V., May, A., and Diener, H.-C. (1999b). Lysine-acetylsalicylic acid in acute migraine attacks. *Eur. Neurol.* 41, 88–93.

Lindskog, U., Ödkvist, L., Noaksson, L., and Wallquist, J. (1999). Benign paroxysmal vertigo in childhood: a long-term follow-up. *Headache* 39, 33–37.

Lipton, R. B., Stewart, W. F., Ryan, R. E., Saper, J., Silberstein, S., and Sheftell, F. (1998). Efficacy and safety of acetaminophen, aspirin, and caffeine in alleviating migraine headache pain—Three double-blind, randomized, placebo-controlled trials. *Arch. Neurol.* 55, 210–217.

Lipton, R. B., Stewart, W. F., Stone, A. M., Lainez, M. J. A., and Sawyer, J. P. C. (2000). Stratified care vs step care strategies for migraine. The Disablility in Strategies of Care (DISC) Study: a randomized trial. *JAMA* 284, 2599–2605.

Ljung, O. (1980). Treatment of migraine with metoprolol. *N. Engl. J. Med.* 303, 664–668.

Lücking, C. H., Oestreich, W., Schmidt, R., and Soyka, D. (1988). Flunarizine vs. propranolol in the prophylaxis of migraine: Two double-blind comparative studies in more than 400 patients. *Cephalalgia* 8, 21–26.

Ludin, H.-P. (1989). Flunarizine and propranolol in the treatment of migraine. *Headache* 29, 218–223.

MacGregor, E. A., Wilkinson, M., and Bancroft, K. (1993). Domperidone plus paracetamol in the treatment of migraine. *Cephalalgia* 13, 124–127.

Marcus, D. A., Scharff, L., and Turk, D. (1999). Longitudinal prospective study of headache during pregnancy and postpartum. *Headache* 39, 625–632.

Markley, H. G., Cheronis, C. D., and Piepho, R. W. (1984). Verapamil in prophylactic therapy of migraine. *Neurology* 34, 973–976.

Martin, G. R. (1997). Pre-clinical pharmacology of zolmitriptan (Zomig (TM); formerly 311C90), a centrally and peripherally acting 5HT1B/1D agonist for migraine. *Cephalalgia* 17, 4–14.

Mathew, N. T., Saper, J. R., Silberstein, S. D., Rankin, L., Markley, H. G., Solomon, S., Rapoport, A. M., Silber, C. J., and Deaton, R. L. (1995). Migraine prophylaxis with divalproex. *Arch. Neurol.* 52, 281–286.

May, A., Gijsman, H. J., Wallnöfer, A., Jones, R., Diener, H. C., and Ferrari, M. D. (1996). Endothelin antagonist bosentan blocks neurogenic inflammation, but is not effective in aborting migraine attacks. *Pain* 67, 375–378.

McGrath, P. J., Humphreys, P., Keene, D., Goodman, J. T., Lascwelles, M. A., Cunningham, S. J., and Forestone, P. (1992). The efficacy and efficiency of self-administered treatment for adolescent migraine. *Pain* 49, 321–324.

McGrath, P. J., and Sorbi, M. J. (1993). Psychological treatment. *In* "The Headaches" (J. Olesen, P. Tfelt-Hansen, and K. M. A. Welch, Eds.), pp. 289–294. Raven, New York.

McQuay, H., Carroll, D., and Moore, A. (1995). Variation in the placebo effect in randomised controlled trials of analgesics: All is as blind as it seems. *Pain* 64, 331–335.

Melchart, D., Linde, K., Fischer, P., White, A., Allais, G., Vickers, A., and Berman, B. (1999). Acupuncture for recurrent headaches: A systematic review of randomized controlled trials. *Cephalalgia* 19, 779–786.

Merikangas, K. R., Fenton, B. T., Cheng, S. H., Stolar, M. J., and Risch, N. (1997). Association between migraine and stroke in a large-scale epidemiological study of the United States. *Arch. Neurol.* 54, 362–368.

Migraine-Nimodipine European Study Group (1989a). European multicenter trial of nimodipine in the prophylaxis of classic migraine (migraine with aura). *Headache* 29, 639–642.

Migraine-Nimodipine European Study Group (1989b). European multicenter trial of nimodipine in the prophylaxis of common migraine (migraine without aura). *Headache* 29, 633–638.

Mikkelsen, B., Pedersen, K. K., and Christiansen, L. V. (1986). Prophylactic treatment of migraine with tolfenamic acid, propranolol and placebo. *Acta Neurol. Scand.* 73, 423–427.

Moskowitz, M. A. (1990). Basic mechanisms in vascular headache. *Neurol. Clin.* 8, 801–815.

Moskowitz, M. A. (1992). Neurogenic versus vascular mechanisms of sumatriptan and ergot alkaloids in migraine. *TiPS* 13, 307–311.

Myllylä, V. V., Havanka, H., Herrala, L., Kangasniemi, P., Rautakorpi, I., Turkka, J., Vapaatalo, H., and Eskerod, O. (1998). Tolfenamic acid rapid release versus sumatriptan in the acute treatment of migraine: Comparable effect in a double-blind, randomized, controlled, prallel-group study. *Headache* 38, 201–207.

Nappi, G., and Johnson, F. N. (2000). The clinical efficacy of zolmitriptan. *Rev. Contemp. Pharmacother.* 11, 99–118.

Neumann, M., Demarez, J. P., Harmey, J. L., Le Bastard, B., and Cauquil, J. (1986). Prevention of migraine attacks through the use of dihydroergotamine. *Int. J. Clin. Pharm. Res.* 6, 11–13.

Newman, L. C., Lipton, R. B., Lay, C. L., and Solomon, S. (1998). A pilot study of oral sumatriptan as intermittent prophylaxis of menstruation-related migraine. *Neurology* 51, 307–309.

O'Quinn, S., Davis, R., Guttermann, D., Pait, G., and Fox, A. (1999). Prospective large-scale study of the tolerability of subcutaneous sumatriptan injection for the acute treatment of migraine. *Cephalalgia* 19, 223–231.

Olesen, J., Tfelt-Hansen, P., and Welch, K. M. A. (1998). The headaches. Raven Press, New York.

Olsson, J. E., Behring, H. C., Forssman, B., Hedman, C., Hedman, G., Johansson, F., Kinnman, J., Palhagen, S. E., Samuelsson, M., and Strandman, E. (1984). Metoprolol and propranolol in migraine prophylaxis: A double-blind multicenter study. *Acta Neurol. Scand.* 70, 160–168.

Ophoff, R. A., Terwindt, G. M., Vergouwe, M. N., van Eijk, R., Oefner, P. J., Hoffman, S. M. G., Lamerdin, J. E., Mohrenweiser, H. W., Bulman, D. E., Ferrari, M., Haan, J., Lindhout, D., van Ommen, G. B., Hofker, M. H., Ferrari, M. D., and Frants, R. R. (1997). Familial hemiplegic migraine and episodic ataxia type-2 are caused by mutations in the Ca2+ channel gene CACNL1A4. *Cell* 87, 543–552.

Paiva, T., Esperanca, P., Marcelino, L., and Assis, G. (1985). A double-blind trial with dihydroergotamine spray in migraine crisis. *Cephalalgia* 5, 140–141.

Peikert, A., Wilimzig, C., and Köhne-Volland, R. (1996). Prophylaxis of migraine with oral magnesium: Results from a prospective, multi-center, placebo-controlled and double-blind randomized study. *Cephalalgia* 16, 257–263.

Pfaffenrath, V., Wessely, P., Meyer, C., Isler, H. R., Evers, S., Grotemeyer, K. H., Taneri, Z., Soyka, D., Göbel, H., and Fischer, M. (1996). Magnesium in the prophylaxis of migraine—a double-blind, placebo-controlled study. *Cephalalgia* 16, 436–440.

Quality Standards Subcommittee of the American Academy of Neurology (1994). Practice parameter: the utility of neuroimaging in the evaluation of headache in patients with normal neurologic examinations. *Neurology* 44, 1353–1354.

Rasmussen, B. K. (1993). Migraine and tension-type headache in a general population: Precipitating factors, female hormones, sleep pattern and relation to lifestyle. *Pain* 53, 65–72.

Reid, G. J., and McGrath, P. J. (1996). Psychological treatments for migraine. *Biomed. Pharmacother.* 50, 58–63.

Rohr, J., and Dufresne, J. (1985). Dihydroergotamine nasal spray for the treatment of migraine attacks: A comparative double-blind crossover study with placebo. *Cephalalgia* 5, 142–143.

Roon, K. I., Olesen, J., Diener, H. C., Ellis, P., Hettiarachchi, J., Poole, P. H., Christianssen, I., Kleinermans, D., Kok, J. G., and Ferrari, M. D. (2000). No acute antimigraine efficacy of CP-122,288, a highly potent inhibitor of neurogenic inflammation: Results of two randomized, double-blind, placebo-controlled clinical trials. *Ann Neurol* 47, 238–241.

Russel, M. B., Rasmussen, B. K., Thornvaldesen, P., and Olesen, J. (1995). Praevalence and sex ratio of the subtypes of migraine. *Int. J. Epidemiol.* 24, 612–618.

Sances, G., Martignoni, E., Fioroni, L., Blandini, F., Facchinetti, F., and Nappi, G. (1990). Naproxen sodium in menstrual migraine prophylaxis: A double-blind placebo controlled study. *Headache* 30, 705–709.

Sargent, J., Solbach, P., Damasio, H., Baumel, B., Corbett, J., Eisner, L., Jessen, B., Kudrow, L., Mathew, N., Medina, J., Saper, J., Vijayan, N., Watson, C., and Alger, J. (1985). A comparison of naproxen sodium to propranolol hydrochloride and a placebo control for the prophylaxis of migraine headache. *Headache* 25, 320–324.

Schoenen, J., Jacquy, J., and Lenaerts, M. (1997). High-dose power riboflavin as a novel prophylactic antimigraine therapy: Results from a double-blind, randomized, placebo-controlled trial. *Cephalalgia* 17, 244.

Scholz, E., Gerber, W. D., Diener, H. C., Langohr, H. D., and Reinecke, M. (1987). Dihydroergotamine versus flunarizine versus nifedipine versus metoprolol versus propranolol in migraine prophylaxis. A comparative study based on time series analysis. *In* "Advances in Headache Research" (F. Clifford Rose, Ed.), pp. 135–145. John Libey, London.

Senter, H. J., and Lieberman, A. N. (1976). Cerebral manifestations of ergotism. Report of a case and review of the literature. *Stroke* 7, 88–92.

Shimell, C. J., Fritz, V. U., and Levien, S. L. (1990). A comparative trial of flunarizine and propranolol in the prevention of migraine. *S. Afr. Med. J.* 77, 75–78.

Silberstein, S. D. (1998). Methysergide. *Cephalalgia* 18, 421–435.

Silberstein, S. D., and Lipton, R. B. (1993). Epidemiology of migraine. *Neuroepidemiology* **12**, 179–194.

Solomon, C. G. D., Steel, M. J. G., and Spaccavento, C. L. J. (1983). Verapamil prophylaxis of migraine. *JAMA* **250**, 2500–2502.

Sorbi, M., and Tellegen, B. (1988). Stress coping in migraine. *Soc. Sci. Med.* **26**, 351–358.

Sorensen, P. S., and the Danish Migraine Study Group (1989). Prophylactic effect of flunarizine versus metoprolol in migraine. *Cephalalgia* **9**, 355–356.

Sorensen, P. S., Larsen, B. H., Rasmussen, M. J. K., Kinge, E., Iversen, H., Alslev, T., Nohr, P., Pedersen, K. K., Schroder, P., Lademann, A., and Olesen, J. (1991). Flunarizine versus metoprolol in migraine prophylaxis: A double-blind, randomized parallel group study of efficacy and tolerability. *Headache* **31**, 650–657.

Stark, S., Spierings, E. L. H., McNiel, S., Putnam, G. P., Bolden-Watson, C. P., and O'Quinn, S. (2000). Naratriptan efficacy in migraineurs who respond poorly to oral sumatriptan. *Headache* **40**, 513–520.

Steiner, T. J., Joseph, R., Hedman, C., and Clifford Rose, F. (1988). Metoprolol in the prophylaxis of migraine: Parallel group comparison with placebo and dose-ranging follow-up. *Headache* **28**, 15–23.

Stellar, S., Ahrens, S., Meibohm, A. R., and Reines, A. S. (1984). Migraine prevention with timolol. A double-blind crossover study. *JAMA* **252**, 2576–2580.

Stensrud, P., and Sjaastad, O. (1980). Comparative trial of Tenormin (atenolol) and Inderal (propranolol) in migraine. *Headache* **20**, 204–207.

Stewart, W., Lipton, R., Whyte, J., Dowson, A., Kolodner, K., Liberman, J., and Sawyer, J. (1999). An international study to assess reliability of the migraine disability assessment (MIDAS) score. *Neurology* **53**, 988–994.

Stewart, W. F., Lipton, R. B., Celentano, D. D., and Reed, M. L. (1992). Prevalence of migraine headache in the United States—Relation to age, race, income, and other sociodemographic factors. *JAMA* **267**, 64–69.

Terwindt, G. M., Ophoff, R. A., Haan, J., Frants, R. R., and Ferrari, M. D. (1996). Familial hemiplegic migraine: A clinical comparison of families linked and unlinked to chromosome 19. *Cephalalgia* **16**, 153–155.

Tfelt-Hansen, P. (1998). Efficacy and adverse events of subcutaneous, oral, and intranasal sumatriptan used for migraine treatment: A systematic review based on number needed to treat. *Cephalalgia* **18**.

Tfelt-Hansen, P., Henry, P., Mulder, L. J., Schaeldewaert, R. G., Schoenen, J., and Chazot, G. (1995). The effectiveness of combined oral lysine acetylsalicylate and metoclopramide compared with oral sumatriptan for migraine. *Lancet* **346**, 923–926.

Tfelt-Hansen, P., and Olesen, J. (1980). Paracetamol (acetaminophen) versus acetylsalicylic acid in migraine. *Eur. Neurol.* **4**, 107–111.

Tfelt-Hansen, P., Saxena, P. R., Dahlöf, C., Pascual, J., Lainez, M., Henry, P., Diener, H. C., Schoenen, J., Ferrari, M. D., and Goadsby, P. J. (2000). Ergotamine in the acute treatment of migraine. A review and European consensus. *Brain* **123**, 9–18.

Tfelt-Hansen, P., Standnes, B., Kangasniemi, P., Hakkarainen, H., and Olesen, J. (1984). Timolol vs. propranolol vs. placebo in common migraine prophylaxis: A double-blind multicenter trial. *Acta. Neurol. Scand.* **69**, 1–8.

Tfelt-Hansen, P., Teall, J., Rodriguez, F., Giacovazzo, M., Paz, J., Malbecq, W., Block, G. A., Reines, S. A., Visser, W. H., on behalf of the Rizatriptan 030 Study Group (1998). Oral rizatriptan versus oral sumatriptan: A direct comparative study in the acute treatment of migraine. *Headache* **38**, 748–755.

The Diclofenac-K/Sumatriptan Migraine Study Group (1999). Acute treatment of migraine attacks: Efficacy and safety of a nonsteroidal antiinflammatory drug, diclofenac-potassium, in comparison to oral sumatriptan and placebo. *Cephalalgia* **19**, 232–240.

The Oral Sumatriptan and Aspirin plus Metoclopramide Comparative Study Group (1992). A study to compare oral sumatriptan with oral aspirin plus oral metoclopramide in the acute treatment of migraine. *Eur. Neurol.* **32**, 177–184.

Tokola, R. A., Kangasniemi, P., Neuvonen, P. J., and Tokola, O. (1984). Tolfenamic acid, metoclopramide, caffeine and their combinations in the treatment of migraine attacks. *Cephalalgia* **4**, 253–263.

Tulunay, F. C., Karan, O., Aydin, N., Culcuoglu, A., and Guvener, A. (1987). Dihydroergotamine nasal spray during migraine attacks—A double-blind crossover study with placebo. *Cephalalgia* **7**, 131–133.

van de Ven, L. L. M., Franke, C. L., Koehler, P. J., on Behalf of the Investigators (1997). Prophylactic treatment of migraine with bisoprolol: A placebo-controlled study. *Cephalalgia* **17**, 596–599.

Vilming, S., Standnes, B., and Hedman, C. (1985). Metoprolol and pizotifen in the prophylactic treatment of classical and common migraine. A double-blind investigation. *Cephalalgia* **5**, 17–23.

Visser, W. H., Jaspers, N., de Vriend, R. H. M., and Ferrari, M. D. (1996). Risk factors for headache recurrence after sumatriptan: a study in 366 migraine patients. *Cephalalgia* **16**, 264–269.

Walach, H., Haeusler, W., Lowes, T., Mussbach, D., Schamell, U., Springer, W., Stritzl, G., Gaus, W., and Haag, G. (1997). Classical homeopathic treatment of chronic headaches. *Cephalalgia* **17**, 119–126.

Weiller, C., May, A., Limmroth, V., Jüptner, M., Kaube, H., van Schayck, R., Coenen, H. H., and Diener, H. C. (1995). Brain stem activation in spontaneous human migraine attacks. *Nat. Med.* **1**, 658–660.

Whitmarsch, T. E., Coleston-Shields, D. M., and Steiner, T. H. (1997). Double-blind randomized placebo-controlled study of homoeopathic prophylaxis of migraine. *Cephalalgia* **17**, 600–604.

Winner, P., Ricalde, O., Le Force, B., Saper, J., and Margui, B. (1996). A double-blind study of subcutaneous dihydroergotamine vs subcutaneous sumatriptan in the treatment of acute migraine. *Arch. Neurol.* **53**, 180–184.

Wörz, R., Reinhardt-Benmalek, B., Grotemeyer, K. H., and Foh, M. (1991). Bisoprolol and metoprolol in the prophylactic treatment of migraine with and without aura—A randomized double-blind cross-over multicenter study. *Cephalalgia* **11** (Suppl 11), 152–153.

Ziegler, D. K., and Ellis, D. J. (1985). Naproxen in prophylaxis of migraine. *Arch. Neurol.* **42**, 582–584.

Ziegler, D. K., Hurwitz, A., and Hassanein, R. S. (1987). Migraine prophylaxis. A comparison of propranolol and amitriptyline. *Arch. Neurol.* **44**, 486–489.

Ziegler, D. K., Hurwitz, A., Preskorn, S., Hassanein, R., and Seim, J. (1993). Propranolol and amitriptyline in prophylaxis of migraine. *Arch. Neurol.* **50**, 825–830.

CHAPTER 2

# Cluster Headache and Paroxysmal Hemicrania

Peter J. Goadsby and Thomas Brandt

## INTRODUCTION

Cluster headache and paroxysmal hemicrania are forms of primary neurovascular headache which feature devastating pain and share some form of cycling, circadian or cirannual, along with prominent activation of the cranial parasympathetic autonomic innervation. From a physiological viewpoint they are fascinating (May and Goadsby, 1999), but for the clinician they represent a considerable opportunity for successful therapeutic intervention. Readers interested in greater detail are referred to monographs by Kudrow (1980), Lance and Goadsby (1998), Olesen and Goadsby (1999), and Sjaastad (1992). It has recently been proposed that the primary headaches with prominent cranial parasympathetic features be regarded together pathophysiologically as the trigeminal-autonomic cephalalgias (Goadsby and Lipton, 1997).

## CLUSTER HEADACHE

### Clinical Aspects

Cluster headache (CH) is a relatively rare very severe episodic primary headache that has been recognized for over 300 years. Patients are often smokers although there is no causal association known. The International Headache Society diagnostic criteria (Headache Classification Committee of The International Headache Society, 1988) define the condition in terms of its episodicity and clinical features (Table I). Cluster headache is characterized by intermittent, repeated, brief attacks of very severe unilateral pain that is usually reported to occur over or behind one eye. There are usually associated autonomic features such as lacrima-tion, nasal congestion, conjunctival injection, and either a full or partial Horner's syndrome. By these criteria each attack may last from 15 min to 3 h and the frequency of attacks varies from one every other day to eight per day. Although not currently part of the formal classification most patients with cluster headache are agitated, moving about or pacing around, in stark contrast to migraine sufferers who do their best to be still. Most patients with cluster headache have them in a bout or *cluster* that may last from 6 weeks to several months and are thus designated episodic cluster headache. Some 10–15% of patients have no substantial breaks and are classified as chronic cluster headache. These can be the most challenging of cases, being frequently resistant to simpler treatments.

It is important to differentiate cluster headache from similar conditions, which most often consist of shorter more frequent attacks, and to be aware of the rare but recognized causes of secondary cluster headache (Table II) as they guide logical investigation.

### Natural Course

Cluster headache is thought to have a prevalence of 0.1% of the population. The age of onset is generally in the third to fourth decade, although patients as young as 5 clearly have the disorder and it may manifest late in life. Cluster headache tends to affect males more than females in a ratio of 3–4 to 1. A family history is uncommon although kindreds with what seems an autosomal dominant pattern of inheritance have been noted (Russell *et al.*, 1996). The condition seems to affect all races. Patients with the episodic form are completely well and untriggerable between bouts. For those in a bout and those with the chronic form, alcohol or nitrates will reliably trigger most patients in 30–60 min. It is generally accepted that treatment does not alter the course of the condition, although data from long-term follow-up is lacking. Remissions periods of years have been observed, and various estimates for the transition from episodic to chronic cluster headache are given, with about 10% being accepted (Kudrow, 1987).

## Principles of Therapy

Cluster headache has been regarded *wrongly* as a vascular headache and has been attributed to an inflammatory process in the cavernous sinus and tributary veins or to a pericarotid pathological process. Given the circadian rhythmicity of attacks and cycling of bouts a

**TABLE I**  Diagnostic Features of Cluster Headache Modified from the International Headache Society (Headache Classification Committee of The International Headache Society, 1988)

Cluster headache has two key forms—
1. **Episodic:** Attacks occur in periods lasting 7 days to 1 year separated by pain-free periods lasting 1 month.
2. **Chronic:** Attacks occur for more than 1 year without remission or with remissions lasting less than 1 month.
— Headache attacks must have each of:
  • severe unilateral orbital, supraorbital, temporal pain lasting 15 min to 3 h;
  • frequency: one every second day to 8 per day;
  • Associated with one of:
    — lacrimation
    — nasal congestion
    — rhinorrhea
    — forehead/facial sweating
    — miosis
    — ptosis
    — eyelid oedema
    — conjunctival injection
— For each of 1 and 2 above no plausible secondary cause is demonstrable (see Table II)

purely vasogenic cause cannot explain the entire picture of cluster headache (Lance and Goadsby, 1998). Indeed MRI imaging reveals no pathological change in the cavernous sinus during active bouts, and studies demonstrating changes provide no differentiation of cluster headache from migraine or paroxysmal hemicrania.

A recent study using positron emission tomography (PET) in acute cluster headache provides a logical basis with which to understand the disorder in terms of brain dysfunction. The PET study observed areas of activation that fell into two broad groups: areas known to be involved in pain processing or responses to pain, such as cingulate and insula cortex and contralateral thalamus, and areas activated specifically in cluster headache but not in other causes of head pain, notably the posterior hypothalamic gray (May et al., 1998). Taken with imaging findings in migraine (see Chapter 1), these data demonstrate that primary headache syndromes share some processing pathways as might be expected, but may be distinguished on a functional neuroanatomical basis by areas of activation specific to the clinical syndrome (Bahra et al., 2001). A most exciting development of this finding has been a parallel study using voxel-based morphometry that has shown a structural change in the posterior hypothalamic gray matter in the same place as the function change seen on PET. It seems likely that Cluster headache is a disorder of the brain with secondary (neurally driven) vascular change.

The biological basis for cluster headache provides an understanding of the medical treatments. Notably, since trigeminovascular activation is seen in both migraine and cluster headache, one might expect that the acute therapies would have overlap, but given that the brain structures involved are so very different, it might be expected that preventative strategies would have differences. It can then readily be understood that the triptans, particularly sumatriptan by injection, would be expected to be useful in acute cluster headache; and it is clearly beneficial, while many key preventatives in

**TABLE II**  Differential Diagnosis of Secondary Cluster-like Headache

| Similar secondary headaches | Secondary cluster headaches[a] | Secondary paroxysmal hemicrania[a] |
| --- | --- | --- |
| Tolosa–Hunt syndrome | Meningioma of the lesser wing of sphenoid | Tolosa–Hunt-like syndrome |
| Maxillary sinusitis | Vertebral artery dissection or aneurysm | Pituitary microadenoma |
| Temporal arteritis | Unilateral cervical cord infarction | Maxillary cyst |
| Raeder's paratrigeminal neuralgia | High cervical meningioma | Cerebral metastases from a parotid epidermoid carcinoma |
| | Head or neck injury | |
| | Pituitary adenoma | |
| | Occipital lobe AVM | |
| | Facial trauma | |
| | Orbitosphenoidal aspergillosis | |
| | Pseudoaneurysm of intracavernous carotid artery | |
| | Lateral medullary infarction | |

[a]Each of these has been reported in the literature.

cluster headache, notably verapamil and lithium, seem not to be useful in migraine.

## Practical Management

The management of CH includes offering advice on general measures to patients, treatment with abortive and preventative agents, and rarely surgery. Contrary to the treatment of migraine, emphasis is often placed on preventing attacks, rather than on acute therapy. This seems appropriate but should not result in under-management of the acute attacks.

### General Measures and Patient Education (C)

Patients should be advised to abstain from taking alcohol during a cluster bout. Otherwise, dietary factors seem to have little importance in CH. Anecdotal evidence suggests that patients should be cautioned against prolonged exposure to volatile substances, such as solvents and oil-based paints. Patients can be advised to avoid afternoon naps, as sleeping sometimes precipitates attacks.

### Abortive Agents

The pain of CH builds up very rapidly to such an excruciating intensity that most oral agents are too slowly absorbed to deal with the pain within a reasonable period of time. The most efficacious abortive agents are those that involve parentral or pulmonary administration.

**Triptans.** Subcutaneous sumatriptan, 6 mg, is the drug of choice in abortive treatment of a cluster attack (***). It has a rapid effect and high response rate (Ekbom and The Sumatriptan Cluster Headache Study Group, 1991). In CH, unlike in migraine, subcutaneous sumatriptan can be prescribed at a frequency of twice daily, on a long-term basis if necessary, without risk of tachyphylaxis or rebound headache (Goadsby, 1994), although oral Sumatriptan used frequently may have some risk for rebound headache. In this era of a cost-conscious medical practice, some practitioners are reluctant to prescribe this relatively expensive drug. We feel that, given the devastating morbidity associated with this excruciating pain syndrome, it is unethical to withhold treatment for cost reasons. The nasal formulation of sumatriptan (20 mg) works in some patients, has been demonstrated to be superior to placebo (van Vliet et al., 2001), and offers a useful choice (**). There is no controlled evidence to support the use of oral sumatriptan in CH. Sumatriptan, 100 mg, three times daily taken prior to an anticipated onset of an attack or at regular times does not prevent the attack (Monstad et al., 1995) and therefore should not be used for CH prophylaxis. Zolmitriptan provides meaningful pain relief after oral administration of 5 mg in the majority of patients with episodic CH, but not in chronic CH (Bahra et al., 2000) (**). However, its efficacy is modest and does not approach the efficacy or speed of subcutaneous sumatriptan or oxygen.

**Oxygen (**).** Inhalation of 100% oxygen, at 7–12 L/min, is rapidly effective in relieving pain in the majority of sufferers (Fogan, 1985). It should be inhaled continuously for 15–20 min via a non-rebreathing facial mask. Patients need to be informed that they should cover any apertures on the face mask.

**Topical Lignocaine (**).** Lignocaine solution, 20–60 mg, given as nasal drops (4–6% lidocaine solution) or a spray deep in the nostril on the painful side, results in mild to moderate relief in some patients, though only a few patients obtain complete pain relief (Costa et al., 2000). Therefore, intranasal lignocaine serves as a useful adjunct to other abortive treatments but is rarely adequate on its own.

**Approaches That Are not Recommended.** Oral or rectal ergotamine is generally too slow in onset to provide meaningful relief in a timely manner. Opiates, nonsteroidal anti-inflammatory drugs, and combination analgesics have no role in the acute management of CH.

### Preventative Treatments

The aim of preventative therapy is to produce a rapid suppression of attacks and to maintain that remission with minimal side effects until the cluster bout is over or for a longer period in patients with chronic cluster headache. Preventative treatments can be divided into short-term preventatives: those suitable for rapidly controlling the attack frequency but not for prolonged use and long-term therapies that are required for prolonged medical management of cluster headache.

**Short-term Prevention.** Patients with either short bouts, perhaps in weeks, or in whom one wishes to quickly control the attack frequency, can benefit from short-term prevention. These medicines are distinguished by the fact that they cannot be used over the long term and thus may require replacement by long-term agents in many patients.

*Corticosteroids.* Corticosteroids are highly efficacious and the most rapid-acting of the preventative agents. However, caution has to be exercised in their use because of the potential for serious side effects, such as osteonecrosis of the femoral head. Treatment should be limited to a short intensive course of 2–3 weeks in tapering doses. We start patients on oral prednisolone 1 mg/kg, to a maximum of 60 mg od, for 5 days and thereafter decrease the dose by 10 mg every 3 days. Unfortunately, relapse almost invariably occurs as the dose is tapered. For this reason, steroids are used as an initial therapy in conjunction with preventatives until the latter are effective.

*Methysergide.* Methysergide is a potent prophylactic agent for the treatment of CH. It is an ideal choice in patients with short cluster bouts lasting less than 4–5 months. Doses up to 12 mg daily can be used if tolerated. Patients are started on 1 mg od and the daily dose then increased by 1 mg every 3 days (in a tds regime) until the daily dose is 5 mg; thereafter, the dose is incremented by 1 mg every 5 days. Prolonged treatment has been associated with fibrotic reactions (retroperitoneal, pulmonary, pleural, and cardiac), though these are rare (1 : 2000). Though occasionally used in chronic CH (CCH), a drug holiday of 1 month after every 6 months of treatment is necessary and neurological supervision entirely appropriate.

*Ergotamine.* Ergotamine is an effective preventative agent that is particularly useful in short-term management of episodic CH (ECH) when attacks occur predictably during the day or at night. Ergotamine, 1–2 mg orally or rectally, can be taken at bedtime or about 1 h before the attack is due. It is rarely suitable for use in CCH. Concomitant use of sumatriptan is contraindicated.

**Long-term Prevention.**  Some patients with either long bouts of episodic cluster headache or chronic cluster headache will require preventative treatment over many months, or even years. Verapamil and lithium are particularly useful in this setting.

*Verapamil.* Verapamil is the preventative drug of choice in both episodic and chronic CH. Clinical experience has clearly demonstrated that higher doses than those used in cardiological indications are needed. Dosages commonly employed range from 240 to 960 mg twice daily in divided doses. Verapamil can cause heart block by slowing conduction in the atrioventricular node. Observing for PR interval prolongation on ECG can monitor potential development of heart block. After performing a baseline ECG, patients are usually started on 80 mg bd and thereafter the total daily dose is increased in increments of 80 mg every 10–14 days. An ECG is performed prior to each increment. The dose is increased until the cluster attacks are suppressed, side effects intervene, or the maximum dose of 960 mg daily is achieved.

*Lithium.* Lithium is an effective agent for CH prophylaxis, though the response is less robust in ECH than in CCH. Renal and thyroid function tests are performed prior to initiation of therapy. Patients are then started on 300 mg bd and the dose titrated upward, aiming for a serum lithium level in the upper part of the therapeutic range. Most patients will benefit from dosages between 600 and 1200 mg daily. The concomitant use of nonsteroidal anti-inflammatory drugs (NSAIDs), diuretics, and carbamazepine is contraindicated.

*Other Drugs.*  Though sodium valproate, pizotifen, topiramate, gabapentin, and melatonin are now being used, they are of as yet unproven efficacy. Of these our experience suggests that topiramate will ultimately prove the most useful in those patients who can tolerate it.

*Surgery.*  This is a last-resort measure in treatment–resistant patients. A number of procedures that interrupt either the trigeminal sensory or autonomic (parasympathetic) pathways can be performed though few are associated with long-lasting results while the side effects can be devastating. Trigeminal procedures include thermocoagulation or sensory root section. The latter is said to be effective in 80% of patients (Kirkpatrick *et al.*, 1993), but there is no long-term follow-up. Certainly some patients continue to have attacks after these procedures and even respond to sumatriptan (Matharu and Goadsby, 2002). It is essential that pharmacological treatment is exploited to the fullest before surgery is undertaken. The most recent suggestion has been that stimulation of the ipsilateral posterior hypothalamic gray matter (Leone *et al.*, 2001) in the region of reported structural change (May *et al.*, 1999) can control the condition. Surgery without expert neurological review in cluster headache is at best ill-advised.

## Paroxysmal Hemicrania

### Clinical Aspects

Paroxysmal hemicrania was first described by Sjaastad *et al.* (1980), who reported eight cases, seven female, with a frequent unilateral severe but short-lasting headache without remission, coining the term chronic paroxysmal hemicrania (CPH). The essential features of paroxysmal hemicrania, as it is now understood, are:

- female preponderance;
- unilateral, usually frontotemporal, very severe pain;
- short-lasting attacks (2–45 min);
- very frequent attacks (usually more than 5 a day);
- marked autonomic features ipsilateral to the pain;
- robust, quick (less than 72 h), excellent response to indomethacin.

The International Headache Society has provided diagnostic criteria for CPH, and its second edition revision recognizes both an episodic form and a chronic form of paroxysmal hemicrania (Table III). If the disorder is suspected, causes listed in Table II should be excluded.

### Natural Course

Paroxysmal hemicrania is predominantly a problem of females in a ratio of about 3:1. It is a rare condition and probably has a prevalence of 0.01% or less, although there are no studies to know this with certainty. The mean daily frequency of attacks varies from 2 to 40 with the pain persisting from a few minutes to 45 min for each attack. The attacks are almost invariably unilateral and have prominent features of cranial parasympathetic autonomic activation. A review of 84 cases showed a history of remission in 35 cases, whereas 49 were chronic. Paroxysmal hemicrania (PH) usually begins in adulthood at the mean age of 34 years with a range of 6 to 81 years. Children with CPH have been reported. The author has seen a 4-year-old child with an otherwise typical indomethacin-sensitive case. Attacks that swap sides have been recorded, as have attacks of autonomic features without pain, just as is known for cluster headache. Otalgia and an interesting sensation of fullness of the external auditory meatus responding to indomethacin has been reported and may have been a form of paroxysmal hemicrania. There has been some interesting speculation from Dodick about extratrigeminal pain in episodic paroxysmal hemicrania, and we doubt that the full clinical dimensions of these syndromes have been defined.

TABLE III    Diagnostic Features of Paroxysmal Hemicrania Modified from the International Headache Society (Headache Classification Committee of The International Headache Society, 1988)

Paroxysmal Hemicrania has two key forms—
1. **Episodic:** Attacks occur in periods lasting 7 days to 1 year separated by pain-free periods lasting 1 month.
2. **Chronic:** Attacks occur for more than 1 year without remission or with remissions lasting less than 1 month.
— Headache attacks must have each of:
  - severe unilateral orbital, supraorbital, temporal pain lasting 15 min to 3 h;
  - frequency: greater than 5/day for half the time;
  - associated with one of:
    - Lacrimation
    - Nasal congestion
    - Rhinorrhea
    - Forehead/facial sweating
    - Miosis
    - Ptosis
    - Eyelid oedema
    - Conjunctival injection
  - Headache is treatable with indomethacin
— For each of 1 and 2 above no plausible secondary cause is demonstrable (Table II)

### Principles of Therapy

It is clear from the phenotype of the acute attack that there is activation of the trigeminal-autonomic reflex (May and Goadsby, 1999). Indeed this has been demonstrated with elevations of the trigeminal marker peptide, calcitonin gene-related peptide (CGRP) and the cranial parasympathetic marker peptide, vasoactive intestinal polypeptide (VIP) (Edvinsson and Goadsby, 1998). There has been speculation that paroxysmal hemicrania is also an inflammatory condition of the cavernous sinus. For the reasons argued in the section on cluster headache, this seems unlikely. Brain imaging studies of acute episodes of paroxysmal hemicrania are keenly awaited.

The main, in fact only clear, treatment of paroxysmal hemicrania is indomethacin.

### Practical Management

**Indomethacin.** Indomethacin is the established treatment of paroxysmal hemicrania, indeed defining the condition. One usually starts at a dose of 25 mg three times daily, if there is no response in 72 h the dose can be increased to 50 mg three times daily. Some patients will require 75 mg three times daily; some will use a single dose daily and some will simply use a suppository at night. We have found the use of the indomethacin injection (100 mg imi) as a practical test of the likely therapeutic response (Antonaci et al., 1998). Generally, in a positive outcome attacks will be blocked for up to 1 day. Many patients require treatment with $H_2$ blockers or $H^+$ pump inhibitors while on indomethacin. There has never been a controlled trial of indomethacin in this condition.

**Other Treatments.** The response of PH to triptans is unclear. We feel it is likely that lack of response as

TABLE IV    Differential Diagnosis of Short-Lasting Headaches

| Feature | Cluster headache | Chronic paroxysmal hemicrania | Episodic paroxysmal hemicrania | SUNCT[a] |
|---|---|---|---|---|
| Gender (M:F) | 4:1 | 1:3 | 1:1 | 8:1 |
| Pain | | | | |
| — Type | Boring | Boring | Boring | Stabbing |
| — Severity | Very | Very | Very | Moderate |
| — Location | Orbital | Orbital | Orbital | Orbital |
| Duration | 15–180 m | 2–45 m | 1–30 m | 15–120 s |
| Frequency | 1–8/day | 1–40/day | 3–30/day | 1/day–30/h |
| Autonomic | + | + | + | + |
| Trigger | Alcohol | Alcohol | Alcohol | Cutaneous |
| Indomethacin | ? | + | + | − |

[a]Shortlasting unilateral neuralgiform headache with conjunctival injection and tearing.

reported may be due to the shortness of attacks and have seen clearly indomethacin-sensitive cases that have responded to sumatriptan by injection. Rofecoxib may be helpful in some patients and a controlled study is needed. Piroxicam has been suggested to be helpful, although not as effective as indomethacin. By analogy with cluster headache, verapamil has been used in CPH, although the response is not spectacular and higher doses require exploration. CPH can coexist with trigeminal neuralgia (PH-Tic syndrome), just as does cluster headache (Cluster-Tic) syndrome, and these require treatment of both problems.

## ACKNOWLEDGMENTS

The work of the author described herein has been supported by the Wellcome Trust and the Migraine Trust.

## REFERENCES

Antonaci, F., Pareja, J. A., Caminero, A. B., and Sjaastad, O. (1998). Chronic paroxysmal hemicrania and hemicrania continua. Parenteral indomethacin: The "Indotest." *Headache* 38, 122–128.

Bahra, A., Gawel, M. J., Hardebo, J.-E., Millson, D., Brean, S. A., and Goadsby, P. J. (2000). Oral zolmitriptan is effective in the acute treatment of cluster headache. *Neurology* 54, 1832–1839.

Bahra, A., Matharu, M. S., Buchel, C., Frackowiak, R. S. J., and Goadsby, P. J. (2001). Brainstem activation specific to migraine headache. *Lancet* 357, 1016–1017.

Costa A., Pucci, E., Antonaci, F., Sances, G., Granella, F., Broich, G., and Nappi, G. (2000). The effect of intranasal cocaine and lidocaine on nitroglycerin-induced attacks in cluster headache. *Cephalalgia* 20, 85–91.

Edvinsson, L., and Goadsby, P. J. (1998). Neuropeptides in headache. *Eur. J. Neurol.* 5, 329–341.

Ekbom, K., The Sumatriptan Cluster Headache Study Group (1991). Treatment of acute cluster headache with sumatriptan. *N. Engl. J. Med.* 325, 322–326.

Fogan, L. (1985). Treatment of cluster headache: A double blind comparison of oxygen vs air inhalation. *Arch. Neurol.* 42, 362–363.

Goadsby, P. J. (1994). Cluster headache and the clinical profile of sumatriptan. *Eur. Neurol.* 34(Suppl.), 35–39.

Goadsby, P. J., and Lipton, R. B. (1997). A review of paroxysmal hemicranias, SUNCT syndrome and other short-lasting headaches with autonomic features, including new cases. *Brain* 120, 193–209.

Headache Classification Committee of The International Headache Society (1988). Classification and diagnostic criteria for headache disorders, cranial neuralgias and facial pain. *Cephalalgia* 8(Suppl. 7), 1–96.

Kirkpatrick, P. J., O'Brien, M., and MacCabe, J. J. (1993). Trigeminal nerve section for chronic migrainous neuralgia. *Br. J. Neurosurg.* 7(5), 483–490.

Kudrow, L. (1980). Cluster headache: Mechanisms and Management. Oxford University Press, Oxford.

Kudrow, L. (1987). Cluster headache. In "Headache: Clinical, Therapeutic, Conceptual and Research Aspects" (J. N. Blau, Ed.). Chapman and Hall, London.

Lance, J. W., and Goadsby, P. J. (1998). "Mechanism and Management of Headache," 6th Ed. Butterworth-Heinemann, London.

Leone, M., Franzini, A., D'Amico, D., Grazzi, L., Rigamonti, A., Usai, S., et al. (2001). Stereotactic electrode implant in inferior posterior hypothalamic gray matter to relieve intractable chronic cluster headache: The first reported case. *Neurology* 56 (Suppl. 3), A218.

Matharu, M. S., and Goadsby, P. J. (2002). Persistence of attacks of cluster headache after trigeminal nerve root section. *Brain* 175, 976–984.

May, A., Ashburner, J., Buchel, C., McGonigle, D. J., Friston, K. J., Frackowiak, R. S. J., and Goadsby, P. J. (1999). Correlation between structural and functional changes in brain in an idiopathic headache syndrome. *Nat. Med.* 5, 836–838.

May, A., Bahra, A., Buchel, C., Frackowiak, R. S. J., and Goadsby, P. J. (1998). Hypothalamic activation in cluster headache attacks. *Lancet* 351, 275–278.

May, A., and Goadsby, P. J. (1999), The trigeminovascular system in humans: Pathophysiological implications for primary headache syndromes of the neural influences on the cerebral circulation. *J. Cereb. Blood Flow Metabol.* 19, 115–127.

Monstad, I., Krabbe, A., Micieli, G., Prusinski, A., Cole, J., Pilgrim, A., and Shevlin, P. (1995). Preemptive oral treatment with sumatriptan during a cluster period. *Headache* 35, 607–613.

Olesen, J., and Goadsby, P. J. (1999). In "Cluster Headache and Related Conditions" (J. Olesen, Ed.), Vol. 9. Oxford University Press, Oxford.

Russell, M. B., Andersson, P. G., and Iselius, L. (1996). Cluster headache is an inherited disorder in some families. *Headache* 36, 608–612.

Sjaastad, O. (1992). Cluster headache syndrome. Saunders, London.

Sjaastad, O., Apfelbaum, R., Caskey, W., Christoffersen, B., Diamond, S., Graham, I., et al. (1980). Chronic paroxysmal hemicrania (CPH). The clinical manifestations. A review. *Uppsala J. Med. Sci.* 31(Suppl.), 27–33.

van Vliet, J. A., Bahra, A., Martin, V., Aurora, S. K., Mathew, N. T., Ferrari, M. D., and Goadsby, P. J. (2001). Intranasal sumatriptan is effective in the treatment of acute cluster headache—A double-blind placebo-controlled crossover study. *Cephalalgia* 21, 270–271.

*Acute and Chronic Pain*

## CHAPTER 3
# Tension-Type Headache

Rigmor Jensen, Jes Olesen, and H. Christoph Diener

Tension-type headache is the most frequent headache disorder and known by almost everyone. Despite the widespread prevalence the pathophysiology behind tension-type headache is widely unknown, and treatment strategies are still widely unspecific.

## CLINICAL ASPECTS

Until 1988, there was no internationally accepted classification of headache disorders and scientific results varied considerably and were practically incomparable due to imprecise classification. With the introduction of the International Headache Classification in 1988 (Headache Classification Committee of the International Headache Society, 1988) tension-type headache is the term designated by The International Headache Society to describe what previously was called tension headache, muscle contraction headache, psychomyogenic headache, stress headache, etc., and diagnostic criteria and subtyping (i.e., episodic vs chronic) were introduced (Table I) (Headache Classification Committee of the International Headache Society, 1988). Distinguishing episodic from chronic tension-type headache and tension-type headache from migraine has practical implications in management strategies. Chronic tension-type headache is often associated with more severe pain and more accompanying symptoms, is often combined with medication overuse, and is less influenced by daily hassles and stress than the episodic form. Studies from specialized clinics led to the conclusion that chronic tension-type headache evolves from migraine, along a continuum of the same disease with evolving manifestations (Leston, 1996). Population studies, however, show a different picture (Göbel *et al.*, 1994; Rasmussen, 1996; Rasmussen *et al.*, 1991; Schwartz *et al.*, 1998). The latter studies show that tension-type headache and migraine differ in gender ratio, age distribution, and

clinical presentation. Therefore, it could be argued that the "continuum theory" is an artifact of referral bias, thereby mimicking a continuum (Rasmussen, 1993, 1996). It is therefore most likely that migraine and tension-type headache are different disorders although they are coexisting in the many patients (Rasmussen, 1993; Rasmussen *et al.*, 1991; Ulrich *et al.*, 1996). Episodes of tension-type headache are more pronounced and frequent in subjects with coexisting migraine than in nonmigraineurs (Rasmussen, 1993; Ulrich *et al.*, 1996). This indicates that migraine can be a precipitating factor to tension-type headache in genetically predisposed individuals. It can be extremely difficult to distinguish between various headache disorders in the severely affected patients even in highly specialized headache clinics. Therefore, a diagnostic headache diary (Russel *et al.*, 1992) and a long-term follow-up are mandatory. A diagnosis of primary headaches as tension-type headache requires exclusion of other organic disorders. The absence of specific and distinguishing features of tension-type headache may explain why physicians, and subsequently patients, question the diagnosis whereas the migraine symptoms are more characteristic. Consequently, paraclinical investigations to exclude other organic disease are more frequently performed in tension-type headache (and probably should be) than in other headaches such as migraine. If an intracranial lesion is suspected on the basis of a clinical history and/or examination a CT or MRI should be performed. At present, there are no other reliable technical investigations that are useful in the differential diagnosis. Therefore, a careful history to uncover depression, anxiety, and other central factors and careful palpation of pericranial muscles and tender insertions to quantify the peripheral factor are extremely important. Furthermore, a general and neurological examination as well as a prospective follow-up using diagnostic headache diaries (Russel *et al.*, 1992)

**TABLE I   Diagnostic Criteria**

Diagnostic criteria for episodic tension-type headache according to the IHS (code 2.1)

A. At least 10 previous headache episodes fulfilling criteria B–D listed below. Number of days with such headache <180/year (<15/month).

B. Headache lasting from 30 min to 7 days.

C. At least two of the following pain characteristics:
1. Pressing/tightening (nonpulsating quality)
2. Mild or moderate severity
3. Bilateral location
4. No aggravation by walking stairs or similar routine physical activity

D. Both of the following:
1. No nausea or vomiting (anorexia may occur)
2. Photophobia and phonophobia are absent, one but not the other is present

E. At least one of the following:
1. History, physical and neurological examinations do not suggest one of the disorders listed in group 5–11 (symptomatic disorders).
2. History and physical and neurological examinations do suggest such disorder, but is ruled out by appropriate investigations.
3. Such disorder is present, but tension-type does not occur for the first time in close temporal relation to the disorder.

Diagnostic criteria for chronic tension-type headache (code 2.2)

A. Average headache frequency >15 days/month or more (≥180 days/year) for 6 months or more fulfilling criteria B–D listed below.

B. At least two of the following pain characteristics:
1. Pressing/tightening (nonpulsating quality)
2. Mild or moderate severity
3. Bilateral location
4. No aggravation by walking stairs or similar routine physical activity

C. Both of the following:
1. No vomiting
2. No more than one of the following: nausea, photophobia, or phonophobia

D. As E. in episodic tension-type headache, above

**TABLE II   Alarming Symptoms where Secondary Headaches Should Be Considered**

| Symptom | |
| --- | --- |
| Late onset of headache (age >40 years) | Morning headache |
| Prior trauma | Explosive vomiting |
| Visual disturbances | Progressive symptoms |
| Sudden onset of headache | Reduced consciousness |
| Other neurological deficits | |

**TABLE III   Differential Diagnosis of Tension-Type Headache with and without Structural Lesions**

| Headache with structural lesion | Headaches without structural lesions |
| --- | --- |
| Frontal sinusitis | Influenza headache |
| Arterial hypertension | Alcohol headache/hangover |
| Tumor | Post-traumatic headache |
| Hydrocephalus | Cough headache |
| Glaucoma | Nitrate headache |
| Pseudotumor cerebri | Food headache |
| Temporal arteritis | Altitude headache |
| Meningitis | Hypoglycemia |
| Subarachnoid hemorrhage | |
| Carotid dissection | |
| Ischemic stroke | |
| Subdural hemorrhage | |
| Cerebral hemorrhage | |

phenonomen complicating the primary headache disorders and may affect more than 50% in some headache clinic populations. Recognizing this condition is of crucial importance, since it has been demonstrated that a short time interval between the onset of drug abuse and first withdrawal is the most important predictor for a favorable long-term outcome (Diener *et al.*, 1989; Schnider *et al.*, 1996).

## NATURAL COURSE

In its milder and infrequent forms, tension-type headache is a nuisance, not a disease, but in its frequent forms, it becomes distressing and socially disturbing like other primary headaches such as migraine or cluster headache (Holroyd *et al.*, 2000). The prevalence of episodic tension-type headache varies considerably from 38.3% in an American study (Schwartz *et al.*, 1998) to 74% in a Danish cross-sectional study (Rasmussen *et al.*, 1991). In contrast, the prevalence of chronic tension-type headache is quite uniform, 2–3% in most studies (Göbel *et al.*, 1994; Rasmussen *et al.*, 1991; Schwartz *et al.*, 1998). The male:female ratio of tension-type headache is 4:5, indicating that, unlike migraine, females are only slightly more affected

are of utmost importance to reach the diagnosis. Neurological deficits like paresis, sensory disturbances, coordination problems, and neuropsychological deficits, e.g., disturbances of orientation and thinking are never seen with tension-type headache (Table II).

Other mostly secondary causes of headache are summarized in Table III. Cervical spondylosis almost never leads to chronic and diffuse headache. The same is true for low blood pressure (which is considered a disease in some European countries). In clinical practice, the most frequent cause for chronic daily headache is chronic analgesic and/or ergotamine or triptan abuse, to which patients may evolve after having presented initially with migraine or episodic/chronic tension-type headache (Diener and Dahlöf, 1999; Gaist *et al.*, 1998; Schnider *et al.*, 1994, 1996). Although the mechanism of drug-induced headache is not clear, it is a very widespread

(Rasmussen *et al.*, 1991). The average age at onset of tension-type headache is higher than in migraine, about 25–30 years, and the prevalence declines with increasing age in cross-sectional epidemiological studies (Rasmussen *et al.*, 1991). The mean duration of tension-type headache has been reported to be 10.3 years in the German population study (Göbel *et al.*, 1994) and 19.9 years in a clinical study (Jensen *et al.*, 1998), illustrating the considerable referral bias and the lifetime consistency in this disorder.

## PATHOPHYSIOLOGY

For decades it has been a matter of debate whether the pain in tension-type headache originates from myofascial tissues or from central mechanisms in the brain (Bendtsen *et al.*, 1996b; Jensen, 1999; Jensen *et al.*, 1998). Clinical and laboratory investigations to substantiate any of these hypothesis are few. Although the pain clinically resembles pain from the myofascial tissues, modern pain physiology indicates that both peripheral and central mechanisms may contribute. The increased tenderness, which is the most pronounced and consistent finding in these patients probably represents activation of peripheral nociceptors (Jensen, 1999), whereas the decreased pain thresholds which have been reported in chronic tension-type headache patients (Jensen, 1999) most likely represents a central misinterpretation. The texture of pericranial, shoulder, and chewing muscles is often altered in tension-type headache with generalized increased consistency (Ashina *et al.*, 1998, 1999c; Sakai *et al.*, 1995). A recent study of the stimulus–response function to mechanical pressure demonstrated for the first time that chronic tension-type headache has a physiological basis and is caused at least partly by qualitative changes in the central processing of sensory information (Bendtsen *et al.*, 1996b). A defect either in the opioid system or in the production of neurotransmitters (Bach *et al.*, 1992; Langemark *et al.*, 1995) has also been suspected as the nociceptive flexor reflex, a spinally organized reflex, is decreased in chronic tension-type headache (Langemark *et al.*, 1993) but no recent studies have confirmed these findings. Studies of neuropeptides and endorphins in these patients have mainly been negative (Ashina *et al.*, 1999a) and only one former study has noted increased metenkephalin in cerebrospinal fluid (Langemark *et al.*, 1995). These various abnormalities may result in or be a function of the disturbed balance between peripheral input and central modulation. The primary eliciting cause and the evolution of pain are, however, still unknown.

Nitric oxide (NO) plays an important role for the pathophysiology of primary headaches including chronic tension-type headache. Thus, a NO synthase inhibitor reduced headache and muscle hardness in a recent study (Ashina *et al.*, 1999d) while the NO donor glyceryl trinitrate (GTN) caused more headache in patients with chronic tension-type headache than in healthy controls (Ashina *et al.*, 1999b). These findings suggest that GTN-induced delayed headache can be used as a valuable human model of tension-type headache and that NO-related central sensitization may be an important factor in the underlying pathophysiology.

Headaches are generally reported to occur in relation to emotional conflict and psychosocial stress but the cause–effect relation is not clear. Stress and mental tension were the most frequently reported precipitating factors but occurred with similar frequency in tension-type headache and migraine (Rasmussen, 1993; Ulrich *et al.*, 1996). These results are in correspondence with the findings of widely normal personality profiles in subjects with episodic tension-type headache, whereas studies of subjects with the chronic form often reveal higher frequency of depression and anxiety (Holroyd *et al.*, 1993; Mitsikostas and Thomas, 1999; Rasmussen, 1992). As in other chronic pain disorders psychological abnormalities in tension-type headache may be viewed as secondary rather than primary (Holroyd *et al.*, 1993), and anxiety and depression are probably comorbid with chronic tension-type headache.

Migraine was previously believed to be a heterogenous syndrome with many causes but is now less likely since the great majority of migraineurs respond to the highly specific 5-HT$_{1B/D}$ agonists. This suggest that migraine mechanisms actually are rather homogenous, despite the numerous different trigger factors and it cannot be excluded that a similar condition exists in tension-type headache. In contrast to migraine, which is an all-or-none phenonomen that runs its course once started, is tension-type headache usually more graded varying from a mild short-lasting episode at one day to a long-lasting moderate to severe pain during another episode. It is still unknown what mechanism can explain such a periodicity in otherwise healthy subjects. Evidence of a genetic predisposition to chronic tension-type headache reflecting in a 3.18-fold increased risk in first-degree relatives compared to the general population has been published (Östergaard *et al.*, 1997; Russell *et al.*, 1999), but the mode of transmission seems complex.

On this basis, it can be concluded that the underlying pain mechanisms in tension-type headache are highly dynamic as tension-type headache represents a wide variety of frequency and intensity, not only between subjects, but also within the individual subject over time. The initiating stimulus may be either a condition of mental stress, unphysiological motor stress, a local myofascial release of irritants, or a combination of these. Secondary to the peripheral stimuli, the supraspinal pain perception structures may become activated, and due to central modulation of the incoming stimuli, a self-limiting process will be the result in most subjects. As chronic tension-type headache usually evolves from the episodic form (Jensen et al., 1998), an effective prevention of this evolution from a peripheral mechanism in the episodic to a central mechanism in chronic tension-type headache will therefore be of major importance in future treatment strategies.

## PRINCIPLES OF THERAPY

### Nonpharmacological Treatment

Various physical treatment modalities such as hot and cold packs, ultrasound and electrical stimulation, improvement of posture, relaxation, and exercise programs have all been used (Table IV) (Carlsson and Jensen, 2000). However, the majority of these modalities are not properly evaluated, and most studies are uncontrolled. In one open-label study, the beneficial long-term effect of physical therapy was excellent (Hammill et al., 1996), whereas a controlled study reported only a minor although significant reduction from 17 to 13 headaches per 4 weeks after 8 weeks of standardized treatment (Jensen and Olesen, 1995). Cervical manipulation has also frequently been recommended (Vernon et al., 1999), but a recent controlled study concluded that there was no significant effect of spinal manipulation in patients with episodic tension-type headache (Bove and Nielsson, 1998). In comparative studies amitriptyline had a better prophylactic effect during active treatment periods whereas the long-term effect was in favor of cervical manipulation. Many studies have investigated various forms of muscle relaxation with or without EMG biofeedback (Bussone et al., 1998). Most former studies have methodological shortcomings but pain intensity and frequency are usually reported to be reduced to 40–60% of their pretreatment values (Bogaards and Kuile, 1994). There is limited evidence for EMG biofeedback being better than simple muscle relaxation alone, whereas the combination of muscle relaxation and EMG biofeedback was favorable in several studies (Bogaards and Kuile, 1994). An extensive meta-analysis concluded that treatment outcome was more affected by patient characteristics than by treatment characteristics, and a better outcome was most pronounced in younger patients with a short disease duration. These results stress the importance of carefully designed studies, including properly classified and well described patients.

Although acupuncture is frequently applied to headache patients only few controlled studies are available. A recent meta-analysis found no evidence that acupuncture is effective in tension-type headache (Melchart et al., 1999).

As tension-type headache frequently is precipitated by stress, relaxation therapy (Jacobson, 1938) and stress management may be helpful in selected cases (Bogaards and Kuile, 1994). Although many links have been made between tension-type headache and personality, mood, and behavioral disorders, especially in the chronic form, the findings are not consistent (Blanchard et al., 1980; Blanchard et al., 1983; Holroyd et al., 1993; Holroyd and Penzien, 1994) and they allow no specific recommendations for management. If depression or anxiety is a significant finding, a specific treatment strategy for these disorders should therefore be initiated.

### Pharmacological Treatment

#### The Acute Episode

Due to the lack of pathophysiological knowledge of tension-type there is no selective or specific therapy. Traditional pharmacological treatment of the acute episode includes simple analgesics, nonsteroidal anti-inflammatory drugs (NSAIDs), and muscle relaxants but the efficacy of these drugs has only rarely been systematically tested using modern-day methodology (Table IV) (Mathew and Schoenen, 2000).

In a placebo-controlled study the effect of solid aspirin versus effervescent aspirin was studied. Both formulations were significantly better than placebo, but there was no significant difference between solid and effervescent aspirin (Langemark and Olesen, 1987). In another placebo-controlled study, it was noted that acetaminophen and aspirin were more effective than placebo, but not different from each other (Peters et al., 1983). However, as the gastric side-effect profile is much better with acetaminophen than with aspirin, acetaminophen may be recommended as the first drug of choice for these mild or moderate headache episodes. Although simple over-the-counter (OTC) drugs are the most commonly used drugs for headache, excessive and frequent use, often combined with caffeine and/or sedatives, should clearly be avoided due to the high risk of drug-induced headache. Therefore, thorough information and an upper daily/weekly limit of such drug consumption are essential to these patients (Mathew and Schoenen, 2000).

The value of NSAIDs in treating tension-type headache is better substantiated in RCTs (randomized controlled trials). Ibuprofen (200 and 400 mg) and 25 mg Ketoprofen are significantly more effective than placebo and at least as effective, but not always superior to aspirin or acetaminophen (Dahlöf and Jacobs, 1996; Packman et al., 2000; Steiner and Lange, 1998;

Van Gerven *et al.*, 1996). When 12.5 or 25 mg of ketoprofen, 200 mg ibuprofen, and 275 mg naproxen sodium were compared to each other no significant difference between these NSAIDs was demonstrated (Lange and Lentz, 1995), whereas 50 mg Ketoprofen was more effective with an early onset compared to 200 mg ibuprofen in another study (Van Gerven *et al.*, 1996). The most recent studies reported a highly significant and early effect of intramuscular injections of 60 mg ketorolac (Harden *et al.*, 1998) compared to placebo and of a new solubilized capsule of 400 mg ibuprofen compared to 1000 mg acetaminophen and placebo in acute episodes of tension-type headache (Packman *et al.*, 2000).

The use of muscle relaxants is mainly on an empirical basis and results are variable (Mathew and Schoenen, 2000). Tizanidine was reported to be effective in the reduction of pain intensity (Fogelholm and Murros, 1992) but these results could not be replicated in larger, recent trial (Murros *et al.*, 2000). On this basis and due to the risk of habituation of most available muscle relaxants, their use cannot be recommended.

### Prophylactic Pharmacological Treatment

Tricyclic antidepressants (TCA) are extensively used (Table V) but the scientific evidence is to some extent contradictory (Mathew and Bendtsen, 2000). In later years the effect of amitriptyline in chronic but not in episodic tension-type headache has been confirmed (Bendtsen *et al.*, 1996a; Diamond and Baltes, 1971). The most extensive study reported a significant 30% reduction in the area under the headache curve during amitriptyline compared with placebo while the specific serotonin reuptake inhibitor citalopram had no significant effect (Bendtsen *et al.*, 1996a). The mechanism of action of amitriptyline in tension-type headache is independent of its antidepressant effect and the effective dosage in headache is usually much lower than used in the treatment of depression, namely 10–75 mg per day (Mathew and Bendtsen, 2000).

Prophylaxis of coexisting headache with more than two to three migraine attacks superimposed on frequent tension-type headache can be performed with amitriptyline (25–75 mg) and metoprolol (100–150 mg). Few other tricyclic antidepressants such as imipramine and buspirone, a 5-HT$_{1A}$ agonist, have also been reported to have a prophylactic effect in chronic tension-type headache (Mitsikostas *et al.*, 1997), but unfortunately these studies never been replicated. The lack of significant effect of the selective serotonine reuptake inhibitors (SSRIs) in nondepressed patients (Bendtsen *et al.*, 1996, Foster and Bafaloukos, 1994; Langemark and Olesen, 1994) indicate that serotonergic mechanisms are not of decisive importance. The effect of SSRIs in patients with coexisting depression and headache appear nevertheless to be favorable and similar to the effect of other antidepressants.

Recently two placebo-controlled studies confirm the prior pilot studies that injections of botulinum toxin type A in pericranial muscles have a prophylactic effect, although the most recent study using smaller doses was negative (Rollnik *et al.*, 2000). Relja *et al.* reported a significant and rather long-lasting decrease in headache intensity in 16 CTH patients (Relja and Korsic, 1999) and Smuts *et al.* reported a significant pronounced reduction in headache days compared to placebo injections in his study of 41 patients (Smuts *et al.*, 1999). The mechanism of action of this botulinum toxin in tension-type headache is not clear but probably related to a transitory but long-lasting interruption of the interaction between peripheral nociception due to muscular overactivity and central pain processing. This treatment strategy may provide important information about the underlying pathophysiology and further studies of the prophylactic effect and its mode of action are certainly needed.

Among future emerging therapies are easily soluble NSAIDs; eventually the selective COX2 inhibitors with fewer gastrointestinal side effects are promising for the acute episode, and for the chronic form the more specific and longer acting nitric oxide synthase inhibitors may have a beneficial effect (Ashina *et al.*, 1999d). Other (nontricyclic) antidepressants should be investigated as amitriptyline is quite effective in chronic tension-type headache but due to numerous side effects more specific and modern antidepressants with fewer side effects would be valuable.

## PRACTICAL MANAGEMENT

First it is important to establish an accurate diagnosis, where the individual headache episode is identified and separated from migraine or a secondary headache, most frequently drug-induced headache. According to the IHS classification (Headache Classification Committee of the International Headache Society, 1988), all different types of headache in an individual patient should be diagnosed, and treatment should then be directed toward the most frequent or most incapacitating type of headache in the patient. Second, avoidance of any possible trigger factors is very important but unfortunately also very often difficult as trigger factors may vary over time not only between but also within patients. One of the most important elements in treating headache patients in general is, however, to take their complaints seriously, show empathy and examine them thoroughly as most have been met with medical

**TABLE IV** Therapy in Episodic Tension-Type Headache

| Nonpharmacological therapy | TR | Pharmacological therapy | TR |
|---|---|---|---|
| Hot or cold packs | * | Aspirin (500–1000 mg) | ** |
| Relaxation | * | Acetaminophen (Paracetamol) (500–1000 mg) | ** |
| Physical therapy | * | Ibuprofen (200–400 mg) | *** |
| Cervical manipulation | * | Ketoprofen (25–50 mg) | ** |
| | | Naproxen (275–550 mg) | ** |

*Treatment based on personal experience, expert opinion, retrospective, or nonrandomized studies.

**Treatment is based on at least one randomized prospective study of reasonable size.

***Treatment recommendations (TR) based on more than one well-designed randomized, placebo-controlled trial or a meta-analysis.

**TABLE V** Therapy in Chronic Tension-Type Headache

| Nonpharmacological therapy | TR | Pharmacological therapy | TR |
|---|---|---|---|
| Cervical manipulation | * | Amitriptyline | *** |
| Relaxation | ** | Clomipramine | * |
| Physiotherapy | * | Doxepin | * |
| Excercise | * | Botulinum toxin A | ** |
| Stress coping | * | | |
| Muscle relaxation biofeedback | * | | |

*Treatment based on personal experience, expert opinion, retrospective, or nonrandomized studies.

**Treatment is based on at least one randomized prospective study of reasonable size.

***Treatment recommendations (TR) based on more than one well-designed randomized, placebo-controlled trial or a meta-analysis.

ignorance and lack of interest for years. The mainstay in treatment of the acute episode is either a standard dose of aspirin or acetaminophen (Table IV). If there is no consistent effect then 200–400 mg ibuprofen or 25–50 mg ketoprofen can be recommended, either as conventional tablets or as soluble capsules, whereas intravenous injections should be restricted to clinical use.

### Treatments No Longer Recommended

Muscle relaxants and tranquilizers should be used neither in acute nor in chronic tension-type headache. Barbiturate-containing drugs have a high risk of medication-overuse headache and addiction. The effect of migraine-specific drugs in tension-type headache is controversial, as a positive effect often is considered in favor of a common etiology for migraine and tension-type headache. Subcutaneous injections of 6 mg sumatriptan had some effect in patients with chronic tension-type headache although not as marked as in migraine, and not considered to be clinically relevant (Brennum *et al.*, 1992). In episodic tension-type headache there was no effect of 100-mg tablets of sumatriptan in one study (Brennum *et al.*, 1996), whereas there was a pronounced effect of 6 mg SC in another study where patients had coexisting migraine (Cady *et al.*, 1997). The effect of newer triptans in tension-type headache have not yet been studied, and until further evidence is available the use of triptans should be restricted to migraine. Combined drugs consisting of analgesics, tranquilizers, and sedatives should clearly be avoided because of the potential of habituation and subsequent analgesic rebound headache phenonomen. As sodium valproate had no effect on chronic tension-type headache in an open labeled study (Lenaerts *et al.*, 1996) and mainly studied in patients with coexisting migraine or drug-induced headache, it can now be con-

cluded that sodium valproate plays no role in the treatment of pure tension-type headache.

## REFERENCES

Ashina, M., Bendtsen, L., Jensen, R., Ekman, R., and Olesen, J. (1999a). Plasma levels of substance P, neuropeptide Y and vasoactive intestinal polypeptide in patients with chronic tension-type headache. *Pain* **83**, 541–547.

Ashina, M., Bendtsen, L., Jensen, R., Hansen, L., Sakai, F., and Olesen, J. (1999b). Possible mechanisms of action of nitric oxide synthase inhibitors in chronis tension-type headache. *Brain* **122**, 1629–1635.

Ashina, M., Bendtsen, L., Jensen, R., Sakai, F., and Olesen, J. (1998). Measurement of muscle hardness: A methodological study. *Cephalalgia* **18**, 106–111.

Ashina, M., Bendtsen, L., Jensen, R., Sakai, F., and Olesen, J. (1999c). Muscle hardness in patients with chronic tension-type headache: Relation to actual headache state. *Pain* **79**, 201–205.

Ashina, M., Lassen, L. H., Bendtsen, L., Jensen, R., and Olesen, J. (1999d). Effect of inhibition of nitric oxide synthase on chronic tension-type headache: A randomised crossover trial. *Lancet* **353**, 287–289.

Bach, F. W., Langemark, M., Secher, N. H., and Olesen, J. (1992). Plasma and cerebrospinal fluid β-endorphin in chronic tension-type headache. *Pain* **51**, 163–168.

Bendtsen, L., Jensen, R., and Olesen, J. (1996a). A non-selective (amitriptyline), but not a selective (citalopram), serotonin reuptake inhibitor is effective in the prophylactic treatment of chronic tension-type headache. *J. Neurol. Neurosurg. Psychiatry* **61**, 285–290.

Bendtsen, L., Jensen, R., and Olesen, J. (1996b). Qualitatively altered nociception in chronic myofacial pain. *Pain* **65**, 259–264.

Blanchard, E., Andrasik, F., Ahles, T., Teders, S., and O'Keefe, D. (1980). Migraine and tension headache: A meta-analytic review. *Behav. Ther.* **11**, 611–631.

Blanchard, E. B., Andrasik, F., Arena, J. G., Neff, D. F., Jurish, S. E., Teders, S. J., Barron, K. D., and Rodichok, L. D. (1983). Nonpharmacologic treatment of chronic headache: Prediction of outcome. *Neurology* **33**, 1596–1603.

Bogaards, M. C., and Kuile, M. M. (1994). Treatment of recurrent tension headache: A metaanalytic review. *Clin. J. Pain.* **10**, 174–190.

Bove, G., and Nielsson, N. (1998). Spinal manipulation in the treatment of episodic tension-type headache. A randomized controlled trial. *JAMA* 280, 1576–1579.

Brennum, J., Brinck, T., Schriver, L., Wanscher, B., Soelberg Sorensen, P., Tfelt-Hansen, P., and Olesen, J. (1996). Sumatriptan has no clinically relevant effect in the treatment of episodic tension-type headache. *Eur. J. Neurol.* 3, 23–28.

Brennum, J., Kjeldsen, M., and Olesen, J. (1992). The 5-HT1-like agonist sumatriptan has a significant effect in chronic tension-type headache. *Cephalalgia* 12, 375–379.

Bussone, G., Grazzi, L., DÁmico, D., Leone, M., and Andrasik, F. (1998). Biofeedback-assisted relaxaton training for young adolescents with tension-type headache: A controlled study. *Cephalalgia* 18, 463–467.

Cady, R. K., Gutterman, D., Saiers, J. A., and Beach, M. E. (1997). Responsiveness of non-IHS-migraine and tension-type headache to sumatriptan. *Cephalalgia* 17, 588–590.

Carlsson, J., and Jensen, R. (2000). Physiotherapy of tension-type headache. *In* "The Headaches" (J. Olesen, P. Tfelt-Hansen, and K. M. L. Welch, Eds.), pp. 651–656. Lippincott, Willimas & Wilkins, Philadelphia.

Carruthers, A., Langtry, J. A. A., Carruthers, J., and Robinson, G. (1999). Improvement of tension-type headache when treating wrinkles with botulinum toxin A injections. *Headache* 39, 662–665.

Dahlöf, C. G. H., and Jacobs, L. D. (1996). Ketoprofen, paracetamol and placebo in the treatment of episodic tension-type headache. *Cephalalgia* 16, 117–123.

Diamond, S., and Baltes, B. J. (1971). Chronic tension headache treated with amitriptyline—A double-blind study. *Headache* 11, 110–116.

Diener, H. C., and Dahlöf, C. G. H. (1999). Headache associated with chronic use of substances. *In* "The Headaches" (J. Olesen, P. Tfelt-Hansen, and K. M. A. Welch, Eds.), 2nd ed., pp 871–878. Lippincott, Williams & Wilkins, Philadelphia.

Diener, H. C., Dichgans, J., Scholz, E., Geiselhart, S., Gerber, W. D., and Bille, A. (1989). Analgesic-induced chronic headache: Long-term results of withdrawal therapy. *J. Neurol.* 236, 9–14.

Fogelholm, R., and Murros, K. (1992). Tizanidine in chronic tension-type headache: A placebo controlled double-blind cross-over study. *Headache* 32, 509–513.

Foster, C. A., and Bafaloukos, J. (1994). Paroxetine on the treatment of chronic daily headache. *Headache* 34, 587–589.

Gaist, D., Tsiropoulus, I., Sindrup, S. H., Hallas, J., Rasmussen, B. K., and Kragstrup, J. (1998). Inappropriate use of sumatriptan: Population based register and interview study. *Br. J. Med.* 316, 1352–1353.

Göbel, H., Petersen-Braun, M., and Soyka, D. (1994). The epidemiology of headache in Germany: A nationwide survey of a representative sample on the basis of the headache classification of the International Headache Society. *Cephalalgia* 14, 79–106.

Haddock, C. K., Rowan, A. B., Andrasik, F., Wilson, P. G., Talcott, G. W., and Stein, R. J. (1997). Home-based behavioral treatments for chronic benign headache: A meta-analysis of controlled trials. *Cephalalgia* 17, 113–118.

Hammill, J. M., Cook, T. M., and Rosecrance, J. C. (1996). Effectiveness of a physical therapy regimen in the treatment of tension-type headache. *Headache* 36, 149–153.

Harden, R. N., Rogers, D., Fink, K., and Gracely, R. H. (1998). Controlled trial of ketorolac in tension-type headache. *Neurology* 50, 507–509.

Headache Classification Committee of the International Headache Society (1988). Classification and diagnostic criteria for headache disorders, cranial neuralgias and facial pain. *Cephalalgia* 8, 1–93.

Holroyd, K. A., France, J. L., Nash, J. M., and Hursey, K. G. (1993). Pain state as artifact in the psychological assessment of recurrent headache sufferers. *Pain* 53, 229–235.

Holroyd, K. A., and Penzien, D. B. (1994). Psychosocial interventions in the management of recurrent headache disorders: Overview and effectiveness. *Behav. Med.* 20, 53–63.

Holroyd, K. A., Stensland, M., Lipchik, G. L., Hill, K. R., O'Donnell, F. S., and Cordingley, G. (2000). Psychosocial correlates and impact of chronic tension-type headache. *Headache* 40, 3–16.

Jacobson, E. (1938). Progressive relaxation. University of Chicago Press, Chicago.

Jensen, R. (1999). Pathophysiological mechanisms of tension-type headache: A review of epidemiological and experimental studies. Thesis. *Cephalalgia* 19, 602–621.

Jensen, R., Bendtsen, L., and Olesen, J. (1998). Muscular factors are of importance in tension-type headache. *Headache* 38, 10–17.

Jensen, R., and Olesen, J. (1995). Is there an effect of physiotherapy in tension-type headache? *Cephalalgia* 15, 152.

Lange, R., and Lentz, R. (1995). Comparison of ketoprofen, ibuprofen and naproxen sodium in the treatment of tension-type headache. *Drugs Exp. Clin. Res.* 3, 89–96.

Langemark, M., Bach, F. W., Ekman, R., and Olesen, J. (1995). Increased cerebrospinal fluid met-enkephalin immunoreactivity in patients with chronic tension-type headache. *Pain* 63, 103–107.

Langemark, M., Bach, F. W., Jensen, T. S., and Olesen, J. (1993). Decreased nociceptive flexion reflex threshold in chronic tension-type headache. *Arch. Neurol.* 50, 1061–1064.

Langemark, M., and Olesen, J. (1987). Effervescent ASA versus solid ASA in the treatment of tension headache. A double-blind, placebo controlled study. *Headache* 27, 90–95.

Langemark, M., and Olesen, J. (1994). Sulpiride and paroxetine in the treatment of chronic tension-type headache. An explanatory double-blind trial. *Headache* 34, 20–24.

Lenaerts, M., Basting, E., Sianard, J., and Schonen, J. (1996). Sodium valproate in severe migraine and tension-type headache: An open study of long-term effect and correlation with blood levels. *Acta. Neurol. Belg.* 96, 126–129.

Leston, J. A. (1996). Migraine and headache are not seperate disorders. *Cephalalgia* 16, 220–223.

Mathew, N., and Bendtsen, L. (2000). Prophylactic pharmacotherapy of tension-type headache. *In* "The Headaches" (J. Olesen, P. Tfelt-Hansen, and K. M. A. Welch, Eds.), pp. 667–673. Lippincott, Williams & Williams, Philadelphia.

Mathew, N., and Schoenen, J. (2000). Acute pharmacotherapy of tension-type headache. *In* "The Headaches" (J. Olesen, P. Tfelt-Hansen, and K. L. A. Welch, Eds.), pp. 661–666. Lippincott, Williams & Williams, Philadelphia.

Melchart, D., Linde, K., Fischer, P., White, A., Allais, G., Vickers, A., and Berman, B. (1999). Acupuncture for recurrent headaches: A systematic review of randomized controlled trials. *Cephalalgia* 19, 779–786.

Mitsikostas, D. D., Gatzonis, S., Thomas, A., and Ilias, A. (1997). Buspirone vs amitriptyline in the treatment of chronic tension-type headache. *Acta. Neurol. Scand.* 96, 247–251.

Mitsikostas, D. D., and Thomas, A. M. (1999). Comorbidity of headache and depressive disorders. *Cephalalgia* 19, 211–217.

Murros, K., Kataja, M., Hedman, C., Havanka, H., Säkö, E., Färkkilä, M., Peltola, J., Keränen, T., for the Finnish Sirdalud Study Group (2000). Modified-release formulation of tizanidine in chronic tension-type headache. *Headache* 40, 633–637.

Östergaard, S., Russell, M. B., Bendtsen, L., and Olesen, J. (1997). Increased familial risk of chronic tension-type headache. *Br. Med. J.* 314, 1092–1093.

Packman, B., Packman, E., Doyle, G., Cooper, S., Ashraf, E., Koronkiewicz, K., and Jayawardena, S. (2000). Solubilized ibuprofen: Evaluation of onset, relief and safety of a novel formulation of the treatment of episodic tension-type headache. *Headache* 40, 561–567.

Peters, B. H., Fraim, C. J., and Masel, B. E. (1983). Comparison of 650 mg aspirin and 1000 mg acetaminophen with each other, and with placebo in moderately severe headache. *Am. J. Med.* **74**, 36–42.

Rasmussen, B. K. (1992). Migraine and tension-type headache in a general population: Psychosocial factors. *Int. J. Epidemiol.* **21**, 1138–1143.

Rasmussen, B. K. (1993). Migraine and tension-type headache in a general population: Precipitating factors, female hormones, sleep pattern and relation to lifestyle. *Pain* **53**, 65–72.

Rasmussen, B. K. (1996). Migraine and tension-type headache are seperate disorders. *Cephalalgia* **16**, 217–220.

Rasmussen, B. K., Jensen, R., Schroll, M., and Olesen, J. (1991). Epidemiology of headache in a general population—A prevalence study. *J. Clin. Epidemiol.* **44**, 1147–1157.

Relja, M., and Korsic, M. (1999). Treatment of tension-type headache by injections of botulinum toxin type A: A double-blind, placebo controlled study. *Neurology* **52**, A203.

Rollnik, J. D., Tanneberger, O., Schubert, M., Schneider, U., and Dengler, R. (2000). Treatment of tension-type headache with botulinum toxin type A: A double-blind, placebo-controlled study. *Headache* **40**, 300–305.

Russel, M., Rasmussen, B. K., Brennum, J. H. I. Jensen, R., and Olesen, J. (1992). Presentation of a new instrument. *Cephalalgia* **12**, 369–374.

Russell, M., Ostergard, S., Bendtsen, L., and Olesen, J. (1999). Familial occurence of chronic tension-type headache. *Cephalalgia* **19**, 207–210.

Sakai, F., Ebihara, S., Akiyama, M., and Horikkawa, M. (1995). Pericranial muscle hardness in tesnion-type headache. A non-invasive method and its clinical application. *Brain* **118**, 523–531.

Saper, J. R., Silberstein, S. D., Lake, A. E., and Winters, M. E. (1994). Double-blind trial of fluoxetine: Chronic daily headache and migraine. *Headache* **34**, 497–502.

Schnider, P., Auli, S., Feucht, M., Mraz, M., Travniczek, A., Zeiler, K., and Wessely, P. (1994). Use and abuse of analgesics in tension-type headache. *Cephalalgia* 162–167.

Schnider, P., Aull, S., Baumgartner, C., Merterer, A., Wöber, C., Zeiler, K., and Wessely, P. (1996). Long-term outcome of patients with headache and druf abuse after inpatient withdrawal: Five year follow-up. *Cephalalgia* **16**, 481–485.

Schwartz, B. S., Stewart, W. F., Simon, D., and Lipton, R. B. (1998). Epidemiology of tension-type headache. *JAMA* **279**, 381–383.

Smuts, J. A., Baker, M. K., Smuts, H. M., Stassen, J. M. R., Rossouw, E., and Barnard, P. W. A. (1999). Prophylactic treatment of chronic tension-type headache using botulinum toxin type A. *Eur. J. Neurol.* **6**, S99–S102.

Steiner, T. J., and Lange, R. (1998). Ketoprofen (25 mg) in the symptomatic treatment of episodic tension-type headache: Double-blind placebo-controlled comparison with acetaminophen (1000 mg). *Cephalalgia* **18**, 38–43.

Ulrich, V., Russel, M. B., Jensen, R., and Olesen, J. (1996). A comparison of tension-type headache in migraineurs and in non-migraineurs: A population based study. *Pain* **67**, 501–506.

Van Gerven, J. M. A., Schoemaker, R. C., Jacobs, L. D., Reints, A., Owersloot van der Meij, M. J., and Hoedemaker, H. G. (1996). Self-medication of a single episode of tension-type headache with ketoprofen, ibuprofen and placebo, home-monitored with an electronic patient diary. *Br. J. Clin. Pharmacol.* **42**, 475–481.

Vernon, H., McDemaid, C. S., and Hagino, C. (1999). Systematic review of randomized clinical trials of complementary/alternative therapies of tension-type and cervicogenic headache. *Complement Ther. Med.* **7**, 142–155.

*Acute and Chronic Pain*

CHAPTER 4
# Medication Overuse Headache

H. Christoph Diener and Ninan Mathew

## CLINICAL ASPECTS

Inappropriate use of headache medication for the treatment of headache episodes may contribute to the development of chronic daily headache that is refractory to medical and nondrug treatments. Physicians experienced in the treatment of migraine and other headaches are well aware that the daily intake of antipyretic or antiinflammatory analgesics, opioids, ergot alkaloids, and "triptans" may result in chronic daily headache. Conversely, if a patient complains of chronic daily headache and takes pain medication every day, or every second day, this headache is most likely to be sustained by the medication and will improve or vanish with abstinence.

It should be noted that almost no experimental work has been done in this field, and the following is based mainly on clinical series describing patients presenting at headache clinics with this problem, with subsequent treatment and follow-up.

### Definition

The International Headache Society defined drug-induced headache originally as follows (Headache Classification Committee of the International Headache Society, 1988).

Headache induced by chronic substance use or exposure

- occurs after daily doses of a substance for >3 months;
- requires a certain minimum dose;
- is chronic (15 days or more a month);
- disappears within 1 month after withdrawal of the substance.

Ergotamine-induced headache

- is preceded by daily ergotamine intake (oral >2 mg, rectal >1 mg);
- is diffuse, pulsating, and distinguished from migraine by absent attack pattern and/or absent associated symptoms.

Analgesics-abuse headache is characterized by one or more of the following:

- >50 g of aspirin a month or equivalent of other mild analgesics;
- >100 tablets a month of analgesics combined with barbiturate or other non-narcotic compounds;
- one or more narcotic analgesic.

Headache from substance withdrawal (chronic use)

- occurs after a high daily dose (specified when possible under each substance) of a substance for >3 months;
- occurs within hours after elimination of the substance;
- is relieved by renewed intake of the substance;
- disappears within 14 days after withdrawal of the substance.

The new classification of the IHS will call this kind of headache medication-overuse headache.

### Epidemiology

Data on prevalence and incidence rates of chronic medication-overuse headache are rare. In a Spanish population-based study, about 1% of the population suffered from daily headache as a result of medication-overuse headache (Castillo *et al.*, 1999). Most headache centers report that between 5 and 10% of the patients they see fulfill the criteria of medication-overuse headache (Granella *et al.*, 1987). Micieli *et al.* observed an incidence of 4.3% in 3000 consecutive headache patients (Micieli *et al.*, 1988). Patients with cluster headache almost never develop medication-overuse headache. A survey in family doctors showed that medication-overuse headache was the third most common cause of headache (Rapoport *et al.*, 1996). Taken together, these studies indicate that medication-overuse headache is a major health problem. This is also true if one considers the side effects of chronic intake of analgesics, ergotamine, and triptans such as chronic kidney failure (combination analgesics), gastrointestinal ulcers (NSAIDs), or ergotism (Andersson, 1988; Mihatsch *et al.*, 1983; Mihatsch and Knüsli, 1982).

## Pathophysiology

Most headache experts agree that patients with migraine and tension-type headache have a higher potential for medication-overuse headache than patients who use drugs for other diseases. Relapses of migraine occur in migraineurs who have been placed on analgesics for other ailments. The association between analgesic overuse and headache has been studied in conditions other than primary headache disorders. Chronic overuse of analgesics does not cause increased headache in nonmigraineurs (Bahra *et al.*, 2000). For example, a group of patients who were consuming fairly large amounts of analgesics regularly for arthritis did not show increased incidence of headache (Lance *et al.*, 1988). The conclusion drawn from various clinical observations and studies is that medication-overuse headache may be restricted to those who are already headache sufferers. The basis for this could either be genetic or due the fact that migraine pain is more severe than joint pain. Different mechanisms probably contribute to the transition from the original headache to medication overuse headache. Psychological factors include the reinforcing properties of pain relief by drug consumption, a very powerful component of positive conditioning. Many patients report taking migraine drugs prophylactically because they are worrying about missing work (or, inevitably, the job) or missing an important social event (dinner, theater, etc.). More importantly, patients often fear an imminent headache and take analgesics or specific migraine drugs prophylactically. They are often instructed by the physicians or by the instructions supplied with the medication to take the migraine drug as early as possible at the start of either the aura or the headache phase of a migraine attack.

Withdrawal headache is an additional factor. Whenever the patient tries to stop or reduce the medication, a worsening of the preexisting headache occurs. Barbiturates are contained in drugs for the treatment of tension-type headache and have a high potency for addiction and cause headache. The psychotropic side effects of analgesic or migraine drugs such as sedation or mild euphoria and their stimulating action may lead to drug dependency. Barbiturates, codeine, other opioids, and caffeine are the most likely substances to have this effect. Caffeine increases vigilance, relieves fatigue, and improves performance and mood (Griffiths and Woodson, 1988a,b). The typical symptoms of caffeine withdrawal such as irritability, nervousness, restlessness, and especially "caffeine withdrawal headache" (Silverman *et al.*, 1992; van Dusseldorp and Katan, 1990), which may last for several days, encourage the patients to continue their abuse. Despite the fact that caffeine may enhance the analgesic action of acetylsalicylic acid and acetaminophen, caffeine-containing combinations should not be used. Similarly, caffeine and meprobamat, the main metabolite of carisoprodol, should be removed from ergotamine-containing formulations.

There are reports on physical dependence on codeine and other opioids in headache patients (Fisher and Glass, 1997; Ziegler, 1994). There are no studies that have investigated the effects of codeine intake over periods as long as 10 years as many headache patients have done. It should be remembered that up to 10% of codeine is metabolized to morphine.

Ergotamine and DHE may certainly lead to physical dependency (Saper and Jones, 1986). Many patients who feel a migraine attack may take ergotamine as prophylactic treatment. The reason for the physical dependency on ergotamine remains obscure. In one study the tyramine-induced mydriasis after ergotamine administration was increased during abuse but not after withdrawal of ergotamine, indicating a central inhibition of pupillary sympathetic activity during abuse (Fanciullacci *et al.*, 1992). Thus a possible CNS effect of ergotamine can be observed after chronic use but not after a single dose of the drug. Other studies investigating the effect of chronic use of ergotamine on the CNS regulation of the autonomic nervous system are needed.

The drugs leading to chronic medication overuse headache vary considerably in the different series depending probably on both selection of patients, (e.g., "pure" ergotamine abusers being reported), and cultural factors. Potentially each component contained in analgesics or drugs for the treatment of migraine attacks can induce headache. This is also true for acetylsalicylic acid and paracetamol (Rapoport *et al.*, 1985). It is, however, difficult to identify a single substance as 90% of patients take more than one compound at a time. Four studies (Baumgartner *et al.*, 1989; Diener *et al.*, 1988; Mathew *et al.*, 1990; Micieli *et al.*, 1988) investigated the frequency of the chemical compounds of drugs used. Combination analgesics containing butalbital (short acting barbiturate), caffeine, aspirin with or without codeine were the leading candidates for medication overuse headache in one study (Mathew *et al.*, 1990). Sumatriptan can also lead to medication overuse headache. This was first observed in patients who abused ergotamine (Catarci *et al.*, 1994; Kaube *et al.*, 1994). Later de novo cases were reported (Gaist *et al.*, 1996, 1998; Pini and Trenti, 1994). Recently, patients who developed drug-induced headache from naratriptan and zolmitriptan were reported (Katsarava *et al.*, 2000; Limmroth *et al.*, 1999). Due to the delay between frequent intake of triptans and the development of medication overuse headache, it is likely that similar cases will be observed in the future with the other recently approved triptans (rizatriptan, eletriptan, frovatriptan, almotriptan). The risk appears to be particularly high in headache patients with a former history of misuse of analgesics and/or ergotamine. Results from headache diaries show that the number of tablets or suppositories taken per day averages 4.9 (range 0.25–25). Patients take on average 2.5–5.8 different pharmacological components simultaneously (range 1–14) (Diener and Dahlöf, 1999). The number of doses per day is much smaller in patients who abuse triptans (Katsarava *et al.*, 2000).

One recent prospective study shows the characteristics of medication overuse headache (Katsarava *et al.*, 2000). Patients who abuse analgesics develop a constant diffuse headache of moderate intensity. They may still suffer from intermittent migraine attacks on top of the daily headache. Most patients with overuse of triptans experience an increase in the frequency of migraine attacks and may experience daily migraine-like headaches. Patients with overuse of ergots usually show a mixture of the two headache characteristics described above.

The withdrawal headache experienced after stopping medication resembles a severe and prolonged migraine attack in patients with migraine as primary headache. Several clinical characteristics are helpful in identifying the occurrence of analgesic rebound headache in patients with primary headache disorders (Mathew *et al.*, 1990). The following are the clinical features of medication overuse headache:

- The headaches are refractory, daily, or nearly daily.
- The headaches occur in a patient with primary headache disorders who use immediate relief medications very frequently, often in excessive quantities.
- The headache itself varies in its severity, type, and location from time to time.
- Physical or intellectual effort may bring on headache. In other words, the threshold for head pain appears to be low.
- Headaches are accompanied by asthenia, nausea and other gastrointestinal symptoms, restlessness, anxiety, irritability, memory problems, difficulty in intellectual concentration, and depression. Those consuming large quantities of ergot derivatives or triptans may exhibit cold extremities, tachycardia, paresthesias, diminished pulse, hypertension, light-headedness, muscle pain of the extremities, weakness of the legs, and depression.
- There is evidence of tolerance to analgesics over time, with patients needing progressively larger doses.
- Withdrawal symptoms are observed when patients are taken off pain medications abruptly.
- Spontaneous improvement of headache occurs on discontinuing the medications.
- Concomitant prophylactic medications are relatively ineffective while the patients are consuming excess amounts of immediate-relief medications.

## NATURAL COURSE

For the purpose of this chapter, a meta-analysis was performed, summarizing 29 studies comprising a total of 2612 patients with chronic medication overuse headache (for details see (Diener and Dahlöf, 1999)). Sixty-five percent of the patients reported migraine as primary headache, 27% of patients reported tension-type headache, and 8% of patients reported mixed or other headaches (e.g., cluster headache). Women were more prone to medication-overuse headache than men (3.5 : 1; 1533 women, 442 men). This ratio is slightly higher than could be expected from the gender differences in frequency of migraine. The mean duration of primary headache was 20.4 years. The mean admitted time of frequent drug intake was 10.3 years and the mean duration of daily headache was 5.9 years. The mean time of frequent drug intake and chronic headache is much shorter for ergotamine and triptans than for analgesics.

The success rate of withdrawal therapy within a time window of 1–6 months is 72.4% (17 studies, $N = 1101$ patients). Success is defined as no headache at all or an improvement of more than 50% in terms of headache days. Three studies had a longer observation period between 9 and 35 months (Baumgartner *et al.*, 1989; Diener *et al.*, 1989; Schnider *et al.*, 1996). The success rates in these studies were 60, 70, and 73%, respectively. A 5-year follow-up study found a relapse rate of 40% (Schnider *et al.*, 1996).

## PRINCIPLES OF THERAPY

Abrupt drug withdrawal is the treatment of choice for mediaction-overuse headache. There are, however, no prospective and randomized trials comparing continuation of drug intake and drug withdrawal. A survey of 22 studies dealing with therapy of drug-induced headache shows that most centers used drug withdrawal as the primary therapy (see Diener and Dahlöf (1999)). Clinical experience indicates that medical and behavioral headache treatment fails, as long as the patient continues to take symptomatic drugs daily. The typical withdrawal symptoms last for 2–10 days (average 3.5 days) and include withdrawal headache, nausea, vomiting, arterial hypotension, tachycardia, sleep disturbances, restlessness, anxiety, and nervousness. The withdrawal phase is much shorter in patients abusing only triptans. Seizures or hallucinations were only rarely observed even in patients abusing barbiturate-containing migraine drugs.

Drug withdrawal is performed differently. Most authors prefer inpatient programs. Hering and Steiner (Hering and Steiner, 1991) abruptly withdrew the offending drugs on an outpatient basis by adequate explanation of the disorder, regular follow-up, and amitriptyline (10 mg at night) and naproxen (500 mg) for relief of headache symptoms. A consensus paper by the German Migraine Society (Haag et al., 1999) recommends outpatient withdrawal for patients who do not take barbiturates or tranquilizers with their analgesics and are highly motivated. Inpatient treatment should be performed in patients who take tranquilizers, codeine, or barbiturates and who failed to withdraw the drugs as outpatients or who have a high depression score.

Treatment recommendations for the acute phase of drug withdrawal vary considerably between the 22 studies mentioned above. They include fluid replacement, analgesics, tranquilizers, neuroleptics, amitriptyline, valproate, intravenous DHE, oxygen, and electrical stimulation. Valproate has been shown to have beneficial effects in the prophylactic treatment of chronic daily headache complicated by excessive analgesic intake (Mathew and Ali, 1991). A recent large open trial showed, that cortisone effectively reduces withdrawal symptoms, including rebound headache (Krymchantowski and Barbosa, 2000). A double-blind study showed a single subcutaneous dose of sumatriptan to be better than placebo in the treatment of ergotamine withdrawal headache but the headache reappeared within 12 h (Diener et al., 1990). An open randomized study indicated that naproxen was better than symptomatic treatment with antiemetics and analgesics (Mathew, 1987). Further double-blind controlled trials are needed.

## PRACTICAL MANAGEMENT

A careful history is necessary in evaluating chronic headache patients. It is very common that these patients take several different substances daily despite the fact that the effect is negligible. The mechanism behind this behavior is merely an attempt to avoid a disabling withdrawal headache. The present and prior use of prescription drugs, nonprescription compounds, and caffeine intake should be recorded. Many patients also abuse other substances, such as tranquilizers, opioids, decongestants, and laxatives. In addition, it is often helpful to let the patient keep a diagnostic headache diary for 1 month in order to actually register headache pattern and drug use (Russel et al., 1992). History and examination should also search for possible complications of regular drug intake, e.g., recurrent gastric ulcers, anemia, and ergotism. A good indicator is the number of physicians consulted by the patient and the number of previous unsuccessful therapies. Thus in one study the headache patients had consulted an average of 5.5 physicians who had prescribed 8.6 different therapies (Diener et al., 1989).

### Inpatient Treatment

A short hospital stay is recommended if medication-overuse headache has lasted more than 5 years or when additional intake of tranquilizers, barbiturates, or opioids exists. It is further indicated in patients with failed outpatient withdrawal or with concomitant depression or anxiety disorder. All pain or headache medication is stopped abruptly. Fluids should be replaced by infusion if frequent vomiting occurs. Vomiting can be treated with antiemetics e.g. metoclopramide or domperidome. The withdrawal headache can be treated with nonsteroidal anti-inflammatory drugs, e.g., naproxen 500 mg bid. In some countries aspirin is available in injectable form and 1000 mg is given every 8 to 12 h. Dihydroergotamin (DHE) 1–2 mg intravenously every 8 h is given if the headache has migrainous features and if the patient has not abused ergotmaine, DHE, or a triptan before (Raskin, 1986; Silberstein et al., 1990; Silberstein and Silberstein, 1992). Prednisone 100 mg on the first day with tampering by 20 mg for the next days is very effective. Symptoms of opiod withdrawal can be treated with clonidine. The initial dose is 0.1–0.2 mg TID. and titered up or down based on withdrawal symptoms (tachycardia, tremot, sleeping disturbances). Some patients may require anxiolytic medication which should be given for no longer than 1 week. Patients need support by the treating physicians and nurses as well as encouragment from family and friends. Behavioral techniques such as relaxion therapy and stress management should be initiated as soon as the withdrawal symptoms fade.

## Outpatient Treatment

Outpatient treatment is advised in patients who take monosubstances or analgesic mixtures not containing barbiturates or codeine. Patients with migraine as the original headache can start prophylactic medication 4 weeks before withdrawal. Beta blockers will improve withdrawal symptoms such as restlessness, tachycardia, or tremor. In patients with chronic tension-type headache a tricylic may be initiated 4 weeks prior to detoxification (e.g., amitriptyline 10 mg, increasing to 25–75 mg at nighttime). Ergots, triptans and nonopiods should be stopped abruptly. Opioids and barbiturates should be withdrawn more slowly depending on the dose and duration of intake. Withdrawal headache after ergots and triptans can be treated with oral or parenteral NSAIDs (e.g., 500 mg naproxen tid for 5–7 days).

## Long-Term Treatment

If more than three migraine attacks per month continues after withdrawal, medical and behavioral prophylaxis should be initiated. The clinical experience shows that many patients respond to prophylactic treatment, e.g., with beta blockers, flunarizine, or valproic acid after drug withdrawal despite the fact that these drugs seemingly had been unsuccessful before. Ergotamine, triptans, and possibly analgesics counteract the action of prophylactic therapy and will not improve drug-induced headache. The same phenomenon can be observed for the action of amitriptyline and behavioral therapy in patients with tension-type headache.

The most important preventive measure is proper instruction and an appropriate surveillance of patients. The migraine patients at risk often have a mixture of migraine and tension-type headaches and should be instructed carefully to use specific antimigraine drugs only for migraine attacks. This point was already stressed in 1951 by Peters and Horton concerning ergotamine abuse; i.e., complications can be avoided if enough time is taken for proper instruction of the patient, so that he or she can distinguish between vasodilating and nondilating headache (Peters and Horton, 1951).

Restricting the dose of ergotamine per attack (4 mg ergotamine), per week (no more than twice per week), and per month (no more than 20 mg ergotamine) is also helpful in avoiding dependency. In a similar way, the number of doses of triptans should be limited per attack and to 12 doses per month. Migraine drugs that contain barbiturates, caffeine, codeine, or tranquilizers as well as mixed analgesics should be avoided. Patients who nonprescription medication should be advised to avoid caffeine combinations. Probably an early start of migraine prophylaxis, either by medical or behavioral treatment, can be a preventive measure to avoid drug-induced headache.

## REFERENCES

Andersson, P. G. (1988). Ergotism—The clinical picture. In "Drug-Induced Headache" (H. C. Diener and M. Wilkinson, Eds.), pp. 16–19. Springer, Heidelberg, New York.

Bahra, A., Walsh, M., Menon, S., and Goadsby, P. J. (2000). Does chronic daily headache arise de novo in association with regular analgesic use? Cephalalgia 20, 294.

Baumgartner, C., Wessely, P., Bingöl, C., Maly, J., and Holzner, F. (1989). Longterm prognosis of analgesic withdrawal in patients with drug-induced headaches. Headache 29, 510–514.

Castillo, J., Munoz, P., Guitera, V., and Pascual, J. (1999). Epidemiology of chronic daily headache in the general population. Headache 39, 190–196.

Catarci, T., Fiacco, F., Argentino, C., Sette, G., and Cerbo, R. (1994). Ergotamine-induced headache can be sustained by sumatriptan daily intake. Cephalalgia 14, 374–375.

Diener, H. C., Bühler, K., Dichgans, J., Geiselhart, S., Gerber, W. D., and Scholz, E. (1988). Analgetikainduzierter Dauerkopfschmerz. Existiert eine kritische Dosis? Arzneimitteltherapie 6, 156–164.

Diener, H. C., and Dahlöf, C. G. H. (1999). Headache associated with chronic use of substances. In "The Headaches" (J. Olesen, P. Tfelt-Hansen, and K. M. A. Welch, Eds.), Lippincott, Williams & Wilkins, Philadelphia. 2nd ed., pp. 871–878.

Diener, H. C., Dichgans, J., Scholz, E., Geiselhart, S., Gerber, W. D., and Bille, A. (1989). Analgesic-induced chronic headache: Long-term results of withdrawal therapy. J. Neurol. 236, 9–14.

Diener, H. C., Haab, J., Peters, C., Ried, S., Dichgans, J., and Pilgrim, A. (1990). Subcutaneous sumatriptan in the treatment of headache during withdrawal from drug-induced headache. Headache 31, 205–209.

Fanciullacci, M., Alessandri, M., Pietrini, U., Briccolani-Bandini, E., and Beatrice, S. (1992). Long-term ergotamine abuse: Effect on adrenergically induced mydriasis. Clin. Pharm. Ther. 51, 302–307.

Fisher, M. A., and Glass, S. (1997). Butorphanol (Stadol): A study in problems of current drug information and control. Neurology 48, 1156–1160.

Gaist, D., Hallas, J., Sindrup, S. H., and Gram, L. F. (1996). Is overuse of sumatriptan a problem? A population-based study. Eur. J. Clin. Pharmacol. 3, 161–165.

Gaist, D., Tsiropoulus, I., Sindrup, S. H., Hallas, J., Rasmussen, B. K., and Kragstrup, J. (1998). Inappropriate use of sumatriptan: Population based register and interview study. Br. J. Med. 316, 1352–1353.

Granella, F., Farina, S., Malferrari, G., and Manzoni, G. C. (1987). Drug abuse in chronic headache: A clinico-epidemiologic study. Cephalalgia 7, 15–19.

Griffiths, R. R., and Woodson, P. P. (1988a). Caffeine physical dependence: a review of human and laboratory animal studies. Psychopharmakology 94, 437–451.

Griffiths, R. R., and Woodson, P. P. (1988b). Reinforcing properties of caffeine: Studies in humans and laboratory animals. Pharmacol. Biochem. Behav. 29, 419–427.

Haag, G., Baar, H., Grotemeyer, K. H., Pfaffenrath, V., Ribbat, M. J., and Diener, H. C. (1999). Prophylaxe und Therapie des medikamenteninduzierten Dauerkopfschmerzes. *Schmerz* **13**, 52–57.

Headache Classification Committee of the International Headache Society (1988). Classification and diagnostic criteria for headache disorders, cranial neuralgias and facial pain. *Cephalalgia* **8**, 1–93.

Hering, R., and Steiner, T. J. (1991). Abrupt outpatient withdrawal of medication in analgesic-abusing migraineurs. *Lancet* **337**, 1442–1443.

Katsarava, Z., Fritsche, G., Diener, H. C., and Limmroth, V. (2000). Drug-induced headache (DIH) following the use of different triptans. *Cephalalgia* **20**, 293.

Kaube, H., May, A., Diener, H. C., and Pfaffenrath, V. (1994). Sumatriptan misuse in daily chronic headache. *Br. Med. J.* **308**, 1573.

Krymchantowski, A. V., and Barbosa, J. S. (2000). Prednisone as initial treatment of analgesic-induced daily headache. *Cephalalgia* **20**, 107–113.

Lance, F., Parkes, C., and Wilkinson, M. (1988). Does analgesic abuse cause headache de novo? *Headache* **38**, 61–62.

Limmroth, V., Kazarawa, S., Fritsche, G., and Diener, H. C. (1999). Headache after frequent use of new 5-HT agonists zolmitriptan and naratriptan. *Lancet* **353**, 378.

Mathew, N. T. (1987). Amelioration of ergotamine withdrawal with naproxen. *Headache* **27**, 130–133.

Mathew, N. T., and Ali, S. (1991). Valproate in the treatment of persistent chronic daily headache. An open labeled study. *Headache* **31**, 71–74.

Mathew, N. T., Kurman, R., and Perez, F. (1990). Drug induced refractory headache—Clinical features and management. *Headache* **30**, 634–638.

Micieli, G., Manzoni, G. C., Granella, F., Martignoni, E., Malferrari, G., and Nappi, G. (1988). Clinical and epidemiological observations on drug abuse in headache patients. *In* "Drug-Induced Headache" (H. C. Diener and M. Wilkinson, Eds.), pp. 20–28. Springer, Berlin Heidelberg New York.

Mihatsch, M. J., Hofer, H. O., Gudat, F., Knuesli, C., Torhorst, J., and Zollinger, H. U. (1983). Capillary sclerosis of the urinary tract and analgesic nephropathy. *Clin. Nephrol.* **20**, 285–301.

Mihatsch, M. J., and Knüsli, C. (1982). Phenacetin abuse and malignant tumors. *Klin Wochenschr* **60**, 1339–1349.

Peters, G. A., and Horton, B. T. (1951). Headache: With special reference to the excessive use of ergotamine preparations and withdrawal effects. *Proc. Staff Meet Mayo Clin.* **26**, 153–161.

Pini, L. A., and Trenti, T. (1994). Case report: Does chronic use of sumatriptan induce dependence? *Headache* **34**, 600–601.

Rapoport, A., Stang, P., Gutterman, D. L., Cady, R., Markley, H., Weeks, R., Saiers, J., and Fox, A. W. (1996). Analgesic rebound headache in clinical practice: Data from a physician survey. *Headache* **36**, 14–19.

Rapoport, A., Weeks, R., and Sheftell, F. (1985). Analgesic rebound headache: Theoretical and practical implications. *In* "Proceedings Second International Headache Congress, Copenhagen" (J. Olesen, P. Tfelt-Hansen, and K. Jensen, Eds.), pp. 448–449. Kopenhagen.

Raskin, N. H. (1986). Repetitive intravenous dihydroergotamine as therapy for intractable migraine. *Neurology* **36**, 995–997.

Russel, M. B., Rasmussen, B. K., Brennum, J., Iversen, H. K., Jensen, R. A., and Olesen, J. (1992). Presentation of a new instrument: The diagnostic headache diary. *Cephalalgia* **12**.

Saper, J. R., and Jones, J. M. (1986). Ergotamine tartrate dependency: Features and possible mechanisms. *Clin. Neuropharmacol.* **9**, 244–256.

Schnider, P., Aull, S., Baumgartner, C., Merterer, A., Wöber, C., Zeiler, K., and Wessely, P. (1996). Long-term outcome of patients with headache and druf abuse after inpatient withdrawal: Five year follow-up. *Cephalalgia* **16**, 481–485.

Silberstein, S. D., Schulman, E. A., and McFaden Hopkins, M. (1990). Repetitive intravenous DHE in the treatment of refractory headache. *Headache* **30**, 334–339.

Silberstein, S. D., and Silberstein, J. R. (1992). Chronic daily headache: Long-term prognosis following inpatient treatment with repetitive i.v. DHE. *Headache* **32**, 439–445.

Silverman, K., Evans, S. M., Strain, E. C., and Griffiths, R. R. (1992). Withdrawal syndrome after the double-blind cessation of caffeine consumption. *N. Engl. J. Med.* **327**, 1109–1114.

van Dusseldorp, M., and Katan, M. B. (1990). Headache caused by caffeine withdrawal among moderate coffee drinkers switched from ordinary to decaffeinated coffee: A 12 week double blind trial. *Br. Med. J.* **300**, 1558–1559.

Ziegler, D. K. (1994). Opiate und opioid use in patients with refractory headache. *Cephalalgia* **14**, 5–10.

## CHAPTER 5
# Other Forms of Headache

Stephen D. Silberstein and Volker Limmroth

## HEMICRANIA CONTINUA

Hemicrania continua is a rare, indomethacin-responsive headache disorder characterized by a continuous, moderate to severe, unilateral headache that varies in intensity, waxing and waning without disappearing completely (Silberstein *et al.*, 1996). While the essential feature of hemicrania continua is unilateral headache, some bilateral or alternating side cases have been reported (Newman *et al.*, 1992; Pasquier *et al.*, 1987). It is frequently associated with jabs and jolts (idiopathic stabbing headache). Exacerbations of pain are often associated with autonomic disturbances, such as ptosis, miosis, tearing, and sweating. Hemicrania continua is not triggered by neck movements, but tender spots in the neck may be present. Some patients may have photophobia, phonophobia, and nausea.

Hemicrania continua almost invariably has a prompt and enduring response to indomethacin. However, the requirement of a therapeutic response to make a diagnosis is problematic, since it excludes the diagnosis of hemicrania continua in patients who were never tried on or who failed to respond to indomethacin. Cases have been described that did not respond to indomethacin but meet the phenotype; for this reason Goadsby and Lipton (1997) have provided an alternate means of diagnosis (Table I).

### Earliest Descriptions

Medina and Diamond (1981) probably were the first authors to describe hemicrania continua (including a response to indomethacin) in a subset of their 54 cluster-headache variant patients. The term hemicrania continua was coined by Sjaastad and Spierings (1984). They reported two patients, a woman, age 63, and a man, age 53, who each developed a strictly unilateral headache that was continuous from onset and absolutely responsive to indomethacin. Boghen and Desaulniers (1983), described a patient with a similar headache that they called "background vascular headache responsive to indomethacin." The patient was a 49-year-old man who had a 20-year history of left-sided headache that rarely radiated to the right side. The headache had a sustained, pressure-like quality and was associated with intermittent jabs of pain. Treatment with 50 mg of indomethacin completely relieved the pain.

Approximately 100 cases of hemicrania continua have been reported, but there is still uncertainty about its clinical features. Atypical features, including bilaterality, alternating-side headaches, and unresponsiveness to indomethacin have been described. Two secondary cases, one a patient with a mesenchymal tumor (Antonaci and Sjaastad, 1992) and another with HIV (Brilla *et al.*, 1998), have been reported. Peres *et al.* (2001) recently reported 34 new cases seen at Jefferson Headache Center. Of all the cases where gender data are available (61 women and 22 men), the female:male ratio is 2.8:1. The sex ratio in the Peres *et al.* (2001) 34 patients (2.4:1 female:male ratio) is similar to the overall ratio.

Hemicrania continua exists in continuous and remitting forms. In the remitting variety, distinct headache phases last weeks to months, with prolonged pain-free remissions (Iordanidis and Sjaastad, 1989; Newman *et al.*, 1993, 1994; Pareja *et al.*, 1990). In the continuous variety, headaches occur on a daily, continuous basis, sometimes for years. The continuous variety can be subclassified into: (1) an evolutive, unremitting form that arises from the remitting form and (2) an unremitting form characterized by continuous headache from the onset (Bordini *et al.*, 1991). Hemicrania continua takes precedence over the diagnosis of other types of chronic daily headache.

TABLE I    Proposed Criteria for Hemicrania Continua (adapted from Goadsby and Lipton, 1997)

A. Headache present for at least 1 month
B. Strictly unilateral headache
C. Pain has all three of the following present:
   1. Continuous but fluctuating
   2. Moderate severity, at least some of the time
   3. Lack of precipitating mechanisms
D. Absolute response to indomethacin or one of the following autonomic features with severe pain exacerbation:[a]
   (a) Conjunctival infection
   (b) Lacrimation
   (c) Nasal congestion
   (d) Rhinorrhea
   (e) Ptosis
   (f) Eyelid edema
E. May have associated stabbing headaches
F. Not attributed to another disorder

[a]Silberstein *et al.* (1996) criteria lack D.2.

Three main categories of associated symptoms occur: autonomic symptoms, "jabs and jolts," and migrainous features. Autonomic symptoms consist of conjunctival injection, tearing, rhinorrhea, nasal stuffiness, eyelid edema, and forehead sweating. They are not as prominent in hemicrania continua as they are in cluster headache and chronic paroxysmal hemicrania. Of 41 cases with available description, autonomic symptoms were reported in 26 patients (63%) and were absent in 15 (37%). The most common symptom described by Peres *et al.* (2001) and in the literature was tearing, but conjunctival injection and ptosis were slightly more frequent in the literature than Peres *et al.* (2001) observed. Autonomic symptoms commonly occur during pain exacerbations rather than during the baseline headache. Symptoms of ocular discomfort, at times premonitory, have been described in patients with hemicrania continua (Pareja, 1999). Some patients report a feeling of sand in the eye, which may be specific for hemicrania continua.

Jabs and jolts syndrome, described by Sjaastad *et al.* (1993) as sharp, knife-like pains, less than 1 min in duration, occurs in patients with tension-type, migraine, and cluster headache and in headache-free individuals. Its prevalence in the general population is 35% (Sjaastad *et al.*, 2001). We found 12 descriptions (26%) of jabs and jolts in the 46 cases where the information could be ascertained. It occurred in 14 (41%) of the patients studied by Peres *et al.* (2001).

Migrainous features are common in hemicrania continua. Twenty-three (50%) of the Peres *et al.* patients had at least one symptom (nausea, vomiting, photophobia, or phonophobia). It was not possible to apply diagnostic criteria for migraine in the previously reported cases because of lack of information. Nausea, vomiting, photophobia, and phonophobia also occur in cluster headache and chronic paroxysmal hemicrania (Bordini *et al.*, 1991).

Associated neurologic signs and symptoms include attacks of hemiparesis with a familial history of hemiplegic migraines (Evers *et al.*, 1999) and unilateral paresthesias (Pasquier *et al.*, 1987). Antonaci reported visual phenomena described as dark spots in a case of hemicrania continua secondary to a mesenchymal tumor (Antonaci and Sjaastad, 1992). Exacerbations are one of the most common features of hemicrania continua, occurring in 74% of patients. Nocturnal exacerbations occur in hemicrania continua and could result in a mistaken diagnosis of cluster or hypnic headache. Fourteen of 46 cases (30%) had nocturnal exacerbations, sometimes more than once, that usually lasted 1–2h. Peres *et al.* found 10 (29.4%) of their patients had nocturnal exacerbations.

## TREATMENT AND PROGNOSIS

In the past, hemicrania continua has been defined by its response to indomethacin, with a dose range of 50 to 300 mg a day. Other successful treatments include ibuprofen, 800 mg (Kumar and Bordiuk, 1991), piroxicam beta-cyclodextrin, 20 to 40 mg a day (Trucco *et al.*, 1992), and rofecoxib, 25 mg (Peres and Zukerman, 2000). Sumatriptan (Antonaci *et al.*, 1998b) and many other analgesics are not effective. Espada *et al.* (1999) reported five men and four women who had hemicrania continua diagnosed using the Goadsby and Lipton diagnostic criteria (Goadsby and Lipton, 1997). Eight of the patients had continuous hemicrania continua and one remitting. The mean age of onset of 53.3 years (range 29 to 69). All of the patients had initial relief with indomethacin (mean daily dose, 94.4 mg; range 50 to 150). Follow-up was possible in eight patients. Three patients were able to discontinue indomethacin after 3, 7, and 15 months, respectively, and remained pain-free. Three patients discontinued treatment because of side effects and had headache recurrence; two of these had relief with aspirin, however. Two other patients continued to take indomethacin with partial relief.

A recently reported option for the diagnosis of indomethacin-responsive headaches is the so-called "indotest" (Antonaci *et al.*, 1998a). Twelve patients with hemicrania continua were given 50 mg of intramuscular indomethacin, and some were given 100 mg on a second day. The time between indomethacin injection and complete pain relief was 73 ± 66 min with the 50-mg injection and 61 ± 56 min with the 100-mg injection. The pain-free period after the 50-mg injection was 13 ± 8 h and 13 ± 10 h after the 100-mg injection. The authors suggest a standard dose of 50 mg of intramuscular indomethacin with observation up to 3 h, since relief occurred in all patients by 2 h.

## DIFFERENTIAL DIAGNOSIS

Hemicrania continua is differentiated from cluster headache and chronic paroxysmal hemicrania primarily by its continuous moderate pain and the lack of autonomic features between the painful exacerbations. Hemicrania continua can be aggravated by a C7 root irritation due to a disc herniation (Sjaastad *et al.*, 1995). A case of a mesenchymal tumor in the sphenoid bone in which the response to indomethacin faded after 2 months has also been reported (Antonaci and Sjaastad, 1992). These cases suggest that escalating doses or loss of indomethacin's efficacy should be treated with suspicion and the patient reevaluated. The condition is seen in noncaucasian populations (Joubert, 1991).

The intermittent form of hemicrania continua has features in common with migraine. Unilateral headache occurs in 60% of migraineurs and never alternates sides in 20% (Sjaastad *et al.*, 1980). Both hemicrania continua and migraine have a female preponderance and a similar age of onset. The distribution of pain in hemicrania continua and migraine are similar, with a preponderance in the forehead and temporal regions. Nausea, photophobia, and phonophobia are common in hemicrania continua and migraine. Migraine and hemicrania continua patients report similar triggers: alcohol, weather changes, exertion, stress, odors, chocolate, bananas, and cheese. Jabs and jolts (idiopathic stabbing headache) is common in migraine and universal in hemicrania continua.

The continuous form of hemicrania continua is in the differential diagnosis of the primary chronic daily headaches, which also includes chronic migraine, chronic tension-type headache, and new daily persistent headache. Chronic tension-type headache is characterized by a relative lack of autonomic symptoms (including nausea, vomiting, and photo- and phonophobia). These are seen in some, but not all, patients with hemicrania continua. Chronic migraine evolves out of episodic migraine headache and may retain many migrainous features, as do many cases of hemicrania continua. Chronic migraine is often seen when there is acute medication overuse, which can occur with hemicrania continua. New daily persistent headache is defined as a new headache of sudden onset that occurs without a clear prior history of migraine; however, it often has migrainous features (Silberstein, 1993). This is similar to the continuous form of hemicrania continua. Chronic tension-type headache, transformed migraine, new daily persistent headache, and hemicrania continua may all be unilateral. The only factor that

separates hemicrania continua is indomethacin responsiveness. This responsiveness does not necessarily imply that a unique anatomic or physiologic defect causes the pain of hemicrania continua. Indomethacin may ameliorate pain in all forms of chronic daily headache, but an exquisite sensitivity to indomethacin independent of the headache illness may indicate a unique pathophysiology for hemicrania continua.

## SHORT-LASTING UNILATERAL NEURALGIFORM HEADACHE WITH CONJUNCTIVAL INJECTION AND TEARING

Short-lasting unilateral neuralgiform headache with conjunctival injection and tearing (SUNCT) is a very rare headache syndrome first described by Sjaastad *et al.* (1978). The clinical picture consists of brief attacks of moderate to severe orbital/periorbital pain accompanied by ipsilateral conjunctival injection, lacrimation, and nasal obstruction or rhinorrhea (Sjaastad *et al.*, 1978, 1989). The long-term temporal pattern is largely unpredictable. Symptomatic periods alternate with spontaneous remissions in an erratic fashion (Table II).

SUNCT patients are usually men, with a male:female gender ratio of 17:2 (Pareja and Sjaastad, 1994). The pain occurs in paroxysms that last between 5 and 300 s (Pareja *et al.*, 1996c), although longer, duller interictal pains have been reported, and two patients have had attacks that lasted up to 2 h (Pareja *et al.*, 1996a). Patients may have as many as 30 episodes an hour, but five to six attacks an hour is the norm. Attack frequency may vary, from as many as 20 attacks a day to once or twice in 1–4 weeks (Sjaastad *et al.*, 1991). Attacks can occur in an almost status-like pattern. A systematic study of attack frequency demonstrated a mean of 28

TABLE II Diagnostic Criteria for Short-Lasting Unilateral Neuralgiform Headache with Conjunctival Injection and Tearing (SUNCT)

A. At least 30 attacks fulfilling B–E, below
B. Attacks of unilateral, moderately severe orbital or temporal stabbing or throbbing pain lasting from 15 to 120 s
C. Attack frequency from 3 to 100/day
D. Pain is associated with at least one of the following signs or symptoms of the affected side with feature 1 being most often present and very prominent:
  1. Conjunctival injection
  2. Lacrimation
  3. Nasal congestion
  4. Rhinorrhea
  5. Ptosis
  6. Eyelid edema
E. Not attributable to another disorder

*Note.* The literature suggests that the most common secondary cause of SUNCT would be a lesion in the posterior fossa.

attacks a day, with a range of 6 to 77 (Pareja *et al.*, 1996b).

In most patients the paroxysms were moderate to severe; they can be very severe but this is less common. The pain is usually described as burning, stabbing (piercing, stinging-like, pricking), or electric. Less frequent descriptions include steady and staccato-like pain. The pain onset is abrupt, the maximum intensity is usually reached within 2–3 s, and the descrescendo phase is usually also abrupt. The pattern of solitary attacks is usually plateau-like, but other patterns, such as repetitive and sawtooth-like, have been described. Attacks can occur at any time of day, and during the worst periods nocturnal attacks can occur. Conjunctival injection and lacrimation are prominent accompaniments. Rhinorrhea or nasal obstruction occur but are not invariably present. The autonomic features can be bilateral but there is a clear symptomatic-side preponderance. Most patients describe precipitating mechanisms. Attacks can be triggered by applying stimuli to the trigger zone just after the preceding paroxysm had finished (Sjaastad *et al.*, 1989) or even during its decrescendo phase. SUNCT can exhibit a clustering pattern. Paroxysms are usually located in the VI division of the trigeminal area, but the precipitating mechanisms are usually located within the V2 and V3 divisions of the trigeminal nerve and also within extratrigeminal areas, such as the neck (Calvo *et al.*, 1997).

Three patients with secondary SUNCT syndromes have been reported. The first two patients had homolateral cerebellopontine angle arteriovenous malformations and the third had a cavernous hemangioma of the brainstem (Morales *et al.*, 1994). A posterior fossa lesion causing otherwise typical SUNCT has also been noted in a patient with human immunodeficiency virus/acquired immunodeficiency syndrome (Graff-Radford, personal communication).

Orbital phlebography is reported to be abnormal in patients with SUNCT, showing a narrowed superior ophthalmic vein homolateral to the pain (Kruszewski, 1992). Forehead sweating is usually increased during bouts of SUNCT (Kruszewski *et al.*, 1993). Pupillary studies are normal (Zhao and Sjaastad, 1993). Since conjunctival injection occurs during SUNCT, it is not surprising that intraocular pressure and corneal temperatures are elevated during attacks. This most likely reflects marked parasympathetic activation with local vasodilation. Similarly, bradycardia in association with attacks of SUNCT may indicate increased parasympathetic outflow (Kruszewski *et al.*, 1992). The parasympathetic manifestations favor a central pathogenesis for SUNCT as a manifestation of the trigeminovascular reflex (Goadsby *et al.*, 1991), rather than a peripheral vasculitic cause. Transcranial Doppler and SPECT studies have not demonstrated convincing changes in the vasomotor activity (Shen and Johnsen, 1994) or cerebral blood flow (Poughias and Aasly, 1995) during attacks of pain.

May *et al.* (1999) described a 71-year-old woman who presented with a short history of episodes of severe left-sided orbital and temporal pain in paroxysms that lasted 60 to 90 s and were accompanied by rhinorrhea, conjunctival injection, and ipsilateral lacrimation of the eye. Her clinical examination and structural imaging were normal, and a clinical diagnosis of SUNCT was made. The patient had a BOLD contrast MRI study in which significant activation was seen in the region of the ipsilateral hypothalamic gray. The region of activation was the same in this patient as has been reported in acute attacks of cluster headache. If the biological mode of posterior hypothalamic activation is correct, the underlying cause for trigeminal autonomic cephalgias (Goadsby and Lipton, 1997) may be similar, and the variation in duration and frequency might be generally dependent on a different disorder of the inferior posterior hypothalamic neurons.

### Treatment

SUNCT is refractory to all treatments, including indomethacin, and procedures that are useful for other short-lasting headaches are not useful in SUNCT (Pareja *et al.*, 1995). Two patients who had a provisional diagnosis of SUNCT and responded to sumatriptan most likely represent a spontaneous remission and are only described in a limited way (Ghose, 1995).

## IDIOPATHIC STABBING HEADACHE

Pareja *et al.* (1996d) studied the clinical features of idiopathic stabbing headache ("jabs and jolts syndrome") in 38 patients who were diagnosed over a 1-year period. Mean age at the onset of symptoms was $47.1 \pm 14.5$ years (SD), and a clear female preponderance was demonstrated (female/male ratio, 6.6). Painful attacks were ultrashort; i.e., virtually all attacks in more than two-thirds of cases lasted only 1 s. The frequency of attacks varied immensely, ranging from one attack per year to 50 attacks daily. The pain paroxysms usually occurred with an irregular or sporadic temporal pattern. The localization of painful attacks was reported frequently as unifocal, usually in the orbital area, but multifocal patterns were also observed, with the attacks frequently changing location from one area to the next. The majority of attacks occurred spontaneously, and accompanying phenomena were reported only rarely. Indomethacin (75 mg daily) had a complete or partial effect in most patients.

The International Headache Society classification for head and face pain uses the term idiopathic stabbing headache for a disorder first described by Lansche (1964) as ophthalmodynia periodica. In 1979, Sjaastad et al. (1979, 1980) used the term "jabs and jolts" syndrome, inspired by the patients' own descriptions; they complained of either short-lasting head pain (jabs) or head pain so marked that it was accompanied by a shock-like feeling caused simply by head movement ("I have the jolts"). This term is now officially acknowledged by the IASP. Raskin and Schwartz (1980) later reported ice-pick-like headache in migraine sufferers. Terms such as "needle-in-the-eye" syndrome and sharp, short-lived head pain syndrome have been used (Mathew, 1980). The International Headache Society diagnostic criteria are (Headache Classification Committee of the International Headache Society, 1988):

(A) Pain is confined to the head and exclusively or predominantly felt in the distribution of the first division of the trigeminal nerve (orbit, temple, and parietal area).
(B) Pain is stabbing in nature and lasts for a fraction of a second. Occurs as single stabs or series of stabs.
(C) It recurs at irregular intervals (hours to days).
(D) Diagnosis depends upon the exclusion of structural changes at the site of pain and in the distribution of the affected cranial nerve.

The head pain in idiopathic stabbing headache may be the most short-lasting of all headaches (1–2 s), followed by trigeminal neuralgia (a few seconds), and short-lasting unilateral neuralgiform headaches with conjunctival injection (SUNCT), with 15 s to 1 min or more (Sjaastad and Kruszewski, 1992; Sjaastad et al., 1989). Sjaastad et al. (2001) carried out an epidemiologic study of headache in a rural parish in southern Norway including 1838 parishioners ranging in age from 18 to 65 (88.6% of the target group) were examined in a structured interview based on a questionnaire. Idiopathic stabbing headache ("jabs and jolts syndrome") was present in >30% of parishioners. Rasmussen and Olesen (1992) assessed the lifetime prevalence of headache disorders in a cross-sectional epidemiologic survey of a representative 25- to 64-year-old in the general population. They classified the headaches on the basis of a clinical interview and a physical and neurologic examination using the operational diagnostic criteria of the International Headache Society. Lifetime prevalence of idiopathic stabbing headaches was 2%, of external compression headache 4%, and of cold stimulus headache 15%. Benign cough headache, benign exertional headache, and headache associated with sexual activity each occurred in 1% of the interviewees.

Idiopathic stabbing headache is rather frequent, occurring in 33/100,000/year in the hospital series. Pareja et al. (1996d) obtained a history of other headaches in more than half of their idiopathic stabbing headache patients. Migraine was the most frequently experienced associated headache. When idiopathic stabbing headache and migraine attacks appeared simultaneously, they frequently did so in the same region. The female preponderance might be explained partly by the fact that the associated headache syndromes have a female preponderance. Most patients reported attacks of 1 s duration, and the mean duration (i.e., 2 s) was consistent with that described previously. The longer duration reported by some patients may reflect an actual phenomenon or be due to an overestimation. In the subgroup of patients with orbital attacks of 5 to 10 s, the differential diagnostic problem is SUNCT. Idiopathic stabbing headache is a primary benign headache. It is a fairly frequent complaint, usually with its onset in the middle or late stages of life. Its main clinical features are (1) brief paroxysms with a unifocal or multifocal location in the head, the pain not infrequently changing localization; (2) marked variability in the frequency of attacks; (3) an irregular or sporadic temporal pattern; (4) a paucity of accompanying phenomena or precipitating mechanism; (5) coexistence with other headaches, with or without a temporal concurrence of both types of pain; (6) female preponderance; and (7) partial responsiveness to indomethacin.

Martins et al. (1995) described six patients who had an identical type of headache, which consisted of short episodes (lasting around one week) of daily attacks of ice-pick-like pain that recurred every minute in the same points of the scalp. The pain was felt outside the cutaneous area of the trigeminal nerve (retroauricular, parietal, and occipital regions). All of the patients had been seen in the emergency department of a general hospital over a period of 7 years because of these acute headaches. None of the patients had a history of migraine. Although the pain is identical to idiopathic stabbing headache, it differs by its temporal profile, its posterior (extratrigeminal) location, and its lack of association with migraine. While the bouts were severe and recurred in two patients, they were all self-limited, benign, and responded promptly to indomethacin. Although idiopathic stabbing headache may occur as a primary entity, it is more likely to appear with another headache type. Lance and Anthony (1971) and Ekbom (1975) described cluster headache patients in whom these pains often herald the end of an attack. Patients with migraine are also particularly subject to sharp, jabbing pains (Lansche, 1964; Raskin and Schwartz, 1980). Drummond and Lance (1984) found that ice-pick pains were coincidental with the site of a patient's usual headache in 40% of migraineurs. Similar head pains have been described in giant cell arteritis (Russell, 1959) and in the cluster variant syndrome (Medina and Diamond, 1981).

Raskin and Schwartz (1980) studied the incidence and clinical characteristics of sharp, jabbing pain about the head in 100 migraineurs and 100 control subjects. Three patients in the control group reported sharp, jabbing head pains at least once annually. Forty-two of the migraineurs admitted to this symptom; 25 reported attacks more than monthly. The pain usually occurred as a single jab (64%), but could also occur as a volley of jabs (12%). The pain was usually unifocal, occurring most often in the temple or orbital regions. The foci were usually ipsilateral when multifocal. Seven of the migraineurs and one control subject reported at least one of the following precipitants to their sharp jabbing pain: light, sudden postural change, physical exertion, transition to dark, and head motion during headache. In seven patients, these ice-pick pains preceded an attack of migraine.

## EXERTIONAL AND COUGH HEADACHES

Although coughing and exertion rarely provoke headache, they can aggravate any type of headache. However, transient, severe head pain upon coughing, sneezing, weight lifting, bending, straining at stool, or stooping defines cough headache. Originally described by Tinel (1932a) in 1932 as "la céphalée à l'effort" and later by Symonds (1956), cough headache mainly affects middle-aged men. It runs its course over a few years and is uncommon in the clinic: at the Mayo Clinic over a 14-year period, Rooke (1968) made only 93 diagnoses. He proposed the broader term benign exertional headache for any headache that is precipitated by exertion, has an acute onset, and is unassociated with structural central nervous system disease, thus combining cough and exertional headache. In a population-based study (Rasmussen et al., 1991), benign cough headache and benign exertional headache each had a prevalence of about 1%.

The most recent classification of these disorders was done by the International Headache Society (Headache Classification Committee of the International Headache Society, 1988). The International Headache Society separates benign cough headache and benign exertional headache since these entities have different clinical features, diagnostic evaluations, and treatment responses (Pascual et al., 1996; Sands et al., 1991).

Headache associated with sexual activity describes bilateral headaches precipitated by masturbation or coitus in the absence of any intracranial disorder. Benign sexual headache is now a well-defined entity, with three types recognized in the International Headache Society classification (Headache Classification Committee of the International Headache Society, 1988). They are described according to the presumed clinical pathophysiologic mechanism. The most frequent, type 2, begins suddenly at the time of orgasm and is thought to be related to hemodynamic changes. It is often associated with exertional headache.

### Benign Cough Headache

Benign cough headache is infrequent. The mean age of onset is 55 years, with a range of 19 to 73 years. It is twice as common in patients over 40 years of age and is four times more common in men than in women (Rooke, 1968). The pain begins immediately (Nick, 1980; Symonds, 1956) or within seconds after coughing, sneezing, or Valsava maneuver (lifting, straining at stool, blowing, crying, or singing) (Nightingale and Williams, 1987; Tinel, 1932a,b). The pain is severe in intensity, with a bursting, explosive, or splitting quality that lasts a few seconds or minutes. The headache is usually bilateral, with maximal pain at the vertex or in the occipital, frontal, or temporal region. Bending the head or lying down may be impossible (Mathew, 1980). The headache is not generally associated with nausea or vomiting and the neurologic examination is usually normal. Vomiting suggests an organic basis for the headache (Sands et al., 1991). As many as 25% of cases have an antecedent respiratory infection (Raskin, 1988; Rooke, 1968; Symonds, 1956). Most patients are pain-free between attacks of head pain, but in some cases the paroxysms are followed by dull, aching pain that may persist for hours; 5 of the 21 patients reported by Symonds had such additional headaches (Symonds, 1956). As these patients often express their complaint as a continuous headache, they should be asked directly about the role of exertion as a trigger factor.

### Symptomatic Cough Headache

Age at onset is significantly lower for symptomatic cough headache than for benign cough headache. Pascual et al. (1996) found that symptomatic cough headache could be precipitated by laughing, weight lifting, or acute body or head postural changes in addition to coughing. Symptomatic cough headache can be caused by hind brain abnormalities, posterior fossa meningiomas, mid-brain cysts, basilar impression, acoustic neurinoma, and brain tumor such as Arnold-Chiari malformation (Raskin, 1978; Rushton and Rooke, 1962; Williams, 1980). In the series of Pascual et al. (1996) headache was the only symptom, at first, of Arnold-Chiari Type I malformation in three patients;

however, all patients had or eventually developed posterior fossa signs or syringomyelia. Besides beginning earlier in life than benign cough headache, symptomatic cough headache does not respond to indomethacin. If a patient does not respond to indomethacin, has posterior fossa signs, or is younger than 50, MRI must be done. Cough headache can be confused with other disorders, such as exertional headache, effort migraine, and coital headache (Ekbom, 1986; Raskin, 1988). In fact, 40% of patients with coital headache of the vascular type had exertional headache, suggesting a relationship between these entities (Silbert *et al.*, 1991).

## Benign Exertional Headache

Benign exertional headache begins almost 40 years earlier than benign cough headache ($P < 0.005$). It is typically throbbing, lasts from 5 min to 24 h, and is provoked by physical exercise. The pain usually begins during exertion, is nonexplosive, and can be either bilateral or unilateral.

## Symptomatic Exertional Headache

Symptomatic exertional headache is usually explosive in onset, severe, and bilateral. Twelve of the Pascual *et al.* (1996) patients presented because of acute headache that coincided with physical exercise. Etiologies included subarachnoid hemorrhage, sinusitis, and brain mass.

## Benign Sexual Headache

Benign sexual headache (Type 2) begins later than benign exertional headache and earlier than benign cough headache. The headache is usually bilateral, but it can be unilateral. It is usually severe and explosive with occasional throbbing and stabbing. Its duration varies, lasting from less than 1 min to 3 h (average ½ h). The frequency of the episodes was directly related to that of sexual intercourse or masturbation. Up to one-third of patients have similar episodes with physical exertion.

## Symptomatic Sexual Headache

Symptomatic sexual headache is an explosive headache that occurs during coitus and is usually symptomatic of a subarachnoid hemorrhage. Rooke (1968) followed 103 patients who had exertional headache but no detectable intracranial disease on initial examination. After 3 or more years of followup, 10 patients subsequently developed organic intracranial lesions. Thirty of the remaining 93 patients had complete headache relief within 5 years. The remainder improved or were headache-free after 10 years. This precomputed tomography-era study emphasizes the importance of careful evaluation for organic disease. Apart from that of Rooke (1968), the largest series of headaches of sudden onset provoked by cough, physical exercise, or sexual excitement was performed by Pascual *et al.* (1996). Benign and symptomatic cases differed in several clinical aspects. Symptomatic cough headache began earlier in life, tended to last longer, and was more frequent than benign cough headache. Chiari type I malformation was the only cause. SAH, sinusitis, and brain metastases were the causes of symptomatic exertional headache.

Sands *et al.* (1991) reviewed 219 cases of exertional and cough headache. The study summarized cases of cough and exertional headache, grouping together etiologically related cases from several selected series. Of the 219 cases, 48 had an identifiable organic etiology. The group with posterior fossa space-occupying lesions includes cases with Arnold–Chiari deformity and hindbrain herniation. Post-traumatic and postcraniotomy cases were also grouped together. From this review, one cannot accurately estimate how many patients with exertional headache have structural disease (Ekbom, 1986; Ibbotson, 1987; Mathew, 1980; Nick, 1980; Nightingale and Williams, 1987; Paulson, 1983; Powell, 1982; Rooke, 1968; Symonds, 1956; Tinel, 1932b; Williams, 1980).

Symptomatic exertional and sexual headaches began later in life and lasted longer than benign exertional and sexual headaches. Male predominance was not present in the symptomatic exertional headache group. Furthermore, all patients with symptomatic headaches had manifestations of meningeal irritation or intracranial hypertension. Patients with subarachnoid bleeding had only had one headache episode. Although neuroradiological studies could be avoided in cases with clinically typical benign sexual or exertional headaches (men around the third decade of life, with a normal examination and short-duration, multiple episodes of pulsating pain that responded to ergotamine or to preventive beta blockers), the remaining patients must have a brain computed tomography (and a cerebrospinal fluid examination if the computed tomography scan is normal).

Benign cough headache and benign exertional headache are separate conditions. Besides the different precipitants (sudden Valsalva maneuvers and sustained physical exercise, respectively), benign cough headache begins later than benign exertional headache. Cough headache starts 43 years later, on average, than exertional headache, and while the youngest patient with benign cough headache in this series was 44 years old, the oldest patient with benign exertional headache was 48. Benign cough headache tended to be shorter than benign exertional headache and the pain quality and response to treatment were different. Benign cough headache was described as sharp or stabbing and responsive to indomethacin, whereas benign exertional headache was pulsating, tended to last longer, and improved with ergotamine or propranolol. It is not uncommon for patients to experience both benign sexual headache and benign exertional headache; this occurred in 31% of patients in the Pascual *et al.* (1996) series.

Other types of headache may be exacerbated by exertion. Severe migraine, postlumbar puncture headache, and, rarely, pseudotumor cerebri may be aggravated by coughing (Silberstein and Marcelis, 1993). Paroxysmal headache may also occur in patients with third ventricular colloid cysts, which may produce intermittent obstruction of cerebrospinal fluid flow through the foramen of Monroe, resulting in abrupt increases of intracranial pressure. Similar headaches have also been experienced by patients who have lateral ventricular tumors, craniopharyngiomas, pinealomas, and tumors of the cerebellum and cerebrum. Pheochromocytomas (Lance and Hinterberger, 1976; Paulson *et al.*, 1979; Thomas *et al.*, 1966) may also cause paroxysmal headache, especially during exercise. In Rooke's series of 303 patients with intracranial lesions, however, none of the 27 patients with unruptured cerebral aneurysm or vascular anomaly complained of exertional headache (Rooke, 1968).

Van den Bergh *et al.* (1987) reported two patients with Arnold-Chiari malformation who had headache paroxysms brought on by coughing, sneezing, and laughing. In Arnold-Chiari malformation, a ball-valve mechanism may be responsible for CSF passing more easily from the spine to the cranium than vice versa. Lumbar CSF pressure waves following a cough occur sooner and rise higher than cisternal pressure waves; the lumbar pressure waves also fall sooner and lower. Therefore, there is a phase during which the lumbar pressure exceeds the cisternal pressure, followed by a phase in which the cisternal pressure is greater than the lumbar pressure; this is aggravated by the ball-valve effect and can produce cough headache (Williams, 1976). Stevens *et al.* (1993) retrospectively studied 141 patients with adult Arnold-Chiari malformation (defined as descent of the hindbrain into the cervical canal, with meningomyelocele absent and hydrocephalus rare). Headache was present in 41 patients and was considered a symptom only if it was exacerbated by head movement, exercise, or coughing.

Outcomes of preoperative cough and posture-related headache showed no relationship to tonsillar descent or any other imaging parameter, including the size of the cisterna magna, yet the latter feature was significantly improved in 62.6% of cases. Posture- and cough-related headache, like drop attacks, are thought to result from intermittent tonsillar impaction in the foramen magnum. Therefore, the lack of association of such features with small or obliterated cisterna magna or low-lying tonsils both pre- and postoperatively suggests that the origin of these symptoms is more complex (Stevens *et al.*, 1993).

The long-term outlook for these patients is favorable. If the headaches are frequent or severe, prophylactic therapy is required, as the short duration of the headaches renders abortive therapy impractical. Some patients respond dramatically to indomethacin in doses of 25 to 150 mg daily (Mathew, 1980). If there is gastrointestinal intolerance to indomethacin, concomitant treatment with misoprostol, sulcralfate, or antacids may be helpful. When indomethacin fails, Raskin (1988) reports that naproxen, ergonovine, and phenelzine are useful, but propranolol is not.

Symonds (1956) performed four lumbar punctures or pneumoencephalograms on each of 21 patients with benign cough headache syndrome. One patient developed a typical postlumbar puncture headache following the lumbar puncture, but her cough headache syndrome remitted for 3 weeks and then recurred. Some patients with benign cough headache respond to lumbar puncture to treat the disorder (Raskin, 1988).

## REFERENCES

Antonaci, F., Pareja, J. A., Caminero, A. B., and Sjaastad, O. (1998a). Chronic paroxysmal hemicrania and hemicrania continua. Parenteral indomethacin: The "indotest." *Headache* 38, 122–128.

Antonaci, F., Pareja, J. A., Caminero, A. B., and Sjaastad, O. (1998b). Chronic paroxysmal hemicrania and hemicrania continua: Lack of efficacy of sumatriptan. *Headache* 38, 197–200.

Antonaci, F., and Sjaastad, O. (1992). Hemicrania continua: A possible symptomatic case, due to mesenchymal tumor. *Funct. Neurol.* 7, 471–474.

Boghen, D., and Desaulniers, N. (1983). Background vascular headache: Relief with indomethacin. *Can. J. Neurol. Sci.* 10, 270–271.

Bordini, C., Antonaci, F., Stovner, L. J., Schrader, H., and Sjaastad, O. (1991). "Hemicrania continua": A clinical review. *Headache* 31, 20–26.

Brilla, R., Evers, S., Soros, P., and Husstedt, I. W. (1998). Hemicrania continua in an HIV-infected outpatient. *Cephalalgia* 18, 287–288.

Calvo, J. F., Tinetti, N., and Leston, J. (1997). SUNCT. The first Argentinian case. *J. Neurol. Sci.* 150, S34.

Drummond, P. D., and Lance, J. W. (1984). Neurovascular disturbances in headache patients. *Clin. Exp. Neurol.* 20, 93–99.

Ekbom, K. (1975). Some observations on pain in cluster headache. *Headache* 14, 219–225.

Ekbom, K. (1986). Cough headache. *In* "Headache (Handbook of Clinical Neurology)" (P. J. Vinkin, G. W. Bruyn, and H. L. Klawans, Eds.), pp. 67–371. Elsevier Science, New York.

Espada, F., Escalza, I., Morales-Asin F., Navas, I., Inignez, C., and Mauri, J. A. (1999). Hemicrania Continua: Nine new cases. *Cephalalgia* 19, 442. [Abstract]

Evers, S., Bahra, A., and Goadsby, P. J. (1999). Coincidence of familial hemiplegic migraine and hemicrania continua? A case report. *Cephalalgia* 19, 533–535.

Ghose, R. R. (1995). SUNCT syndrome. *Med. J. Aust.* 162, 667–668.

Goadsby, P. J., and Lipton, R. B. (1997). A review of paroxysmal hemicranias, SUNCT syndrome and other short-lasting headaches with autonomic features, including new cases. *Brain* 120, 193–209.

Goadsby, P. J., Zagami, A. S., and Lambert, G. A. (1991). Neural processing of craniovascular pain: Synthesis of the central structures involved in migraine. *Headache* 31, 365–371.

Headache Classification Committee of the International Headache Society (1988). Classification and diagnostic criteria for headache disorders, cranial neuralgia, and facial pain. *Cephalalgia* 8, 1–96.

Ibbotson, S. (1987). Weight-lifter's headache. *Br. J. Sports Med.* 3, 138.

Iordanidis, T., and Sjaastad, O. (1989). Hemicrania continua: A case report. *Cephalalgia* 9, 301–303.

Joubert, J. (1991). Hemicrania continua in a black patient—The importance of the non-continuous stage. *Headache* 31, 480–482.

Kruszewski, P. (1992). Short-lasting, unilateral, neuralgiform headache attacks with conjunctival injection and tearing (SUNCT syndrome). V. Orbital phlebography. *Cephalalgia* 12, 387–389.

Kruszewski, P., Sand, T., Shen, J. M., and Sjaastad, O. (1992). Short-lasting, unilateral, neuralgiform headache attacks with conjunctival injection and tearing (SUNCT) syndrome. IV. Respiratory sinus arrhythmia during and outside paroxysms. *Headache* 32, 377–383.

Kruszewski, P., Zhao, J. M., Shen, J. M., and Sjaastad, O. (1993). SUNCT syndrome: Forehead sweating pattern. *Cephalalgia* 13, 108–113.

Kumar, K. L., and Bordiuk, J. D. (1991). Hemicrania continua: A therapeutic dilemma. *Headache* 31, 345. [Letter]

Lance, J. W., and Anthony, M. (1971). Migrainous neuralgia or cluster headache. *J. Neurol. Sci.* 13, 401–414.

Lance, J. W., and Hinterberger, H. (1976). Symptoms of pheochromocytoma with particular reference to headache correlated with catecholamine production. *Arch. Neurol.* 33, 281–288.

Lansche, R. K. (1964). Ophthalmodynia periodica. *Headache* 4, 247–249.

Martins, I. P., Parreira, E., and Costa, I. (1995). Extratrigeminal icepick status. *Headache* 35, 107–110.

Mathew, N. T. (1980). Indomethacin responsive headache syndrome. *Headache* 21, 147–150.

May, A., Bahra, A., Buchel, C., Turner, R., and Goadsby, P. J. (1999). Functional magnetic resonance imaging in spontaneous attacks of SUNCT: Short-lasting neuralgiform headache with conjunctival injection and tearing. *Neurology* 46, 791–794.

Medina, J., and Diamond, S. (1981). Cluster headache variant. Spectrum of a new headache syndrome. *Arch. Neurol.* 38, 705–709.

Morales, F., Mostacero, E., Marta, J., and Sanchez, S. (1994). Vascular malformation of the cerebellopontine angle associated with SUNCT syndrome. *Cephalalgia* 14, 301–302.

Newman, L. C., Lipton, R. B., Russell, M., and Solomon, S. (1992). Hemicrania continua: Attacks may alternate sides. *Headache* 32, 237–238.

Newman, L. C., Lipton, R. B., and Solomon, S. (1993). Episodic paroxysmal hemicrania: 3 New cases and a review of the literature. *Headache* 33, 195–197.

Newman, L. C., Lipton, R. B., and Solomon, S. (1994). Hemicrania continua: Ten new cases and a review of the literature. *Neurology* 44, 2111–2114.

Nick, J. (1980). La céphalée d'effort. A propos d'une série de 43 cases. *Sem. Hop. Paris* 56, 621–628.

Nightingale, S., and Williams, B. (1987). Hindbrain hernia headache. *Lancet* 1, 731–734.

Pareja, J. A. (1999). Hemicrania continua: Ocular discomfort heralding painful attacks. *Funct. Neurol.* 14, 93–95.

Pareja, J. A., Caballero, V., and Sjaastad, O. (1996a). SUNCT syndrome. Status-like pattern. *Headache* 36, 622–624.

Pareja, J. A., Joubert, J., and Sjaastad, O. (1996b). SUNCT syndrome. Atypical temporal patterns. *Headache* 36, 108–110.

Pareja, J. A., Kruszewski, P., and Sjaastad, O. (1995). SUNCT syndrome: Trials of drugs and anesthetic blockades. *Headache* 35, 138–142.

Pareja, J. A., Ming, J. M., Kruszewski, P., Caballero, V., Pamo, M., and Sjaastad, O. (1996c). SUNCT syndrome: Duration, frequency and temporal distribution of attacks. *Headache* 36, 161–165.

Pareja, J. A., Palomo, T., Gorriti, M. A., Pareja, J., and Espejo, J. (1990). "Hemicrania episodica"—A new type of headache or a pre-chronic stage of hemicrania continua? *Headache* 30, 344–346.

Pareja, J. A., Palomo, T., Gorriti, M. A., Pareja, J., Espejo, J., Moron, B., and Trigo, M. (1990). Hemicrania continua. The first Spanish case: A case report. *Cephalalgia* 10, 143–145.

Pareja, J. A., Ruiz, J., de Isla, C., al-Sabbah, H., and Espejo, J. (1996d). Idiopathic stabbing headache (jabs and jolts syndrome). *Cephalalgia* 16, 93–96.

Pareja, J. A., and Sjaastad, O. (1994). SUNCT syndrome in the female. *Headache* 34, 217–220.

Pascual, J., Igessias, F., Oterino, A., and Vazquez-Barquero, A. (1996). Cough, exertional, and sexual headaches: An analysis of 72 benign and symptomatic cases. *Neurology* 46, 1520–1524.

Pasquier, F., Leys, D., and Petit, H. (1987). "Hemicrania continua": The first bilateral case? *Cephalalgia* 7, 169–170.

Paulson, G. W. (1983). Weightlifters headache. *Headache* 23, 193–194.

Paulson, G. W., Zipf, R. E., and Beekman, J. F. (1979). Pheochromocytoma causing exercise-related headache and pulmonary edema. *Ann. Neurol.* 5, 96–99.

Peres, M. F., and Zukerman, E. (2000). Hemicrania continua responsive to rofecoxib. *Cephalalgia* 20, 130–131.

Peres, M. F., Stiles, M. A., Oshinsky, M., and Rozen, T. D. (2001). Remitting form of hemicrania continua with seasonal pattern. *Headache* 41, 592–594.

Poughias, L., and Aasly, J. (1995). SUNCT syndrome: Cerebral SPECT images during attacks. *Headache* 35, 143–145.

Powell, B. (1982). Weight lifter's cephalalgia. *Ann. Emerg. Med.* 11, 449–451.

Raskin, N. (1978). Headaches associated with organic diseases of the nervous system. *Med. Clin. N. Am.* 62, 459–466.

Raskin, N. H. (1988). The indomethacin-responsive syndromes. *In* "Headache" (N. H. Raskin, Ed.) pp. 255–268. Churchill Livingstone, New York.

Raskin, N. H., and Schwartz, R. K. (1980). Icepick-like pain. *Neurology* 30, 203–205.

Rasmussen, B. K., Jensen, R., Schroll, M., and Olesen, J. (1991). Epidemiology of headache in a general population—a prevalence study. *J. Clin. Epidemiol.* 44, 1147–1157.

Rasmussen, K. R., and Olesen, J. (1992). Symptomatic and non-symptomatic headaches in a general population. *Neurology* 42, 1225–1231.

Rooke, E. D. (1968). Benign exertional headache. *Med. Clin. N. Am.* 52, 801–808.

Rushton, J. G., and Rooke, E. D. (1962). Brain tumor headache. *Headache* 2, 147–152.

Russell, R. W. (1959). Giant cell arteritis: A review of 35 cases. *Q. J. Med.* 8, 471–489.

Sands, G. H., Newman, L., and Lipton, R. (1991). Cough, exertional, and other miscellaneous headaches. *Med. Clin. N. Am.* 75, 733–746.

Shen, J. M., and Johnsen, H. J. (1994). SUNCT syndrome: Estimation of cerebral blood flow velocity with transcranial doppler ultrasonography. *Headache* **34**, 25–31.

Silberstein, S. D. (1993). Tension-type and chronic daily headache. *Neurology* **43**, 1644–1649.

Silberstein, S. D., and Marcelis, J. (1993). Headache associated with abnormalities in intracranial structure or pressure. *In* "Wolff's Headache and Other Head Pain" (D. J. Dalessio and S. D. Silberstein, Eds.), pp. 438–461. Oxford University Press, New York.

Silberstein, S. D., Lipton, R. B., and Sliwinski, M. (1996). Classification of daily and near-daily headaches: Field trial of revised IHS criteria. *Neurology* **47**, 871–875.

Silbert, P. L., Edis, R. H., Stewart-Wynne, E. G., and Gubbay, S. S. (1991). Benign vascular sexual headache and exertional headache: Interrelationships and long term prognosis. *J. Neurol. Neurosurg. Psychiatry* **54**, 417–421.

Sjaastad, O., Apfelbaum, R., Caskey, W., and *et al.* (1980). Chronic paroxysmal hemicrania (CPH): The clinical manifestations. A review. *Ups. J. Med. Sci.* **31**, 27–35.

Sjaastad, O., Pettersen, H., and Bakketeig, L. S. (2001). The Vaga study: Epidemiology of headache. I. The prevalence of ultrashort paroxysms. *Cephalalgia* **21**, 207–215.

Sjaastad, O., Egge, K., and Horven, I. (1979). Chronic paroxysmal hemicrania. V. Mechanical precipitation of attacks. *Headache* **19**, 31–36.

Sjaastad, O., Joubert, J., Elsas, T., Bovim, G., and Vincent, M. (1993). Hemicrania continua and cervicogenic headache. Separate headaches or two faces of the same headache? *Funct. Neurol.* **8**, 79–83.

Sjaastad, O., and Kruszewski, P. (1992). Trigeminal neuralgia and SUNCT syndrome: Similarities and differences in the clinical pictures. An overview. *Funct. Neurol.* **7**, 103–107.

Sjaastad, O., Kruszewski, P., Fostad, K., Elsas, T., and Qvigstad, G. (1992). SUNCT syndrome. VII. Ocular and related variables. *Headache* **32**, 489–495.

Sjaastad, O., Russel, D., Horven, I., and Bunaes, U. (1978). Multiple neuralgiform unilateral headache attacks associated with conjunctival injection and appearing in clusters. A nosological problem. *Proc. Scand. Migraine Soc.* **31**.

Sjaastad, O., Saunte, C., and Salvesen, R. (1989). Short-lasting unilateral neuralgiform headache attacks with conjunctival injection, tearing, sweating, and rhinorrhea. *Cephalalgia* **9**, 147–156.

Sjaastad, O., and Spierings, E. L. (1984). "Hemicrania continua": Another headache absolutely responsive to indomethacin. *Cephalalgia* **4**, 65–70.

Sjaastad, O., Stovner, L. J., Stolt-Nielsen, A., Antonaci, F., and Fredriksen, T. A. (1995). CPH and hemicrania continua: Requirements of high indomethacin dosages—An ominous sign? *Headache* **35**, 363–367.

Sjaastad, O., Zhao, J. M., Kruszewski, P., and Stovner, L. J. (1991). Short-lasting unilateral neuralgiform headache attacks with conjunctival injection, tearing, etc. (SUNCT). III. Another Norwegian case. *Headache* **31**, 175–177.

Stevens, J. M., Serva, W. A. D., Kendall, B. E., Valentine, A. R., and Ponsford, J. R. (1993). Chiari malformation in adults: Relation of morphological aspects to clinical features and operative outcome. *J. Neurol. Neurosurg. Psychiatry* **56**, 1072–1077.

Symonds, C. (1956). Cough headache. *Brain* **79**, 557–568.

Thomas, J. E., Rooke, E. D., and Kvale, W. F. (1966). The neurologist's experience with pheochromocytoma: A review of 100 cases. *JAMA* **10**, 100–104.

Tinel, J. (1932a). La cephalee a l'effort. Syndrome de distension douloureuse des veines intracraniennes. *Medicine (Paris)* **13**, 113–118.

Tinel, J. (1932b). Un syndrome d'algie veineuse intracranienne. La cephalee a l'effort. *Prat. Med. Fr.* **13**, 113–119.

Trucco, M., Antonaci, F., and Sandrini, G. (1992). Hemicrania continua: A case responsive to piroxicam-beta-cyclodextrin. *Headache* **32**, 39–40.

Van den Bergh, V., Amery, W. K., and Waelkens, J. (1987). Trigger factors in migraine: A study conducted by the Belgian migraine society. *Headache* **27**, 191–196.

Williams, B. (1976). Cerebrospinal fluid pressure changes in response to coughing. *Brain* **99**, 331–346.

Williams, B. (1980). Cough headache due to craniospinal pressure dissociation. *Arch. Neurol.* **37**, 226–230.

Zhao, J. M., and Sjaastad, O. (1993). SUNCT syndrome. VIII. Pupillary reaction and corneal sensitivity. *Funct. Neurol.* **8**, 409–414.

CHAPTER 6

# Hormone-Sensitive Headache in Women

Stephen D. Silberstein and Helene Massiou

## INTRODUCTION

There is a link between estrogen and progesterone, the female sex hormones, and migraine (Epstein *et al.*, 1975; Goldstein and Chen, 1982). Migraine occurs more frequently in adult women (18%) than in men (6%), although its prevalence is equal, genderwise, in children (Goldstein and Chen, 1982; Waters and O'Connor, 1971). Migraine develops most frequently in the second decade, with the peak incidence occurring with adolescence (Epstein *et al.*, 1975). Menstrually related migraine begins at menarche in a third of affected women (Epstein *et al.*, 1975). Menstrually related migraine occurs mainly at the time of menses in many migrainous women and exclusively with menses (true menstrual migraine [TMM]) in some (Epstein *et al.*, 1975). Menstrual migraine (MM) can be associated with other somatic complaints that arise before and often persist into menses, such as nausea, backache, breast tenderness, and cramps, and, like them, appears to be the result of falling sex hormone levels (American Psychiatric Association, 1994; Waters and O'Connor, 1971).

Migraine often worsens during the first trimester of pregnancy and, although many women become headache-free during the last two trimesters, 25% have no change in their migraine (Lance and Anthony, 1966; Ratinahirana *et al.*, 1990; Somerville, 1972a). MM typically improves with pregnancy, perhaps due to sustained high estrogen levels (Lance and Anthony, 1966; Ratinahirana *et al.*, 1990; Somerville, 1972a). Hormonal replacement with estrogens can exacerbate migraine and oral contraceptives (OCs) can change its character and frequency (Bickerstaff, 1975). Migraine prevalence decreases with advancing age but may either regress or worsen at menopause (Goldstein and Chen, 1982; Neri *et al.*, 1993; Whitty and Hockaday, 1968). Changes in the headache pattern with OC use and during menarche, menstruation, pregnancy, or menopause are related to changes in estrogen levels. These phenomena suggest a relationship between migraine headaches and changes in sex hormone levels (Lundberg, 1986).

## MENSTRUAL MIGRAINE

When migraine occurs before menstruation, there may also be features of premenstrual syndrome (PMS), including depression, anxiety, crying spells, difficulty in thinking, lethargy, backache, nausea, and breast tenderness and swelling (Epstein *et al.*, 1975). Migraine that occurs during menstruation (menstrual migraine [MM]) is often associated with dysmenorrhea and was believed to be frequently longer in duration (Johannes *et al.*, 1995) and refractory to treatment (Solbach *et al.*, 1984). MM is typically not associated with aura, even in patients preselected for migraine with aura.

A 4-month diary study of 79 women screened for migraine with aura was performed in Washington County, Maryland (Johannes *et al.*, 1995). All headache types occurred more frequently during the first 3 days of the menstrual cycle, but this was statistically significant only for migraine without aura (66% higher), despite the fact that the entry criteria required a history of migraine with aura. Attacks of migraine without aura, but not migraine with aura, were more likely to occur 2 days before the onset of menses and on the first 2 days of menses. Migraine headaches were slightly but significantly more painful during the first 2 days of menses.

Most women have increased headache and migraine attacks (usually without aura) at the time of menses. Some women have migraine (usually without aura) only with menses. Menstrual migraine can be defined by looking at attacks that are regularly triggered by menstruation. Attacks that occur only with menstruation, even if infrequent, would be called TMM. Attacks that occur both at menstruation and at other times of the month could be called "menstrually triggered migraine." A frequency indicator (e.g., frequent, ≥70%; common, 35 to 70%; and infrequent, ≤35%) would look at the tightness of this association (Silberstein and Merriam, 1997).

## Pathogenesis

MM occurs at the time of the greatest fluctuation in estrogen levels. Somerville (1972b) reported that MM occurs during or after the simultaneous fall of estrogens and progesterone. Estrogens given premenstrually delay the onset of migraine but not menstruation (Somerville, 1975). In contrast, progesterone administration delays menstruation but does not prevent migraine (Somerville, 1971). Somerville concluded that estrogen withdrawal might trigger migraine attacks in susceptible women. Additional clinical observations support Somerville's conclusion. Women taking combined OCs often develop migraine headache during the steroid-free week and postmenopausal women given depo-estradiol injections frequently develop estrogen-withdrawal migraine.

## Treatment of Menstrual Migraine (Table I)

### Overview

The use of a headache calendar can help establish a relationship between headache and the menstrual cycle and establish whether or not headaches are menstrually triggered migraine. Women who have headaches throughout their menstrual cycle should be treated with reassurance, education, and pharmacologic intervention. Behavioral intervention may be useful in selected instances (Silberstein and Lipton, 1997).

### Acute Treatment

Drugs that are proven effective or are commonly used for the acute treatment of menstrual migraine include nonsteroidal anti-inflammatory drugs (NSAIDs), dihydroergotamine, the triptans, and the combination of aspirin, acetaminophen, and caffeine (AAC) (Silberstein et al., 1999).

Dihydroergotamine, a nonselective 5-HT$_1$ agonist available in parenteral form and as a nasal spray, is

**TABLE I  Preventive Treatment of Menstrual Migraine**

1. Perimenstrual use of standard preventive drugs
2. Perimenstrual use of nonstandard preventive drugs
   - Nonsteroidal antiinflammatory drugs (NSAIDs)
   - Ergotamine and its derivatives
   - Triptans
   - Magnesium
3. Hormonal therapy
   - Estrogens (with or without androgens or progestin)
   - Combined oral contraceptives
   - Synthetic androgens (Danazol)
   - Antiestrogen (Tamoxifen)
   - Medical oophorectomy (GnRH analogs)
4. Dopamine agonists (Bromocriptine)

effective for the treatment of MM (D'Alessandro et al., 1983).

Sumatriptan, zolmitriptan, and rizatriptan, selective 5-HT$_1$ agonists, are as effective for menstrually associated migraine (Solbach and Waymer, 1993) as for non-MM and, in addition, control the nausea and vomiting associated with attacks (DeLignieres et al., 1986).

## Preventive Treatment

The goal of standard continuous preventive therapy is to reduce the frequency, duration, and intensity of attacks.

Women on preventive medication who continue to have MM can increase the dose of their medication prior to their menses. Women who do not use preventive medicine or have migraine exclusively with their menses can just be treated perimenstrually with short-term prophylaxis (Silberstein et al., 2001). Drugs that have been used perimenstrually for short-term prophylaxis include NSAIDs, ergotamine, dihydroergotamine, methysergide, methergine, triptans, and magnesium. NSAIDs in adequate doses can be used preventively 1–2 days before the expected onset of headache and continued for the duration of vulnerability.

Ergotamine and dihydroergotamine can be used prophylactically at the time of menses without significant risk of developing ergot dependence. Ergotamine tartrate, at bedtime or twice a day, is an effective prophylactic agent. Ergotamine in combination with belladonna and phenobarbital (Bellergal™) may be useful in treating other perimenstrual symptoms in addition to headache (Robinson et al., 1977).

Newman et al. used oral sumatriptan (25 mg tid) 2–3 days before the expected headache onset and continued it for a total of five days (Newman et al., 1998). Breakthrough headaches were rare and significantly reduced in severity compared with baseline headache. Newman et al. determined the efficacy of oral naratriptan (1 and

2.5 mg) as short-term prophylactic treatment for MM (Newman *et al.*, 2000). More women treated with naratriptan 1 mg reported absence of MMs across all treated perimenstrual periods and had at least a 50% reduction in MMs compared with placebo-treated women (23% versus 8%; and 61% versus 38%, $P < 0.05$; respectively). More women treated with naratriptan 2.5 mg had a reduced number of MMs and MM days, but the reductions were not significantly different than placebo.

If severe MM cannot be controlled by these measures, then hormonal therapy may be indicated. Successful hormonal or hormonal modulation therapy of MM has been reported with estrogens (DeLignieres *et al.*, 1986) (alone or in combination with progesterone or testosterone) (Magos *et al.*, 1986), combined OCs, synthetic androgens, estrogen modulators and antagonists (Calton and Burnett, 1984), and medical oophorectomy with GnRH analog with or without add-back therapy and prolactin release inhibitors (Thomas *et al.*, 1991). Progesterone alone is not effective (Silberstein and Merriam, 1997).

Pradalier *et al.* found that using the TTS 25 patch from 4 days before to 4 days after menstruation was not as effective as using the TTS 100 patch (Pradalier *et al.*, 1994). Dennerstein *et al.* suggested that a serum estradiol level of 60 to 80 pg/mL is needed during the crucial week to prevent MM (Dennerstein *et al.*, 1978). Pfaffenrath, using TTS 50 patches (TTS = 39 pg/ml), did not find a significant difference from placebo in a placebo-controlled, double-blind trial (Pfaffenrath, 1993). Similarly, Smits *et al.* found minimal benefit of TTS 50 in a placebo-controlled trial, except for patients who had abnormal CNVs and normal ES2 (Smits *et al.*, 1993).

Combinations of estrogens and progestogens or progestogens alone in the form of OCs may be a reasonable approach for some patients with intractable MM, particularly if it is associated with severe dysmenorrhea. Women on OCs who have menstrual-related problems can extend the active OCs for 6 to 12 weeks and delay the menstrually related symptoms.

Neither hysterectomy nor oophorectomy has been proven to be effective in unselected cases in the treatment of migraine. However, medical ovariectomy using GnRH analogs to suppress ovulation is effective in treating patients with refractory PMS (Conn and Crowley, 1991; Muse *et al.*, 1984; Hammarback and Backstrom, 1988). Since GnRH analogs induce hypogonadism, with many of the same short-term and long-term side effects as menopause, treatment is usually limited to 6 months unless replacement estrogens are used (Conn and Crowley, 1991; Pickersgill, 1998). Add-back therapy can be used to limit these side effects and is not usually detrimental to effective GnRH agonist treatment (Pickersgill, 1998). Add-back treatment prevents bone mineral loss and minimizes the adverse effects of hypoestrogenism.

Another strategy that is employed involves a dopamine receptor agonist, used short-term or continuously. Bromocriptine (Parlodel) (Andersch *et al.*, 1978; Andersen *et al.*, 1977; Wentz, 1985), a dopamine D2 receptor agonist, is an inhibitor of prolactin release. A dose of 2.5–5 mg a day during the luteal phase of the menstrual cycle may decrease the premenstrual symptoms of breast engorgement, irritability, and headache. In an open trial (Herzog, 1995), 24 women with severe, disabling MM (occurring within 3 days of menstruation) were treated with continuous bromocriptine, 2.5 mg, three times daily. Seventy-five percent of the women had at least a 25% reduction in headache compared to baseline. Overall headache frequency decreased 72%. None of the patients had less than a 10% increase in headache; 3 could not tolerate bromocriptine, and 3 did not benefit.

## MIGRAINE IN THE MENOPAUSE

At menopause, the permanent cessation of menstruation, sex steroid hormone levels are low and gonadotropin levels are elevated. The average age of menopause is between 51 and 52 years, with a range of 40–60 years. The timing of menopause is both genetically and environmentally influenced (Mishell *et al.*, 1997).

Ovarian function decreases gradually; the time between the onset of menstrual irregularity and menopause is called the perimenopause. In the perimenopausal period, plasma concentrations of follicle-stimulating hormone (FSH) are elevated and are associated with increased serum estradiol levels and urine levels of estrogen conjugates. As the follicles become depleted and inhibin levels remain low, they no longer respond to elevated FSH levels and estradiol levels fall. Replacement with exogenous estrogen does not reduce FSH to premenopausal levels, since FSH release is mainly suppressed by inhibin and inhibin levels remain low postmenopausally. Therefore, measurement of FSH cannot be used to determine whether sufficient exogenous estrogen is being given to produce physiologic replacement (Mishell *et al.*, 1997).

### Hormonal Replacement

The menopause is associated with both early and late symptoms (Utian, 1987a,b). Hot flushes, a vasomotor change, correlate with bursts of activity in hypothalamic pacemaker neurons leading to pulses of GnRH and, thus, lutenizing hormone (LH) (Ravnikar, 1990; Rebar and Spitzer, 1987). Hormonal replacement with estrogens, alone or in combination with progestins, is often used to treat symptoms and prevent osteoporosis (LaRosa, 1995; Shoemaker *et al.*, 1977). Estrogen use may delay the onset and decrease the risk of Alzheimer's disease (Tang *et al.*, 1996) and improve cognition, but this is controversial (Yaffe *et al.*, 1998). Estrogen therapy may decrease the risk of coronary artery disease and hip fracture, but long-term, unopposed estrogen therapy increases the risk of endometrial carcinoma (Grady *et al.*, 1992). The increased endometrial cancer risk can be avoided by adding a progestin to the estrogen regimen for women who have a uterus, probably without reducing the benefit for coronary artery disease (Grady *et al.*, 1992, 1995; Martin and Freeman, 1993; Nablusi *et al.*, 1993). The effect of estrogen replacement therapy on the risk of breast cancer is uncertain, with studies still yielding inconsistent results (Grady *et al.*, 1992, 1995). There appears to be no increased risk with short-term estrogen use, but the risk of breast cancer may increase slightly with long-term use.

The menopause presents a particular set of problems in women for whom estrogen replacement is indicated but leads to a worsening of migraine symptoms. Women should use the lowest dose of estrogen that will relieve vasomotor symptoms, prevent vaginal and urethral epithelial atrophy, maintain the collagen content of the skin, reduce the rate of bone resorption, and prevent acceleration of atherosclerosis. The most commonly used estrogen, Premarin, is a mixture of conjugated estrogens of equine (CEE) origin (Stumpf, 1990). Pure estrones, estradiols, and synthetic ethinyl estradiol are also available (Cedars and Judd, 1987). Estrogen replacement should be physiologic, not pharmacologic (Mishell *et al.*, 1997). The optimal long-term estrogen dose given to asymptomatic women to reduce the risk of osteoporosis and cardiovascular disease (physiologic replacement dose) is 0.625 mg of CEE or estrone sulfate, or 1 mg of micronized estradiol, and probably 0.05 mg transdermally. Higher estrogen doses may be needed for 1 or 2 years to relieve hot flushes.

Estrogens can be taken sequentially for 25 days a month with the addition of a progestational agent on days 16 to 25 to induce bleeding; alternatively, estrogens and progesterone can be taken continuously (Wentz, 1985). Estrogens are available orally or parenterally in the form of injection, vaginal cream, estrogen ring, or transdermal patch (Stumpf, 1990). Experimental implants and transdermal patches provide stable blood estrogen levels (Ravnikar, 1990; Shoemaker *et al.*, 1977; Whitehead *et al.*, 1990). In addition, parenteral administration of estrogens produces fewer hepatic effects and a higher, more physiologic serum estradiol-to-estrone ratio than oral administration (Cedars and Judd, 1987; Studd and Magos, 1987).

The selective estrogen receptor modulators are a new class of drugs that differ from each other and from classic estrogen. Raloxifene, the newest selective estrogen receptor modulator, has estrogen-like effects on bone and on lipid metabolism, but no effect on the breast or uterus (Delmas *et al.*, 1997). It, like estrogen, is associated with venous thrombosis.

Although migraine prevalence decreases with advancing age (Goldstein and Chen, 1982), migraine can either regress or worsen at menopause (Whitty and Hockaday, 1968). Women with prior migraine generally (2/3) improve with physiologic menopause. In contrast, surgical menopause usually (2/3) worsens migraine (Neri *et al.*, 1993).

Headache management can be difficult in women who require hormone replacement therapy for menopausal symptoms but develop headaches as a result of the therapy. Several empirical strategies may be utilized (Table II). Reducing the dose of estrogen or changing the type of estrogen from a conjugated estrogen to pure estradiol, to synthetic ethinyl estradiol, or to a pure estrone may significantly reduce headache. Changing from interrupted to continuous administration may be very effective if the headaches are associated with estrogen withdrawal. Parenteral estrogens, with or without adjunct hormones, can be effective. The estradiol cutaneous patch, which provides a physiologic ratio of estradiol to estrone and a steady-state concentration of estrogen, has been associated anecdotally with fewer headache side effects; however, this has not been proven in any controlled study (Anonymous, 1986; Judd, 1987; Stumpf, 1990). The new selective estrogen receptor modulator, raloxifene, can also be used if a woman requires, but cannot tolerate, nonselective estrogen.

Progestins, used to prevent endometrial hyperplasia, can cause headache in addition to other symptoms of PMS, particularly if used cyclically. Giving a lower dose of a progestin (medroxyprogesterone, 2.5 mg vs 7.5 mg)

---

**TABLE II    Treatment of Hormonal Replacement Headache**

Estrogens
1. Reduce estrogen dose
2. Change estrogen type from conjugated estrogen to pure estradiol to synthetic estrogen to pure estrone
3. Convert from interrupted to continuous dosing
4. Convert from oral to parenteral dosing
5. Add androgens
6. Switch to selective estrogen receptor modulator

Progestin
1. Switch from interrupted (cyclic) to continuous lower dose
2. Change progestin type
3. Change delivery system (PO to vaginal)
4. Discontinue progestin (periodic endometrial biopsy or vaginal ultrasound)

continuously can often control this. Another strategy is to change the type of progestin. Women who received norethindrone had less depression than those who received medroxyprogesterone acetate (Smith *et al.*, 1994). For women with a uterus who have intolerable mental symptoms with progestins, an estrogen-only regimen may be used in conjunction with an annual endometrial biopsy or vaginal ultrasonography to measure the endometrial thickness. If the endometrial echo complex is less than 4 mm, it may not be necessary to perform a biopsy, since the endometrial cancer risk is very low (Mishell *et al.*, 1997).

Another strategy is to use targeted drug delivery. The new adhesive vaginal gel containing micronized progesterone in an emulsion system was designed to maximize progesterone's therapeutic effect on the uterus while minimizing the potential for systemic side effects.

## MIGRAINE ASSOCIATED WITH HORMONAL CONTRACEPTIVE USE

### Introduction

Hormonal contraceptive steroids are available as OCs, subcutaneous implants, depo-injections, and vaginal preparations (in some countries) (Baird and Glasier, 1993). There are three major types of OC formulations: two are combination OCs (COCs) (fixed-dose and phasic combinations) and one is progestin only. The OCs most commonly used in the United States contain combinations of synthetic estrogen (ethinyl estradiol or mestranol) and synthetic progestin. In an attempt to minimize the associated androgenic side effects, the type of synthetic progestin has been changed.

There has been a progressive decrease in the ethinyl estradiol content in combined OCs; most now contain 35 μg or less. Formulations with 50 μg or more of estrogen are now called first-generation OCs. Those with less than 50 μg of estrogen are called second-generation OCs. Formulations with the new progestins, desogestrel, norgestimate, and gestodene, are called third-generation OCs.

### Adverse Events

The most common dose-dependent side effects seen with estrogen OCs are nausea, breast tenderness, fluid retention, and depression. Depression occurs with high, but not low, estrogen (<50 μg) formulations. Physiologic doses of estrogen alone (less than the pharmacologic dose used in OCs) improve mood, while an added progestin provokes depression, irritability, tension, and

fatigue. Weight gain, breakthrough bleeding, nausea, headache, breast tenderness, mood swings, acne, and hirsutism are the most common causes of premature OC discontinuation (Silberstein and Merriam, 1997).

There is persistent controversy concerning OCs and the risk of stroke in migraineurs (Bickerstaff, 1975).

Bousser and Kittner reviewed the relationship between OCs and stroke (Bousser and Kittner, 2000): high estrogen content (≥50 μg) increases the risk of stroke, all stroke subtypes, and stroke death; low estrogen content (<50 μg) carries a very low or no risk of stroke; there are no data on progestogen-only OCs; stroke risk is greatly increased if associated risk factors are present, in particular, hypertension, cigarette smoking, and migraine; OCs, even at low doses, significantly increase the risk of cerebral venous thrombosis, which is further enhanced if congenital thrombophilia is present; and the attributable risk of stroke in young women using OCs is about one per 200,000 woman-years. The contraceptive and noncontraceptive benefits of low-dose OCs vastly outweigh their risks provided that other risk factors are absent or well controlled.

### Contraceptive Use

The older combined OCs can induce, change, or alleviate headache (Bickerstaff, 1975). OCs can trigger the first migraine attack, most often in women with a family history of migraine (Bickerstaff, 1975; Ryan, 1978). Existing migraine may exacerbate and headaches may occur on the days off the OC (Bickerstaff, 1975; Neri *et al.*, 1993; Phillips, 1968; Ryan, 1978). The headache pattern may become more severe and frequent and may be associated with neurologic symptoms (Collaborative Group for the Study of Stroke in Young Women, 1975; Dalton, 1976; Ryan, 1978). In most women, however, the headache pattern does not change, and some women may have a distinct improvement in their headaches (Larrson-Cohn and Lundberg, 1970; Whitty *et al.*, 1966). New onset of migraine usually occurs in the early cycles of OC use, but it can occur after prolonged OC usage (Ryan, 1978). Stopping the OC may not bring immediate headache relief; there may be a delay of ½–1 year or no improvement (Dalton, 1976; Whitty and Hockaday, 1968).

Studies from neurologic or migraine clinics show increased incidence and severity of migraine in users of the older OCs (Carroll, 1971; Dalton, 1976; Phillips, 1968). Four double-blind, placebo-controlled studies (Cullberg, 1972; Goldzieher *et al.*, 1971; Nilsson and Solvell, 1967; Silbergeld *et al.*, 1971) showed no difference in headache incidence between OC and placebo. Both groups had decreasing headache incidence with continued observation. Some uncontrolled studies show an increase in headache frequency in women on OCs (Cullberg *et al.*, 1969; Desrosiers, 1973; Nilsson *et al.*, 1967). Headaches occur during the pill-free interval and may resolve with the use of daily, continuous, COCs (long-cycle treatment). Phasic OCs cannot be used continuously because the variation in steroid levels may cause breakthrough bleeding (Cachrimanidou *et al.*, 1993). Contraception with progestins alone is a hormonal alternative to COCs, but the progestin-only contraceptives have a higher incidence of headache (Darney *et al.*, 1990; Schwallie and Assenzo, 1973). They are the contraceptive of choice for hypertensive women (Baird and Glasier, 1993).

Three types of injectable, long-acting steroid formulations are currently in use for contraception throughout the world. These include depo-medroxyprogesterone acetate, given in a dose of 150 mg every 3 months; norethindrone enanthate, given in a dose of 200 mg every 2 months; and several once-a-month injections of combinations of different progestins and estrogens. Only depo-medroxyprogesterone acetate is currently available in the United States.

Norplant, a system of subdermal implants of capsules made of polydimethylsiloxane (Silastic) that release a steady dose of the progestin, is an effective contraceptive that lasts 5 years. The primary side effects are irregular menstrual bleeding and headaches. Headache, which was the primary reason cited for removal other than menstrual disturbance, occurs in about 5 to 20% of patients.

Women need to be counseled about hormonal contraception, which may generate new headaches or aggravate or even ameliorate preexisting headaches. Women who take OCs must be followed for headache aggravation or the development of neurologic symptoms. Women should not use OCs if they smoke; have uncontrolled hypertension or other cardiovascular disease, such as thromboembolism, thrombophlebitis, stroke, or vasculitis; or have diabetes with retinopathy or nephropathy. Progestins can be used for contraception when estrogens have caused increased headaches or are contraindicated.

Migraine itself may be a risk factor for stroke in women under the age of 45. Migraine with aura may be associated with double the risk compared with migraine without aura. The risk for stroke is higher in women over the age of 35 and OCs should be used with caution in these women, particularly in the presence of any other risk factors. OC use in women under the age of 35 who have migraine without aura is relatively safe. Women with intractable menstrual migraine or a history of headache relief with OCs are particularly good candidates for a trial of OC. OC use in women under the age of 35 who have migraine with typical aura is probably safe, but caution should be exercised. It is an unproven but popular belief that women with prolonged aura or hemiplegic, basilar, or confusional migraine should not use OCs.

The International Headache Society Task Force concluded that, in the absence of migraine aura or other risk factors, there is no contraindication to the use of OCs (Bousser *et al.*, 2000). Women should be counseled and regularly assessed for additional risk factors. Women with migraine who are using COCs and have additional risk factors have a potentially increased risk of ischemic stroke. Risk factors should be identified and evaluated: migraine type, particularly migraine with aura, should be diagnosed; women who have migraine and smoke should stop smoking before starting COCs; other risk factors, such as hypertension and hyperlipidemia, should be treated; nonethinylestradiol methods should be considered for women who are at increased risk of ischemic stroke, particularly those who have multiple risk factors. Progestogen-only hormonal contraceptive use is probably not associated with an increased risk of ischemic stroke. No specific tests need to be undertaken other than those routinely performed or indicated by the patient's history or the presence of specific symptoms, e.g., a patient with a relative who experienced arterial disease when age 45 years or less. Migraine-related symptoms that may necessitate further evaluation and/or cessation of COCs include a new persisting headache, new onset of migraine aura, increased headache frequency or intensity, or the development of unusual aura symptoms, particularly prolonged aura.

## Course of Migraine during and after Pregnancy

Approximately 60 to 70% of migraineurs will improve during pregnancy, while some women who have not had migraine will experience their first migraine headache.

The incidence of miscarriage, toxemia, congenital anomalies, and stillbirth was not increased in a sample of 777 migraine sufferers compared to the national averages or controls (Wainscott and Volans, 1978). Olesen (Olesen *et al.*, 1999) recently found that the odds ratio for having a newborn with a low birth weight was increased (OR 3.0, 95% CI 1.3 to 7.0) for all migraine patients who delivered at term ($n = 115$) compared with the outcome of healthy pregnancies.

Postnatal headache (PNH) occurred in 39% of 71 randomly selected women during their first postpartum week (Stein, 1986). It was most frequent on days 3–6 postpartum and was associated with a past history or a family history of migraine. PNH, while less severe than the patients' typical migraine, was bifrontal, prolonged, and associated with photophobia, nausea, and anorexia.

Most women with migraine improve during pregnancy. Some women have their first attack during pregnancy. Migraine often recurs postpartum and can begin for the first time in general. Despite drug use, migraineurs do not differ from nonmigraineurs in miscarriages, toxemia, congenital anomalies, or stillbirth.

The major concerns in the management of the pregnant patient are the effects of both the medication and the disease on the fetus. Because of the possible risk of injury to the fetus, medication use should be limited; however, it is not contraindicated during pregnancy (Pitkin, 1995). Since migraine usually improves after the first trimester, many women can manage their headaches with this reassurance and nonpharmacologic means of coping, such as ice, massage, and biofeedback (Pitkin, 1995; Silberstein, 1991). Some women, however, will continue to have severe, intractable headaches, sometimes associated with nausea, vomiting, and possible dehydration. Not only are these conditions disruptive to the patient, they may pose a risk to the fetus that is greater than the potential risk of the medications used to treat the pregnant patient (Silberstein, 1991).

Symptomatic treatment, designed to reduce the severity and duration of symptoms, is used to treat an acute headache attack. Individual attacks should be treated with rest, reassurance, and ice packs. For headaches that do not respond to nonpharmacologic treatment, symptomatic drugs are indicated. The NSAIDs, acetaminophen (alone or with codeine), codeine alone, or other opioids can be used during pregnancy (Koren et al., 1998). Aspirin in low intermittent doses is not a significant teratogenic risk, although large doses, especially if given near term, may be associated with maternal and fetal bleeding. Aspirin use should probably be reserved unless there is a definite therapeutic need for it (other than headache). In general, NSAIDs may be safely taken for pain during the first trimester of pregnancy. However, their use should be limited during later pregnancy, as some NSAIDs may constrict or close the fetal ductus arteriosus (Koren et al., 1998). Ergotamine, dihydroergotamine, and sumatriptan should be avoided (Pitkin, 1995).

The associated symptoms of migraine, such as nausea and vomiting, can be as disabling as the headache pain itself. In addition, some medications that are used to treat migraine can produce nausea. Metoclopramide, which decreases the gastric atony seen with migraine and enhances the absorption of coadministered medications, is extremely useful in migraine treatment (Silberstein, 1997). Mild nausea can be treated with phosphorylated carbohydrate solution (emetrol) or doxylamine succinate and vitamin $B_6$ (pyridoxine) (Koren et al., 1998; Silberstein, 1997). More severe nausea may require the use of injections or suppositories. Trimethobenzamide, chlorpromazine, prochlorperazine, and promethazine are available orally, parenterally, and in suppository form and can all be used safely. We frequently use promethazine and prochlorperazine suppositories. Corticosteroids can be utilized occasionally. Some use prednisone in preference to dexamethasone (which crosses the placenta more readily). Domperidone is an antiemetic used outside the United States. In the United Kingdom (MacGregor, 1994) its use is not advised during pregnancy, because of variable embryotoxic effects in animal tests. In France, on the contrary, the product summary indicates no teratogenicity in animals or humans. Minimal amounts are transferred in breast milk.

Severe acute attacks of migraine should be treated aggressively (Rayburn and Lavin, 1986). We start IV fluids for hydration and then use prochlorperazine; 10 mg IV, to control both nausea and head pain. IV opioids or IV corticosteroids can supplement this. This is an extremely effective way of handling status migrainosus during pregnancy.

Preventive therapy is designed to reduce the frequency and severity of headache attacks. Consider prophylaxis when patients experience at least three or four prolonged, severe attacks a month that are particularly incapacitating or unresponsive to symptomatic therapy and may result in dehydration and fetal distress. Beta-adrenergic blockers such as propranolol have been used under these circumstances, although adverse effects, including intrauterine growth retardation, have been reported (Koren et al., 1998; Silberstein, 1997). If the migraine is so severe that drug treatment is essential, the patient should be told of the risks posed by all the drugs that are used. If the patient has a coexistent illness that requires treatment, pick one drug that will treat both disorders. For example, propranolol (Koren et al., 1998) can be used to treat hypertension and migraine while fluoxetine can be used to treat comorbid depression.

If a woman inadvertently takes a drug while she is pregnant or becomes pregnant while taking a drug, determine the dose, timing, and duration of the exposure(s). Ascertain the patient's past and present state of health and the presence of mental retardation or chromosomal abnormalities in the family. Using a reliable source of information (such as TERIS), determine if the drug is a known teratogen (although for many drugs this is not possible) (Blake and Niebyl, 1988; Briggs et al., 1994; Friedman and Polifka, 1994; Little and Gilstrap, 1992).

If the drug is teratogenic or the risk is unknown, have the obstetrician confirm the gestational age by ultrasound. If the exposure occurred during embryogenesis, then high-resolution ultrasound can be performed to determine whether damage to specific organ systems or structures has occurred. If the high-resolution ultrasound is normal, it is reasonable to reassure the patient that the gross fetal structure is normal (within the 90% sensitivity of the study) (Little and Gilstrap, 1992). However, fetal ultrasound cannot exclude minor anomalies or guarantee the birth of a normal child. Delays in achieving developmental milestones, including cognitive development, are potential risks that cannot be predicted or diagnosed prenatally. Have the obstetrician discuss the results of these studies with the mother and the significant other; formal prenatal counseling may be helpful in uncertain cases (Little and Gilstrap, 1992).

## Postpartum Headache

The two most frequently used methods for providing analgesia and anesthesia for labor and cesarean section are spinal and epidural analgesia/anesthesia. Headaches in the parturient are usually attributed to a spinal anesthetic or an accidental dural puncture. After spinal anesthesia, nonpostdural puncture headaches occur in 5 to 16% of patients. Headache occurs in 25% of parturients who have no neuroaxial anesthetic intervention (Ponder, 1999). Maternal cortical vein thrombosis is an unusual complication of pregnancy, with the majority of episodes occurring within the first and third weeks postpartum (Ravindran *et al.*, 1989). Subdural hematoma is a rare complication of a spinal anesthetic, accidental dural puncture complicating an epidural anesthetic (Cohen *et al.*, 1997; Eerola *et al.*, 1981; Jonsson *et al.*, 1983; Pavlin *et al.*, 1979; Edelman and Wingard, 1980). Subdural hematoma is due to the persistent leakage of cerebrospinal fluid through the dural rent. Spontaneous subarachnoid hemorrhage, due to aneurysm or arteriovenous malformation, occurs with an incidence of 1–5 per 10,000 pregnancies, which is higher than the incidence for the general population. Septic meningitis may occur as a result of a break in sterile technique or from contamination with an infectious agent present in blood or other tissue space. Postpartum headaches may be the harbinger of a serious medical event. Misdiagnosis of postdural puncture headache and resultant placement of an epidural blood patch can cloud and confuse further neurologic workups. Placement of an epidural blood patch in a parturient complaining of a headache that is a result of increased intracranial pressure can result in a further exacerbation of symptoms, herniation, or even death (Bader, 1994; Cohen *et al.*, 1997).

## Other Primary Headaches

### Tension-Type Headaches

With a lifetime prevalence of 69% in men and 88% in women, tension-type headache (TTH) is the most common headache type. TTH can begin at any age, but onset during adolescence or young adulthood is most common.

Chronic TTH is approximately twice as frequent in women as it is in men (Rasmussen and Lipton, 2000). In a Danish population-based study (Rasmussen, 1993), menstruation was reported to be a precipitating factor by 39% of women who had TTH more than 14 days a year; 67% of them stated that their headaches were unchanged during pregnancy, and 28% reported that their headaches improved or disappeared. In this population, about half of the migraineurs reported headache improvement during pregnancy. In this study and that of Marcus *et al.* (1999), migraine appears to be more strongly influenced by pregnancy than does TTH. Maggioni *et al.* (1997), however, found that the improvement for migraine and TTH during pregnancy is similar. TTH is reported as a common side effect of oral contraceptives, but its real prevalence in relation to this type of treatment remains unknown. It may be influenced by the dose of ethinyl-estradiol and the type and dose of progestative, but direct comparative data are lacking.

### Cluster Headache

Cluster headaches affect men much more frequently than women. The various case series in the literature suggest male-to-female ratios ranging from 3.7 : 1 to 11 : 6.1, with the average being 6.5 : 1. One study suggested that male predominance is progressively decreasing over the years (Manzoni, 1997). The course of cluster headache did not appear to be modified by menstruation in two studies (Ekbom and Waldenlind, 1981; Manzoni *et al.*, 1988). In one study (Ekbom and Waldenlind, 1981), six of eight women experienced remission during pregnancy, whereas in the other one (Manzoni *et al.*, 1988), there was no change. Both studies found a lower fertility rate in women suffering from cluster headache.

### Chronic Paroxysmal Hemicrania

Chronic paroxysmal hemicrania (CPH) is a rare headache syndrome that predominates in women. Among the 84 patients reviewed in 1989 (Antonaci and Sjaastad, 1989), the sex distribution was 70% women and 30% men. In the same series, CPH started immediately postpartum in 5 of 10 patients, and attacks disappeared during pregnancy in 9 of 10 patients.

Headaches worsened during menstruation in 11 of 15 cases and improved in 4 patients. Oral contraceptives had no influence. There is not sufficient information available regarding the effects of menopause on CPH.

### Hemicrania Continua

Hemicrania continua is a rare, indomethacin-responsive headache with a large female preponderance. In the review of Bordini *et al.* (1991), the female-to-male sex ratio among the 18 reported cases was 5:1. Menstruation was associated with pain exacerbations in two patients.

### Secondary Headaches during Pregnancy

New-onset headache with aura can be due to a symptomatic disorder, such as vasculitis, brain tumor, or occipital arteriovenous malformation (Silberstein *et al.*, 1998). Some disorders that produce headache, such as stroke, cerebral venous thrombosis, eclampsia, and subarachnoid hemorrhage, occur more frequently during pregnancy. Sinusitis, meningitis, and idiopathic intracranial hypertension can present as intractable headache (Silberstein *et al.*, 1998). Subarachnoid hemorrhage can present as a severe bout of acute-onset headache. These symptomatic conditions require neuroimaging and/or a lumbar puncture to diagnose them. Some disorders are more common or occur exclusively during pregnancy and produce headache. These include stroke, cerebral venous thrombosis, eclampsia, subarachnoid hemorrhage, pituitary tumor, and choriocarcinoma (Fox *et al.*, 1990). Idiopathic intracranial hypertension does not occur more commonly than expected during pregnancy.

Eclampsia is a rare complication of pregnancy; it is characterized by hypertension, proteinuria, and edema. Neurologic symptoms include headache, visual disturbances, focal deficits, impaired consciousness, and seizures. Neuroimaging may show ischemia, hemorrhages, and cerebral edema. The risks of arterial cerebral infarction, venous thrombophlebitis, and intracerebral hemorrhage are increased in the 6 weeks after delivery but not during pregnancy (Sharshar *et al.*, 1995). Subarachnoid hemorrhages are mainly due to aneurysmal rupture. The rupture risk increases with time during pregnancy; however, ruptures are rare during delivery and postpartum (Dias and Sekhar, 1990).

Brain tumors, and especially pituitary adenomas, may enlarge during pregnancy (Molitch, 1998; Roelvink *et al.*, 1987). Choriocarcinoma is complicated in 20% of cases by cerebral metastases, which can invade the vessels and provoke ischemic or hemorrhagic strokes (Weir *et al.*, 1978).

## REFERENCES

American Psychiatric Association (1994). "Diagnostic and Statistical Manual of Mental Disorders."

Andersch, B., Hahn, L., Wendestam, C., *et al.* (1978). Treatment of premenstrual syndrome with bromocriptine. *Acta Endocrinol*, 88, 165–174.

Andersen, A. N., Larsen, J. F., Steenstrup, O. R., *et al.* (1977). Effect of bromocriptine on the premenstrual syndrome: A double-blind clinical trial. *Br. J. Obstet. Gynecol.* 84, 370–374.

Anonymous (1986). Transdermal estrogen. *Med. Lett. Drugs Ther.* 28, 119–120.

Antonaci, F., and Sjaastad, O. (1989). Chronic paroxysmal hemicrania (CPH): A review of the clinical manifestations. *Headache* 29, 648–656.

Bader, A. M. (1994). Neurologic and neuromuscular disease. *In* "Obstetric Anesthesia, Principles and Practice" (D. H. Chestnut, ed.), pp. 920–941. Mosby Year Book, St. Louis, MO.

Baird, D. T., and Glasier, A. F. (1993). Hormonal contraception. *N. Engl. J. Med.* 328, 1543–1549.

Bickerstaff, E. R. (1975) "Neurological Complications of Oral contraceptives." Clarendon Press, Oxford.

Blake, D. A., and Niebyl, J. R. (1988). Requirements and limitations in reproductive and teratogenic risk assessment. *In* "Drug Use in Pregnancy" 2nd ed. (J. R. Niebyl, ed.), pp. 1–9, Lea & Febiger, Philadelphia.

Bordini, C., Antonaci, F., Stovner, L. J., *et al.* (1991). Hemicrania Continua—A clinical review. *Headache* 31, 20–26.

Bousser, M. G., Conard, J., Kittner, S., *et al.* (2000). Recommendations on the risk of ischemic stroke associated with use of combined oral contraceptives and hormone replacement therapy in women with migraine (the International Headache Society Task Force on Combined Oral Contraceptives and Hormone Replacement therapy). *Cephalalgia* 20, 155–156.

Bousser, M. G., and Kittner, S. J. (2000). Oral contraceptives and stroke. *Cephalalgia* 20, 183–189.

Briggs, G. G., Freeman, R. K., and Yaffe, S. J. (1994). "Drugs in Pregnancy and Lactation." Williams & Wilkins, Baltimore.

Cachrimanidou, A. C., Hellberg, D., Nilsson, S., *et al.* (1993). Long-interval treatment regimen with a desogestrel-containing oral contraceptive. *Contraception* 48, 205–216.

Calton, G. J., and Burnett, J. W. (1984). Danazol and migraine. *N. Engl. J. Med.* 310, 721–722.

Carroll, J. D. (1971). Migraine and oral contraception. Sandoz Conference Proceeding. Basel, Switzerland, 45–46.

Cedars, M. I., and Judd, H. L. (1987). Nonoral routes of estrogen administration. *Obstet. Gynecol. Clin. N. Am.* 14, 269–298.

Cohen, J. E., Godes, J., and Morales, B. (1997). Postpartum bilateral subdural hematomas following spinal anesthesia: A case report. *Surg. Neurol.* 47, 6–8.

Collaborative Group for the Study of Stroke in Young Women (1975). Oral contraceptives and stroke in young women. *JAMA* 231, 718–722.

Conn, P. M., and Crowley, W. F. (1991). Gonadotropin-releasing hormone and its analogues. *N. Engl. J. Med.* 324, 93–103.

Cullberg, J. (1972). Mood changes and menstrual symptoms with different gestagen/estrogen combinations: A double-blind comparison with a placebo. *Acta Psychiatry Scand.* 236, 259–276.

Cullberg, J., Celli, M. G., and Jonsson, C. O. (1969). Mental and sexual adjustment before and after six months use of an oral contraceptive. *Acta Psychiatry Scand.* 45, 259–276.

D'Alessandro, R., Gamberini, G., Lozito, A., *et al.* (1983). Menstrual migraine, intermittent prophylaxis with a timed-release pharmacological formulation of dihydroergotamine. *Cephalalgia* 15, 158.

Dalton, K. (1976). Migraine and oral contraceptives. *Headache* 15, 247–251.

Darney, P. D., Atkinson, E., Tanner, S., *et al.* (1990). Acceptance and perceptions of Norplant among users in San Francisco, USA. *Stud Fam Plan* 21, 152–160.

DeLignieres, B., Vincens, M., Mauvais-Jarvis, P., *et al.* (1986). Prevention of menstrual migraine by percutaneous estradiol. *Br. Med. J. Clin. Res*, 293, 1540.

Delmas, P. D., Bjarnason, N. H., Mitlak, B. H., *et al.* (1997). Effects of raloxifene on bone mineral density, serum cholesterol concentrations, and uterine endometrium in postmenopausal women. *N. Engl. J. Med.* 337, 1641–1647.

Dennerstein, L., Laby, B., Burrows, G. D., *et al.* (1978). Headache and sex hormone therapy. *Headache* 18, 146–153.

Desrosiers, J. J. J. (1973). Headaches related to contraceptive therapy and their control. *Headache* 13, 117–124.

Dias, M. S., and Sekhar, L. N. (1990). Intracranial hemorrhage from aneurysms and arteriovenous malformations during pregnancy and the puerperium. *Neurology* 27, 855–866.

Edelman, J. D., and Wingard, D. W. (1980). Subdural hematomas after lumbar dural puncture. *Anesthesiology* 52, 166–167.

Eerola, M., Kaukinen, L., and Kaukinen, S. (1981). Fatal brain lesion following spinal anesthesia. *Acta Anesthesiol. Scand.* 25, 115–116.

Ekbom, K., and Waldenlind, E. (1981). Cluster headache in women: Evidence of hypofertility. Headache in relation to menstruation and pregnancy. *Cephalalgia* 1, 167–174.

Epstein, M. T., Hockaday, J. M., and Hockaday, T. D. R. (1975). Migraine and reproductive hormones throughout the menstrual cycle. *Lancet* 1, 543–548.

Fox, M. V., Harms, R. W., and Davis, D. H. (1990). Selected neurologic complications of pregnancy. *Mayo Clin. Proc.* 65, 1595–1618.

Friedman, J. M., and Polifka, J. E. (1994). "Teratogenic Effects of Drugs: A Resource for Clinicians (TERIS)." Johns Hopkins University Press, Baltimore, MD.

Goldstein, M., and Chen, T. C. (1982). The epidemiology of disabling headache. *In* "Advances in Neurology" (M. Critchley, ed.), pp. 377–390. Raven Press, New York.

Goldzieher, J. W., Moses, L. E., Averkin, E., *et al.* (1971). A placebo-controlled double-blind cross-over investigation of the side effects attributed to oral contraceptives. *Fertil. Steril.* 22, 623.

Grady, D., Gebretsadik, T., Kerlikrowske, K., *et al.* (1995). Hormone replacement therapy and endometrial cancer risk: A metaanalysis. *Obstet Gynecol.* 85, 304–313.

Grady, D., Rubin, S. M., Petitti, D. B., *et al.* (1992). Hormone therapy to prevent disease and prolong life in postmenopausal women. *Ann. Intern. Med.* 117, 1016–1037.

Hammarback, S., and Backstrom, T. (1988). Induced anovulation as a treatment of premenstrual tension syndrome: A double-blind cross-over study with GnRH-agonist versus placebo. *Acta Obstet. Gynecol. Scand.* 67, 159–166.

Herzog, A. G. (1995). Continuous bromocriptine therapy in menstrual migraine. *Neurology* 48, 101–102.

Johannes, C. B., Linet, M. S., Stewart, W. F., *et al.* (1995). Relationship of headache to phase of the menstrual cycle among young women: A daily diary study. *Neurology* 45, 1076–1082.

Jonsson, L. O., Einarsson, P., and Olsson, G. L. (1983). Subdural hematoma and spinal anesthesia. *Anesthesia* 38, 144–164.

Judd, H. (1987). Efficacy of transdermal estradiol. *Obstet. Gynecol.* 156, 1326–1331.

Koren, G., Pastuszak, A., and Ito, S. (1998). Drugs in pregnancy. *N. Engl. J. Med.* 338, 1128–1137.

Lance, J. W. and Anthony, M. (1966). Some clinical aspects of migraine. *Arch. Neurol.* 15, 356–361.

LaRosa, J. C. (1995). Has HRT come of age? *Lancet* 345, 76–77.

Larrson-Cohn, U., and Lundberg, P. O. (1970). Headache and treatment with oral contraceptives. *Acta Neurol. Scand.* 46, 267–278.

Little, B. B., and Gilstrap, L. C. (1992). Counseling and evaluation of the drug-exposed pregnant patient. *In* "Drugs and Pregnancy" (LC Gilstrap, Ed.), pp. 23–29. Elsevier, New York.

Lundberg, P. O. (1986). Endocrine headaches. *In* "Handbook of Clinical Neurology (F. C. Rose, ed.), pp. 431–440. Elsevier, New York.

MacGregor, A. (1994). Treatment of migraine during pregnancy. *IHS News Headache* 4, 3–9.

Maggioni, F., Alessi, C., Maggino, T., *et al.* (1997). Headaches during pregnancy. *Cephalalgia* 17, 765–769.

Magos, A. L., Brincat, M., and Studd, J. W. W. (1986). Treatment of the premenstrual syndrome by subcutaneous estradiol implants and cyclical oral noresthisterone: Placebo controlled study. *Br. Med. J. Clin. Res.* 292, 1629–1633.

Manzoni, G. C. (1997). Male preponderance of cluster headache is progressively decreasing over the years. *Headache* 37, 588–589.

Manzoni, G. C., Micieli, G., Granella, F., *et al.* (1988). Cluster headache in women: Clinical findings and relationship with reproductive life. *Cephalalgia* 8, 37–44.

Marcus, D. A., Scharff, L., and Turk, D. (1999). Longitudinal prospective study of headache during pregnancy and postpartum. *Headache* 39, 625–632.

Martin, K. A., and Freeman, M. W. (1993). Postmenopausal hormone replacement therapy. *N. Engl. J. Med.* 328, 1115–1117.

Mishell, D. R., Stenchever, M. A., Droegemueller, W., and Herbst, A. L. (1997). Menopause: Endocrinology, consequences of estrogen deficiency, effects of hormonal replacement therapy, treatment regimens. *In* "Comprehensive gynecology" (D. R. Mishell, M. A. Stenchever, W. Droegemueller, and A. L. Herbst, Eds.), pp. 1159–1198. Mosby, St. Louis, MO.

Molitch, M. E. (1998). Pituitary diseases in pregnancy. *Sem. Perinatol.* 22, 457–470.

Muse, K. N., Cetel, N. S., Fitterman, L. A., *et al.* (1984). The premenstrual syndrome: Effects of "medical ovariectomy." *N. Engl. J. Med.* 311, 1345–1349.

Nablusi, A. A., Folsom, A. R., White, A., *et al.* (1993). Association of hormone replacement therapy with various cardiovascular risk factors in postmenopausal women. *N. Engl. J. Med.* 328, 1069–1075.

Neri, I., Granella, F., Nappi, R. M. G. C., *et al.* (1993). Characteristics of headache at menopause: A clinico-epidemiologic study. *Maturitas* 17, 31–37.

Newman, L. C., Lipton, R. B., Lay, C. L., *et al.* (1998). A pilot study of oral sumatriptan as intermittent prophylaxis of menstruation-related migraine. *Neurology* 51, 307–309.

Newman, L. C., Mannix, L. K., Landy, S. H., Silberstein, S. D., Putnam, G. P., Jobsis, M. M., Batenhorst, A. S., and O'Quinn, S. V. (2000) Naratriptan as prophylaxis for menstrually associated migraine: A randomized, double-blind, placebo-controlled study. Neurology 54, A14. [Abstract].

Nilsson, A., Jacobson, L., and Ingemanson, C. A. (1967). Side-effects of an oral contraceptive with particular attention to mental symptoms and sexual adaptation. *Acta Obstet. Gynecol. Scand.* 46, 537–556.

Nilsson, L., and Solvell, L. (1967). Clinical studies on oral contraceptives: A randomized, double-blind, cross-over study of four different preparations (Anovlar$^R$ mite, Lyndiol$^R$ mite, Ovulen$^R$, and Volidan$^R$). *Acta Obstet. Gynecol. Scand.* 46, 3–31.

Olesen, C., Steffensen, F. H., Sorensen, H. T., *et al.* (1999). Pregnancy outcome following prescription for sumatriptan. *Headache* 40, 20–24.

Pavlin, D. J., McDonald, J. S., Child, B., *et al.* (1979). Acute subdural hematoma: An unusual sequela to lumbar puncture. *Anesthesiology* 51, 338–340.

Pfaffenrath, V. (1993) Efficacy and safety of percutaneous estradiol vs. placebo in menstrual migraine. *Cephalalgia* **13**, 168. [Abstract].

Phillips, B. M. (1968). Oral contraceptive drugs and migraine. *Br. Med. J. Clin. Res.* **2**, 99.

Pickersgill, A. (1998). GnRH agonists and add-back therapy: Is there a perfect combination? *Br. J. Obstet. Gynecol.* **105**, 475–485.

Pitkin, R. M. (1995). Drug treatment of the pregnant woman: The state of the art. "Proceedings from the Food and Drug Administration Conference on Regulated Products and Pregnant Women."

Ponder, T. M. (1999). Differential diagnosis of postdural puncture headache in the parturient. *Clin. Forum Nurse Anesthetists* **10**, 145–154.

Pradalier, A., Vincent, D., Beaulieu, P. H., Baudesson, G., *et al.* (1994). Correlation between estradiol plasma level and therapeutic effect on menstrual migraine. *In* "New Advances in Headache Research" (F. C. Rose, Ed.), pp. 129–132. Smith-Gordon.

Rasmussen, B. K. (1993). Tension-type headaches in a general population: Precipitating factors, female hormones, sleep pattern, and relation to lifestyle. *Pain* **53**, 65–72.

Rasmussen, B. K. and Lipton, R. B. (2000). Epidemiology of tension-type headache. *In* "The Headaches" (J. Olesen, P. Tfelt-Hansen, and K. M. A. Welch, Eds.), pp. 545–550. Lippincott Williams and Wilkins, Baltimore.

Ratinahirana, H., Darbois, Y., and Bousser, M. G. (1990). Migraine and pregnancy: A prospective study in 703 women after delivery. *Neurology* **40**, 437.

Ravindran, R. S., Zandstra, G. C., and Viegas, O. J. (1989). Post-partum headache following regional analgesia: A symptom of cerebral venous thrombosis. *Can. J. Anesth.* **36**, 705–707.

Ravnikar, V. (1990). Physiology and treatment of hot flushes. *Obstet. Gynecol.* **75**, S3–S8.

Rayburn, W. F., and Lavin, J. P. (1986). Drug prescribing for chronic medical disorders during pregnancy: An overview. *Am. J. Obstet. Gynecol.* **155**, 565–569.

Rebar, R. W. and Spitzer, I. B. (1987). The physiology and measurement of hot flushes. *Am. J. Obstet. Gynecol.* **156**, 1284–1288.

Robinson, K., Huntington, K. M., and Wallace, M. G. (1977). Treatment of the premenstrual syndrome. *Br. J. Obstet. Gynecol.* **84**, 784–788.

Roelvink, N. C., Kamphorst, W., VanAlphen, H. A., *et al.* (1987). Pregnancy-related primary brain and spinal tumors. *Arch. Neurol.* **44**, 209–215.

Ryan, R. E. (1978). A controlled study of the effect of oral contraceptives on migraine. *Headache* **17**, 250–252.

Schwallie, P. C., and Assenzo, J. R. (1973). Contraceptive use-efficacy study utilizing medroxyprogesterone acetate administered as an intramuscular injection once every 90 days. *Fertil. Steril.* **24**, 331–339.

Sharshar, T., Lamy, C., and Mas, J. L. (1995). Incidence and causes of strokes associated with pregnancy and puerperium: A study in public hospitals of Ile de France. *Stroke* **26**, 930–936.

Shoemaker, E. S., Forney, J. P., and MacDonald, P. C. (1977). Estrogen treatment of postmenopausal women. *JAMA* **238**, 1524–1530.

Silbergeld, S., Brast, N., and Noble, E. P. (1971). The menstrual cycle: A double-blind study of symptoms, mood and behavior, and biochemical variables using enovid and placebo. *Psychosom. Med.* **33**, 411–428.

Silberstein, S. D. (1991). Appropriate use of abortive medication in headache treatment. *Pain Manage* **4**, 22–28.

Silberstein, S. D. (1997). Migraine and pregnancy. *Neurol. Clin.* **15**, 209–231.

Silberstein, S. D., Armellino, J. J., Hoffman, H. D., *et al.* (1999). Treatment of menstruation-associated migraine with the nonprescription of acetaminophen, aspirin, and caffeine: Results from three randomized, placebo-controlled studies. *Clin. Ther.* **21**, 475–491.

Silberstein, S. D., and Lipton, R. B. (1997). Chronic daily headache. *In* "Headache" (P. J. Goadsby and S. D. Silberstein, Eds.), pp. 201–225. Butterworth-Heinemann, Stoneham, MA.

Silberstein, S. D., Lipton, R. B., and Goadsby, P. J. (1998). Headache in clinical practice. Isis Medical Media.

Silberstein, S. D. and Merriam, G. R. (1997). Sex hormones and headache. *In* "Headache" (P. J. Goadsby and S. D. Silberstein, Eds.), pp. 143–173. Butterworth-Heinemann, Stoneham, MA.

Silberstein, S. D., Saper, J. R., and Freitag, F. (2001). Migraine: Diagnosis and treatment. *In* "Wolff's Headache and Other Head Pain" (S. D. Silberstein, R. B. Lipton, and D. J. Dalessio, Eds.), pp. 121–237. Oxford University Press, Oxford.

Smith, R. N. J., Holland, E. F. N., and Studd, J. W. W. (1994). The symptomatology of progestogen intolerance. *Maturitas* **18**, 87.

Smits, M. G., VanderMeer, Y. G., Pfeil, J. P., *et al.* (1993). Perimenstrual migraine: Effect of estraderm TTS and the value of contingent negative variation and exteroceptive temporalis muscle suppression test. *Headache* **34**, 103–106.

Solbach, M. P., and Waymer, R. S. (1993). Treatment of menstruation-associated migraine headache with subcutaneous sumatriptan. *Obstet. Gynecol.* **82**, 769–772.

Solbach, P., Sargent, J., and Coyne, L. (1984). Menstrual migraine headache: Results of a controlled, experimental, outcome study of nondrug treatments. *Headache* **24**, 75–78.

Somerville, B. W. (1971). The role of progesterone in menstrual migraine. *Neurology* **21**, 853–859.

Somerville, B. W. (1972a). A study of migraine in pregnancy. *Neurology* **22**, 824–828.

Somerville, B. W. (1972b). The role of estradio withdrawal in the etiology of menstrual migraine. *Neurology* **22**, 355–365.

Somerville, B. W. (1975). Estrogen-withdrawal migraine. I. Duration of exposure required and attempted prophylaxis by premenstrual estrogen administration. *Neurology* **25**, 239–244.

Stein, G. S. (1986). Headaches in the first postpartum week and their relationship to migraine. *Headache* **21**, 201–205.

Studd, J. and Magos, A. (1987). Hormone pellet implantation for the menopause and premenstrual syndrome. *Obstet. Gynecol. Clin. N. Am.* **14**, 229–249.

Stumpf, P. G. (1990). Pharmacokinetics of estrogen. *Obstet. Gynecol.* **75**, 9–17.

Tang, M. X., Jacobs, D., Stern, Y., *et al.* (1996). Effect of oestrogen during menopause on risk and age at onset of Alzheimer's disease. *Lancet* **348**, 429–432.

Thomas, E. J., Okuda, K. J., and Thomas, N. M. (1991). The combination of depot gonadotrophin releasing hormone agonist and cyclical hormone replacement therapy for dysfunctional uterine bleeding. *Br. J. Obstet. Gynecol.* **98**, 1155–1159.

Utian, W. H. (1987a). Overview on menopause. *Am. J. Obstet. Gynecol.* **156**, 1280–1283.

Utian, W. H. (1987b). The fate of the untreated menopause. *Obstet. Gynecol. Clin. N. Am.* **14**, 1–11.

Wainscott, G., and Volans, G. N. (1978). The outcome of pregnancy in women suffering from migraine. *Postgrad. Med. J.* **54**, 98–102.

Waters, W. E., and O'Connor, P. J. (1971). Epidemiology of headache and migraine in women. *J. Neurol. Neurosurg. Psychiatry* **34**, 148–153.

Weir, B., MacDonald, N., and Mielke, B. (1978). Intracranial vascular complications of choriocarcinoma. *Neurosurgery* **2**, 138–142.

Wentz, A. C. (1985). Management of the menopause. *In* "Novak's Textbook of Gynecology" (H. W. Jones, A. C. Wentz, and L. S. Burnett, Eds.), pp. 397–442. Williams and Wilkins, Baltimore.

Whitehead, M. I., Hillard, T. C., and Crook, D. (1990). The role and use of progestogens. *Obstet. Gynecol.* **75**, S59–S76.

Whitty, C. W. M., and Hockaday, J. M. (1968). Migraine: A followup study of 92 patients. *Br. Med. J.* **1**, 735–736.

Whitty, C. W. M., Hockaday, J. M., and Whitty, M. M. (1966). The effect of oral contraceptives on migraine. *Lancet* **1**, 856–859.

Yaffe, K., Sawaya, G., Lieberburg, I., *et al.* (1998). Estrogen therapy in postmenopausal women: Effects on cognitive function and dementia. *JAMA* **279**, 688–695.

# CHAPTER 7
# Atypical Facial Pain

Marianne Dieterich and Charlotte Feinmann

## CLINICAL ASPECTS

The term *atypical facial pain* was introduced originally to distinguish trigeminal neuralgia from other facial pain syndromes (Frazier and Russell, 1924) and is used mostly as a residual category for otherwise unclassifiable pain syndromes in the facial region (Weddington and Blazer, 1979). The sparse and inconsistent literature on atypical facial pain may be attributed to vague and partly contradictory definitions, the lack of differentiation from similarly ill-defined syndromes, inconsistent application of the term according to the speciality of the treating physician, and the nonexistence of systematic diagnostic procedures. Only recently the existence of atypical facial pain was "discovered" although chronic orofacial pain affects about 10% of Western populations (Lipton *et al.*, 1993). Inadquate professional recognition and innapropriate management of this pain disorder result in continued patient upset and high economic costs.

The International Headache Society (IHS, 1988) defines atypical facial pain as a persistent facial pain that does not have the characteristics of the cranial neuralgias and is not associated with physical signs or demonstrable organic causes. It is present daily and persists for most or all of the day. It is confined at onset to a limited area on one side of the face and may spread to the upper and lower jaws or other areas of the face or neck. It is deep and poorly localized. The pain is not associated with sensory loss or other physical signs. Laboratory investigations including X-ray of face and jaws do not demonstrate relevant abnormalities.

Although the IHS criteria distinguish atypical pain from temporomandibular joint disease (TMJD), oromandibular dysfunction, anesthesia dolorosa, and thalamic pain (Solomon and Lipton, 1988, 1990), it is not clear how this subdivision is useful. Atypical odontalgia (Reik, 1984) and burning mouth or burning tongue

syndrome (Solomon and Lipton, 1990) are probably subtype forms of atypical facial pain. However, the International Association for the Study of Pain (IASP) does not recognize the term "atypical facial pain," although confusingly it does recognize "atypical odontalgia," it may be more sensible to describe all symptoms as variants of chronic facial pain.

The National Institute of Health (1996) called for validated diagnostic methods to develop a multidisciplinary classification. Dworkin and coworkers (1998) have developed guidelines but they have not been widely used. Recent calls for a "mechanism" based classification (Woolf, 1997) have had little success as the exact "mechanism" of chronic facial pain is unclear. Symptoms without a conventional medical explanation result from a complex interaction of biological, psychological, social, and cultural factors. However, descriptions of disorders tend to be influenced by the background of the specialist assessing the patient. Wessley *et al.* (1999) have suggested that each medical specialty has defined its own syndrome in terms of symptoms that relate to the organ of interest.

Furthermore, intracranial tumors of the trigeminal nerve and ganglion (Fee *et al.*, 1975; Garen *et al.*, 1989; Nijenson *et al.*, 1975), tumors of the cerebellar–pontine angle (e.g., acoustic neuroma; Bullitt *et al.*, 1986; Pradat *et al.*, 1969; Rushton *et al.*, 1959), erosive tumors of the base of the skull, the orbits, and nasopharynx (Bullitt *et al.*, 1986; Paulson, 1977; Ruff *et al.*, 1985), as well as infections of the jaws after previous tooth extraction (Roberts and Person, 1979; Roberts *et al.*, 1984) may all co-occur with atypical facial pain. Thus, exclusion diagnosis requires the measures listed in Table I. The differential diagnosis of atypical facial pain and other well-defined facial neuralgias is described in Chapter 6. In some cases of unilateral facial pain a cervicogenic headache or cluster headache must be excluded (Chapter 2).

TABLE I   Requirements for Exclusion Diagnosis in Atypical Facial Pain

| Examination | Exclusion |
| --- | --- |
| Neurological examination | Sensory loss of the face or neck or other physical signs; anesthesia dolorosa, thalamic pain, cranial polyneuropathy |
| Dental examination | Temporomandibular joint disease, oromandibular dysfunction, infection of the jaw |
| ENT examination | Tumors of the cerebellar–pontine angle, e.g., acoustic neuroma, erosive tumors of the nasopharynx |
| Ophthalmological examination | Erosive tumors of the orbits |
| X-ray of the sinus, skull, orbits, cervical part of the spine in four planes | Erosive tumors of the base of the skull, orbits, nasopharynx, sinus cavernosus, sinus petrosus |
| CT scan (+ skull base) | (See X-ray) |
| MR tomography | (See X-ray) and inflammation and meningiomas of the cavernous sinus and the orbit |

There is no sharply demarcated, stereotyped clinical syndrome, although atypical facial pain is easily recognized. Atypical facial pain is experienced in areas supplied by the fifth and ninth cranial nerves and by the second and third cervical nerves. However, the pain is not confined to the peripheral distribution of the nerves and does not respect the boundaries of their territories. The pain is unilateral in two-thirds and bilateral in one-third of patients. The majority of the latter, however, report that the pain is stronger on one side (Pfaffenrath et al., 1993). In more than half of the patients the pain always occurs on the same side, but sometimes a changing-localization or even multifocal pain is described. It is most commonly situated in the maxillary area and in the lower part of the face and is experienced predominantly as superficial but can also be deep or even both, deep and superficial. Trigger zones or precipitating factors are unusual, but stress as well as cold or weather changes have been cited as aggravating factors. The pain rarely disturbs sleep. In almost 90% the pain is present daily and persists continuously with fluctuating intensity. An attackwise daily pain is the exception. The pain is generally described as drawing, burning, stabbing, pressing and throbbing; almost every patient describes the pain with a variety of three to five different terms. Some of the patients perceived sensations resembling coenesthetic phenomena as are associated with psychosis (Delaney, 1976). Accompanying symptoms like dejection, dysesthesia, and paresthesia are the most predominantly reported symptoms (60%). The appearance of the patients while describing their excruciating pain diverges from the picture of the suffering they are illustrating (Solomon and Lipton, 1988).

Accordingly, there is little work impairment associated with facial pain (Pfaffenrath et al., 1993).

## NATURAL COURSE

The incidence, prevalence, and natural cause of atypical facial pain are not known. Eighty percent of chronic facial pain patients complain of generalized symptoms such as headache, limb pain, irritable bowel, or backache. A prospective general population study showed that patients with other chronic symptoms experience more chronic facial pain (Le Resche et al., 1996). Chronic facial pain predominantly affects women (90%) of 30–60 years of age. Symptom-free phases are common (50%); they occur either spontaneously or while undergoing treatment and last for several weeks to months. As patients with a facial pain syndrome present to doctors and others in a variety of disciplines, they rarely present de novo to the neurologist but have usually had X-rays of sinuses and teeth, removal of a tooth or retained root, multiple dental prostheses, coagulation of the trigeminal nerve, and nasal sinus operations. The average number of ineffective surgical procedures is 3.5 ± 3.0 (range 1–13) (Pfaffenrath et al., 1993). None of the procedures had a long-lasting effect; in contrast, the pain worsened in the majority of cases. Atypical facial pain can initially arise following injuries and operations or infiltrations of the face, teeth, and gums and may persist without any local basis. As it develops, patients often undergo (and encourage) dental operations and operations of the ear as well as injection and infiltration treatment, thereby causing secondary damage to the soft tissue, joints, fasciae, muscles, and neuronal structures of the face, which may complicate the clinical picture or lead to chronic pain. Inappropriate surgical treatment predicts intractable pain (Feinmann et al., 1984). Chronic symptoms and syndromes pose a major challenge to medicine; they are common, frequently persistent, and are associated with frequent disabilty, distress, and unnecessary expenditur on medical resources. In primary and secondary care a substantial proportion of patients complaints are found to be unexplained. This problem is magnified in secondary and tertiary care.

## PRINCIPLES OF THERAPY

The etiology and pathogenesis of atypical facial pain are still unclear. However, many chronic facial pain patients specifically relate the onset of their symptoms to dental treatment. This is important as the role of dental treatment in producing or prolonging pain is unclear. Other reported precipitating factors include life stresses such as chronic ill health or exposure to illness in childhood or having experienced poor parental care. The patients also seem likely to recall episodes of sexual or physical abuse during childhood (Hotopf *et al.*, 1999). Once intiated the patients may inadvertently exacerbate or maitain pain. A proportion of patients completely avoid movement of the jaw, resulting in muscular atrophy and joint stiffness. Others stretch and hyperextend the jaw producing local irritation. Prodding and touching painful areas of the face, teeth, or gums may also irritate already sensitive muscles and nerves. In many patients, psychiatric disorders are associated, particularly anxiety, depression, or phobic disorder. High levels of anxiety related to concerns such as "will the pain ever get better" increase the percerption of pain as does depressed mood. Apart from the biochemical association between pain and depression, depressive symptoms such as loss of interest or fatigue are critical in maintaining a preoccupation with symptoms (Madland *et al.*, 2000). Patients with atypical facial pain are said to have psychological abnormalities (Baile and Myers, 1986; Eversole *et al.*, 1985; Feinmann *et al.*, 1984; Lascelles, 1966; Pfaffenrath *et al.*, 1992; Smith *et al.*, 1969; Violon, 1980; Weddington and Blazer, 1979). Some authors assume depression to be the most frequent cause (Ueda, 1982; Lehmann and Buchholz, 1986), whereas in other cases the depression was present before any symptoms of atypical facial pain were evident (Violon, 1980). In one of the most extensive psychopathological studies a psychiatric illness according to the DSM-III criteria was found in 62% of 121 patients with atypical facial pain, of which, however, only 16.5% suffered from depression (Remick and Blasberg, 1985). Overall, it seems that patients with chronic facial pain are no different from other chronic pain populations and not very different from any group of medical outpatients in terms of psychopathology.

There is substantial evidence that tricyclic antidepressants and rather less monoamine oxidase (MAO) inhibitors can relieve facial pain (Remick and Blasberg, 1985; Sharav *et al.*, 1987; Solomon and Lipton, 1988), due not only to their antidepressant effect but to additional analgesic properties. This explains to a lesser extent why pain relief occurs independent of depressed mood. In addition, the depletion of central serotonin (5-HT) and opioid storage areas along with those of other neurotransmitters could be a common biochemical mechanism underlying both depression and pain (Magni, 1987; Sicuteri, 1981). Accordingly, subcutaneous treatment with sumatriptan (6 mg), a selective 5-HT-like receptor agonist, produced temporary improvement in sensory, affective, and total pain 120 min after injection (Harrison *et al.*, 1997). Most patients, however, described the medication as ineffec-tive overall. On the other hand, the administration of a 5-HT antagonist aggravated atypical facial pain in 40% of patients, whereas a control group showed no reaction at all (Hampf, 1989).

Additional hypotheses about the pathogenesis of atypical facial pain are relevant to treatment: an adjustment of the pain threshold in the trigeminal inflow area, the convergence of meningeal and facial afferents onto trigeminal brainstem neurons (Ellrich *et al.*, 1999), microtraumas that lead to long-term neuronal stimulation and spontaneous activity. For example, an abnormal vessel–nerve contact through an ectatic basilar artery leads to compression of the trigeminal nerve (Martins and Ferro, 1989). Similarly, previous dental extraction was known to cause functional and structural changes in the trigeminal nucleus (Dostrovsky *et al.*, 1982; Globel and Bink, 1977), whereby a peripheral lesion of the nerve or nerve root close to the brainstem could lead to ectopic action potentials and transversally spreading ephaptic activation, which in turn could no longer be suppressed in the nucleus (corresponding to the mechanism of trigeminal neuralgia, Fromm *et al.*, 1984; see Chapter 6). The physician is advised to treat the pathologic transmission of stimuli with drugs in the same manner as anticonvulsives (carbamazepine) are used for trigeminal neuralgia (Martins and Ferro, 1989). These hypotheses agree with recent findings during electrophysiological testing of the trigeminofacial system that blink and jaw reflexes showed abnormalities in 70% of the patients (Jääskeläinen *et al.*, 1999) which indicates a trigeminal dysfunction despite normal MRI scans. A cumulative biological reaction is mounted by an organism in response to acute or chronic noxious stimuli producing many effects in the CNS, including breaking down the blood–brain barrier, changing neuronal function with altered gene expression, and abnormal neurotransmitter production (McEwen, B. S. 1998).

Chronic facial pain may be assumed to be a variant of tension-type headache (Chapter 3) with a dysfunction of central pain-relevant structures (Olesen and Schoenen, 1993). An additional lesion of the peripheral nerve, such as due to a surgical procedure, would then lead to an accentuation and chronification by long-term neuronal stimulation of the trigeminal neurons (Sessle, 1989).

Patients presenting with symptoms of atypical facial pain are often misdiagnosed as having a structural organic disorder and receive inappropriate and irreversible treatment. Eventually they may consider themselves beyond therapeutic help. Referral to a psychiatrist is often seen as the end of the line and such referrals should be handled sympathetically. The fear of being stigmatized as a psychiatric patient is still considerable. Therefore, it is vital to emphasize to patients that their pain is "real" and not imaginary. The concept of "cure" should also be reevaluated by telling the patient that the pain can be alleviated by treatment but may never entirely disappear. A pragmatic treatment approach is recommended in which the patient is counseled in terms of lifestyle and reassured that no serious physical disorder is present. A strong association between chronic pain and adverse life events and other long-term problems indicates the need for conservative forms of management. Chronic pain patients have a tendency to overestimate the intensity of their pain, which usually correlates with the psychopathological abnormality. Possibly this could be a reason for the reputation these patients have as "expert killers" or "doctor hoppers." More significance should be given to behavioral therapy (behavioral techniques and biofeedback; Harrison *et al.*, 1997) with the aim of enabling patients to evaluate and control their own pain experience more realistically and to achieve better stress management. Psychotherapy may relieve psychological symptoms once pain relief has been achieved with an antidepressant.

It is usually important that patients continue working and that the "pain" does not become a central issue in the family. Those who do not respond to reassurance should then be treated with antidepressives in slowly increasing doses. Even in the case of initially successful treatment, relapses are common. One of the most important therapeutic objectives is to discourage patients from unnecessary or indeed detrimental operations (Solomon and Lipton, 1990). Analgesics or opioids should be strictly avoided as they are usually ineffective and likely to provoke drug-induced head and face pain (Pfaffenrath *et al.*, 1992).

## PRACTICAL MANAGEMENT

The question of optimal pharmacological treatment of atypical facial pain is controversial. In the literature, tricyclic antidepressants—e.g., amitriptyline, doxepin and dothiepin, 5-HT antagonists, MAO inhibitors, and sometimes anticonvulsants—are recommended most often (Feinmann, 1988; Feinmann *et al.*, 1984; Hampf, 1989; Loeser, 1985; Moore and Nally, 1975; Solomon and Lipton, 1990). In few cases success has been reported with electrical stimulation of the trigeminal ganglion (Raab *et al.*, 1987; Lazorthes *et al.*, 1987) or application of TENS (Eriksson *et al.*, 1984). No particular type of treatment heals atypical facial pain. One-third of the patients considered none of the therapeutic measures to be successful (Pfaffenrath *et al.*, 1992). One of the most difficult decisions confronting a physician who treats a patient with atypical facial pain is when to discontinue a medication that is keeping a patient free of pain. There are no hard and fast rules for the timing of discontinuation of medication after pain relief has

been achieved. If the patient has been treated for at least 6 months and has been free of pain for 8 weeks, then gradual tapering of medication is appropriate.

### Tricyclic Antidepressants

Patients with atypical facial pain might respond to *amitriptyline* (25–125 mg/day) or *clomipramine* (25–125 mg/day) in separate doses (Table II). Dose-dependent side effects (see Chapter 94) are drowsiness, dryness of the mouth, accommodation disturbances, constipation, and, in elderly patients, arterial hypotension and tachycardia.

### Serotonin-Specific Reuptake Inhibitors

Many patients find the side effects of tricyclics difficult to cope with, the newer serotonin-specific reuptake inhibitors (SSRIs) have fewer side effects overall, only 10% of patients experience side effects compared to 75% who take the older drugs. The SSRIs do not interact with alcohol and they do not cause weight gain. Fluoxetine has been shown to be effective in the management of facial pain (Harrison *et al.*, 1997). The only problem is the greater expense. Unfortunately many practitioners are reluctant to prescribe antidepressants of too low a dose. None of the antidepressants are any more effective than others but they all have different side effects. The most effective management is to fit the side effect profile of the drug to the patient's need, i.e., sedation with the tricyclics, alertness, and weight loss with the SSRIs.

**TABLE II   Drug Treatment of Atypical Facial Pain**

| Drug | Average daily dose (mg) | Maximum daily dose (mg) | Level of evidence |
|---|---|---|---|
| First choice | | | |
|   Amitriptyline | 50–75 | 150 | *** |
|   Clomipramine | 25–125 | 150 | ** |
| Second choice | | | |
|   Fluoxetine | 20–60 | 80 | * |
| Third choice | | | |
|   Carbamazepine | 300–1200 | 2400 | * |
| Not in the UK | | | |
| On a trial basis | | | |
|   Diphenylhydantoin | 300 | 400 | * |
|   Gabapentin | 300–1200 | 2500 | * |
|   Oxcarbazepine | 600–2400 | 3000 | * |

*Treatment based on personal experience, expert opinion, retrospective, or nonrandomized studies.

**Treatment is based on at least one randomized prospective study of reasonable size.

***Treatment recommendations (TR) based on more than one well-designed randomized, placebo-controlled trial or a meta-analysis.

## Antiepileptic Drugs

*Carbamazepine* is the drug of first choice for neuralgias and neuropathic pain. It is only rarely recommended for atypical facial pain, starting with 100 mg twice a day and increasing the dosage by 100 mg per day to a maintenance dosage of 1200 mg/day (Table II). The optimal dose is determined by efficacy and side effects such as sleepiness, nystagmus, dizziness, or unsteady gait. The latter can often be attributed to a rapid dose increase. A dose of 300 to 800 mg tid is sufficient. Long-term treatment demands laboratory controls at 3-month intervals to rule out leukopenia, granulocytopenia, or liver damage. The administration of 300 mg diphenylhydantoin as a single night dose may be attempted. The side effects correspond to those of carbamazepine.

## Treatments No Longer Recommended

*Serotonin agonists* such as sumatriptan are not an appropriate therapeutic option, because it caused only temporary and too little improvement of the pain scores. *Neuroleptics* such as thioridazine (25–75 mg) or MAO inhibitors such as tranylcypromine (20 mg) or phenelzine are no longer recommended. A randomized, controlled trial of high-dose dextromethorphan (NMDA glutamate receptor antagonist) found no analgesic efficacy in pain due to trigeminal neuropathy and anesthesia dolorosa (Gilron *et al.*, 2000). All forms of invasive treatment (surgery, infiltration of teeth, jaw, nasal sinuses, etc.) are contraindicated due to their potential additional damage and inefficacy. Operations on the trigeminal nerve (microvascular decompression, thermocoagulation) do not alleviate pain and may lead among other complaints to anesthesia dolorosa (Hier, 1986). Infrared radiation and neural therapy, autogenic training, acupuncture, massage, psychotherapy and chiropractic maneuvers, electrotherapy, hypnosis, and hydrotherapy rarely have any effect. The danger involved in fashionable diagnoses such as temporomandibular joint syndrome is clearly illustrated in a publication by Reik (1985): Over a period of 6 years, the frequency of this diagnosis on admission increased by 500%, and in four-fifths of these cases unjustified dental operations were performed.

## RELATED SYNDROMES

Temporomandibular joint (TMJ) syndrome (previously described as Costen's syndrome) was distinguished from chronic facial pain by the local, physical findings, muscular tension, particularly in the morning, local pressure pain, odd joint position with restricted jaw movement, and intensification of pain when chewing (Laskin, 1969; Reik and Hale, 1981). In spite of the organic findings, however, many of these patients have a psychological profile similar to that of patients with atypical facial pain, and the courses of both syndromes are confused (Yusuf and Rothwell, 1986). Although, on the one hand, symptoms such as noises in the mandibular joint (39%), pain when opening the mouth (12%), and mouth opening (7%, Agerberg and Carlsson, 1972) occur in the normal population without any indication of illness, considerable mechanical strain in conjunction with certain behavioral patterns lead to a chronic pain syndrome (Madland and Feinmann, 1999; Zarb and Speck, 1985). Physical approaches by an orthodontist such as thermotherapy, ultrasound, special exercises, and particularly occlusal correction, bite planes, and splints are no longer recommended. The NIH held a consensus statement (1996) which stated that there is no evidence linking occlusal abnormalities with facial pain and have recommended only conservative, reversible forms of therapy. In a systematic review of splint and occlusal therapy no evidence was found to support such management (Forssell *et al.*, 1999), echoing the earlier statement from the American Dental Association (1993) that the most important part of treatment was the personality of the therapist.

## REFERENCES

Agerberg G., and Carlsson, G. E. (1972). Functional disorder of the masticatory system. I. Distribution of symptoms according to age and sex judged from investigation by questionnaire. *Acta Dont. Scand.* **30**, 597–613.

Baile, W. F., Jr., and Myers, D. (1986). Psychological and behavioral dynamics in chronic atypical facial pain. *Anesth. Prog.* **33**, 252–257.

Bullitt, E., Tew, J. M., Boyd, J. (1986). Intracranial tumors in patients with facial pain. *J. Neurosurg.* **64**, 865–871.

Delaney, J. F. (1976). Atypical facial pain as a defense against psychosis. *Am. J. Psychiatry* **133**, 1151–1154.

Derbyshire, S. W. G., Jones, A. K., Devani, P., Friston, K. J., Feinmann, C., Harris, M., Pearce, S., Watson, J. D. G., and Frackowiak, R. S. J. (1994). Cerebral responses to pain in patients with atypical facial pain measured by positron emission tomography. *JNNP* **57**, 1166–1172.

Dostrovsky, J., Ball, G. J., and Hu, J. W. (1982). Functional changes associated with partial tooth pulp removal in neurons of the trigeminal spinal tract nucleus and their clinical implications. *In* "Anatomical Physiological and Pharmacological Aspects of Trigeminal Pain" (B. Mathews and R. G. Hill, Eds.), pp. 293–310. Excerpta Medica, Amsterdam.

Drummond, P. D. (1988). Vascular changes in atypical facial pain. *Headache* **28**, 121–123.

Dworkin, S., LeResche (1998). A research classification for TMJ disorders. *Pain* **36**, 281–285.

Ellrich, J., Andersen, O. K., Messlinger, K., and Arendt-Nielsen, L. (1999). Convergence of meningeal and facial afferents onto trigeminal brainstem neurons: An electrophysiological study in rat and man. *Pain* 82, 229–237.

Eriksson M. B., Sjoelund, B. H., and Sundbaerg, G. (1984). Pain relief from peripheral conditioning stimulation in patients with chronic facial pain. *J. Neurosurg.* 61, 149–155.

Eversole, L. R., Stone, C. E., Matheson D., and Kaplan, H. (1985). Psychometric profiles and facial pain. *Oral Surg. Oral Med. Oral Pathol.* 60, 269–274.

Fee, W. E., Jr., Epsy, C. D., and Konrad, H. R. (1975). Trigeminal neurinomas. *Laryngoscope* 85, 371–376.

Feinmann, C. (1988). The contribution of psychiatry toward the understanding of facial pain and headache. *In* "Headache—Problems in Diagnosis and Management" (A. Hopkins, Ed.), pp. 275–304. Saunders, London.

Feinmann, C., Harris, M., and Cawley, R. (1984). Psychogenic facial pain: Presentation and treatment. *Br. Med. J.* 288, 436–438.

Ford, B., Greene, P., and Fahn, S. (1994). Oral and genital tardive pain syndromes. *Neurology* 44, 2115–2119.

Forssell, H, Kalso, E., Koskela, P., Vehmanen, R., Puukka, P., and Alanen, P. (1999). Occlusal treatments in temporomandibular disorders: a qualitative systematic review of randomized controlled trials. *Pain.* 83(3), 549–560.

Frazier C. H., and Russell, E. C. (1924). Neuralgia of the face. An analysis of seven hundred and fifty-four cases with relation to pain and other sensory phenomena before and after operation. *Arch. Neurol. Psychiatry* 11, 557–563.

Fromm, G. H., Terrence, C. F., and Manoon, J. C. (1984). Trigeminal neuralgia. Current concepts regarding etiology and pathogenesis. *Arch. Neurol.* 41, 1204–1207.

Garen, P. D., Powers, J. M., Kings, J. S., and Perot, P. L., Jr. (1989). Intracranial fibroosseous lesion. Case report. *J. Neurosurg.* 70, 475–477.

Gilron, I., Booher, S. L., Rowan, J. S., Smlooer, B., and Max, M. B. (2000). A randomized, controlled trial of high dose dextromethorphan in facial neuralgias. *Neurology* 55, 964–971.

Globel, S., and Bink, J. M. (1977). Degenerative changes in primary axons and in neurons in nucleus caudalis following tooth pulp exstirpation in the cat. *Brain Res.* 132, 347–354.

Goldberg, D. (1996). A dimensional model of common mental disorders. *Br. J. Psychiatry* 168, 44–49.

Hampf, G. (1989). Effect of serotonin antagonists on patients with atypical facial pain. *J. Craniomandib. Disord.* 33, 211–212.

Harrison, S. D., Balawi, S. A., Feinmann, C., and Harris, M. (1997). Atypical facial pain: A double-blind placebo-controlled crossover pilot study of subcutaneous sumatriptan. *Eur. Neuropsychopharmacol.* 7, 83–88.

Hier, D. B. (1986). Headache. *In* "Manual of Neurologic Therapeutics" (M. A. Samuels, Ed.), pp. 15–29. Little, Brown, Boston and Toronto.

Hotopf, M., Majoy, R., Wadsworth, M., and Wessley, S. (1999). Childhood risk for adults with medically unexplained symptoms. Results from a national cohart study. *Am. J. Psych.* 156, 1796–1800.

IASP, Merkey, H., and Bogduk, N. (1994). *In* "Classification of Chronic Pain" (H. Merskey, Ed.), pp. 59–60. IASP Press, Seattle.

IHS—The Headache Classification Committee of the International Headache Society (1988). Classification and diagnostic criteria for headache disorders, cranial neuralgias and facial pain. *Cephalalgia* 8(Suppl. 7), 71–72.

Jääskeläinen, S. K., Forssell, H., and Tenovuo, O. (1999). Electrophysiological testing of the trigeminofacial system: Aid in the diagnosis of atypical facial pain. *Pain* 80, 191–200.

Kaaya, S., Goldberg, D., and Gask, L. (1992). Management of somatic presentations of illness in general medical settings. Evaluation of a new training course for general practitioners. *Med. Ed.* 26, 138–144.

Lascelles, R. G. (1966). Atypical facial pain and depression. *Br. J. Psychiatry* 112, 651–659.

Laskin, D. M. (1969). Etiology of the pain-dysfunction syndrome. *J. Am. Dent. Assoc.* 79, 147–153.

Lazorthes, Y., Armengaud, Y. P., and Da Motta, M. (1987). Chronic stimulation of the Gasserian ganglion for the treatment of atypical facial neuralgia. *PACE Proc. Control Eng.* 10, 257–265.

Lehmann, H. J., and Buchholz, G. (1986). "Atypische Gesichtsneuralgie" oder depressiver Gesichtsschmerz? Diagnostische Aspekte einer gut abgrenzbaren Form der larvierten Depression. *Fortschr. Neurol. Psychiatrie* 54, 154–157

Le Resche, L. (1998). "The Aetiology of Tempoormandibular Joint Pain." IASP Press.

Lesse, S. (1960). Atypical facial pain syndromes: A study of 200 cases. *Arch. Neurol.* 3, 100–101.

Lipton, J. A., Ship, J., Lareche, and Robinson, D. (1993). Estimated prevelance and distribution of reported orofacial pain in the US. *Pain* 124, 115–121.

Loeser, Y. D. (1985). Tic douloreux and atypical facial pain. *J. Can. Dent. Assoc.* 51, 917–923.

MacFarlane, G. T., Hunt, I. M., and Silman, A. J. (2000). Role of mechanical and psychosocial factors in the onset of forearm pain in prospective population based study. *Br. Med. J.* 321, 676.

Madland, G., Feinmann, C., and Newman, S. (2000). Factors associated with anxiety and depression in facial arthromyalgia. *Pain* 84, 225–232.

Magni, G. (1987). On the relationship between chronic pain and depression where there is no organic lesion. *Pain* 31, 1–21.

Martins, I. P., and Ferro, J. M. (1989). Atypical facial pain, ectasia of basilar artery, and baclofen: A case report. *Headache* 29, 581–583.

McEwen, B. S. (1998). Stress, adaptation, and disease. Allostasis and allostatic load. *Ann. NY Acad. Sci.* 840, 33–44.

Moore, D. S., and Nally, F. F. (1975). Atypical facial pain: An analysis of 100 patients with discussion. *J. Can. Dent. Assoc.* 41, 396–401.

NIH (1996). Technology assessment statement management of temporomandibular disorders. Washington, DC: National Institute of Health.

Nijensohn, D. E., Araujo, J. C., and MacCarthy, C. S. (1975). Meningiomas of Meckel's cave. *J. Neurosurg.* 43, 197–202.

Olesen, J., and Schoenen, J. (1993). Synthesis. *In* "The Headaches" (J. Olesen, P. Tfelt-Hansen, and K. M. A. Welch, Eds.), pp. 493–496. Raven, New York.

Paulson, G. W. (1977). Atypical facial pain. *Oral Surg. Oral Med. Oral Pathol.* 43, 338–341.

Pfaffenrath, V., Rath, M., Keeser, W., and Pöumlat;llmann, W. (1992). Atypischer Gesichtsschmerz—die Qualitaøt der IHS-Kriterien und psychometrische Daten. *Nervenarzt* 63, 595–601.

Pfaffenrath, V., Rath, M., Pöumlat; llmann, W., and Keeser, W. (1993). Atypical facial pain—Application of the IHS criteria in a clinical sample. *Cephalalgia* 12(Suppl.), 84–88.

Pradat, P., Guilly, P., David, M., and Metzger, J. (1969). Neuralgie faciale atypique datant de 35 ans: kyste epidermoide lateroprotuberantiel; interet de la tomographie hypocycloide. *Neurochirurgie* 15, 497–502.

Raab, W. H., Kobal, G., Steude, U., Hamburger, C., and Hummel, C. (1987). Die elektrische Stimulation des Ganglion Gasseri bei Patienten mit atypischem Gesichtsschmerz. Klinische Erfahrung und experimentelle Kontrolle durch elektrische Pulpareizung. *Dtsch. Zahnaerztl. Z.* 42, 793–797.

Reik, L., Jr. (1984). Atypical odontalgia.: A localized form of atypical facial pain. *Headache* 24, 222–224.

Reik, L., Jr. (1985). Atypical facial pain: A reappraisal. *Headache* 25, 30–32.

Reik, L., Jr., and Hale, M. (1981). The temporomandibular joint pain-disfunction syndrome: A frequent cause of headache. *Headache* 21, 151–156.

Remick, R. A., and Blasberg, B. (1985). Psychiatric aspects of atypical facial pain. *Can. Dent. Assoc. J.* **51**, 913–916.

Roberts, A. M., and Person, P. (1979). Etiology and treatment of idiopathic trigeminal and atypical facial pain neuralgias. *Oral Surg.* **48**, 298–308.

Roberts, A. M., Person, P., Chandran, N. B., and Hori, J. M. (1984). Further observations on dental parameters of trigeminal and atypical neuralgias. *Oral Surg.* **58**, 121–129.

Ruff, T., Lenis, A., and Diaz, J. A. (1985). Atypical facial pain and orbital cancer. *Arch. Otolaryngol.* **111**, 338–339.

Rushton, J. G., Gibilisco, J. A., and Goldstein, N. P. (1959). Atypical facial pain. *J. Am. Med. Assoc.* **171**, 545–548.

Sessle, B. J. (1989). Neural mechanisms of oral and facial pain. *Otolaryngol. Clin. North Am.* **22**, 1059–1072.

Sharav, Y., Singer, E., and Schmidt, E. (1987). The analgesic effect of amitriptyline on chronic facial pain. *Pain* **31**, 199–209.

Sicuteri, F. (1981). Opioid receptor impairment—Underlying mechanism in "pain diseases"? *Cephalalgia* **1**, 77–82.

Smith, D. P., Pilling, L. F., and Pearson, J. S. (1969). A psychiatic study of atypical facial pain. *Can. Med. Assoc. J.* **100**, 286–291.

Solomon, S., and Lipton, R. B. (1988). Atypical facial pain: A review. *Semin. Neurol.* **8**, 332–338.

Solomon, S., and Lipton, R. B. (1990). Facial pain. *Neurol. Clin.* **8**, 913–928.

Stress, adaptation, and disease. Allostasis and allostatic load. *Ann. N Y Acad. Sci.* **840**, 33–44.

Ueda, N. (1982). Ueber die atypischen Gesichtsschmerzen in der larvierten Depression. *ZWR* **91**, 57–60.

Violon, A. (1980). The onset of facial pain: A psychological study. *Psychother Psychosom.* **34**, 11–16.

Von Knorff, M., Dworkin, S., Le Resche, L., and Kruger, N. (1988). An epidemiologic comparison of pain complaints. *Pain* **32**, 193–199.

Weddington, W. W., and Blazer, D. (1979). Atypical facial pain and trigeminal neuralgia. A comparison study. *Psychosomatics* **20**, 348–356.

Wessley, S., Nimnuan, C., and Sharpe, M. (1999). Functional somatic syndromes: One or many. *Lancet* **354**, 936–939.

Woolf, C. (1997). A mechanism based classification for chronic pain. *Pain* **381**, 138.

Yusuf, H., and Rothwell, P. S. (1986). Temporomandibular joint pain-dysfunction in patients suffering from atypical facial pain. *Br. Dent. J.* **161**, 208–212.

Zarb, G. A., and Speck, J. E. (1985). Die Behandlung der mandibulaøren Dysfunktion. *In* "Physiologie und Pathologie des Kiefergelenks" (G. A. Zarb and G. E. Carlsson, Eds.), pp. 421–447. Quintessenz Verlags-GmbH, Berlin, Chicago, London, Rio de Janeiro, and Tokyo.

CHAPTER 8

# Trigeminal and Glossopharyngeal Neuralgia

G. D. Perkin and Thomas Brandt

## TRIGEMINAL NEURALGIA (TIC DOULOUREUX)

### Clinical Aspects

Trigeminal neuralgia is characterized by attacks of severe, unilateral lancinating pain in the territory of the trigeminal nerve, most commonly its third division. The attacks, which may recur many times a day, last from seconds to minutes. They may occur spontaneously or be stimulus triggered. Long remissions, lasting months or years, are characteristic, particularly in the early stages of the condition. Triggering factors include chewing, speaking, cleaning the teeth, inserting or removing dentures, and facial contact (trigger zones). Idiopathic trigeminal neuralgia (now termed typical trigeminal neuralgia) is due, in many cases, to cross-compression of the nerve by vascular structures close to the nerve entry zone alongside the brainstem (see below).

A pathological condition caused by vascular compression of the root entry/exit zone of the cranial nerves has been designated hyperactive dysfunction syndrome (HDS). Trigeminal neuralgia, hemifacial spasm and glossopharyngeal neuralgia are included under that umbrella, though opinions differ as to how often vascular cross-compression is found when surgery is performed for trigeminal neuralgia. Where combined syndromes occur, vascular cross-compression is even more likely (Kobata *et al.*, 1998). Symptomatic trigeminal neuralgia occurs as a result of other pathological processes affecting the nerve, either within the brainstem or outside it. Features suggesting symptomatic trigeminal neuralgia include young onset (<40 years), atypical pain, onset in the first division, and evidence of neurological deficit on examination. The most common

cause of symptomatic trigeminal neuralgia is multiple sclerosis (MS). It is estimated that 2–4% of patients with trigeminal neuralgia have MS, while 1.5% of MS patients experience the condition.

Trigeminal neuralgia is sometimes the presenting manifestation of MS or occurs during the course of the disease. Bilateral, chronic facial pain occurs in 25–30% of such cases. Numerous other causes of symptomatic trigeminal neuralgia have been described (Table I).

Various investigative procedures have been advocated for the assessment of trigeminal neuralgia. Magnetic resonance imaging is the procedure of choice for visualizing the trigeminal nerve (Majoie *et al.*, 1998). Specialized techniques have been developed to enhance the imaging of the nerve root entry zone, producing a positive yield (in terms of confirming neurovascular contact) of 70–100% in symptomatic nerves and in 6–9% of asymptomatic nerves (Meaney *et al.*, 1995).

In one study, the sensitivity of MRA (in 27 patients with trigeminal neuralgia) was 88.5% but the specificity only 50% if judged by the results of surgical exploration (Boecher-Schwarz *et al.*, 1998). Imaging in cases of MS has usually confirmed the presence of brainstem plaques at a site likely to be affecting the entry zone of sensory fibers (Gass *et al.*, 1997). Vascular cross-compression may still be the mechanism in MS patients, however, with decompression resulting in pain relief (Boecher-Schwarz *et al.*, 1998).

Differential diagnoses of importance for trigeminal neuralgia include glossopharyngeal neuralgia (see below), Raeder's paratrigeminal syndrome, atypical facial pain (see Chapter 7) and the rare auriculotemporal (Frey's) syndrome with gustatory sweating (De Benedittis, 1990). In the SUNCT syndrome, attacks of brief neuralgic-type pain are accompanied by conjunctival injection and tearing. Confusion is possible with

TABLE I    Causes of Symptomatic Trigeminal Neuralgia

Multiple sclerosis
Acoustic neurinoma
Meningeal carcinomatosis
Epidermal tumors
Cavernomas
Fibrous dysplasia
Syringobulbia
Arteriovenous angioma
Persistent primitive trigeminal artery
Cavernous aneurysm
Trigeminal nerve injury
Chiari Type 1

TABLE II    Medical Options (In Year Order of Their Introduction)

Phenytoin
Antihistamines
Carbamazepine
Clonazepam
Baclofen
Valproic acid
L-Baclofen
Oxcarbazepine
Proparacaine (eye drops)
Lamotrigine
Gabapentin

TABLE III    Nonmedical Options

Radiofrequency thermal rhizotomy
Percutaneous microcompression of the Gasserian ganglion
Glycerol rhizotomy
Microvascular decompression
Partial sensory trigeminal rhizotomy
Peripheral neurectomy
Peripheral nerve injection
Stereotactic radiosurgery

trigeminal neuralgia when the latter's paroxysmal pain is accompanied by lachrymation, as reported in 25% of patients in one series (Benoliel and Sharav, 1998).

## Natural Course

Typical trigeminal neuralgia is a disease of the middle-aged and elderly. The peak onset is in the seventh to eighth decades with a female to male ratio of 3:2. The overall incidence in Rochester, MN from 1945 to 1984 was 4.3 per 100,000, 5.9 for women, and 3.4 for men. The annual incidence was age-related—0.2% under 40, 8.9% for 50–60 years of age, 17.5% in those ages 60–70, and approximately 25% in those over 70 (Katusic et al., 1990). Hypertension increases the rise of neurovascular compression and is even more prevalent in those with the combined hyperactive dysfunction syndrome. Trigeminal neuralgia is usually sporadic though; rarely, familial cases have been recorded.

With increased duration of the condition, the intervals between pain attacks become shorter. It has been suggested that untreated, trigeminal neuralgia can progress to atypical features, with a dull, constant background pain, and the development of overt sensory loss (Burchiel and Slavin, 2000). With appropriate medical management, approximately 70% of patients respond initially. The figure decreases with increasing duration of the condition. Figures for the success rate of the various operative procedures for trigeminal neuralgia vary widely. Generally the outcome proves less satisfactory in those patients with atypical features (Zakrzewska et al., 1999). The management of the trigeminal neuralgia associated with MS depends on what mechanism is thought to be responsible for the pain.

## Principles of Therapy

Aberrant, often arteriosclerotically elongated and dilated arteries in the cerebello–pontine angle are thought to cause a segmental demyelination at the junction between the central (oligodendrogial) and peripheral (Schwann cell) myelin sheath (Gardner and Miklos, 1959). Some contact between brainstem arteries and the trigeminal nerve has been found at postmortem in 58% of individuals without a history of trigeminal neuralgia (Hardy and Rholon, 1978), though compression sufficient to cause morphological changes in the root entry zone occurs in only 7% of unselected examinations (Klun and Prestor, 1986).

The pain is believed to originate through ephaptic excitation, that is paroxysmal pathological transmission of impulses between neighboring demyelinated axons. The causes of sex specificity (female predominance) and predominance of the right side (in cases of trigeminal neuralgia) remain unclear. That female predominance becomes even more apparent in patients with the combined hyperactive dysfunction syndrome (Kobata et al., 1998). The medical and nonmedical therapeutic options are listed in Tables II and III.

Drugs influence the condition by their effects on pathological axonal transmission, as for example with phenytoin and carbamazepine. In addition, reduction of Gaba concentrations in the central pain-inhibiting system of the periaqueductal gray matter is likely to be of significance.

Response to carbamazepine in the first few months of treatment reaches 60–80% (Zakrzewska and Patsalos, 1992), higher than that reported for phenytoin. After a period, response rate falls to as low as 25% of patients (Taylor et al., 1981). Oxcarbazepine, a drug related to carbamazepine, may well be as effective, but a double-blind comparison between it and carbamazpine is not available. Baclofen is effective in some cases but less so than carbamazepine. Both valproic acid and clonazepam have been reported to be beneficial in open studies. Lamotrigine appears to be an effective agent perhaps in up to 80% of cases (Canavero and Bonicalzi, 1997). A need to rapidly increase the dose in order to control symptoms can be disadvantageous in the light of the drug's capacity to induce a rash. Gabapentin has also proved beneficial in doses between 900 and 2400 mg/day (Sist et al., 1997). In a study of seven patients with MS complicated by trigeminal neuralgia, resistant to various drug therapies (including carbamazepine in all seven) Gabapentin produced complete relief in six and partial in one. Doses ranged from 900 to 2400 mg daily (Khan, 1995). Proparacaine 0.5% solution instilled into the eye is reported to relieve trigeminal neuralgia for up to 4 months after one to three installations (Vassilouthis, 1994).

Interventional, operative, or radiosurgery therapy is considered if medical management fails. With the exception of microvascular decompression, all methods are selectively destructive, the principal objective being to relieve the pain by blocking trigeminal afferents. In general, the more peripheral the selective destruction, the higher the recurrence rate. Nearly all effective destructive surgery results in sensory deficits in about 20% of cases with incomplete lesions producing correspondingly higher rates.

## Radiofrequency Thermal Rhizotomy

This method of treatment is based on the assumption that radiofrequency lesions can destroy nociceptive fibers (unmyelinated C and poorly myelinated A-Delta fibers) while preserving heavily myelinated fibers (Taha and Tew, 1997). A 20-G needle is inserted through the foramen ovale under fluoroscopic control to reach the retrogasserian part of the nerve. Intravenous anesthesia is used. The final electrode position is determined by the patient's response to electrical stimulation. A preliminary lesion is produced at 60 to 70 for 70 s, and then increasing temperature is used to produce additional lesions, the aim being to produce dense hypalgesia in the relevant area.

Anesthesia dolorosa is seen in less than 4% of cases (Broggi et al., 1990). Dysaesthesiae occur in up to 50% of patients and are influenced by the type of electrode used. Corneal anesthesia occurred in 15% of patients in one series when pain in the ophthalmic division was being treated. 2% of patients developed corneal ulceration and keratitis. Trigeminal motor weakness occurs in about a quarter (Scrivani et al., 1999). Rarely, cerebrovascular accidents, diplopia, carotico-cavernous fistulae, seizures, temporal lobe abscess, and death have been reported. The procedure can be repeated, thereby increasing the risk of affecting other sensory modalities.

A recurrence rate of 12–21% is reported with a 6- to 7-year follow-up (Broggi et al., 1990; Siegfried, 1981; Tew and Keller, 1977), rising to 20–25% of patients over a 10- to 14-year follow-up (Taha et al., 1995). A recently published series, however, found only 26% of subjects were pain-free 11 years after the procedure, although a substantial number of patients in that study were lost to follow-up (Yoon et al., 1999). Outcome is less good in those with atypical symptoms (Yoon et al., 1999; Zakrzewska et al., 1999). Over a period of months after the procedure, investigation of sensory detection thresholds shows an initial recovery of sensitivity to touch followed by heat and pain (Hampf et al., 1990).

## Percutaneous Microcompression of the Gasserian Ganglion

General anesthesia is used. Atropine is sometimes given prophylactically to block the bradycardia that occurs in about two-thirds of subjects. Using fluoroscopic control, a thin-walled 14-G needle is inserted in the foramen ovale. A Fogarty catheter is then advanced to the entrance of Meckel's cave (Brown and Gouda, 1997). The balloon is inflated with Omnipaque to a pressure of 650–950 mmHg for 1–1½ min. The initial success rate ranges from 80 to over 90%, with a relapse rate of 20–50% reported in the first 2 to 5 years (Frank and Fabrizi, 1989; Lichtor and Mullan, 1990; Meglio and Cioni, 1989).

Subarachnoid bleeding sometimes occurs. Numbness is inevitable though it decreases during the first year after the procedure. Dysaesthesiae are uncommon. Herpetic eruptions occur quite frequently after the procedure. Trigeminal weakness, ear pain, and diplopia have been reported. The procedure can be repeated successfully.

## Glycerol Rhizotomy

Under local anesthetic, a standard 20-G spinal needle is directed percutaneously through the foramen ovale, using fluoroscopic control. Cerebrospinal fluid (CSF) flow usually indicates that the needle is in the trigeminal cistern. The cisternal volume is calculated using a nonionic contrast medium, leading to the injection of around 0.2 to 0.4 mL of glycerol (Jho and Lunsford, 1997). Headache, nausea, and vomiting sometimes follow the procedure. Hypertensive or hypotensive reactions (the latter with bradycardia) can appear during the injection. Herpes around the mouth appears in about one-third of patients. Postoperative sensory loss occurs in about 50% and is moderate or dense in 20%. Distressing dysesthesiae are seen.

Comparison has been made with radiofrequency rhizotomy, but not in a controlled fashion (Tan *et al.*, 1995). Complete pain relief occurs in about three-quarters of patients (Jho and Lunsford, 1997). Recurrence rates over 2–10 years lie between 30 and 50%. The procedure can be repeated, but with a higher risk of sensory deficit.

## Microvascular Decompression

Through a retromastoid craniectomy, the trigeminal nerve is examined microsurgically for vascular compression around the nerve root entry zone. Intraoperative monitoring with brainstem auditory evoked potentials (BAEPs) is recommended. The nerve is visualized from the pons to Meckel's cave. The nerve is decompressed of all vessels, whether arterial or venous. Veins are usually coagulated, the arteries are separated with Teflon, cotton gauze, or polyvinyl alcohol foam (Lovely and Jannetta, 1997). In some instances, no cross-compression is found.

Mortality rates have been reported between 0 and 1.7%. The figure for collective data from the literature on 3000 cases is 0.3%. Rarely, brainstem, cerebellar, or hemisphere stroke occurs, usually in the form of a hematoma. Hydrocephalus is sometimes seen. Cranial nerve palsies (IV, VI, VII, and VIII) are encountered. Hearing loss is less likely if intraoperative BAEPs are used. Persistent facial numbness is reported by 18% of patients (Lovely and Jannetta, 1997), but is severe in only 1%. Foreign body granuloma formulation has been reported with the use of cotton gauze (Parmar and Sharr, 1999).

It has been suggested that the procedure has a similar outcome, in terms of pain relief, to balloon microcompression and radiofrequency thermocoagulation, but is superior to glycerol rhizotomy. The percentage of completely pain-free patients lies between 53 and 94 over a mean follow-up between 5 months and 20 years (Broggi *et al.*, 2000). Jannetta (1985) reported that in patients who had not had previous destructive surgery 80% remained pain free. In a recent study, only 10% of patients had relapses during a 17-year follow-up period (Coakham and Moss, 1998). The procedure is successful in patients with MS who do not have brainstem plaques but do have cross-compression at operation (Broggi *et al.*, 2000).

## Partial Sensory Trigeminal Rhizotomy

Retromastoid craniotomy is required for this procedure. It has most often been used where there is no evidence of vascular compression of the nerve or where microvascular decompression has failed and no significant vascular contact found at the time of reoperation (Young and Wilkins, 1993). One-third to one-half of the cross-sectional area of the sensory root is cut about 2 to 5 mm from the pons. Electrophysiological techniques can be used intraoperatively to map the trigeminal nerve root and to allow more accurate placement of the lesion (Stechison *et al.*, 1996).

Two-thirds of patients have a sensory deficit after the procedure, with 20% complaining of dense numbness in one or more divisions of the nerve. Other reported complications include CSF leakage, deafness, and brainstem infarction. Poor outcome, in terms of pain relief, was reported in 30% of patients in one series (Young and Wilkins, 1993). Outcome is less successful if the patient has had previous posterior fossa surgery. Patients with pain involving the third division fare better with partial sectioning of the caudolateral aspect of the sensory root than do patients with pain in the other divisions.

## Peripheral Neurectomy

Supraorbital neurectomy is carried out through an incision in the eyebrow. All branches of the supraorbital and supratrochlear nerves are divided and avulsed. Infraorbital neurectomy is performed via an intraoral incision. The infraorbital nerve is avulsed well into the infraorbital foramen. Sectioning of the inferior alveolar nerve is performed within the inferior alveolar canal.

Loss of facial sensation in the relevant division is inevitable but corneal sensation remains intact. The procedure is usually effective though in one report some patients had continued to use carbamazepine. Over 24 months of follow-up, 15% of patients had pain recurrence in one series.

## Peripheral Nerve Block with Phenol/Glycerol

Although peripheral nerve procedures have been largely dismissed by some authors, based on the high relapse rate, inevitable numbness and frequent unpleasant dysesthesiae, the techniques continue to have their advocates (Wilkinson, 1999). In a recent series using phenol and glycerol, median duration of pain relief was 9 months, with at least partial recovery of facial sensory loss between 3 and 12 months after injection.

## Peripheral Nerve Block with Tetracaine

In a recent study of three patients, peripheral nerve blockade was achieved using a mixture of 4% tetracaine and 0.5% bupivicaine. Pain relief for more than 3 months was achieved. Sensory deficits occurred in one patient but none had discomforting dysesthesiae (Goto *et al.*, 1999).

### Gamma Knife Radiosurgery

Initial attempts to control the pain of trigeminal neuralgia by irradiation concentrated the therapy on the trigeminal ganglion (Leksell, 1983). Subsequently, attention was turned to the proximal trigeminal nerve adjacent to the pons. Using high-resolution MRI, a radiosurgery dose of 70–90 Gy was applied with reports of complete pain relief in 58% of patients over a median follow-up of 18 months (Kondziolka *et al.*, 1996). An increasing literature is reporting on the further use of this technique, its morbidity and its efficacy over a longer period of follow-up.

A report of 120 patients treated for typical trigeminal neuralgia has been recently published (Kondziolka *et al.*, 1998). The mean age was 67 years. All patients had had extensive trials of medication and many had had interventional procedures, including microvascular decompression in 42. Probably 11 of the patients had multiple sclerosis. Treatment was directed at a single 4-mm isocenter placed 2 to 4 mm anterior to the junction of the trigeminal nerve and the pons. The range of maximum dose lay between 70 and 90 Gy. Median follow-up was 18 months. Total or good pain relief (good defined as 50–90% decrease) was obtained in 77% of patients. Median time to response was 4 weeks. Follow-up case ascertainment was 87%. Pain relapse occurred in 10% of patients who had previously had complete relief. Rather poorer responses occurred in the MS patients. Ten percent of patients developed new or increased trigeminal sensory symptoms. Many of the patients were not followed personally. Assumptions regarding the mode of action of the treatment center on possible functional electrophysiological block of ephaptic transmission. Other groups have reported similar findings. Of 57 patients treated by Regis *et al.* (1999), 87% reported cessation of pain, falling to 79% at last follow-up. None of these cases developed new or increased trigeminal paresthesiae or sensory loss. Further information regarding the long-term benefit of this treatment is needed and whether it would be possible to reirradiate those patients with relapse.

### Practical Management

#### Carbamazepine

An initial dose of 200 to 300 mg per day is used, in divided doses. The use of slow-release preparations twice daily lessens peak serum levels and improves tolerance. Most patients respond to doses between 600 and 800 mg per day, though occasionally up to 1600 mg is necessary. Serum concentrations of carbamazepine increase in a linear fashion with dose increases and do not need to be obtained routinely. In some patients, toxic effects reflect levels of the primary metabolite carbamazepine–10,11-epoxide rather than carbamazepine itself.

Dose-dependent effects include dizziness, nausea, vomiting, diplopia, and ataxia. The drug is teratogenic. A rash occurs in about 3% of subjects. Hematological effects, which are uncommon, include neutropenia or pancytopenia. Hyponatremia, usually of mild degree, is recognized to occur, but is seldom symptomatic. Carbamazepine is a potent inducer of liver enzymes and will accelerate metabolism of many other drugs, including the oral contraceptive. Once control has been achieved, the drug can be slowly withdrawn, in case the patient has gone into remission. Withdrawal rates should not exceed 100 mg every 48 h.

#### Oxcarbazepine

Oxcarbazepine is closely linked to carbamazepine. As an anti-convulsant, its dosage ranges from 600 to 3600 mg per day. It is generally better tolerated than carbamazepine though its side effect profile is similar (Zakrzewska and Patsalos, 1989). It is a less potent enzyme inducer than carbamazepine.

#### Phenytoin

Phenytoin is increasingly less often used. In the original study, a dose of 300 mg per day was employed (Bergouignan, 1942). It would seem sensible to employ varying doses according to patient response and serum levels. The drug can be given once daily. Dose-related side effects include drowsiness, nausea, ataxia, dysarthria, and abnormal movements. Gingival hypertrophy is commonplace. Rashes occur, as can coarsening of the facial features, hirsutism, and acne. Lymphadenopathy is a rare complication. The drug can induce folate and vitamin D deficiency. It is also a potent liver enzyme inducer and is teratogenic.

#### Baclofen

Baclofen, perhaps more in the levo than in the racemic form, is effective in some patients (Fromm *et al.*, 1984). Initially 15 mg per day is used, in divided doses increasing thereafter to a maximum of 80–90 mg per day. Dose-related side effects include drowsiness, nausea, vomiting, and ataxia. The drug is less effective than carbamazepine, but may exert a synergistic effect with it. Rapid withdrawal of Baclofen can trigger anxiety, hallucinations, and seizures.

### Clonazepam

The drug is started in a dose of 0.5 mg per day, increasing, if tolerated, to a maintenance dose of 4 to 8 mg per day. Drowsiness, lethargy, and fatigue are major problems and limit the drug's use.

### Valproate Sodium

Only limited data are available on the use of valproate. In one open trial, 600–1200 mg per day was used. Gastric intolerance is common in the form of nausea, vomiting, and anorexia. Weight gain also occurs. Tremor occurs as a dose-related effect. Restlessness, confusion, and irreversible hair loss are encountered. Rarely, hepatotoxicity is seen. The drug sometimes alters platelet function. It is teratogenic. A slow release preparation allows a once or twice daily regime.

### Lamotrigine

Lamotrigine is best introduced slowly, thereby lessening the risk of inducing a rash. The disadvantage is the greater time in achieving adequate serum levels. In an open study, a median dose of 400 mg was used (range 150–600 mg) (Canavero and Bonicalzi, 1997). Side effects include drowsiness, unsteadiness, and a skin rash, probably with a frequency similar to that seen with carbamazepine. The drug is not thought to be an enzyme inducer and is said not to be teratogenic.

### Gabapentin

In an open study, dosage ranged from 900 to 2400 mg per day. The drug is generally well tolerated. It is not an enzyme inducer and does not interact with other drugs. Its use is not recommended in pregnancy.

### Recommendations for Medical Treatment

Most individuals would accept that medical treatment is the initial treatment of choice. Carbamazepine is used first, pressing the dose to tolerance in an attempt to achieve control. Oxcarbazepine has probably a similar level of efficacy and may well become the first alternative to consider if carbamazepine fails. There is insufficient data to make firm recommendations regarding the other drugs detailed. In terms of probable efficacy and known tolerance, second-line drugs include phenytoin, baclofen, valproate, lamotrigine, and Gabapentin. Because of poor tolerance, clonazepam can probably be

considered as a third-line drug. Drug combinations may be worth considering, for example carbamazepine and baclofen.

### Surgical Procedures

It is difficult to make firm statements about the relative merits of the various surgical procedures used in this condition as a result of the near absence of comparative studies. It is generally considered that microvascular decompression, balloon microcompression, and radiofrequency rhizotomy are of comparable efficacy but that all are more effective than glycerol rhizotomy. The picture is now more complicated since the introduction of gamma-knife radiosurgery, whose exact role is not yet finalized.

For younger patients, without any contraindication to surgery, microvascular decompression is probably the first choice. If the procedure fails, or there is no cross-compression at operation, either balloon microcompression or radiofrequency rhizotomy should be considered. If the third division is affected, radiofrequency rhizotomy is preferred to balloon microcompression. If these procedures fail, either partial sensory trigeminal rhizotomy can be considered or the use of gamma-knife radiosurgery.

In the older patient where there are contraindications to craniotomy, either balloon microcompression or radiofrequency rhizotomy should be considered first. If this fails, the choice lies between glycerol rhizotomy or gamma-knife radiosurgery. In the very frail individual, gamma-knife radiosurgery will become the treatment of choice if it proves that its benefit is reasonably well sustained. If that fails, peripheral nerve section or block can be considered, bearing in mind the high recurrence rate and the subsequent incidence of distressing numbness or paresthesiae.

### Treatments No Longer Recommended

- Alcohol injection into peripheral trigeminal fibers;
- alcohol injection into, or excision of, the Gasserian ganglion;
- subtemporal extradural nerve section;
- brainstem tractotomy.

## GLOSSOPHARYNGEAL NEURALGIA (COLLET-SICARD SYNDROME)

### Clinical Aspects

Glossopharyngeal neuralgia is characterized by attacks of severe, unilateral, lancinating pain with characteristics similar to those of trigeminal neuralgia. The pain, however, is experienced in the pharynx and ear, sometimes extending to the throat or tragus. Triggering factors include swallowing (particularly cold liquids), speaking, chewing, coughing, yawning, and blowing the nose. In a variant (Vago-Collet-Sicard Syndrome), the pain originates in the anterior ear and jaw angle and then radiates to the pharynx. In 10% of cases, paroxysms of pain are accompanied by bradycardia or asystole and hypotension leading to syncope (Bruyn, 1986). Glosspharyngeal neuralgia is more often bilateral (25% of cases) than trigeminal neuralgia but does not appear to have an association with multiple sclerosis (Olds et al., 1995).

A trial of pharyngeal anesthesia with cocaine or the use of a nerve block has been advocated for confirmation of the diagnosis, necessitating significant relief of pain for a positive conclusion. The specificity of the procedure has not been properly evaluated (Mairs and Stewart, 1990). Diagnosis is complicated where typical attacks alternate with attacks of trigeminal neuralgia, as happens in 10% of cases (Rushton et al., 1981).

A symptomatic cause of glossopharyngeal neuralgia should be suspected in younger patients (<40 years), if background pain persists between attacks and if focal neurological deficit occurs. Pathologies sometimes responsible include tumors of the posterior cranial fossa and of the throat, nasopharynx, epipharynx, tongue or parotid gland (Chalmers and Olson, 1989; Papay et al., 1989).

The condition, with slightly atypical features (trigger zones in both glossopharyngeal and trigeminal territories), has been described in association with an intrinsic pontine lesion (demonstrated by MRI—nature unknown) (McCarron and Bone, 1999). Following Dandy's observation that the pain could persist after sectioning of the glossopharyngeal nerve alone, but would disappear after sectioning of the adjacent upper two or three rootlets of the vagus nerve, the term vagoglossopharyngeal neuralgia has been coined on the assumption that it is well nigh impossible, on clinical grounds, to decide whether the glossopharyngeal or vagus nerve is relevant in terms of pain genesis (Kondo, 1998).

Differential diagnosis of glossopharyngeal neuralgia (or, perhaps more correctly, vagoglossopharyngeal neuralgia) includes trigeminal neuralgia of the third division (seldom a problem) and geniculate neuralgia. The sensory root of the seventh nerve (nervus intermedius) conveys general sensation from the concha, tympanum, medial surface of the auricle, the external auditory canal, and the soft palate. In clinical practice, geniculate neuralgia and the otitic form of glossopharyngeal neuralgia are impossible to distinguish (Taha and Tew, 1995).

## Natural Course

Glossopharyngeal neuralgia is rare. In a Mayo Clinic review, the annual incidence rate was 0.7 per 100,000. Its incidence ranges from 0.2 to 1.3% of that of trigeminal neuralgia. The overall age-specific incidence rate remains stable until 60, after which the rate rises with a peak incidence between 70 and 79 years. There is a left-sided predominance, particularly in women.

## Principles of Therapy

Glossopharyngeal neuralgia is thought to be caused by vascular compression through arteriosclerotic elongation of the posterior inferior cerebellar artery, vertebral artery, or a persistent hypoglossal artery resulting in segmental demyelination and ephaptic neural transmission.

The drug management is that of trigeminal neuralgia.

## Interventional or Operative Management

### Microvascular Decompression

The procedure is similar to that used for trigeminal neuralgia. The compression, where identified, is usually due to the posterior inferior cerebellar artery. Resolution of symptoms has been reported in 60 to 100% of patients, though series are few and numbers small. In a recent study, follow-up of at least 5 years was reported in 16 patients treated by microvascular decompression (Kondo, 1998). Immediate pain relief occurred in 95%. Complications of the procedure included swallow disturbance and hoarseness. In this series there was one postoperative death. During a mean follow-up period of 11.6 years, there were no recurrences. The authors point out that in the case of cross-compression of the glossopharyngeal nerve, the offending vessels are partly hidden behind the posterolateral sulcus of the medulla oblongata, necessitating greater retraction of the cerebellar hemisphere. Furthermore, the nerves themselves, being smaller than the fifth or seventh nerves, are more easily traumatized.

### Percutaneous Thermocoagulation

This is achieved by a lateral cervical approach through the jugular foramen (Salar et al., 1983). Concomitant vagal injury is a hazard, reduced by the use of CT-guided electrode placement. Persisting dysphoma and dysphonia are recognized.

## Glossopharyngeal Nerve Section

This can be performed in the neck at the level of the jugular foramen or within the posterior cranial fossa. Nerve section produces a high cure rate (providing the upper rootlets of the vagus are included), but cough, altered taste, and dysphagia are reported complications.

## Computer-Tomography Guided Tractotomy

Primary sensory fibers from the VIIth, IXth, and Xth cranial nerves enter the descending tracts of the trigeminal nerve, and all of the nociceptive afferents of the cranial nerves form the descending cranial nociceptive tract. Computer-tomography-guided percutaneous trigeminal tractotomy–nucleotomy is used to produce a destructive lesion in the second-order neurons of the oral pole of the nucleus caudalis. In a series of nine patients with vagoglossopharyngeal or geniculate neuralgia, good or excellent results were obtained over a mean follow-up of 4 years. One patient was lost to follow-up. The procedure had to be repeated in three patients before this outcome was realized. Two patients had temporary ataxia after the procedure (Kanpolat et al., 1998).

## Practical Management

The first choice of treatment is medical, and the drug choice identical to that described for trigeminal neuralgia. If medical treatment fails, interventional or surgical treatment is undertaken. In frail subjects, the choice lies between thermocoagulation and trigeminal tractotomy–nucleotomy. The latter may be preferable. The same procedure might be considered initially for fit individuals, proceeding then to microvascular decompression if the procedure fails. The rarity of this condition makes it unlikely that comparative data on the various interventional techniques will ever be available.

# REFERENCES

Benoliel, R., and Sharav, Y. (1998). Trigeminal neuralgia with lacrimation or SUNCT syndrome? *Cephalalgia* 18, 85–90.

Bergouignan, M. (1942). Cures hereuses de neuralgies essentielles par le diphenyl-hydantoinate de soude. *Rev. Laryngol. Otol. Rhinol.* 63, 34–41.

Boecher-Schwarz, H. G., Bruehl, K., Kessel, G., Guenthner, M., Perneczky, A., and Stoeter, P. (1998). Sensitivity and specificity of MRA in the diagnosis of neurovascular compression in patients with trigeminal neuralgia: A correlation of MRA and surgical findings. *Neuroradiology* 40, 85–95.

Broggi, G., Franzini, A., Lasio, G., Giorgi, C., and Servello, D. (1990). Long-term results of percutaneous retrogasserian thermorhizotomy for "essential" trigeminal neuralgia: Considerations of 1000 consecutive patients. *Neurosurgery* 26, 783–787.

Broggi, G., Ferroli, P., Franzini, A., Servello, D., and Dones, I. (2000). Microvascular decompression for trigeminal neuralgia: Comments on a series of 250 cases, including 10 patients with multiple sclerosis. *J. Neurol. Neurosurg. Psychiatry* 68, 59–64.

Brown, J. A., and Gouda, J. J. (1997). Percutaneous balloon compression of the trigeminal nerve. *Neurosurg. Clin.* 8, 53–62.

Bruyn, G. W. (1968). Glossopharyngeal Neuralgia. *In* "Handbook of Clinical Neurology" (P. J. Vinken, G. W. Bruyn, and H. L. Klawans, Eds.), Vols. 4/48, pp. 459–473. Elsevier, Amsterdam.

Burchiel, K. J., and Slavin, K. V. (2000). On the natural history of trigeminal neuralgia. *Neurosurgery* 46, 152–155.

Canavero, S., and Bonicalzi, V. (1997). Lamotrigine control of trigeminal neuralgia: An expanded study. *J. Neurol.* 244, 527–532.

Chalmers, A. C., and Olson, J. L. (1989). Glossopharyngeal neuralgia with syncope and cervical mass. *Otolaryngol. Head Neck Surg.* 100, 252–255.

Coakham, H. B., and Moss, T. (1998). Microvascular decompression. *J. Neurosurg.* 88, 617–618.

De Benedittis, G. (1990). Auriculotemporal syndrome (Frey's syndrome) presenting as tic douloureux: Report of two cases. *J. Neurosurg.* 72, 955–958.

Frank, F., and Fabrizi, A. P. (1989). Percutaneous surgical treatment of trigeminal neuralgia. *Acta Neurochir. (Wien)* 97, 128–130.

Fromm, G. H., Terrence, C. F., Chatta, A. S., and Glass, J. D. (1984). Baclofen in the treatment of trigeminal neuralgia: Double-blind study and long-term follow-up. *Ann. Neurol.* 15, 240–244.

Gardner, W. J., and Miklos, M. V. (1959). Response of trigeminal neuralgia to decompression of sensory root: Discussion of cause of trigeminal neuralgia. *J. Am. Med. Assoc.* 170, 1773–1776.

Gass, A., Kitchen, N., MacManus, D. G., Moseley, I. F., Hennerici, M. G., and Miller, D. H. (1997). Trigeminal neuralgia in patients with multiple sclerosis: Lesion localization with magnetic resonance imaging. *Neurology* 49, 1142–1144.

Goto, F., Ishizaki, K., Yoshikawa, D., Obata, H., Arii, H., and Terada, M. (1999). The longlasting effects of peripheral nerve blocks for trigeminal neuralgia using a high concentration of tetracaine dissolved in bupivacaine. *Pain* 79, 101–103.

Hampf, G., Bowsher, D., Wells, C., and Miles, J. (1990). Sensory and autonomic measurements in idiopathic trigeminal neuralgia before and after radiofrequency thermo-coagulation: Differentiation from some other causes of facial pain. *Pain* 40, 241–248.

Hardy, D. G., and Rhoton, A. L. (1978). Microsurgical relationships of the superior cerebellar artery and the trigeminal nerve. *J. Neurosurg.* 49, 669–678.

Jannetta, P. J. (1985). Microsurgical management of trigeminal neuralgia. *Arch. Neurol.* 42, 800–801.

Jho, H.-D., and Lunsford, L. D. (1997). Percutaneous retrogasserian glycerol rhizotomy. *Curr. Tech. Results Neurosurg. Clin.* 8, 63–74.

Kanpolat, V., Savas, A., Batay, F., and Sinav, A. (1998). Computer tomography-guided trigeminal tractotomy-nucleotomy in the management of vagoglossopharyngeal and geniculate neuralgias. *Neurosurgery* 43, 484–490.

Katusic, S., Beard, C. M., Bergstrahl, E., and Kurland, L. T. (1990). Incidence and clinical features of trigeminal neuralgia, Rochester, Minnesota, 1945–1984. *Ann. Neurol.* 27, 89–95.

Khan, O. A. (1995). Gabapentin relieves trigeminal neuralgia in multiple sclerosis patients. *Neurology* 51, 611–614.

Klun, B., and Prestor, B. (1986). Microvascular relations of the trigeminal nerve: An anatomical study. *Neurosurgery* 19, 535–539.

Kobata, H., Kondo, A., Iwasaki, K., and Nishioka, T. (1998). Combined hyperactive dysfunction syndrome of the cranial nerves: Trigeminal neuralgia, hemifacial spasm and glosso-pharyngeal neuralgia: 11-Year experience and review. *Neurosurgery* 43, 1351–1362.

Kondo, A. (1998). Follow-up results of using microvascular decompression for treatment of glossopharyngeal neuralgia. *J. Neurosurg.* 88, 221–225.

Kondziolka, D., Lunsford, L. D., Flickinger, J. C., *et al.* (1996). Stereotactic radiosurgery for trigeminal neuralgia: A multi-institutional study using the gamma unit. *J. Neurosurg.* 84, 940–945.

Kondziolka, D., Perez, B., Flickinger, J. C., Habeck, M., and Lunsford, L. D. (1998). Gamma knife radiosurgery for trigeminal neuralgia. Results and expectations. *Arch. Neurol.* 55, 1524–1529.

Leksell, L. (1983). Stereotactic radiosurgery. *J. Neurol. Neurosurg. Pyschiatry* 46, 797–803.

Lichtor, T., and Mullan, J. F. (1990). A 10-year follow-up review of percutaneous microcompression of the trigeminal ganglion. *J. Neurosurg.* 72, 49–54.

Lovely, T. J., and Jannetta, P. J. (1997). Microvascular decompression for trigeminal neuralgia. Surgical technique and long-term results. *Neurosurg. Clin.* 8, 11–29.

Mairs, A. P., and Stewart, J. J. (1990). Surgical treatment of glossopharyngeal neuralgia via the pharyngeal approach. *J. Otolaryngol.* 104, 12–16.

Majoie, C. B. L. M., Hulsmans, F. J. H., Castelijns, J. A., Verbeeten, J. R. B., Tiren, D., Van Beek, E. J. R., Valk, J., and Bosch, D. A. (1998). Symptoms and signs related to the trigeminal nerve: Diagnostic yield of MR imaging. *Radiology* 209, 557–562.

McCarron, M. O., and Bone, I. (1999). Glossopharyngeal neuralgia referred from a pontine lesion. *Cephalalgia* 19, 115–117.

Meaney, J. F. M., Eldridge, P. R., Dunn, L. T., Nixon, T. E., Whitehouse, G. H., and Miles, J. B. (1995). Demonstration of neurovascular compression in trigeminal neuralgia with magnetic resonance imaging. *J. Neurosurg.* 83, 799–805.

Meglio, M., and Cioni, B. (1989). Percutaneous procedures for trigeminal neuralgia: Microcompression versus radiofrequency thermocoagulation: Personal Experience. *Pain* 38, 9–16.

Olds, M. J., Woods, C. I., and Winfield, J. A. (1995). Microvascular decompression in glossopharyngeal neuralgia. *Am. J. Otol.* 16, 326–330.

Papay, F. A., Roberts, J. K., Wegryn, T. L., Gordon, T., and Levine, H. L. (1989). Evaluation of syncope from head and neck cancer. *Laryngoscope* 99, 382–388.

Parmar, D. N., and Sharr, M. M. (1999). Cotton gauze foreign body granuloma following microvascular decompression. *Br. J. Neurosurg.* 13, 87–89.

Regis, J., Bartolomei, F., Metellus, P., Rey, M., Genton, P., Dravet, C., Bureau, M., Semah, F., Gastaut, J. L., Peragut, J. C., and Chauvel, P. (1999). Radiosurgery for trigeminal neuralgia and epilepsy. *Neurosurg. Clin. North Am.* 10, 359–377.

Rushton, J. G., Stevens, J. C., and Miller R. H. (1981). Glossopharyngeal (vagoglossopharyngeal) neuralgia. *Arch. Neurol.* 38, 201–205.

Salar, G., Ori, C., Baratto, V., Iob, I., and Mingrino, S. (1983). Selective percutaneous thermo lesions of the ninth cranial nerve by lateral cervical approach: Report of eight cases. *Surg. Neurol.* 20, 276–279.

Scrivani, S. J., Keith, D. A., Mathews, E. S., and Kaban, L. B. (1999). Percutaneous stereotactic differential radiofrequency thermal rhizotomy for the treatment of trigeminal neuralgia. *J. Oral Maxillofac. Surg.* 57, 104–111.

Siegfried, J. (1981). Percutaneous controlled thermocoagulation of the Gasserian ganglion in trigeminal neuralgia: Experiences with 1000 cases. *In* "The Cranial Nerves" (M. Samii and P. J. Jannetta, Eds.), pp. 322–330. Springer. Berlin/Heidelberg/New York.

Sist, T., Filadora, V., Miner, M., and Lema, M. (1997). Gabapentin for idiopathic trigeminal neuralgia: Report of two cases. *Neurology* 48, 1467.

Stechison, M. T., Molier, A., and Lovely, T. J. (1996). Intraoperative mapping of the trigeminal nerve root: Technique and application in the surgical management of facial pain. *Neurosurgery* 38, 76–82.

Taha, J. M., and Tew, J. M., Jr. (1995). Long-term results of treatment of idiopathic neuralgias of the glossopharyngeal and vagal nerves. *Neurosurgery* 36, 926–931.

Taha, J. M., and Tew, J. M. Jr. (1997). Treatment of trigeminal neuralgia by percutaneous radiofrequency rhizotomy. *Neurosurg. Clin.* 8, 31–52.

Tan, L. K. S., Robinson, S. N., and Chatterjee, S. (1995). Glycerol versus radiofrequency rhizotomy—A comparison of their efficacy in the treatment of trigeminal neuralgia. *Br. J. Neurosurg.* 9, 165–169.

Taylor, J. C., Braver, S., and Esir, M. L. E. (1981). Long-term treatment of trigeminal neuralgia with carbamazepine. *Postgrad. Med. J.* 57, 16–18.

Tew, J. M., and Keller, J. P. (1997). The treatment of trigeminal neuralgia by percutaneous radiofrequency technique. *Clin. Neurosurg.* 24, 557–558.

Vassilouthis, J. (1994). Relief of trigeminal neuralgia by proparacaine. *J. Neurol. Neurosurg. Psychiatry* 57, 121.

Wilkinson, H. A. (1999). Trigeminal nerve peripheral branch phenol/glycerol injections for tic douloureux. *J. Neurosurg.* 90, 828–832.

Yoon, K. B., Wiles, J. R., Miles, J. B., and Nurhikko, T. J. (1999). Long-term outcome of percutaneous thermo-coagulation for trigeminal neuralgia. *Anaesthesia* 54, 798–808.

Young, J. N., and Wilkins, R. H. (1993). Partial sensory trigeminal rhizotomy at the pons for trigeminal neuralgia. *J. Neurosurg.* 79, 680–687.

Zakrzewska, J. M., and Patsalos, P. N. (1989). Oxcarbazepine: A new drug in the management of intractable trigeminal neuralgia. *J. Neurol. Neurosurg. Psychiatry* 52, 472–476.

Zakrzewska, J. M., and Patsalos, P. N. (1992). Drugs used in the management of trigeminal neuralgia. *Oral. Surg. Oral. Med. Oral Pathol.* 74, 439–450.

Zakrzewska, J. M., Jassim, S., and Bulman, J. S. (1999). A prospective, longitudinal study on patients with trigeminal neuralgia who underwent radiofrequency thermocoagulation of the Gasserian ganglion. *Pain* 79, 51–58.

CHAPTER 9

# Postlumbar Puncture Headache Syndrome

Marianne Dieterich and G. D. Perkin

## CLINICAL ASPECTS (DIFFERENTIAL DIAGNOSIS)

The symptoms of the postlumbar puncture headache (PLPH) syndrome are positional, beginning in the upright posture and subsiding or improving when the patient reclines. The headache, which is predominantly occipital but may be frontal, parietal or diffuse, occurs in about 20–40% of subjects after diagnostic lumbar puncture (LP) with a 20- to 22-G needle. Nausea (73%), numbness (60%), neck stiffness (11%), and lower back pain are common accompaniments; blurred vision, photophobia, diminished hearing, tinnitus, and ear pressure occur occasionally (Abouleish *et al.*, 1975; Vilmig and Kloster, 1998). Typically, the symptoms begin within a few hours (mostly 24–48 h) following LP and persist for an average of 4 days. Rarely, transient cranial nerve lesions (most often of the VIth nerve, and nerves I, IX, or X) are induced by LP alone, but more frequently are a toxic effect of the contrast medium used in myelography. Subdural hematoma and defective hearing in the low-frequency range (2%) occur rarely. The latter could be explained by a relative endolymphatic hydrops secondary to a low-pressure state in the perilymphatic space, in turn the consequences of the low CSF pressure.

Sometimes symptoms of PLPH occur spontaneously without a detectable dural defect (spontaneous low-CSF-pressure headache). In these instances, suggested causes include a hyperactive resorption of CSF (Marcelis and Silberstein, 1990) or a spontaneous mechanical dural hole. MRI scans in patients with intracranial hypotension—either after LP or spontaneously—show meningeal enhancement, subdural effusions, and downward brain displacement that resolve spontaneously parallel to the clinical syndrome (Pannullo *et al.*, 1993). Mild or moderate increased enhancement of the meninges is seen in about 80% of the patients with PLPH and is related to the dilatation of meningeal veins (Hannerz *et al.*, 1999; Mittl and Yousem, 1994). In a few patients MRI also shows abnormal dural venous sinus enhancement suggesting compensatory venous expansion to maintain adequate intracranial volume (Bakshi *et al.*, 1999; Fishman and Dillon, 1993).

Differential diagnosis has to exclude other types of headache, for example, cervicogenic headache, migraine without aura, meningitis, subarachnoidal hemorrhage, or complications due to subdural hematoma or intracerebral hemorrhage, especially when the symptoms show an unusual course or persist for more than 14 days.

The most important diagnostic sign is the positional dependence of the symptoms.

## NATURAL COURSE

The natural course of PLPH shows remission in 53% of the patients within 4 days and in 72% within 7 days (Vandam and Dripps, 1956). Rarely symptoms last for several weeks to months. PLPH occurs much less frequently among children under the age of 13 years (Bolder, 1986) and adults above the age of 60 years. Women are affected twice as frequently as men. PLPH syndrome is dependent on the diameter of the needle (Table I), the shape of the needle's tip (e.g., Whitacre or Sprotte's "atraumatic" needle with a closed circular cone and rounded profile is of benefit), and the orientation of the tip during the LP (spreading out, not injuring the longitudinal dural fibers). Within a range of 10–25 mL, the volume of the withdrawn CSF does not influence the incidence of PLPH (Alpers, 1925). Another factor alleged to decrease the incidence is maintaining

TABLE I  Relationship of Incidence to Needle Size of Postlumbar Puncture Headache Syndrome

| Needle size (G) | Incidence of PLPH (%) | |
| --- | --- | --- |
| | Diagnostic LP | Spinal anesthesia |
| 16–19 | About 70 | 10–30 |
| 20–22 | 16–37 | 1–30 |
| 24–27 | 1–25 | 1–25 |
| 29 | 0–3.3 | 0–2 |
| 32 | — | <1 |
| 21, Atraumatic cone | 16 | 3.5 |
| 22, Atraumatic cone | 2.5–6.5 | 0–3.5 |
| 22–24, Atraumatic cone | About 2 | 0–11 |
| 25, Atraumatic cone | — | 0–8.5 |

TABLE II  Level of Evidence in Different Options of Prevention and Treatment

| Prevention | | Treatment | |
| --- | --- | --- | --- |
| Small needle size | *** | Caffeine | ** |
| Atraumatic needle tip | *** | Theophylline | ** |
| Bevel direction parallel (to dural fibers) | *** | Epidural blood patch | ** |
| Replacement of stylet (before needle is withdrawn) | ** | Triptans | — |
| Duration of recumbency | — | | |
| Increased fluids | — | | |

***Treatment recommendations (TR) based on more than one well-designed randomized, placebo-controlled trial or a meta-analysis.
**Treatment is based on at least one randomized prospective study of reasonable size.
*Treatment based on personal experience, expert opinion, retrospective, or nonrandomized studies.

the patient in a head-down position for 30 min after the procedure (Easton, 1979; Smith *et al.*, 1980). Maintaining the patient in this position (prone with the foot of the bed raised 25 cm) for 60 min has been argued to be a means for allowing the dural hole to seal (Raskin, 1990).

## PATHOPHYSIOLOGY

The continued leakage of CSF (5–30 mL/h initially) through the dural (and arachnoidal) defect made by the needle plays a crucial pathophysiological role in triggering the headache. It exceeds the production rate of CSF, thus leading to a state of low spinal fluid volume and pressure and to downward descent of the brain with subsequent stretching of pain-sensitive structures such as dura, nerves, and blood vessels (Kunkle *et al.*, 1943; Wolff, 1972). The IXth and Xth cranial nerves and the upper three cervical nerves below the tentorium cerebelli transmit pain from the suboccipital region and the neck; the second and third divisions of the Vth cranial nerve above the tentorium cerebelli transmit frontal pain (Pickering, 1948; Wolff, 1972). The significant reduction of CSF volume (normally 100–160 mL) induces a decreased resorption, while the production rate remains constant (Pappenheimer *et al.*, 1962; Cutler *et al.*, 1968). In addition, a disturbance in the equilibrium between intravascular and extravascular pressures occurs. Because veins are thin-walled and adjust passively to the pressure in and around them, dilatation follows (Kunkle *et al.*, 1943; Forbes and Nason, 1935), thereby inducing an increase of the brain volume. Such vasodilatation might cause diapedesis of cells and protein into the subarachnoid space and the CSF, explaining the CSF pleocytosis (6–50 cells/μL) and elevated protein (580–1800 mg/L) found in patients with intracranial hypotension (Lipman, 1977). Thus, PLPH syndrome is due to a combination of traction of pain-sensitive structures augmented by tension produced by venous dilatation and increase in brain volume. Confirmation of this hypothesis comes from the fact that the pain can be increased by jugular compression.

Psychogenic factors frequently considered to be of major importance are of subordinate significance (Diener *et al.*, 1985; Dripps and Vandam, 1954). Personality tests disclosed no statistically significant differences between patients with PLPH and controls regarding personality traits (Vilming *et al.*, 1997).

## PRINCIPLES OF THERAPY AND MANAGEMENT

### Prevention

#### Needle

**Needle Size.** There is convincing Class I evidence in anesthesiology series and Class I or II evidence in neurology series for a correlation of needle size and frequency of PLPH (Evans *et al.*, 2000). The correlation between the incidence of PLPH syndrome and the needle size used in LP (16–18 G, 70%; 20–22 G, 35–40%; 24–27 G, 5–12%; Table I) demonstrates that the incidence of the syndrome can be reduced by using small needles (Hilton-Jones *et al.*, 1982; Tourtellotte *et al.*, 1972). In spinal anesthesia, the incidence is lower, for example, for the 20- to 22-G needle 16–18% (Dieterich and Brandt, 1985; Hilton-Jones *et al.*, 1982), which is the result of other factors not applicable to diagnostic LP—the fluid introduced is an anesthetic, more fluid is introduced than removed, postoperative patients usually stay in bed longer and are given analgesics so that headache is not noticed during this period, and the majority of the patients are older with a lower susceptibility. In diagnostic LP, the choice of the needle size is limited by the frequent necessity to remove more than 10 mL CSF and by the desire to measure CSF pressure.

**Needle Design.** Not only is the size of the needle important, but the tip is as well, which should have a lateral opening and a cone-nose shape such as the atraumatic needles of Whitacre or Sprotte (Class I evidence in anesthesiology series). Most of the studies show that this can reduce the incidence of PLPH to 2.5–6.5% in LP with a 22-G needle (Braune and Huffmann, 1992; Engelhardt *et al.*, 1992) and to 16% with a 21-G needle (Strupp *et al.*, 1998), to 6% in myelography (22 G; Kleyweg *et al.*, 1998; Müller *et al.*, 1994), and to 0.02–3.5% in spinal anesthesia (Engelhardt *et al.*, 1992; Halpern and Preston, 1994; Harrison and Langham, 1994). Owing to the higher flexibility of this needle and the blunt top, it is inserted with a sharp short introducer.

**Direction of Bevel.** The direction of insertion should be such that the longitudinally running dural fibers are not cut but separated by introducing the needle bevel at right angles (Class I evidence in anesthesiology). Five studies of patients receiving spinal anesthesia have demonstrated a reduction in the incidence of PLPH by 50% or more when the bevel is parallel rather than perpendicular (Lybecker *et al.*, 1990). A further beneficial effect was observed when the mandrin was first reinserted in the needle and then both were withdrawn together (5% vs 16%; Strupp *et al.*, 1998).

### Postural Maneuvers

Prospective single or double-blind clinical trials have shown that different body positions adopted by the patient after LP (2–24 h bed rest, head horizontal position, 30 min to 4 h prone or supine position, head tilted down at 30°; for 30 min) either only delay the onset of PLPH or fail to influence it at all (Carbaat and van Crevel, 1981; Dieterich and Brandt, 1985; Handler *et al.*, 1982; Hilton-Jones *et al.*, 1982, 1984). Some authors have suggested that the upright position may actually improve or reverse the underlying pathophysiologic derangements, because, in two prospective, randomized clinical studies, the incidence of PLPH was even lower in the subgroup of patients mobilized immediately (Vilming *et al.*, 1988) or at 6 h (Thornberry and Thomas, 1988). Therefore, early mobilization after LP can be justified.

The preventative effects of an increased oral fluid intake after LP or the use of the calcium entry blocker

flunarizine have been disproven in prospective studies (Dieterich and Brandt, 1988; Heide and Diener, 1987).

**Symptomatic Treatment**

Once established, the PLPH syndrome should be treated according to its severity. Current practice and review of the literature suggest that treatment should include postural maneuvers, the administration of caffeine (or theophylline), and an epidural blood patch.

### Posture

For symptomatic treatment of a PLPH syndrome, bed rest for a few days with the head horizontal or even slightly down will provide relief. As a result, the use of analgesics, antiemetics, and sedatives can be reduced or rendered unnecessary. Patients with mild symptoms occurring within about 30 min of mobilization should be able to perform most normal daily activities. When symptoms occur within the first 30 min, patients should be mobilized several times per day (e.g., for meals) to prevent deep vein thrombosis. When postural headache occurs within a few seconds to minutes after mobilization and is severe enough to necessitate lying down in bed in a horizontal position for the rest of the day, patients should sit up several times per day for a short time and be given prophylactic low-dose heparin.

### Medication

In several uncontrolled (Ford *et al.*, 1989; Friedman and Merritt, 1957; Jarvis *et al.*, 1986) and controlled (Sechzer, 1979; Sechzer and Abel, 1978) studies, the efficacy of intravenous caffeine has been established, for example, 500 mg caffeine sodium benzoate given as an intravenous bolus over 2–3 min was beneficial in 75%, a second injection given because of the persistence of symptoms within the next 2 h improved the rate to 85% (Sechzer and Abel, 1978). In 70%, the headache failed to return. Orally administered theophylline, which is pharmacologically similar to caffeine and may increase CSF production, can also improve the patient's discomfort (Feuerstein and Zeides, 1986). Both substances induce intracerebral arterial constriction by blocking adenosine receptors, resulting in a decrease of cerebral blood flow and intracranial pressure (Phillis and DeJong, 1987; Raskin, 1990). If the symptoms persist after 1 g of caffeine, an epidural blood patch is recommended (Raskin, 1988). In patients with severe symptoms or inability to cope with a high oral fluid intake, it is traditional to give isotonic or hypotonic intravenous fluids to increase CSF production. However, their beneficial effect has not yet been adequately documented.

## Epidural Blood Patch

In the case of severe, longer lasting PLPH syndromes, an epidural blood patch has been recommended (Gormley, 1960), which provides prompt and permanent relief of the symptoms in 80–96% of cases without producing major side effects (Crawford, 1980; Olsen, 1987). The epidural injection of 20–30 mL autologous blood can be, although not necessarily, performed in the same interspace as for the LP, but no earlier than 24 h after LP. Afterward, the patient should lie prone for 30–60 min, if possible for 10 min with the head 30° down (Fishman and Dillon, 1997). The epidural blood spreads in all directions and covers an average of four to nine spinal segments, three to six segments above and one to three below the original site of the injection (Griffiths *et al.*, 1993; Szeinfeld *et al.*, 1986), so that it might be advisable to choose a segment below the LP site. Therefore, the injected volume of autologous blood must not be smaller than 15 mL. The occlusion of the dural defect by a gelatinous tamponade has been assumed to be the effect of the blood patch (Abouleish *et al.*, 1975; Ostheimer *et al.*, 1974). Based on a CSF production rate of about 0.35 mL/min and a diminished CSF resorption rate, the diminished CSF volume can be replaced within a few hours (0.5–6 h). For reasons not yet explained, a prophylactic blood patch administered immediately after the LP either fails to prevent the syndrome (Berrettini *et al.*, 1987; Loeser *et al.*, 1978) or reduces it by only about 50% (4–8 mL blood; Heide and Diener, 1990). Speculatively, a partial effect of the epidural blood patch (and epidural saline injections) can be explained by a deactivation of adenosine receptors due to a sudden increase in CSF pressure (Raskin, 1990). Because there is a high failure rate (71% within 24 h versus 4% after 24 h) and the incidence of PLPH is low when small and atraumatic needles are used, the use of a prophylactic or early blood patch cannot be recommended (Baysinger *et al.*, 1986; Heide and Diener, 1990; Olsen, 1987). Complications after epidural blood patching are few and moderate. The most common are transient lower back pain and stiff neck (35%) and tenderness over the LP site. Transient high temperature (5%) and paresthesia of the legs are less frequent, while radicular pain in the legs and acute aseptic meningitis are rare. Severe chronic complications, such as epidural abscess or adhesive arachnoiditis, have not been reported in the literature. Patients with septicemia, lower back infection, disorders of the coagulation system, or those taking anticoagulant medication should be excluded from this procedure.

## Conclusion

In conclusion, therefore, the following therapeutic principles are relevant:

1. Prophylactic aspects in decreasing the incidence of the syndrome by the use of small diameter needles with atraumatic cones.
2. Once PLPH has been established, it can be relieved by lying down, so that bed rest is a sensible therapeutic measure.
3. The use of caffeine or theophylline both of which lead to intracerebral vasoconstriction by blocking brain adenosine receptors.
4. The use of an epidural blood patch forming a gelatinous tamponade over the dural hole and compressing the dural sac.

## TREATMENTS NO LONGER RECOMMENDED

The following treatments are no longer recommended:

- antidiuresis with desmopressine;
- abdominal bandage (to increase the intraabdominal pressure);
- increased oral fluid intake;
- prophylactic flunarizine;
- epidural saline infusions (in 50% effect only temporary).

## REFERENCES

Abouleish, E., dela Vega, S., Blendinger, I., and Tio, T. O. (1975). Long-term follow-up of epidural blood patch. *Anesth. Analg.* (NY) **54**, 459–463.

Alpers, B. J. (1925). Lumbar puncture headache. *Arch. Neurol. Psychiatry* **14**, 806–812.

Bakshi, R., Mechtler, L. L., Kamran, S., Gosy, E., Bates, V. E., Kinkel, P. R., and Kinkel, W. R. (1999). MRI findings in lumbar puncture headache syndrome: Abnormal dural-meningeal and dural venous sinus enhancement. *Clin. Imaging* **23**, 73–76.

Baysinger, C. L., Menk, E. J., Harte, E., and Middaugh, R. (1986). The successful treatment of dural puncture headache after failed epidural blood patch. *Anesth. Analg.* (NY) **65**, 1242–1244.

Berrettini, W. H., Simmons-Alling, S., and Nurnberger, J. I., Jr. (1987). Epidural blood patch does not prevent headache after lumbar puncture. *Lancet* **1**, 856–857.

Bolder, P. M. (1986). Postlumbar puncture headache in pediatric oncology patients. *Anesthesiology* **65**, 696–698.

Braune, H.-J., and Huffmann, G. (1992). A prospective double-blind clinical trial, comparing the sharp Quincke needle (22 G) with an "atraumatic" needle (22 G) in the induction of post-lumbar puncture headache. *Acta Neurol. Scand.* **86**, 50–54.

Carbaat, P. A. T., and van Crevel, H. (1981). Lumbar puncture headache: Controlled study on the preventive effect of 24 hours bed rest. *Lancet* **2**, 1133–1135.

Crawford, J. S. (1980). Experiences with epidural blood patch. *Anaesthesia* **35**, 513–515.

Cutler, R. W. P., Page, L., Galicich, J., and Watters, G. V. (1968). Formation and absorption of cerebrospinal fluid in man. *Brain* **91**, 707–720.

Diener, H. C., Bendig, M., and Hempel, V. (1985). Der postpunktionelle Kopfschmerz. *Fortschritte der Neurologie und Psychiatrie* **53**, 344–349.

Dieterich, M., and Brandt, T. (1985). Is obligatory bed rest after lumbar puncture obsolete? *Eur. Arch. Psychiatry Neurol. Sci.* **235**, 71–75.

Dieterich, M., and Brandt, T. (1988). Incidence of post-lumbar puncture headache is independent of daily fluid intake. *Eur. Arch. Psychiatry Neurol. Sci.* **237**, 194–196.

Dripps, R. D., and Vandam, L. D. (1954). Long-term follow-up of patients who received 10,098 spinal anesthetics. *J. Am. Med. Assoc.* **156**, 1486–1491.

Easton, J. D. (1979). Headache after lumbar puncture. *Lancet* **1**, 974–975.

Engelhardt, A., Oheim, S., and Neundörfer, B. (1992). Post-lumbar puncture headache: Experiences with Sprotte's atraumatic needle. *Cephalalgia* **12**, 259.

Evans, R. W., Armon, C., Frohman, E. M., and Goodin, D. S. (2000). Assessment: Prevention of post-lumbar puncture headaches. Report of the therapeutics and technology assessment subcommittee of the american academy of neurology. *Neurology* **55**, 909–914.

Feuerstein, T. J., and Zeides, A. (1986). Theophylline relieves headache following lumbar puncture. *Klin. Wochenschr.* **64**, 216–218.

Fishman, R. A., and Dillon, W. P. (1993). Dural enhancement and cerebral displacement secondary to intracranial hypotension. *Neurology* **43**, 609–611.

Fishman, R. A., and Dillon, W. P. (1997). Intracranial hypotension. *J. Neurosurg.* **86**, 165.

Forbes, H. S., and Nason, G. I. (1935). The cerebral circulation. *Arch. Neurol. Psychiatry* **34**, 533–547.

Ford, C. C., Ford, D. C., and Koenigsberg, M. D. (1989). A simple treatment of post-lumbar puncture headache. *J. Emerg. Med.* **7**, 29–31.

Friedman, A. P., and Merritt, H. H. (1957). Treatment of headache. *J. Am. Med. Assoc.* **163**, 1111–1117.

Gormley, J. B. (1960). Treatment of postspinal headache. *Anesthesiology* **21**, 565–566.

Griffiths, A. G., Beards, S. C., Jackson, A., and Horseman, E. L. (1993). Visualization of extradural blood patch for post lumbar puncture headache by magnetic resonance imaging. *Br. J. Anaesth.* **70**, 223–225.

Halpern, S., and Preston, N. R. (1994). Postdural puncture headache and spinal needle design. *Metanal. Anesthesiol.* **81**, 1376–1383.

Handler, C. E., Smith, F. R., Perkin, G. D., and Rose, F. C. (1982). Posture and lumbar puncture headache: A controlled trial in 50 patients. *J. R. Soc. Med.* **75**, 404–407.

Hannerz, J., Ericson, K., and Bro Skejo, H. P. (1999). MR imaging with gadolinium in patients with and without post-lumbar puncture headache. *Acta Radiol.* **40**, 135–141.

Harrison, D. A., and Langham, B. T. (1994). Post-dural puncture headache: A comparison of the Sprotte and Yale needles in urological surgery. *Eur. J. Anaesthesiol.* **11**, 325–327.

Heide, W., and Diener, H. C. (1987). The calcium entry blocker flunarizine does not prevent postpuncture headache. *Headache* **27**, 168–169.

Heide, W., and Diener, H. C. (1990). Epidural blood patch reduces the incidence of post lumbar puncture headache. *Headache* **30**, 280–281.

Hilton-Jones, D. (1984). What is postlumbar puncture headache and is it avoidable? *In* "Dilemmas in the Management of the Neuro-

logical Patient," (C. Garfield, Ed.), pp. 144–157. Churchill Livingstone, Edinburgh, London, Melbourne, and New York.

Hilton-Jones, D., Harrad, R. A., Gill, M. W., and Warlow, C. P. (1982). Failure of postural manoeuvres to prevent lumbar puncture headache. *J. Neurol. Neurosurg. Psychiatry* **45**, 743–746.

Jarvis, A. P., Greenawalt, J. W., and Fagraeus, L. (1986). Intravenous caffeine for post-dural puncture headache. *Anesth. Analg. (NY)* **65**, 316–317.

Kleyweg, R. P., Hertzberger, L. I., and Carbaat, P. A. T. (1998). Significant reduction in post-lumbar puncture headache using an atraumatic needle. A double-blind, controlled clinical trial. *Cephalalgia* **18**, 635–637.

Kunkle, E. C., Ray, B. S., and Wolff, H. G. (1943). Experimental studies on headache: Analysis of the headache associated with changes in intracranial pressure. *Arch. Neurol. Psychiatry* **49**, 323–358.

Lipman, I. J. (1977). Primary intracranial hypotension: The syndrome of spontaneous low cerebrospinal fluid pressure with traction headache. *Dis. Nerv. Syst.* **38**, 212–213.

Loeser, E. A., Hill, G. E., Bennett, G. M., and Sederberg, J. H. (1978). Time vs. success rate for epidural blood patch. *Anesthesiology* **49**, 147–148.

Lybecker, H., Moller, J. T., May, O., and Nielsen, H. K. (1990). Incidence and prediction of postdural puncture headache. A prospective study of 2021 spinal anesthesias. *Anesth. Analg.* **70**, 389–394.

Marcelis, J., and Silberstein, S. D. (1990). Spontaneous low cerebrospinal fluid pressure headache. *Headache* **30**, 192–196.

Mittl, R. L., and Yousem, D. M. (1994). Frequency of unexplained meningeal enhancement in the brain after lumbar puncture. *AJNR* **15**, 633–638.

Müller, B., Adelt, K., Reichmann, H., and Toyka, K. (1994). Atraumatic needle reduces the incidence of post-lumbar puncture syndrome. *J. Neurol.* **241**, 376–380.

Olsen, K. S. (1987). Epidural blood patch in the treatment of post-lumbar puncture headache. *Pain* **30**, 293–301.

Ostheimer, G. W., Palahniuk, R. J., and Shnider, S. M. (1974). Epidural blood patch for post lumbar-puncture headache. *Anesthesiology* **41**, 307–308.

Pannullo, S. C., Reich, J. B., Krol, G., Deck, M. D. F., and Posner, J. B. (1993). MRI changes in intracranial hypotension. *Neurology* **43**, 916–926.

Pappenheimer, J. R., Heisey, S. R., Jordan, E. F., and Downer, J. C. (1962). Perfusion of the cerebral ventricular system in unanesthetized goats. *Am. J. Physiol.* **203**, 763–774.

Phillis, J. W., and DeLong, R. E. (1987). An involvement of adenosine in cerebral blood flow regulation during hypercapnia. *Gen. Pharmacol.* **18**, 133–139.

Pickering, G. W. (1948). Lumbar-puncture headache. *Brain* **71**, 271–280.

Raskin, N. H. (1988). "Headache," pp. 290–295. Churchill-Livingstone, New York.

Raskin, N. H. (1990). Lumbar puncture headache: A review. *Headache* **30**, 197–200.

Sechzer, P. H. (1979). Post-spinal anesthesia headache treated with caffeine. Part II: Intracranial vascular distention, a key factor. *Curr. Ther. Res.* **26**, 440–448.

Sechzer, P. H., and Abel, L. (1978). Post-spinal anesthesia headache treated with caffeine. Evaluation with demand method. Part 1. *Curr. Ther. Res.* **24**, 307–312.

Serafini, A. N. (1986). Epidural blood patch: Evaluation of the volume and spread of blood injected into the epidural space. *Anesthesiology* **64**, 820–822.

Smith, F. R., Perkin, G. D., and Rose, F. C. (1980). Posture and headache after lumbar puncture. *Lancet* **1**, 1245.

Strupp, M., Brandt, T., and Müller, A. (1998). Incidence of postlumbar puncture syndrome reduced by reincerting the stylet: A randomized prospective study of 600 patients. *J. Neurol.* **245,** 589–592.

Szeinfeld, M., Ihmeidan, I. H., Moser, M. M., Machado, R., Klose, K. J., Thornberry, E. A., and Thomas, T. A. (1988). Posture and post-spinal headache. A controlled trial in 80 obstetric patients. *Br. J. Anaesth.* **60,** 195–197.

Tourtellotte, W. W., Henderson, W. G., Tucker, R. P., Gilland, O., Walker, J. E., and Kokman, E. (1972). A randomized double-blind clinical trial comparing the 22 versus 26 gauge needle in the production of the post-lumbar puncture syndrome in normal individuals. *Headache* **12,** 73–78.

Vandam, L. D., and Dripps, R. D. (1956). Long-term follow-up of patients who received 10,098 spinal anesthetics. *J. Am. Med. Assoc.* **161,** 586–591.

Vilming, S. T., and Kloster, R. (1998). Pain location and associated symptoms in post-lumbar puncture headache. *Cephalalgia* **18,** 697–703.

Vilming, S. T., Schrader, H., and Monstad, I. (1988). Post-lumbar-puncture headache: The significance of body posture. *Cephalalgia* **8,** 75–78.

Vilming, S. T., Ellertsen, B., Troland, K., Schrader, H., and Monstad, I. (1997). MMPI profiles in post-lumbar puncture headache. *Acta Neurol. Scand.* **95,** 184–188.

Wolff, H. G. (1972). Headache and Other Head Pain, 3rd ed. Oxford Univ. Press, New York.

*Acute and Chronic Pain*

CHAPTER 10
# Whiplash Injury

Matthias Keidel

## CLINICAL ASPECTS

### Definition

Whiplash injury of the cervical spine and surrounding ligaments is typically provoked by a sudden combination of retro- and anteflexion movements of the cervical spine, brought about by an indirect force on, e.g., the patient's trunk (Gay and Abbott, 1953). Thus whiplash injury is characterized by a 'whiplash mechanism' of strain, twisting or flexion–extension force and results in a variety of whiplash-associated disorders (WADs; Spitzer *et al.*, 1995).

The Quebec Task Force (Spitzer *et al.*, 1995) adopted the following definition of whiplash: "Whiplash is an acceleration–deceleration mechanism of engery transfer to the neck. It may result from rearend or side-impact motor vehicle collisions, but can also occur during diving or other mishaps. The impact may result in bony or soft-tissue injuries (whiplash injury), which in turn may lead to a variety of clinical manifestations (whiplash associated disorders)".

Frequent causes are traffic accidents (rear-end and frontal collisions) and, in the case of the newborn, abrupt throwing into and catching from the air or shaking (Hadley *et al.*, 1989). The cervicocranial region and the caudal part of the cervical spine are thereby particularly strained. Accordingly, upper and lower whiplash injuries can be differentiated clinically (Wiesner and Mummenthaler, 1975). The whiplash acceleration can happen in any direction in respect to the body axis. Thus lateral and torsion whiplash injury are possible. Occasionally the whiplash mechanism may be complicated by damage to bony structures, discs, nerve roots, or spinal cord. A classification of whiplash injury including the whiplash-associated disorders proposed by the Quebec Task Force is based on five grades, which correspond to the severity of the injury described

on a clinical–anatomic axis. The classification scheme is given in Table I (Spitzer *et al.*, 1995).

### Pathogenesis

Dramatic accelerations, decelerations, or rotational movements of the body and head with shearing and traction are pathogenic (Keidel and Stude, 2002; Martinez and Garcia, 1968). Mechanical effects upon the neck range from alterations in the curvature and loosening of joints with subluxation and vertebral fracture with dislocation to the rare fatal segment severance, odontoid or occipital condylar avulsion, or base of skull ring fracture (Delank, 1988; Schmidt, 1989). Lesions of the perispinal soft tissue find their cause in straining, extension, or tearing of the connective tissue, vessels, muscle fibers, and anterior cervical soft tissues (oesophagus, larynx, thyroid), sometimes with secondary hemorrhage and tissue swelling. Coincident vascular injuries of the head can give rise to intracranial (epi-/subdural or intracerebral), intramedullary, or vitreous or retinal hemorrhages (*cave*: patients treated with dicumarol; Bauer and Pils, 1984; Birsner and Leask, 1954; Carter and McCormick, 1983; Hadley *et al.*, 1989; Ommaya and Yarnall, 1969). In individual cases thrombotic complications can develop at the site of traumatic intimal dissections (Kessler *et al.*, 1987) of the internal carotid or vertebral arteries (Simeone and Goldberg, 1968). The formation of an aneurysm following intimal dissection is likewise possible (Mokri, 1990; Zagorski and Berlit, 1990). Nervous structures can be directly or indirectly damaged. Functional losses can result, for instance, (otolithic) dizziness following traumatic otoconia divulsion (Brandt, 1991); plexus strain after lateral whiplash injury; radiculopathy after dislocated fracture, traumatic disc prolapse, or foraminal hemorrhage; paraplegia after epidural or intramedullary hematoma,

TABLE I   Clinical Classification of Whiplash-Associated Disorders (Quebec Task Force; Modified from Spitzer *et al.*, 1995)

| Grade | Clinical presentation | Symptoms and signs | Pathology | Time axis |
|-------|----------------------|--------------------|-----------|-----------|
| 0 | No neck complaint(s), no physical sign(s) | | | |
| I | Complaint(s) of neck pain, stiffness, or tenderness, no physical sign(s) | Neck pain, stiffness or tenderness | Microscopic, multimicroscopic lesion, no muscle spasm | Doctor visit >24 h after trauma |
| II | Neck complaint(s) and musculoskeletal sign(s)[a] | Neck pain with nonspecific radiation to the head, face, occipital region, shoulder, and arm from soft-tissues injuries; neck pain with limited range of motion due to muscle spasm | Neck sprain and bleeding around soft-tissue (articular capsules, ligaments, tendons, muscles); muscle spasm secondary to soft-tissue injury | Doctor visit <24 h after trauma |
| III | Neck complaint(s) and neurological signs[b] | Cervicobrachialgia, cervical herniated disc, cervicalgia with headache, headache of cervical origin, cervico-scapulalgia | Injuries to neurologic system by mechanical injury or by irritation secondary to bleeding or inflammation | Doctor visit <24 h after trauma |
| IV | Neck complaint(s) and fracture or dislocation | Limited grades of cervical motion with neurologic symptoms and signs | | |

[a]Musculoskeletal signs include decreased range of motion and point tenderness.

[b]Neurologic signs include decreased or absent deep tendon reflexes, weakness, and/or sensory deficits. Symptoms and disorders that can be manifest in all grades include deafness, dizziness, tinnitus, headache, memory loss, dysphagia, and temporomandibular joint pain.

vertebral body fracture or medullary contusion; brainstem syndrome after contusional brainstem injury or secondary to vertebrobasilar injury or cerebral deficits after subdural hematoma or after infarct following thrombotic internal carotid artery occlusion.

These sequelae are summarized in Table II according to their described pathogenic mechanisms. All these are complications of severe whiplash injuries (grade III or IV) or of serious cervical spinal or head injury, but are not the features of the classical, uncomplicated whiplash injury.

## Symptoms and Signs

In a typical case history the patient with whiplash injury develops a painful stiffness of the neck muscles after an initial pain-free intervall of up to some hours. The nuchal pain may exhibit an interscapular or brachial irradiation and is often accompanied by a headache (Keidel and Ramadan, 2001). In terms of group statistics nearly all patients complain of neck pain (91%) and neck stiffness (89%; averages from eight studies). 68% to 88% claim headache, in half to two-thirds (67%) of whom it is occipital (Balla and Karnaghan, 1987; Keidel *et al.*, 1993; Keidel, 2000). In 77% the headache is of dull-pressing or dragging character. The pain is maximum during afternoon

and evening (Keidel *et al.*, 1993; Keidel and Ramadan, 2001). The existence of a "posttraumatic" migraine following whiplash injury (Jacome, 1986; Weiss *et al.*, 1991) is not supported by Edmeads (1987) and Pearce (1992, 1994).

Other symptoms are claimed with the following frequency: dizziness of varying quality (21–50%; Hildingsson and Toolanen, 1990; Oosterveld *et al.*, 1991; Wiesner and Mumenthaler, 1975; Zenner, 1987), shoulder pains (36–49%), interscapular pains (12–20%), arm complaints (27–39%) (Balla, 1984; Hildingsson and Toolanen, 1990; Hohl, 1974; Keidel 1995; Maimaris *et al.*, 1988; Pearce, 1989; Sturzenegger *et al.*, 1994), and transitory syncope and memory lapses, vegetative symptoms, and depressive-neurasthenic complaints (42–68%; Keidel *et al.*, 1995; Radanov *et al.*, 1991). In the realms of performance and personality, deterioration in motivation and concentration, sleep disorders, or irritability may develop (Ettlin *et al.*, 1992; Keidel *et al.*, 1992; Kischka *et al.*, 1991; Krajewski and Wolff, 1990; Radanov *et al.*, 1993; Wiesner and Mumenthaler, 1975). Visual symptoms occur in a range of 3.9% (Wiesner and Mumenthaler, 1975) to 26% (Burke *et al.*, 1992; Hildingsson and Toolanen, 1990; Keidel, 1995; Oosterveld *et al.*, 1991; Sturzenegger *et al.*, 1994). Disturbances of convergence, accomodation, or fusion are prominent (Burke *et al.*, 1992). Complaints of auditory or swallowing disturbances are rare

TABLE II  Possible Sequelae (Whiplash-Associated Disorders) of Mechanical Damage to the Cervical and Cranial Structures Due to "Whiplash Mechanism" Coincident on Primary Neck Injury (Ordered by Syndrome)

| Cervical | Cervico-cephalic | Cervico-medullary |
|---|---|---|
| Dorsal/ventral cervical-syndrome (cervical spine distortion, overextension of the perispinal soft tissue mantle).<br>*Dorsal:* stiff neck, nuchal pain syndrome (radiating occipital, interscapular, scapular, proximal brachial).<br>*Ventral:* Dysphagia, swelling of the throat.<br>Retropharyngeal oedema/haematoma<br>Perforation/rupture of the oesophagus<br>Retropharyngeal abscess<br>Mediastinitis/sepsis<br>Intra-thyroid haematoma (experimental) | Headaches<br>Vegetative syndrome<br>Neurasthenic syndrome<br>Amnestic syndrome<br>Intracranial haematoma (subdural/-arachnoidal, intracerebral)<br>Carotid thrombosis/occlusion<br>Aneurysm (internal carotid artery)<br>Vertebral thrombosis/occlusion<br>Brain stem syndrome (with cranial nerve deficits; mechanical, vascular)<br>Cerebral contusion (rotational acceleration without direct traumatic contact, helmet impact, e.g., orbital brain with anosmia)<br>Retinal/vitreous bleeding (newborn, small children)<br>Whiplash maculopathy (reversible vitreous body detachment in the central fovea)<br>Bleeding of the inner ear (experimental) | Transversal medullary syndrome (myelocompression, e.g. through structural damage of the cervical spine (fracture, dislocation, kinking), disc rupture/prolapse, dural haematoma).<br>Haematomyelia<br>Syringomyelia<br><br>**Cervico-brachial**<br>Cervical radiculopathy<br>Disturbance of the brachial plexus<br>Thoracic outlet syndrome<br>Upper quadrant syndrome<br>Sympathetic reflex dystrophy |

(Keidel, 1995; Sturzenegger *et al.*, 1994; Wiesner and Mumenthaler, 1975). The common complaints are summarized in Table III (for further survey, see Barnsley *et al.* 1994; Evans, 1992). In the rare complicated whiplash injury (grade III or IV, Table I), cervico-brachial and cervico-medullary complaints can be present, caused by localized involvement of central or peripheral nervous structures such as central, radicular or peripheral pains, paresthesiae, deafness, pareses, or indications of bladder/bowl disturbances (see Table II with respect to accompanying clinical pictures).

## Investigations

Choice of additional investigations is determined by the prior history and examination. For common whiplash injuries grades I and II with cervical syndrome, but without conspicuous neurological signs, only X-ray radiography of the cervical spine is sometimes necessary (recordings in two planes, oblique views, peroral dens representation, and functional X-rays). 65.8% of patients with whiplash injury without osseous cervico-spinal damage have pathological X-ray findings (in 20.6% prevertebral soft tissue swelling; in 20.6% preceding degenerative changes; and in 37% with kinking of the cervical spine; Miles *et al.*, 1988). The frequency of abnormal X-rays is related to their timing. In con-

trast, the incidence of radiographically detectable degenerative changes in intervertebral discs in an asymptomatic normal sample with an average age of 30 years lies at only 6% (Hohl, 1974, 1989), but may be 50% by the age of 50 (also see data of Parmar and Raymakers 1993). The loss of physiological lordosis of the cervical spine is not pathognomic for whiplash injury (Roskamp, 1962; Weir, 1975) and can result from transient emotional tension and muscle spasm. Skeletal scintigraphy does not provide sufficient information of ligamentous or bony lesions. It is not recommended as a diagnostic tool for screening purposes (Hildingsson *et al.*, 1989).

## Investigation of Complications

In early reports no pathological MRI findings were obtained in whiplash injury (Maimaris, 1989; Van Meydam *et al.*, 1986). Recently, in some (severe complicated) cases radiographically occult nonbony lesions such as ruptures of the anterior or posterior longitudinal ligaments, retropharyngeal hematomas or other soft tissue swelling, tearing through the intervertebral disk or separation of the disc from the vertebral end plate could be demonstrated by MR imaging (Davis *et al.*, 1991; Naegele *et al.*, 1992) or postmortem (Taylor and Kakulas, 1991). Such cases represent difficult or missed

**TABLE III**  Frequency of Main Symptoms after Whiplash Injury

| Symptom | Incidence (%) | Characteristics | References |
|---|---|---|---|
| Neck pain | 74–100 | Dull pressing, dragging, "neuralgic" | Hohl (1974); Wiesner and Mumenthaler (1975); Balla (1980); Zenner (1987); Balla and Karnaghan (1987); Maimaris et al. (1988); Pearce (1989); Sturzenegger et al. (1994); Keidel (1995) |
| Neck stiffness | 78–95 | Musculoligamental strain | Hohl (1974); Wiesner and Mumenthaler (1975); Pearce (1989) |
| Headache | 40–97 | Occipital maximum, dull pressing and/or dragging | Hohl (1974); Wiesner and Mumenthaler (1975); Balla (1980); Zenner (1987); Maimaris et al. (1988); Pearce (1989); Sturzenegger et al. (1994); Keidel et al. (1993); Keidel (1995) |
| Shoulder pain | 36–46 | | Hohl (1974); Balla (1984); Maimaris et al. (1988); Pearce (1989); Sturzenegger et al. (1994); Keidel (1995) |
| Interscapular pain | 12–20 | | Hohl (1974); Balla (1984); Maimaris et al. (1988); Pearce (1989); Keidel (1995) |
| Arm complaints | 27–39 | Pain, numbness, feeling of swelling, paraesthesia | Hohl (1974); Balla (1984); Maimaris et al. (1988); Pearce (1989); Keidel (1995) |
| Dizziness | 21–50 | Various character | Wiesner and Mumenthaler (1975); Zenner (1987); Hildingsson and Toolanen (1990); Oosterveld et al. (1991); Keidel (1995) |
| Depressive–neurasthenic symptoms | 3–68 | Cardio-autonomic, orthostatic, depressive, dysphoric, sleep | Wiesner and Mumenthaler (1975); Radanov et al. (1991); Keidel (1995) |
| Neuropsychological disturbances | | Concentration, attention, cognition, visual and verbal memory | Krajewski and Wolff (1990), Kischka et al. (1991), Keidel et al. (1991, 1992); Radanov et al. (1991); Ettlin et al. (1992) |
| Visual symptoms | 4–26 | Convergence, accommodation, fusion, blurred vision | Wiesner and Mumenthaler (1975); Hildingsson and Toolanen (1990); Oosterveld et al. (1991); Burke et al. (1992); Struzenegger et al. (1994); Keidel (1995) |
| Auditory symptoms | 3–21 | Tinnitus, other ear noise, hearing decrease, change in quality | Wiesner and Mumenthaler (1975); Sturzenegger et al. (1994); Keidel (1995) |
| Swallowing disturbances | 7 | Hoarseness, sore throat, floor of the mouth pain | Sturzenegger et al. (1994); Keidel (1995) |
| Lower back pain | 25–37 | | Hohl (1974); Dvorak (1987); Radanov et al. (1991); Keidel (1995) |

clinical diagnoses and are not an indication of the pathology responsible for persisting common whiplash symptoms.

A cervico-brachial syndrome with sensory disturbances or paresis requires investigation of root damage (cervical spine X-ray, spinal CT, myelography, myelo-CT, possibly MRI), plexus involvement (F-threshold, fractionated SEPs, EMG, soft-tissue MRI), "thoracic outlet" syndrome (Doppler sonography, possibly angiography), or medullary pain (spinal CT, MRI; in selected cases functional CT or MRI; Dvorak et al., 1987; Naegele et al., 1992). For cervico-cephalic syndrome with increasing or atypical headache and focal neurology it is wise to exclude at the onset the possibility of unrecognized subdural hematoma, intracerebral bleeding, cerebral insult (CT) or traumatic internal carotid occlusion (Doppler, TCD, Duplex, MRA, pos-

sibly angiography) (Pöllmann et al., 1997; Ramadan and Keidel, 2001). For hearing disorders, dizziness, or swallowing disturbances, laryngoscopy, an audiogram, nystagmography, and labyrinth testing are sensible (Oosterveld et al., 1991). For visual disorders, ophthalmological investigation (e.g., perimetry, funduscopy, assessment of convergence/accomodation amplitudes and fusion range; Burke et al., 1992) is needed in selected patients. Cardio-autonomic disturbances can be quantified by spectral ECG analysis, alteration of brainstem-mediated antinociceptive reflexes by EMG analysis (Keidel et al., 2001), and subjective impediment of performance by repetitive-test and neuropsychological investigations (Ettlin et al., 1992; Keidel et al., 1992, 1994; Radanov et al., 1993). A cervico-medullary syndrome after complicated severe acceleration trauma of the cervical spine (e.g., grade IV) requires urgent spinal imaging and functional diagnosis (plain film cervical spine X-ray, spinal CT, myelography and myelo-CT, spinal MRI, tibial-SEPs).

## NATURAL COURSE

Whiplash injury has apparently increased in frequency since the decree of seat-belt legislation (Allen *et al.*, 1985; Deans *et al.*, 1987). It appears in 15% (Lösvund *et al.*, 1988) to 39% (Hutchinson, 1987) of all reported rear-end automobile accidents and in 6% (Hutchinson, 1987) to 9% (Faverjon *et al.*, 1988) of frontal or side collisions, respectively, yet in merely 4% after rolling (Hutchinson, 1987). Many less severe incidents are possibly not reported, so these frequencies may be falsely high. Peak age is 37 years (Laubichler, 1987; Maimaris *et al.*, 1988; Olsnes, 1989; Wiesner and Mumenthaler, 1975). Women predominate the disorder at 62% (average of nine studies from judicial cases, not selected). Increased litigation may falsely increase the apparent incidence.

A third of recently injured patients are initially asymptomatic (Erdmann, 1973; Laubichler, 1987; Wiesner and Mumenthaler, 1975). The duration of the pain-free interval lies between 4 and 16 h, with a peak at 12 h (Erdmann, 1973). Interval durations up to 48 h are possible (in 6.7% of patients, Wiesner and Mumenthaler, 1975). Only a third of the patients investigated in a study by Maimaris (1988) developed symptoms within the first hour, 50% within 1 day, and 13% even later. Robinson and Cassar-Pullicino (1993) reported the onset of symptoms in 86% immediately or within 24 h. In minor whiplash injury the headache arises on average 5 h after the accident (Keidel *et al.*, 1993). The time span until complaints are maximal can be up to 2 weeks (Erdmann, 1973).

Neck pain and occipital headache recede as a rule within days or weeks (Edmeads, 1987; Keidel, 2000; Keidel *et al.*, 1992, 1993; Keidel and Ramadan, 2001). A half to three-quarters of the patients with neck pain are complaint-free after 3 months (18% within the first week, a further 18% within the first month; 49% in total, Deans *et al.* (1987); 66%, Balla (1988); 71%, Pearce (1989)). In 88% of patients who were complaint-free after 2 years, the symptoms had resolved within the first 2 months (Maimaris *et al.*, 1988). After 6 months the proportion of complaint-free patients rises to 82% (Pearce (1989); 74%, Balla (1988); 57%, Deans *et al.*, (1987)). And 89%, although with occasional residual complaints, are back at work by 6 months (Pearce, 1989). Initial complaints of somatic-vegetative disturbances, depressive mood, and subjective deficits in selective attention, concentration, visual or verbal memory, and cognitive abilities usually recover within 6 months

(Ettlin *et al.*, 1992; Keidel *et al.*, 1992; Radanov *et al.*, 1993). About a third (36%) complained of occasional neck pain and only 6% of continuous pain by 1 year (Deans *et al.*, 1987). Only a quarter of the patients (15%, Pearce (1989); 26.3%, Deans *et al.* (1987)) claim pain for more than 1 year (continuous in 3.7%; intermittent in 22.6%; Deans *et al.* (1987)). In those with pain for more than 2 years there is also interscapular pain in 49% and shoulder pain in 69% of the patients (Maimaris *et al.*, 1988).

Prolonged courses with intermittent residual complaints of up to 13 years are possible (Gargan and Bannister, 1990; Hohl, 1974; Maimaris *et al.*, 1988; Parmar and Raymakers 1993; Robinson and Cassar-Pullicino, 1993; Wiesner and Mumenthaler, 1975). In a long-term review of patients, of which 56% had claimed for compensation, 74% suffered residual neck pain and 33% headache after a mean of 10.3 years following whiplash injury (Gargan and Bannister, 1990). Severe residual symptoms persisted in 12%. In a comparable percentage of 14% neck pain demanding therapy and causing a certain degree of disability persisted over 8 years on average in a medicolegal group of patients (Parmar and Raymakers, 1993). In both studies, after 2 or 3 years the frequency of residual symptoms did not show a major alteration with further passage of time (Gargan and Bannister, 1990; Parmar and Raymakers, 1993).

### Prognostic Factors

A delayed recovery is found in patients with neural or bony damage (grade III or IV). These include occipital headache, distal symptoms such as hand or arm pains, neurological symptoms, interscapular back pain, pathological posttraumatic cervical spine X-ray findings with subluxation, fracture, inversion of cervical lordosis with kyphotic kinking, reduced mobility in a segment of the cervical spine (more importantly bi- or polysegmentally), pre-existing degenerative cervical spine changes, and radiological spondylosis (Hohl, 1974, 1989; Maimaris *et al.*, 1988; Miles *et al.*, 1988; Norris and Watt, 1983; Parmar and Raymakers, 1993).

A six- to eightfold increased incidence of degenerative cervical (intervertebral disc) changes after whiplash injury was reported by Hohl (1989). Comparable observations were made by Watkinson (1990). In his whiplash group followed up for a mean of 10.8 years degenerative spondylosis occurred in 33%, compared with 10% of the control group. However, in a study of Parmar and Raymakers (1993), 8 years after whiplash injury there was no significant (radiological) increase of cervical spondylosis in the patient group compared to the radiological findings of an age-matched asymptomatic cohort analyzed by Friedenberg and Miller (1963). These data are supported by Robinson and Cassar-Pullicino (1993). They found in a retrospective long-term study following a mean time of 13.5 years after whiplash injury that acute neck sprain does not lead to cervical spondylosis or to worsening of pre-existing degenerative changes. Hildingsson and Toolanen (1990) reported that pre-existing degenerative spondylosis or a straight or kyphotic cervical curve did not influence the posttraumatic outcome.

Further factors responsible for delayed recovery are a neck pain onset within 12 h after injury, an initial high intensity of head or neck pain, vegetative complaints, neurasthenic symptoms (e.g., disturbance of concentration or sleep; anxiety), depression, subjective impediment, preexisting tension headache, past head trauma, past neck pain, a marked decrease of passive cervical mobility (especially of inclination), and a higher age (Keidel *et al.*, 1993; Lee *et al.*, 1993; Parmar and Raymakers, 1993; Radanov *et al.*, 1993). Younger, male patients show in contrast a favorable course, as do patients with only mild initial symptoms and no initial hospitalization, with correspondingly shorter duration of treatment. Initial brief unconsciousness, neck stiffness, prevertebral soft tissue swelling, circumstances of the accident, place of seat in the accident car, and the amount of bodywork and seat damage have no influence on the prognosis (Hildingsson and Toolanen, 1990; Hohl, 1974, 1989; Maimaris *et al.*, 1988; Miles *et al.*, 1988; Pennie and Agambar, 1991). The duration of headache following whiplash injury is not related to the time between accident and onset of symptoms, nor to the initial pain frequency per day, the concomitant neck tenderness and resistance to palpation, the intake of analgesics, or gender (Keidel *et al.*, 1993).

The persistence of symptoms longer than 6 months indicates the developement of a chronic posttraumatic syndrome in terms of a "late whiplash syndrome" (Balla, 1980; Ramadan and Keidel, 2001), which widely resembles the postconcussion syndrome observed in contact head traumata (Lidvall, 1974; Pearce, 1993, 1994). It is characterized by neck pain, headache, dizziness, fatigue, sensory hypersensitivity, disturbances in concentration, subjective memory impairment, anxiety, and irritability (Pearce, 1992). There is a wide and controversial discussion that in addition to the factors mentioned above disability may be prolonged or become chronic by posttraumatic neurosis in patients with personality disturbances, neurotic features, or a predisposition to psychoneurotic behavior. Patients are more prone to protracted symptoms if they have an emotional, depressive reaction to the accident, have adverse psychosocial or socio-cultural factors, quest for financial or psychosocial gain, malingering, or are subject to pressure

from lawyers or by a delayed settlement of financial compensation (Balla, 1988; Farbmann, 1973; Gay and Abbott 1953; Gotten, 1956; Hodge, 1971; Lee, 1993; Leopold and Dillon, 1960; Pearce, 1992, 1994; Ritter and Kramer, 1991; Trimble, 1981). Retrospective studies that have investigated the possible relationship between persistence of complaints and the amount of compensation claimed, other damage claims, and the duration of the judicial insurance dispute have produced conflicting results. No relationship between presence or settlement of litigation and relief of symptoms was seen by Hodgson and Grundy (1989), Hohl (1974), Macnab (1964, 1971), Norris and Watt (1983), Pennie and Agambar (1991), Robinson and Cassar-Pullicino (1993), and Schutt and Dohan (1968). A summary of the factors with an influence upon prognosis is given in Table IV.

## PRACTICAL MANAGEMENT

### Principles of Therapy

It has not yet been determined, through controlled studies of complete physical and physiotherapeutic measures, that any assumed therapeutic superiority of early mobilization and of immobilization really does have advantages over and above the untreated course. Treatment in the early phase of whiplash injury of the cervical spine with acute pain is different than in the rehabilitative "late phase" after the disappearence of the cervical pain. The cervical spine is conventionally immobilized in acute therapy. Thereby the stretched joint capsule/ligamentous apparatus and the soft tissue mantle are, in theory, rested (Schlegel, 1968). Supplementary heat application, aimed or increasing blood flow, is used to relieve muscle tension with possible secondary bad posture of the cervical spine, and may aid the resorption of hematomata. However, there are no controlled studies to confirm these putative healing regimens.

Rehabilitation follows immediately. During rehabilitation according to physiotherapeutic treatment rules, the limited post-traumatic function of the cervical spine and muscles are improved or normalized. Compared to just supplying a restraining collar, early application of physiotherapy is claimed to lead to a faster improvement of neck pain and movement (McKinney *et al.*, 1989; Mealey *et al.*, 1986).

### Acute Pain

Pain relief and rest are made possible by wearing a collar. Except for minor injury the 'Philadelphia collar'

TABLE IV    Factors That Influence the Prognosis of the Course of Complaints after Whiplash Injury

| Unfavorable | Favorable | No influence |
|---|---|---|
| Neurological excitatory symptoms or deficits | Few initial symptoms | Initial short unconsciousness |
| Hand or arm pain ("distal symptoms") | Absent hospitalization | Painful neck stiffness |
| Interscapular accentuated back pain | Short inability to work | Prevertebral jugular soft tissue swelling |
| Vegetative, neurasthenic or depressive complaints | Early mobilizing physiotherapy | Accident circumstances |
| Occipital headache | Short duration of therapy | Place of seat in accident car |
| Tension headache or head trauma in the history | Young age | Amount of bodywork and seat damage |
| Pathological X-ray findings of the cervical spine (subluxation, fracture, kyphotic kinking, segment blockade) | Male sex | |
| Degenerative cervical spine changes | | |
| Decrease of passive cervical mobility | | |
| Higher age | | |
| Long-term wearing of a neck collar (>3 months) | | |
| Renewed physiotherapy after remission | | |
| Litigation, compensation and other forensic consequenses? | | |

Note. No weighting lies in the order of the factors.

is preferred by many to other soft foam collars because of its superior stiffness. Care should be paid that the chin is a little raised when putting it on, so that with an optimal transverse position of the intervertebral joints maximal relief in middle and lower regions of the cervical spine region is obtained (Herrmann, 1971; Krämer, 1983b). In general, immobilization of the cervical spine by means of a cervical collar should be done as briefly as possible (if any).

Only with moderate and especially severe injury of the cervical spine (grades II and III; see Table I) is temporary bed rest necessary, because of postural insufficiency and pain in the neck musculature. The abolition of the weight of the head when lying relieves the damaged neck structures from supportive and postural function, thereby alleviating pain. To avoid hospitalization and "somatization tendencies," the shortest duration possible of lying down must be implemented. The same applies to the duration of wearing the collar in order to avoid inactivity, which may weaken the neck musculature. During the painful acute phase 24-h use is necessary. Attendant upon improvement in complaints, a prompt increase in movement during the day is encouraged. Finally the collar is worn only at night. The treatment is finished when pain disappears. It should usually last not longer than 7 days with mild grade I whiplash injury (without neurological excitatory or deficiency symptoms and osseous cervical spine injury). Even with more severe injury (grade III, Table I), early mobilization is desirable, since with longer immobilization secondary damage with sometimes nearly irreversible movement restrictions in the cervical spine region can develop, and a substantially longer course of illness and psychological features may materialize.

Supplementary heat application can help. Dry heat such as infrared light, arc light, warm air, warming bottles, or electric-pillow appliances have all proved as useful as moist heat (hydrocollator pack or other steam packs). They can initially be used daily and later at longer intervals.

The acute drug treatment of the cervicocephalic pain syndrome following whiplash injury includes the prescription of (only temporarily!) analgesics/antirheumatics and/or muscle relaxants. (After exclusion of any intracranial haemorrhage) acetylsalicylic acid (ASS) is recommended as an analgesic in a dose of 500–1000 mg daily, max. 1500 mg daily. Alternatives are paracetamol tablets (or suppositories if available), 3 × 500 mg/die, or diclofenac-sodium-coated tablets or suppositories, 3 × 50 mg/die, or naproxen, 500–1000 mg daily, if necessary with concomitant gastro-protective therapy. Monotherapy is to be preferred and additional doses of analgesics should be avoided. No analgetic combination preparations should be prescribed, because of the high probability of developing a drug-induced headache, which is often misclassified as a post-traumatic headache by the patient. The danger of an analgesic-induced persistent headache increases with a drug intake >4 weeks. This has to be prevented by a controlled and short-lasting prescription of analgetics and by narrow timed re-examinations of the patient.

As a muscle relaxant tetrazepam, 2 × 50 mg/die, can be recommended. Other benzodiazepines (e.g., diazepam, 5 mg nocte) are sometimes useful in sedating and relieving muscle spasm in the acute stages, but should not be continued for more than 2–3 weeks.

None of the therapies have been subject to controlled trials.

## Prolonged Pain

When pain improves (after about 1–4 weeks), *early* active body and postural exercises and isometric contraction exercises can be begun; first of all isometric contraction exercises when supine, then head- and torso-stabilizing exercises while sitting, and finally strengthening exercises of the neck musculature (Herrmann, 1971; Mealey *et al.*, 1986). If pain re-emerges under physical therapy, treatment must be continued at a 'lower' level. With continued improvement of complaints, graduated exercises can be helpful, and strong encouragement is given to return to light work.

In case of the persistence of a post-traumatic (tension-like) headache and some neurasthenic complaints a trial of tricyclic thymoleptics given over at least 8–16 weeks is recommended. The prescription of tricyclic antidepressants as amitriptyline (evening dose of 25–100 mg or 25-0-75 mg daily) is appropriate. Doses of amitriptyline less than 100 mg/die have been shown to have no consistent antidepressant action, though they may have some limited central analgesic effect. As a second choice the prescription of maprotiline (25–75 mg daily), nortriptyline, doxepine (50–100 mg, max. 150 mg daily), imipramine (75–100 mg, max. 150 mg daily), or MAO inhibitors (tranylcypromine, 20–40 mg daily) are sometimes used. Fluoxetine and imipramine are less effective in the treatment of sustained neurasthenic complaints (Langohr *et al.*, 1994). The specialist, who performs the treatment, should be aware of (i) a slow increase of dosage, (ii) a sufficient dosage, (iii) a sufficiently long intake (treatment success can only be evaluated after 8 weeks of intake), and (iv) the contraindications (among others prostatism, glaucoma). An ECG should be done before treatment with tricyclic thymoleptics. MAO blockers should be given only in refractory cases and fully controlled.

Prompt resolution of compensation claims and of accident-related problems and the full supportive guidance of the patient can help to shorten the duration of complaints. The recommendations for drug treatment are adapted from the German Migraine and Headache Society (Keidel *et al.*, 1998). The therapy in the early and late stages of the cervico-cephalic pain syndrome is summarized in Table V.

None of these recommendations is based on controlled trials.

## General Management

In summary, the general drug-free management of the post-traumatic headache together with neck pain in the context of a cervicocephalic pain syndrome following whiplash injury includes: (i) immobilization of the cervical spine by means of a cervical collar as short as possible (if any); (ii) supplementary physical therapy only in the acute phase with dry heat (infrared light, warm air, heat pillows), with moist heat (fango (volcanic mud) packs) or better with cold packs; (iii) physiotherapy with, e.g., passive and active movement of the cervical spine, isometric tension, complex movements, and postural strengthening; (iv) manual treatment attempts for releasing vertebral segmental blockade in selected cases with a strong indication (cave: vertebral dissection); (v) general physical health recommendations (sport, vegetative stabilisation with, e.g., hot and cold showers, brush massage, regular lifestyle); (vi) learning of automobilization of the cervical spine and of muscle relaxation techniques of the shoulder–neck musculature (according to Jacobson), possibly in combination with EMG biofeedback techniques; (vii) in cases of chronification concomitant psychotherapy based (among others) on concepts of cognitive coping strategies for pain and stress; (viii) course dependent approach of an interdisciplinary and multimodal regime of therapy; (ix) empathetic guidance of the patient; (x) explanation of the good prognosis in general; (xi) narrow timed re-examinations of the patient in the acute phase; and (xii) fast solving of accident-related medico-legal problems.

## Treatments No Longer Recommended

In accordance with the principle of rest positioning of overstretched structures there is in the early post-traumatic phase following whiplash injury an absolute contraindication for chiropractic treatment or physio-therapeutic stretching measures such as manual tractions with passive bending, extending, or pulling of the cervical spine. Manipulation in the initial phase leads to amplified irritation of the damaged structures and/or vegetative reactions and is contraindicated. Local injection or infiltration treatments with local anesthetics are in our opinion not necessary if there is sufficient rest and analgesia. The same applies to oral doses of more potent analgesics. The wearing of plaster collars is obsolete with whiplash injury grades I and II. The benefit of local "unguent treatment" is not proven and in necessary cases the systematic antiphlogistic treatment is preferable. Reflex zone massages, neural therapy, acupuncture, acupressure, fresh cells or ozone therapy, and the use of neuroleptics or ergotamine preparations in the post-traumatic headache treatment is nonbeneficial and unnecessary. There is no evidence of effectiveness of antihistamines or steroids, and they are not indicated in whiplash injury. The reported efficacy of intra- or subcutaneous injection of sterile water or of NaCl solu-

TABLE V  Treatment of Whiplash Injury of the Cervical Spine (Cervical Syndrome) in the Acute and Late Phases

| Principle | Recommendation | |
|---|---|---|
| **Acute phase** | | |
| Physical treatment | | |
| Immobilization (whiplash injury of the cervical spine grade I/II, as short as possible, not longer than 1 week) | Cervical collar | |
| Cold/heat application | e.g., Cold pack, hydrocollator pack, infrared light | |
| Drug treatment | | |
| Antiphlogistics (Analgesics) | Acetylsalicylic acid (ASS) | 3 or 4 × 500 mg/die |
| | Alternatives | |
| | Diclofenac-sodium | 3 × 50 mg/die |
| | Paracetamol | 3 × 500 mg/die |
| | Ibuprofen (retard) | 400–600 mg/die |
| | Naproxen | 500–1000 mg/die |
| Muscle relaxants | Tetrazepam | 2 × 50 mg/die |
| | Diazepam | 5 mg nocte (short course) |
| Physiotherapeutic treatment | Isometric stretching exercises | |
| | Movement exercises (passive, active) | |
| | Complex movements | |
| | Strengthening exercises | |
| | Building up posture | |
| | Loosening the musculature | |
| | Enhance morale and distraction from pain | |
| **Prolonged recovery** | | |
| Drug treatment | | |
| Thymoleptics | Amitriptyline | Evening dose of 25–100 mg daily or 25-0-75 mg daily |
| | Alternatives | |
| | 2nd choice | |
| | Doxepine | 50–100 mg daily, max. 150 mg daily |
| | 2nd choice | |
| | Imipramine | 75–100 mg daily, max. 150 mg daily |
| MAO blockers | Tranylcypromine | 20–40 mg daily |
| Muscle relaxation exercises | Functional muscle relaxation, e.g., according to Jacobson | |
| | Return to suitable work | |

*Note.* Adapted from the recommendations of the German Migraine and Headache Society (Keidel *et al.*, 1998).

tion in cervical paravertebral muscles (Byrn *et al.*, 1993; Sand *et al.*, 1992) needs further controlled confirmation by other groups and cannot be recommended at the present stage. The same holds true for subcutaneous injections of sumatriptan (Gawel *et al.*, 1993), which may improve the throbbing pain component in headache following whiplash injury but does not influence the common dull-pressing or dragging (cervico-)cephalic pain.

### Therapy for Special Secondary Illnesses

When there are additional injuries of osseous or nervous structures in the cervico-cephalic, -brachial, or -medullary syndromes, therapy is determined by the specific clinical picture (see Clinical Aspects, above, and Table II). These aspects are referred to in therapy chapters in this book such as dizziness (Chapter 15), radicular syndrome (Chapter 90), compression syndrome of peripheral nerves (Chapters 91 and 92), paraplegia (Chapter 68), subarachnoid hemorrhage (Chapter 36), cerebral hemorrhage (Chapter 35) or cerebral ischemia (Chapters 30 and 33).

### CONCLUSION

Acute soft tissue damage, contusion, and stretching of ligaments and muscles explain the common short-lived symptoms of neck sprains; the prognosis is good and many are pain-free and back at work within 3 months. We have no good objective evidence that explains the protracted complaints of a significant minority of patients with so-called chronic whiplash syndrome. The resistance to, and failure to sustain, improvement to a wide variety of conservative pharmacological and physical therapies is extraordinary in the absence of signs of organic pain-evoking pathology. When such symptoms persist for 1 to 2 years, they tend to continue indefinitely. Mechanical lesions (annular tears, facet joints, avulsion of vertebral end-plates, and subluxations) are not pertinent mechanisms, as shown by many MRI studies. Radanov *et al.* (1991, 1994) found that psychosocial factors at injury did not predict the outcome, though neuroticism correlated with the initial pain intensity. Most such cases are tied up with litigation. This proces and the possibility of large financial rewards may have significant contribution to many otherwise inexplicable symptoms (Pearce, 1992, 1994).

# REFERENCES

Allen, M. J., Barnes, M. R., and Bodiwala, G. G. (1985). The effect of seat belt legislation on injuries sustained by car occupants. *Injury* 16, 471–476.

Balla, J. I. (1980). The late whiplash syndrome. *Aust. NZJ. Surg.* 50, 610–614.

Balla, J. I. (1984). Headaches arising from disorders of the cervical spine. *Headache* 10, 243–267.

Balla, J. I., and Karnaghan, J. (1987). Whiplash headache. *Clin. Exp. Neurol.* 23, 179–182.

Balla, J. I. (1988). Report to the Motor Accidents Board of Victoria on whiplash injuries, 1984. *In* "Headache and Cervical Disorders" (A. Hopkins, Ed.), pp. 256–269. Saunders, London.

Barnsley, L., Lord, S., and Bogduk, N. (1994). Whiplash injury. *Pain* 58, 283–307.

Bauer, G., and Pils, P. (1984). Hirnblutung nach Auffahrunfall. *Unfallheilkunde* 87, 37–39.

Birsner, J. W., and Leask, H. (1954). Retropharyngeal soft tissue swelling due to whiplash injury. *Arch. Surg.* 68, 369–373.

Bosworth, D. M. (1959). Editorial. *J. Bone Jt. Surg.* 41-A, 16.

Brandt, Th. (Ed) (1991). "Vertigo." Springer Verlag, Berlin Heidelberg New York.

Burke, J. P., Orton, H. P., West, J., Strachan, I. M., Hockey, M. S., and Ferguson, D. G. (1992). Whiplash and its effect on the visual system. *Graefe's Arch. Clin. Exp. Ophthalmol.* 230, 335–339.

Byrn, C., Olsson, I., Falkheden, L., Lindh, M., Hoesterey, U., Fogelberg, M., Linder, L. E., and Bunketorp, O. (1993). Subcutaneous sterile water injections for chronic neck and shoulder pain following whiplash injuries. *Lancet* 341, 449–452.

Carter, J. E., and McCormick, A. Q. (1983). Whiplash shaking syndrome: Retinal hemorrhages and computerized axial tomography of the brain. *Child Abuse Neglect* 7, 279–286.

Davis, S. J., Teresi, L. M., Bradley, W. G., Ziemba, M. A., and Bloze, A. E. (1991). Cervical spine hyperextension injuries: MR findings. *Radiology* 180, 245–251.

Deans, G. T., Magalliard, J. N., Kerr, M., and Rutherford, W. H. (1987). Neck sprain—A major cause of disability following car accidents. *Injury* 18, 10–12.

Delank, H. W. (1988). Das Schleudertrauma der HWS, eine neurologische Standortsuche. *Unfallchirurg* 91, 381–387.

Dvorak, J., Valach, L., and Schmid, S. (1987). Verletzungen der Halswirbelsäule in der Schweiz. *Orthopäde* 16, 2–12.

Edmeads, J. (1987). Does the neck play a role in migraine? *In* "Migraine" (J. N. Blau, Ed.), pp. 653–654. Chapman Hall, London.

Erdmann, H. (ed.) (1973). "Die Wirbelsäule in Forschung und Praxis," Band 56, Schleuderverletzung der Halswirbelsäule, Erkennung und Begutachtung. Hippokrates Verlag, Stuttgart.

Ettlin, T. M., Kischka, U., Reichmann, S., Radii, E. W., Heim, S., Wengen, D., and Benson, D. F. (1992). Cerebral symptoms after whiplash injury of the neck: A prospective clinical and neuropsychological study of whiplash injury. *J. Neurol. Neurosurg. Psychiatry* 55, 943–948.

Evans, R. W. (1992). Some observations on whiplash injuries. *Neurol. Clin.* 10, 975–997.

Farbman, A. A. (1973). Neck sprain. *JAMA* 223, 1010–1015.

Faverjon, G., Henry, C., Thomas, C., Tarriere, C., Patel, A., Got, C., and Guillon, F. (1988). Head and neck injuries for belted front occupants involved in real frontal crashes: Patterns and risks. *In* "Proceedings of the 1988 International IRCOBI Conference on the Biomechanics of Impacts, Bergisch Gladbach Sept. 1988" (D. Cesari and A. Charpenne Eds.), 301–317. IRCOBI, 109 Avenue Salvador Allende, 69500 Bron, France.

Fischer, D., and Palleske, H. (1976). Das EEG nach der sogenannten Schleuderverletzung der Halswirbelsäule (zerviko-zephales Beschleunigungstrauma). *Zbl. Neurochir.* 37, 25–35.

Friedenberg, Z. B., and Miller, W. T. (1963). Degenerative disc disease of the cervical spine. *J. Bone Jt. Surg.* 72B, 901.

Gargan, M. F., and Bannister, G. C. (1990). Long-term prognosis of soft-tissue injuries of the neck. *J. Bone Jt. Surg.* 72B, 901–903.

Gawel, M. J., Rothbart, P., and Jacobs, H. (1993). Subcutaneous sumatriptan in the treatment of acute episodes of posttraumatic headache. *Headache* 33, 96–97.

Gay, J. R., and Abbott, K. H. (1953). Common whiplash injuries of the neck. *JAMA* 152, 1698–1704.

Gotten, N. (1956). Survey of one hundred cases of whiplash injury after settlement of litigation. *JAMA* 162, 865–867.

Hadley, M. N., Sonntag, V. K. H., Rekate, H. L., and Murphy, A. (1989). The infant whiplash-shake injury syndrome: a clinical and pathological study. *Neurosurgery* 24, 536–540.

Herrmann, H.-D. (1971). Das Schleudertrauma der Halswirbelsäule; Therapie (II). *Med. Welt.* 22, 1366–1370.

Hildingsson, C., Hietala, S. O., and Toolanen, G. (1989). Scintigraphic findings in acute whiplash injury of the cervical spine. *Injury* 20, 265–266.

Hildingsson, C., and Toolanen, G. (1990). Outcome after soft-tissue injury of the cervical spine. *Acta Orthop. Scand.* 61(4), 357–359.

Hodge, J. R. (1971). The whiplash neurosis. *Psychosomatics* 12, 245–249.

Hodgson, S. P., and Grundy, M. (1989). Whiplash injuries: Their long-term prognosis and its relation to compensation. *Neuro-Orthop* 7, 88–91.

Hohl, M. (1974). Soft-tissue injuries of the neck in automobile accidents. Factors influencing prognosis. *J. Bone. Jt. Surg.* 56-A, 1675–1682.

Hohl, M. (1989). Soft-tissue neck injuries. *Cervical Spine* 2, 436–441.

Hutchinson, T. P. (ed.) (1987). "Road Accident Statistics," Chap 14, 255–266. Rumsby Scientific, Adelaide, South Australia.

Jacome, D. E. (1986). Basilar artery migraine after uncomplicated whiplash injuries. *Headache* 26, 515–516.

Jacome, D. E. (1987). EEG in whiplash: A reappraisal. *Clin. EEG* 18, 41–45.

Keidel, M., Yagüez, L., Wilhelm, H., and Diener, H. C. (1992). Prospektiver Verlauf neuropsychologischer Defizite nach zerviko-zephalem Akzelerationstrauma. *Nervenarzt* 63, 731–740.

Keidel, M., and Diener, H. C. (1993). Headache and acceleration trauma of the cervical spine. *News Headache* 3/3, 1.

Keidel, M., Eisentraut, R., Baume, B., Yagüez, L., and Diener, H. C. (1993). Prospective analysis of acute headache following whiplash injury. *Cephalalgia* 13(Suppl. 13), 177.

Keidel, M., Eisentraut, R., and Diener, H. C. (1993). "Predictors for Prolonged Recovery from Posttraumatic Headache in Whiplash Injury." Seattle: IASP Publications.

Keidel, M. (1995). Der posttraumatische Verlauf nach zerviko-zephaler Beschleunigungsverletzung. Klinische, neurophysiologische und neuropsychologische Aspekte. *In* "Neuroorthopädie VI" (B. Kügelgen, Ed.). Springer, Heidelberg Berlin New York.

Keidel, M., Neu, I. S., Langohr, H. D., and Göbel, H. (1998). Therapie des posttraumatischen Kopfschmerzes nach Schädel-Hirn-Trauma und HWS-Distorsion. Empfehlungen der Deutschen Migräne- und Kopfschmerzgesellschaft. *Schmerz* 12, 350–367.

Keidel, M., and Ramadan, N. (2000). Acute posttraumatic headache. *In*: "The Headaches" (J. Olesen, K. M. A. Welch, and P. Tfelt-Hansen, Eds.), 2nd ed., pp. 765–770. Lippincott-Raven, Philadelphia.

Keidel, M. (2000). Posttraumatic headache. *In* "Drug Treatment of Migraine and other Frequent Headaches" (H. C. Diener, Ed.), Monographs in Clinical Neuroscience, vol. 17, pp. 329–336. Karger, Basel.

Keidel, M., Rieschke, P., Stude, Ph., Eisentraut, R., Schayck, R. van, and Diener, H. C. (2001). Antinociceptive reflex alteration in acute posttraumatic headache following whiplash injury. *Pain* 92, 319–326.

Keidel, M., and Stude, Ph. (2002). Brain Lesion. *In* "Encyclopedia of the Brain." Lippincott-Raven, Philadelphia.

Kessler, Ch., Hipp, M., Langkau, G., Pawlik, G., and Petrovici, J.-N., (1987). Plättchenszintigraphische Befunde bei Carotisthrombosen nach Halswirbelsäulen-Schleudertrauma. *Nervenarzt* 58, 428–431.

Kischka, U., Ettlin, Th., Heim. S., and Schmid, G. (1991). Cerebral symptoms following whiplash injury. *Eur. Neurol.* 31, 136–140.

Krämer, G. (1980). Das zerviko-zephale Beschleunigungstrauma ("HWS-Schleudertrauma") in der Begutachtung. Unter besonderer Berücksichtigung zentralnervöser und psychischer Störungen. *Akt. Neurol.* 7, 211–230.

Krämer, G. (1983). HWS-Schleudertraumen. *Med. Welt.* 34(41), 1134–1140.

Krämer, G. (1983). Therapie neurologischer Störungen nach Schleudertraumen der Halswirbelsäule. *Dtsch. Med. Wschr.* 108, 589–590.

Krajewski, C., and Wolff, H. D. (1990). Psychodiagnostische Untersuchung von HWS-Schleudertrauma-Patienten. *Man. Med.* 28, 35–39.

Langohr, H. D., Keidel, M., Goebel, H., Wallasch, T. M., and Baar, H. (1994). Kopfschmerz nach Schädel-Hirn-Trauma und HWS-Distorsion. Diagnose und Therapie. *In* "Migräne und andere Kopf-und Gesichtsschmerzen. Therapieempfehlungen der Deutschen Migräne- und Kopfschmerzgesellschaft," pp. 49–57. Arcis, München.

Laubichler, W. (1987). Die Problematik einer Begutachtung von Verletzungen der Halswirbelsäule einschließlich cervico-cephalem Beschleunigungstrauma. *Unfallchirurg* 90, 339–346.

Lee, J., Giles, K., and Drummond, P. D. (1993). Psychological disturbances and an exaggerated response to pain in patients with whiplash injury. *J. Psychosomat. Res.* 37, 105–110.

Leopold, R. L., and Dillon, H. (1960). Psychiatric considerations in whiplash injuries of the neck. *PA Med. J.* 63, 385–389.

Lidvall, H. F., Linderoth, B., and Norlin, B. (1974). Causes of the postconcussional syndrome (PCS). *Acta Neurol. Scand. Suppl.* 56, 1–144.

Lövsund, P., Nygren, A., Salen, B., and Tingvall, C. (1988). Neck injuries in rear end collisions among front and rear seat occupants. *In* "Proceedings of the International IRCOBI Conference on the Biomechanics of Impacts, Bergisch Gladbach Sept. 1988" (D. Cesari and A. Charpenne Eds.), pp. 319–325. IRCOBI, 109 Avenue Salvador Allende, 69500 Bron, France.

Macnab, I. (1964). Acceleration injuries of the cervical spine. *J. Bone Jt. Surg.* 46, 1797–1799.

Macnab, I. (1971). The "whiplash syndrome." *Orthop. Clin. North Am.* 2, 389–403.

Maimaris, C., Barnes, M. R., and Allen, M. J. (1988). 'Whiplash injuries' of the neck: A retrospective study. *Injury* 19, 393–396.

Maimaris, C. (1989). Neck sprains after car accidents. *Br. Med. J.* 299, 123.

Martinez, J. L., and Garcia, D. J. (1968). A model for whiplash. *J. Biomech.* 1, 23–32.

McKinney, L. A., Dornan, J. O., and Ryan, M. (1989). The role of physiotherapy in the management of acute neck sprains following road-traffic accidents. *Arch. Emerg. Med.* 6, 27–33.

Mealy, K., Brennan, H., and Fenelon, G. C. C. (1986). Early mobilisation of acute whiplash injuries. *Br. Med. J.* 292, 656–657.

Miles, K. A., Maimaris, C., Finlay, D., and Barnes, M. R. (1988). The incidence and prognostic significance of radiological abnormalities in soft tissue injuries to the cervical spine. *Skeletal Radiol.* 17, 493–496.

Mokri, B. (1990). Traumatic and spontaneous extracranial internal carotid artery dissections. *J. Neurol.* 237, 356–361.

Naegele, M., Koch, W., Kaden, B., Woell, B., and Reiser, M. (1992). Dynamische Funktions-MRT der Halswirbelsäule. *Fortschr. Röntgenstr.* 157/3, 222–228.

Norris, S. H., and Watt, I. (1983). The prognosis of neck injuries resulting from rear-end collisions. *J. Bone Jt. Surg.* 65B, 608–611.

Olsnes, B. T. (1989). Neurobehavioral findings in whiplash patients with long-lasting symptoms. *Acta Neurol. Scand.* 80, 584–588.

Ommaya, A. K., Yarnell, P. (1969). Subdural hematoma after whiplash injury. *Lancet* 2, 237–239.

Oosterveld, W. J., Kortschot, H. W., Kingma, G. G., and Jong, H. A. A. de (1991). Electronystagmographic findings following cervical whiplash injuries. *Acta Otolaryngol.* (Stockh.) 111, 201–205.

Parmar, H. V., and Raymakers, R. (1993). Neck injuries from rear impact road traffic accidents: prognosis in persons seeking compensation. *Injury* 24, 75–78.

Pearce, J. M. S. (1989). Whiplash injury: a reappraisal. *J. Neurol. Neurosurg. Psychiatry* 52, 1329–1331.

Pearce, J. M. S. (1992). Whiplash injury, fact or fiction. *Headache Q. Curr. Treat. Res.* 3, 45–50.

Pearce, J. M. S. (1993). Subtle cerebral lesions in 'chronic whiplash syndrome'? *J. Neurol. Neurosurg. Psychiatry* 56, 1328–1329.

Pearce, J. M. S. (1994). Headache. *J. Neurol. Neurosurg. Psychiatry* 57, 134–143.

Pearce, J. M. S. (1994). Polemics of chronic whiplash injury. *Neurology* 44, 1993–1998.

Pennie, B., and Agambar, L. (1991). Patterns of injury and recovery in whiplash. *Spine* 22, 57–59.

Pöllmann, W., Keidel, M., and Pfaffenrath, V. (1997). Headache and the cervical spine: A critical review. *Cephalalgia* 17, 801–816.

Radanov, B. P., Di Stefano, G., Schnidrig, A., and Ballinari, P. (1991). Role of psychosocial stress in recovery from common whiplash. *Lancet* 338, 712–715.

Radanov, B. P., Sturzenegger, M., Di Stefano, G., Schnidrig, A., and Aljinovic, M. (1993). Factors influencing recovery from headache after common whiplash. *Br. Med. J.* 307, 652–655.

Radanov, B. P., Di Stefano, G., Schnidrig, A., Sturzenegger, M., and Augustiny, K. F. (1993). Cognitive functioning after common whiplash. A controlled follow-up study. *Arch. Neurol.* **50**, 87–91.

Radanov, B. P., Di Stefano, G., Schnidrig, A., and Sturzenegger, M. (1994). Common whiplash: Psychosomatic or somatopsychic? *J. Neurol. Neurosurg. Psychiatry* **57**, 486–490.

Ramadan, N., and Keidel, M. (2000). Chronic posttraumatic headache. *In* "The Headaches" (J. Olesen, K. M. A. Welch, and P. Tfelt-Hansen, Eds.), 2nd ed., pp. 771–780. Lippincott-Raven, Philadelphia.

Ritter, G., and Kramer, J. (Eds) (1991). "Unfallneurose, Rentenneurose, Posttraumatic Stress Disorder." Perimed Verlag, Erlangen.

Robinson, D. D., and Cassar-Pullicino, V. N. (1993). Acute neck sprain after road traffic accident: A long-term clinical and radiological review. *Injury* **24**(2), 79–82.

Roskamp, H. (1962). Angst und Cervikalsyndrom. *Zschr. Psychosom. Med.* **8**, 157–167.

Sand, T., Bovim, G., and Helde, G. (1992). Intracutaneous sterile water injections do not relieve pain in cervicogenic headache. *Acta Neurol. Scand.* **86**, 526–528.

Schlegel, K. F. (1968). Die akuten Schleuderverletzungen der Halswirbelsäule und ihre Behandlung. *Beih. Z. Orthop.* **104**, 265–273.

Schmidt, G. (1989). Zur Biomechanik des Schleudertraumas der Halswirbelsäule. *Versicherungsmed* **4**, 121–125.

Schutt, C. H., and Dohan, F. C. (1968). Neck injury to women in auto accidents. *J. Am. Med. Assoc.* **206**, 2689–2692.

Simeone, F. A., and Goldberg, H. I. (1968). Thrombosis of the vertebral artery from hyperextension injury to the neck. *J. Neurosurg.* **29**, 540–544.

Spitzer *et al.* (1995). Scientific monograph of the Quebec Task Force on whiplash-associated disorders: Redefining "whiplash" and its management. *Spine* **20/8S**, 1S–73S.

Sturzenegger, M., DiStefano, G., Radanov, B. P., and Schnidrig, A. (1994). Presenting symptoms and signs after whiplash injury: the influence of accident mechanisms. *Neurology* **44**, 688–693.

Taylor, J. R., and Kakulas, B. A. (1991). Neck injuries. *Lancet* **338**, 1343.

Trimble, M. R. (ed.) (1981). "Post-Traumatic Neurosis: From Railway Spine to the Whiplash." Wiley, New York.

Van Meydam, K., Sehlen, S., Schlenkhoff, D., Kiricuta, J. C., and Beyer, H. K. (1986). Kernspintomographische Befunde beim Halswirbelsäulentrauma. *Fortschr. Röntgenstr.* **145**, 657–660.

Watkinson, A. F. (1990). Whiplash injury. *Br. Med. J.* **301**, 983.

Weir, D. C. (1975). Röntgenographic signs of cervical spine injury. *Clin. Orthop.* **109**, 9.

Weiss, H. D., Stern, B. J., and Goldberg, J. (1991). Post-traumatic migraine: Chronic migraine precipitated by minor head or neck trauma. *Headache* **31**, 451–456.

Wiesner, H., and Mumenthaler, M. (1975). Schleuderverletzungen der Halswirbelsäule. Eine katamnestische Studie. *Arch. Orthop. Unfall-Chir.* **81**, 13–36.

Yarnell, P. R., and Rossie, G. V. (1988). Minor whiplash head injury with major debilitation. *Brain. Inj.* **2**, 255–258.

Zagorski, A., and Berlit, P. (1990). Das disseziierende Karotisaneurysma – drei Fälle. *Med. Klin.* **85**, 125–128.

Zenner, P. (ed.) (1987). Die Schleuderverletzung der Halswirbelsäule und ihre Begutachtung. Springer Verlag, Berlin, Heidelberg, New York.

CHAPTER 11

# Acute and Chronic Pain

Geoffrey K. Gourlay and H. Christoph Diener

## INTRODUCTION AND TAXONOMY

All of us have experienced pain at some time and therefore understand the difficulty in translating that experience into a language that can be understood by others including our medical advisors. The visual depiction of pain in a myriad of forms has been a subject of interest to artists and sculptors for centuries including the modern-day variant called graphic designers, frequently used by pharmaceutical companies to advertise their drugs and/or equipment to treat pain. The *International Association for the Study of Pain* (IASP), the pre-eminent multidisciplinary international society, established an expert task force to provide a taxonomy and classification of pain terms and conditions that has been instrumental in providing improved communication, particularly between health professionals (Merskey and Bogduk, 1994). According to the IASP definition (currently undergoing minor modification), *pain* is defined *as an unpleasant sensory or emotional experience associated with actual or potential tissue damage or described in terms of such damage.* This definition highlights the fact that pain is an unpleasant experience and although invariably associated with a nociceptive focus, this does not have to be the case. Therefore, the still commonly used but arbitrary broad subdivision of pain patients into psychogenic or organic (or whatever equivalent terms you wish to use) is of little use other than to indicate where the treatment emphasis should be placed. Thus, many factors other than nociception can influence the perception of pain. Consequently, a holistic approach incorporating the assessment and therapeutic strategies of a range of medical specialties together with psychology, physiotherapy, occupational therapy, and social work aspects is most likely to produce the most meaningful gains, particularly in chronic noncancer pain patients (*persistent pain* is the evolving term used to describe this group of patients).

Clearly, significant attention should be given to treating the underlying causes of the pain if that is at all possible as the first intervention.

The subdivision of pain into either *acute* or *chronic pain* is common although somewhat arbitrary and depends on the situation. Acute pain includes postoperative pain, myocardial infarction, injury, or infrequent migraine/other headaches, biliary/renal colic, etc. Pain that persists for longer than the normal time of healing or for longer than 3–6 months (Merskey and Bogduk, 1994) is then regarded as chronic pain, including regular or indeed irregular episodic acute pain conditions. There are some treatment differences between *acute* and *chronic* pain, particularly in the pharmacological algorithms used (*vide infra*), but a return of as much function as possible is a key feature common to both types of pain.

Although further gains are necessary, there have been significant improvements in the treatment of acute pain, both organizationally (acute pain services, promulgated guidelines to treat certain pain states) and pharmacologically (methods of drug delivery [e.g., patient-controlled analgesia for postoperative pain control], drug combinations, routes of administration [spinal drug delivery], etc). Marvelous new drugs that might revolutionize the treatment of pain have been nonexistent.

Complex chronic pain patients can be a very daunting management prospect, frequently best referred to specialist pain management units that are resourced with appropriate multidisciplinary staff able to undertake the comprehensive assessments required as well as formulate and initiate a suitable management plan. The assessment should aim to give a diagnosis wherever possible, including the etiology of the pain (nociceptive, neuropathic, or idiopathic) and other characteristic features (inflammatory processes or malignancy). Other commonly used pain terms used in this context are given in Table I. This assessment should ideally include a

**TABLE I**  IASP List of Pain Terms (Abridged)

| Term | Definition |
| --- | --- |
| Allodynia | Pain due to a stimulus which does not normally provoke pain. |
| Analgesia | Absence of pain in response to a stimulation which would normally be painful. |
| Anesthesia dolorosa | Pain in an area or region which is anesthetic. |
| Causalgia | A syndrome of sustained burning pain, allodynia, and hyperpathia after a traumatic nerve lesion, often combined with vasomotor and sudomotor dysfunction and later trophic changes. |
| Central pain | Pain initiated or caused by a primary lesion or dysfunction in the central nervous system. |
| Dysesthesia | An unpleasant abnormal sensation, whether spontaneous or evoked. Special cases of dysesthesia include hyperalgesia and allodynia. A dysesthesia should always be unpleasant, whereas a paresthesia should not be unpleasant—always specify whether the sensations are spontaneous or evoked. |
| Hyperpathia | A painful syndrome characterized by an abnormally painful reaction to a stimulus, especially a repetitive stimulus, as well as an increased threshold. It may occur with allodynia, hyperesthesia, hyperalgesia, or dysesthesia. |
| Hyperalgesia | An increased response to a stimulus which is normally painful. |
| Hyperesthesia | Increased sensitivity to somatosensory or nociceptive stimulation—there is diminished threshold and increased response to normally recognized stimuli. Includes both allodynia and hyperalgesia. |
| Hypoalgesia | Diminished pain in response to a normally painful stimulus. |
| Hypoesthesia | Decreased sensitivity to stimulation. |
| Neuralgia | Pain in the distribution of a nerve or nerves—common usage, especially in Europe, implies a paroxysmal quality. This term should not be reserved for paroxysmal pains. |
| Neuritis | Inflammation of a nerve or nerves—should only be used if inflammation is though to be present. |
| Neuropathy | A disturbance of function or pathological change in a nerve or nerves. |
| Neuropathic pain | Pain initiated or caused by a primary lesion or dysfunction in the nervous system—may be further subdivided into *central* or *peripheral* neuropathic pain. |
| Nociceptor | A receptor preferentially sensitive to a noxious stimulus or to a stimulus that would become noxious if prolonged. |
| Noxious stimulus | A stimulus which is damaging to normal tissue. |
| Pain | An unpleasant sensory or emotional experience associated with actual or potential tissue damage or described in terms of such damage. |
| Pain threshold | The least experience of pain which a subject can recognize. |
| Pain tolerance level | The greatest level of pain which a subject is prepared to tolerate. |
| Paresthesia | An abnormal sensation, whether spontaneous or evoked. |

*Note.* Adapted from the IASP Classification of Chronic Pain (Merskey and Bogduk, 1994) with permission. This reference provides expanded definitions and additional explanatory notes for many of these terms.

comprehensive psychological and psychiatric evaluation (examining *inter alia*, models of pain behavior, cultural and religious perspectives) as well as that provided by a rehabilitation physician and physiotherapist to provide independent quantification of the degree of physical impairment relative to the observed pain behavior. Finally, an interview with a spouse or significant other is highly recommended as they may be recruited into the management plan, since they may also be a source of negative reinforcement of the pain in some situations. For some patients, the nociceptive focus is no longer present but chronic pain nevertheless persists.

The devised management plan with specific goals, which may be both physical, psychological and vocational (where appropriate) could include any of the following:

(1) evaluation and changes to the pharmacological regimen (*vide infra*);

(2) psychological strategies, which could be group (e.g., cognitive bahavioral programs) and/or individual (hypnosis, pacing issues, anxiety management, or other counseling, sometimes including marital counseling) sessions;

(3) goal-oriented, graded exercise program directed at teaching the patient appropriate skills that he/she may use at home or work rather than passive treatment (rarely should there be the need for greater than 10 sessions with a physiotherapist);

(4) psychiatric counseling where appropriate;

(5) constructive interaction with rehabilitation providers, including insurance case managers if involved (which may also include an ergonomic evaluation of the workplace environment);

(6) implanted pain-relieving devices (pumps or spinal cord stimulators) in some patients;

(7) rarely a recommendation for further surgery.

## NEUROBIOLOGY OF PAIN

The perception of pain is complex and involves sensory, emotional, and behavioral factors; this section will deal with the sensory aspects and the biological basis of the transition from acute to chronic pain (chronification).

The primary afferent nociceptor can be divided into two main categories: C fiber polymodal and A-delta mechanothermal nociceptors. Tissue damage invariably leads to inflammatory processes with the release of calcitonin-gene-related peptide (CGRP), neurokinin A, and substance P. This results in the production and release of bradykinin, histamine, 5-HT, cytokines, NO (nitric oxide), and potassium. The inflammatory "soup" not only leads to sensitization of high-threshold nociceptors but also to the activation of silent nociceptors. Under conditions of chronic pain or chronic inflammation, these receptors have lower thresholds and may respond to non-noxious stimuli.

Damage of peripheral nerves or roots leads to biochemical, morphological, and physiological changes that cause spontaneous pain (Devor, 1994). The ends of damaged nerve fibers start to sprout and exhibit altered response characteristics to various stimuli. This process is partly promoted by the release of nerve growth factor. Demyelination of peripheral nerves, e.g., in diabetic neuropathy, may lead to the generation of ectopic impulses, resulting in the sharp and shooting pain experienced by many patients with polyneuropathy.

The sympathetic nervous system plays an important role in the generation and maintenance of chronic pain (Jänig, 1996). Following nerve damage or trauma, a long-lasting disturbance of sympathetic activity leads to changes in temperature and blood supply, decreased motor function, allodynia, hyperalgesia, and chronic pain. This condition is now called "complex regional pain syndrome." The basis for this development is that increased sympathetic activity can lead to abnormal activity or response characteristics of primary afferent fibers.

Major recent breakthroughs in pain research include the cloning and characterization of a capsaicin receptor, receptors for purines (P2X3), an acid-sensing ion channel, and a tetrodotoxin-resistant sodium channel (Besson, 1999). All afferent inputs terminate in the dorsal horn of the spinal cord. The primary afferent nociceptors terminate in laminae I, II, and V, mostly on the same level, but may also ascend or descend several segments. Second-order neurones in the dorsal spinal chord are well characterized. Nociceptive, high-threshold neurones can be differentiated from wide dynamic range neurones. When wide-range neurones become sensitized, they will respond to tactile stimuli and transmit those as pain signals (allodynia). Many of the neuromodulators and neurotransmitters in dorsal horn neurones have been identified. The major neurotransmitters for nociceptive transmission are the excitatory amino acids glutamate and aspartate. They act on NMDA (N-methyl D-aspartate), AMPA (α-amino-3-hydroxy-5-methyl-4-isoxazole proprionic acid), and metabotrophic glutamate receptors (Coderre, 1993).

NMDA receptors are involved in many phenomena that contribute to the long-term changes seen in chronic pain. These include long-term potentiation, induction of oncogenes, central sensitization and changes in peripheral receptive fields. Long-term potentiation might be the correlate of the so-called "pain memory." NMDA receptor antagonists are able to attenuate long-term potentiation and therefore prevent the chronification of pain.

Central sensitization occurs at the level of the dorsal horn and shows the plasticity of the central nervous system. With repetitive nociceptive stimulation, dorsal horn neuronal activity shows a *wind-up* (i.e., an enhanced temporal summation in nociceptive dorsal horn neurones following repetitive c-fiber afferent stimulation at a constant intensity). The clinical correlate is the zone of "secondary hyperalgesia" due to an expansion of the receptive field size.

Modulation of pain signals happens at the level of the spinal cord. Afferent pain signals initiate inhibitory processes limiting the effects of subsequent pain impulses. This inhibition is mediated by local inhibitory interneurones but also by descending pathways from supraspinal structures. Modulation of pain signals is mediated by opioid, glycine, alpha-adreno-, and GABA receptors.

Pain information is mediated in the spinal cord via spinothalamic, spinoreticular and spinomesencephalic tracts. Second-order neurones terminate in the brainstem, thalamus, and cortex. PET studies have been useful in identifying central structures involved in pain perception (Tölle *et al.*, 1999). These include the sensory, motor, premotor, frontal, and insular cortex (Derbyshire *et al.*, 1997). The affective component of pain is mediated by the anterior cingulate. Descending inhibition arises from the periaqueductal gray matter, locus coeruleus, nucleus raphe magnus, and the hypothalamus. The major modulators of descending inhibition are opioids, 5-HT, GABA, and noradrenaline. Detailed reviews covering the neurobiology of pain transmission have been published (Dickenson, 1996; Carlton and Coggeshall, 1998; Besson, 1999).

## PRINCIPLES OF THERAPY

Philosophically, physicians, and other health care workers must be prepared to promote pain to the status of a symptom worthy of attention in its own right (i.e., the fifth vital sign), rather than just treating the diagnosed condition and hope that pain may miraculously resolve. It is crucial to diagnose the pain condition as accurately as possible and all initial efforts must be directed to treating the cause of the pain with the aim removing or more realistically, particularly for chronic pain, minimizing the nociceptive focus. However, despite these efforts, ongoing pain remains for many patients, not infrequently for the rest of their lives.

It is also important to realize that a thorough mechanistic understanding of the pain complaint will provide the most rational basis for therapy. The continual reassessment of the efficacy and side effects of any prescribed regimen and a willingness to reoptimize therapy is the only way to maximize the treatment of pain. These reassessments would include the adequacy of:

(1) drug doses,
(2) dosing interval,
(3) an examination of drug combinations that may,
   (a) activate different pain pathways (nonsteroidal anti-inflammatory drugs [NSAIDs], opioids, adjuvant analgesic agents, etc.) and
   (b) search for additivity and hopefully (but rarely) synergy where lower doses of usually two drugs have additive (or synergistic) analgesic effects but not adverse effects. Under these circumstances, the combination may result in the same extent or improved analgesia to that seen with higher doses of either drug individually, but a more favorable side effect profile. Higher doses of the combination may also result in superior pain control before side effects become dose limiting,
(4) change drugs in the same class,
(5) change routes of administration where appropriate,
(6) remember that pain is defined as an unpleasant experience and therefore the range of techniques provided by clinical psychologists can provide dramatic improvements in some patients,
(7) fully explore the options that physical therapists can offer.

Referral of difficult chronic pain patients to specialized pain management units for assessment and treatment is highly recommended where the multidisciplinary team approach can provide most of the skills mentioned above in the one unit.

## DRUG THERAPY

### Placebo Response

The placebo response is a factor in the success of all drug therapy, but is particularly significant in analgesic studies where the incidence is estimated to vary between approximately 0 and 75% of patients in individual

studies in both acute (McQuay et al., 1995a) and chronic pain (McQuay et al., 1996) with an average across acute pain studies of 18%. The analgesic response in individual patients can vary between 0 and 100% in both the active and the placebo groups; i.e., for the extreme cases, some patients can report 100% pain relief with placebo or zero pain relief with active drug, although these outcomes are uncommon. Many psychological factors such as conditioning, expectancy (Pollo et al., 2001), desire for pain relief, and distortions in memory may influence the magnitude of placebo response (Price, 2000) because *pain* is defined as an *unpleasant experience*.

### Assessing Efficacy in Analgesic Studies

The strength of evidence supporting a pharmacological or indeed any other intervention can range from a single uncontrolled case report or clinical impression (Level 5 evidence) to a meta-analysis (Level 1 evidence) of blinded randomized clinical trials (McQuay and Moore, 1998a). The evidence-based medicine (EBM) approach has embraced the latter as the *gold standard* to provide the most objective evidence of both desirable (i.e., pain relief) and undesirable (i.e., side effects of varying intensities) effects (Moore, 2000). In recent years, meta-analyses of some of the pharmacological and other interventions have been published and updated, and a web site (http://www.jr2.ox.ac.uk/bandolier/booth/painpag/index.html) provides a current bibliography of this data. Many of these reports have expressed the results as number needed to treat (NNT) for the desirable effect of analgesia and numbers needed to harm (NNH) for adverse events (which can be further subdivided into minor and major side effects, the latter usually involving drug withdrawal). For example, the intervention of a single 10-mg intramuscular morphine dose in the treatment of postoperative pain had a NNT of 2.9. This means that on average, one in three patients (strictly one in 2.9 patients) administered that morphine dose would have achieved the desired analgesic response (defined as at least 50% pain relief) that they would not have with the administration of placebo (McQuay and Moore, 1998b). The NNT data given throughout this chapter for both acute and chronic pain relates to the definition of analgesia given above. A limitation of this approach, particularly for the acute pain area, is that many of the studies examine only single-dose data.

### Acute Pain

Over the past two decades, substantial progress has been made in improving the treatment of acute pain (mainly postoperative pain, but also pain associated with trauma) by re-evaluating routes of administration (intravenous, spinal, etc.), modes of administration (i.e., continuous infusions versus patient initiated bolus doses via patient-controlled analgesia (PCA) pumps), and drug combinations (opioids, NSAIDs, local anesthetics, clonidine, NMDA receptor antagonists, etc.) and by prioritizing pain as an important treatment requirement in its own right (Kehlet, 1999; Plummer, 1999). Nevertheless, postoperative pain remains poorly managed because of the following inadequacies (Plummer, 1999):

(1) education of medical and nursing staff,
(2) organization, and
(3) information given to patients and patient reluctance to demand adequate pain relief.

The establishment of acute pain services in many hospitals has addressed many of these failings. Specific guidelines regarding the optimization of postoperative pain control can be found in the following references: Kehlet (1999), MacIntyre and Ready (2001), NH&MRC (1999), Plummer (1999).

**Chronic Pain**

The steps and philosophy contained in the WHO Analgesic Ladder (Figure 1) provides a logical basis for the pharmacological treatment of chronic cancer and noncancer pain. This approach recognizes that from a population basis, patients can have pain of varying intensities when they consult their doctors or medical specialists and that pain intensity can increase over time, particularly in the case of cancer pain. Thus, the analgesic ladder provides three points of entry and allows progression between steps. Step 1 is the lowest pain intensity and is treated with NSAIDs and the range of the so-called *adjuvant* analgesic drugs (antidepressants, anticonvulsants, membrane-stabilizing agents, etc., *vide infra*). If the pain is of intermediate severity on presentation or if optimized Step 1 drugs no longer provide adequate analgesia, the patient is moved to Step 2, which adds the *weak opioids* (codeine, oxycodone, dextropoxyphene, or perhaps tramadol) to the drugs used at Step 1. Similarly, if the pain is severe, an opioid from Step 3 (morphine, methadone, fentanyl, etc.) is used *in lieu* of a Step 2 opioid. In view of the superior efficacy and predictability of Step 3 opioids, many physicians seriously question the role of Step 2 opioids and recommend going directly to Step 3 from Step 1. However, in some countries Step 3 opioids are difficult if not impossible to obtain; thus Step 2 does have a place under certain circumstances. Some clinicians have proposed an additional step (Step 4) which would include

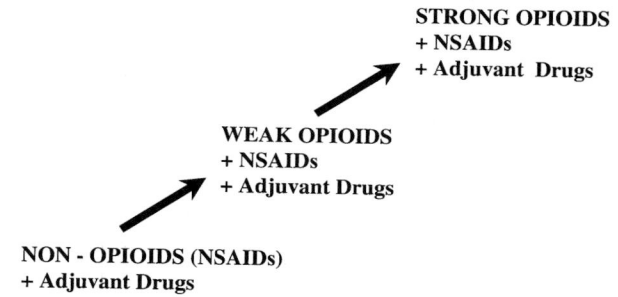

FIGURE 1  WHO analgesic ladder.

spinally administered opioids, neurolytic techniques, or neurosurgical approaches to treat the 15–20% of patients in whom optimized Step 3 drugs can no longer (or never did) provide adequate analgesia without severe side effects, particularly in the treatment of cancer pain. For chronic noncancer pain, the same therapeutic strategy can be employed, but physicians must be very careful about adding opioids to the regimen because of concerns relating to iatrogenic addiction. The following section provides a more detailed consideration of the various drug classes used to treat all types of persistent pain.

*NSAIDs*

The NSAIDs and paracetamol (acetominophen) are commonly used to treat pain, but any possible benefit must be considered in relation to their well-documented propensity for adverse effects and toxicity, particularly upper gastrointestinal PUBs (perforations, ulcers, and bleeds) and renal effects. NSAIDs exert their pharmacological effects by inhibiting prostaglandin formation via action on the enzyme, cyclo-oxygenase (COX). Prostaglandins are one of a number of inflammatory mediators that sensitize nociceptors in inflamed tissues. Two isoforms of this enzyme exist, namely COX–1, which is constitutively expressed in most tissue, and COX–2, which is induced in inflamed tissue by the actions of cytokines and other mediators. The newer NSAIDs are highly specific COX–2 inhibitors and offer the hope of fewer side effects (Kurumbail *et al.*, 1996) because of the localization of the particular isoform only in inflamed tissue. In contrast, the older NSAIDs are either predominantly COX–1 inhibitors or inhibit both isoforms essentially equally and consequently have more widespread actions, resulting in many undesirable pharmacological effects because of the presence of this isoform in most tissue. Table II provides the relative cyclo-oxygenase inhibition (COX–2/COX–1 ratio) for selected NSAIDs—lower ratios represent greater COX–2 selectivity as the data represent the ratio of $IC_{50}$ values (drug concentration to result in 50% inhibition

TABLE II    NSAID's—Pharmacokinetic Parameters, COX-2/COX-1 Selectivity, Dose, and Effect of Food on Absorption

| Drug | Terminal half-life (h) | Recommended oral dose (mg) | Dose frequency (h) | COX-2 / COX-1 ratio[a] | Effect of food on absorption[b] |
|---|---|---|---|---|---|
| Aspirin | 0.2–0.3 | 300–900 | 4–6 | 1a | A |
| Diclofenac | 1.0–2.0 | 25–50 | 8 | 1d | A |
| Ibuprofen | 2–3 | 200–800 | 8 | 1c | A |
| Naproxen | 12–20 | 250–500 | 12 | 1d | B |
| Indomethacin | 2–8 | 50–75 | 8 | 1b | A |
| Piroxicam | 30–60 | 20–30 | 24 | 1c | A |
| Celecoxib | 4–15 | 100–200 | 12 | 2 | C |
| Rofecoxib | 15–20 | 12.5–25 | 24 | 2 | B |
| Paracetamol (acetaminophen) | 1.5–3.0 | 500–1000 (4 g/24 h) | 4–6 | — | A |

[a]COX-2/COX-1 ratio represents a ratio of $IC_{50}$ values [i.e., drug concentration to result in 50% inhibition] of the various drugs listed to differentially inhibit the respective COX enzymes in the human whole-blood assay (Young *et al.*, 1996); the lower the value, the greater the selectivity towards COX-2 inhibition. 1a, ratio >50; 1b, ratio between 10 and 50; 1c, ratio between 1 and 10; 1d, ratio between 0.1–1; 2, ratio <0.01. Different data sets exist for human whole blood assay and indeed other assays indicative of COX activity (e.g., platelet aggregation, serum thromboxane $B_2$). Nevertheless, the relative inhibition of the two COX enzymes by the various NSAIDs is reasonably congruent.

[b]A represents a decrease in the rate of absorption, but no change in oral bioavailability. B represents no change in both the rate of absorption and oral bioavailability. C represents a decrease in the rate of absorption, but an increase in oral bioavailability.

of respective COX activity) in the human whole-blood assay (Young *et al.*, 1996).

NSAIDs have been shown to be very effective in the treatment of acute pain with NNTs as low as 2 for both diclofenac and ibuprofen, values equivalent to 10 mg of intramuscular morphine (McQuay and Moore, 1998c). Systematic reviews in both acute (Moore *et al.*, 1997; Tramer *et al.*, 1998) and chronic pain (Eisenberg *et al.*, 1994; Riedemann *et al.*, 1993; Tramer *et al.*, 1998) have generally failed to identify a particular NSAID that consistently provides superior analgesia. Therefore, physicians often choose a particular NSAID based more on their assessment of the adverse effect profile or pharmacokinetic considerations, than any superior analgesia and these considerations have resulted in a significant market share for the new COX–2 NSAIDs in many markets soon after release. A summary of usual dosing regimens and pharmacokinetic parameters are also given in Table II.

While the conventional oral route is widely used, NSAIDs may be administered by a variety of routes including intravenous, intramuscular, rectal, and topical. At the present time, there are no advantages of the other routes over oral administration (Tramer *et al.*, 1998) and therefore the latter should be used, with the possible exception of topically administered NSAIDs, which are effective (NNT of 3.1) with perhaps a lower incidence of gastrointestinal side effects (Moore *et al.*, 1998).

The major toxicity of the traditional NSAIDs is gastroduodenal damage (Roderick *et al.*, 1993; Stalnikowicz and Rachmilewitz, 1993). The incidence of gastric and duodenal ulcers varied between 18–47% and 2–8%, respectively (Stalnikowicz and Rachmilewitz, 1993). Misoprostol significantly reduced the inci-

dence of gastric ulcers and ranitidine reduced the incidence of duodenal ulcers. The COX–2 inhibitors (celecoxib and rofecoxib) have a clinically significant reduction in the incidence of gastroduodenal ulcers.

### NSAIDs + Opioids

While NSAIDs (or paracetamol) combined with *opioids* are popular in many countries, only modest increases in analgesic activity, if any, have been demonstrated. These formulations, which usually involve low doses of the so-called *weak opioids* (e.g., codeine, propoxyphene, etc.) in a fixed dosage ratio for over-the-counter preparations, have no place in the contemporary pharmacological treatment of pain. The addition of 60 mg codeine to paracetamol (doses ranging from 300 to 1000 mg) was shown to have a modest but nevertheless statistically significant analgesic effect (NNT of only 7.7) compared to paracetamol alone in the treatment of acute pain (Moore *et al.*, 1997)—i.e., the addition of codeine resulted in an analgesic response in only 13% of patients compared to that seen with paracetamol alone. A meta-analysis of NSAIDs alone or in combination with various opioids (either oxycodone, pentazocine, propoxyphene, or codeine) in the treatment of cancer pain (Eisenberg *et al.*, 1994) indicated that the extent of analgesia was the same for both formulation types. These results provide experimental evidence to question the validity of Step 2 in the WHO analgesic ladder as previously proposed.

### Antidepressants

Pain, depression, and anxiety coexist in many chronic pain patients. Antidepressants have a specific analgesia action (in addition to their antidepressant or sedative effects) by activating decending pain inhibitory pathways that use noradrenalin and serotonin as neurotransmitters (Onghena and Van Houdenhove, 1992). There is divided opinion as to the optimal analgesic dose; the systematic review (Onghena and Van Houdenhove, 1992) indicated a similar extent of analgesia was obtained with both normal antidepressant doses and lower than normal doses whereas a positive dose–response relationship was established for amitriptyline in a double-blind, multiple-dose study (McQuay et al., 1993).

Table III shows the NNT (analgesia) and NNH (adverse effects), all relative to placebo, for the use of antidepressants and anticonvulsants in various neuropathic pain states; e.g., diabetic neuropathy, postherpetic neuralgia (Robotham and Petersen, 2001), atypical facial pains, and the so-called *central pain* (McQuay et al., 1996). It is apparent that antidepressants are very effective in the treatment of neuropathic pain in general as indicated in the similarity of the NNT values for the various pain states. However, the incidence of the

usually reported side effects is more variable, depending on the etiology of the neuropathic pain (most frequent for patients with central pain), while approximately 1 in 20 patients could not tolerate the antidepressant (as indicated by the NNH[2] data). This last figure is lower than many clinicians would have expected and may relate to the differences between research studies and clinical practice. Nevertheless patients, particularly older patients, should be started at a lower dose which should be incrementally increased to fully therapeutic doses to minimize the impact of adverse effects.

Antidepressants (with their corresponding NNT value) have varying efficacy in the treatment of chronic pain (e.g., imipramine (3.7), desipramine (3.2), combined tricyclics (3.2), paroxetine (5), and fluoxetine (15.3)). Mianserin was not significantly different from placebo. Thus, SSRIs (selective serotonin reuptake inhibitors) were less effective than tricyclic antidepressants. Nevertheless, the efficacy of the newer SSRIs have not been investigated to the extent of the older antidepressants and they may be more effective than these figures indicate. In three studies, tricyclic antidepressants were more effective than the active control of benzodiazepines.

An older systematic review (Onghena and Van Houdenhove, 1992) of placebo-controlled studies in general chronic noncancer patients suggested the magnitude of the analgesic effect is similar for patients having a predominantly organic or psychogenic basis for their pain, while the magnitude of the analgesic effect was independent of the antidepressant action, the presence of masked depression or the use antidepressants with differing capacities to induce sedation. Consequently, antidepressants with sedation as a side effect should be used in those patients who have poor sleep patterns.

**TABLE III**  Proportions of Patients Experiencing Analgesia and Side Effects Administered Either Antidepressants or Anticonvulsants in Various Neuropathic Pain States

|  | NNT | NNH[1] | NNH[2] |
|---|---|---|---|
| Diabetic neuropathy |  |  |  |
| Antidepressants | 3.0 | 2.8 | 19.1 |
| Anticonvulsants | 2.5 | 3.1 | 20.0 |
| Post herpetic neuralgia |  |  |  |
| Antidepressants | 2.3 | 6 | 19.6 |
| Atypical facial pain |  |  |  |
| Antidepressants | 2.8 | NR | NR |
| Central pain |  |  |  |
| Antidepressants | 1.7 | 2 | 335 |
| Trigeminal neuralgia |  |  |  |
| Anticonvulsants | 2.6 | 3.4 | 24.0 |
| Migraine prophylaxis |  |  |  |
| Anticonvulsants | 1.6 | 2.4 | 39.3 |

*Note.* Data in this table has been compiled from different sources (McQuay et al., 1995b, 1996). NNT represents numbers needed to treat for the analgesic outcome of >50% pain relief. NNT = $1/((IMP_{act}/TOT_{act}) - (IMP_{cont}/TOT_{cont}))$, where $IMP_{act}/TOT_{act}$ represents the experimental event rate (i.e., number of patients achieving the defined analgesic outcome with active drug [$IMP_{act}$ = improved on active] divided by the total number of patients administered the active drug [$TOT_{act}$ = total number of patients given active]) and $IMP_{cont}/TOT_{cont}$ represents the control event rate (i.e., an equivalent fraction, but with the control drug; may be placebo or active control). NNH[1] represent numbers needed to harm for minor (i.e., commonly reported) side effects. The same formula is used, but the data is dichotomized on the basis of the definition of commonly reported adverse events. NNH[2] represent numbers needed to harm for major side effects necessitating drug withdrawal. The same formula is used, but the data is dichotomized on the basis of the definition of major adverse events.

NR = Not recorded.

### Benzodiazepines

Benzodiazepine use should be kept to an absolute minimum in chronic pain patients because of problems of tolerance, dependence and marked withdrawal symptoms when the pharmacological regimen is rationalized. A slowly tapering dose algorithm over a prolonged time is necessary to reduce or eliminate benzodiazepines from the therapeutic regimen. There are sufficient numbers of problem chronic pain patients with benzodiazepine dependency in most pain management units without using these drugs in new pain patients.

### Anticonvulsants

Anticonvulsants have also been used to treat both cancer and chronic noncancer pain, particularly if pain is neuropathic in origin, but not acute nociceptive pain. The more established anticonvulsants (e.g., carbamazepine, phenytoin, clonazepam, and valproate) are still widely used to treat chronic pain. However, more recent interest has focused attention on the newer agents, gabapentin (Robotham *et al.*, 1998), lamotrigine, and pregabalin (a derivative of gabapentin) because they hold the hope of a more favorable side-effect profile, although not necessarily greater analgesia. While the introduction of the established agents into the therapeutic regimen usually follows some variant of the traditional dose escalation titration protocol for these agents in the treatment of epilepsy, gabapentin can be started at therapeutic doses (i.e., 0.9–2.4 g in divided doses). Table III also shows the NNTs for pain relief and NNHs for minor and major adverse effects for anticonvulsants (mainly carbamazepine and phenytoin) in chronic pain (McQuay *et al.*, 1995b). The NNT values for the anticonvulsants are very similar (vary between 1.6 and 2.6) for the various neuropathic pain states examined (Table III). Further, there is a marked similarity in efficacy and side effects between anticonvulsants and antidepressants in the treatment of diabetic neuropathy. However, it should be remembered that this approach represents population-based statistics—therefore, a particular patient may (or may not) respond to antidepressant medication, but not anticonvulsant medication and vice versa.

A direct comparison of the efficacy of amitriptyline and carbamazepine in treating pain following strokes (Leijon and Bovie, 1989) indicated that twice as many patients responded to amitriptyline with a similar incidence of minor and major side effects between the two drugs (McQuay *et al.*, 1995b).

### Opioids

The generally accepted view is that nociceptive pains are opioid sensitive but neuropathic pains **may be** opioid sensitive, but at higher opioid doses. In some chronic pain patients, unacceptable side effects will occur before adequate analgesia is perceived by the patient and the physician must terminate opioid prescription in such cases.

**Acute Pain.** Opioids are very effective if used appropriately in the treatment of acute pain (e.g., postoperative pain, trauma, emergency medical conditions [angina, biliary, and renal cholic, etc.] as previously noted. Potent μ—receptor agonists such as morphine, fentanyl, and hydromorphone should be used preferentially while pethidine (meperidine) and the mixed agonist/antagonist drugs should be avoided because of neurotoxicity and a ceiling effect in analgesia, respectively. Methadone is very effective in the treatment of postoperative pain (Gourlay *et al.*, 1984), but can present difficulties if clinicians and nursing staff do not

appreciate the fact that its pharmacokinetics are substantially different from the other opioids (Table IV). The doses given under the equianalgesic dose column can be taken as initial starting doses for opioid-naive patients. As previously stated, the only way of ensuring effective pain control is to reevaluate the efficacy of any opioid regimen and to titrate to effect by varying the dose, dosing interval, or both. Changing opioid is not usually required in the treatment of acute pain and is invariably done because of an unacceptable side effect profile.

**Chronic Pain.** The difficult clinical question that physicians must ask themselves when treating chronic pain is, *should this patient be prescribed opioid medication*? For patients with cancer pain, this question is much more easily dealt with than for patients with chronic noncancer pain although a disproportionate degree of *opiophobia* still exists with the former. Opioid therapy for chronic noncancer pain should be seen as a last resort when all other attempts to optimize pharmacotherapy and other treatment options have not yielded the desired benefits. While the introduction of opioids into the therapeutic regimen is still regarded as contentious by many clinicians, some national pain societies have formulated position statements and promulgated guidelines (American Pain Society web site, www.ampainsoc.org/advocacy/opioids.htm; Graziotti and Goucke, 1996; Schug *et al.*, 1991) to educate physicians in an attempt to minimize the inappropriate prescription of opioids. Almost all proponents of opioid prescription for noncancer pain recognize that many patients are inappropriately prescribed opioids as part of the pharmacological regimen and that comprehensive educational programs directed toward non-pain-specialist physicians are essential to better target this therapy to suitable patients. The very significant increased opioid prescription rates over the past decade in most countries has been predominantly related to use in the treatment of chronic persistent pain.

The philosophy of the WHO analgesic ladder for the treatment of cancer pain was described earlier (Figure 1). The physician is faced with choices from either Step 2 (*weak opioids*) or Step 3 (*strong opioids*). The Step 2 opioids include codeine, oxycodone, dextropoxyphene, and perhaps tramadol. Codeine is not particularly effective as an analgesic, shows weak binding to μ-opioid receptors (Gourlay, 1999) and probably exerts its action via morphine (*vide infra, Role of Metabolites*).

Oxycodone is considered by some to be a Step 3 opioid because various studies have shown it to be equi- or more potent than morphine. However, recent animal evidence suggests that oxycodone may have additional analgesic activity via interaction at κ receptors (Ross and Smith, 1997).

**TABLE IV** Opioid Pharmacokinetics and the Role of Metabolites

| Opioid | Terminal half-life (h) | Clearance (L/min) | Equianalgesic dose (mg)[a] (IV, IM/oral) | Dosing interval[c] (h) | Oral bioavailabilty (%) | Active metabolites |
|---|---|---|---|---|---|---|
| Morphine | 2–4 | 0.8–1.2 | 10/30–40 | 2–4/12, 24 | 10–50 | M6G |
| Pethidine | 3–4 | 0.6–0.8 | 100/200–300 | 2–4 | 30–60 | Norpethidine |
| Methadone | 6–150 | 0.1–0.3 | 10/2–5[b] | 8–24 | 60–90 | No |
| Fentanyl | 3–7 | 0.7–1.5 | 0.1/0.025–0.05 | 3 Days | <2/90[d] | No |
| Codeine | 3–4 | 0.6–0.9 | 60/120 | 2–4 | 60–90 | Morphine |
| Oxycodone | 2–6 | 0.4–1.1 | 5–10/15 | 3–4/12 | 40–130 | Oxymorphone |
| Hydromorphone | 2–4 | 0.4 | 2/3–4 | 4–6 | 35–80 | No |

[a] The suggested equianalgesic doses represent starting doses or initial conversion dose equivalents. Titration to an effective dose level is essential by closely monitoring the balance between analgesic efficacy and any side effects; clinicians must be prepared to change the dose, dosing interval, or both to achieve the desired outcome. Effective doses may be substantially larger than the figures contained in this table for those patients with severe cancer or noncancer pain—they are a guide for patients with minimal prior recent opioid exposure. The doses on the left-hand side in this column represent parenteral doses; appropriate care should be exercised if physicians elect to administer opioids IV with respect to dose (use conservative doses) and speed of administration. The values on the right-hand side represent equivalent oral doses except for fentanyl where the values are for the transdermal formulation (Durogesic) and the dose is in mg/h (equivalent to 25–50 µg/h).

[b] An unusual situation arises for methadone where the equivalent oral dose is lower than the parenteral dose. This occurs because methadone is a capacity eliminated opioid with a long and variable half-life which results in accumulation of blood methadone concentrations at steady state on prolonged dosing. The acute potency oral dose is 10 mg, but the majority of patients are likely to exhibit sign of toxicity if this IM/oral dose ratio is used except in those taking potent inducers of cytochrome P450 3A4 (e.g., phenytoin, HIV protease inhibitors, fluconazole, etc).

[c] The dosing interval data usually represent immediate release formulations. For morphine and oxycodone, the values on the right of the / represent the dosing interval for various modified release oral formulations—in the case of morphine, formulations are available in different countries with either a 12-h or a 24-h dosing interval. The value for fentanyl represents the usual dosing interval for the transdermal formulation.

[d] The value on the left represents the oral (not buccal) bioavailability, while the value on the right represents the mean transdermal bioavailability.

Tramadol has a potency equivalent to propoxyphene/paracetamol and aspirin/codeine combinations (Moore and McQuay, 1997). The vast majority of drugs are administered as racemates and we are starting to appreciate the complimentary mechanisms of action of the different enantiomers for some analgesic drugs, for example, tramadol and methadone. For tramadol, the d-enantiomer exhibits preferential binding activity at µ-receptors and is a more potent inhibitor of serotonin reuptake while the l-enantiomer is more efficient in blocking noradrenaline uptake (Kovelowski et al., 1998).

Dextroproxyphene has an unacceptable incidence of psychotomimetic side effects and should not be used. Thus, most of the opioids in Step 2 of the WHO analgesic ladder generally provide less predictable pain relief and have a number of problems associated with their use.

Step 3 opioids include morphine, methadone, fentanyl (transdermal formulation as well as the hardened lozenge on a stick formulation for incident pain) pethidine, hydromorphone, levorphanol (in some countries), buprenorphine (in some countries), and the mixed agonist/antagonist group of drugs (i.e., butorphenol, nalbuphine, pentazocine). The opioids from Step 3 generally provide a more predictable outcome and are therefore recommended. Table IV shows the pharmaco-kinetic parameters for selected opioids together with the pharmacological activity of their metabolites. It is evident that the majority of the opioids have a high clearance and a relatively short half-life of between 2 and 6 h, indicating extensive metabolism, predominently in the liver (Gourlay, 1999). Methadone is the exception with a relatively low clearance and long and variable half-life (Gourlay et al., 1986). Methadone has good oral bioavailability and its metabolites are not considered to have analgesic activity (Table IV). Methadone is usually administered as a racemate. While the l-enantiomer is the biologically active species at the µ-receptor (is 50 times more potent than the d-enantiomer as an analgesic in man (Olsen et al., 1977)), both enantiomers are noncompetitive antagonists at the NMDA receptor with an activity similar to that of dextromethorphan (Ebert et al., 1995; Gorman et al., 1997) and therefore would be expected to block the phenomenon of windup (Shimoyama et al., 1997). This is another example where the administration of the racemate may prove to be superior to the opioid active enantiomer because the commonly believed "inactive" enantiomer interacts with a complementary system to possibly produce an enhanced outcome. Thus, some clinicians would suggest that methadone is relatively underulitized in the treatment of both cancer and chronic pain. Nevertheless, methadone is different from other μ-receptor agonists and the novice prescriber can easily get into difficulties of either underdosing (leading to inadequate analgesia) or overdosing (resulting in toxicity) because of the highly variable terminal half-life. It has been suggested that the conversion ratio of oral opioids to methadone at steady state is different (much lower) than the usually accepted acute potency ratios would indicate (Bruera and Neumann, 1999). The clear message to physicians contemplating using methadone is to prescribe it regularly to become familiar with its nuances and they will then find it a valuable addition to their opioid repertoire.

Pethidine (meperidine) use should be consciously minimized as it has no unique advantages and is liable for significant abuse by chronic pain patients. Further, acute pain studies indicate that pethidine appears to be less efficacious than morphine following optimized PCA administration (Plummer, 1999) and can be neurotoxic when administered regularly, particularly to patients with renal insufficiency (vide infra, Role of Metabolites).

Hydromorphone is approximately 5 times more potent than morphine, has a lower clearance, and no metabolites that are thought to have analgesic or antagonist activity (Table IV). Uncontrolled studies suggest some patients experience improved pain relief with hydromorphone. Buprenorphine is usually administered as a sub-lingual formulation in chronic pain patients, but is associated with a high incidence of nausea and vomiting in ambulant patients. Buprenorphine is a partial μ-receptor agonist that binds avidly to the receptor and is not readily displaced by naloxone, indicating that other supportive measures may be necessary in overdose situations.

### Corticosteroids

Corticosteroids (prednisone, prednisolone) have been prescribed for the treatment of a range of inflammatory conditions, sometimes for prolonged periods of time (e.g., treatment of rheumatoid arthritis) because of perceived analgesic effects. Nevertheless, a meta-analysis did not demonstrate any analgesic benefit, only changes in the usual rheumatological parameters (Saag et al., 1996). In cancer pain, steroids significantly improve patient well being by a somewhat diverse range of effects, both direct (e.g., reducing tumor mass) and indirect (e.g., increasing appetite, reducing nausea, etc). Thus, even low-dose corticosteroids have little place in the long-term management of chronic pain and only brief courses should be used when other therapeutic strategies have failed because of the risk of significant side effects (Cushing's syndrome, osteoporosis, diabetes mellitus, etc.).

### Local Anesthetics

Neuropathic pain can respond to the systemic and oral administration of local anesthetic-type drugs that exert their pharmacological effects by blocking sodium channels, but evidence from randomised clinical trials is lacking. Intravenous lignocaine infusions are used in an attempt to predict the success of oral therapy with sodium channel blockers (mexiletine and tocainide); for example, a positive correlation was demonstrated between a favorable outcome of a lignocaine infusion and subsequent oral therapy with mexiletine (Galer et al., 1996). There were no arrhythmias during the infusion (Kalso et al., 1998) and the duration of pain relief can last longer than the expected duration of the local anesthetic effect (up to 8 days in diabetic neuropathy with significant dysesthesia (Kastrup et al., 1987)). Oral mexiletine was effective in the treatment of diabetic neuropathy and peripheral nerve damage (Chabal et al., 1992), but not for dysesthetic pain following spinal cord injury. These drugs were ineffective in the treatment of similar presentations associated with cancer (Kalso et al., 1998). A gradual dose escalation protocol should be followed if these drugs are to be used to minimize the impact of side effects seen with mexiletine therapy.

### Other Agents

Clonidine, NMDA receptor antagonists, baclofen, and a number of topical preparations have been also used to treat pain, particularly if there is a neuropathic component to the presentation.

Clonidine (α2 adrenergic agonist) has been administered by a variety of routes (orally, transdermally, and intrathecally) but sedation limits its use.

Baclofen (a GABA_B a receptor agonist) has been used to treat spasticity in addition to neuropathic pain syndromes by both oral (60–80 mg per day in divided doses) and intrathecal (75–200 μg per day as an infusion) routes. Both sedation and nausea are common side effects.

NMDA receptor antagonists (ketamine and dextromethorphan) administered in combination with opioids have used to treat neuropathic pain (Cherry et al., 1995).

Topical capsaicin cream (Zhang and Li Wan Po, 1994), EMLA (eutectic mixture of local anaesthetics), or a high lignocaine concentration in a gel formulation (Robotham, 1994) have been used to treat neuropathic pains with a strong peripheral component such as postherpetic neuralgia (Rob-otham and Petersen, 2001) and diabetic neuropathy.

## Routes of Administration

The oral route of administration is overwhelmingly preferred in the treatment of both cancer and chronic noncancer pain and many practitioners regard morphine as the opioid drug of choice. The more traditional formulations of immediate release solution or tablets, while very effective if administered *round the clock* in adequate doses, have been relegated to second place in recent years by a variety of sustained-release morphine products (bioequivalence cannot necessarily be assumed) which can be administered either once (every 24 h) or twice daily (Gourlay, 1998; Gourlay *et al.*, 1987). The efficacy and convenience of these formulations makes them ideal for this indication. While most patients with chronic noncancer pain can be treated with doses of less than 100 mg per day and most at doses of less than 200 mg per day, some do require higher doses. For cancer pain, opioiphobia in medical and other related professions is still having an **inappropriate influence** on prescribing patterns; the balance between analgesia and side effects must be the main consideration. Fears of uncontrolled dependency problems and frank addiction (defined as a diverse range of maleficent behaviors aimed at obtaining opioid drugs at all costs) are not practical problems in the vast majority of pain patients. A percentage of patients will show apparent tolerance (i.e., a requirement for an increased opioid dose) which is most likely due to disease progression and may necessitate changing drugs or routes of administration.

A number of other routes are available including transdermal, spinal (intrathecal or epidural), subcutaneous infusions, rectal, and vaginal (although this is rarely used). These alternative routes are considered when patients can no longer obtain sufficient pain relief without unacceptable side effects, have significant nausea and vomiting poorly responsive to optimized antiemetic therapy, or have difficulty swallowing (e.g., dysphagia, psychological reasons). The frequency that these alternative routes of administration are used varies from country to country.

While the ease of implementation probably accounts for the popularity of subcutaneous infusions, avoidance of first-pass hepatic metabolism (associated with oral administration) has also been proposed as a reason for the change in route of administration. Certainly, there is little advantage of subcutaneous infusion over modern modified release oral formulations with respect to the lack of fluctuation in plasma opioid concentration–time curves. Although the morphine metabolite/morphine blood concentration ratios is smaller following subcutaneous infusion than with oral administration (as a consequence of the avoidance of first pass hepatic

metabolism), the relative proportions of the two glucuronide metabolites does not change between the two routes of administration (Faura *et al.*, 1998). Thus, from the pharmacological perspective, we would have to invoke sophisticated dose–response relationships for the various metabolites to successfully argue that subcutaneous infusions result in an improved pharmacodynamic response relative to oral administration. Physicians should be aware of the possibilities of drug incompatibilities and stability if more than one drug is added to syringe drivers.

Fentanyl has desirable physiochemical characteristics (lipophilicity and ionisation at physiological pH) that enable analgesic blood concentrations to be achieved following transdermal administration (Gourlay, 2001). The TTS–fentanyl (Durogesic) systems are used to treat both cancer and chronic pain, but not acute pain because of the inherent fentanyl pharmacokinetic properties of this formulation, namely a long lag phase (lasting 12–18 h in some patients) to effective fentanyl analgesic concentrations following the first dose and a long decay phase following patch removal (Gourlay, 2001). The slowly declining fentanyl concentrations associated with the longer-than-expected decay phase is particularly relevant if the patches are removed because of adverse events which means repeated naloxone doses, possibly a naloxone infusion or other supportive measure may be required. Equivalent analgesia and probably a lower incidence of constipation is associated with TTS–fentanyl administration compared to MS Contin (Ahmedzai and Brooks, 1997). Physicians placing patients on long-term opioid therapy should also pay attention to bowel hygiene.

Any of the opioids mentioned in this chapter could be incorporated into a suppository base and administered rectally/vaginally (Bruera *et al.*, 1995; Gourlay, 1998). The suppository base containing the opioid acts as a reservoir which usually results in a slower onset and perhaps a longer duration of pain relief than immediate-release oral formulations. Further, there can be differing rectal/oral dose ratios as rectally administered opioids can have a lesser extent of hepatic first-pass metabolism compared to equivalent oral doses, depending on the exact anatomical location of the suppository in the rectum (Gourlay, 1998). Therefore, conservative dose ratios should be used and careful dose titration implemented. Many of the oral tablet and surprisingly capsule sustained-release morphine formulations are effective if administered rectally (Gourlay, 1998). Two oral sustained-release morphine formulations (Kapanol, Skennan) consist of pellets contained in a capsule. These formulations can be administered as a sprinkle onto soft food such as ice cream and yogurt for patients who have difficulty in swallowing.

The spinal (i.e., epidural or intrathecal) route of administration has been used to administer opioid drugs in the belief of a more favorable balance between improved pain relief and the incidence and severity of adverse effects due to a selective spinal action. This approach is invasive and could be included in the *Invasive Therapy* section. However, the somewhat arbitrary subdivision involves techniques that are more likely to have significant neurological deficits for the *Invasive Therapies* section. While externalized catheters have been used, even for prolonged periods of time, it is more common to attach the spinal catheter to an implanted drug delivery device (Plummer *et al.*, 1991). This route of administration brings with it additional side effects of a technical as well as medical nature which tends to limit its use to specialized pain management units (Cherry and Gourlay, 1987; Gourlay, 1994). Physicians have a number of choices to make if they intend to use this route of administration including whether to site the catheter in the epidural or intrathecal spaces, the nature of the implanted drug delivery device (intermittent bolus doses, continuous infusion (Gourlay *et al.*, 1991), or both), which opioid will be administered (e.g., hydrophilic (morphine) or lipophilic opioids (fentanyl, sufentanil, of hydromorphone)), which also influences the precise dermatological location of the catheter tip (Cherry and Gourlay, 1987; Gourlay, 1994). Fibrosis around the catheter tip if it is located in the epidural space can lead to poor quality analgesia and intense pain on injection if intermittent bolus doses are administered (Cherry and Gourlay, 1992). This therapeutic modality has been shown to be effective in a range of chronic noncancer pain conditions (Cousins *et al.*, 1988).

### Role of Metabolism

Table IV shows that the metabolites of some opioids possess pharmacological activity, which can either be beneficial or result in significant adverse events. The metabolic conversions involve either cytochrome P450-catalzyed reactions (usually dealkylations and/or hydroxylations) or conjugations (usually with glucuronic acid). For morphine, this involves the formation of morphine-6-glucuronide (has potent analgesic activity) and morphine-3-glucuronide (inactive or perhaps a functional antagonist at other nonopioid receptors) by hepatic conjugation, although the amount of the 3-glucuronide exceeds the 6-glucuronide by a factor of 6–10 (Faura *et al.*, 1998).

Pethidine undergoes hepatic *N*-demethylation to nor-pethidine (nor-meperidine) which accumulates on repeated pethidine administration because of a lower intrinsic clearance. The metabolite causes *p*ethidine-*a*ssociated *n*eurotoxicity (PAN), which varies from unpleasant feelings through myoclonic jerks to grand mal seizures in its most fulminant presentation (Mather and Gourlay, 1984). Renal disease can exacerbate both the incidence and severity of PAN as the parent drug and metabolite undergo significant renal excretion. Pethidine causes the same extent of spasm of the Sphincter of Oddi as other commonly used opioids at analgesic doses although less so at subanalgesic doses (Ruskis, 1982).

A commonly held view is that codeine is a pro-drug and undergoes metabolic activation to morphine via cytochrome P450 isoform 2D6-catalyzed hepatic demethylation (Chen *et al.*, 1988; Eckhardt *et al.*, 1998). Polymorphic distribution of this isoform in Caucasians and the potential for reduced conversion to morphine by concomitantly administered drugs (also metabolised by 2D6) suggests that some patients will not experience adequate analgesia even when prescribed high codeine doses (Eckhardt *et al.*, 1998; Poulsen *et al.*, 1996; Sidrup *et al.*, 1991). It is probable that the same considerations apply to the conversion of oxycodone to oxymorphone.

The major determinant of methadone elimination is the amount of functional cytochrome P450 isoform 3A4 in the liver which means methadone metabolism can be either stimulated [phenytoin, carbamazepine, and rifampicin (Plummer *et al.*, 1988)] or inhibited [fluconazole (Cobb *et al.*, 1998), HIV—1 protease inhibitors (Iribarne *et al.*, 1998), and some SSRI antidepressants] by the concurrent administration of other drugs.

### Other Therapies

### Physical Therapies

These range from the application of heat and cold, TENS (transcutaneous electrical nerve stimulation), acupuncture, specific exercise, and/or graded exercise programs (both with identified goals to be attained).

The application of cold and heat are frequently used to treat acute and chronic pain, particularly of musculoskeletal origin. The popularity of these therapeutic interventions relates to the ease of implementation. Increased tissue metabolism and blood flow (resulting in improved tissue nutrition at a local level) together with possible interruption of the pain/spasm/pain cycle results from the application of heat, which can be subdivided into modalities providing heat to the superficial layers (hot packs, paraffin bath, hydrotherapy, or infrared therapy) or deeper structures (short waves, microwaves, or ultrasound therapy). The application of heat can change the elastic properties of connective tissue and consequently can be combined with other techniques (massage, stretching, and exercise) to alter tendon and muscle length to reduce the perception of pain. Care must be exercised to ensure the temperature is not too high nor the duration of therapy too long; otherwise a significant burn may occur. Other contraindications include application to areas of acute inflammation or trauma, patients with cardiovascular and respiratory insufficiency (particularly if large areas of the body are heated), patients taking anticoagulants or with bleeding disorders, and those patients with significant sensory impairment. The application of cold (via ice packs, vapor sprays, etc.) can have analgesic, anti-inflammatory and antipyretic effects due to vasoconstriction and reduced blood flow, decreased metabolic rate, a reduction in nerve conduction velocity and possibly the reflex secretion of endorphins. Contraindications to cold therapy principally include vascular insufficiency, cold sensitivity, and Reynaud's phenomenon.

TENS is widely used to treat many pain states including acute postoperative pain and labor pain. The popularity of this treatment in part relates to the low adverse event profile which includes cutaneous reactions to the gel as perhaps the most problematic event (frequently overcome by the use of hypoallergenic self-adhesive pads). Difficulty in applying the pads in the correct area (e.g., back pain in patients who have an isolated existence) and social issues (if the pads need to be applied to a visible part of the patient's anatomy) have been identified as minor problems. The Gate Control Theory (Melzack and Wall, 1965) suggests that excitation of large-diameter mylenated Aβ afferent fibers results in stimulation-induced analgesia involving inhibitory, possibly GABAergic, interneurones. Stimulation at high frequency (80–100 Hz) and low intensity recruits Aα/β fibers and results in analgesia or hypoanalgesia for the duration of the stimulation and for periods poststimulation in some patients. Brief periods of stimulation at low frequency (1–10 Hz) and higher intensity probably recruits Aδ fibers, can result in pain and discomfort, and is thought to explain electroacupuncture (Sandkuhler, 2000). A number of sessions are usually required and some patients' report prolonged analgesia. An evidenced-based approach to evaluate the efficacy of TENS in pain states suggests only a marginal effect and is complicated by the fact that most studies are not randomized, placebo TENS-controlled studies. It has proved difficult to have placebo TENS as the patients are clearly physically aware when the TENS unit is active and this is required for pain relief. Different stimulation patterns are sometimes used in an attempt to overcome this criticism.

The traditional form of acupuncture involves placing a series of needles at defined acupuncture points depending on the desired treatment, which are stimulated by either twisting, application of electric current or heat (moxibustion). It is assumed that this modality activates endogenous inhibitory systems. Acupuncture has gained widespread support in some centers despite a highly variable analgesic effect and the incorrect assumption that it is essentially a harmless technique. Deaths have occurred with acupuncture as a consequence of pneumothorax and infection although admittedly at a low rate when compared with the widespread use of this intervention on a global basis. The NNT value of 3 for **nonblinded** studies examining the use of acupuncture in the treatment of low back pain suggests this intervention is efficacious. However, the NNT for acupuncture drops to 13 when only **blinded** studies are included in the analysis, indicating acupuncture is almost ineffective. This shows the value of considering blinded studies when evaluating the efficacy of any intervention, particularly in highly subjective areas such as pain.

There are various assessments of physical performance that are frequently applied to patients with pain (Protas, 1999; Simmonds, 1999), which range from questionnaires, simple measures of endurance (e.g., time taken to walk a certain distance or conversely, distance walked over, say, 5 min), to sophisticated computer aided tests of physical capacity. These approaches must be applied within the framework of a pathophysiological and functional understanding of the pain complaint. Such measures can be very helpful in monitoring progress in rehabilitation programs and increasing range of movement and endurance. The aim is to increase muscle strength and movement control to provide support (particularly in back pain). It is very important that in the therapeutic sense, physiotherapy teaches the means by which the patient actively partakes in their rehabilitation (at home, work, or other situations) rather than numerous visits to the clinic where the patient is the passive recipient of prescribed treatment.

### Psychological and Psychiatric Therapies

The capacity to offer a range of psychological techniques is crucial to the success of treating chronic pain patients. Ideally, all chronic pain patients would have a comprehensive psychological and social assessment which would also involve an interview with the individual(s) closest to the patient. In some cases, a specific neuropsychological assessment can be very helpful in those patients with significant cognitive impairment. A detailed examination of the options in this area is beyond the scope of this text, but *inter alia* includes relaxation training and imagery, hypnosis, CBT (cognitive behavioral therapy), ego strengthening, problem-solving skills to maintain gains and help control setbacks, and sometimes assertiveness training and marital therapy. A systematic review and meta-analysis of either CBT or just behavioral therapy indicated that CBT was significantly more effective (compared to wait-list controls) on all measured domains including pain experience, depression and other mood effects, coping (positive and negative) behaviors, and social role functioning, while CBT produced significantly greater change for pain experience, positive coping measure, and reduced behavioral expression of pain compared to other alternative active treatments (Morley et al., 1999). These approaches can reduce pain (by minimizing anxiety and stress that can increase muscle tension that results in pain) as well as providing individualized coping strategies that may also be used at home and at work (if appropriate). Many of these options are included in *outpatient pain programs* which also aim to optimize drug therapy, increase mobility and promote return to work if appropriate. Some of these techniques are also very important in treating both acute and chronic pain in children, particularly recurrent acute pain frequently experienced by children with cancer (e.g., repeated lumbar punctures and chemotherapy). Psychiatric assessment of chronic pain patients can also be most helpful in providing a contextual basis of the impact of somatization, depression, prior abuse (both sexual [which can be high in female patients, particularly with poorly defined abdominal pain] and physical), psychotic features, dependency traits, and hypochondriasis.

### Invasive Therapies

The role of invasive neurodestructive techniques in contemporary pain management has been progressively diminishing with the increasing use of opioid drugs (particularly by the spinal route) to the extent that many neurodestructive techniques are now rarely performed. While specific comparative outcome studies of opioid therapy versus neurosurgical techniques are rare, opioid therapy is increasingly being favored because it avoids the obligatory neurological deficits associated with neurodestructive techniques. Nevertheless, improved outcome data collection which is occurring slowly will allow a more informed assessment of the role of neurodestructive techniques in the future. Clearly, this type of intervention specifically for the treatment of pain requires comprehensive preprocedure patient assessment along the lines used in multidisciplinary pain management units. Thus, these techniques are outside the expertise of all but the highly experienced specialist practitioner and indeed are rarely implemented by the majority of pain management units.

Diagnostic regional nerve blocks, usually performed with shorter acting local anesthetics (usually lignocaine) while longer acting agents (such as bupivacaine) are used for treatment, are also the province of specialist anesthetists. A combination of corticosteroids and local anesthetics are injected to treat painful trigger points. While regional sympathetic blocks, particularly with guanethidine and similar drugs to treat complex regional pain syndrome (Type 1) are still popular, evolving evidence casts considerable doubt on the efficacy of this technique. Neurolytic solutions such as 10% phenol in radio-opaque dye (for chemical lumbar sympathectomy) or absolute alcohol (for intrathecal alcohol injections) result in irreversible blocks, the selectivity of which depends on the skill of the practitioner in placing the caustic solution in the appropriate anatomical location and minimizing unwanted spread. Other than chemical lumbar sympathectomy (for increasing blood flow in cases of peripheral vascular disease), many of these neurolytic blocks are used very selectively to treat severe cancer pain resistant to optimized pharmacotherapy, usually in the terminal stages of the disease.

Patients with chronic noncancer pain are rarely good candidates for surgical ablative procedures because the pain can frequently reoccur for reasons that are not fully understood, but may *inter alia* include nerve regeneration and central reorganization. Tic douloureux is the exception to the rule where an ablative procedure produces meaningful improvements when medications have failed. Various neurectomy procedures have had variable popularity throughout the past century, but currently are **rarely** performed and are essentially of historical interest only (Loeser, 1999). Dorsal rhizotomy or dorsal root section is associated with a high incidence of deafferentation pain and has essentially been replaced by cordotomy and DREZ (Dorsal Root Entry Zone) lesioning.

Cordotomy or spinothalamic tractotomy can be performed surgically, but more commonly percutaneously (using a radio-frequency electrode) because the latter technique produces superior results in terms of pain relief and adverse events, is quick and can be offered on an outpatient basis or a brief inpatient stay to sicker patients who could not stand the stress of a major surgical procedure. Bilateral cordotomy is almost never performed now because of a high incidence of significant adverse events. Percutaneous cordotomy is restricted to cancer patients and the extent of pain relief can diminish over time, necessitating a repeat procedure. In comparison to the other neurodestructive techniques described here, the DREZ lesion is a relatively recent technique, first proposed by Nashold (Nashold and Ostdahl, 1979). The DREZ procedure involves creating as series of radio-frequency lesions in the dorsal root entry zone of the affected segments. DREZ lesion is most successful for brachial plexus avulsion (produces pain relief in approximately 2/3 patients), but other diagnoses such a phantom and stump pain, postherpetic neuralgia and failed back surgery have success rates of 50, 25, and almost 0%, respectively (Loeser, 1999).

Radio-frequency, glycerol, and balloon compression gangliolysis have all been used to treat tic douloureux with a high success rate that approaches 80% after 1 year for the radio-frequency technique.

A range of other neurosurgical techniques including tractotomies of the brainstem, thalamotomy, hypothalomotomy, cingulotomy, and neocortical procedures are now rarely done because of the unfavorable balance between the extent and duration of pain relief and significant adverse events and are primarily of historical interest. Electrical stimulation of the spinal cord (spinal cord stimulation, SCS) has been used to effectively treat neuropathic pain originating from some peripheral nerve lesions, spinal cord lesions, and spinal root or ganglion lesions (Meyerson and Linderoth, 1999). More recently, SCS has been heavily promoted to treat intractable angina (NYHA class III–IV) after more conventional therapies have failed to produce a satisfactory outcome, but further studies must be done to better characterize patient selection and whether the beneficial effects decrease over time. Centers that specialize in SCS emphasize the need for comprehensive preimplantation evaluation as typical patients usually have chronic multifactorial disease with pain as the major but not only reason for their incapacity (Meyerson and Linderoth, 1999). Trial stimulation with externalized leads is frequently performed prior to implantation of internalized systems because of the high cost of the latter and difficulties in predicting a positive outcome. While numerous reports suggest high success rates, the outcome can be highly variable for poorly understood reasons.

Intracerebral stimulation (ICS) of sensory thalamic nuclei (for some neuropathic pain states), periaqueductal–periventricular gray region (appears to be more selective for nociceptive form of pain) and the motor cortex (only for neuropathic pain) have been proposed. While ICS techniques have been used for some time, they must still be seen as experimental and only performed in centers with extensive experience. Thus, on a worldwide basis, these techniques are not readily available other than in highly specialized centers and therefore are included only for completeness. ICS stimulation of the sensory thalamus in principle targets the same patients considered for SCS and consequently the latter is frequently preferred. A recent meta-analysis indicated that long-term benefit of ICS was seen in between 20 and 80% of patients (Bendok and Levy, 1998). Motor cortex stimulation is essentially an experimental technique that may offer hope in the treatment of central pain states (e.g., poststroke pain) and trigeminal neuropathy (Nguyen et al., 1999).

## PRACTICAL MANAGEMENT

### Neuropathic Pain (Table V)

Neuropathic pain does not respond well to simple analgesics or NSAIDs. Treatment should start with tricyclic antidepressants. Patients who suffer from the sedative side effects can be treated with imipramine. Patients must be informed that side effects may arise immediately and generally improve with time, whereas efficacy develops more slowly. Anticonvulsants are more effective in pain that is triggered or evoked than in the burning and lancinating pain. The dose of carbamazepine should be increased in small steps, particularly in elderly patients. Opioids should be used in patients who do not respond to the first two treatment steps. Tricyclic antidepressant, anticonvulsants and opioids can be combined. Sustained release opioid formulations should be used to decrease the risk of tachyphylaxis and dependency.

### Phantom Pain

Preemptive analgesia to avoid phantom limb pain is effective in animal experiments but has failed so far in humans. Early occurring phantom limb pain can be

**TABLE V    Pragmatic Therapy of Neuropathic Pain**

| Priority | Class of substances | Example | Dose/day | Comments |
|---|---|---|---|---|
| 1 | Antidepressants | Amitriptyline | 25–150 mg | Slow titration, many side effects |
| | | Clomipramine | 25–150 mg | See amitriptyline |
| | | Imipramine | 25–150 mg | Less sedation than amitriptyline |
| 2 | Anticonvulsants | Carbamazepine | 600–1500 mg | Slow titration, best efficacy in trigeminal neuralgia |
| | | Gabapentin | 400–3600 mg | Good tolerability |
| 3 | Local anesthetics | Lidocaine | 5 mg/kg in 60 min IV | Limited efficacy |
| | | Mexiletine | 150 mg max 10 mg/kg | Many side effects |
| 4 | Opioids | See Table IV | See Table IV | Preferable to use sustained-release formulation of your chosen opioid if available, dose regularly |
| 5 | Topical agents | Capsaicin | 0.075% | |
| 6 | NMDA antagonists | Dextromethorphan | 300 mg | Many side effects, poor efficacy |
| 7 | Corticosteroids | Prednisone | 60 mg intrathecal | For chronic postherpetic neuralgia |

**TABLE VI**  Pragmatic Therapy of Complex Regional Pain Syndrome

| Priority | Class of substances/Procedure | Example | Comments |
|---|---|---|---|
| 1 | Physical therapy I | Immobilization/cold packs/contralateral therapy | Immobilization of limbs should be minimized—for severe pain only. Performed by physical therapist |
| 2 | Physical therapy II | Isometric/isotonic training | |
| 3 | Analgesics | NSAIDs | |
| 4 | Psychological therapy | Behavioral therapy | |
| 5 | Antidepressants | Amitriptyline Imipramine | Slow titration |
| 6 | Anticonvulsants | Carbamazepine | Slow titration |
| 7 | Opioids | See Table IV | Preferable to use sustained-release formulation of your chosen opioid if available, dose regularly |
| 8 | Local anesthetics/sympatholytic block | Lidocaine | Evidence suggests lack of effect |
| 9 | Corticosteroids | Prednisone | For short term treatment |

treated with calcitonin infusions (200 units). Triggered sharp pain is treated with either carbamazepine or gabapentin, whereas burning constant pain responds more favorably to tricyclic antidepressants or sustained-release opioids. Local capsaicin may reduce chronic stump pain. Almost all patients require psychological pain therapy including cognitive and behavioral therapy. Transcutaneous electrical nerve stimulation will improve the pain in few patients. The long-term effects of spinal cord stimulation are, in most cases, disappointing.

### Central Pain Following Stroke or Spinal Cord Injury

Central pain following stroke or spinal cord injury is basically treated like chronic neuropathic pain. Lamotrigine is effective in some patients with poststroke pain. Rarely do opioids have any effect in central pain states. In some cases spinal cord stimulation or deep brain stimulation may be indicated.

### Complex Regional Pain Syndrome (Table VI)

This entity has to be treated as early as possible to prevent chronification, where it becomes much more resistant to effective treatment. Pain in the acute phase is treated with NSAIDs and if the analgesic response is poor, sustained-release opioids can be prescribed. Neuropathic pain is treated with tricyclic antidepressants and/or anticonvulsants as previously described. If these procedures fail, sympatholytic therapy (at the level of the stellate ganglion) either by injection of local anesthetics or by small doses of opioids (morphine, 1–2 mg; fentanyl, 50–100 µg) is undertaken in some but by no means all pain units. Immobilization (for brief periods, rarely exceeding a few days and ONLY if the pain is

so severe to restrict active limb movement) and contralateral physical therapy are applied in the acute phase, followed by active isotonic training and sensory sensitasation. In the later phase active physical therapy is indicated. Psychological support, behavioral therapy and social support become more important the longer the pain persists.

## REFERENCES

Ahmedzai, S., and Brooks, D. (1997). Transdermal fentanyl versus sustained—release oral morphine in cancer pain: Efficacy, and quality of life. *J. Pain Symptom Manage.* **13**, 254–261.

Bendok, B., and Levy, R. M. (1998). Brain stimulation for persistent pain management. *In* "Textbook of Stereotactic and Functional Neurosurgery" (P. Gildenberg and R. Tasker, Eds.), pp. 1539–1546. McGraw-Hill, New York.

Besson, M. (1999). The neurobiology of chronic pain. *Lancet* **353**, 1610–1615.

Bruera, E., Watanabe, S., Fainsinger, R. L., Spachynski, K., Suarez-Almazor, M., and Inturrisi, C. (1995). Custom made capsules and suppositories of methadone for patients on high dose opioids for cancer pain. *Pain* **62**, 141–146.

Bruera, E., and Neumann, C. M. (1999). Opioid toxicities: Assessment and management. *In* " Pain 1999—An Updated Review" (M. Max, Ed.), pp. 443–457. IASP Press, Seattle.

Carlton, S. M., and Coggeshall, R. E. (1998). Nociceptive integration: Does it have a peripheral component? *Pain Forum* **7**, 71–78.

Chabal, C., Jacobson, L., Mariano, A., Chaney, E., and Britel, C. (1992). The use of oral mexiletine for the treatment of pain after peripheral nerve injury. *Anesthesiology* **77**, 513–517.

Chen, Z. R., Somogyi, A. A., and Bochner, F. (1988). Polymorphic-O-demethylation of codeine. *Lancet* **2**, 914–915.

Cherry, D. A., and Gourlay, G. K. (1987). The spinal administration of opioids in the treatment of acute and chronic pain: Bolus doses, continuous infusion, intraventricular administration and implanted drug delivery systems. *Palliative Med.* **1**, 89–106.

Cherry, D. A., and Gourlay, G. K. (1992). CT contrast evidence of injectate encapsulation after long-term epidural administration. *Pain* **49**, 369–371.

Cherry, D. A., Plummer, J. L., Gourlay, G. K., Coates, K. R., and Odgers, C. L. (1995). Ketamine as an adjunct to morphine in the treatment of pain. *Pain* 62, 119–121.

Cobb, M. N., Desai, J., Brown, L. S., Zannikos, P. N., and Rainey, P. M. (1998). The effect of fluconazole on the clinical pharmacokinetics of methadone. *Clin. Pharmacol. Ther.* 63, 655–662.

Coderre, T. J. (1993). The role of excitatory amino acid receptors and intracellular messengers in persistent nociception after tissue injury in rats. *Mol. Neurobiol.* 7, 229–246.

Cousins, M. J., Cherry, D. A., and Gourlay, G. K. (1988). Acute and chronic pain: Use of spinal opioids. *In* "Neural Blockade in Clinical Anaesthesia and Management of Pain" (M. J. Cousins and P. O. Bridenbaugh, Eds.), 2nd ed., pp. 955–1029. Lippincott, Philadelphia.

Derbyshire, S. W. G., Jones, A. K. P., Gyulai, F., Clark, S., Townsend, D., and Firestone, L. L. (1997). Pain processing at three levels of noxious stimulation produced differential patterns of central activity. *Pain* 73, 413–445.

Devor, M. (1994). The pathophysiology of damaged peripheral nerves. *In* "Textbook of Pain" (P. D. Wall and R. Melzack, Eds.), pp. 79–100. Churchill-Livingstone, London.

Dickenson, A. H. (1996). Pharmacology of pain transmission and control. *In* " Pain 1996—An Updated Review" (J. N. Cambell, Ed.), pp. 113–121. IASP Press, Seattle.

Ebert, B., Andersen, S., and Krogsgaard-Larsen, P. (1995). Ketobemidone, methadone and pethidine are non-competitive N-methyl-D-aspartate (NMDA) antagonists in the rat cortex and spinal cord. *Neurosci. Lett.* 187, 165–168.

Eckhardt, K., Li, S., Ammon, S., Schanzle, G., Mikus, G., and Eichelbaum, M. (1998). Same incidence of adverse drug events after codeine administration irrespective of the genetically determined differences in morphine formation. *Pain* 76, 27–33.

Eisenberg, E., Berkey, C., Carr, D. B., Mosteller, F., and Chalmers, T. C. (1994). Efficacy and safety of nonsteroidal antiinflammatory drugs for cancer pain: A meta-analysis. *J. Clin. Oncol.* 12, 2756–2765.

Faura, C. C., Collins, S. L., Moore, R. A., and McQuay, H. J. (1998). Systematic review of factors affecting the ratios of morphine and its major metablites. *Pain* 74, 43–53.

Galer, B. S., Harle, J., and Rowbotham, M. (1996). Response to intravenous lignocaine infusion predicts subsequent response to oral mexiletine: A prospective study. *J. Pain Symptom Manage.* 12, 161–167.

Gorman, A. L., Elliott, K. J., and Inturrisi, C. E. (1997). The d- and l-isomers of methadone bind to the non-competitive site on the N-methyl-D-aspartate (NMDA) receptor in rat forebrain and spinal cord. *Neurosci. Lett.* 223, 5–8.

Gourlay, G. K. (1994). Long-term use of opioids in chronic pain patients with non-terminal disease states. *Pain Rev.* 1, 45–59.

Gourlay, G. K. (1998). Sustained relief of chronic pain: Pharmacokinetics of SR morphine. *Clin. Pharmacokinet.* 35, 173–190.

Gourlay, G. K. (1999). Clinical pharmacology of the treatment of chronic non-cancer pain. *In* "Pain 1999—An Updated Review" (M. Max, Ed.), pp. 433–442. IASP Press, Seattle.

Gourlay, G. K. (2001). Treatment of cancer pain with transdermal fentanyl. *Lancet Oncol.* 2, 165–172.

Gourlay, G. K., Willis, R. J., and Wilson, P. R. (1984). Post-operative pain control with methadone: Influence of supplementary methadone doses and blood concentration—response relationships. *Anesthesiology* 61, 19–26.

Gourlay, G. K., Cherry, D. A., and Cousins, M. J. (1986). A comparative study of the efficacy and pharmacokinetics of oral methadone and morphine in the treatment of severe pain in patients with cancer. *Pain* 25, 297–312.

Gourlay, G. K., Plummer, J. L., Cherry, D. A., Onley, M. M., Parish, K. A., Wood, M. M., and Cousins, M. J. (1991). Comparison of intermittent bolus with continuous infusion of epidural morphine in the treatment of severe cancer pain. *Pain* 47, 135–140.

Gourlay, G. K., Cherry, D. A., Onley, M. M., Tordoff, S. G., Conn, D. A., Hood, G. M., and Plummer, J. L. (1997). Pharmacokinetics and pharmacodynamics of twenty-four-hourly Kapanol compared to twelve-hourly MS Contin in the treatment of severe cancer pain. *Pain* 69, 295–302.

Graziotti, P. J., and Goucke, C. R. (1996). The use of oral opioids in patients with chronic non-cancer pain: Management strategies. *Med. J. Aust.* 167, 30–34.

Iribarne, C., Berthou, F., Carlhant, D., Dreano, Y., Picart, D., Lohezic, F., and Riche, C. (1998). Inhibition of methadone and buprenorphine-N-demethylations by three HIV-1 protease inhibitors. *Drug Metab. Dispos.* 26, 257–260.

Jänig, W. (1996). The puzzle of "reflex sympathetic dystrophy": Mechanisms, hypotheses, open questions. *In* "Reflex Sympathetic Dystrophy: A Reappraisal" (W. Jänig and M. Stanton-Hicks, Eds.). Seattle, IASP Press.

Kalso, E., Tramer, M. R., McQuay, H. J., and Moore, R. A. (1998). Systematic local-anaesthetic-type drugs in chronic pain: A systematic review. *Eur. J. Pain* 2, 3–14.

Kastrup, J., Petersen, P., Dejgard, A., Angelo, H. R., and Hilsted, L. (1987). Intravenous lidocaine infusion—A new treatment of chronic painful diabetic neuropathy. *Pain* 28, 69–75.

Kehlet, H. (1999). Controlling acute pain—Role of pre-emptive analgesia, peripheral treatment, and balanced analgesia, and effects on outcome. *In* "Pain 1999—An Updated Review" (M. Max, Ed.), pp. 459–462. IASP Press, Seattle.

Kovelowski, C. J., Raffa, R. B., and Porreca, F. (1998). Tramadol and its enantiomers differentially suppress c-fos-like immunoreactivity in rat brain and spinal cord following acute noxious stimulus. *Eur. J. Pain* 2, 211–219.

Kurumbail, R. G., Stevens, A. M., Gierse, J. K., McDonald, J. J., Stegeman, R. A., Pak, J. Y., Gildehaus, D., Miyashiro, J. M., Penning, T. D., Siebert, K., Isakson, P. C., and Stallings, W. C. (1996). Structural basis for selective inhibition of cyclooxygenase-2 by anti-inflammatory agents. *Nature* 384, 644–648.

Leijon, G., and Boivie, L. (1989). Central post-stroke-pain—A controlled trial of amitriptyline and carbamazepine. *Pain* 36, 27–36.

Loeser, J. D. (1999). Ablative neurosurgery for pain. *In* " Pain 1999—An Updated Review" (M. Max, Ed.), pp. 255–267. IASP Press, Seattle.

MacIntyre, P. E., and Ready, L. B. (2001). "Acute Pain Management: A Practical Guide," 2nd ed. Saunders, London.

Mather, L. E., and Gourlay, G. K. (1984). The biotransformation of opioids: Significance for pain therapy. *In* "Opioid Agonist/Antagonist Drugs in Clinical Practice" (W. S. Nimmo and G. Smith, Eds.), pp. 31–47. Excerpta Medica, Amsterdam.

McQuay, H. J., Carroll, D., and Glynn, C. J. (1993). Dose–response for analgesic effect of amitriptyline in chronic pain. *Anaesthesia* 48, 281–285.

McQuay, H., Carroll, D., and Moore, A. (1995a). Variation in the placebo effect in randomised controlled trials of analgesics: All is as blind as it seems. *Pain* 64, 331–335.

McQuay, H., Carroll, D., Jadad, A. R., Wiffen, P., and Moore, A. (1995b). Anticonvulsant drugs for management of pain: A systematic review. *Br. Med. J.* 311, 1047–1052.

McQuay, H. J., Tramer, M., Nye, B. A., Carroll, D., Wiffen, P. J., and Moore, R. A. (1996). A systematic review of antidepressants in neuropathic pain. *Pain* 68, 217–227.

McQuay, H., and Moore, R. (1998a). Introduction. *In* "An Evidence—Based Resource for Pain Relief," pp. 1–3. Oxford University Press, Oxford.

McQuay, H., and Moore, R. (1998b). Injected morphine in post-operative pain. In "An Evidence—Based Resource for Pain Relief," pp. 118–126. Oxford University Press, Oxford.

McQuay, H., and Moore, R. (1998c). Oral ibuprofen and diclofenac in postoperative pain. In "An Evidence—Based Resource for Pain Relief," pp. 78–93. Oxford University Press, Oxford.

Melzack, R., and Wall, P. D. (1965). Pain mechanisms: A new theory. Science 150, 971–979.

Merskey, H., and Bogduk, N. (1994). "Classification of Chronic Pain: Descriptions of Chronic Pain Syndromes and Definitions of Pain Terms," 2nd ed. IASP Press, Seattle.

Meyerson, B. A., and Linderoth, B. (1999). Electric stimulation of the central nervous system. In " Pain 1999—An Updated Review" (M. Max, Ed.), pp. 269–280. IASP Press, Seattle.

Moore, R. A. (2000). Understanding clinical trials: What have we learned from systematic reviews. In "Progress in Pain Research and Management" (M. Devor, M. C. Robotham, and Z. Wiesenfeld-Hallin, Eds.), vol. 16, pp. 757–770. IASP Press, Seattle.

Moore, R. A., and McQuay, H. J. (1997). Single-patient data mata-analysis of 3453 postoperative patients: Tramadol versus placebo, codeine and combination analgesics. Pain 69, 287–294.

Moore, A., Collins, S., Carroll, D., and McQuay, H. (1997). Paracetamol with and without codeine in acute pain: A quantitative systematic review. Pain 70, 193–201.

Moore, R. A., Tramer, M. R., Carroll, D., Wiffen, P. J., and McQuay, H. J. (1998). Quantitative systematic review of topically applied non-steroidal anti-inflammatory drugs. Br. Med. J. 316, 333–338.

Morley, S., Eccleston, C., and Williams, A. (1999). Systematic review and meta-analysis of randomized controlled trials of cognitive behaviour therapy and behaviour therapy for chronic pain in adults, excluding headache. Pain 80, 1–13.

Nashold, B. S., and Ostdahl, R. H. (1979). Dorsal root entry zone lesions for pain relief. J. Neurosurg. 51, 59–69.

NH & MRC (National Health and Medical Research Council) (1999). "Acute Pain Management: The Scientific Evidence." Commonwealth of Australia, Canberra.

Nguyen, J. P., Lefaucher, J. P., Decq, P., Uchiyama, T., Carpentier, A., Fontaine, D., Brugieres, P., Pollin, B., Feve, A., Rostaing, S., Cesaro, P., and Keravel, Y. (1999). Chronic motor cortex stimulation in the treatment of central and neuropathic pain: Correlations between clinical, electrophysiological and anatomical data. Pain 82, 245–251.

Olsen, G. D., Wendel, H. A., Livermore, J. D., Leger, R. M., Lynn, R. K., and Gerber, N. (1977). Clinical effects and pharmacokinetics of racemic methadone and its optical isomers. Clin. Pharmacol. Ther. 21, 147–157.

Onghena, P., and Van Houdenhove, B. (1992). Antidepressant-induced analgesia in chronic non-malignant pain: A meta-analysis of 39 placebo-controlled studies. Pain 49, 205–219.

Plummer, J. L. (1999). Clinical pharmacology of acute pain: postoperative pain, acute myocardial infarction and migraine. In " Pain 1999—An Updated Review" (M. Max, Ed.), pp. 463–469. IASP Press, Seattle.

Plummer, J. L., Gourlay, G. K., Cherry, D. A., and Cousins M. J. (1988). Estimation of methadone clearance: Application in the management of cancer pain. Pain 33, 313–322.

Plummer, J. L., Cherry, D. A., Cousins, M. J., Gourlay, G. K., Onley, M. M., and Evans, K. H. A. (1991). Long term spinal administration of morphine in cancer and non-cancer pain: A retrospective survey. Pain 44, 215–220.

Pollo, A., Amanzio, M., Arslanian, A., Casadio, C., Maggi, G., and Benedetti, F. (2001). Response expectancies in placebo analgesia and their clinical relevance. Pain 93, 77–84.

Poulsen, L., Brosen, K., Arendt-Nielsen, L., Gram, L. F., Elback, K., and Sindrup, S. H. (1996). Codeine and morphine in extensive and poor metabolizers of sparteine: Pharmacokinetic, analgesic effect and side effects. Eur. J. Clin. Pharmacol. 51, 289–295.

Price, D. D. (2000). Factors that determine the magnitude and presence of placebo analgesia. In "Progress in Pain Research and Management" (M. Devor, M. C. Robotham, and Z. Wiesenfeld-Hallin, Eds.), vol. 16, pp. 1085–1095. IASP Press, Seattle.

Protas, E. J. (1999). Physical activity and low back pain. In " Pain 1999—An Updated Review" (M. Max, Ed.), pp. 145–151. IASP Press, Seattle.

Riedemann, P. J., Bersinic, S., Cuddy, L. J., Torrance, G. W., and Tugwell, P. X. (1993). A study to determine the efficacy and safety of tenoxicam versus piroxicam, diclofenac and indomethacin in patients with osteoarthritis: A meta-analysis. J. Rheumatol. 20, 2095–2103.

Robotham, M. (1994). Topical analgesic agents. In "Pharmacological Approaches to the Treatment of Chronic Pain: New Concepts and Critical Issues" (H. L. Fields and J. C. Liebeskind, Eds.), Progress in Pain Research and Management, vol. 1, pp. 211–227. IASP Press, Seattle.

Robotham, M. C., and Petersen, K. L. (2001). Zoster-associated pain and neural dysfunction. Pain 93, 1–5.

Robotham, M., Harden, N., Stacey, B., Bernstein, P., and Magnus-Miller, L. (1998). Gabapentin for the treatment of postherpetic neuralgia. JAMA 280, 1837–1842.

Roderick, P. J., Wilkes, H. C., and Meade, T. W. (1993). The gastrointestinal toxicity of aspirin: An overview of randomised clinical trials. Br. J. Clin. Pharmacol. 35, 219–226.

Ross, F. B., and Smith, M. T. (1997). The intrinsic antinociceptive effects of oxycodone appear to be κ-opioid receptor mediated. Pain 73, 151–157.

Ruskis, A. F. (1982). Effects of narcotics on gastrointestinal tract, liver and kidneys. In "Narcotic Analgesics in Anesthesiology" (L. M. Kitahata and J. G. Collins, Eds.), pp. 143–156. Williams & Wilkins, Baltimore.

Saag, K. G., Criswell, L. A., Sems, K. M., Nettleman, M. D., and Kolluri, S. (1996). Low-dose corticosteroids in rheumatoid arthritis: A meta-analysis of their moderate-term effectiveness. Arthritis Rheumatism 39, 1818–1825.

Sandkuhler, J. (2000). Long-lasting analgesia following TENS and acupuncture: Spinal mechanisms beyond gate control. In "Progress in Pain Research and Management" (M. Devor, M. C. Robotham, and Z. Wiesenfeld-Hallin, Eds.), vol. 16, pp. 359–369. IASP Press, Seattle.

Schug, S. A., Merry, A. F., and Acland, R. H. (1991). Treatment principles for the use of opioids in pain of non-malignant origin. Drugs 42, 228–32.

Shimoyama, N., Shimoyama, M., Elliott, K. J., and Inturrisi, C. E. (1997). d-Methadone is antinociceptive in the rat formalin test. J. Pharmacol. Exp. Ther. 283, 648–652.

Simmonds, M. (1999). Physical function and physical performance in patients with pain: What are the measures and what do they mean? In "Pain 1999—An Updated Review" (M. Max, Ed.), pp. 127–136. IASP Press, Seattle.

Sindrup, S. H., Brosen, K., Bjerring, P., Arendt-Nielsel, L., Larsen, U., Angelo, H. R., and Gram, L. F. (1991). Codeine increases pain thresholds to copper vapour laser stimuli in extensive but not poor metabolisers of sparteine. Clin. Pharmacol. Ther. 49, 686–693.

Stalnikowicz, R., and Rachmilewitz, D. (1993). NSAID-induced gastroduodenal damage: Is prevention needed. J. Clin. Gastroenterol. 17, 238–243.

Tölle, T. R., Kaufmann, T., Siessmeier, T., Lautenbacher, S., Berthele, A., Munz, F., Zieglgänsberger, W., Willoch, F., Schwaiger, M., Conrad, B., and Bartenstein, P. (1999). Region-specific encoding of sensory and affective components of pain in the human brain: A positron emission tomography correlation analysis. *Ann. Neurol.* **45**, 40–47.

Tramer, M. R., Williams, J. E., Carroll, D., Wiffen, P. J., McQuay, H. J., and Moore, R. A. (1998). Comparing analgesic efficacy of non-steroidal anti-inflammatory drugs given by different routes in acute and chronic pain: A qualitative systematic review. *Acta. Anaesthesiol. Scand.* **42**, 71–79.

Young, J. M., Panah, S., Satchawatcharaphong, C., and Cheung, P. S. (1996). Human whole blood assays for inhibition of prostaglandin G/H synthases-1 and -2 using A23187 and lipopolysaccharide stimulation of thromboxane $B_2$ production. *Inflamm. Res.* **45**, 246–253.

Zhang, W. Y., and Li Wan Po, A. (1994). The effectiveness of topically applied capsaicin: A meta-analysis. *Eur. J. Clin. Pharmacol.* **46**, 517–522.

*Cranial Nerves and Brain Stem*

# CHAPTER 12
# Optic Nerve Lesions

Valérie Biousse and Nancy J. Newman

Damage to an optic nerve classically causes an abnormality in visual sensory function, a relative afferent pupillary defect (if the optic nerve disease is unilateral or asymmetric), and a change in the appearance of the optic nerve. Acquired optic neuropathies can produce any kind of visual field defect, including a central scotoma, cecocentral scotoma, arcuate visual field defect, altitudinal defect, or even a temporal or nasal hemianopic defect. Unless the optic neuropathy is bilateral or the lesion is located near the chiasm, the field defect produced by an optic nerve lesion is always monocular. The visual field defects that occur with various optic neuropathies are not themselves localizing. Rather, it is the pattern of visual loss (rapid vs slow onset, progressive vs stable) and the presence or absence of specific ocular or neurologic signs (relative afferent pupillary defect, acquired color deficit, optic disc swelling, optic disc pallor, optociliary shunt veins, proptosis, ocular motor pareses) that most often allow the examiner to diagnose an optic neuropathy, to localize the pathology along the course of the optic nerve, and to determine its etiology (Table I). The optic disc has only two ways to respond to the many acquired pathologic processes that may affect the optic nerve. It can swell (anterior optic neuropathy) (Table II), or it can remain normal in appearance (posterior or retrobulbar optic neuropathy). If the pathologic process causes irreversible damage to the optic nerve, the disc will eventually become pale 4 to 6 weeks later. Disorders of optic nerve function can result from a wide variety of pathologic processes (Table I). Optic neuropathies are best approached by considering the category of disease and the clinical entities therein (Table I). Once a lesion has been localized to the optic nerve, the next step involves differentiating among possible causes. While at times challenging, this step is critical, since different etiologies carry distinct therapeutic and prognostic implications. While the complete neuro-ophthalmic examination may yield diagnostic clues, historical

factors often provide the most useful information. For example, an acute optic neuropathy in a healthy, 25-year-old white woman will most often be idiopathic demyelinating optic neuritis, while the same presentation in a 60-year-old, hypertensive man will most likely be anterior ischemic optic neuropathy (AION). Examination features in both cases, however, may be remarkably similar (at least initially). Therefore, the tempo of the optic nerve dysfunction (acute, subacute, progressive, etc.), the presence or absence of pain, associated neurologic symptoms and signs, family history, the appearance of the optic disc, and other factors may narrow the differential diagnosis considerably (Table III) (Newman, 1996).

## OPTIC NEURITIS

### Clinical Aspects

Optic neuritis is an inflammatory disorder of the optic nerve that typically occurs in young adults. It is characterized by a subacute, painful loss of central vision that may progress for 7 to 10 days. Visual acuity varies from a mild reduction to severe loss. Pain is usually exacerbated by eye movement and may precede or coincide with visual loss. Color vision is usually impaired out of proportion to visual acuity. The classic visual field defect has been said to be a central scotoma; however, almost any type of visual field defect is possible. The term retrobulbar optic neuritis is applied when the optic nerve appears normal in the acute phase; anterior optic neuritis or papillitis is used to refer to those cases with optic disc swelling. In both cases, temporal pallor of the disc develops 4 to 6 weeks after visual loss. There are multiple causes of inflammatory optic neuritis, including infectious diseases such as syphilis, cat scratch disease, and Lyme disease, and noninfectious inflammation such as sarcoidosis (see below). However, in most cases optic

**TABLE I    The Differential Diagnosis of an Optic Neuropathy**

Inflammatory
    Idiopathic inflammatory optic neuritis (associated with multiple
        sclerosis)
    Systemic inflammatory diseases
    Infectious diseases
Vascular
    Nonarteritic anterior ischemic optic neuropathy
    Nonarteritic posterior ischemic optic neuropathy
    Arteritic anterior ischemic optic neuropathy
    Arteritic posterior ischemic optic neuropathy
Compressive/infiltrative
Paraneoplastic
Toxic/nutritional
Hereditary
Traumatic
Raised intracranial pressure (papilledema)
Glaucoma
Congenitally anomalous

**TABLE II    The Differential Diagnosis of the "Swollen Disc"**

Disc elevation without true swelling
    Optic disc anomalies
        Drusen
        Tilted disc
        Crowded disc
    Optic disc infiltration
    Leber's hereditary optic neuropathy
True disc swelling
    Elevated intracranial pressure (papilledema)
    Inflammatory optic neuropathy
        Demyelinating
        Sarcoidosis
        Infectious
    Vascular optic neuropathy
        Anterior ischemic optic neuropathy
            Nonarteritic
            Arteritic
        Central retinal vein occlusion
    Compressive optic neuropathy
        Neoplastic
            Meningioma
            Hemangioma
        Non-neoplastic
            Thyroid ophthalmopathy
            Orbital inflammatory pseudotumor
    Infiltrative optic neuropathy
        Neoplastic
            Leukemia
            Lymphoma
            Glioma
        Non-neoplastic
            Sarcoidosis
    Toxic/Metabolic/Nutritional deficiencies
    Traumatic optic neuropathy
    Intraocular hypotony

neuritis remains idiopathic or is associated with multiple sclerosis (Table IV). Optic neuritis may also be associated with other primary demyelinating diseases such as **Devic' s disease (neuromyelitis optica)**. Patients with Devic's disease typically present with acute or subacute visual loss in one or both eyes caused by acute optic neuropathy preceded or followed within days or weeks by a transverse or ascending myelopathy. Some authors believe that Devic's disease and multiple sclerosis are variants of the same disease. However, there is no specific treatment for Devic's disease and its prognosis is usually poor. In a small percentage of cases, inflammatory optic neuritis is not associated with a primary demyelinating process in the optic nerve or the central nervous system. Instead, the condition develops in the setting or as the presenting manifestation of an underlying or **systemic infection, vaccination, or systemic inflammatory disease** (Table V). Associated ocular findings such as neuroretinitis (see below), systemic symptoms and signs, and patients' characteristics should direct subsequent evaluation in the appropriate direction. **Neuroretinitis** is a form of optic neuritis characterized ophthalmoscopically by optic disc swelling associated with a macular star figure composed of lipid. This form of optic neuritis is almost never caused by demyelination and occurs most often in the setting of cat-scratch disease or in association with other systemic infectious diseases such as Lyme disease, syphilis, toxoplasmosis, and tuberculosis, as well as with sarcoidosis.

### Natural Course

**Idiopathic demyelinating optic neuritis** is the most common acute optic neuropathy in people under the age of 45. At least two-thirds of patients are women.

Patients typically improve over several weeks, regardless of treatment. The risk of subsequent development of multiple sclerosis after an isolated attack of idiopathic optic neuritis has been estimated as high as 75% at 15 years (Newman, 1996). The Optic Neuritis Treatment Trial (ONTT), a large, multicenter trial designed to evaluate the effects of corticosteroid treatment of acute idiopathic optic neuritis, has provided important information regarding the natural history, prognosis, and treatment of this disorder (Beck *et al.*, 1995; The Optic Neuritis Study Group, 1997). A total of 457 patients between the ages of 18 and 45 years with acute unilateral optic neuritis were examined within 8 days of the onset of visual symptoms. Seventy-five percent of the patients were women, and the mean age was 32 years. Pain, typically exacerbated by eye movement, was reported by 92% of patients. The optic disc was swollen in 35% of cases and normal in 65%. Visual field defects in the central 30° were found frequently, and their configuration included arcuate and altitudinal defects as well as true central scotomas.

**TABLE III**  Clinical Characteristics of Common Optic Neuropathies

| | | Optic neuritis | Aion | Compressive/ Infiltrative | Toxic/Nutritional | Hereditary | Papilledema |
|---|---|---|---|---|---|---|---|
| Age of patients | | Younger | Older (>50 years) | 30–40: meningioma Childhood: glioma | Any age | Younger | Any age |
| Laterality | | Unilateral | Unilateral | Unilateral | Bilateral | Bilateral | Bilateral |
| Visual loss | | Rapidly progressive Acuity rarely spared | Acute Acuity variable | Progressive | Slowly progressive | Subacute/Slowly progressive | Vision preserved until late |
| Pain | | Orbital pain frequent with eye movements | Pain infrequent (except in GCA) | Absent | Absent | Absent | Headache (raised intra-cranial pressure) |
| Color vision | | Commonly abnormal | Commonly spared | Abnormal | Affected early | Abnormal | Preserved until late |
| Visual field | | Central defects | Altitudinal defect | Variable | Ceco-central scotoma | Ceco-central scotoma | Peripheral constriction |
| Optic disc | Acute | Normal (2/3) or disc edema (1/3) | Disc edema, ± segmental Small cup to disc ratio | Variable | Hyperemic | Pseudo-edema in LHON | Disc edema |
| | Late | Temporal pallor | Segmental pallor | Pale | Pale | Pale | Pale with peripapillary changes |
| Visual prognosis | | Good | Variable 40% risk for the other eye within 5 years | Variable | May improve | Poor | Poor if not treated early |
| Systemic diseases | | Risk of development of multiple sclerosis | HTN (51%), DM (24%) GCA to be ruled out | Neurofibromatosis Malignancy | Poor nutrition Peripheral neuropathy | Mitochondrial diseases DIDMOAD syndrome | Any cause of raised intra-cranial pressure |

*Note.* GCA, giant cell arteritis; HTN, hypertension; DM, diabetes mellitus; LHON, Leber's hereditary optic neuropathy; DIDMOAD, diabetes insipidus, diabetes mellitus, optic atrophy, and deafness.

TABLE IV    Recommendations for the Management of a Patient with Idiopathic Optic Neuritis

1) Consider brain MRI to assess the risk of future neurologic events of multiple sclerosis.
2) Investigations such as chest X-ray, blood tests and lumbar puncture are not necessary in evaluating patients with typical clinical features of optic neuritis (young adult with sudden visual loss with progression of symptoms of 1 week or less accompanied by pain on eye movements, with either a swollen or normal optic disc but no more than a minimal vitreous cellular reaction, with no history of systemic disease that might produce optic neuritis), and with visual improvement beginning within 1 month.
3) Treatment with oral prednisone should be avoided.
4) Treatment with intravenous methylprednisolone should be considered if brain MRI demonstrates multiple signal abnormalities consistent with multiple sclerosis, or if the patient wants to recover vision faster (severe visual loss, bilateral optic neuritis).
5) Treatment with interferon Beta-1a should be considered in patients with at least two lesions suggestive of demyelinating disease on brain MRI, in order to reduce the rate of subsequent development of clinically definite multiple sclerosis.

TABLE V    Differential Diagnosis of Inflammatory Optic Neuritis

Optic neuritis associated with demyelinating disease
  Idiopathic optic neuritis
  Multiple sclerosis
  Devic's disease
  Acute disseminated encephalomyelitis (ADEM)
Optic neuritis associated with infectious diseases
  Bacterial infections: syphilis, cat scratch disease (Bartonella henselae), Lyme disease, any bacterial meningitis, mycoplasma pneumoniae, tuberculosis, Whipple's disease
  Viral infections: herpes zoster, herpes simplex, HIV, Epstein-Barr virus, coxsackie virus, adenovirus, cytomegalovirus, hepatitis A and B virus, measles, mumps, rubeola, rubella virus
  Parasitic infections: toxoplasmosis, cysticercosis, toxocariasis, intraocular nematode infection
  Fungal infections: cryptococcosis, aspergillosis, mucormycosis, candidosis, histoplasmosis
  Bee and wasp sting
Postvaccination optic neuritis
  Hepatitis B virus, rabies virus, tetanus toxoid, variola virus, combined smallpox, tetanus, and diphteria vaccine, combined measles, mumps, and rubella vaccine, influenza vaccine, BCG
Optic neuritis associated with other inflammatory disorders
  Sarcoidosis
  Systemic lupus erythematosus
  Polyarteritis nodosa
  Wegener's granulomatosus
  Inflammatory bowel disease
  Other vasculitides

Initial diagnostic evaluation, including ANA and FTA blood tests, chest X-ray, brain magnetic resonance imaging (MRI) and cerebrospinal fluid analysis, rarely revealed an alternative diagnosis, leading the study group to conclude that these tests are not necessary diagnostically in a patient with the typical features of optic neuritis. However, the ONTT found that brain MRI was a powerful predictor of the subsequent risk of multiple sclerosis. Indeed, patients with three or more lesions on MRI had a 5-year risk of multiple sclerosis of 51%, whereas patients with a normal MRI had a risk of 16% (The Optic Neuritis Study Group, 1997). Among those patients with abnormal MRI, other less powerful predicting factors for the development of multiple sclerosis among ONTT patients included prior nonspecific neurologic symptoms and a history of optic neuropathy in the fellow eye (previous optic neuropathy in the study eye was an exclusion criteria and therefore could not be evaluated). Among those patients with normal baseline MRI and no history of neurologic symptoms, positive predictive factors for the development of multiple sclerosis included Caucasian race, family history, female gender, retrobulbar location, pain, and a preceding viral illness. Lack of pain, the presence of optic disc swelling, and mild visual loss were features of the optic neuritis associated with a low risk of multiple sclerosis in patients with normal baseline MRI.

## Practical Management

A practice parameter issued by the American Academy of Neurology provides recommendations regarding the treatment of idiopathic optic neuritis (Kaufman *et al.*, 2000). Patients in the ONTT were randomized to three treatment arms: (1) oral prednisone (1 mg/kg/day) for 14 days; (2) intravenous methylprednisolone (250 mg, 4 times daily) for 3 days, followed by oral prednisone (1 mg/kg/day) for 11 days; (3) oral placebo for 14 days. Compared with the placebo and the oral prednisone regimens, the intravenous therapy provided a more rapid recovery of vision but no long-term benefit. After 1-year follow-up, there were no significant differences among the groups in visual acuity, contrast sensitivity, color vision, or visual field. The regimen of oral prednisone alone not only provided no benefit to vision, but also was associated with an increased rate of new attacks of optic neuritis in both the initially affected and the fellow eyes. Within the first 2 years of follow-up, new attacks of optic neuritis in either eye occurred in 30% of the patients in the oral prednisone group and 14% in the intravenous group.

Visual recovery from optic neuritis, including visual acuity and visual fields, was rapid and occurred within the first month in nearly all patients. At 6 months, 94% of all patients had vision of 20/40 or better and 75% had improved to 20/20 or better. Therefore, the absence of visual improvement within the first month after the onset of treated and untreated optic neuritis, or the worsening of visual symptoms after termination of a course of steroids, should be considered atypical and warranting of further diagnostic investigations. The

ONTT also found that the group receiving the intravenous regimen had a lower rate of development of multiple sclerosis within the first 2 years (7.5%) than did the placebo (16.5%) or oral prednisone (14.7%) groups (Beck *et al.*, 1995). However, this protective effect was no longer appreciable after 3 years: 17.3% of patients treated with intravenous steroids and 20.7% treated with placebo had developed multiple sclerosis (The Optic Neuritis Study Group, 1997).

The recently published CHAMPS trial (Controlled High Risk Subject Avonex Multiple Sclerosis Prevention Study) has suggested that interferon Beta-1a should be initiated immediately following a first episode of idiopathic optic neuritis accompanied by abnormal brain MRI (Jacobs *et al.*, 2000). A total of 383 patients with a first demyelinating event, including 100 patients with a first isolated idiopathic optic neuritis, and at least two clinically silent T2 lesions on the brain MRI were included. All patients received intravenous methylprednisolone according to the ONTT protocol, within 8 days after the visual loss. Patients were randomized to either interferon Beta-1a (30 µg IM once weekly) or to placebo. In this study, interferon Beta-1a treatment of patients with acute idiopathic demyelinating events such as optic neuritis reduced the rate of developing definite multiple sclerosis by 43%. This effect was observed within 6 months and sustained throughout the study (34 months).

Based on the results noted above, several recommendations can be made for the clinician managing patients with acute optic neuritis (Table IV). It is important to remember that these recommendations are only applicable to patients with typical acute optic neuritis seen early in the course of their disease.

# ANTERIOR ISCHEMIC OPTIC NEUROPATHY

## Clinical Aspects

Anterior ischemic optic neuropathy is the most common cause of acute optic nerve disease in the elderly. It results from acute ischemia to the anterior portion of the optic nerve, whose main source of blood supply is from the posterior ciliary circulation. It is characterized by an acute, painless monocular loss of vision that may progress over several hours or days. Examination shows a relative afferent pupillary defect and optic disc swelling, frequently with peripapillary hemorrhages. Gradually the optic disc develops pallor and the edema resolves. The typical visual field defect is altitudinal, especially inferiorly (Hayreh, 1996). Although AION has been associated with a variety of systemic diseases (Hayreh *et al.*, 1994), the most important differential is between arteritic AION, caused by giant cell arteritis (GCA), and nonarteritic AION. Patients with GCA and AION are in danger of catastrophic, irreversible, bilateral total blindness that may be prevented by prompt intervention with corticosteroid therapy. **Diabetic papillopathy** is an atypical form of AION that usually occurs in young patients with insulin-dependent diabetes. It is distinguished from typical nonarteritic AION by the slight degree (or even absence) of visual loss, the frequency of bilateral involvement (50%), and the good visual outcome.

## Natural Course of Arteritic Anterior Ischemic Optic Neuropathy

AION is the most common ophthalmic manifestation of GCA (Gordon and Levin, 1998; Hayreh *et al.*, 1998). Arteritic AION occurs in older patients, typically in their 70s and 80s. It is usually associated with systemic symptoms such as headache, scalp tenderness, jaw claudication, polymyalgia rheumatica, fatigue, and weight loss; however, it can be the only manifestation of the disease in about 25% of patients. Visual loss in arteritic AION is usually severe, with acuities reduced to no light perception, light perception only or hand motion, and may be preceded by recurrent episodes of transient monocular visual loss or transient diplopia. It is often bilateral and can be associated with a concurrent retinal or choroidal infarction. Elevation of the erythrocyte sedimentation rate (ESR) and C-reactive protein (CRP) is highly suggestive of the disease and the diagnosis is proven by finding granulomatous inflammation with giant cells on biopsy of a superficial temporal artery. In the absence of treatment, vision deteriorates and there is a high risk of involvement of the second eye within days or weeks.

## Practical Management of Arteritic Anterior Ischemic Optic Neuropathy

Arteritic AION requires emergency treatment to prevent complete blindness. Systemic corticosteroid therapy should be instituted immediately upon presumed diagnosis and should not be delayed awaiting results of the ESR or temporal artery biopsy. Temporal artery biopsies remain positive even after 2 weeks and probably up to 7 weeks of steroid therapy. The recommended dose for use of oral prednisone after visual loss has occurred in GCA has been in the range of 1 to 2 mg/kg/day for 2 weeks, with gradual tapering over the successive 6 to 18 months. Most neuro-ophthalmologists use large doses of intravenous methylprednisolone (1 to 2 g/day for 3 to 5 days) as routine first therapy for patients who have acute visual loss from GCA, especially when recurrent transient visual loss, complete loss of vision in one eye, or early signs of second eye involvement are apparent. Response of systemic symptoms to steroids is usually rapid and dramatic, with relief of headache and malaise within 24 h. Unfortunately, the chance of recovery of useful vision is poor, especially when there has been loss of light perception, although rare cases of visual recovery have been reported in patients with aggressive intravenous therapy.

## Natural History of Nonarteritic Anterior Ischemic Optic Neuropathy

Nonarteritic AION occurs predominantly in white patients in their 50s and 60s. About 6000 new cases occur annually in the United States. A large multicenter trial, the Ischemic Optic Neuropathy Decompression Trial (IONDT) (1995), has provided important information about nonarteritic AION and its natural history. Compared with patients with arteritic AION, patients with nonarteritic AION may have relatively well-preserved visual acuity, with up to 56% of patients retaining acuity of 20/60 or better. Recurrences in the same eye are rare, but subsequent involvement of the fellow eye is substantial, estimated at about 20% at 5 years. Although nonarteritic AION results from vascular occlusive disease of small vessels supplying the anterior portion of the optic nerve, its exact cause remains unclear. Anatomical factors such as a congenitally small and crowded optic nerve head with a small cup-to-disc ratio (so-called "disc at risk") may contribute to the vascular event (Hayreh, 1996). As a disease of the small vessels, nonarteritic AION is not associated with ipsilateral internal carotid artery stenosis and embolic AION is extremely rare (Hayreh, 1996). In the IONDT, 60% of patients had one or more risk factors thought to be associated with small vessel cerebrovascular disease, including hypertension, diabetes, and cigarette use (IONDT, 1995). Nonarteritic AION may also be precipitated by hypotension or blood loss (Table VI) (Hayreh, 1996; Hayreh *et al.*, 1994).

## Practical Management of Arteritic Anterior Ischemic Optic Neuropathy

There is no established treatment for nonarteritic AION. The clinician's primary role in managing patients with nonarteritic AION is to exclude GCA and to control other factors that might affect the final visual outcome and the subsequent involvement of the second eye. However, as suggested by Hayreh (1994, 1996), the treatment of arterial hypertension with its risk of nocturnal arterial hypotension and iatrogenic AION,

**TABLE VI**    Causes of Anterior Ischemic Neuropathy (AION)

Arteritic AION
  Giant cell arteritis
  Systemic vasculitis other than giant cell arteritis
    Systemic lupus erythematosus
    Peri-arteritis nodosa
    Churg-Strauss syndrome
Nonarteritic AION
  Small crowded disc ("disc at risk")
  Other abnormalities localized to the disc
    Drusen
    Anomalous disc
    Severe papilledema
  Hypotension
    Operative (especially spinal and cardiac surgery)
    Systemic hemorrhage
    Cardiac arrest
    Renal dialysis
  Anemia
  Hypercoagulable disorders
  Radiation optic neuropathy
  Acute intraocular hypertension (during ocular surgery)

should be progressive and moderate. Antiplatelet agents such as aspirin are often prescribed for patients with nonarteritic AION, especially to prevent second eye involvement, but a prospective trial of aspirin in AION has yet to be performed. Other treatments have been tried in patients with nonarteritic AION, including anticoagulants, vasodilators, vasopressors, phenytoin, L-Dopa, corticosteroids, hyperbaric oxygen, and optic nerve sheath decompression surgery. However, all these treatments have proven unsucessful or remain of unproven benefit (IONDT, 1995).

## TOXIC/NUTRITIONAL OPTIC NEUROPATHY

### Clinical Aspects

Toxic and nutritional optic neuropathies generally have similar clinical features, and may often coexist in the same patient. Characteristic clinical features include progressive, symmetrical central visual loss, reduced perception of color, cecocentral scotomas, and ultimately temporal optic disc pallor corresponding to loss of nerve fibers in the papillomacular bundle (Lessell, 1998). There may be a similar underlying mechanism of optic nerve injury in these "metabolic" optic neuropathies, but the pathogenesis remains unknown. Causation in many cases is likely multifactorial. In humans, definitively proven cases of optic nerve damage caused by a single recognized toxic agent or a deficiency in a single identified nutrient are rare. Exceptions include the toxic optic neuropathies associated with medications such as ethambutol, although other factors must influence expression as not all patients treated with even high dosages of these medications will suffer optic nerve damage (Table VII). Perhaps most confusing as regards the overlap of toxic and nutritional optic neuropathies is so-called tobacco–alcohol amblyopia, an often invoked but poorly understood disease. Invoked etiologies include the toxic effects of tobacco smoking alone, the combined effects of tobacco and alcohol, nutritional deficiency, and combined toxic and nutritional influences. Historically, it is particularly likely to occur in adult men who smoke cigars or pipes. Cessation of smoking can result in recovery of vision, as can supplementation with B-complex vitamins, especially hydroxocobalamin, despite continued tobacco use. Nutritional inadequacies may contribute, since B vitamins and sulfur-containing amino acids are essential in the detoxification of cyanide, and tobacco use is associated with lower systemic levels of vitamin $B_{12}$ and folate (Lessell, 1998). Nutritional deficiency or imbalance as a cause of optic neuropathy is best supported by the experimental and clinical data regarding disorders of vitamin $B_{12}$ metabolism. Other vitamins proposed as etiological candidates for nutritional optic neuropathy include thiamin, niacin, and folate. However, a specific vitamin deficiency is rarely identified and a combination of relative deficiencies may contribute to optic nerve dysfunction.

# COMPRESSIVE/INFILTRATIVE OPTIC NEUROPATHY

Compressive lesions may affect the intracranial, prechiasmal optic nerve or the intraorbital optic nerve (Table VIII) (Shults, 1998). Intracranial, intracanalicular, and posterior orbital compressive lesions typically do not produce disc swelling.

**TABLE VII   Suggested Causes of Toxic/Nutritional Optic Neuropathies**

Toxins
  ? Alcohol
  Methanol
  Tobacco
  Ethylene Glycol
  Lead
  Organic solvents
  Amiodarone
  Disulfiram
  Ethambutol
Nutritional
  Vitamin B12 deficiency
  ? Thiamin, niacin, and folate deficiency

# HEREDITARY OPTIC NEUROPATHIES

## Leber's Hereditary Optic Neuropathy

### Clinical Aspects

LHON is a maternally inherited, bilaterally sequential optic neuropathy that occurs predominantly in otherwise healthy young men. Visual acuity typically deteriorates permanently to levels of 20/200 or worse, although spontaneous recovery of vision can occur even years later. Visual fields show central or cecocentral defects. During the acute phase of visual loss, funduscopic appearance may be normal or there may be hyperemia and apparent swelling of the optic disc with dilation and tortuosity of the retinal vasculature. Ultimately, the patient will develop optic nerve pallor and loss of nerve fiber layer, predominantly in the papillomacular bundle (Biousse and Newman, 2001a; 2001b).

While most patients with LHON have optic neuropathy as the sole manifestation of the disease, some also have cardiac conduction abnormalities, minor neurologic abnormalities, or disease clinically indistinguishable from multiple sclerosis. Several "Leber's Plus" pedigrees are described with maternally related members with LHON-like optic neuropathy plus more severe neurological manifestations, including dystonia and basal gangliar lesions, or encephalopathic episodes, spasticity, and psychiatric disturbances.

**TABLE VIII   Causes of Compressive and Infiltrative Optic Neuropathies**

Compressive
  Neoplastic
    Optic nerve sheath meningioma
    Intraorbital tumor (hemangioma, lymphangioma, metastasis, etc.)
    Sphenoid meningioma
    Pituitary tumor
    Craniopharyngioma
  Non-neoplastic
    Thyroid eye disease
    Orbital pseudotumor
    Orbital hemorrhage
    Paget's disease
    Fibrous dysplasia
    Ectatic internal carotid artery
Infiltrative
  Neoplastic
    Optic nerve glioma
    Metastatic carcinoma
    Nasopharyngeal carcinoma and other contiguous tumors
    Lymphoma
    Leukemia
    Meningeal carcinomatosis
  Non-neoplastic
    Sarcoidosis

The inheritance pattern of LHON is maternal, in which only female family members pass on the risk of the disease to all their offspring. Since 1988, several point mutations in the protein-encoding portions of the mitochondrial genome have been proposed as causal in LHON (so-called "primary" mutations). They include the mitochondrial DNA mutations at positions 11778, 3460, and 14484. These mutations can be identified simply and rapidly on any tissue that contains mitochondrial DNA, including whole blood. The 11778 mutation accounts for approximately 31–90% of LHON probands, the 3460 mutation for 8–15%, and the 14484 for 10–15%, depending on patient ethnic origin. Among the LHON primary mutations, the one clear phenotypic distinction is that patients with the 14484 mutation have a substantially greater chance for spontaneous recovery of vision (up to 60%) and better final visual acuities. Good visual outcome strongly correlates with younger age at onset.

### Natural History

The presence of a primary LHON mutation does not guarantee that a patient will lose vision. Indeed, the determinants of expression in LHON are poorly understood, as is the specificity for optic nerve dysfunction. Other contributing genetic factors could include the relative quantity of mutant mitochondrial DNA in individuals and in tissues (heteroplasmy), other mitochondrial DNA abnormalities, or nuclear DNA influences. Internal and external environmental factors may also play a role in expression, including systemic illnesses, nutritional deficiencies, or toxic exposures. Attempts at therapy in LHON have in general proved ineffective. However, it is reasonable to counsel the at-risk patient to avoid agents that may stress mitochondrial function, such as tobacco, alcohol, and environmental toxins (Biousse and Newman, 2001a; 2001b).

### Other Hereditary Optic Neuropathies

They most often continue to be classified according to pedigree analysis. **Autosomal dominant optic atrophy** (ODOA), or Kjer's disease, is clinically characterized by symmetrical, insidious onset of visual loss in the first decade of life, blue-yellow dyschromatopsia, cecocentral scotomas and wedges of temporal disc atrophy (Biousse and Newman, 2001a; 2001b). Visual loss is variable even within families, and ranges from subtle to 20/200. Other neurologic abnormalities are uncommon. The disease is genetically heterogenous and has been linked to chromosomes 3 and 18. Mutation in a nuclear gene on chromosome 3 encoding for a mitochondrial-related protein have recently been identified in patients with ADOA, emphasizing the importance of mitochondrial function in the pathophysiology of hereditary optic nerve dysfunction (Biousse and Newman, 2001a; 2001b). The autosomal recessive optic neuropathies are a less common, more heterogeneous group of disorders. They are frequently associated with systemic and neurologic abnormalities and visual loss is typically severe. One particular phenotype, **Wolfram's syndrome**, or DIDMOAD (diabetes insipidus, diabetes mellitus, optic atrophy, and deafness), may prove, in some cases, to represent a disorder of mitochondrial function caused by defects in either mitochondrial or nuclear DNA (Newman, 1998).

## TRAUMATIC OPTIC NEUROPATHY

### Clinical Aspects

Traumatic optic neuropathy is an uncommon but potentially devastating complication of head injury. It should always be suspected in any patient with evidence of optic nerve dysfunction (e.g., otherwise unexplained decreased visual acuity, RAPD, dyschromatopsia) following head trauma. Most commonly, the optic nerve is injured indirectly, as a result of concussive forces to the head, particularly the forehead. This causes both a mechanical and ischemic insult to the optic nerve, mostly at the level of the optic canal.

### Natural History

Although visual loss may be devastating and permanent, there are many anecdotal reports of spontaneous recovery of vision following optic nerve injury (Steisaper and Goldberg, 1994).

### Practical Management

Management of indirect traumatic optic neuropathy is controversial (Steisaper and Goldberg, 1994). A recent comparative nonrandomized study, the International Optic Nerve Trauma Study compared the visual outcome of indirect traumatic optic neuropathy managed with optic canal decompression, high-dose corticosteroids, or observation (Levin *et al.*, 1999). There was no indication that canal decompression produced better results than high-dose corticosteroids, nor that the timing of treatment with either corticosteroids or surgery were key factors. Therefore, at this time, evaluation and management of indirect traumatic optic neuropathy should include a high index of suspicion in

**TABLE IX** Causes of Raised Intracranial Pressure and Papilledema

Intracranial mass
Hydrocephalus
Meningeal process
Increased venous pressure
Idiopathic intracranial hypertension

the appropriate setting, a CT scan to exclude a retro-orbital hematoma or a bony fragment impinging on the optic nerve (both of which require early surgical intervention), and treatment on an individual patient basis.

## PAPILLEDEMA

Patients with any cause of increased intracranial pressure may develop optic disc swelling (Table IX). This condition is called papilledema. The symptoms and signs in patients with papilledema generally are those typically associated with raised intracranial pressure, including headache, nausea, vomiting, and pulsatile tinnitus. Visual symptoms in such patients include transient visual obscurations and diplopia from sixth nerve palsy. The optic disc swelling in papilledema is usually bilateral and symmetric, but it may be asymmetric or even unilateral. Although early visual loss is insidious and often not recognized by the patient whose peripheral visual field is affected first, chronic papilledema from raised intracranial pressure is a common cause of bilateral optic neuropathy and may lead to optic atrophy and severe visual loss. Careful repeated ophthalmological evaluations, including formal visual fields, are mandatory in patients with papilledema, regardless of its etiology (Corbett and Thomson, 1989).

## GLAUCOMATOUS OPTIC NEUROPATHY

Neurologists often forget that chronic open-angle glaucoma (OAG) is the most common cause of nonacute progressive bilateral optic neuropathy. This disorder is characterized by slowly progressive peripheral visual field loss, associated with elevated intraocular pressure (the normal range is 8–21 mmHg), and cupping of the optic disc. The diagnosis of OAG generally requires documentation of all three findings. Since the visual field defects primarily involve the nasal periphery, central visual acuity is preserved until very late in the course of the disease. A vertically enlarged cup, with nasal displacement of the vessels and absence of pallor of the retained rim of neural tissue, are highly suggestive of glaucoma.

## REFERENCES

Beck, R. W., and Trobe, J. D., for the Optic Neuritis Treatment Group (1995). What have we learned from the Optic Neuritis Treatment Trial. *Ophthalmology* **102**, 1504–1508.

Biousse, V., and Newman, N. J. (2001a). The neuro-ophthalmology of mitochondrial diseases. *Sem. Neurol.* **21**, 275–291.

Biousse, V., and Newman, N. J. (2001b). Hereditary optic neuropathies. *Clin. N. Am. Ophthalmol.* **14**, 547–568.

Cuba Neuropathy Field Investigation Team (1995). Epidemic optic neuropathy in Cuba: Clinical characterization and risk factors. *N. Engl. J. Med.* **338**, 1176–1182.

Corbett, J. J., and Thomson, H. S. (1989). The rational management of idiopathic intracranial hypertension. *Arch. Neurol.* **46**, 1049–1051.

Gordon, L. K., and Levin, L. A. (1998). Visual loss in giant cell arteritis. *JAMA* **280**, 385–386.

Hayreh, S. S., Joos, K. M., Podhajsky, P. A., and Long, C. R. (1994). Systemic diseases associated with nonarteritic anterior ischemic optic neuropathy. *Am. J. Ophthalmol.* **118**, 766–780.

Hayreh, S. S. (1996). Acute ischemic disorders of the optic nerve. Pathogenesis, clinical manifestations, and management. *Ophthalmol. Clin. North Am.* **9**, 407–442.

Hayreh, S. S., Podhajsky, P. A., and Zimmerman, B. (1998). Ocular manifestations of giant cell arteritis. *Am. J. Ophthalmol.* **125**, 509–520.

Ischemic Optic Neuropathy Decompression Trial Research Group (IONDT) (1995). Optic nerve decompression surgery for nonarteritic anterior ischemic optic neuropathy (NAION) is not effective and may be harmful. *JAMA* **273**, 625–632.

Ischemic Optic Neuropathy Decompression Trial Study Group (1996). The clinical profile of nonarteritic anterior ischemic optic neuropathy: Experience of the Ischemic Optic Neuropathy Decompression Trial. *Arch. Ophthalmol.* **114**; 1366–1374.

Jacobs, L. D., Beck, R. W., Simon, J. H., Kinkel, R. P., Brownscheidle, C. M., Murray, T. J., Simonian, N. A., Slasor, P. J., Sandrock, A. W., and the CHAMPS Study Group (2000). Intramuscular interferon-beta-1a therapy initiated during a first demyelinating event in multiple sclerosis. *N. Engl. J. Med.* **343**, 898–904.

Kaufman, D. I., Trobe, J. D., Eggenberger, E. R., and Whitaker, J. N. (2000). Practice parameter: The role of corticosteroids in the management of acute monosymptomatic optic neuritis. Report of the Quality Standards Subcommittee of the American Academy of Neurology. *Neurology* **54**, 2039–2044.

Lessell, S. (1998). Toxic and deficiency optic neuropahies. In "Walsh & Hoyt Clinical Neuro-Ophthalmology" (N. R. Miller and N. J. Newman, Eds.), 5th ed., vol. 1, pp. 663–679. Williams & Wilkins, Baltimore.

Levin, L. A., Beck, R. W., Joseph, M., et al. (1999). The treatment of traumatic optic neuropathy: The International Optic Nerve Trauma Study. *Ophthalmology* **106**, 1268–1277.

Newman, N. J. (1996). Optic neuropathy. *Neurology* **46**, 315–322.

Newman, N. J. (1998). Hereditary optic neuropathies. In "Walsh & Hoyt's Clinical Neuro-Ophthalmology" (N. R. Miller and N. J. Newman, Eds.), 5th ed., vol. 1, pp. 741–773. Williams & Wilkins, Baltimore.

Steinsaper, K. D., and Goldberg, R. A. (1994). Traumatic optic neuropathy. A review. *Surv. Ophthalmol.* **38**, 487–518.

Shults, W. F. (1998). Compressive optic neuropathies. In "Walsh & Hoyt Clinical Neuro-Ophthalmology" (N. R. Miller and N. J. Newman, Eds.), 5th ed., vol. 1, pp. 649–659. Williams & Wilkins, Baltimore.

The Optic Neuritis Study Group (1997). The 5-year risk of multiple sclerosis after optic neuritis: Experience of the Optic Neuritis Treatment Trial. *Neurology* **49**, 1404–1413.

# CHAPTER 13
# Ocular Motor Disorders

Andreas Straube and R. John Leigh

The clinical examination of eye movements presents several advantages to the neurologist. First, because much is now known about their neural substrate, specific abnormalities of eye movements may allow precise topological diagnosis, for example internuclear ophthalmoplegia and lesions of the medial longitudinal fasciculus. Second, abnormal eye movements may be the first sign of an occult disease process, for example opsoclonus with neuroblastoma. Third, the effectiveness of novel treatments can be monitored by measuring specific properties, such as saccade speed in Neimann-Pick type C disease or the gain of the internally triggered saccades in Parkinson's disease. Fourth, the nature of complex defects, such as neglect, can be probed by measuring the ocular motor responses to carefully controlled stimuli. Abnormal eye movements may disrupt vision in several ways. Nystagmus may cause excessive motion of images on the retina, which degrades vision and may cause the illusion of motion of the surroundings (oscillopsia). Inappropriate saccadic eye movements shift the line of sight away from the object of interest and, if repetitive, may interfere with tasks such as reading. Misalignment of the visual axes, so that noncorresponding images fall on each retina, causes diplopia. Although empirical observations have identified most of the currently available therapies, treatments of abnormal eye movements are now increasingly being based on knowledge of the underlying anatomy, physiology and pharmacology and controlled drug trials.

Most of this chapter concerns treatment of nystagmus and its visual consequences. Several fundamental issues are noted at the outset. In health, three main mechanisms act to hold gaze steady so that vision is clear and stable. The first is visual fixation, which has two components: the visual system's ability to detect retinal image slip and program corrective eye movements, and the ability to suppress unwanted eye movements. The second is the vestibulo-ocular reflex, by which the motion detectors of the inner ear promptly initiate eye movements to compensate for head perturbations, especially during locomotion. The third mechanism is the gaze-holding system, which makes it possible to hold the eyes at an eccentric position (such as lateral gaze). Disturbance of any of these systems will cause the eye to drift away from the target and may lead to nystagmus. General strategies to treat abnormal eye movements such as nystagmus include: (1) placing the eye at a position (or vergence angle) in which the oscillations are minimized; (2) optical and electronic methods for negating the visual consequences of the eye movements; (3) procedures for weakening the extraocular muscles; (4) application of auditory or somatosensory stimuli to suppress the nystagmus; (5) drugs that suppress the oscillations but permit normal eye movements.

## NYSTAGMUS

### Central Disorders of the Vestibulo-Ocular Reflex and Vestibular Balance in Yaw, Pitch, and Roll

#### Clinical Aspects

The vestibulo-ocular reflex (VOR) normally generates eye rotations, at short latency, in the same plane as the head rotation that stimulated them. Thus, disorders of the vestibular *periphery* cause nystagmus with a direction that is determined by the pattern of involvement of semicircular canals. Thus, complete, unilateral loss of one labyrinth causes a mixed horizontal–torsional nystagmus that is suppressed by visual fixation (which remains intact). *Central* vestibular disorders may also cause an imbalance, leading to upbeat, downbeat, or torsional nystagmus (see below). Another consequence of central vestibular disease is a change in the size (gain) of the overall response. As a consequence of this, patients complain of oscillopsia during fast head movements. A VOR gain larger than 1 (eye speed exceeds head speed) occurs after a disinhibition of the brainstem circuits responsible for the VOR. Typically this occurs with lesions of the vestibular cerebellum, such as forms of cerebellar degeneration. In some patients with a cerebellar or multisystem atrophy, the VOR gain can also be reduced, but more frequently this is due to a peripheral lesion involving the labyrinth or vestibular nerve which may be caused by inflammation, intoxication, or trauma. An acute unilateral lesion of central vestibular structures, i.e., the vestibular nuclei and vestibular cerebellum (flocculus/paraflocculus, nodulus), causes an acute imbalance between the resting activity of the vestibular nuclei on each side of the brain. If the brainstem mechanism that holds eccentric gaze (the velocity-to-position integrator) is preserved, this difference in the resting activity of the vestibular nuclei causes spontaneous nystagmus that beats in the affected plane. If the gaze-holding mechanism is involved, there is a tonic shift of the neutral zone of the eyes, instead of a nystagmus.

If damage to the vestibular nerve lies close to the vestibular nucleus, it is often difficult to differentiate clinically from a peripheral nerve or labyrinth lesion. Diagnostic hints for a peripheral lesion are a mixed horizontal–torsional spontaneous nystagmus (beating toward the unaffected ear) that suppresses with fixation. In addition, following an unpredictable passive fast head movement ("head thrust") toward the affected ear, a corrective saccade may occur (Halmagyi and Curthoys, 1988).

### Natural Course

In cases with an increased VOR gain due to a cerebellar or multisystem degeneration, the change in gain is often permanent because the adapting structures, such as the cerebellum and the commissural fibers, are involved ("the repair shop is broken"). In cases with isolated lesions of the flocculus there will probably be a partial compensation. In addition to these structural lesions, several central acting drugs can decrease the VOR gain (e.g., benzodiazepine, barbiturates, other hypnotics, and anticonvulsants). In general the effects on the VOR of these drugs are reversible (Remler et al., 1990). Currently, no epidemiological data are available.

### Principles of Therapy and Practical Management

There are only anecdotal reports about the treatment of an increased VOR gain. Thurston et al. (1987) reported a sustained reduction of the VOR gain in a patient with an unclassified system atrophy of the cerebellum during therapy with physostigmine 1 mg po

daily. Furthermore, Tijssen et al. (1985) observed an improvement in visual suppression of the VOR (not a VOR gain change in the proper sense) after the administration of physostigmine, 1 mg, in five patients with a hereditary ataxia. There was also a reduction of the oscillopsia during head movement in the light. On the other hand, Pyykkö et al. (1985) found a decrease in the gain of the VOR and OKN after scopolamine, a central anticholinergic, in normal subjects.

### Down- and Upbeat Nystagmus

#### Clinical Aspects

In a preliminary classification of vestibular brainstem, disorders, Brandt and Dieterich (1993) describe down- and upbeat nystagmus as a central disorder of the vertical VOR. The pathogenesis of these disorders probably reflects certain anatomical and pharmacological differences between the pathways serving central projections of different labyrinthine semicircular canals. For example, the cerebellar flocculus is known to inhibit central projections of the anterior, but not the posterior, semicircular canals. Therefore, lesions of the vestibular cerebellum (such as Chiari malformation) or of the whole cerebellar cortex (e.g., due to paraneoplastic degeneration or genetic disorder of P/Q calcium channels) will remove inhibition of the anterior canals input, the eyes will drift up, and downbeat nystagmus will result. It also seems possible that upbeat nystagmus induced by nicotine may reflect an unmasking of a pre-existing vestibular tonus difference (Pereira et al., 2000).

*Downbeat nystagmus* is the central form of vestibular nystagmus that beats downward in primary gaze position and is normally enhanced on lateral-and-down gaze, or by bending the head backward or downward. Visual fixation has little effect on the amplitude or frequency of the nystagmus, although convergence affects it in some patients. In general the nystagmus is accompanied by a combination of visual and vestibulocerebellar ataxia with a tendency to fall backward (Büchele et al., 1983). Lesions that cause downbeat nystagmus occur in the vestibular cerebellum and underlying medulla (Leigh and Zee, 1999). As noted above, the pathophysiological mechanism for downbeat nystagmus appears to be due to a central imbalance of the vertical VOR, perhaps involving GABA-ergic and cholinergic transmission. The most common cause of downbeat nystagmus is cerebellar degeneration (hereditary, sporadic, or paraneoplastic). Other important causes are Chiari malformation, drug intoxications (especially the anticonvulsants and lithium); multiple sclerosis is an uncommon cause, and a congenital form is rare (Halmagyi et al., 1983).

*Upbeat nystagmus* is a jerk nystagmus beating upward in central gaze position, usually in combination with a disturbance of vertical smooth pursuit. In some patients the upbeat nystagmus changes to a downbeat nystagmus during convergence. Probable causes of upbeat nystagmus are lesions in the ascending pathways from the anterior canals (and/or the otoliths) at the pontomesencephalic or pontomedullary junction, near the perihypoglossal nuclei (Fisher *et al.*, 1983). Most often upbeat nystagmus is seen after medullary lesions (Stahl *et al.*, 2000). The main causes are multiple sclerosis, tumors of the brainstem, Wernicke's encephalopathy, cerebellar degeneration, and intoxications.

### Natural Course

Currently there are no studies of the natural course of down- and upbeat nystagmus. Clinical observations suggest that when downbeat nystagmus persists it may be because the cerebellar structures, which are responsible for the adaptation, are more often involved. In upbeat nystagmus the cerebellum is less often damaged, and therefore a central compensation is more likely to occur than in downbeat nystagmus (Brandt, 1999).

### Principles of Therapy and Practical Management

**Downbeat Nystagmus.** Clinical therapeutic studies are not available. A reduction in downbeat nystagmus and the accompanying oscillopsia was found using the GABA-A agonist clonazepam (3 × 0.5 mg PO daily, Currie and Matsuo, 1986). The authors suggested using a test dose of 1–2 mg to determine the likely long-term benefits. We repeatedly saw a decrease in downbeat nystagmus as well as ataxia during treatment with the GABA-A agonist baclofen (3 × 10 mg PO daily) (Dieterich *et al.*, 1991). Averbuch-Heller *et al.* (1997) reported a modest improvement or deterioration of downbeat nystagmus in patients during treatment with baclofen or gabapentin. An additional observation was that a single intravenous injection of the cholinergic drug physostigmine (Ach-esterase inhibitor) worsened downbeat nystagmus in five patients. This effect was partially reversed in one patient by the anticholinergic drug biperiden, suggesting that there might be some benefit from anticholinergic drugs as has been shown in a double-blind study using intravenous scopolamine (Barton *et al.*, 1994).

In isolated patients with a craniocervical anomaly, a surgical decompression is performed by removing parts of the occipital bone in the region of the foramen magnum (Pedersen *et al.*, 1980; Spooner and Baloh, 1981). Furthermore, there are reports about the effect of biofeedback treatment (Edagawa and Nabatame, personal communication) and, if the nystagmus ceases with convergence, by artificially induced convergence with prism glasses (base out).

**Upbeat Nystagmus.** Treatment with baclofen (3 × 5–10 mg PO daily) resulted in an improvement in several patients (Dieterich *et al.*, 1991). In isolated cases a contact lens device can suppress oscillopsia by artificially stabilizing a visual image on the retina (Leigh *et al.*, 1988; Rushton and Rushton, 1984). This system is useful only during reading or while watching television.

Practically, in symptomatic cases of down- or upbeat nystagmus, a trial of treatment with baclofen, 3 × 5–10 mg PO daily, or gabapentin, 3 × 300–400 mg, should be started.

## Seesaw Nystagmus

### Clinical Aspects

Seesaw nystagmus is a rare pendular or jerklike oscillation that can be classified as a disorder of the VOR in the roll plane (Brandt and Dieterich, 1993). In this form of nystagmus one half cycle consists of the elevation and intorsion of one eye while there is synchronous depression and extorsion of the other eye; during the next half cycle there is a reversal of the vertical and torsional movements. The frequency is normally 4–5 Hz. This disorder has usually been found in patients with lesions in the diencephalic–mesencephalic area (near the chiasm and/or interstitial nucleus of Cajal) (Brandt and Büchele, 1982; Leigh and Zee, 1999). Since there is a dog model of seesaw nystagmus in dogs with genetic defect of the chiasm the loss of crossed visual input seems to be the crucial element in the pathophysiology of seesaw nystagmus (Stahl *et al.*, 2000). Jerky seesaw nystagmus had been proposed to be due to unilateral lesion of the interstitial nucleus of Cajal (INC) with sparing of the adjacent rostral interstitial nucleus of the medial longitudinal fascicles (Halmagyi *et al.*, 1994). However, experimental inactivation of INC has not produced seesaw nystagmus in monkeys (Helmchen *et al.*, 1998).

### Principles of Therapy and Practical Management

There are reports of a beneficial effect of alcohol (1.2 g/kg body weight) in two patients (Frisèn and Wikkelso, 1986; Lepore, 1987) and clonazepam (Carlow, 1986). This observation is similar to that observed in acquired pendular nystagmus, described by Mossman *et al.* (1993). Recently, Averbuch-Heller reported on three patients with a seesaw component to their pendular nystagmus, who improved during treatment with gabapentin, a substance with multiple central-acting mechanisms. GABA-ergic mechanisms are, therefore, implicated and a trial of treatment with GABA-ergic drugs (benzodiazepine, baclofen, gabapentin) should be considered.

## Periodic Alternating Nystagmus

### Clinical Aspects

Periodic alternating nystagmus is a spontaneous, horizontal-beating nystagmus that reverses direction approximately every 2 minutes. The nystagmus amplitude gradually decreases, reverses its direction, and then increases again. The nystagmus amplitude gradually decreases, reverses its direction, and then increases again. During the nystagmus the patients often complain of increasing/decreasing oscillopsia. An animal model for periodic alternating nystagmus is reported in monkey (while the animals are in darkness), following removal of the nodulus and uvula of the vestibulocerebellum. Reported patients with periodic alternating nystagmus commonly have vestibular cerebellar lesions, which also disrupt visual fixation (so that their nystagmus is also present during normal viewing). These observations and animal experiments support the idea that a disinhibition of the GABA-ergic velocity-storage mechanism, which is mediated in the vestibular nuclei, is responsible (Furman et al., 1990; Waespe et al., 1985). The underlying etiologies are craniocervical anomalies (Chiari malformation), multiple sclerosis, tumors of the brainstem, cerebellar atrophies, and posterior circulation stroke. Quite commonly, individuals with congenital nystagmus show an irregular periodic component to their oscillations.

### Natural Course

In general, periodic alternating nystagmus does not improve spontaneously.

### Principles of Therapy and Practical Management

Several reports describe a positive effect of baclofen, a GABA-B agonist, in a dose of 3 × 5–10 mg PO daily (Carlow, 1986; Halmagyi et al., 1980; Isago et al., 1985; Larmande and Larmande, 1983; Nuti et al., 1986). This treatment is less effective in congenital forms, which probably have a different pathogenesis. As an alternative treatment, 5-hydroxytryptophan (500–1000 mg PO daily) can be tried (Larmande and Larmande, 1983). Furthermore, phenothiazine and barbiturates have been found to be effective in single cases (Isago et al., 1985).

Practically, a treatment trial with baclofen (3 × 5–10 mg PO daily) is recommended. The effect usually be-

gins within 3–5 days. The effect of baclofen can diminish with prolonged use, and some cases do not respond at all.

## Acquired Pendular Nystagmus

### Clinical Aspects

Acquired pendular nystagmus (APN) is perhaps the most visually distressing form of nystagmus; oscillopsia and impaired vision are common. Clinically patients can present with nystagmus that is horizontal, vertical, or mixed (i.e., circular, elliptical, or diagonal in trajectory), which can either be monocular or binocular (Gresty et al., 1982; Leigh et al., 1992; Traccis et al., 1990). The frequency of this nystagmus is 2–7 Hz (Zee, 1985), and often (but not in all cases) the nystagmus is associated with a head tremor (not synchronized with the nystagmus), trunk and limb ataxia, or visual impairment. Acquired pendular nystagmus occurs with several disorders of myelin (multiple sclerosis, toluene abuse, Pelizaeus-Merzbacher disease), as a component of the syndrome of oculopalatal tremor (myoclonus), and in Whipple's disease (Leigh and Zee, 1999). More than one mechanism may be responsible for these oscillations. Based on the observation that the nystagmus is often dissociated and that eye movements other than optokinetic nystagmus and voluntary saccades are also disturbed, a lesion in the brainstem near the oculomotor nuclei is suggested (Gresty et al., 1982). Alternatively an instability of the gaze-holding network (neural integrator) has been proposed; this suggestion has received experimental support (Das et al., 2000) and has suggested potential therapies (Stahl et al., 2000).

### Natural Course

The course of APN depends on the etiology, but it is usually a permanent symptom.

### Principles of Therapy and Practical Management

Current approaches for drug treatment of APN rest on recent demonstration of gabaergic, glutaminergic, and cholinergic mechanisms in the medial vestibular nucleus and nucleus prepositus hypoglossi, which play a key role in normal gaze holding (Arnold et al., 1999; Phelan and Gallagher, 1992; Straube et al., 1991).

Some reports state that anticholinergic treatment with trihexyphenidyl (20–40 mg PO daily) is effective (Herishanu and Louzoun, 1986; Jabbari et al., 1987), but in a double-blind study by Leigh et al. (1991) only one of six patients showed improvement from this oral treatment, whereas three patients showed a decrease in nystagmus and improvement of visual acuity during treatment with tridihexethyl chloride (a quaternary anticholinergic that does not cross the blood–brain barrier). In contrast to this, Barton *et al.* (1994) found in a double-blind trial that scopolamine (0.4 mg IV) decreased the nystagmus in all five tested patients with APN. In a further three patients the combination with lidocaine, 100 mg IV, decreased the nystagmus (Ell *et al.*, 1982; Gresty *et al.*, 1982). Recently, Starck *et al.* (1997) saw an improvement in 3 of 10 patients who received a scopolamine patch (containing 1.5 mg scopolamine, released at a rate of 0.5 mg per day). The same authors failed to observe further improvement when scopolamine and mexiletine (400–600 mg PO daily) were given in combination. The most effective substance in their study was memantine, a glutamate antagonist, which improved significantly the nystagmus in all 9 tested patients (15–60 mg PO daily). A further two patients responded to clonazepam (3 × 0.5–1.0 mg PO daily), a GABA-A agonist (Starck *et al.*, 1997). The beneficial effect of GABA-ergic drugs are also reported in two other groups. Traccis *et al.* (1990) showed an improvement in 1 of 3 patients with APN and a cerebellar ataxia due to a multiple sclerosis when treated with isoniazid (800–1000 mg PO daily) and glasses with prisms that induced a convergence. This observation was not confirmed by other investigations (Leigh *et al.*, 1994). Another substance with a GABA-ergic as well as calcium-channel-antagonistic mechanism is gabapentin, which improved the nystagmus (and visual acuity) in a group of 10 of 15 patients substantially (Averbuch-Heller *et al.*, 1997). Mossman *et al.* (1993) saw a decrease of the nystagmus after alcohol consumption, and in an old study (Nathanson *et al.*, 1953) amobarbital sodium (50–150 mg IV) was effective.

In addition to these pharmacological approaches, surgical steps have been tried by changing the insertions of the eye muscles (Cüppers, 1971; Mühlendyck, 1979). More recently, temporary weakening of eye muscles using botulinum toxin injections has been attempted (Leigh *et al.*, 1992). The disadvantages of this method are the variable degree of paresis and diplopia it causes and a possible temporary ptosis.

Practically, treatment should start with memantine in a dosage of 15–60 mg PO or alternatively 3 × 300–400 mg gabapentin daily. If there is no or only a small effect, benzodiazepines like clonazepam (3 × 0.5–1.0 mg PO daily) can be tried. Further possibilities are scopolamine patches or trihexyphenidyl.

## Congenital and Manifest Latent Nystagmus

### Clinical Aspects

Congenital nystagmus is an involuntary eye movement disorder that generally begins at birth or early infancy. Normally the direction of the nystagmus is horizontal and the waveform is variable. Hallmarks of congenital nystagmus are an increase in the nystagmus velocity during attempted fixation, suppression with convergence, apparent inversion of optokinetic nystagmus, and a so-called static neutral or null zone (at which nystagmus is minimal). The presence of "foveation periods"—during which the eye is momentarily still while pointing at the object—is a valuable laboratory confirmation.

Latent nystagmus is another congenital oscillation that occurs in individuals who never developed normal binocular vision. Although often present with both eyes viewing ("manifest latent nystagmus"), it is accentuated by covering one eye; the direction of the nystagmus is then toward the fixating eye. Strabismus is a common associated finding. Congenital nystagmus and manifest latent nystagmus usually require no neurological investigation and so need to be carefully distinguished from acquired forms of nystagmus.

### Natural Course

Both types of nystagmus begin usually at birth or in early infancy. Sometimes the amplitude of the congenital nystagmus decreases with age. Spontaneous improvement is not common.

### Principles of Therapy and Practical Management

**Congenital Nystagmus.** Often treatment is not necessary because of a lack of oscillopsia. The first reported treatment trials were the surgical shifting of the neutral zone (zone of minimal eye oscillations) to the primary position (Anderson, 1953). Another technique, described by Mühlendyck (1979), results in an artificially induced convergence, producing a reduction in the nystagmus. Recently, Atilla *et al.* (1999) described an improvement in visual function as well as decrease in nystagmus intensity due to symmetric recession of the horizontal rectus muscle. To avoid surgery, botulinum toxin injections can be used in some cases (similar to the procedure in strabismus correction). All these forms of therapy have the disadvantage of being invasive and having clinical outcomes that are not always predictable (Abadi and Whittle, 1992). Working with a canine model for congenital nystagmus, Dell'Osso and colleagues (1999) have demonstrated reductions in the nystagmus following a procedure in which the extraocular muscles were detached and then reattached at their insertions on the globe (tenotomy). This procedure is currently being evaluated in a clinical trial, with promising preliminary results. At present, surgically removing and then reattaching the horizontal muscles to the globe (which may interfere with extraocular proprioception) is being evaluated in a controlled trial. Non-surgical treatment trials have included prism glasses to force a convergence reaction (Bagolini, 1978), biofeedback treatment (Mezawa *et al.*, 1990), and drug treatment (5-hydroxytryptophan: Larmande and Pautrizel, 1981; baclofen: Yee *et al.*, 1982). Mezawa *et al.* (1990) were able to decrease the average intensity of the nystagmus by about 40% and increase the foveation time by about 190% by using auditory feedback training. Electrical stimulations or vibration to the forehead or neck decreases congenital nystagmus in some subjects (Sheth *et al.*, 1995).

If necessary a treatment trial should be started with baclofen (3 × 5–10 mg PO daily). If the effect is not satisfactory, biofeedback training or a surgical procedure, usually the shifting of the neutral zone, should be considered.

**Manifest Latent Nystagmus.** Optical or surgical realignment of the eyes decreases this type of nystagmus in some patients (Zubcov *et al.*, 1990).

### Opsoclonus and Ocular Flutter

#### Clinical Aspects

Opsoclonus consists of repetitive bursts of conjugate saccadic oscillations, which have horizontal, vertical, and torsional components. During each burst of these high-frequency oscillations, the movement is continuous, without an intersaccadic interval. These oscillaitons are often triggered by eye closure and convergence; amplitudes range up to 2–15° (overview in Leigh and Zee, 1999). In ocular flutter the same pattern is restricted to the horizontal plane. The ocular symptoms are often accompanied by cerebellar signs, such as gait and limb myoclonus. Pathophysiologically, a functional disturbance of active saccadic suppression by the pontine omnipause neurons is the most probable mechanism. Since histological abnormalities of these neurons has not been shown (Ridley *et al.*, 1987), a functional lesion of the glutaminergic cerebellar afferents to the omnipause cells is a likely cause for their disinhibition. In monkeys bilateral pharmacological lesions of the deep cerebellar nuclei cause an opsoclonus-like eye movement disturbance (peronal observation). Opsoclonus can be observed in encephalitis (postviral, e.g., coxsackie B37; postvaccinal), or as a paraneoplastic symptom (infants, neuroblastoma; adults, carcinoma of the lung, breast, ovary, or uterus). Single case reports describe opsoclonus in hyperosmolar coma, intoxications (antidepressant, morphine, lithium, thallium, haloperidol, cocaine, DDT, and other pesticides), hydrocephalus, and hemorrhages, infarctions, trauma, and tumors of the brainstem (Leopold, 1985).

#### Natural Course

If the opsoclonus is the result of a viral infection or is postvaccinal then the prognosis is usually favorable. If the basic disease is a neuroblastoma, then, even after complete remission of the tumor disease, functional disabling symptoms persist in about 50% of the children (Leopold, 1985). In adults with carcinoma the prognosis may be worse.

#### Principles of Therapy and Practical Management

In addition to therapy for the tumor disease, treatment with ACTH or prednisolone is recommended. The doses are similar to the schedules used in immune-suppressive therapy—initially ACTH 1 mg im daily for 1 week and then a slowly progressive decrement or an equivalent dose of oral prednisolone (Carlow, 1986; Leopold, 1985). Four of five patients with square-wave oscillations, probably a related fixation disturbance, showed an improvement under a therapy with valproic acid (Traccis *et al.*, 1997). In single cases an improvement has been observed during treatment with propranolol (3 × 40–80 mg PO daily), nitrazepam (15–30 mg PO daily), and clonazepam (3 × 0.5–2.0 mg PO daily) (overview in Carlow, 1986; Leopold, 1985). Nausieda *et al.* (1981) reported a dramatic improvement after the administration of 200 mg thiamine IV in one patient. Intravenous immunoglobulin has also been reported to be effective (Pless and Ronthal, 1996).

In cases with a tumor, the therapy should start with ACTH or prednisolone. In all other cases, or if there was no improvement with ACTH or prednisolone, treatment with a single dose of thiamine (200 mg IV) should be given. As the next step propranolol (3 × 40–80 mg PO daily), clonazepam (3 × 0.5–2.0 mg PO daily), or valproic acid (2000 mg PO daily) can be tried.

### Oculopalatal Tremor (Myoclonus)

#### Clinical Aspects

This ocular disorder is characterized by a mainly vertical beating and sometimes also dissociated nystagmus with an atypical waveform (mostly pendular with a frequency of about 3 Hz). In contrast to the acquired pendular fixation nystagmus, oculopalatal myoclonus sometimes occurs with synchronous myoclonic jerks of the soft palate and of other branchial muscles. This condition is caused by pontine tumors, hemorrhages, and trauma or less frequently by inflammation of the brainstem and systemic degenerations that involve the red nucleus, the dentate nucleus, the central tegmental tract, or the inferior olive. The denervation of the inferior olive, with subsequent cholinergic denervation supersensitivity and hypertrophy, due to such lesions involving the dentatoolivary connections is considered to be the crucial mechanism underlying the development of the condition.

### Natural Course

The ocular signs are first observed several months (average 10 months) after the occurrence of the lesion. Their further course depends on the underlying disease process, but usually persists.

### Principles of Therapy and Practical Management

On the basis of the likely pathophysiological mechanism the use of anticholinergic medications have been proposed. Two groups describe an improvement in the signs after the administration of high doses of trihexyphenidyl (20–60 mg PO daily) in six patients (Jabbari et al., 1987; Straube and Büttner, 1991). There are also single case reports about positive effects during treatment with valproic acid (750–2000 mg PO daily), carbamazepine (600–1200 mg PO daily), phenytoin (250–400 mg PO daily), and 5-hydroxytryptophan combined with carbidopa (overview in Carlow, 1986; Straube and Büttner, 1991). We saw an improvement during treatment with valproic acid (3 × 600 mg PO daily) in combination with prisms that forced an upward gaze.

Treatment should start with small doses of trihexyphenidyl (2 mg PO daily), which are slowly increased (about 2 mg every day) until the symptoms are improved or the side effects do not allow a further increase. If this does not sufficiently improve the symptoms, a combination with other drugs (carbamazepine or valproate) should be tried.

Averbuch-Heller and colleagues (1997) reported that gabapentin was effective in some patients with oculopalatal tremor.

## NUCLEAR AND INFRANUCLEAR OCULAR DISORDERS

### Superior Oblique Myokymia

#### Clinical Aspects

Superior oblique myokymia consists of paroxysmal monocular high-frequency oscillations. In the primary gaze position and in abduction these oscillations are mainly torsional, but when the eyes are in adduction the oscillations have a vertical component. The oscillations can be provoked by voluntary eye movements, such as looking down. During such paroxysmal attacks the patients usually complain of oscillopsia. The pathophysiology of this condition is not totally clear. Analogous to the paroxysmal neuralgia of the Vth nerve, a nerve vessel contact involving the IVth nerve (Lee, 1984), or alternatively spontaneous discharges in the IVth nucleus (Hoyt and Keane, 1962) or of the superior oblique muscle, may be responsible (Leigh et al., 1991).

### Natural Course

Spontaneous unpredictable remissions, which can last for days up to years, are typical of superior oblique myokymia.

### Principles of Therapy and Practical Management

Several reports indicate a therapeutic effect of anticonvulsants, especially carbamazepine. Carbamazepine (3–4 × 200–400 mg PO daily) or, less often, phenytoin (250–400 mg PO daily) are recommended (Rosenberg and Glaser, 1983; Susac et al., 1973). Long-term studies about the continued effectiveness of these drugs are not available. Rosenberg and Glaser (1983) described a decrease in the effect of the treatment after 1 month in some patients. Propranolol (10 mg PO daily) was reported as a useful drug in a single case (Tyler and Ruiz, 1990). Benzodiazepine, barbiturates, and ergotamine were without effect (Herzau et al., 1978; Hoyt and Keane, 1962). In chronic cases that did not improve with anticonvulsants, a tenotomy of the superior oblique muscle has been performed (Palmer and Shults, 1984).

Practically, treatment should be started with carbamazepine (3–4 × 200–400 mg PO daily) or phenytoin (250–400 mg PO daily). The side effects and the risk of such therapy are the same as when used for the treatment of trigeminal neuralgia.

### Other Paroxysmal Ocular Motor Disorders

#### Clinical Aspects

There is increasing evidence that, in addition to the well-known superior oblique myokymia, there are other paroxysmal ocular motor disorders. This is a heterogeneous group of diseases with varying symptoms and etiology (neurovascular compression, migraine, hereditary paroxysmal nystagmus, and ataxia). In general, the ocular signs occur acutely and last for seconds (neurovascular compression, paroxysmal ataxia, and nystagmus) to days (migraine, paroxysmal ataxia, and nystagmus). In the case of basilar (vestibular) migraine (Dieterich and Brandt, 1999), considered to be due to a vascular disturbance, the ocular motor signs are usually accompanied by other symptoms. On the other hand, there may be no other symptoms if the oculomotor symptoms result from a neurovascular compression (Straube *et al.*, 1994; overview: Brandt, 1999), ocular neuromyotonia (Frohman and Zee, 1995; Ricker and Mertens, 1970), or some forms of migraine (benign recurrent vertigo) (Dieterich and Brandt, 1999; Slater, 1979). Clinically the ocular signs are expressed as a spontaneous jerk nystagmus, a positional nystagmus, or other complex forms of nystagmus.

In the case of neurovascular compression and ocular neuromyotonia, the neuropathological mechanism may be peripheral ephaptic transmission that takes place in the part of the cranial nerve, which still contains central myelin (derived from oligodendroglia), where the nerve has direct contact with a blood vessel. Another theory is that the pulsation of the blood vessel causes an afferent sensory inflow that then causes a false central response.

### Natural Course

The course depends on the etiology. The ocular symptoms are normally benign and self-limiting.

### Principles of Therapy and Practical Management

**Basilar (Vestibular) Migraine.**   The same prophylactic therapy schedule as in other forms of migraine with aura are indicated, but there are no special therapy studies concerning these subtypes of migraine published. Most experience has been with beta blockers such as propranolol ($3 \times 20$–$80$ mg PO daily) and metroprolol ($1 \times 50$–$200$ mg PO daily). Alternatively, flunarizine or valproic acid are recommended.

**Hereditary Paroxysmal Ataxia and Nystagmus.** Acetazolamide (Baloh and Winder, 1991; Griggs *et al.*, 1978) is the treatment of first choice; alternatively flunarizine ($1 \times 5$–$10$ mg PO daily) can be tried (Boel and Casaer, 1988).

**Neurovascular Compression.**   As a first therapeutic trial, an anticonvulsant (carbamazepine $2$–$4 \times 200$–$600$ mg PO daily; phenytoin $250$–$400$ mg PO daily) should be given; if the symptoms do not cease, a surgical approach may be considered (Jannetta *et al.*, 1984). There are no satisfactory follow-up studies, and the diagnostic criteria are not yet established.

**Ocular Neuromyotonia.**   Most of the reported patients showed at least partial benefit from treatment with carbamazepine, $2 \times 200$–$600$ mg PO daily, or phenytoin, $250$–$400$ mg PO daily (Diaz *et al.*, 1992; Frohman and Zee, 1995; Helmchen *et al.*, 1992).

### Isolated Paresis of Ocular Muscles

#### Clinical Aspects

Acute weakness of an extraocular muscle usually causes diplopia (see Leigh and Zee, 1999, for a review), although the magnitude of diplopia is generally less than the objective degree of disconjugacy (Brandt, 1999). The commonest cause of extraocular muscle weakness is a palsy of the oculomotor, trochlear, or abducens nerves; however, myasthenia gravis should always be considered. Most palsies of the three nerves supplying the extraocular muscles occur in association with hypertension and diabetes (presumed nerve infarction), but compressive aneurysm, tumor, cavernous sinus inflammation, and trauma are other leading causes. Disease affecting the muscles or the surrounding tissues, such as ophthalmic thyroid disease, commonly causes diplopia. Congenital strabismus does not usually cause diplopia.

#### Natural Course

Nerve palsies occurring in association with diabetes and hypertension usually spontaneously recover in 6–12 months. Palsies that persist require repeat investigations to look for an underlying cause, such as tumor. Oculomotor nerve palsies are sometimes complicated by aberrant regeneration of fibers. Diplopia due to myasthenia is sometimes refractory to general treatments for this condition.

#### Principles of Therapy and Practical Management

Initially, to avoid the disturbing double vision, one eye should be patched. Alternating the patched eye may stimulate recovery mechanisms, when the weak eye views. Some patients may benefit from training themselves to try to fixate in the pulling direction of the paretic muscle. If there is no recovery after a period of 1.5–2 years, large misalignments of the eyes can be decreased surgically. Another approach has been to induce temporary paralysis of the antagonistic muscle with botulinum toxin injections (Dunn *et al.*, 1986; Scott, 1980). Smaller degrees of misalignment can often be corrected by wearing spectacles fitted with prisms.

## REFERENCES

Abadi, R. V., and Whittle, J. (1992). Surgery and compensatory head postures in congenital nystagmus. A longitudinal study. *Arch. Ophthalmol.* **110**, 632–635.

Anderson, J. R. (1953). Causes and treatment of congenital eccentric nystagmus. *Br. J. Ophthalmol.* **37**, 267–281.

Arnold, D. B., Robinson, D. A., and Leigh, R. J. (1999). Nystagmus induced by pharmacological inactivation of the brainstem ocular motor integrator in monkey. *Vision Res.* **39**, 4286–4295.

Atilla, H., Erkam, N., and Işikcelik, Y. (1999). Surgical treatment in nystagmus. *Eye* 13, 11–15.

Averbruch-Heller, L., Tusa, R. J., Fuhry, L., Rottach, K. G., Ganser, G. L., Heide, W., Büttner, U., and Leigh, R. J. (1997). A double-blind controlled study of gabapentin and baclofen as treatment for acquired nystagmus. *Ann. Neurol.* 41, 818–825.

Bagolini, B. (1978). Orthoptic and prismatic treatment of congenital nystagmus. In "Strabismus (R. D. Reinecke, Ed.), pp. 191–201. Grune & Stratton, New York.

Baloh, R. W., and Spooner, J. W. (1981). Downbeat nystagmus. A type of central vestibular nystagmus. *Neurology* 31, 304–310.

Baloh, R. W., and Winder, A. (1991). Acetazolamide-responsive vestibulocerebellar syndrome: Clincial and oculographic features. *Neurology* 41, 429–433.

Barton, J. J. S., Huaman, A. G., and Sharpe, J. A. (1994). Muscarinic antagonists in the treatment of acquired pendular and downbeat nystagmus: A double-blind, randomized trial of three intravenous drugs. *Ann. Neurol.* 35, 319–325.

Boel, M., and Casaer, P. (1988). Familial periodic ataxia responsive to flunarizine. *Neuropediatrics* 19, 218–220.

Brandt, T. (1999). "Vertigo. Its Multisensory Syndromes," 2nd ed. Springer-Verlag, London.

Brandt, T., and Dieterich, M. (1993). Preliminary classification of vestibular brain-stem disorders. In "Brain-Stem Localization and Function" (L. R. Caplan and H. C. Hopf, Eds.), pp. 79–91. Springer-Verlag, Berlin and Heidelberg.

Brandt, T., and Büchele, W. (1982). "Augenbewegungsstörungen." Fischer, Stuttgart.

Büchele, W., Brandt, T., and Degner, D. (1983). Ataxia and oscillopsia in downbeat-nystagmus vertigo syndrome. *Adv. Oto-RhinoLaryngol.* 30, 291–297.

Carlow, T. J. (1986). Medical treatment of nystagmus and ocular motor disorders. *Int. Ophthalmol. Clin.* 26, 251–264.

Cüppers, C. (1971). Probleme der operativen Therapie des oculären Nystagmus. *Klin. Monatsbl. Augenheilkd.* 159, 145.

Currie, J., and Matsuo, V. (1986). The use of clonazepam in the treatment of nystagmus induced oscillopsia. *Ophthalmology* 93, 924–932.

Das, V. E., Oruganti, P., Kramer, P. D., and Leigh, R. J. (2000). Experimental tests of a neural-network model for ocular oscillations caused by disease of central myelin. *Exp. Brain Res.* 133, 189–197.

Dell'Osso, L. F., Hertle, R. W., Williams, R. W., and Jacobs, J. B. (1999). A new surgery for congenital nystagmus: effects of tenotomy on an achiasmatic canine and the role of extraocular proprioception. *J AAPOS* 3, 166–182.

Diaz, J. M., Urban, E. S., Schiffman, J. S., and Peterson, A. C. (1992). Post-irradiation neuromyotonia affecting trigeminal nerve distribution: An unusual presentation. *Neurology* 42, 1102–1104.

Dieterich, M., and Brandt, T. (1999). Episodic vertigo related to migraine (90 cases): Vestibular migraine? *J. Neurol.* 246, 883–892.

Dieterich, M., Straube, A., Brandt, T., Paulus, W., and Büttner, U. (1991). The effects of baclofen and cholinergic drugs on upbeat and downbeat nsytagmus. *J. Neurol. Neurosurg. Psychiatry* 54, 627–632.

Dunn, W. J., Arnold, A. C., and O'Connor, P. S. (1986). Botulinum toxin for the treatment of dysthyroid ocular myopathy. *Ophthalmology* 93, 470–475.

Ell, J., Gresty, M., Chambers, B. R., and Frindley, L. (1982). Acquired pendular nsytagmus: Characteristics, pathophysiology and pharmacological modification. In "Physiological and Pathological Aspects of Eye Movements" (A. Roucoux and M. Crommeilinck, Eds.), pp. 89–98. Junk, The Hague, Boston, and London.

Fisher, A., Gresty, M., Chambers, B., and Rudge, P. (1983). Primary position upbeating nystagmus: A variety of central positional nystagmus. *Brain* 106, 949–964.

Frisèn, L., and Wikkelso, C. (1986). Posttraumatic seesaw nystagmus abolished by ethanol ingestion. *Neurology* 36, 841–844.

Frohman, E. M., and Zee, D. S. (1995). Ocular Neuromyotonia: Clinical features, physiological mechanisms, and response to therapy. *Ann. Neurol.* 37, 620–626.

Furman, J. M. R., Wall, C., and Pang, D. (1990). Vestibular function in periodic alternating nystagmus. *Brain* 113, 1425–1439.

Gresty, M., Ell, J. J., and Findley, L. J. (1982). Acquired pendular nystagmus: Its characteristics, localising value and pathophysiology. *J. Neurol. Neurosurg. Psychiatry* 45, 431–439.

Griggs, R. C., Moxley, R. T., La France, R. A., and McQuillen, J. (1978). Hereditary paroxysmal ataxia: Response to acetazolamide. *Neurology* 28, 1259–1264.

Halmagyi, M. G., Aw, S. T., Dehaene, I., Curthoys, I. S., and Todd, M. J. (1994). Jerk-waveform see-saw nystagmus due to unilateral meso-diencephalic lesion. *Brain* 117, 789–803.

Halmagyi, M. G., and Curthoys, I. S. (1988). A clinical sign of canal paresis. *Arch. Neurol.* 45, 737–739.

Halmagyi, M. G., Rudge, P., and Gresty, M. A. (1980). Treatment of periodic alternating nystagmus. *Ann. Neurol.* 8, 609–611.

Halmagyi, M. G., Rudge, P., Gresty, M. A., and Sanders, M. D. (1983). Downbeating nystagmus. A review of 62 cases. *Arch. Neurol.* 40, 777–784.

Helmchen, C., Dieterich, M., Straube, A., and Büttner, U. (1992). Abduzensneuromyotonie mit partieller Okulomotoriusparese. *Nervenarzt* 63, 625–629.

Helmchen, C., Rambold, H., Fuhry, L., and Büttner, U. (1998). Deficits in vertical and torsional eye movements after uni- and bilateral muscimol inactivation of the interstitial nucleus of Cajal of the alert monkey. *Exp. Brain Res.* 119, 436–452.

Herishanu, Y., and Louzoun, Z. (1986). Trihexyphenidyl treatment of vertical pendular nystagmus. *Neurology* 36, 82–84.

Herzau, V., Körner, F., Kommerell, G., and Friedel, B. (1978). Obliquus superior myokymie: Eine klinische und elektromyographische studie. In "Augenbewegungsstörungen. Neurophysiologie und Klinik" (G. Kommerell, Ed.), pp. 81–90. Springer-Verlag (Bergmann), Munich.

Hoyt, W. F., and Keane, J. R. (1962). Superior oblique myokymia: Report and discussion of five cases of benign intermittent uniocular microtremor. *Arch. Ophthalmol.* 84, 461–467.

Isago, H., Tsuboya, R., and Kataura, A. (1985). A case of periodic alternating nystagmus: With special reference to the efficacy of baclofen treatment. *Auris Nasus Larynx* 12, 15–21.

Jabbari, B., Rosenberg, M., Scherokman, B., Gunderson, C. H., McBurney, J. W., and McClintock, W. (1987). Effectivness of trihexyphenidyl against pendular nystagmus and palatal myoclonus: Evidence of cholinergic dysfunction. *Movement Disorders* 2, 93–98.

Jannetta, P. J., Moller, M. D., and Moller, A. R. (1984). Disabling positional vertigo. *N. Engl. J. Med.* 310, 1700–1705.

Larmande, P., and Larmande, A. (1983). Action du baclofene sur le nystagmus alternant peridoque. *Bull. Mem. Soc. Fr. Ophtalmol.* 94, 390–393.

Larmande, P., and Pautrizel, B. (1981). Traitement du nsytagmus congénital par 5-hydroxytryptophane. *Presse Med.* 10, 3166.

Lee, J. P. (1984). Superior oblique myokymia: A possible etiologic factor. *Arch. Ophthalmol.* 102, 1178–1179.

Leigh, R. J., Averbuch-Heller, L., Tomsak, R. L., Remler, B. F., Yaniglos, S. S., and Dell'Oso, L. F. (1994). Treatment of abnormal eye movements that impair vision: Strategies based on current concepts of physiology and pharmacology. *Ann. Neurol.* 36, 129–141.

Leigh, R. J., Burnstine, T. H., Ruff, R. L., and Kasmer, R. J. (1991). The effect of anticholinergic agents upon acquired nystagmus: A double-blind study of trihexyphenidyl and tridihexethyl chloride. *Neurology* 41, 1737–1741.

Leigh, R. J., Rushton, D. N., Thurston, S. E., Hertle, R. W., and Yaniglos, S. S. (1988). Effects of retinal image stabilization in acquired nystagmus due to neurologic disease. *Neurology* **38**, 122–127.

Leigh, R. J., Tomsak, R. L., Grant, M. P., Remler, B. F., Yaniglos, S. S., Lystad, L., and Dell'Osso, L. F. (1992). Effectiveness of botulinum toxin administered to abolish acquired nystagmus. *Ann. Neurol.* **32**, 633–642.

Leigh, R. J., Tomsak, R. L., Seidman, S. H., and Dell'Osso, L. F. (1991). Superior oblique myokymia. Quantitative characteristics of the eye movements in three patients. *Arch. Ophthalmol.* **109**, 1710–1713.

Leigh, R. J., and Zee, D. S. (1999). "The Neurology of Eye Movements," 3rd ed. Oxford University Press, New York.

Leopold, H. C. (1985). Opsoklonus- und Myoklonie-Syndrom. Klinische und elektronystagmographische Befunde mit Verlaufsstudien. *Fortschr. Neurol. Psychiatrie* **53**, 42–54.

Lepore, F. E. (1987). Ethanol-induced resolution of pathologic nystagmus. *Neurology* **37**, 877.

Mezawa, M., Ishikawa, S., and Ukai, K. (1990). Changes in waveform of congenital nystagmus associated with biofeedback treatment. *Br. J. Ophthalmol.* **74**, 472–476.

Mossman, S. S., Bronstein, A. M., Rudge, P., and Gresty, M. A. (1993). Acquired pendular nystagmus suppressed by alcohol. *NeuroOphthalmology* **13**, 99–106.

Mühlendyck, H. (1979). Therapeutische Möglichkeiten bei Nystagmuspatienten mit guter Binokularfunktion und Abnahme der Nystagmusintensität in der Nähe. *Schielen* **11**, 133.

Nathanson, M., Bergman, P. S., and Bender, M. B. (1953). Visual disturbances as the result of nystagmus on direct forward gaze. Effect of Amobarbital Sodium. *Arch. Neurol. Psychiatry* **69**, 427–435.

Nausieda, P. A., Tanner, C. M., and Weiner, W. J. (1981). Opsoclonic cerebellopathy. A paraneoplastic syndrome responsive to thiamine. *Arch. Neurol.* **38**, 780–782.

Nuti, D., Ciacci, G., Giannini, F., Rossi, A., and Frederico, A. (1986). Aperiodic alternating nystagmus: Report of two cases and treatment by baclofen. *Ital. J. Neurol. Sci.* **7**, 453–459.

Optican, L. M., and Zee, D. S. (1984). A hypothetical explanation of congenital nystagmus. *Biol. Cybernetics* **50**, 119–134.

Palmer, E. A., and Shults, W. T. (1984). Superior oblique myokymia: Preliminary results of surgical treatment. *J. Pediatr. Ophthalmol. Strabismus* **21**, 91–101.

Pedersen, R. A., Troost, B. T., Abel, L. A., and Zorub, D. (1980). Intermittent down beat nystagmus and oscillopsia reversed by suboccipital craniectomy. *Neurology* **30**, 1232–1242.

Pereira, C. B., Strupp, M., Eggert, T., Straube, A., and Brandt, T. (2000). Nicotine-induced nystagmus: Three-dimensional analysis and dependence on head position. *Neurology* **55**, 1563–1565.

Phelan, K. D., and Gallagher, J. P. (1992). Direct muscarinic and nicotinic receptor mediated excitation of rat medial vestibular nucleus neurons in vitro. *Synapse* **10**, 349–358.

Pless, M., and Ronthal, M. (1996). Treatment of opsoclonus-myoclonus with high-dose intravenous immunoglobulin. *Neurology* **46**, 583–584.

Pyykkö, I., Schalén, J., and Matsuoka, I. (1985). Transdermally administered scopolamine vs. dimenhydrinate. II. Effect on different types of nystagmus. *Acta Oto-Laryngol* **99**, 597–604.

Remler, B. F., Leigh, J., Osorio, I., and Tomsak, R. L. (1990). The characteristics and mechanisms of visual disturbance associated with anticonvulsant therapy. *Neurology* **40**, 791–796.

Ricker, V. K., and Mertens, H. G. (1970). Okuäre Neuromyotonie. *Klin. Monatsbl. Augenheilkd.* **156**, 837–842.

Ridley, A., Kennard, C., Scholtz, C. L., Büttner-Ennever, J. A., Summers, B., and Turnbull, A. (1987). Omnipause neurons in two cases of opsoclonus associated with oat cell carcinoma of the lung. *Brain* **110**, 1699–1709.

Rosenberg, M. I., and Glaser, J. S. (1983). Superior oblique myokymia. *Ann. Neurol.* **13**, 667–669.

Rushton, D. N., and Rushton, R. H. (1984). An optical method for approximate stabilization of vision of the real world. *J. Physiol. (London)* **357**, 3P.

Scott, A. B. (1980). Botulinum toxin injection into extraocular muscles as an alternative to stabismus surgery. *Ophthalmology* **87**, 1044–1049.

Sheth, N. V., Dell'Osso, L. F., Leigh, R. J., Van Doren, C. L., and Peckham, H. P. (1995). The effects of afferent stimulation on congenital nystagmus foveation periods. *Vision Res.* **35**, 2371–2382.

Slater, R. (1979). Benign recurrent vertigo. *J. Neurol. Neurosurg. Psychiatry* **42**, 363–367.

Spooner, J. W., and Baloh, R. W. (1981). Arnold-Chiari malformation. Improvement in eye movements after surgical treatment. *Brain* **104**, 51–60.

Stahl, J. S., Averbuch-Heller, L., and Leigh, R. J. (2000). Acquired Nystagmus. *Arch. Ophthalmol.* **118**, 544–549.

Starck, M., Albrecht, H., Pöllmann, W., Straube, A., and Dieterich, M. (1997). Drug therapy of acquired nystagmus in multiple sklerose. *J. Neurol.* **244**, 9–16.

Straube, A., and Büttner, U. (1991). Medikamentöse Therapie supranukleärer Augenbewegungsstörungen. *Nervenarzt* **62**, 212–220.

Straube, A., Büttner, U., and Brandt, T. (1994). Recurrent attacks with skew deviation, torsional nystagmus and contraction of the left frontalis muscle. *Neurology*, **44**, 177–178.

Straube, A., Kurzan, R., and Büttner, U. (1991). Differential effects of bicuculline and muscimol microinjections into the vestibular nuclei on simian eye movements. *Exp. Brain Res.* **86**, 347–358.

Susac, J. O., Smith, J. L., and Schatz, N. J. (1973). Superior oblique myokymia. *Arch. Neurol.* **29**, 432–434.

Thurston, S. E., Leigh, J. R., Abel, L. A., and Dell'Osso, L. (1987). Hyperactive vestibulo-ocular reflex in cerebellar degeneration: Pathogenesis and treatment. *Neurology* **37**, 53–57.

Tijssen, C. C., Endtz, L. J., and Goor, C. (1985). The influence of physostigmine on visual-vestibular interaction in hereditary ataxias. *J. Neurol. Neurosurg. Psychiatry* **48**, 977–981.

Traccis, S., Marras, M. A., Puliga, M. V., Ruiu, M. C., Masala, P. G., Carboni, A., Aiello, I., Pugliatti, M., and Rosati, G. (1997). Square-wave jerks and squware-wave oscillations: Treatment with valproic acid. *Neuro-ophthalmology* **18**, 51–58.

Traccis, S., Rosati, G., Monaco, M. F., Aiello, I., and Agnetti, V. (1990). Successful treatment of acquired pendular elliptical nystagmus in multiple sclerosis with isoniazid and base-out prisms. *Neurology* **40**, 492–494.

Tyler, R. D., and Ruiz, R. S. (1990). Propranolol in the treatment of superior oblique myokymia. *Arch. Ophthalmol.* **108**, 175–176.

Waespe, W., Cohen, B., and Raphan, T. (1985). Dynamic modification of the vestibuloocular reflex by the nodulus and uvula. *Science* **228**, 199–202.

Yee, R. D., Baloh, R. W., and Honrubia, V. (1982). Effect of baclofen on congenital nystagmus. *In* "Functional Basis of Ocular Motility" (G. Lennerstrand, D. S. Zee, and E. Keller, Eds.), pp. 151–158. Pergamon, Oxford.

Zee, D. S. (1985). Mechanisms of nystagmus. *Am. J. Otolaryngol.* (*Suppl.*) 30–34.

Zubcov, A. A., Reinecke, R. D., Gottlob, I., Manley, D. R., and Calhoun, J. H. (1990). Treatment of manifest latent nystagmus. *Am. J. Ophthalmol.* **110**, 160–167.

CHAPTER 14

# Idiopathic (Bell's) Facial Palsy

Amparo Gutierrez and Austin J. Sumner

## CLINICAL ASPECTS (DIFFERENTIAL DIAGNOSIS)

Facial nerve compromise is the most commonly encountered cranial neuropathy. Clearly the idiopathic or Bell's palsy form is the most common cause of infranuclear seventh nerve dysfunction. Idiopathic facial nerve paralysis (IFNP) is defined as a facial paralysis of unknown origin; thus it is a diagnosis of exclusion (Table I). Most patients with Bell's palsy display a rather sudden onset of unilateral partial or complete facial muscle paralysis with no suggestion of central nervous system or posterior fossa/ear disease. Approximately 60% of patients have a preceding viral illness (Katusic *et al.*, 1986). Associated complaints in decreasing order of frequency include numbness or pain in front or behind the ear, alterations in taste, hyperacusis, facial numbness ipsilateral to the side of weakness and decreased tearing in the affected eye.

Physical examination will reveal some degree of weakness involving muscles of facial expression and platysmal. Facial muscles pull to the opposite side on smiling. Saliva and food may collect on the paralyzed side, and on attempting to close the eye the eyeball may be diverted upward and inward (Bell's phenomenon). A lesion proximal to the geniculate ganglion results in decreased tearing in the affected eye. If the chorda tympani is affected there may be a decrease in salivation and impaired taste in the anterior two-thirds of the tongue. The clinician must carefully examine the patient for the presence of any masses in the head or neck region, signs of ear vesicles, infection, or a symptom complex compatible with posterior fossa involvement. A thorough examination in combination with electrical and imaging studies should always be pursued when a patient suspected of having Bell's palsy has an atypical presentation or clinical course.

## NATURAL COURSE

The incidence of Bell's palsy is 20 to 30 per 100,000 persons. Pregnancy, diabetes, and hypertension have all been associated with an increased incidence of Bell's palsy (Dumitru, 1995). It affects men and women almost equally, peak incidence is between the ages of 10 and 40. Onset of facial paralysis has an equal distribution between right and left sides (Devriese *et al.*, 1990). Approximately 2–9% of patients will have a recurrent palsy affecting the same or opposite side (Proctor *et al.*, 1976). Full recovery can be expected to occur in approximately 80–90% of patients. Several factors may serve as indicators of a poor outcome, age over 60 years, untreated hypertension, nonear pain, and abnormal submandibular salivary flow rate. Recovery usually begins within 3 weeks of onset in the majority of patients. Patients that do not improve early are often left with residual deficits; there is little hope of recovery past 1 year. Ten to 15% of patients are left with residual unilateral weakness or secondary deformities such as synkinesis, tearing, or contracture.

Electrophysiological testing has played an important role in determining the extent of facial nerve injury when there is early, complete paralysis of the facial muscles. Testing is best performed on the fifth day after onset or within 2 weeks. Various techniques have been used, facial motor nerve conduction, blink reflex, electromyography, nerve excitability test, maximal stimulation test, and magnetic stimulation. The best technique appears to be facial motor nerve conduction studies. The facial motor nerve is routinely evaluated for (1) side-to-side comparison of compound motor action potential (CMAP) amplitudes and (2) facial nerve CMAP onset latency. CMAP side-to-side comparisons have been the best studied and seem to hold the best prognostic values. May studied 273 patients with IFNP.

TABLE I  Etiology of Facial Nerve Paralysis

Idiopathic
Trauma
Herpes zoster cephalicus (Ramsay–Hunt syndrome)
Demyelinating disorders
 • Multiple sclerosis, Guillian-Barré, CIDP
Tumors
 • Benign: schwannomas, neurofibromas, glomus jugulare,
   hemangiomas
 • Malignant: parotid tumors, basal cell carcinoma, temporal
   bone metastasis from breast, lungs, kidneys, squamous cell
   carcinoma of middle/external ear
Granulomas
 • Sarcoidosis, Melkersson–Rosenthal syndrome
Birth (congenital/acquired)
Infection
 • Syphyllis, tuberculosis, Lyme, leprosy, HIV
Endocrinological
 • Diabetes mellitus

He showed that patients with a 25% or greater sparing of CMAP amplitude when compared to the unaffected side had a 98% chance of a satisfactory recovery (May et al., 1983). Less than an 11% CMAP amplitude side-to-side comparison usually predicted a poor outcome.

Additional testing which can be performed includes the Schirmer's test for decreased lacrimation, stapedius reflex for hyperacusis, sialometry for salivation, and electrogustometry for abnormal taste on the anterior two-thirds of the tongue. Despite the relative popularity of these tests, they are subject to a number of shortcomings, and their utility with respect to localization and prognostication has been questioned (Ekstran, 1979). Magnetic resonance imagining when performed often shows gadolinium enhancement of the facial nerve, this may last for several months but does not appear to play a prognostic role (Jonsson et al., 1995). The most frequently enhancing segments are usually the geniculate ganglion and intracanalicular segment (Saatci et al., 1996). Enhancement occurs secondary to a breakdown in the blood–nerve barrier and may signify the presence of the initial demyelinating insult.

## PRINCIPLES OF THERAPY

Idiopathic facial nerve palsy is a self-limiting, non-life-threatening, and spontaneously remitting disorder in the majority of cases. Histological studies of the facial nerve during the acute stage of Bell's palsy, either at autopsy or during surgery, have shown signs of edema, perivascular perineurial lymphocytic and macrophage infiltration of the nerve, an increase in the axon:myelin surface ratio (thinning of myelin), and a decrease in the total fiber count (Podvinec, 1984) (Prescott, 1988). Edema of the nerve confined within the facial canal

also contributes to the constriction of nerve fibers and subsequent production of conduction slowing or block (neurapraxia). Prolonged or increased constriction within the canal leads to Wallerian degeneration and more severe injury (axonotmesis). Therefore there is a wide spectrum of injury in Bell's palsy, from neurapraxia to axonotmesis.

A variety of viral etiologies have been proposed for idiopathic facial nerve palsy, including varicella–zoster virus, cytomegalovirus or adenovirus, mononucleosis, Coxsackie, influenza, polio, and mumps. Current evidence points to herpes simplex virus-1 as a very common causative agent. A recent study found herpes simplex virus type 1 (HSV-1) by polymerase chain reaction of endoneurial fluid in 77% of patients with Bell's palsy (Murkakami et al., 1996). Whether the presence of HSV-1 represents a new infection or a reactivation of the latent virus within the geniculate ganglion is not known.

Although more than 150 years have passed since Sir Charles Bell established that the facial muscles are under the control of a separate cranial nerve, the treatment of IFNP still remains controversial. Medical treatment should be directed at a causative agent and the inflammatory event. Various treatment modalities have been employed in IFNP surgical decompression, electrophysiotherapy, physical therapy, biofeedback, injection of the stylomastoid foramen with steroids, intravenous dextran, intravenous steroids, oral nicotinic acid, and acyclovir. The most widely used treatment has been oral steroids. Determining the effectiveness of steroids in a disease such as Bell's palsy, where there is such a high, rapid remission rate, has been difficult. A large number of prospective and retrospective studies of the use of steroids in the treatment of Bell's palsy have been performed (Adour et al., 1972; May et al., 1976; Prescott, 1988). Only one randomized, double-blind placebo-controlled study seemed to show that patients with Bell's palsy benefited from early treatment with steroids (Austin et al., 1993). The difficulty with this study is that 29% of participants did not participate in the follow-up.

The hypothesis that HSV-1 contributes to the pathogenesis of Bell's palsy has gained increasing acceptance since 1975. This has paved the way for using antiviral treatment in Bell's palsy. Adour performed a placebo-controlled study in 119 patients, comparing acyclovir, an antiviral agent, with placebo (Adour et al., 1996). Patients received either acyclovir 2000 mg/day (400 mg five times daily) or placebo. All patients received prednisone. The number of patients with good recovery was significantly higher in those receiving the acyclovir and prednisone. The incidence of sequelae was nonsignificant in the two groups but there appeared to be a favorable trend. Similar findings were seen in a second trial

(Ramos-Macias *et al.*, 1992). A subsequent prospective, controlled, randomized study was performed comparing prednisone to acyclovir in 101 patients with Bell's palsy (Diego *et al.*, 1998) In this study the prednisone-treated group had a more favorable recovery than the acyclovir group. This study did not have a placebo group nor was it blinded. Thus, overall, the current state of affairs points to the need of further studies to clarify the appropriate treatment protocol.

## PRACTICAL MANAGEMENT

### Prednisone/Acyclovir

Treatment with steroids is modest at best, they should be started within seven days of onset of facial paralysis. The protocol most widely used is prednisone 1 mg/kg per day (up to 60 mg) for five days; if paralysis is incomplete, taper the prednisone over five days. If the paralysis is complete, then treatment is continued for 10 days and then tapered over a further 5 days. Treatment should not be stopped abruptly due to the potential for rebound edema. Incidences of side effects with steroids are minimal. Steroids should be avoided in patients with contraindications to their use. Acyclovir is used in conjunction with prednisone (2000 mg/day given in divided doses) in order to minimize gastrointestinal irritation. Patients need to be screened for renal or hepatic dysfunction prior to initiation of treatment. An alternative medication valacylovir, an ester of acyclovir, which has higher absorption and serum levels than acyclovir can be utilized at 1 g, orally, three times per day.

### Eye Care

Lubricating eye drops during the day and an ointment-like petrolatum at night is important in preventing corneal ulcerations. Taping the eye shut or the use of an eye patch may also be warranted.

### Eyelid Surgery

This procedure is reserved for patients with severe palsy or with poor recovery. Suture tarsorrhaphy is effective but limits vision. Gold weighting of the upper eyelid provides an alternative therapy, and can be removed as the facial nerve regenerates.

### Surgical Decompression of the Nerve

The main role for surgical exploration is in patients where the diagnosis of Bell's palsy is in doubt. Surgery is indicated for the following reasons: (1) history of trauma, (2) clinical and radiographic evidence suggestive of an infiltrating lesion, (3) facial paralysis progresses over more than 1 month, (4) presence of other cranial nerve deficits, and (5) no recovery seen at 1 year. **Surgical exploration of patients who have not recovered after 1 year reveals atrophy and fibrosis of the facial nerve. In these cases surgical resection and reanimation may be warranted.** Facial reanimation techniques include end-to-end anastamosis of the facial nerve. Potential nerves to be used in grafting are the sural, hypoglossal, or greater auricular nerve (May and Klein, 1991).

### Botulinum Toxin Injections

Disfiguring synkinesis can occur as a result of facial nerve palsy. The use of botulism toxin has been useful in amelorating facial synkinesis in this setting (Roggenkamper *et al.*, 1994).

## TREATMENTS NO LONGER RECOMMENDED

Recognizing that Bell's palsy is a neuritis should make surgical treatment moot in the acute phase (Knox *et al.*, 1998).

## REFERENCES

Adour, K. K., Ruboyianes, J. M., Von Doersten, P. G., Byl, F. M., Trent, C. S., Quesenberry, C. P., and Hitchcock, T. (1996). Bell's palsy treatment with acyclovir and prednisone alone: A double blind, randomized, controlled trial. *Ann. Otol. Rhinol. Laryngol.* **105**, 371–378.

Adour, K. K., Wingerd, J., Douglas, M. A., Bell, N., Manning, J. J., and Hurley, J. P. (1972). Prednisone for idiopathic facial paralysis (Bell's palsy). *N. Engl. J. Med.* **287**, 1268–1272.

Austin, J. R., Peskind, S. P., Austin, S. G., and Rice, D. H. (1993). Idiopathic facial nerve paralysis: A randomized double blind controlled study of placebo versus prednisone. *Laryngoscope* **103**, 1326–1333.

Devriese, P. P., Schumacher, T., Scheide, A., and De Jingk, R. (1990). Incidence, prognosis and recovery of Bell's palsy. A survey of 1000 patients (1974–1983). *Clin. Otolaryngol.* **15**, 15–27.

Diego, J. I., Prim, M. P., De Sarria, M. J., Madero, R., and Gavilan, J. (1998). Idiopathic Facial Paralysis: A randomized, prospective, and controlled study using single-dose prednisone versus acyclovir three times daily. *Laryngoscope* **108**, 573–575.

Dumitru, D. (1995). "Electrodiagnostic Medicine," pp. 700–740. Hanley & Belfus.

Ekstran, T. (1979). Bell's palsy: Prognostic accuracy of case history, sialometry, and taste impairment. *Clin. Otolaryngol.* **4**, 183–196.

Jonsson, L., Tien, R., Engstrom, M., and Thuomas, K. (1995). GD-DPTA enhanced MRI in Bell's palsy and herpes zoster oticus: An overview and implications for future studies. *Acta Otolaryngol. (Stockh.)* **115**, 577–584.

Katusic, S. K., Beard, C. M., Wiederholt, W. C., Bergstralh, E. J., and Kurlan, L. T. (1986). Incidence, clinical features, and prognosis in Bell's palsy. *Ann. Neurol.* **20**, 622–627.

Knox, G. W. (1998). Treatment controversy in Bell's palsy. *Arch. Otolaryngol. Head Neck Surg.* **124**, 821–824.

May, M., Blumenthal, F., and Klein, S. R. (1983). Acute Bell's palsy: Prognostic value of evoked electromyography, maximal stimulation, and other electrical tests. *Am. J. Otol.* **5**, 1–7.

May, M., and Hardin, W. B. (1977). Facial nerve palsy: Interpretation of neurological findings. *Trans. Am. Acad. Ophthalmol. Otolaryngol.* **84**, 710–722.

May, M., and Klein, S. R. (1991). Differiential diagnosis of facial nerve palsy. *Otolaryngol. Clin. North Am.* **24**, 613–645.

May, M., Wette, R., Hardin, W. B., and Sullivan, J. (1976). The use of steriods in Bell's palsy: A prospective controlled study. *Laryngoscope* **86**, 1111–1122.

Murkakami, S., Mizobuchi, M., Nakashiro, Y., Takashi, D., Hato, N., and Yanagihara, N. (1996). Bell palsy and herpes simplex virus: Identification of viral DNA in endoneurial fluid and muscle. *Ann. Intern. Med.* **124**, 27–30.

Podvinec, M. (1984). Facial nerve disorders: Anatomical, histological and clinical aspects. *Adv. Otorhinolaryngol.* **32**, 124–193.

Prescott, C. A. (1998). Idiopathic facial nerve palsy (The effect of Treatment with Steroids). *J. Laryngol. Otol.* **102**, 403–407.

Proctor, B., Corgill, D. A., and Proudd, G. (1976). The pathology of Bell's palsy. *Trans Am. Acad. Ophthalmol. Otolaryngol.* **82**, 70–80.

Ramos-Macias, A., De Miquel Martinez, I., Martin Sanchez, A. M., Gomez Gonzalez, J. L., and Martin Galan, A. (1992). Incorporacion del aciclovir en el tratamiento de la paralisis periferica. Un estudio de 45 casos. *Acta Otorrinolaringol. Esp.* **43**, 117–120.

Roggenkamper, P., Laskawi, R., Damenz, W., and Baetz, A. (1994). Orbicular synkinesis after facial paralysis: Treatment with botulism toxin. *Doc. Ophthalmol.* **86**, 395–402.

Saatci, I., Sahinturk, F., Sennaroglu, L., Boyvat, F., Gursel, B., and Besim, A. (1996). MRI of the facial nerve in idiopathic facial palsy. *Eur. Radiol.* **6**(5), 631–636.

Yanigahara, N., Mori, H., Maurimauri, H., Kozawa, T., Nakamura, K., and Kita, M. (1984). Bell's palsy, nonrecurrent versus and unilateral versus bilateral. *Arch. Otolaryngol.* **110**, 374–377.

CHAPTER 15

# Dizziness and Vertigo

Thomas Brandt

Vertigo is an unpleasant distortion of static gravitational orientation or an erroneous perception of motion of either the sufferer or the environment. It is not a well-defined disease entity but rather the outcome of many pathological processes causing a mismatch between the visual, vestibular, and somatosensory systems, all of which subserve both static and dynamic spatial orientation.

Vestibular syndromes are commonly characterized by a combination of phenomena involving perceived vertigo, nystagmus, ataxia, and nausea. These four manifestations correlate with different aspects of vestibular function and emanate from different sites within the central nervous system.

- The vertigo itself results from a disturbance of cortical spatial orientation.
- Nystagmus is secondary to a direction-specific imbalance in the vestibulo-ocular reflex, which activates brainstem ocular motor neuronal circuitry.
- Vestibular ataxia and postural imbalance are caused by inappropriate or abnormal activation of mono- and polysynaptic vestibulospinal pathways.
- The unpleasant autonomic responses with nausea, vomiting, and anxiety travel along ascending (anxiety) and descending vestibulo-autonomic pathways to activate the medullary vomiting center.

Vertigo, dizziness, and disequilibrium are common complaints of patients of all ages, particularly the elderly. As presenting symptoms, they occur in 5–10% of all patients seen by general practitioners and 10–20% of all patients seen by neurologists and otolaryngologists. The clinical spectrum of vertigo is broad, extending from vestibular rotatory vertigo with nausea and vomiting to presyncope light-headedness, from drug intoxication to hypoglycemic dizziness, from visual vertigo to phobias and panic attacks, and from motion sickness to height vertigo. Appropriate preventions and treatments differ for different types of dizziness and vertigo.

The prevailing good prognosis of vertigo should be emphasized, because

(1) many forms of vertigo have a benign cause and are characterized by spontaneous recovery of vestibular function or central compensation of a peripheral or central vestibular tone imbalance;
(2) most forms of vertigo can be effectively relieved by pharmacological treatment (Table I), physical therapy (Table II), surgery (Table III), or psychotherapy (Brandt, 1999).

## GENERAL THERAPEUTIC APPROACHES

There is no common treatment, and vestibular suppressants (Table IV) provide only symptomatic relief of vertigo and nausea. A specific therapeutic approach thus requires recognition of the various particular pathomechanisms involved (Baloh and Halmagyi, 1996; Bronstein *et al.*, 1996; Brandt, 1999). Such therapy can include causative, symptomatic, or preventive approaches.

### Antivertiginous and Antiemetic Drugs

A variety of drugs used for symptomatic relief of vertigo and nausea (Table IV) have the major side effect of general sedation (Foster and Baloh, 1996). Vestibular suppressants, including anticholinergics, antihistamines, and benzodiazepines, provide symptomatic relief of distressing symptoms by downregulating vestibular excitability. Antiemetics preferably control nausea and vomiting by acting on the medullary vomiting center, the

**TABLE I    Pharmacologic Therapies for Vertigo**

| Therapy | Vertigo |
|---|---|
| Vestibular suppressants | Symptomatic relief of nausea (in acute peripheral and vestibular nuclei lesions), prevention of motion sickness |
| Antiepileptic drugs | Vestibular epilepsy, vestibular paroxysmia (disabling positional vertigo), paroxysmal dysarthria and ataxia in MS, other central vestibular paroxysms, superior oblique myokymia |
| Beta-receptor blockers | Basilar migraine (vestibular migraine; benign recurrent vertigo) |
| Betahistine | Menière's disease |
| Antibiotics | Infections of the ear and temporal bone |
| Ototoxic antibiotics | Menière's disease (Menière's drop attacks) |
| Corticosteroids | Vestibular neuritis, autoimmune inner ear disease |
| Baclofen | Downbeat or upbeat nystagmus or vertigo |
| Acetazolamide | Familial episodic ataxia or vertigo |

chemoreceptor trigger zone, or the gastrointestinal tract itself. Vestibular suppressants are often acetylcholine and histamine antagonists, which act by competitive inhibition at muscarinic receptors in the vestibular nuclei, their most likely site of action. Vestibular suppression by benzodiazepines is best explained by their $GABA_A$ agonistic effect, because GABA is the major neuroinhibitory transmitter for vestibular neurons. Antiemetics are effective mainly due to their dopamine ($D_2$) antagonist properties, but some antiemetics also have muscarinergic or antihistaminic ($H_1$) properties that may assist in vestibular suppression as well. Primary vestibular suppressants such as scopolamine also effectively suppress vomiting by virtue of their muscarinergic action. Antiemetics are more selective in action. They are primarily used to control nausea and vomiting; for treatment of severe vertigo with nausea,

they are often combined with antivertiginous drugs (Foster and Baloh, 1996).

There are only four clear indications for the use of antivertiginous (vestibular suppressants) and antiemetic drugs to control vertigo, nausea, and vomiting:

(1) to prevent nausea due to acute peripheral vestibulopathy (for the first 1–3 days or as long as nausea lasts),
(2) to prevent severe vertigo and nausea due to acute brainstem or archicerebellar lesions near the vestibular nuclei,
(3) to prevent severe vertigo attacks recurring on a frequent basis, and
(4) to prevent motion sickness.

For the first two of these conditions, fast-acting compounds with vestibular and general sedation should be

**TABLE II    Physical Therapies for Vertigo**

| Therapy | Vertigo |
|---|---|
| Deliberate maneuvers | Benign paroxysmal positioning vertigo |
| Vestibular exercises | Vestibular rehabilitation, central compensation of acute vestibular loss, habituation for prevention of motion sickness, improvement of balance skills (e.g., in the elderly) |
| Physical therapy (neck collar) | Cervical vertigo (fiction or reality?) |

**TABLE III    Surgical Interventions for Vertigo**

| Surgery | Causes of vertigo |
|---|---|
| Surgical decompression of eighth nerve | Tumor (vestibular schwannoma) or cyst |
| Neurovascular decompression | Vestibular paroxysmia (disabling positional vertigo) |
| Ampullary nerve section or canal plugging | Benign paroxysmal positioning vertigo (exceptional cases only), superior canal dehiscence (plugging or resurfacing) |
| Endolymphatic shunt | Menière's disease (questionable) |
| Vestibular nerve section or labyrinthectomy | Intractable Menière's disease (exceptional cases only) |
| Surgical patching | Perilymph fistula |
| Surgical decompression of vertebral artery | Rotational vertebral artery occlusion syndrome |

TABLE IV   Commonly Used Antivertiginous and Antiemetic Drugs

| Drug | Dosage | Action |
|------|--------|--------|
| Anticholinergics | | Muscarine antagonist |
|   Scopolamine (Transderm Scop) | 0.6 mg PO q 4–6 h or transdermal patch: 1 q 3 days | |
| Antihistamines | | |
|   Dimenhydrinate (Dramamine) | 50 mg PO q 4–6 h or IM q 4–6 h or 100 mg suppository q 8–10 h | Histamine (H₁) antagonist Muscarine antagonist |
|   Meclizine (Antivert, Bonine) | 25 mg PO q 4–6 h | Histamine (H₁) antagonist Muscarine antagonist |
|   Promethazine (Phenergan) | 15 or 50 mg PO q 4–6 h or IM q 4–6 h or suppository q 4–6 h | Histamine (H₁) antagonist Muscarine antagonist Dopamine (D₂) antagonist |
| Phenothiazine | | |
|   Prochlorperazine (Compazine) | 5 or 10 mg PO q 4–6 h or IM q 6 h or 25 mg suppository q 12 h | Muscarine antagonist Dopamine (D₂) antagonist |
| Butyrophenone | | |
|   Droperidol (Inapsine) | 2.5 or 5 mg IM q 12 h | Muscarine antagonist Dopamine (D₂) antagonist |
| Benzodiazepines | | |
|   Diazepam (Valium) | 5 or 10 mg PO bid–qid IM q 4–6 h or IV q 4–6 h | GABAₐ agonist |
|   Clonazepam (Klonopin) | 0.5 mg PO tid | GABAₐ agonist |

preferably administered, e.g., diazepam or promethazine combined with dimenhydrinate if nausea and vomiting are exceptionally severe. These drugs should not be given after nausea has disappeared, because they prolong the time course of central compensation of an acute vestibular tone imbalance.

Antivertiginous and antiemetic drugs are not indicated for patients suffering from chronic dizziness. A prophylactic treatment with vestibular suppressants, e.g., scopolamine or dimenhydrinate, is justified only in exceptional situations of rare patients who have frequent and severe vertigo attacks. In severe cases of benign paroxysmal positioning vertigo it may become necessary to control nausea and vomiting when performing physical liberatory maneuvers. It is our own experience that severe central positioning vomiting is best controlled by benzodiazepines rather than antiemetics or typical vestibular suppressants (Arbusow *et al.*, 1998). Scopolamine administered transdermally as Transderm Scop provides a continuous blood level over a 3-day period and effectively prevents motion sickness. The selection of vestibular suppressants and antiemetic drugs should take account of the fact that those that reach a peak effect 7–9 h after ingestion (Manning *et al.*, 1992) are ineffective for treating short vertigo attacks. Other effective drugs can be expected to be developed from compounds that interfere with the presynaptic histamine receptor $H_3$ or GABAₐ receptors.

## Specific Drug Treatment for Vestibular Disorders

Corticosteroids and virostatics are currently being tested in prospective studies to determine their value as treatment for vestibular neuritis, which is most likely caused by viral inflammation of the vestibular ganglion. Paroxysmal vestibular syndromes resulting from pathological excitation rather than loss of function due to a lesion can effectively be treated by antiepileptics such as carbamazepine or phenytoin. These conditions include vestibular epilepsy, vestibular paroxysmia due to neurovascular cross-compression, or paroxysmal ataxia in multiple sclerosis. Familial episodic ataxia I and II are highly responsive to acetazolamide. Episodic vertigo that occurs in the form of an aura with basilar or "vestibular" migraine can be prevented by administering beta-receptor blockers.

Some reliable pharmacological therapies are available for abnormal vestibular and nonvestibular eye movements, which override fixation and thus cause oscillopsia and impair vision (see Chapter 13; Leigh *et al.*, 1994; Leigh and Ramat, 1999). The GABA_B agonist baclofen suppresses periodic alternating nystagmus in patients (Halmagyi *et al.*, 1980) and animals with experimental lesions of the nodulus and uvula. The GABAergic anticonvulsant gabapentine (Averbuch-Heller *et al.*, 1997) and the glutamate antagonist memantine (Starck *et al.*, 1997) also effectively suppress acquired pendular nystagmus. Baclofen provides an effective treatment of some patients with downbeat or upbeat nystagmus (Dieterich *et al.*, 1991); occasional patients will also respond to gabapentine (Averbuch-Heller *et al.*, 1997).

### Surgical Treatment

Surgical procedures for the treatment of dizzy patients primarily involve otolaryngologists but also to a minor extent neurosurgeons. There is no doubt that surgery is an appropriate therapy, e.g., for cerebellopontine angle tumors, an infratentorial cavernoma, or for patching a perilymph fistula, such as a dehiscence of the superior semicircular canal (Minor *et al.*, 1998). The same holds for the rotational vertebral artery syndrome, because of the danger of vertebral artery occlusion or embolism (Strupp *et al.*, 2000). In these cases, vertigo may be part of the clinical syndrome, but the indication for surgery is based mainly on the impending risk of brain and cranial nerve damage. Indications for surgical interventions based only on the goal of controlling recurrent or chronic vertigo are rare and should always be considered second choice after conservative management has failed. The multiple procedures can be classified as

- nondestructive
  — decompression of the eighth nerve (vestibular schwannoma, cerebello-pontine angle cyst)
  — neurovascular decompression of the eighth nerve (vestibular paroxysmia)
  — endolymphatic shunt in Menière's disease
  — surgical patching of perilymph fistulas
- selectively destructive
  — retrolabyrinthine or middle fossa vestibular nerve section in intractable Menière's disease
  — semicircular canal plugging or ampulary nerve section in intractable benign paroxysmal positioning vertigo
- destructive
  — oval window or transmastoid labyrinthectomy
  — translabyrinthine vestibular nerve section
  — laser labyrinthectomy.

### Vestibular Compensation and Substitution: Vestibular Exercises and Physical Therapy for Vestibular Rehabilitation

Vestibular exercises are performed either to promote central habituation so as to prevent motion sickness or to readjust vestibulo-ocular and vestibulo-spinal reflexes as a form of retraining for exceptional patient populations (vestibular compensation).

Animal experiments have shown that exercise may facilitate vestibular compensation (Fetter and Zee,

1988; Igarashi, 1986; Lacour *et al.*, 1976). The special role of visual input has been convincingly demonstrated. Furthermore, animal experiments suggest that there is a critical period for functional recovery that is crucial for achieving either optimal or minimal repair (Xerri and Lacour, 1980). The few available clinical control studies provide evidence that physical therapy is superior to general conditioning exercises (Herdman *et al.*, 2000). Such physical therapy helps patients with chronic dizziness (Horak *et al.*, 1992) and those after resection of an acoustic neuroma recover balance earlier than if not treated (Herdman *et al.*, 1995); similarly patients after acute unilateral vestibular neuritis exhibit normalization of postural sway within a significantly shorter time course than the control group (Strupp *et al.*, 1998).

Vestibular compensation is no "simple or single" process. It consists of multiple processes for perceptual, vestibulo-ocular, and vestibulo-spinal readjustment, which have different time courses at different sites in the brain and the spinal cord (Brandt *et al.*, 1997; Curthoys and Halmagyi, 1994; Dieringer, 1995). Therefore, vestibular rehabiliation should incorporate different exercises that involve eye, head, and body movements and the monitoring of patients' progress separately for the different perceptual, ocular motor, and postural vestibular functions. A study on the efficacy of vestibular exercises for compensation and substitution after an acute unilateral partial vestibular loss (cases of vestibular neuritis without recovery of peripheral function during the training phase) found that only postural balance was significantly facilitated, not the time course of recovery of ocular torsion or tilts of perceived vertical as measured in degrees (Strupp *et al.*, 1998). Recovery from bilateral labyrinthine loss was also demonstrated in animal experiments (Igarashi *et al.*, 1988) and in patients with chronic bilateral vestibular deficits (Krebs *et al.*, 1993). In such cases recovery takes place more slowly and incompletely, leaving permanent instability during intensified balance tasks and in darkness as well as rapid head movements while walking. Up to now no controlled studies have focused on the effects of physical exercise on the rehabilitation of patients with central vestibular disorders. However, the patients' rapid recovery, e.g., from lateropulsion in Wallenberg's syndrome, when they become mobilized and are able to perform intensive physical therapy seems to support the efficacy of exercises, but provides no proof.

Vestibular compensation is less perfect than generally believed. For instance, after acute unilateral vestibular deafferentation, which occurs in vestibular neuritis, the process of normalization is impressive for *static* conditions in the absence of head motion: the initial rotatory vertigo, spontaneous nystagmus, and postural imbalance subside. Compensation is, however, less impressive for *dynamic* conditions, especially when the vestibular system is exposed to high-frequency head accelerations (Curthoys and Halmagyi, 1994; Halmagyi *et al.*, 1990). The dynamic disequilibrium, i.e., VOR asymmetry, causes oscillopsia, the illusory movement of the environment due to excessive slip of images upon the retina during fast head movements or walking, because after uni- and bilateral peripheral vestibular lesions the VOR cannot generate fast compensatory eye rotations during high-frequency head rotations. The dynamic vestibular tone imbalance can be detected clinically by provoking a directional head-shaking nystagmus (Hain *et al.*, 1987) or by bedside testing of the VOR with rapid head rotation (Halmagyi and Curthoys, 1988). Thus, the so-called simple and complete vestibular compensation for peripheral deficits is only a legend.

Pharmacological and metabolic studies suggest that the process of compensation can be retarded by alcohol, phenobarbitol, chlorpromazine, diazepam, and ACTH antagonists, whereas caffeine, amphetamines, and steroids may accelerate it. Drug-induced modification of vestibular compensation continues to attract research interest, although without convincing proof (Curthoys, 2000). Gingko biloba extract, for example, has been reported to accelerate compensation, but according to a recent study in guinea pigs it seems that the vehicle carrying the extract was the responsible agent (Schlatter *et al.*, 1999). Vestibular compensation is usually considered a central "repair mechanism for a vestibular tone imbalance secondary to a peripheral vestibular loss. However, central compensation is also possible for central vestibular tone imbalance. It is still poorly understood which central vestibular syndromes can be compensated and which cannot. Upbeat and downbeat nystagmus may serve as an example. Acquired upbeat nystagmus is rarely permanent, whereas acquired downbeat nystagmus may be permanent.

# VESTIBULAR NERVE AND LABYRINTHINE DISORDERS

## Benign Paroxysmal Positioning Vertigo (BPPV; Frequent Posterior Semicircular Canal Type)

### Clinical Aspects

BPPV is the most common cause of vertigo, particularly in the elderly. It is a mechanical disorder of the inner ear in which precipitating positioning of the head causes an abnormal stimulation, usually of the posterior semicircular canal of the ear undermost and, less frequently, of the horizontal or the anterior semicircular canal. Definite diagnostic criteria for BPPV are based

on the time history of the burst of rotational vertigo, associated with the typical positioning nystagmus (for posterior canal BPPV):

(1) latency: vertigo and nystagmus begin one or more seconds after the head is tilted toward the affected ear and increase in severity to a maximum;
(2) duration less than 1 min: nystagmus gradually reduces after 10–40 s and ultimately abates even with maintenance of the precipitating head position;
(3) linear-rotatory nystagmus: the nystagmus is best seen with Frenzel's glasses, which prevent suppression by fixation. The nystagmus is linear-rotatory, with the fast phase beating toward the ear undermost or upward when the gaze is directed toward the ear uppermost;
(4) reversal: when the patient returns to a seated position, the vertigo and the nystagmus may recur less violently in the opposite direction;
(5) fatigability: constant repetition of this maneuver will result in ever-lessening symptoms.

### Natural Course

The age of onset of BPPV ranges from adolescence to old age and in the idiopathic group exhibits a peak incidence in the sixth and seventh decades; about 30% of the elderly above 70 years of age have had BPPV. There is a striking preponderance of females, exceeding a ratio of 2 : 1 in the idiopathic group (Baloh *et al.*, 1987), whereas the sexes are about equally distributed in the post-traumatic and postviral neurolabyrinthitis forms.

In the early stages BPPV is usually experienced on awakening in the morning rather than on first lying down. Large series of patients (Baloh *et al.*, 1987; Katsarkas and Kirkham, 1978) support the common clinical finding that the following conditions figure in the etiology of BPPV: head (labyrinthine) trauma, viral neural labyrinthitis, vertebro-basilar ischemia, post-surgery (ear and general), prolonged bedrest due to unrelated diseases, and most often idiopathic aging. In a series of 240 patients described by Baloh *et al.* (1987), the origin was idiopathic in about half of the cases. In the remainder, the most commonly identified causes were head trauma (17%) and vestibular neuritis (15%). We found that, of a total of 104 patients with unilateral BPPV, 12% had suffered from vestibular neuritis days or years previously. In about 10%, BPPV is bilateral (mostly asymmetrical), in particular in posttraumatic cases.

The natural history of BPPV is considered benign because of its spontaneous recovery within weeks or months, but in about 30% of patients, the condition persists when untreated and in another 20–30% it recurs after variable periods. Among 50% of our patients, the condition lasted more than 4 weeks before a diagnosis was established and more than 6 months in 10% of patients.

### Principles of Therapy

Schuknecht (1969) hypothesized that heavy debris settles on the cupula ("cupulolithiasis") of the canal, transforming it from a transducer of angular acceleration into a transducer of linear acceleration. There is now general acceptance than the debris floats freely within the endolymph of the canal ("canalolithiasis") (Brandt and Steddin, 1993; Epley, 1992; Parnes and McClure, 1991). The debris, consisting of particles possibly detached from the otolith, congeals to form a free-floating clot (plug); since the clot is heavier than the endolymph, it will always gravitate to the most dependent part of the canal during head position changes that alter the angle of the canal's plane relative to gravity (Figures 1 and 2). Analogous to a plunger, the clot induces bidirectional (push or pull) forces on the cupula,

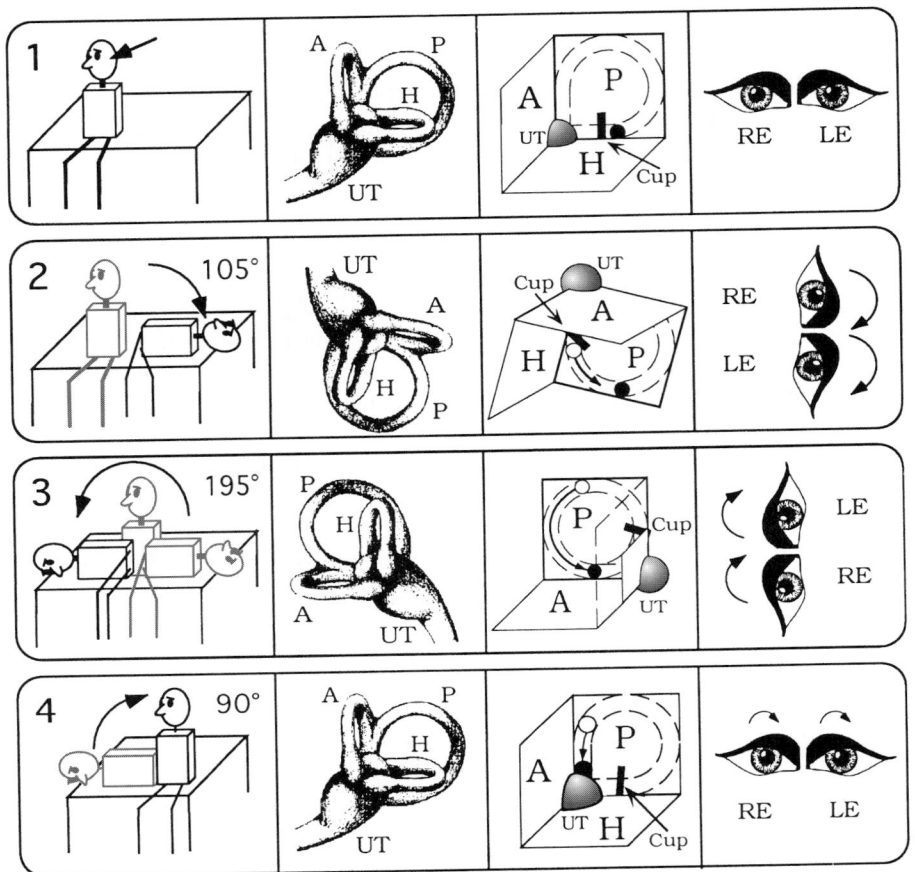

**FIGURE 1** Schematic drawing of the Semont liberatory maneuver in a patient with typical BPPV of the left ear. Boxes from left to right: position of body and head, position of labyrinth in space, position and movement of the clot in the posterior canal and resulting cupula deflection, and direction of the rotatory nystagmus. The clot is depicted as an open circle within the canal; a black circle represents the final resting position of the clot. (1) In the sitting position, the head is turned horizontally 45° to the unaffected ear. The clot, which is heavier than endolymph, settles at the base of the left posterior semicircular canal. (2) The patient is tilted approximately 105° toward the left (affected) ear. The head position change, relative to gravity, causes the clot to gravitate to the lowermost part of the canal and the cupula to deflect downward, inducing BPPV with rotatory nystagmus beating toward the undermost ear. The patient maintains this position for 3 min. (3) The patient is turned approximately 195° with the nose down, causing the clot to move toward the exit of the canal. The endolymphatic flow again deflects the cupula such that the nystagmus beats toward the left ear, now uppermost. The patient remains in this position for 3 min. (4) The patient is slowly moved to the sitting position; this causes the clot to enter the utricular cavity. A, P, and H, anterior, posterior, and horizontal semicircular canals, respectively; Cup, cupula; UT, utricular cavity; RE, right eye; and LE, left eye. Reprinted with permission from Brandt *et al.* (1994a).

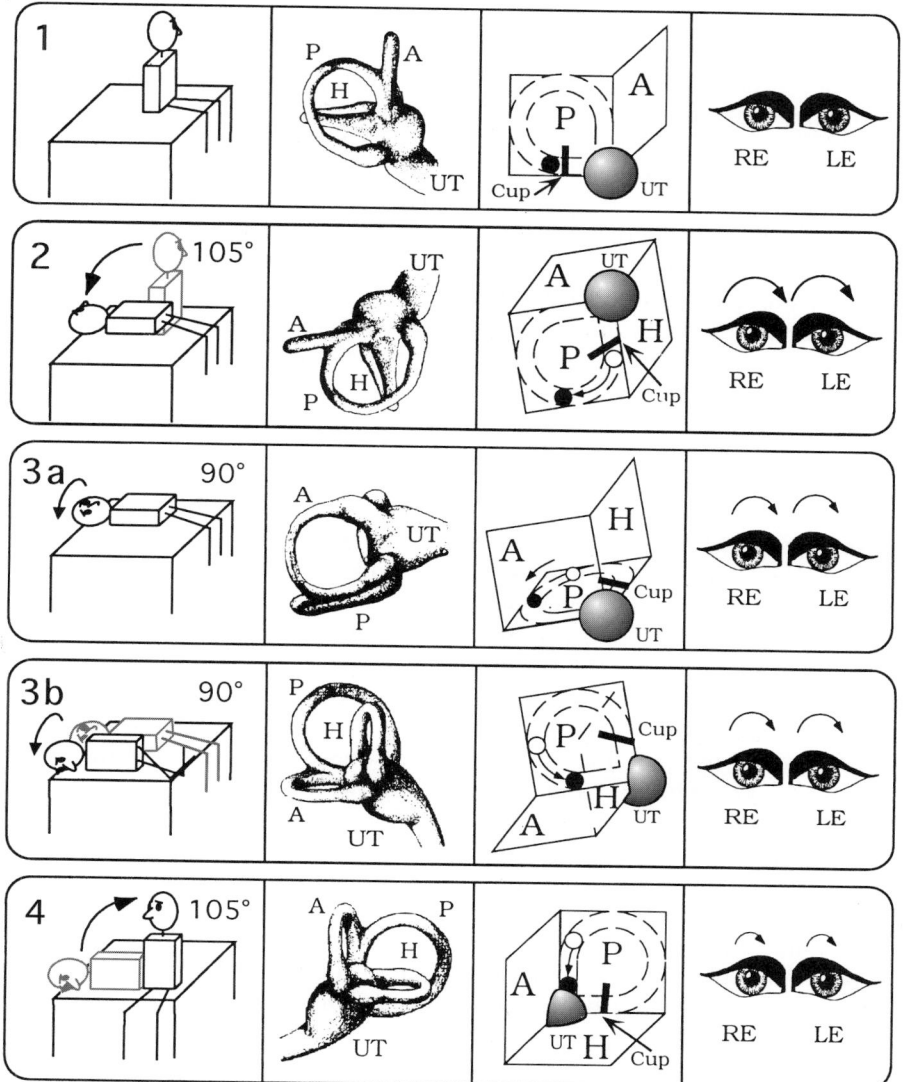

**FIGURE 2** Schematic drawing of modified Epley liberatory maneuver. Patient characteristics and abbreviations are as in the legend to Figure 1. (1) In the sitting position, the head is turned horizontally 45° to the affected (left) ear. (2) The patient is tilted approximately 105° backward into a slight head-hanging position, causing the clot to move in the canal, deflecting the cupula downward, and inducing the BPPV attack. The patient remains in this position for 3 min. (3a) The head is turned 90° to the unaffected ear, now undermost, and (see 3b) the head and trunk continue turning another 90° to the right, causing the clot to move toward the exit of the canal. The patient remains in this position for 3 min. The positioning nystagmus beating toward the affected (uppermost) ear in positions 3a and 3b indicates effective therapy. (4) The patient is moved to the sitting position. Reprinted with permission from Brandt *et al.* (1994a).

thereby triggering the BPPV attack. Canalolithiasis explains all features of BPPV (Brandt and Steddin, 1993): latency, short duration, fatigability, changes in direction of nystagmus with changes in head position, and efficacy of physical therapy.

In 1980, we proposed the first effective physical therapy (positional exercises) for BPPV, based on the assumption that cupulolithiasis was the underlying mechanism (Brandt and Daroff, 1980a). The exercises were a sequence of rapid lateral head–trunk tilts, repeated serially to promote loosening and, ultimately, dispersion of the debris toward the utricular cavity. In 1988, Semont *et al.* introduced a single liberatory maneuver (Fig. 1) and Epley, in 1992, proposed a variation, later modified by Herdman *et al.* (1993). With canalolithiasis as the established mechanism of BPPV, we can now explain the efficacy of the therapies by anatomic and physical principles.

## Practical Management

**Physical Therapy.** If performed properly, all three forms of therapy (Brandt–Daroff exercises and Semont and Epley's liberatory maneuvers) are effective in BPPV patients (Herdman, 1990; Herdman *et al.*, 1993).

For the Brandt–Daroff exercises, we instruct patients to sit; to then move rapidly into the challenging position to induce the correct plane-specific stimulation of the posterior semicircular canal; to remain in the position until the evoked vertigo subsides, or for at least 30 s; and then to sit up for 30 s before assuming the opposite head-down position for an additional 30 s.

Figure 1 illustrates the Semont maneuver in a patient with typical (posterior canal) left-sided BPPV. The clot causes no deflection of the cupula in the upright position. When the patient is quickly tilted toward the affected left ear with a 45° head rotation to the right (moving the left posterior canal to a plane corresponding with the plane of the head tilt), the clot gravitates toward the lower part of the canal, causing the cupula to deflect downward (ampullofugal) and triggering a typical BPPV attack. These events explain the latency of a few seconds (the time needed for the clot-induced endolymph flow to develop by gravitational force), the ineffectiveness of a very slow positioning movement (the clot would then slowly gravitate along the undermost wall of the canal without plugging the canal and deflecting the cupula), and the short duration of the positional vertigo/nystagmus (the cupula deflection ends when the clot reaches its lowest position in the canal (Brandt and Steddin, 1993). If the patient is swung toward the right side with the nose down, the clot will gravitate downward, causing stimulation of the posterior canal of the affected left ear (now uppermost). The patient is then slowly moved to the upright position; the clot will gravitate downward through the common crus of the posterior and anterior canals and enter the utricular cavity, where it becomes harmless. Semont *et al.* (1988) recommended having the patient maintain the upright position for 48 h after the liberation, but we have not found this to be necessary.

Figure 2 illustrates the Epley maneuver (Epley, 1992) as modified by Herdman *et al.* (1993) in a patient with typical posterior canal (left-sided) BPPV. The clot causes no deflection of the cupula in the upright position with the head turned horizontally 45° to the affected ear. When the patient is quickly tilted backward into a slight head-hanging position, the clot gravitates downward in the posterior canal, deflecting the cupula downward and inducing a BPPV attack. Rotation of the head and trunk toward the unaffected right ear causes further movement of the clot downward (ampullofugal) toward the exit of the canal, resulting in positioning vertigo and nystagmus toward the affected (now uppermost) ear. The final uprighting of the patient causes the clot to enter the utricular cavity, and it becomes harmless.

The process illustrated in Figures 1 and 2 explain the seemingly paradoxical observation in the literature that the final liberatory positioning with the affected ear uppermost (Figure 1, panel 3) induces nystagmus that beats toward that ear. The upward direction of the nystagmus induced by the final positioning provides reasonable certainty that the clot has exited the canal and the patient will be free of symptoms ("liberated"). If the nystagmus does not beat up toward the affected ear, the clot is probably still inside the canal; if the nystagmus beats downward toward the unaffected ear, the clot must have moved toward the cupula, causing an ampullopetal deflection (Figure 3). In either situation, the procedure should be repeated. If the nystagmus fails to beat upward during the second procedure and the BPPV persists, we reschedule a return visit for the same maneuver. If the second session fails, we try a different liberatory maneuver (i.e., modified Epley, if we first used Semont, and vice versa). If those liberatory maneuvers fail, we prescribe Brandt–Daroff exercises.

Following effective liberation, approximately 10–20% will experience a recurrence of attacks in the first

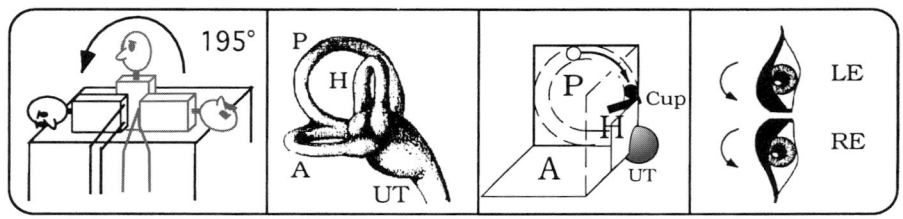

**FIGURE 3** Schematic drawing of an ineffective Semont liberatory maneuver to be compared with Fig. 1, panel 3. After the patient is tilted to the right, the clot migrates back toward the cupula. Endolymph flow causes an ampullopetal cupula deflection with the nystagmus beating downward to the unaffected ear. This probably indicates a failure of the liberatory maneuver.

2 weeks. The recurrences may be due to re-entry of the clot to the posterior canal from the utricular cavity and should be treated with the same maneuver that induced resolution of the initial episode.

**Surgical Therapy.** Only a few individuals do not respond well to physical therapy and ultimately require surgery. Here, singular neurectomy, a technically highly demanding surgical approach with a significant risk of failure and inadvertent loss of hearing, has been replaced by nonampullary plugging of the posterior semicircular canal (Pace-Balzan and Rutka, 1991; Parnes and McClure, 1991). Nonampullary plugging seems to constitute a safe and effective alternative to singular neurectomy for the small group of patients with physically intractable BPPV.

### Treatments No Longer Recommended

Owing to the pathomechanism of the condition, medical treatment of BPPV with antivertiginous drugs cannot be recommended. The only indication for these drugs is excessive nausea accompanying the BPPV in rare patients who therefore refuse the appropriate therapy.

## Horizontal BPPV (Rare Horizontal Semicircular Canal Type, h-BPPV)

h-BPPV accounts for about 10–20% of all patients presenting with BPPV). It may be combined with posterior-BPPV (p-BPPV) of the same or the contralateral ear. Transitions between h-BPPV and p-BPPV are also possible, particularly as a result of therapeutic positioning maneuvers. Whereas typical h-BPPV is caused by canalolithiasis, atypical h-BPPV may occur with ageotropic positioning nystagmus caused by cupulolithiasis. The etiological factors are the same as in p-BPPV.

### Clinical Aspects

The clinical characteristics of the horizontal semicircular canal variant of BPPV are easy to distinguish from those of posterior canal BPPV (Baloh *et al.*, 1993). The diagnosis of h-BPPV is based on the following features (Strupp *et al.*, 1995):

- The patient has a history of brief episodes of vertigo, usually induced by rolling the head from side to side while supine.
- Positioning testing reveals a linear horizontal nystagmus toward the undermost ear (geotropic) when the head of the supine patient is rapidly turned from side to side around the longitudinal *z*-axis (barrel roll).

- Horizontal positioning nystagmus with the head turned to either side beats stronger toward the affected ear; when this position is maintained, nystagmus often reverses its direction.
- Positioning nystagmus in h-BPPV exhibits short latencies (<5 s) and lasts longer (20–60 s) than in p-BPPV.
- h-BPPV rarely fatigues with repetitive positioning maneuvers.
- About one-third of the patients show moderate, horizontal semicircular paresis during caloric irrigation of the affected ear.
- Positioning vertigo attacks are often more severe than in p-BPPV and more frequently associated with nausea. As distinct from p-BPPV, attacks are not elicited by the patient getting in or out of bed, bending over, or extending the neck.

### Natural Course

h-BPPV is reported to occur from the third to the ninth decade; the mean age of manifestation is in the sixth decade (De la Meilleure *et al.*, 1996; Nuti *et al.*, 1996). There is a slight preponderance of cases in females. The right ear is affected as frequently as the left. The onset of the condition is usually abrupt as is its spontaneous remission after days to weeks. There are rare persistent courses which last for months to years. We share the experience of Baloh *et al.* (1993) and De la Meilleure *et al.* (1996): about half of the patients have a few or multiple relapses of the same condition. This is also true for the h-BPPV variant of cupulolithiasis.

### Principles of Therapy

Several features of h-BPPV remain unclear and are still a subject of speculation. For instance, why does canalolithiasis of the horizontal semicircular canal occur despite the fact that debris leave the canal on a simple head or body tilt from one side to the other, as is often performed while lying in bed? We suspect that the following two conditions must be fulfilled for the debris to remain in the canal (Steddin and Brandt, 1996):

(1) The diameter of the congealed debris must be greater than that of the bottleneck-like narrowing of the distal branch of the canal (Curthoys and Oman, 1987).
(2) The configuration of the debris, which congeal in the canal, must be so stable that the clot does not break into pieces small enough to pass the bottleneck.

If the fatigability of symptoms on repetitive testing is explained by transient dissolving of the debris, then nonfatigability, which is frequently seen in h-BPPV, may prove the above assumption. Consequently, as long as there is no fatigue of vertigo and nystagmus in h-BPPV, maneuvers intended to sluice the debris out of the canal will have minor success. Lempert and Tiel-Wilck (1996) described a single 270° "barbecue rotation" toward the unaffected side, which was performed in rapid steps of 90° at 30-s intervals. Baloh (1994) suggested a 360° rotation around the yaw axis in four quick steps of 90° with the initial motion toward the healthy side; each position was held for about 1 min.

Vannucchi *et al.* (1997) compared the therapeutic results obtained by maintaining a prolonged position on the healthy side (35 patients) with repetitive head shaking in a supine position (24 patients) and no therapy (15 patients). More than 90% of the patients treated with prolonged position recovered within 3 days, although 6 of 35 patients subsequently developed p-BPPV (which responded successfully to repositioning maneuvers). The rationale was obviously that the "heavy particles" in the nonampullary arm of the horizontal canal gradually moved out when the patient maintained a prolonged position on the side of the nonaffected ear. This maneuver was not effective if performed only for a duration of 10–20 min; it was intended to be maintained for up to 12 h. In those who failed to respond, the Brandt–Daroff exercises (Brandt and Daroff, 1980a) were performed and within a matter of days led to full recovery. This agrees with the report of Baloh *et al.* (1993) and our own experience with such patients. Nuti *et al.* (1998) compared the effects of "barbecue rotation" versus forced prolonged position in a large group of 92 patients with h-BPPV. Forced prolonged position was successful in more than 70%, while the "barbecue rotation" was slightly less successful but had more immediate results.

Thus, for the time being, we first try "barbecue rotation" and propose that prolonged bedrest with the head turned toward the unaffected ear be maintained for up to 12 h. If this is still unsuccessful after 2 days, we advise the patients to perform the Brandt–Daroff exercises. Both physical therapies can be performed at home and do not require the presence of a physical therapist.

## Vestibular Neuritis

### Clinical Aspects

The chief symptom of the common unilateral labrynthine lesion, also known as vestibular neuritis, is acute onset of prolonged severe rotational vertigo associated with spontaneous nystagmus, postural imbalance, and nausea, without concomitant auditory dysfunction. Caloric testing invariably shows ipsilateral hyporespon-siveness (horizontal semicircular canal paresis). In vestibular neuritis, the fast phase of the rotational spontaneous nystagmus as well as the initial perception of apparent body motion are directed away from the side where the lesion is, and the postural reactions initiated by vestibulo-spinal reflexes are usually opposite to the direction of vertigo. These result in both the Romberg fall and past pointing toward the side of the lesion. The nystagmus is typically reduced in amplitude by fixation and enhanced by eye closure or Frenzel's lenses.

### Natural Course

Vestibular neuritis is a common cause of vertigo, and the age of onset peaks during the sixth decade (slightly more frequent among women than men). Its annual incidence is about 3.5/100,000 population (Sekitani *et al.*, 1993). Occasionally onset is preceded by short attacks of rotational vertigo a couple of days beforehand. Vertigo and the postural imbalance subside slowly over a period of 1–2 weeks, and all distressing symptoms abate when the tone is re-equalized by either peripheral restitution (50–70% of patients) or central compensation. In 30–50% of cases, there remains a distinct non- or underresponsiveness to thermic irrigation of the horizontal semicircular canal (Okinaka *et al.*, 1993; Ohbayashi *et al.*, 1993). In these patients, a high-frequency defect of the vestibulo-ocular reflex persists, which causes slight oscillopsia when the head is vigorously moved. Relapses of vestibular neuritis are possible but rare.

### Principles of Therapy

Epidemic occurrences and a few autopsy studies that exhibited cell degeneration of one or more vestibular nerve trunks (Schuknecht and Kitamura, 1981) support viral vestibular nerve site origin (Arbusow *et al.*, 1999), similar to that of Bell's palsy or sudden unilateral hearing loss. The possibility that the three semicircular canals and the otoliths (utricular and saccular) may be separately involved in partial labyrinthine lesions is suggested by the occasional observation of an acute unilateral vestibulopathy (horizontal canal paresis) and a benign paroxysmal positioning vertigo simultaneously in the same ear of a patient (Büchele and Brandt, 1988).

There is a distinct lack of systematic treatment studies with antiviral substances such as adeninearabinoside, acyclovir, or interferon. However, it has been shown that the combination of acyclovir and steroids significantly improves the outcome of Bell's palsy (Adour *et al.*, 1996), which most probably has the same pathogenesis. To date, two studies have reported that corticosteroids have a beneficial effect on the course of vestibular neuritis. The study of Ariyasu *et al.* (1990) included 20 randomly selected patients and was double-blind, prospective, placebo-controlled, and crossover; nevertheless, uncertainties about the precise diagnosis remained ("acute vestibular vertigo"). While the second study by Ohbayashi *et al.* (1993) had no clearly prospective design, it also reported that corticosteroids facilitate early recovery from vertigo and nystagmus.

Initially, the vestibular cerebellum helps to reduce the tone imbalance by diminishing spontaneous activity of the vestibular nuclei neurons and thus downregulates vestibular sensitivity. The actual repair mechanism—the central compensation develops over a period of weeks—is obviously produced by both activation and sprouting of commissural fibers between the vestibular nuclei. Here, the inferior vermis is evidently more important in terms of vestibulo-spinal compensation, whereas the archicerebellar flocculus, uvula, and nodulus are more important for vestibulo-ocular motor adaptation.

Readjustment of the vestibular reflexes, which act on eye and body muscles, requires sensory feedback from the sensory mismatch elicited by voluntary movements. Therefore, on the basis of our current knowledge of vestibular physiology, continued management of vestibular neuritis should consist of vestibular exercises that promote central compensation (Strupp *et al.*, 1998; see also p. 142).

### Practical Management

During the first 1–3 days when nausea is prominent, vestibular sedatives such as dimenhydrinate (50–100 mg every 6 h) or scopolamine (0.6 mg) can be administered parenterally for symptomatic relief, with the major side effect being general sedation. As soon as the patient is free of vomiting, this treatment should be stopped and replaced by a graduated vestibular exercise program (supervised by a physiotherapist) to promote central compensation.

Corticosteroids in the acute phase will most likely improve the condition.

### Treatments No Longer Recommended

Measures that stimulate blood flow are still commonly practiced but remain ineffective in vestibular neuritis.

### Related Clinical Syndromes

Typical initial symptoms of herpes zoster oticus (Ramsay-Hunt syndrome) are burning pain and the eruption of small blisters, in which case acyclovir is indicated. Toxic accompanying labyrinthitis in the case of middle-ear infections is treated with antibiotics,

whereas acute purulent labyrinthitis also requires surgery and drainage. Tuberculous labyrinthitis is more often a complication of tuberculous meningitis than of tuberculous middle-ear infection. Inherited syphilitic labyrinthitis has its peak onset in the fifth to sixth decade. Likewise, borelliosis can also cause acute vertigo (Rosenhall *et al.*, 1988). Cogan's syndrome, an autoimmune disease with interstitial keratitis and audiovestibular symptoms (mainly hearing disorders), is found predominantly among young adults and may respond positively (sometimes only temporarily) to corticosteroids or, like other autoimmune disorders of the inner ear, to a combination of cyclophosphamide and prednisolone (McCabe, 1989; Vollertsen *et al.*, 1986). Acute unilateral labyrinthine dysfuntion can also be caused by ischemia in the case of labyrinth infarcts (Gradt and Baloh, 1989) as well as by venous obstruction in the case of hyperviscosity syndrome. Reduction of blood viscosity is an effective form of treatment for all symptoms of hyperviscosity syndrome, including vertigo (Andrews *et al.*, 1988).

## Menière's Disease

### Clinical Aspects

The typical syndrome is characterized by fluctuating hearing loss, tinnitus, and prolonged decrescendo nystagmus-vertigo attacks for hours. This is accompanied by a horizontally rotating nystagmus, postural imbalance as well as nausea and vomiting. Most single attacks have no premonitory symptoms or recognizable precipitating factors. There is no connection to any particular time of the day and they even occur during sleep. Approximately one-third of patients report an increase in tinnitus and hearing loss, as well as an apparent fullness of the ear, which precedes the vertigo attack as a kind of aura. During the attack, most patients experience an increase in the severity of tinnitus and hearing loss. At the beginning, particularly with monosymptomatic forms of the condition and in the vertigo-free interval, diagnosis is difficult and clinical function tests may not be sufficiently informative. Later on, sustained progressive loss of hearing is the most important diagnostic measure.

Vestibular drop attacks may occur in the earlier or later stages of Menière's disease (Baloh *et al.*, 1990) with a reflex-like vestibulo-spinal loss of postural tone.

### Natural Course

Menière's disease is undoubtedly overdiagnosed. The usual age of onset is in the fourth to sixth decades, and there is a slight preponderance of females. Incidence of

Menière's disease was calculated to be 46/100,000 in a Swedish population if the cochlear type (with only fluctuating hearing loss) was not included (Stahle *et al.*, 1978). It rarely occurs during childhood (Sad and Yaniv, 1984). Genetic predisposing factors are suggested by the frequently positive family history.

In the course of the disease, tinnitus and fluctuating hearing loss sometimes announce the first vertigo attack years in advance since the hydrops normally begins in the inferior part of the labyrinth, the cochlear duct, with initial ruptures of Reissner's membrane within the helicotrema. In this case, fullness of the ear is a characteristic sensation. Pure vestibular attacks without cochlear symptoms are comparatively rare.

Menière's disease usually begins in one ear with an irregularly increasing and fortunately, after some years, decreasing frequency of attacks. The major reduction of hearing occurs within the first few years of the disease; the same is true for the reduction in vestibular function as measured by caloric testing. The longer one follows up patients with Menière's disease, the greater is the percentage found to develop bilateral disease (Morrison, 1986). In the earlier stages, when symptoms have been present for up to 2 years, about 15% of the cases are bilateral while, after one to two decades, 30–60% are bilateral. There is agreement on the relatively benign natural cause of the disease, with a spontaneous remission rate of about 80% in 5–10 years (Friberg *et al.*, 1984). It is most likely that spontaneous remission of Meniere's attacks is due to permanent fistulization between the endo- and perilymph, allowing a continuous asymptomatic leakage of excessive endolymph.

### Principles of Therapy

Menière's disease arises from endolymphatic hydrops and periodic ruptures in the membranes separating endolymph from perilymph, which give rise to intermediate potassium palsy of vestibular nerve fibers. Contrary to earlier hypotheses, endolymphatic hydrops does not develop as a consequence of increased endolymph production or due to malfunction of the membrane that stabilizes concentrations of electrolytes and osmolarity. Impaired resorption of endolymph by the endolymphatic sac is the major cause of hydrops or obliteration of the endolymphatic duct with interruption of longitudinal endolymph circulation. An endolymph hydrops can also arise asymptomatically. Etiology may be inflammatory (labyrinthitis), traumatic, or accidental (Schuknecht and Gulya, 1983).

The management of Menière's disease has four aims:

(1) to treat the acute attack,
(2) to prevent further attacks,
(3) to improve and/or preserve hearing and vestibular function, and
(4) to prevent the development of bilateral Menière's disease.

To date, conservative and surgical procedures have proven to be effective for only the first two aims. There is much confusion in the literature about which therapy is effective and which is the treatment of first choice. No other vestibular disorder has been the subject of such a large number of articles (~1500 between 1966 and 1996) which nevertheless leave unanswered the questions of etiology, pathomechanism, and effective treatment of the condition.

According to a recent review of the literature (Class and van de Heyning, 2000), only diuretics and betahistine have a proven effect in double-blind studies on the long-term control of vertigo, but no medical therapy has a proven effect on hearing or long-term evolution of the disease.

The effect of diuretics (acetazolamide) on experimental hydrops in guinea pigs was confined to the period of administration, and neither the administration nor the nonadministration of the drugs had any effect on the extent of cochlear sensory and neural atrophy in animals (Shinkawa and Kimura, 1986). Urea, thiazide diuretics, and acetazolamide have been recommended for chronic treatment of Menière's disease. In a double-blind, placebo-controlled, crossover study, a combination of triamterene and hydrochlorothiazide effectively controlled vertigo (Van Deelen and Huizing, 1986). This combination again had no positive long-term effect on hearing, which confirms an earlier report by Klockhoff *et al.* (1974) on long-term results with chlorthalidone.

The histamine derivative betahistine has been recommended as the drug of first choice (Le Pére, 1967). A 1-year prospective double-blind study concluded that this treatment is preferable to no treatment (Meyer, 1985). While it acts to improve the microcirculation of the stria vascularis as a partial $H_1$ receptor agonist, it also has inhibitory effects on polysynaptic vestibular nucleus neurons. Other studies support the view that betahistine is significantly more effective than placebo (Oosterveld, 1984) and even more effective than the diuretic hydrochlorothiazide (Petermann and Mulch, 1982) or flunarizine (Fraysse *et al.*, 1991). None of these studies document an improvement in hearing loss. To prevent attacks, betahistine should be administered continuously for 6–12 months.

### Intratympanic Gentamicin Therapy

"Functional labyrinthectomy" can be performed by applying ototoxic aminoglycosides (gentamicin or streptomycin). In the 1980s it was thought that selective damage of the dark cells of the secretory epithelium was possible (thereby improving endolymphatic hydrops) before significantly affecting vestibular and cochlea function. Intratympanic instillations were stopped when daily audiograms or a check of spontaneous nystagmus by using Frenzel's glasses indicated a beginning end-organ deafferentation. Since then indications and recommendations for intratympanic gentamicin therapy (Graham and Goldsmith, 1994; Halmagyi *et al.*, 1994; Hellström and Ödkvist, 1994) have changed, especially when Magnusson and Padoan (1991) observed that the onset of ototoxic effects was delayed by a few days to 1 week after gentamicin instillation.

It is most likely that the route of transport of gentamicin from the middle to the inner ear is through the round window membrane to the perilymphatic space and from there to the hair cells in the endolymphatic space (Bergenius and Ödkvist, 1996). Ototoxicity is probably caused by a reversible transduction channel blocking (Kroese *et al.*, 1989) and by damage to mitochondria due to excessive mitochondrial superoxide production that leads to cell death (Hutchin and Cortopassi, 1994).

These multistage mechanisms of gentamicin ototoxicity are consistent with findings that the functional deficit was reversible at an early stage and became irreversible at a late stage (Halmagyi *et al.*, 1994). Application of excessive gentamicin can cause unnecessary, inadvertent damage to the inner ear receptors, including the cochlear hair cells. *Low-dose treatment—* which does not even diminish or abolish caloric responses of the treated ear—has also been demonstrated to be effective (Driscoll *et al.*, 1997; Murofushi *et al.*, 1997; Yamazaki *et al.*, 1991) and *is therefore recommended as the standard procedure.*

### Practical Management

The acute attack is self-limiting and subsides within a few hours (rarely less than 1 h or more than 1 day) in a slow decrescendo. Head movements or rapid changes in head position should be restricted because of cross-coupled accelerations and positional vertigo. If nausea is prominent, diazepam or vestibular sedatives such as dimenhydrinate (50 mg), perphenazine (4 mg), promethazine hydrochloride (25 mg), or scopolamine (0.6 mg) can be administered parenterally for symptomatic relief.

Pragmatic therapy includes

— Vestibular sedatives such as dimenhydrinate or scopolamine effectively reduce vertigo and nausea in the acute attack.

— Betahistine is the drug of first choice for preventing vertigo attacks (8–16 mg/day for 6–12 months).

— Diuretics (hydrochlorothiazide, 20–50 mg/day, or acetazolamide, 500 mg/day) are considered second choice for preventing vertigo attacks.

— Combinations of betahistine and diuretics may be tried if single-drug treatment fails.

— Intratympanic gentamicin is the treatment of first choice for preventing vertigo attacks or drop attacks in rare patients with intractable and frequent attacks lasting for more than 6–12 months and nonserviceable hearing in the affected ear. All the reported experience with this kind of treatment indicates that one injection per week (1–2 mL with concentrations less than 30 mg/mL) on an outpatient basis could be recommended in order to better monitor the delayed ototoxic effects.

— Selective vestibular neurectomy is considered second choice in the same exceptional patients or first choice in patients with moderate hearing loss.

Surgical and other procedures destructive to the peripheral labyrinth or vestibular nerve can successfully stop attacks of vertigo but do not improve hearing. For patients in whom more conservative procedures have failed, selective destructive surgical techniques such as middle-fossa vestibular nerve section or ultrasonic or cryo-surgical vestibular destruction, have been proposed with the intention of preserving serviceable hearing function. Since this surgical approach does not affect the hydrops pathomechanism and, therefore, does not prevent ongoing fluctuating hearing loss, there is an obvious recent tendency for it to be less frequently considered. Particularly in elderly patients, ablative surgical procedures may cause long-lasting postural imbalance, owing to the reduced capability of central compensation of the postoperative tone imbalance. Surgical fistualization in various parts of the membraneous labyrinth has been used in animal experiments, and cochlear endolymphatic shunt operation is a solution which Schuknecht and Bartley (1985) reported to have a success rate of 70% of the cases being relieved of vertigo.

### Treatments No Longer Recommended

Changing views on the pathogenesis have prompted the development of a variety of procedures. The existence of a large number of different therapies, each defended fiercely by its advocates, usually indicates that there is no demonstrably effective therapy available. Dietetic programs including restriction of salt, water, alcohol, nicotine, caffeine are as useless in treating the disease as are physical exercise, avoidance of exposure to low temperatures, or use of subatmospheric pressure chambers. Stellate ganglion blocks, vasoactive agents, tranquillizers, neuroleptics, and lithium have been employed under the questionable assumption that it is possible to diminish endolymphatic hydrops. All these procedures have been fashionable therapies at one time or another, but—with the possible exception of diuretics and betahistine—there has never been proof of their efficacy in controlled prospective studies.

## Perilymph Fistula

### Clinical Aspects

Perilymph fistulas may lead to episodic vertigo and sensorineural hearing loss owing to the pathologic elasticity of the bony labyrinth, usually at the round or oval window. Some fistulas appear as solely otolithic vertigo with periodic head motion intolerance, gait ataxia, and distressing tilt sensations on head tilt. Even though some typical clinical vascular and pressure fistula tests are useful, a definite diagnosis can be made only by exploratory tympanotomy, which sometimes demonstrates pumping of the stapes when the patients perform a Valsalva maneuver. The combination of vertigo, eye movements, and postural instability (either semicircular canal or otolith type) induced by high-intensity auditory stimulation in congenitally deaf patients and those with labyrinthine fistulas (e.g., luxated stapes footplate) is known as the Tullio phenomenon.

### Natural Course

A history of trauma (or physical exertion), with subsequent vertigo and hearing loss, and a positional nystagmus of short latency but long duration, which is not as violent as in benign paroxysmal positioning vertigo, should lead the physician to suspect a fistula. Most often, these fistulas heal spontaneously.

### Principles of Therapy

In the acute case, medical treatment is universally recommended because of the mostly benign spontaneous course and because the results of surgical intervention are not encouraging (Singleton et al., 1978).

### Practical Management

Medical care consists of absolute bed rest with head elevation for 5 to 10 days; avoidance of straining, sneezing, or coughing; and use of stool softeners. If symptoms persist for 3 to 4 weeks, exploratory tympanotomy is indicated under local anesthesia, with the Vasalva maneuver or gentle palpation of the footplate to make the leak apparent. For patching the fistula, perichondrial graft from the tragus is superior to fat.

### Fistula of the Anterior Semicircular Canal

A new syndrome has been described by Minor et al. (1998): a dehiscence of the bone overlying the anterior semicircular canal. The clinical manifestation of this syndrome was sound- and/or pressure-induced dizziness with vertical and rotatory oscillopsia or diplopia associated with vertical–torsional eye movements typical for excitation or inhibition of the anterior semicircular canal of the affected ear. The bony dehiscence can be identified on high-resolution temporal bone CT scan. About 1–2% of people fail to develop a normal thickness of bone overlying the superior canal. They are predisposed to the disruption of this thin layer of bone by head trauma or by changes in intracranial pressure (Minor, 2000).

Disabling disequilibrium prompted Minor and coworkers to plug the affected canal via the middle cranial fossa approach or to resurface the area of dehiscence. The symptoms improved after the plugging or resurfacing procedure (Minor, 2000; Minor et al., 1998).

## Vestibular Paroxysmia (Disabling Positioning Vertigo)

### Clinical Aspects

Neurovascular cross-compression of the root entry zone of the V, VII, and VIIIth cranial nerves can elicit distressing attacks of trigeminal neuralgia, hemifacial spasm, and glossopharyngeal neuralgia. The syndrome of neurovascular compression of the VIIIth cranial nerve was first described by Jannetta and later termed "disabling positional vertigo" (Jannetta et al., 1984; Moller et al., 1986), a description for a heterogeneous collection of signs and symptoms, which can even mimic classic symptoms or Menière's disease (Moller, 1988) but is not a reliably diagnosable disease entity. The lack of a well-defined syndrome and of a reliable diagnostic test make it difficult for the nonsurgical clinician to believe in this disease. We rely on our own preliminary experience with this interesting and treatable condition of episodic vertigo (vestibular paroxysmia) which has escaped notice for so long. In analogy to trigeminal neuralgia, we propose basing the diagnosis on five characteristic features (Brandt and Dieterich, 1994b):

(1) short and frequent attacks of rotational or to-and-fro vertigo lasting from seconds to minutes,
(2) attacks frequently dependent on particular head positions and modification of the duration of the attack by changing head position,
(3) hypacusias and/or tinnitus permanently or during attack,

(4) auditory or vestibular deficits measurable by neurophysiological methods,

(5) positive response to antiepileptic drugs (carbamazepine).

Transitions from conduction block to ectopic discharges may rarely occur during compression of the eighth nerve, which results in episodes of vestibular hypofunction and paroxysmal vestibular excitation (Arbusow *et al.*, 1998).

### Natural Course

Incidence and prevalence are not known for this condition. The age range was 25 to 67 years in our few patients with a mean age of 51 and a mean duration of the condition before diagnosis of 7 years.

### Principles of Therapy

Neurovascular cross-compression can cause local demyelinization of the root entry zone of the VIIIth nerve. Ephaptic transmission between bare axons and/or central hyperactivity initiated and maintained by the peripheral compression are the suggested mechanisms. Analogous to trigeminal neuralgia, antiepileptic drugs are the first choice of medical treatment of the condition (Brandt and Dieterich, 1994b) before surgical microvascular decompression (Møller, 1991) is contemplated.

### Practical Management

Carbamazepine should be tried at an initial dosage of $3 \times 100$ mg daily, increasing to a maximum of $2 \times 400$ mg daily within 10 days. In a first study, all patients responded promptly and significantly, even to the low initial dosage (Brandt and Dieterich, 1994b). Eight of 11 became symptom-free, and only 3 reported infrequent residual attacks. Efficacy of treatment has been followed up for 6 years in one patient and for 3 years in five others. In two patients carbamazepine had to be replaced by phenytoin because of allergic reactions. As phenytoin appeared less efficient, we tried pimozide, which gave complete relief.

The efficacy of microvascular decompression in patients suspected of suffering from "disabling positioning vertigo" has been reported with variable success. In most of the patients (Jannetta *et al.*, 1984; Møller *et al.*, 1986), it is the lack of a pathognomonic sign or test for neurovascular compression of the VIIIth nerve which explains the critical reserve of the nonsurgical clinician. Too many patients would be selected for an unnecessary and risky operation.

# CENTRAL VESTIBULAR VERTIGO

Central vestibular vertigo syndromes are caused, in the majority of cases, by dysfunction induced by a lesion, but a small proportion result from pathological excitation of various structures, ranging from the vestibular nuclei to the vestibular cortex (vestibular epilepsy, paroxysmal dysarthria, and ataxia in multiple sclerosis). Most of the patients have an intra-axial, paramedian infratentorial lesion, especially of the pontomedullary tegmentum, involving vestibular nuclei connections to oculomotor and vestibulo-cerebellar structures (Table I). A classification of vestibular brainstem syndromes has been proposed according to the three major functional planes of action of the vestibulo-ocular reflex (VOR; Brandt, 1999):

(1) disorders of the VOR in horizontal (yaw) plane: pseudovestibular neuritis (lacunar AICA or PICA infarction; multiple sclerosis plaques);

(2) disorders of VOR in sagittal (pitch) plane: downbeat and upbeat nystagmus/vertigo;

(3) disorders of VOR in frontal (roll) plane: ocular tilt reaction, lateropulsion, ocular skew-torsion, tilt of perceived vertical.

In this context, disorders of the VOR are not limited to those affecting eye movements but include those affecting the vestibulo-spinal reflexes and perception because VOR, perception, and control of posture are intimately and inextricably connected (Leigh and Brandt, 1993). Central vestibular syndromes are the cause of vertigo in up to 25% of patients whose main complaint is vertigo. The proportion is greater if one includes all the cerebellar and brainstem disorders in which vertigo is a secondary symptom.

Disorders of the VOR in yaw indicate a unilateral pontomedullary lesion involving the vestibular nuclei or the VIIIth nerve. The natural course is a gradual recovery within days to weeks by central compensation of the central vestibular tone imbalance facilitated by vestibular exercises. The same is true for vestibular syndromes secondary to a tone imbalance in the roll plane. This is secondary to lesions of the graviceptive pathways, which travel from the vestibular nuclei, crossing midline at pontine level, to the contralateral medial longitudinal fasciculus to reach the oculomotor nuclei and the interstitial nucleus of Cajal (Brandt and Dieterich, 1994a). Typical clinical signs are that of an ocular tilt reaction, skew deviation, ocular torsion, and tilts of perceived vertical which recover spontaneously or are facilitated by visual–vestibular exercises. Disorders of the VOR in pitch plane indicate bilateral midline lesions of the medulla or the pontomesencephalic junction. Baclofen may effectively suppress downbeat as well as upbeat

**TABLE V   Central Vestibular Syndromes**

| Site | Syndrome | Mechanism/Etiology |
|---|---|---|
| Vestibular cortex (multisensory) | Vestibular epilepsy | Vestibular seizures are auras (simple or complex partial multisensory seizures) |
| | Volvular epilepsy | Sensorimotor "vestibular" rotatory seizures with walking in small circles |
| | Nonepileptic cortical vertigo | Rare rotatory vertigo in acute lesions of the parieto-insular vestibular cortex |
| | Spatial hemineglect (contraversive) | Multisensory horizontal deviation of spatial attention with (right) parietal or (frontal?) cortex lesions |
| | Transient room-tilt illusions | Paroxysmal or transient mismatch of visual and vestibular 3-D spatial coordinate maps in vestibular brain stem, parietal or frontal cortex lesions |
| | Tilt of perceived vertical with body lateropulsion (mostly contraversive) | Vestibular tone imbalance in roll with acute lesions of the parieto-insular vestibular cortex |
| Thalamus | Thalamic astasia | Dorsolateral vestibular thalamic lesions |
| | Tilt of perceived vertical (ipsiversive or contraversive) with body lateropulsion | Vestibular tone imbalance in roll |
| Mesodiencephalic brain stem | Ocular tilt reaction (contraversive; ipsiversive if paroxysmal) | Vestibular tone imbalance in roll (integrator-OTR with INC lesions) |
| | Torsional nystagmus (ipsiversive or contraversive) | Ipsiversive in INC lesions, contraversive in riMLF lesions |
| Mesencephalic brain stem | Skew torsion (contraversive) | Vestibular tone imbalance in roll with MLF lesions |
| | Upbeat nystagmus | Vestibular tone imbalance in pitch in bilateral brachium conjunctivum lesions |
| Ponto-medullary brain stem | Tilt of perceived vertical lateropulsion ocular tilt reaction | Vestibular tone imbalance in roll with medial and/or superior vestibular nuclei lesions |
| | Pseudo "vestibular neuritis" | Lacunar infarction or MS plaque at the root entry zone of the eighth nerve |
| | Downbeat nystagmus | Vestibular tone imbalance in pitch |
| | Transient room-tilt illusion | Acute severe vestibular tone imbalance in roll or pitch |
| | Paroxysmal room-tilt illusion in MS | Transversally spreading ephaptic axonal activity |
| | Paroxysmal dysarthria/ataxia in MS | Transversally spreading ephaptic axonal activity |
| | Paroxysmal vertigo evoked by lateral gaze | Vestibular nuclei lesion? |
| Medulla | Upbeat nystagmus | Vestibular tone imbalance in pitch? (nucleus prepositus hypoglossi) |
| Vestibular cerebellum | Downbeat nystagmus | Vestibular tone imbalance in pitch caused by bilateral flocculus lesions (disinhibition) |
| | Positional downbeat nystagmus | Disinhibited otolith–canal interaction in nodulus lesions? |
| | Familial episodic ataxia (EA1 with myokymia and EA2 with vertigo) | EA1, autosomally dominant inherited potassium channelopathy; EA2, autosomally dominant inherited calcium channelopathy |
| | Encephalitis with predominant vertigo | Viral infection of cerebellum |
| | Epidemic vertigo | Viral infection of cerebellum |

nystagmus with distressing oscillopsia and postural imbalance (Dieterich *et al.*, 1991; see also Chapter 13).

## Vestibular Epilepsy

### Clinical Aspects

In vestibular epilepsy the patient experiences a sudden rotational or linear vertigo accompanied by a contraversive body or head and eye rotation. The symptom lasts only several seconds and may be associated with mild nausea but rarely with vomiting. Tinnitus, often unilateral, and contralateral paresthesias may precede or accompany the vertigo. Contraversive epileptic nystagmus may occur. An aura, temporal lobe seizures, or absences may also be experienced as episodic vertigo by the patient. Vertiginous epilepsy should be distinguished from vestibulogenic epilepsy. This is a variety of sensory-evoked epilepsy, either partial complex or grand mal, induced by peripheral labyrinthine stimulation (caloric irrigation or spinning).

## Principles of Therapy

Vestibular seizures are caused by focal discharges from either the temporal lobe or the parietal association cortex (Schneider *et al.*, 1968), both of which receive vestibular projections from the thalamus. Several distinct and separate areas of the parietal and temporal cortex have been identified in animal studies as receiving vestibular afferents, such as area 2v at the tip of the intraparietal sulcus, area 3aV in the central sulcus, the parieto-insular vestibular cortex at the posterior end of the insula, and area 7 in the inferior parietal lobule (Brandt *et al.*, 1994b; Grüsser *et al.*, 1990a,b; Leigh, 1994). Our knowledge of vestibular cortex function in humans is less precise, derived mainly from stimulation experiments reported anecdotally in the older literature (Foerster, 1936; Penfield and Jasper, 1954). Although electrophysiological and cytoarchitectonic data in animals demonstrate several multisensory areas rather than a single primary vestibular cortex, the parieto-insular vestibular cortex seems to represent the integration center of the multisensory vestibular cortex areas within the parietal lobe.

### Practical Management

Anticonvulsive therapy is the same as that for other focal or partial complex seizures, e.g., carbamazepine (600–1200 mg daily) or phenytoin (200–500 mg daily) (see Chapter 20).

### Related Clinical Syndromes

Paroxysmal dysarthria and ataxia: The combination of nonepileptic paroxysmal dysarthria, vertigo and ataxia is a well-known manifestation of multiple sclerosis. The suggested mechanism of the attacks is a transversally spreading ephaptic activation of adjacent axons within a partially demyelinated lesion in fiber tracts of the pontine tegmentum involving the brachium conjunctivum (Osterman and Westerberg, 1975). Carbamazepine is the most effective treatment (Espir and Millac, 1970) and results in complete disappearance of the attacks in most cases. If the patient shows allergic reactions, phenytoin can be administered as an alternative. Carbamazepine is also the first choice of drug for superior oblique myokymia, a rare ocular form of visual dizziness which manifests in oscillopsia due to myokymia of the superior oblique eye muscle.

Paroxysms of vertigo and ataxia have also been described as familial episodic ataxia (Donat and Auger, 1979), an autosomal dominant potassium channel (EA$_1$) or calcium channel (EA$_2$) disorder. Acetazolamide, a drug for prevention of periodic paralysis, is also effective in preventing familial episodic ataxia (Griggs and Nutt, 1995).

## Basilar Migraine/Vestibular Migraine

### Clinical Aspects

Sudden attacks of basilar migraine occur predominantly in adolescent girls (Bickerstaff, 1961). Vertigo, nystagmus, and ataxia are key symptoms of the aura lasting from minutes to 1 h and are frequently combined with other signs of dysfunction within the basilar and posterior cerebral artery territory: dysarthria, perioral and distal limb paresthesias, drop attacks, and scotomata or visual hallucinations as well as loss of consciousness or global amnesia. Accompanying, but usually subsequent, headache is predominantly occipital.

The term vestibular migraine describes episodic vertigo that can be related to migraine as the most probable pathomechanism, but it does not fulfill the criteria of the International Headache Society for basilar migraine (Dieterich and Brandt, 1999). Monosymptomatic audiovestibular attacks may last from seconds to hours without associated headache in one-third of the patients. Two-thirds of these patients show mild central ocular motor signs in the symptom-free interval. Preventive medication with metoprolol or flunarizine is most effective. In a recent study, Neuhauser *et al.* (2001) substantiated the epidemiologic association of migraine and vertigo. They found that migrainous vertigo affects a significant proportion of patients in both dizziness and headache clinics, thus confirming that "it is time for more attention to migrainous vertigo" (Stahl and Daroff, 2000).

### Natural Course

Typically there is a family history of migraine; attacks exhibit an obvious relation to the menstrual cycle; additional classic migraine attacks may occur in the same patient, and with increasing age, basilar migraine tends to become a rare event. There is, however, a second peak of onset of this condition in the fifth to sixth decades, and vestibular migraine may manifest any time throughout life.

### Practical Management

See management of migraine (Chapter 1).

### Related Clinical Syndromes

Benign paroxysmal vertigo of childhood (Basser, 1964), paroxysmal torticollis in infancy, and benign recurrent vertigo in adults (Slater, 1979) are also regarded as typical manifestations of migraine, even without the apparently obligatory headache. Benign paroxysmal vertigo in childhood affects children between 1 and 4 years. The attacks last only a matter of seconds to minutes and spontaneous remission of the disorder occurs within a few years.

### Traumatic Vertigo

Traumatic vertigo is among the most frequent seque-lae associated with head and neck injuries and baro-trauma. Central vestibular vertigo syndromes caused by brainstem dysfunction (concussion, hemorrhage) and the classic paroxysmal positioning vertigo are well recognized. The peripheral end organs and the vestibu-lar nerves may also be affected by temporal bone fractures or hemorrhages into the endolymphatic and perilymphatic space (e.g., perilymph fistulas). Patients often describe their post-traumatic vertigo as a head motion intolerance and unsteadiness of gait similar to walking on pillows. Since these symptoms resemble those of otolith dysfunction (Gresty et al., 1992), one might speculate that this vulnerable accelerometer is affected by trauma (Brandt and Daroff, 1980b). Otoconia is easily dislodged by linear acceleration, resulting in unequal loads on the macula beds and a tone imbalance between the two otoliths (Brandt, 1999). Vestibular exercises are effective in promoting and stabilizing central compensation of the peripheral tone imbalance. Vertigo resulting from barotrauma is mostly of peripheral labyrinthine origin (alternobaric vertigo, round or oval window fistulas) or part of the decompression sickness.

## PSYCHOGENIC VERTIGO

Vertigo is a frequent symptom of psychiatric illness, in particular anxiety, depression, and personality dis-orders, but less frequently psychoses.

### Phobic Postural Vertigo

#### Clinical Aspects

Phobic postural vertigo (Brandt, 1996; Brandt et al., 1994c) has been described as a syndrome that is dis-tinguishable from agoraphobia, acrophobia, and the pseudo-agarophobic syndrome "space phobia" (Marks, 1981). Closely related to stance and locomotion, it is characterized by a combination of nonrotational vertigo with subjective postural and gait instability (mainly in patients with an obsessive–compulsive personality).

The monosymptomatic disturbance of balance mani-fests with superimposed attacks that occur with and without recognizable provoking factors in the same patient and are experienced with and without accom-panying excess anxiety, misleading both patient and physician to a false diagnosis of organic disease.

The diagnosis of phobic postural vertigo is based mainly on the following six characteristic features (Brandt et al., 1994c):

(1) Dizziness and subjective disturbance of balance in the upright posture and during gait despite normal clinical balance tests;
(2) postural vertigo described as fluctuating unsteadi-ness, often taking the form of attacks or sometimes the perception of illusory body perturbation for mere fractions of seconds;
(3) anxiety and distressing vegetative symptoms accom-panying or subsequent to the vertigo; elicited by direct questioning, although most patients expe-rience vertigo attacks both with and without excess anxiety;
(4) vertigo attacks that can occur spontaneously but upon specific questioning are found to be almost invariably associated with particular constellations of perceptual stimuli (bridges, staircases, empty rooms, streets, driving a car) or social situations (department store, restaurant, concert, meeting, reception) from which the patients have difficulty withdrawing and which they recognize as provoking factors. There is a tendency for rapid conditioning, generalization, and avoidance behavior to develop;
(5) typically, an obsessive–compulsive-type personality in patients often found to have affective lability and mild (reactive) depression;
(6) frequently, the onset of the condition following periods of particular stress or after the patient has experienced an illness, usually a vestibular disorder (about 20%).

#### Natural Course

Phobic postural vertigo (PPV) has become the second most common cause of vertigo in our dizziness unit since we began diagnosing it as a clinical entity. Of 3574 consecutive neurological inpatients and outpatients presenting with vertigo, 620 (17.3%) were found to suffer from benign paroxysmal positioning vertigo and 510 (14.3%) from PPV, whereas less that 10% were diagnosed as having other, well-known disorders such as Ménière's syndrome or vestibular neuritis. The peak age of onset of the condition is between 30 and 50 years with a mean of 41 years, and there is a slight preponderance among females.

## Principles of Therapy

Attempts have been made to classify phobic postural vertigo as a panic disorder. We doubt the usefulness of this because panic disorder is not a distinct diagnostic entity, and phobic postural vertigo occurs with and without excess anxiety in the same patient.

We believe that PPV is a disorder of space constancy due to uncoupling of the efference copy (Brandt, 1991). The voluntary impulse for initiation of a movement is simultaneously accompanied by an appropriate efference copy to make identification possible. It has been suggested (v. Holst and Mittelstaedt, 1950) that the efference copy is used to recalibrate the perceptual systems on the basis of previous experience in such a way that the incoming sensory information is interpreted as arising from movement of the observer relative to a stationary environment. Such decoupling would cause the sensory effects of normal postural adjustments (of which we are usually unconscious) to be interpreted as arising from motion of the surroundings. If there is no efference copy (e.g., if we move our eyeball by placing a finger on the eyelid), we see illusory movements of the environment: oscillopsia. The description of the sensation of vertigo in phobic patients (involving involuntary body fluctuations and the occasional perception of individual head movements as disturbing external acceleration) can be explained by a transient uncoupling of efference and efference copy, leading to a mismatch between anticipated and actual motion. Healthy people can experience mild sensations of vertigo without simultaneous anxiety if they are in a state of exhaustion, when differences between voluntary head movements and involuntary fluctuations become blurred. In the phobic patient, this partial uncoupling could be caused by a constant anxious controlling and checking of balance regulation. The way could thus be paved for the perception of sensorimotor adjustment, which would otherwise occur unconsciously, by means of learned and automatically recalled programs for different patterns of activation of the muscles necessary to maintain an upright position. The analysis of postural sway during upright stance revealed a significant increase in sway activity in the 3.53- to 8-Hz frequency band in patients with PPV; however, this increase did not impair their objective postural stability. An increase in higher frequency sway activity may simply reflect a change in postural strategy rather than a sensorimotor dysfunction. The patients' conscious control of stance may augment the coactivation of anti-gravity muscles, a strategy applied by normal subjects when performing demanding balancing tasks (Krafczyk et al., 1999). The more difficult the balance task is, the better the balance performance. During a very difficult task, i.e., tandem stance on foam rubber with the eyes closed, body sway

activity and sway path values did not differ between patients and controls (Querner et al., 2000).

Our therapeutic regimen consists mainly of freeing the patients from their fear of an occult organic disease by providing them with a detailed explanation of the mechanism that causes, and the factors that provoke, phobic postural vertigo attacks. A first follow-up of 78 patients with PPV revealed a favorable course in the majority, despite only short-term suggestive and behavioral therapy of one to three sessions (Brandt et al., 1994c). This might be further improved by additional pharmacological or behavioral therapies not yet proven in controlled prospective studies. In their follow-up, Kapfhammer et al. (1997) emphasized the existence of significant psychological problems at follow-up, despite the considerable improvement in the complaints of vertigo.

## Practical Management

Our short-term suggestive and behavioral therapy is usually achieved in two to three sessions. We do not subject the patients to long-term psychotherapy. We attempt to guide the patient by suggestion, assuring him or her during the discussion that the nature of the disease is known and that self-controlled therapy is possible. We advise the patient against dwelling on the illness too much (decoupling of catastrophic thoughts), provided the obsessional symptoms are not too severe. We recommend a self-controlled desensitization—within the context of behavioral therapy—by repeated exposure to situations that evoke the patient's vertigo. We advocate not overly strenuous physical activity in order to improve the patient's sense of diminished fitness. Some patients may require additional medical treatment with antidepressants such as imipramine.

# PHYSIOLOGICAL (STIMULATION) VERTIGO SYNDROMES

## Motion Sickness

### Clinical Aspects

The full picture of acute motion sickness develops after initial symptoms such as dizziness, physical discomfort, tiredness, periodic yawning, and pallor. An increase in facial pallor is followed by cold sweating, increased salivation, oversensitivity to smells, pains in the back of the head, and feelings of pressure in the upper abdomen. Finally, the central symptoms of nausea, retching, and vomiting develop with motor incoordination, loss of drive and concentration, apathy, and fear of impending doom (Money, 1970).

Recognized subforms of motion sickness (kinetosis) are carsickness (visual-vestibular conflict), seasickness (unfamiliar complex linear and angular accelerations at low frequency), simulator sickness (optokinetic motion sickness), and spacesickness (conflict of sensory input from the otoliths, semicircular canals, and the visual system during active head movements in microgravity).

### Natural Course

Each individual can be made motion sick, although there is considerable interindividual variation to susceptibility. The incidence of motion sickness at sea varies from less than 1% to almost 100% (Money, 1970). During an Atlantic crossing at moderate turbulence, about 25% of passengers succumb to motion sickness within the first 3 days, whereas seasickness among surviving air crew in life rafts is reported to be as high as 60%. It is well known that susceptibility is greater in females than in males and that incidence of motion sickness decreases with increasing age. Infants below the age of 2 are highly resistant to motion sickness because they use their visual system only to a limited extent for dynamic spatial orientation and are therefore subject to fewer visual–vestibular perception conflicts (Brandt et al., 1976). Children older than 2 years, with a peak of around 10–12 years, are more susceptible than adults. Loss of vestibular function produces absolute resistance, whereas blindness does not give protection against motion sickness.

Motion sickness is an acute disorder with spontaneous remission within hours to a day after cessation of the inducing motion. If the stimulation lasts (ship, spacecraft), a recovery occurs within 3 days through central habituation (Figure 4).

### Principles of Therapy

Motion sickness is generated either by unfamiliar accelerations to which a person has therefore not adapted or by an intersensory mismatch involving conflicting vestibular and visual stimuli. The mismatch arises, for example, when the multisensory consequences of being a passenger in a moving vehicle or of moving actively do not match the expected patterns which have been calibrated by prior experience of active locomotion (Brandt, 1999; Dichgans and Brandt, 1978; Probst et al., 1982; Reason, 1978; Figure 4).

The most effective therapeutic measure in terms of physical prevention is habituation through intermittent exposure to the critical stimuli. This habituation is only temporary and acceleration-specific; i.e., resistance to seasickness does not protect against airsickness.

If resistance is not achieved using "vestibular training," the avoidance of head movements or the use of head supports during acceleration in vehicles in the actual situation of stimulation helps to reduce the severity of motion sickness because additional complex accelerations (cross-coupled accelerations, Coriolis effects) are avoided. In enclosed vehicles or when a passenger is reading on the back seat of a car, motion sickness is predominantly caused by body oscillations, when vestibular acceleration is in disagreement with the visual information of no movement. This type of motion sickness can be significantly reduced by adequate visual control of the vehicle's movements in comparison with the visual stimulus condition "eyes closed." However, susceptibility to motion sickness in vehicles considerably increases if the field of view is filled with largely stationary contrasts (Dichgans and Brandt, 1978; Probst et al., 1982).

Anti-motion-sickness drugs such as dimenhydrinate or scopolamine inhibit spontaneous activity and the excitability of vestibular nuclei neurons in the case of body accelerations, which reduces susceptibility to motion sickness.

Some researchers reported that autogenic-feedback training to control autonomic responses during "Coriolis stimulations" reduced the incidence of motion sickness to some extent (Cowings and Toscano, 1982; Toscano and Cowings, 1982). Subsequent studies, however, have not confirmed this finding (Jozsvai and Pigeau, 1996).

### Practical Management

The options for physical prevention are listed in Table VI. Effective medical prevention is possible with anti-motion sickness drugs such as scopolamine (0.6 mg) or dimenhydrinate (50–100 mg). Doubling the recommended single dose leads to a marked increase in central sedative side effects with no significant improvement of its efficacy against motion sickness (Wood et al., 1996). In cases of exceptionally strong motion sickness, combinations of either D-amphetamine sulfate and L-scopolamine hydrobromide or promethazine with D-amphetamine provide far better protection than any single drug without increasing distressing side effects (Wood and Graybiel, 1970).

### Treatments No Longer Recommended

Hypnotics are not recommended, since vomiting due to motion sickness interrupts sleep.

## Height Vertigo (Acrophobia)

### Clinical Aspects

Height vertigo, a visually induced syndrome commonly experienced on top of high structures, is

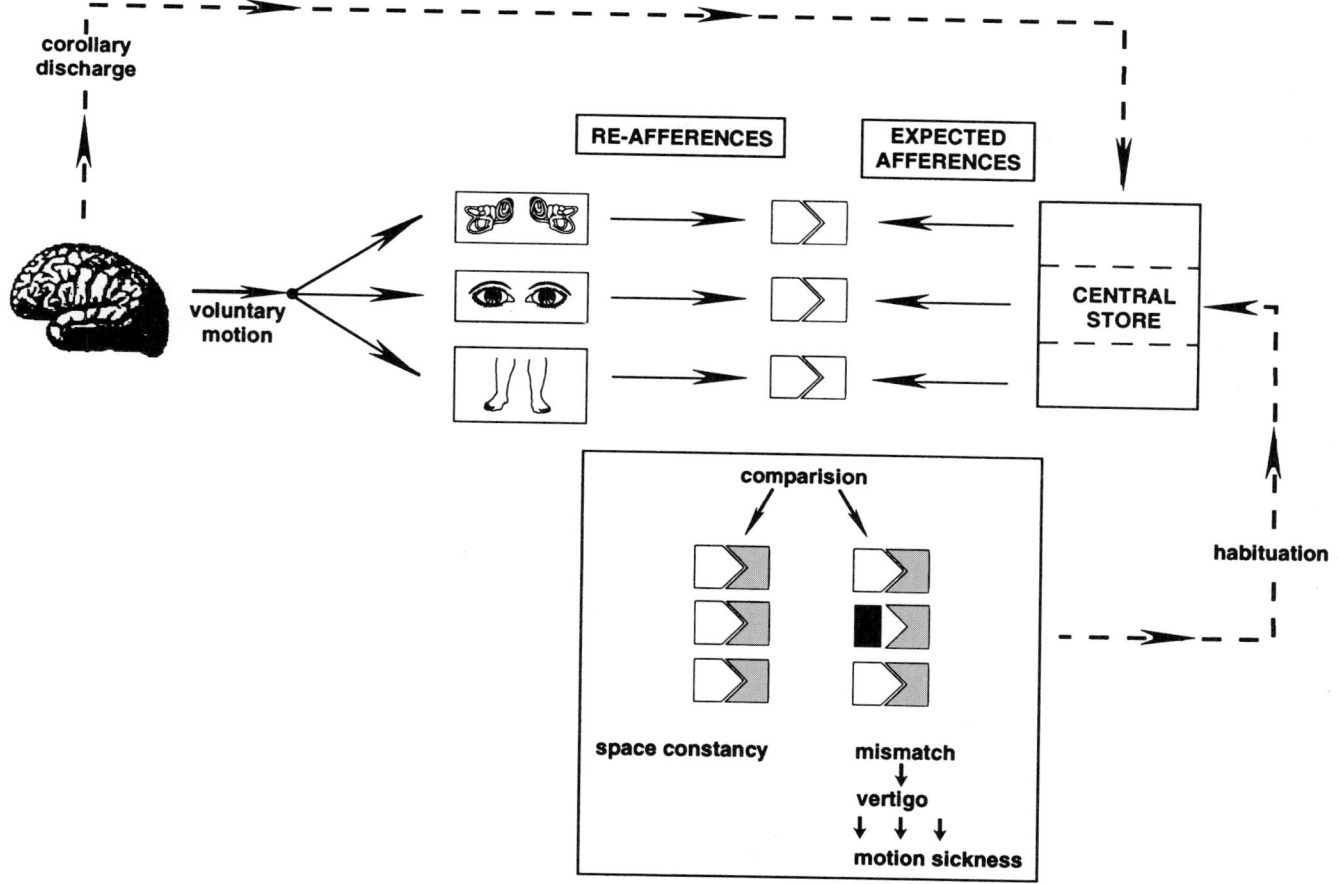

**FIGURE 4** Schematic diagram of the sensory conflict or the neural mismatch concept of vertigo and motion sickness. An active movement leads to stimulation of the sensory organs, whose messages are compared with a multisensory pattern of expectation calibrated by earlier experience of motions (central store). The pattern of expectation is either prepared by the efference copy signal, which is emitted parallel to and simultaneously with the motion impulse, or by vestibular excitation during passive transportation in vehicles. If concurrent sensory stimulation and the pattern of expectation are in agreement, self-motion is perceived while "space constancy" is maintained. If, for example, there is no appropriate visual report of motion, as a result of the field of view being filled with stationary environmental contrasts (reading in car), a sensory mismatch occurs. With repeated stimulation, motion sickness is induced through summation; the repeated stimulation leads to a rearrangement of the stored pattern of expectation in a few days. An acute unilateral labyrinthine loss causes vertigo, because the self-motion sensation induced by the vestibular tone imbalance is contradicted by vision and the somatosensors. Reprinted with permission from Brandt, "Vertigo: Its Multisensory Syndromes, 2nd Edition", Fig. 1.2, p. 5, 1999.

manifested by a subjective instability of posture and locomotion, coupled with a fear of falling and vegetative symptoms. Neurotic acrophobia results when physiological height vertigo induces a conditioned phobic reaction characterized by a dissociation between the objective and the subjective risk of falling. Although the acrophobic is normally aware of this dissociation, he cannot overcome his avoidance behavior.

### Natural Course

Along with many animal species, humans are subject to a largely genetically determined depth avoidance, known as *visual cliff behavior* (Walk *et al.*, 1957). Height vertigo is physiological and must be distinguished from pathological acrophobia. When the subject looks down, height vertigo manifests itself after a delay of several seconds and quickly remits after termination of the situation that induced the stimulus. Patients with labyrinth and balance dysfunction and alcoholics are more susceptible. Habituation to height vertigo may occur through repeated exposure, as is observed in steeplejacks, roof workers, and tightrope artists, who achieve a remarkable degree of postural balance with seeming insensitivity to height.

**TABLE VI**   Physical Prevention of Motion Sickness

| Strategy | Mechanisms |
| --- | --- |
| Chronic | |
|   Vestibular training: repetitive exposure to stimuli during active head movements | Acceleration-specific central habituation |
|   Simulator training | Using the benefit of positive visual–vestibular habituation transfer |
| Acute | |
|   Head fixation | Avoidance of additional acceleration and cross-coupling (e.g., Coriolis effects) |
|   Head position(relative to gravity vector) | Using the benefit of the natural habituation for head $z$-axis accelerations |
|     Ship: supine | |
|     Car: supine, head in driving direction | |
|     Helicopter: sitting | |
|   Counterbalance of vehicle-induced body motion | |
|   Visual control of vehicle motion. If not possible, eyes closed | Avoidance of visual-vestibular mismatch |

### Principles of Therapy

In the textbooks, height vertigo has generally been attributed to psychopathological processes, such as neurotic acrophobia, and psychoanalytical attempts have been made to explain its pathogenesis. However, there is a simple geometrical explanation for physiological height vertigo being a "distance vertigo" caused by visual destabilization of postural balance when the distance between the observer's eye and the nearest visible stationary contrasts becomes critically large (Bles *et al.*, 1980; Brandt *et al.*, 1980). Thus, a visually induced postural imbalance causes a "physiological vertigo syndrome" which, without question, may be contaminated by additional cognitive and psychological factors, mainly responsible for height anxiety and the vegetative symptoms.

### Practical Management

Knowledge of the stimulus characteristics required for optimal visual contribution to postural balance dictates practical advice on how to reduce physiological height vertigo in susceptible subjects (Brandt, 1991):

1. One should avoid a free upright stance in critical situations at high altitudes. This is done intuitively when grasping for a stationary framework or leaning against a wall for support.

2. When looking down, one should obtain stationary cues from nearby contrasts in the peripheral visual field. Visual stabilization of posture is then maintained by the retinal periphery, while the central retina mainly serves egocentric recognition and pursuit of objects. Staring at moving objects such as clouds increases the danger of falling because additional postural destabilization through linearvection may be induced. One

should avoid long exposure times because height vertigo usually takes several seconds to develop. Looking through binoculars is dangerous because it restricts the visual field and introduces an unusual, and therefore unadapted, magnification factor of retinal slip.

3. Body and head position should be adjusted to the gravitational vector because vision will receive a relatively greater sensorial weight (which is undesirable) if the otoliths are displaced beyond their optimal working range by extreme head tilt (head extension vertigo). It may also be true, based on other observations, that the feet should be firmly planted on an "earth-horizontal" surface.

Psychotherapy of acrophobia is dominated by behavioral approaches that can be classified as either systematic or *in vivo* desensitization procedures. Systematic desensitization is based on the construction of a graduated hierarchy of anxiety-provoking visual scenes, which, after a training phase in muscle relaxation, are subsequently visualized by the patient while in a relaxed state. Although therapist or self-directed desensitization has been widely used, *in vivo* desensitization is more effective in the treatment of phobia.

*In vivo* desensitization techniques aim to reduce avoidance behavior or anxiety by increasing contact with real life rather than the patient imaging provoking stimuli. Successive approximation to the feared situation is supported by instructions and reinforcement. Contact desensitization stresses the advantage of joint participation and physical contact with the therapist model (participant modeling). Flooding, an alternative technique, also relies on exposure to real phobic stimuli but without a graduated approach. It is based on getting the patient into the feared situation and maintaining the exposure for prolonged periods of time.

Drugs used for symptomatic relief from panic attacks in phobic patients are either tranquilizers or antidepressants such as imipramine. The long-term course of untreated anxiety neurosis, however, indicates that during a 5-year interval most children and 50% of adults have either recovered from their phobias or substantially improved (Agras *et al.*, 1972; Noyes *et al.*, 1980).

# REFERENCES

Adour, K. K., Ruboyianes, J. M., Von Doersten, P. G., Byl, F. M., Trent, C. S., Quesenberry, C. P., Jr., Hitchcock, T. (1996). Bell's palsy treatment with acyclovir and prednisone compared with prednisone alone: A double-blind, randomized, controlled trial. *Ann. Otol. Rhinol. Laryngol.* **105**, 371–378.

Agras, W. S., Chapin, H. N., and Oliveau, D. C. (1972). The natural history of phobia. *Arch. Gen. Psychiatry* **26**, 315–317.

Andrews, J. C., Hoover, L. A., Lees, R. S., and Honrubia, V. (1988). Vertigo in the hyperviscosity syndrome. *Otolaryngol. Head Neck Surg.* **98**, 144–149.

Arbusow, V., Schulz, P., Strupp, M., Dieterich, M., von Reinhardshoettner, A., Rauch, E., and Brandt, T. (1999). Distribution of herpes simplex virus type 1 in human geniculate and vestibular ganglia: Implications for vestibular neuritis. *Ann. Neurol.* **46**, 416–419.

Arbusow, V., Strupp, M., and Brandt, T. (1998). Amiodarone-induced severe prolonged head-positional vertigo and vomiting. *Neurology* **51**, 917.

Ariyasu, L., Byl, F. M., Sprague, M. S., and Adour, K. K. (1990). The beneficial effect of methylprednisolone in acute vestibular vertigo. *Arch. Otolaryng. Head Neck Surg.* **116**, 700–703.

Averbuch-Heller, L., Tusa, R. J., Fuhry, L. *et al.* (1997). A double-blind controlled study of gabapentin and baclofen as treatment for acquired nystagmus. *Ann. Neurol.* **41**, 818–825.

Baloh, R. W. (1994). Reply to the letter by Lempert: Horizontal benign positional vertigo. *Neurology* **44**, 2214.

Baloh, R. M., and Halmagyi, G. M. (Eds.) (1996). "Disorders of the Vestibular System." New York: Oxford University Press.

Baloh, R. W., Jacobson, K., and Honrubia, V. (1993). Horizontal semicircular canal variant of benign positional vertigo. *Neurology* **43**, 2542–2549.

Baloh, R. W., Jacobsen, B. A., and Winder, T. (1990). Drop attacks with Ménière's syndrome. *Ann. Neurol.* **28**, 384–387.

Baloh, R. W., Honrubia, V., and Jacobson, K. (1987). Benign positional vertigo. *Neurology* **37**, 371–378.

Basser, L. S. (1964). Benign paroxysmal vertigo of childhood. *Brain* **87**, 141.

Bergenius, J., Ödkvist, L. M. (1996). Transtympanic aminoglycoside treatment in Ménière's disease. *In* "Disorders of the Vestibular System" (R. W. Baloh and G. M. Halmagyi, Eds.) pp. 575–582. Oxford University Press, New York.

Bickerstaff, E. R. (1961). Basilar artery migraine. *Lancet* **1**, 15–17.

Bles, W., Kapteyn, T. S., Brandt T., and Arnold, F. (1980). The mechanism of physiological height vertigo. II. Posturography. *Acta Otolaryngol.* **89**, 534–540.

Brandt, T. (1999). "Vertigo: Its Multisensory Syndromes," 2nd ed. Springer-Verlag, London.

Brandt, T. (1996). Phobic postural vertigo. *Neurology* **46**, 515–519.

Brandt, T. (1991). Man in motion. Historical and clinical aspects of vestibular function. *Brain* **114**, 2159–2174.

Brandt, T., and Dieterich, M. (1994a). Vestibular syndromes in the roll plane: Topographic diagnosis from brainstem to cortex. *Ann. Neurol.* **36**, 337–347.

Brandt, T., and Dieterich, M. (1994b). Vestibular paroxysmia: Vascular compression of the eighth nerve? *Lancet.* **1**, 798–799.

Brandt, T., and Daroff, R. B. (1980a). Physical therapy for benign paroxysmal positional vertigo. *Arch. Otolaryngol.* **106**, 484–485.

Brandt, T., and Daroff, R. B. (1980b). The multisensory physiological and pathological vertigo syndromes. *Ann. Neurol.* **7**, 195–203.

Brandt, T., and Steddin, S. (1993). Current view of the mechanism of benign paroxysmal positioning vertigo: Cupulolithiasis or canalolithiasis? *J. Vestib. Res.* **3**, 373–382.

Brandt, T., Strupp, M., Arbusow, V., and Dieringer, N. (1997). Plasticity of the vestibular system: Central compensation and sensory substitution for vestibular deficits. *Ann. Neurol.* **73**, 297–309.

Brandt, T., Steddin, S., and Daroff, R. B. (1994a). Therapy for benign paroxysmal positioning vertigo, revisited. *Neurology* **44**, 796–800.

Brandt, T., Dieterich, M., and Danek, A. (1994b). Vestibular cortex lesions affect the perception of verticality. *Ann. Neurol.* **35**, 403–412.

Brandt, T., Huppert, D., and Dieterich, M. (1994c). Phobic postural vertigo, a first follow-up. *J Neurol.* **241**, 191–195.

Brandt, T., Arnold, F., Bles, W., and Kapteyn, T. S. (1980). The mechanism of physiological height vertigo. I. Theoretical approach and psychophysics. *Acta Otolaryngol.* **89**, 513–523.

Brandt, T., Wenzel, D., and Dichgans, J. (1976). Die Entwicklung der visuellen Stabilisation des aufrechten Standes beim Kind: Ein Reifezeichen in der Kinderneurologie. *Arch. Psychiat. Nervenkr.* **223**, 1–13.

Bronstein, A. M., Brandt, T., and Woollacott, M. (1996). Clinical disorders of balance, posture and gait. Arnold, London.

Büchele, W., and Brandt, T. (1988). Vestibular neuritis—A horizontal semicircular canal paresis? *Adv. Oto-Rhino-Laryng.* **42**, 157–161.

Claes, J., and van de Heyning, P. H. (2000). A review of medical treatment for Ménière's disease. *Acta Otolaryngol. (Suppl.)* **544**, 34–39.

Cowings, P. S., and Toscano, W. B. (1982). The relationship of motion sickness susceptibility to learned autonomic control for symptom suppression. *Aviat. Space Environ. Med.* **53**, 570–575.

Curthoys, I. S. (2000). Vestibular compensation and substitution. *Curr. Opin. Neurol.* **13**, 27–30.

Curthoys, I. S., and Halmagyi, G. M. (1994). Vestibular compensation: A review of the oculomotor, neural, and clinical consequences of unilateral vestibular loss. *J. Vestib. Res.* **5**, 67–107.

Curthoys, J. S., and Oman, C. M. (1987). Dimensions of the horizontal semicircular duct ampulla and utricle in human. *Acta Otolaryngol. (Stockh.)* **103**, 254–261.

De la Meilleure, G., Dehaene, I., Depondt, M., Damman, W., Crevits, L., and Vanhooren, G. (1996). Benign paroxysmal positional vertigo of the horizontal canal. *J. Neurol. Neurosurg. Psychiatry* **60**, 68–71.

Dichgans, J., and Brandt, T. (1978). Visual-vestibular interaction: Effects on self-motion perception and postural control. *In* "Handbook of Sensory Physiology" (R. Held, H. W. Leibowitz, and H. L. Teuber, Eds.), vol VIII, Perception, pp. 755–804. Springer, Berlin Heidelberg New York.

Dieringer, N. (1995). "Vestibular compensation": Neural plasticity and its relations to functional recovery after labyrinthine lesions in frogs and other vertebrates. *Prog. Neurobiol.* **46**, 97–129.

Dieterich, M., and Brandt, T. (1999). Episodic vertigo related to migraine (90 cases): Vestibular migraine? *J. Neurol.* **246**, 883–889.

Dieterich, M., Straube, A., Brandt, T., Paulus, W., and Büttner, U. (1991). The effects of baclofen and cholinergic drugs on upbeat and downbeat nystagmus. *J. Neurol. Neurosurg. Psychiatry* **54**, 627–132.

Donat, J. R., and Auger, R. (1979). Familial periodic ataxia. *Arch. Neurol.* **36**, 568–569.

Dirscoll, C. L. W., Kasperbauer, J. L., Facer, G. W. Harner, S. G., and Beatty, C. W. (1997). Low-dose intratympanic gentamicin and the treatment of Meniere's disease: Preliminary results. *Laryngoscope* **107**, 83–89.

Epley, J. M. (1992). The canalith repositioning procedure: For treatment of benign paroxysmal positioning vertigo. *Otolaryngol. Head Neck Surg.* **107**, 299–304.

Espir, M. L. E., and Millac, P. (1970). Treatment of paroxysmal disorders in multiple sclerosis with carbamazepine (Tegretol). *J. Neurol. Neurosurg. Psychiatry* **33**, 528–531.

Fetter, M., and Zee, D. S. (1988). Recovery from unilateral labyrinthectomy in rhesus monkeys. *J. Neurophysiol.* **59**, 370–393.

Foerster, O. (1936). Sensible kortikale Felder. *In* "Handbuch der Neurologie" (O. Bumke and O. Foerster, Eds.), Vol. VI, pp. 358–449. Springer, Berlin.

Foster, C., Baloh, R. W. (1996). Drug therapy for vertigo. *In* "Disorders of the Vestibular System" (R. W. Baloh and G. M. Halmagyi, Eds.), pp. 541–550. Oxford University Press, New York, Oxford.

Fraysse, B., Bebear, J. P., Dubreuil, C., Berges, C., Dauman, R. (1991). Betahistine dihydrochloride versus flunarizine. A double-blind study on recurrent vertigo with or without cochlear syndrome typical of Meniere's disease. *Acta Otolaryngol. (Stockh.) (Suppl.)* **490**, 1–10.

Friberg, U., Stahle, J., and Svedberg, A. (1984). The natural cause of Meniere's disease. *Acta Otolaryngol. (Stockh.) (Suppl.)* **406**, 72–77.

Gradt, A., and Baloh, R. W. (1989). Vertigo of vascular origin. Clinical and electro-nystagmographic features in 84 cases. *Arch. Neurol.* **46**, 281–284.

Graham, M. D., and Goldsmith, M. M. (1994). Labyrinthectomy. Indications and surgical technique. *Otolaryngol. Clin. North Am.* **27**, 325–335.

Gresty, M. A., Bronstein, A. M., Brandt, T., and Dieterich, M. (1992). Neurology of otolith function: Peripheral and central disorders. *Brain* **115**, 647–673.

Griggs, R. C., and Nutt, J. G. (1995). Episodic ataxias as channelopathies. *Ann. Neurol.* **37**, 285–286.

Grüsser, O.-J., Pause, M., and Schreiter, U. (1990a). Localization and responses of neurons in the parieto-insular vestibular cortex of awake monkeys (Macaca fascicularis). *J. Physiol.* **430**, 537–557.

Grüsser, O.-J., Pause, M., and Schreiter, U. (1990b). Vestibular neurons in the parieto-insular cortex of monkeys (Macaca fascicularis): Visual and neck receptor responses. *J. Physiol.* **430**, 559–583.

Hain, T. C., Fetter, M., and Zee, D. S. (1987). Head-shaking nystagmus in patients with unilateral peripheral vestibular lesions. *Am. J. Otolaryngol.* **8**, 36–47.

Halmagyi, G. M., and Curthoys, I. S. (1988). A clinical sign of canal paresis. *Arch. Neurol.* **45**, 737–739.

Halmagyi, G. M., Fattore, C. M., Curthoys, I. S., and Wade, S. (1994). Gentamicin vestibulotoxicity. *Otolaryngol. Head Neck Surg.* **111**, 571–574.

Halmagyi, G. M., Curthoys, I. S., Cremer, P. D., Henderson, C. J., Todd, M. J., Staples, M. J., and D'Cruz, D. M. (1990). The human horizontal vestibulo-ocular reflex in response to high-acceleration stimulation before and after unilateral vestibular neurectomy. *Exp. Brain Res.* **81**, 479–490.

Halmagyi, G. M., Rudge, P., and Gresty, M. A. (1980). Treatment of periodic alternating nystagmus. *Ann. Neurol.* **8**, 609–611.

Hellström, S., and Ödkvist, L. (1994). Pharmacologic labyrinthectomy. *Otolaryngol. Clin. North Am.* **27**, 307–315.

Herdman, S. J. (1990). Treatment of benign paroxysmal positional vertigo. *Phys. Ther.* **10**, 381–388.

Herdman, S. J., Blatt, P. J., and Schubert, M. C. (2000). Vestibular rehabilitation of patients with vestibular hypofunction or with benign paroxysmal positional vertigo. *Curr. Opin. Neurol.* **13**, 39–43.

Herdman, S. J., Clendaniel, R. A., Mattox, D. E., Holiday, M., and Niparko, J. K. (1995). Vestibular adaptation exercises and recovery: Acute stage following acoustic neuroma resection. *Otolaryngol. Head Neck Surg.* **113**, 71–77.

Herdman, S. J., Tusa, R. J., Zee, D. S., Proctor, L. R., and Mattox, D. E. (1993). Single treatment approaches to benign paroxysmal positional vertigo. *Arch. Otolaryngol. Head Neck Surg.* **119**, 450–454.

Holst, E. V., and Mittelstaedt, H. (1950). Das Reafferenzprinzip (Wechselwirkungen zwischen Zentralnervensystem und Peripherie). *Naturwissenschaften* **37**, 464–476.

Horak, F. B., Jones-Rycewicz, C., Black, O., and Shumway-Cook, A. (1992). Effects of vestibular rehabilitation on dizziness and imbalance. *Otolaryngol. Head Neck Surg.* **106**, 175–180.

Hutchin, T., and Cortopassi, G. (1994). Proposed molecular and cellular mechanism for aminoglycoside ototoxicity. *Antimicrob. Agents Chemother.* **38**, 2517–2520.

Igarashi, M. (1986). Compensation for peripheral vestibular disturbances—Animal studies. *In* "Disorders of Posture and Gait" (W. Bles and T. Brandt, Eds.), 337–351. Elsevier, Amsterdam.

Igarashi, M., Ishikawa, K., Ishii, M., and Yamane, H. (1988). Physical exercise and balance compensation after total ablation of vestibular organs. *Prog. Brain Res.* **76**, 395–401.

Jannetta, P. J., Møller, M. B., and Møller, A. R. (1984). Disabling positional vertigo. *N. Engl. J. Med.* **310**, 1700–1705.

Jozsvai, E. E., and Pigeau, R. A. (1996). The effect of autogenic training and biofeedback on motion sickness tolerance. *Aviat. Space Environ. Med.* **67**, 963–968.

Kapfhammer, H. P., Mayer, C., Hock, U., Huppert, D., Dieterich, M., and Brandt, T. (1997). Course of illness in phobic postural vertigo. *Acta Neurol. Scand.* **95**, 23–28.

Katsarkas, A., and Kirkham, T. (1978). Paroxysmal positional vertigo. A study of 255 cases. *J. Laryngol.* **7**, 320–330.

Klockhoff, I., Lindblom, U., and Stahle, J. (1974). Diuretic treatment of Meniere's disease: long term results with clorthalidone. *Arch. Otolaryngol.* **100**, 262–265.

Krafczyk, S., Schlamp, V., Dieterich, M., Haberhauer, P., and Brandt, T. (1999). Increased body sway at 3.5–8 Hz in patients with phobic postural vertigo. *Neurosci. Lett.* **259**, 149–152.

Krebs, D. E., Gill-Body, K. M., Riley, P. O., and Parker, S. W. (1993). Double-blind placebo controlled trial of rehabilitation for bilateral vestibular hypofunction: Preliminary report. *Otolaryngol. Head Neck Surg.* **109**, 735–741.

Kroese, A. B. A., Das, A., and Hudspeth, A. J. (1989). Blockage of transduction channels of hair cells in bullfrog's sacculus by aminoglycoside antibiotics. *Hear Res.* **37**, 203–218.

Lacour, M., Roll, J. R., and Apaix, M. (1976). Modifications and development of spinal reflexes in the alert baboon (Papio papio) following unilateral vestibular neurectomy. *Brain Rev.* **113**, 255–269.

Leigh, R. J. (1994). Human vestibular cortex. *Ann. Neurol.* **35**, 383–384.

Leigh, R. J., and Ramat, S. (1999). Neuropharmacologic aspects of the ocular motor system and the treatment of abnormal eye movements. *Curr. Opin. Neurol.* **12**, 21–27.

Leigh, J., and Brandt, T. (1993). A reevaluation of the vestibulo-ocular reflex: New ideas of its purpose, properties, neural substrate, and disorders. *Neurology* **43**, 1288–1295.

Leigh, R. J., Averbuch-Heller, L., Tomsak, R. L., Remler, B. F., Yaniglos, S. S., and Dell'Osso, L. F. (1994). Treatment of abnormal eye movements that impair vision: Strategies based on current concepts of physiology and pharmacology. *Ann. Neurol.* **36**, 129–141.

Lempert, T., and Tiel-Wilck, K. (1996). A positional maneuver for treatment of horizontal-canal benign positional vertigo. *Laryngoscope* **106**, 476–478.

Le Père, D. M. (1967). Evaluation of a new symptomatic treatment for Menière's disease. *Clin. Med.* **74,** 63–64.

McCabe, B. F. (1989). Autoimmune inner ear disease. *Am. J. Otol.* **10,** 196–197.

McClure, J. A. (1985). Horizontal canal BPPV. *J. Otolaryngol.* **14,** 30–35.

Magnusson, M., and Padoan, S. (1991). Delayed onset of ototoxic effect of gentamicin in treatment of Menière's disease. *Acta Otolaryngol. (Stockh.)* **111,** 671–676.

Manning, C., Scandale, L., Manning, E. J., and Gengo, F. M. (1992). Central nervous system effects of meclicine and dimenhydrinate: Evidence of acute tolerance to antihistamines. *J. Clin. Pharmacol.* **32,** 996–1002.

Marks, E. M. (1991). "Phobia": A pseudo-agoraphobic syndrome. *J. Neurol. Neurosurg. Psychiatry* **44,** 387–391.

Meyer, E. D. (1985). Zur Behandlung des Morbus Menière mit Betahistindimesilat (Aequamen)—Doppelblindstudie gegen Placebo (Cross-over). *Laryngol. Rhinol. Otol.* **64,** 269–272.

Minor, L. B. (2000). Superior canal dehiscence syndrome. *Am. J. Otol.* **21,** 8–19.

Minor, L. B., Solomon, D., Zinreich, J. S. and Zee, D. S. (1998). Sound- and/or pressure-induced vertigo due to dehiscence of the superior semicircular canal. *Arch. Otolaryngol. Head Neck Surg.* **124,** 249–258.

Møller, M. B. (1998). Controversy in Menière's disease: Result of microvascular decompression of the eighth nerve. *Ann. J. Otol.* **9,** 60–63.

Møller, M. B., Møller, A. R., Jannetta, P. J., and Sekhar, L. N. (1986). Diagnosis and surgical treatment of disabling positional vertigo. *J. Neurosurg.* **64,** 21–28.

Møller, A. R. (1991). The cranial nerve vascular compression syndrome. I. a review of treatment. *Acta Neurochir.* **113,** 18–23.

Money, K. E. (1970). Motion sickness. *Physiol. Rev.* **50,** 1–39.

Morrison, A. V. (1986). Predictive test for Meniere's disease. *Am. J. Otol.* **7,** 5–10.

Murofushi, T., Halmagyi, G. M., and Yavor, R. A. (1997). Intratympanic gentamicin in Menière's disease: Results of therapy. *Am. J. Otol.* **18,** 52–57.

Neuhauser, H., Leopold, M., von Brevern, M., Arnold, G., and Lempert, T. (2001). The interrelations of migraine, vertigo, and migrainous vertigo. *Neurology* **56,** 436–441.

Noyes, R., Clancy, J., Hoenk, P. R., and Slymen, D. J. (1980). The prognosis of anxiety neurosis. *Arch Gen Psychiatry* **37,** 173–178.

Nuti, D., Agus, G., Barbiere, M.-T., and Passali, D. (1998). The management of horizontal-canal paroxysmal positional vertigo. *Acta Otolaryngol. (Stockh.)* **118,** 455–460.

Nuti, D., Vannucchi, P., and Pagnini, P. (1996). Benign paroxysmal positional vertigo of the horizontal canal: A form of canalolithiasis with variable clinical features. *J. Vestib. Res.* **6,** 173–184.

Ohbayashi, S., Oda, M., Yamamoto, M., Urano, M., Harada, K., Horikoshi, H., Orihara, H., and Kitsuda, C. (1993). Recovery of the vestibular function after vestibular neuronitis. *Acta Otolaryngol. (Stockh.) (Suppl.)* **503,** 31–34.

Okinaka, Y., Sekitani, T., Okazaki, H., Miura, M., and Tahara, T. (1993). Progress of caloric response of vestibular neuronitis. *Acta Otolaryngol. (Stockh.) (Suppl.)* **503,** 18–22.

Oosterveld, W. J. (1984). Betahistine dihydrochloride in the treatment of vertigo of peripheral vestibular origin. A double-blind placebo-controlled study. *J. Laryngol. Otol.* **98,** 37–41.

Osterman, P. O., and Westerberg, C.-E. (1975). Paroxysmal attacks in multiple sclerosis. *Brain* **98,** 189–202.

Pace-Balzan, A., and Rutka, J. A. (1991). Non-ampullary plugging of the posterior semicircular canal for benign paroxysmal positional vertigo. *J. Laryngol. Otol.* **105,** 901–906.

Parnes, L. S., and McClure, J. A. (1991). Posterior semicircular canal occlusion in the normal hearing ear. *Otolaryngol. Head Neck Surg.* **104,** 52–57.

Penfield, W., and Jasper, H. (1954). Epilepsy and the functional anatomy of the human brain. Little Brown, Boston.

Petermann, W., and Mulch, G. (1982). Zur Langzeittherapie des Morbus Meniere. Betahistin-dihydrochlorid und Hydrochlorothiazid im Wirkungsvergleich. *Fortschr. Med.* **100,** 431–435.

Probst, Th., Krafczyk, S., Büchele, W., and Brandt, T. (1982). Visuelle Prevention der Bewegungskrankheit im Auto. *Arch. Psychiat. Nervenkr.* **231,** 409–421.

Querner, V., Krafczyk, S., Dieterich, M., and Brandt, T. (2000). Patients with somatoform phobic postural vertigo: The more difficult the balance task, the better the balance performance. *Neurosci. Lett.* **285,** 21–24.

Reason, J. T. (1978). Motion sickness adaptation: A neural mismatch model. *J. R. Soc. Med.* **71,** 819–829.

Rosenhall, U., Hanner, P., and Kaijser, B. (1988). Borellia infection and vertigo. *Acta Otolaryngol. (Stockh.)* **106,** 111–116.

Sad, J., and Yaniv, E. (1984). Menière's disease in infants. *Acta Otolaryngol. (Stockh.)* **97,** 33–37.

Schlatter, M., Kerr, D. R., Smith, P. F., and Darlington, C. L. (1999). Evidence that the ginkgo biloba extract, Egb 761, neither accelerates nor enhances the rapid compensation of the static symptoms of unilateral vestibular deafferentation in guinea pig. *J. Vestib. Res.* **9,** 111–118.

Schneider, R. C., Calhaun, H. D., and Crosby, E. C. (1968). Vertigo and rotational movement in cortical and subcortical lesions. *J. Neurol. Sci.* **6,** 493–516.

Schuknecht, H. F. (1969). Cupulolithiasis. *Arch Otolaryngol.* **90,** 765–778.

Schuknecht, H. F., and Bartley, M. (1985). Cochlear endolymphatic shunt for Menière's disease. *Ann. J. Otol. Suppl.* 20–22.

Schuknecht, H. F., and Gulya, A. J. (1983). Endolymphatic hydrops: an overview and classification. *Ann. Otol. Rhinol. Larnygol.* **92,** 1–20.

Schuknecht, H. F., and Kitamura, K. (1981). Vestibular neuritis. *Ann. Otol. Rhinol. Laryngol. (Suppl.)* **90,** 1–19.

Sekitani, T., Imate, Y., Noguchi, T., and Inokuma, T. (1993). Vestibular neuronitis: Epidemological survey by questionnaire in Japan. *Acta Otolaryngol. (Stockh.) (Suppl.)* **503,** 9–12.

Semont, A., Freyss, G., and Vitte, E. (1988). Curing the BPPV with a liberatory maneuver. *Adv. Otorhinolaryngol.* **42,** 290–293.

Shinkawa, H., and Kimura, R. S. (1986). Effect of diuretics on endolymphatic hydrops. *Acta Otolaryngol. (Stockh.)* **101,** 43–52.

Singleton, G. T., Post, K. N., Karlan, M. S., and Bock, D. G. (1978). Perilymph fistulas: Diagnostic criteria and therapy. *Ann. Otol. Rhinol. Laryngol.* **87,** 797–803.

Slater, R. (1979). Benign recurrent vertigo. *J. Neurol. Neurosurg. Psychiatry* **42,** 363–367.

Stahl, J. S., and Daroff, R. B. (2001). Time for more attention to migrainous vertigo? *Neurology* **56,** 428–429.

Stahle, J., Stahle, Ch., and Arenberg, I. K. (1978). Incidence of Menière's disease. *Arch. Otolaryngol.* **104,** 99–102.

Starck, M., Albrecht, H., Pöllman, W., Straube, A., and Dieterich, M. (1997). Drug therapy for acquired pendular nystagmus in multiple sclerosis. *J. Neurol.* **244,** 9–16.

Steddin, S., and Brandt, T. (1996). Horizontal canal benign paroxysmal positioning vertigo (h-BPPV): Transition of canalolithiasis to cupulolithiasis. *Ann. Neurol.* **40,** 918–922.

Strupp, M., Planck, J. H., Arbusow, V., Steiger, H.-J., Brückmann, H., and Brandt, T. (2000). Rotational vertebral artery occlusion syndrome with vertigo due to "labyrinthine excitation." *Neurology* **54,** 1376–1379.

Strupp, M., Arbusow, V., Maag, K. P., Gall, C., and Brandt, T. (1988). Vestibular exercises improve central vestibulo-spinal compensation after an acute unilateral peripheral vestibular lesion; A prospective clinical study. *Neurology* **51**, 838–844.

Strupp, M., Brandt, T., and Steddin, S. (1955). Horizontal canal benign paroxysmal positioning vertigo: Reversible ipsilateral caloric hypoexcitability caused by canalolithiasis? *Neurology* **45**, 2072–2076.

Toscano, W. V., and Cowings, P. S. (1982). Reducing motion sickness: A comparison of autogenic-feedback training and an alternative cognitive task. *Aviat. Space Environ. Med.* **53**, 449–453.

Van Deelen, G. W., and Huizing, E. H. (1986). Use of a diuretic (dyazide) in the treatment of Meniere's disease. A double-blind cross-over placebo-controlled study. *ORL J. Otorhinolaryngol. Relat. Spec.* **48**, 287–292.

Vannucchi, P., Giaonnoni, B., and Pagnini, P. (1997). Treatment of horizontal semicircular canal benign paroxysmal positional vertigo. *J. Vestib. Res.* **7**, 1–6.

Vollertsen, R. S., McDonald, T. J., Younge, B. R., Banks, P. M., Stanson, A. W., and Ilstrup, D. M. (1986). Cogan's syndrome: 18

Cases and a review of the literature. *Mayo Clin. Proc.* **61**, 344–361.

Walk, R. D., Gibson, E. J., and Tighe, T. J. (1957). Behaviour of light- and dark-raised rats on a visual cliff. *Science* **126**, 80–81.

Wood, C. D., and Graybiel, A. (1970). Evaluation of anti-motion sickness drugs: A new effective remedy revealed. *Aerospace Med.* **41**, 932–933.

Wood, C. D., Graybiel, A., and Kennedy, R. S. (1996). Comparison of effectiveness of some antimotion sickness drugs using recommended and larger than recommended doses as tested in the slow rotation room. *Aerospace Med.* **37**, 259–262.

Xerri, C., and Lacour, M. (1980). Compensation des déficits posturaux et cinétiques après neurectomie vestibulaire unilatérale chez le chat. *Acta Otolaryngol. (Stockh.)* **90**, 414–424.

Yamazaki, T., Hayashi, M., and Komatsuzaki, A. (1991). Intratympanic gentamicin therapy for Meniere's disease placed by a tubal catheter with systematic isosorbide. *Acta Otolaryngol. (Stockh.) (Suppl.)* **481**, 613–616.

CHAPTER 16

# Tinnitus

Ulrich Büttner and Alan H. Lockwood

Tinnitus is the false perception of tones and noises without an appropriate external stimulus. It is a common phenomenon. According to investigations in England, 17% of the population without noise-induced trauma had tinnitus that lasted more than 5 min (Coles, 1987). The American Tinnitus Association estimates 50 million Americans have experienced the disorder.

Tinnitus occurs largely in adults and becomes more frequent in old age. About 1% of the population suffers from tinnitus with men more often affected than women. Acute tinnitus stops spontaneously in 60–80% of the cases. The chances of spontaneous recovery are small when the tinnitus lasts more than 3 months (chronic tinnitus). Most patients (85%) with continuous or intermittent tinnitus are relatively untroubled by the phenomenon (compensated tinnitus). For the remaining 15%, the disorder may be a source of substantial disability (decompensated tinnitus). This is usually manifested by anxiety, inattentiveness, insomnia, depression, and/or other relatively nonspecific complaints. Some patients with depression have suicidal thoughts (2%). However, suicidal attempts without a history of depression are rare (Lewis *et al.*, 1994).

Clinically, it is important to distinguish between subjective and objective tinnitus. In patients with objective tinnitus, the sounds are real, and may be heard with a stethoscope placed over the external auditory meatus or adjacent to the ear. Thus, objective tinnitus is not a phantom perception.

## OBJECTIVE TINNITUS

A distinction is made between the various noise types with etiological indications.

### Souffle during Inspiration/Expiration

#### *Etiology*

This condition is due to an abnormally wide tuba auditiva eustachii, often caused by considerable weight loss.

#### *Treatment*

In individual cases a tube in the tuba Eustachii can lead to a narrowing (Robinson and Hazell, 1989).

### Series of Sharp Clicks (Lasting Several Seconds to Minutes)

#### *Etiology*

This is called palatal myoclonus. It is a form of segmental myoclonus and is characterized by rhythmic contractions of the soft palate. There are often synchronous contractions of other voluntary muscles of the pharynx, larynx, face, neck, eyes, and occasionally the muscles of respiration. It is usually caused by a lesion in the pathways connecting the inferior olivary nucleus and the dentate nucleus. These lesions are usually seen on MRI scans where hypertrophy of the inferior olive may also be evident.

#### *Therapy*

The disorder, once established, usually persists and is resistant to treatment. However, there are reports of responses to tetrabenazine, 5-HTP, trihexaphenadyl, and carbamazepine (Williams *et al.*, 1998). Botulinum toxin injections may be tried in selected cases (Bryce and Morrison, 1998; Saeed and Brookes, 1993). The tensor veli palatini or the stapedius muscle can be transected surgically (Badia *et al.*, 1994).

### Vascular Bruit (Pulsatile Tinnitus)

#### *Etiology*

Vascular bruit is caused by (1) vascular stenoses or malformations of the head and neck, such as carotid stenosis and dural arteriovenous fistulas of the sigmoid or rarely other locations, or (2) vascular tumors of the petrous bone such as glomus tumor.

#### *Treatment*

Carotid stenoses are treated by endovascular dilatation or stenting or by open surgery. Dural arteriovenous fistulas are eliminted by transvenous or transarterial embolization or by surgical methods (Olteanu-Nerbe *et al.*, 1997). With dural fistulae more than 80% of the patients can achieve clinical resolution. The therapeutic outcome is better in less severe cases (Shah *et al.*, 1999). Glomus tumors usually require surgical removal, possibly after transarterial embolization to reduce surgical blood loss.

### Continual Swishing Noise (Disappears when Pressure Is Applied to the Jugular Vein)

#### *Etiology*

This has a venous origin, such as stenosis or external compression of the sigmoid sinus, high jugular bulb, aneurysm of the sigmoid sinus, or rarely intracranial hypertension.

#### *Treatment*

If compression by a tumor is identified as the cause, treatment depends on the nature of the neoplasm. The most common tumor, meningioma, requires surgery. Radiosurgery, an alternative method for small meningiomas using convergent X-ray beams does commonly not result in a dramatic shrinkage of the tumor volume and is therefore not a primary consideration in cases with a disturbing noise. High located jugular bulbs leading to Menière's disease associated with a pulsatile component can be lowered surgically (Couloigner *et al.*, 1999). For other venous anomalies, venous obliteration by endovascular technique can be considered. The effect and tolerance of treatment can be tested by balloon test occlusion beforehand.

### Spontaneous Otoacoustic Emissions (SOAE)

#### *Etiology*

Rarely, patients with SOAE will hear these sounds and complain of tinnitus (Penner, 1992). SOAEs are produced by movement of the cilia of the outer hair cells. While many normal people have SOAEs, most are unaware of their presence. These sounds are usually inaudible to others, but they are occasionally loud enough to be heard by an examiner. SOAEs can be detected by inserting very small microphones into the external auditory meatus and examining the spectra of recorded sounds.

## SUBJECTIVE TINNITUS

### Subjective Tinnitus with Rotational Vertigo

See information on Meniére's disease (Chapter 15).

### Subjective Tinnitus with Hearing Defects

#### *Otosclerosis (Mostly Low-Frequency Tinnitus)*

Tinnitus subsides after surgery only in 30–50% of cases (House, 1989). It even might worsen.

#### *Sudden Hearing Loss*

ENT physicians use low-molecular-weight dextran, pentoxiphyllin, and nicotinic acid, although their success rate has not been established by controlled studies. Therapeutic effects have been reported with prednisolone (Lamm, 1995).

#### *Head Trauma (High Frequency, Ringing Tinnitus)*

Tinnitus after head trauma usually lasts only a couple of hours but can persist for years with inner ear lesions. There is no known treatment.

#### *Noise-Induced Trauma as a Result of Acute or Chronic Exposure*

After acute exposure therapeutic effects for cortisone in combination with hyperbaric oxygen have been reported (Lamm, 1995).

#### *Acoustic Neuroma*

Tinnitus is often continuous and high-frequency, on one side. It can precede the hearing defect. About 4% of the neuromas present initially with tinnitus. Treatment options include surgery, after which tinnitus may persist (see below).

### Otototoxic Medication

Certain aminoglycosides (especially canamycine, streptomycine, neomycine, bycomycine), heavy metals, and cisplatine can cause tinnitus. Treatment involves withdrawl from medication. Permanent deficits may persist.

## Subjective Tinnitus without Hearing Defects

### Clinical Aspects

Approximately only 10–20% of patients who complain of tinnitus do not suffer from concurrent hypacusis. Causing factors for tinnitus are often emotionally disturbing situations. About 10% of the patients complain about hyperacusis before the onset of tinnitus (Hazell and Sheldrake, 1992). Irrespective of its origin, tinnitus frequently causes major disturbance, particularly in quiet surroundings and at night (Fichter and Goebel, 1996). Patients react by becoming depressed and, in some cases, suicidal. Sleep is disturbed and patients complain about impaired memory and lack of concentration.

Hearing loss, particularly in the high frequencies in the worse ear (often maximal at or about 4.0 kHz) is the most important risk factor or predictor for the development of sustained subjective tinnitus (Davis and Refaie, 2000). Some patients with tinnitus have hearing thesholds that are reportedly in the normal range. In these individuals, it is usually not possible to determine whether hearing thresholds have shifted toward the impaired range. Since thresholds are measured in dB, a logarithmic scale, substantial differences in sound energy may translate into a relatively small shift in thresholds.

The exact location (central or peripheral) of the lesion is not known. An altered relation between the activity of outer and inner hair cells in the cochlea has been discussed (Jastreboff, 1990). Since tinnitus still persists after severance of the acoustic nerve, a central accentuation after primary peripheral lesion is likely (House and Brackmann, 1981). In the rat, abnormal activity has been recorded in the inferior colliculus with salicylate levels causing tinnitus (Jastreboff and Sasaki, 1994). This abnormal activity can be influenced by lidocaine. Lidocaine also affected the abnormally increased perfusion in the auditory cortex of a patient with chronic tinnitus using the SPECT method (Staffen et al., 1999). In patients who could affect their tinnitus loudness by oral facial movements, PET studies suggest that the tinnitus

experience arises in the central auditory system and not the cochlea (Lockwood et al., 1998). In these patients, who all had unilateral tinnitus, changes in neural activity that paralleled changes in the loudness of tinnitus were found unilaterally in auditory cortical sites. This unilaterality contrasted with bilateral activation of auditory cortex that was associated with tonal stimuli delivered to the ear that the patients said their tinnitus was heard. Since cochlear activation was associated with bilateral activation whereas tinnitus was associated with unilateral activity, the investigators deduced that tinnitus was not of cochlear origin. In this same group of patients, who all had sensorineural hearing loss, tonal stimuli activated a more extensive portion of auditory cortex than was activated by identical stimuli delivered to control subjects with normal hearing. This expansion of cortical sites responsive to the stimuli demonstrated plastic changes in the central auditory system. This led to the hypothesis that tinnitus is due to plastic transformations of the central auditory system due to deafferentation, a condition that may be analogous to phantom pain sensations that may follow somatosensory deafferentation.

In analogy to trigeminal neuralgia, the possibility of ephaptic stimulation of neighboring, partially demyelinized axons through vascular loop compression of the nerve entry zone at the brainstem is discussed. In this cases a therapeutic effect of carbamazepine should be expected, although this could not be confirmed by a double-blind study against placebo in a random selection of patients (Hulshof and Vermaj, 1985). In individual cases with paroxysmal tinnitus due to a meningeoma (Espir et al., 1997) or with paroxysmal vertigo and tinnitus (Brandt and Dieterich, 1994), effective treatment has been achieved. Also surgery (microvascular decompression) has been performed in some patients (Møller et al., 1993).

### Etiology

This type of tinnitus can be caused by medication (many of the drugs can also lead to hearing defects), such as quinine, salicylate (dose dependent, reversible after 2–3 days), indometacine, carbamazepine, propranolol, levodopa, aminophyllin, coffeine, tetracycline, doxycycline, or salbutamol (for detailed review, see Brown et al., 1981). Most cases remain etiologically unclear.

### Natural Course

Drug-induced tinnitus disappears when the drug in question is withdrawn. Ototoxic drugs are exceptions to this rule. These drugs damage the cochlea and cause hearing loss that, in turn, leads to tinnitus. As a rule, if the etiology is not clear, the tinnitus subsides spontaneously after a matter of weeks or months, recurring only during times of fatigue and stress. Courses of highly irritative tinnitus over many years are not unusual. For only 8–13% of the patients with chronic tinnitus the tinnitus increases with time. As a rule after 1½ years the tinnitus gets less loud and is no longer of central importance.

### Practical Treatment

A conclusive and generally accepted form of treatment is not known. However, the following approaches should be given careful consideration, which sometimes prove successful also after various attempts. They are almost exclusively prescribed by the ENT physician or at least implemented under her/his guidance.

**Counseling, Behavioral Therapy.**   In many instances simple psychotherapeutic measures improve the situation for the often highly disturbed, insecure, and suffering patients. They range from simple tinnitus counselling (Preyer and Booth, 1995) to relaxation programs (Goebel, 1992; Lindberg *et al.*, 1989) for patients in specialized hospitals.

**Masking.**   Many tinnitus patients perceive some relief when other sources of noise drown out the tinnitus. In straightforward cases, this can be achieved by listening to the radio before falling asleep. Sometimes also wide frequency noise is used in this context. Hearing aids often lead to suppression of tinnitus by intensifying audibility (House, 1989).

**Tinnitus Retraining Therapy.**   This a relatively new form of therapy for patients with tinnitus, hyperacusis, or both (Jastreboff and Jastreboff, 2000; Mattox *et al.*, 1997). This is a multidisciplinary approach to the patient utilizing didactic informational sessions, counseling, and the use of a device that presents constant low-level noise to the affected ear. Although the practitioners of TRT report 70–80% success rates, the treatment has the disadvantage of being lengthy, requiring up to 2 years. TRT is available in only a few centers that specialize in tinnitus, and therapists require specialized training.

**Cochlear Implants.**   Cochlear implants can lead to tinnitus reduction (Hazell *et al.*, 1993; Souliere *et al.*, 1992). This therapy can be used only in deaf patients, since electrical stimulation damages the healthy cochlea (Portmann *et al.*, 1979).

**Vascular Decompression.**   A microvascular decompression at the root entry zone of the acoustic nerve in the brainstem (in analogy to trigeminus neuralgia) has been performed with the Gardner–Jannetta method (Møller *et al.*, 1993). Careful indication is mandatory. About half of the patients report some improvement. The surgery can deteriorate hearing, since the acoustic nerve is very vulnerable. The general lethality for the Gardner–Janetta operation is 1%.

### Ineffective or Obsolete

**Drug Therapy.**   Although there are numerous anecdotal responses to a variety of medications, a completely reliable and generally accepted form of treatment is not known (Dobie, 1999). The U.S. Food and Drug Administration has not approved any drug for the treatment of tinnitus. Although infusions of lidocaine reduce the loudness of tinnitus in most patients, this treatment has not been pursued due to considerable side effects. The same applies for oral analog tocanide HCl (Dobie, 1999). Lidocaine applications by means of local iontophoresis showed no effects (Dobie, 1999).

A specific effect in tinnitus treatment has not been established for the following drugs: carbamazepine (Hulshof and Vermeij, 1985), cinnarizine, clonazepam, cyclandelate (a vasodolating agent) (Hester *et al.*, 1998), flunarizine, nicotinic acid and derivatives, meclofenoxate, oxazepam, sulpuride, pentoxyphyllin, vitamine A, Gingko (von Wedel *et al.*, 1995), baclofen (Møller, 1997; Westerberg *et al.*, 1996), betahistine, lamotrigine (Simpson *et al.*, 1999), melatonin (Rosenberg *et al.*, 1998), caraverine (a quinoxaline derivative) (Domeisen *et al.*, 1998). A positive effect of sodium valproate (400 mg daily) in individual cases needs further examination (Menkes and Larson, 1998).

**Biofeedback.**   Either no or insignificant effects were achieved with biofeedback (Tonkin, 1985; Dobie, 1999).

**Surgical Measures.**   Surgery is not indicated for treatment of tinnitus alone. If for other reasons (e.g., tumor) a nerve must be severed, tinnitus is abated in only 45% of patients and either remains or becomes even worse in 55% (House and Brackmann, 1981; Parving *et al.*, 1992; Pulec, 1995).

**Miscellaneous.**   Acupuncture (Park *et al.*, 2000; Vilholm *et al.*, 1998) and hyperbaric oxygen have not been shown to be successful (Lenarz *et al.*, 1999).

## REFERENCES

Badia, L., Parikh, A., and Brookes, G. B. (1994). Management of middle ear myoclonus. *J. Laryngol. Otol.* **108**, 380–382.

Brandt, T., and Dieterich, M. (1994). Vestibular paroxysmia: Vascular compression of the eighth nerve? *Lancet* 1, 798–799.

Bryce, G. E., and Morrison, M. D. (1998). Botulinum toxin treatment of essential palatal myoclonus tinnitus. *J. Otolaryngol.* 27, 213–216.

Brown, D. R., Penny, J. E., Henley, C. M., Hodges, K. B., Kupetz, S. A., Glenn, D. W., and Jobe, P. C. (1981). Ototoxic drugs and noise. *In* "Tinnitus" (D. Evered and G. Lawrenson, Eds.), Ciba Foundation Symposium 85, pp. 151–171. Pitman, London.

Coles, R. R. A. (1987). Epidemiology of tinnitus. *In* "Tinnitus" (J. W. P. Hazell, Ed.), pp. 46–70. Churchill-Livingstone, Edinburgh.

Couloigner, V., Grayeli, A. B., Boccara, D., Julien, N., and Sterkers, O. (1999). Surgical treatment of the high jugular bulb in patients with Meniere's disease and pulsatile tinnitus. *Eur. Arch. Otorhinolaryngol* 256, 224–229.

Davis, A., and Refaie, A. E. (2000). Epidemiology of tinnitus. *In* "Tinnitus Handbook" (R. S. Tyler, Ed.), pp. 1–24. Singular, San Diego, CA.

Dobie, R. A. (1999). A review of randomized clinical trials in tinnitus. *Laryngoscope* 109, 1202–1211.

Domeisen, H., Hotz, M. A., and Häusler, R. (1998). Caroverine in tinnitus treatment. *Acta Otolaryngol. (Stockh.)* 118, 606–607.

Espir, M., Illingworth, R., Ceranic, B., and Luxon, L. (1997). Paroxysmal tinnitus due to a meningioma in the cerebellopontine angle. *J. Neurol. Neurosurg. Psychiatry* 62, 401–403.

Fichter, M., and Goebel, G. (1996). Psychosomatische Aspekte des chronischen komplexen Tinnitus. *Deutsches Ärzteblatt.* 93, A1771–A1776.

Goebel, G. (1992). "Ohrgeräusche: Psychosomatische Aspekte des komplexen chronischen Tinnitus: Vorkommen, Auswirkungen, Diagnostik und Therapie." Quintessenz Verl., München.

Hazell, J. W. (1990). Tinnitus III: The practical management of sensorineural tinnitus. *J. Otolaryngol.* 19, 11–18.

Hazell, J. W., Jastreboff, P. J., Meerton, L. E., and Conway, M. J. (1993). Electrical tinnitus suppression: frequency dependence of effects. *Audiology* 32, 68–77.

Hazell, J. W. P., and Sheldrake, J. B. (1992). Hyperacusis and tinnitus. *In* "Proceedings, IVth International Tinnitus Seminar Bordeaux, France," pp. 245–248. Kugler, Amsterdam/New York.

Hester, O., Theilman, G., Green, W., and Jones, R. O. (1998). Cyclandelate in the management of tinnitus: A randomized, placebo-controlled study. *Otolaryngol. Head Neck Surg.* 118, 329–332.

House, J. W. (1989). Therapies for tinnitus. *Ann. J. Otol.* 10, 163–165.

House, J. W., and Brackmann, D. E. (1981). Tinnitus: Surgical treatment. *In* "Tinnitus" (D. Evere and G. Lawrenson, Eds.), Ciba Foundation Symposium 85, pp. 204–216. Pitman, London.

Hulshof, J. H., and Vermeij, P. (1985). The value of carbamazepine in the treatment of tinnitus. *ORL* 47, 262–266.

Jastreboff, P. J. (1990). Phantom auditory perception (tinnitus): Mechanisms of generation and perception. *Neurosci. Res.* 8, 221–254.

Jastreboff, P. J., and Jastreboff, M. M. (2000). Tinnitus Retraining Therapy (TRT) as a method for treatment of tinnitus and hyperacusis patients. *J. Am. Acad. Audiol.* 11, 162–177.

Jastreboff, P. J., and Sasaki, C. T. (1994). An animal model of tinnitus: A decade of development. *Ann. J. Otol.* 15, 19–27.

Jastreboff, P. J. (1995). A neuropsychological approach to tinnitus theory and practice. Whurr, London.

Lamm, K. (1995). Rationale Grundlagen einer Innenohrtherapie. *Otorhinolaryngol. Nova* 5, 153–160.

Lenarz, T. *et al.* (Eds) (1999). Tinnitus. *HNO* 47, 14–18.

Lewis, J. E., Stephens, S. D., and McKenna, L. (1994). Tinnitus and Suicide. *Clin Otolaryngol* 19, 50–54.

Lindberg, P., Scott, B., Melin, L., and Lyttkeus, L. (1989). The psychological treatment of tinnitus: an experimental evaluation. *Behav. Res. Ther.* 27: 593–603.

Lockwood, A. H., Salvi, R. J., Coad, M. L., Towsley, M. L., Wack, D. S., and Murphy, B. W. (1998). The functional neuroanatomy of tinnitus. Evidence for limbic system links and neural plasticity. *Neurology* 50, 114–120.

Mattox, D. E., Jastreboff, P., and Gray, W. (1997). Tinnitus Habituation Therapy: The University of Maryland Tinnitus and Hyperacusis Center Experience. *Int. Tinnitus J.* 3, 31–32.

Menkes, D. B., and Larson, P. M. (1998). Sodium valproate for tinnitus. *J. Neurol. Neurosurg. Psychiatry* 65, 803.

Møller, M. B., Møller, A. R., Jannetta, P. J., and Jho, H. D. (1993). Vascular decompression surgery for severe tinnitus: Selection criteria and results. *Laryngoscope* 103, 421–427.

Møller, A. R. (1997). A double-blind placebo-controlled trial of baclofen in the treatment of tinnitus. *Am. J. Otol.* 18, 268–269. [Comment]

Olteanu-Nerbe, V., Uhl, E., Steiger, H.-J., Yousry, T., and Reulen, H.-J. (1997). Dural arteriovenous fistulas including the transverse and sigmoid sinuses: Results of treatment in 30 cases. *Acta Neurochir. (Wien)* 139, 307–318.

Park, J., White, A. R., and Ernst, E. (2000). Efficacy of acupuncture as a treatment for tinnitus. *Arch. Otolaryngol. Head Neck Surg.* 126, 489–492.

Parving, A., Tos, M., Thomsen, J., Míller, H., and Buchwald, C. (1992). Some aspects of life quality after surgery for acoustic neuroma. *Arch. Otolaryngol. Head Neck Surg.* 118, 1061–1064.

Penner, M. J. (1992). Linking spontaneous otoacoustic emissions and tinnitus. *Br. J. Audiol* 26, 115–123.

Portman, M., Cazals, Y., Negrevergne, M., and Aran, J. M. (1979). Temporary tinnitus suppression in man through electrical stimulation of the cochlea. *Acta Otolaryngol.* 87, 294–299.

Preyer, S., and Booth, F. (1995). Tinnitusmodelle zur Verwendung bei der Tinnituscounsellingtherapie des chronischen Tinnitus. *HNO* 43, 338–351.

Pulec, J. L. (1995). Cochlear nerve section for intractable tinnitus. *Ear Nose Throat J.* 74, 470–476.

Robinson, P. J., and Hazell, J. W. (1989). Patulous eustachian tube syndrome: The relationship with sensorineural hearing loss. Treatment by eustachian tube diathermy. *J. Laryngol. Otol.* 103, 739–742.

Rosenberg, S. I., Silverstein, H., Rowan, P. T., and Olds, M. J. (1998). Effect of melatonin on tinnitus. *Laryngoscope* 108, 305–310.

Saeed, S. R., and Brookes, G. B. (1993). The use of clostridium botulinum toxin in palatal myoclonus. A preliminary report. *J. Laryngol. Otol.* 107, 208–210.

Shah, S. B., Lalwani, A. K., and Dowd, C. F. (1999). Transverse/sigmoid sinus dural arteriovenous fistulas presenting as pulsatile tinnitus. *Laryngoscope* 109, 54–58.

Simpson, J. J., Gilbert, A. M., Weiner, G. M., and Davies, W. E. (1999). The assessment of lamotrigine, an antiepileptic drug, in the treatment of tinnitus. *Am. J. Otol.* 20, 627–631.

Souliere, C. R. Jr., Kileny, P. R., Zwolan, T. A., and Kemink, J. I. (1992). Tinnitus suppression following cochlear implantation. A multifactorial investigation. *Arch. Otolaryngol. Head Neck Surg.* 118, 1291–1297.

Staffen, W., Biesinger, E., Trinka, E., and Ladurner, G. (1999). The effect of lidocaine on chronic tinnitus: A quantitative cerebral perfusion study. *Audiology* 38, 53–57.

Tonkin, J. P. (1985). Tinnitus: Clinical approaches to management. *Curr. Ther.* 37–40.

Vilholm, O. J., Møller, K., and Jorgensen, K. (1998). Effect of traditional Chinese acupuncture on severe tinnitus: A double-blind, placebo-controlled, clinical investigation with open therapeutic control. *Br. J. Audiol.* **32,** 197–204.

v. Wedel, H., v. Wedel, U. C., Streppel, M., and Walger, M. (1997). Zur Effektivität partieller und kompletter apparativer Maskierung beim chronischen Tinnitus. *HNO* **45,** 690–694.

Westerberg, B. D., Roberson, J. B. Jr., and Stach, B. A. (1996). A double-blind placebo-controlled trial of baclofen in the treatment of tinnitus. *Am. J. Otol.* **17,** 896–903.

Williams, A., Goodenberger, D., and Calne, D. B. (1998). Palatal myoclonus following Herpes Zoster ameliorated by 5-hydroxytryptophan and Carbidopa. *Neurology* **28,** 358.

CHAPTER 17

# Anosmia, Ageusia, and Other Disorders of Chemosensation

Richard L. Doty and Steven M. Bromley

## INTRODUCTION

Until they malfunction, the chemical senses of taste and smell often go unappreciated by laymen and medical practitioners alike. These senses, however, are very important to everyday life, as they determine the flavor of foods and beverages, provide a vast array of aesthetic pleasures, and aid in the detection of environmental hazards, including leaking natural gas, spoiled food, and myriad of other air- and water-borne toxins and pollutants. From the perspective of the neurologist, alterations in smell and taste function can be the first clinical sign of an underlying neurological disorder, including brain neoplasm, drug or environmental toxicity, and—in the case of olfaction—Alzheimer's disease, epilepsy, idiopathic Parkinson's disease, multiple sclerosis, and schizophrenia. It goes without saying that the early diagnosis of such disorders is essential for optimal medical intervention to thwart or delay further symptom development or progression.

In this chapter we provide the nomenclature used to characterize disorders of tasting and smelling, examine a number of commonly encountered chemosensory pathologies, and describe up-to-date means for quantitatively assessing, managing, and treating taste and smell disorders.

## CLINICAL ASPECTS

### Definitions of Chemosensory Disorders

Disorders of the chemical senses are varied, ranging from phantom sensations that appear in the absence of any obvious stimuli (termed *phantosmias* for smells and *phantogeusias* for tastes) to altered or reduced sensations in response to modality-appropriate stimuli. *Anosmia* reflects the inability to perceive odors, whereas a*geusia* reflects the inability to perceive tastants (e.g., sweet, sour, bitter, or salty tasting substances). Such losses can be for all modality-specific stimuli (termed *total anosmia* for odors or *total ageusia* for tastes) or for just some such stimuli (termed *partial anosmia* for odors or *partial ageusia* for tastes). Lessened function for odorants or tastants are termed *hyposmia* (microsmia) or *hypogeusia* (microgeusia), respectively, and can be further subdivided, on the basis of quantitative testing, into mild, moderate, and severe categories. *Hypergeusia* or *hyperosmia* reflects abnormally heighted taste or smell sensations, whereas taste or smell *agnosia* reflects the inability to recognize a taste or smell sensation even though sensory processing, language, and general intellectual functions are essentially intact, as in some stroke patients. Distorted smell sensations are termed *dysosmias* or *parosmias*, whereas distorted taste sensations are termed *dysgeusias* or *parageusias*. Gustatory dysfunction is more often than not regional, depending upon which taste bud field or taste nerve is damaged. Although *presbyosmia* or *presbygeusia* are sometimes used to describe smell or taste losses due to aging, these terms are less specific than those noted above and are laden, by definition, with the notion that it is age *per se* that is causing the age-related deficit.

### Measurement of Chemosensory Disorders

Many patients are inaccurate in describing their chemosensory function; some are unaware of a chemosensory deficit, whereas others overstate the nature of their problem. It is common, for example, for both the patient and physician to ascribe "taste loss" simply to taste bud dysfunction, when often this reflects diminished retronasal activation of the olfactory receptors during deglutition (Burdach and Doty, 1991). Thus, most flavor sensations (e.g., lemon, peanut, coffee, cola, meat sauce, chocolate, vanilla, and strawberry) are dependent upon such retronasal stimulation, not upon the taste system. Taste buds mediate sweet, sour, salty, and bitter sensations, as well as possibly the sensations of "metallic," "chalky," and "umami" produced, respectively, by iron salts, calcium salts, and sodium-based substances like monosodium glutamate. Hence, it is incumbent upon the neurologist to employ sensitive and modern means for quantitatively assessing chemosensory function in the office setting. Such testing is essential to (a) verify the validity of the patient's complaint, (b) establish the nature and the degree of the problem, (c) accurately monitor changes in function over time (including those resulting from therapeutic interventions), (d) detect malingering, and (e) determine appropriate compensation for disability. In the past, the more astute neurologists have tested smell and taste by simply asking the patient to identify several crude odorants (e.g., peppermint, coffee grounds, or tobacco) or tastants (e.g., swishing and expectorating an aqueous solution of sugar water). Unfortunately, this qualitative type of testing can lead to the wrong conclusions (e.g., patients have difficulty identifying odors and some tastants without response alternatives), can be easily faked by malingerers, lacks reliability, and has no normative reference.

In the case of olfaction, a number of standardized and practical psychophysical tests have been developed in the past two decades (for review, see Doty, 2001). The most widely used is the microencapsulated 40-odor University of Pennsylvania Smell Identification Test (or UPSIT, commercially known as the Smell Identification Test), which is available in English, Spanish, French, and German versions (Doty, 1995). This reliable test (test–rest $r = 0.94$) can be self-administered either unilaterally or bilaterally in 10 to 15 min and scored in less than 1 min by nonmedical personnel. In addition to providing the neurologist with a percentile ranking of a patient's performance relative to age- and sex-referenced controls, an absolute determination of normosmia, mild microsmia, moderate microsmia, severe microsmia, anosmia, or probable malingering can be made. To date, this test has been administered to approximately 200,000 patients and used in hundreds of published medical and scientific studies. Although local electrial potentials can be recorded from the olfactory epithelium (termed the electro-olfactogram or EOG) and evoked potentials can be obtained from chemical stimulation of the olfactory system, these measures are not generally practical. For example, the EOG can be recorded only in limited regions of the epithelium, thereby not pro-

viding an accurate assessment of the functioning of the entire epithelium. Because anesthesia cannot be used, electrodes are often not tolerated by patients, limiting the number of persons from which such potentials can be obtained. The measurement of evoked potentials requires long test sessions (due, in part, to the necessity to have long intertrial intervals to avoid adaptation) and expensive equipment necessary to provide square-wave pulses of odorants to the olfactory region without evoking intranasal somatosensory sensations and adaptation (for review, see Doty and Kobal, 1995).

Taste testing is performed using either chemical or electrical stimuli. Regional taste testing allows for assessment of deficits in one or more of the paired nerves that innervate the taste buds; whole-mouth testing is generally insensitive to major deficits in one or more of these nerves, reducing its usefulness in the neurological examination. In the case of chemical testing, an examiner can present set concentrations of liquid tastants (e.g., sucrose, citric acid, caffeine, and sodium chloride) to specified lingual regions (in some cases in comparison with blank trials). After presentation of the tastants, the same regions are rinsed with water. Most stimulus-presentation procedures (e.g., those using pipettes) confine testing to lingual surfaces, although Q-tips and other devices have been used to assess function on the palate. Accurate chemical testing requires rinsing the mouth and expectoration after each stimulus presentation, and can be quite time consuming. For example, if responsiveness to each of the four basic taste qualities is to be made on left and right anterior (CN VII) and posterior (CN IX) tongue regions, 16 trials (4 tastants × 4 tongue regions) are needed to present a single stimulus for each quality. Since multiple stimuli are required to produce reliable responses, the number of trials increases considerably.

Electrical assessment of the taste system is more practical than chemical assessment. To perform electrical testing, an electrogustometer must be employed. Such a device allows for the easy presentation of brief (e.g., 0.5 s) microamp currents to small regions of the tongue for known durations (Frank and Smith, 1991). Neither stimulus preparation nor rinsing are required. While electrogustometric thresholds and suprathreshold measures of stimulation intensity are easy to assess, the relationship of such measures to those obtained using chemical stimuli is not well established and no specific taste qualities are usually evoked. Nevertheless, at low-current levels electrical stimuli have been shown to activate only taste afferents, not trigeminal (CN V) afferents. Unfortunately, sound normative data based upon forced-choice testing paradigms are still generally lacking for such stimuli.

On forced-choice psychophysical tests, such as the UPSIT, malingering appears as the reporting of fewer incorrect responses than expected on the basis of chance performance, as would be expected of an anosmic. Since, in the case of the UPSIT, there are 4 response alternatives for each item and the patients must provide an answer even if no smell is perceived, 25% of the items, on average, should be correctly identified by chance alone (i.e., 10 of 40). A sampling distribution exists around this expected probability, and empirical data are available on this point (Doty *et al.*, 1984b). The theoretical probability of a true anosmic having an UPSIT score <6 is less than 5 in 100. The theoretical probability of a true anosmic scoring zero on the UPSIT is less than 1 in 100,000 (Doty, 1995). In general, if a patient scores within the probable malingering region of UPSIT scores, the UPSIT should be administered again to confirm the apparent avoidance of correct responses. Multiplication of the two probabilities is then used to establish the statistical likelihood of a true deficit independent of malingering.

## Natural Course of Olfactory Disorders

Olfactory disturbance results from numerous etiologies and affects patients of all ages (Bromley, 2000; Deems *et al.*, 1991; Schiffman, 1997). Although precise data on the incidence and prevalence of olfactory dysfunction in the general population are not available, it appears that about 1–2% of the population under the age of 65 years suffers from smell or taste dysfunction. The prevalence and magnitude of olfactory loss in older persons, however, is staggering. About half of the population between 65 and 80 years of age appears to suffer from significant decrements in the ability to smell. This figure rises to nearly 75% of those over the age of 80 years (Doty *et al.*, 1984a). Such smell loss adversely affects the quality of life of most older persons, influencing nutrition, appetite, and, in some cases, even general immunity and defensive responses to illnesses.

Anosmia or hyposmia usually is due to one of three general causes: (a) nasal passage obstructions that produce *conductive or transport impairments* (e.g., by chronic rhinosinusitis, polyposis, excessive mucus secretion, etc.), (b) injury to the olfactory neuroepithelium resulting in *sensorineural impairment* (e.g., by viruses, airborne toxins, etc.), and (c) injury to central nervous system structures causing *central olfactory neural impairment* (e.g., tumors, masses impinging on the olfactory tract, etc.). These categories, however, are not mutually exclusive. For example, chronic rhinosinusitis can produce damage to the olfactory receptors in addition to blocking airflow, and altered receptor function can, over time, lead to degeneration within the olfactory bulb, a central structure.

As shown in Table I, there are many disorders that have been associated with olfactory disturbances. Most chronic cases of anosmia or hyposmia are a result of upper respiratory infections, head trauma, and nasal and paranasal sinus disease. Typically, these disorders produce long-lasting or permanent damage to the olfactory epithelium (Deems *et al.*, 1991). Other reasons for olfactory impairment include intranasal neoplasms (e.g., inverting papillomas, hemangiomas, and esthesioneuroblastomas), intracranial tumors or lesions (e.g., olfactory groove meningiomas, frontal lobe gliomas), exposure to airborne toxins (including cigarette smoke), iatrogenic interventions (e.g., septoplasty, rhinoplasty, turbinectomy, laryngectomy, radiation therapy, medications), psychiatric disorders, and various endocrine and metabolic disorders.

Most dysosmias—distorted smell sensations—reflect ongoing degenerative or regenerative processes within the olfactory epithelium and can remit over time. Many patients report that prior to onset of their anosmia, they experienced a period of weeks or months of steady or intermittent dysosmia. Usually some smell function is present during the dysosmic period, as measured psychophysically. On rare occasions dysosmias present as aura-like hallucinations that may reflect central dysfunction, e.g., temporal lobe seizure activity. In many such cases, no seizure activity can be documented by EEG, and no evidence of CNS lesions or tumors is apparent. Dysosmias can also occur in a number of psychiatric disturbances which usually are diagnosed on other grounds (e.g., psychosis). Some dysosmias are medication-induced, while others can arise from purulent nasal secretions in rhinosinusitis or from exhalations in halitosis or uremia.

Hyperosmia is unusual, with documented cases of abnormally heightened smell function being exceptionally rare. Although one laboratory has reported that untreated adrenal cortical insufficiency produces hyperosmia, this finding has yet to be confirmed and animal studies find no evidence for hypersensitivity following adrenalectomy (Doty *et al.*, 1991). Hyperosmia has been claimed to accompany such controversial syndromes as multiple chemical sensitivity, but the limited data available fail to support this notion (Doty *et al.*, 1988a). In some cases, olfactory epileptic auras, typically generated from the amygdala and other medial temporal lobe structures, may manifest as perceived increased sensitivity to odors. However, most patients with long-term epilepsy and intractable seizure activity, such as candidates for temporal lobe resection, are hyposmic (West and Doty, 1995).

Seizure activity within the mesial temporal lobe (commonly the amygdala or hippocampus), and rarely the orbitofrontal cortex of the frontal lobe, can produce an olfactory epileptic aura, although auras may also occur

**TABLE I**   Reported Agents, Diseases, Drugs, Interventions, and Other Etiologic Categories Associated in the Medical or Toxicological Literature with Olfactory Dysfunction

**Air pollutants and industrial dusts**
  Acetone
  Acids (e.g., sulfuric)
  Ashes
  Benzene
  Benzol
  Butyl acetate
  Cadmium
  Carbon disulfide
  Cement
  Chalk
  Chlorine
  Chromium
  Coke/coal
  Cotton
  Cresol
  Ethyl acetate
  Ethyl and methyl acrylate
  Flour
  Formaldehyde
  Grain
  Hydrazine
  Hydrogen selenide
  Hydrogen sulfide
  Iron carboxyl
  Lead
  Nickel
  Nitrous gases
  Paint solvents
  Paper
  Pepper
  Peppermint oil
  Phosphorus oxychloride
  Potash
  Silicone dioxide
  Spices
  Trichloroethylene

**Drugs**
  Adrenal steroids (chronic use)
  Amino acids (excess)
    Cysteine
    Histidine
  Analgesics
    Antipyrine
  Anesthetics, local
    Cocaine HCl
    Procaine HCl
    Tetracaine HCl
  Anticancer agents (e.g., methotrexate)
  Antihistamines (e.g., Chlorpheniramine malate)
  Antimicrobials
    Griseofulvin
    Lincomycin
    Macrolides
    Neomycin
    Pencillins
    Streptomycin
    Tetracyclines
    Tyrothricin
  Antirheumatics
    Mercury/gold salts
    D-Penicillamine

  Antithyroids
    Methimazole
    Propylthiouracil
    Thiouracil
  Antivirals
  Cardiovascular/hypertensives
  Gastric medications
    Cimetidine
  Hyperlipoproteinemia medications
    Artovastatin Calcium (Lipitor)
    Cholestyramine
    Clofibrate
  Intranasal saline solutions with
    Acetylcholine
    Acetyl, β-methylcholine
    Menthol
    Strychnine
    Zinc sulfate
  Local Vasoconstrictors
  Opiates
    Codeine
    Hydromophone HCl
    Morphine
  Psychopharmaceuticals (e.g., LSD, psilocybin)
  Sympathomimetics
    Amphetamine sulfate
    Fenbutrazate HCl
    Phenmetrazine theoclate

**Endocrine/metabolic**
  Addison's disease
  Congenital adreanl hyperplasia
  Cushing's syndrome
  Diabetes mellitus
  Froelich's syndrome
  Gigantism
  Hypergonadotropic hypogonadism
  Hypothyroidism
  Kallmann's syndrome
  Pregnancy
  Panhypopituitarism
  Pseudohypoparathyroidism
  Sjögren's syndrome
  Turner's syndrome

**Infections—viral/bacterial**
  Acquired immunodeficiency syndrome (AIDS)
  Acute viral rhinitis
  Bacterial rhinosinusitis
  Bronchiectasis
  Fungal
  Influenza
  Rickettsial
  Microfilarial

**Lesions of the nose/airway blockage**
  Adenoid hypertrophy
  Allergic rhinitis
    Perennial
    Seasonal
  Atrophic rhinitis
  Chronic inflammatory rhinitis
  Hypertrophic rhinitis

*continues*

**TABLE I** *continued*

Nasal Polyposis
Rhinitis medicamentosa
Structural abnormality
   Deviated septum
   Weakness of alae nasi
Vasomotor rhinitis

**Medical interventions**
Adrenalectomy
Anesthesia
Anterior craniotomy
Arteriography
Chemotherapy
Frontal lobe resection
Gastrectomy
Hemodialysis
Hypophysectomy
Influenza vaccination
Laryngectomy
Oophorectomy
Paranasal sinus exenteration
Radiation therapy
Rhinoplasty
Temporal lobe resection
Thyroidectomy

**Neoplasms—intracranial**
Frontal lobe gliomas and other tumors
Midline cranial tumors
   Parasagital meningiomas
   Tumors of the corpus callosum
Olfactory groove/cribriform plate meningiomas
Osteomas
Paraoptic chiasma tumors
   Aneurysms
   Craniopharyngioma
   Pituitary tumors (esp. adenomas)
   Suprasellar cholesteatoma
   Suprasellar meningioma
Temporal lobe tumors

**Neoplasms—intranasal**
Neuro-olfactory tumors
   Esthesioepithelioma
   Esthesioneuroblastoma
   Esthesioneurocytoma
   Esthesioneuroepithelioma
Other benign or malignant nasal tumors
   Adenocarcinoma
   Leukemic infiltration
   Nasopharyngeal tumors with extension
   Neurofibroma
   Paranasal tumors with extension
   Schwannoma

**Neoplasms—extranasal and extracranial**
Breast
Gastrointestinal tract
Laryngeal
Lung
Ovary
Testicular

**Neurologic**
Amyotrophic Lateral Sclerosis
Alzheimer's disease
Cerebral abscess (esp. frontal or ethmoidal regions)
Down's syndrome
Familial dysautonomia
Guam ALS/PD/Dementia
Head trauma
Huntington's disease
Hydrocephalus
Korsakoff's psychosis
Migraine
Meningitis
Multiple sclerosis
Myesthenia gravis
Paget's disease
Parkinson's disease
Refsum's syndrome
Restless leg syndrome
Syphilis
Syringomyelia
Temporal lobe epilepsy
   Hamartomas
   Mesial temporal sclerosis
   Scars/previous infarcts
Vascular insufficiency/anoxia
   Small multiple cerebrovascular accidents
   Subclavian steal syndrome
   Transient ischemic attacks

**Nutritional/metabolic**
Abetalipoproteinemia
Chronic alcoholism
Chronic renal failure
Cirrhosis of liver
Gout
Protein calorie malnutrition
Total parenteral nutrition w/o adequate replacement
Trace metal deficiencies
   Copper
   Zinc
Whipple's disease
Vitamin deficiency
   Vitamin A
   Vitamin $B_6$
   Vitamin $B_{12}$

**Pulmonary**
Chronic obstructive pulmonary disease

**Psychiatric**
Anorexia nervosa (severe stage)
Attention deficit disorder
Depressive disorders
Hysteria
Malingering
Olfactory reference syndrome
Schizophrenia
Schizotypy
Seasonal affective disorder

in migraineurs as an associated predome of a migraine headache. Because of the *de novo* nature of olfactory epileptic auras, they are frequently called "hallucinations." By definition, an olfactory epileptic aura—constituting about ~1% of all auras—is not a warning that a seizure is coming, but rather it is a partial simple seizure, and it is suggestive of a focal epileptiform lesion (Acharya *et al.*, 1998). Hence, such a sensation need not be followed by additional seizure activity. In general, olfactory auras are unpleasant (e.g., feces, rotting fruit, vomitus), although in rare instances pleasant auras also occur (West and Doty, 1995). For example, a patient with an aneurysm at the bifurcation of the right middle cerebral artery compressing the right orbitofrontal cortex experienced, prior to the onset of complex partial seizures, pleasant smelling auras seeming to come from behind the right ear (Mizobuchi *et al.*, 1999). Intraoperative spikes were recorded from the right orbitofrontal region, superior temporal gyrus, and uncus during the aneurysm clipping. The authors suggested that the perception of the odor coming from a specific direction (i.e., from the right) was being mediated by the orbitofrontal cortex.

Olfactory dysfunction is a hallmark of a number of common neurodegenerative disorders, including Alzheimer's disease (AD), idiopathic Parkinson's disease (PD), Huntington's disease (HD), alcoholic Korsakoff's syndrome (KS), Pick's disease (PD), the parkinsonian dementia complex of Guam (PDG), amyotrophic lateral sclerosis (ALS), and multiple sclerosis (MS) (Doty, 1991; Doty *et al.*, 1998; Moberg *et al.*, 1997). Indeed, olfactory dysfunction may be the first clinical sign of AD and idiopathic PD (Doty *et al.*, 1987, 1988b). In the case of PD, the smell loss is more prevalent than tremor, a cardinal sign of the disorder. Such loss is bilateral and unrelated to disease stage, duration of illness, use of anti-parkinsonism medications, and severity of the symptoms, such as tremor, rigidity, bradykinesia, or gait disturbance (Doty *et al.*, 1992b). Importantly, a number of other movement-related neurological disorders are *unaccompanied* by smell loss, including corticobasal degeneration (Wenning *et al.*, 1995), progressive supranuclear palsy (Doty *et al.*, 1993; Wenning *et al.*, 1995), 1-methyl-4-phenyl-1,2,3,6-tetrahydropyridine (MPTP) -induced parkinsonism (Doty *et al.*, 1992a), and essential tremor (Busenbark *et al.*, 1992). Such findings suggest that assessment of smell function can aid in the differential diagnosis of several neurodegenerative disorders, especially those that present with similar motoric clinical signs. McCaffrey *et al.* (2000) have recently shown that even a three-item self-administered smell identification test (the Pocket Smell Test) discriminates better between patients with AD and depression than the widely used 30-item Mini-Mental State Examination (MMSE).

Several studies report that smell testing is useful in identifying persons at risk for later significant cognitive decline or AD. For example, Graves *et al.* (1999) administered a 12-item version of the UPSIT, termed the Brief Smell Identification Test or B-SIT, and several cognitive tests to 1985 Japanese-American people around the age of 60 years. Two years later 1604 of these people were retested. About two-thirds of the follow-up participants were genotyped for apolipoprotein E (apoE). Low B-SIT scores in conjunction with one or more APOE-ε4 alleles were associated with a very high risk of subsequent cognitive decline, and smell testing was better than a global cognitive test in identifying persons who later came to exhibit such decline. More recently, Devanand *et al.* (2000) administered the UPSIT to 90 outpatients with mild cognitive impairment and to matched controls at 6-month intervals over a several-year time period. Lower UPSIT scores were found in patients with mild cognitive impairment than in the controls. Most importantly, patients with UPSIT scores less than 34 were more likely to develop AD than the controls. Low UPSIT scores, in conjunction with lack of awareness of olfactory dysfunction on the part of the patients, was predictive of the time of AD development. UPSIT scores ranging from 30 to 35 exhibited moderate to strong sensitivity and specificity for diagnosis of AD at follow-up.

Although not classically considered a neurodegenerative disease, smell loss is present in patients with schizophrenia. Unlike other neuropsychological measures, UPSIT scores appear to be correlated with disease duration, suggesting a possible degenerative component to this disorder in olfaction-related pathways (Moberg *et al.*, 1997). Interestingly, recent MRI studies have found the olfactory bulbs and tracts of patients with schizophrenia to be markedly smaller than those of controls (Turetsky *et al.*, 2000).

### Natural Course of Taste Disorders

As in the case of olfaction, gustatory problems are widespread. A 1998 national survey suggests that more than 1.1 million Americans suffer from taste problems (Hoffman *et al.*, 1998). Total whole-mouth loss of taste function (total ageusia) is generally believed to be rare, since peripheral damage would have to involve all of the taste nerves and taste bud fields. Thus, when present, total taste loss is usually ascribed to central events (e.g., ischemia, pharmacological agents, metabolic disturbance). Although there is evidence of an age-related decline in whole-mouth taste function, the degree of loss is not comparable to the degree of loss observed in olfaction. Regional losses, however, are likely very common in older people. For example, sensitivity to NaCl was measured in one study on the tongue tip and 3 cm posterior to the tongue tip in 12 young (20–29 years of age) and 12 elderly (70–79 years of age) subjects. The young subjects were more sensitive to NaCl on the tongue tip than on the more posterior stimulation site and exhibited, at both tongue loci, in increase in detection performance as the stimulus concentration increased. The elderly subjects, who would be generally expected to exhibit moderate whole-mouth taste deficits, performed at chance levels (Matsuda and Doty, 1995).

Taste perception can be altered in a number of ways, including (a) the release of foul-tasting materials into the oral cavity (e.g., from oral infections, gingivitis, sialadenitis), (b) the decrease of movement of tastants into the taste buds (e.g., from blocked or damaged taste pores, excessive oral dryness), (c) damage to the taste buds proper (e.g., from caustic burns, allergic reactions to chemicals or oral products), (d) injury to the taste nerves (e.g., as a result of post-viral Bell's palsy, dental or surgical procedures), and (e) damage to central nervous system taste pathways (e.g., from tumors, epilepsy, infarcts). Some dysgeusias arise from the use of different metals in fillings or bridges within the mouth that set up subtle electrical currents that traverse segments of the oral cavity.

Some forms of *hyper*geusia and dysgeusia may arise from selective damage or alterations to the taste nerves. Thus, anesthetizing one chorda tympani nerve increases the perceived intensity of bitter substances, such as quinine, applied to taste fields innervated by the contralateral glossopharyngeal nerve (Yanagisawa *et al.*, 1998). Such anesthetization decreases the perceived intensity of NaCl applied to an area innervated by the ipsilateral glossopharyngeal nerve. Anesthetization of both chorda tympani nerves intensified the taste of quinine and diminished the taste of NaCl bilaterally in the posterior lingual regions innervated by CN IX. A phantom taste, usually localized to the posterior tongue contralateral to the anesthesia, appeared in the absence of stimulation in ~40% of their subjects. This phantom taste disappeared when its region of origin was anesthetized. The implication of these findings is that when individual taste nerves incur damage, other taste nerves may be activated—perhaps by the release of central inhibition—in ways that induce dysgeusic or hypergeusic sensations.

Be it as it may, most persons are, in fact, unaware of regional losses of taste function, and even patients who have had one chorda tympani nerve severed during middle ear surgery rarely complain of taste loss. This lack of awareness stems, in part, from the redundancy of the multiple taste nerves, as well as possibly neural compensatory mechanisms associated with the aforementioned release of central inhibition (Kveton and Bartoshuk, 1994).

While major head injury causes around 15% of patients to exhibit olfactory deficits, most typically total bilateral anosmia, fewer than 1% of persons with such injury exhibit ageusia to sweet, sour, salty, or bitter taste qualities (Deems *et al.*, 1991; Sumner, 1967). Injuries that affect the middle ear (e.g., basilar temporal bone fractures) can impair chorda tympani nerve-mediated taste function unilaterally, as well alter salivary secretion. Trauma to the lingual nerve (a branch of the mandibular division of CN $V_3$ that also carries CN VII taste fibers) around the jaw and tongue can also compromise taste in some cases.

Bell's palsy, a common cause of ipsilateral taste loss of the anterior two-thirds of the tongue, also causes unilateral facial weakness, hyperacusis, and impairment of the taste–salivary reflex. Return of taste function within the first 2 weeks of onset of Bell's palsy is reportedly associated with complete recovery from facial paresis; however, impairment of taste for longer than 2 weeks suggests a poor prognosis for rapid return of full strength.

Peripheral neoplasms and mass lesions can cause taste loss or distortion. For example, while squamous cell carcinoma of the mucous membranes of the upper aerodigestive tract can interfere with taste by direct destruction of receptors, mass lesions affecting the course of CN VII, IX, and X (e.g., acoustic neuromas, facial nerve schwannomas) may cause impairment through neural compression and impaired conduction. Cancer-related morbidity and treatment (e.g., malnutrition, SIADH, chemo-therapy, radiation therapy) can also lead to taste dysfunction. For example, Panayiotou *et al.* (1995) described three patients presenting with a persistent sweet dysgeusia who were diagnosed as having small cell carcinoma of the lung. In each case, hyponatremia secondary to inappropriate secretion of antidiuretic hormone was present. Resolution of the dysgeusia paralleled an increase in serum sodium concentration after water restriction alone. The authors concluded that the dysgeusia was likely due to the hyponatremia rather than other factors (e.g., tumor, antiduretic factor, medications, chemotherapy).

Taste loss and dysgeusia, as well as salivary dysfunction, can follow chemotherapy and radiotherapy, significantly impacting the quality of life and leading to decreased appetite, enjoyment from eating, weight loss, and nutritional deficits. Symptoms typically begin early in the course of treatment. Post-treatment recovery can be prolonged, with return to normal function taking months and, in some instances, even years (Bartoshuk, 1990). Radiation therapy of the head and neck can directly damage taste cells and buds, taste nerve fibers, and salivary glands, with the attendant compromise of food transport, altered levels of proteins potentially involved in taste transduction, and promotion of opportunistic oral infections (e.g., oral candidiasis) (Conger, 1973; Della Fera *et al.*, 1995).

Central nervous system lesions, as seen with stroke and multiple sclerosis, can cause taste disturbance. Susceptible regions of the brainstem include the nucleus tractus solitarius and the pontine tegmentum, involving both gustatory lemnisci. An ipsilateral multiple sclerosis plaque at the midpontine tegmentum was recently associated with hemiageusia (Combarros *et al.*, 2000). A small hemorrhage in the right tegmentum of the middle pons was reported to produce ageusia to all taste qualities on the right side of the tongue (Kojima and Hirano, 1999). Recently, three cases of focal ischemic lesions in the brainstem have been described that produced ipsilateral hemiageusia (Lee *et al.*, 1998).

Lateralized infarcts of the thalamus and infarcts in the coronal radiata have been associated with contralateral hypogeusia and/or dysgeusia (reflecting the crossed taste pathways at this level of the nervous system) (Fujikane *et al.*, 1999). Onoda and Ikeda (1999) reviewed 15 cases of unilateral gustatory impairment due to central infarcts, noting that the pattern of central injuries implies that the gustatory pathways in humans ascend ipsilaterally from the solitary tract of the medulla to the pons, and then cross superiorly in the midbrain to reach the contralateral thalamus. Complete loss of function does not occur from unilateral lesions above the thalamus, presumably because of the multiple areas involved in higher order taste processing.

A variety of surgical interventions have been linked to taste dysfunction. The chorda tympani nerve is at risk from surgical procedures that involve the middle ear, given its course between the malleus and the incus along the surface of the medial tympanic membrane. This nerve is often stretched or sectioned during tympanoplasty, mastoidectomy, and stapedectomy, in some cases producing long-lasting symptoms. In one study of 116 surgical patients, 78% of patients with bilateral section and 32% of patients with unilateral section of the chorda tympani reported persistent adverse gustatory symptoms (Bull, 1965). CN IX can be damaged during tonsillectomy, bronchoscopy, or laryngoscopy (Arnhold-Schneider and Bernemann, 1987; Donati *et al.*, 1991; Ohtuka *et al.*, 1994), reflecting the proximity of its lingual branch to the muscle layer of the palatine tonsillar bed (Ohtsuka *et al.*, 1994). Surgical treatment for snoring (e.g., uvulopalatoplasty), as well as such surgery-related procedures as intubation and employment of a laryngeal mask, have all been associated with taste loss or alteration (Evers *et al.*, 1999; Ostergaard *et al.*, 1997; Walker and Gopalsami, 1996). Recently, Shafer *et al.* (1999) tested taste function in 17 patients prior to third molar surgery and at 1 and 7 months after surgery. A ~15% average reduction of perceived inten-

sity was observed 1 month after surgery for NaCl, citric acid, and quininine. The ability identify the quality of NaCl was decreased after the extraction. Normal intensity perception for citric acid was still not present 6 months after the surgery. This study suggests that gustatory deficits (a) are common after third molar extraction, (b) persist for relatively long periods of time, and (c) are associated with the depth of impaction.

The use of medications is among the most common causes of taste disturbance. In extreme cases such effects can influence palatability of foods to the extent that body weight is compromised and depression ensues. Repetitive use of some oral topical agents, including hydrogen peroxide or steroids, can adversely alter taste. A number of medications (e.g., anticholinergics, antidepressants, antihistamines) cause dryness within the mouth and can produce hyperviscous saliva—physical conditions which may result in some individuals in decrements in taste acuity. Among drugs that are known to produce taste-related side effects are antiproliferative agents, lipid reducing drugs, antihypertensive drugs, diuretics, antifungal agents, antirheumatic drugs, antibiotic drugs, and drugs with sulfhydryl groups, such as penicillamine and captopril (Ackerman and Kasbekar, 1997; Doty *et al.*, 1991). It is noteworthy that, according to the Physician's Desk Reference (PDR), approximately one-fourth of all cardiac medications employed in the United States (including antilipemic agents, adrenergic blockers, ACE inhibitors, angiotensin II antagonists, calcium channel blockers, vasodilators, anticoagulants, antiarrhymics, and various diuretics) exhibit potential side effects of "altered taste," "bad taste," "bitter taste," or "metalic taste." The physician should be aware that the onset of gustatory symptoms for some agents can take weeks or even months. For example, in a study of 87 patients experiencing taste loss as a result of the antifungal agent terbinafine, the average latency from the first use of the drug to the experience of taste loss was 35 days. Recovery after drug cessation took several months (Stricker *et al.*, 1996). A common side effect of vinblastine (a mitotic inhibitor) is loss of taste (Kellokumpu-Lehtinen *et al.*, 1989). Rats injected with this agent exhibit a marked depression in preferences for sweet-tasting solutions (i.e., glucose and saccharine water) (Beidler and Smith, 1991). Altered taste is a frequent complaint of patients receiving neoadjuvant bleomycin, cisplatin, and methotrexate therapy (Duhra and Foulds, 1988; Fetting *et al.*, 1985; Greene *et al.*, 1994).

Numerous other etiologies for taste dysfunction have been described in the literature, including hypothyroidism, renal disease, liver disease, myasthenia gravis, Guillain–Barre syndrome, diabetes mellitus, and familial dysautonomia (a genetic disorder with lack of taste buds and papillae). Idiopathic dysgeusia has been associated with blood transfusions (Erick, 1996). Like olfactory epileptic auras, gustatory symptoms have all been reported in association with epileptic seizures. Examples of taste sensations that have been reported in such cases include "peculiar," "rotten," "sweet," "like a cigarette," "like rotten apples," and "like vomitus" (for review, see West and Doty, 1995). Some of these "tastes," however, likely represent smell sensations that are wrongly categorized as tastes by both the patients and their physicians.

## PRINCIPLES OF THERAPY

### Clinical Evaluation

The evaluation of a patient with a chemosensory disturbance should include, in addition to quantitative assessment described above, a complete medical history, followed by a thorough physical examination of the head and neck and appropriate brain and rhinosinus imaging.

### History

In the history, the examiner should obtain specifics regarding the nature, time of onset, duration and pattern of fluctuations, if any, of the patient's chemosensory symptoms to aid in establishing causality. For example, *sudden olfactory loss* suggests the possibility of head trauma, infection, or ischemia. *Gradual loss* can reflect not only accumulations of damage within the olfactory epithelium from viruses, pollutants and other insults, but the development of progressive obstructive lesions or tumors. *Intermittent loss* can signify an intranasal inflammatory process. Careful documentation of the types of stimuli that can or cannot be distinguished helps to differentiate between retronasal CN I flavor loss and true taste-bud-mediated gustatory loss. In making this distinction, it is useful to inquire whether the patient can detect the saltiness of potato chips, pretzels, or salted nuts; the sourness of vinegar, pickles, or lemons; the sweetness of sugar, soda, cookies, or ice cream; and the bitterness of coffee, beer, or tonic water. A true taste-bud-mediated dysfunction is suspected if a patient reports a problem in detecting any of these stimuli.

Questions regarding epistaxis, discharge (clear, purulent, or bloody), nasal obstruction, allergies and somatic symptoms, including headache or irritation, may have localizing value. Discovery of antecedent events, such as head trauma, viral upper respiratory infections, allergies, toxic exposures, and iatrogenic (e.g., operative)

interventions are critical, as is the identification of underlying medical conditions known to have potential chemosensory consequences (e.g., renal failure, liver disease, hypothyroidism, diabetes, or dementia). A review of the history of tobacco, alcohol, or recreational drug use may also provide clues to etiology (e.g., chronic alcoholism in the context of Wernicke and Korsakoff syndromes). A detailed assessment of the medications being used prior to and during the onset of the dysfunction should be performed.

Delayed puberty in association with anosmia (with or without midline craniofacial abnormalities, deafness, and renal anomalies) suggests the possibility of Kallmann's syndrome or one of its variants. A family history of smell dysfunction suggests the possibility of a genetic etiology, although most persons exhibit smell loss in their later years. Subtle signs of central tumors, dementia, parkinsonism, and seizure activity (e.g., automatisms, occurrence of black-outs, auras, and déjà vu) should be sought in both the history and physical exam. Although the examiner should be aware of pending litigation and the possibility of malingering, care must be taken not to assume malintent on the part of the patient and objective reliance on sensory tests and physical findings must be maintained.

Present or past problems with chewing, salivation, swallowing, oral pain or burning, dryness of the mouth, periodontal disease, speech articulation, bruxism, or foul breath odor should be determined. Questioning concerning diet, oral habits, stomach problems, and possible problems with digestion or acid reflux can aid in establishing if acid reflux into the oral cavity, which can irritate or damage posterior taste buds, is involved. Documentation of dental work or exposure to radiation prior to or at the time of symptom onset should be noted, as should hearing or balance problems, since past or current ear infections or surgery can alter chorda tympani nerve function.

### Physical Examination, Laboratory Tests, and Medical Imaging

For most patients complaining of olfactory problems, neurologic and otolaryngologic assessment is warranted. Visual acuity, visual field, and optic disc examinations are of value in detecting possible intracranial mass lesions that produce increased intracranial pressure (papilledema) and optic atrophy (e.g., the Foster Kennedy syndrome). A thorough nasal endoscopic evaluation should pay close attention to the olfactory meatal area. The color, swelling, surface texture, inflammation, exudate, erosion, ulceration, epithelial metaplasia, and atrophy of the nasal mucosa should be assessed. Involvement of the osteomeatal complex is suggested by the presence of mucopus below the Eustacian tube orifice, whereas mucopus above this orifice suggests posterior ethmoid and/or sphenoid sinus disease. Airflow to the receptor region can be compromised by masses, polyps, and adhesions of the turbinates to the septum. Unusual crusting, dryness, or spaciousness, as is seen in atrophic rhinitis, suggests atrophy of the lamina propria. Allergy can often be inferred from a pale mucous membrane, usually reflecting edema within the lamina propria. Metaplasia within the epithelium, in addition to swelling, inflammation, exudates, erosion, and ulceration, can be caused by industrial or environmental pollutants.

Altered CN VII, IX, and X function, independent of taste, can shed light on whether and at what location a gustatory dysfunction may be present (e.g., abnormal facial musculature, swallowing, salivation, gag reflex, voice production). Evidence of abnormal epithelial color, or visual signs of scarring, inflammation, or atrophy of lingual papillae, should be sought. Palpation of the tongue should be used to rule out neoplastic lesions deep in the tongue's musculature. Assessment of the condition of the teeth and gums is important, since gingivitis or pyorrhea may produce exudates that cause or contribute to dysgeusic symptoms. The nature and integrity of the fillings, bridges and other dental work should be examined (as noted earlier, dissimilar metals can induce small electrical currents that, in turn, produce abnormal oral sensations). In idiopathic cases, neuroimaging should be obtained to rule out CNS tumors or lesions.

Photography of the lingual surface under relatively low power magnification, after staining the tongue with dark food dye, allows for counting or better visualizing selected classes of papillae, although this procedure is best confined to centers such as ours with experience in this area. In some cases where local pathology is suspected, biopsies of circumvallate or fungiform papillae can be obtained to determine the presence of specific pathology.

Some laboratory tests (e.g., blood serum evaluation) may help in identifying underlying medical conditions suggested by history and physical examination, such as allergy, infection, kidney disease, diabetes mellitus, hypothyroidism, nutritional deficiencies (e.g., $B_6$, $B_{12}$), and liver disease. Despite the fact that biopsy of the olfactory epithelium is possible, interpretation is hindered by the same sampling issues that limit the EOG measurement, most notably the fact that metaplasia of respiratory-like epithelium occurs throughout the olfactory epithelia of even persons with no olfactory problems.

Medical imaging is essential for understanding the basis of some smell and taste disturbances. Magnetic resonance imaging (MRI) is most appropriate for evaluating soft tissue (e.g., olfactory bulbs, tracts, and cortical parenchyma). Computed tomography (CT) is the method of choice to assess sinonasal tract inflammatory disorders and is superior to MRI in the evaluation of bony structures adjacent to the olfactory pathways (e.g., ethmoid, cribriform plate). Coronal CT scans are valuable in evaluating paranasal anatomy. Functional MRI (fMRI), positron emission tomography (PET), and single-proton emission computed tomography (SPECT), despite their research potential, have limited usefulness in routine patient evaluation. Plain radiographs have substantial limitations and should not be used unless other imaging means are unavailable.

## PRACTICAL MANAGEMENT

In most chemosensory disorders, the likelihood for at least some spontaneous recovery seems to be better for patients with not-too-severe loss. In the case of olfaction, a more optimistic prognosis would be expected in patients with UPSIT scores above 25 than in those with scores less than 25. This presumably reflects less extensive damage to the basal cell layer of the epithelia from which other cell types arise and possibly less fibrosis around the foramina of the cribriform plate, thereby allowing for some subsequent regeneration and reconnection to the bulb. Quantitative testing provides a basis for assessing prognosis and helps to place the patient's problem into general perspective. Thus, it is therapeutic for half of the older persons tested to learn that while their smell function is not what it used to be, it still falls above the average (i.e., median) of their peer group. Fortunately, the taste nerves and buds are relatively resilient, and spontaneous resolution of many taste problems—particularly dysgeusias—largely resolve with time (Deems et al., 1996).

Meaningful treatments are available for many conductive disorders of olfaction, although complete restoration of function is not always possible. Inflammatory elements of nasal and sinus disease can usually be treated medically. Surgery or medical treatment can be effective in mitigating nasal obstruction secondary to polyposis, intranasal tumors, and distorted intranasal architecture. Allergy management, topical and systemic corticosteroid therapies, antibiotic therapy, and functional endoscopic sinus surgery have all restored olfactory function in some patients. A brief course of systemic steroid therapy can be useful in distinguishing between conductive and sensorineural olfactory loss, as patients with the former will usually respond positively to the treatment. However, longer term systemic steroid administration, even at relatively low doses, is not advised (Scott, 1989). Topical nasal steroids are often ineffectual in altering smell dysfunction because of lack of penetrance of the steroid into the higher recesses of the nose, a problem that can be overcome to some degree by administering the spray or drops in the head-down Moffett's position.

Management of olfactory dysfunction caused by *sensorineural* damage is challenging. Spontaneous recovery in head trauma patients, when it occurs, typically occurs within a few months of the injury, usually as a result of resorption of hematomas or decreases in inflammation. Although longer-term improvement can occur on rare occasion, this is believed to occur mainly in patients with initial borderline loss (Doty *et al.*, 1997). Cessation of tobacco smoking can result in improvement in olfactory function and flavor sensation over time (Frye *et al.*, 1990). Resection of central nervous system tumors that impinge on the olfactory bulbs and tracts, or of epileptogenic tissue within the temporal or frontal lobes, can sometimes be performed in such a manner that restores at least some normal olfactory function, or mitigates disturbing dysosmias. Chitanondh (1966), for example, reported successful treatment of seven patients with seizure disorder, olfactory hallucinations, and psychiatric problems by stereotactic amygdalotomy. Intractable unilateral chronic dysosmia of many years' duration that significantly impairs health and daily functioning is sometimes amenable to surgical intervention (e.g., ablation of regions of the olfactory epithelium or, less conservatively, olfactory bulb removal) (Leopold *et al.*, 1991). In the majority of dysosmic cases, however, spontaneous resolution occurs. Only rarely is significant smell loss present in dysosmia, supporting the notion that at least some olfactory system function is needed for expression of most forms of dysosmia.

Antifungal and antibiotic treatments have been reported to be useful in some dysgeusic cases, although double blind studies of the efficacy of such treatments are lacking. Some salty or bitter dysgeusias reportedly respond to chlorhexidine mouth wash, conceivably as a result of its strong positive charge (Helms *et al.*, 1995). In cases of taste loss secondary to hypothyroidism, thyroxin replacement therapy reportedly returns taste sensitivity to normal levels (Mattes and Kare, 1986). Taste and smell disorders secondary to drug therapy can sometimes be reversed by discontinuing the drug, by employing alternative agents, or by altering drug dose. It should be reiterated, however, that a number of pharmacological agents seem to induce long-term alterations in taste that may take months to disappear after discontinuance of the drug.

Proper oral hygiene and routine dental care is of paramount importance for proper chemosensory function. In disease states where excessive dryness of the oral cavity exists, a physician can provide artificial saliva (e.g., Xerolube) to improve comfort and mucosal function. In some cases of ageusia or hypogeusia, adding a flavor enhancer like monosodium glutamate to foods increases palatability and encourages intake.

## TREATMENTS NO LONGER RECOMMENDED

There is no compelling evidence that zinc and vitamin therapies work except in some rare cases where frank zinc or vitamin deficiencies exist and neurological impairment is not marked.

## ACKNOWLEDGMENTS

This paper was supported, in part, by Grants PO1 DC 00161, RO1 DC 04278, RO1 DC 02974, and RO1 AG 27496 from the National Institutes of Health, Bethesda, MD (R. L. Doty, Principal Investigator).

## REFERENCES

Acharya, V., Acharya, J., and Luders, H. (1998). Olfactory epileptic auras. *Neurology* 51, 56–61.

Ackerman, B. H., and Kasbekar, N. (1997). Disturbances of taste and smell induced by drugs. *Pharmacotherapy* 17, 482–496.

Arnhold-Schneider, M., and Bernemann, D. (1987). Uber die Haufigkeit von Geschmacksstorungen nach Tonsillektomie. *HNO* 35, 195–198.

Bartoshuk, L. M. (1990). Chemosensory alterations and cancer therapies. *NCI Monographs* 179–184.

Beidler, L. M., and Smith, J. C. (1991). Effects of radiation therapy and drugs on cell turnover and taste. *In* "Smell and Taste in Health and Disease" (T. V. Getchell, R. L. Doty, L. M. Bartoshuk and J. B. Snow, Jr. Eds.) pp. 753–763. Raven Press, New York.

Bromley, S. M. (2000). Smell and taste disorders: A primary care approach. *Am. Fam. Phys.* 61, 427–436.

Bull, T. R. (1965). Taste and the chorda tympani. *J. Laryngol. Otol.* 79, 479–493.

Burdach, K., and Doty, R. L. (1987). Retronasal flavor perception: Influences of mouth movements, swallowing and spitting. *Physiol. Behav.* 41, 353–356.

Busenbark, K. L., Huber, S. I., Greer, G., Pahwa, R., and Koller, W. C. (1992). Olfactory function in essential tremor. *Neurology* 42, 1631–1632.

Chambers, K. C., and Bernstein, I. L. (1995). Conditioned flavor aversions. *In* "Handbook of Olfaction and Gustation" (R. L. Doty, Ed.), pp. 745–773. Dekker, New York.

Chitanondh, H. (1966). Stereotaxic amygdalotomy in the treatment of olfactory seizures and psychiatric disorders with olfactory hallucination. *Confin. Neurol.* 27, 181–196.

Combarros, O., Sanchez-Juan, P., Berciano, J., and De, P. C. (2000). Hemiageusia from an ipsilateral multiple sclerosis plaque at the midpontine tegmentum. *J. Neurol. Neurosurg. Psychiatry* 68, 796.

Conger, A. D. (1973). Loss and recovery of taste acuity in patients irradiated to the oral cavity. *Radiat. Res.* 53, 338–347.

Deems, D. A., Doty, R. L., Settle, R. G., Moore-Gillon, V., Shaman, P., Mester, A. F., Kimmelman, C. P., Brightman, V. J., and Snow, J. B., Jr. (1991). Smell and taste disorders, a study of 750 patients from the University of Pennsylvania Smell and Taste Center. *Arch. Otolaryngol. Head Neck Surg.* 117, 519–528.

Deems, D. A., Yen, D. M., Kreshak, A., and Doty, R. L. (1996). Spontaneous resolution of dysgeusia. *Arch. Otolaryngol. Head Neck Surg.* **122,** 961–963.

Della Fera, M. A., Mott, A. E., and Frank, M. E. (1995). Iatrogenic causes of taste disorders: Radiation therapy, surgery, and medication. *In* "Handbook of Olfaction and Gustation" (R. L. Doty, Ed.), pp. 785–791. Dekker, New York.

Devanand, D. P., Michaels-Marston, K. S., Liu, X., Pelton, G. H., Padilla, M., Marder, K., Bell, K., Stern, Y., and Mayeux, R. (2000). Olfactory deficits in patients with mild cognitive impairment predict Alzheimer's disease at follow-up. *Am. J. Psychiatry* **157,** 1399–1405.

Donati, F., Pfammatter, J. P., Mauderli, M., and Vassella, F. (1991). Neurologische Komplikationen nach Tonsillektomie. *Schweiz. Med. Wochenschr.* **121,** 1612–1617.

Doty, R. L. (1991). Olfactory dysfunction in neurogenerative disorders. *In* "Smell and Taste in Health and Disease" (T. V. Getchell, R. L. Doty, L. M. Bartoshuk, and J. B. Snow, Jr., Eds.), pp. 735–751. Raven Press, New York.

Doty, R. L. (1995). "The Smell Identification Test TM Administration Manual," 3rd ed. Sensonics, Haddon Heights, NJ.

Doty, R. L. (2001). Olfaction. *Annu. Rev. Psychol.* **52,** 423–452.

Doty, R. L., Bartoshuk, L. M., and Snow, J. B., Jr. (1991). Causes of olfactory and gustatory disorders. *In* "Smell and Taste in Health and Disease" (T. V. Getchell, R. L. Doty, L. M. Bartoshuk, and J. B. Snow, Jr., Eds.), pp. 449–461. Raven Press, New York.

Doty, R. L., Deems, D. A., Frye, R. E., Pelberg, R., and Shapiro, A. (1988a). Olfactory sensitivity, nasal resistance, and autonomic function in patients with multiple chemical sensitivities. *Arch. Otolaryngol. Head Neck Surg.* **114,** 1422–1427.

Doty, R. L., Deems, D. A., and Stellar, S. (1988b). Olfactory dysfunction in parkinsonism: A general deficit unrelated to neurologic signs, disease stage, or disease duration. *Neurology* **38,** 1237–1244.

Doty, R. L., Golbe, L. I., McKeown, D. A., Stern, M. B., Lehrach, C. M., and Crawford, D. (1993). Olfactory testing differentiates between progressive supranuclear palsy and idiopathic Parkinson's disease. *Neurology* **43,** 962–965.

Doty, R. L., and Kobal, G. (1995). Current trends in the measurement of olfactory function. *In* "Handbook of Olfaction and Gustation" (R. L. Doty, Ed.), pp. 191–225. Marcel Dekker, New York.

Doty, R. L., Li, C., Mannon, L. J., and Yousem, D. M. (1998). Olfactory dysfunction in multiple sclerosis: Relation to plaque load in inferior frontal and temporal lobes. *Ann. NY Acad. Sci.* **855,** 781–786.

Doty, R. L., Reyes, P. F., and Gregor, T. (1987). Presence of both odor identification and detection deficits in Alzheimer's disease. *Brain Res. Bull.* **18,** 597–600.

Doty, R. L., Risser, J. M., and Brosvic, G. M. (1991). Influence of adrenalectomy on the odor detection performance of rats. *Physiol. Behav.* **49,** 1273–1277.

Doty, R. L., Shaman, P., Applebaum, S. L., Giberson, R., Siksorski, L., and Rosenberg, L. (1984a). Smell identification ability: Changes with age. *Science* **226,** 1441–1443.

Doty, R. L., Shaman, P., and Dann, M. (1984b). Development of the University of Pennsylvania Smell Identification Test: A standardized microencapsulated test of olfactory function. *Physiol. Behav.* **32,** 489–502.

Doty, R. L., Singh, A., Tetrude, J., and Langston, J. W. (1992a). Lack of olfactory dysfunction in MPTP-induced parkinsonism. *Ann. Neurol.* **32,** 97–100.

Doty, R. L., Stern, M. B., Pfeiffer, C., Gollomp, S. M., and Hurtig, H. I. (1992b). Bilateral olfactory dysfunction in early stage treated and untreated idiopathic Parkinson's disease. *J. Neurol. Neurosurg. Psychiatry* **55,** 138–142.

Doty, R. L., Yousem, D. M., Pham, L. T., Kreshak, A. A., Geckle, R., and Lee, W. W. (1997). Olfactory dysfunction in patients with head trauma. *Arch. Neurol.* **54,** 1131–1140.

Duhra, P., and Foulds, I. S. (1988). Methotrexate-induced impairment of taste acuity. *Clin. Exp. Dermatol.* **13,** 126–127.

Erick, M. (1996). Idiopathic dysgeusia associated with blood transfusion: A case report. *J. Am. Diet Assoc.* **96,** 450. [Letter]

Evers, K. A., Eindhoven, G. B., and Wierda, J. M. (1999). Transient nerve damage following intubation for trans-sphenoidal hypophysectomy. *Can. J. Anesth.* **46,** 1143–1145.

Fetting, J. H., Wilcox, P. M., Sheidler, V. R., Enterline, J. P., Donehower, R. C., and Grochow, L. B. (1985). Tastes associated with parenteral chemotherapy for breast cancer. *Cancer Treat. Reports.* **69,** 1249–1251.

Frank, M. E., and Smith, D. V. (1991). Electrogustometry: A simple way to test taste. *In* "Smell and Taste in Health and Disease" (T. V. Getchell, R. L. Doty, L. M. Bartoshuk, and J. B. Snow, Jr., Eds.), pp. 503–514. Raven Press, New York.

Frye, R. E., Schwartz, B. S., and Doty, R. L. (1990). Dose-related effects of cigarette smoking on olfactory function. *JAMA* **263,** 1233–1236.

Fujikane, M., Itoh, M., Nakazawa, M., Yamaguchi, Y., Hirata, K., and Tsudo, N. (1999). [Cerebral infarction accompanied by dysgeusia—A clinical study on the gustatory pathway in the CNS]. *Rinsho Shinkeigaku—Clin. Neurol.* **39,** 771–774. [in Japanese]

Graves, A. B., Bowen, J. D., Rajaram, L., McCormick, W. C., McCurry, S. M., Schellenberg, G. D., and Larson, E. B. (1999). Impaired olfaction as a marker for cognitive decline: Interaction with apolipoprotein E epsilon4 status. *Neurology* **53,** 1480–1487.

Greene, D., Nail, L. M., Fieler, V. K., Dudgeon, D., and Jones, L. S. (1994). A comparison of patient-reported side effects among three chemotherapy regimens for breast cancer. *Cancer Practice* **2,** 57–62.

Helms, J. A., Della-Fera, M. A., Mott, A. E., and Frank, M. E. (1995). Effects of chlorhexidine on human taste perception. *Arch. Oral Biol.* **40,** 913–920.

Hoffman, H. J., Ishii, E. K., and MacTurk, R. H. (1998). Age-related changes in the prevalence of smell/taste problems among the United States adult population. Results of the 1994 disability supplement to the National Health Interview Survey (NHIS). *Ann. N Y Acad. Sci.* **855,** 716–722.

Kellokumpu-Lehtinen, P., Nordman, E., and Toivanen, A. (1989). Combined inteferon and vinblastine treatment of advanced melanoma: evaluation of the treatment results and the effects of the treatment on immunological functions. *Cancer Immunol. Immunother.* **28,** 213–217.

Kojima, Y., and Hirano, T. (1999). [A case of gustatory disturbance caused by ipsilateral pontine hemorrhage]. *Rinsho Shinkeigaku—Clin. Neurol.* **39,** 979–981. [in Japanese]

Kveton, J. F., and Bartoshuk, L. M. (1994). The effect of unilateral chorda tympani damage on taste. *Laryngoscope* **104,** 25–29.

Lee, B. C., Hwang, S. H., Rison, R., and Chang, G. Y. (1998). Central pathway of taste: Clinical and MRI study. *Eur. Neurol.* **39,** 200–203.

Leopold, D. A., Schwob, J. E., Youngentob, S. L., Hornung, D. E., Wright, H. N., and Mozell, M. M. (1991). Successful treatment of phantosmia with preservation of olfaction. *Arch. Otolaryngol. Head Neck Surg.* **117,** 1402–1406.

Matsuda, T., and Doty, R. L. (1995). Regional taste sensitivity to NaCl: Relationship to subject age, tongue locus and area of stimulation. *Chem. Senses* **20,** 283–290.

Mattes, R. D., and Kare, M. R. (1986). Gustatory sequelae of alimentary disorders. *Digest Dis.* **4,** 129–138.

McCaffrey, R. J., Duff, K., and Solomon, G. S. (2000). Olfactory dysfunction discriminates probable Alzheimer's dementia from major depression: A cross validation and extension. *J. Neuropsychiatry Clin. Neurosci.* **12,** 29–33.

Mizobuchi, M., Ito, N., Tanaka, C., Sako, K., Sumi, Y., and Sasaki, T. (1999). Unidirectional olfactory hallucination associated with ipsilateral unruptured intracranial aneurysm. *Epilepsia* 40, 516–519.

Moberg, P. J., Doty, R. L., Turetsky, B. I., Arnold, S. E., Mahr, R. N., Gur, R. C., Bilker, W., and Gur, R. E. (1997). Olfactory identification deficits in schizophrenia: correlation with duration of illness. *Am. J. Psychiatry* 154, 1016–1018.

Ohtsuka, K., Tomita, H., and Murakami, G. (1994). [Anatomical study of the tonsillar bed: the topographical relationship between the palatine tonsil and the lingual branch of the glossopharyngeal nerve]. *Nippon Jibiinkoka Gakkai Kaiho [J. Oto-Rhino-Laryngol. Soc. Jpn.]* 97, 1481–1493. [in Japanese]

Ohtuka, K., Tomita, H., Yamauchi, Y., and Kitagoh, H. (1994). [Taste disturbance after tonsillectomy]. *Nippon Jibiinkoka Gakkai Kaiho [J. Oto-Rhino-Laryngol. Soc. Jpn.]* 97, 1079–1088. [in Japanese]

Onoda, K., and Ikeda, M. (1999). Gustatory disturbance due to cerebrovascular disorder. *Laryngoscope* 109, 123–128.

Ostergaard, M., Kristensen, B. B., and Mogensen, T. S. (1997). [Reduced sense of taste as a complication of the laryngeal mask use]. *Ugeskrift for Laeger* 159, 6835–6836. [in Danish]

Panayiotou, H., Small, S. C., Hunter, J. H., and Culpepper, R. M. (1995). Sweet taste (dysgeusia). The first symptom of hyponatremia in small cell carcinoma of the lung. *Arch. Intern. Med.* 155, 1325–1328.

Scott, A. E. (1989). Caution urged in treating "steroid dependent anosmia." *Arch. Otolaryngol. Head Neck Surg.* 115, 109–110.

Shafer, D. M., Frank, M. E., Gent, J. F., and Fischer, M. E. (1999). Gustatory function after third molar extraction. *Oral Surg. Oral Med. Oral Pathol. Oral Radiol. Endodont.* 87, 419–428.

Schiffman, S. S. (1997). Taste and smell losses in normal aging and disease. *JAMA* 278, 1357–1362.

Stricker, B. H., Van, R. M., Sturkenboom, M. C., and Ottervanger, J. P. (1996). Taste loss to terbinafine: A case-control study of potential risk factors. *Br. J. Clin. Pharmacol.* 42, 313–318.

Sumner, D. (1967). Post-traumatic ageusia. *Brain* 90, 187–202.

Turetsky, B. I., Moberg, P. J., Yousem, D. M., Doty, R. L., Arnold, S. E., and Gur, R. I. (2000). Olfactory bulb volume is reduced in patients with schizophrenia. *Am. J. Psychiatry* 157, 828–830.

Walker, R. P., and Gopalsami, C. (1996). Laser-assisted uvulopalatoplasty: Postoperative complications. *Laryngoscope* 106, 834–838.

Wenning, G. K., Shephard, B., Hawkes, C., Petruckevitch, A., Lees, A., and Quinn, N. (1995). Olfactory function in atypical parkinsonian syndromes. *Acta Neurol. Scand.* 91, 247–250.

West, S. E., and Doty, R. L. (1995). Influence of epilepsy and temporal lobe resection on olfactory function. *Epilepsia* 36, 531–542.

Yanagisawa, K., Bartoshuk, L. M., Catalanotto, F. A., Karrer, T. A., and Kveton, J. F. (1998). Anesthesia of the chorda tympani nerve and taste phantoms. *Physiol. Behav.* 63, 329–335.

## CHAPTER 18
# Hiccup

Michael Fetter and Christopher Kennard

## CLINICAL ASPECTS

Hiccup is a forceful, involuntary inspiration commonly experienced by fetuses and newborns and to a lesser extent by children and adults. It usually presents in otherwise healthy subjects as an annoyance for short periods (for minutes to hours after excitement, heavy meals, or ingestion of cold drinks and especially alcohol). The condition usually resolves spontaneously or with simple folk remedies and does not require medical attention. In contrast, prolonged hiccups (weeks to months) is a rare but disabling condition that can induce fatigue, incapacitation, exhaustion, depression, weight loss, and sleep deprivation. A wide variety of pathologic conditions can cause chronic hiccup (Table I); the most common causes are located in the gastrointestinal tract, with gastro-esophageal reflux as the most important. Detailed medical history and physical examination will often guide diagnostic investigation (i.e., abdominal ultrasound, chest CT and brain MRI scan, endoscopy including pH monitoring, and manometry) (Launois *et al.*, 1993; Marsot-Dupuch *et al.*, 1995; Rousseau, 1995).

The only comprehensive collection of patients with hiccups lasting for more than 48 h has been published by Souadjian and Cain (1968). This study consisted of 181 men, but only 39 women. It was also surprising that an organic cause of the hiccup could be found in 93% of the men, but only in 8% of the women.

Hiccup consists of a periodic, synchronous, myoclonic contraction of the diaphragm and inspiratory (external) intercostal musculature. Usually glottal closure follows in 35 ms, stopping the air exchange abruptly and producing the typical "hic" (Newsom Davis, 1970). In comparison to the normal respiratory excursions, inspiration is accelerated and shortened and expiration is inhibited. There is no respiratory effect. Hiccups are most likely to occur during the period of maximal inspiration because lung inflation inhibits vagal mucosal and laryngotracheal afferents, which are known to inhibit hiccups (Salem *et al.*, 1967). In contrast to the protection reflexes like vomiting or coughing, hiccup has no physiological relevance. Three hypotheses regarding the pathophysiology of hiccups have so far been proposed.

### Hiccup as a Referred Gastrointestinal (Vagal) Reflex

Hiccup could be due to a referred gastrointestinal reflex elicited by abdominal inputs. The afferent pathway consists of the phrenic nerve, vagal nerve, cervical plexus (C2–C4), and sympathetic fibers of the lower thoracic segments (Th6–Th12), and the efferent pathway consists of the motor part of the phrenic nerve (diaphragm), vagus nerve (musculature of the larynx and esophagus), and sympathetic fibers of the cervical (C5–C7) and thoracic segments (Th1–Th11) innervating the auxiliary respiratory musculature (Newsom Davis, 1970). Despite the involvement of numerous neural elements, there is a well-coordinated contraction. This suggests supraspinal, polysynaptic control. The center of coordination is unknown, but based on clinical experience it is conceivable that it is located in the medulla oblongata.

### Hiccup as Myoclonus

Hiccups could be a subcortical equivalent of myoclonus possibly generated at the level of the pons or medulla oblongata, with disturbances in the so-called myoclonic triangle (Guillain–Mollaret): inferior olive, dentate nucleus, red nucleus. However, there is no correlation of hiccup with respiratory myoclonus, palatal myoclonus, or cranial nerve myorhythmia.

TABLE I    Etiology of Hiccups

Central nervous system
  Diffuse disorders
    Increased intracranial pressure
    Meningoencephalitis
    Sarcoidosis
    Metabolic disorders (e.g., diabetic coma, hepatic coma, uremic coma, Addison's disease, hyponatraemia)
    Drugs and intoxication (e.g., alcohol, cortisone, short-acting narcotics, opioids, barbiturates, diazepam, muscle relaxants, nonsteroidal anti-inflammatory drugs, non-imipraminic antidepressants, dopaminergic antiparkinsonians, beta-lactams, macrolides, fluoroquinolones, digitalic compounds, methohexitone, chlordiazepoxide, nicethamide, midazolam, cyclophosphamide, cefotetan, perphenazine)
  Focal lesions
    Structural lesions of the brain stem (e.g., Arnold–Chiari malformations)
    Brain stem infarction, brain stem hemorrhage (e.g., Wallenberg's syndrome)
    Tumors of the brain stem and medulla oblongata (tuberculoma, giant aneurysm of the upper vertebral, basilar, or posterior inferior cerebellar artery)
    Multiple sclerosis, encephalitis of the brain stem
    Head trauma with brain stem concussion
    Focal epilepsy (myoclonus)

Peripheral lesions
  Cervical
    Nerve root compression C4 (disc prolapse)
    Tumors of the neck (e.g., hyperplasia or neoplasm of the thyroid gland, lymphomas)
  Thoracic
    Tumors of the mediastinum (e.g., lymphomas, tuberculoma)
    Lymphogranulomatosis or sarcoidosis
    Cancer of the esophagus or the lung
    Hiatus hernia, diverticulosis of the esophagus, obstruction of the esophagus, reflux esophagitis, esophagitis due to infection (e.g., herpes)
    Laryngo-bronchitis, pneumonia, pleuritis, pericarditis, mediastinitis
    Coronary insufficiency, myocardial infarction
  Abdominal
    Gastric stasis, gastritis, ulcers of the stomach or the duodenum
    Ileus, peritonitis, subphrenic abscess
    Pancreatitis
    Diseases of the billiary system
    Tumors of the abdominal organs (e.g., cancer of the stomach, the pancreas, or the liver)
Other causes
  Glaucoma
Psychogenic causes

## Hiccup as a Primitive Motor Pattern

Hiccup seems to be an important primitive motor pattern (it is very common *in utero* and in newborn infants and babies) that allows the fetus to produce maximal inspiration movements without the risk of inhaling amniotic fluid (Fuller, 1990). It is suppressed postnatally by slowly developing inhibitory mechanisms. With severe peripheral stimulation or a lesion of the central inhibitory mechanisms hiccups can reoccur.

Evidence from animal experiments suggests that these inhibitory mechanisms are mediated through GABA(B) receptors within central connections of the hiccup reflex arc. In cats a hiccup-like reflex can be elicited by electrical stimulation to a limited area within the medullary reticular formation, the so-called hiccup-evoking site (HES). The hiccup-like response can rapidly be suppressed after microinjection of baclofen into HES, indicating that HES has GABA(B) receptors. Retrograde labeling suggests that the nucleus raphe magnus may be most likely the source of the GABAergic inhibitory inputs to the hiccup reflex arc (Oshima *et al.*, 1998). These experimental findings support the growing use of baclofen as a therapy of first choice.

## NATURAL COURSE

In most cases the cause of hiccups remains unclear. Usually hiccups stop within minutes to a few hours or with simple folk remedies. There are, however, reports of patients in whom hiccups persisted for years.

This symptom is of special interest for anesthesiologists, because of the risk of hyperventilation in patients with impaired glottic closure who are intubated. Hiccup can also cause severe difficulties during artificial respiration.

## PRINCIPLES OF THERAPY

If a cause can be identified (see Table I) and adequate treatment is given, the hiccups usually subside. Intermittently, symptomatic measures have to be undertaken. The principle on which the so-called folk remedies are based is probably the alteration of vagal afferents by stimulation of the dorsal part of the throat (e.g., swallowing dry sugar). The same principle is proposed for stimulation of the throat with a catheter (Salem *et al.*, 1967). Similarly, the effect of local anesthetics on the throat is probably due to modification of the relevant input.

The use of gastrointestinal medication (metoclopramide, spasmolytics) is based on the assumption of a gastrointestinal reflex as the cause of hiccups. Anticonvulsant drugs and drugs influencing serotonin metabolism (5-hydroxytryptophan, serotonin reuptake inhibitors) are used, based on the assumption that it is a form of myoclonus. The evidence for the use of baclofen has already been described. Other medications are simply based on empirical results, and for most of them no controlled studies are available.

## PRACTICAL MANAGEMENT

### Folk Remedies

Numerous folk remedies to treat periodic hiccups have been suggested: swallowing granulated sugar (Engleman *et al.*, 1971), drinking lemon juice or vinegar, stopping breathing after deep inspiration, stopping breathing and reclining the head, dry swallowing several times, placing pressure on both eyes, stimulating the phrenic nerves by pinching the dorsal edge of the sternomastoid muscle, pulling the tongue, provoking sneezing (by pepper, etc.), inhalating smoke, or changing the breathing pattern from costo-clavicular to costo-abdominal breathing. In some anesthetized patients hiccup can be stopped by brief or continuous increase of the insufflation pressure (Saitto *et al.*, 1982).

### Noninvasive Therapy

If the preceding measures do not stop hiccup, non- or semi-invasive procedures can be tried, such as noninvasive phrenic nerve stimulation (Aravot *et al.*, 1989), digital rectal massage (Odeh *et al.*, 1990), mechanic stimulation of the dorsal part of the throat with a nasogastric tube (works in some patients even under anesthesia) (Salem *et al.*, 1967) or an endoscope (Beda *et al.*, 1993). Local anesthetic applied to the mucosa of nose and throat is sometimes helpful. Similarly, the inhalation of 10–15% $CO_2$ for 3 to 5 min or the reinhalation of expired air via a plastic bag can stop hiccups.

### Pharmacological Therapy

If hiccups persist despite these measures, a pharmacological treatment should be tried. Several groups of drugs are available (Table II). Drugs of first choice are metoclopramide (Williamson and Macintyre, 1977), bromopride, domperidone, and, since recently, also baclofen (Bhalotra, 1990; Burke *et al.*, 1988; Fodstad and Nilsson, 1993; Guelaud *et al.*, 1995; Lance and Bassil, 1989; Ramirez and Graham, 1992; Yaqoob *et al.*, 1989). In idiopathic chronic hiccup a combination of cisapride, omeprazole, and baclofen has been suggested as therapy of choice (Petroianu *et al.*, 1998). The same authors also showed that an add-on of gabapentin or exchange of baclofen with gabapentin can be effective if the combination therapy still fails (Petroianu *et al.*, 2000). Extrapyramidal side effects can occur with

TABLE II  Medical Treatment of Hiccups (Bold: Drugs of First Choice)

| Substance | IV | IM | Oral |
|---|---|---|---|
| Gastrointestinal substances | | | |
| **Metoclopramide** | 10–20 mg | 10–20 mg | 3 × 10 mg |
| **Domperidone** | — | — | 3 × 20 mg |
| **Bromopride** | — | — | 3 × 10 mg |
| Atropine | 0.5–1 mg | — | — |
| Scopolamine | 20 mg | — | 3 × 10 mg |
| Biperiden | 5 mg | — | — |
| Cimetidine | 200–400 mg | — | 4 × 200 mg |
| Ranitidine | — | — | 2 × 150 mg |
| Cisapride | — | — | 3 × 5–10 mg |
| Omeprazole | — | — | 10–20 mg |
| Antacids | | | |
| Anticovulsant drugs | | | |
| Carbamazepine | — | — | 2 × 400 mg |
| Valproic acid | — | — | 3 × 600 mg |
| Diphenylhydantoin | 750 mg | — | 3 × 100 mg |
| Gabapentin | — | — | 900–1200 mg |
| Psychopharmacological drugs | | | |
| Chlorpromazine | 50 mg | — | — |
| Promethazine | 50 mg | 50 mg | — |
| Triflupromazine | 10 mg | 20 mg | 70 mg supp. |
| Promazine | 50–100 mg | 50–100 mg | 50–100 mg |
| Amitriptyline | 50 mg | 50 mg | 75–150 mg |
| Haloperidol | 2.5–5 mg | 2.5–5 mg | 3 × 5 mg |
| Various substances | | | |
| **Baclofen** | — | — | Up to 3 × 25 mg |
| Nifedipine | — | — | 3 × 20 mg |
| Sulfonamide | — | — | 3 × 1 g |
| 5-hydroxytryptophan | — | — | Up to 2000 mg |
| Amantadine | 1–3 × 200 mg | — | 2–3 × 100 mg |

metoclopramide, bromopride, and domperidone, such as oral and pharyngeal dyskinesia, which can be interrupted by biperiden (5 mg, IV). Spasmolytics and parasympatholytics (atropine, scopolamine, biperiden) and antacids also act in the gastrointestinal tract. As drugs of second choice anticonvulsants can be used (carbamazepine, valproic acid, phenytoin), the serotonin precursor 5-hydroxytryptophan (slowly increasing dosage up to 2000 mg/day because of gastrointestinal side effects) or psychopharmacological drugs (chlorpromazine, promethazine, triflupromazine, promazine, amitriptyline, droperidol, haloperidol) (Jacobson *et al.*, 1981; Stalnikowicz *et al.*, 1986; Williamson and Macintyre, 1977). Amantadine has also been reported to be helpful in single cases (Askenasy *et al.*, 1988). Recently, nifedipine has been advocated to stop hiccups in cases in which all other medications failed (Lipps *et al.*, 1990; Mukhopadhyay *et al.*, 1986).

Hiccups that persist despite medication for more than a week or recommence are always a symptom of an organic disease (most of the time in the esophagus [Bizec *et al.*, 1995]) and call for appropriate investigations.

## Invasive Therapy

If hiccups persist and the patient is progressively exhausted more invasive measures have to be considered, such as a cervical epidural block (Sato *et al.*, 1993) or the use of breathing pacemakers with implanted electrodes, which control the diaphragm by electrical stimulation of the phrenic nerve (Dobelle, 1999; Okuda *et al.*, 1998). Another possibility is the temporary or permanent elimination of the phrenic nerve by instillation of a local anesthetic, surgical crush, or section of the nerve above the anterior scalenus muscle, which denervates the diaphragm (Lanz and Dick, 1983). Often only a lesion of the left nerve is required to stop the hiccups. These procedures, however, are debated (Petroianu, 1998).

There is also a case report of intractable idiopathic hiccups treated by microvascular decompression of the vagus nerve (Johnson, 1993). The success of microvascular decompression has been documented for other conditions, such as trigeminal neuralgia and hemifacial spasm, which are characterized by hyperactive dysfunction due to neurovascular contact. In this case report the vagus nerve was separated from the posterior inferior cerebellar artery by inserting a Teflon pledget between the nerve and the vessel, thereby eliminating the neurovascular contact. This procedure immediately stopped the hiccups in this patient, but they reoccurred after 1 year. A second operation showed that the Teflon pledget had fallen out of place. Once the contact of the nerve with the artery was eliminated by wrapping the artery with a tuft of Teflon, the hiccups stopped again. The patient remained free of hiccups for the remainder of the observation period (3 years). The author concludes that patients with intractable idiopathic hiccups who fail medical therapy should be considered for microvascular decompression of the vagus nerve.

## REFERENCES

Aravot, D. J., Wright, G., Rees, A., Maiwand, O. M., and Garland, M. H. (1989). Non-invasive phrenic nerve stimulation for intractable hiccups. *Lancet* 8670, 1047.

Askenasy, J. J., Boiangiu, M., and Davidovitch, S. (1988). Persistent hiccup cured by amantadine. *N. Engl. J. Med.* 318, 711.

Beda, B. Y., Niamkey, E. K., Ouattara, D., Diallo, A. D., Adom, A. H., Djakoure, S., Yoboue, L., Yangni-Angate, Y., Kadjo, K., and Toutou, T. (1993). Stopping persistent hiccups in the adult by endoscopic maneuver. *Ann. Gastroenterol. Hepatol.* 29, 11–13.

Bhalotra, R. (1990). Baclofen therapy for intractable hiccoughs. *J. Clin. Gastroenterol.* 12, 122.

Bizec, J. L., Launois, S., Bolgert, F., Lamas, G., Chollet, R., and Derenne, J. P. (1995). Hiccups in adults. *Rev. Mal. Respir.* 12, 219–229.

Burke, A. M., White, A. B., and Brill, N. (1988). Baclofen for intractable hiccups. *N. Engl. J. Med.* 319, 1354.

Dobelle, W. H. (1999). Use of breathing pacemakers to suppress intractable hiccups of up to thirteen years duration. *ASAIO J.* 45, 524–525.

Engleman, E. G., Lankton, J., and Lankton, B. (1971). Granulated sugar as a treatment for hiccups in conscious patients. *N. Engl. J. Med.* 285, 1489.

Fodstad, H., and Nilsson, S. (1993). Intractable hiccup: A diagnostic and therapeutic challenge. *Br. J. Neurosurg.* 7, 255–260.

Fuller, G. N. (1990). Hiccups and human purpose. *Nature* 343, 420.

Guelaud, C., Similowski, T., Bizec, J. L., Cabane, J., Whitelaw, W. A., and Derenne, J. P. (1995). Baclofen therapy for chronic hiccup. *Eur. Respir. J.* 8, 235–237.

Jacobson, P. L., Messenheimer, T., and Farmer, W. (1981). Treatment of intractable hiccups with valproic acid. *Neurology* 31, 1458.

Johnson, D. L. (1993). Intractable hiccups: Treatment by microvascular decompression of the vagus nerve. *J. Neurosurg.* 78, 813–816.

Lance, J. W., and Bassil, G. T. (1989). Familial intractable hiccup relieved by baclofen. *Lancet* 8657, 276–277.

Lanz, E., and Dick, W. (1983). Phrenicus-Blockade bei therapieresistentem Schluckauf. *Dtsch. Med. Wochenschr.* 108, 1854–1855.

Launois, S., Bizec, J. L., Whitelaw, W. A., Cabane, J., and Derenne, J. P. (1993). Hiccup in adults: An overview. *Eur. Respir. J.* 6, 563–575.

Lipps, D. C., Jabbari, B., Mitchell, M. H., and Daigh, J. D. J. (1990). Nifedipine for intractable hiccups. *Neurology* 40, 531–532.

Marsot-Dupuch, K., Bousson, V., Cabane, J., and Tubiana, J. M. (1995). Intractable hiccups: The role of cerebral MR in cases without systemic cause. *Am. J. Neuroradiol.* 16, 2093–2100.

Mukhopadhyay, P., Osman, M. R., Wajima, F., and Wallace, T. J. (1986). Nifedipine for intractable hiccups. *N. Engl. J. Med.* 314, 1256.

Newsom Davis, J. (1970). An experimental study of hiccup. *Brain* 93, 851–872.

Odeh, M., Bassan, H., and Oliven, A. (1990). Termination of intractable hiccups with rectal massage. *J. Intern. Med.* 272, 145–146.

Okuda, Y., Kitajima, T., and Asai, T. (1998). Use of a nerve stimulator for phrenic nerve block in treatment of hiccups. *Anesthesiology* 88, 525–527.

Oshima, T., Sakamoto, M., Tatsuta, H., and Arita, H. (1998). GABAergic inhibition of hiccup-like reflex induced by electrical stimulation in medulla of cats. *Neurosci. Res.* 30, 287–293.

Petroianu, G. (1998). Idiopathic chronic hiccup (ICH): Phrenic nerve block is not the way to go. *Anesthesiology* 89, 1284–1285.

Petroianu, G., Hein, G., Petroianu, A., Bergler, W., and Rufer, R. (1998). ETICS Study: Empirical therapy of idiopathic chronic singultus. *Z. Gastroenterol.* 36, 559–566.

Petroianu, G., Hein, G., Stegmeier-Petroianu, A., Bergler, W., and Rufer, R. (2000). Gabapentin "add-on therapy" for idiopathic chronic hiccup (ICH). *J. Clin. Gastroenterol.* 30, 321–324.

Ramirez, F. C., and Graham, D. Y. (1992). Treatment of intractable hiccup with baclofen: Results of a double-blind randomized, controlled cross-over study. *Am. J. Gastroenterol.* 87, 1789–1791.

Rousseau, P. (1995). Hiccups. *South Med. J.* 88, 175–181.

Saitto, C., Gristina, G., and Cosmi, E. V. (1982). Treatment of hiccups by continuous positive airway pressure (CPAP) in anesthetised subjects. *Anesthesiology* 57, 345.

Salem, M. R., Baraka, A., Rattenborg, C. C., and Holady, D. A. (1967). Treatment of hiccups by pharyngeal stimulation in anaesthetized and conscious subjects. *J. Am. Med. Assoc.* **202**, 126–130.

Sato, S., Asakura, N., Endo, T., and Naito, H. (1993). Cervical epidural block can relieve postoperative intractable hiccups. *Anesthesiology* **78**, 1184–1186.

Souadjian, J. V., and Cain, J. C. (1968). Intractable hiccup. Etiologic factors in 220 cases. *Postgrad. Med.* **43**, 72–77.

Stalnikowicz, R., Fich, A., and Troudart, T. (1986). Amitriptyline for intractable hiccups. *N. Engl. J. Med.* **315**, 64–65.

Williamson, B. W. A., and Macintyre, I. M. C. (1977). Management of intractable hiccup. *Br. Med. J.* **2**, 501–503.

Yaqoob, M., Prabhu, P., and Ahmad, R. (1989). Baclofen for intractable hiccups. *Lancet* **8662**, 562–563.

CHAPTER 19

# Sleep and Its Disorders

Christian Guilleminault and Ilonka Eisensehr

## NORMAL SLEEP

Sleep consists of two different states, rapid eye movement (REM) and non-REM (NREM) sleep. REM sleep is defined by the presence of an EEG pattern of low-voltage fast EEG associated with occurrence of rapid eye movements, isolated or in bursts, and postural relaxation, i.e., muscle atonia. The muscle atonia is interrupted by bursts of muscle tone, leading, at times, to movement jerks. These jerks and bursts of eye movements are called "phasic events," occurring on a background of tonic muscle inhibition (called "tonic REM sleep"). Electrophysiologic studies performed on cats have shown that the phasic events are associated with bursts of waves recorded simultaneously in the pons, lateral geniculate, and occipital lobe (Jouvet, 1972; Laurent *et al.*, 1974). These waves called "ponto-geniculo-occipital (PGO) spikes" or waves subdivided "phasic" from "tonic" REM sleep. The term REM sleep has a number of synonyms such as "desynchronized (D) sleep," "dream sleep," "paradoxical sleep," and, in infants, "active sleep."

NREM sleep, also called "synchronized sleep," and, in infants, "quiet sleep" has been subdivided in four sleep stages. Stage 1 is seen at sleep onset and is defined by low voltage and mixed frequency (2 to 7 Hertz), with absence of rapid eye movements and presence of muscle tone. Vertex sharp waves may be seen, and slow eye movements are often present. Stage 2 is scored when 12 to 14 Hertz sleep spindles and/or K complexes are present against a background activity of relatively low-voltage mixed EEG frequencies. Stage 3 is scored when a moderate amount (20 to 50%) of high-amplitude ($75\,\mu v$ or greater), slow-wave (0.5–3.5 Hz) EEG activity is seen. Stage 4 is defined by the presence of a predominance (greater than 50%) of high-amplitude, slow-wave activity. Stages 3 and 4 combined are often called slow-wave sleep (SWS).

A healthy young adult has the simplest pattern of sleep, regardless of gender. Nocturnal sleep is associated with a regular pattern with a regular reoccurrence of "sleep cycles." Individuals first enter NREM sleep with stage 1 NREM, which lasts only 1 to 7 min. This stage may be interrupted by wakefulness. Stage 2 NREM sleep follows this short transition from wakefulness and continues for 10 to 25 min. Progressive gradual appearance of high-voltage, slow-wave signals switch to stage 3, followed by stage 4, which last between 20 and 40 min. A brief switch to stage 2 may precede the occurrence of REM sleep. The first REM period is short, lasting between 4 and 8 min. REM sleep often ends with a brief body movement, and a new sleep cycle begins. NREM and REM sleep continue to alternate throughout the night in a cyclical fashion. As night progresses each sleep cycle differs slightly from the preceding one. The average length of the first sleep cycle is around 90 min and is about 100 to 120 min from the second to the fourth cycle (usually the last one). The last sleep cycle is usually the longest. The organization of NREM and REM sleep also changes with each cycle. There is a predominance of SWS during the first two sleep cycles and a predominance of REM sleep in the last two cycles. SWS is rare to nonexistent in the last cycle, while REM sleep is the longest (Figure 1).

Sleep patterns vary with age. Sleep develops during infancy. In the neonatal period, REM sleep usually represents more than 50% of total sleep time. This percentage declines rapidly to around 25% during the prepubertal period and then stabilizes around 20% until old age (Roffwarg *et al.*, 1966). SWS is usually of maximum duration during childhood and declines progressively in adulthood with low levels attained around 50 to 55 years of age (Feinberg and Floyd, 1979). There may be a sex difference in this decline with men presenting a more notable decline (Fukuda *et al.*, 1999).

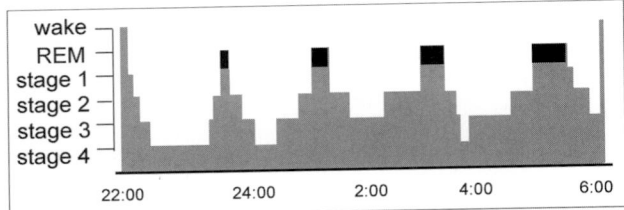

**FIGURE 1** The hypnogram of a healthy young adult. Night sleep consists of four to five sleep cycles, each beginning with light sleep, going then through slow-wave sleep, and ending with REM sleep. Over the course of the night, slow-wave sleep decreases and REM sleep increases.

## TESTS

The investigation of sleep and its pathologies is based on clinical investigations and different types of sleep monitoring.

### Polysomnography

#### Technique

Polysomnography is a term coined to describe the monitoring during sleep of many biologic variables. The utility of this monitoring becomes obvious when investigation of controls for vital functions are found to be different during the three states of alertness: wakefulness and NREM and REM sleep. Most of these functions are under the control of the autonomic nervous system (ANS). Sympathetic and parasympathetic activities are very much dependent upon the state of alertness, and the balance between these activities is state dependent. This recording is different from basic electroencephalographic (EEG) monitoring, but uses the 10–20 international electrode placement system for EEG recording. The EEG electrodes used are $C_3$, $C_4$, and occipital leads ($O_1$, $O_2$), and it is recommended to add a frontal electrode, most commonly $F_z$. These electrodes refer to the opposite side of the head ($A_1$, $A_2$). Occipital leads may be a bipolar montage: referred to each other. Electrodes for recording the electro-oculogram (EOG) are applied to the skin at the outer canthus (or lateral temporal corner) of both eyes. The electromyogram (EMG) from the chin muscles is monitored with two electrodes placed under the chin. The minimum requirements for identification of sleep states and stages are three channels of EEG, two of EOG and one of EMG. An electrocardiographic (ECG) channel is mandatory. It is usually the derivation $V_2$. In a standard montage, leg muscles and respiration are monitored. The electrode placement for leg EMG is on the skin above right and left anterior tibialis muscles. Respiration may be monitored with different means. The most common montage comprises a nasal cannula/pressure transducer system, a mouth thermistor, a neck microphone, a thoracic and an abdominal band, and a pulse-oximeter; all of this equipment is commercially available. It is possible, depending on the goal of the test, to measure $CO_2$ or other variables. The most common additions are esophageal pressure ($P_{es}$) measurement, systematically done in some laboratories, esophageal pH, skin impedance, pulse transit time (PTT), and more EEG leads. The signals obtained from these sensors are amplified by nonspecific amplifiers (i.e., Grass amplifiers) or generic amplifiers related to the computerized sleep system on which the signals are monitored and scored. There are many commercially available sleep systems. A minimum of 16 channels must be available and commonly 32 channels are available but many sleep laboratories, for economic reasons, monitor two beds simultaneously on their 32-channel system. Sleep monitoring must include video recording of the sleeping patient, which can be synchronized with the polysomnographic recording. This is especially important for evaluation of nocturnal movement disorders. The frequency at which signals are scanned is an important variable. Usually the minimum is 128 Hz/channel. But this will be too low if one wants to perform specific analyses such as fast Fourier transformation (FFT) of the ECG that may allow investigations of the high frequencies and low frequencies of the ECG reported to indicate sympathetic and parasympathetic activities during sleep. A sampling frequency of 500 Hz is the minimum required for appropriate and meaningful data collection.

Most of the variables collected during sleep are qualitative or semiquantitative at best. For example, thermistors used to collect airflow measure only temperature and have a very large margin of error (up to 1000%). Nasal cannula/pressure transducer systems are semiquantitative sensors, but nonlinear in the low range and the "flow limitation" is still a visual and subjective interpretation with commercially available systems, despite the fact that this type of airflow monitoring is more sensitive to nasal flow limitation than thermistors. The combined nasal/oral cannula systems commercially available are not satisfactory, and a mouth thermistor gives better results (Norman *et al.*, 1997). Usage of end-tidal $CO_2$ measurement to investigate abnormal breathing results in many more errors than the combined nasal cannula/pressure transducer system and mouth thermistor to date (Weese-Mayer *et al.*, 2000).

Polysomnography can be performed using portable equipment. The lower the number of signals monitored, the lower the amount of information that will be collected and the greater the chance of ignoring pathology in different cases. Portable equipment has the advantage of allowing monitoring in hospital beds and at home. The disadvantages are related to loss of signals and the need to perform a repeat study. There is a relationship between the number of sensors and their invasiveness and the amount of sleep disturbance induced. A one-night polysomnography is not sufficient to thoroughly study sleep. To investigate duration of sleep, or effects of drugs on sleep, monitoring should be obtained during several successive nights, to avoid the disturbances related to the "first night effect" (Mosko *et al.*, 1988) and those related to equipment invasiveness.

### Scoring of the Polysomnogram

Sleep is usually scored by 20- or 30-s segments. The scoring is performed by one of two international scoring systems: (a) Rechtschaffen and Kales international manual (Rechtschaffen and Kales, 1968). The manual gives specific instructions on how to score sleep states and stages. It is based on the 20- or 30-s "epoch." This scoring system was developed to score the sleep of normal subjects and not of patients. (b) The American Sleep Disorders Association Atlas Task Force (1993). This scoring system was developed to complement the international Atlas and to be more helpful when looking at pathology. It examines smaller epochs (3 s and longer) and abnormal events. Both atlases are used simultaneously to study pathology. Use of EEG cyclic alternating pattern (CAP) analyses has been proposed to study arousal (Terzano *et al.*, 1985, 1988). Scoring may be performed, after visual scoring, with computer-based analyses of the EEG. The computer-based analyses trying to simulate Rechtschaffen and Kales Atlas in patients have given poor results, but analyses based on more computer appropriate approaches have been helpful. This includes quantitative power spectral analyses of EEG with FFT used to identify specific EEG frequency bands.

Two other polygraphic tests are commonly used: multiple sleep latency test (MSLT) and maintenance of wakefulness test (MWT). These tests are based on polygraphic monitoring of sleep and wakefulness and must be preceded by polysomnography. A minimum total nocturnal sleep time in the preceding night has to be demonstrated for these two tests to have validity.

### Multiple Sleep Latency Test

This procedure is designed to evaluate: (1) the complaint of excessive daytime sleepiness by quantifying the time required for falling asleep and (2) the possibility of specific disorders such as narcolepsy by checking for abnormally short latencies to REM sleep (Association of Sleep Disorders Centers Task Force on Daytime

Sleepiness, 1986). Polygraphic variables include central and occipital EEG, chin muscle EMG, EOG, and ECG. Patients are monitored for five 20-min periods in a dark, quiet, temperature-controlled environment, dressed in street clothes, without the effect of alcohol, caffeine, or other stimulant drugs. The use of sleep-related medications should be stopped for 10 to 15 days before the test, depending on the duration of the drug's action. The first nap is scheduled between 90 and 150 min after the end of nocturnal sleep, and the remaining naps are at 2-h intervals. Sleep onset is defined as any of the following: the first three consecutive epochs of stage 1, any single epoch of stages 2, 3, or 4, or REM sleep. Sleep offset is defined as two consecutive epochs of wakefulness after sleep onset. Sleep is scored using the Rechtschaffen and Kales instructional criteria with a 30-s epoch. If sleep occurs during the 20-min test, the subject is allowed to sleep for the following 15 min. The observation of the occurrence of REM sleep during 15 min of sleep is called "presence of a sleep onset REM sleep period." Between naps it is imperative that the technician observes the patient and keeps him as alert as possible.

### Maintenance of Wakefulness Test

This is an alternative to the MSLT (Mitler *et al.*, 1982). It requires that subjects sit in a comfortable chair/armchair, reclining at a 45° angle, in the dark, with eyes closed in street clothes. The instructions given are different from those of the MSLT, it is requested that the subject avoid falling asleep and remain awake for 20 min (some have used 30 and 40 min, but the test has been validated for 20 min only). Testing periods recur every 2 h and the same rules as in the MSLT apply. Latencies to sleep onset and to REM sleep are determined.

There have been some variants to scoring these tests and some have used presence of one epoch of stage 1 NREM sleep to call sleep onset.

Based on data base obtained by the usage of these tests, a mean latency of 5 min or less at the MSLT and of 10 min or less at the MWT have been considered as indications of sleepiness (Richardson *et al.*, 1978), but further research has shown that this limit of 5 min is too low, and most clinical settings will consider a mean sleep latency of 8 min or below as pathological. A mean sleep latency above 10 min at both tests has been considered as normal alertness in adults. Presence of occurrence of REM sleep during 15 min of sleep is considered abnormal if it occurs at least in two of the five naps. Despite the wide usage of these tests, their specificity and sensitivity have not been determined, and clinical judgment must be applied. Van den Hoed *et al.* (1981), when determining a cut-off point of 8 min at MSLT, have shown that some narcoleptic–cataplectic patients may have to be monitored up to 4 successive days to show two or more sleep onset REM periods. A positive test indicates sleepiness on the day of the test. These tests may be associated with long-term (8 to 15 days) recording of movements using an actigraph placed on the nondominant arm that may show activity during long-term monitoring. Subjective rating, validated scales have been commonly given with the above tests. All of items have come under criticism but give information on the subjective rating of alertness. The most commonly used are the Epworth Sleepiness Scale, the fatigue scales, the Karolinska scale, and the Stanford sleepiness scale.

## SLEEP DISORDERS (Table I)

### Disorders of Initiating and Maintaining Sleep (DIMS)

#### Psychiatric Insomnia

**Clinical Features.** DIMS may be associated with psychiatric disorders and it is one of the symptoms of psychiatric syndromes. Twenty-one percent of individuals reporting severe insomnia also report psychiatric symptoms approximating the DSM-III criteria of major depression and almost 50% demonstrate high scores on a well-validated anxiety scale (Mellinger *et al.*, 1985). The major affective disorders are commonly associated with either sleep-onset insomnia and short sleep as in mania or early morning awakenings with short REM sleep latency as in depression (Kupfer and Reynolds, 1983). The short REM sleep latency after sleep onset has been considered a marker of depression. Of people with chronic insomnia, 26% receive a principal diagnosis of personality disorder (Tan *et al.*, 1984). Patients with dementia have a disturbance of the sleep–wake cycle, which produces episodes of wakefulness during nocturnal sleep and periods of lowered alertness and sleep during the day.

**Treatment.** Therapy should focus on the underlying psychiatric disorder. In case of dementia chronobiological approaches and sleep hygiene (Table II) may be helpful: regular bedtimes and rise times, limiting of daytime napping, maximizing daytime activity, and increasing exposure to sunlight or bright light during the day.

TABLE I   Summary of Sleep Disorders

| Diagnosis | Symptoms | Typical PSG features | Therapy |
|---|---|---|---|
| **DIMS** | | | |
| Psychiatric insomnia | Sleep onset insomnia, shortened sleep duration, early morning awakening | Short REM sleep latency, long sleep latency | Treatment of underlying psychiatric disorder (depression, mania, etc.) |
| Psychophysiologic insomnia | Sleep onset insomnia, sleep maintenance insomnia | Long sleep latency, sleep fragmentation | Hypnotics (Table III), behavior modification with stimulus control, sleep hygiene, recommendations (Table II) |
| Drug and alcohol use | Rebound insomnia after withdrawal, anxiety dreams | REM rebound, short sleep duration | Gradual withdrawal, supportive psychotherapy |
| Periodic limb movement disorder | Insomnia | Periodic myoclonus of the tibial anterior muscles | Pergolide (0.25–0.75 mg), pramipexol (0.125–0.75 mg), L-dopa + benserazid (or carbidopa) (62.5–375 mg), clonazepam (0.5–1 mg) |
| Sleep walking | Disorientation, confusion upon awakening | Complex behavior out of SWS initiated with synchronous delta waves | Locate mattress on the floor, bedrooms on the ground floor, lock windows, remove anything from the bedroom that could be potentially hazardous, consider side effects of prescribed medications such as benzodiazepines, carbamazepine |
| Confusion arousals | Disorientation, confusion, anxiety, tachycardia, tachypnea, sweat, harmful behavior upon awakening | Confused complex behavior out of NREM sleep | See sleep walking |
| REM sleep behavior disorder | Acting out violent dreams, self injury and injury to the bedpartner | Absence of REM sleep atonia | Clonazepam (0.5–1.5 mg) 30 min before bed time, appropriate treatment of Parkinson's disease |

*continues*

**TABLE I** *continued*

| Diagnosis | Symptoms | Typical PSG features | Therapy |
|---|---|---|---|
| Bruxism | Teeth damages, morning pain of the temporomandibular muscles | Rhythmic activity of temporomandibular muscles, arthrosis of temporomandibular joint | Rubber device ("mouth guard") |
| **EDS** | | | |
| Sleep deprivation | Cognitive impairment, disordered mood, physical fatigue, mental drowsiness | Short sleep period | Sleep hygiene recommendations (Table II) |
| Obstructive sleep apnea syndrome | Hypertension, nocturia, impotence, gastro-esophageal reflux, depression | >5 Obstructive apneas/ hypopneas per hour of total sleep time | Nasal continuous positive airway pressure, otolaryngological surgery, maxillo-mandibular and orthodontics, weight loss, in mild forms: theophylline (serum level: 11 µg/mL), acetazolamide: 350 mg at bedtime |
| Upper airway resistance syndrome | Sleep fragmentation | Arousals, repetitive increase in oesophageal pressure | See obstructive sleep apnea syndrome |
| Central sleep apnea syndrome | Sleep fragmentation | >5 Central sleep apneas per hour of total sleep time | See obstructive sleep apnea syndrome and: oxygen (2–3 L/min), nasal bilevel positive airway pressure |
| Narcolepsy | Sleep attacks, cataplexy, automatic behavior, hypnagogic and hypnapompic hallucinations, sleep paralysis, nocturnal sleep disruption | Sleep onset REM periods in >2 MSLT naps, reduced REM latency in the PSG | Regularly scheduled brief naps, modafinil (200–400 mg), methylphenidate: 10–60 mg, mazindol: 4–8 mg, metamphetamine: 20–25 mg, dextroamphetamine: 5–60 mg, SSRIs ortricyclics for cataplexy (i.e., fluoxetine: 20–60 mg, clomipramine: 50–150 mg) |
| Idiopathic CNS hypersomnia | Excessive daytime sleepiness, naps not refreshing, reduced sleep latency | Sleep latency <10 min in the MSLT | Modafinil (200–400 mg), methylphenidate: 20–60 mg, mazindol: 4–8 mg, metamphetamine: 20–25 mg, dextroamphetamine: 5–60 mg, |
| Psychiatric disorders and excessive daytime sleepiness | Atypical depression | Low sleep efficiency with long time in bed | MAO inhibitors, bright light therapy |
| CNS disorders and excessive daytime sleepiness | Brain lesions | Sleep latency <10 min in the MSLT | Stimulants |
| Circadian disorders | Desynchronization of the physiologic cycle with the major sleep period or daily schedule | Early awakening, long sleep latencies | Bright light therapy, sleep hygiene recommendations (Table II), melatonin |

*Note.* CNS, central nervous system; DIMS, disorders of initiating and maintaining sleep; EDS, excessive daytime sleepiness; MSLT, multiple sleep latency test; REM, rapid eye movement; SSRI, serotonin reuptake inhibitor.

## Psychophysiological Insomnia

**Clinical Features.** Psychophysiological insomnia is a common cause of DIMS. It can be transient and situational or persistent. In the former it involves brief periods of disturbed sleep, often induced by acute emotional arousal or conflict. It does not last longer than 3 weeks. The latter is defined as sleep onset and sleep maintenance insomnia. It is the result of reinforcing factors of chronic, somatized tension and negative conditioning to sleep. Patients often consider themselves as light sleepers and often have multiple somatic complaints (particularly tension headache, palpitation, and low back pain). The majority of these patients are tense and anxious.

**Treatment.** Patients with transient psychophysiological insomnia may be treated with hypnotics (Table III). In those cases in which the patients know ahead of time that they will take their medication (e.g., on the night before an important examination), almost any prescription of hypnotic seems acceptable (Table III). In the more frequent case in which the insomniac always attempts to fall asleep first without a hypnotic, but occasionally gives up, when sleep does not occur within 1 or 2 h, a rapid-acting hypnotic with a short half-life may be preferable. Patients with persistent psychophysiological insomnia respond to nonpharmacological approaches that include behavior modification with stimulus control, sleep restriction, and sleep hygiene (Table II) recommendations and morning light treatment with appropriate protocol (Bootzin *et al.*, 1991; Glovinsky and Spielman, 1991; Hauri, 1981). Sleep curtailment to a few hours per night is often used until a person learns again to use the time in bed for sleeping. In stimulus control therapy, persons with insomnia are asked to get out of bed whenever they cannot sleep. Once out of bed, the subject must use dim light, absolutely avoid watching TV or using a computer due to the bright light alerting effect, and engage in quiet activities leading to sleep, such as reading a nonexciting book, or listening to music. This is done to break the learned association between the bedroom environment and frustration. Insomnia group therapy is an approach that reinforces the above treatments. Relaxation and muscle biofeedback techniques, reverse conditioning, have also been successful in helping these patients. Hypnotics should not be prescribed during the time that behavioral treatment is administered (Hauri and Wisbey, 1993).

### Drug and Alcohol Use

**Clinical Features.** A common cause of DIMS is drug and alcohol use. It is linked to withdrawal from a CNS depressant after long-term use. Very common is the chronic usage of sleep medications. Alcohol promotes breathing disturbances during sleep (Stradling and Crosby, 1991). Caffeine increases sleep onset latency and amount of REM sleep and decreases REM latency and SWS (Karacan *et al.*, 1977). Tolerance to caffeine develops earliest at the fourth day of use (Zwyghuizen-Doorenbos *et al.*, 1988). Nicotine has an arousing effect with high dosages and a sedative effect with lower amounts. One study showed that chronic smokers took significantly longer to fall asleep than nonsmokers and improved immediately when smoking was withdrawn (Soldatos *et al.*, 1980). Stimulants initially prolong the REM and sleep onset latency and decrease the total sleep time (Post *et al.*, 1974). During the first week after withdrawal of stimulants there is commonly a short sleep onset and a short REM latency with increased total sleep time, which is followed by insomnia and reduced SWS for some weeks (Watson *et al.*, 1972). All hypnotics, independently of pharmaceutical industry claims, disrupt sleep architecture with long-term usage (long-term is defined as more than 3 weeks). Patients will often have an acute rebound when they stop taking their hypnotic agent. They may develop rebound insomnia, including during the early morning hours with very short-acting hypnotics (e.g., zoliplone, zolpidem). Sometimes REM rebound can be seen associated with anxiety dreams (Ciraulo *et al.*, 1991; Salzman, 1990).

**Treatment.** To avoid the problems mentioned above, withdrawal from these medications should be gradual, with supportive psychotherapy during the withdrawal. Nonpharmacological treatments should be installed during or just before withdrawal.

### Periodic Limb Movements during Sleep (PLMS)

**Clinical Features.** Two syndromes that may be seen in association can lead to nocturnal sleep disruption:

**TABLE II    Sleep Hygiene Instructions**

| | |
|---|---|
| I | Avoid naps 8 h after arising (exception: sleep disorder for which naps are beneficial). |
| II | Restrict sleep period to average number of hours you have slept in the previous week. |
| III | Do not stay in bed but get up after prolonged wake phases during the night. |
| IV | Keep a regular time out of bed. |
| V | Do not go to bed hungry. |
| VI | Exercise on a regular basis during the day. |
| VII | Do not smoke after 7 p.m. or drink caffeine after 10 a.m., and avoid severe use of alcoholic beverages. |
| VIII | Keep your room dark, quiet, well ventilated, and at a comfortable temperature throughout the night. |

**TABLE III    Hypnotics**

| Classification | Example | Influences on sleep |
|---|---|---|
| Long-acting (half-life > 5 h) | Flunitrazepam, diazepam | Reduce sleep latency and REM sleep, increase total sleep time, EDS the day after, withdrawal: disorders of initiating and maintaining sleep, REM rebound, nightmares |
| Middle long-acting (half-life: 3–5 h) | Zopiclone, lormetazepam, | First half of the night: reduced sleep latency, last third of the night: fragmented sleep, REM rebound |
| Short-acting (<5 h) | Zolpidem tartrate, zaleplon | Reduced sleep latency, may induce rebound insomnia and fragmentation of sleep at the end of the night, risk of increased intake with the second dose during the night with zaleplon |

restless legs and periodic limb movement syndromes. The complaint is one of difficulty initiating sleep and awakening after sleep onset. These syndromes will lead to arousals. They may lead to conditioning and difficulty in falling asleep again with a mixed presentation. Restless legs syndrome is described extensively in another chapter of this book. Periodic limb movements initially described by Symmonds under the name of "nocturnal myoclonus," were further investigated by Lugaresi and colleagues (1972), who reported that the repetitive, short bursts of muscle tone involving the anterior tibialis muscles and small components of the big toe similar to a spontaneous Babinski sign were not an epileptic phenomenon. Sometimes a complete Babinski maneuver with flexion and extension of the leg can be seen. Guilleminault et al. (1975) described the existence of periodic limb movement syndrome in isolation and its association with complaints of insomnia. There is controversy concerning the role of the isolated periodic limb movements in sleep. No change in sleep structure with or without periodic leg movements was found in one study (Karadeniz et al., 2000), but the issue is more linked to the presence or absence of an association of the movement with a short EEG arousal. Scoring of periodic limb movements with arousal during polysomnography is critical for the decision to pharmacologically treat the syndrome. When appropriately justified, treatment has eliminated complaints of disrupted nocturnal sleep and daytime fatigue.

**Treatment.** PLMS should be treated only in patients complaining about insomnia or excessive daytime sleepiness (EDS). If there exists an additional specific sleep disorder associated with insomnia or EDS in the same patient, this sleep disorder should be treated first. The treatment of choice for PLMS is a dopamine agonist. The combination L-Dopa/Carbidopa (Benserazid, Sinemet) has the inconvenience of being rapidly eliminated and not sufficiently long acting to cover the entire night (Trenkwalder et al., 1995). It may lead to a rebound phenomenon, or "activation" (Guilleminault et al., 1993). Pramipexole (0.25 mg) at bedtime is considered the drug of choice (Montplaisir et al., 1999). Pergolide given at bedtime and at low dosage (0.25–0.75 mg) (Wetter et al., 1999) and Ropinirole or other dopamine agonists have been shown to be helpful. Magnesium (12.4 mmol/day) and valproate (125–250 mg/day) were shown to reduce periodic limb movements in open trials (Ehrenberg et al., 2000; Hornyak et al., 1998). Clonazepam 0.5 to 1 mg can help by maintaining sleep and decreasing, but often not eliminating, the nocturnal jerks (Montagna et al., 1984); codeine derivatives have also shown positive results (Walters et al., 1993). Other possible therapies for PLMS include carbamazepine (Telstad et al., 1984) and gabapentine (Mellick and Mellick, 1996).

## The Parasomnias

They are undesirable phenomena during sleep which include sleep talking, sleep walking, sleep terror, nightmare, REM behavior disorder, and bruxism. Some are seen during NREM sleep; others are seen during REM sleep (anxiety dreams, REM behavior disorder), and some are seen in both states such as bruxism. Often, the differential diagnosis is a hypermotoric seizure indicating frontal lobe epilepsy. But epilepsy during sleep is the underlying cause of a parasomnia in about 1% of the patients. The most common diagnostic problem is the distinction between NREM sleep parasomnia and a REM behavior disorder. This distinction may be difficult, as both may exist in the same individual. Somnambulism, REM behavior disorder, and confusion arousal may lead to violent behavior during sleep with harm to patient and others.

### NREM Sleep Parasomnias

**Clinical Features.** Sleep walking and sleep terrors most frequently occur during the first one-third of nocturnal sleep, out of SWS with the abnormal behavior preceded by bursts of hypersynchronic delta (Kales et al., 1979; Nino-Murcia and Dement, 1987). Often the sleep prior the parasomnia is disrupted by short arousals. During the event the individual is confused and disoriented. The clinical presentation often is one of great anxiety with tachycardia, tachypnea, and sweat. The surrounding is threatening and bizarre and intervention by a third party may lead to behavior of escape or defense against a felt threat with exchange of blows or usage of any instrument, including knife or gun, if available. Patients often have a family history of somnambulism (Kales et al., 1980). Cardiovascular drugs may be responsible for nightmares, sleep terrors, and sleep walking episodes. Other medications, including tricyclics and benzodiazepines, may trigger sleepwalking but to a lesser degree than antiarrhythmic medications do (Huapaya, 1979a,b; Luchins et al., 1978; Risberg et al., 1975). Febrile illness may enhance NREM sleep nightmares, sleep terrors, and sleep walking (Dorus, 1979; Nadel, 1981). Tourettes syndrome, which includes a decrease in REM sleep and an increase in SWS, has been reported to be associated with sleep walking and sleep terrors (Barabas et al., 1984). Breathing disorders during sleep and severe PLMS may be associated with sleepwalking.

**Treatment.** Treatment of NREM sleep parasomnia should lead to avoidance of self-inflicted injuries. Subjects should sleep on a mattress on the floor, which limits fast walking movement, sleep in sleeping bags, in rooms located on the ground floor without furniture and windows covered by heavy drapes. Treatment of sleep disorders associated with the parasomnia, particularly sleep-disordered breathing, should be a priority. If an anxiety disorder is present, anxiolytic and psychoactive therapy should be prescribed.

### REM Sleep Parasomnia

**Clinical Features.** Anxiety dreams occur in REM sleep and may lead to screaming and awakenings from terrifying nightmares with difficulty to fall asleep again. REM sleep behavior is associated with absence of the normal REM sleep atonia. The behavior may involve acting out the dream (from running in bed, eating fish, driving a car) with performance of movements associated with the activity performed in the dream. REM behavior disorder may lead to violence on the unsuspecting sleeping bed partner if the dream is active or violent. REM behavior disorder can be seen without any other neurological disorders. It has preceded Parkinson's disease by up to 10 years (Schenck et al., 1996). It is commonly observed in patients with established Parkinson's disease, multiple system atrophy, Lewy body disease, Alzheimer's dementia, specific brainstem lesions, corticobasal degenerations, progressive supranuclear palsy, and narcolepsy and is associated with HLADQw1 (see Noachtar and Eisensehr (2000) for review). Serotonin reuptake inhibitors, selegiline, alcohol, and stimulants were reported to induce REM sleep behavior disorder (see Noachtar and Eisensehr (2000) for review). Reduced striatal dopamine transporters have recently been shown in patients with idiopathic REM behavior disorder (Eisensehr et al., 2000). During polysomnography, the usual muscle atonia seen during REM sleep is absent in a large amount of REM epochs. The EEG and the eye movements are those of REM sleep. The acting out does not occur with each dream period and is variable in its expression.

**Treatment.** REM behavior disorder has responded very well to clonazepam, 0.5 to 1.0 mg administered 30 min before bedtime (Schenck and Mahowald, 1990). Possible symptomatic reasons for REM behavior disorder should be eliminated: i.e., alcohol abuse, serotonin reuptake inhibitor, selegiline, stimulants. Appropriate treatment of Parkinson's disease during sleep (prescription of a long-acting dopamine agonist at bedtime) has also had positive effects (Tan et al., 1996).

### NREM/REM Sleep Parasomnia

**Clinical Features.** Bruxism is an oral habit characterized by a rhythmic activity of the temporo-mandibular muscles that causes a forced contact between dental surfaces during sleep. It is the third most frequent parasomnia in the general population (Glaros, 1981). It is most frequently associated with sleep disordered breathing. Anxiety disorder and other parasomnia were the next two most common associations with bruxism (Ohayon et al., 2001).

**Treatment.** Treatment for bruxism commonly includes a rubber device, worn over the teeth at night, called a mouth guard (Glaros and Rao, 1977).

## Syndromes of Sleepiness

### Insufficient Sleep

**Clinical Features.** The most common cause of daytime sleepiness is insufficient sleep, which may reflect poor sleep hygiene (behaviors impacting sleep, Table II) or self-imposed or socially dictated sleep deprivation. Busy people tend to regard sleep as a bank from which time can be borrowed as necessary to allow them to accomplish more by prolonging wakefulness. Thus, a sleep-debt is accumulated over time. If the sleep debt is not repaid in sleep, per se, some other currency must be used—this usually takes the form of daytime dysfunction and may include cognitive impairment, disordered mood, physical fatigue, or mental drowsiness.

**Treatment.** Adequate sleep hygiene (Table II) should be recommended.

### Sleep-Related Breathing Disorders

**Clinical Features.** Sleep-related breathing disorders (SRBDs) represent a very common (but underdiagnosed) cause of sleep fragmentation, and hence, daytime sleepiness. These disorders are a consequence of abnormal anatomy (crowding of the airway) superimposed on normal sleep physiology (reduction in muscle tone). During inspiration, pressure in the airway becomes negative relative to atmospheric pressure. If the airway pressure reaches a certain threshold, some degree of airway collapse and narrowing will occur. While the brain is awake, sufficient tone is maintained in the muscles of the airway to avoid significant airway collapse. With sleep onset, however, muscle tone diminishes and the soft tissues of the airway become more compliant. With sleep onset, heavy snorers show a higher supraglottic impedance than normals (Stoohs and Guilleminault, 1991). If the airway pressure reaches the critical threshold, complete or partial airway collapse will occur. Any factor, such as anatomic crowding of the airway, which contributes to airway pressure negativity, will tend to facilitate airway narrowing during sleep. Obstructive sleep apnea (OSA) was the first described SRBD. It usually, but not invariably, occurs in individuals, who snore. In OSA, complete collapse of the airway occurs during sleep; this leads to episodes of cessation of breathing, which last for more than 10 s, lead to blood oxygen desaturation, and tend to be associated with fragmented sleep. After hypoxemia or increased mechanical effort brief (usually just a few seconds) arousals occur in the brain. Coinciding with the arousal, muscle tone is immediately restored to waking levels; airway muscles tighten and stiffen, the airway expands, and breathing resumes. As the brain returns to sleep and muscle tone once again declines, the apneic process tends to recur, frequently throughout the night. Increased risks of developing hypertension, cardiac arrhythmia, myocardial infarction, and stroke have been assumed for patients with OSA (Hla *et al.*, 1994; Hung *et al.*, 1990; Lavie *et al.*, 2000; Palomäki *et al.*, 1992; Peppard *et al.*, 2000; Ohayon *et al.*, 2000; Motta and Guilleminault, 1985). Other associated symptoms and health problems include nocturia, impotence, headache, gastro-esophageal reflux, and depression (Shepard, 1994). When OSA is associated with daytime sleepiness, the disorder is called obstructive sleep apnea syndrome (OSAS). Prevalence of OSAS has been calculated from a cohort to be 4% for men and 2% for women between the ages of 30 and 60 years in the United States (Young, 1993) and 2% in a representative sample of the U.K. population (Ohayon *et al.*, 1997). Most studies have included mostly Caucasians; prevalence of OSAS in other groups is not well known. OSAS has important negative effects including reduced productivity, cognitive dysfunction, irritability, judgment error, and increased accident rates (Wu and Yan-Go, 1996). Less dramatic SRBDs occur: there may be partial airway collapse during sleep, leading to episodes of reduced airflow (termed "hypopneas"), short of frank apnea. An even more subtle form of SRBD, called upper airway resistance syndrome (UARS), has been described (Guilleminault *et al.*, 1993). In UARS, episodes of increased airway resistance occur during sleep, due to mechanisms similar to those described in OSA. Although oxygen desaturation and significant limitation of airflow do not occur in UARS, breathing is maintained at the expense of increased respiratory effort and resultant sleep fragmentation. Unless special instrumentation is used to monitor esophageal pressure (a sensitive indicator of respiratory effort), this disorder may be overlooked. Patients with UARS develop the same daytime consequences as those with OSA.

**Treatment.** The most common treatment for patients with OSA, which is also effective in some cases of central sleep apnea (apnea without any indication of body effort and without airway collapse) is long-term nasal continuous positive airway pressure (nCPAP). nCPAP alone or with supplemental oxygen is adjusted to the patient's specific breathing disorder in the sleep laboratory. Mask fitting and leaks become common problems with a pressure requirement of 12 or more cmH$_2$O. Thus, it is important to use the minimal required pressure to achieve an open airway throughout sleep. Nasal congestion is a common side effect of nCPAP therapy. Intranasal vasoconstrictors or steroids before going to sleep should be applied, particularly with history of atopy. Most patients with chronic nasal congestion affecting their ability to tolerate nCPAP benefit from humidification of the air applied through the nCPAP mask, which can be offered by a portable vaporizer or is integrated within the nCPAP machine. At the beginning of nCPAP therapy, patients with severe sleep apnea will have long periods of REM and stage 4 NREM sleep (Issa and Sullivan, 1986). The duration and intensity of these rebound episodes decrease quickly after the first night and last about 1 week. The loss of sleepiness after nCPAP is applied to patients with OSA can be detected by improved results on practical performance tests, such as driving simulator (Findley *et al.*, 1989). Increased weight usually aggravates OSA. Therefore, weight loss should be recommended to overweight patients (Guilleminault, 1989). Dental appliances that increase the upper airway size are available for treatment of mild forms of OSA (Lowe, 1994). Surgical therapy may be suggested to selected patients. Surgeries address anatomical reduction of the upper airway lumen noted during wakefulness. They consist of surgeries on soft tissues or cartilages located in the naso-pharynx (septoplasty, turbinectomy, radio frequency reduction of turbinates, adenoidectomy), in the oro-pharynx (tonsillectomy, uvulo-palato-pharyngo-plasty, uvulo-flap surgery), in the hypopharynx (exceptionally posterior glossectomy—a dangerous choice—radio-frequency reduction on base of tongue), and/or different osteotomies: genio-tubercle advancement with or without hyoidectomy and resuspension, maxillo-mandibular osteotomies with mid-face advancement, and distraction osteogenesis (antero-posterior, or lateral on mandibule, maxilla or both).

Much more unfrequently, Theophylline applied to a serum level of 11 µg/mL in patients with heart failure and acetazolamide, 350 mg, before bedtime have been shown to reduce respiratory events in mild to moderate sleep apnea (Inoue *et al.*, 1999).

### Narcolepsy

**Clinical Features.** The prevalence of narcolepsy is estimated to be 0.03 to 0.05%. Onset of symptoms may occur at virtually any age, but peak onset occurs in adolescence (with a secondary peak in the fourth decade). EDS is usually the presenting symptom, but some patients also have associated features including cataplexy (sudden, transient loss of muscle tone, usually in response to an emotional stimulus), sleep paralysis, hypnagogic hallucinations, and disrupted nocturnal sleep. The daytime sleepiness often becomes irresistible, leading to episodes of sleep at inappropriate times, so-called "sleep attacks." Periods of automatic behavior may also occur, a reflection of brief intrusions of sleep ("microsleeps") into the drowsy state. The clinical expression of the disorder likely depends upon an interplay between one or more genetic factors and environmental triggers. A genetic basis is suggested by association with specific HLA alleles and increased prevalence in first-degree relatives (40 times that in the general population) (Guilleminault et al., 1989). That genetic factors alone are insufficient to explain the disorder is supported by the infrequent familial cases and the low concordance rate (25–30%) of narcolepsy between identical twins (Partinen et al., 1994).

Honda (1986) first described the associations between narcolepsy and HLA DR15 and HLA DQ6 in the Japanese population. The association has been confirmed in 96% of Caucasians with narcolepsy (Mignot et al., 1994). The incidence of HLA-DR15 varies among ethnic groups, with a lower incidence in African-Americans (African-Blacks have not been studied). HLA-DQ6 (and more particularly DQB1-0602) is a more sensitive marker for narcolepsy across the ethnic groups. DQB1-0602 also occurs more often in narcoleptics with cataplexy (76%) than in those without cataplexy (41%) (Mignot et al., 1994). The strong HLA associations suggest that an autoimmune process may be involved. Overall, HLA typing is of limited usefulness in diagnosing narcolepsy, because the subtypes of interest often occur in normal individuals, and HLA associations are strongest in individuals with cataplexy (who also pose the least diagnostic difficulty); however, HLA negativity might cause one to question the diagnosis. In 1998, de Lecea et al. reported finding a hypothalamic-specific mRNA encoding the precursor of a pair of peptides homologous to secretin. They named the peptides hypocretin 1 and hypocretin 2 (Hcrt 1 and Hcrt 2) to denote their hypothalamic specificity and their resemblance to secretin. The Stanford group found a mutation in the hypocretin 2 receptor gene in narcoleptic dogs (Lin et al., 1999). Around the same time, researchers in Texas observed narcoleptic-like behavior in prehypocretin knockout mice (Chemelli et al., 1999). Nishino et al. (2000) hypothesized that human narcolepsy involves a disruption in hypocretin neurotransmission. They measured CSF hypocretin in nine narcoleptic patients with cataplexy and eight controls. Hypocretin 1 was detected in all control samples (250–285 pg/mL). In seven of nine patients, hypocretin levels were below the limits of detection of the assay (<40 pg/mL). Several investigators have tried to quantify plevels of hypocretin in the plasma; unfortunately, they never tested the assay against CSF findings in a blind fashion. The peptide with the currently available assays is at the limits of detection, and the current publications on levels of hypocretin in the plasma in narcoleptics and OSAS patients are most probably erroneous, due to the lack of appropriate validation of the assay (what has been measured is a "hypocretin-like immunological" reaction). More sensitive assays will need to be developed and are unavailable at this time (July 2001). The only valid results at this time are those measured in the CSF. Very recently, pathologic studies of brains of narcoleptics (compared to those of age-matched controls) have shown absence of hypocretin neurons in the hypothalamus (Peyron et al., 2000). Along with the clinical history, polysomnographic studies are essential in confirming the diagnosis of narcolepsy. Overnight recordings often demonstrate shortening of REM sleep latency, as well as disruption of sleep architecture with brief awakenings. MSLT reults are abnormal, with mean sleep latencies usually less than 5 min and the occurrence of two or more sleep onset REM sleep periods (SOREMPs).

**Treatment.**    Presently, the treatment of narcolepsy consists of support, medications for cataplexy, generally of the tricyclic or serotonin reuptake blocker, such as fluoxetine (with norepinephrine active metabolite), antidepressant classes, and stimulant medications for EDS such as methylphenidate, mazindol, amphetamines, and modafinil (Nishino and Mignot, 1997). Recently, gamma-hydroxybutyrate has been used with success in treating both EDS and cataplexy in narcoleptic patients (Mamelak et al., 1986). Our initial treatment consists of modafinil, 300 to 400 mg, taken in two divided dosages (morning and lunch) for daytime sleepiness, and fluoxetine, 20 to 40 mg, in the morning for cataplexy. If sleepiness persists modafinil may be increased to a maximum of 600 mg or may be combined with methyl phenidate administered in divided doses up to 40 mg. None of the current medication controls completely sleepiness. Experimental investigations have shown that gamma-hydroxybutyrate between 6 and 9 g taken for 1 month in two divided dosages during the night, leads to the best improvement at the multiple sleep latency test compared to modafinil and methylphenidate, but gamma-hydroxy butyrate is still considered an experimental drug in many countries at this time. Regularly scheduled (3×/day), brief naps are also beneficial (Mullington and Broughton, 1993). Narcoleptic patients often report temporary improvement in symptoms after naps of 10–15 min.

### Idiopathic CNS Hypersomnia

**Clinical Features.**    This disorder is characterized by EDS, but without sleep attacks, cataplexy, or nocturnal sleep disruption (Billiard *et al.*, 1998). It is believed to be less common than narcolepsy, but prevalence is difficult to determine, because strict diagnostic criteria are lacking, and no specific diagnostic marker is currently available. Onset of symptoms occurs in adolescence or early adulthood. Etiology of the disorder is not known, although viral illness may herald the onset of sleepiness in a subset of patients. Familial cases are known to occur, with increased frequency of HLA-Cw2. HLA typing in sporadic cases is unrevealing. Patients with idiopathic CNS hypersomnia report increased total sleep times. No amount of sleep ameliorates the EDS, and naps are generally not refreshing. Occupational and social functioning may be severely impacted by sleepiness. Polysomnography usually reveals shortened initial sleep latency, increased total sleep time and normal sleep architecture. Mean sleep latency on MSLT is usually reduced, often in the 6- to 10-min range, but SOREMPS are not seen.

**Treatment.**    Treatment is often less than satisfactory and includes stimulant medication and lifestyle modifications. The treatment of sleepiness consists primarily of administration of stimulant medications, such as methylphenidate, mazindol, amphetamines, and modafinil.

### Psychiatric Disorders and EDS

**Clinical Features.**    Psychiatric disorders are often associated with disrupted sleep. This is especially true of depression. While the majority of depressed patients with sleep disruption have insomnia, some of them have EDS. This subset of patients is often diagnosed with "atypical" depression or depression with the DSM-IV atypical features (which includes mood reactivity, increased appetite, leaden "paralysis," and rejection sensitivity along with hypersomnia) (diagnostic and statistical manual of mental disorders: DSM-IV-TR). These patients are thought to respond better to MAO inhibitors and possibly norepinephrine reuptake inhibitors than other types of antidepressants. There are also patients who might be said to have "psychogenic hypersomnia" (Vgontzas *et al.*, 2000). Generally, they are young adults who complain of EDS and have MSLT mean sleep latencies in the 7- to 10-min range. Overnight studies demonstrate long times in bed and poor sleep efficiency (ratio of total sleep time to total time in bed). Some of these patients often develop symptoms after a prolonged period of stress or following a period of disrupted sleep.

**Treatment.**    Patients respond to stress management, to improved sleep hygiene (Table II) with reduction of time in bed, and to a reduced sleep duration. Exposure to bright light immediately after arising in the morning (using commercially available light boxes) has also been found to be useful.

### Nervous System Disorders and EDS

**Clinical Features.**    EDS is a clinical feature of many encephalopathies. These disorders often present with other symptoms and signs, but EDS may dominate the picture, particularly in chronic cases. Structural brain lesions, including strokes, tumors, cysts, abscesses, hematomas, vascular malformations, and multiple sclerosis plaques are known to produce EDS. Somnolence may result from direct involvement of discrete brain regions (especially the brainstem reticular formation or midline diencephalic structures) or because of effects on sleep continuity (for example, nocturnal seizure activity or secondary SRBD). EDS has been reported following encephalitis or head trauma. Even post-traumatic narcolepsy with cataplexy has been described (Francisco and Ivanhoe, 1996). Patients with neuromuscular disorders may also develop EDS because of SRBD (Chokroverty, 2001) or PLMS (Yokota *et al.*, 1991).

**Treatment.**    If there is no adequate treatment option of the underlying neurological disease, stimulants can be used to improve EDS. If there is a specific sleep disorder, such as SRBD in neuromuscular disorders, this may be treated adequately.

## Circadian Disorders

The normal circadian cycle, regulated by the suprachiasmatic nucleus (SCN) of the hypothalamus, is a major determinate of alertness or sleepiness across the 24-h period (Turek, 1998). The cycle is entrained by factors such as physical activity and, especially, environmental light. If this physiologic cycle becomes desynchronized with the major sleep period or with the daily schedule, EDS may develop. Transient situations of this sort, such as jet lag, pose no diagnostic difficulty, but more chronic conditions may be overlooked if a careful sleep history is not taken.

### Delayed Sleep Phase Syndrome (DSPS)

**Clinical Features.**    Delayed sleep phase, i.e., a circadian driven tendency for the major sleep period to begin and end at later times, is common during puberty and may be associated with hormonal changes occurring at that time (Weitzman *et al.*, 1981). The sleep of patients DSPS is normal in quality and architecture, but it occurs at times that conflict with societal dictates and that the patients may find problematic. In DSPS, the shifting of the major sleep period causes disruption of "normal" activities and often conflict within the family. School performance typically suffers, particularly in morning classes, when the individual is in a state of suboptimal alertness. DSPS may be mistaken for insomnia, as the patient may simply complain of difficulty with sleep initiation. Often, a psychological or psychiatric component (including disorders of personality) is part of the picture. Treatment programs may be doomed to failure unless this aspect is addressed.

**Treatment.** Treatment strategies include sleep hygiene recommendations (Table II) and bright light applied at the beginning of the wake time.

### Advance Sleep Phase Syndrome (ASPS)

**Clinical Features.** Another chronic circadian disorder, known as advanced sleep phase syndrome (ASPS) (Wagner, 1996) involves the shift of the major sleep period to an earlier time; this condition often occurs in geriatric individuals. The sleep of patients ASPS is normal in quality and architecture, but it occurs at times which conflict with societal dictates and which the patients may find problematic. The patient with ASPS may be diagnosed with depression because of a complaint of early awakening. A careful history, perhaps supplemented with sleep diaries or actigraphy (a portable method for monitoring motor activity over time) usually eliminates any diagnostic uncertainty.

**Treatment.** Treatment strategies include sleep hygiene recommendations (Table II) and bright light applied at the beginning of the wake time.

### Other Circadian Rhythm Disorders

**Clinical Features.** EDS is a common problem for shift workers, who tend to have reduced total sleep times per 24-h period, in addition to their circadian disruptions (Akerstedt *et al.*, 1987). Much less common are patients with a "non-24-hour sleep–wake syndrome" (International Classification of sleep disorders, revised). Most of these individuals are blind and lack effective input to the SCN from the optic nerves. Therefore, they are unable to entrain the circadian clock with environmental light. Their circadian clock behaves as in "free-running" conditions and the major sleep period tends to be progressively delayed each day. Over time, the major sleep period will work its way around the 24-h clock; the patient will, thus report symptoms that vary according to the synchrony of the circadian clock and the 24-h clock. This disorder may rarely occur in patients with structural abnormalities involving the hypothalamus.

**Treatment.** Treatment strategies include sleep hygiene recommendations (Table II) and bright light applied at the beginning of the wake time. Melatonin has had very positive results in fixing the "free running" presentation of blind subjects (Sack *et al.*, 2000). It is mandatory to find at which stage of the free running cycle the subject is before beginning administration of the drug, usually administered near 20:00 h orally.

## CONCLUSION

Although progress has been made in understanding sleep disorders, there is still much to learn. Most patients with troubling (and often debilitating) symptoms can be helped with proper diagnosis and treatment.

## REFERENCES

Akerstedt, T., Torsvall, L., and Gillberg, M. (1987). Sleepiness in shift work. A review with emphasis on continuous monitoring of EEG and EOG. *Chronobiol. Int.* **4**, 129–140.

Association of sleep disorders centers task force on daytime sleepiness (Carskadon MA, chairman). (1986). Guidelines for the Multiple Sleep Latency Test (MSLT): A standard measure of sleepiness. *Sleep* **9**, 519–524.

Barabas, G., Mathews, W. S., and Ferrari, M. (1984). Somnambulism in children with Tourette's syndrome. *Dev. Med. Child Neurol.* **26**, 457–460.

Billiard, M., Merle, C., Carlander, B., Ondze, B., Alvarez, D., and Besset, A. (1998). Idiopathic hypersomnia. Review. *Psychiatry Clin. Neurosci.* **52**, 125–129.

Bootzin, R. R., Epstein, D., and Wood, J. M. (1991). Stimulus control instructions. *In* "Case Studies in Insomnia" (P. Hauri, Ed.). Plenum, New York.

Chemelli, R. M., Willie, J. T., Sinton, C. M., Elmquist, J. K., Scammell, T., Lee, C., Richardson, J. A., Williams, S. C., Xiong, Y., Kisanuki, Y., Fitch, T. E., Nakazato, M., Hammer, R. E., Saper, C. B., and Yanagisawa, M. (1999). Narcolepsy in orexin knock out mice: Molecular genetics of sleep regulation. *Cell* **98**, 437–451.

Chokroverty, S. (2001). Sleep-disordered breathing in neuromuscular disorders: A condition in search of recognition. *Muscle Nerve* **24**, 451–455.

Ciraulo, D. A., Sands, B. F., Shader, R. I., and Greenblatt, D. J. (1991). Anxiolytics. *In* "Clinical manual of chemical dependence" (D. A. Ciraulo and R. I. Shader, Eds.), pp. 135–174. American Psychiatric Press, Washington, DC.

de Lecea, L., Kilduff, T. S., Peyron, C., Gao, X., Foye, P. E., Danielson, P. E., Fukuhara, C., Battenberg, E. L., Gautvik, V. T., Bartlett, F. S. 2nd, Frankel, W. N., van den Pol, A. N., Bloom, F. E., Gautvik, K. M., and Sutcliffe, J. G. (1998). The hypocretins: Hypothalamus-specific peptides with neuroexcitatory activity. *Proc. Natl. Acad. Sci. USA* **95**, 322–327.

Dorus, E. (1979). Sleep walking and febrile illness. *Am. J. Psychiatry* **136**, 620. [Letter]

Ehrenberg, B. L., Eisensehr, I., Corbett, K. E., Crowley, P. F., and Walters, A. S. (2000). Valproate for sleep consolidation in periodic limb movement disorder. *J. Clin. Psychopharmacol.* **20**, 574–578.

Eisensehr, I., Linke, R., Noachtar, S., Schwarz, J., Gildehaus, F. J., and Tatsch, K. (2000). Reduced striatal dopamine transporters in idiopathic REM sleep behavior disorder. *Brain* 123, 155–1160.

Feinberg, I., and Floyd, T. C. (1979). Systematic trends across the night in human sleep cycles. *Psychophysiology* 16, 283–291.

Findley, L. J., Fabrizio, M. J., Knight, H., Norcross, B. B., LaForte, A. J., and Suratt, P. M. (1989). Driving simulator performance in patients with sleep apnea. *Am. Rev. Respir. Dis.* 140, 529–530.

Francisco, G. E., and Ivanhoe, C. B. (1996). Successful treatment of post-traumatic narcolepsy with methylphenidate: A case report. *Am. J. Phys. Med. Rehabil.* 75, 63–65.

Fukuda, N., Honma, H., Kohsaka, M., Kobayashi, R., Sakakibara, S., Kohsaka, S., and Koyama, T. (1999). Gender difference of slow wave sleep in middle aged and elderly subjects. *Psychiatry Clin. Neurosci.* 53, 151–153.

Glaros, A. G. (1981). Incidence of diurnal and nocturnal bruxism. *J. Prosthet. Dent.* 45, 545–549.

Glaros, A. G., and Rao, S. M. (1977). Bruxism. A critical review. *Psychol. Bull.* 84, 767–781.

Glovinsky, R. P., and Spielman, A. J. (1991). Sleep restriction therapy. *In* "Case Studies in Insomnia" (P. Hauri, Ed.), pp. 19–23. Plenum, New York.

Guilleminault, C., Cetel, M., and Philip, P. (1993). Dopaminergic treatment of Restless Legs and rebound phenomenon. *Neurology* 43, 445–446.

Guilleminault, C. (1989). Weight loss in sleep apnea. *Chest* 96, 703–704.

Guilleminault, C., Mignot, E., and Grumet, F. C. (1989). Familial patterns of narcolepsy. *Lancet* 2, 1376–1379.

Guilleminault, C., Raynal, D., Weitzman, E. D., and Dement, W. C. (1975). Sleep-related periodic myoclonus in patients complaining of insomnia. *Trans. Am. Neurol. Assoc.* 100, 19–22.

Guilleminault, C., Stoohs, R., Clerk, A., Cetel, M., and Maistros, P. (1993). A cause of excessive daytime sleepiness. The upper airway resistance syndrome. *Chest* 104, 781–787.

Hauri, P. (1981). Treating psychophysiologic insomnia with biofeedback. *Arch Gen Psychiatry* 38, 752–758.

Hauri, P., and Wisbey, J. (1993). Can we mix behavioral therapy with hypnotics. *Sleep Res.* 22, 207. [Abstract]

Hla, K. M., Young, T. B., Bidwell, T., Palta, M., Skatrud, J. B., and Dempsey, J. (1994). Sleep apnea and hypertension. A population based study. *Ann. Int. Med.* 120, 382–388.

Honda, Y. (1986). Clinical features of narcolepsy: Japanese experiences. *In* "HLA in Narcolepsy" (Y. Honda, T. Juji, Eds.), pp. 24–57. Springer, Berlin, Heidelberg, Tokyo, New York.

Hornyak, M., Voderholzer, U., Hohagen, F., Berger, M., and Riemann, D. (1998). Magnesium therapy for periodic leg movements-related insomnia and restless legs syndrome: an open pilot study. *Sleep* 21, 501–505.

Huapaya, L. (1979). Seven cases of somnambulism induced by drugs. *Am. J. Psychiatry* 36, 985.

Huapaya, L. (1979). Somnambulism and bedtime medication. *Am. J. Psychiatry* 36, 1207.

Hung, J., Whitford, E. G., Parsons, R. W., and Hillman, D. R. (1990). Association of sleep apnoea with myocardial infarction in men. *Lancet* 336, 261–264.

Inoue, Y., Takata, K., Sakamoto, I., Hazama, H., and Kawahara, R. (1999). Clinical efficacy and indication of acetazolamide treatment on sleep apnea syndrome. *Psychiatry Clin. Neurosci.* 53, 321–322.

Issa, F. G., and Sullivan, C. E. (1986). The immediate effects of continuous positive airway pressure treatment on sleep patterns in patients with obstructive sleep apnea syndrome. *Electroencephalogr. Clin. Neurophysiol.* 63, 10–17.

Jouvet, M. (1972). The role of monoamines and acetylcholine-containing neurons in the regulation of the sleep-wake cycle. *Ergebn. Physiol.* 64, 166–307.

Kales, A., Soldatos, C. R., Bixler, E. O., Ladda, R. L., Charney, D. S., Weber, G., and Schweitzer, P. K. (1980). Hereditary factors in sleep-walking and night terrors. *Br. J. Psychiatry* 137, 111–118.

Kales, J. D., Kales, A., Soldatos, C. R, Chamberlin, K., and Martin, E. D. (1979). Sleep walking and night terrors related to febrile illness. *Am. J. Psychiatry* 136, 1214–1215.

Karacan, I., Thornby, J. I., Anch, M., Booth, G. H., Williams, R. L., and Salis, P. J. (1977). Dose-related sleep disturbances induced by coffee and caffeine. *Clin. Pharmacol. Ther.* 20, 682–688.

Karadeniz, D., Ondze, B., Besset, A., and Billiard, M. (2000). Are periodic leg movements during sleep (PLMS) responsible for sleep disruption in insomnia patients? *Eur. J. Neurol.* 7, 331–336.

Kupfer, D. J., and Reynolds, C. F. (1983). Neurophysiologic studies of depression: State of the art. *In* "The Origins of Depression: Current Concepts and Approaches" (A. Berlin, Ed.), pp. 235–252. Springer, New York.

Laurent, J. P., Cespuglio, R., and Jouvet, M. (1974). Delimitation des voies ascendantes de l'activite ponto-geniculo-occipitale chez le chat. *Brain Res.* 65, 29–52.

Lavie, P., Herer, P., and Hoffstein, V. (2000). Obstructive sleep apnoea syndrome as a risk factor for hypertension: Population study. *Br. Med. J.* 320, 479–482.

Lin, L., Faraco, J., Li, R., Kadotani, H., Rogers, W., Lin, X., Qiu, X., de Jong, P. J., Nishino, S., and Mignot, E. (1999). The sleep disorder canine narcolepsy is caused by a mutation in the hypocretin (orexin) receptor 2 gene. *Cell* 98, 365–376.

Lowe, A. A. (1994). Dental appliances for the treatment of snoring and obstructive sleep apnea. *In* "Principles and Practice of Sleep Medicine" (M. H. Kryger, T. Roth, and W. C. Dement, Eds.), pp. 722–735. Saunders, Philadelphia.

Luchins, D. J., Sherwood, P. M., Gillin, J. C., Mendelson, W. B., and Wyatt, R. J. (1978). Filicide during psychotropic-induced somnambulism. A case report. *Am. J. Psychiatry* 135, 1404–1405.

Lugaresi, E., Coccagna, G., Mantovani, M., and Lebrun, R. (1972). Some periodic phenomena arising during drowsiness and sleep in man. *Electroencephalogr. Clin. Neurophysiol.* 32, 701–705.

Mamelak, M., Scharf, M. B., and Woods, M. (1986). Treatment of narcolepsy with gamma-hydroxybutyrate. A review of clinical and sleep laboratory findings. *Sleep* 9, 285–289.

Mellick, G. A., and Mellick, L. B. (1996). Management of restless legs syndrome with gabapentin. *Sleep* 19, 224–226.

Mellinger, G. D., Balter, M. B., and Uhlenhuth, E. H. (1985). Insomnia and ist treatment. *Arch. Gen. Psychiatry* 42, 225–232.

Mignot, E., Lin, X., Arrigoni, J., Macaubas, C., Olive, F., Hallmayer, J., Underhill, P., Guilleminault, C., Dement, W. C., and Grumet, F. C. (1994). DQB1*0602 and DQA1*0102 (DQ1) are better markers than DR2 for narcolepsy in Caucasian and black Americans. *Sleep* 17, 60–67.

Mitler, M. M., Gujavarty, K. S., Sampson, M. G., and Brownman, C. P. (1982). Multiple daytime nap approaches to evaluating the sleepy patient. *Sleep* 5, 5119–5127.

Montagna, P., de Bianchi, L. S., Succoni, M., Cirignotta, F., and Lugaresi, E. (1984). Clonazepam and vibration in restless legs syndrome. *Acta Neurol. Scand.* 69, 428–430.

Montplaisir, J., Nicolas, A., Denesle, R., and Gomez-Mancilla, B. (1999). Restless legs syndrome improved by pramipexole: A double-blind randomized trial. *Neurology* 52, 938–943.

Mosko, S. S., Dickel, M. J., and Ashurst, J. (1988). Night-to-night variability in sleep apnea and sleep related periodic leg movements in the elderly. *Sleep* 11, 340–348.

Motta, J., and Guilleminault, C. (1985). Cardiac dysfunction during sleep. *Ann. Clin. Res.* 17, 190–198.

Mullington, J., and Broughton, R. (1993). Scheduled naps in the management of daytime sleepiness in narcolepsy-cataplexy. *Sleep* 16, 444–456.

Nadel, C. (1981). Somnambulism, bedtime medication and overeating. *Br. J. Psychiatry* **139**, 79–83. [Letter]

Nino-Murcia, G., and Dement, W. C. (1987). Psychophysiological and psychopharmacological aspects of somnambulism and night terrors in children. *In* "Psychopharmacology: The Third Generation of Progress" (H. Y. Metzer, Ed.), pp. 873–879. Raven Press. New York.

Nishino, S., and Mignot, E. (1997). Pharmacological aspects of human and canine narcolepsy. Review. *Prog. Neurobiol.* **52**, 27–78.

Nishino, S., Ripley, B., Overeem, S., Lammers, G. J., and Mignot, E. (2000). Hypocretin (orexin) deficiency in human narcolepsy. *Lancet* **355**, 39–40.

Noachtar, S., and Eisensehr, I. (2000). REM Sleep Behavior Disorder. Review. *Nervenarzt* **71**, 802–806.

Norman, R. G., Ahmed, M. M., Walsleben, J. A., and Rapoprt, D. M. (1997). Detection of respiratory events during NPSG: Nasal cannula/pressure sensor versus thermistor. *Sleep* **20**, 1175–1184.

Ohayon, M. M., Guilleminault, C., Priest, R. G., Zulley, J., and Smirne, S. (2000). Is sleep-disordered breathing an independent risk factor for hypertension in the general population (13,057 subjects)? *J. Psychosomat. Res.* **48**, 6.

Ohayon, M. M., Guilleminault, C., Priest, R. G., and Caulet, M. (1997). Snoring and breathing pauses during sleep: Telephone interview survey of a United Kingdom population sample. *Br. Med. J.* **314**, 860–863.

Ohayon, M. M., Li, K. K., and Guilleminault, C. (2001). Risk factors for sleep bruxism in the general population. *Chest* **119**, 53–61.

Palomäki, H., Partinen, M., and Erkinjuntti, T. (1992). Snoring, sleep apnea syndrome and stroke. *Neurology* **42**, 75–82.

Partinen, M., Hublin, C., Kaprio, J., Koskenvuo, M., and Guilleminault, C. (1994). Twin studies in narcolepsy. *Sleep* **17**, 13–16.

Peppard, E. P., Young, T., Palta, M., and Skatrud, J. (2000). Prospective study of the association between sleep-disordered breathing and hypertension. *N. Engl. J. Med.* **342**, 1378–1384.

Peyron, C., Faraco, J., Rogers, W., Ripley, B., Overeem, S., Charnay, Y., Nevsimalova, S., Aldrich, M., Reynolds, D., Albin, R., Li, R., Hungs, M., Pedrazzoli, M., Padigaru, M., Kucherlapati, M., Fan, J., Maki, R., Lammers, G. J., Bouras, C., Kucherlapati, R., Nishino, S., and Mignot, E. (2000). A mutation in a case of early onset narcolepsy and a generalized absence of hypocretin peptides in human narcoleptic brains. *Nat. Med.* **6**, 991–997.

Post, R. M., Gillin, J. C., Wyatt, R. J., and Goodwin, F. K. (1974). The effect of orally administered cocaine on sleep of depressed patients. *Psychopharmacologia* **37**, 59–66.

Rechtschaffen, A., and Kales, A., (Ed) (1968). A manual of standardized terminology, techniques and scoring system for sleep stages of human subjects. U.S. Department of Health, Education, and Welfare, Public Health Service-National Institutes of Health, National Institute of Neurological Diseases and Blindness, Neurological Information Network. Bethesda, MD.

Richardson, G. S., Carskadon, M. A., Flagg, W., van den Hoed, J., Dement, W. C., and Mitler, M. M. (1978). Excessive daytime sleepiness in man: Multiple sleep latency measurement in narcoleptic and control subjects. *Electroencephalogr. Clin. Neurophysiol.* **45**, 621–627.

Risberg, A. M., Risberg, J., and Elquist, D. (1975). Effects of diryrazine and methaqualone on the sleep pattern in normal man. *Eur. J. Clin. Pharmacol.* **8**, 227–231.

Roffwarg, H. P., Muzio, J. N., and Dement, W. C. (1966). Ontogenetic development of the human sleep-dream cycle. *Science* **152**, 604–619.

Sack, R. L., Brandes, R. W., Kendall, A. R., and Lewy, A. J. (2000). Entrainment of free-running circadian rhythms by melatonin in blind people. *N. Engl. J. Med.* **343**, 1070–1077.

Salzman, C. (Chairperson) (1990). Benzodiazepine dependence, toxicity and abuse: A Task Force Report. American Psychiatric Association, Washington DC.

Schenck, C., Bundline, S. R., and Mahowald, M. W. (1996). Delayed emergence of a parkinsonian disorder in 38% of 29 older men initially diagnosed with idiopathic rapid eye movement behavior disorder. *Neurology* **46**, 422–425.

Schenck, C. H., and Mahowald, M. W. (1990). Polysomnographic, neurologic, psychiatric, and clinical outcome report on 70 consecutive cases with the REM sleep behavior disorder (RBD): Sustained clonazepam efficacy in 89.5% of 57 treated patients. *Cleve. Clin. J. Med.* **57**, 10–24.

Shepard, J. W. (1994). Cardiorespiratory changes in obstructive sleep apnea. *In* "Principles and Practice of Sleep Medicine" (M. H. Kryger, T. Roth, and W. C. Dement, Eds.), pp. 667–677. Saunders, Philadelphia.

Soldatos, C. R., Kales, J. D., Scharf, M. B., Bixler, E. O., and Kales, A. (1980). Cigarette smoking associated with sleep difficulty. *Science* **207**, 551–553.

Stoohs, R., and Guilleminault, C. (1991). Snoring during NREM sleep: Respiratory timing, esophageal pressure and EEG arousal. *Respir. Physiol.* **85**, 151–167.

Stradling, J. R., and Crosby, J. H. (1991). Predictors and prevalence of obstructive sleep apnea and snoring in 1001 middle aged men. *Thorax* **46**, 85–90.

Symonds, C. P. (1953). Nocturnal myoclonus *J. Neurol. Neurosurg. Psychiatry* **16**, 166–171.

Tan, A., Salgado, M., and Fahn, S. (1996). Rapid eye movement sleep behavior disorder preceding Parkinson's disease with therapeutic response to levodopa. *Mov. Dis.* **11**, 214–216.

Tan, T., Kales, J. D., Kales, A., Soldatos, C. R., and Bixler, E. O. (1984). Biopsychobehavioral correlates of insomnia. IV. Diagnoses based on DSM-III. *Am. J. Psychiatry* **141**, 356–362.

Telstad, W., Sorensen, O., Larsen, S., Lillevold, P. E., Stensrud, P., and Nyberg-Hansen, R. (1984). Treatment of the restless legs syndrome with carbamazepine: A double blind study. *Br. Med. J.* **288**, 444–446.

Terzano, M. G., Mancia, D., Salati, M. R., Costani, G., Decembrino, A., and Parrino, L. (1985). The cyclic alternating pattern as a physiologic component of normal NREM sleep. *Sleep* **8**, 137–145.

Terzano, M. G., Parrino, L., and Spaggiari, M. C. (1988). The cyclic alternating pattern sequences in the dynamic organization of sleep. *Electroenceph. Clin. Neurophysiol.* **69**, 437–447.

The Atlas Task Force (1992). EEG arousals: Scoring rules and examples: A preliminary report from the Sleep Disorders Atlas Task Force of the American Sleep Disorders Association. *Sleep* **15**, 173–184.

The Atlas Task Force (1993). Recording and scoring leg movements. *Sleep* **16**, 748–759.

Trenkwalder, C., Stiasny, K., Pollmaecher, T., Wetter, T. C., Schwarz, J., Kohnen, R., Krueger, H. P., Kazenwadel, J., Ramm, S., Kuenzel, M., and Oertel, W. H. (1995). L-Dopa therapy of uremic and idiopathic restless legs syndrome: A double-blind, crossover trial. *Sleep* **18**, 681–688.

Turek, F. W. (1998). Circadian rhythms. Review. *Horm. Res.* **49**, 109–113.

Van den Hoed, J., Kraemer, H., Guilleminault, C., Zarcone, V. P. Jr., Miles, L. E., Dement, W. C., and Mitler, M. M. (1981). Disorders of excessive daytime somnolence: Polygraphic and clinical data for 100 patients. *Sleep* **4**, 23–37.

Vgontzas, A. N., Bixler, E. O., Kales, A., Criley, C., and Vela-Bueno, A. (2000). Differences in nocturnal and daytime sleep between primary and psychiatric hypersomnia: diagnostic and treatment implications. *Psychosomat. Med.* **62**, 220–226.

Wagner, D. R. (1996). Disorders of the circadian sleep–wake cycle. Review. *Neurol. Clin.* **14**, 651–670.

Walters, A. S., Wagner, M. L., Hening, W. A., Grasing, K., Mills, R., Chokroverty, S., and Kavey, N. (1993). Successful treatment of the idiopathic restless legs syndrome in a randomized double-blind trial of oxycodone versus placebo. *Sleep* **16**, 327–332.

Watson, R., Hartmann, E., and Schildkraut, J. J. (1972). Amphetamine withdrawal: Affective state, sleep patterns, and MHPG excretion. *Am. J. Psychiatry* **129**, 39–45.

Weese-Mayer, D. E., Corwin, M. J., Peucker, M. R., Di Fiore, J. M., Hufford, D. R., Tinsley, L. R., Neuman, M. R., Martin, R. J., Brooks, L. J., Davidson Ward, S. L., Lister, G., and Willinger, M. (2000). Comparison of apnea identified by respiratory inductance plethysmography with that detected by end-tidal $CO_2$ or thermistor. The CHIME Study Group. *Am. J. Respir. Crit. Care Med.* **162**, 471–480.

Weitzman, E. D., Czeisler, C. A., Coleman, R. M., Spielman, A. J., Zimmerman, J. C., Dement, W., Richardson, G., and Pollak, C. P. (1981). Delayed sleep phase syndrome. A chronobiological disorder with sleep-onset insomnia. *Arch. Gen. Psychiatry* **38**, 737–746.

Wetter, T. C., Stiasny, K., Winkelmann, J., Buhlinger, A., Brandenburg, U., Penzel, T., Medori, R., Rubin, M., Oertel, W. H., and Trenkwalder, C. (1999). A randomized controlled study of pergolide in patients with restless legs syndrome. *Neurology* **52**, 944–950.

Wu, H., and Yan-Go, F. (1996). Self-reported automobile accidents involving patients with obstructive sleep apnea. *Neurology* **46**, 1254–1257.

Yokota, T., Hirose, K., Tanabe, H., and Tsukagoshi, H. (1991). Sleep-related periodic leg movements (nocturnal myoclonus) due to spinal cord lesion. *J. Neurol. Sci.* **104**, 13–18.

Young, T. (1993). Analytic epidemiology studies of sleep disordered breathing-what explains the gender difference in sleep disordered breathing? *Sleep* **8**, 1–2.

Zwyghuizen-Doorenbos, A., Roehrs, T., Lamphere, J., Zorick, F., and Roth, T. (1988). Increased daytime sleepiness enhances ethanol's sedative effects. *Neuropsychopharmacology* **1**, 279–286.

## CHAPTER 20
# The Epilepsies

Samden D. Lhatoo and Josemir W. A. S. Sander

Epilepsy may be defined as "the occurrence of transient paroxysms of excessive or uncontrolled discharges of neurons, which may be caused by a number of different etiologies, leading to epileptic seizures." The term "epilepsy" encompasses a number of seizure disorders that only have in common a tendency for the patient to have recurrent epileptic attacks. A seizure (also termed "ictus" or "ictal event") is an event which is discernible to the person experiencing the seizure or an observer. It is a stereotypic event in which an individual's awareness of his surroundings may be impaired and behavior altered. Seizures frequently have a sudden onset and usually cease spontaneously. They are commonly brief, lasting from seconds to minutes, and are often followed by a period of drowsiness and confusion (the postictal period). The clinical manifestations of seizures may take many different forms, varying from patient to patient and reflecting the functions of the cortical tissues in which the excessive discharge arises and to which it spreads. By convention, it is necessary for seizures to be recurrent and unprovoked to constitute epilepsy; by definition a single attack is not sufficient to make this diagnosis, even though most people having one seizure will have further attacks.

## PATHOPHYSIOLOGY

Hughlings Jackson suggested epilepsy was a recurrent, disorderly discharge of the nerve tissue (Jackson JH 1873), and this prescient statement has since been confirmed physiologically. Classically, two major physiological patterns of epileptic discharges have been recognized: those arising from focal cortical disturbances (with focal EEG discharges comprising spikes, sharp waves, or spike/wave) and those characterized by immediate synchronous discharge in both hemispheres (with synchronous spike/wave on the EEG). The origin of synchronous spike/wave discharges is still a matter of considerable debate; some authorities propose a corticothalamic basis and others, cortical generation. In the latter proposition, it is a faster speed of propagation which sets the generalized spike/wave discharges apart from the discharges of focal epilepsy.

At a cellular neuronal level the seizure threshold is set by influences causing inhibitory and excitatory postsynaptic potentials (IPSPs and EPSPs). Membrane depolarization results in an action potential that is propagated and leads to repetitive firing of surrounding individual neurons. In acute experimental animal models of epilepsy a neuron that is functionally converted to an epileptic state will exhibit an all-or-none reaction and a paroxysmal depolarization shift that is more prolonged than the physiological excitatory postsynaptic potential. This results in burst firing and, if adjacent neurons are recruited in sufficient numbers, a seizure. The cellular electrophysiology of human epilepsy and the molecular basis for the paroxysmal depolarization shift and burst firing have been reviewed in depth by Meldrum (1990) and Schwartzkroin (1994). The past decade has seen an important increase in the understanding of putative mechanisms underlying neuronal hyperexcitability, which may lead to different types of seizure disorders.

Several single gene disorders have been described (Table I), where each involved gene encodes an ion channel gated either by voltage or by a neurotransmitter. These disorders account for only a small proportion of all cases of epilepsy and these are called "channelopathies" which cause episodic or paroxysmal disturbances that manifest as seizures in a manner similar to those seen in other paroxysmal ion channel disorders such as the episodic paralyses, hemiplegic migraine, episodic ataxia, and ventricular arrhythmias. In other causes of epilepsy following neuronal injury such as hippocampal sclerosis, synaptic reorganization due to

TABLE I    Single Gene Mutations in Epilepsy

| Gene | Locus | Epilepsy | Reference |
|------|-------|----------|-----------|
| CHRNA4 | 20q | ADNFLE | (Steinlein, and others 1995) |
| CHRNB2 | 1q | ADNFLE | (De Fusco, and others 2000) |
| KCNQ2 | 20q | BFNC | (Singh, and others 1998) |
| KCNQ3 | 8q | BFNC | (Charlier, and others 1998) |
| SCN1B | 19q | GEFS+ | (Wallace, and others 1998) |
| SCN1A | 2q | GEFS+ | (Escayg, and others 2000) |
| GABRG2 | 5q | GEFS+? | (Baulac S, and others 2001) |
| GABRG2 | 5q | FS/CAE | (Wallace RH, and others 2001) |

*Note.* ADNFLE, autosomal dominant nocturnal frontal lobe epilepsy; BFNC, benign familial neonatal convulsions; GEFS+, generalized epilepsy with febrile seizures plus; CAE, childhood absence epilepsy; FS, febrile seizures. CHRNA4, gene for α-4 subunit of the ligand gated neuronal nicotinic acetylcholine receptor; CHRNB2, gene for β-2 subunit of the nicotinic acetylcholine receptor; KCNQ2, KCNQ3, voltage-gated neuronal potassium channel genes; SCN1B, SCN1A, genes for β-1 and α-1 subunits, respectively, for the voltage-gated neuronal sodium channel; GABRG2, genes for the γ-2 subunit of the GABA–A receptor.

mossy fiber sprouting may contribute to epileptogenesis by forming new excitatory circuits between dentate granule cells in the hippocampus (McNamara, 1999). Some experimental studies have shown however, that this model may be too simplistic and mossy fiber sprouting may not be crucial to epileptogenesis. The role of excitatory glutamatergic and inhibitory GABA-ergic transmission in hippocampal sclerosis is also not completely clear although changes in glutamate and GABA receptors may favor predominant excitation (Dalby and Mody, 2001). Brain tumors may predispose to epilepsy by mechanisms such as selective loss of surrounding inhibitory interneurons, loss of GABA receptors, alteration of glutamate, and calcium regulation by local hemosiderin deposition and even mechanical distortion. Similar mechanisms may be involved in epileptogenesis in head injuries. In cortical dysgenesis, there may be an imbalance between GABA-ergic inhibitory interneurons and excitatory neurons along with abnormalities of synaptic connections and function.

## THE GENETICS OF EPILEPSY

The concordance for epilepsy in monozygotic twins is 62% compared to 18% in dizygotic twins which suggests that there is a strong genetic influence in seizure disorders (Berkovic *et al.*, 1998). In idiopathic generalized epilepsy, the figures for monozygotic and dizygotic twins are 76% for monozygotic twins and 33% for dizygotic twins. As well as mendelian disorders such as neurofibromatosis where epilepsy is part of the phenotype, genetic mutations with a mendelian inheritance have also been described in epilepsies where seizures are

TABLE II    International Classification of Seizure Type[a]

I.  Partial seizures
   A. Simple partial seizures
     1. With motor signs
     2. With somatosensory or special sensory hallucinations
     3. With autonomic symptoms and signs
     4. With psychic symptoms
   B. Complex partial seizures
     1. Simple partial onset followed by impairment of consciousness
     2. With impaired consciousness at onset
   C. Partial seizures evolving to secondary generalized seizures
     1. Simple partial seizures evolving to generalized ones
     2. Complex partial seizures evolving to generalized ones
II. Generalized seizures
   A. Absence seizures
     1. Typical
     2. Atypical
   B. Myoclonic seizures
   C. Clonic seizures
   D. Tonic seizures
   E. Tonic–clonic seizures
   F. Atonic seizures
III. Unclassified epileptic seizures

[a]Commission on Classification (1981).

the main disorder (Table I). The majority of seizure disorders with a genetic basis, however, are likely to be due to mutiple gene disorders with complex inheritance. Epilepsy genetics are currently the focus of much study and advances in our understanding of this fascinating field are likely to impact on all aspects of the management of seizure disorders (Berkovic and Scheffer, 1999), including drug treatment. In medically intractable epilepsy for example, increased expression of P glycoprotein, encoded by the human multidrug resistance gene-1, has been shown in the brains of some patients (Sisodiya, and others 2001).

## CLASSIFICATION OF SEIZURE TYPE

By international convention, seizures are currently divided fundamentally into two main types: partial and generalized (Table II).

Partial or focal seizures have clinical or EEG evidence of local cortical onset (usually in a restricted part of one hemisphere) and may spread to other parts of the brain (often in both hemispheres) during the course of a seizure. Partial seizures are subclassified into three groups: simple partial seizures; complex partial seizures; and partial seizures that secondarily generalize (Table III). The distinction between simple and complex partial seizures depends on whether or not consciousness is impaired, and this may be difficult to detect clinically (Table IV). Furthermore, simple partial seizures often evolve into complex partial seizures, and both may

**TABLE III  Partial Epileptic Seizures**

| | |
|---|---|
| Simple partial seizures (focal) | |
| Consciousness | Unaffected |
| Age group | Any age |
| Duration | Seconds to minutes |
| Symptoms | Depend upon localization of focus (see Table IV); no postictal confusion |
| Ictal EEG | Contralateral epileptiform discharges; in many instances no ictal abnormalities in scalp recordings |
| Special features | An aura is a simple partial seizure |
| Complex partial seizures (psychomotor) | |
| Consciousness | Impaired |
| Age group | Any group |
| Duration | Minutes |
| Symptoms | Depend upon localization of focus (see Table IV); postictal confusion |
| Ictal EEG | Unilateral or frequently bilateral epileptiform discharges, diffuse or focal |
| Secondarily generalized seizures | |
| Initial symptoms | Depend on seizure type (simple or complex partial) and localization of epileptic focus |
| Subsequent symptoms | Generalized tonic–clonic convulsions |

become secondary generalized. Partial seizures can also be divided according to anatomical localization (Table IV), although there is a considerable overlap of symptoms partly due to the rapid spread of partial seizures from one cortical region to another. The commonest partial seizures are of temporal lobe origin.

In contrast, generalized seizures have no clinical or EEG evidence of local onset but simultaneously involve all or a large part of both hemispheres from the onset of the seizure. Various clinical and EEG generalized seizure patterns exist, of which the tonic–clonic (grand mal) seizure is the most common (Table V).

## CLASSIFICATION OF EPILEPSY SYNDROMES

A categorization of epilepsy according to seizure type provides only limited information, and similar seizure types occur in widely differing epilepsies. Therefore, a more synoptic classification of the epilepsies and epilepsy syndromes has been developed that accounts for other clinical aspects as well as seizure type. The classification of epilepsy syndromes maintains the dichotomy between partial and generalized seizures, which are then subdivided by etiology into symptomatic (where there is evidence for underlying brain disorder), idiopathic (where there is evidence for a genetic basis), and cryptogenic (where there is no evidence for either). Within these categories, other clinical features such as age of onset are used to differentiate the epilepsies, and the classification is comprehensive but inevitably complex (Table VI).

The different epilepsy syndromes may differ in their associated clinical features, their natural histories and prognosis, their response to standard medical and surgical treatment, and their relapse rate after withdrawal

**TABLE IV  Symptoms of Partial Seizures with Different Focus Localization[a]**

| | |
|---|---|
| Frontal lobe seizures | |
| Usual pattern | Simple partial, complex partial, with or without secondary generalization |
| Symptoms | Adversive movements of the head; prominent motor manifestations (especially in the legs); motor automatisms; vocalization; bizarre attacks, which often appear hysterical |
| Frequency | Several times daily |
| Duration | Seconds; sudden beginning despite impairment of consciousness; very little postictal confusion |
| Parietal lobe seizures | |
| Usual pattern | Simple partial with or without secondary generalization |
| Symptoms | Sensory or motor symptoms (Jacksonian march); rarely painful sensations |
| Temporal lobe seizures | |
| Usual pattern | Simple partial (aura in 80%), complex partial, with or without secondary generalization |
| Symptoms | Epigastric rising sensation; olfactory and gustatory hallucinations; deja vu, jamais vu, oral and other primitive automatisms; visual hallucinations; autonomic changes |
| Frequency | Typically several per month, rarely many a day |
| Duration | Typically 2–5 min postictal confusion, gradual recovery |
| Occipital lobe seizures | |
| Usual pattern | Simple partial with or without secondary generalization |
| Symptoms | Unformed simple visual phenomena (sparks, flashes, phosphenes), eye deviation, forced blinking, oculoclonic activity |

[a]Gram (1990), reprinted with permission.

TABLE V  Generalized Epileptic Seizures[a]

| | |
|---|---|
| **Absences** | |
| Age of onset | Children and adolescents |
| Duration | <30 s |
| Symptoms | Sudden loss of consciousness accompanied by staring and blinking; long absences may be accompanied by automatisms, e.g., lip smacking, chewing, fiddling, fumbling; immediate regain of consciousness |
| EEG | Bilateral regular 3 (2–4) Hz spike waves |
| **Myoclonic seizures** | |
| Age of onset | Children and juveniles |
| Duration | 1–5 s |
| Symptoms | Brief jerks in arms or legs; often occur in a series |
| EEG | Polyspike/waves, spike/waves, or sharp and slow waves |
| **Atonic seizures(astatic)** | |
| Age of onset | Infants and children |
| Duration | Few seconds |
| Symptoms | Sudden loss of muscle tone often causing severe head injuries (fall) |
| EEG | Polyspike/waves, flattening, or low-voltage fast activity during the seizures |
| **Primary generalized tonic–clonic seizures (grand mal)** | |
| Age of onset | Any age |
| Duration | 1–3 min |
| Symptoms | Initial cry (sometimes); loss of muscle tone (fall); respiratory arrest; cyanosis; tonic convulsions; clonic convulsions; possible tongue biting and urinary incontinence; relaxation followed by a deep sleep |
| EEG | Bilateral regular synchronous spike/wave-discharges, ictal EEG often obscured by muscular artifacts |

[a]Gram (1990), reprinted with permission.

of medication in patients with complete seizure control. For instance, the idiopathic generalized epilepsy syndromes respond specifically to treatment with valproate with complete control of seizures, but the response to drug treatment of similar seizure types (e.g., myoclonic seizures, absence seizures) in other epilepsy syndromes is not nearly as good. The differing syndromes also show different susceptibilities to external provocation of seizures (photic stimulation or insomnia).

Undoubtedly, as our understanding of the genetics of epilepsy increases, the syndromic classification of the epilepsies is likely to change to incorporate these advances and a revised classification system is very likely in the near future.

## FREQUENCY AND ETIOLOGY OF EPILEPSY

The annual incidence of epilepsy is between 50 and 120 cases per 100,000 persons, and the prevalence between 5 and 10/1000 persons in the general population. Its incidence may be somewhat higher in developing countries but most prevalence estimates are very similar in developed as well as developing countries (Sander and Shorvon, 1996). In a general population of 1 million persons, there are about 5000 persons with active epilepsy and about 500 new cases each year. Of the active cases, about 20% have frequent seizures (Hart and Shorvon 1995a, b), and about 30% have some additional neurological or psychiatric handicap. Age

has a marked effect on incidence, and the age-related incidence of epilepsy follows a U-shaped pattern, falling from the highest incidence in the first 2 years to low levels in mid-life and rising again to peak at 70 or 80 years (Hauser et al., 1983; MacDonald et al., 2000). Epilepsy has a wide range of causes, and indeed almost all gray-matter diseases can result in seizures (Tables VII and VIII). The most important factor influencing the range of causes is age. In the neonatal period the commonest and most important causes are ischemic–hypoxic encephalopathy, cerebral infection, hypoglycemia, hypocalcemia, and other metabolic, genetic, or congenital conditions. In later childhood the commonest causes are genetic, metabolic, or congenital, and the majority of patients in adulthood and late adulthood have partial epilepsy secondary to stroke, brain tumor, trauma, alcohol abuse, or acute infectious illness.

## DIFFERENTIAL DIAGNOSIS OF EPILEPSY

Epilepsy can be confused with any other transient alteration of neurological function (Table IX) and misdiagnosis is often a result of incomplete history taking as well as misinterpretation of EEG investigations. Up to a quarter of patients seen in secondary and tertiary care clinics with possible epilepsy may not have true seizure disorders (Smith et al., 1999). The commonest problem is their differentiation from syncope and psychogenic attacks. Conversely, epilepsy is diagnosed infrequently (<2%) in patients suspected of having

TABLE VI  International Classification of Epilepsies and Epilepsy Syndromes and Related Seizure Disorders[a]

1. Localization-related (local, focal, partial) epilepsies and syndromes
    1.1. Idiopathic (with age-related onset)
        Benign childhood epilepsy with centro-temporal spikes
        Childhood epilepsy with occipital paroxysms
        Primary reading epilepsy
    1.2. Symptomatic
        Chronic progressive epilepsia partialis continua
        Syndromes characterized by seizures with specific modes of precipitation
        Temporal lobe epilepsies
        Frontal lobe epilepsies
        Parietal lobe epilepsies
        Occipital lobe epilepsies
    1.3. Cryptogenic
2. Generalized epilepsies and syndromes
    2.1. Idiopathic (with age-related onset)
        Benign neonatal familial convulsions
        Benign myoclonic epilepsy in infancy
        Childhood absence epilepsy
        Juvenile absence epilepsy
        Juvenile myoclonic epilepsy
        Epilepsy with grand mal seizures
        Other generalized idiopathic epilepsies
        Epilepsies with seizures precipitated by specific modes of activation
    2.2. Cryptogenic or symptomatic
        West syndrome (infantile spasms)
        Lennox–Gastaut syndrome
        Epilepsy with myoclonic–astatic seizures
        Epilepsy with myoclonic absences
    2.3. Symptomatic
        2.3.1. Nonspecific etiology
            Early myoclonic encephalopathy
            Early infantile epileptic encephalopathy with burst suppression
            Other symptomatic generalized epilepsies
        2.3.2. Specific syndromes
            Epilepsy complicating other disease states
3. Epilepsies and syndrome undetermined whether focal or generalized
    3.1. With both generalized and focal seizures
        Neonatal seizures
        Severe myoclonic epilepsy of infancy
        Epilepsy with continuous spike waves during slow wave sleep
        Acquired epileptic aphasia (Landau-Kleffner syndrome)
        Other undetermined epilepsies
    3.2. Without unequivocal generalized or focal features
4. Special syndromes
    4.1. Situation-related seizures
        Febrile convulsions
        Isolated seizures or isolated status epilepticus
        Seizures occurring only with acute metabolic or toxic events

[a]Commission on Classification and Terminology of the International League against Epilepsy (1989).

recurrent syncope in tertiary referral autonomic clinics (Mathias *et al.*, 2001).

## THE INVESTIGATION OF EPILEPSY

Epilepsy is a fundamentally clinical diagnosis but the presence of epileptiform discharges on EEG is useful supportive information. The role of EEG, however, should be to answer specific diagnostic, classificatory and therapy decision-related questions such as differentiating between partial and primary generalized epilepsy, to look for evidence of photosensitivity or non-convulsive status epilepticus, etc; an equivocal report often leads to difficulties. For example, only 40% of patients with epilepsy have an abnormal interictal, wake

TABLE VII     Causes of Epilepsy in Children and Adults

Genetic and congenital abnormalities
Anoxic and perinatal injury
Trauma
Cerebral infection
Cerebrovascular disease
Cerebral tumor
Cerebral degenerative disorders
Cerebral inflammatory and immunological disorders
Toxins and drug withdrawal
Metabolic disorders

EEG. This figure improves to 80% with sleep EEGs and sleep deprivation only marginally improves the yield.

Bilateral synchronous spike/wave discharges are the hallmark of idiopathic generalized epilepsy and help differentiate these generalized absence attacks from the absence of complex partial seizures. Similarly, bilateral synchronous spike/wave discharges help in the differentiation of the syndrome of grand mal on awakening (a form of idiopathic generalized epilepsy) from cryptogenic or partial onset tonic–clonic seizures. Focal paroxysmal EEG findings are occasionally found in patients without a history of epilepsy and occur in the interictal scalp recordings of less than half of those with partial epilepsy. The diagnosis of symptomatic generalized epilepsy can be supported by bilateral irregular slow spike/wave discharges or hypsarrhythmia in infants with infantile spasms. For screening of brain lesions, a focal slow-wave EEG may be useful, although modern neuro-imaging has largely superseded EEG in this role (Bronen, 1992). The EEG plays a minor role in defining prognosis or response to drug treatment (except perhaps for the quantification of spike/wave discharges over a prolonged period) and should not be routinely used for these purposes.

MRI scanning is superior to CT for the diagnosis of most structural causes of epilepsy. In particular, MRI can demonstrate hippocampal sclerosis, cortical dysgenesis, and small foreign tissue lesions (e.g., cavernous hemangiomata), which are largely invisible on CT. Hippocampal sclerosis is a common cause of surgically treatable temporal lobe epilepsy. The affected

TABLE VIII     Causes of Neonatal Seizures

Ischemic–hypoxic encephalopathy
Intracerebral hemorrhage
Cerebral infarction
Cerebral infections
Hypoglycemia
Hypocalcemia
Other metabolic disorders
Other genetic and hereditary causes
Toxic exposure and drug withdrawal
Congenital cerebral malformations

TABLE IX     Attacks That May Be Mistaken for Epilepsy

Syncope
  *Reflex syncope*
    Psychogenic
    Micturition syncope
    Cough syncope
    Valsalva maneuver
  *Cardiac syncope*
    Dysrhythmias (heart block, tachycardias)
    Valvular disease (especially aortic stenosis)
    Cardiomyopathies
    Shunts
    Displaced cardiac pacemaker
  *Perfusion failure*
    Hypovolemia
    Autonomic failure
Psychogenic attacks
  Pseudoseizures
  Panic attacks
  Hyperventilation
  Anxiety with derealization/depersonalization
  Tantrum
Transient ischemic attacks
Migraine
Narcolepsy, cataplexy
Hypoglycemia
Drop attacks in the elderly
Pheochromocytoma
Deliberate simulation

hippocampus is atrophic on T1-weighted images and hyperintense on T2-weighted images. Volumetric T1-weighted MRI, with thin-slices and on-line reformatting, and with three-dimensional rendering, is the preferable method for detecting cortical dysgenesis. CT scan remains useful only for emergency evaluation in a restless patient (and for young children, where imaging under an anesthetic is not indicated), where MR is not available for initial screening, and as a confirmation that MR signal void is due to calcium in small cortical lesions. Functional imaging with SPECT or PET is occasionally a useful complementary investigation for the presurgical evaluation of patients with contradictory or noncontributory MRI and EEG findings. EEG triggered functional MRI which combines the spatial and temporal attributes of both investigations is currently at a developmental stage and may enhance the currently available range of noninvasive investigations (Krakow *et al.*, 1999).

Serum antiepileptic drug monitoring is useful in assessing toxicity, compliance, and efficacy and is limited to only a few antiepileptic drugs (Glauser and Pippenger, 2000). This is an often over-requested investigation, however, and only a quarter of requests may be appropriate (Schoenenberger *et al.*, 1995). Given the expense and inconvenience of serum level monitoring, it is remarkable that no consistent evidence demonstrates an improved therapeutic outcome from its use

(Jannuzzi *et al.*, 2000). Properly used serum level monitoring, however, is useful for the detection of poor drug compliance, for the evaluation of side effects, for the evaluation of drug interactions, and for cases where pharmacokinetic changes are extreme or unusual. Serum level monitoring is most useful in patients treated with phenytoin, as the kinetics of phenytoin are non-linear, the serum level/dose relationship is unpredictable, and levels are often changed by alterations of concomitant medication. Serum level monitoring should not replace clinical judgment, however, and dose titration and reduction should be based primarily on the clinical response, not laboratory data. A common example of the misuse of serum level monitoring is the reduction of dosage in a patient who has improved seizure control and no signs of clinical side effects, although his serum drug concentrations are in the so-called toxic range. A second example is, conversely, the increase of dosage in a patient with already complete seizure control, because the clinical chemist's so-called therapeutic range has not been reached.

## THE NATURAL COURSE AND PROGNOSIS OF EPILEPSY

The natural history of untreated epilepsy is difficult to assess because effective drug treatment has been available for over 100 years, since the introduction of bromides in 1857. Nevertheless, the widely shared view that epilepsy is a chronic and uncontrollable condition in over 80% of all cases, as concluded by Rodin in 1968 based on hospital and institutional studies, has now been revised because of the evidence from longitudinal population-based studies of newly diagnosed patients with epilepsy (Sander, 1993). Such studies avoid the selection bias of hospital and clinic investigations of chronic epilepsy.

The U.K. National General Practice Study of Epilepsy (NGPSE), which is the largest population-based, prospective study of newly diagnosed epilepsy, found that of 792 patients, 71% achieved a 5-year remission within 9 years of the initial seizure (Cockerell *et al.*, 1995, 1997). Therefore, the long-term prognosis for most patients with epilepsy is good. Interestingly, multivariate analysis of these patients showed that the single most important baseline factor in determining prognosis was the number of seizures in the first 6 months after enrolment to the study. The chance of entering 1-year remission and 5-year remission from seizures was 95 and 47%, respectively, for patients who had 2 seizures in the first 6 months compared with 75 and 24%, respectively, in those who had 10 or more seizures (MacDonald *et al.*, 2000). The subgroup of patients who will have chronic, difficult-to-treat epilepsy can be identified early in the course of the epilepsy, and appropriate medical or surgical treatment should in general be applied early in these cases.

In one partly retrospective study, after 20 years' follow-up, as many as 65% of newly diagnosed patients with epilepsy were without seizures for at least 5 years. About 50% of these patients successfully stopped drug treatment, and a further 20% were seizure-free on medication. Only 30% continued to have seizures (Annegers *et al.*, 1979). In another population-based study, only 19% of patients reported a seizure in the preceding 2 years when studied 15 years after the onset of epilepsy (Goodridge and Shorvon, 1983). Only 38% of patients were still taking antiepileptic drugs. In both studies, 42% (Annegers *et al.*, 1979) and 47% (Goodridge and Shorvon, 1983) of the patients had become seizure-free within 1 year of diagnosis.

The epilepsy syndromes with the best prognosis are those with idiopathic partial epilepsies such as patients with benign rolandic epilepsy, with rare generalized tonic–clonic seizures, or with primary generalized epilepsies. Patients with the poorest prognosis are those with evidence of focal or diffuse cerebral disease, often associated with cognitive or behavioral disturbances, early onset of epilepsy, partial or mixed seizure types, progressive neurological or cerebral disorders, and those with the symptomatic generalized epilepsy syndromes such as Lennox–Gastaut syndrome or West syndrome. The duration of uncontrolled epilepsy is critical; the longer the seizures continue after the onset of the drug therapy, the worse is the ultimate prognosis. The number of seizures prior to onset of treatment, however, appears not to be critical for treatment response (Placencia *et al.*, 1993).

Recurrence after a first seizure is common. In the NGPSE, 60% of patients had a recurrence within the first 6 months after a first seizure, 67% within 12 months, and 78% within 36 months (Hart *et al.*, 1990). This is higher than previously thought, largely because earlier studies ascertained patients only after a significant interval from the first seizure, and most patients will already have had a second seizure within a few weeks of the first attack. The fact that the recurrence rate is high renders rather invalid the traditional distinction between a single seizure (not epilepsy) and two or more seizures (epilepsy).

The overall mortality of people with epilepsy is about two to three times that of persons without epilepsy (Lhatoo *et al.*, 2001). This mortality is partly due to the influence of the underlying condition, but may also be directly related to the epilepsy, particularly in patients with chronic epilepsy. Some deaths occur in accidents or in status, but about one-third of recorded deaths in chronic epilepsy are sudden without obvious explanation. Most of these deaths (designated SUDEP, sudden unexpected death in epilepsy) occur in young adults (Jay and Leestma, 1981). The estimated incidence of sudden death is about 1:200–1:1000 patients with epilepsy, with the highest rate in young adults, those with frequent seizures, those with tonic–clonic seizures, and possibly those with learning disabilities (Nashef et al., 1995). Other suspected risk factors for SUDEP include male gender, refractory seizures, remote symptomatic epilepsy, treatment noncompliance, recent head injury, and alcohol, although most of these were inferred from studies of selected populations without adequate controls (Lhatoo et al., 1999). A recent case–control study suggested that relative risk for SUDEP was increased to 10.16 (95%CI 2.94, 35.18) in patients with more than 50 seizures per year compared with patients with two or less seizures per year (Nilsson et al., 1999). The exact pathogenetic mechanisms of SUDEP are unknown, however, and hypotheses include obstructive apnea, central apnea, and cardiac dysrhythmias. Thus seizure control as well as adequate supervision of the patients with refractory epilepsy appear likely to be important preventive measures (Langan et al., 2000).

## THE TREATMENT OF EPILEPSY

A number of mechanisms may contribute to epileptogenicity in the human brain (Schwartzkroin PA 1994), but the pathophysiology of many specific epilepsy syndromes is currently not well understood. Despite general agreement that antiepileptic drugs such as valproate and ethosuximide, which are effective against absence seizures, share a common molecular action (that is, blockage of calcium ion currents as assessed experimentally in thalamic neurons (Rogawski and Porter, 1990)), drugs that are effective against partial seizures show a variety of different molecular actions. These include blockage of sodium ion currents in neuronal membranes by phenytoin, carbamazepine, valproate, lamotrigine, oxcarbazepine, and topiramate and enhancement of GABA-ergic inhibition by vigabatrin, valproate, benzodiazepines, phenobarbital, felbamate, and gabapentin. Several clinically effective antiepileptic drugs cause a reduction in glutamatergic excitatory neurotransmission, for example, lamotrigine and felbamate. As a consequence the choice of an individual antiepileptic drug is usually not determined by the putative mechanism of action. Instead, efficacy against specific seizures or syndromes determines the choice of drug for first-line single-drug treatment and adjunctive therapy in refractory seizures or syndromes. Adequate evidence for efficacy can be provided only by controlled comparative clinical trials. In this chapter, only agents that have been marketed will be discussed, and drugs commonly used but without adequate evidence for efficacy will be characterized as such.

### Starting Treatment

Antiepileptic drug treatment has been advocated as prophylaxis before the onset of seizures in patients with high risk of developing epilepsy; for instance, after traumatic brain injury or bacterial brain abscess. There is, however, no firm evidence that this prevents the later occurrence of epilepsy nor that it renders the subsequent epilepsy more treatable (Temkin et al., 1999; Temkin, 2001).

Drug treatment following a single tonic–clonic seizure is effective in preventing relapse (Musicco et al., 1993) but has not been shown to affect prognosis (Musicco et al., 1997). Nevertheless, drug treatment is often recommended following a single seizure, especially in high-risk categories such as symptomatic epilepsy. Practice varies in this situation, but clearly, preventing a further seizure provides psychosocial as well as medical benefits in certain cases. Moreover, given the high recurrence rate of single seizures, it is probably entirely reasonable to start treatment.

It is commonly recommended that drug treatment be started when two or more tonic–clonic seizures have been clinically diagnosed. This clinical practice is supported by data (albeit in selected populations and uncontrolled) that the recurrence rate increases to 60% in patients with two seizures compared to 27% in patients with one seizure (Hauser and Kurland, 1975). Factors that may modify the usual practice are precipitation by alcohol, drugs, reflex stimuli, the probability of poor drug compliance, and the attitude of the patient or carers. Prospective randomized trials are needed to compare early and delayed treatment in patients with untreated seizures.

### Choice of First-Line Antiepileptic Drug

The choice of the appropriate drug is based in practice largely on seizure type, efficacy, toxicity, availability, handling of the drug, and despite calls for rationalization of therapy, fashion, or tradition. Valproate appears to be most effective for idiopathic generalized epilepsies, especially absence seizures or juvenile myoclonic epilepsy and, despite the lack of comparative treatment data, for the tonic clonic seizures of primary generalized epilepsy (Table X). Partial seizures are probably best treated by carbamazepine (Marson et al., 2000). Comparative toxicity is an important factor determining the choice of drugs. Most

**TABLE X**    Choice of Antiepileptic Drugs

|  | First-line drugs | Second-line drugs |
|---|---|---|
| Generalized epilepsy | | |
| Idiopathic | | |
| Simple absence | VPA | ESM, LTG, BZPs |
| Juvenile myoclonic | VPA | LTG, PHB |
| Tonic clonic | VPA | LTG, TPM, LVT, CBZ, PHB, PHT, VGB |
| Symptomatic | | |
| Partial epilepsy | CBZ, VPA | LTG, GBP, TOP, LVT, OXC, TGB, VGB |
| Generalised epilepsy | VPA, CBZ | LTG, TOP, LVT, OXC, VGB |
| Unclassified epilepsy | VPA, CBZ | LTG, GBP, TOP, LVT, TGB, OXC, VGB |

*Note.* VPA, sodium valproate; CBZ, carbamazepine; OXC, oxcarbazepine; PHT, phenytoin; PHB, phenobarbital; ESM, ethosuximide; BZPs, benzodiazepines; LTG, lamotrigine; GBP, gabapentin; TOP, topiramate; LVT, levetiracetam; vigabatrin, VGB; TGB, tiagabine.

authorities would agree that carbamazepine is easier to handle and causes fewer cognitive disturbances than phenytoin, which is compromised by its nonlinear kinetics and dose-related toxicity, requiring plasma drug monitoring. For most patients the choice is between valproate, lamotrigine, carbamazepine, and phenytoin. Once the drug is chosen, its daily dose should be increased in increments until complete seizure control is achieved or side effects preclude further dose increments. In the majority of patients complete seizure control will be achieved on such monotherapy. Whether such therapy is curative or simply suppressive (controlling seizures until natural remission occurs) is uncertain.

## Stopping Antiepileptic Drugs

Up to 70% of patients in whom epilepsy is diagnosed will enter a prolonged remission of their seizures within 1 or 2 years of starting treatment. In such patients therapy may be safely discontinued after 2–3 years. The decision to withdraw therapy is a personal one, and relapse after withdrawal occurs in a substantial minority of cases (Lhatoo and Sander, 2000). For some patients it is of overriding importance to remain free of seizures; for others concern about the consequences of long-term antiepileptic drug treatment, for instance, for contraception and pregnancy, will influence the decision. Concern about the subtle cognitive impairment induced by chronic antiepileptic drug exposure especially in children has recently been emphasized, and a case can be made for early discontinuation of therapy

**TABLE XI**    Stopping Antiepileptic Treatment

| Factors associated with a low relapse rate (20–60%) | Factors associated with a high relapse rate (>60%) |
|---|---|
| Childhood epilepsy | Late-onset epilepsy |
| Absence seizures alone | Partial epilepsy |
| Few generalized tonic–clonic seizures | Cerebral disorder |
| No cerebral disorder | Long duration of uncontrolled epilepsy |
| Short duration of epilepsy | Abnormal EEG during dose reduction |
| Normal EEG | Juvenile myoclonic epilepsy |
| | More than one type of seizures |
| | Poor compliance |

in some patients. However, in one study, a significant improvement attributable to antiepileptic drug withdrawal has been reported in only one of a battery of cognitive tests (psychomotor speed), suggesting that the impact of antiepileptic drug treatment on higher-order cognitive function is rather limited (Aldenkamp *et al.*, 1993). In a controlled study, the features found to influence the risk of recurrence are shown in Table XI (Medical Research Council Antiepileptic Drug Withdrawal Study Group 1991). The epilepsy syndrome is an important factor that determines the rate of relapse after withdrawal. Between 75 and 100% of patients with juvenile myoclonic epilepsy withdrawn from treatment develop a relapse (Shinnar *et al.*, 1994; Janz and Durner, 1997), whereas in childhood absence epilepsy, the figure is much smaller. The time taken to achieve complete seizure control, adversely affects the rate of relapse on withdrawal. Conversely, long periods of remission are associated with smaller rates of subsequent relapse on cessation of therapy.

## Established and Novel Antiepileptic Drugs

Antiepileptic drugs are universally preferred treatment for the great majority of patients with epilepsy. Initial single first-line drug therapy will lead to remission in approximately 70% (Goodridge and Shorvon, 1983). In addition, single or multiple drug therapy with first- or second-line agents including new antiepileptic drugs are the mainstay of treatment for the remaining 30% of patients with epilepsy unresponsive to first-line drugs. New drugs, however, produce long-term seizure remission in less than 15% of patients with chronic epilepsy and less than one-third of such patients continue with new drugs such as lamotrigine, topiramate, gabapentin, and levetiracetam beyond 5 years despite lack of seizure remission (Krakow *et al.*, 2001; Lhatoo *et al.*, 2000). Surgical treatment is valuable but suitable for only a small proportion of patients with drug-resistant epilepsy.

**TABLE XII    Pharmacokinetics of Antiepileptic Drugs**

| Drug | Absorption (bioavail-ability) | Distribution volume (L/kg) | Protein binding (% bound) | Elimination half-life (h) | Route of elimination | Comments |
|---|---|---|---|---|---|---|
| Carbamazepine | Slow (75–85%) | 0.8–1.6 | 70–78 (single) 8–24 (chronic) | 24–45 | Hepatic metabolism; CBZ 10, 11 epoxide active metabolite | Enzyme inducer; auto-induction of metabolism |
| Clobazam | Rapid (80–90%) | 0.7–1.6 | 87–90 | 10–30 | Hepatic metabolism; -des-methyl -clobazam active metabolite | Tolerance; withdrawal exacerbations |
| Clonazepam | Rapid (80–90%) | 2.1–4.3 | 80–90 | 30–40 | Hepatic metabolism | Sedative; tolerance; withdrawal exacerbations |
| Ethosuximide | Rapid (90–95%) | 0.6–0.9 | 0 | 20–60 | Hepatic metabolism, 25% excreted unchanged | More rapid clearance in children |
| Phenobarbital | Slow (95–100%) | 0.51–0.57 | 48–54 | 72–144 | Hepatic metabolism, 25% excreted unchanged | Enzyme inducer; sedative |
| Phenytoin | Slow (85–95%) | 0.5–0.7 | 90–93 | 9–40 | Saturable, hepatic metabolism | Enzyme inducer; elimination half-life concentration dependent |
| Primidone | Rapid (90–100%) | 0.4–0.8 | 20–30 | 4–12 | Hepatic metabolism, active metabolites; 40% excreted unchanged | Enzyme inducer; phenobarbital a major metabolite |
| Valproate | Rapid (100%) | 0.09–0.17 | 88–92 | 7–17 | Hepatic metabolism; active metabolites | Enzyme inhibitor; concentration-dependent protein binding |

**TABLE XIII    Prescribing Antiepileptic Drugs in Adults**

| Drug | Indications | Starting dose | Common daily dose | Maintenance range | Dosage interval | Target range |
|---|---|---|---|---|---|---|
| Carbamazepine | Partial and generalized tonic–clonic seizures | 100–200 mg twice a day | 800 mg | 400–2000 mg | Two to three times a day | 4–10 mg/L; (17–42 mmol/L) |
| Clobazam | Adjunctive therapy for refractory epilepsy | 10 mg at night | 20 mg | 10–40 mg | Once or twice a day | None |
| Clonazepam | Myoclonic and generalized tonic–clonic seizures; Status epilepticus | 0.5–1 mg | 4 mg | 2–8 mg | Once or twice a day | None |
| Ethosuximide | Absence seizures | 500 mg | 1000 mg | 500–2000 mg | Once or twice a day | 40–100 mg/L[a]; (283–708 μmol/L) |
| Phenobarbital | Partial and generalized tonic–clonic, clonic, and tonic seizures; prophylaxis of febrile convulsions; resistant status epilepticus | 30–60 mg | 120 mg | 60–240 mg | Once or twice a day | 10–40 mg/L[a]; (40–172 μmol/L) |
| Phenytoin | Partial and generalized tonic–clonic seizures; status epilepticus | 100–200 mg | 300 mg | 100–700 mg | Once or twice a day | 10–20 mg/L; (49–80 μmol/L) |
| Primidone | Partial and generalized tonic–clonic seizures | 125–250 mg | 500 mg | 250–1500 mg | Once or twice a day | Monitor derived phenobarbital level |
| Valproate | Primary generalized epilepsies; partial seizures; prophylaxis of febrile convulsions | 200 mg twice a day | 1000 mg | 400–3000 mg | Twice a day | 50–100 mg/L[b]; (347–693 μmol/L) |

[a]Target range is unhelpful.
[b]Twice a day with controlled-release formulation.

The optimum use of antiepileptic drugs is governed by an appreciation of the clinical pharmacokinetics of antiepileptic drugs (Table XII). The indications and the starting maintenance dose are summarized in Table XIII. The comparative merit of an antiepileptic drug is also determined by the side effect profile (Table XIV) and other features (Table XV).

## Benzodiazepines

The relative frequency of side effects and the development of tolerance to benzodiazepines such as diazepam or clonazepam (Ko *et al.*, 1997), render them unsuitable as regularly prescribed oral treatment for epilepsy except as a last resort. The main disadvantages are sedation, cognitive impairment, muscle fatigue, ataxia and visual disturbances. Withdrawal symptoms occur with both clonazepam and clobazam and can result not only in seizure exacerbation, but frank psychiatric disturbances such as anxiety and psychosis. Clobazam, however, can be a useful adjunctive drug for patients with refractory symptomatic partial seizures and its anxiolytic properties are of added benefit. It may be particularly useful in nonconvulsive status and benign childhood partial epilepsies. Hypersensitivity reactions do not occur. Intermittent administration of clobazam may prevent or reverse tolerance.

**TABLE XIV**   Adverse Reactions of Marketed Antiepileptic Drugs[a]

| Carbamazepine | Valproate | Phenytoin | Phenobarbital | Primidone | Ethosuximide | Clonazepam/ clobazam |
|---|---|---|---|---|---|---|
| *Dose-related effects* | | | | | | |
| Diplopia | Anorexia | Anorexia | Fatigue | Nausea | Anorexia | Fatigue |
| Dizziness | Dyspepsia | Dyspepsia | Listlessness | Vomiting | Nausea | Dizziness |
| Nystagmus | Nausea | Nausea | Tiredness | Drowsiness | Vomiting | Nystagmus |
| Drowsiness | Vomiting | Vomiting | Depression | Weakness | Singultus | Ataxia |
| Ataxia | Hair loss | Aggression | Insomnia | Dizziness | Agitation | Irritability[b] |
| Nausea | Rash | Ataxia | Distractability | Diplopia | Drowsiness | Aggression |
| Hyponatremia | Peripheral edema | Cognitive | Aggression | Nystagmus | Headache | Hyperkinesia |
| Hypocalcemia | Weight gain | impairment | Poor memory | Ataxia | Lethargy | Hypersalivation |
| Orofacial | Drowsiness | Depression | Decreased libido | Personality change | Parkinsonism | Bronchorrhoea |
| dyskinesia | Tremor | Drowsiness | Impotence | Neonatal | Psychotic episode | Weight gain |
| Cardiac | | Headache | Neonatal | hemorrhage | | Muscle |
| arrhythmia | | Nystagmus | hemorrhage | Decreased libido | | weakness |
| | | Gum hypertrophy | Hypocalcemia | Impotence | | Psychotic |
| | | Coarse facies | Osteomalacia | Hypocalcemia | | episode |
| | | Hirsutism | | Osteomalacia | | |
| | | Megaloblastic | | Megaloblastic | | |
| | | anemia | | anemia | | |
| | | Hyperglycemia | | Neonatal | | |
| | | Hypocalcemia | | hemorrhage | | |
| | | Osteomalacia | | | | |
| | | Neonatal | | | | |
| | | hemorrhage | | | | |
| *Idiosyncratic effects* | | | | | | |
| Agranulocytosis | Acute pancreatitis | Blood dyscrasias | Maculopapular | Rash | Rash | |
| Aplastic anemia | Hepatoxicity | Lupuslike | rash | Agranulocytosis | Erythema | |
| Hepatotoxicity | Thrombocytopenia | syndrome | Exfoliation | Thrombocytopenia | multiforme | |
| Photosensitivity | Hyperammonemia | Reduced serum | Toxic epidermal | Lupuslike | Stevens–Johnson | |
| Stevens–Johnson | Stupor | IgA | necrolysis | syndrome | syndrome | |
| syndrome | Encephalopathy | Pseudolymphoma | Hepatotoxicity | Frozen shoulder | Lupuslike | |
| Lupuslike | Teratogenicity | Peripheral | Dupytren's | Dupytren's | syndrome | |
| syndrome | | neuropathy | contracture | contracture | Agranulocytosis | |
| Morbilliform rash | | Rash | Frozen shoulder | Teratogenicity | Aplastic anemia | |
| Thrombocytopenia | | Stevens–Johnson | Teratogenicity | | | |
| Pseudolymphoma | | syndrome | | | | |
| | | Hepatoxicity | | | | |
| | | Teratogenicity | | | | |

[a]Brodie (1990), with permission.
[b]In children.

**TABLE XV  Strengths and Weaknesses of Marketed Antiepileptic Drugs[a]**

| | Strengths | Weaknesses |
|---|---|---|
| Carbamazepine | Good first-line efficacy for tonic–clonic and partial seizures | Not effective for absence or myoclonic seizures (may aggravate); rashes in 5–10%; drug interaction, enzyme inducer; CNS side effects |
| Clobazam/ Clonazepam | Second-line efficacy as add-on drug for primary and secondary tonic–clonic seizures, partial seizures and myoclonic seizures; no enzyme induction; no drug monitoring | Tolerance; withdrawal; cognitive disturbances; depression |
| Ethosuximide | Good first-line efficacy for absence seizures<br>No drug monitoring<br>No enzyme induction | No efficacy for tonic–clonic seizures or partial seizures; gastrointestinal side effects; psychotic reactions |
| Felbamate | Second-line efficacy as a single or add-on drug for refractory partial seizures<br><br>Second-line efficacy as add-on drug for refractory Lennox-Gastaut syndrome | Aplastic anemia and hepatic failure (both side effects have severely limited suitability in every day practice); CNS-side effects insomnia; headache; weight loss; relevant drug interaction |
| Gabapentin | Second-line efficacy as add-on drug for refractory partial seizures<br>No drug monitoring<br>No enzyme induction | No efficacy for absence or myoclonic seizures; CNS side effects; potential human carcinogen |
| Lamotrigine | Efficacy as single or add-on drug for refractory and newly diagnosed partial and generalized seizures | Rashes in 5–10%; relevant drug interaction |
| Levetiracetam | Second-line efficacy as add-on drug for refractory partial Seizures<br>No drug monitoring<br>No enzyme induction | Newest marketed drug; limited data on use in generalised epilepsies |
| Oxcarbazepine | Second-line efficacy as single or add-on drug for refractory and newly diagnosed partial seizures | No efficacy for absence or myoclonic seizures CNS side effects; hyponatremia; drug interaction |
| Phenobarbital/ Primidone | Efficacy for tonic–clonic partial and myoclonic seizures<br>Low cost<br>No drug monitoring | Cognitive and behavioral side effects; enzyme inducer; drug interaction |
| Phenytoin | Good first-line efficacy for tonic–clonic and partial seizures | Not effective for absence or mycolonic seizures<br>Rashes in 5–10%; enzyme inducer; CNS side effects; cognitive disorders, depression; nonlinear kinetics may require plasma monitoring; cosmetic changes |
| Tiagabine | Second-line efficacy as add-on drug for refractory partial seizures<br>No drug monitoring<br>No enzyme induction | Limited data on use in generalised epilepsies |
| Topiramate | Highly effective against partial and secondarily generalized seizures;<br>Excellent pharmacokinetics<br>No enzyme induction<br>No drug monitoring | Tends to cause neurological side effects at high doses and if titrated too rapidly;<br>Renal calculi<br>Teratogenicity in animals |
| Valproate | Good first-line efficacy for all types of seizures<br>No enzyme induction<br>Seldom rashes | Hepatotoxicity and teratogenicity<br>CNS side effects; weight gain |
| Vigabatrin | Efficacious as add-on drug for refractory partial seizures<br>No enzyme induction<br>No drug monitoring<br>Effective in West's syndrome | No efficacy for absence or myoclonic seizures<br>CNS side effects depression, psychotic episodes; weight gain; tolerance<br>Visual field defects most common serious side effect limiting usage in adult patients |
| Zonisamide | Good second-line efficacy as add-on drug for refractory partial seizures and refractory myoclonic seizures | CNS side effects; sedation; renal calculi |

[a]Brodie (1990), with permission.

## Carbamazepine

Carbamazepine is a first-line drug for the initial treatment of children and adults with tonic–clonic seizures and partial seizures (Mattson, 1997). Generally well tolerated, it is highly effective and easy to use. The strengths and weaknesses of carbamazepine are reviewed in Table XV. Carbamazepine must be started at a low dose (100 mg in an adult) to keep development of transient neurotoxicity at a minimum. The dose can be increased every 4 weeks to a maintenance level that controls the seizures. Even when taking this cautious approach, some patients develop diplopia, nausea, dizziness, and headache on initiation of therapy, although these side effects usually resolve. The introduction of controlled-release preparations reduced the toxicity associated with peak plasma concentrations of carbamazepine (Aldenkamp et al., 1987). Fluid retention may limit the use of carbamazepine in the elderly or those with cardiac failure. In addition, carbamazepine may aggravate bradycardia in patients with heart disease. Hyponatremia is rarely symptomatic but occasionally leads to confusion, peripheral edema, and an increase in the number of seizures (Brodie, 1990).

Unfortunately, carbamazepine is a potent enzyme inducer, reducing the effectiveness of several drugs, such as oral contraceptives or steroids given for treatment of asthma, haloperidol, theophylline, and warfarin. Conversely, other drugs inhibit its metabolism with resultant neurotoxicity; these include cimetidine, danazol, dextropropoxyphene, diltiazem, erythromycin, isoniazid, verapamil, and viloxazine. Interactions of carbamazepine with other antiepileptic drugs are common. Carbamazepine increases the clearance of ethosuximide, sodium valproate, clobazam, clonazepam, lamotrigine, vigabatrin, felbamate, and gabapentin. Inhibition or enzyme induction is seen with phenobarbital, phenytoin, or primidone with usually small but unpredictable changes in the plasma concentration of these drugs.

## Valproate

Valproate is a first-line drug for the initial treatment of patients with childhood absence seizures, juvenile myoclonic epilepsy and the tonic clonic seizures of primary generalized epilepsy (Rowan, 1997). Some controlled comparative trials suggest that the efficacy of valproate against partial seizures is similar to that of carbamazepine and phenytoin, especially in patients with predominant secondarily generalized tonic clonic seizures (Mattson et al., 1985), although it has been suggested that carbamazepine may be better (Marson et al., 2000).The primary strengths and weaknesses of valproate are reviewed in Table XV. Common side effects include tremor, weight gain, ankle swelling, and usually minor loss of hair. Cognitive impairment is sometimes seen. Encephalopathy has been occasionally reported, possibly due to hyperammonemia, which is a common result of valproate therapy. Rare cases of fatal hepatotoxicity have occurred, especially in handicapped infants during polytherapy, possibly related to a genetically determined metabolic disorder. Adult hepatotoxicity is even rarer and less than 30 patients have been reported in the past two decades (Konig et al., 1999). The use of valproate during pregnancy is associated

with an increased risk of neural tube defects; and a fetal antiepileptic drug syndrome has also been observed. Studies have linked valproate therapy with the development of polycystic ovarian disease, although this is controversial (Isojarvi et al., 1993; Herzog and Schachter, 2001).

Valproate is the only first-line drug that is not an enzyme inducer, and this is especially advantageous when treating the elderly or others with coexisting medical disorders. Valproate mildly inhibits the oxidative metabolism of other antiepileptic drugs. This is rarely of clinical relevance except when prescribed with lamotrigine. Valproate profoundly inhibits the metabolism of lamotrigine, leading to greatly elevated serum levels. This interaction results also in an increased incidence of lamotrigine-induced rashes (Brodie, 1992). Because of this, in patients medicated with valproate, lamotrigine therapy should be initiated at a very low dose of 25 mg every second day for the first 2 weeks. Conversely, felbamate inhibits the oxidative metabolism of valproate, requiring a reduction of dose by 20–30% when felbamate is added (Schmidt, 1994). The full pharmacological action of valproate may take several weeks to develop after steady-state concentrations have been reached. Valproate monitoring is not recommended as there is no relationship between clinical effects and the plasma concentration.

## Ethosuximide

Ethosuximide is still used for the treatment of absence seizures in children and adults (Bromfield EB 1997). In children below the age of 3 and those at risk for valproate-induced hepatoxicity it is the first-choice drug. Ethosuximide is not generally effective against tonic–clonic seizures. Ethosuximide is usually started at a dose of 250–500 mg/day with weekly increments. Plasma drug monitoring is not usually required, although it can be useful. The most important side effects are sedation, malaise, nausea, vomiting, and rarely psychotic episodes with paranoid ideation. Ethosuximide-induced headache and abdominal pain can be severe in individual patients. Ethosuximide does not inhibit or induce the metabolism of other drugs but its clearance is reduced by valproate.

## Phenobarbital

Phenobarbital is as effective as carbamazepine or phenytoin for treatment of tonic–clonic seizures (Heller et al., 1990). One controlled study found it less effective than carbamazepine against partial seizures (Mattson et al., 1985) but others did not (Heller et al., 1990). In patients controlled at low daily doses, phenobarbital is a very cost-effective and nontoxic medication and is used especially in countries with low health-care budgets.

The main disadvantage of phenobarbital is its potential to cause cognitive side effects (adverse effects on cognition and behavior, fatigue, insomnia, hyperactivity, especially in mentally retarded children, and aggressive behavior). A difficult withdrawal syndrome is often encountered, and tolerance may occur but is rarely a significant problem.

### Phenytoin

Phenytoin is as effective as carbamazepine for the treatment of partial seizures and tonic–clonic seizures (Browne, 1997). However, because of its greater potential to cause side effects, it is now not used as first-line therapy in many countries. The commonest side effects include cosmetic changes (e.g., gingival hyperplasia, acne, hirsutism, and possibly facial coarsening) and neuropsychiatric disturbance (e.g., depression, sedation, and fatigue). Phenytoin at common plasma concentrations has been shown to result in slightly more cognitive disturbances than carbamazepine, when used as a single-drug treatment (Aldenkamp et al., 1994). The nonlinear pharmacokinetics of phenytoin can also result in large increases in plasma concentrations with even small dose increments when the enzymatic monooxygenase metabolism is saturated. Conversely and no less troublesome, the circulating concentration may fall abruptly even with modest dose reduction. The starting dose of phenytoin is 100–200 mg in adults, and the dose can be increased by weekly increments. If seizures continue, the daily dose should be adjusted with the assistance of plasma concentrations. At concentrations below 40 μmol/L, 100-mg increments and above 40 mol/l, 25 mg increments are recommended. Above plasma concentrations of 80–100 μmol/L, neurotoxicity becomes a problem in most patients.

In many, but not all patients diplopia, ataxia, tremor, and nausea are easily recognized, but in others the signs and symptoms are mistaken as psychogenic. Complete seizure control may be observed with plasma concentrations as low as 10 μmol/L and above 100 μmol/L. Conversely, individual patients have only responded at much higher plasma concentrations and remained without signs of clinical neurotoxicity despite plasma concentrations of 160 μmol/L. Phenytoin is a potent enzyme inducer, similar to carbamazepine, and can be expected to exert similar interactions on other drugs such as antiepileptic drugs, anticoagulants, steroids, cyclosporin, and oral contraceptives. Conversely, the metabolism of phenytoin can be inhibited by enzyme inducers such as allopurinol, chloramphenicol, cimetidine, imipramine, isoniazid, metronidazol, phenothiazine, and sulfonamides.

### Primidone

Primidone is a barbiturate that is largely metabolized to phenobarbital, and its effects are very similar to those of phenobarbital (Bourgeois, 1997). An idiosyncratic reaction comprising severe dizziness and nausea can be produced when the drug is taken for the first time, and primidone should therefore be started at a very low dose (125 mg on alternate days) increased at intervals of every other week to an initial maintenance dose of 500 mg in adults. Because of a renal overflow mechanism at a certain plasma concentration, fatal suicides have not been reported with primidone overdose alone, in contrast to phenobarbital. Crystalluria can be observed in patients with primidone overdose.

### Newer Antiepileptic Drugs

In addition to established antiepileptic drugs, the following novel compounds have been registered in recent years in European and American countries: felbamate, gabapentin, lamotrigine, oxcarbazepine, topiramate, tiagabine, levetiracetam, fosphenytoin, and vigabatrin; and in Japan, zonisamide. The pharmacokinetics of the newly marketed antiepileptic drugs (Table XVI), their indications (Table XVII) and current side-effect profiles (Table XVIII) are briefly summarized.

### Felbamate

Felbamate is effective as add-on and as monotherapy in refractory partial epilepsy (Theodore, 1997). The molecular action of felbamate involves both enhancement of GABA-ergic inhibition and attenuation of glutamatergic excitation (Rho et al., 1994). The drug was licensed in Europe and in the United States in 1994 and then cases of aplastic anemia and hepatotoxicity (some fatal) were reported. Both these idiosyncratic side effects appear within the first 6 months of therapy, particularly in patients on polytherapy. The drug is now used in very refractory cases only.

### Fosphenytoin

Fosphenytoin (Browne, 1997), a prodrug of phenytoin, is converted into phenytoin by plasma phosphatase enzymes. It is water-soluble and can be administered both intravenously as well as intramuscularly. Fosphenytoin doses are expressed as phenytoin equivalents or PE and therefore 100 PE of fosphenytoin is equal to 100 mg of phenytoin. Free phenytoin levels after fosphenytoin infusion are achieved three times faster than with phenytoin and can therefore be infused at 150 PE/minute as compared to 50 mg/minute, with the same

**TABLE XVI**  Pharmacokinetics of Newly Marketed Antiepileptic Drugs[a]

| Drug | Absorption (bioavailability) | Distribution volume (L/kg) | Protein bound (%) | Elimination half-life (h) | Route of elimination | Comments |
|---|---|---|---|---|---|---|
| Felbamate | Rapid (>90%) | 0.7–0.8 | 22–25 | 2–6 | 90% Renal | Felbamate inhibits elimination of valproate, carbamazepine-epoxide, phenytoin, and phenobarbital |
| Gabapentin | Rapid (51–72%) | 0.57 | 0 | 5–7 | >90% Renal | No relevant drug interaction, extent of absorption diminished with increasing dose |
| Lamotrigine | Rapid (98%) | 0.9–1.3 | 55 | 12 h in enzyme induced Patients; 25 h in patients on valproate | Hepatic metabolism 5%, excreted unchanged | Lamotrigine elimination inhibited by valproate and increased by enzyme-inducing drugs |
| Levetiracetam | Rapid (100%) | 0.5–0.7 | 0 | 6–11 | >90% Renal excretion | Inconsistent effects on phenytoin levels |
| Oxcarbazepine | Rapid (95%) | 0.3–0.8 (MHD) | 38 (MHD) | 8–10 (MHD) | Hepatic metabolism to active compound 10-monohydroxy-carbamazepine (MHD); renal excretion 95% | Enzyme inducer, mild |
| Topiramate | Rapid (100%) | 0.6–1.1 | 15 | 18–23 | Not metabolized. Excreted unchanged renally | Not an enzyme inducer; no significant interactions |
| Tiagabine | Rapid (96%) | 1.0 | 96 | 4–8 | Hepatic metabolism | Not an enzyme inducer; may reduce valproate levels |
| Vigabratin | Rapid | 0.8 | 0 | 5–7 | Not metabolized in humans, renal excretion 70% | No relevant drug interaction |
| Zonisamide | Rapid | 1.4–1.9 | 50 | 27–36 | Hepatic metabolism; urinary excretion | |

[a] Partly adapted from Brodie (1990).

**TABLE XVII**  Prescribing Newly Marketed Antiepileptic Drugs in Adults

| Drug | Indications | Starting dose | Common daily dose | Dosage interval |
|---|---|---|---|---|
| Felbamate | Adjunctive therapy for refractory partial seizures (because of toxicity only in highly selected cases); adjunctive therapy for refractory Lennox–Gastaut syndrome | 600 mg | 1800–2400 | Three or four times a day |
| Gabapentin | Adjunctive therapy for refractory partial seizures | 600 mg | 1800–4200 mg | Three times a day |
| Lamotrigine | Adjunctive and single-drug therapy for refractory and newly diagnosed partial and generalized seizures | 25 mg every other day for patients on valproate, 25 mg per day in other patients | 100–200 mg for patients on valproate, 200–400 mg for other patients | Twice a day |
| Levetiracetam | Adjunctive therapy for refractory partial seizures | 1000 mg | 1000–3000 mg | Twice a day |
| Oxcarbazepine | Adjunctive drug therapy for refractory partial seizures; single-drug therapy for newly diagnosed partial seizures | 300 mg | 1500 mg | Three times daily |
| Tiagabine | Adjunctive therapy for for refractory partial seizures | 15 mg | 15–45 mg | Twice/thrice a day |
| Topiramate | Adjunctive therapy for refractory partial seizures | 50 mg | 400–600 mg | Twice a day |
| Vigabatrin | Adjunctive therapy for refractory partial seizures | 500 mg | 2–3 mg | Twice a day |
| Zonisamide | Adjunctive therapy for refractory partial and myoclonic seizures | 100 mg | 300–400 mg | Twice a day |

TABLE XVIII    Adverse Reactions to Newly Marketed Antiepileptic Drugs

| Felbamate | Gabapentin | Lamotrigine | Oxcarbazepine | Vigabatrin | Topiramate | Zonisamide | Tiagabine | Levetiracetam |
|---|---|---|---|---|---|---|---|---|
| | | | | *Dose-related effects* | | | | |
| Nausea | Somnolence | Dizziness | Drowsiness | Somnolence | Dizziness | Drowsiness | Dizziness | Dizziness |
| Anorexia | Dizziness | Headache | Dizziness | Fatigue | Nystagmus | Ataxia | Asthenia | Infection |
| Dizziness | Ataxia | Ataxia | Headache | Irritability | Ataxia | Loss of appetite | Nervousness | Asthenia |
| Vomiting | Fatigue | Nausea | Ataxia | Dizziness | Headache | Slowing of | Tremor | Somnolence |
| Insomnia | Nystagmus | Diplopia | Diplopia | Headache | Sedation | thought | Diarrhoea | Headache |
| Weight loss | Headache | Vomiting | Nystagmus | Weight gain | Paresthesia | Dysphasia | Depression | |
| Diplopia | Tremor | Asthenia | Dysarthria | Encephalopathy | Asthenia | Headache | Somnolence | |
| Somnolence | Nausea | Somnolence | Nausea | Psychotic episodes | Confusion | Weight loss | Headaches | |
| Headache | Vomiting | Depression | Vomiting | Depression | Agitation | Leukopenia | | |
| Dyspepsia | Rhinitis | Insomnia | Diarrhea | Manic–affective | Anorexia | Abnormal LFT | | |
| | Diplopia | Hyperkinesia | Poor appetite | disorder | Nausea | Renal calculi | | |
| | Weight gain | Leukopenia | Parkinsonlike | Visual field | Diarrhea | Paresthesia | | |
| | Tantrums | Purpura | syndrome | defects | Abdominal | Psychotic | | |
| | Aggression | Anemia | Hyponatremia | | pain | episodes | | |
| | Hyperactivity | | | | | | | |
| | | | | *Idiosyncratic effects* | | | | |
| Hypersensitivity | Rash[a] | Rash | Rash | Rash | Renal | Rash | Rash | Rash |
| syndromes | | Stevens– | | | calculi | Stevens– | | |
| Stevens–Johnson | | Johnson | | | | Johnson | | |
| syndrome | | syndrome | | | | syndrome | | |
| Liver failure | | Lyell | | | | | | |
| Aplastic | | syndrome | | | | | | |
| anemia | | | | | | | | |

[a]Causal relationship uncertain.

attendant risks of sudden hypotension, arrhythmias and respiratory depression. There have been no proper controlled trials reported so far although it is likely that it will occupy a position similar to phenytoin in the drug armamentarium, although cost may be a factor in prescribing.

## Gabapentin

Gabapentin is a mild antiepileptic drug useful as add-on treatment of refractory partial seizures. Clinical trials showed that 22% of patients with partial seizures will have at least a 50% reduction in seizure frequency (Chadwick, 1994). Gabapentin is not effective for treatment of idiopathic generalized epilepsy. As with standard antiepileptic drugs, secondarily generalized tonic–clonic seizures may respond better than simple or complex partial seizures. The side effects of gabapentin are usually mild and include diplopia, ataxia, fatigue, and headache (Wong and Lhatoo, 2000). Male rats developed acinar pancreatic tumors in a dose-related fashion, and gabapentin has therefore been termed a potential human carcinogen with an acceptable low risk (Chadwick, 1994). An advantage is the lack of interaction of gabapentin with other drugs, which makes plasma drug monitoring unnecessary. It is, however, known to cause seizure exacerbation in some cases and is not particularly effective in patients with severe epilepsy.

## Lamotrigine

Lamotrigine is effective as monotherapy of newly diagnosed partial seizures and generalized tonic–clonic seizures and as add-on treatment of refractory partial seizures (Binnie, 1997). Clinical trials showed a reduction of at least 50% in about 24% of patients, with up to 5% of patients becoming seizure-free at doses of 100–300 mg/day. Lamotrigine is also effective for add-on treatment of refractory generalized epilepsy with atypical absences, atonic seizures, and myoclonic seizures. Lamotrigine is a triazine derivative and, like phenytoin or carbamazepine, causes an extended inactivation of neuronal membrane sodium channels, thus inhibiting repetitive discharges in experimental models. The pathological release of glutamate, a major excitatory neurotransmitter in human brain, is also blocked. The elimination of lamotrigine is accelerated by enzyme-inducing antiepileptic drugs, such as carbamazepine, phenytoin, and phenobarbital, and inhibited by valproate. Even in monotherapy, the initiation of lamotrigine therapy should be at a very low dose (25 mg/day) to avoid the development of a rash (Brodie, 1990) in up to 5% of patients, particularly younger age groups and those on concurrent valproate. The rash may be severe, amounting in some cases to a Stevens–Johnson syndrome. Other side effects include sedation, dizziness, diplopia, and ataxia, but generally the drug is well tolerated (Wong and Lhatoo, 2000).

## Levetiracetam

Levetiracetam is the most recently marketed antiepileptic drug, a pyrrolidone derivative whose mode of action remains unknown (Shorvon SD, 2000). It is well tolerated and the most frequent central nervous system adverse events in trials were dizziness, asthenia, and somnolence. An increased frequency of upper respiratory tract "infections" were noted but these were not treatment limiting. Placebo-controlled trials in refractory partial epilepsy have shown 50% seizure reduction in up to 33% on 1000 mg and 40% on 3000 mg compared to 11% in the placebo groups (Cereghino et al., 2000). It also appears to be efficacious as monotherapy. There are no formal trials examining its usefulness in the primary generalized epilepsies although it has been shown to be effective in eliminating photoparoxysmal responses on EEG, and has been successfully used in juvenile myoclonic epilepsy and myoclonic jerks. It is a promising compound and likely to be useful in the treatment of refractory epilepsy.

## Oxcarbazepine

Oxcarbazepine was developed as a structural variant of carbamazepine (Gram, 1997). It is a pro-drug, as its 10-hydroxymetabolite is responsible for the antiepileptic effect. When compared to carbamazepine, oxcarbazepine has shown similar efficacy as an add-on drug for refractory partial seizures and as a first-line agent in previously untreated patients with tonic–clonic and partial seizures (Grant and Faulds, 1992). Common side effects include drowsiness, dizziness, ataxia, headache, and hyponatremia. The hyponatremia is more marked than with carbamazepine and occasionally leads to confusion and increase of seizures. Other side effects include rashes, diarrhea, nausea, vomiting, and anorexia (Wong and Lhatoo, 2000). The enzyme-inducing activity of oxcarbazepine is probably limited to the P450 IIIa isoenzyme of the cytochrome P450 complex; nevertheless, interactions occur, for instance with oral contraceptives.

## Tiagabine

Tiagabine is another new drug with mild to moderate efficacy in seizure control. Its proposed mechanism of action is by inhibiting glial cell GABA reuptake. It is used as adjunctive therapy in partial seizures with or without secondary generalization (Leppick, 1995). Results from controlled trials have shown that up to one-third of patients on tiagabine achieve a 50% reduction in seizure frequency, although complete remission from seizures is an infrequent occurrence. The com-

monest adverse events related to therapy are central nervous system related and consist of sedation, tremor, headache, mental slowing, tiredness, and dizziness. Confusion, irritability, and depression may occur (Wong and Lhatoo, 2000). Increases in seizure frequency and episodes of nonconvulsive status have also been reported. So far, no life-threatening idiosyncratic reactions have been encountered. Use in pregnancy is not recommended although no teratogenicity has been reported in humans.

## Topiramate

Topiramate (Kramer and Reife, 1997) is a sulfamate-substituted D-fructose, a naturally occurring monosaccharide. Its antiepileptic action was discovered during a search for antidiabetic drugs. It has a variety of mechanisms of action, the major being its effect on the voltage-dependent Na channels in the neuronal membrane. It affects glutaminergic (via the AMPA receptor) and GABA-ergic (at the GABA receptor) transmission. The drug also has carbonic anhydrase action. Animal experimentation showed topiramate to be highly effective against a wide variety of experimental epilepsy models, and the experimental profile suggests a wide spectrum of activity in human epilepsy. Animal toxicology showed only mild effects, but the drug was shown to have teratogenicity at high doses. The drug has excellent pharmacokinetic properties and there are no active metabolites. In randomized clinical trials, topiramate has been shown to be highly effective against partial and secondarily generalized seizures. Topiramate can cause predominately neurological side effects, particularly at high dosage and if titrated too fast (Lhatoo and Walker, 1999). These adverse effects include headache, sedation, asthenia, and confusion. Loss of weight is common, and the drug also causes renal stones in less than 2% of patients (Lhatoo and Walker, 1999). There are no serious recorded idiosyncratic reactions, and there is no hematological toxicity. To lessen the incidence of treatment limiting side effects, the drug needs to be started as low as 15 mg daily or every alternate day, increasing by weekly or 2 weekly increments of 25 mg.

## Vigabatrin

Vigabatrin is effective as add-on treatment of refractory partial seizures (Ben-Menachem and French 1997). In clinical trials, the addition of 1–3 g/day led in about 46% of cases to a reduction of at least 50%, with 5% of patients becoming seizure-free. Vigabatrin is not effective in patients with idiopathic generalized epilepsy. The antiepileptic effect of vigabatrin is due to the irreversible inhibition of GABA-transaminase that leads to a permanent severalfold rise in GABA, the major inhibitory neurotransmitter in human brain. The most common side effects are transient drowsiness and weight gain, which may be unacceptable in some patients, and less often, depression and manic–depressive disorder, confusion, and psychotic episodes. Recent, well documented descriptions of visual field defects in patients on long-term therapy with the drug have drastically limited its use. The visual field defects, which may be asymptomatic, may occur in up to half of all patients and are thought to be as a result of GABA-mediated retinal toxicity. It can lead to blindness and when vigabatrin is used, therefore, visual fields need to be carefully monitored (Wong and Lhatoo, 2000). Rare cases of encephalopathy with stupor and an increase in seizures have been reported.

### Zonisamide

Zonisamide is an antiepileptic drug that is very effective as add-on treatment for refractory partial seizures and refractory myoclonic seizures (Seino and Tsugutaka, 1997). Clinical trials suggest that about 50% of patients with refractory partial seizures will have a 50% or more reduction in seizures after the introduction of zonisamide. In the United States, renal calculi were recorded (in fewer than 20 patients) and this resulted in the termination of its clinical development. In Japan, however, the drug continues to be used, and renal calculi have not proven a problem. The common side effects of zonisamide include sedation and fatigue, but zonisamide is generally well tolerated (Wong and Lhatoo, 2000).

## REFRACTORY EPILEPSY

Epilepsy can be successfully treated in approximately 70% of patients with single-drug therapy with first-line agents. The remaining patients have continuing seizures, and some will not respond adequately to treatment with any of the currently available antiepileptic drugs (Sander, 1993). Refractory partial seizures may be difficult to treat compared to the primary generalized epilepsies. The tonic–clonic, myoclonic, atonic, and tonic seizures in symptomatic generalized epilepsy (e.g., West syndrome, early myoclonic encephalopathy, severe myoclonic epilepsy in infancy, and Lennox–Gastaut syndrome) are unlikely to respond to any conventional drug therapy, and the prognosis for seizure control in such cases is very poor.

When assessing a patient who is apparently unresponsive to single-drug therapy with a first-line agent, it is first critically important to reevaluate the diagnosis. A significant number of such patients have been mis-diagnosed, and the "seizures" are due to psychogenic disturbance or syncope or less commonly to other mechanisms (Table IX). When epilepsy is confirmed, the seizure type and syndrome should be defined, if necessary by video-EEG monitoring. The etiology of the epilepsy should be identified, and magnetic resonance imaging is indicated in most such patients to rule out surgically amenable lesions. Even patients with apparently refractory idiopathic generalized epilepsy may turn out to have an underlying metabolic or progressive cause. A detailed history of previous treatment should be taken, and those drugs not previously taken should be tried. Compliance should be assessed; poor compliance is not uncommonly related to poor instruction, memory problems, and covert alcoholism in some patients. Assessing the serum levels of some drugs is important, and increasing the daily dose from suboptimal to clinically effective levels may produce improvement or even complete seizure control. Adjusting lifestyle will sometimes improve seizure control, especially those with idiopathic generalized epilepsy where fatigue or alcohol is often an important precipitant.

Where single-drug therapy has failed, a second drug should be added, and the dose of the first drug is often usefully reduced, lowering the risk of side effects when transferring to polytherapy (Table XIX). The addition of a second drug followed by the withdrawal of the first has been shown to succeed in about one-third of patients with refractory partial epilepsy (Schmidt and Richter K 1986). The benefit of two-drug therapy in failures of single-drug regimens has been clinically evaluated, albeit in an uncontrolled fashion. A reduction of the number of seizures by more than 75% was reported in 13% of the patients exposed to a second standard drug (phenytoin, carbamazepine, phenobarbital, or primidone) when another of the four drugs had failed before. In that study, 37% of patients had at least 50% fewer seizures during two-drug therapy (Schmidt, 1982). In another clinical observation, 32 of 82 patients with refractory partial seizures had fewer seizures when a second drug was added and 11% became seizure-free (Mattson et al., 1985). In a retrospective analysis of two-drug therapy in failures of single-drug treatment, the number of patients with side effects did not, however, increase when a second drug was added (Table XIX). In summary, two-drug therapy has been shown in uncontrolled observations to be effective in about 20–30% of patients with refractory partial seizures or tonic clonic seizures. The choice of first-line and second-line add-on drugs, established or new, depends on seizure type and syndrome (Table X). Where therapy has failed, most patients will benefit most from a low to moderate dose regimen of the individually most effective agent, preferably in single-drug therapy, which provides a measure of seizure control but does not expose

**TABLE XIX** Transfer from Single- to Two-Drug Therapy[a]

| | Single-drug therapy | Two-drug therapy |
|---|---|---|
| Number of patients | 68 | 68 |
| Number of patients with seizure reduction of 75% or more | 0 | 15 |
| Median seizure frequency per month | 2.3 ($P \geq 0.015$) | 1.3 |
| Number of patients with side effects during both treatment regimens | 26 | 26 |
| Number of patients with side effects only during single-drug therapy | 16 | 0 |
| Number of patients with side effects only during two-drug therapy | 0 | 13 |
| Sum of patients with side effects | 42 | 39 |

[a]Schmidt and Richter (1986), copyright Lippincott Williams & Wilkins.

the patient to chronic side effects. Finally, the potential for surgical treatment of the epilepsy should be evaluated in suitable patients within 2–5 years of failure of optimal antiepileptic drug treatment.

## THE SURGICAL TREATMENT OF EPILEPSY

For a proportion of the 20–30% of patients who have refractory epilepsy, surgery provides a realistic means of attaining seizure freedom with minimum morbidity (Engel, 1997). Unfortunately, shortage of surgical centres and lack of awareness of this modality of treatment result in a much smaller number being treated than require it.

Several different types of surgical procedures are used in the treatment of epilepsy (Table XX). Some involve curative resection, such as anterior temporal lobectomy for hippocampal sclerosis, whereas others involve palliative resection, such as corpus callosotomy and hemispherectomy. Many centers are currently evaluating nonresective curative surgery such as radiosurgical techniques. Anterior temporal lobectomy for hippocampal sclerosis for example, a condition which has fairly bleak prognosis for seizure control with medication alone, is particularly effective and produces long-term complete seizure remission in 70–90% or more of patients who are carefully screened for the procedure. In a further 20%, seizure frequency is significantly ameliorated. Despite the difficulties inherent in the undertaking of randomized, controlled trials in surgery, a recent study randomized patients with refractory temporal lobe epilepsy to either temporal lobectomy (40 patients) or medical treatment (40 patients). At 1 year, 58% of patients in the surgical group were free of seizures impairing awareness, compared to 8% in the medical group ($P < 0.001$). Complete seizure freedom (including auras) occurred in 38% compared to 3% ($P < 0.001$). Although the figures for seizure freedom are lower than

**TABLE XX** Surgical Procedures in the Treatment of Epilepsy

Temporal lobe surgery
  "En bloc" anterior temporal lobe resection
  Selective amygdalohippocampectomy
  Tailored resection/lesionectomy
  Radiosurgery
Extratemporal resection
Corpus callosotomy
Hemispherectomy
Multiple subpial transection
Vagal nerve stimulation

is the average in published literature, the superiority of surgery to continued medical treatment is easily established. Significant morbidity in the form of postoperative memory impairment occurred in 5% (Wiebe S, and others 2001).

Careful preoperative evaluation that includes neuropsychological testing and neuropsychiatric assessments in addition to standard investigations such as ictal EEG and epilepsy protocol MRI scans to define the epileptogenic zone are essential. Functional imaging with functional MRI (fMRI), single-photon emission computed tomography (SPECT), and positron emission tomography (PET) may have a role in some patients where lesion localization is difficult. Surgical patients with the following characteristics may have a particularly good chance of achieving postsurgical seizure remission:

1. temporal lobe epilepsy with childhood onset;
2. no secondary generalized seizures;
3. history of childhood febrile convulsions;
4. MRI evidence of unilateral hippocampal sclerosis or other lesion confined to the temporal lobe;
5. unilateral epileptiform discharges on EEG confined to the anterior temporal lobe.

Characteristics that may predict less benefit from surgery include:

1. dual or multiple cerebral lesions;
2. marked intellectual dysfunction;
3. interictal psychiatric disturbances;
4. multiple seizure types;
5. interictal EEG evidence of generalized or extratemporal epileptiform discharges;
6. normal MRI even after appropriate epilepsy protocol scans.

Morbidity and mortality are infrequent in centers that regularly perform these procedures and in most centers, serious morbidity is less than 5% and mortality less than 0.5%. Some of the possible adverse events associated with temporal lobe surgery are listed (Table XXI).

TABLE XXI    Potential Complications of Temporal Lobe Surgery

| | |
|---|---|
| Visual field defects | |
|    Minor | 20% |
|    Major | 3% |
| Speech disturbances (usually temporary) | <1% |
| Memory deficits (usually temporary) | <1% |
| Diplopia (usually temporary) | <1% |
| Hemiparesis | 1% |
| Scalp infections | <1% |
| Aseptic meningitis | <1% |
| Subdural collections | <1% |

## Vagal Nerve Stimulation

This is best considered a palliative form of treatment and a surgical adjunct in patients with refractory partial seizures where resective or ablative surgery is not possible. It works on the principle that electrical stimulation of the vagus nerve may modify or even halt seizure activity. Proposed mechanisms of action based on animal studies include vagally mediated release of GABA and glycine, transient "desynchronization" of cortical rhythm, and influences on the reticular activating system. Observations on patients undergoing carotid surgery have suggested that vagal stimulation may produce EEG changes. PET scanning studies have indicated that vagal stimulation may produce alterations in cerebral blood flow.

The device, which is called the Neuro Cybernetic Prosthesis (NCP) system, consists of a pulse generator implanted in the left upper chest wall, a bipolar lead with electrodes coiled around the left vagus nerve, a programming "wand" with specialized IBM-computer-compatible software for customizing the generator, and hand-held magnets for activating the generator. The generator emits a current output at 30 Hz signal frequency with a 500-μs pulse width for 30 s on and 5 min off cycles. The magnets allow activation at any time in an attempt to abort a seizure (Schachter 1998).

In two randomized, double blind, multicenter studies in Europe and North America that compared high-stimulation (as above) and low-stimulation protocols, patients who had six or more partial seizures per month treated with one to three AEDs were studied over a 12-week treatment period. A third of the high-stimulation group had 50% seizure reduction compared to a fifth in the low-stimulation group. Interestingly, about a quarter of the patients receiving high stimulation in both trials noted a worsening of their seizure disorder. Complete seizure remission has not yet been reported despite more than 3500 insertions worldwide.

Adverse effects are mainly related to the surgical procedure and include left vocal cord palsy, lower facial weakness and fluid accumulation in the subcutaneous generator pocket. Other effects include pain, cough, dysphonia, dyspnoea, and vomiting.

## FEBRILE SEIZURES

Febrile seizures occur in 3–4% of children under the age of 5 years. The use of antipyretics and tepid sponging is often advocated for temperature control although these have not been proven effective (Uhari et al., 1995). Similarly, there is little evidence to suggest that specific treatment to prevent recurrent febrile seizures reduces the risk of later developing afebrile seizures. Although fashionable for a time, the intermittent use of phenobarbital at the time of fever has been shown to be ineffective (Wolf et al., 1977) as has daily and continuous treatment with phenobarbitone or valproate (Newton RW 1988; McKinlay I and Newton R 1989). Other studies have suggested that rectal or oral diazepam given during a febrile illness may be effective in preventing recurrences (Rosman et al., 1993; Knudsen, 1991), but may not influence subsequent development of habitual seizures or affect cognitive and motor ability (Knudsen et al., 1996). Thus prophylaxis is probably best advised only where frequent, usually prolonged seizures have occurred in a very young child. Parents or physicians often have questions regarding the continuation of routine childhood immunizations in children who have experienced a febrile seizure. Seizures following childhood immunizations are not different from any other febrile seizure (Hirtz et al., 1983).

The long-term prognosis for children with febrile seizures is better than previously assumed, and large-scale population-based studies have shown that long-term sequelae (neurological deficit, epilepsy, mental retardation, or behavioral problems) only rarely follow febrile seizures (MacDonald et al., 1999).

## EPILEPSY AND CONTRACEPTION, FERTILITY, AND PREGNANCY

### Fertility

Reduced fertility may be more common in women with epilepsy than previously thought. A study in the United Kingdom found fertility to be lower among women with treated epilepsy, with an overall rate of 47.1 live births per 1000 women ages 15–44 per year (42.3–52.2), compared with a national rate of 62.6 in the same age group. Fertility rates are therefore 33% higher in women in the general population without epilepsy (Wallace et al., 1998). Potential explanations include medical factors such as genetic influences and the adverse effects of antiepileptic drugs; social factors such as low marriage rates, stigmatization, and the avoidance of pregnancy because of the potential teratogenicity of antiepileptic drugs.

## Contraception

Antiepileptic drugs may reduce the effectiveness of oral contraception, and breakthrough bleeding is a useful warning sign that the contraceptive dose is too low. This is a problem with enzyme-inducing drugs such as carbamazepine, phenytoin, phenobarbital, and primidone, although other mechanisms may also exist. In view of this potential, it is usual to avoid the low-dose estrogen contraceptives and recommend therapy with medium- or high-dose compounds to avoid ineffective contraception. Valproate is the only first-line antiepileptic drug that does not have the potential to interact with the oral contraceptive pill. No interactions have been observed with the newer drugs with the exceptions of topiramate and oxcarbazepine, although clinical experience is limited. Breakthrough bleeding has been noted with the use of oxcarbazepine. There is no evidence that oral contraceptives influence the plasma concentrations of antiepileptic drugs or seizure control.

## Prepregnancy Counseling and Teratogenicity

The prevalence of epilepsy in women of childbearing age is 0.31–1.0% (Schmidt, 1993), and epilepsy is the most common neurological disorder in pregnant women. Pregnancy should not be discouraged in women with epilepsy as was often the practice in the past (Commission on Genetics, 1993). The complex effects of antiepileptic drugs on the course and the outcome of pregnancy require detailed discussion and counseling. Although there is no doubt that the offspring of women with epilepsy taking antiepileptic drugs have an increased risk of minor or major malformations (Delgado-Escueta *et al.*, 1992), the exact contribution of antiepileptic drug intake during pregnancy to this higher risk has been controversial. A recent study found an odds ratio of 2.8 (95% confidence interval 1.1–5.1) for malformations in infants of mothers on one anticonvulsant compared to controls who were mothers without epilepsy or anticonvulsant treatment (Holmes *et al.*, 2001). The odds ratio was similarly high (4.2; 95% confidence interval 1.1–5.1) in infants born to mothers on more than one anticonvulsant compared to controls. Infants whose mothers had epilepsy but were not on medication did not have a higher risk than controls, suggesting a primary role for drugs in malformations in children of mothers with epilepsy. Many experts would agree, however, that both the severity of maternal epilepsy and other genetic and social factors also influence the risk of malformation (Steegers-Theunissen *et al.*, 1994).

In general the risk of uncontrolled epilepsy is equal to or greater than the risk of drug-induced teratogenesis, and drug treatment is usually continued with the minimum effective regimen. Prior to pregnancy, an attempt should be made to change therapy to the lowest effective dosage of the most suitable single drug. In selected cases, however, the gradual withdrawal of all antiepileptic drugs is appropriate prior to the pregnancy (Table XXII). The data on the comparative teratogenicity of the major drugs is not based on prospective controlled trials but on surveys that are inherently potentially biased. Despite a relative lack of definitive epidemiological studies, a meta-analysis of pregnancies with mothers taking antiepileptic drugs suggests an approximately two- to threefold increased risk of major malformations (Table XXIII). There are also specific risks with different antiepileptics. Exposure to valproate is associated with a twofold incidence of neural tube defects when compared to other antiepileptic drugs (Lindhout and Schmidt, 1986), although neural tube defects have also been associated (with less evidence) with the exposure to carbamazepine during pregnancy (Rosa, 1991). A survey of congenital abnormalities in one Dutch center confirmed that valproate was associated with a higher rate of open spina bifida compared

**TABLE XXII    Management of Epilepsy during Pregnancy**

Prior to pregnancy
1. Reassess and confirm the diagnosis of epilepsy
2. Re-evaluate the continued need for antiepileptic drug therapy
3. Transfer to single-drug therapy, if possible
4. Treat with the lowest individual effective daily dose
5. Change antiepileptic drugs, if necessary
6. Prescribe folic acid supplementation
7. Reassure of the altogether benign course of pregnancy, delivery, and its good outcome in over 90% of the cases and on the benign prognosis of epilepsy, provided good drug compliance is maintained

Pregnancy
1. Counsel on the value of good drug compliance
2. Monthly neurological examinations including drug monitoring
3. Increase daily dose of antiepileptic drug, if necessary, preferably based on clinical criteria; i.e., seizure control and side effects
4. Reassure about the good course and outcome of pregnancy
5. Change antiepileptic drugs, if necessary
6. Individualize obstetric monitoring, including determination of serum alpha-fetoprotein and ultrasound examination, if necessary
7. Counsel on continued drug compliance during delivery and on breast feeding, which is recommended

Puerperium
1. Lower the daily dose of antiepileptic drugs in case of side effects
2. Maintain a good night's sleep despite breast feeding during the day
3. Pediatric monitoring, if necessary

TABLE XXIII

Pregnancy and epilepsy
  Risks for the mother
    Pregnancy
      Vaginal bleeding
      Protracted course of delivery
    Epilepsy
      No dramatic change
      Avoid poor compliance
      Provide reassurance
  Risks for the child
    Lower postnatal Apgar scores
    Increased risk for asphyxia
    Lower incidence of physiological icterus
    Lower vitamin K, replacement required 1 mg/kg
    Postnatal sedation vs. hyperexcitability
    Breast feeding recommended despite associated risks
Major malformations associated with antiepileptic drug intake
  Cardiovascular malformations
  Cheilo- or palatoschisis, or both
  Skeletal abnormalities (clubfoot, hip dislocation, etc.)
  Central nervous system (micro-, hydro-, anencephalus,
    meningomyelocele)
  Gastrointestinal malformations (intestinal atresia, diaphragmatic
    hernia, etc.)
  Urogenital malformations (hypospadia, etc.)
  Others
Minor anomalies associated with antiepileptic drug intake
  Epicanthus
  Hypertelorism
  Ptosis
  Prominent metopic ridge
  Short, low-bridged nose
  Long philtrum
  Wide mouth with full lips
  Low-set or abnormal ears
  Distal digital hypoplasia
  Webbed neck
  Hirsutism or low hairline

to other antiepileptic drugs (5.4% vs 0%; $P = >0.0003$). Interestingly, women whose offspring had open spina bifida received a higher daily dose of valproate per mg/kg body weight and showed higher plasma concentrations, suggesting a dose–response relationship (Omtzigt et al., 1992). Phenytoin can also cause a range of specific teratogenic effects (e.g., cleft palate, harelip, neuroblastoma) and other drugs such as the benzodiazepines, phenobarbital, and primidone have also been associated with major and minor malformations with uncertain frequency. Early prenatal screening for some defects (including spina bifida) with ultrasound, alphafetoprotein estimations, and in some cases amniocentesis is possible, and if termination of pregnancy is carried out, the risks of a live birth with spina bifida is, of course, considerably reduced. There is very little clinical evidence of the use in pregnancy of the novel antiepileptic drugs such as lamotrigine, gabapentin, vigabatrin, or felbamate. These show no preclinical evidence of teratogenicity but in general should be avoided in pregnancy. The incidence of neural tube defects can be reduced by the administration of folic acid before and in the earliest stages of pregnancy and should be given to all women prior to conception (Table XXII). Although conclusive data in women with epilepsy is not available, folic acid supplementation of 4 mg before conception prevented nearly three-quarters of such abnormalities in women without epilepsy (MRC Vitamin Study Research Group, 1991). Earlier concerns that folic acid supplementation may increase the number of seizures are not well founded. From a clinical perspective, reassurance that epilepsy only rarely influences the course of pregnancy and that a good outcome of pregnancy is seen in approximately 90% of all cases is a most important part of pregnancy counseling. The risk of minor (approximately 5–10%) or major malformations (approximately 2–5.5%) and the risk of epilepsy (slightly increased in most cases) need to be put into a careful but reassuring perspective prior to pregnancy.

## Pregnancy

The management of epilepsy in pregnancy requires attention to certain specific issues (Table XXII). The total plasma concentration of all the major antiepileptic drugs may decrease during the course of pregnancy for various metabolic and hormonal reasons. The higher free concentration that is not protein bound may partly compensate for a lower total concentration. In some situations, measurement of free rather than total serum concentrations are appropriate, and changes in total plasma concentrations therefore should not necessarily prompt revision of the daily dose in asymptomatic patients. Poor drug compliance is common in pregnancy, encouraged sometimes by anxiety about teratogenic effects, and a major cause of increased seizure frequency during pregnancy. Valproate, phenytoin, and carbamazepine should be given by divided doses in view of experimental data, suggesting a greater potential for teratogenic effects at high peak plasma concentrations (Delgado-Escueta et al., 1992).

The extent to which seizures result in an adverse outcome of pregnancy is not well known. However, not surprisingly, tonic–clonic seizures result in changes in fetal heart rate and potentially in hypoxia. The effect of other types of seizures, where hypoxia is less likely (for instance, absence seizure, myoclonic seizures, or complex partial seizures), is not known. Continuous ultrasound monitoring to assess fetal well-being throughout the pregnancy is advisable. In cases of valproate or carbamazepine exposure and a family history of neural tube defects, amniocentesis for alphafetoprotein analysis and acetylcholinesterase electrophoresis should also be considered.

There is a slightly higher risk of obstetric complications for the mother during pregnancy. Vaginal bleeding is more common, the rate of caesarian section and forceps or complicated delivery is higher, and all women should be delivered in a hospital. Women with epilepsy more often have a protracted course of delivery (Table XXIII). Seizures are common during the delivery period (occurring overall in 1% of patients with epilepsy), and it is sometimes appropriate to increase antiepileptic drug dosage temporarily at the time of delivery.

### Neonatal Period

Following delivery, plasma antiepileptic drug concentrations of the mother often rise, and clinical toxicity at this time is a not uncommon but preventable occurrence. Sleep deprivation should be minimized. Oral vitamin K prophylaxis for the baby with 2 mg at birth, at the end of the first week, and at the fourth week are recommended for prevention of late hemorrhagic disease of the newborn (Delgado-Escueta *et al.*, 1992). Parenteral vitamin K prophylaxis is controversial because of concerns that it may cause childhood cancer (Golding *et al.*, 1992). Breastfeeding should be encouraged as in a mother who does not have epilepsy, because of its general benefits for mother and child. Antiepileptic drugs taken by the mother enter breast milk to a variable degree: valproate 5–10%, phenytoin 30%, phenobarbital 40%, carbamazepine 45%, primidone 60%, and ethosuximide 90% (Delgado-Escueta *et al.*, 1992). Only maternal barbiturates at high dosage are likely to cause sedation in the newborn, and if sedation does occur breastfeeding should be reduced or discontinued. Occasionally, symptoms of drug withdrawal occur in the offspring following birth or the discontinuation of breastfeeding. If the mother is not seizure-free, measures should be considered to prevent seizure-related accidents to the neonate. Finally, the mother should be reassured that the delay in length and weight gain in the first postnatal months seen in the offspring of some women with epilepsy does not seem to affect weight at a later stage and that there is no reliable evidence that prenatal exposure to antiepileptic drugs causes permanent cognitive impairment (Gaily *et al.*, 1990).

## STATUS EPILEPTICUS

Status epilepticus is defined as a fixed epileptic state in which repeated or continuous seizures occur for more than 30 min (Shorvon, 1994). There are a variety of

**TABLE XXIV    Classification of Status Epilepticus**

Status epilepticus confined to the neonatal period
  Neonatal status epilepticus
  Status epilepticus in neonatal epilepsy syndromes
Status epilepticus confined to infancy and childhood
  Infantile spasm (West syndrome)
  Febrile status epilepticus
  Status epilepticus in childhood myoclonic syndromes
  Status epilepticus in benign childhood partial epilepsy syndromes
  Electrical status epilepticus during slow wave sleep (ESES)
  Syndrome of acquired epileptic aphasia
Status epilepticus occurring in childhood and adult life
  Tonic–clonic status epilepticus
  Absence status epilepticus
  Epilepsia partialis continua (EPC)
  Myoclonic status epilepticus in coma
  Specific forms of status epilepticus in mental retardation
  Myoclonic status epilepticus in other epilepsy syndromes
  Simple nonconvulsive partial status
  Complex partial status
Status epilepticus confined to late adult life
  De novo absence status of late onset

*Note.* Adapted from Shorvon (1994), copyright Cambridge University Press.

types of status epilepticus (Table XXIV), and treatment varies according to seizure type and syndrome. Psychogenic status is also not uncommon, and its recognition is important to avoid invasive therapy. Prominent features of pseudostatus are bizarre motor activity, poor response to treatment, a fluctuating pattern of seizures, lack of metabolic consequences, and lack of exhaustion (although iatrogenic sedation may occur).

The commonest form is tonic–clonic status, and immediate aggressive emergency treatment is essential once the diagnosis of status epilepticus is firmly established (Shorvon SD 1994); inadequate therapy results in an increased morbidity and mortality. The treatment of status should be considered in stages. Status is often preceded by a premonitory stage, when serial seizures occur and when impending status can be predicted; immediate therapy at this stage will often prevent the evolution to full status. Once status has developed, it is usual to divide this into an early phase and an established phase, during which subanesthetic therapy is given, and then a refractory stage when general anesthesia is necessary. The common causes of tonic–clonic status are shown in Table XXV and emergency measures to be taken in Table XXVI. The drug management of tonic–clonic status varies from center to center, but a standard staged regime is given in Table XXVII. Intravenous therapy carries significant risk, especially of hypotension or respiratory or cardiocirculatory arrest. These are particular problems of benzodiazepine therapy, but can occur with any of the major antiepileptic drugs, especially after prolonged infusion. Phenytoin may also cause cardiac arrhythmias and nystagmus and ataxia and

**TABLE XXV    Causes of Tonic–Clonic Status Epilepticus**[a]

Background of epilepsy
    Poor anticonvulsant compliance
    Recent dose reduction or discontinuation
    Alcohol withdrawal
    Pseudostatus
Presenting de novo
    Cerebrovascular disease
    Meningoencephalitis
    Acute head injury
    Cerebral tumor
    Brain abscess
    Metabolic disorders (e.g., renal failure, hypoglycemia,
        hyponatremia, post-cardiac arrest, hepatic encephalopathy)
    Drug overdose (e.g., tricyclic antidepressant, phenothiazine,
        theophylline, isoniazid, cocaine)
    Inflammatory arteritis (e.g., systemic lupus erythematosus)

[a]Brodie (1990).

should not be given faster than 50 mg/min. Following emergency treatment, oral maintenance via gastric tube should be initiated. In a previously untreated patient, an oral loading dose of phenytoin would probably be preferred in most centers. A current phenytoin loading dosage regime (in milligrams) is 0.65 weight (kg) × 15 µg/mL current concentration (Winter and Tozer, 1986), followed by 5–7 mg/kg the next day guided by plasma monitoring. Medical complications are not uncommon during prolonged status and can interfere with drug therapy (Table XXVIII). The outcome of status is largely dependent on the underlying neurological cause. In about 90% of cases, the condition can be controlled, but a failure to respond to emergency

**TABLE XXVI    Status Epilepticus: Emergency Management of Tonic–Clonic Status**

Stage 1 (0–10 min)
    Assess cardiorespiratory function
    Secure airway, administer oxygen, resuscitate
Stage 2 (0–60 min)
    Antiepileptic drug therapy
    Regular monitoring
    Intravenous lines
    Draw 50–100 mL blood for emergency investigations
    Glucose (50 mL of 50% solution) or thiamine (250 mg as HPIV
        parenterovite) where appropriate
    Treat acidosis where appropriate
Stage 3 (0–60/90 min)
    Etiology
    Pressor therapy where appropriate
    Treat other medical complications
Stage 4 (30–90 min)
    Intensive care
    Seizure and EEG monitoring
    Intracranial pressure monitoring where appropriate
    Maintenance of antiepileptic therapy

*Note.* Adapted from Shorvon (1994), copyright Cambridge University Press.

**TABLE XXVII    Emergency Antiepileptic Drug Regimen for Tonic–Clonic Status Epilepticus in Newly Presenting Adult Patients**

Premonitory Stage
    Lorazepam 4 mg iv, repeat once after 10 min. if necessary
or
    Diazepam 10–20 mg give IV or rectally, repeated once after
        15 min; later if status continues to threaten, at a rate not
        exceeding 2–5 mg/min
If seizures continue, treat as follows:
    Early status
        Lorazepam (IV) 4 mg bolus, repeated after 10 min (rate not
            critical)
If seizures continue 30 min after first injection, treat as follows:
    Established status
        Phenobarbital bolus of 10 mg/kg at a rate of 100 mg/min (e.g.,
            about 700 mg over 6 min in an adult)
or
        Phenytoin infusion at a dose of 18 mg/kg at a rate of 50 mg/min
            (e.g., 1000 mg in 20 min; with diazepam, if not already given,
            at 10 mg)
If seizures continue 30–60 min or longer, treat as follows:
    Refractory status
        Propofol 2 mg/kg, then 5–10 mg/kg/hour or:
        Thiopentone 100–250 mg, then 3–5 mg/kg/h or:
        Midazolam 5–10 mg, then 0.05–0.4 mg/kg/h
Anesthetic continued for 12–24 h after the last clinical or
    electrographic seizure, then dose tapered

*Note.* Adapted from Shorvon (1994), with the permission of Cambridge University Press.

**TABLE XXVIII    Medical Complications in Tonic–Clonic Status Epilepticus**

Cerebral
    Hypoxic or metabolic cerebral damage
    Seizure-induced cerebral damage
    Cerebral edema and raised intracranial pressure
    Cerebral venous thrombosis
    Cerebral hemorrhage and infarction
Cardiovascular, respiratory and autonomic
    Hypotension
    Hypertension
    Cardiac failure, tachy- and bradyarrhythmia, arrest
    Cardiogenic shock
    Respiratory failure
    Disturbances of respiratory rate and rhythm, apnea
    Pulmonary edema, hypertension, embolism
    Hyperpyrexia
    Sweating, hypersecretion, tracheobroncheal obstruction
    Peripheral ischemia
Metabolic
    Dehydration
    Electrolyte disturbance (especially hyponatremia, hyperkalemia,
        hypoglycemia)
    Acute renal failure (especially acute tubular necrosis)
    Acute hepatic failure
    Acute pancreatitis
Other
    Disseminated intravascular coagulopathy or multiorgan failure
    Rhabdomyolysis
    Fractures
    Infections (especially pulmonary, skin, urinary)
    Thrombophlebitis, dermal injury

*Note.* Adapted from Shorvon (1994), with the permission of Cambridge University Press.

**TABLE XXIX** Common Reasons for the Failure of Emergency Drug Therapy to Control Seizures in Status Epilepticus

---

Inadequate emergency antiepileptic drug therapy (especially the administration of drugs at too low a dose)

Failure to initiate maintenance antiepileptic drug therapy (seizures will recur as effect of emergency drug treatment wears off)

Failure to identify and control the underlying cause

Failure to identify or control hypoxia, hypotension, cardiorespiratory failure, or metabolic disturbance

Failure to identify or control other medical complications

Misdiagnosis (especially pseudo status)

---

*Note.* Adapted from Shorvon (1994), with the permission of Cambridge University Press.

therapy is often due to potentially reversible or preventable factors (Table XXIX).

# REFERENCES

Aldenkamp, A. P., Alpherts, W. C. J., Blennow, G., Elmquist, D., Heijbel, J., Nilsson, H. L., Sanstedt, P., Tonnby, B., Wahlander, L., and Wosse, E. (1993). Withdrawal of antiepileptic medication in children—Effects on cognitive function: The multicenter Holmfrid study. *Neurology* 43, 41–50.

Aldenkamp, A. P., Alpherts, W. C. J., Diepman, J., van t Slot, B., Overweg, J., and Vermeulen, J. (1994). Cognitive side effects of phenytoin compared with carbamazepine in patients with localisation related epilepsy. *Epilepsy Res.* 19, 37–43.

Aldenkamp, A. P., Alpherts, W. C. J., Moerland, M. C., Ottevanger, N., and Van Parys, J. A. P. (1987). Controlled release carbamazepine: Cognitive side effects in patients with epilepsy. *Epilepsia* 28, 507–514.

Annegers, J. F., Hauser, W. A., and Elveback, L. R. (1979). Remission of seizures and relapse in patients with epilepsy. *Epilepsia* 20, 729–737.

Baulac, S., Huberfeld, G., Gourfinkel-An, I., Mitropoulou, G., Beranger, A., Prud'homme, J., Baulac, M., Brice, A., Bruzzone, R., and LeGuern, E. (2001). First genetic evidence of GABA$_A$ receptor dysfunction in epilepsy: A mutation in the γ2-subunit gene. *Nat Genet* 28, 46–48.

Ben-Menachem, E., and French, J. A. (1997). Vigabatrin. *In* "Epilepsy: A Comprehensive Textbook" (J. Engel, Jr. and T. A. Pedley, Eds.), pp. 1609–1618. Lippincott-Raven, Philadelphia.

Berkovic, S. F., Howell, R. A., and Hay, D. A. (1998). Epilepsy in twins: Genetics of the major epilepsy syndromes. *Ann. Neurol.* 43, 435–445.

Berkovic, S. F., and Scheffer, I. E. (1999). Genetics of the epilepsies. *Curr. Opin. Neurol.* 12, 177–182.

Binnie, C. D. (1997). Lamotrigine. *In* "Epilepsy: A Comprehensive Textbook" (J. Engel, Jr. and T. A. Pedley, Eds.), pp. 1531–1540. Lippincott-Raven, Philadelphia.

Bourgeois, B. F. (1997). Primidone. *In* "Epilepsy: A Comprehensive Textbook" (J. Engel, Jr. and T. A. Pedley, Eds.), pp. 1581–1590. Lippincott-Raven, Philadelphia.

Brodie, M. J. (1990). "Established Anticonvulsants and Treatment of Refractory Epilepsy", *Lancet, A Lancet Review: Epilepsy*, pp. 17–20.

Brodie, M. J. (1992). Lamotrigine. *Lancet* 339, 1397–1400.

Bromfield, E. B. (1997). Ethosuximide and other succinimides. *In* "Epilepsy: A Comprehensive Textbook" (J. Engel, Jr. and T. A. Pedley, Eds.), pp. 1503–1508. Lippincott-Raven, Philadelphia.

Bronen, R. A. (1992). Epilepsy: The role of MR imaging. *Am. J. Resonance* 159, 1165–1174.

Browne, T. R. (1997). Fosphenytoin (Cerebyx). *Clin. Neuropharmacol.* 20, 1–12.

Browne, T. R. (1997). Phenytoin and other hydantoins. *In* "Epilepsy: A Comprehensive Textbook" (J. Engel, Jr. and T. A. Pedley, Eds.), pp 1557–1580. Lippincott-Raven, Philadelphia.

Cereghino, J. J., Biton, V., Abou-Khalil, B., Dreifuss, F., Gauer, L. J., and Leppick, I. E. (2000). Levetiracetam for partial seizures: Results of a double blind, randomised clinical trial. *Neurology* 55, 236–242.

Chadwick, D. W. (1994). Gabapentin. *Lancet* 343, 89–91.

Charlier, C., Singh, N. A., Ryan, S. G., Lewis, T. B., Reus, B. E., Leach, R. J., and Leppert, M. (1998). A pore mutation in a novel KQT-like potassium channel gene in an idiopathic epilepsy family. *Nat. Genet.* 18(1), 53–55.

Cockerell, O. C., Johnson, A. L., Sander, J. W., Hart, Y. M., and Shorvon, S. D. (1995). Remission of epilepsy: Results from the National General Practice Study of Epilepsy. *Lancet* 346(8968), 140–144.

Cockerell, O. C., Johnson, A. L., Sander, J. W., and Shorvon, S. D. (1997). Prognosis of epilepsy: A review and further analysis of the first nine years of the British National General Practice Study of Epilepsy, a prospective population-based study. *Epilepsia* 38(1), 31–46.

Commission on Classification and Terminology of the International League against Epilepsy. (1989). Proposal for revised classification of epilepsies and epileptic syndromes. *Epilepsia* 30, 389–399.

Commission on Genetics ILAE. (1993). Guidelines for the care of women of childbearing age with epilepsy. *Epilepsia* 34, 588–589.

Dalby, N. O., and Mody, I. (2001). The process of epileptogenesis: a pathophysiological approach. *Curr. Opin. Neurol.* 14, 187–192.

De Fusco, M., Becchetti, A., Patrignani, A., Annesi, G., Gambardella, A., Quattrone, A., Ballabio, A., Wanke, E., and Casari, G. (2000). The nicotinic receptor beta2 subunit is mutant in nocturnal frontal lobe epilepsy. *Nat. Genet.* 26(3), 275–276.

Delgado-Escueta, A. V., Janz, D., and Beck-Mannagetta, G. (1992). Pregnancy and teratogenesis in epilepsy. *Neurology* 42(Suppl. 5), 1–160.

Engel, J. Jr. (1997). Update on the surgical treatment of the epilepsies. *Neurology* 43, 1612–1617.

Escayg, A., MacDonald, B. T., Meisler, M. H., Baulac, S., Huberfeld, An., Brice, A., Leguern, E., Moulard, B., Chaigne, D., and others (2000). Mutations of SCN1A, encoding a neuronal sodium channel, in two families with GEFS + 2. *Nat. Genet.* 24(4), 343–345.

Gaily, E. K., Granstrom, M. L., Hiilesmaa, V. K., and Bardy, A. H. (1990). Head circumference in children of epileptic mothers: Contributions of drug exposure and genetic background. *Epilepsy Res.* 5, 217–222.

Glauser, T. A., and Pippenger, C. E. (2000). Controversies in blood-level monitoring: Reexamining its role in the treatment of epilepsy. *Epilepsia* 41(Suppl. 8), S6–S15.

Golding, J., Greenwood, R., Birmingham, K., and Mott, M. (1992). Childhood cancer, intramuscular Vitamin K, and pethidine given during labour. *Br. Med. J.* 305, 341–346.

Goodridge, D. M., and Shorvon, S. D. (1983). Epileptic seizures in a population of 6000. I. Demography, diagnosis and classification, and role of the hospital services. *Br. Med. J. Clin. Res. Ed.* 287(6393), 641–644.

Goodridge, D. M., and Shorvon, S. D. (1983). Epileptic seizures in a population of 6000. II. Treatment and prognosis. *Br. Med. J. Clin. Res. Ed.* 287(6393), 645–647.

Gram, L. (1990). "Epileptic Seizures and Syndromes. A Lancet Review: Epilepsy," pp. 7–10.

Gram, L. (1997). Oxcarbazepine. *In* "Epilepsy: A Comprehensive Textbook" (J. Engel, Jr. and T. A. Pedley, Eds.), pp. 1541–1546. Lippincott-Raven, Philadelphia.

Grant, S. M., and Faulds, D. (1992). Oxcarbazepine: A review of its pharmacology and therapeutic potential in epilepsy, trigeminal neuralgia and affective disorders. *Drugs* **43**, 873–888.

Hart, Y. M., Sander, J. W., Johnson, A. L., and Shorvon, S. D. (1990). National General Practice Study of Epilepsy: Recurrence after a first seizure. *Lancet* **336**(8726), 1271–1274.

Hart, Y. M., and Shorvon, S. D. (1995a). The nature of epilepsy in the general population. I. Characteristics of patients receiving medication for epilepsy. *Epilepsy Res.* **21**(1), 43–49.

Hart, Y. M., and Shorvon, S. D. (1995b). The nature of epilepsy in the general population. II. Medical care. *Epilepsy Res.* **21**(1), 51–58.

Hauser, W. A., Annegers, J. F., and Andersen, V. E. (1983). Epidemiology and genetics of epilepsy. *In* "Epilepsy" (A. A. Ward, J. Penry, and D. Purpura, Eds.), pp. 274–275. Raven Press, New York.

Hauser, W. A., and Kurland, L. T. (1975). The epidemiology of epilepsy in Rochester, Minnesota 1935 through 1967. *Epilepsia* **16**, 1–66.

Heller, A. J., Chesterman, P., Elwes, R. D., Crawford, D., Chadwick, D. W., Johnson, A. L., and Reynolds, E. H. (1990). Monotherapy for newly diagnosed epilepsy: A comparative trial and prognostic evaluation. *J. Neurol.* **237**, 27–28.

Herzog, A. G., and Schachter, S. C. (2001). Valproate and the polycystic ovarian syndrome: Final thoughts. *Epilepsia* **42**, 311–315.

Hirtz, D. G., Nelson, K. B., and Ellenberg, J. H. (1983). Seizures following childhood immunizations. *J. Pediatr.* **102**, 14–18.

Holmes, L. B., Harvey, E. A., Coull, B. A., Huntington, K. B., Khoshbin, S., Hayes, A. M., and Ryan, L. M. (2001). The teratogenicity of anticonvulsant drugs. *N. Engl. J. Med.* **344**, 1132–1138.

Isojarvi, J. I. T., Laatikainen, T. J., Pakarinen, A. J., Juntunen, K. T. S., and Millyla, V. V. (1993). Polycystic ovaries and hyperandrogenism in women taking valproate for epilepsy. *N. Engl. J. Med.* **329**, 1383–1388.

Jackson, J. H. (1873). On the anatomical, physiological and pathological investigation of epilepsies. West Riding Lunatic Asylum Medical Reports 3. *In* "Selected Writings of John Hughlings Jackson" (J. Taylor, Ed.), pp. 90–91. Hodder and Stoughton, London.

Jannuzzi, G., Cian, P., Fattore, C., Gatti, G., Bartoli, A., Monaco, F., and Perucca, E. (2000). A multicenter randomized controlled trial on the clinical impact of therapeutic drug monitoring in patients with newly diagnosed epilepsy. *Epilepsia* **41**(2), 222–230.

Janz, D., and Durner, M., (1997). Juvenile myoclonic epilepsy. *In* "Epilepsy: A Comprehensive Textbook" (J. Engel, Jr. and T. A. Pedley, Eds.), pp. 2389–2400. Lippincott-Raven, Philadelphia.

Jay, G. W., and Leestma, J. E. (1981). Sudden death in epilepsy: A comprehensive review of the literature and proposed mechanisms. *Acta Neurol. Scand.* **63**(Suppl. 82), 1–66.

Knudsen, F. U. (1991). Intermittent diazepam prophylaxis in febrile convulsions. *Acta Neurol. Scand.* (Suppl.), **83**, 1–24.

Knudsen, F. U., Paerregaard, A., Andersen, R., and Andersen, J. (1996). Long-term outcome of proplylaxis for febrile convulsions. *Arch. Dis. Child.* **74**, 13–18.

Ko, D. Y., Rho, J. M., DeGiorgio, C. M., and Sato, S. (1997). Benzodiazepines. *In* "Epilepsy: A Comprehensive Textbook" (J. Engel, Jr. and T. A. Pedley, Eds.), pp. 1475–1490. Lippincott-Raven, Philadelphia.

Konig, S. A., Schenk, M., Sick, C., Holm, E., Heubner, C., Weiss, A., Konig, I., and Hehlmann, R. (1999). Fatal liver failure associated with valproate therapy in a patient with Friedreich's disease: Review of valproate hepatotoxicity in adults. *Epilepsia* **40**(7), 1036–1040.

Krakow, K., Walker, M. C., Otoul, C., and Sander, J. W. A. S. (2001). Long-term continuation of levetiracetam in patients with refractory epilepsy. *Neurology* **56**, 1772–1774.

Krakow, K., Woermann, F. G., Symms, M. R., Allen, P. J., Lemieux, L., Barker, G. J., Duncan, J. S., and Fish, D. R. (1999). EEG-triggered functional MRI of interictal epileptiform activity in patients with partial seizures. *Brain* **122**(9), 1679–1688.

Kramer, L. D., and Reife, R. A. (1997). Topiramate. *In* "Epilepsy: A Comprehensive Textbook" (J. Engel, Jr. and T. A. Pedley, Eds.), pp. 1593–1598. Lippincott-Raven, Philadelphia.

Langan, Y., Nashef, L., and Sander, J. W. A. S. (2000). Sudden unexpected death in epilepsy: A series of witnessed deaths. *J. Neurol. Neurosurg. Psychiatry* **68**, 211–213.

Leppick, I. E. (1995). Tiagabine: The safety landscape. *Epilepsia* **36**(Suppl. 6), S10–S13.

Lhatoo, S. D., Johnson, A. L., Goodridge, D. M., MacDonald, B. K., Sander, J. W. A. S., and Shorvon, S. D. (2001). Mortality in epilepsy in the first 11 to 14 years after diagnosis: Multivariate analysis of a long-term, prospective, population-based cohort. *Ann. Neurol.* **49**, 336–344.

Lhatoo, S. D., Langan, Y., and Sander, J. W. A. S. (1999). Sudden unexpected death in epilepsy. *Postgrad. Med. J.* **75**, 706–709.

Lhatoo, S. D., and Sander, J. W. A. S. (2000). Stopping drug therapy in epilepsy. *Curr. Pharm. Des.* **6**(8), 861–863.

Lhatoo, S. D., and Walker, M. C. (1999). The safety and adverse event profile of topiramate. *Rev. Contemp. Pharmacother.* **10**(3), 185–191.

Lhatoo, S. D., Wong, I. C. K., Polizzi, G., and Sander, J. W. A. S. (2000). Long-term retention rates of lamotrigine, gabapentin, and topiramate in chronic epilepsy. *Epilepsia* **41**(12), 1592–1596.

Lindhout, D., and Schmidt, D. (1986). In-utero exposure to valproate and neural tube defects. *Lancet* **1**(8494), 1392–1393.

MacDonald, B. K., Cockerell, O. C., Sander, J. W. A. S., and Shorvon, S. D. (2000). The incidence and lifetime prevalence of neurological disorders in a prospective community-based study in the UK. *Brain* **123**, 663–664.

MacDonald, B. K., Johnson, A. L., Goodridge, D. M., Cockerell, O. C., Sander, J. W. A. S., and Shorvon, S. D. (2000). Factors predicting prognosis of epilepsy after presentation with seizures. *Ann. Neurol.* **48**, 833–841.

MacDonald, B. K., Johnson, A. L., Sander, J. W. A. S., and Shorvon, S. D. (1999). Febrile convulsions in 220 children—Neurological sequelae at 12 years follow-up. *Eur. Neurol.* **41**, 179–186.

Marson, A. G., Williamson, P. R., Hutton, J. L., Clough, J. E., and Chadwick, D. W. (2000). Carbamazepine versus valproate monotherapy for epilepsy. *Cochrane Database Syst. Rev.* **3**, CD001030.

Mathias, C. J., Deguchi, K., and Schatz, I. (2001). Observations on recurrent syncope and presyncope in 641 patients. *Lancet* **357**(9253), 348–353.

Mattson, R. H. (1997). Carbamazepine. *In* "Epilepsy: A Comprehensive Textbook" (J. Engel, Jr. and T. A. Pedley, Eds.), pp. 1491–1502. Lippincott-Raven, Philadelphia.

Mattson, R. H., Cramer, J. A., and Collins, J. F., *et al.* (1985). Comparison of carbamezepine, phenobarbital, phenytoin and primidone in partial and secondary generalised tonic clonic seizures. *N. Engl. J. Med.* **313**, 145–151.

McKinlay, I., and Newton, R. (1989). Intention to treat febrile convulsions with rectal diazepam, valproate, or phenobarbitone. *Dev. Med. Child. Neurol.* **31**, 617–625.

McNamara, J. O. (1999). Emerging insights into the genesis of epilepsy. *Nature* **399**, 15–22.

Medical Research Council Antiepileptic Drug Withdrawal Study Group (1991). Randomised study of antiepileptic drug withdrawal in patients in remission. *Lancet* **337**, 1175–1180.

Meldrum, B. S. (1990). Anatomy, physiology and pathology of epilepsy. Epilepsy: A Lancet Review. *Lancet*, 11–14.

MRC Vitamin Study Research Group. (1991). Prevention of neural tube defects: Results of the MRC vitamin study. *Lancet* **338**, 132–137.

Musicco, M., Beghi, E., Bordo, B., Viani, F., Hauser, W. A., Nicolosi, A., Fratiglioni, L., Bogliun, G., Aloisi, P., Cerone, G., and others.

(1993). Randomized clinical trial on the efficacy of antiepileptic drugs in reducing the risk of relapse after a first unprovoked tonic-clonic seizure. *Neurology* 43(3 I), 478–483.

Musicco, M., Beghi, E., Solari, A., and Viani, F. (1997). Treatment of first tonic-clonic seizure does not improve the prognosis of epilepsy. *Neurology* 49(4), 991–998.

Nashef, L., Fish, D. R., Garner, S., Sander, J. W., and Shorvon, S. D. (1995). Sudden death in epilepsy: A study of incidence in a young cohort with epilepsy and learning difficulty. *Epilepsia* 36(12), 1187–1194.

Newton, R. W. (1988). Randomized controlled trials of phenobarbitone and valproate in febrile convulsions. *Arch. Dis. Child.* 63, 1189–1191.

Nilsson, L., Farahmand, B. Y., Persson, P. G., and Tomson, T. (1999). Risk factors for sudden unexpected death in epilepsy: A case control study. *Lancet* 353, 888–893.

Omtzigt, J. G., Los, F. J., and Grobbee, D. E., *et al.* (1992). The risk of spina bifida aperta after first trimester exposure to sodium valproate in a prenatal cohort. *Neurology* 42(4 Suppl. 5), 119–125.

Placencia, M., Sander, J. W., Shorvon, S. D., Roman, M., Alarcon, F., Bimos, C., and Cascante, S. (1993). Mar. Antiepileptic drug treatment in a community health care setting in northern Ecuador: A prospective 12-month assessment. *Epilepsy Res.* 14(3), 237–244.

Rho, J. M., Donevan, S. D., and Rogawski, M. A. (1994). Mechanism of action of the anticonvulsant felbamate: Opposing effects on N-Methyl-D-Aspartate and aminobutyric acid A receptors. *Ann. Neurol.* 35, 229–234.

Rogawski, M. A., and Porter, R. J. (1990). Antiepileptic drugs: Pharmacological mechanism and clinical efficacy with consideration of promising developmental stage compounds. *Pharmacol. Rev.* 42, 223–286.

Rosa, W. F. (1991). Spina bifida in infants of women treated with carbamazepine during pregnancy. *N. Engl. J. Med.* 324, 674–677.

Rosman, N. P., Colton, T., Labazzo, J., Gilbert, P. L., Gardella, N. B., and Kaye, E. M., *et al.* (1993). A controlled trial of diazepam administered during febrile illnesses to prevent recurrence of febrile seizures. *N. Engl. J. Med.* 329, 79–84.

Rowan, A. J. (1997). Valproate. *In* "Epilepsy: A Comprehensive Textbook" (J. Engel, Jr. and T. A. Pedley, Eds.), pp. 1599–1608. Lippincott-Raven, Philadelphia.

Sander, J. W. A. S. (1993). Some aspects of prognosis in the epilepsies: A review. *Epilepsia* 34, 1007–1016.

Sander, J. W. A. S., and Shorvon, S. D. (1996). Epidemiology of the epilepsies. *J. Neurol. Neurosurg. Psychiatry* 61(433), 443.

Schachter, S. C. (1998). Vagus nerve stimulation. *Epilepsia* 39, 677–686.

Schmidt, D. (1982). Two antiepileptic drugs for intractable epilepsy with complex partial seizures. *J. Neurol. Neurosurg. Psychiatry* 45, 1119–1124.

Schmidt, D. (1993). Epilepsy in women. *In* "A Textbook of Epilepsy" (J. Laidlaw, A. Richens, and D. W. Chadwick, Eds.), pp. 637–644. Churchill Livingstone, Edinburgh.

Schmidt, D. (1994). Felbamate: Successful development of a new compound for the treatment of epilepsy. *Epilepsia* 34(Suppl. 7), S30–S33.

Schmidt, D., and Richter, K. (1986). Alternative single anticonvulsant drug therapy for refractory epilepsy. *Ann. Neurol.* 19, 85–87.

Schoenenberger, R. A., Tanasijevic, M. J., Jha, A., and Bates, D. W. (1995). Appropriateness of antiepileptic drug level monitoring. *JAMA* 274(20), 1622–1666.

Schwartzkroin, P. A. (1994). Cellular electrophysiology of human epilepsy. *Epilepsy Res.* 17, 185–192.

Seino, M., and Tsugutaka, I. (1997). Zonisamide. *In* "Epilepsy: A Comprehensive Textbook" (J. Engel, Jr. and T. A. Pedley, Eds.), pp. 1619–1626. Lippincott-Raven, Philadelphia.

Shinnar, S., Berg, A. T., Moshe, S. L., Kang, H., O'Dell, C., Alemany, Goldensohn, E. S., and Hauser, W. A. (1994). Discontinuing antiepileptic drugs in children with epilepsy: A prospective study. *Ann. Neurol.* 35(5), 534–545.

Shorvon, S. D. (1994). "Status Epilepticus: Its Clinical Features and Treatment in Children and Adults." University Press, Cambridge.

Shorvon, S. D. (2000). Oxcarbazepine: A review. *Seizure* 9, 75–79.

Singh, N. A., Charlier, C., Stauffer, D., DuPont, B. R., Leach, R. J., Melis, R., Ronen, G. M., Bjerre, I., Quattlebaum, T., and others. (1998). A novel potassium channel gene, KCNQ2, is mutated in an inherited epilepsy of newborns. *Nat. Genet.* 18(1), 25–29.

Sisodiya, S. M., Lin, W.-R., Squier, M. V., and Thom, M. (2001). Multidrug-resistance protein 1 in focal cortical dysplasia. *Lancet* 357(9249), 42–43.

Smith, D., Defalla, B. A., and Chadwick, D. W. (1999). The misdiagnosis of epilepsy and the management of refractory epilepsy in a specialist clinic. *Mont. J. Assoc. Physicians* 92(1), 15–23.

Steegers-Theunissen, R. P. M., Renier, W. O., Borm, G. F., Thomas, C. M. G., and Merkus, H. M. W., *et al.* (1994). Factors influencing the risk of abnormal pregnancy outcome in epileptic women: A multicentre prospective study. *Epilepsy Res.* 18, 261–270.

Steinlein, O. K., Mulley, J. C., Propping, P., Wallace, R. H., Phillips, H. A., Sutherland, G. R., Scheffer, I. E., and Berkovic, S. F. (1995). A missense mutation in the neuronal nicotinic acetylcholine receptor alpha4 subunit is associated with autosomal dominant nocturnal frontal lobe epilepsy. *Nat. Genet.* 11(2), 201–203.

Temkin, N. R. (2001). Antiepileptogenesis and seizure prevention trials with antiepileptic drugs: Meta-analysis of controlled trials. *Epilepsia* 42, 515–524.

Temkin, N. R., Dikmen, S. S., Anderson, G. D., Wilensky, A. J., Holmes, M. D., Cohen, W., Newell, D. W., Nelson, P., Awan, A., Winn, H. R. (1999). Valproate therapy for prevention of posttraumatic seizures: A randomized trial. *J. Neurosurg.* 91, 593–600.

Theodore, W. H. (1997). Felbamate. *In* "Epilepsy: A Comprehensive Textbook" (J. Engel, Jr. and T. A. Pedley, Eds.), pp. 1509–1514. Lippincott-Raven, Philadelphia.

Uhari, M., Rantala, H., Vainionpaa, M., and Kurttila, B. M. (1995). Effect of acetaminophen and of low intermittent doses of diazepam on prevention of recurrences of febrile seizures. *J. Pediatr.* 126, 991–995.

Wallace, R. H., Marini, C., Petrou, S., Harkin, L. A., Bowser, D. N., Panchal, R. G., Williams, D. A., Sutherland, G. R., Mulley, J. C., Scheffer, I. E., and others. (2001). Mutant GABA_A receptor γ2-subunit in childhood absence epilepsy and febrile seizures. *Nat. Genet.* 28, 49–52.

Wallace, H., Shorvon, S., and Tallis, R. (1998). Age-specific incidence and prevalence rates of treated epilepsy in an unselected population of 2,052,922 and age-specific fertility rates of women with epilepsy. *Lancet* 352(9145), 1970–1973.

Wallace, R. H., Wang, D. W., Singh, R., Scheffer, I. E., George, A. L. J., Phillips, H. A., Saar, K., Reis, A., Johnson, E. W., and others. (1998). Febrile seizures and generalized epilepsy associated with a mutation in the Na+-channel beta1 subunit gene SCN1B. *Nat. Genet.* 19(4), 366–370.

Wiebe, S., Blume, W. T., Girvin, J. P., and Eliasziw, M. (2001). A randomized, controlled trial of surgery for temporal-lobe epilepsy. *N. Engl. J. Med.* 345, 311–318.

Winter, M. D., and Tozer, T. N. (1986). Phenytoin. *In* "Applied Pharmacokinetics" (W. E. Evans, J. J. Schontag, and W. J. Jusko, Eds.), pp. 439–539. Applied Therapeutics, Spokane.

Wolf, S. M., Carr, A., David, D. C., Davidson, S., Dale, E., Forsythe, A., Goldenberg, E. D., Hanson, R., Lulejian, G. A., Nelson, M. A., and others. (1977). The value of phenobarbital in the child who has had a single febrile seizure: A controlled prospective study. *Pediatrics* 59, 378–385.

Wong, I. C. K., and Lhatoo, S. D. (2000). Adverse reactions to new anticonvulsant drugs. *Drug Saf.* 23, 35–56.

*Cognitive and Behavioral Disorders*

# CHAPTER 21
# Surgical Therapy of Epilepsy

Soheyl Noachtar, Peter A. Winkler, and Hans Otto Lüders

## CLINICAL ASPECTS

The objective of surgical treatment of epilepsy is improvement in the psychosocial status of patients with drug-resistant epileptic seizures by completely controlling or at least significantly reducing seizure occurrence without causing incapacitating neurological or neuropsychological deficits.

Confirmation of the diagnosis of epilepsy and its medical intractability is the essential prerequisite for epilepsy surgery. After excluding nonepileptic events such as psychogenic pseudoseizures, the clinician must establish that adequate drug trials, including verification of compliance, have been performed. Criteria to determine medical intractability vary somewhat, but the points listed in Table I are usually accepted as the minimum. Most epileptologists feel that adequate trials of two to three drugs of first choice in monotherapy are a minimum requirement to establish medical intractability (Bourgeois, 1991) (see Chapter 20 in this volume). The efficacy of an antiepileptic drug should be determined by increasing the daily dose until either seizure control is achieved or unacceptable side effects occur. The recommended "therapeutic range" as a criterion for drug failure is no longer acceptable because some patients will become seizure free at serum drug levels above the "therapeutic range" without having toxic side effects (Hermanns *et al.*, 1996; Lesser *et al.*, 1984). Some authors also recommend testing drug combinations, but side effects are often increased despite little or no improvement in seizure control (Schmidt, 1982). Consideration of surgical therapy in appropriate candidates should not be excessively delayed by repeated trials with adjunctive drugs. The rationale of early surgery is based on the notion that a therapeutic intervention early in the course of epilepsy will avoid devastating psychosocial effect of chronic epilepsy and, therefore, improve the long-term quality of life outcome. If acceptable seizure control is not achieved a patient should be referred to an epilepsy surgery center for further evaluation (Table II).

## NATURAL COURSE

The prevalence of epilepsy has been assessed at 0.4 to 0.8% (Hauser and Hesdorffer, 1990). Whereas about 70–80% will enter 5-year remission (Annegers *et al.*, 1979), approximately 20–30% develop chronic intractable epilepsy (Sillanpää, 2000). There is some evidence that a good prognosis is related to the rapidity with which seizures are brought under control. Poor prognostic factors include a high number of seizures, a long duration of epilepsy, partial seizures, psycho-social handicap, and psychiatric and neurological deficits. Patients who do not respond sufficiently to medications have psychosocial problems and are subject to cognitive and sedative drug effects. About 70,000 persons in the United States might be eligible for surgical therapy (Hauser, 1992).

The concern that recurrent seizures may have detrimental effects on the brain similar to kindling and secondary epileptogenesis (Morrell, 1985) has led to a bias toward early surgical intervention (Moshe and Shinnar, 1993). Whether the concept of secondary epileptogenesis, which is derived from animal studies, has clinical implications for human epilepsy remains controversial (Goldensohn, 1984; Lüders, 2001). In any case, early surgical intervention in children with focal epilepsies offers the possibility of preventing the psychosocial morbidity and neuropsychological decline that are linked to intractable epilepsy (Lindsay *et al.*, 1984).

**TABLE I**  Definition of Medical Intractability

No acceptable seizure control despite
— Two or three drugs of first choice
— Appropriate monotherapy trials

## PRINCIPLES OF THERAPY

The results of epilepsy surgery depend on how well the epileptogenic zone (region of cortex that can generate seizures) can be identified and how completely it can be removed without resecting functional essential cortex (Awad *et al.*, 1991). If complete resection is not feasible, partial resection of epileptogenic cerebral tissue may also be worthwhile in selected cases, although the results are less favorable (Wyllie *et al.*, 1987). Another approach has been interruption of pathways of seizure spread, which was the rationale for resection of the anterior temporal lobe in isolated patients with posterior temporal or extratemporal seizure onset, although results were relatively unfavorable when compared to complete resection of the epileptogenic zone (Fish *et al.*, 1991).

Electroencephalography is the most specific method to define epileptogenic cortex. Interictal epileptiform discharges, particularly if consistent over time, can provide useful information (Ojemann, 1987). Ictal EEG video recording is, however, considered by many to be critical in localizing the epileptogenic zone. A careful analysis of the first clinical signs and symptoms of a seizure and of the evolution of the seizure symptomatology can also provide important clues (Noachtar, 2000; Lüders and Noachtar, 2000, 2001). One must keep in mind, however, that often an epileptic seizure arises from a "silent" region of cortex and would remain asymptomatic unless it spreads to "eloquent" cortex such as primary motor, primary sensory, or supplementary sensorimotor areas (Lüders and Awad, 1991). Invasive EEG techniques including stereotactically implanted multicontact depth electrodes or subdural strips and grids have been developed to further define the epileptogenic zone in selected patients (Phase II in Figure 1).

**TABLE II**  Candidacy for Considering Epilepsy Surgery

— Confirmed diagnosis of epilepsy
— Medical intractability
— Disabling seizures
— Resectable focus (except callosotomy candidates, vagus nerve stimulation, and deep brain stimulation)
— Motivated patient
— No progressive underlying cause (except Rasmussen's encephalitis)
— High probability that better seizure control will improve quality of life

Although surface EEG recordings are less sensitive than invasive studies, they provide the best overview and therefore the most efficient way to define the approximate localization of the epileptogenic zone. Invasive electrodes are subject to sampling errors if misplaced and should be used only after exhaustive non-invasive evaluations have (1) failed to localize the epileptogenic zone and (2) led to a testable hypothesis regarding this localization. Invasive EEG studies are associated with additional risks (Van Buren, 1987) that are justifiable only if there is a good chance of obtaining useful and essential localizing information.

Electrical stimulation of the cortex can be performed either intraoperatively or extraoperatively and is useful to delineate the limits of cortical resection when the epileptogenic zone is adjacent to or overlapping functional cortex (Lüders *et al.*, 1988).

Structural imaging with high-resolution MRI frequently provides essential information. If a structural lesion is found and its location is consistent with clinical and EEG data regarding the epileptogenic zone, removal of the lesion alone may be sufficient to control seizures (Cascino *et al.*, 1992). In other patients, removal of additional cortex, identifiable with intraoperative or extraoperative invasive recording may be necessary. This is frequently the case in patients with epilepsy caused by cortical dysplasia, posttraumatic lesions, and cerebrovascular insults. Definition of the anatomic relationship of epileptogenic cortex to a lesion is a crucial issue in the evaluation of patients considered for epilepsy surgery. Space-occupying progressive brain lesions that themselves are reasons for surgery are considered elsewhere in this volume (see Chapters 62–64 in this volume).

Interictally, epileptogenic zones particularly in the temporal lobe are frequently associated with reduced regional cerebral metabolism that can be detected by [$^{18}$F]Deoxyglucose PET (Engel *et al.*, 1990). Interictal SPECT may show a corresponding area of reduced cerebral blood flow, but is less sensitive than PET, and may provide misleading information. Ictal SPECT studies particularly in temporal lobe epilepsy, however, are more reliable and show a region of increased cerebral perfusion (Newton *et al.*, 1992). In extratemporal epilepsies, ictal SPECT localizes better in patients with frontal seizure onset than in patients with parietooccipital seizure onset (Noachtar *et al.*, 1998).

Neuropsychological testing is an integral part of the evaluation. In some patients, selective deficits may provide confirmatory information regarding location of abnormal cortex. More importantly, this testing may provide information about the patient's preoperative cognitive functioning which is helpful for counseling about the risks of cognitive deficits after surgery and for planning postsurgical rehabilitation. Although seizure

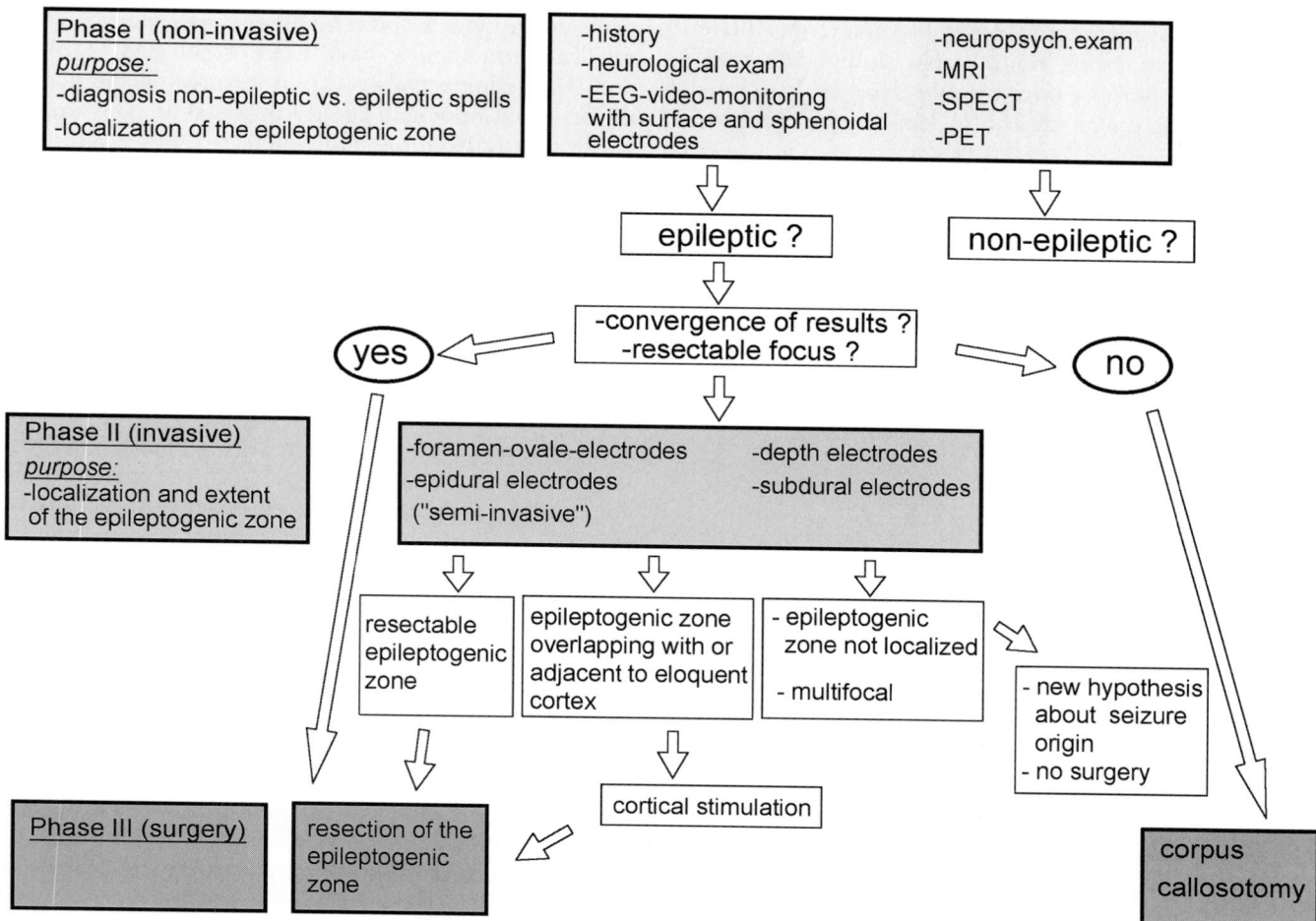

**FIGURE 1** Presurgical evaluation of epilepsies.

freedom is the obvious objective of epilepsy surgery, the gain of seizure control must be weighed against the attendant risks of cognitive deficits that may be associated with epilepsy surgery. In adults an IQ below 70 has been considered a poor prognostic factor for resective epilepsy surgery since it usually indicates diffuse brain damage which not infrequently is associated with a widespread epileptogenic zone. However, a resectable epileptogenic zone can sometimes be identified in such individuals, and its resection may favorably affect quality of life. In addition, patients with Lennox–Gastaut syndrome, who usually have IQs below 70, may qualify for callosotomy. Moreover, in selected infants whose developmental delay appears to result at least in part from the effects of seizures, this delay may itself be a reason to refer a child with secondary generalized or severe focal epilepsy for surgical evaluation (Chugani *et al.*, 1990).

An additional role of neuropsychology is in the performance and interpretation of the intracarotid amobarbital (Wada) test, which is used preoperatively for lateralization of language and memory. This is still commonly performed despite continuing controversy regarding its ability to prevent postoperative deficits by altering the area of resection.

Convergence of data from structural (MRI) and functional studies (EEG, SPECT, PET, and Wada test/ neuropsychology) is crucial for planning epilepsy surgery. Conflicting results among these tests may be resolved by invasive monitoring and stimulation or may in some patients lead to a decision not to operate, either because the epileptogenic zone cannot be adequately localized or because the risks of removing it are unacceptable (see Figure 1).

## PRACTICAL MANAGEMENT

Presurgical evaluation should (1) select those patients who could benefit from surgery and (2) exclude those whom surgery would not help and may harm. For this purpose a highly specialized multidisciplinary team is required. Table II lists the selection criteria for considering a patient for epilepsy surgery. Presurgical evaluation of epilepsy patients consists of three consecutive phases, the first being noninvasive studies (Phase I, Figure 1). The purpose of this noninvasive evaluation is to establish the diagnosis and to localize the epilepto-

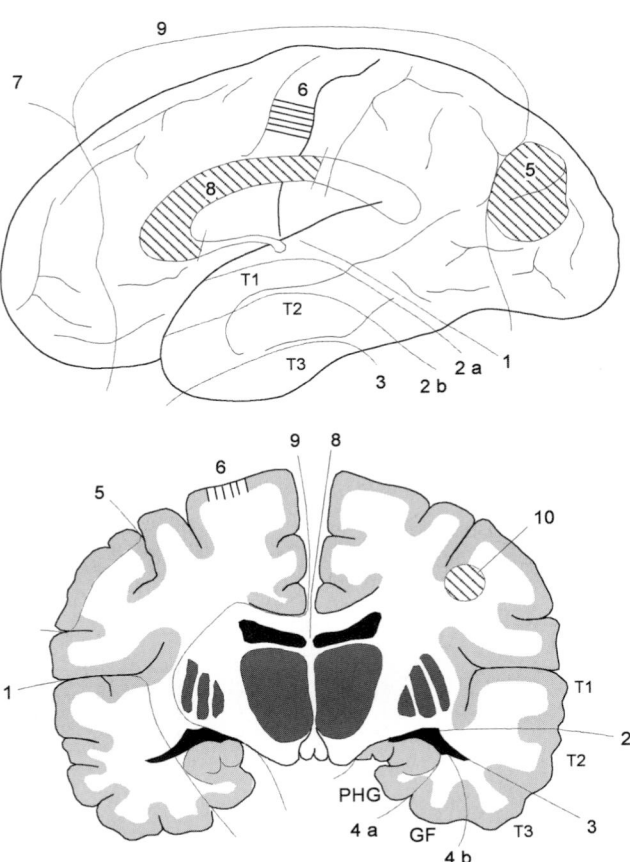

**FIGURE 2**  Surgical procedures in epilepsy surgery (modified from Stodieck, 1993).

---

T1 = superior temporal gyrus, T2 = medial temporal gyrus, T3 = inferior temporal gyrus, FG = fusiform gyrus, PHG = parahippocampal gyrus, CC = corpus callosum

**Temporal**
1. Lateral temporal resection                                      (Bailey and Gibbs, 1951)
2. "En-bloc" resection                                               (Falconer, 1971)
   a. Nondominant hemisphere
   b. Dominant hemisphere
3. Basal temporal resection                                        (Shimizu et al., 1990)
4. Selective amygdalohippocampectomy          (Niemeyer, 1958; Yasargil et al., 1985)

   a. "Superselective"
   b. Including parahippocampal gyrus and fusiform gyrus

**Extratemporal**
5. Cortical resection                                   (Olivier, 1992)
6. Multiple subpial transections               (Morrell et al., 1989)
7. Lobectomy, multilobe resection           (Rasmussen, 1963, 1987)
8. Anterior 2/3 callosotomy                     (Wilson et al., 1982)
9. Functional hemispherectomy              (Rasmussen, 1983)
10. Stereotactic lesionectomy                   (Kelly, 1986)

genic zone. Once a correct diagnosis of epilepsy and its medical intractability have been established (Tables I and II), the impact of recurrent seizures on the quality of life of an individual should be assessed. This impact varies greatly from one individual to another, although certain generalizations can be made. One aura per month, for example, would probably not interfere with a normal lifestyle and would not justify the risk of surgery; the same number of seizures with loss of consciousness or falling, however, probably would. This concept of disabling seizures, however, is very difficult to apply in early infancy and childhood. In this age group stagnation or regression of cognitive development is the crucial factor (Wyllie, 1991).

The therapeutic procedure in epilepsy surgery depends on the location and extent of the epileptogenic zone. Many patients referred for epilepsy surgery have seizures arising from the temporal lobe, which may be divided into the mesial structures and the lateral neocortex. In mesial temporal lobe epilepsies the limits of the epileptogenic zone are defined by anatomical structures and are usually restricted to the hippocampus, the amygdala, and the parahippocampal gyrus (Figure 2). Epilepsy surgery in these patients aims at removal of these structures. In regions other than the mesial temporal lobe the localization of the epileptogenic zone is more variable and usually more difficult to define (Lüders and Awad, 1991).

Concordance of noninvasive results implicating a resectable focus allows proceeding to epilepsy surgery (Phase III) based on noninvasive studies only; otherwise, Phase II, namely, an invasive evaluation, is needed. These investigations are indicated and justified only if noninvasive studies are inconclusive or reveal discrepant results, but still support a testable hypothesis of a resectable focus. Under these circumstances, properly placed invasive electrodes frequently provide useful additional information about the localization and extent of the epileptogenic zone. If Phase I reveals multifocality or diffuse epileptogenicity, the patient does not qualify for resective surgery, but may be a candidate for corpus callosum section (see below).

Figure 2 shows different surgical procedures commonly used in epilepsy surgery. Outcome and classification of outcome are given in Tables III and IV, respectively. Morbidity and mortality of surgical treatment of epilepsy are shown in Table V.

## MESIAL TEMPORAL EPILEPSIES

In many patients with intractable unilateral mesial temporal lobe epilepsies, "en-bloc" anterior temporal resection (Figure 2) can be performed on the basis of only noninvasive studies if clinical, EEG, and MRI data

**TABLE III**  Outcome in Epilepsy Surgery

| | Anterior temporal lobectomy | | Neocortical resections | | Neocortical lesionectomy | | Hemispherectomy | | Callosotomy | |
|---|---|---|---|---|---|---|---|---|---|---|
| | n | % | n | % | n | % | n | % | n | % |
| Seizure-free | 2429 | 67.9 | 363 | 45.1 | 195 | 66.6 | 128 | 67.4 | 43 | 7.6 |
| Improved | 860 | 24.0 | 283 | 35.2 | 63 | 21.5 | 40 | 21.1 | 343 | 60.9 |
| Not improved | 290 | 8.1 | 159 | 19.8 | 35 | 11.9 | 22 | 11.6 | 177 | 31.4 |
| Total | 3579 | 100 | 805 | 100 | 293 | 100 | 190 | 100 | 563 | 100 |

Modified with permission from Engel *et al.*, 1993.

**TABLE IV**  Proposal for a New Classification of Outcome with Respect to Epileptic Seizures following Epilepsy Surgery[a]

| Outcome classification[b] | Definition |
|---|---|
| 1[c] | Completely seizure free; no auras[d] |
| 2 | Only auras; no other seizures |
| 3[e] | One to three seizure days per year[g]; ± auras |
| 4[f] | Four seizure days per year to 50% reduction of baseline seizure-days[g], ± auras |
| 5[h] | Less than 50% reduction of baseline seizure-days to 100% increase of baseline seizure-days; ± auras |
| 6 | More than 100% increase of baseline seizure-days; ± auras |

From Weiser *et al.* (2001), reprinted with permission by the journal *Epilepsia*.

[a]Seizures during the first month postsurgery are not counted. These may be related to the surgery ("neighborhood-seizures") and do not predict long-term outcome.

[b]Seizure outcome class is determined for each year at yearly intervals following the date of surgery. Patients often change from one classification to another from year to year. The "last available aoutcome" can be indicated in a separate column with the specification that a minimal follow-up of 1 year postsurgery is required. It should further be specified by mean ± standard deviation, and minimal and maximal follow-up. It is understood that the last available year hs to be taken into consideration. For example, a patient had surgery on September 5, 1992 and is followed to end of July 1998. The period that counts for the "last available outcome" would then be September 5, 1996, to September 5, 1997. (See also footnote to Table 3.) Only where the length of the follow-up period is critical, the last available outcome can be calculated for the last 12-month period (i.e., in this example, July 31 1997, to July 31, 1998). It should be indicated whether the year prior to the last surgical anniversary or the last 12-month period is taken into account.

[c]In the year-by-year reporting, class 1 should contain a subgroup "completely seizure free since surgery; no auras" (see Table 3 and Fig. 1). It is recommended that those patients in class 1 without AEDs also be separately reported.

[d]An aura is defined as a simple partial seizure that is not observable by witnesses (i.e., pure subjective experience), and does not affect the patient's function. Auras should be counted only if they are of short duration and are similiar or identical to the auras the patient experienced prior to surgery. If the history is not clear, auras should not be counted. Postoperatively, a large number of patients are anxious and may report subjective feelings that are not epileptic.

[e]Class 3 is included because it is quite common for patients to have "rare seizures" postoperatively. These are often nocturnal and often tonic-clonic. It is practical experience that some patients have these rare seizures on certain, usually provocative-occasions, and usually can cope with fairly well.

[f]In classes 4, 5, and 6, the change of seizure-free days with reference to baseline should further be detailed in at least 10% subcategories, or (better) in absolute numbers to allow for cumulative data reporting. For plain statistical reasons, classes 4, 5 and 6 can be replaced by an absolute scale denoting the percentage improvement (classes 4 and 5) or worsening (classes 5 and 6) in seizure days frequency above/below baseline, in 10% increments/decrements.

[g]The number of "baseline seizure-days" is calculated by determining the seizure-day frequency during the 12 months prior to surgery, with correction for the effects of AED reduction during diagnosic evaluation. We admit that this classification adequately works only in patients with chronic epilepsy who have had seizures for more than 12 months. Exceptions are, however, patients who have an outcome falling into class 1 to 3. For patients who had a seizure onset less than 12 months before surgery and who have an outcome outside classes 1, 2 and 3, no baseline is available. In such rare instances where the seizure onset is less than 12 months and where the patients do not fall into outcome classes 1–3, the baseline can be approximated but should be properly documented. For instance, in a patient who had a seizure onset that began nine months before surgery, with one seizure day in the first month, two in the second, three in the third, building up to several a week before surgery, the last month before surgery could be used as baseline, but this has to be documented.

[h]Class 5 means that the surgery either did not significantly improve or significantly worsen the clinical situation of the patient. After a long discussion and on the advice of statisticians, we decided to use 100% increase of baseline seizure-days (vs. 50%) on the argument that for mathematical and biological reasons it is justified to assume that seizure frequencies and their changes after surgery have a normal log distribution (logarhythmische Normalverteilung). This implies that a 50% decrease is equivalent to a 100% increase.

TABLE V   Morbidity and Mortality in Epilepsy Surgery[a]

| | Anterior temporal lobectomy | Extratemporal resections | Hemispherectomy | Callosotomy |
|---|---|---|---|---|
| Number of patients | 1911 | 432 | 84 | 130 |
| Morbidity[b] | 5.1% | 5.8% | 16.7%[c] | 9.2% |
| Mortality | 0.5% | 0.0% | 3.6%[c] | 1.5% |

[a]From Van Buren, 1987, copyright Lippincott Williams & Wilkins.

[b]Complications like infection and hemorrhage, hemiparesis, hemianopsia, aphasia, acute hydrocephalus, memory deficits, split brain syndrome, and psychic problems.

[c]In most of these patients anatomical hemispherectomies, which bear more risks than functional hemispherectomies, were performed.

are convergent (Sperling et al., 1992; Winkler et al., 1999). Noninvasive EEG monitoring includes sphenoidal electrodes, which are particularly sensitive to mesial temporal foci (Morris et al., 1989). In these cases MRI frequently demonstrates mesial hippocampal sclerosis (Jackson et al., 1990). Mesial temporal resections have a favorable outcome with regard to seizure control (Table IV). Good results depend on the completeness of removal of mesial temporal structures (Nayel et al., 1991). Amygdalohippocampectomy is a surgical option in these patients (Wieser and Yasargil, 1982) (Figure 2). Patients considered for dominant temporal lobectomy are at risk for verbal memory disturbances, particularly if their preoperative mesial temporal structures are not atrophic and memory skills are above normal (Chelune et al., 1991). There is some evidence that in such cases more selective resections such as amygdalohippocampectomy result in less cognitive deficits (Foldvary, personal communication).

Twenty to fifty percent of patients with medically intractable temporal lobe epilepsies will present with independent bitemporal interictal epileptiform EEG abnormalities and nonlateralized ictal surface recordings. In a considerable number of these patients invasive EEG recordings will demonstrate the origin of all or most seizures in one temporal lobe and lead to successful epilepsy surgery (So et al., 1989). Depth electrodes in both temporal lobes are considered the most sensitive technique in these cases, but foramen-ovale electrodes provide an alternative semiinvasive recording technique with less risk (Wieser et al., 1985).

## EXTRAMESIAL TEMPORAL EPILEPSIES

In extramesial temporal neocortical epilepsies the limits of resection for epilepsy surgery are more difficult to define, because the limits of the epileptogenic zones are more variable. Neocortical epilepsies are divided into lesional and nonlesional. In the former, MRI can guide epilepsy surgery. Frequently the epileptogenic

zone lies in the vicinity of the structural lesion. Invasive studies may be needed to define the epileptogenic zone. If surface EEG and imaging studies do not localize the focus with sufficient precision to guide invasive studies, semiinvasive EEG recordings may be helpful, either in providing localizing information or in guiding placement of invasive electrodes (Figure 1) (Noachtar, 2001). Improvement of imaging techniques such as MRI, PET and ictal SPECT, however, leads to less use of semiinvasive electrodes (Noachtar, 2001). If the presumed epileptogenic zone lies near eloquent cortex such as speech area or primary motor cortex, its relationship to functional cortex must be defined by intraoperative or extraoperative recording and stimulation techniques (Lüders et al., 1987) (Figure 1). Preoperative localization of subdural electrodes is improved with three-dimensional MRI images showing the exact relation of the subdural electrodes to the cortex relief (Winkler et al., 2000). Based on invasive ictal recordings and results of electrical stimulation of the cortex, epilepsy surgery can be tailored individually by resecting the epileptogenic zone and leaving functionally intact cortex.

The results of lesional epilepsy surgery are better because epilepsy surgery can be based on data from structural imaging as well as functional tests (EEG, SPECT, and PET), whereas in nonlesional epilepsy surgery must rely on functional tests only (Van Ness, 1991). Surgical techniques for extratemporal surgery are shown in Figure 2. Stereotactic guided lesionectomies, which have the advantage of allowing resection of deep-seated lesions involving eloquent cortex with relatively low surgical morbidity, have been reported to achieve similar good results (56% Class I and 74% Class I and Class II, Table IV) (Cascino et al., 1992).

Patients with intractable seizures and severe hemispheric damage often benefit from hemispherectomy. Most patients qualifying for hemispherectomy have medically intractable motor seizures arising from one hemisphere. With rare exceptions only children with a severe hemiparesis and no remaining useful hand function qualify for this surgery. The essential evaluation consists of clinical data and imaging and noninvasive EEG video studies. For hemispherectomy to be considered, these studies should show that seizures arise only or almost exclusively from the damaged hemisphere, whereas the other hemisphere is relatively intact. A hemispherectomy is indicated only in patients who do not qualify for a more limited resection. Anatomic hemispherectomy, which refers to removal of the cortex of one whole hemisphere including the mesial temporal structures, was abandoned because of long-term complications, particularly superficial cerebral hemosiderosis which was almost invariably fatal. This unacceptable complication led to the development of the "functional hemispherectomy," which consists of total disconnection of one hemisphere but only a very limited resection of brain tissue (the frontal and occipital lobes actually remain in place) (Rasmussen, 1983) (Figure 2).

The results with respect to seizure outcome are as good as in temporal lobe surgery with 2/3 to 3/4 of the patients becoming seizure free (Table III). The risk of surgery must be weighed against the devastating and life-threatening effect of seizures and the chance for a marked developmental progress in these infants and young children. If a hemispherectomy of the dominant side is considered, careful investigation of language representation is required. In young children, the intracarotid amobarbital test is difficult to perform to establish speech dominance. Several studies indicate, however, that language will almost invariably transfer to the contralateral hemisphere if the disease starts before 6 years of age. The exact age limit is not yet known, but even after the age of 6 years language shift may still occur, although speech in these children is more likely to be impaired (Taylor, 1991). In progressive hemispheric syndromes like Rasmussen encephalitis (Villemure et al., 1991), early hemispherectomy, even before the neurological deficit is maximal, may be considered to prevent intellectual deterioration.

The technique of multiple subpial transections was introduced as a surgical option for cases in which the epileptogenic zone lies in functional essential cortex and cannot be resected (Morrell et al., 1989). The rationale is based on the concept that disruption of horizontal cortical interconnections with sparing of vertically oriented fibers will reduce seizure spread without causing functional deficits. This technique is generally considered a palliative technique although some authors have reported good results (Vasquez et al., 1993).

## GENERALIZED EPILEPSIES

Candidates for corpus callosotomy include patients with medically intractable symptomatic generalized epilepsies and multiple-seizure types including tonic, atonic, generalized tonic–clonic seizures, absences, and, less frequently, focal seizures. Most of these patients also have mental retardation and are categorized as having Lennox–Gastaut syndrome. Noninvasive EEG video recordings and imaging studies are sufficient in the evaluation of these patients. The rationale is based on the disruption of seizure spread from one hemisphere to the other.

Tonic and atonic seizures leading to injuries seem to respond better to callosotomy than other seizure types (Gates et al., 1987). Because some patients after complete callosotomy develop a disconnection syndrome, a staged approach has been developed. An anterior 2/3 callosotomy is performed first. If this fails to improve seizure control, complete disconnection can be performed in a second step usually without producing a disabling disconnection syndrome (Wilson et al., 1982) (Figure 2). Focal EEG discharges with secondary bilateral synchrony and focal lesions in imaging studies have been considered good prognostic signs for callosotomy (although resective surgery should first be considered in these patients), whereas severe mental retardation has been reported to be a poor prognostic factor (Spencer et al., 1988). At present the results of different groups regarding the effect of callosotomy particularly on absences and generalized tonic–clonic seizures are controversial and prospective studies are needed (Blume et al., 1993).

Postoperative neuropsychological complications such as leg weakness, speech difficulty, or aggressive behavior may occur but usually improve sufficiently within several days to weeks so that they no longer interfere with the daily lives of these patients. Further, most patients actually show a postoperative neuropsychological improvement, probably due to reduced seizure frequency and lower anticonvulsant levels. It has been suggested that patients whose language and hand dominance do not correspond (crossed dominance) may be greatly dependent on transfer of language through the corpus callosum, raising the concern that postoperative speech problems may occur. Intracarotid amobarbital test may be indicated in this situation. Corpus callosotomy was performed frequently in most major surgery of epilepsy centers in the early 1990s. However, relatively poor results have led most centers to abandon the technique or only perform it infrequently.

## ALTERNATIVE SURGICAL TECHNIQUES

Vagus nerve stimulation is a new technique and the first controlled studies have been published recently. This procedure has been shown to significantly reduce seizures in controlled studies to an extent similar to newly marketed anitepileptic drugs. Interestingly, further reduction of seizure frequency may occur with long-term treatment even after 1 year (Ben-Menachem *et al.*, 1999). Overall, vagal nerve stimulation is a viable treatment for medically refractory patients who are not candidates for resective epilepsy surgery. However, only selected patients achieve seizure control that significantly affects their quality of life and seizure freedom is extremely rare.

Gamma knife radio surgery has been shown to be very effective in a study of patients with mesial temporal sclerosis, but it takes about 10 months after radiation to achieve seizure control (Regis *et al.*, 1999). A recent study suggests that gamma knife surgery can be used to treat epilepsy caused by cavernomas located in highly functional areas, particularly the regions near the central sulcus, whereas cavernomas in mesial temporal regions did not respond well (Bartolomei *et al.*, 1999).

There is a renewed interest in deep brain stimulation for treatment of epilepsy. The anterior thalamic nucleus has been a target since the 1960s, but very little controlled data are available (Fisher *et al.*, 1992). Electrical stimulation of the centromedian thalamic nucleus has been reported to reduce seizure frequency in patients with medically intractable generalized seizures and epilepsia partialis continua (Velasco *et al.*, 1989), but these results were not confirmed by a controlled study (Fisher *et al.*, 1992). Similarly, the use of electrical stimulation of the cerebellum, first performed in the 1970s remains controversial. A controlled study failed to show reduction of seizure frequency (Wright *et al.*, 1985). The subthalamic nucleus which is a good target for treatment of parkinson's disease recently attracted the interest of epileptologists (Benabid *et al.*, 2000). Despite theoretic advantages, electrical stimulation is still far from being an established therapeutic technique.

## TREATMENTS NO LONGER RECOMMENDED

Stereotactic ablations of the Field of Forel (Jinnai and Mukawa, 1970) and stereotactic amygdalotomy, which initially was developed as a surgical therapy for behavioral problems, are not considered effective surgical treatments for epilepsies (Vaernet, 1972).

## REFERENCES

Annegers, J. F., Hauser, W. A., and Elveback, L. R. (1979). Remission of seizures and relaps in patients with epilepsy. *Epilepsia* 20, 729–737.

Awad, I. A., Rosenfeld, J., Ahl, H., Hahn, J. F., and Lüders, H. (1991). Intractable epilepsy and structural lesions of the brain: Mapping, resection strategies and seizure outcome. *Epilepsia* 32, 179–186.

Bailey, P., and Gibbs, F. A. (1951). The surgical treatment of psychomotor epilepsy. *JAMA* 145, 365–370.

Bartolomei, F., Regis, J., Kida, Y., Kobayashi, T., Vladyka, V., Liscak, R., Forster, D. M., Kemeny, A., Shrotner, O., and Pendl, G. (1999). Gamma knife radiosurgery for epilepsy associated with cavernous hemangiomas: A retrospective study of 49 cases. *Stereotact. Funct. Neurosurg. (Suppl.)* 72, 1, 22–28.

Benabid, A. L., Koudsie, A., Pollak, P., Kahane, P., Chabardes, S., Hirsch, E., Marescaux, C., and Benazzouz, A. (2000). Future prospects of brain stimulation. *Neurol. Res.* 22, 237–246.

Ben-Menachem, E., Hellstrom, K., Waldton, C., and Augustinsson, L. E. (1999). Evaluation of refractory epilepsy treated with vagus nerve timulation for up to 5 years. *Neurology* 52, 1265–1267.

Blume, W. T., Aicardi, J., and Dreifuss, F. (1993). Syndromes not amenable to resective surgery. *In* "Surgical Treatment of the Epilepsies," 2nd ed. (J. Engel, Jr., Ed.), pp. 103–118. Raven Press, New York.

Bourgeois, B. F. D. (1991). General concepts of medical intractability. *In* "Epilepsy Surgery" (H.-O. Lüders, Ed.), pp. 77–81. Raven Press, New York.

Cascino, G. D., Kelly, P. J., Sharbrough, R. W., Hulihan, J. F., Hirschorn, K. A., and Trenerry, M. R. (1992). Long-term follow-up of stereotactic lesionectomy in partial epilepsy: Predictive factors and electroencephalographic results. *Epilepsia* 33, 639–644.

Chelune, G. J., Naugle, R. I., Lüders, H., and Awad, I. A. (1991). Prediction of cognitive change as a function of preoperative ability level among temporal lobectomy patients at six months follow-up. *Neurology* 41, 399–404.

Chugani, H. T., Shields, W. D., Shewmon, D. A., Olson, D. M., Phelps, M. E., and Peacock, W. J. (1990). Infantile spasms: I. PET identifies focal cortical dysgenesis in cryptogenic cases for surgical treatment. *Ann. Neurol.* 27, 406–413.

Engel, J., Jr. (1993). Appendix II. Presurgical evaluation protocols. *In* "Surgical Treatment of the Epilepsies," 2nd ed. (J. Engel, Jr., Ed.), pp. 707–750. Raven Press, New York.

Engel, J., Jr., Henry, T. R., Risinger, M. W., Mazziota, J. C., Sutherling, W. W., Levesque, M. F., and Phelps, M. E. (1990). Presurgical evaluation for partial epilepsy: Relative contribution of chronic depth recordings versus FDG-PET and scalp-sphenoidal ictal EEG. *Neurology* 40, 1670–1677.

Engel, J. Jr., Van Ness, P. C., Rasmussen, T. B., and Ojemann, L. M. (1993). Outcome with respect to epileptic seizures. *In* "Surgical Treatment of the Epilepsies," 2nd ed. (J. Engel, Jr., Ed.), pp. 609–621. Raven Press, New York.

Falconer, M. A. (1971). Anterior temporal lobectomy for epilepsy. *In* "Operative Surgery," Vol. 14, "Neurosurgery" (V. Logue, Ed.), Butterworths, London. pp. 142–149.

Fish, D. R., Andermann, F., and Olivier, A. (1991). Complex partial seizures and small temporal or extratemporal structural lesions: Surgical management. *Neurology* 41, 1781–1784.

Fisher, R. S., Uematsu, S., Krauss, G. L., Cysyk, B. J., McPherson, R., Lesser, R. P., Gordon, B., Schwerdt, P., and Rise, M. (1992). Placebo-controlled pilot study of centromedian thalamic stimulation in treatment of intractable seizures. *Epilepsia* 33, 841–851.

Fisher, R. S., Uthman, B. M., Ramsay, R. E., Penry, J. K., Morrel, F., Velasco, F., Velasco, M., Wilder, B. J., Whistler, W. W., Krauss, G. L., and Davis, R. (1993). Alternative surgical techniques for epilepsy. *In* "Surgical Treatment of the Epilepsies," 2nd ed. (J. Engel, Ed.), pp. 549–564. Raven Press, New York.

Gates, J. R., Rosenfeld, W. E., Maxwell, R. E., and Lyons, R. E. (1987). Response of multiple seizure types to corpus callosum section. *Epilepsia* 28, 28–334.

Goldensohn, E. S. (1984). The relevance of secondary epileptogenesis to the treatment of epilepsy: Kindling and the mirror focus. *Epilepsia* 25, S156–S173.

Hauser, W. A. (1992). The natural history of drug resistant epilepsy: Epidemiologic considerations. *In* "Surgical Treatment of Epilepsy. Epilepsy Research," Suppl. 5, (W. H. Theodore, Ed.), pp. 25–28. Elsevier, Amsterdam.

Hauser, W. A., and Hesdorffer, D. C. (1990). "Epilepsy: Frequency, Causes, and Consequences," pp. 1–52. Demos Press, New York.

Hermanns, G., Noachtar, S., Tuxhorn, I., Holthausen, H., Ebner, A., and Wolf, P. (1996). Systematic testing of medical intractability for carbamazepine, phenytoin and phenobarbital or primidone in monotherapy in patients considered for epilepsy surgery. *Epilepsia* 37, 675–679.

Jackson, G., Berkovic, S., Tress, B., Kalnins, R., Fabinyi, G., and Bladin, P. (1990). Hippocampal sclerosis may be reliably detected by MRI. *Neurology* 40, 1869–1875.

Jinnai, D., and Mukawa, J. (1970). Forel-H-tomy for the treatment of epilepsy. *Confin. Neurol.* 32, 307–315.

Kelly, P. J. (1986). Computer-assisted stereotaxis: New approaches for the management of intracranial intra-axial tumors. *Neurology* 36, 535–541.

Lesser, R. P., Pippenger, C. E., Lüders, H., and Dinner, D. S. (1984). High-dose monotherapy in treament of intractable seizures. *Neurology* 34, 707–711.

Lindsay, J., Glaser, G., Richards, P., and Ounsted, C. (1984). Developmental aspects of focal epilepsies of childhood treated by neurosurgery. *Dev. Med. Child Neurol.* 26, 574–587.

Lüders, H. O. (2001). Clinical evidence for secondary epileptogenesis. *In* "Brain Plasticity and Epilepsy" (J. Engel, Jr., P. A. Schwartzkroin, S. Moshe, and D. H. Lowenstein, Eds.), pp. 469–480. Academic Press, San Diego.

Lüders, H. O., and Awad, I. A. (1991). Conceptual considerations. *In* "Epilepsy Surgery" (H. O. Lüders, Ed.), pp. 51–62. Raven Press, New York.

Lüders, H., Lesser, R. P., Dinner, D. S., Morris, H. H., III, Hahn, J. F., Friedman, L., and Skipper, G. (1987). Commentary: Intracranial electrical stimulation with subdural electrodes. *In* "The Surgical Treatment of the Epilepsies" (J. Engel, Jr., Ed.), pp. 297–321. Raven Press, New York.

Lüders, H., Lesser, R. P., Dinner, D. S., Morris, H. H., Wyllie, E., and Godoy, J. (1988). Localization of cortical function: New information from extraoperative monitoring in patients with epilepsy. *Epilepsia (Suppl.)* 2, S56–S65.

Lüders, H. O., and Noachtar, S. (Eds.) (2000). "Epileptic Seizures: Pathophysiology and Clinical Semiology." Churchill Livingstone, New York.

Lüders, H. O., and Noachtar, S. (2001). "Atlas of Epileptic Seizures and Syndromes." Saunders, Philadelphia.

Morrell, F. (1985). Secondary epileptogenesis in man. *Arch. Neurol.* 42, 318–335.

Morrell, F., Whisler, W. W., and Bleck, T. P. (1989). Multiple subpial transections: A new approach to the surgical treatment of focal epilepsy. *J. Neurosurg.* 70, 231–239.

Morris, H. H., III, Kanner, A., Lüders, H., Murphy, D., Dinner, D. S., Wyllie, E., and Kotagal, P. (1989). Can sharp waves localized at the sphenoidal electrode accurately identify a mesial-temporal epileptogenic focus? *Epilepsia* 30, 532–539.

Moshe, S. L., and Shinnar, S. (1993). Early intervention. *In* "Surgical Treatment of the Epilepsies," 2nd ed. (J. Engel, Jr., Ed.), pp. 123–132. Raven Press, New York.

Nayel, M. H., Awad, I. A., and Lüders, H. (1991). Extent of mesiobasal resection determines outcome after temporal lobectomy for unilateral temporal lobe seizure focus. *Neurosurgery* 29, 55–61.

Newton, M. R., Berkovic, S. F., Austin, M. C., Rowe, C. C., McCay, W. J., and Bladin, P. F. (1992). A postictal switch in blood flow distribution characterizes human temporal lobe seizures. *J. Neurol. Neurosurg. Psychiatry* 55, 891–894.

Niemeyer, P. (1958). The transventricular amygdalohippocampectomy in temporal lobe epilepsy. *In* "Temporal Lobe Epilepsy" (M. Baldwin and P. Bailey, Eds.), pp. 461–482. Thomas, Springfield, IL.

Noachtar, S. (2000). Seizure semiology. *In* "Epilepsy: Comprehensive Review and Case Discussions" (H. O. Lüders, Ed.), pp. 127–140. Dunitz, London.

Noachtar, S. (2001). Epidural electrodes. *In* "Epilepsy Surgery," 2nd ed. (H. O. Lüders and Y. Comair, Eds.), pp. 585–591. Lippincott Williams & Wilkins, Philadelphia.

Noachtar, S., Arnold, S., Yousry, T. A., Bartenstein, P., Werhahn, K. J., and Tatsch, K. (1998). Ictal technetium-99m ethyl cysteinate dimer single-photon emission computed tomography findings and propagation of epileptic seizure activity in patients with extratemporal epilepsies. *Eur. J. Nucl. Med.* 25, 166–172.

Ojemann, G. A. (1987). Surgical therapy for medical intractable epilepsy. *J. Neurosurg.* 66, 489–499.

Olivier, A. (1992). Extratemporal cortical resections: Principles and methods. *In* "Epilepsy Surgery" (H. O. Lüders, Ed.), pp. 559–568. Raven Press, New York.

Rasmussen, T. (1963). Surgical therapy of frontal lobe epilepsy. *Epilepsia* 4, 181–198.

Rasmussen, T. (1983). Hemispherectomy for seizures revisited. *Can. J. Neurol. Sci.* 10, 71–78.

Rasmussen, T. (1987). Cortical resection for multilobe epileptogenic lesions. *In* "Presurgical Evaluation of Epileptics" (H. G. Wieser and C. E. Elger, Eds.), pp. 344–351. Springer, Berlin.

Regis, J., Bartolomei, F., Rey, M., Genton, P., Dravet, C., Semah, F., Gastaut, J. L., Chauvel, P., and Peragut, J. C. (1999). Gamma knife surgery for mesial temporal lobe epilepsy. *Epilepsia* 40, 1551–1556.

Schmidt, D. (1982). Two antiepileptic drugs for intractable epilepsy with complex-partial seizures. *J. Neurol. Neurosurg. Psychiatry* 45, 1119–1124.

Shimizu, H., Suzuki, I., and Ishijima, B. (1990). Zygomatic approach for resection of mesial temporal epileptic focus. *Neurosurgery* 25, 798–801.

Sillanpää, M. (2000). Long-term outcome of Epilepsy. *Epileptic Disord.* 2, 79–88.

So, N., Gloor, P., Quesney, L. F., Jones-Gotman, M., Olivier, A., and Andermann, F. (1989). Depth electrode investigations in patients with bitemporal epileptiform abnormalities. *Ann. Neurol.* 25, 423–431.

Spencer, S. S., Spencer, D. D., Williamson, P. D., Sass, K., Novelly, R. A., and Mattson, R. H. (1988). Corpus callosotomy for epilepsy: I. Seizure effects. *Neurology* 38, 19–24.

Sperling, M. R., O'Connor, M. J., Saykin, A. J., Phillips, C. A., Morell, M. J., Bridgman, P. A., French, J. A., and Gonatas, N. (1992). A non-invasive protocol for anterior temporal lobectomy. *Neurology* 42, 416–422.

Stodieck, S. R. G. (1993). Pharmakoresistente epilepsien. *In* "Therapie und Verlauf Neurologischer Erkrankungen," 2nd ed. (T. Brandt, J. Dichgans, and H. C. Diener, Eds.), pp. 219–231. Kohlhammer, Stuttgart.

Taylor, L. P. (1991). Neuropsychological assessment of patients with chronic encephalitis. *In* "Chronic Encephalitis and Epilepsy: Rasmussen's Syndrome" (F. Andermann, Ed.), pp. 111–121. Butterworths, Boston.

Van Buren, J. M. (1987). Complications of surgical procedures in the diagnosis and treatment of epilepsy. *In* "Surgical Treatment of the Epilepsies" (J. Engel, Jr., Ed.), pp. 465–475. Raven Press, New York.

Vaernet, K. (1972). Stereotaxic amygdalotomy in temporal lobe epilepsy. *Confin. Neurol.* 34, 176–183.

Van Ness, P. C. (1991). Surgical outcome for neocortical (extrahippocampal) focal epilepsy. *In* "Epilepsy Surgery" (H.-O. Lüders, Ed.), pp. 613–624. Raven Press, New York.

Vasquez, B., Devinsly, O., Perrine, K., Luciano, D. J., and Dogali, M. (1993). Multiple subpial transections of language areas: Postoperative function and seizure control. *Neurology* **43**, 997S. [Abstract].

Velasco, M., Velasco, F., Velasco, A. L., Luján, M., and Vázquez del Mercado, J. (1989). Epileptiform EEG activities of the centromedian thalamic nuclei in patients with intractable partial motor, complex partial, and generalized seizures. *Epilepsia* **30**, 295–306.

Villemure, J. G., Andermann, F., and Rasmussen, T. B. (1991). Hemispherectomy for the treament of epilepsy due to chronic encephalitis. *In* "Chronic Encephalitis and Epilepsy: Rasmussen's Syndrome" (F. Andermann, Ed.), pp. 235–241. Butterworth–Heinemann, Boston.

Wieser, H. G., Blume, W., Fish, D., Goldensohn, E., Hufnagel, A., King, D., Sperling, M. R., and Lüders, H. O. (2001). Proposal for a new classification outcome with respect to epileptic seizures following epilepsy surgery. *Epilepsia* **42**, 282–286.

Wieser, H. G., Elger, C. E., and Stodieck, S. R. G. (1985). The "foramen-ovale-electrode": A new recording method for the preoperative evaluation of patients suffering from mesio-basal temporal lobe epilepsy. *Electroencephalogr. Clin. Neurophysiol.* **61**, 314–322.

Wieser, H. G., and Yasargil, M. G. (1982). Selective amygdalohippocampectomy as a surgical treatment of mesiobasal limbic epilepsy. *Surg. Neurol.* **17**, 445–457.

Wilson, D. H., Reeves, A., and Gazzaniga, M. (1982). Central commissurotomy for intractable generalized epilepsy: Series two. *Neurology* **32**, 687–697.

Winkler, P. A., Herzog, C., Henkel, A., Arnold, A., Werhahn, K. J., Yousry, T. A., Uttner, I., Ilmberger, J., Tatsch, K., Weis, S., Bartenstein, P., and Noachtar, S. (1999). Non-invasive protocol for epilepsy surgery of focal epilepsies. *Nervenarzt.* **70**, 1088–1093.

Winkler, P. A., Vollmar, C., Krishnan, K. G., Pfluger, T., Bruckmann, H., and Noachtar, S. (2000). Usefulness of 3-D reconstructed images of the human cerebral cortex for localization of subdural electrodes in epilepsy surgery. *Epilepsy Res.* **41**, 169–178.

Wright, G. D., McLellan, D. L., and Brice, J. G. (1985). A double-blind trial of chronic cerebellar stimulation in twelve patients with severe epilepsy. *J. Neurol. Neurosurg. Psychiatry* **47**, 769–774.

Wyllie, E. (1991). Candidacy for epilepsy surgery: Special considerations in children. *In* "Epilepsy Surgery" (H.-O. Lüders, Ed.), pp. 127–130. Raven Press, New York.

Wyllie, E., Lüders, H., Morris, H. H. III, Lesser, R. P., Dinner, D. S., Hahn, J., Estes, M. L., Rothner, A. D., Erenberg, G., Cruse, R., and Friedman, D. (1987). Clinical outcome after complete or partial cortical resection for intractable epilepsy. *Neurology* **37**, 1634–1641.

Yasargil, M. G., Teddy, P. G., and Roth, P. (1985). Selective amygdalohippocampectomy. Operative anatomy and surgical technique. *In* "Advances and Technical Standards in Neurosurgery" (L. Symon, Ed.), Vol. 12, pp. 93–123. Springer–Verlag, Vienna.

*Cognitive and Behavioral Disorders*

# CHAPTER 22
# Dysarthria and Dysphonia

Hermann Ackermann and T. A. T. Hughes

Besides inborn structural anomalies of the vocal tract (e.g., cleft palate) and deteriorations of language acquisition (e.g., developmental apraxia of speech), a variety of medical conditions arising in adolescence or adulthood that affect the peripheral speech apparatus (e.g., surgical resections, muscular dystrophy, myasthenia gravis), the cranial nerves (e.g., polyneuritis), or the central nervous system (e.g., closed head trauma, cerebrovascular disorders, degenerative diseases) may disrupt articulation and/or phonation. This chapter is restricted to the therapy for speech and voice impairments subsequent to adult disorders of the nervous or muscular system. Only the principles of the available treatment regimens are outlined to help the physician in planning rehabilitation.

## CLINICAL ASPECTS

### Terminology

As a rule, the speech sounds of the various human languages are produced by laryngeal and supralaryngeal modulation of the expiratory air flow (Kent, 1997). The "speech apparatus," thus, comprises three components: respiration, phonation, and articulation. Neuromuscular disorders often compromise all three functional domains. Depending upon lesion site and severity, dysfunctions restricted to a single subsystem of the vocal tract may, however, emerge such as hypophonia subsequent to hemorrhage of the left putamen or isolated velopharyngeal incompetency due to injury of the caudal cranial nerves.

Damage to the anterior perisylvian cortex or the rostral insula of the language-dominant hemisphere may give rise to a fairly consistent syndrome of speech motor deficits, including slowed speaking rate, effortful and imprecise sound production, and altered prosodic qual-

ities (*aphemia or apraxia of speech*). This variant of communication disorders is considered a higher-level dysfunction, reflecting impaired programming/planning of vocal tract movements (see Chapter 25). Besides rostral peri-/intrasylvian areas, various further cerebral structures contribute to speech motor control: the motor cortex, basal ganglia, cerebellum, and brainstem nuclei. Apart from apraxia of speech, the notion *dysarthria* is applied to all speech disorders arising from disrupted neuromotor control, including even abnormalities subsequent to dysfunctions of single cranial nerves (Darley *et al.*, 1975). As an exception, *dysphonia* denotes deficits restricted to the laryngeal level even in cases in which the underlying pathology is thought to be supranuclear as, for example, in spastic dysphonia.

### Syndromes of Dysarthria

Most bulbar nuclei relevant for speech production receive corticofugal tracts from the motor cortex of both cerebral hemispheres. As a rule, therefore, persistent speech and voice disorders are associated with bilateral lesions of the precentral gyrus or corticobulbar pathways. The respective communication disorders (*spastic dysarthria*) are characterized by slowed articulatory gestures of reduced amplitude, hypernasality due to insufficient velar elevation, tongue retraction, increased constriction of the pharynx, and hyperadduction of the shortened vocal folds. In its extreme, anarthria and/or aphonia may emerge. Unilateral damage to the motor cortex or the respective efferent tracts causes, if at all, only slight and transient dysarthria.

Dysfunctions of the cranial lower motor neurons as well as muscular disorders involving the speech apparatus may give rise to a syndrome called *flaccid dysarthria*, characterized, among others, by insufficient elevation of the velum, resulting in hypernasality and articulatory imprecision, and laryngeal dysfunctions such as breathy voice quality.

A variety of speech and vocal abnormalities including reduced loudness, monopitch, breathy and harsh voice quality, tremulous speech, and articulatory imprecision or even "speech freezing" due to undershooting orofacial movements may emerge in Parkinson's disease (*hypokinetic dysarthria*). In contrast to corticobular and cerebellar dysfunctions, parkinsonian dysarthrics show mostly normal or even an accelerated speech rate (Ackermann, 1999).

Most likely, *ataxic dysarthria* of cerebellar origin results from damage to the paramedian regions of the superior hemispheres. This syndrome is characterized, at the perceptual level, by slowed speech tempo, sometimes "scanning" rhythm, and articulatory imprecision. Fluctuations of pitch, including voice tremor, and loudness have been observed. Kinematic analyses have revealed slowed articulatory movements in these subjects (Ackermann and Hertrich, 2000).

### Neurogenic Dysphonias

*Spasmodic dysphonia* is characterized by a strained and harsh voice quality, low volume and pitch, voice tremor, and irregularly distributed stoppages as well as catches of the voice. It is now widely accepted that this disorder represents a variant of focal dystonia (*laryngeal dystonia*).

Parkinson's disease, cerebellar disorders, spasmodic dysphonia, and amyotrophic lateral sclerosis also may give rise to vocal tremor. In addition, quivering voice may be a component of the syndrome of essential tremor or even represent an isolated sign (*essential vocal tremor*). The bandwidth of frequencies in essential voice tremor extends from 4 to 6 Hz. Rhythmic contractions of the cricothyroid and rectus abdominis muscles seem to be the underlying pathophysiological mechanism (see Blitzer *et al.*, 1992).

### Acquired Dysfluencies

A variety of CNS disorders such as traumatic brain injury, ischemic lesions of either cerebral hemisphere, extrapyramidal dysfunctions, dementia of the Alzheimer type, and motor neuron diseases may rarely give rise to acquired dysfluencies. Palilalia (compulsive repetition of mostly utterance-final words or phrases) occurs in a variety of CNS dysfunctions such as postencephalitic parkinsonism, pseudobulbar palsy, and Pick's disease (for a review, see Ackermann, 1999).

## PRINCIPLES OF THERAPY

### Overview

Four approaches to dysarthria therapy are available: behavioral treatment techniques (drill, exercises), instrumental aids including prosthetic and augmentative devices, medication, and surgical procedures. In the absence of medical or surgical intervention acting upon the underlying disease, "compensated intelligibility" rather than "normal speech" must be considered the goal of dysarthria therapy.

### Behavioral Therapy (Including Biofeedback Techniques)

The concept of drill is fundamental to behavioral treatment of dysarthrias. Traditional approaches within this domain aim at the formation of response contingencies to stimulus presentation in order to establish new skills, compensation strategies, or adjustments. More recently, behavioral techniques for the therapy of articulatory disorders have been extended by so-called pragmatic approaches. Rather than articulatory impairment, treatment focuses on the efficacy of verbal communication in various contexts. Thus, patients and their relatives are instructed, for example, to alter the communication environment, to modify the length of utterances, to enhance self-monitoring, and to ensure the orientation of a listener to the topic at hand in order to improve communication (e.g., Kearns and Simmons, 1988).

In severe dysarthria, behavioral treatment begins with exercises on individual components of the speech apparatus or even with nonspeech movements (e.g., Netsell and Rosenbek, 1985). Postural adjustments such as lying supine or monitoring the amount of air inhaled as well as the evenness of exhalation may improve respiratory support. Articulation drill relies on individualized word lists that improve existing and facilitate deficient motor capabilities. Once a patient is able to combine several differentiated sounds, training proceeds to the level of utterances. At later stages of treatment or in less impaired individuals, exercises aim at the simultaneous coordinated activity of all functional components, such as learning to change stress and rhythm. Behavioral strategies can be supported by biofeedback procedures based on either acoustic measures, such as vocal intensity or pitch, or electromyographic activity obtained from various muscles or kinematic parameters.

Ramig and co-workers introduced a therapy program for Parkinson's disease that focuses on vocal loudness (Lee Silverman Voice Treatment; Ramig *et al.*, 1995):

patients are trained to increase vocal fold adduction in maximum effort tasks and to generalize the effects on louder speech. Besides significant increases of speech intensity and fundamental frequency variability, this approach also achieves sustained improvement of disordered articulation or speech tempo.

## Specific Instrumental Aids

*Accelerated speech tempo* as observed, for example, in Parkinson's disease can impair the intelligibility of speech utterances. *Pacing boards* segmented into several sections by raised dividers may help patients to better control their speech rate. The patient moves a finger along the board and produces one syllable per each segment. As a less inconvenient alternative, *delayed auditory feedback* can be used to reduce speech tempo.

*Lowered voice volume* represents a significant problem especially in parkinsonian patients. Most healthy individuals show a consistent increase of loudness during speech when exposed to masking noise (*Lombard effect*). Patients with Parkinson's disease exhibit marked improvement of vocal intensity under these conditions as well. As an alternative *voice amplifiers* can be used.

*Velopharyngeal insufficiency* (reduced or missing elevation of the velum giving rise to inadequate nasal emission) may significantly impair intelligibility of speech utterances due to decreased intraoral pressure. *Dental prostheses*, either palatal lifts or pharyngeal bulbs, can improve this condition. Patients with velopharyngeal insufficiency due to lesions of the upper as well as the lower motor neurons or even myasthenia gravis have been reported to benefit from application of these devices.

Involuntary movements of orofacial structures (*orofacial dyskinesia*) may interfere with speech production. Immobilization of the mandible by means of an *occlusal splint* or a *bite raiser* might be helpful for these patients.

*Nonverbal systems of communication* must be considered in patients with *persistent anarthria* or *unintelligible speech* if improvement of articulatory or phonatory functions cannot be expected. Patients with sufficient motor capabilities of at least one hand may use a typewriter or computer keyboard. In cases of insufficient manual control, presentation of pictures, words, or letters by means of boards or electronic devices represents an alternative. Dependent upon residual motor functions, an adequate "indicating mode" has to be selected. Patients able to point with a finger or a headstick or to activate a series of switches can indicate themselves message components displayed on a board or screen ("direct-selection strategy"). More disabled patients transmit information by signaling "yes" or "no" in response to pictorial or verbal material presented by another person ("scanning strategy"; for further details, see McNeil, 1997).

## Medication

For several neurological diseases giving rise to speech and voice impairments (myasthenia gravis, Parkinson's disease, essential tremor), drug therapy is available. However, therapeutic efficacy of dopaminergic medication on parkinsonian dysarthria has been questioned. Conceivably, L-dopa and dopamine agonists exert a differential influence on the various speech and voice deficits in this disease. Moreover, essential voice tremor seems to be less susceptible to drug therapy than essential tremor of the hands (see Ackermann, 1999).

Injection of botulinum toxin into the laryngeal muscles has been shown to significantly reduce the symptoms of spastic dysphonia. This procedure either aims at unilateral or slight bilateral paresis of the vocal folds. Application of botulinum toxin, therefore, must be restricted to spastic dysphonia with hyperadduction of the folds. The therapeutic effects start within hours to several days after injection. Breathy voice quality has been observed as a side effect of botulinum toxin treatment; however, severe complications have not been reported so far (Blitzer *et al.*, 1992).

## Surgery

Unilateral section of the recurrent laryngeal nerve represents an alternative to botulinum injection in patients with spasmodic dysphonia. The reported numbers of successfully treated patients in terms of effective conversational voices extend from about 30 to 80%. Recurring spasticity can be corrected by laser thinning of the paralyzed vocal fold via direct laryngoscopy. Injection of Teflon or similar materials into the posterior pharyngeal walls producing an anteriorly projecting bulge may improve velopharyngeal insufficiency as well. Furthermore, surgical procedures introducing posterior pharyngeal flaps have been recommended in these cases as well (see Ackermann, 1999).

# EVALUATION OF TREATMENT EFFICACY

So far, the effectiveness of dysarthria therapy has been evaluated most in parkinsonian patients. The first investigations during the 1960s and 1970s failed to document carryover or maintenance of treatment-related changes during follow-up once intervention had been discontinued. Therefore, transfer to daily life is considered a significant problem of behavioral treatment in this disease. A recent Cochrane review identified three studies comparing speech therapy in parkinsonian subjects with placebo and two studies considering two different forms of intervention (Deane *et al.*, 2000). These data indicate significant improvement of voice intensity, pitch contol, and intelligibility in response to specific treatment. Lasting effects seem to be bound, however, to intensive daily interventions across several weeks.

A number of single- and multiple-case studies demonstrated improved intelligibility and enhanced speech motor functions following adjustment of instrumental aids such as a palatal lift or delayed auditory feedback. Since these procedures may pose considerable demands on compliance and cooperation, patients must be carefully selected in order to achieve long-term benefits.

At least during the first months after traumatic brain damage, spontaneous recovery of neurological impairments can be expected to some extent. It is difficult, therefore, to estimate the contribution of therapy on any observed improvement during this period. A recent review of therapy for speech disorders due to non-progressive brain damage found no trials that were of a quality high enough to be included in a systematic review (Sellars *et al.*, 2001). However, various case studies documented the beneficial effects of speech exercises concomitant with instrumental aids, even long after the recognized period of spontaneous recovery in subjects with strokes or closed head injury. Thus, it must be recognized that behavioral treatment supported by instrumental aids may enable dysarthric patients to speak more intelligibly (for further details, see McNeil, 1997).

## PRACTICAL MANAGEMENT

As a rule, treatment should be tailored to each patient according to the profile of deficits in the various subsystems of the speech apparatus. Therefore, detailed auditory-perceptual evaluation, eventually assisted by instrumental investigations such as acoustic analysis,

electroglottography, fiber-endoscopic laryngoscopy, or kinematic techniques, is required prior to therapy. In addition to severity and type of speech or voice deficits, the therapeutic approach must consider motor disabilities outside the speech domain, intellectual capabilities, and personal motivation of the patient as well as the expected course of the underlying disease. Effective therapy may entail, as suggested by data on parkinsonian patients, intensive treatment lasting several hours daily. The ability of the patient to cope with these demands merits consideration.

## REFERENCES

Ackermann, H. (1999). Acquired disorders of articulation: Classification and intervention. *In* "Concise Encyclopedia of Language Pathology" (F. Fabbro, Ed.), pp. 261–268. Elsevier, Amsterdam.

Ackermann, H., and Hertrich, I. (2000). The contribution of the cerebellum to speech processing. *J. Neuroling.* **13**, 95–116.

Blitzer, A., Brin, M. F., Sasaki, C. T., Fahn, S., and Harris, K. S., Eds. (1992). "Neurologic Disorders of the Larynx." Thieme, New York, NY.

Darley, F. L., Aronson, A. E., and Brown, J. R. (1975). "Motor Speech Disorders." Saunders, Philadelphia, PA.

Deane, K. H. O., Whurr, R., Playford, E. D., Ben-Shlomo, Y., and Clarke, C. E. (2000). Cochrane systematic reviews of speech and language therapy for speech disorders in Parkinson's disease. *Movement Disord.* **15**, 170 (Suppl. 3).

Kearns, K. P., and Simmons, N. N. (1988). Motor speech disorders: The dysarthrias and apraxia of speech. *In* "Handbook of Speech-Language Pathology and Audiology" (N. J. Lass, I. V. McReynolds, J. L. Northern, and D. E. Yoder, Eds.), pp. 592–621. Dekker, Toronto.

Kent, R. D. (1997). "The Speech Sciences." Singular Press, San Diego, CA.

McNeil, M. R., Ed. (1997). "Clinical Management of Sensorimotor Speech Disorders." Thieme, New York, NY.

Netsell, R., and Rosenbek, J. C. (1985). Treating the dysarthrias. *In* "Speech and Language Evaluation in Neurology: Adult Disorders" (J. K. Darby, Ed.), pp. 363–392. Grune & Stratton, Orlando, FL.

Ramig, L. O., Countryman, S., Thompson, L. L., and Horii, Y. (1995). Comparison of two forms of intensive speech treatment for Parkinson disease. *J. Speech Hear. Res.* **38**, 1232–1251.

Sellars, C., Hughes, T., and Langhorne, P. (2001). Speech and language therapy for dysarthria due to non-progressive brain damage. *In* "The Cochrane Library," Vol. 2. Update Software, Oxford.

CHAPTER 23
# Dysphagia

T. A. T. Hughes and Hermann Ackermann

## CLINICAL ASPECTS

### Normal Swallowing

A bolus is moved from the mouth to the esophagus by an elegant pressure driven process which is a considerable challenge even to the undamaged nervous system (Miller, 1982). An understanding of the important elements of this process allows the effects of common neurological deficits to be anticipated.

If the bolus placed in the mouth needs alteration it is manipulated by the tongue and chewed. Sensory information from the oral cavity allows the consistency and shape of the bolus to be appreciated. This sensory information coupled with fine control of the tongue, lips, and suprahyoid musculature produces a tailored swallow, voluntarily initiated, in which the duration of the component parts changes according to the size and consistency of the bolus (Lazarus *et al.*, 1993).

When suitably prepared the bolus is pushed toward the pharynx by the tongue. To prevent nasal regurgitation the palate rises to close off the nasopharynx. To prevent entry of bolus into the airway the larynx is closed by its sphincteric action during which the true cords, the false cords, and the arytenoids become apposed, and the hood-like epiglottis covers the laryngeal vestibule.

The mechanism that opens the upper esophagus is of great importance. The main muscle of the upper esophagus, the cricopharyngeus, has to first relax but its opening is dependent on the upward and forward movement of the hyoid and larynx. This movement pulls open the relaxed sphincter and the associated drop in pressure exerts a downward pull on the bolus. With the nasopharynx and larynx closed, the bolus is driven through the pharynx to the esophagus both by this hypopharyngeal suction and the following wave of peristalsis. With the bolus safely dispatched, the involved musculature relaxes allowing the larynx to reopen and return to its original position and the tongue and pharyngeal wall to part.

Respiration is necessarily arrested during swallowing and typically expiration precedes and follows the deglutition; apnea promotes the egress of bolus remnants from the laryngeal inlet immediately before and after the swallow.

The determinants of swallowing capacity in health are age, sex, and height (or weight). Using a simple timed test of swallowing the average volume per swallow (ml) and the swallowing capacity (ml.s.$^{-1}$) fall with age and are greater in men than in women (Hughes and Wiles, 1996). Age-related changes in swallowing function have also been described in studies of average bolus volume and in videofluoroscopy studies (Logemann, 1990).

### Abnormal Swallowing

Most neurological conditions have the potential to affect the speed and amplitude of the individual movements of swallowing and/or to disrupt their integration.

## COMPLICATIONS OF ABNORMAL SWALLOWING

### Nutritional

If swallowing becomes too much of a chore and so slow that intake is compromised, nutritional failure may develop and lead to the complications of malnutrition including lethargy, pressure sores, and reduced physical activity.

## Respiratory

The respiratory complications stem from the common aerodigestive tract in humans; the pharynx is both an airway and a "foodway" (Wiles, 1991). Potential complications include upper airway obstruction due to bolus obstruction of the larynx or the complications of occlusion of distal airways including lobar collapse and infection. The recurrent coughing and choking that may plague patients as they struggle with swallowing failure, even if not accompanied by other respiratory complications, may discourage further attempts at oral feeding.

## Psychosocial

The coughing and spluttering of the dysphagic patient makes communal eating difficult and may discourage the sufferer from persisting in their efforts to feed by mouth; drooling due to ineffective lip closure or poor head posture compounds the situation.

## MECHANISMS OF ABNORMAL SWALLOWING

The consequences of abnormal swallowing differ markedly according to the underlying disease, its severity, and its speed of onset. However, some generalizations about the different mechanisms of dysphagia can inform more detailed assessments of individual patients.

### Upper Motor Neuron Syndromes

In conditions affecting mainly the supranuclear and corticobulbar fibers the fine control of facial and lingual movement is compromised causing impaired oral control, poor preparation of the bolus, and unpredictable bolus presentation to the pharynx. However, although the timing and preparation may be clumsy the larynx and suprahyoid musculature are not denervated, and crude untailored swallows can occur due to reflex closure and movement of the larynx. At its simplest this is an upper airway reflex that protects the airway by closing the larynx and then pulling the bolus away from the laryngeal inlet into the upper esophagus.

### Lower Motor Neuron Syndromes

The nature of the swallowing problem in upper motor neuron diseases contrasts sharply with qualitative lower cranial nerve palsies which lead to denervation of the bulbar structures. Unilateral 10th nerve lesions are the best example; the larynx is unable to shut because the true and false cords and the arytenoids on one side are paralyzed. Thus during both tailored (voluntary) and crude (reflex) swallowing the larynx will not close. This makes aspiration of bolus into the larynx inevitable, particularly in the acute stages. In this regard there is little difference in the voluntary and reflex bulbar functions for qualitative lower motor neuron lesions, whereas in upper motor neuron disease there is some disassociation.

In peripheral neuromuscular diseases causing weakness but not complete paralysis of the involved structures, swallowing is not unhinged by a loss of fine upper motor neuron control or by a complete breakdown in the lower motor neuron circuitry; instead all stages are affected, to a degree determined mainly by the severity of the weakness.

## MECHANISMS OF AIRWAY DEFENSE

If bolus enters the airway there are a number of ways of restoring its patency. If the pharyngeal airway is occluded gagging promotes the egress of bolus from the pharynx to the mouth and, if ineffective, may be accompanied or followed by coughing, retching, or vomiting. Prompt alteration of posture and immediate and powerful coughing are perhaps the most effective mechanisms of airway defence. If these defence mechanisms fail and vigorous blows to the back do not dislodge the offending bolus, the Heimlich maneuver may be required. All of these mechanisms serve to move the bolus rostrally. Obviously swallowing itself, as mentioned above, is also an effective way of removing (caudally) a bolus approaching the laryngeal vestibule.

## FAILURE OF ORAL SUSTENANCE AND DYSPHAGIA

Although it is essential to understand the neurology of the swallowing process and the mechanisms of dysphagia, in practice there are often many other significant problems that compromise feeding more than the bulbar impairments. For a given defect in the swallowing process problems with inappropriate food consistency, maintenance of posture, motivation, insight, respiratory function, respiratory control, appetite, vertigo, nausea, visual disturbance, and the attitude and availability of care givers may contribute as much or more to failure of oral feeding as the defects in the mere swallowing process. Using the term *failure of oral sustenance* rather than dysphagia to describe patients who cannot be fed and watered by mouth encourages a more holistic approach and puts the neurology of the bulbar dysfunction into a relevant clinical context. This is crucial when addressing the broader issues in the neurological patient who is developing respiratory or nutritional complications of oral feeding.

## SPECIFIC DISEASES

A very large number of neurological diseases are complicated by dysphagia but for illustrative purposes we shall consider just four very broad categories: stroke, motor neuron disease, diseases resulting in myopathic weakness, and extrapyramidal and cerebellar diseases.

### Stroke

In patients with lateral medullary syndrome, infarction of the nucleus ambiguus causes a unilateral paresis of the larynx, palate, and pharynx and involvement of neighboring areas may result in intraoral and pharyngeal sensory loss and some cerebellar impairments; these deficits, particularly the first, unhinge swallowing. Voluntary and reflex movements of the vocal cords are lost because the nuclear lesion produces a lower motor neuron paresis of the sphincteric action of the larynx that makes laryngeal closure impossible.

This contrasts with cortical strokes where the lower motor neuron circuitry of the bulbar structures is preserved and in theory crude reflex untailored swallows are still possible soon after the event. This may account for the resolution of most dysphagia associated with hemispheric stroke within 2 weeks (Barer, 1989), in contrast to lateral medullary infarcts in which typically the dysphagia is prolonged. The dysphagia associated with cortical stroke has been the subject of elegant studies using transcortical magnetic stimulation (Hamdy et al., 1997). These suggest that recovery of swallowing is associated with increased cortical activity on the unaffected side revealing a previously undescribed plasticity of cortical involvement in swallowing function.

In relation to failure of oral sustenance there are a number of other relevant deficits in stroke, apart from the purely bulbar-related deficits, all of which are of sudden onset and leave the patient with no time to compensate. In lateral medullary syndrome the sudden onset of nausea, vertigo, unsteadiness, weakness, and loss of postural control in combination with impaired coughing (due to the vocal cord paresis) and the likely loss of appetite may contribute as much or more to failure of oral feeding as the swallowing problem. Similarly in hemisphere stroke the sudden onset, and psychological

chaos, of significant weakness and sensory loss in combination with cognitive and language problems may make oral feeding seem improbable to the patient, at least in the early stages.

### Motor Neuron Disease

The variable upper and lower motor neuron involvement in motor neuron disease produces different patterns of bulbar dysfunction.

In patients with mainly upper motor neuron involvement the profound slowing of tongue movements impairs oral control of the bolus but the intact lower motor neuron circuitry allows the larynx to close and move reflexly. In addition, reflex coughing and the palatal and pharyngeal reflexes may be exaggerated due to loss of descending inhibition; this enhances airway protection in what is otherwise precarious swallowing. Such patients are often profoundly dysarthric because of the slow tongue movements and may be become anarthric and aphagic but are still able to maintain their own airway due to the preservation of reflex airway protective mechanisms.

In predominantly lower motor neuron forms of the disease there can be profound weakness of the face and tongue but because fine control and speed of movement is relatively preserved oral control of the bolus is possible and presentation to the pharynx is less clumsy than in upper motor neuron forms of the disease. However, gradual denervation of the larynx and respiratory muscles produces weakness that impairs laryngeal closure, and when aspiration does occur the posture changes and coughing are less effective in maintaining the airways.

### Myopathies

In myopathies the central and peripheral neural circuits are intact but weakness may lead to failure of oral sustenance. The strength required to chew, and to pull open the relaxed upper esophageal sphincter, means that weakness may preclude eating anything but soft or pureed food and if respiratory muscle weakness is also present any aspiration may lead to the early onset of complications.

If the myopathy is of very gradual onset, such as is seen in oculopharyngeal dystrophy and mitochondrial myopathies, patients learn how to compensate for their bulbar dysfunction and may maintain nutrition and hydration without significant respiratory complications until an advanced stage of the disease. Such patients may show an *apparent* dramatic worsening of bulbar and respiratory function following trivial medical problems due to the lack of any reserve of function.

### Extrapyramidal and Cerebellar Diseases

The frequency with which swallowing problems are seen in these conditions attests to the importance of coordination and timing of movement to accomplish adequate oral preparation and smooth transit of the bolus.

The rigidity and bradykinesia of parkinsonism impairs oral control of the bolus and the reduction in amplitude of movement of the hyoid and larynx may cause the apparent cricopharyngeal abnormalities reported on videofluoroscopy. On this background other features of the extrapyramidal diseases including the side effects of medication, changes in posture, oromandibular and/or cervical dystonias, vocal cord abductor palsies (in multi-system atrophy), and cognitive changes further threaten the success of oral sustenance.

While severe swallowing problems are usually not a feature of early cerebellar disease, even when the patient is obviously dysarthric, in the late stages the incoordination of the tongue makes oral preparation difficult and potentially hazardous. Unless there is an accompanying neuropathy or myopathy, however, the lower motor neuron circuitry of the bulbar reflexes and respiratory system are preserved and therefore the airway defences may be adequate to prevent respiratory complications until the very late stages of the disease. It is common to come across patients with advanced cerebellar disease whose tongue control is so poor as to make them unintelligible but they are still able to take food of an appropriate consistency by mouth.

## THERAPY AND MANAGEMENT

### Behavioral Treatments

The approach to patients with swallowing problems differs depending on how acute the problem is but the principles are similar. The effectiveness of different interventions for dysphagia after stroke has been the subject of a systematic review and the results suggest the need for further research (Bath *et al.*, 2000). Alteration of the consistency of the diet is the first treatment option and if the disease is subacute or chronic in onset patients often establish this for themselves. As a general rule foods that require chewing, such as meat, biscuits, and tough vegetables, represent a considerable challenge to oral preparation and propulsion. At the other end of the spectrum of consistency thin liquids are a stringent test

of oral control (rather than preparation) and laryngeal closure; the less viscous a fluid the more likely it is to seep out of the back of the mouth and penetrate the larynx. For this reason soft foods or pureed foods, which do not require a lot of preparation or propulsion and which will not leap into the airway, have the dietary consistencies that many patients find the easiest to cope with.

In diseases that are slowly progressive, patients and their care givers usually discover the most appropriate consistency of food by trial and error but, in dysphagia of acute onset, patients cannot experiment with different consistencies and they need professional guidance. It is contentious whether there are clinical or radiological indices of swallowing function that predict the volume and consistency of oral intake which an individual will be able to tolerate and in practice there may be no alternative to observation by an informed witness over a representative time period. Videofluoroscopy has been used extensively to look for symptomatic and silent aspiration of bolus into the airway during swallowing and is undoubtedly an effective way of visualizing this phenomena. However in view of the many factors that contribute to the success or failure of oral sustenance, particularly respiratory function and the airway defence mechanisms, it is questionable whether the presence, or absence, of aspiration during this investigation answers the relevant clinical questions more reliably than careful clinical observation during the graded introduction of oral feeding. Obviously if a trial of oral intake is judged to be too risky because of an altered consciousness level, poor cooperation, fluctuating vigilance, or progressive or obviously severe bulbar and respiratory failure, then the safest option is to feed via an alternative route and review the case after a suitable time interval.

A number of strategies used by speech and language therapists may allow patients to compensate for underlying impairments (Logemann, 1983). Head turning toward, or tilting the head away from, the side of a 10th or 12th nerve lesion may promote pharyngeal transit and reduce post swallow pharyngeal residue. Instructing patients to deliberately breathe out before and after each swallow may reduce the risk of significant aspiration, a technique know as a supraglottic swallow. Videofluoroscopy may be used to demonstrate to patients the effect of swallowing maneuvers on the passage of the bolus.

While most patients should sit up for feeding, the effect of gravity on the bolus in the mouth may be a disadvantage if oral control is very impaired; some patients find it easier to be fed in different postures including side lying (Drake *et al.*, 1997). If patients need to be fed the importance of the feeder appreciating the preferred pace and style cannot be over emphasized.

## Medical Treatments

The underlying pathology requires treatment and obvious improvements may obviate the need for further consideration of the swallowing problem, e.g., appropriate treatments for myasthenia gravis. In susceptible patients a number of medications may lead to a decompensation of bulbar and respiratory function and withdrawal of benzodiazepines, anticholinergic preparations, and major tranquilizers (and their distant relatives) should always be considered as a simple but potentially beneficial intervention.

If the underlying disease is not responsive to treatment, conventional medical treatments are limited. There are a number of surgical interventions that may lead to improved swallowing, enhanced airway protection, or both. The timing of intervention and the selection of appropriate patients is crucial to the success of surgery because the potential beneficiaries often have other progressive disabilities. In patients in whom cricopharyngeal dysfunction is suspected, cricopharyngeal myotomy has been tried with variable success (Bucholz, 1995). The unpredictable response may be related to the severity of accompanying neurological deficits or because the apparent dysfunction is due to reduced hyolaryngeal movement rather than to a primary disorder of the cricopharynx. Some operations provide effective airway protection at the expense of the subject losing their voice and therefore are considered only in exceptional circumstances (Baredes, 1988).

Drooling may be troublesome and we have found Robinul (glycopyrronium bromide) to be very effective; other options include atropine tablets or transdermal anticholinergic preparations.

If oral feeding has failed a nasogastric tube can be considered. In cooperative patients this may be an effective way of delivering feed and some patients learn to insert their own tube intermittently at meal times (Logemann, 1983). In a small randomized study, patients judged to have a signficant swallowing problem of >4 weeks duration were more likely to be effectively fed through a gastrostomy than a nasogastric tube although the complication rate was higher (Park *et al.*, 1992). Subsequent studies have confirmed that if long-term failure of oral sustenance is predicted gastrostomy

represents the appropriate technique (Norton *et al.*, 1996). The timing of insertion and the identification of suitable patients merits further study in all neurological diseases if the morbidity and mortality of the procedure is to be reduced (Raha and Woodhouse, 1994).

## REFERENCES

Baredes, S. (1988). Surgical management of swallowing disorders. *Otolaryngol. Clin. North Am.* 21, 711–720.

Barer, D. H. (1989). The natural history and functional consequences of dysphagia after hemispheric stroke. *J. Neurol. Neurosurg. Psychiatry* 52, 236–241.

Bath, P. M., Bath, F. J., and Smithard, D. G. (2000). Interventions for dysphagia in acute stroke. *Cochrane Database Syst. Rev.* (2), CD000323.

Bucholz, D. W. (1995). Cricopharyngeal myotomy may be effective treatment for selected patients with neurogenic oropharyngeal dysphagia. 10(4), 255–258.

Drake, W., O'Donoghue, S., Bartram, C., *et al.* (1997). Eating in side-lying facilitates rehabilitation in neurogenic dysphagia. *Brain Inj.* 11, 137–142.

Hamdy, S., Aziz, Q., Rothwell, J. C., Crone, R., Hughes, D., Tallis, R. C., and Thompson, D. G. (1997). Explaining oropharyngeal dysphagia after unilateral hemispheric stroke. *Lancet* 350, 636–692.

Hughes, T. A. T., and Wiles, C. M. (1996). Clinical measurement of swallowing in health and in neurogenic dysphagia *Q. J. Med.* 89, 109–116.

Lazarus, C. L., Logemann, J. A., Rademaker, A. W., Kahrilas, P. J., Pajak, T., Lazar, R., and Halper, A. (1993). Effects of bolus volume, viscosity and repeated swallows in nonstroke subjects and stroke patients. *Arch. Phys. Med. Rehab.* 74, 1066–1070.

Logemann, J. A. (1983). "Evaluation and Treatment of Swallowing Disorders." College-Hill, San Diego.

Logemann, J. A. (1990). Effects of aging on the swallowing mechanism. *Otolaryngol. Clin. North Am.* 23(6), 1045–1056.

Miller, A. J. (1982). Deglutition. *Physiol. Rev.* 62,129–184.

Norton, B., Homer-Ward, M., Donnelly, M. T., Long, R. G., and Holmes, G. K. (1996). A randomised prospective comparison of percutaneous endoscopic gastrostomy and nasogastric tube feeding after acute dysphagic stroke. *Br. Med. J.* 312, 13–16.

Park, P. H. R., Allison, A. C., Lang, J., Spence, E., Morris, A. J., Danesh, B. J. Z., Russell, R. I., and Mills, P. R. (1992). Randomised comparison of percutaneous endoscopic gastrostomy and nasogastric tube feeding in patients with persisting neurological dysphagia. *Br. Med. J.* 304, 1046–1049.

Raha, S. K., and Woodhouse, K.(1994) The use of percutaneous endoscopic gastrostomy (PEG) in 161 consecutive elderly patients. *Age Aging* 23,162–163.

Wiles, C. M. (1991). Neurogenic dysphagia. *J. Neurol. Neurosurg. Psychiatry* 54, 1037–1039.

## CHAPTER 24
# Disorders of Higher Visual Function

Josef Zihl and Christopher Kennard

About 20–40% of patients with acquired cerebrovascular or traumatic brain injury have higher visual disorders (Hier *et al.*, 1983; Uzzell *et al.*, 1988). The term "higher visual disorders" denotes visual deficits following injury to cortical brain structures. Since adequate vision is a crucial prerequisite for performance in various cognitive domains as well as for rehabilitation outcome and vocational success (Patel *et al.*, 2000; Reding and Potes, 1988), the detailed examination of cerebral visual disorders and their treatment are an integral part of the work of neurological and neuropsychological institutions.

It is now generally accepted that the posterior brain is a patchwork of more than 30 functionally specialized visual areas, arranged in a modular form and with flexible networks to subserve complex visual abilities, e.g., visual spatial orientation and visual recognition. Despite this functional segregation, *isolated* visual disorders are only rarely observed, since focal brain injury is seldom limited to one single visual cortical area (for reviews, see Cowey, 1994; Zeki, 1993). Thus, the typical consequence of posterior brain injury is an association of various visual disorders.

Under the aspects of diagnostics, the higher visual disorders caused by brain injury can be classified into two main groups: (1) disorders of basic visual functions, e.g., visual field and visual acuity, and (2) disorders of complex visual abilities, e.g., visual cognition. (For disorders of visual space perception, spatial cognition, and visual spatial attention, see Chapter 26).

Although the neuroscience of vision is a major research topic, research devoted to the study of treatment has attracted interest only in the last few years. When evaluating studies on the treatment of patients with cerebral visual deficits, one must balance the various methodological approaches available against their possible benefits for the patient. Therefore, one must consider very carefully the efficacy of a treatment procedure for reducing the degree of a patient's visual handicap. Not every procedure that shows positive results represents a clinically valid and useful method of treatment, which also fulfills the criteria of ecological validity and cost effectiveness. In the end, the critical question for the patient is not whether visual function will return or if it can be compensated for, but whether he will be able to successfully cope with the demands of daily life.

## CEREBRAL BLINDNESS

### Clinical Aspects

Unilateral injury to the retrogeniculate visual pathway, in particular the striate cortex, leads to hemianopias (loss of vision in one hemifield), quadranopias (loss of vision in one quadrant), and scotomata (islands of blindness occurring mainly in the paracentral visual field region), which correspond retinotopically to the injured area and are homonymous (i.e., involve both eyes). After bilateral retrogeniculate injury, corresponding portions in either visual hemifield are affected; typical forms are bilateral hemianopia ("tunnel vision"), bilateral upper or lower quadranopia (also called upper or lower hemianopia), and bilateral paracentral scotomata. Complete destruction of both retrogeniculate pathways results in total cerebral blindness. Central scotoma, i.e., the absolute or relative loss of vision in the foveal region of both visual fields, results from bilateral injury either to the occipital pole, where the central visual field is represented, or to the portion of the visual pathway containing the foveal fiber connections to the striate cortex. Chronic cerebral hypoxia is probably the most frequent cause of this visual field disorder, but demyelinating diseases, for example, multiple sclerosis, can also induce it. Patients with central scotoma have a variety of visual impairments. These depend on the size of the scotoma, which may be surrounded by an amblyopic zone, and the associated visual disorders. As a rule, visual acuity and form vision are severely reduced, and visual recognition, reading, and spatial localization pose difficulties (Zihl, 2000).

In a group of 714 patients with retrogeniculate injury of cerebrovascular (61.5%) or traumatic (14.6%) origin (the rest underwent surgical intervention for an occipital tumor or had hypoxia), 88.8% showed unilateral field defects. In this group, homonymous hemianopia was by far the most frequent type of field loss (65.2%), followed by quadranopia (16.1%) and paracentral scotomata (7.7%). Eleven percent of patients showed depressed light vision and loss of color and form vision in the affected hemifield ("homonymous hemiamblyopia"). The fovea was always spared in patients with unilateral field defects, but the degree of parafoveal sparing varied; sparing was less than 5° in the majority of patients (74.4%; Zihl, 2000).

Apart from anopia (loss of all visual functions) or amblyopia (degraded vision) in the corresponding hemifields, hemianopic field sections have also been reported to be dedicated to only one visual function: for color (hemiachromatopsia; Paulson et al., 1994), spatial contrast (Rowe and Sarkies, 1998), form (Frassinetti et al., 1999), and movement (Plant et al., 1993; Schenk and Zihl, 1997). Thus, segregation of function includes the representation of the visual field also in extrastriate visual cortical areas.

Homonymous visual field defects (e.g., hemianopia or hemiachromatopsia) may also occur in focal posterior atrophy ("posterior dementia"; Caselli, 2000; Freedman and Costa, 1992).

## Natural Course

The spontaneous recovery of vision in patients with partial cerebral blindness has been frequently reported. A follow-up study using quantitative perimetry in 225 patients with homonymous hemianopia after occipital stroke showed that spontaneous recovery occurred in 16% within 12 weeks. Those patients with field sparing >10° also showed more recovery (3°–24°; mean, 7°) than patients with field sparing <5° (1°–6°; mean, 3°; Zihl, 1994). Only four patients had a complete recovery (about 2%).

Recovery from complete cerebral blindness occurred in about one-third (27%) of 111 patients (Aldrich et al., 1987; Bergmann, 1957; Gloning et al., 1962; Symonds and Mackenzie, 1957). Only four patients (6%) had complete recovery; in the rest recovery ranged from the return of "crude" light vision to the return of partial visual field portions with or without color and form vision. Recovery normally took place within 8–12

weeks after the onset of blindness; however, in some patients it took up to 2 years.

The recovery of vision seems to proceed in stages, during which different visual functions return, one after the other. The detection of (moving or flickering) light stimuli is followed by the perception of forms. At this stage forms often seem "fuzzy" or are perceived as "vague impressions" characterized as "foggy vision." Visual acuity may be reduced to the ability to count fingers, and colors may appear pale or "dirty." Vision may eventually improve to a degree that allows the patient to recognize objects and faces and to visually navigate at least within familiar surroundings. As a rule, vision returns in the central field region; only four cases of a persistent loss of central visual field but some degree of recovery in the field periphery have been reported (Symonds and Mackenzie, 1957). The extent of recovery was found to be inversely related to age, a history of diabetes or hypertension, and the presence of cognitive or language impairment. Likewise, patients with cortical blindness after spontaneous stroke have a poor recovery (Aldrich et al., 1987). Functional patterns and visual evoked potentials (VEPs) have been used to evaluate blindness and recovery of patients with complete cerebral blindness. VEPs are usually impaired in terms of increased latencies or reduced amplitudes of P1, or both. However, VEPs may also be normal in these patients (Aldrich et al., 1987; Celesia et al., 1980), suggesting that cerebral blindness, and recovery from it, cannot be reliably evaluated by VEP recordings. Interestingly patients who spontaneously recover from blindness show residual metabolism in the striate cortex (Bosley et al., 1987; Wunderlich et al., 2000), which extrastriate visual cortical, parietal, and frontal activation may enhance (Rausch et al., 2000). This indicates that damage to the striate cortex was incomplete and therefore not irreversible. Thus, even a severely reduced striate cortex activity is sufficient for the generation of a pattern evoked visual potential but not for vision.

Recovery from central scotoma appears to be rather limited except in patients with multiple sclerosis. Eventually they may fully recover; however, the central field loss is typically of prechiasmatic origin (Slamovits et al., 1991).

Transient cortical blindness may occur during uni- or bilateral focal seizure activity of the primary generalized type (Joseph and Louis, 1995); be associated with preeclampsia and eclampsia (Cunningham et al., 1995); appear as a manifestation of hypomagnesemia secondary to cisplatin therapy (Al-Tweigeri et al., 1999); and arise after the failure of free tissue transfer (Tuchler et al., 1995), cervical trauma (Vaccaro et al., 1998), reversible occipital white matter abnormalities (Meyer et al., 1998), or vertebral angiography due to arterial spasm (Jackson et al., 1995). As a rule, recovery from blindness in these cases takes place within a few days or weeks.

## Principles of Therapy and Practical Management

### Recovery of the Visual Field

Two main approaches have been followed to improve the visual performance in patients with homonymous visual field disorders: (1) procedures to recover visual field portions from scotoma and (2) procedures to establish or improve compensatory strategies for overcoming the visual field defect. It has been shown in monkeys (Cowey and Weiskrantz, 1963; Mohler and Wurtz, 1977) and humans (Zihl and von Cramon, 1985) that systematic practice and training of saccadic eye movements to localize targets presented in the (perimetrically) blind field region can lead to a partial reduction of the size of the scotoma, provided the visual field loss is not due to irreversible brain damage. These results have been confirmed (Kasten *et al.*, 1998; Kerkhoff *et al.*, 1992a,b; Van der Wildt and Bergsma, 1997). Although some reports have proven the effect of visual field enlargement on field-dependent visual abilities (oculomotor scanning, reading), it is still an open question whether this approach can (and should) be considered useful in routine rehabilitation. Furthermore, we still lack a valid and reliable measure that indicates whether a patient has a good or poor prognosis for visual field recovery and the degree of recovery to be expected. Therefore, this approach is mainly of theoretical interest from the point of view of the plasticity of the occipital brain. From a pragmatic point of view, we need methods that focus on the main consequences of homonymous visual field defects and that can be easily applied to all patients requiring treatment for field loss (Zihl, 1988).

### Oculomotor Compensation

The methods developed for the compensation of visual field defects focus on the recovery of effective visual exploration of the visual environment and on improvement of reading performance. Hemianopic patients may show short-term improvements in visual scanning (Zangenmeister *et al.*, 1995), but this adaptation may be restricted to familiar stimulus conditions. The majority of patients (about 80%) report difficulties gaining a quick and complete overview, orienting themselves visually especially in unfamiliar surroundings, avoiding obstacles, and reading even several weeks after the onset of the visual field loss. In fact, about 60% of patients show scanning difficulties and about 80% impaired reading (Zihl, 2000); these difficulties are mirrored in the eye movement patterns (Pambakian *et al.*, 2000; Zihl, 1995a). There is no obvious left–right difference with respect to visual scanning; with respect to reading, however, patients with right-sided field loss show greater impairment because of the left-to-right orthography. Interestingly about 15% of patients acquire effective compensation strategies without any treatment; these patients have an injury restricted to the optic radiation and/or the striate cortex (Zihl, 1995a,b). Although visual exploration and reading may not be as fast as before brain injury, these patients do not need rehabilitation; within a few weeks they can successfully cope with daily life.

**Visual Exploration.** The first step in improving patients' visual exploration ability is to enlarge the amplitude of their saccadic eye movements. Larger eye movements enable them to cover the entire visual world by a few gaze shifts and thus gain information about its spatial structure. This information is required to visually guide oculomotor scanning movements in a systematic way through the actual scene. Since most of the patients also have difficulties orienting themselves visually, i.e., with the visual spatial guidance of their scanning eye movements, their scanning strategy is next improved with visual search tasks. This schedule of treatment caused a significant improvement in visual exploration (expressed as field of view) and visual scanning (expressed as errorless searching performance) in a group of 126 patients with hemianopia (Zihl, 1994) and in 19 patients with bilateral homonymous visual field loss (Zihl, 2000). Eye movement recordings showed that after practice patients utilized a much more systematic and spatially organized scanning strategy, characterized by significantly fewer saccades and fixations. Furthermore, the improvement persisted when they were tested 6 weeks after the training. As regards the subjective complaints of patients after training, only about 15% of patients still reported "slowness" of vision, but otherwise no further difficulties. Similar results have been reported by Kerkhoff *et al.* (1992b) who used a nearly identical method. The amount of practice (4–8 sessions for enlarging saccadic eye movements; 7–15 sessions for improving the searching strategy) is acceptable; it is important to carry out at least one session (45–60 min) per day and not to interrupt the period of practice. It must be emphasized that while patients with additional injury to the occipital white matter or to occipitoparietal cortical areas may require more sessions, they may eventually not benefit to the same extent. The explanation for this outcome is that the posterior parietal cortex, one of the critical structures underlying the visual spatial guidance of the scan path, is either also injured or disconnected from its occipital afferents (Zihl and Hebel, 1997).

**Reading.** As with visual exploration, the main aim in treating "hemianopic dyslexia" is to compensate for the parafoveal visual field loss by oculomotor adaptation. Especially patients with sparing of the visual field by less than 5° have difficulties reading: their reading speed is decreased and their reading often inaccurate. The method of treatment is to basically reorganize eye movements during reading. Briefly, patients with left-sided field loss have to shift their gaze first of all to the beginning of the line and then to the first letter of every single word in the line, whereas patients with right-sided field loss first shift their gaze to the end of a word. Thus, in both instances, patients are instructed to intentionally perceive the whole word before reading it. There are different ways to present text material: words can either move from left to right (Kerkhoff et al., 1992a; Zihl, 1988), or can be presented tachistoscopically (Zihl, 2000); both procedures require a PC. After treatment, patients (n = 96) showed a significant increase in the reading speed and a decrease in errors. Eye movement recordings showed that the improvement was mainly the result of adaptation of the eye movement pattern. After practice there were fewer fixations, larger saccadic jumps, and shorter fixation periods (Zihl, 1995b).

Using a similar method of treatment, Kerkhoff et al. (1992a) found comparable effects in a group of 56 patients. Although the postinjury time differed in the two studies (22 weeks in Zihl's study and 44 weeks in Kerkhoff's), improvement was observed within a few weeks, indicating that it was not just spontaneous adaptation. Interestingly patients with right-sided field loss required more training sessions (mean, 9; range, 7–14; 30-min duration each) than did patients with left-sided field loss (mean, 7; range, 4–11). However, more sessions may be required if additional disorders (e.g., visual disorientation or attentional deficits) are present. Follow-up testing for up to 2 years revealed either further improvement or at least maintenance of the post-treatment performance level.

The treatment of patients with bilateral homonymous field loss is similar to that of patients with unilateral defects, but more training sessions are required to achieve a sufficient degree of compensation. For patients with central scotoma, systematic practice of fixation and saccadic localization followed by oculomotor scanning training; discrimination of patterns and forms; and eventually visual recognition of objects, faces, scenes, and words have been found useful (Zihl, 2000).

### Head Shifts

Hemianopic patients are often instructed to shift their head toward the affected side and to keep it there to overcome the field loss. However, head movements usually follow eye movements, and head displacement depends on the amplitude and temporal course of the saccadic movement (Uemura et al., 1980). Thus a reversal of this order during eye–head coordination for locating and fixating objects does not contribute to gaze shifts in hemianopic patients (Zangenmeister et al., 1982). It even has a detrimental effect on visual exploration (Kerkhoff et al., 1992a).

### Optical Aids

Optical aids (mirror spectacles or prism systems) to compensate for the loss of vision in one hemifield have been reported to have some positive results (e.g., Rossi et al., 1990). Lee and Perez (1999) studied the possibility of increasing awareness of the peripheral visual field in nine selected patients with complete hemianopia by having them wear spectacles with ground-in sectorial prisms. Seven patients reported subjective improvement in activities of daily living scores, but this improvement was not assessed objectively. Using a high-power prism segment with 12 patients, Peli (2000) also reported similarly positive results. However, it is still unclear whether hemianopic patients also benefit in functional terms by wearing prism glasses. Further studies are required to test their usefulness under defined everyday conditions.

### "Blindsight"

When forced to guess, patients can detect and localize light targets presented in their blind field region, although they deny having seen anything (for a comprehensive review, see Weiskrantz, 1996). It has also been reported that practice has considerable effects on the detection and location of stimuli in the hemianopic field (Bridgeman and Staggs, 1982; Zihl, 1980). Patients with "blindsight" performance often report less visual difficulties in everyday life and more confidence in their visual orientation. This reduction in their visual handicap can be explained in terms of more frequent and larger gaze shifts as a by-product of practice with blindsight. Furthermore, after damage to the striate cortex, residual vision can be unconsciously used to detect (and locate) a light stimulus (Scharli et al., 1999; Wessinger et al., 1999) and therefore serve as the basis for eliciting orienting eye movements toward the hemianopic side.

## SPATIAL CONTRAST SENSITIVITY, VISUAL ADAPTATION, AND COLOR VISION

### Clinical Aspects and Natural Course

Spatial contrast sensitivity can be impaired after either uni- or bilateral retrogeniculate injury. Although recovery has been observed in single cases after therapy for brain tumors (e.g., Bodis-Wollner, 1972), spontaneous recovery does not seem to be the rule (Hess *et al.*, 1990).

Visual adaptation is impaired not only when the peripheral visual system is affected but also after retrogeniculate injury. Patients may exhibit reduced light or dark adaptation, or both. This visual disorder appears to be irreversible, since it persists for years (Zihl and Kerkhoff, 1990). The ability to tolerate strong light intensity may also be impaired after minor head injury (Waddell and Gronwall, 1984).

Color vision can be impaired either in one hemifield (and partly the fovea) after left- or right-sided injury to the lingual and fusiform gyri or in the entire hemifield (cerebral dyschromatopsia); it can be completely lost (cerebral achromatopsia) when these structures are damaged bilaterally. Some patients also exhibit impaired selective color constancy (Kennard *et al.*, 1995; Ruttiger *et al.*, 1999). Interestingly, imagery for colors can be preserved in cerebral achromatopsia (e.g., Bartolomeo *et al.*, 1997). Spontaneous recovery of color vision has been observed in patients with total cerebral blindness (see earlier). Complete spontaneous recovery of color vision within 6 months has been reported in a patient who had had carbon monoxide poisoning (Fine and Parker, 1996). Patients with cerebral achromatopsia due to bilateral strokes failed to exhibit recovery of color vision over a period of 6 years (Pearlman *et al.*, 1979).

Impairments in contrast and color vision have also been reported in patients with Parkinson's and Alzheimer's disease (for reviews, see Harris, 1998, and Mendez *et al.*, 1990). Birch *et al.* (1998) reported color vision deficiencies in 17 of 57 (29.8%) Parkinson patients; Mendola *et al.* (1995) found impaired contrast and pattern vision in about half of their 77 subjects with Alzheimer's disease.

## Principles of Therapy and Practical Management

Three patients with posterior brain injury have provided preliminary evidence that spatial contrast sensitivity can be improved with systematic practice and is associated with an increase in visual acuity and reading performance (Zihl, 2000). It must still be proven if this procedure is a useful clinical tool.

For the remediation of impaired visual adaptation, sunglasses are recommended for patients with defective light adaptation, and the variable adjustment of illumination, especially when reading, for patients with impaired dark adaptation.

The discrimination of color hues in cases of severely impaired color vision can, in principle, be improved by systematic practice (Mendola and Corkin, 1999).

Patients also seem to benefit from specific training involving the selection of the odd-one-out in an array of colors. A distinct reduction of error rates was observed in the Farnsworth Munsell test and improved color naming performance after practice (Zihl, 2000). Although this procedure is mainly experimental, it may represent an interesting approach to remediate cerebral color vision deficiency.

## VISUAL RECOGNITION

### Clinical Aspects

Visual agnosic disorders are impairments of complex visual capacities that are based on both visual-perceptual and visual-cognitive abilities and their interactions. Visual agnosia refers to the difficulty or inability of patients to visually identify familiar stimuli despite (sufficiently) preserved visual functions, cognitive capacities, and language. Recognition is preserved in another modality.

Visual agnosia, in its pure form, is a rare condition; its incidence may be in the range of 1–3% (Gloning *et al.*, 1968). Visual *object* agnosia refers to the difficulty of identifying and recognizing objects, *prosopagnosia* denotes the loss of the ability to visually recognize familiar faces (including one's own), *topographical agnosia* is often used to describe various difficulties of geographical orientation (see Chapter 26), and letter agnosia (pure alexia) refers to the difficulty of identifying individual letters and/or making words out of letters ("letter-by-letter reading"). In some cases, category-specific agnosias have been reported, but usually patients exhibit visual agnosias in more than one visual category (for a comprehensive review, see Grüsser and Landis, 1991). A typical site of injury causing visual agnosia is the region of the posterior and medial temporooccipital cortex, either bilateral or on the right side. The case of alexia is an exception; here the lesion site is the angular gyrus on the left side.

Recognition of facial expressions of emotion (for example, fearful as opposed to happy faces), can be impaired after injury to the amygdala (Young *et al.*, 1995), indicating that this structure plays an important role in the assignment of emotional ratings to facial expressions.

Visual agnosic disorders have also been reported in patients with posterior cortical atrophy. These disorders typically show a progression in severity (Caselli, 2000).

### Natural Course

Only a few case reports exist on recovery from visual agnosia. Adler (1950) found only minimal recovery within a 5-year period in a patient with visual agnosia due to carbon monoxide poisoning. Sparr *et al.* (1991) examined the same patient again about 40 years later, at which time the patient's status was more or less unchanged. In a review of the literature, Sparr *et al.* pointed out that carbon monoxide poisoning induces visual agnosia with a very poor prognosis. Kertesz (1979) followed up a patient who had had a traumatic head injury and showed visual agnosia that persisted for more than 10 years. Wilson and Davidoff (1993) examined a patient 10 years after she had sustained a severe head injury in an accident. They found that visual agnosia for real objects had resolved and identification of photographs (objects, faces), drawings, and words had improved. In a patient with prosopagnosia, Bruyer *et al.* (1983) found no evidence of improvement over a 1-year period of observation.

It appears that limited spontaneous recovery from visual agnosia can take place, except in those patients who also have mental deterioration which prevents the development and use of spared visual recognition procedures, context information, and nonvisual cues for functional adaptation. Wapner *et al.* (1978) reported, for example, the case of an agnosic artist who regained the ability to draw. The reappearance of this capacity may be due to the use of spared visual mental imagery, which can be observed in some patients who have severe visual agnosia (e.g., Chatterje and Southwood, 1995).

### Principles of Therapy and Practical Management

The first evidence for successful treatment of visual agnosic patients was reported by Poppelreuter (1917/1990), who found an improvement in visual recognition in three patients. The conventional way of ameliorating the patients' capacity to recognize objects and people is to teach them to use context information and to reorganize their visual perception with intact kinesthetic information (e.g., Tanemura, 1999). In a pilot study we (Zihl, 2000) attempted to reestablish and improve the prerequisites for identifying visual material in two patients with severe visual agnosia for objects, faces, scenes, and letters. Practice procedures included complete processing of stimulus information, selection of relevant features, i.e., features characteristic for a particular object, and the development and use of cognitive strategies to supervise and control visual identification. Visual materials included objects (food, personal hygiene articles, clothing), animals, faces, scenes, letters, and words. Both patients benefited from systematic practice, but one patient remained severely prosopagnosic. Although this outcome provides only preliminary evidence, it appears that visual agnosic patients may

learn to perform quite complex visual cognitive activities and may regain, at least in part, the ability to visually recognize stimuli. However, it must be added that both patients required quite a lot of training (416 and 940 sessions, respectively).

## VISUAL ILLUSIONS AND HALLUCINATIONS

### Clinical Features

Visual *illusions* are misperceptions of external objects and, therefore, represent abnormal processing of incoming visual information. The image may be altered in size (micropsia, macropsia), shape (dysmorphopsia or metamorphopsia), position (telopsia, allaesthesia), number (polyopia), color, movement, or persistence (visual perseveration). Visual *hallucinations* are visual perceptions that are not based on incoming visual information, which lead the subject to see something that is not evident to others in the same environment. These sensations may be unformed, consisting of flashes, light (colored or white), lines, and simple shapes, or be complex hallucinations involving faces, persons, numbers, animals, or detailed scenes.

If patients are aware of the illusory character of these visual phenomena they are referred to as "pseudohallucinations." Interestingly, the sites of "production" of visual illusions and visual hallucinations are the visual cortical areas that normally process the particular stimulus category. Visual spatial illusions and hallucinations arise from activities along the occipitoparietal (dorsal) stream, while illusions and hallucinations regarding colors, objects, faces, and scenes are "produced" along the occipitotemporal (ventral) stream (Ffytche *et al.*, 1998; Manford and Andermann, 1998). It is assumed that spontaneous brain activity is the basis for these visual phenomena, either as a consequence of pathological neuronal activity, as for example, in dural arteriovenous malformations, migraine, epilepsy, after drug intake, or because a given visual cortical area is deprived of its normal input due to deafferentiation (Grüsser and Landis, 1991).

Patients with Alzheimer's disease (e.g., Chapman *et al.*, 1999; Wilson *et al.*, 2000) and with dementia with Lewis bodies have also been reported to have visual hallucinations (Imamura *et al.*, 1999). Variables associated with the occurrence of visual hallucinations included older age, female sex, eye pathologies with decreased visual acuity, and severity of cognitive impairment. Holroyd and Seldonkeller (1995) found visual hallucinations in 18.4% of 98 patients with probable Alzheimer's disease. Patients with Parkinson's disease may also exhibit visual hallucinations. They are typically dopaminergic drug-induced but may also be facilitated by visual deficits, comorbid psychotic illness, or cognitive decline (Goetz *et al.*, 1998; Klein *et al.*, 1997). As regards prevalence, Fenelon *et al.* (2000) reported visual hallucinations in 39.8% of 216 patients with Parkinson's disease.

## Natural Course

Visual illusions appear to be a highly transient symptom after brain injury; they usually persist for seconds or minutes and disappear within a few days (Gloning *et al.*, 1968). In single cases, however, they may continue for several weeks or even months (Jacobs, 1980; Kawamura *et al.*, 1987). Similarly, visual hallucinations of the simple form are usually transient in nature, lasting some weeks but they can also persist for years. Their onset may be immediate after brain injury or after a delay of several days (Lance, 1976). Colored patterns usually appear days before a stroke occurs and persist for several hours or days, maximally 2 weeks (Kölmel, 1984). Complex visual hallucinations show an onset latency of 1–6 days and may be experienced by the patient usually over 5–15 days or in single cases up to several weeks (Kölmel, 1985). Their course depends on the underlying pathophysiological mechanism. If due to epileptic phenomena, the hallucinations are brief in duration; if due to relative deafferentiation caused by lesions in the anterior visual pathway, they may persist for days or weeks. When due to delerium, drug intoxication, and withdrawal states, the hallucinations may persist for as long as the responsible drug or disordered conscious state is maintained.

## REFERENCES

Adler, A. (1950). Course and outcome of visual agnosia. *J. Nerv. Ment. Dis.* 3, 41–51.

Aldrich, M. S., Alessi, A. G., Beck, R. W., and Gilman, S. (1987). Cortical blindness: Etiology, diagnosis, and prognosis. *Ann. Neurol.* 21, 149–158.

Al-Tweigeri, T., Magliocco, A. M., and DeCoteau, J. F. (1999). Cortical blindness as a manifestation of hypomagnesemia secondary to cisplatin therapy: Case report and review of literature. *Gynecol. Oncol.* 72, 120–122.

Bartolomeo, P., Bacoudlevi, A. C., and Denes, G. (1997). Preserved imagery for colours in a patient with cerebral achromatopsia. *Cortex* 33, 369–378.

Bergmann, P. S. (1957). Cerebral blindness. *Arch. Neurol. Psychiatry* 78, 568–584.

Birch, J., Kolle, R. U., Kunkel, M., Paulus, W., and Upadhyay, P. (1998). Acquired colour deficiency in patients with Parkinson's disease. *Vision Res.* 38, 3421–3426.

Bodis-Wollner, I. (1972). Visual acuity and contrast sensitivity in patients with cerebral lesions. *Science* 178, 769–771.

Bosley, T. M., Dann, R., Silver, F. L., Alavi, A., Kushner, M., Chawluck, J. B., Savi, P. J., Sergott, R. C., Schatz, N. J., and Raivich, M. (1987). Recovery of vision after ischemic lesions: Positron emission tomography. *Ann. Neurol.* 21, 444–450.

Bridgeman, B., and Staggs, D. (1982). Plasticity in human blindsight. *Vision Res.* 22, 1199–1203.

Bruyer, R., Laterre, C., Seron, X., Feyereisen, P., Strypstein, E., Pierrard, E., and Rectem, D. (1983). A case of prosopagnosia with some preserved covered remembrance of familiar faces. *Brain Cogn.* 2, 257–284.

Caselli, R. J. (2000). Visual syndromes as the presenting feature of degenerative brain disease. *Semin. Neurol.* 20, 139–144.

Celesia, G. G., Archer, C. R., Kuroiwa, Y., and Goldfader, P. R. (1980). Visual function of the extrageniculo-calcarine system in man: Relationship of cortical blindness. *Arch. Neurol.* 37, 704–706.

Chapman, F. M., Dickinson, J., McKeith, I., and Ballard, C. (1999). Association among visual hallucinations, visual acuity, and specific eye pathologies in Alzheimer's disease: Treatment complications. *Am. J. Psychiatry* 156, 1983–1985.

Chatterje, A., and Southwood, M. H. (1995). Cortical blindness and visual imagery. *Neurology* 45, 2189–2195.

Cowey, A. (1994). Cortical visual areas and the neurobiology of higher visual processes. *In* "The Neuropsychology of High-Level Vision" (M. Farah and G. Ratcliff, eds.), pp. 3–31. Erlbaum, Hillsdale, NJ/Hove, UK.

Cowey, A., and Weiskrantz, L. (1963). A perimetric study of visual field defects in monkeys. *Q. J. Exp. Psychol.* 15, 91–115.

Cunningham, F. G., Fernandez, C. O., and Hernandez, C. (1995). Blindness associated with preeclampsia and eclampsia. *Am. J. Obstet. Gynecol.* 172, 1291–1298.

Fenelon, G., Mahieux, F., and Ziegler, M. (2000). Hallucinations in Parkinson's disease: Prevalence, phenomenology and risk factors. *Brain* 123, 733–745.

Ffytche, D. H., Howard, R. J., Brammer, M. J., David, A., Woodruff, P., and Williams, S. (1998). The anatomy of conscious vision: An fMRI study of visual hallucinations. *Nat. Neurosci.* 1, 738–742.

Fine, R. D., and Parker, G. D. (1996). Disturbance of central vision after carbon monoxide poisoning. *Aust. N. Z. J. Ophthalmol.* 24, 137–141.

Frassinetti, F., Nichelli, P., and di Pellegrino, G. (1999). Selective horizontal dysmetropsia following prestriate lesion. *Brain* 122, 339–350.

Freedman, L., and Costa, L. (1992). Pure alexia and right hemiachromatopsia in posterior dementia. *J. Neurol. Neurosurg. Psychiatry* 55, 500–502.

Gloning, I., Gloning, K., and Hoff, H. (1968). "Neuropsychological Symptoms and Syndromes in Lesions of the Occipital Lobe and Adjacent Areas." Gauthier–Villars, Paris.

Gloning, I., Gloning, K., and Tschabischer, H. (1962). Die occipitale Blindheit auf vaskulärer Basis. *Albrecht von Graefes Arch. Klin. Ophthalmol.* 165, 138–177.

Goetz, C. G., Vogel, C., Tanner, C. M., and Stebbins, G. T. (1998). Early dopaminergic drug-induced hallucinations in parkinsonian patients. *Neurology* 51, 811–814.

Grüsser, O.-J., and Landis, Th. (1991). "Visual Agnosias and Other Disturbances of Visual Perception and Cognition." CRC Press, Boca Raton.

Harris, J. (1998). Vision in Parkinson's disease—What are the deficits and what are their origins. *Neuro-Ophthalmology* 19, 113–135.

Hess, R. F., Zihl, J., Pointer, S., and Schmid, C. (1990). The contrast sensitivity deficit in cases with cerebral lesions. *Clin. Vision Sci.* 5, 203–215.

Hier, D. B., Mondlock, J., and Caplan, L. R. (1983). Behavioral abnormalities after right hemisphere stroke. *Neurology* 33, 337–344.

Holroyd, S., and Seldonkeller, A. (1995). A study of visual hallucinations in Alzheimer's disease. *Am. J. Geriatr. Psychiatry* 3, 198–205.

Imamura, T., Ishii, K., Hirono, N., Hashimoto, M., Tanimukai, S., Kazuai, H., Hanihara, T., Sasaki, M., and Mori, E. (1999). Visual hallucinations and regional cerebral metabolism in dementia with Lewy bodies (DLB). *Neuroreport* **10**, 1903–1907.

Jackson, A., Stewart, G., Wood, A., and Gillespie, J. E. (1995). Transient global amnesia and cortical blindness after vertebral angiography—Further evidence for the role of arterial spasm. *Am. J. Neuroradiol.* **16 (Suppl. S)**, 955–959.

Jacobs, L. (1980). Visual allesthesia. *Neurology* **30**, 1059–1063.

Joseph, J. M., and Louis, S. (1995). Transient ictal cortical blindness during middle age—A case report and review of the literature. *J. Neuro-Ophthalmology* **15**, 39–42.

Kasten, E., Wust, S., Behrensbaumann, W., and Sabel, B. A. (1998). Computer-based training for the treatment of partial blindness. *Nat. Med.* **4**, 1083–1087.

Kawamura, M., Hirayama, K., Shinohara, Y., Watanabe, Y., and Sugishita, M. (1987). Alloesthesia. *Brain* **110**, 225–236.

Kennard, C., Lawden, M., Morland, A. B., and Ruddock, K. H. (1995). Colour identification and colour constancy are impaired in a patient with incomplete achromatopsia associated with prestriate cortical lesions. *Proc. R. Soc. London B* **260**, 169–175.

Kerkhoff, G., Münsinger, U., Eberle-Strauss, G., and Stögerer, E. (1992a). Rehabilitation of hemianopic dyslexia in patients with postgeniculate field disorders. *Neuropsychol. Rehab.* **2**, 21–42.

Kerkhoff, G., Münsinger, U., Haaf, E., Eberle-Strauss, G., and Stögerer, E. (1992b). Rehabilitation of homonymous scotomata in patients with postgeniculate damage of the visual system: Saccadic compensation training. *Restor. Neurol. Neurosci.* **4**, 245–254.

Kertesz, A. (1979). Visual agnosia: The dual deficit of perception and recognition. *Cortex* **15**, 403–419.

Klein, C., Kömpf, D., Pulkowski, U., Moser, A., and Vieregge, P. (1997). A study of visual hallucinations in patients with Parkinson's disease. *J. Neurol.* **244**, 371–377.

Kölmel, H. W. (1984). Coloured patterns in hemianopic fields. *Brain* **107**, 155–167.

Kölmel, H. W. (1985). Complex visual hallucinations in the hemianopic field. *J. Neurol. Neurosurg. Psychiatry* **48**, 29–38.

Lance, J. W. (1976). Simple formed hallucinations confined to the area of a specific visual field defect. *Brain* **99**, 719–734.

Lee, A. G., and Perez, A. M. (1999). Improving awareness of peripheral visual field using sectorial prism. *J. Am. Optom. Assoc.* **70**, 624–628.

Manford, M., and Andermann, F. (1998). Complex visual hallucinations: Clinical and neurobiological insights. *Brain* **121**, 1819–1840.

Mendez, M. F., Tomsak, R. L., and Remler, B. (1990). Disorders of the visual system in Alzheimer's disease. *J. Clin. Neuro-ophthalmol.* **10**, 62–69.

Mendola, J. D., and Corkin, S. (1999). Visual discrimination and attention after bilateral temporal lobe lesions: A case study. *Neuropsychologia* **37**, 91–102.

Mendola, J. D., Croningolomb, A., Corkin, S., and Growdon, J. H. (1995). Prevalence of visual deficits in Alzheimer's disease. *Optometry Visual Sci.* **72**, 155–167.

Meyer, M. A., Galloway, G., and Khan, S. (1998). Transient blindness associated with reversible occipital white matter abnormalities—Two patients studied by MR, CT, and $^{18}$F-FDG PET imaging. *J. Neuroimag.* **8**, 240–242.

Mohler, C. W., and Wurtz, R. H. (1977). Role of striate cortex and superior colliculus in visual guidance of saccadic eye movements in monkeys. *J. Neurophysiol.* **40**, 74–94.

Pambakian, A. L. M., Wooding, D. S., Patel, N., Morland, A. B., Kennard, C., and Mannan, S. K. (2000). Scanning the visual world: A study of patients with homonymous hemianopia. *J. Neurol. Neurosurg. Psychiatry* **69**, 751–759.

Patel, A. T., Duncan, P. W., Lai, S. M., and Studenski, S. (2000). The relation between impairments and functional outcomes poststroke. *Arch. Phys. Med. Rehabil.* **81**, 1357–1363.

Paulson, H. L., Galetta, S. L., Grossman, M., and Alavi, A. (1994). Hemiachromatopsia of unilateral occipitotemporal infarcts. *Am. J. Ophthalmol.* **118**, 518–523.

Pearlman, A. L., Birch, J., and Meadows, J. C. (1979). Cerebral color blindness. An acquired defect in hue discrimination. *Ann. Neurol.* **5**, 253–261.

Peli, E. (2000). Field expansion for homonymous hemianopia by optically induced peripheral exotropia. *Optometry Visual Sci.* **77**, 453–464.

Plant, G. T., Laxer, K. D., Barbaro, N. M., Schiffman, J. S., and Nakayama, K. (1993). Impaired visual motion perception in the contralateral hemifield following unilateral posterior cerebral lesions in humans. *Brain* **116**, 1303–1335.

Poppelreuter, W. (1917/1990). "Disturbances of Lower and Higher Visual Capacities Caused by Occipital Damage." Translated by J. Zihl and L. Weiskrantz. Oxford Univ. Press (Clarendon), Oxford.

Rausch, M., Widdig, W., Eysel, U. T., Penner, I. K., and Tegenthoff, M. (2000). Enhanced responsiveness of human extrastriate areas to photic stimulation in patients with severely reduced vision. *Exp. Brain Res.* **135**, 34–40.

Reding, M. J., and Potes, E. (1988). Rehabilitation outcome following initial unilateral hemispheric stroke: Life table analysis approach. *Stroke* **19**, 1354–1358.

Rossi, P. W., Kheyfets, S., and Reding, M. (1990). Fresnal lens improve visual perception in stroke patients with homonymous hemianopia or unilateral visual neglect. *Neurology* **40**, 1587–1599.

Rowe, F. J., and Sarkies, N. J. (1998). Assessment of visual function in idiopathic intracranial hypertension—A prospective study. *Eye* **12**, 111–118.

Ruttiger, L., Braun, D. I., Gegenfurtner, K. R., Petersen, D. Sconle, P., and Sharpe L. T. (1999). Selective color constancy deficits after circumscribed unilateral brain lesions. *J. Neurosci.* **19**, 3094–3106.

Scharli, H., Harman, A. M., and Hogben, J. H. (1999). Residual vision in a subject with damaged visual cortex. *J. Cogn. Neurosci.* **11**, 502–510.

Schenk, T., and Zihl, J. (1997). Visual motion perception after brain damage: I. Deficits in global motion perception. *Neuropsychologia* **35**, 1289–1297.

Slamovits, T. L., Rosen, C. E., Cheng, K. P., and Striph, G. G. (1991). Visual recovery in patients with optic neuritis and visual loss to no light perception. *Am. J. Ophthalmol.* **111**, 209–214.

Sparr, S. A., Jay, M., Drislane, F. W., and Venna, N. (1991). An historic case of visual agnosia revisited after 40 years. *Brain* **114**, 789–800.

Symonds, C., and Mackenzie, I. (1957). Bilateral loss of vision from cerebral infarction. *Brain* **80**, 415–455.

Tanemura, R. (1999). Awareness in apraxia and agnosia. *Top. Stroke Rehab.* **6**, 33–42.

Tuchler, R. E., Lyos, A. T., Rainey, A. M., and Anous, M. M. (1995). Cortical blindness after a failed free tissue transfer—A case report and review of the literature. *Ann. Plastic Surg.* **34**, 431–434.

Uemura, T., Arai, Y., and Shimazaki, C. (1980). Eye–head coordination during lateral gaze in normal subjects. *Acta Otolaryngol.* **90**, 191–198.

Uzzell, B. P., Dolinskas, C. A., and Langfitt, T. W. (1988). Visual field defects in relation to head injury severity. *Arch. Neurol.* **45**, 420–424.

Vaccaro, A. R., Urban, W. C., and Aiken, R. D. (1998). Delayed cortical blindness and recurrent quadriplegia after cervical trauma. *J. Spinal Disord.* **11**, 553–559.

Van der Wildt, G. J., and Bergsma, D. P. (1997). Visual field enlargement by neuropsychological training of a hemianopsia patient. *Doc. Ophthalmol.* **93**, 277–292.

Waddell, P. A., and Gronwall, D. M. A. (1984). Sensitivity to light and sound following minor head injury. *Acta Neurol. Scand.* **69**, 270–276.

Wapner, W., Judd, T., and Gardner, H. (1978). Visual agnosia in an artist. *Cortex* **14**, 343–364.

Weiskrantz, L. (1996). Blindsight revisited. *Curr. Opin. Neurobiol.* **6**, 215–220.

Wessinger, C. M., Fendrich, R., and Gazzaniga, M. S. (1999). Variability of residual vision in hemianopic subjects. *Rest. Neurol. Neurosci.* **15**, 243–253.

Wilson, B., and Davidoff, J. (1993). Partial recovery from visual object agnosia: A 10 year follow-up study. *Cortex* **29**, 529–542.

Wilson, R. S., Gilley, D. W., Bennett, D. A., Beckett, L. A., and Evans, D. A. (2000). Hallucinations, delusions, and cognitive decline in Alzheimer's disease. *J. Neurol. Neurosurg. Psychiatry* **69**, 172–177.

Wunderlich, G., Suchan, B., Herzog, H., Hömberg, V., and Seitz, R. J. (2000). Visual hallucinations in recovery from cortical blindness: Imaging correlates. *Arch. Neurol.* **57**, 561–565.

Young, A. W., Aggleton, J. P., Hellawell, D. J., Johnson, M., Broks, P., and Hanley, J. R. (1995). Face processing impairments after amygdalotomy. *Brain* **118**, 15–24.

Zangenmeister, W. H., Meienberg, O., Stark, L., and Hoyt, W. F. (1982). Eye–head coordination in homonymous hemianopia. *J. Neurol.* **226**, 243–254.

Zangenmeister, W. H., Oechsner, U., and Freska, C. (1995). Short-term adaptation of eye movements in patients with visual hemifield defects indicates high level control of human scanpath. *Optometry Visual Sci.* **72**, 467–477.

Zeki, S. (1993). "A Vision of the Brain." Blackwell Scientific, Oxford.

Zihl, J. (1980). "Blindsight": Improvement of visually guided eye movements by systematic practice in patients with cerebral blindness. *Neuropsychologia* **18**, 71–77.

Zihl, J. (1988). Sehen. *In* "Neuropsychologische Rehabilitation" (D. von Cramon and J. Zihl, Eds.), pp. 105–131. Springer-Verlag, Berlin/Heidelberg New York.

Zihl, J. (1994). Rehabilitation of visual impairments in patients with brain damage. *In* "Low Vision. Research and New Developments in Rehabilitation" (A. C. Kooijman, P. L. Looijestijn, J. A. Welling, and G. J. van der Wildt, Eds.), pp. 287–295. IOS Press, Amsterdam/Oxford.

Zihl, J. (1995a). Visual scanning behavior in patients with homonymous hemianopia. *Neuropsychologia* **33**, 287–303.

Zihl, J. (1995b). Eye movement patterns in hemianopic dyslexia. *Brain* **118**, 891–912.

Zihl, J. (2000). "Rehabilitation of Visual Disorders after Brain Injury." Psychology Press, Hove, UK.

Zihl, J., and Hebel, N. (1997). Patterns of oculomotor scanning in patients with unilateral posterior parietal or frontal lobe damage. *Neuropsychologia* **35**, 893–906.

Zihl, J., and von Cramon, D. (1985). Visual field recovery from scotoma in patients with postgeniculate damage. *Brain* **108**, 335–365.

Zihl, J., and Kerkhoff, G. (1990). Foveal photopic and scotopic adaptation in patients with brain damage. *Clin. Vision Sci.* **5**, 185–195.

## CHAPTER 25
# Aphasia

David Glenn Clark and Jeffrey L. Cummings

## CLINICAL ASPECTS

### Definition

Damage to regions of the brain devoted to language processing results in a loss of the ability to interpret or express thoughts in the form of language. In most adults the regions that are most vital for symbolic communication are located in the perisylvian region of the left cerebral hemisphere. Depending on the size and location of the damaged area there may be preferential loss of the capability to express or to comprehend spoken or written language. The set of clinical presentations denoting any acquired disorder of language is labeled with the general term **aphasia**. Aphasia must be distinguished from abnormalities of the motor or sensory systems that are utilized by, but are outside of, the language network. Thus, dysarthria—which includes disorders of articulation that may result from spasticity, from lesions of cranial nerves or their nuclei, or even from loss of teeth—must be considered separately. Similarly, aphasia excludes various sensory disturbances that may lead to failure of communication, including some focal cerebral lesions that affect auditory or visual perception. At the border between aphasic and non-aphasic deficits are certain uncommon perceptual deficits that are specific for linguistic sensory information, such as "pure word deafness" and "pure word blindness." Most investigators classify these, as well as other forms of alexia and the various agraphias, under the general classification of aphasia or aphasia-related syndromes. It is important to distinguish between disturbances of the language network and those of other higher cortical modules, including those mediating memory, attention, and executive function. Despite these distinctions, aphasia frequently coexists with one or more motor, sensory, or cognitive abnormalities.

### Classical Aphasiology and Cognitive Neurolinguistics

Linguists describe the phenomena of language in terms of features that are universal among all human languages: semantics, phonology, syntax, and morphology (Fromkin, 2000). The term **semantics** pertains to the meanings that are assigned to words. From a neurological perspective a word is endowed with meaning only when it is linked to a set of sensory associations that are established through the speaker's experience with the world. Neural representations of concrete objects have a distributed location in the cortex that results from the speaker's experience with the object. During lexical access, these arbitrary associations activate neurons within areas of association cortex that are specific to the category of the object being named (Damasio *et al.*, 1996). Thus, a lesion of the sensory association cortex in the inferior temporal region and temporo-occipito-parietal junction can disrupt the ability to access the names of tools and other items that can be manipulated with the hand. This may result from disconnection of sensory cortex associated with the hand (in the inferior parietal lobe) from an area in which the sensory information converges prior to activating a symbolic representation. Deficits produced by such lesions are out of proportion to impairment in assigning appropriate names to animals or familiar human faces. The contribution of various regions of cortex to categorical object naming has been demonstrated with functional neuroimaging and lesion analysis studies. The neural representations of words that serve other grammatical functions (such as verbs) may be linked to areas of association cortex that are separate from or overlapping those related to object naming (Perani *et al.*, 1999). Wernicke's area, which is a poorly delineated region in the posterior part of the superior temporal gyrus and adjacent parietal lobe, comprises a critical nexus for

*Neurological Disorders: Course and Treatment, Second Edition*

establishing the connections between word forms and the concepts they represent. A lesion in this area can disturb the ability to link semantic associations with the appropriate word form. This results in the classical syndrome of Wernicke's aphasia, in which both comprehension and self-expression are impaired by loss of the lexical–semantic interface. Aphasias resulting from lesions in the vicinity of Wernicke's area are associated with **semantic paraphasic errors**, in which the speaker substitutes a semantically related word for the target word (such as "table" instead of "chair"). Lesions in Wernicke's area can cause disruption in normal phonological sequencing, which often results in the generation of **neologisms**. These differ greatly from the target word and may render output unintelligible.

**Phonology** relates to the sound patterns that make up words. These patterns are understood as programs of motor activity that lead to the expression of a spoken symbol. Access to these motor programs is mediated by the lower portion of the third left frontal gyrus, anterior to the operculum. Paraphasic errors are rare with lesions in this region, but when they occur, they are more likely to be **phonemic**, rather than semantic; that is, the speaker substitutes an incorrect phoneme in an otherwise correct sequence, such as saying "cuff" when trying to pronounce "cup." Semantic paraphasic errors that occur in the setting of Broca's aphasia are likely to be followed by an immediate negation by the speaker (Goodglass, 1993).

The term **syntax** refers to the integrated process of placing appropriate words in a coherent order and of parsing linguistic input according to the accepted rules of order. **Morphology** is related to syntax, but focuses on the precise rules that guide the formation of words by combining meaningful linguistic units called morphemes. These include **free morphemes**, such as the word "book," and **bound morphemes**, such as the prefix "re-" or the suffix "–able." The normal use of syntax and morphology is dependent largely on portions of Broca's area. Some patients with lesions in Broca's area have preserved capacity for expressing nouns and verbs, but develop nonfluent, **agrammatic** speech characterized by difficulty placing content words in all but the simplest syntactic arrangements, usually without appropriate morphological affixes. The language output is typically sparse and halting. This contrasts sharply with **paragrammatism**, a syntactic abnormality that is characteristic of Wernicke's aphasia, in which output is fluent and patients use function words and inflections incorrectly. In addition, sentences may be blended together nonsensically and at times nouns and verbs will occupy the wrong syntactic positions within a sentence.

Precise generalizations can be made regarding the syntax of all human languages; these rules are the result of a set of features that are common to all normal human brains. Such rules reflect the nature of neuronal interactions in language cortex and certain subcortical nuclei. None of the key features of language are due to the action of a single, well-delineated region of the brain. Thus, abnormalities of syntax or phonology may result from lesions in a number of locations, including both Broca's and Wernicke's regions. Normal language comprehension and expression result from processes that take place in parallel between Wernicke's and Broca's areas. Many other regions of the language-dominant hemisphere act in concert with these epicenters to link symbolic neural representations with semantic concepts, auditory or visual word forms, or the motor programs for producing spoken or written words and sentences.

The term **pragmatics** refers to complex aspects of discourse and the subtle rules that underlie the normal sociocultural use of language. The elements of pragmatics are difficult to quantify and may be impacted by many neurologic lesions that do not have a direct effect on the other more readily definable features of language. Nevertheless, pragmatics are an essential part of communication. Normal speakers exercise a number of skills in order to use language that is appropriate to the context, such as estimating the appropriate distance from their conversational partner, taking turns speaking and giving statements and questions their respective tonal qualities. Some studies suggest that all brain-damaged individuals exhibit some degree of pragmatic impairment (Newhoff and Apel, 1997).

### Cortical Lesions Affecting Language

The most well recognized aphasic syndromes result from lesions of the left perisylvian cerebral cortex. Broca's aphasia results from lesions anterior to the rolandic fissure, in the inferior left frontal gyrus, surrounding frontal areas and subjacent subcortical structures (Damasio, 2000). There is often associated weakness of the right face and upper extremity. The language defect is characterized by slow, effortful speech with loss of normal melodic modulation and a reduction in the number of words per utterance. Patients generally produce nouns with less difficulty than verbs and other words. The classic syndrome also consists of agrammatism, in which normal canonical word order may be violated and the patient misuses or omits inflections and grammatical function words. Some patients exhibit **phonetic disintegration**, in which the temporal alignment of phonemes is distorted, leading to misarticulation of individual phonemes. Confrontational naming and repetition are also impaired. Although patients usually understand routine conversation quite well, auditory discrimination of individual phonemes may be impaired. Many patients have difficulty comprehending syntactically complex sentences, especially those that can be semantically reversed. For example, a patient with Broca's aphasia may interpret the sentence, "The boy was pushed by the girl," as meaning that the boy pushed the girl. This is due to interpretation based on word order and failure to take into account the reversal of thematic roles that results from the verb construction and prepositional phrase. Typically the syndrome of Broca's aphasia results from a lesion involving the superior division of the left middle cerebral artery, with infarction extending to subcortical regions and the frontal lobe. Infarction limited to Broca's area causes transient mutism; recovery is often nearly complete apart from residual dysarthria.

Lesions resulting in Wernicke's aphasia lie posterior to the rolandic fissure and usually involve the auditory association cortex of the left superior temporal gyrus, as well as other surrounding regions. Hemiparesis is not present to aid in the recognition of Wernicke's aphasia. There is defective repetition of sentences. Defective naming and phonemic and neologistic paraphasic errors may render output incomprehensible. Patients have difficulty comprehending sentences that they hear. Language output is fluent in the sense that patients speak effortlessly, the normal melodic intonation of language is present, and the number of words per utterance is normal or increased. The syndrome of Wernicke's aphasia results most often from occlusion of the inferior division of the middle cerebral artery, resulting in infarction of the temporal and parietal lobes.

**Conduction aphasia** may result from lesions of the supramarginal gyrus or from combined damage to the primary auditory cortices, insula, and underlying white matter. Patients have difficulty naming objects to confrontation and are unable to repeat sentences *verbatim*. They produce frequent phonemic paraphasias, but otherwise can express themselves intelligibly and can comprehend simple sentences without difficulty. The only common motor finding is right facial weakness; sensory abnormalities of the right hand are commonly present.

Very large left hemisphere lesions may cause **global aphasia**, a severe disruption of all features of language. Such lesions are usually the result of large infarctions in the left middle cerebral artery territory and involve the regions associated with Broca's and Wernicke's aphasias, as well as the insula and basal ganglia. These patients may retain the capacity to say a few words or sentences, but often make repetitive stereotyped utterances. The use of some expletives may also be preserved.

Occasionally patients can present with aphasic syndromes that resemble the classical syndromes of Broca's, Wernicke's, or global aphasia except that repetition is spared (Goodglass, 1993). These are termed **transcortical** motor, sensory, and mixed aphasias, respectively. Transcortical motor aphasia results from small lesions of the mesial frontal lobe or underlying white matter, causing disconnection of language areas from areas important in volition. The result is an impairment of the generation of spontaneous utterances. The preservation of areas critical to linguistic processing allows virtually flawless repetition. Transcortical sensory aphasia usually results from watershed zone infarction of the temporoparietal area in the dominant hemisphere. The lesion is slightly posterior to Wernicke's area, and outside the perisylvian region usually associated with language. Apart from having spared capacity for repetition, these patients evidence less paragrammatism than those with typical Wernicke's aphasia.

Larger, more widespread lesions have sometimes led to an aphasic syndrome with severe impairments in comprehension and expression but spared repetition. Geschwind *et al.* (1968) described such a case and referred to it as "isolation of the speech area," since the widespread cerebral lesions were found to spare the left perisylvian cortex at autopsy. Others have called it **transcortical mixed aphasia**. This patient demonstrated a classic feature of the disorder, called **completion phenomenon:** a tendency to respond to half-spoken cliches by completing them. For example, when an examiner said, "Ask me no questions," she replied, "Tell me no lies," but never appeared to comprehend what was being said. She also demonstrated preserved capacity to sing along with records and even to learn words to new songs. Some patients have preserved automatic speech, including the ability to count or to recite memorized prayers or the alphabet. Considering the very large left hemisphere infarcts that have sometimes led to this syndrome, some authors have speculated that the preservation of repetition may be mediated by the right hemisphere (Grossi *et al.*, 1991).

## Subcortical Aphasic Syndromes

The basal ganglia and thalamus play roles in the normal use of language. Cortical and subcortical regions are integrated into interactive systems: prefrontal cortex projects to the striatum as the first segment of circuits that eventually influence thalamic nuclei. These circuits are completed by thalamic afferents back to cerebral cortex. Other areas of cortex have reciprocal connections to thalamic nuclei, including parietal and superior temporal regions. Current models of large-scale neurocognitive networks include the basal ganglia and thalamus. The striatum is believed to integrate, compare or synchronize computations as they occur in epicenters of each network (Mesulam, 1998). Lesions of the internal capsule and striatum that interrupt the subcortical circuit connecting the dorsolateral frontal cortex and dorsolateral caudate affect the generative aspects of language that rely on normal executive function (Mega and Alexander, 1994). These patients suffer from anomia during conversation and have deficits in confrontational testing; speech is often dysarthric and hypophonic. Language is usually fluent and grammatical but patients can have difficulty with novel syntactic constructions and may develop echolalia. Prolonged latencies, perseverations, and occasional bizarre content also have been noted. Repetition, reading, and single-word comprehension are usually intact, and phonemic paraphasic errors are rare.

At least two thalamic aphasic syndromes have been described (Graff-Radford and Damasio, 1984). One of these is associated with lesions of the ventrolateral and ventral anterior nuclei of the dominant thalamus. These patients have impaired naming, comprehension, reading, and writing. Although language is characterized as fluent, utterances are sparse and speech may be dysarthric, hypophonic, and dysprosodic. Patients also tend to produce semantic paraphasic errors and perseverations. The ventrolateral nucleus receives input primarily from basal ganglia and sends projections to primary motor cortex, including Broca's area. The ventral anterior nucleus also receives input from basal ganglia, but has output primarily to the supplementary motor area. A second aphasic syndrome is associated with lesions of the pulvinar and lateral posterior nuclei. It is usually associated with normal fluency or abnormally increased speech output, termed **logorrhea**. Neologistic paraphasic errors may occur. This language disorder is generally transient and is almost always associated with other neurocognitive deficits, usually in the domains of verbal and nonverbal memory, attention, affect, and motivation. The pulvinar has reciprocal interaction with Wernicke's area and the lateral posterior nucleus is associated with the inferior parietal lobule.

## Etiologies

Vascular disease is the most common etiology of aphasia, particularly thromboembolic stroke in the middle cerebral artery territory. Intraparenchymal hemorrhage, aneurysmal rupture, and hypotensive, watershed infarction may also lead to aphasic disturbances. Stroke accounts for 85% of aphasias (Reinvang, 1984). Recovery from aphasia due to stroke depends on a number of factors (Kertesz and McCabe, 1977). Younger patients demonstrate more complete recovery than do older patients. Broca's aphasia generally has a better prognosis than do other syndromes, but the overall pattern of recovery is more variable. A broad spectrum of variability is also seen in recovery from aphasia due to CNS hemorrhage. Aphasia that is more severe at onset is associated with a poor prognosis compared to aphasia that is initially mild.

Trauma, especially closed-head injury, is the second most common cause of aphasia. It generally has a better prognosis than aphasia due to vascular insult, but this may be due to the influence of the young age of many patients with head trauma (Levin, 1991). Traumatic aphasia is usually fluent with anomia characterized by frequent semantic paraphasias and circumlocutions. Comprehension and repetition are relatively spared.

Any mass lesion of the left cerebral hemisphere may produce aphasia. Those that grow slowly and are not associated with large amounts of edema may produce more subtle deficits and progress more insidiously. Aphasia may not be the presenting feature, since masses are often associated with signs of increased intracranial pressure such as headache, nausea, and vomiting. Neoplastic mass lesions that may produce aphasia include metastases, gliomas, and meningiomas, among others. Masses of infectious etiology may produce fever, weight loss, and night sweats in addition to elevated intracranial pressure and focal neurologic signs. Infectious masses include tuberculomas, syphilitic gummas and abscesses caused by other types of bacteria, protozoans such as toxoplasma or amoebae, parasitic diseases such as cysticercosis, paragonimiasis, and echinococcosis, and fungal infections such as cryptococcosis.

Infectious diseases that do not produce mass lesions may result in aphasia. These include herpes simplex encephalitis and other encephalitides (Ku *et al.*, 1996). Progressive multifocal leukoencephalopathy may produce aphasia by undercutting areas of cortex integral to language function (Singer *et al.*, 1994).

Some forms of epilepsy can manifest as aphasia. Children may rarely acquire aphasia due to Landau–Kleffner syndrome, a form of epilepsy associated with paroxysmal temporal lobe discharges on EEG that are activated during sleep (Roger *et al.*, 1993). Patients usually present before the age of 6 years with auditory verbal agnosia and reduced or absent verbal expression. The deficits may progress in a gradual or stepwise manner. Hyperkinesia and personality disturbances are frequently associated and 70% of patients have coexistent epileptic seizures. While the prognosis of the epilepsy is favorable, that of the aphasia is poor, with only 40–50% of patients going on to lead normal social and professional lives. In adults, aphasia may rarely be the sole manifestation of partial status epilepticus (Toledo *et al.*, 2000).

Dementing diseases that affect primarily cerebral cortex are often associated with language impairment. Alzheimer's disease has been noted to present with aphasia and in one study of hospitalized patients with Alzheimer's disease all were found to have some degree of language impairment (Kertesz *et al.*, 1986). The language impairment followed a characteristic evolution through several syndromes, beginning with anomia on confrontational tasks. There follows a progression through disturbances best classified as transcortical sensory and Wernicke's aphasia and then finally degeneration to a syndrome with mutism or echolalia and palilalia. Comprehension is impaired for both spoken and written language and discourse production is characterized by frequent communication of irrelevant information (Bayles, 1993). The first case of frontotemporal dementia described by Pick presented with a syndrome similar to transcortical sensory aphasia (Pick, 1892). Infrequently, language impairment is the only debilitating feature of a progressive dementia at presentation. If this condition of isolated language disturbance persists for 2 or more years, the patient can be classified as having primary progressive aphasia (Mesulam, 2001). Primary progressive aphasia may manifest as a fluent or nonfluent aphasia. The latter often progresses to include other elements of frontotemporal dementia, such as personality change and disinhibition. In contrast, the former may develop into semantic dementia, with defects of visual processing or other evidence of dilapidation of semantic concepts, or conceivably, impaired access to them (Snowden *et al.*, 1992). The underlying histopathology of both syndromes is variable, but that of Pick's disease is the most common specific finding. Because these patients retain a capacity and eagerness for cooperation they give valuable insights into normal and pathological language processing by providing multiple, sequential perspectives of language dissolution. Ongoing therapy can improve communication but the prognosis is frequently poor for recovery of communication skills.

## ASSESSMENT

The bedside evaluation of language includes testing of fluency in spontaneous speech, auditory comprehension, repetition, naming, reading, and writing. Naming can be assessed by asking the patient to identify common objects, such as a wristwatch or a pen. More subtle deficits of naming may be elicited by asking for the names of parts of objects, such as the "band" of the watch, or of body parts such as fingernails, knuckles, or eyebrows. Tests of generative naming are sensitive to mild anomia. These require the patient to generate as many words in specific categories as possible in a short period of time, usually 60 s. The words must meet specified criteria, such as "names of animals" or "starting with the letter F." Most normal individuals can gener-

ate at least 12 animal names or 10 F-words in 1 min. If more detailed information is required, the clinician may administer the Boston Naming Test, a widely used set of simple drawings that the patient is asked to name at the bedside.

The assessment of fluency is valuable for coarse localization of lesions producing aphasia, with nonfluent forms typically resulting from lesions anterior to the Rolandic fissure and fluent forms arising from more posterior lesions (Goodglass, 1993). Aphasic patients tend to fall into fluent and nonfluent categories based on the number of words per utterance group. Nonfluent patients generally produce four or fewer words per utterance, while fluent patients readily produce five or more. Fluent aphasias are also characterized by reduced semantic content of their speech and by paraphasic errors.

Apart from casual conversation, the examiner may evaluate auditory comprehension by having the patient comply with a graded sequence of questions and commands. Initially, the patient is required to answer simple questions with yes/no answers, such as "Are you lying in bed?" or "Is the TV on?" This can be advanced to more syntactically complex questions. Many severely impaired patients retain the ability to follow one-step, midline commands, such as "Close your eyes," but fail to follow commands that require use of the extremities or left/right distinctions or that have multiple steps. In order to test the patient's understanding of names of items, spatial relationships, and action words it is often useful to place a number of common items on a tray table and give commands related to the items (e.g., "Touch the pen with the comb"). Still more sensitive is the Token Test (De Renzi and Vignolo, 1962), which makes use of colored circular and rectangular tokens as stimuli. This deprives the patient of some of the implicit cues that may be present if common objects are used; it can be administered fairly quickly at the bedside. However, performance on the Token Test can be affected by nonlinguistic factors, such as attention, working memory, anxiety, or color perception.

The examiner can assess repetition rapidly using a graded set of stimuli. Severely impaired patients may have difficulty repeating even short phrases consisting only of high-frequency words, like "This is it." Longer phrases, such as "They heard him speak on the radio last night" or "The judge was impressed by the evidence," add a degree of difficulty. Patients with very subtle deficits of repetition may have trouble only with very unusual combinations of words like "No ifs, ands, or buts," "The phantom soared over the foggy heath," or various tongue twisters.

Reading is assessed with a short passage of prose from a newspaper or novel and should always be accompanied by questions testing reading comprehension.

Writing should be screened by asking the patient to write a sentence spontaneously. If the patient is not able to perform this task it may be useful to test his or her ability to write the letters of the alphabet, write to dictation, or copy written material.

If bedside screening leads the clinician to believe the patient is aphasic, a number of test batteries are available for more detailed language assessment. Two of the most widely used English language tests are the Boston Diagnostic Aphasia Examination (BDAE) and the Western Aphasia Battery (WAB), which is a more recent modification of the BDAE. Both of these place the patient's disorder into one of the syndromic categories discussed above.

Classification of aphasia by syndrome has been scrutinized. Goodglass (1993) defends use of syndrome labels as a means by which clinicians can communicate sets of aphasic features to one another rapidly. Depending on the clinician's experience the syndrome label may provide important information regarding the localization or etiology of the disorder. However, he points out that each syndrome is a mixture of features from different language domains and that the degree to which each domain is affected is highly variable. Caplan (1992) proposes a more descriptive means of categorizing aphasias, in which 27 subtests are used to assess an array of language "modules" individually. The Psycholinguistic Assessment of Language Processing in Aphasia (PALPA) is a commercially available battery of 60 subtests that follows a similar philosophy, with the goal of providing clinicians a means to define aphasia in terms of distinct processing deficits (Kay et al., 1996). Hypotheses for further research may then be formulated and tested. The proposed psycholinguistic modules and their putative functional relationships are depicted. One criticism of the PALPA is that there is no fixed method of administration, and examiners are not meant to give all 60 subtests to each patient. This raises questions regarding its reliability and validity, features that have been well established for other psychometric aphasia tests, such as the WAB (Shewan and Kertesz, 1980).

The examination pursued depends on the clinician's goal. Many simply want to know the nature of the language impairment in order to focus treatment on the patient's deficits. In this case a practical assessment of the patient's language aptitude may provide the most relevant information. A modality-oriented test such as the Porch Index of Communicative Ability (Porch, 1981) or a test oriented toward general communication skills, such as Communicative Abilities in Daily Living (Holland, 1980), may be the most appropriate. Most localization-based uses of language assessment are based on principles incorporated in the BDAE and WAB. Research on the effectiveness of aphasia treatments may require a test with strict psychometric standardization, such as the WAB. PALPA may have a use, particularly when the clinician desires extremely detailed information regarding a patient's deficits and spared capacities, especially for the purpose of designing psycholinguistic experiments.

## PRINCIPLES OF THERAPY

Therapy for aphasia is complex; patients improve regardless of whether therapy is given (Kertesz and McCabe, 1977) and there have been few well-controlled studies. Nonetheless, it is now generally accepted that therapy is beneficial (Basso et al., 1979; Holland et al., 1996; Wertz et al., 1986) and that programs of therapy must be selected based on the needs of the individual patient.

### Reactivation of Linguistic Functions

Many patients with aphasia have some spared aspects of language function and therapies are designed accordingly. It is uncertain whether treatment is a form of re-education or leads to reorganization of language-specific parts of the brain or reallocation of other resources in the brain to assume the function of language. Shuell et al. (1955) argued that language function is not destroyed by aphasia and can be reactivated through therapy. They also proposed that aphasia was a general disturbance of language with varying degrees of severity. Much of traditional speech therapy is based on the technique that Shuell and others promoted, which consisted of intense, persistent auditory stimulation and induction of responses by the patient. The traditional view is that repeated, frequent auditory stimulation leads not only to improvements in comprehension but also to generalized improvement in articulation, reading, word-finding, and writing. At least one large study has shown the stimulation approach to be beneficial for patients who begin treatment within 6 months of onset of aphasia (Basso et al., 1979).

Despite the measurable benefits of traditional therapy, the quest for more effective treatment strategies has continued. A general approach has been to focus early efforts on sharpening preserved aspects of language function, gradually expanding therapy to address the restoration of other skills. For example, many patients with expressive deficits are able to use expletives and to sing songs, even pronouncing the lyrics. This may be due to the right hemisphere's role in nonpropositional communication. Melodic Intonation Therapy (Sparks and Holland, 1976) is a technique developed to expand this capability; the therapist trains the patient to articulate phrases by intoning each syllable within a limited musical scale of approximately five half-steps. It has been shown to benefit patients who scored poorly on the articulatory agility scale of the BDAE and who have frontal lobe lesions in the region of Broca's area (Naeser and Helm-Estabrooks, 1985).

Patients with deficits of comprehension have some preserved capacity for phoneme discrimination. Naeser et al. (1986) sought to turn this skill to their patients' advantage through the technique of Sentence Level Auditory Comprehension (SLAC) therapy. Patients are first trained to discern between words that differ with regard to only one consonant (e.g., pill/sill). They are then required to make the same discriminations as the words are used in sentences of increasing complexity. In a study of chronically aphasic patients, five of seven showed statistically significant improvement in Token Test scores. The study did not address the issue of functional improvement in daily living. An advantage of this therapy is that it was administered with a special tape player that read the sentences from magnetic tape on the stimulus cards. Thus, the patients were able to practice without a therapist.

Patients in whom fluency is impaired have improved with the Helm Elicited Language Program for Syntax Stimulation (HELPSS; Helm-Estabrooks and Ramsberger, 1986). Patients are read brief stories and invited to repeat the final sentence. Later, the patients are required to provide the final sentence from memory. The investigators found significant improvements in spontaneous morpheme counts and on scores of the Northwestern Syntax Screening Test.

Perseveration can be an integral component of aphasic symptoms and has been noted to play a role in semantic and phonemic paraphasic errors, and neologisms. Treatment of Aphasic Perseveration (TAP) is a set of strategies devised to improve fluency by reducing perseveration. The therapist explains the problem of perseveration and encourages the patient to make a conscious effort to overcome it. Further strategies include repetition of stimuli after a 5- to 10-s delay, use of tactile, orthographic, phonemic or gestural cues, speaking in unison with the therapist, singing, and use of Melodic Intonation Therapy. Authors of the technique described three patients in whom perseveration was a dominant barrier to recovery and demonstrated the effectiveness of TAP in each patient by crossing over twice between TAP and traditional speech therapy (Helm-Estabrooks et al., 1987).

Katz and Wertz (1997) reported a technique of computer-provided reading treatment tested on patients that had been aphasic for more than 1 year. An advantage to the treatment is that no clinician needs to be present during sessions. Patients who received the reading therapy for 26 weeks had significant improvements in scores on the Porch Index of Communicative Ability (especially on the Verbal and Pantomime subtests) and on the WAB Aphasia Quotient and repetition subtest.

Language-Oriented Treatment (LOT) is a battery of specifically targeted therapies devised by Shewan and Bandur (1994). The authors cite current neurolinguistic theories as the underlying basis for the approach. Therapists address language disorders by systematically focusing on five modalities of the communication system: auditory processing, visual processing, gestural communication, oral expression, and graphic expression. Patients can be trained with tasks of increasing complexity within each modality. The therapist has freedom to tailor therapy to each patient's needs. In a comparison study between patients who chose to have treatment and patients who chose to undergo no treatment, those who were treated were found to have significantly higher improvements in Language Quotient (LQ), which is a composite of the oral and written subtests of the WAB.

Other language-oriented learning methods can be devised according to the specific needs of patients. This approach can have the dual objective of rehabilitating patients and of giving support to neurolinguistic theory. Thompson et al. (1997) reported treatment of two patients with agrammatic aphasia and lesions in Broca's area. Therapy was focused on a single type of syntactic operation. The investigators noted improvement not only in the element of syntax that was the focus of therapy, but also in a theoretically related syntactic operation. The effects of treatment did not generalize to theoretically unrelated syntactic transformations.

## Pragmatic Therapy

In some cases patients are so severely affected that language functions are lost due to destruction of the brain structures necessary for their execution. There is evidence that even patients who are unable to use language at all retain the capacity for symbolic thought and can be taught to communicate with an alternate set of symbols. Velletri-Glass et al. (1973) have reported success teaching globally aphasic patients the use of a symbol set that had originally been taught to chimpanzees. These symbols consisted of various shapes cut out of colored paper. In a similar approach, patients with global aphasia who did not improve after 6 months of conventional treatment were taught to communicate by arranging Blissymbols (Johanssen-Horbach et al., 1985). These are simple, iconic drawings on 3″ × 3″ cards, which even hemiparetic patients can carry and manipulate fairly easily. Globally aphasic patients also learned to utilize symbols that represented abstract function words. Similar systems have been devised using a computer with a graphical user interface. One example is Computer-based Visual Communication (C-VIC). The computer is less portable than the deck of cards, but allows for more rapid communication, since the search time for each card is shortened (Steele *et al.*, 1989).

Communication can be enhanced by use of manual signs. This is made difficult by the fact that many patients with severe aphasia have a right hemiparesis, left upper extremity apraxia, and severe deficits of comprehension for both spoken and written language. Attempts have been made to train aphasic patients with forms of American Sign Language (ASL) and Native American Hand Talk (Amer-Ind) that are adapted for single-hand use. Amer-Ind is generally felt to be more concrete than ASL, simpler with regard to its spatial descriptiveness, and more readily utilized with one hand.

Visual Action Therapy (VAT) is a nonverbal treatment program in which patients are taught to produce symbolic gestures that represent objects. The gestures are not intended to be an independent symbolic code. Rather, it is hoped that patients will learn a skill for pantomime that they can generalize to objects that are not in the training set. Globally aphasic recipients of VAT improved on the gestural-pantomime and auditory comprehension subtests of the Porch Index of Communicative Ability (Helm-Estbrooks *et al.*, 1982).

Some pragmatic techniques focus on enhancing the patient's ability to communicate by whatever means available. A particularly useful method is PACE (Promoting Aphasics' Communicative Effectiveness) therapy (Davis and Wilcox, 1985). This method consists of an interactive discussion between the patient and the therapist, with each taking turns describing scenes depicted on cards (Thompson, 1994). The task can be varied by using sets of cards that require the expression of actions, nouns or spatial relationships. Patients are encouraged to use any means to get their point across, including pantomime, and the technique has been combined with the use of Amer-Ind (Rao, 1994). Aten *et al.* (1982) described Functional Communication Treatment (FCT), a technique that is similar to PACE therapy in that adequacy of communication is emphasized over linguistic competence. Therapists using FCT focus their efforts on the communication of information that has daily importance, such as ordering meals in restaurants, giving important demographic or biographical information, or discussing entertainment such as sports or television. A group of patients with chronic (more than six months duration), nonfluent, agrammatic aphasia had significant improvement in scores on the Communicative Abilities in Daily Living (CADL) scores, but not on the Porch Index of Communicative Abilities (PICA).

## Pharmacotherapy

The Persian philosopher and physician, ibn Sina, is said to have recommended the ingestion of cashew for aphasia, as well as "for virtually all psychiatric and neurological afflictions" (Albert *et al.*, 1988; Sarno, 1991). Many continue to hope for a less labor-intensive method of treating language disorders than those currently available. Based on the possibility that much of the functional impairment of aphasia results from damage to neurons that rely on certain neurotransmitters, some investigators have sought a pharmacological means of improving language function in aphasic patients. A number of studies have been performed to evaluate pharmacotherapy aimed at the dopaminergic and noradrenergic systems (Small, 1994) but have been difficult to interpret due to inconsistent anatomical and functional classification of aphasia, lack of neuroimaging, lack of placebo controls and other study design flaws. It seems feasible, however, that drug therapy may be a useful adjunct to other forms of language therapy.

Normal function of the frontal lobes relies on dopaminergic activity and a few reports have documented improvement in patients following treatment with bromocryptine, usually in the setting of nonfluent aphasia of moderate severity (Sabe *et al.*, 1992) or of transcortical motor aphasia (Raymer *et al.*, 2001). However, a subsequent randomized, double-blind, placebo-controlled trial of bromocryptine in a series of seven patients showed no statistically significant benefit of the drug (Sabe *et al.*, 1995). The authors concluded that the benefits seen in their previous open-label study may have been due to a practice effect and emphasized the need for well-designed, controlled studies.

Norepinephrine may enhance recovery from nervous system lesions in animals (Boyeson and Feeney, 1990). Dextroamphetamine exerts its action by triggering release of norepinephrine and preventing its reuptake from the nerve terminal. There is evidence from animal studies that dextroamphetamine may enhance neural sprouting and synaptogenesis after experimental infarction (Stroemer *et al.*, 1998). In humans, dextroamphetamine improves recovery from hemiparesis after stroke (Walker-Batson, 1995). A subsequent double-blind, placebo-controlled study of 21 patients with moderate to severe aphasia was performed. Patients received 5 weeks of speech therapy, with each session preceded by administration of placebo or a 10-mg dose of dextroamphetamine. Patients who received dextroamphetamine had significantly more improvement in PICA scores than patients who received placebo. However, when the patients were re-evaluated after 6 months, the difference between the two groups was not statistically significant (Walker-Batson *et al.*, 2001).

Piracetam is a γ-amino butyric acid (GABA) derivative that is devoid of GABAergic activity or antagonism. It is approved in Europe for use as a "nootropic" agent that is believed to enhance cognitive functions such as learning and memory through facilitation of cholinergic and excitatory amino acid neurotransmission. A number of studies has been performed evaluating the efficacy of piracetam in promoting recovery from aphasia. The earliest of these studies was not limited to aphasia, but evaluated 67 aphasic speakers of German. All of these patients received 12 weeks of speech therapy, along with placebo or 4.8 g piracetam. Global scores on the Aachen Aphasia Test (AAT) were significantly more improved in the treated group at 12 weeks. No significant difference was found in scores on any of the individual subtests. Differences in AAT scores did not reach statistical significance at 24-week follow-up (Enderby *et al.*, 1994). Two subsequent studies have shown similar improvements in AAT scores associated with piracetam use; both were carried out for only 6 weeks. Thus, neither addresses the question of whether the measured differences lead to a persistent benefit (Huber *et al.*, 1997; Kessler *et al.*, 2000).

## A Practical Approach to Management

Therapy must be tailored according to each patient's current stage of recovery, severity, pattern of deficits, and neurological and psychological condition. Table I contains a list of the treatments discussed in this chapter.

**TABLE I**

| General treatments | Language-oriented treatment |
|---|---|
| | Shuell's stimulation approach |
| Perseveration | Treatment of aphasic perseveration |
| Articulation | Melodic intonation therapy |
| Comprehension | Sentence level auditory comprehension |
| Agrammatism | Helm elicited language program for syntax stimulation |
| | NP- and *wh*-movement training[a] |
| Reading | Computer provided reading treatment |
| Severe nonfluency | Blissymbols |
| | Amer-Ind |
| | American Sign Language |
| | C-VIC[b] |
| | Visual action therapy |
| Practical therapies | Promoting aphasics' communicative effectiveness |
| | Functional communication treatment |
| | Group therapy |

[a]NP stands for "noun phrase." *wh* refers to question words, such as "what" and "where."
[b]C-VIC stands for "computer-assisted visual communication."

In the acute stage, an attempt should be made to enhance spontaneous recovery through stimulation techniques and exposure to a variety of communicative activities. Unwanted positive symptoms, such as perseveration, should be suppressed through behavioral therapy. The opportunity for therapeutic intervention in the acute period may be compromised by the patient's medical condition.

Once the patient is medically stable, the exact linguistic deficits should be determined through standard test batteries. The therapist should consider the use of supplemental testing to ascertain the details of the patient's particular aphasic problems and aspects of language performance that are spared. Focused, language-oriented therapy can then be undertaken to address the patient's needs. A collection of treatment methods that have proven efficient for specific deficits are discussed above. The effectiveness of all other therapy may depend critically on comprehension, and disorders of comprehension should be treated first. Severe nonfluency also may have an impact on subsequent treatment and should be addressed early. Language deficits associated with Alzheimer's disease may respond partially to treatment with cholinesterase inhibitors. In the future, pharmacologic adjuncts to therapy may prove useful during the subacute phase after stroke. It may become necessary to coordinate dosage and therapy schedules so that each form of therapy is maximally enhanced by the drug.

Improvements made from language-specific therapy can then be consolidated through pragmatic techniques, such as PACE therapy. Group therapy is effective for improving functional communication, and can reduce costs. Patients who fail to improve despite directed language therapy may benefit from the use of an alternate symbol set, such as Amer-Ind or C-VIC.

Psychosocial factors play a role in recovery. Psychotherapeutic support or therapeutic groups with patients and their spouses or relatives should be planned early. The patient's major conversational partners should be "coached" in methods of communication with the patient (Holland, 1991). Depression is a common complication of anterior aphasias, and can retard therapeutic efforts if not recognized and treated promptly.

## THE FUTURE OF APHASIA TREATMENT

The progress of treatment for aphasia rests on discerning the optimal use of all available treatments and on discovering new ones. Future treatments may address the underlying cause (stroke, degenerative disease, *etc.*) or the communication deficit itself. Ideally there will be constructed a foundation of clinical evidence from which a detailed treatment plan can be designed for patients with any conceivable language deficit. This will require ongoing improvements in assessment techniques, carefully designed clinical trials, and rigorous correlation of clinical features with outcomes. The potential benefit of amphetamine and piracetam must be ascertained, and studies of other drugs must be undertaken. Aphasic patients may eventually receive a cocktail or series of medications. Small (2000) raises the issue of the use of neurotrophins and other chemical modulators, as well as stem cell or neuronal transplantation into the area of infarction. In the absence of a means to trigger regrowth or repair of neural tissue, investigators in a broad range of fields have given attention to the possibility of interfacing the brain with computers or other machines (www.engin.umich.edu/dbi/ and www.neuroprosthesis.org/project.htm). Although the initial thrust of this effort has been focused on the guidance of prosthetic limbs, orthoses also could be conceived for the remediation of language or other cognitive modalities.

# REFERENCES

Albert, M. L., Bachman, D. L., Morgan, A., and Helm-Estabrooks, N. (1988). Pharmacotherapy for aphasia. *Neurology* 38(6), 877–879.

Albert, M. L., Sparks, R. W., and Helm, N. A. (1973). Melodic intonation therapy. *Arch. Neurol.* 29, 130–131.

Aten, J. L., Calugiuri, M. P., and Holland, A. L. (1982). The efficacy of Functional Communication Therapy for chronic aphasic patients. *J. Speech Hearing Disord.* 47, 93–96.

Basso, A., Capitani, E., and Vignolo, L. A. (1979). Influence of rehabilitation on language skills in aphasic patients: A controlled study. *Arch. Neurol.* 36(4), 190–196.

Bayles, K. A. (1993). Pathology in language behavior in dementia. *In* "Linguistic Disorders and Pathologies" (G. Blanken, J. Dittmann, H. Grimm, J. C. Marshall, and C. W. Wallesch, Eds.), pp. 388–408. de Gruyter, Berlin.

Boyeson, M., and Feeney, D. M. (1990). Intraventricular norepinephrine facilitates motor recovery following sensorimotor cortex injury. *Pharmacol. Biochem. Behav.* 35, 497–501.

Caplan, D. (1992). "Language: Structure, Processing and Disorders." MIT Press, Cambridge, MA.

Damasio, A. R., and Damasio, H. (2000). Aphasia and the neural basis of language. *In* "Principles of Behavioral and Cognitive Neurology" (M. M. Mesulam, Ed.), pp. 294–315. Oxford University Press, New York, NY.

Damasio, H., Grabowski, T. J., Tranel, D., Hichwa, R. D., and Damasio, A. R. (1996). A neural basis for lexical retrieval. *Nature* 380(6574), 499–505.

Davis, G. A., and Wilcox, M. J. (1985). "Adult Aphasia Rehabilitation." College Hill, San Diego.

De Renzi, E., and Vignolo, L. (1962). The Token Test: A sensitive test to detect receptive disturbances in aphasia. *Brain* 85, 665–678.

Enderby, P., Broeckx, J., Hospers, W., Schildermans, F., and Deberdt, W. (1994). Effect of piracetam on recovery and rehabilitation after stroke: A double-blind, placebo-controlled study. *Clin. Neuropharmacol.* 17(4), 320–331.

Fromkin, V. A. (ed.) (2000). "Linguistics: An Introduction to Linguistic Theory." Blackwell, Malden, MA.

Geschwind, N., Quadfasel, F. A., and Segarra, J. M. (1968). Isolation of the speech area. *Neuropsychologia* 6, 327–340.

Goodglass, H. (1993). "Understanding Aphasia." Academic Press, San Diego, CA.

Graff-Radford, N., and Damasio, A. R. (1984). Disturbances of speech and language associated with thalamic dysfunction. *Sem. Neurol.* 4(2), 162–168.

Grossi, D., Trojana, L., Chiacchio, L., Soricelli, A., Mansi, L., Postiglione, A., and Salvatore, M. (1991). Mixed transcortical aphasia: Clinical features and neuroanatomical correlates. *Eur. Neurol.* 31(4), 204–211.

Helm-Estbrooks, N., Fitzpatrick, P. M., and Barresi, B. (1982). Visual action therapy for global aphasia. *J. Speech Hearing Disord.* 47, 385–389.

Helm-Estabrooks, N., Emery, P., and Albert, M. L. (1987). Treatment of Aphasic Perseveration (TAP) program. A new approach to aphasia therapy. *Arch. Neurol.* 44, 1253–1255.

Helm-Estabrooks, N., and Ramsberger, G. (1986). Treatment of agrammatism in long-term Broca's aphasia. *Br. J. Disord. Commun.* 21, 39–45.

Holland, A. L. (1980). "Communicative Abilities in Daily Living." University Park Press, Baltimore, MD.

Holland, A. L., Fromm, D. S., DeRuyter, F., and Stein, M. (1996). Treatment efficacy: Aphasia. *J. Speech and Hearing Res.* 39, S27–S36.

Holland, A. L. (1991). Pragmatic aspects of intervention in aphasia. *J. Neurolinguistics* 6, 197–211.

Huber, W., Willmes, K., Poeck, K., Van Vleyman, B., and Deberdt, W. (1997). Piracetam as an adjuvant to language therapy for aphasia: A randomized double-blind placebo-controlled pilot study. *Arch. Phys. Med. Rehabil.* 78, 245–250.

Johanssen-Horbach, H., Cegla, B., Mager, U., and Schempp, B. (1985). Treatment of chronic global aphasia with a nonverbal communication system. *Brain Language* 24, 74–82.

Katz, R. C., and Wertz, R. T. (1997). The efficacy of computer-provided reading therapy for chronic aphasic adults. *J. Speech Language Hear. Res.* 40(3), 493–507.

Kay, J., Lesser, R., and Coltheart, M. (1996). Psycholinguistic assessment of language processing in aphasia (PALPA): An introduction. *Aphasiology* 10(2), 159–215.

Kertesz, A., and McCabe, P. (1977). Recovery patterns and prognosis in aphasia. *Brain* 100, 1–18.

Kertesz, A., Appell, J., and Fisman, M. (1986). The dissolution of language in Alzheimer's disease. *Can. J. Neurol. Sci.* 13, 415–418.

Kessler, J., Thiel, A., Karbe, H., and Heiss, W. D. (2000). Piracetam improves activated blood flow and facilitates rehabilitation of post-stroke aphasic patients. *Stroke* 31, 2112–2116.

Ku, A., Lachman, E. A., and Nagler, W. (1996). Selective language aphasia from herpes simplex encephalitis. *Pediatr. Neurol.* 15, 169–171.

Levin, H. S. (1991). Aphasia after head injury. *In* "Acquired Aphasia" (M. T. Sarno, Ed.), 2nd ed. Academic Press, San Diego, CA.

Mega, M., and Alexander, M. P. (1994). Subcortical aphasia: The core profile of lenticulostriate infarction. *Neurology* 44, 1824–1829.

Mesulam, M. M. (1998). From sensation to cognition. *Brain* 121, 1013–1052.

Mesulam, M. M. (2001). Primary progressive aphasia. *Ann. Neurol.* 49, 425–432.

Naeser, M. A., and Helm-Estabrooks, N. (1985). CT scan lesion localization and response to MIT with nonfluent aphasia cases. *Cortex* 21(2), 203–223.

Naeser, M. A., Haas, G., Mazurski, P., and Laughlin, S. (1986). Sentence Level Auditory Comprehension treatment program for aphasic adults. *Arch. Phys. Med. Rehabil.* **67**(6), 393–399.

Newhoff, M., and Apel, K. (1997). Impairments in pragmatics. *In* "Aphasia and Related Neurogenic Language Disorders" (L. L. LaPointe, Ed.), 2nd ed. Thieme, New York, NY.

Perani, D., Cappa, S. F., Schnur, T., Tettamanti, M., Collina, S., Rosa, M. M., and Fazio, F. (1999). The neural correlates of verb and noun processing. A PET study. *Brain* **122**, 2337–2344.

Pick, A. (1892). As translated by WC Schoene. On the relation between aphasia and senile atrophy of the brain. *In* "Neurological Classics in Modern Translation," (D. A. Rottenberg and F. H. Hochberg, Eds.), pp. 35–40. Hafner Press, New York, NY.

Porch, B. E. (1981). "The Porch Index of Communicative Ability." Consulting Psychologists Press, Palo Alto, CA.

Rao, P. R. (1994). Use of Amer-Ind code by persons with aphasia. *In* "Language Intervention Strategies in Adult Aphasia" (R. Chapey, Ed.), pp. 358–367. Williams & Wilkins, Baltimore, MD.

Raymer, A. M., Bandy, D., Adair, J. C., Schwartz, R. L., Williamson, D. J. G., Rothi, L. J. G., and Heilman, K. Effects of bromocryptine in a patient with crossed nonfluent aphasia: A case report. *Arch. Phys. Med. Rehab.* **82**, 139–144.

Reinvang, I. (1984). The natural history of aphasia. *Adv. Neurol.* **42**, 13–22.

Roger, J., Genton, P., Bureau, M., and Dravet, C. (1993). Less common epileptic syndromes. *In* "The Treatment of Epilepsy Syndromes. Principles and Practice" (E. Wyllie, Ed.), pp. 624–635. Lea & Febiger, Philadelphia, PA.

Sabe, L., Leiguarda, R., and Starkstein, S. E. (1992). An open-label trial of bromocryptine in non-fluent aphasia. *Neurology* **42**(8), 1637–1638.

Sabe, L., Salvarezza, F., Cuerva, A. G., Leiguarda, R., and Starkstein, S. (1995). A randomized, double-blind, placebo-controlled study of bromocryptine in nonfluent aphasia. *Neurology* **45**, 2272–2274.

Sarno, M. T. (1991). Recovery and rehabilitation in aphasia. *In* "Acquired Aphasia, Second Edition" (M. T. Sarno, Ed.), pp. 521–582. Academic Press, San Diego, CA.

Shewan, C. M., and Bandur, D. L. (1994). Language-oriented treatment: A psycholinguistic approach to aphasia. *In* "Language Intervention Strategies in Adult Aphasia" (R. Chapey, Ed.), pp. 184–206. Williams & Wilkins, Baltimore, MD.

Shewan, C. M., and Kertesz, A. (1980). Reliability and validity characteristics of the Western Aphasia Battery (WAB). *J. Speech Hear. Disord.* **45**(3), 308–324.

Shuell, H., Carroll, V., and Street, B. S. (1955). Clinical treatment of aphasia. *J. Speech Hear. Disord.* **20**, 43–53.

Singer, E. J., Singer, P., Tomiyasu, U., Licht, E., Fahy-Chandon, B., and Tourtellotte, W. W. (1994). AIDS presenting as progressive multifocal leukoencephalopathy with clinical response to zidovudine. *Acta Neurol. Scand.* **90**(6), 443–447.

Small, S. L. (1994). Pharmacotherapy of aphasia. A critical review. *Stroke* **25**, 1282–1289.

Small, S. L. (2000). The future of aphasia treatment. *Brain Language* **71**, 227–232.

Snowden, J. S., Neary, D., Mann, D. M. A., Goulding, P. J., and Testa, H. J. (1992). Progressive language disorder due to lobar atrophy. *Ann. Neurol.* **31**, 174–183.

Sparks, R. W., and Holland, A. L. (1976). Method: Melodic intonation therapy for aphasia. *J. Speech Hear. Disord.* **41**, 287–297.

Steele, R. D., Weinrich, M., Wertz, R. T., and Kleczewska, M. K. (1989). Computer-based visual communication in aphasia. *Neuropsychologia* **27**(4), 409–426.

Stroemer, R. P., Kent, T. A., and Hulsebosch, C. E. (1998). Enhanced neocortical neural sprouting, synaptogenesis, and behavioral recovery with D-amphetamine therapy after neocortical infarction in rats. *Stroke* **29**(11), 2381–2393.

Thompson, C. K. (1994). Treatment of nonfluent Broca's aphasia. *In* "Language Intervention Strategies in Adult Aphasia" (R. Chapey, Ed.), pp. 407–428. Williams & Wilkins, Baltimore, MD.

Thompson, C. K., Shapiro, L. P., Ballard, K. J., Jacobs, B. J., Schneider, S. S., and Tait, M. E. (1997). Training and generalized production of *wh*- and NP- movement structures in agrammatic aphasia. *J. Speech, Language Hear. Res.* **40**(2), 228–244.

Toledo, J. C., Minagar, A., and Lowe, M. R. (2000). Persisting aphasia as the sole manifestation of partial status epilepticus. *Clin. Neurol. Neurosurg.* **102**, 144–148.

Velletri-Glass, A., Gazzaniga, M. S., and Premack, D. (1973). Artificial language training in global aphasics. *Neuropsychologia* **11**, 95–103.

Walker-Batson, D., Smith, P., Curtis, S., Unwin, H., and Greenlee, R. (1995). Amphetamine paired with physical therapy accelerates motor recovery after stroke. *Stroke* **26**, 2254–2259.

Walker-Batson, D., Curtis, S., Natarajan, R., Ford, J., Dronkers, N., Salmeron, E., Lai, J., and Unwin, H. (2001). A double-blind, placebo-controlled study of the use of amphetamine in the treatment of aphasia. *Stroke* **32**, 2093–2098.

Wertz, R. T., Weiss, D. G., Aten, J. L., Brookshire, R. H., García-Buñuel, L., Holland, A. L., Kurtzke, J. F., Lapointe, L. L., Milianti, F. J., Brannegan, R., Greenbaum, H., Marshall, R. C., Vogel, D., Carter, J., Barnes, N. S., and Goodman, R. (1986). Comparison of clinic, home and deferred language treatment of aphasia. *Arch. Neurol.* **43**(7), 653–658.

## CHAPTER 26
# Disorders of Spatial Orientation

Hans-Otto Karnath and Josef Zihl

Orienting in space requires not only visual spatial abilities, for example visual localization, but also spatial cognition, and spatial attention. A further spatial capacity is to use incoming or stored visuo-spatial information for construction (visuoconstructive abilities). Disturbances of these processes are typically observed after occipito-parietal and posterior parietal lobe injury, with right-hemisphere damage more frequently causing deficits (for comprehensive reviews, see Benton and Tranel, 1993; Grüsser and Landis, 1991). Disorders of spatial orientation further may involve the ability to correctly perceive upright orientation of the body or the ability to visually or tactilely explore space or may affect awareness for one side of space.

## VISUAL SPACE PERCEPTION

### Clinical Aspects

Disorders of *visual spatial localization* can occur in the hemifield contralateral to the damaged hemisphere and are typically associated with defective saccadic localization accuracy. Patients with bilateral posterior brain damage and disturbance of visual localization in the entire visual field report difficulties in visually guided activities because accurate fixating of, reaching for, and grasping of objects, as well as reading, writing, and drawing, are impaired (Postma *et al.*, 2000; Zihl, 2000). Systematic contralesional shifts in the *visual vertical and horizontal axes* have been observed after lesions of both hemispheres but more frequently after right-sided damage (Brandt *et al.*, 1994; Lütgehetmann and Stäbler, 1992). Deviations of the visual vertical can also occur after thalamic stroke (VPL, VPM, pulvinar) (Anastasopoulos and Bronstein, 1999; Dieterich and Brandt, 1993), in Parkinson's disease (Bronstein *et al.*, 1996), and in multiple sclerosis (Jackson *et al.*, 1995), proba-

bly because of disruption of visual, vestibular, and somatosensory information. Shifts in the subjective midline (*straight-ahead direction*), which were frequently assessed using line bisection tasks, have been found in patients with occipital and posterior parietal damage and are typically associated with hemianopia or visual neglect. In patients with hemianopia, the shifts are typically toward the affected hemifield (i.e., leftward in the case of left-sided hemianopia; e.g., Barton and Black, 1998) and are associated with corresponding fixation behavior (Barton *et al.*, 1998). In contrast, patients with left-sided visual neglect may show right-sided displacement. Interestingly, in patients with a combination of visual neglect and hemianopia no significant shift has been observed (Ferber and Karnath, 1999), indicating that both occipital and parietal lobe structures may contribute to the visual straight-ahead direction. It has been argued that the contralesional bias in hemianopic patients may reflect either an adaptive contralateral attentional gradient or a nonveridical spatial representation within the remaining normal field (Barton *et al.*, 1998). However, since the displacement of the straight-ahead direction and the degree of spontaneous compensation of the hemianopia appear unrelated, the displacement may rather represent a genuine visual–spatial disorder. Furthermore, no relationship exists between the extent of visual field sparing in hemianopia and the amount of displacement, indicating that this spatial disorder is not caused by the visual field defect (Zihl and von Cramon, 1986). As a consequence of shifts in the main visual spatial axes, patients may have difficulty in drawing, writing, and copying and keeping the straight-ahead direction in walking or in guiding a wheelchair.

Defective *depth perception* and *stereopsis* have been reported in patients with uni- or bilateral posterior damage, with unilateral damage causing more moderate deficits (De Renzi, 1982; Miller *et al.*, 1999). Depth perception deficits may affect localization in terms of under- or overestimation of distances and may be associated with micropsia or macropsia. In its severest form, patients may completely lose the capacity to see differences in depth; in front of a flight of stairs they may report a "number of straight lines on the floor" or may complain that everything they see appears flat to them, "as in a picture or photograph" (Zihl and von Cramon, 1986).

Patients with impaired *visual spatial orientation* have difficulties in orienting themselves in a scene or a stimulus array and may "get lost" in reading and picture viewing. These difficulties are mirrored in their oculomotor scanning patterns: they appear spatially incoherent and disorganized, without any evidence for an adaptation of the scanning path to the spatial configuration of the scene. As a consequence, oculomotor scanning is very time-consuming and is characterized by a high number of fixations and fixation repetitions, indicating that (transient) storing of executed fixation positions is also impaired (Zihl, 2000; Zihl and Hebel, 1997). A higher-order form of visual spatial disorientation refers to impairment or loss of *spatial knowledge* of the actual or virtual (mental) environment, i.e., geographic orientation in the real world, in maps, or both. This disorder is known as topographical agnosia, topographagnosia, or environmental agnosia (Grüsser and Landis, 1991). Patients with this syndrome have difficulties orienting themselves in their familiar environment, have problems learning new spatial routes, and find it difficult to find their way using a map. The site of injury in these patients is typically temporo-parietal or temporo-occipital in the right hemisphere, including the posterior hippocampal gyrus. Environmental agnosia may also be present in patients with Alzheimer's disease (Kaida *et al.*, 1998; Mendez *et al.*, 1990).

## Natural Course

Meerwaldt (1983) found evidence for the recovery of perception of visual spatial axes within 6 months in a group of 17 patients who had right occipital infarction. Hier *et al.* (1983) reported recovery from visuospatial deficits within 15 weeks after stroke in about 70% of 41 patients. Unfortunately, visual-spatial abilities were not tested in either of these studies; it remains unclear, therefore, whether the return of visual spatial performance was of any behavioral significance.

## Principles of Therapy and Practical Management

Therapy of visual spatial disorders usually includes practice with position and distance estimation and block design (Diller *et al.*, 1974; Kerkhoff, 1998; Weinberg *et al.*, 1979; Young *et al.*, 1983). Although improved ADL scores were reported after practice (Weinberg *et al.*, 1979; Langdon and Thompson, 2000), it remains unclear whether this effect is specific to the therapy or is in part also due to an increased patient awareness of the problem and would therefore indicate improved coping.

Quantitative observations on the effect of systematic treatment have been reported by Lütgehetmann and Stäbler (1992). Using a personal computer based program to improve the adjustment of the visual vertical and horizontal axes and visual localization and discrimination of line length, line orientation, and line bisection, they found significant improvements at least in single patients. Unfortunately, no information is available concerning the behavioral significance of these improvements. Further quantitative observations on the effect of systematic practice have been reported in a single case approach (Zihl, 2000). The patient, a 48-year-old woman, had bilateral posterior parietal lobe damage and showed, as the main consequence, a severe impairment in visual space perception. When examined 7 months postinjury, there was no visual neglect and only a mild lower hemiamblyopia. She had, however, marked difficulties with visually guided oculomotor and hand motor activities, including fixation, reaching, and grasping; in fact, she never knew where an object was and where she herself had looked. The severity of her disorder is best described as "space blindness"; she was unable, for example, to find food on her plate or a glass of wine in front of her, to find the appropriate parts of her face she wanted to paint, and to find her way through the ward of the hospital. Her comment described vividly her condition: "I can see many things around me, and can also recognize some of them, but when I try to look accurately at them, I cannot find them, and have no idea, where they could be." Her eye movements resembled her difficulties; she showed highly inaccurate fixation responses and completely lost her way when scanning a picture or reading a word. Her visually guided pointing responses were also highly inaccurate, and she used mainly tactile cues to find objects. Training procedures consisted of visual localization of objects by oculomotor and hand motor responses and visual search for single and multiple objects. After 46 sessions (45 min each) she reported an improvement in vision in her everyday life, for example, in reaching for a door handle, in picking up food with a fork, in reaching for a glass of wine without overturning it, with putting on make-up using a mirror, and in finding her way through the hospital. This improvement in daily activities was paralleled by an increase in the accuracy of fixation, the spatial organization of the scanpath, and visually guided reaching and grasping performance. Furthermore, her performance in visually recognizing objects and in reading also improved, without further practice, indicating that the severe impairment in visual space perception had secondarily also affected object recognition and reading.

## VISUOCONSTRUCTIVE ABILITIES

### Clinical Aspects

We summarize under this heading impairments in the ability to construct a copy of a visually presented model or of its mental representation (i.e., from memory) by means of assembling blocks or by drawing. [Alternatively, the term "constructional apraxia" is often used, although there is no relation to apraxia. To avoid confusion, this term should no longer be used.] Typically, patients have difficulties with the length, size, and orientation of forms or form elements, with the spatial position of elements within a larger figure or a given spatial framework, and with the three- and two-dimensionality of objects (for a review, see Grossi and Trojano, 1999). It is still an open question whether impaired visual constructive abilities can be explained by visual spatial perceptual deficits or are independent visuo-motor deficits (e.g., Trojano and Grossi, 1998), because visuoconstructive disorders are often present in association with visual spatial perceptual deficits or with executive impairments. Patients with visuoconstructive disorders have typically had brain damage to the parietal lobe, with right-sided damage causing more profound constructive deficits than left-sided damage. Patients with right-sided or bilateral parietal lobe damage may also show left-sided neglect or Balint syndrome. Impaired visuoconstructive abilities have also been reported after bacterial meningitis (Merkelbach et al., 2000), following infarction in the right basal ganglia (Lazar et al., 1995), after callosal infarction (Marangolo et al., 1998), in Huntington's disease (Gomeztortosa et al., 1996), in the early stage of Alzheimer's disease (Kalman et al., 1995; Mendez et al., 1990), and in dementia with Lewy bodies (DLB) (Mori et al., 2000).

### Natural Course and Principles of Therapy and Practical Management

There are no reports available on spontaneous recovery of visuoconstructive abilities. Systematic practice with block design (Weinberg et al., 1979; Young et al., 1983) has been found to have a positive effect on visuoconstructive abilities, which shows also some generalization to everyday life activities.

## BALINT SYNDROME

### Clinical Aspects

Patients with Balint's syndrome typically show an inability to perceive the visual field as a whole, because their field of attention is severely restricted, and to perceive more than one stimulus at a time within the spared field of attention. In addition, they have difficulties shifting their gaze voluntarily or on command (*oculomotor apraxia or psychic paralysis of gaze*). The final feature that makes up the triad of symptoms is *optic, or visuomotor, ataxia*, an inability to direct movement of an extremity using visual guidance. As a consequence, visually guided oculomotor and hand motor activities, visuoconstructive abilities, visual orientation, and recognition, including reading, are severely impaired (Ghika et al., 1998). Patients with this syndrome may also have severe visual disorientation and defective depth perception, either secondary to the attentional and oculomotor deficits or as associated visual impairments. Spatial localization deficits may be present not only for visual, but also for auditory stimuli (Phan et al., 2000). The inability to process more than one stimulus at a time may refer to spatial and object features, but may also be more pronounced for one or the other of these stimulus categories. According to the two-visual-pathway model (dorsal or "where"-, and ventral or "what"-pathway; Ungerleider and Haxby, 1994), difficulties with the processing of multiple stimulus locations (between-object representation) are also referred to as "dorsal," and difficulties with the processing of multiple stimulus properties (within-object representation) as "ventral" simultanagnosia (Karnath et al., 2000a; Robertson et al., 1997). Balint's syndrome typically results from bilateral parieto-occipital lobe damage, including occipitofrontal and parietofrontal fiber connections (Grüsser and Landis, 1991), bilateral posterior cortical atrophy and Alzheimer's disease (Galton et al., 2000; Mendez et al., 1990; Ross et al., 1996), Jakob-Creutzfeld disease, and corticobasal ganglionic degeneration (Mendez, 2000). Migraine aura may also contain features of Balint's syndrome (Shah and Nafee, 1999). The frequency of Balint's syndrome after stroke is only in the range of 2% (Gloning et al., 1968), but it amounts to about 30% in degenerative diseases (Rizzo, 1993).

### Natural Course and Principles of Therapy and Practical Management

Allison et al. (1969) reported a patient with a minor form of Balint's syndrome, who after about 4 years showed good spontaneous recovery concerning oculomotor scanning and visual–spatial orientation, while reading and simultaneous processing of multiple stimuli remained impaired. Montero *et al.* (1982) described good recovery in three patients within 2–3 months, while in one patient vision improved slowly over a period of 5 years. Trivelli *et al.* (1996) reported spontaneous recovery of recognition strategies in a single patient with ventral simultanagnosia; Perez *et al.* (1996) used intensive verbal cueing and "organizational strategies" and found subsequent improvement in visual recognition, reaching, and scanning. In three patients who had severe Balint's syndrome, we (Zihl, 2000) used initiation of gaze shifts and increase of their frequency both on command and intentionally, localization of single and multiple visual stimuli by oculomotor and hand motor responses, visual scanning of scenes with low and high stimulus densities, and visual search tasks to enlarge their restricted field of attention and improve their oculomotor scanning and reaching activities. After intensive practice up to 3 months (daily: three to five sessions of a 30-min training each) we found enlargement of the fields of attention and search, an increase in the frequency of exploratory eye movements and improved oculomotor scanning. These improvements also led to a better overview and an improvement in visual orientation and object and scene identification. However, patients still required more time to scan the actual environment, had difficulties in complex and unfamiliar environments, and reading was still impossible. Nevertheless, it appears worth trying to treat patients with Balint's syndrome, because visual orientation in familiar surroundings is an important prerequisite for regaining—at least in part—an independent life.

## SPATIAL NEGLECT

### Clinical Features

Spatial neglect is a laterized disorder of space-related behavior in patients with brain damage that describes a characteristic failure to explore the side of space contralateral to the lesion and to react or respond to stimuli or subjects located on this side. In the acute stage of the disease, such patients behave as if one side of the surrounding space had ceased to exist. The patients are not aware that they have this disorder.

The investigation of a large sample of 602 patients with acute stroke admitted during a 1.5-year period from a community-based catchment area revealed spatial neglect in 23% of all subjects (Pedersen *et al.*, 1997). Of these neglect patients, 85% showed a right hemisphere lesion. Compared to controls, these lesions were larger in size and affected more frequently cortical regions. Within the right hemisphere, cortical lesions

center on the superior temporal gyrus (Karnath *et al.*, 2001b). Subcortically, right-sided lesions of the caudate nucleus (Caplan *et al.*, 1990; Kumral *et al.*, 1999), the putamen (Karnath *et al.*, 2002a), and the thalamic pulvinar (Karnath *et al.*, 2002a) may lead to spatial neglect.

The typical clinical behavior of patients with spatial neglect in the acute stage is a spontaneous deviation of the head and the eyes toward the ipsilesional side. They orient toward that side when addressed from the front or the left and ignore contralesionally located people or objects. Neglect of stimuli can be observed in visual, auditory, and tactile modalities. Beside these characteristic manifestations of spontaneous behavior, patients with spatial neglect typically disregard targets located on the contralesional side in bedside tests such as the letter or bells cancellation tasks or neglect left-sided objects when copying a scene or when completing a clock-face. Searching for objects in peripersonal space, the patients show a shift of their ocular and tactile exploratory movements toward the side of the lesion (Karnath *et al.*, 1998; Karnath and Perenin, 1998); the movements concentrate on the ipsilesional side while the contralesional side is disregarded. Patients with spatial neglect may also show neglect of their contralesional extremities. The left arm or leg may appear hemiplegic but can be moved when the examiner directs the patient's attention to that arm or leg. Patients also may appear hemianopic but perceive contralesionally presented visual stimuli when explicitly instructed by the investigator that such stimuli will appear on their left side. This forced cueing is only transiently effective; patients do not adopt the compensatory shift to the contralesional side in the absence of forced requirements.

### Natural Course

Summarizing the follow-up observations on 125 patients (Campbell and Oxbury, 1976; Colombo *et al.*, 1982; Hier *et al.*, 1983; Levine *et al.*, 1986; Stone *et al.*, 1992), about 75% of patients recover from spatial neglect within 6 months, but the symptoms of neglect in the remainder may persist (Zarit and Kahn, 1974; Zoccolotti *et al.*, 1989) and may cause severe disabilities in everyday life (Denes *et al.*, 1982; Fullerton *et al.*, 1986; Kinsella and Ford, 1985; Stone *et al.*, 1993). The severity of spatial neglect and the course of recovery depend on lesion size and presence of premorbid cerebral atrophy (Levine *et al.*, 1986).

Spatial neglect was found to be associated with diminished sensory-motor and cognitive abilities, lower functioning in activities of daily living, a longer rehabilitation period, and a lower rate of discharge to independent living (Kalra *et al.*, 1997; Katz *et al.*, 1999; Kinsella and Ford, 1985; Pedersen *et al.*, 1997). The severity of spatial neglect, together with the initial degree of paralysis and the patient's age, were found to be significant predictors of functional activity in daily living (Katz *et al.*, 1999; Stone *et al.*, 1993). However, investigation of 602 acute stroke patients revealed that spatial neglect has no independent influence on the length of rehabilitation, the functional outcome, and discharge to independent living when the neurological and functional status at admission, aphasia, orientation, anosognosia, age, gender, prior stroke, and comorbidity were included in the analysis (Pedersen *et al.*, 1997).

It is important to note that the incidence of spatial neglect as well as the percentage of recovered patients strongly depend on the measures used to diagnose and follow up spatial neglect. Patients may perform normally on some neglect tests but may remain impaired on others (Halligan and Marshall, 1992). Thus, the pattern of recovery may be different for different behavioral deficits of spatial neglect. In the recovered stage of the disorder, neglect of contralesional stimuli often is no longer observed in the patients' spontaneous behavior but occur in situations in which two stimuli appear simultaneously. In these situations, the patients orient toward the stimulus on the ipsilesional side and neglect that on the contralesional side (Karnath, 1988).

## Principles of Therapy and Practical Management

### Active Orienting to the Contralesional Side

Since the main symptom of spatial neglect is defective exploration of contralesional space, many interventions to treat the disorder are based on active orienting to the contralesional side. These procedures aim to enhance the field of exploration toward the contralesional side by training the patient to perform eye or hand movements to the impaired side. Thereby, visual and tactile scanning is improved and compensatory search strategies utilized.

The first major research based on this idea was carried out by Weinberg and colleagues (1977). These authors used texts that were specially designed using graded visual cues. This material should treat the pathologically biased visual exploratory behavior of the patients during reading of this material inducing more frequent head and eye turning to the contralesional, left side. Their work has suggested positive effects to scanning that were maintained even 1 year post-treatment. Generalization effects to activities of daily living were not investigated.

More specific training procedures to treat the pathological exploratory behavior of the neglect patients have been carried out successfully by presenting scanning material on large visual screens via slide projection or video beam (Antonucci *et al.*, 1995; Kerkhoff, 1998; Kerkhoff *et al.*, 1992; Pizzamiglio *et al.*, 1992). In these studies, patients were trained to find contralesionally located stimuli by systematically scanning the whole scene.

In addition to this procedure, Kerkhoff and coworkers (1992, 1998) included systematic saccade training to fixed and nonfixed targets on the contralesional side, trained the coordination of eye and head movements during visual search, and later transferred such strategies to more natural situations relevant in everyday life. The therapy led to significant enlargement of the visual search field, a reduction of search time, and better reading performance. About 2 years after discharge with no further training, the extent of the visual search field was retested and was found unchanged.

Beside training of systematic scanning on a large display, Pizzamiglio *et al.* (1992) included reading, copying, and figure description in their treatment program. All procedures were carried out to stimulate the patient to actively and sequentially scan various parts of the visual field. The program was found to be effective in reducing spatial neglect. Also, the authors showed an extension of exploratory scanning to functional situations similar to those of real life. The effect of training remained stable when assessed at follow-up several months later. A later study of this group explicitly studied the effect of the training on the recovery of motor and functional impairment (Paolucci *et al.*, 1996). It revealed that after 2 months of rehabilitation, the group with neglect training showed higher functional recovery than the group with only general cognitive intervention.

A new technique to carry out exploration trainings was suggested by Wiart *et al.* (1997). The authors used a specific device that associates exploration not only with active eye and head movements but also with active trunk rotation toward the contralesional side. In comparison to a group of neglect patients that obtained physiotherapy and occupational therapy only, significant improvement of neglect symptoms and of activities of daily living were found in patients with acute and chronic spatial neglect who had this training. The results were maintained at follow-up, 1 month after the end of the treatment.

### Limb Activation and Contralesional Cueing

While the above training procedures demand active orienting of the eyes, head, trunk, and hands toward the contralesional side, Robertson and North (1992, 1993) used motoric activation of the left-sided limbs when stationary located in the left hemispace. In one patient, the authors showed that irrelevant active but not passive movements of the left fingers or leg led to a reduction of neglected targets in a cancellation task. A further patient showed significant improvement in several neglect and self-care tasks with this procedure (Wilson et al., 2000). In a larger group of acute stroke patients, the technique was found to shorten the length of hospital stay (Kalra et al., 1997).

Recently, Cubelli et al. (1999) replicated part of the experiments conducted by Robertson and North to verify the presence of the beneficial effect of left hand movements in left hemispace. Ten consecutive patients with right brain damage and spatial neglect were investigated. In only one patient, the authors found a significant decrease of neglect symptoms when irrelevant movements were performed with the left hand in the left hemispace. Likewise, Brown et al. (1999) observed that such movements do not produce consistent improvements of contralateral neglect and do not improve leftward saccades in an overt orienting task.

Robertson and North (1992, 1993) reported that reduction of contralesional neglect with left-sided movements was not observed when the same patient performed finger movements with the left hand in the right hemispace. This suggests that the limb activation effect is due to a motorically induced cueing effect. This conclusion is compatible with similar observations reported by Halligan et al. (1991) and suggests that limb activation does not induce lasting recovery. It has repeatedly been shown that even forced cueing is only transiently effective. Patients do not adopt the shift to the contralesional side in the absence of forced requirements. Cueing procedures using visual stimuli (Halligan et al., 1992; Riddoch and Humphreys, 1983), transcutaneous electrical stimulation (Karnath, 1995; Pizzamiglio et al., 1996), or verbal instruction to attend to the neglected side (Karnath, 1988) did not show lasting reduction of contralesional neglect.

### Neck Muscle Vibration

Recent neurophysiological findings showed that the brain integrates visual, auditory, tactile, vestibular, and proprioceptive input for higher-order representations of space in nonretinal, egocentric and allocentric, frames of reference for appropriate spatial behavior (Andersen et al., 1993; for review see Thier and Karnath, 1997). On this basis it was speculated whether in neglect patients the transformation of this polysensory input might be working with a systematic ipsilesional error, thus leading to a rightward orientation of patients and left-sided neglect (Karnath, 1994, 1997). In line with this assumption, several studies have demonstrated a substantial, though transient, reduction of neglect symptoms by different sensory stimulations via the peripheral pathways contributing to these higher-order representations of space, such as vestibular stimulation (Rubens, 1985), optokinetic stimulation (Pizzamiglio et al., 1990), and neck proprioceptive stimulation by muscle vibration (Karnath et al., 1993).

Substantial recovery of spatial neglect outlasting the duration of the application recently has been demonstrated for neck muscle vibration (Ferber et al., 1998; Schindler et al., in press). The latter study evaluated the long-term efficacy of combined vibration and exploration treatment with that of visual exploration training alone. In the combination treatment, the patients performed the same visual exploration training while the contralesional neck muscles were vibrated. The authors observed a specific and lasting reduction of neglect symptoms with neck muscle vibration which was superior to visual exploration training alone. The reduction in the visual modality transferred to the tactile modality with a concomitant improvement in activities of daily living. At follow-up testing 2 months after discharge, improvements were found unchanged. Since neck muscle vibration is noninvasive, has no side effects, and is technically easy to apply, it may become a useful additional tool in the rehabilitation of spatial neglect.

### Eye Patching and Prism Exposure

Another technique that also is easy to apply is the use of glasses with half-field patches. Patches obscuring the right visual half-field of both eyes were found to significantly reduce spatial neglect after three month of daily application (Beis et al., 1999). Analyzing eye movements during reading aloud a series of letters, the authors found significant improvement in the time spent on the left side of the display and the number of saccades directed to that side. Moreover, they found an improvement of the patients' activities of daily living (e.g., transfer bed, toileting, dressing, etc). Covering the right half-field on both eyes appears to influence foveal perception and seems to positively affect the mechanisms controlling voluntary orienting of gaze. In contrast, monocular patching of the whole right eye did not produce a consistent reduction of neglect (Beis et al., 1999; Walker et al., 1996).

A recent study suggested that prism adaptation to a rightward optical deviation also may lead to improvement of neglect symptoms (Rossetti et al., 1998). Prismatic lenses creating an optical shift of 10° to the right were exposed while making pointing responses to visual left- or right-sided targets for 2 to 5 min. After this short period of prism exposure, the patients with left-sided neglect showed a compensatory shift in the direction opposite the rightward optical deviation and improvement of left-sided neglect symptoms in different tests. The authors reported that this improvement lasted for at least 2 h after prism removal. Hence, future studies may show that prism exposure is an effective rehabilitation technique to promote lasting cortical reorganization processes of impaired right-hemispheric functions.

## PUSHER SYNDROME

### Clinical Features

While most patients with hemiparesis have good trunk balance soon after stroke, some patients have the peculiar behavior of using the non-affected arm or leg to push away actively from the nonparalyzed side. Without assistance, this contraversive pushing leads to loss of postural balance and falling toward the paralyzed side. When sitting or standing, these patients actively lean toward the hemiparetic side and resist any attempt to correct passively their tilted body posture. They use the nonparetic arm to resist actively against attempts of passive correction toward the earth-vertical upright orientation and report the impression of lateral instability and fear of falling toward the nonparalyzed side. In contrast, these patients show no fear when their active pushing leads to an unstable, tilted body position toward the paretic, contralesional side. Davies (1985) termed this behavior the "Pusher syndrome." A systematic investigation in a large sample of 327 acute stroke patients with hemiparesis found the disorder in 10.4% (Pedersen et al., 1996).

A helpful tool to clinically assess contraversive pushing is the standardized "Scale for Contraversive Pushing (SCP)" (Karnath et al., 2000b, 2001a). The SCP assesses (a) symmetry of spontaneous posture while sitting and standing, (b) use of the arm and/or the leg to extend the area of physical contact to the ground while sitting and standing, and (c) resistance to passive correction of posture while sitting and standing.

The disorder typically is associated with unilateral lesions of the left or the right posterolateral thalamus (Karnath et al., 2000c), leading to an altered perception of the body's orientation in relation to gravity. Pusher patients experience their body as oriented "upright" when actually tilted about 20° to the ipsilesional side (Karnath et al., 2000b). Surprisingly, these patients show undisturbed processing of visual and vestibular inputs determining visual vertical. Thus, in contrast to their disturbed perception of upright body posture, orientation perception of the visual world is unaffected. In other words, pusher patients have a severe tilt of body posture, in spite of normal visual–vestibular functioning.

### Natural Course

On admission, patients with contraversive pushing are more severely impaired concerning level of consciousness, paresis of upper and lower extremities, ability to walk, and show lower initial ADL function than hemiparetic patients without contraversive pushing (Pedersen et al., 1996). While this study revealed no evidence for a regular combination of contraversive pushing with other neuropsychological deficits such as spatial neglect, anosognosia, aphasia, or apraxia, a recent investigation found spatial neglect in 80% of right brain-damaged pusher patients and aphasia in all pusher patients investigated with left brain damage (Karnath et al., 2000c).

Contraversive pushing has a good prognosis (Karnath et al., 2002b) and does not seem to negatively influence the functional outcome of rehabilitation. The gain in ADL function and the discharge rate to nursing homes did not differ from hemiparetic patients without contraversive pushing. However, patients with contraversive pushing needed 3.6 weeks (= 63%) longer to reach the same functional outcome level compared to hemiparetic patients without contraversive pushing (Pedersen et al., 1996). Specific strategies to treat the disorder thus aim to shorten this period and enable earlier discharge.

### Principles of Therapy and Practical Management

Contraversive pushing prevents patients from learning to transfer weight to the nonhemiparetic side. They use the non-affected leg to push away actively from the nonparalyzed side. Thus, while standing or walking, the nonhemiparetic leg cannot be used as the standing leg. However, the latter is the key for patients with hemiparesis to re-learn these abilities.

The leading aspect for physiotherapy thus is to enable the pusher patient to transfer weight to the nonhemiparetic side. Davies (1985) suggested various manipulations such as using a back slab to hold the patient's paretic leg in extension or to carry out activities which automatically elicit the desired upright posture without the patient being dependent upon verbal instructions and feedback from the therapist. For example, such activities can be hitting of a balloon with the nonparetic hand or any other activity which requires the patient to reach forward and upward with the sound hand. Also, when kicking a football with the paretic foot, the patient transfers his/her weight spontaneously over to the nonparetic leg.

On the basis of the recently discovered mechanism underlying contraversive pushing (Karnath et al., 2000b), one may speculate about new approaches for therapy. The study revealed that pusher patients were able to align correctly their longitudinal body axis to earth-vertical when they could visually explore the structured surroundings. This indicates that seeing the upright orientation of the surrounding objects, persons, etc., helped their impaired graviceptive system. However, in contrast to these findings, pusher patients' spontaneous orientation of the body in daily life activities shows the characteristic tilt due to their active pushing away from the nonparetic side. Apparently, visual input does not suffice to control automatically and continuously upright body posture in pusher patients. It seems to help them only under certain constrained conditions in which their attention is explicitly drawn to the structured surroundings (Karnath *et al.*, 2000b).

Nevertheless, Karnath *et al.* (2001a) speculated that for the purpose of rehabilitation this preserved ability to align the body axis to earth-vertical with the help of visual cues might be a helpful new tool. Although the patients are not spontaneously able to use the visual input to control upright body posture, this might become possible when training procedures apply this ability for conscious strategies of posture control in these subjects.

# REFERENCES

Allison, R. S., Hurwitz, L. J., White, G. J., and Wilmot, T. J. (1969). A follow-up study of a patient with Balint's syndrome. *Neuropsychologia* 7, 319–333.

Anastasopoulos, D., and Bronstein, A. M. (1999). A case of thalamic syndrome: Somatosensory influences on visual orientation. *J. Neurol. Neurosurg. Psychiatry* 67, 390–394.

Andersen, R. A., Snyder, L. H., Li, C.-S., and Stricanne, B. (1993). Coordinate transformations in the representation of spatial information. *Curr. Opin. Neurobiol.* 3, 171–176.

Antonucci, G., Guariglia, C., Judica, A., Magnotti, L., Poalucci, S., Pizzamiglio, L., and Zoccolotti, P. (1995). Effectiveness of neglect rehabilitation in a randomized group study. *J. Clin. Exp. Neuropsychol.* 17, 383–389.

Barton, J. J. S., and Black, S. E. (1998). Line bisection in hemianopia. *J. Neurol. Neurosurg. Psychiatry* 64, 660–662.

Barton, J. J. S., Behrmann, M., and Black, S. (1998). Ocular search during line bisection—The effects of hemi-neglect and hemianopia. *Brain* 121, 1117–1131.

Beis, J.-M., André, J.-M., Baumgarten, A., and Challier, B. (1999). Eye patching in unilateral spatial neglect: efficacy of two methods. *Arch. Phys. Med. Rehabil.* 80, 71–76.

Benton, A. L., and Tranel, D. (1993). Visuoperceptual, visuospatial, and visuoconstructive disorders. *In* "Clinical Neuropsychology" (K. M. Heilman and E. Valenstein, Eds.), 3rd ed., pp. 165–213. Oxford University Press, New York.

Brandt, T., Dieterich, M., and Danek, A. (1994). Vestibular cortex lesions affect the perception of verticality. *Ann. Neurol.* 35, 403–412.

Bronstein, A. M., Yardley, L., Moore, A. P., and Cleeves, L. (1996). Visually and posturally mediated tilt illusion in Parkinson's disease and in labyrinthine defective subjects. *Neurology* 47, 651–656.

Brown, V., Walker, R., Gray, C., and Findlay, J. M. (1999). Limb activation and the rehabilitation of unilateral neglect: Evidence of task-specific effects. *Neurocase* 5, 129–142.

Campbell, D. C., and Oxbury, J. M. (1976). Recovery from unilateral visuo-spatial neglect? *Cortex* 12, 303–312.

Caplan, L. R., Schmahmann, J. D., Kase, C. S., Feldmann, E., Baquis, G., Greenberg, J. P., Gorelik, P. B., Helgason, C., and Hier, D. B. (1990). Caudate infarcts. *Arch. Neurol.* 47, 133–143.

Colombo, A., De Renzi, E., and Faglioni, P. (1982). The time course of visual hemi-inattention. *Arch. Psychiatr. Nervenkrankh.* 231, 539–546.

Cubelli, R., Paganelli, N., Achilli, D., and Pedrizzi, S. (1999). Is one hand always better than two? A replication study. *Neurocase* 5, 143–151.

Davies, P. M. (1985). "Steps to Follow. A Guide to the Treatment of Adult Hemiplegia." Springer, New York.

De Renzi, E. (1982). "Disorders of Space Exploration and Cognition." Wiley, Chicester.

Denes, G., Semenza, C., Stoppa, E., and Lis, A. (1982). Unilateral spatial neglect and recovery from hemiplegia—A follow-up study. *Brain* 105, 543–552.

Dieterich, M., and Brandt, T. (1993). Thalamic infarctions: Differential effects on vestibular function in the roll plane (35 patients). *Neurology* 43, 1732–1740.

Diller, L., Ben-Yishay, Y., Gerstman, L. J., Goodkin, R., and Gordon, W. (1974). "Studies in Cognition and Rehabilitation in Hemiplegia," Rehabilitation Monograph 50. New York Institute of Rehabilitation Medicine, New York.

Ferber, S., and Karnath, H.-O. (1999). Parietal and occipital lobe contribution to perception of straight ahead orientation. *J. Neurol. Neurosurg. Psychiatry* 67, 572–578.

Ferber, S., Bahlo, S., Ackermann, H., and Karnath, H.-O. (1998). Treatment of neglect with vibration of neck muscles?—A case study. *Neurol. Rehabil.* 4, 21–24. [in German]

Fullerton, J., McSherry, D., and Stout, M. (1986). Albert's test: A neglected test of perceptual neglect. *Lancet* I, 430–432.

Galton, C. J., Patterson, K., Xuereb, J. H., and Hodges, J. R. (2000). Atypical and typical presentations of Alzheimer's disease: A clinical, neuropsychological, neuroimaging and pathological study of 13 cases. *Brain* 123, 484–498.

Ghika, J., Ghika-Schmid, F., and Bogousslavsky, J. (1998). Parietal motor syndrome: A clinical description in 32 patients in the acute phase of pure parietal stroke studied prospectively. *Clin. Neurol. Neurosurg.* 100, 271–282.

Gloning, I., Gloning, K., and Hoff, H. (1968). "Neuropsychological Symptoms and Syndromes in Lesions of the Occipital Lobe and Adjacent Areas." Gauthier-Villars, Paris.

Gomeztortosa, E., Delbarrio, A., Barroso, T., and Ruiz, P. J. G. (1996). Visual processing disorders in patients with Huntington's disease and asymptomatic carriers. *J. Neurology* 243, 286–292.

Grossi, D., and Trojano, L. (1999). Constructional apraxia. *In* "Handbook of Clinical and Experimental Neuropsychology" (G. Denes and L. Pizzamiglio, Eds.), pp. 441–450. Psychology Press, Hove, UK.

Grüsser, O.-J., and Landis, Th. (1991). "Visual Agnosias and Other Disturbances of Visual Perception and Cognition." CRC Press, Boca Raton.

Halligan, P. W., and Marshall, J. C. (1992). Left visuo-spatial neglect: A meaningless entity? *Cortex* 28, 525–535.

Halligan, P. W., Manning, L., Marshall, J. C. (1991). Hemispheric activation vs spatio-motor cueing in visual neglect: A case study. *Neuropsychologia* 29, 165–176.

Halligan, P. W., Donegan, C. A., and Marshall, J. C. (1992). When is a cue not a cue? On the intractability of visuospatial neglect. *Neuropsychol. Rehabil.* 2, 283–293.

Hier, D. B., Mondlock, J., and Caplan, L. R. (1983). Recovery of behavioural abnormalities after right hemisphere stroke. *Neurology* 33, 345–350.

Jackson, R. T., Epstein, C. M., and Delaune, W. R. (1995). Abnormalities in posturography and estimations of visual vertical and horizontal in multiple sclerosis. *Am. J. Otol.* 16, 88–93.

Kaida, K., Takeda, K., Nagata, N., and Kamakura, K. (1998). Alzheimer's disease with asymmetric parietal lobe atrophy—A case report. *J. Neurol. Sci.* **160,** 96–99.

Kalman, J., Magloczky, E., and Janka, Z. (1995). Disturbed visuospatial orientation in the early stage of Alzheimer's dementia. *Arch. Gerontol. Geriatr.* **21,** 27–34.

Kalra, L., Perez, I., Gupta, S., and Wittink, M. (1997). The influence of visual neglect on stroke rehabilitation. *Stroke* **28,** 1386–1391.

Karnath, H.-O. (1988). Deficits of attention in acute and recovered visual hemi-neglect. *Neuropsychologia* **26,** 27–43.

Karnath, H.-O. (1994). Disturbed coordinate transformation in the neural representation of space as the crucial mechanism leading to neglect. *Neuropsychol. Rehabil.* **4,** 147–150.

Karnath, H.-O. (1995). Transcutaneous electrical stimulation and vibration of neck muscles in neglect. *Exp Brain Res* **105,** 321–324.

Karnath, H.-O. (1997). Spatial orientation and the representation of space with parietal lobe lesions. *Phil. Trans. R. Soc. B* **352,** 1411–1419.

Karnath, H.-O., and Perenin, M.-T. (1998). Tactile exploration of peripersonal space in patients with neglect. *NeuroReport* **9,** 2273–2277.

Karnath, H.-O., Christ, K., and Hartje, W. (1993). Decrease of contralateral neglect by neck muscle vibration and spatial orientation of trunk midline. *Brain* **116,** 383–396.

Karnath, H.-O., Niemeier, M., and Dichgans, J. (1998). Space exploration in neglect. *Brain* **121,** 2357–2367.

Karnath, H.-O., Ferber, S., Rorden, C., and Driver, J. (2000a). The fate of global information in dorsal simultanagnosia. *Neurocase* **6,** 295–306.

Karnath, H.-O., Ferber, S., and Dichgans, J. (2000b). The origin of contraversive pushing: Evidence for a second graviceptive system in humans. *Neurology* **55,** 1298–1304.

Karnath, H.-O., Ferber, S., and Dichgans, J. (2000c). The neural representation of postural control in humans. *Proc. Natl. Acad. Sci. USA* **97,** 13931–13936.

Karnath, H.-O., Brötz, D., and Götz, A. (2001a). Clinical symptoms, origin, and therapy of the "pusher syndrome". *Nervenarzt* **72,** 86–92.

Karnath, H.-O., Ferber, S., and Himmelbach, M. (2001b). Spatial awareness is a temporal not a posterior parietal lobe function. *Nature,* **411,** 950–953.

Karnath, H.-O., Himmelbach, M., and Rorden, C. (2002a). The subcortical anatomy of human spatial neglect: Putamen, caudate nucleus, and pulvinar. *Brain* **125,** 350–360.

Karnath, H.-O., Johannsen, L., Broetz, D., Ferber, S., and Dichgans, J. (2002b). Prognosis of contraversive pushing. *J. Neurol.* in press.

Katz, N., Hartman-Maeir, A., Ring, H., and Soroker, N. (1999). Functional disability and rehabilitation outcome in right hemisphere damaged patients with and without unilateral spatial neglect. *Arch. Phys. Med. Rehabil.* **80,** 379–384.

Kerkhoff, G. (1998). Rehabilitation of visuospatial cognition and visual exploration in neglect: A cross-over study. *Restorative Neurol. Neurosci.* **12,** 27–40.

Kerkhoff, G., Münssinger, U., Haaf, E., Eberle-Strauss, G., and Stögerer, E. (1992). Rehabilitation of homonymous scotomata in patients with postgeniculate damage of the visual system: saccadic compensation training. *Restorative Neurol. Neurosci.* **4,** 245–254.

Kinsella, G., and Ford, B. (1985). Hemi-inattention and the recovery patterns of stroke patients. *Int. Rehabil. Med.* **7,** 102–106.

Kumral, E., Evyapan, D., and Balkir, K. (1999). Acute caudate vascular lesions. *Stroke* **30,** 100–108.

Langdon, D. W., and Thompson, A. J. (2000). Relation of impairment to everyday competence in visual disorientation syndrome: Evidence from a single case study. *Arch. Phys. Med. Rehabil.* **81,** 686–691.

Lazar, R. M., Weiner, M., Wald, H. S., and Kula, R. W. (1995). Visuoconstructive deficit following infarction in the right basal ganglia—A case report and some experimental data. *Arch. Clin. Neuropsychol.* **10,** 543–553.

Levine, D. N., Warach, J. D., Benowitz, L., and Calvanio, R. (1986). Left spatial neglect: Effects of lesion size and premorbid brain atrophy on severity and recovery following right cerebral infarction. *Neurology* **36,** 362–366.

Lütgehetmann, R., and Stäbler, M. (1992). Deficiences of visual spatial orientation: Diagnostic and therapy of brain damaged patients. *Z. Neuropsychologie,* **3,** 130–142. [in German]

Marangolo, P., De Renzi, E., Di Pace, E., Ciurli, P., and Castriota-Skandenberg, A. (1998). Let not thy left hand know thy right hand knoweth. The case of a patient with an infarct involving the callosal pathways. *Brain* **121,** 1459–1467.

Meerwaldt, J. D. (1983). Spatial disorientation in right-hemisphere infarction: A study of the speed of recovery. *J. Neurol. Neurosurg. Psychiatry* **46,** 426–429.

Mendez, M. F. (2000). Corticobasal ganglionic degeneration with Balint's syndrome. *J. Neurosychiatry Clin. Sci.* **12,** 273–275.

Mendez, M. F., Tomsak, R. L., and Remler, B. (1990). Disorders of the visual system in Alzheimer's disease. *J. Clin. Neuroophthalmol.* **10,** 62–69.

Merkelbach, S., Sittinger, H., Schweizer, J., and Muller, M. (2000). Cognitive outcome after bacterial meningitis. *Acta Neurol. Scand.* **102,** 118–123.

Miller, L. J., Mittenberg, S., Carey, V. M., McMorrow, M. A., Kushner, T. E., and Weinstein, J. M. (1999). Astereopsis caused by traumatic brain injury. *Arch. Clin. Neuropsychol.* **14,** 537–543.

Montero, J., Pena, J., Genis, D., Rubio, F., Peres-Serra, J., and Barraquer-Bordas, L. (1982). Balint's syndrome. *Acta Neurol. Belg.* **82,** 270–280.

Mori, E., Shimomura, T., Fujimori, M., Hirono, N., Imamura, T., Hashimoto, M., Tanimuka, S., Kazui, H., and Hanihara, T. (2000). Visuoperceptual impairment in dementia with Lewy bodies. *Arch. Neurol.* **57,** 489–493.

Paolucci, S., Antonucci, G., Guariglia, C., Magnotti, L., Pizzamiglio, L., and Zoccolotti, P. (1996). Facilitatory effect of neglect rehabilitation on the recovery of left hemiplegic stroke patients: A cross-over study. *J. Neurol.* **243,** 308–314.

Pedersen, P. M., Wandel, A., Jorgensen, H. S., Nakayama, H., Raaschou, H. O., and Olsen, T. S. (1996). Ipsilateral pushing in stroke: Incidence, relation to neuropsychological symptoms, and impact on rehabilitation. The Copenhagen stroke study. *Arch. Phys. Med. Rehabil.* **77,** 25–28.

Pedersen, P. M., Jorgensen, H. S., Nakayama, H., Raaschou, H. O., and Olsen, T. S. (1997). Hemineglect in acute stroke—Incidence and prognostic implications. *Am. J. Phys. Med. Rehabil.* **76,** 122–127.

Perez, F. M., Tunkel, R. S., Lachmann, E. A., and Nagler, W. (1996). Balint's syndrome arising from bilateral posterior cortical atrophy or infarction—Rehabilitation strategies and their limitation. *Disabil. Rehabil.* **18,** 300–304.

Phan, M. L., Schendel, K. L., Recanzone, G. H., and Robertson, L. C. (2000). Auditory and visual spatial localization deficits following bilateral parietal lobe lesions in a patient with Balint's syndrome. *J. Cognit. Neurosci.* **12,** 583–600.

Pizzamiglio, L., Frasca, R., Guariglia, C., Incoccia, C., and Antonucci, G. (1990). Effect of optokinetic stimulation in patients with visual neglect. *Cortex* **26,** 535–540.

Pizzamiglio, L., Antonucci, G., Judica, A., Montenero, P., Razzano, C., and Zoccolotti, P. (1992). Cognitive Rehabilitation of the hemineglect disorder in chronic patients with unilateral right brain damage. *J. Clin. Exp. Neuropsychol.* **14,** 901–923.

Pizzamiglio, L., Vallar, G., and Magnotti, L. (1996). Transcutaneous electrical stimulation of the neck muscles and hemineglect rehabilitation. *Restor. Neurol. Neurosci.* **10,** 197–203.

Postma, A., Sterken, Y., de Vries, L., and de Haan, E. H. E. (2000). Spatial localization in patients with unilateral posterior left or right hemisphere lesions. *Exp. Brain Res.* **134**, 220–227.

Riddoch, M. J., and Humphreys, G. W. (1983). The effect of cueing on unilateral neglect. *Neuropsychologia* **21**, 589–599.

Rizzo, A. (1993). "Balint's syndrome" and associated visuospatial disorders. *Baillieres Clin. Neurol.* **2**, 415–437.

Robertson, I. H., and North, N. (1992). Spatio-motor cueing in unilateral neglect: The role of hemispace, hand and motor activation. *Neuropsychologia* **30**, 553–563.

Robertson, I. H., and North, N. (1993). Active and passive activation of left limbs: Influence on visual and sensory neglect. *Neuropsychologia* **31**, 293–300.

Robertson, L., Treisman, A., Friedmanhill, S., and Grabowecky, M. (1997). The interaction of spatial and object pathways—Evidence from Balint's syndrome. *J. Cognit. Neurosci.* **9**, 295–317.

Ross, S. K., Graham, N., Stuartgreen, L., Prins, M., Xuereb, J., Patterson, K., and Hodges, J. R. (1996). Progressive biparietal atrophy—An atypical presentation of Alzheimer's diesease. *J. Neurol. Neurosurg. Psychiatry* **61**, 388–395.

Rossetti, Y., Rode, G., Pisella, L., Farné, A., Ling, L., Boisson, D., and Perenin, M.-T. (1998). Prism adaptation to a rightward optical deviation rehabilitates left hemispatial neglect. *Nature* **395**, 166–169.

Rubens, A. B. (1985). Caloric stimulation and unilateral visual neglect. *Neurology* **35**, 1019–1024.

Schindler, I., Kerkhoff, G., Karnath, H.-O., Keller, I., and Goldenberg, G. (in press). Neck muscle vibration induces lasting recovery in spatial neglect. *J. Neurol. Neurosurg. Psychiatry.*

Shah, P. A., and Nafee, A. (1999). Migraine aura masquerading as Balint's syndrome. *J. Neurol. Neurosurg. Psychiatry* **67**, 554–555.

Stone, S. P., Patel, P., Greenwood, R. J., and Halligan, P. W. (1992). Measuring visual neglect in acute stroke and predicting its recovery: The visual neglect recovery index. *J. Neurol. Neurosurg. Psychiatry* **55**, 431–436.

Stone, S. P., Patel, P., and Greenwood, R. J. (1993). Selection of acute stroke patients for treatment of visual neglect. *J. Neurol. Neurosurg. Psychiatry* **56**, 463–466.

Thier, P., and Karnath, H.-O. (1997). "Parietal Lobe Contributions to Orientation in 3D Space." Springer, Heidelberg.

Trivelli, C., Turnbull, O., and della Sala, S. (1996). Recovery of object recognition in a case of simultanagnosia. *Appl. Neuropsychol.* **3**, 166–173.

Trojano, L., and Grossi, D. (1998). "Pure" constructional apraxia: A cognitive analysis of a single case. *Behav. Neurol.* **11**, 43–49.

Ungerleider, L. G., and Haxby, J. V. (1994). "What" and "where" in the human brain. *Curr. Opin. Neurobiol.* **4**, 157–165.

Walker, R., Young, A. W., and Lincoln, N. B. (1996). Eye patching and the rehabilitation of visual neglect. *Neuropsychol. Rehabil.* **6**, 219–231.

Weinberg, J., Diller, L., Gordon, W. A., Gerstman, L. J., Lieberman, A., Lakin, P., Hodges, G., and Ezrachi, O. (1977). Visual scanning training effect on reading-related tasks in aquired right brain damage. *Arch. Phys. Med. Rehabil.* **58**, 479–486.

Weinberg, J., Diller, L., Gordon, W. A., Gerstman, L. J., Lieberman, A., Lakin, P., Hodges, G., and Ezrachi, O. (1979). Training sensory awareness and spatial organisation in people with right brain damage. *Arch. Phys. Med. Rehabil.* **60**, 491–496.

Wiart, L., Bon Saint Côme, A., Debelleix, X., Petit, H., Joseph, P. A., Mazaux, J. M., and Barat, M. (1997). Unilateral neglect syndrome rehabilitation by trunk rotation and scanning training. *Arch. Phys. Med. Rehabil.* **78**, 424–429.

Wilson, F. C., Manly, T., Coyle, D., and Robertson, I. H. (2000). The effect of contralesional limb activation training and sustained attention training for self-care programmes in unilateral spatial neglect. *Restor. Neurol. Neurosci.* **16**, 1–4.

Young, G. C., Collins, D., and Hren, M. (1983). Effect of pairing scanning training with block design training in the remediation of perceptual problems in left hemiplegics. *J. Clin. Neuropsychol.* **5**, 201–212.

Zarit, S. H., and Kahn, R. L. (1974). Impairment and adaptation in chronic disabilities: Spatial inattention. *J. Nerv. Mental Dis.* **159**, 63–72.

Zihl, J. (2000). "Rehabilitation of Visual Disorders after Brain Injury." Psychology Press, Hove, UK.

Zihl, J., and von Cramon, D. (1986). "Zerebrale Sehstörungen." Kohlhammer-Verlag, Stuttgart.

Zihl, J., and Hebel, N. (1997). Patterns of oculomotor scanning in patients with unilateral posterior parietal or frontal lobe damage. *Neuropsychologia* **35**, 893–906.

Zoccolotti, P., Antonucci, G., Judica, A., Montenero, P., Pizzamiglio, L., and Razzano, C. (1989). Incidence and evolution of the hemineglect disorder in chronic patients with unilateral right brain damage. *Intern. J. Neurosci.* **47**, 209–216.

## CHAPTER 27
# Memory: Disturbances and Therapy

Hans J. Markowitsch

Current theories of memory and memory disorders, as they apply to human neurology, are described and discussed. The many patient groups that possibly require memory-related therapeutic interventions are outlined and an example of a patient, successfully treated for severe amnesia, is given. Variables, such as personality dimensions and the etiologies of memory disturbances, are presented and evaluated. The importance of a careful analysis of memory and other cognitive functions is emphasized. Among the possible therapeutic interventions, internal and external memory training techniques are outlined especially and it is stressed that therapeutic interventions usually must be individualized and should be based on proper emphatic and motivational grounds.

## INTRODUCTION

Memory disorders are among the most common consequences of brain damage and a great number of possible causes for memory impairments are described in the neurological and psychiatric literature. Table I gives an overview of the principle disease categories leading to memory disturbances. Qualitatively and quantitatively there is, of course, a variety of memory problems which range from short-lasting ones as in transient global amnesia (which are per definition less than 1 day in duration) to permanent amnesias after bilateral damage to major bottleneck structures implicated in information encoding (Markowitsch, 2000). In cases where permanent tissue damage exists, the possibilities for memory rehabilitation are much more reduced than under conditions of psychic illnesses or toxic conditions, which vary widely in duration interindividually (leaving aside the many interindividual differences that exist anyway). Nevertheless, memory problems belong to the most frequent concomitants of brain damage and

are also found—in benign as well as in pathological forms—as a deterioration accompanying aging.

While in previous decades, memory disturbances were considered mainly along quantitative dimensions with bilaterally damaged patients being affected most severely ("global amnesic syndrome"), and right unilaterally damaged ones being least affected, this picture became much more diversified with description of various forms of memory and the finding that amnesic patients have preserved capacities, usually lying, for example, in immediate (short-term, on-line, or working) memory and in those forms of information retrieval which do not require conscious monitoring ("implicit forms of memory").

## THE DIVERSITIES OF MEMORY AND THEIR CONSEQUENCES FOR TREATMENT

Memory is not an entity, though its disturbance in certain brain-damaged patients was previously equated with the "global amnesic syndrome," meaning—as the expression implies—a total inability to acquire new information long term with or without considerable disturbances in retrieving old (long-term stored) memories. This view was challenged by more sophisticated, diversified testing of memory abilities and by theoretical formulations of possible memory subdivisions.

The postulation of more than one memory system soon implied as well that not all subsystems might be affected to the same degree by a given organic or psychic disorder. As theorists further postulated a hierarchy of memory subsystems with information propagating from the lower to the higher, it could be inferred that some forms of memory might be preserved in a given individual, while others might not. This view corresponded with clinical evidence showing preserved as well as impaired memory functions in a given patient.

TABLE I    Overview of Patient Groups Who Usually Demonstrate Memory Disorders

| Etiology | Common lesion sites |
| --- | --- |
| Traumas, closed head injury | Temporal pole, orbitofrontal and prefrontal cortex |
| Intra-cranial tumors | Medial and anterior thalamus, medial temporal lobe, posterior cingulate cortex |
| Cerebral infarcts, ruptured aneurysms | Medial temporal lobe, limbic nuclei of the thalamus (posterior cerebral artery), orbitofrontal cortex, basal forebrain (anterior communicating artery) |
| Avitaminoses (e.g., B$_1$ deficiency) | Limbic thalamus, mammillary bodies (e.g., Korsakoff's disease) |
| Viral infections (e.g., herpes simplex encephalitis) | Hippocampal region, limbic and paralimbic cortex |
| Neurotoxin exposure | Hippocampal formation |
| Temporal lobe epilepsy | Hippocampus and medial temporal lobe |
| Anoxia or hypoxia (e.g., after a heart attack or drowning) | Hippocampus (CA1 sector) |
| Degenerative diseases of the CNS (e.g., Alzheimer's or Pick's diseases) | Temporal and other association cortices |
| Drugs (e.g., anticholinergics, benzodiazepines) | Limbic system |
| Electroconvulsive therapy | Probably limbic system |
| Transient global amnesia | Probably medial temporal lobe and/or medial thalamus (when an etiology is found) |
| Paraneoplastic limbic encephalitis | Limbic structure in medial temporal lobe |
| Dissociative disorders | Amnesia possibly due to hormonal changes in the brain |
| Mnestic block syndrome | Massive glucocorticoid release leading to a disruption in the memory processing pathways of the medial temporal lobe |

A dichotomy of preserved versus impaired memories even held for the distinction between short-term (lasting seconds to a few minutes at most) and long-term memories. In rare patients there may be drastically impaired short-term memory together with largely preserved long-term-memory abilities (e.g., Markowitsch *et al.*, 1999), while for the majority of memory disturbed patients the reverse is true—a preserved, sometimes even above-average short-term-memory together with a profound long-term-memory impairment (e.g., Markowitsch *et al.*, 1993b).

For those patients with long-term-memory impairment there are still two dimensions along which they may or may not be disturbed. One is the time domain and the other that of contents. The time domain refers to the possibilities of disturbed information retrieval or disturbed information acquisition: Patients may become unable to retrieve certain memories which they had acquired a long time prior to brain damage or the occurrence of a psychic alteration ("retrograde amnesia"), or they may become unable to successfully acquire new information long term ("anterograde amnesia") (Figure 1). Some patients can recall information but cannot locate the memories in correct chronological time. They do not know when events occurred. The contents

domain refers to the possibility that only one of several long-term-memory systems may be impaired. In order to specify this, I will shortly introduce current memory subdivisions.

First, however, I will shortly discuss the possible distinctions between anterograde and retrograde amnesia. In former times a close connection between the inability to acquire new information long term and an impaired access to previously stored old information was taken for granted; more recent research results show that the two may exist as distinct entities. There are patients with selective retrograde amnesia who nevertheless can learn and keep in mind new information and there are patients with the reverse pattern—an inability to acquire new information long term combined with preserved old memories (Markowitsch, 2000).

### The Long-Term-Memory Systems

The distinction into several long-term-memory systems is of great importance for memory therapy, as (a) these memory systems are represented in different neural nets, as (b) they may be selectively disturbed, as

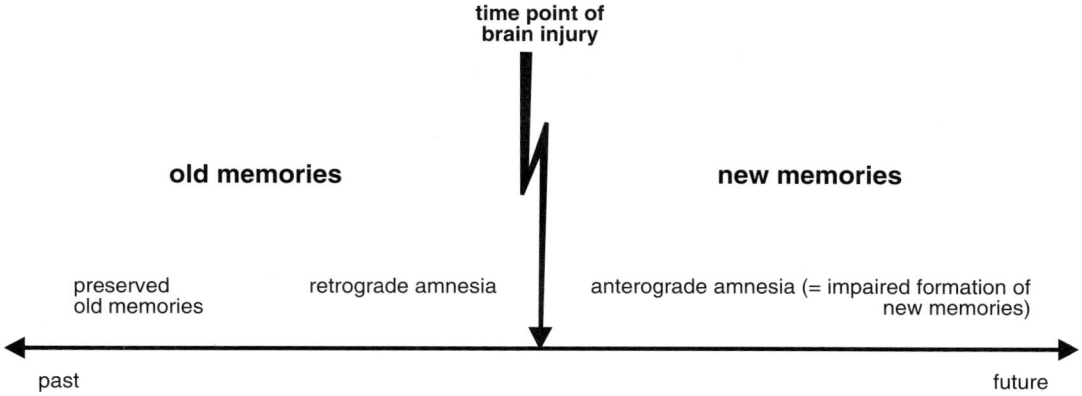

**FIGURE 1**  Possible consequences of brain injury on old and new memories.

they are probably arranged in a hierarchical manner with some being dependent on the proper action of others (Tulving and Markowitsch, 1998), and as (c) it may—after brain damage—be possible to *learn* to shift from one system to another (Ewald, 1997; Thöne, 1996; Thöne and Glisky, 1995) or at least to be able to rely for information retrieval on different memory systems (for details see the SPI-model in Tulving, 1995). Table II and Figure 2 give the principal long-term-memory systems and describe their characteristics.

It is of importance to note (a) that the different memory systems depend on different brain structures or networks and (b) that a hierarchy or gradation exists between them. Furthermore, the existence of several forms of hemispheric asymmetries needs to be kept in mind: First, there is the more general, long established one: the left hemisphere being the analytic one, which principally processes verbal information, and the right hemisphere being for global analysis, processing non-verbal information. The right hemisphere is also more prone to processing emotional information. Second, there are some more recently postulated asymmetries: Tulving and coworkers (1994) proposed with their HERA-model a hemispheric asymmetry between encoding and retrieval of information with encoding being processed by the left and retrieval of episodic memories by the right hemisphere. Related results were obtained in patients with left versus right fronto-temporal damage who after brain damage became unable to retrieve either episodic events (right hemispheric damage) or semantic facts (left hemispheric damage).

With respect to the engagement of different brain structures, the episodic memory system is the most complex, and hierarchically the highest one, depending on limbic and prefrontal structures. Semantic memory is probably much less dependent on prefrontal regions. The procedural memory system relies on portions of the

**TABLE II   Content-Dependent Divisions of Memory**

| Episodic memory Memory for episodes | Context-embedded autobiographical memory allowing mental time traveling |
|---|---|
| Declarative memory Memory for facts Knowledge system | Storage of context-free information |
| Procedural memory | Memory for skills, rules, sequences |
| Priming | Higher likelihood of reidentifying previously perceived stimuli |
| Other forms | Lower forms of memory such as classical conditioning or sensitization |

basal ganglia and possibly the cerebellum, and the priming system most likely processes information directly from the peripheral sensory organs over the respective thalamic relay nuclei to cortical structures. Consequently, even very focal damage can interfere with at least one memory system.

## GROUPS OF PATIENTS REQUIRING MEMORY-RELATED THERAPEUTIC INTERVENTIONS

The overview of patient groups with memory disorders, given in Table I, shows how many patients in principle may require memory treatment. Treatment is, however, dependent on a number of factors—personality dimensions as well as brain damage (or etiology) related variables make individually tailored or at least group-specific interventions necessary.

The patients requiring most severe observations and treatment are those injured by trauma. Head-injured patients often have reduced alertness and consequently require intensive medical care prior to psychological

## MEMORY

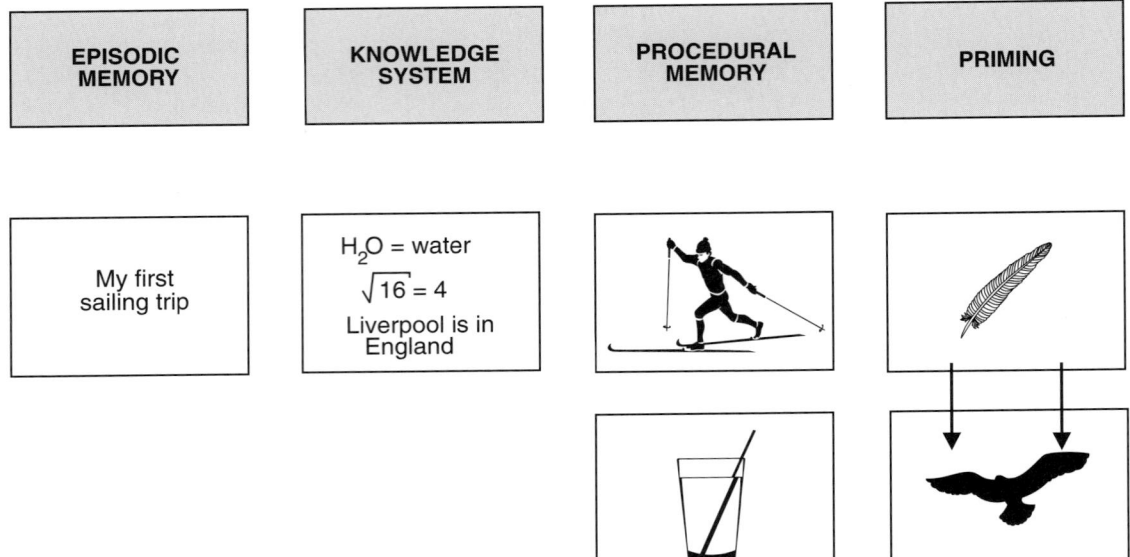

| EPISODIC MEMORY | KNOWLEDGE SYSTEM | PROCEDURAL MEMORY | PRIMING |

**FIGURE 2** The principal content-dependent subdivisions of long-term memory with examples. Episodic memory is context-embedded and allows a time travel. The knowledge system consists of context-free facts. Procedural memory is based on skills, primarily motor skills. Priming refers to a higher likeliness of identifying stimuli which were perceived unconsciously during a previous occasion.

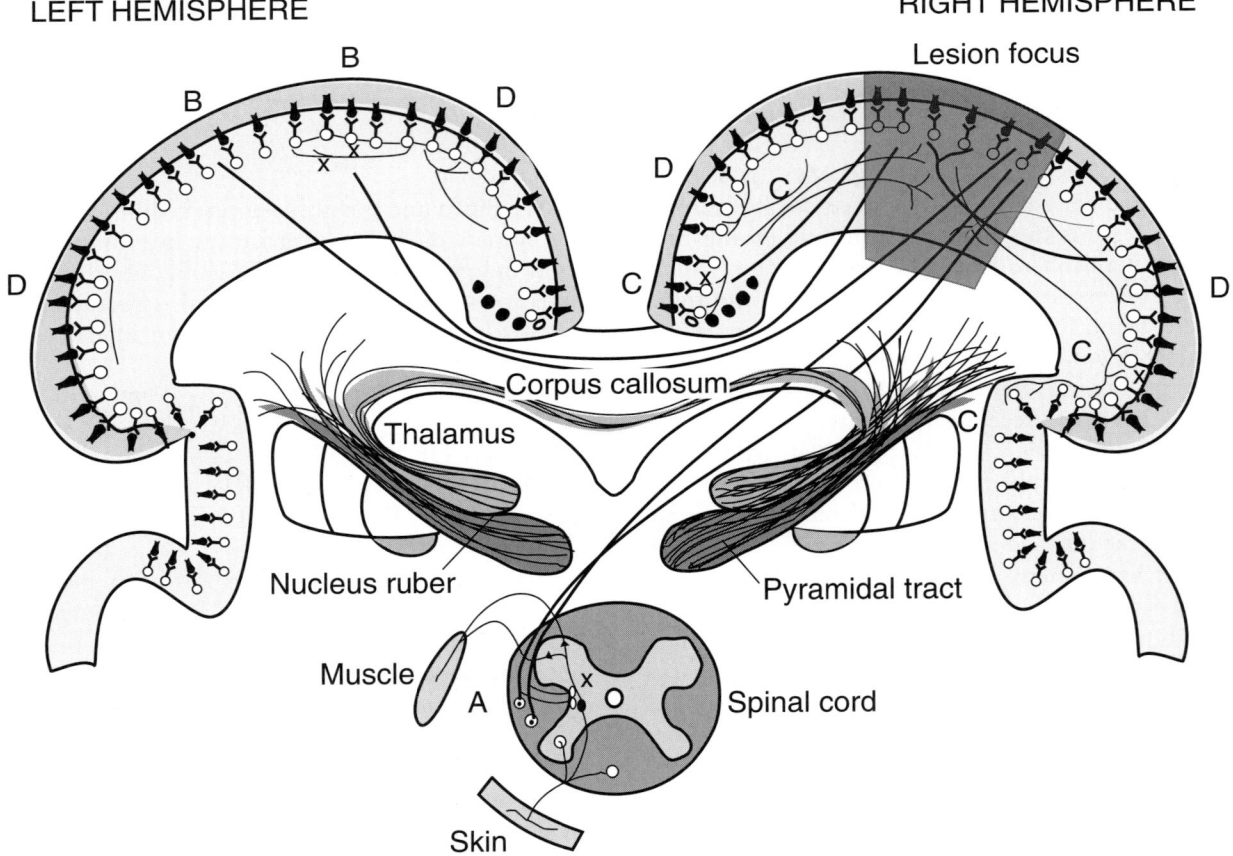

**FIGURE 3** Von Monakow's (1914) diagrammatical illustration of his diaschisis model. Damage to an area in the cerebrum ("Lesion focus") leads to activity changes in widespread brain regions, which, however, follow distinct pathways. According to von Monakow these are the pathways labeled A, B, C, and D, whereby pathway A characterizes cortico-spinal diaschisis, pathway B commissural diaschisis, pathway C associative diaschisis, and pathway D central pathways of diaschisis. x, Point of attack of diaschisis.

diagnosis and treatment. The duration and depth of coma is predictive for later memory problems, a long-known fact (Russell, 1935). A number of factors contribute to this poor outcome, among them the severity and spread of neural damage, the most frequently affected targets (which correspond to those processing both anterograde and retrograde memory; Markowitsch, 2000), and the suddenness in the change from a normally working to a damaged brain.

Already nearly 100 years ago, von Monakow (1914) created his diaschisis model with which he predicted that shock-based brain damage will have a number of immediate consequences on the brain and consequently on behavior. A few of these changes will remain, while others may weaken or disappear over time (Figure 3). Barbizet (1970) gave an example of a patient with trauma-caused major memory disturbances. He created a time line to show short-term and long-term changes after traumatic brain injury (Figure 4).

## Personality Dimensions

There are both traits and states, which influence recovery from memory disturbances. Examples for traits or personality dimensions are the patients' age or intelligence, their degree of introversion or extraversion, and their lifestyles. Motivation and mood are examples of states. The ability to attend and concentrate on external stimuli is another example of dimensions influencing the mnestic performance of an individual. It is consequently necessary to assess a number of variables in addition to memory in order to establish the prerequisites for successful memory therapy. As traits may exaggerate after brain injury, it is of special importance to shape the conditions surrounding a therapeutic set in a way optimal for a patient's needs.

An example, demonstrating the complex changes which may accompany memory impairments, is given below.

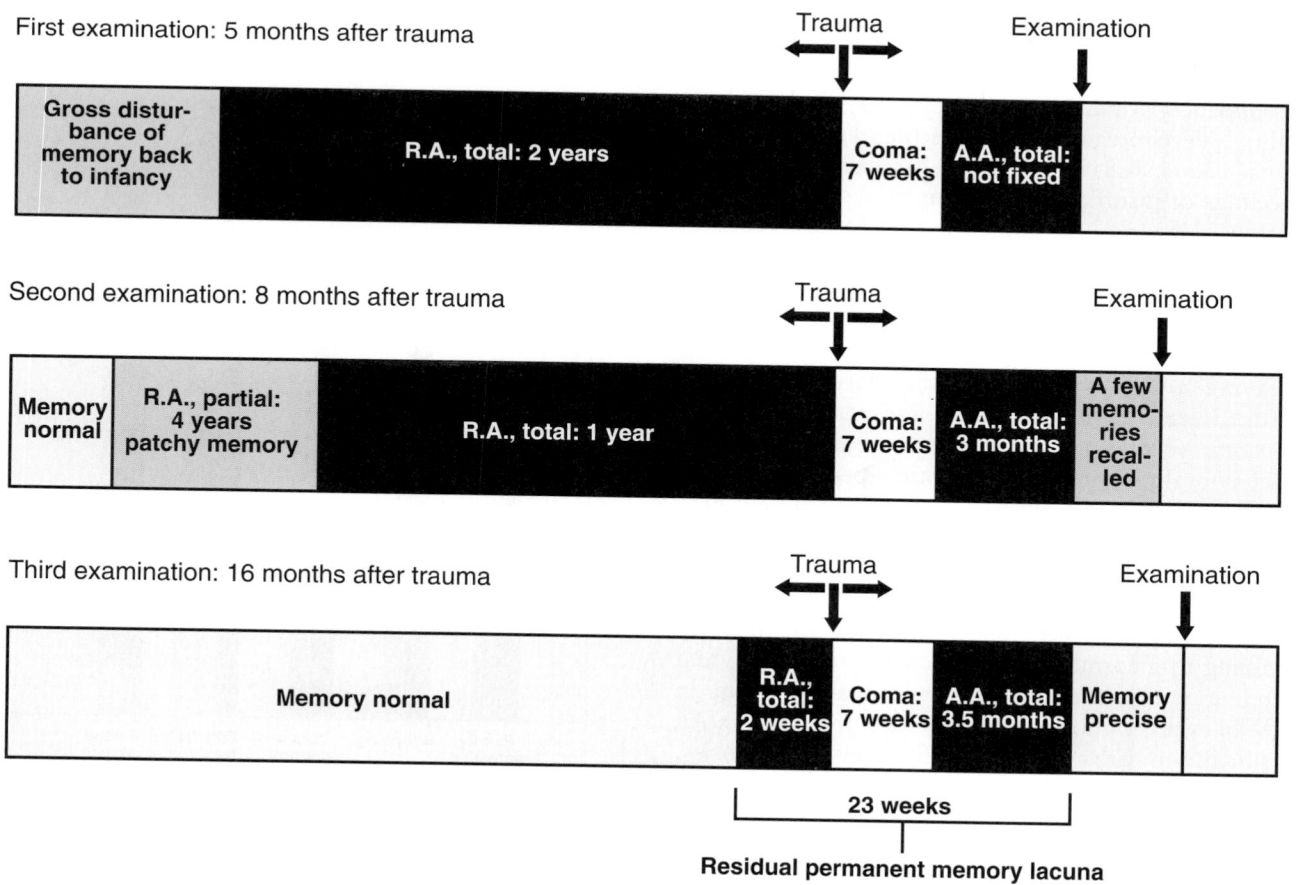

**FIGURE 4**   The time-line: Development and changes of post-traumatic amnesia across time (after Fig. 9 of Barbizet, 1970).

### BOX 1
### Example of a Patient Demonstrating Various Personality and Cognitive Changes in Addition to Memory Disturbances after Brain Damage

Calabrese and Markowitsch (1995) studied and treated a 54-year-old man after diffuse hypoxic brain damage due to cardiac decompensation. The primary aim of treatment was to allow him to live on his own and to be able to remember, plan, and execute his ideas. This aim was achieved during altogether 1½ years.

**General Description of the Patient.** From the beginning of diagnosis and therapeutic treatment, the former trade union activist and technical worker was conscious and fully oriented, friendly, and rational in his rapport. With respect to his mood, he talked in a reduced voice (though he stated that he had been accustomed to talk before large audiences) and was somewhat depressed in his mood. The patient's drive and his abilities to concentrate and to memorize were severely reduced. Retrograde memory was spotty. He was quite critical toward himself with respect to his organic brain syndrome, stating for example: "If I still would try to fulfill my former tasks, I would run around as a fool" and that his reduced self-confidence would be "paralyzing his whole personality." Psychoreactive elements (depression, anxious uncertainty, loss of self-esteem) became apparent. His feelings of insufficiency led him to avoid social situations. He strictly avoided meeting former acquaintances. In order to assure not to forget some objects or simple processes (like having switched off the light before leaving his house) he developed some "control strategies" (like turning back several times before leaving) which over time became nearly compulsive. Adjustment to his cognitive failures clearly had become substantial and had to be cotreated. Thus, behaviorally oriented psychotherapeutic elements were employed in parallel to cognitive training procedures, in order to alleviate the corollary symptoms. First, a hierarchical list of social situations in which he felt "uncomfortable" was established, beginning with short "indirect" social interactions, such as talking to a former colleague on the telephone, and ending with situations like participating again in social events (following an invitation for a party). All critical behaviors fell within five major categories, namely family activities, shopping, conversations with former friends and colleagues, official negotiations, and public speech (Figure 5). From discussing various situations described by the patient as insur-

mountable, together with the practical problems which could be encountered in these social interactions, it became evident that he underestimated his actual social competence. Indeed, his solutions for some hypothesized problems were theoretically correct and he was also straightforward in verbal management of them. Nonetheless he hesitated to bring himself into such situations, because of his (already premorbidly existent) pretension to master social interactions perfectly. Thus, it was decided to face the respective situations "*in vivo*," starting from the least embarrassing interaction (e.g., telephone call) and continuing to more demanding ones. The stepwise successful mastering of these situations should help foster a feeling of personal value.

As in the most brain-damaged patients, attentional difficulties were particularly prominent in this patient. He was unable to sustain heightened attention over time. His initial speed in simple reaction tasks was also reduced as was his ability to attend to

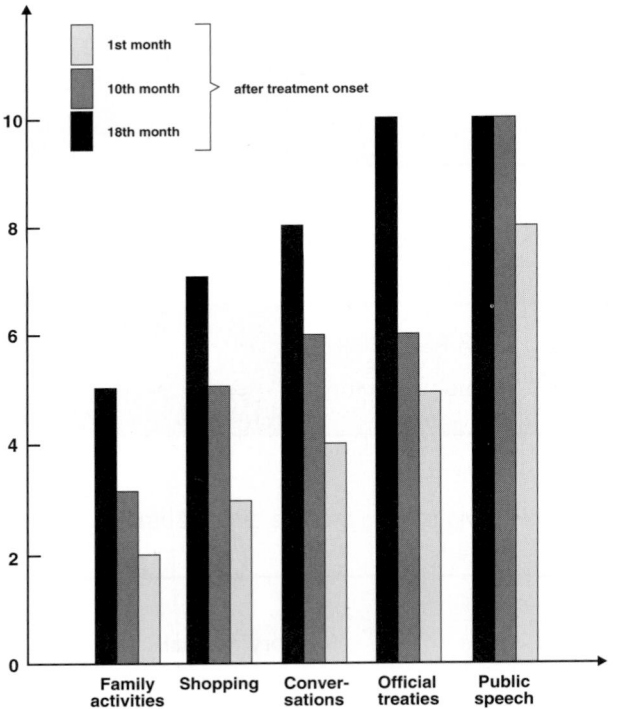

FIGURE 5  The patient's rating of his feelings in social situations at the beginning of treatment and 9 and 17 months later. High numbers indicate situations regarded as very demanding, anxiety-provoking, and/or problematic.

multiple-tasks situations. This limitation in cognitive speed and capacity is also shown in his initial values on the Trail-making test and in the backward reproduction condition of the digit and block spans (see Table III).

Before starting with cognitive remediation, for about 6 weeks a stepwise approach to those "real-life-situations," which the patient had described as critical, was undertaken. This was accompanied by careful preparations (description of the scene, participants, etc.) and exercises (imagining typical embarrassing situations) which helped to adjust the patient to the respective situations prospectively and to resolve his self-doubting attitude.

**Treatment.** It was decided to retrain attentional functions first, since an alleviation of the capacities of attention and concentration is needed as a prerequisite for proper consolidation of mnemonic strategies. These functions were trained with a commercial hardware–software program device (Vienna Testing System; Brickenkamp, 1986), applying tasks which involve visual scanning and require visual information processing under time pressure. The symbol-digit-modalities subtest from the WAIS and the Attention/Concentration Index from the

Wechsler Memory Scale—revised (WMS-r) were used as targets. The control measure was a delayed name reproduction task where the patient had to memorize family names. While for the control function the patient had even a drop in his performance, the symbol-digit task and also the attention/concentration index improved gradually, although without reaching statistical significance during the intervention phase. The attention/concentration training was continued during the following rehabilitation period. The fundamentals for this training involving reinforced practice were adopted from Ben-Yishay et al. (1979, 1985).

The memory training program included training for remembering names and appointments and new routes and verbal information as present in item lists.

**Outcome.** The patient improved considerably in most of the trained functions (cf. Table III). He also improved greatly in his capacity to deal with social situations (Figure 5). With every aim that the patient accomplished successfully, his depressed mood gradually diminished. At the beginning of cognitive remediation therapy, he had regained substantial self-confidence and a positive outlook regarding his personal future. This was also reflected by a substantial reduction in his depression score.

**Reflections.** Helping the patient to face and to cope with his social reality and to deal with his actual possibilities, is not only a "noncognitive" component of neurorehabilitative efforts, but turns out to be a prerequisite for the restoration of cognitive strategies over the long term. Since psychotherapy remains an individual-specific process, a differentiation of the contribution of socio-behavioral-therapeutic and cognitive strategies is impossible, at least on the basis of a single-case approach. Both components act in a synergistic manner and serve as premises for the attainment of specific goals.

The results in this initially severely memory-impaired patient suggest that specific, prolonged training is effective long term and generalizes to everyday situations. While it was impossible to exclude training-independent recovery processes for this patient, it can be considered quite unlikely that the patient would have succeeded in gaining his present cognitive status without this or similar treatments. Especially the combined application of psychosocial (psychotherapeutic), attention, and cognitive training should lead to profound long-term improvements.

TABLE III  Pre-treatment and Post-treatment Results of a Patient with Primary Memory Problems in Tests of Intelligence, Attention, and Memory (Patient BK of Calabrese and Markowitsch (1995))

|  | Pre-treatment | Post-treatment (after 1½ years) |
|---|---|---|
| Intelligence |  |  |
| General information (GI) | 48[a] | 52 |
| Similarities (SI) | 52 | 55 |
| Picture completion (PC) | 51 | 50 |
| Block design (BD) | 47 | 57 |
| Attention[b] |  |  |
| Trail-making test A | 25% | 50% |
| Trail-making test B | 10% | 25% |
| d2 | 16% | 35% |
| Short-term memory |  |  |
| Corsi block test (forward) | 4 | 5 |
| Corsi block test (backward) | 3 | 5 |
| Digit span (forward) | 5 | 5 |
| Digit span (backward) | 4 | 5 |
| Auditory verbal learning test |  |  |
| Total (1–5) | 21 | 36 |
| Rey-Osterrieth figure |  |  |
| Copy | 32 | 35 |
| Reproduction after 5 min | 14 | 22 |
| Everyday memory test | 47 | 112 |

[a] T score.
[b] Percentiles.

## Etiology

Severity of memory disorders and consequences for treatment are very dependent on the etiology of the underlying disease or disease process. Abrupt injuries shall be distinguished from those with a gradual onset and continuing progress. The first category is usually of more severe immediate consequences and even may result in an initial overshoot of symptoms. The diaschisis concept, mentioned above, provides a model for such diseases, for which brain infarcts and traumatic injuries are examples. For the second category slow-growing tumors and so-called degenerative diseases are examples. While some brain diseases with gradual progression are benign or can at least be treated successfully, the majority are progressive and up to now untreatable. Among these are age-related diseases such as Alzheimer's and Parkinson's disease.

In animal research a huge literature exists, modeling the consequences of abrupt versus slowly developing lesions (Butters et al., 1971; Markowitsch et al., 1985). Usually, abrupt lesions in humans are compared to one-stage lesions of the total region in question, while slowly developing ones in humans are compared to serial or several-stage lesions in animals. Serial lesions frequently lead to minor and transient deficits, while one-stage lesions may cause lasting and widespread deterioration of cognitive functions. Comparing the results of the two approaches to brain damage suggests that it is not so much the locus and extent of the damaged tissue, but the way or manner in which the damage was caused, which accounts for the behavioral deterioration.

Plasticity and recovery of function consequently depend significantly on the etiology of damage (Azari and Seitz, 2000; Kolb and Whishaw, 1998; Markowitsch and Calabrese, 1996; Robertson and Murre, 1999). Consequently, the etiology has to be considered carefully as a variable predicting functional outcome short term and long term.

## MEMORY ASSESSMENT

As other intellectual functions, memory is not an isolated intellectual entity. Instead, memories are embedded in the basic intellectual capacity of an individual, represented above all by his or her intelligence. Furthermore, the abilities to attend and concentrate are of considerable importance for memorization. An individual's motivation and mood determine performance, especially under demanding conditions (see Personality Dimensions, above) (Benedict, 1989). Memory in itself is also not an entity, but must be divided into systems along the dimensions of time and contents.

Amnesia for a long time was equated with the global amnesic syndrome, which meant—as the expression implies—a total inability of acquiring new information long-term with or without considerable disturbances in retrieving old (long-term stored) memories. This view was challenged by more sophisticated, diversified testing of memory abilities and by theoretical formulations of possible memory subdivisions (cf. The Long-Term-Memory Systems, above).

While in a few cases, amnesia for episodic or declarative memories may be permanent (Markowitsch et al., 1993a,b; Scoville and Milner, 1957), the majority of brain damaged patients have more limited memory disturbances, both with respect to content and duration (cf. Fig. 4). Sometimes, only material- or modality-specific memory impairments are seen (Markowitsch, 2000). From this it follows that not enough weight can be laid on a proper and detailed memory assessment as well as on careful repetitions of the memory status at different time points after initial memory impairment (and—usually—brain damage). Table IV contains examples of functions and tests.

While memory improvement, as estimated from improvement in memory tests, is not a primary source for mnestic progress in everyday situations, it nevertheless provides index scores of the principal available cognitive capacities.

## THERAPEUTIC POSSIBILITIES AND APPROACHES

These should be subdivided into more general, supportive steps and actions and direct or indirect interventions, especially training of memory functions. Furthermore, they need to be divided into immediate and prolonged actions.

Immediately after brain damage, usually self-assuring and stabilizing measures are needed, including affective affirmation and other forms of social care and support (Johnson, 1998); Kendall and Terry (1996) used the expression "psychosocial adjustment" in this context. Only later, detailed memory and further cognitive assessment can be done and thereafter various forms of direct training. For many patients this must include general forms of psychotherapy in addition to specific treatment of memory rehabilitation (Ben-Yishay et al., 1979, 1985; Fraser and Clemmons, 2000; Robertson and Murre, 1999). Of course, premorbid personality factors contribute substantially to the postmorbid recovery process (Symonds, 1937; Tate, 1998). Symonds (1937) had phrased this "it is not only the kind of injury that matters, but the kind of head" (p. 1092).

TABLE IV    Suggested Range of Cognitive Functions to Be Assessed after Brain Damage Leading to Memory Deterioration

| Function | Examples of tests |
|---|---|
| Intelligence | Wechsler Adult Intelligence Test—revised |
| Attention and concentration | Trail Making Test |
| Language | Aachen Aphasia Test, Token test |
| Mood and affect | Beck Depression Inventory |
| Memory in general | Wechsler Memory Scale III |
| Everyday memory | Rivermead Behavioural Memory Test |
| Short-term memory, working memory | Corsi block span, digit span backward |
| Concept formation, cognitive flexibility, reasoning, ability to calculate | Wisconsin Card Sorting Test, Tower of Hanoi, Fibonacci series, transcoding tests, simple calculations |
| Anterograde memory, verbal | Pair associate learning, California Verbal Learning Test, Learning of new facts |
| Anterograde memory, nonverbal | Rey-Osterrieth Figure |
| Recall vs recognition memory | Story recall, Face recognition test (words, objects), Recurring figures test |
| Emotional vs neutral memory | Emotional vs neutral stories and photographs |
| Procedural memory | Mirror image reading or writing |
| Priming | Incomplete Pictures Test, word stem completion |
| Retrograde memory, factual | Famous faces test, Famous scenes test, Famous names test, Semantic knowledge test |
| Retrograde memory, autobiographic | Autobiographic Memory Interview, Individual episodic remembrance of owned objects |

*Note.* Description of tests can be found in Lezak (1995), Spreen and Strauss (1991), Kroll *et al.* (1997), and Markowitsch *et al.* (1993a, b, 2000).

## Aims of Intervention

Optimally, a functional restitution should be the aim of all memory therapy. As this can be achieved only rarely, other, less ambitious aims have been formulated. One is the compensation of disturbed memory processes by using previously neglected strategies of information processing. After the rise of differentiating memory in subsystems, various strategies of using possibly undisturbed memory subsystems, when other ones are disturbed, became prevalent.

The method of "vanishing cues," in which, for example, a word is presented first in complete letters and later in training is reduced until only one (the first) letter remains, became prominent and aimed to benefit from implicit over explicit learning. The idea was based on the observed hyperspecificity in patients with traditional forms of amnesia (i.e., with episodic anterograde amnesia) (e.g., Butters *et al.*, 1993; Glisky *et al.*, 1986a). Glisky *et al.* (1986b), for example, trained the hyperspecific knowledge of learning to use a computer. Thöne (1996, 2000) later criticized the value of this method by pointing out that amnesic patients in general profit more from the rest capacity of their explicit memory system than from narrow-range implicit memory systems.

This had also been found in a direct comparison of implicit and explicit training procedures in Korsakoff patients (Thöne and Markowitsch, 1995). In their work, Thöne and Markowitsch selected a task in which new information had to be associated with known names and famous faces. This design was chosen because implicit memory processes can occur more readily when the patients can build on already existing knowledge (Schacter, 1987). The authors compared deep conceptual with flat perceptual learning and explicit with implicit recall, the combination of conceptual processing and explicit recall was most successful for this kind of information acquisition.

Other therapeutic approaches intend to reach an undisturbed, smoothly running everyday memory functioning as possible. Within this approach, improvement of metacognitive knowledge ("learning to learn") is a primary aim (Unverhau, 1994). It has to be kept in mind that even in everyday situations very different forms of memory are required: Persons and their names have to be remembered, events, loci, and news have to be stored, passing time has to be monitored and partitioned, and future dates, intentions, and plans have to be kept in mind.

Taken together, there are methods that serve to reduce memory demands, methods aimed at improving memory components, and methods used to improve metacognitive knowledge and aspects of problem solving (Schuri, 1998).

## Training of Memory Functions

When starting memory training, it is important that the patient gain insight into the advantage of the training procedures and of their usefulness for him or her. Training must be performed in a way that reveals successes easily and directly. There are various approaches applied and the preferred method depends on the patient's educational background, age, and the locus, extent, and etiology of brain damage. It may also be advisable to use different methods for training different types of memory functions.

Conventional memory training strategies can be criticized because they frequently rely on artificial material and lack the transfer to everyday situations. Furthermore, the proposed techniques—for example, the PQRST-technique (cf. Table V)—are very time-consuming and require considerable effort in attention and concentration. It is therefore necessary to train patients *situations* in which the application of such memory techniques is worth the effort. As non-brain-damaged subjects, we usually acquire information *incidentally* and only rarely decide consciously that we wish to encode the perceived material (Thöne, 2000). All of these variables hinder an efficient, successful use of memory techniques in patients.

A principle distinction is that between internally and externally guided memory training techniques. Internal techniques are used to improve the patient's cognitive capacity, while external techniques merely intend to support the patient so that his or her deficits are less obvious and she or he may lead a more normal life.

### Internal Memory Training Techniques

Table IV provides an overview of commonly applied internal memory techniques. A frequently used form is so-called face–name association learning as it has everyday relevance and as it is trainable with many variants—suited to the deficit and the abilities of the patient (Herholz *et al.*, 2001; Thöne, 1996; Thöne and Glisky, 1995). Visual imagery is usually highly effective for this training (Downes *et al.*, 1997). Vogel *et al.* (1987) showed that visual imagery even allows acquisition and retention of complex paired associations in amnesic patients, while without visual imagery acquisition was negligible (Figure 6).

As mentioned above, the success rate with different mnemonic technique seems to differ more generally, with the implicit method of vanishing cues usually not leading to more general memory improvements. A possible cause may be that patients may make errors when applying this technique and that such errors are as resistant to extinction as the correct responses (Baddeley and Wilson, 1997). Therefore the technique

of "errorless learning" was propagated for training memory to amnesic patients (Baddeley and Wilson, 1997; Squires *et al.*, 1997). Furthermore, it was found that the spacing of time between learning and recall appeared to be of importance in amnesics (Landauer and Bjork, 1978; Schacter *et al.*, 1985). This playing with time intervals had a tradition that goes back to attempts to optimize delayed alternation learning in frontal-lobe damaged nonhuman primates (Pribram and Tubbs, 1967).

### External Memory Training Techniques

Recent microtechnological progress allows the use of numerous external memory aids and these may provide efficient help to amnesics of diverse background, though especially the more sophisticated ones are still quite costly and usually involve particular engagement from the providing and monitoring side. Still, the patients need to first learn about the items and understand their features (Kapur, 1995), or, as Sohlberg and Mateer (1989) put it, "acquisition," "application," and "adaptation" are the three prominent features for successful use of external memory aids.

Written lists or the knot in the handkerchief represent the early times and the simple approaches, notepads and pagers the more advanced ones, and satellite-monitored control and guiding systems the latest stage in this development (Goldstein *et al.*, 1996; Kapur, 1995; Squires *et al.*, 1996; Zencius *et al.*, 1990). In fact, electronic organizers nowadays are used by normal individuals to a great extent as well. In a chapter in Kapur (1995) many such external memory aids are explained and more recently, the use of neuropagers and satellite-controlled guiding systems have been evaluated (Wilson *et al.*, 1997).

Wilson *et al.* (1997) studied 15 patients with neurological damage and everyday memory problems (the onset of damage being $\frac{1}{2}$ to 13 years earlier) with respect to their memory performance before, during, and after using a neuropager. The progress reminded them for instance to take medicine, to prepare lunch packets, to take certain items with them, and to use a diary. Prior to the use of a neuropager, 37% of the acts were performed; during its use 85% were performed and thereafter 74%. However, the epochs of measures were of unequal length and were disproportionally short for the epoch after ending the use of a neuropager (prior, $2 \times 5$ weeks; during, 12 weeks; after, 3 weeks). The costs of a neuropager were considered low, amounting to about £50 including its rental and respective costs for the controller of the neuropager.

Especially geriatric patients with severe memory problems may profit from external memory aids. In geriatric wards color codes may be used to mark, for instance, different levels of the building or room doors;

**TABLE V**  Internal Memory Training Techniques (after Ewald, 1997)

| Method | Characteristics |
|---|---|
| Method of Loci | Oldest known mnemonic technique—the system of places, named after the Greek poet Simonides (556–467 B.C). The patient selects a place well-known to him or her (e.g., living room) and imagines the objects present therein in a fixed spatial sequence. The chosen loci or places should be vivid, making them distinct and easy to remember. The subject pairs each object with an item he or she wishes to memorize, building thereby a self-generated image of the two. The image can be funny, surreal, grotesque, or realistic (Vogel *et al.*, 1987). Recall occurs via a sequential reconstruction of the objects in the locus or place. |
| Chaining | Chaining means the building of connections between words, objects, or characteristics in order to remember them. For example, by creating a sentence which contains the items to be remembered (e.g., bread, jam, ketchup, roses). Each item is used as trigger for the next item (enhancing its retrieval), similarly as individual members of a chain. |
| Key word technique | The key word technique combines the mnemotechnics "substitution" and "visual association." A phonetically related word is used as key for memorizing another word (to be learned anew); a bridge is built via visual association or imagery. Example: The Spanish word 'perro' for dog is learned via a similarly sounding word, for instance "pear." "Pear" then serves as a phonetic link, and, in a second step—the imagery link—a visual image is formed, containing a pear and a dog (e.g., a dog eating a pear). When now hearing "perro," the phonetic linkage will serve as a reminder for the key word "pear," and the visual image of the pear will trigger the picture of the dog eating it. |
| Rhyming, imagery, and association | These techniques may be used especially for learning numbers (phone numbers, house numbers, dates). A basic rhyming list has to be learned by heart and can be used repeatedly for memorizing numbers. |
| Word associations | The word association method is similar to the method of loci, with the difference that associated nouns function as loci. An own list of word associations is created and learned by heart, and later terms are associated with this list (imagery technique). In order to recall the associated terms, the list is retrieved. Example: (a) first base, (b) lion, (c) zebra, (d) grass. The first base serves as the starting point from which on the created imagery-story begins (e.g., a lion sitting on this first base, lions like to eat zebras, and so on). |
| Substitution | Complex, difficult-to-remember information is substituted by a simply memorizable term. In order to remember the altitude of Mount Yako (6052 m), the mountain can be remembered as having so many seconds as one minute and so many weeks as one year. The method of substitution may also be used for abstract words, which are substituted by visualizable (imaginable) ones and decoded via a "peg system" (see below) (e.g., alphabetic mnemonics). |
| Phonetic substitution | For terms that cannot be transformed into visual images at all, phonetically closely related terms are created, which then allow a visual association. |
| "Peg"-word technique | Peg-word lists are best described as numbered pegs on the wall of memory. Peg words serve as hooks where data to be stored can be attached to a mental filing system consisting of a series of prememorized nouns of a self-generated peg-word list, facilitating the sequential storage of information. Peg-word lists can be generated for dates, phone numbers, or names, and they can include concrete nouns or sounds that are later transformed into words. Peg words can also stand for the letters of the alphabet, representing specific words or intonations to store names and abreviations. Examples: a, aid, ale; b, beetle, beef, beer. It is of importance that these peg words are visualizable. Information to be remembered needs to be transferred according to the peg-word list. |
| Cueing via first letters | First-letter cueing is a very common technique for memorizing. A special set of pegs can be created, to be used for one set of information. Examples are acronyms such as "HOMES," standing for Huron, Ontario, Michigan, Erie and Superior (the Great Lakes). Or, if the first letters do not allow to create meaningful words, sentences or phrases may be used ("Richard of York gains battles in vain", standing for the correct sequence of colors in the visual spectrum: red, orange, yellow, green, blue, indigo, violet) (everyday examples: AIDS, UN, EU). |
| PQRST method | PQRST stands as acronym for Preview, Question, Read, State, Test and was developed as study technique used to remember the contents of written material. (*Preview*: scanning through the text or material; *Question*: asking oneself important questions about the text; *Read*: a second, and this time thorough, reading of the text in order to answer the previously stated questions; *State*: formulating of answers (if answers to key questions remain unclear → re-reading of the text; *Test*: testing in intervals whether one still remembers the text's contents well. |
| Remembering faces and names | The technique was conceived to memorize people's names by forming associations between a face and a name. The basic idea is the association of a person's name with someone or something already familiar to the individual. After hearing and understanding the name of a person, the subject must try to recall another, already known person with the same name. An association has to be created, connecting the person just met with the person having the similar name, by associating features of the introduced person with aspects of the known person. The recall of the newly introduced person's name is concluded by remembering the initial association which consequently evokes the respective name. |

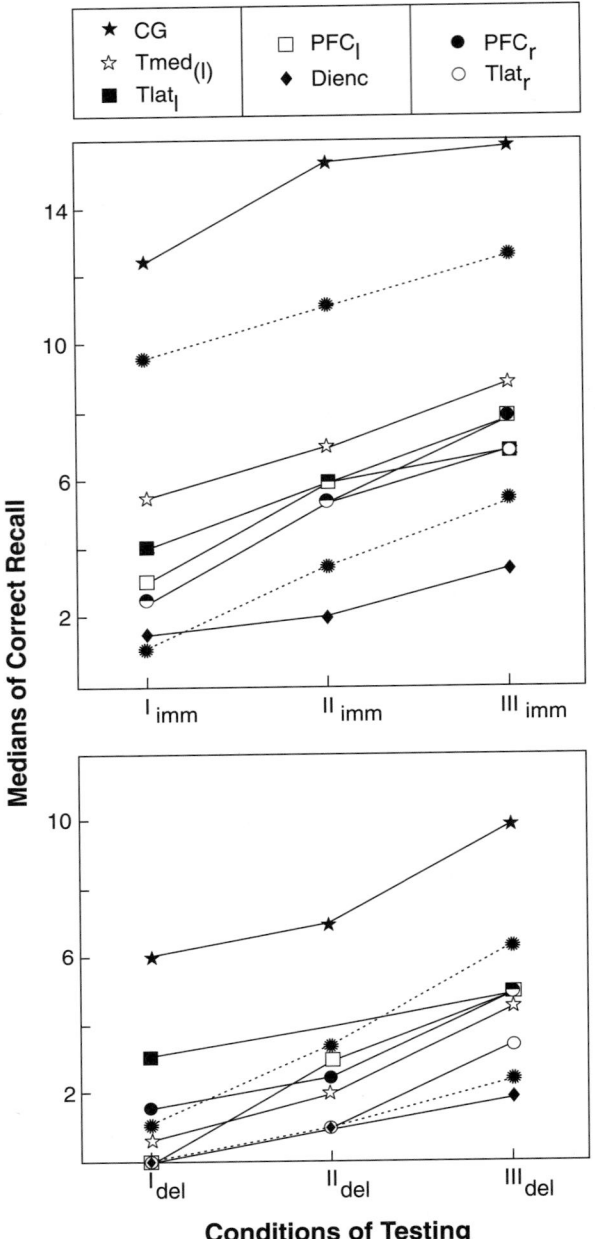

**FIGURE 6** Comparison of memory retrieval in 36 patients, divided into six groups, and 13 control subjects (CG) in two lists of paired-associate learning. The patients had right (r) or left (l) prefrontal cortical (PFC) damage, right or left anterior lateral temporal (Tlat) or (largely left-hemispheric), medial temporal lobe (Tmed₁), or diencephalic (Dienc) brain damage. The patients and the control group were tested under conditions of immediate (imm) (top) and delayed (del) (bottom) recall (48 h delay). Recall condition I was free recall, condition II was with visual imagery, and condition III with visual imagery and cued recall. The results demonstrate that visual imagery and providing cues increase retrieval substantially. Even severely brain-damaged individuals encode and store much more information than they recall under demanding conditions. (Data from Fig. 2 of Vogel et al., 1987).

colored lines may help to guide movement from one place (e.g., the own room) to another (e.g., a meeting room) (Harris, 1992; Wilson, 1992). Similarly, easily visible calendars, time plans, alarm clocks, or tables may be of great help for orientation and memory. Reality-orientation training is especially apt at guiding demented and confused patients (Folsom, 1968; Ford, 1996). The clinic staff may be included in the routines of training (Wilson and Moffat, 1992).

A long tradition used tokens, in which patients are either rewarded with a token when they remember specific things, events, or acts or "punished" by token withdrawal when they fail to remember the respective acts, events, or things. Token techniques, as well as other forms of training may rely not only on direct memory improvement, but can include the training of abilities to attend and concentrate, problem-solving abilities, and meta-memory training. Especially frontal lobe damage may require the training of such additional abilities. Training of meta-memory may allow the patient to evaluate his or her abilities, thereby providing a reality-oriented adaptation after brain damage and self-initiative in applying strategies. The patient learns when and where to use newly learned strategies and learns to attend to his or her own behavior and its consequences.

Of principal importance for all training concepts is that it comes to a generalization of learned information and learned behavioral acts and strategies. I stress that therapists must have adequate training themselves (Mozzoni and Bailey, 1996), that they must establish ways of measuring for a reliable and valid control of therapeutic strategies (Dobkin, 1990; Lindgren et al., 1997; Teasdale et al., 1997; Trexler et al., 1994), and that therapists need to specify predictors of functional recovery (Jørgensen et al., 1997; Woischneck et al., 1997). Patients with traumatic brain damage profit, for example, usually much less from compensatory memory training than patients with other etiologies of their brain damage (Prigatano et al., 1993).

A frequently discussed issue is the use of group or individual training. Especially under the aspect of cost and therapist availability, group training has, compared to individual training, the advantage that patients are not isolated with their problems; they can compare their deficits and training progress, exchange ideas, and give hints to each other. Group therapy helps to exchange and compare memory strategies and group training allows the therapist to compare progress between patients. The many interindividual differences and the superimposed differences in etiology and lesion locus and extent (Markowitsch and Calabrese, 1996), however, limit its application and suggest that group therapy is best used as a supplement to individually shaped memory training.

## Medical Support for Memory Amelioration

Long-term memory amelioration in focally brain-damaged individuals is principally performed within the fields of behavior and psychotherapy. Pharmacological interventions are usually supportive at best (e.g., to enhance attention and concentration: Sarter *et al.*, 1997; Sahakian *et al.*, 1993), though there is a wide variety of approaches (Caplan, 1997; Curran, 2000a,b; Foster *et al.*, 1998; Lynch, 1998; Wessel *et al.*, 1997). When it comes to degenerative diseases of the brain, there are, however, a number of approaches, particularly for Parkinson's (e.g., Rinne *et al.*, 2000) and Alzheimer's patients (Nordberg, 1996; Wilcock, 1997). As with other brain-damaged patients, pharmacological interventions cannot reestablish memory in degenerative disorders, but they can (re-)activate the brain's metabolism (Alexander *et al.*, 1997; Hirano *et al.*, 1998; Yasuno *et al.*, 1998) and thereby—if appropriately canalized—support psychological interventions.

## More General Strategies to Cope with Memory Disturbances

Generally, reducing the demands in everyday life and the increased use of external resources may help to cope with memory disturbances. As much as possible, the environment should be adapted to fit with the patient's capabilities. Reality-orientation-training may be of help, as are all the above-mentioned internal and external memory training strategies. Furthermore, problem solving strategies and meta-cognitive training are useful. Finally, probably the most effective help is given by caring relatives and friends who provide a positive motivational background.

## CONCLUSIONS

Memory cannot be trained as in body building, though physical training and somatic efforts may be of some more general advantage (leading, e.g., to an increased release of endorphins) (e.g., Ide and Secher, 2000). Up to now, we have relied mostly on psychological and psychotherapeutic approaches to improve memory after brain damage. Neurogenesis and other plastic changes within neuronal tissue up to now still remain largely speculative when it comes to human tissue (Gould *et al.*, 1999; Magavi *et al.*, 2000), though there is, of course, convincing evidence for functional recovery, for example, from strokes (Azari and Seitz, 2000). Robot-aided neurorehabilitation (Krebs *et al.*, 1998), implant techniques (Dobelle *et al.*, 1976), and other bio-technical approaches have their greatest advantages in guiding or substituting sensory and motor functions, but much less so (up to now) on the memory level.

It has been found repeatedly that a proper emphatic and motivational basis and support by friends and relatives is of greatest help for memory improvement after brain damage and that methods of drill and practice can be much less recommended for memory training than they can for retraining of sensory and motor functions (Chervinsky *et al.*, 1998; Johnson, 1998). Individualized training of memory functions cannot be overestimated (Calabrese and Markowitsch, 1995; Kendall and Terry, 1996; Robertson and Murre, 1999).

Finally, we need not only behavioral measures for functional recovery, but also brain measures. Here, modern functional neuroimaging provides a great perspective (Herholz *et al.*, 2001; Markowitsch *et al.*, 2000; Wykes, 1998). Markowitsch *et al.* (2000), for example, showed in a patient with severe anterograde and retrograde episodic amnesia that this amnesia correlated with reduced glucose metabolism in memory-processing regions of his brain and that various forms of memory therapy after 1 year re-established normal cerebral glucose metabolism and led to major improvements in neuropsychological memory tests.

## ACKNOWLEDGMENT

While writing this chapter, my research was supported by the German Research Council (D.F.G.) (Grants Ma 795/26, 27, 29, 30).

## REFERENCES

Alexander, G. E., Furey, M. L., Grady, C. L., Pietrini, P., Brady, D. R., Mentis, M. J., and Schapiro, M. B. (1997). Association of premorbid intellectual function with cerebral metabolism in Alzheimer's disease: Implications for the cognitive reserve hypothesis. *Am. J. Psychiatry* **154**, 165–172.

Azari, N. P., and Seitz, R. J. (2000). Brain plasticity and recovery from stroke. *Am. Scientist* **88**, 426–431.

Baddeley, A., and Wilson, B. A. (1997). When implicit memory fails: Amnesia and the problem of error elimination. *Neuropsychologia* **32**, 53–68.

Barbizet, J. (1970). "Human Memory and its Pathology." Freeman, San Francisco, CA.

Benedict, R. H. B. (1989). The effectiveness of cognitive remediation strategies for victims of traumatic head-injury: A review of the literature. *Clin. Psychol. Rev.* **9**, 605–626.

Ben-Yishai, Y., Rattock, J., and Diller, L. A. (1979). A remedial "module" for the systematic amelioration of basic attentional disturbances in head trauma patients. *In* "Working Approaches to Remediation of Cognitive Deficits in Brain Damaged Persons." New York University Medical Center, New York.

Ben-Yishay, Y., Rattock, J., Lakin, P., Piasetsky, E. B., Ross, B., Silver, S., Zide, E., and Ezrachi, O. (1985). Neuropsychological rehabilitation: Quest for a holistic approach. *Sem. Neurol.* **5**, 252–259.

Brickenkamp, R. (1986). "Handbuch apparativer Verfahren in der Psychologie." Hogrefe, Göttingen.

Butters, M. A., Glisky, E. L., and Schacter, D. L. (1993). Transfer of new learning in memory-impaired patients. *J. Clin. Exp. Neuropsychol.* 15, 219–230.

Butters, N., Pandya, D., Sanders, K., and Dye, P. (1971). Behavioral deficits in monkeys after selective lesions within the middle third of sulcus principalis. *J. Comp. Physiol. Psychol.* 76, 8–14.

Calabrese, P., and Markowitsch, H. J. (1995). Recovery of mnestic functions after hypoxic brain damage. *Int. J. Rehabil. Health* 1, 247–260.

Caplan, L. R. (1997). New therapies for stroke. *Arch. Neurol.* 54, 1222–1224.

Chervinsky, A. B., Ommaya, A. K., de Jonge, M., Spector, J., Schwab, K., and Salazar, A. M. (1998). Motivation for traumatic brain injury rehabilitation questionnaire (MOT-Q): Reliability, factor analysis, and relationship to MMPI-2 variables. *Arch. Clin. Neuropsychol.* 13, 433–446.

Curran, H. V. (2000a). Psychopharmacological approaches to human memory. *In* "The New Cognitive Neurosciences" (M. S. Gazzaniga, Ed.), 2nd ed., pp. 797–804. MIT Press, Cambridge, MA.

Curran, H. V. (2000b). Psychopharmacological perspectives on memory. *In* "The Oxford Handbook of Memory" (E. Tulving and F. I. M Craik, Eds.), pp. 539–556. Oxford University Press, New York.

Dobelle, W. H., Mladejovsky, M. G., Evans, J. R., Roberts, T. S., and Girvin, J. P. (1976). "Braille" reading by a blind volunteer by visual cortex stimulation. *Nature* 259, 111–112.

Dobkin, B. H. (1990). Focused stroke rehabilitation programs do not improve outcome. *Arch. Neurol.* 46, 701–703.

Downes, J. J., Kalla, T., Davies, A. D. M., Flynn, A., Ali, H., and Mayes, A. R. (1997). The pre-exposure technique: A novel method for enhancing the effects of imagery in face-name association learning. *Neuropsychol. Rehabil.* 7, 195–214.

Ewald, K. (1997). "Computer Assisted Mnemonic Strategy Acquisition and Tailored Memory Training Approaches: A Study with Brain Injured Individuals." Logos Verlag, Berlin.

Folsom, J. C. (1968). Reality orientation for the elderly patient. *J. Ger. Psychiatry* 1, 291–307.

Ford, S. (1996). Cognitive rehabilitation of patients with dementia: A review of the effectiveness of learning mnemonics. *Int. J. Rehab. Health* 2, 277–283.

Foster, J. K., Lidder, P. G., and Sünram, S. I. (1998). Glucose and memory: fractionation of enhancement effects? *Psychopharmacology* 137, 259–270.

Fraser, R. T., and Clemmons, D. C. (Eds.) (2000). "Traumatic Brain Injury Rehabilitation." CRC Press, Boca Raton.

Glisky, E. L., Schacter, D. L., and Tulving, E. (1986a). Learning and retention of computer-related vocabulary in amnesic patients: Method of vanishing cues. *J. Clin. Exp. Neuropsychol.* 8, 292–312.

Glisky, E. L., Schacter, D. L., and Tulving, E. (1986b). Computer learning by memory-impaired patients—Acquisition and retention of complex knowledge. *Neuropsychologia* 24, 313–328.

Goldstein, G., Beers, S. R., Longmore, S., and McCue, M. (1996). Efficacy of memory training: A technological extension and replication. *Clin. Neuropsychol.* 10, 66–72.

Gould, E., Reeves, A. J., Fallah, M., Tanapat, P., Gross, C. G., and Fuchs, E. (1999). Hippocampal neurogenesis in adult old world primates. *Proc. Natl. Acad. Sci. USA* 96, 5263–5267.

Harris, J. E. (1992). Ways to help memory. *In* "Clinical Management of Memory Problems" (B. Wilson and N. Moffat, Eds.), pp. 56–85. Chapman and Hall, London.

Herholz, K., Kessler, J., Ehlen, P., Lenz, O., Kalbe, E., and Markowitsch, H. J. (2001). The role of prefrontal cortex, pre-cuneus, and cerebellum during face-name association learning. *Neuropsychologia* 39, 643–650.

Hirano, N., Mori, E., Ishii, K., Imamura, T., Shimamura, T., Tanimukai, S., Kazui, H., Hashimoto, M., Yamashita, H., and Sasaki, M. (1998). Regional metabolism: Associations with dyscalculia in Alzheimer's disease. *J. Neurol. Neurosurg. Psychiatry* 65, 913–916.

Ide, K., and Secher, N. H. (2000). Cerebral blood flow and metabolism during exercise. *Progr. Neurobiol.* 61, 397–414.

Johnson, R. (1998). How do people get back to work after severe head injury? A 10 year follow-up study. *Neuropsychol. Rehabil.* 8, 61–79.

Jørgensen, H. S., Nakayama, H., Reith, J., Raschou, H. O., and Olsen, T. S. (1997). Stroke recurrence: Predictors, severity, and prognosis. the Copenhagen stroke study. *Neurology* 48, 891–895.

Kapur, N. (1995). Memory aids in the rehabilitation of memory disordered patients. *In* "Handbook of Memory Disorders" (A. D. Baddeley, B. A. Wilson, and F. N. Watts, Eds.), pp. 533–556. Wiley, Chichester.

Kendall, E., and Terry, D. J. (1996). Psychosocial adjustment following closed head injury: A model for understanding individual differences and predicting outcome. *Neuropsychol. Rehabil.* 6, 101–132.

Kolb, B., and Whishaw, I. Q. (1998). Brain plasticity and behavior. *Ann. Rev. Psychol.* 49, 43–64.

Krebs, H. I., Hogan, N., Aisen, M. L., and Volpe, B. T. (1998). Robot-aided neurorehabilitation. *IEEE Transact. Rehabil. Eng.* 6, 75–87.

Kroll, N., Markowitsch, H. J., Knight, R., and von Cramon, D. Y. (1997). Retrieval of old memories—The temporo-frontal hypothesis. *Brain* 120, 1377–1399.

Landauer, T. K., and Bjork, R. A. (1978). Optimum rehearsal patterns and name learning. *In* "Practical Aspects of Memory" (M. M. Gruneberg, P. E. Morris, and R. N. Sykes, Eds.), pp. 625–632. Academic Press, London.

Lezak, M. D. (1995). "Neuropsychological Assessment", 3rd ed. Oxford University Press, New York.

Lindgren, M., Hagstadius, S., Abjörnsson, G., and Ørbaek, P. (1997). Neuropsychological rehabilitation of patients with organic solvent-induced chronic toxic encephalopathy. A pilot study. *Neuropsychol. Rehab.* 7, 1–22.

Lynch, G. (1998). Memory and the brain: Unexpected chemistries and a new pharmacology. *Neurobiol. Learn. Mem.* 70, 82–100.

Magavi, S. S., Leavitt, B. R., and Macklis, J. D. (2000). Induction of neurogenesis in the neocortex of adult mice. *Nature* 405, 951–955.

Markowitsch, H. J. (2000). Memory and amnesia. *In* "Principles of Cognitive and Behavioral Neurology" (M.-M. Mesulam, Ed.), pp. 257–293. Oxford Univ. Press, New York.

Markowitsch, H. J., and Calabrese, P. (1996). Commonalities and discrepancies in the relationship between behavioural outcome and the results of neuroimaging in brain-damaged patients. *Behav. Neurol.* 9, 45–55.

Markowitsch, H. J., Kessler, J., and Streicher, M. (1985). Consequences of serial cortical, hippocampal, and thalamic lesions and of different lengths of overtraining on the acquisition and retention of learning tasks. *Behav. Neurosci.* 99, 233–256.

Markowitsch, H. J., Calabrese, P., Liess, J., Haupts, M., Durwen, H. F., and Gehlen, W. (1993a). Retrograde amnesia after traumatic injury of the temporo-frontal cortex. *J. Neurol. Neurosurg. Psychiatry* 56, 988–992.

Markowitsch, H. J., von Cramon, D. Y., and Schuri, U. (1993b). Mnestic performance profile of a bilateral diencephalic infarct patient with preserved intelligence and severe amnesic disturbances. *J. Clin. Exp. Neuropsychol.* 15, 627–652.

Markowitsch, H. J., Kalbe, E., Kessler, J., von Stockhausen, H.-M., Ghaemi, M., and Heiss, W.-D. (1999). Short-term memory deficit after focal parietal damage. *J. Clin. Exp. Neuropsychol.* 21, 784–796.

Markowitsch, H. J., Kessler, J., Weber-Luxenburger, G., Van der Ven, C., and Heiss, W.-D. (2000). Neuroimaging and behavioral correlates of recovery from "mnestic block syndrome" and other cognitive deteriorations. *Neuropsychiatry Neuropsychol. Behav. Neurol.* 13, 60–66.

Monakow, C. von. (1914). "Die Lokalisation im Grosshirn und der Abbau der Funktion durch kortikale Herde." Bergmann, Wiesbaden.

Mozzoni, M. P., and Bailey, J. S. (1996). Improving training methods in brain injury rehabilitation. *J. Head Trauma Rehabil.* 11, 1–17.

Nordberg, A. (1996). Pharmacological treatment of cognitive dysfunction in dementia disorders. *Acta Neurol. Scand.* 168, 87–92.

Pribram, K. H., and Tubbs, W. E. (1967). Short-term memory, parsing and the primate frontal cortex. *Science* 156, 1765–1767.

Prigatano, G. P., Amin, K., and Jaramillo, K. (1993). Memory performance and use of a compensation after traumatic brain injury. *Neuropsychol. Rehabil.* 3, 53–62.

Rinne, J. O., Portin, R., Ruottinen, H., Nurmi, E., Bergman, J., Haaparanta, M., and Solin, O. (2000). Cognitive impairment and the brain dopaminergic system in Parkinson disease. *Arch. Neurology* 57, 470–475.

Robertson, I. H., and Murre, J. M. J. (1999). Rehabilitation of brain damage: Brain plasticity and principles of guided recovery. *Psychol. Bull.* 125, 544–575.

Russell, W. R. (1935). Amnesia following head injuries. *Lancet* 2, 762–763.

Sahakian, B. J., Owen, A. M., Morant, N. J., Crockford, H. A., Crooks, M., Hill, K., and Levy, F. (1993). Further analysis of the cognitive effects of tetrahydroaminoacridine (THA) in Alzheimer's disease: Assessment of attentional and mnemonic function using CANTAB. *Psychopharmacology* 110, 395–410.

Sarter, M., Bruno, J. P., and Himmelheber, A. M. (1997). Cortical acetylcholine and attention: Neuropharmacological and cognitive principles directing treatment strategies for cognitive disorders. *In* "Pharmacological Treatment of Alzheimer's Disease: Molecular and Neurobiological Foundations" (E. Brioni and M. W. Decker, Eds.), pp. 105–128. Wiley-Liss, New York.

Schacter, D. L. (1987). Implicit memory: History and current status. *J. Exp. Psychol. Learn. Mem. Cogn.* 13, 501–518.

Schacter, D. L., Rich, S. A., and Stampp, M. S. (1985). Remediation of memory disorders: Experimental evaluation of the spaced-retrieval technique. *J. Clin. Exp. Neuropsychol.* 7, 79–96.

Schuri, U. (1998). Gedächtnisstörungen: Intervention. *In* "Lehrbuch Klinische Psychologie-Psychotherapie" (U. Baumann and M. Perrez, Eds.), pp. 593–603. Huber, Bern.

Scoville, W. B., and Milner, B. (1957). Loss of recent memory after bilateral hippocampal lesions. *J. Neurol. Neurosurg. Psychiatry* 20, 11–21.

Sohlberg, M. M., and Mateer, C. A. (1989). Training use of compensatory memory books: A three stage behavioral approach. *J. Clin. Exp. Neuropsychol.* 11, 871–887.

Spreen, E., and Strauss, O. (1991). "A Compendium of Neuropsychological Tests." Oxford University Press, Oxford.

Squires, E. J., Hunkin, N. M., and Parkin, A. J. (1996). Memory notebook training in a case of severe amnesia: Generalising from paired associate learning to real life. *Neuropsychol. Rehab.* 6, 55–65.

Squires, E. J., Hunkin, N. M., and Parkin, A. J. (1997). Errorless learning of novel associations in amnesia. *Neuropsychologia* 35, 1103–1111.

Symonds, C. P. (1937). Mental disorder following head injury. *Proc. R. Soc. Med.* 30, 1081–1094.

Tate, R. L. (1998). "It is not only the kind of injury that matters, but the kind of head": The Contribution of premorbid psychosocial factors to rehabilitation outcomes after severe traumatic brain injury. *Neuropsychol. Rehabil.* 8, 1–18.

Teasdale, T. W., Skovdahl, Hansen, H., Gade, A., and Christensen, A.-L. (1997). Neuropsychological test scores before and after brain-injury in relation to return to employment. *Neuropsychol. Rehabil.* 7, 23–42.

Thöne, A. I. T. (1996). Memory rehabilitation—Recent developments and future directions. *Restor. Neurol. Neurosci.* 9, 125–140, 1996.

Thöne, A. I. T. (2000). Neuropsychologische Therapie bei Patienten mit schwerer Amnesie: Möglichkeiten und Grenzen. *In* "Aktuelle Aspekte der Neurorehabilitation" (I. Daum and W. Widdig, Eds.), pp. 160–184. Pabst-Science, Lengerich.

Thöne, A. I. T., and Glisky, E. L. (1995). Learning of name-face associations in memory impaired patients: A comparison of different training procedures. *J. Int. Neuropsychol. Soc.* 1, 29–38.

Thöne, A. I. T., and Markowitsch, H. J. (1995). Möglichkeiten des Erwerbs neuer Informationen bei chronisch alkoholabhängigen Patienten mit Gedächtnisstörungen. Ein Vergleich verschiedener Trainingsstrategien. *Verhaltensmed. Heute* 5, 59–63.

Trexler, L. E., Webb, P. M., and Zappala, G. (1994). Strategic aspects of neuropsychological rehabilitation. *In* "Brain Injury and Neuropsychological Rehabilitation" (A.-L. Christensen, and B. P. Uzzell, Eds.), pp. 99–121. Lawrence Erlbaum, Hillsdale, NJ.

Tulving, E. (1995). Organization of memory: Quo vadis? *In* "The Cognitive Neurosciences" (M. S. Gazzaniga, Ed.), pp. 839–847. MIT Press, Cambridge, MA.

Tulving, E., and Markowitsch, H. J. (1998). Episodic and declarative memory: Role of the hippocampus. *Hippocampus* 8, 198–204.

Tulving, E., Kapur, S., Craik, F. I. M., Moscovitch, M., and Houle, S. (1994). Hemispheric encoding/Retrieval asymmetry in episodic memory: Positron emission tomography findings. *Proc. Natl. Acad. Sci. USA* 91, 2016–2020.

Unverhau, S. (1994). Strategien der Gedächtnistherapie bei neurologischen Erkrankungen. *In* "Neurologie und Gedächtnis" (M. Haupts, H. F. Durwen, W. Gehlen, and H. J. Markowitsch, Eds.), pp. 106–120. Huber, Bern.

Vogel, C. C., Markowitsch, H. J., Hempel, U., and Hackenberg, P. (1987). Verbal memory in brain damaged patients under different conditions of retrieval aids. A study of frontal, temporal, and diencephalic damaged subjects. *Int. J. Neurosci.* 33, 237–256.

Wessel, K., Langenberger, K., Nitschke, M. F., and Kömpf, D. (1997). Double-blind crosscover study with physostigmine in patients with degenerative cerebellar diseases. *Archs. Neurol.* 54, 397–400.

Wilcock, G. K. (1997). Pharmacological and non-pharmacological therapeutic interventions in Alzheimer's disease. *Clinician* 15, 30–37.

Wilson, B. A. (1992). Memory therapy in practice. *In* "Handbook of Behavior Therapy and Psychological Science" (B. A. Wilson, and N. Moffat, Eds.), pp. 227–252. Pergamon Press, Elmsford.

Wilson, B. A., Evans, J. J., Emslie, H., and Malinek, V. (1997). Evaluation of neuropage: A new memory aid. *J. Neurol. Neurosurg. Psychiatry* 63, 113–115.

Wilson, B. A., and Moffat, N. (1992). The development of group memory therapy. *In* "Clinical Management of Memory Problems" (B. A. Wilson, and N. Moffat, Eds.), pp. 243–273. Chapman and Hall, London.

Woischneck, D., Firsching, R., Rückert, N., Hussein, S., Heissler, H., Aumüller, E., and Dietz, H. (1997). Clinical predictors of the psychosocial long term outcome after brain injury. *Neurol. Res.* 19, 305–310.

Wykes, T. (1998). What are we changing with neurocognitive rehabilitation? Illustrations from two single cases of changes in neuropsychological performance and brain systems as measured by SPECT. *Schiz. Res.* **34**, 77–86.

Yasuno, F., Imamura, T., Hirono, N., Ishii, K., Sasaki, M., Ikejiri, Y., Hashimoto, M., Shimamura, T., Yamashita, H., and Mori, E. (1998). Age at onset and regional cerebral glucose metabolism in Alzheimer's disease. *Dementia Geriatr. Cognit. Disord.* **9**, 63–67.

Zencius, A., Wesolowski, M. D., and Burke, W. H. (1990). A comparison of four memory strategies with traumatically brain-injured clients. *Brain Injury* **4**, 33–38.

*Cognitive and Behavioral Disorders*

# CHAPTER 28
# Delirium

Marshal Folstein and Louis R. Caplan

## CLINICAL ASPECTS

### Definition and Classification

Delirium is a syndrome characterized by altered consciousness and multiple cognitive impairments including deterioration in the capacity to direct and sustain attention. The domain of consciousness, which varies from alertness to stupor, is distinguished from cognitive domains such as memory and attention, which depend on consciousness. Demented patients are fully alert but severely cognitively impaired. Patients with mania or schizophrenia cannot direct or sustain attention and yet have normal levels of consciousness.

The current definition of delirium in the ICD and DSM systems derives from the work of Bonhoeffer (1974). Bonhoeffer described several syndromes, delirium, stupor, clouded states, and twilight states, that he characterized as exogenous reaction types, which he differentiated from schizophrenia and mania. The term *delirium* encompasses all of these syndromes. Validation of the concept that a single syndrome was the product of many exogenous causes is found in the report of Wolff and Curran (1935), who described among 110 hospitalized patients a core syndrome that was caused by a variety of etiologies. The characteristic features were alteration of consciousness, cognitive impairment, and prominent noncognitive symptoms such as anxiety, depression, delusions, and hallucinations. Patients were hyperactive or agitated, hypoactive or somnolent, or had both hypoactive and hyperactive periods.

Twilight and postictal states that occur after temporal lobe epilepsy are characterized by inaccessibility, marked slowing of cognition, and intermittently unpredictable and occasionally violent behavior and will not be considered further in this chapter.

The unity of the syndrome of delirium is still debated depending on the primacy of etiology in the classification scheme. The Bonhoeffer syndromes share the common characeristic feature of altered consciousness which can be caused by different etiologies which affect different parts of the brain. Many neurologists reserve the term *delirium* for the hyperactive form. This view is articulated by Adams and Victor (1989), who consider delirium of the delirium tremens type to be a separate disorder from a condition that they called *acute confusional state* and that others referred to as *metabolic encephalopathy*. According to this view, delirium (a hyperactive agitated state) follows withdrawal from alcohol, benzodiazepines, and barbiturates and often develops after ingestion of amphetamine, L-dopa, or steroids; occurs after head trauma; and can be seen during emergence from anesthesia. Delirious patients are vigilant and overactive, have prominent hallucinations and delusions, and suffer sleep-cycle disruptions as well as autonomic instability. On the other hand, the acute confusional state is seen in drug or alcohol intoxication, metabolic failure, infections, and electrolyte imbalance. These patients are quiet, slow, and somnolent and only rarely have hallucinations, but do have tachycardia and fever. Many patients, however, are underactive or overactive at different times in the course of the same illness.

Two sets of diagnostic criteria are used worldwide. The *Diagnostic and Statistical Manual of the American Psychiatric Association* and the *International Classification of Disease* (ICD) are similar but not identical and so are presented in Tables I and II.

### Prevalence, Causes, and Differential Diagnosis

The prevalence of delirium in the community is less than 1% in adults and increases with age (Folstein *et al.*, 1991). Men and women are equally affected. The prevalence of delirium in clinical samples varies from 14

*Neurological Disorders: Course and Treatment, Second Edition*

**TABLE I   DSM-IV Criteria for Delirium**

A. Disturbance of consciousness (i.e., reduced clarity of awareness of the environment) with reduced ability to focus or sustain shift of attention.
B. A change in cognition (such as memory deficit, disorientation, language disturbance) or the development of a perceptual disturbance that is not better accounted for by a pre-existing, established, or evolving dementia.
C. The disturbance develops over a short period of time (usually hours to days) and tends to fluctuate during the course of the day.
D. There is evidence from the history, physician examination, or laboratory findings that the disturbance is caused by the direct physiological consequences of a general medical condition.

**TABLE II   ICD 10 Diagnostic Guidelines for Delirium**

For definite diagnosis, symptoms should be present in each of the following areas:
A. Impairment of consciousness and attention
B. Global disturbance of cognition with perceptual distortions, illusions and hallucinations, impairment of abstract thinking and comprehension with or without delusions, but typically with some degree of incoherence, impairment of immediate recall and of recent memory, disorientation for time as well as in more severe cases for place and person
C. Psychomotor disturbances
D. Disturbance of sleep–wake cycle
E. Emotional disturbances (e.g., depression, anxiety, irritability, euphoria, apathy, or wondering perplexity)
The category includes acute brain syndrome, acute confusional state, acute infective psychosis, acute organic reaction, acute psycho-organic syndrome.

to 75% (Francis *et al.*, 1990; Kolbeinsson and Jonsson, 1993; Koponen and Riekkinen, 1993; Murray *et al.*, 1993; Rabins and Folstein, 1982; Rockwood, 1993). The risks for delirium include metabolic, toxic, inflammatory, and infectious disorders. In a community sample most had diabetes, epilepsy, or were taking many prescribed drugs (Folstein *et al.*, 1991). Delirium is quite common after surgery (Craven, 1990; Williams-Russo *et al.*, 1992). However, a surprising number of delirious patients had structural brain disease that was most often due to strokes.

A growing number of clinical reports show that certain stroke locations cause delirium, although none of these studies used structured interviews to exclude other conditions that must be differentiated from delirium. For example, Peroutka *et al.* (1982) reported the case of an elderly woman with a small right parietal infarct who developed prominent delusions and hallucinations in clear consciousness with a minimum of cognitive decline and a very mild left hemiparesis. The symptoms responded to a low dose of haloperidol and remitted as the CT image of the stroke normalized. Lesions of the temporal lobes can result in an agitated delirium (Caplan *et al.*, 1986). The most common stroke lesion causing a hyperactive agitated delirium is embolism to the inferior division of the middle cerebral artery (MCA). When this occurs on the left side a Wernicke-type aphasia dominates the clinical picture. Right-sided inferior division MCA occlusion causes a left hemianopia, often upper quadrantic, and diminished ability to draw and copy and other visuo-spatial dysfunction. Weakness and sensory loss are usually not present. Hemorrhages and large infarcts, especially involving the right cerebral hemisphere, can also cause an agitated confusional state with poor concentration and attention (Schmidley and Messing, 1984). The critical structures involved in these cases were the medial temporal lobes and not the parietal lobe as was once thought. Inflammatory lesions such as herpes simplex and limbic (paraneoplastic) encephalitis are also accom-

panied by similar agitation and delirium, usually with accompanying seizures and prominent amnesia. Hyperphagia and hypersexuality may be accompanying features.

Acute lesions of the lingual and fusiform gyri can also lead to a hyperactive delirious state. The most common cause is embolism to the rostral basilar artery (Caplan, 1980; Chaves and Caplan, 2001; Hornstein *et al.*, 1962; Medina *et al.*, 1974; Mehler, 1989), causing infarction of the bilateral temporo-occipital lobes in the distribution of the posterior cerebral arteries (PCAs). Accompanying features are cortical blindness and inability to make new memories (Caplan, 1980). Again, motor and sensory abnormalities are often absent. Infarction in the PCA territory on one side can also produce a similar syndrome if the fusiform and/or lingual gyrus is involved (Devinsky *et al.*, 1988). Posteriorly placed hematomas as occur in head injuries and cerebral amyloid angiopathy can cause a similar picture usually with hemianopia.

Lesions of the cingulate gyrus can also cause agitation, hyperactivity, and acute psychosis (Amyes and Nielsen, 1955; Faris, 1969). A wide variety of lesions, including multiple sclerosis, tumors, infarcts, and hemorrhages, in this region can cause delirium. Delirium and agitation often with aggressiveness can also follow lesions of the orbital surface of the frontal lobes. This is most often found after head injury that causes subfrontal contusions.

Infarcts and hemorrhages involving the caudate nucleus (Caplan *et al.*, 1990; Chung *et al.*, 2001; Mendez *et al.*, 1989) and the thalamus (Barth *et al.*, 2001; Bogousslavsky and Caplan, 1993) can also cause agitation and delirium, often alternating with apathetic, inactive states—abulia. Portions of the frontal lobe, especially the anterior perforated substance, the caudate nucleus, and the medial thalamus, have strong projections to limbic cortex, possibly explaining the similarity of signs.

The variety of lesions that can cause delirium suggests a widely distributed integrated system or that delirium has multiple underlying mechanisms. Fisher (1983) posited that an integrated system that subserves all impulse to action includes mesencephalic—frontal connections that share circuitry with the reticular activating system and the thalamus. Lesions affecting posterior brain structures more often cause an agitated delirium while lesions of anterior structures more often cause akinesia, apathy, and abulia.

The signs and symptoms of delirium are presented here as a product of the clinical methods used to elicit them. The diagnosis of delirium requires two steps. The first step is recognition on the basis of mental state examination and history of a change in the level of consciousness and other features of the mental state. The second step is determination of the cause by means of history, physical examination, and laboratory studies.

The history concerning the delirious patient may reveal previous medical illnesses such as stroke, diabetes, and epilepsy; visual impairment; multiple drug intake including alcohol and psychoactive drugs as well as drugs with anticholinergic potential such as eye drops, cimetadine, antihistamines, and others; and episodes of delirium during previous illnesses (DePaulo et al., 1982; Folstein, 1983; Folstein et al., 1984; Inouye et al., 1990, 1993; Kulick and Wilbur, 1982; Marcantonio et al., 1994). The family history of delirious patients is not well studied. The history of the present illness of delirium reveals that the onset of delirium occurs over a period of days to hours. The symptoms and signs of delirium fluctuate greatly. Patients can appear relatively well one minute and minutes later be impaired, only to improve hours later.

Most important in the history is whether the onset was very abrupt, suggesting a stroke, or gradual, suggesting a more insidious onset of a metabolic, toxic, or inflammatory disorder. Accompanying systemic symptoms, especially headache and fever, are important, as well as neurologic symptoms, especially visual loss and altered memory, because they suggest focal brain lesions (Rabins and Folstein, 1982).

## Examination

### The Mental State Examination in Delirium Is Very Important and Includes a Number of Essential Parts (Folstein 1983; Folstein et al., 1984)

**Appearance and Behavior.** The delirious patient inhabits his or her own world, is drowsy and quiet or overactive and restless, or is unable to focus attention on the examiner. The level of consciousness can be drowsy and hypoactive, beclouded and approaching stupor, or vigilant and highly distractible as in delirium tremens (Ross et al., 1991). Abnormalities of consiousness fluctuate from hour to hour and are reversible if the underlying condition is corrected. Alertness and accessibility can be reliably rated by analogue scales (Anthony et al., 1982, 1985). Most patients are found in bed or sitting, perhaps because ataxia is common. They pick at the bedsheets or mutter to themselves. They appear to be unable to persist with motor tasks. Perseveration is also commonly found.

Many abnormal behaviors of delirious patients derive from their abnormal moods and perceptions. Some patients barricade themselves in a room or jump out a window because they believe they are about to be killed. Other patients can become aggressive against the caregiving staff because they misinterpret their intentions. Delirious patients are often slowed down in their responsiveness and reaction, and perception times are slow (Pauker and Folstein, 1978). Sleep disorders are the rule and may become a cause of distress to patients. Patients appear drowsy or desiring sleep but do not sleep in normal cycles during their delirium. In fact, the offset of delirium is signaled by the appearance of a normal sleep cycle. Appetite and thirst are not studied, but delirious patients often neglect to eat and drink.

**Talk.** Speech is often quiet and slurred but may be rambling and incoherent. Paraphasic errors are uncommon but comprehension of spoken language is poor. Patients do not follow complex commands, perhaps because of their inattention. They may have difficulty reading, writing, and copying designs (Chedru and Geschwind, 1972).

**Mood.** Most delirious patients are frightened (Wolf and Curran, 1935). Depression can also be present and lead to suicidal behavior. Obsessions are not usually present, but other phenomena secondary to fears are common. Patients often fear their caregivers and become frightened because of their misinterpretation of the environment.

**Delusions and Hallucinations.** Hallucinations occur mostly in active hypervigilant patients rather than in the somnolent form of delirium. Visual hallucinations are most frequent but auditory and tactile hallucinations also occur and are often followed by delusional explanations. Sensory misinterpretation and distortion are common (Wolf and Curran, 1935).

### Cognitive Examination

Disorientation to time and place and defects in attention, concentration, and memory are the rule. Other executive functions such as planning or set changing are impaired. Delirious patients are unable to quickly register events and thus are unable to remember them.

Delirious patients have anterograde and retrograde amnesia. Even after recovery the events occurring during the delirium are not remembered although islands of memory can be preserved. Memory functions are also very important, but difficult to test; hiding objects, especially money, and later asking its whereabouts is often effective. It is of interest that the delusions and hallucinations which occur are often remembered during and after the delirium. The patients are unable to perform tasks that require complex motor patterns such as writing. The documentation of handwriting and the construction of complex figures with matchsticks or copying designs are useful outcome measures of recovery or deterioration.

Especially important are signs of temporal lobe, frontal lobe, and occipital lobe dysfunction. Vision and the visual fields must be tested. Having the patient scan the room and describe what is seen and looking at pictures or a newspaper are very good screening tests for visual field defects and for visual neglect.

Cognitive screening tests such as the Mini-mental State Examination (MMS) are often abnormal but MMS scores above 23 are found in 30% of patients. The MMS items most affected are orientation, serial sevens, and recall. Most delirious patients cannot copy the interlocking pentagons. Tests such as the mental alteration test in which patients are asked to count to 10 and say the alphabet to j and then alternate letters and numbers such as a1, b2, c3, or the trails b test, or the digit symbol test are sensitive indicators of delirium (Jones *et al.*, 1993).

The confusion assessment method (CAM) specifies four criteria: acute onset and fluctuating course, inattention, disorganized thinking, and altered level of consciousness. The CAM algorithm for diagnosis of delirium required the presence of both the first and the second criteria and of either the third or the fourth criterion (Inouye *et al.*, 1990).

### Physical and Neurological Examination

Tachycardia and fever are frequent (Rabins and Folstein, 1982). In addition the delirious patient often has tremor, myoclonus, asterixis, and gait disorders. All varieties of motor abnormalities can be seen including decerebrate rigidity. Asymmetrical motor, sensory, and visual signs suggest focal brain lesions as the cause of the delirium and the need for brain imaging to clarify the cause.

### Laboratory and Imaging Examinations

Laboratory tests often help identify the the cause of the delirium. Hypoalbuminemia (Dicksonn, 1991;

Levkoff *et al.*, 1988), leukocytosis, elevated anticholinergic substances (Tune *et al.*, 1981, 1986, 1992, 1993, 2000), and diffuse slowing of the EEG are common (Engel and Romano, 1959). Blood screens for other drugs and toxins should routinely be performed when the etiology is unknown. Tests of thyroid, liver, and renal functions and serum $Ca^{2+}$ are also important. Neuroimaging-CT or MRI are very important in patients whose history and careful neurological examination suggest focal brain lesions. Since the patients are usually restless and agitated, sedation may be necessary to accomplish a diagnostic imaging procedure. If a brain infarct is suspected by the risk factor profile, suddenness of onset, and focal signs, then consideration should be given to performing a magnetic resonance angiogram (MRA) at the same time as the MRI (and under the same sedative). Cardiac-origin embolism is a very frequent cause of MCA and PCA infarcts with delirium. Since cardiac lesions can be lethal especially in hyperactive delirious patients, evaluation of the heart is usually important and should be planned early in the course.

Engel and Romano (1959) described diffuse slowing of the EEG in delirium. The slowing of the background is sometimes in the range of normal, as when an individual with a normal background frequency of 12 per second slows to 8 per second in the course of a delirium. Serial tracings throughout the course are needed to document these changes and demonstrate a return to the normal faster frequency after recovery. EEG distinguishes delirium from schizophrenia or depression because in these conditions the pattern is normal. Because delirium is often superimposed on Alzheimer's disease and other causes of dementia that can also have EEG slowing, the finding is nonspecific. Recent studies of quantitated EEG suggest that this modality might be a useful means of distinguishing slowing due to delirium from slowing due to dementing conditions (Jacobson *et al.*, 1993a,b).

Blood and serological tests may also help clarify the cause of the delirium. Recent reports of hypoalbuminemia in delirious patients suggest that drug transport and nutritional factors might be important mechanisms in the cause of delirium. For example, Levkoff *et al.* (1988) found that urinary tract infection, low albumin, increased WBC, and proteinuria were significant risk factors for the development of delirium. Dicksohn (1991) reported high rates of hypoalbuminemia in delirious patients referred for psychiatric consultation. Since toxic blood levels of prescribed and illegal drugs are common, serum levels should be evaluated in every delirious patient. For example, Tune and colleages (1981, 1986, 1992, 1993, 2000), in a series of studies spanning two decades, showed that delirium in the ICU, in a nursing home, and after electroconvulsive therapy (EST) was often associated with elevated levels of anticholinergic substances that could be detected by a radioreceptor assay. Others (Mondimore *et al.*, 1983; Rovner *et al.*, 1988) have corroborated that anticholinergic substance levels are elevated in patients with delirium.

The CSF may show abnormalities of transmitter metabolites including MOPEG and HIAA in cases of delirium due to alcohol withdrawal.

## NATURAL COURSE AND PROGNOSIS

The course and prognosis of delirium varies with the cause but usually it is of acute or subacute onset, developing over hours to days and persisting an average of 6 days (DePaulo *et al.*, 1982). The mental state typically fluctuates from hour to hour. The prognosis for the mental state is good in that all abnormalities can return to normal if the cause of the delirium can be treated. In practice the prognosis is guarded especially in the elderly because the mortality rates are high and many are left in a dependent state and often require institutional care: 30–65% of elderly delirious patients do not fully recover. Fatality rates vary from 3 to 30% (Cole and Primeau, 1993).

## PRINCIPLES AND THERAPY

### Prevention

Delirium can be prevented by the judicious use of drugs that can cause delirium and the control of underlying medical conditions such as diabetes, epilepsy, infection, malnutrition and electrolyte imbalance, and endocrine disturbance. Alcohol and benzodiazepine withdrawal syndromes can be prevented by the appropriate tapering of drugs and the use of replacement therapies that block the same receptors as in the use of benzodiazepines to treat alcohol withdrawal or the use of physostigmine to treat anticholinergic excess or flumazenil to treat benzodiazepine overdose (Milam and Bennett, 1987). Control of known factors that predict the development of delirium can prevent its occurrence. Inouye *et al.* (1999, 2000) showed that interventions that effectively prevented delirium consisted of standardized protocols for the management of six risk factors for delirium: cognitive impairment, sleep deprivation, immobility, visual impairment, hearing impairment, and dehydration.

### Treatment

When delirium cannot be prevented it must be treated by rational empirical and empathic methods. The rational treatment of delirium is based on treatment of the etiology. Empirical treatment is the use of drugs and other interventions demonstrated to be effective in clinical trials. Empathic methods are used to support the patient and family through the severe medical illnesses associated with delirium and to interpret the patient's abnormal behaviors to the staff so that delirious behaviors are not seen as purposeful uncooperativeness.

Rational treatment is based on treatment of the underlying cause of the delirium. Although delirium is often due to multiple disorders, many patients become delirious because of the accumulation of anticholinergic effects of prescribed drugs (Tune *et al.*, 1981, 1986, 1992, 1993, 2000). For many years a pattern of anticholinergic delirium has been recognized. Clinically, anticholinergic delirium is characterized by drowsiness, restlessness, hallucinations, dysarthria, ataxia, stupor, coma, dry mouth, and dry skin. The pupils are large and the pulse is rapid. Urinary retention is common. The syndrome is treated by increasing acetylcholine at the synapse by physostigmine, which is an anticholinesterase and thus blocks the breakdown of endogenous acetylcholine. There are many reports that physostigmine is useful in the treatment of delirium due to many drugs.

The treatment of delirium due to alcohol or benzodiazepine withdrawal is accomplished by the use of benzodiazepines, which are gradually tapered over the course of days to weeks. There are no data indicating that these drugs shorten the course of withdrawal once it has begun, but their use can prevent the transition of mild withdrawal from alcohol to delirium tremens if the drugs are initiated early in the withdrawal period. Furthermore the use of benzodiazepines can prevent the necessity of physical restraints, which should lead to faster mobilization of the patient and thus fewer complications of prolonged bedrest. Oversedation, however, should be avoided because decreased consciousness predisposes to pneumonia, aspiration, and a variety of other medical complications. Furthermore oversedated patients cannot cooperate well for historical details and examination so that medical and neurological conditions are more difficult to recognize and treat. Recall that delirium is not a disease: delirium is a syndrome that accompanies many medical, neurological, and toxic conditions and effective treatment depends on recognition of the cause in the individual patient.

## PRACTICAL MANAGEMENT

Maintenance of nutritional and fluid intake and preservation of mobility and muscular tone form the foundation of empirical treatment. The high prevalence of hypoalbuminemia in delirium suggests that profound disturbances in nutrient supply and transport are the rule. Meticulous attention to the state of the patient's intake of nutrients, vitamins, and fluids is critical.

Physical activity to prevent decubiti, pneumonia, or deep vein thrombosis is useful. Although delirious patients are often ataxic and need assistance in sitting and walking, early rehabilitation out of bed is desirable if the patient is not hypotensive. The sooner the patient can be made to regulate his or her own activity and intake, the easier will be the care.

Patients report that physical restraints are extremely unpleasant and their use is often misinterpreted by delirious patients. For this reason and also to encourage physical activity, restraints should be avoided unless absolutely necessary to protect the life of the patient or staff. Deaths of delirious patients struggling against restraints have been reported (O'Halloran and Lewman, 1993). Since autonomic instability is the rule in delirium, patients should be maintained in a calm, supportive environment (Emson, 1994; Mirchandani et al., 1994).

Patients in distress from delusions and hallucinations should be provided symptomatic pharmacological relief. Patients with quiet delirium seldom require any additional medication. Many reports document the efficacy of haloperidol with or without benzodiazepine as an effective short-term treatment for hallucinating or anxious delirious patients. Haloperidol given orally or intravenously is widely used for managing delirium in the postoperative period. However, treatment with haloperidol is often followed by complications of extrapyramidal symptoms and drowsiness especially when high doses are used; therefore, treatment should be brief and discontinued after the delirium has cleared. Haloperidol may take days to weeks to clear and some patients treated with this drug feel stiff, inert, and apathetic for weeks even after the drug is discontinued. Risperidone offers similar antianxiety effects without causing as severe extrapyramidal effects, although side effects depend heavily on the dose and duration of treatment (Miller et al., 1998; Rosebush and Mazurek, 1999). Prolongation of the Q-T interval has been reported after intravenous haloperidol (Adams, 1988; Adams et al., 1986; Fernandez et al., 1988; Gelfand et al., 1992; Menza et al., 1988; Metzger and Friedman, 1993; Moulaert, 1989; Sanders and Stern, 1993; Sanders et al., 1991). Furthermore, Feeny et al. (1982) suggest that, during experimental stroke in mice, recovery is delayed for weeks even after a single dose of treatment with haloperidol.

Management of the external environment is important. The patient must be repeatedly oriented and told who each new person is entering his or her room because sensory misinterpretations are common. The patient must be told that he or she is delirious and not going crazy. Patients should also be reassured that they will recover their usual mental state.

The presence of familiar people such as family is helpful. But family members must be taught the symptoms of delirium and reassured that recovery is possible. Lights during the day and dim lighting at night are useful. Avoid inappropriate stimulation such as overhead speakers, pinging monitors, and middle-of-the-night intrusions. Music, television, and radio can be useful in some patients but misinterpreted by others.

Empathy or putting oneself in the shoes of the patient is needed to understand the content of the symptoms and also to provide reassurance.

Delirious patients live in a present state filled with false ideas and bewildering perceptions. Caregivers must understand that the patient may not remember what he or she is told, often does not understand instructions, and cannot make complicated decisions. Because the patient is cognitively impaired, some caregivers believe that they can be treated without regard to feelings and reactions. In fact, the delirious patient—like the demented patient—is acutely responsive to his or her relationship with anyone nearby. The highest levels of professional respect must be maintained if the patient is not to think he or she is being abused.

The patient's delusions and hallucinations are often filled with characters and situations from the past. A knowledge of the life story of the patient is essential if a patient's experiences while delirious are to be appreciated by the staff. Disorientation and difficulty in thinking impairs patients' ability to make meaningful clinical decisions. Cognitively impaired patients should not be challenged with medical decisions but, rather, family members with power of attorney or guardianship should be used to help the patient make decisions.

The family and friends are often frightened by the patient's mental state and must be educated about the symptoms of delirium and its causes as well as its prognosis. The family and significant others often can be enlisted to help with the care of the patient and thus reduce staff time and costs. Professional caregivers must understand the patient's mental state and not misinterpret actions as intentional.

## TREATMENTS NO LONGER USED

Physical restraint and sedatives are seldom used. Physical restraints are discussed above. Many sedative drugs such as bromides, barbiturates, and paraldehyde have been replaced by short-acting barbiturates such as lorazepam or neuroleptics such as risperidol that can be given orally, intramuscularly, or intravenously if needed. The acutely aggressive and disoriented patient can be calmed within 10 min with a single injection of 1 to 5 mg of haloperidol and 2 mg of lorazepam. The use of any sedative drug in the delirious patient must be closely monitored to avoid the complications of oversedation, including aspiration, hypotension, respiratory arrest, and hypostatic pneumonia due to prolonged immobility.

# REFERENCES

Adams, F. (1988). Emergency intravenous sedation of the delirious, medically ill patient. *J. Clin. Psychiatry* 49(Suppl.), 22–27.

Adams, F., Fernandez, F., and Anderson, B. S. (1986). Emergency pharmacotherapy of delirium in the critically ill cancer patient. *Psychosomatics* 27(Suppl. 1), 33–38.

Adams, R., and Victor, M. (1989). Delirium and other acute confusional states. *In* "Principles of Neurology," pp. 323–334. McGraw-Hill, New York.

Amyes, E. W., and Nielsen, J. M. (1955). Clinical study of vascular lesions of the anterior cingulate region. *Bull. Los Angeles Neurol. Soc.* 20, 112–130.

Anthony, J. C., LeResche, L., Niaz, U., Von Korff, M. R., and Folstein, M. F. (1982). Limits of the Mini-Mental State as a screening test for dementia and delirium among hospital patients. *Psychol. Med.* 12, 397–408.

Anthony, J. C., LeResche, L. A., Von Korff, M. R., Niaz, U., and Folstein, M. F. (1985). Screening for delirium on a general medical ward: The tachistoscope and a global accessibility rating. *Gen. Hosp. Psychiatry* 7, 36–42.

Barth, A., Bogousslavsky, J., and Caplan, L. R. (2001). Thalamic infarcts and hemorrhages. *In* "Stroke Syndromes" (J. Bogousslavsky and L. R. Caplan, Eds.), 2nd ed. pp. 461–468. Cambridge University Press, Cambridge.

Bogousslavsky, J., and Caplan, L. R., (1993). Vertebrobasilar occlusive disease: Selected clinical aspects. 4. Thalamic infarcts. *Cerebrovasc. Dis.* 3, 193–205.

Bonhoeffer, C. (1974). Exogenous psychoses. *In* "Themes and Variations in European Psychiatry" (S. Hirsch and M. Sheppard, Eds.), pp. 47–63.

Caplan, L. R. (1980). Top of the basilar syndrome: Selected clinical aspects. *Neurology* 30, 72–79.

Caplan, L. R., Kelly, M., and Kase, C. S. *et al.* (1986). Infarcts of the inferior division of the right middle cerebral artery: Mirror image of Wernicke's aphasia. *Neurology* 36, 1015–1020.

Caplan, L. R., Schmahman, J., Kase, C. S. *et al.* (1990). Caudate infarcts. *Arch. Neurol.* 47, 133–143.

Chaves, C. J., and Caplan, L. R. (2001). Posterior cerebral artery. *In* "Stroke Syndromes" (J. Bogousslavsky and L. R. Caplan, Eds.), 2nd ed., pp. 479–489. Cambridge University Press, Cambridge.

Chedru, F., and Geschwind, N. (1972). Writing disturbances in acute confusional states. *Neuropsychologia* 10, 343–353.

Chung, C.-S., Lee, H.-S., and Caplan, L. R. (2001). Caudate infarcts and hemorrhages. *In* "Stroke Syndromes" (J. Bogousslavsky and L. R. Caplan, Eds.), 2nd ed., pp. 469–478. Cambridge University Press, Cambridge.

Cole, M. G., and Primeau, F. J. (1993). Prognosis of delirium in elderly hospital patients. *Can. Med. Assoc. J.* 149, 41–46.

Craven, J. L. (1990). Postoperative organic mental syndromes in lung transplant recipients. *J. Heart. Transplant.* 9, 129–312.

DePaulo, J. R., Jr., Folstein, M. F., and Correa, E. I. (1982). The course of delirium due to lithium intoxication. *J. Clin. Psychiatry* 11, 447–449.

Devinsky, O., Bear, D., and Volpe, B. T. (1988). Confusional states following posterior cerebral artery infarction. *Arch. Neurol.* 45, 160–163.

Dicksonn, L. R. (1991). Hypoalbuminemia in delirium. *Psychosomatics* 32, 317–323.

Emson, H. E. (1994). Death in a restraint jacket from mechanical asphyxia. *Can. Med. Assoc. J.* 151, 985–1157.

Engel, G. L., and Romano, J. (1959). A syndrome of cerebral insufficiency. *J. Chron. Dis.* 9, 260–277.

Faris, A. A. (1969). Limbic system infarction. *Neurology* 19, 91–96.

Feeny, D. M., Gonzales, A., and Law, W. A. (1982). Amphetamine, haloperidol and experience interact to affect rate of recovery after motor cortex injury. *Science* 217, 855–857.

Fernandez, F., Holmes, V. F., Adams, F., and Kavanaugh, J. J. (1988). Treatment of severe, refractory agitation with a haloperidol drip. *J. Clin. Psychiatry* 49, 239–241.

Fisher, C. M. (1983). Abulia minor vs. agitated behavior. *Clin. Neurosurg.* 31, 9–31.

Folstein, M. F. (1983). Delirium. *Curr. Ther.* 898–900.

Folstein, M. F., Bassett, S. S., Romanoski, A. J., and Nestadt, G. (1991). The epidemiology of delirium in the community: The Eastern Baltimore Mental Health Survey. *Int. Psychogeriatr.* 3, 169–176.

Folstein, M. F., Fetting, J. H., Lobo, A., Niaz, U., and Capozzoli, K. D. (1984). Cognitive assessment of cancer patients. *Cancer* 53(Suppl. 10), 2250–2257.

Francis, J., Martin, D., and Kapoor, W. N. (1990). A prospective study of delirium in hospitalized elderly. *JAMA* 263(8), 1097–1101.

Gelfand, S. B., Indelicato, J., and Benjamin, J. (1992). Using intravenous haloperidol to control delirium. *Hosp. Community Psychiatry* 43, 215.

Hornstein, S., Chamberlain, W., and Conomy, J. (1962). Infarction of the fusiform and clacarine regions: Agitated delirium and hemianopsia. *Trans. Am. Neurol. Assoc.* 92, 357–367.

Inouye, S. (2000). Prevention of delirium in hospitalized older patients: Risk factors and targetted intervention startegies. *Ann. Med.* 32, 257–263.

Inouye, S. K., van Dyke, C. H. *et al.* (1990). Clarifying confusion: The confusion assessment method. A new method for detection of delirium. *Ann. Intern. Med.* 113, 941–948.

Inouye, S. K., Viscoli, C. M., Horwitz, R. I., Hurst, L. D., and Tinetti, M. E. (1993). A predictive model for delirium in hospitalized elderly medical patients based on admission characteristics. *Ann. Intern. Med.* 119, 474–481.

Inouye, S. K., Bogardus, S. T. Jr., Charpentier, P. A., Leo-Summers, L., Acampora, D., Holford, T. R., and Cooney, L. M., Jr. (1999). A multicomponent intervention to prevent delirium in hospitalized older patients. *N. Engl. J. Med.* 340, 669–676.

Jacobson, S. A., Leuchter, A. F., and Walter, D. O. (1993a). Conventional and quantitative EEG in the diagnosis of delirium among the elderly. *J. Neurol. Neurosurg. Psychiatry* 56, 153–158.

Jacobson, S. A., Leuchter, A. F., Walter, D. O., and Weiner, H. (1993b). Serial quantitative EEG among elderly subjects with delirium. *Biol. Psychiatry* 34, 135–140.

Jones, B. N., Teng, E. L., Folstein, M., and Harrison, K. (1993). A new bedside test of cognition for patients with HIV infection. *Ann. Int. Med.* 119, 1001–1004.

Kolbeinsson, H., and Jonsson, A. (1993). Delirium and dementia in acute medical admissions of elderly patients in Iceland. *Acta Psychiatry Scand.* 87, 123–127.

Koponen, H. J., and Riekkinen, P. J. (1993). A prospective study of delirium in elderly patients admitted to a psychiatric hospital. *Psychol. Med.* 23, 103–109.

Kulik, A. V., and Wilbur, R. (1982). Delirium and stereotypy from anticholinergic antiparkinson drugs. *Prog. Neuropsychopharmacol. Biol. Psychiatry* 6, 75–82.

Levkoff, S. E., Saran, C., Cleary, P. D., Gallop, J., and Phillips, R. S. (1988). Identification of factors associated with the diagnosis of delirium in elderly hospitalized patients. *J. Am. Geriatr. Soc.* **36**, 1099–1104.

Marcantonio, E. R., Goldman L., Mangione, C. M., Ludwig, L. E., *et al.* (1994). A clinical prediction rule for delirium after elective noncardiac surgery. *JAMA* **271**, 134–139.

Medina, J., Rubino, F., and Ross, E. (1974). Agitated delirium caused by infarction of the hippocampal formation and fusiform and lingual gyri: A case report. *Neurology* **24**, 1181–1183.

Mehler, M. (1989). The rostral basilar artery syndrome: Diagnosis, etiology, prognosis. *Neurology* **39**, 9–16.

Mendez, M. F., Adams, N. L., and Lewandowski, K. S. (1989). Neurobehavioral changes associated with caudate lesions. *Neurology* **39**, 349–354.

Menza, M. A., Murray, G. B., Holmes, V. F., and Rafuls, W. A. (1988). Controlled study of extrapyramidal reactions in the management of delirious, medically ill patients: Intravenous haloperidol versus intravenous haloperidol plus benzodiazepines. *Heart Lung.* **17**, 238–241.

Metzger, E., and Friedman, R. (1993). Prolongation of the corrected QT and torsades de pointes cardiac arrhythmia associated with intravenous haloperidol in the medically ill. *J. Clin. Psychopharmacol.* **13**, 128–132.

Milam, S. B., and Bennett, C. R. (1987). Physostigmine reversal of drug-induced paradoxical excitement. *Int. J. Oral. Maxillofac. Surg.* **16**, 190–193.

Miller, C. H., Mohr, F., Umbricht, D., Woerner, M., Fleischacker, W. W., and Lieberman, J. A. (1998). The prevalence of acute extrapyramidal signs and symptoms in patients treated with clozapine, risperidone, and conventional antipsychotics. *J. Clin. Psychiatry* **59**, 69–75.

Mimura, M., Namiki, A., Kishi, R., Ikeda, T. *et al.* (1992). Central cholinergic action produces antagonism to ketamine anesthesia. *Acta Anaesthesiol. Scand.* **36**, 460–462.

Mirchandani, H. G., Rorke, L. B., Sekula-Perlman, A., and Hood, I. C. (1994). Cocaine-induced agitated delirium, forceful struggle, and minor head injury. A further definition of sudden death during restraint. *Am. J. Forensic Med. Pathol.* **15**, 95–99.

Mondimore, F. M., Damloumji, N., Folstein, M. F., and Tune, L. (1983). Post-ECT confusional states associated with elevated serum anticholinergic levels. *Am. J. Psychiatry* **140**, 930–931.

Moulaert, P. (1989). Treatment of acute nonspecific delirium with i.v. haloperidol in surgical intensive care patients. *Acta Anaesthesiol. Belg.* **40**, 183–186.

Murray, A. M., Levkoff, S. E., Wetle, T. T., Beckett, L. *et al.* (1993). Acute delirium and functional decline in the hospitalized elderly patient. *J. Gerontol.* **48**, M181–M186.

O'Halloran, R. L., and Lewman, L. V. (1993). Restraint asphyxiation in excited delirium. *Am. J. Forensic Med. Pathol.* **14**, 289–295.

Pauker, N. E., and Folstein, M. F. (1978). The clinical utility of the hand-held tachistoscope. *J. Nerv. Ment. Dis.* **166**, 126–129.

Peroutka, S. J., Sohmer, B. H., Kumar, A. J., Folstein, M. F., and Robinson, R. G. (1982). Hallucinations and delusions following a right temporoparietooccipital infarction. *Johns Hopkins Med. J.* **151**, 181–185.

Rabins, P. V., and Folstein, M. F. (1982). Delirium and dementia: Diagnostic criteria and fatality rates. *Br. J. Psychiatry* **140**, 149–153.

Rockwood, K. (1993). The occurrence and duration of symptoms in elderly patients with delirium. *J. Gerontol.* **48**, M162–M166.

Rosebush, P. I., and Mazurek, M. F. (1999). Neurologic side effects in neuroleptic-naïve patients treated with haloperidol or risperidone. *Neurology* **52**, 782–785.

Ross, C. A., Peyser, C. E., Shapiro, I., and Folstein, M. F. (1991). Delirium: Phenomenologic and etiologic subtypes. *Int. Psychogeriatr.* **3**, 135–147.

Rovner, B. W., David, A., Lucas-Blaustein, M. J., Conklin, B., Filipp, L., and Tune, L. E. (1988). Self-care capacity and anticholinergic drug levels in nursing home patients. *Am. J. Psychiatry* **145**, 107–109.

Sanders, K. M., and Stern, T. A. (1993). Management of delirium associated with use of the intra-aortic balloon pump. *Am. J. Crit. Care.* **2**, 371–377.

Sanders, K. M., Murray, G. B., and Cassem, N. H. (1991). High-dose intravenous haloperidol for agitated delirium in a cardiac patient on intra-aortic balloon pump. *J. Clin. Psychopharmacol.* **11**, 146–147. [Letter]

Schmidley, J., and Messing, R. (1984). Agitated confusional state with right hemisphere infarctions. *Stroke* **15**, 883–885.

Tune, L. E. (2000). Serum anticholinergic activity levels and delirium in the elderly. *Semin. Clin. Neuropsychiatry* **5**, 149–153.

Tune, L. E., and Folstein, M. F. (1986). Post-operative delirium. *Adv. Psychosom. Med.* **15**, 51–68.

Tune, L. E., Carr, S., Cooper, T., Klug, B., and Golinger, R. C. (1993). Association of anticholinergic activity of prescribed medications with postoperative delirium. *J. Neuropsychiatry Clin. Neurosci.* **5**, 208–210.

Tune, L. E., Carr, S., Hoag, E., and Cooper, T. (1992). Anticholinergic effects of drugs commonly prescribed for the elderly: Potential means for assessing risk of delirium. *Am. J. Psychiatry* **149**, 1393–1394.

Tune, L. E., Damlouji, N., Holland, A., Gardiner, T., Folstein, M. F., and Coyle, J. (1981). Association of postoperative delirium with raised serum levels of anticholinergic drugs. *Lancet* **2**(8248), 651–652.

Williams-Russo, P., Urquhart, B. L., Sharrock, N. E., and Charlson, M. E. (1992). Post-operative delirium: Predictors and prognosis in elderly orthopedic patients. *J. Am. Geriatr. Soc.* **40**, 759–767.

Wolf, H. G., and Curran, D. (1935). Nature of delirium in allied states: The dysergastic reaction. *Arch. Neurol. Psych.* **33**, 175–215.

CHAPTER 29

# Diagnosis and Treatment of Dementia

Margaret M. Swanberg and Jeffrey L. Cummings

## INTRODUCTION

Dementia, as defined in the *Diagnostic and Statistical Manual of Mental Disorders*, 4th edition (DSM-IV) (American Psychiatric Association, 1994), is a clinical syndrome characterized by progressive memory decline and impairment in at least one other cognitive domain. The cognitive changes must be acquired and of sufficient severity to compromise social or occupational function. A different means of operationally characterizing dementia was proposed by Benson and Cummings, who defined it as an acquired and persistent impairment of intellectual function with deficits in at least three of the following five cognitive domains: language, memory, visuospatial skills, personality and behavior, and other cognition (abstraction, calculation, executive function, judgement, etc.) (Cummings *et al.*, 1980). Alternate definitions of the dementia syndrome also have been proffered. Most identify a group of patients with similar cognitive features.

Dementia is mostly a disorder of the elderly and the incidence rises an estimated 5% per every 5 years after the age of 65. These figures appear even more alarming when considering that the elderly population continues to increase dramatically. It is estimated that by the year 2030 people over the age of 65 will comprise close to 20% of the U.S. population (Schoenberg, 1986), and the corresponding dementia population will be large.

The most common dementing illnesses, Alzheimer's disease (AD), dementia with Lewy bodies (DLB), frontotemporal degeneration (FTD), vascular dementia (VaD), and hydrocephalic dementia make up close to 99% of all dementia syndromes with neurological etiologies. The diagnosis and treatment of these more common conditions will be reviewed below. The dementia syndrome of depression and dementia associated with toxic–metabolic or infectious causes are not considered in detail in this chapter.

## CLINICAL FINDINGS OF THE DEMENTIA SYNDROMES

The most important causes of dementia to consider are those that could potentially be reversible, leading to stabilization or reversal of the dementia syndrome. These causes are rare, but do merit investigation (Clarfield, 1988; Massoud *et al.*, 2000; Weytingh *et al.*, 1995). Table I lists the potentially reversible causes of dementia.

Required laboratory evaluations include screens for $B_{12}$ deficiency and thyroid dysfunction (Knopman *et al.*, 2001). If the patient has risk factors for syphilis, prior infection or lives in a region of the United States where syphilis is common, then the syphilis serology is justified. Figures 1 and 2 depict a decision tree algorithm for the diagnosis of dementia and the specific dementia syndromes. Metabolic abnormalities may cause dementia or more commonly exacerbate cognitive impairment from other dementias.

Structural imaging with either computerized tomography (CT) or magnetic resonance imaging (MRI) is recommended as a standard part of any dementia evaluation (Cummings, 2000a; De Carli, 2001; Knopman, 2001). Structural imaging is important to detect lesions such as strokes, tumors, vascular malformations, and hydrocephalus, which may contribute to or actually cause cognitive impairment. Structural imaging also reveals ischemic changes that may cause dementia or co-occur with degenerative disorders.

The use of positron emission tomography (PET) and single-photon emission computerized tomography

TABLE I    Reversible Causes of Dementia

Infectious
    Neurosyphillis
    Neuroborelliosis
Metabolic/endocrine
    Hypothyroidism
    B12 deficiency
    Hepatic disease
    Renal disease
    Wilson's disease
Trauma/toxins
    Neoplasm
    Subdural hematoma
    Head trauma
    Alcohol abuse
    Dementia pugilistica
Other
    CADASIL[a]
    Fabry's disease
    Multiple sclerosis
    Kuf's disease
    Depression
    Normal pressure hydrocephalus

[a]CADASIL, Cerebral Autosomal Dominant Arteriopathy with Subcortical Infarcts and Leukoariosis.

(SPECT) is not recommended as a routine part of the dementia assessment (Knopman, 2001). There may be a role for PET or SPECT as an adjunct to clinical diagnosis, especially in confusing or unclear cases, but its routine use requires further study.

Although there is a large body of evidence addressing the role of APOE alleles in AD, causative mutations in AD (Berg et al., 1998; Mayeux et al., 1998) and tau mutations on chromosome 17 in familial cases of FTD (Chow et al., 1999; Hutton, 2001; Poorkaj et al., 1998), and mutations associated with familial prion diseases, the routine use of genetic screens for either the diagnosis of AD, FTD, or Creutzfeldt-Jakob (CJD) is not recommended in the diagnosis of dementia syndromes. CSF studies have also not been found to be helpful as a standard screen in the assessment of dementia with the exception of the 14-3-3 protein to aid in the diagnosis of CJD (Knopman et al., 2001). These studies may be considered in specific clinical circumstances.

# CLINICAL FINDINGS IN SPECIFIC DEMENTIA SYNDROMES

## Alzheimer's Disease

Alzheimer's disease affects 2.5–4 million Americans and, with an estimated prevalence approaching almost 50% in those over age 85, it is the most common of all dementing illnesses. Onset of AD is most common after age 60, although onset in the 40s and 50s may be seen, especially among familial cases. The core clinical syndrome of AD is relatively stereotyped and includes severe impairments in recent memory, mild to moderate language deficits, apraxia, agnosia, and visuospatial disturbances including environmental disorientation. There is substantial variation in the degree of impairment in these domains in individual patients. Neuropsychiatric symptoms including depression, delusions, misidentification syndromes, and hallucinations are also common with varied presentation among individual patients. The neurologic examination is typically normal in the early stages of AD (except for mental status examination), although with progression of the disease neurological abnormalities including extrapyramidal symptoms (EPS), gait disorder, primitive reflexes, incontinence, and seizures can occur.

The National Institute of Neurologic and Communicative Disorders and Stroke–Alzheimer's Disease and Related Disorders Association (NINCDS-ADRDA) Work Group (McKhann et al., 1984) and the DSM-IV (APA, 1997) definitions present the most widely used clinical criteria for the diagnosis of AD. The NINCDS-ADRDA criteria have been divided into definitions of definite, probable, and possible AD. Definite AD requires histopathologic evidence either from autopsy or from biopsy samples plus a clinical diagnosis of probable AD. Probable AD consists of deficits in two or more cognitive domains (including memory) that are progressive in nature. These criteria specify that the cognitive complaints must be present for at least 6 months, progressive, and cannot be explained by other systemic disorders or brain diseases. To make the diagnosis of possible AD, a deficit in only one sphere need be present or there may be another brain disorder or systemic illnesses present, but not believed to be the cause of the dementia. In addition, disturbances of consciousness must be excluded to make a diagnosis of AD. Both sets of criteria have been validated extensively in the literature with average sensitivity of 81%; however, specificity averages 70% (Blacker et al., 1994; Holmes et al., 1999; Lopez et al., 1999). Table II summarizes the NINCDS-ADRDA criteria.

## Clinical Course

The typical presentation of AD is that of an insidious onset of progressive declines in memory, language, and praxis, with an average life expectancy after diagnosis of between 8 and 10 years (Jost and Grossberg, 1995). Higher education and greater occupational attainment may predict later disease onset; however, disease progression may be more rapid (Stern et al., 1999). The memory deficit is most marked for recent events and is generally not helped by prompting or cuing. However,

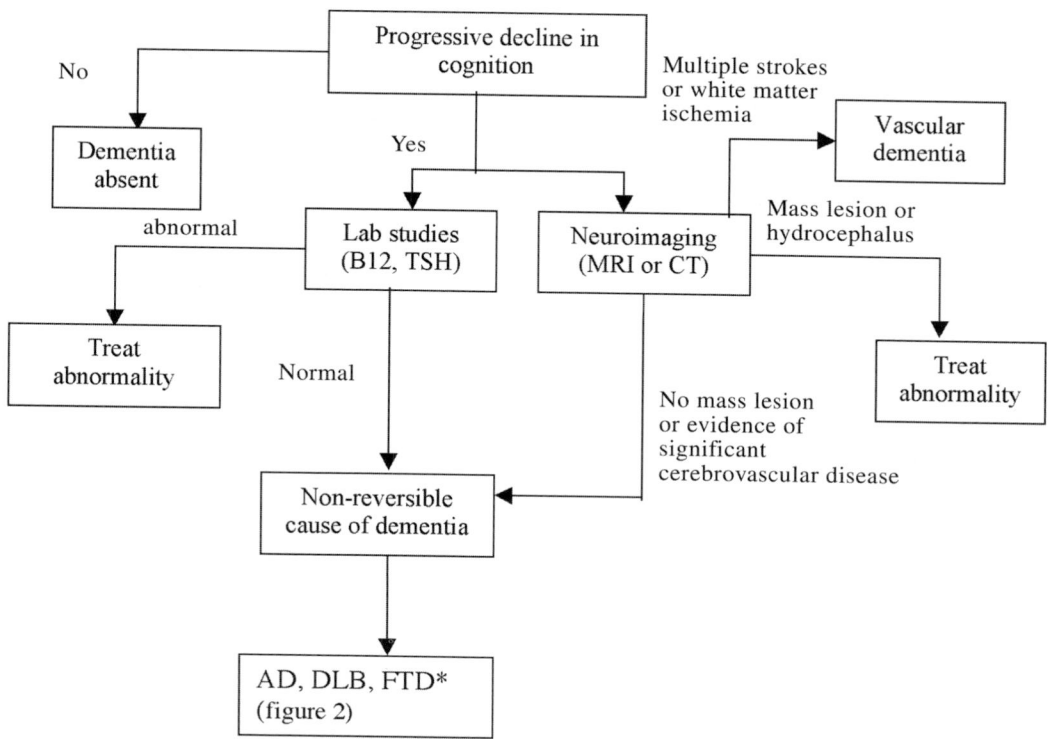

**FIGURE 1** Decision tree for the diagnosis of dementia. AD, Alzheimer's disease; DLB, dementia with Lewy bodies; FTD, frontotemporal dementia.

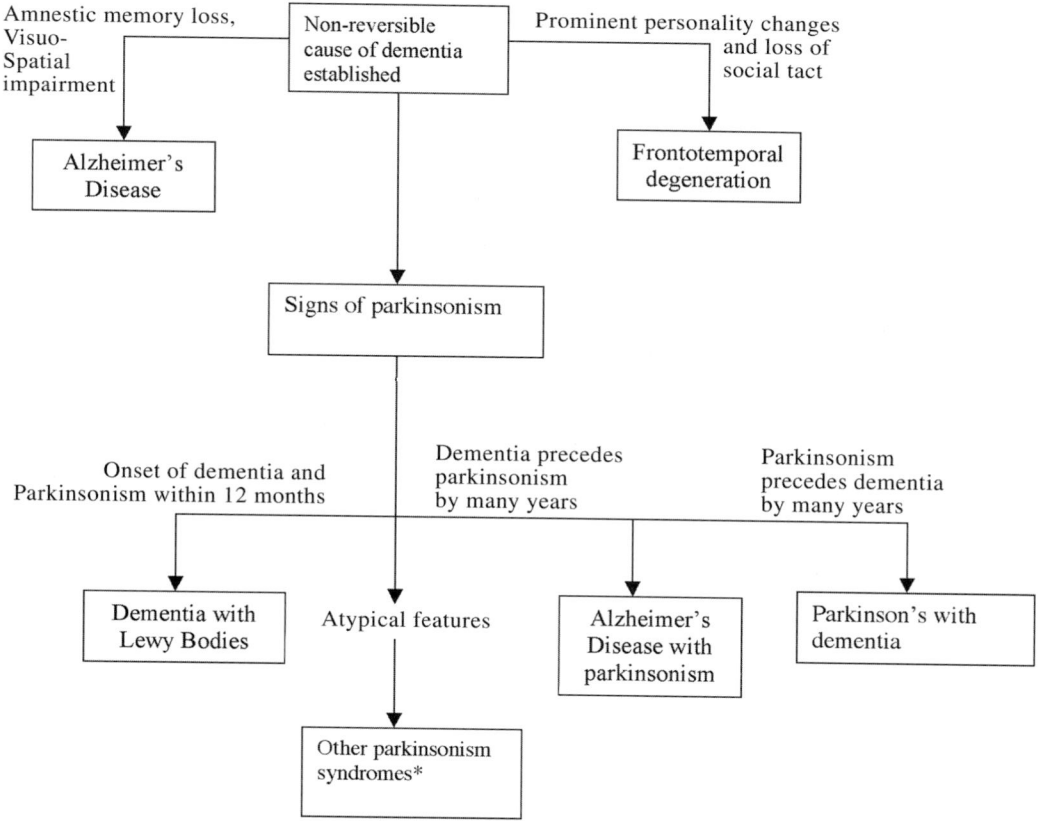

**FIGURE 2** Decision tree for major causes of dementia. *Progressive supranuclear palsy, corticobasal ganglionic degeneration, multiple systems atrophy.

TABLE II   NINCDS-ADRDA Criteria for AD

Definite AD
  Clinical criteria for probably AD
  Histopathologic evidence of AD
Probable AD
  Dementia established by clinical examination and mental status
    questionnaire
  Dementia confirmed by neuropsychological testing
  Deficits in two or more areas of cognition
  Progressive worsening of memory and other cognition functions
  No disturbance of consciousness
  Absence of systemic disorders or brain diseases that could
    produce a dementia syndrome
  Onset between 40–90
Possible AD
  Presence of systemic disorder or brain disease capable of
    producing a dementia but not thought to be the cause of the
    dementia
  Gradual progressive decline in only one intellectual function
    without any other identifiable cause
Unlikely AD
  Sudden onset
  Focal neurologic findings
  Seizures or gait disturbance early in the course of illness

*Note.* Adapted from McKhann *et al.* (1984).

with earlier diagnosis, mild AD cases are frequently seen with what is typically called a "retrieval" memory impairment, compared to the "encoding" believed to be the hallmark of AD memory loss. Remote memories are better preserved until later in the disease course.

Language changes are described as beginning with dysnomia followed by what is best described as a transcortical-type sensory aphasia with impaired comprehension and relatively preserved repetition ability (Cummings *et al.*, 1985). Word-finding difficulty and cicumlocuitous speech are also characteristic in the mild to moderate stages of disease. End-stage disease is marked by muteness with loss of all ability to verbally communicate. There is evidence suggesting that younger onset patients have more marked language impairment and perhaps faster disease progression.

Alzheimer's disease can have varied presentations including a "frontal variant" characterized by predominant deficits in executive function and insight with relative preseveration of memory until later in the disease (Johnson *et al.*, 1999) and language predominant presentations. (Galton *et al.*, 2000; Li *et al.*, 2000).

Common neuropsychiatric features including delusions and depression are prevalent throughout all disease stages but are more common in moderate and severe disease states.

## Pathologic Findings

There are five major neuropathologic findings in the brains of AD patients: neuritic plaques, neurofibrillary tangles, amyloid angiopathy, granuolovacuolar degeneration, and neuronal loss. The tangles are composed of paired helical filaments and are most prominent in the pyramidal neurons of the neocortex, the hippocampus, the amygdala, the locus ceruleus, and the raphe nuclei in the brainstem. The plaques have an amyloid core and are found in the cortex and the hippocampus. Different pathologic criteria are employed to aid in the postmortem diagnosis of AD. The varying criteria focus on either the extent of plaques or tangles in the various regions of the AD brain (Braak, 1991).

## Dementia with Lewy Bodies

Dementia with Lewy bodies, originally described by Japanese investigators (Kosaka *et al.*, 1984), is a relatively newly recognized dementia syndrome. With a prevalence of between 10 and 30% of late-onset dementias this disorder is now believed to be the second most common form of dementia following AD (Del Ser, 2000; McKeith *et al.*, 1999a). Over the past decade recognition and knowledge of DLB has grown. Initially thought to be a variant of AD, consensus criteria for the clinical and pathologic diagnosis of DLB now make it a distinct syndrome if not a separate disease. The average course and treatment response of DLB also help to distinguish this syndrome from AD.

The diagnosis of DLB relies on the use of consensus criteria proposed by the Consortium on DLB International Workshop in 1996 and revised in 1999 (McKeith *et al.*, 1996, 1999b). The consensus criteria are divided into central, core, and supportive features. The central feature is that of cognitive impairment sufficient to interfere with daily, social, or occupational functioning. Core features are spontaneous motor features of parkinsonism, visual hallucinations, and fluctuating cognition. Only two of these core features are essential for a diagnosis of probable DLB. Falls, systematized delusions in other modalities, syncope, alterations of consciousness, and neuroleptic sensitivity comprise the supportive features. These criteria have been shown to have acceptable sensitivity and specificity (Luis *et al.*, 1997; Mega *et al.*, 1996). The duration of illness is usually shorter than that of AD with average survival from time of diagnosis of 5–7 years (Olichney *et al.*, 1998). The prominent neuropsychiatric symptoms may lead to earlier and more frequent nursing home placement (Ballard *et al.*, 1999; Morris *et al.*, 2001). These consensus criteria are summarized in Table III.

## Clinical Course

With an average life expectancy shorter than typical AD, DLB is a more rapidly progressive dementing

**TABLE III**  Diagnostic Criteria for Dementia with Lewy Bodies

Central feature
  Progressive decline in cognition sufficient to interfere with normal
    or social or occupational functioning
  Prominent or persistent memory impairment is not necessary in
    the early stages, but is present with progression
Core features (two of three required)
  Fluctuating cognition with pronounced variations in attention
    and alertness
  Recurrent and usually well formed and detailed visual
    hallucinations
  Spontaneous motor features of Parkisonism
Supportive features
  Repeated falls
  Syncope
  Transient loss of consciousness
  Systematized delusions
  Neuroleptic sensitivity
  Hallucinations in other modalities
Unlikely DLB
  Cerebrovascular disease as evidenced by physical exam or
    neuroimaging
  Evidence of any physical illness or other brain disorder, sufficient
    to account for the clinical picture

*Note.* Adapted from McKeith *et al.* (1996).

disorder. In contrast to PD where the motor features precede the cognitive changes by several years, the dementia and motor features typically occur within 12 months of each other in DLB. This nearly simultaneous presentation also helps to distinguish DLB from AD, where the dementia precedes the motor features by several years. The relative lack of the typical PD resting tremor and the more frequent occurrence of myoclonus also help to distinguish these two disease entities (Louis *et al.*, 1997).

Depression is common but has not consistently been shown to be more prevalent in DLB compared to AD. Visual hallucinations are of a different character and duration in DLB compared to AD and do help to distinguish the two entities clinically. Hallucinations are more vivid and complex in DLB, composed of scenes of people or animals that are normal in size. These hallucinations are more likely to be of multiple types and more severe in contrast to AD. Persecutory delusions are one of the hallmarks in AD compared to more complex and bizarre delusions in DLB; misidentifications are also more prevalent in DLB (Ballard *et al.*, 2001; Hirono and Cummings, 1999). Neuropsychiatric features become more prominent as the disorder progresses.

### Pathologic Findings

The pathologic criteria for DLB are divided into those that are essential for the diagnosis and those that are associated but not required. The only essential feature

is the Lewy body (LB), which can be classified as "brainstem" or "cortical." The brainstem LB is a spherical intracytoplasmic eosinophilic neuronal inclusion body. It is typically seen in conditions such as Parkinson's disease. Cortical LBs are less-well-defined, spherical inclusions seen in cortical neurons. Cortical LBs can be identified with ubiquitin and alpha-synuclein stains. Associated pathologic features include Lewy-related neurites, plaques, neurofibrillary tangles, and microvacuolation (McKeith *et al.*, 1996).

### Frontotemporal Degeneration

Pick's disease was first described by Arnold Pick in 1892. He reported a case series of six patients with predominant language impairment and temporal lobe atrophy at autopsy. Alzheimer later described the microscopic findings of argyrophillic inclusions (Pick bodies) and swollen cells (Pick cells). Compared to AD, the brains of Pick's dementia show abrupt demarcation between affected and unaffected brain with relative preserved hippocampi. Over the past several years the syndrome of Pick's disease has been incorporated into the larger category of frontotemporal degeneration.

### Clinical Course

Classification of FTD utilizes the Lund-Manchester Criteria (Lund-Manchester Groups, 1994) as revised in 1998 by Neary and colleagues (Neary, 1998). These criteria emphasize behavioral changes including disinhibition, social inappropriateness, dietary changes, and changes in affect. Cognitive changes manifested by reduced verbal output and executive dysfunction with preserved posterior functions also are important in the diagnosis. Unlike patients with AD, those with FTD may present with symptoms associated with motor neuron disease and a full neurologic evaluation is indicated (Bak and Hodges, 1999; Rakowicz and Hodges, 1998). In addition, to further distinguish this syndrome from AD is the younger age of onset, generally before age 65. Table IV summarizes these criteria.

There are three principal variants of FTD: frontal variant FTD, semantic dementia, and nonfluent progressive aphasia. The frontal variant is characterized by combinations of disinhibition, poor impulse control, antisocial behavior, and stereotypies (associated with orbitofrontal cortex dysfunction), apathy (reflecting mesial frontal dysfunction), and abnormalities in planning and organization (related to the dorsolateral prefrontal cortex) (Bozeat *et al.*, 2001; Hodges *et al.*, 2001; Miller *et al.*, 1997). Semantic dementia typically presents with loss of memory for words, impaired comprehension, and the use of substitute words or phrases

**TABLE IV** Diagnostic Criteria for Frontotemporal Degeneration

Core diagnostic features
  Social behavioral changes
    Poor hygiene
    Lack of social tact
    Disinhibition
  Oral/dietary changes
    Hyperorality
    Food fixations
    Excessive tobacco or alcohol use
  Perseveration
    Mental rigidity
    Aberrant motor behavior
    Environmental dependency
  Anxiety/affect
    Somatic preoccupation
    Depression, suicidal ideation
    Apathy
Cognitive changes
  Language/speech
    Progressive reduction in verbal output
    Stereotypies of words phrases or themes
  Executive dysfunction
    Sequencing defects
    Defects in cognitive strategy
  Preserved posterior functions
    Preserved visuospatial ability
    Preserved calculations
Supportive features
  Onset before age 65

*Note.* Adapted from Lund-Manchester criteria (1994).

**TABLE V** NINCDS-AIREN Criteria for Vascular Dementia

A. Dementia present
B1. Vascular lesion(s) on brain imaging
B2. Focal neurologic signs
C. Relation between A and B with either
  1. Onset within 3 months of stroke
  2. Abrupt cognitive deterioration
  3. Fluctuating stepwise progression

*Note.* Adapted from Roman *et al.* (1993).

(Garrard *et al.*, 1997; Hodges *et al.*, 1992). Patients present with complaints of decreased speech output or word finding difficulty. Comprehension is relatively preserved, however as the disease progresses, problems do arise. Primary progressive aphasia is characterized by a slowly progressive nonfluent aphasia with relative preservation of other cognitive abilities until the final phase of the disorder.

### Pathologic Findings

There is selective atophy of the frontal and temporal lobes. Microscopically, there is neuronal loss, "ballooned cells," gliosis, and mild spongiosis appearing in the superficial portions of the cortex. Stains for the tau protein are frequently positive. The classic Pick body, not seen in all cases of FTD, is an intraneuronal argyrophilic inclusion that also stains positive for tau.

## Vascular Dementia

Varying sources estimate that VaD accounts for 20% of all dementias (Desmond, 1996). Dementia due to cerebrovascular disease may result from the effects of multiple brain infarcts involving large volumes of brain

tissue, from "strategic" infarcts that damage areas that play a key role in memory and other cognitive processes, or from subcortical lacunar infarcts and ischemic white matter damage associated with small vessel disease. There are several different diagnostic criteria for VaD, the most widely used being the National Institute of Neurological Disorders and Stroke–Association Internationale pour la Recherche et l'Enseignement en Neurosciences (NINDS-AIREN) criteria (Roman *et al.*, 1993). These are summarized in Table V. The diagnosis of VaD remains challenging because of the frequent co-occurrence of vascular dementia and AD. A recent evaluation of the various criteria found low intercriteria reliability further complicating the diagnosis (Chui *et al.*, 2000).

### Clinical Findings

The typical findings of VaD vary and depend on the areas affected by the vascular changes and the underlying pathology. Causes of VaD include macrovascular causes as well as multiple large infarcts and strategic single infarcts as well as microvascular causes as well as small vessel disease, hypoperfusion (from various etiologies), and hemorrhage (Nyenhuis and Gorelick, 1998). Focal neurologic features are common and can include hemiparesis, aphasia, dysarthria, ataxia, visual loss, and hemisensory loss. Gait changes, abnormalities of tone, and "lower half" parkinsonism are common. Cognitive findings include greater deficits in attention, concentration, and executive functioning compared to AD (Kertesz and Clydesdale, 1994; McPherson and Cummings, 1996). Vascular dementia patients also usually demonstrate more psychomotor slowing compared to AD. Additional neuropsychiatric features include greater incidence of depression (Cummings *et al.*, 1987), behavioral retardation, and anxiety (Sultzer *et al.*, 1993).

### Clinical Course

The course of VaD varies with etiology. The criteria for multi-infarct dementia require a stepwise progression of deficits that coincide with the occurrence of strokes (Hachinski *et al.*, 1974). Strategic infarct dementia may remain stable or improved; small-vessel disease is typically progressive.

### Pathologic Findings

The amount of infarcted brain tissue (in milliliters) previously was thought to correlate most strongly with the presence of VaD. More recent attention has focused on the changes in the small vessels and the cerebral white matter as an important causative factor of vascular dementia syndrome. Microvascular damage associated with subcortical white matter changes and microinfarction has been correlated with a history of dementia. Pathologic changes include arteriosclerosis, demyelination, gliosis, and nerve cell loss in the deep nuclei (Esiri *et al.*, 1997; Love *et al.*, 2000; Rosenberg *et al.*, 2001).

### Other Dementia Syndromes

Other less common causes of dementia include diseases associated with extrapyramidal features. Parkinson's disease (PD) may be a cause of dementia and can generally be distinguished from both AD and DLB based on the timing of the extrapyramidal and cognitive features. Parkinson's disease is associated with rest tremor, bradykinesia, rigidity, and a good response to levodopa. If there is early impairment of vertical eye movements, early loss of postural stability with frequent falls, and axial rigidity along with dementia, progressive supranuclear palsy is likely. Cortical basal ganglia degeneration (CBGD) may cause dementia, often in association with unilateral apraxia, rigidity, and alien hand syndromes (Riley, 1990). The cognitive features of these dementias typically include psychomotor slowing, impaired mental flexibility, impaired attention, and "retrieval type" memory deficits (characterized by inability to spontaneously recall a list of items in the presence of good ability to pick the items out of a list that includes distractor items).

Rapidly progressing dementia (often with progression to severe dementia and death within 1 year of onset) may be caused by the spongiform encephalopathies, including CJD. This syndrome is often associated with myoclonus, an abnormal EEG, and the presence of 14-3-3 protein in the CSF. Paraneoplastic syndromes such as limbic encephalitis also may cause rapidly progressing dementia.

## PRINCIPLES OF THERAPY

The treatment of any dementing illness requires first a search for reversible causes and then appropriate treatment of these. When a reversible condition cannot be found, the physician is faced with the challenging task of relaying this information to concerned family members and the patient. Fortunately, in recent years the physician has been provided an arsenal of potential therapies that may benefit both patient and caregiver. These therapies address both the cognitive deficits as well as the behavioral management of the more common dementing illnesses. In addition numerous interventions for the caregiver are now incorporated and include respite care, support groups, and individual therapy.

## PRACTICAL MANAGEMENT

### Alzheimer's Disease

#### Cognition Enhancing and Disease-Modifying Therapies

**Cholinesterase Inhibitors.** The cholinesterase inhibitors (ChE-Is) are the only agents approved by the Food and Drug Administration (FDA) for the treatment of AD. All other agents are used on an off-label basis. There are four ChE-Is available for use: tacrine (Cognex), donepezil (Aricept), rivastigmine (Exelon), and galantamine (Reminyl).

Neuronal loss in the basal forebrain nuclei results in reduced production of choline acetyltransferase. The loss of this enzyme leads to a reduced capacity for the synthesis of acetylcholine and a severe widespread cholinergic deficiency. Cholinesterase inhibitors increase cholinergic activity by inhibition of acetylcholinesterase, the enzyme that hydrolyzes acetylcholine in the synapse. By inhibiting this enzyme, more acetylcholine is available to stimulate the postsynaptic cholinergic receptors (Francis *et al.*, 1999).

The FDA requires demonstrated benefit on a measure of cognition and a global scale to approve a new drug for use in dementia. The ChE-Is as a class have shown modest improvement in cognition as measured by the Alzheimer's Disease Assessment Scale—Cognitive portion (ADAS-Cog) (Talwalker *et al.*, 1996). The most frequently employed global measure is the Clinical Interview-Based Impression of Change with Caregiver Input (CIBIC+) (Joffr *et al.*, 2000). Approximately 25% of those on ChE-I therapy were shown to have a clinically appreciable improvement and approximately another 50% remained stable without evidence of cognitive decline for 6–12 months (Corey-Bloom *et al.*, 1998; Raskind *et al.*, 2000; Rogers *et al.*, 1998a,b). Table VI summarizes the use of the ChE-Is.

Gastrointestinal side effects including nausea, vomiting, and diarrhea are the most frequently reported in all of the ChE-I trials. Many of these side effects are related

TABLE VI    Cholinesterase Therapy for Alzheimer's

| Drug name | Dose (mg) | Frequency of dose | Special mention |
| --- | --- | --- | --- |
| Donepezil (Aricept) | 10 | Once daily | Ease of administration, more rapid titration |
| Rivastigmine (Exelon) | 6–12 | Twice daily | Higher incidence of side effects, monitor weight loss |
| Galantamine (Reminyl) | 16–24 | Twice daily | Requires slow titration |

to the speed of titration and may be minimized with slower increases and administration with food. Dosage adjustments should be made approximately every 4–6 weeks as tolerated. Special precautions should be taken when using tacrine due its hepatotoxicity; liver function tests should be monitored every other week during the titration phase and every 3 months thereafter (Davis et al., 1992). Rivastigmine has been associated with weight loss and body weight should be monitored in those using this agent (Corey-Bloom et al., 1998).

**Anti-oxidant Treatment.** Anti-oxidants have been investigated as a potential therapy for AD. The most widely studied are vitamin E and selegiline. A randomized, double-blind, placebo-controlled trial examined the use of vitamin E, selegiline, or both in patients with AD. The endpoints were progression to severe dementia, loss of ability to perform ADLs, institutionalization, or death. Either vitamin E or selegiline delayed nursing home placement as well as the combined endpoints; the use of both agents simultaneously did not confer additional benefit (Sano et al., 1997).

**Other Therapies.** Other agents such as anti-inflammatory drugs are still undergoing clinical trials and are not recommended for routine use. Estrogen replacement therapy when administered in usual doses does not benefit postmenopausal women with AD. In recent years, the use of complimentary alternative medications (CAM) has become more popular in the United States. Many patients and caregivers are exploring these options for the treatment of AD. Gingko biloba is the most widely studied and trials comparing this agent to placebo have shown very modest improvement in cognitive measures (Kanowski et al., 1996; LeBars et al., 1997). Caution should be used when taking gingko in conjunction with anti-platelet agents or anticoagulants due to the increased risk of bleeding (Rowin and Lewis, 1996; Matthews, 1998). Huperzine A, another neutraceutical, has shown some benefit in limited studies

and is believed to have anti-cholinesterase activity (Ved et al., 1997). There is no role established for the use of ginseng in the treatment of AD.

### Pharmacotherapy for Behavioral Disorders

There are numerous neuropsychiatric features associated with AD including agitation, psychosis, depression, anxiety, and insomnia (Chung and Cummings, 1998). Not all symptoms require pharmacologic therapy and nonpharmacologic strategies should be tried first. When medication is necessary, the physician should choose the agent based on the target symptom.

Psychosis is manifested by delusions of theft, infidelity, or misidentification and occasionally by hallucinations. Neuroleptic agents and the atypical antipsychotics have demonstrated efficacy and tolerability in the treatment of agitation and psychosis in demented individuals regardless of severity of illness (DeDeyn and Katz, 2000; deDeyn et al., 1999; Katz et al., 1999; Street et al., 1998; Tariot et al., 2001; Teri et al., 2000). These agents also have been shown to be efficacious for the treatment of agitation in AD. While both classes of agents are effective, the atypical antipsychotics (risperidone, olanzapine and quetiapine) appear to be better tolerated (Doody et al., 2001b). Table VII summarizes the use of neuroleptics and atypical antipsychotics.

There have been case series and placebo-controlled trials evaluating the efficacy of anticonvulsants for the treatment of agitation in AD. The most widely studied anticonvulsants are carbamazepine and valproate. (Raskind, 1999; Tariot et al., 1994, 1998, 2001) Both of these agents have benefit; however, routine use may be limited secondary to side effects, rash, hepatotoxicity, hematologic abnormalities, and altered electrolytes,

TABLE VII    Behavioral Pharmacotherapy for Alzheimer's Disease

| Symptom | Drug | Dose (mg/day) |
| --- | --- | --- |
| Agitation/psychosis | Risperidone | 0.5–1.5 |
| | Olanzapine | 2.5–10 |
| | Quetiapine | 100–400 |
| | Haloperidol | 0.5–3 |
| | Carbamazepine | 200–1200 |
| | Divalproex | 250–3000 |
| Depression | Citalopram | 10–30 |
| | Sertraline | 50–200 |
| | Fluoextine | 20–80 |
| | Nortriptyline | 50–100 |
| Sleep Disturbances | Trazodone | 25–150 |
| | Zolpidem | 5–10 |
| | Temazepam | 15–30 |
| | Zaloplon | 5–20 |

which are more common in the elderly population. The recently published American Academy of Neurology (AAN) practice parameter on the treatment of AD does not recommend routine use of anticonvulsants in the treatment of agitation (Doody *et al.*, 2001b). The newer anticonvulsants have not been studied in a controlled fashion; therefore, recommendations regarding their use cannot be made.

Several classes of antidepressants have been studied for the treatment of depression in AD. The selective serotonin reuptake inhibitors (SSRIs) and the tricyclic antidepressants have demonstrated efficacy in the treatment of depression in general populations; however, in the treatment of AD, the SSRIs with their lower incidence of anticholinergic activity may offer a more favorable side effect profile (Lykestsos *et al.*, 2000; Nyth *et al.*, 1992; Tarango *et al.*, 1997). There may be a role for the use of trazodone in demented patients especially in those with sleep disturbances and agitation (Sultzer *et al.*, 1997; Teri, 2000).

ChE-Is in addition to their cognitive benefits have also been shown to improve behaviors in AD (Cummings, 2000b; Raskind *et al.*, 2000). These agents can provide relief in patients who have hallucinations and have been shown to improve apathy. Other symptoms may improve but the appropriate clinical trials necessary to establish the behavioral efficacy of these agents have not been conducted.

## Future Directions

Research into the development of new treatment strategies for AD is advancing rapidly. On the horizon are newer ChE-Is, agents such as memantine that target N-methyl-D-aspartate (NMDA) receptors mediating glutamate excitotoxicity, and neurotrophic agents (Emilien *et al.*, 2000; Grundman and Thal, 2000). In addition development of inhibitors of β- and γ-secretases, which are involved in the cascade that creates β-amyloid, as well as anti-amyloid immunization strategies are active research areas.

One of the more exciting potential therapies for the treatment and possible prevention of AD is the vaccine. Use of this vaccine in the mouse model of AD has shown that mice with AD-type pathology could be immunized against β-amyloid. These transgenic mice develop many of the neuropathologic markers of AD including amyloid plaques. Vaccination of young mice prevented the development of plaques. Treatment of the older mice reduced the extent and progression of the plaques. The brains of the immunized mice also had fewer dystrophic neurites that contribute to the development of the neuritic plaques (Janus *et al.*, 2000; Morgan *et al.*, 2000; Schenk *et al.*, 1999).

## Dementia with Lewy Bodies

The consensus criteria for DLB emphasize cognitive decline, signs of parkinsonism, and hallucinations. Each of these symptoms are troubling to the patient and caregiver. Appropriate treatment of DLB encompasses therapy for each of these problems, with some overlap between them.

### Cholinesterase Inhibitor Therapy

The cholinergic deficit in DLB has been found to be greater than that in AD (Langlais *et al.*, 1993; Perry *et al.*, 1994; Shiozaki *et al.*, 1999). Based on this evidence, the ChE-Is have been studied in this patient population. Tacrine was the first ChE-I studied and this literature showed a modest benefit in DLB. Donepezil has been more widely studied; however, double-blind controlled studies are lacking. The putative beneficial effect is based on case series and case reports (Rojas-Fernandez, 2001; Shea *et al.*, 1998). These studies have reported an improvement on the MMSE of up to four points. Neuropsychiatric features including hallucinations, delusions, and agitation also have been reported to improve following donepezil therapy (Kaufer *et al.*, 1998; Lanctot and Herrmann, 2000; Skjerve and Nygaard, 2000). Donepezil is generally well tolerated in DLB patients with only 3 of the 20 reported patients discontinuing therapy secondary to side effects.

McKeith and colleagues conducted a randomized placebo controlled trial of rivastigmine in DLB patients and found that this agent was efficacious (McKeith *et al.*, 2000). Improvement was noted in global measures, cognitive measures, and neuropsychiatric measures to include approximately 63% of treated patients showing at least a 30% improvement on the Neuropsychiatric Inventory (Cummings *et al.*, 1994). Rivastigmine was well tolerated in DLB with the majority of side effects being gastrointestinal (nausea, anorexia, diarrhea). Mild worsening of parkinsonism has been reported with all of the ChE-Is but is uncommon.

### Treatment of Motor Symptoms

Parkinsonism is generally mild in DLB; however, these symptoms at times merit therapy. Sinemet has shown limited efficacy in the treatment of DLB, potentially related to intolerability and neuropsychiatric side effects (Byrne *et al.*, 1989; Hely *et al.*, 1996; Louis *et al.*, 1995, 1997). Dopamine agonists used in the treatment of Parkinson's disease can exacerbate or induce hallucinations and psychosis, which are already prominent in DLB. This combination of disease symptoms and drug side effects makes it difficult to titrate Sinemet to typical therapeutic doses. There have been limited reports that tacrine therapy may have some benefit for symptoms of parkinsonism which may be particularly helpful considering these agents also improve the neuropsychiatric symptoms (Hutchinson and Fazzini, 1996).

### Treatment of Neuropsychiatric Features

Characteristic features of DLB that aid in distinguishing this disorder from AD are the presence of visual hallucinations and the frequency of delusions (Ballard et al., 2001; Hirono and Cummings, 1999). These symptoms are disturbing to both the patient and the caregiver. Treatment can be problematic because of neuroleptic sensitivity. Conventional neuroleptics should not be used in this patient population because of exacerbation of extrapyramidal symptoms (EPS). Several studies have examined the use of atypical antipsychotics (risperidone, olanzapine, and clozapine) in DLB. These results are conflicting with some literature suggesting that these agents are well tolerated and other reports that they induce worsening of EPS and possible neuroleptic malignant syndrome (Burke et al., 1998; Chacko et al., 1993; Herrmann et al., 1998; Sechi et al., 2000). A post hoc analysis of olanzapine in DLB patients suggests acceptable tolerability and efficacy when using the 5- and 10-mg doses; higher doses have not been shown to be more effective and may potentially be deleterious (Cummings et al., 2002). Quetiapine and ziprasidone have recently become available and studies using these agents are not available.

### Frontotemporal Degeneration

FTD is characterized by prominent changes in personality, impaired reasoning, and lack of social tact. Misdiagnosis is common delaying appropriate management. Treatment of FTD centers around neurotransmitter deficits involving both serotonin and dopamine. There is essentially no involvement of the cholinergic system in FTD, making use of the ChE-Is unlikely to be effective (Francis et al., 1993).

Decreased serotonin binding has been found in the frontal lobes of patients with FTD, and it has been suggested that frontal lobe dysfunction in FTD patients is linked to changes in the serotonin system (Sparks and Markesbery, 1991; Sparks et al., 1994). There have been case reports and small case series utilizing SSRIs for symptomatic relief of many of the behavioral symptoms in FTD. Fluvoxamine has been found to be effective for hyperphagia (Hope, 1991), fluoxetine for depression (Hoehn-Saric et al., 1994), and paroxetine for agitation, delusions, and depression. An open-label trial examining sertraline, fluoxetine, and paroxetine found im-

proved disinhibition, depression, hyperphagia, and compulsions (Swartz et al., 1997). Sertraline and citalopram appear to be the best tolerated, with few reported side effects (Hoehn-Saric et al., 1994; Lauterbach, 1994; Swartz et al., 1997).

Treatment using dopamine blockers and agonists is controversial. CSF studies of dopamine in patients with FTD have found that dopamine release is unaffected and others have found dopamine metabolites to be lowered (Francis et al., 1993). Bromocriptine has been found to ameliorate perseveration and improve overall functioning following traumatic brain injury involving the frontal lobes (Imamura et al., 1998; McDowell et al., 1998). Studies of the long-term use of bromocriptine and other dopamine agonists in FTD are warranted.

Aggressive behaviors can be very troublesome to families and pharmacotherapy is often required. Small doses of atypical antipsychotics have been found to be modestly effective (Perry and Miller, 2001). The anticonvulsants carbamazepine and valproate have shown some efficacy in treating agitation and aggression in a variety of syndromes and may be efficacious in FTD (Fava, 1997).

### Vascular Dementia

Treatment of VaD is based on the association of cognitive impairment with cerebral arteriosclerosis. Management of cerebrovascular risk factors is appropriate for treatment of VaD. Several population-based studies assessed the role of hypertension and its management in the prevention of cognitive impairment. Results have varied between showing a strong association between uncontrolled hypertension contributing to cognitive decline (Launer, et al., 2000; Skoog, 1999) to no association (Elias, 1995; Ferrucci, 1996; Guo, 1997).

Antiplatelet agents have proved to be effective in the secondary prevention of stroke. These agents also have been studied to determine efficacy in prevention of VaD. These agents are widely prescribed in an attempt to prevent VaD or to slow its progression; however, controlled trials are limited. There have been a few controlled trials assessing the role of aspirin in the treatment of VaD. A few studies have suggested a benefit to aspirin therapy in VaD (Henderson et al., 1997; Meyer et al., 1989), while others have been inconclusive (Williams et al., 1988).

There may be an associated cholinergic deficit in VaD, particularly in the mixed VaD–AD type, suggesting potential benefit from treatment with ChE-Is. A case series of six patients treated with donepezil demonstrated stable cognition over 6 months and caregivers reported improved patient activity, engagement, and self-care (Mendez *et al.*, 1999). Recently, a randomized, double-blind, placebo trial found galantamine improved cognition, ADLs, and behavior. Further trials are underway. (Erkinjunti *et al.*, 2002).

Although data regarding the utility of anti-platelet agents and antihypertensives in prevention or treatment of VaD are limited, these agents should continue to be prescribed to VaD patients for prevention of heart disease and stroke, where they have proven efficacy.

Other approaches to treating or preventing VaD include agents such as pentoxifylline and propentofylline. In initial studies these two agents have shown some promise for use as neuroprotective therapies (European Pentoxifylline Multi-Infarct Dementia Study Group, 1996; Marcusson *et al.*, 1997). Limited success has been seen with calcium channel antagonists (Pantoni *et al.*, 1996). Other classes of agents are being considered or investigated.

### Hydrocephalic Dementia

Hydrocephalic dementia is a relatively uncommon dementia syndrome accounting for approximately 1% of all dementia diagnoses. Treatment options for the cognitive changes secondary to hydrocephalus are limited. Shunting procedures for normal pressure hydrocephalus are believed to improve gait abnormalities; however, their utility for improving cognition has been variable. A study involving 40 patients who underwent shunting examined the prognosis of dementia. Neuropsychologic testing was performed prior to and 12 months after the shunting procedure. Cognition improved in 16 and remained the same or deteriorated in 24 (Thomsen *et al.*, 1986). A more recent study examined functional outcome including improvement in gait disturbances, ADLs, and cognition. Improvement was seen in all measures at the 3-month-postoperation evaluation; however, deterioration in all measures at 36 months was noted when compared to the 3-month data (Malm *et al.*, 2000). Shunt-related complications including malfunction, subdural hematomas, hygromas, and death occurred in 24% of patients. Further study for management of hydrocephalus and the related cognitive changes is needed.

## THERAPIES NO LONGER RECOMMENDED

### Tacrine

Tacrine (Cognex) was the first ChE-I therapy approved for the treatment of AD. It demonstrated modest benefit on cognitive and global impression scales. This agent requires slow titration over a several-week period with many patients being unable to toler-

ate the drug in "effective doses." A common side effect reported was hepatoxicity with close to 25% of patients on this agent having potentially clinically significant elevations in liver transaminase levels (Gracon *et al.*, 1998). Generally, these elevations are reversible with cessation of the drug; however, the more rapid titration, better tolerability, and lack of hepatoxicity associated with the other agents in this class, make tacrine a less than desirable choice.

### Estrogen

Epidemiologic studies suggest that estrogen hormonal therapy may improve cognition or delay the onset of AD in postmenopausal women (Henderson *et al.*, 2000; Kawas *et al.*, 1997). Controlled clinical trials examining the use of estrogen in postmenopausal women with mild to moderate AD have shown that the use of estrogen did not provide any benefit over placebo (Mulnard *et al.*, 2000). Unless clinical trials demonstrate a clear benefit of hormonal therapy in women with AD, the routine use of estrogen in this population is discouraged.

### Hydergine

Hydergine is an ergot alkaloid preparation used for several years for the treatment of memory loss. Early clinical trials suggested a very mild benefit in "senile dementia"; however standardized cognitive scales were infrequently used to allow quantitative measurements of improvement, in addition, subject selection criteria were not uniform among the studies. More recent trials examining the use of this compound in patients who meet NINCDS-ADRDA criteria for probable AD using standardized assessment scales have failed to find a significant benefit (Schneider and Olin, 1995; Thompson *et al.*, 1991). With the advent of cholinesterase inhibitor therapy, use of ergot compounds is no longer recommended.

## REFERENCES

American Psychiatric Association (APA) (1994). "Diagnostic and Statistical Manual of Mental Disorders," 4th ed. American Psychiatric Association, Washington, DC.

Bak, T., and Hodges, J. R. (1999). Cognition, language and behavior in motor neurone disease: Evidence of frontotemporal dementia. *Dement. Geriatr. Cogn. Disord.* **10**(Suppl. 1), 29–32.

Ballard, C., Holmes, C., McKeith, I., *et al.* (1999). Psychiatric morbidity in dementia with Lewy Bodies: A prospective clinical and neuropathological comparative study with Alzheimer's disease. *Am. J. Psychiatry* **156**, 1039–1045.

Ballard, C. G., O'Brien, J., Swann, A., *et al.* (2001). The natural history of psychosis and depression in dementia with Lewy bodies and Alzheimer's disease: Persistence and new cases over 1 year of follow-up. *J. Clin. Psychiatry* **62**, 46–49.

Berg, L., McKeel, D. W., Miller, J. P., *et al.* (1998). Clinicopathologica studies in cognitively healthy aging and Alzheimer's disease: Relation of histologic markers to dementia severity, age, sex, and apolipoprotein E genotype. *Arch. Neurol.* **55**, 326–335.

Blacker, D., Albert, M. S., Bassett, S. S., *et al.* (1994). Reliability and validity of NINCDS-ADRDA criteria for Alzheimer's Disease. The National Institute of Mental Health Genetic Initiative. *Arch. Neurol.* **51**, 1198–1204.

Bozeat, S., Gregory, C. A., Lambon, R., *et al.* (2001). Which neuropsychiatric and behavioural features distinguish frontal and temporal variants of frontotemporal dementia from Alzheimer's disease? *J. Neurol. Neurosurg. Psychiatry* **69**, 178–186.

Braak, H., and Braak, E. (1991). Neuropathological staging of Alzheimer-related changes. *Acta Neuropathol.* **82**, 239–259.

Burke, W. J., Pfeiffer, R. F., and McComb, R. D. (1998). Neuroleptic sensitivity to clozapine in dementia with Lewy Bodies. *J. Neuropsychiatry Clin. Neursci.* **10**, 227–229.

Byrne, E., Lennox, G., Lowe, J., *et al.* (1989). Diffuse Lewy body disease: Clinical features in 15 cases. *J. Neurol. Neurosurg. Psychiatry* **52**, 709–717.

Chacko, R. C., Hurley, R. A., and Jankovic, J. (1993). Clozapine use in diffuse Lewy body disease. *J. Neuropsychiatry Clin. Neurosci.* **5**, 206–208.

Chow, T. W., Miller, B. L., Hayashi, V. N., *et al.* (1999). Inheritance of frontotemporal dementia. *Arch. Neurol.* **56**, 817–822.

Chui, H. C., Mack, W., Jagust, W. J., *et al.* (2000). Clinical criteria for the diagnosis of vascular dementia. A multicenter study of comparability and interrater reliability. *Arch. Neurol.* **57**, 191–196.

Chung, J. A., and Cummings, J. L. (1998). Neurobehavioral and neuropsychiatric symptoms in Alzheimer's disease: Characteristics and treatment. *Neurol. Clin.* **18**, 829–846.

Clarfield, A. M. (1988). The reversible dementias: Do they reverse? *Ann. Intern. Med.* **109**, 476–486.

Corey-Bloom, J., Anand, R., Veach, J., *et al.* (1998). A randomized trial evaluating the efficacy and safety of ENA 713 (rivastigmine), a new acetylcholinesterase inhibitor, in patients with mild to moderately severe Alzheimer's disease. *Int. J. Geriatr. Psychopharmacol.* **1**, 55–65.

Cummings, J. L., Benson, D. F., LoVerme, S. (1980). Reversible dementia. *JAMA* **243**, 2424–2439.

Cummings, J. L., Benson, D. F., Hill, M. A., *et al.* (1985). Aphasia in dementia of the Alzheimer type. *Neurology* **35**, 394–397.

Cummings, J. L., Mega, M., Gray, K., *et al.* (1994). The neuropsychiatry inventory: Comprehensive assessment of psychopathology in dementia. *Neurology* **44**, 2308–2314.

Cummings. J. L. (2000a). Neuroimaging in the dementia assessment: Is it necessary? *J. Am. Geriatr. Soc.* **48**, 1345–1346.

Cummings. J. L. (2000b). Cholinesterase inhibitors: A new class of psychotropic compounds. *Am. J. Psychiatry* **157**, 4–16.

Cummings, J. L., Miller, B., Hill, M. A., and Neshkeo, R. (1987). Neuropsychiatric aspects of multi-infarct dementia and dementia of the Alzheimer type. *Arch. Neurol.* **44**, 389–393.

Cummings, J. L., Street, J., Masterman, D., *et al.* (2002). Efficacy of olanzapine in the treatment of psychosis in dementia with Lewy bodies. *Dement. Geriatr. Cog. Disord.* **13**, 67–73.

Davis, K. L., Thal, L. J., Gamzu, E. R., *et al.* (1992). A double-blind, placebo-controlled multicenter study of tacrine for Alzheimer's disease. *N. Engl. J. Med.* **327**, 1253–1259.

De Carli, C. (2001). The role of neuroimaging in dementia. *Clin. Geriatr. Med.* **17**, 255–279.

de Deyn, P., Rabheru, K., Rasmussen, A., *et al.* (1999). A randomized trial of risperidone, placebo, and haloperidol for behavioral symptoms of dementia. *Neurology* **53**, 946–955.

de Deyn, P., and Katz, I. R. (2000). Control of aggression and agitation in patients with dementia: Efficacy and safety of risperidone. *Int. J. Geriatr. Psychiatry* **15**, S14–S23.

Del Ser, T., McKeith, I., Anand, R., *et al.* (2000). Dementia with Lewy Bodies: Findings from an international multicenter study. *Int. J. Geriatr. Psychiatry* **15**, 1034–1045.

Desmond, D. W. (1996). Vascular dementia: A construct in evolution. *Cerebrovasc. Brain Metab. Rev.* **8**, 296–325.

Doody, R. S., Geldmacher, D. S., Gordon, B., Perdomo, C. A., and Pratt, R. D. (2001a). Open-label, multicenter, phase 3 extension study of the safety and efficacy of donepezil in patients with Alzheimer Disease. *Arch. Neurol.* **58**, 427–433.

Doody, R. S., Stevens, J. C., Cummings, J. L., *et al.* (2001b). Practice parameter: management of dementia (an evidence-based review). Report of the Quality Standards Subcommittee of the American Academy of Neurology. *Neurology* **56**, 1154–1166.

Elias, M. F., D'Agostino, R. B., Elias, P. K., and Wolf, P. A. (1995). Neuropsychological test performance, cognitive functioning, blood pressure, and age: The Framingham Heart Study **21**, 369–391.

Emilien, G., Beyreuther, K., Masters, C. L., *et al.* (2000). Prospects for pharmacologic intervention in Alzheimer's disease. *Arch. Neurol.* **57**, 454–459.

Erkinjunti, T., Kurz, A., Gauthier, S., Bullock, R., and Damarju, C. V. (2002). Efficacy of galantamine in probable vascular dementia and Alzheimer's disease combines with cerebrovascular disease. *Lancet* **359**, 1283–1290.

Esiri, M. M., Wilcock, G. K., and Morris, J. H. (1997). Neuropathologic assessment of the lesions of significance in vascular dementia. *J. Neurol. Neurosurg. Psychiatry* **63**, 749–753.

European Pentoxifylline Multi-Infarct Dementia Study (1996). *Eur. Neurol.* **36**, 315–321.

Farlow, M. R., Hake, A., Messina, J., Hartman, R., Veach, J., and Anand, R. (2001). Response of patients with Alzheimer disease to rivastigmine treatment is predicted by the rate of disease progression. *Arch. Neurol.* **58**, 417–422.

Fava, M. (1997). Psychopharmacologic treatment of pathologic aggression. *Psychiatr. Clin. North Am.* **20**, 427–451.

Ferrucci, L., Guralnik, J., Salive, M. E., *et al.* (1996). Cognitive impairment and risk of stroke in the older population. *J. Am. Geriatr. Soc.* **44**, 237–241.

Francis, P., Holmes, C., Webster, M., *et al.* (1993). Preliminary neurochemical findings in non-Alzheimer dementia due to lobar atrophy. *Dementia* **4**, 172–177.

Francis, P. T., Palmer, A. M., Snape, M., *et al.* (1999). The cholinergic hypothesis of Alzheimer's disease: A review of progress. *J. Neurol. Neurosurg. Psychiatry* **66**, 137–147.

Galton, C. J., Patterson, K., Hodges, J. R., *et al.* (2000). Atypical and typical presentations of Alzheimer's disease: A clinical, neuropsychological, neuroimaging, and pathological study of 13 cases. *Brain* **123**, 484–498.

Garrard, P., Perry, R., and Hodges, J. R. (1997). Disorders of semantic memory. *J. Neurol. Neurosurg. Psychiatry* **62**, 431–435.

Grundman, M., and Thal, L. J. (2000). Treatment of Alzheimer's disease. *Neurol. Clin.* **18**, 807–823.

Gracon, S. I., Knapp, M. J., Berghoff, W. G., Pierce, M., DeJong, R., Symons, J., Dombey, S. L., and Kraemer, D. (1998). Safety of tacrine: Clinical trials, treatment IND, and post marketing experience. *Alzheimer's Dis. Assoc. Disord.* **12**, 93–101.

Guo, Z., Fratiglioni, L., Winbad, B., and Viitanen, M. (1997). Blood pressure and performance on the mini-mental state examination in the very old. Cross-sectional and longitudinal data from the Kungsholmen Project. *Am. J. Epidemiol.* **145**, 1106–1113.

Hachinski, V. C., Lassen, N. A., and Marshall, J. (1974). Multi-infarct dementia. A cause of mental deteriorioration in the elderly. *Lancet* **2**, 207–210.

Hely, M. A., Reid, W. G. J., and Morris, J. G. L. (1996). Diffuse Lewy body disease: Clinical features in nine cases without co-existent Alzheimer's disease. *J. Neurol. Neurosurg. Psychiatry* **60**, 531–538.

Henderson, A. S., Jorm, A. F., Korten, A. E., *et al.* (1997). Aspirin, Anti-inflammatory drugs and risk of dementia. *Int. J. Geriatr. Psychiatry* **12**, 926–930.

Henderson, V. W., Paganini-Hill, A., Miller, B. L. *et al.* (2000). Estrogen for Alzheimer's disease in women: Randomized, double-blind, placebo-controlled trial. *Neurology* **54**, 295–301.

Herrmann, N., Rivard, M., Flynn, M., *et al.* (1998). Risperidone for the treatment of behavioral disturbances in dementia: A case series. *J. Neuropsychiatry Clin. Neurosci.* **10**, 220–223.

Hirono, N., and Cummings, J. L. (1999). Neuropsychiatric aspects of dementia with Lewy bodies. *Curr. Psychiatry Rep.* **1**, 85–92.

Hodges, J. R., Patterson, K., Oxbury, S., *et al.* (1992). Semantic dementia, progressive fluent aphasia with temporal lobe atrophy. *Brain* **115**, 1783–1806.

Hodges, J. R. (2001). Frontotemporal dementia (Pick's disease): Clinical features and assessment. *Neurology* **56**(Suppl. 4), S6–S10.

Hoehn-Saric, R., Lipsey, J. R., and McLeod, D. R. (1994). Apathy and indifference in patients on fluvoxamine and fluoxetine. *J. Clin. Psychiatry* **55**, 137–141.

Holmes, C., Cairns, N., Lantos, P., *et al.* (1999). Validity of current clinical criteria for Alzheimer's Disease, vascular dementia and dementia with Lewy Bodies. *Br. J. Psychiatry* **174**, 45–50.

Hope, R. A. (1991). Hyperphagia in dementia: Fluvoxamine takes the biscuit. *J. Neurol. Neurosurg. Psychiatry* **54**, 88.

Hutchinson, M., and Fazzini, E. (1996). Cholinesterase inhibition in Parkinson's disease. *J. Neurol. Neurosurg. Psychiatry* **61**, 324–325.

Hutton, M. (2001). Missense and splice site mutations in tau associated with FTDP-17: Multiple pathogenic mechanisms. *Neurology* **56**(11), S21–S25.

Imamura, T., Takanashi, M., Hattori, N., *et al.* (1998). Bromocriptine treatment for perseveration in demented patients. *Alzheimer Dis. Assoc. Disord.* **12**, 109–113.

Janus, C., Pearson, J., McLaurin, J., *et al.* (2000). AB peptide immunization reduces behavioral impairment and plaques in a model of Alzheimer's disease. *Nature* **408**, 979–982.

Joffr, C., Graham, J., and Rockwood, K. (2000). Qualitative analysis of the clinician interview-based impression of change (plus): Methodological issues and implications for clinical research. *Int. Psychogeriatr.* **12**, 403–413.

Johnson, J. K., Head, E., Kim, R., *et al.* (1999). Clinical and pathological evidence for a frontal variant of Alzheimer disease. *Arch. Neurol.* **56**, 1233–1239.

Jost, B. C., and Grossberg, G. T. (1995). The Natural History of Alzheimer's Disease: A brain bank Study. *J. Am. Geriatr. Soc.* **43**, 1248–1255.

Kanowski, S., Herrmann, W. M., Stephan, K., Wierich, W., and Horr, R. (1996). Proof of efficacy of the ginkgo biloba special extract Egb 761 in outpatients suffering from mild to moderate primary degenerative dementia of the Alzheimer type or multi-infarct dementia. *Pharmacopsychiatry* **29**, 47–56.

Katz, I. R., Jeste, D. V., Mintzer, J. E., *et al.* (1999). Comparison of risperidone and placebo for psychosis and behavioral disturbances associated with dementia: A randomized, double-blind trial. Risperidone Study Group. *J. Clin. Psychiatry* **60**, 107–115.

Kaufer, D. I., Catt, K. E., DeKosky, S. T., *et al.* (1998). Dementia with Lewy bodies: Response of delirium-like features to donepezil. *Neurology* **51**, 1512.

Kawas, C., Resnick, S., and Morrison, A. (1997). A prospective study of estrogen replacement therapy and the risk of developing Alzheimer's disease: The Baltimore Longitudinal Study of Aging. *Neurology* **48**, 1517–1521.

Kertesz, A., and Clydesdale, S. (1994). Neuropsychological deficits in vascular dementia vs. Alzheimer's disease. Frontal lobe deficits prominent in vascular dementia. *Arch. Neurol.* **S1**, 1226–1231.

Knopman, D. S. (2001). Practice parameter: Diagnosis of dementia (an evidence-based review). Report of the Quality Standards Subcommittee of the American Academy of Neurology. *Neurology* **56**, 1143–1153.

Kosaka, K., Yoshimura, M., Ikeda, K. (1984). Diffuse type of Lewy body disease: Progressive Dementia with abundant cortical Lewy bodies and senile changes of varying degree. A new disease? *Clin. Neuropathol.* **3**, 185–192.

Lanctot, K., and Herrmann, N. (2000). Donepezil for behavioral disorders associated with Lewy bodies. A case series. *Int. J. Geriatr. Psychiatry* **15**, 338–345.

Langlais, P. J., Thal, L., Hansen, L., *et al.* (1993). Neurotransmitters in basal ganglia and cortex of Alzheimer's disease with and without Lewy bodies. *Neurology* **43**, 1927–1934.

Launer, L. J., Ross, G. W., Pehovitch, H., Foley, D., White, L. R., and Haulick, R. J. (2000). Midlife blood pressure and dementia: The Honolulu–Asia aging study. *Neurobiol. Aging* **21**, 49–55.

Lauterbach, E. C. (1994). Reversible intermittent rhythmic myoclonus with Fluoxetine in presumed Pick's disease. *Mov. Disord.* **9**, 343–346.

Le Bars, P. L., Katz, M. M., Berman, N., Itil, T. M., Freedman, A. M., *et al.* (1997). A placebo-controlled, double-blind, randomized trial of an extract of ginkgo biloba for dementia. *JAMA* **278**, 1327–1332.

Li, F., Iseki, E., Kato, M., *et al.* (2000). An autopsy case of Alzheimer's disease presenting with primary progressive aphasia: A clinico-pathological and immunohistochemistry study. *Neuropathology* **20**, 239–245.

Lopez, O. L., Litvan, I., Catt, K. E., *et al.* (1999). Accuracy of four clinical diagnostic criteia for the diagnosis of neurodegenerative dementias. *Neurology* **53**, 1292–1299.

Lopez, O. L., Becker, J. T., Klunk, W., *et al.* (2000). Research evaluation and diagnosis of probable Alzheimer's disease over the last two decades. *Neurology* **55**, 1854–1862.

Louis, E. D., Goldman, J. E., Powers, J. M., *et al.* (1995). Parkinsonian features of eight pathologically diagnosed cases of diffuse Lewy body disease. *Movement Disord.* **10**, 188–194.

Louis, E. D., Klatka, L., Liu, Y., *et al.* (1997). Comparison of extrapyramidal features in 31 pathologically confirmed cases of Parkinson's disease. *Neurology* **48**, 376–380.

Love, S., Barber, R., and Wilcock, G. K. (2000). Neuronal death in brain infarcts in man. *Neuropathol. Appl. Neurobiol.* **26**, 55–66.

Luis, C. A., Barker, W. W., Gajaraj, K., *et al.* (1997). Sensitivity and specificity of three clnical criteria for dementia with Lewy bodies in an autopsy verified sample. *Int. J. Geriatr. Psychiatry* **14**, 526–533.

The Lund and Manchester Groups (1994). Clinical and neuropathological criteria for frontotemporal dementia. *J. Neurol. Neurosurg. Psychiatry* **57**, 416–418.

Lykestos, C. G., Sheppard, J., Steele, C. D., *et al.* (2000). Randomized, placebo-controlled, double-blind clinical trial of sertraline in the treatment of depression complicating Alzheimer's disease: Initial results from the depression in Alzheimer's disease study. *Am. J. Psychiatry* **157**, 1686–1689.

Malm, J., Kristensen, B., Koskinen, L. O., *et al.* (2000). Three-year survival and functinal outcome of patients with idiopathic adult hydrocephalus syndrome. *Neurology* **55**, 576–578.

Marcusson, J., Rother, M., Kittner, B., Rossner, M., Smith, R. J., Babic, T., Moller, H. J., and Labs, K. H. (1997). A 12-month randomized, placebo-controlled trial of propentofylline in patients with dementia according to DSM-IIIR. The European Propentofylline Study Group. *Dement. Geriatr. Cogn. Disord.* **8**, 320–328.

Massoud, F., Devi, G., and Moroney, J. T. (2000). The role of routine laboratory studies and neuroimaging in the diagnosis of dementia: A clinicopathological study. *J. Am. Geriatr. Soc.* 48, 1204–1210.

Matthews, M. K. Jr. (1998). Association of Ginkgo biloba with intracerebral hemorrhage. *Neurology* 50, 1933–1934.

Mayeux, R., Saunders, A. M., Shea, S., *et al.* (1998). Utility of the apolipoprotein E genotype in the diagnosis of Alzheimer's disease. Alzheimer's Disease Centers Consortium on apolipoprotein E and Alzheimer's disease. *N. Engl. J. Med.* 338, 506–511.

McDowell, S., Whyte, J., and D'Esposito, M. (1998). Differential effect of a domaninergic agonist on prefrontal function in traumatic brain injury patients. *Brain* 121, 1155–1164.

McKeith, I., Galasko, D., Kosaka, K., *et al.* (1996). Consensus guidelines for the clinical and pathologic diagnosis of dementia with Lewy Bodies (DLB): Report of the consortium on DLB international workshop. *Neurology* 47, 1113–1124.

McKeith, I., O'Brien, J., and Ballard, C. (1999a). Diagnosing dementia with Lewy Bodies. *Lancet* 354, 1227–1228.

McKeith, I., Perry, E., and Perry, R. (1999b). Report of the second dementia with Lewy body international workshop: Diagnosis and treatment. *Neurology* 53, 902–905.

McKeith, I., Del Ser, T., Spano, P., *et al.* (2000). Efficacy of rivastigmine in dementia with Lewy bodies: A randomized, double-blind, palcebo-controlled international study. *Lancet* 356, 2031–2036.

McKhann, G., Drachman, D., Folstein, M., *et al.* (1984). Clinical diagnosis of Alzheimer's Disease: Report of the NINCDS-ADRDA Work Group under the auspices of Department of Health and Human services Task Force on Alzheimer's Disease. *Neurology* 34, 939–944.

McPherson, S. E., and Cummings, J. L. (1996). Neuropsychological aspects of vascular dementia. *Brain Cog.* 31, 269–282.

Mega, M. S., Masterman, D. L., and Benson, D. F. (1996). Dementia with Lewy bodies: Reliability and validity of clinical and pathologic criteria. *Neurology* 47, 1403–1409.

Mendez, M. F., Younesi, F. L., and Perryman, K. M. (1999). Use of donepesil for vascular dementia: Preliminary clinical experience. *J. Neuropsychiatry Clin. Neurosci.* 11, 268–269.

Meyer, J. S., Rogers, R. L., Lotfi, J., *et al.* (1989). Randomized clinical trial of daily aspirin therapy in Multi-infarct Dementia, A pilot study. *J. Am. Geriatr. Soc.* 37, 549–555.

Miller, B. L., Darby, A., Benson, D. F., *et al.* (1997). Aggressive, socially disruptive and antisocial behaviour associated with frontotemporal dementia. *Br. J. Psychiatry* 170, 150–155.

Morgan, D., Diamond, D. M., Gottschall, P. E., *et al.* (2000). AB peptide vaccination prevents memory loss in an animal model of Alzheimer's disease. *Nature* 408, 982–985.

Morris, S. K., Olichney, J. M., and Corey-Bloom, J. (2001). Psychosis in dementia with Lewy Bodies. *Sem. Clin. Neuropsychiatry* 70, 157–164.

Mulnard, R. A., Cotman, C. W., Kawas, C., *et al.* (2000). Estrogen replacement therapy for treatment of mild to moderate Alzheimer's disease: A randomized controlled trial: Alzheimer's Disease Cooperative Study. *J. Am. Med. Assoc.* 283, 1007–1015.

Neary, D., Snowden, J. S., Gustafson, L., *et al.* (1998). Frontotemporal lobar degeneration: A consensus on clinical diagnostic criteria. *Neurology* 51, 1546–1554.

Nyenhuis, D. L., and Gorelick, P. B. (1998). Vascular dementia: A contemporary Review of epidemiology, diagnosis, prevention, and treatment. *J. Am. Geriatr. Soc.* 46, 1437–1448.

Nyth, A., gottfires, C., Lyby, K., *et al.* (1992). A controlled multicenter clinical study of citalopram and placebo in elderly depressed patients with and without concomitant dementia. *Acta Psychiatr. Scand.* 86, 138–145.

Olichney, J. M., Galasko, D., Salmon, D. P., *et al.* (1998). Cognitive decline is faster in Lewy body variant than in Alzheimer's disease. *Neurology* 51, 351–357.

Pantoni, L., Carosi, M., Amigoni, S., Mascalchi, M., and Inzitari, D. (1996). A preliminary open trial with nimodipine in patients with cognitive impairment and leukoariosis. *Clin. Neuropharmacol.* 19, 497–506.

Perry, E. K., Haroutunian, V., Perry, R. H., *et al.* (1994). Neocortical cholinergic activities differentiate Lewy body dementia from classical Alzheimer's disease. *NeuroRreport* 5, 747–749.

Perry, R. J., and Miller, B. L. (2001). Behavior and treatment in frontotemporal dementia. *Neurology* 56(Suppl. 4), S46–S51.

Poorkaj, P., Bird, T. D., Wijsman, E., *et al.* (1998). Tau is a candidate gene for chromosome 17 frontotemporal dementia. *Ann. Neurol.* 43, 815–825.

Rakowicz, Z., and Hodges, J. R. (1998). Dementia and aphasia in motor neurone disease: An under recognized association. *J. Neurol. Neurosurg. Psychiatry* 65, 881–889.

Raskind, M. A. (1999). Evaluation and management of aggressive behavior in the elderly demented patient. *J. Clin. Psychiatry* 60, 45–49.

Raskind, M. A., Peskind, E. R., Wessel, T., *et al.* (2000). Galantamine USA-1 Study Group. Galantamine in AD: A 6-month randomized placebo controlled trial with a 6-month extension. *Neurology* 54, 2261–2268.

Riley, D. E. (1990). Cortical-basal ganglionic degeneration. *Neurology* 40, 1203–1212.

Rogers, S. L., Farlow, M. R., and Doody, R. S. (1998a). A 24-week, double-blind, placebo-controlled trial of donepezil in patients with Alzheimer's disease. *Neurology* 50, 136–145.

Rogers, S. L., and Friedhoff, L. T. (1998b). Long-term efficacy and safety of donepezil in the treatment of Alzheimer's disease: An interim analysis of the results of a US multicentre open label extension study. *Eur. Neuropsychopharmacol.* 8, 67–75.

Rojas-Fernandez, C. (2001). Successful use of donepezil for the treatment of dementia with Lewy bodies. *Ann. Pharmacother.* 35, 202–205.

Roman, G. C., Tatemichi, T. K., Erkinjuntti, T., *et al.* (1993). Vascular dementia: diagnostic criteria for research studies: Report of the NINCDS-AIREN international workshop. *Neurology* 43, 250–260.

Rosenberg, G. A., Sullivan, N., and Esiri, M. M. (2001). White matter damage is associated with matrix metalloproteinases in vascular dementia. *Stroke* 32, 1162–1168.

Rowin, J., and Lewis, S. L. (1996). Spontaneous bilateral subdural hematomas associated with chronic Ginkgo biloba ingestion. *Neurology* 46, 1775–1776.

Sano, M., Ernesto, C., Thomas, R. G., *et al.* (1997). A controlled trial of selegiline, alpha-tocopherol, or both as treatment for Alzheimer's disease. *N. Engl. J. Med.* 336, 1216–1222.

Schenk, D., Barbour, R., Dunn, W., *et al.* (1999). Immunization with amyloid-beta attenuates Alzheimer-disease-like pathology in the PDAPP mouse. *Nature* 400, 173–177.

Schneider, L. S., and Olin, J. T. (1995). Overview of clinical trials of hydergine in dementia. *Arch. Neurol.* 51, 787–798.

Schoenberg, B. S. (1986). Epidemiology of Alzheimer's disease and other dementing illnesses. *J. Chronic Dis.* 39, 1095–1104.

Sechi, G., Agnetti, V., Masuri, R., *et al.* (2000). Risperidone, neuroleptic malignant syndrome and proable dementia with Lewy bodies. *Prog. Neuro-Psychopharmacol. Biol.* 24, 1043–1051.

Shea, C., MacKnight, C., and Rockwood, K. (1998). Donepezil for treatment of dementia with Lewy bodies: A case series of nine patients. *Int. Psychogeriatr.* 10, 229–238.

Shiozaki, K., Iseki, E., Uchiyama, H., *et al.* (1999). Alterations of muscarinin acetylcholine receptor subtypes in diffuse Lewy body disease: Relation to Alzheimer's disease. *J. Neurol. Neurosurg. Psychiatry* 67, 209–213.

Skjerve, A., and Nygaard, H. (2000). Improvement in sundowning in dementia with Lewy bodies after treatment with donepezil. *Int. J. Geriatry. Psychiatry* **15**, 1147–1151.

Skoog, I., Gorelick, P. B., Winblad, B., *et al.* (1999). Prevention of vascular dementia. *Alz. Dis. Assoc. Disord.* **13**(Suppl. 13), S131–S139.

Sparks, D. L., and Markesbery, W. R. (1991). Altered serotonergic and cholinergic synaptic markers in Pick's disease. *Arch. Neurol.* **48**, 796–799.

Sparks, D. L., Danner, F. W., Davis, D. G., *et al.* (1994). Neurochemical and histopathologic alterations characteristic of Pick's disease in a non-demented individual. *J. Neuropathol. Exp. Neurol.* **53**, 37–42.

Stern, Y., Albert, S., Tang, M. X., *et al.* (1999). Rate of memory decline in AD is related to education and occupation: Cognitive reserve? *Neurology* **53**, 1942–1947.

Street, J., Mitan, S., Tamura, R., *et al.* (1998). Olanzapine in the treatment of psychosis and behavioral disturbances associated with Alzheimer's disease. *Eur. J. Neurol.* **5**, S39.

Sultzer, D. L., Levin, H. S., Mahler, M. E., and Cummings, J. L. (1993). A comparison of psychiatric symptoms in vascular dementia and Alzheimer's disease. *Am. J. Psychiatry.* **150**, 1806–1212.

Sultzer, D. L., Gray, K. F., Gunay, I., *et al.* (1997). A double-blind comparison of trazodone and haloperidol for treatment of agitation in patients with dementia. *Am. J. Geriatr. Psychiatry* **5**, 60–69.

Swartz, J. R., Miller, B. L., Lesser, I. M., and Darby, A. L. (1997). Frontotemporal dementia: Treatment response to serotonin selective reuptake inhibitors. *J. Clin. Psychiatry* **58**, 212–216.

Swartz, J. R., Miller, B. L., Lesser, I. M., *et al.* (1997). Frontotemporal dementia: Treatment response to serotonin selective reuptake inhibitors. *J. Clin. Psychiatry* **58**, 212–216.

Talwalker, S., Overall, J. E., Srirama, M. K., *et al.* (1996). Cardinal features of cognitive dysfunction in Alzheimer's disease: A factor analytic study of the Alzheimer's Disease Assessment Scale. *Neurology* **9**, 39–46.

Tarango, F., Lykestos, C., Mangone, C., *et al.* (1997). A double-blind, randomized, fixed-dose trial of fluoxetine vs. amitryptyline in the treatment of major depression complicating Alzheimer's disease. *Psychosomatics* **38**, 246–252.

Tariot, P. N., Erb, R., Leibovici, A., *et al.* (1994). Carbamazepine treatment of agitation in nursing home patients with dementia: A preliminary study. *J. Am. Geriat. Soc.* **42**, 1160–1166.

Tariot, P. N., Erb, R., Podgorski, C. A., *et al.* (1998). Efficacy and tolerability of carbamazepine for agitation and aggression in dementia. *Am. J. Psychiatry* **155**, 54–61.

Tariot, P. N., Ryan, J. M., Porsteinsson, A. P., *et al.* (2001). Alzheimer's Disease and dementia: Pharmacologic therapy for behavioral symptoms of Alzheimer's disease. *Clin. Geriatr. Med.* **17**, 359–376.

Teri, L., Logsdon, R. G., Peskind, E., *et al.* (2000). Treatment of agitation in AD. A randomized, placebo-controlled clinical trial. *Neurology* **55**, 1271–1278.

Thompsen, T. L. 2nd, Filley, C. M., Mitchell, W. D., *et al.* (1990). Lack of efficacy of hydergine in patients with Alzheimer's disease. *N. Engl. J. Med.* **323**, 445–458.

Thomsen, A. M., Brgesen, S. E., Gjerris, F., *et al.* (1986). Prognosis of dementia in normal-pressure hydrocephalus after a shunt operation. *Ann. Neurol.* **20**, 304–310.

Ved, H. S., Koenig, M. L., Dave, J. R., and Doctor, B. P. (1997). Huperzine A, a potential therapeutic Agent for dementia, reduces neuronal cell death caused by glutamate. *Neuroreport* **8**, 963–968.

Weytingh, M. D., Bossuyt, P. M., and van Crevel, H. (1995). Reversible dementia: More than 10% or less than 1%? A quantitative review. *J. Neurology* **242**, 466–471.

White, L., Petrovitch, H., Ross, G. W., *et al.* (1996). Prevalence of dementia in older Japanese-American men in Hawaii: The Honolulu-Asia Aging Study. *JAMA* **276**, 955–960.

Williams, P. S., Rands, G., Orrel, M., *et al.* (1988). Aspirin for vascular dementia. *Cochrane Libr.* **2001**, 2.

CHAPTER 30
# Acute Ischemic Stroke

Louis R. Caplan and Werner Hacke

## CLINICAL ASPECTS

### Definitions and Subtype Mechanisms

In acute ischemic stroke a portion of the brain is acutely deprived of sufficient blood flow. The brain requires about 50 cc/mL/min of blood that contains an adequate level of oxygen and sugar to function normally. The need for blood supply differs largely between white and gray matter, and is also different in different parts of the brain. Ischemia can be subdivided into three different mechanisms: decreased systemic perfusion, thrombosis, and embolism. In the category of decreased systemic perfusion diminished flow to brain tissue is due to low systemic perfusion pressure. The most common causes are cardiac pump failure (most often due to myocardial infarction or arrhythmia) and systemic hypotension (due to blood loss or hypovolemia). These general causes of hypoperfusion conditions lead to global brain ischemia especially in border zone or so-called watershed regions at the periphery of the major vascular-supply territories, the cortical layer of the hemispheres, and the basal ganglia and can cause selective neuronal ischemia. Systemic hypoperfusion does not cause a focal acute ischemic stroke and will not be considered further in this chapter.

By convention, thrombosis refers to blockage of blood flow due to a localized *in situ* occlusive process within one or more blood vessels. The lumen of the artery is narrowed or occluded by an alteration in the vessel wall or by superimposed clot formation. The commonest vascular pathology is atherosclerosis, in which fibrous and muscular tissues overgrow in the subintima, and fatty materials form plaques that can encroach on the lumen. Next, platelets adhere to plaque crevices and form clumps that serve as nidi for the deposition of fibrin, thrombin, and clot (Caplan, 2000). Plaque rupture allowing contact of the gruel of an atherosclerotic plaque with tissue factor and other substances within the blood can activate the coagulation cascade, leading to formation of an occlusive thrombus. Atherosclerosis affects chiefly the larger extracranial and intracranial vessels. Occasionally, clot forms within the lumen because of a primary hematological problem, such as polycythemia, thrombocytosis, or a systemic hypercoagulable state. The smaller penetrating intracranial vessels are more often damaged by hypertension than by atherosclerotic processes. In such cases, increased arterial tension leads to hypertrophy of the media and deposition of fibrinoid material into the vessel wall, a process that gradually encroaches on the already small lumen, a process sometimes referred to as lipohyalinosis. Atheromatous plaques, often referred to as micoatheromas, can obstruct the orifices of penetrating arteries.

Less common vascular pathologies leading to obstruction include fibromuscular dysplasia, an overgrowth of medial and intimal elements that compromises vessel contractility and luminal size; arteritis, especially of the Takayasu or giant-cell type; dissection of the vessel wall, often with luminal or extraluminal clot temporarily obstructing the vessel; and hemorrhage into a plaque, leading to acute or chronic compromise of the vascular lumen. At times, the focal vascular abnormality is a functional change in the contractility of blood vessels, a so-called vasospasm. Intense focal vasoconstriction can lead to decreased blood flow and thrombosis. Dilation of blood vessels also alters local blood flow, and clots often form in dilated segments.

In embolism, material formed elsewhere within the vascular system lodges in a vessel and blocks blood flow in the recipient artery. In contrast to thrombosis, embolic luminal blockage is not due to a localized process originating within the blocked vessel. The material arises proximally, most commonly from the heart; from major arteries such as the aorta, carotid, and vertebral arteries; and from systemic veins. Brain embolism and its treatment is discussed in Chapter 33.

All three mechanisms of ischemia lead to temporary or permanent tissue injury. Permanent injury is termed infarction. Capillaries or other vessels within the ischemic tissue may also be injured, so that reperfusion can lead to leakage of blood into the ischemic tissue, a hemorrhagic infarct, or a parenchymatous hematoma within the region of infarction. The extent of brain damage depends on the location and duration of the poor perfusion and the ability of collateral vessels to perfuse the tissues at risk.

### Clinical Diagnosis

The clinician attempts to answer the "what" (the stroke mechanism subtype) and the "where" (the anatomy of the brain and vascular lesions) questions at the bedside. Different clinical data are used to answer these two quite different questions. In determining stroke mechanism, the following clinical bedside data available from the history are helpful: (1) ecology—the past and present personal and family illnesses of the patient; (2) the presence and nature of past strokes or transient ischemic attacks (TIAs); (3) the time of onset of the symptoms and activity at the time of onset; (4) the temporal course and progression of the neurological symptoms; (5) accompanying symptoms such as headache, vomiting, and decreased level of consciousness.

Findings from the general physical examination may uncover disorders not known from the history and so add to the data used for diagnosing stroke mechanism. Elevated blood pressure, cardiac enlargement or murmurs, and vascular bruits are examples of physical findings that influence identification of the stroke mechanism.

Diagnosis of stroke location is made using different data such as analysis of the neurological symptoms and their distribution and the findings on neurological examination. Clinicians should estimate the probability of various vascular mechanisms and the location of the likely brain and vascular lesions. These hypotheses can then be explored by imaging and other laboratory diagnostic techniques.

### Laboratory and Imaging Diagnosis

During the past two decades there has been a revolution in technology able to define quickly, safely, and accurately stroke pathophysoiology and the cardiac and cerebral vascular lesions that cause acute strokes in individuals patients. Advanced brain imaging with computed tomography (CT); magnetic resonance imaging (MRI); and newer MR modalities including fluid attenuating inversion recovery (FLAIR) images; diffusion-weighted (dwMRI), perfusion-weighted (pwMRI), and functional MRI using the BOLD-effect; and MR spectroscopy show clinicians the localization, severity, and potential reversibility of ischemia. Vascular lesions can be defined using spiral CT angiography, MR angiography, and extracranial and transcranial ultrasound. Cardiac and aortic sources of embolic strokes are now better studied using transesophageal echocardiography. More sophisticated hematological testing now gives new insights into the role of altered coagulability in causing or contributing to thrombo-embolism. Clinicians can now recognize the key important data elements needed to logically choose treatment for patients with acute brain ischemia: (1) the nature, location, and severity of cardiac and cerebrovascular lesions; (2) the mechanism by which these lesions cause ischemia–hypoperfusion? embolism? functional changes such as vasoconstriction? (3) the cellular and serological components of the blood, that relate to coagulability, viscosity, and blood flow; and (4) the state of the brain—normal, reversibly ischemic ("stunned"), or infarcted.

All patients suspected of having acute brain ischemia should have brain and vascular studies and hematological screening tests (complete blood count, platelet count, prothrombin time, activated partial thromboplastin time (APTT)). Preferably the imaging should be performed on one machine: CT scan and CTA, or MRI and MRA. Neck and transcranial ultrasound could be combined with CT or MRI to yield both brain images and vascular data.

With these diagnostic advances have come new treatments, new ideas about treatment, and more and new information about old treatments.

## NATURAL COURSE

Brain ischemia can develop suddenly and the deficit be maximal at or very near the onset of symptoms. This clinical course is typical of brain embolism although in about 20% of patients there may be further fluctuations or stepwise worsenings or improvements within the 24–48 h after onset (Caplan, 1999). In some patients with brain embolism there is sudden amelioration of a severe clinical deficit, the so-called "spectacular shrinking deficit." Patients with *in situ* occlusive thrombosis often have had TIAs affecting the same vascular territory in the hours, days, weeks, and months preceding the acute brain ischemic event. The neurological symptoms and signs may fluctuate between normal and abnormal and by severity of abnormality. There may be stepwise or gradual increases in neurological symptoms and signs. The clinical deficit usually evolves during a longer period of time in patients with *in situ* occlusion of the large extracranial and intracranial arteries than in those patients who have penetrating artery occlusive disease.

In most patients progression, if it occurs, develops within the first 3–6 h after stroke onset. Progression of ischemia is much less common after the first day. However, some patients with penetrating artery disease who develop lacunar infarcts worsen gradually over 3–7 days. Brain edema, bleeding into areas of infarction, and adverse systemic factors such as infection and hypovolemia can lead to worsening during the days and weeks after stroke and should not be confused with recurrent stroke. The occurrence of subsequent strokes depends heavily on the stroke mechanism and the presence and nature of the causative cardiac, cerebrovascular, and hematological causes of the acute ischemic stroke (Bogousslavsky and Yamamoto, 1998).

## PRINCIPLES OF THERAPY

### General Aims of Treatment

There are four general goals of treatment of patients with acute brain ischemia: (1) achieve the best physiological state possible for the patient, (2) minimize brain infarction during the acute period, (3) prevent brain and systemic complications, and (4) begin preventive strategies to prevent subsequent strokes and other vascular lesions.

Stroke patients often are affected by concurrent diseases and conditions. Diabetes and fever are frequent and need to be corrected. There may be cardiac failure, fluid deficiency, pre-existing pulmonary disease with low $O_2$ saturation, that need treatment and may, if untreated, worsen the patients prognosis.

Ischemia by definition means inadequate delivery of blood containing required nutrients. Maximizing blood flow to ischemic zones is clearly important. Clinicians should ask themselves a number of questions related to minimizing acute brain infarction. Can areas of vascular blockage in large arteries be opened or circumvented by medical or surgical treatments (reperfusion)? Can local perfusion through the microcirculation supplying the ischemic zone be improved? Occlusive thrombosis and thromboembolism are important in the great majority of patients with acute ischemic stroke. Can the coagulation system be altered to diminish the development of white platelet–fibrin clots and red thrombin-dependent clots? Metabolic changes within the ischemic zone are important in causing cell death. Can the brain be made more resistant to ischemia by manipulating its physical–chemical environment (neuroprotection)?

Edema and raised ICP can potentiate nerve-cell damage. Can they be controlled?

Sometimes, the brain infarct that occurs during a stroke is not the only medical problem that patients, families, and doctors must battle. Strokes, like many other serious medical illnesses, can be followed by a host of other problems, usually referred to as secondary complications. These complications can sometimes cause neurological deterioration; in other instances, patients feel worse, and the deterioration is falsely attributed to worsening of the stroke. Most of the complications that occur after stroke are medical and not neurological. Complications are very common. Complications may occur during the hospitalization for acute stroke or may develop during rehabilitation and neurological recovery.

Complications can be very serious and can cause death. Brain edema, cardiac abnormalities, and pulmonary embolism dominate during the first week. Pneumonia, urinary tract infections, bedsores, phlebothrombosis and pulmonary embolism, gastrointestinal bleeding, contractures, falls, osteopenia, and depression do occur during the first week but continue to be problems during recovery and even after patients have left the hospital. Randomized trials, analyses, and meta-analyses have all shown that units dedicated solely to the care of stroke patients decrease mortality and morbidity among stroke patients. One very important function of stroke units and stroke teams is to systematically pursue measures to monitor complications and to prevent their occurrence.

Stroke is most often caused by a systemic disorder, atherosclerosis. Cerebrovascular lesions are often accompanied by coronary artery and peripheral limb artery athero-occlusive lesions. Preventive strategies should be begun during the first week to recognize and treat risk factors that are likely to cause future heart and brain ischemia and brain hemorrhage.

### Reperfusion

#### Thrombolysis

Patients with acute ischemic stroke, especially for those who present within the first hours after the onset of neurological symptoms, should be evaluated for application of thrombolytic agents in countries where these agents are approved for use in stroke patients. The first experiments with thrombolytic agents in ischemic stroke were conducted in the 1960s using streptokinase (Sloan, 1987). Meyer *et al.* randomized 73 patients with worsening stroke to anticoagulants versus intravenous streptokinase. The time of stroke onset was within 3 days for either arm of the study. Not surprisingly 10 of the enrolled patients died and some of the patients had significant brain hemorrhages (del Zoppo, 1988; Meyer et al., 1964, 1965). Thrombolytic agents were introduced again in the 1980s for the treatment of coronary artery disease. Inspired by encouraging results in cardiovascular disease, pilot trials using lytic agents were planned and performed (del Zoppo, 1988; Pessin et al., 1995; Sloan, 1987). Later, three trials of intravenous streptokinase were all stopped due to unacceptably high intracerebral hemorrhage rates and excess mortality (Donan et al., 1995; Multicenter Acute Stroke Trial, 1995, 1996). The NINDS trial was a positive, randomized-controlled trial for the intravenous use of rt-PA within 3 h of symptom onset. In addition, there is one positive trial for intra-arterial use of pro-urokinase within 6 h. Intravenous rt-PA is the only route of thrombolysis now approved in the United States, Canada, and Germany for the treatment of acute ischemic stroke. There are now many reports of the effectiveness and risks of intravenous (Albers et al., 2000; Anderson et al., 1998; Chiu et al., 1997; Clark et al., 1999; Cruz-Flores et al., 1998; Dafer et al., 1998; del Zoppo et al., 1992, 1998; Grond et al., 1998a,b; Hacke et al., 1995, 1998; Hanson et al., 1998; Jansen et al., 1986; Mori et al., 1992; Multicenter Acute Stroke Trial—Italy, 1995; Multicenter Acute Stroke Trial—Europe, 1996; National Institute of Neurological Disorders and Stroke, 1995; Steiner et al., 1998; Tanne et al., 2000; Trouillas et al., 1998; von Kummer and Hacke, 1992; von Kummer et al., 1991; Wang et al., 2000; Wolpert et al., 1993; Yamaguchi et al., 1993, 1995), intra-arterial (Barnwell et al., 1994; Becker et al., 1996; Berg-Dammer et al., 1996; Casto et al., 1996; Cross et al., 1997; del Zoppo et al., 1998a,b; Furlan et al., 1999; Gonner et al., 1998; Hacke et al., 1998; Matsumoto and Satoh, 1991; Mitchell et al., 1997; Mobius et al., 1991; Mori et al., 1988, 1991; Suarez et al., 1999; Theron et al., 1989; Wijdicks et al., 1997; Zeumer et al., 1983, 1993), and combined intravenous and intra-arterial (Lewandowsky et al., 1999) thrombolysis used to treat acute ischemic stroke patients.

**Intravenous Thrombolysis.** The first reported large multicenter randomized trial of intravenous thrombolysis was the European Cooperative Acute Stroke Study (ECASS) (Hacke et al., 1995). In the ECASS, 620 patients with acute hemispheral strokes were recruited in 75 hospitals in 14 European countries. A total of 313 patients were randomized to receive rt-PA (1.1 mg/kg) and 307 patients were randomized to the placebo group. Treatment was given within 6 h of the onset of symptoms of brain ischemia. Patients with hemorrhage or major early infarct signs (diffuse hemispheral swelling, parenchymal hypodensity, effacement of cerebral sulci in more than one-third of the MCA territory) on the initial CT scans were excluded. CT scans were read on site. An independent blinded CT scan reading panel then retrospectively reviewed the CT scans and determined protocol violations of the CT scan entry criteria. There were many protocol deviations in this study, mostly because of failure at local centers to recognize CT abnormalities that should have excluded patients. One hundred nine patients were excluded after review at the coordinating center, 52 of them because of the presence of extended early infarct signs on CT; 66 patients in the rt-PA group and 43 in the placebo-treated group had to be excluded. Analysis of the results involved both an intention-to-treat analysis and analyzing only those patients in the target population (per-protocol-analysis, that is those patients who did not have protocol violations). In the target population, there was a significantly better outcome in the rt-PA-treated patients as measured by the modified Rankin Scale scores at 90 days. Stay in the hospital was significantly shorter in rt-PA-treated patients and neurological outcome was better. Intracerebral hemorrhages and mortality were higher in the rt-PA treated patients but these differences were not statistically significant. Large parenchymal hematomas were more often found in patients treated with rt-PA. Patients treated within 3 h did better after rt-PA than controls and those treated with rt-PA between 3 and 6 h (Steiner et al., 1998).

The ECASS study showed that treatment of patients with large early infarct signs on CT scan (exceeding 33% of the MCA territory) was particularly dangerous. Initial analysis of CT scans at local hospitals was often unreliable. Two hemorrhages and many early infarcts were missed by local physicians. The mortality and brain hemorrhage rate among patients with protocol violations randomized to the rt-PA group was extremely high, 33.3 and 40%, respectively. Among 52 patients included in the study despite major early infarct signs, 40% died. Among the 31 patients with major early infarct signs on CT treated with rt-PA, 15 (48.4%) died, mostly during the first week (Steiner et al., 1998).

The NINDS trial was the next large randomized stroke trial to be published and differed in the following ways: a strict 3-h window for treatment, using 0.9 mg/kg of rt-PA, a smaller dose than used in ECASS, tighter guidelines for control of hypertension, and the lack of exclusion of patients due to large regions of early brain ischemia seen on CT scan (National Institute of Neurological Disorders, 1995). As in ECASS, no vascular imaging was required for entrance into the study. The trial was divided into two parts. Part I was rt-PA vs placebo to determine whether treatment with rt-PA would result in an improvement of four points on the NIH stroke scale within 24 h. Part II was rt-PA versus placebo using clinical scales measured at 3 months after treatment. No statistical significance between medication and placebo was found in the first part of the study.

The results in Part II showed a statically significant difference between the rt-PA and the placebo groups. There was a 1.7 odds ratio (95% confidence 1.2–2.6) for a favorable outcome at 3 months in the rt-PA group (National Institute of Neurological Disorders, 1995). Mortality at 3 months in the rt-PA group was 17%, while it was 21% in the placebo group. Patients in the rt-PA group were 30% more likely to have minor or no disability at 3 months. Symptomatic hemorrhages were more common in the rt-PA group 6.4% vs 0.6%, particularly for patients who had more severe deficits and those over 75 years of age (National Institute of Neurological Disorders, 1995).

During the summer of 1996, about ½ year after the publication of the NINDS rt-PA Study Group, the FDA approved the use of rt-PA for treatment of acute stroke patients when the drug was given within the first 3 h. Subsequent published treatment protocols adopted by committees of the American Heart Association (Adams et al., 1996), the American Academy of Neurology (Quality Standards Subcommittee of the American Academy of Neurology, 1996), and the European Stroke Initiative (EUSI) recommended intravenous administration of rt-PA (0.9 mg/kg—maximum of 90 mg) given in a 10% bolus followed by an infusion lasting 60 min to patients within 3 h of onset of ischemic stroke. The recommendations stipulated that a CT scan done before the infusion should not show major infarction, mass effect, edema, or hemorrhage. Neither the NINDS rt-PA study group nor the American Heart Association nor the American Academy of Neurology committees required or suggested vascular tests before thrombolytic treatment.

In the ECASS II trial, investigators treated 800 patients from Europe, Australia, and New Zealand with rt-PA or placebo within 6 hours of stroke onset (Hacke et al., 1998). The dose of rt-PA was the same as that used in the NINDS trial rather than the higher dose used in ECASS I. Patients with major infarcts on CT scan were excluded but vascular imaging was not performed before treatment. Guidelines for control of hypertension were more explicit than in ECASS I. In ECASS II, 36.6% of patients in the placebo-treated group had favorable outcomes—a much better result than in the ECASS I and NINDS trials. Among the rt-PA-treated group, 40.3% had favorable outcomes; this result was not statistically significantly different from that of the placebo-treated group (Hacke et al., 1998). Treatment results and frequency of hemorrhages were similar in the 0 to 3 h and 3 to 6 h treatment groups. Symptomatic hemorrhages developed in 8.8% of the rt-PA-treated patients.

ATLANTIS was a trial that followed NINDS in order to expand the 3-h time window. Starting with a 3 to 6 h window, this was quickly changed to a 3 to 5 h time window. Recruitment of patients took a long time, and finally, there was no difference in outcome between rtPA- and placebo-treated patients. Symptomatic hemorrhages, however, remained higher in the rt-PA group (Clark et al., 1999).

STARS (Standard Treatment with Altepase to Reverse Stroke) evaluated 389 patients (Albers et al., 2000). There was a symptomatic hemorrhage rate of 3.3% with 7 deaths and an asymptomatic hemorrhage rate of 8.2% in this study. Rankin scales at 30 days were used as an outcome measure. Among rt-PA-treated patients, 35% had very favorable outcomes (Rankin 0–1) and 43% had functional improvement (Rankin 1–2). In multivariate analysis, this study found that patients with more favorable outcomes had lower NIH scores, were under age 85, had lower initial mean arterial pressures, and fewer early CT signs (Albers et al., 1999).

Similar results have been reported by a Canadian Phase 4 study and several local or oligocenter series from the United States and Europe (Anderson et al., 1998; Chiu et al., 1997; Cruz-Flores et al., 1988; Dafer et al., 1998; Grond et al., 1998; Hanson et al., 1998; Tanne et al., 2000; Trouillas et al., 1998; Wang et al., 2000). Given the results of these studies and a series of smaller follow-up studies that reproduced a low hemorrhage rate with similar recovery rates, IV rt-PA was recommended for patients with acute stroke treated within 3 h of symptom onset (Tanne et al., 2000).

Recent meta-analyses show a time-dependent efficacy of rtPA. The earlier treatment starts, the more likely it is to be effective. However, in selected patients a positive effect of rtPA can be found even up to 6 h.

In seven clinical studies, the vascular lesions were defined by angiography but the thrombolytic drug was delivered intravenously (del Zoppo et al., 1992; Jansen et al., 1986; Mori et al., 1992; von Kummer and Hacke, 1992; von Kummer et al., 1991; Wolpert et al., 1993; Yamaguchi et al., 1993, 1995). In these studies, angiographic catheters were left in situ so that angiography could be performed both before and after intravenous delivery of the drug. Table I shows the results of these studies. Only two of these studies (Yamaguchi et al., 1993, 1995), had control patients who were not given a thrombolytic drug. The thrombolytic drug, rt-PA, was given within 6 h, except in one study (Mori et al., 1992), that had an 8-h limit. Among the total of 370 patients in these observational series treated with rt-PA, one-third of the arteries treated showed significant recanalization as compared to only 5% of 58 control arteries. MCA branch occlusions recanalized best followed by occlusions of the superior and inferior divisions of the MCA. Mainstem MCA occlusions recanalized less often than branch and division MCA lesions. ICA occlusions seldom recanalized and there were no recanalizations when both the ICA and MCA were occluded (von Kummer and Hacke, 1992). Very few patients with

**TABLE I**   Intravenous rt-PA—Angiographic Studies

| Author | Time (h) | Total patients treated | Arteries recanalized/total | | | Hemorrhage | Outcome |
|---|---|---|---|---|---|---|---|
| | | | ICA | MCA | BA | | |
| Yamaguchi, 1991 | <6 | 51 tPA<br>47 *controls* | 10/47 tPA<br>*2/46* | | 0 | 24HI, 4PH<br>*22HI, 5PH* | 37% Good;<br>*22% controls*<br>*good* |
| Yamaguchi, 1995[a] | <6 | 121 tPA | | 28/121 | | 11HI&PH | MCA > ICA |
| Mori, 1992 | <6 | 19 tPA,<br>12 *controls* | 1/6<br>*0/4* | 7/13<br>*1/8* | 0 | 8HI, 1PH<br>*4HI, 1PH* | tPA > *controls* |
| Von Kummer and Hacke, 1992 | <6 | 32 tPA | 1/11 | 10/21 | 0 | 9HI, 3PH | 44% Good |
| Von Kummer *et al.*, 1991 | <6 | 27 tPA | 3/10 | 9/17 | 1/5 | 6HI, 0PH | 44% Good |
| Jansen *et al.*, 1995[b] | <6 | 8 tPA, 8U | 2/16 | | | 1PH | 19% Good |
| Del Zoppo *et al.*, 1992, 1993 | <8 | 104 (93[c]) tPA | 2/23 | 12/34 mst<br>14/26 M2<br>29/44 br | 0/1 | 21HI, 11PH | No data |
| Totals | | 370<br>59 *Controls* | 128/389 (33%) tPA<br>*3/58 (5%) Controls* | | 16% | 58HI, 20PH<br>*26HI, 6PH* | |

*Note.* HI, hemorrhagic infarct; PH, parenchymatous hemorrhage; mst, mainstem MCA; M2, MCA divisions; br, branch.
[a]Only patients with ICA/MCA emboli included.
[b]Only intracranial ICA occlusions.
[c]Ninety-three of the 104 completed rt-PA therapy according to protocol; recanalization data presented as arteries not as patients.

documented basilar artery occlusions were given intravenous rt-PA and only 1/6 recanalized. In another study, Grond et al, reported favorable results of treatment in 10 of 12 patients with acute vertebrobasilar territory ischemia given intravenous rt-PA followed by heparin but the occlusive vascular lesions were not documented before treatment (Grond *et al.*, 1988). Embolic occlusions recanalized more often than *in situ* thrombosis of atherostenotic arteries. Recanalization was better when there was angiographic evidence of good collateral circulation prior to administration of rt-PA.

**Intra-Arterial Thrombolysis.** A number of uncontrolled trials reported the results of the effectiveness of intrarterial thrombolysis using different agents at different onset times and in different vascular territories. Combining the results of these angiographically monitored trials, 65% of patients had effective recanalization using endovascular techniques. Mainstem MCA occlusions responded best and ICA occlusions responded worse (Caplan, 2000). Sixty-nine percent of basilar artery occlusions showed recanalization. In these studies, all patients received therapy within 24 h and 42% had good outcomes while 18.5% had hemorrahagic complications (Barnwell *et al.*, 1994; Becker *et al.*, 1996; Berg-Dammer *et al.*, 1996; Casto *et al.*, 1996; Cross *et al.*, 1997; del Zoppo *et al.*, 1988; Gonner *et al.*, 1998; Hacke *et al.*, 1988a; Jansen *et al.*, 1983; Matsumoto and Satoh, 1991; Mitchell *et al.*, 1997; Mobius *et al.*, 1991; Mori *et al.*, 1988, 1991; Suarez *et al.*, 1999; Theron *et al.*, 1989; Wijdicks *et al.*, 1997; Zeumer *et al.*, 1986, 1993). The extent of reperfusion of blood after lysis generally depended on the patho-physiology of the stroke and the location of the occluded artery. Table II shows the results of these studies. Higher survival rates were regularly associated with recanalization.

PROACT I (Prolyse in Acute Cerebral Thromboembolism Trial) was the first randomized double-blind controlled trial of the therapeutic effectiveness of intra-arterial thrombolysis with pro-urokinase although numerous uncontrolled studies had been performed (del Zoppo *et al.*, 1998). PROACT I compared the use of locally injected, intra-arterial pro-urokinase (r-proUK) versus heparin within 6 h of onset in patients with a radiographically proven MCA occlusion (del Zoppo *et al.*, 1998). The aim of this study was only to compare the recanalization rate of pro-urokinase versus placebo The interventional radiologist was not allowed to mechanically disturb the clot and was instructed to inject medication at the proximal end of the thrombus. There was successful recanalization in 15/26 (57.7%) of patients treated with medication while 2/14 (14.3%) recanalized with heparin alone. The study had too few patients to properly address outcome but the 90-day cumulative mortality was 26.9% in pro-urokinase group and 42.9% in the placebo group ($2P = 0.48$) (del Zoppo *et al.*, 1998).

PROACT II was a more extensive study (Furlan *et al.*, 1999). The entrance criteria were similar to those in PROACT I. While it was an open label study, the follow-up was blinded to medication versus placebo. Among 474 patients who had angiograms, 142 (30%) had no MCA occlusions. Forty percent of patients in the treatment group had slight or no neurologic disability at day 90, defined as a Rankin scale of 2, while this was

**TABLE II** Intra-Arterial Thrombolytic Studies

| Author | Drug | Total | Reperfusion of arteries | | | Hem | "Good outcome" |
|---|---|---|---|---|---|---|---|
| | | | ICA | MCA | Basilar | | |
| Del Zoppo et al., 1988 | U/S | 19 | 7/8 | 9/11 | | 4 | 17 (89%) |
| Hacke et al., 1988 | U/S | 43 | | | 19/43 | 4 | 13 (30%) |
| Zeumer et al., 1993 | U/tPA | 59 | 29/31 | | 28/28 | 8 | 17 (29%) |
| Mori et al., 1988 | U | 22 | | 8/22 | | 4 | 11 (50%) |
| Mori, 1991 | UPA | 44 | 1/8 | 13/31 | 2/5 | 10 | 16 (36%) |
| Mobius et al., 1991 | U/S | 18 | | | 14/18 | | 10 (56%) |
| Matsumoto and Sato, 1991 | U/tPA | 93 | 21/36 | 28/41 | 9/16 | 25 | 30 (32%) |
| Theron et al., 1989 | S/U | 12 | 3/3 | 9/9 | | 3 | 11 (92%) |
| Zeumer et al., 1983 | S | 5 | | | 4/5 | | 3 (60%) |
| Casto et al., 1996 | U | 12 | | 8/8 | 3/4 | 3 | 5 (42%) |
| Barnwell et al., 1994 | U | 12 | 2/2 | 5/8 | 2/2 | 3 | 9 (75%) |
| Berg-Dammer et al., 1996[a] | S/U | 14 | | 13/14 | | | 9 (64%) |
| Mitchell et al., 1997 | U | 16 | | | 13/16 | 2 | 7 (44%) |
| Jansen et al., 1995 | U/tPA | 16 | 2/16 | | | 3 | 1 (6%) |
| Becker et al., 1996 | U | 12 | | | 10/13[a] | 2 | 3 (25%) |
| Cross et al., 1997 | U | 20 | | | 11/20 | 7 | 4 (20%) |
| Wijdicks et al., 1997 | U | 9 | | | 7/9 | 1 | 5 (56%) |
| Gonner et al., 1998 | U | 43 | 1/9 | 13/23 | 5/10 | 8 | 26 (60%) |
| Suarez et al., 1999 | | | | | | 8 | |
| Totals | | 523 | 39/87 | 143/216 | 127/189 | 95 | 245 (47%) |
| recanalized | | 338 (65%) | 45% | 62% | 67% | 18% | |

*Note.* U, urokinase; S, streptokinase; U-PA, urokinase tPA; Hem, hemorrhagic changes.
[a] 13 Occlusions/12 patients; 2 basilar, 10 bilat. vertebral artery.

true of only 25% of the control group ($P = 0.04$). The mortality rate was similar in the treatment (25%) and placebo (27%) groups. The symptomatic hemorrhage rate was much higher in the treatment group (10%) compared to 2% in the placebo group ($P = 0.04$) The study showed favorable recanalization rates in the treatment group 66% versus 18% in control groups ($P < 0.001$) (Furlan et al., 1999).

One trial considered the safety and feasibility of combined intravenous and local intra-arterial delivery of rt-PA in patients treated within 3 hours of symptom onset (Lewandowsky et al., 1999). Thirty-five patients were assigned randomly, 17 into an intravenous (IV) followed by intra-arterial (IA) group and 18 were treated with IV placebo and then IA rt-PA. Occlusions were found on angiography in 22 patients (63%). Recanalization was better in the IV/IA group. There were no parenchymal brain hematomas. Systemic bleeding was more common in the IV/IA group. There were no statistically significant differences in outcomes between the two groups. More deaths (5, 29%) occurred in the IV/IA group compared to 1 (5%) in the placebo/IA group (Lewandowsky et al., 1999).

The indications, timing, and patient selection rules for administration of thrombolytic agents remain controversial (Caplan et al., 1997). The present authors and others (Albers, 1999) believe that in the near future the present USA recommendations for intravenous thrombolysis (Adams et al., 1996; Quality Standards Committee, 1996) should be modified to take into consideration the results of vascular imaging and modern CT and MRI brain imaging protocols which give useful information about the presence and size of infarction and the region that is underperfused as well as the presence and location of arterial occlusions.

### Surgery

Although carotid artery surgery has proven effective in preventing future strokes in patients who have had recent brain or ocular ischemia ipsilateral to severe carotid artery stenosis (>70% luminal narrowing) (Barnett et al., 1998; Biller et al., 1998; European Carotid Surgery, 1998; MRC European Carotid Surgery, 1991; North American Symptomatic Carotid Endarterectomy, 1991), carotid endarterectomy during an acute stroke carries a considerable risk of increased ischemia and brain hemorrhage. In isolated occasional circumstances emergency carotid endarterectomy during active brain ischemia can reverse neurological signs and is a consideration when the patient is worsening despite maximal medical therapy. Similarly surgery on the subclavian and vertebral arteries in the neck can be successful in patients with posterior circulation transient ischemia and minor strokes (Caplan, 1996; Kieffer et al., 1996; Spetzler et al., 1987) but ordinarily has not been performed for acute vertebro-basilar territory strokes.

In many early anecdotal reports surgeons noted the effectiveness of extracranial (EC) to intracranial (IC) artery bypass, using the superficial temporal artery as the usual donor artery and an MCA branch as the recipient artery. A large, randomized study of the effectiveness of such EC-IC bypasses, however, proved beyond reasonable doubt that the surgery as it was customarily performed at that time had no benefit, and in some circumstances, operated patients fared worse than patients treated medically (EC-IC Bypass Study Group, 1985). The patients in this series had surgery about 6 weeks after stroke, a timing employed to prevent reperfusion hemorrhage in regions where the capillaries and arterioles might be ischemic and vulnerable to leakage. Earlier bypass surgery and surgery using larger proximal recipient arteries are untested alternative options in selected patients. Posterior circulation bypass has not been studied in trials but has been essentially superceded by thrombolysis and angioplasty with stenting.

### Angioplasty and Stenting

During the past 25 years, interventional neuroradiology techniques have developed into an important therapeutic alternative for treating patients with a variety of cerebrovascular conditions. Initially, interventional techniques using coils, catheters, balloons, glues, and other devices were applied mostly to treatment of patients with intracranial aneurysms and vascular malformations. Neuroradiologists and other trained interventionalists use catheters and balloons to mechanically split plaques and dilate stenotic coronary and peripheral limb arteries. Mechanical vascular dilation has also been used to treat fibromuscular dysplasia. Carotid and vertebral artery angioplasty for stenotic disease of the neck arteries was first performed in patients considered to be poor candidates for surgical endarterectomy. More recently carotid artery angioplasty has begun to be performed much more widely. Series of patients with carotid artery (Eckert et al., 1996; Kachel et al., 1991; Porta et al., 1991) and with subclavian or vertebral artery stenosis in the neck (Higashida et al., 1986; Malek et al., 1999) have been reported. Randomized secondary prevention trials are now being performed to compare the relative effectiveness and complication rate of angioplasty vs carotid artery surgery in patients with severe carotid and vertebral artery stenosis.

Intracranial angioplasty is effective in dilating vasoconstricted intracranial arteries after SAH. Percutaneous transluminal angioplasty using intra-arterial balloons has also occasionally been performed for stenosis of intracranial arteries in patients with TIAs or minor strokes, especially the MCA (Alazzaz et al.,

2000; Takis et al., 1997) and the intracranial vertebral and basilar arteries (Alazzaz et al., 2000; Gomez et al., 2000; Mori et al., 2000; Takis et al., 1997). Dissections, vasospasm, and blockage of penetrating arteries during balloon insufflation are important complications of intracranial angioplasty. Angioplasty of intracranial arteries at the sites of the origins of important penetrating arteries carries a risk of infarction in the territorial supply of these penetrators. For this reason, angioplasty of the mainstem MCA at the site of the lenticulostriate arterial origins and the basilar artery have the highest risk of complications. Thrombosis can also complicate intracranial angioplasty. More recently emergent angioplasty with stenting has been used to treat patients with acute strokes due to MCA occlusions as an alternative to thrombolysis (Nakano et al., 1998; Suh et al., 1999). This treatment has been successful but clots occasionally migrated to occlude more distal branches of the MCA and rethrombosis was a problem (Nakano et al., 1998). The technology used to treat extracranial and intracranial stenoses by interventional techniques is improving rapidly and stents are now used more widely. In the future angioplasty will undoubtedly prove to be a more commonly used strategy to treat patients with occlusive cerebrovascular disease who have acute strokes and recurrent brain ischemia.

### Prevention of Clot Formation, Propagation, and Embolism

Formation of thrombi depend on local vascular injury or roughening, the number of platelets and their activation, and the presence of serum coagulant and anticoagulant substances. In general, so-called "red clots," fibrin-dependent thrombi, tend to form in regions where there is low flow or stagnation, whereas smaller so-called "white platelet clots" adhere to roughened places in faster-moving streams of blood (Deykin, 1967; Weksler, 1985). Standard anticoagulants, such as heparin (and low-molecular-weight heparin and heparinoids) and warfarin, are posited to be more useful in preventing red clots, and antiplatelet agglutinating agents are thought to be better at preventing white platelet plugs. Heparin and warfarin should be most effective in patients with occlusive disease of veins and large arteries and in patients with cardiac disorders that predispose to cardiac-origin thromboembolism, whereas agents that decrease platelet aggregation might have an advantage in patients who have plaque disease without severe stenosis. In this chapter we will discuss only management of acute stroke. Prevention of subsequent strokes will be covered in Chapter X. The two types of medicines that have been used acutely are aspirin and heparins (unfractionated heparin, low-molecular-weight heparin, and heparinoids). Warfarin is customarily given after acute heparinization and other antiplatelet drugs (dipyridamole, ticlopidine, clopidogrel)have not as yet been studied in trials of acute stroke treatment.

Polycythemia and thrombocytosis increase the probability of clot formation. Hemodilution decreases the hematocrit and so reduces viscosity and thrombotic tendencies in patients with polycythemia. Thrombocytosis can also cause a clotting tendency and usually accompanies hematological proliferative disorders that require specific therapy.

### Heparin, Low-Molecular-Weight Heparin, and Heparinoids

Heparin has been used for cerebrovascular disease since the 1940s. It is a heterogeneous mixture of sulfated mucopolysaccharides containing multiple active proteins ranging in size form 3000–37,500 Da. It is known to bind and enhance the activity of Anti-Thrombin III, which in turn binds to thrombin and neutralizes it. It also decreases hyperlipidemia, antagonizes thromboplastin, inactivates factor Xa, and reduces fibrin production (Caplan, 1986; Hirsh et al., 1998; Llinas and Caplan, 2001).

Heparin is customarily used in patients as a prelude to oral anticoagulation. Individuals with previously undiagnosed protein C and S deficiency can develop warfarin-induced skin necrosis if not pretreated with heparin (Hirsh, 1991). The evidence from trials that IV heparin helps in the acute setting to decrease propagation of clot or to retard secondary embolism is meagre. Multiple small, poorly controlled trials showed benefit with IV heparin in TIA and stroke in evolution (Baker et al., 1962; Brust, 1977; Carter, 1961; Fisher, 1961; Llinas and Caplan, 2001). Genton et al. reviewed the results of 11 heparin studies and concluded that there was evidence of benefit in patients with cardiac risk for embolization specifically with rheumatic and ischemic heart disease (Genton et al., 1977). Ramirez et al. (1986) noted that 81% of patients treated with heparin had good recoveries and that the medication reduced fluctuations. Duke et al., however, found no benefit for stroke progression in patients with "partial stable strokes" (Duke et al., 1986). In neither trial was the underlying cause of the stroke determined or controlled (Duke et al., 1986; Ramirez et al., 1986).

More recent trials in acute ischemic stroke patients have most often used low-molecular-weight heparin or heparinoids. Low-molecular-weight heparin is a fractionated form of heparin that includes only molecules from 1000 to 5000 Da. Low-molecular-weight heparin has less ability to bind to and inactivate thrombin but retains the ability to inactivate factor Xa (Hirsh et al.,

1998). Three trials studied the efficacy of low-molecular-weight heparin in acute stroke (Hommel et al., 1998; Kay et al., 1995; The Publication Committee, 1998). These trials were analyzed according to stroke etiology and stroke subtype. The Nadroparin low-molecular-weight heparin trial divided 308 patients into placebo and high- and low-dose groups (Kay et al., 1995). The high dose was the dose used for DVT prophylaxis and the low dose was half this dose. Patients were given the medication within 48 h of symptom onset. The study showed significant reduction of risk of death or dependency at 6 months (45% treatment vs 52% low dose vs 65% placebo, $P = 0.005$) but no significant difference at 3 months (Kay et al., 1995). The number of patients was too small to determine if specific stroke subtypes or etiologies did better than others. Given the high rate of intracranial disease among this population, it was felt that the low-molecular-weight heparin was effective in the treatment of stroke caused by intracranial disease. In a larger trial of 750 European patients treated with fraxiparine, this heparinoid showed no significant benefit (Hommel et al., 1998).

The TOAST trial (Trial of ORG 101072 in Acute Stroke Treatment) enrolled 1281 patients and gave them ORG 10172 (a heparinoid compound) or placebo beginning within 24 h of symptom onset (The Publication Committee, 1998). Patients were given continuous intravenous infusion over 7 days, aiming to keep anti-factor Xa activity from 0.6 to 0.8 antifactor Xa units/mL. There was an increased proportion of "very favorable outcome" at 7 days in those given heparinoid. For all types of stroke, patients had better outcomes, with 68% of patients having favorable outcomes versus 54.7% treated with placebo ($P = 0.04$). Before subtype etiological analysis, the heparinoid showed no statistically significant benefit over placebo in primary or secondary outcome measures. Major hemorrhage was more common in the treatment group: 32 patients in the treatment group and 10 in the control group, a statistically significant difference ($P < 0.005$) (The Publication Committee, 1998).

In TOAST trial subanalyses, the heparinoid was found to be effective in treating patients with large artery occlusive disease (Adams et al., 1999). Forty-three percent of patients with large artery disease had favorable outcomes versus 29.1% who received placebo ($P = 0.02$). Patients with large artery disease had significantly fewer recurrences during the time of the infusion than placebo, 6% vs 11% (Trial of ORG 101072 in Acute Stroke Treatment). The only patients in the TOAST trial who had sufficient vascular imaging for analysis were those patients who had occlusive ICA disease in the neck. In patients with greater than 50% internal carotid artery stenosis, at 7 days, 53.8% (64/119) vs 38% (41/108) had favorable outcomes

($P = 0.0239$), and 27.7% (33/119) vs 17.6% (41/108) had very favorable outcomes when ORG 10172 vs placebo was compared (Adams *et al.*, 1999; The Publication Committee, 1998).

The IST (International Stroke Study) treated patients within 48 h of stroke onset for 14 days with placebo, 300 mg of aspirin, heparin 5000 U bid, heparin 12,500 U bid, aspirin plus heparin 5000 U bid, or aspirin plus heparin 12,500 U bid (International Stroke Trial, 1997). Heparin was given subcutaneously and monitoring of coagulation times was optional. Some patients had no brain imaging. The patients in the aspirin group were treated with aspirin for 14 days. There was a nonsignificant trend toward fewer deaths within 14 days among patients treated with heparin compared with those who did not receive heparin (9% vs 9.3%). Heparin-treated patients had fewer recurrent ischemic strokes (2.9% vs 3.8%) and fewer pulmonary emboli (0.5% vs 0.8%) but more hemorrhagic strokes (1.2% vs 0.4%) and extracranial bleeding than those patients not treated with heparin. At 6 months, the same percentage of patients (62.9%) were dead or dependent whether or not they were treated with heparin (International Stroke Trial, 1998).

The use of heparin in patients with cardiac origin embolism is discussed in more detail in Chapter 33 on brain embolism. The use of heparin in patients with cardio-embolic stroke is based mostly on the theory that since anticoagualtion with coumadin is protective, the use of heparin must also be protective. The risk for recurrent cardio-embolic events varies in several studies and depends heavily on the nature of the cardiac source. The risk of embolism recurrence is estimated at between 2 and 22% during the first 2 weeks, making a cumulative risk of approximately 1% a day (Cerebral Embolism Study Group, 1984; Cerebral Embolism Task Force, 1986). In retrospective analyses of treatment of patients with cardio-embolic stroke, heparin reduces the rate of recurrence during the first 2 weeks after cardiogenic embolization (Cerebral Embolism Study Group, 1983, 1984; Furlan *et al.*, 1982). Heparin reduces the risk from 15% without heparin to 5% with anticoagulation with heparin (Cerebral Embolism Study Group, 1983, 1984; Furlan *et al.*, 1982). In one small prospective trial, the risk reduction went from 10 to 0% when heparin was given immediately (Cerebral Embolism Study Group, 1983), but the results did not meet statistical significance. In the TOAST trial the rate of recurrent embolization from a cardiac source was 2.8% of ORG 10172-treated patients versus 7.3% placebo-treated patients (The Publication Committee, 1998). The rate of re-embolization was so low that statistical significance was not achieved. Patients with a high risk of recurrent cardiac-origin embolism were most likely not entered by TOAST investigators.

## Aspirin and Other Agents That Decrease Platelet Aggregation

In the IST trial some patients were treated with aspirin 300 mg/day with or without subcutaneous low- or high-dose heparin (International Stroke Trial, 1997). The aspirin-alone group had a statistically significant decrease in recurrent stroke within the 14 days following the acute event, 2.8% with aspirin and 3.9% with those patients not given ASA ($2P < 0.001$). There was also a significant reduction in the rate of death and nonfatal stroke for aspirin versus placebo, 11.3% versus 12.4% ($2P = 0.02$). The major extracranial hemorrhage rate in the aspirin group was 1.1% compared with 0.6% in the placebo group ($2P = 0.0004$) although there was no significant increase in intracranial hemorrhagic events (International Stroke Trial, 1997).

The CAST trial (The Chinese Acute Stroke Trial) investigated aspirin use alone in the treatment of acute stroke (CAST, 1997). More than 20,000 patients were enrolled and were given 160 mg of aspirin acutely (within 48 h) and for 4 weeks versus placebo. More than 2500 patients (12% did not have CT scans or other brain imaging) were included. There were statistically significant reductions in rates of death within the 4 weeks for ASA versus placebo, 3.3% versus 3.9%, and a statistically significant reduction of rates of nonfatal stroke or death during hospitalization, ASA 5.3% versus placebo 5.9%. A statistically significant reduction in stroke recurrence was shown with ASA 1.6% versus placebo 2.1%. The intracranial hemorrhage rate was not statistically significantly elevated. As in the IST trial, there was a statistically significant increase in extracranial hemorrhages with aspirin 0.8% compared to placebo 0.6% (CAST, 1997). Unfortunately as with many of the megatrials, there was no assessment for stroke subtype or etiology and no vascular and scant cardiac testing was performed (CAST, 1997).

Other antiplatelet aggregants (dipyridamole, ticlopidine, clopidogrel, combined aspirin, and modified-release dipyridamole) have not been tested as yet in randomized trials of patients with acute stroke. They have been tested as prophylactic agents for secondary stroke prevention. Newer drugs that are antagonists of the glycoprotein platelet llb/llla complex give promise of even more effective inhibition of platelet functions. The platelet glycoprotein llb/llla complex is the site of binding to adhesive proteins including fibrinogen. Binding to fibrinogen stimulates platelet aggregation and adhesion to blood vessels. To date GP llb/llla inhibitors have been used acutely and intravenously but agents that can be used orally and chronically are now being tested. Since these GP llb/llla inhibitors are antagonists of the final common pathway of platelet adhesion with fibrinogen, they give promise of being the most effective agents to prevent white clot formation.

## Maximizing Blood Flow

Increasing collateral circulation and perfusion in the ischemic-zone capillary bed may improve ischemia.

Vasodilating agents have long been prescribed to increase blood flow. Physicians attempted to dilate brain arteries by having patients breathe air with high $CO_2$ content. Other agents include cyclandelate, isoxsuprine, hydergine, papaverine, and nicotinic acid. Unfortunately, there is little information regarding the effect and utility of vasodilating agents on CBF in patients with TIAs or ischemic strokes (Caplan, 1979). Abnormal or even paradoxical responses of the cerebral circulation in stroke patients might theoretically render vasodilator treatment ineffective or even harmful (Caplan, 1979). Cerebral arteries have a scarcity of medial elastic fibers and are less responsive than systemic vessels to vasodilator stimuli. Vasodilator agents produce more systemic than cerebral vasodilation and thus could lead to hypotension or globally decreased CBF. Arteries within nonischemic regions should retain the ability to vasodilate while arteries within ischemic zones could be damaged by ischemia sufficiently to lose their ability to dilate. In that circumstance vasodilating drugs could result in increased flow into nonischemic areas, creating a type of steal away from the ischemic areas.

Calcium-channel-blocking agents have been studied to determine whether they would improve function in patients with ischemic stroke and SAH. Calcium has a number of actions that could influence the outcome of ischemia, including promoting vasoconstriction by effects on vascular smooth muscle, altering coagulation, killing cells when extracellular $Ca^{2+}$ passes through cell membranes into the intracellular compartment, and lowering blood pressure (Brunwald, 1982; Gorelick and Caplan, 1985). A Dutch trial of oral nimodipine in ischemic stroke patients found that the drug reduced mortality in men and had a beneficial effect in reducing clinical deficits in patients with moderate and severe neurological abnormalities (Gelmers et al., 1988). However, more recent trials of the effectiveness of nimodipine in patients with ischemic stroke have had disappointing results (American Nimodipine Study Group, 1992; TRUST, 1990).

Volume expansion is another method used to attempt to augment cerebral blood flow and increase perfusion within the microcirculation of the brain. Physicians have attempted to increase blood volume by simply increasing fluid intake and by using various volume expanders. Albumin, plasma, and solutions of colloids and crystalloids have all been used. Most often the expansion

is in serum volume, thus effectively diluting the more viscous erythrocyte portion of the blood (Heros and Korosue, 1989). The two major determinants of blood viscosity within brain vessels are fibrinogen and Hct. Lowering the Hct by hemodilution could reduce whole blood viscosity and increase blood flow (Thomas, 1985; Thomas et al., 1997). The optimal Hct for blood flow and preservation of oxygen transport is around 33% (Wood and Kee, 1985). Rapid hemodilution with reduction of the Hct theoretically could significantly improve blood flow to ischemic zones. Hemodilution may be isovolemic (that is, blood replaced with equivalent fluid volume), or replacement could be more (hypervolemic) or less (hypovolemic) than the original blood volume. Plasma, Ringer's solution, or colloid solutions such as dextran 40 or hydroxyethyl starch are usually used for fluid replacement. The most frequently used colloidal solutions are dextrans and starches. Dextrans are polysaccharide molecules made by the action of bacteria on sucrose. The most commonly used solution (10% dextran solution in normal saline or 5% dextrose) is called "dextran 40" and contains molecules ranging from 10,000 to 80,000 molecular weight (average 40,000). Infusion causes rapid volume expansion but rapid urinary excretion of the dextran molecules of <50,000 molecular weight leads to an osmotic diuresis with reduction in plasma volume. Dextran also coats RBCs, platelets, and endothelium, which is posited to decrease blood viscosity and prevent cellular aggregation and so improve microcirculatory flow. Dextran has a potential antithrombotic action by its hemodiluting effect, erythrocyte coating, and decreased platelet aggregation. Another frequently used colloidal solution is hydroxyethyl starch (HES). Volume is important to maintain during hemodilution because either hypovolemia with excessively reduced volume or hypervolemia with potential cardiac overload and brain edema can be harmful.

Another means of reducing viscosity and also potentially altering blood coagulability is the use of substances that reduce fibrinogen levels. Ancrod, a substance derived from the purified protein fraction of venom from the Malayan pit viper, has a thrombin-like enzymatic effect. Ancrod selectively acts on fibrinogen and inhibits formation of cross-linked fibrin. Ancrod can reduce fibrinogen levels to about 100 mg and has been studied for its utility in patients with brain ischemia (Liu et al., 1998; Olinger et al., 1988; Sherman et al., 2000). In one study, at onset, the mean plasma fibrinogen level was 385 mg/dL (Olinger et al., 1988). After IV infusions of ancrod, D-dimer levels rose, indicating clot lysis, and fibrinogen levels fell to a mean of 116 mg/dL at 6 h and 52 mg/dL at 24 h after start of treatment (Olinger et al., 1988). In the Stroke Treatment with Ancrod Trial (STAT), favorable functional status was achieved in more patients in the ancrod-treated group than in the placebo group (42.2% vs 34.4%, $P = 0.04$) and there were fewer severely disabled stroke patients in the ancrod-treated group (11.8% vs 19.8%, $P = 0.01$) (Sherman *et al.*, 2000). Mortality was not significantly different in the placebo and Ancrod groups but there was a trend for more hemorrhages in the Ancrod group (Sherman *et al.*, 2000). Unfortunately, a European 6-h ancrod trial (ESTAT) did not replicate the results and was terminated due to excess complication rates.

Another method of rapidly reducing blood fibrinogen levels and whole blood viscosity is to use plasmapheresis techniques. Investigators in Austria have developed a technique that they call heparin-induced extracorporeal low-density-lipoprotein precipitation (HELP) (Schuff-Werner *et al.*, 1989; Walzl *et al.*, 1993). In this phoresis system, blood is removed from a cubital vein and passed through a filter which separates the cellular components from the plasma. Isovolemic acetate buffer and heparin are added to the plasma. Fibrinogen, LDL cholesterol, and triglycerides are removed by this process. The blood is then reinfused into the cubital vein on the opposite side. This HELP treatment reduces fibrinogen levels, lowers whole blood viscosity at high and low shear rates, lowers plasma viscosity, and reduces red cell transit time (Schuff-Werner *et al.*, 1989; Walzl *et al.*, 1993). This treatment has been used in patients with acute brain ischemia to increase cerebral blood flow and can also be used repeatedly in patients with microvascular occlusive disease with vascular dementia who have high fibrinogen levels. There are no randomized trials to support this methodology.

Omega-3 fatty acids, especially eicosopentanoic acid (EPA), can also lower blood fibrinogen levels (Radack *et al.*, 1989). In a preliminary study, EPA also reduced blood viscosity, especially in those patients with high baseline viscosities. EPA and omega-3 fatty acids, plentiful in various fish oils, also reduce platelet aggregability, in addition to their effects on fibrinogen and viscosity. The low frequency of atherosclerosis in Eskimos has been attributed to diets rich in fish and omega-3 fatty acids. To date, there have been no trials of these substances in stroke prevention or treatment. Atromid, ticlopidine, and pentoxyphilline also have some fibrinogen-lowering effects.

## Increasing the Brain's Resistance to Ischemia (Neuroprotection)

Neuronal death depends on multiple factors (Garcia, 1995). Important are level of activity (the more work that goes on, the more fuel is needed), presence of local metabolites such as lactic acid and oxygen-free radicals,

temperature of the system (at low temperatures, there is less metabolism and less need for fuel), integrity of the neuronal cell membranes, and influx of calcium into cells and the extracellular-to-intracellular gradient for calcium. Preliminary evidence from experimental whole-brain ischemia in young animals indicates that hyperglycemia makes the brain more vulnerable to ischemia. Sugar increases metabolism and leads to the production of lactic acid. Acidosis can be destructive to brain tissue (Siesjo and Smith, 1997). Neurotransmitters, especially glutamate, released at sites of ischemia might overexcite neurons and cause toxic damage, increasing the effects of the initial ischemia (Choi, 1995, 1997). This theory is often referred to as the excitotoxin hypothesis and has stimulated much research concerning neurotransmitters and ischemia.

These observations on factors that influence brain vulnerability to anoxic and ischemic insults have stimulated interest in many possible therapeutic regimens. Treatments aimed at reducing cell damage are usually now referred to as neuroprotection. Unfortunately, although a variety of neuroprotective strategies have shown effectiveness in laboratory animals, especially when given before or soon after development of brain ischemia, none to date has shown definite benefits in randomized trials of human patients with strokes (Devuyst and Bogousslavsky; Steiner and Hacke, 1998; Venti *et al.*, 2000; Wahlgren, 1997; Warach *et al.*, 2000). Neuroprotective therapy might be most effective when combined with measures to promote reperfusion such as thrombolytic treatment (Steiner and Hacke, 1998).

Although early studies suggested that hyperglycemia, with or without diabetes mellitus, might aggravate ischemic brain damage, more recent reviews indicate that the issues are complex (Helgason, 1988). Studies of glycosylated hemoglobin now allow an estimation of the adequacy of blood-glucose control in diabetics. Elevated fasting blood-sugar levels, especially when accompanied by high levels of glycosylated hemoglobin, do predict poor outcome (Adams *et al.*, 1988; Helgason, 1988). Others have related high glucose levels in stroke patients to a stress response (Woo *et al.*, 1988, 1990); Large hemorrhages and infarcts induce catecholamine secretion, which in turn increases the WBC count and the blood-sugar levels. Reduction of elevated blood sugar levels by insulin might decrease brain lactate production in ischemic zones and thereby reduce necrosis, but this idea remains an unproven hypothesis. At present, it seems prudent to avoid hyperglycemia and to treat very high blood-sugar levels with insulin. There are insufficient data to warrant reducing normal blood sugars to hypoglycemic levels.

Barbiturate anesthesia (Black *et al.*, 1978; Safar, 1980) and hypothermia (Diringer, 1999; Ginsberg,

1997; Schwab *et al.*, 1998) are neuroprotective strategies that reduce cerebral energy metabolism and thereby reduce the brain's requirements for fuel, oxygen, and blood. Unfortunately, each can lead to circulatory changes and can complicate the examination and management of stroke patients.

At present, hypotheses and theories abound and far outweigh the data, but investigation of putative neuroprotectant substances may prove very fruitful in the future. Unfortunately trials in human stroke patients have not always been well designed to show an effect of the various therapies. Ideally, neuroprotective agents would be posited to work when given early to patients who had stunned brain at risk for infarction. Patients with lacunar infarcts and those that already have important brain infarction are very unlikely to respond. When main arteries are occluded, the agent used may not reach the stunned brain region in sufficient concentration to be effective. Most of the trials have given the therapeutic agents too late and have not required sufficient brain and vascular imaging to limit treatment to those expected to respond.

## Treatment of Ischemic Brain Edema

Edema formation is common after ischemic stroke. Whenever possible, patients who develop brain edema and other stroke complications should be managed in dedicated stroke units (Indredavik *et al.*, 1997; Kaste *et al.*, 1995; Ronning and Gulvog, 1998; Stroke Unit, 1997). In patients with large infarcts, edema may be so prominent that it causes herniation. This condition is found with large MCA territory infarcts (malignant MCA infarcts) and space occupying cerebellar infarcts. Treatment of ischemic brain edema is either (1) medical, (2) mechanical, or (3) surgical. Medical therapy comprises treatment of fever, lowering of extremely high blood pressure, and administering barbiturates, osmotherapeutics, or Tris buffer (Brott *et al.*, 1994; Diringer *et al.*, 1988; Hacke *et al.*, 2000; Ringleb *et al.*, 1998). Among the mechanical measures, head elevation and facilitating venous outflow are most important. When signs of increased intracranial pressure develop, the first treatment is infusion of osmotically active substances, either intravenous glycerol (four infusions of 250 mL of 10% glycerol given over 30–60 min) or mannitol (25–50 g given over 20–30 min every 3–6 h) or glycerol by mouth (4 × 50 mL of 10% glycerol) (Hacke *et al.*, 2000). Replacement fluids should be given to maintain the serum osmolality in the 300–320 mOsm/L range. Hypotonic and glucose-containing solutions should be avoided.

Some centers may use barbiturates (short-acting agents such as thiopental given as a bolus with ICP and EEG monitoring). Most patients will be ventilated, and hyperventilation may reduce increased intracerebral pressure (ICP) for a short time. All the above mentioned measures are short-lived and lose their efficacy rapidly. Surgical interventions include ventricular drainage and suboccipital craniectomy for cerebellar infarcts and hemicraniectomy for malignant MCA infarcts. With hemicraniectomy, mortality may be lowered from 80% in untreated patients to about 20% (Rieke *et al.*, 1995; Schwab *et al.*, 1998). However, treatment must be instituted before first signs of herniation occur. Randomized data to support this approach are not available as yet.

Recently Schwab and colleagues showed that moderate hypothermia could reduce intracranial pressure and improve mortality in 25 patients with large space-taking middle cerebral artery territory infarcts (Schwab *et al.*, 1998). They maintained body core temperature at 33–34°C for 48–72 h using a cooling blanket, cool ventilator air, and fanning. Muscle-paralyzing drugs and narcotics were also given. The mortality in this study was 44%, less than the usual 80%, and survivors had a relatively favorable outcome with a mean Bartel Index score of 70 (Schwab *et al.*, 1998).

Decompressive surgery should be considered in patients with very large life-threatening cerebral hemisphere infarctions (Rieke *et al.*, 1995; Schwab *et al.*, 1998). This usually consists of a wide hemicraniectomy. To be most effective this must be done early in the course of illness (Schwab *et al.*, 1998). In patients with large space-taking cerebellar infarcts, surgical decompression of infarcted cerebellar tissue can be life saving (Rieke *et al.*, 1993).

## Stroke Units and Stroke Services

Within the past two decades, many hospitals have employed units and teams that are dedicated to the care of stroke patients. Many general and specialty hospitals have begun to designate such units. These units vary in personnel and emphasis. Some are administered under internal medicine specialists, some by neurologists, and others by physical medicine and rehabilitation specialists. Some emphasize very acute stroke care and monitoring of acute stroke patients, and these units are excellent resources for clinical research and clinical stroke trials that study acute-phase treatments and practices. Other stroke units emphasize early rehabilitation. Some units have access to the most advanced diagnostic technology. All stroke units have in common specialized nursing care, protocols and practices to prevent complications of stroke, and education of stroke patients and their families and significant others. Improved mortality and morbidity in Europe during the

5 years between the ECASS I (Hacke *et al.*, 1995) and the ECASS II (Hacke *et al.*, 1998) trials is mostly due to experience and the development of stroke units and teams in the European centers that participated in these trials. Treatment in stroke units reduces mortality, decreases stroke morbidity, and allows more patients to remain independent and to return to their homes. Randomized trials and systematic reviews and meta-analyses of trials all have shown a considerable benefit for acute stroke units (Indredavik *et al.*, 1997; Kaste *et al.*, 1995; Ronning and Gulvog, 1998; Stroke Unit, 1997a,b).

Stroke is a complex disease. Care should include stroke and atherosclerosis risk assessment and stroke prevention strategies; rapid and efficient clinical evaluation and diagnosis, rapid complete brain and vascular imaging studies and blood tests; management of blood pressure and fluid balance; medical and/or surgical treatment of the process causing the acute stroke; institution early of rehabilitation techniques; surveilance and preventive strategies and treatment to prevent the common stroke complications such as aspiration, phlebothrombosis, urinary and pulmonary infections, and bedsores; and teaching of the patients and their families about their specific problem and about stroke in general. These functions are undoubtedly best performed in dedicated stroke units. Stroke units and stroke rehabilitation services all have in common a focus on stroke patients, a multidisciplinary team approach, and emphasis on early treatment and comprehensive rehabilitation.

## PRACTICAL MANAGEMENT

### Evaluation

The evaluation should be hypothesis driven. Questions to be answered in sequence include:

1. Does the patient have a stroke or possibly a stroke mimic (postseizure state, hyperglycemia, hysterical paralysis, etc.)?
2. What type of stroke and stroke subtype? Ischemic (thrombotic, embolic, or systemic hypoperfusion) or hemorrhagic (subarachnoid or intracerebral hemorrhage)?
3. What part of the brain is involved?
4. What blood vessels that supply the abnormal brain region are most likely affected and by what pathology?
5. Was the brain ischemia transient or instead is the brain reversibly ischemic ("stunned") or infarcted?
6. Does the patient have important concomitant conditions?

The evaluation should be rapid but not hasty. Remember that time elapsed equals brain tissue lost. Systematic checklists and protocols ensure thoroughness of the evaluation and assist rapid throughput of the patient.

### History and Examination

The history and general and neurological examinations are of cardinal importance even in this age of technological advances. The physician should question the patient and others who can give reliable information about cardiovascular and other risk factors for stroke, prior TIAs and strokes, time of onset of neurological symptoms, activity at onset, clinical course since onset, known important medical illnesses and surgeries, and neurological and other symptoms since onset. The general examination should focus on cardiovascular findings. Abnormalities of the pulse rate and regularity, blood pressure, cardiac size, sounds, and murmurs should be noted. Temperature should be accurately determined and recorded. Craniocervical pulses should be palpated and listened to for bruits.

A rapid and efficient neurological examination should include the state of alertness; abnormalities of language, attention, memory, and behavior; cranial nerve abnormalities, loss of strength, coordination, and sensation; abnormalities and asymmetries of the deep tendon reflexes and plantar responses; and altered gait. Quantitative estimate of severity of the neurological deficit should be documented by a commonly used stroke scale (NIH stroke scale, Scandinavian Stroke Scale, etc.) and the degree of functional deficit and disability should be estimated using a modified Rankin score and Barthel index score whenever feasible.

### Blood Testing

Routine blood testing should include a complete blood count (hemoglobin, hematocrit, white blood count and differential, platelet count) prothrombin time, partial thromboplastin time, and measures of serum glucose, electrolytes, calcium, and renal function.

### Cardiac Evaluation

Cardiac evaluation including an EKG and cardiac auscultation are important. In some patients an echocardiogram and chest X-ray should be performed within 24 h. Measurement of oxygen saturation by pulse oximetry is often helpful.

### Brain Imaging and Vascular Tests

Whenever possible each patient should have brain imaging (either CT or MRI) and vascular studies (MRA,

CTA or neck and transcranial ultrasound (TCD)) to identify nonstroke lesions and brain hemorrhages, to define the presence, location and size of brain infarcts, and to determine the nature, location, and severity of intracranial and extracranial vascular lesions. When echo-planar MR studies are available, diffusion-weighted MRI along with susceptibility images and MRA can provide quickly and safely the necessary information to guide treatment.

### General ("Medical") Treatment

Treatment of acute stroke patients includes care of cardiac and respiratory abnormalities, fluid and metabolic management, blood pressure and blood sugar control, and management of fever and infections. Monitoring of blood pressure, cardiac rate and rhythm, and pulse oximetry are important especially when abnormalities are found on the initial examination. The intensity of treatment needed clearly depends on the nature, presence, and severity of abnormalities found.

#### Pulmonary Function and Airway Protection

Patients with reduced consciousness, congestive heart failure, pneumonia, chronic lung disease may be hypoxemic. Analysis of blood gases, and transcutaneous oximetry can detect and quantify hypoxia and hypercarbia. In hypoxemic patients administer 2–4 L of $O_2$/min via a nasal tube. Drugs that treat bronchial spasm are sometimes needed. Endotracheal intubation is sometimes necessary if there is a pathological breathing pattern, the patient is stuporous or comatose, there is severe hypoxemia or hypercarbia, or if the patient is at high risk of aspiration.

#### Blood Pressure Management and Cardiac Care

Every stroke patient should have an initial electrocardiogram. Arrythmias may require treatment. Patients with severe stroke deficits and those who are hemodynamically unstable should have monitoring of their cardiac rhythm. Cardiac output should be optimized with maintenance of a high normal blood pressure and a normal heart rate and rhythm when possible and avoidance of fluid volume depletion. In stroke patients that are hemodynamically unstable, central venous pressure should be maintained at 8–10 cm $H_2O$ (Hacke *et al.*, 2000).

Blood pressure should not be lowered unless it is extremely high (Brott *et al.*, 1994; Hacke *et al.*, 2000; Ringleb *et al.*, 1998). For BP >220 systolic and 120–140 diastolic, the following regimens are suggested: captopril 6.25–12.5 orally, or labetolol 5–20 mg IV, or ura-

padil 10–50 mg IV followed by 4–8 mg/h IV, or clonidine 0.15–0.3 mg IV or SC, or dihydralazine 5 mg IV + metropolol 10 mg (Brott *et al.*, 1994; Hacke *et al.*, 2000). If the diastolic pressure exceeds 140 mm Hg then rapid reduction of pressure is warranted. Nitroglycerin 5 mg IV followed by 1–4 mg/h IV or sodium nitroprusside 1–2 mg are recommended (Brott *et al.*, 1994; Hacke *et al.*, 2000).

#### Blood Glucose Management

Concentrated carbohydrate solutions should not be given until the blood glucose is known. Insulin titration should be begun if the blood glucose level exceeds 200 mg/dL (10 mmol/L) (Hacke *et al.*, 2000). In patients who are hypoglycemic 10–20% glucose solution should be infused usually through a central venous line (Hacke *et al.*, 2000).

#### Fever and Infection Management

Antipyretics should be given to patients with fever (>37.5°C). Dehydration should be avoided and infections (most often pneumonia and urinary tract infections) should be treated with appropriate antibiotics.

#### Fluid and Electrolyte Management

Plasma volume contraction should be avoided. In patients treated with intravenous fluids, electrolytes should be monitored daily. Excess fluid intake can cause pulmonary edema and cardiac failure and can enhance brain edema. In patients who have raised intracranial pressure, aim at a slightly negative fluid balance (about 500 mL negative balance daily). When insulin is administered, potassium requirements are often increased. When large volumes of fluid need to be replaced, solutions with high osmolality are used. When infused substances are irritative to veins, a central venous line should be used for infusions.

Hyponatremia can develop in patients because of inappropriate secretion of ADH or excess release of atrial naturetic factor. Inappropriate ADH is managed best by fluid restriction and, in some patients, administration of hypertonic saline. Normovolemia should be maintained when excess atrial naturetic factor secretion is suspected (Diringer *et al.*, 1988; Hacke *et al.*, 2000).

### Reperfusion

For management to be effective and safe, evaluation and treatment must occur quickly and efficiently. Whenever possible hospitals should have stroke teams and stroke protocols that assure rapid throughput and rapid access to the needed technology and rapid assessment of the results of brain and vascular imaging.

When patients are seen during the first few hours after the onset of acute brain ischemia, reperfusion should be considered. Thrombolysis and other means of reperfusion are best performed by physicians with considerable training and experience in evaluating and treating stroke patients and patients with other neurological conditions. The main factors involved in the decision on whether or not to attempt reperfusion are: (1) the time since the onset of neurological symptoms (Generally reperfusion is considered during the first 6 h after symptom onset.), (2) the severity of the clinical deficit (The more severe the deficit, the less likely that reperfusion will be helpful.), (3) the presence and extent of brain infarction (The larger the area of infarction, the less likely that reperfusion will be helpful and the more likely that it will cause significant brain hemorrhage and edema.), and (4) the presence and location of thromboembolic occlusion of cervical and cerebral arteries that supply the region of ischemia/infarction.

Brain imaging is an essential prerequisite before a decision on thrombolysis can be made. MRI with diffusion-weighted images and susceptibility imaging can be performed in a short time. MR imaging is probably superior to CT in defining acute infarction and will suffice to exclude hemorrhage when interpreted by physicians experienced in MRI interpretation in acute stroke. CT can separate reliably hemorrhage from ischemia but is not as effective as MR in defining the extent of infarction during the first few hours after onset. When brain imaging shows hemorrhage thrombolysis or other forms of reperfusion are not considered. When brain imaging shows extensive brain infarction (e.g., >1/3 of the MCA territory), thrombolysis carries a very high risk.

When the technology for vascular testing is available and the investigations can be performed accurately and quickly then we advise that vascular imaging be performed before deciding on reperfusion. Vascular testing can include: MRA when MRI is performed, CT angiography (CTA) when a helical CT scanner is used, or extracranial and transcranial ultrasound. Ultrasound can be performed as an adjunct to either CT or MRI. When no occlusion is present we do not use thrombolytic agents. When an intracranial branch occlusion is present, and treatment can be given within 3 h of symptom onset we advocate intravenous administration of rtPA (0.9 mg/kg—maximum of 90 mg) given in a 10% bolus followed by an infusion lasting 60 min following the recommendations of the comittees of the American Heart Association (Adams *et al.*, 1998) and American Academy of Neurology (Quality Standards Subcommittee, 1996). When the arterial occlusion involves the neck or intracranial ICA or the basilar

artery, and/or more than 3 h has transpired and a trained neurological interventionalist is available, then we advocate intra-arterial administration of thrombolytic agent. At times, in patients in whom the occlusive lesion has developed *in situ* (and is not embolic), angioplasty with stenting may be needed to maintain patency after thrombolysis.

When advanced technology is not available and patients are seen within the first 3 h after onset then we advocate following the guidelines of the AAN and AHA for treatment of acute stroke patients.

Mechanical methods of reperfusion including lasers, pressure or vacuum systems, and angioplasty with stenting might be used especially in patients in whom thrombolytic drugs are contraindicated.

### Antithrombotic Treatment

Every acute stroke patient should be placed on some type of antithrombotic treatment, i.e., agents that decrease platelet functions or standard anticoagulants such as heparin, heparinoids, or warfarin.

A strategy that we now advocate until the results of randomized trials are available that include patients with well defined vascular lesions is to use platelet antiaggregants when there is no obstruction to flow but ischemia is most likely due to the process of platelet plugs and small white or white and red thrombi. Heparin (or low-molecular-weight heparin or heparinoids) and warfarin are reserved for situations of severe stenosis or clots within the heart. These standard anticoagulants are also used for 3 to 4 weeks in patients with a recent *in situ* occlusion within a large artery. This method is strictly hypothetical, and results to date are anecdotal and have not been tested scientifically.

Table III reviews present recommendations for the use of platelet antiaggregants and anticoagulants. The timing of the use of antithrombotic substances is important. When thrombolytic agents are considered it is best to hold antithrombotic drugs until after thrombolysis, or until the decision not to use thrombolytics. In patients in whom the acute stroke is due to embolism (see also chapter on brain embolism) the timing of the use of heparin (or low-molecular-weight heparin or heparinoids) is important.

## TREATMENTS NO LONGER RECOMMENDED

Acute surgical exploration to remove clots in recipient arteries in patients with brain embolism has been superceded by thrombolysis and other interventional

**TABLE III  Present Recommended Use of Platelet Aggregants and Anticoagulants**

**Heparin** (standard dose) (short-term, 2–4 weeks)
Usually given by constant IV infusion, keeping aPTT between 60 and 100 s (1.5–2 × control aPTT)
1. Immediate therapy of definite cardiac-origin brain embolism (large cerebral infarct, hypertension, bacterial endocarditis, or sepsis would delay or contraindicate this use)
2. For patients with severe stenosis or occlusion of the ICA origin, ICA siphon, MCA, VA, or basilar artery, with less than a severe clinical deficit—treatment then could be shifted to warfarin or surgery

**Heparin** (subcutaneous mini dose)
For prophylaxis of deep-vein occlusion in patients immobilized by stroke (unless contraindicated)

**Warfarin**
Usually overlapped with heparin keeping INR between 2.0–3.0
1. Long-term (greater than 3 months) in patients with cardiogenic brain embolization and rheumatic heart disease, atrial fibrillation with large atria or prior cerebral embolism, prosthetic valves, and some hypercoagulable states
2. Long-term (greater than 3 months) in patients with severe stenosis of the ICA origin, ICA siphon, MCA stem, VA, basilar artery—used until studies show artery has been occluded for at least 3 weeks
3. Shorter-term (3–6 weeks) in patients with recent occlusion of the ICA, MCA, VA, or basilar arteries

**Platelet antiaggregants** (aspirin, clopidogrel, combined aspirin–dipyridamole)
1. For patients with plaque disease of the extracranial and intracranial arteries without severe stenosis
2. For patients with polycythemia or thrombocytosis and related ischemic attacks

techniques and is no longer recommended. Extracranial to intracranial bypass is also no longer recommended for patients with acute brain ischemia.

# REFERENCES

Adams, H. P., Brott, T. G., Furlan, A. J. et al. (1996). Use of thrombolytic drugs. A supplement to the guidelines for the management of patients with acute ischemic stroke. A statement for Health Care Professionals from a special writing group of the Stroke Council American Heart Association. *Stroke* 27, 1711–1718.

Adams, H. P., Olinger, C. P., Marler, J. R. et al. (1988). Comparison of admission serum glucose concentration with neurologic outcome in cerebral infarction. *Stroke* 19, 455–458.

Adams, H. P. Jr, Bendixen, B. H., Leira, E. et al. (1999). Antithrombotic treatment of ischemic stroke among patients with occlusion or severe stenosis of the internal carotid artery: A report of the Trial of Org 10172 in acute stroke treatment (TOAST). *Neurology* 53, 122–131.

Alazzaz, A., Thornton, J., Aletich, V. A., Debrun, G. M., Ausman, J. I., and Charbel, F. (2000). Intracranial percutaneous transluminal angioplasty for arteriosclerotic stenosis. *Arch. Neurol.* 57, 1625–1630.

Albers, G. (1999). Expanding the window for thrombolytic therapy in acute stroke. The potential role of acute MRI for patient selection. *Stroke* 30, 2230–2237.

Albers, C. W., Bates, V. E., Clark, W. M., Bell, R., Verro, P., and Hamilton, S. A. (2000). Intravenous tissue-type plasminogen activator for treatment of acute stroke: The Standard Treatment with Alteplase to Reverse Stroke (STARS) Study. *JAMA* 283(9), 1145–1150.

American Nimodipine Study Group (1992). Clinical trial of nimodipine in acute ischemic stroke. *Stroke* 23, 3–8.

Anderson, A., Smith, D. B., and Hughs, R. L. (1998). The Colorado stroke network experience with intravenous rt-PA in acute ischemic stroke. *Neurology* 50, A157. [Abstract]

Baker, R. N., Brownward, J. A., Fang, H. C. et al. (1962). Anticoagulant therapy in cerbral infarction. *Neurology* 12, 823–835.

Barnett, H. J. M., Taylor, D. W., Eliasziw, M. et al. for the North American Symptomatic Carotid Endarterectomy Trial Collaborators (1998). Benefit of carotid endarterectomy in patients with symptomatic moderate or severe stenosis. *N. Engl. J. Med.* 339, 1415–1425.

Barnwell, S. L., Clark, W. M., Nguyen et al. (1994). Safety and efficacy of delayed intra-arterial urokinase therapy with mechanical clot disruption for thromboembolic stroke. *Am. J. Neurol. Res.* 15, 1817–1822.

Becker, K. J., Monsein, L. H., Ulatowski, J. et al. (1996). Intraarterial thrombolysis in vertebrobasilar occlusion. *Am. J. Neurol. Res.* 17, 255–262.

Berg-Dammer, E., Henkes, H., Nahser, H. C., and Kuhne, D. (1996). Thromboembolic occlusion of the middle cerebral artery due to angiography and endovascular procedures: Safety and efficacy of local intra-arterial fibrinolysis. *Cerebrovasc. Dis.* 6, 222–230.

Biller, J., Feinberg, W. M., Castaldo, J. E. et al. (1998). Guidelines for carotid endarterectomy. A statement for healthcare professionals from a special writing group of the Stroke Council, American Heart Association. *Stroke* 29, 554–562.

Black, K., Weidler, J., Jallad, N. et al. (1978). Delayed pentobarbital therapy of acute focal cerebral ischemia. *Stroke* 9, 245–251.

Braunwald, E. (1982). Mechanism of action of calcium-channel blocking agents. *N. Engl. J. Med.* 307, 1618–1627.

Brott, T., Fieschi, C., and Hacke, W. (1994). General therapy of acute ischemic stroke. *In* W. Hacke, D. F. Hanley, K. Einhaupl, and T. P. Bleck Eds., pp. 554–577. Neurocritical care, Berlin, Springer Verlag, 1994.

Brust, J. C. M. (1977). Transient ischemic attacks: Natural history and anticoagulation. *Neurology* 27, 701–707.

Caplan, L. R. (1979). Use of vasodilating drugs for cerebral symptomatology. *In* "Drug Therapy Reviews" (R. Miller and D. Greenblatt, Eds.), pp. 305–317. Elsevier-North Holland, Amsterdam.

Caplan, L. R. (1986). Anticogulation of cerebral ischemia. *Clin. Neuropharamocol.* 9, 399–414.

Caplan, L. R. (1996). "Posterior circulation disease: Clinical findings, diagnosis, and management." Blackwell Science, Boston.

Caplan, L. R. (1999). Brain embolism. *In* "Clinical Neurocardiology" (L. R. Caplan, J. W. Hurst, and M. I. Chimowitz, Eds.), pp. 35–185. Dekker, New York.

Caplan, L. R. (2000). "Caplan's Stroke, a Clinical Approach," 3rd ed. Butterworth-Heinemann, Boston.

Caplan, L. R., Mohr, J. P., Kistler, J. P., and Koroshetz, W. (1997). Thrombolysis-not a panacea for ischemic stroke. *N. Engl. J. Med.* **337**, 1309–1310, 1313.

Carter, A. B. (1961). Anticoagulant treatment in progressing stroke. *Br. Med. J.* **2**, 70–73.

CAST (Chinese Acute Aspirin Trial) Collaborative Group (1997). CAST: Randomized-placebo controlled trial of early aspirin use in 20,000 patients with acute ischemic stroke. *Lancet* **349**, 1641–1649.

Casto, L., Caverni, L., Canerlingo, M. *et al.* (1996). Intra-arterial thrombolysis in acute ischaemic stroke: experience with a super-selective catheter embedded in the clot. *J. Neurol. Neurosurg. Psychiatry* **60**, 667–670.

Cerebral embolism study group (1983). Immediate anticoagulation of embolic stroke: A randomized trial. *Stroke* **14**, 668–676.

Cerebral embolism study group (1984). Immediate anticoagulation of embolic stroke: Brain hemorrhage and management options. *Stroke* **15**, 779–789.

Cerebral embolism task force (1986). Cardiogenic brain embolism. *Arch. Neurol.* **43**, 71–84.

Chiu, D., Krieger, D., Villar-Cordova, C. *et al.* (1997). Intravenous tissue plasminogen activator for acute ischemic stroke feasibility, safety and efficacy in the first year of clinical practice. *Stroke* **29**, 18–22.

Choi, D. W. (1995). Excitotoxicity and stroke. *In* "Brain Ischemia, Basic Concepts and Clinical Relevance" (L. R. Caplan, Ed.), pp. 29–36. Springer-Verlag, London.

Choi, D. W. (1997). The excitotoxic Concept. *In* "Primer on Cerebrovascular Diseases" (K. M. A. Welch, L. R. Caplan, D. J. Reis, B. K. Siesjo, and B. Weir, Eds.), pp. 187–190. Academic Press, San Diego.

Clark, W. M., Wissman, S., Albers, G. W., Jhamandas, J. H., Madden, K. P., and Hamilton, S. (1999). Recombinant tissue-type plasminogen activator (Altepase) for ischemic stroke 3 to 5 hours after symptom onset. The ATLANTIS study: A randomized controlled trial. Altepase Thrombolysis for Acute Non-interventional Therapy in Acute Stroke. *JAMA* **282**(21), 2019–2026.

Cross, L., D. T., Moran, C. J., Akins, P. T., Angtuaco, E., and Diringer, M. N. (1997). Relationship between clot location and outcome after basilar artery thrombosis. *Am. J. Neurol.* **18**, 1221–1228.

Cruz-Flores, S., Thompson, D. W., Banet, G. *et al.* (1988). Intravenous thrombolysis in acute ischemic stroke: Preliminary experience with tissue plasminogen activator. *Neurology* **50**, A113–A114. [Abstract]

Dafer, R. M., Tiejen, G. E., and Korsnack, A. (1998). Experience with rt-PA at a small medical center. *Neurology* **50**, A115. [Abstract]

del Zoppo, G., Ferbert, A., Otis, S. *et al.* (1988). Local intra-arterial fibrinolytic therapy in acute carotid territory stroke. A pilot study. *Stroke* **19**, 307–313.

del Zoppo, G. J. (1988). Thrombolytic therapy in cerebrovascular disease. *Stroke* **19**, 1174–1179.

del Zoppo, G. J., Higashida, R. T., Furlan, A. J. *et al.* (1998b). PROACT: A phase II randomized trial of recombinant pro-urokinase by direct arterial delivery in acute middle cerebral artery stroke. PROACT Investigators. Prolyse in Acute Cerebral Thromboembolism. *Stroke* **29**, 4–11.

del Zoppo, G. J., Poeck, K., Pessin, M. S. *et al.* (1992). Recombinant tissue plasminogen activator in acute thrombotic and embolic stroke. *Ann. Neurol.* **32**, 78–86.

Devuyst, G., and Bogousslavsky, J. (1999). Recent progress in drug treatment for acute stroke. *J. Neurol. Neurosurg. Psychiatry.*

Deykin, D. (1967). Thrombogenesis. *N. Engl. J. Med.* **276**, 622–628.

Diringer, M., Ladenson, P., Stern, B., Schleimer, J., and Hanley, D. (1988). Plasma atrial naturetic factor and subarachnoid hemorrhage. *Stroke* **19**, 1119–1124.

Diringer, M. N. (1999). Hypothermia: An old idea comes of age. *Neurol. Network Commentary* **3**, 120–126.

Donnan, G. A., Davis, S. M., Chambers, B. R. *et al.* (1995). Trials of streptokinase in severe acute ischemic stroke. *Lancet* **345**, 578–579.

Duke, R. J., Bloch, R. F., Alexander, G. G. *et al.* (1986). Intravenous heparin for the prevention of stroke progression in acute partial stable stroke: A randomized controlled trial. *Ann. Intern. Med.* **105**, 825–828.

EC-IC Bypass Study Group (1985). Failure of the extracranial-intracranial arterial bypass to reduce the risk of ischemic stroke. *N. Engl. J. Med.* **313**, 1191–1200.

Eckert, B., Zanella, F. E., Thie, A., Steinmetz, J., and Zeumer, H. (1996). Angioplasty of the internal carotid artery: Results, complications, and follow-up of 61 cases. *Cerebrovasc. Dis.* **6**, 97–105.

European Carotid Surgery Trialists Collaborative Group (1998). Randomised trial of endarterectomy for recently symptomatic carotid stenosis: Final results of the MRC European Carotid Surgery Trial (ECST). *Lancet* **351**, 1379–1387.

Fisher, C. M. (1961). Anticoagulant therapy in cerebral thrombosis and cerebral embolism: A national cooperative study: Interim report. *Neurology* **11**, 132–138.

Furlan, A. J., Cavalier, S. J., Hobbs, R. E. *et al.* (1982). Hemorrhage and anticoagulation after non-septic embolic brain infarction. *Neurology* **32**, 280–282.

Furlan, A. J., Higashida, R. T., Wechsler, L. *et al.* (1999). A randomized trial of intra-arterial pro-urokinase for acute ischemic stroke of less than 6 hours duration due to middle cerebral artery occlusion. PROACT Investigators. prolyse in Acute Cerebral thromboembolism. *JAMA* **282**, 2003–2011.

Garcia, J. (1995). Mechanisms of cell death in ischemia. *In* "Brain Ischemia, Basic Concepts and Clinical Relevance" (L. R. Caplan, Ed.), pp. 7–18. Springer-Verlag, London.

Gelmers, H. J., Gorter, K., deWeerdt, C. *et al.* (1988). A controlled trial of nimodipine in acute ischemic stroke. *N. Engl. J. Med.* **318**, 203–207.

Genton, E., Barnett, H. J. M., Fields, *et al.* (1977). Study group of antithrombotic therapy. Report of the joint committee on stroke resources. XIV. Cerebral ischemia. The role of thrombosis andantithrombotic therapy. *Stroke* **8**, 150–175.

Ginsberg, M. D. (1997). Hypothermic neuroprotection in cerebral ischemia. *In* "Primer on Cerebrovascular Diseases" (K. M. A. Welch, L. R. Caplan, D. J. Reis, B. K. Siesjo, and B. Weir, Eds.), pp. 272–275. Academic Press, San Diego.

Gomez, C. R., Misra, V. K., Liu, M. W. *et al.* (2000). Elective stenting of symptomatic basilar artery stenosis. *Stroke* **31**, 95–99.

Gonner, F., Remonda, L., Mattle, H. *et al.* (1998). Local intra-arterial thrombolysis in acute ischemic stroke. *Stroke* **29**, 1894–1900.

Gorelick, P. B., and Caplan, L. R. (1985). Calcium, hypercalcemia and stroke. *Curr. Concepts Cerebrovasc. Dis. (Stroke)* **20**, 13–17.

Grond, M., Rudolf, J., Schmulling, S. *et al.* (1998). Can NINDS results be transferred into daily routine? *Stroke* **29**, 288. [Abstract]

Grond, M., Rudolf, J., Schmulling, S., Stenzel, C., Neveling, M., and Heiss, W.-D. (1998). Early intravenous thrombolysis with rt-PA in vertebrobasilar stroke. *Arch. Neurol.* **55**, 466–469.

Hacke, W., Kaste, M., Fieschi, C. *et al.* (1995). Intravenous thrombolysis with recombinant tissue plasminogen activator for acute hemispheric stroke. The European Cooperative Acute Stroke Study (ECASS). *JAMA* **274**, 1017–1025.

Hacke, W., Kaste, M., Fieschi, C. *et al.* (1998). Randomized double-blind placebo controlled trial of thrombolytic therapy with intravenous alteplase in acute ischemic stroke (ECASS II). *Lancet* **352**, 1245–1251.

Hacke, W., Kaste, M., Olsen, T. S., Bogousslavsky, J., and Orgogozo, J.-M. for the European Stroke Initiative (EUSI) Executive Committee (2000). Acute treatment of ischemic stroke. *Cerebrovasc. Dis.* 10 (Suppl. 3), 22–33.

Hacke, W., Zeumer, H., Ferbert, A., Bruckmann, H., and del Zoppo, G. (1988a). Intra-arterial thrombolytic therapy improves outcome in patients with acute vertebrobasilar occlusive disease. *Stroke* 19, 1216–1222.

Hanson, S. K., Brauer, D. J., Anderson, D. C. *et al.* (1998). Stroke treatment in the community (STIC): Intravenous rt-PA in clinical practice. *Neurology* 50, A155–A156. [Abstract]

Helgason, C. (1988). Blood glucose and stroke. *Stroke* 19, 1049–1053.

Heros, R. C., and Korosue, K. (1989). Hemodilution for cerebral ischemia. *Stroke* 20, 423–427.

Higashida, R. T., Hieshima, G. B., Tsai, F. Y., Bentson, J. R., and Halbach, V. V. (1986). Percutaneous transluminal angioplasty of the subclavian and vertebral areries. *Acta Radiol. Suppl.* 369, 124–126.

Hirsh, J. (1991) Heparin. *N. Engl. J. Med.* 324, 1565–1574.

Hirsh, J., Warkentin, T. E., Raschke, *et al.* (1998). Heparin and low molecular weight heparin: Mechanism of action, pharamocokinetics, dosing considerations, monitoring, efficacy, and safety. *Chest* 114(Suppl.), 489s–510s.

Hommel, M., for the FISS bis Investigators (1998). Fraxiparine in ischemic stroke study (FISS bis). *Cerebrovasc. Dis.* 8 (Suppl. 4), A64. [Abstract]

Indredavik, B., Slordahl, S. A., Bakke, F., Rokseth, R., and Haheim, L. L. (1997). Stroke unit treatment. Long term effects. *Stroke* 28, 1861–1866.

International Stroke Trial Collaborative Group (1997). The International Stroke Trial: A randomized trial of aspirin, subcutaneous heparin, both or neither among 19,435 patients with acute ischemic stroke. *Lancet* 349, 1569–1581.

Jansen, O., von Kummer, R., Forsting, M., Hacke, W., and Sartor, K. (1986). Thrombolytic therapy in acute occlusion of the intracranial internal carotid artery bifurcation. *Am. J. Neurol. Res.* 16, 1977–1986.

Kachel, R., Basche, S., Heerklotz, I., Grossman, K., and Endler, S. (1991). Percutaneous transluminal angioplasty (PTA) of supraaortic arteries especially the internal carotid artery. *Neuroradiology* 33, 191–194.

Kay, R., Wong, K. S., Yu, Y. L. *et al.* (1995). Low molecular weight heparin for the treatment of acute ischemic stroke. *N. Engl. J. Med.* 333, 1588–1593.

Kaste, M., Palmomaki, H., and Sarna, S. (1995). Where and how should elderly stroke patients be treated? a randomized trial. *Stroke* 26, 249–253.

Kieffer, E., Koskas, F., Bahnini, A., Ruotolo, C., and Rancurel, G. (1996). Long-term results after reconstruction of the cervical vertebral artery. *In* "Cerebrovascular ischaemia—Investigation & management." (L. R. Caplan, E. G. Shifrin, A. N. Nicolaides, and W. S. Moore, Eds.), pp. 617–625. Med-Orion, London.

Kobayashi, S., Hirai, A., Terano, T. *et al.* (1981). Reduction in blood viscosity by eicosopentaenoic acid. *Lancet* 2, 197.

Lewandowski, C., Frankel, M., Tomsick, T. A. *et al.* (1999). Combined intravenous and intra-arterial r-TPA versus intra-arterial therapy of acute ischemic stroke. Emergency Management of Stroke (EMS) Bridging Trial. *Stroke* 30, 2598–2605.

Liu, M., Counsell, C., Wardlaw, J., and Sandercock, P. (1998). A systematic review of randomized evidence for fibrinogen-depleting agents in acute ischemic stroke. *J. Stroke Cerebrovasc. Dis.* 7, 63–69.

Llinas, R., and Caplan, L. R. (2001). Evidence-based treatment of patients with ischemic cerebrovascular disease. *Neurol. Clin.* 19, 1.

Malek, A. M., Higashida, R. T., Phatouros, C. C. *et al.* (1999). Treatment of posterior circulation ischemia with extracranial percuta-

neous balloon angioplasty with stent placement. *Stroke* 30, 2073–2085.

Matsumoto, K., and Satoh, K. (1991). Topical intraarterial urokinase infusion for acute stroke. *In* "Thrombolytic Therapy in Acute Ischemic Stroke" (W. Hacke, G. J. del Zoppo, and M. Hirshberg, Eds.), pp. 207–212. Springer-Verlag, Heidelberg.

Meyer, J. S., Gilroy, J., and Barnhart, M. E. *et al.* (1964). Anticoagulants plus streptokinase therapy in progressing stroke. *JAMA* 189, 373.

Meyer, J. S., Gilroy, J., Barnhart, M. E., and Johnson, J. F. (1965). Therapeutic Thrombolysis, in cerebral thromboembolism: Randomized Evaluation of Streptokinase. *In* "Cerebral Vascular Disease, 4th Princeton Conference" (C. Millikan, R. Siekert, and J. P. Whisnant, Eds.), pp. 200–213. Grune&Stratton, New York.

Mitchell, P. J., Gerraty, R. P., Donnan, G. A. *et al.* (1997). Thrombolysis in the vertebrobasilar circulation: The Australian Urokinase Stroke Trial. *Cerebrovasc. Dis.* 7, 94–99.

Mobius, E., Berg-Dammer, E., Kuhne, D., and Nahser, H. C. (1991). Local thrombolytic therapy in acute basilar artery occlusion. Experience with 18 patients. *In* "Thrombolytic Therapy in Acute Ischemic Stroke" (W. Hacke, G. J. del Zoppo, and M. Hirschberg, Eds.), pp. 213–215. Springer-Verlag, Heidelberg.

Mori, E. (1991). Fibrinolytic recanalization therapy in acute cerebrovascular thromboembolism. *In* "Thrombolytic Therapy in Acute Ischemic Stroke" (W. Hacke, G. J. del Zoppo, and M. Hirschberg, Eds.), pp. 137–145. Springer-Verlag, Heidelberg.

Mori, T., Kazita, K., Chokyu, K., Mima, T., and Mori, K. (2000). Short-term arteriographic and clinical outcome after cerebral angioplasty and stenting for intracranial vertebrobasilar and carotid atherosclerotic occlusive disease. *Am. J. Neurol. Res.* 21, 249–254.

Mori, E., Tabuchi, M., Yoshida, T., and Yamadori, A. (1988). Intracarotid urokinase with thromboembolic occlusion of the middle cerebral artery. *Stroke* 19, 808–812.

Mori, E., Yoneda, Y., Tabuchi, M. *et al.* (1992). Intravenous recombinant tissue plasminogen activator in acute carotid artery territory stroke. *Neurology* 42, 976–982.

MRC European Carotid Surgery Trial (1991). Interim results for symptomatic patients with severe (70–99%) or with mild (0–29%) carotid stenosis. *Lancet* 337, 1235–1243.

Multicenter Acute Stroke Trial—Europe Study Group (1996). Thrombolytic therapy with streptokinase in acute ischemic stroke. *N. Engl. J. Med.* 335, 145–150.

Multicenter Acute Stroke Trial—Italy (MAST-I) Group (1995). Randomized controlled trial of streptokinase, aspirin, and combination of both in treatment of acute ishaemic stroke. *Lancet* 346, 1509–1514.

Nakano, S., Yokogami, K., Ohta, H., Yano, T., and Ohnishi, T. (1998). Direct percutaneous transluminal angioplasty for acute middle cerebral artery occlusion. *Am. J. Neurol. Res.* 19, 767–772.

National Institute of Neurological Disorders and Stroke rt-PA Study Group (1995). Tissue plasminogen activator for acute ischemic stroke. *N. Engl. J. Med.* 333, 1581–1587.

North American Symptomatic Carotid Endarterectomy Trial Collaborators (1991). Beneficial effect of carotid endarterectomy in symptomatic patients with high-grade carotid stenosis. *N. Engl. J. Med.* 325, 445–453.

Olinger, C. P., Brott, T. G., Barsan, T. G. *et al.* (1988). Use of ancrod in acute or progressing ischemic cerebral infarction. *Ann. Emerg. Med.* 17, 1208–1209.

Pessin, M. S., del Zoppo, G. J., and Furlan, A. J. (1995). "Thrombolytic Treatment in Acute Stroke: Review and Update of Selected Topics in Cerebrovascular Disease, 19th Princeton Conference, 1994," pp. 409–418. Butterworth-Heinemann, Boston.

Porta, M., Munari, L. M., Belloni, G., Moschini, L., and Bonaldi, G. (1991). Percutaneous angioplasty of atherosclerotic carotid arteries. *Cerebrovasc. Dis.* **1**, 265–272.

Quality Standards Subcommittee of the American Academy of Neurology (1996). Practice advisory: Thrombolytic therapy for acute ischemic stroke—Summary statement. *Neurology* **47**, 835–839.

Radack, K., Deck, C., and Huster, G. (1989). Dietary supplementation with low-dose fish oils lowers fibrinogen levels: A randomized double-blind controlled study. *Ann. Int. Med.* **111**, 757–758.

Ramirez-Lassepas, M., Quinones, M. R., and Nino, H. H. (1986). Treatment of acute ischemic stroke: Open trial with continuous infusion heparinzation. *Arch. Neurol.* **42**, 386–390.

Rieke, K., Krieger, D., Adams, H.-P., Aschoff, A., Meyding-Lamade, U., and Hacke, W. (1993). Therapeutic strategies in space-occupying cerebellar infarction based on clinical, neuroradiological, and neurophysiological data. *Cerebrovasc. Dis.* **3**, 45–55.

Rieke, K., Schwab, S., Krieger, D., von Kummer, R., Aschoff, A., Schuchardt, V., and Hacke, W. (1995). Decompressive surgery in space-occupying hemispheric infarction. Results of an open prospective trial. *Crit. Care Med.* **23**, 1576–1587.

Ringleb, P. A., Bertram, Keller, E., and Hacke, W. (1998). Hypertension in patients with cerebrovascular accident. To treat or not to treat? *Nephrol. Dial. Transplant* **13**, 2179–2181.

Ronning, O. M., and Gulvog, B. (1998). Outcome of subacute stroke rehabilitation. A randomized controlled trial. *Stroke* **29**, 779–784.

Safar, P. (1980). Amelioration of post ischemic brain damage with barbiturates. *Curr. Concepts Cerebrovasc. Dis. (Stroke)* **15**, 1–5.

Schuff-Werner, P., Schutz, E., Seyde, W. C. *et al.* (1989). Improved hemorheology associated with a reduction in plasma fibrinoigen and LDL in patients being treated by heparin-inducedextracorporeal LDL precipitation (HELP) *Eur. J. Clin. Invest.* **19**, 30–37.

Schwab, S., Schwarz, S., Spranger, M., Keller, E., Bertram, M., and Hacke, W. (1998). Moderate hypothermia in the treatment of patients with severe middle cerebral artery infarction. *Stroke* **29**, 2461–2466.

Schwab, S., Steiner, T., Aschoff, A., Schwarz, S., Steiner, H. H., and Hacke, W. (1998). Hemicraniectomy in malignant MCA infarction. A prospective trial of 63 patients in support of early intervention. *Stroke* **29**, 188–1894.

Sherman, D. G., Atkinson, R. P., Chippendale, T. *et al.* for the STAT participants (2000). Intravenous ancrod for treatment of acute ischemic stroke. The STAT Study: A randomized controlled trial. *JAMA* **283**, 2395–2403.

Siesjo, B., and Smith, M.-L. (1997). Mechanism of acidosis-related damage. *In* "Primer on Cerebrovascular Diseases" (K. M. A. Welch, L. R. Caplan, D. J. Reis, B. K. Siesjo, and B. Weir, Eds.), pp. 223–226. Academic Press, San Diego.

Sloan, M. A. (1987). Thrombolysis and stroke, past and future. *Arch. Neurol.* **44**, 748–768.

Spetzler, R. F., Hadley, M. N., Martin, N. A. *et al.* (1987). Vertebrobasilar insufficiency. I. microsurgical treatment of extracranial vertebrobasilar disease. *J. Neurosurg.* **66**, 648–661.

Steiner, T., Bluhmki, E., Kaste, M. *et al.* (1998). The ECASS 3-hour cohort. Secondary analysis of ECASS data by time stratification. *Cerebrovasc. Dis.* **8**, 198–203.

Steiner, T., and Hacke, W. (1998). Combination therapy with neuroprotectants and thrombolytics in acute ischemic stroke. *Eur. Neurol.* **40**, 1–8.

Stroke Unit Trialistsí Collaboration (1997a). Collaborative systematic review of the randomized trials of organized in-patient (stroke unit) care after stroke. *Br. Med. J.* **314**, 1151–1159.

Stroke Unit Trialistsí Collaboration (1997b). How do stroke units improve patient outcomes? a collaborative systematic review of the randomized trials. *Stroke* **28**, 2139–2144.

Suarez, J. I., Sunshine, J. L., Tarr, R. *et al.* (1999). Predictors of clinical improvement, angiographic recanalization, and intracranial hemorrhage after intra-arterial thrombolysis for acute ischemic stroke. *Stroke* **30**, 2094–2100.

Suh, D. C., Sung, K.-B., Cho, Y. S. *et al.* (1999). Transluminal angioplasty for middle cerebral artery stenosis in patients with acute ischemic stroke. *Am. J. Neurol. Res.* **20**, 553–558.

Takis, C., Kwan, E., Pessin, M. S., Jacobs, D. H, and Caplan, L. R. (1997). Intracranial angioplasty: Experience and complications. *Am. J. Neurol. Res.* **18**, 1661–1668.

Tanne, D., Gorman, M. J., Bates, V. E., Kasner, *et al.* (2000). Intravenous tissue plasminogen activator for acute stroke in patients aged 80 years and older: The tPA stroke survey experience. *Stroke* **31**(2), 370–375.

The publication committee for the Trial of Org 10172 in Acute Stroke Treatment (TOAST) Investigators (1998). Low molecular weight heparinoid, ORG 10172 (Danaparoid) an outcome after acute ischemic stroke. A randomized controlled trial. *JAMA* **279**, 1265–1272.

Theron, J., Courtheoux, P., Casasco, A. *et al.* (1989). Local intra-arterial fibrinolysis in the carotid territory. *Am. J. Neurol. Res.* **10**, 753–765.

Thomas, D. J. (1985). Hemodilution in acute stroke. *Stroke* **16**, 763–764.

Thomas, D. J., duBoulay, G. H., Marshall, J. *et al.* (1997). Effect of haemotocrit on cerebral blood flow in man. *Lancet* **2**, 941–943.

Trouillas, P., Nighoghossian, N., Derex, *et al.* (1998). Thrombolysis with intravenous rt-PA in a series of 100 cases of acute carotid territory stroke. *Stroke* **29**, 2529–2540.

TRUST study group (1990). Randomized, double-blind placebo-controlled trial of nimodipine in acute stroke. *Lancet* **336**, 1205–1209.

Venti, M., Parnetti, L., Silvestrelli, and Gallai, V. (2000). Role of neuroprotective drugs in acute ischemic stroke. *Cerebrovasc. Dis.* **10** (suppl. 4), 24–26.

von Kummer, R., Forsting, M., Sartor, K., and Hacke, W. (1991). Intravenous plasminogen activator in acute stroke in Thrombolytic therapy in acute ischemic stroke. W. Hacke, G. J. del Zoppo, and M. Hirschberg, (Eds.), pp. 161–167. Springer-Verlag, Heidelberg.

von Kummer, R., and Hacke, W. (1992). Safety and efficacy of intravenous tissue plasminogen activator and heparin in acute middle cerebral artery stroke. *Stroke* **23**, 646–652.

Wahlgren, N. G. (1997). Neuroprotectants in late clinical development—A status report. *Cerebrovasc. Dis.* **7** (Suppl. 2), 13–17.

Walzl, M., Lechner, H., Walzl, B., and Schied, G. (1993). Improved neurological recovery of cerebral infarctions after plasmapheretic reduction of lipids and fibrinogen. *Stroke* **24**, 1447–1451.

Wang, D. Z., Rose, J. A., Honings, D. S., Garwachi, D. J. *et al.* (2000). Treating acute stroke patients with intravenous tPA. the OSF stroke network experience. *Stroke* **31**, 77–81.

Warach, S., Pettigrew, L. C., Dashe, J. F. *et al.* (2000). Effect of Citicholine on ischemic lesions as measured by diffusion-weighted magnetic resonance imaging. *Ann. Neurol.* **48**, 713–722.

Weksler, B. (1985). Antithrombotic therapies in the management of cerebral ischemia. *In* "Cerebrovascular Diseases: Proceedings of the Fourteenth Princeton Conference" (F. Plum and W. Pulsinelli Eds.), pp. 211–223. Raven Press, New York.

Wijdicks, E. F. M., Nichols, D. A., Thielen, K. R. *et al.* (1997). Intra-arterial thrombolysis in acute basilar artery thromboembolism: The initial Mayo Clinic experience. *Mayo Clin. Proc.* **72**, 1005–1013.

Wolpert, S. M., Bruckmann, H., Greenlee, R., Wechsler, L., Pessin, M. S., del Zoppo, G. J., and the rt-PA Acute Stroke Study Group (1993). *Am. J. Nerol. Res.* **14**, 3–13.

Woo, E., Lam, C. W. K., Kay, R. *et al.* (1990). The influence of hyperglycemia and diabetes mellitus on immediate and 3-month morbidity and mortality after acute stroke. *Arch. Neurol.* **47**, 1174–1177.

Woo, E., Ma, J. T. C., Robinson, J. D. *et al.* (1988). Hyperglycemia is a stress response in acute stroke. *Stroke* **19**, 1359–1364.

Wood, J. H., and Kee, D. B. (1985). Hemorrheology of the cerebral circulation in stroke. *Stroke* **16**, 765–772.

Yamaguchi, T., Hayakawa, T., and Kikuchi, H. (1993). For the Japanese Thrombolytic Study Group, Intravenous tissue plasminogen activator in acute thromboembolic stroke: A placebo-controlled double-blind trial. *In* "Thrombolytic therapy in acute ischemic stroke II" (G. J. del Zoppo, E. Mori, and W. Hacke, Eds.), pp. 59–65. Berlin, Springer-Verlag.

Yamaguchi, T., Kikuchi, H., and Hayakawa, T., for the Japanese Thrombolysis Study Group (1995). Clinical efficacy and safety of intravenous tissue plasminogen activator in acute embolic stroke: A randomized double-blind, dose-comparison study of duteplase.

*In* "Thrombolytic Therapy in Acute Ischemic Stroke" (T. Yamaguchi, E. Mori, K. Minematsu, and G. J. del Zoppo, Eds.), pp. 223–229. Springer-Verlag, Tokyo.

Yamamamoto, H., and Bogousslavsky, J. (1998). Mechanism of second and further strokes. *J. Neurol. Neurosurg. Psychiatry* **64**, 771–776.

Zeumer, H., Freitag, H.-J., Zanella, F., Thie, A., and Arning, C. (1993). Local intra-arterial fibrinolytic therapy in patients with stroke: Urokinase versus recombinant tissue plasminogen activator (r-tPA). *Neuroradiology* **35**, 159–162.

Zeumer, H., Hacke, W., and Ringelstein, E. B. (1983). Local intra-arterial thrombolysis in vertebrobasilar thromboembolic disease. *Am. J. Neurol. Res.* **4**, 401–404.

*Cerebrovascular Disorders*

## CHAPTER 31
# Prevention of Stroke

H. Christoph Diener and Gregory W. Albers

## PRIMARY PREVENTION

### Clinical Aspects

In the context of this chapter primary prevention of stroke means to prevent ischemic strokes or transient ischemic attacks (TIA) in patients who have not yet had a cerebrovascular event. These patients might be healthy, might suffer from vascular risk factors, or might already have had another vascular event such as myocardial infarction (MI) or angina pectoris. Secondary prevention is covered in the second part of this chapter.

### Natural Course

Depending on the region of the world between 100 and 700/100,000 persons per year will suffer a first stroke. Between 50 and 200/100,000 will have a recurrent event. Stroke rates are highest in the Eastern European countries (Khaw, 1996). Risk factors for stroke are well known. Many of them are modifiable or potentially modifiable (Table I). Patients with atrial fibrillation have a five times increased stroke risk (Narayan *et al.*, 1997). Stroke risk increases with age.

### Principles of Therapy

#### Identification and Treatment of Vascular Risk Factors

The most important risk factor for stroke is hypertension. Stroke risk doubles for every 7.5 mm Hg increase in blood pressure (MacMahon *et al.*, 1990). Several placebo-controlled trials showed that antihypertensive treatment reduces stroke risk by 38–42% (Collins *et al.*, 1990). This is also true for the elderly

and patients with isolated systolic hypertension (SHEP Cooperative Research Group, 1991; Staessen *et al.*, 1997). Early trials were performed with diuretics and beta blockers. Similar efficacy for stroke prevention was shown for the ACE inhibitors ramipril (The Heart Outcomes Prevention Evaluation Study Investigators, 2000) and captopril (Hansson *et al.*, 1999b) and the calcium channel blocker nitrendipin (Staessen, *et al.*, 1997). The STOP-2 trial (Hansson *et al.*, 1999a) compared conventional antihypertensive drugs (atenolol, metoprolol, pindolol, hydrochlorothiazide plus amiloride) with newer drugs (enalapril, lisinopril, felodipine, iradipine) and found no difference in stroke reduction. The lower limit for the reduction of blood pressure has not been clearly defined. The HOT trial showed that the lower the blood pressure, the lower the stroke risk (Hansson *et al.*, 1998). This trial also indicated that the benefit of angiotensin-converting enzyme (ACE) inhibitors (ramipril) on the reduction of stroke, MI, and death is only partly dependent on their blood pressure lowering effect. Not all drugs that lower blood pressure may prevent strokes. The alpha blocker doxazosin was clearly inferior to a diuretic (The ALLHAT Officers and Coordinators for the ALLHAT Collaborative Research Group, 2000). Long-acting nifedipine (Brown *et al.*, 2000) is not superior to diuretics for stroke prevention. Diltiazem decreases stroke risk more than diuretics, beta blockers, or a combination of both (Hansson *et al.*, 2000).

Smoking increases the stroke risk by a factor of 1.5 (Wolf *et al.*, 1988). There are no prospective randomized trials on smoking cessation and stroke risk. Observational studies indicate that the stroke risk decreases after smoking cessation and that it takes 2 to 5 years after a smoker quits to reach the stroke risk of a nonsmoker (Bonita *et al.*, 1986; Kawachi *et al.*, 1993).

Increased cholesterol and triglycerides are important risk factors for myocardial infarction but play a lesser

**TABLE I** Risk Factors for Stroke

| Nonmodifiable risk factors | Modifiable risk factors | Potentially modifiable risk factors | Modification not investigated in large trials |
|---|---|---|---|
| Age | Hypertension | Diabetes mellitus | Patent foramen ovale |
| Gender | Atrial fibrillation | Obesity | Alcohol dependence |
| Race | Other cardiac diseases | Physical inactivity | Drug abuse |
| Genetic disposition | Cigarette smoking | Cardiac failure | Chronic infection |
| | Hyperlipidemia | | Migraine |
| | | | Hyperhomocysteinemia |
| | | | Hypercoagulability |
| | | | Antiphospholipid antibodies |
| | | | Contraceptives |

role in stroke (Prospective Studies Collaboration, 1995). Several studies in patients with coronary heart disease treated with hydroxymethyl glutaryl coenzyme A (HMG CoA) reductase inhibitors (so called "statins") showed a significant decrease in TIA and stroke rates (The Long-Term Intervention with Pravastatin in Ischemic Disease (LIPID) Study Group, 1998; The Scandinavian Simvastatin Survival Study, 1994; White et al., 2000). This has not been shown for dietary interventions and fibrates. The decrease in stroke risk is independent from the decrease in cholesterol plasma levels (Blauw et al., 1997). This may indicate neuroprotective properties of the statins or an effect on plaque stabalization. Diabetes mellitus is a major risk factor for stroke. Large intervention trials showed that strict control of blood glucose will decrease the likelihood of microvascular complications such as nephropathy, retinopathy, and polyneuropathy but has only a minor impact on the risk of stroke (Adler et al., 2000; Diabetes Control and Complications Trial Research Group, 1993; Stratton et al., 2000). Additional risk factors are physical inactivity (Sacco et al., 1998; Wannamethee and Shaper, 1992) and obesity (Goldberg et al., 1995). Regular physical activity reduces stroke risk (Lee et al., 1999). Low-dose oral contraceptives are not clearly associated with an increased stroke risk (Schwartz et al., 1998). However, a recently published meta-analysis found a relative risk of 1.93 (95%; CI, 1.35–2.74) for low-estrogen preparations in population-based studies that controlled for smoking and hypertension (Gillum et al., 2000). Chronic alcoholism increases stroke risk, whereas small amounts of alcohol reduce the risk of ischemic stroke (Berger et al., 1999; Easton, 1995). Increased levels of homocysteine are a risk factor for stroke (Perry et al., 1995; Ridker et al., 1999). It is unknown whether treatment with folic acid or vitamins decreases stroke risk.

### Nonrheumatic Atrial Fibrillation

In general, patients with nonrheumatic atrial fibrillation have a stroke risk of approximately 5%/year.

However, the risk of stroke increases substantially in patients with specific risk factors. The most important risk factors are previous stroke/TIA, history of hypertension, poor LV function, age >75 years, rheumatic mitral valve disease, and prosthetic heart valves. The benefit of oral anticoagulation in patients with atrial fibrillation (AF) was studied in six large randomized trials. A meta-anlysis of these studies (Hart et al., 1998; Report of the Quality Standards Subcommittee of the American Academy of Neurology, 1998) showed that anticoagulation with warfarin achieving an international normalized ratio (INR) between 2 and 3 was associated with a 60–70% relative risk reduction for ischemic stroke (Table II). INR values above the therapeutic range (2.0–3.0) lead to bleeding complications and INR values below 2.0 are substantially less effective in preventing ischemic stroke. Four of the randomized trials investigated the efficacy of aspirin in AF patients. Compared to placebo aspirin reduced stroke risk on average by 20% (the same range as in secondary prevention of stroke).

### Primary Prevention in Patients with Other Cardiac Diseases

In patients with rheumatic valvular heart disease (Salem et al., 1998) or after valve replacement with a mechanical heart valve long-term anticogulation is highly effective (Cannegieter et al., 1995; Stein et al., 1998). The estimated annual stroke risk for patients with mechanical valves is 1–4% and for patients with bioprothetic valves is 0.2–2.9%. An INR in the range of 2.5 to 3.5 is most effective in reducing the risk of embolic events in patients with mechanical valves and has a low rate of bleeding complications. In bioprosthetic valves in the mitral position anticoagulation is recommended for the first 3 months followed by antiplatelet drugs thereafter. Stroke can be a complication of an acute myocardial infarction and occurs in about 2.5% of patients with acute MI in the first 4 weeks. Anticoagulation is recommended following MI

TABLE II  Primary Prevention of Stroke

| Condition | Recommended prevention |
|---|---|
| Asymptomatic | Treatment of risk factors |
| Idiopathic atrial fibrillation; age, <65 years | None or aspirin |
| Atrial fibrillation; no vascular risk factors; age, 65–75 years | Aspirin, 300–325 mg*** |
| Atrial fibrillation; at least one vascular risk factor (hypertension, diabetes mellitus, cardiac failure, smoking) | Oral anticoagulation INR, 2–3; age, >75 years; INR, 2.0*** |
| Atrial fibrillation; contraindications for anticoagulation (e.g., cerebral microangiopathy) | Aspirin, 300–325 mg |
| ICA stenosis, >60% | Treatment of risk factors; consider surgery for selected patients** |

Note. Aspirin = acetylsalicyclic acid; INR = international normalized ratio.
**: One prospective randomized study.
***: More than one prospective, randomized, controlled study.

in patients with atrial fibrillation or poor left ventricular function. Patent foramen ovale (PFO) appears to be a risk factor for stroke in young people, however, the stroke recurrence rate is very low. The risk–benefit ratio of antithrombotic therapy in asymptomatic patients with PFOs has not been addressed.

### Antiplatelet Drugs

Two large prospective trials investigated the use of acetylsalicyclic acid (aspirin) in primary prevention in male physicians. The open British trial included 5139 physicians and treated half of them with 500 mg aspirin per day (Peto et al., 1988). There was no difference in the incidence of stroke. The Physicians Health Study (Steering Committee of the Physicians Health Study Research Group, 1989) was a randomized, double-blind trial with 22,071 male physicians who were treated with 325 mg aspirin every other day or placebo. The trial showed a 44% relative risk reduction for myocardial infarction but no risk reduction for stroke. There was a tendency for an increased number of cerebral hemorrhages with the use of aspirin. The Nurses Health Study, performed in women, also was unable to show a benefit of aspirin in stroke prevention (Iso et al., 1999). A recent meta-analysis of five large trials included 250,251 persons. Doses of aspirin varied between 75 and 650 mg/day. Annual mortality ranged between 0.4 and 3%. The stroke rate with placebo was only 0.3%/year. Across all studies relative stroke risk increased by 8% which was mainly due to an increased number of cerebral hemorrhages (Hart et al., 2000). Other antiplatelet drugs have not been investigated for primary stroke prevention.

### Carotid Endarterectomy

The role of carotid endarterectomy in patients with asymptomatic carotid stenosis is hotly debated (Barnett et al., 1996). The annual incidence of ischemic strokes

in patients with asymptomatic carotid stenosis >50% ranges from 0.2 to 3.3% (Hennerici et al., 1987; Norris et al., 1991). The average incidence of cerebral infarcts ipsilateral to a carotid stenosis is only 0.5% (Rosa, 1990). The risk is increased in patients with a high degree of internal carotid artery stenosis (Norris et al., 1991), in patients with multiple vessel disease including both carotid and the vertebral arteries, and in patients with progression of stenosis (Mess et al., 1990).

The first prospective study on 406 patients with carotid stenosis ranging from 50 to 90% did not show a positive effect of an operative treatment (The Casanova Study Group, 1991). The VA study involving 444 men found a reduced incidence of TIA in the operated patients but no difference in stroke rates or mortality (Hobson et al., 1993). The Asymptomatic Carotid Endarterectomy Study (ACAS) enrolled 1662 patients with >60% stenosis of the internal carotid artery (ICA) and randomized them to best medical treatment alone or to best medical treatment plus carotid endarterectomy. Mean follow-up was 2.7 years. Concerning the projected 5-year risk of stroke, there was a 53% relative risk reduction in favor of the operation (Executive Committee for the Asymptomatic Carotid Atherosclerosis, 1995). The relative risk reduction seems impressive, but the absolute risk reduction was only about 1% per year. In a subgroup analysis there was no benefit of endarterectomy in women, and severe strokes leading to permanent disability were not prevented. A recently published subgroup analysis of the ACAS Study focused on the 163 patients who had a contralateral carotid occlusion at the time they were entered in the study. Surprisingly, the medically managed patients with contralateral occlusions had a very low stroke rate and there appeared to be no benefit of surgery for this subgroup (Baker et al., 2000). Another concern is whether surgeons in a community practice will match the low complication rates of the ACAS trial. The largest randomized trial, the Asymptomatic Carotid Sugery

Trial (ACST) is still ongoing (Halliday for the Steering Committee and for the Collaborators, 1994). In addition some randomized trials comparing endarterectomy and carotid stenting will include asymptomatic patients. Results are not yet available.

### Practical Management

Patients with hypertension should be treated with antihypertensive drugs. Hypertension is defined as systolic blood pressure >140 mm Hg and diastolic blood pressure >90 mm Hg. Patients with clinical manifestations of vascular disease (diabetes mellitus, cardiac failure, coronary heart disease) should maintain systolic blood pressure <140 mm Hg or lower. Patients with myocardial infarction or angina pectoris should be treated with a beta blocker. Patients with diabetes mellitus and incipient or manifest nephropathy have the highest benefit with ACE inhibitors. This is also true for patients with cardiac failure. Patients with isolated systolic hypertension can be treated with diuretics or a long-acting calcium antagonist. There is no well-defined lower limit for lowering blood pressure. The limiting factors are tolerability and the incidence of orthostatic hypotension.

Neurologists should council all smokers to quit and offer support with pharmacological therapies (nicotine patch, nicotine gum, SSRI, buproprion) and nonpharmacological interventions (behavioral therapy, support groups). Patients with a prior history of angina pectoris or myocardial infarction and high cholesterol levels should be treated with a statin. Placebo-controlled trials with stroke as an endpoint are only available for simvastatin and pravastatin. LDL cholesterol levels should be <160 mg/dL in patients with one vascular risk factor and <130 mg/dL if two or more risk factors exist. HDL cholesterol levels should be >35 mg/dL and triglycerides <200 mg/dL. Diabetes should be treated with diet, antidiabetic drugs, insulin when required, and regular physical activity.

Patients with atrial fibrillation and vascular risk factors, e.g., hypertension, cardiac failure, age >75 years should receive oral anticoagulation with an INR range between 2.0 and 3.0. An INR of 2.0 may be optimal for patients >75 years. Patients with lone atrial fibrillation below age 65 should not receive anticoagulation. Patients >65 years without other risk factors can be treated with aspirin 300–325 mg/day. Aspirin is also used in patients with contraindications for warfarin. Most contraindications, however, apply for both warfarin and aspirin (severe hepatic disease, high bleeding risk, bleeding ulcers, chronic alcohol abuse, frequent falls, dementia, uncontrolled hypertension).

Operation of asymptomatic ICA stenosis should only be considered for patients with stenosis >60% who are good surgical candidates and can be operated on by a surgeon who has a complication rate below 5%.

### Treatments No Longer Recommended

- Aspirin is not recommended for primary prevention. The possible benefit in reducing ischemic stroke is counterbalanced by an increase in cerebral hemorrhages.
- A fixed combination of aspirin (325 mg per day) and warfarin (INR 1.2–1.5) is not recommended for stroke prevention in patients with atrial fibrillation (Stroke Prevention in Atrial Fibrillation Investigators, 1996).
- CEA for asymptomatic stenosis by a surgeon with a complication rate >5%.
- Anticoagulation in patients with mitral valve prolapse (Gilon et al., 1999).
- Treatment of patent foramen ovale in asymptomatic persons.

## SECONDARY PREVENTION

### Clinical Aspects

Secondary prevention is defined as treatments or procedures to prevent another ischemic cerebral event after a TIA or a first stroke.

### Natural Course

Among the 80–85% of patients who survive a stroke, the risk of recurrent stroke is between 5 and 15% in the first year with the highest risk in the first few weeks after the acute event. The risk of a recurrent stroke or a stroke following a TIA is higher in patients with multiple vascular risk factors and patients who in addition to a TIA or stroke already have coronary heart disease diabetes or peripheral arterial disease. Other risk factors include age >60 years, symptom duration >10 min, and weakness or speech impairment occurring during the TIA (Johnston, 2000). The risk is higher after TIA in the cerebral circulation than following amaurosis fugax (The Amaurosis Fugax Study Group, 1990).

### Principles of Therapy

Treatment should be based on the nature, location, and severity of the causative lesion. The causes can be vascular (stenosis, plaques, occlusion of extracranial and intracranial large arteries, or disease of small penetrating arteries), cardiac (causing embolism or systemic hypoperfusion) or hematologic causing hypercoagulability. The mechanism of ischemia should be clarified (hemodynamic versus embolic). All patients with TIAs and strokes should be evaluated for the underlying causes of the ischemia and appropriate treatment aimed at the etiology. Unfortunately in the past, few systematic studies of brain ischemia based on etiology were performed. Exceptions are studies of treatment for patients with atrial fibrillation and carotid stenosis which is discussed in detail herein.

There are four different types of prophylaxis for patients with cerebral ischemia (Table III).

- Treatment of vascular risk factors.
- Modification of coagulation in order to prevent the formation, propagation, and embolization of white platelet–fibrin thrombi and red erythrocyte–fibrin thrombi. The commonest drugs used are those that affect platelet functions (e.g., aspirin, ticlopidine, clopidogrel, aspirin plus dipyridamole, GPIIb/IIIa antagonists) and standard anticoagulants (heparin, heparinoids, warfarin). The standard anticoagulants may be more effective at preventing red clots while the drugs that affect platelet function may be more active against white thrombi.
- Reperfusion. These treatments aim to prevent occlusion of brain-supplying arteries with severe occlusive changes but not thrombosed. Reperfusion can be obtained by direct surgery (endarterectomy) or stenting with or without angioplasty.
- Augmentation of brain blood flow. This can be performed by increasing blood volume and/or blood pressure and by using agents that increase flow in the affected or collateral arteries.

Herein we concentrate the discussion on drugs and other treatments that have been systematically tested in controlled trials.

### Treatment of Risk Factors

As shown previously, there exist many studies on the control of risk factors and the influence on stroke. There are almost no controlled trials on risk factor modifica-

TABLE III Principles of Prophylaxis in TIA or after Ischemic Stroke

| Mechanism of ischemia | Treatment or prophylaxis |
|---|---|
| Platelet–fibrin aggregates (e.g., arteriosclerotic plaque) | Antiplatelet drugs |
| Erythrocyte–fibrin thrombi (e.g., cardiac origin) | Anticoagulation |
| ICA stenosis, >70% | Endarterectomy; stenting plus antiplatelet drugs |
| Hemodynamic | Elevation of blood pressure |

tion in secondary prevention. The INDANA group performed a meta-analysis of all trials investigating the influence of antihypertensive treatment on recurrent strokes in patients with TIA (12%) or stroke (88%). Nine placebo-controlled studies included 6772 patients (The INDANA Project Collaborators, 1997). The average follow-up period was 1.8 years (much shorter than in the primary prevention trials). The number of strokes was 237 in patients treated with active drug and 270 in placebo-treated patients. This resulted in a relative risk of 72% with a 95% confidence interval between 61 and 85%. Nonsignificant risk reductions were observed for mortality (14%) and myocardial infarctions (12%). Recently, the PROGRESS trial demonstrated that aggressive blood pressure reduction following stroke with perindopril plus indapamid can substantially reduce stroke recurrence (PROGRESS Collaborative Group, 2001). The use of statins for stroke prevention has only been investigated in primary prevention trials (Blauw et al., 1997; Plehn et al., 1999; Rosendorff, 1998; The Long-Term Intervention with Pravastatin in Ischemic Disease (LIPID) Study Group, 1998; The Scandinavian Simvastatin Survival Study, 1994; White et al., 2000). Two large ongoing trials investigate the use of statins in primary and secondary prevention (Shepherd et al., 1999), and the SPARCL trial. A placebo-controlled trial with gemfibrozil found a decrease in cardiac mortality but not in stroke incidence (Rubins et al., 1999). Treatment of diabetes mellitus was investigated mainly in primary prevention (Diabetes Control and Complications Trial Research Group, 1993). Prospective trials investigating decrease in homocysteine or Lp(a)-lipoprotein levels and the use of vitamin E are under way.

### Antiplatelet Drugs

Antiplatelet therapy has become the mainstay of secondary stroke prevention. Recurrence risk after TIA or ischemic stroke ranges from 5 to 20% per year (Wilterdink and Easton, 1992). Numerous trials and meta-analyses have left little doubt that antiplatelet therapy effectively reduces stroke risk in patients with prior stroke or TIA (Antiplatelet Trialists Collaboration, 1988, 1994; Patrono and Roth, 1996). Nevertheless, controversies exist. Foremost among them is the debate over the optimal antiplatelet regimen. The majority of research in secondary stroke prevention supports the clinical value of acetylsalicylic acid (aspirin).

Early studies demonstrated that antiplatelet therapy significantly reduces the risk of fatal and nonfatal stroke among patients with prior stroke or TIA (American–Canadian Co-operative Study Group, 1985). The second 1994 meta-analysis by the Antiplatelet Trialists' Collaboration concluded that antiplatelet therapy

TABLE IV    Probable or Definite Ischemic Stroke, Survival to End of Study; Data from
Antithrombotic Trialists' Collaboration, 2002

| | | Probable or definite ischemic stroke | | |
| Category of trial | No. of trials with data | Antiplatelet | Adjusted controls | Percent odds reduction (SD) |
| --- | --- | --- | --- | --- |
| Prior MI | 12 | 57/5.476 (1.04%) | 82/5,507 (1.49%) | 31% (14) |
| Acute MI | 15 | 37/8,821 (0.43%) | 65/8,830 (0.75%) | 42% (15) |
| Prior stroke/TIA | 21 | 768/9,553 (8.04%) | 1012/9.610 (10.53%) | 25% (5) |
| Other high risk | 140 | 155/4,498 (3.45%) | 235/4,529 (5.19%) | 37% (9) |

reduced the risk of nonfatal stroke by 23% in patients with a history of stroke or TIA (Antiplatelet Trialists Collaboration, 1994). When high-risk patients (defined as those with acute myocardial infarction, prior myocardial infarction, prior stroke, or TIA) were combined with the low-risk group (defined as patients who are candidates for primary prevention), the overall reduction in risk of nonfatal stroke was 25%, which was highly significant ($2P < 0.00001$). Among high-risk patients, the reduction in risk of stroke was 37% (Antithrombotic Trialists Collaboration, 2002, Table IV). However, these findings may not accurately reflect the potential benefits of antiplatelet therapy, because most of the early stroke-prevention studies enrolled patients without regard to the pathophysiology of their stroke. For example, in patients with atrial fibrillation, warfarin has been shown to be twice as effective as aspirin (Hart and Halperin, 1994). Similarly, in patients with TIA or minor stroke with ipsilateral internal carotid artery stenosis greater than 70%, endarterectomy has yielded a 60% reduction in risk of stroke (European Carotid Surgery Trialists' Collaborative Group, 1998; North American Symptomatic Carotid Endarterectomy Trial Collaborators, 1991). If these patients had been excluded from meta-analyses, a more significant effect for antiplatelet therapy may have emerged.

Relative risk reduction also may have been skewed by limiting inclusion criteria to patients with minor stroke or TIA. With the notable exceptions of ESPS-1 (The ESPS Group, 1987), ESPS-2 (Diener et al., 1997), and the Canadian American Ticlopidine Study (Gent et al., 1989), which enrolled patients regardless of stroke severity, most major trials excluded high-risk patients (e.g., survivors of moderate-to-severe strokes). Yet these are the patients who would be expected to benefit most from antiplatelet therapy. In fact, the Antiplatelet Trialists determined that 37 vascular events per 1000 high-risk patients would be avoided if the patients were treated with antiplatelet therapy for 33 months (Antiplatelet Trialists Collaboration, 1994).

Clinically, the most important concern with any therapy is the individual patient's outcome. Decreased morbidity and mortality are clinical goals of stroke prevention that must be considered in the context of the patient's quality of life. Fatal stroke is clearly an undesirable outcome, and nonfatal strokes can lead to severe, permanent disability with devastating consequences for patients, their families, and society. However, very few trials have stratified nonfatal stroke on the basis of disability. The Antiplatelet Trialists identified only 14 trials of high-risk patients that provided information on subsequent disability. Antiplatelet therapy was associated with a 24% reduction in disabling or fatal stroke, and a 17% reduction in nondisabling stroke (Antiplatelet Trialists Collaboration, 1994). This meta-analysis indicated that the benefits of antiplatelet therapy extend to the patient's quality of life by reducing disability and its consequences. With one exception, however, no single trial in patients with TIA or stroke has shown that antiplatelet therapy reduces all cause or vascular mortality. ESPS1 was the only trial that showed a reduced mortality with a combination of aspirin and dipyridamole (The ESPS Group, 1987).

Patients with an acute ischemic stroke have a 3–5% risk of an immediate second stroke within the first 4 weeks after the event. Two large trials, the International Stroke Trial (IST) (International Stroke Trial Collaborative Group, 1997) and a Chinese study (CAST (Chinese Acute Stroke Trial) Collaborative Group, 1997) investigated, whether the early use of aspirin can decrease the number of recurrent strokes. In analogy to MI, the studies also investigated whether mortality can be reduced by aspirin.

IST was a randomized open trial in 19,435 stroke patients recruited in 467 hospitals in 36 countries. The study included patients within 48 h of stroke onset. In a factorial design patients were treated with 300 mg aspirin or no aspirin and a combination of heparin or no heparin. Here only the aspirin results are reported. Mortality within the first 2 weeks in the aspirin group

($n = 872$; 9.0%) was lower than that in the control group ($n = 909$; 9.4%). The difference, however, was not significant. The number of recurrent strokes within 14 days was lower with aspirin (2.8%) versus control (3.9%). The number of cerebral hemorrhages was not increased. Hemorrhagic complications requiring blood transfusions or surgery were more frequent in the aspirin group. IST had many methodological shortcomings (not blinded, no placebo-control, low number of early CTs). The number of patients was large enough to compensate in part for this and the results can be transfered to clinical practice. CAST was a randomized placebo-controlled trial recruiting 21,106 chinese patients within 48 h of an acute stroke. In contrast to IST almost all patients were investigated by CT to exclude cerebral hemorrhage. The oberservation period was 4 weeks. The aspirin dose was 160 mg. Mortality was significantly reduced in the aspirin group ($n = 343$, 3.3% aspirin; $n = 398$, 3.9% placebo; $P = 0.04$). Early recurrent ischemic strokes were reduced in the aspirin group (1.6% versus 2.1%), but hemorrhagic strokes were more frequent in the aspirin group (1.1% vs 0.9%).

A meta-analysis of both trials including 20,207 patients treated with either 300 or 160 mg aspirin and 20,190 patients treated with placebo or no aspirin showed 7/1000 treated patients had fewer recurrent ischemic strokes, but 2/1000 had additional cerebral hemorrhages. Mortality was reduced by 5/1000. These numbers just achieved significance. While reduction of recurrent stroke and mortality of 0.5% is small to the individual, the world-wide public health importance of using aspirin in acute stroke is tremendous. Despite its small expense, aspirin remains underutilized.

In conclusion, aspirin in doses between 160 and 300 mg results in a modest decrease in recurrent ischemic strokes and mortality in patients with acute ischemic strokes. The downside is a small increase in bleeding complications including cerebral hemorrhages.

The use of aspirin in the secondary prevention of stroke was studied in 10 placebo-controlled trials of different doses of aspirin and placebo. Only 4 trials were large enough to have appropriate power to answer the question of whether aspirin is effective (Diener *et al.*, 1996; ESPS-Working Group, 1992; The SALT Collaborative Group, 1991; UK-TIA Study Group, 1991). In summary, aspirin reduced the relative risk of the combined endpoint stroke, MI, and vascular death by 13%. The absolute risk reduction was 3%. This modest effect explains why other antiplatelet drugs such as ticlopidine (Gent, *et al.*, 1989; Hass *et al.*, 1989) and clopidogrel (CAPRIE Steering Committee, 1996) were investigated.

The preferred dose of aspirin for the secondary prevention of TIA or stroke was hotly debated for many years. Many North Americans favored high doses (>900 mg) (Dyken, 1993; Dyken *et al.*, 1992) whereas most Europeans used lower doses (Patrono and Roth, 1996). Tijssen performed a meta-analysis including more than 9000 patients from 10 studies (reported in Diener, 1998). The studies were categorized according to dose ranges (see Table V). There was no relationship between the dose of aspirin and risk reduction. The same result was achieved by an independent meta-analysis by Algra *et al.* (Algra and van Gijn, 1996; Algra and van Gijn, 1999). The American FDA ruled that the recommended dose of aspirin for secondary prevention of stroke is 50 to 325 mg (Department of Health and Human Services, and Food and Drug Administration, 1998).

A major side effect of aspirin therapy is bleeding. The rate of severe hemorrhages—hemorrhages that require the patient to have a transfusion or to be hospitalized—has not been found to be dose-dependent. In the Dutch TIA trial (The Dutch TIA Trial Study Group, 1991) the effect of 30 mg aspirin was compared to that of 283 mg in 3000 patients. Major bleeding complications occurred in 3.4% of patients in the medium-dose group versus 2.6% in the low-dose group. In the UK-TIA trial (UK-TIA Study Group, 1991) the number of cerebral hemorrhagic strokes was identical (0.9%) in the 300-mg and 1200-mg dose groups, whereas gastrointestinal hemorrhages were slightly higher in the high-dose population (4.8% vs 3.1%). The incidence of all-site bleedings from four studies using low to medium doses of aspirin is shown in Table VI. There is, however,

TABLE V  Dose of Aspirin and Relative Risk of Vascular Events in Relation to Placebo

| Dose of aspirin (mg) | Relative risk | 95% Confidence interval |
|---|---|---|
| 1000–1300 | 0.87 | 0.76–0.98 |
| 300 | 0.91 | 0.76–0.98 |
| 50–75 | 0.86 | 0.77–0.97 |
| Overall | 0.87 | 0.80–0.94 |

TABLE VI  Bleeding Risk in Different Studies for Stroke Prevention

| Study | Treatment | Incidence of all-site bleeding (%) |
|---|---|---|
| ESPS-2 (Diener *et al.*, 1996) | 50 mg aspirin | 8.2 |
| SALT (The SALT Collaborative Group, 1991) | 75 mg aspirin | 7.2 |
| Dutch TIA (The Dutch TIA Trial Study Group, 1991) | 282 mg aspirin | 8.7 |
| CAPRIE (CAPRIE Steering Committee, 1996) | 325 mg aspirin | 9.3 |

definitely a relationship between the aspirin dose and gastrointestinal side effects. The results of the UK-TIA trial (UK-TIA Study Group, 1991) show that upper gastrointestinal symptoms occurred in 41.5% in the group given 1200 mg aspirin, 31.4% in the group given 300 mg aspirin, and 25.7% in the placebo group. This trend was the same in all the other trials. A formal meta-analysis is impossible because the single trials used different definitions and reporting systems for gastrointestinal side effects.

Ticlopidine inhibits platelet aggregation by directly altering platelet membranes, independent of any effect on prostaglandins. The efficacy of ticlopidine versus placebo was studied in the Canadian-American Ticlopidine Study (Gent et al., 1989). The trial included 1072 stroke patients who were treated either with 250 mg ticlopidine b.i.d. or placebo. The relative risk reduction for recurrent stroke, myocardial infarction, and death was 24.1% (P = 0.023) in patients who received 250 mg ticlopidine twice daily. In the on-treatment group, stroke was reduced by 33.5% in favor of ticlopidine. In the intention-to-treat analysis, other outcome events, such as stroke, myocardial infarction, or death, or all strokes were also reduced by ticlopidine but the difference did not reach significance. There was no significant difference between ticlopidine and placebo concerning vascular death and death of any cause. Another multicenter trial compared a daily dose of 500 mg ticlopidine and 1300 mg daily of aspirin in 3069 patients with TIA or minor stroke (Hass et al., 1989). In the first year, the event rates clearly diverged, favoring ticlopidine over aspirin. At 3 years, however, the rate of nonfatal stroke or death from all causes was 17% for ticlopidine and 19% for aspirin, a result of borderline statistical significance (P = 0.048). This study found a statistically significant 21% reduction (P = 0.024) in fatal or nonfatal stroke risk at 3 years in patients who received ticlopidine versus aspirin. The relative risk reduction of the combined outcome stroke, myocardial infarction, or vascular death was reduced by 9% in favor of ticlopidine, which was not statistically significant.

The overall incidence of adverse events with ticlopidine and aspirin was 62.3% versus 53.2%, respectively. Diarrhea was the most common adverse event, reported in 20% of ticlopidine patients and 10% of the aspirin patients. The incidence of dyspepsia and bleeding was similar for the two treatment groups. Severe neutropenia with ticlopidine occurred in about 0.9% of patients. Because of the potential for significant adverse events, such as neutropenia, blood counts are recommended during the first 3 months of ticlopidine therapy. More recently many more cases of thrombotic thrombocytopenic purpura have been reported with the intake of ticlopidine (Bennett et al., 1998; Page et al., 1991). This side effect does not only occur in the first 3 months of treatment. At present, because of its adverse effects profile and the availability of alternative antiplatelet agents, ticlopidine is rarely used.

Clopidogrel is an antiplatelet agent that is chemically related to ticlopidine. Clopidogrel is a new thienopyridine derivative that acts by selectively and irreversibly inhibiting binding of adenosine diphosphate (ADP) to its platelet receptor and affects the ADP-dependent activation of the GPIIb–IIIa complex, which is the major receptor of available fibrinogen. A pivotal randomized, blinded, international trial, Clopidogrel versus Aspirin in Patients at Risk of Ischaemic Events (CAPRIE), examined the relative safety and efficacy of daily doses of 75 mg clopidogrel versus 325 mg aspirin in nearly 20,000 patients with stroke, myocardial infarction, or peripheral arterial disease (CAPRIE Steering Committee, 1996). The results of the trial showed that clopidogrel was more effective than aspirin in preventing a combined endpoint of ischemic stroke, myocardial infarction, or vascular death: The trialists found a significant 8.7% relative-risk reduction (P = 0.043) for clopidogrel versus aspirin. Although the CAPRIE trial was not powered to detect treatment differences within patient subgroups, a subgroup analysis, which was not part of the original design, was performed. A test of heterogeneity revealed that there was not an equal effect on treatment outcome of the three clinical subgroups (P = 0.042). Overall, when the results for these subgroups were examined, there was no significant difference between clopidogrel and aspirin in patients with stroke or myocardial infarction. There was, however, a significant benefit favoring clopidogrel in patients with peripheral arterial disease. Specifically, the relative-risk reduction for stroke patients was 7.3% (P = 0.26) in favor of clopidogrel. For patients with myocardial infarction, there was a relative-risk increase of 3.7% (P = 0.66) in favor of aspirin. In comparison, patients with peripheral arterial disease had a significant risk reduction of 23.8% (P = 0.0028) in favor of clopidogrel. The combination of aspirin (75 mg) plus clopidogrel versus clopidogrel in patients after TIA and stroke and high risk is being investigated in the ongoing MATCH trial.

Regarding safety, side effects of clopidogrel and aspirin were similar. Although clopidogrel was not compared with its relative, ticlopidine, the CAPRIE trialists note that the newer agent has a more favorable safety profile than ticlopidine: neutropenia occurs in only 0.1% of clopidogrel-treated patients. In the CAPRIE trial, the frequency of severe rash was higher with clopidogrel than with aspirin. Recently a few cases of thrombotic thrombocytopenic purpura were described within the first 2 weeks of clopidogrel intake (Bennett et al., 2000).

Dipyridamole is an antiplatelet agent that inhibits cyclic nucleotide phosphodiesterase and blocks of the uptake of adenosine (Patrono *et al.*, 1998). The first European Stroke Prevention Study (ESPS-1), which was reported in 1987, included 2500 patients with stroke or TIA (The ESPS Group, 1987). Patients received high-dose aspirin (990 mg) plus dipyridamole (225 mg) daily or placebo. The combination antiplatelet regimen reduced the 2-year incidence of stroke and all-cause mortality by 33.5%. Fatal and nonfatal stroke risk was reduced by 38.1%, representing a greater risk reduction than ever reported with monotherapy for secondary stroke prevention. Four trials investigated whether the combination of dipyridamole plus aspirin is superior to aspirin alone. Three trials with patient numbers of less than 1000 were negative (American–Canadian Co-operative Study Group, 1985; Bousser *et al.*, 1983; Guiraud-Chaumeil *et al.*, 1982). They might have lacked the statistical power to detect a difference. The largest trial was ESPS-2 (Diener *et al.*, 1996). In light of the aspirin-related side effects observed in ESPS-1, the question of appropriate aspirin dose was a paramount issue for the second trial. Further, the study sought to clarify the relative contributions of aspirin and dipyridamole in prevention of secondary stroke.

ESPS-2 analyzed 6602 stroke or TIA patients who were randomly assigned to four treatment arms: placebo, aspirin alone (25 mg twice daily), extended-release dipyridamole alone (200 mg twice daily), or aspirin (25 mg twice daily) plus extended-release dipyridamole (200 mg twice daily). These dosages were chosen on the basis of a double-blind, randomized study that showed the additive effect of low-dose aspirin and dipyridamole on the inhibition of *ex vivo* platelet aggregation in healthy volunteers (Müller *et al.*, 1990). The low aspirin dose of 25 mg twice daily had not been tested previously for clinical endpoints. Therefore, the trial included a placebo arm to test if low-dose aspirin was superior to placebo in secondary prevention of stroke.

The trial confirmed the additive effects of aspirin and dipyridamole. The aspirin-plus-dipyridamole regimen of ESPS-2 produced a statistically significant 37% reduction ($P = 0.001$) in risk of fatal or nonfatal stroke over 2 years compared with placebo, similar to the risk reduction of the earlier ESPS-1 trial. ESPS-2 also was the first trial to show that 25 mg aspirin twice daily is clinically effective in secondary stroke prevention. When given as monotherapy versus placebo, each agent in the combination was associated with a statistically significant reduction in stroke risk: 18% ($P = 0.013$) for aspirin and 16% ($P = 0.039$) for dipyridamole. Neither aspirin nor dipyridamole or the combination reduced mortality (Diener *et al.*, 1997).

The overall incidence of adverse events in the ESPS-2 trial was approximately the same for treatment arms and reduced in frequency after the first 30 days. About one-third of all patients had mostly mild or insignificant recurring events—either GI disturbances (including diarrhea) or headache. The incidence of anysite bleeding was 8.7% for the aspirin–dipyridamole combination, 4.7% for dipyridamole alone, 8.2% for aspirin alone, and 4.5% for placebo. Headache was the most common reason for discontinuation of drug intake, occurring in 8.1% of patients in the aspirin–dipyridamole combination arm, 8.0% in the dipyridamole-alone group, 1.9% in the aspirin-alone arm, and 2.4% in the placebo group.

### GPIIb/IIIa Antagonists

GPIIb/IIIa antagonists given iv are able to reduce early mortality in patients with myocardial infarctions (Topol *et al.*, 1999). They failed when given orally for the long-term prevention of cardiac events. The same seems to be true for the secondary prevention of stroke (Topol *et al.*, 2000).

### Anticoagulants

The role of anticoagulants or aspirin as secondary prevention in patients with atrial fibrillation was investigated in a randomized trial (EAFT Group, 1993). Anticoagulation resulted in a 70% risk reduction, whereas aspirin decreased the stroke risk by 15%. The SPIRIT study assessed the efficacy of oral anticoagulation for secondary prevention of noncardioembolic stroke and compared high-intensity oral anticoagulation (INR, 3.0–4.5) with aspirin 30 mg per day (The Stroke Prevention in Reversible Ischemia Trial (SPIRIT) Study Group, 1997) in 1316 patients with TIA or minor ischemic stroke. The study was stopped prematurely by the Safety Committee because of a significant increase in the rate of major bleeding complications including cerebral hemorrhages in the anticoagulation group. The Warfarin–Aspirin Recurrent Stroke Study (WARSS) compared a lower INR of 1.4 to 2.8 with 325 mg aspirin. No difference in the rate of recurrent stroke or death was detected (Mohr *et al.*, 2002). The Warfarin–Aspirin Symptomatic Intracranial Disease (WASID) study was a nonrandomized retrospective trial in patients with 50–99% stenosis of an intracranial artery. Sixty-three patients received aspirin and 88 were anticoagulated with warfarin (Chimowitz *et al.*, 1995). There was a significantly lower incidence of vascular events in the warfarin-treated group but an increased incidence of major hemorrhagic complications. A larger prospective randomized trial is under way.

### Carotid Endarterectomy or Stenting

The efficacy of carotid endarterectomy for stroke prevention was proven by two large randomized trials.

The North American Symptomatic Carotid Endarterectomy Trial (NASCET) (Barnett *et al.*, 1998; North American Symptomatic Carotid Endarterectomy Trial Collaborators, 1991) randomized patients with symptomatic stenosis (70–99%) and moderate stenosis (30–69%). Degree of stenosis was evaluated by angiography. The European Carotid Surgery Trial (European Carotid Surgery Trialists' Collaborative Group, 1991; European Carotid Surgery Trialists' Collaborative Group, 1998) also included patients with <30% stenosis. Taken together the two trials showed that surgery resulted in a 60–80% relative risk reduction for ipsilateral stroke. No benefit for carotid endarterectomy was seen with less than 50% stenosis. Benefit in the 50–69% stenosis group was very small. Predictors for a significant risk reduction with carotid endarterectomy are male sex, severe stenosis, ulceration within the stenosis, and hemispheric versus retinal ischemia (Rothwell *et al.*, 1999). Surgical benefit is only maintained with a low perioperative complication rate. The 30-day complication rate in the two trials was between 5.8 and 7%. A large randomized trial showed that lower doses of aspirin (81 or 325 mg) are preferable to higher doses (650 or 1300 mg) in patients who undergo carotid endarterectomy (Taylor *et al.*, 1999). Angioplasty and stenting is presently being investigated in several randomized trials. Smaller trials and case series indicate that the complication rate may be identical to that of surgery (Eckert *et al.*, 1996; Ferguson and Ferguson, 1996; Golledge *et al.*, 2000; CAVATAS Investigators, 2001). Limited long-term data on the incidence of restenosis are not available; however, a recent study of carotid angioplasty in patients who were not good surgical candidates demonstrated a relatively high short-term rate of restenosis or occlusion (Malek *et al.*, 2000).

**Practical Management** (Table VII)

In patients with TIA or ischemic stroke of noncardiac origin antiplatelet drugs are able to decrease the risk of stroke by 11–15% and the risk of stroke, MI, and vascular death by 15–22%. Aspirin leads to a moderate but significant reduction of stroke, MI and vascular death in patients with TIA and ischemic stroke. Low doses are as effective as high doses but are better tolerated in terms of gastrointestinal side effects. The recommended aspirin dose therefore is between 50 and 325 mg. Bleeding complications increase at higher doses but also occur with the lowest doses. The combination of aspirin (2 × 25 mg) with slow release dipyridamole (2 × 200 mg) is superior to aspirin alone for stroke prevention but it is more expensive. Depending on cost considerations, this agent may be used as either a first or a second choice therapy. Ticlopidine is effective in secondary stroke prevention in patients with TIA and stroke. For some endpoints it is superior to aspirin. Due to its side effect profile (neutropenia, TTP) and the availability of safer alternatives, ticlopidine is not recommended as an option for initial therapy. Clopidogrel has a better safety

**TABLE VII**  Secondary Prevention of Stroke after TIA or Ischemic Stroke with Antiplatelet Drugs, Anticoagulants, or Carotid Endarterectomy

| Condition | Secondary prevention | RRR versus placebo or control group | RRR versus aspirin | Cost/benefit ratio (high +++) |
|---|---|---|---|---|
| TIA or ischemic stroke; ICA stenosis | Aspirin, 50–300 mg*** | 18–22% | | +++ |
| <70% or no stenosis; cardiac source | Ticlopidine, 2 × 250 mg*** | 33% | 21% | + |
| of embolism excluded | Aspirin, 50 mg, plus DP, 400 mg** | 37% | 23% | ++ |
| Ischemic stroke plus MI or PAD | Clopidogrel, 75 mg** | | 9% | + |
| Contraindications for or intolerability of | Ticlopidine, 2 × 250 mg | | 21% | + |
| aspirin | Clopidogrel, 75 mg | | 9% | + |
| Contraindications for or intolerability of aspirin, ticlopidine, and clopidogrel | Dipyridamole, 2 × 200 mg | 18% | | + |
| Repeated TIA despite aspirin | Aspirin, 50 mg, plus DP, 400 mg | n.d. | n.d. | ? |
| | Clopidogrel plus aspirin | n.d. | n.d. | ? |
| | Anticoagulation INR 2–3 | n.d. | n.d. | ? |
| Stenosis of the ICA >70%; TIA or minor stroke ipsilateral to the stenosis | Endarterectomy plus 75–325 mg aspirin*** or inclusion in trial comparing operation with stenting | 65% | | + |
| High degree stenosis intracranially or vertebral artery | Stenting and/or angioplasty | n.d. | n.d. | ? |
| Cardiac source of embolism | Anticoagulation INR, 3–4, 5** | 70% | 40% | +++ |
| Cardiac source of embolism; contraindications for warfarin | Aspirin, 325 mg | 21% | | +++ |

*Note.* Abbreviations used: Aspirin, Acetylsalicylic acid; DP, Dipyridamole; RRR, relative risk reduction for recurrent stroke; ICA, Internal carotid artery; MI, myocardial infarction; PAD, peripheral arterial disease; n.d., not done.
**: One prospective randomized study.
***: More than one prospective, randomized, controlled study.

profile than ticlopidine. Although not investigated in patients with TIA, clopidogrel should also be effective in these patients assuming the same pathophysiology as in patients with stroke. Clopidogrel is typically used as a second line treatment for stroke prevention in patients who fail aspirin or a first choice for patients who are unable to take aspirin. Some experts use clopidogrel as a first choice for patients with stroke or TIA who have concomitant peripheral arterial disease or myocardial infarction.

A frequent clinical problem is patients who are already on aspirin because of coronary heart disease or a prior cerebral ischemic event and suffer a first or recurrent TIA or stroke. No single clinical trial has investigated this problem. Therefore all recommendations given are not evidence based. Possible strategies include to continue aspirin, add dipyridamole, add clopidogrel, or switch to ticlopidine or clopidogrel or anticoagulation with an INR of 2.0–3.0. The combination of low-dose warfarin plus aspirin has never been studied in the secondary prevention of stroke.

In patients with a high-risk cardiac source of embolism, anticoagulation is recommended with an INR of 2.0–3.0. At the present time, anticoagulation with an INR between 3.0 and 4.5 cannot be recommended in patients with TIA or stroke. Anticoagulation with an INR between 3.0 and 4.5 carries too high a bleeding risk. Whether anticoagulation with lower INRs is safe and effective is not yet known. Treatment of vascular risk factors should also be performed in secondary stroke prevention.

Carotid endarterectomy is recommended in patients with severe (>70%) symptomatic stenosis of the internal carotid stenosis if the complication rate is <6%. Selected patients with stenosis between 50 and 69% may also benefit. These patients should preferentially be included in randomized trials comparing operation with stenting. Relative contraindications are severe disabling stroke, time interval between event and surgery >6 months, uncontrolled hypertension, and short life expectancy. Although adequate efficacy data are not available, stenting with or without angioplasty may be considered in symptomatic distal carotid stenosis and vertebral as well as basilar artery stenosis (Terada *et al.*, 1996). The optimal antithrombotic therapy prior to, during, and after angioplasty or stenting was not investigated in randomized trials. In analogy to cardiology (Bertrand *et al.*, 2000) many interventional neuroradiologists use the combination of clopidogrel plus aspirin.

## Treatment No Longer Recommended

- Extra-intracranial bypass surgery
- GPIIa/IIIb antagonists
- Carotid surgery in ICA stenosis <50%

- Anticoagulation with INR 3–4.5 (INRs of 2.5–3.5 are recommended for patients with certain types of mechanical heart valves)

## REFERENCES

Adler, A. I., Stratton, I. M., Neil, H. A. W., Yudkin, J. S., Matthews, D. R., Cull, C. A., Wright, A. D., Turner, R. C., and Holman, R. R., on behalf of the UK Prospective Diabetes Study Group (2000). Association of systolic blood pressure with macrovascular and microvascular complications of type 2 diabetes (UKPDS 36): Prospective observational study. *Br. Med. J.* 321, 412–419.

Algra, A., and van Gijn, J. (1996). Aspirin at any dose above 30 mg offers only modest protection after cerebral ischaemia. *J. Neurol. Neurosurg. Psychiatry* 60, 197–199.

Algra, A., and van Gijn, J. (1999). Cumulative meta-analysis of aspirin efficacy after cerebral ischaemia of arterial origin. *J. Neurol. Neurosurg. Psychiatry* 65, 255.

American–Canadian Co-operative Study Group (1985). Persantin-aspirin in cerebral ischemia: II. Endpoint results. *Stroke* 16, 406–415.

Antiplatelet Trialists Collaboration (1988). Secondary prevention of vascular disease by prolonged antiplatelet treatment. *Br. Med. J.* 296, 320–331.

Antiplatelet Trialists Collaboration (1994). Collaborative overview of randomised trials of antiplatelet therapy: I. Prevention of death, myocardial infarction, and stroke by prolonged antiplatelet therapy in various categories of patients. *Br. Med. J.* 308, 81–106.

Antithrombotic Trialists Collaboration (2002). Collaborative meta-analysis of randomised trials of antiplatelet therapy for prevention of death, myocardial infarction, and stroke in high risk patients. *Br. Med. J.* 324, 71–86.

Baker, W. H., Howard, V. J., Howard, G., and Toole, J. F., for the ACAS Investigators (2000). Effect of contralateral occlusion on long-term efficacy of endarterectomy in the asymptomatic carotid atherosclerosis study (ACAS). *Stroke* 31, 2330–2334.

Barnett, H., Eliasziw, M., Meldrum, H., and Taylor, D. (1996). Do the facts and figures warrant a 10-fold increase in the performance of carotid endarterectomy on asymptomatic patients? *Neurology* 46, 603–608.

Barnett, H. J., Taylor, D. W., Eliasziw, M., Fox, A. J., Ferguson, G. G., Haynes, R. B., Rankin, R. N., Clagett, G. P., Hachinski, V. C., Sackett, D. L., Thorpe, K. E., and Meldrum, H. E., for the North American Symptomatic Carotid Endarterectomy Trial Collaborators (1998). Benefit of carotid endarterectomy in patients with symptomatic moderate or severe stenosis. *N. Engl. J. Med.* 339, 1415–1425.

Bennett, C. L., Connors, J. M., Carwile, J. M., Moake, J. L., Bell, W. R., Tarantolo, S. R., McCarthy, L. J., Sarode, R., Hatfield, A. J., Feldman, M. D., Davidson, C. J., and Tsai, H. M. (2000). Thrombotic thrombocytopenic purpura associated with clopidogrel. *N. Engl. J. Med.* 342, 1773–1777.

Bennett, C. L., Kiss, J. E., Weinberg, P. D., Pinevich, A. J., Green, D., Kwaan, H. C., and Feldman, M. D. (1998). Thrombotic thrombocytopenic purpura after stenting and ticlopidine. *Lancet* 352, 1036–1037.

Berger, K., Ajani, U. A., Kase, C. S., Gaziano, J. M., Nurning, J. E., Glynn, R. J., and Hennekens, C. H. (1999). Light-to-moderate alcohol consumption and the risk of stroke among U.S. male physicians. *N. Engl. Med.* 345.

Bertrand, M. E., Rupprecht, H. J., Urban, P., and Gershlick, A. H., for the CLASSICS Investigators (2000). Double-blind study of the safety of clopidogrel with and without a loading dose in combination with aspirin after coronary stenting: The clopidogrel aspirin stent international cooperative study (CLASSICS). *Circulation* **102**, 624–629.

Blauw, G. J., Lagaay, A. M., Smelt, A. H. M., and Westendorp, R. G. J. (1997). Stroke, statins, and cholesterol: A meta-analysis of randomized, placebo-controlled, double-blind, double-blind trials with HMG–CoA reductase inhibitors. *Stroke* **28**, 946–950.

Bonita, R., Scragg, R., Stewart, A., Jackson, R., and Beaglehole, R. (1986). Cigarette smoking and risk of premature stroke in men and women. *Br. Med. J.* **293**, 6–8.

Bousser, M. G., Eschwege, E., Haguenau, M., Lefauconier, J. M., Thibult, N., Touboul, D., and Touboul, P. J. (1983). "A.I.C.L.A." controlled trial of aspirin and dipyridamole in the secondary prevention of atherothrombotic cerebral ischemia. *Stroke* **13**, 5–14.

Brown, M. J., Palmer, C. R., Castaigne, A., Leeuw de, P. W., Mancia, G., Rosenthal, T., and Ruilope, L. M. (2000). Morbidity and mortality in patients randomised in double-blind treatment with a long-acting calcium-channel blocker or diuretic in the international nifedipine GITS study: Intervention as a goal in hypertension treatment (INSIGHT). *Lancet* **356**, 366–372.

Cannegieter, S., Rosendaal, F., Wintzen, A., van der Meer, F., Vandenbroucke, J., and Briet, E. (1995). Optimal oral anticoagulant therapy in patients with mechanical heart valves. *N. Engl. J. Med.* **333**, 11–17.

CAPRIE Steering Committee (1996). A randomised, blinded, trial of clopidogrel versus aspirin in patients at risk of ischaemic events (CAPRIE). *Lancet* **348**, 1329–1339.

CAVATAS Investigators (2001). Endovascular versus surgical treatment in patients with carotid stenosis in the Carotid and Vertebral Artery Transluminal Angioplasty Study (CAVATAS): a randomised trial. *Lancet* **357**, 1729–1737.

CAST (Chinese Acute Stroke Trial) Collaborative Group (1997). CAST: Randomized placebo-controlled trial of early aspirin use in 20.000 patients with acute ischaemic stroke. *Lancet* **349**, 1641–1649.

Chimowitz, M., Kokkinos, J., Strong, J., Brown, M., Levine, S., Silliman, S., Pessin, M., Weichel, E., Sila, C., Furlan, A., Kargman, D., Sacco, R., Wityk, R., Ford, G., and Fayad, P. (1995). The warfarin–aspirin symptomatic intracranial disease study. *Neurology* **45**, 1488–1493.

Collins, R., Peto, R., MacMahon, S., Hebert, P., Fiebach, N. H., Eberlein, K. A., Godwin, J., Qizilbash, N., Taylor, J. O., and Hennekens, C. H. (1990). Blood pressure, stroke and coronary heart disease. Part 2, Short-term reductions in blood pressure: Overview of randomized drug trials in their epidemiological context. *Lancet* **335**, 827–838.

Department of Health and Human Services, and Food and Drug Administration (1998). Internal analgesic, antipyretic, and antirheumatic drug products for over-the counter human use: Final rule for professional labeling of aspirin, buffered aspirin and aspirin in combination with antacid drug products. *Federal Register* **63**, 56802–56819.

Diabetes Control and Complications Trial Research Group (1993). The effect of intensive treatment of diabetes on the development and progression of long-term complications in insulin-dependent diabetes mellitus. *N. Engl. J. Med.* **329**, 977–986.

Diener, H. C. (1998). Use of acetylsalicyclic acid in the secondary prevention of stroke. *Int. J. Clin. Pract. (Suppl.)* **97**, 12–15.

Diener, H. C., Cuhna, L., Forbes, C., Sivenius, J., Smets, P., and Lowenthal, A. (1996). European Stroke Prevention Study 2: Dipyridamole and acetylsalicylic acid in the secondary prevention of stroke. *J. Neurol. Sci.* **143**, 1–13.

Diener, H. C., Forbes, C., Riekkinen, P. J., Sivenius, J., Smets, P., Lowenthal, A., and the ESPS Group (1997). European Stroke Prevention Study 2: Efficacy and safety data. *J. Neurol. Sci.* **151**, S1–S77.

Dyken, M. L. (1993). Controversies in stroke: Past and present. The Willis Lecture. *Stroke* **24**, 1251–1258.

Dyken, M. L., Barnett, H. J. M., Easton, J. D., Fields, W. S., Fuster, V., Hachsinski, V., Norris, J. W., and Sherman, D. G. (1992). Low-dose aspirin and stroke: "It ain't necessarily so." *Stroke* **23**, 1395–1399.

EAFT Group (1993). Secondary prevention in non-rheumatic atrial fibrillation after transient ischaemic attack or minor stroke. *Lancet* **342**, 1255–1262.

Easton, J. D. (1995). Does alcohol cause or prevent stroke? *Cerebrovasc. Dis.* **5**, 375–380.

Eckert, B., Zanella, F. E., Thie, A., Steinmetz, J., and Zeumer, H. (1996). Angioplasty of the internal carotid artery: Results, complications and follow-up in 61 cases. *Cerebrovasc. Dis.* **6**, 97–105.

ESPS-Working Group (1992). Second European stroke prevention study. *J. Neurol.* **239**, 299–301.

European Carotid Surgery Trialists' Collaborative Group (1991). MRC European carotid surgery trial: Interim results for symptomatic patients with severe carotid stenosis and with mild carotid stenosis. *Lancet* **337**, 1235–1243.

European Carotid Surgery Trialists' Collaborative Group (1998). Randomised trial of endarterectomy for recently symptomatic carotid stenosis: Final results of the MRC European Carotid Surgery Trial (ECST). *Lancet* **351**, 1379–1387.

Executive Committee for the Asymptomatic Carotid Atherosclerosis (1995). Endarterectomy for asymptomatic carotid artery stenosis. *JAMA* **273**, 1421–1428.

Ferguson, R. D. G., and Ferguson, J. G. (1996). Carotid angioplasty. *Arch. Neurol.* **53**, 696–698.

Gent, M., Blakely, J. A., Easton, J. D., Ellis, D. J., Hachinski, V. C., Harbison, J. W., Panak, E., Roberts, R. S., Sicurella, J., and Turpie, A. G. G. (1989). The Canadian American ticlopidine study (CATS) in thromboembolic stroke. *Lancet* **i**, 1215–1220.

Gillum, L. A., Mamidipudi, S. K., and Johnston, S. C. (2000). Ischemic stroke risk with oral contraceptives. *JAMA* **284**, 72–78.

Gilon, D., Bounanno, F. S., Joffe, M. M., Leavitt, M., Marshall, J. E., Kistler, J. P., and Levine, R. A. (1999). Lack of evidence of an association between mitral-valve prolapse and stroke in young patients. *N. Engl. J. Med.* **341**, 8–13.

Goldberg, R. J., Burchfiel, C. M., Benfante, R., Chiu, D., Reed, D. M., and Yano, K. (1995). Lifestyle and biological factors associated with atherosclerotic disease in middle-aged men: 20 year findings from the Honolulu heart programme. *Arch. Int. Med.* **155**, 686–694.

Golledge, J., Mitchell, A., Greenhalgh, R. M., and Davies, A. H. (2000). Systematic comparision of the early outcome of angioplasty and endarterectomy for symptomatic carotid artery disease. *Stroke* **31**, 1439–1443.

Guiraud-Chaumeil, B., Rascol, A., David, J., Boneu, B., Clanet, M., and Bierme, R. (1982). Prevention des recidives des accidents vasculaires cerebraux ischemiques par les anti-agregants plaquettaires. *Rev. Neurol. (Paris)* **138**, 367–385.

Halliday, A. W., for the Steering Committee and for the Collaborators (1994). The Asymptomatic Carotid Surgery Trial (ACST): Rationale and design. *Eur. J. Vasc. Surg.* **8**, 703–710.

Hansson, L., Hedner, T., Lund-Johansen, P., Kjeldsen, S. E., Lindholm, L. H., Syvertsen, J. O., Lanke, J., Faire de, U., Dahlöf, B., and Karlberg, B. E., for the NORDIL study group (2000). Randomised trial of effects of calcium antagonists compared with diuretics and beta-blockers on cardiovascular morbidity and mortality in hypertension: The nordic diltiazem (NORDIL) study. *Lancet* **356**, 359–365.

Hansson, L., Lindholm, L. H., Ekbom, T., Dahlöf, B., Lanke, J., Scherstén, B., Wester, P. O., and Hedner, T. (1999a). Randomised trial of old and new antihypertensive drugs in elderly patients:

Cardiovascular mortality and morbidity the Swedish Trial in old patients with hypertension-2 study. *Lancet* **354**, 1751–1756.

Hansson, L., Lindholm, L. H., Niskanen, L., Lahnke, J., Hedner, T., Niklason, A., Luomanmäki, K., Dahlöf, B., de Faire, U., Mörlin, C., Karlberg, B. E., Wester, P. O., and Björck, J. E., for the Captopril Prevention Project (CAPPP) Study Group (1999b). Effect of angiotensin-converting-enzyme inhibition compared with conventional therapy on cardiovascular morbidity and mortality in hypertension: The Captorpil Prevention Project (CAPPP) randomised trial. *Lancet* **353**, 611–616.

Hansson, L., Zanchetti, A., Carruthers, S. G., Dahlöf, B., Elmfeld, D., Julius, S., Menard, J., Rahn, K. H., Wedel, H., and Westerling, S., for the HOT Study Group (1998). Effects of intensive blood-pressure lowering and low-dose aspirin in patients with hypertension: Principal results of the Hypertension Optimal Treatment (HOT) randomised trial. *Lancet* **351**, 1755–1762.

Hart, R. G., and Halperin, J. L. (1994). Atrial fibrillation and stroke. Revisiting the dilemmas. *Stroke* **25**, 1337–11341.

Hart, R. G., Halperin, J. L., McBride, R., Benavente, O., Man-Son-Hing, M., and Kronmal, R. A. (2000). Aspirin for the primary prevention of stroke and other major vascular events: Meta-analysis and hypotheses. *Arch. Neurol.* **57**, 326–332.

Hart, R. G., Sherman, D. G., Easton, J. D., and Cairns, J. A. (1998). Prevention of stroke in patients with nonvalvular atrial fibrillation. *Neurology* **51**, 674–681.

Hass, W. K., Easton, J. D., Adams, H. P., Pryse-Phillips, W., Molony, B. A., Anderson, S., and Kamm, B. (1989). A randomized trial comparing ticlopidine hydrochloride with aspirin for the prevention of stroke. *N. Engl. J. Med.* **321**, 501–507.

Hennerici, M., Hülsbömer, H., Hefter, H., Lammerts, D., and Rautenberg, W. (1987). Natural history of asymptomatic extracranial arterial disease: Results of a long-term prospective study. *Brain* **110**, 777–791.

Hobson, R. W., Weiss, D. G., Fields, W. S., Goldstone, J., Moore, W. S., Towne, J. B., Wright, C. B., and the Veterans Affairs Cooperative Study Group (1993). Efficacy of carotid endarterectomy for asymptomatic carotid stenosis. *N. Engl. J. Med.* **328**, 221–227.

International Stroke Trial Collaborative Group (1997). The International Stroke Trial (IST): A randomised trial of aspirin, subcutaneous heparin, both, or neither among 19.435 patients with acute ischaemic stroke. *Lancet* **349**, 1564–1565.

Iso, H., Hennekens, C. H., Stampfer, M. J., Rexrode, K. M., Colditz, G. A., Speizer, F. E., Willett, W. C., and Mannson, J. E. (1999). Prospective study of aspirin use and risk of stroke in women. *Stroke* **30**, 1764–1771.

Johnston, S. C., Gress, D. R., Warren, S., Sidney, S. (2000). Short-term prognosis after emergency department diagnosis of TIA. *JAMA* **284**, 2901–2906.

Kawachi, I., Colditz, G. A., Stampfer, M. J., Willett, W. C., Manson, J. E., Rosner, B., Speizer, F. E., and Hennekens, C. H. (1993). Smoking cessation and decreased risk of stroke in women. *JAMA* **269**, 232–236.

Khaw, K. T. (1996). Epidemiology of stroke. *J. Neurol. Neurosurg. Psychiatry* **61**, 333–338.

Lee, I. M., Hennekens, C. H., Berger, K., Buring, J. E., and Manson, J. E. (1999). Exercise and risk of stroke in male physicians. *Stroke* **30**, 1–6.

MacMahon, S., Peto, R., Cutler, J., Collins, R., Sorlie, P., Neaton, J., Abbott, R., Godwin, J., Dyer, A., and Stamler, J. (1990). Blood pressure, stroke and coronary heart diesease. Part 1. Prolonged differences in blood pressure: Prospective observational studies corrected for the regression dilution bias. *Lancet* **335**, 765–774.

Malek, A. M., Higashida, R. T., Phatouros, C. C., Lempert, T. E., Meyers, P. M., Smith, W. S., Dowd, C. F., and Halbach, V. V. (2000). Stent angioplasty for cervical carotid artery stenosis in high-risk symptomatic NASCET-ineligible patients. *Stroke* **12**, 3029–3033.

Mess, W., Rautenberg, W., Sitzer, M., Dudek, M., Diehl, R., and Hennerici, M. (1990). Asymptomatic extracranial arterial stenosis: How to predict stroke? *J. Neurol.* **237**, 157.

Mohr, J. P., Thompson, J. L., Lazar, R. M. *et al.* Warfarin-Aspirin Recurrent Stroke Study Group (2001). A comparison of warfarin and aspirin for the prevention of recurrent ischemic stroke. *N. Engl. J. Med.* **345**(20), 1444–1451.

Müller, T. H., Su, C. A., Weisenberger, H., Brickl, R., Nehmiz, G., and Eisert, W. G. (1990). Dipyridamole alone or combined with low dose acetylsalicyclic acid inhibits platelet aggregation in human whole blood ex vivo. *Br. J. Clin. Pharmacel.* **30**, 179–186.

Narayan, S. M., Cain, M. E., and Smith, J. M. (1997). Atrial fibrillation. *Lancet* **350**, 943–950.

Norris, J. W., Zhu, C. Z., Bornstein, N. M., and Chambers, B. R. (1991). Vascular risks of asymptomatic carotid stenosis. *Stroke* **22**, 1485–1490.

North American Symptomatic Carotid Endarterectomy Trial Collaborators (1991). Beneficial effect of carotid endarterectomy in symptomatic patients with high-grade carotid stenosis. *N. Engl. J. Med.* **325**, 445–453.

Page, Y., Tardy, B., Zeni, F., Comtet, C., Terrana, R., and Bertrand, J. C. (1991). Thrombotic thrombocytopenic purpura related to ticlopidine. *Lancet* **i**, 774–776.

Patrono, C., Coller, B., Dalen, J. E., Fuster, V., Gent, M., Harker, L. A., Hirsh, J., and Roth, G. (1998). Platelet-active drugs: The relationships among dose, effectiveness, and side effects. *Chest* **114** (*Suppl.*), 470S–488S.

Patrono, C., and Roth, G. J. (1996). Aspirin in ischemic cerebrovascular disease. *Stroke* **27**, 756–760.

Perry, I., Refsum, H., Morris, R., Ebrahim, S., Ueland, P., and Shaper, A. (1995). Prospective study of serum total homocysteine concentration and risk of stroke in middle-aged british men. *Lancet* **346**, 1395–1398.

Peto, R., Gray, R., Collins, R., Wheatley, K., Hennekens, C., Jamrozik, K., Warlow, C., Hafner, B., Thompson, E., Norton, S., Gilliland, J., and Doll, R. (1988). Randomised trial of prophylactic daily aspirin in British male doctors. *Br. Med. J.* **296**, 313–316.

Plehn, J. F., Davis, B. R., Sacks, F. M., Rouleau, J. L., Pfeffer, M. A., Bernstein, V., Cuddy, E., Moyé, L. A., Piller, L., Rutherford, J., Simpson, L. M., and Braunwald, E., for the CARE Investigators (1999). Reduction of stroke incidence after myocardial infarction with pravastatin—The Cholesterol and Recurrent Events (CARE) Study. *Circulation* **99**, 216–223.

PROGRESS Collaborative Group (2001). Randomised trial of a perindorpril-based blood-pressure-lowering regimen among 6105 individuals with previous stroke or transient ischaemic attack. *Lancet* **358**, 1033–1041.

Prospective Studies Collaboration (1995). Cholesterol, diastolic blood pressure, and stroke: 13,000 strokes in 450,000 peoples in 45 prospective cohorts. *Lancet* **346**, 1647–1653.

Report of the Quality Standards Subcommittee of the American Academy of Neurology (1998). Practice parameter—Stroke prevention in patients with nonvalvular atrial fibrillation. *Neurology* **51**, 671–673.

Ridker, P. M., Manson, J. E., Buring, J. E., Shih, J., Matias, M., and Hennekes, C. H. (1999). Homocysteine and risk of cardiovascular disease among postmenopausal women. *JAMA* **281**, 1817–1821.

Rosa, A. (1990). Should certain carotid artery stenoses be operated? *Rev. Neurol. (Paris)* **146**, 319–329.

Rosendorff, C. (1998). Statins for preventions of stroke. *Lancet* **351**, 1002–1003.

Rothwell, P. M., and Warlow, C. P., on behalf of the European Carotid Surgery Trialists' Collaborative Group (1999). Prediction of benefit from carotid endarterectomy in individual patients: A risk-modelling study. *Lancet* **353**, 2105–2110.

Rubins, H. B., Robins, S. J., Collins, D., Fye, C. L., Anderson, J. W., Elam, M. B., Faas, F. H., Linares, E., Schaefer, E. J., Schectman, G., Wilt, T. J., and Wittes, J., for the Veterans Affairs High-Density Lipoprotein Cholesterol Intervention Trial Study Group (1999). Gemfibrozil for the secondary prevention of coronary heart disease in men with low levels of high-density lipoprotein cholesterol. *N. Engl. J. Med.* **341**, 410–418.

Sacco, R. L., Gan, R., Boden-Albala, B., Lin, I., Kargman, D. E., Hauser, W. A., Shea, S., and Paik, M. C. (1998). Leisure-time physical activity and ischemic stroke risk—The Northern Manhattan Stroke Study. *Stroke* **29**, 380–387.

Salem, D. N., Levine, H. J., Pauker, S. G., Eckman, M. H., and Daudelin, D. H. (1998). Antithrombotic therapy in valvular heart disease. *Chest* **114** (*Suppl.*), 590S–601S.

Schwartz, S. M., Petitti, D. B., Siscovick, D. S., Longstreth, W. T., Sidney, S., Raghunathan, T. E., Quesenberry, C. P., and Kelaghan, J. (1998). Stroke and use of low-dose oral contraceptives in young women—A pooled analysis of two U.S. studies. *Stroke* **29**, 2277–2284.

SHEP Cooperative Research Group (1991). Prevention of stroke by antihypertensive drug treatment in older persons with isolated systolic hypertension. *JAMA* **265**, 3255–3264.

Shepherd, J., Blauw, G. J., Murphy, M. B., Cobbe, S. M., Bollen, E. L. E. M., Buckley, B. M., Ford, I., Jukema, J. W., Hyland, M., Gaw, A., Lagaay, A. M., Perry, I. J., Macfarlane, P. W., Meinders, A. E., Sweeney, B. J., Packard, C. J., Westendorp, R. G. J., Twomey, C., and Stott, D. J., on behalf of the PROSPER Study Group (1999). The design of a prospective study of pravastatin in the elderly at risk (PROSPER). *Am. J. Cardiol.* **84**, 1192–1197.

Staessen, J. A., Fagard, R., Thijs, L., Celis, H., Arabizde, G. G., Birkenhäger, W. H., Bulpitt, C. J., de Leeuw, P. W., Dollery, C. T., Fletcher, A. E., Forette, F., Leonetti, G., Nachev, C., O'Brien, E. T., Rosenfeld, J., Rodicio, J. L., Tuomilehto, J., and Zanchetti, A., for the Systolic Hypertension in Europe (Syst-Eur) Trial Investigators (1997). Randomised double-blind comparison of placebo and active treatment for older patients with isolated systolic hypertension. *Lancet* **350**, 757–764.

Steering Committee of the Physicians Health Study Research Group (1989). Final report on the aspirin component of the ongoing physicians health study. *N. Engl. J. Med.* **321**, 129–135.

Stein, P. D., Alpert, J. S., Dalen, J. E., Horstkotte, D., and Turpie, A. G. G. (1998). Antithrombotic therapy in patients with mechanical and biological prosthetic heart valves. *Chest* **114** (*Suppl.*), 602S–610S.

Stratton, I. M., Adler, A. I., Neil, H. A. W., Matthews, D. R., Manley, S. E., Cull, C. A., Hadden, D., Turner, R. C., and Holman, R. R., on behalf of the U.K. Prospective Diabetes Study Group (2000). Association of glycaemia with macrovascular and microvascular complications of type 2 diabetes (UKPDS 35): Prospective observational study. *Br. Med. J.* **321**, 405–412.

Stroke Prevention in Atrial Fibrillation Investigators (1996). Adjusted dose warfarin versus low intensity, fixed dose warfarin plus aspirin for high-risk patients with atrial fibrillation: Stroke Prevention in Atrial Fibrillation III Randomised Clinical Trial. *Lancet* **348**.

Taylor, D. W., Barnett, H. J. M., Haynes, R. B., Ferguson, G. G., Sackett, D. L., Thorpe, K. E., Simard, D., Silver, F. L., Hachinski, V., Clagett, G. P., Barnes, R., and Spence, J. D., for the ASA and Carotid Endarterectomy (ACE) Trial Collaborators (1999). Low-dose and high-dose acetylsalicylic acid for patients undergoing carotid endarterectomy: A randomised controlled trial. *Lancet* **353**, 2179–2184.

Terada, T., Higashida, R., Halbach van, V., Dowd, C., Nakai, E., Yokote, H., Itakura, T., and Hieshima, G. (1996). Transluminal angioplasty for arteriosclerotic disease of the distal vertebral and basilar arteries. *J. Neurol. Neurosurg. Psychiatry* **60**, 377–381.

The ALLHAT Officers and Coordinators for the ALLHAT Collaborative Research Group (2000). Major cardiovascular events in hypertensive patients randomized to doxazosin versus chlorthalidone: The Antihypertensive and Lipid-lowering treatment to prevent Heart Attack Trial (ALLHAT). *JAMA* **283**, 1967–1975.

The Amaurosis Fugax Study Group (1990). Current management of amaurosis fugax. *Stroke* **21**, 201–208.

The Casanova Study Group (1991). Carotid surgery versus medical therapy in asymptomatic carotid stenosis. *Stroke* **22**, 1229–1235.

The Dutch TIA Trial Study Group (1991). A comparison of two doses of aspirin (30 mg vs. 283 mg a day) in patients after a transient ischemic attack or minor ischemic stroke. *N. Engl. J. Med.* **325**, 1261–1266.

The ESPS Group (1987). The European Stroke Prevention Study (ESPS). Principal end-points. *Lancet* **ii**, 1351–1354.

The Heart Outcomes Prevention Evaluation Study Investigators (2000). Effects of an angiotensin-converting-enzyme inhibitor, ramipril, on cardiovascular events in high-risk patients. *New. Engl. J. Med.* **342**, 145–153.

The INDANA Project Collaborators (1997). Effect of antihypertensive treatment in patients having already suffered from stroke: Gathering the evidence. *Stroke* **28**, 2557–2562.

The Long-Term Intervention with Pravastatin in Ischemic Disease (LIPID) Study Group (1998). Prevention of cardiovascular events and death with pravastatin in patients with coronary heart disease and a broad range of initial cholersterol levels. *N. Engl. J. Med.* **339**, 1349–1357.

The SALT Collaborative Group (1991). Swedish aspirin low-dose trial (SALT) of 75 mg aspirin as secondary prophylaxis after cerebrovascular ischaemic events. *Lancet* **338**, 1345–1349.

The Scandinavian Simvastatin Survival Study (1994). Randomised trial of cholesterol lowering in 4444 patients with coronary heart disease: The Scandinavian Simvastatin Survival Study (4S). *Lancet* **344**, 1383–1389.

The Stroke Prevention in Reversible Ischemia Trial (SPIRIT) Study Group (1997). A randomized trial of anticoagulants versus aspirin after cerebral ischemia of presumed arterial origin. *Ann. Neurol.* **42**, 857–865.

Topol, E. J., Byzova, T. V., and Plow, E. F. (1999). Platelet GPIIb-IIIa blockers. *Lancet* **353**, 227–231.

Topol, E. J., Eston, J. D., Amarenco, P., Calif, R., Harrington, R., Graffagnino, C., Davis, S., Diener, H. C., Fergurson, J., Fitzgerald, D., Shuaib, A., Koudstaal, P. J., Theroux, P., Van De Werf, F., Willerson, J. T., Chan, R., Samuels, R., Ilson, B., and Granett, J. (2000). Design of the blockade of the glycoprotein IIb/IIIa receptor to avoid vascular occlusion (BRAVO) trial. *Am. Heart. J.* **139**, 927–933.

UK-TIA Study Group (1991). The United Kingdom Transient Ischaemic Attack (UK-TIA) Aspirin Trial: Final results. *J. Neurol. Neurosurg. Psychiatry* **54**, 1044–1054.

Wannamethee, G., and Shaper, A. G. (1992). Physical activity and stroke in British middle aged man. *Brit. Med. J.* **304**, 597–601.

White, H. D., Simes, R. J., Anderson, N. E., Hankey, G. J., Watson, J. D. G., Hunt, D., Colquhoun, D. M., Glasziou, P., MacMahon, S., Kirby, A. C., West, M. J., and Tonkin, A. M. (2000). Pravastatin therapy and the risk of stroke. *N. Engl. J. Med.* **343**, 317–326.

Wilterdink, J. L., and Easton, J. D. (1992). Vascular event rates in patients with atherosclerotic cerebrovascular disease. *Arch. Neurol.* **49**, 857–863.

Wolf, P. A., D'Agostino, R. B., Kannel, W. B., Bonita, R., and Belanger, A. J. (1988). Cigarette smoking as a risk factor for stroke. *JAMA* **259**, 1025–1029.

CHAPTER 32

# Microangiopathic Diseases of the Brain

Martin Dichgans

## CLINICAL ASPECTS

Microangiopathic diseases of the brain (cerebral microangiopathies) affect blood vessels with a diameter below 500 μm. Most of these disorders predominantly affect the arteries. However, in some conditions capillaries and venules may be involved as well. Cerebral microangiopathies account for at least 20% of strokes and are probably the most common cause of transient ischemic attacks (TIAs) (Kappelle *et al.*, 1991). Furthermore, microangiopathies are an important cause of vascular dementia.

Microangiopathy may be suspected based on the clinical syndrome (e.g., a classic lacunar syndrome) and neuroimaging findings such as an acute small subcortical infarct in the presence of diffuse white matter signal abnormalities. However, there is no common diagnostic marker for all diseases affecting small blood vessels of the brain.

Microangiopathies may be separated into two groups depending on whether a specific underlying condition can be identified (Table I). Specific conditions may be suspected based on an early age at onset, a positive family history, clinical features (e.g., migraine with aura in CADASIL), neuroimaging (e.g., microbleeds on gradient echo images in cerebral amyloid angiopathy) or laboratory tests (e.g., low platelet counts and hemolytic anemia in thrombotic thrombocytopenic purpura). Evidence for a specific underlying condition may also come from additional extracerebral manifestations (e.g., retinal symptoms and hearing loss in Susac syndrome). Dissecting specific conditions may have implications for therapeutic decisions and genetic counseling.

In the majority of cases no specific cause will be found. Many of these patients have chronic arterial hypertension, which has been identified as the second most important risk factor for cerebral small vessel disease following age. In this patient group, blood pressure control is the most important preventive treatment.

## MICROANGIOPATHIES WITHOUT A SPECIFIC UNDERLYING CONDITION (CEREBRAL SMALL VESSEL DISEASE, SMALL ARTERY OCCLUSIVE DISEASE)

### Clinical Aspects

In most patients with cerebral microangiopathy no specific underlying condition can be identified. Many of these patients have chronic arterial hypertension, which is commonly accepted as the second most important risk factor for cerebral small vessel disease following age.

The clinical presentation of small vessel disease (SVD) is highly variable. Patients may present with a classic lacunar syndrome (pure motor stroke, pure sensory stroke, sensorimotor stroke, ataxic hemiparesis, dysarthria—clumsy hand syndrome) or other lacunar syndromes such as brainstem syndromes. Lenticular and thalamic infarcts may cause neuropsychological deficits including aphasia, neglect, apraxia, confusional states, memory deficits, and frontal lobe type abnormalities. In some patients with SVD subcortical dementia is the predominant manifestation. Additional manifestations include extrapyramidal symptoms (Parkinson-like without resting tremor, hemichorea–hemiballismus) and pseudobulbar palsy with labile laughing and crying. Some patients develop seizures.

Neuroimaging reveals a broad spectrum of changes. In the situation of an acute neurologic deficit (e.g.,

**TABLE I** Classification of Cerebral Microangiopathies

I. Microangiopathies without a specific underlying condition ("cerebral small vessel disease")
   Commonly accepted risk factors are
      Age
      Hypertension
   Possible associations that have not been proven
      Diabetes
      Fibrinogen
      Homocysteine
      Smoking
      Polycythemia
      Sleep apnea
      Cardiac arrhythmia
      Hypotension
      Infections (syphilis, tuberculosis, Lyme disease, neurocysticercosis)
      Pseudoxanthoma elasticum
II. Specific causes of cerebral microangiopathy
   a. Without extracerebral manifestations
      CADASIL
      Cerebral amyloid angiopathy (CAA)
      Granulomatos angiitis (see Chapter 38)
   b. With extracerebral manifestations
      Susac syndrome
      Thrombotic thrombocytopenic purpura
      Fabry disease
      Cerebroretinal vasculopathy (CRV) and hereditary endotheliopathy with retinopathy nephropathy and stroke (HERNS)
      Vasculitis (see Chapter 38)
      Antiphospholipid antibody syndrome (see Chapter 38)
      Radiation therapy (Rottenberg et al., 1977; Shanley, 1995)
      Intravascular lymphomatosis (Baumann et al., 2000; Williams et al., 1998)

a lacunar syndrome), diffusion-weighted (DW) MRI may show an acute ischemic lesion (hyperintense on DW images). In addition, there may be evidence for previous small infarcts (cystic lesions that are isointense to cerebrospinal fluid (CSF)) in locations typical for lacunes (basal ganglia, thalamus, internal capsule, pons). Most patients also have multiple confluent white matter signal abnormalities (hyperintense on T2-weighted images, hypodense on CT scans). White matter hyperintensities (WMH) are not at all specific for cerebral SVD. In SVD WMH tend to spare the subcortical U fibers. Gradient echo images may show microbleeds in subcortical nuclei (Roob and Fazekas, 2000). MR angiography is not contributory in establishing the diagnosis of SVD but may allow identification of other arterial pathologies causing lacunar syndromes such as stenosis of the middle cerebral artery trunk (Lyrer et al., 1997). Using transcranial color-coded duplex sonography a prolonged arteriovenous cerebral transit time has been shown in patients with vascular dementia due to SVD (Puls et al., 1999).

Blood vessels in these patients show various changes inluding increased tortuousity, lipid deposition, disorganization of the vessel wall, disappearance of vascular smooth muscle cells, lipohyalinosis, fibrinoid necrosis, intramural hemorrhages, and microaneurysms. Most of these changes have been linked to chronic arterial hypertension.

## Natural Course

Cerebral SVD affects both sexes with a similar frequency. The annual incidence rate of lacunar infarction has been reported to be 13.4/100,000 persons (Sacco et al., 1991). Mean age at onset for lacunar stroke is between 65 and 70 years with a very broad range. Survival in subcortical infarcts due to SVD is better than in ischemic stroke in general. In a large population (mean age, 70 years) with a first lacunar infarct, the 1-year survival rate was 97% (confidence interval, 84–100%) (Sacco et al., 1991). Functional outcome regarding physical independence is also better than that in stroke in general. In most patients with a first-ever lacunar infarct there is a good initial recovery of stroke-related impairment although deterioration may occur (Nakamura et al., 1999; Samuelsson et al., 1996). The annual recurrence rate is between 4.7 and 10% (Gandolfo et al., 1986; Sacco et al., 1991). In a recent follow-up study of patients with a first-ever lacunar infarct 11% developed dementia within a 3-year period (Samuelsson et al., 1996). Apart from acute worsening, symptoms of SVD may evolve in a slowly progressive manner. Periods of stabilization and sometimes temporary improvement are often present. Advanced stages of SVD correspond to the syndrome of "Binswanger's disease" ("subcortical arteriosclerotic encephalopathy"), which is clinically characterized by pyramidal tract signs, extrapyramidal symptoms, and subcortical dementia with abulia, gait imbalance, pseudobulbar signs, and urinary incontinence (Caplan, 1995).

## Principles of Therapy

The pivotal role of arterial hypertension in the pathogenesis of small vessel disease has been firmly established. In longitudinal studies, blood pressure has been shown to be an important predictor of white matter hyperintensity progression (Schmidt et al., 1999; Yamamoto et al., 1998). Consequently, the most important preventive treatment is blood pressure control, even though there have been no prospective studies that have systematically investigated the effects of such intervention.

Some authors have reported increased fibrinogen levels and increased blood viscosity in patients with SVD (Caplan, 1995; Schneider *et al.*, 1987). Also, several studies have linked fibrinogen levels with stroke, stroke recurrence, and white matter hyperintensities (Coull *et al.*, 1991; Grotta *et al.*, 1985; Longstreth *et al.*, 1996; Schmidt *et al.*, 1999; Wilhelmsen *et al.*, 1984). Hyperlipidemia, macroglobulinemia, and hyperglobulinemia may also affect viscosity. Increased viscosity could reduce cerebral blood flow at a microcirculatory level, particularly in the presence of thickened and narrowed blood vessels. Therefore, lowering blood viscosity might be beneficial in patients with SVD (Walzl *et al.*, 1993). Some authors suggested phlebotomy when the hematocrit is >45%. However, these therapies have not been systematically evaluated.

Some studies have posed a role for other risk factors including diabetes and hyperlipidemia (Table I). A role of these and other potential risk factors has not been firmly established in SVD but when present they should be controlled.

Marked white matter changes are associated with a significantly increased risk for phenprocoumon-induced intracranial hemorrhages (The Stroke Prevention in Reversible Ischemia Trial (SPIRIT) Study Group, 1997). This risk is further increased by an age above 65 years and arterial hypertension. Therefore, extensive white matter (WM) changes represent a relative contraindication for oral anticoagulation.

## Practical Management

Consistent treatment of arterial hypertension is the most important preventive treatment. Acute strokes are treated following the principles outlined in Chapter 30. During the first days following an acute stroke, diastolic blood pressure values up to 120 mm Hg and systolic values up to 200 mm Hg may be tolerated. Management of SVD includes treatment of symptoms such as depression, urinary incontinence, and sleep disturbances. Emotional lability may respond to selective serotonin reuptake inhibitors (e.g., a single evening dose of 100 mg fluvoxamine; Iannaccone and Ferini-Strambi, 1996). Regular fluid uptake is important, particularly in elderly and demented individuals.

## Treatments No Longer Recommended

Nootropics are no longer recommended. Extensive white matter abnormalities represent a relative contraindication for oral anticoagulation, particularly in hypertensive individuals and in those above 65 years of age.

# SPECIFIC CAUSES OF CEREBRAL MICROANGIOPATHY

## Cadasil

### Clinical Aspects

Cerebral autosomal dominant arteriopathy with subcortical infarcts and leukoencephalopathy (CADASIL) is an inherited nonarteriosclerotic, nonamyloid angiopathy due to mutations in *Notch3* (chromosome 19p13; Dichgans *et al.*, 1998; Joutel *et al.*, 1996). Recurrent ischemic episodes (TIAs and strokes) and progressive cognitive deficits are the predominant clinical manifestations (Chabriat *et al.*, 1995; Dichgans *et al.*, 1998). Migraine with aura is an early and characteristic feature that occurs in up to 40% of the cases. Additional manifestations include psychiatric symptoms (in particular mood disorders) and—less often—epileptic seizures (Abbott *et al.*, 1986). Magnetic resonance imaging reveals diffuse white matter abnormalities and small subcortical infarcts. T2-signal hyperintensities in the temporopolar white matter with involvement of subcortical U fibers is a frequent and characteristic finding (Auer *et al.*, 2001) that may help to differentiate CADASIL from sporadic SVD. Lumbar puncture and conventional angiography are not contributory. In fact, conventional angiography should be avoided because of a high rate of complications (Dichgans and Petersen, 1997). Vascular risk factors are usually absent. In many cases there is a family history of stroke or dementia compatible with an autosomal dominant trait. However, sporadic cases due to neomutations have been described. The underlying vascular lesion is a unique nonamyloid angiopathy involving small arteries (100–400 μm) and capillaries primarily in the brain but also in other organs. The diagnosis may therefore be established by skin biopsy (Mayer *et al.*, 1999). Ultrastructural examination reveals characteristic granular deposits within the vascular basal membrane. When properly conducted (ultrastructural examination of at least two arterioles), the sensitivity of this method is very high. Genetic testing is possible (see below).

### Natural Course

The prevalence of CADASIL is unknown although a large number of families have now been reported from all over the world (more than 150 families in Germany). CADASIL patients carry mutations in *Notch3*. Inheritance is autosomal dominant. The phenotype greatly varies both between and within families. Some gene carriers may remain clinically asymptomatic until their sixties. Migraine with aura is an early symptom (manifestation usually before the age of 40). Mean age at onset for ischemic symptoms is 46 years with a range from about 30 to 70 years (Dichgans *et al.*, 1998). The penetrance of MRI white matter signal hyperintensities is complete by about age 35 (Chabriat *et al.*, 1995). Cognitive deficits evolve in a slowly progressive fashion with additional stepwise worsening. The overall course is relentlessly progressive. Advanced stages are characterized by spastic tetraparesis and subcortical dementia with pseudobulbar palsy, gait disturbance, and urinary incontinence. At age 60 years, about half of the individuals are unable to walk without assistance. Mean age at death is 61 years. Median survival times are 64 years in men and 69 years in women. A major cause of death is pneumonia due to aspiration (swallowing deficits).

Genetic counseling in affected individuals and those at risk of being affected should adhere to the principles outlined in Chapter 107. Because of the large size of the *Notch3* gene mutational screening may be quite laborious. Once a mutation has been identified in the index case, genetic testing of additional family members is straightforward.

### Principles of Therapy

There is no known treatment for the disease. So far, there have been no systematic trials in this condition. PET, MR, and doppler sonography studies have indicated a reduction in cerebral blood flow and a reduced vasomotor response to vasodilating stimuli (Chabriat *et al.*, 2000; Pfefferkorn *et al.*, 2001). Theoretically, therapies aiming at increasing CBF (e.g., by lowering blood viscosity) could be of benefit. However, such therapies have not been investigated systematically. Azetazolamide (250 mg per day) has been reported to improve migraine in one patient treated with this agent (Weller *et al.*, 1998).

### Practical Management

Regular fluid uptake is important, particularly in elderly and demented individuals. Acute strokes are treated according to the principles outlined in Chapter 30. Management focuses on the control of symptoms such as headache, mood disturbances, urinary incontinence, and sleep disturbance. Forced crying and laughing may respond to selective serotonin reuptake inhibitors (e.g., a single evening dose of 100 mg fluvoxamine; Iannaccone and Ferini-Strambi, 1996). When oral hydration and feeding become insufficient, the patient should receive additional tube feeding (see Chapter 30).

## Cerebral Amyloid Angiopathy

### Clinical Aspects

Cerebral amyloid angiopathy (CAA) refers to the deposition of amyloid (usually β-amyloid) in the walls of cortical and leptomeningeal blood vessels. Apart from amyloid deposition, vessel walls show a number of changes such as cracking between single layers, microaneurysm formation, and fibrinoid necrosis. Rupture of such structurally weakened arteries eventually results in cerebral hemorrhage. CAA accounts for a significant proportion of spontaneous lobar hemorrhage in the elderly and is an important cause of warfarin-associated lobar hemorrhage (Rosand *et al.*, 2000). CAA-related hemorrhage, like vascular amyloid itself, favors the cerebral cortex and corticosubcortical or lobar regions (Greenberg, 1998). In contrast, brain regions characteristic of hypertensive hemorrhages, such as the basal ganglia, thalamus, pons, and cerebellum are usually spared.

The main clinical manifestations of CAA are (i) recurrent or multiple lobar hemorrhages, (ii) transient neurologic deficits, and (iii) dementia with leukoencephalopathy. Cognitive deficits may evolve in a slowly progressive fashion. Neuroimaging may show hemorrhages and white matter signal abnormalities. Gradientecho MRI is helpful in establishing the presence of small previous hemorrhages in corticosubcortical regions. A clinical diagnosis of "probable CAA" may be made in patients older than 60 years who have multiple lobar hemorrhages without other causes of hemorrhage. A definite diagnosis can only be made at autopsy or when tissue becomes available (e.g., by evacuation of an acute hematoma or cortical biopsy).

### Natural Course

CAA is an age-related phenomenon. In a large autopsy study the estimated prevalence of moderate to severe CAA was 2.3% for individuals ages 65 to 74 years and 12.1% for ages 85 and above. Moreover, CAA is particularly common in Alzheimer's disease. The apolipoprotein E ε4 and ε2 alleles are associated with an increased risk and earlier age of first hemorrhage. However, they are neither sensitive nor specific for cerebral amyloid angiopathy. CAA occurs in association with Down's syndrome. Rare monogenic variants with an autosomal dominant mode of transmission and early age at onset have been described in single families mostly from Europe and the U.S. These conditions are caused by point mutations in the *amyloid precursor protein* (*APP*) gene, the *cystatin* C gene, and the BRI gene (Kalaria, 2001).

## Principles of Therapy

So far there is no specific treatment for CAA. Whereas the biology of vascular damage in CAA appears to be distinct from other types of hemorrhage, there is no evidence for differences in the behavior of the acute hematoma.

## Practical Management

Acute treatment may necessitate surgical intervention. Preventive treatment is limited to withdrawal of anticoagulants or antiplatelet agents. In patients with CAA and a valid indication for anticoagulation the high mortality of phenprocoumon-associated hemorrhage represents a strong argument against this agent. Consistent blood pressure control in hypertensive individuals is strongly recommended.

## Susac Syndrome

### Clinical Aspects

Susac syndrome (retinocochleocerebral vasculopathy) is a distinctive microangiopathy involving the brain, retina, and cochlea (Petty *et al.*, 1998; Susac *et al.*, 1979). Clinically, the syndrome consists of a triad of encephalopathy, retinopathy, and auditory symptoms.

Neurologic symptoms are diffuse and multifocal and progress during the course of the disease. They include headache (in more than 50% of the cases, often migraine-like, usually early during the course), focal neurologic deficits, cognitive deficits, mood disturbance, abulia, sphincter disturbance, and—less often—seizures. Magnetic resonance imaging reveals diffuse and multiple small foci of increased signal on T2-weighted images in subcortical white matter (including the corpus callosum), basal ganglia, thalamus, pons, and cerebellum. In many cases there is some enhancement of lesions upon administration of gadolinium. Cranial CT is usually normal except for atrophy in later stages of the disease. In single patients cerebral angiography has shown "branch occlusion" but it is normal in at least 90% of the cases. CSF examination may show signs of a mild inflammatory reponse. The most consistent CSF finding is an increased protein content (usually below 300 mg/dL). In some cases there is a mild lymphocytic pleocytosis with less than 20 cells/mm³. Oligoclonal bands are usually absent.

Retinal manifestation typically consists in segmental loss of vision. Ophthalmoscopic examination reveals retinal infarcts and retinal arteriolar narrowing and occlusions. Fluorescein angiography shows branch occlusions and vessel leakage. The veins are not involved. Vestibulocochlear manifestations include hearing loss (uni- or bilateral) as well as tinnitus and vertigo. Audiometry demonstrates sensorineural hearing loss, usually in the low- to mid-frequency ranges, which is attributed to infarction of the cochlear apex in the distribution of small end arteries. Caloric testing may demonstrate vestibular dysfunction.

### Natural History

More than 80% of the patients are women. Onset is usually before the age of 40. In most cases the onset is acute or subacute. In most cases the disease starts with symptoms in only one or two of the three systems. Typically, but not always, symptoms in the remaining systems follow within a few months. In most cases the course is fluctuating and monophasic with a fulminant phase that lasts from months to a few years. The functional outcome appears to be related to the clinical deficits at the most active phase of the disease. Many patients remain permanently disabled and are unable to return to school or work.

### Principles of Therapy

The etiology of the arteriopathy is unknown. Because of the small number of patients reported, the apparent monophasic course of the disease, and the limited success of the various therapies tested no firm conclusions on the best treatment can be drawn. Despite the fact that there is no convincing evidence for an autoimmune process or vasculitis most authors have used corticosteroids or immunosuppressants and about 90% of the patients appear to have shown improvement at some point while receiving this treatment (Petty *et al.*, 1998). In some patients new symptoms developed when the dose of corticosteroids was tapered (Petty *et al.*, 1998; Susac *et al.*, 1979). Antiplatelet and anticoagulant therapy have repeatedly been used but most patients developed new symptoms during such treatment. Single patients receiving plasmapheresis, intravenous immunoglobulins, or a combination of nimodipine and aspirin have shown improvement (Petty *et al.*, 1998; Wildemann *et al.*, 1996). However, these reports allow no firm conclusions. Hyperbaric oxygen has been reported to reverse visual field and acuity changes in a patient with recurrent retinal symptoms (Li *et al.*, 1996).

### Practical Management

Corticosteroids given orally or intravenously may be the best agent to start with. Immunosuppressants (cyclophosphamide or azathioprine) can be added. However, the potential benefits and risks of these therapies must be weighed against the prognostic uncertainties. Aspirin (30 to 100 mg per day) can be given either alone or in combination with nimodipine (90 mg per day). Retinal artery branch occlusions may be treated with hyperbaric oxygen. Some patients require hearing aids.

## Thrombotic Thrombocytopenic Purpura

### Clinical Aspects

Thrombotic thrombocytopenic purpura (TTP) is a potentially fatal disease characterized by widespread platelet thrombi in arterioles and capillaries (Moake, 1998; Moschcowitz, 1924). The disease is caused by a constitutive or acquired deficiency of von Willebrandt factor-cleaving protease.

TTP is characterized by a pentad of thrombocytopenia, microangiopathic hemolytic anemia, neurological symptoms, renal failure, and fever. These signs occur variably, depending on the number and sites of the arteriolar lesions. The most frequent neurologic manifestations are mental status changes, headache, coma, seizures, and focal neurologic deficits (hemiparesis, aphasia, visual field deficits; Amorosi and Ultmann, 1966; O'Brien and Sibley, 1958). T2-weighted images may reveal multiple hyperintense foci in the brainstem, thalamus, basal ganglia, white matter, or cortical gray matter. Some of these signal abnormalities are reversible whereas others correspond to true infarcts. Less frequent, territorial infarctions have been reported. Laboratory investigations show fragmentation of erythrocytes and thrombocytopenia. Anemia often parallels thrombocytopenia. Together with the serum level of LDH these parameters can be used to estimate the severity of the disorder. Neurological symptoms and renal failure are usually seen only when the platelet count is significantly diminished (<30,000/mm³). Proteinuria and moderate elevation of blood urea nitrogen is considered an early sign of renal manifestation. Coagulation tests are usually normal. The diagnosis may be established by demonstrating a reduced or absent activity of von Willebrand factor-cleaving protease or by inhibitory IgG autoantibodies against the protease (Furlan et al., 1998; Tsai and Lian, 1998).

### Natural History

TTP may occur as an acquired or—less frequent—familial condition. Idiopathic cases occur at a rate of 3.7 per year per million persons. TTP has been associated with a number of drugs (Gordon and Kwaan, 1997). Among cases reported in ticlopidine-treated patients, 95% occurred after 2 to 12 weeks of treatment, whereas 91% of the cases among clopidogrel-treated patients occurred within 2 weeks after initiation of treatment (Bennett et al., 1998, 2000). TTP has further been associated with various situations such as infection, bone marrow transplantation, cancer, chemotherapy, and pregnancy. In most patients onset is acute. Mean age at onset in a large study of 108 patients was 38 years (range, 16–77 years) (Bell et al., 1991). Prior to the introduction of plasma exchange the mortality rate of TTP was almost 100%. In promptly treated cases the mortality ranges from 10 to 20%. Relapses may occur, particularly in familial cases. Most relapses occur within 30 days of diagnosis (Bell et al., 1991).

### Principles of Therapy

Patients with nonfamilial TTP have an acquired deficiency of von Willebrand factor-cleaving protease due to inhibitory antibodies against this protease (Tsai and Lian, 1998). Patients with familial TTP have a complete deficiency in the absence of inhibitory antibodies (Furlan et al., 1998). In patients with chronic relapsing TTP the processing of von Willebrand factor has successfully been restored by the infusion of fresh frozen plasma or cryosupernatant which contain the cleavage protease that is missing in patients with chronic relapsing TTP (Moake, 1998). In acute single episodes of TTP, autoantibodies interfere with the activity of the protease. As shown by a number of studies most of these patients require additional plasmapheresis, probably to remove autoantibodies and large multimers of von Willebrand factor.

In the so far largest interventional study on TTP–hemolytic uremic syndrome, Bell and coworkers used the following treatment regimen (Bell et al., 1991): patients with minimal symptoms and no CNS symptoms received 200 mg of prednisone a day. Patients with rapid clinical deterioration (no improvement after 48 h of prednisone) and patients presenting with CNS symptoms and rapid decline of platelet counts and hematocrit values received prednisone plus plasma exchange. Corticosteroids alone were effective in 28 of 30 patients. Relapses occurred in 2 of 30 patients treated with steroids alone and in 67 of 78 patients who received plasma exchange. Ten of 108 patients died.

A variety of other therapeutic approaches have been tried in an uncontrolled fashion with limited success. Such treatments included immunosuppressive agents, immunoadsorption with a protein A–Sepharose column, immunoglobulins, treatment with vincristine, and splenectomy.

### Practical Management

Early recognition and diagnosis is essential. Based on a large uncontrolled study (Bell et al., 1991) the following treatment regimen may be proposed: If the patient has minimal symptoms and is free of CNS symptoms (except for mild headache), corticosteroids (200 mg of oral prednisone or 200 mg of intravenous prednisolone per day) should be given. If the patient has moderate-to-severe symptoms with changes in mental status or other neurologic manifestations, or if there is rapid clinical deterioration (hematocrit, < 0.2; platelet count, <10,000/mm³; serum LDH, >600 U/l; or creatinine, >5.0 mg/dl), the patient should receive 200 mg of prednisolone per day intravenously, plasmapheresis, and plasma exchange. If there is clinical improvement, and laboratory values are normal or nearly normal for 2 successive days, plasma exchange may be discontinued. Corticosteroids may be rapidly reduced to 60 mg per day and then reduced more slowly (5 mg/day) (for details see, Bell *et al.*, 1991).

## Fabry Disease

### Clinical Aspects

Fabry disease (FD) is an X-linked recessive lysosomal storage disorder caused by a deficiency of α-galactosidase A. The deficiency of this enzyme results in intracellular accumulation of globotriaosylceramide (Gb3) in different areas of the body, but predominantly in vascular endothelial and smooth muscle cells. The main clinical manifestations of FD are listed in Table II.

Cerebrovascular symptoms are mediated both by small artery involvement and large artery disease (Crutchfield *et al.*, 1998; Grewal, 1994; Mitsias and Levine, 1996). Small artery disease is characterized by progressive occlusion of blood vessels, secondary to deposition of Gb3 within the vessel wall. The most frequent sign of large artery disease is dolichoectasia of the basilar and vertebral arteries. Large infarcts involving the cortex may result from artery-to-artery embolism from ectatic vessels. In fact, there is evidence for endothelium and leukocyte activation consistent with a prothrombotic state in FD (DeGraba *et al.*, 2000). Some patients may develop intracranial hemorrhage due probably to vessel degeneration and uncontrolled hypertension from renal failure.

The diagnosis of FD may be suspected based on the typical skin changes, acroparesthesias, hypohydrosis, corneal opacities, and lenticular lesions. The diagnosis is usually confirmed by skin biopsy (lipid inclusions in the vascular endothelium) and biochemical studies.

### Natural History

The incidence of Fabry disease is 1:117,000 live births. Patients have mutations in the α-galactosidase A gene with almost every family having its own mutation. Men are predominantly affected but female carriers may have similar symptoms. Angiokeratomas usually appear in childhood. Painful small fiber neuropathy is often the first sign that brings patients to neurological attention. Pain crises may be induced by fever, hot weather, exercise, or fatigue. In most cases they become less frequent in the third and fourth decades of life. Renal insufficiency often occurs in the fourth or fifth decade. Mean age at onset for cerebrovascular symptoms is around 34 years (range, 16–70 years). In hemizygous patients the penetrance for cerebrovascular involvement on MRI is complete by about age 54. Death usually occurs from systemic complications, in particular, renal failure, cardiac insufficiency, and stroke.

### Principles of Therapy

Enzyme replacement therapy is under investigation. Only recently sufficient quantities of purified α-galactosidase A have become available through genetic engineering. In a small study on 10 FD patients a single infusion of α-galactosidase A produced by cultured fibroblasts was shown to be safe and reduce the levels of Gb3 in blood, liver, and renal tubular epithelial cells (Schiffmann *et al.*, 2000b). In a consecutive double-blind, placebo-controlled study on 26 hemizygous men, infusions were given every 2 weeks for a total of 6 months (Schiffmann *et al.*, 2000a). Significant effects were seen regarding pain, creatinine clearance, insulin clearance, glomerular histology, and levels of Gb3 in plasma and urinary sediment. The authors concluded that enzyme replacement therapy with α-galactosidase

**TABLE II**   Clinical Manifestations of Fabry Disease

| | |
|---|---|
| Skin: | Angiokeratoma corporis diffusum (clustered nonblanching angiectases primarily over the lower part of the trunk, buttocks, and scrotum), Hypohydrosis |
| Painful neuropathy: | Intermittent crises of agonizing burning pain in the hands and feet. Since the neuropathy primarily affects small nerve fibers, electromyography and nerve conduction velocities may be normal. |
| Cardiac involvement: | Cardiac hypertrophy, arrhythmia, conduction abnormalities, valvular insufficiency, coronary artery disease leading to myocardial infarction |
| Renal: | Proteinuria, progressive renal insufficiency |
| Cerebrovascular: | Small artery disease |
| | Large artery disease |
| Eye: | Corneal dystrophy |
| | Posterior capsular cataract |
| | Excessively tortuous retinal blood vessels |
| Others: | Progressive sensorineural hearing loss, postprandial intestinal cramps, diarrhea, achalasia |

A is safe and likely to improve the prognosis of Fabry disease (Brady and Schiffmann, 2000). Renal transplantation replaces the missing enzyme and may produce some regression of nonrenal manifestations. Strategies aiming at inhibiting the synthesis of Gb3 are under investigation (Abe *et al.*, 2000).

### Practical Management

Painful paresthesias may be treated with carbamazepine (400–800 mg per day), phenytoin (300 mg per day), gabapentin, or lamotrigine. There is no specific treatment for strokes in FD. Renal transplantation and chronic hemodialysis may be lifesaving. Enzyme replacement therapy seems to improve the prognosis of FD. This treatment is starting to become available at specialized centers.

### Cerebroretinal Vasculopathy and Hereditary Endotheliopathy with Retinopathy, Nephropathy, and Stroke (HERNS)

Cerebroretinal vasculopathy (CRV) is a rare condition characterized by a microangiopathy of the brain and retina in conjunction with a cerebral pseudotumor typically located within the frontoparietal white matter (Grand *et al.*, 1988).

Clinical symptoms include progressive visual loss, migraine-like headache, seizures, focal neurologic deficits, and progressive cognitive worsening. The few families reported so far suggest an autosomal dominant pattern of inheritance. Small intracerebral vessels exhibit amorphous thickening of their walls with adventitial fibrosis. Some of the vessels become necrotic and show thrombosis.

Jen and co-workers described a large family in which multiple members from three generations had developed a syndrome consisting of retinopathy, nephropathy, and stroke (Jen *et al.*, 1997). In most cases the disease had started with visual disturbances ("blind spots") followed by focal neurologic deficits within a few years. On brain magnetic resonance images these individuals were shown to have multiple white matter signal hyperintensities. In the acute stage they had contrast-enhancing lesions with surrounding vasogenic edema most commonly in frontoparietal regions. Fluorescein angiograms demonstrated capillary obliteration with telangiectatic microaneurysms. Ultrastructural examination of microvessels from different organs including the brain, kidney, and skin revealed multilayered capillary basal membranes. Based on these findings the authors suspected a primary endothelial injury and coined the acronym HERNS. Although renal insufficiency has not been reported as a feature of CRV, evidence exists that CRV and HERNS may represent variants of the same condition (Ophoff *et al.*, 2001). Corticosteroids have been advocated for the treatment of edema in the acute stage. Surgical resection of the pseudotumor has not helped those who underwent this procedure.

## REFERENCES

Abbott, R. D., Yin, Y., Reed, D. M., and Yano, K. (1986). Risk of stroke in male cigarette smokers. *N. Engl. J. Med.* **315**, 717–720.

Abe, A., Gregory, S., Lee, L., Killen, P. D., Brady, R. O., Kulkarni, A., and Shayman, J. A. (2000). Reduction of globotriaosylceramide in Fabry disease mice by substrate deprivation. *J. Clin. Invest.* **105**, 1563–1571.

Amorosi, E. and Ultmann, J. (1966). Thrombotic thrombocytopenic purpura: Report of 16 cases and review of the literature. *Medicine* **45**, 139–159.

Auer, D. P., Putz, B., Gossl, C., Elbel, G. K., Gasser, T., and Dichgans, M. (2001). Differential lesion patterns in CADASIL and sporadic subcortical arteriosclerotic encephalopathy: MR imaging study with statistical parametric group comparison. *Radiology* **218**, 443–451.

Baumann, T. P., Hurwitz, N., Karamitopolou-Diamantis, E., Probst, A., Herrmann, R., and Steck, A. J. (2000). Diagnosis and treatment of intravascular lymphomatosis. *Arch. Neurol.* **57**, 374–377.

Bell, W. R., Braine, H. G., Ness, P. M., and Kickler, T. S. (1991). Improved survival in thrombotic thrombocytopenic purpura-hemolytic uremic syndrome: Clinical experience in 108 patients. *N. Engl. J. Med.* **325**, 398–403.

Bennett, C. L., Connors, J. M., Carwile, J. M., Moake, J. L., Bell, W. R., Tarantolo, S. R., McCarthy, L. J., Sarode, R., Hatfield, A. J., Feldman, M. D., Davidson, C. J., and Tsai, H. M. (2000). Thrombotic thrombocytopenic purpura associated with clopidogrel. *N. Engl. J. Med.* **342**, 1773–1777.

Bennett, C. L., Weinberg, P. D., Rozenberg-Ben-Dror, K., Yarnold, P. R., Kwaan, H. C., and Green, D. (1998). Thrombotic thrombocytopenic purpura associated with ticlopidine: A review of 60 cases. *Ann. Intern. Med.* **128**, 541–544.

Brady, R. O. and Schiffmann, R. (2000). Clinical features of and recent advances in therapy for fabry disease. *JAMA* **284**, 2771–2775.

Caplan, L. R. (1995). Binswanger's disease revisited. *Neurology* **45**, 626–633.

Chabriat, H., Pappata, S., Ostergaard, L., Clark, C. A., Pachot-Clouard, M., Vahedi, K., Jobert, A., Le Bihan, D., and Bousser, M. G. (2000). Cerebral hemodynamics in CADASIL before and after acetazolamide challenge assessed with MRI bolus tracking. *Stroke* **31**, 1904–1912.

Chabriat, H., Vahedi, K., Iba-Zizen, M. T., Joutel, A., Nibbio, A., Nagy, T. G., Krebs, M. O., Julien, J., Dubois, B., and Ducrocq, X. (1995). Clinical spectrum of CADASIL: A study of 7 families. Cerebral autosomal dominant arteriopathy with subcortical infarcts and leukoencephalopathy. *Lancet* **346**, 934–939.

Coull, B. M., Beamer, N., de Garmo, P., Sexton, G., Nordt, F., Knox, R., and Seaman, G. V. (1991). Chronic blood hyperviscosity in subjects with acute stroke, transient ischemic attack, and risk factors for stroke. *Stroke* **22**, 162–168.

Crutchfield, K. E., Patronas, N. J., Dambrosia, J. M., Frei, K. P., Banerjee, T. K., Barton, N. W., and Schiffmann, R. (1998). Quantitative analysis of cerebral vasculopathy in patients with Fabry disease. *Neurology* **50**, 1746–1749.

DeGraba, T., Azhar, S., Dignat-George, F., Brown, E., Boutiere, B., Altarescu, G., McCarron, R., and Schiffmann, R. (2000). Profile of endothelial and leukocyte activation in Fabry patients. *Ann. Neurol.* 47, 229–233.

Dichgans, M., Mayer, M., Uttner, I., Bruning, R., Muller-Hocker, J., Rungger, G., Ebke, M., Klockgether, T., and Gasser, T. (1998). The phenotypic spectrum of CADASIL: Clinical findings in 102 cases. *Ann. Neurol.* 44, 731–739.

Dichgans, M., and Petersen, D. (1997). Angiographic complications in CADASIL. *Lancet* 349, 776–777.

Furlan, M., Robles, R., Galbusera, M., Remuzzi, G., Kyrle, P. A., Brenner, B., Krause, M., Scharrer, I., Aumann, V., Mittler, U., Solenthaler, M., and Lammle, B. (1998). von Willebrand factor-cleaving protease in thrombotic thrombocytopenic purpura and the hemolytic–uremic syndrome. *N. Engl. J. Med.* 339, 1578–1584.

Gandolfo, C., Moretti, C., Dall'Agata, D., Primavera, A., Brusa, G., and Loeb, C. (1986). Long-term prognosis of patients with lacunar syndromes. *Acta. Neurol. Scand.* 74, 224–229.

Gordon, L. I., and Kwaan, H. C. (1997). *Ca. Semin. Hematol.* 34, 140–147.

Grand, M. G., Kaine, J., Fulling, K., Atkinson, J., Dowton, S. B., Farber, M., Craver, J., and Rice, K. (1988). Cerebroretinal vasculopathy: A new hereditary syndrome. *Ophthalmology* 95, 649–659.

Greenberg, S. M. (1998). Cerebral amyloid angiopathy: Prospects for clinical diagnosis and treatment. *Neurology* 51, 690–694.

Grewal, R. P. (1994). Stroke in Fabry's disease. *J. Neurol.* 241, 153–156.

Grotta, J. C., Pettigrew, L. C., Allen, S., Tonnesen, A., Yatsu, F. M., Gray, J., and Spydell, J. (1985). Baseline hemodynamic state and response to hemodilution in patients with acute cerebral ischemia. *Stroke* 16, 790–795.

Iannaccone, S., and Ferini-Strambi, L. (1996). Pharmacologic treatment of emotional lability. *Clin. Neuropharmacol.* 19, 532–535.

Jen, J., Cohen, A. H., Yue, Q., Stout, J. T., Vinters, H. V., Nelson, S., and Baloh, R. W. (1997). Hereditary endotheliopathy with retinopathy, nephropathy, and stroke (HERNS). *Neurology* 49, 1322–1330.

Joutel, A., Corpechot, C., Ducros, A., Vahedi, K., Chabriat, H., Mouton, P., Alamowitch, S., Domenga, V., Cecillion, M., Marechal, E., Maciazek, J., Vayssiere, C., Cruaud, C., Cabanis, E. A., Ruchoux, M. M., Weissenbach, J., Bach, J. F., Bousser, M. G., and Tournier-Lasserve, E. (1996). Notch3 mutations in CADASIL: a hereditary adult-onset condition causing stroke and dementia. *Nature* 383, 707–710.

Kalaria, R. N. (2001). Advances in molecular genetics and pathology of cerebrovascular disorders. *TRENDS in Neurosciences* 24, 392–400.

Kappelle, L. J., van Latum, J. C., Koudstaal, P. J., and van Gijn, J. (1991). Transient ischaemic attacks and small-vessel disease. Dutch TIA Study Group. *Lancet* 337, 339–341.

Li, H. K., Dejean, B. J., and Tang, R. A. (1996). Reversal of visual loss with hyperbaric oxygen treatment in a patient with Susac syndrome. *Ophthalmology* 103, 2091–2098.

Longstreth, W. T., Manolio, T. A., Arnold, A., Burke, G. L., Bryan, N., Jungreis, C. A., Enright, P. L., O'Leary, D., and Fried, L. (1996). Clinical correlates of white matter findings on cranial magnetic resonance imaging of 3301 elderly people. The Cardiovascular Health Study. *Stroke* 27, 1274–1282.

Lyrer, P. A., Engelter, S., Radu, E. W., and Steck, A. J. (1997). Cerebral infarcts related to isolated middle cerebral artery stenosis. *Stroke* 28, 1022–1027.

Mayer, M., Straube, A., Bruening, R., Uttner, I., Pongratz, D., Gasser, T., Dichgans, M., and Muller-Hocker, J. (1999). Muscle and skin biopsies are a sensitive diagnostic tool in the diagnosis of CADASIL. *J. Neurol.* 246, 526–532.

Mitsias, P., and Levine, S. R. (1996). Cerebrovascular complications of Fabry's disease. *Ann. Neurol.* 40, 8–17.

Moake, J. L. (1998). Moschcowitz, multimers, and metalloprotease. *N. Engl. J. Med.* 339, 1629–1631.

Moschcowitz, E. (1924). Hyaline thrombosis of the terminal arterioles and capillaries: A hitherto undescribed disease. *Poc. N. Y. Pathol. Soc.* 24, 21–24.

Nakamura, K., Saku, Y., Ibayashi, S., and Fujishima, M. (1999). Progressive motor deficits in lacunar infarction. *Neurology* 52, 29–33.

O'Brien, J. L., and Sibley, W. A. (1958). Neurologic manifestations of thrombotic thrombocytopenic purpura. *Neurology* 8, 44–63.

Ophoff, R. A., DeYoung, J., Service, S. K., Joosse, M., Caffo, N. A., Sandkuijl, L. A., Terwindt, G. M., Haan, J., van den Maagdenberg, A. M., Jen, J., Baloh, R. W., Barilla-LaBarca, M. L., Saccone, N. L., Atkinson, J. P., Ferrari, M. D., Freimer, N. B., and Frants, R. R. (2001). Hereditary vascular retinopathy, cerebroretinal vasculopathy, and hereditary endotheliopathy with retinopathy, nephropathy, and stroke map to a single locus on chromosome 3p21.1–p21.3. *Am. J. Hum. Genet.* 69(2), 447–453.

Petty, G. W., Engel, A. G., Younge, B. R., Duffy, J., Yanagihara, T., Lucchinetti, C. F., Bartleson, J. D., Parisi, J. E., Kasperbauer, J. L., and Rodriguez, M. (1998). Retinocochleocerebral vasculopathy. *Medicine (Baltimore)* 77, 12–40.

Pfefferkorn, T., von Stuckrat-Barre, S., Herzog, J., Gasser, T., Hamann, G., and Dichgans, M. (2001). Reduced cerebrovascular C02 reactivity in CADASIL: A transcranial doppler sonography study. *Stroke* 32, 17–21.

Puls, I., Hauck, K., Demuth, K., Horowski, A., Schliesser, M., Dorfler, P., Scheel, P., Toyka, K. V., Reiners, K., Schoning, M., and Becker, G. (1999). Diagnostic impact of cerebral transit time in the identification of microangiopathy in dementia: A transcranial ultrasound study. *Stroke* 30, 2291–2295.

Roob, G., and Fazekas, F. (2000). Magnetic resonance imaging of cerebral microbleeds. *Curr. Opin. Neurol.* 13, 69–73.

Rosand, J., Hylek, E. M., O'Donnell, H. C., and Greenberg, S. M. (2000). Warfarin-associated hemorrhage and cerebral amyloid angiopathy: A genetic and pathologic study. *Neurology* 55, 947–951.

Rottenberg, D. A., Chernik, N. L., Deck, M. D., Ellis, F., and Posner, J. B. (1977). Cerebral necrosis following radiotherapy of extracranial neoplasms. *Ann. Neurol.* 1, 339–357.

Sacco, S. E., Whisnant, J. P., Broderick, J. P., Phillips, S. J., and O'Fallon, W. M. (1991). Epidemiological characteristics of lacunar infarcts in a population. *Stroke* 22, 1236–1241.

Samuelsson, M., Soderfeldt, B., and Olsson, G. B. (1996). Functional outcome in patients with lacunar infarction. *Stroke* 27, 842–846.

Schiffmann, R., Kopp, J., Austin, H., Moore, D. F., Sabnis, S., Weibel, T., Balow, J. E., and Brady, R. O. (2000a). Efficacy and safety of enzyme replacement therapy in Fabry disease demonstrated by a double-blind placebo controlled trial. *Am. J. Med. Genet.* 67, 38.

Schiffmann, R., Murray, G. J., Treco, D., Daniel, P., Sellos-Moura, M., Myers, M., Quirk, J. M., Zirzow, G. C., Borowski, M., Loveday, K., Anderson, T., Gillespie, F., Oliver, K. L., Jeffries, N. O., Doo, E., Liang, T. J., Kreps, C., Gunter, K., Frei, K., Crutchfield, K., Selden, R. F., and Brady, R. O. (2000b). Infusion of α-galactosidase A reduces tissue globotriaosylceramide storage in patients with Fabry disease. *Proc. Natl. Acad. Sci. USA* 97, 365–370.

Schmidt, R., Fazekas, F., Kapeller, P., Schmidt, H., and Hartung, H. P. (1999). MRI white matter hyperintensities: Three-year follow-up of the Austrian Stroke Prevention Study. *Neurology* 53, 132–139.

Schneider, R., Ringelstein, E. B., Zeumer, H., Kiesewetter, H., and Jung, F. (1987). The role of plasma hyperviscosity in subcortical arteriosclerotic encephalopathy (Binswanger's disease). *J. Neurol.* 234, 67–73.

Shanley, D. J. (1995). Mineralizing microangiopathy: CT and MRI. *Neuroradiology* **37**, 331–333.

Susac, J. O., Hardman, J. M., and Selhorst, J. B. (1979). Microangiopathy of the brain and retina. *Neurology* **29**, 313–316.

The Stroke Prevention in Reversible Ischemia Trial (SPIRIT) Study Group (1997). A randomized trial of anticoagulants versus aspirin after cerebral ischemia of presumed arterial origin. *Ann. Neurol.* **42**, 857–865.

Tsai, H. M., and Lian, E. C. (1998). Antibodies to von Willebrand factor-cleaving protease in acute thrombotic thrombocytopenic purpura. *N. Engl. J. Med.* **339**, 1585–1594.

Walzl, M., Lechner, H., Walzl, B., and Schied, G. (1993). Improved neurological recovery of cerebral infarctions after plasmapheretic reduction of lipids and fibrinogen. Stroke **24**, 1447–1451.

Weller, M., Dichgans, J., and Klockgether, T. (1998). Acetazolamide-responsive migraine in CADASIL. *Neurology* **50**, 1505.

Wildemann, B., Schulin, C., Storch-Hagenlocher, B., Hacke, W., Dithmar, S., Kirchhof, K., Jansen, O., and Breitbart, A. (1996). Susac's syndrome: Improvement with combined antiplatelet and calcium antagonist therapy. *Stroke* **27**, 149–151.

Wilhelmsen, L., Svardsudd, K., Korsan-Bengtsen, K., Larsson, B., Welin, L., and Tibblin, G. (1984). Fibrinogen as a risk factor for stroke and myocardial infarction. *N. Engl. J. Med.* **311**, 501–505.

Williams, R. L., Meltzer, C. C., Smirniotopoulos, J. G., Fukui, M. B., and Inman, M. (1998). Cerebral MR imaging in intravascular lymphomatosis. *Am. J. Neuroradiol.* **19**, 427–431.

Yamamoto, Y., Akiguchi, I., Oiwa, K., Hayashi, M., and Kimura, J. (1998). Adverse effect of nighttime blood pressure on the outcome of lacunar infarct patients. *Stroke* **29**, 570–576.

*Cerebrovascular Disorders*

CHAPTER 33
# Brain Embolism

Louis R. Caplan and Werner Hacke

## CLINICAL FINDINGS

### Definition and Pathophysiology

Brain embolism occurs when material, usually a thrombus, formed or introduced in one part of the vascular system travels to an artery that supplies brain tissue. There are three main components: the recipient artery that receives the material, the embolic material itself, and the donor source where the material originated.

Donor sources include the heart, the aorta, and the neck arteries and their intracranial arterial branches proximal to the recipient artery. The most common cardiac sources are atrial fibrillation and other atrial arrythmias, myocardial ischemic pathology, and valvular heart diseases. However, less common cardiac pathologies such as myocardiopathies, cardiac tumors, and septal defects within the heart can also lead to brain embolism. Aortic atheromas especially those that are large, protuberant, and mobile are prone to embolize to the brain and systemic arteries. The most common disease of arteries that leads to embolism is atherosclerosis. Materials from plaques and thrombi form within arteries in the neck and proximal intracranial arteries and embolize distally. Occasionally tumor particles, fat, and air and foreign particles introduced into veins or arteries embolize to the brain.

### Clinical Symptoms and Signs

The clinical findings nearly always relate to blockage, temporary or persistent, of the recipient artery. When emboli are released into the circulation, they travel distally until they lodge in an extracranial or intracranial brain artery. Unlike thrombi that are formed locally in areas of atherosclerosis or other endothelial damage, emboli are loosely adherent to vascular walls where they first land. An embolus sometimes precipitates a vaso-constrictive response in the recipient artery. Blockage of the recipient artery causes a sudden decrease in blood flow and pressure in the artery distal to the blockage. The brain tissue supplied by the artery becomes ischemic. These changes in the supply zone of the recipient artery quickly promote the development of collateral blood flow. Collateral blood flow develops very rapidly and is often sufficient to maintain adequate perfusion of the brain.

Sudden blockage of a brain-feeding artery often leads to symptoms of dysfunction of the area of brain supplied. When collateral blood flow develops, the symptoms may stabilize or improve. The nonadherent embolic particles often do not remain at their initial recipient site but fragment and move distally. Sequential arteriograms, or even sequential injections during the same arteriogram, can show passage of emboli from their initial resting place (Caplan, 1999; Dalal *et al.*, 1965; Fisher and Perlman, 1967; Liebeskind *et al.*, 1971). Distal movement of emboli often occurs without causing new symptoms since smaller fragments may pass through the system without causing further ischemic damage. Alternatively, the embolus might block a distal branch. If collateral circulation to the supply area of that branch is not adequate, then further brain ischemic symptoms and brain infarction develop. The nature of the neurologic symptoms depends entirely on the artery that is blocked and the size and location of the resultant brain ischemia and infarction.

### Diagnosis

#### Recipient Artery-Related Findings

Evidence for brain embolism relates to the location, type, and size of brain infarcts, and the location and nature of abnormalities found within recipient arteries. Brain imaging, computed tomography (CT), and magnetic resonance imaging (MRI), provide information about brain infarcts, and magnetic resonance angiography (MRA), CT angiography (CTA), transcranial Doppler ultrasound (TCD), and contrast angiography are technologies that provide information about recipient and donor arteries.

When embolism causes brain infarction, the infarcts often become hemorrhagic (Caplan, 1999; Fisher and Adams, 1951, 1987). Obstruction of a brain-supplying artery causes ischemia to neurons as well as ischemic damage to the blood vessels within the area of ischemia. When the obstructing embolus moves distally, the previously ischemic region is reperfused with blood. The damaged capillaries and arterioles within that region are no longer competent and blood leaks into the surrounding infarcted tissue. The essential cause of hemorrhagic infarction is reperfusion of previously ischemic tissue. The other mechanism that causes hemorrhagic infarction besides embolism is systemic hypoperfusion. After cardiac arrest or shock, the reinstitution of effective circulation after a prolonged period of brain hypoperfusion can lead to hemorrhage within border-zone infarcts. Hemorrhagic changes are extremely common in patients with brain embolism (Caplan, 1999; Okada et al., 1989; Yamaguchi et al., 1984).

About 80% of emboli that arise from the heart go into the anterior (carotid artery) circulation, equally divided between the two sides. The remaining 20% of emboli go to the posterior (vertebral and basilar arteries) circulation, a rate roughly equal to the proportion of blood supply entering the vertebrobasilar arteries (Bogousslavsky et al., 1988; Caplan, 1980, 1999; Caplan et al., 1983). The recipient artery destination depends on the size and nature of the embolic particles. Calcific particles from heart valves or mitral annular calcifications are less mobile and adapt less well to the shape of their recipient arteries than red (erythrocyte–fibrin) and white (platelet–fibrin) thrombi. The circulating bloodstream seems to be able to bypass obstructing cholesterol crystal emboli, especially in the retinal arteries.

Within the anterior and posterior circulations there are predilection sites for the destination of embolic particles. Large emboli entering a common carotid artery could become lodged in the common or internal carotid artery, especially if atheromatous plaques had already narrowed the lumens of these vessels. If the emboli were able to pass through the carotid arteries in the neck, the next common lodging place is the intracranial bifurcation of the internal carotid arteries (ICAs) into the anterior cerebral (ACAs) and middle cerebral (MCAs) arteries. Bifurcations are common

resting places for emboli. Emboli that pass through the carotid intracranial bifurcations most often go into the MCAs and their branches. Emboli seldom go into the penetrating artery (lenticulostriate arteries) branches of the MCAs or the penetrators from the ACAs because these vessels originate at about a 90° angle from the parent arteries.

Embolism into the MCAs causes a variety of different patterns of infarction. Obstruction of the mainstem MCA before the lenticulostriate branches can cause a large infarct that encompasses the entire MCA territory including the deep basal ganglia and internal capsule as well as the cerebral cortex and subcortical white matter of both the suprasylvian and infrasylvian MCA territories. In some patients an embolus blocks the intracranial carotid artery causing infarction of the anterior cerebral artery (ACA) territory as well as the entire MCA territory. In young patients when the mainstem MCA is blocked, the rapid development of collateral circulation over the convexity of the brain often leads to sparing of the superficial territory of the MCA. The lenticulostriate branches are blocked by the clot in the mainstem MCA and collateral circulation to the deep MCA territory is poor. The resultant infarct is limited to the basal ganglia and surrounding cerebral white matter and is usually referred to as a striato-capsular infarct. Passage of an embolus into the superior division of the MCA leads to a cortical/subcortical infarct in the region of the suprasylvian convexity. Embolism to the the inferior division leads to an infarct limited to the temporal and inferior parietal lobes below the sylvian fissure. When an embolus rests first in the mainstem MCA and then travels to one of the divisional branches, infarction involves the deep territory and cortex above or below the sylvian fissure. Small emboli block cortical branches and cause small cortical/subcortical infarcts involving one or several gyri. Occasionally emboli block the anterior cerebral artery or its distal branches. This causes an infarct in the paramedian area of one frontal lobe.

Emboli entering the posterior circulation can block the vertebral arteries in the neck or intracranially. Emboli able to pass through the intracranial vertebral arteries (ICVAs) will usually be able to pass through the proximal and middle portions of the basilar artery which are wider than the ICVAs. The basilar artery becomes narrower as it courses cranially. Emboli often block the distal basilar artery bifurcation ("top of the basilar") or one of its branches. The most frequent brain areas infarcted are the posterior inferior portion of the cerebellum in the territory of the posterior inferior cerebellar artery (PICA) branch of the ICVA, the superior surface of the cerebellum in the territory of the superior cerebellar artery, and

**TABLE I** Topography of Infarcts in the Lausanne Stroke Registry in Patients with Potential Cardiac Sources of Embolism

| | |
|---|---|
| Anterior circulation | 213 (70%) |
| Global MCA | 33 (11%) |
| Superior division MCA | 60 (20%) |
| Inferior division MCA | 54 (18%) |
| Deep subcortical | 56 (18%) |
| Anterior Cerebral Artery (ACA) | 9 (3%) |
| ACA and MCA together | 1 (0.3%) |
| Posterior circulation | 69 (23%) |
| Brainstem | 18 (6%) |
| Thalamus (deep PCA) | 12 (4%) |
| Superficial PCA | 21 (7%) |
| Superficial and deep PCA | 3 (1%) |
| Cerebellum | 10 (3%) |

Bogousslavsky *et al.*, 1991, copyright Lippincott Williams & Wilkins.

the thalamic and hemispheral territories of the PCAs (Caplan, 1980, 1996, 1999). Table I shows the most frequent sites of infarction in patients with potential cardiac sources of embolism in the Lausanne Stroke registry (Bogousslavsky *et al.*, 1991).

Emboli of cardiac origin are often larger than those arising in the proximal arteries and so the infarcts are, on average, larger than artery-to-artery infarcts (Lodder *et al.*, 1986; Timsit *et al.*, 1993). Another important feature of cardiac origin embolism is development of multiple cortical/subcortical infarcts in multiple vascular territories within both carotid circulations and the posterior circulation, especially in the absence of severe proximal arterial occlusive lesions. Emboli arising from the aorta probably share the same patterns as that found in cardiac-origin embolism although recipient sites of aortic-origin emboli have not been extensively studied. Emboli that arise from proximal arteries go only into branches of that artery. Repeated embolism into one MCA suggests an intrinsic lesion of the carotid artery on that side. Proximal arterial disease often induces circulatory changes with increased collateral circulation. The pre-existence of collateral circulation might limit the size of intra-arterial embolic infarcts when compared to cardiac and aortic origin embolism in which there is no such pre-event adaptation.

The distribution of infarcts according to superficial and deep intracranial territories also differs between patients with cardiac and intra-arterial sources of embolism. Superficial and deep infarcts were much more common in patients with cardiac sources of embolism (Timsit *et al.*, 1993). Large emboli more often block the mainstem MCAs and PCAs before their penetrating artery branches, leading to infarcts that are both deep and superficial. In contrast purely superficial infarcts are more common in patients with intra-arterial embolism (Timsit *et al.*, 1993).

A number of features related to the recipient arteries themselves suggest the presence of embolism but not its source. These include blockage of superficial branches of intracranial arteries, a filling defect within an artery, abrupt sharply demarcated occlusion of an intracranial artery that does not harbor an atherosclerotic lesion, absence of vascular occlusion proximal to a superficial or superficial and deep brain infarct, and movement or disappearance of an obstructing vascular occlusive lesion on subsequent vascular studies. TCD is also helpful in showing high-intensity transient signals (HITSs) that represent emboli that pass through intracranial arteries (Daffertshofer *et al.*, 1996; Markus, 1993; Markus and Harrison, 1995; Sliwka *et al.*, 1997; Tong and Albers, 1995).

### Donor Sources

Detection of a potential donor source of embolism provides circumstantial evidence that brain ischemia might be due to embolism. Cardiac sources are elicited by careful history of cardiac symptoms, electrocardiography, cardiac rhythm monitoring, and echocardiography. Transesophageal echocardiography is also able to provide images of the aorta and show potential aortic sources of embolism. The ascending aorta can also be insonated using a Duplex ultrasound probe placed in the right supraclavicular fossa and the arch and proximal descending thoracic aorta can be imaged using a left supraclavicular probe (Weinberger *et al.*, 1998). Most plaques are located in the curvature of the arch from the distal ascending aorta to the proximal descending aorta, regions well shown using B-mode ultrasound (Weinberger *et al.*, 1998). Potential donor sources of embolism in the neck and major intracranial arteries can be imaged using MRA, CTA, and extracranial and transcranial ultrasound.

## NATURAL COURSE

The most common and characteristic time course in patients with embolism to a brain artery is the very sudden onset of neurologic symptoms and signs that are maximal at onset. After an embolus blocks the recipient artery, collateral circulation begins to develop and some improvement occurs. The breakup and distal movement of the embolus strongly effects the subsequent clinical course. Movement of emboli is most common during the first 48 h after symptom onset (Caplan, 1999; Mohr *et al.*, 1978). Transcranial Doppler monitoring of intracranial arteries after onset of embolic strokes also shows a high frequency of emboli passage. The rate of spontaneous recanalization during the first 24 h is high. Angiographic studies show that up to 20% of MCA emboli recanalize spontaneously within 6 h (Furlan *et al.*, 1999).

Movement of emboli before the development of irreversible brain damage allows reperfusion of previously ischemic brain tissue and is usually accompanied by clinical improvement in symptoms and signs. However, in some patients the embolus or its fragments block an important distal branch leading to further ischemia and worsening of symptoms. For example, a patient with an embolus to the left mainstem MCA might have the sudden onset of right hemiplegia, hemisensory loss, and aphasia. When the embolus passes and the lenticulostriate arteries supplying the internal capsule and basal ganglia regions are reperfused, the hemiparesis might improve. Improved cortical blood flow might be accompanied by improvement in speech. If the embolus passed into the inferior division of the MCA supplying the temporal lobe and occluded the artery, the patient might then have the further onset of a fluent Wernicke-type aphasia. When there is further worsening after initial improvement in patients with embolism, the worsening usually occurs in a single step and nearly always occurs during the first 48 h. Multiple stepwise worsenings, gradual smooth worsening, and delayed worsenings are unusual clinical courses in patients with brain embolism. Late worsening after 48 h should raise suspicion of hemorrhage into the area of infarction since hemorrhagic transformation often occurs between days 2 to 7 after stroke onset. Some patients with very large brain infarcts develop brain swelling, usually during the period of 18 h to 4 days after the onset of the stroke.

Another pattern characteristic of brain embolism has been called spectacular shrinking deficit by Mohr *et al.* (1986). This term describes sudden, complete or nearly complete clearing of a sudden onset severe neurologic deficit. Most often the patient has had a mainstem MCA or basilar artery embolus which rapidly passed. The abrupt onset of a severe neurologic deficit with rapid improvement is nearly always due to embolism (Minematsu *et al.*, 1992).

Temporary deficits that qualify as transient ischemic attacks (TIAs) also occur in some patients with brain embolism. In patients with arterial sources of emboli, the attacks are always in the supply territory of the affected artery. In patients with cardiogenic or aortic origin embolism, when TIAs occur they are usually random and in different vascular territories.

## PRINCIPLES OF THERAPY

The treatment of patients with brain embolism has two basic goals: (1) treatment of the acute embolic event and (2) prophylactic treatment to prevent further embolism.

### Treatment of the Acute Event of Brain Embolism

The goals of treatment of the acute ischemic event are minimization of brain ischemic injury and reperfusion whenever possible.

#### Reperfusion

**Thrombolysis.** The most important predictor of recovery from brain embolism is whether ischemic brain is reperfused and, if so, how quickly. Reperfusion occurs when an occlusive embolus passes, either spontaneously or after treatment. Augmentation of perfusion through collateral circulation to an ischemic brain region also promotes salvage of neurons. Many emboli pass spontaneously, usually within the first 48 h after the onset of neurological symptoms. Vascular studies (CTA, MRA, TCD) can determine the presence of blockage of major intracranial and extracranial arteries by emboli. If the patient is seen and evaluated soon after stroke onset and embolic occlusions are identified by vascular imaging then an attempt at thrombolysis should be made unless there are contraindications or the patient already has a major brain infarct.

The timing of thrombolysis is important if brain tissue is to be saved: "*time = brain.*" Salvagability of ischemic brain tissue varies considerably from patient to patient. After an embolic occlusion, there are three zones of brain tissue supplied by the recipient recently occluded artery. One zone, usually at the core center of blood supply, soon becomes irreversibly damaged (infarcted). Brain tissue at the very periphery of supply often remains normal since adequate blood is usually supplied through adjacent collateral channels. Between these two zones lies brain tissue that is in a state somewhere between infarcted and normal. This tissue has inadequate blood supply to function but is not irreversibly damaged. This state has been called "stunned brain" or penumbral tissue. Infarcted brain tissue contains cells that are dead and blood vessels that may have also been damaged by ischemia. The major danger of thrombolysis (and of spontaneous reperfusion) is that reperfusion of these damaged vessels in the infarcted zone could cause major bleeding into ischemic brain tissue.

The decision on whether or not to pursue thrombolysis should rest not only on the time that has expired since symptom onset but also on the presence of viable penumbral tissue that can be salvaged and the extent of brain that is already infarcted. The extent of infarction determines the risk of treatment; the presence and size of penumbral, stunned tissue determines the potential benefit of thrombolysis that accomplishes reperfusion. The newer magnetic resonance techniques of diffusion-weighted and perfusion MR scans performed with echo-planer equipment, when coupled with MRA, should give clinicians a quantitative estimate of these factors (Albers, 1999; Sorenson *et al.*, 1996; Warach *et al.*, 1992). Alternatively, clinicians estimate the extent of normal, infarcted, and stunned brain supplied by the occluded artery by using brain imaging (CT and T2-weighted MRI scans), vascular studies (CTA, MRA, TCD, angiography), and neurological examination. If the patient has a severe neurological deficit and a large infarct is present on brain scans, then much of the brain is infarcted and there is little to gain by thrombolysis, which carries a substantial risk of hemorrhage in this circumstance. However, if the patient has a severe neurological deficit and brain scans are normal, then there could be considerable stunned, salvageable brain which could potentially be restored to function by successful thrombolysis.

Thrombolytic drugs can be given either intravenously or intra-arterially. Each has advantages and disadvantages (Furlan *et al.*, 1999; Pessin *et al.*, 1995). Intravenous therapy can be administered quickly and needs no special training. The amount of drug that reaches large obstructed arteries is, however, more limited than intra-arterial infusion of drug, which can deliver the drug locally within the obstructing clot. Intra-arterial therapy requires a trained interventionalist. Arteriography is ordinarily required before, during, and after intra-arterial thrombolysis. This delays treatment. The major advantage of intra-arterial therapy is that the interventionalist can physically manipulate the clot, a process that facilitates thrombolysis, and can perform angioplasty at the same sitting if necessary. Usually less drug is used during intra-arterial therapy and the rate of hemorrhagic complications is lower than with intravenous therapy (Pessin *et al.*, 1995).

There is considerable data from reports of patient treated with either intravenous or intra-arterial therapy after their occlusive lesions have been shown by angiography. There is also data from randomized trials concerning outcome in patients treated intravenously in whom no vascular studies were required prior to treatment. The results of these trials are shown in Chapter 30 on acute ischemic stroke treatment. Embolic occlusions respond better than thrombi formed locally in vessels that have severe atherostenosis. Freshly formed thromboemboli lyse more often than older clots. Thrombi that block the extracranial or intracranial ICA rarely respond to intravenous thrombolysis. An especially important and common situation is thrombosis of the ICA in the neck which has caused neurological deficit by an embolus breaking off from the neck thrombus and embolizing intracranially to the MCA. Intra-

venous thrombolysis in this situation is ineffective because the drug does not reach the MCA clot because of proximal obstruction. An interventionalist may be able to pass a catheter through the clot in the neck and then manipulate the catheter to and within the MCA clot in order to deliver the thrombolytic drug into the MCA clot. After lysing the intracranial clot, the catheter can be maneuvered back into the neck in order to lyse the neck clot and, if needed, perform an angioplasty with stenting of the atherostenotic ICA disease. MCA branch occlusions seem to lyse best with intravenous therapy. To date very few patients with basilar artery thromboemboli have been studied after intravenous thrombolysis but intra-arterial therapy of patients with basilar artery occlusion is often effective, especially if the occlusion is embolic.

The most commonly used thrombolytic drugs for stroke patients are recombinant tissue plasminogen activator (rt-PA) and urokinase. Streptokinase has had an unacceptable rate of bleeding complications when used for stroke (Donnan *et al.*, 1995). For details see Chapter 30.

**Mechanical Clot Lysis or Removal.**   Mechanical clot lysis and removal for example using laser energy or a vacuum type extraction device has begun to be used experimentally and may have a bright future. These treatments would be introduced through the vascular system by trained interventionalists. To date there have been no published trials of such treatments.

**Angioplasty and Stenting.**   When an embolus arises from a region of atherosclerotic narrowing, angioplasty with or without stenting might be required to prevent acute rethrombosis after clot lysis. In some patients direct angioplasty might be effective without thrombolysis (Nakano *et al.*, 1998).

**Surgery.**   Reperfusion can also be accomplished by surgical embolectomy but this procedure has been superceded by thrombolysis and mechanical thrombus removal by endovascular means. Direct endarterectomy of stenotic neck arteries can also be performed acutely but carries a risk of reperfusion hemorrhage in the presence of acute brain infarction and so this procedure is seldom performed in patients with embolism arising from neck arteries. A surgical bypass can be created, connecting one vessel to another beyond an intracranial embolic obstruction. An artery can be directly sewn to another, or a venous conduit can be interposed. Many anecdotal reports noted the effectiveness of extracranial (EC) to intracranial (IC) artery bypass, using the superficial temporal artery as the donor artery and an MCA branch as the recipient artery. A large, randomized study of the effectiveness of such EC-IC bypasses, however, proved beyond reasonable doubt that the surgery as it was customarily performed had no benefit, and in some circumstances, operated patients fared worse than patients treated medically (EC-IC Bypass Study Group, 1985). The patients in this series had surgery about 6 weeks after stroke to prevent reperfusion hemorrhage in regions where the capillaries and arterioles might be ischemic and vulnerable to leakage. Might bypass procedures earlier in the course, although more risky, be more effective?

### Prevention of Propagation and Embolization of Existent Thromboemboli

The most common embolic materials are thrombi formed on the surface of endothelial lesions that involve the heart, aorta, or major extracranial and intracranial arteries. In general, so-called "red clots" (erythrocyte–fibrin thrombi) tend to form in regions where there is low flow or stagnation, whereas smaller so-called white platelet clots adhere to roughened places in faster-moving streams of blood (Deykin, 1967; Weksler, 1985). Standard anticoagulants, such as heparin (and low-molecular-weight heparin and heparinoids) and warfarin, theoretically should be more useful in preventing red clots, and antiplatelet agglutinating agents should be better at preventing white platelet plugs. Heparin and warfarin should work best in occlusive disease of veins and large arteries and in cardiac disorders that predispose to cardiac-origin thromboembolism, whereas agents that decrease platelet aggregation might have an advantage in plaque disease without severe stenosis.

Heparin is also often prescribed as a treatment for patients with acute thromboembolism. The posited purpose of heparinization is to prevent propagation of red erythrocyte–fibrin thrombi and breakoff of the tail of existing thrombi and so prevent embolization. In addition, heparin and, later, warfarin are used to prevent further thromboembolism after the acute event. As far as is known, heparin does not actually lyse existing thrombi, although cardiac clots often disappear during heparin treatment. The decision on whether or not to prescribe heparin acutely to prevent the next thromboembolic stroke depends on weighing the risk of acute reembolization vs the risk of hemorrhage related to heparin therapy. In patients with lesions with high rates of re-embolization, e.g., mitral stenosis with atrial fibrillation, atrial fibrillation (especially with large atria and atrial thrombi), and acute myocardial infarction with mural thrombi, then acute heparinization is probably warranted. In patients with a low risk of acute rembolization, such as chronic atrial fibrillation or mitral annulus calcification, heparin can be withheld during the acute period. If the patient has a large brain infarct, then the risk of brain hemorrhage after heparin is higher than when there is no brain infarct or a small brain infarct.

There is little data from trials that tests the effectiveness of heparin in preventing acute worsening in patients with thromboembolism verified by modern neuroimaging. In the TOAST trial, a heparinoid was effective in a subgroup of patients with large artery occlusive disease (Publications Committee, 1998). In this study, the low-molecular-weight heparinoid ORG 10172 was given within 24 h of the onset of symptoms of an acute ischemic stroke (Publications Committee, 1998). This heparinoid was given by continuous intravenous infusion for 7 days and the dose was adjusted after 24 h to maintain the antiXa factor activity at 0.6–0.8 antifactor Xa units/mL. ORG 10172 (Danaparoid) is a mixture of glycosaminoglycans isolated from porcine intestinal mucosa. The antifactor Xa activity of Danaparoid is attributed to its heparin sulfate component, which has a high affinity for antithrombin III.

Although Danaparoid treatment was not effective in the entire group of patients with ischemic stroke, there was effectiveness in the group of patients diagnosed as having large artery atherosclerosis, most of whom likely had embolism of red clot from their atherosclerotic lesions. In the TOAST trial, in patients classified as having large artery atherosclerosis, heparinoid reduced the number of recurrences of stroke during the 7 days of infusion, and the rates of favorable and very favorable outcomes were significantly higher in patients given heparinoid when compared with placebo. Sixty-eight percent of patients with large artery atherosclerosis treated with Danaparoid had favorable outcome vs 54.7% treated with placebo ($P = 0.04$); 43% of patients with large artery atherosclerosis treated with Danaparoid had very favorable outcomes vs 29.1% treated with placebo ($P = 0.02$). Recurrent strokes developed in 6% of Danaparoid-treated patients with large artery atherosclerosis vs 11% of those treated with placebo and 2.8% of patients with cardioembolism treated with Danaparoid had recurrent strokes vs 7.3% of patients treated with placebo. Because of the small numbers the figures for recurrent strokes did not meet statistical significance (Publications Committee, 1998). Danaparoid was effective among the group of patients with large-artery atherosclerosis who had severe internal carotid artery stenosis in the neck (>50% luminal narrowing or occlusion) (Bandixen et al., 1998). This was the only subgroup in the TOAST study in whom all of the patients had vascular tests that defined the large artery lesions. Significantly more patients with severe carotid artery disease had favorable and very favorable outcomes among patients treated with heparinoid.

There is data about the effectiveness of warfarin in patients with some cardiac lesions. This will be presented in the section below on prophylaxis against future thromboembolism. Theoretically, aspirin, clopidogrel, and aspirin combined with dipyridamole or clopidogrel could prevent the new formation of white platelet–fibrin clots and so reduce brain ischemia when the embolus consists of white clots. Although aspirin has been used in the acute treatment of brain ischemia there is no vascular data available in the patients in these trials. Gp2b/3a inhibitors are now being used in patients with acute vascular occlusive lesions since sometimes, e.g., after carotid surgery, the intracranial embolus consists of a carpet of white platelet–fibrin clots.

## Treatment of Brain Edema and Mass Effect

Brain edema often develops early in patients with embolic strokes. The edema allows recognition of ischemic tissue on diffusion-weighted MR scans. Ischemic edema can be intracellular (so-called cytotoxic edema or dry edema) or exist mostly in the extracellular spaces and connective tissue (vasogenic edema or wet edema). Brain edema that lies in the interstices outside of cells might be posited to respond to osmotic diuretics such as mannitol and glycerol, or to corticosteroids. However, studies have shown that these agents are not very effective in series of stroke patients with large brain infarcts or hemorrhages. Most edema is probably within cells and indicates that the cells are sick. Restitution of the normal metabolic functions of these cells is likely to be more therapeutic than so-called anti-edema agents. There are probably some patients, mostly young patients, who quickly develop a great deal of vasogenic, extracellular brain edema with the consequences of increased intracranial pressure and displacement of brain compartments. In these patients a therapeutic trial of osmotic agents and/or corticosteroids is probably warranted since the situation is often desperate.

The usual medical treatments of brain edema shrink mostly uninfarcted brain. These medical treatments—barbiturates, hypothermia, corticosteroids, hyperventilation, osmotic diuretics (mannitol, glycerol, hypertonic dextrose), and furosemide—are not very effective in patients with severe brain swelling. In some patients with massive brain swelling and increased intracranial pressure, removal of the skull overlying the side of the infarct (hemicraniectomy) can be life-saving but some patients may be left with severe neurological residual deficits (Schwab *et al.*, 1996, 1998). Some patients make extraordinary recoveries after hemicraniotomies and survive with very little neurologic deficit. Induced hypothermia of 33°C has also been reported beneficial for large-space-occupying infarcts (Diringer, 1999; Schwab *et al.*, 1998).

## Prevention of Cell Death of Ischemic Neurons

Experimental studies have begun to dissect the cellular mechanisms and causes of brain cell death in relation to ischemic injuries (Garcia, 1995). Clinicians have now begun to give drugs in an effort to make the brain more resistant to ischemia. This type of therapy is usually called neuroprotective treatment. Neuroprotective therapy attempts to ameliorate the cellular metabolic consequences of ischemic injury (Devuyst and Bogousslavsky, 1999; Fisher, 1995; Oral and Fisher, 1996; Venti *et al.*, 2000). In ischemic brain, calcium moves into cells via calcium-conducting channels and excitotoxic amino acids, free oxygen radicals, and leukocytes attracted to the ischemic zone promote cytotoxic injury to neurons (Garcia, 1995). Alterations in growth factors and gene expression also contribute to programmed cell death. Knowledge of the mechanisms of cell injury have led to the development of many neuroprotective agents now in various stages of development and trials. These agents have very diverse posited modes of action and varied strategies for neuronal protection (Devuyst and Bogousslavsky, 1999; Fisher, 1995; Oral and Fisher, 1996; Venti *et al.*, 2000). Physicians hope that these agents can mitigate ischemic damage and delay neuronal death long enough to allow reperfusion. These agents are in various development and trial stages. None has, as yet, been definitively shown to be effective in patients with brain ischemia.

## Prophylaxis against Recurrent Thromboembolism

### Drug Treatments

No single drug is effective in preventing embolization of all of the different materials that embolize. Drugs have rather specific effects depending mostly on the nature of the embolic material. Table II from Caplan (1991) shows the variety of different materials that can embolize from the heart, aorta, and arterial donor sources. In only a few circumstances has the effectiveness of either antiplatelet or anticoagulant drugs been analyzed in randomized trials which defined the source of emboli.

**Substances That Alter Coagulation.** Standard anticoagulants (heparin, heparinoids, low-molecular-weight heparins, warfarin) are posited to be most effective in situations that promote red thrombus formation. Red thrombi are most likely to form in the heart in patients with atrial fibrillation and sick sinus syndrome; in myocardial infarction, myocardial aneurysms, and hypokinetic and akinetic zones; in the ventricles in patients with low ejection fractions and congestive heart failure; in patients with myocarditis

**TABLE II**   Types of Embolic Materials

| Cardiac origin | Aortic origin | Arterial origin |
| --- | --- | --- |
| Red thrombi | Red thrombi | Red thrombi |
| White platelet–fibrin aggregates | White platelet–fibrin aggregates | White platelet–fibrin aggregates |
| Marantic endocarditis vegetations | Cholesterol crystals | Cholesterol crystals |
| Bacteria from endocarditis vegetations | Atheromatous plaque debris | Atheromatous plaque debris |
| Calcium (valves andmitral annulus calcification [MAC]) | Calcium from vascular calcification | Calcium from vascular calcification |
| Myxoma and fibroelastoma fragments | | Air |
| | | Fat |
| | | Tumor cells and tumor mucin |
| | | Talc or cellulose (microcrystalline) from injected drugs |

*Note.* Modified from Caplan (1991).

and myocardiopathies; in the leg and systemic veins in patients with paradoxical embolism; and in patients with valvular heart diseases and those who have prosthetic valves. Red thrombi also are often present in patients with protruding mobile aortic atheromas that are larger than 4 mm and in large recently occluded or very stenotic arteries.

Data about the effectiveness of anticoagulants is most robust in regard to prophylaxis against brain embolism in patients with atrial fibrillation who do not have valvular heart disease. Table III from (Caplan, 1999) reviews the six major randomized trials of anticoagulants in patients with atrial fibrillation (Boston Area Anticoagulation Trial, 1990; EAFT, 1993, 1995; Petersen *et al.*, 1989; Stroke Prevention, 1991, 1994, 1996, 1997). All trials showed a consistent and considerable risk reduction for stroke in patients treated with warfarin. The European Atrial Fibrillation Trial (EAFT) Study Group addressed specifically the question of the optimal level of anticoagulation by reviewing the results of their own trial (EAFT, 1995). No treatment effect was found with anticoagulation responses below INRs of 2.0. The rate of thromboembolic events was lowest at INRs from 2 to 3.9; most major hemorrhages occurred at INRs of 5.0 and above. The EAFT group recommended a target of 3.0 with a range from 2 to 5.0 (EAFT, 1995). Fixed-dose warfarin with a target of 1.3 to 1.5 was not as effective as standard adjusted-dose

warfarin at an average INR of 2.4, even when aspirin, 325 mg/day, was added to the low fixed-dose warfarin in another study (Stroke Prevention, 1996). Warfarin is about 50% more effective than aspirin in reducing the rate of stroke in patients with atrial fibrillation without valvular disease (Albers, 1994). The rate of intracranial hemorrhages and major bleeding episodes was low in all groups in all trials except in warfarin-treated patients over 75 in the SPAF II study. In that study, 7 of 197 warfarin-treated older patients had intracranial hemorrhages which were most often fatal (Stroke Prevention, 1996). Aspirin is also an effective treatment; in aggregate, aspirin conveys a stroke risk reduction of 20 to 25% with no clear relationship to aspirin dose. The SPAF III investigators enrolled 892 patients with atrial fibrillation who had none of the four prespecified risk factors: (1) recent congestive heart failure or left ventricular fractional shortening, <25%; (2) previous thromboembolism; (3) systolic blood pressure >160 mm Hg; and (4) women >75 years of age (Stroke Prevention, 1997). The rate of ischemic stroke (2%/year) and disabling ischemic stroke (0.8%/year) were low and were just above the rate of major bleeding with aspirin therapy (0.5%/year) in this study (Stroke Prevention, 1997).

Patients with clinical and/or echocardiographic risk factors should be treated with warfarin with an INR maintained between 2 and 4 (target 2.6 to 3.0) In elderly patients (>75 years) the risk of serious hemorrhage is higher. Koudstaal, in discussing treatment of patients >80 years old, said it well: "very elderly atrial fibrillation patients have most to win but also most to lose from anticoagulation treatment (Koudstaal, 1995). Treatment should be decided on an individual basis with the clinician weighing the risk of stroke without warfarin versus the risk of important hemorrhage on warfarin treatment. Koudstaal noted that there was no good reason to deny anticoagulant treatment to an 85-year-old patient with a recent stroke due to atrial fibrillation if that patient had no major risk factors for hemorrhage and was expected to be compliant with testing and treatment (Koudstaal, 1995).

Trial data on the effectiveness of anticoagulants for prevention of embolism in patients with other cardiac conditions that would be expected to promote red clot formation is meager. Most strokes that occur in patients with acute myocardial infarction are caused by embolization of thrombi formed in the left ventricle and atrium. The stroke occurrence with and without anticoagulation was studied in three trials of large numbers of patients with myocardial infarcts (Drapkin and Mersky, 1972; Hart, 1987; Medical Research Council, 1969; Veterans Cooperative Study, 1973). The data from these trials is summarized in Table IV. Among a total of 3562 patients with myocardial infarcts, 2.9%

**TABLE III**    Trials of Prophylactic Therapy in Patients with Atrial Fibrillation without Valvular Disease

| Trial | Design | Results |
|---|---|---|
| Copenhagen AFASKA (Petersen et al., 1989) | 1007 Patients; mean age 73; coumadin (INR 2.8–4.2) vs aspirin (75 mg/day) vs placebo. | Thromboemboli (stroke, TIA, systemic embolism) coumadin 2%/year; aspirin 5.5%/year; placebo 5.5%/year. |
| BAATAF (Boston Area Anticoagulation Trial, 1990) | 628 Patients; mean age 68; coumadin (INR 1.5–2.7) vs other medical Rx (could include aspirin). | Coumadin 2 strokes (0.4%/year); Control 13 (3%/year). No benefit of aspirin (8 of 13 strokes in controls on aspirin); 2 hemorrhages, 1 each group. |
| SPAF (Stroke Prevention, 1991) | 1330 Patients; mean age 67; warfarin-eligible patients randomized to warfarin (INR 2–3.5), aspirin (325 mg/day), or placebo. Warfarin-ineligible patients randomized to aspirin or placebo. | Warfarin 2.3%/year vs 7.4%/year placebo; stroke in warfarin-ineligible aspirin group 3.6%/year vs 6.3% in placebo group. Major bleeding 1.5, 1.4, 1.6% in warfarin, aspirin, and placebo groups, respectively. |
| EAFT (1993) | 1007 Patients, mean age 73; warfarin-eligible patients randomized to warfarin (INR 2.5–4), aspirin (300 mg), or placebo. Warfarin-ineligible to aspirin or placebo. | Strokes in 8% of 225 in warfarin group, 15% of 404 in aspirin group, 19% of 378 in placebo group. Major bleeding 2.8%/year warfarin group and 0.9%/year aspirin group. |
| SPAF II (Stroke Prevention, 1994) | 1100 Patients; mean age 69.6; warfarin (INR 2–4.5) vs aspirin (325 mg/day compared in patients <75 and patients >75. | 715 Patients <75; ischemic stroke and systemic embolism 1.3%/year warfarin vs 1.9%/year aspirin; major hemorrhage 0.9%/year aspirin, 1.7%/year warfarin; 385 > 75; ischemic stroke and systemic embolism 3.6%/year warfarin, 4.8%/year aspirin; major bleeds 4.2% warfarin,[a] 1.6% aspirin. |
| SPAF III (Stroke Prevention, 1996) | 1044 Patients with one or more risk factors; mean age 72; low-intensity fixed-dose warfarin (INR 1.2–1.5) plus aspirin (325mg/day vs adjusted-dose warfarin (INR 2–3). | INR 1.3 fixed-dose warfarin vs INR target 2.4 adjusted group. Ischemic stroke and systemic embolism in 7.9% of fixed-dose and aspirin vs 1.9% in adjusted dose group. |
| SPAF III (Stroke Prevention, 1997) | 892 Patients with posited low risk were given 325 mg aspirin. | The rate of ischemic stroke was low (2%/year) and disabling ischemic stroke only 0.8%/year. The rate of major bleeding was 0.5%/year. |

*Note.* Reprinted from Caplan (1999), by courtesy of Marcel Dekker, Inc. AFASAK, Copenhagen Atrial Fibrillation Aspirin Anticoagulation Study. BAATAF, Boston Area Anticoagulation Trial for Atrial Fibrillation. SPAF, Stroke Prevention in Atrial Fibrillation Study. EAFT, European Atrial Fibrillation Trial.
[a]71% of intracranial hemorrhages fatal; 29% had residual deficit.

of patients not anticoagulated had strokes vs 1.2% of those who were anticoagulated had strokes (Hart, 1987). These trials were carried out in the 1969–1973 period when the intensity of anticoagulation was much higher than that used now. Despite the high intensity of anticoagulation, the complication rates attributable to anticoagulants were low. Minor bleeding occurred in 7%,

major bleeds in 1.5%, and intracranial hemorrhages in 0.05%; there were no fatal hemorrhages (Drapkin and Mersky, 1972; Hart, 1987; Medical Research Council, 1969; Veterans Cooperative Study, 1973).

The data about the effectiveness of anticoagulants in patients with valvular heart disease come predominantly from studies in patients with rheumatic mitral stenosis performed between 1951 and 1975 (Adams *et al.*, 1974; Carter, 1957, 1965; Fleming and Bailey, 1971). The frequency of embolism in patients with mitral stenosis ranges in various series between 10% (Szekely, 1964) and 20% (Coulshed *et al.*, 1970; Keen and Leveaux, 1958). About 50 to 75% of emboli detected clinically involve the brain. Although embolism does occur in patients with mitral stenosis who are in normal sinus rhythm, the development of atrial fibrillation greatly increases the risk of embolism. Daley *et al.* found a mitral valve lesion in 97%, atrial fibrillation in 90%, and either a mitral valve lesion or atrial fibrillation in 100% of 194 patients with rheumatic heart disease and

**TABLE IV**    Acute Myocardial Infarction, Stroke Occurrence, and Anticoagulation

| Trials | No. of patients | Strokes (no anticoagulants) (%) | Strokes (anticoagulated) (%) |
|---|---|---|---|
| Veterans[59] | 999 | 3.8 | 0.8 |
| MRC (UK)[60] | 1427 | 2.5 | 1.1 |
| Bronx Hospital[61] | 1136 | 2.3 | 1.7 |
| **Totals** | 3562 | 2.9 | 1.2 |

*Note.* Modified from Hart (1987), with permission from Springer-Verlag, London.

systemic embolism (Daley *et al.*, 1951). The frequency of embolism is about seven times higher in patients with atrial fibrillation than in those with sinus rhythm (Caplan, 1999). A third of recurrences of embolism occur during the first month, and two-thirds of recurrences occur during the first year after the onset of atrial fibrillation (Szekely, 1964).

Observational studies show that anticoagulants clearly reduce the frequency of recurrent embolism. Fleming and Bailey reported only five recurrent embolic episodes among 217 patients with mitral stenosis treated with anticoagulants during a period of 9½ years, a rate of 0.8% per patient treatment year (Fleming and Bailey, 1971). Adams and colleagues studied the effect of anticoagulants on mortality among 84 patients with mitral valve disease and atrial fibrillation who had survived brain embolism (Adams *et al.*, 1974). Two weeks after their strokes, half the patients were given anticoagulants (phenindione at first, later warfarin). The reduction in mortality was dramatic, especially during the first 6 months after brain embolism (Adams *et al.*, 1974). Carter also showed that anticoagulant treatment improved the prognosis in patients with brain embolism in terms of immediate outcome, recurrences, and late survival (Carter, 1957). Unfortunately, bleeding was a very significant problem during the time of these studies in patients on long-term anticoagulants because the intensity of treatment was much higher than now used. The longer that patients were on anticoagulants, the higher the rate of serious hemorrhage often into the cranial cavity.

Embolism is an important complication in patients with both mechanical and bioprosthetic valves. The frequency of major embolism in patients with mechanical valves is estimated to be about 4% per year if no antithrombotic therapy is used (Cannegiester *et al.*, 1994). This figure is reduced to about 2% per year by medications that decrease platelet aggregation and to 1% per year with warfarin anticoagulation (Cannegiester *et al.*, 1994).

The effect of anticoagulants and antiplatelet aggregants on the recurrence rate of stroke in patients with PFOs has occasionally been studied (Bogousslavsky *et al.*, 1996; French Study Group, 1995; Hanna *et al.*, 1994). In one small study, no recurrences ocurred among 15 patients taking aspirin (6), warfarin (7), or after surgical closure (2) (Cannegiester *et al.*, 1994). Bogousslavsky and colleagues studied stroke recurrence among 140 consecutive patients who had PFOs and brain ischemic events (Bogousslavsky, *et al.*, 1996). One-fourth of the patients also had atrial septal aneurysms. During a mean follow-up period of 3 years, the stroke or death rate was 2.4% per year. Only 8 patients had a recurrent brain infarct (1.9% per year) (Bogousslavsky *et al.*, 1996). Ninety-two (66%) took

aspirin, 250 mg/day, while 37 (26%) were given anticoagulants and 11 (8%) had surgical closure of the PFO within 12 weeks of the stroke after being treated with anticoagulants. There was no significant difference in the effect of any of the treatments on recurrence. The relatively low rate of recurrence contrasted with the severity of the initial stroke which left disabling effects in half the patients (Bogousslavsky *et al.*, 1996). In a French multicenter study, 132 patients with PFOs and/or atrial septal aneurysms and cryptogenic stroke were followed for an average of 22.6 months (French Study Group, 1995). The recurrence rate was approximately 2 to 3% at 2 years and was higher in patients with both PFOs and atrial septal aneurysms. Recurrences ocurred in 4 patients who were taking antiplatelet agents and in one patient treated with anticoagulants (French Study Group, 1995).

Although there are no trial data, anticoagulants are widely accepted as effectively reducing the frequency of brain embolism in patients with sick sinus syndrome, myocardial contractile abnormalities, patients with low ejection fractions and congestive heart failure, and patients with myocardiopathies, myocardial aneurysms, cardiac thrombi, and regions of reduced ventricular function.

Various expert groups have published recommendations for intensity of anticoagulation (Hirsh *et al.*, 1989; Poller, 1991; Yamaguchi *et al.*, 2000). Two intensities of anticoagulation are usually recommended: a less intense range (INR 2.0–3.0, roughly equivalent to a PT of 1.3–1.5) and a more intense range (INR 3.0–4.5, about equivalent to a PT of 1.5–2.0 times control) (Hirsh *et al.*, 1989). The higher intensity range has more risk of bleeding. A large Dutch trial (SPIRIT) that compared the effectiveness of aspirin vs oral anticoagulation (INR target range 3.0–4.5) was prematurely stopped after an interim analysis showed an unacceptable rate of hemorrhage in the anticoagulant-treated group (Stroke Prevention, 1997). The frequency of bleeding increased by a factor of 1.43 for each 0.5-unit increase in the INR (Stroke Prevention, 1997). The Atrial Fibrillation Investigators analyzed the results of five trials of anticoagulation in patients with atrial fibrillation (INR targets 1.4–4.2) and recommended a target range of 2.0 to 3.0 as having the best benefit/risk ratio (Atrial Fibrillation, 1994; Hart, 1997). Hylek and colleagues also analyzed the results of anticoagulation in atrial fibrillation trials and found that the rate of stroke increased when the INR fell below 2.0 and that optimal protection ocurred in patients with INRs between 2.0 and 3.0 (Hylek *et al.*, 1996). No further protection was attributable to INRs above 3.0 (Hylek *et al.*, 1996). Yamaguchi *et al.* found that INRs below 1.6 were not very effective and that the optimal therapeutic range was 1.6–2.3 (Yamaguchi *et al.*, 2000).

We have already cited some data about the effect of antiplatelet agents in patients with atrial fibrillation and valvular disease. There are scant data about the effectiveness of antiplatelet drugs in preventing embolism in patients with well defined donor sources for brain embolism. In patients with prosthetic valves degenerative abnormalities in the valves leads to the formation of white platelet–fibrin thrombi (Caplan, 1999). Prosthetic foreign materials also activates the coagulation cascade, promoting the formation of red erythrocyte–fibrin clots. The combination of antiplatelet agents and warfarin is probably more effective in preventing thromboembolism than anticoagulants alone.

There is a lack of data about the effectiveness of antiplatelet agents in preventing thromboembolism in patients with disorders that are posited to promote the formation of white platelet–fibrin thrombi. In some conditions fibrotic changes develop in the heart valves and the endocardium. Fibrous valve thickening, often with grossly visible vegetations containing mixtures of platelets and fibrin, are found on the heart valves and adjacent endocardium in patients who have no evidence of either rheumatic fever or bacterial endocarditis. The first detailed description of such lesions was by Emanuel Libman and Benjamin Sachs, who reported four patients studied clinically and pathologically of an "atypical verrucous endocarditis" (Caplan, 1999). Libman–Sacks endocardiits develops in patients with systemic lupus erythematosis (SLE). Valvular and fibrous endocardial lesions that have identical distribution and histology as that found in SLE occurring in patients with the antiphospholipid antibody syndrome and in patients with cancer and other debilitating diseases called marantic endocardiits or nonbacterial thrombotic endocardiits (NBTE). Other patients have fibrin strands visible by echocardiography. Antiplatelet agents are posited to be more effective in preventing embolism in patients with these disorders than anticoagulants but there are no data from trials or observational studies that confirm this hypothesis.

Antiplatelet agents have shown some effect in preventing recurrent brain ischemia in large studies of patients with TIAs and minor strokes. Unfortunately the nature of the vascular lesions and whether or not the strokes were caused by embolism has not been defined in these studies. Antiplatelet agents are posited to decrease the occurrence of embolism in patients with nonstenosing aortic and arterial plaques. They might also be effective alone or with warfarin in patients with severe arterial stenosis but there is no therapeutic data about this point.

**Antiarrythmic Agents.** Cardiac arrythmias, especially atrial fibrillation and the sick sinus syndrome, cause brain embolism because red thrombi form within the inefficiently contracting cardiac atria. Defibrillation and the application of agents such as quinidine, procainamide, flecainide, and amiodarone reduce the frequency of brain embolism if they are able to obtain and maintain normal sinus rhythm. Anticoagulation is usually given prior to conversion to normal sinus rhythm to prevent embolization of thrombi that have already formed in the heart.

**Antibiotics.** Valvular vegetations in patients with infective endocarditis are composed of platelets, fibrin, erythrocytes, and inflammatory cells attached to damaged endothelium of native and prosthetic valves. Bacteria and occasionally fungi are enmeshed within the fibrinous material often deep within the vegetations, explaining why antibiotics sometimes have difficulty sterilizing the lesions. The most important treatment that prevents embolization of material from vegetations is the rapid introduction of specific antimicrobial drugs (Caplan, 1999; Hart *et al.*, 1990; Salgado *et al.*, 1989). Most neurologic complications occur before or near the time of diagnosis and initial antibiotic treatment. Recurrent strokes do occur after bacteriologic cure but rarely. In one series, among 147 patients discharged from the hospital after treatment of infective endocarditis, 15 developed strokes after discharge; all except one of the stroke patients had prosthetic valve endocarditis (Salgado *et al.*, 1989). Strokes in this series occurred long after discharge (median 22 months) and were better explained by recurrence of endocarditis, complications of anticoagulants, or the presence of noninfective disease of the prosthetic valves rather than cerebrovascular complications of the original endocarditic episode (Salgado *et al.*, 1989). Experts agree that native valve endocarditis is not an indication for anticoagulation even when brain or systemic embolism has occurred. Controversy surrounds the issue of maintenance of anticoagulation in patients who have mechanical valve endocarditis but most clinicians favor cautious continuation of anticoagulants unless a brain hemorrhage develops. When a hemorrhagic infarct or brain hemorrhage develops, anticoagulation with warfarin is usually stopped temporarily. In patients with a major risk of recurrent embolization it may be safe to use heparin beginning soon after the hemorrhage is discovered and later switch back to warfarin.

### Surgery and Interventional Radiological Treatment

Cardiac surgery in patients with infective endocarditis is performed to debride or replace infected valves for cardiac indications—mostly heart failure related to valve dysfunction, lack of control of infection, valve infection with fungal or other virulent organism not controllable by antimicrobial drugs, and valve or chordae tendinae rupture (Caplan, 1999). Surgery is the only effective treatment in preventing embolism from cardiac tumors such as myxomas and fibroelastomas.

Surgical therapy of congenital cardiac defects including closure of atrial septal defects and patent foramen ovale is generally considered effective in preventing brain embolism in patients with these disorders. In one study, Lausanne investigators found no recurrence of stroke during an average follow-up of 2 years among 30 patients who had suture closure of PFOs during cardiopulmonary bypass surgery (Devuyst et al., 1996). No patients were given antiaggregants or anticoagulants after surgery. There were no serious surgical complications. After surgery, two patients had interatrial shunting as determined by TCD and TEE but the shunts were much smaller than before surgery (Devuyst et al., 1996). PFOs are now also being closed using an umbrella-like device placed through a catheter. Bridges and colleagues reported the results of transcatheter closure of PFOs among 36 patients with PFOs, of whom half had multiple ischemic events (Bridges et al., 1982). Twelve patients had had ischemic events while taking warfarin before trancatheter closure. In 28 (78%) the closure was complete and another 5 had nearly complete closure. No patient had a recurrent stroke during 8.4 months of follow-up but 4 patients had transient ischemic events (Bridges et al., 1982).

Studies have now conclusively shown that endarterectomy is more effective than medical therapy in patients with severe carotid artery occlusive disease who have symptoms of brain or retinal ischemia ipsilateral to the atherosclerotic occlusive lesion (Barnett et al., 1998; Biller et al., 1998; ECST, 1998; MRC, 1991; North American Symptomatic, 1991). Angioplasty with or without placement of stents is now also being widely applied to treat occlusive extracranial and intracranial stenotic lesions in order to prevent embolism and hypoperfusion related to these lesions. The effectiveness of this treatment compared to surgery or to drugs that alter coagulation has not been systematically studied to date.

## PRACTICAL MANAGEMENT

### Treatment of the Acute Brain and Vascular Lesion (Recipient Artery)

#### Attempting to Open Occluded Arteries (Reperfusion)

When the patient with brain embolism is seen soon after neurological symptom onset, the first consideration is whether or not the patient is a candidate for reperfusion. The key factors that determine the advisability of reperfusion are: (1) time since symptom onset,

(2) the presence location and size of brain infarction, and (3) the presence, nature, and location of an occlusive thromboembolus within recipient and donor extracranial and intracranial arteries supplying the region of brain ischemia. The most useful method of assessing vascular occlusion and brain infarction and its size relative to the region of diminished brain perfusion, is modern MRI using echo planar equipment and the techniques of diffusion-weighted and perfusion MRI and MRA. The DWI lesion allows a reasonable estimate of the presence, location, and size of infarction. An estimate of the residual at-risk tissue is obtained from determining the presence and location of any occluded artery on MRA and the region underperfused on perfusion-weighted MRI. The brain tissue underperfused that is supplied by the occluded artery minus the region already infarcted is the tissue still at risk for further infarction (DWI and PDI mismatch).

When there is no occlusive thromboembolism detected within the large extracranial and intracranial arteries significant extension of infarction is quite rare. When the region of infarction is large or the at-risk tissue volume is small, then reperfusion is unwise. CT can also be used to estimate the same factors. Infarction is shown on an unenhanced scan and films after contrast can produce a CTA showing vascular occlusions; brain scan films after contrast can show the region of relative underperfusion. New helical multislice CT technology also can offer perfusion-CT assessment.

Doppler ultrasound can also be used to detect and monitor the presence and location of vascular occlusion. The severity of the neurological deficit also correlates with the relative size of brain infarction. The more severe the deficit, the more likely there is a sizable brain infarct, and the less likely that reperfusion would be effective. Time alone is not a reliable guide but the more time that has expired since symptom onset, the more futile are reperfusion strategies.

Having said this, we nevertheless recommend following the guidelines published by the comittees of the American Heart Association and the American Academy of Neurology (Adams et al., 1996; Quality Standards, 1996) which are using intravenous rt-PA, 0.9 mg/kg (maximum of 90 mg), given in a 10% bolus followed by an infusion lasting 60 min when the patient is seen within 3 h of symptom onset without the need to show either vessel occlusion or a zone of salvagable tissue. If there is an occlusive intracranial thrombus shown by MRA, CTA, or TCD and there is considerable at-risk brain tissue and the size of the already visible brain infarction on plain CT is relatively small, the decision to use thrombolysis may be easier. When only CT is available, then we advocate following the published guidelines (Adams et al., 1996; Quality

Standards, 1996) when the patient is seen within the first 3 h after symptom onset and the clinical history and examination favor an embolic mechanism of stroke. If the patient is seen after 3 h, the estimated infarct size is not too large, and there is still sizable at-risk brain tissue present, intravenous rtPA treatment according to the ECASS II protocol (Hacke *et al.*, 1998) is an option; however, the patient and his relatives need to be informed that this is still an experimental treatment that is not yet approved by the authorities. The same holds true if there is an occlusion that involves the intracranial ICA, mainstem MCA, or the intracranial vertebral or basilar arteries and angiography followed by intra-arterial thrombolysis or mechanical clot lysis is performed.

The decision on whether to pursue thrombolysis or any other reperfusion strategy must include the patient and their significant others and must carefully weigh the benefit/risk ratio of treatment. Thrombolysis is also discussed in more detail in Chapter 30 on acute ischemic stroke. Mechanical clot lysis and acute angioplasty with stenting are also alternative strategies available and depend on the availability of an experienced interventionalist and the clinical situation.

### Maximizing Blood Flow

Managing blood pressure, blood volume, and cardiac output can help optimize cerebral blood flow to underperfused penumbral brain tissue. Cerebral blood flow (CBF) increases with rising blood pressure until the blood pressure becomes very high, approaching the malignant range. For this reason, surgeons often give intravenous agents such as phenylephrine to raise blood pressure just prior to clamping the ICA during an endarterectomy. During the acute period of brain ischemia caused by embolism, it is unwise to lower the systemic pressure unless it is extremely high, for example, above 200/120 Torr. The patient's symptoms, signs, and neurological function are better guides to the appropriateness of any antihypertensive treatment than is the measured blood pressure.

Blood volume also affects perfusion pressure and cerebral blood flow. Some patients who are not able to eat normally will become dehydrated and relatively hemoconcentrated. Other factors such as vomiting, restrictions on eating because of concern for aspiration, or simply frequent concentrated diagnostic testing occupying patients at meal times all contribute to reduced fluid intake during the early hours and days after onset of embolic strokes. In general, blood volume, especially plasma volume, should be maintained high. Fluids must often be given intravenously or by nasogastric tube. Care, however, must be taken to avoid fluid overload and the complications of cardiac failure and brain edema. Careful monitoring of cardiac and brain function should accompany therapeutic attempts to augment fluid volume.

Some patients have anoxic–ischemic brain damage due to cardiac malfunction; in others, strong cardiac pumping function helps maximize CBF. Attention to cardiac rhythm and pump function is important, especially during the acute, fragile period of brain ischemia. Cardiac output can sometimes be improved by the use of digitalis or vasodilators, use of pacemakers or medications to treat slow rhythms and heart block, adjustment of already prescribed drugs such as digitalis and diuretics, correction of abnormal serum potassium and calcium ion levels, and control of tachyrhythmias. Cardiac-ejection fractions and output can be studied and monitored noninvasively by echocardiography.

### Acute Anticoagulation

Heparin is sometimes prescribed as a treatment for patients with acute thromboembolism. The posited purpose of heparinization is to prevent propagation of thrombi and breakoff of the tail of existing thrombi and so prevent embolization. As far as is known, heparin does not actually lyse existing thrombi, although cardiac clots often can no longer be imaged in some patients during heparin treatment. The decision on whether or not to prescribe heparin acutely to prevent the next thromboembolic stroke depends on weighing the risk of acute reembolization vs the risk of hemorrhage related to heparin therapy (Hacke, 2000; Phan *et al.*, 2000). In patients with lesions with high rates of re-embolization, e.g., mitral stenosis with atrial fibrillation, atrial fibrillation especially with large left atria and atrial thrombi, acute myocardial infarction with mural thrombi, then acute heparinization may be used, although there are no convincing data from clinical trials to support this decision. In patients with a low risk of acute re-embolization, such as chronic atrial fibrillation or mitral annulus calcification, heparin should be withheld during the acute period. If the patient has a large brain infarct, then the risk of brain hemorrhage after heparin is higher than when there is no brain infarct or a small brain infarct. Poorly controlled hypertension, thrombocytopenia, and other bleeding diathesis present high risks for heparinization. Heparin is often used after thrombolysis to maintain arterial patency. The European Stroke Initiative (EUSI) also recommends the early use of IV heparin under the following conditions: (1) arterial dissection, (2) protein-S-, -C insufficiency, APC-resistance, (3) high risk cardio-embolic source, and (4) high-graded ICA stenosis with planned early reconstruction (Hacke *et al.*, 2000).

## Treatment of Brain Edema and Increased Intracranial Pressure

In most patients with small or moderate sized infarcts increased intracranial pressure and extensive brain edema is not a problem. Some patients, mostly young men and women quickly develop a great deal of vasogenic, extracellular brain edema that causes increased intracranial pressure and displacement and herniations of brain compartments. In these patients a therapeutic trial of osmotic agents and/or corticosteroids is warranted since the situation is often desperate. Intubation with hyperventilation may be needed in those with reduced level of consciousness. Hyperventilation reduces $CO_2$ content in the blood and causes vasoconstriction which in turn decreases the amount of blood within the cranium and reduces intracranial pressure at least temporarily. In some patients who develop massive brain swelling and increased intracranial pressure leading to herniation, removal of the skull overlying the side of the infarct (hemicraniectomy) can be life-saving but some patients treated in this way may be left with severe neurological residual deficits (Schwab et al., 1996, 1998). Some patients make extraordinary recoveries after hemicraniotomies and survive with very little neurologic deficit. For space-occupying cerebellar infarcts, ventriculostomy with suboccipital craniectomy is recommended, although the data base is smaller than that for hemicraniectomy for space-occupying MCA infarcts. Recently induced hypothermia was shown to reduce mortality among patients with the so-called malignant MCA syndrome (Diringer, 1999; Schwab et al., 1998).

## Prophylaxis—Prevention of Re-embolization

### Cardiac Source of Embolism

All patients with embolization that arises from the heart should be given some form of antithrombotic treatment to prevent embolism. In general patients with disorders that predispose to red clot formation should receive prophylactic warfarin anticoagulation. Patients with disorders that predispose to white clot formation should be treated with antiplatelet aggregants. Some conditions predispose to both white and red clots and patients with these disorders should receive both antiplatelet aggregants and warfarin anticoagulation.

**Arrythmias.** Most patients with atrial fibrillation should receive warfarin prophylaxis. Some patients with low risk of embolization can be treated with aspirin or another antiplatelet aggregant (clopidogrel, 75 mg qd, or aspirin combined with modified release dipyridamole, 25/200 bid). We advise following the recommendations

**TABLE V**  Modified Recommendations for Antithrombotic Treatment of Patients with Atrial Fibrillation by a Committee of the Fifth American College of Chest Physicians Conference on Antithrombotic Therapy

| Age | Risk factors[a] | Recommendation |
|---|---|---|
| <65 Years | Absent | Aspirin |
|  | present | warfarin (INR target 2.5, 2–3 range) |
| 65–75 Years | absent | aspirin or warfarin |
|  | present | warfarin (INR target 2.5, 2–3 range) |
| >75 Years | all patients | warfarin (INR target 2.5, 2–3 range) |

Laupacis et al., 1998, with permission.
[a]Prior TIA, systemic embolism or stroke, poor left ventricular function, rheumatic mitral valve disease, prosthetic heart valve.

of a committee of the Fifth American College of Chest Physicians Conference on Antithrombotic Therapy (Laupacis et al., 1998), which are shown in Table V. Patients with sick sinus syndrome who have large left atria or other emboligenic risk factors also should be treated prophylactically with warfarin (INR target 2.5, range 2.0–3.0). We also follow the committee's recommendation that effective oral anticoagulant therapy (INR target 2.5, range 2–3) be given for 3 weeks before elective cardioversion of patients who have been in atrial fibrillation for more than 48 h (Laupacis et al., 1998).

**Rheumatic Heart Disease.**  We recommend long-term warfarin (INR target 2.5, range 2.0–3.0) in patients with rheumatic mitral valve disease who have a history of systemic embolism or who have paroxysmal or chronic atrial fibrillation or sinus rhythm with left atrial diameter >5.5 cm (Salem et al., 1998). If recurrent embolism occurs despite adequate warfarin therapy we suggest adding aspirin or dipyridamole, 400 mg/day, or clopidogrel, 75 mg/day. There is no need to anticoagulate patients with rheumatic aortic valve disease unless they also have mitral valve disease or have atrial fibrillation or systemic embolism (Salem et al., 1998).

**Prosthetic Heart Valves.**  Anticoagulation is recommended for all patients with prosthetic heart valves. Patients with bioprosthetic valves are most often treated for 3 months after surgery using a target INR of 2.0 to 3.0 (Caplan, 1999). Oral anticoagulant therapy reduces the frequency of embolism in patients with mechanical valve prostheses; the intensity of anticoagulation in these patients is usually higher than that used with bioprosthetic valves (INR of 2.5–4.0). Patients with caged-ball prostheses may require a higher intensity of anticoagulation than those with bileaflet disk valves and those with single-tilting disk valves (Vongpatanasin et al., 1996). Adding antiplatelet medications can further reduce the frequency of embolism. Aspirin (100 mg/day) added to warfarin (INR target 3.0 to 4.5) in patients with mechanical prosthetic valves or tissue valves and atrial fibrillation, reduces mortality and the frequency of embolization (Turpie *et al.*, 1993). Patients with bioprosthetic valves are most often treated for 3 months with anticoagulants (INR target 2.5, range 2.0–3.0). If patients also have atrial fibrillation or left atrial thrombi are found at surgery or systemic embolism has occurred, then the recommendation is that warfarin be continued long-term using the same target range (Stein *et al.*, 1998). Otherwise, aspirin or another antiplatelet aggregant is begun at 3 months when warfarin is discontinued. The risk of bleeding is greatly increased during combined high-intensity anticoagulation with aspirin therapy. Aspirin and lower intensity anticoagulation (INR 2.0 to 3.0) is probably as effective as high-intensity anticoagulation and has less risk of serious bleeding. Since prosthetic valves induce formation of both white platelet–fibrin and red erythrocyte–fibrin clots, the use of combined antiplatelet aggregant and anticoagulant therapy makes sense. Pregnant women with prosthetic valves should be treated with heparin or low-molecular-weight heparin during pregnancy since the incidence of thromboemboli is increased while pregnant and also in the puerperium. A group of physicians at the Mayo Clinic reviewed the literature concerning the intensity of anticoagulation in patients with prosthetic heart valves and made recommendations about treatment (Tiede *et al.*, 1998). We suggest following their recommendations, which are listed in Table VI.

**Other Valve Disorders.** Patients with mitral annulus calcification and mitral valve prolapse are ordinarily not treated with warfarin unless they have evidence of systemic embolism, thrombi are shown along the valve, the left atrium is greatly enlarged, or atrial fibrillation is present (Salem *et al.*, 1998). Antibiotics are the most important treatment in patients with infective endocarditis. If the infective endocarditis is engrafted upon

prosthetic valves and the patient has been taking warfarin it can be continued. Embolism in patients with infective endocarditis is not an indication for anticoagulation unless a condition other than the valve infection is the cause of the embolic event. The prophylactic treatment of patients with fibrous noninfective vegetations and with fibrous strands is not clear. We recommend antiplatelet aggregants (aspirin 100–325 mg qd, or clopidogrel 75 mg qd, or combined aspirin/dipyridamole 25/200 bid).

**Myocardial Ischemia and Cardiomyopathies.** Patients with acute myocardial infarcts are often treated for 3 months with warfarin to prevent embolization (Cairns *et al.*, 1998). Some patients with acute myocardial infarcts develop myocardial aneurysms or hypokinetic or akinetic zones within the left ventricle that can become the source of thromboemboli. These patients should be maintained on anticoagulants especially if they have had an embolic event. Patients with congestive heart failure, very low ejection fractions, thrombi seen on echocardiography, and atrial fibrillation should be maintained on long-term warfarin anticoagulation (INR target 2.5, range 2.0–3.0) (Cairns *et al.*, 1998). Similarly, patients with myocardiopathies and low ejection fractions should be maintained on warfarin anticoagulation especially if they have congestive heart failure or atrial fibrillation.

Paradoxical embolism through interatrial septal defects and patent foramen ovales is much more frequently recognized now than in the past. Optimal prophylactic therapy is uncertain. Although theoretically, warfarin should be more effective in preventing red clot formation, the results of preliminary studies show no important differences in outcome in patients treated prophylactically with either aspirin or warfarin (Bogousslavsky *et al.*, 1996; French Study Group, 1995; Hanna *et al.*, 1994). Surgical closure prevents re-embolization. Percutaneous procedures to close the

**TABLE VI** Recommended Prophylactic Treatment for Patients with Prosthetic Heart Valves

| Prosthetic valve | Target INR | Aspirin (mg/day) |
|---|---|---|
| Aortic valve | | |
| Newer generation mechanical bileaflet | 2.5 | 81 |
| All other mechanical valves | 3.0 | 81 |
| Bioprosthetic valves with | 2.5–3 Months | 325[a] |
| thromboembolic risk factors | 2.5 Indefinite time | 81 |
| Mitral valve | | |
| First generation tilting disc valves | 3.5 | 81 |
| All other mechanical valves | 3.0 | 81 |
| Bioprosthetic valves | 2.5 Indefinite time | 81 |
| Low thromboembolic or high bleed risk | 2.5 3–6 Months | 325[a] |

Modified from Tiede, D. J., Nishimura *et al.* (1998), with permission.
[a]Treatment continued even after coumadin stopped.

atrial communications are promising and will likely prove in the near future to be the preferred treatment. We usually prescribe warfarin (INR target 2.5, range 2.0–3.0) until the defect is repaired percutaneously.

### Aortic Source Embolism

Recent reports indicate that the aorta is a much more frequent source of embolism than was previously recognized. There are no randomized trials or even reports of the success of one or another treatment in preventing embolism in patients with aortic plaques. Protruding mobile plaques >5 mm in length pose a considerable risk of embolism. These lesions usually contain red thrombus material. We recommend warfarin prophylaxis (INR target 2.5, range 2.0–3.0) for patients with these lesions. Antiplatelet aggregants (aspirin, 100–325 mg qd, or clopidogrel, 75 mg qd, or combined aspirin/dipyridamole, 25/200 bid) are recommended for patients with flat or irregular plaques that are smaller and nonmobile. Surgery is recommended for repeated embolism despite these treatments. Perhaps in the future it may be possible to treat these plaques using interventional radiological techniques.

### Arterial Source Embolism

The large extracranial and intracranial arteries are a common source of discharge of embolic material into the distal portions and branches of these arteries.

We recommend warfarin treatment (INR target 2.5, range 2.0–3.0) for 4–6 weeks for patients with complete occlusions of large extracranial (carotid, vertebral) and intracranial (MCA, intracranial vertebral, basilar) arteries who have important residual brain tissue at-risk for further infarction (that is, they have not already developed infarction of most of the territory supplied by the occluded artery). There is at present no data from randomized trials regarding treatment of patients with intracranial artery stenosis, although a trial comparing antiplatelets with warfarin is in progress at this writing. One of the authors (WH) uses antiplatelet aggregants and gives anticoagulants only in selected patients, especially those who have failed antiplatelet aggregants. LRC recommends long-term warfarin anticoagulation (INR target 2.5, range 2.0–3.0) for patients with severe stenosis of extracranial and intracranial arteries who will not have surgical or interventional radiologic treatment. The anticoagulants are continued as long as the severe stenosis is present. If the artery occludes then anticoagulation is continued for 4–6 weeks after recognized occlusion, after which the patient is placed on platelet antiaggregants. Patients with nonstenosing plaques are maintained on platelet antiaggregants (aspirin, 100–325 mg qd, or clopidogrel, 75 mg qd, or combined aspirin/dipyridamole, 25/200 bid).

### Prevention of Complications and Risk-Factor Management

During the acute phase of stroke treatment and thereafter, attention must be given to prevention of the common complications of stroke and toward recognition and management of stroke risk factors whenever possible. Management is no different for patients with embolic strokes than for any other cause of acute brain ischemia. The strategies used are discussed in Chapter 30 on acute ischemic stroke.

## TREATMENTS NO LONGER RECOMMENDED

Acute surgical exterpation of clots in recipient arteries has been superceded by thrombolysis and other interventional techniques and is no longer recommended. Extracranial to intracranial bypass is also no longer recommended for patients with acute brain ischemia due to embolism. Ticlopidine, an antiplatelet aggregant, has frequencies of hematological complications, such as thrombotic thrombocytopenic purpura, leukopenia, and anemia, that exceed clopidogrel, a substance from the same class that is structurally and functionally quite close to ticlopidine. Ticlopidine offers no important benefits compared with clopidogrel. We suggest using clopidogrel in the situations that formerly prompted ticlopidine prescription. Patients who have been taking ticlopidine for some time can safely be continued on this drug.

## REFERENCES

Adams, G. F., Merrett, J. D., Hutchinson, W. M., and Pollock, A. M. (1974). Cerebral embolism and mitral stenosis: Survival with and without anticoagulants. *J. Neurol. Neurosurg. Psychiatry* 37, 378–383.
Adams, H. P., Brott, T. G., Furlan, A. J. *et al.* (1996). Use of thrombolytic drugs. A supplement to the guidelines for the management of patients with acute ischemic stroke. A statement for Health Care Professionals from a special writing group of the Stroke Council American Heart Association. *Stroke* 27, 1711–1718.
Albers, G. (1994). Atrial fibrillation and stroke. Three new studies, three remaining questions. *Arch. Intern. Med.* 154, 1443–1448.
Albers, G. W. (1999). Expanding the window for thrombolytic therapy in acute stroke. The potential role of acute MRI for patient selection. *Stroke* 30, 2230–2237.
Atrial Fibrillation Investigators (1994). Risk factors for stroke and efficacy of antithrombotic therapy in atrial fibrillation: Analysis of pooled data from 5 randomized clinical trials. *Arch. Intern. Med.* 154, 1949–1957.

Barnett, H. J. M., Taylor, D. W., Eliasziw, M. *et al.*, for the North American Symptomatic Carotid Endarterectomy Trial Collaborators (1998). Benefit of carotid endarterectomy in patients with symptomatic moderate or severe stenosis. *N. Engl. J. Med.* **339**, 1415–1425.

Bendixen, B. H., Adams, H. P., Leira, E. C. *et al.* (1998). Responses to treatment with a low molecular weight heparinoid or placebo among patients with acute ischemic stroke secondary to large artery atherosclerosis. *Neurology* **50**, A345.

Biller, J., Feinberg, W. M., Castaldo, J. E. *et al.* (1998). Guidelines for carotid endarterectomy. A statement for healthcare professionals from a special writing group of the Stroke Council, American Heart Association. *Stroke* **29**, 554–562.

Bogousslavsky, J., Cachin, C., Regli, F. *et al.* (1991). Cardiac sources of embolism and cerebral infarction. Clinical consequences and vascular concomitants. *Neurology* **41**, 855–859.

Bogousslavsky, J., Garazi, S., Jeanrenaud, X. *et al.* (1996). Stroke recurrence in patients with patent foramen ovale: The Lausanne study. *Neurology* **46**, 1301–1305.

Bogousslavsky, J., van Melle, G., and Regli, F. (1988). The Lausanne Stroke Registry: Analysis of 1000 consecutive patients with first stroke. *Stroke* **19**, 1083–1092.

Boston Area Anticoagulation Trial for Atrial Fibrillation Investigators (1990). The effect of low-dose warfarin on the risk of stroke in patients with nonrheumatic atrial fibrillation. *N. Engl. J. Med.* **323**, 1505–1511.

Bridges, N. D., Hellensbrand, W., Catson, L. *et al.* (1982). Transcatheter closure of patent foramen ovale after presumed paradoxical embolism. *Circulation* **86**, 1902–190.

Cairns, J. A., Theroux, P., Lewis, H. D., Eskowitz, M., Meade, T. W., and Sutton, G. C. (1998). Antithrombotic agents in coronary artery disease. *Chest* **114**, 611S–633S.

Cannegieter, S. C., Rosendaal, F. R., and Briet, E. (1994). Thromboembolic and bleeding complications in patients with mechanical heart valve prostheses. *Circulation* **89**, 635–641.

Caplan, L. R. (1980). Top of the basilar syndrome: selected clinical aspects. *Neurology* **30**, 72–79.

Caplan, L. R. (1991). Of birds and nests and brain emboli. *Rev. Neurol.* **147**, 265–273.

Caplan, L. R. (1996). "Posterior Circulation Disease. Clinical Findings, Diagnosis, and Management." Blackwell Science, Boston.

Caplan, L. R. (1999). Brain embolism. *In* (L. R. Caplan, J. W. Hurst, and M. I. Chimowitz, Eds.), pp. 35–185. Dekker, New York.

Caplan, L. R., Hier, D. B., DíCruz, I. (1983). Cerebral embolism in the Michael Reese Stroke Registry. *Stroke* **14**, 530–536.

Carter, A. B. (1965). Prognosis of cerebral embolism. *Lancet* **2**, 514–519.

Carter, A. B. (1957). The immediate treatment of cerebral embolism. *Q. J. Med.* **26**, 335–347.

Coulshed, N., Epstein, E. J., McKendrick, C. S. *et al.* (1970). Systemic embolism in mitral valve disease. *Br. Med. J.* **32**, 26–34.

Daffertshofer, M., Ries, S., Schminke, U., and Hennerici, M. (1996). High-intensity transient signals in patients with cerebral ischemia. *Stroke* **27**, 1844–1849.

Dalal, P., Shah, P., Sheth, S. *et al.* (1965). Cerebral embolism: Angiographic observations on spontaneous clot lysis. *Lancet* **1**, 61–64.

Daley, R., Mattingly, T. W., Holt, C. L. *et al.* (1951). Systemic arterial embolism in rheumatic heart disease. *Am. Heart J.* **42**, 566–581.

Devuyst, G., and Bogousslavsky, J. (1999). Recent progress in drug treatment for acute stroke. *J. Neurol. Neurosurg. Psychiatry.* **67**, 420–425.

Devuyst, G., Bogousslavsky, J., Ruchat, P. *et al.* (1996). Prognosis after stroke followed by surgical closure of patent foramen ovale: A prospective follow-up study with brain MRI and simultaneous transesophageal and transcranial Doppler ultrasound. *Neurology* **47**, 1162–1166.

Deykin, D. (1967). Thrombogenesis. *N. Engl. J. Med.* **276**, 622–628.

Diringer, M. N. (1999). Hypothermia: An old idea comes of age. *Neurology Network Commentary* **3**, 120–126.

Donnan, G. A., Davis, S. M., Chambers, B. R. *et al.* (1995). Trials of streptokinase in severe acute ischemic stroke. *Lancet* **345**, 578–579.

Drapkin, A., and Mersky, C. (1972). Anticoagulant therapy after acute myocardial infarction: Relation of therapeutic benefit to patientís age, sex, and severity of infarction. *JAMA* **222**, 541–549.

EC-IC Bypass Study Group (1985). Failure of the extracranial-intracranial arterial bypass to reduce the risk of ischemic stroke. *N. Engl. J. Med.* **313**, 1191–1200.

European Atrial Fibrillation Trial (EAFT) Study Group (1993). Secondary prevention in non-rheumatic atrial fibrillation after transient ischaemic attack or minor stroke. *Lancet* **342**, 1255–1262.

European Atrial Fibrillation Trial (EAFT) Study Group (1995). Optimal oral anticoagulation therapy in patients with non-rheumatic atrial fibrillation and recent cerebral ischemia. *N. Engl. J. Med.* **333**, 5–10.

European Carotid Surgery Trialists Collaborative (ECST) Group (1998). Randomised trial of endarterectomy for recently symptomatic carotid stenosis: Final results of the MRC European Carotid Surgery Trial (ECST). *Lancet* **351**, 1379–1387.

Fisher, C. M., and Adams, R. D. (1951). Observations on brain embolism with special reference to the mechanism of hemorrhagic infarction. *J. Neuropathol. Exp. Neurol.* **10**, 92–94.

Fisher, C. M., and Adams, R. D. (1987). Observations on brain embolism with special reference to hemorrhagic infarction. *In* "The Heart and Stroke" (A. J. Furlan, Ed.), pp. 17–36. Springer-Verlag, London.

Fisher, C. M., and Perlman, A. (1967). The nonsudden onset of cerebral embolism. *Neurology* **17**, 1025–1032.

Fisher, M. (1995). Prophylactic neuroprotection. *In* "Stroke Therapy" (M. Fisher, Ed.), pp. 233–245. Butterworth-Heinemann, Boston.

Fleming, H. A., and Bailey, S. M. (1971). Mitral valve disease, systemic embolism and anticoagulants. *Postgrad. Med. J.* **47**, 599–604.

French Study Group on Patent Foramen Ovale and Atrial Septal Aneurysm (1995). Recurrent cerebrovascular events in patients with patent foramen ovale or atrial septal aneurysms and cryptogenic stroke or TIA. *Am. Heart J.* **130**, 1083–1088.

Furlan, A. J., Higashida, R. T., Wechsler, L. *et al.* (1999). A randomized trial of intra-arterial pro-urokinase for acute ischemic stroke of less than 6 hours duration due to middle cerebral artery occlusion. PROACT Investigators. Prolyse in Acute Cerebral thromboembolism. *JAMA* **282**, 2003–2011.

Garcia, J. H. (1995). Mechanisms of cell death in ischemia. *In* "Brain Ischemia. Basic Concepts and Clinical Relevance" (L. R. Caplan, Ed.), pp. 7–18. Springer-Verlag, London.

Hacke, W. (2000). The dilemma of reinstituting anticoagulation for patients with cardioembolic sources and intracranial hemorrhage. How wide is the strait between Skylla and Karybdis? *Arch. Neurol.* **57**, 1682–1684.

Hacke, W., Kaste, M., Fieschi, C. *et al.* (1998). Randomized double-blind placebo controlled trial of thrombolytic therapy with intravenous alteplase in acute ischemic stroke (ECASS II). *Lancet* **352**, 1245–1251.

Hacke, W., Kaste, M., Olsen, T. S., Bogousslavsky, J., and Orgogozo, J.-M., for the European Stroke Initiative (EUSI) Executive Committee (2000). Acute treatment of ischemic stroke. *Cerebrovasc. Dis.* **10** (Suppl. 3), 22–33.

Hanna, J. P., Sun, J. P., Furlan, A. J. *et al.* (1994). Patent foramen ovale and brain infarct. Echocardiographic predictors, recurrence, and prevention. *Stroke* **25**, 782–786.

Hart, R. G. (1987). Prevention and treatment of cardioembolic stroke. *In* "The Heart and Stroke: Exploring Mutual Cerebrovascular and

Cardiovascular Issues" (A. J. Furlan, Ed.), pp. 117–138. Springer-Verlag, London.

Hart, R. G. (1997). Oral anticoagulation for secondary prevention of stroke. *Cerebrovasc. Dis.* 7 (Suppl. 6), 24–29.

Hart, R. G., Foster, J. W., Luther, M. F., and Kanter, M. C. (1990). Stroke in infective endocarditis. *Stroke* 21, 695–700.

Hirsh, J., Poller, L., Deykin, D. *et al.* (1989). Optimal therapeutic range for oral anticoagulants. *Chest* 95 (Suppl.), 5S–11S.

Hylek, E. M., Skates, S. J., Sheehan, M. A., and Singer, D. E. (1996). An analysis of the lowest effective intensity of prophylactic anticoiagulation for patients with nonrheumatic atrial fibrillation. *N. Engl. J. Med.* 335, 540–546.

Keen, G., and Leveaux, V. M. (1958). Prognosis of cerebral embolism in rheumatic heart disease. *Br. Med. J.* 2, 91–92.

Koudstaal, P. J. (1995). Anticoagulation in very elderly patients (>80 years) with stroke and atrial fibrillation. *Cerebrovasc. Dis.* 5, 8–9.

Laupacis, A., Albers, G., Dalen, J., Dunn, M. I., Jacobson, A. K., and Singer, D. E. (1998). Antithrombotic therapy in atrial fibrillation. *Chest* 114, 579S–589S.

Liebeskind, A., Chinichian, A., and Schechter, M. (1971). The moving embolus seen during cerebral angiography. *Stroke* 2, 440–443.

Lodder, J., Krijne-Kubat, B., and Broekman, J. (1986). Cerebral hemorrhagic infarction at autopsy: Cardiac embolic cause and the relationship to the cause of death. *Stroke* 17, 626–629.

Markus, H. S. (1993). Transcranial Doppler detection of circulating cerebral emboli, a review. *Stroke* 24, 1246–1250.

Markus, H. S., and Harrison, M. J. (1995). Microembolic signal detection using ultrasound. *Stroke* 26, 1517–1519.

Medical Research Council (1969). Assessment of short term anticoagulant administration after cardiac infarction. *Br. Med. J.* 1, 335–342.

Minematsu, K., Yamaguchi, T., and Omae, T. (1992). Spectacular shrinking deficit: Rapid recovery from a major hemispheric syndrome by migration of an embolus. *Neurology* 42, 157–162.

Mohr, J. P., Caplan, L. R., Melski, J. W. *et al.* (1978). The Harvard Cooperative Stroke Registry: A propective registry. *Neurology* 28, 754–762.

Mohr, J. P., Gautier, J. C., Hier, D. B., and Stein, R. W. (1986). Middle Cerebral artery in Stroke, pathophysiology, diagnosis, and management (H. J. M. Barnett, J. P. Mohr, B. M. Stein, and F. M. Yatsu, Eds.), vol. 1, pp. 377–450. Churchill-Livingstone, New York.

MRC European Carotid Surgery Trial (1991). Interim results for symptomatic patients with severe (70–99%) or with mild (0–29%) carotid stenosis. *Lancet* 337, 1235–1243.

Nakano, S., Yokogami, K., Ohta, H., Yano, T., and Ohnishi, T. (1998). Direct percutaneous transluminal angioplasty for acute middle cerebral artery occlusion. *Am. J. Neurol. Res.* 19, 767–772.

North American Symptomatic Carotid Endarterectomy Trial Collaborators (1991). Beneficial effect of carotid endarterectomy in symptomatic patients with high-grade carotid stenosis. *N. Engl. J. Med.* 325, 445–453.

Okada, Y., Yamaguchi, T., Minematsu, K. *et al.* (1989). Hemorrhagic transformation in cerebral embolism. *Stroke* 20, 598–603.

Onal, M. Z., and Fisher, M. (1996). Thrombolytic and cytoprotective therapies for acute ischemic stroke: A clinical overview. *Drugs Today* 32, 573–592.

Pessin, M. S., del Zoppo, G. J., and Furlan, A. J. (1995). Thrombolytic treatment in acute stroke: Review and update of selective topics. *In* "Cerebrovascular Diseases, Nineteenth Princeton Stroke Conference" (M. Moskowitz and L. R. Caplan, Eds.), pp. 409–418. Butterworth-Heinemann, Boston.

Petersen, P., Godtfredsen, J., Boysen, G. *et al.* (1989). Placebo-controlled, randomized trial of warfarin and aspirin for prevention of thromboembolic complications in chronic atrial fibrillation: The Copenhagen AFASAK study. *Lancet* 1, 175–179.

Phan, T. G., Koh, M., and Wijdicks, E. F. M. (2000). Safety of discontinuation of anticoagulation in patients with intracranial hemorrhage at high thromboembolic risk. *Arch. Neurol.* 57, 1710–1713.

Poller, L. (1991). The effect of low-dose warfarin on the risk of stroke in patients with nonrheumatic atrial fibrillation. *N. Engl. J. Med.* 325, 129–130.

Publications Committee for the Trial of ORG 10172 in Acute Stroke Treatment (TOAST) Investigators (1998). Low molecular weight heparinoid, ORG 10172 (Danaparoid), and outcome after acute ischemic stroke. A randomized controlled trial. *JAMA* 279, 1265–1272.

Quality Standards Subcommittee of the American Academy of Neurology (1996). Practice advisory: Thrombolytic therapy for acute ischemic stroke—Summary statement. *Neurology* 47, 835–839.

Salem, D. N., Levine, H. J., Pauker, S. G., Eckman, M. H., and Daudelin, D. H. (1998). Antithrombotic therapy in valvular heart disease. *Chest* 114, 590S–601S.

Salgado, A. V., Furlan, A. J., Keys, T. F., Nichols, T. R., and Beck, G. J. (1989). Neurologic complications of endocarditis: A 12-year experience. *Neurology* 39, 173–178.

Schwab, S., Rieke, K., Aschoff, A. *et al.* (1996). Hemicraniotomy in space-occupying hemisopheric infarction: Useful early intervention or desperate activism. *Cerebrovasc. Dis.* 6, 325–329.

Schwab, S., Schwarz, S., Spranger, M., Keller, E., Bertram, M., and Hacke, W. (1998). Moderate hypothermia in the treatment of patients with severe middle cerebral artery infarction. *Stroke* 29, 2461–2466.

Schwab, S., Steiner, T., Aschoff, A. *et al.* (1998). Early hemicraniectomy in patients with complete middle cerebral artery infarction. *Stroke* 29, 1888–1893.

Sliwka, U., Lingnau, A., Stohlmann, W.-D. *et al.* (1997). Prevalence and time course of microembolic signals in patients with acute strokes, a prospective study. *Stroke* 28, 358–363.

Sorenson, A. G., Buonanno, F. S., Gonzales, R. G. *et al.* (1996). Hyperacute stroke: Evaluation with combined multisection diffusion-weighted and hemodynamically-weighted echo-planar MR imaging. *Radiology* 199, 391–401.

Stein, P. D., Alpert, J. S., Dalen, J. E., Horstkotte, D., and Turpie, A. G. G. (1998). Antithrombotic therapy in patients with mechanical and bioprosthetic heart valves. *Chest* 114, 602S–610S.

Stroke Prevention in Atrial Fibrillation Investigators (1991). The stroke prevention in atrial fibrillation study: Final results. *Circulation* 84, 527–539.

Stroke Prevention in Atrial Fibrillation Investigators (1994). Warfarin versus aspirin for prevention of thromboembolism in atrial fibrillation: Stroke Prevention in Atrial Fibrillation II Study. *Lancet* 343, 687–691.

Stroke Prevention in Atrial Fibrillation Investigators (1996). Adjusted-dose warfarin versus low-intensity, fixed-dose warfarin plus aspirin for high-risk patients with atrial fibrillation: Stroke Prevention in Atrial Fibrillation III randomised clinical trial. *Lancet* 348, 633–638.

Stroke Prevention in Atrial Fibrillation Investigators (1997). Prospective identification of patients with nonvalvular atrial fibrillation at low risk of stroke during treatment with aspirin: Stroke Prevention in Atrial Fibrillation III Study. *Circulation* 96 (Suppl.), I-281. [Abstract]

Stroke Prevention in Reversible Ischemia Trial (SPIRIT) Study Group (1997). A randomized trial of anticoagulants versus aspirin after cerebral ischemia of presumed arterial origin. *Ann. Neurol.* 42, 857–865.

Szekely, P. (1964). Systemic embolism and anticoagulant prophylaxis in rheumatic heart disease. *Br. Med. J.* 1, 1209–1212.

Tiede, D., Nishimura, R. A., Gastineau, D. *et al.* (1998). Modern management of prosthetic valve anticoagulation. *Mayo Clin. Proc.* **73**, 665–680.

Timsit, S. G., Sacco, R. L., Mohr, J. P. *et al.* (1993). Brain infarction severity differs according to cardiac or arterial embolic source. *Neurology* **43**, 728–733.

Tong, D. C., and Albers, G. W. (1995). Transcranial Doppler-detected microemboli in patients with acute stroke. *Stroke* **26**, 1588–1592.

Turpie, A. G. G., Gent, M., Laupacis, A. *et al.* (1993). A comparison of aspirin with placebo in patients treated with warfarin after heart-valve replacement. *N. Engl. J. Med.* **329**, 524–529.

Venti, M., Parnetti, L., Silvestrelli, G., and Gallai, V. (2000). Role of Neuroprotective drugs in acute ischemic stroke. *Cerebrovasc. Dis.* **10** (Suppl. 4), 24–26.

Veterans Cooperative Study (1973). Anticoagulants in acute myocardial infarction: Results of a cooperative trial. *JAMA* **225**, 724–729.

Vongpatanasin, W., Hillis, D., and Lange, R. A. (1996). Prosthetic heart valves. *N. Engl. J. Med.* **335**, 407–416.

Warach, S., Chien, D., Li, W. *et al.* (1992). Fast magnetic resonance diffusion-weighted imaging of acute human stroke. *Neurology* **42**, 1717–1723.

Weinberger, J., Azhar, S., Danisi, F., Hayes, R., and Goldman, M. (1998). A new noninvasive technique for imaging atherosclerotic plaque in the aortic arch of stroke patients by transcutaneous real-time B-mode ultrasonography. *Stroke* **29**, 673–676.

Weksler, B. (1985). Antithrombotic therapies in the management of cerebral ischemia. *In* "Cerebrovascular Diseases: Proceedings of the Fourteenth Princeton Conference" (F. Plum and W. Pulsinelli, Eds.), pp. 211–223. Raven Press, New York.

Yamaguchi, T., for Japanese NVAF-embolism Secondary Prevention Cooperative Study Group (2000). Optimal intensity of warfarin therapy for secondary prevention of stroke in patients with non-valvular atrial fibrillation. A multicenter prospective randomized trial. *Stroke* **31**, 817–821.

Yamaguchi, T., Minematsu, K., Choki, J. I., and Ikeda, M. (1984). Clinical and neuroradiological analysis of thrombotic and embolic cerebral infarction. *Jpn. Circ.* **48**, 50–58.

## CHAPTER 34
# Spinal Cord Ischemia

Louis R. Caplan

## CLINICAL ASPECTS

When contrasted to brain ischemia, spinal cord strokes are rare and caused by more diverse etiologies. The rarity of spinal-cord strokes and the relative lack of accessibility of the spinal cord and its vascular system to imaging during life have long been barriers to the understanding of spinal-cord vascular diseases. The spinal cord and its vascular system are seldom examined in detail at necropsy. When thoroughly sought, spinal ischemic lesions are found. In London, Ontario, Canada, a systematic search for examples of spinal cord ischemia found at postmortem showed 52 cases among 1200 consecutive necropsies (4%) (Buchan and Barnett, 1986; Vinters and Gilbert, 1979).

### Spinal Cord Vascular Anatomy

Clinicians will best understand and visualize the cord circulatory system by first focusing on the blood supply of a spinal-cord segment in axial section (Figure 34.1) (Caplan, 2000). A single anterior spinal artery courses in the ventral midline, running rostrally from the spinomedullary junction at the foramen magnum, caudally to the tip of the spinal cord, the filum terminale. Paired smaller posterior spinal arteries are located on the dorsal surface of the spinal cord that commonly form a plexus of small vessels. The anterior spinal artery gives off deep branches, which course along the ventral sulcus and then branch as they reach the central gray matter to supply left and right branches to the anterior horn regions on each side (Caplan, 2000; Gillilan, 1958; Mawad *et al.*, 1990). Lateral circumferential arteries and their penetrators are directed laterally from the midline anterior spinal artery as it travels dorsally within the ventral sulcus to supply the ventral white matter, tips of the anterior horns, and the pyramidal

tracts (Caplan, 2000; Gillilan, 1958; Mawad *et al.*, 1990). This pattern of blood supply is very comparable to that found in the brainstem, where paramedian penetrators and short and long circumferential arteries branch from the vertebral and basilar arteries. The posterior spinal artery plexus gives off penetrating branches to the posterior columns and posterior gray horns (Buchan and Barnett, 1986; Caplan, 2000; Gillilan, 1958; Mawad *et al.*, 1990). The regions of supply of the anterior and posterior spinal arteries are shaded diagrammatically on Figure 1. The area between the two zones of supply in the central portion of the spinal cord has often been called the "border zone" or "watershed region of supply" (the white zone between the two shaded areas on Figure 1) (Hogan and Romanul, 1966).

The anterior spinal artery supply originates from 5 to 10 usually single, unpaired radicular arteries, which feed into the anterior spinal artery at various levels (Figure 2). The most rostral supply comes from the intracranial vertebral arteries. Each intracranial vertebral artery gives off a ramus from its distal segment that joins with that of the contralateral vertebral artery to form a midline anterior spinal artery that feeds the medulla and descends in the midline through the foramen magnum to supply the rostral cervical spinal cord. Branches from the thyrocervical and costocervical branches of the subclavian arteries and the extracranial vertebral arteries feed into the spinal cord at the cervical enlargement.

The thoracic portion of the anterior spinal artery is fed by radicular branches of the deep cervical and intercostal arteries and by branches of the aorta. Blood supply is most marginal in the upper thoracic region ($T_2$–$T_4$), and this region has been called a spinal-cord watershed region. The largest supply artery usually arises in the lower thoracic or upper lumbar segments, most often between $T_9$ and $T_{12}$, but can arise as low as $L_2$. This artery is known as the artery of Adamkiewicz;

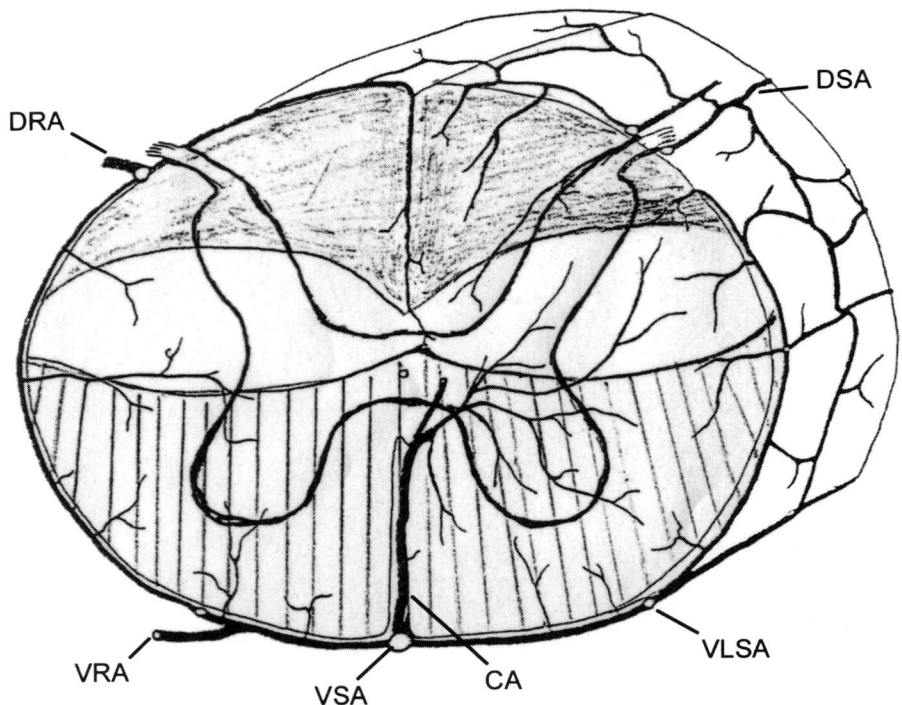

**FIGURE 1** Axial section of the spinal cord showing the arterial circulation. The area supplied by the posterior spinal artery supply is shaded gray. The anterior arterial supply is designated by vertically oriented lines. The white area between is an estimate of the border-zone region between the anterior and the posterior arterial supplies. CA, central artery; DRA, dorsal radicular arte; DSA, dorsal spinal artery; VLSA, ventrolateral spinalartery; VRA, ventral radicular artery; VSA, ventral spinal artery. Modified from Austin (1961).

it usually arises from the left and supplies the lumbar enlargement of the spinal cord. The sacral cord and cauda equina are supplied anteriorly from the hypogastric or obturator arteries.

In contrast to the supply of the anterior spinal artery, there are many more posterior radicular arteries, which enter along nerve roots from each side at nearly every spinal level, to supply the plexus of vessels that lie on the posterior cord surface (Caplan, 2000; Gillilan, 1958). Other segmental arteries arise from the vertebral arteries and from the aorta and iliac arteries to supply the paraspinous structures and then end on the anterior and posterior nerve roots without supplying the spinal cord or penetrating the dura mater. These vessels are often the origin of spinal vascular malformations and fistulas (Heros, 1988).

The venous spinal-cord anatomy has also been clarified (Gillilan, 1970). Similar to the arterial supply, there are anterior and posterior venous drainage systems. Radicular veins are plentiful and drain into the paravertebral and intravertebral plexus into the pelvic veins. The posterior portions of the spinal cord drain into a large midline posterior vein. The anterior and posterior venous system forms an extensive network that encircles the spinal cord. Venous spinal-cord infarc-

tion is probably more common than venous brain infarcts and is most often due to mechanical compression, infection, and inflammation, which obliterate the veins, and to dural arteriovenous fistulas, which cause increased venous pressure. Venous hypertension is an important contributor to spinal cord infarction in patients with spinal dural fistulas and other vascular malformations.

### Causes of Spinal Cord Ischemia

The etiologies of spinal cord ischemia can be considered under seven quite different subcategories: (1) spinal dural arteriovenous fistulas (Heros, 1988; Caplan, 1991; Bradac *et al.*, 1993; Hurst *et al.*, 1995; Hemphill *et al.*, 1998; DeChiro *et al.*, 1971; Rosenblum *et al.*, 1987; Gilbertson *et al.*, 1995; Aminoff and Logue, 1974; Teal *et al.*, 1992; Criscuolo *et al.*, 1989; Larsson *et al.*, 1991; Bowen *et al.*, 1995); (2) diseases of the aorta, especially surgical procedures involving aortic clamping and aortic aneurysms and aortic dissections (Mawad *et al.*, 1990; Hogan and Romanul, 1966; Cheshire *et al.*, 1996; Ross, 1985; Rockman *et al.*, 2001; Herrick and Mills, 1971); (3) other usual causes

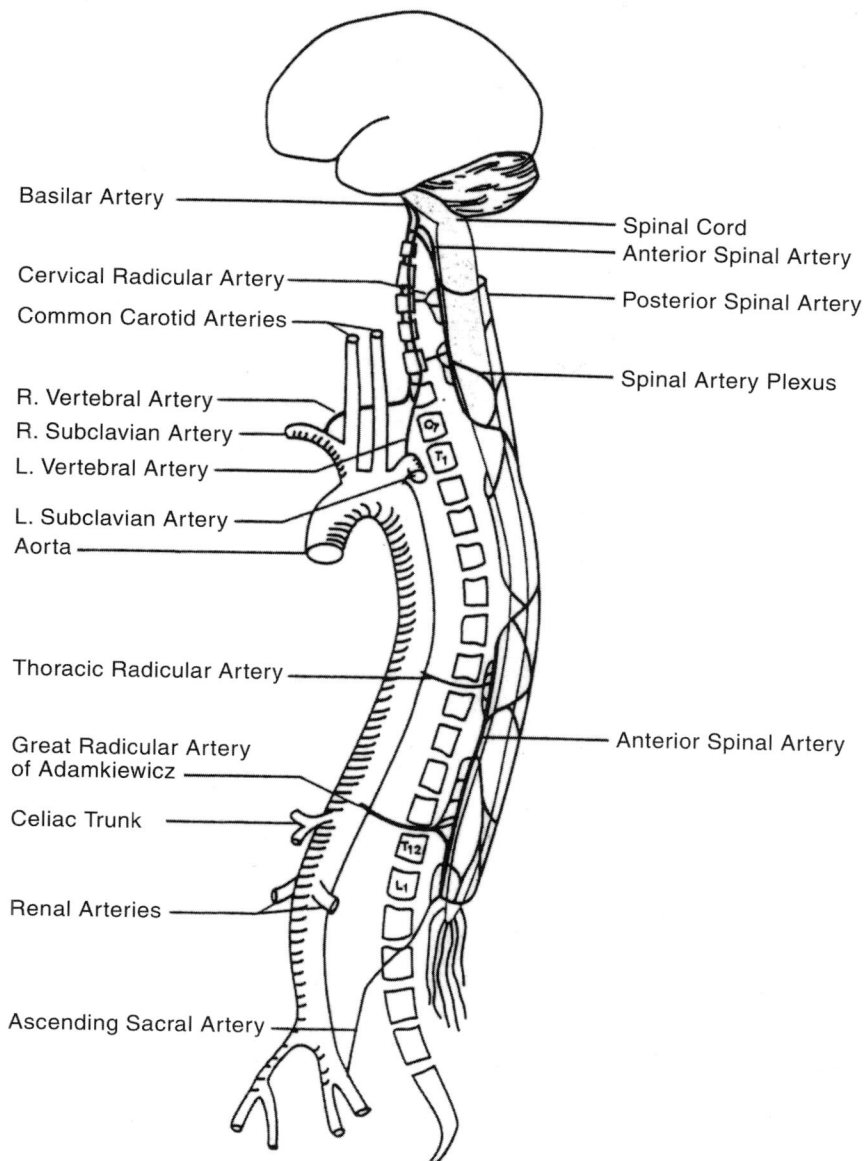

**FIGURE 2** Segmental arterial supply of the spinal cord. The anterior radicular artery issues from the posterior ramus of a lumbar intercostal artery. Reprinted from Cheshire *et al.* (1996), copyright Lippincott Williams & Wilkins.

of stroke, e.g., embolism from cardiac disorders; (4) cartilaginous disc embolism (Caplan, 1991; Srigley *et al.*, 1981; Mikulis *et al.*, 1992; Kestle *et al.*, 1989); (5) hypoperfusion—cord ischemia due to systemic hypotension or shock (Vinters and Gilbert, 1979); (6) venous spinal cord infarcts (Tosi *et al.*, 1996; Freyaldenhoven and Mrak, 2001; Roa *et al.*, 1982); and (7) unusual vascular processes such as vasculitis due to infectious agents (meningococci, syphyllis, tuberculosis, schistosomiasis, etc.) and inflammatory diseases, arachnoiditis, and infarction due to drug use and hypercoagulability (Kim *et al.*, 1984; Hughes, 1971; O'Farrell *et al.*, 2000;

Caplan *et al.*, 1990; Rothman and Nelson, 1980; Caccamo *et al.*, 1992).

### Spinal Dural Arteriovenous Fistulas

Contrary to the situation within the cranium, spinal vascular malformations most often present with ischemia rather than bleeding. So-called Type I malformations, often referred to as "dural," derive their blood supply from arteries located in the dural sleeves of spinal roots (Heros, 1988; Bradac *et al.*, 1993). The small niduses which contain the arteriovenous communications are fed by dural branches of radicular arteries. The arteriovenous fistulas drain intradurally by one or more arterialized, enlarged, usually tortuous veins, which course on the dorsal surface of the spinal cord, usually above but occasionally below the feeding vessels. The dural feeding arteries do not participate in the blood supply of the spinal cord. These lesions occur predominantly in men (4:1 ratio) between 40 and 70 years of age, and they involve mostly the lower thoracic and lumbosacral segments. In one series, 26 of 27 spinal dural fistulas were fed by arteries in the thoracic or lumbar regions, and the remaining fistula was sacral (Rosenblum *et al.*, 1987). In another series of 12 dural fistulas, 8 were located between $T_8$ and $T_{12}$, 2 were at $S_1$, and 1 each was at $L_1$ and $L_5$ (Bradac *et al.*, 1993). There is usually only one feeder, but at times two, or at most three, arterial feeders have been found (Heros, 1988). Dural fisulas are usually described as having "low-flow" because angiography results only in very slow, low—volume filling of the lesions. Cutaneous angiomas are not seen, and bruits are not audible.

The most frequent presentation of dural malformations is that of progressive neurological worsening, often with acute deteriorations (Bradac *et al.*, 1993). Dural fisulas usually do not cause subarachnoid bleeding except when the lesions are cervical (Hemphill *et al.*, 1998). Cervical fistulas that cause subarachnoid bleeding are fed by the vertebral artery and are often located near the cervico-medullary junction. Pain is present in about 40% of patients and can be radicular, sometimes mimicking sciatic pain. Symptoms often worsen after exercise. Transient symptoms of spinal cord dysfunction are common. One patient had two episodes of leg paralysis and numbness that occurred while she was driving a car (Teal *et al.*, 1992). Her husband had to grab the wheel and foot controls to avert a crash. Soon, strength and feeling returned. Another patient had a Brown–Séquard distribution attack while in the hospital that lasted several hours (Teal *et al.*, 1992). Spinal transient ischemic attacks (TIAs) are quite rare in all other causes of spinal cord strokes. Activity worsened symptoms in 19 of 27 (27%) patients in the National Institutes of Health (NIH) series (Rosenblum *et al.*, 1987). These lesions probably cause symptoms because of venous hypertension and occasional thrombosis of the venous drainage system.

Some dural fistulas are located in the paravertebral region but drain into epidural veins and often also into the intradural venous system (McCreary *et al.*, 2000). The enlarged epidural veins can cause a compressive myelopathy. The venous hypertension that results from these paravertebral fistulas and dural fistulas can cause venous hypertension. In order to perfuse the spinal cord, arterial inflow pressure must exceed venous pressure. The low-pressure inflow and increased venous pressure found in patients with dural fistulas often lead to intermittent insufficiency of perfusion.

### Aortic Diseases and Procedures

Disease of the aorta is undoubtedly the most common recognized cause of spinal-cord infarction. Most often, paraplegia is recognized after repair of thoracic and abdominal aortic aneurysms (Mawad *et al.*, 1990; Hogan and Romanul, 1966; Cheshire *et al.*, 1996; Ross, 1985; Rockman *et al.*, 2001; Herrick and Mills, 1971). During repair of aortic aneurysms, flow through radicular supply arteries to the anterior spinal artery is compromised. When the thoracic cord is involved, usually a flaccid paraplegia is noted directly after surgery, with incontinence and a thoracic sensory level. Later, the lower limbs become spastic. When the lumbar spinal cord is involved, a conus medullaris infarct develops with hypotonia, wasting and areflexia of the legs, loss of sphincter function, and variable loss of touch and pinprick sensation in the lower limbs, perineum, and lower abdomen.

Similar findings are noted in patients with unruptured aneurysms, dissections of the aorta, traumatic rupture of the aorta, thromboembolic aortic occlusions, and ulcerative aortic plaque disease. Thrombi and plaques can obstruct the orifices of radicular spinal arteries. Dissections can tear or interrupt the orifices of spinal-cord-feeding arteries, sometimes over a long rostro-caudal area. Cholesterol crystals and other plaque materials can embolize into spinal arteries and can block branches. In some patients, spinal-cord ischemia develops gradually and insidiously and can be misdiagnosed as motor-neuron disease or diabetic amyotrophy because of selective ischemia involving the anterior horns and sometimes the pyramidal tracts (Herrick and Mills, 1971). In contrast to brain ischemic strokes, TIAs are quite unusual, but do occur (Cheshire *et al.*, 1996). Often the posterior columns are spared, and vibration sense is retained.

### Cartilaginous Disc Embolism

Cartilaginous material from intervertebral discs can somehow invade the spinal arteries and veins and can cause devastating spinal-cord strokes (Caplan, 1991; Srigley *et al.*, 1981; Mikulis *et al.*, 1992; Kestle *et al.*, 1989; Tosi *et al.*, 1996; Freyaldenhoven and Mrak, 2001). Most reported cases are cervical and involve young women. Some reported patients were pregnant, puerperal, or took oral contraceptives. Minor trauma, sudden neck motion, exercise, and lifting are commonly mentioned as precipitants. The first symptom is usually pain in the neck or upper back or radicular pain. Then a rapidly progressive, sometimes asymmetric, spinal cord syndrome with quadriparesis develops. Syringomyelia-like dissociated sensory loss with loss of pain and temperature but preserved touch sensation may be found in the upper limbs or cape region. Once paralysis develops the abnormalities are usually persistent. The same syndrome can affect the lumbar spinal cord and can cause conus medullaris infarction (Caplan, 1991). Undoubtedly, cartilaginous disc embolism occurs more often than is presently diagnosed and is an important cause of spinal-cord infarction.

### Hypoxic–Ischemic Spinal Cord Ischemia

Spinal-cord ischemia may develop during shock or cardiac arrest. Infarction is most likely to affect the thoracic spinal cord between the $T_4$ and $T_8$ segments, the most vulnerable region of the spinal cord (Cheshire et al., 1996). Nearly always, the spinal cord signs and symptoms are overshadowed by brain hypoxic–ischemic injury. Deeply comatose patients remain hypotonic and areflexic because of accompanying spinal cord ischemia. Very rarely, a pure spinal syndrome complicates systemic hypoperfusion.

### Venous Spinal Cord Infarction

Venous infarction is also an important cause of spinal-cord infarction. The infarcts may remain bland (Roa et al., 1982; Kim et al., 1984) or become frankly hemorrhagic (Caplan, 1991; Hughes, 1971). Venous infarcts are usually attributable to one of three different mechanisms: spinal dural fistulas, coagulopathies with venous thrombosis, or mechanical compression of veins by epidural mass lesions (Roa et al., 1982; Kim et al., 1984).

### Other Less Common Causes

Infection and inflammation of the meninges covering the spinal cord can spread to the spinal arteries, causing acute spinal-cord strokes. The phenomenon is similar to Heubner's arteritis found in the brain in the presence of tuberculosis and syphilis. These two disorders, as well as fungal infections and Lyme borreliosis, probably now account for the vast majority of infectious spinal arteritis. Schistosomiasis is an important cause of spinal-cord infarction in some parts of the world. Occasionally acute bacterial meningitis can also cause spinal cord infarction (O'Farrell et al., 2000). Almost invariably, lumbar puncture with spinal fluid analysis provides the clue to this problem.

Chronic adhesive arachnoiditis from any cause can also cause scarring and obliteration of spinal penetrating arteries and ischemic necrosis of the central portion of the spinal cord (Caplan et al., 1990). The clinical findings are very similar to syringomyelia except that any level of the spinal cord can be involved. Arachnoidal scarring can be due to trauma, hemorrhage, or infection. The signs and symptoms develop gradually, sometimes years after the spinal injury, bleed, or meningitis (Caplan et al., 1990).

Embolism can cause spinal-cord infarction but does so only rarely. Bacterial endocarditis, atrial myxomas and nonbacterial thrombotic (marantic) endocarditis are disorders in which small particles can embolize to the spinal cord. Hypercoagulable states such as sickle cell anemia (Rothman and Nelson, 1980) and granulomatous angiitis (Caccamo et al., 1992) are other rare causes of spinal-cord infarction. Spinal-cord strokes have been reported after the use of heroin, especially when there has been a long hiatus between the last heroin use and the present injection or insufflation (McCreary et al., 2000; Ell et al., 1981). Aberrant or unusual anatomical structures can occasionally block arterial feeders to the spinal cord, causing ischemia (Rogopoulos et al., 2000).

### Diagnosis and Imaging

The clinical diagnosis of spinal cord ischemia can be difficult. In patients with the acute onset of symptoms and signs that indicate spinal-cord dysfunction, the main differential diagnostic considerations are spinal cord compression, inflammatory disorders causing myelitis, and infarction. When acute spinal-cord dysfunction follows aortic procedures or aortic dissection, then recognition of the vascular etiology is clear. Most spinal-cord infarcts begin abruptly and are maximal at or near onset; however, infarction due to spinal dural fistulas can be preceded by TIAs and the course is often characterized by gradually progressive or stepwise increases in deficits. These fistulas are most often found in older individuals, most often men. Other less common causes of spinal-cord infarction can occur at any age and seldom have distinctive features except for known underlying diseases, e.g., endocarditis, hypercoagulable states, systemic infections, etc.

Myelography and CT can show cord swelling and compression, but these techniques do not image the internal structure of the spinal cord. Regions of spinal-cord softening can now be imaged on newer-generation MRI scanners (Mawad et al., 1990; Larsson et al., 1991; Goyal et al., 1991; Yuh et al., 1992). High-intensity signals are often seen on T2-weighted images. Diffusion-weighted images can show even early infarcts not visible on T2-weighted images (Gass et al., 2000). Contrast enhancement on T1-weighted images is a common finding in patients with spinal dural fistulas and may be related to venous hypertension with an abnormal blood–cord barrier (Larsson et al., 1991; Bowen et al., 1995).

Abnormal bone signals are also shown in some patients since arteries supplying the spinal cord also nourish the vertebral bodies (Rogopoulos et al., 2000). The bony abnormalities are best seen on T2-weighted images and consist of increased intensity near the end-plates and in the deep medullary portion of the vertebral bodies (Rogopoulos et al., 2000). At times there is a triangular region of increased intensity within the vertebral body that is quite characteristic and is located adjacent to the level of the spinal-cord infarct.

Spinal dural fisulas themselves are often not well seen by standard MRI, but increased T2 signal in the cord and spinal cord edema are common. MRI may show hyperintensity on T2-weighted sequences, enhancement on T1-weighted images, spinal cord edema, and prominent perimedullary vessels. Some patients with dural fistulas show hypointensity in the peripheral portion of the spinal cord on T2-weighted images (Hurst and Grossman, 2000). The hypointensity is probably best explained by slow flow of blood related to the venous hypertension (Hurst and Grossman, 2000). Dilated tortuous epidural veins are sometimes visible on MR sequences and at angiography (Gilbertson et al., 1995). Myelography is an important diagnostic test because usually, coiled enlarged serpiginous veins are visible along the dorsal cord surface. Often patients must be turned during myelography, so that films are taken with the patients lying supine on their backs to show the abnormal veins. Selective spinal angiography in expert, experienced hands often shows the feeding arteries, but sometimes the feeding arteries cannot be opacified (Gilbertson et al., 1995). MR angiography using fast three-dimensional contrast enhancement usually shows the abnormal dorsal surface veins (Bowen et al., 1995; Mascalchi et al., 1997; Binkert et al., 1999) and phase display after contrast injection can show the direction of flow within the epidural veins and so indicate the likely location of arterial feeders (Mascalchi et al., 1997). MR angiography may show a draining medullary vein within a neural foramen (Binkert et al., 1999) and occasionally also shows the feeding artery (Mascalchi et al., 1997; Binkert et al., 1999). Transesophageal echocardiography may show aortic plaques in patients who present with acute paraplegia (DeChiro et al., 1971). In patients with cartilaginous disc embolism myelography and other studies usually fail to show herniated discs with compression of the cord or nerve roots. Infarction is usually bland but can be hemorrhagic.

## NATURAL COURSE

Spinal cord deficits in patients with spinal dural fistulas usually progressively worsen if the lesion is untreated, and most patients become unable to walk within 5 years of the onset of weakness. Patients who develop spinal cord infarction after surgical procedures or manipulation of the aorta usually have a maximal deficit after awakening from anesthesia but may improve somewhat during the ensuing days and weeks. Spinal-cord infarction caused by fibrocartilaginous embolism usually has a painful onset and then progression during hours or days (Tosi et al., 1996; Freyaldenhoven and Mrak, 2001). Unlike the situation in patients with brain ischemia, except in the special situation of spinal dural fisulas, recurrence of spinal ischemia is rare.

The severity of the neurological signs and electrophysiological tests (in the form of somatosensory evoked responses from various locations in the upper and lower limbs) are useful in predicting outcome of ischemic spinal-cord lesions (Iseli et al., 1999; Jacobs et al., 2000). Monitoring of somatosensory evoked responses is also used during surgery on the aorta to indicate whether there is spinal-cord dysfunction while the patient is anesthetized (Jacobs et al., 2000).

## PRINCIPLES OF THERAPY

In patients with dural arteriovenous fistulas the goal of treatment is to prevent progressive spinal cord damage by eradicating the fistula. The interventional and surgical treatment of spinal dural fistulas is discussed in Chapter 41. In others causes of spinal cord ischemia the insult is mostly a single event, often procedure-related. The goal is prevention of the primary event rather than prophylaxis for a recurrence and minimization of the neurological deficit once it has developed.

During procedures in which the aorta will be clamped every attempt should be made to identify the major feeding arteries supplying the spinal cord and avoiding their compromise during the procedure. The artery of Adamkiewicz originates at variable levels. Also if there are areas of severe atherosclerotic plaques, these regions should also be avoided during clamping since atherosclerotic debris could be mobilized and embolize to the spinal cord, lower limbs, and kidneys. Many surgeons monitor somatosensory potentials using electrophysiological techniques during surgery in order to identify early spinal-cord ischemia.

A variety of maneuvers and pharmacological treatments have been postulated to minimize spinal-cord ischemic damage but, as yet, none has been satisfactorily proven to be effective or generally adopted. These strategies have been used during procedures on the aorta to prevent or minimize damage and sometimes also applied after spinal-cord ischemia from other causes.

1. Regional hypothermia. Cooling can be performed by regional perfusion of the epidural or intrathecal space (Cambria and Davison, 2000; Tabayashi and Motoyoshi, 2001). Hypothermia has been used in brain ischemia to reduce the demands for oxygen and nutrients and so reduce ischemic damage.

2. Selective perfusion of segmental arteries (Ueda et al., 2000) or retrograde perfusion through spinal-cord veins (Follis et al., 1999) has been postulated to maintain perfusion during aortic procedures that involve aortic clamping. Segmental arteries can be perfused through catheters attached to a distal bypass circuit (Ueda et al., 2000).

3. Cerebrospinal fluid drainage. Drainage of CSF has been used to attempt to decompress the spinal cord after spinal-cord infarction (Azizzadeh et al., 2000; Tiesenhausen et al., 2000). Cord ischemia can cause swelling and spinal cord edema, raising CSF pressure and perhaps contributing to compression of feeding and draining vessels which could secondarily compound the ischemic damage.

4. Pharmacological "neuroprotective" agents. A variety of agents have been tried in experimental animals in an attempt to make the spinal cord more resistant to ischemia (De Haan, 2000; Kazama et al., 2001; Lips et al., 2000). These agents include pentobarbital, riluzole, ketamine, a 21-aminosteroid, naloxone, gangliosides, tirilizad mesylate, and others. Sometimes these agents are used in combinations and sometimes they are used with hypothermia.

5. High-dose methylprednisolone. Coricosteroids have been proven to be effective if given soon after traumatic spinal-cord injury (Bracken et al., 1992, 1997). These agents have also been given after spinal-cord infarction in order to diminish spinal-cord edema. Theoretically mannitol, glycerol and other osmotic agents might have a similar therapeutic effect.

Omental transposition has been used to try to augment spinal-cord blood flow just as it has been used in patients with brain ischemia, but the results in patients with spinal-cord injury have not been promising (Duffill et al., 2001).

Clearly rehabilitation techniques are very important in patients with spinal-cord infarcts. Sphincter training, mobilization of the limbs, prevention of phlebothrombosis, and treatment of spasticity are crucial, as they are in all patients who have spinal-cord damage of any cause. Locomotor training can be effective in patients with some residual function in the lower extremities (Wirz et al., 2001).

## PRACTICAL MANAGEMENT

At present there are no recognized effective strategies that reliably minimize spinal-cord damage caused by ischemia after infarction has occurred. Every attempt should be made to prevent ischemia during aortic surgical repair and interventional radiological procedures on the aorta. Every attempt should be made to identify regions of severe atherosclerosis within the aorta and the location of major feeding arteries to the spinal cord either before surgery or during surgery before aortic clamping. Monitoring of somatosensory-evoked responses during surgery should allow recognition of early spinal-cord dysfunction. If spinal-cord dysfunction is shown, then every attempt should be made to maximize spinal-cord blood flow. Spinal-cord cooling, high-dose methylprednisolone, and CSF drainage are reasonable procedures that could be used in patients with spinal-cord infarction, although none of these measures has been tested in randomized therapeutic trials.

In older individuals, especially those with transient or progressive spinal-cord signs, every attempt should be made to identify spinal dural arteriovenous fistulas and repair them.

## TREATMENTS NO LONGER RECOMMENDED

Transplantation of omentum to the spinal cord has been shown to be generally ineffective and should not be attempted.

## REFERENCES

Aminoff, M. J., and Logue, V. (1974). Clinical features of spinal vascular malformations. *Brain* 97, 197–210.

Austin, G. (1961). The Spinal Cord. Charles C. Thomas Publisher, Ltd., Springfield, Illinois.

Azizzadeh, A., Huynh, T. T., Miller, C. C., Safi, H. J., and Miller, C. C. 3rd. (2000). Reversal of twice-delayed neurologic deficits with cerebrospinal fluid drainage after thoracoabdominal aneurysm repair: A case report and plea for a national database collection. *J. Vasc. Surg.* 31, 592–598.

Binkert, C. A., Kollias, S. S., and Valvanis, A. (1999). Spinal cord vascular disease: Characterization with fast three-dimensional contrast-enhanced MR angiography. *Am. J. Neuroradiol.* 20, 1785–1793.

Bowen, B. C., Fraser, K., Kochan, J. P., Pattany, P. M., Green, B. A., and Quencer, R. M. (1995). Spinal Dural Arteriovenous fistulas: Evaluation with MR angiography. *Am. J. Neuroradiol.* 16, 2029–2043.

Bracken, M. B., Shepard, M. J., and Collins, W. F. et al. (1992). Methylprednisolone or naloxone treatment after acute spinal cord injury: 1-Year follow-up data: Results of the Second national Acute Spinal Cord Injury Study. *J. Neurosurg.* 76, 23–31.

Bracken, M. B., Shepard, M. J., and Holford, T. R. *et al.* (1997). Administration of methylprednisolone for 24 or 48 hours or tirilizad mesylate for 48 hours in the treatment of acute spinal cord injury: Results of the Third National Acute Spinal Cord Injury Randomized Controlled Trial. National Acute Spinal Cord Injury Study. *J. Am. Med. Assoc.* **277,** 1597–1604.

Bradac, G. B., Daniele, D., Riva, A., Bracchi, M., Stura, G., Riccio, A., and Pagni, C. A. (1993). Spinal dural arteriovenous fistulas: An underestimated cause of myelopathy. *Eur. Neurol.* **34,** 87–94.

Buchan, A. M., and Barnett, H. J. M. (1986). Infarction of the spinal cord. *In* "Stroke: Pathophysiology, Diagnosis and Management" (H. J. M. Barnett, J. P. Mohr, B. Stern, and F. Yatsu, Eds.), pp. 707–719. Churchill-Livingstone, New York.

Caccamo, D. V., Garcia, J. H., and Ho, K-l. (1992). Isolated granulomatous angiitis of the spinal cord. *Ann. Neurol.* **32,** 580–582.

Cambria, R. P., and Davison, J. K. (2000). Regional hypothermia with epidural cooling for spinal cord protection during thoracoabdominal aneurysm repair. *Sem. Vasc. Surg.* **13,** 315–324.

Caplan, L. R. (1991). Case records of the Massachusetts General Hospital: Case 5—1991. *N. Engl. J. Med.* **324,** 322–332.

Caplan, L. R. (2000). Spinal cord strokes. *In* "Caplan's Stroke, a Clinical Approach (L. R. Caplan, Ed.), 3rd ed., pp. 435–443. Boston.

Caplan, L. R., Noronha, A., and Amico, L. (1990). Syringomyelia and arachnoiditis. *J. Neurol. Neurosurg. Psychiatry* **53,** 106–113.

Cheshire, W. P., Santos, C. S., Massey, E. W., and Howard, J. F. (1996). Spinal cord infarction: Etiology and outcome. *Neurology* **47,** 321–330.

Criscuolo, G. R., Oldfield, E. H., and Doppman, J. L. (1989). Reversible acute and subacute myelopathy in patients with dural arteriovenous fistulas. *J. Neurosurg.* **70,** 354–359.

DeChiro, G., Doppman, J. L., and Ommaya, A. K. (1971). Radiology of spinal cord arteriovenous malformations. *Prog. Neurol. Surg.* **4,** 329–354.

De Haan, P. (2000). Pharmacologic adjuncts to protect the spinal cord during transient ischemia. *Sem. Vasc. Surg.* **13,** 264–271.

Duffill, J., Buckley, J., Lang, D., Neil-Dwyer, G., McGinn, F., and Wade, D. (2001). Prospective study of omental transposition in patients with chronic spinal injury. *J. Neurol. Neurosurg. Psychiatry* **71,** 73–80.

Ell, J. J., Uttley, D., and Silver, J. R. (1981). Acute myelopathy in association with heroin addiction. *J. Neurol. Neurosurg. Psychiatry* **44,** 448–450.

Follis, F., Dragan, R., Blisard, K. S., Hartshorne, M., Temes, T., Pett, S. B., and Wernly, J. A. (1999). Retrograde perfusion of the spinal cord during aortic crossclamping: Initial observations in the swine model. *J. Thorac. Cardiovasc. Surg.* **118,** 597–603.

Freyaldenhoven, T. E., and Mrak, R. E. (2001). Fibrocartilaginous embolization. *Neurology* **56,** 1354.

Gass, A., Back, T., Behrens, S., and Maras, A. (2000). MRI of spinal cord infarction. *Neurology* **54,** 2195.

Gilbertson, J. R., Miller, G. M., Goldman, M. S., and Marsh, W. R. (1995). Spinal dural arteriovenous fistulas: MR and myelographic findings. *Am. J. Neuroradiol.* **16,** 2049–2057.

Gillilan, L. (1958). The arterial blood supply of the human spinal cord. *J. Comp. Neurol.* **110,** 75–103.

Gillilan, L. (1970). Veins of the spinal cord: Anatomic details— Suggested clinical applications. *Neurology* **20,** 860–868.

Goyal, M., Willinsky, R., Montanera, W., and terBrugge, K. (1999). Paravertebral arteriovenous malformations with epidural drainage: Clinical spectrum, imaging features, and results of treatment. *Am. J. Neuroradiol.* **20,** 749–755.

Hemphill, J. C. III, Smith, W. S., and Halbach, V. V. (1998). Neurologic manifestations of spinal epidural arteriovenous malformations. *Neurology* **50,** 817–819.

Heros, R. (1988). Arteriovenous malformations of the spinal cord. *In* "Surgical Management of Cerebrovascular Disease" (R. G.

Ojemann, R. C. Heros, R. M. Crowell, Eds.), 2nd ed., pp. 451–466. Williams & Wilkins, Baltimore.

Herrick, M. K., and Mills, P. E. (1971). Infarction of spinal cord. Two cases of selective gray matter involvement secondary to asymptomatic aortic disease. *Arch. Neurology* **24,** 228–241.

Hogan, E. L., and Romanul, F. C. A. (1966). Spinal cord infarction occurring during insertion of aortic graft. *Neurol.* **16,** 67–74.

Hughes, J. T. (1971). Venous infarction of the spinal cord. *Neurology* **21,** 794–800.

Hurst, R. W., and Grossman, R. I. (2000). Peripheral spinal cord hypointensity on T2-weighted MR images: A reliable imaging sign of venous hypertensive myelopathy. *Am. J. Neuroradiol.* **21,** 781–786.

Hurst, R. W., Kenyon. L. C., Lavi, E., Raps, E. C., and Marcotte, P. (1995). Spinal dural arteriovenous fistula: The pathology of venous hypertensive myelopathy. *Neurology* **45,** 1309–1313.

Iseli, E., Cavigelli, A., Dietz, V., and Curt, A. (1999). Prognosis and recovery in ischaemic and traumatic spinal cord injury: Clinical and electrophysiological evaluation. *J. Neurol. Neurosurg. Psychiatry* **67,** 567–571.

Jacobs, M. J., Meylaerts, S. A., de Haan, P., de Mol, B. A., and Kalkman, C. J. (2000). Assessment of spinal cord ischemia by means of evoked potential monitoring during thoracoabdominal aortic surgery. *Sem. Vasc. Surg.* **13,** 299–307.

Kazama, S., Miyoshi, Y., Nie, M., Imai, H., Lin, Z. B., Kurata, A., and Machii, M. (2001). Protection of the spinal cord with pentobarbital and hypothermia. *Ann. Thorac. Surg.* **71,** 1591–1595.

Kestle, J. R. W., Resch, L., Tator, C. H., and Kucharczyk, W. (1989). Intervertebral disc embolization resulting in spinal cord infarction. *J. Neurosurg.* **71,** 938–941.

Kim, R. C., Smith, H. R., Henbest, M. L., and Choi, B. H. (1984). Nonhemorrhagic venous infarction of the spinal cord. *Ann. Neurol.* **15,** 379–385.

Larsson, E-M., Desai, P., Hardin, C. W., Story, J., and Jinkins, J. R. (1991). Venous infarction of the spinal cord resulting from dural arteriovenous fistula: MR imaging findins. *Am. J. Neuroradiol.* **12,** 739–743.

Lips, J., de Haan, P., Bodewits, P., Vanicky, I., Dzoljic, M., Jacobs, M. J., and Kalkman, C. J. (2000). Neuroprotective effects of riluzole and ketamine during transient spinal cord ischemia in the rabbit. *Anesthesiology* **93,** 1303–1311.

Mascalchi, M., Quillici, N., Ferrito, Mangiafico, S., Scazzeri, F., Torselli, P., Petruzzi, P., Cosottini, M., Tessa, C., and Bartolozzi, C. (1997). Identification of the feeding arteries of spinal vascular lesions with phase-contrast MR angiography with three-dimensional acquisition and phase display. *Am. J. Neuroradiol.* **18,** 351–358.

Mawad, M. E., Rivera, V., and Crawford, S. *et al.* (1990). Spinal cord ischemia after resection of thoracoabdominal aortic aneurysms: MR findings in 24 patients. *Am. J. Neuroradiol.* **11,** 987–991.

McCreary, M., Emerman, C., Hanna. J., and Simon, J. (2000). Acute myelopathy following intranasal insufflation of heroin: A case report. *Neurology* **55,** 316–317.

Mikulis, D. J., Ogilvy, C. S., McKee, A., Davis, K. R., and Ojeman, R. G. (1992). Spinal cord infarction and fibrocartilaginous emboli. *Am. J. Neuroradiol.* **13,** 155–160.

O'Farrell, R., Thornton, J., Brennan, P., Brett, F., and Cunningham, A. J. (2000). Spinal cord infarction and tetraplegia—Rare complication of meningococcal meningitis. *Br. J. Anaesth.* **84,** 514–517.

Roa, K. R., Donnenfeld, H., Chusid, J. G., and Valdez, S. (1982). Acute myelopathy secondary to spinal venous thrombosis. *J. Neurol. Sci.* **56,** 107–113.

Rockman, C. B., Riles, T. S., and Landis, R. (2001). Lower extremity paraplegia subsequent to endovascular management of abdominal aortic aneurysms. *J. Vasc. Surg.* **33,** 178–180.

Rogopoulos, A., Benchimol, D., Paquis, P., Mahagne, M. H., and Bourgeon, A. (2000). Lumbar artery compression by the diaphragmatic crus: A new etiology for spinal cord ischemia. *Ann. Neurol.* **48**, 261–264.

Rosenblum, B., Oldfield, E. H., Doppman, J. L., and DiChiro, G. (1987). Spinal arteriovenous malformations: A comparison of dural arteriovenous fistulas and intradural AVMs in 81 patients. *J. Neurosurg.* **67**, 795–802.

Ross, R. T. (1985). Spinal cord infarction in diseases and surgery of the aorta. *Can. J. Neurol. Sci.* **12**, 289–295.

Rothman, S. M., and Nelson, J. S. (1980). Spinal cord infarction in a patient with sickle cell anemia. *Neurology* **30**, 1072–1076.

Srigley, J. R., Lambert, C. D., Bilbao, J. M., and Pritzker, K. P. (1981). Spinal cord infarction secondary to intervertebral disc embolism. *Ann. Neurol.* **9**, 296–301.

Tabayashi, K., and Motoyoshi, N. (2001). As originally published in 1993: Protection from postischemic spinal cord injury by perfusion cooling of the epidural space. Updated in 2001. *Ann. Thorac. Surg.* **71**, 1063–1064.

Teal, P. A., Wityk, R. J., Rosengart, A., and Caplan, L. R. (1992). Spinal TIAs—A clue to the presence of spinal dural AVMs. *Neurology* **142**(Suppl. 3), 341.

Tiesenhausen, K., Amann, W., Koch, G., Hausegger, K. A., and Oberwalder, P. (2000). Cerebrospinal fluid drainage to reverse paraplegia after endovascular thoracic aortic aneurysm repair. *J. Endovasc. Ther.* **7**, 132–135.

Tosi, L., Rigoli, G., and Beltramello, A. (1996). Fibrocartilaginous embolism of the spinal cord: A clinical and pathogenetic reconsideration. *J. Neurol. Neurosurg. Psychiatry* **60**, 55–60.

Ueda, T., Shimizu, H., Mori, A., Kashima, I., Moro, K., and Kawada, S. (2000). Selective perfusion of segmental arteries in patients undergoing thoracoabdominal aortic surgery. *Ann. Thorac. Surg.* **70**, 38–43.

Vinters, H. V., and Gilbert, J. J. (1979). Hypoxic myelopathy. Canadian Journal of Neurological Science **6**, 380.

Wirz, M., Colombo, G., and Dietz, V. (2001). Long term effects of locomotor training in spinal humans. *J. Neurol. Neurosurg. Psychiatry* **71**, 93–96.

Yuh, W. T., Marsh, E. E., Wang, A. K., Russell, J. W., Chiang, F., Koci, T. M., and Ryals, T. J. (1992). MR imaging of spinal cord and vertebral body infarction. *Am. J. Neuroradiol.* **13**, 145–154.

*Cerebrovascular Disorders*

CHAPTER 35

# Therapy of Intracerebral Hemorrhage

Joao A. Gomes, Carlos S. Kase, and Louis R. Caplan

## CLINICAL ASPECTS

Intracerebral hemorrhage (ICH) is due to rupture of small arteries located within the substance of the brain, and its main mechanism is hypertension. The affected vessels are perforating branches of 50 to 200 μm in diameter in deep, basal ganglionic hemorrhages (Fisher, 1971). In the case of superficially located lobar hematomas, larger cortico-subcortical or leptomeningeal arteries are the site of rupture.

A number of mechanisms other than hypertension are known to cause ICH. These include cerebral amyloid angiopathy (CAA), vascular malformations (arteriovenous malformations, cavernous angiomas), brain tumors, anticoagulant and fibrinolytic agents, sympathomimetic drugs, and vasculitis. Most of these nonhypertensive causes of ICH lead to hematomas in lobar locations, whereas hypertensive ICH predominates in the deep portions of the hemispheres, in the basal ganglia and thalamus.

### Features Related to Increased Intracranial Pressure

Pathologically, an ICH is a round or ovoid mass of blood in the brain parenchyma that destroys and displaces adjacent tissues. Before reaching its final size, the hematoma has progressively increased in size over periods of minutes to hours (Broderick *et al.*, 1990), and the resulting mass, if large, causes an increase in intracranial pressure (ICP). Some small hematomas may remain unchanged, while others attain intermediate and large sizes. The increased ICP accompanying large hematomas is responsible for a set of symptoms that frequently occur in ICH, regardless of location. They include headache, vomiting, and a decline in the level of consciousness. These symptoms reflect a combination of the general effects of increased ICP with several potential local mechanical factors related to the hematoma. The latter include stretching of blood vessels and meninges, impingement on cerebral areas that control alertness, obstruction of cerebrospinal fluid (CSF) pathways with resultant hydrocephalus, and extension of the hemorrhage into the ventricular cavity.

### Features Specific to Hemorrhage Location

The location of the ICH in the brain determines the neurologic findings on examination. Hematomas in the basal ganglia (putamen) usually present with severe contralateral hemiplegia, hemisensory loss, supratentorial horizontal gaze palsy, and homonymous hemianopia, along with aphasia in dominant hemisphere lesions or hemineglect in nondominant hemisphere hemorrhages (Hier *et al.*, 1977). Thalamic hemorrhages have a similar clinical picture, except for the prominence of oculomotor deficits (paresis of upward gaze, small nonreactive pupils, poor convergence) that reflect the pressure effects of the thalamic hematoma on the tectum of the midbrain (Fisher, 1961). A more benign presentation characterizes basal ganglia hematomas that originate in the head of the caudate nucleus; the small parenchymal component of the lesion causes either slight transient hemiparesis or no motor deficit at all, the main clinical features resulting from early entry of blood into the adjacent ventricular system, often mimicking subarachnoid hemorrhage (SAH) from a ruptured aneurysm (Stein *et al.*, 1984). Lobar hemorrhages are characterized by a combination of findings that depend on the specific lobe involved (Ropper and Davis, 1980; Weisberg, 1985) (Table I).

**TABLE I** Lobar Hemorrhage: Clinical Features by Location

| Location | Clinical findings |
|---|---|
| Frontal, superior | Bifrontal headache, contralateral leg weakness or hemiparesis |
| Frontal, inferior | Hemiparesis, hemisensory loss, horizontal gaze palsy to the side opposite to the hemiparesis |
| Temporal, posterior | Retroauricular headache, Wernicke aphasia or neglect, homonymous hemianopia |
| Parietal, lateral | Unilateral temple headache, hemiparesis and hemisensory loss, homonymous hemianopia, aphasia or hemineglect |
| Occipital | Ocular/periocular headache, homonymous hemianopia, dysgraphia/dyslexia, "alexia without agraphia" |

Cerebellar hemorrhage presents with sudden onset of inability to stand and walk, with ipsilateral ataxia, at times combined with ipsilateral pontine cranial nerve involvement (infratentorial horizontal gaze palsy, peripheral facial palsy, trigeminal sensory loss) (Fisher, 1961; Fisher et al., 1965; Ott et al., 1974). In instances of massive, bilateral pontine hemorrhage, the clinical picture includes coma, quadriplegia, decerebrate rigidity, pinpoint reactive pupils, bilateral infratentorial horizontal gaze paralysis, respiratory rhythm abnormalities, and hyperthermia (Fisher, 1961). Partial, predominantly unilateral tegmental pontine hemorrhages produce less severe asymmetric syndromes characterized by ipsilateral oculomotor abnormalities (the "one-and-a-half" syndrome), hemiataxia, and facial and trigeminal involvement, along with contralateral hemiparesis and hemisensory loss (Caplan and Goodwin, 1982; Kase et al., 1980).

## NATURAL COURSE

Intracerebral hemorrhage accounts for approximately 10% of strokes in Caucasian/Black mixed populations in the United States, while in Asia, especially in Japan, it is responsible for as many as 25% of strokes. In the United States each year between 37,000 and 52,400 individuals have an intracerebral hemorrhage (Quereshi et al., 2001). The incidence of intracerebral hemorrhage has great geographical variability: 6 per 100,000 in men and 7 per 100,000 in women over age 35 in Rochester, Minnesota (Furlan et al., 1979); 12.6/100,000 for men, 15.1/100,000 for women, 13.5/100,000 for Whites, and 17.5/100,000 for Blacks in Cincinnati, Ohio (Broderick et al., 1992); 70/100,000 in women and 220/100,000 in men in Hisayama, Japan (Ueda et al., 1988). The racial differences in ICH incidence in the United States consistently reveal higher

rates among Blacks. In the greater Cincinnati area, Blacks had an overall age- and sex-adjusted incidence that was 1.4-fold greater than in Whites, and in the age group below 75 years, the incidence was 2.3 times higher in Blacks (Broderick et al., 1992). These differences are likely the result of the significantly increased prevalence and poor control of hypertension in the Black population. The reasons for the strikingly high incidence of ICH in Asians are not clear; however, dietary habits of high salt consumption, poorly controlled hypertension, and generally low serum cholesterol levels may be implicated.

Risk factors for ICH include advancing age, Black race, hypertension, cigarette smoking, alcohol use, and low serum cholesterol levels. Other factors possibly associated with an increased frequency of ICH are sympathomimetic drug use, anticoagulant or fibrinolytic treatment, aspirin use, exposure to extremely low ambient temperature, and migraine (Ueshima et al., 1980; Wolf, 1994). The role of hypertension as a cause of ICH is favored by its high prevalence in patients admitted with ICH, as well as its frequent association with cardiomegaly by chest X-rays and/or electrocardiogram (ECG). However, in some populations the frequency of hypertension among patients with ICH has only reached figures close to 50% (Brott et al., 1986), suggesting that other factors operate in a substantial proportion of cases. Among these, cigarette smoking has been associated with a 2.5 greater incidence of "hemorrhagic stroke" (ICH and SAH) in Hawaiian men of Japanese ancestry, independently from the effect of age, diastolic blood pressure, serum cholesterol, alcohol consumption, hematocrit, and body mass index (Abbott et al., 1986). Intracerebral hemorrhage has also been shown to relate, in a dose-dependent manner, to alcohol use (Donahue et al., 1986). The age-adjusted estimated relative-risk of ICH in drinkers, as compared to nondrinkers, was 2.1 for light drinkers (1–14 oz/month), 2.4 for moderate drinkers (15–39 oz/month), and 4.0 for heavy drinkers (>40 oz/month). Low serum cholesterol levels (<160 md/dL) have been related to an increased incidence of ICH in Japanese, Hawaiians of Japanese ancestry, and Caucasians (Jacobs et al., 1992; Kagan et al., 1980; Ueshima et al., 1980). This effect of low serum cholesterol levels was particularly strong when it was associated with diastolic blood pressure greater than 90 mm Hg (Kagan et al., 1980).

A decline in the incidence of ICH has been documented in various populations in the last decades (Aurell and Hood, 1964; Furlan et al., 1979; Tanaka et al., 1981; Ueda et al., 1988). This has occurred largely as a result of improved detection and treatment of hypertension, as well as changes in dietary habits in some populations (Shimamoto et al., 1989).

The mortality from ICH varies greatly depending on the type of series analyzed. In pre-CT series, which relied on autopsy data and on clinically diagnosed large hemorrhages, mortality figures ranged from 58 to 92% (Furlan et al., 1979; McKissock et al., 1961). Since the introduction of CT, the ability to detect small hemorrhages with better prognosis (that in pre-CT days would have been diagnosed as cerebral infarcts) (Caplan, 1994; Drury et al., 1984; Rowe et al., 1988) has resulted in an overall lower mortality of 20 to 56% (Bogousslavsky et al., 1988; Silver et al., 1984).

The data on long-term prognosis of ICH are sparse, but they suggest a relatively good functional prognosis in survivors and a low frequency of ICH recurrence (Fieschi et al., 1988; Fogelholm et al., 1992; Franke et al., 1992; Steiner et al., 1984), although the latter point has been disputed by a recent study showing a recurrence rate of 5.4% (Bae et al., 1999). After surviving an episode of ICH, patients are found to have the same mortality rates as age- and sex-matched members of the general population (Fogelholm et al., 1992). Similarly, Franke et al. (1992) found that ICH survivors had equal mortality rates and functional levels after 1 year in comparison to matched survivors of cerebral infarction. In contrast with the sparsity of data on long-term outcome of ICH, abundant information has been gathered on predictive factors of outcome during the acute phase. Factors that are highly predictive of mortality after ICH include decreased level of consciousness (measured as low Glasgow Coma Score [GCS] of 3–8) and large hematoma size (Tuhrim et al., 1988). Other factors related to poor outcome have included advancing age, amount of alcohol consumed (Juvela, 1995), presence of hydrocephalus (Diringer et al., 1998; Phan et al., 2000), and blood pressure on admission (Terayama et al., 1997). Pulse pressure, gaze palsy, motor deficit, $pO_2$, ECG changes, admission blood sugar, IVH, and neurological deterioration after onset have been shown to be less consistent predictors of outcome (Kase and Crowell, 1994; Tuhrim et al., 1988).

## PRINCIPLES OF THERAPY

Small hematomas that do not cause headache or vomiting and do not increase ICP have an excellent prognosis without treatment. The issues in these patients involve more contolling factors that increase bleeding and discovering the mechanism so as to prevent recurrent ICH. Massive hematomas which render patients quickly comatose have a dire prognosis regardless of therapy. Intermediate and large hematomas that increase ICP are those that lend themselves most to medical and surgical therapy. Especially important is treatment of patients whose hematomas are initially intermediate or large who develop increasing neurological signs and decreased level of consciousness while under observation.

The first concern with a patient who presents with acute ICH is to assure stability of vital signs, management of hypertension if necessary and, especially, airway patency and protection. The latter is best achieved with early endotracheal intubation (see below). The next step is the neurological evaluation to assess the severity of the neurological deficits, localize the ICH, and often suggest its mechanism. The clinical findings are complemented by CT, which is virtually 100% reliable in the diagnosis of acute ICH (Dul and Drayer, 1994). The next step is to attempt to control factors that increase bleeding.

### Control of Bleeding

#### Reversal of Anticoagulation

An emergency issue in ICH is the management of patients who have bled while on anticoagulant treatment, as the correction of the iatrogenic coagulopathy is mandatory (Hart et al., 1995). Normalization of a prolonged International Normalized Ratio (INR) in patients with warfarin-associated ICH can be achieved with the use of vitamin K. Although higher doses may be needed when a subcutaneous route is used, this method of administration is generally considered safer since IV administration may be complicated by anaphylactic reactions (Raj et al., 1999). Despite the effectiveness of vitamin K in normalizing a prolonged INR, it takes several hours before its effect becomes clinically significant, and the use of fresh frozen plasma (FFP) is required for immediate reversal. However, because of the variable content of clotting factors in FFP (particularly factor IX) and the effects of dilution, clotting factor concentrates might be superior for rapid and complete reversal of anticoagulation (Makris et al., 1997). The risk of viral contamination and thrombotic side effects limit the use of clotting factor concentrates. The development of recombinant coagulation factors will eventually eliminate the concern for viral contamination (Erhardtsen et al., 1998).

Oftentimes physicians are faced with the dilemma of reversing anticoagulation in patients who develop an ICH and who had been treated with oral anticoagulation for conditions associated with a high thromboembolic risk (i.e. prosthetic heart valves or atrial fibrillation). It has been increasingly recognized that antithrombotic therapy-associated ICH most likely represents unmasking of an underlying vasculopathy that would have otherwise caused a subclinical hemorrhage in the absence of anticoagulation (Hart *et al.*, 1995). In this regard, the association between CAA and warfarin-related ICH has been well documented by means of determination of genetic risk factors (APOE alleles) and neuropathologic studies. The APOE alpha2 allele has been found to be over-represented in patients with warfarin-related ICH of lobar location (Rosand *et al.*, 2000). These authors posited that this finding might be due to an association between the alpha2 allele and the particular vasculopathic processes (i.e., fibrinoid necrosis) that facilitate extravasation of blood.

Although no universal guidelines for the management of these patients can be adopted based on the current knowledge of this condition, there are now data to support the discontinuation of warfarin for up to 2 weeks with still low probability of embolic events (Ananthasubramanian *et al.*, 2001; Butler and Tait, 1998; Phan *et al.*, 2000). The risk of such an event in a patient with a prosthetic heart valve has been estimated as between 1 in 1300 and 1 in 240 during the first week without anticoagulation (Butler and Tait, 1998). After 1–2 weeks, it seems advisable to restart anticoagulation, especially since early recurrence of ICH seems to be rare (Phan *et al.*, 2000), and onset of anticoagulation as early as 1 week after ICH has been done safely (Butler and Tait, 2000).

For patients who bleed while being treated with IV heparin, reversal of the activated plasma thromboplastin time (aPTT) with protamine sulfate is standard treatment. Both the time elapsed since administration and the amount of heparin given have to be considered to calculate the total dose of protamine needed for reversal.

There is scarce information regarding the optimal treatment of ICH following the administration of thrombolytic agents. As a general principle, several factors should guide the management of thrombolysis-related ICH, including: (1) location and size of the hematoma, (2) interval between drug administration and hemorrhage onset, (3) the thrombolytic drug used, and (4) the likelihood of neurological deterioration or death (Greenberg *et al.*, 1996). Cryoprecipitate, FFP, fibrinogen and platelets have all been used to treat this dreaded complication of thrombolytic therapy. Similarly, the use of aminocaproic acid has been advocated in the past; however, its current role remains controversial as it increases the risk of deep venous thrombosis and a clear benefit from its use has not been unequivocally shown.

Finally, in a thromboembolic rabbit stroke model, the combination of a matrix metalloproteinase and rt-PA significantly attenuated the rate of hemorrhage development without decreasing the rate of thrombolysis. This therapeutic measure holds promise as an effective means of reducing the risk of thrombolysis-related

ICH, but further validation in human studies is required (Lapchak *et al.*, 2000).

### Evaluation of Vascular Malformations That Could Have Caused Bleeding

Another consideration after acute presentation with ICH relates to the potential of a ruptured aneurysm, arteriovenous malformation (AVM), or cavernous angioma as the cause of the hemorrhage. A ruptured aneurysm is suggested by the location of the ICH in atypical areas, such as the vicinities of the Sylvian fissure or interhemispheric fissure, where bleeding from middle cerebral artery or anterior communicating artery aneurysms, respectively, occur (Vermeulen and van Gijn, 1990). In addition, a component of local or diffuse SAH in association with ICH should raise suspicion of a ruptured aneurysm or AVM. In these instances, strong consideration should be given to cerebral angiography for diagnosis of the nature of the bleeding lesion, in particular if surgical drainage of the ICH is being contemplated, as preoperative knowledge of the type, size, and location of such lesions may help prevent intraoperative rerupture and, ideally, allow definitive treatment.

### Control of Severe Hypertension

Very high blood pressure acts to increase leakage of blood into the brain and increases the size of acute hematomas. The elevated blood pressure also increases ICP. It is essential to closely monitor BP and treat severe hypertension in patients with acute ICH. Increased ICP causes an increased venous pressure in the cranium. In order to perfuse brain tissue, the arterial pressure must exceed the venous pressure creating an arterial–venous gradient. The blood pressure should be lowered to slightly hypertensive ranges but not to normal pressure in order to maintain the arterial–venous gradient and perfuse the brain adequately.

Unfortunately, no specific guidelines exist as to what values of hypertension require aggressive treatment, since correction of any level of hypertension is not indicated. Furthermore, overly aggressive BP control carries the risk of hypotension, which in turn can result in hypoxia, with further neurologic damage and increase in ICP. This is particularly the case in previously hypertensive patients, whose autoregulatory curve is shifted to the right, making them susceptible to ischemic brain damage if the BP is brought to the normotensive level.

It is generally accepted that levels of hypertension greater than 120 mm Hg diastolic or a mean arterial pressure (MAP) greater than 130 mm Hg should be treated (Broderick *et al.*, 1999; Diringer, 1993). In patients with an ICP monitor, the cerebral perfusion pressure should be kept >70 mm Hg and pressors should be given if the systolic blood pressure drops below 90 mm Hg. The treatment of hypertension in this setting should ideally exclude agents that are cerebral vasodilators, as they are likely to further increase ICP. This includes agents such as nitroprusside, nitroglycerin, hydralazine, verapamil, and nicardipine. Although its vasodilator effect makes it have an unpredictable effect on ICP, nitroprusside is often used in instances of severe hypertension because of its rapid and predictable effect on BP and ease of administration and titration. The same potential for increasing ICP applies to hydralazine, which is at times used as an oral dose of 50 to 100 mg twice a day, in order to maintain lower BP levels. A more useful agent is the α- and β-adrenergic blocker labetalol, which can be administered intravenously, produces minimal changes in heart rate and cardiac output, causes no rebound hypertension when discontinued, and does not produce tachyphylaxis (MacCarthy and Bloomfield, 1983). When given to patients with intracranial hemorrhage as an intravenous bolus of 5–25 mg, labetalol produces a rapid (within 5 to 15 min) drop in systolic and diastolic BP, in the order of 10 to 20% (Patel et al., 1993). These doses are not associated with hypotension, changes in mental status, or worsening in clinical status.

## Management of Increased Intracranial Pressure

Intracranial pressure remains relatively constant under normal conditions, at a level of 10 to 15 mm Hg. With increases in size of any of the components that determine ICP (brain, blood, and CSF volume), a compensatory attempt at maintaining ICP constant takes place through adjustments in the size of the other, unaffected intracranial components. These include migration of CSF from the intracranial compartment into the spinal subarachnoid space, decreased production of CSF, and movement of venous blood into the systemic circulation. When these compensatory mechanisms are exhausted a rise in ICP occurs, with the potential for major changes in cerebral hemodynamics, as ICP directly impacts on cerebral perfusion pressure (CPP). Changes in CPP, with levels below 50 mm Hg, are in turn responsible for brain ischemia and the potential for further deterioration in neurological function.

Increased ICP has a major impact on morbidity and mortality after ICH. A persistent elevation of ICP above 20 mm Hg, that is not responsive to maneuvers aimed at lowering ICP, is associated with a grim prognosis for survival (Diringer, 1993). For this reason, the management of increased ICP after ICH is essential, and it includes two main aspects, the prevention of elevations in ICP and the treatment of elevated ICP. A number of factors that operate in patients with acute ICH can contribute to increases in ICP. These include: hypertension, hypoxia, seizures, hyperthermia, and elevations of intrathoracic pressure.

### General Problems That Relate to ICP

**Hypertension.** We have already commented on hypertension as a factor that promotes continued bleeding. Hypertension if uncontrolled also has a deleterious effect on ICP. In the setting of acute stroke, including ICH, there is loss of cerebral autoregulation, and cerebral blood flow (CBF) becomes dependent on the systemic blood pressure (BP). Under these conditions, systemic hypertension results in increased CPP and CBF, which in turn increase cerebral blood volume (CBV) and promote the development of brain edema, with worsening intracranial hypertension (Ropper, 1993).

**Hypoxia.** Hypoxia, defined as $pO_2$ <60 mm Hg or $pCO_2$ >50 mm Hg, increases CBF and CBV in the presence of poor intracranial compliance, resulting in raised ICP. This makes adequate oxygenation mandatory in patients with ICH, and the first step toward this goal in comatose patients is early endotracheal intubation. This not only assures adequate oxygenation ($pO_2$ of 100 to 150 mm Hg), but also protects the airway against aspiration in lethargic or comatose patients. Intubation also facilitates pulmonary toilet, another essential measure to assure adequate oxygenation and prevention of pulmonary infections. Intubation needs to be performed rapidly and safely, keeping in mind that tracheal stimulation results in increased ICP. In order to block such effect, intubation can be performed after the intravenous administration of either a very short-acting barbiturate or lidocaine (Diringer, 1993). Other pharmacologic agents that are helpful while performing intubation include atropine, midazolam, propofol, and succinylcholine (Broderick et al., 1999).

**Seizures.** Seizures occur in 10 to 15% of patients with ICH in general and in 15 to 35% of those with lobar ICH (Caplan, 1994). Although seizures are more likely to occur at the onset of ICH, their subsequent occurrence can be associated with transient increases in CBF, CBV, and ICP, thus requiring urgent treatment. Diazepam, phosphenytoin, and phenytoin are the preferred agents to use. In order to stop an ongoing generalized seizure, IV diazepam is used, followed by IV loading with phenytoin. This results in therapeutic blood levels within 1 h of administration due to the rapid distribution of the drug after IV loading. A maintenance dose of phenytoin is then administered daily. Fosphenytoin, a water-soluble formulation of phenytoin, is associated with less serious side effects and is considered safer than and equally effective as phenytoin. In the event of seizure recurrence despite therapeutic doses of phenytoin, phenobarbital loading is required. This agent has the disadvantage of potentially affecting the level of alertness, impairing the ability to fully monitor the patient's neurological status. However, recurrent seizures and, especially, status epilepticus, may require the use of anesthetic doses of thiopental in cases of failure of the other two agents. Thiopental has the additional disadvantage of producing hypotension, requiring close monitoring of BP during its administration.

**Hyperthermia.**    Hyperthermia results in an increased metabolic rate (5 to 7% per each centigrade degree) and CBF, with increased ICP. Thus, fever in patients with ICH needs to be treated promptly, with cooling blankets and non-aspirin-containing antipyretics such as acetaminophen. Antibiotic treatment of infections should be instituted early.

The occurrence of extreme degrees of hyperthermia (43 to 45°C) is an ominous sign in patients with ICH, indicating hypothalamic damage and generally occurring shortly before death. It is a particularly common occurrence in patients with massive, bilateral pontine hemorrhage.

**Elevations of Intrathoracic Pressure.**    Intrathorathic pressure elevations frequently occur in patients with ICH, as a result of endotracheal suctioning and coughing, use of positive end-expiratory pressure (PEEP), and chest therapy. These measures can all result in impaired venous return with elevation of ICP. A simple measure in these instances is to elevate the patient's head 30° in order to facilitate venous return. In addition, hyperventilation with 100% oxygen before and after suctioning is recommended, along with the IV use of either thiopental or lidocaine or in order to block increased ICP in response to these maneuvers (Diringer, 1993).

### Specific Measures to Reduce Increased Intracranial Pressure

These measures are instituted with the purpose of rapidly decreasing ICP, in order to avoid its potential effects on cerebral function through either decreased CPP or shifting of intracranial contents. The proven therapies for this purpose include hyperventilation, osmotic diuretics, and high-dose intravenous barbiturates.

**Hyperventilation.**    Hyperventilation reduces ICP promptly and reliably by decreasing arterial and CSF $pCO_2$, the latter resulting in a rise in CSF pH, which is responsible for vasoconstriction (Koehler and Traystman, 1982). This produces a virtually immediate decrease in CBV and ICP that may reach its peak after about 30 min (Ropper, 1993). Reductions in arterial $pCO_2$ of 5 to 10 mmHg can result in acute reductions of ICP of 25 to 30% (James et al., 1977). The most effective levels of therapeutic hyperventilation are achieved with reductions in arterial $pCO_2$ to 30–35 mmHg. The effects of controlled hyperventilation are, however, transient, as the pH in the CSF tends to normalize despite continuing arterial hypocarbia. For this reason, therapeutic hyperventilation is most useful in the hours that follow its institution, its benefits gradually declining over time. After a stable reduction in ICP has been achieved, generally after 24 to 72 h, it is prudent to wean patients from hyperventilation gradually, over periods of 4 to 24 h, since its abrupt discontinuation carries a risk of rebound increase in ICP (Diringer, 1993; Broderick et al., 1999). The inability to reduce increased ICP with adequate levels of therapeutic hyperventilation is a poor prognostic sign, and is usually associated with similarly poor responses to other measures used to reduce ICP (Ropper, 1993).

Patients with deteriorating levels of consciousness and suspected high ICP should be considered for invasive ICP monitoring, particularly if the Glasgow Coma Scale (GCS) is <9 (Broderick et al., 1999).

**Osmotic Diuretics.**    These agents dehydrate the brain by causing an osmotic gradient between the intravascular space and the intracellular space, promoting movement of water out of the brain. The aim of the therapy is thus to create a state of serum hyperosmolality of ≤310 mOsm/L (Broderick et al., 1999), which can be achieved by the use of osmotic diuretics such as mannitol, as well as with loop diuretics, agents that are at times used in combination.

Mannitol produces a rapid reduction of ICP, its effect being apparent within 10 to 20 min from an IV bolus injection (Mendelow et al., 1985), before a sizable diuretic response occurs, suggesting that its initial effect may be independent from diuresis. In this regard, experimental data indicate that a rapid decrease in blood viscosity accompanied by vasoconstriction follows the administration of a bolus of 1 g/kg of mannitol, changes that result in a 28% decrease in ICP, an effect that is maximal 10 min after the injection (Muizelaar et al., 1983). These data suggest that factors other than diuresis account for the rapid response of ICP to mannitol injections, at least initially.

A potential side-effect of mannitol is a rebound increase in ICP following its use for several hours or its discontinuation, on account of entry of mannitol to areas of injured brain with abnormally permeable blood–brain barrier. A concern was that, in patients with active intracerebral bleeding, the mannitol could enter the hematoma and its osmotic effects could cause imbibation of fluid into the hematoma, thus increasing the size and mass effect of the hematoma. However, the clinical significance of such effect is uncertain (Ropper, 1993), and experimental data have failed to show an increase in brain water in damaged cerebral hemispheres of animals treated with IV mannitol (Berger et al., 1994). This occasionally observed rebound effect of mannitol on ICP is thought to be related to premature discontinuation of therapy, rather than to a pharmacologic effect of this agent. More pertinent concerns with the use of mannitol are the risks of wide variations in blood volume, electrolyte imbalances, and renal failure. Following IV infusions of mannitol, a rapid increase in intravascular volume can result in both cardiac decompensation and dilutional hyponatremia, whereas continued use and diuresis may lead to volume contraction, hypernatremia, and hypokalemia. These effects in turn can result in renal failure, which can be prevented by close monitoring of fluid status and electrolyte balance, including proper fluid replacement with normal saline (which has an osmolality of 310 mOsm/L), monitoring of central venous pressure, and adjustments in electrolyte administration (Diringer, 1993).

The value of alternative hypertonic solutions for acute reduction of ICP has not been as clearly established as with mannitol. In experimental studies, Berger et al. (1994) have shown a potential benefit of the hypertonic/hyperoncotic combination of 7.2% NaCl and 10% Dextran 60 on ICP, and its use in humans has resulted in rapid restoration of BP and cardiac output after severe hemorrhagic shock (Holcroft et al., 1987).

The use of loop diuretics, such as furosemide (10 mg every 2–8 h), is followed by a gradual increase in serum osmolality and reduction in ICP, an effect that can be enhanced by fluid replacement with hypertonic solutions. The loop diuretics can be particularly helpful in combination with mannitol in patients whose serum hyperosmolality cannot be maintained with the latter agent alone, and further use is associated with congestive heart failure (Ropper, 1993).

The potential for a rebound increase in ICP after discontinuation of osmotic diuretics calls for slow correction of serum hyperosmolality once the desired effect of these agents has been achieved. This gradual correction of serum osmolality should take 1 to 2 weeks (Diringer, 1993).

**High-Dose Intravenous Barbiturates.** Intravenous barbiturates produce a reduction in brain metabolism, CBV, and ICP (Marshall et al., 1978) and may act as free radical scavengers as well (Broderick et al., 1999).

The experience with thiopental at a dose of 1 to 5 mg/kg indicates that a clinically significant reduction in ICP follows shortly after injection, thus being a useful agent for the prevention of transient increases in ICP in maneuvers such as endotracheal intubation (Diringer, 1993; Ropper, 1993). However, its value as a more sustained treatment of intracranial hypertension is less clear. This is primarily the result of its substantial risk of inducing hypotension, as well as its marked effects in neurological function, which make the neurologic examination unavailable as a way of monitoring clinical progress. High-dose IV barbiturates have also been associated with a predisposition to infections (Broderick et al., 1999).

**Other Therapies.** A commonly prescribed drug in ICH is dexamethasone, given with the purpose of reducing brain edema around the acute hemorrhage. Although corticosteroids have proven value in reducing cerebral edema, especially surrounding cerebral metastases (Patchell and Posner, 1985), their role in acute ICH is still controversial. Their widespread use has not been supported when they have been tested in controlled clinical trials. Poungvarin et al. (1987) conducted a randomized clinical trial of dexamethasone treatment in acute ICH, and found no differences in mortality between the drug and placebo groups after 21 days from onset, with a higher frequency of complications in the drug group, leading to early termination of the study. A recently published guidelines for the management of ICH by the American Heart Association discourages the use of steroids given the lack of proven efficacy and potentially serious side effects (Broderick et al., 1999).

The use of neuromuscular paralysis results in reductions of ICP by preventing increases in intrathoracic pressure associated with coughing, straining, and "bucking" the ventilator (Ropper, 1993). Although their use is justified under specific circumstances (such as during endotracheal intubation), their more prolonged use has the disadvantage of producing paralysis, including the extraocular movements, rendering the neurological examination unavailable for assessing the patient's clinical course and response to treatment.

The removal of CSF via a ventricular catheter is an effective way of rapidly reducing ICP, and it is particularly useful in conditions associated with hydrocephalus, such as thalamic and cerebellar hemorrhage. Although ventriculostomy catheters carry the potential risks of hemorrhage at the time of insertion and infection after prolonged use, their value in patients with acute ICH with hydrocephalus is well established. Infection risks are minimized by replacing catheters after 7 days, and monitoring the CSF for onset of pleocytosis.

## Use of Continuous Intracranial Pressure Monitoring

Several techniques for ICP monitoring have become available in recent years, and they are used in the intensive care management of some patients with neurological disorders, especially head trauma. Although the value and, especially, the impact on outcome of such intensive monitoring of ICP in acute stroke patients has not been clearly established, it has advanced the understanding of the course of ICP in patients with acute ICH.

Recordings of ICP in ICH patients by Janny *et al.* (1978) showed the highest levels of ICP early in the course, with subsequent decline in ICP until reaching normal values over periods of 20 to 30 days. The observations of Papo *et al.* (1979) suggested good correlation between measured levels of ICP and level of consciousness and outcome at the two extremes of ICP, either normal or markedly elevated (>30 mm Hg), but the correlation was poor for intermediate levels of ICP. These authors suggested that continuous ICP monitoring is a valuable tool in the management of patients with ICH, especially as an indicator of the need for surgical drainage of the hematoma in patients whose elevated ICP fails to respond to maximal medical measures. This view was supported by the study of Ropper and King (1984), who monitored 10 comatose patients with supratentorial ICH and found no survivors among the four patients whose ICP remained >20 mm Hg despite maximal "medical" therapy (hyperventilation, mannitol, dexamethasone, high-dose intravenous barbiturates). On the other hand, the three patients who had similar ICP recordings, but in addition had surgical hematoma evacuation survived.

Although continuous ICP monitoring provides useful information to guide some aspects of the management of ICH, its value in modifying clinical outcome has not been clearly documented, and it needs to be further evaluated by comparing similar groups of patients treated with and without this technique.

## Choice between Surgical and Nonsurgical Treatment

This aspect of the management of ICH is one of the most controversial ones, as a result of a paucity of scientific data, despite a number of studies that have attempted to address this important issue. Most of the studies have had methodological flaws, the most common being the nonrandom allocation of treatment, while others have been biased by either comparing patients with different degrees of clinical severity or performing delayed surgery, the latter thus selecting patients with better natural survival rates.

## Clinical Trials

The first randomized clinical trial that compared surgical with nonsurgical treatment, conducted before the introduction of CT, was reported by McKissock *et al.* in 1961. Although limited by the lack of brain imaging of the ICH (thus not including important prognostic variables such as hematoma size and precise location), this study included a substantial number of patients (180), and its main conclusion of a lack of benefit for surgical drainage of hematomas, in terms of both mortality and functional outcome (Table II), has been corroborated by modern, CT-based studies.

Following the study of McKissock *et al.* (1961), a number of uncontrolled studies failed to provide reliable data to help decide between surgical and nonsurgical treatment of ICH, but added useful information on the subject: Cuatico *et al.* (1965) and Luessenhop *et al.* (1967) observed a more favorable outcome in operated patients with lobar hemorrhages than in those with basal ganglionic hemorrhages; Paillas and Alliez (1973) reported better results in surgical patients treated after a delay of "five to ten days" from ICH onset, a fact that takes advantage of the improved prognosis of patients who survive the critical initial days after ICH onset (Silver *et al.*, 1984); Kaneko *et al.* (1977) obtained good results (mortality of 8%) in patients with putaminal hemorrhage who had early surgical drainage of the hematoma, within 7 h from onset.

After the introduction of CT scan, studies have benefited from a better definition of the anatomical characteristics of the hematomas, thus allowing the comparison of patients with similar prognosis, that in addition can be stratified according to both clinical and CT features. Kanaya *et al.* (1980) found no benefit for surgery for patients who were at either end of the clinical spectrum, either alert or comatose. For those in the intermediate levels of depressed consciousness ("stuporous" and "semicomatose without herniation signs"), both survival and functional outcome were better for the surgical group. These authors concluded that patients with putaminal hemorrhage who are alert or somnolent should be treated nonsurgically, the same

TABLE II  Outcome after Surgical versus Conservative Treatment of Intracerebral Hemorrhage, Randomized Clinical Trial[a]

| Outcome | Conservative (n = 91) | Surgery (n = 89) |
|---|---|---|
| Full recovery | 11 (12%) | 10 (11%) |
| Partial disability | 20 (22%) | 8 (9%) |
| Total disability | 14 (15%) | 13 (15%) |
| Dead | 46 (51%) | 58 (65%) |

[a]Reprinted with permission from McKissock *et al.* (1961).

TABLE III Cumulative Time from Hemorrhage to Death, Correlated with Type of Treatment[a]

| | Treatment | | Deaths | |
| Time after ICH | Nonsurgical (n = 26) | Surgical (n = 26) | n | % |
|---|---|---|---|---|
| 0–4 Days | 5 | 4 | 9 | 17 |
| 1 Week | 8 | 8 | 16 | 31 |
| 1 Month | 10 | 10 | 20 | 38 |
| 6 Months | 10 | 12 | 22 | 42 |
| 12 Months | 10 | 13 | 23 | 44 |
| 2 Years | 11 | 14 | 25 | 48 |

[a]Reprinted with permission from Juvela et al. (1989).

TABLE IV Cerebellar Hemorrhage: Surgical Mortality According to Preoperative Mental Status[a]

| | No. | Died | No. | % |
|---|---|---|---|---|
| Alert | 2 | 0 | 0 | |
| Drowsy | 10 | 2 | 20 | 17 |
| Stuporous | 4 | 2 | 50 | |
| Comatose | 12 | 10 | 83 | 75 |

[a]Modified from Ott et al. (1974), with permission from the American Medical Association.

as comatose patients, whereas those with intermediate levels of depressed consciousness should be considered candidates for surgical hematoma drainage. Similar encouraging results were reported by Kaneko et al. (1983), who operated early (within 7 hours of onset) on patients in "stupor" (GCS scores of 10–12) or "semi-coma" (GCS scores of 6–9), with hematomas of >20–30 cc, with more than 5 mm of midline shift: surgical mortality was 7%, and 89% of the survivors were ambulatory 6 months after surgery.

The above results stimulated Batjer et al. (1990) to conduct a randomized trial of surgical and non-surgical therapy in putaminal hemorrhage. The patients were randomly assigned to three treatment groups: (1) "best medical management" (BMM), which included hyperventilation, osmotic and loop diuretics, and dexamethasone; (2) best medical management plus ICP monitoring by ventriculostomy, allowing CSF drainage for maintenance of ICP below 20 mm Hg; (3) surgical drainage of the hematoma. After entry of 21 patients into the trial, an interim analysis showed dismal results in all treatment groups, with an overall mortality or survival in vegetative state of 71%, only four patients (19%) living independently at home. This led to termination of the trial. The small numbers did not allow for statistical comparisons between the treatment groups.

In a randomized, prospective trial, Juvela et al. (1989) included 26 surgically treated and 26 non-surgically treated patients with supratentorial ICH who had a GCS <9 and/or severe hemiparesis or aphasia and who were admited within 24 h of ICH onset. The mortality rates did not differ between the groups, both acutely and in long-term follow-up (Table IV). They concluded, based on an overall mortality of 42% at 6 months and a poor quality of life in survivors, that the treatment of supratentorial ICH should be nonsurgical.

In 1998, Morgenstern et al. reported the results of a single-center pilot feasibility study of early hematoma removal (open craniotomy within 12 h of documented symptom onset vs best medical therapy). In the 34 patients that comprised the randomized series of this prospective trial, a nonsignificant trend for better survival and functional outcome in those who underwent early surgery was found. The surgical and medical groups, however, differed significantly in certain baseline characteristics (i.e., race and location of hemorrhage). Although this study was not powered to detect a difference in outcome between surgery and medical therapy, it demonstrated the feasibility and safety of conducting a randomized trial of early surgery for patients with spontaneous supratentorial ICH.

Similarly, Zuccarello et al. (1999), conducted a randomized feasibility study of early surgery for ICH. Twenty patients were randomized and assigned to either surgical intervention or best medical therapy. Surgery was performed within 24 h of symptom onset and allowed open craniotomy or CT-guided stereotactic placement of a catheter with concomitant instillation of urokinase to facilitate clot removal. No statistically significant difference was found between the two groups for the primary endpoint (Glasgow Outcome Scale >3) at 3 months; however, the 3-month NIHSS score significantly favored the surgically treated group. This study, too, given the small number of patients enrolled, showed baseline differences in anatomical location as well as in intraventricular extension of the hemorrhage between the two groups. Besides demonstrating the feasibility of a larger randomized study of early surgery in ICH involving both academic and community centers, the authors also provided an estimation of the sample size needed for such a trial (107 patients per group).

Recently Morganstern et al. (2001) reported the results among patients with intracerebral hematomas operated on within 4 h after onset of symptoms. The study was stopped after an interim analysis of the first 11 patients. Postoperative rebleeding occurred in 40% of the patients treated within 4 h (with 75% mortality) compared to 12% in those treated after 12 h. This study showed that ultra-early sutgical decompression carries a risk of rebleeding and may not be advisable.

A large, multicenter, randomized trial aiming to recruit 1000 patients (Surgical Trial in Intracerebral Hemorrhage (STICH)) is under way. Some, however, noted that the time to randomization (seeking surgery within the first 3–4 hours) as well as the standardization of minimally invasive surgical techniques need to be improved if favorable, reliable results are to be expected (Gregson et al., 2000).

Despite the lack of fully adequate studies, certain facts regarding the surgical/nonsurgical choice of treatment in ICH have been established: patients with either small (<10 cc) hematomas and minimally depressed level of consciousness (somnolent), or large hematomas (>60 cc) and severely depressed level of consciousness (lethargic or comatose) do not benefit from surgery on account of the naturally good outcome in the former and the universally poor outcome in the latter. In patients with intermediate levels of clinical involvement and hematoma size, the data are still inconclusive. Patients with intermediate size hematomas who have a progressive decline in the level of consciousness have a poor prognosis, and thus should be considered for surgical drainage if the lesion is accessible. The timing of surgical drainage is also an important consideration. During the first hours the hematoma is liquid and readily suctioned during direct surgical drainage or stereotactically through a burr hole. Later the hematoma becomes a solid, firm clot that cannot be suctioned readily, and more extensive corticotomy is needed for removal. Later, after 7 to 10 days, liquefaction occurs and the hematoma can again be easily drained. Thus, early or later surgery is technically easier.

### Surgery for Hematomas in Specific Locations

Despite the lack of definite information regarding the value of surgical treatment in ICH in general, there are several situations in which clinical experience and the results of small, nonrandomized clinical series have provided useful guidelines for the choice of therapy. These include hemorrhages of lobar, thalamic, and cerebellar location.

**Lobar Hemorrhage.**    The superficial, more accessible location of hematomas of lobar location results in their being surgically treated more often than deep, basal ganglionic hematomas (Broderick et al., 1994). This applies particularly to young patients with medium size hematomas, in the 10 to 50 cc range, since those with either small (<10 cc) or large (>50 to 85 cc) hematomas do not appear to benefit from surgical treatment (Broderick et al., 1999; Garde et al., 1983; Kase et al., 1982; Volpin et al., 1984). Surgery is particularly appropriate in patients with medium-size hematomas who present with an obtunded or lethargic level of con-

sciousness, or who deteriorate to those levels after admission, a change that is often accompanied by CT evidence of increased mass effect and midline shift due to hematoma enlargement. An additional value of surgical treatment in patients with lobar ICH is the resection of a lesion with potential for rebleeding, such as an AVM, or a cavernous angioma, which are often the cause of such hemorrhages (Wakai et al., 1992).

The decisions between surgical and nonsurgical treatment in lobar hematomas, although facilitated by clinical and CT volumetric data, are still based on the findings of the individual patient. The value of each form of therapy will be determined only by a clinical trial involving randomization of comparable groups of patients to one or the other form of treatment.

**Thalamic Hemorrhage.**    Thalamic hemorrhages are located so deep in the hemisphere that direct surgical approach is associated with a high risk of added damage to unaffected brain tissue located superficial to the hemorrhage. For this reason, direct evacuation of thalamic hemorrhages is very seldom considered. However, some surgical approaches are at times justified, in view of its tendency to lead to ventricular extension of the hemorrhage and hydrocephalus. In these instances, the association with hydrocephalus can worsen the overall neurological impairment by further decreasing the level of alertness, along with potential accentuation of oculomotor abnormalities, especially Parinaud's syndrome (Chung et al., 1996; Weisberg, 1986). Management of hydrocephalus with urgent ventriculostomy can at times produce a dramatic reversal of the oculomotor and consciousness abnormalities (Waga et al., 1979).

Although direct hematoma evacuation is not recommended in thalamic hemorrhage, ventriculostomy should always be considered in patients with hydrocephalus. In addition, the current development of stereotactic techniques for evacuation of deep ICHs (see below), which minimize the surgical trauma to unaffected superficial tissues, may lead to consideration of direct evacuation of thalamic hematomas in the near future.

**Cerebellar Hemorrhage.**    Patients with cerebellar hemorrhages usually have a good prognosis if the hematoma is small, i.e., 1 to 2 cm in diameter as measured on CT, since they do not require surgical treatment and their functional prognosis is generally excellent (Little et al., 1978). However, larger hematomas and, especially, those associated with clinical signs of brainstem compression and CT signs of hydrocephalus and effacement of the quadrigeminal cistern (Taneda et al., 1987) often need surgical treatment, as their nonsurgical management almost invariably leads to a fatal outcome. Surgical treatment often involves urgent ventriculostomy for relief of hydrocephalus, followed by posterior fossa craniotomy and drainage of the cerebellar hematoma.

The surgical decision in patients with cerebellar hemorrhage is, however, often difficult, because patients may not show the clinical or CT features indicative of the need for surgery upon presentation, but they may develop later in the course, at times precipitously, not always allowing for timely surgical intervention (Fisher *et al.*, 1965; Ott *et al.*, 1974). Because of the notorious tendency of cerebellar hematomas to produce sudden clinical deterioration with signs of brainstem compression, or even sudden apnea, at times after a stable course under hospital observation, the tendency is to consider these patients for surgical treatment early in their course. This is particularly the case since it is well established that the preoperative level of consciousness is a major determinant of outcome (Table IV). Ideally, since surgical mortality is markedly lower in patients who are alert or drowsy preoperatively (17%), the surgical decision needs to be made at this stage, as preoperative deterioration to stupor or coma carries a marked increase in risk of a poor outcome (75%). The clinical signs of brainstem compression, and CT findings of effacement of the fourth ventricle, hydrocephalus, and quadrigeminal cistern obliteration are all taken into consideration in arriving at a decision for surgical treatment of cerebellar hemorrhage.

## Newer Surgical Approaches

New techniques are being developed for drainage of intracerebral hematomas with the least amount of associated surgical trauma. These include the use of stereotactic and endoscopic techniques for the evacuation of deep hematomas.

### Stereotactic Drainage of Intracerebral Hematomas

This technique was pioneered by Backlund and Holst (1978), who introduced a new stereotactic instrument for the evacuation of ICHs, later used primarily by Japanese surgeons, who have used it to drain intracerebral hematomas under local anesthesia. The procedure includes either simple hematoma aspiration and irrigation (Tanikawa *et al.*, 1985; Tanizaki *et al.*, 1985), or aspiration after liquefying the hematoma with local instillations of a fibrinolytic agent (urokinase in the past, t-PA currently), under CT guidance (Matsumoto and Hondo, 1984; Niizuma *et al.*, 1985). The fibrinolytic agent most commonly used in the past was urokinase, as instillations of 6000 IU (in a 3–5 cc volume of saline) into the hematoma, followed by aspiration of its lique-

fied portions, a procedure that is repeated every 6 to 12 h through a silicone tube left in place in the cavity of the hematoma. Since the removal of urokinase from the U.S. market in 1999, the agent used for this purpose has been rt-PA.

Preliminary results with this stereotactic approach have been encouraging in patients with putaminal, thalamic, and lobar hematomas, in whom 50 to 80% of the hematoma can be gradually removed. Although experience with this technique is still limited, it shows promise by offering lower mortality and better functional outcome in comparison with conventional craniotomy for hematoma evacuation (Matsumoto and Hondo, 1984). Its main role may be in the surgical treatment of deeply located hematomas (thalamic, putaminal), as well as in patients who are poor surgical risks for craniotomy under general anesthesia.

A recent pilot series that assessed the feasibility and safety of stereotactic CT-guided aspiration and thrombolysis with urokinase in 12 patients was reported by Montes *et al.* (2000). Three patients (all with lobar hematomas) had full recovery at 3 months and a similar number of deaths were recorded (all in patients who had declined resuscitative measures), before hospital discharge. None of the patients with ganglionic hemorrhage were independent at 6 months. The authors also claim to have made the procedure less cumbersome by placing the catheter under real-time CT guidance, which avoided delays and the risks of general anesthesia and report the successful use of t-PA as the thrombolytic agent in three patients.

In another randomized feasibility study of surgical evacuation of ICH within 24 h, the four patients who underwent stereotactic surgery (three with putaminal and one with lobar hemorrhage) were independent at 3 months (Zuccarello *et al.*, 1999).

MRI-guided stereotactic aspiration of ICH has also been successfully performed. Tyler *et al.* (2000) resorted to this technique in 10 patients with thalamic or basal ganglionic hematomas who were operated on within 1 to 34 days after hemorrhage. In two cases local instillation of rt-PA was used. No complications or rebleeding were observed, and they reported improvement in BP control, language function, and cognition.

### Endoscopic Drainage of Intracerebral Hematomas

The technique of endoscopic drainage of ICHs was reported by Auer *et al.* in 1989. It involved the introduction of a "neuroendoscope" into the hematoma by an ultrasound-assisted stereotactic technique, with removal of the hematoma under visual control via a miniature video camera attached to the endoscope. Patients treated had putaminal, lobar, and thalamic hemorrhages.

In a meta-analysis (Prasad *et al.*, 1997) of the four randomized clinical trials comparing surgery vs best medical treatment, reanalysis of the data reported by Auer *et al.* found an 18% improvement in outcome. Furthermore, the analysis showed that open craniotomy was associated with a 13% rate of treatment failure (Prasad, 1997).

### Ultra-Early, Minimally Invasive, Thrombolytic-Assisted Evacuation of Intracerebral Hematomas

The 30-day mortality rate of ICH has consistently been shown to be around 45–50% in most series, with about half of the deaths typically occurring early in the course, usually within the first two days. The volume of the parenchymal hemorrhage, as calculated by CT scan using the ellipsoid method, is the strongest predictor of survival regardless of location, being more favorable in patients with volumes of <30 cc and rather dismal in those who harbor a hemorrhage >60 cc (Broderick *et al.*, 1993). Early enlargement of hematomas is a common occurrence that is associated with clinical deterioration and up to 40% of the patients may show an increase of more than one-third in the size of the hemorrhage during the first 24 h (Brott *et al.*, 1997; Morganstern *et al.*, 2001).

Development of brain edema has also been implicated in increasing mortality following an ICH. Among the factors that determine edema formation, the oncotic pressure exerted by extravasated plasma proteins seems to be a decisive one by promoting very early (within 1 h) perihematomal edema accumulation. Furthermore, animal models have shown the role of thrombin in increasing water content in gray matter within the first 24 h following an ICH and the concept of "blood toxicity" is gaining wide acceptance as it has been shown that the activation of the coagulation cascade with subsequent clot formation are necessary steps for the development of both immediate (1 h) and prolonged (24 h) perihematomal edema (Xi *et al.*, 1998). In this regard, a pilot trial is being conducted to evaluate the safety and feasibility of using argatroban, a selective thrombin inhibitor, in patients with ICH as a means of reducing edema formation. Such therapy has been tested in four patients and preliminary data suggest less disability with decreased perihematomal fluid accumulation (Hamada *et al.*, 2000).

There is accumulating evidence that a variety of factors might ultimately prove to be essential for a good outcome following surgical intervention for ICH. Among these, a minimally invasive surgical approach, either by the use of a neuroendoscope or stereotactically guided catheter aspiration, resulting in less trauma and tissue reaction, might prove to be beneficial. Indeed, these views are supported by the recent meta-analysis of the four randomized trials of surgery in ICH performed by Prasad *et al.* (1997).

Other factors that might impact surgical outcome include ultra-early (<4 h) clot aspiration and the concurrent use of thrombolytic agents to aid in clot liquefaction, facilitating its removal and decreasing the volume of residual clot. Besides the evidence discussed above, thrombolytic-assisted ultra-early aspiration of the clot is further supported by the fact that hematoma evacuation becomes more technically difficult with time, a phenomenon probably related to the loss of intrinsic plasminogen and stabilization of the fibrin molecule rendering the clot more resistant to proteolysis (Zeumer *et al.*, 1993). Also, while only 25–30% of the clot is successfully aspirated without thrombolytic treatment, removal of 60–70% has been achieved with the use of urokinase (Montes *et al.*, 2000). Finally, further support is derived from an animal study which demonstrated that removal of more than 70% of the hematoma volume was associated with a significant (77%) decrease in perihematoma edema with preservation of the blood–brain barrier in a porcine model of ultra-early hematoma aspiration following treatment with t-PA (Wagner *et al.*, 1999). The validity of this approach should be tested in a randomized clinical trial of early treatment of ICH by this technique in comparison with conventional craniotomy and best medical therapy.

## PRACTICAL MANAGEMENT

### Control of Bleeding

#### Treatment of Bleeding Diathesis

In patients on warfarin anticoagulants, anticoagulation can be reversed using vitamin K (5 to 25 mg subcutaneously or IV). Although higher doses are required when the subcutaneous route is used, this method of administration is safer since IV administration may be complicated by anaphylactic reactions (Raj *et al.*, 1999). Fresh frozen plasma acts more quickly. The usual dose range is 15 to 20 mL/kg at a rate of one unit every 45 to 60 min (Broderick *et al.*, 1999).

Protamine sulfate should be used for patients who bleed while being treated with IV heparin. The time elapsed since administration and the amount of heparin given are considered to calculate the total dose of protamine needed for reversal. If the bleeding occurs immediately while the infusion of IV heparin is ongoing, 1–1.5 mg of protamine should be used for every 100 units of heparin. If treatment with protamine sulfate is given after 30–60 min of heparin discontinuation, 0.5–0.75 mg for every 100 units of heparin previously infused should suffice; after 60–120 min of heparin discontinuation 0.375 to 0.5 mg/100 U of heparin is required, and finally 0.25 to 0.375 mg/100 U heparin is given if the length of heparin discontinuation is >120 min (Lacy, 1998). Serious side effects may be associated with the use of protamine sulfate, including hypotension, bradycardia, and hypersensitivity reactions, especially following rapid administration. Doses greater than 100 mg may cause paradoxical anticoagulation and hemorrhage.

### Control of Severe Hypertension

Levels of hypertension greater than 120 mm Hg diastolic or a mean arterial pressure (MAP) greater than 130 mm Hg should be treated in patients with acute hematomas (Broderick *et al.*, 1999; Diringer, 1993). In patients with an ICP monitor, the cerebral perfusion pressure should be kept at >70 mm Hg and pressors should be given if the systolic blood pressure drops below 90 mm Hg. We suggest using the α- and β-adrenergic blocker labetalol, which can be administered intravenously, produces minimal changes in heart rate and cardiac output, causes no rebound hypertension when discontinued, and does not produce tachyphylaxis (MacCarthy and Bloomfield, 1983). When given to patients with intracranial hemorrhage as an intravenous bolus of 5–25 mg, labetalol produces a rapid (within 5 to 15 min) drop in systolic and diastolic BP, in the order of 10 to 20% (Patel *et al.*, 1993). These doses are not associated with hypotension, changes in mental status, or worsening in clinical status. A common schedule of administration of labetalol is with 10 to 40 mg IV boluses every 10 min as needed for BP control, to a maximum of 160 mg, or as a continuous drip at a dose of 2–8 mg/min.

Hydralazine is another acceptable agent and is used as an oral dose of 50 to 100 mg twice a day. Other treatments for severe hypertension are esmolol (500 μg/kg as a loading dose followed by 50–200 μg/kg/min) and enalapril (0.625–1.2 mg Q 6 h on a prn basis). Finally, the availability of parenteral clonidine in some countries makes it yet another option, for either subcutaneous (0.075 mg) or intravenous use. The approaches to the treatment of hypertension in acute stroke in an American and a European center are listed in Table V. Note that the two institutions not only use different agents, but also different levels of hypertension to trigger therapeutic intervention, highlighting the point that strict rules based on scientific data are not available in regard to treatment of hypertension in acute stroke.

### Control of Increased ICP

Blood pressure control has been discussed above. For intubation in patients with decreased alertness, very

**TABLE V** Antihypertensive Treatment in Acute Stroke[a]

| | |
|---|---|
| **Cincinnati regimen** | |
| 1. Systolic BP 180–230 mm Hg, and/or diastolic BP <120 | Do not treat BP |
| 2. Diastolic BP >140 mm Hg, systolic BP slightly increased on repeated measurements 5 min apart | Sodium nitroprusside 2 repeated micragram/kg/min, may be with double dose after 3–5 min |
| 3. Systolic BP >230 mm Hg or diastolic BP 120–140 mm Hg or both on repeated measurements 20 min apart sublingual | a. Labetalol 10 mg IV may be repeated with double doses after 10 min as needed to maximum of 160 mg<br>b. Nifedipine 10 mg |
| **Heidelberg regimen** | |
| 1. Systolic BP <200 mm Hg, diastolic BP <120 mm Hg | Do not treat |
| 2. Diastolic BP >120 mm Hg, or systolic BP slightly increased on repeated measurements 15 min apart | a. Nitroglycerin, 5 mg IV or 10 mg PO<br>b. Sodium nitroprusside, rarely |
| 3. Systolic BP >220 mm Hg or diastolic BP 110–120 mm Hg, or both on repeated measurements 15 min apart | a. Nifedipine 10 mg sublingually<br>b. Clonidine 0.075 SC<br>c. Uripidil 12.5 mg IV |

[a]Reprinted from Brott *et al.* (1994), with permission from Springer-Verlag. BP, blood pressure; IV, intravenously; PO, orally; SC, subcutaneously.

short-acting barbiturates such as thiopental (1.0 to 1.5 mg/kg) or lidocaine (1 to 2 mg/kg) (Diringer, 1993) are used. Seizures increase ICP. In order to stop an ongoing generalized seizure, IV doses of 5 mg of diazepam are used, followed by IV loading with phenytoin at a dose of 15–18 mg/kg, given at a rate no faster than 50 mg/min, while monitoring the ECG for heart block (prolongation of P–R interval) and BP for hypotension. This results in therapeutic blood levels within 1 h of administration due to the rapid distribution of the drug after IV loading. A maintenance dose of 300 mg is generally adequate for the average-size adult, but precise dose titration depends on the measurement of phenytoin blood levels, aiming at a therapeutic level of 10–20 μg/mL. Fosphenytoin, a water-soluble formulation of phenytoin, is associated with less serious side effects and is safer and equally effective compared with phenytoin. In the event of seizure recurrence despite therapeutic doses of phenytoin, phenobarbital loading is required. This agent has the disadvantage of potentially affecting the level of alertness, impairing the ability to fully monitor the patient's neurological status. However, recurrent seizures and, especially, status epilepticus, may require the use of anesthetic doses of thiopental in cases of failure of the other two agents. Thiopental has the additional disadvantage of producing hypotension, requiring close monitoring of BP during its administration.

Since suctioning of secretions can increase ICP, hyperventilation with 100% oxygen before and after suctioning is recommended, along with the IV use of either lidocaine (1 mg/kg) or thiopental (0.5 to 1.0 mg/kg), in order to block increased ICP (Diringer, 1993).

Mannitol (20%) is often used as an osmotic diuretic in patients with space-taking hematomas at a dose of 0.25 to 0.5 g/kg every 4 h, with twice-a-day measurement of serum osmolality and usually not exceeding five days of therapy (Broderick *et al.*, 1999).

Intracranial pressure monitors are sometimes used to guide treatment.

## Surgery

Because hematomas represent so-called benign masses, the logical treatment for life-threatening lesions is surgical drainage. The following factors should be considered in deciding on surgical therapy: size, location, mass effect and drainage patterns, etiology, timing, and clinical course.

### Size

Hematomas over 3 cm in their widest diameter have a higher mortality and more delayed recovery than smaller lesions. Thus, the larger the lesion is on CT or MRI, the more logical would be surgical drainage.

### Location

Lobar and cerebellar hematomas are more accessible surgically. Although putaminal hemorrhages can be drained through the sylvian fissures and insular cortex, large left basal ganglionic hemorrhages usually leave patients aphasic and dependent, so treatment is less aggressive than for right-sided lesions. Cerebellar ICH can cause respiratory arrest without preceding gradual deterioration of neurological function or alertness, and surgical removal of a portion of the cerebellum often leaves no important residual handicap so that the threshold for recommending surgery for cerebellar hematomas is lower than for other lesions of comparable size. Cerebellar, lobar, and right putaminal hemorrhages are most accessible to surgical drainage.

### Mass Effect and Drainage Patterns

The size of the hematoma is not the only factor that determines mass effect. Older patients may have sufficient preexisting atrophy so that their cranial cavities can accommodate a sizable hematoma without a criti-

cal rise in ICP or shift in intracranial compartments. Some lesions have significant surrounding edema, while others have relatively little. Hydrocephalus can add to the increased mass effect. Compression of the third or lateral ventricle, shift of the midline, or uncal herniation favor surgical decompression. In posterior fossa ICH, displacement of the fourth ventricle and effacement of the ambient, cerebellopontine, and other cisterns favors surgery. Does the lesion drain? Entry into the ventricles or superficially into the subarachnoid space CSF may spontaneously decompress the lesion. Surgical drainage would be indicated more strongly for lesions with greater mass effect and no spontaneous decompression.

### Cause

Hematomas caused by amyloid angiopathy may tend to bleed even after surgical drainage because of the fragility of the blood vessels. Similarly, hemorrhages in patients with anticoagulant-related or other bleeding disorders will also continue to bleed unless the coagulopathy is reversed before surgery. When operating on hematomas caused by vascular malformations, ideally, surgeons would like to remove the malformations while also draining the hematoma. The threshold for surgical treatment is most favorable for vascular malformations, moderately so for accessible lesions caused by hypertension, and least favorable for CAA and for ICH due to a bleeding diathesis.

### Timing

During the first 24 to 36 h, hematomas are still at least partly liquid and can be more easily drained. However, very early (first 4–6 h) rebleeding is common. Later, hematomas solidify and become technically more difficult to drain. After 7 to 10 days, blood begins to be absorbed, and again the lesion becomes softer. Thus, ideally, for technical reasons, drainage should occur either early (6–24 h) or after 7 to 10 days. In general, if the patient has survived the first week, improvement occurs as edema subsides, so there seems little argument for late drainage except for concurrent removal of a vascular malformation. Some have wondered whether late surgery (1–2 weeks) would speed recovery, but this argument is unsupported by data.

### Clinical Course

The most important factor to consider is whether the patient is improving, stable, or worsening. Patients who deteriorate and show a decreased level of consciousness to severe lethargy or stupor have a poor outlook for recovery. In patients with putaminal hemorrhage, other poor prognostic signs include the development of ipsilateral pupillary dilation, ipsilateral extensor plantar response, or an ipsilateral conjugate-gaze paresis; these signs are indicative of midline shift or early brainstem compression. In patients with cerebellar hemorrhage, the development of bilateral extensor plantar responses is a poor prognostic sign. In deteriorating patients with accessible lesions, surgery should not be delayed if medical decompression is not quickly beneficial.

The particular means of surgical decompression depends to a great extent on the neurosurgeon and the technology available. Stereotactic drainage, often with the additional use of thrombolytic agents to liquefy hematomas will undoubtedly increase in future decades.

## TREATMENTS NO LONGER RECOMMENDED

In the past, dexamethasone and other corticosteroids were often given to reduce brain edema around acute hemorrhages. Although corticosteroids have proven value in reducing cerebral edema, especially surrounding cerebral metastases (Patchell and Posner, 1985), their role in acute ICH is still unclear. Their widespread use has not been supported when they have been tested in controlled clinical trials. Recently published guidelines for the management of ICH by the American Heart Association discourage the use of steroids given the lack of proven efficacy and potentially serious side effects (Broderick et al., 1999).

## REFERENCES

Abbott, R. D., Yin, Y., Reed, D. M., and Yano, K. (1986). Risk of stroke in male cigarrete smokers. N. Engl. J. Med. 315, 717–720.

Ananthasubramaniam, K., Beattie, J. N., Rosman, H., Jayam, V., and Borzak, S. (2001). How safely and for how long can warfarin therapy be withheld in prosthetic heart valves patients hospitalized with a major hemorrhage? Chest 119, 478–484.

Auer, L. M., Deinsberger, W., Niederkorn, K., Gell, G., Kleinert, R., Schneider, G., Holzer, P., Bone, G., Mokry, M., Körner, E., Kleinert, G., and Hanusch, S. (1989). Endoscopic surgery versus medical treatment for spontaneous intracerebral hematoma: A randomized study. J. Neurosurg. 70, 530–535.

Aurell, M., and Hood, B. (1964). Cerebral hemorrhage in a population after a decade of active antihypertensive treatment. Acta Med. Scand. 176, 377–383.

Backlund, E.-O., and von Holst, H. (1978). Controlled subtotal evacuation of intracerebral hematomas by stereotactic technique. Surg. Neurol. 9, 99–101.

Bae, H., Jeong, D., Doh, J., Lee, K., Yun, I., and Byun, B. (1999). Recurrence of bleeding in patients with hypertensive intracerebral hemorrhage. Cerebrovasc. Dis. 9, 102–108.

Batjer, H. H., Reisch, J. S., Allen, B. C., Plaizier, L. J., and Su, C. J. (1990). Failure of surgery to improve outcome in hypertensive putaminal hemorrhage: A prospective randomized trial. Arch. Neurol. 47, 1103–1106.

Berger, S., Schurer, L., Hartl, R., Deisbock, T., Dautermann, C., Murr, R., Messmer, K., and Baethmann, A. (1994). 7.2% NaCl/10% Dextran 60 versus 20% mannitol for treatment of intracranial hypertension. Acta Neurochir. 60, 494–498.

Bogousslavsky, J., Van Melle, G., and Regli, F. (1988). The Lausanne Stroke Registry: Analysis of 1000 consecutive patients with first stroke. Stroke 19, 1083–1092.

Broderick, J. P., Adams, H. P., Barsan, W., Feinberg, W., Feldman, E., Grotta, J., Kase, C., Krieger, D., Mayberg, M., Tilley, B., Zabramski, J. M., and Zuccarello, M. (1999). Guidelines for the management of spontaneous intracerebral hemorrhage. A statement for healthcare professionals from a special writing group of the stroke council, American Heart Association. Stroke 30, 905–915.

Broderick, J., Brott, T., Tomsick, T., Tew, J., Duldner, J., and Huster, G. (1994). Management of intracerebral hemorrhage in a large metropolitan population. Neurosurgery 34, 882–887.

Broderick, J. P., Brott, T., Duldner, J. E., Tomsick, T., and Huster, G. (1993). Volume of Intracerebral Hemorrhage. A powerful and easy-to-use predictor of 30-day mortality. Stroke 24, 987–993.

Broderick, J. P., Brott, T. G., Tomsick, T., Barsan, W., and Spilker, J. (1990). Ultra-early evaluation of intracerebral hemorrhage. J. Neurosurg. 72, 195–199.

Broderick, J. P., Brott, T., Tomsick T., Huster, G., and Miller, R. (1992). The risk of subarachnoid and intracerebral hemorrhages in blacks as compared with whites. N. Engl. J. Med. 326, 733–736.

Brott, T., Broderick, J., Kothari, R., Barsan, W., Tomsick, T., Sauerbeck, L., Spilker, J., Duldner, J., and Khoury, J. (1997). Early hemorrhage growth in patients with intracerebral hemorrhage. Stroke 28, 1–5.

Brott, T., Fieschi, C., and Hacke, W. (1994). General therapy of acute ischemic stroke. In "Neurocritical Care" (W. Hacke, D. F. Hanley, K. M. Einhäupl, T. P. Bleck, M. N. Diringer, and A. Ropper, Eds.), p. 563. Springer-Verlag, Heidelberg.

Brott, T., Thalinger, K., and Hertzberg, V. (1986). Hypertension as a risk factor for spontaneous intracerebral hemorrhage. Stroke 17, 1078–1083.

Butler, A. C., and Tait, R. C. (1998). Restarting anticoagulation in prosthetic heart valve patients after intracranial hemorrhage: A 2-year follow-up. Br. J. Haem. 103, 1064–1066.

Caplan, L. R. (1994). General symptoms and signs. In "Intracerebral Hemorrhage" (C. S. Kase and L. R. Caplan, Eds.), pp. 31–43. Butterworth-Heinemann, Boston.

Caplan, L. R., and Goodwin, J. A. (1982). Lateral tegmental brainstem hemorrhages. Neurology 32, 252–260.

Chung, C.-S., Caplan, L. R., Han, W., Pessin, M. S., Lee, K.-H., and Kim, J.-M. (1996). Thalamic haemorrhage. Brain 119, 1973–1986.

Cuatico, W., Adib, S., and Gaston, P. (1965). Spontaneous intracerebral hematomas: A surgical appraisal. J. Neurosurg. 22, 569–575.

De Reuck, J., De Bleecker, J., and Reyntjens, K. (1989). Steroid treatment in primary intracerebral hemorrhage. Acta Neurol. Belg. 89, 7–11.

Diringer, M. N. (1993). Intracerebral hemorrhage: Pathophysiology and management. Crit. Care Med. 21, 1591–1603.

Diringer, M. N., Edwards, D. F., and Zazulia, A. R. (1998). Hydrocephalus: A previously unrecognized predictor of poor outcome from supratentorial intracerebral hemorrhage. Stroke 29, 1352–1357.

Donahue, R. P., Abbott, R. D., Reed, D. M., and Yano, K. (1986). Alcohol and hemorrhagic stroke: The Honolulu Heart Program. JAMA 255, 2311–2314.

Drury, I., Whisnant, J. P., and Garraway, W. M. (1984). Primary intracerebral hemorrhage: Impact of CT on incidence. Neurology 34, 653–657.

Dul, K., and Drayer, B. P. (1994). CT and MR imaging of intracerebral hemorrhage. In "Intracerebral Hemorrhage" (C. S. Kase and L. R. Caplan, Eds.), pp. 73–93. Butterworth-Heinemann, Boston.

Erhardtsen, E., Nony, P., Dechavanne, M. et al. (1998). The effect of recombinant factor VIIa (NovoSevn) in healthy volunteers receiving acenocoumarol to International Normalized Ratio above 2.0. Blood Coagul. Fibrinol. 9, 741–748.

Fieschi, C., Carolei, A., Fiorelli, M., Argentino, C., Bozzao, L., Fazio, C., Salvetti, M., and Bastianello, S. (1988). Changing prognosis of primary intracerebral hemorrhage: Results of a clinical and computed tomographic follow-up study of 104 patients. Stroke 19, 192–195.

Fisher, C. M. (1961). Clinical syndromes in cerebral hemorrhage. In "Pathogenesis and Treatment of Cerebrovascular Disease" (W. S. Fields, Ed.), pp. 318–342. Charles C. Thomas, Springfield, IL.

Fisher, C. M. (1971). Pathological observations in hypertensive cerebral hemorrhage. J. Neuropathol. Exp. Neurol. 30, 536–550.

Fisher, C. M., Picard, E. H., Polak, A., Dalal, P., and Ojemann, R. G. (1965). Acute hypertensive cerebellar hemorrhage: Diagnosis and surgical treatment. J. Nerv. Ment. Dis. 140, 38–57.

Fogelholm, R., Nuutila, M., and Vuorela, A.-L. (1992). Primary intracerebral hemorrhage in the Jyvaskyla region, Central Finland 1985–89: Incidence, case fatality rate, and functional outcome. J. Neurol. Neurosurg. Psychiatry 55, 546–552.

Franke, C. L., van Swieten, J. C., Algra, A., and van Gijn, J. (1992). Prognostic factors in patients with intracerebral haematoma. J. Neurol. Neurosurg. Psychiatry 55, 653–657.

Furlan, A. J., Whisnant, J. P., and Elveback, L. R. (1979). The decreasing incidence of primary intracerebral hemorrhage: A population study. Ann. Neurol. 5, 367–373.

Garde, A., Bohmer, G., Selden, B., Neiman, J. (1983). 100 cases of spontaneous intracerebral haematoma: Diagnosis, treatment and prognosis. Eur. Neurol. 22, 161–172.

Greenberg, M. K., Alter, M., Ashwal, S. et al. (1996). Practice Advisory: Thrombolytic therapy for acute ischemic stroke—Summary statement: Report of the quality Standards Subcommittee of the American Academy of Neurology. Neurology 47, 835–839.

Gregson, B. A., Mendelow, A. D., Fernandes, H., Pearson, A. J., and Siddique, M. S. (2000). Letters to the Editor: Surgery for intracerebral hemorrhage. Stroke 31, 791.

Hamada, R., and Matsuoka, H. (2000). Antithrombin therapy for intracranial hemorrhage. Stroke 31, 794–795. [Letter to the editor]

Hart, R. G., Boop, B. S., and Anderson, D. C. (1995). Oral anticoagulants and intracranial hemorrhage. Facts and hypotheses. Stroke 26, 1471–1477.

Hier, D. B., Davis, K. R., Richardson, E. P., and Mohr, J. P. (1977). Hypertensive putaminal hemorrhage. Ann. Neurol. 1, 152–159.

Holcroft, J. W., Vassar, M. J., Turner, J. E., Derlet, R. W., and Kramer, G. C. (1987). 3% NaC and 7.5% NaCl/Dextran 70 in the resuscitation of severely injured patients. Ann. Surg. 206, 279–287.

Jacobs, D., Blackburn, H., Higgins, M., Reed, D., Iso, H., McMillan, G., Neaton, J., Nelson, T., Potter, J., Rifkind, B., Rossouw, J., Shekelle, R., and Yusuf, S. (1992). Report of the conference on low blood cholesterol: Mortality associations. Circulation 86, 1046–1060.

James, H. E., Langfitt, T. W., Kumar, V. S., and Ghostine, S. Y. (1977). Treatment of intracranial hypertension: Analysis of 105 consecutive continuous recordings of intracranial pressure. Acta Neurochir. 36, 189–200.

Janny, P., Colnet, G., Georget, A.-M., and Chazal, J. (1978). Intracranial pressure with intracerebral hemorrhages. Surg. Neurol. 10, 371–375.

Juvela, S. (1995). Risk factors for impaired outcome after spontaneous intracerebral hemorrhage. Arch. Neurol. 52, 1193–1200.

Juvela, S., Heiskanen, O., Poranen, A., Valtonen, S., Kuurne, T., Kaste, M., and Troupp, H. (1989). The treatment of spontaneous intracerebral hemorrhage: A prospective randomized trial of surgical and conservative treatment. J. Neurosurg. 70, 755–758.

Kagan, A., Popper, J. S., and Rhoads, G. G. (1980). Factors related to stroke incidence in Hawaii Japanese men: The Honolulu Heart Study. Stroke 11, 14–21.

Kanaya, H., Yukawa, H., Itoh, Z., Kutsuzawa, H., Kagawa, M., Kanno, T., Kuwabara, T., Mizukami, M., Araki, G., and Irino, T. (1980). Grading and the indications for treatment in ICH of the basal ganglia (cooperative study in Japan). In "Spontaneous Intracerebral Haematomas: Advances in Diagnosis and Therapy" (H. W. Pia, et al., Eds.), pp. 268–274. Springer-Verlag, Heidelberg.

Kaneko, M., Koba, T., and Yokoyama, T. (1977). Early surgical treatment for hypertensive intracerebral hemorrhage. J. Neurosurg. 46, 579–583.

Kaneko, M., Tanaka, K., Shimada, T., Sato, K., and Uemura, K. (1983). Long-term evaluation of ultra-early operation for hypertensive intracerebral hemorrhage in 100 cases. J. Neurosurg. 58, 838–842.

Kase, C. S., and Crowell, R. M. (1994). Prognosis and treatment of patients with intracerebral hemorrhage. In "Intracerebral Hemorrhage" (C. S. Kase, and L. R. Caplan, Eds.), pp. 467–489. Butterworth-Heinemann, Boston.

Kase, C. S., Maulsby, G. O., and Mohr, J. P. (1980). Partial pontine hematomas. Neurology 30, 652–655.

Kase, C. S., Williams, J. P., Wyatt, D. A., and Mohr, J. P. (1982). Lobar intracerebral hematomas: Clinical and CT analysis of 22 cases. Neurology 32, 1146–1150.

Koehler, R. C., and Traystman, R. J. (1982). Bicarbonate ion modulation of cerebral blood flow during hypoxia and hypercapnia. Am. J. Physiol. 243, H33–H40.

Lacy, C. F., Armstrong, L. L., Ingrim, N. B., and Lance, L. L. (1998). Protamine sulfate. In "Drug Information Handbook," pp. 1069–1070. Lexi-Comp Inc, Ohio.

Lapchak, P. A., Chapman, D. F., and Zivin, J. A. (2000). Metalloproteinase inhibition reduces thrombolytic (tissue plasminogen activator)-induced hemorrhage after thromboembolic stroke. Stroke 31, 3034–3040.

Little, J. R., Tubman, D. E., and Ethier, R. (1978). Cerebellar hemorrhage in adults: Diagnosis by computerized tomography. J. Neurosurg. 48, 575–579.

Luessenhop, A. J., Shevlin, W. A., Ferrero, A. A., McCullough, D. C., and Barone, B. M. (1967). Surgical management of primary intracerebral hemorrhage. J. Neurosurg. 27, 419–427.

MacCarthy, E. P., and Bloomfield, S. S. (1983). Labetalol: A review of its pharmacology, pharmacokinetics, clinical uses and adverse effects. Pharmacotherapy 3, 193–219.

Makris, M., Greaves, M., Phillips, W., Kitchen, S., Rosendaal, F. R., and Preston, F. E. (1997). Emergency oral anticoagulant reversal: The relative efficacy of fresh frozen plasma and clotting factor concentrate on correction of the coagulopathy. Thromb. Haemost. 77, 477–480.

Marshall, L. F., Shapiro, H. M., Rauscher. A., and Kaufman, N. M. (1978). Pentobarbital therapy for intracranial hypertension in metabolic coma. Reye's syndrome. Crit. Care Med. 6, 1–5.

Marshall, L. F., Smith, R. W., Rauscher, A., and Shapiro, H. M. (1978). Mannitol dose requirements in brain-injured patients. J. Neurosurg. 48, 169–172.

Matsumoto, K., and Hondo, H. (1984). CT-guided stereotaxic evacuation of hypertensive intracerebral hematomas. J. Neurosurg. 61, 440–448.

McKissock, W., Richardson, A., and Taylor, J. (1961). Primary intracerebral haemorrhage: A controlled trial of surgical and conservative treatment in 180 unselected cases. *Lancet* **2**, 221–226.

Mendelow, A. D., Teasdale, G. M., Russel, T., Flood, J., Patterson, J., and Murray, G. D. (1985). Effect of mannitol on cerebral blood flow and cerebral perfusion pressure in human head injury. *J. Neurosurg.* **63**, 43–48.

Montes, J. M., Wong, J. H., Fayad, P. B., and Awad, I. A. (2000). Stereotactic computed tomography-guided aspiration and thrombolysis of intracerebral hematoma: Protocol and preliminary experience. *Stroke* **31**, 834–840.

Morganstern, L. B., Demchuk, A. M., Kim, D. H., Frankowski, R. F., and Grotta, J. C. (2001) Rebleeding leads to poor outcome in ultra-early craniotomy for intracerebral hemorrhage. *Neurology* **56**, 1294–1299.

Muizelaar, J. P., Wei, E. P., Kontos, H. E., and Becker, D. P. (1983). Mannitol causes compensatory cerebral vasoconstriction and vasodilation in response to blood viscosity changes. *J. Neurosurg.* **59**, 822–828.

Niizuma, H., Otsuki, T., Johkura, H., Nakazato, N., and Suzuki, J. (1985). CT-guided stereotactic aspiration of intracerebral hematoma: Result of a hematoma-lysis method using urokinase. *Appl. Neurophysiol.* **48**, 427–430.

Ott, K. H., Kase, C. S., Ojemann, R. G., and Mohr. J. P. (1974). Cerebellar hemorrhage: Diagnosis and treatment. A review of 56 cases. *Arch. Neurol.* **31**(3), 160–167.

Paillas, J. E., and Alliez, B. (1973). Surgical treatment of spontaneous intracerebral hemorrhage: Immediate and long-term results in 250 cases. *J. Neurosurg.* **39**, 145–151.

Papo, I., Janny, P., Caruselli, G., Colnet, G., and Luongo, A. (1979). Intracranial pressure time course in primary intracerebral hemorrhage. *Neurosurgery* **4**, 504–511.

Patchell, R. A., and Posner, J. B. (1985). Neurologic complications of systemic cancer. *Neurol. Clin.* **3**, 729–750.

Patel, R. V., Kertland, H. R., Jahns, B. E., Zarowitz, B. J., Mlynarek, M. E., and Fagan, S. C. (1993). Labetalol: Response and safety in critically ill hemorrhagic stroke patients. *Ann. Pharmacother.* **27**, 180–181.

Phan, T. G., Koh, M., and Wijdicks, E. F. M. (2000). Safety of discontinuation of anticoagulation in patients with intracranial hemorrhage at high thromboembolic risk. *Arch. Neurol.* **57**, 1710–1713.

Poungvarin, N., Bhoopat, W., Viriyavejakul, A., Rodprasert, P., Buranasiri, P., Sukondhabhant, S., Hensley, M. J., and Strom, B. L. (1987). Effects of dexamethasone in primary supratentorial intracerebral hemorrhage. *N. Engl. J. Med.* **316**, 1229–1233.

Prasad, K., Browman G., Srivastava, A., and Menon, G. (1997). Surgery in primary supratentorial intracerebral hematoma: A meta-analysis of randomized trials. *Acta Neurol. Scand.* **95**, 103–110.

Quereshi, A. I., Tuhrim, S., Broderick, J. P., Batjer, H. H., Hondo, H., and Hanley, D. F. (2001) Spontaneous intracerebral hemorrhage. *N. Engl. J. Med.* **344**, 1450–1460.

Raj, G., Kumar, R., and Mckinney, W. P. (1997). Time course of reversal of anticoagulation effect of warfarin by intravenous and subcutaneous phytonadione. *Arch. Intern. Med.* **1999**, 2721–2724.

Ropper, A. H. (1993). Treatment of intracranial hypertension. *In* "Neurological and Neurosurgical Intensive Care" (A. H. Ropper, Ed.), 3rd ed., pp. 29–52. Raven Press, New York.

Ropper, A. H., and Davis, K. R. (1980). Lobar cerebral hemorrhages: Acute clinical syndromes in 26 cases. *Ann. Neurol.* **8**, 141–147.

Ropper, A. H., and King, R, B. (1984). Intracranial pressure monitoring in comatose patients with cerebral hemorrhage. *Arch. Neurol.* **41**, 725–728.

Rosand, J., Hylek, E. M., O'Donnell, H. C., and Greenberg, S. M. (2000). Warfarin-associated hemorrhage and cerebral amyloid angiopathy. A genetic and pathologic study. *Neurology* **55**, 947–951.

Rowe, C. C., Donnan, G. A., and Bladin, P. F. (1988). Intracerebral hemorrhage: Incidence and use of computed tomography. *Br. Med. J.* **297**, 1177–1178.

Shimamoto, T., Komachi, Y., Inada, H., Doi, M., Iso, H., Sato, S., Kitamura, A., Iida, M., Konishi, M., Nakanishi, N., Terao, A., Naito, Y., and Kojima, S. (1989). Trends for coronary heart disease and stroke and their risk factors in Japan. *Circulation* **79**, 503–515.

Silver, F. L., Norris, J. W., Lewis, A. J., and Hachinski, V. C. (1984). Early mortality following stroke: A prospective review. *Stroke* **15**, 492–496.

Stein, R. W., Kase, C. S., Hier, D. B., Caplan, L. R., Mohr, J. P., Hemmati, M., and Henderson, K. (1984). Caudate hemorrhage. *Neurology* **34**, 1549–1554.

Steiner, I., Gomori, J. M., and Melamed, E. (1984). The prognostic value of the CT scan in conservatively treated patients with intracerebral hematoma. *Stroke* **15**, 279–282.

Tanaka, H., Ueda, Y., Date, C., Baba, T., Yamashita, H., Hayashi, M., Shoji, H., Owada, K., Baba, K. I., Shibuya, M., Kon, T., and Detels, R. (1981). Incidence of stroke in Shibata, Japan: 1976–1978. *Stroke* **12**, 460–466.

Taneda, M., Hayakawa, T., and Mogami, H. (1987). Primary cerebellar hemorrhage: Quadrigeminal cistern obliteration on CT scans as a predictor of outcome. *J. Neurosurg.* **67**, 545–552.

Tanikawa, T., Amano, K., Kawamura, H., Kawabatake, H., Notani, M., Iseki, H., Shiwaku, T., Nagao, T., Iwata, Y., Taira, T., Umezawa, Y., Shimizu, T., and Kitamura, K. (1985). CT-guided stereotactic surgery for evacuation of hypertensive intracerebral hematoma. *Appl. Neurophysiol.* **48**, 431–439.

Tanizaki, Y., Sugita, K., Toriyama, T., and Hokama, M. (1985). New CT-guided stereotactic apparatus and clinical experience with intracerebral hematomas. *Appl. Neurophysiol.* **48**, 11–17.

Terayama, Y., Tanahashi, N., Fukuuchi, Y., and Gotoh, F. (1997). Prognostic value of admission blood pressure in patients with intracerebral hemorrhage. *Stroke* **28**, 1185–1188.

Tuhrim, S., Dambrosia, J. M., Price, T. R., Mohr, J. P., Wolf, P. A., Heyman, A., and Kase, C. S. (1988). Prediction of intracerebral hemorrhage survival. *Ann. Neurol.* **24**, 258–263.

Tyler, D., and Mandybur, G. (1999). Interventional MRI-guided stereotactic aspiration of acute/subacute intracerebral hematomas. *Stereotact. Funct. Neurosurg.* **72**, 129–135.

Ueda, K., Hasuo, Y., Kiyohara, Y., Wada, J., Kawano, H., Kato, I., Fujii, I., Yanai, T., Omae, T., and Fujishima, M. (1988). Intracerebral hemorrhage in a Japanese community, Hisayama: Incidence, changing pattern during long-term follow-up, and related factors. *Stroke* **19**, 48–52.

Ueshima, H., Iida, M., Shimamoto, T., Konishi, M., Tsujioka, K., Tanigaki, M., Nakanishi, N., Ozawa, H., Kojima, S., and Komachi, Y. (1980). Multivariate analysis of risk factor for stroke: Eight year follow-up study of farming villages in Akita, Japan. *Prev. Med.* **9**, 722–740.

Vermeulen, M., and van Gijn, J. (1990). The diagnosis of subarachnoid haemorrhage. *J. Neurol. Neurosurg. Psychiatry* **53**, 365–372.

Volpin, L., Cervellini, P., Colombo, F., Zanusso, M., and Benedetti, A. (1984). Spontaneous intracerebral hematomas: A new proposal about the usefulness and limits of surgical treatment. *Neurosurgery* **15**, 663–666.

Waga, S., Okada, M., and Yamamoto, Y. (1979). Reversibility of Parinaud syndrome in thalamic hemorrhage. *Neurology* **29**, 407–409.

Wagner, K. R., Xi, G., Hua, Y., Zuccarello, M., DE Courten-Myers, G. M., Broderick, J. P., and Brott, T. G. (1999). Ultraearly clot aspiration after lysis with tissue plasminogen activator in a porcine model of intracerebral hemorrhage: Edema reduction and blood–brain barrier protection. *J. Neurosurg.* **90**, 491–498.

Wakai, S., Kumakura, N., and Nagai, M. (1992). Lobar intracerebral hemorrhage: A clinical, radiographic, and pathological study of 29 consecutive operated cases with negative angiography. *J. Neurosurg.* **76**, 231–238.

Weisberg, L. A. (1985). Subcortical lobar intracerebral haemorrhage: Clinical-computed tomographic correlations. *J. Neurol. Neurosurg. Psychiatry* **48**, 1078–1084.

Weisberg, L. A. (1986). Thalamic hemorrhage: Clinical-CT correlations. *Neurology* **36**, 1382–1386.

Wolf, P. A. (1994). Epidemiology of intracerebral hemorrhage. *In* "Intracerebral Hemorrhage" (C. S. Kase, and L. R. Caplan, Eds.), pp. 21–30. Butterworth-Heinemann.

Xi, G., Wagner, K. R., Keep, R. F., Hua, Y., de Courten-Myers, G. M., Broderick, J. P., Brott, T. G., and Hoff, J. T. (1998). Role of blood clot formation on early edema development after experimental intracerebral hemorrhage. *Stroke* **29**, 2580–2586.

Zeumer, H., Freitag, H. J., and Thie, A. (1993). Local intra-arterial fibrinilytic therapy in patients with stroke: Uokinase versus recombinant tissue plasminogen activator. *Neuroradiology* **35**, 159–162.

Zuccarello, M., Brott, T., Derex, L., Kothari, R., Sauerbeck, L., Tew, J., Van Loveren, H., Yeh, H., Tomsick, T., Pancioli, A., Khoury, J., and Broderick, J. P. (1999). Early Surgical Treatment for Supratentorial Intracerebral Hemorrhage. A randomized Feasibility Study. *Stroke* **30**, 1833–1839.

*Cerebrovascular Disorders*

## CHAPTER 36
# Subarachnoid Hemorrhage

J. van Gijn

## CLINICAL ASPECTS

The clinical hallmark of subarachnoid hemorrhage (SAH) is a history of sudden, unusually severe headache. A period of unresponsiveness longer than 1 h occurs in almost half the patients, and focal signs develop at the same time of the headache or soon afterward in one-third of patients (Hop *et al.*, 1999; Linn *et al.*, 1998). In patients in whom headache is the only symptom, it may be difficult to recognize the seriousness of the underlying condition. Only half the patients with aneurysmal rupture describe the onset as instantaneous; the other half describe it as coming on in seconds to even a few minutes (Linn *et al.*, 1998). Another problem is that in the group of patients whose headache came on within a split second, innocuous forms of headache outnumber subarachnoid hemorrhage by 10 to 1 (Linn *et al.*, 1994). Vomiting occurs in 70% of patients with aneurysmal rupture but also in 43% of patients with innocuous thunderclap headache, and preceding bouts of similar headaches are recalled in 20% of patients with aneurysmal rupture and 15% of patients with innocuous thunderclap headache (Linn *et al.*, 1998). Neck stiffness is a common sign in SAH of any cause but it takes hours to develop and therefore it cannot be used to exclude the diagnosis if a patient is seen soon after the onset of the headache; it does not occur if patients are in deep coma. Subhyaloid hemorrhages require experience with fundoscopy; they occur in approximately 17% of patients, at least of those who reach hospital alive (Frizzell *et al.*, 1997; Pfausler *et al.*, 1996).

The lack of clinical features that distinguish reliably and at an early stage between SAH and innocuous types of sudden headache necessitates a brief consultation in a hospital for all patients with an episode of severe headache that comes on within minutes. Such an approach serves the patient's best interests and is also cost-effective. The discomfort and cost of referring the 90% of patients with innocuous headache is outweighed by avoidance of a misdiagnosis in the other 10% (Tolias *et al.*, 1996).

It is even more difficult to suspect aneurysmal rupture if other symptoms prevail over the headache, such as in patients who are initially unresponsive, or in patients presenting with a seizure or a confusional state, or if there is an associated head trauma. Epileptic seizures at the onset of aneurysmal SAH occur in approximately 6–17% of patients (Hart *et al.*, 1981; Pinto *et al.*, 1996; Rhoney *et al.*, 2000; Sarner and Rose, 1967). Of course the majority of patients with *de novo* epilepsy above age 25 will have underlying conditions other than subarachnoid hemorrhage, but the diagnosis should be suspected if the postictal headache is unusually severe. One to two percent of patients with subarachnoid hemorrhage present with an acute confusional state, and in most such patients a history of sudden headache is lacking (Reijneveld *et al.*, 2000). The differential diagnosis of acute confusional state is extensive, and subarachnoid hemorrhage accounts for at most a few percent of all patients seen in an emergency ward because of an acute confusional state (Benbadis *et al.*, 1994).

## NATURAL COURSE

### Epidemiological Aspects (for Aneurysmal Subarachnoid Hemorrhage)

The incidence of subarachnoid hemorrhage has remained stable over the past three decades. In a meta-analysis of relevant studies, the pooled incidence rate was 10.5 per 100,000 person-years (Linn *et al.*, 1996). There seems to be a decline across time, but this is caused by diagnostic bias. That more recent studies report lower incidence rates than older studies can be entirely explained by the increasing proportion of patients investigated with CT scanning. In a virtual study where CT is applied to all patients, the incidence is calculated to be 5.6 per 100,000 patient-years (Linn et al., 1996); this is only slightly lower than the incidence of 6.9 published later for a study spanning a 30-year period of the population in Olmsted, Minnesota (Menghini et al., 1998). The average age of patients with subarachnoid hemorrhage is substantially lower than for other types of stroke, peaking in the sixth decade (Lanzino et al., 1996; Longstreth et al., 1993).

Gender, race, and region have a marked influence on the incidence of subarachnoid hemorrhage. Women have a 1.6 times (95% confidence interval [CI]: 1.5 B 2.3) higher risk than men (Linn et al., 1996), and black people a 2.1 times (95% CI 1.3–3.6) higher risk than whites (Broderick et al., 1992). In Finland and Japan the incidence rates are much higher than in other parts of the world.

Genetic factors may predispose to subarachnoid hemorrhage. Between 5 and 20% of patients with subarachnoid hemorrhage have a positive family history (Schievink, 1997). First-degree relatives of patients with subarachnoid hemorrhage have a threefold to sevenfold increased risk of being struck by the same disease (Bromberg et al., 1995; De Braekeleer et al., 1996; Gaist et al., 2000; Schievink et al., 1995; Wang et al., 1995). In second-degree relatives the incidence of subarachnoid hemorrhage is similar to that in the general population (Bromberg et al., 1995).

The occurrence of subarachnoid hemorrhage is also associated with specific heritable disorders of connective tissue, but these patients account for only a minority of all patients with subarachnoid hemorrhage. Even though autosomal dominant polycystic kidney disease (ADPKD) is the most common heritable disorder associated with subarachnoid hemorrhage, it is found in only 2% of all patients with subarachnoid hemorrhage (Schievink et al., 1992). Other genetically determined disorders that have been associated with subarachnoid hemorrhage are Ehlers-Danlos disease IV, and neurofibromatosis type 1, but these associations are weaker than between ADPKD and aneurysms, and they are seldom found in patients with subarachnoid hemorrhage (Pepin et al., 2000; Schievink et al., 1994).

Of modifiable risk factors for subarachnoid hemorrhage, only smoking, hypertension, and heavy drinking emerged as significant risk factors in a systematic review of 8 longitudinal and 10 case–control studies, with odds ratios in the order of two or three (Teunissen et al., 1996). For the use of oral contraceptives the risk was significantly increased in another meta-analysis published (relative risk 1.42; 95%CI 1.12 B 1.80) (Johnston et al., 1998a); the risks were not clear for hormone

replacement therapy or an increased level of plasma cholesterol (Teunissen et al., 1996).

On a population basis, attributable risks have been estimated as follows: drinking alcohol 100 to 299 g/week accounts for 11% of the cases of SAH, drinking alcohol 300 g/wk or more for 21%, and smoking for 20%. An additional 17% can be attributed to hypertension, 11% to a positive family history for SAH, and 0.3% to ADPKD (Ruigrok et al., 2001).

### Outcome

Case fatality ranges between 32 and 67% in population-based studies from 1960 onward that reported on it. The weighted average is 51%. Of patients who survive the hemorrhage, approximately one-third remain dependent (Hop et al., 1997). Recovery to an independent state does not necessarily mean that outcome is good. In a study on quality of life in patients after subarachnoid hemorrhage, only 9 of 48 (19%; 95% CI 9 to 33%) patients who were independent 4 months after the hemorrhage had no significant reduction in quality of life (Hop et al., 1998a). One year to 18 months after the hemorrhage patients may have considerably improved in functional terms, but in at least half of the survivors the quality of life is still reduced (Hackett et al., 2000; Hop et al., 2001). The improvement in the first year and a half shows that long-term follow up is essential in studies on effectiveness of new treatment strategies on functional outcome after subarachnoid hemorrhage. All in all, only a small minority of all patients with subarachnoid hemorrhage have a truly good outcome. The relatively young age at which subarachnoid hemorrhage occurs and the poor outcome together explain why the loss of years of potential life before age 65 from subarachnoid hemorrhage is comparable with that of ischemic stroke (Johnston et al., 1998b).

## CAUSES OF SUBARACHNOID HEMORRHAGE

The great majority (85%) of subarachnoid hemorrhages is attributable to saccular aneurysms (Kassell et al., 1990; van Gijn et al., 1980a; Velthuis et al., 1998). Two thirds of the remainder (10% of the total) are caused by nonaneurysmal subarachnoid hemorrhage, and the remaining 5% by a variety of rare conditions (Table I).

### Saccular Aneurysms

Saccular aneurysms are not congenital, but develop during the course of life. They almost never occur

**TABLE I**  Causes of Subarachnoid Hemorrhage

| Cause | Frequency | Site of blood on CT | Characteristic features |
|---|---|---|---|
| Ruptured aneurysm | 85% | Basal cisterns or none | |
| Nonaneurysmal perimesencephalic hemorrhage | 10% | Basal cisterns | Pattern of hemorrhage on CT |
| Rare conditions | 5% | | |
| — Arterial dissection (transmural) | | Basal cisterns | Preceding neck trauma or pain; lower cranial nerve palsy |
| — Cerebral arteriovenous malformation | | Superficial | Vascular lesion often visible on CT |
| — Dural arteriovenous fistula | | Basal cisterns | History of skull fracture |
| — Vascular lesions around the spinal cord | | Basal cisterns or none | Pain in lower part of neck or in back; radicular pain or cord deficit |
| — Septic aneurysm | | Usually superficial | History; preceding fever or malaise |
| — Pituitary apoplexy | | Usually none | Visual or oculomotor deficits; adenoma on CT |
| — Cocaine abuse | | Basal cisterns or superficial | History |
| — Trauma (without visible brain contusion) | | Basal cisterns or superficial | History |

in neonates and are only rarely found in children (Heiskanen, 1986). In those exceptional cases there is usually a specific underlying cause for the aneurysm, such as trauma, infection, or connective-tissue disorder (Ferry *et al.*, 1974; Stehbens, 1982). The frequency in which saccular aneurysms are found in the general population can be estimated only from postmortem studies and depends on the definition and the diligence with which the search for unruptured aneurysms has been performed. In a systematic overview of studies reporting the prevalence of intracranial aneurysms in patients studied for reasons other than subarachnoid hemorrhage, 23 studies were identified, totaling 56,304 patients; 6685 (12%) of these patients were from 15 angiography studies (Rinkel *et al.*, 1989). The prevalence was lowest in retrospective autopsy studies (0.4%) and highest in prospective angiography studies (6%), especially in patients with autosomal polycystic kidney disease, familial predisposition, or atherosclerosis.

A role of acquired changes in the arterial wall is likely because hypertension, smoking and alcohol abuse are risk factors for subarachnoid hemorrhage, in general (Teunissen *et al.*, 1996). It may well be the influence of these factors that leads to local thickening of the intimal layer ("intimal pads") in the arterial wall, distal and proximal to a branching site, changes that some investigators regard as the earliest stage in the formation of aneurysms. The formation of these pads, in which the intimal layer is inelastic, may cause increased strain in the more elastic portions of the vessel wall (Crompton, 1966). Also abnormalities of structural proteins of the extracellular matrix have been identified in the arterial wall at a distance from the aneurysm itself (Chyatte *et al.*, 1990).

Saccular aneurysms arise almost invariably at sites of arterial branching, usually at the base of the brain, either on the circle of Willis itself or at a nearby branching site. They occur most commonly on the anterior communicating artery, the middle cerebral artery (at the point where the main stem divides into its branches), and on the carotid artery (mostly at the origin of the posterior communicating artery, and sometimes at the origin of other branches or at the terminal bifurcation). Less common locations are the tip of the basilar artery, the origin of the posterior inferior cerebellar artery from the vertebral artery, the pericallosal artery, or nonbranching sites (Ogawa *et al.*, 2000).

Some neoplastic conditions may lead to the formation of aneurysms: cerebellar hemangioblastoma (Guzman *et al.*, 1999) or metastasis from systemic carcinoma (Gliemroth *et al.*, 1999). Iatrogenic causes include radiation therapy (Jensen *et al.*, 1997), acrylate applied externally for microvascular decompression (Tokuda *et al.*, 1998), and operation for a superficial temporal artery–middle cerebral artery bypass, with the aneurysm at the site of the anastomosis (Sasaki *et al.*, 1996).

## Nonaneurysmal Perimesencephalic Hemorrhage

Perimesencephalic hemorrhage constitutes approximately 10% of all episodes of subarachnoid hemorrhage and two-thirds of those with a normal angiogram (Farrés *et al.*, 1992; Ferbert *et al.*, 1992; Kitahara *et al.*, 1993; Pinto *et al.*, 1993; van Gijn *et al.*, 1985a; Vermeer *et al.*, 1997). In this radiologically distinct and strikingly harmless variety of subarachnoid hemorrhage, the extravasated blood is confined to the cisterns around the midbrain, and the center of the bleeding is immediately anterior to the midbrain (Rinkel *et al.*, 1991a; Schwartz *et al.*, 1996; van Gijn *et al.*, 1985a). In some cases, the only evidence of blood is found anterior to the pons (Zentner *et al.*, 1996). For this reason some have proposed the term "pretruncal hemorrhage" (Schievink *et al.*, 1997), but in other patients the blood is found mainly in the ambient cistern or only in the quadrigeminal cistern (Rinkel *et al.*, 1995; Schwartz and Mayer, 2000; van Gijn *et al.*, 1985a). There is no extension of the hemorrhage to the lateral Sylvian fissures or to the anterior part of the interhemispheric fissure. Some sedimentation of blood in the posterior horns of the lateral ventricles may occur, but frank intraventricular hemorrhage or extension of the hemorrhage into the brain parenchyma indicates arterial hemorrhage and rules out this particular condition (Rinkel *et al.*, 1991a). This disease entity is defined only by the characteristic distribution of the extravasated blood on brain CT, in combination with the absence of an underlying aneurysm.

Perimesencephalic hemorrhage can occur in any patient over the age of 20 years, but most patients are in their sixth decade, as with aneurysmal hemorrhage. A history of hypertension was obtained more often than expected in only a single study (Canhao *et al.*, 1999), and not in another (Rinkel *et al.*, 1991b).

Clinically, there is little to distinguish idiopathic perimesencephalic hemorrhage from aneurysmal hemorrhage. In one-third of the patients, strenuous activities immediately precede the onset of symptoms, a proportion similar to that found in aneurysmal hemorrhage (Linn *et al.*, 1998; van Gijn *et al.*, 1985a). The headache onset is often more gradual (minutes rather than seconds) than with aneurysmal hemorrhage (Linn *et al.*, 1998; van Gijn *et al.*, 1985a), but the predictive value of this feature is poor. Loss of consciousness and focal symptoms are exceptional, and then only transient; a seizure at onset virtually rules out the diagnosis (Linn *et al.*, 1998). On admission, all patients are in fact in perfect clinical condition, apart from their headache (van Gijn *et al.*, 1985a; Rinkel *et al.*, 1991b). Transient amnesia is found in about one-third, and is associated with enlargement of the temporal horns on the initial CT scan (Hop *et al.*, 1998b). Typically, the early course is uneventful: rebleeds and delayed cerebral ischemia simply do not occur. Approximately 20% of patients have enlarged lateral ventricles on their admission brain CT scan, associated with extravasation of blood in all perimesencephalic cisterns, which probably causes blockage of the cerebrospinal fluid (CSF) circulation at the tentorial hiatus (Rinkel *et al.*, 1992). Only few have symptoms from this ventricular dilatation, and even then an excellent outcome can be anticipated (Rinkel *et al.*, 1990a,b). The period of convalescence is short, and almost invariably patients are able to resume their previous work and other activities (Brilstra *et al.*, 1997; Rinkel *et al.*, 1990b). Rebleeds after the hospital period have not been documented thus far (Canhao *et al.*, 1995; Rinkel *et al.*, 1991b), and the quality of life in the long term is excellent (Brilstra *et al.*, 1997).

A perimesencephalic pattern of hemorrhage may occasionally (in 2.5–5% of cases) be caused by rupture of a posterior fossa aneurysm (Pinto *et al.*, 1993; Rinkel *et al.*, 1991a; Van Calenbergh *et al.*, 1993). The chance of finding an aneurysm in 5% of patients has to be weighed against the risks of complications from angiography imposed upon the remaining 95% of patients. In recent years, CTA has been studied as method to confirm or exclude the presence of an aneurysm in patients with a perimesencephalic pattern of hemorrhage on CT. In a prospectively collected series of 40 patients with either a perimesencephalic hemorrhage or a posterior circulation aneurysm in whom CTA and conventional angiography were performed, radiologists detected an aneurysm in 16 patients and no aneurysm in the remaining 24 patients. These findings were confirmed after reading the angiograms (Velthuis *et al.*, 1999a). A formal decision analysis based on these observations indicated that a strategy where CTA is performed and not followed by conventional angiography if negative results in a better utility than a strategy where CTA is followed by conventional angiography or if all patients are initially investigated by conventional angiography (Ruigrok *et al.*, 2000).

### Rare Causes of Subarachnoid Hemorrhage

Table I lists the unusual causes of subarachnoid hemorrhage that together constitute about 5% of all episodes, with some clinical and radiological features that may help to establish the diagnosis.

Trauma and spontaneous SAH are sometimes difficult to disentangle. Patients may be found alone after having been beaten up in a brawl or hit by a drunken driver who made away, without external wounds to indicate an accident, while a decreased level of consciousness or retrograde amnesia make it impossible to obtain a history. The presence of neck stiffness and subarachnoid blood on the CT scan will cause the patient to be evaluated for SAH. Conversely, patients may cause an accident while riding a bicycle or driving a car at time of the aneurysmal rupture. Meticulous reconstruction of traffic or sports accidents may therefore be rewarding, especially in patients with disproportionate headache or neck stiffness.

## INVESTIGATIONS

### Brain Scanning (CT, MRI)

If subarachnoid hemorrhage is suspected, CT scanning is the first-line investigation because of the characteristically hyperdense appearance of extravasated blood in the basal cisterns. The pattern of hemorrhage often suggests the location of an underlying aneurysm (van Gijn et al., 1980b), although with variable degrees of certainty (Van der Jagt et al., 1999). A false-positive diagnosis of subarachnoid hemorrhage on CT is possible in the presence of generalized brain edema, with or without brain death, which causes venous congestion in the subarachnoid space and in this way may mimic SAH (Avrahami et al., 1998). The CT scan should be carefully scrutinized because small amounts of subarachnoid blood may easily be overlooked. If after a thorough review no blood is found, aneurysmal subarachnoid hemorrhage cannot be excluded. Even if CT scanning is performed within 12 h after the hemorrhage and with a modern CT machine, studies are negative in about 2% of patients with subarachnoid hemorrhage (van der Wee et al., 1995).

Brain CT may also help in distinguishing primary SAH from traumatic brain injury, but the aneurysmal pattern of hemorrhage is not always immediately appreciated in patients admitted with trauma (Vos et al., 2000). If trauma is the cause of SAH, the blood is usually confined to the superficial sulci at the convexity of the brain, adjacent to a fracture or to an intracerebral contusion, which findings dispel any lingering concern about the possibility of a ruptured aneurysm. Nevertheless, patients with basal–frontal contusions may show a pattern of hemorrhage resembling that of a ruptured anterior communicating artery aneurysm (Sakas et al., 1995), and in patients with blood confined to the Sylvian fissure or ambient cistern it may also be difficult to distinguish trauma from aneurysmal rupture by the pattern of hemorrhage alone (Rinkel et al., 1993). In patients with direct trauma to the neck or with head injury associated with vigorous neck movement, the accident can immediately be followed by massive hemorrhage into the basal cisterns resulting from a tear or even a complete rupture of one of the arteries of the posterior circulation, which is often rapidly fatal (Dowling and Corry, 1988; Harland et al., 1983).

MR imaging with fluid attenuated inversion recovery (FLAIR) techniques demonstrates subarachnoid hemorrhage in the acute phase as reliably as CT (Noguchi et al., 1995), but MR is impracticable because the facilities are less readily available than CT scans, and restless patients cannot be studied unless anesthesia is given. After a few days, however, MR imaging is increasingly superior to CT in detecting extravasated blood, up to 40 days later (Noguchi et al., 1997; Ogawa et al., 1995). This makes MRI a unique method for identifying the site of the hemorrhage in patients with a negative CT scan but a positive lumbar puncture (see below), such as those who are not referred until 1 or 2 weeks after symptom onset (Renowden et al., 1994).

## Lumbar Puncture

Lumbar puncture is still an indispensable step in the exclusion of subarachnoid hemorrhage in patients with a convincing history and negative brain imaging. Lumbar puncture should not be carried out rashly, or without some background knowledge. A first rule is that at least 6 and preferably 12 h should have elapsed between the onset of headache and the spinal tap. The delay is essential, because if there are red cells in the CSF sufficient lysis will have taken place during that time for bilirubin and oxyhemoglobin to have formed (Vermeulen et al., 1990). The pigments give the CSF a yellow tinge after centrifugation (xanthochromia), a critical feature in the distinction from a traumatic tap; the pigments are invariably detectable until at least 2 weeks later (Vermeulen et al., 1989). The "three tube test" (a decrease in red cells in consecutive tubes) is notoriously unreliable, and a false-positive diagnosis of subarachnoid hemorrhage can be almost as invalidating as a missed one. Spinning down the blood-stained CSF should be done immediately, otherwise oxyhemoglobin will form in vitro. If the supernatant seems crystal clear, the specimen should be stored in darkness until absence of blood pigments is confirmed by spectrophotometry (Vermeulen and van Gijn, 1990). Although the sensitivity and specificity of spectrophotometry have not yet been confirmed in a series of patients with suspected SAH and a negative CT scan (Beetham et al., 1998), it is the best technique currently available.

Keeping patients in an emergency department or admitting them to hospital until 6–12 h after symptom onset may be a practical problem. Yet we see no alternative, until a scientifically sound method has been devised to distinguish a traumatic tap reliably from blood that was previously present. Even the smoothest puncture can end in a vein. Immediately proceeding with CT or MR angiography in all patients with blood-stained CSF is not a good idea, because a small (<5 mm) aneurysm may well be coincidental and should be left untreated. Conversely, a negative study may not take away all concerns that have been raised—not only with the patients themselves but also with advisors of insurance companies.

## The Search for a Ruptured Aneurysm: Is Catheter Angiography Still Necessary?

The gold standard for detecting aneurysms is conventional angiography, but this procedure can be time consuming and it is not innocuous. A systematic review of three prospective studies in which patients with SAH were distinguished form other indications for catheter angiography found a complication rate (transient or permanent) of 1.8% (Cloft *et al.*, 1999). At any rate, the aneurysm may rerupture during the procedure, in 1–2% overall (Hayakawa *et al.*, 1978; Koenig *et al.*, 1979; Saitoh *et al.*, 1995). The rupture rate in the 6-h period after angiography has been estimated at 5% (Saitoh *et al.* 1995), which is higher than the expected rate.

Other imaging modalities are MR angiography (MRA) and CT angiography (CTA). MRA is safe but less suitable in the acute stage, because then patients are often restless or need extensive monitoring (Anzalone *et al.*, 1995). A review of studies comparing MRA and intra-arterial angiography in patients with recent subarachnoid hemorrhage, under blinded-reader conditions, showed a sensitivity in the range of 69–100% for detecting at least one aneurysm per patient. For the detection of all aneurysms the sensitivity is 70–97%, with specificity in the range 75–100% (Wardlaw and White, 2000). Despite its limitations but thanks to its noninvasive nature, MRA is a feasible tool for detecting aneurysms in relatives of patients with subarachnoid hemorrhage (Kojima *et al.*, 1998; Raaymakers *et al.*, 1999; Ronkainen *et al.*, 1995). The technique is still evolving, with improvements such as 3D imaging of aneurysms (Metens *et al.*, 2000).

CT angiography is based on the technique of spiral CT. It can easily be obtained immediately after the noncontrast CT on which the diagnosis is first made. It is minimally invasive, because it does not require intra-arterial catheterization. Compared with MRA it involves radiation and it requires injection of iodine-based contrast, but it is much simpler to perform, especially in ill patients. After the data acquisition, which can be done within 1 min, postprocessing techniques are needed to produce an angiogram-like display. The most practical procedure for daily routine is cine review of the axial source images combined with maximum intensity projection (MIP) of a limited volume of interest (Velthuis *et al.*, 1997). In addition, MIP images derived from CTA can be rotated and studied on a computer screen at every conceivable angle, which is a great advantage over the limited views with conventional angiography.

The sensitivity of CTA (compared with catheter angiography) is 85–98%, in the same range as that of MRA (Alberico *et al.*, 1995; Hope *et al.*, 1996; Wardlaw and White, 2000). On the other hand, CTA sometimes detects aneurysms that were missed with conventional angiography (Hashimoto *et al.*, 2000). In a study in which CTA and conventional angiography were compared in 80 patients with subarachnoid hemorrhage, neurosurgeons rated CT angiograms as equal or superior to conventional angiography in 83% (95% CI 73–90%) of 87 aneurysms (Velthuis *et al.*, 1998). It is not surprising, therefore, that an increasing proportion of patients with a ruptured aneurysm is successfully operated with CTA as the only imaging method (Anderson *et al.*, 1999; Velthuis *et al.*, 1999b). There is no doubt that catheter angiography is on the way out for the pretreatment assessment of cerebral aneurysms, as the techniques of CTA and MRA are still improving and as neurosurgeons and interventional radiologists are growing familiar with it.

## PRINCIPLES OF TREATMENT AND PROGNOSTIC FEATURES (ANEURYSMAL HEMORRHAGE)

In the following sections it shall be assumed that the cause of SAH is an aneurysm, unless specifically indicated otherwise. General principles of nursing and medical care are outlined in Table II.

The three baseline variables most closely related to poor outcome in aneurysmal SAH are the neurological condition of the patient on admission, age, and the amount of subarachnoid blood on the in initial CT scan (Hijdra *et al.*, 1988; Kassell *et al.*, 1990). Of these three prognostic features the neurological condition of the patient on admission, particularly the level of consciousness, is the most important determinant (Hijdra *et al.*, 1988).

Several grading systems have been developed for this initial assessment, in most cases consisting of approximately five categories of severity, in hierarchical order. No single system has gained worldwide acceptance, but until the 1990s the most widely used scales were those of Hunt and Hess (1968), and of Botterell, either in the original version (Botterell *et al.*, 1956) or in a modified version (Nishioka, 1966). The constituent features of these grading systems are not only the level of consciousness, but also headache, neck stiffness, and focal neurological deficit. Unfortunately, these more or less traditional systems are neither valid nor reliable. Headache and neck stiffness are very poor predictors of outcome in their own right; the construction of these grading scales attributes equal weight to the presence of an impaired level of consciousness, a focal deficit, or both, the actual grade depending on the severity; both these features are classified in vague terms. In view of the overlapping and equivocal terminology it is not surprising that a formal study of observer variability demonstrated large inconsistencies when the same patients were grade by different physicians, on either the Hunt and Hess scale or the Nishioka–Botterell scale (Lindsay *et al.*, 1982).

Classification into a few levels of the sum score of the Glasgow Coma Scale, which consists of eye-opening, motor response, and verbal response (Teasdale and Jennett, 1974), proved more reliable than any of the

**TABLE II** Recommendations for Nursing and General Management of Patients with SAH

Nursing
- Continuous observation (Glasgow Coma Scale, temperature, ECG monitoring, pupils, any focal deficits)*

Nutrition

*Oral route*
- Only with intact cough and swallowing reflexes*
- Keep stools soft by adequate fluid intake and by restriction of milk content; if necessary add laxatives*

*Nasogastric tube*
- Deflate endotracheal cuff (if present) on insertion*
- Confirm proper placement by X-ray*
- Begin with samll test feeds of 5% dextrose*
- Prevent aspiration by feeding in sitting position and by checking gastric residue every hour*
- Tablets should be crushed and flushed down (phenytoin levels will not be adequate in conventional doses)*

Blood pressure
- Do not treat hypertension unless there is clinical or laboratory evidence of progressive end organ damage*

Fluids and electrolytes
- Intravenous line mandatory*
- Give at least 3 L/day (normal saline, 0.9%)*
- Insert an indwelling bladder catheter if patient is incontinent*
- Compensate for a negative fluid balance and for fever*
- Monitoring of electrolytes (and leukocyte count), at least every other day*

Pain
- Start with paracetamol and/or dextropropoxyphene; avoid aspirin*
- Midazolam can be used if pain is accompanied by anxiety (5 mg via infusion pump)*
- For severe pain, use codeine or, as a last resort, morphine*

Prevention of deep vein thrombosis and pulmonary embolism
- Apply compression stockings or intermittent compression by pneumatic devices***

Prevention of delayed cerebral ischemia
- Give intravenous nimodipine, 60 mg every 4 h**

---

*Retrospective or non-randomized study(ies), or empirical recommendation without scientific proof.
**One prospective randomized study.
***More than one prospective, randomized, placebo-controlled study.

previous systems used to classify the degree of wakefulness (Lindsay *et al.*, 1983). The prognostic value is not the same for all elements of the Glasgow Coma Scale; for example, that a patient is disoriented rather than alert has stronger implications for outcome than losing a point on the dimensions "best motor response" or "eye opening" (Hirai *et al.*, 1996). A committee of the World Federation of Neurological Surgeons (WFNS) has proposed a new grading scale of five levels, essentially based on the GCS, with focal deficit making up one extra level for patients with a GCS score of 14 or 13. In other words, the WFNS Scale takes account of the fact that a focal neurological deficit in patients with SAH rarely occurs with a normal level of consciousness and assumes that the presence or absence of such a deficit does not add much to the prognosis in patients with a GCS score of 12 or less (Drake *et al.*, 1988). No formal studies of the validity and reliability of the WFNS Scale have yet been undertaken, but at least its core is made up by the well-validated GCS.

It is often tacitly assumed that in patients with ruptured aneurysms the initial clinical condition is related only to the impact of the first hemorrhage. This is incorrect, as some complications such as early rebleeding or acute hydrocephalus can occur within hours of the original rupture. Particularly, the presence of acute hydrocephalus may be sadly overlooked if the telltale history of increasing drowsiness in the first few hours after the bleed is not properly interpreted (van Gijn *et al.*, 1985b). Regardless of clinical condition, patients should be investigated and should be treated according to the problems that are identified (see Poor Clinical Condition on Admission, below).

## PRACTICAL MANAGEMENT

### Poor Clinical Condition on Admission

A decreased level of consciousness, with the initial hemorrhage or after early rebleeding, may be caused by intracerebral hematoma, subdural hematoma or hydrocephalus. Only by exclusion it should be assumed that the cause is global brain damage as a result of high intracranial pressure and subsequent ischemia at the time of hemorrhage.

### Early Rebleeding

In the first few hours after the initial hemorrhage, up to 15% of patients have a sudden episode of clinical deterioration that suggests rebleeding (Fujii *et al.*, 1996; Hijdra *et al.*, 1987; Kassell and Torner, 1983; Ohkuma *et al.*, 2001). As such sudden episodes often occur before the first CT scan, or even before admission to the hospital, a definite diagnosis is difficult and the true frequency of rebleeding on the first day is inevitably underestimated.

A common question in this situation is whether patients with a rebleed should be resuscitated and artificially ventilated if respiratory arrest occurs: in the series of 39 patients with a CT-confirmed rebleed, 14 had initial respiratory abnormalities that called for assisted ventilation. Spontaneous respiration returned within 1 h in 8 of these 14 patients, and in 3 more between 1 and 24 h (Hijdra *et al.*, 1984). Whether or not the patient would regain spontaneous respiration could not be predicted from the anatomical site of hemorrhage on CT, the initial presence or absence of brainstem reflexes or the type of respiratory disorder (Hijdra *et al.*, 1984). Many patients with initial apnea who were

**TABLE III   Management of Rebleeding**

- In case of respiratory arrest, resuscitate and ventilate; within hours either spontaneous respiration will return or all other brain stem functions will be lost*
- Repeat CT scan
- Consider emergency clipping or coiling of aneurysm after recovery, as the majority will rebleed again, with a high case fatality; any intracerebral hematoma can be removed at the same time*

*Retrospective or non-randomized study(ies).

successfully resuscitated later died from subsequent complications, but survival without brain damage is possible even after respiratory arrest. After resuscitation, it will usually become clear within a matter of hours whether the patient will indeed survive the episode or whether dysfunction of the brainstem will persist (Table III).

### Intracerebral Hematoma

Intraparenchymal hematomas occur in up to 30% of patients with ruptured aneurysms (van Gijn and van Dongen, 1982). Not surprisingly, the average outcome is worse than in patients with purely subarachnoid blood (Hauerberg et al., 1994). When a large hematoma is the most likely cause of the poor condition on admission, immediate evacuation of the hematoma should be seriously considered (with simultaneous clipping of the aneurysm if it can be identified). In such cases the aneurysm will often have been demonstrated only by MR angiography of CT angiography. Surgical treatment may not only be lifesaving in patients with impending transtentorial herniation, particularly with temporal hematomas, but may even result in independent survival, according to uncontrolled reports (Brandt et al., 1987; Suzuki et al., 2000) and a small randomized study (Heiskanen et al., 1988).

### Acute Subdural Hematoma

An acute subdural hematoma is usually associated with recurrent aneurysmal rupture but can also occur with the initial hemorrhage. It may be life threatening and in those cases immediate evacuation is called for (O'Sullivan et al., 1994).

### Acute Hydrocephalus

Gradual obtundation within 24 h of hemorrhage, sometimes accompanied by slow pupillary responses to light and downward deviation of the eyes, is fairly characteristic for acute hydrocephalus (van Gijn et al., 1985b; Rinkel et al., 1990a). If the diagnosis is confirmed by CT this can be a reason for early ventricular drainage, although some patients improve spontaneously in the first 24 h (Hasan et al., 1989a).

Acute hydrocephalus with large amounts of intraventricular blood is often associated with a poor clinical condition from the outset. If such patients are left alone, more than 90% have a poor outcome (Roos et al., 1995). An indirect comparison of observational studies suggests that insertion of an external ventricular catheter is not very helpful in these patients, but that a strategy where such drainage is combined with fibrinolysis through the drain results in a good outcome in half the patients (Nieuwkamp et al., 2000). This needs to be contirmed in studies with concurrent, randomized controls. The same applies to the strategy of surgically removing a large clot from the fourth ventricle (Lagares et al., 2001).

### Global Cerebral Ischemia

Not all patients who arrive in a moribund state can be salvaged, because irreversible brain damage may have occurred immediately after aneurysm rupture. In a consecutive series of 31 patients who died on the first day, 9 had a potentially treatable supratentorial hematoma. Sixteen others showed dysfunction of the brainstem, associated with massive intraventricular hemorrhage on CT, including distension of the fourth ventricle with blood; in nine cases this occurred together with an intracerebral hematoma. In the six remaining patients, however, neither a supratentorial hematoma nor intraventricular hemorrhage could explain the progressive dysfunction of the brainstem and the fatal outcome, the CT scan showing no abnormality other than subarachnoid blood (Hijdra et al., 1982). The most likely explanation is a prolonged period of global cerebral ischemia at the time of hemorrhage, as a result of the pressure in the cerebrospinal fluid spaces being elevated to the level of that in the arteries, for as long as a few minutes. This is quite distinct from delayed ischemia, which is focal or multifocal (see below). Such an immediate and potentially lethal arrest of the circulation to the brain is indeed suggested by autopsy evidence and by the recording of intracranial pressure or transcranial Doppler sonography at the time of recurrent aneurysmal hemorrhage (Grote et al., 1988; Smith 1963).

### Prevention of Rebleeding

We mentioned above that early rebleeding, within hours of the initial hemorrhage, occurs in at least 15% of patients. At present it is virtually impossible to prevent this from happening, but medical or surgical intervention can prevent recurrent hemorrhages occurring later. In patients who survive the first day, the risk of rebleeding is more or less evenly distributed over the next 4 weeks, although there may be a second peak early in the third week (Hijdra *et al.*, 1987). Given that the proportion of patients who eventually rebled was 32% in a consecutive series of patients not treated with antifibrinolytic agents but in whom one-third of the patients had undergone aneurysm clipping around day 12, the total risk of rebleeding without medical or surgical intervention in the 4 weeks after the first day can be estimated at 35–40% (Hijdra *et al.*, 1987). Between 4 weeks and 6 months after the hemorrhage the risk of rebleeding gradually decreases, from the initial level of 1–2%/day to a constant level of about 3%/year (Winn *et al.*, 1977).

### Drug Treatment

Medical treatment for prevention of rebleeding has not yet been successful; treatment with antifibrinolytic agents does reduce the rebleed rate, but fails to improve overall outcome. A systematic review of antifibrinolytic agents included eight trials published before 2000 that met predefined inclusion criteria and totaled 937 patients (Roos *et al.*, 1999). By far the largest study was a Dutch–Scottish trial (Vermeulen *et al.*, 1984). In this meta-analysis antifibrinolytic treatment did not provide any evidence of benefit on outcome. The risk of rebleeding was significantly reduced by antifibrinolytic therapy, but this was offset by a similar increase of the risk of secondary cerebral ischemia. In other words, antifibrinolytic drugs work, but they do not help. All trials in this meta-analysis had been performed before the 1990s, at a time when prevention or treatment of secondary cerebral ischemia had yet to be developed. Therefore a new clinical trial on antifibrinolytic therapy was undertaken in which all 492 patients were maximally protected against ischemia by means of calcium antagonists and normovolemia. Tranexamic acid again significantly reduced the rate of rebleeding, but the overall outcome was not different between the two groups, mainly because of cerebral ischemia (Roos *et al.*, 2000).

A pilot study has been performed with activated factor VII to promote clot formation at the site of rupture, but the safety of this treatment has not yet been established (Pickard *et al.*, 2000).

### Operative Clipping of the Aneurysm

Surgical obliteration of the aneurysm has been the mainstay of treatment for decades. Until the 1980s this was deferred until day 10–12, because of the many complications with earlier operations. Since then, many neurosurgeons have adopted a policy of early clipping of the aneurysm, i.e., within 3 days of the initial bleed. The main rationale, of course, is optimal prevention of rebleeding. The theoretical advantages of early operation have not yet been proven by systematic studies, which is an uncomfortable reflection. In the only randomized trial of the timing of operation performed so far, 216 patients were allocated to operation within 3 days, after seven days or in the intermediate period (Öhman and Heiskanen, 1989). The outcome tended to be better after early than after intermediate or late operation, but as the difference was not statistically significant, a disadvantage could not be excluded. The same result—no difference in outcome after early or late operation—emerged from a review of observational studies (de Gans *et al.*, 2002). The U.S. study found the worst outcome in patients operated on between day 7 and day 10 after the initial hemorrhage. This disadvantageous period for performing the operation in the second week after SAH coincides with the peak time of cerebral ischemia (Hijdra *et al.*, 1986), and of cerebral vasospasm (Weir *et al.*, 1978), both phenomena being most common from day 4 to day 12.

### Endovascular Treatment ("Coiling")

Since the introduction of controlled detachable coils for the endosaccular packing of aneurysms (Guglielmi *et al.*, 1992), endovascular embolization is increasingly used. In some institutions endovascular embolization is even the preferred method of treatment (Cognard *et al.*, 1997). It is mostly performed under general anesthesia to ensure complete immobilization of the patient, but not always (Qureshi *et al.*, 2001).

Numerous observational studies have published complication rates, occlusion rates, and short-term follow-up results. These have been summarized, up to March 1997, in a systematic review of 48 eligible studies about 1383 patients; in 900 of these the aneurysm had ruptured (Brilstra *et al.*, 1999). Permanent complications of the procedure occurred in 3.7% of 1256 patients in whom this was recorded (95% CI, 2.7-4.9%). A greater than 90% occlusion of the aneurysm was achieved in almost 90% of patients. The most frequent complication was procedure-related ischemia, even if patients are treated with heparin; the second most frequent complication is aneurysm perforation, which occurs in 2–4% of patients (Levy *et al.*, 2001; Sluzewski *et al.*, 2001). Relatively many of the aneurysms treated with controlled detachable coils are located at the basilar artery (Tateshima *et al.*, 2000; Uda *et al.*, 2001). Other suitable sites are the carotid artery and the anterior communicating artery. Pericallosal arteries are difficult to reach (Menovsky *et al.*, 2002); these aneurysms constitute thus far only 2% of all aneurysms treated with controlled detachable coils. Another problematic site is the trifurcation of the middle cerebral artery (6% of all aneurysms treated with controlled detachable coils), because often one or more of the branches originate from the aneurysm. If the aneurysm is wide-necked or fusiform, endovascular occlusion can be combined with surgical remodeling (Hoh *et al.*, 2001), or with balloon-assisted remodeling (Aletich *et al.*, 2000; Malek *et al.*, 2000).

Indirect comparisons between endovascular and surgical treatment are inappropriate, if only because there are so many differences in study design, patients, and aneurysms. Moreover, recanalization may occur after a coiling procedure, especially if a portion of the neck has remained open (Hayakawa *et al.*, 2000; Thornton *et al.*, 2002). Even after apparently complete coiling, rerupture may occur (Manabe *et al.*, 1998), and the long-term rates of rebleeding after endovascular coiling still need to be established. A first report from a single center (Oxford, U.K.) in which more than 300 patients had been followed up after aneurysm embolization for a median period of almost 2 years, showed rebleeding rates of 0.8% in the first year, 0.6% in the second year, and 2.4% in the third year, with no rebleeding in subsequent years (Byrne *et al.*, 1999). On the other hand, it should not be assumed that surgical treatment is always definitive: in a retrospective review of postoperative angiograms in a series of 66 patients with ruptured aneurysms and 12 additional aneurysms, all treated by surgical clipping, 8% of patients showed aneurysms with a residual lumen or aneurysms that were previously undetected (Macdonald *et al.*, 1993).

An analysis in which the factor "confounding by indication" was eliminated as much as possible suggested that coiling provides a better outcome than surgical treatment (Johnston, 2000). Nevertheless, controlled trials are urgently needed in patients with aneurysms for which it is uncertain whether surgical clipping or endovascular coiling should be the preferred treatment. A first such study, although a small one (109 patients), found no difference in outcome at 3 or 12 months between the surgical group and the endovascular group (Koivisto *et al.*, 2000; Vanninen *et al.*, 1999).

## Prevention of Secondary Cerebral Ischemia

Delayed cerebral ischemia occurs mainly in the first or second week after aneurysmal subarachnoid hemorrhage, in up to one-third of patients, depending on case mix and timing of operation (Hijdra *et al.*, 1986). Despite many years of intensive research, the pathogenesis of secondary cerebral ischemia following subarachnoid hemorrhage has not been elucidated. It is a generally held belief that after the hemorrhage a thus-

far-unidentified factor is released in the subarachnoid space, which induces vasoconstriction and thereby secondary ischemia.

A much cited study from Boston (of only 41 patients in total) postulates a close relationship between the location of subarachnoid blood and the "thickness" of the clot on the one hand and the occurrence of vasospasm and delayed cerebral ischemia on the other (Kistler *et al.*, 1983). Several observations argue against this popular notion. First, the presence of subarachnoid blood, though a powerful predictor of delayed cerebral ischemia, is not in itself a sufficient factor for the development of secondary ischemia. After all, secondary ischemia does not occur in patients with a perimesencephalic (nonaneurysmal) subarachnoid hemorrhage (Rinkel *et al.*, 1991b), and it is rare in patients with subarachnoid hemorrhage secondary to intracerebral hematoma or a ruptured arteriovenous malformation. Second, in larger series of patients than in the Boston study the site of delayed cerebral ischemia does not correspond with the distribution or even the side of subarachnoid blood (Brouwers *et al.*, 1992; Hop *et al.*, 1996). The method of quantifying local amounts of subarachnoid blood in these later studies proved reliable between observers (Hijdra *et al.*, 1990), whereas the Boston method (canonized as the Fisher scale, after the last author), is associated with wide interobserver variation (Svensson *et al.*, 1996). Third, many patients with vasospasm never develop secondary ischemia. These observations collectively suggest that not only the presence of subarachnoid blood per se but rather the combination with other factors such as a ruptured artery determines whether and where secondary ischemia will develop.

Despite the lack of pathophysiological insight, some progress has been made in the prevention of secondary ischemia after aneurysmal SAH by changes in general medical care (notably increased fluid intake and avoidance of antihypertensive drugs) as well as by specific drug treatment. Transcranial Doppler sonography may suggest impending cerebral ischemia by means of the increased blood flow velocity from arterial narrowing in the middle cerebral artery or in the posterior circulation, but there is considerable overlap with patients who do not develop ischemia (Lysakowski *et al.*, 2001; Sloan *et al.*, 1989, 1994). One reason is that narrowing in distal branches of the middle cerebral artery often escapes detection (Okada *et al.*, 1999). In comparison with angiographic studies only velocities above 200 mL/min in the middle cerebral artery are reasonably accurate in indicating vasospasm, but the sensitiviy of TCD monotoring is low (Lysakowski *et al.*, 2001; Vora *et al.*, 1999). Even then, impaired autoregulation, which probably is a crucial factor in the development of delayed cerebral ischemia, usually precedes arterial narrowing

(Lam *et al.*, 2000; Lang *et al.*, 2001; Rätsep and Asser, 2001).

## Management of Blood Pressure

Whether hypertension should be controlled is a difficult issue in patients with SAH, especially if the blood pressure rises above 200/110 mm Hg. Following intracranial hemorrhage, the range between the upper and the lower limits of the autoregulation of cerebral blood flow becomes more narrow, which makes the perfusion of brain more dependent on arterial blood pressure (Kaneko *et al.*, 1983). Consequently, aggressive treatment of surges of blood pressure entails a definite risk of ischemia in areas with loss of autoregulation. The rationalistic approach is therefore to advise against treating hypertension following aneurysmal rupture. The empirical evidence for this advice is sparse, but tends to support the avoidance of antihypertensive drugs. In an American cooperative Study conducted between 1963 and 1970, 1005 patients with ruptured aneurysms were randomized between four treatment modalities; one arm consisted of drug-induced lowering of the blood pressure, another of bed rest alone (the other two arms were surgical: carotid ligation and intracranial surgery). In the intention-to-treat analysis, antihypertensive drugs failed to reduce either case fatality or the rate of rebleeding within the first 6 months after the initial event; on-treatment analysis suggested that induced hypotension did decrease the rate of rebleeding in comparison with bed rest, but not the case fatality (Torner *et al.*, 1981). It should be kept in mind, however, that the diagnosis of rebleeding had to be made in the pre-CT era and was therefore probably inaccurate. An observational study from the 1980s, in which all events had been documented by means of serial CT scanning, compared patients in whom hypertension had been newly treated with normotensive controls; the rate of rebleeding was lower but the rate of cerebral infraction was higher than in untreated patients, despite the blood pressures being, on average, still higher than in the controls (Wijdicks *et al.*, 1990).

All this suggests that hypertension after SAH is a compensatory phenomenon, at least to some extent and one that should not be interfered with. In keeping with this, a further observational study suggested that the combined strategy of avoiding antihypertensive medication and increasing fluid intake may decrease the risk of cerebral infarction (Hasan *et al.*, 1989a).

It seems best to reserve antihypertensive drugs (other than those the patients were on already) for patients with extreme elevations of blood pressure as well as evidence of rapidly progressive end organ damage. Such damage can be diagnosed from either clinical signs (e.g., new retinopathy, heart failure, etc.) or laboratory evidence (e.g., signs of left ventricular failure on chest X-ray, proteinuria, or oliguria with a rapid rise of creatinine levels).

## Management of Fluid Balance and Electrolytes

Fluid management in SAH is important to prevent a reduction in plasma volume, which may contribute to the development of cerebral ischemia. Nevertheless, the arguments for a liberal (some might say aggressive) regimen of fluid administration are indirect. In approximately one-third of the patients plasma volume drops by more than 10% within the preoperative period, which is significantly associated with a negative sodium balance; in other words, there is loss of sodium as well as of water (Hasan *et al.*, 1990; Wijdicks *et al.*, 1985a). Moreover, fluid restriction in patients with hyponatremia is associated with an increased risk of cerebral ischemia (Wijdicks *et al.*, 1985b); such a regimen was applied in the past because hyponatremia was erroneously attributed to water retention, via inappropriate secretion of antidiuretic hormone. Two nonrandomized studies with historical controls suggested that a daily intake of at least 3 L of saline (against 1.5–2.0 L in the past) is associated with a lower rate of delayed cerebral ischemia and a better overall outcome (Hasan *et al.*, 1989a; Vermeij *et al.*, 1998).

Despite the incomplete evidence, it seems reasonable to prevent hypovolemia. We favor giving 2.5–3.5 L/day of normal saline, unless contraindicated by signs of impending cardiac failure. Nevertheless, it appears that many patients need a daily fluid intake of 4–6 L (sometimes as much as 10 L) to balance the production of urine plus estimated insensible losses (via perspiration and expired air). Fluid requirements may be guided by recording of central venous pressure (directly measured value should be above 8 mm Hg) or pulmonary wedge pressures (to be kept above 7 mm Hg), but frequent calculation of fluid balance (four times per day until approximately day 10) is the main measure for estimating how much fluid should be given. Fluid intake should be increased proportionally in patients with fever, from whatever cause.

## Calcium Antagonists

Initially, the rationale for the use of calcium antagonists in the prevention or treatment of secondary ischemia was based on the assumption that these drugs reduce the frequency of vasospasm by counteracting the influx of calcium in the vascular smooth-muscle cell. This anti-vasospastic effect of calcium antagonists was confirmed by many *in vitro* studies with intracranial arteries and also by *in vivo* assessments of arterial lumen changes after experimental subarachnoid hemorrhage.

Clinical trials have been undertaken with three types of calcium antagonists: nimodipine, nicardipine and AT877, of which nimodipine is the most extensively studied and used. A recent systematic review of all randomized controlled trials on calcium antagonists in patients with subarachnoid hemorrhage, showed a significant reduction in frequency of poor outcome, which resulted from a reduction in the frequency of secondary ischemia (Feigin et al., 2000). When the different agents are analyzed separately, the nimodipine trials showed a significant reduction in the frequency of poor outcome, but the nicardipine and AT877 trials did not. On the other hand, nicardipine and AT877 significantly reduce the frequency of vasospasm, whereas the nimodipine trials showed only a trend toward reduction of vasospasm, despite a larger number of patients included. In brief, administration of nimodipine improves outcome in patients with subarachnoid hemorrhage, but it is uncertain whether nimodipine acts through neuroprotection, through reducing the frequency of vasospasm, or both. Nicardipine and AT877 definitely reduce the frequency of vasospasm, but the effect on overall outcome remains unproved, which again underlines the weak relation between vasospasm and outcome.

The practical implications are that the regimen employed in the dominant nimodipine trial (60 mg orally every 4 h, to be continued for 3 weeks) is currently regarded as the standard treatment in patients with aneurysmal subarachnoid hemorrhage. If the patient is unable to swallow, the tablets should be crushed and washed down a nasogastric tube with normal saline. Yet the entire evidence about efficacy and dosage of nimodipine hinges on a single, large clinical trial (Pickard et al., 1989). Because the results might be affected by unpublished negative trials, the benefits of nimodipine cannot be regarded as being beyond all reasonable doubt.

### Neuroprotective Drugs Other Than Calcium Antagonists

Tirilazad has been studied in four randomized, controlled trials, totaling more than 3500 patients (Haley et al., 1997; Kassell et al., 1996; Lanzino et al., 1999a,b). This drug belongs to the category of 21 amino steroids that inhibit iron-dependent lipid peroxidation. The only beneficial effect on overall outcome was seen in a single subgroup of a single trial, i.e., those treated with 6 mg/kg/day (two other groups received 0.2 mg/kg/day or 2 mg/kg/day) (Kassell et al., 1996). This possible benefit could not be reproduced in the corresponding subgroup from a parallel trial (Haley et al., 1997), nor in two further trials with an even higher dose (15 mg/kg/day) in women (Lanzion et al., 1999b); the

gender distinction was made because in the first two trials women had seemed to respond less than men to tirilazad mesylate.

A single trial with another hydroxyl radical scavenger, N'-propylenedinico-tinamide (nicaraven) in 162 patients showed a decreased rate of delayed cerebral ischemia but not of poor outcome at 3 months after SAH (Asano et al., 1996). Curiously enough, the reverse was found a trial of 286 patients with ebselen, a selenoorganic compound with antioxidant activity through a glutathione peroxidase-like action: improved outcome at 3 months after SAH, but without any reduction in the frequency of delayed ischemia (Saito et al., 1998).

### Aspirin and Other Antiplatelet Agents

Several studies have found that blood platelets are activated from day 3 after subarachnoid hemorrhage, mostly through increased levels of thromboxane $B_2$, the stable metabolite of thromboxane $A_2$, a substance which promotes platelet aggregation and vasoconstriction (Juvela et al., 1990; Ohkuma et al., 1991; Vinge et al., 1988). The practical question is whether interventions aimed at counteracting platelet activation are therapeutically useful. A retrospective analysis of 242 patients who had survived the first 4 days after SAH showed that patients who had used salicylates before their hemorrhage (as detected by history and urine screening) had a significantly decreased risk of delayed cerebral ischemia, with or without permanent deficits (relative risk 0.04, 95% CI: 0.18–0.93) (Juvela, 1995). As early as in 1982 a first clinical trial was done, which failed to show benefit from aspirin (Mendelow et al., 1982), but the number of patients was small (53), unoperated patients were also included, and all were treated with tranexamic acid, which increases the risk of ischemia (see above). There is a need for a prospective and randomized study of salicylates or other antiplatelet drugs as a preventive measure against delayed cerebral ischemia, preferably after clipping of the aneurysm to avoid rebleeding being precipitated by the antiplatelet and so antihemostatic action. A pilot study of aspirin after early operation in 50 patients has shown that this treatment is feasible and probably safe (Hop et al., 2000).

Four antiplatelet agents other than aspirin have been tested in separate trials of patients with subarachnoid hemorrhage: dipyridamole (100 mg/day orally or 10 mg/day intravenously), in 320 patients (Shaw et al., 1985); the thromboxane $A_2$ synthetase inhibitor nizofenone (10 mg/day intravenously), in 77 patients (Saito et al., 1983); the thromboxane $A_2$ synthetase inhibitor cataclot (1 fg/kg/min intravenously), in 24 patients (Tokiyoshi et al., 1991); and the experimental antiplatelet agent OKY-46, 160 or 800 mg orally, in 256 patients (Suzuki *et al.*, 1989). In a systematic overview of these four trials and the two aspirin trials mentioned above, the rate of poor outcome was not significantly different between patients treated with antiplatelet agents and controls (unpublished review).

### Other Strategies to Prevent Delayed Cerebral Ischemia

Prophylactic volume expansion in patients with aneurysmal subarachnoid hemorrhage has been applied in four small randomized trials. In one of these, with only 30 patients, the treatment was started preoperatively; the rate of ischemic episodes significantly decreased, but no information was given on long-term outcome (Rosenwasser *et al.*, 1983). A second study, involving 32 patients, applied hypervolemia together with induced hypertension and hemodilution but again this regimen did not result in a detectable improvement of outcome (Egge *et al.*, 2001). Two other studies randomized patients after aneurysm clipping, but reported only physiological surrogate measures and not functional outcome (Lennihan *et al.*, 2000; Mayer *et al.*, 1998).

Observational studies suggest an adverse effect of intraoperative hypotension (Chang *et al.*, 2000), but there is no evidence that avoidance of even moderate hypotension during aneurysm surgery reduces the risk of delayed ischemia.

Calcitonin-gene-related peptide is a potent vasodilatator, but in a randomized clinical trial, no effect of this drug was found (European CGRP in Subarachnoid Haemorrhage Study Group, 1992). Another strategy aimed at reducing the frequency of vasospasm is lysis of the intracisternal blood clot with intrathecally administered recombinant tissue plasminogen activator, but a clinical trial in 100 patients failed to show a reduction in the rate of secondary ischemia or improvement in outcome (Findlay *et al.*, 1995).

Safety studies have been performed with magnesium sulfate (Boet *et al.*, 2000) and with a novel endothelin receptor antagonist (Shaw *et al.*, 2000); efficacy studies are still under way.

Prophylactic transluminal balloon angioplasty has been advocated (Muizelaar *et al.*, 1999), but there are no controlled studies to support this procedure.

### Treatment of Delayed Cerebral Ischemia (Table IV)

Once delayed cerbral ischemia has occurred, despite all preventive measures mentioned above, treatment with hypervolemia, hemodilution, and induced hypertension, the so-called triple-H therapy, has become widely used, although evidence from clinical trials is still

**TABLE IV** Management of Cerebral Ischemia

- Immediately administer 500 ml of plasma expander*
- Insert subclavian vein catheter or pulmonary arterial balloon catheter; maintain central venous pressure between 8–12 mm Hg, or pulmonary wedge pressure between 14–18 mm Hg*
- Keep arterial pressure 20–40 mm Hg above baseline values*
- Maintain fluid intake with at least 3 L of normal saline (0.9%) per 24 h*
- Correct hyponatremia, if severe*

*Retrospective or non-randomized study(ies), or empirical recommendation without scientific proof.

lacking. Since the 1960s, induced hypertension has been used to combat ischemic deficits in patients with SAH (Farhat *et al.*, 1967; Kosnik and Hunt 1976). Later, induced hypertension was often combined with volume expansion. In a series of patients with progressive neurological deterioration and angiographically confirmed vasospasm, the deficits could be permanently reversed in 43 of 58 cased (Kassell *et al.*, 1982). In 16 patients who had responded to this treatment, the neurological deficits recurred when the blood pressure transiently dropped, but again resolved when the pressure increased. The most plausible explanation for these phenomena is a defect of cerebral autoregulation that makes the perfusion of the brain passively dependent on the systemic blood pressure. The risks of deliberately increasing the arterial pressure and plasma volume include rebleeding of an unclipped aneurysm, increased cerebral edema or hemorrhagic transformation in areas of infarction (Amin-Hanjani *et al.*, 1999), myocardial infarction and congestive heart failure.

Only few centers have experience with the endovascular approach in the treatment of symptomatic vasospasm after SAH (Bejjani *et al.*, 1998; Eskridge *et al.*, 1998; Firlik *et al.*, 1997; Higashida *et al.*, 1989; Newell *et al.*, 1989; Nichols *et al.*, 1994). Most of these reports document sustained improvement in more than half of the cases (the total numbers were 10–20 in each of the first four studies, 31 and 50 in the two most recent ones), but the series were uncontrolled and evidently there must be publication bias. Less favorable experiences have followed (Polin *et al.*, 2000). Rebleeding can be precipitated by this procedure, even after the aneurysm has been clipped (Newell *et al.*, 1989; Linskey *et al.*, 1991). Hyperperfusion injury has also been reported (Schoser *et al.*, 1997). In view of the risks, the high costs and the lack of controlled trials, transluminal angioplasty should presently be regarded as a strictly experimental procedure.

The same caveat applies to uncontrolled reports of improvement of ischemic deficits after intra-arterial infusion of papaverine, following superselective catheterization (Elliott *et al.*, 1998; Fandino *et al.*, 1998; Kaku *et al.*, 1992); moreover, not all these case studies

**TABLE V   Management of Acute Hydrocephalus**

- Consider diagnosis if level of consciousness gradually deteriorates, particularly on the first day after the bleed
- Repeat the CT scan and compare the bicaudate index with that on any previous scan
- *Spontaneous improvement* occurs within 24 hours in 50% of the patients (except those with massive intraventricular hemorrhage); take action if patient further deteriorates or fails to improve within 24 hours*
- *Lumbar punctures are reasonably* safe if there is no brain shift, and effective in about 50% of the patients who have no intraventricular obstruction*
- *External drainage* of the ventricles is very effective in restoring the level of consciousness, but carries a high risk of rebleeding (consider emergency clipping or coiling at the same time), and of infection (this may to some degree be prevented by prophylactic antibiotics or subcutaneous tunnelling)*

*Retrospective or non-randomized study(ies).

are positive (Polin *et al.*, 1998). Of the related compound milrinone favorable effects on arterial diameter have been reported but not yet on individuals (Arakawa *et al.*, 2001).

## Management of Acute Hydrocephalus (Table V)

In several series of patients admitted within 3 days of SAH and who were not selected for aneurysm surgery, the frequency of acute hydrocephalus (defined as a bicaudate index above the 95th percentile for age was consistently around 20% (hasan *et al.*, 1991; Milhorat, 1987; van Gijn *et al.*, 1985b). Of those, only 10–28% had a normal level of consciousness, but approximately one-third of that subgroup still deteriorated in the next few days (Hasan *et al.*, 1989a).

The classical presentation of acute hydrocephalus is that of a patient who is alert immediately after the initial hemorrhage, but who in the next few hours becomes increasingly drowsy, to the point that the patient only moans and localizes to pain. Nonetheless only 50% of all patients with acute hydrocephalus present in this way (van Gijn *et al.*, 1985b). In the other 50%, consciousness is impaired from the onset, or the course is unknown because the patient was alone at the time of hemorrhage. If the patient is admitted very early and secondary deterioration occurs because of hydrocephalus, serial investigation by CT may show that the level of consciousness correlates more or less inversely with the width of the lateral ventricles (Rinkel *et al.*, 1990a). When different patients are compared within one series; however, the relationship between the level of consciousness and the degree of ventricular dilatation is rather erratic (Hasan *et al.*, 1989a; van Gijn *et al.*, 1985b). Ocular signs do not always accompany obtun-

dation as a result of acute hydrocephalus; they help to corroborate but not to exclude the diagnosis. In a prospective study in which 30 of 34 patients with acute hydrocephalus had an impaired level of consciousness, 9 of these 30 had small, nonreactive pupils, and 4 of these 9 also showed persistent downward deviation of the eyes, with otherwise intact brainstem reflexes (van Gijn *et al.*, 1985b). These eye signs reflect dilatation of the proximal part of the aqueduct, which causes dysfunction of the pretectal area (Swash, 1974). All nine patients with nonreactive pupils had a relative ventricular size of more than 1.20 and were in coma; i.e., they did not open their eyes, did not obey commands and did not utter words.

Intraventricular blood was found to be associated with acute hydrocephalus in all studies that addressed this question, at least if the ventricles contained enough blood to suggest direct intraventricular hemorrhage rather than just sedimentation in the posterior horns, which reflects diffusion of red blood cells throughout the cerebrospinal fluid spaces (Graff-Radford *et al.*, 1989; Hasan *et al.*, 1991; Mehta *et al.*, 1996; van Gijn *et al.*, 1985b). On the other hand, not all patients with acute hydrocephalus have intraventricular blood (the proportions varied between 35 and 65% in these studies), and it is an erroneous notion that acute hydrocephalus is, by definition, the result of intraventricular obstruction. In some patients, it is probable that clots obstructing the tentorial hiatus are partly or wholly responsible (Hasan *et al.*, 1992; Rinkel *et al.*, 1992).

### Wait and See

A policy of wait and see for 24 h is eminently justified in patients with dilated ventricles who are alert, because only about one-third of them will become symptomatic in the next few days (Hasan *et al.*, 1989a). Postponing interventions for 1 day can be rewarding even if the level of consciousness is decreased. The reason is that spontaneous improvement within this period has been documented in approximately 50% of the patients (7 of 13) with acute hydrocephalus who were only drowsy, and also in almost 50% of the patients (19 of 43) who had a Glasgow Coma Score of 12/14 or worse but no massive intraventricular hemorrhage (Hasan *et al.*, 1989a). On the other hand, it is not always easy to make a definitive decision on the need for surgical measures even after 1 day has elapsed, because patients may temporarily improve to some extent but then reach a plateau phase or again deteriorate; such fluctuations were encountered in about one-third of the cases (Hasan *et al.*, 1989a). Any further deterioration in the level of consciousness warrants active intervention.

## Lumbar Puncture

Lumbar puncture has been suggested as a therapeutic measure a long time ago (Kolluri *et al.*, 1984), but formal studies are scarce. In a prospective but uncontrolled study, Hasan *et al.*, treated 17 patients in this way; they had acute hydrocephalus with neither a hematoma nor gross intraventricular hemorrhage, i.e., less than complete filling of the third of fourth ventricle (Hasan *et al.*, 1991). Between one and seven spinal taps per patient were performed in the first 10 days, the number depending on the rate of improvement; each time a maximum of 20 mL of cerebrospinal fluid was removed, the aim being a closing pressure of 15 cm $H_2O$. Five of the 17 patients had a decreased level of consciousness on admission, and in them the lumbar punctures were started immediately; in 11 of the remaining 12 patients deterioration occurred a few days after admission, and the last patient had a fluctuating level of consciousness. Of the total of 17 patients, 12 showed initial improvement: 6 fully recovered, 2 showed incomplete improvement but fully recovered after insertion of an internal shunt, and 4 patients died of other complications several days after the lumbar punctures had been started. Of the 5 remaining patients in whom lumbar puncture had no effect, 2 recovered after an internal shunt and 3 died of other complications. The rate of rebleeding (2 of the 17 patients) was similar to what might have been expected, but of course the numbers were small (Hasan *et al.*, 1991).

Until controlled trials are available (and we think these are still needed and ethically justifiable) the tentative conclusion is that lumbar puncture seems a safe and reasonably effective way of treating those forms of acute hydrocephalus that are not obviously caused by intraventricular obstruction.

## External Ventricular Drainage

External drainage of the cerebral ventricles by a catheter inserted through a burr hole is, in many centers, the most common method of treating acute hydrocephalus. Internal drainage, to the right atrium or the peritoneal cavity, is rarely considered in the first few days because the blood in the cerebrospinal fluid will almost inevitably block the shunt system. After insertion of an external catheter, the improvement is usually rapid and sometimes dramatic (Hasan *et al.*, 1989a; van Gijn *et al.*, 1985b).

Unfortunately other problems tend to intervene soon, particularly rebleeding and ventriculitis. Rebleeding after insertion of an external drain occurs significantly more often than in patients with acute hydrocephalus who are not shunted or in patients without hydrocephalus (Hasan *et al.*, 1989a; Kawai

*et al.*, 1997; Papo *et al.*, 1984; Paré *et al.*, 1992; Raimondi and Torres, 1973; van Gijn *et al.*, 1985b). This complication did not occur in only a single published series (Mehta *et al.*, 1996). None of the studies had a randomized design for assessing the effect of shunt insertion, and it is possible that the development of acute hydrocephalus is associated with a more severely disrupted aneurysm which is more prone to rebleeding as part of its natural history. Yet it seems plausible that the increased risk of aneurysmal rebleeding in patients undergoing ventricular drainage results from a rise in aneurysmal transmural pressure, since intracranial pressure is rapidly lowered by ventricular drainage. It has been suggested that an excess of rebleeding can be prevented by keeping the pressure of the cerebrospinal fluid between 15 and 25 mm Hg (Pickard, 1984). Disappointingly, this had not been borne out by subsequent experience (Hasan *et al.*, 1989a). Others advocate capitalization on the expected improvement after external drainage by performing early aneurysm surgery at the same time (Milhorat, 1987). This is clearly another issue where a randomized controlled trial is needed.

Ventriculitis is a frequent complication after external drainage, especially if drainage is continued for more than 3 days. Of 31 patients treated with external drainage, 17 had died after 3 months and infection contributed to the death of 5 (Hasan *et al.*, 1989a). The use of prophylactic antibiotics or a long subcutaneous tunnel has been advocated but these measures have not been subjected to a controlled study.

## SYSTEMIC COMPLICATIONS

Neurologists and neurosurgeons are regularly faced by non-neurological complications in patients with aneurysmal SAH that need prompt therapeutic intervention (Gruber *et al.*, 1999). Forty percent had at least one life-threatening complication in a series of 451 patients enrolled in a clinical trial of nicardipine (Solenski *et al.*, 1995), and a similar proportion of medical complications (37%) may account for deaths (Gruber *et al.*, 1998). Hyponatremia is the most common of these but several other systemic disorders may cause secondary deterioration.

### Hyponatremia (Table VI)

Hyponatremia, with or without intravascular volume change, is the most common electrolyte disturbance following aneurysmal rupture. The frequency depends on the cutoff point; if defined as a sodium level of 134 mmol or less on at least 2 consecutive days, it occurs in about one-third of patients (Hasan *et al.*,

**TABLE VI**  Management of Hyponatremia

- Almost invariably caused by sodium *depletion*, not by sodium dilution (SIADH)
- Associated hypovolaemia increases risk of delayed cerebral ischaemia
- Give isotonic saline (with or without plasma expander) or a mixture of glucose and saline; no free water*
- If necessary, add fludrocortisone acetate, 400 mg/day in two doses, orally or intravenously**
- Keep central venous pressure between 8 and 12 mm Hg, or pulmonary capillary wedge pressure between 14 and 18 mm Hg*

*Retrospective or non-randomized study(ies), or empirical recommendation without scientific proof.
**One prospective randomized study.

1989a). It develops most commonly between the second and the 10th day (Wijdicks *et al.*, 1985a).

The clinical manifestations of hyponatremia include an impaired level of consciousness, asterixis, hemiparesis, seizures, and coma (Arieff *et al.*, 1976). These usually do not occur until the plasma sodium is less than 125 mmol/L, but irritability, restlessness, and confusion can result from a rapid decline of sodium, particularly if the downward trend continues over a few days. Sodium levels below 100 mmol/L almost always give rise to seizures and, rarely, ventricular tachycardia or fibrillation (Arieff, 1986). But the most dreaded complication of hyponatremia in patients with SAH is the precipitation of delayed ischemia, through associated hypovolemia.

In general, the causes of hyponatremia vary according to the patient's volume status. In most, but not all, cases, the total body sodium has remained constant but the water content of the extracellular volume is at fault; therefore, hyponatremia can be best classified according to the extracellular volume status. After the syndrome of inappropriate secretion of antidiuretic hormone (SIADH) had been described in the 1950s (Bartter *et al.*, 1967), hyponatremia in SAH has often been incorrectly attributed to this syndrome (Doczi *et al.*, 1981). In SIADH there is a continuing secretion of ADH, not appropriate to changes of plasma volume and osmolality. The extracellular volume increases and by the expansion of the intravascular component of this volume a *dilutional hyponatremia* ensues. Natriuresis takes place because the volume expansion increases the glomerular filtration rate and inhibits the secretion of aldosterone. Balance studies have shown that the degree of natriuresis is relatively small and approximately equals intake. The high concentration of urinary sodium in SIADH can be simply explained by the fact that the sodium intake must be excreted in a small volume of urine.

In contrast, hyponatremia after SAH results from excessive natriuresis, or *cerebral salt wasting* (Harrigan,

1996). A prospective study of 21 patients demonstrated that the plasma volume decreaed in most patients who developed hyponatremia and that this was preceded by a negative sodium balance in all instances (Wijdicks *et al.*, 1985a). Plasma volume considerably decreased, even in some patients with normal sodium levels, usually as a result of excessive natriuresis. Serum vasopressin levels were increased or normal on admission, but had decreased by the time hyponatremia occurred. It is clear that the volume status in these patients is extremely important.

A possible contributing factor in the development of hyponatremia may be hydrocephalus, particularly enlargement of the third ventricle (van Gijn *et al.*, 1985b; Wijdicks *et al.*, 1988). This relationship was independent of the amount of cisternal blood or the location of the ruptured aneurysm, though others found it especially with aneurysms of the anterior communicating artery (Sayama *et al.*, 2000), but is partly dependent on the amount of intraventricular blood. Mechanical pressure on the hypothalamus can perhaps disturb sodium and water homeostasis. Three substances have been identified as being related to natriuresis and possibly act as intermediary factors: a digoxin-like substance (Wijdicks *et al.*, 1987), atrial natriuretic factor (Diringer *et al.*, 1988; Rosenfeld *et al.*, 1989; Wijdicks *et al.*, 1991, 1997), and brain natriuretic peptide (Berendes *et al.*, 1997; Wijdicks *et al.*, 1997; Tomida *et al.*, 1998).

Correction of hyponatremia in SAH is truly a problem of correcting volume depletion. Acute symptomatic hyponatremia is rare and requires urgent treatment with hypertonic saline (1.8% or even 3%). On the other hand, over-rapid infusion of sodium may precipitate myelinolysis in the pons and the white matter of the hemispheres. A retrospective survey suggests the maximal rate of correction should be by 12 mmol/L per day (Sterns *et al.*, 1986), but others maintain that more rapid correction is safe as long as the sodium level does not exceed 126 mmol/L during the first 24 h (Ayus *et al.*, 1987). A mild degree of hyponatremia (125–134 mmol/L) is usually well tolerated, self-limiting, and need not be treated in itself. Hyponatremia in patients with evidence of a negative fluid balance or excessive natriuresis is corrected with saline (0.9%; sodium concentration 150 mmol/L) or with a mixture of glucose and saline. In persistent hyponatremia administration of fludrocortisone may be effective (Hasan *et al.*, 1989c).

### Disorders of Heart Rhythm

Aneurysmal rupture is commonly associated with cardiac arrhythmias and electrocardiographic (ECG) abnormalities. This is one of the reasons why patients with SAH may be initially misdiagnosed as acute myocardial infarction and admitted to coronary care units. Cardiogenic shock may occur, usually in combination with pulmonary edema. A usual explanation is sustained sympathetic stimulation, a notion supported by massively elevated plasma levels of norepinephrine (Naredi *et al.*, 2000). There may be structural damage to the myocardium as well, which is often evident on echocardiograms (Mayer *et al.*, 1999; Pollick *et al.*, 1988). The histological features of myocardial damage are contraction bands, focal myocardial necrosis and subendocardial ischemia. A more unconventional theory is the existence of an arrhythmogenic center in the right insular cortex (Hirashima *et al.*, 2001).

The most common ECG abnormalities in SAH are ST- and T-segment changes, prominent U waves, QT prolongation, and sinus arrhythmias (Brouwers *et al.*, 1989; Marion *et al.*, 1986; Stober *et al.*, 1988). Life-threatening arrhythmias such as ventricular fibrillation or "torsade de pointe" may be seen on 24-h monitoring, but are extremely rare (Andreoli *et al.*, 1987; Carruth *et al.*, 1980; Hijdra *et al.*, 1984). A striking finding in a large series of patients investigated by serial ECGs was that every patient had at least one abnormal ECG (Brouwers *et al.*, 1989). Virtually all ECG abnormalities changed to other abnormalities, without any consistent order, and then disappeared, in an observation period of 10 days. Some patients had ECG changes that closely resembled those in acute myocardial infarction; these spontaneously disappeared the following day without any change in neurological or cardiac condition. Potentially dangerous arrhythmias or actual death from cardiac failure did not occur in this limited series. In contrast, others found runs of ventricular tachycardia in 20% of those monitored and prolonged QT intervals in 60% (Estanol Vidal *et al.*, 1979). Prolonged QT intervals often represent delayed ventricular repolarization and predispose to ventricular arrhythmias. The prognostic value of these ECG changes is unclear, but in patients with SAH they are more important as indicators of severe intracranial disease than as predictors of potentially serious cardiac complications (Brouwers *et al.*, 1989; Manninen *et al.*, 1995).

Generally, severe ventricular arrhythmias are of short duration. Beta-blockade has been proposed as preventive treatment aimed at lowering the sympathetic tone. In patients with head injury a double-blind, randomized study found that beta blockers reduced catecholamine-induced cardiac necrosis, but not in-hospital case fatality (Cruickshank *et al.*, 1987). In patients with SAH, routine administration of beta blockers is not warranted until there is evidence of improved overall outcome; the net benefits may be disappointing because beta blockers also lower blood pressure. Controlled clinical trials of cardioprotective agents in SAH are scarce, but are less urgent than those aimed at the neurological complications.

### Neurogenic Pulmonary Oedema

Neurogenic pulmonary edema is a dramatic and dangerous complication (Wijdicks and Bonel, 1998). After SAH it usually has an extremely rapid onset, within hours, although a delayed course has been reported (Fisher and Aboul Nasr, 1979). What triggers pulmonary edema is unclear. Elevated intracranial pressure can lead to a massive sympathetic discharge mediated by the anterior hypothalamus. Systemic hypertension often co-exists, but pulmonary edema may present or even worsen during the lowering of systemic or pulmonary pressures. It has also been suggested that increased sympathetic activity leads to generalized vasoconstriction, including the pulmonary vasculature, and that direct damage to endothelial cells may result in increased permeability (Theodore and Robin, 1976).

Fortunately, pulmonary edema is not common, occurring in less than 10% of patients with SAH. It is related to the severity of aneurysmal hemorrhage and is rarely seen in patients with a normal level of consciousness (Weir, 1978). The typical clinical picture consists of unexpected dyspnea, cyanosis, and production of pink and frothy sputum. Many patients are pale, sweat excessively, and are hypertensive. A chest X-ray usually demonstrates impressive pulmonary edema which may disappear in a matter of hours following positive end-expiratory pressure ventilation (Wauchob *et al.*, 1984). Diuretics are often used as standard therapy and it has been claimed that labetalol or chlorpromazine are also beneficial (Wohns *et al.*, 1985; Harrier, 1988). A problem is that liberal administration of fluids is beneficial for brain perfusion but may delay recovery of pulmonary edema and hence impair brain oxygenation, and that positive end-expiratory pressure ventilation increases intracranial pressure.

Neurogenic pulmonary edema may be complicated by reversible decompensation of the left ventricle, clinically manifested by sudden hypotension following initially elevated blood pressures, transient lactic acidosis, mild elevation of the creatine kinase MB fraction, and varied ECG changes during the first day, followed by widespread and persistent T-wave inversion (Mayer *et al.*, 1994; Parr *et al.*, 1996). In these cases of pulmonary edema with secondary cardiac injury, treatment with pressor agents may be indicated.

## REFERENCES

Alberico, R. A., Patel, M., Casey, S., Jacobs, B., Maguire, W., and Decker, R. (1995). Evaluation of the circle of Willis with three-dimensional CT angiography in patients with suspected intracranial aneurysms. *Am. J. Neuroradiol.* 16, 1571–1578.

Aletich, V. A., Debrun, G. M., Misra, M., Charbel, F., and Ausman, J. I. (2000). The remodeling technique of balloon-assisted Guglielmi detachable coil placement in wide-necked aneurysms: Experience at the University of Illionis at Chicago. *Journal of Neurosurgery* 93, 388–396.

Amin-Hanjani, S., Schwartz, R. B., Sathi, S., and Stieg, P. E. (1999). Hypertensive encephalopathy as a complication of hyperdynamic therapy for vasospasm: Report of two cases. *Neurosurgery* 44, 1113–1116.

Anderson, G. B., Steinke, D. E., Petruk, K. C., Ashforth, R., and Findlay, J. M. (1999). Computed tomographic angiography versus digital subtraction angiography for the disgnosis and early teatment of ruptured intracranial aneurysms. *Neurosurgery* 45, 1315–1320.

Andreoli, A., di Pasquale, G., Pinelli, G., Grazi, P., Tognetti, F., and Testa, C. (1987). Subarachnoid hemorrhage: Frequency and severity of cardiac arrhythmias. A survey of 70 cases studied in the acute phase. *Stroke* 18, 558–564.

Anzalone, N., Triulzi, F., and Scotti, G. (1995). Acute subarachnoid haemorrhage: 3D time-of-flight MR angiography versus intra-arterial digital angiography. *Neuroradiology* 37, 257–261.

Arakawa, Y., Kikuta, K., Hojo, M., Goto, Y., Ishii, A., and Yamagata, S. (2001). Milrinone for the treatment of cerebral vasospasm after subarachnoid hemorrhage: Report of seven cases. *Neurosurgery* 48, 723–728.

Arieff, A. I. (1986). Hyponatremia, convulsions, respiratory arrest, and permanent brain damage after elective surgery in healthy women. *N. Engl. J. Med.* 314, 1529–1535.

Arieff, A. I., Llach, F., and Massry, S. G. (1976). Neurological manifestations and morbidity of hyponatremia: Correlation with brain water and electrolytes. *Medicine (Baltimore)* 55, 121–129.

Asano, T., Takakura, K., Sano, K., Kikuchi, H., Nagai, H., Saito, I., Tamura, A., Ochiai, C., and Sasaki, T. (1996). Effects of a hydroxyl radical scavenger on delayed ischemic neurological deficits following aneurysmal subarachnoid hemorrhage: Results of a multicenter, placebo-controlled double-blind trial. *J. Neurosurg.* 84, 792–803.

Avrahami, E., Katz, R., Rabin, A., and Friedman, V. (1998). CT diagnosis of non-traumatic subarachnoid haemorrhage in patients with brain edema. *Eur. J. Radiol.* 28, 222–225.

Ayus, J. C., Krothapalli, R. K., and Arieff, A. I. (1987). Treatment of symptomatic hyponatremia and its relation to brain damage. A prospective study. *N. Engl. J. Med.* 317, 1190–1195.

Bartter, F. C., and Schwarz, W. B. (1967). The syndrome of inappropriate secretion of antdiuretic hormone. *Am. J. Med.* 42, 790–806.

Beetham, R., Fahie-Wilson, M. N., and Park, D. (1998). What is the role of CSF spectrophotometry in the diagnosis of subarachnoid haemorrhage? *Ann. Clin. Biochem.* 35, 1–4.

Bejjani, G. K., Bank, W. O., Olan, W. J., and Sekhar, L. N. (1998). The efficacy and safety of angioplasty for cerebral vasospasm after subarachnoid hemorrhage. *Neurosurgery* 42, 979–986.

Benbadis, S. R., Sila, C. A., and Cristea, R. L. (1994). Mental status changes and stroke. *J. Gen. Intern. Med.* 9, 485–487.

Berendes, E., Walter, M., Cullen, P., Prien, T., Van Aken, H., Horsthemke, J., Schulte, M., Von Wild, K., and Scherer, R. (1997). Secretion of brain natriuretic peptide in patients with aneurysmal subarachnoid haemorrhage. *Lancet* 349, 245–249.

Boet, R., and Mee, E. (2000). Magnesium sulfate in the management of patients with Fisher Grade 3 subarachnoid hemorrhage: A pilot study. *Neurosurgery* 47, 602–606.

Botterell, E. H., Lougheed, W. M., Scott, J. W., and Vandewater, S. L. (1956). Hypothermia, and interruption of carotid, or carotid and vertebral circulation, in the surgical management of intracranial aneurysms. *J. Neurosurg.* 13, 1–42.

Brandt, L., Sonesson, B., Ljunggren, B., and Såveland, H. (1987). Ruptured middle cerebral artery aneurysm with intracerebral hemorrhage in younger patients appearing moribund: Emergency operation? *Neurosurgery* 20, 925–929.

Brilstra, E. H., Hop, J. W., and Rinkel, G. J. E. (1997). Quality of life after perimesencephalic haemorrhage. *J. Neurol. Neurosurg. Psychiatry* 63, 382–384.

Brilstra, E. H., Rinkel, G. J. E., van der Graaf, Y., van Rooij, W. J. J., and Algra, A. (1999). Treatment of intracranial aneurysms by embolization with coils—A systematic review. *Stroke* 30, 470–476.

Broderick, J. P., Brott, T., Tomsick, T., Huster, G., and Miller, R. (1992). The risk of subarachnoid and intracerebral hemorrhages in blacks as compared with whites. *N. Engl. J. Med.* 326, 733–736.

Bromberg, J. E. C., Rinkel, G. J. E., Algra, A., Greebe, P., van Duyn, C. M., Hasan, D., Limburg, M., ter Berg, H. W. M., Wijdicks, E. F. M., an van Gijn, J. (1995). Subarachnoid haemorrhage in first and second degree relatives of patients with subarachnoid haemorrhage. *Br. Med. J.* 311, 288–289.

Brouwers, P. J. A. M., Wijdicks, E. F. M., Hasan, D., Vermeulen, M., Wever, E. F., Frericks, H., and van Gijn, J. (1989). Serial electrocardiographic recording in aneurysmal subarachnoid hemorrhage. *Stroke* 20, 1162–1167.

Brouwers, P. J. A. M., Wijdicks, E. F. M., and van Gijn, J. (1992). Infarction after aneurysm rupture does not depend on distribution or clearance rate of blood. *Stroke* 23, 374–379.

Byrne, J. V., Sohn, N. J., and Molyneux, A. J. (1999). Five-year experience in using coil embolization for ruptured intracranial aneurysms: Outcomes and incidence of late rebleeding. *J. Neurosurg.* 90, 656–663.

Canhao, P., Falcao, F., Melo, T., Ferro, H., and Ferro, J. (1999). Vascular risk factors for perimesencephalic nonaneurysmal subarachnoid hemorrhage. *J. Neurol.* 246, 492–496.

Canhao, P., Ferro, J. M., Pinto, A. M., Melo, T. P., and Campos, J. G. (1995). Perimesencephalic and nonperimesencephalic subarachnoid haemorrhages with negative angiograms. *Acta Neurochir. (Wien)* 132, 14–19.

Carruth, J. E., and Silverman, M. E. (1980). Torsade de pointe atypical ventricular tachycardia complicating subarachnoid hemorrhage. *Chest* 78, 886–888.

Chang, H. S., Hongo, K., and Nakagawa, H. (2000). Adverse effects of limited hypotensive anesthesia on the outcome of patients with subarachnoid hemorrhage. *J. Neurosurg.* 92, 971–975.

Chyatte, D., Reilly, J., and Tilson, M. D. (1990). Morphometric analysis of reticular and elastin fibers in the cerebral arteries of patients with intracranial aneurysms. *Neurosurgery* 26, 939–943.

Cloft, H. J., Joseph, G. J., and Dion, J. E. (1999). Risk of cerebral angiography in patients with subarachnoid hemorrhage, cerebral aneurysm, and arteriovenous malformation—A meta-analysis. *Stroke* 30, 317–320.

Cognard, C., Pierot, L., Boulin, A., Weill, A., Toevi, M., Castaings, L., Rey, A., and Moret, J. (1997). Intracranial aneurysms: Endovascular treatement with mechanical detachable spirals in 60 aneurysms. *Radiology* 202, 783–792.

Crompton, M. R. (1966). The pathogenesis of cerebral aneurysms. *Brain* 89, 797–814.

Cruickshank, J. M., Neil-Dwyer, G., Degaute, J. P., Hayes, Y., Kuurne, T., Kytta, J., Vincent, J. L., Carruthers, M. E., and Patel, S. (1987). Reduction of stress/catecholamine-induced cardiac necrosis by beta 1-selective blockade. *Lancet* 2, 585–589.

De Braekeleer, M., Pérusse, L., Cantin, L., Bouchard, J. M., and Mathieu, J. (1996). A study of inbreeding and kinship in intracranial aneurysms in the Saguenay Lac-Saint-Jean region (Quebec, Canada). *Ann. Hum. Genet.* 60, 99–104.

de Gans, K., Nieuwkamp, D. J., Rinkel, G. J. E., and Algra, A. (2002). Timing of aneurysm surgery in subarachnoid hemorrhage: A systematic review of the literature. *Neurosurgery* 50, 336–340.

Diringer, M., Ladenson, P. W., Stern, B. J., Schleimer, J., and Hanley, D. F. (1988). Plasma atrial natriuretic factor and subarachnoid hemorrhage. *Stroke* **19**, 1119–1124.

Doczi, T., Bende, J., Huszka, E., and Kiss, J. (1981). Syndrome of inappropriate secretion of antidiuretic hormone after subarachnoid hemorrhage. *Neurosurgery* **9**, 394–397.

Dowling, G., and Curry, B. (1988). Traumatic basal subarachnoid hemorrhage. Report of six cases and review of the literature. *Am. J. Forensic Med. Pathol.* **9**, 23–31.

Drake, C. G., Hunt, W. E., Sano, K., Kassell, N., Teasdale, G., Pertuiset, B., and Devilliers, J. C. (1988). Report of World Federation of Neurological Surgeons Committee on a Universal Subarachnoid Hemorrhage Grading Scale. *J. Neurosurg.* **68**, 985–986.

Egge, A., Waterloo, K., Sjoholm, H., Solberg, T., Ingebrigtsen, T., and Romner, B. (2001). Prophylactic hyperdynamic postoperative fluid therapy after aneurysmal subarachnoid hemorrhage: A clinical, prospective, randomized, controlled study. *Neurosurgery* **49**, 593–605.

Elliott, J. P., Newell, D. W., Lam, D. J., Eskridge, J. M., Douville, C. M., Le Roux, P. D., Lewis, D. H., Mayberg, M. R., Grady, M. S., and Winn, R. (1998). Comparison of balloon angioplasty and papaverine infusion for the treatment of vasospasm following aneurysmal subarachnoid hemorrhage. *J. Neurosurg.* **88**, 277–284.

Eskridge, J. M., McAuliffe, W., Song, J. K., Deliganis, A. V., Newell, D. W., Lewis, D. H., Mayberg, M. R., and Winn, H. R. (1998). Balloon angioplasty for the treatment of vasospasm: Results of first 50 cases. *Neurosurgery* **42**, 510–516.

Estanol Vidal, B., Badui Dergal, E., Cesarman, E., Marin San Martin, O., Loyo, M., Vargas Lugo, B., and Perez Ortega, R. (1979). Cardiac arrhythmias associated with subarachnoid hemorrhage: Prospective study. *Neurosurgery* **5**, 675–680.

European CCRP in Subarachnoid Haemorrhage Study Group (1992). Effect of calcitonin-gene-related peptide in patients with delayed postoperative cerebral ischaemia after aneurysmal subarachnoid haemorrhage. *Lancet* **339**, 831–834.

Fandino, J., Kaku, Y., Schuknecht, B., Valavanis, A., and Yonekawa, Y. (1998). Improvement of cerebral oxygenation patterns and metabolic validation of super-selective intraarterial infusion of papaverine for the treatment of cerebral vasospasm. *J. Neurosurg.* **89**, 93–100.

Farhat, S. M., and Schneider, R. C. (1967). Observations on the effect of systemic blood pressure on intracranial circulation in patients with cerebrovascular insufficiency. *J. Neurosurg.* **27**, 441–445.

Farrés, M. T., Ferraz Leite, H., Schindler, E., and Mühlbauer, M. (1992). Spontaneous subarachnoid hemorrhage with negative angiography: CT fingings. *J. Comput. Assist. Tomogr.* **16**, 534–537.

Feigin, V. L., Rinkel, G. J. E., Algra, A., Vermeulen, M., and van Gijn, J. (2000). "Calcium Antagonists for Aneurysmal Subarachnoid Haemorrhage (Cochrane Review)," Issue 3, The Cochrane Library, Update Software, Oxford.

Ferbert, A., Hubo, I., and Biniek, R. (1992). Non-traumatic subarachnoid hemorrhage with normal angiogram. Long-term follow-up and CT predictors of complications. *J. Neurol. Sci.* **107**, 14–18.

Ferry, P. C., Kerber, C., Peterson, D., and Gallo, A. A., Jr. (1974). Arteriectasis, subarachnoid hemorrhage in a three-month-old infant. *Neurology* **24**, 494–500.

Findlay, J. M., Kassell, N. F., Weir, B. K. A., Haley, E. C., Jr., Kongable, G., Germanson, T., Truskowski, L., Alves, W. M., Holness, R. O., Knuckey, N. W., Yonas, H., Steinberg, G. K., West, M., Winn, H. R., and Ferguson, G. (1995). A randomized trial of intraoperative, intracisternal tissue plasminogen activator for the prevention of vasospasm. *Neurosurgery* **37**, 168–178.

Firlik, A. D., Kaufamann, A. M., Jungreis, C. A., and Yonas, H. (1997). Effect of transluminal angioplasty on cerebral blood flow

in the managment of symptomatic vasospasm following aneurysmal subarachnoid hemorrhage. *J. Neurosurg.* **86**, 830–839.

Fisher, A., and Aboul Nasr, H. T. (1979). Delayed nonfatal pulmonary edema following subarachnoid hemorrhage. Case report. *J. Neurosurg.* **51**, 856–859.

Frizzell, R. T., Kuhn, F., Morris, R., Quinn, C., and Fisher, W. S., III (1997). Screening for ocular hemorrhages in patients with ruptured cerebral aneurysms: A prospective study of 99 patients. *Neurosurgery* **41**, 529–533.

Fujii, Y., Takeuchi, S., Sasaki, O., Minakawa, T., Koike, T., and Tanaka, R. (1996). Ultra-early rebleeding in spontaneous subarachnoid hemorrhage. *J. Neurosurg.* **84**, 35–42.

Gaist, K., Vaeth, M., Tsiropoulos, I., Christensen, K., Corder, E., Olsen, J., and Sorensen, H. T. (2000). Risk of subarachnoid haemorrhage in first degree relatives of patients with subarachnoid haemorrhage: Follow up study based on national registries in Denmark. *Br. Med. J.* **320**, 141–145.

Gliemroth, J., Nowak, G., Kehler, U., Arnold, H., and Gaebel, C. (1999). Neoplastic cerebral aneurysm from metastatic lung adenocarcinoma associated with cerebral thrombosis and recurrent subarachnoid haemorrhage. *J. Neurol. Neurosurg. Psychiatry* **66**, 246–247.

Graff-Radford, N. R., Torner, J., Adams, H. P., Jr., and Kassell, N. F. (1989). Factors associated with hydrocephalus after subarachnoid hemorrhage. A report of the Cooperative Aneurysm Study. *Arch. Neurol.* **46**, 744–752.

Grote, E., and Hassler, W. (1988). The critical first minutes after subarachnoid hemorrhage. *Neurosurgery* **22**, 654–661.

Gruber, A., Reinprecht, A., Görzer, H., Fridrich, P., Czech, T., Illievich, U. M., and Richling, B. (1998). Pulmonary function and radiographic abnormalities related to neurological outcome after aneurysmal subarachnoid hemorrhage. *J. Neurosurg.* **88**, 28–37.

Gruber, A., Reinprecht, A., Illievich, U. M., Fitzgerald, R., Dietrich, W., Czech, T., and Richling, B. (1999). Extracerebral organ dysfunction and neurologic outcome after aneurysmal subarachnoid hemorrhage. *Crit. Care Med.* **27**, 505–514.

Guglielmi, G., Vinuela, F., Duckwiler, G., Dion, J., Lylyk, P., Berenstein, A., Strother, C., Graves, V., Halbach, V., Nichols, D., *et al.* (1992). Endovascular treatment of posterior circulation aneurysms by electrothrombosis using electrically detachable coils. *J. Neurosurg.* **77**, 515–524.

Guzman, R., and Grady, M. S. (1999). An intracranial aneurysm on the feeding artery of a cerebellar hemangioblastoma—Case report. *J. Neurosurg.* **91**, 136–138.

Hackett, M. L., Anderson, C. S., and ACROSS, G. (2000). Health outcomes 1 year after subarachnoid hemorrhage—An international population-based study. *Neurology* **55**, 658–662.

Haley, E. C. Jr., Kassell, N. F., Apperson-Hansen C., Maile, M. H., and Alves, W. M. (1997). A randomized, double-blind, vehicle-controlled trial of tirilazad mesylate in patients with aneurysmal subarachnoid hemorrhage: A cooperative study in North America. *J. Neurosurg.* **86**, 467–474.

Harland, W. A., Pitts, J. F., and Watson, A. A. (1983). Subarachnoid haemorrhage due to upper cervical trauma. *J. Clin. Pathol.* **36**, 1335–1341.

Harrier, H. D. (1988). Use of labetalol in trauma. *Crit. Care Med.* **16**, 1159–1160.

Harrigan, M. R. (1996). Cerebral salt wasting syndrome: A review. *Neurosurgery* **38**, 152–160.

Hart, R. G., Byer, J. A., Slaughter, J. R., Hewett, J. E., and Easton, J. D. (1981). Occurrence and implications of seizures in subarachnoid hemorrhage due to ruptured intracranial aneurysms. *Neurosurgery* **8**, 417–421.

Hasan D., Lindsay, K. W., and Vermeulen, M. (1991). Treatment of acute hydrocephalus after subarachnoid hemorrhage with serial lumbar puncture. *Stroke* **22**, 190–194.

Hasan D., Lindsay, K. W., Wijdicks, E. F. M., Murray, G. D., Brouwers, P. J. A. M., Bakker, W. H., van Gijn J., and Vermeulen, M. (1989c). Effect of fludrocortisone acetate in patients with subarachnoid hemorrhage. *Stroke* **20**, 1156–1161.

Hasan, D. and Tanghe, H. L. (1992). Distribution of cisternal blood in patients with acute hydrocephalus after subarachnoid hemorrhage. *Ann. Neurol.* **31**, 374–378.

Hasan, D., Vermeulen, M., Wijdicks, E. F. M., Hijdra, A., and van Gijn, J. (1989a). Effect of fluid intake and antihypertensive treatment on cerebral ischemia after subarachnoid hemorrhage. *Stroke* **20**, 1511–1515.

Hasan, D., Vermeulen, M., Wijdicks, E. F. M., Hijdra, A., and van Gijn, J. (1989a). Management problems in acute hydrocephalus after subarachnoid hemorrhage. *Stroke* **20**, 747–753.

Hasan, D., Wijdicks, E. F. M., and Vermeulen, M. (1990). Hyponatremia is associated with cerebral ischemia in patients with aneurysmal subarachnoid hemorrhage. *Ann. Neurol.* **27**, 106–108.

Hashimoto, H., Iida, J., Hironaka, Y., Okada, M., and Sakaki, T. (2000). Use of spiral computerized tomography angiography in patients with subarachnoid hemorrhage in whom subtraction angiography did not reveal cerebral aneurysms. *J. Neurosurg.* **92**, 278–283.

Hauerberg, J., Eskesen, V., and Rosenorn, J. (1994). The prognostic significance of intracerebral haematoma as shown on CT scanning after aneurysmal subarachnoid haemorrhage. *Br. J. Neurosurg.* **8**, 333–339.

Hayakawa, I., Watanabe, T., Tsuchida, T., and Sasaki, A. (1978). Perangiographic rupture of intracranial aneurysms. *Neuroradiology* **16**, 293–295.

Hayakawa, M., Murayama, Y., Duckwiler, G. R., Gobin, Y. P., Guglielmi, G., and Viñuela, F. (2000). Natural history of the neck remnant of a cerebral aneurysm treated with the Guglielmi detachable coil system. *J. Neurosurg.* **93**, 561–568.

Heiskanen, O. (1986). Risks of surgery for unrupted intracranial aneurysms. *J. Neurosurg.* **65**, 451–453.

Heiskanen, O., Poranen, A., Kuurne, T., Valtonen, S., and Kaste, M. (1988). Acute surgery for intracerebral haematomas caused by rupture of an intracranial arterial aneurysm. A prospective randomized study. *Acta Neurochir.* (*Wien*) **90**, 81–83.

Higashida, R. T., Halbach, V. V., Cahan, L. D., Brant Zawadzki, M., barnwell, S., Dowd, C., and Hieshima, G. B. (1989). Transluminal angioplasty for treatment of intracranial arterial vasospasm. *J. Neurosurg.* **71**, 648–653.

Hijdra, A., Brouwers, P. J., Vermeulen, M., and van Gijn, J. (1990). Grading the amount of blood on computed tomograms after subarachnoid hemorrhage. *Stroke* **21**, 1156–1161.

Hijdra, A. and van Gijn, J. (1982). Early death from rupture of an intracranial aneurysm. *J. Neurosurg.* **57**, 765–768.

Hijdra, A., van Gijn, J., Nagelkerke, N. J., Vermeulen, M., and van Crevel, H. (1988). Prediction of delayed cerebral ischemia, rebleeding, and outcome after aneurysmal subarachnoid hemorrhage. *Stroke* **19**, 1250–1256.

Hijdra, A., Van Gijn, J., Stefanko, S., van Dongen, K. J., Vermeulen, M., and van Crevel, H. (1986). Delayed cerebral ischemia after aneurysmal subarachnoid hemorrhage: clinicoanatomic correlations. *Neurology* **36**, 329–333.

Hijdra, A., vermeulen, M., van Gijn, J., and van Crevel, H. (1984). Respiratory arrest in subarachnoid hemorrhage. *Neurology* **34**, 1501–1503.

Hijdra, A., Vermeulen, M., van Gijn, J., and van Crevel, H. (1987). Rerupture of intracranial aneurysms: a clinicoanatomic study. *J. Neurosurg.* **67**, 29–33.

Hirai, S., Ono, J., and Yamaura, A. (1996). Clinical grading and outcome after early surgery in aneurysmal subarachnoid hemorrhage. *Neurosurgery* **39**, 441–446.

Hirashima, Y., Takashima, S., Matsumura, N., Kurimoto, M., Origasa, H., and Endo, S. (2001). Right Sylvian fissure subarachnoid hemorrhage has electrocardiographic consequences. *Stroke* **32**, 2278–2281.

Hoh, B. L., Putman, C. M., Budzik, R. F., Carter, B. S., and Ogilvy, C. S. (2001). Combined surgical and endovascular techniques of flow alteration to treat fusiform and complex wide-necked intracranial aneurysms that are unsuitable for clipping or coil embolization. *J. Neurosurg.* **95**, 24–35.

Hop, J. W., Brilstra, E. H., and Rinkel, G. J. E. (1998b). Transient amnesia after perimesencephalic haemorrhage: The role of enlarged temporal horns. *J. Neurol. Neurosurg. Psychiatry* **65**, 590–593.

Hop, J. W. and Rinkel, G. J. E. (1996). Secondary ischemia after subarachnoid hemorrhage. *Cerebrovasc. Dis.* **6**, 264–265.

Hop, J. W., Rinkel, G. J. E., Algra, A., Berkelbach van der Sprenkel, J. W., and van Gijn, J. (2000). Randomized pilot trial of postoperative aspirin in subarachnoid hemorrhage. *Neurology* **54**, 872–878.

Hop, J. W., Rinkel, G. J. E., Algra, A., and van Gijn, J. (1997). Case–fatality rates and functional outcome after subarachnoid hemorrhage—A systematic review. *Stroke* **28**, 660–664.

Hop, J. W., Rinkel, G. J. E., Algra, A., and van Gijn, J. (1998a). Quality of life in patients and partners after aneurysmal subarachnoid hemorrhage. *Stroke* **29**, 798–804.

Hop, J. W., Rinkel, G. J. E., Algra, A., and van Gijn, J. (1999). Initial loss of consciousness and risk of delayed cerebral ischemia after aneurysmal subarachnoid hemorrhage. *Stroke* **30**, 2268–2271.

Hop, J. W., Rinkel, G. J. E., Algra, A., and van Gijn, J. (2001). Changes in functional outcome and quality of life in patients and caregivers after aneurysmal subarachnoid hemorrhage. *J. Neurosurg.* **95**, 957–963.

Hope, J. K., Wilson, J. L., and Thomson, F. J. (1996). Three-dimensional CT angiography in the detection and characterization of intracranial berry aneurysms. *Am. J. Neuroradiol.* **17**, 437–445.

Hunt, W. E. and Hess, R. M. (1968). Surgical risk as related to time of intervention in the repair of intracranial aneurysms. *J. Neurosurg.* **28**, 14–20.

Jensen, F. K. and Wagner, A. (1997). Intracranial aneurysm following radiation therapy for medulloblastoma—A case report and review of the literature. *Acta Radiol.* **38**, 37–42.

Johnston, S. C. (2000). Combining ecological and individual variables to reduce confounding by indication: Case study—subarachnoid hemorrhage treatment. *J. Clin. Epidemiol.* **53**, 1236–1241.

Johnston, S. C., Colford, J. M., Jr., and Gress, D. R. (1998a). Oral contraceptives and the risk of subarachnoid hemorrhage—A meta-analysis. *Neurology* **51**, 411–418.

Johnston, S. C., Selvin, S., and Gress, D. R. (1998b). The burden, trends, and demographics of mortality from subarachnoid hemorrhage. *Neurology* **50**, 1413–1418.

Juvela, S. (1995). Aspirin and delayed cerebral ischemia after aneurysmal subarachnoid hemorrhage. *J. Neurosurg.* **82**, 945–952.

Juvela, S., Kaste, M., and Hillbom, M. (1990). Platelet thromboxane release afer subarachnoid hemorrhage and surgery. *Stroke* **21**, 566–571.

Kaku, Y., Yonekawa, Y., Tsukahara, T., and Kazekawa, K. (1992). Superselective intraarterial infusion of papaverine for the treatment of cerebral vasospasm after subarachnoid hemorrhage. *J. Neurosurg.* **77**, 842–847.

Kaneko, T., Sawada, T., and Niimi, T. (1983). Lower limit of blood pressure in treatment of acute hypertensive intracranial hemorrhage (AHCH). *J. Cereb. Blood Flow Metab.* **3** (**Suppl. 1**), S51–S52.

Kassell, N. F., Haley, E. C., Jr., Apperson-Hansen, C., Stat, M., Alves, W. M., Dorsch, N. W., Fabinyi, G., Matheson, J., Reilly, P., Siu, K., Stokes, B., Stuart, G., Koos, W., Calliauw, L., Selosse, P., Astrup,

J., Gjerris, F., Mendelow, A. D., Castel, J. P., Christiaens, J. L., Cophignon, J., Keravel, Y., Lagarrigue, J., and Mourier, K. (1996). Randomized, double-blind, vehicle-controlled trial of tirilazad mesylate in patients with aneurysmal subarachnoid hemorrhage: A cooperative study in Europe, Australia, and New Zealand. *J. Neurosurg.* **84,** 221–228.

Kassell, N. F., Peerless, S. J., Durward, Q. J., Beck, D. W., Drake, C. G., and Adams, H. P., Jr. (1982). Treatment of ischemic deficits from vasospasm with intravascular volume expansion and induced arterial hypertension. *Neurosurgery* **11,** 337–343.

Kassell, N. F. and Torner, J. C. (1983). Aneurysmal rebleeding: A preliminary report from the Cooperative Aneurysm Study. *Neurosurgery* **13,** 479–481.

Kassell, N. F., Torner, J. C., Haley, E. C., Jr., Jane, J. A., Adams, H. P., Jr., and Kongable, G. L. (1990). The International Cooperative Study on the Timing of Aneurysm Surgery. 1. Overall management results. *J. Neurosurg.* **73,** 18–36.

Kawai, K., Nagashima, H., Narita, K., Nakagomi, T., Nakayama, H., Tamura, A., and Sano, K. (1997). Efficacy and risk of ventricular drainage in cases of grade V subarachnoid hemorrhage. *Neurol. Res.* **19,** 649–653.

Kistler, J. P., Crowell, R. M., Davis, K. R., Heros, R., Ojemann, R. G., Zervas, T., and Fisher, C. M. (1983). The relation of cerebral vasospasm to the extent and location of subarachnoid blood visualized by CT scan: A prospective study. *Neurology* **33,** 424–436.

Kitahara, T., Ohwada, T., Tokiwa, K., Kurata, A., Miyasaka, Y., Yada, K., and Kan, S. (1993). [Clinical study in patients with perimesencephalic subarachnoid hemorrhage of unknown etiology]. *No. Shinkei. Geka.* **21,** 903–908.

Koenig, G. H., Marshall, W. H., Jr., Poole, G. J., and Kramer, R. A. (1979). Rupture of intracranial aneurysms during cerebral angiography: Report of ten cases and review of the literature. *Neurosurgery* **5,** 314–324.

Koivisto, T., Vanninen, R., Hurskainen, H., Saari, T., Hernesniemi, J., and Vapalahti, M. (2000). Outcomes of early endovascular versus surgical treatment of ruptured cerebral aneurysms—A prospective randomized study. *Stroke* **31,** 2369–2377.

Kojima, M., Nagasawa, S., Lee, Y. E., Takeichi, Y., Tsuda, E., and Mabuchi, N. (1998). Asymptomatic familial cerebral aneurysms. *Neurosurgery* **43,** 776–781.

Kolluri, V. R. and Sengupta, R. P. (1984). Symptomatic hydrocephalus following aneurysmal subarachnoid hemorrhage. *Surg. Neurol.* **21,** 402–404.

Kosnik, E. J., and Hunt, W. E. (1976). Postoperative hypertension in the management of patients with intracranial arterial aneurysms. *Journal of Neurosurgery* **45,** 148–154.

Lagares, A., Putman, C. M., and Ogilvy, C. S. (2001). Posterior fossa decompression and clot evacuation for fourth ventricle hemorrhage after aneurysmal rupture: Case report. *Neurosurgery* **49,** 208–211.

Lam, J. M. K., Smielewski, P., Czosnyka, M., Pickard, J. D., and Kirkpatrick, P. J. (2000). Predicting delayed ischemic deficits after aneurysmal subarachnoid hemorrhage using a transient hyperemic response test of cerebral autoregulation. *Neurosurgery* **47,** 819–825.

Lang, E. W., Diehl, R. R., and Mehdorn, H. M. (2001). Cerebral autoregulation testing after aneurysmal subarachnoid hemorrhage: The phase relationship between arterial blood pressure and cerebral blood flow velocity. *Crit. Care Med.* **29,** 158–163.

Lanzino, G., and Kassell, N. F. (1999b). Double-blind, randomized, vehicle-controlled study of high-dose tirilazad mesylate in women with aneurysmal subarachnoid hemorrhage. Part II. A cooperative study in North America. *J. Neurosurg.* **90,** 1018–1024.

Lanzino, G., Kassell, N. F., Dorsch, N. W. C., Pasqualin, A., Brandt, L., Schmiedek, P., Truskowsk, L. L., and Alves, W. M. (1999a).

Double-blind, randomized, vehicle-controlled study of high-dose tirilazad mesylate in women with aneurysmal subarachnoid hemorrhage. Part, I. A cooperative study in Europe, Australia, New Zealand, and South Africa. *J. Neurosurg.* **90,** 1011–1017.

Lanzino, G., Kassell, N. F., Germanson, T. P., Kongable, G. L., Truskowski, L. L., Torner, J. C., Jane, J. A., Spetzler, R. F., Zabramski, J., Culicchia, F., Carter, L. P., Feinberg, W., Urbina, C., Lopez, L., Brown, D., Tallman, D., Selman, W. R., Harrington, F., Warf, B., Barnett, G. H., Little, J., Palmer, J., Campbell, R. L., and Shapiro, S. (1996). Age and outcome after aneurysmal subarachnoid hemorrhage: Why do older patients fare worse. *J. Neurosurg.* **85,** 410–418.

Lennihan, L., Mayer, S.A, Fink, M. E., Beckford, A., Paik, M. C., Zhang, H., Wu, Y. C., Klebanoff, L. M., Raps, E. C., and Solomon, R. A. (2000). Effect of hypervolemic therapy on cerebral blood flow after subarachnoid hemorrhage: A randomized controlled trial. *Stroke* **31,** 383–391.

Levy, E., Koebbe, C. J., Horowitz, M. B., Jungreis, C. A., Pride, G. L., Dutton, K., Kassam, A., and Purdy, P. D. (2001). Rupture of intracranial aneurysms during endovascular coiling: Management and outcomes. *Neurosurgery* **49,** 807–811.

Lindsay, K. W., Teasdale, G., Knill Jones, R. P., and Murray, L. (1982). Observer variability in grading patients with subarachnoid hemorrhage. *J. Neurosurg.* **56,** 628–633.

Lindsay, K. W., Teasdale, G. M., and Knill Jones, R. P. (1983). Observer variability in assessing the clinical features of subarachnoid hemorrhage. *J. Neurosurg.* **58,** 57–62.

Linn, F. H. H., Rinkel, G. J. E., Algra, A., and van Gijn, J. (1996). Incidence of subarachnoid hemorrhage—Role of region, year, and rate of computed tomography: A meta-analysis. *Stroke* **27,** 625–629.

Linn, F. H. H., Rinkel, G. J. E., Algra, A., and van Gijn, J. (1998). Headache characteristics in subarachnoid haemorrhage and benign thunderclap headache. *J. Neurol. Neurosurg. Psychiatry* **65,** 79–793.

Linn, F. H. H., Wijdicks, E. F. M., van der Graaf, Y., Weerdesteyn-van Vliet, F. A., Bartelds, A. I., and van Gijn, J. (1994). Prospective study of sentinel headache in aneurysmal subarachnoid haemorrhage. *Lancet* **344,** 590–593.

Linskey, M. E., Horton, J. A., Rao, G. R., and Yonas, H. (1991). Fatal rupture of the intracranial carotid artery during transluminal angioplasty for vasospasm induced by subarachnoid hemorrhage. Case report. *J. Neurosurg.* **74,** 985–990.

Longstreth, W. T., Jr., Nelson, L. M., Koepsell, T. D., and van Belle, G. (1993). Clinical course of spontaneous subarachnoid hemorrhage: A population-based study in King County, Washington. *Neurology* **43,** 712–718.

Lysakowski, C., Walder, B., Costanza, M. C., and Tramér, M. R. (2001). Transcranial Doppler versus angiography in patients with vasospasm due to a ruptured cerebral aneurysm—A systematic review. *Stroke* **32,** 2292–2298.

Macdonald, R. L., Wallace, M. C., and Kestle, J. R. (1993). Role of angiography following aneurysm surgery. *J. Neurosurg.* **79,** 826–832.

Malek, A. M., Halbach, V. V., Phatouros, C. C., Lempert, T. E., Meyers, P. M., Dowd, C. F., and Higashida, R. T. (2000). Balloon-assist technique for endovascular coil embolization of geometrically difficult intracranial aneurysms. *Neurosurgery* **46,** 1397–1406.

Manabe, H., Fujita, S., Hatayama, T., Suzuki, s., an dYagihashi, S. (1998). Rerupture of coil-embolized aneurysm during long-term observation—Case report. *J. Neurosurg.* **88,** 1096–1098.

Manninen, P. H., Ayra, B., Gelb, A. W., and Pelz, D. (1995). Association between electrocardiographic abnormalities and intracranial blood in patients following acute subarachnoid hemorrhage. *J. Neurosurg. Anesthesiol.* **7,** 12–16.

Marion, D. W., Segal, R., and Thompson, M. E. (1986). Subarachnoid hemorrhage and the heart. *Neurosurgery* 18, 101–106.

Mayer, S. A., Fink, M. E., Homma, S., Sherman, D., LiMandri, G., Lennihan, L., Solomon, R. A., Klebanoff, L. M., Beckford, A., and Raps, E. C. (1994). Cardiac injury associated with neurogenic pulmonary edema following subarachnoid hemorrhage. *Neurology* 44, 815–820.

Mayer, S. A., Lin, J., Homma, S., Solomon, R. A., Lennihan, L., Sherman, D., Fink, M. E., Beckford, A., and Klebanoff, L. M. (1999). Myocardial injury and left ventricular performance after subarachnoid hemorrhage. *Stroke* 30, 780–786.

Mayer, S. A., Solomon, R. A., Fink, M. E., Lennihan, L., Stern, L., Beckford, A., Thomas, C. E., and Klebanoff, L. M. (1988). Effect of 5% albumin solution on sodium balance and blood volume after subarachnoid hemorrhage. *Neurosurgery* 42, 759–767.

Mehta, V., Holness, R. O., Connolly, K., Walling, S., and Hall, R. (1996). Acute hydrocephalus following aneurysmal subarachnoid hemorrhage. *Can. J. Neurol. Sci.* 23, 40–45.

Mendelow, A. D., Stockdill, G., Steers, A. J., Hayes, J., and Gillingham, F. J. (1982). Double-blind trial of aspirin in patient receiving tranexamic acid for subarachnoid hemorrhage. *Acta Neurochir. (Wien)* 62, 195–202.

Menghini, V. V., Brown, R. D., Jr., Sicks, J. D., O'Fallon, W. M., and Wiebers, D. O. (1998). Incidence and prevalence of intracranial aneurysms and hemorrhage in Olmsted County, Minnesota, 1965 to 1995. *Neurology* 51, 405–411.

Menovsky, T., Van Rooij, W. J., Sluzewski, M., and Wijnalda, D. (2002). Coiling of ruptured pericallosal artery aneurysms. *Neurosurgery* 50, 11–14.

Metens, T., Rio, F., Balériaux, D., Roger, T., David, P., and Rodesch, G. (2000). Intracranial aneurysms: Detection with gadolinium-enhanced dynamic three-dimensional MR angiography—Initial results. *Radiology* 216, 39–46.

Milhorat, T. H. (1987). Acute hydrocephalus afer aneurysmal subarachnoid hemorrhage. *Neurosurgery* 20, 15–20.

Muizelaar, J. P., Zwienenberg, M., Rudisill, N. A., and Hecht, S. T. (1999). The prophylactic use of transluminal balloon angioplasty in patients with Fisher Grade 3 subarachnoid hemorrhage: A pilot study. *J. Neurosurg.* 91, 51–58.

Naredi, S., Lambert, G., Edén, E., Zäll, S., Runnerstam, M., Rydenhag, B., and Friberg, P. (2000). Increased sympathetic nervous activity in patients with nontraumatic subarachnoid hemorrhage. *Stroke* 31, 901–906.

Newell, D. W., Eskridge, J. M., Mayberg, M. R., Grady, M. S., and Winn, H. R. (1989). Angioplasty for the treatment of symptomatic vasospasm following subarachnoid hemorrhage. *J. Neurosurg.* 71, 654–660.

Nichols, D. A., Meyer, F. B., Piepgras, D. G., and Smith, P. L. (1994). Endovascular treatment of intracranial aneurysms. *Mayo Clin. Proc.* 69, 272–285.

Nieuwkamp, D. J., de Gans, K., Rinkel, G. J., and Algra, A. (2000). Treatment and outcome of severe intraventricular extension in patients with subarachnoid or intracerebral hemorrhage: A systematic review of the literature. *J. Neurol.* 247, 117–121.

Nishioka, H. (1966). Report on the cooperative study of intracranial aneurysms and subarachnoid hemorrhage. Section VII. I. Evaluation of the conservative management of ruptured intracranial aneurysms. *J. Neurosurg.* 25, 574–592.

Noguchi, K., Ogawa, T., Inugami, A., Toyoshima, H., Sugawara, S., Hatazawa, J., Fujita, H., Shimosegawa, E., Kanno, I., Okudera, T., Uemura, K., and Yasui, N. (1995). Acute subarachnoid hemorrhage: MR imaging with fluid-attenuated inversion recovery pulse sequences. *Radiology* 196, 773–777.

Noguchi, K., Ogawa, T., Seto, H., Inugami, A., Hadeishi, H., Fujita, H., Hatazawa, J., Shimosegawa, E., Okudera, T., and Uemura, K. (1997). Subacute and chronic subarachnoid hemorrhage: Diagnosis with fluid-attenuated inversion-recovery MR imaging. *Radiology* 203, 257–262.

O'Sullivan, M. G., Whyman, M., Steers, J. W., Whittle, I. R., and Miller, J. D. (1994). Acute subdural haematoma secondary to ruptured intracranial aneurysm: Diagnosis and management. *Br. J. Neurosurg.* 8, 439–445.

Ogawa, A., Suzuki, M., and Ogasawara, K. (2000). Aneurysms at nonbranching sites in the supraclinoid portion of the internal carotid artery: Internal carotid artery trunk aneurysms. *Neurosurgery* 47, 578–583.

Ogawa, T., Inugami, A., Fujita, H., Hatazawa, J., Shimosegawa, E., Noguchi, K., Okudera, T., Kanno, I., Uemura, K., Suzuki, A., and Yasui, N. (1995). MR diagnosis of subacute and chronic subarachnoid hemorrhage: Comparison with CT. *Am. J. Roentgenol.* 165, 1257–1262.

Ohkuma, H., Suzuki, S., Kimura, M., and Sobata, E. (1991). Role of platelet function in symptomatic cerebral vasospasm following aneurysmal subarachnoid hemorrhage. *Stroke* 22, 854–859.

Ohkuma, H., Tsurutani, H., and Suzuki, S. (2001). Incidence and significance of early aneurysmal rebleeding before neurosurgical or neurological management. *Stroke* 32, 1176–1180.

Okada, Y., Shima, T., Nishida, M., Yamane, K., Hatayama, T., Yamanaka, C., and Yoshida, A. (1999). Comparison of transcranial Doppler investigation of aneurysmal vasospasm with digital subtraction angiographic and clinical findings. *Neurosurgery* 45, 443–449.

Öhman, J., and Heiskanen, O. (1989). Timing of operation for ruptured supratentorial aneurysms: A prospective randomized study. *J. Neurosurg.* 70, 55–60.

Papo, I., Bodosi, M., Merei, T. F., and Luongo, A. (1984). [Hydrocephalus following subarachnoid hemorrhage] L'hydrocephalie apres hemorragie sous-arachnoidienne. *Neurochirurgie* 30, 159–164.

Paré, L., Delfino, R., and Leblanc, R. (1992). The relationship of ventricular drainage to aneurysmal rebleeding. *J. Neurosurg.* 76, 422–427.

Parr, M. J., Finfer, S. R., and Morgan, M. K. (1996). Reversibel cardiogenic shock complicating subarachnoid haemorrhage. *Br. Med. J.* 313, 681–683.

Pepin, M., Schwarze, U., Superti-Furga, A., and Byers, P. H. (2000). Clinical and genetic features of Ehlers-Danlos syndrome type IV, the vascular type [see comments]. *N. Engl. J. Med.* 342, 673–680.

Pfausler, B., Belcl, R., Metzler, R., Mohsenipour, I., and Schmutzhard, E. (1996). Terson's syndrome in spontaneous subarachnoid hemorrhage: A prospective study in 60 consecutive patients. *J. Neurosurg.* 85, 392–394.

Pickard, J. D. (1984). Early posthaemorrhagic hydrocephalus. *Br. Med. J.* 289, 569–570.

Pickard, J. D., Kirkpatrick, P. J., Melsen, T., Andreasen, R. B., Gelling, L., Fryer, T., Matthews, J., Minhas, P., Hutchinson, P. J. A., Mennon, d., Downey, S. P., Kendall, I., Clark, J., Carpenter, T. A., Williams, E., and Persson, L. (2000). Potential role of NovoSeven (R) in the prevention of rebleeding following aneurysmal subarachnoid haemorrhage. *Blood Coagulation & Fibrinolysis* 11, S117–S120.

Pickard, J. D., Murray, G. D., Illingworth, R., Shaw, M. D., Teasdale, G. M., Foy, P. M., Humphrey, P. R., Lang, D. A., Nelson, R., Richards, P.h, and *et al.* (1989). Effect of oral nimodipine on cerebral infarction and outcome after subarachnoid haemorrhage: British aneurysm nimodipine trial. *Br. Med. J.* 298, 636–642.

Pinto, A. N., Canhao, P., and Ferro, J. M. (1996). Seizures at the onset of subarachnoid haemorrhage. *J. Neurol.* 243, 161–164.

Pinto, A. N., Ferro, J. M., Canhao, P., and Campos, J. (1993). How often is a perimesencephalic subarachnoid haemorrhage CT pattern caused by ruptured aneurysms? *Acta Neurochir. (Wien)* 124, 79–81.

Polin, R. S., Coenen, V. A., Hansen, C. A., Shin, P., Baskaya, M. K., Nanda, A., and Kassell, N. F. (2000). Efficacy of transluminal angioplasty for the management of symptomatic cerebral vasospasm following aneurysmal subarachnoid hemorrhage. *J. Neurosurg.* **92,** 284–290.

Polin, R. S., Hansen, C. A., German, P., Chadduck, J. B., and Kassell, N. F. (1998). Intra-arterially administered papaverine for the treatment of symptomatic cerebral vasospasm. *Neurosurgery* **42,** 1256–1264.

Pollick, C., Cujec, B., Parker, S., and Tator, C. (1988). Left ventricular wall motion abnormalities in subarachnoid hemorrhage: An echocardiographic study. *J. Am. Coll. Cardiol.* **12,** 600–605.

Qureshi, A. I., Suri, M. F. K., Khan, J., Kim, S. H., Fessler, R. D., Ringer, A. J., Guterman, L. R., and Hopkins, L. N. (2001). Endovascular treatment of intracranial aneurysms by using Guglielmi detachable coils in awake patients: Safety and feasibility. *J. Neurosurg.* **94,** 880–885.

Raaymakers, T. W. M., Buys, P. C., Verbeeten, B., Jr., Ramos, L. M. P., Witkamp, T. D., Hulsmans, F. J., Mali, W. P. T. M., Algra, A., Bonsel, G. J., Bossuyt, P. M. M., Vonk, C. M., Buskens, E., Limburg, M., van Gijn, J., Gorissen, A., Greebe, P., Albrecht, K. W., Tulleken, C. A. F., and Rinkel, G. J. E. (1999). MR angiography as a screening tool for intracranial aneurysms: Feasibility, test characteristics, and interobserver agreement. *Am. J. Roentgenol.* **173,** 1469–1475.

Raimondi, A. J. and Torres, H. (1973). Acute hydrocephalus as a complication of subarachnoid hemorrhage. *Surg. Neurol.* **1,** 23–26.

Rätsep, T. and Asser, T. (2001). Cerebral hemodynamic impairment after aneurysmal subarachnoid hemorrhage as evaluated using transcranial Doppler ultrasonography: Relationship to delayed cerebral ischemia and clinical outcome. *J. Neurosurg.* **95,** 393–401.

Reijneveld, J. C., Wermer, M., Boonman, Z., van Gijn, J., and Rinkel, G. J. E. (2000). Acute confusional state as presenting feature in aneurysmal subarachnoid hemorrhage: frequency and characteristics. *J. Neurol.* **247,** 112–116.

Renowden, S. A., Molyneux, A. J., Anslow, P., and Byrne, J. V. (1994). The value of MRI in angiogram-negative intracranial haemorrhage. *Neuroradiology* **36,** 422–425.

Rhoney, D. H., Tipps, L. B., Murry, K. R., Basham, M. C., Michael, D. B., and Coplin, W. M. (2000). Anticonvulsant prophylaxis and timing of seizures after aneurysmal subarachnoid hemorrhage. *Neurology* **55,** 258–265.

Rinkel, G. J. E., Djibuti, M., Algra, A., and van Gijn, J. (1998). Prevalence and risk of rupture of intracranial aneurysms—A systematic review. *Stroke* **29,** 251–256.

Rinkel, G. J. E. and van Gijn, J. (1995). Perimesencephalic haemorrhage in the quadrigeminal cistern. *Cerebrovasc. Dis.* **5,** 312–313.

Rinkel, G. J. E., van Gijn, J., and Wijdicks, E. F. M. (1993). Subarachnoid hemorrhage without detectable aneurysm. A review of the causes. *Stroke* **24,** 1403–1409.

Rinkel, G. J. E., Wijdicks, E. F. M., Ramos, L. M. P., and van Gijn, J. (1990a). Progression of acute hydrocephalus in subarachnoid haemorrhage: A case report documented by serial CT scanning. *J. Neurol. Neurosurg. Psychiatry* **53,** 354–355.

Rinkel, G. J. E., Wijdicks, E. F. M., Vermeulen, M., Hageman, L. M., Tans, J. T., and van Gijn, J. (1990b). Outcome in perimesencephalic (nonaneurysmal) subarachnoid hemorrhage: A follow-up study in 37 patients. *Neurology* **40,** 1130–1132.

Rinkel, G. J. E., Wijdicks, E. F. M., Vermeulen, M., Hasan, D., Brouwers, P. J. A. M., and van Gijn, J. (1991b). The clinical course of perimesencephalic nonaneurysmal subarachnoid hemorrhage. *Ann. Neurol.* **29,** 463–468.

Rinkel, G. J. E., Wijdicks, E. F. M., Vermeulen, M., Ramos, L. M. P., Tanghe, H. L., Hasan, D., Meiners, L. C., and van Gijn, J. (1991a). Nonaneurysmal perimesencephalic subarachnoid hemorrhage: CT and MR patterns that differ from aneurysmal rupture. *Am. J. Neuroradiol.* **12,** 829–834.

Rinkel, G. J. E., Wijdicks, E. F. M., Vermeulen, M., Tans, J. T. J., Hasan, D., and van Gijn, J. (1992). Acute hydrocephalus in nonaneurysmal perimesencephalic hemorrhage: Evidence of CSF block at the tentorial hiatus. *Neurology* **42,** 1805–1807.

Ronkainen, A., Puranen, M. I., Hernesniemi, J. A., Vanninen, R. L., Partanen, P. L., Saari, J. T., Vainio, P. A., and Ryynanen, M. (1995). Intracranial aneurysms: MR angiographic screening in 400 asymptomatic individuals with increased familial risk. *Radiology* **195,** 35–40.

Roos, Y. B. W. E. M. and for the STAR Study Group (2000). Antifibrinolytic treatment in subarachnoid hemorrhage—A randomized placebo-controlled trial. *Neurology* **54,** 77–82.

Roos, Y. B. W. E. M., Hasan, D., and Vermeulen, M. (1995). Outcome in patients with large intraventricular haemorrhages: A volumetric study. *J. Neurol. Neurosurg. Psychiatry* **58,** 622–624.

Roos, Y. B. W. E. M., Rinkel, G. J. E., Vermeulen, M., Algra, A., and van Gijn, J. (2000). "Antifibrinolytic Treatment in Aneurysmal Subarachnoid Haemorrhage," Issue 3. The Cochrane Library, Update Software, Oxford.

Rosenfeld, J. V., Barnett, G. H., Sila, C. A., Little, J. R., Bravo, E. L., and Beck, G. J. (1989). The effect of subarachnoid hemorrhage on blood and CSF atrial natriuretic factor. *J. Neurosurg.* **71,** 32–37.

Rosenwasser, R. H., Delgado, T. E., Buchheit, W. A., and Freed, M. H. (1983). Control of hypertension and prophylaxis against vasospasm in cases of subarachnoid hemorrhage: A preliminary report. *Neurosurgery* **12,** 658–661.

Ruigrok, Y. M., Buskens, E., and Rinkel, G. J. E. (2001). Attributable risk of common and rare determinants of subarachnoid hemorrhage. *Stroke* **32,** 1173–1175.

Ruigrok, Y. M., Rinkel, G. J. E., Buskens, E., Velthuis, B. K., and van Gijn, J. (2000). Perimesencephalic hemorrhage and CT angiography—A decision analysis. *Stroke* **31,** 2976–2983.

Saito, I., Asano, T., Ochiai, C., Takakura, K., Tamura, A., and Sano, K. (1983). A double-blind clinical evaluation of the effect of Nizofenone (Y-9179) on delayed ischemic neurological deficits following aneurysmal rupture. *Neurol. Res.* **5,** 29–47.

Saito, I., Asano, T., Sano, K., Takakura, K., Abe, H., Yoshimoto, T., Kikuchi, H., Ohta, T., and Ishibashi, S. (1998). Neuroprotective effect of an antioxidant, ebselen, in patients with delayed neurological deficits after aneurysmal subarachnoid hemorrhage. *Neurosurgery* **42,** 269–277.

Saitoh, H., Hayakawa, K., Nishimura, K., Okuno, Y., Teraura, T., Yumitori, K., and Okumura, A. (1995). Rerupture of cerebral aneurysms during angiography. *Am. J. Neuroradiol.* **16,** 539–542.

Sakas, D. E., Dias, L. S., and Beale, D. (1995). Subarachnoid hemorrhage presenting as head injury. *Br. Med. J.* **310,** 1186–1187.

Sarner, M., and Rose, F. C. (1967). Clinical presentation of ruptured intracranial aneurysm. *J. Neurol. Neurosurg. Psychiatry* **30,** 67–70.

Sasaki, T., Kodama, N., and Itokawa, H. (1996). Aneurysm formation and rupture at the site of anastomosis following bypass surgery. *J. Neurosurg.* **85,** 500–502.

Sayama, T., Inamura, T., Matsushima, T., Inoha, S., Inoue, T., and Fukui, M. (2000). High incidence of hyponatremia in patients with ruptured anterior communicating artery aneurysms. *Neurol. Res.* **22,** 151–155.

Schievink, W. I. (1997). Genetics of intracranial aneurysms. *Neurosurgery* **40,** 651–662.

Schievink, W. I., Schaid, D. J., Michels, V. V., and Piepgras, D. G. (1995). Familial aneurysmal subarachnoid hemorrhage: A community-based study. *J. Neurosurg.* **83,** 426–429.

Schievink, W. I., Schaid, D. J., Rogers, H. M., Piepgras, D. G., and Michels, V. V. (1994). On the inheritance of intracranial aneurysms. *Stroke* 25, 2028–2037.

Schievink, W. I., Torres, V. E., Piepgras, D. G., and Wiebers, D. O. (1992). Saccular intracranial aneurysms in autosomal dominant polycystic kidney disease. *J. Am. Soc. Nephrol.* 3, 88–95.

Schievink, W. I., and Wijdicks, E. F. M. (1997). Pretruncal subarachnoid hemorrhage: An anatomically correct description of the perimesencephalic subarachnoid hemorrhage. *Stroke* 28, 2572–2572.

Schoser, B. G., Heesen, C., Eckert, B., and Thie, A. (1997). Cerebral hyperperfusion injury after percutaneous transluminal angioplasty of extracranial arteries. *J. Neurol.* 244, 101–104.

Schwartz, T. H. and Mayer, S. A. (2000). Quadrigeminal variant of perimesencephalic nonaneurysmal subarachnoid hemorrhage. *Neurosurgery* 46, 584–588.

Schwartz, T. H. and Solomon, R. A. (1996). Perimesencephalic nonaneurysmal subarachnoid hemorrhage: Review of the literature. *Neurosurgery* 39, 433–440.

Shaw, M. D., Foy, P. M., Conway, M., Pickard, J. d., Maloney, P., Spillane, J. A., and Chadwick, D. W. (1985). Dipyridamole and postoperative ischemic deficits in aneurysmal subarachnoid hemorrhage. *J. Neurosurg.* 63, 699–703.

Shaw, M. D. M., Vermeulen, M., Murray, G. D., Pickard, J. D., Bell, B. A., and Teasdale, G. M. (2000). Efficacy and safety of the endothelin$_{A/B}$ receptor antagonist TAK-044 in treating subarachnoid hemorrhage: A report by the Steering Committee on behalf of the UK/Netherlands/Eire TAK-044 Subarachnoid Haemorrhage Study Group. *J. Neurosurg.* 93, 992–997.

Sloan, M. A., Burch, C. M., Wozniak, M. A., Rothman, M. I., Rigamonti, D., Permutt, T., and Numaguchi, Y. (1994). Transcranial Doppler detection of vertebrobasilar vasospasm following subarachnoid hemorrhage. *Stroke* 25, 2187–2197.

Sloan, M. A., Haley, E. C., Jr., Kassell, N. F., Henry, M. L., Stewart, S. R., Beskin, R. R., Sevilla, E. A., and Torner, J. C. (1989). Sensitivity and specificity of transcranial Doppler ultrasonography in the diagnosis of vasospasm following subarachnoid hemorrhage. *Neurology* 39, 1514–1518.

Sluzewski, M., Bosch, J. A., Van Rooij, W. j., Nijssen, P. C. G., and Wijnalda, D. (2001). Rupture of intracranial aneurysms during treatment with Guglielmi detachable coils: Incidence, outcome, and risk factors. *J. Neurosurg.* 94, 238–240.

Smith, B. (1963). Cerebral pathology in subarachnoid haemorrhage. *J. Neurol. Neurosurg. Psychiatry* 26, 67–70.

Solenski, N. J., Haley, E. C., Jr., Kassell, N. F., Kongable, G., Germanson, T., Truskowski, L., Torner, J. C., and Particip Multictr Coop Aneurysm Study (1995). Medical complications of aneurysmal subarachnoid hemorrhage: A report of the multicenter, cooperative aneurysm study. *Crit. Care Med.* 23, 1007–1017.

Stehbens, W. E. (1982). Intracranial berry aneurysms in infancy. *Surg. Neurol.* 18, 58–60.

Sterns, R. H., Riggs, J. E., and Schochet, S. S. Jr. (1986). Osmotic demyelination syndrome following correcting of hyponatremia. *N. Engl. J. Med.* 314, 1535–1542.

Stober, T., Anstatt, T., Sen, S., Schimrigk, K., and Jager, H. (1988). Cardiac arrhythmias in subarachnoid haemorrhage. *Acta Neurochir. (Wien)* 93, 37–44.

Suzuki, M., Otawara, Y., Doi, M., Ogasawara, K., and Ogawa, A. (2000). Neurological grades of patients with poor-grade subarachnoid hemorrhage improve after short-term pretreatment. *Neurosurgery* 47, 1098–1104.

Suzuki, S., Sano, K., Handa, H., Asano, T., Tamura, A., Yonekawa, Y., Ono, H., Tachibana, N., and Hanaoka, K. (1989). Clinical study of OKY-046, a thromboxane synthetase inhibitor, in prevention of cerebral vasospasms and delayed cerebral ischaemic symptoms after subarachnoid haemorrhage due to aneurysmal rupture: A randomized double-blind study. *Neurol. Res.* 11, 79–88.

Svensson, E., Starmark, J. E., Ekholm, S., Von Essen, C., and Johansson, A. (1996). Analysis of interobserver disagreement in the assessment of subarachnoid blood and acute hydrocephalus on CT scans. *Neurol. Res.* 18, 487–494.

Swash, M. (1974). Periaqueductal dysfunction (the Sylvian aqueduct syndrome): A sign of hydrocephalus? *J. Neurol. Neurosurg. Psychiatry* 37, 21–26.

Tateshima, S., Murayama, Y., Gobin, Y. P., Duckwiler, G. R., Guglielmi, G., and Viñuela, F. (2000). Endovascular treatment of basilar tip aneurysms using Guglielmi detachabel coils: Anatomic and clinical outcomes in 73 patients from a single institution. *Neurosurgery* 47, 1332–1339.

Teasdale, G., and Jennett, B. (1974). Assessment of coma and impaired consciousness—A practical scale. *Lancet* 2, 81–84.

Teunissen, L. L., Rinkel, G. J. E., Algra, A., and van Gijn, J. (1996). Risk factors for subarachnoid hemorrhage—A systematic review. *Stroke* 27, 544–549.

Theodore, J. and Robin, E. D. (1976). Speculations on neurogenic pulmonary edema (NPE). *Am. Rev. Respir. Dis.* 113, 405–411.

Thornton, J., Debrun, G. M., Aletich, V. A., Bashir, Q., Charbel, F. T., and Ausman, J. (2002). Follow-up angiography of intracranial aneurysms treated with endovascular placement of Guglielmi detachable coils. *Neurosurgery* 50, 239–249.

Tokiyoshi, K., Ohnishi, T., and Nii, Y. (1991). Efficacy and toxicity of thromboxane synthetase inhibitor for cerebral vasospasm after subarachnoid hemorrhage. *Surg. Neurol.* 36, 112–118.

Tokuda, Y., Inagawa, T., Takechi, A., and Inokuchi, F. (1998). Ruptured de novo aneurysm induced by ethyl 2-cyanoacrylate: Case report. *Neurosurgery* 43, 626–628.

Tolias, C. M. and Choksey, M. S. (1996). Will increased awareness among physicians of the significance of sudden agonizing headache affect the outcome of subarachnoid hemorrhage? Coventry and Warwickshire Study: Audit of subarachnoid hemorrhage (Establishing historical controls), hypothesis, campaign layout, and cost estimation. *Stroke* 27, 807–812.

Tomida, M., Muraki, M., Uemura, K., and Yamasaki, K. (1998). Plasma concentrations of brain natriuretic peptide in patients with subarachnoid hemorrhage. *Stroke* 29, 1584–1587.

Torner, J. C., Nibbelink, D. W., and Burmeister, L. F. (1981). Statistical comparisons of end results of a randomised treatment study. *In* (A. L. Sahs, D. W. Nibbelink, J. C. Torner Eds.), "Aneurysmal Subarachnoid Hemorrhage—Report of the Cooperative Study" pp. 249–275. Urban & Schwartzenberg, Baltimore.

Uda, K., Murayama, Y., Gobin, Y. P., Duckwiler, G. R., and Viñuela, F. (2001). Endovascular treatment of basilar artery trunk aneurysms with Guglielmi detachable coils: Clinical experience with 41 aneurysms in 39 patients. *J. Neurosurg.* 95, 624–632.

Van Calenbergh, F., Plets, C., Goffin, J., and Velghe, L. (1993). Nonaneurysmal subarachnoid hemorrhage: Prevalence of perimesencephalic hemorrhage in a consecutive series. *Surg. Neurol.* 39, 320–323.

Van der Jagt, M., Hasan, D., Bijvoet, H. W. C., Pieterman, H., Dippel, D. W. J., Vermeij, F. H., and Avezaat, C. J. J. (1999). Validity of prediction of the site of ruptured intracranial aneurysms with CT. *Neurology* 52, 34–39.

van der Wee, N., rinkel, G. J. E., Hasan, D., and van Gijn, J. (1995). Detection of subarachnoid haemorrhage on early CT: Is lumbar puncture still needed after a negative scan? *J. Neurol. Neurosurg. Psychiatry* 58, 357–359.

van Gijn, J., Hijdra, A., Hijdicks, E. F. M., Vermeulen, M., and van Crevel, H. (1985b). Acute hydrocephalus after aneurysmal subarachnoid hemorrhage. *J. Neurosurg.* 63, 355–362.

van Gijn, J., and van Dongen, K. J. (1980b). Computed tomography in the diagnosis of subarachnoid haemorrhage and ruptured aneurysm. *Clin. Neurol. Neurosurg.* 82, 11–24.

van Gijn, J., and van Dongen, K. J. (1980a). Computerized tomography in subarachnoid hemorrhage: Difference between patients with and without an aneurysm on angiography. *Neurology* 30, 538–539.

van Gijn, J., and van Dongen, K. J. (1982). The time course of aneurysmal haemorrhage on computed tomograms. *Neuroradiology* 23, 153–156.

van Gijn, J., van Dongen, K. J., Vermeulen, M., and Hijdra, A. (1985a). Perimesencephalic hemorrhage: A nonaneurysmal and benign form of subarachnoid hemorrhage. *Neurology* 35, 493–497.

Vanninen, R., Koivisto, T., Saari, T., Hernesniemi, J., and Vapalahti, M. (1999). Ruptured intracranial aneurysms: Acute endovascular treatment with electrolytically detachable coils—A [prospective randomized study. *Radiology* 211, 325–336.

Velthuis, B. K., Rinkel, G. J. E., Ramos, L. M. P., Witkamp, T. D., Berkelbach van der Sprenkel, J. W., Vandertop, W. P., and Van leeuwen, M. S. (1998). Subarachnoid hemorrhage: Aneurysm detection and preoperative evaluation with CT angiography. *Radiology* 208, 423–430.

Velthuis, B. K., Rinkel, G. J. E., Ramos, L. M. P., Witkamp, T. D., and Van Leeuwen, M. S. (1999a). Perimesencephalic hemorrhage—Exclusion of vertebrobasilar aneurysms with CT angiography. *Stroke* 30, 1103–1109.

Velthuis, B. K., Van Leeuwen, M. S., Witkamp, T. D., Boomstra, S., Ramos, L. M., and Rinkel, G. J. E. (1997). CT angiography: Source images and postprocessing techniques in the detection of cerebral aneurysms. *Am. J. Roentgenol.* 169, 1411–1417.

Velthuis, B. K., Van Leeuwen, M. S., Witkamp, T. D., Ramos, L. M. P., Berkelbach van der Sprenkel, J. W., and Rinkel, G. J. E. (1999b). Computerized tomography angiography in patients with subarachnoid hemorrhage: From aneurysm detection to treatment without conventional angiography. *J. Neurosurg.* 91, 761–767.

Vermeer, S. E., Rinkel, G. J. E., and Algra, A. (1997). Circadian fluctuations in onset of subarachnoid hemorrhage—New data on aneurysmal and perimesencephalic hemorrhage and a systematic review. *Stroke* 28, 805–808.

Vermeij, F. H., Hasan, D., Bijvoet, H. W. C., and Avezaat, C. J. J. (1998). Impact of medical treatment on the outcome of patients after aneurysmal subarachnoid hemorrhage. *Stroke* 29, 924–930.

Vermeulen, M., Hasan, D., Blijenberg, B. G., Hijdra, A., and van Gijn, J. (1989). Xanthochromia after subarachnoid haemorrhage needs no revisitation, *J. Neurol. Neurosurg. Psychiatry* 52, 826–828.

Vermeulen, M., Lindsay, K. W., Murray, G. D., Cheah, F., Hijdra, A., Muizelaar, J. P., Schannong, M., Teasdale, G. M., van Crevel, H., and van Gijn, J. (1984). Antifibrinnolytic treatment in subarachnoid hemorrhage. *N. Engl. J. Med.* 311, 432–437.

Vermeulen, M., and van Gijn, J. (1990). The diagnosis of subarachnoid haemorrhage. *J. Neurol. Neurosurg. Psychiatry* 53, 365–372.

Vinge, E., Brandt, L., Ljunggren, B., and Andersson, K. E. (1988). Thromboxane B2 levels in serum during continuous administration of nimodipine to patients with aneurysmal subarachnoid hemorrhage. *Stroke* 19, 644–647.

Vora, Y. Y. Suarez-Almazor, M., Steinke, D. E., Martin, M. L., and Findlay, J. M. (1999). Role of transcranial Doppler monitoring in the diagnosis of cerebral vasospasm after subarachnoid hemorrhage. *Neurosurgery* 44, 1237–1247.

Vos, P. E., Zwienenberg, M., O'Hannian, K. L., and Muizelaar, J. P. (2000). Subarachnoid haemorrhage following rupture of an ophthalmic artery aneurysm presenting as traumatic brain injury. *Clin. Neurol. Neurosurg.* 102, 29–32.

Wang, P. S., Longstreth, W. T., Jr., and Koepsell, T. D. (1995). Subarachnoid hemorrhage and family history. A population-based case-control study. *Arch. Neurol.* 52, 202–204.

Wardlaw, J. M., and White, P. M. (2000). The detection and management of unruptured intracranial aneurysms. *Brain* 123, 205–221.

Wauchob, T. D., Brooks, R. J., and Harrison, K. M. (1984). Neurogenic pulmonary oedema. *Anaesthesia* 39, 529–534.

Weir, B., Grace, M., Hansen, J., and Rothberg, C. (1978). Time course of vasospasm in man. *J. Neurosurg.* 48, 173–178.

Weir, B. K. (1978). Pulmonary edema following fatal aneurysm rupture. *J. Neurosurg.* 49, 502–507.

Wijdicks, E. F. M. and Borel, C. O. (1998). Respiratory management in acute neurologic illness. *Neurology* 50, 11–20.

Wijdicks, E. F. M., Ropper, A. H., Hunnicutt, E. J., Richardson, G. S., and Nathanson, J. A. (1991). Atrial natriuretic factor and salt wasting after aneurysmal subarachnoid hemorrhage. *Stroke* 22, 1519–1524.

Wijdicks, E. F. M., Schievink, W. I., and Burnett, J. C., Jr. (1997). Natriuretic peptide system and endothelin in aneurysmal subarachnoid hemorrhage. *J. Neurosurg.* 87, 275–280.

Wijdicks, E. F. M., van Dongen, K. J., van Gijn, J., Hijdra, A., and Vermeulen, M. (1988). Enlargement of the third ventricle and hyponatraemia in aneurysmal subarachnoid haemorrhage. *J. Neurol. Neurosurg. Psychiatry* 51, 516–520.

Wijdicks, E. F. M., Vermeulen, M., Hijdra, A., and van Gijn, J. (1985b). Hyponatremia and cerebral infarction in patients with ruptured intracranial aneurysms: Is fluid restriction harmful? *Ann. Neurol.* 17, 137–140.

Wijdicks, E. F. M., Vermeulen, M., Murray, G. D., Hijdra, A., and van Gijn, J. (1990). The effects of treating hypertension following aneurysmal subarachnoid hemorrhage. *Clin. Neurol. Neurosurg.* 92, 111–117.

Wijdicks, E. F. M., Vermeulen, M., ten Haaf, J. A., Hijdra, A., Bakker, W. H., and van Gijn, J. (1985a). Volume depletion and natriuresis in patients with a ruptured intracranial aneurysm. *Ann. Neurol.* 18, 211–216.

Wijdicks, E. F. M., Vermeulen, M., van Brummelen, P., den Boer, N. C., and van Gijn, J. (1987). Digoxin-like immunoreactive substance in patients with aneurysmal subarachnoid haemorrhage. *Br. Med. J.* 294, 729–732.

Winn, H. R., Richardson, A. E., and Jane, J. A. (1977). The long-term prognosis in untreated cerebral aneurysms. I. The incidence of late hemorrhage in cerebral aneurysm: A 10-year evaluation of 364 patients. *Ann. Neurol.* 1, 358–370.

Wohns, R. N., Tamas, L., Pierce, K. R., and Howe, J. F. (1985). Chlorpromazing treatment for neurogenic pulmonary edema. *Crit. Care Med.* 13, 210–211.

Zentner, J., Solymosi, L., and Lorenz,, M. (1996). Subarachnoid hemorrhage of unknown etiology. *Neurol. Res.* 18, 220–226.

*Cerebrovascular Disorders*

CHAPTER 37

# Cerebral Venous and Sinus Thrombosis

Michael Strupp, Karl M. Einhäupl, and
Marie-Germaine Bousser

## CLINICAL FEATURES

Cerebral venous and sinus thrombosis (CVST) is a disorder with strikingly diverse manifestations that range from isolated headache to psychiatric symptoms and may mimic a host of other neurological conditions. Its mode of onset (acute—like subarachnoid hemorrhage—or weeklong progressive development), clinical course, and appearance in cranial computer tomography and magnetic resonance imaging vary greatly. Thus, CVST still poses a diagnostic and therapeutic challenge and should be considered in almost any brain syndrome. It is reasonable to distinguish between nonseptic and septic CVST. Septic CVST occurs far less frequently and almost always in patients with bacterial cranial infections, e.g., otitis, mastoiditis, and sinusitis. Its clinical picture may vary, its prognosis is worse than that of nonseptic CVST, and it requires a different treatment.

The frequency of nonseptic CVST is certainly underestimated, as it is often overlooked, especially due to the variety of its presenting symptoms. Isolated involvement of a sinus often leads to increased intracranial pressure without localizing signs or symptoms, whereas additional extension of the thrombus into cortical veins or the involvement of multiple dural sinuses leads to focal neurological deficits. The neurological symptoms mainly depend on the sinus (superior sagittal, lateral, or cavernous) or veins (cortical or deep) involved (see Table I), but clinicoanatomical correlation is not always possible, and often multiple sinuses are involved. In contrast to arterial stroke, the mode of onset is subacute in 65–70% of patients with CVST (Ameri and Bousser, 1992), and the clinical course often fluctuates. During pregnancy and puerperium, however, the onset is often acute (Cantu and Barinagarrementeria, 1993). The

development of intracerebral hemorrhage (ICH) (Figure 1) in one-third of the patients may lead to acute focal neurological deficits or their acute worsening.

Headache, seizures, focal neurological deficits, altered consciousness, and papilledema, which can present in isolation or in association, are the most frequent signs. The leading symptom of CVST is headache in 70–90% of all cases (in most cases severe, persisting, unilateral, or bilateral); it precedes the appearance of focal neurological deficits (Ameri and Bousser, 1994). Focal or more generalized disturbances, such as seizures, appear in 35–50%, motor deficits in 35–60%, disorders of consciousness in 30–60%, papilledema in 25–50%, and cognitive and behavioral symptoms in about 25%. Of all patients, 17–40% present with isolated intracranial hypertension mimicking benign intracranial hypertension (BIH) as the clinical manifestation of CVST. A primary and extended thrombosis of a large sinus leads to more generalized neurological disorders (headache, disorders of consciousness), whereas patients with isolated cortical vein thrombosis have more focal neurological symptoms. Although rare, thrombosis of deep cerebral veins may appear as diencephalic edema mimicking a tumor or thalamic hemorrhage. All in all, differential diagnosis must include ischemic or hemorrhagic stroke, abscess, tumor, encephalitis, metabolic encephalopathy, BIH, or, e.g., in cavernous sinus thrombosis, Tolosa–Hunt syndrome, dysthyroid ophthalmopathy, or orbital cellulitis.

The current "gold standard" for diagnosing and following-up CVST is no longer angiography but magnetic resonance imaging (MRI) in combination with magnetic resonance angiography (MRA) (Villringer et al., 1989; Vogl et al., 1994a), since it has the ability to visualize flowing and stagnant blood. In the T1- and

**TABLE I**  Neurological Signs and Symptoms of CVST

| Sinus/vein involved | Clinical presentation | Remarks |
| --- | --- | --- |
| Superior sagittal sinus thrombosis | Headache, papilledema, intracranial hypertension<br>Generalized or focal seizures<br>Motor deficits<br>Psychiatric symptoms | The most frequently involved sinus |
| Lateral sinus thrombosis | Headache, ear or neck pain, fever, draining ear, sepsis "otogenic intracranial hypertension"; rarely focal neurological deficits due to associated intracranial abscess or thrombosis of superior petrosal sinus (e.g., hemianopia) or jugular bulb or vein to hypoglossal canal (N. IX–XII involvement) | Most often complication of otitis media, mastoiditis, chronic ear infection or cholesteatoma; right lateral sinus usually dominant lateral sinus, therefore more likely to become symptomatic |
| Cavernous sinus thrombosis | Headache, proptosis, chemosis, edema of eyelids and conjunctiva, diplopia, N. III, IV, VI ophthalmoplegia, visual disturbances or loss, involvement of N. V, rarely pituitary insufficiency or amnestic syndrome | Often due to an infection of the face or ethmoid, sphenoid, or maxillary sinuses; in most cases bilateral symptoms and signs; proof of diagnosis often difficult |
| Cortical vein thrombosis | Frontal lobe, central paresis of the leg(s) and proximal arm (often bilateral), bladder disturbances; parietal lobe, sensory loss, especially astereognosis; vein of Labbé, hemiparesis, and/or dysphasia or hemianopia; internal occipital lobe, homonymous hemianopia | Rarely isolated involvement of a single vein, often associated with sinus thrombosis and therefore bilateral |
| Deep venous thrombosis | Involvement of diencephalon, basal ganglia, and/or mesencephalon; vein of Galen, often intraventricular hemorrhage or hemorrhagic infarction of thalamus and basal ganglia | Rarely in adults; if treated early also good prognosis |
| Cerebellar vein | Brainstem and cerebellar signs and increased intracranial pressure | Should be considered a differential diagnosis of acute cerebellar or brainstem syndromes |
| Spinal cord venous thrombosis | Painful sensorimotor deficits with bladder and bowel disturbances | Often located centrally within the spinal cord |

*Note.* Neurological signs and symptoms of CVST often depend on the sinus (superior sagittal, lateral, or cavernous) or veins (cortical or deep) involved, but a clinicoanatomical correlation is not always possible, because in many cases multiple sinuses are involved, and also indirect, nonlocalizing signs develop.

T2-weighted images and spin-echo (the latter to differentiate between anatomical variants and a thrombosed sinus) as well as gradient echo images, the appearance of a thrombosed vein or sinus varies with the stage of thrombosis (as shown in Table II), depending on the products of hemoglobin and their location. Oxyhemoglobin, which appears in the first minutes to hours, is (most often) isointense on T1-weighted images and hypointense on T2-weighted images. Nevertheless, due to local edema the thrombus may appear hyperintense in the acute phase. After day 2 intracellular methemoglobin increases the signal intensity on T1-weighted images. Extracellular methemoglobin (>2 weeks after onset of CVST) leads to hyperintense signals in T1- and T2-weighted images (Figure 2). Phase contrast images, which are important for demonstrating flowing blood, show a high correlation with conventional angiography (Nadel *et al.*, 1991). To differentiate flow-artifacts from a thrombus, time-of-flight (TOF) techniques are useful. Several weeks after onset the diagnosis of CVST becomes more difficult with MRI (Isensee *et al.*, 1994), and the thrombus may appear as a contrast-enhancing structure.

Parenchymal T2 hyperintensities may reflect venous infarction or edema (Yuh *et al.*, 1994). It was shown that in CVST—in contrast to arterial stroke—signal changes in diffusion-weighted MRI are less prominent and normalize earlier during venous infarction, despite marked signal hyperintensities in the fast fluid-attenuated inversion recovery sequences. This suggests that prominent or proceeding vasogenic edema is associated with CVST (Corvol *et al.*, 1998). In perfusion imaging CVST is characterized by increased regional cerebral blood volume and slowed bolus passage but almost no perfusion deficit, which would indicate venous congestion; therefore, it may help to differentiate arterial from venous stroke (Keller *et al.*, 1999).

Cranial computed tomography (CT) is usually the first emergency investigation performed; however, in up to 30% of cases, especially the early ones, it is normal (Ameri and Bousser, 1992; Bousser, 2000). Nonenhanced CT can detect nonspecific changes such as brain swelling, localized hypodense or hyperdense areas (due to bleedings, Figure 1), and rarely, spontaneously hyperdense thrombosed sinus ("cord sign"), usually

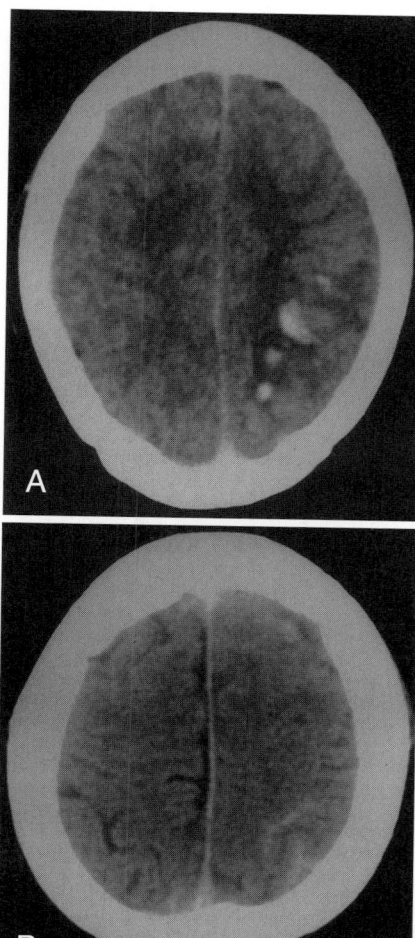

**FIGURE 1** Two different forms of bleeding in CVST (**A**) multiple parietal subcortical (intraparenchymal) left-sided bleeding with bilateral swelling; (**B**) left-sided fronto-parietal (subarachnoidal) bleeding within the gyri, like a subarachnoid hemorrhage. Reprinted with permission from Dr. K. Seelos, Department of Neuroradiology, University of Munich, Germany.

seen during the first 4–6 days after onset of CVST (Anxionnat *et al.*, 1994), especially the suggestive dense triangle ("dense or filled delta sign") (Ford and Sarwar, 1981). Contrast-enhanced CT may reveal gyral or ring enhancement (in areas of venous infarction), tentorial enhancement, and in about 20% the "empty triangle or delta sign," i.e., a central hypodensity surrounded by a margin of contrast enhancement. However, this sign is not seen on early scans (first 3–4 days). In addition to MRI and MRA, helical CT venography is indicated in the very early (before day 5) or late stages (after 6 weeks), visualizing thrombosed sinuses as filling defects, or whenever MRI shows equivocal signs.

Some rare cases of isolated cortical vein thrombosis may still require conventional cerebral angiography to establish the final diagnosis of venous thrombosis. Indirect signs (dilated collateral veins with corkscrew appearance, delayed venous emptying, collateral circulation) are more important for establishing the diagnosis in conventional angiography than the nonvisualization of a thrombosed sinus or vein.

Cerebrospinal fluid (CSF) abnormalities (found in only 50%) such as moderate lymphocytic or mixed pleocytosis (25%), elevated protein (16%), subarachnoid blood (9%), and xanthochromia are nondiagnostic. The CSF opening pressure is in most cases elevated, and it is assumed that the pressure is correlated with the involvement of the sinuses (Martin and Enevoldson, 1996), but this is also not specific. Lumbar puncture may be necessary to exclude meningitis or to remove CSF when vision is threatened.

Once CVST is recognized, the next diagnostic step is to determine its etiology from among over 100 causes that have been identified (or assumed, since a causal relationship has not always been proven). They encompass all causes of deep-vein leg thrombosis and numerous local causes such as infections, trauma, tumors, and phakomatoses (see Table III; for a review see Ameri and Bousser, 1992; Bousser, 2000). Coagulopathies were identified in about 50% of children and in 10 to 25% of adults with CVST. In particular, Factor V Leiden mutation, anticardiolipin antibodies, and prothrombin gene mutation are independent risk factors for CVST. Hormonal changes, i.e., oral contraceptives especially when there is a concomitant coagulopathy or during pregnancy (marked reduction of free protein S levels during the third trimester) and puerperium often cause CVST. In developing countries CVST related to pregnancy is still a serious problem. In India, for instance, it accounts for 25% of maternal deaths and is the most common cause of stroke in young women (Bansal *et al.*, 1980).

Despite an extensive evaluation, however, the etiology remains unknown in 20–25% of cases. Especially these patients require prolonged follow-up with repeated and detailed consultations and examinations.

## CLINICAL COURSE

There are still many open questions in CVST, e.g., incidence, natural history, morbidity, and mortality. The data available are mainly based on retrospective studies and case reports originating from heterogeneous patient groups. Further studies are required.

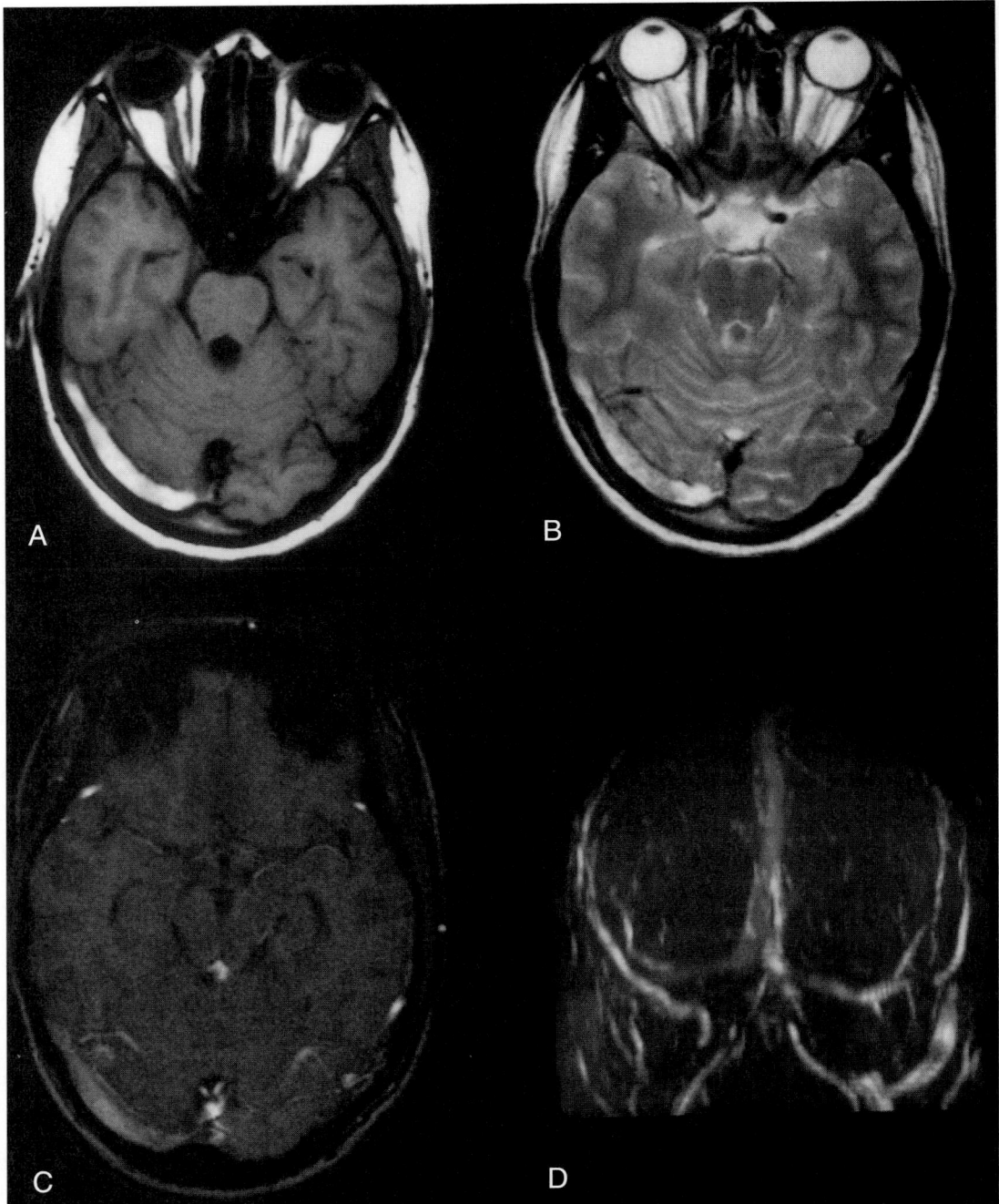

**FIGURE 2** Magnetic resonance imaging and venous magnetic resonance angiography of CVST in a 24-year-old female, 6 days after symptom onset with subacute holocephalic headache and right-sided abducens paresis, showing a complete thrombosis of the right lateral and sagittal sinus. (**A**) In the T1-weighted imaging the right lateral sinus is hyperintense, indicating a subacute thrombosis. (**B**) In the T2-weighted imaging there is an intermediate signal intensity in the middle part of the lateral sinus and a hyperintense signal in the other parts, and there is no flow void (compare with the hypointense signal in the straight sinus due to flow void). (**C**) Flow-sensitive 2D-fast low angle shot-sequence (FLASH): in the right lateral sinus the thrombus appears hyperintense (due to the "shining-through phenomenon"), whereas the straight sinus is more hyperintense due the venous blood flow. (**D**) Three-dimensional reconstruction of the flow-sensitive 2d-FLASH sequence: the thrombosed sinus as well as the other sinuses appear hyperintense, which may be a pitfall; i.e., correct diagnosis is possible only after consideration of all sequences in this case. Reprinted with permission from Dr. K. Seelos, Department of Neuroradiology, University of Munich, Germany.

**TABLE II**  Signal Characteristics of CSVT and Their Time Course in T1- and T2-Weighted and Gradient Echo Sequences as Well as Venous MRA and Contrast-Enhanced MRI

| | Acute phase (ca. days 1 to 5) | Subacute phase (ca. days 5 to 15) | Chronic phase (>several weeks) |
|---|---|---|---|
| Changes of hemoglobin over time and dominating form of hemoglobin | Oxyhemoglobin suddenly becomes deoxyhemoglobin and later methemoglobin; further, due to disruption of the erythrocyte membrane intracellular methemoglobin becomes extracellular | Extracellular methemoglobin | Fibrous tissue within the sinus or partially or completely recanalized sinus |
| T1-weighted images | Isointense | Hyperintense (first in T1-, later in T2-weighted images) | Isointense, increasing signal inhomogeneity |
| Flow void | No flow void of the thrombosed sinus or vein; increased flow void in the collateral veins | No flow void of the thrombosed sinus or vein; increased flow void in the collateral veins | In case of recanalization, flow void |
| T2-weighted images | Hypointense (→ isointense) (due to edema hyperintense signal also possible) | (Isointense →) Hyperintense (magnetic fields ≤1 T hyperintense; magnetic fields >1 T initially hypointense, later hyperintense) | Hyperintense, increasing signal inhomogeneity |
| Gradient echo Venous MRA (e.g., with time-of-flight sequence; scans with two orientations necessary) | Isointense No flow in the thrombosed sinus or vein | Hyperintense No flow in the thrombosed sinus or vein or central area of nonenhancing intensity, surrounded by an enhanced rim (e.g., "negative delta sign") or evidence for recanalization | Isointense No flow in the thrombosed sinus or vein or evidence of recanalization |
| Contrast-enhanced MRI | No intravenous contrast enhancement (e.g., "empty triangle sign" of CT) | No intravenous contrast enhancement (e.g., "empty triangle sign" of CT); dilated collateral veins | Thrombus may show contrast enhancement. In case of no recanalization dilated collateral veins |

*Note.* The appearance of CVST on MRI depends on the parameters of the sequences used, the velocity of blood flow, and the orientation of the vessels in relation to the images. Differentiation between CVST, anatomical variants, or artifacts is often difficult. Diagnosis is easy, if the signal within the sinus/vein is hyperintense on T1- and T2-weighted images (Isensee *et al.*, 1994; Nadel *et al.*, 1991; Vogl *et al.*, 1993; 1994b; Yuh *et al.*, 1994).

One of the most puzzling aspects of CVST is its diversity in onset, course, and outcome. For example, reported mortality due to CVST ranges from 5 to 80%. Initially and for almost a century CVST was almost exclusively diagnosed at autopsy and therefore assumed always lethal. Early angiographic studies reported a mortality of 30 to 80% (Hahn *et al.*, 1954). This previous assumption that CVST is a rare and severe disease with a poor prognosis had to be revised in light of more recent clinical trials (Ameri and Bousser, 1992; Bousser, 1991; de Bruijn and Stam, 1999; Einhäupl *et al.*, 1991). These trials reported a more favorable outcome, mainly as a result of earlier and better diagnosis (especially by MRI) and treatment with anticoagulants. In the future CVST will be recognized even more often due to the increasing availability of noninvasive neuroimaging techniques, especially when patients are not under the care of neurologists.

Recent retrospective studies and two prospective trials of CVST (the latter with a total of 20 (Einhäupl *et al.*, 1991) and 59 patients (de Bruijn and Stam, 1999) reported a mortality rate between 5 and 30% in patients not on anticoagulation therapy and 5–15% in patients treated with unfractioned or low-molecular-weight heparin (Ameri and Bousser, 1992; Bousser *et al.*, 1985; Cantu and Barinagarrementeria, 1993; Daif *et al.*, 1995; de Bruijn and Stam, 1999; Einhäupl *et al.*, 1991; Preter *et al.*, 1996). The main causes of death were brain herniation as a sequela of raised intracranial pressure due to brain edema or hemorrhagic infarction, status epilepticus, or pulmonary embolism. The factors of a poor prognosis are advanced age, coma, involvement of the deep or cerebellar venous system, severely raised intracranial pressure, large hemorrhagic infarcts, infectious (see below) or malignant etiologies, rapid progression of symptoms, and intercurrent complications, such as uncontrolled seizures or pulmonary embolism. Although the outcome is much better than in arterial stroke, it is still unpredictable: whereas deeply comatose or hemiplegic patients may recover completely without

**TABLE III** Potential Etiologies and Predisposing Conditions of CVST

| Potential etiologies | Examples | Comments |
|---|---|---|
| Coagulation and hematologic disorders (hereditary or acquired) | AT-III, protein C, and protein S deficiencies, factor V-Leiden, mutation of prothrombin gene, circulating anticoagulants, disseminated intravascular coagulation, heparin- or heparinoid-induced thrombocythemia (HIT); red blood cell disorders: polycythemia, posthemorrhagic anemia, sickle cell disease; congenital thrombophilias; thrombocythemia (primary or secondary) | Requires careful evaluation with multiple tests |
| Infective causes | Local<br>    Intracranial infection: abscess, empyema, meningitis<br>    Regional infections: otitis, tonsilitis, sinusitis, stomatitis, skin<br>General<br>    Bacterial: septicemia, endocarditis, typhoid, tuberculosis<br>    Viral: measles, hepatitis, encephalitis, herpes, HIV, CMV<br>    Parasitic: malaria, trichinosis<br>    Fungal: aspergillosis | Often affects the lateral sinuses |
| Vasculitis/connective tissue diseases/collagenoses | Digestive: Crohn's disease, ulcerative colitis<br>Connective tissue: systemic lupus erythematodes, temporal arteritis, Wegner's granulomatosis, Sjögren's syndrome<br>Venous thromboembolic disease, Hughes-Stovin syndrome<br>Others: Behçet's syndrome, sarcoidosis | Careful evaluation |
| Hormonal changes | Gyneco-obstetric: pregnancy and puerperium<br>Oral contraceptives (estrogens, progestogens) | Especially important in developing countries |
| Internal disorders | Cardiac: congenital heart disease, cardiac insufficiency, pacemaker<br>General: cirrhosis, nephrotic syndrome, dehydratation, hyperhomocysteinaemia | |
| Local, noninfective | Head trauma, tumors, e.g., meningeoma, surgery, lumbar puncture | |
| Neoplasia | Local: meningioma, metastasis, glomus tumor<br>Systemic: any visceral carcinoma, lymphoma, leukemia, carcinoid | |
| Medication | Androgens, danazol, epsilon-aminocaproic acid, heparin (-induced thrombocytopenia, HIT), L-asparaginase, oral contraceptives, progestogens, androgen therapy, danazol | Especially important with concomitant coagulation disorders |
| Iatrogenic | Local: neurosurgical operation, infusions into the internal jugular vein, neck dissection and radiation<br>General: any surgery with or without deep vein thrombosis | |
| Phakomatoses | Sturge–Weber syndrome | |

any sequelae, patients presenting with initially isolated headache may suddenly worsen and die. The prognosis of CVST during pregnancy and puerperium is reported to be better (Cantu and Barinagarrementeria, 1993); these authors suggested that this is due to "more limited and transient occlusion, with rapid sinovenous recanalization by spontaneous thrombolysis or development of collaterals."

Several follow-up studies have shown that survivors usually have complete or almost complete recovery. Minor to moderate neurological deficits were found in 5 to 50% of survivors (for reference, see Ameri and Bousser 1992). Thirty percent of patients with a hemorrhagic infarct were not able to resume work, whereas 80% of those with a nonhemorrhagic venous edema and infarcts did.

Clinical recovery results from the combination of collateral channels and recanalization. In single cases good clinical outcome may occur even without recanalization. A long-term follow-up study (mean 12.1 years) of 40 patients with CVST revealed a positive correlation between clinical outcome and recanalization (Strupp *et al.*, 2002). Complete recanalization was achieved in 21 (52%), partial recanalization in 12 (30%), and no recanalization in 7 of the 40 patients (18%). Eleven patients showed slight focal neurological deficits (mainly paresthesia, incomplete cranial nerve palsy, or mild aphasia); 6 had chronic fluctuating headache that worsened during physical activity. Focal neurological deficits or headache were observed in 7 of the 21 patients with complete recanalization (33%), in 4 of the 12 patients with partial recanalization (33%), and even in 6 of the 7 patients with no recanalization (85%) ($p < 0.005$ complete vs no recanalization). Parenchymal lesions were identified in 14 of 40 patients; 9 of these 14 had focal neurological deficits or headache, whereas only 8 of 26 of the patients without parenchymal lesions had such deficits.

The long-term clinical outcome of CVST is favorable. Although recurrence is rare, it usually occurs within the first 3 months after the initial manifestations of CVST, mainly in patients with a previously undiagnosed underlying cause, such as Behçet's disease (Preter *et al.*, 1996).

# THERAPEUTIC PRINCIPLES

Treatment of nonseptic CVST consists of (a) anticoagulants to prevent the thrombus from spreading intracerebrally (and also from causing pulmonary embolism), (b) in rare cases fibrinolytic therapy or mechanical thrombectomy, (c) specific therapy of the underlying disease (Table III), and (d) general therapy, e.g., of seizures, headache, and elevated intracranial pressure.

## Anticoagulation

### Unfractionated Heparin

Animal experiments have shown that venous occlusion causes an increase of capillary pressure, the primary important factor in the pathophysiology of CVST (Frerichs *et al.*, 1994). This, for instance, explains the edema and the high frequency of hemorrhagic infarcts (in about 30% of patients) caused by diapedesis of erythrocytes. The rationale for using anticoagulants is to limit the spread of thrombus, especially into cortical veins. Therefore, early treatment should prevent onset of focal neurological deficits and seizures. There is no proven evidence that anticoagulants may also promote dissolution, but they can prevent the reocclusion of recanalized (due to endogenous lysis) vessels. Capillary pressure decreases when adequate collaterals develop and/or recanalization occurs, thus preventing the development of diapedesis bleeding and promoting clinical recovery. In animal experiments it has been shown that heparin reduces intracapillary pressure (Einhäupl *et al.*, 1990). All findings indicate that heparin is theoretically a useful drug.

Notwithstanding, anticoagulation has long been disapproved of in CVST, because of the high rate of *spontaneous* hemorrhagic infarcts and the fear of increased risk of heparin-induced bleeding (Hacker and May, 1969; Gettelfinger and Kokmen, 1977; Sigsbee *et al.*, 1979; Yerby and Bailey, 1980). Heparin-induced bleeding, however, is pathophysiologically unlikely in CVST, and, indeed, several studies (see also below), e.g., a prospective nonrandomized study on the safety of heparin, showed that ICH was less frequent in patients on heparin therapy than in those without. Heparin

therapy even proved beneficial when started or continued in patients who had already had a hemorrhagic infarct. Nevertheless, it should be noted that some authors (Gettelfinger and Kokmen, 1977) have reported isolated cases of ICH during heparin therapy.

Although there have been only two prospective, randomized, controlled trials of heparin for the treatment of CVST (see below), and no consensus among neurologists has been reached so far, "heparin remains the first-line treatment (...) because of its efficacy, safety, and feasibility" (Bousser, 1999). It is recommended for all patients with CVST, especially those who deteriorate despite general therapy and/or develop concomitant pelvic or deep vein thrombosis, unless there is a definite medical contraindication.

A placebo-controlled, double-blind clinical trial showed a significant advantage of intravenous dose-adjusted unfractionated heparin therapy in patients with CVST (Einhäupl *et al.*, 1991). This study had to be stopped after 20 patients had participated because of the dramatic difference between the two groups. There were no deaths and eight complete recoveries in the treatment group (after 90 days). By contrast, three patients died and only one recovered completely in the control group. An additional retrospective analysis of 102 patients showed the safety of heparin—even in the presence of intracranial bleeding—as well as its effectiveness: of the 27 patients with ICH treated with heparin 4 died (15%) compared with 9 of 13 (69%) who did not receive heparin. Heparin therapy is useful even in advanced cases of CVST, since a fluctuating weeklong clinical course usually indicates an ongoing thrombotic process.

### Low-Molecular-Weight Heparin

Recently, a randomized placebo-controlled trial of anticoagulant treatment with low-molecular-weight heparin (LMWH; subcutaneous nadroparin, 180 anti-Xa units/kg per day) was performed. In the final analysis, 59 patients received either LMWH or placebo for 3 weeks, followed by 10 weeks of oral anticoagulation with warfarin (INR 2.5–3.5) in those allocated to LMWH (de Bruijn and Stam, 1999). Patients with ICH caused by CVST were also included. After 3 weeks, 6 of 30 patients (20%) in the LMWH group and 7 of 29 patients (24%) in the placebo group were in a poor state (defined as death or Barthel Index score of <15). After 12 weeks, 4 of 30 patients (13%) in the LMWH group and 6 of 29 (21%) in the placebo group had a poor outcome (defined as death or Oxford Handicap Score of ≥3). Although no new symptomatic ICHs occurred, all 6 deaths were of patients who had had ICH before treatment. One patient in the LMWH group had a major gastrointestinal hemorrhage, and one patient in the placebo group died from clinically suspected pulmonary embolism. The authors concluded that patients with CVST treated with anticoagulants (LMWH followed by oral anticoagulation) had a favorable outcome more often than controls, but the difference was not statistically significant. Moreover, anticoagulation proved to be safe, even in patients with ICH.

## Endovascular Thrombolysis

Endovascular thrombolysis (with urokinase, streptokinase, or recombinant tissue plasminogen activator (rtPA)) is theoretically an attractive method for dissolving the thrombus. It leads to recanalization and thereby reduces the increased intracapillary pressure. Total lysis of the clot in sinus thrombosis was documented in animal experiments 1 h after tPA was administered in seven of eight animals; one animal retained partial clot fragments (Alexander et al., 1990).

Although the first patients were treated with urokinase as early as 1971, there are still no prospective, randomized, controlled trials of endovascular thrombolysis. A few uncontrolled case series and a retrospective analysis on the effects of urokinase, used initially, on small numbers of patients were published (Di Rocco et al., 1981; Horowitz et al., 1995; Smith et al., 1994; Tsai et al., 1992; total of 42 patients). More recently two studies with rtPA in combination with heparin were published (Frey et al., 1999; Kim and Suh, 1997; total of 21 patients; see below). Recombinant tissue plasminogen activator has several advantages over urokinase or streptokinase, which might also decrease the risk of ICH: (1) when bound to fibrin, lytic properties of rtPA increase 400-fold (Hoylaerts et al., 1982); (2) its half-life is short (7–8 min); (3) it avoids plasminemia; and (4) it produces the lowest level of fibrinogen degradation products. (It has to be mentioned that rtPA has a high affinity for fibrin; i.e., it may be clot selective, but this is not true for the high doses used for lysis.) All in all, theoretically rtPA seems to be the drug of choice for endovascular thrombolysis.

Kim and Suh treated nine patients for CVST over a 2-year period with rtPA. They placed a microcatheter directly into the thrombus in the dural sinus via the transfemoral route. Thrombolysis was initiated by rapid injection of 10 mg rtPA over 10 min, followed by a continuous infusion of 50 mg within 3 h, and then a continuous infusion at 5 mg per h, until complete thrombolysis was achieved or a total dose of 100 mg per day was reached. If complete recanalization did not occur at 100 mg per day, repeat thrombolysis was tried on the following day. In all nine patients successful recanalization with improvement of the symptoms was achieved. The time required for complete thrombolysis varied from 8 to 43 h. The total dose of rtPA ranged from 50 to 300 mg. In one patient a small intrapelvic hemorrhage and in another oozing at a femoral puncture site occurred. The authors concluded that direct fibrinolytic therapy with rtPA is safe, fast, and effective for treating dural sinus thrombosis.

Frey and coworkers (1999) included in their study 12 patients with symptoms of 1 to 40 days' duration; in addition, 4 had subtle pretreatment hemorrhagic venous infarction, 2 had hemorrhagic infarction, 1 small parenchymal hemorrhage, and 5 no focal lesion. A loading dose of rtPA was instilled throughout the clot at 1 mg/cm, followed by continuous intrathrombus infusion at 1 to 2 mg/h. Intravenous heparin was infused concomitantly. Flow was restored completely in six patients and partially in three, with a mean rtPA dose of 46 mg (range, 23 to 128 mg) at a mean time of 29 h (range, 13 to 77 h). It took less time to restore flow than it did with urokinase (an average of 71 h has been reported for 29 documented patients). Symptoms improved in these nine patients together with flow restoration. In three patients flow could not be restored. In two of them, hemorrhaging worsened, and treatment was shortened after the initial rtPA dose (in one patient the hematoma had to be evacuated). The authors concluded from their experience that intrathrombus rtPA in conjunction with intravenous heparin in patients with CVST is encouraging, but that "this therapy should probably be regarded as unsafe in patients with obvious hemorrhage." All in all, local thrombolysis appears to restore flow more frequently and rapidly than heparin alone, but a better clinical outcome has not been proven, and—as mentioned above—the risk of hemorrhage is greater, at least when there was already an ICH prior to treatment with rtPA (Frey et al., 1999). It is important to note that the patients were simultaneously treated with heparin; thus, the effects of heparin and rtPA could not be separated.

Therefore, on the basis of the trials published so far, we recommend the following. (1) Endovascular thrombolysis is not the treatment of first choice. (2) It would, however, be indicated if symptoms worsen despite heparin and etiologic and symptomatic treatment (see below). The most frequent cause of progression is insufficient anticoagulation, because it is extremely rare to observe deterioration in patients who receive proper anticoagulation. (3) Endovascular thrombolysis requires an experienced team that includes an interventional neuroradiologist. (4) Further studies are required to determine the optimal dose of rtPA (ranging from 23 to 300 mg in the two studies mentioned above), proper methods of administration, its safety, indication, and efficacy.

## Mechanical Endovascular and Surgical Thrombectomy

In individual patients, who did not respond to anticoagulation, mechanical endovascular (Dowd *et al.*, 1999; Novak *et al.*, 2000; Opatowsky *et al.*, 1999; Scarrow *et al.*, 1999)—in one case with balloon dilation and stent deployment (Malek *et al.*, 1999)—or surgical thrombectomy (Ekseth *et al.*, 1998; Kourtopoulos *et al.*, 1994) of CVST have been performed. Both procedures are regularly used in combination with thrombolysis. For instance, a 54-year-old patient with progressive neurological symptoms, who became comatose and hemiplegic 8 days after presentation, was treated with a hydrodynamic thrombectomy catheter ("AngioJet rheolytic thrombectomy catheter"), followed by intrasinus urokinase thrombolytic therapy for 2 days. Thrombolysis and nearly total neurologic recovery were achieved (Dowd *et al.*, 1999). A 13-year-old patient with multiple-sinus thrombosis refractory to anticoagulant therapy, dural arteriovenous malformations, and chronic intracranial hypertension underwent a transvenous mechanical recanalization, balloon angioplasty, and stent deployment in the occipital sinus and successfully achieved sustained venous outflow. At the 12-month follow-up examination the patient was still neurologically intact (Malek *et al.*, 1999). The published data report on only very few patients (with neurological deterioration despite anticoagulant) who were treated surgically with open thrombectomy in combination with local thrombolysis; all survived (Ekseth *et al.*, 1998; Kourtopoulos *et al.*, 1994).

Mechanical endovascular maceration of the clot and surgical thrombectomy, however, are still experimental in character. They play a very limited role in the treatment of CVST and are generally provided to patients with an otherwise deleterious course in highly specialized centers.

## PRAGMATIC THERAPY

### Dose-Adjusted Intravenous Unfractionated Heparin Treatment

Intravenous heparin treatment with an initial bolus of 5000 U should be initiated immediately after the diagnosis is established, even if a hemorrhagic infarct already exists. Continuous treatment is started with 20 U/kg/h, i.e., usually 1000–1400 U per h. The dosage is subsequently adjusted to a therapeutic target activated partial thromboplastin time (aPTT) of two- to threefold (i.e., 80–100 s) of the reference value, usually by an increase of the heparin dose of 100–200 U every 6–8 h;

this should occur within the first 24 h. Full-dose heparin treatment must be used, so that aPTT is at least doubled (80–100 s). Once a steady dosage schedule is established, daily monitoring is sufficient. Due to the short half-life of intravenous heparin (0.5–2 h, depending on the dosage) it should be administered continuously by an intravenous infusion pump system which leads to steady-state plasma levels. Intravenous heparin therapy is easy to control. Activated PTT may normalize within 1 h after infusion is stopped, which may be necessary if complications occur or surgery is necessary. Although CVST is an unpredictable and potentially lethal disorder, there are very few real contraindications for heparin, for example, gastrointestinal arterial bleeding or known heparin-induced thrombocytopenia (HIT) in the patient's history. Required daily heparin dosage varies significantly among patients, in adults between 30,000 and 60,000 U (average 35,000 U).

The administration of initially very high heparin doses (>2000–3000 U per h) to reach target aPTT level, so-called heparin resistance, may indicate an antithrombin (AT) deficiency, which makes it necessary to substitute AT. Apart from the very rare hereditary AT deficiency (about 0.02%), heparin resistance is mainly acquired either due to hepatic cirrhosis, nephrotic syndrome, disseminated intravascular coagulation (AT concentration <25%), or accelerated clearance, as in massive thrombosis (both of which necessitate the substitution of AT).

Relevant side effects of heparin are bleeding and HIT. Heparin-induced thrombocytopenia is characterized by strong IgG-mediated platelet activation with a substantial risk of thrombotic complications. The frequency of HIT has remained unclear, but is assumed to range between 1 and 5%. Usually, HIT is defined as a >50 T/L platelet count fall that begins after 5–10 days of heparin treatment. In patients who have been exposed previously to heparin, the platelet count fall may occur within 24 h of exposure to heparin. At the diagnosis of HIT, heparin titration has to be continued. Patients with HIT who require anticoagulation should receive lepirudin (Refludan; initially an intravenous bolus of 0.4 mg/kg/h, subsequently 0.15 mg/kg as a continuous infusion; dosage adaptation by measuring the aPTT ratio, which should be 1.5–2.5) (Greinacher *et al.*, 2000).

Heparin therapy should be continued until remission of the acute stage of CVST (normalizing level of consciousness or remission of mental confusion, improvement of headache and focal neurological deficits), i.e., in most cases for 10 to 20 days. Afterward anticoagulation therapy is slowly switched over to oral anticoagulation.

## Adjustment to Oral Anticoagulation

After the acute period of CVST—in most cases 1–3 weeks after diagnosis—treatment with oral anticoagulants (warfarin (Coumadin, 5 mg per tablet) or phenprocoumon (Marcumar, 3 mg per tablet)) should be initiated concomitantly with full-dose heparin, i.e., overlapping until the "international normalized ratio" (INR) of the prothrombin time is within its therapeutic range (see below).

Due to the half-lives of the vitamin K-dependent coagulation factors (II, VII, IX, X), which range from 6 to 60 h, a few days are required before an effect of oral anticoagulation is observed (peak effect usually after 3 to 5 days). The usual initial dose of warfarin is 10 mg daily orally for 2 days; subsequent maintenance doses range from 3 to 9 mg/day, given as a single dose at the same time each day (preferably in the evening). If phenprocoumon is used, the initial dose is 9 to 12 mg/day on day 1, 9 mg/day on day 2; on the following days dosage depends on the INR (maintenance doses usually range from 0.5 to 4.5 mg/day).

To adjust oral anticoagulation therapy, the INR should be determined before treatment and daily or every second day after initiation until it is well stabilized. Subsequently greater but regular intervals (under steady-state condition in 2–4 weeks) are sufficient. Several therapy studies on *extracerebral* venous thrombosis have shown that an INR of 2.5–3.5 is sufficient (for ref. see Martindale, 2000). We also recommend this for CVST, although there have been no relevant studies. Patients should have a registration card on which the daily dosage and the INR results are noted.

If the patient's clinical status deteriorates, heparin therapy must be resumed without termination of oral anticoagulation, because deterioration is usually due to ineffective anticoagulation during oral anticoagulation adjustment. However, if clinical deterioration persists, investigations including brain and vascular scanning should be performed to determine the cause.

If CVST occurs during pregnancy, oral anticoagulants must be avoided because of their ability to pass the placenta and their teratogenic effects. In these cases anticoagulation should be continued with heparin or preferably LMWH (due to the lower risk of HIT), both of which do not cross the placenta and have been associated with neither fetal malformations nor increased fetal mortality or prematurity (for reference, see Majerus et al. (1995)). The drug should, however, be discontinued 24 h before delivery to minimize the risk of postpartum bleeding.

It is recommended that oral anticoagulation therapy should be continued for 6 months in patients in whom the underlying cause of thrombosis was not found, although there are no studies on such duration of treatment. For *extracerebral* venous thrombosis of unknown etiology or proven pulmonary embolism, 6 weeks to 3 months are recommended. Lifelong treatment should be considered in patients with persisting coagulation disorders (see Table III). Moreover, the duration of oral anticoagulation depends on the treatment of the known underlying cause such as Behçet's syndrome or hyperglobulia. In CVST steroids in combination with anticoagulation should be given for an underlying autoimmune disorder despite their possible thrombogenic effect.

Since sudden discontinuation may theoretically result in rebound hypercoagulability with the risk of rethrombosis, the dosage of oral anticoagulants should be slowly reduced after a period of long-term treatment. Regular follow-up visits are required after termination of anticoagulation, and patients must be informed about early signs (headache) that indicate a possible relapse.

## Treatment of Seizures

Seizures occur in 30–50% of all patients with CVST and can be followed by transient hemiparesis (Todd's paresis), lasting several hours or days. Anticonvulsants should not be given prophylactically; they should be restricted to only those patients with seizures (Ameri and Bousser, 1992). For acute treatment phenytoin is the drug of choice. The oral loading dose is 15 mg/kg in three doses at 1-h intervals, which gives low therapeutic levels after 8 to 12 h. Intravenously loading of 15 mg/kg (at maximum 50 mg/min) gives immediate therapeutic levels. Effective concentrations of phenytoin (usually >10 μg/mL) should be achieved within 4–6 h, because a series of seizures frequently occurs in patients with CVST. The maintenance dose of 4–7 mg/kg per day (i.e., in most cases 200 to 300 mg per day) should be given orally. It should be noted that plasma drug concentration increases disproportionally as dosage is increased. We continue the treatment for 3 months, and if no further seizures occur, the dosage is gradually reduced and then stopped. The development of persistent epilepsy due to CVST is rare.

## Analgesic Treatment

Nonopioid analgesics, such as acetaminophen (benuron, 500 to 1000 mg tid), or in severe cases "weak" opioid analgesics, such as tramadol (50 mg tid up to 400 mg per day) can be used to treat (the usually severe) headache. These drugs may be given early after symptom onset and continuously to treat pain effectively. Since the combination of antiplatelet drugs such as acetylsalicylic acid and heparin poses a high risk of hemorrhage, they should be avoided (see Table II).

### Treatment of Elevated Intracranial Pressure

Although brain swelling is observed on CT scans in about 50% of all patients with CVST, antiedematous treatment is necessary in only 20%. Therapy follows the general principles for treating raised intracranial pressure (see Chapter 56). In patients with isolated intracranial hypertension and threatened vision due to papilledema, we favor a lumbar puncture to remove CSF before starting heparin. This is often followed by rapid improvement of headache and disappearance of visual disturbances. The use of steroids to treat increased intracranial pressure is controversial, since steroids may inhibit fibrinolysis. However, in a study with 19 patients with Behçet's syndrome who received anticoagulants and steroids, none died, 25% had persisting neurological deficits, and there was no recurrence (follow-up interval: 3 years) (Wechsler *et al.*, 1992, 1993). This study concluded that steroids should be considered for the treatment of the underlying disorder as well as for increased intracranial pressure.

### Counseling of Patients with CVST Related to Oral Contraceptives and Pregnancy

Oral contraceptives are a risk factor for CVST, especially if there is a concomitant coagulation disorder (see Table III); this is also true for the "third-generation pills" (de Bruijn *et al.*, 1998). Patients who develop CVST while on oral contraceptives should be urged to stop taking the pill.

The question of whether these women should be warned against a further pregnancy is still difficult to answer. To assess whether subsequent pregnancies increase the risk of recurrent stroke and CVST the French Study Group on Stroke in Pregnancy recently analyzed their occurrence in 373 women with arterial ischemic stroke and in 68 with CVST; 10 of the cases of CVST were related to pregnancy or puerperium. During a mean follow-up of 5 years 89 women with previous arterial stroke and 26 with previous CVST became pregnant and delivered their children. No recurrent CVST occurred (Lamy *et al.*, 2000). In another study there was no recurrence in 12 pregnancies of seven women with previous CVST (the majority was treated prophylactically with LMWH) (Preter *et al.*, 1996). Six of our own patients had no relapse of CVST during pregnancy. All in all, a previous CVST—even during pregnancy or puerperium—does not seem to be a contraindication to a subsequent pregnancy. There is,

however, no consensus about treatment during pregnancy and the postpartum period. If there is an additional coagulation disorder or another underlying cause, continuous treatment with LMWH is necessary. If the etiology remains unclear, LMWH should be given during the last trimester, be discontinued 24 h before delivery, subsequently started again, and given 4 weeks after delivery.

Since there is evidence that the high rate of CVST related to pregnancy in developing countries, especially Mexico and India, may also be related to high homocysteine levels that predispose to thrombosis, it is recommended to supplement folate.

## SEPTIC CVST

Since the introduction of antibiotics there has been a considerable decline in septic CVST, at least in the developed countries. Nowadays only about 10% of CVST are related to infection. It is, however, still the most common cause of cavernous sinus thrombosis and an important etiology of lateral and petrosal sinus thrombosis. Septic CVST is due to propagation from infections of the ethmoid, sphenoid air cells, or ear or to hematogenous spread from the nose, face, ear, or neck. The cavernous sinus is frequently affected since it receives blood from the face, paranasal sinuses, and fossa pterygoidea; the lateral and petrosal sinuses are affected by infections of the ear. Purulent meningitis is sometimes followed by septic CVST, which predominantly affects the superior sagittal sinus or superficial cortical veins. A more distant focus is an exceptional cause (Southwick *et al.*, 1986). The underlying infection is most often caused by *Streptococcus pneumoniae*, *Staphylococcus aureus*, Hemophilus species, Proteus species, *Escherichia coli*, or anaerobes. Septic CVST is usually accompanied by fever, leukocytosis, and a distinct CSF pleocytosis (granulocytes), due to an associated meningitis.

Septic cavernous sinus thrombosis is characterized by chemosis, exophthalmos, and painful ophthalmoplegia (due to lesions of the III, IV, and VIth cranial nerves); initially it is unilateral but frequently becomes bilateral. Extension to other sinuses or to the intracavernous portion of the carotid artery is rare, but often dramatic. Septic thrombosis of posterior sinuses often extends to other sinuses and veins of the brain, causing the same clinical features as noninfective CVST. Isolated septic cortical vein thrombosis without the sinus being affected is extremely rare (DiNubile *et al.*, 1990). Clinical symptoms (seizures and hemiparesis) will often be attributed to purulent meningitis, due to the accompanying or dominating meningeal syndrome. Therefore, septic thrombosis of the lateral and superior sagittal sinuses is often overlooked; diagnosis can be established only if MRI and MRA, CT, or—in rare cases—angiography are performed.

The prognosis of septic CVST is worse than that of nonseptic CVST, e.g., the mortality was almost 80% in patients with complete septic thrombosis of the superior sagittal sinus (Southwick *et al.*, 1986), about 50% if the cortical veins were affected (DiNubile *et al.*, 1990)— these patients were not treated with heparin—and about

30% in patients with septic cavernous sinus thrombosis (Bharucha *et al.*, 1996).

Therapeutic principles for septic CVST include early systemic administration of antibiotics (prior to the return of culture reports), surgical removal of the infectious focus, and anticoagulation with full-dose heparin.

As long as the pathogen has not been identified from the assumed focus, cerebrospinal fluid or blood samples, adults (without evidence of a nosocomial infection or a

**TABLE IV    Treatment Strategies for CVST**

(a) Nonseptic CVST, including general therapy

| | Generic (drug) | Aim and monitoring of the treatment | Dosage | Duration of treatment |
|---|---|---|---|---|
| Acute phase: Dose-adjusted heparin | Heparin** (Calciparin, Liquemin, Thrombophob) | Increase of aPTT to 80–100 (120) s, control of aPTT, initially every 6 h, then every 12 to 24 h | Intially 5000 U as a bolus, then 1000–1600 U per hour via infusion pump system | Usually 2–3 weeks until patient in a stable condition |
| Subacute phase: Oral anticoagulation[a] | Warfarin* (Coumadin, 5 mg per tablet) Phenprocoumon (Marcumar, 3 mg per tablet) | INR 2.5–3.5, regular controls, initially once a day; if stable, every second or fourth week | Warfarin  Days 1 and 2, 10 mg/day; ≥day 3, depending on INR; maintenance dosage, 3 to 9 mg/day  Phenprocoumon  Day 1, 9 to 12 mg/day; day 2, 9 mg/day; ≥day 3, depending on INR; maintenance dosage, 0.5 to 4.5 mg/day | Usually 6 months; lifelong if, for instance, persisting coagulation disorder was found |
| Headache, Minor | Nonopioid analgesics Acetaminophen (Benuron) | Reduction or relief of pain | 500 to 1000 mg tid | |
| Severe | "Weak opioid analgesics" Tramadol | | 50 mg tid orally or subcutaneously up to 400 mg/day | |
| Epileptic seizures | Phenytoin | Effective and early treatment to prevent series of seizures or status epilepticus | Initially  500–1000 mg phenytoin intravenously over 4 to 6 h, up to 1500 mg/day (infusion rate <50 mg/min)  Subsequently  200 or 300 mg/day orally | Only in those patients with seizures, not prophylactically; duration of treatment, if no further seizures occur: 3 months |

(b) Septic CVST

| Surgery | Medication | |
|---|---|---|
| | Antibiotics[b] | Anticoagulation |
| Removal of the infectious focus | Pathogen unknown  Adults, without nosocomial infection: combination of  1. Third generation cephalosporin, e.g., ceftriaxone (4 g per day) or cefotaxime 2 g every 8 h) plus  2. flucloxacillin (2 g every 4 h) or fosfomycin (5 g every 8 h)  Adults with nosocomial infection: combination of  Meropenem (2 g every 8 h) and vancomycin (0.5 g every 6 h) or  Ceftazidine (2 g every 8 h) and vancomycin (0.5 g every 6 h) | As above in the acute phase |

[a]In case of contraindications (e.g., pregnancy) anticoagulant treatment should be low molecular-weight heparin (e.g., subcutaneous nadroparin, 90 anti-Xa units/kg every 12 h).
[b]After identification of pathogen, treatment according to the antibiogram.

compromised immune system) should be administered a combination of two antibiotics directed against the pathogens found in common infections of the face, neck, or ear, and therefore, most likely associated with septic CVST (*S. pneumoniae*, meningococci, *Haemophilus influenzae*, and *S. aureus*): (1) a third-generation cephalosporin (such as ceftriaxone (4 g per day intravenously) or cefotaxime (2 g every 8 h intravenously)) and (2) flucloxacillin (2 g every 4 h) or fosfomycin (5 g every 8 h) (Table IV). If anaerobic infection from the teeth or paranasal sinus is suspected, metronidazole should be added. If a nosocomial infection is suspected, including Gram-negative Enterobacteriaceae and *Pseudomonas aeruginosa*, a combination of meropenem (2 g every 8 h) and vancomycin (0.5 g every 6 h) or ceftazidine (2 g every 8 h) and vancomycin (0.5 g every 6 h) is recommended. Later antibiotics are selected according to the bacterial susceptibility tests in the cultures.

The effect of heparin in septic CVST has not yet been systematically investigated. Levine and coworkers (Levine *et al.*, 1988) reported reduced morbidity but no effect on mortality in 104 retrospectively studied patients with septic cavernous sinus thrombosis who received a combined therapy of antibiotics and heparin compared with only antibiotic treatment. A retrospective analysis by Southwick and coworkers (1986) suggested that treatment with heparin may reduce mortality in cases of septic cavernous sinus-thrombosis. Most authors favor the use of heparin in septic CVST as well as in nonseptic cases in order to prevent the extension of the thrombosis.

## INEFFECTIVE OR OBSOLETE

Antiplatelet drugs are not recommended, since they are less effective than anticoagulation for treating venous thrombosis. The value of general rheological measures such as low-molecular-weight dextran, hydroxylethyl starch, or albumin has not been investigated systematically. Theoretically, they may increase the central venous pressure and thereby further decrease the venous flow. Moreover, dextrin should be considered with reservation due to its antiplatelet effect, and a combined application of dextran and heparin may pose a higher risk of hemorrhage. Restriction of volume intake and forced diuresis for treatment of brain edema must be avoided, since these measures can cause an additional deterioration of blood viscosity.

## REFERENCES

Alexander, L. F., Yamamoto, Y., Ayoubi, S., al Mefty, O., and Smith, R. R. (1990). Efficacy of tissue plasminogen activator in the lysis of thrombosis of the cerebral venous sinus. *Neurosurgery* **26**, 559–564.

Ameri, A., and Bousser, M. G. (1992). Cerebral venous thrombosis. *Neurol. Clin.* **10**, 87–111.

Ameri, A., and Bousser, M. G. (1994). Cerebral venous thromboses. Clinical diagnosis. *Ann. Radiol. Paris* **37**, 101–107.

Anxionnat, R., Blanchet, B., Dormont, D., Bracard, S., Chiras, J., Maillard, S., Louail, C., Moret, C., Braun, M., Roland, J., *et al.* (1994). Present status of computerized tomography and angiography in the diagnosis of cerebral thrombophlebitis cavernous sinus thrombosis excluded. *J. Neuroradiol.* **21**, 59–71.

Bansal, B. C., Gupta, R. R., and Prakash, C. (1980). Stroke during pregnancy and puerperium in young females below the age of 40 years as a result of cerebral venous/venous sinus thrombosis. *Jpn. Heart J.* **21**, 171–183.

Bharucha, N. E., Bharucha, E. P., and Bhabba, S. K. (1996). Bacterial infections. *In* "Neurology in Clinical Practice" (W. Bradley, R. Daroff, G. Fenichel, and C. Marsden, Eds.), pp. 1181–1243. Butterworth-Heinemann, Boston.

Bousser, M. G. (1991). Cerebral venous thrombosis. Report of 76 cases. *J. Mal. Vasc.* **16**, 249–254.

Bousser, M. G. (1999). Cerebral venous thrombosis: Nothing, heparin, or local thrombolysis? *Stroke* **30**, 481–483. [Editorial; Comment].

Bousser, M. G. (2000). Cerebral venous thrombosis: Diagnosis and management. *J. Neurol.* **247**, 252–258.

Bousser, M. G., Chiras, J., Bories, J., and Castaigne, P. (1985). Cerebral venous thrombosis—A review of 38 cases. *Stroke* **16**, 199–213.

Cantu, C., and Barinagarrementeria, F. (1993). Cerebral venous thrombosis associated with pregnancy and puerperium. Review of 67 cases. *Stroke* **24**, 1880–1884.

Corvol, J. C., Oppenheim, C., Manai, R., Logak, M., Dormont, D., Samson, Y., Marsault, C., and Rancurel, G. (1998). Diffusion-weighted magnetic resonance imaging in a case of cerebral venous thrombosis. *Stroke* **29**, 2649–2652.

Daif, A., Awada, A., al Rajeh, S., Abduljabbar, M., al Tahan, A. R., Obeid, T., and Malibary, T. (1995). Cerebral venous thrombosis in adults. A study of 40 cases from Saudi Arabia. *Stroke* **26**, 1193–1195.

de Bruijn, S. F., and Stam, J. (1999). Randomized, placebo-controlled trial of anticoagulant treatment with low-molecular-weight heparin for cerebral sinus thrombosis. *Stroke* **30**, 484–488.

de Bruijn, S. F., Stam, J., Koopman, M. M., and Vandenbroucke, J. P. (1998). Case-control study of risk of cerebral sinus thrombosis in oral contraceptive users and in [correction of who are] carriers of hereditary prothrombotic conditions. The Cerebral Venous Sinus Thrombosis Study Group. *Br. Med. J.* **316**, 589–592.

Di Rocco, C., Iannelli, A., Leone, G., Moschini, M., and Valori, V. M. (1981). Heparin-urokinase treatment in aseptic dural sinus thrombosis. *Arch. Neurol.* **38**, 431–435.

DiNubile, M. J., Boom, W. H., and Southwick, F. S. (1990). Septic cortical thrombophlebitis. *J. Infect. Dis.* **161**, 1216–1220.

Dowd, C. F., Malek, A. M., Phatouros, C. C., and Hemphill, J. C., III (1999). Application of a rheolytic thrombectomy device in the treatment of dural sinus thrombosis: A new technique. *Am. J. Neuroradiol.* **20**, 568–570.

Einhäupl, K. H., Villringer, A., Haberl, R. L., Pfister, H. W., Deckert, M., Steinhoff, H., and Schmiedek, P. (1990). Clinical spectrum of sinus venous thrombosis. *In* "Cerebral Sinus Thrombosis: Experimental and Clinical Aspects" (K. Einhäupl, O. Kempski, and A. Baethmann, Eds.), pp. 149–155. Plenum, New York, London.

Einhäupl, K. M., Villringer, A., Meister, W., Mehraein, S., Garner, C., Pellkofer, M., Haberl, R. L., Pfister, H. W., and Schmiedek, P. (1991). Heparin treatment in sinus venous thrombosis. *Lancet* **338**, 597–600.

Ekseth, K., Bostrom, S., and Vegfors, M. (1998). Reversibility of severe sagittal sinus thrombosis with open surgical thrombectomy combined with local infusion of tissue plasminogen activator: Technical case report. *Neurosurgery* 43, 960–965.

Ford, K., and Sarwar, M. (1981). Computed tomography of dural sinus thrombosis. *Am. J. Neuroradiol.* 2, 539–543.

Frerichs, K. U., Deckert, M., Kempski, O., Schurer, L., Einhäupl, K., and Baethmann, A. (1994). Cerebral sinus and venous thrombosis in rats induces long-term deficits in brain function and morphology—Evidence for a cytotoxic genesis. *J. Cereb. Blood Flow Metab* 14, 289–300.

Frey, J. L., Muro, G. J., McDougall, C. G., Dean, B. L., and Jahnke, H. K. (1999). Cerebral venous thrombosis: Combined intrathrombus rtPA and intravenous heparin. *Stroke* 30, 489–494.

Gettelfinger, D. M., and Kokmen, E. (1977). Superior Sagittal Sinus Thrombosis. *Arch. Neurol.* 34, 2–6.

Greinacher, A., Eichler, P., Lubenow, N., Kwasny, H., and Luz, M. (2000). Heparin-induced thrombocytopenia with thromboembolic complications: Meta-analysis of 2 prospective trials to assess the value of parenteral treatment with lepirudin and its therapeutic aPTT range. *Blood* 96, 846–851.

Hacker, H., and May, B. (2001). Zur Behandlung der Hirnvenenthrombose. *Nervenarzt* 40, 440–443.

Hahn, T. (1954). Die Elektroenzephalographie bei zerebralen Thrombophlebitiden und Thrombosen. *Schweiz. Arch. Neurol. Psychiatr.* 73, 57–99.

Horowitz, M., Purdy, P., Unwin, H., Carstens, G., III, Greenlee, R., Hise, J., Kopitnik, T., Batjer, H., Rollins, N., and Samson, D. (1995). Treatment of dural sinus thrombosis using selective catheterization and urokinase. *Ann. Neurol.* 38, 58–67.

Hoylaerts, M., Rijken, D. C., Lijnen, H. R., and Collen, D. (1982). Kinetics of the activation of plasminogen by human tissue plasminogen activator. Role of fibrin. *J. Biol. Chem.* 257, 2912–2919.

Isensee, C., Reul, J., and Thron, A. (1994). Magnetic resonance imaging of thrombosed dural sinuses. *Stroke* 25, 29–34.

Keller, E., Flacke, S., Urbach, H., and Schild, H. H. (1999). Diffusion- and perfusion-weighted magnetic resonance imaging in deep cerebral venous thrombosis. *Stroke* 30, 1144–1146.

Kim, S. Y., and Suh, J. H. (1997). Direct endovascular thrombolytic therapy for dural sinus thrombosis: Infusion of alteplase. *Am. J. Neuroradiol.* 18, 639–645.

Kourtopoulos, H., Christie, M., and Rath, B. (1994). Open thrombectomy combined with thrombolysis in massive intracranial sinus thrombosis. *Acta Neurochir. Wien.* 128, 171–173.

Lamy, C., Hamon, J. B., Coste, J., and Mas, J. L. (2000). Ischemic stroke in young women: Risk of recurrence during subsequent pregnancies. French Study Group on Stroke in Pregnancy. *Neurology* 55, 269–274.

Levine, S. R., Twyman, R. E., and Gilman, S. (1988). The role of anticoagulation in cavernous sinus thrombosis. *Neurology* 38, 517–522.

Majerus, P., Broze, G., Miletich, J., and Tollefsen, D. (1995). Anticoagulant, thrombolytic, and antiplatelet drugs. *In* "Goodman & Gilman's—The Pharmacological Basis of Therapeutics" (J. Hardman and L. Limbird, Eds.), pp. 1341–1359. McGraw-Hill, New York.

Malek, A. M., Higashida, R. T., Balousek, P. A., Phatouros, C. C., Smith, W. S., Dowd, C. F., and Halbach, V. V. (1999). Endovascular recanalization with balloon angioplasty and stenting of an occluded occipital sinus for treatment of intracranial venous hypertension: Technical case report. *Neurosurgery* 44, 896–901.

Martin, P. J., and Enevoldson, T. P. (1996). Cerebral venous thrombosis. *Postgrad. Med. J.* 72, 72–76.

Martindale (2000). "*Martindale—The Extra Pharmacopoeia*," 30 ed. The Pharmaceutical Press, London.

Nadel, L., Braun, I. F., Muizelaar, J. P., and Laine, F. J. (1991). Tumoral thrombosis of cerebral venous sinuses: Preoperative diagnosis using magnetic resonance phase imaging. *Surg. Neurol.* 35, 189–195.

Novak, Z., Coldwell, D. M., and Brega, K. E. (2000). Selective infusion of urokinase and thrombectomy in the treatment of acute cerebral sinus thrombosis. *Am. J. Neuroradiol.* 21, 143–145.

Opatowsky, M. J., Morris, P. P., Regan, J. D., Mewborne, J. D., and Wilson, J. A. (1999). Rapid thrombectomy of superior sagittal sinus and transverse sinus thrombosis with a rheolytic catheter device. *Am. J. Neuroradiol.* 20, 414–417.

Preter, M., Tzourio, C., Ameri, A., and Bousser, M. G. (1996). Long-term prognosis in cerebral venous thrombosis. Follow-up of 77 patients. *Stroke* 27, 243–246.

Scarrow, A. M., Williams, R. L., Jungreis, C. A., Yonas, H., and Scarrow, M. R. (1999). Removal of a thrombus from the sigmoid and transverse sinuses with a rheolytic thrombectomy catheter. *Am. J. Neuroradiol.* 20, 1467–1469.

Sigsbee, B., Deck, M. D., and Posner, J. B. (1979). Nonmetastatic superior sagittal sinus thrombosis complicating systemic cancer. *Neurology* 29, 139–146.

Smith, T. P., Higashida, R. T., Barnwell, S. L., Halbach, V. V., Dowd, C. F., Fraser, K. W., Teitelbaum, G. P., and Hieshima, G. B. (1994). Treatment of dural sinus thrombosis by urokinase infusion. *Am. J. Neuroradiol.* 15, 801–807.

Southwick, F. S., Richardson, E. P. J., and Swartz, M. N. (1986). Septic thrombosis of the dural venous sinuses. *Medicine Baltimore* 65, 82–106.

Strupp, M., Covi, M., Seelos, K., Dichgans, M., and Brandt, T. (2002). Cerebral venous thrombosis: Correlation between recanalization and clinical outcome—A long-term follow-up of 40 patients. *J. Neurol.* (in press).

Tsai, F. Y., Higashida, R. T., Matovich, V., and Alfieri, K. (1992). Acute thrombosis of the intracranial dural sinus: Direct thrombolytic treatment. *Am. J. Neuroradiol.* 13, 1137–1141.

Villringer, A., Seiderer, M., Bauer, W. M., Laub, G., Haberl, R. L., and Einhäupl, K. M. (1989). Diagnosis of superior sagittal sinus thrombosis by three-dimensional magnetic resonance flow imaging. *Lancet* 1, 1086–1087.

Vogl, T. J., Bergman, C., Villringer, A., Einhäupl, K., Lissner, J., and Felix, R. (1994a). Dural sinus thrombosis: Value of venous MR angiography for diagnosis and follow-up. *Am. J. Roentgenol.* 162, 1191–1198.

Vogl, T. J., Bergmann, C. U., Villringer, A., Einhäupl, K. M., Balzer, J. O., Steinhoff, H., and Felix, R. (1993). [Venous MR angiography for the primary diagnosis and follow-up of sinus venous thrombosis. The correlation with the clinical picture and DSA]. *Rofo Fortschr. Geb. Rontgenstr. Neuen Bildgeb. Verfahr.* 159, 78–85.

Vogl, T. J., Hoffmann, Y., Muhler, A., and Felix, R. (1994b). Contrast medium enhanced MR angiography. *Radiologe* 34, 423–429.

Wechsler, B., Dell'Isola, B., Vidailhet, M., Dormont, D., Piette, J. C., Bletry, O., and Godeau, P. (1993). MRI in 31 patients with Behçet's disease and neurological involvement: Prospective study with clinical correlation. *J. Neurol. Neurosurg. Psychiatry* 56, 793–798.

Wechsler, B., Vidailhet, M., Piette, J. C., Bousser, M. G., Dell, I. B., Bletry, O., and Godeau, P. (1992). Cerebral venous thrombosis in Behçet's disease: Clinical study and long-term follow-up of 25 cases. *Neurology* 42, 614–618.

Yerby, M. S., and Bailey, G. M. (1980). Superior sagittal sinus thrombosis 10 years after surgery for ulcerative colitis. *Stroke* 11, 294–296.

Yuh, W. T., Simonson, T. M., Wang, A. M., Koci, T. M., Tali, E. T., Fisher, D. J., Simon, J. H., Jinkins, J. R., and Tsai, F. (1994). Venous sinus occlusive disease: MR findings. *Am. J. Neuroradiol.* 15, 309–316.

## CHAPTER 38
# The Vasculitides

Patricia M. Moore and Arno Villringer

## DEFINITION AND CLASSIFICATION

The vasculitides are a collection of diseases sharing the central feature of tissue injury due to inflammation of blood vessels. Clinical features, acuity of presentation, chronicity, or recurrences of inflammation depend upon the etiology, immunopathogenesis, and target organ systems. Peripheral and/or central nervous system features comprise a prominent part of many of these disorders (Moore and Richardson, 1998). The classification of the vasculitides is evolving. There is no simple way to group the vasculitides that have neurological manifestations; however, one that is useful when discussing the diseases in terms of therapeutic approaches appears in Table I.

The *idiopathic systemic vasculitides* are noted for the constitutional features of systemic inflammation, a target group of affected organs, and requirement of biopsy for diagnosis. The epidemiology of these disorders varies among different regions of the world although in Europe, the United States, and Canada the incidence approaches 4/100,000 (Scott and Watts, 2000; Watts *et al.*, 2000). Overall, giant cell arteritis is most frequent. Wegener's, microscopic polyarteritis, and Churg–Strauss share a middle incidence. Polyarteritis nodosa is now considered more unusual. The *secondary vasculitides* are undoubtedly the largest group and most frequently encountered of all the vasculitides. This group illustrates the wide variety and frequency of identifiable causes of vasculitis as well as the spectrum of clinical manifestations. In the third group, *focal vasculitides*, are several types of vasculitis sufficiently restricted in their vascular territories or clinical presentation that they are properly a distinctive group. Examples are: isolated angiitis of the central nervous system

(CNS) (a rare disorder characterized by recurrent inflammation of blood vessels within the dural reflections), Behcet's disease (a small-vessel vasculitis most frequently encountered in the Middle East and South America), and Takayasu's arteritis (a large-vessel vasculitis in which most of the neurological clinical manifestations present during the stage of scarring rather than during a stage of active inflammation). The last group of disorders, *immune-mediated vasculopathies*, is not truly inflammatory; thus the term vasculitis would be a misnomer, but there is an immune component to the vessel damage. This is important because early recognition and therapy directed at the immune cascade likely minimize accrued damage. Increasingly, therapies to minimize disease progression in disorders including transplant vasculopathy, SLE cerebrovascular disease, and, possibly, atherosclerosis utilize early anti-inflammatory or immunosuppressive agents.

## GENERAL CLINICAL ASPECTS OF NEUROLOGICAL SYMPTOMS IN VASCULITIDES

In vasculitis, the predominant mediator of tissue injury is ischemia. Depending on the size, distribution, and density of the vessels involved, the clinical features vary. In the central nervous system, encephalopathies, hemorrhages both subarachnoid and intraparenchymal, thrombotic strokes, seizures, cranial neuropathies, and myelopathies occur. In the peripheral nervous system, mononeuropathy multiplex may suggest a vasculitis although polyneuropathies occur as frequently. Radiculopathies, plexopathies, and cutaneous neuropathies also occur.

TABLE I  A Classification of Vasculitis

Idiopathic systemic vasculitides
  Wegener's granulomatosis
  Churg–Strauss angiitis
  Polyarteritis nodosa
  Microscopic polyarteritis
Secondary vasculitides
  Hypersensitivity vasculitis
  Infection associated vasculitis
    Hepatitis C
    Hepatitis B
    HIV
    Herpes Zoster Varicella
    Bacterial
    Aspergillus
    Mucormycoses
    Treponemes
  Drug associated vasculitis
  Neoplasia associated vasculitis
    Hodgkin's
  Connective tissue disease associated vasculitis
    Rheumatoid vasculitis
  Kawasaki disease
Focal or restricted vasculitides
  Isolated angiitis of the CNS
  Takayasu's arteritis
  Behcet's disease
  Cogan's syndrome
  Eale's disease
Immune-mediated vasculopathies
  Transplant vasculopathy
  SLE vasculopathy
  Susak vasculopathy–microangiopahy of the brain, ear, and retina

## GENERAL PRINCIPLES OF DIAGNOSIS

Diagnosis of vasculitis depends on a combination of clinical, radiographic, and pathologic features. The gold standard in diagnosis is confirmation of vasculitis in a biopsy specimen. Brain biopsy not only identifies vasculitis as the cause of symptoms but also, critically at times, determines a possible infectious cause for the vasculitis (Hurst and Grossman, 1994; Parisi and Moore, 1994). Angiography may suggest the diagnosis but no abnormalities are pathognomonic for vasculitis; nor do angiographically defined abnormalities distinguish among the different etiologies or types of vasculitis (Christophe et al., 1999; Vassallo et al., 1999). In addition, vasculitis may not be identified in an angiogram particularly when only small vessels are affected. Magnetic resonance angiography is a promising tool but the spatial resolution is not yet the equivalent of conventional cerebral angiography. Cranial MRI and CT delineate well tissue damage secondary to vasculitis but they do not detect early vasculitis, predict histological features, or distinguish underlying causes.

## PATHOPHYSIOLOGY

Inflammatory cells binding to and passing through the blood vessel wall are a vital physiological process central to infection control and tumor surveillance functions of the immune system. The underlying sequential expression of cell surface molecules in response to proinflammatory stimuli is normally a well-regulated, self-limited process. Under a variety of circumstances persistent or recurrent inflammation in the blood vessel damages the structure or the function of the vessel and tissue ischemia occurs. Multiple factors determine the susceptibility to and development of vasculitis. Inherited determinants increase the risk but are, themselves, insufficient to induce disease. Infections are documented causes of many types of vasculitis and likely important modulators and triggers of other cases (Calli et al., 1999; Dawson and Starkebaum, 1999; Heckmann et al., 1999; Kalita et al., 1999; Nau and Bruck, 2002; Podlecka et al., 1998; Sorimachi et al., 2001; Stone et al., 1998). There are defined mechanisms of vasculitis associated with a specific agent: infection of the endothelial cells themselves, overexpression of proinflammatory cytokines and adhesion molecules within the vessel walls, activation of monocytes, and circulating immune complexes. In other disorders, a spectrum of infections suggests a role for superantigen activation of autoreactive T- and B-lymphocytes. Both the acute responses and the longer-term tissue responses to individual infectious agents vary considerably among patients, likely due to genetically controlled immune response factors. A classic example of this is the two polarized responses to infections with leprosy, tuberculoid and lepromatous.

Immunopathogenic mechanisms involved in vascular inflammation include immune complex deposition, T lymphocyte–endothelial interactions, granulomatous processes, and autoantibody-mediated changes in the blood vessel wall. Tissue injury in the vasculitides ultimately results from diminished blood flow beyond the ability of other blood vessels or the tissue to adapt. Reduced blood flow accompanies inflammation and develops from (1) obstruction of the lumen by inflammatory cells, (2) increased local thrombosis, and (3) vasospasm (Moore, 2000).

### Immune Complex-Mediated Vasculitis

Experience with the mechanisms of immune complex-mediated vasculitis in the experimental models of (1) serum sickness and (2) Arthus reaction provides much of our current information. In serum sickness, circulating antibody–antigen complexes, with certain immunochemical properties, deposit in the blood vessels. There, trapped immune complexes activate complement, inducing an influx of polymorphonuclear leukocytes and macrophages. These cells mediate local inflammation, damage, and necrosis to the vessel wall. Thrombosis, occlusion, and hemorrhage often ensue. The Arthus reaction is characterized by a localized, in situ, formation of immune complexes inducing a similar local vascular injury. A combination of systemic and local immune complex damage may also occur when an antigen that initially circulated is subsequently trapped in the vessel wall where antibodies react with it. The prototype clinical entity with this pathogenic mechanism is hypersensitivity vasculitis. In these patients, circulating immune complexes occur. Immunoglobulin and complement can be demonstrated in biopsied cutaneous lesions. Immune complex mechanisms may also play a role in polyarteritis nodosa (PAN); since half of the patients have circulating immune complexes and one-third have hepatitis B antigenemia (Guillevin et al., 1995b). Furthermore, vasculitis in patients with rheumatoid arthritis, systemic lupus erythematosus, and cryoglobinemia may follow these pathogenic principles. Immune complex-induced vasculitis may also occur as a secondary vasculitis in various viral infections such as human Epstein–Barr virus infection and serum hepatitis, as well as in nonviral infections such as with Neisseria gonococcus and N. meningococcus. Some medications and toxins (such as sulfonamide) but not others (amphetamine) may be associated with immune complex-mediated vasculitis.

## Antibody-Mediated Vasculitis

Autoantibodies are associated with vasculitis in several conditions. In Kawasaki's diseases, the autoantibodies appear to be pathogenic. Kawasaki's disease is an acute childhood panvasculitis with endothelial necrosis. High circulating levels of interleukins, interleukin la (IL-1), and tumor necrosis factor (TNF) induce neoantigens on the surface of endothelial cells. Anti-endothelial antibodies react with these modified cells causing lysis of the endothelium, increased coagulation, and thrombosis with secondary inflammation (Leung et al., 1998). The pathogenesis of anti-neutrophil cytoplasmic antibodies (ANCAs) is not as clear. However, antibodies to one of the antigenic substrates of ANCA, proteinase 3, may react with neutrophils, causing a respiratory burst and degranulation. The release of lysosomal enzymes potentially injures the vessel wall and incites inflammation. Alternately, binding of the ANCA antibodies to the endothelium has also been described as a mechanism of vascular injury.

## T-Cell-Mediated Vasculitis

In cell-mediated mechanisms, the interaction between T cells and endothelial cells (or smooth muscle or microglial cells) may be the crucial process in initiating local vasculitis. Tissue-specific antigens on endothelial cells may provide an explanation for organ-specific vasculitis such as isolated CNS angiitis (Moore, 1989) or peripheral nervous system vasculitis. Recruitment of inflammation within the vessel wall results from the release of cytokines that may initially come from either the T cell or the endothelium. Often interferon released from T cells induces adhesion molecule expression on endothelial cell as well as release of cytokines by the endothelial cells. These cytokines then recruit non-antigen-specific T cells to the site. This mutually stimulatory interaction of lymphocytes and endothelial cells is normally highly regulated. In the vasculitides, regulation may be defective either because of persistent antigen stimulation or an abnormality in the regulatory process.

## Nonspecific Vascular Injury

Vascular inflammation may also result from nonspecific vascular injury. The types of injury associated with inflammation range from atherosclerotic lesions to acute pressor responses. It is likely that cytokines and platelet adhesions play a role in these lesions. Damaged vessels may also expose previously hidden antigens that serve as a nidus for secondary vascular inflammation. Atherosclerosis and malignant hypertension may work this way. Some toxins such as amphetamine also appear to result in vasculitis through this mechanism. Infections such as equine viral encephalitis and rickettsial disease directly injure endothelial cells and may thus initiate inflammation.

## GENERAL PRINCIPLES OF TREATMENT

Therapy of vasculitis varies with the suspected cause and pathogenic mechanism. However, underlying principals remain: (1) remove the underlying cause of inflammation, (2) reduce the inflammatory cells and pathogenic mediators by the method with least short- and long-term toxicity, and (3) restore the balance of healing within the vessel. Treatment strategies used for vasculitis in its many forms have been difficult to analyze objectively for reasons including: (1) The natural history remains variable among populations of patients. (2) The rarity of many disorders complicates recruitment in randomized, controlled studies. (3) The nature of immunosuppressive therapy, both the obvious side effects and the need for dosage adjustments to achieve desired effects, confounds blinded studies. (4) The features of treatment failure resemble both complications of therapy and long-term manifestations of disease. Nonetheless, some recent studies are applying controlled techniques in some types of vasculitis.

Complications of therapy for vasculitis can be divided into those that appear acutely and those that occur over a longer term. *Infection* continues to be a major source of morbidity and mortality in patients with vasculitis. Recognition and treatment of infections in these patients are particularly difficult because the spectrum of potential pathogens is extensive and the clinical manifestations of infection mimic those of the underlying disease. Less acutely but equally troublesome are the burden of chronic disease, the high incidence of relapse, and the longer-term toxic effects of therapy.

Many of the pharmacological agents used to treat immune-mediated disease and chronic inflammation are drugs used for various diseases over the past three to five decades. As the mortality for many of the vasculitides has decreased, the morbidity of the medications becomes more apparent. For instance, glucocorticoid therapy is effective in treating acute vascular inflammation but contributes to some of the long-term damage in the vessel wall as well as reducing resistance to numerous infections. Current studies focus on learning new ways to use these medications with greater effectiveness and fewer side effects. Older agents are used in new ways that restrict their action (more selective), restrict site (smaller distribution), or modify the dosage or route. Newer medications and strategies also prove effective in patients with vasculitis resistant to traditional therapies (Calabrese, 1997; Gross, 1999; Langford and Sneller, 1997; Thomas-Golbanov and Sridharan, 2001). Tables II and III show agents currently useful in specific disorders and several specific regimens.

## Agents Used in the Treatment of Vasculitis

The role of *anti-platelet agents* in acute disease is not clear. Given the prominent scarring and early atherosclerosis reported in patients with long-term survival from systemic vasculitis, maintenance with anti-platelet agents would appear to be useful, although long-term studies have not been conducted. *Aspirin* is a central part of the acute treatment of Kawasaki's disease. We recommend daily aspirin to patients after therapy for isolated angiitis of the CNS and systemic vasculitides.

*Nonsteroidal anti-inflammatory drugs (NSAIDS)* have both anti-inflammatory and analgesic actions. They act by inhibiting the cyclooxygenase activities of prostaglandin endoperoxidase H (PGH) synthases that block the biosynthesis of prostanoids including prostaglandins and thromboxanes. The clinical efficacy and side effects of these medications depend on their relative potency against two isoenzymes, PGH synthases 1 and 2 (also called COX 1 and 2, respectively), which are encoded by separate genes. Briefly, PGH synthase 1 (COX 1) is constitutively expressed in all tissues particularly the gastrointestinal tract cells, which use PGH 1 to produce prostaglandins needed for intracellular housekeeping functions. PGH 2 (COX 2) is produced largely in endothelial cells, fibroblasts, and macrophages in response to cytokines, growth factors, and tumor promoters. Most NSAIDS including aspirin, indomethacin, ibuprofen, and naproxen are more effective at inhibiting COX 1 than COX 2; this is apparent in the prominent gastrointestinal side effects of these medications.

**TABLE II**    Currently Recommended Medications in Specific Vasculitides

| Disease/natural history | Recommended therapy | Small series |
|---|---|---|
| Wegener's granulomatosis | Cyclophosphamide/prednisone, trimethoprine-sulfamethoxazol | Etaneracept |
| PAN | Cyclophosphamide/prednisone | |
| Churg–Strauss | Cyclophosphamide/prednisone | Interferon |
| Temporal arteritis | Prednisone | Prednisone/methotrexate |
| | | Inflixamab |
| Isolated angiitis of the CNS | Cyclophosphamide/prednisone | |
| Takayasu's | Anti-platelet, anti-hypertensive, prednisone/cyclophosphamide | MMF |
| | | Surgery |
| | | Interferon |
| Behcet's | Azathioprine | Thalidomide |
| | Chlorambucil | Methotrexate |
| Secondary vasculitides | | Lamivudine |
| Viral | Interferon, vidurbarine, plasmapheresis | |
| Other infectious | Appropriate agent to treat underlying infection | |
| Kawasaki | Intravenous immunoglobulin, aspirin | |
| Toxic | Remove cause | |

**TABLE III** Current Treatment Regimens for Specific Vasculitides

| | |
|---|---|
| Isolated angiitis of the CNS | *Prednisone*: 40–60 mg/day orally at initiation; taper to 40 mg/day by end of first 4 weeks; taper in 5 mg increments every week to 15 mg; maintain 15 mg/day of prednisone for 6–9 months; reevaluate clinical features and taper 2.5 mg/day every 2 weeks until therapy is withdrawn <br> *Cyclophosphamide*: 500 mg intravenously every 3–4 weeks with 2 L of fluid for 9 months; after clinical reevaluation cyclophosphamide dosage may be discontinued or reduced by half the dosage times for two additional treatments <br> Alternative: <br> Prednisone as above with oral cyclophosphamide 1–2 mg/kg/day; if documented relapses occur with either regimen, therapy should be repeated with oral cyclophosphamide and prednisone for 1 year |
| Wegener's granulomatosis | *Prednisolone*: Days 1–3, 0.5 g intravenously; days 4–14, 1 mg/kg/day orally. By day 15 taper steroids with a reduction of 10 mg/week to a dosage of 30 mg/day. Then change tapering to 5 mg/week until a dosage of 15 mg/day. From there tapering is 2.5 mg/week. Tapering is varied according to clinical features but after 6 months the dosage of prednisolone should be 12.5 mg/day. <br> *Cyclophosphamide*: 0.75 g/m2 intravenously every 4 weeks <br> or <br> *Cyclophosphamide*: 2 mg/kg/day for 1 year |
| Hepatitis B-related PAN | *Plasmapheresis*: 3 exchanges a week for 3 weeks; then 2 exchanges a week for 2 weeks; then one to two times a week according to clinical situation <br> *Prednisone*: 1 mg/kg/day for 1 week; then taper over 1 week and stop <br> *Vidabarine* intravenously 15 mg/kg/day for 1 week; then 7.5 mg/kg/day for 2 weeks or <br> *Interferon-b*: 3 million units a week for a maximum of 1 year <br> Alternative: <br> *Interferon-b*: 3 million units a week for a maximum of 1 year <br> *Lamivudin* <br> *Prednisone*: 1 mg/kg/day for 1 week |

Selective COX-2 inhibitors are newer medications that reduce the inducible form of cyclooxygenase (COX2), which is the major source of prostaglandin in the inflammatory response. Two agents, rofecoxib (Vioxx) and celecoxib (Cerebrex), are selective COX 2 inhibitors and permit higher dosage of medication targeted at inducible enzymes with fewer side effects. Of note, the COX 2 inhibitors do not affect platelet function. This is an advantage if bleeding is a potential complication. However, many physicians accustomed to using a nonsteroidal for both its anti-inflammatory and anti-platelet actions need to remember to add a specific anti-platelet agent if needed.

*Glucocorticoids (CSs)* are small lipophilic molecules that circulate in the blood mostly bound to CS binding globulin and albumin. Free CS molecules traverse cell membranes and interact with highly conserved CS receptors that reside in the cytosol of all nucleated cells. The mobility of the molecule and the wide distribution of its receptor are responsible for its pleomorphic effects. Occupancy of glucocorticoid and mineralocorticoid receptors throughout the body (including in the hippocampus and hypothalamus) determines the net activity of many pathways. Both inadequate and excessive (endogenous or pharmacological) levels of cortisol damage the host. The immunomodulatory and anti-inflammatory actions of CS result from several actions: (1) A change in traffic patterns of leukocytes, which impedes leukocyte accumulation, (2) reduced expression of pro-inflammatory cytokines (IL-1, IL-2, IL-6, IFN, and TFN), thus inhibiting T-cell proliferation and T-cell-dependent immunity. The anti-inflammatory effects of CS are probably the greatest, or at least most discernible, on macrophages where they inhibit cytokine gene transcription, PGH synthase 2 (COX 2)-mediated production of prostanoids, and nitric oxide synthetase. These effects reduce vasodilation. Adverse events are considerable and do not necessarily correlate with the length of treatment. Infections remain a major complication of CS therapy (Hamuryudan *et al.*, 1998; Sneller, 1998; Stuck *et al.*, 1989). Other devastating conditions induced or accelerated by corticosteroid therapy are osteoporosis and aseptic necrosis. Recent investigations provide guidelines for minimizing osteoporosis during CS therapy which include exercise, vitamin D and calcium supplementation, sodium restriction as well as the indications for calcitonin, biphosphonates, sodium fluoride, and gonadal hormone replacement (American College of Rheumatology Task Force 1996; Lukert and Raisz, 1990; Morand, 2000; Reid, 2000). Other serious side effects of CS therapy include diabetes mellitus, hypertension, and cataract formation. CS may contribute to the chronic vasculopathy through hypertension, diabetes, effects on the shuttle of fatty acids in the vessel wall, and the increased distribution of B adrenergic receptors.

Despite the caveats, corticosteroid therapy is indicated to reduce inflammation in any disorder in which the effects of the disease outweigh the considerable long-term side effects of the medications. In some inflammatory diseases we have no alternative agents to prednisone, or the equivalent, but the side effects should be clear to the patient and physician. Temporal arteritis is an example of a vasculitis in which no alternative agent is equally effective, at least in preventing blindness. Temporal arteritis, which is a systemic inflammatory disease, overlaps with a milder disorder, polymyalgia rheumatica. Distinguishing between the two is important because the latter responds to a lower dosage of prednisone and does not carry the risk of blindness. In other disorders such as PAN and SLE, CS therapy is strongly suspected of a major contribution to the development of chronic vasculopathy (Lukert and Raisz, 1990; Thomas *et al.*, 2002). Studies continue to investigate medications for their ability to spare or reduce the use of CS.

The dosages of CS preparations are numerous, and all are completely and rapidly absorbed orally. Prednisone is the least expensive corticosteroid and is available in many dose sizes. After absorption it is converted to prednisolone. Methylprednisolone (4 mg is equivalent to 5 mg prednisone) has a somewhat longer half-life and is much more expensive. Dexamethasone, even more expensive and with a longer half-life, does not bind to CS binding globulin and variably crosses the blood–brain barrier (BBB). Intravenous administration is often prescribed although the rationale is not clear unless a patient cannot tolerate any oral medication. The immunological impact is considered to be high at a dose above the equivalent of 30 mg prednisone/day, medium at a dose of 10–30 mg/day, and low at a dose <10 mg/day. The rationale for very high dosages is not apparent based on definable immunological or inflammatory effects (Rota *et al.*, 1992; Wilcke and Davis, 1982).

*Cyclophosphamide* is a cytotoxic alkylating agent introduced nearly 40 years ago for the treatment of malignancies. As an anti-metabolite, it causes interstrand and intrastrand DNA crosslinks that result in dysfunction of the DNA template. These intracellular biochemical reactions lead to cell death, largely through apoptotic mechanisms. Cyclophosphamide exerts its effect throughout the cell cycle but the greatest action is during cell division. Cyclophosphamide's effect on clonal expansion of lymphocytes underlies its use in immune-mediated disorders. The rapidity of its effect in some vasculitides suggests an additional mechanism, possibly on endothelial cells, but this has not been established.

Cyclophosphamide is used in the vasculitis of rheumatoid arthritis, Wegener's granulomatosis, PAN, isolated angiitis of the CNS, and autoimmune processes without a CNS vasculitis such as SLE and polymyositis. Patients usually tolerate cyclophosphamide well, although it must be used with care and scrupulous attention to hydration (>2 L fluid a day) to avoid hemorrhagic cystitis and potential bladder malignancies. Some patients develop prominent nausea, requiring antiemetic therapy. Infrequently, allergic reactions, teratogenicity, and carcinogenicity (0.1–1% with long term use) occur. Rarely, we encounter cardiac, pulmonary, hepatic disorders or syndrome of inappropriate antidiuretic hormone (SIADH). In the dosages used for treatment of vasculitis (smaller than those used to treat cancers), the major side effects include infection, nausea, gonadal failure, and hemorrhagic cystitis. The actual dosage of medication varies with the specific disease and will likely continue to be modified as a result of ongoing studies. Cyclophosphamide is prescribed either orally (1.5–2 mg/kg/day) or intravenously (500–1500 mg every 3–4 weeks).

### Combination Therapy
### Cyclophosphamide/Prednisone

The systemic vasculitides including Wegener's granulomatosis, Churg–Strauss angiitis, microscopic polyarteritis, and PAN, as well as systemic vasculitis complicating rheumatoid arthritis and isolated angiitis are usually successfully treated with a combination of cyclophosphamide and a glucocorticoid. With the reduction in disease mortality come the newer questions: how can we reduce disease morbidity, morbidity from the medications, and the relapse rate? There have been several prospective randomized trials of the treatment of systemic vasculitis with several dosages and routes of cyclophosphamide (Abgrall *et al.*, 2001; de Groot *et al.*, 2001a; Gayraud *et al.*, 2001). It is difficult to compare these studies because they neither measure the same outcomes nor do they agree on the escalation therapy for treatment failures. One study comparing induction of remission, frequency of remission, number of relapses, and complications of therapy (particularly infection, bladder related side effects, and malignancies) found pulse cyclophosphamide/prednisolone to be slightly less effective but less toxic than a continuous oral regimen of prednisolone and cyclophosphamide (with the latter replaced after three months with azathioprine) (Adu *et al.*, 1997). More recent treatment regimens stratify patients by predicted course so that brief or intermittent therapy is used for patients with a better prognosis and more aggressive therapy given to patients at a higher risk of early death (de Groot and Gross, 1998; Haubitz *et al.*, 1998; Langford *et al.*, 1999). Newer methods of therapy continue to be explored (Langford and Sneller, 1997).

Several combination regimens are effective in inducing remission. After choosing a starting dosage, these medications are titrated to clinical response and side effects. Two regimens that have been effective in systemic vasculitis and isolated angiitis of the CNS (although control trials of the latter have not been possible because of the rarity of the disease) are: (1) Intravenous cyclophosphamide 500 mg every 3 weeks for 1 year with oral prednisone, initially 40–60 mg/day for several weeks and then tapered to 20 mg for 6 months and maintained at 10–15 mg for at least 6 months prior to slow taper. After discontinuing medication, patients need to carry identification that they have been adrenally suppressed and should receive a stress dose of prednisone equivalent in the event of a medical emergency. (2) Oral cyclophosphamide 100–150 mg per day for 1 year (with reduction of dosage if the neutrophil count falls below 1500 mm$^3$) with oral prednisone. One recent article describes the successful treatment of patients with isolated angiitis of the CNS and amyloid angiopathy with cyclophosphamide alone (Fountain and Lopes, 1999). The major clinical problem in the management of isolated angiitis of the CNS is accurate diagnosis, particularly distinguishing it from the numerous infectious and neoplastic causes of similar clinical features.

*Methotrexate* is a potent folate antagonist that has been utilized successfully in rheumatoid arthritis and psoriasis. It has recently been investigated in the treatment of Takaysu's arteritis (Stuck *et al.*, 1989), Wegener's granulomatosis (Langford *et al.*, 1997, 1999), and neuropsychiatric Behcet's disease (Hirohata *et al.*, 1998).

*Chlorambucil* is a cell-cycle-nonspecific agent from the alkylating class of anticancer drugs. Among the vasculitides, it may be useful in Behcet's disease. Short-term, high-dose chlorambucil was used in five patients with intractable sympathetic ophthalmia and six patients with severe Behcet's disease. Total cumulative doses ranged from 306 m to 4.2 g with duration of therapy of no longer than 36 weeks (most cases were treated for less than 24 weeks). Adverse effects include myelosuppression that may result in leukopenia and thrombocytopenia. Hepatotoxicity, pulmonary fibrosis, bronchopulmonary dysplasia, sterility, peripheral neuropathy, seizures, and acute pneumonitis are known side effects. There is also a risk for developing a malignancy (Luqmani *et al.*, 1990; Nichols *et al.*, 2001; Tessler and Jennings, 1990).

## Newer Pharmacological Agents for Vasculitis

*Mycophenolate mofetil (MFF)* is the morpholinoethyl ester prodrug of mycophenolic acid (MPA). MFF is an immunosuppressant drug that acts by impairment of de novo purine synthesis. This drug is relatively selective for lymphocytes and inhibits antibody production by B

cells more than other immunosuppressants. The FDA recently approved MMF for the prevention of rejection in renal transplantation. There has been reported success with this medication in the maintenance phase of systemic vasculitis after traditional induction with cyclophosphamide/prednisone as well as in Takayasu's syndrome (Daina *et al.*, 1999; Haubitz and de Groot, 2002; Waiser *et al.*, 1999). If the low side effect profile continues, this may decrease the long-term morbidity of traditional vasculitis therapies.

*Leflunomide* is an isoxazole derivative structurally unrelated to other immunosuppressant medications. It diminishes pyrimidine nucleosides and may inhibit NFκ-B activation (Herrmann *et al.*, 2000; Manna *et al.*, 2000).

*Thalidomide* appears to have a specific inhibitory activity on TNF-α production. Based on recent studies indicating efficacy in erythema nodosa leprosum, acquired immunodeficiency syndrome (AIDS), graft-versus-host disease (GVHD), and Behcet's disease, thalidomide is undergoing a reevaluation of its utility in vasculitis (Hamuryudan *et al.*, 1998; Sakane and Takeno, 2000). Side effects include peripheral neuropathy, drowsiness/somnolence, orthostatic hypotension, and teratogenicity.

## Biological Modifiers

Biologic modifiers encompass a variety of substances that diminish or enhance biologic pathways. They include monoclonal antibodies, cytokines, receptor antagonists, gene promotors, and other agents almost invariably cloned, sequenced, and synthesized. As a group, a major attraction is their utilization of natural pathways and their ability to target selected aspects of an immune response. A limitation of their use in chronic diseases is the short duration of their effect. When used repeatedly, the development of host antibodies may limit their efficacy.

Interferons as a class possess antitumor, antiviral, and immunomodulating activity. They are species-restricted and receptor-dependent. Binding to specific cell-surface receptors is required for activity. Interferon is a primary therapeutic agent in the treatment of viral vasculitis, most specifically that associated with hepatitis c viremia (Komocsi *et al.*, 2002). There is reported success with Churg–Strauss angiitis (Alpsoy *et al.*, 2002; Gross, 1999; Kosar *et al.*, 1999; Nichols *et al.*, 2001; O'Duffy *et al.*, 1998; Tatsis *et al.*, 1998). Of note, interferon-γ has also been suspected of either inducing or exacerbating a vasculitis. Characteristic side effects of interferon-γ are flu-like symptoms and gastrointestinal disturbances, but alopecia, neutropenia, thrombocytopenia, and liver enzyme elevations also occur. In addition, the neurological, psychiatric, and autoimmune side effects of interferon therapy may restrict its use.

## Agents That Interfere with Activity of Tumor Necrosis Factor-α (TNF-α)

Tumor necrosis factor (TNF) promotes inflammation by binding to receptors on a variety of cells, stimulating them to release an array of other inflammatory mediators. Soluble TNF receptors are cleaved from the membrane portions of the cell complex and appear to function as a natural regulator of TNF-α.

*Etanercept* is an expressed protein which contains two chains of a soluble TNF receptor linked by the Fc portion of an immunoglobulin molecule (TNFR:Fc = soluble TNF receptor (p75):Fc fusion protein). The molecule functions by absorbing excess circulating TNF-α and keeping it from binding to its natural receptor. Several studies have indicated efficacy and safety in patients with rheumatoid arthritis (Stone *et al.*, 2001).

*Infliximab* is a chimeric (part mouse, part human) monoclonal antibody to TNF. This also reduces inflammation and is currently indicated for Crohn's disease. An issue in using nonhuman monoclonal antibodies as therapy is the development of host antibodies to injected monoclonal antibody. Its main side effects are hypersensitivity reactions, urticaria, dyspnea, hypertension, fever, rash, headache, and polyarthralgias (Airo *et al.*, 2002).

*Gene therapy* is on the frontier of treatment of vascular disease and possesses the possibility to diminish some of the chronic changes associated with proliferative responses (Harari *et al.*, 1999).

## Interventional Procedures/Maneuvers

*Plasmapheresis* may be effective therapy in vasculitis when circulating autoantibodies or immune complexes contribute to the pathogenesis of the vasculitis. It is most effective in essential mixed cryoglobulinemia but also may be useful in Kawasaki's syndrome. A controlled study of CS alone compared with CS and plasmapheresis did not demonstrate improved short-term or long-term outcome in patients with PAN or Churg–Strauss (Guillevin *et al.*, 1995a). Except for patients with refractory systemic vasculitis, it is not recommended for routine or initial treatment. Both the complications of vascular access and the high cost limit its utility. The main side effects are hypotension and occasional blood-clotting abnormalities. Typically, six exchanges are performed over the first 2 weeks, which are reduced to weekly exchanges and then monthly. An immunosuppressant such as azathioprine is usually started to prevent rebound inflammation and to maintain the therapeutic effect.

*Intravenous immunoglobulin* is proven effective in Kawasaki's disease where it is the treatment of choice (Burns, 2001; Tse *et al.*, 2002). It is not a prominent therapy of other vasculitides, although occasional successes have been reported. No controlled studies have been performed. Side effects include fever, chills, hypotension, nausea, abdominal pain, headache, dizziness, encephalopathy, and rarely anaphylaxis (Mathy *et al.*, 1998). The dosage varies. In one study a single intravenous dose of 2 g/kg body weight was more effective than 400 mg/kg over 4 days in Kawasaki's syndrome (Michelfelder and Shim, 2002).

*Surgery*, despite its many potential benefits, is currently demonstrably beneficial only in Takayasu's arteritis. Because the CNS manifestations result from obliterative and hypertensive vascular changes occurring during the disease's fibrotic stage, bypass procedures have occasionally proved advantageous. More recently angioplasty, grafts of the subclavian artery, to

---

**TABLE IV   Wegener's Granulomatosis**

**Definition and classification:** Wegener's granulomatosis is a disease of the upper and lower respiratory tracts, kidneys, and vasculature characterized by necrotizing granuloma and vasculitis. The typical history is one of upper respiratory tract symptoms, including epistaxis and hemoptisis; deafness and multiple cranial neuropathies suggest diagnosis (Cotch *et al.*, 1996; de Groot *et al.*, 2001b; Hoffman *et al.*, 1992; Langford and Sneller, 1997; Leavitt *et al.*, 1988; Nishino *et al.*, 1993). Renal disease is frequent.

**Clinical aspects and natural course:** Prominent systemic inflammation – sedimentation rate invariably elevated during active disease; ANCA autoantibodies characteristic but unlikely to be marker of disease activity; thrombocytosis, anemia, and leukocytosis frequent.

**Pathogenesis and pathology**

*Etiology and pathogenesis:* Unknown. The early stages may be infectious. The later course of vasculitis is associated with autoantibodies to cytoplasmic components of neutrophils and monocytes (cANCAs).

*Pathology:* Granulomatous lesions of upper and lower respiratory tract, focal segmental glomerulitis, and necrotizing vasculitis are abundant. Neurological abnormalities may result from (1) local extension of granulomas in the sinuses to the anterior and middle cranial fossa resulting in multiple cranial neuropathies or (2) vasculitis with tissue ischemia.

**Practical management:** Untreated the diseases has a 2-year mortality of 93% with an average survival time of 5 months. With current combination therapy 90% of patients can be brought into remission with this regimen. Should relapse occur, reinstitution of immunosuppressive treatment is usually effective.

**Combination (cyclophosphamide/prednisone therapy)** is usually but not invariably effective. The goals now are to limit side effects of aggressive treatment and several studies compare different routes of therapy or switching to a potentially less toxic immunosuppressant when a remission is induced.

**TABLE V    Churg Strauss Syndrome**

**Definition and classification:** a necrotizing vasculitis characterized by a history of allergic diathesis, invariable lung involvement and peripheral eosinophilia (Abu-Shakra *et al.*, 1994; Gross, 2002; Guillevin *et al.*, 1999).

**Clinical aspects and natural course:** Often a history of atopy or asthma. Constitutional symptoms are frequent. Eosinophilia invariable. Anti-neutrophil cytoplasmic antibodies variable. Triad for diagnosis: asthma, eosinophilia, and vasculitis in two extrapulmonary organs. Heart, PNS, and CNS affected in about two-thirds of patients.

**Pathogenesis and pathology**

*Pathogenesis:* Unclear but eosinophil mediated vascular injury appears likely.

*Pathology (Churg and Strauss, 1951):* Necrotizing vasculitis usually with fibrinoid necrosis; the predominant cellular infiltrate is eosinophilic.

**Practical management:** Prednisone alone may be effective in some patients with mild disease. Combination prednisone/cyclophosphamide therapy is usual. Interferon-α is effective is some cases.

---

repair areas of damaged wall has been described (Gu and Wang, 2001; Ziyal *et al.*, 1999).

## SPECIFIC VASCULITIDES

### Idiopathic Systemic Vasculitides

A brief review of the clinical features of polyarteritis nodosa, microscopic polyangiitis, Wegener's granulo-

**TABLE VI    Polyarteritis Nodosa**

**Definition and classification:** Necrotizing arteritis of mainly small and medium-sized muscular arteries, affecting multiple organs notably sparing the lungs (Adu *et al.*, 1997; Calabrese *et al.*, 1995; Gaston *et al.*, 1988; Gayraud *et al.*, 2001; Gordon *et al.*, 1993).

**Clinical aspects and natural course:** Evidence of systemic inflammation usually present, laboratory evidence of elevated sedimentation rate, anti-nuclear antibodies variable organs involved: kidney, 85%; heart, 76%; PNS, 67%; liver, 62%; gastrointestinal, 50%; skin, 50%; CNS, 40%.

**Pathogenesis and pathology**

*Pathogenesis:* Circulating immune complexes are detectable in 50% patients; association with hepatitis B and C suggests an infectious etiology.

*Pathology:* Necrotizing pan-arteritis, predominant cellular infiltrate is neutrophilic.

**Diagnosis:** Constitutional symptoms and multiple organ system involvement, mononeuropathy or mononeuropathy multiplex, sudden onset of severe hypertension. Angiography of hepatic, renal, and mesenteric circulation. Biopsy of symptomatic organs, such as muscle or skin or sural nerve.

**Prognosis/practical management:** Untreated the 5-year survival rate is 13%, 40–50% die in the first 3 months. With combination prednisone/cyclophosphamide therapy long-term remission, 5-year survival rates of 80% have been achieved. Encephalopathy, seizures, and neurological symptoms may improve significantly.

---

**TABLE VII    Microscopic Polyarteritis**

**Definition and classification:** A small-vessel vasculitis, prominent renal involvement, and circulating ANCA antibodies. Microscopic polyangiitis is the most common cause for pulmonary–renal vasculitic syndrome (Jennette *et al.*, 2001; Lhote *et al.*, 1996).

**Clinical features:** Pauci-immune necrotizing and crescentic glomerulonephritis, and hemorrhagic pulmonary capillaritis are common in patients with microscopic polyangiitis.

**Pathogenesis and pathology**

*Pathogenesis:* Unknown but cANCA are characteristic.

*Pathology:* Absence or paucity of immunoglobulin localization in vessel walls; small vessels most prominently affected. Occasionally arterioles spared. Pathologic features similar to Wegener's and Churg Straus angiitis.

**Practical management:** Combination cyclophosphamide/prednisone is effective.

---

matosis, and Churg–Strauss angiitis are displayed in Tables IV–VII.

## Secondary Vasculitides

Diagnosis and treatment of the most frequent secondary vasculitis, hypersensitivity vasculitis is shown in Table VIII. However, many cases of vasculitis secondary

**TABLE VIII    Hypersensitivity Vasculitis**

**Definition and classification:** Hypersensitivity vasculitis is a group of diseases characterized by a cutaneous venulitis. Several distinctive types exist although the nomenclature is evolving. Palpable purpura associated with medications or infections (hypersentivity vasculitis), serum sickness, Henoch–Schönlein–purpura (HSP), and cryoglobulinemia occur most frequently. Neurological involvement is distinctly unusual although subarachnoid hemorrhage is reported in HSP and plexopathies are notable in serum sickness (Koutkia *et al.*, 2001; Sais *et al.*, 1998; Vidaller and Sais, 2001).

**Clinical aspects and natural course**

*Epidemiology:* Very frequent. Usually resolves spontaneously.

**Pathogenesis and pathology**

*Pathogenesis:* Immune complex deposition in the vessel wall, leading to activation of complement, subsequent attraction of polymorphonuclear leukocytes, and subsequent destruction of vascular structures by proteolytic enzymes. The pathogenic experimental models for this disease are Arthus reaction and serum sickness.

*Pathology:* Leucocytoclastic vasculitis, an acute necrotizing process with a predominantly neutrophilic infiltration of all three layers of vessel walls, accompanied by nuclear debris (leukocytoclasis). During the progression of inflammation, the relative amount of neutrophils and mononuclear cells and complement activation probably varies.

**Diagnosis:** Clinical findings with skin biopsy.

**Practical management:** Remove underlying antigen or treat the infection. If there is renal or systemic involvement, corticosteroids are usually effective. The prognosis for resolution of skin lesions is excellent.

to infection are more complicated than a cutaneous venulitis. *Vasculitis secondary to infections* is both frequent and clinically important. The number of patients with vasculitis from secondary causes far exceeds those with a primary, idiopathic vasculitis. A high index of suspicion enables a clinician to properly diagnose and institute therapy promptly for a vasculitis secondary to infection. A spectrum of infections demonstrably associated with prominent vascular inflammation in the central nervous system includes bacteria, fungi, viruses, and protozoa. Inflammation of the vasculature may develop from extension from neighboring foci or because of direct invasion of pathogenic microorganisms in vessel walls. Mechanisms of infection-mediated inflammation include toxic disruption of the endothelium, immune-mediated cytolysis of infected endothelia, and immune complex (antibody–infectious organism)-mediated inflammation.

Infections present in many cases of polyarteritis nodosa include hepatitis C and hepatitis B; treatment regimens are shown in Table III. Lamivudin, a nucleoside analog and retroviral transcriptase inhibitor, appears effective in some cases (Lau *et al.*, 2002; Wicki *et al.*, 1999). Vasculitis in the CNS due to infection may be more difficult to diagnose. Strokes associated with viral infection usually demonstrate more pleomorphic features. Histologically, inflammation and necrosis of the cerebral blood vessel wall certainly appear with viruses such as herpes simplex, herpes zoster, cytomegalovirus, toxoplasmosis, human immunodeficiency virus, and varicella. However, viruses such as herpes zoster may also cause vasoocclusive disease without inflammatory changes.

## Focal or Restricted Vasculitides

Giant cell arteritis (temporal arteritis) is a systemic vasculitis with clinically restricted features described elsewhere in this book. Isolated angiitis of the CNS, Behcet's disease, and Takayasu's arteritis are shown in Table IX–XI.

---

**TABLE IX    Isolated Angiitis of the CNS**

**Definition and classification:** Isolated angiitis of the CNS (IAC) is an idiopathic vasculitis occurring within the dural reflections. It is also referred to as granulomatous angiitis of the CNS and primary angiitis of the CNS. Careful diagnosis is important because vasculitis secondary to Hodgkin's disease, infection, and toxins may have similar presentations (Chu *et al.*, 1998; Kumar *et al.*, 1997; Lanthier *et al.*, 2001; Moore, 1989; Riemer *et al.*, 1999; Scolding *et al.*, 1997).

**Clinical aspects and natural course**

*Symptoms:* Typically, encephalopathy with headache and multifocal neurological signs. Isolated strikes, cranial neuropathies, and myelopathies are less frequent. Visceral symptoms and signs are absent.

*Blood sedimentation rate:* Almost invariably normal. Autoantibodies such as antinuclear antibodies (ANAs), rheumatoid factor, and ANCA are conspicuously absent. The cerebrospinal fluid may be normal (about 50% of patient or show a mild pleocytosis or elevated protein.

**Pathogenesis and pathology**

*Pathogenesis:* The etiology is unknown. The presence of a prominent cellular infiltrate in the absence of deposited immunoglobulin suggests a cell-mediated process.

*Pathology:* Typically a mononuclear cell infiltrate. Granulomas are a variable histological finding.

**Diagnosis**

Angiography and biopsy are the mainstays of diagnosis. Angiography typically reveals multiple localized areas of narrowing, occlusion or vessels, delayed emptying, and anastomotic channels. Leptomeningeal and wedge cortical biopsy at temporal tip of nondominant hemisphere or stereotactic biopsy of edge of MRI identified lesions is appropriate. Biopsy should not only confirm the diagnosis of vasculitis but also exclude other potential reasons for vasculitis or vasculopathy, in particular infections and neoplasia.

**Practical management/prognosis**

*Natural course and prognosis if untreated:* Prognosis in IAC is recurrent and usually fatal if untreated. Combination prednisone/cyclophosphamide is effective and results in long-term cure or remission >85% of patients. It is not clear if prednisone alone is effective but in biopsy-proven cases this appears to be inadequate treatment.

---

**TABLE X    Takayasu's Disease**

**Definition and classification:** Takayasu's disease is a large vessel giant cell arteritis with predilection for the aortic arch (Desiron and Zeaiter, 2000; Fraga and Medina, 2002; Hoffmann *et al.*, 2000; Sabbadini *et al.*, 2001; Wang *et al.*, 2002).

**Clinical aspects and natural course:** Occurs most frequently in Asian and Oriental populations.

**Pathogenesis and pathology**

*Pathogenesis:* Unknown. The neurological effects appear to occur during the vasoocclusive phase exacerbated by the hypertensive vasculopathy.

*Pathology:* (1) Acute inflammatory infiltrate, (2) intermediate stage with partially necrotic media and revascularization, and (3) sclerosing stage with scarring throughout the arterial wall. Common sequelae are weakening of artery with aneurysm and progressive stenosis culminating in occlusion. The arterial ostium at its site of origin from the aorta is stenotic.

**Diagnosis:** Usually history of systemic inflammation over months, myalgias, arthralgias, fatigue, fever, and weight loss. Confirmation by angiography of aortic arch and the descending aorta.

**Practical management:** Many of the clinical abnormalities appear after the inflammatory phase, so it is not clear if immunosuppressive is effective. Some recent studies suggest immunosuppression diminishes progression of aortopathy. Vascular bypass surgery may be beneficial for neurological abnormalities although the surgery may be technically difficult. It is not known if immune intervention benefits neurological features.

**TABLE XI    Behcet's Disease**

**Definition and classification:** Behcet's disease is an inflammatory disorder characterized by relapsing uveitis and genital and oral ulcers as well as arthralgias (Akman-Demir *et al.*, 1996; Fresko *et al.*, 1999; Nakamura *et al.*, 2002).

**Clinical aspects and natural course:** Associated with HLA-B5 in Japan and the Mediterranean nations. Usually chronic mild disorder. An infectious etiology is suspected but not proven. The disease is generally painful, but most often benign; characterized by recurrences over a number of years. Neurological and posterior uveal tract lesions usually indicate a poorer prognosis. Neurological involvement occurs in 10–29% of patients and includes meningoencephalitis, pseudotumor cerebri, venous thrombosis, and focal signs or brain stem signs. CSF pleocytosis is present in 86% of cases when the CSF was examined.

**Pathogenesis and pathology**

*Pathogenesis:* Uncertain.

*Pathology:* Inflammation with lymphocytes and plasma cells in and around small vessels; perivascular focus of cellular loss and demyelination.

**Diagnosis:** Complete form: at least three criteria including oral aphthae. Inflammatory bowel disease, systemic lupus erythematosus, Reiter's syndrome, orogenital herpesvirus infection must be excluded.

**Practical management**

Permanent remission of symptoms has not been reported. Corticosteroids, azathioprine, chlorambucil, and, more recently, thalidomide are among the reportedly useful treatments.

**TABLE XII    Systemic Lupus Erythematosus**

**Definition and classification:** Multisystem inflammatory disorder characterized by persistently circulating, distinctive autoantibodies, and immune complexes. Cutaneous, renal, hematological, musculoskeletal, and neurological abnormalities dominate the clinical features. The tempo of the disease ranges from mild and indolent to acute and fulminant.

**Clinical aspects and natural course:** Neurological, psychiatric, and behavioral abnormalities (NP-SLE) occur in 40–75% of patients during the course of disease. Stroke and hemorrhage may occur in up to 15% of patients (Moore, 1997a,b; Welch *et al.*, 1997).

**Pathogenesis and pathology**

*Pathogenesis:* Antibodies to brain cells, neuroendocrine changes, effects of inflammatory mediators and cytokines, and vascular disease are all implicated.

*Pathology:* Histological features are scanty compared with clinical abnormalities. A bland degeneration of cerebral blood vessels without evidence of vasculitis occurs. Also present is microvascular occlusive disease. A range of processes may contribute, including chronic immune complex disease, emboli from the heart, and coagulopathies.

**Practical management:** In the absence of acute inflammation the benefits of corticosteroids are doubtful and their contribution to early atherosclerosis in patients with SLE increases the risk of this therapy. Anti-platelet agents and low-dosage immunosuppressive therapy may minimize or prevent some recurrence. Plasmapheresis has a role in acute neurological events.

**TABLE XIII    Rare Vasculitic Syndromes with Neurological Involvement**

| Disease | Characteristics | Neurological symptoms | Treatment of neurological symptoms |
|---|---|---|---|
| Cogan's syndrome | Inflammation of eye and ear associated with a systemic vasculitis.<br>Most frequent in young adults.<br>Meniere-like episodes.<br>Interstitial keratitis, iritis, conjunctivitis, uveitis, vitritis, scleritis.<br>Coronary vasculitis reported in 10%. | Cerebral infarction association with vasculitis reported but unusual (Hammer *et al.*, 1994; Karni *et al.*, 1991) | Mild eye inflammation: topical corticosteroids.<br>Hearing loss: oral prednisone.<br>Systemic vasculitis: combination therapy (corticosteroids with cyclophosphamide or cyclosporinA). |
| Lymphomatoid granulomatosis | Most recently classified with T cell lymphomas. | | |
| Eales's disease | Retinal vasculitis, frequent in young men of the Middle East and India.<br>Recurrent vitreous hemorrhages induce specks, floaters, cobwebs, curtains, and blurred vision. | Meningitis.<br>Ischemic infarcts.<br>Myelopathy (Gordon *et al.*, 1988; Katz *et al.*, 1991). | Local treatment for the eye includes photocoagulation and vitrectomy. Immunosuppression of uncertain benefit. |
| Kawasaki disease (mucocutaneous lymph node syndrome) | Antibody mediated pan vasculitis of young childhood, particularly prevalent in Japan. Histologically, a panvasculitis with endothelial necrosis.<br>High fever, conjunctivitis, polymorphous exanthema, cervical lymphadenopathy, and desquamation of the fingertips. Coronary arteries are involved in approximately one-quarter of patients. | Aseptic meningitis<br>Seizures<br>Stroke | Current therapeutic recommendation is a single dose of immunoglobulins (2 g/kg, iv) combined with aspirin. |

## Immune Mediated Vasculopathies

*Connective tissue diseases* are systemic inflammatory diseases in which one of the target systems is the vasculature. True inflammation of the CNS vasculature occurs occasionally, particularly in rheumatoid arthritis vasculitis and possibly Sjogren's disease. The more frequent vascular complication in the CNS is a bland or degenerative vasculopathy as is seen in SLE.

*Sjogren's disease* is a systemic disorder manifest by sicca complex and a variety of extraglandular manifestations. Both central and peripheral nervous system abnormalities occur although the histology is not convincingly vasculitis. An alternative, distinctive neuropathy in Sjogren's is not vasculitis but a dorsal root ganglionitis. These patients present with a sensory neuropathy and ataxia usually associated with autonomic insufficiency. Similarly, in the nonspecific systemic inflammatory diseases such as associated with cryoglobulins, a mononeuropathy multiplex, polyneuropathy, or autonomic neuropathy may occur associated with vascular inflammation.

*Systemic lupus erythematosus (SLE)* is a multisystemic inflammatory disease characterized by a loss of self-tolerance. Autoantibodies are a hallmark of systemic lupus erythematosus. The pathogenesis of clinical features varies. Immune complex deposition with subsequent inflammation is the principal pathogenesis of the renal and cutaneous disease. Anemia, leukopenia, thrombocytopenia, and some coagulopathies likely result from direct antibody-mediated effects. Table XII summarizes some of the features.

There are other types of vasculitis that are encountered even less frequently than those discussed above. Some of these are shown in Table XIII.

# REFERENCES

American College of Rheumatology Task Force on Osteoporosis Guidelines. Recommendations for the prevention and treatment of glucocorticoid-induced osteoporosis (1996). *Arthritis Rheum.* **39**, 1791–1801.

Abgrall, S., Mouthon, L., Cohen, P. *et al.* (2001). Localized neurological necrotizing vasculitides. Three cases with isolated mononeuritis multiplex. *J. Rheumatol.* **28**, 631–633.

Abu-Shakra, M., Smythe, H., Lewtas, J., Badley, E., Weber, D., and Keystone, E. (1994). Outcome of polyarteritis nodosa and Churg–Strauss syndrome. An analysis of twenty-five patients. *Arthritis Rheum.* **37**, 1798–1803.

Adu, D., Pall, A., Luqmani, R. A. *et al.* (1997). Controlled trial of pulse versus continuous prednisolone and cyclophosphamide in the treatment of systemic vasculitis. *Q. J. Med.* **90**, 401–409.

Airo, P., Antonioli, C. M., Vianelli, M., and Toniati, P. (2002). Antitumour necrosis factor treatment with infliximab in a case of giant cell arteritis resistant to steroid and immunosuppressive drugs. *Rheumatology (Oxford)* **41**, 347–349.

Akman-Demir, G., Baykan-Kurt, B., Serdaroglu, P. *et al.* (1996). Seven-year follow-up of neurologic involvement in Behcet syndrome. *Arch. Neurol.* **53**, 691–694.

Alpsoy, E., Durusoy, C., Yilmaz, E. *et al.* (2002). Interferon alfa-2a in the treatment of Behcet disease: A randomized placebo-controlled and double-blind study. *Arch. Dermatol.* **138**, 467–471.

Burns, J. C. (2001). Kawasaki disease. *Adv. Pediatr.* **48**, 157–177.

Calabrese, L. H. (1997). Therapy of systemic vasculitis. *Neurol. Clin.* **15**, 973–991.

Calabrese, L. H., Hoffman, G. S., and Guillevin, L. (1995). Therapy of resistant systemic necrotizing vasculitis. Polyarteritis, Churg–Strauss syndrome, Wegener's granulomatosis, and hypersensitivity vasculitis group disorders. *Rheum. Dis. Clin. North Am.* **21**, 41–57.

Calli, C., Savas, R., Parildar, M., Pekindil, G., Alper, H., and Yunten, N. (1999). Isolated pontine infarction due to rhinocerebral mucormycosis. *Neuroradiology* **41**, 179–181.

Christophe, C., Azzi, N., Bouche, B., Dan, B., Levivier, M., and Ferster, A. (1999). Magnetic resonance imaging and angiography in cerebral fungal vasculitis. *Neuropediatrics* **30**, 218–220.

Chu, C. T., Gray, L., Goldstein, L. B., and Hulette, C. M. (1998). Diagnosis of intracranial vasculitis: a multi-disciplinary approach. *J. Neuropathol. Exp. Neurol.* **57**, 30–38.

Churg, J., and Strauss, L. (1957). Allergic granulomatosis, allergic angiitis, and periarteritis nodosa. *Am. J. Path.* **27**, 227–301.

Cotch, M. F., Hoffman, G. S., Yerg, D. E., Kaufman, G. I., Targonski, P., and Kaslow, R. A. (1996). The epidemiology of Wegener's granulomatosis. Estimates of the five-year period prevalence, annual mortality, and geographic disease distribution from population-based data sources. *Arthritis Rheum.* **39**, 87–92.

Daina, E., Schieppati, A., and Remuzzi, G. (1999). Mycophenolate mofetil for the treatment of Takayasu arteritis: report of three cases. *Ann. Intern. Med.* **130**, 422–426.

Dawson, T. M., and Starkebaum, G. (1999). Isolated central nervous system vasculitis associated with hepatitis C infection. *J. Rheumatol.* **26**, 2273–2276.

de Groot, K., Adu, D., and Savage, C. O. (2001a). The value of pulse cyclophosphamide in ANCA-associated vasculitis: Meta-analysis and critical review. *Nephrol. Dial. Transplant.* **16**, 2018–2027.

de Groot, K., and Gross, W. L. (1998). Wegener's granulomatosis: Disease course, assessment of activity and extent and treatment. *Lupus* **7**, 285–291.

de Groot, K., Schmidt, D. K., Arlt, A. C., Gross, W. L., and Reinhold-Keller, E. (2001b). Standardized neurologic evaluations of 128 patients with Wegener granulomatosis. *Arch. Neurol.* **58**, 1215–1221.

Desiron, O., and Zeaiter, R. (2000). Takayasu's arteritis. *Acta Chir. Belg.* **100**, 1–6.

Fountain, N. B., and Lopes, M. B. (1999). Control of primary angiitis of the CNS associated with cerebral amyloid angiopathy by cyclophosphamide alone. *Neurology* **52**, 660–662.

Fraga, A., and Medina, F. (2002). Takayasu's arteritis. *Curr. Rheumatol. Rep.* **4**, 30–38.

Fresko, I., Yurdakul, S., Hamuryudan, V. *et al.* (1999). The management of Behcet's syndrome. *Ann. Med. Interne (Paris)* **150**, 576–581.

Gaston, J. S., Scott, D. G., and Bacon, P. A. (1988). Acute confusional state and hyponatraemia due to inappropriate antidiuretic hormone secretion in polyarteritis nodosa. *Ann. Rheum. Dis.* **47**, 428–430.

Gayraud, M., Guillevin, L., Le Toumelin, P. *et al.* (2001). Long-term followup of polyarteritis nodosa, microscopic polyangiitis, and Churg–Strauss syndrome: Analysis of four prospective trials including 278 patients. *Arthritis Rheum.* **44**, 666–675.

Gordon, M., Luqmani, R. A., Adu, D. *et al.* (1993). Necrotizing vasculitis–relapse despite cytotoxic therapy. *Adv. Exp. Med. Biol.* 336, 477–481.

Gordon, M. F., Coyle, P. K., and Golub, B. (1988). Eales' disease presenting as stroke in the young adult. *Ann. Neurol.* 24, 264–266.

Gross, W. L. (1999). New concepts in treatment protocols for severe systemic vasculitis. *Curr. Opin. Rheumatol.* 11, 41–46.

Gross, W. L. (2002). Churg–Strauss syndrome: update on recent developments. *Curr. Opin. Rheumatol.* 14, 11–14.

Gu, Y. O., and Wang, Z. G. (2001). Surgical treatment of cerebral ischaemia caused by cervical arterial lesions due to Takayasu's arteritis: preliminary results of 49 cases. *ANZ. J. Surg.* 71, 89–92.

Guillevin, L., Cohen, P., Gayraud, M., Lhote, F., Jarrousse, B., and Casassus, P. (1999). Churg–Strauss syndrome. Clinical study and long-term follow-up of 96 patients. *Medicine (Baltimore)* 78, 26–37.

Guillevin, L., Lhote, F., Cohen, P. *et al.* (1995a). Corticosteroids plus pulse cyclophosphamide and plasma exchanges versus corticosteroids plus pulse cyclophosphamide alone in the treatment of polyarteritis nodosa and Churg–Strauss syndrome patients with factors predicting poor prognosis. A prospective, randomized trial in sixty-two patients. *Arthritis Rheum.* 38, 1638–1645.

Guillevin, L., Lhote, F., Cohen, P. *et al.* (1995b). Polyarteritis nodosa related to hepatitis B virus. A prospective study with long-term observation of 41 patients. *Medicine (Baltimore)* 74, 238–253.

Hammer, M., Witte, T., Mugge, A. *et al.* (1994). Complicated Cogan's syndrome with aortic insufficiency and coronary stenosis. *J. Rheumatol.* 21, 552–555.

Hamuryudan, V., Mat, C., Saip, S. *et al.* (1998). Thalidomide in the treatment of the mucocutaneous lesions of the Behcet syndrome. A randomized, double-blind, placebo-controlled trial. *Ann. Intern. Med.* 128, 443–450.

Harari, O. A., Wickham, T. J., Stocker, C. J. *et al.* (1999). Targeting an adenoviral gene vector to cytokine-activated vascular endothelium via E-selectin. *Gene Ther.* 6, 801–807.

Haubitz, M., and de Groot, K. (2002). Tolerance of mycophenolate mofetil in end-stage renal disease patients with ANCA-associated vasculitis. *Clin. Nephrol.* 57, 421–424.

Haubitz, M., Ehlerding, C., Kamino, K., Koch, K. M., and Brunkhorst, R. (1998). Reduced gonadal toxicity after i.v. cyclophosphamide administration in patients with nonmalignant diseases. *Clin. Nephrol.* 49, 19–23.

Heckmann, J. G., Kayser, C., Heuss, D., Manger, B., Blum, H. E., and Neundorfer, B. (1999). Neurological manifestations of chronic hepatitis C. *J. Neurol.* 246, 486–491.

Herrmann, M. L., Schleyerbach, R., and Kirschbaum, B. J. (2000). Leflunomide: An immunomodulatory drug for the treatment of rheumatoid arthritis and other autoimmune diseases. *Immunopharmacology* 47, 273–289.

Hirohata, S., Suda, H., and Hashimoto, T. (1998). Low-dose weekly methotrexate for progressive neuropsychiatric manifestations in Behcet's disease. *J. Neurol. Sci.* 159, 181–185.

Hoffman, G. S., Kerr, G. S., Leavitt, R. Y. *et al.* (1992). Wegener granulomatosis: an analysis of 158 patients. *Ann. Intern. Med.* 116, 488–498.

Hoffmann, M., Corr, P., and Robbs, J. (2000). Cerebrovascular findings in Takayasu disease. *J. Neuroimaging* 10, 84–90.

Hurst, R. W. and Grossman, R. I. (1999). Neuroradiology of central nervous system vasculitis. *Semin. Neurol.* 14, 320–340.

Jennette, J. C., Thomas, D. B., and Falk, R. J. (2001). Microscopic polyangiitis (microscopic polyarteritis). *Semin. Diagn. Pathol.* 18, 3–13.

Kalita, J., Bansal, R., Ayagiri, A., and Misra, U. K. (1999). Midbrain infarction: a rare presentation of cryptococcal meningitis. *Clin. Neurol. Neurosurg.* 101, 23–25.

Karni, A., Sadeh, M., Blatt, I., and Goldhammer, Y. (1991). Cogan's syndrome complicated by lacunar brain infarcts. *J. Neurol. Neurosurg. Psychiatry* 54, 169–171.

Katz, B., Wheeler, D., Weinreb, R. N., and Swenson, M. R. (1991). Eales' disease with central nervous system infarction. *Ann. Ophthalmol.* 23, 460–463.

Komocsi, A., Lamprecht, P., Csernok, E. *et al.* (2002). Peripheral blood and granuloma CD4(+)CD28(−) T cells are a major source of interferon-gamma and tumor necrosis factor-alpha in Wegener's granulomatosis. *Am. J. Pathol.* 160, 1717–1724.

Kosar, A., Haznedaroglu, S., Karaaslan, Y. *et al.* (1999). Effects of interferon-alpha2a treatment on serum levels of tumor necrosis factor-alpha, tumor necrosis factor-alpha2 receptor, interleukin-2, interleukin-2 receptor, and E-selectin in Behcet's disease. *Rheumatol. Int.* 19, 11–14.

Koutkia, P., Mylonakis, E., Rounds, S., and Erickson, A. (2001). Leucocytoclastic vasculitis: an update for the clinician. *Scand. J. Rheumatol.* 30, 315–322.

Kumar, R., Wijdicks, E. F., Brown, R. D., Jr., Parisi, J. E., and Hammond, C. A. (1997). Isolated angiitis of the CNS presenting as subarachnoid haemorrhage. *J. Neurol. Neurosurg. Psychiatry* 62, 649–651.

Langford, C. A., and Sneller, M. C. (1997). New developments in the treatment of Wegener's granulomatosis, polyarteritis nodosa, microscopic polyangiitis, and Churg–Strauss syndrome. *Curr. Opin. Rheumatol.* 9, 26–30.

Langford, C. A., Sneller, M. C., and Hoffman, G. S. (1997). Methotrexate use in systemic vasculitis. *Rheum. Dis. Clin. North Am.* 23, 841–853.

Langford, C. A., Talar-Williams, C., Barron, K. S., and Sneller, M. C. (1999). A staged approach to the treatment of Wegener's granulomatosis: Induction of remission with glucocroticoids and daily cyclophosphamide switching to methotrexate for remission maintenance. *Arthritis Rheum.* 42, 2666–2673.

Lanthier, S., Lortie, A., Michaud, J. Laxer, R., Jay, V., and de Veber, G. (2001). Isolated angiitis of the CNS in children. *Neurology* 56, 837–842.

Lau, C. F., Hui, P. K., Chan, W. M. *et al.* (2002). Hepatitis B associated fulminant polyarteritis nodosa: Successful treatment with pulse cyclophosphamide, prednisolone and lamivudine following emergency surgery. *Eur. J. Gastroenterol. Hepatol.* 14, 563–566.

Leavitt, R. Y., Hoffman, G. S., and Fauci, A. S. (1988). The role of trimethoprim/sulfamethoxazole in the treatment of Wegener's granulomatosis. *Arthritis Rheum.* 31, 1073–1074.

Leung, D. Y., Schlievert, P. M., and Meissner, H. C. (1998). The immunopathogenesis and management of Kawasaki syndrome. *Arthritis Rheum.* 41, 1538–1547.

Lhote, F., Cohen, P., Genereau, T., Gayraud, M., and Guillevin, L. (1996). Microscopic polyangiitis: Clinical aspects and treatment. *Ann. Med. Interne (Paris)* 147, 165–177.

Lukert, B. P., and Raisz, L. G. (1990). Glucocorticoid-induced osteoporosis: pathogenesis and management. *Ann. Intern. Med.* 112, 352–364.

Luqmani R. A., Palmer, R. G., and Bacon, P. A. (1990). Azathioprine, cyclophosphamide and chlorambucil. *Baillieres Clin. Rheumatol.* 4, 595–619.

Manna, S. K., Mukhopadhyay, A., and Aggarwal, B. B. (2000). Leflunomide suppresses TNF-induced cellular responses: Effects on NF-kappa B, activator protein-1, c-Jun N-terminal protein kinase, and apoptosis. *J. Immunol.* 165, 5962–5969.

Mathy, I., Gille, M., Van Raemdonck, F., Delbecq, J., and Depre, A. (1998). Neurological complications of intravenous immunoglobulin (IVIg) therapy: An illustrative case of acute encephalopathy following IVIg therapy and a review of the literature. *Acta Neurol. Belg.* 98, 347–351.

Michelfelder, E. C., Shim, D. (2002). Kawasaki disease: Current therapeutic perspectives. 4, 341–350.

Moore, P. M. (1989). Diagnosis and management of isolated angiitis of the central nervous system. *Neurology* **39**, 167–173.

Moore, P. M. (1997a). Autoantibodies to nervous system tissue in human and murine systemic lupus erythematosus. *Ann. N. Y. Acad. Sci.* **823**, 289–299.

Moore, P. M. (1997b). Neuropsychiatric systemic lupus erythematosus. Stress, stroke, and seizures. *Ann. N. Y. Acad. Sci.* **823**, 1–17.

Moore, P. M. (2000). Vasculitis of the central nervous system. *Curr. Rheumatol. Rep.* **2**, 376–382.

Moore, P. M., and Richardson, B. (1998). Neurology of the vasculitides and connective tissue disease. *J. Neurol. Neurosurg. Psychiatry* **65**, 10–22.

Morand, E. F. (2000). Corticosteroids in the treatment of rheumatologic diseases. *Curr. Opin. Rheumatol.* **12**, 171–177.

Nakamura, T., Takahashi, K., and Kishi, S. (2002). Optic nerve involvement in neuro-Behcet's disease. *Jpn. J. Ophthalmol.* **46**, 100–102.

Nau, R., and Bruck, W. (2002). Neuronal injury in bacterial meningitis: mechanisms and implications for therapy. *Trends Neurosci.* **25**, 38–45.

Nichols, J. C., Ince, A., Akduman, L., and Mann, E. S. (2001). Interferon-alpha 2a treatment of neuro-Behcet disease. *J. Neuroophthalmol.* **21**, 109–111.

Nishino, H., Rubino, F. A., DeRemee, R. A., Swanson, J. W., and Parisi, J. E. (1993). Neurological involvement in Wegener's granulomatosis: An analysis of 324 consecutive patients at the Mayo Clinic. *Ann. Neurol.* **33**, 4–9.

O'Duffy, J. D., Calamia, K., Cohen, S. *et al.* (1998). Interferon-alpha treatment of Behcet's disease. *J. Rheumatol.* **25**, 1938–1944.

Parisi, J. E., and Moore, P. M. (1994). The role of biopsy in vasculitis of the central nervous system. *Semin. Neurol.* **14**, 341–348.

Podlecka, A., Dziewulska, D., and Rafalowska, J. (1998). Vascular changes in tuberculous meningoencephalitis. *Folia Neuropathol.* **36**, 235–237.

Reid, I. R. (2000). Preventing glucocorticoid-induced osteoporosis. *Z. Rheumatol.* **59**(Suppl. 2), II/97–II102.

Riemer, G., Lamszus, K., Zschaber, R., Freitag, H. J., Eggers, C., and Pfeiffer, G. (1999). Isolated angiitis of the central nervous system: Lack of inflammation after long-term treatment. *Neurology* **52**, 196–199.

Rota, S., Rambaldi, A., Gaspari, F. *et al.* (1992). Methylprednisolone dosage effects on peripheral lymphocyte subpopulations and eicosanoid synthesis. *Kidney Int.* **42**, 981–990.

Sabbadini, M. G., Bozzolo, E., Baldissera, E., and Bellone, M. (2001). Takayasu's arteritis: Therapeutic strategies. *J. Nephrol.* **14**, 525–531.

Sais, G., Vidaller, A., Jucgla, A., Servitje, O., Condom, E., and Peyri, J. (1998). Prognostic factors in leukocytoclastic vasculitis: A clinicopathologic study of 160 patients. *Arch. Dermatol.* **134**, 309–315.

Sakane, T., and Takeno, M. (2000). Novel approaches to Behcet's disease. *Expert Opin. Invest. Drugs* **9**, 1993–2005.

Scolding, N. J., Jayne, D. R., Zajicek, J. P., Meyer, P. A., Wraight, E. P., and Lockwood, C. M. (1997). Cerebral vasculitis—Recognition, diagnosis and management. *Q. J. Med.* **90**, 61–73.

Scott, D. G., and Watts, R. A. (2000). Systemic vasculitis: Epidemiology, classification and environmental factors. *Ann. Rheum. Dis.* **59**, 161–163.

Sneller, M. C. (1998). Evaluation, treatment, and prophylaxis of

infections complicating systemic vasculitis. *Curr. Opin. Rheumatol.* **10**, 38–44.

Sorimachi, T., Kamada, K., Ozawa, T., and Takeuchi, S. (2001). Basilar artery vasculitis secondary to sphenoid sinusitis—Case report. *Neurol. Med. Chir. (Tokyo)* **41**, 454–457.

Stone, J. H., Pomper, M. G., and Hellmann, D. B. (1998). Histoplasmosis mimicking vasculitis of the central nervous system. *J. Rheumatol.* **25**, 1644–1648.

Stone, J. H., Uhlfelder, M. L., Hellmann, D. B., Crook, S., Bedocs, N. M., and Hoffman, G. S. (2001). Etanercept combined with conventional treatment in Wegener's granulomatosis: A six-month open-label trial to evaluate safety. *Arthritis Rheum.* **44**, 1149–1154.

Stuck, A. E. Minder, C. E., and Fery, F. J. (1989). Risk of infectious complications in patients taking glucocorticosteroids. *Rev. Infect. Dis.* **11**, 954–963.

Tatsis, E., Schnabel, A., and Gross, W. L. (1998). Interferon-alpha treatment of four patients with the Churg–Strauss syndrome. *Ann. Intern. Med.* **129**, 370–374.

Tessler, H. H., and Jennings, T. (1990). High-dose short-term chlorambucil for intractable sympathetic ophthalmia and Behcet's disease. *Br. J. Ophthalmol.* **74**, 353–357.

Thomas-Golbanov, C., and Sridharan, S. (2001). Novel therapies in vasculitis. *Expert. Opin. Invest. Drugs* **10**, 1279–1289.

Thomas, G. N., Tam, L. S. Tomlinson, B., and Li, E. K. (2002). Accelerated atherosclerosis in patients with systemic lupus erythematosus: A review of the causes and possible prevention. *Hong Kong Med. J.* **8**, 26–32.

Tse, S. M., Silverman, E. D., McCrindle, B. W., and Yeung, R. S. (2002). Early treatment with intravenous immunoglobulin in patients with Kawasaki disease. *J. Pediatr.* **140**, 450–455.

Vassallo, R., Remstein, E. D., Parisi, J. E., Huston, J., III, Brown, R. D., Jr. (1999). Multiple cerebral infarctions from nonbacterial thrombotic endocarditis mimicking cerebral vasculitis. *Mayo Clin. Proc.* **74**, 798–802.

Vidaller, A., and Sais, G. (2001). Cutaneous leukocytoclastic vasculitis: The dynamic nature of the infiltrate and the expression of adhesion molecules. *J. Cutan. Pathol.* **28**, 327–329.

Waiser, J., Budde, K., Braasch, E., and Neumayer, H. H. (1999). Treatment of acute c-ANCA-positive vasculitis with mycophenolate mofetil. *Am. J. Kidney Dis.* **34**, e9.

Wang, Z., Shen, L., Yu, J. *et al.* (2002). Management of cerebral ischemia due to Takayasu's arteritis. *Clin. Med. J. (Engl.)* **115**, 342–346.

Watts, R. A., Lane, S. E., Bentham, G., and Scott, D. G. (2000). Epidemiology of systemic vasculitis: a ten-year study in the United Kingdom. *Arthritis Rheum.* **43**, 414–419.

Welch, K. M., Nagesh, V., Boska, M., and Moore, P. M. (1997). Detection of cerebral ischemia in systemic lupus erythematosus by magnetic resonance techniques. *Ann. N. Y. Acad. Sci.* **823**, 120–131.

Wicki, J., Olivieri, J., Pizzolato, G. *et al.* (1999). Successful treatment of polyarteritis nodosa related to hepatitis B virus with a combination of lamivudine and interferon alpha. *Rheumatology (Oxford)* **38**, 183–185.

Wilcke, J. R., and Davis, L. E. (1982). Review of glucocorticoid pharmacology. *Vet. Clin. North Am. Small Anim. Pract.* **12**, 3–17.

Ziyal, I. M., Sekhar, L. N., Chandrasekar, K., and Bank, W. O. (1999). Vertebral artery to common carotid artery bypass in Takayasu's disease with delayed cerebral ischemia. *Acta Neurochir. (Wien)* **141**, 655–659.

CHAPTER 39

# Giant Cell Arteritis and Polymyalgia Rheumatica

James J. Corbett and A. Melms

## CLINICAL ASPECTS

Giant cell arteritis (GCA) is the most common vasculitic syndrome, with a high prevalence in elderly patients (Hunder *et al.*, 1990; Huston *et al.*, 1978). Due to the variety of presenting symptoms such as localized headaches or rheumatic complaints, cranial or temporal arteritis (TA, Horton's arteritis) and polymyalgia rheumatica (PMR) are addressed as individual clinical entities. Over the course, major symptoms may change. Hence, the coexistence and overlap of symptoms suggest that GCA and PMR are two variants of the same disease. In this vein, approximately 50% of patients with biopsy-proven GCA complain of rheumatic symptoms (Huston *et al.*, 1978) and likewise, patients with PMR have asymptomatic temporal arteritis in as many as 50% of biopsies (Bengtsson and Malmvall, 1981). Neurologic disease in patients with GCA or PMR may appear in the guise of a large variety of disorders suggesting primary ophthalmologic conditions (visual disturbances, diplopia, scotoma, amaurosis) (Hayreh *et al.*, 1998a,b), pharyngeal symptoms (sore throat, dysphagia, jaw claudication), dementia, or depression (Caselli, 1990) (Table I). The onset may be acute or subacute and symptoms often blow up in a few days with throbbing headaches, scalp tenderness, visual disturbances, myalgia, and malaise. Systemic inflammatory disease with low-grade fever and evidence of weight loss is reflected by moderate to marked elevation of the erythrocyte-sedimentation rate (ESR). Mixed among these symptomatic patients are a group of occult patients whose diagnosis is devilishly difficult (Desmet *et al.*, 1990; Hayreh *et al.*, 1998a; Healey and Wilske, 1980). This constellation of findings makes GCA and PMR a likely diagnosis in many elderly patients, one that needs prompt action. Despite numerous laboratory studies performed in this context the markedly elevated ESR is still the best single laboratory parameter available in suspected cases and most useful in monitoring therapy. It should be remembered, however, that up to 10% of patients will have a normal ESR. A set of diagnostic criteria for GCA has been proposed by the American Rheumatology Association (Hunder *et al.*, 1990) (Table II).

### Giant Cell Arteritis

Giant cell arteritis is a segmental granulomatous inflammation of medium-sized and larger arteries that frequently affects branches of the aortic arch. In the neurologic setting, the most prominent manifestation is cranial arteritis, most often as temporal arteritis (Horton's arteritis). A typical, but not invariable, finding is tenderness, induration, and decreased pulse of the temporal arteries on palpation. Vertebral and internal carotid arteries may be affected in some patients (Thielen *et al.*, 1998) but the intracranial segments are spared. Anecdotal case reports have shown GCA of the mesenteric and coronary arteries and in the aorta itself. Aneurysmal rupture of the aorta occurs rarely. The most frequent causes of vasculitis-related death are cerebral infarction and myocardial infarction.

As a cardinal feature of temporal arteritis (TA), approximately two-thirds of the patients complain of recent onset of persistent throbbing pain in the scalp and neck pain (Hayreh *et al.*, 1997). Patients may also complain of painful jaw claudication, pain in the tongue, sore throat, or dysphagia (Table I). Although quite rare, these symptoms may be accompanied by ischemic infarcts and necrosis of the scalp, the tongue, lips, or throat.

TABLE I   Symptoms of Giant Cell Arteritis in 166 Patients

| Symptom | Patients with symptom (%) | Patients in whom it was initial symptom |
|---|---|---|
| Headache | 72 | 33 |
| Polymyalgia rheumatica | 58 | 25 |
| Malaise, fatigue | 56 | 20 |
| Jaw claudication | 40 | 4 |
| Fever | 35 | 11 |
| Cough | 17 | 8 |
| Neuropathy | 14 | 0 |
| Sore throat, dysphagia | 11 | 2 |
| Amaurosis fugax | 10 | 2 |
| Permanent visual loss | 8 | 3 |
| Claudication of limbs | 8 | 0 |
| TIA/stroke | 7 | 0 |
| Neuro-otologic disorder | 7 | 0 |
| Scintillating scotoma | 5 | 0 |
| Tongue claudication | 4 | 0 |
| Depression | 3 | 0.6 |
| Diplopia | 2 | 0 |
| Tongue numbness | 2 | 0 |
| Myelopathy | 0.6 | 0 |

Note. Some patients had more than one symptom.

TABLE II   Criteria for Giant Cell Arteritis (Modified from American College of Rheumatology 1990; Hunder *et al.*, 1990)

Age 50 years or greater at onset of symptoms
New onset of localized headache
Tenderness or decreased pulse of temporal artery
Elevated erythrocyte sedimentation rate 50 mm/h or greater
Temporal artery biopsy showing necrotizing arteritis with
  predominance of mononuclear cell infiltrates or granulomatous
  process with multinucleated giant cells

Note. The presence of three or more of these five criteria is associated with a sensitivity of 93.5% and a specificity of 91.2%.

Amaurosis fugax and the presence of retinal nerve fiber layer infarcts reflect retinal and choroidal ischemia and imminent and almost invariably irreversible loss of vision in one or both eyes. This is an emergency situation requiring prompt diagnostic and therapeutic action. Frequently, the disease affects both eyes and therapeutic intervention is directed at saving the remaining eye. Ophthalmoscopic signs of anterior ischemic optic neuropathy (disc swelling) may precede visual loss by 1 or 2 days.

Therapy must be initiated when diagnosis is suspected and must not be delayed pending the results of the biopsy. Biopsy of the superficial temporal artery or less commonly the occipital artery may be performed after initiation of steroid therapy (Achkar *et al.*, 1994; Guevara *et al.*, 1998). Due to the segmental nature of the inflammatory foci (skip lesions) the biopsy should be generous (about 2 inches) and multiple histopathological sections should be examined (Kattah *et al.*, 1999). Biopsy should be performed bilaterally if necessary. Pathological findings can be expected in 70 to 90% (Bengtsson *et al.*, 1981; Delecoeuillerie *et al.*, 1988; McDonnell *et al.*, 1986).

In some atypical cases magnetic resonance angiography or PET (Turlakow *et al.*, 2001) may help to guide diagnosis. Doppler sonography of branches of the superficial temporal artery may disclose segmental stenosis and help to find the artery in severe stenosis and thus may be useful for selecting the site for biopsy (Schmidt *et al.*, 1997).

The characteristic histologic appearance is mononuclear cell infiltrates throughout the vessel wall disrupting the internal elastic membrane. There are activated CD4$^+$ T-lymphocytes, macrophages, and multinucleated giant cells, although the latter are not required to establish the diagnosis of GCA. Autoantibodies against neutrophil cytoplasmic antigens (ANCA, anti-proteinase 3, anti-myeloperoxidase) which are useful in the diagnosis and classification of vasculitic syndromes are negative in GCA and PMR. Areas of healed arteritic scarring have the same diagnostic significance as active arteritis (McDonnell *et al.*, 1986).

### Polymyalgia Rheumatica

Polymyalgia rheumatica (PMR) is diagnosed two to four times more frequently than GCA. The major symptoms are rheumatic complaints of the limb girdles and joint stiffness reminiscent of rheumatoid arthritis. In contrast to rheumatoid arthritis mobility of these joints is not impaired. Laboratory studies disclose systemic inflammation almost always with markedly elevated ESR. Muscle enzymes are normal; rheumatoid factors or antinuclear antibodies suggesting rheumatoid arthritis or collagen disease are generally negative. Likewise, there are no inflammatory infiltrates in muscle biopsy or myopathic abnormalities in electromyography indicating polymyositis. Approximately 8% of patients presenting with GCA/PMR initially do not complain of typical symptoms (Strachan *et al.*, 1980; Healey and Wilske, 1980; Desmet *et al.*, 1990). Patients with PMR who have minimal rheumatic complaints but present with fever, weight loss, hypochromic microcytic anemia, or mental status changes are a special diagnostic challenge. Hence, correct diagnosis leading to PMR/GCR is made by exclusion. Atypical clinical presentation at onset (sometimes called occult giant cell arteritis) may mimic occult infections, B-cell lymphomas and other malignancies, other rheumatic disorders, muscle diseases, fibromyalgia, or psychogenic rheumatic complaints and may delay the final diagnosis and adequate

**TABLE III** Laboratory Findings in GCA and PMR

Elevated erythrocyte sedimentation rate (ESR) in 70–80% of patients typically more than 60 mm n.W. in 1 h with the Westergren method; *ESR may be normal (less than 30 mm/h) in 3 to 30% of elderly patients*

Slight anemia (mean hemoglobin 11.7″ 1.6 g/dL)

Slight thrombocytosis (mean platelet count 427″ 116 × 10³/μL)

Slight elevation of serum aspartate aminotransferase and alkaline phosphatase (in 15%)

Elevation of C-reactive protein, haptoglobulin, fibrinogen elevation of α-1 and α-2 globulin fractions.

Raised complement components C3 and C4.

Raised levels of interleukin 6

Normal values for creatinine kinase and other muscle enzymes

No specific autoantibodies; ANCA negative

Normal electromyogram

Normal muscle biopsy

**TABLE IV** Differential Diagnosis in Polymyalgia Rheumatica

| Diagnosis | Clinical and laboratory findings |
| --- | --- |
| Primary fibromyalgia | No evidence for systemic inflammation (ESR, C-reactive protein normal) |
| Myofascial pain | Tender and trigger points on palpation |
| Rheumatoid arthritis | Rheumatoid factor, joint X-rays; chronic erosive arthropathy; chronic symmetric synovitis |
| Para- or postinfectious myalgia | Serology for viral infections (e.g., hepatitis, rubella, influenza) or bacterial infectious (e.g., Yersinia, Salmonella, Borrelia) |
| Bacterial endocarditis | Blood cultures, echocardiography |
| Paraneoplastic syndromes | Screening for tumor |
| Polymyositis | Muscle weakness; elevated muscle enzymes; EMG findings; muscle biopsy |
| Plasmocytoma amyloidosis | Immunoelectrophoresis; Bence–Jones proteins; bone X-rays; bone marrow biopsy |
| Thyroid disease | Thyroid hormones, autoantibodies to thyroid antigens |

treatment. Biopsy of a "clinically uninvolved" temporal artery or a therapeutic trial of steroids with dramatic or persistent response may resolve the diagnostic problem in some patients. A positive temporal artery biopsy limits the diagnostic evaluation for suspected malignancy and provides a tissue diagnosis and rationale for adequate long-term steroid therapy (Hall and Hunder, 1984). The differential diagnoses of the PMR syndrome are summarized in Table IV.

## NATURAL COURSE

GCA and PMR are diseases of the elderly. The prevalence in persons age 50 years or greater was estimated to be 133 cases per 100,000. In this age group, incidence of GCA was reported to be 18.3 per 100,000 (Women 25, Men 9.4; Nordburg and Bengtsson, 1990) and incidence of PMR was 53.7 per 100,000 (Chuang *et al.*, 1982). Incidence is slightly higher in women (2:1 to 3:1). On diagnosis, the vast majority of patients are in their sixth or seventh decade of life. Both GCA and PMR appear to be long-lasting but eventually self-limiting diseases. The natural course as documented before the introduction of steroid therapy was relapsing and remitting along with frequent complications over 5 to 7 years (Barber, 1957). Ischemic strokes and emboli were, and still are, major causes of death. Today, when one receives appropriate therapy, life expectancy is not significantly reduced. Approximately 20% of patients lost vision in one eye before introduction of steroid therapy. Today this disabling event still occurs in 0.5 to 3%. Other symptoms show dramatic improvement within a very few days after onset of steroids. Such rapid improvement may be taken as diagnostic of the disease and lack of symptomatic improvement is strong evidence in favor of another cause of an elevated

sedimentation rate. Complications and relapses can be avoided by adequate therapeutic regimen adjustment of dosage, although the course of the disease does not appear to be shortened the same way. The result of the biopsy does not affect prognosis or outcome (Vilaseca *et al.*, 1987) but is reassuring when steroid complications occur as a justification for the continued use of steroids.

## PRINCIPLES OF THERAPY

Immunopathology in GCA shows cell-mediated damage to the vessel wall involving CD4⁺ T cells presumably targeting elastin. The etiology of the chronic inflammatory process is not known. However, these findings are the rationale for anti-inflammatory and immunosuppressive treatment with steroids. Although several laboratory studies are abnormal in GCA/PMR including ESR, C-reactive protein (CRP), interleukin 6, and others indicating an acute phase reaction, there is no specific laboratory test available to unequivocally establish these syndromes. Generally, ESR and CRP and history of symptoms are the mainstay of diagnosis and follow-up.

## PRACTICAL MANAGEMENT

In every suspected case with CGA or PMR, steroid therapy should be initiated immediately without delay to avoid sudden loss of vision, which can occur both in patients with giant cell arteritis and in those with PMR. Patients who present with very recent loss of vision should be given a trial with high-dose steroids intravenously (e.g., ranging from 2 mg per kg body weight to 1000 mg prednisolone equivalent). This may save some function in the affected eye in occasional cases. However, ischemic damage and visual loss is almost invariably irreversible (Hayreh, 2000). In ischemic ophthalmopathy, anticoagulants may be tried but their efficacy is not proved. In GCA, steroids should be initiated with 1 mg per kg body weight of prednisone per day. In PMR without severe complaints, 20 to 40 mg prednisone per day is generally sufficient. Prompt relief of signs and symptoms within a few days on steroids is diagnostic especially if, for some reason, biopsy of the temporal artery has not been performed or was reported negative. After symptoms subside, steroids may be gradually reduced by 5 mg every other week. Generally, maintenance with low steroid dosage (5 to 10 mg prednisone per day) is recommended for 18 to 24 months. Alternating steroid regimen has been found to be ineffective (Hunder et al., 1975). Treatment failure and flare-up of symptoms is most often a consequence of tapering steroids too rapidly, although occasionally it may be necessary to augment steroid treatment with methotrexate or dapsone (Evans et al., 1994; Jover et al., 2001; Rienitz et al., 1988). The reappearance of clinical symptoms and elevation of the ESR strongly suggest a relapse and require immediate increase of dosage of at least 20 mg per day until these symptoms subside. Patients with insufficient anti-inflammatory treatment are at continued risk for visual loss. Patients should be informed about the prolonged time needed for steroid treatment to obtain optimal compliance. Occasionally, patients require a very low dosage of steroids (2.5 to 5 mg prednisone) over several years (Kyle and Hazleman, 1990). Patients on long-term steroids require prophylactic treatment and observation for osteoporosis and associated fractures, gastric ulcers, and the development of diabetes and cataracts. Such side effects occur less frequently in patients whose dosage can be lowered for maintenance. After long-term steroids have been stopped, patients should continue to be followed closely for 6 to 12 months to recognize recurrence of arteritic symptoms although recurrence with visual symptoms has been reported as late as 7 years later (Cullen, 1972).

## ALTERNATIVES IN THERAPY

There are studies underway to determine the usefulness of other cytotoxic drugs in sparing high-dose steroid use (Ghanchi and Dutton, 1997). Recently, the combination of prednisone in combination with methotrexate or placebo was investigated in a randomized, double-blind study over 2 years (Jover et al., 2001). Patients treated with prednisone and MTX (10 mg once weekly) had less relapses and the cumulative dose of prednisone was lower than in those patients receiving predisone and placebo. Side effects were similar in both groups; however, it seems premature for a general recommenation of this regime until more data are available. Previously, azathioprin has been reported to reduce the dosage of steroids in stabilized disease (De Silva and Hazleman, 1986), but due to its slow mode of action it is not well suited with regard to imminent loss of vision.

## REFERENCES

Achkar, A. A., Lie, J. T., Hunder, G. G., O'Fallon, W. M., and Gabriel, S. E. (1994). How does previous corticosteroid treatment affect the biopsy findings in giant cell (temporal) arteritis? Ann. Int. Med. 120, 987–992.

Barber, H. S. (1957). Myalgic syndrome with constitutional effects: Polymyalgia rheumatica. Ann. Rheum. Dis. 16, 230–237.

Bengtsson, B. A., and Malmvall, B. E. (1981). The epidemiology of giant cell arteritis including temporal arteritis and polymyalgia rheumatica. Arthritis Rheum. 24, 899–904

Caselli, R. J. (1990). Giant cell (temporal) arteritis: A treatable cause of multi-infarct dementia. Neurology 40, 753–755.

Caselli, R. J., Hunder, G. G., and Whisnant, J. P. (1988). Neurologic disease in biopsy-proven giant cell (temporal) arteritis. Neurology 38, 352–359.

Caselli, R. J., Daube, J. R., Hunder, G. G., and Whisnant, J. P. (1988). Peripheral neuropathic syndromes in giant cell (temporal) arteritis. Neurology 38, 685–689.

Chuang, T. Y., Hunder, G. G., Ilstrup, D. M., and Kurland, L. T. (1982). Polymyalgia rheumatica: A ten year epidemiologic and clinical study. Ann. Int. Med. 97, 672–680.

Cullen, J. F. (1972). Temporal arteritis. Occurrence of ocular complications 7 years after diagnosis. Br. J. Ophthalmol. 56, 584–588.

Delecoeuillerie, G., Joly, P., Cohen de Lara, A., and Paolaggi, J. B. (1988). Polymyalgia rheumatica and temporal arteritis: A retrospective analysis of prognostic features and different corticosteroid regimens (11 year survey of 210 patients). Ann. Rheum. Dis. 47, 733–739.

De Silva, M., and Hazleman, B. L. (1986). Azathioprine in giant cell arteritis/polymyalgia rheumatica: A double blind study. Ann. Rheum. Dis. 45, 136–138.

Desmet, G. D., Knockaert, D. C., and Bobbaers, H. J. (1990). Temporal arteritis: The silent presentation and delay in diagnosis. J. Intern. Med. 227, 237–240.

Evans, J. M., Butts, K. P., and Hunder, G. G. (1994). Persistent giant cell arteritis despite corticosteroid treatment. Mayo Clin. Proc. 69, 1060–1061.

Ghanchi, F. D., and Dutton, G. N. (1997). Current concepts in giant cell (temporal) arteritis. Survey Ophthalmol. 42, 99–123.

Guevara, R. A., Newman, N. J., and Grossniklaus, H. E. (1998). Positive temporal artery biopsy 6 months after prednisone treatment. Arch. Ophthalmol. 116, 1252–1253.

Hall, S., and Hunder, G. G. (1984). Is temporal artery biopsy prudent? Mayo Clin. Proc. 59, 793–796.

Hayreh, S. S. (2000). Steroid therapy for visual loss in patients with giant-cell arteritis. Lancet 355, 1572–1573.

Hayreh, S. S., Podhajsky, P. A., Raman, R., and Zimmerman, B. (1997). Giant cell arteritis: Validity and reliability of various diagnostic criteria. *Am. J. Ophthalmol.* **123**, 285–295.

Hayreh, S. S., Podhajsky, P. A., and Zimmerman, B. (1998a). Ocular manifestations of giant cell arteritis. *Am. J. Ophthalmol.* **125**, 509–520.

Hayreh, S. S., Podhajsky, P. A., and Zimmerman, B. (1998b). Occult giant cell arteritis: Ocular manifestations. *Am. J. Ophthalmol.* **125**, 521–526.

Healey, L. A., and Wilske, K. R. (1980). Presentation of occult giant arteritis. *Arthritis Rheum.* **23**, 651–653.

Hunder, G. G., Bloch, D. A., Michel, B. A., Stevens, M. B., Arend, W. P., Calabrese, L. H., Edworthy, S. M., Fauci, A. S., Leavitt, R. Y., Lie, J. T., Lightfoot, R. W., Masi, A. T., McShane, D. J., Mills, J. A., Wallace, S. L., and Zvaifler, N. J. (1990). The American College of Rheumatology 1990 criteria for the classification of giant cell arteritis. *Arthritis Rheum.* **33**, 1122–1128.

Hunder, G. G., Sheps, S. G., Allen, G. L., and Joyce, J. W. (1975). Daily and alternate day regimens in treatment of giant cell arteritis: Comparison in a prospective study. *Ann. Int. Med.* **82**, 613–618.

Huston, K. A., Hunder, G. G., Lie, J. T., Kennedy, R. H., and Elveback, L. R. (1978). Temporal arteritis: A 25-year epidemiologic, clinical, and pathologic study. *Ann. Int. Med.* **88**, 162–167.

Jover, J. A., Hernandez-Garcia, C., Morado, I. C., Vargas, E., Banares, A., and Fernandez-Gutierrez, B. (2001). Combined treatment of giant cell arteritis with methotrexate and prednison. *Ann. Intern. Med.* **134**, 106–114.

Kattah, J. C., Mejico, L., Chrousos, G. A. *et al.* (1999). Pathologic findings in a steroid responsive optic nerve infarct in giant cell arteritis. *Neurology* **53**, 177–180.

Kyle, V., and Hazlemann, B. L. (1990). Stopping steroids in polymyalgia rheumatica and giant cell arteritis. *Br. Med. J.* **300**, 344–345.

McDonnell, P. J., Moore, G. W., Miller, N. R., Hutchings, G. M., and Green, W. R. (1986). Temporal arteritis: A clinicopathologic study. *Ophthalmology* **93**, 518–530.

Nordburg. E., and Bengtsson, B. A. (1990). Epidemiology of biopsy-proven giant cell arteritis. *J. Intern. Med.* **227**, 233–236.

Reinitz, E., and Aversa, A. (1988). Long-term treatment of temporal arteritis with dapsone. *Am. J. Med.* **85**, 456–457.

Schmidt, W. A., Kraft, H. E., Vorpahl, K., Volker, L., and Gromnica-Ihle, E. J. (1997). Color duplex ultrasonography in the diagnosis of temporal arteritis. *N. Engl. J. Med.* **337**, 1336–1342.

Strachan, R. W., How, J., and Bewsher, P. D. (1980). Masked giant cell arteritis. *Lancet* **1**, 194–196.

Thielen, K. R., Wydicks, E. F. M., and Nichols, D. A. (1998). Giant cell (temporal) arteritis: Involvement of the vertebral and internal carotid arteries. *Mayo Clin. Proc.* **73**, 444–446.

Turlakow, A., Yeung, H. W. D., Pui, J., Macapinlac, H., Liebovitz, E., Rusch, V., Goy, A., and Larson, S. M. (2001). Fludeoxyglucose positron emission tomography in the diagnosis of giant cell arteritis. *Arch. Intern. Med.* **161**, 1003–1007.

Vilaseca, J., Gonzales, A., Cid, M. C., Lopez-Vivancos, J., and Ortega, A. (1987). Clinical usefulness of temporal artery biopsy. *Ann. Rheum. Dis.* **46**, 282–285.

CHAPTER 40

# Interventional Neuroradiology in the Brain, Head, and Neck Region

T. Yousry and R. T. Higashida

## HEAD AND NECK TUMORS

By decreasing the neo-vascularity of head and neck tumors, surgical excision is facilitated with better visibility and shorter operative times, and in addition, the decreased blood loss can potentially avoid blood transfusion. Tumors for which preoperative embolization is common include meningiomas, glomus tumors, and juvenile nasopharyngeal angiofibromas.

### Meningioma

Most meningiomas are benign and therefore curable lesions, for which the best available treatment is complete surgical resection. The goal of interventional neuroradiologists is to facilitate this by selecting patients for preoperative devascularization of as much of the tumor as possible (Nelson *et al.*, 1994). This is best achieved with polyvinal alcohol (PVA) particulate embolization, using the smallest particles to achieve better penetration of the tumor vessels, compatible with the lowest risk of cranial nerve palsy which increases with smaller particles, in the particular vessels being treated (Wakhloo *et al.*, 1993). PVA of 150–250 µm are a reasonable compromise of these two conflicting goals, although prolonged embolization of dilute 50- to 150-µm PVA will cause greater tumor necrosis.

Convexity meningiomas frequently have bilateral supply from both middle meningeal arteries and superficial temporal arteries, all of which should be sequentially embolized. Due to such extensive supply, it is even more important to protect the distal normal arterial territory from small particulate embolization to avoid scalp necrosis. This is best achieved by superselective catheterization, but if multiple proximal branches contribute blood supply to the tumor, the more distal normal vessels can be protected by embolization with large gelfoam pledgets or coils, thus redirecting flow and the smaller particles to the tumor vessels. Although some controversy exists as to the value of embolization of meningiomas involving the cerebral convexity, preoperative embolization certainly is of value in more complex presentations, including giant hypervascular meningiomas, meningiomas exhibiting malignant or angioblastic characteristics, and those involving the skull base, scalp, or critical vascular structures (Nelson *et al.*, 1994). Complications are uncommon and only marginally higher than seen with diagnostic angiography and consist of cranial nerve palsy and brain infarction (Halbach *et al.*, 1992).

### Juvenile Nasopharyngeal Angiofibroma

Juvenile nasopharyngeal angiofibroma is a particularly vascular benign hamartoma of young men, and during surgical resection, significant blood loss was common before the advent of routine preoperative embolization. Both CT and MRI are necessary in defining the anatomical extent of the lesion, in particular the presence of intracranial and parapharyngeal extension, which will give an indication of the likely vessels involved, as well as crucial information for surgical planning. Typical supply is predominantly from terminal branches of the internal maxillary artery, usually the ascending palatine and sphenopalatine arteries.

The ascending pharyngeal, facial, and internal carotid arteries are the other commonly encountered vessels of supply, increasing in frequency with larger lesions. Particles of PVA are directed to the branches of the internal maxillary artery, and ascending pharyngeal artery using superselective catheter placement. During this intervention one must remain aware of the numerous anastomoses (in particular with the internal carotid, ophthalmic and vertebral arteries) (Halbach *et al.*, 1992) and cranial nerve supply, and choose an agent appropriate to the site and anatomy. PVA particles of 150–250 μm are ideal for penetration of the vascular bed of this lesion, while 300- to 600-μm particles are safer in the presence of known intracranial anastomoses or cranial nerve supply. Liquid agents are usually never used. The timing of the procedure should be at least 24 h and probably not more than 72 h before surgery. Reductions in duration of surgery, intraoperative blood loss, and transfusions have been shown (Economou *et al.*, 1988; Roberson *et al.*, 1979) and preoperative embolization is indicated for most patients with juvenile nasopharyngeal angiofibromas.

## Paraganglioma

Paragangliomas constitute the bulk of the remaining hypervascular tumors of the head and neck commonly managed with embolization, always as a preoperative adjunct and never as definitive treatment. Glomus tympanicum, jugular, jugulo-tympanic, vagal, and carotid body tumors are all potential candidates for embolization. If there is evidence that the lesions are secreting vasoactive agents, preoperative alpha blockade should be considered, but this constitutes a minority of lesions. Typical arteries of supply are the ascending pharyngeal (APA), posterior auricular, and occipital arteries. The agent of choice is PVA particles (150–300 or 300–600 μm) injected from a microcatheter in a superselective position, with particular care in the ascending pharyngeal artery due to the jugular foramen and hypoglossal canal cranial nerve supply, and occipital artery due to the vertebral artery anastomoses of varying size.

Mutiple vessels of supply are usually encountered in carotid body tumors, requiring sequential selective catheterization and embolization. Occasionally there will be significant contribution from small branches of the internal carotid artery, and in general these should not be embolized when small and numerous. Their presence should be indicated to the surgeon. Glomus jugular tumors are usually supplied by the ascending pharyngeal artery and only a few other branches (depending upon size), and complete angiographic devascularization can be achieved. As in embolization elsewhere, complete angiographic "occlusion" corresponds to minimal bleeding if smaller particles are used. Larger particles can achieve a similar angiographic result and yet cause significant intraoperative bleeding. Effectiveness of this technique has been shown in large series of cases (Hurst, 1996; Valavanis, 1986; Zanella and Valavanis, 1994).

## EPISTAXIS

### Clinical Aspects

Intractable epistaxis is uncommon, and when encountered especially if recurrent, than an underlying abnormality should be suspected. Severe hypertension is most frequently encountered, but disorders of coagulation and hereditary hemorrhagic telangiectases (Osler-Weber Rendu disease) can also cause nose bleeds.

### Principles of Therapy

Embolization has been applied to epistaxis for the past 30 years and is effective in both anterior and posterior epistaxis. Merland *et al.* (1980) reported a 94% success rate in 54 patients treated with embolization. In general, success rates of 95–100% are achievable, with low complications, although Elahi *et al.* (1995) reported 96% success and 6% "major neurological complications."

Patients with hereditary hemorrhagic telangiectasia have recurrent and often severe epistaxis, with most episodes manageable by nonsurgical means. If intractable, embolization is very effective in controlling the bleeding, although it may not achieve alteration of long-term bleeding patterns. Embolization is always preferable to surgical ligation, which not only tends to be ineffective, but removes vascular access for endovascular treatments.

Post-traumatic epistaxis can indicate an underlying false aneurysm, either of the external carotid branches, or more importantly of internal carotid artery bleeding into the sphenoid sinus. These cases constitute an emergency and should be studied with a complete diagnostic angiogram as soon as possible. If a false aneurysm of the internal carotid artery is found as the cause of massive epistaxis, immediate occlusion using detachable balloons is indicated. Attempts to preserve the internal carotid artery are unlikely to succeed, and fatal hemorrhage may occur during the attempt.

The goal in the usual refractory epistaxis, in which preliminary angiography has not shown a false aneurysm, is to stop bleeding. This can be achieved by decreasing the perfusion pressure of the mucosal arteries, and hence occlusion at the level of the mucosa or even more proximal can be effective. If angiography has not indicated a specific bleeding site, then bilateral embolization of the terminal branches of the internal maxillary arteries is indicated. Selective embolization of the sphenopalatine artery can be achieved, but the greater palatine and posterior superior alveolar arteries should also be embolized, and flow embolization from the terminal internal maxillary artery may best achieve this without inducing spasm. Depending on the site, bilateral superior labial embolization may be required. The contribution from the anterior ethmoidal arteries is not amenable to safe embolization.

The agent most commonly used is PVA particles (250–500 or 500–700 μm), which is associated with a very low incidence of cranial nerve ischemia, stroke, or local tissue necrosis. Knowledge of normal anastomoses within the intracranial circulation, recognition of aberrant anatomy, and avoidance of spasm in the external carotid artery are the keys to successful treatment.

# INTRACRANIAL ANEURYSMS

## Clinical Aspects

### Signs and Symptoms

Intracranial aneurysms may present as incidental findings or with focal neurologic signs, mass effect on cerebral or brainstem structures, compression of cranial nerves, or ischemic/embolic phenomena. Most are asymptomatic until they rupture, causing subarachnoid hemorrhage (SAH) and a *uniquely severe headache*, which may be accompanied by nausea, vomiting, and loss of consciousness. The concept of a "warning leak" is emphasized in many reports of SAH, and nearly 40% of patients presenting with SAH report a similar headache in the preceding 2 weeks. Recognition of such cases could allow timely investigation and referral to an endovascular or neurosurgical center for early management. Although headache is frequent in patients who visit primary care providers, SAH is common in the group presenting with *sudden severe headache* (occurring in up to 25% of cases). Clinical grade predicts outcome, and there is a major opportunity for the primary care provider to improve outcome by early diagnosis. Misdiagnosis affects outcome. Among patients with the correct diagnosis of good grade SAH, 91% had good outcomes, while only 53% had similar outcomes if incorrectly diagnosed (Le Roux and Winn, 1998; Mitchell and Tress, 1999).

### Diagnostic Evaluation

**Computed Tomography.** Investigation should start with *computed tomography* (CT), which is the first imaging modality of choice. CT should detect 90–95% of cases of acute SAH. CT angiography can also detect the aneurysms directly. CT angiography may demonstrate aneurysms as small as 2–3 mm with sensitivities of 77–97% and specificities of 87–100% (Hope *et al.*, 1996). This modality of imaging may be useful when patients with identified unruptured aneurysms are followed conservatively, in patients with partially clipped aneurysms, or in those who have undergone treatment

with endovascular coil techniques (Dorsch *et al.*, 1995; Vieco *et al.*, 1995, 1996).

**Lumbar Puncture.** In all cases with strong clinical suspicion of an acute SAH and a negative CT head scan *lumbar puncture* is mandatory. CSF must be examined for red blood cells and xanthochromia to confirm or exclude the diagnosis. Xanthochromia is detected in all patients with SAH between 12 h and 2 weeks and is still detectable in 70% after 3 weeks.

**Magnetic Resonance Imaging.** Using the appropriate sequences, it is possible to detect SAH in MRI with a high sensitivity and specificity (Wiesmann *et al.*, 2001). MR angiography (MRA) is also useful as a screening modality, particularly for aneurysms greater than 3–5 mm (Maeder *et al.*, 1996; Ronkainen *et al.*, 1995). MRA/CTA can provide important additional information, particularly in anatomically complex sites or large aneurysms, but cannot replace the temporal and spatial resolution of digital subtraction angiography (DSA). But MRA/CTA can be used as a low-risk screening tool with a good sensitivity for aneurysms of >4 mm in a high-risk population such as patients with a familial history of SAH and autosomal dominant polycystic kidney disease.

**Digital Subtraction Angiography.** Intra-arterial catheter angiography continues to be the "gold standard" in the diagnostic evaluation of intracranial aneurysms. Transcatheter studies provide the most information about small perforating vessels and define spatial resolution better than other imaging modalities (Setton *et al.*, 1996; Tu *et al.*, 1996). These capacities are further increased by the development of newer techniques, such as rotational 3D angiography. Since as many as 20% of patients with one aneurysm have multiple aneurysms, selective catheterization should be performed of both internal carotid arteries and the intracranial segments of both vertebral arteries as well to visualize the origins of both posterior/inferior cerebellar arteries (Setton *et al.*, 1996).

## Natural Course

### Epidemiology

The frequency of detection of intracranial aneurysms ranges from 0.2 to 9.9%, with a mean of 5%. Angiographic series detects aneurysms in 0.65–1% of the population, and these are likely to be larger and more significant than smaller aneurysms detected in selected postmortem series. The majority of intracranial aneurysms are berry or saccular aneurysms located in the anterior circulation (85%), arise at branch points of the circle of Willis, and are multiple in 20–30%. SAH is a major clinical problem, with an annual incidence of 11 per 100,000, higher than many other neurological disorders, including brain tumors and multiple sclerosis. Approximately 5–15% of stroke cases are secondary to ruptured saccular aneurysms (Graves *et al.*, 1992).

More women than men are affected, the difference most pronounced in the postmenopausal age group. Of patients who have aneurysmal SAH, 50–60% die, 12% before reaching the hospital. Survivors have a 35–50% chance of long-term disability, and half of patients assessed as having a good outcome are unable to return to previous employment and have residual neuropsychologic and cognitive impairment. The outcome is determined by the extent of the initial hemorrhage, grade at presentation, repeat hemorrhage, cerebral vasospasm, and hydrocephalus (Le Roux and Winn, 1998; Menghini *et al.*, 1998; Mitchell and Tress, 1999; Schievink, 1997).

Lack of randomized trials or population-based studies has made the true natural history of unruptured intracranial aneurysms difficult to determine. Accumulating evidence suggests an influence of aneurysm size on the risk of rupture in patients with unruptured aneurysms, with larger lesions more likely to bleed. Although small aneurysms have low rupture rates, it remains unclear that a precise size exists below which the risk of rupture is negligible (the "critical size"). The International Study of Unruptured Intracranial Aneurysms (ISUIA, 1998) retrospectively studied the frequency of rupture among 1449 patients with known unruptured intracranial aneurysms. Aneurysms less than 10 mm in diameter, on anterior circulation arteries, and in patients with no past history of SAH had rupture rates considerably lower than previous studies indicated. Early longitudinal studies estimate the annual risk of rupture to be 1–2%, with at least half of these ruptures being fatal. There is an increased risk of rupture for large, posterior circulation and posterior communicating artery aneurysms.

### Pathogenesis

Genetic factors are implicated in the pathogenesis of intracranial aneurysms. Familial occurrence has long been recognized, with 7–20% of patients with aneurysmal SAH having a first- or second-degree relative with a known aneurysm (Raaymakers *et al.*, 1998; Ronkainen *et al.*, 1997). Many connective tissue disorders have been associated with intracranial aneurysms, including autosomal dominant polycystic kidney disease (ADPCKD), Neurofibromatosis type 1, Ehlers Danlos syndrome, and Marfans syndrome. Among these, autosomal dominant polycystic kidney disease is most

common. Familial aneurysms rupture at an earlier age, may be smaller when they rupture, and are more often followed by development of new aneurysms (Schievink, 1997).

Modifiable acquired risk factors associated with the formation of aneurysms include atherosclerosis, cigarette smoking, hypertension, and elevated cholesterol. Cigarette smoking, alcohol intake, and hypertension have an association with SAH (Le Roux and Winn, 1998; Schievink, 1997).

### Principles of Therapy

#### Unruptured Aneurysms

There is a discrepancy between the prevalence of intracranial aneurysms and the incidence of SAH, indicating that most aneurysms do not rupture. Unfortunately, no reliable tools exist to determine which aneurysms will rupture. Morbidity and mortality associated with treating unruptured intracranial aneurysms are much less than in ruptured aneurysms. Yet, the results of the International Study of Unruptured Intracranial Aneurysms (ISUIA, 1998) prompted the authors to conclude that it appears unlikely that surgery will reduce the rates of disability and death in patients with unruptured intracranial aneurysms smaller than 10 mm in diameter and no history of SAH. In those in whom rupture did occur during follow-up, 83% died. Small asymptomatic aneurysms (<10 mm) on anterior circulation vessels, with no past history of SAH, especially if in sites considered low risk for SAH (cavernous ICA) should not be treated. Conversely unruptured aneurysms, which are symptomatic, greater than 10 mm diameter, or in patients with a history of SAH, would appear to be cost effective for treatment (Johnston *et al.*, 1999).

**Surgery.** Morbidity and mortality associated with surgical treatment of unruptured intracranial aneurysms varies with experience, site, and size of the aneurysm (Mitchell and Tress, 1999). To date, no reports in the literature are randomized or have cohort controls. The morbidity and mortality associated with surgical treatment of unruptured aneurysms remain significant, with reported mortality of 0–7.7% and morbidity of 0–30% (King *et al.*, 1994; Raaymakers *et al.*, 1998; Solomon *et al.*, 1996). If cognitive defects are also assessed, rates of postsurgical combined morbidity and mortality in unruptured aneurysms of up to 16% were reported in one prospective study (ISUIA, 1998).

**Endovascular Therapy.** Currently, endosaccular occlusion of intracranial aneurysms is performed using the electrolytically detachable Guglielmi detachable coil system (GDC, Target Therapeutics, Fremont, CA) (Figure 1) (Casasco *et al.*, 1993; Graves *et al.*, 1995; Guglielmi *et al.*, 1991a,b; Halbach *et al.*, 1994; Mawad and Klucznik, 1995). This device is currently approved in the United States and Canada and involves braided platinum microwires of different sizes and variable lengths, which can form complex shapes when deployed within the aneurysm sac. There are now additional endovascular detachable coils now available in Europe and the U.S., including those from Cordis Neurovascular Systems (Miami, FL). To date, more than 100,000 patients have been treated with detachable coils worldwide. Published reports of early clinical and angiographic results have been promising, but the long-term efficacy of the coiling method is as yet unproven. Malisch (1997) reported a consecutive series of 100 patients with mid-term (1–2 years) clinical results with follow-up from 2 to 6 years, average 3.5 years. Of concern is the mid-term incidence of post-coil embolization hemorrhage in patients with large aneurysms (4% incidence of rebleeding) and in giant aneurysms (33% incidence). An additional study by Eskridge and Song (1998) evaluated endosaccular occlusion in 150 basilar tip aneurysms, which are the results of the Food and Drug Administration Multi-Center Clinical Trial. In this group, 83 patients had ruptured aneurysms and 67 had unruptured basilar tip aneurysms. The rebleeding rate for treated ruptured aneurysms was up to 3.3% and the bleeding rate for unruptured aneurysms was up to 4.1%. However, Johnston *et al.* (1999) reported a decreased morbidity and mortality of endovascular versus surgical management of unruptured intracranial aneurysms. Their results clearly favor the endovascular approach, especially in the posterior cerebral circulation.

### Ruptured Aneurysms

The obliteration of the ruptured aneurysm is an important component of management of the patient with acute SAH (Figure 2). Recurrent hemorrhage occurs in 20–30% of patients in the first 2 weeks of the initial bleed, with significant morbidity (Kassell *et al.*, 1990). Microvascular surgical clipping of the aneurysm is the "gold standard" against which other techniques must be measured. In particular, the long-term clinical efficacy of clipping is well documented with low rates of subsequent SAH or aneurysm regrowth. Late angiographic follow-up of surgically treated aneurysms was presented by David (1999), demonstrating only 2 recurrent aneurysms in 135 cases with no residual aneurysm postclipping. Of 12 cases with known residual aneurysm rests, most appeared stable but a subset with broad necks showed higher risks of SAH.

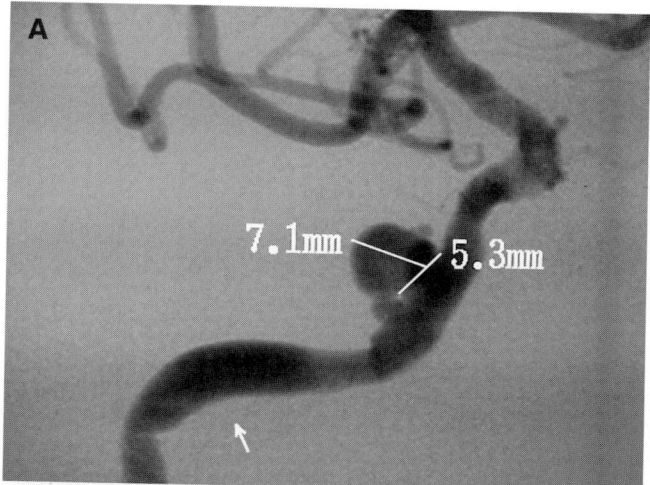

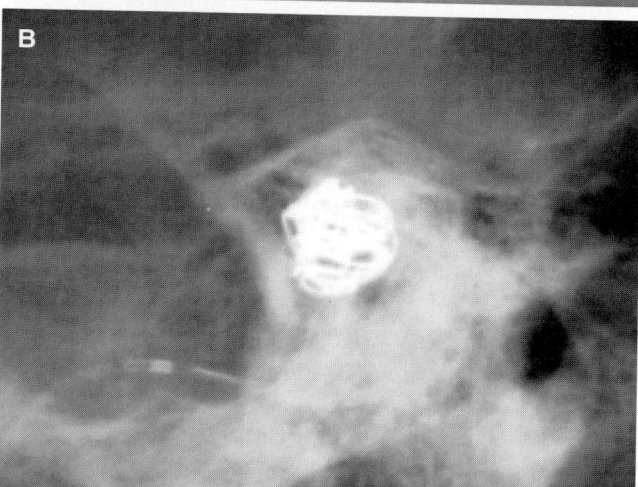

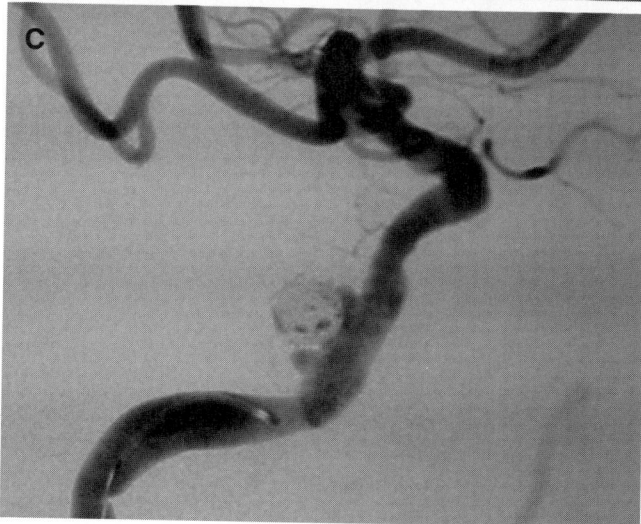

**FIGURE 1** Precavernous aneurysm treated with GDC. (**A**) Carotid angiogram showing large irregular precavernous aneurysm. (**B**) GDC placed into aneurysm. (**C**) Post-treatment showing occlusion of aneurysm and preservation of normal internal carotid artery.

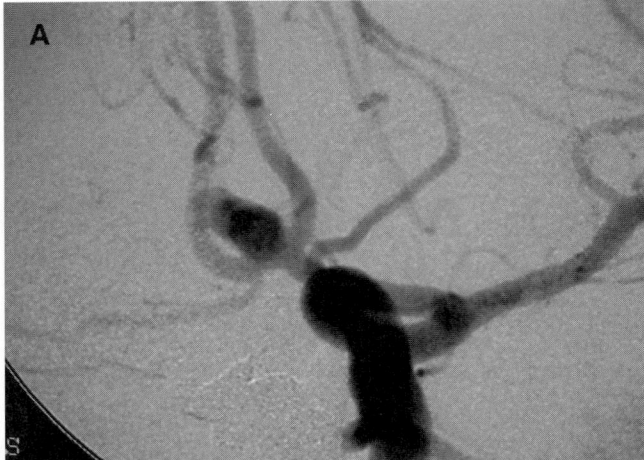

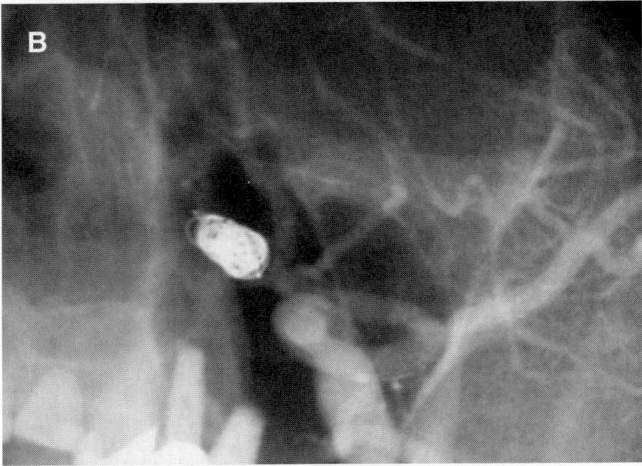

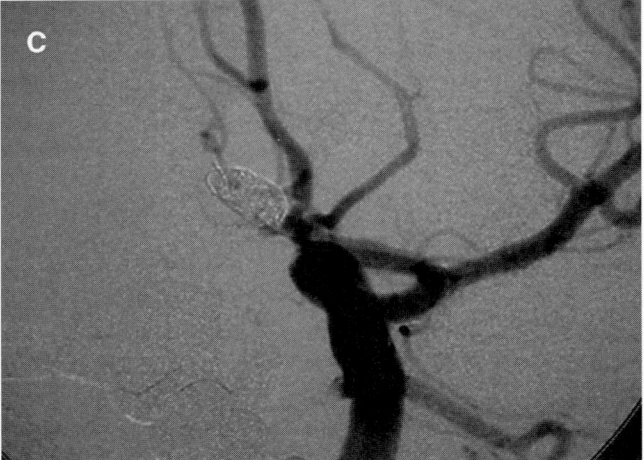

FIGURE 2  A 57-year-old man who presented with a grade 1 acute subarachnoid hemorrhage. Diagnosed with 4 × 10-mm irregular anterior communicating artery aneurysm. (A) Pretreatment showing anterior communicating artery aneurysm. (B) GDC placed into aneurysm. (C) Post-treatment showing occlusion of aneurysm.

Surgery is associated with much higher morbidity and mortality in patients with ruptured rather than unruptured aneurysms. The outcome following SAH may be more determined by the initial hemorrhage and subsequent hydrocephalus or vasospasm than the technique used to manage the aneurysm. However, the endovascular technique can be applied in poor-grade post-SAH patients with little increased morbidity associated with the procedure, and vasospasm can be treated with balloon angioplasty or papaverine injection, which is a potent vasodilator, at the same procedure. While there may be improved outcomes with endovascular management of ruptured aneurysms, this may be less than seen in unruptured aneurysms due to the direct effects of the SAH.

### Aneurysm Morphology

Not all aneurysms are geometrically suited to endovascular coil treatment, and the best results are in small aneurysms with small necks, absolute neck size of <4 mm, aneurysms with dome-to-neck ratios of >2, and sidewall aneurysms rather than terminal aneurysms (Debrun *et al.*, 1998; Vinuela *et al.*, 1997). The number of aneurysms potentially treatable by the endovascular coil technique is increasing with new technological developments such as intracranial stents, which provide a scaffold with which to contain the coil, the balloon remodeling technique, softer smaller coils, and three-dimensional coils (Higashida, 1997; Lefkowitz *et al.*, 1999; Moret *et al.*, 1997). Giant aneurysms remain difficult to successfully treat, while preserving the parent vessel, with current endovascular or microsurgical techniques.

Surgery and endovascular repair can be complementary, and it is possible to create a neck with a clip and to complete the obliteration of the aneurysm with coils, to clip an aneurysm in which coiling fails, and to coil an aneurysm in which surgery has failed (Fraser *et al.*, 1994; Gurian *et al.*, 1995; Hacein-Bey *et al.*, 1998).

There will remain patients with aneurysms that cannot be obliterated using current surgical or endovascular techniques with vessel preservation. These may be treated by parent vessel occlusion, a useful technique that with appropriate selection can allow palliation of symptoms due to mass effect, and decrease the chance of SAH (Hacein-Bey *et al.*, 1998; Higashida *et al.*, 1989b, 1991; Hodes *et al.*, 1991; Taki *et al.*, 1992; Vazquez-Anon *et al.*, 1992). Surgical or endovascular occlusion can be performed, with the use of an extracranial to intracranial bypass graft if trial occlusion or anatomy of the circle of Willis suggests a high likelihood of cerebral hypoperfusion and stroke (Hacein-Bey *et al.*, 1998; Taki *et al.*, 1992; Vazquez-Anon *et al.*, 1992). Morbidity is lower with an intact circle of Willis, posterior communicating artery diameters of >1 mm, no clinical or perfusion deficit during trial balloon occlusion, and endovascular versus surgical occlusion. Endovascular balloon occlusion close to the site of the aneurysm has less thrombus burden in the internal carotid artery stump to permit distal embolism formation.

## Practical Management

### Equipment and Technique

All procedures should be performed using *digital subtraction angiography*, capable of high resolution, road mapping, and preferably biplane image acquisition.

**Perioperative Medication.** *Anticoagulation* should be used in all cases of aneurysm treatment, in ruptured and unruptured aneurysms. The timing, degree, method, monitoring, and duration of anticoagulation differ between different endovascular therapy units. No randomized trial has resolved these issues. We typically begin anticoagulation with a heparin bolus of 5000 units, tailored to the patients body habitus (70 units/kg). The bolus is administered at the start of catheterization in unruptured aneurysms and after microcatheter placement in ruptured aneurysms. Anticoagulation is monitored by hourly activated clotting times (ACTs), and maintained usually by regular intravenous bolus injections rather than continuous infusion. In most instances, heparin is not reversed, and clotting is allowed to return gradually to normal. Heparin is continued for 1–2 days only in cases of thromboembolic complications requiring treatment, or in patients assessed at unusually high risk of embolization, such as giant aneurysms with a very broad neck.

All procedures are performed with the patients under general anesthesia. This is essential to ensure a still patient throughout the procedure and to facilitate immediate management of any complication with either endovascular or surgical rescue. Dense packing with coils should be the goal of therapy, and subsequent packing of neck remnants is usually harder than at the initial treatment. In poor grade SAH patients it is sometimes expedient to not densely pack the body and fundus, providing protection from immediate rebleed with minimal morbidity and allowing elective surgical or repeat packing when the patient is in better medical condition. Follow-up angiography is usually scheduled at 6 months, 1 year, and/or 2 years after treatment.

*Intracranial stents* have become increasingly useful for the treatment of wide-necked intracranial aneurysms and extracranial pseudoaneurysms. A stent has been deployed in the basilar artery across the wide neck of a mid-basilar aneurysm to serve as an endovascular scaffold with subsequent placement of a microcatheter between the interstices of the stent (Higashida *et al.*, 1997). Guglielmi detachable coils (GDCs) were then deployed within the otherwise untreatable aneurysm and the stent precluded herniation of the coils within the parent vessel lumen. Such stent-assisted coil embolization of aneurysms has since been reported for the treatment of carotid (Perez-Cruet *et al.*, 1997) and vertebral artery aneurysms (Lylyk *et al.*, 1998; Sekhon *et al.*, 1998) and has also been used successfully to treat petrous internal carotid dissecting pseudoaneurysms (Mericle *et al.*, 1998). There have also been numerous reports on the effect of stent deployment on alteration of intra-aneurysmal hemodynamics with eventual thrombosis, even in the absence of coil embolization (Lieber *et al.*, 1997; Wakhloo *et al.*, 1998). Future advances in stent technology including a lower-profile design with greater flexibility and specifically tailored strut porosity for intracranial use is expected to expand the use of stents in the treatment of cerebral aneurysms.

### Complications

*Thromboembolism* remains the most common major complication of endovascular coil treatment, either from thrombus formation around the catheter system used (despite heparinised pressure infusions), from local atherosclerosis, or from within the aneurysm. The process of coil thrombosis involves inducing thrombus within the aneurysm and coil mass. Packing further coils into the lumen can potentially displace thrombus or aggregated platelets. Heparinization remains the mainstay of prevention together with usual care of pressurized flushing solutions. Injections within the aneurysm ("aneurysmogram") should be avoided in most cases. The use of abciximab is a potent means of preventing platelet aggregation, but is long-lasting and should be avoided in acute SAH. If periprocedure thrombosis is recognized, management should depend upon the site and likely significance of the occlusion. Minor thrombi can be treated with the administration of antiplatelet medications such as acetylsalicylic acid. Occlusion of major vessels may require local infusion of Urokinase or rtPA as indicated. This has been safely performed even in the presence of a ruptured aneurysm, as long as low doses distal to the aneurysm can be administered directly into the clot. It remains a hazardous procedure, and the risks of hemorrhage must be weighed against the risk of the deficit likely to result from arterial occlusion.

*Rupture of the aneurysm* is most likely to occur with small aneurysms presenting with acute SAH. When recognized, heparinization must be reversed with Protamine, the coil deployed (a coil should never be removed if it is suspected to lie partially outside the aneurysm), and further coils deployed as rapidly as possible to densely pack the aneurysm. Emergency intraventricular drain placement in the angiography laboratory can be lifesaving.

*Residual aneurysm rests* after postsurgical clipping have been shown to pose a significant risk of subsequent growth and rupture (Drake and Vanderkinden, 1967; Lin *et al.*, 1989), although there may be a subset that do not pose as great a risk (David *et al.*, 1999). While incomplete GDC obliteration of a recently ruptured aneurysm has been shown to markedly reduce the chance of rehemorrhage in the short term, rehemorrhage rates have been reported on later follow-up. These are approximately 1–5% in smaller aneurysm rests, but up to 33% has been reported in giant ruptured aneurysms with large unpacked areas (Boccardi *et al.*, 1998; Casasco and George, 1998; Christoforidis and Valavanis, 1998; Debrun *et al.*, 1998; Eskridge and Song, 1998; Kuether *et al.*, 1998; Lefkowitz *et al.*, 1999; Malisch *et al.*, 1997, 1998; McDougall *et al.*, 1996; Vinuela *et al.*, 1997). It is reasonable to pursue complete obliteration as the goal of therapy, to retreat residual rests if large, but very small "dog-ear" remnants may not pose the same risk of hemorrhage and should be followed. If they remain stable, treatment can be deferred.

**Vasospasm.** The mechanism and pathophysiology for cerebral vasospasm is still poorly understood. Previous reports indicate that spasm is due to direct endothelial damage, myointimal cellular proliferation, inflammatory changes, and/or a reduction in smooth muscle contractility within the vessel wall (Smith *et al.*, 1985).

Cerebral vasospasm following SAH is a leading source of delayed cerebral ischemia following intracranial aneurysmal rupture and is the single most important cause of death and disability for survivors (Awad *et al.*, 1987; Higashida *et al.*, 1989a; Origitano *et al.*, 1990). It is estimated that 6000–10,000 potential patients per year may benefit from revascularization therapy for arterial spasm (Kassel *et al.*, 1985). Vasospasm can be treated by intra-arterial papaverine infusion, balloon dilatation, or by a combination of both. The first reports of balloon angioplasty for cerebral arterial vasospasm were from Zubkov and Nikiforov in 1984. They reported successful treatment of 33 patients involving 105 vascular territories, with good clinical outcome in the majority of patients. Higashida (1992) reported on 28 patients involving 99 vascular territories in which 19 patients (66.7%) showed clinical and neurological improvement following angioplasty. Eskridge (1998) reported on 50 patients

with clinical evidence of vasospasm-induced ischemia, and 28 (61%) showed sustained neurological improvement within 72 h of angioplasty.

Indications for utilizing these techniques include patients who are clinically symptomatic, have failed maximal medical therapy, have angiographic stenosis correlating with their symptoms or increase in the flow velocity in the intracerebral arteries by >30% from baseline, and who do not have evidence of massive cerebral infarction or cerebral hemorrhage. Acute hemorrhage, within 6 h of presentation, is a relative contraindication, since patients undergoing therapy need to be systemically anticoagulated during the procedure to avoid thromboembolic complications from the balloon dilatation procedure or microcatheter.

PTA is performed while the patient is anticoagulated and utilizes a specifically designed, low-profile compliant balloon mounted on an atraumatic flexible catheter. The balloon catheter is deployed via a guide catheter positioned in the extracranial internal carotid or vertebral artery. Upon inflation within the intracranial vessel to be treated, typically the intradural portion of the supraclinoid internal carotid artery, the balloon is gently inflated under digital roadmap technique, with a dilute mixture of radio-opaque contrast material, in short durations to enable intermittent cerebral perfusion. More distally, the M1 segment of the middle cerebral artery is the most commonly treated vessel. On rare occasions, and in particular when one of the anterior cerebral arteries is dominant, the A1 segment is amenable to angioplasty as well.

Selective intra-arterial papaverine injection has been used via a microcatheter in symptomatic spastic vessels, with doses of 150–300 mg in 100 mL of normal saline, over 30 min (Clouston *et al.*, 1995; Eskridge *et al.*, 1998; Morgan *et al.*, 1995). This appears safe (although there are reports of transient elevation of intracranial pressure, transient monocular blindness, and of significant systemic hypotension) and efficacious in early vasospasm, and it has the benefit of dilating vessels beyond the reach of balloon catheters, but is of short duration, usually <4–6 h. Papaverine-treated vessels showed an average decrease in transcranial Doppler TCD velocity of 20% on postprocedure day 1, which by postprocedure day 2 was no longer significantly different from pretreatment velocities. In contrast, vessels treated with angioplasty showed a 45% lower mean TCD velocity, which remained sustained on postprocedure day 2. Eskridge (1998) showed a similar longlasting effect by angiographic, clinical, and TCD velocity criteria, with only 1 of 170 treated vessel segments requiring repeat angioplasty. A recent study by Elliott (1998) also indicates that there is a longer-lasting benefit with percutaneous transluminal balloon angioplasty (PTA) than with intraluminal infusion of papaverine. PTA of intracranial vessel segments using a compliant microballoon catheter has established its efficacy and relative safety in the treatment of SAH-related intracranial vasospasm which has failed maximal hypertensive, hypervolemic, and hemodilutional therapy.

The main complications include vessel rupture (usually fatal; less common if stiffer angioplasty balloon catheters are avoided), dissection (gentle dilatation, undersized compliant balloon), stroke from distal embolization (minimized with full periprocedural anticoagulation with heparin), and rehemorrhage if the ruptured aneurysm is not treated, and therefore in general it is preferred only to treat postaneurysm clipping or coiling, in order to ensure that aneurysms do not rebleed following balloon angioplasty for vasospasm.

## Conclusion

Treatment of unruptured aneurysms with endovascular coils has a low complication rate of around 3%–8% with minimal mortality. All recent reports have indicated improving outcomes due to a combination of technological advances and increasing experience. In selected patients, up to 70% of aneurysms can be completely obliterated, but 20% of aneurysms can regrow a neck remnant usually due to compaction of the coils (Ausman, 1997). The neck remnants pose a risk of delayed SAH, and if growing or large should be treated with further coiling, or less commonly surgical clipping. The procedural morbidity and mortality in ruptured aneurysms depend on grade, but seem to be markedly less than for surgery. There are centers in which 80% of all aneurysms are currently treated with endovascular coil techniques.

The task force of the Stroke Council of the American Heart Association published the following recommendations with respect to aneurysm treatment (Mayberg *et al.*, 1994):

(1) Treatment of small incidental intracavernous internal carotid artery (ICA) aneurysms is not indicated. For large symptomatic intracavernous aneurysms, treatment decisions should be individualized based on patient age, severity and progression of symptoms, and treatment alternatives.

(2) Symptomatic intradural aneurysms of all sizes should be considered for treatment, with relative urgency for treatment of acutely symptomatic aneurysms. Symptomatic large or giant aneurysms carry higher surgical risks requiring careful analysis of the individualized patient and aneurysmal risks and surgeon expertise.

(3) Coexisting or remaining aneurysms of all sizes in patients with SAH due to another treated aneurysm appear to carry a higher risk for future hemorrhage than similar sized aneurysms without a prior SAH history and warrant considerations for treatment. Aneurysms located at the basilar apex and posterior cerebral artery locations carry a somewhat higher risk for rupture and should be given greater consideration for treatment. Such decisions must consider the patient's age, existing medical and neurologic condition, and relative risks of repair for the remaining aneurysm(s). If a decision is made for observation, regular re-evaluation with selective contrast angiography or CTA/MRA is appropriate on an annual or semiannual basis or, if symptoms of concern should arise, is indicated looking for changes in aneurysmal size.

(4) Special consideration should be given to young patients and those with a positive family history for SAH. Small aneurysms with daughter sac formation may deserve special consideration. In aneurysms managed conservatively, periodic follow-up imaging is appropriate and necessary if specific symptoms arise. If changes in aneurysmal size are observed, this should lead to special consideration for treatment.

(5) Concerning patients with small (4–9 mm) unruptured aneurysms and no history of SAH, although recent data from the ISUIA support conservative management, other studies suggest that treatment should be offered. Currently there is no generally accepted consensus on their optimal management, and treatment should be individualized.

The best strategy, until definitive studies of long-term efficacy are completed, is to have a multidisciplinary approach discussing each patient, and making a decision based on aneurysm size, location, geometry, patient clinical grade, presence or absence of vasospasm, vascular access, local experience and patient preference. If geometrically favorable, endovascular treatment is the preferred treatment modality for posterior circulation aneurysms, partially extradural aneurysms, paraclinoid aneurysms, and for ruptured aneurysms regardless of site if the patient is in bad clinical grade. Significant issues remain, particularly which unruptured aneurysms should be treated (ISUIA, 1998; Johnston *et al.*, 1999).

## CEREBRAL AVMS

### Clinical Aspects

#### Signs and Symptoms

Intracranial AVMs are typically diagnosed prior to age 40 but can occasionally be seen in the elderly.

Clinically they present in over 50% with intracranial hemorrhage (Brown *et al.*, 1997) most often intracerebral hemorrhage (ICH) but SAH and intraventricular hemorrhage (IVH) can also occur. In approximately 20–67% they present with seizures (Brown *et al.*, 1988; Wilkins, 1985). Other presentations include headaches in 15%, focal neurological deficit in less than 5% of cases, and occasionally pulsatile tinnitus. In children under the age of 2, presentations can include congestive heart failure, large head due to hydrocephalus, and seizures. Vascular malformation-related steal phenomena causing focal neurologic deficits by altering perfusion in the tissue in the region of the AVM are rare (Brown *et al.*, 1988; Mast *et al.*, 1995).

### Diagnostic Evaluation

**Computed Tomography.** On noncontrast CT brain scans, AVMs may reveal calcifications and hypodense areas, enhancing after application of contrast media (Kuman *et al.*, 1984).

**Magnetic Resonance Imaging.** MRI is very sensitive, showing an inhomogeneous signal void on T1- and T2-weighted sequences, commonly with hemosiderin, suggesting prior hemorrhage (Huston *et al.*, 1991; Kucharczyk *et al.*, 1985). MRI can also provide crucial information about the exact anatomic location of the AVM, required in some classification concepts (Valavanis and Yasargil, 1998).

**Digital Subtraction Angiography.** DSA is the "gold standard" for identifying arterial and venous anatomy. It is the only method that can provide a reliable estimation of the hemodynamics of the AVM, type of feeding arteries, associated aneurysms, and size and angioarchitecture of the nidus (Valavanis and Yasargil, 1998).

### Natural Course

The incidence and prevalance of intracranial vascular malformations are uncertain. Autopsy data suggest that vascular malformations are about 1/7th as common as intracranial saccular aneurysms and there is an overall frequency of detection of AVMs in approximately 4.3% of the population (McCormick *et al.*, 1976; Michelson *et al.*, 1978). In another autopsy series, AVMs were detected in 1.4% of cases; 12.2% of which were symptomatic (Olivecrona *et al.*, 1948; The AVM Study Group, 1999). Population-based data are limited regarding intracranial vascular malformations. The annual incidence of AVMs has been estimated to be between 0.01–0.001% (Brown *et al.*, 1996; Jessurun

*et al.*, 1993). Regardless of the initial clinical presentation, the annual bleeding rate is thought to be up to 4%, with an annual mortality rate of 1% and a morbidity rate of 1.7% (Ondra *et al.*, 1990). The risk of recurrent intracranial hemorrhage is elevated for a short period of time following the first hemorrhage.

During the first year after initial hemorrhage, the risk of recurrence ranges between 6, 18, and even 33% (Forster *et al.*, 1972; Fults and Kelly, 1984; Graf *et al.*, 1983; Mast *et al.*, 1997) and decreases to 11.3% in subsequent years (Mast *et al.*, 1997). The risk of recurrent hemorrhage may be even higher, up to 25% in the first year after the second hemorrhage (Forster *et al.*, 1972). Angiographic features that seem to correlate with a higher bleeding incidence are AVM-associated flow-related aneurysms, pseudoaneurysms, stenosis or occlusion of draining veins, or of veins located in the deep parts of the brain or in the posterior fossa (Valavanis and Yasargil, 1998). The postulation that small and micro-AVMs are at higher bleeding risk than larger AVMs is controversial. Recent reports suggest that there is no direct correlation between the size of the AVM and the frequency of hemorrhage (Valavanis and Yasargil, 1998).

### Principles of Therapy

The primary goal of treatment of a brain AVM is to prevent new or recurrent hemorrhage. Other goals include improving or stabilizing neurological deficits, treating intractable epilepsy, and reducing the severity and frequency of chronic headaches (Valavanis and Yasargil, 1998). To assist in the decision-making process, various grading systems have been developed with the Spetzler and Martin (1986) system gaining the largest acceptance. This system is based on criteria directly related to anticipated surgical difficulties. Scores range between one and five, with one point given for a lesion less than 3 cm, two points for a lesion between 3 and 6 cm, and three points for a lesion greater than 6 cm. Location within eloquent cortex provides an additional point as does deep venous drainage. The score is calculated by summing the points for each category.

When this system was retrospectively applied by the authors, lesions graded I–III were found to have low treatment-associated morbidity. However, grade IV had 31.2% and grade V had 50% new treatment-associated morbidity. In addition, grade V lesions had 29.9% and grade IV 16.7% permanent deficit. This led the authors to recommend surgery for all grades I and II lesions. Grade III lesions should be treated on a case-by-case basis; however, they usually recommend surgery for both symptomatic and asymptomatic patients. Grades IV and V lesions require a multidisciplinary approach with individual analysis. The problem with the Spetzler–Martin grading system is that it does not correlate with the anticipated endovascular difficulties that differ significantly from surgical ones. It is therefore of limited use in this context (Valavanis and Yasargil, 1998).

In general, the approach to treating a patient with an AVM is a multidisciplinary approach that includes the neuroradiologist, neurosurgeon, and radiotherapist. The specific goals of endovascular therapy should be determined prior to the start of the procedure and will be either (1) curative, (2) preoperative or preradiosurgical, or (3) palliative by targeted embolization to eliminate the angioarchitecturally weak elements or vascular elements responsible for venous hypertension or tissue hypoperfusion (Valavanis and Yasargil, 1998). Complete cure of an AVM is achieved by embolization alone in only 10–20%, with one group reporting occlusion rates of 41% (Valavanis and Yasargil, 1998) (Figure 3). Preoperative embolization is the most frequently applied form of embolization. Its goal is either to convert an inoperatable AVM into an operable AVM or to facilitate removal of operable but complex AVMs. This can be achieved by: (a) a reduction of the size of the nidus, in cases of large AVMs with conversion into multiple small compartments; (b) occlusion of deep feeding arteries; (c) obliteration of intranidal aneurysms or other vascular cavities; and (d) occlusion of arteriovenous (AV) fistulae (Valavanis and Yasargil, 1998). Preradiosurgical embolization is gaining increasing attention. Its goal is to make AVMs treatable by radiosurgery. This is achieved by reducing the size of the nidus to less than 3 cm, as smaller volumes have a higher cure rate with less morbidity; occlusion of intranidal aneurysms or other vascular cavities; occlusion of large AV fistulae; and occlusion of the dural supply to the AVM (Valavanis and Yasargil, 1998). Palliative embolization is employed in large or giant AVMs that cannot be completely obliterated by embolization and cannot be operated. The goals of palliative embolization should include a decrease in the probability of bleeding by weak angioarchitectural elements such as flow-related aneurysms, pseudoaneurysms, and intranidal venous varices proximal to an obvious obstruction of venous drainage.

## Practical Management

### Equipment and Technique

All procedures should be performed using digital subtraction angiography (*DSA*), capable of high resolution, road mapping, and preferably biplane image acquisition.

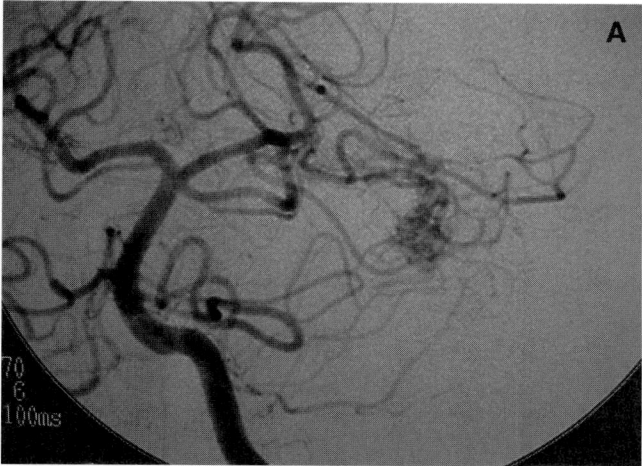

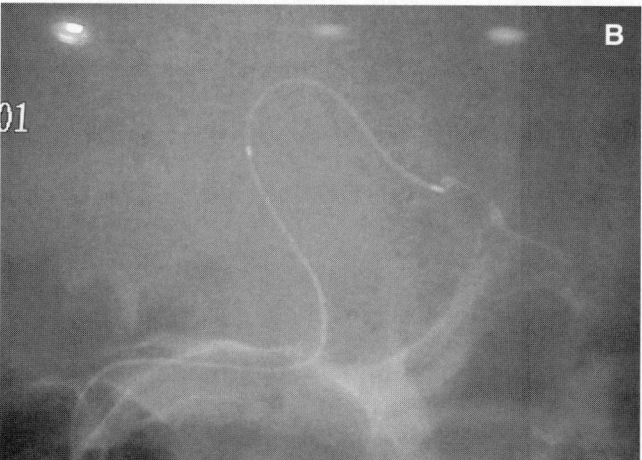

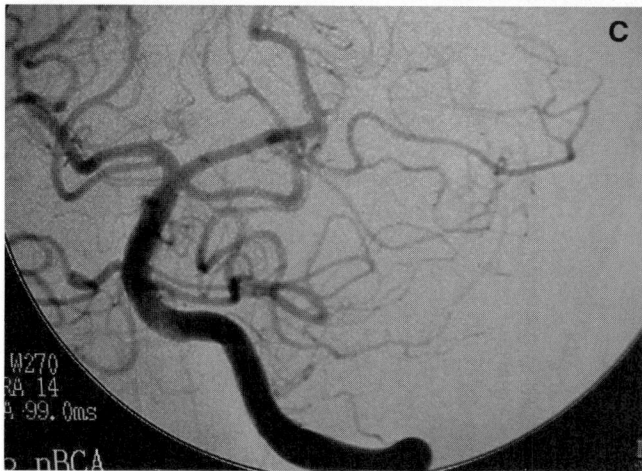

FIGURE 3 Cerebellar AVM embolization. (A) Vertebral angiogram frontal view shows a small cerebellar AVM, supplied by the superior cerebellar artery, arising from the basilar artery. (B) Selective catheterization of the left superior cerebellar artery. (C) Postembolization of the AVM with NBCA liquid tissue adhesive shows complete obliteration of the AVM following treatment.

**Anticoagulation.** Systemic anticoagulation is often used during intracranial embolization procedures at many centers. The disadvantages for using anticoagulation are that, if a hemorrhage occurs during the procedure, the bleeding complications will be more severe. However, it is well known that thromboembolic complications increase with catheter size and duration of the procedure (Formanek *et al.*, 1970). Since most procedures require relatively large catheters and long procedure times, systemic anticoagulation is routinely used at most centers. Several centers have reported large numbers of cases performed without systemic heparinization, with only local perfusion of heparinized saline flushes between coaxial catheters and guidewires, and have reported no difficulty from catheter-induced thrombosis. Conversely, large numbers of cases have been performed in patients who recently had intracranial hemorrhage and no additional hemorrhagic complications were noted with systemic anticoagulation (Barnwell *et al.*, 1993). Considerations for premedication with corticosteroids, anticonvulsants, aspirin, calcium channel blockers, and antibiotics are made, but none have rigorous support for their use.

**Anesthesia.** Endovascular embolization of AVMs can be performed using local anesthesia or under general anesthesia. An important argument in deciding is the question of performing superselective injections of sodium amylbarbital, with or without concomitant EEG monitoring. The principle is that if vessels supply normal functioning cerebral cortex, their presence will be disclosed by a neurological deficit after the barbiturate agent, and embolization will be avoided at that particular site of the microcirculation, thus avoiding a potential permanent neurologic deficit (Paiva *et al.*, 1995; Peters *et al.*, 1993). Others who do not routinely perform provocative testing argue that good superselective angiography is more important in deciding whether to embolize an AVM (Valavanis, 1996) and that in the presence of a high flow lesion the agent injected is likely to bypass the normal lower flow vessels ("sump effect") giving a false sense of security. As the embolization proceeds, the hemodynamics change and embolization of hitherto silent vessels and cortical territories can occur.

The advantages of general anesthesia are:

1. Better working conditions by immobilizing the patient, elimination of patient anxiety and its associated cardiovascular reactions;
2. Prolongation of the duration of individual sessions and reduction of the number of sessions required;
3. Safer and faster management of complications (Valavanis and Yasargil, 1998).

## Complications

Mortality rates during embolization have been reported to be 1.08% or less and neurological morbidity rates of 2–5% (Purdy *et al.*, 1991; Rosenwasser *et al.*, 1998; Spetzler *et al.*, 1987; Valavanis and Yasargil, 1998; Vinuela *et al.*, 1991).

*Procedural complications* have been shown to relate to the experience of the operator, and have also decreased as newer catheters and agents have become available. Published figures from experienced operators, using modern equipment and materials are of 1.6% mortality, 1.4% severe deficits, 5.6% mild deficits, and 10.9% transient deficits (Berenstein, 1989). The complications from combined surgical and endovascular treatment are probably less than of each if used alone in larger malformations, but in smaller superficial lesions this may not be true (Deruty *et al.*, 1996; Nakstad *et al.*, 1994). The most serious procedural complications are brain ischemia and hemorrhage.

Ischemic complications occurring in the course of catheterization and navigation are related to dissection, blood clot formation around the catheter, or thromboemboli (Valavanis and Yasargil, 1998). Ischemic complications occurring in conjunction with the embolization are due to occlusion of normal arteries. Experience leading to optimal selection and use of the embolic agents and delivery system will decrease these complications. Ischemia related to thromboembolism can be managed with judicious use of local thrombolytics, abciximab, and heparinization; embolization with embolic agents cannot usually be remedied and general supportive measures and management of cerebral infarction should be instituted (oxygen, antiplatelet agents, neuroprotective drugs).

Hemorrhagic complications can be caused by mechanical perforation in the course of catheterization or may result from hemodynamic changes. If perforation occurs within the nidus or adjacent feeding vessel, the catheter or wire should *not* be removed (Halbach *et al.*, 1991). Treatment of these perforations include immediate reversal of anticoagulation, placement of coils across the perforation site, temporary inflation of a balloon across the perforation, and the institution of medical therapy such as anticonvulsant administration. Recognition of the perforation by the catheter tip is also critical since the catheter should be left in place for placement of a coil across the perforation site as the tip is withdrawn into the arterial lumen.

*Postprocedural complications* occur in the early post-embolization period (<72 h) (Valavanis and Yasargil, 1998). Hemorrhage occurring in this period always results from hemodynamic changes, usually by compromise of venous drainage (Valavanis and Yasargil,

1998). It may be difficult to diagnose the underlying cause of neurological deterioration secondary to thrombosis of veins, i.e., mass effect versus venous overload with resultant acute venous hypertension. Wilson and Hieshima (1993) suspect early neurological deterioration within the first 24 h postembolization or post-surgical resection of the larger high-flow AVMs is probably the result of venous thrombosis or stasis leading to venous overload and associated cerebral edema. Late deterioration (>72 h) is probably secondary to ongoing delayed thrombosis of critically draining veins and may be minimized by optimal volume expansion, avoiding hemoconcentration, and pre-embolization administration of 8 mg of Decadron, continued with 4 mg doses every 6 h with a steroid taper over a period of 1 week. This applies more to very large and complex AVMs in which such mass effect or propagating venous thrombosis may be anticipated.

Normal perfusion pressure breakthrough is thought to be the cause of either immediate or early deterioration (Spetzler et al., 1978, 1987), although it was not observed in other large series (Valavanis and Yasargil, 1998). It is assumed that staged embolization can avoid or minimize this cause of post-treatment neurological deterioration (Spetzler et al., 1987).

## Conclusion

All cerebral parenchymal AVMs in which treatment is indicated, should be discussed by a team comprising specialists from multiple disciplines. Treatment modalities available are often combined, and must be tailored to the individual patient. Embolization can be the definitive modality, but is often used more as an adjunct that precedes other therapeutic modalities, often making their application possible, especially in complex AVMs. Simple, superficial, and small AVMs may be treated by all three modalities with equal success rates.

## CEREBRAL DURAL ARTERIOVENOUS FISTULAS (DAVF)

### Clinical Aspects

#### Signs and Symptoms

**Dural Arteriovenous Fistulas (DAVFs).** Patients may be asymptomatic apart from tinnitus, and, depending on the angioarchitecture, the risk of cerebral venous complications leading to hemorrhage can be estimated to be low or high. Symptoms are related to the specific site of the abnormality, and can range from minor symptoms to disabling neurological deficits. Patients with DAVFs of the transverse and the sigmoid sinus tend to present with pulse synchronous tinnitus and may have an audible bruit. Anterior cranial fossa DAVFs are an uncommon site of fistulas, but tend to present with frontal lobe hemorrhage(s) (Awad et al., 1990; Halbach et al., 1987; Malik et al., 1984). Their importance is in recognizing the pattern of hemorrhage.

**Carotid Cavernous Sinus Fistulas (CCF).** Carotid-cavernous sinus fistulas cause a marked increase in pressure within the cavernous sinus, and increased flow through the superior ophthalmic vein and/or the inferior petrosal sinus. The direction of venous flow determines the clinical presentation, conveniently divided into ocular and central. Ocular symptoms include proptosis, chemosis, glaucoma, venous hemorrhages and pallor of the retina, and papillaedema. Neurologic findings of cerebral venous drainage reflect elevated venous pressures and flow and include cerebral dysfunction due to elevated intracranial pressure, venous hemorrhages, infarction, and bruit (Halbach et al., 1987a). The higher the flow the more severe the symptoms in the majority of cases, but a cerebral venous pattern of drainage carries a higher risk of major morbidity independent of flow rates.

#### Diagnostic Evaluation

**CT of DAVF.** Usually, only secondary changes such as intracerebral hematoma and venous congestion can be detected on CT.

**CCF.** An enlarged superior ophthalmic vein (SOV), as well as an enlarged cavernous sinus, can be visualized. Secondary changes such as intracerebral hematoma and venous congestion are easily detected, if they occur.

**MRI of DAVF, CCF.** MRA can show high signal intensity in the dural sinus, reflecting arterialized high flow shunting in the venous system, but MRA cannot exclude the diagnosis. Additional secondary changes can be easily detected, if present.

**DSA of DAVF, CCF.** An angiographic assessment is always necessary to establish the diagnosis, to detect high-risk features such as cerebral venous drainage, and to plan the therapeutic procedure.

In traumatic or direct carotid cavernous sinus fistulas (CCFs), it is important to define the site and size of defects in the internal carotid artery and the venous drainage pattern. Injection of the vertebral artery while compressing the ipsilateral carotid, or injecting the ipsilateral carotid artery while simultaneously compressing it, are techniques that can help characterize the anatomy. If the anatomy is such that carotid occlusion is likely or possible, elective trial carotid occlusion should be performed.

## Natural Course

### Epidemiology

DAVFs as well as dural CCFs typically occur in older women. An exception is the anterior cranial fossa DAVFs, which are characterized by a strong male predominance (87%) (Halbach *et al.*, 1992). Transverse and sigmoid sinus DAVFs are the second most common site after dural CCFs. Superior sagittal and inferior petrosal sinus DAVFs are less common. Spontaneous and postangiographic occlusion has been reported in some dural CCFs. (Halbach *et al.*, 1987).

### Pathogenesis

DAVFs are considered to be acquired. A suggested etiology of DAVFs is prior dural sinus thrombosis, with recanalization associated with the formation of a fistulous connection. Sinus thrombosis can also occur in patients with DAVFs. They are also reported to follow trauma, infection, recent surgery, and vascular disease.

**CCF.** Carotid-cavernous sinus fistulas are spontaneous or acquired communications between the carotid artery and the cavernous sinus, and are classified into direct and indirect depending upon the pattern of arterial supply. Cavernous DAVFs refer to indirect arterial supply. These consist of an abnormal communication between dural branches of the internal or external carotid arteries and the cavernous sinus, functioning as arteriovenous fistulas (the connections are nearly always multiple). Cavernous DAVFs are also related to sinus thrombosis. In contrast, direct or traumatic CCFs are related to trauma or to the rupture of a cavernous aneurysm. Direct or traumatic CCFs are high-flow arteriovenous shunts from the internal carotid artery to the cavernous sinus, they are invariably symptomatic, almost never spontaneously occlude, and usually require treatment. This contrasts with cavernous DAVFs, which may have a immediate spontaneous closure rate postangiography and may have minor symptoms not requiring treatment.

## Principles of Therapy

### DAVF

The importance of the pattern of venous drainage in predicting intracranial hemorrhage (Awad *et al.*, 1990; Lasjaunias *et al.*, 1986; Malik *et al.*, 1984) has been emphasized, and several classifications have been proposed based on venous anatomy. Cognard (1995) classified dural AVFs into five types:

I. located in the main dural sinus, with antegrade flow
II. in the main dural sinus, with poor antegrade flow
    IIa. with reflux into the sinus
    IIb. reflux into cortical veins alone
    IIa + b. both of above
III. direct cortical venous drainage without venous ectasia
IV. direct cortical venous drainage with venous ectasia
V. spinal venous drainage.

Patients with type I DAVFs have a benign course. Patients with type IIa DAVFs present with intracranial hypertension in 20% of cases, and patients with IIb DAVFs present with hemorrhage in 10% of cases. Hemorrhage was present in 40% of cases of type III dural AVFs and 65% of type IV DAVFs. Type V produced progressive myelopathy in 50% of cases. Therefore, patients with types I and IIa DAVFs may need no treatment at all, or subtotal treatment (arterial embolization) may be adequate, and they may not need to be treated with sinus occlusion. In types IIb DAVFs and above, treatment is indicated and can include sinus occlusion to achieve complete cure.

Treatment strategies include carotid–jugular compression and transarterial embolization either alone or combined with transvenous occlusion of the involved sinus. The latter combination has the highest success rate, but should be performed only after careful analysis of the venous drainage of the entire intracranial contents indicates an intact contralateral pathway, preferably with evidence of long-standing venous hypertension having already redirected flow to the contralateral side.

Anterior cranial fossa DAVFs can be embolized via the ophthalmic artery, but complete permanent obliteration may be associated with higher risk than surgery.

### Carotid Cavernous Sinus Fistulas (CCF)

CCF are classified by Barrow *et al.*, (1985) into four groups

A. direct communication between intracavernous ICA and cavernous sinus;
B. supply from the meningeal branches of the ICA alone;

C. supply from meningeal branches of the ECA alone;
D. mixed supply from the ICA/ECA meningeal branches.

Most cases are type D when detailed angiography is performed. This classification does not differentiate type D with minor versus dominant supply from the internal carotid artery (ICA), a distinction with significant impact on the approach and results of treatment, as the expected success of transarterial embolization alone is higher when the dominant supply is from the external carotid artery (ECA). Immediate management with a combined transvenous and transarterial treatment is then favored. The pattern of venous drainage determines the risk of intracranial hemorrhage and should be considered in all management decisions.

The aim of management is to prevent visual loss, ophthalmoplegia, cerebral venous infarction, and hemorrhage, and to achieve complete closure of the abnormal connection between the arteries and veins. All patients with cerebral venous drainage, and those with persistent ocular symptoms or decreased visual acuity should be treated. These lesions can spontaneously thrombose or occlude following angiography or carotico-jugular compression.

*Carotico-jugular compression* consists of instructing the patient to compress the carotid artery and jugular vein just below the angle of the mandible with the contralateral hand, for 1 min, two to three times every hour when awake, which can continue for several weeks. It is important to observe the patient to recognize a transient ischemic event, a contraindication to the procedure. Closure rates of up to 10%–30% have been reported. Whether the closure of the fistula results from the maneuver or reflects the natural history cannot be estimated.

*Transarterial embolization* alone has reported success up to 78%, although in most series considerably less is achieved. Superselective embolization of individual arterial feeders, and embolization with 150- to 250 μm polyvinyl alcohol particles is the first choice, with low chance of cranial nerve palsy. Permanent agents and alcohol should be avoided in all but the most refractory high risk fistulas. The risks of occluding the branches from the internal carotid artery are higher, and in general should be avoided in favor of combining treatment with transvenous embolization or allowing time for delayed thrombosis after the initial embolization.

The highest chance of success comes with combined transarterial and transvenous embolization (Halbach et al., 1992).

*Transvenous embolization* is another treatment method. The cavernous sinus is easily catheterized from a femoral vein approach by either the ipsilateral or contralateral inferior petrosal sinus (IPS) in the majority of patients, allowing direct deposition of microcoils (typically fibered platinum of 2–4 mm diameter and 10–60 mm length) (Figure 4). Even in the absence of significant opacification during angiography, the IPS can usually be found. The anatomy of drainage of the IPS varies from a classic large single venous sinus to a plexus of small channels and can join the internal jugular vein above, at, or below the skull base. Of these patterns, there is clearly a subset in whom a transvenous approach via the IPS to the cavernous sinus is precluded. This group of patients, and those with persistent but slowed flow after combined embolization (usually from the many compartments of the cavernous sinus) are instructed in the technique of carotico-jugular compression. Some of these will completely occlude during follow-up, and in the remaining symptomatic group, a surgical approach to the superior ophthalmic vein allows further fibered platinum coils to be placed via a microcatheter into the anterior patent part of the sinus (Gioulekas et al., 1997; Tress et al., 1983).

Close angiographic and clinical follow-up is essential, being aware that the bruit can alter in pitch or volume, or even disappear in the presence of persistent and potentially high-risk fistulas. There remains a small number of patients in whom despite complete obliteration of the fistula, there is paradoxical (usually transient) worsening of ocular symptoms (Sergott et al., 1987), probably due to poor venous collaterals after occlusion of the cavernous sinus and adjacent portion of the superior ophthalmic vein. This is usually self-limited.

### Direct Carotid Cavernous Sinus Fistulas

Transarterial balloon embolization with flow-directed detachable balloons is the optimal method in direct CCFs (Figure 5) (Debrun et al., 1988; Desal et al., 1997; Higashida et al., 1989c).

Current endovascular techniques rarely result in parent vessel occlusion, and complete obliteration of the fistula can be achieved in >90% of cases. The majority of the remaining cases are cured after repeated or transvenous approaches, with an overall success rate (Debrun et al., 1988) of over 95%, although occasionally requiring more than one procedure.

In cases previously treated by surgical internal carotid ligation, now with recurrent symptoms, direct carotid puncture can allow access and successful treatment (Halbach et al., 1989). Transvenous embolization is an alternative when the arterial route (direct puncture or transfemoral) fails (Halbach et al., 1988a).

Follow-up angiography is recommended in 3–6 months, to assess for residual fistulas, which can be asymptomatic, or the development of a false aneurysm. If the patient is asymptomatic and the false aneurysm

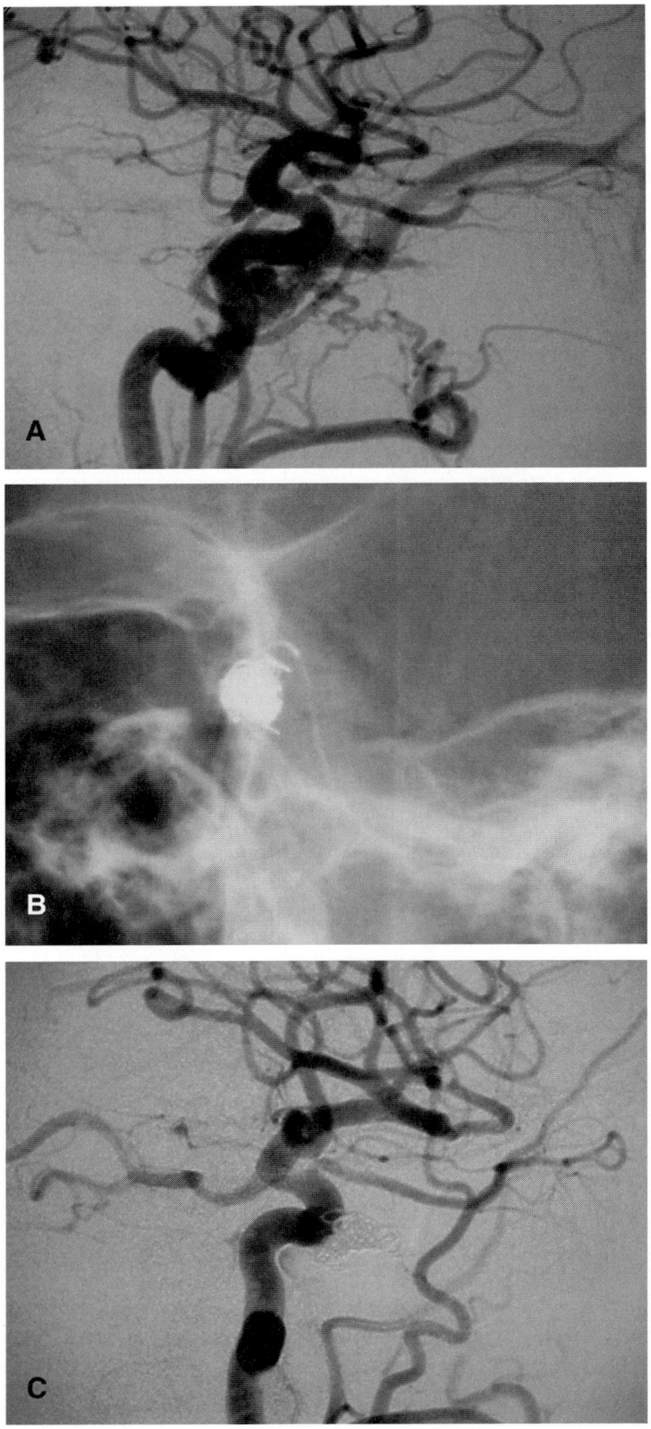

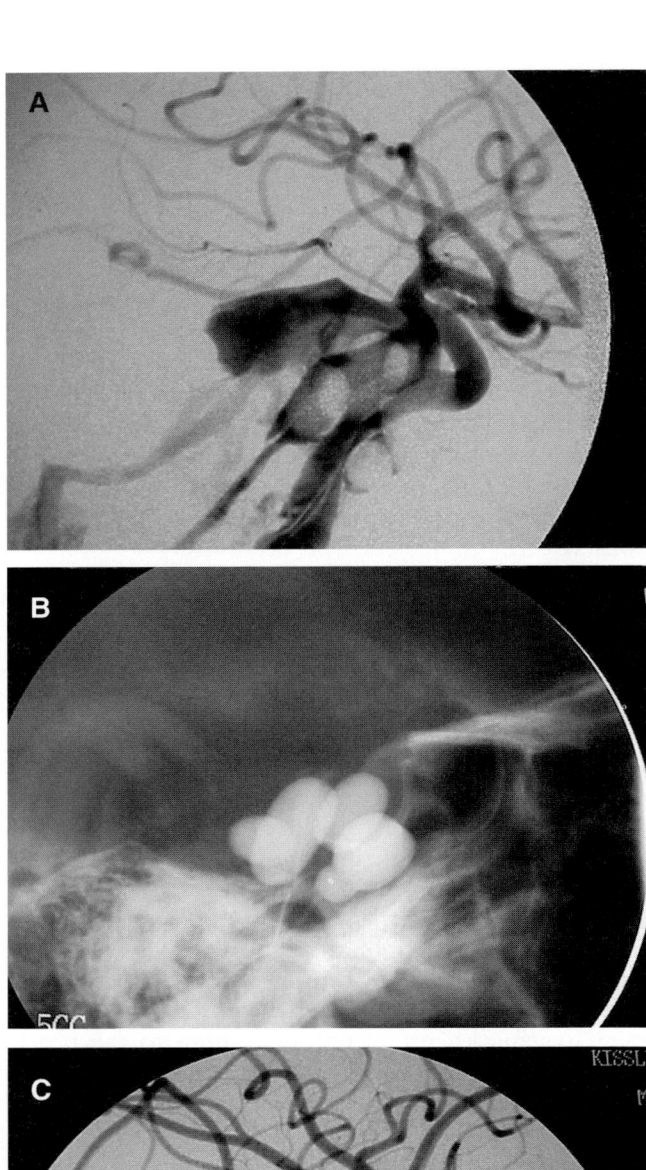

**FIGURE 4** Dural carotid cavernous sinus fistula. (**A**) Pretreatment showing left sided dural carotid cavernous sinus fistula, supplied by meningeal, dural branches of the cavernous internal carotid artery with shunting of blood flow to the cavernous sinus. (**B**) Coils placed into the left cavernous sinus, from a transfemoral venous route, via the left internal jugular vein, up the inferior petrosal sinus, and into the cavernous sinus. (**C**) Postembolization showing complete occlusion of the dural carotid cavernous sinus fistula, with complete cure.

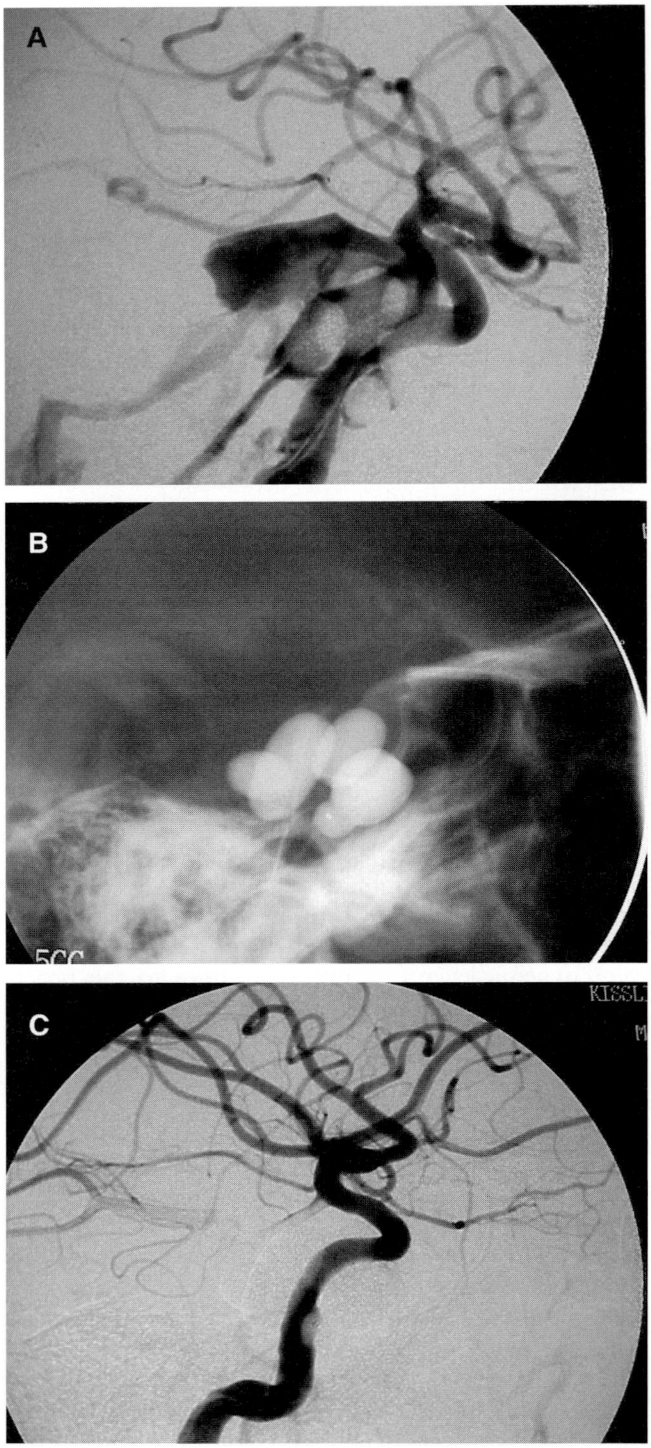

**FIGURE 5** Direct traumatic carotid cavernous sinus fistula. (**A**) Traumatic carotid cavernous sinus fistula, lateral view angiogram, shows rapid shunting of blood from the carotid artery directly into the cavernous sinus. (**B**) Five detachable silicone balloons, placed into the cavernous sinus for occlusion. (**C**) Postembolization angiogram showing complete occlusion of the traumatic carotid cavernous sinus fistula.

is small, treatment is not performed, but in all other cases of false aneurysm further treatment may be necessary. Although usually in patients with a wide neck, treatment with further balloons or detachable coils is preferred initially, aiming at vessel preservation.

## Practical Management

The symptoms and angiographic features will determine the need and mode of treatment.

### Anticoagulation

Systemic anticoagulation is routinely used at most centers. Other centers only use local perfusion of heparinized saline flushes between coaxial catheters and guidewires.

### Anesthesia

Arterial endovascular embolization of DAVFs and CCFS can be performed using local or general anesthesia. However, transvenous embolization should be performed under general anesthesia, the main reason being the unpredictable length of the sessions, which would significantly affect the compliance of patients.

## Complications

### Arterial Embolization

Similar to AVMs, ischemic complications occurring in the course of catheterization and navigation are related to dissection, blood clot formation around the catheter, or thromboemboli (Valavanis and Yasargil, 1998). Ischemic complications occurring in conjunction with the embolization are due to occlusion of normal arteries. Ischemia related to embolic agents can also occur during embolization in the ECA when anastomoses between the ECA and the ICA are patent.

### Venous Embolization

The main complication when closing a dural sinus is the danger of increased venous pressure, leading to venous hypertension and possible hemorrhage. The increased venous pressure can result from the failure of regular veins to drain adequately after the closure of the sinus or from a pouch remnant in the sinus still draining the fistula. The pouch connects the fistula to other veins that suddenly must drain a larger volume than before. The sinus being partially occluded becomes inaccessible from an endovascualr approach.

## Carotid and Vertebral Artery–Jugular Fistulas

Carotid and vertebral artery jugular fistulas are usually post-traumatic, but spontaneous and presumed congenital fistulas are also known to occur. Local symptoms may occur such as pain, tinnitus is common, neurological symptoms uncommon, and a bruit is often audible on examination. The fistulas are typically high flow, often with a very short "fistula" separating the artery and vein. If long-standing, patients can develop a "cirsoid" aneurysm, with radiating draining veins of markedly increased caliber. Usually these fistulas are considered low risk of neurological deficit. However, bleeding can be massive, but when superficial can be controlled by local pressure. Local tissue necrosis can follow from mass effect and decreased tissue perfusion. Treatment is not always required and this condition is rarely considered an emergency and can be discussed at length with the patient.

When treatment is indicated, the goal is obliteration of the fistula with preservation of the parent artery. Detachable balloons can best achieve this; occasionally coils may be necessary but care must be taken to avoid distal migration of the embolic agent used. If coils migrate, they can usually be retrieved from the heart or proximal pulmonary arteries; if distal in the pulmonary arteries they often are asymptomatic. If the length of the fistula precludes coils or balloons, covered stents can be considered, usually from the arterial but also from the venous side. When all other options fail, obliteration can be achieved with sacrifice of the parent artery; detachable balloons are preferred. It is important to occlude the site of the fistula or to trap the fistula to avoid having it fill in a retrograde fashion or from local anastomoses.

Success in treating vertebral artery–jugular vein fistulas is achieved in more than 90% of cases (Beaujeux et al., 1992; Halbach et al., 1988a), and less so in the case of cirsoid aneurysms in which combined surgical excision may be necessary (Barnwell et al., 1989).

## ENDOVASCULAR THERAPY OF OCCLUSIVE EXTRACRANIAL VASCULAR DISEASE

### Natural Course

Atherosclerotic disease as a cause for stroke represents a major source of mortality in North America and Europe and is associated with high medical and social cost. Currently there are more than 700,000 cases of new or recurrent strokes per year (American Heart Association, 1993), of which 20–30% are attributed to atherosclerosis of the carotid bifurcation and proximal internal carotid artery. More than 150,000 surgical endarterectomies are performed annually for atherosclerotic carotid lesions in the United States alone (Strandness, 1994), with good evidence from prospective randomized trials that this is more effective than best medical therapy alone in stroke prevention in symptomatic (North American Symptomatic Carotid Endarterectomy Trial Collaborators, 1991) and asymptomatic (Executive Committee for the Asymptomatic Carotid Atherosclerosis Study, ACAS, 1995) patients with severe carotid stenoses.

## Principles of Therapy

In 1964, Dotter and Judkins first reported the technique of percutaneous transluminal angioplasty (PTA) for treatment of high-grade atherosclerotic lesions of the peripheral arteries. In the early 1980s, PTA for symptomatic vascular lesions involving the brachiocephalic arteries were initially reported (Bockenheimer et al., 1983; Motarjeme et al., 1982). PTA for cerebral vessels was delayed when compared to other regions of the body due to the more difficult technical problems of access and more importantly because of the perceived potential complications of inducing stroke by fragmentation of debris from the angioplasty site. As larger case series of cerebral angioplasty were reported, the safety, efficacy, and long-term success of the procedure were shown to be comparable to surgical alternatives. Successful results of cervical and intracranial PTA involving patients with atherosclerosis, fibromuscular dysplasia, radiation induced fibrosis, acute dissection, vasculitis, and postsurgical restenosis from intimal hyperplasia have also been described (Courtheoux et al., 1985; Hodgins et al., 1982; Tsai et al., 1986).

The endovascular approach now provides greater therapeutic options for patients particularly at high risk from surgery, who are symptomatic despite maximal medical therapy. Such lesions include severe atherosclerosis in patients with significant comorbidity, tandem intracranial stenosis, distal lesions of the cervical carotid, contralateral carotid occlusion, and recurrent stenosis of vessels having previously undergone endarterectomy.

A review by Kachel (1996) summarizing the cumulative experience of balloon angioplasty of the carotid, vertebral, subclavian, and innominate arteries, described 1971 cases with an overall technical success rate of 94.6%. The mortality varied by territory from 0 to 2.1%, with overall morbidity of 0.9% and minor technical complications of 0–6.3%. The author's experience included percutaneous angioplasty of 74 symptomatic carotid lesions with a technical success of 93%, 1 major stroke (1.4%), 2 minor complications (2.7%), and no reported restenosis at 70-month follow-up. Higashida et al. (1993a) reported angioplasty of 256

extracranial vessels with greater than 70% stenosis with 1 major stroke (0.4%), 5 transient ischemic events (2%), no mortality, and 15 cases (6%) of restenosis at 6–12 months follow-up.

A prospective analytical study of 29 patients having undergone carotid angioplasty for severe symptomatic ipsilateral carotid stenosis by NASCET criteria was reported by Schoser (1998). Neurological and ultrasonographic follow-up (mean of 33 months) revealed that 78% of patients had no further neurological sequellae, 10% had a single episode of ipsilateral transient ischemia or amaurosis fugax, and 7% had recurrent episodes, with no patient developing a stroke. Fifty percent of treated vessels remained normal with ultrasound (<50% stenosis), 40% with mild stenosis (50–70%), and 10% with severe stenosis (>70%). Angioplasty alone can achieve reasonable long-term patency rates.

## Carotid Artery

### Angioplasty and Stenting

Yadav et al. (1996) reported their experience in balloon angioplasty and stenting of carotid restenosis in a series of 22 patients who had previously undergone carotid endarterectomy. They were able to successfully decrease the stenosis by 79% with a morbidity of 4% (a minor stroke in a single patient) and no restenosis (>50%) at 6 months follow-up. Given the high risk associated with reoperation on previous endarterectomized lesions, which have been reported to have operative complication rates of 10.5% (Piepgras et al., 1986), balloon angioplasty and stenting was concluded to be an attractive alternative.

Diethrich (1996) reported on a study of 110 nonconsecutive patients (79 males; mean age 72 years, range 45 to 85), either symptomatic with stenoses of 70% or greater or asymptomatic with greater than 75% stenoses, treated with stent assisted angioplasty. Lesions meeting the treatment criteria were in the proximal common, mid-common, distal common, internal, and external carotid arteries. In 110 patients (117 arteries) intended for treatment, 109 (99.0%) were successfully treated. There were seven strokes (two major, five reversible) and five transient ischemic attacks (TIAs). Clinical success at 30 days was 89%, and, over a mean 7.6-month follow-up, no new neurological symptoms developed.

Roubin (1996) reported on the results of 146 procedures in 74 patients, with only one technical failure. One in-hospital death occurred; there were two major strokes and seven minor strokes. The restenosis rate at 6 months was less than 5%, and only one TIA was reported. No strokes or deaths occurred during the follow-up period. Vitek *et al.* (1999) reported having treated a total of 445 vessels in 404 patients using stent-assisted angioplasty with predilatation, stent placement, and postdeployment dilatation. Patients were pretreated with aspirin and ticlopidine and were not maintained under intravenous anticoagulation following the procedure. They reported on the treatment of 40 patients with contralateral carotid occlusion and 70 patients with postendarterectomy restenosis. Overall technical success was 98%, with a 30-day mortality/morbidity and death rate of 1.9% (0.7% neurological and 1.2% systemic), 0.7% risk of major stroke, and 5.8% minor stroke. The authors determined their annual risk of minor stroke and showed a steady decline from 7.2% in 1994–1995 to 4.4 and 2.2% in each successive year. The decrease in complication rate with experience illustrates the fact that stent-assisted angioplasty is still early, yet appears to have short-term risk comparable to that reported for carotid endarterectomy. Six-month follow-up obtained in 80% of patients demonstrated stenosis greater than 50% in 5% of patients. Vitek *et al.* (1999) reported clinical follow-up in 95% of patients in the same series of patients with two neurological deaths and one major and three minor strokes for a freedom of any stroke of 92% and freedom from disabling stroke or death of 98%, both at 2 years.

In addition, the feasibility and safety of percutaneous carotid angioplasty and elective stenting was further evaluated prospectively in a consecutive series of 107 high-risk patients (Yadav *et al.*, 1997). One hundred twenty-six carotid arteries with significant stenosis were treated. This series represented a high-risk subset that included patients with previous ipsilateral endarterectomy and severe medical comorbidity. There were seven minor strokes, two major strokes, and one death during the initial hospitalization and first 30 days after the procedure. For the combined end point of all strokes and death, the incidence was 7.9% of vessels treated (9.3% of patients treated). For ipsilateral major stroke and death, the incidence was 1.6%. There were no strokes during the follow-up period. Of 81 asymptomatic patients in whom angiographic follow-up was available, four (4.9%) had restenoses managed with repeat intervention.

Theron (1996) reported on angioplasty for carotid artery stenosis in 259 patients. Of note was the use of cerebral protection (triple coaxial catheter) in 136 cases of atherosclerotic stenosis in the internal carotid artery or in the carotid bifurcation. A stent was placed in 69 patients when images obtained immediately after angioplasty showed signs of dissection or insufficient arterial opening. No procedure-related complications occurred in the 71 cases of nonatherosclerotic stenosis and in the

14 cases of proximal carotid artery and siphon atherosclerotic stenosis. Among the 38 patients who underwent angioplasty without cerebral protection, dissection occurred in 2 (5%) and embolic complication occurred in 3 (8%) during the procedure. Among 136 patients in whom cerebral protection was used, no embolic complications occurred during angioplasty, and only two (1%) occurred during or after stent placement when protection was not possible. No residual dissection flaps were found after stents were placed, and the restenosis rate decreased from 16 to 4%, with the use of stents.

Mathur *et al.* (1998) reported their analysis of multiple factors that portend a higher risk of complication in stent-assisted angioplasty of the carotid artery in 271 vessels of 231 patients. Their patients constituted a high-risk subset who had coronary artery disease (71%), bilateral carotid disease (39%), and contralateral carotid occlusion (12%). The treated vessels had undergone previous endarterectomy (22%), contained ulcerated plaques (24%), or were calcified lesions (32%). Only 14% of these patients would have been eligible to undergo endarterectomy by NASCET criteria. The rate of minor stroke was 6.2% and that of major stroke 0.7% during the first 30 days after and including the procedure. Furthermore, the rate of any stroke for the NASCET eligible subset was 2.7% for the same time interval. Multivariate analysis showed that advanced age and long or multiple stenoses were independent predictors of procedural-related stroke in carotid stent-assisted angioplasty.

Henry (1998) attempted carotid angioplasty in 174 arteries (163 patients; 126 men; mean age, 71 ± 10 years, range 47 to 93). The majority (65%) were asymptomatic. Most (82%) carotid arteries were treated without cerebral protection; stents (primarily Palmaz and Wallstent) were deployed routinely in all cases. Immediate technical success was achieved in 173 of 174 (99.4%); eight (4.6%) neurological complications occurred in the periprocedural period: three transient ischemic attacks, two minor strokes, and three major strokes. Two major complications developed despite cerebral protection. There were no deaths or myocardial infarctions and only three cervical access site hematomas. Over a mean 12.7 months follow-up, no ipsilateral neurological complications were seen. There were four (2.3%) restenoses, for primary and secondary patencies at 3 years of 96 and 99%, respectively.

A more recent report in 1998, based on surveys and literature review, indicated that the total number of endovascular carotid stent procedures performed worldwide included 2048 cases, with a technical success of 98.6% (Wholey *et al.*, 1998). Overall, there were 63

(3.08%) minor strokes, 27 (1.32%) major strokes, and 28 (1.37%) deaths within a 30-day postprocedure period. Restenosis rates of carotid stenting have been 4.80% at 6 months. Mathias (1999) reported their own experience of angioplasty and stent placement for atherosclerotic internal carotid artery stenosis. In 633 patients, 799 ICA stenoses were treated; 70% of them were symptomatic and 30% asymptomatic. In 99% of the patients, the stenoses could be decreased with a reduction of the degree of stenosis from 82% to 12%. Transient neurological deficits occurred in 5% and permanent deficits in 2.7% of patients with decreasing morbidity incidence over the years. The reported 5-year patency rate was 91.6%.

Fifty-one patients with severe coexisting carotid and symptomatic coronary artery occlusive disease successfully underwent staged or simultaneous coronary angioplasty and carotid stenting (Al-Mubarak et al., 1999). One pericardial effusion and two minor strokes with full recovery occurred in the hospital, but no major neurologic events, myocardial infarction, or deaths were observed, and no repeat revascularization was required within a 30-day follow-up.

Thus far, the overall results are promising, with most series published in this early phase of the technique having major stroke or death rates on the order of those reported in larger multi-institutional studies of carotid endarterectomy, especially considering the inclusion of high-risk patients, with reported surgical morbidity rates of up to 18% (Piepgras et al., 1986; Sundt et al., 1994). The risk of procedural stroke or death from endarterectomy in NASCET (1991) was 5.8% and in ACAS (1995) 2.3%. In addition, they also compare favorably to the retrospective analysis of 3111 carotid endarterectomies performed by Sundt (1994) given that a large proportion of the patients undergoing stent-assisted angioplasty constitute Sundt Classes III and IV. There are some exceptions to this general trend. A comparison of carotid stent-assisted angioplasty (107 patients) and endarterectomy (166 patients) by Jordan (1998) reported an early minor stroke rate of 6.6% for the former and 0.6% for the latter, and a combined major stroke/death rate of 9.7% for the former and 0.9% for the latter. The authors conclude that PTA carries a higher neurological risk and requires more monitoring than CEA in the treatment of patients with carotid artery stenosis. Extensive experience with carotid and cerebral angiography, and with interventional procedures, is a key requirement to achieving a low technical complication rate. In experienced hands, results from multiple centers suggest that carotid stent-assisted angioplasty is now an acceptable alternative to carotid endarterectomy in select patient subsets, which are known to constitute very high risk for surgical complications (Figure 6).

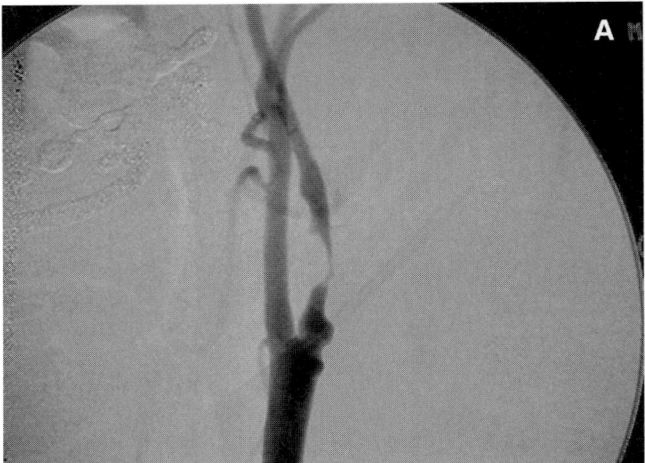

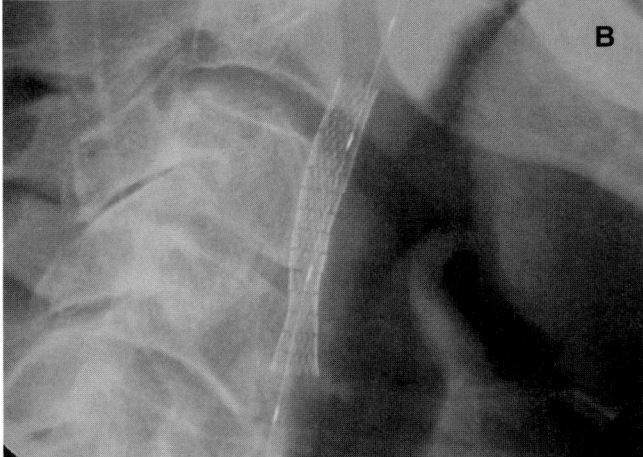

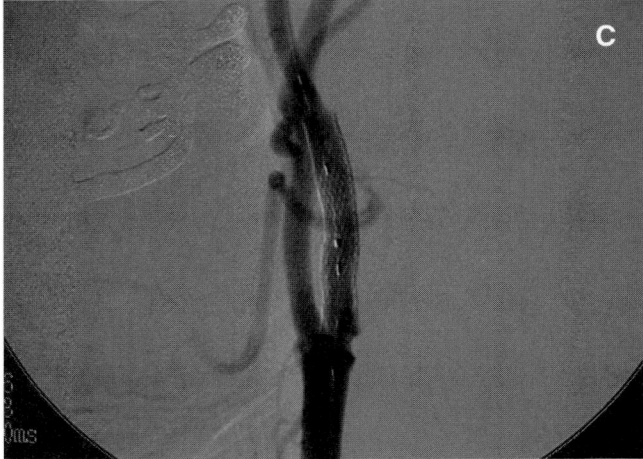

FIGURE 6   Left cervical carotid artery stenosis treated with stent. (A) Left cervical high-grade internal carotid artery stenosis in a patient with recurrent transient ischemic attacks in spite of being on antiplatelet medical therapy. (B) Self expanding nitinol stent placed across stenosis. (C) Poststent deployment shows wide patency of internal carotid artery.

## Practical Management

Ultrasound has been shown to be sufficiently sensitive to detect stenosis when compared to angiography in a series of 170 stent placements in 119 patients (Robbin *et al.*, 1997). Ultrasound did not fail to detect any significant stenosis and was promising in its ability to assess intrastent intimal hyperplasia.

The patient is first subjected to a complete and thorough angiographic evaluation, using a high-quality digital subtraction unit to determine the location and severity of stenosis, and the adequacy of collateral blood supply to the affected territory. Following the study the patient undergoes measurement of a baseline activated clotting time (ACT) and receives an initial weight-based (70 units/kg) intravenous bolus of heparin followed by a post-heparin ACT determination to achieve an ACT value equal to or greater than 2.5 times the baseline value (>250 s). The patient then receives either an hourly dose equal to half the initial bolus or is placed on a heparin drip of 15–20 u/kg/h. Patients are administered enteric-coated aspirin (325 mg qd) and either ticlopidine (Ticlid 250 mg bid) or clopidogrel (Plavix 75 mg qd) starting 3 days prior to the procedure. Following the procedure, the patient is kept on daily aspirin indefinitely and on ticlopidine or clopidogrel for at least 6 weeks. The role of glycoprotein IIb/IIIa inhibitors, which have been shown to decrease mortality and morbidity in a number of coronary stent studies (Tcheng, 1996), remains to be defined in carotid and vertebral angioplasty and stenting.

Patients who have not undergone prior CEA are administered atropine 0.5–1 mg intravenously immediately prior to balloon dilatation of the carotid artery to blunt any parasympathetic discharge.

## Stent Designs

Stents applied in extracranial carotid atherosclerosis involve the use of the self-expanding WallStent or the newer Nitinol Stents (Cordis/Johnson & Johnson; AVE/Medtronics; Guidant). The newer nitinol stents are self-expandable designed with significantly lower radial force, and with a greater metal surface area coverage with finer and more numerous interstices. Its compliant design and the consequently lower radial force can be a hindrance in severely calcified lesions, but the benefits outweigh the limitations when contrasted with noncompliant stents, and this has been a preferred stent over recent years.

## Complications

Potential complications include vessel occlusion, perforation, dissection, spasm, thromboemboli, occlusion of adjacent vessels, transient ischemic episodes, and stroke.

Angioplasty of atherosclerotic lesions has been reported to induce the release of multiple emboli including atheroma, cholesterol crystals, thrombus, and platelet aggregates (Bladin *et al.*, 1998; Crawley *et al.*, 1997; Muller *et al.*, 1998; Theron *et al.*, 1996). Embolization of microparticles has also been shown during open endarterectomy and has been shown to correlate with complex plaque morphology (Gaunt *et al.*, 1996) and with clinical postoperative cerebral ischemia (Levi *et al.*, 1997). Analysis of 14 patients undergoing percutaneous angioplasty and 14 undergoing endarterectomy with shunt placement revealed that endarterectomy resulted in significantly greater total occlusion time (337 s vs 26 s), but a lower count of microembolic signals (52 vs 202 events) then seen in angioplasty, although neither parameters were predictive of later neurologic events (Crawley *et al.*, 1997). It is unclear whether stent placement concomitant with angioplasty may help decrease the number of microemboli by trapping them under the metal interstices, or whether primary stenting would decrease emboli compared to secondary stenting. The problem of distal embolization during balloon dilatation of atherosclerotic stenoses has engendered interest in various methods of protection. Distal protection using a specially designed triple-coaxial catheter has been described by Théron (1990). Distal embolic complications were found in 3 of 38 patients undergoing angioplasty without (8%) and in none of 136 undergoing angioplasty (0%) with distal protection (Theron *et al.*, 1996). A number of commercial devices are now under development to provide distal protection as a means of decreasing the thromboembolic burden associated with angioplasty.

## Other Complications

These can be separated (Dorros, 1996) into those related to *arterial access* including hematoma, retroperitoneal hemorrhage, pseudoaneurysm, arteriovenous fistula, arterial thrombosis, groin infection; *catheterization* including arterial dissection, embolism of air or thrombus, vessel perforation, tear or rupture; *contrast media*, including allergic reaction, hypotension, and acute renal failure; and to *angioplasty and stent placement*, including distal embolization, pseudoaneurysm formation, stent infection and dissection. In a recent meta-analysis, angiography-associated risk of combined and transient neurological complications ranged from 0.8 to 3.0% for patients undergoing the procedure for subarachnoid hemorrhage or for transient ischemic attack/stroke, respectively. The rate of permanent neurological complication was found to be very low at

0.07% for patients presenting with subarachnoid hemorrhage (Cloft et al., 1999). An excellent account of the multitude of technical complications relating to carotid stent-assisted angioplasty has recently been published by Théron (1998). The authors described the type and mechanism of complications encountered, as well as precautionary measures needed for avoidance of such pitfalls.

### Current Indications

Angioplasty and stenting in the extracranial carotid artery has become a viable alternative to surgical carotid endarterectomy in certain subsets of very high-risk patients. Long-term clinical and angiographic follow-up as well as randomized prospective trials will help determine the role of percutaneous angioplasty with stenting in the treatment of carotid stenosis.

## Extracranial Vertebral Artery

### Stenosis

The frequency and severity of atherosclerosis in the vertebral arteries is variably reported as equal to or somewhat less than in the carotid arteries and about equal to that of the cerebral arteries. Most reports identify the origin or first segment of the vertebral artery to be the most frequent site of stenosis (Moufarrij et al., 1984). The morphological and pathological characteristics of atherosclerotic plaque of the vertebral artery differ from internal carotid artery disease, being more frequently annular and concentric, fibrous, and smooth, with a low incidence of ulceration or intramural hemorrhage (Hass et al., 1968; Imparato et al., 1981). Atherosclerotic plaque at the vertebral artery origin may also involve the adjacent subclavian artery. This has important implications for endovascular therapy, with particulate embolization during endovascular manipulations likely to be less frequent than encountered with carotid plaque, but increased vessel recoil could be anticipated to cause decreased long-term patency rates.

### Natural History

Data is available to assist in determining the prognosis of stenotic disease of the posterior circulation. However, there is no large prospective trial of treatment modalities stratified by degree of stenosis comparable to NASCET, and consequently much controversy over natural history and treatment exists. In general it would seem that intracranial vertebral artery disease has a worse prognosis than extracranial, with basilar artery disease worse than vertebral. The exact pathophysiological basis for stroke in this setting is often not determined, and can include hypoperfusion, artery to artery embolization, and arterial thrombosis.

Moufarrij (1984) reported on 96 patients with greater than or equal to 50% unilateral vertebral artery stenosis, followed up for an average of 4.6 years. None of the patients had definite vertebro-basilar transient ischemic attacks (VB TIAs), and the only two patients (2%) who sustained a brainstem infarction had fatal strokes and both were known to have basilar artery stenosis in addition to their vertebral artery (VA) stenosis. They concluded that vertebral artery stenosis is most frequently located at the VA origin (93%), and is associated with a low incidence of brainstem infarction.

In a subsequent study, 44 patients with greater than or equal to 50% stenosis of a distal vertebral artery and/or basilar artery were followed up for an average of 6.1 years. Seven patients (16%) had definite, and three patients possible VB TIAs. Eight patients (18%) sustained a stroke, five of which were in the VB territory. The observed stroke rate was 17 times the expected rate for a matched normal population. Distal vertebrobasilar occlusive disease appears to carry a higher risk for brainstem ischemia (Moufarrij et al., 1986).

In another study, 80 patients with either occlusion or high-grade stenosis involving the V1 segment of the vertebral artery were followed (Wityk et al., 1998). Hypertension, cigarette smoking, and coronary artery disease were common risk factors. Occlusive disease involving the V1 segment of the vertebral artery was common in patients with posterior circulation ischemia, but was often associated with other potential mechanisms of stroke. Occlusive disease of the V1 segment was the primary mechanism of ischemia in 9% of patients, but in other cases was associated with severe intracranial occlusive disease of the posterior circulation, evidence of artery-to-artery embolism, and proximal vertebral arterial dissection. The WASID investigators (1998) concluded that patients with symptomatic intracranial vertebral artery or basilar artery stenosis were at high risk of stroke, MI, or sudden death. They reported on 68 patients with symptomatic stenosis of the intracranial posterior circulation of >50% severity, treated with warfarin (n = 42) or aspirin (n = 12, followed for a median time period of 13.8 months). During follow-up, 16% of patients had nonfatal strokes, and 6% had fatal strokes. The calculated stroke rate for patients with intracranial vertebral stenosis was 13.7 per 100 patient-years of follow-up.

### Principles of Therapy

Balloon angioplasty for the treatment of vertebral artery occlusive disease was first reported in 1981

(Motarjeme *et al.*, 1981). Over the following decade, numerous investigators reported favorable short-term results using angioplasty (Brückmann *et al.*, 1986; Courtheoux *et al.*, 1985; Higashida *et al.*, 1986, 1993b; Kachel *et al.*, 1987; Motarjeme *et al.*, 1982; Theron *et al.*, 1984). Higashida (1993b) reported treating 34 proximal vertebral artery atherosclerotic lesions in 33 patients presenting with vertebro-basilar ischemic symptoms or stroke in whom medical therapy had failed. Technical success (<30% residual stenosis) was achieved in all cases. They reported a 9% incidence of transient intraprocedural complications and no permanent complications. Follow-up radiography revealed a 9% incidence of restenosis within 5 months. Clinical follow up at 12 months disclosed improvement of symptoms in all but one patient who died from a rupture of an unrelated contralateral vertebral artery aneurysm. In Kachel's (1996) review of the endovascular literature up to 1996, 268 vertebral balloon angioplasty procedures had been attempted with an overall technical success rate of 95.1%, morbidity rate of 0.7%, minor complication of 3.3%, and no procedure-related mortality. Endovascular therapy of extracranial vertebral artery strictures has evolved from simple balloon angioplasty to stent-supported angioplasty principally to overcome the problems of elastic recoil and early restenosis (Feldmann *et al.*, 1996; Motarjeme, 1996; Storey *et al.*, 1996). Malek *et al.* (1999b) treated 21 patients with symptomatic occlusive subclavian (*n* = 13) or vertebral artery stenosis (*n* = 8) refractory to medical therapy using stents. Technical success was achieved in all cases with the mean degree of overall stenosis reduced from 75% to 4.5%. There were no periprocedural vertebro-basilar strokes. One patient (4.8%) experienced a periprocedural transient ischemic attack. At a mean clinical

follow-up of 21 months, 90.5% of surviving patients had resolution or improvement of symptoms. One patient died from a carotid artery territory stroke.

Despite the limited published data, stent-supported angioplasty is emerging as a viable therapeutic option for medically refractory vertebral artery occlusive disease offering improved rates of technical success and better long-term patency over balloon angioplasty alone. Preference is for stent-assisted angioplasty, either primary or secondary dependent upon the degree of stenosis (Figure 7).

### Practical Management

The general preoperative medication, angiographic study, and guide catheter placement outlined for ulcerated internal carotid disease apply to vertebral angioplasty and stent insertion. Once a guide catheter is in position in the subclavian artery, the stenosis is crossed using a microcatheter and wire, and an exchange wire left across the lesion. If ostial, the wire is placed in the distal cervical vertebral artery, and if a distal vertebral stenosis is present, then the wire is usually left in the posterior cerebral artery. Appropriate monitoring of the position of the tip of the wire is necessary to ensure it does not move forward during catheter exchange, to minimize the chance of perforation or dissection.

### Subclavian and Innominate Artery

Recent experience in the percutaneous angioplasty of subclavian artery for aortoarteritis (32 vessels) and atherosclerosis (23 vessels) has been performed by Tyagin (1998) with a 92.8% success in stenotic lesions, and a

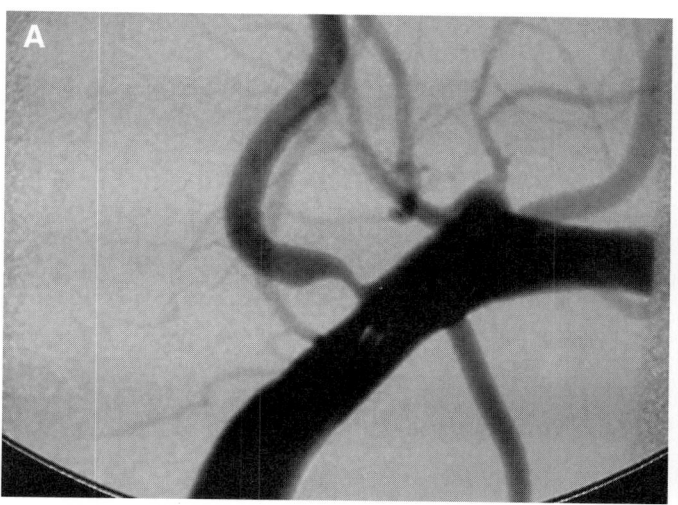
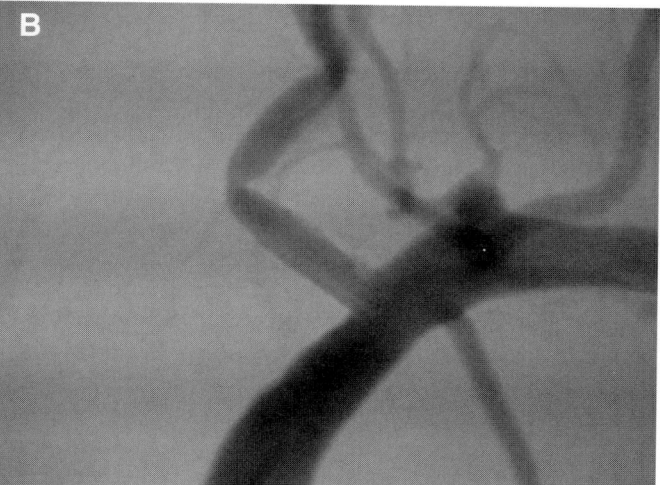

**FIGURE 7**  Vertebral artery stent. (**A**) A 60-year-old female with vertebrobasilar insufficiency symptoms, and an occluded contralateral vertebral artery. Left proximal vertebral artery shows high-grade stenosis. (**B**) Post-stent deployment shows widely patent left proximal vertebral artery.

60% success in recannalizing totally occluded arteries. The average stenosis decreased from 88.7% for atherosclerosis and 89% for aortoarteritis, to 15.5% and 8.3%, respectively, following percutaneous angioplasty without neurological complications and with good long-term symptomatic relief (3–120 months). Stent-assisted angioplasty is appropriate, associated with low morbidity, and seems effective (Phatouros *et al.*, 1999). The technical and procedural elements do not significantly differ from those discussed in the preceding sections.

## ENDOVASCULAR THERAPY OF OCCLUSIVE INTRACRANIAL VASCULAR DISEASE

### Intracranial Cerebral Arteries: Angioplasty and Stenting

#### Clinical Aspects

Atherosclerotic stenosis of intracranial vessels can damage brain tissue by causing either a thromboembolic territorial infarct or a hemodynamic infarct. The estimated risk of stroke in this setting varies widely from 10% to 46% per year, independent of medical therapy (Bogousslavsky *et al.*, 1986; Chimowitz *et al.*, 1995; Marzewski *et al.*, 1982; The EC/IC Bypass Study Group, 1985; WASID, 1998; Wechsler *et al.*, 1986). Intracranial arterial disease (IAD) is an independent risk factor for subsequent stroke in medically treated patients with symptomatic extracranial internal carotid artery stenosis (Kapelle *et al.*, 1999). Treatment of such a stenosis depends on its patho-physiological effects. If the stenosis results in thromboemboli, antiplatelet or anticoagulant medication is prescribed. If the stenosis causes hemodynamic abnormalities, these effects may be attenuated by treatments designed to improve the patient's cardiac output and raise blood pressure although this is a controversial procedure, when used on regular basis (Helgason, 1997; Kwiatkowski and Libman, 1997). However, in some patients, the signs and symptoms of stenosis recur or progress despite medication. Decreasing the risk of stroke in patients with stenosis of intracranial arteries presents challenges.

Although bypass surgery may be appropriate to treat stenotic portions of the artery in the anterior circulation, bypass surgery for stenosis in the posterior intracranial circulation is associated with significantly higher risk of complications (Hopkins and Budny, 1989). PTA has been found effective and associated with low morbidity and mortality when used to treat atherosclerotic stenosis in the anterior intracranial circulation (Brückmann *et al.*, 1986; Higashida *et al.*,

1996) and to relieve stenosis in the posterior intracranial circulation (Clark *et al.*, 1995; Higashida *et al.*, 1991, 1993b; Nahser *et al.*, 2000; Takis *et al.*, 1997).

#### Natural Course

Studies of the natural history of stenosis of the intracranial vertebral and basilar arteries show that this condition poses a high risk for brainstem ischemia (Acheson *et al.*, 1971).

#### Principles of Therapy

**Cerebral Angioplasty.** The first intracranial angioplasty (of a basilar artery) was performed through surgical exposure of the extracranial vertebral artery at the base of the skull and involved resection of the arch of the atlas (Sundt *et al.*, 1980). With the development of new instruments the invasiveness of this procedure was reduced considerably, but it was still associated with mortality of 37.5% (Higashida *et al.*, 1993a). Further technical improvements resulted in a decisive decrease in mortality (Clark *et al.*, 1995; Higashida *et al.*, 1996; Houdart *et al.*, 1996; Mori *et al.*, 1997; Takis *et al.*, 1997). The rate of permanent neurologic deficits in larger series reported to date has ranged between 5% (Nahser *et al.*, 2000) and 33% (Takis *et al.*, 1997).

**Intracranial Stents.** Intracranial atherosclerotic stenoses of the carotid and vertebral arteries and their branches can be treated with stent placement following or concurrently with angioplasty (Figures 8 and 9). This treatment is reserved for lesions that are symptomatic despite maximal medical therapy, have either had a mechanical complication of angioplasty, or are considered at increased chance of poor outcome from balloon angioplasty alone. In the carotid artery, stent placement has been performed or reported in the intracranial petrous, cavernous, proximal supraclinoid segment, and middle cerebral arteries. In the posterior circulation, stents have been deployed in the intracranial V4 segment and in the mid-segment of the basilar artery (Malek *et al.*, 1999a; Nakayama *et al.*, 1998; Phatouros *et al.*, 1999). Compromise of perforating vessels supplying the brainstem remains a factor, which may limit the use of mid-basilar stents, although in the experience to date, there have been no instances of perforator occlusion.

Intimal dissection is not unexpected following plaque rupture by balloon angioplasty, and is anticipated to heal over time. Stent placement can be used as a bailout measure for treatment of a flow-limiting angioplasty-

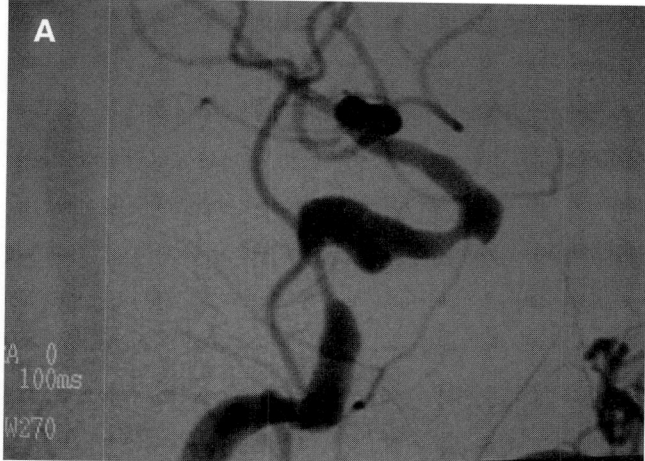

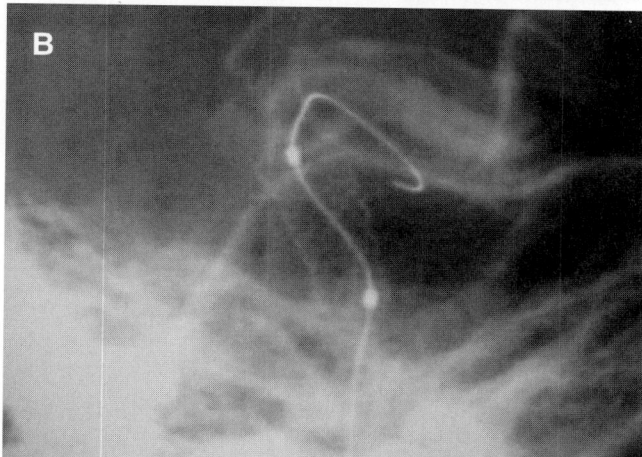

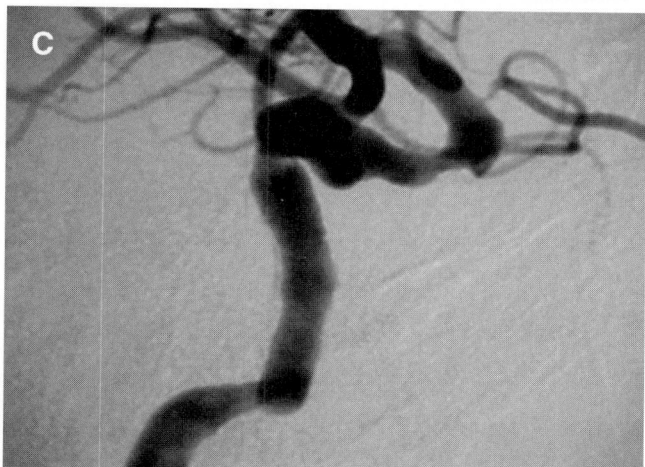

**FIGURE 8** Precavernous carotid stenosis treated with stent. (A) High-grade precavernous internal carotid artery stenosis in a patient with prior stroke from this stenosis. (B) Stent and balloon across high-grade stenosis. (C) Post-stent shows wide patency of precavernous stenosis.

induced intimal dissection. Intracranial stent-assisted angioplasty decreases the incidence of vessel recoil following conventional angioplasty.

Stent-assisted angioplasty is preferred for all focal high-grade and eccentric lesions to minimize the chance of acute vessel closure or delayed thromboembolic complications related to a dissection flap. Predilatation is uncommonly performed, and the balloon and stent diameters are carefully chosen to be the same or 0.1–0.2 mm less than the diameter of the adjacent normal vessel. The stents used are second-generation low-profile balloon-premounted coronary stents (*GFX*, S540 AVE; *GR-2*, Cook; *MultiLink*, Guidant) of 2.5 to 4.0 mm diameter.

Connors and Wojak (1999) reported routine use of ReoPro® with a very low incidence of ischemic stroke since adopting this protocol. ReoPro should be considered on a case-by-case basis for any perceived high-risk factors for platelet aggregation.

### Complications

Treatment of severe stenosis in the intracranial arteries entails a greater risk of complications because of the delicate vascular structure and the lack of media and muscular layers. Of these, the most important complications are stroke due to distal embolization of intraluminal thrombus, dissection, vessel rupture (usually fatal for vessels within the subarachnoid space), and reperfusion injury (Schoser *et al.*, 1997). Many groups therefore limit the treatment of intracranial atherosclerotic lesions to patients who have failed maximal medical therapy and are left with no options short of high-risk surgical bypass or revascularization procedures. As with all emerging indications, a very close clinical and angiographic follow-up is advocated in order to better define appropriate indications and drawbacks of this therapy in the future.

## Intra-arterial Thrombolytic Therapy

### Natural History

Standard medical management of ischemic stroke results in severe neurologic deficit or death in many patients. The 30-day and 5-year mortality rates for stroke in the carotid territory are 17% and 40%, respectively (Chambers *et al.*, 1987). This is higher in the subset with complete occlusion of the proximal middle cerebral artery, and Saito (1987) have reported up to 78% dead or severely disabled with only 9% having a good outcome. Major vertebro-basilar occlusion with acute severe stroke has a particularly poor prognosis, with up to 80% mortality.

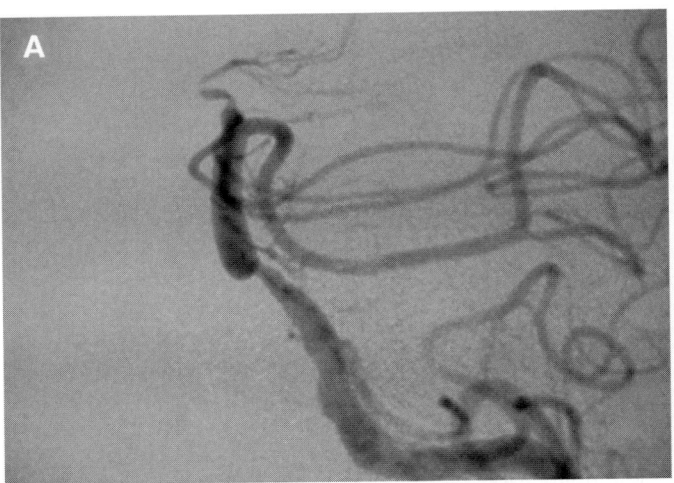
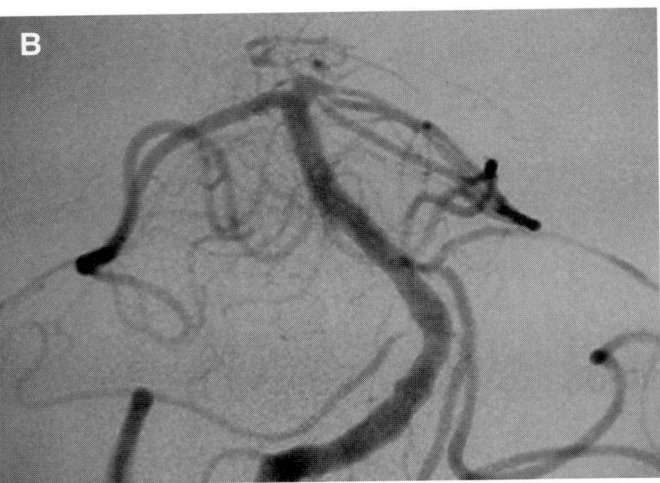

FIGURE 9    Mid-basilar stenosis treated with stent. (A) High-grade mid-basilar artery stenosis, in a patient with posterior fossa TIAs. (B) Following stent deployement, with an AVE S670 stent, the basilar artery is now widely patent.

### Principles of Therapy

The goal of emergency therapy for acute ischemic stroke is preservation of tissue that is not irreversibly damaged (the "ischemic penumbra") to minimize the subsequent clinical deficit. Neuroprotective agents and reperfusion are possible means of achieving these goals. No large randomized prospective trial has shown significant improvement in outcome from the use of neuroprotective agents. Several multicenter randomized trials evaluating systemic (intravenous) administration of thrombolytic agents have been conducted (Donnan *et al.*, 1995a,b; Donnen, 1993; NINDS, 1995; Hacke *et al.*, 1995, 1996), and only the National Institute of Neurological Disorders and Stroke (NINDS) Tissue Plasminogen Activator Stroke Trial demonstrated an acceptably low rate of intracranial hemorrhage and significant efficacy (NINDS, 1995).There was a 30% relative increase in patients with good outcomes in the tPA group compared with the placebo group. This was despite symptomatic intracranial hemorrhage being more common in the tPA group (6.4% vs 0.6%) and comparable overall mortality. tPA received U.S. Food and Drug Administration approval for intravenous administration within 3 h of onset of stroke symptoms in 1996.

Early recanalization correlates with improved outcome in acute stroke patients (Wardlaw and Warlow, 1992), especially in patients with good collateral blood flow and no major early signs of infarction on CT (Figure 10). Recanalization rates with IA thrombolysis are superior to IV thrombolyis for major cerebrovascular occlusions (average 70% recanalization for IA thrombolysis compared to 34% for IV thrombolysis). The differences in recanalization rates are most appar-

ent with large vessel occlusions such as the internal carotid artery (ICA); the supraclinioid carotid, proximal middle cerebral artery and anterior cerebral arteries ("carotid T occlusion"); and the proximal segment of the middle cerebral artery (Furlan *et al.*, 1999).

Published series of intra-arterial thrombolysis in the carotid circulation show good outcomes in similar proportions of patients to those in the NINDS intravenous thrombolysis trial, and comparable to slightly increased rates of hemorrhage. This was achieved despite significantly worse disability at entry and delayed time to treatment in one series (Jahan *et al.*, 1999).

Randomized multicenter trials of local intracranial intra-arterial thrombolysis include the Prolyse for Acute Cerebral Thromboembolism (PROACT) I and II trials (del Zoppo and Higashida *et al.*, 1998; Furlan and Higashida *et al.*, 1999). PROACT I demonstrated the safety of intraarterial recombinant pro-Urokinase (Prolyse). In the PROACT II trial, patients with proximal middle cerebral artery occlusions who presented within 6 h were randomized to receive either intravenous saline or recombinanat pro-Urokinase (Abbott Pharmaceuticals, Chicago, IL) intra-arterially at the site of occlusion for up to 2 h. After confirmation of appropriate occlusion, a microcatheter was placed at or into the intracranial thrombus and pro-Urokinase was infused. This procedure was shown to significantly increase the rate of reperfusion. PROACT II demonstrated a 60% relative neurological improvement for pro-Urokinase therapy in comparison to placebo, no significant increase in morbidity or mortality and a greater rate of recanalization.

Direct thrombolytic therapy has been associated with mortality of 31–68%; however, this has been lower in more recent reports of case series (Becker *et al.*, 1996; Hacke *et al.*, 1995; Higashida *et al.*, 1994a; Mitchell,

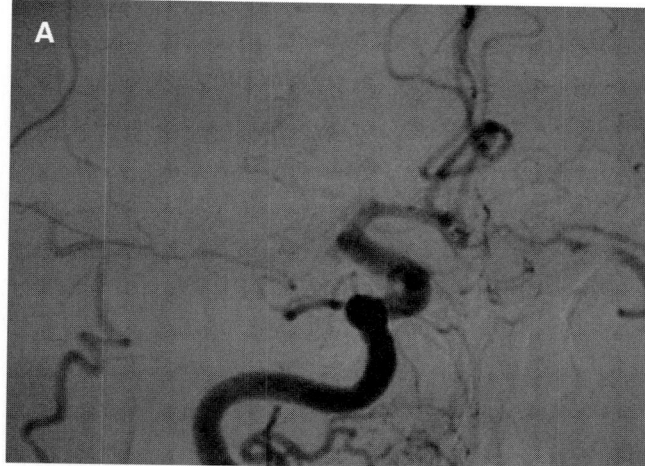

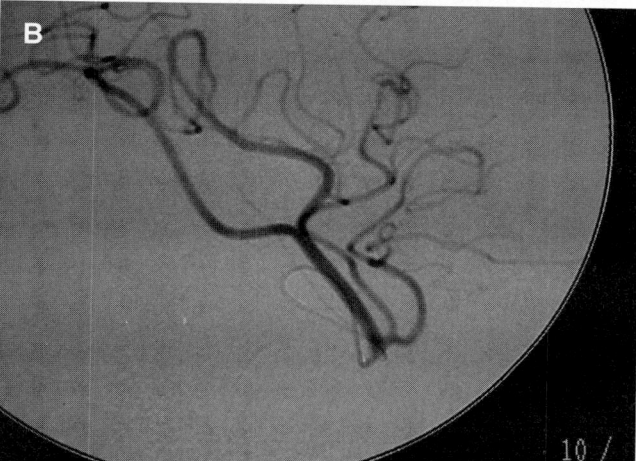

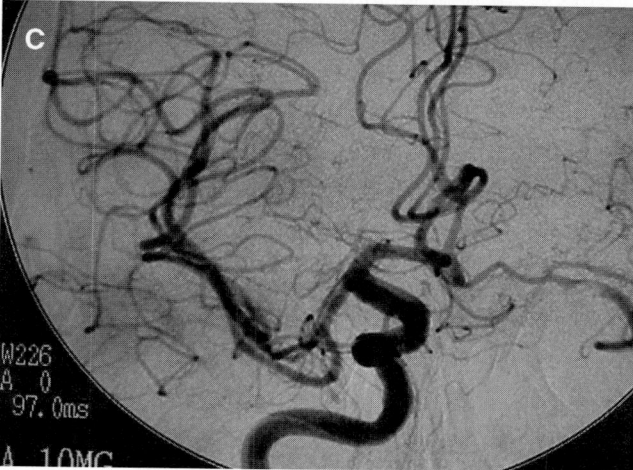

FIGURE 10 Acute occlusion of right middle cerebral artery by an acute embolus. (A) Carotid angiogram showing acute occlusion of the right M1 segment of the middle cerebral artery. (B) Superselective catheter placement during thrombolysis into the right middle cerebral artery. (C) Post-thrombolysis angiogram now shows patency with good blood flow of the right middle cerebral artery.

1997; Zeumer *et al.*, 1993). Of significance, however, is that overall outcome in all these series is markedly better than the natural history of the disease. Case series of intra-arterial thrombolysis have been reported in the vertebrobasilar circulation, but no large prospective randomized trial has been concluded. A wider therapeutic window is recognized in the vertebrobasilar territory, with good outcomes being reported up to 24 h after initial symptom onset (Mitchell, 1997).

### Practical Management

**Intra-arterial Lysis.** Case selection is crucial to achieve good outcomes in ischemic stroke patients when using thrombolysis. A noncontrast CT scan must be obtained before any therapy to exclude other pathologies (tumor) and exclude intracranial hemorrhage or large area of hypodensity. A poor outcome and higher incidence of hemorrhagic complications follows thrombolytic therapy in patients having hypodensity in greater than one-third of the territory of the middle cerebral artery within 6 h of symptom onset (von Kummer *et al.*, 1997). The presence of coma or tetraparesis for several hours also portends a poor prognosis, despite recanalization in patients presenting with vertebrobasilar thrombosis and stroke (Brandt *et al.*, 1996; Hacke *et al.*, 1988).

Such symptoms do not preclude survival, however, and recovery has been documented after successful recanalization in such patients (Brandt *et al.*, 1996; Wijdicks and Scott, 1997). Better collateral blood flow is correlated with improved response to thrombolysis and with longer tolerance of ischemia (Cross *et al.*, 1997). CT is very useful in selecting patients with the potential to do well, but our ability to identify these patients should improve further with perfusion CT techniques, and with the greater availability of perfusion/diffusion capable MRI to differentiate infarcted from ischemic tissue and identify those in whom reperfusion has already occurred.

Currently for intra-arterial thrombolytic treatment many interventional stroke centers are utilizing urokinase (Abbott Laboratories, Abbott Park, IL) in the dose range of 25,000 to 50,000 units over 5- to 10-min intervals at the rate of 250,000–500,000 units per hour. Urokinase is available in 250,000-unit vials and is mixed with 10 mL of sterile water, yielding a concentration of 25,000 units/mL. After each infusion of 125,000–250,000 units, a superselective angiogram is performed and if there is partial clot dissolution, the catheter is advanced into the remaining thrombus where additional thrombolysis is performed. As the thrombus is dissolved, the catheter is then directed into the more distal branches of the intracranial circulation, so that the majority of the drug enters the occluded artery and is not washed preferentially into adjacent open blood vessels. Patients should be treated in a relatively short period of time from symptom onset. Currently, therapy is directed at treating patients within 4–6 h of symptom onset. Treatment should not be denied patients with vertebrobasilar thrombosis even out to 24 h if other factors are favorable.

There is no standard heparin regimen established for IA thrombolysis in acute stroke. PROACT I (del Zoppo and Higashida et al., 1998) reported a 27% rate of symptomatic brain hemorrhage when a conventional heparin regimen (100 U/kg bolus, 1000 U per hour for 4 h) was employed with IA r-proUK. Subsequently a standard low-dose heparin regimen was used (2000 U bolus, 500 U per hour for 4 h) which reduced the symptomatic brain hemorrhage rate with IA r-proUK to 7% in PROACT I and 10% in PROACT II. This dose of heparin does not prolong the activated partial thromboplastin time (aPTT) in most patients. Based on the PROACT I and II trials, many neurointerventionalists now use the low-dose heparin regimen during IA thrombolysis.

The potent glycoprotein IIb/IIIa platelet inhibitor abciximab (ReoPro®, Eli Lilly & Co., Indianapolis, IN) has been used successfully instead of heparin in patients undergoing acute or elective cerebrovascular interventions (Adams et al., 1999). Coronary doses of abciximab appear to be safe in patients with acute ischemic stroke up to 24 h after onset (Wallace et al., 1997). Glycoprotein IIb/IIIa agents may be most efficacious when the risk of acute reocclusion is great such as with basilar artery atherothrombosis. The safety and efficacy of IIb/IIIa agents in patients with embolic occlusion of cerebral vessels, which is the usual cause of middle cerebral artery occlusion, is less clear. It is typically given as an intravenous bolus of 0.25 mg/kg followed by continuous infusion of 10 μg/min for 12 h.

**Intracranial Angioplasty and Stents.** Angioplasty of the cerebral arteries for acute occlusion has been reported to be of benefit in treatment of acute middle cerebral artery occlusion following a failed thrombolysis attempt (Ueda et al., 1998). The underlying mechanism remains unclear. It is possible that angioplasty may treat an underlying stenotic lesion. It is hypothesized that the increased flow resulting from the caliber improvement by the dilatation may increase the endogenous production of endothelial nitric oxide and prostacyclin, both of which have vasodilatory and platelet-inhibitory effects (Berthiaume et al., 1992; Tsao et al., 1995), and increase the release and production of t-PA. A host of endothelial-derived factors and cytokines have been shown to be regulated by prevailing flow and associated shear stress conditions (Malek et al., 1994). The contribution of platelet aggregation to restenosis following successful thrombolysis has been

recently confirmed by successful treatment of this condition using the glycoprotein IIb/IIIa inhibitor abciximab (Wallace et al., 1997). There has also been report of a case of recalcitrant basilar thrombosis despite thrombolysis and angioplasty, which responded to intracranial mid-basilar stent deployment (Phatouros et al., 1999). A similar benefit of angioplasty, when combined with suboptimal thrombolysis was also shown in three cases of intracranial vertebrobasilar occlusion (Nakayama et al., 1998), and in four cases by Yokote (1998).

### Conclusion

Intra-arterial stroke thrombolysis is continuing to evolve. Mechanical clot removal, new catheter techniques, and new adjunctive antithrombotic agents should improve the degree, speed, and safety of IA recanalization. Patient selection and the treatment window will be better defined by using technologies such as MRA with perfusion–diffusion MRI. Comprehensive management utilizing all available tools including IV thrombolysis and cytoprotective agents will be necessary to achieve improved patient outcomes, but only when timely arrival of patients to medical care occurs. Application in the vertebrobasilar circulation is supported by the particularly poor prognosis and improved results shown in recent series. Patients in whom there is carotid territory stroke, and a relative contraindication for systemic thrombolysis, or who fall outside a 3-h time window may benefit from direct IA thrombolysis.

## Intravenous Thrombolytic Therapy for Dural Sinus Thrombosis

### Clinical Aspects

Dural venous sinus occlusion can be asymptomatic, associated with minor symptoms of headache, or with focal neurological deficits and decreased consciousness.

### Natural History

The mortality and morbidity is varied and is difficult to predict. Of 77 patients age 18 to 77 with long-term follow up (Preter et al., 1996) 85% had no neurological sequelae, 14% remained neurologically impaired, recurrent seizures occurred in 14% of patients; and 12% had a second cerebral venous thrombosis (CVT), and noncerebral thrombotic events occurred in 14%. The authors concluded that CVT has an essentially good long-term prognosis, although with the occurrence of a second CVT or another thrombotic episode in 20% of patients, a minority of patients will need long-term anticoagulation.

### Diagnostic Imaging

Tsai (1995) attempted to identify those patients at increased risk of poor outcome by using MRI and direct dural sinus pressure measurement. Acute dural sinus thrombosis leads to distinct stages of parenchymal changes, the severity of which depends on the degree of venous congestion, which, in turn, is closely related to the intradural sinus pressure. They identified five distinct stages of brain parenchymal changes, each stage correlating with increasing intradural sinus pressure. The pressures measured ranged from 20 to 50 mm Hg, and brain parenchymal changes were reversible up to stage III if thrombolytic treatment was performed. All stage V patients died.

Diagnosis can be difficult owing to anatomical variation, poor visualization of normal cerebral veins, and the presence of intraluminal filling defects in normal dural sinuses. Review and comparison of interventional studies (angiography and venography) and cross-sectional imaging (MRI, MRA, CT) is essential to reach a correct diagnosis, and facilitate management plans.

### Principles of Therapy

**Medical Therapy.** The outcome in the majority of cases of dural sinus thrombosis managed with anticoagulation or supportive measures has been excellent, and as a result more aggressive therapy has been reserved for clinical deterioration despite anticoagulation and in patients with raised intracranial pressure and multiple dural sinus occlusions.

**Transvenous Thrombolysis.** Access is usually from femoral venous catheterization, following diagnostic evaluation with full cerebral angiography and study of the venous phase of the arterial injection. In the dural sinuses, either bolus infusions are used, or continuous infusions through the catheter at rates between 20,000 and 150,000 units per hour of Urokinase have been reported. If a continuous infusion is used, it can be continued overnight. Angiography is performed intermittently throughout the procedure to document progression of clot lysis, and to readjust catheter placement within the thrombus to ensure maximum infusion directly into the occluded segments. The catheter will easily pass through recent thrombus, despite "complete" occlusion on imaging and if catheterization is difficult there is usually a history of preexisting chronic organized thrombosis of the dural sinus. Heparin has usually been started before the interventional procedure

and may be continued even in the presence of small amounts of hemorrhage. Thrombolysis is more effective in the presence of heparin, as rethrombosis and extension of thrombus into other veins is prevented. If enlarging venous hemorrhages are found while the patient is on heparin, definitive trial-based data are unavailable to guide decisions, but most investigators would consider stopping heparin, particularly if a mechanical device was available and could be used to clear the sinuses. Excellent results have been reported in some cases. (Smith *et al.*, 1994; Tsai *et al.*, 1992, 1995). Mechanical devices have also been used in dural sinus thrombosis. More often as adjunctive techniques to drug-based thrombolysis, but also as the sole technique when there are contraindications to fibrinolytic drugs. The large volume of thrombus compared to arterial occlusions and the more frequent associated hemorrhage are ideal situations for the use of mechanical clot disruption. The Possis AngioJet mechanical thrombectomy device (Possis Medical, Minneapolis, MN) is effective in removing large amounts of thrombus, and, if continuing improvements allow more reliable and safe access to the more distal portions of the dural sinuses, it may become the preferred mode of treatment.

### Conclusions

A significant proportion of patients, many young and otherwise healthy, with dural sinus or cortical venous thrombosis have major morbidity or die despite therapy with anticoagulation. Patients deemed at higher risk can be identified, especially those with extensive occlusion with compromised remaining venous outflow on angiography; MR indicating major parenchymal changes; elevated intrasinus pressure at venography; and failed anticoagulant therapy. Combinations of chemical and mechanical thrombolysis are now available, and should be used in these higher risk patients. If outcomes can be shown to be improved by this strategy, and devices improve, application to larger numbers of patients under a prospective trial may be appropriate

## REFERENCES

Acheson, J., and Hutchinson, E. C. (1971). The natural history of focal cerebral vascular disease. *Am. J. Med.* **157**, 15–25.

Adams, H. P., Bendixen, B. H., Leira, E., Chang, K. C., Davis, P. H., Woolson, R. F., Clarke, W. R., and Hausen, M. D. (1999). Antithrombotic treatment of ischaemic stroke among patients with occlusion or severe stenosis of the internal carotid artery. A report on the Trial of Org 10172 in Acute Stroke Treatment (TOAST). *Neurology* **13**, 122–125.

Al-Mubarak, N., Roubin, G. S., Liu, M. W., Dean, L. S., Gomez, C. R., Iyers, S. S., and Vitek, J. J. (1999). Early results of percutaneous intervention for severe coexisting carotid and coronary artery disease. *Am. J. Cardiol.* **84**, 600–602, A9.

American Heart Association (1993). "Heart and Stroke Facts Statistics." Dallas, Texas.

Ausman, J. L. (1997). The future of neurovascular surgery. Part I. Intracranial aneurysms. *Surg. Neurol.* **48**, 98–100.

Awad, I. A., Carter, L. P., Spetzler, R. F., Medina, R., and Williams, F. C., Jr. (1987). Clinical vasospasm after subarachnoid haemorrhage: Response to hypervolemic hemodilution and arterial hypertension. *Stroke* **18**, 365–372.

Awad, I. A., Little, J. R., Akarawi, W. P., and Ahl, J. (1990). Intracranial dural AVMs: Factors predisposing to an aggressive neurological course. *J. Neurosurg.* **72**, 839–850.

Barnwell, S. L., Dowd, C. F., Higashida, R. T. *et al.* (1993). Endovascular therapy for cerebral arteriovenous malformations. *In* "Brain Surgery Complication Avoidance and Management" (M. L. J. Apuzzo, Ed.), Vol. 1, pp. 1225–1242. Churchill Livingstone, London.

Barnwell, S. L., Halbach, V. V., Higashida, R. T., Hieshima, G. B., and Wilson, C. B. (1989). Compex dural arteriovenous fistulas. Results of combined endovascular and neurosurgical treatment in 16 patients. *J. Neurosurg.* **71**, 352–358.

Barrow, D. L., Spector, R. H., Braun, I. F., Landman, J. A., Tindall, S. C., and Tindall, G. T. (1985). Classification and treatment of spontaneous carotid sinus fistula. *J. Neurosurg.* **62**, 248–256.

Beaujeux, R. L., Reizine, D. C., Casasco, A., Aymard, A., Rufenacht, D., Khayata, M. H., Riche, M. C., and Merland, J. J. (1992). Endovascular treatment of vertebral arteriovenous fistulas. *Radiology* **183**, 361–367.

Becker, K. J., Monsein, L. H., Ulatowski, J., Mirski, M., Williams, M., and Hanley, D. F. (1996). Intra-arterial thrombolysis in vertebrobasilar occlusion. *Am. J. Neurol. Res.* **131**, 421–433.

Berenstein, A., Choi, J. S., Kupersmith, M. *et al.* (1989). Complications of endovascular embolization in 182 patients with cerebral AVMS. *Am. J. Neuroradiology* **57**, 5.

Berthiaume, F., and Frangos, J. A. (1992). Flow-induced prostacylin production is mediated by a pertussis toxin-sensitive G-protein. *FEBS Lett.* **308**, 2863–2870.

Bladin, C. F., Bingham, L., Grigg, L., Yapanis, A. G., Gerraty, R., and Davis, S. M. (1998). Transcranial Doppler detection of microemboli during percutaneous transluminal coronary angioplasty. *Stroke* **29**, 2367–2370.

Boccardi, A., Branca, V., Valvassori, L., and Scialfa, G. (1998). Endovascular treatment with GDSs: Results in 100 patients. *J. Neurosurg. Sci.* **42**, 127–129.

Bockenheimer, S. A., and Mathias, K. (1983). Percutaneous transluminal angioplasty in arteriosclerotic internal carotid artery stenosis. *Am. J. Nerol. Res.* **4**, 791–792.

Bogousslavsky, J., Barnett, H. J., Fox, A. J., Hachinski, V. C., and Taylor, W. (1986). Atherosclerotic disease of the middle cerebral artery. *Stroke* **17**, 1112–1120.

Brandt, T., Grau, A. J., and Hacke, W. (1996). Severe stroke. *Bailliers. Clin. Neurol.* **5**, 515–541.

Brown, R. D., Jr., Wiebers, D. O., Forbes, G., O'Fallon, W. M., Piepgras, D. G., Marsh, W. R., and Maciunas, R. J. (1988). The natural history of unruptured intracranial arteriovenous malformations. *J. Neurosurg.* **68**, 352–357.

Brown, R. D., Jr., Wiebers, D. O., Torner, J. C., and O'Fallon, W. M. (1996). Incidence and prevalence of intracranial vascular malformations in Olmstead County, Minnesota, 1965 to 1992. *Neurology* **46**, 949–952.

Brown, R. D., Jr., Wiebers, D. O., Torner, J. C., and O'Fallon, W. M. (1997). Frequency of intracranial hemorrhage as presenting and subtype analysis: A population based study of intracranial vascular malformations in Olmsted County, Minnesota. *J. Neurosurg.* **85**, 29–32.

Brückmann, H., Ringelstein, E. B., and Zeumer, H. (1986). Percutaneous transluminal angioplasty of the vertebral artery: A therapeutic alternative to operative reconstruction of proximal vertebral artery stenoses. *J. Neurol.* **233**, 336–339.

Casasco, A. E., Aymard, A., Gobin, Y. P., Houdart, E., Rogopoulos, A., George, B., Hodes, J. E., Cophignou, J., and Merland, J. J. (1993). Selective endovascular treatment of 71 intracranial aneurysms with platinum coils. *J. Neurosurg.* **79**, 3–10.

Casasco, A., and George, B. (1998). Endovascular treatment of saccular intracranial aneurysm. *J. Neurosurg. Sci.* **42**, 125–126.

Chambers, B. R., Norris, J. W., Shurvell, B. L., and Hachinski, V. C. (1987). Prognosis of acute stroke. *Neurology* **37**, 221–225.

Chimowitz, M. I., Kokkinos, J., Strong, J., Brown, M. B., Levine, S. R., Silliman, S., Pessin, M. S. Weichel, E., Sila, C. A., Furlan, A. J., *et al.* (1995). The warfarin-aspirin symptomatic intracranial disease study. *Neurology* **45**, 1488–1493.

Christoforidis, G., and Valavanis, A. (1998). Clinical experience with endovascular treatment of aneurysms using Guglielmi detachable coils. *Crit. Rev. Neurosurg.* **8**, 295–309.

Clark, W. M., Barnwell, S. L., Nesbit, G., O'Neill, O. R., Wynn, M. L., and Coull, B. M. (1995). Safety and efficacy of percutaneous transluminal angioplasty for intracranial atherosclerotic stenosis. *Stroke* **26**, 1200–1204.

Cloft, H. J., Joseph, G. J., and Dion, J. E. (1999). Risk of cerebral angiography in patients with subarachnoid hemorrhage, cerebral aneurysm, and arteriovenous malformation: A meta-analysis. *Stroke* **30**, 317–320.

Clouston, J. E., Numaguchi, Y., Zoarski, G. H., Aldrich, E. F., Simard, J. M., and Zitnay, K. M. (1995). Intraarterial papaverine infusion for cerebral vasospasm after subarachnoid hemorrhage. *Am. J. Neurol. Res.* **16**, 27–38.

Cognard, C., Gobin, Y. P., Pierot, L., Baily, A. L., Houdart, E., Casasco, A., Chiras, J., and Merland, J. J. (1995). Cerebral dural AV fistulas: Clinical and angiographic correlation with a revised classification of venous drainage. *Radiology* **194**, 671–80.

Connors, J. J., 3rd, and Wojak, J. C. (1999). Percutaneous transluminal angioplasty for intracranial atherosclerotic lesions: Evolution of technique and short-term results. *J. Neurosurg.* **91**, 415–423.

Courtheoux, P., Tournade, A., Theron, J., Henriet, J. P., Maiza, D., Derlon, J. M., Pelouze, G., and Evrard, C. (1985). Transcutaneous angioplasty of vertebral artery atheromatous ostial stricture. *Neuroradiology* **27**, 259–264.

Crawley, F., Clifton, A., Buckenham, T., Loosemore, T., Taylor, R. S., and Brown, M. M. (1997). Comparison of hemodynamic cerebral ischemia and microembolic signals detected during carotid endarterectomy and carotid angioplasty. *Stroke* **28**, 2460–2464.

Cross, D. T. 3rd, Moran, C. J., Atkins, P. T., Angtuaco, E. E., and Diringer, M. N. (1997). Relationship between clot location and outcome after basilar artery thrombolysis. *Am. J. Neurol. Res.* **18**, 1221–1228.

David, C. A., Vishteh, A. G., Spetzler, R. F., Lemde, M., Lawton, M. T., and Partovi, S. (1999). Late angiographic follow-up review of surgically treated aneurysms. *J. Neurosurg.* **91**, 396–401.

Debrun, G. M. (1998). The future of aneurysm surgery. *Surg. Neurol.* **49**, 583.

Debrun, G. M., Aletich, V. A., Kehri, P., Misra, M., Ausman, J. I., and Charbel, F. (1998). Selection of cerebral aneurysms for treatment using Guglielmi detachable coils: The preliminary University of Illinois at Chicago experience. *Neurosurgery* **43**, 1281–1295, discussion 1296–1297.

Debrun, G. M., Aletich, V. A., Kehri, P., Misra, M., Ausman, J. I., Charbel, F., and Shownkeen, H. (1998). Aneurysm geometry: An important criterion in selecting patients for Gugliemi detachable coiling. *Neurol. Med. Chir.* **38**, 1–20.

Debrun, G. M., Vinuela, F. V., and Fox, A. J. (1982). Aspirin and systemic heparinization in diagnostic and interventional neuroradiology. *Am. J. Neurol. Res.* **3**, 337–340.

Debrun, G. M., Vinuela, F., Fox, A. J., Davis, K. R., and Ahn, H. S. (1988). Indications for treatment and classification of 132 carotid-cavernous fistulae. *Neurosurgery* 22, 285–289.

Del Zoppo, G. J., Higashida, R. T., Furlan, A. J., Pessin, M. S., Rowley, H. A., and Gent, M. (1998). PROACT: A phase II randomised trial; of recombinant prourokinase by direct arterial delivery in acute middle cerebral artery stroke. PROACT Investigators. Prolyse in Acute Cerebral Thromboembolism. *Stroke* 29, 4–11.

Deruty, R., Pelisson-Guyotat, I., Amat, D., Mottolese, C., Bascouler-gue, Y., Turjman, F., and Gerard, J. P. (1996). Complications after multidisciplinary treatment of cerebral arteriovenous malformations. *Acta Neurochir. (Wien)* 138, 119–131.

Desal, H., Leaute, F., Auffray-Calvier, E., Martin, S., Guillon, B., Robert, R., and De Kersaint-Gilly, A. (1997). Direct carotid-cavernous fistula clinical, radiologic and therapeutic studies. Apropos of 49 cases. *J. Neurorad.* 24, 141–154.

Diethrich, E. B., Ndiaye, M., and Reid, D. B. (1996). Stenting in the carotid artery: Initial experience in 110 patients. *J. Endovasc. Surg.* 3, 42–62.

Donnan, G. A., Davis, S. M., Chambers, B. R., Gates, P. C., Hankey, G. J., McNeil, J. J., Rosen, D., Stewart-Wynne, E. G., and Tuck, R. R. (1995a). Trials of streptokinase in acute ischaemic stroke. *Lancet* 345, 578–579.

Donnan, G. A., Hommel, M., Davis, S. M., and McNeil, J. J. (1995b). Streptokinase in acute stroke. Steering Committees of the ASK and MAST-E trials. Australian Streptokinase Trial. *Lancet* 346, 56.

Donnen, G. A. (1993). Lifesaving for stroke. *Lancet* 342, 383–384.

Dorros, G. (1996). Complications associated with extracranial carotid artery interventions. *J. Endovasc. Surg.* 3, 166–170.

Dorsch, N. W., Young, N., Kingston, R. J., and Compton, J. S. (1995). Early experience with spiral CT in the diagnosis of intracranial aneurysms. *Neurosurgery* 36, 230–236.

Dotter, C. T., and Judkins, M. P. (1964). Transluminal treatment of arteriosclerotic obstruction: Description of a new technique and preliminary report of its application. *Circulation* 30, 654–670.

Drake, C. G., and Vanderkinden, R. G. (1967). The late consequences of incomplete surgical treatment of cerebral aneurysms. *J. Neurosurg.* 27, 226–238.

Economou, T. S., Abemayor, E., and Ward, P. H. (1988). Juvenile nasopharyngeal angiofibroma: An update of the UCLA experience, 1960–1985. *Laryngoscope* 98, 170–125.

Elahi, M. M., Parnes, L. S., Fox, A. J., Pelz, D. M., and Lee, D. H. (1995). Therapeutic embolisation in the treatment of intractable epistaxis. *Arch. Otolaryngol. Head Neck Surg.* 121, 65–69.

Elliott, J. P., Newell, D. W., Lam, D. J., Eskridge, J. M., Douville, C. M., Le Roux, P. D., Lewis, D. H., Mayberg, M. R., Grady, M. S., and Winn, H. R. (1998). Comparison of balloon angioplasty and papaverine infusion for the treatment of vasospasm following aneurysmal subarachnoid hemorrhage. *J. Neurosurg.* 88, 277–284.

Eskridge, J. M., McAuliffe, W., Song, J. K., Deliganis, A. V., Newell, D. W., Lewis, D. H., Mayberg, M. R., and Winn, H. R. (1998). Balloon angioplasty for the treatment of vasospasm: results of first 50 cases. *Neurosurgery* 42, 510–516, discussion 516–517.

Eskridge, J. M., and Song, J. K. (1998). Endovascular embolisation of 150 basilar tip aneurysms with Guglielmi detachable coils: Results of the Food and Drug Administration multicentre clinical trial. *J. Neurosurg.* 89, 81–86.

Executive Committee for the Asymptomatic Carotid Atherosclerosis Study (1995). Endarterectomy for asymptomatic carotid artery stenosis. *JAMA* 273, 1421–1428.

Feldman, R. L., Trigg, L., Gaudier, J., and Galat, J. (1996). Use of coronary Palmaz-Schatz stent in the percutaneous treatment of an intracranial carotid artery stenosis. *Cathet. Cardiovasc. Diagn.* 84, 883–887.

Formanek, G., Frech, R., and Amplatz, K. (1970). Arterial thrombus formation during clinical percutaneous catheterization. *Circulation* 41, 833–839.

Forster, D. M. C., Steiner, L., and Hakanson, S. (1972). Arteriovenous malformations of the brain. A long-term clinical study. *J. Neurosurg.* 37, 562–570.

Fraser, K. W., Halbach, V. V., Teitelbaum, G. P., Smith, T. P., Higashida, R. T., Dowd, C. F., Wilson, C. B., and Hieshima, G. B. (1994). Endovascular platinum coil embolisation of incompletely surgically clipped aneurysms. *Surg. Neurol.* 41, 4–8.

Fults, D., and Kelly, D. L. (1984). Natural history of arteriovenous malformations of the brain: A clinical study. *Neurosurgery* 15, 658–662.

Furlan, A., Higashida, R., Wechsler, L., Gent, M., Rowley, H., Kase, C., Pessin, M., Ahuja, A., Callahan, F., Clark, W. M., Silver, F., and Rivera, F. (1999). Intra-arterial prourokinase for acute ischaemic stroke. The PROACT II study: A randomized cotrolled trial. Prolyse in Acute Cerebral Thromboembolism. *JAMA* 282, 2003–2011.

Gaunt, M. E., Brown, L., Hartshorne, T., Bell, P. R., and Naylor, A. R. (1996). Unstable carotid plaques: Preoperative identification and association with intraoperative embolisation detected by transcranial Doppler. *Eur. J. Vasc. Endovasc. Surg.* 11, 78–82.

Gioulekas, J., Mitchell, P., Tress, B., and McNab, A. A. (1997). Embolisation of carotid cavernous fistulas via the superior ophthalmic vein. *Aust. N. Z. J. Ophthalmol.* 25, 47–53.

Graf, C. J., Perret, G. E., and Torner, J. C. (1983). Bleeding from cerebral arteriovenous malformations as part of their natural history. *J. Neurosurg.* 58, 331–337.

Graves, V. B., Strother, C. M., Duff, T. A., and Perl, J. 2nd (1995). Early treatment of ruptured aneurysms with Guglielmi detachable coils: Effect on subsequent bleeding. *Neurosurgery* 37, 640–647, discussion 647–648.

Graves, V. B., Strother, C. M., Partington, C. R., and Rappe, A. (1992). Flow dynamics of lateral carotid artery aneurysm and their effects on coils and balloons: And experimental study in dogs. *Am. J. Neurol. Res.* 13, 189–196.

Guglielmi, G., Vinuela, F., Dion, J., and Duckwiler, G. (1991a). Electrothrombosis of saccular aneurysms via endovascular approach. 2. Preliminary clinical experience. *J. Neurosurg.* 75, 8–14.

Guglielmi, G., Vinuela, F., Sepetka, I., and Macellari, V. (1991b). Electrothrombosis of saccular aneurysms via endovascular approach. 1. Electrochemical basis, technique, and experimental results. *J. Neurosurg.* 75, 1–7.

Gurian, J. H., Martin, N. A., King, W. A., Duckwiler, G. R., Guglielmi, G., and Vinuela, F. (1995). Neurosurgical management of cerebral aneurysms following unsuccessful or incomplete endovascular embolisation. *J. Neurosurg.* 83, 843–853.

Hacein-Bey, L., Connolly, E. S. Jr, Mayer, S. A., Young, W. L., Pile-Spellman, J., and Solomon, R. A. (1998). Complex intracranial aneurysms: Combined operative and endovascular approaches. *Neurosurgery* 43, 1304–1312.

Hacke, W., Kaste, M., Fieschi, C., Toni, D., Lesaffre, E., von Kummer, R., Boysen, G., Bluhmki, E., Hoxter, G., Mahagne, M. H., *et al.* (1995). Intravenous thrombolysis with recombinant tissue plasminogen activator for acute hemispheric stroke. The European Cooperative Acute Stroke Study. *JAMA* 274, 1017–1025.

Halbach, V. V., Hieshima, G. B., Higashida, R. T., and Reicher, M. (1987a). Carotid cavernous fistulae: Indications for urgent treatment. *Amer. J. Roentgenology* 149, 587–593.

Halbach, V. V., Higashida, R. T., and Hieshima, G. B. (1992). Endovascular therapy of head and neck tumors. In "Endovascular therapy in the Central Nervous System" (F. Vinuela, Ed.), pp. 17–28. Raven Press, New York.

Halbach, V. V., Higashida, R. T., and Heishina, G. B. (1992). Endovasclular therapy of dural fistulas. In "Endovascular Therapy of the Central Nervous System" (F. Vinuela, Ed.), pp. 29–50. Raven Press, New York.

Halbach, V. V., Higashida, R. T., Dowd, C. F., Barnwell, S. L., Fraser,

K. W., Smith, T. P., Teitelbaum, G. P., and Hieshima, G. B. (1994). The efficacy of endosaccular aneurysm occlusion in alleviating neurological deficits produced by mass effect. *J. Neurosurg.* **80**, 659–666.

Halbach, V. V., Higashida, R. T., Dowd, C. F., Barnwell, S. L., and Hieshima, G. B. (1991). Management of vascular perforations that occur during neurointerventional procedures. *Am. J. Neurol. Res.* **12**, 319–327.

Halbach, V. V., Higashida, R. T., and Hieshima, G. B. (1988a). Treatment of vertebral arteriovenous fistulas. *Amer. J. Roentgenology* **150**, 405–412.

Halbach, V. V., Higashida, R. T., Hieshima, G. B., Hardin, C. W., Dowd, C. F., and Barnwell, S. F. (1989). Tansarterial occlusion of solitary intracerebral arteriovenous fistulas. *Am. J. Neurol. Res.* **10**, 747–752.

Halbach, V. V., Higashida, R. T., Hieshima, G. B., Hardin, C. W., and Yang, P. J. (1988b). Transvenous embolisation of direct carotid cavernous fistulas. *Am. J. Neurol. Res.* **9**, 741–747.

Halbach, V. V., Higashida, R. T., Hieshima, G. B., and Norman, D. (1987b). Normal perfusion pressure breakthrough occurring during treatment of carotid and vertebral fistulas. *Am. J. Neurol. Res.* **8**, 751–756.

Halbach, V. V., Higashida, R. T., Yang, P., Barnwell, S. L., Wilson, C. B., and Hieshima, G. B. (1988c). Preoperative balloon occlusion of arteriovenous malformations. *Neurosurgery* **22**, 301–307.

Halbach, V. V., Higashida, R. T., Yeng, P. J. *et al.* (1988). Transvenous embolization of direct carotid cavernous sinus fistulas. *Am. J. Neuroradiology* **9**, 741–747.

Hass, W. K., Fields, W. S., North, R. R., Kircheff, I. I., Chase, N. E., and Bauer, R. B. (1968). Joint study of extracranial arterial occlusion. II. Arteriography, techniques, sites, and complications. *JAMA* **203**, 961–968.

Helgason, C. M. (1997). Mechanisms of antiplatelet agents and the prevention of stroke. *In* "Primer on Cerebrovascular Diseases" (K. M. A. Welch, L. R. Caplan, D. J. Reis, B. K. Siesjö, and B. Weir, Eds), pp. 712–716. Academic Press, San Diego.

Henry, M., Amor, M., Masson, I., *et al* (1998). Angioplasty and stenting of the extracranial carotid arteries. *J. Endovasc. Surg.* **5**, 293–304.

Higashida, R. T. (1997a). Retrievable versus non-retrievable coils. *Am. J. Neurol. Res.* **18**, 1390.

Higashida, R. T., Halbach, V. V., Barnwell, S. L., Dowd, C. F., and Hieshima, G. B. (1994a). Thrombolytic therapy in acute stroke. *J Endovasc Surg* **1**, 4–15.

Higashida, R. T., Halbach, V. V., Cahan, L. D., Brant-Zawadzkim, M., Barnwell, S., Dowd, C., and Hieshima, G. B. (1989a). Transluminal angioplasty for treatment of intracranial arterial vasospasm. *J. Neurosurg.* **71**, 648–653.

Higashida, R. T., Halbach, V. V., Cahan, L. D., Hieshima, G. B., and Konishi, Y. (1989b). Detachable balloon embolization therapy of posterior circulation intracranial aneurysms. *J. Neurosurg.* **71**, 512–519.

Higashida, R. T., Halbach, V. V., Dowd, C. F., Dormandy, B., Bell, J., and Hieshima, G. B. (1992). Intravascular balloon dilatation therapy for intracranial arterial vasospasm: Patient selection, technique, and clinical results. *Neurosurg. Rev.* **15**, 89–95.

Higashida, R. T., Halbach, V. V., Tsai, F. Y., Dowd, C. F., and Hieshima, G. B. (1994b). Interventional neurovascular techniques for cerebral revascularization in the treatment of stroke. *Amer. J. Roentgenology* **163**, 793–800.

Higashida, R. T., Halbach, V. V., Tsai, F. Y., Norman, D., Pribram, H. F., Mehringer, C. M., and Hieshima, G. B. (1989c). Interventional treatment of traumatic carotid and vertebral artery lesions: Results in 234 cases. *Amer. J. Roentgenology* **153**, 577–582.

Higashida, R. T., Hieshima, G. B., Tsai, F. Y., Bentson, J. R., and Halbach, V. V. (1986). Percutaneous transluminal angioplasty of the subclavian and vertebral arteries. *Acta Radiol. Suppl.* **369**, 124–126.

Higashida, R. T., Hieshima, G. B., and Halbach, V. V. (1991). Advances in the treatment of complex cerebrovascular disorders by interventional neurovascular techniques. *Circulation* **83**, I-196–I-206.

Higashida, R. T., Smith, W., Gress, D., Urwin, R., Dowd, C. F., Balousek, P. A., and Halbach, V. V. (1997b). Intravascular stent and endovascular coil placement for a ruptured fusiform aneurysm of the basilar artery. Case report and review of the literature. *J. Neurosurg.* **87**, 944–949.

Higashida, R. T., Tsai, F. Y., Halbach, V. V., Dowd, C. F., and Hieshima, G. B. (1993a). Cerebral percutaneous transluminal angioplasty. *Heart Dis. Stroke* **2**, 497–502.

Higashida, R. T., Tsai, F. Y., Halbach, V. V., Dowd, C. F., and Hieshima, G. B. (1996). Transluminal angioplasty, thrombolysis, and stenting for extracranial and intracranial cerebral vascular disease. *J. Intervent. Cardiol.* **9**, 245–255.

Higashida, R. T., Tsai, F. Y., Halbach, V. V., Dowd, C. F., Smith, T., Fraser, K., and Hieshima, G. B. (1993b). Transluminal angioplasty for atherosclerotic disease of the vertebral and basilar arteries. *J. Neurosurg.* **78**, 192–198.

Hodes, J. E., Aymard, A., Gobin, Y. P., Rufenacht, D., Bien, S., Reizine, D., Gaston, A., and Merland, J. J. (1991). Endovascular occlusion of intracranial vessels for curative treatment of unclippable aneurysms: report of 16 cases. *J. Neurosurg.* **75**, 694–701.

Hodgins, G. W., and Dutton, J. W. (1982). Subclavian and carotid angioplasties for Takayasu's arteritis. *J. Can. Assoc. Radiol.* **33**, 205–207.

Hope, J. K. A., Wilson, J. L., and Thomson, F. J. (1996). Three dimensional CT angiography in the detection and characterization of berry aneurysms. *Am. J. Neurol. Res.* **17**, 439–445.

Hopkins, L. N., and Budny, J. L. (1989). Complications of intracranial bypass for vertebrobasilar insufficiency. *J. Neurosurg.* **70**, 207–211.

Houdart, E., Ricolfi, F., Brugières, P., Antoine, J. C., and Gaston, A. (1996). Percutaneous transluminal angioplasty of atherosclerotic basilar artery stenosis. *Neuroradiology* **38**, 383–385.

Hurst, R. W. (1996). Interventional neuroradiology of the head and neck. *Neuroimaging Clin. N. Am.* **6**, 473–495.

Huston, J., Rufenacht, D. A., Ehman, R. L., and Wiebers, D. O. (1991). Intracranial aneurysms and vascular malformations: Comparison of time-of-flight and phase-contrast MR angiography. *Radiology* **181**, 721–730.

Imparato, A. M., Riles, T. S., and Kim, G. E. (1981). Cervical vertebral angioplasty for brain stem ischemia. *Surgery* **90**, 842–852.

Investigators TiSoUIA (1998). Unruptured intracranial aneurysms—Risks of rupture and risks of surgical intervention. *N. Engl. J. Med.* **339**, 1725–1733.

Jahan, R., Duckwiler, G. R., Kidwell, C. S., Sayre, J. W., Gobin, Y. P., Villablanca, J. P., Sawer, J., Starkman, S., Martin, N., and Vinuela, F. (1999). Intra-arterial thrombolysis for treatment of acute stroke: Experience in 26 patients with long-term follow-up. *Am. J. Neurol. Res.* **20**, 1291–1299.

Jessurun, G. A. J., Kamphuis, D. J., van der Zande, F. H. R., and Nossent, J. C. (1993). Cerebral arteriovenous malformations in the Netherland Antilles. *Clin. Neurol. Neurosurg.* **95**, 193–198.

Johnston, S. C., Gress, D. R., and Kahn, J. G. (1999). Which unruptured cerebral aneurysms should be treated? A cost-utility analysis. *Neurology* **52**, 1806–1815.

Jordon, W. D., Voellinger, D. C., Fisher, W. S. *et al.* (1998). A comparison of carotid angioplasty with stenting versus endarterectomy with regional anesthesia. *J. Vasc. Surg.* **28**(3), 397–402.

Kachel, R. (1996). Results of balloon angioplasty in the carotid arteries. *J. Endovasc. Surg.* **3**, 22–30.

Kachel, R., Endert, G., Basche, S., Grossmann, K., and Glaser. F. H. (1987). Percutaneous transluminal angioplasty (dilatation) of carotid, vertebral, and innominate artery stenoses. *Cardiovasc. Intervent. Radiol.* **10**, 142–146.

Kappelle, L. J., Eliasziw, M., Fox, A. J., Sharpe, B. L., and Barnett, H. J. (1999). Importance of intracranial atherosclerotic disease in patients with symptomatic stenosis of the internal carotid artery. The North American Symptomatic Carotid Endarterectomy Trail. *Stroke* **30**, 282–286.

Kassell, N. F., Torner, J. C., Haley, E. C., Jr., Jane, J. A., Adams, H. P., and Kongable, G. L. (1990a). The International Cooperative Study on the Timing of Aneurysm Surgery. I. Overall management results. *J. Neurosurg.* **73**, 18–36.

Kassell, N. F., Torner, J. C., Jane, J. A., Haley, E. C. Jr, and Adams, H. P. (1990b). The International Cooperative Study on the Timing of Aneurysm Surgery. II. Surgical results. *J. Neurosurg.* **73**, 37–47.

King, J. T., Jr., Berlin, J. A., and Flamm, E. S. (1994). Morbidity and mortality from elective surgery for asymptomatic, unruptured, intracranial aneurysms: A meta-analysis. *J. Neurosurg.* **81**, 837–842.

Kucharczyk, W., Lemme-Pleghos, L., Uske, A., Brant-Zawadzki, M., Dooms, G., and Norman, D. (1985). Intracranial vascular malformations: MR and CT imaging. *Radiology* **56**, 383–389.

Kuether, T. A., Nesbit, G. M., and Barnwell, S. M. (1998). Clinical and angiographic outcome, with treatment data, for patients with cerebral aneurysms treated with Guglielmi detachable coils: A single-centre experience. *Neurosurgery* **43**, 1016–1025.

Kuman, A. J., Fox, A., Vinuela, F., and Rosenbaum, A. E. (1984). Revisited old and new findings in unruptured larger arteriovenous malformations of the brain. *J. Comput. Assist. Tomogr.* **8**, 648–655.

Kwiatkowski, T. G., and Libman, R. B. (1987). Emergency strategies. *In* "Primer on Cerebrovascular Diseases" (K. M. A. Welch, L. R. Caplan, D. J. Reis, B. K. Siesjö, and B. Weir, Eds.), pp. 671–675. Academic Press, San Diego.

Lasjaunias, P., Chiu, M., TerBrugge, K., Tolia, A., Hurth, M., and Bernstein, M. (1986). Neurological manifestations of intracranial dural AVMs. *J. Neurosurg.* **64**, 724–730.

Le Roux, P. D., and Winn, H. R. (1998). Management of the ruptured aneurysm. *Neurosurg. Clin. N. Am.* **9**, 525–540.

Lefkowitz, M. A., Gobin, Y. P., Akiba, Y., Duckwiler, G. R., Murayama, Y., Guglielmi, G., Martin, N. A., and Vinuela, F. (1999). Balloon-assisted GDC embolisation of wide-necked aneurysms. II. Clinical results. *Neurosurgery* **45**, 531–537, discussion 537–538.

Levi, C. R., O'Malley, H. M., Fell, G., Roberts, A. K., Hoare, M. C., Royle, J. P., Chan, A., Beiles, B. C., Chambers, B. R., Bladin, C. F., and Donnan, G. A. (1997). Transcranial Doppler detected cerebral microembolism following carotid endarterectomy. High microembolic signal loads predict postoperative cerebral ischaemia. *Brain* **120**, 621–629.

Lieber, B. B., Stancampiano, A. P., and Wakhloo, A. K. (1997). Attenuation of hemodynamics in aneurysm models by stenting influence of stent porosity. *Ann. Biomed. Eng.* **25**, 460–469.

Lin, T., Fox, A. J., and Drake, C. G. (1989). Regrowth of aneurysm sacs from residual neck following aneurysm clipping. *J. Neurosurg.* **70**, 556–560.

Lylyk, P.,Ceratto, R., Hurvitz, D., and Basso, A. (1998). Treatment of a vertebral dissecting aneurysm with stents and coils: Technique case report. *Neurosurgery* **43**, 385–388.

Maeder, P. P., Meuli, R. A., and de Tribolet, N. (1996). Three-dimensional volume rendering for magnetic resonance angiography in the screening and preoperative workup of intracranial aneurysms. *J. Neurosurg.* **85**, 1050–1055.

Malek, A. M., and Izumo, S. (1994). Molecular aspects of signal transduction of shear stress in the endothelial cell. (Editorial). *J. Hypertension* **12**, 989–999.

Malek, A. K., Rowinski, O., Ostrowski, T., Hilgertner, L., Januszewicz, M., and Szostek, M. (1995). Transcutaneous balloon angioplasty in the treatment of subclavian steal syndrome. Characteristics of vertebral basal flow with transcranial Doppler technique. *Pol. Tyg. Lek.* **50**, 19–22.

Malek, A. M., Higashida, R. T., Halbach, V. V., Phatouros, C. C., Meyers, P. M., and Dowd, C. F. (1999a). Tandem intracranial stent deployment for treatment of an iatrogenic, flow-limiting, basilar artery dissection: technical case report. *Neurosurgery* **45**, 919–924.

Malek, A. M., Higashida, R. T., Phatouros, C. C., Lempert, T. E., Meyers, P. M., Gress, D. R., Dowd, C. F., and Halbach, V. V. (1999b). Treatment of posterior circulation ischemia with extracranial percutaneous balloon angioplasty and stent placement. *Stroke* **30**, 2073–2085.

Malik, G. M., Pearce, J. E., Ausman, J. I., and Mehta, B. (1984). Dural arteriovenous malformations and intracranial haemorrhage. *Neurosurgery* **15**, 332–339.

Malisch, T. W., Guglielmi, G., Vinuela, F., Duckwiler, G., Gobin, Y. P., Martin, N. A., and Frazee, J. G. (1997). Intracranial aneurysms treated with the Guglielmi detachable coil: Midterm clinical results in a consecutive series of 100 patients. *J. Neurosurg.* **87**, 176–183.

Marzewski, D. J., Furlan, A. J., St. Louis, P., Little, J. R., Modic, M. T., and Williams, G. (1982). Intracranial internal carotid artery stenosis: longterm prognosis. *Stroke* **13**, 821–824.

Mast, H., Mohr, J. P., Osipov, A., Pile-Spellman, J., Marshall, R. S., Lazar, R. M., Stein, B. M., and Young, W. L. (1995). "Steal" is an unestablished mechanism for the clinical presentation of cerebral arteriovenous malformations. *Stroke* **26**, 1215–1220.

Mast, H., Young, W. L., Koennecke, H-C., Sciacca, R. R., Osipov, A., Pile-Spellman, J., Hacein-Bey, L., Duong, H., Stein, B. M., and Mohr, J. P. (1997). Risk of spontaneous haemorrhage after diagnosis of cerebral arteriovenous malformation. *Lancet* **350**, 1065–1068.

Mathias, K., Jäger, H., Sahl, H., Hennigs, S., and Gissler, H. M. (1999). Interventional treatment of arteriosclerotic carotid stenosis. *Radiology* **39**, 125–134.

Mathur, A., Dorros, G., Iyer, S. S., Vitek, J. J., Yadav, S. S., and Roubin, G. S. (1997). Palmaz stent compression in patients following carotid artery stenting. *Cathet. Cardiovasc. Diagn.* **41**, 137–140.

Mathur, A., Roubin, G. S., Iyer, S. S., Piasonboon, C., Liu, M. W., Gomez, C. R., Yadav, J. S., Chastain, H. D., Fox, L. M., Dean, L. S., and Vitek, J. J. (1998). Predictors of stroke complicating carotid artery stenting. *Circulation* **97**, 1239–1245.

Mawad, M. E., and Klucznik, R. P. (1995). Giant serpentine aneurysms: radiographic features and endovascular treatment. *Am. J. Neurol. Res.* **16**, 1053–1060.

Mayberg, M. R., Batjer, H. H., Dacey, R. *et al.* (1994). Guidelines for the management of aneurysmal subarachnoid hemorrhage. *Stroke* **25**, 2315–2328.

McCormick, W. F., and Schochet, S. S., Jr. (1976). "Atlas of Cerebrovascular Disease." Saunders, Philadelphia.

McDougall, C. G., Halbach, V. V., Dowd, C. F., Higashida, R. T., Larsen, D. W., and Hieshima, G. B. (1996). Endovascular treatment of basilar tip aneurysms using electrolytically detachable coils. *J Neurosurg.* **84**, 393–399.

Menghini, V. V., Brown, R. D. Jr., Sicks, J. D., O'Fallon, W. M., and Wiebers, D. O. (1998). Incidence and prevalence of intracranial aneurysms and haemorrhage in Olmsted County, Minnesota, 1965 to 1995. *Neurology* **51**, 405–411.

Mericle, R. A., Lanzino, G., Wakhloo, A. K., Guterman, L. R., and Hopkins, L. N. (1998). Stenting and secondary coiling of intracranial internal carotid artery aneurysm: Technical case report. *Neurosurgery* **43**, 1229–1234.

Merland, J. J., Melki, J. P., Chiras, J. *et al.* (1980). Place of embolization in the treatment of severe epistaxis. *Laryngoscope* **90**, 1694.

Michelson, W. J. (1978). Natural history and pathophysiology of arteriovenous malformations. *Clin. Neurosurg.* **26**, 307–313.

Mitchell, P. J. (1997). Interventional neuroradiology techniques in neurovascular disease. *Br. J. Hosp. Med.* **58**, 8–14.

Mitchell, P. J., and Tress, B. M. (1999). Management of cerebral aneurysms: Current best practice. *Med. J. Aust.* **171**, 121–122.

Moret, J., Cognard, C., Weill, A., Castaings, L., and Rey, A. (1997). Reconstruction technique in the treatment of wide-neck intracranial aneurysms. Long-term angiographic and clinical results. Apropos of 56 cases. *J. Neuroradiol.* **24**, 30–44.

Morgan, M. K., Day, M. J., Little, N., Grinnell, V., and Sorby, W. (1995). The use of intraarterial papaverine in the management of vasospasm complicating arteriovenous malformation resection. Report of two cases. *J. Neurosurg.* **82**, 296–299.

Mori, T., Mori, K., Fukuoka, M., Arisawa, M., and Honda, S. (1997). Percutaneous transluminal cerebral angioplasty: Serial angiographic follow-up after successful dilatation. *Neuroradiology* **39**, 111–116.

Motarjeme, A. (1996). Percutaneous transluminal angioplasty of supra-aortic vessels. *J. Endovasc. Surg.* **3**, 171–181.

Motarjeme, A., Keifer, J. W., and Zuska, A. J. (1981). Percutaneous transluminal angioplasty of the vertebral arteries. *Radiology* **139**, 715–717.

Motarjeme, A., Keifer, J. W., and Zuska, A. J. (1982). Percutaneous transluminal angioplasty of the brachiocephalic arteries. *Amer. J. Roentgenology* **138**, 457–462.

Moufarrij, N. A., Little, J. R., Furlan, A. J., Leatherman, J. R., and Williams, G. W. (1986). Basilar and distal vertebral artery stenosis: Long-term follow-up. *Stroke* **17**, 938–942.

Moufarrij, N. A., Little, J. R., Furlan, A. J., Williams, G., and Marzewski, D. J. (1984). Vertebral artery stenosis: Long-term follow-up. *Stroke* **15**, 260–263.

Muller, M., Behnke, S., Walter, P., Omlor, G., and Schimrigk, K. (1998). Microembolic signals and intraoperative stroke in carotid endarterectomy. *Acta Neurol. Scand.* **97**, 110–117.

Nahser, H. C., Henkes, H., Weber, W., Berg-Dammer, E., Yousry, T. A., and Kuhne, D. (2000). Intracranial vertebrobasilar stenosis: Angioplasty and follow-up. *Am. J. Neurol. Res.* **21**, 1293–1301.

Nakayama, T., Tanaka, K., Kaneko, M., Yokoyama, T., and Uemura, K. (1998). Thrombolysis and angioplasty for acute occlusion of intracranial vertebrobasilar arteries. Report of three cases. *J. Neurosurg.* **88**, 919–922.

Nakstad, P. H., and Nornes, H. (1994). Superselective angiography, embolisation and surgery in treatment of AVMs of the brain. *Neuroradiology* **36**, 410–413.

The National Institute of Neurological Disorders and Stroke rt-PA Stroke Study Group. (1995). Tissue plasminogen for acute ischemic stroke. *N. Eng. J. Med.* **333**, 1581–1587.

Nelson, P. K., Setton, A., Choi, I. S., Ransohoff, J., and Berenstein, A. (1994). Current status of interventional neuroradiology in the management of meningiomas. *Neurosurg. Clin. N. Am.* **5**, 235–259.

North American Symptomatic Carotid Endarterectomy Trial Collaborators (1991). Beneficial effect of carotid endarterectomy in symptomatic patients with high-grade carotid stenosis. *N. Engl. J. Med.* **325**, 445–453.

Olivecrona, H., and Riives, J. (1948). Arteriovenous aneurysms of the brain: their diagnosis and treatment. *Arch. Neurol. Psychiatry* **59**, 567–603.

Ondra, S. L., Troupp, H., George, U. E. D., and Schwab, K. (1990). The natural history of symptomatic arteriovenous malformations of the brain: A 24 year follow-up assessment. *J. Neurosurg.* **73**, 387–391.

Origitano, T. C., Wascher, T. M., Reichman, O. H., and Anderson, D. E. (1990). Sustained increased cerebral blood flow with prophylactic hypertensive hypervolemic hemodilution ("triple-H" therapy) after subarachnoid haemorrhage. *Neurosurgery* **27**, 729–739, discussion 739–740.

Paiva, T., Campos, J., Baeta, E., Gomes, L. B., Martins, I. P., and Parreira, E. (1995). EEG monitoring during endovascular embolisation of cerebral arteriovenous malformations. *Electroencephalogr. Clin. Neurophysiol.* **95**, 3–13.

Perez-Cruet, M. J., Patwardhan, R. V., Mawad, M. E., and Rose, J. E. (1997). Treatment of dissecting pseudo-aneurysms of the cervical internal carotid artery using a wall stent and detachable coils: Case report. *Neurosurgery* **40**, 622–625.

Peters, K. R., Quisling, R. G., Gilmore, R., Mickle, P., and Kuperus, J. H. (1993). Intraarterial use of sodium methohexital for provocative testing during brain embolotherapy. *Am. J. Neurol. Res.* **14**, 171–174.

Phatouros, C. C., Higashida, R. T., Malek, A. M., Smith, W. S., Mully, T. W., De Armond, S. J., Dowd, C. F., and Halbach, V. V. (1999). Endovascular stenting of an acutely thrombosed basilar artery: Technical case report and review of the literature. *Neurosurgery* **44**, 667–673.

Piepgras, D. G., Sundt, T. M., Marsh, W. R. *et al.* (1986). Recurrent carotid stenosis. Results and complications of 57 operations. *Ann. Surgery* **203**(2), 205–213.

Preter, M., Tzourio, C., Ameri, A., Bousser, M. G. (1996). Long-term prognosis in cerebral venous thrombosis. Follow-up of 77 patients. *Stroke* **27**, 243–246.

Purdy, P. D., Batjer, H. H., and Samson, D. (1991). Management of hemorrhagic complications from preoperative embolization of arteriovenous malformations. *J. Neurosurg.* **74**, 205–211.

Raaymakers, T. W., Rinkel, G. J., Limburg, M., and Algra, A. (1998). Mortality and morbidity of surgery for unruptured intracranial aneurysms: A meta-analysis. *Stroke* **29**, 1531–1538.

Robbin, M. L., Lockhart, M. E., Weber, T. M., Vitek, J. J., Smith, J. K., Yadav, J., Mathur, A., Iyer, S. S., and Roubin, G. S. (1997). Carotid artery stents: Early and intermediate follow-up with Doppler ultrasound. *Radiology* **205**, 749–756.

Roberson, G. H., and Reardon, E. J. (1979). Angiography and embolisation of the internal maxillary artery for posterior epistaxis. *Arch. Otolaryngol.* **105**, 333–337.

Ronkainen, A., Herniesmi, J., Puranen, M., Niemitukia, L., Vanninen, R., Ryynanen, M., Kuivaniemi, H., and Tromp, G. (1997). Familial intracranial aneurysms. *Lancet* **349**, 380–384.

Ronkainen, A., Puranen, M. I., Hernesniemi, J. A., Vanninen, R. L., Partanen, P. L., Saari, J. T., Vainio, P. A., and Ryynanen, H. (1995). Intracranial aneurysms: MR angiographic screening in 400 asymptomatic individuals with increased familial risk. *Radiology* **195**, 35–40.

Rosenwasser, R. H., Thomas, J. E., Gannon, P. M., Bell, R., and Jamieson, D. (1998). "Current Strategies for the Management of Cerebral Arteriovenous Malformations." American Association of Neurological Surgeons.

Roubin, G. S., Yadav, S., Iyer, S. S., and Vitek, J. (1996). Carotid stent supported angioplasty: A neurovascular intervetnion to prevent stroke. *Am. J. Cardiol.* **78**, 8–12.

Saito, I., Segawa, H., Shiokawa, Y., Taniguchi, M., and Tutsumi, K. (1987). Middle cerebral artery occlusion: Correlation of computed tomography and angiography with clinical outcome. *Stroke* **18**, 863–868.

Schievink, W. I. (1997). Intracranial aneurysms. *N. Eng. J. Med.* **336**, 28–40.

Schoser, B. G., Heesen, C., Eckert, B., and Thie, A. (1997). Cerebral hyperperfusion inury after percutaneous transluminal angioplasty of extracranial arteries. *J. Neurol.* **244**, 101–104.

Schoser, B. G., Becker, V. U., Eckert, B., Zeumer, H., and Thie, A.

(1998). Clinical and ultrasonic long-term results of percutaneous transluminal carotid angioplasty. A prospective follow-up of 30 carotid angioplasties. *Cerebrovasc. Dis.* **8**, 38–41.

Sekhon, L. H., Morgan, M. K., Sorby, W., and Grinnell, V. (1998). Combined endovascular stent implantation and endosaccular coil placement for the treatment of a wide-necked vertebral artery aneurysm: Technical case report. *Neurosurgery* **43**, 380–383, discussion 384.

Sergott, R. C., Grossman, R. I., Savino, P. J., Bosley, T. M., and Schatz, N. J. (1987). The syndrome of paradoxical worsening of dural cavernous sinus arteriovenous malformations. *Ophthalmology* **94**, 205–212.

Setton, A., Davis, A. J., Bose, A., Nelson, P. K., and Berenstein, A. (1996). Angiography of cerebral aneurysms. *Neuroimaging Clin. N. Am.* **6**, 705–738.

Smith, C. M., Higashida, R. T., Barnwell, S. L., Halbach, V. V., Dowd, C. F., Fraser, K. W., Teitelbaum, G. P., and Hieshima, G. B. (1994). Treatment of dural sinus thrombosis by urokinase infusion. *Am. J. Neurol. Res.* **15**, 801–807.

Smith, R. R., Clower, B. R., Grotendorst, G. M., Yabuno, N., and Cruse, J. M. (1985). Arterial wall changes in early human vasospasm. *Neurosurgery* **16**, 171–176.

Solomon, R. A., Mayer, S. A., and Tarmey, J. J. (1996). Relationship between the volume of craniotomies for cerebral aneurysm performed at New York state hospitals and in-hospital mortality. *Stroke* **27**, 13–17.

Spetzler, R. F., and Martin, N. A. (1986). A proposed grading system for arteriovenous malformations. *J. Neurosurg.* **65**, 476–483.

Spetzler, R. F., Martin, N. A., Carter, L. P., Flom, R. A., Raudzens, P. A., and Wilkinson, E. (1987). Surgical management of large AVMs by staged embolization and operative excision. *J. Neurosurg.* **67**, 17–28.

Spetzler, R. F., Wilson, C. B., Weinstein, P., Mehdorn, M., Townsend, J., and Telles, D. (1978). Normal perfusion breakthrough theory. *In* "Congress of Neurological Surgeons Clinical Surgery," pp. 651–672. Williams & Wilkins, Baltimore.

Storey, G. S., Marks, M. P., Dake, M., Norbash, A. M., and Steinberg, G. K. (1996). Vertebral artery stenting following percutaneous transluminal angioplasty. Technical note. *J. Neurosurg.* **84**, 883–887.

Strandness, D. E. (1994). *In* "Vascular Diseases: Surgical and Interventional Therapy." (D. E. Strandness and A. van Breda, Eds.), pp. 643–650. Churchill Livingstone, New York.

Sundt, T. M., Smith, H. C., Campbell, J. K., Vliestra, R. E., Cucchiera, R. F., and Stanson, A. W. (1980). Transluminal angioplasty for basilar artery stenosis. *Mayo Clin. Proc.* **55**, 673–680.

Sundt, T. M. J., Meyer, F. B., Piepgras, D. G., *et al.* (1994). Risk factors and operative results. *In* "Sundt's Occlusive Cerebrovascular Disease" (F. B. Meyer, Ed.), 2 ed., pp. 241–247. Saunders, Philadelphia.

Taki, W., Nishi, S., Yamashita, K., Sadatoh, A., Nakahara, I., Kikuchi, H., and Iwata, H. (1992). Selection and combination of various endovascular techniques in the treatment of giant aneurysms. *J. Neurosurg.* **77**, 37–42.

Takis, C., Kwan, E. S., Pessin, M. S., Jacobs, D. H., and Caplan, L. R. (1997). Intracranial angioplasty: Experience and complications. *Am. J. Neurol. Res.* **18**, 1661–1668.

Tcheng, J. E. (1996). Glycoprotein IIb/IIIa receptor inhibitors: Putting the EPIC, IMPACT II, RESTORE, and EPILOG trials into perspective. *Am. J. Cardiol.* **78**, 35–40.

The Arteriovenous Malformation (AVM) Study Group (1999). Arteriovenous malformations of the brain in adults. *N. Engl. J. Med.* **340**, 1812–1818.

The EC/IC Bypass Study Group (1985). Failure of extracranial-intracranial arterial bypass to reduce the risk of ischemic stroke.

Results of an international randomized trial. *N. Engl. J. Med.* **313**, 1191–200.

The Warfarin-Aspirin Symptomatic Intracranial Disease (WASID) Study Group (1998). Prognosis of patients with symptomatic vertebral or basilar artery stenosis. *Stroke* **29**, 1389–1392.

Theron, J. (1996). Cerebral protection during carotid angioplasty. *J. Endovasc. Surg.* **3**, 484–486.

Theron, J., Courtheoux, P., Alachkar, F., Bouvard, G., and Maiza, D. (1990). New triple coaxial catheter system for carotid angioplasty with cerebral protection. *Am. J. Neurol. Res.* **11**, 869–874, discussion 875–877.

Theron, J., Courtheoux, P., Henriet, J. P., Pelouze, G., Derlon, J. M., and Maiza, D. (1984). Angioplasty of supraaortic arteries. *J. Neuroradiol.* **11**, 187–200.

Theron, J. G., Payelle, G. G., Coskun, O., Huet, H. F., and Guimaraens, L. (1996). Carotid artery stenosis: Treatment with protected balloon angioplasty and stent placement. *Radiology* **201**, 627–636.

Théron, J. G. L., Oguzman, C., et al. (1998). Complications of carotid angioplasty and stenting. *Neurosurg. Focus* **5**, 4.

Tress, B. M., Thompson, K. R., Klug, G. L., Mee, R. R., and Crawford, B. (1983). Management of carotid-cavernous fistulas by surgery combined with interventional radiology. Report of 2 cases. *J. Neurosurg.* **59**, 1076–1081.

Tsai, F. Y., Higashida, R. T., Matovich, V., and Alfieri, K. (1992). Acute thrombosis of the intracranial dural sinus: direct thrombolytic treatment. *Am. J. Neurol. Res.* **13**, 1137–1141.

Tsai, F. Y., Matovich, V., Hieshima, G., Higashida, R. T., Shah, D. G., Ashraf, A., and Pribram, H. F. (1986). Percutaneous transluminal angioplasty of the carotid artery. *Am. J. Neurol. Res.* **7**, 349–358.

Tsai, F. Y., Wang, A. M., Matovich, V. B., Lavin, M., Berberian, B., Simonson, T. M., and Yuh, W. T. (1995). MR staging of acute dural sinus thrombosis: Correlation with venous pressure measurements and implications for treatment and prognosis. *Am. J. Neurol. Res.* **16**, 1021–1029.

Tsao, P. S., Lewis, N. P., Alpert, S., and Cooke, J. P. (1995). Exposure to shear stress alters endothelial adhesiveness. Role of nitric oxide. *Circulation* **92**, 3513–3519.

Tu, R. K., Cohen, W. A., Maravilla, K. R., Bush, W. H., Patel, N. H., Eskridge, J., and Winn, H. R. (1996). Digital subtraction rotational angiography for aneurysms of the intracranial anterior circulation: injection method and optimization. *Am. J. Neurol. Res.* **17**, 1127–1136.

Tyagi, S., Rao, B. H., and Arora, R. (1998). Transfemoral placement of an endovascular stent-graft for a traumatic subclavian arteriovenous fistula. *Indian Heart J.* **50**, 443–445.

Ueda, T., Sakaki, S., Nochide, I., Kumon, Y., Kohno, K., and Ohta, S. (1998). Angioplasty after intra-arterial thrombolysis for acute occlusion of intracranial arteries. *Stroke* **29**, 2568–2574.

Valavanis, A. (1986). Preoperative embolisation of the head and neck: Indications, patient selection, goals and precautions. *Am. J. Neurol. Res.* **7**, 943–952.

Valavanis, A. (1996). The role of angiography in the evaluation of cerebral vascular malformations. *Neuroimaging Clin. N. Am.* **6**, 679–704.

Valavanis, A., and Yasargil, M. G. (1998). The endovascular treatment of brain arteriovenous malformations. *Adv. Tech. Stand. Neurosurg.* **24**, 131–214.

Vazquez-Anon, V., Aymard, A., Gobin, Y. P., Casasco, A., Rufenacht, D., Khayata, M. H., Abizanda, E., Redondo, A., and Merland, J. J. (1992). Balloon occlusion of the internal carotid artery in 40 cases of giant intracavernous aneurysm: Technical aspects, cerebral monitoring and results. *Neuroradiology* **34**, 245–251.

Vieco, P. T. (1998). CT angiography of the intracranial circulation. *Neuroimaging N. Clin. Am.* **8**, 577–592.

Vieco, P. T., Morin, E. E. D., and Gross, C. E. (1996). CT angiography in the examination of patients with aneurysm clips. *Am. J. Neurol. Res.* **17**, 455–457.

Vieco, P. T., Shuman, W. P., Alsofrom, G. F., and Gross, C. E. (1995). Detection of circle of Willis aneurysms in patients with acute subarachnoid hemorrhage: A comparison of CT angiography and digital subtraction angiography. *Amer. J. Roentgenology* **165**, 425–430.

Vinuela, F., Dion, J. E., Duckwiler, G., Martin, N. A., Lylyk, P., Fox, A., Pelz, D., Drake, C. G., Girvin, J. J., and Debrun, G. (1991). Combined endovascular embolization and surgery in the management of cerebral arteriovenous malformations: Experience with 101 cases. *J. Neurosurg.* **75**, 856–864.

Vinuela, F., Duckwiler, G., and Mawad, M. (1997). GDC embolisation of acute intracranial aneurysm: Perioperative anatomical and clinical outcome in 403 patients. *J. Neurosurg.* **86**, 475–482.

Vitek, J., Roubin, G., and Iyer, S. (1999). "Immediate and Late Outcome Of Carotid Angioplasty with Stenting," Joint Section Meeting, *AANS/CNS/ASITN*. Nashville.

Von Kummer, R., Allen, K. L., Holle, R., Bozzao, L., Bastianello, S., Manelfo, C., Bluhmki, E., Ringleb, P., Meier, D. H., and Hacke, W. (1997). Acute stroke: Usefulness of early CT findings before thrombolytic therapy. *Radiology* **205**, 327–333.

Wakhloo, A. K., Juengling, F. D., Van Velthoven, D. S., Schumacher, M., Henrig, J., and Schwechheimer, K. (1993). Extended preoperative polyvinyl alcohol microembolisation of intracranial meningiomas: Assessment of two embolisation techniques. *Am. J. Neurol. Res.* **14**, 571–582.

Wallace, R. C., Furlan, A. J., Moliterno, D. J., Stevens, G. H., Masaryk, T. J., and Perl, J., 2nd (1997). Basilar artery rethrombosis: Successful treatment with platelet glycoprotein IIB/IIIA receptor inhibitor. *Am. J. Neurol. Res.* **18**, 1257–1260.

Wardlaw, J. M., and Warlow, C. P. (1992). Thrombolysis in acute ischaemic stroke: Does it work? *Stroke* **23**, 1826–1829.

Wechsler, L. R., Kistler, J. P., Davis, K. R., and Kaminski, M. J. (1986). The prognosis of carotid siphon stenosis. *Stroke*. **17**, 714–718.

Wholey, M. H., Wholey, M., Bergeron, P., Dietrich, E. B., Henry, M., Laborde, J. C., Mathias, K., Myla, S., Roubin, G. S., Shawl, F.,

Theron, J. G., and Yadav, J. S. (1998). Current global status of carotid artery stent placement. *Cathet. Cardiovasc. Diagn.* **44**, 1–6.

Wiesmann, M., Mayer, T. E., Yousry, I., Hannan, G. F., and Bruckmann, H. (2001). Detection of hyperacute parenchymal haemorrhage of the brain using echo-planar T2*-weighted and diffusion-weighted MRI. *Eur. Radiol.* **11**, 849–853.

Wijdicks, E. F., and Scott, J. P. (1997). Causes and outcome of mechanical ventilation in patients with hemispheric ischaemic stroke. *Mayo Clin. Proc.* **72**, 210–213.

Wilkins, R. H. (1985). The natural history of intracranial vascular malformations: A review. *Neurosurgery* **16**, 421–430.

Wilson, C. B., and Hieshima, G. (1993). Occlusive hyperemia: A new way to think about an old problem. *J. Neurosurg.* **29**, 333–353.

Wityk, R. J., Chang, H. M., Rosengart, A., Han, W. C., DeWitt, L. D., Pessin, M. S., and Caplan, L. R. (1998). Proximal extracranial vertebral artery disease in the New England Medical Center Posterior Circulation Registry. *Arch. Neurol.* **55**, 470–478.

Yadav, J. S., Roubin, G. S., Iyer, S., Vitek, J., King, P., Jordan, W. D., and Fisher, W. S. (1997). Elective stenting of the extracranial carotid arteries. *Circulation* **95**, 376–381.

Yadav, J. S., Roubin, G. S., King, P., Iyer, S., and Vitek, J. (1996). Angioplasty and stenting for restenosis after carotid endarterectomy. Initial experience. *Stroke* **27**, 2075–2079.

Yokote, H., Terada, T., Ryujin, K., Konoshita, Y., Tsuura, M., Nakai, E., Kamei, I., Moriwaki, H., Hayashi, S., and Ikatura, T. (1998). Percutaneous transluminal angioplasty for intracranial arteriosclerotic lesions. *Neuroradiology* **40**, 590–596.

Zanella, F. E., and Valavanis, A. (1994). Interventional neuroradiology of lesions of the skull base. *Neuroimaging Clin. N. Am.* **4**, 619–637.

Zeman, R. K., Silverman, P. M., Vieco, P. T., and Costello, P. (1995). CT angiography. *Amer. J. Roentgenology* **165**, 1079–1088.

Zeumer, H., Fiertag, H. J., Zanella, F. *et al.* (1993). Local intraarterial fibrinolytic therapy in patients with stroke: Urokinase vs. rtPA. *Neuroradiology* **35**, 159–162.

Zubkov, Y. N., Nikiforov, B. M., and Shustin, V. A. (1984). Balloon catheter technique for dilatation of constricted cerebral arteries after aneurysmal SAH. *Acta Neurochir.* **70**, 65–79.

CHAPTER 41

# Vascular Malformations and Interventional Neuroradiology of the Spinal Cord

Armin K. Thron and Louis R. Caplan

## INTRODUCTION

Vascular malformations of the spinal canal and its meninges are rare diseases that lead to damage of the spinal cord if not treated. Depending on the type of vascular malformation, hemorrhage, circulatory disorders, or space-occupying effects may be the cause of damage. Nonspecific initial symptoms can make an early diagnosis difficult. In cases of clinical suspicion, MRI is the most relevant imaging procedure in the early diagnostic evaluation. Selective spinal angiography is required to define the nature of the lesion and to decide about the appropriate therapy, which may be endovascular, interventional, neurosurgical, combined, or attentive.

## CLASSIFICATION (Table I)

Spinal vascular malformations should be classified similar to those of the brain, differentiating between the true inborn lesions.

- **Arteriovenous malformations (AVMs)**
  with different morphology and hemodynamics of the nidus
  — Fistulous type 1–3 (perimedullary fistula)
  — Glomerular type
  — Juvenile Type
- **Cavernomas**
- **Capillary teleangiectases**
  and the acquired lesions
- **Dural arteriovenous fistulas**

## SPINAL DURAL AV FISTULA

Spinal dural fistulas are the most frequent arteriovenous malformations of the spinal canal and its meninges. There is a striking similarity to AV fistulas (AVFs) of the cranial dura mater. According to König *et al.* (1989) there is a ratio of 20:1 between dural fistulas and real intramedullary arteriovenous angiomas. Usually the disease becomes symptomatic in adults over 40 years of age. Men exceed women at a ratio between 5:1 and 7:1 (Berenstein and Lasjaunias, 1992; Merland *et al.*, 1980; Rosenblum *et al.*, 1987; Symon *et al.*, 1984; Thron, 1988).

Kendall and Logue (1977) were the first to recognize that this disease, which had formerly been described as so-called retromedullary angiomas, in fact is explained by an AV fistula situated within the dura mater. This was confirmed by Merland *et al.* (1980) who pointed out the angiographical characteristics of the disease in detail. Figure 1 shows the distribution of the shunt level in 114 personal cases diagnosed and treated between 1984 and 2000.

The arteriovenous shunts are located inside the dura mater close to the spinal nerve roots. The arterial blood enters a radicular vein where it passes the dura. Flow is directed to the superficial veins of the spinal cord. Medullary arteries do not participate in the supply of the fistula; it is supplied by radiculomeningeal arteries, that is, arteries supplying the segmental nerve roots and dura. They are branches of the segmental vessels that persist as intercostal or lumbar arteries in the thoracolumbar region. Depending on the location of the fistula, the supply may also come from the sacral or

**TABLE I  Classification, Onset, Development, and Therapy of Spinal AVMs and AVFs**

| Type | Onset age | Development | Localization | Involved vessels | Therapy |
|---|---|---|---|---|---|
| Dural AVF | >50 years | Subacute/chronic | Thoracolumbar >Sacral. >Craniocerv. | Mening. artery Spinal vein | Surgery Embolization with glue |
| AVM | | | | | |
| Perimedullary | >20 years | Rapid | Thoracolumbar cone(!) | Medullary artery Spinal vein | Embolization with coils, balloon, or surgery |
| Fistula type 1–3 | <40 years | Progressive | | | |
| Glomus type | >20 years | Lengthy | Entire spinal cord | AVM nidus fed by spinal cord arteries and veins | Embolization and/or surgery |
| Juvenile type | >20 years | Progressive | | | |

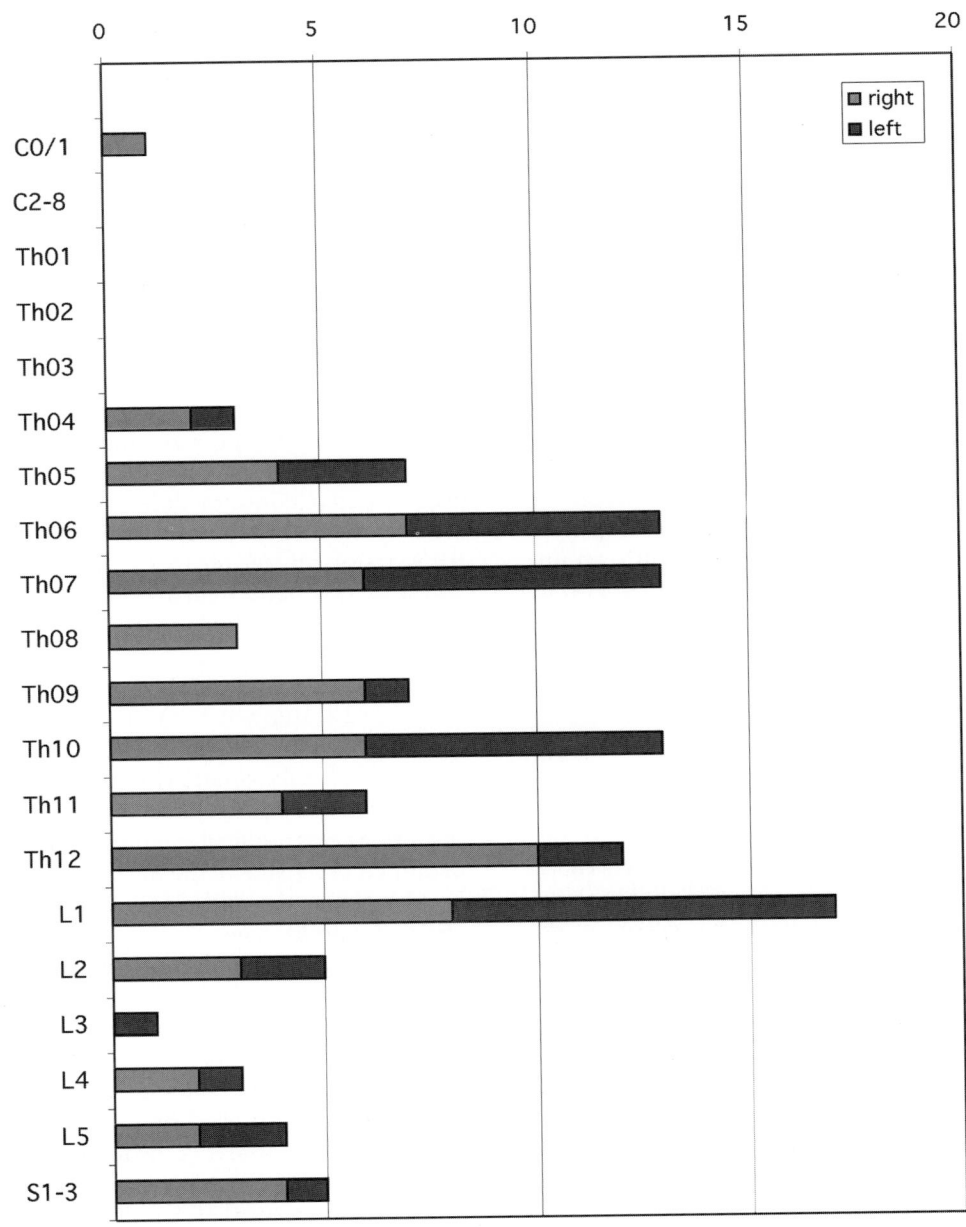

**FIGURE 1**  Segmental distribution of the av-shunt in 114 personal cases of SDAVF diagnosed between 1984 and 2000.

hypogastric arteries (Djindjian, 1970; Heindel et al., 1975; Picard et al., 1969; Stein et al., 1972). In addition, intracranial dural fistulas have been described to drain to spinal cord veins and evoke the same symptoms as spinal dural fistulas (Mahagne et al., 1992; Woimant et al., 1982; Wrobel et al., 1988). The cause of the fistulas is unknown.

## SYMPTOMATOLOGY

The symptoms are slowly progressive and consist of sensory loss and paraparesis that is independent of the location of the fistula (König et al., 1989; Morgan and Marsh, 1989; Schrader et al., 1989). Frequent accompanying symptoms are pain in the back or irradiating pains in the legs. An acute onset of the disease (11%) or a progressive development interrupted by intermediate remissions (11%) are the rare forms (according to Symon et al., 1984). Between 1980 and 1990 when the clinical findings and the course of this disease were not well known and diagnosis depended on myelograms or MR images of first generation machines almost all patients had additional sphincter disturbances at the time of diagnosis. (Hassler et al., 1989; Thron, 1988). The disease affects men more often than women, in their middle or older age. Because the arteriovenous shunts are located within the dura, subarachnoid hemorrhage is not a typical symptom. In our patients it occurred in 1 of 114 cases.

Intervertebral disk disease as well as intraspinal space-occupying lesions are among the differential diagnoses that are excluded.

The functional abnormalities or disturbances within the spinal cord are caused by venous congestion due to arterialization of the medullary veins (Aminoff and Logue, 1974; Hassler et al., 1989; Thron, 1988). In the year 1926, Foix and Alajouanine described a syndrome that they called subacute necrotizing myelitis. The subtitle was "Myelite Centrale Angéio-hypertrophique à Évolution Progressive: Paraplegie Amyotrophique Lentement Ascendante, d'abord Spasmodique, puis Flasque, S'accompagnant de Dissociation Albumino-cytologique." Included in this report is an account of the fate of a man of 29 years of age—still young, given that he had a dural fistula—who, within 2.5 years after the onset of the symptoms, died of the secondary effects of his finally flaccid and complete paraplegia.

Onset and development of the disease, CSF findings, and macroscopic aspects of the spinal cord in this initial case description as well as in classical cases reported later on (Jellinger et al., 1968; Scholz and Manuelidis, 1951; Wechsler, 1964; Wyburn-Mason, 1943) provide

evidence that this disease is the result of dural AV fistulas. (Thron et al., 1987; Criscuolo et al., 1989). Perimedullary fistulas type 1 may present with a similar congestive myelopathy, but in this case the involvement of enlarged spinal cord arteries should be recognized histologically.

## THERAPY

The goal of every form of therapy must be to relieve the spinal cord veins from the blood shunted through the fistula. Surgical occlusion of the intrathecal vein that receives the blood from the arteriovenous shunt is a technically simple and safe intervention. Except for fistulas, located in the dura of the sacrum, surgical intervention (nearly) always results in occlusion of the fistula (Huffmann et al., 1995; Mourier et al., 1989; Reinges et al., 2001; Symon et al., 1984).

Alternatively, the same goal can be achieved by injection of a tissue adhesive (histoacryl) after superselective catheterization of the feeding radiculomeningeal artery. A problem, initially not recognized, was that the liquid embolization material must reach and occlude the proximal segment of the draining vein (Figure 3) in order to prevent collateral reestablishment of flow through the numerous intradural capillaries (Figure 2). If this succeeds, surgical removal is no longer necessary. If the endovascular access does not result in complete occlusion, surgical removal must follow. According to the results of the largest published series, endovascular occlusion was successful in about two-thirds of cases (40 out of 63), whereas 23 fistulas could not be embolized completely or not at all because of technically unfeasible selective catheterization or because of medullary arteries arising from the same intercostal artery (Merland et al., 1986; Mourier et al., 1989).

Due to the slowly progressive, sometimes acutely deteriorating, course and the lack of specific symptoms, definitive angiographic diagnosis and adequate therapy come late for many of the patients. Therefore, the problems of the spinal dural AVFs are mainly problems of diagnosis and not so much problems of alternative treatment modalities. The result of therapy depends on the neurological deficits before treatment. Disconnecting the vein from the shunt succeeds in stopping the clinical course of the constantly progressing deficit. Frequently, an improvement of the existing motor and—to a lesser degree—sensory deficits follows, whereas sphincter disturbances often do not improve. Spectacular improvements after therapy are possible but exceptional (Barth et al., 1984; Behrens and Thron, 1999; Criscuolo et al., 1989; König et al., 1989; Mourier et al., 1989; Symon et al., 1984).

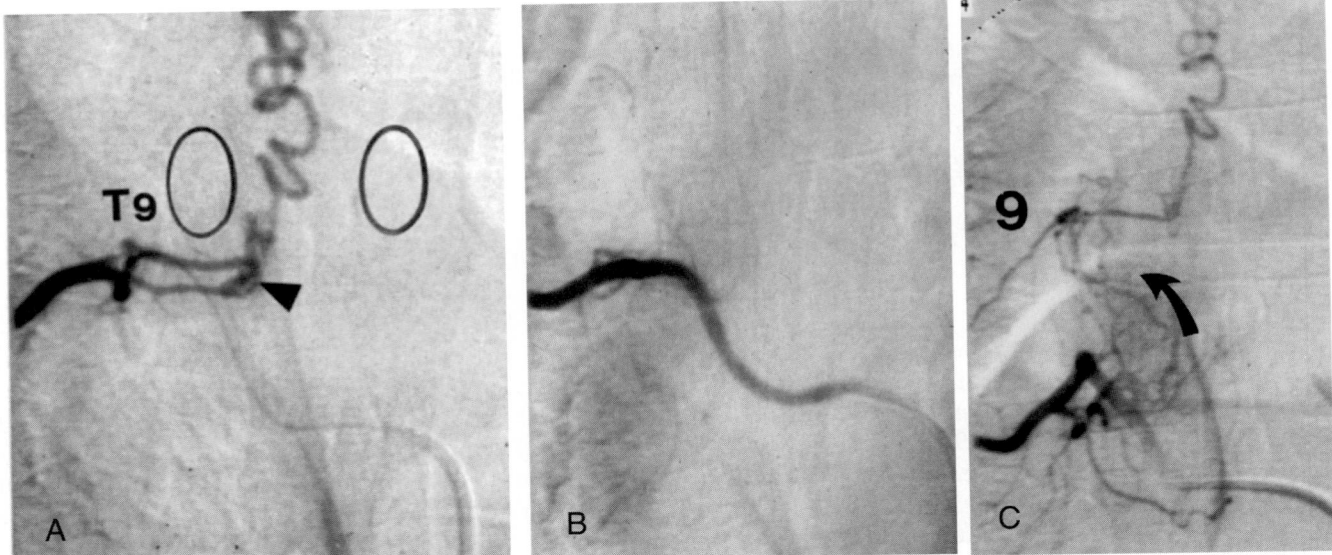

**FIGURE 2**    Spinal dural AVF at T9 level. The shunt (arrowhead) is supplied by two radiculomeningeal branches (**A**). Following embolization with glue the fistula seems to be occluded (**B**). Control angiography after 3 months demonstrates persistence of a part of the AV communication, now supplied by collateral vessels from the T10 level (**C**) and from other neighboring arteries (not shown). The liquid embolization material obviously did not enter the proximal part of the draining vein. Clinically transient improvement was followed by deterioration of symptoms.

## ARTERIOVENOUS MALFORMATIONS

### AVM of Perimedullary Fistula Type (1–3)

These are direct arteriovenous shunts that are located on the ventral or dorsal surface of the spinal cord or the conus medullaris, usually in the thoracolumbar area, occasionally thoracic, and rarely cervical (Heros *et al.*, 1986). Their location thus is intradural, intramedullary, or extramedullary. They are always supplied by spinal cord vessels, either by the anterior spinal artery (ventrally) or by a posterolateral artery (dorsally), depending on their location. They drain into spinal cord veins. Drainage may ascend up to the foramen magnum or into the posterior fossa.

Djindjian *et al.* in 1977, were the first to describe perimedullary fistulas. Merland and his group distinguished three types, according to the vessel size, the shunt volume, and the drainage pattern (Gueguen *et al.*, 1987; Merland *et al.*, 1980; Mourier *et al.*, 1993; Ricolfi *et al.*, 1991).

Perimedullary fistulas become symptomatic in early to middle adulthood with rapidly progressing ascending motor and sensory signs, accompanied by sphincter abnormalities. Spinal subarachnoid hemorrhage may occur, due to the intradural location. In most cases, diagnosis is established only many years after the onset of symptoms.

The type 1 perimedullary fistula is a simple, small AV fistula that is fed by a long, thin anterior or postero-lateral spinal artery, which is only minimally dilated. The fistula is small, and the shunt volume is low. The artery as well as the vein are not significantly dilated. Because the feeding arteries are spinal cord vessels, embolization has to be performed as close as possible to the fistula. Usually this is not possible due to the length of the undilated feeder, so that surgical excision is the therapy of choice if the fistula is accessible to open surgery (Figure 4) (Mourier *et al.*, 1993). This is often not the case in a ventral location. Embolization with particles may be tried only in selected cases of a ventral type 1 perimedullary fistula.

The type 2 perimedullary fistula is a middle-sized fistula that is fed by one or two markedly dilated arteries. The angioarchitecture of these fistulas usually allows superselective catheterization, so that an elective occlusion of the fistula can be performed. If this endovascular therapy failed to occlude the lesion completely, a surgical excision of the partially occluded AVF can be done (seven of nine patients in the largest series) (Mourier *et al.*, 1993).

The type 3 perimedullary fistula is a large AV fistula with multiple feeding arteries of large caliber and a high or very high flow volume. The draining veins are dilated and tortuous. Balloon occlusion after hyperselective catheterization has been performed in the past (Mourier

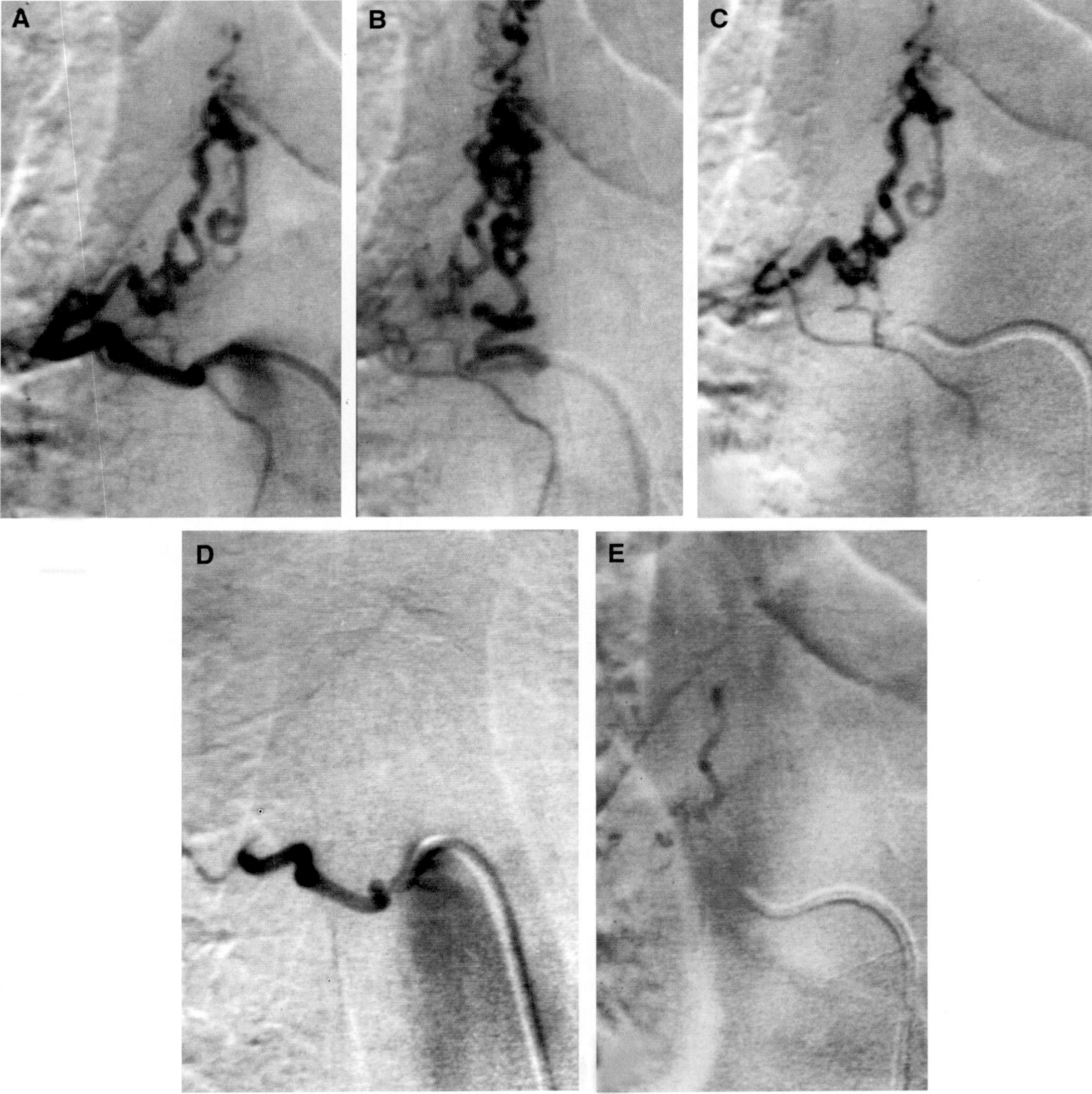

**FIGURE 3**  Spinal dural AVF with high shunt volume (**A** and **B**). Following selective catheterization (**C**), injection of glue occluded the feeding vessel (**D**) and the proximal part of the intradural vein. The radiopaque cast in this vessel is shown in **E**. In this situation no recurrence is expected.

*et al.*, 1993; Richè *et al.*, 1983a,b; Ricolfi *et al.*, 1991). It will perhaps be replaced by coil embolization in most cases. Surgery is not an equivalent alternative because of the large number of vessels, the high blood flow velocity, and the often ventral location of the fistula.

## AVM of Glomerular Type

This frequent type of AVM is characterized by an angioma nidus like in most cerebral AVMs. It may be located superficially on the surface of the spinal cord or deep within the cord parenchyma or may extend to both compartments. Due to the numerous anastomoses

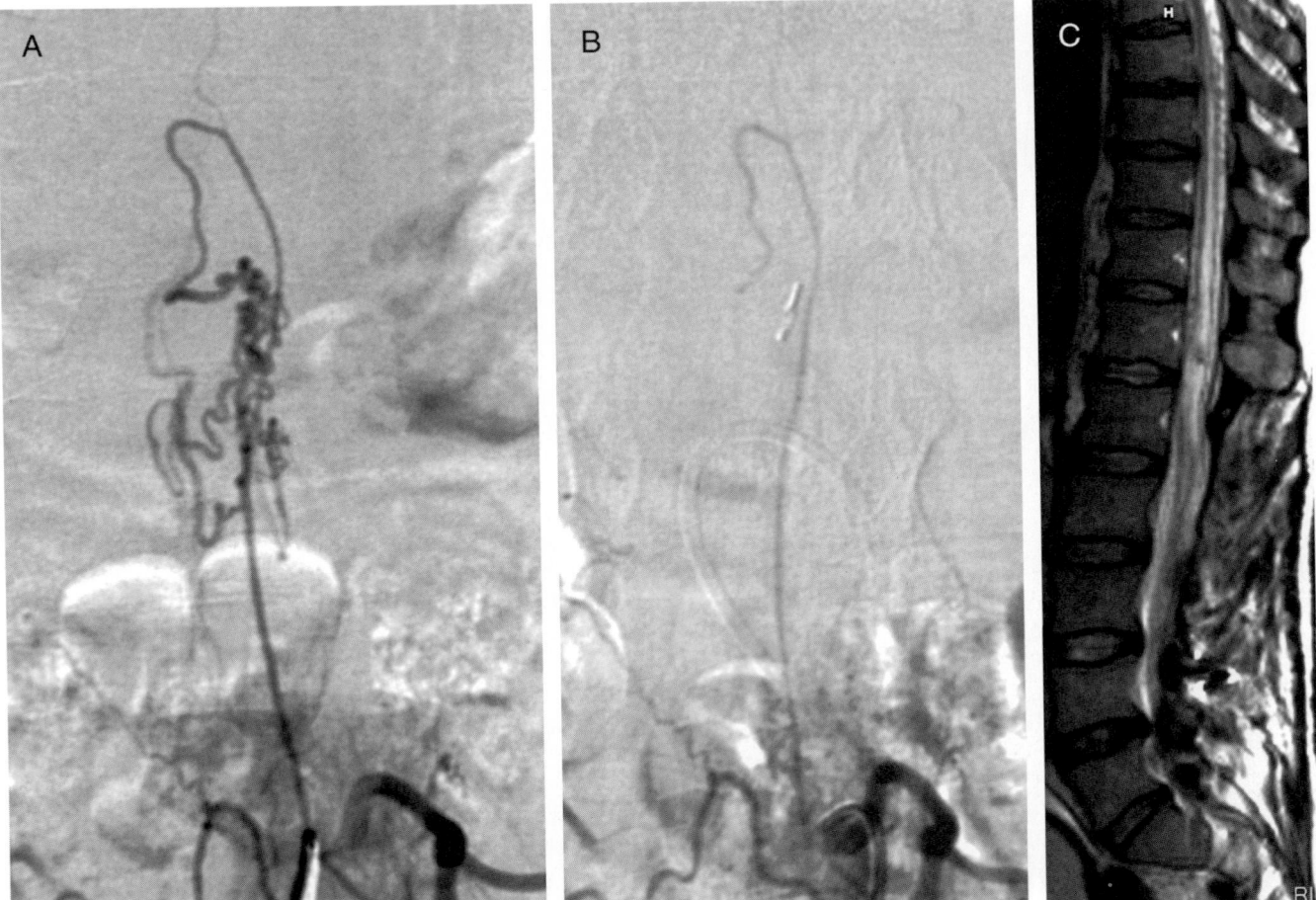

**FIGURE 4**    Spinal AVM of perimedullary fistula type 1. The transition of arterial to venous side (arrowhead) is characterized by the difference in caliber and course of the involved vessels (**A**). If endovascular approach to the fistula point is not possible, surgical dysconnection of the shunt (**B**) is the treatment of choice. Similar to SDAVF, venous congestion leads to an extensive centromedullary edema of the spinal cord with hyperintensity in T2-weighted images on MRI (**C**).

between the spinal cord arteries, the nidus is always supplied by several arteries or branches derived from the anterior or posterior spinal arteries. The AVM drains into the spinal cord veins.

These malformations become symptomatic in children and adolescents. They usually evolve over a long time, and acute worsenings can be followed by marked improvements. According to their possible location at every level of the spinal cord, their symptomatology may differ. Spinal subarachnoid hemorrhage is one of the common symptoms of intramedullary AVMs.

Due to the intramedullary location, surgical removal is only occasionally possible, and endovascular therapy has become an established alternative. It has been carried out since the 1960s (Bentson *et al.*, 1978; Berenstein *et al.*, 1984; Berenstein and Lasjaunias, 1992; Biondi *et al.*, 1990; Djindjian, 1975; Djindjian *et al.*, 1973a,b; Djindjian and Merland, 1978; Doppmann

*et al.*, 1968, 1971; Hall *et al.*, 1989; Horton *et al.*, 1986; Newton and Adams, 1968; Rodesch *et al.*, 1992; Spetzler *et al.*, 1989). For endovascular occlusion of the AVM, many kinds of corpuscular as well as liquid embolization materials have been used. They can be reduced to two different types of embolization materials: particles and glue. Generally, the corpuscular embolization materials only achieve a reduction of flow through the AVM. But with a position of the catheter immediately before or in the nidus of the malformation it is also possible to fill up the angioma with particles until complete and definite occlusion is achieved (Figures 3 and 4). In other cases particle embolization entails a reduction of steal and a lower ischemic risk for the spinal cord. The venous pressure is reduced, and the consequences of the venous hypertension may recede. The risk of hemorrhage can be reduced or annuled by these means. If particles are used that are bigger than the vessels feeding the spinal cord but smaller than those feeding the malformation, risks for the spinal cord can be avoided. As the particles cause no inflammatory reactions, secondary damage of the spinal cord by tumefaction and edema is not to be expected. The disadvantage of the particles is that they may not be definitive embolization materials. Revascularization of the AVM and recanalization of the embolized vessels occur. This is especially true for dural AV fistulas, which should not undergo this type of embolization in the meninges of the spinal cord. Although even definitive cure in patients with superficial or deep glomerular AVMs can be archieved, patients should be kept under clinical and angiographical supervision in order to repeat and accomplish therapy.

The main representative of the corpuscular embolization materials is polyvinyl alcohol (PVA), which is available in different calibrated grain sizes. Embolization with particles is performed after superselective catheterization, preferably via the posterolateral arteries (Figure 4). The disadvantage of revascularization of embolized AV malformations was thought to be avoided by the use of liquid glues.

The prospective advantage of glues is a durable occlusion of the embolized regions and a durable cure of the malformation if one succeeds in eliminating it completely endovascularly. A disadvantage may arise by possibly a higher rate of complications (Theron et al., 1986) caused by the occlusion of normal vessels and the induced inflammatory reaction. Comparison of the results of all published cases is not possible because in the time before the recognition of spinal dural fistulas these were mistaken for angiomas of the spinal cord. This consequently led to a result concerning the treatment of intramedullary AV malformations that was too positive, because prognosis after embolization is often good after the treatment of dural fistulas.

Merland and his group (1980, 1986) presented good results after particle embolization with stabilization of the clinical condition and avoidance of postoperative hemorrhage in all patients. Long-term results were evaluated in 35 patients (repeated treatments with 158 procedures) with thoracic AVMs and a follow-up extending from 1 to 15 years (average 6 years) (Biondi et al., 1990). They found clinical improvement in 63% of their patients compared to the conditions at the beginning of their treatment. Seven patients (20%) had a treatment-related deterioration. A durable complete elimination could be achieved in none of the cases. Théron et al. (1986) reported an improved endovascular treatment with complete AVM occlusion in all of five cases using a new embolization technique with calibrated PVA particles of 150–250 μm in size. In two follow-up angiograms obtained, no revascularization was observed. We share his experience that complete or almost complete and durable occlusion of spinal glomerular AVMs can be achieved by the relatively safe particle embolization (Figures 5–7). A small residual or revascularized part of the malformation can be re-embolized or surgically resected (Latchaw et al., 1980). This is our strategy if the angioarchitecture of the malformation requires a combined approach for complete elimination (Figure 5).

Berenstein and Lasjaunias (1992) and Rodesch et al. (1992) report on their experience with glue embolization of spinal AVMs performed in 38 of 47 patients. They attained a complete endovascular elimination in 53%. The complication rate amounts to 10.6% of permanent and 10.6% of transient deteriorations. Compared to the natural history (Aminoff and Logue, 1974), endovascular treatment offers a distinct improvement in prognosis.

The treatment strategy in symptomatic cases of spinal AVMs must be based on a thorough evaluation of the AVM angioarchitecture of a given case. In a center with experience in surgical and endovascular treatment of this disease the possible treatment options and their risks must be compared. Progress in the development of catheters and embolization materials may change a given concept rapidly. The decision to take an endovascular, microneurosurgical, combined, or attentive approach should be taken up in an interdisciplinary discussion.

## Cavernomas

This type of venous low-flow malformation is composed of large vascular channels without intervening neural tissue. Grossly, cavernomas appear as well-circumscribed mulberry-shaped dark blue or brown lesions. Histologically thrombosed and calcified parts of the malformation are frequently observed. The surrounding parenchyma shows gliosis with deposition of hemosiderin and ferritin. Symptoms occur by bleeding either inside the malformation or into the parenchyma or CSF. Any region of the spinal cord can be involved.

Development and severity of symptoms depend on the location of the lesion and of the size of the hemorrhage. Zevgaridis et al. (1999) describe a clinical course with acute deterioration in 26%, with gradual or intermittent worsening in 30% and a slowly progressive development of symptoms in 41% of their cases. According to Jellinger (1978) cavernomas account for 5–15% of spinal vascular malformations. Most cavernomas become symptomatic in the third or fourth decade of life. The incidence of bleeding events is not well known, it is thought to be much lower than that of spinal AVMs (Zevgaridis et al., 1999).

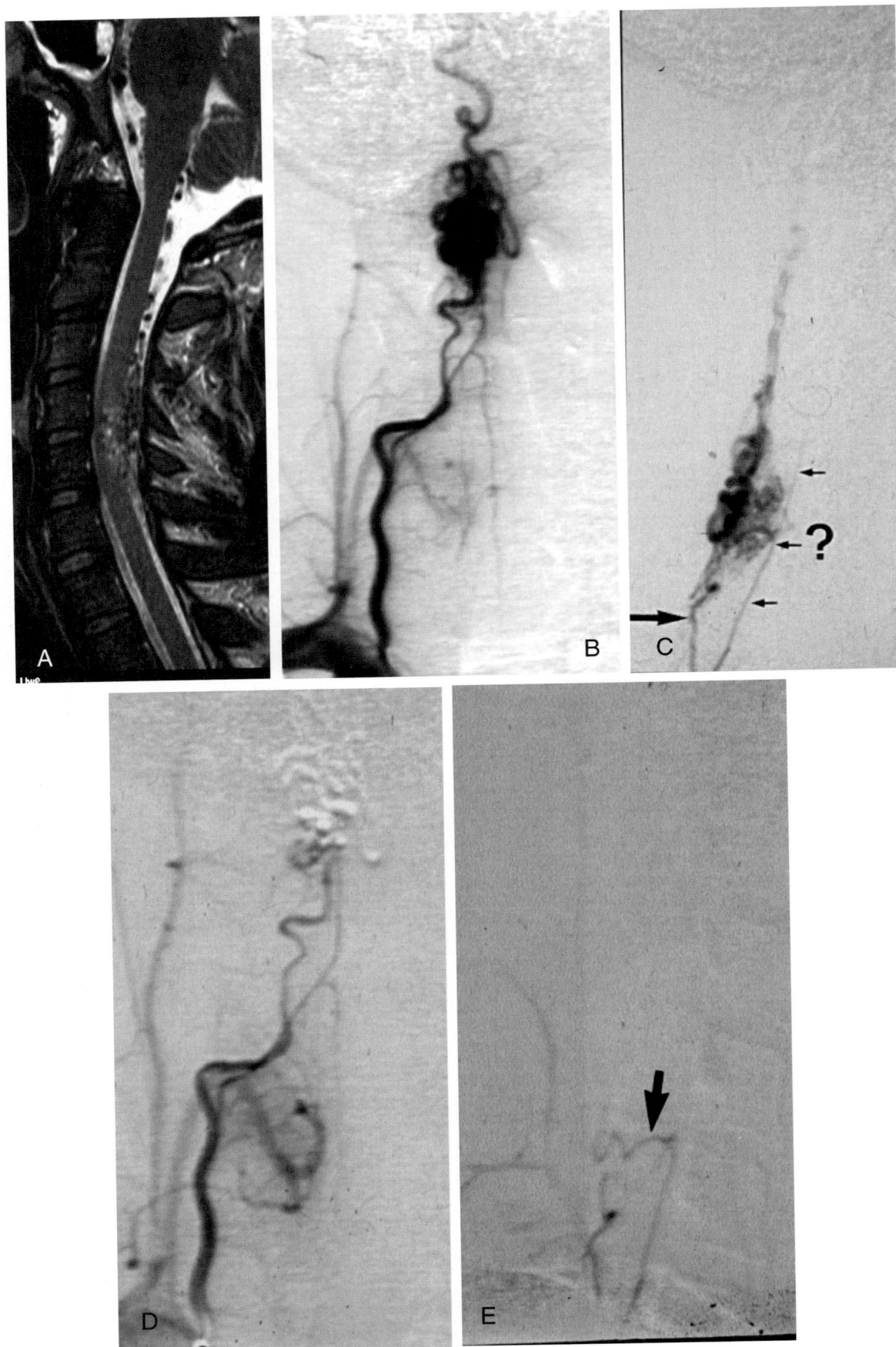

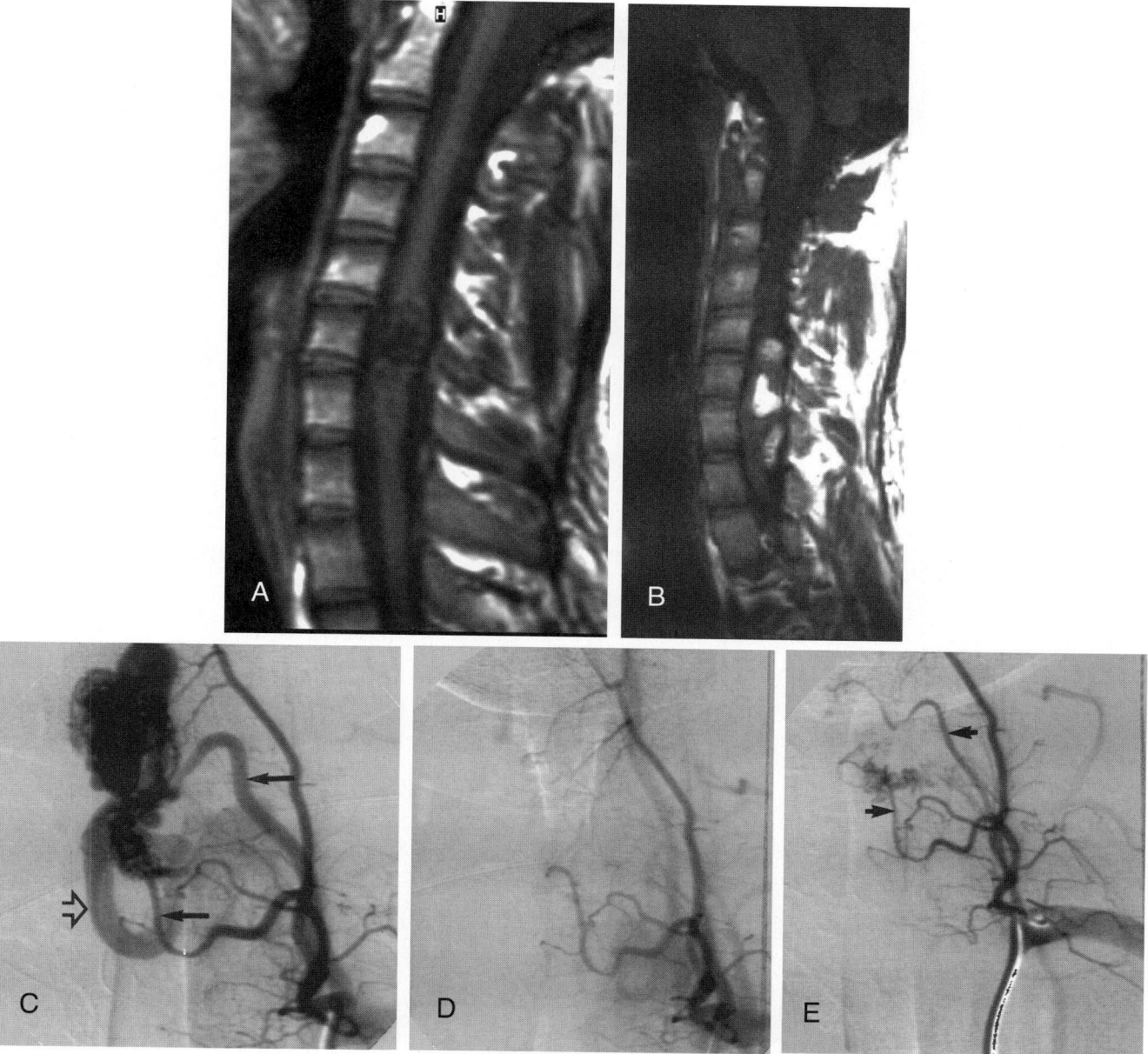

FIGURE 6   Intramedullary spinal AVM at low cervical level. Sagittal T1-weighted image before (A) and 2 months after treatment (B). The large malformation was supplied by several feeders from various levels. The most prominent feeders, which were selected for embolization, originate from the thyrocervical trunk (C, closed arrows). One big draining vein opacifies early (open arrow). The immediate postembolization angiogram (D) suggests complete occlusion. Control DSA (E) and MRI (B) after 2 months show only a small AVM remnant and extensive thrombosis of the large draining veins. The additional feeders that were not embolized had closed together with the nidus. The patient's symptoms improved.

FIGURE 5   Spinal AVM of glomerular type in the cervical spinal cord. Sagittal T2-weighted MRI (A). DSA in a.p. (B) and lateral view (C) before embolization. The main supply comes from a posterior radiculomedullary feeder (large arrow). The contribution from the anterior spinal artery (small arrows, question mark) is much lower. After embolization of the AVM nidus with PVA particles (150–250 μm) through the posterior feeder, there is only a very faint and circumscribed AVM opacification in the a.p. view of the angiogram (D). It is caused by the residual supply from the anterior axis through a central artery (arrow) which is now well demonstrated in the lateral view (E).

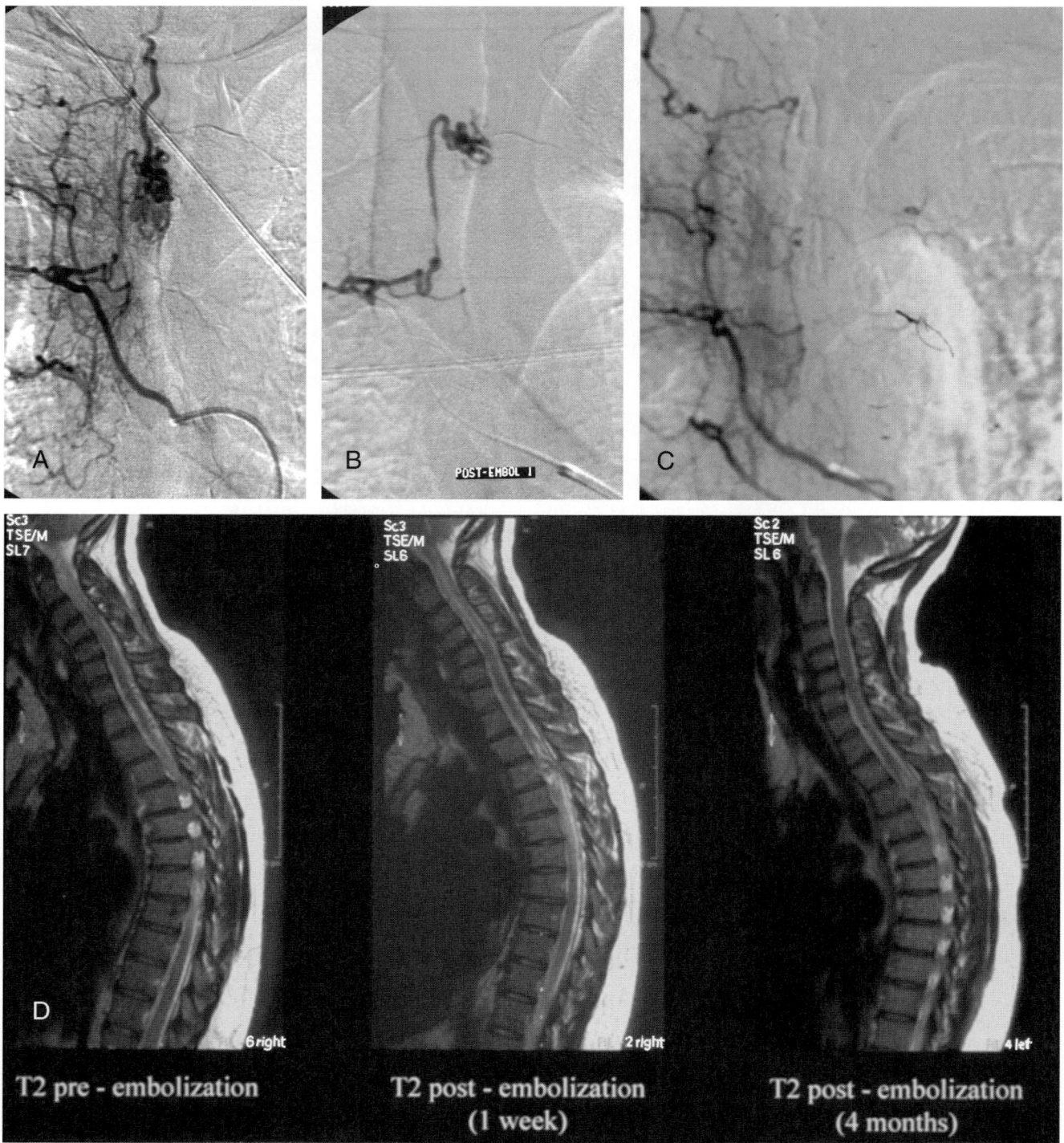

**FIGURE 7**    Intramedullary spinal AVM at Th 3–4 level. The anterior as well as two posterior radiculomedullary arteries from both sides supplied the AVM nidus. The right posterior vessel (**A**) was selected for embolization of the nidus (**B**). Control angiography of all initial feeders after 4 months did not show any residual AVM (**C**). MRI findings on sagittal T2-weighted images before, 1 week after, and 4 months after treatment are shown in **D**. Transient swelling and hyperintensity of the cervical spinal cord was probably due to thrombosis of spinal cord veins and was the reason for a transient deterioration of the neurological deficits. After 4 months the MRI findings had normalized and the patient was almost free of symptoms.

MRI, especially with the presently available resolution, allows diagnosis of even small cavernomas within the spinal cord. The typical MRI finding is a mixed signal (hyper-, iso-, and hypointense) in the center of the lesion in T1- and T2-weighted images ("popcorn-like") that is surrounded by a low-signal hemosiderin/ferritin rim (Fontaine *et al.*, 1988). Only in rare cases does the lesion enhance following injection of contrast medium. Angiographically, cavernomas are occult lesions.

As far as treatment is concerned an endovascular approach to cavernomas is not possible and radiotherapy is not recommended. Therefore, surgical removal is an option, at least in symptomatic cases.

### Capillary Teleangiectases

They form a cluster of capillary-sized blood vessels within CNS parenchyma. They are frequently encountered in the brainstem. They are mostly incidental findings but may cause differential diagnostic problems.

On MRI the lesion is invisible in T1-weighted images without contrast but enhances strongly as an area of small tubular structures. In T2-weighted images the lesion may be invisible or appear as an area of faint hyperintesity without any space-occupying effect (Küker *et al.*, 2000). Capillary teleangiectases should not be mistaken for gliomas.

## EVALUATION

In the rare cases of vascular malformations of the spinal cord, treatment in symptomatic cases offers an improvement in prognosis compared to the natural history. Treatment should be performed in centers with endovascular and surgical experience especially in patients with an AVM. For the perimedullary fistula type 1 and for the spinal cord cavernoma, surgery is the only possible therapy. For the other types of AV malformations the endovascular approach has become an alternative treatment option of first choice. The problems of the spinal dural AV fistulas are mainly in reaching an early diagnosis and not so much problems of the alternative treatment modalities surgery versus glue embolization.

## REFERENCES

Aminoff, M. J., and Logue, V. (1974). The prognosis of patients with spinal vascular malformations. *Brain* **97**, 211.

Barth, M. O., Chiras, J., Kose, M., Vega Molina, J., and Bories, J. (1984). Resultat de l'embolisation des fistules artério-veineuses durales rachiediennes a drainage veineux péri-medullaire. *Neurochirurgie* **30**, 381–386.

Behrens, S., and Thron, A. (1999). Long-term follow-up and outcome in patients treated for spinal dural arteriovenous fistula. *J. Neurol.* **246**, 181–185.

Bentson, J., Rand, R., Calcaterra, T., and Lasjaunias, P. (1978). Unexpected complications following therapeutic embolisation. *Neuroradiology* **16**, 420–423.

Berenstein, A., and Lasjaunias, P. (1992). "Surgical Neuroangiography V: Endovascular Treatment of Spine and Spinal Cord Lesions." Springer-Verlag, Berlin/New York.

Berenstein, A., Young, W., Ransohoff, J., Benjamin, V., and Merkin, H. (1984). Somatosensory evoked potentials during spinal angiography and therapeutic transvascula embolization. *J. Neurosurg.* **60**, 777.

Biondi, A., Merland, J. J., Reizine, D., Aymard, A., Hodes, J. E., Lecoz, P., and Rey, A. (1990). Embolization with particles in thoracic intramedullary arteriovenous malformations: Long-term angiographic and clinical results. *Radiology* **177**, 651.

Criscuolo, G. R., Oldfield, E. H., and Doppman, J. L. (1989). Reversible acute and subacute myelopathy in patients with dural arteriovenous fistulas. *J. Neurosurg.* **70**, 354–359.

Djindjian, R. (1970). "Angiography of the Spinal Cord." University Park Press, Baltimore.

Djindjian, R. (1975). Embolization of angiomas of the spinal cord. *Surg. Neurol.* **4**, 411–420.

Djindjian, R., Cophignon, J., Rey, A., *et al.* (1973a). Superselective arteriographic embolisation by the femoral route in neuroradiology: Study of 50 cases. II. Embolization in vertebromedullary pathology. *Neuroradiology* **6**, 132–142.

Djindjian, R., Cophignon, J., Theron, J., *et al.* (1973b). Superselective arteriographic embolisation by the femoral route in neuroradiology: Study of 50 cases. I. Technique, indications, complications. *Neuroradiology* **6**, 20–26.

Djindjian, M., Djindjian, R., Rey, A., Hurth, M., and Houdart, R. (1977). Intradural extra-medullary spinal arterio-venous malformation fed by the anterior spinal artery. *Surg. Neurol.* **8**, 85–94.

Djindjian, R., and Merland, J. J. (1978). Place de l'embolisation dans le traitement des malformations arterio-veineuse medullaires: A propos de 38 cas. *Neuroradiology* **16**, 428–429.

Doppman, J. I., DiChiro, G., and Ommaya, A. K. (1968). Obliteration of spinal cord arteriovenous malformation by percutaneous embolization. *Lancet* **1**, 577.

Doppman, J. I., DiChiro, G., and Ommaya, A. K. (1971). Percutaneous embolization of spinal cord arterio-venous malformations. *J. Neurosurg.* **34**, 48–55.

Foix, C., and Alajouanine, T. (1926). La myélite necrotique subaigue. Myélite centrale angéio-hypertrophique a evolution progressive: Paraplegie amyotrophique lentement ascendante, d'abord spasmodique, puis flasque, s'accompagnant de dissociation albuminocytologique. *Rev. Neurol.* **33**, 1.

Fontaine, S., Melanson, D., Cosgrove, D. R., and Bertrand, G. (1988). Cavernous hemangiomas of the spinal cord: MR imaging. *Radiology* **166**, 839–841.

Gueguen, B., Merland, J. J., Riche, M. C., and Rey, A. (1987). Vascular malformation of the spinal cord: Intrathecal perimedullary arteriovenous fistulas fed by medullary arteries. *Neurology* **37**, 969–979.

Hall, W. A., Oldfield, E. H., and Doppmann, J. L. (1989). Recanalization of spinal arteriovenous malformations following embolization. *J. Neurosurg.* **70**, 714.

Hassler, W., Thron, A., and Grote, E. H. (1989). Hemodynamics of spinal dural arteriovenous fistulas. *J. Neurosurg.* **70**, 360–370.

Heindel, C. C., Dugger, G. S., and Guinto, F. C. (1975). Spinal arteriovenous malformation with hypogastric blood supply. *J. Neurosurg.* **42**, 462–464.

Heros, R. C., Debrun, G. M., Ojemann, R. G., Lasjaunias, P. L., and Naessens, P. J. (1986). Direct spinal arteriovenous fistula: A new type of spinal (AVM). *J. Neurosurg.* **64**, 134–139.

Horton, J. A., Latchaw, R. E., Gold, L. H., and Pang, D. (1986). Embolization of intramedullary arteriovenous malformations of the spinal cord. *Am. J. Neurol. Res.* **7**, 113–118.

Huffmann, B. C., Gilsbach, J. M., and Thron, A. (1995). Spinal dural arteriovenous Fistulas: A plea for neurosurgical treatment. *Acta Neurochir. (Wien)* **135**, 44–51.

Jellinger, K. (1978). "Pathology of Spinal Vascular Malformations and Tumors in Spinal Angiomas: Advances in Diagnosis and Therapy" (H. W. Pia and R. Djindjian, Eds.). Springer-Verlag, New York.

Jellinger, K., Minauf, M., Garzuly, F., and Neumayer, E. (1968). Angiodysgenetische nekrotisierende Myelopathie (Bericht über 7 Fälle). *Arch. Psychiat. Nervenkrankh.* **211**, 377–404.

Kendall, B. E., and Logue, V. (1977). Spinal epidural angiomatous malformation draining into intrathecal veins. *Neuroradiology* **13**, 181.

König, E., Thron, A., Schrader, V., and Dichgans, J. (1989). Spinal arteriovenous malformations and fistulae: Clinical, neuroradiological and neurophysiological findings. *J. Neurol.* **236**, 260–266.

Küker, W., Nacimiento, W., Block, F., and Thron, A. (2000). Presumed capillary teleangiectasia of the pons: MRI and follow-up. *Eur. Radiol.* **10**, 945–950.

Latchaw, R. E., Harris, R. D., Chou, S. N., and Gold, L. H. A. (1980). Combined embolization and operation in the treatment of cervical arteriovenous malformations. *Neurosurgery* **6**, 131–137.

Mahagne, M. H., Rogopoulos, A., Paquis, P. H., *et al.* (1992). Fistules artérioveineuses durales intracraniennes a drainage veineux médullaire. *Rev. Neurol.* **148**, 789–792.

Merland, J. J., Assouline, E., Rüfenacht, D., Guimaraens, L., and Laurent, A. (1986). Dural spinal artériovenous fistulae draining into medullary veins: Clinical and radiological results of treatment (embolization and surgery) in 56 cases. *In* "Neuroradiology 1985/1986" (J. Valk, Ed.), pp 283–294. Elsevier, New York.

Merland, J. J., Riché, M. C., and Chiras, J. (1980). Les fistules artérioveineus intracanalaires, extramédullaires á drainage veineux medullaire. *J. Neuroradiol.* **7**, 271–320.

Morgan, M. K., and Marsh, W. R. (1989). Management of spinal dural arteriovenous malformations. *J. Neurosurg.* **70**, 832–836.

Mourier, K. L, Gelbert, F., Rey, A., Assouline, E., George, B., Reizine, D., Merland, J. J., and Cophignon, J. (1989). Spinal dural arteriovenous malformations with perimedullary drainage. *Acta Neurochir. (Wien)* **100**, 136–141.

Mourier, F., Gobin, Y. P., George, B., Loc, G., and Merland, J. J. (1993). Intradural perimedullary arteriovenous fistulae: Results of surgical and endovascular treatment in a series of 35 cases. *Neurosurgery* **32**, 885–891.

Newton, T. H., and Adams, J. E. (1968). Angiographic demonstration andnonsurgical embolization of spinal cord angioma. *Radiology* **91**, 873.

Picard, L., Verc, P., Renard, M., *et al.* (1969). Aspects radioanatomiques des angiomes médullaires. *Neuro.-Chir.* **15**, 519–528.

Reinges, H. T., Thron, A., Mull, M., Huffmann, B., and Gilsbach, J. M. (2001). Dural arteriovenous fistulae at the foramen magnum. *J. Neurol.* **248**, 945–950.

Riché, M. C., Melki, J. P., and Merland, J. J. (1983a). Embolization of spinal cord vascular malformations via the anterior spinal artery. *Am. J. Neurol. Res.* **4**, 378–381.

Riché, M. C., Scialfa, G., Gueguen, B., and Merland, J. J. (1983b). Giant extramedullary arteriovenous fistula supplied by the anterior spinal artery: Treatment by detachable balloons. *Am. J. Neurol. Res.* **4**, 391–394.

Ricolfi, F., Gobin, Y., Aymard, A., Brunelle, F., and Merland, J. J. (1991). Endovascular treatment of giant perimedullary spinal arteriovenous fistulae (14 cases). *Proc. XIV Symp. Neuroradiologicum.*

Rodesch, G., Lasjaunias, P., and Berenstein, A. (1992). Embolization of spinal cord arteriovenous malformations. *Rivista di Neuroradiologia* **5**(Suppl. 2), 67–92.

Rosenblum, B., Oldfield, E. H., Doppman, J. L., *et al.* (1987). Spinal arteriovenous malformations: A comparison of dural arteriovenous fistulas and intradural AVM's in 81 patients. *J. Neurosurg.* **67**, 795.

Scholz, W., and Manuelidis, E. E. (1951). Myélite nécrotique (Foix-Alajouanine)-Angiodysgenetische nekrotisierende Myelopathie. *Dtsch. Z. Nervenheilkd.* **165**, 56–71.

Schrader, V., König, E., Thron, A., and Dichgans, J. (1989). Neurophysiological characteristics of spinal arteriovenous malformations. *Electromyogr. Clin. Neurophysiol.* **29**, 169–177.

Spetzler, R. F., Zabramski, J. M., and Flom, R. A. (1989). Management of juvenile spinal AVM's by emolization and operative excision. *J. Neurosurg.* **70**, 628.

Stein, S., Ommaya, A. K., Doppman, J. L., and Di Chiro, G. (1972). Arteriovenous malformation of the cauda equina with arterial supply from branches of the internal iliac arteries. *J. Neurosurg.* **36**, 649–651.

Symon, L., Kuyama, H., and Kendall, B. (1984). Dural arteriovenous malformations of the spine. *J. Neurosurg.* **60**, 238–247.

Theron, J., Cosgrove, R., and Melanson, D. (1986). Spinal arteriovenous malformations: Advances in therapeutic embolization. *Radiology* **158**, 163.

Thron, A. (1988). "Vascular Anatomy of the Spinal Cord: Neuroradiological Investigations and Clinical Syndromes." Springer-Verlag, Berlin/New York.

Thron, A., König, E., Pfeiffer, P., *et al.* (1987). Dural vascular anomalies of the spine—An important cause of progressive radiculomyelopathy. *In* "Stroke and Microcirculation" (J. Cervos Navarro and R. Ferszt, Eds.). Raven Press, New York.

Wechsler, W. (1964). 1st die angiodysgenetische nekrotisierende Myelopathie (Foix-Alajouaninesche Krankheit) eine Mißbildung oder eine Mißbildungs-krankheit? *Arch. Psychiat. Nervenkrankh.* **206**, 131–145.

Woimant, A., Merland, J. J., Riché, M. C., *et al.* (1982). Syndrome bulbo-medullaire en rapport avec une fistule arterioveineuse menigee du sinus lateral a drainage veineux médullaire. *Rev. Neurol.* **132**, 559–566.

Wrobel, C. J., Oldfield, E. H., Di Chiro, G., Tarlov, E. C., Baker, R. A., and Doppman, J. L. (1988). Myelopathy due to intracranial dural arteriovenous fistulas draining intrathecally into spinal medullary veins: Report of three cases. *J. Neurosurg.* **69**, 934–939.

Wyburn-Mason, R. (1943) "The vascular abnormalities and tumors of the spinal cord and its membranes." Klimpton, London.

Zevgaridis, D., Medele, R., Hamburger, C., Steiger, H.-J., and Reulen, H.-J. (1999). Cavernous haemangiomas of the spinal cord: A review of 117 cases. *Acta Neurochir. (Wien)* **141**, 237–245.

*Infections and Inflammatory Diseases*

## CHAPTER 42
# Bacterial Infections

Hans-Walter Pfister and T. P. Bleck

## BACTERIAL MENINGITIS

### Clinical Aspects

The clinical hallmarks of bacterial meningitis are stiff neck, headache, fever, photophobia, malaise, vomiting, alteration of consciousness, confusion, irritability, and rarely acute psychosis. Meningismus is maintained in the majority of the patients with altered consciousness even if comatose.

Diagnosis of bacterial (purulent) meningitis is established by

(a) identification of the bacterial microorganism in the cerebrospinal fluid (CSF) by culture, microscopy of a Gram-stained smear, or antigen detection using latex particle agglutination test, and/or

(b) elevated CSF cell count of more than 1000 white blood cells/μL and CSF consisting of more than 60% polymorphonuclear leukocytes, elevated protein content, and a CSF:blood glucose ratio of less than 0.3.

Bacterial microorganisms are detectable in the CSF in 70–80% of the patients by at least one of the above-mentioned methods. The most common etiologic agents of bacterial meningitis are *Neisseria meningitidis* and *Streptococcus pneumoniae* (Table I).

The likelihood of a positive finding by CSF microscopy or culture decreases by 30 to 40% if antibiotic therapy is administered before admission. In nearly half of the patients with bacterial meningitis positive blood cultures can be obtained. Total protein content in the CSF is typically elevated to levels above 120 mg/dL. The determination of lactate in the CSF has been reported to be a useful additional test. At a very early stage of the disease approximately 10 to 20% of the patients with bacterial meningitis have a cell count of less than 1000 cells/μL (Durand *et al.*, 1993). Apart from an early disease stage, a cell count of less than 1000 cells/μL may be found in partially treated bacterial meningitis; overwhelming bacterial meningeal infection, most often due to pneumococci ("apurulent bacterial meningitis"; Felgenhauer and Kober, 1985); and immunosuppression and leukopenia.

Certain bacterial pathogens such as *Listeria monocytogenes* may either induce a purulent CSF or a mixed granulocytic/lymphocytic pleocytosis with a cell count of less than 1000/μL. White cell count in blood, C-reactive protein and erythrocyte sedimentation rate are usually elevated in patients with bacterial meningitis. Recently, elevated plasma procalcitonin levels were found in patients with bacterial but not viral meningitis (Gendrel *et al.*, 1997; Viallon *et al.*, 1999). These data suggest that the measurement of plasma procalcitonin might be of value in the differential diagnosis of meningitis due to either bacteria or viruses. However, further studies are warranted to clarify the definite role of this parameter.

The presence of a petechial-purpura rash at physical examination is highly suspicious of *N. meningitidis* infection or, more rarely, *Staphylococcus aureus* or pneumococcal infection. A recent study showed that a petechial like rash was present in 75% of 255 patients with acute meningococcal meningitis (Andersen *et al.*, 1997). About 10–15% of meningococcal infections exhibit an overwhelming course, resulting in the Waterhouse–Friderichsen syndrome, i.e., fulminating meningococcal sepsis. This is characterized by high fever, large petechial hemorrhages in the skin and mucous membranes (purpura fulminans), hypotension, disseminated intravascular coagulation, and multiorgan failure. The presence of peripheral cutaneous emboli (Osler's spots), especially located at the fingers and toes, in a patient with bacterial meningitis raises suspicion of underlying infective endocarditis. Herpes simplex infection of the lips—occasionally very severe forms

TABLE I    Typical Microorganisms Related to the Patient's Age

| Age | Typical microorganisms |
|---|---|
| <1 Month | Gram-negative Enterobacteriaceae (*Escherichia coli*, Klebsiella, Enterobacter, Proteus, *Pseudomonas aeruginosa*) |
| | Streptococci, in particular group B streptococci (*S. agalactiae*) |
| | Listeria monocytogenes |
| 1 Months–6 years | *Neisseria meningitidis* |
| | *Streptococcus pneumoniae* |
| | Other streptococci, staphylococci, Listeria, Gram-negative Enterobacteriaceae (incl. *P. aeruginosa*) (*Haemophilus influenzae*[a]) |
| 6 Years–15 years | *N. meningitidis* |
| | *S. pneumoniae* |
| | Other streptococci, *H. influenzae*, Listeria, staphylococci, Gram-negative Enterobacteriaceae (incl. *P. aeruginosa*) |
| >15 Years | *S. pneumoniae* |
| | *N. meningitidis* |
| | Other streptococci, *H. influenzae*, Listeria, staphylococci, Gram-negative Enterobacteriaceae (incl. *P. aeruginosa*) |

[a]Incidence has declined since the introduction of *H. influenzae* type b vaccine.

TABLE II    Major Differential Diagnoses of Bacterial Meningitis

— Viral meningitis/encephalitis (e.g., herpes simplex encephalitis)
— Tuberculous meningitis
— Fungal meningitis (in particular cryptococcal meningitis)
— Carcinomatous meningitis
— Naegleria meningoencephalitis
— Mollaret meningitis
— Rickettsial meningitis
— Parameningeal purulent infectious foci (brain abscess, epidural abscess, subdural empyema)
— Subarachnoid hemorrhage
— Sinus-/venous thrombosis
— Tumor in the posterior fossa

are seen—may develop in approximately 10% of the patients during the acute stage of the disease.

Focal neurologic signs (e.g., hemi- or tetraparesis, ataxia, aphasia, visual field defects) are detectable in approximately 10–15% of the patients. Seizures occur in 20 to 40% of patients, in particular in pneumococcal meningitis rather than in meningococcal disease. Cranial nerve palsies, usually of the sixth, third, seventh, or eighth cranial nerve, are found in approximately 10% of the patients. Papilloedema is very rare in bacterial meningitis. Hearing impairment is a well-known sequela of acute bacterial meningitis in children (Dodge *et al.*, 1984; Lebel *et al.*, 1988). The overall incidence of hearing impairment is about 10% (Fortnum, 1992). However, the incidence of hearing loss was as high as 31% in pneumococcal meningitis as determined by electric response audiometry and conventional tests (Dodge *et al.*, 1984). A recent study showed that high-resolution MRI can visualize and differentiate the sites of auditory and vestibular involvement in adult patients with hearing loss as a complication of bacterial meningitis (Dichgans *et al.*, 1999).

Cranial CT (or MRI) may identify the following in patients with bacterial meningitis: parameningeal infectious foci (bone window technique), e.g., sinusitis, mastoiditis; intracranial free air due to a dural leak; brain abscess or subdural empyema (leading secondarily to bacterial meningitis); meningeal and ventricular ependymal enhancement; basal purulent exudate; and cerebral complications of bacterial meningitis, e.g., hydrocephalus, brain edema with small ventricles, infarction, signs of venous sinus thrombosis.

MRI is superior to CT for detecting of parenchymal ischemic changes. Cerebrovascular involvement is usually detected by MRI (including diffusion weighted imaging) and MR angiography. In addition, MRI may reveal septic sinus thrombosis. Transcranial Doppler sonography may be useful in diagnosing and monitoring the involvement of great arteries at the base of the brain during the acute phase of bacterial meningitis (Haring *et al.*, 1993).

The major *differential diagnoses* of acute bacterial meningitis are summarized in Table II (Pfister and Roos, 1994).

### Natural Course

The worldwide incidence of bacterial meningitis is estimated at 1.2 million cases per year and that of meningococcal meningitis at 500,000 cases per year; overall mortality rates are approximately 10–15% (WHO, 2000). The incidence of pneumococcal meningitis is estimated at 1–2 cases/100,000 persons per year but may reach up to 20/100,000 persons per year in devleoping countries. Bacterial meningitis is much more common in developing countries and in specific geographic areas, such as the meningitis belt of Africa, where there is an estimated incidence of 70 cases per 100,000 persons per year.

The epidemiologic features of bacterial meningitis in the United States in 1995—5 years after the *Haemophilus influenzae* type B conjugate vaccines were licensed for routine immunization of infants—were recently reviewed by Schuchat *et al.* (1997). Because of a 94% reduction in the number of cases of *H. influenzae* meningitis, *S. pneumoniae* and *N. meningitidis* have become the major pathogens causing bacterial meningitis, and meningitis is now a disease predominantly of adults rather than of infants and young children (Philipps and Simor, 1998; Schuchat *et al.*, 1997). *Listeria monocytogenes* is a microorganism involved in bacterial meningitis, especially in neonates, the elderly, and immunocompromised patients (see Table I). Studies from Europe and North America have reported incidences of *L. monocytogenes* of 3–10% among all episodes of community-acquired acute bacterial meningitis among adults. Gram-negative Enterobacteriaceae cause 10% of the overall cases of bacterial meningitis; however, they are the etiologic agents in 60 to 70% of all cases of meningitis following a neurosurgical procedure and are a common cause of meningitis in the elderly and adults debilitated by chronic illness. Anaerobic microorganisms are responsible for less than 1% of the cases. Mixed infection with more than one bacterial microorganism is found in 1% of meningitis cases, in particular in patients with immunosuppression, skull fractures, a previous neurosurgical procedure, or parameningeal infectious foci (e.g., sinusitis, otitis).

An important epidemiologic trend is the worldwide increase in infections with antibiotic-resistant strains of *S. pneumoniae*. The resistance of *S. pneumoniae* is mediated by alterations in the penicillin-binding proteins involved in the synthesis of bacterial cell walls (Tomasz, 1997). Infections with antibiotic-resistant strains of *S. pneumoniae* have emerged as a major problem in Spain, Hungary, South Africa, Asia, and the United States (Schuchat *et al.*, 1997). In 1995, more than one-third of cases of pneumococcal meningitis in the USA were caused by organisms that were not susceptible to penicillin (Schuchat *et al.*, 1997). Likewise, in 1996, 55.8% of the *S. pneumoniae* strains isolated from CSF in France were penicillin-resistant (Doit *et al.*, 1998).

With increasing prevalence of penicillin-resistant *S. pneumoniae* alternative antibiotic strategies for pneumococcal meningitis (e.g., initial combination of vancomycin plus ceftriaxone or rifampin) are necessary. Furthermore, penicillin-resistant strains of *N. meningitidis* are also emerging in some regions of the world (e.g., Spain, South Africa).

The spectrum of meningeal microorganisms causing bacterial meningitis is mainly dependent on the age of the patient, predisposing factors, and underlying diseases (Tables I and III).

More than half of adult patients with bacterial meningitis have underlying risk factors, including parameningeal infections (e.g., otitis, sinusitis, mastoiditis, brain abscess, subdural empyema), a recent neurosurgical procedure, a history of a previous head trauma with or without a dural fistula, a remote infectious focus (e.g., pneumonia, infective endocarditis, psoas abscess), immunodeficient states (diabetes mellitus, chronic alcoholism, therapy with immunosuppressive agents, acquired immunodeficiency syndrome, previous splenectomy), or malignancy. Bacterial meningitis following lumbar puncture or peridural anesthesia is observed very rarely. Recurrent bacterial meningitis may be observed in patients with congenital defects (e.g., meningomyelocele, dermal sinus), previous head trauma (e.g., skull fracture, dural leak), parameningeal foci (e.g., chronic mastoiditis), and immunologic defects (e.g., splenectomy, AIDS). Among more than 800 patients with *Listeria* meningitis/meningoencephalitis hematologic malignancy and kidney transplantation were the leading predisposing factors, but one-third of patients had no underlying diseases recognized (Mylonakis *et al.*, 1998).

**TABLE III** Bacterial Microorganisms and Predisposing Factors of Bacterial Meningitis

| Predisposing factors | Probable microorganisms |
|---|---|
| Sinusitis, mastoiditis, otitis media | *Streptococcus pneumoniae*, staphylococci |
| Traumatic head injury, dural fistual | *S. pneumoniae, Staphylococcus aureus*, Gram-negative Enterobacteriaceae |
| Nosocomial meningitis (e.g., after neurosurgical procedure) | Gram-negative Enterobacteriaceae, *Pseudomonas aeruginosa*, staphylococci |
| Ventriculitis due to external ventricular drainage, shunt infection | *Staphylococcus epidermidis, Staphylococcus aureus*, Gram-negative Enterobacteriaceae |
| Pneumonia | *S. pneumoniae*, other streptococci |
| Endocarditis | *S. aureus, S. pneumoniae*, other streptococci |
| Recurrent meningitis | *S. pneumoniae*, other streptococci, *Haemophilus influenzae* |
| Petechial rash | *N. meningitidis*, staphylococci, *S. pneumoniae* |
| Immunosuppression (e.g., chronic alcoholism, diabetes mellitus, immunosuppressive therapy) | *Listeria monocytogenes*, Gram-negative Enterobacteriaceae, staphylococci, *S. pneumoniae* |
| Splenectomy | *S. pneumoniae* |
| Intravenous drug abuse | Staphylococci, *P. aeruginosa* |

Clinical symptoms and signs of bacterial meningitis usually develop rapidly progressive within several hours. However, meningitis may also have a more subacute presentation, evolving over 24 to 48 h. Appropriate antibiotic therapy usually leads to clinical improvement within several days. Predictors for an unfavorable course of bacterial meningitis include apurulent bacterial meningitis, i.e., high bacterial density in the CSF associated with low cell count; age older than 40 years; underlying or concomitant diseases (e.g., splenectomy or infective endocarditis); type of bacterial microorganism (e.g., Gram-negative bacilli or pneumococci); and delayed antimicrobial therapy.

Mortality rates of meningitis due to *N. meningitidis* is 3 to 10%, due to *H. influenzae* type b less than 7%, and in Gram-negative meningitis 10 to 20%. However, despite further progress in antimicrobial therapy and improvements in intensive care medicine, the mortality rate of meningitis due to pneumococcus, the organism most often responsible for bacterial meningitis in adults, has remained relatively unchanged during the last decades and is still unacceptably high (20 to 30%) (Swartz, 1984; Durand *et al.*, 1993; Pfister *et al.*, 1993; Sigurdardottir *et al.*, 1997). Neurological and neuropsychological sequelae are found overall in (10 to) 30% of patients with bacterial meningitis. In a long-term follow-up study in infants and children with bacterial meningitis 14% were shown to have persistent deficits, sensorineural hearing being most often reported (Pomeroy *et al.*, 1990). Seven percent had one or more late seizures not associated with fever.

A major cause of mortality and sequelae in bacterial meningitis are cerebral and systemic complications arising during the acute phase of the disease. As shown at autopsy in adults with bacterial meningitis the lethal outcome was caused primarily by central nervous complications in one-third of the patients, systemic complications in one-third, and by both in the other third (Pfister *et al.*, 1993).

### Central Nervous System Complications

The major central nervous system complications included angiographically documented cerebrovascular involvement, brain edema, and hydrocephalus (Table IV).

Cerebrovascular complications of bacterial meningitis may include arterial narrowing of the major arteries at the base of the brain, in particular affecting the supraclinoid portion of the internal carotid artery (vasospasm); vessel wall irregularities, focal dilations, and occlusions of the distal branches of the middle cerebral artery (vasculitis); and thrombosis of the sagittal superior sinus and cortical veins (Igarashi *et al.*, 1984; Pfister *et al.*, 1992, 1993; Southwick, 1995). Cerebrovascular involvement may lead to infarction with severe irreversible cerebral damage and to an increase of intracranial pressure (ICP) due to cytotoxic edema. In addition, elevation of ICP in patients with bacterial meningitis may result from increased intracranial blood volume due either to disturbed cerebrovascular autoregulation or to septic sinus venous thrombosis (Paulson *et al.*, 1974; Southwick, 1995). There is a risk of cortical necrosis when cerebral perfusion pressure (defined as the difference between systemic mean arterial blood pressure and ICP) decreases as a result of increased ICP. Both vasogenic and cytotoxic *brain edema* may be present in bacterial meningitis (Dodge and Swartz, 1965). In addition, interstitial edema may occur due to transependymal movement of CSF from the ventricular system into the surrounding brain parenchyma as a consequence of obstructive hydrocephalus.

Other central nervous system complications that may arise during bacterial meningitis include subdural effu-

TABLE IV    Frequency of Neurological Complications in One Prospective and Two Retrospective Studies of Bacterial Meningitis[a]

| | Pfister *et al.* (1993) (*n* = 86) | Durand *et al.* (1993) (*n* = 493 episodes) | Dodge und Swartz (1965) (*n* = 156)[b] |
|---|---|---|---|
| Cerebrovascular involvement (%) | 15[c] | 5[d] | 20[b] |
| Cerebral edema (%) | 14[e] | 4[d] | 3[f] |
| Cerebral herniation (autopsy confirmed) (%) | 8 | 3 | 6 |
| Seizures (%) | 24 | 23 | 31 |
| Hydrocephalus (%) | 12[e] | 11[d] | ND |
| Mortality (%) | 19 | 25 | 20 |

*Note.* ND, not documented.
[a]After Chang and Bleck (1997), copyright Marcel Dekker, Inc.
[b]Diagnosis was made by focal neurological deficits on physical examination.
[c]Angiographically documented.
[d]Diagnosis was made by CT scan in 87 of 122 adults treated for community-acquired meningitis.
[e]Diagnosis was made by CT scan.
[f]Diagnosis was made only in patients who died with autopsy confirmation.

sion (in particular in infants of less than 18 months of age), brain abscess and subdural empyema (infrequently a secondary complication of bacterial meningitis), and rarely central diabetes insipidus, cerebral salt wasting syndrome, and the syndrome of inappropriate secretion of antidiuretic hormone and spinal complications (vasculitis, myelitis, syringomyelia) (Durand et al., 1993; Kastenbauer et al., 2001; Roos, 1997).

### Systemic Complications

The major systemic complications during meningitis include septic shock, adult respiratory distress syndrome, and disseminated intravascular coagulation (Pfister et al., 1993). Less frequently observed are septic or reactive arthritis, rhabdomyolysis, pancreatitis, and septic panophthalmitis (Pfister et al., 1993; Spataro and Marone, 1993). In addition, typical complications that may arise during intensive care therapy are pneumonia, thrombosis of deep veins of the extremities, pulmonary embolism, pharmacogenic or alcohol withdrawal syndrome, electrolyte disturbances, and side effects of drug or surgical therapy.

### Temporal Profile of Major Cerebral Complications

Brain edema noted on CT is an early feature in the course of bacterial meningitis and is predominantly observed during the first 3 days after onset of disease (Pfister et al., 1993). Hydrocephalus is usually observed during the first 2 weeks of disease and may evolve even later, several weeks to months after the onset of the disease. However, hydrocephalus may already be present in the first days of the disease and may provoke a life-threatening condition due to increased ICP and cerebral herniation. Cerebrovascular complications, septic shock, and disseminated intravascular coagulation are diagnosed mainly during the first week of the disease.

## Principles of Therapy

Animal models (in particular of the rat, rabbit, and mouse) have substantially improved our understanding of the complex pathophysiological mechanisms of bacterial meningitis (Koedel and Pfister, 1999a). In recent years it has become clear that meningitis-associated brain injury is not simply dependent on the presence of viable bacteria, but rather occurs as a consequence of host reactions to bacterial components (Frei et al., 1993; Koedel and Pfister, 1999b; Nau et al., 1999; Scheld et al., 1980; Täuber et al., 1985; Tureen et al., 1990; Weber et al., 1995). Once bacteria have entered the subarachnoid space, it appears that subcapsular bacterial

surface components (e.g., cell wall components of Gram-positive bacteria and lipopolysaccharide of Gram-negative bacteria) induce the local production of several inflammatory mediators within the CSF including platelet-activating factor, and certain cytokines, e.g., interleukin-1β, interleukin-6, interleukin-8, and tumor necrosis factor α (Quagliarello and Scheld, 1993). The cytokines may be produced by monocytes, macrophages, brain astrocytes, and microglial cells. The action of these cytokines may lead to an increased expression of adhesion molecules, both on the endothelium and on neutrophils. These processes result in an enhanced adhesion of leukocytes to the cerebral endothelium and finally in transmigration of leukocytes into the subarachnoid space. Leukocytes in the subarachnoid space are thought to be more harmful than beneficial. The major defense function of the leukocytes, namely phagocytosis of microorganisms, is inefficient in the CSF due to low concentrations of anticapsular antibodies and complement (Scheld, 1984). As the disease progresses, alterations of the permeability of the blood–brain barrier and vasogenic edema, hydrocephalus, increase in ICP, loss of cerebrovascular autoregulation, and a reduction of cerebral blood flow may occur with the risk of secondary cerebral brain damage.

There is now substantial evidence that much of the oxidative injury associated by simultaneous production of superoxide and nitric oxide is mediated by the strong oxidant peroxynitrite (Koedel and Pfister, 1999b). Reactive oxygen species (ROS) and peroxynitrite can be cytotoxic via a number of independent mechanisms. Their cytotoxic effects include upregulation of caspases and matrix metalloproteinases, initiation of lipid peroxidation, and induction of DNA single strand breakage, activating poly(ADP-ribose) polymerase (PARP). Agents that interfere with the production of these mediators represent novel, therapeutic strategies to limit meningitis-associated brain damage and, thus, to improve the outcome of this serious disease.

### General Management of a Patient with Suspected Bacterial Meningitis

In a patient with suspected acute bacterial meningitis urgent lumbar puncture (plus blood culture) is indicated after clinical examination. Then, empiric antibiotic therapy is initiated followed by a cranial CT scan (see Fig. 1).

Any delay in initiating treatment should be avoided. Patients who are unconscious and have focal neurologic deficits, receive an initial antibiotic dose immediately after drawing a single blood culture, prior to any other diagnostic procedure. CT scanning and CSF examination should be performed afterward as soon as possible.

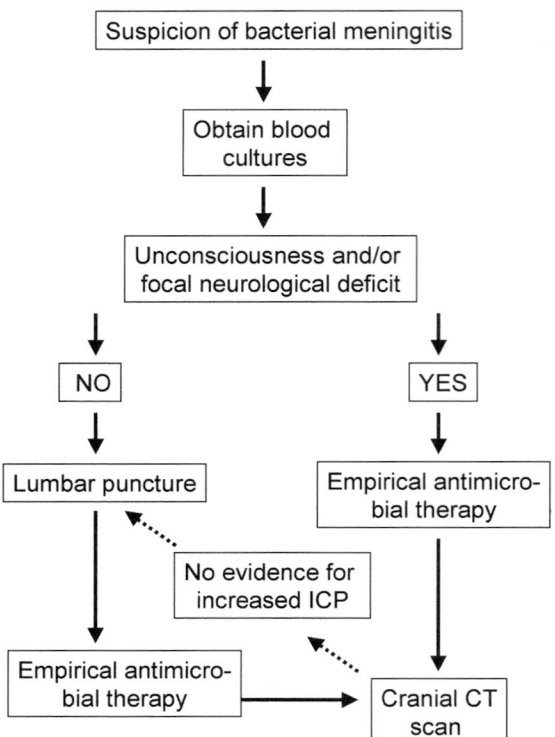

**FIGURE 1** Algorithm for initial management of a patient with acute bacterial meningitis.

Contraindications to lumbar puncture are clinical signs of cerebral herniation (e.g., an unilaterally dilated and unreactive pupil, decerebrate movements) and a focal mass lesion (e.g., large, space-occupying brain abscess) or severe brain edema on CT scan. If there are CT findings indicating mild brain edema and raised ICP, hyperosmolar agents (e.g., 125 mL 20% mannitol) may be infused intravenously just before the lumbar puncture. However, there is no scientific proof of the efficacy of this type of management in this situation. The presence of a parameningeal infectious focus such as sinusitis or mastoiditis should also be investigated by CT, including the bone window technique. In addition, clinical examination by an otolaryngologist should be performed. If a parameningeal focus (e.g., otitis, mastoiditis, sinusitis) is identified as a possible origin of bacterial meningitis, surgical intervention is required as soon as possible. If the patient's clinical condition does not improve despite antibiotic therapy, the possibility of complications of bacterial meningitis should be investigated (e.g., repeated CT or MRI scanning) and additional sources of infection sought (e.g., persistent infectious focus in the paranasal sinuses). Importantly, the sensitivity of the causative pathogen against the antibiotics used must be confirmed by *in vitro* testing (including determination of minimal inhibition concentrations), and antibiotic coverage must be adjusted to the sensitivity results. If the causative organism has not been isolated, broaden-

ing of the antibiotic coverage should be considered in patients who fail to respond to the initial therapy.

### Antibiotic Therapy

If the bacterial microorganism causing meningitis is known, the antibiotic of choice will have high activity against the pathogen, appropriate blood-CSF penetration resulting in adequate CSF-antibiotic concentrations, uninterrupted antibiotic activity within purulent and acidic CSF, and a relatively low incidence of adverse effects. Experimental and clinical studies have shown that the best therapeutic response is achieved with CSF-antibiotic concentrations that exceed by 10- to 20-fold the minimal bactericidal concentration *in vitro* for the particular organism by (Scheld, 1989). If treatment has to be started without microbiological confirmation, empirical therapy is initiated with respect to the patient's age, predisposing factors, underlying diseases, and the most likely meningeal pathogens (Table III). After initiating antibiotic therapy, daily lumbar punctures are usually performed until the CSF becomes sterile. The CSF usually becomes sterile within 24 to 48 h after beginning therapy.

### Adjunctive Therapy

Since the lysis of bacteria by antibiotics induces the release of cell wall components, which in turn initiate the inflammatory cascade, it is hoped that improvement of adjunctive therapy to treat complications will be a promising therapeutic approach.

**Corticosteroids.** Corticosteroids were shown to have beneficial effects in animal models of bacterial meningitis: they decreased ICP, reduced the degree of brain edema and meningeal inflammation, improved disturbances in CSF hydrodynamics, and prevented changes in cerebral blood flow (Pfister *et al.*, 1990; Scheld *et al.*, 1980; Syrogiannopoulos *et al.*, 1987; Täuber *et al.*, 1985). In prospective, placebo-controlled clinical studies children with bacterial meningitis (most of the patients suffered from *H. influenzae* type b meningitis) dexamethasone was reported to reduce the incidence of bilateral hearing loss (Lebel *et al.*, 1988, **) and neurological sequelae (Odio *et al.*, 1991, **). A Swiss study reported a tendency for dexamethasone to be superior to placebo (Schaad *et al.*, 1993, **). It was shown that a 4-day regimen of dexamethasone was equally effective against bacterial meningitis in children as was a 2-day dexamethasone regimen (Syrogiannopoulos *et al.*, 1994, **). Furthermore, in a retrospective data analysis of children with pneumococcal meningitis neurological sequelae were less frequently detected in children treated with dexamethasone than in those treated with antibiotics alone

(Kennedy *et al.*, 1991\*\*). In an open randomized Egyptian study it was shown that dexamethasone may reduce mortality of pneumococcal meningitis in adults (Girgis *et al.*, 1989\*\*).

In a recent meta-analysis the data of 11 randomized clinical trials since 1988 using dexamethasone as adjunctive therapy in bacterial meningitis were evaluated (McIntyre *et al.*, 1997\*\*\*). In *H. influenzae* meningitis in children, dexamethasone reduced severe hearing loss overall. In pneumococcal meningitis, only studies in which dexamethasone was given early suggested protection, which was significant for severe hearing loss and approached significance for any neurological or hearing deficit. Outcomes were similar in studies that used 2 days rather than more than 2 days of dexamethasone therapy.

Recently, the clinical benefit of early adjunctive dexamethasone therapy was investigated in adults with bacterial meningitis in France and Switzerland (Thomas *et al.*, 1999\*\*). Unfortunately, the study had to be stopped prematurely, because of a new national recommendation of experts to use a third-generation cephalosporin and vancomycin as a result of the increasing rate of penicillin-resistant *S. pneumoniae* in France. The difference of rate of cured patients without any neurologic sequelae was not statistically significant between the dexamethasone group and the placebo group.

In the spring of 2001, 280 adult patients have been included in the multicenter, placebo-controlled, double-blind, European Dexamethasone in Bacterial Meningitis Study (principal investigator: Jan de Gans, Amsterdam\*\*). The calculated number of patients in this trial is 300. Thus, the beneficial effect of dexamethasone in bacterial meningitis in adults has not yet been scientifically proven. However, this ongoing study offers the hope to get further information on the role of dexamethasone in adults with bacterial meningitis.

When using a short-term therapy with dexamethasone (e.g., $3 \times 8$ mg/day IV over 2 or 4 days) the risk of a delayed CSF sterilization and the number of serious side effects are likely to be low. There is no proven effect of dexamethasone on established cerebral arterial complications. We do not currently know whether early dexamethasone will be able to prevent cerebrovascular arterial complications.

Corticosteroids are not recommended for the therapy of meningitis following infective endocarditis or in newborns with bacterial meningitis. Experimental or clinical data have not proven the efficacy of dexamethasone in meningococcal meningitis.

**Other Adjunctive Therapeutic Approaches.** Other adjunctive therapeutic approaches include treatment of increased ICP, septic sinus venous thrombosis, epileptic seizures, and the correction of electrolyte disturbances. Anticoagulation of septic sinus venous thrombosis in bacterial meningitis is controversial. There are no prospective controlled clinical studies investigating anticoagulation in the treatment of septic sinus venous thrombosis. Despite the advent of antibiotic therapy, the mortality rate of septic cortical thrombophlebitis and thrombosis of the superior sagittal sinus is still 50–80% (DiNubile *et al.*, 1990; Southwick *et al.*, 1986). Since the outcome with antibiotic therapy alone has been unsatisfactory, anticoagulation with dose-adjusted intravenous heparin may be performed in patients with MRI or angiographically proven septic sinus venous thrombosis.

**Experimental Approaches.** A variety of potential therapeutic agents, which may limit inflammation of the subarachnoid space, have been investigated in animal models of bacterial meningitis and have shown beneficial effects. These anti-inflammatory agents include nonsteroidal anti-inflammatory drugs (e.g., indomethacine), 21-aminosteroids, pentoxifylline, antagonists of leukocyte-endothelial cell adhesion molecules, cytokine antibodies, platelet-activating-factor receptor antagonists, free radical scavengers, nitric oxide synthase inhibitors, inhibitors of PARP, matrix metalloproteinases, and caspases (Braun *et al.*, 1999; Koedel and Pfister, 1999b; Leib *et al.*, 2000). These agents have not yet been investigated in humans, but it appears their effect in the experimental situation is promising. Further studies are needed to clarify whether some of these approaches may be applied clinical practice.

Vasospasm of the large arteries at the base of the brain resembles vasospasm following subarachnoid hemorrhage. In these patients, hypervolemic hypertensive therapy or nimodipine therapy (e.g., 1–2 mg/h IV in adults) should be considered; however, these approaches have not been investigated systematically in bacterial meningitis.

The experimental procedure of CSF-filtration in patients with bacterial meningitis frequently associated with an increased ICP should not be performed.

Recently, it was shown in an open-label study that protein C replacement therapy in severe meningococcal septicemia was associated with a reduction in predicted morbidity and mortality (White *et al.*, 2000). This beneficial effect may reflect both the anticoagulant and anti-inflammatory properties of the PC pathway.

**Practical Management**

*Empiric Antibiotic Therapy (Choice of Antibiotics)*

Recommendations for initial empiric antibiotic therapy for bacterial meningitis according to the patient's age and clinical condition are given in Table V. The most common bacterial pathogens in neonates with bacterial meningitis are Gram-negative Enterobac-

**TABLE V**  Initial Empiric Antibiotic Therapy for Bacterial Meningitis in Newborns, Children, and Adults

| Age groups | Recommended antibiotic regimen |
|---|---|
| Neonates | Cefotaxime plus ampicillin |
| Infants and children | Third-generation cephalosporin[a] |
| Adults | |
|   Healthy, immunocompetent, community-acquired | Third-generation cephalosporin[a] plus ampicillin[b] |
|   Nosocomial (e.g., postneurosurgical or recent head injury) | Vancomycin plus ceftazidime (or meropenem) |
|   Immunocompromised, elderly | Ceftazidime plus ampicillin |
|   Shunt-related meningitis | Vancomycin plus ceftazidime (or meropenem) |

[a]Ceftriaxone or cefotaxime.
[b]In areas with high rates of resistant pneumococcal strains, initial empiric therapy should include two antibiotics, ceftriaxone, and either rifampin or vancomycin.

teriaceae, *Streptococcus agalactiae*, and *L. monocytogenes* (Table I). Therefore, an empiric therapy in this age group consisting of cefotaxime plus ampicillin is often recommended; alternatively some authors recommend ampicillin combined with an aminoglycoside. As third-generation cephalosporins are usually inactive *in vivo* against *L. monocytogenes*, monotherapy with a cephalosporin is not recommended despite laboratory data that may suggest sensitivity to these agents (Khayr *et al.*, 1992). In infants and children (over 2 months of age) initial empiric antibiotic therapy usually includes the administration of a third-generation cephalosporin (e.g., ceftriaxone or cefotaxime). Initial antibiotic treatment of healthy, immunocompetent adults for community-acquired bacterial meningitis, most often caused by pneumococci or meningococci, consists of a third-generation cephalosporin (e.g., ceftriaxone or cefotaxime). It was recommended that in areas with high rates of resistant pneumococcal strains, initial empiric therapy should include two antibiotics, ceftriaxone and either rifampin or vancomycin. CSF isolates of pneumococci and meningococci should be tested for penicillin and ampicillin susceptibility (Table VI).

**TABLE VI**  Antibiotic Treatment of Bacterial Meningitis (Known Pathogen[a])

| Causative organism | Drug of choice | Alternatives |
|---|---|---|
| *Neisseria meningitidis* | Penicillin G | Ceftriaxone (or cefotaxime), ampicillin, rifampin |
| *Streptococcus pneumoniae*, penicillin-susceptible | Penicillin G | Ceftriaxone (or cefotaxime) |
| *S. pneumoniae*, penicillin-tolerant (MIC 0.1–1 µg/mL) | Ceftriaxone (or cefotaxime) | Meropenem, cefepime |
| *S. pneumoniae*, penicillin-resistant (MIC > 1 µg/mL) | Ceftriaxone + Vancomycin or Ceftriaxone + rifampin | Meropenem, cefepime |
| *Haemophilus influenzae* | Ceftriaxone (or cefotaxime) | Ampicillin plus chloramphenicol |
| Group B streptococci | Penicillin G (+ gentamicin) | Ceftriaxone, ampicillin (plus gentamicin) vancomycin |
| Gram-negative Enterobacteriaceae (e.g. *Klebsiella*, *E. coli*, Proteus) | Ceftriaxone (or cefotaxime) plus aminoglycoside | Meropenem, cefepime |
| *Pseudomonas aeruginosa* | Ceftazidime plus aminoglycoside | Meropenem, cefepime |
| Staphylococci (methicillin-susceptible) | Fosfomycin or nafcillin | Rifampin, vancomycin (or flucloxacillin) |
| Staphylococci (methicillin-resistant) | Vancomycin | Trimethoprim-sulfamethoxazole, rifampin |
| *Listeria monocytogenes* | Ampicillin plus gentamicin | Trimethoprim-sulfamethoxazole, meropenem |
| *Bacteroides fragilis* | Metronidazole | Meropenem, clindamycin |

[a]Treatment according to the susceptibility tests.

For penicillin-resistant meningococci or relatively penicillin-resistant pneumococci a third-generation cephalosporin is used, and for highly resistant pneumococci vancomycin therapy is often needed.

A combination of vancomycin plus ceftazidime (or meropenem) is recommended for a patient with bacterial meningitis following a recent head trauma or a neurosurgical procedure. Empirical antibiotic therapy of ventriculitis associated with an external intraventricular drainage device or a ventriculoperitoneal shunt infection should cover *Staphylococcus epidermidis*, *S. aureus*, and Gram-negative Enterobacteriaceae; therefore a combination of a third-generation cephalosporin (or meropenem) plus vancomycin is usually recommended. When bacterial meningitis develops following infection of a ventriculoperitoneal shunt, the shunt often needs to be removed. A temporary external ventricular drainage device should be inserted to control hydrocephalus while the infection is being treated. In cases of severe staphylococcal ventriculitis vancomycin may be given intraventricularly. If ventriculitis is caused by Gram-negative Enterobacteriaceae susceptible to aminoglycosides, then gentamicin may be given intraventricularly in addition to intravenous administration. Immunocompromised adults should be treated with a third-generation cephalosporin plus ampicillin to cover *L. monocytogenes*, *S. pneumoniae*, and Gram-negative Enterobacteriaceae.

### Dosing Recommendations of Antibiotics

The daily dosages and dosing intervals of antibiotics commonly used in therapy for bacterial meningitis in children and adults are summarized in Table VII.

### Duration of Antibiotic Therapy

Recommendations on how long bacterial meningitis should be treated with antibiotics are given in Table VIII.

### Patient Isolation

Patients with clinically suspected meningococcal meningitis (e.g., petechial rash, Gram-negative diplococci on Gram's stained smear of the CSF) should be isolated for the first 24 h after initiation of antibiotic therapy. Patients with bacterial meningitis due to other bacterial microorganisms do not require isolation.

### Adjunctive Therapy

**Dexamethasone.** The usage of dexamethasone is recommended in *H. influenzae* meningitis in children, e.g., 0.15 mg/kg body weight every 6 h for 4 days (***). The American Academy of Pediatrics recommends consideration of dexamethasone therapy in infants and children 2 months of age and older with proven or suspected bacterial meningitis on the basis of CSF examination, Gram-stained smear, or antigen test results (American Academy of Pediatrics, 1990).

In the clinical situation of an adult patient with meningitis and microscopic detection of Gram-positive

**TABLE VII** Antibiotics Commonly Used in the Treatment of Bacterial Meningitis

| Antibiotic | Daily dose and dosing interval | |
| --- | --- | --- |
| | Adults | Children |
| Penicillin G | 20–30 × 10⁶ U/day (every 4 h) | 500,000 U/kg/day (every 4 h) |
| Ampicillin | 12–15 g/day (every 4 h) | 300 mg/kg/day (every 4 h) |
| Cefotaxime | 9–12 g/day (every 8 h) | 200 mg/kg/day (every 6 h) |
| Ceftazidime | 6 g/day (every 8 h) | 150–200 mg/kg/day (every 8 h) |
| Ceftriaxone | 4 g/day (every 24 h) | 100 mg/kg/day[a] (every 24 h) |
| Meropenem | 6 g/day (every 8 h) | 120 mg/kg/day (every 8 h) |
| Fosfomycin | 15 g/day (every 8 h) | 200–300 mg/kg/day (every 8 h) |
| Gentamicin[b,c] | 240–360 mg/day (every 24 h) | 5 mg/kg/day (every 24 h) |
| Trimethoprim (TMP) + Sulfamethoxazole (SMX) | 480 mg/day TMP + 2.4 g/day SMX (every 8 h) | 160–320 mg/day TMP + 0.8–1.6 g/day SMX (every 8 h) |
| Rifampin | 600–1200 mg/day (every 24 h) | 10 mg/kg/day (max. 450 mg/day) (every 24 h) |
| Vancomycin[c] | 2 g/day (every 6 h) | 50 mg/kg/day (every 6 h) |

*Note.* Reduction of dose of the antibiotics (with the exception of ceftriaxone) in case of renal failure is necessary.
[a]100 mg/kg on the first day, then 75 mg/kg/day.
[b]Dose in case of intraventricular administration: children 1–2 mg/day, adults 5–10 mg/day.
[c]Determination of the serum concentrations are required; recommended peak levels 1 h after intravenous administration: gentamicin and tobracinin 5–10 µg/mL; vancomycin 25 µg/mL.

**TABLE VIII** Recommendations of the Duration of Antibiotic Therapy in Uncomplicated Cases of Bacterial Meningitis[a]

| Bacterial pathogen | Recommended duration of antibiotics (days)[b] |
| --- | --- |
| *Neisseria meningitidis* | 7(–10) |
| *Streptococcus pneumoniae* | (10–)14 |
| *Haemophilus influenzae* | 7(–10) |
| *Listeria monocytogenes* | 14–21 |
| Group B streptococci | 14–21 |
| Gram-negative Enterobacteriaceae | 21 |

[a]Modified after Quagliarello and Scheld (1997), copyright Massachusetts Medical Society.
[b]Dependent on the severity of the disease and the complications.

diplococci in a Gram-stained smear of the CSF (indicating the suspicion of pneumococcal infection), dexamethasone may be given immediately before intravenous antibiotic therapy, e.g., 8 mg IV every 8 h for 2–4 days, if there are no contraindications. A beneficial effect of dexamethasone might be expected especially in apurulent bacterial meningitis, often associated with *S. pneumoniae* infection. The institution of lytic antibiotics for this condition may release a flood of cell wall components. Thus ideally, the first dexamethasone dose should be administered several minutes before the first antibiotic dose in order to achieve maximal inhibition of the inflammatory cascade, which is initiated by antibiotic-induced bacteriolysis. The concomitant use of an intravenous $H_2$ receptor antagonist is recommended to prevent gastrointestinal bleeding.

It is recommended that in areas with high rates of resistant pneumococcal strains, initial empiric therapy should include two antibiotics, ceftriaxone and either rifampin or vancomycin (see Empiric Antibiotic Therapy, above). When dexamethasone is used in this situation, ceftriaxone combined with rifampin is preferred, since the penetration of vancomycin (but not of rifampin) into the CSF was reduced with dexamethasone therapy in experimental pneumococcal meningitis, resulting in a delay in CSF sterilization (Paris *et al.*, 1994) .

**Other Adjunctive Therapeutic Approaches.** Treatment of increased ICP includes elevation of the head of the bed to 30°, hyperventilation to maintain a $pCO_2$ concentration between 32 and 35 mm Hg, and the administration of hyperosmolar agents, e.g., mannitol (4 to 6 × 125 mL/day 20% mannitol IV). Stuporous or comatose patients may benefit from a careful ICP measurement and monitoring device. If there is CT evidence of hydrocephalus, ventricular drainage or serial CT scans, depending on the level of consciousness of the patient, should be performed. Anticoagulation with dose-adjusted intravenous heparin (doubling of partial thromboplastin time) may be used in patients with MRI or angiographically proven septic sinus venous thrombosis. Anticonvulsants are given to treat seizures, e.g., rapid intravenous fosphenytoin administration (e.g., 20 mg/kg, no faster than 150 mg/min in adults). Subdural effusion (sterile) usually spontaneously resolves and does not require surgical therapy. CT-guided stereotactic aspiration (and drainage) is recommended for cases of subdural empyema.

### Chemoprophylaxis

Eradication of bacterial pathogens from the nasopharynx by chemoprophylaxis may prevent secondary cases of *H. influenzae* meningitis and meningococcal meningitis.

**Meningococcal Meningitis.** Prophylaxis for meningococcal meningitis is recommended for all people sleeping in the same household and engaging in saliva-exchanging contacts. Rifampin is the drug most often recommended for chemoprophylaxis of meningococcal meningitis (Table IX); alternative drugs are ceftriaxone and ciprofloxacin. In the United States, a single 500-mg oral dose of ciprofloxacin is recommended for adults requiring prophylaxis (Control and Prevention of Meningococcocal Disease, 1997).

*Haemophilus Influenzae* **Meningitis.** Chemoprophylaxis with rifampin is recommended for households in which there is at least one child (other than the index case) under 48 months of age.

### Immunoprophylaxis

The *H. influenzae* type b conjugate vaccine series (bacterial polysacharide conjugated to protein) should begin at age 2 to 6 months. There is no effective vaccine against meningococci of the serogroup B, which accounts for the majority of cases of meningococcal infection in Europe. In the United States, serogroup C accounts for the majority of cases of meningococcal meningitis. Immunization with the quadrivalent vaccine (serogroups A, C, Y, and W 135) is currently recommended for certain high-risk groups, including patients with terminal complement component deficiency or dysfunction, asplenic patients, and travellers to areas with hyperendemic or epidemic meningococcal disease (e.g., Nigeria, Cameroon). Candidates for *S. pneumoniae*

**TABLE IX   Chemoprophylaxis of Meningococcal Meningitis**[a]

| Drug and age group | Dosage |
|---|---|
| Rifampin[b] | |
| Adults | 600 mg every 12 h for 2 days po |
| Children ≥1 month | 10 mg/kg every 12 h for 2 days po |
| Children <1 month | 5 mg/kg every 12 h for 2 days po |
| Ciprofloxacin[c] | |
| Adults | 500 mg given in a single dose po |
| Children | — |
| Ceftriaxone | |
| Adults and children ≥15 years | 250 mg given in a single intramuscular dose |
| Children <15 years | 125 mg given in a single intramuscular dose |

[a]Modified after Rosenstein *et al.* (2000), copyright Massachusetts Medical Society.
[b]Should not be prescribed for pregnant women.
[c]Ciprofloxacin is not remmended for persons <18 years or for pregnant or lactating women.

TABLE X  Other Bacterial Infections of the Central Nervous System

| Disease (bacterial pathogen) | Epidemiology, clinical features | Diagnostic procedures | Therapy, Prognosis |
|---|---|---|---|
| CNS listeriosis (*Listeria monocytogenes*) | Distributed worldwide, ≈3.6 cases per million population per year; epidemic outbreaks may occur; ingestion of contaminated food (e.g., soft cheese, vegetables, unpasteurized milk); neurological manifestations: acute (or subacute) meningitis/meningoencephalitis (≈90%), brain stem encephalitis (≈5–10%), brain abscess or spinal cord abscess (<5%); 70% of the patients are immunocompromised. | CSF: Gram-positive rods on Gram's stain, positive culture in blood and CSF (≈70%), CSF cell count median 600 cells/μL, only rarely >5000 cells/μL, frequently mixed granulocytic-lymphocytic pleocytosis; CT (MRI) may reveal brain swelling, hydrocephalus, brain stem lesions. | Ampicillin 6 × 2 g/day IV for (2–) 3 weeks plus gentamicin 1 × 240–360 mg/day iv for 7 (–10) days. Alternative drug: co-trimoxazole; overall mortality of CNS listeriosis 25–30%, in immunocompromised patients up to 60%. |
| Neurobrucellosis (*Brucella* species, e.g., *Brucella melitensis, Brucella abortus, Brucella suis*) | Zoonosis, ≈500,000 cases worldwide each year, endemic in Middle Eastern countries (e.g., Saudi Arabia, Kuwait) and in the Mediterranean region (e.g., Spain, Portugal); brucella are ingested via unpasteurized milk or milk products, or transmitted by direct contact with infected animals (e.g., sheep, goats, cattle, pigs, camels); neurobrucellosis in 2–5% of infected patients:<br>— subacute or chronic meningitis<br>— acute meningoencephalitis, rarely myelitis<br>— radiculoneuritis (cranial neuropathy)<br>— cerebrovascular involvement (vasculitis, spasm, mycotic aneurysm). | Elevated ELISA- IgM- and IgG-antibody titers against Brucella in serum and CSF (alternative: agglutination titers >1:160), positive CSF- or blood cultures in <50% of the patients; CSF: lymphocytic pleocytosis (<500 cells/μL), protein elevation (up to 500 mg/dL) and reduced CSF/serum glucose ratio. | Doxycycline (200 mg/day po) *plus* rifampin (600 mg/day po) *plus* ceftriaxone (1 × 2 g/day IV) or co-trimoxazole (e.g., 320 mg daily TMP dosage IV) for at least 4 weeks, then according to clinical and CSF response therapy with doxycycline (200 mg/day po) plus rifampin (600 mg/day po) for 3–6 months. Corticosteroids (e.g., dexamethasone, initially 3 × 8 mg/day IV) may be administered during the acute stage of the disease. Major complications: infective endocarditis, intracerebral hemorrhage from mycotic aneurysm, stroke; overall mortality of untreated brucellosis is 2%, complete recovery >50% of treated patients. |
| Whipple's disease (*Tropheryma whippelii*) | Extremely rare disorder, usually in middle-aged men; clinical features: migratory polyarthritis, chronic diarrhea, malabsorption, weight loss, mesenteric lymphadenopathy, abdominal pain. Neurological manifestations in 5–10% of the patients: subacute or chronic granulomatous encephalitis (dementia, supranuclear ophthalmoplegia, oculomasticatory and oculofacial–skeletal myorhythmias, hypothalamic dysfunction, ataxia, seizures), rarely meningitis, radiculoneuritis, myositis. Differential diagnoses: progressive supranuclear palsy, Creutzfeldt–Jakob disease. | Detection of *Tropheryma whippelii* by PCR in the CSF and the jejunal mucous membrane, PAS-positive macrophages (sickle form particles containing cells) in the jejunal mucous membrane, in the CSF (rarely) and in brain tissue (biopsy); CSF cell count (5–900 cells/μL) and protein content (47–158 mg/dL) are elevated in 50% of the patients; CT (MRI) may reveal hypodense parenchymal areas and granulomas in or near the upper brain stem, hydrocephalus. | Co-trimoxazole (e.g., 320 mg daily IV during the first 2 to 4 weeks, then po); alternative drugs: ceftriaxone (2 g/day IV), doxycycline (200 mg/day po), maintainance therapy for 1 to 2 years to prevent relapses, e.g., doxycycline (100 mg/day po) or co-trimoxazole (e.g., 320 mg TMP daily po) or cefixime (2 × 200 mg/day po); stop of progression or clinical improvement in 40–50% of treated patients. |

*continues*

**TABLE X** *Continued*

| Disease (bacterial pathogen) | Epidemiology, clinical features | Diagnostic procedures | Therapy, Prognosis |
|---|---|---|---|
| Actinomycosis *Actinomyces* species (e.g., *Actinomyces israelii*) | CNS actinomycosis extremely rare; these anaerobic gram-positive bacilli are normal inhabitants of the mouth; neurological manifestations in <5% of the patients suffering from cervicofacial, thoracic or abdominal actinomycosis: brain abscess (2/3 of CNS-actinomycosis), subdural empyema, epidural abscess, spinal epidural abscess, meningitis. | Microscopic and cultural detection of actinomyces in brain abscess material (usually mixed flora) or the CSF (rarely). CSF may be purulent or exhibit a mixed granulocytic/lymphocytic pleocytosis. | Penicillin G 4 × 5 million units/day IV for 4 to 6 weeks followed by oral phenoxy-penicillin, 3–6 million units/day for 6 to 12 months to prevent relapses. Alternative drugs: ceftriaxone, doxycycline, macrolides. Surgical interventions for abscess or empyema are frequently necessary. 75% of the treated patients survive, sequelae are common, relapses may occur. |
| Nocardiosis *Nocardia* species (e.g., *Nocardia asteroides, Nocardia farcinica, Nocardia brasiliensis*) | CNS nocardiosis is very rare; opportunistic infection in immunocompromised patients, *Nocardia* inhabit the soil; hematogeneous spread from a pulmonary infection. Neurological manifestations in 20 to 30% of patients with nocardiosis: most commonly brain abscess, usually multiple, very rarely meningitis in the absence of brain abscess. | Detection of bacteria by microscopy (Gram-positive acid-fast bacteria, Ziehl–Neelsen stain) and culture of abscess material or bronchial secretion; CSF may be purulent or reveal a mixed granulocytic/lymphocytic pleocytosis. | Co-trimoxazole (480 mg/day IV TMP dosage) for 6 to 8 weeks, followed by oral administration for 6 to 12 months (e.g., 160–320 TMP/day po) to prevent relapses. Alternative drugs: meropenem, ceftriaxone, vancomycin; surgical interventions for brain abscess; prognosis of CNS nocardiosis is poor; overall mortality ≈50%, 90% if multiple brain abscesses are present. |
| Rickettsial diseases Rocky Mountain spotted fever (RMSF) (*Rickettsia rickettsii*) | *R. rickettsii* is transmitted by ticks. The average annual incidence of Rocky Mountain spotted fever (caused by *R. rickettsii*) during 1993–1996 in the United States was 2.2 cases per million persons; the south Atlantic region accounted for the largest proportion of confirmed cases; classic triad: fever, rash (macular, petechiae), history of tick bite. Encephalitis, meningitis ≈25% of cases. | Serological diagnosis: detection of specific IgM-and IgG antibodies, e.g., by indirect immunofluorescence, EIA, immunoblot. CSF in patients with neurological abnormalities: often mild pleocytosis <200 cells/μL, variable combination of mononuclear and polymorphonuclear leukocytes, protein elevation in 30–50%. | Doxycycline 200 mg/day po for 10 (–14) days. Case fatality rate of Rocky Mountain Spotted Fever was highest among persons 70 years of age and older (9.0%) and lowest among adults 40–49 years of age (0.6%). Pathological hallmark: endangiitis of small and medium vessels. |
| Mediterranean Spotted Fever, syn. tick typhus, Boutonneuse fever, Marseilles fever (*Rickettsia conorii*) | *R. conorii* is transmitted by ticks. Geographic distribution: Mediterranean basin, Southern Africa and Asia (India). Clinical hallmark: acute fever, eschar ("tache noir," at the site of inoculation), rash (macular, papular), regional lymphadenitis; meningitis or meningoencephalitis may occur. | Serological diagnosis: detection of IgM-and IgG antibodies, e.g., by indirect immunofluorescence. CSF may show lymphocytic pleocytosis. | Doxycycline 200 mg/day po for 10 (–14) days. Alternative: ciprofloxacin. Mortality rate <1%. |

| Disease | Clinical features | Diagnosis | Therapy |
|---|---|---|---|
| Q fever (*Coxiella burnetii*) | Worldwide distribution (in Europe in particular in Great Britain, South Eastern parts); humans acquire *C. burnetii* from infected animals (e.g., sheep, goats, cattle), probably by inhalation. Rinder), Clinical hallmarks: Flu-like illness with atypical pneumonia and hepatic involvement; neurological manifestations in 2–7%: aseptic meningitis/meningoencephalitis, cerebellar ataxia. | Serological diagnosis: detection of specific IgM and IgG antibodies (e.g., EIA, immunofluorescence test, complement fixation). CSF may show lymphocytic pleocytosis. | Doxycycline 200 mg/day po for 10 (–14) days (at least 3 days after the patient is afebrile). Alternative: rifampin 600 mg/day po. Acute Q fever is typically a self-limited illness. Mortality rate <1%; lethal courses in case of endocarditis. |
| Human ehrlichiosis (*Ehrlichia species*) | Human ehrlichiosis is a tick-borne disease. Two forms have been differentiated in the US: human monocytic ehrlichiosis—HME—(caused by *Ehrlichia chaffeensis*) and human granulocytic ehrlichiosis—HGE—(caused by the *HGE agens*). Human ehrlichiosis have also been reported in Europe. Clinical hallmarks are <br>— acute fever, headache, nausea, myalgia, arthralgia (exanthema in 30% of the patients) <br>— leukopenia, thrombopenia, anemia <br>— mildly elevated hepatic transaminase activity. <br>Mild lymphocytic meningitis occurs in ≈10% of the patients with HME, is very rarely in patients with HGE. | Serological diagnosis: Antibody detection (immunofluorescence, seroconversion), detection of typical intracytoplasmic inclusion bodies (morulae) in blood (in monocytes in case of HME and in granulocytes in case of HGE), detection of Ehrlichia by PCR. | Doxycycline 200 mg/day po for 10–14 days (at least 3 days after the patient is afebrile). Alternative: rifampin 600 mg/day po. Usually mild monophasic course, the risk for serious disease or death increases with advancing age and delayed onset of therapy; rarely fatal cases (multiorgan failure) of old and immunosuppressed patients have been reported. |
| Mycoplasmosis (*Mycoplasma pneumoniae*) | *M. pneumoniae* is responsible for ≈ 10–20% of cases of community-acquired pneumonia; clinical hallmark: atypical pneumonia (unilateral interstitial infiltrates on chest films), followed by neurological complications: meningoencephalitis, cerebellar ataxia, transverse myelitis, meningitis, peripheral neuropathy, Guillain–Barré syndrome, mononeuritis multiplex, myositis, cerebral infarction. | Presence of specific IgM antibodies in serum or fourfold rise in complement fixation titers; Mycoplasma specific intrathecal antibody production, detection of Mycoplasma by PCR in the CSF; presence of cold agglutinins (titer > 1:64) in ≈50% of the patients; MRI may reveal periventricular white matter lesions or intramedullary lesions (in case of myelitis). | Erythromycin (4 × 500 mg/day IV) for 2 to 3 weeks; alternatively: doxycycline (200 mg/day po) or clarithromycin (2 × 500 mg/day po). Mycoplasmas have no cell wall and are therefore resistant to penicillins and cephalosporins. In severe cases (e.g., myelitis with paraparesis) corticosteroids (e.g., 500 mg methylprednisolone/day iv for 5 days) may be administered. Usually complete recovery, sequelae in ≈10–15% of the patients. Rarely lethal outcome. |

*Notes.* CSF, cerebrospinal fluid; TMP, trimethoprim-sulfamethoxazole; PAS, periodic-acid-Schiff; EIA, enzyme-linked immunosorbent assay.

Clinical data are adapted from Akdeniz *et al.* (1998); AlDeeb *et al.* (1989); Bakken and Dumler (2000); Bencina *et al.* (2000); Dobbins (1995); Dumler and Bakken (1998); Ezpeleta *et al.* (1999); Fingerle *et al.* (1999); Grant *et al.* (1997); Gucuyener *et al.* (2000); Herbay *et al.* (1997); Koskiniemi (1993); Lerner (1996); Lotric-Furlan *et al.* (1998); Louis (1996); Mylonakis *et al.* (1998); Raoult *et al.* (2000a,b); Ratnaike (2000); Relman *et al.* (1992); Spach *et al.* (1993); Sakai *et al.* (1999); Socan *et al.* (2001); Treadwell *et al.* (2000).

immunoprophylaxis with the pneumococcal vaccine (bacterial polysaccharides of 23 pneumococcal types) are patients with asplenic states (splenectomy, sickle cell anemia) or chronic debilitating diseases (e.g., diabetes, congestive heart failure) or HIV.

## OTHER BACTERIAL INFECTIONS

Epidemiology, clinical features, diagnostic procedures, prognosis, and therapeutic approaches for the following bacterial disorders, which may be complicated by neurological manifestations, are summarized in Table X: Listeriosis, neurobrucellosis, Whipple's disease, actinomycosis, nocardiosis, mycoplasmosis, rickettsial diseases and ehrlichiosis.

## REFERENCES

Al Deeb, S. M., Yaqub, B. A., Sharif, H. S., and Phadke J. G. (1989). Neurobrucellosis: Clinical characteristicst, diagnosis, and outcome. *Neurology* 39, 498–501.

American Academy of Pediatrics—Commitee on Infectious Diseases (1990). Dexamethasone therapy for bacterial meningitis in infants and children. *Pediatrics* 86, 130–133.

Andersen, J., Backer, V., Voldsgrard, P., Skinhoj, P., and Wandall, J. H. (Members of the Copenhagen Meningitis Study Group) (1997). Acute meningococcal meningitis: Analysis of features of the disease according to the age of 255 patients. *J. Infect.* 34, 227–235.

Bakken, J. S., and Dumler, J. S. (2000). Human granulocytic ehrlichiosis. *Clin. Infect. Dis.* 31, 554–560.

Bencina, D., Dovc, P., Mueller-Premru, M., Avsic-Zupanc, T., Socan, M., Beovic, B., Arnex, M., and Narat, M. (2000). Intrathecal synthesis of specific antibodies in patients with invasion of the central nervous system by Mycoplasma pneumoniae. *Eur. J. Clin. Microbiol. Infect. Dis.* 19, 521–530.

Braun, J. S., Novak, R., Herzog, K. H., Bodmer, S. M., Cleveland, J. L., and Tuomanen, E. I. (1999). Neuroprotection by a caspase inhibitor in acute bacterial meningitis. *Nat. Med.* 5, 298–302.

Chang, C. W. J., and Bleck, T. P. (1997). Acute management of neurological complications. *In* "Central Nervous Infectious Diseases and Therapy" (K. L. Roos, Ed.), Dekker, New York, pp. 691–709.

Control and Prevention of Meningococcal Disease: Recommendations of the Advisory Committee on Immunization Practices (ACIP) (1997). *Morb. Mortal. Wkly. Rep.* 46(RR-5), 1–51.

Dichgans, M., Jäger, L., Mayer, T. *et al.* (1999). Bacterial meningitis in adults: demonstration of inner ear involvement using high-resolution MRI. *Neurology* 52, 1003–1009.

DiNubile, M. J., Boom, W. H., and Southwick, F. S. (1990). Septic cortical thrombophebitis. *J. Infect. Dis.* 161, 1216–1220.

Dobbins, W. O. (1995). The diagnosis of Whipple's disease. *N. Engl. J. Med.* 332, 390–392.

Dodge, P. R., Davis, H., Feigin, R. D., Holmes, S. J., Kaplan, S. L., Jubelirer, D. P., Stechenberg, B. W., and Hirsh, S. K. (1984). Prospective evaluation of hearing impairment as a sequela of acute bacterial meningitis. *N. Engl. J. Med.* 311, 869–874.

Dodge, P. R., and Swartz, M. N. (1965). Bacterial meningitis: A review of selected aspects. *N. Engl. J. Med.* 272, 725–731, 1003–1010.

Doit, C., Bourrillon, A., and Bingen, E. (1998). Childhood bacterial meningitis-bacterial epidemiology and antibiotic resistance. Presse Medicale 1998; 27, 1177–1182. DiNubile, M. J., Boom, W. H.,

and Southwick, F. S. (1990). Septic cortical thrombophlebitis. *J. Infect. Dis.* 161, 1216–1220.

Dumler, J. S., and Bakken, J. S. (1998). Human ehrlichioses: Newly recognized infections transmitted by ticks. *Annu. Rev. Med.* 49, 201–213.

Durand, M. L., Calderwood, S. B., Weber, D. J., Miller, S. I., Southwick, F. S., Caviness, V. S., Jr., and Swartz, M. N. (1993). Acute bacterial meningitis. A review of 493 episodes. *N. Engl. J. Med.* 328, 21–28.

Ezpeleta, D., Munoz-Blanco, J. L., Tabernero, C., and Gimenez-Roldan, S. (1999). Neurological complications of Mediterranean boutonneuse fever. Presentation of a case of acute encephalomeningomyelitis and review of the literature. *Neurologica* 14, 38–42.

Felgenhauer, K., and Kober, D. (1985). Apurulent bacterial meningitis (compartmental leucopenia in purulent meningitis). *J. Neurol.* 232, 157–161.

Fingerle, V., Goodman, J. L., Johnson, R. C., Kurtti, T. J., Munderloh, U. G., and Wilske, B. (1999). Epidemiologic aspects of human granulocytic Ehrlichiosis in southern Germany. *Wien Klin. Wochenschr.* 111, 1000–1004.

Fortnum, H. M. (1992). Hearing impairment after bacterial meningitis: A review. *Arch. Dis. Child.* 67, 1128–1133.

Frei, K., Nadal, D., Pfister, H. W., and Fontana, A. (1993). Listeria meningitis: Identification of a cerebrospinal fluid inhibitor of macrophage listericidal function as interleukin 10. *J. Exp. Med.* 178, 1255–1261.

Gendrel, D., Raymond, J., Assicot, M. *et al.* (1997). Measurement of procalcitonin levels in children with bacterial and viral meningitis. *Clin. Infect. Dis.* 24, 1240–1242.

Girgis, N. I., Farid, Z., Mikhail, I. A., Farrag, I., Sultan, Y., and Kilpatrick, M. E. (1989). Dexamethasone treatment for bacterial meningitis in children and adults. *Pediatr. Infect. Dis. J.* 8, 848–851.

Grant, A. C., Hunter, S., and Partin, W. C. (1997). A case of acute monocytic ehrlichiosis with prominent neurological signs. *Neurology* 48, 1619–1623.

Gucuyener, K., Simsek, F., Yilmaz , O., and Serdaroglu, A. (2000). Methyl-prednisolone in neurologic complications of Mycoplasma pneumonia. *Ind. J Pediatr.* 67, 467–469.

Haring, H. P., Rötzer, H. K., Reindl, H., Berek, K., Kampfl, A., Pfausler, B., and Schmutzhard, E. (1993). Time course of cerebral blood flow velocity in central nervous system infections. *Arch. Neurol.* 50, 98–101.

Herbay von, A., Ditton, H. J., Schuhmacher, F., and Maiwald, M. (1997). Whipple's disease: Staging and monitoring by cytology and polymerase chain reaction analysis of cerebrospinal fluid. *Gastroenterology* 113, 434–441.

Igarashi, M., Gilmartin, R. C., Gerald, B., Wilburn, F., and Jabbour, J. T. (1984). Cerebral arteritis and bacterial meningitis. *Arch. Neurol.* 41, 531–535.

Kastenbauer, S., Winkler, F., Fesl, G., Schiel, X., Ostermann, H., Yousry, T. A., and Pfister, H. W. (2001). Acute severe spinal cord dysfunction in bacterial meningitis in adults: MRI findings suggest extensive myelitis. *Arch. Neurol.* 58, 806–810.

Khayr, W. F., Cherubin, C. E., and Bleck, T. P. (1992). Listeriosis: Review of a protean disease. *Infect. Dis. Clin. Pract.* 1, 291–298.

Kennedy, W. A., Hoyt, M. J., and McCracken, G. H., Jr. (1991). The role of corticosteroid therapy in children with pneumococcal meningitis. *Am. J. Dis. Child.* 145, 1374–1378.

Koedel, U., and Pfister, H. W. (1999a). Models of experimental meningitis. Role and limitations. *Infect. Dis. Clin. North Am.* 13, 549–577.

Koedel, U., and Pfister, H. W. (1999b). Oxidative stress in bacterial meningitis. *Brain Pathol.* 9, 57–67.

Koskiniemi, M. (1993). CNS manifestations associated with Mycoplasma pneumoniae infections: Summary of cases at the University of Helsinki and review. *Clin. Inf. Dis.* 17, 52–57.

Lebel, M. H., Freji, B. J., Syrogiannopoulos, G. A., Chrane, D. F., Hoyt, M. J., Stewart, S. M., Kennard, B. D., Olsen, K. D., and McCracken, G. H., Jr. (1988). Dexamethasone therapy for bacterial meningitis: Results of two double-blind, placebo-controlled trials. *N. Engl. J. Med.* **319,** 964–971.

Leib, S. L., Leppert, D., Clements, J., and Täuber, M. G. (2000). Matrix metalloproteinases contribute to brain damage in experimental pneumococcal meningitis. *Infect. Immunol.* **68,** 615–620.

Lerner, P. I. (1996). Nocardiosis. *Clin. Infect. Dis.* **22,** 891–903.

Lotric-Furlan, S., Petrovec, M., Zupanc, T. A. *et al.* (1998). Human granulocytic Ehrlichiosis in Europe: Clinical and laboratory findings for four patients from Slovenia. *Clin. Infect. Dis.* **27,** 424–428.

Louis, E. D., Lynch, T., Kaufmann, P., Fahn, S., and Odel, J. (1996). Diagnostic guidelines in central nervous system Whipple's disease. *Ann. Neurol.* **40,** 561–568.

McIntyre, P. B., Berkey, C. S., King, S. *et al.* (1997). Dexamethasone as adjunctive therapy in bacterial meningitis. *JAMA* **278,** 925–931.

Mylonakis, E., Hohmann, E. L., and Calderwoud, S. B. (1998). Central nervous system infection with *Listeria monocytogenes*. 33 Years' experience at a general hospital and review of 776 episodes from the literature. *Medicine* **77,** 313–336.

Nau, R., Wellmer, A., Soto, A., Koch, K., Schneider, O., Schmidt, H., Gerber, J., Michel, U., and Brück, W. (1999). Rifampin reduces early mortality in experimental Streptococcus pneumoniae meningitis. *J. Infect. Dis.* **179,** 1557–1560.

Odio, C. M., Faingezicht, I., Paris, M., Nassar, M., Baltodano, A., Rogers, J., Saez-Llorens, X., Olsen, K. D., and McCracken, G. H., Jr. (1991). The beneficial effects of early dexamethasone administration in infants and children with bacterial meningitis. *N. Engl. J. Med.* **324,** 1525–1531.

Paris, M. M., Hickey, S. M., Uscher, M. I., Shelton, S., Olsen, K. D., and McCracken, G. H., Jr. (1994). Effect of dexamethasone on therapy of experimental penicillin- and cephalosporin-resistant pneumococcal meningitis. *Antimicrob. Agents Chemother.* **38,** 1320–1324.

Paulson, O. B., Brodersen, P., Hansen, E. L., and Kristensen, H. S. (1974). Regional cerebral blood flow, cerebral metabolic rate of oxygen, and cerebrospinal fluid acid–base variables in patients with acute meningitis and with encephalitis. *Acta Med. Scand.* **196,** 191–198.

Pfister, H. W., Borasio, G. D., Dirnagl, U., Bauer, M., and Einhäupl, K. M. (1992). Cerebrovascular complications of bacterial meningitis in adults. *Neurology* **42,** 1497–1504.

Pfister, H. W., Feiden, W., and Einhäupl, K. M. (1993). Spectrum of complications during bacterial meningitis in adults. *Arch. Neurol.* **50,** 575–581.

Pfister, H. W., Koedel, U., Haberl, R., Dirnagl, U., Feiden, W., Ruckdeschel, G., and Einhäupl, K. M. (1990). Microvascular changes during the early phase of experimental bacterial meningitis. *J. Cereb. Blood Flow Metab.* **10,** 914–922.

Pfister, H. W., and Roos, K. L. (1994). Bacterial meningitis. *In* "Neurocritical Care" (W. Hacke, D. F. Hanley, K. M. Einhäupl, T. P. Bleck, and M. N. Diringer, Eds.), pp. 377–397. Springer Verlag, Berlin.

Phillips, E. J., and Simor, A. E. (1998). Bacterial meningitis in children and adults-changes in community-acquired disease may affect patient care. *Postgrad. Med.* **103,** 102.

Pomeroy, S. L., Holmes, S. J., Dodge, P. R., and Feigin, R. D. (1990). Seizures and other neurologic sequelae of bacterial meningitis in children. *N. Engl. J. Med.* **323,** 1651–1657.

Quagliarello, V. J., and Scheld, W. M. (1993). New perspectives on bacterial meningitis. *Clin. Infect. Dis.* **17,** 603–610.

Quagliarello, V. J., and Scheld, W. M. (1997). Treatment of bacterial meningitis. *N. Engl. J. Med.* **336,** 708–716.

Raoult, D., Birg, M. L., La Scola, B. *et al.* (2000). Cultivation of the bacillus of Whipple's disease. *N. Engl. J. Med.* **342,** 620–625.

Raoult, D., Tissot-Dupont, H., Foucoult, C., Gouvernet, J., Fournier, P. E., Bernit, E., Stein, A., Nesri, M., Harle, J. R., and Weiller, P. J. (2000). Q fever 1985–1998. Clinical and epidemiologic features of 1383 infections. *Medicine* **79,** 109–123.

Ratnaike, R. N. (2000). Whipple's disease. *Postgrad. Med. J.* **76,** 760–766.

Relman, D. A., Schmidt, T. M., MacDermott, R. P., and Falkow, S. (1992). Identification of the uncultured bacillus of Whipple's disease. *N. Engl. J. Med.* **327,** 298–301.

Roos, K. L. (1997). Bacterial meningitis. *In* "Central Nervous Infectious Diseases and Therapy" (K. L. Roos, Ed.), pp. 99–126. Dekker, New York.

Rosenstein, N. E., Perkins, B. A., Stephens, D. S., Popovic, T., and Hughes, J. M. (2001). Meningococcal disease. *N. Engl. J. Med.* **344,** 1378–1388.

Sakai, C., Takagi, T., and Satoh, Y. (1999). Nocardia asteroides pneumonia, subcutaneous abscess and meningitis in a patient with advanced malignant lymphoma: Successful treatment based on in vitro antimicrobial susceptibility. *Intern. Med.* **38,** 683–686.

Schaad, U. B., Lips, U., Gnehm, H. E., Blumberg, A., Heinzer, I., and Joanna Wedgwood for the Swiss Meningitis Study Group (1993). Dexamethasone therapy for bacterial meningitis in children. *Lancet* **342,** 457–461.

Scheld, W. M. (1984). Bacterial meningitis in the patient at risk: Intrinsic risk factors and host defense mechanisms. *Am. J. Med.* **76**(5A), 193–207.

Scheld, W. M. (1989). Drug delivery to the central nervous system: General principles and relevance to therapy for infections of the central nervous system. *Rev. Infect. Dis.* **11**(Suppl.7), 1669–1690.

Scheld, W. M., Dacey, R. G., Winn, H. R., Welsh, J. E., Jane, J. A., and Sande, M. A. (1980). Cerebrospinal fluid outflow resistance in rabbits with experimental meningitis. *J. Clin. Invest.* **66,** 243–253.

Schuchat, A., Robinson, K., Wenger, J. D., *et al.* (1997). Bacterial meningitis in the United States in 1995. Active Surveillance Team. *N. Engl. J. Med.* **337,** 970–976.

Sigurdardottir, B., Bjornsson, O. M., Jonsdottir, K. E., Erlendsdottir, H., and Gudmundsson, S. (1997). Acute bacterial meningitis in adults—A 20-year overview. *Arch. Intern. Med.* **157,** 425–430.

Socan, M., Ravnik, I., Bencina, D., Dovc, P., Zakotnik, B., and Jazbec, J. (2001). Neurological symptoms in patients whose cerebrospinal fluid is culture- and/or polymerase chain reaction-positive for Mycoplasma pneumoniae. *Clin. Infect. Dis.* **32,** E31–E35.

Southwick, F. S. (1995). Septic thrombophlebitis of major dural venous sinuses. *Curr. Clin. Trop. Infect. Dis.* **15,** 179–203.

Southwick, F. S., Richardson, E. P., Jr, and Swartz, M. N. (1986). Septic thrombosis of the dural venous sinuses. *Medicine* **65,** 82–106.

Spach, D. H., Liles, W. C., Campbell, G. L., Quick, R. E., Anderson, Jr., D. E., and Fritsche, T. R. (1993). Tick-borne diseases in the United States. *N. Engl. J. Med.* **329,** 936–947.

Spataro, V., and Marone, C. (1993). Rhabdomyolysis associated with bacteremia due to Streptococcus pneumoniae: Case report and review. *Clin. Infect. Dis.* **17,** 1063–1064.

Swartz, M. N. (1984). Bacterial meningitis. More involved than just the meninges. *N. Engl. J. Med.* **311,** 912–914.

Syrogiannopoulos, G. A., Lourida, A. N., Theodoridou, M. C., Pappas, I. G., Babilis, G. C., Economidis, J. J., Zoumboulakis, D. J., Beratis, N. G., and Matsaniotis (1994). Dexamethasone therapy for bacterial meningitis in children: 2- versus 4-day regimen. *J. Infect. Dis.* **169,** 853–858.

Syrogiannopoulos, G. A., Olsen, K. D., Reisch, J. S., and McCracken, G. H., Jr. (1987). Dexamethasone in the treatment of experimental Haemophilus influenzae type b meningitis. *J. Infect. Dis.* **155,** 213–219.

Täuber, M. G., Khayam-Bashi, H., and Sande, M. A. (1985). Effects of ampicillin and corticosteroids on brain water content, cerebrospinal fluid pressure, and cerebrospinal fluid lactate levels in experimental pneumococcal meningitis. *J. Infect. Dis.* **151**, 528–534.

Thomas, R., Le Tulzo, Y., Bouget, J. *et al.* (1999). Trial of dexamethasone treatment for severe bacterial meningitis in adults. Adult Meningitis Study Group. *Intensive Care Med.* **25**, 475–480.

Tomasz, A. (1997). Antibiotic resistance in *Streptococcus pneumoniae*. *Clin. Infect. Dis.* **24**, S85–S88.

Treadwell, T. A., Holman, R. C., Clarke, M. J., Krebs, J. W., Paddock, C. D., and Cilds, J. E. (2000). Rocky Mountain spotted fever in the United States, 1993–1996. *Am. J. Trop. Med. Hyg.* **63**, 21–26.

Tureen, J. H., Dworkin, R. J., Kennedy, S. L., Sachdeva , M., and Sande, M. A. (1990). Loss of cerebrovascular autoregulation in experimental meningitis in rabbits. *J. Clin. Invest.* **85**, 577–581.

Viallon, A., Zeni, F., Lambert, C. *et al.* (1999). High sensitivity and specificity of serum procalcitonin levels in adults with bacterial meningiti*s. Clin. Infect. Dis.* **28**, 1313–1316.

Weber, J. R., Angstwurm, K., Bürger, W., Einhäupl, K. M., Dirnagl, U. (1995). Anti ICAM-1 (CD54) monoclonal antibody reduces inflammatory changes in experimental bacterial meningitis. *J. Neuroimmunol.* **63**, 63–68.

White, B., Livingstone, W., Murphy, C., Hodgson, A., Rafferty, M., and Smith, O. W. (2000). An open-label study of the role of adjuvant hemostatic support with protein C replacement therapy in purpura fulminans-associated meningococcemia. *Blood* **96**, 3719–3724.

WHO (2000). Control of epidemic meningococcal disease. "WHO Practical Guidelines," 2nd ed. Internet WHO/EMC Web Site.

## CHAPTER 43
# Brain and Spinal Cord Abscess

Stefan Kastenbauer, Hans-Walter Pfister, and W. Michael Scheld

## BRAIN ABSCESS

### Clinical Aspects

#### Definition

Brain abscess is a focal, intracerebral infection that begins as a localized area of cerebritis and develops into a collection of pus surrounded by a well-vascularized capsule.

#### Etiology

Brain abscesses can be classified according to the route of infection: contiguous spread of an adjacent infection, hematogenous seeding from a distant focus, and direct inoculation of pathogens or facilitation of their entry into the brain by trauma or neurosurgical procedures.

Frequent contiguous infections predisposing for brain abscess are otitis media or mastoiditis (37% of cases in six recent large studies; Alderson *et al.*, 1981; Arseni and Ciurea, 1981; Chun *et al.*, 1986; O'Donoghue *et al.*, 1992; Nielsen *et al.*, 1982; Yang, 1981) and paranasal sinusitis (8% of cases, e.g., frontal sinusitis, ethmoiditis, pansinusitis, and, rarely, sphenoid sinusitis). Infrequent causative adjacent infections are those of the teeth (3% of cases, e.g., tooth extractions, caries, periodontal disease), the skull (osteomyelitis), or the scalp. Very rarely, a brain abscess can result from an infection within the subarachnoid space (meningitis). The site of the primary infectious focus often determines the location of the abscess. Thus, otogenic abscesses are frequently located in the cerebellum or temporal lobe. Abscesses due to paranasal sinusitis or odontogenic infections are often located in the frontal lobe.

Hematogenous spread (19% of cases) frequently results in multiple brain abscesses that are often located in the distribution area of the middle cerebral artery with a predilection for the gray matter–white matter junctions, where the blood supply is relatively reduced. In these cases, the most common primary sources are the lung (in particular lung abscesses, but also bronchiectasis, pneumonia, empyema, cystic fibrosis, and pulmonary arteriovenous malformations) and the heart (infective endocarditis and congenital cyanotic heart disease; the latter accounts for more than 60% of brain abscesses in children). Intraabdominal sepsis, urinary tract infections, wound and skin infections, or osteomyelitis can also cause a brain abscess via the hematogenous route. The increased susceptibility to brain abscess in patients with congenital cyanotic heart disease or pulmonary arteriovenous malformations is probably due to the right-to-left shunt bypassing the filtering pulmonary capillary bed.

The risk of developing a deep brain infection following a head trauma or a neurosurgical procedure (13% of cases) is increased by retained bone fragments or foreign bodies, contamination of the wound (e.g., with soil), and cranial wound complications, in particular cerebral spinal fluid (CSF) fistulae (Bayston *et al.*, 2000). Brain abscesses following closed skull fractures are rare; they are frequently associated with previously unrecognized dural tears or residual intracerebral bone fragments or debris. Rarely, intracranial pressure monitors and CSF shunt systems have been reported to cause brain abscesses (Gower *et al.*, 1990). In a considerable proportion of patients with brain abscesses, the route of infection remains unclear (15% of cases).

The most common bacterial pathogens isolated from intracerebral pus are given in Table I. The spectrum of etiologic pathogens varies with the underlying conditions (Table II). For example, following craniotomy, a considerable proportion of deep brain infections is due to methicillin-resistant *Staphylococcus aureus* and *Staphylococcus epidermidis* (19.4 and 4.8% in one

TABLE I Bacteriological Results of Abscess Pus Cultures from 368 Patients Reported in Six Recent Studies from Switzerland (Seydoux and Francioli, 1992), the United States (Chun *et al.*, 1986), Great Britain (O'Donoghue *et al.*, 1992; Alderson *et al.*, 1981), China (Yang and Zhao, 1993), and Sweden (Svanteson *et al.*, 1988)

| Etiologic pathogen | Frequency (%) |
|---|---|
| Streptococci | 51.6 |
| Unspecified anaerobic streptococci | 18.5 |
| Unspecified aerobic streptococci | 17.2 |
| *Streptococcus milleri* | 10.7 |
| Unspecified microaerophilic streptococci | 2.9 |
| *Streptococcus pneumoniae* | 2.3 |
| Bacteroidaceae | 36.4 |
| *Bacteroides* species[a] | 30.2 |
| *Fusobacterium* species | 6.2 |
| Enterobacteriaceae | 27.6 |
| *Proteus* species[b] | 17.9 |
| *Escherichia coli* | 8.4 |
| *Klebsiella pneumoniae* | 1.3 |
| Staphylococci | 16.6 |
| *Staphylococcus aureus* | 15.6 |
| *Staphylococcus epidermidis* | 1.0 |
| Other bacteria | |
| *Haemophilus* species | 6.5 |
| *Peptostreptococcus* species | 4.5 |
| *Eubacterium* species | 4.2 |
| *Propionibacterium* acnes | 2.9 |
| *Peptococcus* species | 2.3 |

*Note.* Cultures were positive in 308 cases (83.%). A single organism was detected in 68% of these patients and 32% had mixed cultures. The percentages refer to the total number of positive cultures. Due to the presence of mixed cultures, the total of the percentage numbers exceeds 100%.
[a] Mostly *B. fragilis* or *B. melaninogenicus*.
[b] Mostly *P. mirabilis* or *P. fragilis*.

study; Korinek, 1997). After solid organ or bone marrow transplantation, brain abscesses are mostly due to *Aspergillus*, *Candida*, and *Toxoplasma* species, or, rarely, to *Listeria monocytogenes* (Hagensee *et al.*, 1994; Maschke *et al.*, 1999; Selby *et al.*, 1997).

### Location

Solitary brain abscesses were observed in 79% and multiple abscesses in 21% of cases in five large studies from the CT era (Chun *et al.*, 1986; Miller *et al.*, 1988; O'Donoghue *et al.*, 1992; Seydoux and Francioli, 1992; Yang and Zhao, 1993). They were most frequently (the total exceeds 100% due to multiple abscesses) located in the frontal (38%), temporal (25%), and parietal (24%) lobes, followed by the occipital lobe (11%), cerebellum (7%), pituitary fossa (1%), and basal ganglia (<1%). The brainstem also is a very uncommon location of brain abscesses.

### Diagnosis

Clinical signs and symptoms of brain abscess are summarized in Table III. The wide frequency ranges in the different studies make clear that the clinical diagnosis of brain abscess can be difficult. For example, only 50% or fewer of the patients present with the typical triad of brain abscess: headache, focal neurologic deficit, and fever. The diagnosis of brain abscess is usually made by cranial postcontrast CT or MRI. During the early phase of cerebritis, the CT shows a hypodense area with slight or absent contrast enhancement (Enzmann *et al.*,

TABLE II Most Probable Microbial Flora of Brain Abscesses According to the Underlying Condition or Source of Infection

| Underlying condition/source of infection | Major pathogens |
|---|---|
| Paranasal sinusitis | Streptococci (in particular, *S. milleri*), *Staphylococcus aureus*, *Haemophilus* species, *Bacteroides* species |
| Otogenic infection | Streptococci, *Bacteroides* species, enterobacteria (in particular, *Proteus* species), *Pseudomonas* species, *Haemophilus* species |
| Odontogenic infection | Streptococci, staphylococci, *Bacteroides* species, *Fusobacterium* species, *Actinomyces* species, *Actinobacillus* species |
| Bacterial endocarditis | *Staphylococcus aureus*, viridans streptococci |
| Pulmonary infection (abscess, empyema, bronchiectasis) | Streptococci, staphylococci, *Bacteroides* species, *Fusobacterium* species, enterobacteria |
| Right-to-left shunt (congenital cyanotic heart disease, pulmonary AVM) | Streptococci, staphylococci, *Peptostreptococcus* species |
| Penetrating trauma or neurosurgery | *Staphylococcus aureus*, *Staphylococcus epidermidis*, streptococci, enterobacteria, *Clostridium* species |
| Unknown | Streptococci, *Peptostreptococcus* species, *Bacteroides* species, *Haemophilus* species, staphylococci |
| Immunosuppression: bone marrow or solid organ transplantation | *Aspergillus* species, *Candida* species, *Nocardia* species, *Toxoplasma* gondii |
| AIDS | *Toxoplasma gondii*, *Cryptococcus neoformans*, *Listeria monocytogenes*, *Mycobacterium* species, *Candida* species, *Aspergillus* species |

**TABLE III** Presenting Signs and Symptoms of Brain Abscess (after Chun *et al.*, 1986; Nielsen *et al.*, 1982; O'Donoghue *et al.*, 1992; Svanteson *et al.*, 1988; Yang, 1981; Yang and Zhao, 1993)

| Sign or symptom | Frequency range (%) | Mean (%) |
| --- | --- | --- |
| Headache | 64–97 | 82 |
| Vomiting | 35–85 | 55 |
| Altered consciousness | 28–91 | 54 |
| Fever | 32–62 | 52 |
| Papilloedema | 9–56 | 39 |
| Hemiparesis | 23–44 | 33 |
| Seizures | 13–35 | 25 |
| Neck stiffness | 5–41 | 21 |

1979; Falcone and Post, 2000). Within several days, a ring-like contrast enhancement may become visible during the late cerebritis stage. The expanse of cerebritis, cerebral edema, and mass effect are greatest at this stage. The capsule stage is characterized by a hypodense (necrotic) center of the lesion; a ring enhancement; and a regression of cerebritis, perifocal edema, and mass effect. The MRI or CT scan may also reveal the cause of the intracerebral infection (e.g., sinusitis, otitis, mastoiditis, fracture, osseous dehiscence, or intracranial foreign body) or its complications (e.g., hydrocephalus, ventricular rupture, or, very rarely, hemorrhage into the abscess cavity). A chest X-ray or a CT of the chest should be performed in order to exclude a pulmonary infection or arteriovenous malformation. Valvular vegetations (endocarditis) or a right-to-left shunt should be looked for by auscultation and echocardiography. Tooth infections should be ruled out clinically and by dental (panoramic) radiography. The blood leukocyte counts are elevated in 30–57% (mean, 47%) of cases (Alderson *et al.*, 1981; Chun *et al.*, 1986; O'Donoghue *et al.*, 1992; Seydoux and Francioli, 1992; Svanteson *et al.*, 1988; Yang and Zhao, 1993), the erythrocyte sedimentation rate is accelerated in 48–65% (mean, 59%) of cases (Mampalam and Rosenblum, 1988; Svanteson *et al.*, 1988; Yang and Zhao, 1993), and the serum C-reactive protein value is increased in 77–90% (mean, 82%) of cases (Grimstad *et al.*, 1992; Hirschberg and Bosnes, 1987; Jamjoom, 1996). Blood cultures (aerobic and anaerobic) should be performed although they are rarely positive (only 11% of cases in one study; Chun *et al.*, 1986). A lumbar puncture should not be performed in patients with brain abscess, because it carries a considerable risk of cerebral herniation (1–3% of patients; Nielsen *et al.*, 1982; Yang, 1981) and the diagnostic yield is low: CSF cultures are positive in <10% (mean, 6%) (Chun *et al.*, 1986; Mampalam and Rosenblum, 1988; Nielsen *et al.*, 1982; Seydoux and Francioli, 1992) and other CSF parameters are altered unspecifically (the CSF cell count can range from zero

to more than 100,000 cells/μl, the protein level from normal to more than 500 mg/dl, and CSF glucose can be normal or severely depressed; Chun *et al.*, 1986; Seydoux and Francioli, 1992). The most important diagnostic measure is the abscess aspiration. Purulent material should be examined with Gram's stain and bacterial cultures (aerobic and anaerobic, positive in 69–97% of cases; mean, 84%; Alderson *et al.*, 1981; Chun *et al.*, 1986; O'Donoghue *et al.*, 1992; Seydoux and Francioli, 1992; Svanteson *et al.*, 1988; Yang and Zhao, 1993). On indication, fungal, mycobacterial, and protozoal cultures should also be carried out.

### Differential Diagnosis

The most important clinical differential diagnoses of brain abscess are meningoencephalitis, subdural empyema or epidural abscess, and septic sinus or venous thrombosis. Radiologically, primary brain tumors, metastasis, ischemic infarction, resolving hematoma, thrombosed aneurysm, arteriovenous malformation, tumefactive multiple sclerosis, sarcoidosis, and lymphoma should be considered (Falcone and Post, 2000). Nuclear medicine techniques, such as positron emission tomography (PET), indium-111-labeled leukocyte scintigraphy, or 99mTc-hexamethylpropyleneamine oxime (99mTc-HMPAO) scintigraphy have been evaluated for the discrimination of brain abscesses from other cystic cerebral lesions (Grimstad *et al.*, 1992; Pierce *et al.*, 1995; Schmidt *et al.*, 1990). Also, modern MRI methods, such as diffusion-weighted MRI (Kim *et al.*, 1997) and, in particular, proton MR spectroscopy may be helpful for the identification of brain abscesses (Kim *et al.*, 1998; Shukla-Dave *et al.*, 2001; Yen *et al.*, 1995).

In patients with AIDS, the most common cause of intracranial mass lesions is toxoplasmosis; the most important differential diagnosis is primary CNS lymphoma (PCNSL), followed by brain abscesses due to pathogens other than *Toxoplasma*, metastatic tumors, and cerebrovascular disease (American Academy of Neurology, 1998). The following procedures have been proposed for the evaluation and treatment of intracranial mass lesions in AIDS patients (American Academy of Neurology, 1998): (a) large lesions with mass effect threatening impending herniation should undergo open biopsy with decompression; (b) when available, thallium-201 single photon emission computed tomography can be performed (Tl-201 SPECT, limited sensitivity but good specificity for PCNSL); (c) empirical treatment for toxoplasmosis should be instituted in all other cases, except when a single intracranial mass lesion accompanies negative serology for toxoplasmosis; (d) in the latter cases, a stereotactic biopsy is warranted; (e) clinical or radiographic worsening in patients treated presumptively for toxoplasmosis is an indication for a stereotactic biopsy; (f) in HIV-infected children, proceeding directly to stereotactic biopsy may be considered, since toxoplasmosis is rare in HIV-infected children.

## Natural Course

In the general population, the incidence of brain abscess is estimated at 0.3–1.3 per 100,000 people per year (McClelland *et al.*, 1978; Miller *et al.*, 1988; Nicolosi *et al.*, 1991) with a male:female ratio of 2:1 to 3:1 (Mampalam and Rosenblum, 1988; Nielsen *et al.*, 1982; Yang, 1981; Yang and Zhao, 1993). In certain patient groups, the risk of developing a brain abscess is markedly increased, e.g., 1.7% of patients with acute or chronic sinusitis (Clayman *et al.*, 1991), 0.14% of patients with chronic otitis media (Kangsanarak *et al.*, 1993), 3% of penetrating craniocerebral trauma victims (Rish *et al.*, 1981), 0.58% of craniotomized patients (Korinek, 1997), 0.61 to 4% of transplant recipients (Maschke *et al.*, 1999; Selby *et al.*, 1997), 2% of patients with congenital heart disease (Fischbein *et al.*, 1974), 3 to 37% of patients with pulmonary arteriovenous malformations (Faughnan *et al.*, 2000; Swanson *et al.*, 1999), and up to 5% of patients with acute infective endocarditis (Kanter and Hart, 1991; Le Cam *et al.*, 1984; Pruitt *et al.*, 1978; Salgado *et al.*, 1989) were reported to develop a brain abscess. Therefore, preventive measures are particularly important for these patients at risk (see section on prophylaxis of brain abscess).

Improved diagnostic and therapeutic means (cranial CT and MRI, microbiological methods, surgical, medical, and supportive therapy) have reduced the mortality from 40–60% prior to the antibiotic era to 6–24% (mean, 13%) in recent studies (Mampalam and Rosenblum, 1988; O'Donoghue *et al.*, 1992; Seydoux and Francioli, 1992; Yang and Zhao, 1993). Patients with a very young or old age; reduced level of consciousness on admission; rapidly progressive disease; multiple, large, or deeply located abscesses; or intraventricular rupture of brain abscess have been reported to have a worse prognosis (Chun *et al.*, 1986; Mampalam and Rosenblum, 1988; Seydoux and Francioli, 1992; Takeshita *et al.*, 2001; Yang and Zhao, 1993). Long-term complications of brain abscess include seizures (10 or 16% in two recent studies; Mampalam and Rosenblum, 1988; Seydoux and Francioli, 1992), focal neurological deficits (e.g., 18% of the survivors in one long-term follow-up study had paresis or aphasia; Nielsen *et al.*, 1983), and intellectual impairment (affecting 18% of the survivors in the beforementioned study; Nielsen *et al.*, 1983).

## Principles of Therapy

The treatment of brain abscess usually includes

1. Surgery of brain abscess,
2. Surgery of the primary underlying infectious focus (if detectable),
3. Medical therapy (parenteral antibiotics, anticonvulsants, antiedema therapy).

Randomized controlled studies of therapies for brain abscess do not exist. Treatment recommendations are therefore based on retrospective analyses and clinical experience.

### Surgery of Brain Abscess

The reasons for surgery of a brain abscess are to decrease cerebral mass, to confirm the diagnosis, and to obtain purulent material for culture. The administration of preoperative antibiotics probably reduces the yield of positive cultures (Mampalam and Rosenblum, 1988; Yang and Zhao, 1993). However, antimicrobial therapy can only be withheld until surgery, if the operation can be performed within a short time (hours) and the abscess does not show a significant mass effect (risk of herniation).

The method of choice is CT-guided stereotactical aspiration through a burr hole. Generally, it causes less tissue damage than open craniotomy and excision and is more precise than CT-guided "freehand" burr hole aspiration. Thus, stereotactical burr hole aspiration allows the treatment of deep-seated abscesses and abscesses at critical locations, where a high precision is necessary. It can be performed under local anesthesia. The aspirate should be smeared immediately and cultures (aerobic and anaerobic) should be performed. Often, postaspiration external drainage of the abscess through an intracavitary catheter is recommended for abscesses larger than 3 cm in diameter; in our hospital, the abscess cavity is irrigated daily with 10 ml sterile saline, and when the aspirate is clear, the drain is removed (usually after 3 to 5 days). In patients with abnormal results of clotting studies or thrombocytopenia, open craniotomy and excision may be preferred, because it allows for a better control of intracranial bleeding. Open craniotomy and total excision of the abscess may frequently become necessary in patients with abscesses for whom (repeat) aspiration has failed, brain abscess secondary to foreign bodies, or fungal (or nocardial) abscesses. If the brain abscess is the result of a head trauma, surgical intervention by excision is usually required to perform appropriate toilet, debridement, removal of bone fragments or hair, and closure of dural defects. An infected bone flap (a rare finding in brain abscess surgery) requires removal and 6 months later replacement by cranioplasty.

For patients with multiple brain abscesses, it has been recommended that abscesses larger than 2.5 cm in diameter or with significant mass effect should be stereotactically aspirated (or excised) (Mamelak et al., 1995). If the abscesses are all smaller than 2.5 cm in diameter, then the largest and/or most accessible lesion may be aspirated for diagnostic purposes (Mamelak et al., 1995). Further surgical drainage is recommended if an abscess enlarges after a 2-week interval or fails to diminish in size after 3–4 weeks of antibiotics (Mamelak et al., 1995).

External ventricular drainage of a hydrocephalus is particularly often necessary in patients with basal ganglia or thalamic abscesses (75% of patients in one study; Lutz et al., 1994) or cerebellar abscesses (79.2% of cases in a large series; Nadvi et al., 1997).

In case of intraventricular rupture of brain abscesses (it occurred in 14.5 or 27.3% of patients and was fatal in 79 or 39.4% in two studies; Nielsen et al., 1982; Takeshita et al., 2001), an aggressive approach seems appropriate in view of its dismal prognosis. Open craniotomy with debridement of the abscess cavity, lavage of the ventricular system, ventricular drainage, intrathecal gentamicin, and intravenous administration of appropriate antibiotics is one strategy (Zeidman et al., 1995). Recently, the reaspiration of the abscess, followed by ventricular drainage and also intravenous and intrathecal antibiotic therapy, was propagated (Takeshita et al., 2001).

## Surgical Treatment of the Primary Underlying Infectious Focus

The focus of primary infection (e.g., otitis, sinusitis, bronchiectasis) should be surgically treated as soon as possible. However, surgery of the brain abscess has top priority in case of clinical neurological deterioration.

## Medical Therapy

**Antibiotics.** Antibiotics used for the treatment of brain abscess should be administered intravenously initially, be active against the pathogens that are likely in a given clinical scenario, penetrate into the abscess fluid (and into the site of the underlying infection) in adequate concentrations, and have bactericidal activity (Townsend and Scheld, 1998). A sufficient penetration into brain abscesses has been demonstrated for high-dose penicillin, ampicillin, cefuroxime, chloramphenicol, cotrimoxazole, ceftazidime, moxalactam, and metronidazole (Infection in Neurosurgery Working Party of the British Society for Antimicrobial Chemotherapy, 2000; Mathisen and Johnson, 1997).

The data on ceftriaxone, semisynthetic penicillins (e.g., flucloxacillin, oxacillin, and nafcillin), imipenem (meropenem should be preferred due to its lower neurotoxicity), vancomycin, and quinolones (caution to lowered seizure thresholds) are limited, because only smaller studies or case reports have demonstrated their efficacy for the treatment of brain abscess (Mathisen and Johnson, 1997). For the recommended empirical antimicrobial therapy and dosages please see Tables IV and V. Recipients of bone marrow or solid organ transplants should be treated initially for fungi, *Nocardia*, and *Toxoplasma* with a combination of amphotericin

**TABLE IV**  Empiric Antimicrobial Therapy of Brain Abscess

| Underlying condition | Recommended antimicrobial regimen |
|---|---|
| Community acquired (immunocompetent patients) | Third-generation cephalosporin[a,b] + metronidazole |
| Postoperative or posttraumatic | Meropenem + vancomycin |

[a]Ceftriaxone or cefotaxime; for *Pseudomonas* species ceftazidime or cefepime.
[b]If staphylococci are suspected: plus vancomycin (alternatives: fosfomycin, rifampicin, flucloxacillin, nafcillin).

**TABLE V**  Recommended Dosages for Antimicrobial Therapy of Brain or Spinal Cord Abscesses in Adults (Dosing May Need Adjustment in Patients with Underlying Renal or Liver Disease)

| Antimicrobial agent | Dosage per day |
|---|---|
| Amphotericin B | 0.6–1 mg/kg iv[a] |
| Liposomal amphotericin B | 5 mg/kg iv |
| Cefepime | $3 \times 2$ g iv |
| Cefotaxime | $3 \times 3$ (–4) g iv |
| Ceftazidime | $3 \times 2$ g iv |
| Ceftriaxone | $1 \times$ (2–) 4 g iv |
| Clindamycin | $4 \times 600$ mg iv |
| Flucloxacillin or nafcillin | $6 \times 2$ g iv |
| Fosfomycin | $3 \times 5$ g iv |
| Itraconazole | $2 \times 200$–400 mg po[b] |
| Meropenem | $3 \times 2$ g iv |
| Metronidazole | $3 \times 500$ mg iv |
| Pyrimethamine + sulfadiazine | 25–50 mg po[c] 6–8 g po |
| Rifampicin | $1 \times 10$ mg/kg iv (maximum 600 mg) |
| Trimethoprim + sulfamethoxazole | $3 \times 160$ mg iv + $3 \times 800$ mg iv |
| Vancomycin | $2(-3) \times 1$ g iv[d] |

[a]Dosages of up to 1.5 mg/kg/day may be used for aspergillosis or mucormycosis.
[b]Possible alternative to amphotericin B (Sanchez et al., 1995); poor penetration of the blood–CSF barrier.
[c]Loading dose, 50–200 mg po.
[d]Need to monitor serum concentrations.

B and trimethoprim–sulfamethoxazole. AIDS patients with suspected toxoplasmic encephalitis should receive pyrimethamine plus sulfadiazine or pyrimethamine plus clindamycin. After culture results, the therapy should be modified according to the antibiogram. The appropriate duration of antimicrobial therapy for brain abscess remains unclear. Usually, 6–8 weeks of intravenous therapy is recommended, provided that the pathogens are susceptible, a clear clinical and radiographic response is observed, and the treatment is well-tolerated by the patient. Some favor an additional 2- to 3-month course of oral antimicrobial therapy (e.g., trimethoprim–sulfamethoxazole) to prevent a relapse; clinical evidence for that recommendation, however, is lacking.

**Anticonvulsants.** Seizures are frequent complications of brain abscesses (on admission, seizures are reported in 20–50% of patients; Ciurea et al., 1999; Nielsen et al., 1982; Wong et al., 1989; Yang and Zhao, 1993). They should be treated with phenytoin or carbamazepine. Patients without seizures but with epileptic discharges on the EEG should also be treated with anticonvulsants. However, there is no evidence to support seizure prophylaxis for all patients. Patients with seizures or EEG abnormalities during the acute disease should be treated for at least 6–12 months; if seizures do not recur, the EEG is normal, and the cranial CT shows only residual abscess, but not active enhancing lesions, the antiepileptic medication is then slowly withdrawn.

**Corticosteroids.** Retrospective studies on adjunctive corticosteroid therapy for brain abscess failed to show a beneficial effect on outcome (Chun et al., 1986; Seydoux and Francioli, 1992). Experimental animal studies suggest that corticosteroids may reduce the edema around a brain abscess (Wallenfang et al., 1980) but impair the bacterial clearance and capsule formation (Neuwelt et al., 1984; Quartey et al., 1976; Rosenblum et al., 1986) and the penetration of polar antibiotics, such as benzylpenicillin, into the infected brain tissue (but not that of more lipophilic substances, such as metronidazole) (Kourtopoulos et al., 1983). The use of corticosteroids (e.g., initially $3 \times 8$ mg iv dexamethasone/day) should therefore be restricted to patients with a progressive neurological deterioration or impending cerebral herniation and radiological evidence of significant cerebral edema and mass effect. The corticosteroids should be tapered off over a few days after the patient's condition has stabilized.

**Conservative Treatment.** Medical therapy alone can be considered in patients with lesion in the cerebritis stage (which are much more likely to respond to antibiotic therapy alone due to lack of a capsule) or small (less than 2–3 cm in diameter) or surgically inaccessible abscesses (in particular brainstem abscesses; Carpenter, 1994), as well as in patients with a stable neurological but poor medical condition (Mathisen and Johnson, 1997; Rosenblum et al., 1986). However, surgery should be performed if the patient shows neurological deterioration or serial CTs show that the abscess does not decrease in size within 4 weeks or even grows (Rosenblum et al., 1986).

### Prophylaxis of Brain Abscess

Primary or secondary prevention of brain abscesses is particularly important in patients at risk. Preventive measures include the adequate management of acute and chronic sinusitis or otitis media (Clayman et al., 1991; Kangsanarak et al., 1993), antibiotic prophylaxis for bacteremic procedures or corrective surgery for patients with cyanotic congenital heart disease (Fischbein et al., 1974), or antibiotic prophylaxis and occlusion (transcatheter embolotherapy) of the larger arteriovenous malformations in patients with diffuse pulmonary arteriovenous malformations (Faughnan et al., 2000). For penetrating craniocerebral injuries, broad-spectrum antibiotic prophylaxis should be given as soon as possible after the injury and be continued for 5 days after debridement of the scalp wound and of the intracranial injury, e.g., $3 \times 1.2$ g iv amoxicillin + clavulanic acid/day, or $3 \times 750$ mg iv cefuroxime/day (1.5 g loading dose) + $3 \times 500$ mg/day iv metronidazole (Bayston et al., 2000). A meta-analysis of the efficacy of prophylactic antibiotics for craniotomy reported an about fourfold reduction of the superficial and deep wound infection rate by antibiotic prophylaxis (e.g., a single dose of 2 g iv cefuroxime with the induction of anesthesia; a second dose should be given if the operation lasts longer than 6 h) (Barker, 1994). In order to avoid toxoplasmic encephalitis in HIV-positive persons, the prevention of exposure is particularly important for individuals who lack IgG antibody to *Toxoplasma*. *Toxoplasma*-seropositive patients with a CD4+ T lymphocyte count below 100/μl should be adminstered propylaxis for toxoplasmic encephalitis (1 DS tablet trimethoprim–sulfamethoxazole (TMP–SMZ)/day) (U.S. Public Health Service (USPHS) and Infectious Diseases Society of America (IDSA), 1999). For secondary prevention, 25–75 mg/day pyrimethamine po + 500–1000 mg/day sulfadiazine po + 10–25 mg/day leucovorin po are recommended (U.S. Public Health Service (USPHS) and Infectious Diseases Society of America (IDSA), 1999).

# SUBDURAL EMPYEMA

## Clinical Aspects

### Definition

Subdural empyema is a suppurative inflammation in the subdural space, i.e., between the dura mater and the arachnoid. It is referred to as empyema because the subdural compartment is a preexisting normal space.

### Etiology

In a recent series of 699 South African patients with subdural empyema (Nathoo et al., 1999b), underlying infections were paranasal sinusitis in 67%, bacterial meningitis in 10% (almost exclusively in infants), ear infections in 9%, trauma in 8%, and oral infections in <1% of cases. The remaining 5% were attributed to diabetes mellitus, alcoholism, sepsis, ventriculoperitoneal shunt or postsurgical infections, and infections of the scalp or lung. In recent studies from the developed world, the distribution differs only with respect to a higher proportion of postoperative cases (15–30%; Dill et al., 1995; Hlavin et al., 1994). Subdural empyemas most frequently spread diffusely over the cerebral hemispheres or along the falx cerebri. If the empyema follows frontal sinusitis, then it is usually located over the frontal lobes; empyemas complicating otitis are frequently located over the temporal or occipital lobe. Infratentorial empyema is a rarity (Nathoo et al., 1997; Pathak et al., 1990). The most frequent microorganisms involved are aerobic and anaerobic streptococci (in particular S. milleri), staphylococci (in particular postoperative cases, caution to methicillin-resistant staphylococci; Hlavin et al., 1994; Korinek, 1997), and Gram-negative Enterobacteriaceae (in particular, Proteus and Pseudomonas species and Escherichia coli). In the South African series, multiple organisms were detected in 14.1% and the cultures remained sterile in 17.6% of patients (Nathoo et al., 1999b). With appropriate culturing techniques, anaerobic bacteria may be detected in up to 40% of cases (Dill et al., 1995).

### Diagnosis

The clinical signs and symptoms of subdural empyema are fever (77%), meningismus (74%), periorbital edema (43%), seizures (39%), subgaleal abscess (Pott's puffy tumor, 33%), headache (32%), hemiparesis (25.5%), eyelid abscess (12%), and signs of tentorial herniation (4.7%) (Dill et al., 1995). More than 50% of the patients show a decreased level of consciousness (Bok and Peter, 1993). The diagnostic method of choice is contrast-enhanced cranial CT or MRI. The CT typically shows a hypodense or isodense subdural collection with rim enhancement. MRI is more sensitive than CT for the diagnosis of subdural empyema (Campbell and Zimmerman, 1998). The subdural collection typically is discretely hyperintense on T1-weighted images and strongly hyperintense on T2-weighted images compared with the ventricular CSF (Shapiro, 1997). Lumbar puncture should be avoided in established subdural empyema due to a considerable risk of herniation; furthermore, CSF findings are usually unspecific. Blood cultures should be drawn for aerobic and anaerobic cultures. The erythrocyte sedimentation rate and the blood leukocyte count have been reported to be invariably elevated (Bok and Peter, 1993). Complications of subdural empyema are epidural (15.3%) or brain (6%) abscesses, ostemyelitis (10.0%), cerebral infarction (2.4%), and status epilepticus (1.1%) (Nathoo et al., 1999b). Also, septic venous or sinus thrombosis, brain edema, purulent meningitis, and hyponatremia can develop.

### Differential Diagnosis

Differential diagnoses include focal encephalitis, purulent meningitis, epidural or brain abscess, septic sinus or venous thrombosis, chronic subdural hematomas (more hyperintense on MRI than empyemas), and benign subdural effusions (isointense to CSF on MRI).

## Natural Course

Subdural empyema is observed less frequently than brain abscess and accounts for approximately 15% of all intracranial infections (Nathoo et al., 1999b). It often has an acute course because the inflammation can spread quickly in the subdural space. Therefore, it must be considered a neurological–neurosurgical emergency. If left untreated, subdural empyema is fatal within days or weeks. Improved diagnostic methods and antibiotic and surgical therapy have reduced the mortality to 8 to 12% (Dill et al., 1995; Nathoo et al., 1999b; Pathak et al., 1990). A depressed level of consciousness on admission, a large extent of the empyema, older patient age, and a delayed diagnosis were reported to be associated with an adverse outcome (Bannister et al., 1981; Dill et al., 1995; Hitchcock and Andreadis, 1964; Kaufman et al., 1983; Mauser et al., 1987; Miller et al., 1987). A good outcome (Glasgow Outcome Scale 4 or 5) can be expected in approximately 80% of patients (Nathoo et al., 1999b). Postoperative seizures were reported to ocur in 14.7% (Nathoo et al., 1999b). Older studies reported higher incidences of late seizures (64%), hemipareses

(17%), and speech disorders (6%) in survivors of subdural empyema (Cowie and Williams, 1983).

## Management

The principles of therapy are similar to those of brain abscess: immediate surgical drainage, surgical cure of the underlying infectious source, and medical therapy (antimicrobial therapy; antiedematous and antiepileptic therapy on indication). Randomized controlled studies on therapy of subdural empyema are not available. There is no general agreement on whether craniotomy or burr hole aspiration should be preferred. Earlier studies observed lower mortality rates with craniotomy (Bannister et al., 1981). Studies from the CT era, however, showed similar outcomes of patients treated with burr holes alone, burrholes and small craniectomies, or craniotomy (Bok and Peter, 1993). In our hospital, drainage through multiple burr holes is the standard method; drains are left in the subdural space, they are irrigated daily with 10 ml sterile saline, and when the aspirate is clear, the drains are removed (usually after 3 to 5 days). Burr hole aspiration is usually easier early in the course of the empyema, when the pus tends to be more fluid; later, loculations sometimes make repeated burr hole aspirations or a craniotomy necessary. Empyema fluid should be submitted for Gram stain and culture (aerobic and anaerobic). The medical therapy is that of brain abscess. For the empiric antimicrobial therapy see Tables IV and V. After culture results, the therapy should be modified according to the antibiogram. Antimicrobial therapy should be continued for at least 3 weeks. Cranial CT or MRI should be regularly performed in order to detect postoperative recurrences.

## EPIDURAL ABSCESS

### Clinical Aspects

#### Definition

A cranial epidural abscess is a suppurative inflammation that develops between the dura and the inner table of the skull. It is referred to as an abscess because the epidural space does not exist under normal conditions, when the dura is closely attached to the skull.

#### Etiology

Major causes were paranasal sinusitis (65%), mastoiditis (20%), and trauma (6%) in a recent South African series (Nathoo et al., 1999a). Older studies from developed countries also observed a predominance of otorhinogenic sources. However, in a recent study from the United States, 66% of epidural abscesses and/or subdural empyemas occurred after a craniotomy (Hlavin et al., 1994). Streptococci (in particular S. milleri) and Proteus species are dominant in otorhinogenic epidural abscesses, but staphylococci (including Staphylococcus epidermidis and methicillin-resistant Staphylococcus aureus) are the leading pathogens in postoperative epidural abscesses (Hlavin et al., 1994; Nathoo et al., 1999a).

### Diagnosis

Clinical signs are similar to those of subdural empyema (Nathoo et al., 1999a). Conditions associated with epidural abscess are osteomyelitis of the calvarium, subdural empyema, meningitis, septic sinus or venous thrombosis, cerebritis, and brain abscess. The MRI can detect the extracerebral fluid collection (hyperintense compared with ventricular CSF) with a sensitivity higher than that of the CT, because there are no bony artifacts adjacent to the skull (Shapiro, 1997). The dura can be seen as a thin line separating the abscess from the brain parenchyma.

### Treatment

Surgical drainage can be achieved by formal craniotomy, drainage through burr holes, or an extended craniectomy. Underlying infections should be cured surgically. The medical therapy is that of brain abscess. For the empirical antimicrobial therapy, see Tables IV and V. After culture results, the therapy should be modified according to the antibiogram. The prognosis of epidural abscess is usually good: a complete recovery (Glasgow Outcome Scale 5) can be expected in approximately 95% of patients (Nathoo et al., 1999a).

## SPINAL EPIDURAL ABSCESS

### Clinical Aspects

#### Definition

Spinal epidural abscess is a suppurative inflammation in the epidural space between the dura mater and the vertebral periosteum. Among spinal abscesses, epidural abscesses are observed much more frequently than subdural or intramedullary abscesses.

#### Etiology

Spinal epidural abscesses can result from contiguous spread from an adjacent infectious focus, hematogenous dissemination, or spinal instrumentation. Immunocompromising conditions are frequent among patients with spinal epidural abscesses, e.g., in a recent meta-analysis of 915 patients, 15% were diabetics, 9% were intravenous drug abusers, 5% were chronic alcoholics, 2% were on corticosteroids, and 2% had cancer (Reihsaus *et al.*, 2000). Fourty-four percent had associated infections that were regarded as the source of infection via contiguous or hematogenous spread (Reihsaus *et al.*, 2000). Skin infections (e.g., abscesses, furuncles, or paronychia) were present in 15%, vertebral osteomyelitis and/or discitis in 7%, pulmonary or mediastinal infections in 5%, and sepsis in 5% of patients. Trauma, in particular spinal trauma, preceded the abscess development in 10% of cases. In these patients, it is believed that a small spinal hematoma or other injured tissue may be a nidus for metastatic spread during transient bacteremia. Epidural anesthesia caused 5%, extraspinal operations 5%, spinal operations 3%, and paravertebral injections 1% of spinal epidural abscesses.

The principal etiologic agent of spinal epidural abscess is *S. aureus*. It accounted for 73% of the cases reviewed in the recent meta-analysis (Reihsaus *et al.*, 2000). Coagulase-negative staphylococci caused an additional 4% of abscesses. Streptococci were cultured in 7% (in particular *S. viridans* and *S. pneumoniae*) and *E. coli* in 2% of patients. Mixed bacterial infections were rare (only 3%). Fungi (in particular *Aspergillus fumigatus*) accounted for 1.5% and parasites (*Echinococcus granularis* and *Dracunulus midinensis*, which is endemic in Africa) for 0.35% of cases.

### Location

Thirty-five percent of the abscesses were located in the thoracic, 30% in the lumbar or lumbosacral, and 19% in the cervical epidural space; 7% each were in the cervicothoracic or thoracolumbar region. Abscesses extending from the cervical to the lumbar epidural space were observed in 1% (Reihsaus *et al.*, 2000). In more than 70% cases, spinal epidural abscesses develop dorsally to the spinal cord, because ventrally the dura mater is tightly adherent to the vertebral bodies and to the ligaments. Anterior epidural abscesses usually result from contiguous spread from vertebral osteomyelitis or another contiguous site of infection. They are often confined to one or two segments. In the dorsal epidural compartment, however, there is no barrier to the rostral or caudal spread of the suppuration. Therefore, dorsal epidural abscesses often extend over several segments.

### Diagnosis

According to the classical descriptions of the various stages of spinal epidural abscess (Heusner, 1948;

Rankin and Flothow, 1946) (the percentages refer to Reihsaus *et al.*, 2000), initial clinical findings are severe back pain (71% of cases) associated with fever (66%) and local tenderness (17%). Then, signs of spinal irritation (20%) may develop, such as meningismus or radicular pain. As the disease progresses, neurological deficits appear, such as muscle weakness (26%), fecal or urinary incontinence (24%), and sensory deficits (13%). The progression to severe paraparesis or -plegia (31%) or tetraparesis or -plegia (3%) can then occur rapidly. The neurological symptoms and signs usually develop during a period of several days up to several weeks. Surgical decompressive intervention most often reveals purulent material in the acute forms of the disease and organized granulation and fibrous tissue in the chronic forms.

The diagnostic method of choice is the contrast-enhanced MRI. MRI shows a fluid collection that is hyperintense on T2-weighted images and hypointense on T1-weighted images. The dura can often be seen as a thin line (hypointense on T2-weighted images) between pus and cerebrospinal fluid. After adminstration of contrast media, homogenous or peripheral enhancement of the lesion can be seen. The MR images should also be examined carefully for adjacent infectious lesions, such as vertebral osteomyelitis or discitis and paravertebral or psoas abscesses. Alternatively, myelography and postmyelography CT may be used. This combination can only show the accompanying extradural mass lesion with its attending effects on CSF flow. The inflammatory tissue is shown only indirectly.

Blood cultures are particularly valuable in the management of patients with spinal epidural abscesses, because they are positive in the majority of cases and the etiologic pathogens are almost always identical to those identified in the abscess pus. Seventy-eight percent of patients were reported to have blood leukocyte counts above 10,000/µl and 94% showed an erythrocyte sedimentation rate above 20 mm in the first hour (Reihsaus *et al.*, 2000). Cerebrospinal fluid usually reveals a mixed pleocytosis (up to several hundred leukocytes/µl) and an elevated total protein content (up to several hundred mg/dl). However, lumbar puncture should be avoided if a spinal epidural abscess is suspected, because the needle may inadvertently carry infectious material into the subarachnoid space.

### Differential Diagnosis

Differential diagnoses mainly include intervertebral disc prolapse, spinal tumors, transverse myelitis, epidural hematoma, vertebral osteomyelitis, spinal subdural or intramedullary abscess, and spinal artery syndromes. The clinical picture of spinal epidural abscess can also be mimicked by acute spinal cord dysfunction due to myelitis during bacterial meningitis (Kastenbauer *et al.*, 2001).

## Natural Course

Spinal epidural abscesses are diagnosed in approximately 0.2 to 0.8 cases per 10,000 admissions to large tertiary care centers (Baker *et al.*, 1975; Hlavin *et al.*, 1990). There is a male predominance (male : female ratio = 1 : 0.56; Reihsaus *et al.*, 2000). Spinal epidural abscesses have been reported for all age groups (from 10 days to 87 years old; Reihsaus *et al.*, 2000). However, they are very rare in children (Auletta and John, 2001). The complications of spinal epidural abscess are spinal cord or nerve root compression, myelomalacia secondary to vascular involvement (vasculitis, venous congestion, and thrombosis), purulent meningitis, and septic shock. Until the 1930s, most patients with spinal epidural abscess died (Reihsaus *et al.*, 2000). With the introduction of laminectomy as standard surgical treatment and the advent of antibacterial chemotherapy, the mortality dropped to less than 50%. During the last 3 decades, the mortality rate of spinal epidural abscess was relatively constant at 15%. Complete recovery can be achieved in 40–50% of cases. Minor neurologic deficits are present in 20–30% and severe paresis or paralysis is present in 10–20% (Reihsaus *et al.*, 2000). The chance of complete recovery depends on early diagnosis and treatment. It is most likely in those patients with no neurological deficit or with a deficit that has developed within 24h before diagnosis and treatment. Complete recovery is unlikely in patients with weakness or paralysis for more than 36–48h (Reihsaus *et al.*, 2000).

## Principles of Therapy

Spinal epidural abscess is a neurological–neurosurgical emergency. The neurological deterioration to severe spinal cord dysfunction can occur in just a few hours, making the diagnosis and treatment imparative. The treatment usually consists of decompressive surgery and drainage of the abscess, eradication of the primary underlying infectious focus (if detectable), and parenteral antibiotic therapy.

### Surgery

Posterior epidural abscesses are usually treated by laminectomy, removal of pus or granulation tissue, and postoperative irrigation with sterile saline through extradural drains for several days. For the treatment of anterior epidural abscesses, an anterior approach may be necessary. Adjacent sources of infection should also be cured surgically (e.g., vertebrectomy for severe osteomyelitis with vertebral body destruction).

### Antibiotics

Parenteral antibiotic therapy should be directed against the etiologic agents most likely to be involved: *S. aureus*, coagulase-negative staphylococci, streptococci, and Gram-negative rods. A combination of a third-generation cephalosporin (e.g., ceftriaxone) with another antibiotic showing antistaphylococcal activity (e.g., rifampicin, nafcillin, or flucloxacillin) is recommended (for dosages see Table V). If methicillin-resistant staphylococci are involved, vancomycin should be used. Thus, a nosocomial spinal epidural abscess may be initially treated with vancomycin plus meropenem. When the etiologic agent has been identified in cultures, the antibiotic regimen should be modified according to the antibiogram. A treatment duration of 3–4 weeks is usually recommended; this period should be extended to 6–8 weeks if osteomyelitis is present.

### Corticosteroids

The administration of corticosteroids (e.g., iv dexamethasone $3 \times 8$ mg/day in adults) is controversial. Their benefit has not been proven and controlled studies do not exist.

### Conservative Treatment

Nonsurgical management using antibiotic therapy alone may be considered in selected patients

— without significant neurological deficit,
— with complete paralysis for more than three days, or
— in poor medical condition with a high surgical risk.

Identification of the causative pathogen may be attempted by CT-guided needle biopsy in conservatively treated patients with negative blood cultures. Conservatively treated patients without significant neurological deficit need particularly close neurological monitoring in a hospital with facilities for emergency decompressive surgery. Retrospective analyses suggest that the outcome is comparable to that of surgically treated patients, if these guidelines are followed (Grieve *et al.*, 2000; Leys *et al.*, 1985; Wheeler *et al.*, 1992). In the future, MRI criteria may also be helpful when selecting patients for conservative treatment. A recent study suggested that a good outcome was associated with (Tung *et al.*, 1999)

— abscesses shorter than 3 cm cephalocaudally,

— less than 50% narrowing of the sagittal diameter of the spinal canal, and

— homogenous contrast-enhancement (suggestive of phlegmonous or granulomatous tissue, opposed to the peripheral enhancement of a frank abscess containing pus).

The authors therefore suggested that only patients fulfilling al least two of these criteria should be treated conservatively. However, definite conclusions can only be drawn from controlled studies, which are still lacking.

## SPINAL SUBDURAL EMPYEMA

Spinal subdural empyema is rare. Only approximately 50 cases are on record (Bartels *et al.*, 1992; Levy *et al.*, 1993; Ozates *et al.*, 2000; Schneider and Givens, 1998). It is most often caused by hematogenous spread from a remote infectious focus with *S. aureus*, streptococci, and Gram-negative rods being the most frequent etiologic agents. Direct extension of a contiguous infection is infrequent but iatrogenic cases (e.g., following lumbar puncture or discography) are relatively common (Bartels *et al.*, 1992). The signs and symptoms are similar to those of spinal epidural abscess. Diagnosis is made by MRI or postmyelography CT. The antibiotic therapy is that of spinal epidural abscess; rapid surgical decompression and drainage has been recommended for all cases (Bartels *et al.*, 1992).

## SPINAL INTRAMEDULLARY ABSCESS

Only approximately 100 cases of intramedullary abscesses have been published (Bartels *et al.*, 1995; Byrne *et al.*, 1994; Chan and Gold, 1998; Desai *et al.*, 1999). The majority of intramedullary abscesses are cryptogenic in origin (Chan and Gold, 1998). Approximately 25% of infections spread through a dermal sinus tract (Chan and Gold, 1998). Other causes are hematogenous or iatrogenic infections. The clinical presentation is similar to that of spinal epidural or subdural abscesses (back and/or radicular pain, paresis, sensory loss, and sphincter disturbances). However, fever (<50%) is less frequently reported (Chan and Gold, 1998). Abscence of fever may be one reason why intramedullary abscesses are often initially misdiagnosed as spinal tumor or transverse myelitis. The diagnosis of intramedullary abscess is best made by MRI. Similar to the management of brain abscess, the therapy of intramedullary abscess usually includes immediate decompression/drainage (myelotomy or CT-guided needle aspiration) and antibiotic therapy (Chan and Gold, 1998). For the initial empirical antibiotic therapy, vancomycin, a third-generation cephalosporin, and metronidazole should be given in case of a dorsal midline skin lesion, which suggests infection by *S. aureus* or *epidermidis*, Enterobacticeae, or anaerobes through a dermal sinus tract (Chan and Gold, 1998). Following neurosurgical procedures, *Pseudomonas aeroginosa* must be considered, too; in these cases, a combination of vancomycin, ceftazidime, and metronidazole is recommended (Chan and Gold, 1998). A substantial proportion of cryptogenic intramedullary abscesses are due to *Listeria monocytogenes*; therefore, high-dose ampicillin should then be included in the antibiotic regimen (Chan and Gold, 1998). The primary focus of infection can be a guide for the initial empirical therapy of metastatic intramedullary abscesses.

## REFERENCES

Alderson, D., Strong, A. J., Ingham, H. R., and Selkon, J. B. (1981). Fifteen-year review of the mortality of brain abscess. *Neurosurgery* 8, 1–6.

American Academy of Neurology (1998). Evaluation and management of intracranial mass lesions in AIDS: Report of the Quality Standards Subcommittee of the American Academy of Neurology. *Neurology* 50, 21–26.

Arseni, C., and Ciurea, A. V. (1981). Brain abscesses: Comments with reference to 810 cases. *Zentralbl. Neurochir.* 42, 77–83.

Auletta, J. J., and John, C. C. (2001). Spinal epidural abscesses in children: A 15-year experience and review of the literature. *Clin. Infect. Dis.* 32, 9–16.

Baker, A. S., Ojemann, R. G., Swartz, M. N., and Richardson, E. P. J. (1975). Spinal epidural abscess. *N. Engl. J. Med.* 293, 463–468.

Bannister, G., Williams, B., and Smith, S. (1981). Treatment of subdural empyema. *J. Neurosurg.* 55, 82–88.

Barker, F. G. (1994). Efficacy of prophylactic antibiotics for craniotomy: A meta-analysis. *Neurosurgery* 35, 484–490.

Bartels, R. H., de Jong, T. R., and Grotenhuis, J. A. (1992). Spinal subdural abscess: Case report. *J. Neurosurg.* 76, 307–311.

Bartels, R. H., Gonera, E. G., van der Spek, J. A., Thijssen, H. O., Mullaart, R. A., and Gabreels, F. J. (1995). Intramedullary spinal cord abscess: A case report. *Spine* 20, 1199–1204.

Bayston, R., de Louvois, J., Brown, E. M., Johnston, R. A., Lees, P., and Pople, I. K. (2000). Use of antibiotics in penetrating craniocerebral injuries: "Infection in Neurosurgery," Working Party of British Society for Antimicrobial Chemotherapy. *Lancet* 355, 1813–1817.

Bok, A. P., and Peter, J. C. (1993). Subdural empyema: Burr holes or craniotomy? A retrospective computerized tomography-era analysis of treatment in 90 cases. *J. Neurosurg.* 78, 574–578.

Byrne, R. W., von Roenn, K. A., and Whisler, W. W. (1994). Intramedullary abscess: A report of two cases and a review of the literature. *Neurosurgery* 35, 321–326.

Campbell, B. G., and Zimmerman, R. D. (1998). Emergency magnetic resonance of the brain. *Top. Magn. Reson. Imaging* 9, 208–227.

Carpenter, J. L. (1994). Brain stem abscesses: Cure with medical therapy, case report, and review. *Clin. Infect. Dis.* 18, 219–226.

Chan, C. T., and Gold, W. L. (1998). Intramedullary abscess of the spinal cord in the antibiotic era: Clinical features, microbial etiologies, trends in pathogenesis, and outcomes. *Clin. Infect. Dis.* 27, 619–626.

Chun, C. H., Johnson, J. D., Hofstetter, M., and Raff, M. J. (1986). Brain abscess: A study of 45 consecutive cases. *Medicine (Baltimore)* 65, 415–431.

Ciurea, A. V., Stoica, F., Vasilescu, G., and Nuteanu, L. (1999). Neurosurgical management of brain abscesses in children. *Child. Nerv. Syst.* 15, 309–317.

Clayman, G. L., Adams, G. L., Paugh, D. R., and Koopmann, C. F. (1991). Intracranial complications of paranasal sinusitis: A combined institutional review. *Laryngoscope* 101, 234–239.

Cowie, R., and Williams, B. (1983). Late seizures and morbidity after subdural empyema. *J. Neurosurg.* 58, 569–573.

Desai, K. I., Muzumdar, D. P., and Goel, A. (1999). Holocord intramedullary abscess: An unusual case with review of literature. *Spinal Cord* 37, 866–870.

Dill, S. R., Cobbs, C. G., and McDonald, C. K. (1995). Subdural empyema: Analysis of 32 cases and review. *Clin. Infect. Dis.* 20, 372–386.

Enzmann, D. R., Britt, R. H., and Yeager, A. S. (1979). Experimental brain abscess evolution: Computed tomographic and neuropathologic Correlation. *Radiology* 133, 113–122.

Falcone, S., and Post, M. J. (2000). Encephalitis, cerebritis, and brain abscess: Pathophysiology and imaging findings. *Neuroimaging Clin. North Am.* 10, 333–353.

Faughnan, M. E., Lui, Y. W., Wirth, J. A., Pugash, R. A., Redelmeier, D. A., Hyland, R. H., and White, R. I. (2000). Diffuse pulmonary arteriovenous malformations: Characteristics and prognosis. *Chest* 117, 31–38.

Fischbein, C. A., Rosenthal, A., Fischer, E. G., Nadas, A. S., and Welch, K. (1974). Risk factors of brain abscess in patients with congenital heart disease. *Am. J. Cardiol.* 34, 97–102.

Gower, D. J., Horton, D., and Pollay, M. (1990). Shunt-related brain abscess and ascending shunt infection. *J. Child Neurol.* 5, 318–320.

Grieve, J. P., Ashwood, N., O'Neill, K. S., and Moore, A. J. (2000). A retrospective study of surgical and conservative treatment for spinal extradural abscess. *Eur. Spine. J.* 9, 67–71.

Grimstad, I. A., Hirschberg, H., and Rootwelt, K. (1992). 99mTc-hexamethylpropyleneamine oxime leukocyte scintigraphy and C-reactive protein levels in the differential diagnosis of brain abscesses. *J. Neurosurg.* 77, 732–736.

Hagensee, M. E., Bauwens, J. E., Kjos, B., and Bowden, R. A. (1994). Brain abscess following marrow transplantation: Experience at the Fred Hutchinson Cancer Research Center, 1984–1992. *Clin. Infect. Dis.* 19, 402–408.

Heusner, A. P. (1948). Nontuberculous spinal epidural infections. *N. Engl. J. Med.* 239, 845–854.

Hirschberg, H., and Bosnes, V. (1987). C-reactive protein levels in the differential diagnosis of brain abscesses. *J. Neurosurg.* 67, 358–360.

Hitchcock, E., and Andreadis, A. (1964). Subdural empyema: A review of 29 patients. *J. Neurol. Neurosurg. Psychiatry* 27, 422–434.

Hlavin, M. L., Kaminski, H. J., Fenstermaker, R. A., and White, R. J. (1994). Intracranial suppuration: A modern decade of postoperative subdural empyema and epidural abscess. *Neurosurgery* 34, 974–980.

Hlavin, M. L., Kaminski, H. J., Ross, J. S., and Ganz, E. (1990). Spinal epidural abscess: A ten-year perspective. *Neurosurgery* 27, 177–184.

Infection in Neurosurgery Working Party of the British Society for Antimicrobial Chemotherapy (2000). The rational use of antibiotics in the treatment of brain abscess. *Br. J. Neurosurg.* 14, 525–530.

Jamjoom, A. B. (1996). Short course antimicrobial therapy in intracranial abscess. *Acta Neurochir. (Wien)* 138, 835–839.

Kangsanarak, J., Fooanant, S., Ruckphaopunt, K., Navacharoen, N., and Teotrakul, S. (1993). Extracranial and intracranial complications of suppurative otitis media: Report of 102 cases. *J. Laryngol. Otol.* 107, 999–1004.

Kanter, M. C., and Hart, R. G. (1991). Neurologic complications of infective endocarditis. *Neurology* 41, 1015–1020.

Kastenbauer, S., Winkler, F., Fesl, G., Schiel, X., Ostermann, H., Yousry, T. A., and Pfister, H. W. (2001). Acute severe spinal cord dysfunction in bacterial meningitis in adults: MRI findings suggest extensive myelitis. *Arch. Neurol.* 58, 806–810.

Kaufman, D. M., Litman, N., and Miller, M. H. (1983). Sinusitis: Induced subdural empyema. *Neurology* 33, 123–132.

Kim, S. H., Chang, K. H., Song, I. C., Han, M. H., Kim, H. C., Kang, H. S., and Han, M. C. (1997). Brain abscess and brain tumor: Discrimination with in vivo H-1 MR spectroscopy. *Radiology* 204, 239–245.

Kim, Y. J., Chang, K. H., Song, I. C., Kim, H. D., Seong, S. O., Kim, Y. H., and Han, M. H. (1998). Brain abscess and necrotic or cystic brain tumor: Discrimination with signal intensity on diffusion-weighted MR imaging. *Am. J. Roentgenol.* 171, 1487–1490.

Korinek, A. M. (1997). Risk factors for neurosurgical site infections after craniotomy: A prospective multicenter study of 2944 patients. *Neurosurgery* 41, 1073–1079.

Kourtopoulos, H., Holm, S. E., and Norrby, S. R. (1983). The influence of steroids on the penetration of antibiotics into brain tissue and brain abscesses: An experimental study in rats. *J. Antimicrob. Chemother.* 11, 245–249.

Le Cam, B., Guivarch, G., Boles, J. M., Garre, M., and Cartier, F. (1984). Neurologic complications in a group of 86 bacterial endocarditis. *Eur. Heart. J.* 5(Suppl. C), 97–100.

Levy, M. L., Wieder, B. H., Schneider, J., Zee, C. S., and Weiss, M. H. (1993). Subdural empyema of the cervical spine: Clinicopathological correlates and magnetic resonance imaging: Report of three cases. *J. Neurosurg.* 79, 929–935.

Leys, D., Lesoin, F., Viaud, C., Pasquier, F., Rousseaux, M., Jomin, M., and Petit, H. (1985). Decreased morbidity from acute bacterial spinal epidural abscesses using computed tomography and non-surgical treatment in selected patients. *Ann. Neurol.* 17, 350–355.

Lutz, T. W., Landolt, H., Wasner, M., and Gratzl, O. (1994). Diagnosis and management of abscesses in the basal ganglia and thalamus: A survey. *Acta Neurochir. (Wien)* 127, 91–98.

Mamelak, A. N., Mampalam, T. J., Obana, W. G., and Rosenblum, M. L. (1995). Improved management of multiple brain abscesses: A combined surgical and medical approach. *Neurosurgery* 36, 76–85.

Mampalam, T. J., and Rosenblum, M. L. (1988). Trends in the management of bacterial brain abscesses: A review of 102 cases over 17 years. *Neurosurgery* 23, 451–458.

Maschke, M., Dietrich, U., Prumbaum, M., Kastrup, O., Turowski, B., Schaefer, U. W., and Diener, H. C. (1999). Opportunistic CNS infection after bone marrow transplantation. *Bone Marrow Transplant.* 23, 1167–1176.

Mathisen, G. E., and Johnson, J. P. (1997). Brain abscess. *Clin. Infect. Dis.* 25, 763–779.

Mauser, H. W., Van Houwelingen, H. C., and Tulleken, C. A. (1987). Factors affecting the outcome in subdural empyema. *J. Neurol. Neurosurg. Psychiatry* 50, 1136–1141.

McClelland, C. J., Craig, B. F., and Crockard, H. A. (1978). Brain abscesses in Northern Ireland: A 30 year community review. *J. Neurol. Neurosurg. Psychiatry* 41, 1043–1047.

Miller, E. S., Dias, P. S., and Uttley, D. (1987). Management of subdural empyema: A series of 24 cases. *J. Neurol. Neurosurg. Psychiatry* 50, 1415–1418.

Miller, E. S., Dias, P. S., and Uttley, D. (1988). CT scanning in the management of intracranial abscess: A review of 100 cases. *Br. J. Neurosurg.* 2, 439–446.

Nadvi, S. S., Parboosing, R., and van, D. J. (1997). Cerebellar abscess: The significance of cerebrospinal fluid diversion. *Neurosurgery* **41**, 61–66.

Nathoo, N., Nadvi, S. S., and Van, D. J. (1997). Infratentorial empyema: Analysis of 22 cases. *Neurosurgery* **41**, 1263–1268.

Nathoo, N., Nadvi, S. S., and Van, D. J. (1999a). Cranial extradural empyema in the era of computed tomography: A review of 82 cases. *Neurosurgery* **44**, 748–753.

Nathoo, N., Nadvi, S. S., Van, D. J., and Gouws, E. (1999b). Intracranial subdural empyemas in the era of computed tomography: A review of 699 cases. *Neurosurgery* **44**, 529–535.

Neuwelt, E. A., Lawrence, M. S., and Blank, N. K. (1984). Effect of gentamicin and dexamethasone on the natural history of the rat *Escherichia coli* brain abscess model with histopathological correlation. *Neurosurgery* **15**, 475–483.

Nicolosi, A., Hauser, W. A., Musicco, M., and Kurland, L. T. (1991). Incidence and prognosis of brain abscess in a defined population: Olmsted County, Minnesota, 1935–1981. *Neuroepidemiology* **10**, 122–131.

Nielsen, H., Gyldensted, C., and Harmsen, A. (1982). Cerebral abscess: Aetiology and pathogenesis, symptoms, diagnosis and treatment. A review of 200 cases from 1935–1976. *Acta Neurol. Scand.* **65**, 609–622.

Nielsen, H., Harmsen, A., and Gyldensted, C. (1983). Cerebral abscess: A long-term follow-up. *Acta Neurol. Scand.* **67**, 330–337.

O'Donoghue, M. A., Green, H. T., and Shaw, M. D. (1992). Cerebral abscess on Merseyside 1980–1988. *J. Infect.* **25**, 163–172.

Ozates, M., Ozkan, U., Kemaloglu, S., Hosoglu, S., and Sari, I. (2000). Spinal subdural tuberculous abscess. *Spinal Cord* **38**, 56–58.

Pathak, A., Sharma, B. S., Mathuriya, S. N., Khosla, V. K., Khandelwal, N., and Kak, V. K. (1990). Controversies in the management of subdural empyema: A study of 41 cases with review of literature. *Acta Neurochir. (Wien)* **102**, 25–32.

Pierce, M. A., Johnson, M. D., Maciunas, R. J., Murray, M. J., Allen, G. S., Harbison, M. A., Creasy, J. L., and Kessler, R. M. (1995). Evaluating contrast-enhancing brain lesions in patients with AIDS by using positron emission tomography. *Ann. Intern. Med.* **123**, 594–598.

Pruitt, A. A., Rubin, R. H., Karchmer, A. W., and Duncan, G. W. (1978). Neurologic complications of bacterial endocarditis. *Medicine (Baltimore)* **57**, 329–343.

Quartey, G. R., Johnston, J. A., and Rozdilsky, B. (1976). Decadron in the treatment of cerebral abscess: An experimental study. *J. Neurosurg.* **45**, 301–310.

Rankin, R. M., and Flothow, P. G. (1946). Pyogenic infection of the spinal epidural space. *West. J. Surg. Obstet. Gynecol.* **54**, 320–323.

Reihsaus, E., Waldbaur, H., and Seeling, W. (2000). Spinal epidural abscess: A meta-analysis of 915 patients. *Neurosurg. Rev.* **23**, 175–204.

Rish, B. L., Caveness, W. F., Dillon, J. D., Kistler, J. P., Mohr, J. P., and Weiss, G. H. (1981). Analysis of brain abscess after penetrating craniocerebral injuries in Vietnam. *Neurosurgery* **9**, 535–541.

Rosenblum, M. L., Mampalam, T. J., and Pons, V. G. (1986). Controversies in the management of brain abscesses. *Clin. Neurosurg.* **33**, 603–632.

Salgado, A. V., Furlan, A. J., Keys, T. F., Nichols, T. R., and Beck, G. J. (1989). Neurologic complications of endocarditis: A 12-year experience. *Neurology* **39**, 173–178.

Sanchez, C., Mauri, E., Dalmau, D., Quintana, S., Aparicio, A., and Garau, J. (1995). Treatment of cerebral Aspergillosis with itraconazole: Do high doses improve the prognosis? *Clin. Infect. Dis.* **21**, 1485–1487.

Schmidt, K. G., Rasmussen, J. W., Frederiksen, P. B., Kock-Jensen, C., and Pedersen, N. T. (1990). Indium-111-granulocyte scintigraphy in brain abscess diagnosis: Limitations and pitfalls. *J. Nucl. Med.* **31**, 1121–1127.

Schneider, P., and Givens, T. G. (1998). Spinal subdural abscess in a pediatric patient: A case report and review of the literature. *Pediatr. Emerg. Care* **14**, 22–23.

Selby, R., Ramirez, C. B., Singh, R., Kleopoulos, I., Kusne, S., Starzl, T. E., and Fung, J. (1997). Brain abscess in solid organ transplant recipients receiving cyclosporine-based immunosuppression. *Arch. Surg.* **132**, 304–310.

Seydoux, C., and Francioli, P. (1992). Bacterial brain abscesses: Factors influencing mortality and sequelae. *Clin. Infect. Dis.* **15**, 394–401.

Shapiro, S. (1997). Cranial epidural abscess and cranial subdural empyema. *In* "Central Nervous System Infectious Diseases and Therapy" (K. L. Roos, Ed.), pp. 499–506. Dekker, New York.

Shukla-Dave, A., Gupta, R. K., Roy, R., Husain, N., Paul, L., Venkatesh, S. K., Rashid, M. R., Chhabra, D. K., and Husain, M. (2001). Prospective evaluation of in vivo proton MR spectroscopy in differentiation of similar appearing intracranial cystic lesions. *Magn. Reson. Imaging* **19**, 103–110.

Svanteson, B., Nordstrom, C. H., and Rausing, A. (1988). Nontraumatic brain abscess: Epidemiology, clinical symptoms and therapeutic results. *Acta Neurochir. (Wien)* **94**, 57–65.

Swanson, K. L., Prakash, U. B., and Stanson, A. W. (1999). Pulmonary arteriovenous fistulas: Mayo Clinic experience, 1982–1997. *Mayo Clin. Proc.* **74**, 671–680.

Takeshita, M., Kawamata, T., Izawa, M., and Hori, T. (2001). Prodromal signs and clinical factors influencing outcome in patients with intraventricular rupture of purulent brain abscess. *Neurosurgery* **48**, 310–316.

Townsend, G. C., and Scheld, W. M. (1998). Infections of the central nervous system. *Adv. Intern. Med.* **43**, 403–447.

Tung, G. A., Yim, J. W., Mermel, L. A., Philip, L., and Rogg, J. M. (1999). Spinal epidural abscess: Correlation between MRI findings and outcome. *Neuroradiology* **41**, 904–909.

U.S. Public Health Service (USPHS) and Infectious Diseases Society of America (IDSA) (1999). 1999 USPHS/IDSA guidelines for the prevention of opportunistic infections in persons infected with human immunodeficiency virus. *Morb. Mortal. Wkly. Rep.* **48**, 1–82.

Wallenfang, T., Bohl, J., and Kretzschmar, K. (1980). Evolution of brain abscess in cats formation of capsule and resolution of brain edema. *Neurosurg. Rev.* **3**, 101–111.

Wheeler, D., Keiser, P., Rigamonti, D., and Keay, S. (1992). Medical management of spinal epidural abscesses: Case report and review. *Clin. Infect. Dis.* **15**, 22–27.

Wong, T. T., Lee, L. S., Wang, H. S., Shen, E. Y., Jaw, W. C., Chiang, C. H., Chi, C. S., Hung, K. L., Liou, W. Y., and Shen, Y. Z. (1989). Brain abscesses in children—A cooperative study of 83 cases. *Child. Nerv. Syst.* **5**, 19–24.

Yang, S. Y. (1981). Brain abscess: A review of 400 cases. *J. Neurosurg.* **55**, 794–799.

Yang, S. Y., and Zhao, C. S. (1993). Review of 140 patients with brain abscess. *Surg. Neurol.* **39**, 290–296.

Yen, P. T., Chan, S. T., and Huang, T. S. (1995). Brain abscess: With special reference to otolaryngologic sources of infection. *Otolaryngol. Head Neck Surg.* **113**, 15–22.

Zeidman, S. M., Geisler, F. H., and Olivi, A. (1995). Intraventricular rupture of a purulent brain abscess: Case report. *Neurosurgery* **36**, 189–193.

## CHAPTER 44
# Tuberculous Meningitis

Karen L. Roos and Hans-Walter Pfister

## TUBERCULOUS MENINGITIS

Tuberculous meningitis presents as either a subacute or a chronic meningitis, characterized by fever, headache, night sweats, and malaise, or a fulminant meningoencephalitis with coma, raised intracranial pressure, seizure activity, and stroke. In children, tuberculous meningitis occurs during primary infection with *Mycobacterium tuberculosis*. In adults, tuberculous meningitis does not occur during the course of primary infection, but rather from endogenous reactivation of infection in caseous tuberculous foci adjacent to the subarachnoid space that developed during hematogenous spread of tubercle bacilli in the course of an earlier primary infection, usually pulmonary tuberculosis. Meningitis is the result of the discharge of bacilli and tuberculous antigens into the subarachnoid space. It is not known what causes the growth and rupture of caseous lesions, but immunological mechanisms most likely have a role. This is supported by the evidence that the known risk factors for tuberculous meningitis, including age, malnutrition, alcoholism, diabetes mellitus, chronic corticosteroid therapy, and HIV infection, all affect cell-mediated immunity.

In developed countries, the most common presentation of tuberculous meningitis is low-grade fever, unrelenting headache, anorexia, night sweats, malaise, and stiff neck. The differential diagnosis of tuberculous meningitis includes those diseases that cause this combination of symptoms with a cerebrospinal fluid (CSF) lymphocytic pleocytosis and a mildly decreased glucose concentration. The differential diagnosis mainly includes fungal meningitis, viral meningitis, neurosarcoidosis, neurosyphilis, and leptomeningeal carcinomatosis (Table I). The characteristic CSF abnormalities in tuberculous meningitis are an elevated opening pressure, a lymphocytic pleocytosis (10–500 cells/mm$^3$), a decreased glucose concentration (20–40 mg/dL), and an elevated protein concentration (Table II). Early in infection, the CSF may have a predominance of polymorphonuclear leukocytes, but lymphocytes become the predominant cell type within a few days. Spinal fluid should be examined by acid-fast stains and cultures and molecular diagnostic techniques for *M. tuberculosis* DNA. The last tube of fluid collected from lumbar puncture is the best tube to send for an acid-fast bacilli (AFB) smear. A pellicle may form in the CSF in tuberculous meningitis or a cobweb-like clot on the surface of the fluid. Tubercle bacilli are best demonstrated in a smear of the clot or sediment. Positive smears are reported in 10–40% of cases and are dependent on the amount of time and the expertise of the person examining the stain for acid-fast bacilli and the amount of CSF that is sampled. The growth of the organism is very slow, so cultures take weeks to be positive. Cultures are reported to be positive in 25–75% of cases of tuberculous meningitis, requiring 3 to 6 weeks for growth to be detectable (Leonard and Des Prez, 1990; Traub *et al.*, 1984). The polymerase chain reaction technique has yet to be perfected for the detection of *M. tuberculosis* DNA in CSF, but it is available in most laboratories. False-negative results are due to too few mycobacteria in CSF in the early stages of infection. False-positive results are due to cross-contamination with amplified DNA products in the laboratory (Lin *et al.*, 1995). The diagnosis of tuberculous meningitis is supported by the detection of substances in CSF that are unique to the organism, i.e., tuberculostearic acid. The CSF chloride concentration and the bromide partition test have been disappointing in differentiating tuberculous meningitis from other types of lymphocytic meningitis. The CSF chloride concentration may be decreased in tuberculous meningitis. However, this is a nonspecific abnormality because the CSF chloride concentration typically decreases as the CSF protein concentration increases. The bromide partition test measures the ratio of serum/CSF bromide

**TABLE I** Differential Diagnosis of Headache, Low-Grade Fever, and CSF Lymphocytic Pleocytosis

Bacteria
  *Mycobacterium tuberculosis*
  *Listeria monocytogenes*
  *Mycoplasma pneumoniae*
Spirochetes
  *Treponema pallidum*
  *Borrelia burgdorferi*
Viruses
  Enteroviruses
  Herpes-simplex virus-2
  Arthropod-borne viruses
  Epstein–Barr virus
  HIV
Fungi
  *Cryptococcus neoformans*
  *Histoplasma capsulatum*
  *Coccidioides immitis*
Other
  Sarcoidosis
  Leptomeningeal carcinomatosis (rarely: primary diffuse
    leptomeningeal gliomatosis)
  Behcet's disease
  Cerebral vasculitis (e.g., Wegener's granulomatosis)
  Medications (drug-induced meningitis)

**TABLE II** Cerebrospinal Fluid in Tuberculous Meningitis

1. Elevated opening pressure
2. Lymphocytic pleocytosis (usually 10–500 cells/mm$^3$)
3. Decreased glucose concentration (usually 20–40 mg/dl)
4. Increased protein concentration
5. Local IgA synthesis
6. Acid-fast stain and tuberculous culture
7. PCR for mycobacterial DNA
8. Tuberculostearic acid
9. Decreased CSF chloride concentration
10. Increased adenosine deaminase level
11. Low bromide partition ratio

24 h after the oral or intravenous administration of $^{84}$Br-labeled ammonium bromide (Wiggelinkhuizen and Mann, 1980). In patients with tuberculous meningitis, the ratio of serum/CSF bromide will be low. A low bromide partition ratio is, however, a nonspecific abnormality and will occur in any process that disrupts the blood–brain barrier permeability.

The diagnosis of tuberculous meningitis, in the absence of demonstrating the organism in smear or culture, is supported by finding a site of extrameningeal tuberculous infection. A chest radiograph should be obtained as part of the diagnostic evaluation of tuberculous meningitis. The lungs are the most common site of initial infection. The classic primary complex (Ghon complex) refers to Anton Ghon's observation from autopsy specimens that the primary lesion of tuberculosis is in the lung with secondary infection in the tracheobronchial lymph nodes (Ober, 1983). In addition to the "primary complex," chest radiographic abnormalities suggestive of pulmonary tuberculosis are hilar adenopathy, a miliary pattern, upper lobe infiltrate, and lobar consolidation (Kent *et al.*, 1993). The Mantoux intradermal tuberculin skin test is helpful when positive. The test is interpreted 48 h after placement and is considered positive if the amount of induration is greater than 5 mm. The test may, however, be falsely negative even in the absence of immunosuppression and in association with a positive reaction to the common antigens used to determine anergy. More often, however, patients have global anergy to all of the common control skin test antigens.

Hydrocephalus is the most common abnormality on neuroimaging and often progresses during the course of the disease. As with meningitis of any etiology, there may be meningeal enhancement, especially of the basilar leptomeninges, post contrast administration. The internal carotids, middle cerebral arteries, and medial striate and thalamoperforating arteries are the most common vessels to be narrowed or occluded. Patients with tuberculous meningitis may also develop tuberculomas during therapy (Figure 1).

The clinical presentation, outcome, and CSF profile of tuberculous meningitis is not modified by HIV infection (Verdon *et al.*, 1996). HIV-infected persons with tuberculosis are, however, at increased risk for meningitis (Berenguer *et al.*, 1992). Approximately 50% of HIV-infected persons with tuberculous meningitis will have pulmonary infiltrates. The presence of peripheral, intrathoracic, and intraabdominal adenopathy is particularly common in tuberculosis associated with HIV infection, and acid-fast smears from these sites can help to confirm the diagnosis (Berenguer *et al.*, 1992). Intracerebral mass lesions, specifically tuberculous abscesses, are more frequent in HIV-infected patients, but do not appear to affect mortality (Berenguer *et al.*, 1992; Dube *et al.*, 1992; Whiteman *et al.*, 1995).

The combination of symptoms of headache and low-grade fever with a shift in CSF white blood cells from a polymorphonuclear to a lymphocytic predominance, a mildly decreased glucose concentration and an increasing protein concentration, and/or the development of hydrocephalus is highly suggestive of tuberculous meningitis. Although a subacute meningitis syndrome is the most common presentation of tuberculous meningitis in developed countries, cranial nerve palsies or an alteration in the level of consciousness may also be part of the initial presentation. The British Medical Research Council divides the clinical course of tuberculous meningitis into three stages (Medical Research Council, 1948). In the first stage, the level of consciousness is normal, but patients are irritable and

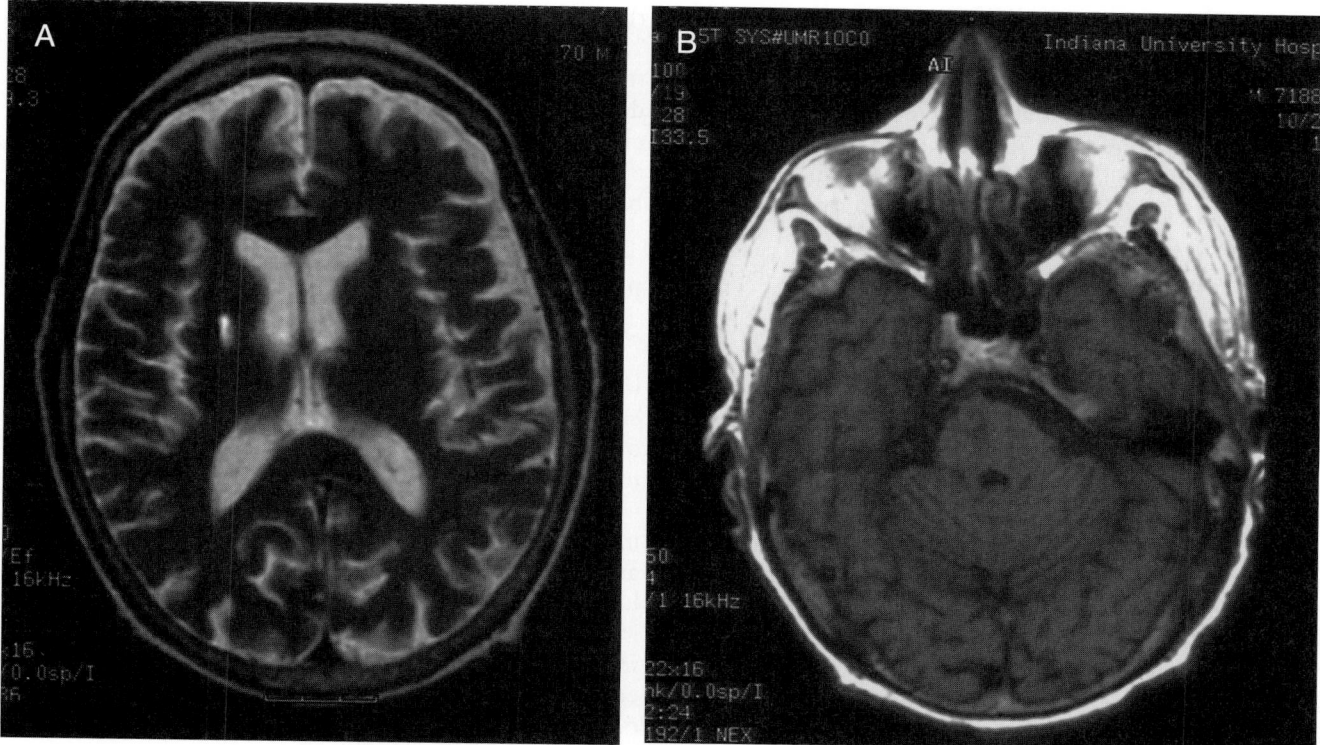

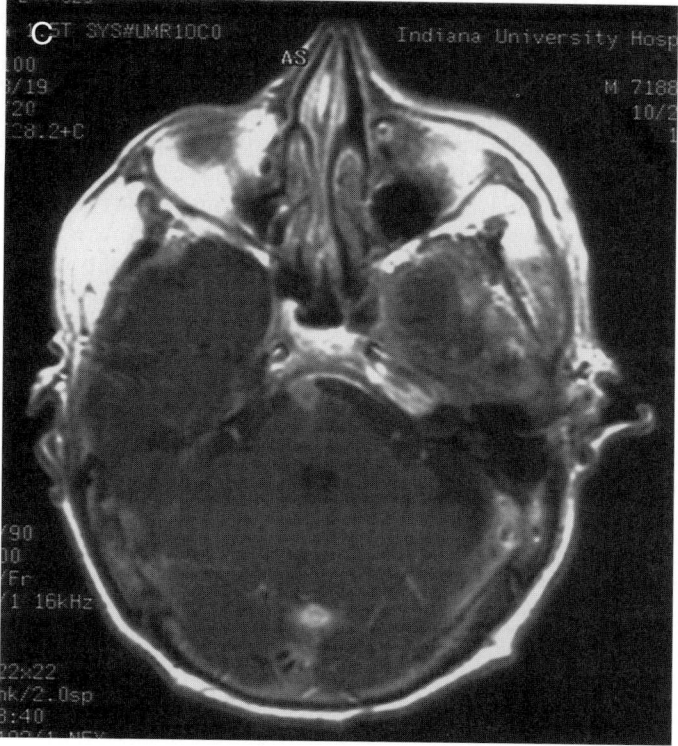

**FIGURE 1**    T2-weighted cranial MR scan (A) demonstrating right posterior limb internal capsule infarction in a patient with tuberculous meningitis. T1-weighted cranial MR scan pre (B) and post (C) gadolinium demonstrating a tuberculoma in the right prepontine cistern that arose during therapy.

febrile. There are no focal neurological signs. Stage 2 is characterized by confusion, but not coma; there may be headache, vomiting, focal neurological deficits such as hemiparesis, and a single cranial nerve palsy. In stage 3, the patient is febrile and stuporous or comatose, may have focal neurological deficits from ischemic infarction, focal or generalized seizures, and multiple cranial nerve palsies.

## NATURAL COURSE

A person becomes infected with M. tuberculosis when they inhale aerosolized droplet nuclei containing tubercle bacilli. Not all infected persons have the same risk of developing pulmonary or CNS disease. An immunocompetent adult with untreated M. tuberculosis infection has a 5–10% lifetime risk of disease. The risk is greatest in the first 2 to 3 years after infection. Infants and children have a 40% risk of developing disease within 1 to 2 years of infection (Starke, 1999).

Approximately 10% of immunocompetent persons with tuberculosis develop CNS disease (Udani et al., 1971). The neurological complications of tuberculous meningitis are the result of a hypersensitivity reaction to the discharge of tubercle bacilli and tuberculous antigens into the subarachnoid space (Table III). This leads to the production of a purulent exudate that obstructs the flow of CSF through the basilar cisterns and the resorption of CSF by the arachnoid granulations, surrounds the cranial nerves, and surrounds and infiltrates the arterial walls of the blood vessels at the base of the brain.

The result is communicating and obstructive hydrocephalus, ischemic and hemorrhagic infarctions, and cranial nerve palsies. Untreated tuberculous meningitis is universally fatal. Factors associated with a poor outcome include advanced stage of disease at the time of hospital admission, delay in starting treatment, extreme age, associated chronic medical illness, CSF protein concentration of 3 g/L, and cerebral infarction

TABLE III   Complications during the Course of Tuberculous Meningitis

- Hydrocephalus
- Cerebral vasculitis
- Tuberculoma
- Tuberculous abscess
- Hyponatraemia, syndrome of inappropriate ADH secretion
- Spinal involvement
  — Myeloradiculitis
  — Arachnoiditis
  — Vasculitis with infarctions
  — Tuberculoma
  — Postmeningitic syringomyelia
- Tuberculous encephalopathy (only in children)

(Haas et al., 1977; Kent et al., 1993; Ogawa et al., 1987; Verdon et al., 1996). There is a nonfatal form of tuberculous meningitis termed serous meningitis that develops when a tuberculous focus adjacent to the subarachnoid space causes a lymphocytic reaction without actually seeding the subarachnoid space with tubercle bacilli (Starke, 1999).

## PRACTICAL MANAGEMENT

The most important rule of management of tuberculous meningitis is to initiate therapy based on a strong clinical suspicion while awaiting the results of cultures of CSF as outcome is affected by the clinical stage at the time therapy is initiated.

Outbreaks of multidrug-resistant tuberculosis are increasingly reported to the Centers for Disease Control and Prevention. The development of resistant strains has been attributed to a number of factors, the most important of which is inadequate treatment due to an irregular drug supply, inappropriate regimens, or poor compliance (Pablos-Mendez et al., 1998). The risk of developing a resistant strain is decreased by the use of more than one drug. There have been no large controlled clinical trials to evaluate the efficacy of one multidrug regimen against another for tuberculous meningitis. Present recommendations are that treatment of tuberculous meningitis in adults be initiated with a combination of isoniazid (300 mg/day), rifampin (10 mg/kg/day, up to 600 mg/day), and pyrazinamide (25–35 mg/kg/day) for 2 months, followed by isoniazid and rifampin for an additional 7 to 10 months (Table IV). Pyridoxine in a dose of 50 mg daily is added to this regimen to avoid peripheral neuropathy due to isoniazid-induced pyridoxine deficiency. Monthly liver function studies should be obtained in adults receiving isoniazid and rifampin. The risk of hepatotoxicity from isoniazid is approximately 1%. This risk doubles with the addition of rifampin (Leonard and Des Prez, 1990). In patients with a high probability of a drug-resistant strain of M. tuberculosis, therapy is initiated with a combination of isoniazid, rifampin, pyrazinamide, and either streptomycin (1 g daily by intramuscular injection) or ethambutol (15–25 mg/kg/day) for 2 months. This is followed by isoniazid and rifampin to complete a 9- to 18-month course of therapy. If the organism is resistant to one of these drugs, the results of susceptibility studies dictate which drugs are used. Treatment of drug-resistant tuberculosis must include at least two bactericidal drugs to which the organism is susceptible. The chief toxicity of ethambutol is optic neuritis which occurs in as many as 3% of patients receiving 25 mg/kg/day, but is very rare at a dose of 15 mg/kg/day (Leonard and Des Prez, 1990). If ethambutol is required for long-term therapy because the organism is resistant to isoniazid, ethambutol is given in a dose of 25 mg/kg/day for 1 to 2 months and 15 mg/kg/day thereafter. Streptomycin is administered in a dose of 1 g/day for the first 6 to 8 weeks, and then 1 g/day in a biweekly regimen thereafter (Molavi and LeFrock, 1985). Recommended treatment regimens are the same for HIV-positive and HIV-negative individuals, except that rifabutin is sometimes substituted for rifampin in HIV-infected individuals. In treating HIV-infected individuals with tuberculous meningitis, it should be kept in mind that the rifamycins interact with protease inhibitors, resulting in decreased activity of protease inhibitors (Garcia-Monco, 1999). Table V lists the adverse effects of antituberculous agents and the recommended course of action.

For children with tuberculous meningitis, the American Academy of Pediatrics recommends initiating therapy for 2 months with a combination of isoniazid (10 mg/kg/day), rifampin (10–20 mg/kg/day), pyrazinamide (20–40 mg/kg/day), and streptomycin (20–40 mg/kg/day by intramuscular injection; Committee on Infectious Diseases, 1992). Streptomycin is used in initial therapy until the results of drug susceptibility test results are known. For patients who may have acquired tuberculosis in geographic areas where resistance to streptomycin is common, capreomycin (15–30 mg/kg/day) or kanamycin (15–30 mg/kg/day) may be used instead of streptomycin. After 2 months of four-drug therapy, isoniazid and rifampin are administered daily or twice weekly (isoniazid, 20–40 mg/kg/dose; rifampin, 10–20 mg/kg/dose) for 10 months (Committee on Infectious Diseases, 1992). The administration of pyridoxine is not generally recommended for children taking isoniazid (Committee on Infectious Diseases, 1992).

It is generally not recommended that CSF white blood cell count or protein and glucose concentrations be

followed to assess response to therapy. The CSF was reexamined in 58 patients with tuberculous meningitis after the initiation of therapy at monthly intervals until the CSF profile returned to normal. The CSF culture for *M. tuberculosis* was negative after 1 month of therapy in all patients. The CSF glucose concentration returned to normal in 23 of 40 cases after 1 month of treatment. In 12 patients, it took 2 months for the CSF glucose concentration to return to normal. In one patient, it took 4 months, but this patient had an uneventful clinical recovery within the first 2 weeks of therapy. The CSF protein concentration returned to a normal value within 6 months in only 2 of 45 cases. In 6 cases, it took longer than 12 months for the protein concentration to return to normal. The CSF cell count decreased by 50% from the initial value within the first month of therapy in 43 of 45 cases, but was still elevated at 6 months in 16 cases. In seven cases the cell count was still abnormal after 24 months of therapy (Kent *et al.*, 1993).

## Adjunctive Therapy

The consensus of opinion is that corticosteroid therapy improves neurological outcome and decreases mortality in patients with stage 2 tuberculous meningitis. Early investigations into the effect of corticosteroid therapy on *M. tuberculosis* infection in animal models demonstrated that corticosteroids enhanced the virulence of *M. tuberculosis* when antituberculous therapy was not administered concomitantly. Anecdotal experience with humans was similar. Corticosteroid therapy without antituberculous therapy was perilous. Coadministration of corticosteroid therapy with antituberculous therapy improved outcome (Des Prez and Organick, 1958; Dooley *et al.*, 1997; Shubin *et al.*, 1959). There have been a number of prospective, randomized double blind and nonblinded clinical trials investigating the efficacy of corticosteroids on tuberculous meningitis. The difficulty in interpreting the results of these studies is that the diagnosis of tuberculous meningitis was not always proven by positive CSF culture or acid-fast smear. For example, Schoeman *et al.* (1997) report the results of corticosteroid therapy on outcome in 141 children with "tuberculous meningitis"; however, *M. tuberculosis* was isolated from the CSF in only 23 patients. In 33 patients, *M. tuberculosis* was isolated from gastric aspirate. In the remaining children, the diagnosis of tuberculous meningitis was based on the history and CSF abnormalities in addition to two of the following: positive Mantoux test, chest radiographic abnormalities suggestive of tuberculosis, and acute hydrocephalus with basilar enhancement on CT scanning. Kumarvelu *et al.* (1994) reported the results of a randomized controlled trial of dexamethasone in tuberculous meningitis in 47 patients. The diagnosis of

**TABLE IV** Antimicrobial Therapy of Tuberculous Meningitis

| Antituberculous agent | Adult dose | Pediatric dose |
|---|---|---|
| Isoniazid | 300 mg/day single oral dose | 10 mg/kg/day |
| Rifampin | 10 mg/kg/day (<50 kg, 450 mg/day) (>50 kg, 600 mg/day) Maximum daily dose—600 mg | 10–20 mg/kg/day Maximum daily dose—500 mg |
| Pyrazinamide | 25–35 mg/kg/day (<50 kg, 1.5 g/day; 51–75 kg, 2.0 g/day; >75 kg, 2.5 g/day) | 20–40 mg/kg/day |
| Pyridoxine | 50 mg/day | — |
| Streptomycin | 1 g/day im | 20–40 mg/kg/day im |
| Ethambutol | 15–25 mg/kg/day | 15–25 mg/kg/day |

"probable tuberculous meningitis" was made if at least three of the following criteria were present: (1) fever, headache, and neck stiffness with or without seizures or altered sensorium for 2 weeks or more; (2) CSF lymphocytic pleocytosis (>20 cells/mm$^3$), elevated protein concentration, decreased glucose concentration, negative bacterial and fungal cultures, and negative cytology; (3) contrast enhanced CT abnormalities of tuberculous meningitis; and (4) clinical, radiological, or histological evidence of extracranial tuberculosis. In both of these treatment trials, the patients probably had tuberculous meningitis, but they did not definitively have tuberculous meningitis because they did not have a positive CSF acid-fast smear, *M. tuberculosis* culture, or evidence of tuberculostearic acid in CSF. However, the inclusion of the results of these studies in determining the efficacy of corticosteroid therapy seems reasonable due to the strict criteria described above. Both of these clinical trials demonstrated efficacy for corticosteroid therapy in tuberculous meningitis.

There are also differences in the severity of disease and in corticosteroid regimens in clinical trials evaluating the efficacy of steroid therapy in tuberculous meningitis. Dexamethasone is used in some clinical trials, prednisone in others. The steroid regimen and severity of disease between treatment groups also vary between treatment trials.

Girgis *et al.* (1991) performed a prospective, randomized but nonblinded study on adjunctive corticosteroid therapy in tuberculous meningitis. The diagnosis was established by culturing *M. tuberculosis* from CSF. There were 160 culture positive patients. They included severity of disease in their analysis. Seventy-five patients received dexamethasone and antituberculous chemotherapy. Dexamethasone was given intramuscularly (12 mg/day to adults and 8 mg/day to children weighing less than 25 kg) for 3 weeks and then tapered over the next 3 weeks. Eighty-five patients received antituberculous chemotherapy alone. The patients in both groups were similar in terms of age, duration of symptoms prior to hospitalization, and level of consciousness and cranial nerve deficits on admission. A significant reduction in mortality was observed in those patients who were drowsy on admission and were treated with corticosteroids compared to those patients who were drowsy on admission and treated with antituberculous therapy only (four of 27 [15%] compared to 14 of 35 [40%], $P < 0.04$). There was no significant difference in mortality between the two treatment groups for those admitted in a coma. During therapy, significantly fewer complications developed in the corticosteroid-treated group, and they had fewer permanent sequelae at 2-year follow-up. The CSF abnormalities (WBC count and glucose and protein concentrations) normalized faster in the corticosteroid-treated group.

**TABLE V**   Major Adverse Effects of Antituberculous Agents

| Antituberculous agent | Adverse effect | Recommended action |
|---|---|---|
| Isoniazid | Hepatotoxicity | Monthly LFTs |
| | Peripheral neuropathy | Pyridoxine |
| Rifampin | Hepatotoxicity | Monthly LFTs |
| | Teratogenic | Avoid during pregnancy |
| | Turns urine orange-brownish color | Warn patient this will happen |
| Pyrazinamide | Hepatotoxicity | Monthly LFTs |
| | Hyperuricemia | Monitor uric acid |
| Streptomycin | Nephrotoxicity | Monthly creatinine, will need to decrease dose in renal insufficiency |
| | Vestibulotoxicity | Reassure patient this will improve post therapy |
| Ethambutol | Optic neuritis | Monthly eye exams |

Meningeal inflammation enhances the penetration of some antimicrobial agents. The argument is often made that corticosteroids will decrease the penetration of antituberculous agents into the cerebrospinal fluid by decreasing meningeal inflammation. Isoniazid and pyrazinamide penetrate the CSF barrier regardless of the degree of meningeal inflammation. Rifampin penetrates the CSF barrier more readily when inflammation is present and streptomycin penetrates poorly in general. Kaojarern *et al.* (1991) measured the concentrations of oral isoniazid, pyrazinamide, rifampin, and intramuscular streptomycin in CSF at various intervals in patients with tuberculous meningitis receiving only antituberculous therapy and in patients with tuberculous meningitis receiving both antituberculous agents and corticosteroids. The mean concentrations of all four drugs in CSF in the two groups of patients at each interval and the total mean were not significantly different, except that the total CSF mean concentration of streptomycin in those receiving steroids was higher than that in the control group ($P < 0.01$). They concluded that there was no effect of corticosteroid therapy on the penetration of these four drugs into the CSF.

Dooley *et al.* (1997) reviewed the literature on adjunctive corticosteroid therapy for tuberculous meningitis from 1966 to 1996 using Medline and concluded that adjunctive corticosteroid therapy had a significant benefit in improving survival and reducing sequelae for patients with stage 2 disease (drowsiness, single cranial nerve palsy, or hemiparesis), less benefit in early disease or late disease (coma), and more benefit with longer regimens (4 weeks to months). Dexamethasone, 8–12 mg/day, or a prednisone equivalent tapered over 6 to 8 weeks is recommended. The Infectious Disease Society of America endorses the use of corticosteroids in the therapy of tuberculous meningitis (McGowan *et al.*, 1992).

## Surgical Management

Hydrocephalus should be treated with a ventriculostomy. Persistent hydrocephalus may require placement of a ventriculoperitoneal shunt.

## REFERENCES

Berenguer, J., Moreno, S., Laguna, F., Vicente, T., et al. (1992). Tuberculous meningitis in patients infected with the human immunodeficiency virus. N. Engl. J. Med. 326, 668–672.

Committee on Infectious Diseases. American Academy of Pediatrics (1992). Chemotherapy for tuberculosis in infants and children. Pediatrics 89, 161–165.

Des Prez, R. M., and Organick, A. (1958). Corticotropin and corticosteroid therapy in tuberculosis. Arch. Intern. Med. 101, 1129–1142.

Dooley, D. P., Carpenter, J. L., and Rademacher, S. (1997). Adjunctive corticosteroid therapy for tuberculosis: A critical reappraisal of the literature. Clin. Infect. Dis. 25, 872–887.

Dube, M. P., Holtom, P. D., and Larsen, R. A. (1992). Tuberculous meningitis in patients with and without human immunodeficiency virus infection. Am. J. Med. 93, 520–524.

Garcia-Monco, J. C. (1999). Central nervous system tuberculosis. Neurol. Clin. 17, 737–759.

Girgis, N. I., Farid, Z., Kilpatrick, M. E., Sultan, Y., and Mikhail, I. A. (1991). Dexamethasone adjunctive treatment for tuberculous meningitis. Pediatr. Infect. Dis. J. 10, 179–183.

Haas, E. J., Madhavan, T., Quinn, E. L., Cox, F., Fisher, E., and Burch, K. (1977). Tuberculous meningitis in an urban general hospital. Arch. Intern. Med. 137, 1518–1521.

Kaojarern, S., Supmonchai, K., Phuapradit, P., Mokkhavesa, C., and Krittiyanunt, S. (1991). Effect of steroids on cerebrospinal fluid penetration of antituberculous drugs in tuberculous meningitis. Clin. Pharmacol. Ther. 49, 6–12.

Kent, S. J., Crowe, S. M., Yung, A., Lucas, C. R., and Mijch, A. M. (1993). Tuberculous meningitis: A 30-year review. Clin. Infect. Dis. 17, 987–994.

Kumarvelu, S., Prasad, K., Khosla, A., Behari, M., and Ahuja, G. K. (1994). Randomized controlled trial of dexamethasone in tuberculous meningitis. Tubercle Lung Dis. 75(3), 203–207.

Leonard, J. M., and Des Prez, R. M. (1990). Tuberculous meningitis. Infect. Dis. Clin. North Am. 4, 769–787.

Lin, J. J., Harn, H. J., Hsu, Y. D., and Tsao, W. L. (1995). Rapid diagnosis of tuberculous meningitis by polymerase chain reaction assay of cerebrospinal fluid. J. Neurol. 242, 147–152.

McGowan, J. E., Chesney, P. J., Crossley, K. B., and LaForce, F. M. (1992). Guidelines for the use of systemic glucocorticosteroids in the management of selected infections. J. Infect. Dis. 165, 1–13.

Medical Research Council. Streptomycin in Tuberculosis Trials Committee (1948).

Streptomycin treatment of tuberculous meningitis. Lancet 1, 582–596.

Molavi, A., and LeFrock, J. L. (1985). Tuberculous meningitis. Med. Clin. North Am. 69, 315–331.

Ober, W. B. (1983). Ghon but not forgotten: Anton Ghon and his complex. Pathol. Annu. 18, 79–85.

Ogawa, S. K., Smith, M. A., Brennessel, D. J., and Lowy, F. D. (1987). Tuberculous meningitis in an urban medical center. Medicine 66, 317–326.

Pablos-Mendez, A., Raviglione, M. C., Laszlo, A., Binkin, N., et al. (1998). Global surveillance for antituberculosis-drug resistance 1994–1997. N. Engl. J. Med. 338, 1641–1649.

Schoeman, J. F., Van Zyl, L. E., Laubscher, J. A., and Donald, P. R. (1997). Effect of corticosteroid on intracranial pressure, computed tomographic findings, and clinical outcome in young children with tuberculous meningitis. Pediatrics 99, 226–231.

Shubin, H., Lambert, R. E., Heiken, C. A., Sokmensuer, A., and Glaskin, A. (1959). Steroid therapy and tuberculosis. JAMA 170, 1885–1890.

Starke, J. R. (1999). Tuberculosis of the central nervous system in children. Semin. Pediatr. Neurol. 6, 1–16.

Traub, M., Colchester, A. C. F., Kingsley, D. P. E., and Swash, M. (1984). Tuberculosis of the central nervous system. Q. J. Med. 53, 81–100.

Udani, P. M., Parekh, U. C., and Dastur, D. K. (1971). Neurological and related syndromes in CNS tuberculosis: Clinical features and pathogenesis. J. Neurol. Sci. 14, 341–357.

Verdon, R., Chevret, S., Laissy, J. P., and Wolff, M. (1996). Tuberculous meningitis in adults: review of 48 cases. Clin. Infect. Dis. 22, 982–988.

Whiteman, M., Espinoza, L., Post, M. J. D., Bell, M. D., and Falcone, S. (1995). Central nervous system tuberculosis in HIV-infected patients: Clinical and radiographic findings. AJNR 16, 1319–1327.

Wiggelinkhuizen, J., and Mann, M. (1980). The radioactive bromide partition test in the diagnosis of tuberculous meningitis in children. J. Pediatr. 97, 843–847.

CHAPTER 45

# Neurosarcoidosis

Barney J. Stern

## CLINICAL ASPECTS

Sarcoidosis is a disease of unknown etiology characterized pathologically by noncaseating granulomas. The diagnosis is most secure if inflammation can be documented in multiple organ systems (Johns and Michele, 1999).

### Clinical Features

Sarcoidosis most commonly affects intrathoracic structures (87% of patients), followed by lymph node, skin, and ocular disease (15–28% of patients). Neurologic disease develops in 5% of patients with sarcoidosis. The neurologic manifestations of sarcoidosis can be categorized as outlined in Table I. Selected differential diagnostic considerations are found in Table II.

Patients with known systemic sarcoidosis who develop an illness compatible with the neurologic manifestations of sarcoidosis should be evaluated to reasonably exclude other disease entities, particularly infection and neoplasia. If the patient does not respond as expected to treatment for neurosarcoidosis, the diagnosis should be questioned and a more extensive diagnostic evaluation pursued.

If a patient without documented sarcoidosis develops an illness compatible with neurosarcoidosis, the patient should be evaluated for multisystem inflammation. Because corticosteroid therapy can mask systemic disease, treatment should be postponed, if possible, while the diagnosis is pursued. Sarcoidosis most frequently affects the lungs, lymph nodes, skin, and eyes. Ocular findings include lacrimal gland inflammation, conjunctival nodules, iritis, uveitis, retinal lesions (vascular sheathing, granulomas, vascular occlusions, and chorioretinitis), and optic disc pathology (edema, nodules, granulomas, and atrophy). Systemic disease can often be documented if a comprehensive approach is followed (Table III).

Patients with possible CNS disease should be questioned about abnormal smell or taste perception, because these symptoms may represent nasal or olfactory problems. Disordered neuroendocrinologic or hypothalamic function can be reflected by symptoms of abnormal menses, libido, potency, breast discharge, thirst, body temperature, sleep, and appetite. Individuals with CNS symptoms other than transient cranial nerve palsies or aseptic meningitis should have a thorough endocrinologic evaluation with specific attention to hypothalamic hypothyroidism, hypocortisolism, and hypogonadism.

### Diagnostic Aids

If a patient has no history of documented sarcoidosis, histologic support for a diagnosis of sarcoidosis can be pursued following leads obtained from the patient's general medical evaluation. Therefore, consideration should be given to a skin or lymph node biopsy of suspicious sites. Transbronchial, nasal mucosal, conjunctival, lacrimal gland, liver, and muscle biopsies are also to be considered. A gallium scan or whole body fluorodeoxyglucose positron emission tomographic (FDG PET) scan can highlight foci of inflammation that might be targeted for biopsy.

The preferred imaging modality to evaluate CNS sarcoidosis is a gadolinium-enhanced MRI. Enhancement probably reflects a breakdown of the blood–brain barrier and implies active inflammation. MRI can also show enlargement of the optic nerves and ventricular system, as well as spinal cord abnormalities.

Cerebrospinal fluid (CSF) analysis can show an elevated pressure, increased total protein, hypoglycorrhachia, and a predominantly mononuclear pleocytosis.

**TABLE I    Neurologic Manifestations of Sarcoidosis**[a]

| Clinical manifestations | Approximate frequency (%) |
|---|---|
| Cranial neuropathy | 50–75 |
| Facial palsy | 25–50 |
| Meningeal disease | 10–20 |
| Aseptic meningitis | |
| Mass hydrocephalus | 10 |
| Parenchymal disease | 50 |
| Brain | |
| Endocrinopathy | 10–15 |
| Mass lesion(s) | 5–10 |
| Encephalopathy/vasculopathy | 5–10 |
| Seizures | 5–10 |
| Vegetative dysfunction | |
| Spinal canal | |
| Extramedullary or intramedullary disease | |
| Cauda equina syndrome | |
| Neuropathy | 15 |
| Axonal neuropathy | |
| Mononeuropathy | |
| Mononeuropathy multiplex | |
| Sensorimotor | |
| Sensory | |
| Motor | |
| Demyelinating neuropathy | |

[a]Adapted from Stern (1992a) and Stern and Schonfeld (1992), with permission.

**TABLE II    Other Selected Diagnostic Considerations**[a]

Multiple sclerosis
Sjogren's syndrome
Systemic lupus erythematosus
Neurosyphilis
Neuroborreliosis
HIV infection
Behçet's syndrome
Vogt–Koyanagi–Harada disease
Toxoplasmosis
Brucellosis
Whipple's disease
Lymphoma, including primary CNS lymphoma
Germ cell tumors
Craniopharyngioma
Isolated angiitis of the CNS
Lymphocytic hypophysitis
Pachymeningitis
Low CSF pressure/volume meningeal enhancement
AIDS-related diffuse infiltrative lymphocytosis syndrome

[a]Adapted from Stern (1992a), with permission.

The IgG index can be elevated and oligoclonal bands detected. The CSF angiotensin-converting enzyme assay is occasionally elevated, though reliable normative data are lacking; specificity is poor (Dale and O'Brien, 1999). None of these abnormalities are pathognomonic for CNS sarcoidosis; infection or neoplasia must always be considered.

Visual, auditory, or somatosensory evoked potentials can be abnormal, especially if there is optic or VIIIth nerve or spinal cord disease. Occasional patients with CNS sarcoidosis have evoked potential abnormalities even without overt clinical involvement at the appropriate site.

Because CNS tissue is rarely examined pathologically, the physician should always question a diagnosis of CNS sarcoidosis. Patients can be classified as having possible, probable, or definite CNS sarcoidosis (Zajicek et al., 1999) based on the certainty of the diagnosis of multisystem sarcoidosis, the pattern of CNS disease, and the response to therapy.

Possible. The clinical syndrome and neurodiagnostic evaluation are suggestive of neurosarcoidosis. CNS infection and malignancy have not been rigorously excluded or there is no pathologic confirmation of systemic sarcoidosis.

Probable. The clinical syndrome and neurodiagnostic evaluation are suggestive of neurosarcoidosis and alternate diagnoses have been excluded, expecially infection and malignancy. There is pathologic evidence of systemic sarcoidosis.

Definite. The clinical presentation is suggestive of neurosarcoidosis, other possible diagnoses are excluded, and there is the presence of supportive nervous system pathology.

"Possible" neurosarcoidosis is the most difficult scenario to deal with, especially if the patient is being treated for presumptive neurosarcoidosis, is not responding well, and alternate treatment strategies are being explored. The use of aggressive immune suppression strategies employing agents other than corticosteroids for patients with "possible" neurosarcoidosis should probably be avoided. On the other hand, patients with "possible" neurosarcoidosis who are

**TABLE III    The Search for Systemic Sarcoidosis**[a]

Serum angiotensin-converting enzyme
Serum calcium
Chest X-ray
Pulmonary function tests including diffusing capacity
Ophthalmologic examination
Endoscopic nasal examination
Whole body gallium scan
24-h Urinary calcium excretion
Anergy screen
Muscle magnetic resonance imaging
Whole body FDG PET scan

[a]Adapted from Stern (1992a), with permission.

doing well with corticosteroid treatment over 1 or 2 years usually do not have an occult infection or malignancy, thereby making a diagnosis of neurosarcoidosis more tenable.

Nerve conduction studies can define peripheral nerve disease as being either primarily axonal or demyelinating. Entrapment neuropathies can be delineated. Electromyography can define the presence of a myopathy or help characterize the pattern of peripheral nerve dysfunction.

## NATURAL COURSE

### Epidemiology

The prevalence of sarcoidosis is on the order of 60 per 100,000 population. The incidence is approximately 11 per 100,000 population, although in certain groups the incidence can be several times greater. Sarcoidosis can occur at any age although the likelihood is greatest in the third and fourth decades. Although there are some indications that sarcoidosis develops most commonly in women, overall there seems to be no sex predilection. Sarcoidosis does seem to occur with greater likelihood in some families but as yet there is no well-defined genetic pattern. In the United States, there is an increased incidence of sarcoidosis in blacks compared to whites; the disease also seems to be more severe in blacks. There are no definite provoking or exacerbating factors reliably associated with sarcoidosis.

Neurologic manifestations develop in approximately 5% of patients with sarcoidosis. Neurologic disease is the presenting feature of sarcoidosis in roughly one-half of patients destined to develop neurosarcoidosis. Furthermore, three-quarters of patients develop their neurologic problems within 2 years of becoming ill with sarcoidosis. The approximate frequency of the various neurologic complications is presented in Table I. One-third to one-half of patients with neurosarcoidosis have more than one neurologic manifestation. Only rarely do patients have isolated neurosarcoidosis.

### Clinical Course

Because most patients with neurosarcoidosis are treated with immunosuppressive agents, it is difficult to define the natural history of the disorder (Stern, 1992b). On the other hand, it is not at all certain that treatment changes the ultimate outcome of the disease, although therapy can certainly improve patients in the short term. Two-thirds of patients have a monophasic neurologic illness, with the remainder having a chronically progressive or remitting–relapsing course. The former

patients usually have an isolated cranial neuropathy or an episode of aseptic meningitis. The latter patients often have CNS parenchymal disease, hydrocephalus, multiple cranial neuropathies (especially involving cranial nerves II and VIII), peripheral neuropathy, or myopathy. Even in severely ill or impaired patients, the inflammatory process can subside over time. However, patients with remitting–relapsing or progressive disease can become severely incapacitated. Patients with CNS parenchymal disease or hydrocephalus are at the highest risk of death, either from the inflammatory process itself or from complications of therapy. Mortality of neurosarcoidosis is approximately 10%.

## PRINCIPLES OF THERAPY

### Pathophysiology

The cause of sarcoidosis is unknown. Activated CD4 lymphocytes congregate at sites of disease activity and there is a dominant T helper 1 (Th1) response with a cytokine profile including interleukins 2 and 12 and interferon γ. (Moller, 1999; Agostini et al., 2000; Muller-Quernheim, 1998) Monocytes or macrophages or both form granulomas. The inflammatory cells secrete a large array of cytokines including tumor necrosis factor. Irreversible fibrosis can develop. Furthermore, small foci of ischemic necrosis can be found, probably as a consequence of vascular compromise from perivascular inflammation.

### Therapy*

There are no rigorous studies defining the optimal therapy of neurosarcoidosis (Figure 1). Most authorities recommend corticosteroids for the initial treatment of neurosarcoidosis. Corticosteroids antagonize inflammation in many ways including decreasing the availability of interleukins 1, 2, and 6, interferon-gamma, and tumor necrosis factor (Boumpas et al., 1993). Other immunosuppressive agents and radiation therapy have been used to treat neurosarcoidosis, but, again, there are no controlled trials to guide usage.

## PRACTICAL MANAGEMENT

### Peripheral Facial Nerve Palsy

The most common neurologic manifestation of neurosarcoidosis is a peripheral facial nerve palsy. Treatment usually consists of prednisone 0.5–1 mg/kg/day (or 40–60 mg/day) for 1 week, followed by a tapering dose

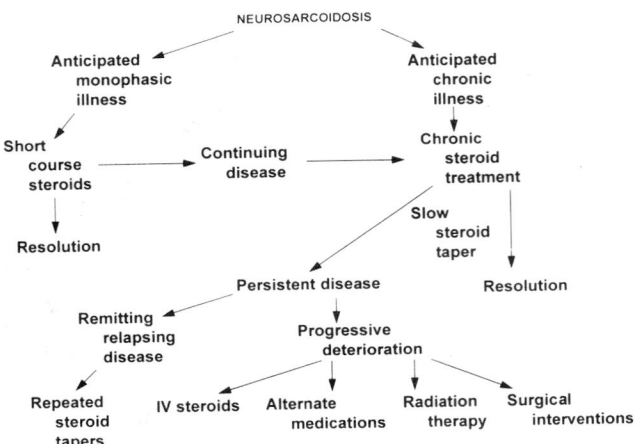

**FIGURE 1** A treatment approach to the patient with neuro-sarcoidosis.

of prednisone during the second week. General supportive care, as used for any patient with a peripheral facial nerve palsy, should be provided, with special attention being directed at protecting the eye. Occasionally a patient who was improving will relapse as the dose of prednisone is decreased; the dose can be increased for a week and a slower taper attempted.

## Other Cranial Nerve Palsies and Aseptic Meningitis

Patients with other cranial neuropathies and aseptic meningitis can be managed with a corticosteroid protocol similar to that described for a peripheral facial nerve palsy. However, often more than 2 weeks of therapy is needed. In particular, patients with optic neuropathy or VIIIth cranial nerve dysfunction can need prolonged, aggressive therapy, as discussed later.

## Peripheral Neuropathy and Myopathy

Patients with a peripheral neuropathy or myopathy can respond to a several-week course of corticosteroid, usually beginning with prednisone, 0.5–1.0 mg/kg/day. Here too, prolonged treatment may be indicated.

## Hydrocephalus

Asymptomatic ventricular enlargement probably does not require therapy. Mild, symptomatic hydrocephalus can respond to corticosteroid treatment; prolonged therapy is often appropriate. Unfortunately, patients can evolve from having mild hydrocephalus to severe, life-threatening disease quite rapidly. Patients

and careproviders should be educated as to the symptoms of progressive hydrocephalus and know how to obtain emergency care. At times, high-dose intravenous corticosteroid therapy can stabilize a patient with life-threatening disease, although usually surgical intervention with a ventricular drain or ventriculoperitoneal shunt is necessary.

## CNS Parenchymal Disease

Patients with a diffuse encephalopathy/vasculopathy or CNS mass lesion can improve with corticosteroid therapy. Treatment rarely alleviates neuroendocrine dysfunction or vegetative symptoms. Seizures most commonly occur in patients with brain parenchymal disease or hydrocephalus. Control of seizures is usually not difficult with standard antiseizure medications if the underlying inflammatory process can be controlled (Krumholz et al., 1991).

## Chronic Corticosteroid Therapy

Treatment for CNS parenchymal disease or other severe manifestations of neurosarcoidosis is usually begun with prednisone, 1.0–1.5 mg/kg/day. Particularly ill patients are started on the higher doses. Patients often require prolonged treatment and are at risk not only from the underlying disease but also from the adverse effects of therapy. As a general principle, prednisone should be tapered very slowly; rapid decreases often lead to patient worsening. The patient can be managed at a high dose for 2–4 weeks, as the clinical response is assessed. The prednisone dose can then be tapered by 5-mg decrements every 2 weeks as the patient is monitored. Patients tend to worsen at a prednisone dose of approximately 10 mg/day, though each individual will define a unique *floor* below which deterioration is likely to occur.

The patient who arrives at a dose of prednisone of 10 mg/day and is doing well should be evaluated for evidence of ongoing inflammation. For the patient with CNS parenchymal disease, this usually means obtaining an enhanced MRI; enhancement suggests active disease and a further decrease in prednisone dose may lead to a clinical exacerbation. (Lexa and Grossman, 1994; Christoforidis et al., 1999; Dumas et al., 2000) Other manifestations of neurosarcoidosis can be evaluated for subclinical deterioration on an individualized basis, for instance, evaluating nerve conduction studies or a serum creatine kinase level. However, persistent CSF abnormalities are usually not an indication for continuing high-dose corticosteroid treatment because patients can remain very functional despite an abnormal CSF and efforts to *normalize* the CSF often necessitate intense immunosuppression. If the disease appears quiescent, the prednisone dose can be tapered by 1 mg every 2–4 weeks. Depending on clinical circumstances, a reevaluation for subclinical inflammation may be indicated again at a dose of prednisone of 5 mg/day.

If, at any time during the tapering process, the patient has a clinical relapse, the prednisone dose can be doubled and the patient observed for 2–4 weeks, before another tapering cycle is attempted. If the patient worsens at a prednisone dose of less than 10 mg/day, the dose can be raised to 10–20 mg/day rather than a more elevated level. Patients may require multiple cycles of corticosteroid tapering. However, it is usually worthwhile to attempt to keep the patient on the lowest possible dose of corticosteroid so as to minimize corticosteroid side effects. Also, the disease can become quiescent and without constantly striving to achieve the lowest possible level of immunosuppression, spontaneous improvement may not be evident. However, patients can relapse months or years after being maintained for a long period on a minimal dose of prednisone, or even after the prednisone has been discontinued for an extended interval.

Daily dosing of prednisone is usually superior to alternate day therapy. However, if a patient has been clinically stable for a long time on a modest daily prednisone dose, an attempt can be made to wean the patient slowly onto alternate day therapy.

### Intravenous Corticosteroid Therapy

Critically ill patients may respond to high-dose corticosteroid therapy (Soucek *et al.*, 1993). Aggressive therapy can also clarify whether a patient on a modest corticosteroid dose can incrementally improve. Methylprednisolone, 20 mg/kg/day, intravenously for 3 days followed by prednisone, 1.0–1.5 mg/kg/day, for 2–4 weeks can be used. The clinical response should be closely monitored and predefined measures of improvement noted. At times traditional measures of neurologic function are not adequate; the results of formal psychometric tests or a timed walk can be used to judge clinical response.

### Alternate Medications

Experience with alternate therapies for neurosarcoidosis is limited and firm guidelines are not available. Indications for the use of alternate treatments include the need to avoid corticosteroids as initial therapy, serious adverse effects from chronic corticosteroid therapy, and disease activity despite *optimal* corticosteroid therapy.

Alternate modalities that have been used to treat sarcoidosis include azathioprine, methotrexate, cyclosporine, cyclophosphamide, chlorambucil, hydroxychloroquine, pentoxifylline, thalidomide, and irradiation. (Stern *et al.*, 1992; Agbogu *et al.*, 1995; Baughman and Lower, 1997; Lower *et al.*, 1997; Sharma, 1998; Zabel *et al.*, 1997; Marques *et al.*, 1999; Moller, 1999; Lee and Koblenzer, 1998; Rousseau *et al.*, 1998) The medications are best used in combination with modest doses of corticosteroids. An attempt is made to stabilize the patient on corticosteroid therapy, if feasible, and the alternate agent is introduced and the dose adjusted by observing the clinical response and monitoring adverse effects. As soon as the dose of the alternate medication is optimal, and probably no longer than 2–3 months after introducing the multidrug regimen, the corticosteroid dose should be tapered, as outlined earlier. If the patient has been taking more than 1.0 mg/kg/day of prednisone, the dose should be brought to this range as quickly as possible (for instance, by 5-mg decrements every week) and a slower taper can then be started. Rarely is it possible to eliminate prednisone entirely; typically prednisone at 10–20 mg/day (approximately 0.1 mg/kg/day) is chronically required. Fortunately, patients often feel much better on this modest prednisone dose compared to the levels they had been accustomed to.

Decisions as to which alternate therapy should be used first currently rely on clinical judgment (Abgobu *et al.*, 1995). Issues of cost and patient compliance should be considered as well as the spectrum of adverse effects of the various agents. For instance, cyclosporine should be avoided in patients with poorly controlled hypertension or renal insufficiency. Azathioprine, methotrexate, and chlorambucil are associated with liver toxicity, and methotrexate, cyclophosphamide, and chlorambucil can cause pulmonary fibrosis. All of these alternate medications except cyclosporine are implicated in causing malignancies, an issue to be considered in young patients with neurosarcoidosis. Hydroxychloroquine has ocular toxicity.

A reasonable first-line agent is azathioprine. This drug can be started at a dose of 50 mg/day for 1 week and, if this is tolerated, the dose can be progressively increased to 2–3 mg/kg/day. Blood and liver profiles should be monitored; the white blood cell count should be lowered to approximately 3500/mm$^3$ or the total lymphocyte count should be lowered to approximately 1000/mm$^3$ in patients on corticosteroids. An increase in the red blood cell mean corpuscular volume to above 100 is also indicative of an adequate azathioprine effect. Some 10% of patients develop an acute idiosyncratic reaction with fever and abnormal liver function necessitating discontinuation of the drug. The optimal dosing of the other immunosuppressive medications is not certain; a literature review is appropriate prior to the use of these agents to evaluate treatment options. Infliximab, a monoclonal antibody directed at tumor necrosis factor, may be useful for refractory disease.

## Irradiation

Patients with severe CNS sarcoidosis who cannot tolerate alternate immunosuppressive agents or who have failed to respond to two alternate medications should be considered for CNS irradiation (Ahmad *et al.*, 1992). A total dose of approximately 20 Gy should be considered; occasionally higher doses can be used. Total nodal radiation has been administered in refractory patients (Ahmad *et al.*, 1992). Patients can stabilize, though ultimately continuing immunosuppressive therapy is frequently required.

## Surgical Considerations

Surgical consultation should be considered in several scenarios. First, without biopsy confirmation of affected neural tissue, the diagnosis of neurosarcoidosis always remains somewhat in doubt. Though it is easy to biopsy nerve or muscle for diagnosis, CNS tissue is rarely obtained for pathologic examination. If a patient with presumptive CNS sarcoidosis is not responding to treatment as anticipated, consideration should be given to biopsy to exclude other diagnoses and attempt to confirm the clinical suspicion of sarcoidosis (Cheng *et al.*, 1994; Edmonds *et al.*, 1992; Fidler *et al.*, 1993).

### Mass Lesion

Patients without known systemic sarcoidosis who develop a brain or spinal cord mass are usually biopsied. If pathologic examination suggests noncaseating granulomas, appropriate cultures should be obtained but surgical excision of the mass should probably be avoided because surgery is rarely curative and patients have been known to deteriorate from surgical excision of a granulomatous mass. Importantly, granulomas can surround some malignancies; therefore, adequate biopsy samples should be taken (Peeples *et al.*, 1991). Patients should be evaluated for systemic sarcoidosis, although corticosteroid therapy can mask systemic disease.

If a patient with known sarcoidosis develops a CNS mass, an empiric trial of corticosteroid treatment is appropriate, especially if infection and malignancy can be reasonably excluded by CSF examination. If the patient does not respond to corticosteroid therapy, a biopsy should be considered. In either event, if a mass progressively enlarges despite corticosteroid therapy,

surgical exploration should be pursued to evaluate the possibility of a malignancy (Peeples *et al.*, 1991).

### Hydrocephalus

Patients with symptomatic hydrocephalus not rapidly responsive to corticosteroid therapy or life-threatening hydrocephalus should be considered for ventricular shunting. Shunt placement is not without risk, however, because shunt obstruction occurs frequently in the setting of brain and CSF inflammation and the presence of a foreign object in the CNS of an immunosuppressed host invites infection. Ventricles or parts of ventricles can become *trapped*, necessitating challenging neurosurgical strategies.

## General Supportive Care

Patients should be closely monitored for adverse effects of therapy. Rehabilitation services should be used as appropriate. Formal exercise and nutritional programs are often beneficial. Depression responds to standard treatments.

### Endocrinopathies

Endocrinologic deficiencies should be defined and treated. In particular, hypothalamic hypothyroidism should be sought. Because patients are often on chronic, low-dose corticosteroid therapy, supplemental corticosteroids should be given during intercurrent illnesses or surgery.

### Osteoporosis

Patients are at particular risk for osteoporosis (Montemurro *et al.*, 1990). Screening for osteoporosis should be done early and repeated periodically, perhaps on an annual basis. Hypogonadism should be treated and the lowest possible dose of corticosteroid used. Because sarcoidosis itself is associated with hypercalcemia and hypercalciuria, treatment of osteoporosis can present quite a challenge. It is best to work closely with a physician expert in the management of osteoporosis, since the field is rapidly evolving. Deflazacort, a corticosteroid with a low propensity for causing osteoporosis, is effective in treating sarcoidosis (Rizzato *et al.*, 1991).

## Pregnancy

There is little information on the effects of pregnancy on neurosarcoidosis. Conception can be difficult if there is an underlying endocrinopathy. Immunosuppressive agents should be used sparingly and in consultation with an obstetrician. Patients with a tendency to

**TABLE IV** Selected Long-Term Complications of Neurosarcoidosis[a]

Infection
    Cryptococcal meningitis
    Tuberculous meningitis
    Toxoplasmosis
    Progressive multifocal leukoencephalopathy
    *Listeria monocytogenes*
Spinal epidural lipomatosis
Corticosteroid myopathy
Lymphoma
Inclusion body myositis

[a]Adapted from Stern (1992a), with permission.

increased intracranial pressure might be at risk during labor and delivery. In general, sarcoidosis often becomes quiescent during pregnancy, although the disease can flare up after birth.

## Long-Term Complications

Patients with refractory neurosarcoidosis not only are predisposed to the primary effects of the inflammatory process but also are at risk of the long-term complications of treatment. If a patient is not doing well, not only should the diagnosis of sarcoidosis be questioned but also the patient should be evaluated for intercurrent complications, as outlined in Table IV.

## REFERENCES

Agbogu, B. N., Stern, B. J., Sewell, C., and Yang, G. (1995). Therapeutic considerations in patients with refractory neurosarcoidosis. *Arch. Neurol.* **52**, 875–879.

Agostini, C., Adami, F., and Semenzato, G. (2000). New pathogenetic insights into the sarcoid granuloma. *Curr. Opin. Rheumatol.* **12**, 71–76.

Ahmad, K., Kim, Y. H., Spitzer, A. R., Gupta, A., Han, I. H., Herskovic, A., and Sakr, W. A. (1992). Total nodal radiation in progressive sarcoidosis. *Am. J. Clin. Oncol.* **15**, 311–313.

Baughman, R. P., and Lower, E. E. (1997). Steroid-sparing alternative treatments for sarcoidosis. *Clin. Chest Med.* **18**, 853–864.

Boumpas, D. T., Chrousos, G. P., Wilder, R. L., Cupps, T. R., and Balow, J. E. (1993). Glucocorticoid therapy for immune-mediated diseases: Basic and clinical correlates. *Ann. Intern. Med.* **119**, 1198–1208.

Cheng, T. M., O'Neill, B. P., Scheithauer, B. W., and Piepgras, D. G. (1994). Chronic meningitis: The role of meningeal or cortical biopsy. *Neurosurgery* **34**, 590–596.

Christoforidis, G. A., Spickler, E. M., Recio, M. V., and Mehta, B. M. (1999). MR of CNS sarcoidosis: Correlation of imaging features to clinical symptoms and response to treatment. *Am. J. Neuroradiol.* **20**, 655–669.

Dale, J. C., and O'Brien, J. F. (1999). Determination of angiotensin-converting enzyme levels in cerebrospinal fluid is not a useful test for the diagnosis of neurosarcoidosis. *Mayo Clin. Proc.* **74**, 535.

Dumas, J. L., Valeyre, D., Chapelon-Abric, C., Belin, C., Piette, J. C., Tandjaoui-Lambiotte, H., Brauner, M., and Goldlust, D. (2000). Central nervous system sarcoidosis: Follow-up MR imaging during steroid therapy. *Radiology* **214**, 411–420.

Edmonds, L. C., Stubbs, S. E., and Ryu, J. H. (1992). Syphilis: A disease to exclude in diagnosing sarcoidosis. *Mayo Clin. Proc.* **67**, 37–41.

Fidler, H. M., Rook, G. A., Johnson, N. M., and McFadden, J. (1993). Mycobacterium tuberculosis DNA in tissue affected by sarcoidosis. *Br. Med. J.* **306**, 546–549.

Johns, C. J., and Michele, T. M. (1999). The clinical management of sarcoidosis. A 50-year experience at the Johns Hopkins Hospital. *Medicine* **78**, 65–111.

Krumholz, A., Stern, B. J., and Stern, E. G. (1991). Clinical implications of seizures in neurosarcoidosis. *Arch. Neurol.* **48**, 842–844.

Lee, J. B., and Koblenzer, P. S. (1998). Disfiguring cutaneous manifestations of sarcoidosis treated with thalidomide: A case report. *J. Am. Acad. Dermatol.* **39**, 835–838.

Lexa, F. J., and Grossman, R. I. (1994). MR of sarcoidosis in the head and spine: Spectrum of manifestations and radiographic response to steroid therapy. *Am. J. Neuroradiol.* **15**, 973–982.

Lower, E. E., Broderick, J. P., Brott, T. G., and Baughman, R. P. (1997). Diagnosis and management of neurological sarcoidosis. *Arch. Intern. Med.* **157**, 1864–1868.

Marques, L. J., Zheng, L., Poulakis, N., Guzman, J., and Costabel, U. (1999). Pentoxifylline inhibits TNF-alpha production from human alveolar macrophages. *Am. J. Resp. Crit. Care Med.* **159**, 508–511.

Moller, D. R. (1999). Cells and cytokines involved in the pathogenesis of sarcoidosis. *Sarcoidosis Vasculitis Diffuse Lung Dis.* **16**, 24–31.

Montemurro, L., Fraioli, P., Riboldi, A., Delpiano, S., Zanni, D., and Rizzato, G. (1990). Bone loss in prednisone treated sarcoidosis: A two-year follow-up. *Ann. Ital. Med. Int.* **5**, 164–168.

Muller-Quernheim, J. (1998). Sarcoidosis: Immunopathogenetic concepts and their clinical application. *Eur. Respir. J.* **12**, 716–738.

Peeples, D. M., Stern, B. J., Jiji, V., and Sahni, K. S. (1991). Germ cell tumors masquerading as central nervous system sarcoidosis. *Arch. Neurol.* **48**, 554–556.

Rizzato, G., Fraioli, P., and Montemurro, L. (1991). Long-term therapy with deflazacort in chronic sarcoidosis. *Chest* **99**, 301–309.

Rousseau, L., Beylot-Berry, M., Doutre, M. S., and Beylot, C. (1998). Cutaneous sarcoidosis successfully treated with low doses of thalidomide. *Arch. Dermatol.* **134**, 1045–1046.

Sharma, O. P. (1998). Effectiveness of chloroquine and hydroxychloroquine in treating selected patients with sarcoidosis with neurological involvement. *Arch. Neurol.* **55**, 1248.

Soucek, D., Prior, C., Luef, G., Birbamer, G., and Bauer, G. (1993). Successful treatment of spinal sarcoidosis by high-dose intravenous methylprednisolone. *Clin. Neuropharmacol.* **16**, 464–467.

Stern, B. J. (1992a). Neurosarcoidosis. *Neurol. Chron.* **2**, 1–6.

Stern, B. J. (1992b). Neurosarcoidosis. *In* "Prognosis of Neurological Disorders" (R. W. Evans, D. S. Baskin, and F. M. Yatsu, Eds.), pp. 535–539. Oxford Univ. Press, New York.

Stern, B. J., and Schonfeld, S. A. (1992). Neurosarcoidosis. *In* "Metabolic Brain Dysfunction in Systemic Disorders" (A. I. Arieff and R. C. Griggs, Eds.), pp. 289–312. Little, Brown and Company Boston.

Stern, B. J., Schonfeld, S. A., Sewell, C., Krumholz, A., Scott, P., and Belendiuk, G. (1992). The treatment of neurosarcoidosis with cyclosporine. *Arch. Neurol.* **49**, 1065–1072.

Zabel, P., Entzian, P., Dalhoff, K., and Schlaak, M. (1997). Pentoxifylline in treatment of sarcoidosis. *Am. J. Resp. Crit. Care Med.* **155**, 1665–1669.

Zajicek, J. P., Scolding, N. J., Foster, O., Rovaris, M., Evanson, J., Moseley, I. F., Scadding, J. W., Thompson, E. J., Chamoun, V., Miller, D. H., McDonald, W. I., and Mitchell, D. (1999). Central nervous system sarcoidosis—diagnosis and management. *Q. J. Med.* **92**, 103–117.

CHAPTER 46

# Neurosyphilis

H. Prange

## CLINICAL ASPECTS

Although neurosyphilis has become relatively rare in the last decade, it still cannot be clinically ignored. Reports from the United States and other countries show that the number of patients with this disease (often combined with acquired immunodeficiency syndrome—AIDS) is increasing (Hook and Marra, 1992; Rowland, 1995). The number of new infections with *Treponema pallidum*, the causative agent of syphilis, rose in some metropolitan areas (e.g., St. Petersburg) to 300 per 100,000 inhabitants. The principal risk group consists of males between the age of 25 and 30. Heterosexual drug addicts form a prominent subgroup.

In the pathogenesis of treponemal infection, syphilis has been divided into three clinical stages that have interim phases of latency. The first stage, primary syphilis, is characterized by a local, circumscribed skin lesion, called a chancre or syphilitic ulcer. The ulcer develops at the site of infection, becoming apparent 10 days to 10 weeks after initial exposure (usually a sexual contact). At this time treponemes in the primary lesion can be visualized with dark-field or fluorescence antibody microscopy.

It is thought that 60–90% of the bacteria evade the local host defence. Their hematogenous and lymphogenous dissemination leads to secondary syphilis, unless an early treatment takes place. Fever, malaise, generalized lymphadenopathy, and dermatological symptoms, such as macular, papular, annular, or follicular rash, most often on the palms and soles, are its typical manifestations. The mucous membranes, scalp (alopecia), eyes, joints, and kidneys may also be involved. About one third of those who are infected show abnormal cerebrospinal fluid (CSF) findings, e.g., moderate pleocytosis, mild glycorrhachia (increased

lactate), and a slight rise in total protein. Patients may be asymptomatic, have a headache, and/or present with meningitic symptoms. As a rule, the meningeal symptomatology is minor in most cases; however, lesions of the cranial nerves (especially VIII, VII, and III) may ensue. Whereas treponemes can be readily identified on dark-field microscopic examination of skin lesions, they are merely occasionally detected in CSF and the anterior chamber of the eye.

At the secondary stage, up to one-third of those infected develop a meningeal reaction (in the older literature called "meningeal catarrh") with pleocytosis, which is asymptomatic or slight in the majority of these patients. *Treponema pallidum* was demonstrated in the CSF of 15 to 30% of patients with no other CSF abnormalities (Chesney and Kemp, 1924). Antibiotic treatment resolves the symptoms of secondary syphilis. However, clinical manifestations of early syphilis also disappear without specific therapy.

A transition to latency can occur. The Oslo study of the natural course of untreated syphilis (Clark and Danbolt, 1964) revealed that 25% of patients with latent syphilis had recurrent mucocutaneous (infectious) lesions, usually within the first year. Mainly for this reason, the Centers for Disease Control (CDC) defined a *T. pallidum* infection lasting less than 1 year as early latent (infectious) syphilis and more than 1 year as late latent (noninfectious) syphilis. Latency may last 2–10 years, after which about a third of untreated patients develop signs of tertiary syphilis. This stage manifests itself as gummatous syphilis (granulomatous reactions in different organs; approx. 15%), cardiovascular syphilis (10%) or neurosyphilis (5–10%). During the last decades, gummatous and cardiovascular syphilis have become rare, probably due to screening programs and inadvertent cotherapy with antibiotics given for other illnesses.

**TABLE I** CDC Indications for CSF Analysis in Patients with Syphilis

- Neurologic, ophthalmologic, otic, or psychiatric signs or symptoms
- HIV infection and latent syphilis
- Evidence of tertiary-stage gummatous or cardiovascular disease
- Treatment failure

## Syphilitic Manifestations in the Central Nervous System

Syphilitic involvement of central nervous structures may occur at all stages of the disease. Patients with acute syphilitic meningitis in the primosecondary stage may present with headache, nausea, vomiting, and stiff neck, at times combined with cranial nerve alterations (predominantly nerves II, III, VII, and VIII), but usually without fever. Laboratory indicators of inflammation such as increased sedimentation rate or C-reactive protein are generally absent, as are intrathecally synthesized antibodies against *T. pallidum*. Acute syphilitic meningitis seldom leads to acute hydrocephalus or spinal cord lesions.

Some patients in late latency do not have any complaints, but slight CNS dysfunctions are detected by neuroophthalmologic and neurootologic examinations, evoked potentials, imaging procedures, or lumbar puncture (Table 1). The term "asymptomatic neurosyphilis" for this group of patients dates back to the early twenties. At that time, it was found that patients in whom CSF abnormalities of the secondary stage persist during the whole phase of latency are at risk of developing neurosyphilis (Moore and Hopkins, 1930). Typical of these patients are slight CSF abnormalities such as low pleocytosis and *de novo* IgG synthesis or oligoclonal IgG bands.

"Classic" CNS manifestations of tertiary syphilis are meningovascular neurosyphilis (syphilis cerebrospinalis), general paresis, and tabes dorsalis. These clinical syndromes may overlap appearing, for example, as taboparesis. Syphilitic amyotrophy, spastic spinal paralysis, cervical hyperplastic pachymeningitis, and cerebral gumma are very infrequent.

*Meningovascular neurosyphilis* usually occurs 4–7 years after infection. The inflammatory process takes place subacutely in meningeal and vascular tissues of the CNS, in some cases mainly the cerebral tissues, in others mainly the spinal tissues. The manifestation of meningovascular neurosyphilis is extraordinarily variable (Hutchinson called this form "the great imitator," 1863). Early symptoms are vision impairment of different origins (50%), vertigo (35%), mono- or hemipareses with apoplectiform onset (20%), and headache (20%) as well as hearing loss, gait disturbances, and

behavioral deviations (20% each). Later typical symptom combinations develop, such as lesions of the brain stem/cranial nerves (25%), hemisyndromes including homonymous anopia (20%), alterations of acoustic and vestibular functions (13%), syndrome of chronic meningitis (10%), spinal syndromes (10%), epilepsy (8%), or organic brain syndrome (8%).

Contrary to the focal ischemic changes due to syphilitic inflammation of cerebral blood vessels typical of meningovascular syphilis, *parenchymatous neurosyphilis* is characterized by diffuse, progressive, and irreversible neuronal degeneration (Roos, 1992). *Tabetic neurosyphilis* (tabes dorsalis; progressive locomotor ataxia) is a well-defined syndrome, characterized by episodic lightning pains, absence of ankle and knee reflexes, abnormal pupils, hyperflexibility of the hip joints, gait disturbances, atonic (deafferentiated) bladder, and optic atrophy. The most frequent initial symptoms in 26 of our patients were lightning pains (54%), impairment of vision (27%), "locomotor" ataxia (27%), gastric and visceral crises often resulting in subsequent laparatomy (15%), paresthesias (15%), vertigo (15%), and behavioral abnormalities (5%). Later the chief signs were reflex abnormalities (100%), impaired vibratory sense (81%), abnormal pupils (mainly the Argyll Robertson phenomenon; 66%), gait disturbances (50%), optic atrophy (45%), Charcot's joints (32%), and ocular palsy (4%). In advanced cases of tabes dorsalis, patients complain of hearing impairment, rectal incontinence, impotence, and perforating ulcerations of the extremities. The condition may arrest spontaneously or be stopped in its progression by antibiotic treatment; however, the lightning pains and ataxia often continue.

*Paretic neurosyphilis* (general paresis, dementia paralytica) is a chronic form of progressive meningoencephalitis which has a peak incidence 10 to 20 years after the primary infection. Initially, there are slight behavioral changes, subtle deteriorations of cognitive functions, dysarthria, complaints of headaches, and vertigo. As the disease progresses, organic brain syndrome with confusion or psychotic episodes, abnormal pupils, tremor of the tongue and hands, epileptic fits, eventually speech impairment, and quadriparesis as well as loss of bowel and bladder control develop. If untreated, general paresis is fatal in 3 to 5 years. The administration of penicillin in advanced cases may stop the inflammation in the brain, but cerebral functions will not be restored.

## Diagnostic Procedures

In the early stages of syphilis, *T. pallidum* can be demonstrated in exudates from lesions and also occasionally in CSF with fluorescent or dark-field microscopy. Sustained *in vitro* cultivation of *T. pallidum* has not been successful so far. Although the detection of treponemal DNA in clinical specimens by polymerase chain reaction (PCR) was considered to have great promise in the early nineties (Hook and Marra, 1992), it subsequently became apparent that PCR is not a sensitive method for diagnosis of late stages of syphilis (Gordon *et al.*, 1994). Thus, serologic tests have retained priority for the diagnosis of secondary and tertiary syphilis (Figure 1).

Two types of serologic tests are used:

(1) quantitative nontreponemal or reagin tests, such as the Venereal Disease Research Laboratory (VDRL) test, the rapid plasma reagin (RPR) test, or the cardiolipin complement fixation reaction (a further development of the Wassermann test);

(2) specific treponemal tests, namely, the fluorescent treponemal–antibody absorption (FTA–ABS) test, the *T. pallidum* hemagglutionation assay (TPHA), or the *T. pallidum* particle agglutination assay (TPPA).

The combined usage of these tests varies in different countries. In the United States, quantitative nontreponemal tests are used for screening; in the case of a positive ("reactive") result, a specific treponemal test, e.g., FTA–ABS test, is employed for confirmation (Rowland, 1995). Many European laboratories prefer the TPHA or TPPA for screening. A positive TPHA, TPPA, or FTA–ABS test in plasma has high specificity for the diagnosis but not to determine the disease activity, because these tests may remain positive for years after successful treatment. In contrast, nontreponemal tests become negative within 6 to 12 months after treatment. Therefore, they are applicable for assessing disease activity unless antibiotics were given during the preceding 12 months. However, nontreponemal tests can be negative in patients with advanced stages of tertiary syphilis or with impaired immunity. Therefore, a positive test result in late syphilis indicates activity of the pathological process, but a negative result does not exclude it. For this reason, some centers assess disease activity by the presence of antibodies of the IgM type. Tests available for this purpose that have high reliability are the 19S-(IgM)FTA–ABS test and *T.p.* IgM ELISA. The detection of treponemal IgM antibodies by these tests indicates disease activity, provided the patient received no antibiotic treatment in the preceding months.

Quantitative antitreponemal antibody testing and CSF analysis are necessary to verify a diagnosis of neurosyphilis. The procedure of choice for CSF antibody studies remains a subject of debate. Many laboratories in the United States will not perform antitreponemal tests of the CSF because standardization is lacking and because small amounts of serum contaminating the CSF may produce a false-positive test. In some European

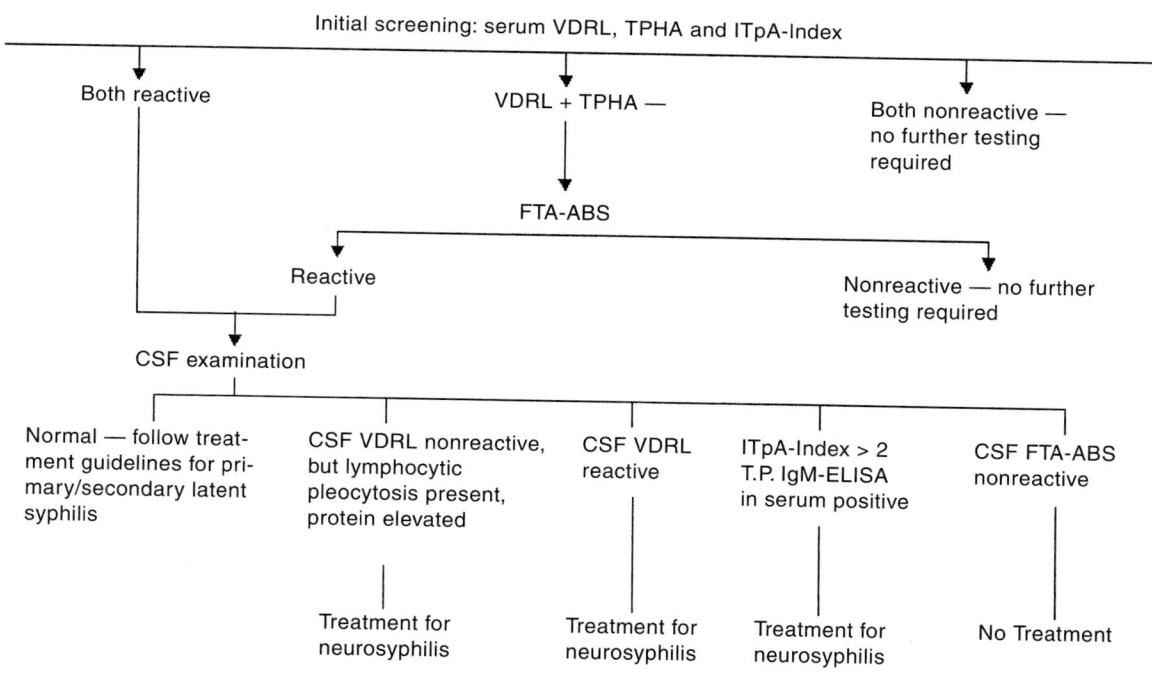

FIGURE 1    Laboratory diagnosis of central nervous system syphilis.

countries, the TPHA or TPPA, given in titer steps, is routinely used for CSF. Further CSF analysis includes cell count, determination of total protein, albumin, and IgG (IgM and IgA) concentrations, and isoelectric focusing for oligoclonal IgG bands. Albumin and immunoglobulin determination as well as oligoclonal antibody detection should be performed simultaneously in serum and CSF samples to obtain CSF/serum quotients. The IgG production index is a measure of intrathecal immunoglobulin synthesis and is calculated from the albumin and IgG quotients as follows:

$$IgG\ index = \frac{IgG_{CSF}/IgG_{serum}}{albu\min_{CSF}/albu\min_{serum}}.$$

The upper limit of the normal range is about 0.69. Indices exceeding this limit indicate intrathecal IgG production due to a local humoral immune response within the CNS. To verify the syphilitic cause of the latter, local production of antitreponemal antibodies must be proven. For this purpose, the ITpA (intrathecally produced *T. pallidum*-specific antibodies) index seems most suitable. It is calculated with the formula

$$ITpA\ index = \frac{TPHA_{CSF}/Total\ IgG_{CSF}}{TPHA_{serum}/Total\ IgG_{serum}}.$$

As antibodies of the IgM type cross the blood/CSF barrier in minute quantities due to their size, the titer of TPHA approximately reflects the rate of antitreponemal IgG antibody production in each compartment. In late stages of syphilis (late latency, tertiary syphilis), an ITpA index >2 suggests probable syphilitic CNS involvement; an ITpA >3 proves neurosyphilis with a specificity of 100 and a sensitivity of 84% (Prange *et al.*, 1983; Prange and Bobis-Seidenschwanz, 1994). Because the ITpA index remains positive in many cases of tabetic and paretic neurosyphilis throughout the patient's life, it cannot be considered a criterion of disease activity.

The activity of the syphilitic CNS process can be estimated from the CSF pleocytosis and proof of antitreponemal IgM antibodies in *serum* (*T.p.* IgM ELISA, 19S-(IgM)FTA–ABS test). However, about one third of patients with untreated neurosyphilis do not have pleocytosis. Therefore, other indicators of disease activity have been sought. A positive CSF-VDRL test, regardless of previous antibiotic therapy, and in special cases (e.g., coinfection with HIV) a positive CSF-PCR (Horowitz *et al.*, 1994) may be taken as evidence of disease activity. The IgG index does not constitute a reliable criterion of activity, but it is well suited for long-term monitoring of the therapeutic response, as it falls in a logarithmically

linear fashion after effective therapy. Six to twelve months after antibiotic therapy, CSF pleocytosis should have faded, the treponemal IgM antibodies should have disappeared from the serum, and the CSF-VDRL test should be negative.

Other diagnostic procedures for patients suspected of having neurosyphilis are crainal CT or MRT, EEG, chest X-ray, and ophthalmologic and otologic examinations, as well as evoked potentials to establish the functional state of the CNS, to detect serious complications, and to exclude other CNS diseases relevant for differential diagnosis. These include tuberculous and cryptococcal meningitis, neurolisteriosis, neuroborreliosis, diffuse CNS toxoplasmosis, some viral meningoencephalitidis, central nervous system sarcoidosis, cerebral vasculitis, and carcinomatous meningeosis.

### Neurosyphilis and HIV Infection

Since syphilis and human immunodeficiency virus (HIV) spread in similar risk groups and often through gential lesions, coinfections are quite frequent. It is known (Gordon *et al.*, 1994) that patients coinfected with syphilis and HIV may show an accelerated and more aggressive course and develop complications normally associated with late syphilis. In addition, there is evidence that coexistent HIV infection can alter the response of patients with early syphilis to benzathine penicillin G therapy (Hook and Marra, 1992). Observations (ours and Fox *et al.*, 2000) that meningovascular neurosyphilis can develop after antibiotic treatment of early syphilis suggests that neurosyphilis may be a more common complication of syphilis in HIV patients. Retrospective studies indicate that the currently recommended neurosyphilis therapy with penicillin G fails in 23 to 60% of HIV-infected patients (Marra *et al.*, 2000). Coinfection of syphilis and HIV may pose problems of discrimination between a syphilitic or and HIV origin of CNS symptoms. Both HIV and syphilis may cause cognitive deficits and neurologic abnormalities as well as CSF pleocytosis, higher immunoglobulin production rates, and/or elevated CSF protein concentrations. The CSF-VDRL test is sensitive in only 70% of HIV-infected subjects. Thus, a negative result does not exclude syphilitic CNS involvement. Conversely, false-positive nontreponemal reagin tests may occur in earlier stages of HIV infection, probably due to polyclonal B-cell activation (Hook and Marra, 1992). The CSF-PCR has also been recently shown to be of limited usefulness for diagnosing neurosyphilis in HIV-infected patients (Marra *et al.*, 1996). These facts suggest that a certain diagnostic dilemma is posed by coinfected patients who present with CNS symptoms and pathologic CSF findings. Theoretically, the determination of intrathecal antibody synthesis against HIV and *T. pallidum* could resolve this dilemma and make it feasible to discriminate among syphilitic, HIV, and a simultaneously syphilitic and HIV related origin of the inflammatory CNS process. We have shown in our laboratory for more than a decade that this procedure is practicable (Felgenhauer *et al.*, 1990; Reiber and Lange, 1991); however, a larger prospective study has not yet been conducted to clinically evaluate this point. The practical question is whether special syphilis treatment schedules are required for HIV-infected patients. This question reinforces the need for careful follow-up after treatment so that retreatment can be given when needed. For patients who appear unlikely to comply, intravenous penicillin administration with doses up to 30 millions IU/day may be adequate. A recent study comparing high-dose penicillin with iv ceftriaxone (2 g/day) revealed a slight advantage for ceftriaxone, but the differences between the study groups were not strong enough to justify abandoning high-dose iv penicillin in HIV-infected patients (Marra *et al.*, 2000).

## NATURAL COURSE

Only one-third of the people exposed develop primary syphilis and have a painless chancre at the site of inoculation. Even if untreated, the ulcer will heal within 3 to 6 weeks, leaving a faint scar. In 10 to 40% of the infected patients, the syphilitic process remains restricted to the chancre, and a spontaneous cure takes place.

In the rest of the inoculated patients infection spreads from the primary site: treponemes enter the bloodstream and the lymphatics and may infect any organ. Four to sixteen weeks after the syphilitic ulcer appeared, patients develop symptoms and signs of secondary syphilis. The most frequent sign at this stage is condylomata at the anal and genital regions. CNS Symptoms may appear. The symptomatology of secondary syphilis spontaneously fades after 3 to 12 weeks; within the following 12 months there are recurrences in 25% of the patients. A period of latency follows, during which there are no clinical manifestations of syphilis, but the serologic indicators of (nontreated) syphilis persist. Without treatment, one-third of these patients develop a tertiary syphilis, characterized by signs of cardiovascular and neurologic dysfunctions resulting from inflammatory reactions of heart, vessel walls, and CNS structures (meninges and cerebral vasculatur). Tertiary syphilis may also be manifested by granulomatous reactions of different organs, e.g., skin, bones, or internal organs. After inoculation 4 to 40 years may elapse. The approximate time course of the clinical manifestations of early syphilis and neurosyphilis is given in Figure 2.

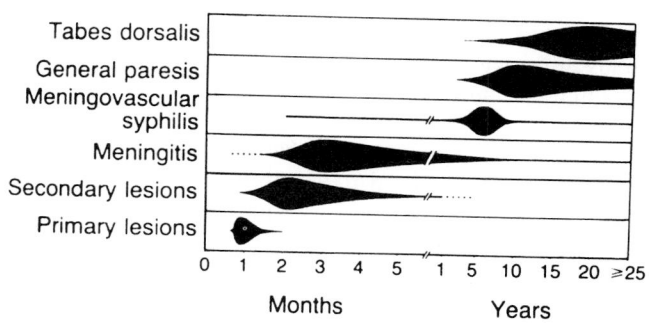

**FIGURE 2** Approximate time course of the clinical manifestations of early syphilis and neurosyphilis. Shaded areas corresponding to each syndrome represent the approximate proportion of patients with the syndrome specified and do not indicate the proportion of all patients with syphilis who have that syndrome (Reprinted with permission from Hook and Marra, 1992, copyright Massachusetts Medical Society. All rights reserved).

## PRINCIPLES OF THERAPY

### Pathogenesis

In the last 20 years *T. pallidum* has been found to have a certain neurotropism for the peripheral and central nervous systems. Electron microscopic examinations of the syphilitic chancre demonstrated that treponemes enter peripheral nerve endings around the ulcer (Secher *et al.*, 1982). They can invade the central nervous system even in the primary stage, often inducing an inflammatory reaction within the CSF (Wile and Hasley, 1921; Engel *et al.*, 1980; Lukehart *et al.*, 1988). CNS invasion is thought to take place by hematogenous propagation, but the early presence of treponemes in peripheral nerves may point to axonal transport as an alternative route to the central nervous system. Treponemes were also detected in CSF samples taken of patients with primary syphilis whose CSF was otherwise normal.

CSF examinations of patients with secondary syphilis revealed pleocytosis in 25 to 30%, increased protein in 30 to 40%, and intrathecal IgG production in 15 to 20%. The majority of the patients were asymptomatic or complained of mild headaches (Mills, 1927; Löwhagen *et al.*, 1983). Obviously, treponemes have already invaded central nervous structures by the end of the secondary stage in many patients. The microorganisms apparently disappear spontaneously even without special therapy. Patients in whom *T. pallidum* persist in the CNS are at special risk of developing neurosyphilis, unless they have received specific or inadvertently effective antibiotic treatment. Lymphocytes and other mononuclear cells infiltrate the meninges and cranial nerves. When the inflammation involves small meningeal vessels and endothelial proliferation ensues, occlusion may cause ischemic necrosis of the brain and spinal cord. If the process continues to progress chronic meningitis, vasculitis, neuronal degeneration, demyelination (mainly of the spinal cord), ependymal granulation, and astrocyte proliferation with gliosis result.

Clinical observations of patients with coinfection of HIV and syphilis found that treponemal CNS invasion may occur even during the stage of late latency. For decades, it has been a subject of debate whether or not antibiotics can eradicate treponemes at this late stage of the disease. This highlights the necessity of long-term controls after treatment.

## Pharmacology and Pharmacokinetics

In the last 50 years of clinical experience it has become clear that *T. pallidum* is susceptible to penicillin, other beta-lactams, macrolides, and tetracyclines. It was difficult to determine *T. pallidum*'s susceptibility to the respective antibiotic, since it cannot be cultivated by standard microbiological techniques. Factors governing the therapeutic efficacy of antibacterial agents against the organism include (1) the concentration that can be achieved at the site of infection, (2) the replication time of *T. pallidum* in the involved tissue, and (3) the support of the tissue defence mechanisms at the site of infection.

Experimental data showed that serum penicillin concentrations of 0.1 mg/ml are maximally treponemicidal, but lower levels will also result in cure when given for longer periods (Rein, 1981). In a seminal article composed on behalf of the World Health Organization (WHO), Idsoe *et al.* (1972) stated that a serum penicillin concentration exceeding 0.018 mg/ml (or 0.03 IU/ml) for 7 to 10 days, without interruption for more than 24 to 30 h, is the standard goal of therapy for early syphilis. Later this therapeutic level of 0.018 mg/ml was extended to include cerebrospinal fluid in cases of neurosyphilis. A series of pharmacokinetic studies on patients with neurosyphilis followed which showed that the low-dosage standard treatment with penicillin was insufficient to achieve such CSF concentrations. This was possible only with high doses of 20 to 30 IU penicillin G per day given in three to six single doses (Volles and Ritter, 1974; Mohr *et al.*, 1976).

A second point to be considered is *T. pallidum*'s slow replication, which *in vitro* can take up to 30 to 33 h. Treponemal cell division may be much slower in humans, especially during the latent period or tertiary syphilis. As most antibiotics act only during active microbial metabolism, prolonged treatment is presumed to be necessary.

The third relevant factor in the efficacy of antibacterial treatment is the as yet unclear role of the defence mechanisms in the CNS. Central nervous system was long considered an immunologically privileged space for

microbial pathogens. *T. pallidum* is thought to employ special strategies to elude tissue defence mechanisms (e.g., masking with substances derived from the host), a fact that emphasises the necessity of high-dose systemic antibiotics to eradicate the pathogen.

*In vitro* concentrations that immobilize 50% of the treponemes are 0.002 μg penicillin G per milliliter, 0.07 micrograms amoxicillin per millilitre, and 0.01 μg ceftriaxone/ml (Korting *et al.*, 1986). These concentrations should be exceeded tenfold *in vivo*. Doses of ceftriaxone of 2 g/day (initial dose 4 g) are adequate to achieve corresponding CSF concentrations (Nau *et al.*, 1993). Clinical studies have confirmed the efficacy of ceftriaxone against neurosyphilis (Marra *et al.*, 2000). The compound is an attractive candidate due to its long half-life and ability to sufficiently penetrate the CSF. Its activity against *T. pallidum* is probably not better than that of high-dose penicillin, but once daily administration and applicability in penicillin-allergic patients are advantages. Chloramphenicol and doxycycline have also been administered to small groups of patients with neurosyphilis; however, definite recommendations cannot yet be made on the basis of the available data (Quinn and Bender, 1988; Clinical Effectiveness Group, 1999).

## Therapy for Neurosyphilis

### Antibiotics

The current recommendations of the CDC for the treatment of latent and tertiary stage syphilis are given in Table II. Recent data confirm that modifications of the recommended antibiotic regimen are necessary for the following reasons: 1) Procain penicillin G, 2.4 (as well 4.8) million IU daily by intramuscular injection, plus probenecid, 500 mg orally 4 times daily for 10 to 14 days, fail to reach treponemocidal levels in CSF of many patients (Goh *et al.*, 1984); 2) since benzathine penicillin G (2.4 million IU), which was recommended for early and latent syphilis, also does not achieve treponemocidal concentrations in the CSF, this penicillin preparation is not suitable for the treatment of asymptomatic neurosyphilis. The *first-line therapy* for neurosyphilis is therefore aqueous crystalline penicillin G, 18 to 24 million IU/day (3 to 4 million IU every 4 hours) by intravenous infusion for at least 14 days. This regimen should be administered to patients with symptomatic or asymptomatic neurosyphilis as well as in those coinfected with HIV.

Some European centers administer 10 million IU penicillin G intravenously three times daily. We found that this regimen was successful; the penicillin concentrations in CSF were always treponemicidal and retreatments were not necessary.

TABLE II    CDC Recommended Therapy for Latent- and Tertiary-Stage Syphilis

| Medication | Route | Dosage | Duration |
|---|---|---|---|
| Early latency | | | |
|   Penicillin G benzathine | Intramuscular | 2.4 million IU | Once |
|   Doxycycline | Oral | 100 mg bid | 2 Weeks |
|   Tetracycline | Oral | 500 mg qid | 2 Weeks |
| Late or unknown period of latency | | | |
|   Penicillin G benzathine | Intramuscular | 2.4 million IU/wk | 3 Weeks |
|   Doxycycline | Oral | 100 mg bid | 4 Weeks |
|   Tetracycline | Oral | 500 mg qid | 4 Weeks |
| Tertiary stage/Neurologic symptomatology | | | |
|   Aqueous penicillin G | Intravenous | 3–4 million IU/q4h | 10–14 Days |
|   Penicillin G procaine | Intramuscular | 2.4 million IU qd | 10–14 Days |
|   + probenecid | Oral | 500 mg qid | 10–14 Days |

An *alternative regimen* is ceftriaxone (2 g iv once daily) with an initial dose of 4 g given for 10 to 14 days (Marra *et al.*, 2000). These regimens are recommended for patients with CNS symptoms and/or pathological CSF findings, irrespective of the stage of the syphilis.

As a *third line therapy*, doxycycline can be administered to patients who are allergic to beta-lactam antibiotics. The recommended dose is 200 mg orally four times daily for 28 days (Clinical Effectiveness Group, 1999). Doxycycline is contraindicated during pregnancy and breast feeding.

Another effective antibiotic is chloramphenicol. As the efficacy of this compound has not yet been tested in a controlled study, it should be used only in special cases in which the aforementioned antibiotics are contraindicated.

### Monitoring Therapeutic Success

After successful treatment CSF pleocytosis will fade within a few weeks or months, and specific serum IgM antibodies will disappear within 6 to 12 months. In exceptional cases it may take 2 to 3 years for the 19S-(IgM)FTA-ABS test to become negative. The reappearance of specific serum IgM antibodies after antibiotic therapy suggests that relapse or reinfection has occured. After effective therapy, the VDRL should decrease fourfold in 3 months and no longer be positive after a year. Intrathecally procuced IgG (according to the IgG index) may persist for years; however, its amount declines exponentially. The TPHA and FTA–ABS tests should not be used to determine response to treatment, since these tests remain positive for long periods of time despite effective treatment. The same is true for the ITpA index. Patients who have positive nontreponemal tests (VDRL) should be checked serologically every 6 months until these are negative. Patients with involvement of the CNS, regardless of the stage of syphilis, should undergo repeated CSF examinations every 6 months until the cell count is normal. Treatment failure may be indicated by clinical progression, increase in nontreponemal test titers by two or more dilutions, and the failure of CSF pleocytosis to resolve.

If 12 to 24 h after initiation of therapy systemic constitutional symptoms emerge (fever, headache, myalgia, malaise, tachycardia, hyper- or hypotension, peripheral leukocytosis, and lymphopenia) or neurologic signs worsen (including convulsions), the Jarisch-Herxheimer reaction (JHR) must be considered. While this reaction is not infrequent in the early stages of syphilis, its incidence in neruosyphilis is quoted at 1 to 2%. Treatment of JHR is symptomatic and should include antipyretic measures and monitoring of the cardiovascular functions. Administration of the given antibiotic should not be discontinued.

### Other Therapies

Epileptic fits should be treated according to the general guidelines for anticonvulsive therapy. The lightning pains that are mostly unresponsive to analgetics may tentatively be treated by carbamazepine, gabapentin, amitryptilin, or flupirtine. Hydrocephalus, a rare complication of neurosyphilis, sometimes requires ventricular shunting. The use of corticosteroids (along with antibiotics) for neurosyphilis must be considered experimental, since no data are yet available on its efficacy.

## REFERENCES

Centers for Disease Control and Prevention (1998). 1998 Guidelines for treatment of sexually transmitted disease. *Morb. Morbid. Wkly. Rep.* **47**, 1–111.

Clinical Effectiveness Group (Association of Genitourinary Medicine and the Medical Society for the Study of Venereal Diseases) (1999).

National guideline for the management of late syphilis. Sex Transm Inf 75 (Suppl) S34–S37.

Chesney, A. M., and Kemp, J. E. (1924). Incidence of Spirochaeta pallida in cerebrospinal fluid during early stage of syphilis. *JAMA* **83**, 1725–1728.

Clark, E. G., and Danbolt, N. (1964). The Oslo study of the natural course of untreated syphilis. *Med. Clin. North Am.* **48**, 613–623.

Engel. S., Krause, H., and Metz, D. (1980). Liquorbefunde bei Frühsyphilis. *Dermatol. Monatsschr.* **166**, 223–228.

Felgenhauer, K., Lüer, W., Weber, T. *et al.* (1990). CSF changes in chronic HIV encephalitis and opportunistic infections of the brain. *In* "Advances in Neruology" (J. S. Chopra *et al.*, eds.), pp. 229–237. Elsevier Science, Amsterdam.

Fox, P. A., Hawkin, D. A., and Dawson, D. (2000). Dementia following an acute presentation of menigovascular neurosyphilis in an HIV-1-positive patient. *AIDS* **14**, 2062–2063.

Goh, B. T., Smith, G. W., Samarasinghe, L. *et al.* (1984). Penicillin concentrations in serum and cerebrospinal fluid after intramuscular injection of aqueous procaine penicillin 0.6 MU with and without probenecid. *Br. J. Vener. Dis.* **60**, 371–373.

Gordon, S. M., Eaton, M. E., George, R. *et al.* (1994). The response of symptomatic neurosyphilis to high-dose intravenous penicillin G in patients with human immunodeficiency virus infection. *N. Engl. J. Med.* **331**, 1469–1473.

Hook, E. W., and Marra, C. (1992). Acquired syphilis in adults. *N. Engl. J. Med.* **326**, 1060–1069.

Horowitz, H. W., Valsamis, M. P., Whicher, V. *et al.* (1994). Brief report: Cerebral syphilitic gumma confirmed by polymerase chain reaction in a man with HIV infection. *N. Engl. J. Med.* **331**, 1488–1491.

Idsoe, O., Guthe, T., and Willcox, R. R. (1972). Penicillin in the treatment of syphilis: The experience of three decades. *Bull. World Health Org.* **47** (Suppl. 1), 1–68.

Korting, H. C., Walther, D., Riethmüller, U. *et al.* (1986). Comparative in vivo susceptibility of Treponema pallidum to cefti-zoxime, ceftriaxone and penicillin G. *Chemotherapy* **32**, 352–355.

Löwhagen, G. B., Andersson, M., Blomstrand, C. *et al.* (1983). Central nervous system involvement in early syphilis. *Acta Derm. Venereol. (Stockholm)* **63**, 409–417.

Lukehart, S. A., Hook, E. W., Baker-Zander, S. A. *et al.* (1988). Invasion of the central nervous system by Treponema pallidum: Implication for diagnosis and treatment. *Ann. Intern. Med.* **109**, 855–862.

Marra, C. M., Gary, D. W., Kuypers *et al.* (1996). Diagnosis of neurosyphilis in patients infected with human immunodeficiency virus type 1. *J. Infect. Dis.* **174**, 219–221.

Marra, C. M., Boutin, P., McArthur, J. C. *et al.* (2000). A pilot study evaluating cetriaxone and penicillin G as treatment agents for neurosyphilis in human immunodeficiency virus-infected individuals. *Clin. Infect. Dis.* **30**, 540–544.

Mills, C. H. (1927). Routine examination of the cerebrospinal fluid in syphilis. *Br. Med. J.* **2**, 527–532.

Mohr, J. A., Griffiths, W., Jackson, R. *et al.* (1976). Neurosyphilis and penicillin levels in cerebrospinal fluid. *JAMA* **236**, 2208–2209.

Moore, J. E., and Hopkins, H. H. (1930). Asymptomatic neurosyphilis. VI. The prognosis of early and late asymptomatic neurosyphilis. *JAMA* **95**, 1637–1641.

Nau, R., Prange, H. W., Muth, P. *et al.* (1983). Passage of cefotaxime and ceftriaxone into the cerebrospinal fluid of patients with uninflamed meninges. *Antimicrob. Agents Chemother.* **27**, 1518–1524.

Prange, H., Moskophidis, M., Schipper H. I. *et al.* (1983). Relationship between neurological features and intrathecal synthesis of IgG antibodies to Treponema pallidum in untreated and treated human neurosyphilis. *J. Neurol.* **230**, 241–252.

Prange, H. W., and Robis-Seidenschwanz, I. (1994/95). Zur Evaluierung serologischer Aktivitätskriterien bei Neurosyphilis. *Verh. Dtsch. Ges. Neurol. Bd.* **8**, 789–791.

Quinn, T. C., and Bender, B. (1988). Sexually transmitted diseases. *In* "The Principles and Practice of Medicine" (A. M. Harvey, R. J. Johns, McKusick *et al.*, Eds.), 22nd ed., pp. 661–663. Appleton & Lange, Norwalk, C. T.

Reiber, H., and Lange, P. (1991). Quantification of virus-specific antibodies in cerebrospinal fluid and serum. *Clin. Chem.* **37**, 1153–1160.

Rein, M. F. (1981). Treatment of neurosyphilis. *JAMA* **246**, 2613.

Roose, K. L. (1992). Neurosyphilis. *Semin. Neurol.* **112**, 209–212.

Rowland, L. P. (1995). Spriochete infections: Neurosyphilis. *In* "Merritt's Textbook of Neurology" (L. P. Rowland, ed.), 9th ed., pp. 200–208. Williams & Wilkins, Baltimore.

Secher, L., Weismann, K., and Kobayasi, T. (1982). Treponema pallidum in peripheral nerve tissue of syphilitic chancres. *Acta Derm. Venereol. (Stockholm)* **62**, 407–411.

Volles, E., and Ritter, G. (1974). Zur Penizillin-Behandlung der Neurolues unter pharmakokinetischem Aspekt. *Z. Neurol.* **206**, 235–242.

Wile, U. J., and Hasley, C. K. (1921). Involvement of nervous system during primary stage of syphilis. *JAMA* **76**, 8–9.

CHAPTER 47

# Lyme Neuroborreliosis, Leptospirosis, and Relapsing Fever

Klaus Hansen and Hans-Walter Pfister

## LYME NEUROBORRELIOSIS

### Clinical Aspects

The spirochete *Borrelia burgdorferi* is the etiologic agent of Lyme borreliosis (Lyme disease), a multisystem disorder characterized by several clinical stages (Steere, 2001). *Borrelia burgdorferi* is transmitted to humans by ticks belonging to the *Ixodes ricinus* complex. The responsible vector is, in Europe, *I. ricinus*, in Asia, *I. persulcatus*, and, in the endemic areas of North America, the eastern and midwestern states, *I. scapularis*. Lyme borreliosis is a zoonosis and the main reservoirs of *B. burgdorferi* are wild mammals, especially mice and deer. Dermatologic and neurologic manifestations have been known in Europe since 1883 and 1922, respectively. In the late 1970s the Yale rheumatologist A. C. Steere recognized an endemic tick-borne multisystem disease also comprising arthritis and called the whole entity, due to the location (Lyme, CT), Lyme disease. Burgdorfer *et al.* (1982) eventually isolated spirochetes from ticks collected in the endemic area of this disease. Thus the new spirochete species was called *B. burgdorferi*. Molecular analysis has led to the delineation of several but so far only three known human pathogenic *B. burgdorferi* species: *B. burgdorferi sensu stricto (ss)*, *B. afzelii*, and *B. garinii* (Baranton *et al.*, 1992). All *B. burgdorferi* strains isolated in North America belonged to *B. burgdorferi ss*, whereas all three species have been isolated in Europe. Besides genomic typing, *B. burgdorferi* strains may be grouped according to their reactivity to monoclonal antibodies to outer surface protein (Osp) (Wilske *et al.*, 1993). European isolates of all three genospecies exhibit a marked intraspecies antigenic heterogeneity, whereas North American strains are more homogeneous. This may

have important implications for clinical differences seen, the applicability of diagnostic tests, and vaccine efficacy (Wang *et al.*, 1999; Baranton *et al.*, 2001). In Europe genotyping of *B. burgdorferi* CSF isolates and of *B. burgdorferi* DNA amplified from CSF has shown a predominance of *B. garinii*, suggesting an organotropism of this species for the nervous system. *Borrelia afzelii*, on the other hand, is primarily associated with cutaneous manifestations (Busch *et al.*, 1996; Wang *et al.*, 1999; Lebech *et al.*, 1999).

The clinical spectrum of Lyme borreliosis includes dermatologic, neurologic, cardiac, and rheumatologic manifestations. Like syphilis the disease can be characterized by three stages: stage I (early localized infection), stage II (early disseminated infection), and stage III (late or chronic infection). During the natural course of the disease not all stages and clinical conditions become manifest. Some, e.g., erythema migrans, may go unnoticed; however, as important is that the natural course of the infection seems directed by different pathogenicty and organotropism of the *B. burgdorferi* species as well as by genetically determined host factors. Neurologic conditions may occur during all stages and are referred to as Lyme neuroborreliosis. Currently widely used diagnostic criteria for this condition in Europe are summarized in Table I.

### Stage I

Erythema migrans is a well demarcated either annular or homogeneous skin rash. It spreads in a centrifugal manner and develops usually 1–3 weeks, rarely up to 3 months, after a tick bite. In most patients it spontaneously resolves within a period of weeks to months. Concomitant flu-like constitutional symptoms including fatigue, headache, arthralgias, myalgias, and fever are

**TABLE I**  Diagnostic Clinical Criteria of Active Lyme Neuroborreliosis (LNB)

| Diagnosis of active neuroborreliosis | Neurological disorder | Obligatory laboratory/clinical findings |
|---|---|---|
| Confirmed | Early stage 2 LNB<br>   Painful lymphocytic meningoradiculitis<br>   Cranial neuritis especially: seventh, sixth, third<br>     (rarely other)<br>   Multiplex type radicular limb paresis<br>   Subacute meningitis: mainly children<br>   Meningoencephalitis<br>   Myelo-radiculitis<br>   Meningovascular involvement with lacunar infarction | Lymphocytic pleocytosis in CSF[a]<br>B. burgdorferi (Bb)-specific intrathecal IgM and or IgG synthesis[b]<br>  (begins in untreated patients in the 2nd week after onset and<br>  is detectable in ~75% by week 3 and in >99% by 8 weeks)<br>Negative blood VDRL test |
| | Chronic stage 3 LNB<br>   Disease duration >6 months<br>   Progressive encephalomyelitis<br>   Progressive myelopathy<br>   Chronic meningitis<br>   Meningovascular involvement with lacunar infarction | Lymphocytic pleocytosis in CSF[a]<br>B. burgdorferi-specific intrathecal IgG synthesis[b]<br>Negative blood VDRL test |
| |    ACA-associated distal axonal polyneuropathy | Concomittant acrodermatitis chronica atrophicans (ACA). In this<br>  condition there are no CSF changes |
| Probable | Early stage 2 LNB<br>   As specified above | Lymphocytic pleocytosis in CSF[a]<br>Disease duration <8 weeks otherwise Bb-specific CSF antibodies<br>  will be present[b]<br>Positive B. burgdorferi-specific IgG/IgM serum antibody test and<br>  especially a significant IgM or IgG titer fall/increase on<br>  consecutive sera are supportive<br>History of erythema migrans within 4–6 months prior to onset of<br>  neurological symptoms is supportive;<br>Exclusion of other causes |
| | Chronic stage 3 LNB<br>   This diagnosis will always be able to be confirmed | |

[a]Usually associated with elevated total protein content, elevated CSF/serum albumin ratio, and the presence of oligoclonal IgG bands in the CSF.
[b]Intrathecally produced specific antibodies are determined by investigating simultaneously drawn samples of cerebrospinal fluid (CSF) and serum comparing the relative amount of Bb-specific IgM/IgG in CSF versus serum expressed as an index or ratio. The finding of a positive Bb-specific intrathecal antibody synthesis without CSF pleocytosis indicates previous LNB.

significantly more frequent in North America than in Europe. Cerebrospinal fluid (CSF) examination during stage I does not show inflammatory changes.

### Stage II

Approximately 5–10% of persons infected with *B. burgdorferi* will if not treated with antibiotics develop neurologic manifestations within 2 to 6 weeks after a tick bite (range, 1 to 18 weeks) (Hansen and Lebech, 1992). Of these individuals 25 and 50% can recall a tick bite or an erythema migrans, respectively. In patients admitted with neuroborreliosis 10–20% still present remnants of erythema migrans. The most common neurologic manifestation (60–80%) during stage II in European patients is lymphocytic meningoradiculitis (Bannwarth syndrome) Hansen and Lebech, 1992; Oschmann *et al.*, 1998). The hallmark of Bannwarth syndrome is the triad of radicular pain, peripheral paresis, especially of the facial nerve, and concomitant lymphocytic CSF pleocytosis (10–1000 cells/µl). Despite of CSF inflammation signs of meningism is usually lacking. Motor and sensory signs follow a multiplex distribution. 25% of the patients only present with painful radiculitis without focal motor signs. Other neurologic manifestations especially in children and adolescents during stage II are lymphocytic meningitis with headache and moderate meningismus (Christen *et al.*, 1994). Less than 5% of the patients reveal subtle myelitis and even less encephalitis. In European patients with neuroborreliosis concomitant extra neural manifestations of stage II (myocarditis, arthritis) are rare <2% (Hansen and Lebech, 1992).

### Stage III

Late or chronic neuroborreliosis constitutes 1–2% of all neuroborreliosis cases and is defined by continuous disease and CSF inflammation of more than 6 months duration (Hansen and Lebech, 1992). Chronic neuroborreliosis may present as chronic meningitis, as a progressive encephalomyelitis, a slowly progressive myelopathy or as in neurosyphilis with meningovascular involvement leading to mainly lacunar infarction (Ackermann *et al.*, 1985; Kohler *et al.*, 1988; Krüger *et al.*, 1991). Most of the patients suffer significant general symptoms such as weight loss, headache, and malaise. All patients show CSF inflammation and *B. burgdorferi*-specific intrathecal IgG production. Chronic neuroborreliosis as reported in Europe seems very rare in North America (Logigian *et al.*, 1990).

A chronic distal and axonal polyneuropathy may be associated with the characteristic late cutaneous *B. burgdorferi* manifestation acrodermatitis chronica athrophicans (ACA). It was studied by Hopf *et al.* (1975), and later, after the discovery of *B. burgdorferi*, by Kristoferitsch *et al.* (1987) and Kindstrand *et al.* (1997). The ACA neuropathy is mild, predominantly sensory, and it usually follows the extension of the skin lesions. In North America a mild distal neuropathy without other Lyme-borreliosis-defining manifestations and without CSF changes has been reported (Halperin *et al.*, 1990; Logigian *et al.*, 1992). The ethiological association to *B. burgdorferi*, however, remains controversial and the condition is not a working diagnosis in Europe. Similarly entities, such as "post-Lyme-disease syndrome" and Lyme encephalopathy, are often referred to as chronic neuroborreliosis (Halperin *et al.*, 1990; Kaplan *et al.*, 1992). However, also in these conditions the causation by *B. burgdorferi* remains to be proven (Seltzer *et al.*, 2000) and a recent controlled treatment trial failed to demonstrate antibiotic efficacy in these conditions (Klempner *et al.*, 2001).

## Differential Diagnoses

Differential diagnoses of Lyme neuroborreliosis with the predominant syndrome of meningitis include viral meningitis, carcinomatous meningitis, neurosarcoidosis, fungal meningitis, tuberculous meningitis, Mollaret meningitis, and other spirochetal infections such as syphilis. Peripheral facial palsy requires the diagnostic differentiation from Guillain-Barré syndrome, Miller-Fisher syndrome, and especially idiopathic facial palsy (Bell's palsy). The intense radicular pain of Bannwarth's syndrome often leads to initial suspicion of a herniated disk, brachial or lumbar plexopathy, herpes zoster radiculitis, or diseases of the spine. Chronic neuroborreliosis may be confused with primary progressive spinal multiple sclerosis (MS), neurosarcoidosis, and stroke-like syndromes. The detection of a specific intrathecal antibody production against *B. burgdorferi* is crucial for differentiation from these disorders. There is no association of neuroborreliosis and MS. The MRI in chronic neuroborreliosis may show meningeal contrast enhance-

ment and lesions compatible with ischemia, but does not show for MS characteristic periventricular white matter lesions or lesions in the corpus callosum.

## Microbiological Diagnosis

Isolation of the spirochete from CSF in patients with neuroborreliosis is rarely successful (<10%) (Karlsson *et al.*, 1990). It requires a difficult, time consuming culture technique and is thus not useful for routine diagnosis. Assays for *B. burgdorferi* antigen detection in CSF have been reported (Coyle *et al.*, 1990), but their diagnostic performance was never confirmed. The introduction of polymerase chain reaction (PCR) did neither live up to the expectations. Specific detection of *B. burgdorferi* DNA in CSF is possible, but the yield was low, ~20–25% according to reports with comparable study populations (Lebech *et al.*, 1992 and 1999; Amouriaux *et al.*, 1993; Christen *et al.*, 1995; Nocton *et al.*, 1996) and thus considerably lower than measurement of the *B. burgdorferi*-specific intrathecal antibody synthesis (Hansen and Lebech, 1991). PCR results reported in the literature vary greatly; and no comparative studies with parallel examination of samples in different laboratories have been performed. A variety of different protocols regarding target genes, primer pairs, extraction procedures, sample volumes, and detection methods have been used. The most likely explanations for the different PCR results on CSF are differences in the selection of patients, PCR methodology, and extremely low levels of organisms present in the CSF. At the moment no general recommendations can be given with respect to optimal sample preparation, target genes, or PCR and detection methods. PCR may be useful in the early phase of neuroborreliosis (<2 weeks) when antibody tests often are negative and PCR as culture is more likely to be positive than later (Lebech *et al.*, 1999).

At present, serological methods are the primary laboratory tools for diagnosis of Lyme borreliosis. The serological tests most often used are enzyme-linked immunosorbent assay (ELISA), indirect immunofluorescent assay (IFA), and Western blotting (WB) (Steere, 2001). The frequency of positive serologic results depends on the disease duration and the type of clinical manifestation. In stage I, 20 to 50% of the patients have developed antibody titers against *B. burgdorferi*, predominantly IgM-antibodies. In stage II the number of seropositive individuals increases from 60 to 100% as all will be seropositive by 10–12 weeks after onset of neurological symptoms. In stage III all patients have a positive IgG blood test (Hansen, 1994). The presence of elevated titers of serum IgG *B. burgdorferi* antibodies is indicative only of previous exposure and it is important to recognize that *B. burgdorferi* IgG antibodies are frequently detectable in asymptomatic individuals. The mere presence of a *B. burgdorferi* antibody response does not establish the presence of active infection nor that coexisting neurological signs and symptoms are due to LNB. In Munich 3.7% of more than 3000 consecutive, unselected neurological patients had elevated serum IgG and/or IgM antibodies against *B. burgdorferi*. This figure reflects the local seroprevalence and is expected considering the usual 98% specificity cutoff level to which antibody assays ought to be adjusted. The increased seroprevalence in highly exposed persons and the lack of standardization of Lyme serology in general adds to the difficulties in serodiagnosis Lyme borreliosis. No antibody test including WB is able to discriminate between active infection or past exposure (Blaauw *et al.*, 1999). Due to these limitations, the very high test frequency nowadays (~15–30 tests/1000/year) where tests are ordered without a high a priory probability of Lyme borreliosis, serology in general has a very low predictive value of a positive test, which probably is less than 20%. Considering the incidence rate of true Lyme borreliosis and the mentioned limitations of the diagnostic methods the number of tests done currently is too high.

Due to the possibility of *B. burgdorferi*-specific antibody detection in CSF the situation for the neurologist is highly advantageous regarding other manifestations of Lyme borreliosis. A *B. burgdorferi*-specific intrathecal antibody response is very rarely found by coincidence. A positive CSF test with concomitant CSF inflammation has a predictive value of >95% (Hansen, 1994). These two CSF parameters are currently the best laboratory evidence of active neuroborreliosis. The *B. burgdorferi*-specific intrathecal antibody response becomes detectable during the second week after onset of neurological symptoms and is present in all untreated neuroborreliosis patients by 2 months.

The method for measuring agent-specific antibodies in CSF must allow to determine the specific intrathecal antibody synthesis, thus discriminating leakage of specific serum antibodies through an inflamed blood–CSF barrier. Several safe techniques have been reported (Stiernstedt *et al.*, 1985; Wilske *et al.*, 1986; Hansen and Lebech, 1991; Kaiser and Lucking, 1993). They may differ in labor intensiveness, their ability to detect *B. burgdorferi*-specific IgG and IgM separately, sample volume needed, and the expression of the test result; however, essentially they perform similar. In early stage II neuroborreliosis, measuring *B. burgdorferi*-specific IgM production is as important as IgG measurement. As in neurosyphilis, the intrathecal anti-*B. burgdorferi* antibody response may, after recovery/therapy, persist for many years and does not, without concomitant CSF inflammation, indicate persistence of the spirochete inside the CNS (Krüger *et al.*, 1989; Hammers-Berggren *et al.*, 1993).

Western blotting, which has been recommended as a confirmatory serological test, is often difficult to interpret, due to the high number of proteins in the spirochete (>100), several being cross-reactive, several having similar molecular weights, thus leading to an electrophoretical comigration. Furthermore, the enormous strain-dependent variability in antigen band pattern has at least in Europe made the use of a confirmatory WB problematic (Robertson *et al.*, 2000). In North America and in some laboratories in Germany a confirmatory WB is widely used in the so-called two-step approach (Nichol *et al.*, 1998). The usefulness of this approach in Europe has, however, been questioned (Blaauw *et al.*, 1999). The diagnostic sensitivity and specificity of ELISAs have been significantly improved by using purified immunodominant native or recombinant antigens (Hansen *et al.*, 1988; Wilske *et al.*, 1998; Mathiesen *et al.*, 1998; Kaiser and Rauer, 1999).

### Natural History

Conditions now attributed to *B. burgdorferi* have been described in the European literature since 1883 and in North America erythema migrans was first reported in 1969. Lyme borreliosis is thus not a new disease and has most likely not been recognized as one etiological entity earlier due to its benign character and usually self-limiting course.

The endemicity depends on a complex ecological interaction comprising the climate, the presence of Ixodes ticks, and the reservoir hosts, especially small rodents and deer (Steere, 2001). Man is an incidental host in this zoonosis and is seemingly also one of the few species to which *B. burgdorferi* is pathogenic. Tick infection rates vary considerably from 0 to 40%, depending on the region, the season, and the developmental stages of the Ixodes ticks. *Ixodes* larvae are practically never infected due to lack of vertical or transovarial transmission. Nymphal infection rates of 10–15% are reported from several endemic areas and are usually considerably lower than in adult ticks. However, due to their abundance, nymphal ticks are the primary cause of the human infections.

The season-dependent activity of Ixodes ticks explains that *Borrelia* infections including neuroborreliosis show a typical seasonal incidence and onset between June and December. Chronic neuroborreliosis may, however, be observed throughout the year.

Population-based incidence figures for neuroborreliosis are only available from Europe. Early neuroborreliosis according to the case definitions in Table I has an incidence of 30–40 cases/$10^6$/year, whereas chronic neuroborreliosis is rare (<0.4/$10^6$/year; Hansen and Lebech, 1992; Christen *et al.*, 1994; Berglund *et al.*,

1995). The natural course of a *B. burgdorferi* infection is well described in the European literature where erythema migrans and lymphocytic meningoradiculitis (Bannwarth's syndrome) were well recognized conditions for many decades, although the etiology and thus antibiotic therapy was unknown. Among individuals with untreated erythema migrans, 5–10% develop neuroborreliosis within 2–4 weeks. The favorable self-limiting outcome without evidence of chronic disease activity in large series of patients with untreated lymphocytic meningoradiculitis after tick bite was reported in the 1960s through the 1980s (Krüger *et al.*, 1989). Permanent neurological deficits like a residual facial palsy should not to be confused with active disease. Although neuroborreliosis may be disabling, fatality due to neuroborreliosis is extremely rare.

Probably due to the species-related organotropism and to genetic host factors only few patients with neuroborreliosis (<2%) show concomitant signs of other organ manifestations as myocarditis and arthritis (Hansen and Lebech, 1992).

Ixodes ticks may carry other pathogens such as *Babesia*, *Ehrlichia*, and virus, e.g., TBE (tickborn encephalitis virus). Coinfections of *Borrelia* and TBE are rare but do occur (Kaiser *et al.*, 1997). Coinfections, especially with *Ehrlichia*, have received considerable attention in recent years; however, evidence was mainly by serology (Fingerle *et al.*, 1997; Lebech *et al.*, 1998). There are no studies reporting on the frequency and clinical significance of this coinfection in terms of comorbidity and different clinical presentation.

## Principles of Therapy

### Pathogenesis

*B. garinii* seems especially associated with neuroborreliosis. However, so far no molecular characterization of the organotrpism is reported. First symptoms of neuroborreliosis occur 2–4 weeks after erythema migrans. The spread to the nervous system may be hematogenous but a transneural spread from the site of entry along the peripheral nerves to the nerve roots, meninges and subarachnoid space is likely. Evidence for the latter is the significant topographical association of the location of a preceding erythema migrans or tick bite and the site of first neurological sign (radicular pain and paresis) (Kristoferitsch, 1991; Hansen and Lebech, 1992). The entire genome of *B. burgdorferi* has now been sequenced (Fraser *et al.*, 1997). It contains no homologous of systems that specify in the secretion of toxins. Known virulence factors of *B. burgdorferi* are the unique outer surface lipoproteins, which, due to their variable expression, help the spirochete evade the host immune response and

seem involved in the attachment to mammalian cells. The spirochetal spread through tissue is facilitated by the surface binding of human plasminogen and its activators (Coleman *et al.*, 1997). Surface proteins do facilitate the attachment to endothelium and extracellular matrix proteins (Guo *et al.*, 1998). *B. burgdorferi* is not an intracellular pathogen, an important fact also regarding accessibility to antibiotic agents. The histopathological tissue reaction comprises a lymphocytic infiltrate with mainly T but also B cells, plasma cells and macrophages. Recently it was reported that T cells from ACA lesions produced a restricted cytokine profile with low γINF level (Mullegger *et al.*, 2000). As γINF acts proinflammmatory this could explain the chronicity of this disorder. *B. burgdorferi* is primarily a human pathogen. Animal models have always required immunomodulation. The only animal model for neuroborreliosis was established in monkeys (Cadavid *et al.*, 2000; Pachner *et al.*, 2001). CSF showed pleocytosis, *B. burgdorferi* infiltrated the leptomeninges, motor-sensory nerve roots but not the brain parenchyma. Human autopsy studies are sparse (Kuntzer *et al.*, 1991), but are regarding the location of inflammatory changes in alignment with the animal studies. Autopsy reports have furthermore confirmed the meningovascular involvement in neuroborreliosis. These findings correspond to clinical experiences of a lymphocytic meningoradiculitis, that meningovascular disease occurs, and that encephalitis is not a feature of neuroborreliosis. Peripheral nerves may be involved, especially when associated with ACA (Hopf, 1975; Kristoferitsch *et al.*, 1988; Kindstrand *et al.*, 1997). Sural nerve biopsies in patients with ACA and in a few with lymphocytic meningoradiculitis (Meier, 1988) have shown perivascular lymphocytic infiltration and vessel obliteration/thrombosis in the perineurium. Signs of necrotizing vasculitis, however, were not reported. In accordance herewith the neuropathy in neuroborreliosis is axonal.

The existence of a "post-Lyme-disease syndrome" (late Lyme encephalopathy, chronic fatigue, and fibromyalgia-like syndromes) is controversial (Seltzer *et al.*, 2000). A recent antibiotic trial in such patients failed to improve the condition (Klemperer *et al.*, 2001). Thus the possibility of a diagnostic failure rather then a therapeutic failure, as often presumed, should be emphasized.

### Therapy

The recommendations of antibiotic therapy are summarized in Table II. A number of *in vitro* susceptibility studies have been reported. However, due to the culture conditions of *B. burgdorferi* (5–7 days at 35°C) and differences in the stability of antibiotic agents under

**TABLE II**  Therapy of Lyme-Borreliosis

| | | |
|---|---|---|
| Stage I early localized infection | | |
| erythema migrans | | |
|   Doxycycline orally | $1 \times 100$ mg/day orally | 14 Days |
|   Penicillin V orally | $3 \times 1$ million units/day | 14 Days |
|   Amoxicillin orally | $3 \times 500$ mg/day | 14 Days |
| Stage II early disseminated infection | | |
| including neuroborreliosis | | |
|   Penicillin G IV | $4 \times 5 \times 10^6$ units/day | 14 Days |
|   Ceftriaxone IV[a] | $1 \times 2$ G/day | 14 Days |
|   Doxycycline orally | Day 1, $2 \times 200$ mg/day | 14 Days |
| | Days 2–14, $2 \times 100$ mg/day | |
| Stage III late infection including chronic | | |
| neuroborreliosis and ACA neuropathy[b] | | |
|   Penicillin G IV | $4 \times 5 \times 10^6$ units/day | 14 Days[c] |
|   Ceftriaxone IV[a] | $1 \times 2$ g/day | 14 Days[c] |
|   Doxycycline orally | Day 1, $2 \times 200$ mg/day | 14 Days |
| | Days 2–14, $2 \times 100$ mg/day | |

[a]Or other third-generation cephalosporin.
[b]There are no controlled studies on the antibiotic efficacy in patients with chronic neuroborreliosis or ACA neuropathy.
[c]Some authors recommend 3–4 weeks of antibiotic therapy; however, no trial evidence favors such a recommendation.

these conditions MIC or MBC values differ and do not correspond to clinical experience. They are not always reflecting the true susceptibility of the organism. This has led to considerable misinterpretations. MBC values for penicillin were for example considerably higher than for ceftriaxone, a finding easily explained by the instability and fast degradation of penicillin during culture. Penicillin resistance of *B. burgdorferi* was claimed but never shown. Antibiotics as sulfamethizol and aminoglycosides are ineffective and are even used in *B. burgdorferi* culture tubes to avoid contamination. The principle drugs recommended against Lyme borreliosis are β-lactams and tetracyclines. Macrolides, although effective *in vitro*, seem less effective *in vivo* and may lead to treatment failures (Hansen *et al.*, 1992; Luft *et al.*, 1996).

Whereas early local disease is treated orally, neuroborreliosis, except for oral doxycycline, needs IV therapy. Early studies in patients with neuroborreliosis either uncontrolled or with historic control groups showed the efficacy of high-dose IV penicillin (Sköldenberg *et al.*, 1983; Steere *et al.*, 1983). This therapy led to resolution of general symptoms, radicular pains, and headaches; however, there was no difference in the resolution of motor deficits like facial palsy.

Prospective randomized trials in patients with neuroborreliosis have later confirmed this finding and furthermore shown the equivalence of penicillin, ceftriaxone and doxycycline (Pfister *et al.*, 1989, 1991; Kohlhepp *et al.*, 1989; Karlsson *et al.*, 1994, 1996). Due to the lipophilicity of doxycycline and thus a high penetration of the blood–CSF barrier and the blood–brain

barrier it is presently the only efficient oral drug for neuroborreliosis as shown by Karlsson *et al.*, (1994, 1996) in a prospective randomized comparison with high-dose IV penicillin. Thus penicillin, ceftriaxone and doxycycline should be regarded as equivalent in efficacy.

The evaluation of different treatment regimens for neuroborreliosis is, however, hampered by the generally favorable outcome even without therapy. Minor differences in efficacy may thus require much higher numbers of patients to be enrolled. However, from our own and published experience over the past 18 years, treatment failures in neuroborreliosis are if at all very rare. Oral doxycycline or IV ceftriaxone given once a day is, however, advantageous versus IV penicillin as patients may be treated on an outpatient basis. Due to the risks of dental discoloration doxycycline as other tetracyclines should not be used in children under 8–10 years.

Additional steroid treatment may accelerate relief of radicular pain, but is not recommended in the treatment of neuroborreliosis.

No trials have addressed a possible difference in efficacy in early and late neuroborreliosis. However, based on the literature and our experience, patients with chronic neuroborreliosis respond clinically and regarding CSF inflammation to the same regimens.

Treatment duration of 10–14 days was derived empirically. Several therapy recommendations in reviews or textbooks do recommend treatment courses up to 4 weeks or even longer. However, no conclusive drug trial has ever shown that 3–4 weeks of relevant therapy was superior to 14 days. Thus, there is no evidence for longer treatment courses.

## Monitoring Therapeutic Efficacy

The experience with neuroborreliosis over the past two decades has shown that treatment failures are very rare. Thus, the earlier recommendation of repeated CSF examination is relative. Antibiotic treatment leads to a more rapid normalization of CSF inflammation.

Oligoclonal IgG bands and *Borrelia*-specific intrathecal antibody synthesis may as in neurosyphilis persist for years and should thus not be considered evidence of active disease in treated patients. Similarly, consecutive measurement of serum antibody to *B. burgdorferi* after therapy is not useful for monitoring treatment efficacy.

The most reliable indicator of treatment response is the resolution of neurological symptoms and normalization of CSF pleocytosis. In stage II neuroborreliosis CSF pleocytosis will disappear gradually by 1–3 months, whereas in chronic neuroborreliosis this may take more than 6 months.

A Jarish Herxheimer reaction may occur within the first day after therapy onset. It is, however, compared to neurosyphilis, rare and mild (Hansen and Lebech, 1992).

## Practical Management

### Diagnosis

Neuroborreliosis is primarily a clinical diagnosis and should follow the guidelines summarized in Table I. Neuroborreliosis presents in most patients with a typical pattern (time course, constellation of signs, and typical CSF abnormalities). Atypical cases are rare.

CSF examination is obligatory in every patient suspected of neuroborreliosis. A normal CSF questions the diagnosis of active neuroborreliosis. In early neuroborreliosis, *B. burgdorferi* antibody test in serum may still be negative. A positive blood test, however, only indicates previous or recent exposure to *B. burgdorferi* and is thus of no help. The best indicator of active neuroborreliosis is CSF inflammation with lymphocytic pleocytosis and a *B. burgdorferi*-specific intrathecal antibody response. The latter is usually detectable by 2–3 weeks after onset of neurological symptoms and latest by 8 weeks. A normal CSF and no *B. burgdorferi*-specific intrathecal antibody production in patients with disease durations of more than 2 months highly questions the diagnosis of neuroborreliosis except for a peripheral neuropathy in association with ACA.

### Therapy

Active neuroborreliosis is a treatable condition. Patients should receive a 14-day course of antibiotic therapy following the recommendations in Table II. There is no place for steroids treating NB.

### Follow-Up

Treatment failures are if ever rare. Suspicion of relapses should lead to a repeated CSF examination for persistence of CSF pleocytosis. Antibody measurement in serum and CSF is not suitable for follow-up due to frequent antibody persistence.

### Prevention

The most important measure is awareness of the possibility to contract a tick bite in endemic areas, self-inspection following outdoor activities and to remove infested ticks as early as possible. Early removal of ticks may prevent transmission. Experimental studies indicate that it takes 12–24 h to establish a *B. burgdorferi* infection (Piesman *et al.*, 1987). Prophylactic antibiotic therapy after tick bite has been studied, is controversial (Shapiro, 2001), and is in our opinion not recommended.

Recently a Lyme disease vaccine has been developed, approved, and since 1999 marketed in North America. The vaccine is based on recombinant OspA in adjuvant. After three injections the vaccine yielded a 76% protection considering erythema migrans. Prevention of neuroborreliosis should be expected but was not evidenced in the vaccine trials (Sigal *et al.*, 1998; Steere *et al.*, 1998), where ~10,000 individuals received active vaccine. The vaccine-induced immune response wanes quickly and booster injections are needed every 1–3 years to maintain protection. Due to the considerable genomic heterogeneity of *B. burgdorferi* strains in Europe the cross-protection of the vaccine is low and insufficient. The vaccine is thus not recommended in Europe.

## LEPTOSPIROSIS

### Clinical Aspects and Natural Course

Leptospirosis is a worldwide distributed zoonosis caused by the pathogenic species *Leptospira interrogans*, the most common serotypes being *L. icterohaemorrhagiae* and *L. canicola*. Humans acquire the disease by contact with infected urine or soil, water, or other fomites contaminated with infected urine. Reservoirs of *Leptospira* are wild (rats, mice) and domestic (dogs, cattle, and swine) animals. An occupational risk arises among farmers and agricultural workers when cattle and other livestock become infected (Ferguson, 1993). The incidence of clinical manifestations peaks from July to October. Mild anicteric febrile disease is the most common manifestation of leptospirosis and accounts for more than 90% of cases. The incubation period is 2 to 26 days. The disease typically has a biphasic course. In the first leptospiremic phase, which lasts about 4 to 7 days, unspecific flu-like symptoms including headache, gastrointestinal symptoms, and myalgias are frequently reported. A characteristic finding on clinical examination is conjunctival suffusion which is observed in 30 to 85% of the patients. After an afebrile interval of 1 to 3 days, the second stage of the disease (immune or leptospiruric phase) follows. A common complication during this stage is lymphocytic meningitis with pleocytosis (up to 500 cells/µL) and moderate elevation of total protein content in the CSF (Coursin et al., 2000). Inflammatory CSF changes are detected in 70 to 90% of the patients in whom lumbar puncture is performed. However, only about half of the patients are clinically symptomatic. On rare occations during this second stage, encephalitis, myelitis, polyneuritis, uveitis, or endocarditis may develop. Weil's disease denotes severe icteric forms of the disease and accounts for about 5 to 10% of leptospirosis cases; renal insufficiency, thrombopenia, and capillary leakage syndrome may occur (Vieira et al., 1999; Coursin et al., 2000).

Diagnosis can be established by the detection of leptospira (e.g., by PCR, direct immunofluorescence, culture) from blood and CSF (during leptospiremic phase) and from the urine (during immune phase). Nevertheless, diagnosis is usually based on serology. The serological test most often used is the microscopic agglutination test.

Most patients with leptospirosis recover well within 2 to 4 weeks even if they have severe disease. Clinical relapses are very rare. Death among patients with leptospirosis in the United States range from about 2 to 10%. Mortality usually results from renal (acute tubular necrosis) rather than hepatic failure. Predictors for an unfavorable clinical course are age over 60 years and severe disease with jaundice.

## Therapy

Doxycycline (200 mg/day IV or orally) is recommended for leptospirosis. Alternatively, penicillin G (4 × 5 mg Mega/day IV) or third-generation cephalosporins (e.g., ceftriaxone 1 × 2 g/day IV or cefotaxime 3 × 2 g/day IV) may be used. The recommended treatment duration is about 7 to 10 days. Several hours after onset of antibiotic therapy, a Jarisch–Herxheimer reaction consisting of fever, chills, headache and myalgia may occur. It is thought that the course of the disease can be positively influenced by antibiotics provided that antibiotic therapy is started within the first 4 days after disease.

# RELAPSING FEVER

## Clinical Features and Natural Course

Epidemic louse-borne relapsing fever (etiologic agent B. recurrentis) is mainly observed in Central and East Africa (e.g., Ethiopia) and occasionally in Asia and South America (Sundnes and Haimanot, 1993). The pathogens are tansmitted by the human body louse Pediculus humanus. Endemic tick-borne relapsing fever is caused by several different borreliae species (e.g., B. duttoni). Endemic relapsing fever is transmitted by various species of Ornithodoros ticks (Spach et al., 1993; Colebunders et al., 1993). Cases of endemic relapsing fever have been reported from North America, Africa, Europe, Asia and South America; in the United States, endemic tick-borne relapsing fever has been reported almost exclusively west of the Mississippi River, in particular the mountainous West (B. hermsii) and the Southwest (B. turicatae) (Cadavid and Barbour, 1998).

After an incubation period of about 7 days (range 4 to 18) the disease begins abruptly with high fever (39–40°C), headache, chills, somnolence, photophobia, myalgia, arthralgia, and malaise. Facultative concomitant signs are rash, hepatosplenomegalia, lymphadenopathy, jaundice, and iridocyclitis. Neurological manifestations include meningitis, neuritis, and rarely encephalitis or radiculomyelitis and are seen in 10–30% of cases. CSF shows mild to moderate lymphocytic pleocytosis (20–700 cells/µL) and elevation of total protein content. Borreliae may be isolated from the CSF in about 10% of patients. Neurological complications are more common in endemic tick-borne relapsing fever than in epidemic louse-borne relapsing fever (Cadavid and Barbour, 1998). Diagnosis is usually made during fever attack by the detection of borreliae on thick and thin blood smears (Giemsa-stain, darkfield microscopy).

The relapsing fever diseases are characterized by recurrent fever attacks. The first fever attack ends abruptly after 3 to 6 days with the risk of hypotension and shock. After an afebrile interval of about 8 days (range 3 to 36) recurrent fever occurs. Duration and intensity of symptoms decrease with each relapse. Without treatment louse-borne relapsing fever has typically 1 to 2 relapses (up to 5), while tick-borne relapsing fever is characterized by 3 to 5 relapses (up to 13).

In 95% of treated patients, relapsing fever resolves completely. Mortality of untreated patients with louse-borne relapsing fever is about 40% and with tick-borne relapsing fever about 5%. The most frequent causes of death are myocarditis, shock, and hepatic failure.

Differential diagnoses include malaria, typhus, brucellosis, 5-days-fever, leptospirosis and dengue-fever.

## Therapy

Tetracyclines are usually recommended for the treatment of relapsing fever (e.g., 200 mg doxycycline/day orally for 10 days), although rare failures have been reported (Liles and Spach, 1993; Estanislao and Pachner, 1999). Alternatively, ceftriaxone (1 × 2 g/day IV) or erythromycin (4 × 500 mg/day IV) can be used. Within several hours after initiation of antibiotic therapy, a Jarisch–Herxheimer reaction (fever, chills, tachycardia, hypotension, rarely intensification of skin lesions) occurs in up to one-third of patients (Seboxa and Rahlenbeck, 1995). The pathophysiological mechanisms of Jarisch–Herxheimer reaction seem to be related to cytokine relase or endogeneous opioids (Negussie *et al.*, 1992). Strategies to diminish Jarisch–Herxheimer reaction (e.g., low initial antibiotic dose, administration of antiinflammatory agents) were inconclusively successful.

# REFERENCES

Ackermann, R., Gollmer, E., and Rehse-Kupper, B. (1985). [Progressive Borrelia encephalomyelitis. Chronic manifestation of erythema chronicum migrans disease of the nervous system]. *Dtsch. Med. Wochenschr.* **110**, 1039–1042.

Amouriaux, P., Assous, M., Margarita, D., Baranton, G., and Saint, G. I. (1993). Polymerase chain reaction with the 30-kb circular plasmid of Borrelia burgdorferi B31 as a target for detection of the Lyme borreliosis agents in cerebrospinal fluid. *Res. Microbiol.* **144**, 211–219.

Baranton, G., Postic, D., Saint, G. I., Boerlin, P., Piffaretti, J. C., Assous, M., and Grimont, P. A. (1992). Delineation of Borrelia burgdorferi sensu stricto, Borrelia garinii sp. nov., and group VS461 associated with Lyme borreliosis. *Int. J. Syst. Bacteriol.* **42**, 378–383.

Baranton, G., Seinost, G., Theodore, G., Postic, D., and Dykhuizen, D. (2001). Distinct levels of genetic diversity of Borrelia burgdorferi are associated with different aspects of pathogenicity. *Res. Microbiol.* **152**, 149–156.

Berglund, J., Eitrem, R., Ornstein, K., Lindberg, A., Ringer, A., Elmrud, H., Carlsson, M., Runehagen, A., Svanborg, C., and Norrby, R. (1995). An epidemiologic study of Lyme disease in southern Sweden. *N. Engl. J. Med.* **333**, 1319–1327.

Blaauw, A. A., Van Loon, A. M., Schellekens, J. F., and Bijlsma, J. W. (1999). Clinical evaluation of guidelines and two-test approach for lyme disease. *Rheumatology. (Oxford)* **38**, 1121–1126.

Burgdorfer, W., Barbour, A. G., Hayes, S. F., Benach, J. L., Grunwaldt, E., and Davis, J. P. (1982). Lyme disease-a tick-borne spirochetosis? *Science* **216**, 1317–1319.

Busch, U., Hizo-Teufel, C., Boehmer, R., Fingerle, V., Nitschko, H., Wilske, B., and Preac-Mursic, V. (1996). Three species of Borrelia burgdorferi sensu lato (B. burgdorferi sensu stricto, B afzelii, and B. garinii) identified from cerebrospinal fluid isolates by pulsed-field gel electrophoresis and PCR. *J. Clin. Microbiol.* **34**, 1072–1078.

Cadavid, D., and Barbour, A. G. (1998). Neuroborreliosis during relapsing fever: review of the clinical manifestations, pathology, and treatment of infections in humans and experimental animals. *Clin. Infect. Dis.* **26**, 151–164.

Cadavid, D., O'neill, T., Schaefer, H., and Pachner, A. R. (2000). Localization of Borrelia burgdorferi in the nervous system and other organs in a nonhuman primate model of lyme disease. *Lab. Invest.* **80**, 1043–1054.

Christen, H. J., Eiffert, H., Ohlenbusch, A., and Hanefeld, F. (1995). Evaluation of the polymerase chain reaction for the detection of Borrelia burgdorferi in cerebrospinal fluid of children with acute peripheral facial palsy. *Eur. J. Pediatr.* **154**, 374–377.

Christen, H. J., Hanefeld, F., Eiffert, H., and Thomssen, R. (1993). Epidemiology and clinical manifestations of Lyme borreliosis in childhood. A prospective multicentre study with special regard to neuroborreliosis. *Acta Paediatr. Suppl.* **386**, 1–75.

Colebunders, R., De Serrano, P., Van Gompel, A., Wynants, H., Blot, K., Van Den, E. E., and Van Den, E. J. (1993). Imported relapsing fever in European tourists. *Scand. J. Infect. Dis.* **25**, 533–536.

Coleman, J. L., Gebbia, J. A., Piesman, J., Degen, J. L., Bugge, T. H., and Benach, J. L. (1997). Plasminogen is required for efficient dissemination of B. burgdorferi in ticks and for enhancement of spirochetemia in mice. *Cell* **89**, 1111–1119.

Coursin, D. B., Updike, S. J., and Maki, D. G. (2000). Massive rhabdomyolysis and multiple organ dysfunction syndrome caused by leptospirosis. *Intens. Care Med.* **26**, 808–812.

Coyle, P. K., Schutzer, S. E., Belman, A. L., Krupp, L. B., and Golightly, M. G. (1990). Cerebrospinal fluid immune complexes in patients exposed to Borrelia burgdorferi: Detection of Borrelia-specific and -nonspecific complexes. *Ann. Neurol.* **28**, 739–744.

Estanislao, L. B., and Pachner, A. R. (1999). Spirochetal infection of the nervous system. *Neurol. Clin.* **17**, 783–800.

Ferguson, I. R. (1993). Leptospirosis surveillance: 1990–1992. *Commun. Dis. Rep. CDR Rev.* **3**, R47–R48.

Fingerle, V., Goodman, J. L., Johnson, R. C., Kurtti, T. J., Munderloh, U. G., and Wilske, B. (1997). Human granulocytic ehrlichiosis in southern Germany: Increased seroprevalence in high-risk groups. *J. Clin. Microbiol.* **35**, 3244–3247.

Fraser, C. M., Casjens, S., Huang, W. M., Sutton, G. G., Clayton, R., Lathigra, R., White, O., Ketchum, K. A., Dodson, R., Hickey, E. K., Gwinn, M., Dougherty, B., Tomb, J. F., Fleischmann, R. D., Richardson, D., Peterson, J., Kerlavage, A. R., Quackenbush, J., Salzberg, S., Hanson, M., van Vugt, R., Palmer, N., Adams, M. D., Gocayne, J., Weidman, J., Utterback, T., Watthey, L., McDonald, L., Artiach, P., Bowman, C., Garland, S., Fujii, C., Cotton, M. D., Horst, K., Roberts, K., Hatch, B., Smith, H. O., and Venter, J. C. (1997). Genomic sequence of a Lyme disease spirochaete, Borrelia burgdorferi. *Nature* **390**, 580–586.

Guo, B. P., Brown, E. L., Dorward, D. W., Rosenberg, L. C., and Hook, M. (1998). Decorin-binding adhesins from Borrelia burgdorferi. *Mol. Microbiol.* **30**, 711–723.

Halperin, J., Luft, B. J., Volkman, D. J., and Dattwyler, R. J. (1990). Lyme neuroborreliosis. Peripheral nervous system manifestations. *Brain* **113 (Pt. 4)**, 1207–1221.

Halperin, J. J., Krupp, L. B., Golightly, M. G., and Volkman, D. J. (1990). Lyme borreliosis-associated encephalopathy. *Neurology* **40**, 1340–1343.

Hammers-Berggren, S., Hansen, K., Lebech, A. M., and Karlsson, M. (1993). Borrelia burgdorferi-specific intrathecal antibody production in neuroborreliosis: A follow-up study. *Neurology* **43**, 169–175.

Hansen, K. (1994). Lyme neuroborreliosis: Improvements of the laboratory diagnosis and a survey of epidemiological and clinical features in Denmark 1985–1990. *Acta Neurol. Scand. Suppl* **151**, 1–44.

Hansen, K., Hindersson, P., and Pedersen, N. S. (1988). Measurement of antibodies to the Borrelia burgdorferi flagellum improves serodiagnosis in Lyme disease. *J. Clin. Microbiol.* **26**, 338–346.

Hansen, K., Hovmark, A., Lebech, A. M., Lebech, K., Olsson, I., Halkier-Sorensen, L., Olsson, E., and Asbrink, E. (1992).

Roxithromycin in Lyme borreliosis: Discrepant results of an in vitro and in vivo animal susceptibility study and a clinical trial in patients with erythema migrans. *Acta Derm. Venereol.* **72**, 297–300.

Hansen, K., and Lebech, A. M. (1991). Lyme neuroborreliosis: A new sensitive diagnostic assay for intrathecal synthesis of Borrelia burgdorferi—specific immunoglobulin G, A, and M. *Ann. Neurol.* **30**, 197–205.

Hansen, K., and Lebech, A. M. (1992). The clinical and epidemiological profile of Lyme neuroborreliosis in Denmark 1985–1990. A prospective study of 187 patients with Borrelia burgdorferi specific intrathecal antibody production. *Brain* **115 (Pt. 2)**, 399–423.

Hopf, H. C. (1975). Peripheral neuropathy in acrodermatitis chronica atrophicans (Herxheimer). *J. Neurol. Neurosurg. Psychiatry* **38**, 452–458.

Kaiser, R., Kern, A., Kampa, D., and Neumann-Haefelin, D. (1997). Prevalence of antibodies to Borrelia burgdorferi and tick-borne encephalitis virus in an endemic region in southern Germany. *Zentralbl. Bakteriol.* **286**, 534–541.

Kaiser, R., and Lucking, C. H. (1993). Intrathecal synthesis of specific antibodies in neuroborreliosis. Comparison of different ELISA techniques and calculation methods. *J. Neurol. Sci.* **118**, 64–72.

Kaiser, R., and Rauer, S. (1999). Advantage of recombinant borrelial proteins for serodiagnosis of neuroborreliosis. *J. Med. Microbiol.* **48**, 5–10.

Kaplan, R. F., Meadows, M. E., Vincent, L. C., Logigian, E. L., and Steere, A. C. (1992). Memory impairment and depression in patients with Lyme encephalopathy: Comparison with fibromyalgia and nonpsychotically depressed patients. *Neurology* **42**, 1263–1267.

Karlsson, M., Hammers, S., Nilsson-Ehle, I., Malmborg, A. S., and Wretlind, B., (1996). Concentrations of doxycycline and penicillin G in sera and cerebrospinal fluid of patients treated for neuroborreliosis. *Antimicrob. Agents Chemother.* **40**, 1104–1107.

Karlsson, M., Hovind-Hougen, K., Svenungsson, B., and Stiernstedt, G. (1990). Cultivation and characterization of spirochetes from cerebrospinal fluid of patients with Lyme borreliosis. *J. Clin. Microbiol.* **28**, 473–479.

Karlsson, M., Hammers-Berggren, S., Lindquist, L., Stiernstedt, G., and Svenungsson, B. (1994). Comparison of intravenous penicillin G and oral doxycycline for treatment of Lyme neuroborreliosis. *Neurology* **44**, 1203–1207.

Kindstrand, E., Nilsson, B. Y., Hovmark, A., Pirskanen, R., and Asbrink, E. (1997). Peripheral neuropathy in acrodermatitis chronica atrophic. *Acta Neurol. Scand.* **95**, 338–345.

Klempner, M. S., Hu, L. T., Evans, J., Schmid, C. H., Johnson, G. M., Trevino, R. P., Norton, D., Levy, L., Wall, D., Mccall, J., Kosinski, M., and Weinstein, A. (2001). Two controlled trials of antibiotic treatment in patients with persistent symptoms and a history of Lyme disease. *N. Engl. J. Med.* **345**, 85–92.

Kohler, J., Kern, U., Kasper, J., Rhese-Kupper, B., and Thoden, U. (1988). Chronic central nervous system involvement in Lyme borreliosis. *Neurology* **38**, 863–867.

Kohlhepp, W., Oschmann, P., and Mertens, H. G. (1989). Treatment of Lyme borreliosis. Randomized comparison of doxycycline and penicillin G. *J. Neurol.* **236**, 464–469.

Kristoferitsch, W. (1991). Neurological manifestations of Lyme borreliosis: clinical definition and differential diagnosis. *Scand. J. Infect. Dis. Suppl* **77**, 64–73.

Kristoferitsch, W., Sluga, E., Graf, M., Partsch, H., Neumann, R., Stanek, G., and Budka, H. (1988). Neuropathy associated with acrodermatitis chronica atrophicans. Clinical and morphological features. *Ann. N. Y. Acad. Sci.* **539**, 35–45.

Kruger, H., Heim, E., Schuknecht, B., and Scholz, S. (1991). Acute and chronic neuroborreliosis with and without CNS involvement:

A clinical, MRI, and HLA study of 27 cases. *J. Neurol.* **238**, 271–280.

Kruger, H., Reuss, K., Pulz, M., Rohrbach, E., Pflughaupt, K. W., Martin, R., and Mertens, H. G. (1989). Meningoradiculitis and encephalomyelitis due to Borrelia burgdorferi: A follow-up study of 72 patients over 27 years. *J. Neurol.* **236**, 322–328.

Kuntzer, T., Bogousslavsky, J., Miklossy, J., Steck, A. J., Janzer, R., and Regli, F. (1991). Borrelia rhombencephalomyelopathy. *Arch. Neurol.* **48**, 832–836.

Lebech, A. M., and Hansen, K. (1992). Detection of Borrelia burgdorferi DNA in urine samples and cerebrospinal fluid samples from patients with early and late Lyme neuroborreliosis by polymerase chain reaction. *J. Clin. Microbiol.* **30**, 1646–1653.

Lebech, A. M., Hansen, K., Pancholi, P., Sloan, L. M., Magera, J. M., and Persing, D. H. (1998). Immunoserologic evidence of Human Granulocytic Ehrlichiosis in Danish patients with Lyme neuroborreliosis. *Scand. J. Infect. Dis.* **30**, 173–176.

Lebech, A. M., Hansen, K., Rutledge, B. J., Kolbert, C. P., Rys, P. N., and Persing, D. H. (1998). Diagnostic detection and direct genotyping of Borrelia burgdorferi by polymerase chain reaction in cerebrospinal fluid in Lyme neuroborreliosis. *Mol. Diag.* **3**, 131–141.

Liles, W. C., and Spach, D. H. (1993). Late relapse of tick-borne relapsing fever following treatment with doxycycline. *West J. Med.* **158**, 200.

Logigian, E. L., Kaplan, R. F., and Steere, A. C. (1990). Chronic neurologic manifestations of Lyme disease. *N. Engl. J. Med.* **323**, 1438–1444.

Logigian, E. L., and Steeer, A. C. (1992). Clinical and electrophysiologic findings in chronic neuropathy of Lyme disease. *Neurology* **42**, 303–311.

Luft, B. J., Dattwyler, R. J., Johnson, R. C., Luger, S. W., Bosler, E. M., Rahn, D. W., Masters, E. J., Grunwaldt, E., and Gadgil, S. D. (1996). Azithromycin compared with amoxicillin in the treatment of erythema migrans. A double-blind, randomized, controlled trial. *Ann. Intern. Med.* **124**, 785–791.

Mathiesen, M. J., Christiansen, M., Hansen, K., Holm, A., Asbrink, E., and Theisen, M. (1998). Peptide-based OspC enzyme-linked immunosorbent assay for serodiagnosis of Lyme borreliosis. *J. Clin. Microbiol.* **36**, 3474–3479.

Meier, C., Grahmann, F., Engelhardt, A., and Dumas, M. (1989). Peripheral nerve disorders in Lyme-Borreliosis. Nerve biopsy studies from eight cases. *Acta Neuropathol. (Berl.)* **79**, 271–278.

Mullegger, R. R., Mchugh, G., Ruthazer, R., Binder, B., Kerl, H., and Steere, A. C. (2000). Differential expression of cytokine mRNA in skin specimens from patients with erythema migrans or acrodermatitis chronica atrophicans. *J. Invest Dermatol.* **115**, 1115–1123.

Negussie, Y., Remick, D. G., Deforge, L. E., Kunkel, S. L., Eynon, A., and Griffin, G. E. (1992). Detection of plasma tumor necrosis factor, interleukins 6, and 8 during the Jarisch-Herxheimer Reaction of relapsing fever. *J. Exp. Med.* **175**, 1207–1212.

Nichol, G., Dennis, D. T., Steere, A. C., Lightfoot, R., Wells, G., Shea, B., and Tugwell, P. (1998). Test-treatment strategies for patients suspected of having Lyme disease: A cost-effectiveness analysis. *Ann. Intern. Med.* **128**, 37–48.

Nocton, J. J., Bloom, B. J., Rutledge, B. J., Persing, D. H., Logigian. E. L., Schmid, C. H., and Steere, A. C. (1996). Detection of Borrelia burgdorferi DNA by polymerase chain reaction in cerebrospinal fluid in Lyme neuroborreliosis. *J. Infect. Dis.* **174**, 623–627.

Oschmann, P., Dorndorf, W., Hornig, C., Schafer, C., Wellensiek, H. J., and Pflughaupt, K. W. (1998). Stages and syndromes of neuroborreliosis. *J. Neurol.* **245**, 262–272.

Pachner, A. R., Cadavid, D., Shu, G., Dail, D., Pachner, S., Hodzic, E., and Barthold, S. W. (2001). Central and peripheral nervous system infection, immunity, and inflammation in the NHP model of Lyme borreliosis. *Ann. Neurol.* **50**, 330–338.

Pfister, H. W., Preac-Mursic, V., Wilske, B., and Einhaupl, K. M. (1989). Cefotaxime vs penicillin G for acute neurologic manifestations in Lyme borreliosis. A prospective randomized study. *Arch. Neurol.* **46**, 1190–1194.

Pfister, H. W., Preac-Mursic, V., Wilske, B., Schielke, E., Sorgel, F., and Einhaupl, K. M. (1991). Randomized comparison of ceftriaxone and cefotaxime in Lyme neuroborreliosis. *J. Infect. Dis.* **163**, 311–318.

Piesman, J., Mather, T. N., Sinsky, R. J., and Spielman, A. (1987). Duration of tick attachment and Borrelia burgdorferi transmission. *J. Clin. Microbiol.* **25**, 557–558.

Robertson, J., Guy, E., Andrews, N., Wilske, B., Anda, P., Granstrom, M., Hauser, U., Moosmann, Y., Sambri, V., Schellekens, J., Stanek, G., and Gray, J. (2000). A European multicenter study of immunoblotting in serodiagnosis of lyme borreliosis. *J. Clin. Microbiol.* **38**, 2097–2102.

Seboxa, T., and Rahlenbeck, S. I. (1995). Treatment of louse-borne relapsing fever with low dose penicillin or tetracycline: A clinical trial. *Scand. J. Infect. Dis.* **27**, 29–31.

Seltzer, E. G., Gerber, M. A., Cartter, M. L., Freudigman, K., and Shapiro, E. D. (2000). Long-term outcomes of persons with Lyme disease. *JAMA* **283**, 609–616.

Shapiro, E. D. (2001). Doxycycline for tick bites—Not for everyone. *N. Engl. J. Med.* **345**, 133–134.

Sigal, L. H., Zahradnik, J. M., Lavin, P., Patella, S. J., Bryant, G., Haselby, R., Hilton, E., Kunkel, M., Adler-Klein, D., Doherty, T., Evans, J., Molloy, P. J., Seidner, A. L., Sabetta, J. R., Simon, H. J., Klempner, M. S., Mays, J., Marks. D., and Malawista, S. E. (1998). A vaccine consisting of recombinant Borrelia burgdorferi outer-surface protein A to prevent Lyme disease. Recombinant Outer-Surface Protein A Lyme Disease Vaccine Study Consortium. *N. Engl. J. Med.* **339**, 216–222.

Skoldenberg, B., Stiernstedt, G., Garde, A., Kolmodin, G., Carlstrom, A., and Nord, C. E. (1983). Chronic meningitis caused by a penicillin-sensitive microorganism? *Lancet* **2**, 75–78.

Spach, D. H., Liles, W. C., Campbell, G. L., Quick, R. E., Anderson, D. E., Jr., and Fritsche, T. R. (1993). Tick-borne diseases in the United States. *N. Engl. J. Med.* **329**, 936–947.

Steere, A. C. (2001). Lyme disease. *N. Engl. J. Med.* **345**, 115–125.

Steere, A. C., Pachner, A. R., and Malawista, S. E. (1983). Neurologic abnormalities of Lyme disease: Successful treatment with high-dose intravenous penicillin. *Ann. Intern. Med.* **99**, 767–772.

Steere, A. C., Sikand, V. K., Meurice, F., Parenti, D. L., Fikrig, E., Schoen, R. T., Nowakowski, J., Schmid, C. H., Laukamp, S., Buscarino, C., and Krause, D. S. (1998). Vaccination against Lyme disease with recombinant Borrelia burgdorferi outer-surface lipoprotein A with adjuvant. Lyme Disease Vaccine Study Group. *N. Engl. J. Med.* **339**, 209–215.

Stiernstedt, G. T., Granstrom, M., Hederstedt, B., and Skoldenberg, B. (1985). Diagnosis of spirochetal meningitis by enzyme-linked immunosorbent assay and indirect immunofluorescence assay in serum and cerebrospinal fluid. *J. Clin. Microbiol.* **21**, 819–825.

Sundnes, K. O., and Haimanot, A. T. (1993). Epidemic of louse-borne relapsing fever in Ethiopia. *Lancet* **342**, 1213–1215.

Vieira, A., Barros, M. S., Valente, C., Trindade, L., Faria, M. J., and Freitas, F. (1999). [Human leptospirosis. A short review concerning a caseload]. *Acta Med. Port.* **12**, 331–340.

Wang, G., Van Dam, A. P., Schwartz, I., and Dankert, J. (1999). Molecular typing of Borrelia burgdorferi sensu lato: Taxonomic, epidemiological, and clinical implications. *Clin. Microbiol. Rev.* **12**, 633–653.

Watt, G., and Warrell, D. A. (1995). Leptospirosis and the Jarisch-Herxheimer reaction. *Clin. Infect. Dis.* **20**, 1437–1438.

Wilske, B., Habermann, C., Fingerle, V., Hillenbrand, B., Jauris-Heipke, S., Lehnert, G., Pradel, I., Rossler, D., and Schulte-Spechtel, U. (1999). An improved recombinant IgG immunoblot for serodiagnosis of Lyme borreliosis. *Med. Microbiol. Immunol. (Berl.)* **188**, 139–144.

Wilske, B., Preac-Mursic, V., Gobel, U. B., Graf, B., Jauris, S., Soutschek, E., Schwab, E., and Zumstein, G. (1993). An OspA serotyping system for Borrelia burgdorferi based on reactivity with monoclonal antibodies and OspA sequence analysis. *J. Clin. Microbiol.* **31**, 340–350.

Wilske, B., Schierz, G., Preac-Mursic, V., Von Busch, K., Kuhbeck, R., Pfister, H. W., and Einhaupl, K. (1986). Intrathecal production of specific antibodies against Borrelia burgdorferi in patients with lymphocytic meningoradiculitis (Bannwarth's syndrome). *J. Infect. Dis.* **153**, 304–314.

## CHAPTER 48
# Parasitic Infections

Raad A. Shakir and Martin Rösener

Parasitic diseases can be counted among the most common diseases of humans. The nervous system may be affected by parasites, protozoa, and helminths. The burden of disease as well as the diagnostic and therapeutic challenges are substantial and in many parts of the world overwhelming.

The protozoa have a worldwide distribution and can multiply in humans. They can cause severe illness, especially in patients with the acquired immunodeficiency syndrome (AIDS).

Helminths affecting the nervous system can lead to major problems because of their size, transmissibility, and challenge to the host immune system.

Treatment of parasitic diseases is made difficult by the lack of easy access and at times toxicity of existing chemotherapeutic agents in addition to the emergence of resistant strains (White, 2000).

Major parasitic infections such as malaria, neurocysticercosis, amebiasis, African and American trypanosomiasis, leishmaniasis, lymphatic filiariasis, and schistosomiasis are dealt with in Chapter 52 and those in patients with AIDS are considered in Chapter 51.

In this chapter other protozoan and parasitic diseases are discussed.

## PROTOZOAN INFECTIONS

### Toxoplasmosis

#### Clinical Features and Natural Course

Toxoplasmosis is caused by *Toxoplasma gondii*, an intracellular protozoan. Because of its presence in migratory birds, it has a worldwide distribution. Cats harboring the intestinal infection are associated with the majority of transmitted disease. Most human infections are acquired orally by ingesting oocytes contained in food products, or cysts contained in raw meat or soil. Seroprevalence rates in the adult general population are approximately 50–90% in Europe and 20% in the United States. Transmission also occurs across the placenta, by blood transfusion, and by organ transplantation.

Ingested *T. gondii* penetrate the bowel mucous membrane and spread hematogeneously to muscles or the central nervous system. Acute infection is usually subclinical in immunocompetent individuals or causes a mild illness with fever, lymphadenopathy, and splenomegaly. There is usually complete clinical recovery within several weeks, followed by a chronic latent stage. In immunosuppressed individuals (e.g., those with AIDS, malignancy, organ transplantation recipients, or patients on immunosuppression), reactivation of latent infection rather than newly acquired infection may lead to a focal encephalitis and abscess formation or diffuse meningoencephalitis with necrosis and perivascular inflammatory infiltrates (Navia *et al.*, 1986). Rarely, toxoplasmosis may lead to encephalomyelitis, myositis, myocarditis, or pneumonia. In non-AIDS patients, diffuse meningoencephalitis with meningeal signs rather than focal signs are more frequently observed. Neurological symptoms and signs such as headaches, focal deficits, seizures, and altered consciousness evolve within several days in the majority of patients.

Cranial CT and MRI usually reveal multiple (rarely solitary) space-occupying lesions. They appear hypodense on CT and as areas of increased signal intensity on T2-weighted scans on MRI. They are usually located in the white matter of the cerebral hemispheres or in the basal ganglia, with ring-like contrast enhancement. Cerebrospinal fluid (CSF) usually shows a mild to moderate lymphocytic pleocytosis with elevation of the total protein content; in some the CSF may be normal. Microscopic detection of *T. gondii* utilizing Giemsa stain of centrifuged CSF or isolation by the use of tissue culture is rarely positive. PCR of the CSF in patients with encephalitis may contain *T. gondii* DNA. Serological investigations (e.g., complement fixation, Sabin–Feldman test, indirect immunofluorescence test, hemagglutination) are usually of little diagnostic relevance, because the sero-prevalence in the general population is high. Low-serum IgG antibody titers and the absence of an increase in antibodies are usually noted in immunosuppressed patients.

Histological examination and immuno-histochemistry of brain lesions obtained through CT-guided stereotactic brain biopsy may be considered if serology and CSF examination are not diagnostic and when the focal lesion is easily accessible. In AIDS patients brain biopsy should be considered, if anti-toxoplasmic therapy has not produced clinical or radiological improvement within 2 weeks.

Differential diagnoses include herpes simplex virus encephalitis, Epstein–Barr virus, and cytomegalovirus infection, fungal encephalitis, bacterial brain abscess, tuberculosis, progressive multifocal leukoencephalopathy, and brain tumor (e.g., lymphoma). Approximately 80–90% of patients respond to chemotherapy clinically and radiologically. The course of toxoplasma encephalitis is frequently complicated by concomitant opportunistic infections in immunocompromised patients.

In approximately 40% of the children **congenital toxoplasmosis** leads to clinical symptoms and signs such as chorioretinitis (70%), microcephaly (20%), disseminated intracranial calcifications (35%), obstructive hydrocephalus (20%), and epilepsy (30–40%). In addition, anemia, exanthema, pneumonia, and hepatosplenomegaly with jaundice may occur. Infants may be born without evidence of infection and symptoms develop later. The mortality rate is about 10%. Survivors frequently have severe neurological sequelae, including psychomotor and mental retardation, epilepsy, spastic paresis, deafness, and visual disturbances. Differential diagnosis of the neonatal disease include human cytomegalovirus, herpes, rubella, syphilis, *Escherichia coli* meningitis, sepsis, and erythroblastosis fetalis (Jones *et al.*, 2001).

### Therapy

Medical therapy for central nervous system toxoplasmosis include pyrimethamine, sulfadiazine and folinic acid, which does not inhibit the action of pyrimethamine on *T. gondii*. To prevent relapses, immunocompetent patients should be treated with anti-toxoplasmic agents for at least 4–6 weeks after resolution of clinical symptoms and negative CT findings. In patients with persistent immunosuppression including AIDS, life-long maintenance therapy is required. Newborns with congenital infection are usually treated with

pyrimethamine, sulfadiazine, and folinic acid for 12 months. Pregnant women with acute toxoplasmosis should be treated until delivery to prevent the risk of fetal infection. Monotherapy with a sulfonamide is usually recommended, because pyrimethamine is contraindicated during pregnancy and spiramycin has poor placental penetration. The role of a number of other antimicrobial agents including clindamycin, clarithromycine, and azithromycin in the treatment of toxoplasmosis has yet to be determined (Georgiev, 1994). Details of anti-toxoplasmic treatment are listed in Table I. Response to therapy is judged by clinical response, serial serology, radiology, ophthalmological, and CSF examinations.

### Prevention

General hygiene measures, especially during pregnancy, are recommended. In addition, the use of gloves when handling cat litter boxes or potentially contaminated soil and the thorough cooking of meat should be observed.

## HELMINTHIC INFECTIONS

### Echinococcosis (Hydatid Disease)

### Clinical Features and Natural Course

Human echinococcosis is caused by the larvae of *Echinococcus granulosus*, *E. multilocularis*, or rarely *E. vogeli*. The ova of *Echinococcus* are ingested after contact with infected animals (e.g., dog, cat, sheep, fox) or through contaminated food. Ingested eggs hatch, releasing larvae, which penetrate the intestinal mucosa and are carried to the liver or other sites, where their cystic development begins. The liver is predominantly involved; secondary pulmonary hydatid cysts may

---

**TABLE I    Therapy of Toxoplasmosis**

1. Therapy of acute CNS toxoplasmosis
   Pyrimethamine, 75 mg daily for 3 days, then 25 mg daily; plus sulfadiazine, 500 mg four times a day; plus folinic acid, 15 mg/day.
   Duration of therapy: at least 4–6 weeks after remission of clinical symptoms, in patients with AIDS for life.
2. Therapy of pregnant patients with acute toxoplasmosis
   Sulfadiazine, 500 mg four times a day till delivery.
3. Therapy of infected neonates
   Sulfadiazine, 50 mg/kg twice daily for 12 months; plus pyrimethamine 2 mg/kg/day for 2 days, then 1 mg/kg/day for 2–6 months, then 1 mg/kg/day three times a week for additional 6 months.

develop. Brain involvement occurs in 1–4% of patients, leading to intracerebral, or very rarely intraventricular, subdural, or intraspinal cysts.

*Echinococcus granulosus* is distributed worldwide, especially in South America, Australia, northern and eastern Africa, and the Mediterranean region. It is characterized by a slowly expanding growth of cysts (80% solitary cysts). Infection with *E. alveolaris* (*multilocularis*), which occurs in middle Europe (e.g., Austria, Switzerland, Germany), northern United States, Canada, Japan, Turkey, and Iran, is characterized by a proliferative, multicystic growth. Cysts can occasionally metastasize. *Echinococcus vogeli* is found in bush and domestic dogs of the South American tropics.

Onset of clinical symptoms varies from 2 to 20 years following infection, with signs and symptoms of cystic lesions of the liver and lung (abdominal pain, hepatomegaly, or cough), or rupture of the cyst (acute abdomen, dyspnoea, hemoptysis, or hypersensitivity reaction). CNS cysts present as seizures or space-occupying lesions. Other presentations include cord compression caused by hydatid bone disease with or without vertebral collapse.

Head CT scans reveal cystic space-occupying lesions with a diameter of up to several centimeters. Blood and CSF eosinophilia and elevation of IgE in serum as well as serological investigations (e.g., indirect hemagglutination test, immunofluorescence test, ELISA) are diagnostically helpful.

### Therapy

The treatment of choice is total surgical removal of the cysts without rupture, because release of cystic material carries the risk of anaphylactic reaction and dissemination of contents. The cysts of *E. multilocularis* lack a capsule; they infiltrate surrounding tissue, and therefore cannot be totally excised. Only 20–40% are operable with a high mortality in inoperable cases of 90% within 10 years (Schantz, 1985). Benzoimidazole derivates (albendazole, mebendazole) proved to be effective for the treatment of human hydatid disease (Al Karawi *et al.*, 1990; Teggi *et al.*, 1993). Medical therapy with albendazole (15 mg/kg/day, for 4–6 weeks; alternatively, mebendazole 50 mg/day) is recommended prior to operation, if surgery is not possible, and if the cysts are only subtotally removed. The drug is given for at least 3 months; serial imaging with MRI or CT controls is needed. Patients who have undergone subtotal operation for cysts of *E. multilocularis* usually require lifelong maintenance therapy because of the high risk of recurrence. Side effects of albendazole and mebendazole include elevation of hepatic enzymes, abdominal pain, headache, alopecia, and urticaria.

## Diphyllobothriasis

### Clinical Features and Natural Course

Diphyllobothriasis, or fish tapeworm disease, is caused by cestodes of *Diphyllobathrium latum*. Infection is acquired by ingestion of larvae in flesh, roe, or liver of raw or undercooked fish. *Diphyllobathrium latum* is a common cestode in people of northern Europe. Most infections are asymptomatic or cause transient diarrhea, headache, or malaise. Since *D. latum* absorbs large quantities of vitamin B$_{12}$, it may cause features of B$_{12}$ deficiency with a megaloblastic anemia or neurological sequelae in 2% of infected patients. The clinical picture consists of paresthesiae, pareses, diminished position and vibration sense, bladder disturbances, and cognitive deficits. Subclinical vitamin B$_{12}$ deficiency might be dangerous for patients undergoing nitrous oxide anesthesia. Diagnosis is based on finding proglottids or eggs in the stool.

### Therapy

Treatment consists of vitamin B$_{12}$ supplementation and the use of albendazole or mebendazole as discussed for hydatid cysts (Schantz, 1996).

## Toxocariasis

### Clinical Features and Natural Course

*Toxocara canis*, the roundworm of dogs, has a worldwide distribution. Humans contract *Toxocara canis* after inadvertent ingestion of eggs from material contaminated with the feces of dogs. Young children with histories of pica are at risk. Infective larvae hatch from eggs in the small intestine, penetrate the intestinal wall, and migrate. They may migrate to any portion of the brain or spinal cord. Encephalitis, epileptic seizures, neurological deficits including cognitive and physical symptoms may occur. Vasculitic lesions such as ischemic infarcts, inflammation, and eosinophilic granulomas are reported.

The diagnosis should be considered in young children with CSF and peripheral eosinophilia, and an encephalitic illness. Diagnosis may be confirmed by serologic tests such as ELISA or Western blot using larval excretory–secretory antigens (Yamasaki *et al.*, 2000).

### Therapy

Albendazole 10 mg/kg daily for 5 days seems to be superior to thiabendazole, 25 mg/kg twice daily for 5

**TABLE II    Rare Parasitic Infections of the Central Nervous System**

| Disease, pathogen | Distribution, transmission | Symptoms and signs | Clinical notes | Treatment |
|---|---|---|---|---|
| Sparganosis, second stage larva of tapeworm of the genus *Spirometra* | Worldwide, most common in East, Southeast Asia, East Africa; ingesting contaminated water or infected fish, snakes or amphibians | Penetration of larvae into the CNS causes seizures, focal signs, bland or hemorrhagic infarcts | Serial CT or MRI scans demonstrate mobility of parasite | Surgical excision; Spargana are resistant to praziquantel, other antihelminths, and gamma irradiation |
| Angiostrongyliasis, *Angiostrongylus cantonensis* (rat lungworm) | Southeast Asia, Papua New Guinea, Pacific, Australia; ingestion of infected snails, slugs, undercooked prawns, or crab | Headache, nausea, vomiting, stiff neck, paresthesias, impaired vision, cranial neuropathies, generalized weakness, muscle twitching, seizures, pathological reflexes, papilledema, depressed consciousness, coma | CSF pressure elevated, eosinophilic meningitis, WBC count 150–2000/$\mu$L, 10–90% eosinophils, larvae sometimes seen, ELISA of CSF and serum sometimes positive | Usually self-limiting, spinal tap improves headache, analgesics, sedatives, corticosteroids |
| Gnathostomiasis, *Gnathostoma spinigerum* | East Asia, South Asia, Israel; ingestion of undercooked or fermented freshwater fish, reptiles, birds that have ingested *Gnathostoma* larvae | Eosinophilic meningoencephalitis or radiculomyelitis with headache, cranial nerve palsies, focal sensory or motor deficits, depressed consciousness, segmental girdle pain, paraparesis | Eosinophilic pleocytosis, hemorrhagic or xanthochromic CSF, CT or MRI may show enhancing lesions | Albendazole 400–800 mg daily for 21 days, corticosteroids |
| Strongyloidiasis, *Strongyloides stercoralis* | Worldwide; infection of humans by skin penetration of filariform larvae | Polymicrobial or eosinophilic meningitis, cerebritis, chronic meningoencephalitis | Diagnosis by detection of eggs, adult females or larval stages in the stool | Albendazol 400 mg once or twice daily for 3 days, thiabendazole, 25 mg/kg twice daily for 3 days, ivermectin 200 mg/kg single dose, in immunodeficient patients thiabendazole 50 mg/kg twice daily for 2–4 weeks |
| Ascariasis, *Ascaris* | Tropical and subtropical regions of Asia, Africa, Europe, and the Americas; ingestion of eggs in contaminated soil | Epileptogenic foci by ectopic migration | Diagnosis by detection of worms or eggs in stool | Albendazol 400 mg single dose, mebendazole 2 × 100 mg for three days |
| Paragonimiasis, *Paragonimus* species | East and South Asia, West Africa, South and Central America; consumption of larval metacercaria in raw or undercooked freshwater crustaceans or water plants | Acute or fluctuating meningoencephalitis or myelitis, formation of cysts, abscesses or granulomas produces seizures or focal cerebral signs, embolisation may produce hemorrhage or infarction, chronic forms cause space occupying lesions, raised intracranial pressure, papilledema, optic atrophy, or mental retardation | CT or MRI demonstrate multiple ring enhancing cysts, calcification, dilated ventricles, CSF shows lymphocytic pleocytosis, diagnosis by serological and intradermal tests, detection of eggs in sputum | Praziquantel 25 mg/kg three times per day for three days |

*Note.* Data adapted from Bia and Barry (1986), Kim *et al.* (1996), and Schmutzhard *et al.* (1988), and Siddiqui and Berk (2001).

days (Sturchler *et al.*, 1989). Mebendazole, 100 mg three times daily for 7 days might be of value too. Corticosteroids may reduce the inflammatory complications of *T. canis*.

## Trichinosis

### Clinical Features and Natural Course

Trichinosis is a widely distributed zoonotic disease caused by the nematode *Trichinella spp.* In addition to the classical agent *T. spiralis* (found worldwide in many carnivorous and omnivorous animals), four other species of Trichinella are now recognized: *T. pseudospiralis* (mammals and birds worldwide), *T. nativa* (Arctic bears), *T. nelsoni* (African predators and scavengers), and *T. britovi* (carnivores of Europe and western Asia). Humans are accidentally infected when eating poorly processed meat of these carnivorous animals (or eating food contaminated with such meat). Mild infections may be asymptomatic. Intestinal invasion can be accompanied by gastrointestinal symptoms (diarrhea, abdominal pain, vomiting). Larval migration into muscle tissues usually occurs about 1 week after infection, can cause periorbital and facial edema, conjunctivitis, fever, myalgias, splinter hemorrhages, rashes, and blood eosinophilia. CNS inflammatory infiltration and damage may result from larval migration and vascular occlusion, or from the effects of toxic parasite antigens, or eosinophil infiltration. CNS involvement causes meningoencephalitis presenting with seizures or delirium.

### Diagnosis

Blood eosinophilia, elevated serum IgE levels, and elevated muscle enzymes are present in symptomatic patients. The definitive test is muscle biopsy. Larvae can be seen, when fresh muscle is pressed between glass slides and examined microscopically. Antibody detection is possible between 2 and 4 weeks after infection by bentonite or latex tests, IFA, or ELISA.

### Therapy

Oral mebendazole (200 mg twice daily) and thiabendazole (25 mg/kg twice daily) for 10 days is active against enteric stages and muscle tenderness (Watt *et al.*, 2000). For disseminated infection, treatment is directed toward the larvae and subsequent immune reactions. Severe allergic reactions are treated with prednisolone, initially at 60 mg per day and tapered over a period of 2–3 weeks. Patients with mild infections recover with bed rest, antipyretics, and analgesics.

## OTHER PARASITIC INFECTIONS OF THE CENTRAL NERVOUS SYSTEM

Major clinical features, course, and therapy of rare parasitic diseases such as sparganosis, angiostrongyliasis, gnathostomiasis, strongyloidiasis, ascariasis, and paragonimiasis are shown in Table II.

## REFERENCES

Al Karawi, M. A., el Shiekh Mohamed, A. R., and Yasawry, M. (1990). Advances in diagnosis and management of hydatid disease. *Hepatogastroenterology* 37, 327–331.

Bia, F. J., and Barry, M. (1986). Parasitic infections of the central nervous system. *Neurol. Clin.* 4, 171–206.

Georgiev, V. S. (1994). Management of toxoplasmosis. *Drugs* 48, 179–188.

Jones, J. L., Lopez, A., Wilson, M., Schulkin, J., and Gibbs, R. (2001). Congenital toxoplasmosis: A review. *Obstet Gynecol. Surv.* 56, 296–305.

Kim, D. G., Paek, S. H., Chang, K. H., Wang, K. C., Jung, H. W., Kim, H. J., Chi, J. G., Choi, K. S., and Han, D. H. (1996). Cerebral sparganosis: Clinical manifestations, treatment, and outcome. *J. Neurosurg.* 85, 1066–1071.

Navia, B. A., Petito, C. K., Gold, J. W. M., Cho, E. S., Jordan, B. D., and Price, R. W. (1986). Cerebral toxoplasmosis complicating the acquired immune deficiency syndrome: Clinical and neuropathological findings in 27 patients. *Ann. Neurol.* 19, 224–238.

Schantz, P. M. (1985). Effective medical treatment for hydatid disease? *J. Am. Med. Assoc.* 253, 2095–2097.

Schantz, P. M. (1996). Tapeworms (cestodiasis). *Gastroenterol. Clin. North Am.* 25, 637–653.

Schmutzhard, E., Boongird, P., and Vejjajiva, A. (1988). Eosinophilic meningitis and radiculomyelitis in Thailand, caused by CNS invasion of Gnathostoma spinigerum and Angiostrongylus cantonensis. *J. Neurol. Neurosurg. Psychiatry* 51, 80–87.

Siddiqui, A. A., and Berk, S. L. (2001). Diagnosis of *Strongyloides stercoralis* infection. *Clin. Infect. Dis.* 33, 1040–1047.

Sturchler, D., Schubarth, P., Gualzata, M., Gottstein, B., and Oettli, A. (1989). Thiabendazole vs. albendazole in treatment of toxocariasis: A clinical trial. *Ann. Trop. Med. Parasitol.* 83, 473–478.

Teggi, A., Lastilla, M. G., and DeRosa, F. (1993). Therapy of human hydatid disease with mebendazole and albendazole. Antimicrob. *Agents Chemother.* 37, 1679–1684.

Watt, G., Saisorn, S., Jongsakul, K., Sakolvaree, Y., and Chaicumpa, W. (2000). Blinded, placebo-controlled trial of antiparasitic drugs for trichinosis myositis. *J. Infect. Dis.* 182, 371–374.

White, A. J. (2000). The disappearing arsenal of antiparasitic drugs. *N. Engl. J. Med.* 343, 1273–1274.

Yamasaki, H., Araki, K., Lim, P. K., Zasmy, N., Mak, J. W., Taib, R., and Aoki, T. (2000). Development of a highly specific recombinant Toxocara canis second-stage larva excretory–secretory antigen for immunodiagnosis of human toxocariasis. *J. Clin. Microbiol.* 38, 1409–1413.

CHAPTER 49

# Acute Viral Infections of the Central Nervous System

R. Malessa and Kenneth L. Tyler

Most viral infections of the central nervous system (CNS) occur as a consequence of systemic viral infection. Viruses spread to the CNS either through the bloodstream (hematogenous spread) or through nerves (neural spread). Hematogenous spread occurs with viruses such as the togaviruses, which are responsible for the majority of cases of epidemic encephalitis and for some herpesviruses including cytomegalovirus (CMV) and Epstein–Barr virus (EBV). Conversely, neural spread is critical to pathogenesis of encephalitis caused by herpes simplex viruses (HSVs), rabies, and possibly polio. According to their clinical features and pathogenesis viral CNS infections can be classified into:

(1) Acute viral infections (most common)
(2) Chronic infections (human immunodeficiency virus (HIV), human T-cell lymphotrophic viruses (HTLV), CMV, rubella, lymphocytic choriomeningitis virus (LCMV)
(3) Latent infections (herpesviruses)
(4) Transmissible neurodegenerative diseases due to viruses subacute sclerosing panencephalitis (SSPE), and progressive multifocal leukoencephalitis (PML).

This chapter focuses on acute viral infections of the CNS and their clinical features, diagnosis, and treatment. HIV infection and its neurological complications are covered in a separate chapter. The clinical aspects, diagnosis and treatment of herpes zoster (shingles) is covered in the chapter on inflammatory and infectious polyneuropathy. Neurodegenerative diseases caused by prions (e.g., Creutzfeldt–Jakob disease) are also covered separately.

## CLINICAL ASPECTS

The clinical features of infections of the CNS caused by different viruses are rarely specific enough to allow for a specific pathogen to be identified based on clinical features alone. Determination of intrathecal synthesis of virus-specific antibodies, culture of viruses from the CSF or in rare cases brain tissue, or amplification of viral nucleic acid by the polymerase chain reaction technique (PCR) are crucial for sensitive, specific, and rapid diagnosis of viral CNS infections. A causative viral agent can confidently be identified in the majority of cases of acute meningitis. However, as many as two-thirds of the cases of acute encephalitis remain of uncertain etiology despite comprehensive investigation (Mateos-Mora and Ratzan, 1990).

### Meningitis

Viral meningitis can be defined as a viral infection of the cranial and spinal subarachnoid space. Viral meningitis is at least twice as common as virus encephalitis or meningoencephalitis (Evans, 1976). The core clinical features of acute viral meningitis include fever, headache, and signs of meningeal irritation (e.g., nuchal rigidity). Additional symptoms such as nausea and vomiting, photophobia, fatigue, somnolence, and irritability may occur. Non-CNS manifestations should be carefully searched for as they may provide important diagnostic clues. For example, the presence of typical genital lesions may suggest HSV-2 infection, parotitis is seen with mumps and less commonly with enteroviruses and

LCMV, the appropriate skin rash may suggest measles, enteroviral infection, or the presence of Varicella–Zoster virus infection (see Tyler, 1984, for review). Focal neurological symptoms are rare and indicate involvement of the brain parenchyma. Mild radicular signs and isolated transient cranial nerve dysfunction may accompany viral meningitis.

In both Europe and North America, the most commonly identified pathogens include enteroviruses (Coxsackie A, B, and ECHO) (50–90% of diagnosed cases), mumps (10–20% of cases in nonimmunized individuals), herpesviruses (5–10%), HIV, and arboviruses (variable). Lymphocytic choriomeningitis virus (LCMV) may be underdiagnosed as a cause of meningitis, and in older series accounted for 10–15% of cases.

Cases of viral meningitis typically last less than 14 days and generally have a benign prognosis (90% of cases). In approximately 10% of cases, the illness is protracted, but residual neurological sequelae are infrequent and death is exceedingly rare. Hydrocephalus may occur in cases with concurrent ependymitis.

## Encephalitis

Viral encephalitis is an infection of the brain tissue, often combined with signs of meningeal (meningoencephalitis) or spinal cord (encephalomyelitis) infection. Patients with encephalitis often show a brief prodromal stage characterized by fever, headache with or without nuchal rigidity, photophobia, malaise, nausea and vomiting. This is typically followed by a variable combination of altered mental status, focal neurological deficits (e.g., hemiparesis, aphasia), and seizures. Many patients develop progressive impairment of consciousness (lethargy, stupor, coma) and signs of increased intracranial pressure.

Illness lasts from 2 weeks up to several months. In the era before effective antiviral agents were available, mortality was generally around 10%, although mortality rates as high as 70–80% occurred with certain pathogens (e.g., herpes simplex encephalitis, HSE). Mortality associated with acyclovir-treated HSE is currently below 20%, although residual neurologic deficits are still frequent (>50%). The incidence of HSE is about 2.3 cases/$10^6$ inhabitants/year (Najioullah et al., 2000). Approximately one-fifth of HSE patients have mild or atypical disease without focal findings and only slow progression in the absence of antiviral therapy (Fodor et al., 1998). Herpes simplex virus can also produce monophasic or recurrent episodes of brainstem encephalitis (Tyler et al., 1995).

The most important pathogens causing acute meningoencephalitis in North America and Europe are HSV-1, arboviruses, enteroviruses, measles and mumps (in nonimmunized individuals), and more rarely LCMV. Ongoing studies, exemplified by the California Encephalitis Project, suggest that even with comprehensive serological and PCR testing, up to two-thirds of cases of acute encephalitis remain of unknown etiology.

## Myelitis

The most common pathogens responsible for acute viral myelitis are Coxsackie A and B viruses, ECHO viruses, and Varicella–Zoster virus (VZV). In Europe, tick-borne encephalitis viruses, including Central European Encephalitis virus, an antigenic variant of Russian spring–summer encephalitis virus, may also cause this syndrome. Both HSV and CMV are important causes of acute myelitis in HIV-infected individuals.

It may be difficult to distinguish between an acute viral infection with myelin damage due to ongoing viral replication and a parainfectious immune-mediated process. Careful determination of the prior history is essential for differential diagnosis. A preceding illness or an immunization 2–4 weeks earlier followed by a symptom-free interval prior to the onset of the neurologic deficit suggests a parainfectious etiology. Laboratory tests may be helpful. Virus may be cultured from CSF or viral genome amplified by PCR in cases of acute viral infection, but these tests are typically negative in parainfectious myelitis. In acute disseminated encephalomyelitis (ADEM), magnetic resonance imaging may demonstrate multifocal white matter lesions with increased T2 signal intensity without or with uniform gadolinium enhancement. Neuromyelitis optica, the combination of severe myelitis and optic neuritis, may be a manifestation of ADEM, systemic lupus erythematosus or multiple sclerosis (Andersen, 2000).

Poliomyelitis, which affects primarily the motor neurons of the anterior horns of the spinal cord and in the brainstem, has become rare since the introduction and widespread use in the United States and Europe of either the live attenuated oral polio vaccine (OPV, e.g., Sabin) or the inactivated polio vaccine (IPV, e.g., Salk). Rare cases (<1 case/2,000,000 vaccine doses) still occur in recipients of the live oral vaccine or in their nonimmune contacts. Paralytic polio in vaccine recipients tends to occur in young children (<4 years), within 1–3 weeks of vaccination. By contrast, cases of paralytic polio in contacts of vaccinees typically involve adults, and occur 3–4 weeks after vaccination. The patients usually present with a viral syndrome (fever, headache, and nausea) followed by muscle pain associated with the development over several days of an asymmetrical flaccid paresis. Proximal muscles are involved more commonly than distal ones, and the legs are more frequently affected than the arms.

Viral infections of the dorsal root ganglia may spread to the spinal cord, leading to myelitis (especially in herpes zoster).

## Diagnostic Features

Examination of the **cerebrospinal fluid (CSF)** is essential to the diagnosis of virtually all viral infections of the CNS. The typical profile in viral infections is a lymphocytic pleocytosis with cell counts of less than 500 cells/mm$^3$. In 5–10% of cases the CSF cell count exceeds 500 cells/mm$^3$. Pleocytosis of this magnitude may be encountered with LCMV, eastern and California encephalitis viruses (EEV, CEV), and occasionally mumps infection (Tyler, 1984). In some infections at their initial stage a polymorphonuclear pleocytosis may be transiently present. Repeat CSF examination 12–24 h after the initial tap will usually show a progressing conversion to lymphocytic pleocytosis. Persisting polymorphonuclear pleocytosis should always suggest the possibility of bacterial infection, although it may be encountered with viral meningitis due to certain Echoviruses (e.g., E9) and EEV. CSF glucose levels are normal (>50% of blood glucose) in 90% of cases, although they may be depressed in some patients with meningitis due to LCMV or mumps. The presence of hypoglycorrhachia should always prompt a careful search for non-viral etiologies of meningitis or encephalitis such as tuberculosis, fungal infections, neoplastic meningitis and sarcoidosis. CSF total protein is typically mildly elevated, but almost never exceeds 800 mg/dL. Such findings are typical, but not entirely specific, as a lymphocytic-normal glucose CSF profile can also be found in parainfectious encephalomyelitis, partially treated bacterial meningitis, in association with parameningeal infections, with parasitic diseases, and early in fungal or tuberculous meningitis. It has been suggested that determination of serum C-reactive protein and procalcitonin levels may be useful in distinguishing bacterial meningitis from viral meningitis in children (Hatherill *et al.*, 1999; Sormunen *et al.*, 1999).

The **EEG** is usually abnormal in cases of viral encephalitis. The most common finding is diffuse slowing, although focal abnormalities may be found in 60–80% of cases. EEG changes may suggest a specific diagnosis (e.g., periodic bilaterally symmetrical, synchronous high-voltage stereotypic slow-wave complexes in subacute sclerosing panencephalitis). Repetitive sharp slow-wave discharges (pseudoperiodic complexes) at 1- to 5-s intervals localized to the temporal regions may occur in HSE, although the absence of EEG changes does not exclude the diagnosis. In early Creutzfeldt–Jakob disease (CJD), the EEG shows disorganization of background rhythms and generalized slowing. As the disease progresses periodic di- or triphasic sharp waves appear at a frequency of 0.5–2/s. These may occur in temporal proximity to clinically observed myoclonic jerks. Typical EEG abnormalities of the type described have been reported in 75–95% of CJD cases, and their absence after 3 or 4 months of illness should prompt consideration of other diagnostic possibilities.

**Computer tomography (CT) and magnetic resonance imaging (MRI)** may be helpful in supporting the diagnosis of viral encephalitis or parainfectious encephalomyelitis and in excluding nonviral causes of encephalitis or encephalopathy. Many patients with HSE have areas of increased signal on T2-weighted images in the frontal and temporal areas. These abnormalities can be detected within days of the onset of illness and may be present when the CT scan is normal or shows only minor abnormalities. Although this pattern is suggestive of HSE it is not diagnostic.

Patients with parainfectious encephalomyelitis may have multiple white matter lesions with increased T2 and decreased T1 signal intensity. During the first 1–2 months following onset of illness these lesions often enhance with gadolinium, reflecting local breakdown of the blood brain barrier. The presence of multiple lesions all enhancing with gadolinium, and resolving together, without the appearance of new lesions, is extremely helpful in differentiating acute disseminated encephalomyelitis (ADEM) from multiple sclerosis. Diffusion-weighted MRI may be superior in detecting early encephalitic changes and serves as a valuable adjunct to conventional MRI (Tsuchiya *et al.*, 1999, 2000). Spect and PET can be used to detect associated perfusion changes and vasculitic involvement (Wakamoto *et al.*, 2000).

Both culture and serology remain important diagnostic tools for identifying the specific etiology of viral CNS infections. Appropriate sources for viral cultures are shown in (Table I). Serologic studies typically require evidence of a fourfold or greater rise in antibody titer against a specific pathogen between acute specimens and convalescent specimens obtained 2–4 weeks after onset of illness. The time required for development of a diagnostic antibody response makes serologic studies useful primarily for retrospective confirmation of a diagnosis, and limit their utility in the acute setting. Obtaining paired serum and CSF antibody titers may increase the specificity of diagnosis. An increased CSF/serum antibody ratio for a specific virus, in the absence of a similarly elevated CSF/serum albumin ratio, is indicative of intrathecal antibody synthesis and suggests infection by that virus. Detection of virus-specific IgM in serum or CSF generally indicates acute infection.

The **polymerase chain reaction (PCR)** allows rapid and sensitive detection of even minute quantities of viral

TABLE I   Virus Isolation and Identification[a]

| | Sputum | Urine | CSF | Blood | Stool | Other | Serology |
|---|---|---|---|---|---|---|---|
| Adeno virus | ++ | − | + | − | + | | ± |
| CEE virus1 | − | − | ± | + | − | | ++ |
| LCM virus2 | − | − | ++ | + | − | | + |
| Herpes simplex virus | − | − | − | − | − | Brain biopsy | + |
| Varicella-zoster virus | − | − | + | − | − | Vesicle scrapings | ± |
| Epstein–Barr virus | + | − | ± | + | − | | ++ |
| Cytomegalovirus | + | + | − | ± | − | | + |
| Influenza virus | ++ | − | − | + | − | | ± |
| Mumps virus | ++ | ++ | ++ | ± | − | | + |
| Measles virus | ++ | + | − | + | − | | + |
| Poliomyelitis virus | + | − | − | − | ++ | | + |
| Coxsackie −A + −B virus | ++ | − | ++ | − | ++ | | + |
| ECHO virus | − | − | + | − | ++ | | + |
| Smallpox virus | + | − | − | ± | − | Vesicle scrapings | + |
| Rubella virus | + | + | + | ± | + | | + |
| Rabies virus | + | − | − | − | − | Brain biopsy | + |

*Note.* (++), very useful; (+), useful; (−), unlikely to be helpful; 1 CEE, Central European Encephalitis; 2 LCM, lymphocytic choriomeningitis.
[a]Modified from Tyler (1984).

DNA or RNA in HSV-I, HSV-II, VZV, EBV, CMV, HIV, enteroviral, and measles infections of the CNS (Cinque *et al.*, 1992; Goswami *et al.*, 1991; Mustafa *et al.*, 1993). In the case of herpes simplex virus and enteroviruses, CSF-PCR is highly sensitive (95–100%) and specific (97–100%). Studies concerning the sensitivity and specificity of CSF PCR in the diagnosis of acute infections other than those caused by herpes simplex virus and enteroviruses is limited. However, in one series a patient with a positive CSF-PCR result was 88 times as likely to have a definite viral infection of the CNS as a patient with a negative PCR result (Jeffery *et al.*, 1997). CSF-PCR has become the diagnostic procedure of choice.

Viral meningoencephalitis must be distinguished from bacterial meningitis, from thrombosis of cerebral veins, from meningism due to other conditions (e.g., subarachnoid hemorrhage), or from metabolic and toxic–allergic encephalopathies (Table II). Parameningeal infections (abscess, empyema) must also be included in the differential diagnosis (Reik and Barwick, 1990). The differential between an acute infectious process and a parainfectious immune-mediated encephalomyelitis may be extremely difficult. The latter develop as the result of an antigenic challenge by a viral protein, due to either viral infection or vaccination. These disorders usually appear 1–3 weeks after infection with viruses including measles (probability 1:1000), chickenpox (1:2500), herpes zoster, mumps and rubella (1:6000), influenza, and Epstein–Barr virus or after vaccinations against smallpox or rabies. The risk of postvaccinal encephalomyelitis has been markedly reduced with modern vaccines that avoid the use of

neural tissues in vaccine development and virus culture. Such postinfectious and postvaccinal diseases often begin with acute fever and multifocal neurological dysfunction, frequently accompanied by seizures. Numerous, small, hyperintense foci can be detected in T2-weighted magnetic resonance images; pathological examinations reveal such foci to result from perivascular inflammation with demyelination. Changes in the CSF do not differ from those typical of acute viral encephalitis (Kerkar and Molayi, 1990). Steroids are frequently used in the treatment of postinfectious immune-mediated CNS infections, although controlled trials documenting their efficacy are lacking.

TABLE II   Differential Diagnosis of Viral Encephalitis

| Infections | |
|---|---|
| Bacterial | Brain abscess, endocarditis, tuberculosis, syphilis, mycoplasma pneumoniae infection |
| Fungal | Cryptococcosis, candidiasis |
| Parasitic | Malaria, toxoplasmosis, cysticercosis, trichinosis |
| Toxic–allergic encephalopathy | Heavy metals, nonsteroidal antiinflammatory agents (ibuprofen, naproxen), aspirin, sulfadiazine, trimethoprim, isoniazid, azathioprine, cystosine arabinoside, barbiturates |
| Metabolic encephalopathy | Electrolyte imbalance, diabetic coma or hypoglycemia, acute porphyria, pheochromocytoma |
| Encephalopathy in systemic illnesses | Sarcoidosis, collagen diseases |

# NATURAL COURSE

Tables III and IV summarize the clinical features and the course of most frequent viral infections of the central nervous system.

# PRINCIPLES OF THERAPY

Immunization against viruses producing neurologic disease remains a vital aspect of preventive therapy. Among the important vaccines currently employed for this purpose are live and inactivated poliomyelitis vaccines, live measles, mumps, rubella and Varicella–Zoster vaccines. Vaccines against Japanese encephalitis, yellow fever virus, influenza, hepatitis B, rabies virus, and adenovirus types 4 and 7 are also available and should be used when specifically indicated.

Hyperimmune immunoglobulins, which contain high titers of antibody against a specific virus, may be useful in both therapy and prevention of certain viral infections (Table V) (Berlit, 1989). Among the currently available immunoglobulins are preparations directed against rabies, Varicella–Zoster virus, hepatitis B virus and CMV. Rabies hyperimmune serum is used, in combination with rabies vaccine, in postexposure prophylaxis of individuals exposed to rabid animals (Table VI). Varicella–Zoster hyperimmune globulin may be useful in preventing dissemination of varicella infection in immunocompromised patients with shingles, or in pre-

**TABLE III**  Acute Viral Infections of the CNS

| Group | Type | Neurologic manifestation | Special features | Clinical course |
|---|---|---|---|---|
| Adenoviruses | Adeno virus | Meningitis, meningoencephalitis | Acute febrile pharyngitis, conjunctivitis, epidemic keratoconjunctivitis | Sometimes severe illness in infants |
| Arboviruses | CEE (Central European encephalitis) virus | Meningitis, meningoencephalitis (tick-borne encephalitis) | Influenzalike symptoms; often biphasic course | 75% meningoencephalitis, fatality rate 0.8–2%; isolated meningitis in 25%; poliomyelitislike in 10% with a fatality rate 20% |
| Arenaviruses | LCM (lymphocytic choriomeningitis) virus | Meningoencephalitis and -myelitis, transmitted by mice and hamsters | Influenzalike illness | Protracted course, weeks up to months, residual deficits infrequent; coma and death infrequent (2.5%) |
| Herpesviruses | HSV type 1 (herpes simplex virus) | temporal lobe encephalitis | Prodromal stage, hemiparesis, aphasia, focal epilepsy | If untreated, fatality rate up to 80%, <21%, if treated |
| | VZV (varicella-zoster virus) | VZV encephalitis/cerebellitis or cerebral vasculitis | Zoster ophthalmicus | Fatality rate of encephalitis 5–10%, residual neurological deficits up to 20% |
| | | VZV ganglionitis, radiculitis, infrequently myelitis; encephalitis in immunodeficiency | Herpes zoster exanthema | 10% postherpetic neuralgia in general, but in about 50% of the elderly (>60 years) |
| | EBV (Epstein–Barr virus) | Meningitis, encephalitis (brain stem encephalitis), cerebellitis, polyneuritis | Infectious mononucleosis, fever, lymphadenopathy | Total recovery in 80–90%, fatality rate 2–5% |
| | CMV (cytomegalovirus) | Meningoencephalitis, polyneuritis, myelitis, (often reactivation in immunodeficiency/transplant recipients) | Lymphomonocytosis, often hepatitis, myocarditis, pneumonia | No clear clinical pattern, insidious or subacute mental deterioration, focal signs (supra- and subtentorial) |
| Myxoviruses | Influenza-A, -B virus | Encephalitis, encephalomyelitis (parainfectious) | Influenza, bronchitis, pneumonia, myalgia, exanthema | Often benign, but also severe and progressive encephalitis; overall fatality rate <10% |

*continues*

**TABLE III**    *Continued*

| Group | Type | Neurologic manifestation | Special features | Clinical course |
|---|---|---|---|---|
| | Mumps virus | Meningitis, meningoencephalitis (sometimes prior to parotitis) | Parotitis, orchitis, pancreatitis, oophoritis | Often short-lasting and benign; sometimes hydrocephalus (ependymitis) |
| | Measles virus | Parainfectious encephalitis | Biphasic course, measles exanthema, sometimes bronchopneumonia | Bad prognosis in case of coma or seizures, fatality rate 20%, neurologic sequelae in about 30–40% |
| | Parainfluenza virus | Meningitis | Bronchitis, pneumonia | Often benign |
| Picornaviruses (enteroviruses) | Poliomyelitis virus types 1–3 | Poliomyelitis | Biphasic course with catarrhal prodrome followed by paralytic polio | Overall fatality rate about 10% (30% in subjects >40 yr), residual paresis in about 30% |
| | Coxsackie virus -A | Meningitis, infrequently encephalitis | Herpangina, influenzalike illness | Benign meningitis, duration up to 10 days |
| | Coxsackie virus -B | Meningitis, infrequently encephalitis | Pleurodynia, myo- and pericarditis | |
| | ECHO viruses | Meningitis, infrequently meningomyelitis | Gastrointestinal symptoms, influenzalike illness | Benign meningitis |
| Smallpox virus | Smallpox virus | Encephalo–myelitis, almost always parainfectious | Cyclical course, initial phase (fever, exanthema), macular–papular–vesicular–pustular rash, and scabbing stages | Cerebral involvement in 2.7% |
| Rubella virus | Rubella virus | Encephalitis (parainfectious) | Rubella exanthema, nuchal lymphadenopathy | Fatality rate in encephalitis about 20% |
| Rhabdoviruses | Rabies virus | Brain stem encephalitis | Prodromal phase with headache, fever; period of excitement, hydrophobia, paralysis, death | Fatal outcome in 95–100% |

**TABLE IV**    **Chronic Viral Infections of the CNS**

| Agent | Disease | Special clinical features |
|---|---|---|
| Conventional | | |
| Measles virus | SSPE (subacute sclerosing panencephalitis) | Frequency: 1/1 mio. Occurs in childhood<br>Stage 1: personality change, intellectual deterioration<br>Stage 2: focal neurological deficits, myoclonic seizures, ataxia, visual impairment<br>Stage 3: impairment of consciousness, coma, fatal outcome within months or a few years |
| Rubella virus | PRP (progressive rubella panencephalitis) | Like SSPE, but prolonged course; exceedingly rare disease, slowly progressive intellectual deterioration, dementia, fatal outcome |
| JC- or SV40- PML virus | PML (progressive multifocal leukoencephalopathy) | Subacute demyelinating leukoencephalopathy in immune-compromised hosts, relatively frequent complication in AIDS (Chapter 51) |
| HIV (human immundeficiency virus) | AIDS (acquired immune deficiency syndrome) | Encephalopathy (HIV-1-associated minor cognitive or motor disorders and AIDS dementia complex) and vacuolar myelopathy/multinucleated cell myelitis are frequent as well as toxoplasma encephalitis, cryptococcal meningitis, PML, and cerebral lymphoma (Chapter 51) |
| Unconventional (prion diseases) | Kuru | Endemic in Papua, New Guinea, transmitted by cannibalism, insidious onset, progressive course with ataxia, tremor, myoclous, rigidity, dementia, death within 9–24 months |
| | CJD (Creutzfeldt–Jacob disease) | Distribution worldwide, mean age 50–60 years, familial cases up to 15%. Stage 1: personality changes, depression. Stage 2: myoclous, spasticity, visual impairment, ataxia, extrapyramidal signs. Stage 3: dementia, rigidity, vegetative state, death within months (maximum 2 years) |
| | Gerstmann–Sträussler–Scheinker disease | Exceedingly rare disease; mutations of the PRNP-gene on chromosome 20p are linked to GSS development; predominantly familial, (e.g., cerebellar dysfunction, hyporeflexia, slowly progressive dementia), mean duration: 5 years |

venting development of disease in susceptible nonimmune individuals exposed to Varicella–Zoster virus (Stevens and Marigan, 1980). CMV hyperimmune globulin may be of value in treatment and prophylaxis of CMV infections. Hyperimmune globulins or intravenous immunoglobulin preparations may also be of value in the treatment of chronic enteroviral meningitis in infants and children with immunodeficiency. Infants (<1 year) and children with immune dysfunction may benefit from prophylactic immunization with human immunoglobulins if they have been exposed to measles.

Prophylactic immunization is also effective after exposure to the Central European encephalitis virus in approximately 60% (results of a survey in Vienna, Austria). Considering the relatively low rate of infection subsequent to a single tick bite in endemic regions (between 1 : 500 and 1 : 5000), prophylactic immunization should be carried out only under special conditions (discussed later in this chapter).

The value of hyperimmune globulins in other settings has not been conclusively established. To date there are no reliable reports on positive effects of adjuvant intravenous, intramuscular, or intrathecal therapy with polyvalent immunoglobulins in patients whose immune system is intact. Both intravenous immunoglobulin and steroid therapy have been reported to be of benefit in individual cases of ADEM, but controlled clinical trials are lacking.

The treatment of elevated intracranial pressure in viral encephalitis is discussed later in this chapter.

**TABLE V** Prophylaxis and Therapy of Viral Infections with Immunoglobulin (Schumacher, 1986)

| Prophylaxis | Therapy |
|---|---|
| a. CMV prophylaxis in cases of polytransfusion and immune deficit | Ig iv, 500 mg IgG for each transfusion |
| b. CMV prophylaxis after bone marrow transplantation | CMV-hyperimmunoglobulin iv |
| c. Prophylaxis of generalized varicella zoster infection in hypo- (a-) gammaglobulinemia and normal T-cell function | Ig iv, >10 g IgG or HSV-hyperimmunoglobulin im |
| d. Persons in contact with smallpox patients | Smallpox virus–hyper-immunoglobulin im |
| e. Persons in contact with rabies-infected animals | Rabies hyperimmunoglobulin im and active immunization (Table VI) |
| f. In case of multiple (>5), simultaneous tick bites in a region of CEE endemia | CEE-hyperimmunoglobulin im |

*Note.* Ig iv, polyvalent immunoglobulin preparation; CMV, Cytomegalovirus; CEE, Central European encephalitis.

**TABLE VI** Guide for Postexposure Treatment of Rabies[a]

| Category | Type of contact with a suspect or confirmed rabid domestic or wild[b] animal or an animal unavailable for observation | Recommended treatment |
|---|---|---|
| I | Touching or feeding of animals, licks on intact skin | None, if reliable case history is available |
| II | Nibbling of uncovered skin, minor scratches or abrasions without bleeding, licks on broken skin | Administer vaccine immediately.[c,d] Stop treatment if animal remains healthy throughout an observation period of 10 days or if an animal is killed humanely and found to be negative for rabies by appropriate laboratory techniques |
| III | Single or multiple transdermal bites or scratches, contamination of mucous membrane with saliva (i.e., licks) | Administer rabies immunoglobulin and vaccine immediately.[c,d] Stop treatment if animal remains healthy throughout an observation period[e] of 10 days or if an animal is killed humanely and found to be negative for rabies by appropriate laboratory techniques |

[a]Modified from the WHO Expert Committee on Rabies (1992).
[b]Exposure to rodents, rabbits, and hares seldom, if ever, requires specific antirabies treatment.
[c]If an apparently healthy dog or cat in or from a low risk area is placed under observation, the situation may warrant delaying initiation of treatment.
[d]Tissue-culture or purified duck-embryo vaccines of potency at least 2.5 IU per dose should be applied to the following intramuscular schedule: one dose of vaccine should be administered on days 0, 3, 7, 14, and 30. All injections must be given into the deltoid region or, in small children, into the anterolateral area of the thigh muscle. Vaccine should never be administered in the gluteal region. An intradermal schedule or an abbreviated multisite intramuscular schedule may be used alternatively. The latter may be particularly effective when postexposure treatment does not include administration of rabies immunoglobulin (for detailed information; WHO Expert Committee on Rabies, 1992). Two kinds of rabies immunoglobulin (RIG) may be used: human rabies immunoglobulin (HRIG) and equine rabies immunoglobulin (ERIG). A skin test must be performed prior to the administration of ERIG. As much as possible of the recommended dose should be delivered to the wound site, the rest is given intramuscularly (into the gluteal region) and followed by a complete course of vaccine.
[e]This observation period applies only to dogs and cats. Except in the case of threatened or endangered species, other domestic and wild animals suspected as rabid should be killed humanely and its tissues examined using appropriate laboratory techniques.

## Antiviral Therapy (See Chapter on Human Immunodeficiency Virus Infection for a Discussion of Antiretroviral Drugs)

Nucleoside analogs (acyclovir, ganciclovir, vidarabine, famciclovir, and valacyclovir) act by competitive inhibition of viral DNA polymerase and early termination of DNA chain synthesis. Since the activity of these agents is dependent upon their phosphorylation by virally encoded thymidine kinases, they are selectively activated within virus-infected cells. Clinically important DNA viruses suppressed by nucleoside analogs belong to the herpes group: herpes simplex types 1 and 2 (HSV-1, HSV-2), Epstein–Barr virus (EBV), Varicella–Zoster virus (VZV) and cytomegalovirus (CMV). **Acyclovir** (Zovirax) is a deoxyguanosine analog that is effective against HSV-1, HSV-2, VZV, and EBV. Acyclovir is ineffective in the treatment of CMV infection because it is not well phosphorylated in CMV-infected cells. Acyclovir-resistant viruses, with altered thymidine kinase or DNA polymerase enzymes have been isolated from clinical specimens. Clinically significant virulent acyclovir-resistant viruses have emerged with increasing frequency in isolates from immunocompromised patients including those with AIDS. Acyclovir is eliminated primarily via the kidneys (approximately 70%) with a half-life of 2–3 h in patients with normal renal function. CSF concentrations of acyclovir are approximately 50% of coincident plasma concentrations. The bioavailability after oral administration is small (20%) (de Miranda and Blum, 1983). Since acyclovir acts relative specifically in infected cells (preferential phosphorylation by viral thymidine kinase), it has only few side affects. Transiently elevated levels of creatinine (5–25%) and neurotoxicity in patients who had received bone marrow transplants (4% with episodes of confusion, tremor or seizures) have been observed.

The efficacy of acyclovir has been shown in herpes simplex encephalitis, the most common cause of sporadic viral encephalitis in the United States (O'Brien and Campoli-Richards, 1989; Whitley, 1988a). In a large, prospective, double-blind, randomized study the fatality rate was lowered to 19%. Fifty percent of the patients who had participated in the study were able to return to their own daily routines within 6 months (Sköldenberg *et al.*, 1984). For therapy of herpes encephalitis, a dosage of 10 mg/kg body weight IV at 8-h intervals for at least 14 days is recommended. In patients with an immune deficit or in case of insufficient response higher doses or more prolonged therapy may be required.

**Vidarabine** (adenine-arabinoside) is effective primarily against HSV-1, HSV-2, and VZV. It has largely been replaced in clinical practice by acyclovir (Sköldenberg *et al.*, 1984; Whitley *et al.*, 1986). Vidarabine may still be of value in the treatment of rare cases of HSE which relapse after initial therapy with acyclovir. *In vivo*, vidarabine is rapidly metabolized to less active intermediates. The half-life of acyclovir is about 3 to 4 h and approximately 60% of the substance is eliminated via the kidneys (dosage reduction required in case of renal insufficiency). Since vidarabine is not very soluble it requires a substantial diluent volume for intravenous infusions (15 mg/kg/day as a 12-h infusion). Vidarabine concentrations in CSF are approximately 30–50% of serum values. Side effects of vidarabine are relatively mild. They include nausea, diarrhea, and vomiting. At higher doses (20 mg/kg), neurological complications may occur and include tremor, ataxia, psychoses, and seizures.

Another deoxyguanosine nucleoside analog, **ganciclovir** (Cytovene) is effective against viruses of the herpes group and other DNA viruses. Ganciclovir has become the drug of choice for treating severe CMV infections in both immunocompetent and immunocompromised patients (e.g., patients with bone marrow or organ transplants, or AIDS) (Cohen *et al.*, 1993; Dieterich *et al.*, 1993; Faulds and Heel, 1990). Dose-limiting toxicity includes bone marrow suppression resulting in neutropenia (<1000 cells/mm³ in 40%) and thrombocytopenia (<50,000 platelets/mm³ in 20%). Reported CNS toxicity includes headache, altered mental status, and seizures. Immunocompetent patients may respond to acute therapy, and do not require maintenance therapy. In immunocompromised patients, disease frequently recurs if therapy is discontinued.

**Famciclovir** (Famvir) is converted following oral administration to penciclovir. The mechanism of action of penciclovir is similar to that of acyclovir. Like acyclovir, penciclovir must be phosphorylated by a virally encoded thymidine kinase and then by cellular kinases into its active triphosphate form. The efficacy and side-effect profile of famciclovir are similar to those of acyclovir.

**Valacyclovir** (Valtrex) is the valyl ester of acyclovir. It acts as a prodrug that is converted into acyclovir. Oral bioavailability is 3–5 times greater than with acyclovir. Pharmacokinetic studies suggest that a dose of 2000 mg qid produces similar serum concentration curves as an intravenous dose of 10 mg/kg of acyclovir given every 8 h. A dose of 1000 mg qid is equivalent to an intravenous acyclovir dose of 5 mg/kg given every 8 h, and a dose of 250 mg qid is equivalent to an oral acyclovir dose of 800 mg given five times a day. The efficacy and side effect profiles of valacyclovir are the same of those of acyclovir.

The antiviral drug **foscarnet** (Foscavir) is a pyrophosphate analog, which inhibits the DNA polymerases of HSV-1, HSV-2, CMV, EBV, and VZV (Crumpacker, 1992). AIDS patients with CMV retinitis respond equally well (response in 60–90%) to treatment with foscarnet or ganciclovir (Fletcher, 1992). Patients treated with foscarnet, however, survive longer than those receiving ganciclovir (AIDS Research Group, AIDS Clinical Trials Group (1992). In AIDS-related CMV retinitis maintenance therapy is always needed to prevent a relapse. Foscarnet may also be of value in the treatment of acyclovir-resistant HSV or VZV isolates, and ganciclovir-resistant isolates of CMV. The major dose-limiting side effect is nephrotoxicity, which can progress to acute renal failure. Adequate hydration is essential during foscarnet therapy and reduces the risk of nephrotoxicity (Jacobson, 1992). Metabolic derangements are common and include hypo- and hypercalcemia, hypo- and hyperphosphatemia, and hypokalemia. CNS toxicity occurs in up to 10% of patients and may include headache, tremor, seizures, and alterations in mental status.

**Cidofovir** (Vistide) is a monophosphate nucleoside analog of deoxycytidine (dCTP) that inhibits dCTP's incorporation into viral DNA. The drug has been used clinically in the treatment of HIV-associated CMV infections, notably retinitis. No detailed trials of the efficacy of cidofovir in the treatment of CMV-associated CNS infections are currently available. As a result, its use should be confined to patients who have failed to respond to therapy with ganciclovir or foscarnet or both. The usual dose is 5 mg/kg IV administered over 1 h and given once weekly for 2 weeks followed by a maintenance dose of 5 mg/kg IV every other week. Patients should be hydrated with 1 L of 0.9% normal saline prior to each cidofovir dose. A second liter of fluid can be administered immediately following each dose of drug. Probenecid should be given 3 h prior to each cidofovir dose (2 g) and at 2 and 8 h postdose (1 g each). Cidofovir has significant dose-dependent nephrotoxicity and dosage needs to be adjusted in patients with renal impairment. Patients receiving cidofovir need to have careful monitoring of Bun, creatinine, and of urine proteinuria and leucocytosis. Additional toxicity includes hematological complications (neutropenia, thrombocytopenia, and anemia) (10–20%) and ophthalmological complications including iritis and decreased intraocular pressures. Patients receiving cidofovir should have careful monitoring of IOP, visual fields and acuity, and complete blood counts and differential. Patients can also experience gastrointestinal side effects, rashes and alopecia, respiratory problems, and neuromuscular side effects (weakness, paresthesias).

**Amantadine** (Symmetrel) **and rimantadine hydrochloride** (Flumadine): These tricyclic amines suppress the uncoating of influenza A virus, and thereby inhibit its replication. Both amantadine and rimantadine are effective in the treatment and prophylaxis of influenza A infection. A seasonal 6-week course of 100–200 mg/days of amantadine or rimantadine is >75% effective in preventing influenza A infection. Both amantadine and rimantadine are also effective in ameliorating the course of influenza A infection if given within 24–48 h of onset of symptoms (Kubar et al., 1989; Little et al., 1978).

**Ribavirin** (Virazole) is a synthetic guanosine analog which has been licensed in the United States for the treatment of respiratory syncytial virus infections, but is also suppresses the synthesis the synthesis of nucleic acids of other DNA and RNA viruses. Ribavirin has no effect in HIV infection (Spanish Ribavirin Trial Group 1991; Spector et al., 1989).

**Interferons** are glycoproteins that produce a variety of biological effects. In addition to antiproliferative and immunmodulating effects, interferons show antiviral properties. Endogenous interferons are produced and released early in the course of viral infections. They bind to specific cell receptors, thereby initiating the production of antiviral proteins. Different natural interferons (human α-, β-, and γ-interferon) can be derived from cultures of leukocytes, fibroblasts, and T-lymphocytes after specific (e.g., viral) stimulation. For pharmacological use interferons are synthesized with recombinant DNA techniques. Their biological effects are strictly species-specific.

Interferon beta has been shown *in vitro* to be an effective suppressor of over 100 different kinds of viruses. Controlled clinical studies on its clinical antiviral efficacy in CNS infections are still lacking. Anecdotal reports suggest that β-interferon may have positive effects in cases of severe HSE, although these should be viewed with caution until controlled clinical trials are performed. CSF concentrations of β-interferon are only 2–3% of coincident serum levels, limiting its potential utility for treatment of CNS viral infections. Therapy is further complicated by β-interferon short serum half-life, which requires that therapy be given by continuous infusion. Clinical side effects include influenza-like symptoms of fever, fatigue, myalgia, and headache. Additional toxicity includes anemia, neutropenia, thrombocytopenia, and coagulopathies.

### General Measures

It is important to consider primary and secondary complications of viral CNS infections in order to provide early and adequate treatment. Primary complications include **abnormal ventilation**. Respiratory function must be closely monitored. Adequate oxygenation, prevention of pulmonary infection and aspiration are essential. Hypercapnia has to be avoided. **Increased intracranial pressure** results from cytotoxic brain edema, and may result in transtentorial herniation.

Specific therapy is discussed below. **Seizures** are a frequent complication of viral encephalitis and prophylactic therapy is advisable in most circumstances.

Secondary complications of viral CNS infections are common to acutely ill, chronically immobilized patients. They include development of pressure sores and decubitus ulcers, contractures, aspiration pneumonia, deep venous thromboses, gastrointestinal bleeding, urinary infections complicating in-dwelling catheters, and both local and systemic infection from in-dwelling venous, central, and arterial lines.

Artificial ventilation may be indicated, if:

- the respiratory rate is >35 breaths/min,
- the vital capacity is below 15 mL/kg body weight,
- the $pO_2$ levels falls below 65 mm Hg,
- the $pCO_2$ levels are above 55 mm Hg (except for patients with chronic hypercapnia),
  — for airway protection in comatose or obtunded patients,
  — for control of elevated ICP.

### Intracranial Pressure Elevation

Treatment of increased intracranial pressure includes appropriate positioning of the patient (elevation of the head of the bed to 30°–45°, avoidance of jugular vein compression by lateral rotation or anteflexion of the head and neck). Patients should be intubated and hyperventilated to maintain the $pCO_2$ at 25–30 mm Hg. Fluids should be restricted, and hypotonic fluids avoided. Hyperosmotic agents should be administered (e.g., mannitol 1.0 to 1.5 g/kg IV given over 15–30 min at 4- to 6-hour intervals).

Administration of a corticosteroids is no longer recommended in viral CNS infection because it may enhance the spread of the virus and has not been proven to effectively reduce cytotoxic brain edema (see also Chapters 55 and 56).

### Seizure Prophylaxis and Treatment

Seizures in viral encephalitis may be difficult to treat. Anti-convulsive drugs such as phenytoin should be given. If required, a more rapid anti-convulsive effect can be achieved by IV administration of clonazepam or diazepam. Use of these agents can result in severe sedation, respiratory depression, and hypotension. They should generally be employed only in intubated patients in an intensive care unit setting. Prophylactic use of anticonvulsants is recommended in cases of focal encephalitis or in the presence of frequent or paroxysmal sharp activity on the EEG.

## PRAGMATIC THERAPY

### Herpes Simplex Encephalitis (HSE)

Successful treatment of HSE depends upon rapid diagnosis and initiation of adequate therapy. Because of minimal toxicity, acyclovir should be given even empirically as soon as HSV encephalitis is suspected on the basis of a combination of clinical and laboratory features that may include:

1. an influenza-like prodromal illness followed by typical features of encephalitis;
2. Wernicke's aphasia and/or focal seizures with secondary generalization;
3. EEG changes indicating temporal lobe dysfunction (especially periodic sharp wave complexes) often combined with slowed background activity;
4. CSF lymphocytic pleocytosis (cell counts usually not in excess of 300/mm³) and slight CSF protein elevation with a normal CSF glucose;
5. normal CT normal during the first 3 days after the onset of neurological symptoms, then hypodense lesions appearing in the frontotemporal regions;
6. T2-weighted MR images that may display hyperintensities in the temporal lobe as early as 2 days after onset of symptoms;
7. PCR of CSF that allows rapid, noninvasive diagnosis in very early stages of HSV-encephalitis. CSF PCR amplification of HSV-specific DNA provides the most sensitive and specific diagnostic test currently available for HSE. PCR becomes positive early after onset of symptoms and in most cases remains positive for 2 weeks or longer. Results are generally available within a few days.

Temporal lobe biopsy may be considered in cases with atypical manifestations, insufficient response to treatment, or when reliable CSF PCR studies are not available. In the era when MRI was not widely available, biopsies in suspected HSE were reported to show non-HSV-related treatable diseases in up to 10% of cases (Whitley *et al.*, 1989). This result will undoubtedly become significantly lower with the widespread use of MRI and the availability of CSF PCR. Brain biopsy should only be performed at institutions with documented low levels of complications with the procedure.

### Treatment Schedules

1. **Acyclovir** (Zovirax) 10 mg/kg/every 8 h for a minimum of 14 days as IV infusion in 100 mL saline (minimum infusion time of 1 h) (Table VII).**

TABLE VII  Adjustment of Acyclovir Dosage Interval in Patients with Renal Insufficiency

| Creatinine clearance (ml/min) | Dosage interval |
| --- | --- |
| >50 | Every 8 hr |
| 50–25 | Every 12 hr |
| 25–10 | Every 24 hr |
| <10 | Half dosage every 24 hr |

2. In cases of allergic reactions to acyclovir, or acyclovir-resistant HSV, **vidarabine** can be administered as slow IV infusion of 15 mg/kg/day for 14 days (Whitley, 1986). Because the drug must be diluted, high fluid volumes may cause difficulties in patients with significant brain edema or mass lesions.**

3. Although serum levels achievable with oral valacyclovir can approximate those reached with intravenous acyclovir, the efficacy of oral anti-herpesviral drugs in the treatment of herpes simplex encephalitis is unknown and they should not be considered first-line therapy for HSE.

4. Prophylaxis against deep venous thrombosis can be achieved with low doses of heparin (e.g., 5000 IU/every 12 h subcutaneously) or through the use of pneumatic venous compression devices.

5. Anti-convulsive therapy (e.g., phenytoin 300 mg/day IV by slow infusion).

### Herpes Simplex Meningitis

There have been no controlled clinical trials of the efficacy of acyclovir or related drugs in the treatment of either primary or recurrent herpes simplex meningitis. Based on isolated case reports, treatment may reduce the duration and severity of disease. Patients who are acutely ill or have severe symptoms can be treated with intravenous acyclovir (5–10 mg/kg IV every 8 h). High-dose regimens of oral drugs including acyclovir (800 mg five times/day), famciclovir (500 mg tid), or valacyclovir (1000 mg tid) may be employed in less ill patients. Patients with recurrent episodes of herpes simplex meningitis may have milder or less frequent episodes if treated prophylactically with oral acyclovir (400 mg bid), famciclovir (250 mg bid), or valacyclovir (500 mg qd).**

### Herpes Zoster Myelitis and Encephalitis

The treatment of VZV infections is similar to that for HSE. The treatment of herpes zoster (shingles) is discussed in the chapter on inflammatory and infectious polyneuropathies.

### Central European Encephalitis (CEE)

To date there is no specific regimen or antiviral drug for the treatment of CEE. Postexposition prophylaxis should be considered only if simultaneous tick bites (>5) are detected in areas in which the virus is endemic. In central Europe endemic regions are located south of the river Main, along rivers in Baden-Wurttemberg, Bavaria, and in Austria. Passive immunization can be accomplished with immunoglobulin derived from donor plasma (FSME-Bulin) according to the following schedule:

0–48 h after a tick bite: 0.1 mL/kg body weight IM
48–96 h after a tick bite: 0.2 mL/kg body weight IM

Immunization more than 4 days after exposure is contraindicated. Protection is achieved within 24 h and lasts approximately 4 weeks.

The fatality rate of CEE in Germany is about 3–4%, but is considerably higher in other endemic regions in Europe.

### CMV Infection

Fifty to 60% of the healthy adult population are seropositive for CMV. CMV is an uncommon cause of neurologic disease in immunocompetent adults. However, CMV is an important pathogen in immuno-compromised individuals including organ and bone marrow transplant recipients and patients with HIV infection. In immunocompromised individuals, CMV may produce severe encephalitis, myelitis, or radiculomyelitis. These may occur as isolated manifestations, or in association with known CMV infection outside the nervous system including CMV retinitis, colitis, or pneumonitis.

**Acute therapy:**
  Ganciclovir (5 mg/kg/every 12 h IV) or foscarnet (60 mg/kg/every 8 h IV).
**Maintenance therapy:**
  Ganciclovir (5 mg/kg/day 6 or 7×/week) or foscarnet (90–120 mg/kg/day).**

Combination therapy with ganciclovir and foscarnet or therapy with cidofovir should be considered in patients who fail to respond to treatment with ganciclovir or foscarnet monotherapy.*

## Rabies

Infection with rabies causes fatal encephalitis with the brunt of the injury in the brainstem and limbic regions. The few reported survivors have received prophylactic therapy prior to the onset of neurologic symptoms. Virus is typically transmitted by the bite of an infected animal, although under special circumstances other forms of transmission, including inhalation of aerosolized virus, have been reported. Elimination of rabies in domestic canine populations through widespread vaccination programs has substantially reduced human rabies infection in developed countries.

The prodromal stage of rabies infection includes headache, fever, paresthesias, and pain at the inoculation site. This is followed by hydro- and aerophobia (50–90%), paralysis (20%), seizures (10%), and death (100%). CT and MRI scans are generally negative; however, in single cases brainstem and hypothalamic–pituitary abnormalities on MRI can be seen (Pleasure and Fischbein, 2000).

In cases of potential rabies exposure prophylaxis and treatment should be instituted following WHO guidelines (see Table VII), regardless of the interval from exposure. Vaccine schedules differ depending on the type and potency of vaccine employed.

## Enteroviral Infection

An experimental antiviral agent, pleconaril (ViroPharma VP 63843), is currently undergoing phase III clinical trials in the treatment of enteroviral meningitis. Preliminary studies suggest that the drug is both safe and effective. Pleconaril acts by intercalating into the viral outer capsid at the receptor binding site and thereby preventing virus attachment to host cells.

## REFERENCES

AIDS Research Group, in collaboration with the AIDS Clinical Trials Group (1992). Mortality in patients with the acquired immunodeficiency syndrome treated with either foscarnet or ganciclovir for cytomegalovirus retinitis. N. Engl. J. Med. 326, 213–220.

Andersen, O. (2000). Myelitis. Curr. Opin. Neurol. 13, 311–316.

Berlit, P. (1989). Immunoglobulin therapy in neurologic diseases. Klin. Wochenschr. 67, 967–970.

Boe, J., Solberg, C. O., and Saeter, T. (1965). Corticosteroid treatment for acute meningoencephalitis: A retrospective study of 346 cases. Br. Med. J. 1, 1094–1095.

Booss, J., and Esiri, M. M. (1986). "Viral Encephalitis—Pathology, Diagnosis and Management." Blackwell, Oxford, London, and Edinburgh.

Cinque, P., Vago, L., Brytting, M., Castagna, A., Accordini, A., Sundqvist, V. A., Zanchetta, N., Monforte, A. D., Wahren, B., and Lazzarin, A. (1992). Cytomegalovirus infection of the central nervous system in patients with AIDS: Diagnosis by DNA amplification from cerebrospinal fluid. J. Infect. Dis. 166, 1408–1411.

Cohen, B. A., McArthur, J. C., Grohman, S., Patterson, B., and Glass, J. D. (1993). Neurologic prognosis of cytomegalovirus polyradiculomyelopathy in AIDS. Neurology 43, 493–499.

Crumpacker, C. S. (1992). Mechanism of action of foscarnet against viral polymerases. Am. J. Med. 92, 3.

de Miranda, P., and Blum, M. R. (1983). Pharmacokinetics of acyclovir after intravenous and oral administration. J. Antimicrob. Chemother. 12(Suppl. B), 29–37.

Deray, G., Martinez, E., Katlama, C., Levaltier, B., Beaufils, H., Danis, M., Rozenheim, M., Baumelou, A., Dohin, E., Gentilini, M., et al. (1989). Foscarnet nephrotoxicity: Mechanism, incidence and prevention. Am. J. Nephrol. 9, 316–321.

Dieterich, D. T., Kotler, D. P., Busch, D. F., Crumpacker, C., Du Mond, C., Dearmand, B., and Buhles, W. (1993). Ganciclovir treatment of cytomegalovirus colitis in AIDS: A randomized, double-blind, placebo-controlled multicenter study. J. Infect. Dis. 167, 278–282.

Evans, A. S. (Ed.) (1976). "Viral Infections of Humans. Epidemiology and Control." Wiley, London, New York, Sydney, and Toronto.

Faulds, D., and Heel, R. C. (1990). Ganciclovir. A review of its antiviral activity, pharmacokinetic properties and therapeutic efficacy in cytomegalovirus infections. Drugs 39, 597–638.

Fletcher, C. V. (1992). Treatment of herpes virus infections in HIV-infected individuals. Ann. Pharmacother. 26, 955–962.

Fodor, P. A., Levin, M. J., Weinberg, A., Sandberg, E., Sylman, J., and Tyler, K. L. (1998). Atypical herpes simplex virus encephalitis diagnosed by PCR amplification of viral DNA from CSF. Neurology 51, 554–559.

Goswami, K. K., Miller, R. F., Harrison, M. J., Hamel, D. J., Daniels, R. S., and Tedder, R. S. (1991). Expression of HIV-1 in the cerebrospinal fluid detected by the polymerase chain reaction and its correlation with central nervous system disease. AIDS 5, 797–803.

Hatherill, M., Tibby, S. M., Sykes, K., Turner, C., and Murdoch, I. A. (1999). Diagnostic markers of infection: Comparison of procalcitonin with C reactive protein and leucocyte count. Arch. Dis. Child. 81, 417–421.

Hosoya, M., Honzumi, K., Sato, M., Katayose, M., Kato, K., and Suzuki, H. (1998). Application of PCR for various neurotropic viruses on the diagnosis of viral meningitis. J. Clin. Virol. 11, 117–124.

Jacobson, M. (1992). A review of the toxicities of foscarnet. J. Acquired Immune Defic. Syndr. 5, 11–17.

Jaffe, M., Srugo, I., Tiroh, E., Colin, A. A., and Tal, Y. (1989). The ameliorating effect of lumbar puncture in viral meningitis. Am. J. Dis. Child. 143, 682–685.

Jeffery, K. J., Read, S. J., Peto, T. E., Mayon-White, R. T., and Bangham, C. R. (1997). Diagnosis of viral infections of the central nervous system: Clinical interpretation of PCR results. Lancet 349, 313–317.

Kerkar, S., and Molavi, A. (1990). Postinfection complications of the central nervous system. In "Infections of the Nervous System" (D. Schlossberg, Ed.), pp. 135–142. Springer-Verlag, New York, Berlin, and Heidelberg.

Kubar, O. L., Brjantseva, E. A., Nikitina, L. E., and Zlydnikov, D. M. (1989). The importance of virus drug-resistance in the treatment of influenza with rimantadine. Antiviral Res. 11, 313–315.

Levy, R. M., Bredesen, D. E., and Rosenblum, M. L. (1985). Neurological manifestations of the acquired immunodeficiency syndrome (AIDS): Experience at UCSF and review of the literature. J. Neurosurg. 62, 475–495.

Little, J. W., Hall, W. J., Douglas, R. G., Jr., Mudholkar, G. S., Speers, D. M., and Patel, K. (1978). Attenuation of airway hyperreactivity by amantadine in natural influenza A infection. Am. Rev. Respir. Dis. 118, 295–303.

Mateos-Mora, M., and Ratzan, K. R. (1990). Acute viral encephalitis. In "Infections of the Nervous System" (D. Schlossberg, Ed.), pp. 105–134. Springer-Verlag, New York, Berlin, and Heidelberg.

McKendall, R. R. (1982). Pharmacology of antiviral chemotherapeutic agents useful in human viral infections of the nervous system. Clin. Neuropharmacol. 5, 115–129.

Meyers, J. D., Lescynski, J., Zaja, J. A., Fluornoy, N., Newton, B., Snydman, D. R., Wright, G. G., Levin, M. J., and Thomas, E. D. (1983). Prevention of cytomegalovirus infection by cytomegalovirus immunoglobulin after marrow transplantation. Ann. Intern. Med. 98, 442–446.

Mustafa, M. M., Weitman, S. D., Winick, N. J., Bellini, W. J., Timmons, C. F., and Siegel, J. D. (1993). Subacute measles encephalitis in the young immunocompromised host: Report of two cases diagnosed by polymerase chain reaction and treated with ribavirin and review of the literature. Clin. Infect. Dis. 16, 654–660.

Najioullah, F., Bosshard, S., Thouvenot, D., Boibieux, A., Menager, B., Biron, F., Aymard, M., and Lina, B. (2000). Diagnosis and surveillance of herpes simplex virus infection of the central nervous system. J. Med. Virol. 61, 468–473.

O'Brien, J. J., and Campoli-Richards, D. M. (1989). Acyclovir. An updated review of its antiviral activity, pharmacokinetic properties and therapeutic efficacy. Drugs 37, 233–309.

Pleasure, S. J., and Fischbein, N. J. (2000). Correlation of clinical and neuroimaging findings in a case of rabies encephalitis. Arch. Neurol. 57, 1765–1769.

Prange, H. W., Hacke, W., and Felgenhauer, K. (1985). Diagnostik und Therapie der Herpes simplex-Enzephalitis. Aktuel. Neurol. 12, 217–225.

Reik, L., Jr., and Barwick, M. C. (1990). Noninfectious causes of acute CNS inflammation. In Infections of the Nervous System (D. Schlossberg, Ed.), pp. 73–89. Springer-Verlag, New York, Berlin, and Heidelberg.

Sande, M. A., and Mandell, G. L. (1985). Antimicrobial agents, antifungal and antiviral agents. In "The Pharmacological Basis of Therapeutics" (L. S. Goodman, A. G. Gilman, T. W. Rall, and F. Murad, Eds.), 7th ed., pp. 1219–1239. Macmillan, New York, Toronto, and London.

Schorre, W. (1979). Die Infektionskrankheiten des Nervensystems. Urban & Schwarzenberg, München, Wien and Baltimore, Maryland.

Schroth, G., Gawehn, J., Thron, A., Vallbracht, A., and Voigt, K. (1987). Early diagnosis of herpes simplex encephalitis by MRI. Neurology 37, 179–183.

Schumacher, K. (1986). Therapie mit Immunoglobulinen. Dtsch. Med. Wochenschr. 111, 550–556.

Sköldenberg, B., Alestig, K., Burman, L., Forkman, A., Lövgren, K. et al. (1984). Acyclovir versus vidarabine in herpes simplex encephalitis. Lancet 2, 707–711.

Sormunen, P., Kallio, M. J., Kilpi, T., and Peltola, H. (1999). C-reactive protein is useful in distinguishing Gram stain-negative bacterial meningitis from viral meningitis in children. Pediatrics 134, 725–729.

Spanish Ribavirin Trial Group (1991). Comparison of ribavirin and placebo in CDC group III human immunodeficiency virus infection. Lancet 338, 6–9.

Spector, S. A., Kennedy, C., McCutchan, J. A., Bozzette, S. A., Straube, R. G., Connor, J. D., and Richman, D. D. (1989). The antiviral effect of zidovudine and ribavirin in clinical trials and the use of p24 antigen levels as a virologic marker. J. Infect. Dis. 159, 822–828.

Stevens, D. A., and Marigan, T. C. (1980). Zoster immune globulin prophylaxis of disseminated zoster in compromised hosts. A randomized trial. Arch. Intern. Med. 140, 52–54.

Tsuchiya, K., Katase, S., Yoshino, A., and Hachiya, J. (1999). Diffusion-weighted MR imaging of encephalitis. Am. J. Roentgenol. 173, 1097–1099.

Tsuchiya, K., Katase, S., Yoshino, A., and Hachiya, J. (2000). MRI of influenza encephalopathy in children: Value of diffusion-weighted imaging. J. comput. Assist. Tomogr. 24, 303–307.

Tyler, K. L. (1984). Diagnosis and management of acute viral encephalitis. Semin. Neurol. 4, 480–489.

Tyler, K. L., Tedder, D. G., Yamamoto, L. J., Klapper, J. A., Ashley, R., Lichtenstein, K. A., and Levin, M. J. (1995). Recurrent brainstem encephalitis associated with herpes simplex virus type 1 DNA in cerebrospinal fluid. Neurology 45, 2246–2250.

Tyms, A. S., Taylor, D. L., and Parkin, J. M. (1989). Cytomegalovirus and the acquired immunodeficiency syndrome. J. Antimicrob. Chemother. 23(Suppl. A), 89–105.

Wakamoto, H., Ohta, M., Nakano, N., and Kunisue, K. (2000). SPECT in focal enterovirus encephalitis: Evidence for local cerebral vasculitis. Pediatr. Neurol. 23, 429–431.

Warrell, M. J., White, N. J., Looareesuwan, S., Phillips, R. E., Suntharasamai, P., and Chanthavanich, P. (1989). Failure of interferon alfa and ribavirin in rabies encephalitis. Br. Med. J. 299, 830–833.

Whitley, R. I., Cobbs, C. G., Alford, C. A., Jr., Soong, S. J., Hirsch, M. S., Connor, J. D., Corey, L., Hanley, D. F., Levin, M., and Powell, D. A. (1989). Diseases that mimic herpes simplex encephalitis. Diagnosis, presentation, and outcome. NIAID Collaborative Antiviral Study Group. J. Am. Med. Assoc. 262, 234–239.

Whitley, R. J. (1988a). Herpes simplex infections of the central nervous system. A review. Am. J. Med. 85(Suppl. 2A), 61–67.

Whitley, R. J. (1988b). Editorial: The frustrations treating herpes simplex virus infections of the central nervous system. J. Am. Med. Assoc. 259, 1067.

Whitley, R. J., Soong, S. J., Dolin, R., Galasso, G. J., Chien, L. T., and Alford, C. A., NIAID Collaborative Antiviral Study Group (1977). Adenine arabinoside therapy of biopsy-proven herpes simplex encephalitis. N. Engl. J. Med. 297, 289–294.

Whitley, R. J., Alford, C. A., Hirsch, M. S., Schooley, R. T., Luby, J. P., Aoki, F. Y., Hanley, D., Nahmias, A. J., and Soong, S. J. (1986). Vidarabine versus acyclovir therapy in herpes simplex encephalitis. N. Engl. J. Med. 314, 144–149.

World Health Organization (1992). "Prevention of Rabies in Humans and Annex 2. WHO Expert Committee on Rabies," Eighth Report, pp. 21–25 and 53–56.

CHAPTER 50

# Fungal Infections of the Central Nervous System

Martin Rösener and Geraint Fuller

Fungi are found throughout the world, with distributions influenced by climate, geography, host exposure, and immunity. Fungal infections of the central nervous system (CNS) are now diagnosed with increasing frequency because of the extended use of immunosuppressive or cytostatic treatments, broad-spectrum antibiotics, CNS invasive procedures, and the acquired immunodeficiency syndrome (AIDS) epidemic.

Fungi are eukaryotic organisms with polysaccharide-based cell walls. On the basis of their reproductive habits, they are categorized as yeasts or molds. Fungi cause disease in humans either by direct tissue invasion or by noninvasive means, such as allergic mechanisms or toxin production. A well-known mycotoxin is ergotamine, a product of the rye ergot. *Cryptococcus neoformans*, *Coccidioides immitis*, *Paracoccidioides brasiliensis*, *Histoplasma capsulatum*, and *Blastomyces dermatitidis* are known as pathogenic, or endemic, fungi for their ability to cause invasive disease in normal hosts. *Candida albicans*, *Aspergillus fumigatus*, and *Zygomycetes rhizopus* (*Mucor*) are considered opportunistic fungi and are almost exclusively pathogens of immunocompromised patients. Infection of the CNS usually occurs as a result of hematogenous spread and less frequently by direct invasion, for example from sinusitis or otitis. Fungal infections of the CNS in human immunodeficiency virus infections are considered in Chapter 51.

## CLINICAL SIGNS AND SYMPTOMS

Most patients present with a subacute meningitis developing over 1 to 4 weeks associated with a low-grade fever, malaise, altered mental state, and lethargy. Patients may later develop focal signs, seizures, and multiple cranial nerve palsies. The patient may have other systemic symptoms that can be useful in the diagnosis of the source of the fungal infection. Other organ involvement may represent other sites of hematogenous spread of the fungus and can provide a more accessible site from which to characterize the fungus, either by biopsy or culture. Particular sites of infection are the lungs, sinuses, kidney, joints, and bone.

Brain imaging is useful to look for focal abnormalities, hydrocephalus, or sinus diesase. MRI is more sensitive than CT. CSF examination finds a lymphocytic pleocytosis. Eosinophilia suggests coccidiodal meningitis. CSF glucose is usually depressed, and CSF protein elevated, sometimes markedly. CSF oligoclonal bands are usually present. CSF Gram stain is usually not helpful, India ink may identify *C. neoformans*. CSF culture is infrequently positive (*vide infra*). Repeat lumbar puncture, taking large volumes, may increase diagnostic yield. Because of the difficulty isolating the fungi, serological tests or identification of the fungus from other organs is important in making the diagnosis. Clues as to other organ involvement usually arise in the history or general physical and ophthalmological examination. Chest x-rays, sinus imaging, bone scans, and urine examination may all contribute.

The differential diagnosis of chronic menigitis includes tuberculosis, neurosyphilis, brucellosis, actinomycetes infections, protozoal infections, meningeal carcinomatosis or lymphoma, benign lymphocytic meningitis, paramenigeal bacterial infection, the uveomeningitic syndromes (sarcoid, Behçet's disease, Vogt-Koyanagi-Harada syndrome), granulomatous angiitis, collagen vascular disorders, hemosiderosis, resolving subarachnoid hemorrhage, subdural hematoma, and drug reactions.

## Parenchymal Syndromes

Patients may also present with the clinical signs of an intracranial mass. The evolution of these is subacute, with headache and localizing signs, seizures, decrements in mentation or consciousness, and evidence of increased intracranial pressure, with or without fever. Fungal CNS infections cause diffuse granulomatous infiltrates, solitary abscesses, multiple microabscesses or macroabscesses, or vasculitic disease, spanning several neuropathological categories of meningeal and parenchymal brain or spinal cord disease. Often there is an overlap between meningeal, vasculitic, and parenchymal disease. These pathological processes can result in acute syndromes, such as stroke, epidural spinal cord compression, or cauda equina syndromes. One specific syndrome is the rhinocerebral syndrome.

### Rhinocerebral Syndrome

The rhinocerebral syndrome is a progressive fungal infection affecting the sinuses, orbit, and brain (DeShazo *et al.*, 1997) most commonly related to Zygomycetes or *Aspergillus* infections. An acute, invasive fungal sinusitis, beginning with nasal stuffiness and discharge, is followed by the development of a painless, necrotic black palatal or nasal septal ulcer, and rapid spread of infection to contiguous soft tissue, orbit, skull, and intracranial sites. Spread of infection is signalled by face pain and headache, fever, orbital celluitis, proptosis, eye movement abnormalities, orbital apex syndrome, or cavernous sinus or internal carotid artery thrombosis. The syndrome is seen in diabetics, particulary with ketoacidosis, leukemic patients who have been neutropenic for long periods of time and receive broad-spectrum antibiotics, organ transplant recipients, patients receiving deferoxamine chelation therapy for hemochromatosis, individuals with malnutition or lactic acidosis, and, rarely, in previously healthy people. The differential diagnosis includes aggressive orbital tumors and bacterial sinus or periorbital infection with septic venous thrombosis.

## CANDIDIASIS

*Candida* species colonize the normal human respiratory, gastrointestinal, and genitourinary tracts. However, in immunodeficient patients *Candida* species are important causes of nosocomial infections. *C. albicans* is the most common cause of CNS infection. Infections usually occur in patients with malignancies, severe disease or debilitation, patients taking corticosteroids or broad-spectrum antibiotics, transplant recipients, and postoperative neurosurgery patients (Nguyen and Yu, 1995). Infection of the CNS occurs as chronic meningitis or a brain abscess and is almost always a result of hematogenous spread from other organs, mainly from the gastrointestinal tract. Cerebral involvement with multiple small (<2 mm) subcortical abscesses is frequently overlooked because of lack of meningeal signs and rare focal signs (Pendlebury *et al.*, 1989). Diagnosis can be made by finding *Candida* on direct microscopy of CSF, by CSF culture, or by CSF *C. albicans* antigen enzyme-linked immunosorbent assay. CSF culture with resistance testing should always be carried out to allow logical antimycotic therapy. Despite antimycotic treatment, death or incomplete recovery occurs in 10%–20% of patients.

## CRYPTOCOCCOSIS

*Cryptococcus* is the most common cause of clinically recognized fungal meningitis and meningocerebral syndromes. Cryptococcosis is caused by *C. neoformans*, whose natural habitat is pigeon droppings. Infection is acquired by inhalation of fungus into the lungs. Hematogenous spread leads to CNS involvement. The clinical signs are bitemporal headaches, nausea, staggering gait, dementia, irritability, confusion, blindness, and sometimes deafness. They are caused by a meningoencephalitis with microgranulomas that can be shown on MRI (Cochius *et al.*, 1989). Diagnosis is made by identifying cryptococci on an India ink smear of CSF sediment, on culture, or by cryptococcal antigen detection by latex agglutination. Approximately 50% of CSF India ink preparations and 75% of CSF culture results are positive in cases of cryptococcal meningitis. Without appropriate therapy, CNS infection is invariably fatal, with a survival period from 2 weeks to several years from the onset of symptoms.

## ASPERGILLOSIS

*A. fumigatus* is the most common pathogenic form of *Aspergillus* species and is ubiquitous in the environment. Inhalation of *Aspergillus* spores is extremely common, but invasion of lung tissue is almost entirely confined to immunosuppressed individuals. Hematogenous spread may lead to cerebral abscess formation and to infiltration of vessel walls, causing arterial occlusion and hemorrhagic infarctions. Meningeal signs are rare. Invasive *Aspergillus* sinusitis is one cause of the rhinocerebral syndrome. Detection of *Aspergillus* by microscopic examination, antigen detection, or culture is difficult. There is currently no reliable serodiagnosic marker for aspergillosis infection. Diagnosis rests on demonstration of tissue invasion on biopsy (Sanchez and Noskin, 1996). Without appropriate therapy, patients survive 2–8 months from the onset of symptoms.

## HISTOPLASMOSIS

*H. capsulatum*, the cause of histoplasmosis, rarely leads to cerebral mycosis. After inhalation and primary infection of the lungs, hematogenous spread may lead to CNS involvement in 10%–20% of patients, usually presenting as a chronic meningoencephalitis and rarely as granulomas or abscesses. Detection by microscopic examination or culture of CSF is rarely achieved. Blood or bone marrow culture results may be positive. Otherwise, diagnosis depends on detection of antibodies by complement fixation. However, the sensitivity and specificity of serological tests are not high. Antigen detection methods may provide a means of following treatment response. Histoplasma meningitis is difficult to treat, with frequent relapses and cures in only approximately 50% of patients.

## COCCIDIOIDOMYOSIS

*C. immitis* is a highly infectious fungus that occurs in the arid zones of the Western and Southwestern United States and similar geological areas of Central and South America. In most cases chronic pulmonary disease that follows inhalation of fungal spores is self-limiting. Approximately 1% of patients have disseminated disease. Immunosuppression poses a higher risk on dissemination. Osteoarticular disease, including lytic skull and vertebra lesions, is seen in approximately one third of patients with disseminated disease. Meningitis usually occurs within 6 months of symptomatic or asymptomatic primary infection and may be the initial manifestation of disease. The clinical picture is nonspecific and characterized by fever, weakness, behavioral changes, mental slowness, ataxia, vomiting, seizures, cranial nerve palsies, or focal deficits. CSF culture is positive in less than 50% of meningitis cases; 70% of patients have positive complement-fixing antibodies in the CSF.

## PARACOCCIDIOIDOMYCOSIS

Paracoccidioidomycosis, endemic in Central and South America, particularly in Brazil, usually a self-limited pulmonary infection, is caused by *P. brasiliensis*. Rarely, disseminated disease with cerebral and cerebellar masses has been recognized.

## BLASTOMYCOSIS

Blastomycosis, caused by *B. dermatitidis*, mostly occurs in North America. The primary disease is a mild respiratory infection after inhalation of the fungus. In immunosuppressed patients, especially patients with AIDS, CNS disease occurs as intracranial or spinal abscesses, meningitis, or osteolytic infection of vertebrae.

Diagnosis is difficult because direct visualization or culture is rarely achieved. Serodiagnosis remains problematic (Areno *et al.*, 1997).

## ZYCOMYCOSIS

There are two orders of Zygomycetes-containing organisms that cause human disease, the Mucorales and the Entomophthorales. Most human illness is caused by the Mucorales, *Rhizopus* spp. Human disease with these organisms occurs predominantly in tropical regions, with transmission occurring by implantation of spores from minor trauma such as insect bites or by inhalation of spores into the sinuses. Human zygomycosis generally occurs in immunocompromised hosts as opportunistic infections. Host risk factors include diabetes mellitus, neutropenia, sustained immunosuppressive therapy, chronic prednisone use, iron chelation therapy, broad-spectrum antibiotic use, severe malnutrition, and primary breakdown in the integrity of the cutaneous barrier such as trauma, surgical wounds, needlesticks, or burns. Zygomycosis occurs only rarely in immunocompetent hosts. The disease manifestations reflect the mode of transmission, with rhinocerebral and pulmonary diseases being the most common manifestations. The Mucorales are associated with angioinvasive disease, often leading to thrombosis, infarction of involved tissues, and tissue destruction mediated by a number of fungal proteases, lipases, and mycotoxins. Serodiagnostic tests for mucormycosis remain insensitive (Sanchez and Noskin, 1996). Overall, mortality from rhinocerebral mucormycosis is greater than 50%.

## PRINCIPLES OF THERAPY

Polyene macrolide antimycotics (amphotericin B), flucytosine, and triazoles (fluconazole, itraconazol) are of value in the treatment of CNS mycoses. Miconazole and ketoconazole currently have no role in the primary treatment of serious CNS fungal infections. No universal consensus as to optimal regimen, total dose, or total duration of therapy exists. All antimycotic agents have a small therapeutic range. Therefore, severe side effects

TABLE I  Pharmacological Properties of Antifungal Agents

| Property | Amphotericin B | Flucytosine | Fluconazole | Itraconazole |
|---|---|---|---|---|
| Oral bioavailibility (%) | <5 | >80 | >80 | >70 |
| Protein binding (%) | 91–95 | 4 | 11 | >99 |
| Peak plasma concentration (μg/ml) | 1.2–2.0 | 30–45 | 10.2 | 0.2–0.4 |
| Dose (mg) | 50 IV | 2000 PO | 200 PO | 200 PO |
| Time to peak plasma concentration (h) | — | 2 | 2–4 | 4–5 |
| Terminal elimination half-life | 15 days | 3–6 h | 22–31 h | 24–42 h |
| Unchanged drug in urine (%) | 3 | >75 | 80 | <1 |
| CSF concentration (%) | 2–4 | >75 | >70 | <1 |

Modified from Como and Dismukes (1994).

are to be expected. Table I shows the most important pharmacological data, and Table II lists the main adverse effects.

**Amphotericin B** is the treatment of choice for most fungal infections of the CNS, particularly those that are life-threatening. It is a polyene macrolide antibiotic, which combines with sterol in the fungal cytoplasmic membrane, increasing membrane permeability. Within the bloodstream, the drug first binds to plasma lipoproteins and then transfers to tissues throughout the body. Penetration into the CSF is poor, being independent of the grade of inflammation of the meninges. Intrathecal administration (maximal 0.5 mg/injection) is necessary if IV therapy fails or if after a complete course of IV

therapy disease recurs. Severe illness and immunosuppression also necessitate intrathecal administration. Intracranial distribution of amphotericin B is poor after lumbar administration because of arachnoid adhesions, and intraventricular administration by way of an Ommaya reservoir is a more logical alternative. Complications of intraventricular administration include shunt infections with *S. epidermidis*, shunt occlusions, ventriculitides, encephalopathies, and seizures.

Three new lipid formulations of amphotericin B are now available, amphotericin B lipid complex, amphotericin B colloidal dispersion, and liposomal amphotericin B. These newer formulations are substantially more expensive but allow patients to receive higher

TABLE II  Adverse Effects and Contraindications of Antifungal Drugs

| Organ or system | Amphotericin B | Flucytosine | Fluconazole | Itraconazole |
|---|---|---|---|---|
| Gastrointestinal tract | Nausea, vomiting, anorexia | Nausea, vomiting (5% of patients), diarrhea, abdominal pain | Nausea, vomiting (<5% of patients) | Nausea, vomiting (<10% of patients) |
| Skin | — | Rash | Rash, Stevens-Johnson syndrome | Pruritus, rash |
| Liver | — | Asymptomatic liver-enzyme elevation (7%), hepatitis (rare) | Asymptomatic liver-enzyme elevation (1–7%), hepatitis (rare) | Asymptomatic liver-enzyme elevation (1–5%), hepatitis (rare) |
| Bone marrow | Anemia | Anemia (less common), leukopenia, thrombocytopenia | — | — |
| Kidney | Azotemia (80%), renal tubular acidosis, hypokalemia, hypomagnesemia | — | — | — |
| Endocrine system | — | — | — | Hypokalemia, hypertension, edema, impotence (rare) |
| Other | Thrombophlebitis, headache, fever, chills | Headache, confusion, hair loss | Headache, seizure | Headache, dizziness |
| Contraindications | Severe liver or renal failure | — | Severe liver failure, children <16 years of age | Children <18 years of age |
| Pregnancy | Under careful consideration of risks | Under careful consideration of risks | Contraindicated | Contraindicated |

Modified from Como and Dismukes (1994).

doses for longer periods of time with decreased renal toxicity than conventional amphotericin B (Robinson and Nahata, 1999).

Because of the high cost, their indication should be limited to patients

- who do not respond to conventional amphotericin B
- who show nephrotoxic side effects with conventional amphotericin B
- in whom conventional amphotericin B is contraindicated because of renal failure.

Allergic reaction to the liposomal component of liposomal amphotericin B has been described (Cesaro *et al.*, 1999).

**Flucytosine** is converted within the fungal cell to the antimetabolite 5-fluorouracil by the enzyme cytosinedesaminase. Flucytosine is well absorbed from the gastrointestinal tract even in the presence of food. Drug resistance appears rather rapidly when flucytosine is used alone. For this reason the drug should only be used in conjunction with amphotericin B, which also permits a lower dose of amphotericin B. In the treatment of cryptococcosis these two drugs have a synergistic action; amphotericin B increases membrane permeability, allowing better penetration of flucytosine into the fungal cell. This combination should always be the first line of treatment for cryptococcosis. Unfortu-

nately, there is a failure rate of 30% (Sugar *et al.*, 1990). The same combination should be used in aspergillosis and candidiasis until resistance tests have been completed.

**Fluconazole**, a triazole, interferes with ergosterol synthesis in the fungal cell membrane by inhibition of a cytochrome P-450–dependent demethylase. It is generally well tolerated and has good CNS penetration. It is used to treat coccidioidal meningitis. In addition, in patients with cryptococcal meningitis, fluconazole is often used after induction with amphotericin B and flucytosine. Fluconazole can be administered orally and intravenously.

**Itraconazole** is a triazole, which like the other azoles acts by inhibition of ergosterol synthesis in the fungal cell wall. It is effective for CNS histoplasmosis and blastomycosis after induction with amphotericin B.

## PRACTICAL MANAGEMENT

Before starting treatment, the pathogen should always be cultured and when possible its pattern of resistance determined. The standard therapy of different fungal infections of the CNS in immunosuppressed and immunocompetent patients is given in Table III. In this section the method of administration and dosage schedules will be considered.

**TABLE III  Therapy of Cerebral Mycoses**

| Mycosis | Immunocompetent host | Immunosuppressed host without AIDS |
|---|---|---|
| Cryptococcosis | AMB, 0.3–0.5 mg/kg/day, with 5-FC, 150 mg/kg/day (in four divided doses) for 6 weeks or AMB, 0.3–0.5 mg/kg/day for 10 weeks | AMB, 0.7 mg/kg/day, with 5-FC, 150 mg/kg/day (in four divided doses) for at least 6 weeks |
| Histoplasmosis | AMB, 500–1000 mg total until stable, followed by ITRA, 400 mg/day for 6 months | AMB, 1000 mg total until stable, followed by ITRA, 400 mg/day for 6 months |
| Coccidioidomycosis | FLU, 400–800 mg/day for 12 months or longer or AMB 0.3–0.5 mg/kg/day, followed by FLU, 400–800 mg/day for at least 12 months | AMB 0.3–0.5 mg/kg/day, followed by FLU, 400–800 mg/day for life |
| Aspergillosis | AMB, 0.7–1.5 mg/kg/day, followed by ITRA, 400–800 mg/day | AMB, 1.5 mg/kg/day, followed by ITRA, 400–800 mg/day |
| Blastomycosis | AMB, 2000–3000 mg total until stable, followed by ITRA, 400 mg/day for 6 months | AMB, 2000–3000 mg total until stable, followed by ITRA, 400 mg/day for 6–12 months |
| Candidiasis | AMB, 0.3–0.5 mg/kg/day, with 5-FC, 150 mg/kg/day (in four divided doses) for 4–6 weeks | AMB, 0.3–0.5 mg/kg/day, with 5-FC, 150 mg/kg/day (in four divided doses) for 4–6 weeks |
| Zygormycosis | AMB 1.5 mg/kg/day | AMB, 1.5 mg/kg/day |

AMB, amphotericin B; 5-FC, 5-fluorocytosine (flucytosine); FLU, fluconazole; ITRA, itraconazole.

**TABLE IV    Dose Escalation with Amphotericin B**

Standard dose
Day 1: test dose of 1 mg amphotericin B in 20 ml 5% glucose
   solution IV; when tolerated: 5 mg in 500 ml 5% glucose solution
   IV 4 hours later.
Day 2: 10 mg in 500 ml 5% glucose solution IV.
Daily elevation by 5 mg until the level dose is reached. Then
   administration every second day.

A rapidly escalating dose for immunosuppressed or severely
   threatened patients
Day 1: test dose of 1 mg amphotericin B in 20 ml 5% glucose
   solution IV; when tolerated: 0.2 mg/kg body weight in 500 ml 5%
   glucose solution IV 4 hours later.
Day 2: 0.4 mg/kg body weight in 500 ml 5% glucose solution IV.
When treatment is continued with flucytosine, administration of
   amphotericin B should be on alternate days, when given as a
   monotherapy daily escalation of 0.2 mg/kg body weight daily
   until the plateau dose of up to 1.5 mg/kg is reached.

## AMPHOTERICIN B

Amphotericin B is introduced in an escalating dosage, and details of this are given in Table IV. To minimize adverse effects, 500 mg of paracetamol or 500 mg of metamizole should be given before each intravenous infusion: After the dose escalation, the final dose of amphotericin B plateaus at about 0.7–1.5 mg/kg body weight every second day if used alone or at 0.3–0.5 mg/kg body weight every second day if given with flucytosine.

## PREPARATION OF SOLUTIONS FOR IV INFUSION

Every solution should be prepared immediately before administration.

Concentration must not exceed 0.1 mg amphotericin B/ml 5% glucose solution.

The dry powder (amphotericin B and desoxycholat) is disolved in 10 ml of water for injection without preservatives. The necessary amount of the clear suspension is added to a 5% glucose solution (250 or 500 ml). Salts or other drugs (except heparin) must not be added to the infusion solutions, because the colloidal drug will precipitate.

The infusion and the infusion system must be protected from light (aluminium foil).

Amphotericin B is extremely irritant, so when the solution is administered by means of a peripheral venous line, 1000 IU heparin can be added to the infusion to prevent formation of venous thrombosis.

The infusion should last 4–6 hours; the rate of administration should be adjusted according to the side effects.

Depending on the response, amphotericin B monotherapy should be continued for 6–8 weeks or 4 weeks after the last positive culture.

The cumulative dose of a full course of amphotericin B is 1500–2000 mg. It should not exceed 4000 mg because of the risk of an irreversible nephropathy.

## INTRATHECAL APPLICATION OF AMPHOTERICIN B

Indications for the additional intrathecal application are as follows:

- Systemic therapy failed (no clearcut improvement after 4 weeks of treatment).
- Relapse after a full course of systemic treatment.
- Patient is moribund at the beginning of therapy.
- Severe immunosuppression.

Amphotericin B can be given by means of a lumbar puncture or intraventricularly with an Ommaya reservoir. In the case of hydrocephalus caused by cisternal adhesions intraventricular administration is compulsory.

## PREPARATION OF SOLUTION FOR INTRATHECAL INJECTION

- The solution must be prepared directly before administration.
- The concentration is 0.25 mg amphotericin B/ml 5% glucose solution.
- After a test dose of 0.025 mg (0.10 ml solution), intrathecal injection is given every second day with 0.025 mg more amphotericin B until a level dose of 0.50 mg is reached. After this injection, 0.50 mg is given twice weekly.

## LABORATORY CONTROLS

- Twice weekly: blood count, reticulocytes, sodium, potassium, creatinine, urea, ALT, AST, GGT, alkaline phosphatase, bilirubin, and urine tests.
- Once weekly: creatinine clearance.
- Once weekly: CSF for cell count, protein, glucose, lactate, culture, and antigen. With intrathecal therapy, every CSF sample should be examined.
- Fortnightly: CT scans to detect adhesions that could lead to occlusive hydrocephalus.
- With acute renal insufficiency, therapy must be discontinued for 2–3 days and continued with a lowered dose of 0.2–0.4 mg/kg body weight.

**Flucytosine** should be used in combination with another antifungal agent only. The daily dosage is 150 mg/kg body weight divided into three or four doses (e.g., 50 mg/kg body weight every 8 hours polly or IV).

With renal insufficiency, the dose must be reduced:

- With a creatinine 2.0–3.5 mg/dl (GFR 40–25 ml/min) to 50 mg/kg body weight every 12 hours.
- With a creatinine 3.5–6.0 mg/dl (GFR 20–10 ml/min) to 50 mg/kg body weight every 24 hours.
- With a creatinine >6.0 mg/dl (GFR <10 ml/min) to 50 mg/kg body weight every 48 hours.

Determination of serum levels is recommended. Dosage should be adjusted to reach therapeutic serum levels of 60–80 µg/ml.

Resistance tests should be undertaken before therapy and every 2 weeks thereafter.

**Fluconazole** is the first-line treatment for immunosuppressed, leukopenic and anemic patients. The initial dose is 400 mg/day for 3 days and then 200 mg/day PO or IV for a treatment period of 10 days–3 months. Fluconazole is used in combination with other antifungal agents, for example, with amphotericin B and flucytosine in the treatment of cryptococcosis. Fluconazole provides a useful prophylaxis of recurrent candidiasis (200 mg every second day for 1 week) and cryptococcosis (200 mg/day for 3 weeks), especially in patients with AIDS.

Weekly controls of liver enzymes and creatinine are recommended.

**Itraconazole** is given orally. Because low gastric pH facilitates resoption, antacids, $H_2$ blocker, or blocker of proton pump should be given 2 hours apart from itraconazole. The usual dosage is 200–400 mg/day.

## TREATMENT OF COMPLICATIONS OF FUNGAL INFECTION

Obstructive **hydrocephalus** is the most common complication of fungal CNS disease, occurring in 10% of cases. Internal shunt systems should be withheld until CSF culture is negative because of the high complication rate of internal shunt systems, with fungal spread, shunt sepsis, or shunt occlusion (Young *et al.*, 1985). Urgent decompression can be achieved by using an external ventricular drain.

Parenchymous infections with **abscesses** and **granulomas** that exert a mass effect can be caused by *Aspergillus*, *Cryptococcus*, *Candida*, and *Histoplasma*. They should be treated like bacterial brain abscesses. Single abscesses or granulomas in a suitable site should be resected either openly or, if possible, by use of a stereotactic procedure (Goodman and Coffey, 1989). This relieves the mass effect and provides material for culture and histological examination. If the pathogen is known before surgery, then the maximal tolerable antimycotic treatment should be given for 48 hours before operating.

Multiple abscesses and granulomas are treated medically, and the treatment is followed by repeated CT scans. Mortality can be reduced almost by half with maximal medical therapy and early neurosurgical intervention (from 64%–39%), according to Young *et al.*, 1985.

## REFERENCES

Areno, J., Campbell, G. J., and George, R. (1997). Diagnosis of blastomycosis. *Semin. Respir. Infect.* **12**, 252–262.

Cesaro, S., Calore, E., Messina, C., and Zanesco, L. (1999). Allergic reaction to the liposomal component of liposomal amphotericin B. *Support Care Cancer.* **7**, 284–286.

Cochius, J. I., Burns, R. J., and Willoughby, J. O. (1989). CNS cryptococcosis: unusual aspects. *Clin. Exp. Neurol.* **26**, 183–191.

Como, J. A., and Dismukes, W. E. (1994). Oral azole drugs as systemic antifungal therapy. *N. Engl. J. Med.* **330**, 263–272.

DeShazo, R., Chapin, K., and Swain, R. (1997). Fungal sinusitis. *N. Engl. J. Med.* **337**, 254–259.

Goodman, M. L., and Coffey, R. J. (1989). Stereotactic drainage of *Aspergillus* brain abscess with long-term survival: case report and review. *Neurosurgery* **24**, 96–99.

Nguyen, M., and Yu, V. (1995). Meningitis caused by *Candida* species: an emerging problem in neurosurgical patients. *Clin. Infect. Dis.* **21**, 323–327.

Pendlebury, W. W., Perl, D. P., and Munoz, D. G. (1989). Multiple microabscesses in the central nervous system: a clinicopathologic study. *J. Neuropathol. Exp. Neurol.* **48**, 290–300.

Robinson, R. F., and Nahata, M. C. (1999). A comparative review of conventional and lipid formulations of amphotericin B. *J. Clin. Pharm. Ther.* **24**, 249–257.

Sanchez, J., and Noskin, G. (1996). Recent advances in the management of opportunistic fungal infections. *Compr. Ther.* **22**, 703–712.

Sugar, A. M., Stern, J. J., and Dupont, B. (1990). Overview: treatment of cryptococcal meningitis. *Rev. Infect. Dis.* **12** (**Suppl 3**), S338–348.

Young, R. F., Gade, G., and Grinnell, V. (1985). Surgical treatment for fungal infections in the central nervous system. *J. Neurosurg.* **63**, 371–381.

CHAPTER 51

# Human Immunodeficiency Virus Infections—Neurological Manifestations

Hadi Manji and Roberto Guiloff

## INTRODUCTION

Acquired immunodeficiency syndrome (AIDS) was described in 1981 (Gottleib *et al.*, 1981; Masur *et al.*, 1981). It is caused by the retrovirus human immuno-deficiency virus type 1 (HIV-1). AIDS may also develop by infection with the antigenically different West African variant HIV type 2 (HIV-2), whose nucleic acid sequences are 40% homologous to those of HIV-1. The exact role of HIV-2 in the pathogenesis of AIDS-associated neurological disease is yet to be determined—HIV-2 related disorders will not be discussed here. Both HIV-1 and HIV-2 are members of the lentivirus (or slow virus) family of retroviruses. Characteristically, these slow viruses have a long incubation period and result in persistent infections. Other members include visna virus, bovine immunodeficiency virus, and feline immunodeficiency virus. HIV-1 affects the immune system by infecting a variety of cells, including T-helper lymphocytes, macrophages, promyelocytes, fibroblasts, and Langerhan's cells (Levy, 1990). Cell-mediated immunity is compromised by the loss and function of T-lymphocytes. There is also evidence of involvement of the humoral immune response with, for example, hypergammaglobulinemia resulting from HIV-induced polyclonal B-cell activation. This may present clinically with autoimmune disorders such as thrombocytopenia purpura (Levy, 1990).

Infection of a CD4-positive cell occurs when the Gp120 viral envelope protein interacts with a conserved CD4 binding site. Additional coreceptors, known as chemokines, are also necessary for this process. Some viruses bind preferentially to the monocyte/macrophage series using the CCR5 chemokine receptor (the so-called macrophage tropic virus), and others to the CXCR4 receptor found on lymphocytes (the lymphotrophic virus). Viruses may evolve from one to the other type. It has become evident that rather than lying dormant after the initial infection, active viral replication occurs continuously within the lymph nodes during the asymptomatic phase. Thus, by a process of sustained attrition, the CD4 lymphocyte count gradually decreases over a variable period of years, eventually leading to immunological deficiency.

## CLINICAL ASPECTS

### Diagnosis of HIV Infection and AIDS

The diagnosis of HIV-1 infection relies on the detection of HIV antibodies by the ELISA method (for screening) and Western blot as a confirmatory test. Specific HIV antibodies are usually detectable 4 weeks to 3 months after infection (the window period). The P24 antigen may be detected in the serum after infection and before the appearance of antibodies. Determination of the immunological profile includes analyses of lymphocyte subsets with absolute CD4 and CD8 cell counts and CD4/CD8 ratios. The CD4 count represents the extent of the damage to the immune system. More recently, with the introduction of polymerase chain reaction (PCR) techniques, measurement of the HIV RNA viral load has been shown to be a useful, independent marker for the risk of disease progression. The two variables, the CD4 count and the HIV RNA plasma load, are now used to make decisions regarding the starting of antiretroviral drugs, as well as monitoring the response to treatment of the complicated regimens since the introduction of the highly active antiretroviral combination therapies (HAART). It should, however, be noted that it is not clear whether the increased numbers of CD4 cells as a result of HAART are all functionally competent.

## HIV and the Nervous System

HIV-1 infects the nervous system early in the course of the infection. The evidence for this lies in the variety of seroconverting neurological presentations that have been described, as well as the cerebrospinal fluid abnormalities that may be found in asymptomatic HIV-infected individuals (McArthur *et al.*, 1988). The neurological complications attributed to HIV itself include HIV dementia, a vacuolar myelopathy, a polymyositis-like syndrome, and a range of peripheral neuropathic disorders, the most common being a distal sensory peripheral neuropathy. In addition, neurological complications also result from opportunistic infections and tumors such as toxoplasmosis and primary CNS lymphoma (PCNSL) (Figure 1).

When considering the differential diagnosis of a neurological presentation in an HIV-infected patient, the clinician needs to know the immunological status. This includes clinical clues such as the presence of, for example, oral candidiasis, which would suggest a significant degree of immunosuppression, and the CD4 count. In HIV-positive immunocompetent individuals, neurological disorders are usually unrelated to the HIV infection except during seroconversion. In the later immunosuppressed stages, the CD4 count is a useful guide to the underlying etiology (Crowe *et al.*, 1991). For example, cytomegalovirus complications such as encephalitis, polyradiculopathy, and retinitis occur at CD counts around 50 cells/μL.

Multiple CNS complications may occur simultaneously or in rapid sequence. Furthermore, the clinical presentation may be the product of interactions between infectious, nutritional, toxic, and metabolic factors. It should be noted that HIV-infected patients may not present with florid signs and symptoms, because the inflammatory processes may become muted—only one third of patients with cryptococcal meningitis present with the classical signs of meningism (Chuck and Sande, 1989). Even with the help of sophisticated diagnostic tools, some patients continue to remain a diagnostic challenge.

### Neurological Complications Caused by HIV-1

**Neurological Seroconversion Illness.** A symptomatic acute retroviral infection occurs in up to 90% of patients. In approximately 10%, this is a neurological syndrome which includes an acute aseptic meningitis or a meningoencephalitis (Carne *et al.*, 1985). The clinical presentation and CSF findings are indistinguishable from those seen in immunocompetent individuals caused by other viruses. Other, albeit rare, descriptions have included myelitis (Denning *et al.*, 1987), cauda equina syndrome (Zeman and Donaghy, 1991), painful brachial neuritis—Parsonage Turner syndrome (Calabrese *et al.*, 1989), rhabdomyolysis (Mahe *et al.*, 1989), Guillain-Barré syndrome, and most recently a case of acute disseminated encephalomyelitis (Narciso *et al.*, 2001).

| | Seroconversion | Asymptomatic | ARC/AIDS |
|---|---|---|---|
| HIV dementia | (meningoencephalitis) | | ------------------- |
| HIV myelopathy | (myelitis) | | ------------------- |
| HIV related distal sensory polyneuropathy (DSPN) | | | ------------------- |
| HIV myopathy | (polymyositis) | (polymyositis) | ------------------- (Zidovudine related) |
| Toxoplasmosis | | | ------------------- |
| Cryptococcal meningitis | | | ------------------- |
| Primary CNS lymphoma | | | |
| PML | | | ------------------- |

FIGURE 1    Timing of neurological complications of HIV-1 infection and AIDS.

**HIV-1–Associated Dementia (HAD) (HIV-1 Encephalopathy, HIV-Associated Cognitive Motor Complex, AIDS Dementia Complex).** This subcortical dementia occurs in up to 15% of AIDS patients with an incidence of 7% (McArthur *et al.*, 1993). Risk factors for the development of HAD include increasing age, low hemoglobin, low CD4 count, and a high viral load. Childs *et al.* (1999), using data from the Multicentre AIDS Cohort Study (MACS), found the risk of dementia was 8.5 times higher in individuals with an HIV RNA load greater than 30,000 copies/μL compared with those with a viral load less than 3000 copies/μL. Those with CD4 counts less than 200 cells/μL had a 3.5-fold risk of dementia compared with those with a CD4 count greater than 500 cells/μL.

Symptoms include impaired attention and concentration, memory loss, affective disorders, apathy, mental slowing, and lethargy. Motor involvement can vary from a minor unsteadiness of gait caused by posterior column impairment to a severe spastic paraparesis caused by the associated vacuolar myelopathy (Working Group of the American Academy of Neurology AIDS Task Force, 1991). Focal neurological signs are rare, although ocular motor disturbances and cerebellar ataxia have been reported (Hamed *et al.*, 1988; Pfister *et al.*, 1989).

The progression of the disorder is variable, with some patients remaining stable for up to 12 months and with others declining rapidly. Predictive markers of rapid progression in one study were patients whose risk factor for HIV infection was intravenous drug use and a low CD4 count at presentation of HIV dementia (Bowman *et al.*, 1998).

Cerebrospinal abnormalities, such as a mild pleocytosis and an elevated protein level, may occur but are nonspecific. Although there is a correlation between the severity of dementia and the CSF viral load, there is too much overlap for this measurement to be a diagnostic marker of HAD (Ellis *et al.*, 1997; McArthur *et al.*, 1997).

Cranial CT shows evidence of atrophy in some patients. T2-weighted MRI additionally shows diffuse or patchy high signal in the white matter. Neuropsychological tests reveal deficiencies in the domains of psychomotor speed, information processing, attention, and verbal and nonverbal memory.

Pathologically the hallmark of HIVD is the presence of multinucleated giant cells, which contain HIV DNA. However, the spectrum of histopathological changes is broad and has been categorized as follows: HIV leukoencephalopathy, diffuse poliodystrophy, HIV encephalitis, vacuolar leukoencephalopathy, and cerebral vasculitis (Budka *et al.*, 1991).

The differential diagnosis of HAD in the early stages is extensive and includes depression, recreational drug use, medication side effects, or the effects of metabolic derangement such as hypoxia caused by pneumonia. In the later stages, the important differentials include progressive multifocal leukoencephalopathy (PML) and CMV encephalitis (Table I).

Because there are no "gold standard" investigations to confirm a diagnosis of HAD, investigations such as brain imaging and CSF examination are necessary to exclude the other causes of cognitive impairment such as toxoplasmosis, cryptococcal meningitis, and neurosyphilis.

Viral invasion occurs as a result of infected macrophages crossing the blood–brain barrier—the so-called Trojan Horse theory (Ho *et al.*, 1987). Although there is little evidence to suggest that neurons are infected per se, productive HIV-1 has been demonstrated within macrophages and the microglial cell lines (Brinkman *et al.*, 1992; Epstein and Gendleman, 1993). Nonproductive HIV infection of astrocytes has also been identified (Takahashi *et al.*, 1996). There is poor correlation between the numbers of productively infected cells with the nervous system and both the clinical and pathological severity that occurs within the brains of HIV-infected individuals (Glass *et al.*, 1993). However, a strong correlation exists between the numbers of activated macrophages, their neurotoxic products, and the clinical severity of the dementia (Wesselingh *et al.*, 1993). These neurotoxic products include the HIV-1 coat protein gp 120 (Toggas *et al.*, 1993) and cytokines such as tumor necrosis factor and interleukin-1.

**HIV-1 Vacuolar Myelopathy.** This condition is characterized, in the typical case, by a relatively slowly progressive spastic paraparesis and posterior column signs in the lower limbs. Urgency and frequency, as well as incontinence of urine, occur frequently. There is no sensory level. Pathologically, there is vacuolar degenera-

TABLE I  Differential Diagnosis: HIV Dementia (HIVD), CMV Encephalitis, and PML

|  | HIVD | CMV | PML |
|---|---|---|---|
| Clinical | Memory, slowing, gait problems | Delirium brainstem signs; CMV elsewhere | Focal signs, dementia |
| Time course | Months | Days, weeks | Weeks, months |
| CD4 | <200 | <50 | <100 |
| MRI | Atrophy diffuse WML | Periventriculitis, enhancing mass lesions, brainstem | Subcortical WML focal, nonenhancing, no mass effect |
| CSF | Nonspecific | PCR + CMV 90% | PCR + JCV 80% |

tion of the white matter in the thoracic spinal cord in a distribution similar to that seen in subacute degeneration caused by vitamin $B_{12}$ deficiency. The vacuoles are related to intramyelinic dilations (Petitio *et al.*, 1985). This degeneration is not specific to AIDS and has been described in other immunodeficient states (Kamin and Petito, 1991).

The differential diagnoses include cord compression caused by, for example, lymphoma, vitamin $B_{12}$ deficiency, or neurosyphilis. Coinfection with HTLV-1, which has a similar mode of transmission to HIV-1, also needs to be considered. Viral myelitis caused by herpes zoster, cytomegalovirus, and mass lesions in the spinal cord from toxoplasmosis evolve much more rapidly (Guiloff and Tan, 1992; Malessa, 1991).

**Peripheral Nerve Syndromes.** A range of peripheral nerve disorders caused by a variety of pathological mechanisms occur within the context of HIV infection (see Table II). Acute inflammatory demyelinating neuropathy (Guillain-Barré syndrome) occurs during seroconversion or in the early asymptomatic stages. This is clinically similar to that found in seronegative individuals, except that a significant pleocytosis may be evident (Cornblath *et al.*, 1987). The trigger may be the HIV in a similar fashion to *Campylobacter jejuni* and CMV infection, accounting for a large proportion of Guillain-Barré episodes in the general population. There is some debate as to whether the incidence of Guillain-Barré is higher than expected in the HIV population (Barohn *et al.*, 1993; Guiloff and Fuller, 1992).

Mononeuritis multiplex caused by HIV-related vasculitis occurs rarely. In one retrospective series from a tertiary referral center only 12 cases were reported over a decade. The patients presented at all stages of their HIV illness, including AIDS. Most presented in a similar fashion to immunocompetent individuals, with the common peroneal nerve being the most frequently affected (Manji *et al.*, 1997). There have been anecdotal reports of vasculitis in peripheral nerves caused by HIV presenting as a painful distal sensory neuropathy (Bradley and Verma, 1996). The differential diagnosis includes a mononeuritis multiplex caused by CMV (Said *et al.*, 1991) and lymphomatous infiltration of peripheral nerves.

The most common peripheral nerve disorder in HIV infection is distal sensory peripheral neuropathy (DSPN). The prevalence rates have been estimated at between 10% and 35% (Cornblath and McArthur, 1988; So *et al.*, 1988). In one autopsy series 95% of patients dying of AIDS had histopathological abnormalities in their sural nerves (de la Monte *et al.*, 1988).

Risk factors for the development of DSPN include the HIV RNA viral load and to a lesser extent the CD4 count. Childs *et al.* (1999), using data from the Multicentre AIDS cohort Study (MACS) found that individuals with an HIV RNA load >10,000 copies/ml had a

**Table II  Peripheral Nerve Disorders in HIV Infection**

| HIV associated | CDC stages |
| --- | --- |
| Distal sensory peripheral neuropathy | AIDS |
| Inflammatory demyelinating neuropathy (acute and chronic) | Seroconversion, asymptomatic |
| Mononeuritis multiplex | All stages |
| DILS[a] | Asymptomatic, AIDS |
| Sensory neuronopathy (dorsal root ganglionitis) | Asymptomatic |
| Motor neuropathy/neuronopathy | AIDS |
| Brachial plexopathy | Seroconversion |
| Lumbar polyradiculopathy | Seroconversion |
| Autonomic neuropathy | AIDS |
| Other infections | |
|     CMV—mononeuritis multiplex, lumbar polyradiculopathy | AIDS |
|     Herpes zoster radiculopathy | All stages |
| Tumors | |
|     Non-Hodgkin's lymphoma—infiltrative radiculopathy, plexopathy | AIDS |
| Toxic neuropathy | |
|     Nucleoside reverse transcriptase inhibitors (ddI, ddC, d4T) | |
|     Thalidomide | |
|     Isoniazid | |
|     Vincristine | |
|     Metronidazole | |
|     Dapsone | |

[a]DILS = Diffuse infiltrative lymphocytosis syndrome.

2.3-fold greater risk of a neuropathy developing than those with <500 copies/ml. The implication being that the HIV viral load somehow "drives" the neuropathic process. Patients with a CD4 count of 750 cells/mm$^3$ had a 1.4-fold greater risk than those with a CD4 count of >750 cells/mm$^3$.

The clinical presentation is complaints of paresthesias on the dorsum and soles of the feet (So et al., 1988). A significant proportion of patients also complain of painful dysesthesias and hyperpathia. There is little or no weakness, and the upper limbs are usually symptom free. Abnormal neurological signs include depressed or absent ankle reflexes and impaired sensation to vibration, pain, and temperature. Neurophysiological tests may be normal or reveal a mild axonal neuropathy. Thermal thresholds are abnormal, implying small fiber dysfunction.

In histological studies, the usual finding is one of a dying-back axonopathy with secondary demyelination (de la Monte et al., 1988). Increased numbers of macrophages are found in the peripheral nerves and dorsal root ganglia. There is loss of both myelinated and unmyelinated nerve fibers. HIV RNA is rarely isolated form the nerve biopsy specimens. The underlying pathophysiology is postulated to be due to the release of cytokines such as tumor necrosis factor-alpha, resulting in neuronal damage in a similar fashion to that found in HIV dementia.

An uncommon form of HIV-associated neuropathy that resembles DSPN and is amenable to treatment with corticosteroids and antiretroviral drug therapy is the diffuse infiltrative lymphocytosis syndrome (DILS) (Gherardi et al., 1998). This multisystem disorder resembles Sjögren's syndrome, although anti Ro /SS-A and anti La/SS-B antibodies are absent. Patients have salivary gland enlargement, xerostomia, keratoconjunctivitis sicca, uveitis, and lymphocytic pulmonary, gastrointestinal and renal infiltrates. The peripheral nerve manifestations include a painful sensorimotor neuropathy, which may be symmetrical or asymmetrical. The CD4 count is variable, but the CD8 cell counts are invariably elevated (Moulignier et al., 1997).

Nerve biopsy specimens from patients with DILS show a marked angiocentric infiltration with polyclonal CD8 cells, and therefore are not lymphomatous, with no evidence of a necrotizing arteritis. In addition, the HIV proviral load in whole nerve extracts is 100,000-fold greater than that found in other HIV-related neuropathies. Because the blood viral load is low, this would support the notion of local HIV replication (Gherardi et al., 1998).

The true frequency and clinical features of autonomic neuropathy in HIV patients have been poorly established. In part, this is due to intercurrent conditions such as diarrhea, malnutrition, anemia, and CMV adrenali-tis. In addition, the vast range of drugs that patients are prescribed confound the interpretation of autonomic function tests. It has been suggested that autonomic denervation in the jejunal mucosa may contribute to chronic diarrhea in HIV disease (Batman et al., 1991). Pathological changes in sympathetic ganglia have been found at autopsy (Chimelli and Scaravelli, 1991). These autonomic disturbances may play a role in the sudden death that occasionally occurs in this group of patients.

There are increasing reports of lower motor neuron syndromes in HIV-infected patients. Two case reports describe a reversal of these syndromes with HAART therapy (MacGowan et al., 2001; Nishio et al., 2001).

**Myopathy.** An immunologically mediated polymyositis may occur at seroconversion or during the asymptomatic phases of the disease. As with non-HIV patients, the subacute illness manifests with proximal weakness and myalgia, an elevated creatinine kinase level, and myopathic changes on EMG studies (Dalakas et al., 1986). The disorder is usually steroid responsive. The inflammatory infiltrates found on muscle biopsy are similar to HIV-negative polymyositis, except there is a reduction in CD4-positive cells (Illa et al., 1991). However, other workers report similar patients with little or no evidence of inflammation on biopsy (Simpson and Bender, 1988).

Zidovudine, a nucleoside reverse transcriptase inhibitor (NRTI), introduced in 1986, causes a mitochondrial myopathy as evidenced by the presence of ragged red fibers on muscle biopsy (Dalakas et al., 1990). Furthermore, muscle mitochondrial DNA levels were found to be significantly depleted in AIDS patients with a zidovudine-related myopathy (Arnaudo et al., 1991). However, controversy ensued after Simpson and coworkers found no significant differences in clinical, biochemical, neurophysiological, or histopathological criteria between the zidovudine and nonzidovudine groups (Morgello et al., 1997; Simpson and Bender, 1988; Simpson et al., 1993).

It is nevertheless clear that zidovudine has a myopathic effect, because some patients benefit from stopping the drug. As with the other toxic side effects such as peripheral neuropathy caused by drugs of the same class, some authors describe the myopathy to be a cumulative dose-dependent effect (Mhiri et al., 1991). Zidovudine-related myopathic side effects seem to have become less prevalent since the standard dosage of the drug has been reduced from 1000–1500 mg/day to 600 mg/day.

### Neurological Complications Caused by Opportunistic Infections

Although *Pneumocystis carinii* pneumonia (PCP), *Mycobacterium avium* complex (MAC), and cytomega-

| | Frequency |
|---|---|
| HIV-1–related disorders | |
| Aseptic meningitis | 1%–6% |
| Acute meningoencephalitis | 1%–2% |
| HIV-1 dementia | 10%–15% |
| Myelopathy | 6%–10% |
| Peripheral neuropathy | 15% |
| Myopathy | 5% |
| Opportunistic infections | |
| Toxoplasma encephalitis | 5%–20% |
| Cryptococcal meningitis | 2%–13% |
| PML | 2% |
| CMV encephalitis | 1%[†] |
| CMV polyradiculopathy | 1% |
| Malignancies | |
| Primary CNS lymphoma | 2%–3% |
| Metastatic lymphoma | 2%–3% |
| Vascular complications | 1%–4% |

(Ischemic CNS infarcts and hemorrhages caused by HIV-1 vasculopathy, thrombotic microangiopathy, endocarditis, vasculitis)

*Because of HAART-significant declines in incidence have occurred.

[†]CMV found in 40% of brains of patients dying with AIDS.

lovirus infections are the most common opportunistic infections encountered in AIDS patients, it was clear from the onset of the epidemic that the nervous system was frequently involved. Approximately 20% of AIDS patients from a central London cohort had neurological opportunistic infections during their illness; this represented 40% of patients with neurological complications (Guiloff, 1989). Table III outlines the frequency of the main neurological complications associated with HIV-1 infection. Since the advent of the highly active antiretroviral therapies, the incidence has been dramatically reduced, at least in areas of the world where these drugs are available.

**Toxoplasmosis.** Infection by *Toxoplasma gondii* is the most frequent cause of an acute or subacute focal encephalopathy in AIDS. The history is usually short (less than 3 weeks), with focal deficits such as hemiparesis, aphasia, visual field defects, or extrapyramidal signs and movement disorders such as choreoathetosis. Headache, fever, and confusion are frequent. A presentation with a diffuse encephalopathy with confusion or coma but no focal signs is well documented (Gray *et al.*, 1989; Wijdicks *et al.*, 1991). The absolute CD4 count is usually <200 cells/mm$^3$. Cranial CT scans demonstrate single or multiple focal lesions with mass effect, which may or may not enhance with contrast. These usually have a predilection for the grey–white matter interface and the basal ganglia. MRI is a more sensitive modality of imaging and may reveal more lesions

than CT, particularly those in the posterior fossa. Single enhancing lesions on MRI, particularly those infiltrating the ventricles, are more likely to be lymphoma (Ciricillo and Rosenblum, 1990).

Cerebral toxoplasmosis is, in most AIDS cases, due to a reactivation of latent cysts as the CD4 count falls <200 cells/μL. The background seroprevalence varies around the world—in France, for example, the seroprevalence rate for *Toxoplasma gondii* is 90% compared with less than 50% in the United Kingdom, reflecting dietary habits because the oocytes are spread in undercooked meat. Spread also occurs by means of cat feces. Thus, negative serological findings make the diagnosis of toxoplasmosis less likely. However, 16% of individuals with toxoplasma encephalitis diagnosed on the basis of tissue biopsy or a successful response to therapy have been reported to be seronegative (Porter and Sande, 1992; Raffi *et al.*, 1997). A further issue to be considered is the reported loss of seropositivity in a minority of patients as their HIV disease progresses (Renold *et al.*, 1992).

The examination of cerebrospinal fluid in patients with mass lesions is clearly not feasible in a large proportion of patients. Using PCR to detect *T. gondii* in the CSF has a reported sensitivity of 53% with a specificity of 99% (Weber, 1999).

**Cryptococcal Meningitis.** The ubiquitous *Cryptococcus neoformans* is particularly found in the excreta from birds, especially pigeons. Pulmonary infection occurs by inhalation. It is often asymptomatic until hematogenous dissemination leads to infection of other organs. The patients present with a history between 1 and 4 weeks of headache, fever, nausea, and vomiting. Only one third of patients have the classic features of meningism—photophobia, neck stiffness, and a positive Kernig's sign (Zuger *et al.*, 1986). Clinical evidence of involvement outside the nervous system is present in 20% of cases. These are mainly pulmonary infiltrates, urinary tract infection, and skin lesions. Brain imaging is usually normal, although occasionally the MRI scan may show dilated Virchow-Robin spaces or basal contrast enhancement with gadolinium. Cystlike structures (gelatinous pseudocysts) and choroidal ependymal granulomas can sometimes be seen (Andreula *et al.*, 1993).

At lumbar puncture, the opening pressure is frequently elevated (Grabyill *et al.*, 2000). The CSF may show a pleocytosis, an elevation of protein, and a reduced glucose, but all the cytochemical markers may also be normal. The diagnosis is established by the detection of cryptococcal antigen in the CSF in 95% of cases. India ink staining of the CSF is positive in 75% of cases, with culture being the "gold standard." The serum cryptococcal antigen is usually positive and provides a useful screening method in patients with mild

Table IV   WHO/CDC Classification of HIV

| Laboratory classification | | | Clinical category | | |
|---|---|---|---|---|---|
| Absolute CD4 count (/μl) | or | total lymphocyte count (μl) | A | B | C |
| 1. >500 | | >2000 | A1 | B1 | C1 |
| 2. >200–499 | | 1000–1999 | A2 | B2 | C2 |
| 3. <200 | | <1000 | A3 | B3 | C3 |

*Clinical category A* (asymptomatic disease): acute infection with HIV, persistent generalized lymphadenopathy (PGL), asymptomatic.

*Clinical category B* (symptomatic disease): any symptomatic conditions not included in category C (e.g., bacterial infections, candidosis (oral or vulvovaginal) for >1 month, cervical dysplasia or carcinoma, constitutional symptoms, oral hairy leukoplakia, two distinct episodes of herpes zoster or herpes zoster involving more than one dermatome, idiopathic thrombocytopenic purpura, *Mycobacterium tuberculosis*, peripheral neuropathy) (previously known as AIDS-related complex).

*Clinical category C*: any condition meeting the 1987 WHO/CDC case definition for AIDS.

symptoms in whom the diagnosis is being considered. If positive, it is necessary to confirm the finding in the CSF.

**Cytomegalovirus.**   More than 90% of HIV-infected individuals have serological evidence of CMV infection. Postmortem, CMV can be isolated in the brains of one third of patients dying with AIDS (Petito *et al.*, 1986). It frequently coexists with other infections such as toxoplasmosis and HIV dementia. Clinical CMV disorders present within a setting of severe immunodeficiency with CD4 counts <50 cells/μL.

CMV encephalitis or meningoencephalitis presents acutely or subacutely over a period of weeks, with confusion and seizures often associated with focal abnormalities such as cranial nerve palsies, as well as brainstem and cerebellar signs (Holland *et al.*, 1994; Kalayjian *et al.*, 1993). HIV dementia, the main differential diagnosis, presents over months, with focal neurological signs being the exception rather than the rule. Most patients will have a history of/or coexistent CMV disease in another organ system such as the lungs, gastrointestinal system, retina, or the adrenal glands.

Imaging studies with CT and MRI may be normal (Post *et al.*, 1986). Diffuse periventricular enhancement (Arribas *et al.*, 1996), low-density areas, and occasional enhancing cortical nodules resembling mass lesions have been reported (Huang *et al.*, 1997). The CSF may be normal or show minor nonspecific abnormalities of cell count and biochemistry. The diagnosis is confirmed by the detection of CMV-DNA by use of PCR analysis. The sensitivity for this technique is reported to be approximately 75%, with a specificity of 95% (Weber, 1999). Because systemic infection with CMV is common, blood isolation of the virus antigen or by CMV-DNA by PCR is not helpful diagnostically.

The main neuropathological features are ventriculoencephalitis, focal parenchymal necrosis, and micro-

nodular encephalitis. The latter is characterized by the presence of microglial nodules, which consist of astrocytes surrounding CMV inclusion-bearing cells (Setinek *et al.*, 1995; Vinters *et al.*, 1989).

Within the peripheral nervous system, CMV may cause an ascending lumbosacral polyradiculopathy. Patients present with back pain followed by the progressive development of a flaccid paraparesis with sensory loss and sphincter disturbance (Miller *et al.*, 1990; So and Olney, 1994). Imaging studies, which are essential to exclude compressive lesions caused by a central disc protrusion and lymphoma, are usually normal but may reveal nodular thickening of the nerve roots. Other differential diagnoses to be considered include a syphilitic and herpetic polyradiculopathy.

The CSF typically shows a polymorphonuclear leukocytosis, which is unusual in a viral infection (So *et al.*, 1994). CMV can be detected in the CSF by PCR techniques.

CMV has been incriminated in the pathogenesis of a painful peripheral neuropathy, which is similar to the distal symmetrical peripheral neuropathy caused by HIV (Fuller *et al.*, 1993). CMV-related mononeuritis multiplex has also been described (Fuller *et al.*, 1993; Said *et al.*, 1991). The diagnosis is usually made on nerve biopsy with CMV inclusions within Schwann cells associated with polymorphonuclear infiltrates and necrosis.

CMV is the most common cause of blindness in AIDS patients with very low CD4 counts. Before the introduction of the combination retroviral therapies, approximately 20% of patients with HIV infection were affected. The clinical presentation varies from being asymptomatic to complaints of floaters, loss of peripheral vision, or if the lesions are centered around the macula—poor visual acuity. On fundoscopy, typically there is a perivascular yellow-white infiltrate with retinal hemorrhages. The differential diagnosis

**TABLE V  CDC Definition of AIDS in Patients with Laboratory Evidence of HIV Infection**

A definitive diagnosis of:
1. Bacterial infections, multiple or recurrent (any combination of at least two within a 2-year period, in a child <13 years of age
2. Disseminated coccidioidomycosis
3. HIV dementia (also termed HIV dementia, AIDS dementia, or subacute encephalitis caused by HIV)
4. Disseminated histoplasmosis
5. Isopsoriasis with diarrhea for >1 month
6. Kaposi's sarcoma
7. Lymphoma of the brain (primary)
8. Non-Hodgkin's lymphoma: diffuse, undifferentiated B-cell type, or unknown phenotype
9. Disseminated mycobacterial disease (other than *Mycobacterium tuberculosis*)
10. Extrapulmonary *Mycobacterium tuberculosis* involving at least one site outside the lungs
11. Recurrent *Salmonella* (nontyphoid) septicemia
12. HIV wasting syndrome (emaciation, slim "disease")

*or a presumptive diagnosis of*
1. Candidosis of the esophagus
2. Cytomegalovirus retinitis with loss of vision
3. Kaposi's sarcoma
4. Disseminated mycobacterial disease (acid-fast bacilli with species not identified by culture)
5. *Pneumocystis carinii* pneumonia
6. Toxoplasmosis of the brain

In 1992, the definition of AIDS was modified by the addition of three clinical conditions in the presence of HIV infection—cervical cancer, two episodes of bacterial pneumonia in 12 months, and pulmonary tuberculosis. In 1993, in the United States, the CDC extended the definition of AIDS to include all persons with a CD4 count $<200 \times 10^6$ cells/L, irrespective of the presence or absence of an indicator disease. This definition has not been accepted in the United Kingdom or Europe.

includes retinal disease caused by toxoplasmosis, lymphoma, syphilis, and herpes zoster and herpes simplex.

**JC Virus.** Before the AIDS epidemic, progressive multifocal leukoencephalopathy (PML) was a rare condition found mainly in patients with lymphoproliferative disorders, granulomatous disorders such as sarcoidosis, and patients on immunosuppressive drugs. HIV-associated cases now account for 85% of cases. Before HAART, the incidence in AIDS patients was 4%. This acquired demyelinating disorder is caused by the JC virus—a member of the genus *Polyomavirus* in the Papovaviridae family. In the general population, 80%–90% of individuals have been exposed to the virus as a banal viral upper respiratory infection. It is reactivated as the immune system function declines (Manji and Miller, 2001).

The clinical presentation of PML involves a progressive deficit, which includes hemiparesis, hemianopia, or ataxia (Gillespie *et al.*, 1991). Cortical involvement presents with dysphasia, seizures, and occasionally dementia with focal signs (Sweeney *et al.*, 1994). By contrast with other causes of focal lesions in the brain of HIV-infected individuals, toxoplasmosis, and primary CNS lymphoma, there are usually no

symptoms of systemic infection or raised intracranial pressure.

Cranial CT shows hypodense lesions. Typically, MRI will reveal large single or multiple lesions involving the white matter, with scalloping at the grey–white interface in the parieto-occipital and frontal lobes. The posterior fossa, thalamus, and basal ganglia are less often involved. The affected areas are low signal on T1-weighted sequences and hyperintense on T2-weighted images. Generally there is little mass effect or enhancement (von Einsidel *et al.*, 1993).

In the early days of the AIDS epidemic, brain biopsy was the only method to confirm a diagnosis of PML. The typical histological features are areas of demyelination with enlarged oligodendrocyte nuclei and inclusion particles, which are shown to be JC viral particles by in situ hybridization techniques. Infected astrocytes with enlarged bizzare nuclei resembling neoplasia are also present.

Nowadays, however, CSF JCV-DNA using PCR has a sensitivity of 75%, with a specificity of 90%–99% (Cinque *et al.*, 1996). The false-negative rate of 25% means that a negative result does not exclude the diagnosis, and it may be necessary to perform repeat lumbar punctures, and if these are also negative, to consider a stereotactic brain biopsy.

**TABLE VI  HAART Drug Toxicities**

| Drug | Toxicity |
| --- | --- |
| 1. Nucleoside reverse transcriptase inhibitors (NRTIs) | |
| **Class associated | Lactic acidosis, hepatic steatosis, lipodystrophy |
| *Drug specific | |
| Zidovudine | Myelosuppression, myopathy |
| Stavudine (d4T) | Peripheral neuropathy |
| Zalcitabine (ddC) | Peripheral neuropathy |
| Didanosine (ddI) | Peripheral neuropathy, pancreatitis |
| Lamivudine (3TC) | Peripheral neuropathy? |
| 2. Nonnucleoside reverse transcriptase inhibitors (NNRTIs) | |
| Efavirenz | Dysphoria, mood changes, vivid dreams, hypercholesterolemia |
| Nevirapine | Hepatitis, Stevens—Johnson syndrome |
| 3. Protease inhibitors (PI) | |
| **Class associated | Lipodystrophy, hyperlipidemia, diabetes mellitus |
| *Drug specific | |
| Nelfinavir | Diarrhea |
| Indinivir | Nephrolithiasis |
| Ritonivir | Perioral dysesthesia Flushing, diarrhea |
| Amprenivir | Diarrhea |
| Lopinivir | Diarrhea |

**Herpes Zoster and Herpes Simplex.** Because the risk of herpes varicella zoster reactivation increases as cell-mediated immunity wanes, for example, in elderly patients or those with cancer, it is not surprising that there is a high incidence of shingles, which is often multidermatomal, in the HIV population. A low CD4 count is predictive of a major zoster neurological complication, as is trigeminal or multidermatomal involvement (Glesby *et al.*, 1995; Veenstra *et al.*, 1996). Although the neurological complications generally occur coincident with the vesicular rash, this is not necessarily always the case. In AIDS patients the following complications have been described: a progressive encephalitis, a granulomatous cerebral vasculitis usually in association with trigeminal nerve zoster, meningomyeloradiculitis, and optic nerve involvement with a retrobulbar neuritis and retinal necrosis.

Although mucocutaneous syndromes caused by herpes simplex 1 and 2 are common, the neurological complications attributed to these viruses are uncommon. These include the typical temporal lobe encephalitis encountered in non-HIV practice but also a milder, less acute meningoencephalitis (Tan *et al.*, 1993). There

are cases reported of a combined encephalitis caused by herpes simplex and CMV (Miller *et al.*, 1995; Vago *et al.*, 1996). Other reported complications include a myeloradiculitis and retinal necrosis.

As with other viral infections such as CMV, the use of PCR techniques on CSF has made a great impact on being able to diagnose both varicella zoster and simplex neurological complications antemortem and, consequently, make an attempt at a rational therapeutic strategy.

**Mycobacterium Tuberculosis.** In the developing countries, infection with *M. tuberculosis* has always been a major problem. With the onset of the HIV pandemic, this has been magnified. In the United States, the numbers of reported cases was on the decline from 1953–1984, after which numbers have been on the increase, in part because of the HIV pandemic. HIV-infected patients are more likely to have active disease than noninfected individuals. In addition, extrapulmonary tuberculosis is more common in those patients who are HIV positive, particularly those with advanced immunosuppression.

Meningitis is the most common neurological presentation, often occurring at CD4 counts much higher than, for example, cryptococcal meningitis, implying greater virulence. A helpful clinical marker is finding concomitant pulmonary tuberculosis on chest x-ray in between 50% and 75% of cases (Berenguer *et al.*, 1992; Yechoor *et al.*, 1996).

The clinical presentation is similar to that found in non-HIV patients, except for a higher incidence of intracerebral mass lesions (Bishburg *et al.*, 1986). Cerebrospinal fluid findings are similar to non-HIV cases, with a moderate pleocytosis (100–300 cells/µL), an elevated protein, and a low glucose. It has been suggested that the culture-positive rate may be lower in HIV patients. The use of PCR in diagnosis of tuberculous meningitis remains, to date, controversial. Cranial CT scans are abnormal in up to 70% of cases and may reveal any of the following features: hydrocephalus, meningeal enhancement, or mass lesions (Berengeur *et al.*, 1992).

Central nervous system mass lesions caused by *M. tuberculosis* (tuberculoma or tuberculous brain abscess) are difficult to distinguish from the other common mass lesions in this group of patients—toxoplasmosis and primary CNS lymphoma.

As a clinical point, the issue of concomitant infection with more than one organism or dual pathological conditions should be considered in patients who do not, or only partially, respond to treatment. Cases of both CNS lymphoma and tuberculoma have been reported. This picture is complicated by the well-described clinical observation that tuberculous brain lesions may

**TABLE VII    Clinical Features, Course, and Therapy for HIV-1–Associated Neurological Complications**

---

*HIV-1 dementia*
  Clinical features.
    Early: Symptoms of poor concentration, memory difficulty, mental slowing, clumsiness. Signs of impaired fine finger movements, impaired tandem gait, impaired saccadic eye movements.
    Late: Apathy, seizures, and severe cognitive impairment; signs of pyramidal disturbance in the legs caused by associated vacuolar myelopathy, frontal release signs, myoclonus.
  Blood tests: Vitamin $B_{12}$, thyroid function, syphilis serology; CD4 count usually below 200 cells/µL.
  CSF: Mild nonspecific abnormalities with lymphocytosis and elevated protein level. CSF HIV RNA load correlates with dementia severity but sensitivity and specificity poor for diagnosis.
  CT/MRI: Cerebral atrophy; T2-weighted images show diffuse white matter abnormalities.
  Neuropsychological tests: Abnormalities in the domains of psychomotor speed, attention, frontal lobe dysfunction, verbal and nonverbal memory.
  Clinical course: Some patients have minor cognitive deficits. HIV dementia (10%–15%)—variable progression. Rapid progression associated with IVDU and lower CD4 count at diagnosis.
  Therapy: HAART. Only one placebo-controlled trial showed benefit with zidovudine, 2000 mg. Zidovudine, stavudine, abacavir, and nevirapine have best CSF penetration and should be included in any regimen.

*HIV-1 associated vacuolar myelopathy*
  Clinical features: Spastic paraparesis with bladder and bowel involvement. No sensory level.
  Blood: Vitamin $B_{12}$ level, HTLV-1, syphilis serology.
  CSF: Exclude herpes simplex and herpes zoster with PCR DNA.
  MRI: Usually normal. May show atrophy or rarely high signal on T2.
  Clinical course: Slowly progressive or static. Associated with HIV dementia.
  Therapy: HAART. As for HIV dementia. Consider oral methionine 3 g/day.
    Symptomatic—baclofen for spasticity, oxybutynin for detrusor instability.

*Distal sensory peripheral neuropathy (DSPN)*
  Clinical features: Numbness, burning, paresthesias, contact hypersensitivity. Little or no weakness. Impaired pain and temperature sensation. Depressed or absent ankle jerks.
  Blood: glucose, renal function, vitamin $B_{12}$. CD4 count < than 200 cells/ul.
  Nerve conduction: Large fiber tests may show an axonal neuropathy but may be normal if only small fibers affected. Thermal thresholds abnormal.
  Nerve biopsy: Only necessary if unusual features present to exclude vasculitis or demyelinating neuropathy. Axonal loss with secondary demyelination. Few inflammatory cells around endoneurial blood vessels and in endoneurium.
  Clinical course: May be slowly progressive or static. Possible regression with HAART.
  Therapy: Stop any neurotoxic drug—ddI, ddC, d4T.
    Symptomatic—gabapentin up to 1200 mg TDS; lamotrigine, 25 mg/day to 300 mg; amitriptylin start 10 mg/night; sodium valproate 1200–2000 mg/day.
Acute or chronic demyelinating neuropathy
  Clinical features; Rare. Symmetrical motor and sensory neuropathy—may start in the upper limbs.
  CSF: May show an elevated cell count >50 cells/µL compared with non-HIV patients and elevated protein level.
  NCT/EMG: Slowed nerve conduction with block, prolonged distal motor latencies, and f waves.
  Clinical course: Acute (Guillain-Barré syndrome)—rapidly progressive over 4 weeks. CIDP progressive/relapsing remitting >8 weeks.
  Therapy: IV immunoglobulin, 0.4 g/kg/day for 5 days. For CIDP—corticosteroids are an alternative.

*Mononeuritis multiplex (vasculitis)*
  Clinical features: Multifocal sensory and motor deficits affecting peripheral nerves—common peroneal nerve, radial nerve. May also present as a symmetrical sensorimotor neuropathy.
  Blood: Vasculitis screen, CD4/CD8 ratio (DILS), hepatitis B and C serology. Consider PMP 22 deletion (HLPP).
  NCT/EMG: Patchy involvement of sensory and motor nerves.
  Nerve biopsy: Epineural and endoneural perivascular inflammatory cells or necrotizing vasculitis. If CMV associated—polymorphonuclear cells and CMV "owl eyes inclusions." DILS—perivascular CD8 infiltrates.
  Clinical course: Acute or subacute onset—progressive.
  Therapy: HIV related and DILS—antiretroviral drugs and corticosteroids. If CMV related—ganciclovir, foscarnet, or cidofovir.

*HIV-related myopathy*
  Clinical features: Slowly progressive proximal myopathy involving arms, legs, neck flexors with myalgia, fatigue, and dysphagia. Zidovudine myopathy—variable presentation from myalgias only to slowly progressive myopathy affecting proximal leg muscles including gluteal muscles.
  Blood: CPK, HTLV1
  EMG: Myopathic changes—polyphasic motor units, positive sharp waves.
  Muscle biopsy: myonecrosis and endomysial inflammatory infiltrates. In zidovudine myopathy, ragged red fibers with myonecrosis and a milder inflammatory infiltrate.
  Clinical course: Subacute onset. Zidovudine myopathy may develop over months usually in patients on high doses >600 mg/day over a prolonged period.
  Therapy: If on zidovudine, consider stopping the drug. Corticosteroids.

paradoxically enlarge in the early stages of treatment. One South African study documented dual infection of *M. tuberculosis* with *Streptococcus pneumoniae, Cryptococcus neoformans*, or *Treponema pallidum* in up to 20% (Silber, 1998).

**Neurosyphilis.**  Positive syphilis serological results are reported in 46%–63% of HIV-seropositive individuals (Berger, 1991; Sindrup *et al.*, 1986). Clinical neurosyphilis is far less frequent, with one series reporting 1.8% of asymptomatic and 2.5% of neurologically symptomatic HIV-seropositive individuals are affected (Berger, 1991).

The clinical spectrum of neurological presentations is similar to that in non HIV patients—in the secondary stages patients may present with acute syphilitic meningitis. In the tertiary stages, early neurological complications include strokelike episodes caused by meningovascular syphilis resulting in endarteritis. Later, patients have general paralysis of the insane (GPI), tabes dorsalis, or syphilitic gummas as mass lesions.

Within the context of HIV infection, a number of factors compound the difficulties in diagnosis and treatment. As a result of the depressed cellular immunity, it seems that infection with *Treponema pallidum* may be more aggressive and may present in an atypical fashion (Lanska *et al.*, 1988; Muscher *et al.*, 1990). There are anecdotal reports of rapid progression to neurosyphilis and relapses despite treatment with adequate doses of benzathine penicillin (Berry *et al.*, 1987; Muscher *et al.*, 1990). This is plausable, because chronic HIV meningitis may increase the risk of developing neurosyphilis. However, most HIV-infected individuals will present with typical signs and serological test results.

Serological testing may also be affected by the B-cell polyclonal activation resulting in false-positive tests (Drabick and Tramont, 1990). Conversely, Haas (1990) showed that treponemal serological tests may revert to negative with progressive HIV disease. From a practical point, it should be noted that it is not possible to use the CSF cytochemical parameters as markers for active neurosyphilis, because HIV itself may result in similar changes.

**Miscellaneous Infections.**  A large number of other infrequent opportunistic infections of the nervous system have been reported in AIDS patients. These include protozoa-like Acanthamoeba, *Trypanosoma cruzi* (Silva *et al.*, 1999), and *Pneumocystis carinii*; fungi such as *Histoplasma capsulatum, Coccidioides immitis, Nocardia asteroides*, and *Candida albicans*; bacterial infections described rarely include *Listeria monocytogenes* and metazoa-like *Strongyloides stercoralis* (Guillof and Tan, 1992).

### Neurological Malignancies in HIV

**Primary Central Nervous System Lymphoma (PCNSL).**   In children infected with HIV, PCNSL is the most common cause of mass lesions; in adults, it is second only to toxoplasmosis. Histologically, this is a high-grade B-cell lymphoma. The Epstein-Barr virus can be isolated from tissue specimens and is purported to have a causal role in the development of the tumor.

The clinical presentation is similar to that with toxoplasmosis with focal neurological signs, such as hemiparesis and seizures, in association with symptoms and signs of raised intracranial pressure. The CT and MRI findings may also be indistinguishable from those caused by toxoplasmosis, although single lesions on MRI and those that adhere to the ventricular wall are more likely to be due to lymphoma.[201] Thallium SPECT scans, where available, may have a role in trying to distinguish between lymphoma, which lights up as a "hot spot", and infective processes such as toxoplasmosis and tuberculoma, which fail to take up the radioisotope (Miller *et al.*, 1998).

Although the reported figures for the sensitivity and specificity for the detection of EBV DNA by PCR techniques in PCNSL have been high, 87% and 98% respectively, in practice it is difficult to justify a lumbar puncture in patients with mass lesions and raised intracranial pressure. A recent report suggests that levels of a soluble glycoprotein, CD23, that is expressed on mature B cells are significantly higher in the CSF of patients with PCNSL and non-Hodgkin's lymphoma with brain involvement than in patients with non-Hodgkin's lymphoma without brain involvement, toxoplasmosis, PML, and HIV dementia (Bossolasco *et al.*, 2001).

**Non-Hodgkin's Lymphoma with Nervous System Involvement.**   Non-Hodgkin's lymphoma is one of the most common malignancies associated with HIV infection. Like PCNL, it is an aggressive, high-grade B-cell tumor. Extranodal disease is usually evident at presentation.

The neurological presentation is with leptomeningeal disease with cranial nerve palsies and spinal root involvement. Extension into the epidural space may result in spinal cord compression or a cauda equina syndrome. Peripherally, there may be infiltration into the brachial or lumbosacral plexuses or into the peripheral nerves.

Imaging studies show meningeal enhancement with contrast. Repeated lumbar punctures for cytological examination may be necessary to reveal the malignant lymphomatous cells.

**Kaposi's Sarcoma.**  Involvement of the nervous system is exceptional. There are few recorded cases of

**TABLE VIII   Clinical Features, Course, and Therapy of Neurological Complications of HIV-1 Infection (Opportunistic Infections and Tumors)**

*Toxoplasmic encephalitis (TE)*
    Clinical features: Focal neurological deficits, 70%; headache, 51%; confusion, 50%; fever, 40%; seizures, 30%.
    Blood: More than 95% serology positive, because TE a reactivation. IgG negative 3%–17%. Risk of development of TE if serology positive
        30%–50%, primary prophylaxis recommended if serology positive and CD4 count less than 100/μL—trimethoprim–sulfamethoxazole or
        dapsone 50 mg + pyrimethamine, 50 mg daily, + folinic acid, 25 mg once weekly.
    CSF (possible only if no significant mass effect): Normal, 14%; tachyzoites, 6%; pleocytosis, 50%; high protein, 79%; low glucose, 17%.
    CT/MRI: Single 30% or multiple 65% hypodense lesion(s). Ringlike, nodular, or other contrast enhancement 95%. Mass effect and edema.
        Repeat imaging after 2 weeks, even if clinical improvement, in case of dual pathology.
    Clinical course: Acute or subacute presentation (1–2 weeks). Clinical response in more than 90% by week 2. Without maintenance therapy,
        relapse in 80%, 10%–33% with maintenance therapy.
    Acute-phase therapy: Pyrimethamine loading dose 100 mg PO for 3 days, then 50 mg/day PO + sulfadiazine, 4–8 g/day PO/IV; + folinic acid,
        15 mg/day PO. Clindamycin, 600 mg qds PO/IV, can be used instead of sulfadiazine. Second-line treatment: pyrimethamine, 50 mg/day PO,
        + atovaquone, 750 mg qds PO. Treatment for 6 weeks.
    Maintenance therapy: Pyrimethamine, 25–50 mg/day + sulfadiazine, 2 g/day, + folinic acid, 10 mg/day. Clindamycin, 600 mg/day. In case of
        relapse, reinstate acute regimen.

*Cryptococcal meningitis (CM)*
    Clinical features: Headaches, 80%; fever, 75%; meningeal signs, 35%; photophobia, 25%; nausea and vomiting, 50%; impaired mental
        status, 30%; seizures, 5%.
    Blood: Serum cryptococcal antigen positive in 98%. A useful screening test but not conclusive, especially in relapses.
    CSF: Opening pressure >200 cm CSF, 66%; cell count normal, 50%; protein >40 mg/dl, 50%; cryptococcal antigen positive, 95%; India ink
        staining positive, 75%; culture positive, >85%.
    CT/MRI: Cryptococcomas rare; hydrocephalus, 9%.
    Clinical course: Subacute (2–4 weeks) or acute onset. Response to therapy 70%–85%. Prognostic markers for poor outcome include altered
        mental status, relapse episode, hyponatremia, CSF white cell count <20 cell/μL, CSF cryptococcal antigen >1:1024. Relapse rate under
        maintenance therapy: fluconazole, 200 mg/day, 2%; amphotericin, 1 mg/kg/week IV, 18%.
    Acute phase therapy: Amphotericin B, 0.4–1.0 mg/kg/day IV (usually by means of a central line) ± flucytosine, 150 mg/kg/day PO. In milder
        cases, 400 mg IV treatment for 4–6 weeks or until CSF culture negative.
        Repeat lumbar punctures or lumbar drain necessary for raised ICP.
    Maintenance therapy: Fluconazole, 200 mg/day PO.

*Primary CNS lymphoma*
    Clinical features: Similar to toxoplasmosis, but no fever. Hemiparesis, 38%; seizures, 23%; confusion and memory loss, 57%; cranial nerve
        deficits, 17%.
    CSF (rarely possible because of mass effect): PCR positive for Epstein-Barr virus.
    CT/MRI: Single or multiple lesions with contrast enhancement and associated cerebral edema. Difficult to differentiate from toxoplasmosis,
        but single lesion on MRI more likely to be lymphoma as is a lesion adjacent to the ventricular wall.
    Diagnosis: Failed empirical treatment for toxoplasmosis for at least 2 weeks, stereotactic brain biopsy.
    Clinical course: Progressive course that may be rapid. Stabilization after radiotherapy and corticosteroid treatment. Prognosis poor.
    Acute phase therapy: Whole-brain radiotherapy, 4000 rad over 3 weeks + dexamethasone, 16–24 mg/day.
    Maintenance therapy: Dexamethasone, if steroid responsive.

*Progressive multifocal leukoencephalopathy*
    Clinical features: Focal or multifocal neurological deficits without raised intracranial pressure. Motor function abnormality, 67%; mental
        status change, 66%; hemiparesis, 39%; cerebellar problems, 34%; language disorder, 31%.
    Blood: Serology for JC virus unhelpful for diagnosis, because 90% of population positive.
    CSF: Normal, 64%. JCV detection by PCR sensitivity, 75%; specificity 90%–99%.
    CT/MRI: Hypodense lesions on CT. MRI lesions low signal on T1-weighted sequence and hyperintense on T2 involving mainly white matter,
        usually in the parieto-occipital and frontal lobes. Contrast enhancement and mass effect rare.
    Clinical course: Subacute progression but occasional long-term survivors described. Increased survival since HAART.
    Acute phase therapy: Reconstitution with HAART. Consider treatment with cidofovir (open-labeled studies only): cidofovir, 5 mg/kg IV day 1
        and day 8, then 5 mg/kg IV every 2 weeks. Requires pretreatment with probenacid and adequate hydration. Alternative consider alpha-
        interferon (one open-labeled study only)—interferon 2b 3 million units SC daily or 5 million units three times a week.
    Maintenance therapy: HAART lifelong. Cidofovir and alpha-interferon unknown.

*CMV encephalitis and polyradiculopathy* (CMV retinitis often present)
    Clinical features:
        Encephalitis: A diffuse encephalopathy with confusion; seizures, but may also develop brainstem and cerebellar signs and symptoms.
        Polyradiculopathy: Lumbosacral polyradiculopathy with flaccid paraparesis, sphincter involvement, and back pain.
    Blood: Unhelpful in diagnosis since seroprevalence in HIV patients 100%.
    CSF: Pleocytosis, 30%. Increased polymorphonuclear count in CSF. CMV-DNA by PCR sensitivity 79%–100%.
    CT/MRI: Encephalitis: periventricular enhancement but this can occur in lymphoma. In polyradiculopathy, the MRI is usually normal but
        may show nodular thickening of nerve roots.

*continues*

**TABLE VIII**  *Continued*

Clinical course:
  Encephalitis: Poor prognosis, although some improvement with treatment.
  Polyradiculopathy: Without treatment, poor prognosis. With treatment, increased survival with functional improvement in some cases.
  Acute phase therapy: Ganciclovir, 5 mg/kg IV bd for 2–3 weeks or foscarnet, 60–70 mg/kg IV tds. Some evidence that a combination of both may be more effective. HAART.
  Maintenance therapy: Ganciclovir, 5 mg/kg od IV or foscarnet 60–120 mg/kg od IV.

*Herpes simplex virus 1, 2 and herpes varicella zoster*
  Clinical features: Infrequent focal or diffuse encephalitis or myeloradiculitis. Skin lesions not obligatory. HVZ associated with a vasculopathy affecting small and large vessels.
  CSF: Isolation of viral DNA by PCR
  Clinical course: Acute or subacute onset, progressive. Dual infection with HSV 1 and CMV may occur.
  Acute phase therapy: Aciclovir, 15–30 mg/kg/day IV for at least 10 days. In cases in which coinfection with CMV, consider ganciclovir or foscarnet.
  Maintenance therapy: Aciclovir, 200–400 mg ods PO.

*Neurosyphilis*
  Clinical features: Secondary syphilis—acute syphilitic meningitis with meningism, cranial nerve palsies, mainly II, VII, and VIII. Tertiary neurosyphilis includes meningovascular syphilis (headaches, malaise, and focal deficits caused by vasculitic strokes); general paresis, tabes dorsalis, and syphilitic gummas (granulomas).
  Blood: VDRL and FTA positive but false negatives and positives occur.
  CSF: Pleocytosis with elevated protein caused by both HIV and syphilis. Reactive CSF VDRL confirms neurosyphilis, but a negative test does not exclude it. A nonreactive CSF FTA-ABS excludes active neurosyphilis.
  Clinical course: Subacute, slowly progressive strokelike episodes in meningovascular syphilis. Occasionally, myeloradiculitis. Relapse possible even after adequate therapy.
  Acute phase therapy: Penicillin G (aqueous), 24–48 mU/day IV for at least 14–21 days. Careful follow-up with repeat serological tests and lumbar puncture recommended to document fall in VDRL titers and CSF parameters.

brain metastases, which may be hemorrhagic, and local infiltration into plexuses and peripheral nerves.

## Cerebrovascular Disease

In contrast to the infectious complications of HIV, seldom caused by cerebrovascular disease are relatively seldom studied. This is, in part, due to the wide range of confounding variables that may predispose to stroke, such as intravenous and recreational drug use, infections and tumors that may have strokelike presentations, infections such as herpes varicella zoster, which may result in an ipsilateral arteritis and contralateral stroke, and meningovascular syphilis and tuberculosis, both of which can be associated with a cerebral vasculitis. Cardiogenic causes of embolic stroke in this group may be due to myocarditis with congestive cardiomyopathy caused by HIV, nonbacterial thrombotic (marantic) endocarditis in the preterminal phases, and infective endocarditis in intravenous drug users.

Although autopsy studies have documented a high prevalence of cerebrovascular disease, stroke as a clinical presentation is rare (Pinto, 1996). A retrospective, hospital-based study of cerebrovascular disease in young HIV-infected black Africans in South Africa found rates similar to those in HIV-seronegative controls. However, there was an increased incidence of large-vessel cryptogenic stroke (Hoffman *et al.*, 2000). The authors postulate that prothrombotic abnormalities could be playing an important role. However, formal cerebral angiography, the most sensitive imaging modality for detection of vasculitis, was not performed.

Various studies have documented prothrombotic coagulation abnormalities in HIV infection-elevated levels of protein S (Bissuel *et al.*, 1992; Stahl *et al.*, 1993), elevated levels of anticardiolipin antibody (Stimmler *et al.*, 1989), and heparin cofactor II deficiency (Toulon *et al.*, 1993).

There is some evidence that HIV itself may be responsible for a vasculopathy; an autopsy study of cerebral infarction in AIDS patients describes a vasculopathy, not a vasculitis, characterized by hyaline small-vessel thickening, vessel wall mineralization, and perivascular inflammatory infiltrates. Of 10 patients with this pathological finding, however, only one had a transient ischemic attack, and none clinically developed a stroke. All were severely immunocompromised (Connor *et al.*, 2000).

A study, using transcranial Doppler ultrasonography and intravenous acetozolamide to increase cerebral blood flow by means of dilatation of small arterioles in HIV patients, who were neurologically asymptomatic with a mean CD4 count of 217/μL, found that both base-line blood flow and reactivity to acetozolamide were reduced (Brilla *et al.*, 1999). Similar findings are reported in patients with SLE and CNS involvement. This is further evidence suggestive of an HIV-related vasculopathy that needs further investigation.

## NATURAL COURSE

In 1981, cases of previously rare disorders of *P. carinii* pneumonia and Kaposi's sarcoma were identified in young homosexual men in San Francisco, Los Angeles, and New York. By the following year, other risk groups, intravenous drug users and hemophiliacs, were identified. Twenty years later, the pandemic is causing devastation worldwide. The United Nations program on AIDS (UNAIDS) estimated that by the end of 2000 there were 36.1 million people living with HIV/AIDS (34.7 million adults and 1.4 million children). There are approximately 16,000 new infections per day. Currently, 95% of all infections occur in developing countries, with sub-Saharan Africa and South East Asia taking the brunt. By 1999, UNAIDS estimated that there was a total of 13.2 million AIDS orphans.

Worldwide, a global summary of HIV transmission breaks down as follows: blood transfusion, 3%–5%; perinatal, 5%–10%; vaginal sexual intercourse, 60%–70%; anal sexual intercourse, 5%–10%; intravenous drug use 5%–10%; and needle stick injuries, <0.01% (Adler, 2001).

Whereas in the developing world, HIV is spread mainly by heterosexual intercourse, in the developed world such as the United States (in June 1999) 48% of cases were seen in men who have sex with men, 26% in intravenous drug users, 10% by heterosexual contact, 6% in men who have sex with men and use IV drugs, and 2% in blood recipients and hemophiliacs. In some European countries, such as Spain and Italy, the most common mode of infection is through intravenous drug use.

The major transmission route is by sexual contact, with unprotected receptive anal intercourse being the most effective way of transmission. It is thought that probably <1% of HIV transmission occurs during a single act of unprotected intercourse between asymptomatically infected men and women (Perriens and Piot, 1993). Contaminated blood products, in particular when derived from pooled plasma, have caused epidemics among hemophiliacs.

Worldwide, perinatal transmission is a major health problem. In 1999, 470,000 children died from HIV infection, and another 570,000 became infected. HIV transmission occurs *in utero*, during delivery, or through breast feeding. Maternal viral load is the best predictor of risk.

After seroconversion, which in 50%–90% of patients present with a glandular fever-like syndrome or acute aseptic meningitis, there follows an asymptomatic phase that may last months or years. The prodromal stage of full-blown AIDS is heralded by the lymphadenopathy syndrome (LAS) or the AIDS-related complex (ARC) with chronic lymphadenopathy, weight loss, fever, and diarrhea. AIDS is characterized by the development of opportunistic infections such as *P. carinii* pneumonia or cerebral toxoplasmosis, malignancies such as Kaposi's sarcoma, B-cell lymphoma, or invasive cervical cancer in women. Occasionally, neurological disorders caused by HIV itself such as HIV dementia may be the AIDS-defining illness (Tables I and II).

In untreated HIV-infected individuals, 51% will have developed AIDS within 10 years. Variables associated with rapid progression include those with a symptomatic seroconverting illness, older age group at diagnosis, and those who received a large innoculum of virus, for example, through a contaminated transfusion in which the donor had a high viral load. Laboratory indices that are used as surrogate markers are the CD4 lymphocyte count, for example, in untreated patients with a CD4 count of 500/μL, the risk of AIDS developing at 1 year is 1% compared with 20% in a patients with a CD4 count around 200/μL (Brettle, 2001). A fall in CD4 cell count, particularly if the decline is rapid, is associated with a higher risk of progression. More recently, the viral load measurement has been shown to be an independent marker for the risk of progression. For example, in untreated individuals with an HIV RNA load of 3,000–14,000/μL, the risk of AIDS over a 6-year period is 17%. With a viral load of 110,000, this risk rises to 80% (Brettle, 2001).

Before the introduction of effective antiretroviral therapy, after a clinical diagnosis of AIDS, the 1-year survival rate was 50% with a 5-year survival of 0%. The introduction of HAART has had a dramatic impact on AIDS-related morbidity and mortality in the developed world. This is related to the significant decrease in the incidence of opportunistic infections, which are the main cause of death in patients with AIDS. Over a 3-year period from 1995–1997, death rates in patients with CD4 counts <100 cells/μL in the United States fell from 29.4/100 person-years to 8.8/100 person-years (Palella *et al.*, 1998). In the same period, the combined incidence of opportunistic infections (*P. carinii* pneumonia, *M. avium* complex, and cytomegalovirus) fell from 21.9/100 person-years to 3.7/100 person-years.

### Neurological Manifestations

The timing of neurological complications of HIV infection is shown in Figure 1. Acute aseptic meningitis is seen at seroconversion. A chronic meningitic picture with cranial nerve palsies, especially affecting the fifth, seventh, and eighth nerves, may present during the asymptomatic phases.

The most common neurological manifestation encountered in HIV/AIDS patients, distal sensory peripheral neuropathy, only occurs in the later AIDS stages of the infection. With the advent of the newer antiretroviral treatments, although the incidence of HIV-related complications, such as HIV dementia, has declined, that of neuropathy has continued to increase. This is, in part, due to drugs such as didanosine (ddI), zalcitabine (ddC), and stavudine (d4T), which can all cause peripheral neuropathy.

HIV dementia occurs during the advanced stages of AIDS. The mean CD4 count in one study was 109 cells/µL at the time of diagnosis (Portegies *et al.*, 1989). The associated vacuolar myelopathy also occurs only during the advanced stages of AIDS.

Toxoplasmosis, cryptococcal meningitis, primary CNS lymphoma, and PML all occur in the later immunosuppressed stages of HIV infection. In an analysis of 115 patients with AIDS-related toxoplasmosis, the median CD4 count at presentation was 50 cells/µL (Porter and Sande, 1992). Background seroprevalence rates for *T. gondii* vary geographically. It is in the region of 90% in France and 50% in the United Kingdom. The estimated risk of an HIV-infected individual who is seropositive developing cerebral toxoplasmosis is about 28% (Grant *et al.*, 1990). Primary prophylaxis with trimethoprim-sulfamethoxazole, 980 mg/day, is recommended in seropositive patients with a CD4 count <100 cells/µL (Kovacs and Masur, 2000). In acute cases, more than 90% of patients had a clinical or radiological response by day 14 of antitoxoplasmosis therapy. Maintenance therapy after the acute episode is recommended, although with the newer antiretroviral drugs, it may be feasible to discontinue prophylaxis after a sustained response in CD4 counts. At present the data are too limited to make definitive recommendations.

Cryptococcal meningitis is the most common fungal infection of the nervous system encountered in AIDS. The causative organism, *C. neoformans*, is less virulent than, for example, *M. tuberculosis*, and usually occurs when the CD4 count is <100 cells/µL. Other concomitant infections, notably *P. carinii* pneumonia, are found in 15%–35% of patients. The acute mortality of cryptococcal meningitis during induction therapy is 10%–25%; the 12-month survival rate among all patients is 30%–60% (Powderly, 1993). Poor prognostic markers include hyponatremia, extrameningeal cryptococcus, an altered mental state at presentation, a CSF cryptococcal antigen titer of >1:1054, and a CSF white cell count of <20 cells/µL (Chuck and Sande, 1989; Saag *et al.*, 1992). In patients with AIDS, complete cure is unlikely; the goal therefore is to control the infection. After initiation of treatment with amphotericin B or fluconazole, the duration of positive CSF culture ranges from 15–41 days. Approximately 25% of patients treated with these drugs show clinical improvement without a complete mycological response

(Powderly, 1993; Saag *et al.*, 1992). This is known as the quiescent phase, whose natural history is unknown. Maintenance therapy for cryptococcal meningitis is necessary, because relapse rates of 50%–60% and a shorter life expectancy have been reported for patients without such therapy. As with toxoplasmosis, the data are limited as to the issue of stopping secondary prophylaxis in patients with improved immune function after HAART.

The natural history of PML in patients with AIDS is variable, making it difficult to interpret the results of treatment protocols in uncontrolled studies. In general, however, the median survival figures range between 6 and 9 months before the introduction of HAART. Berger *et al.*, studying a group of HIV-associated cases of PML, found the following factors to be predictive for survival periods of more than 12 months: PML as the AIDS-defining illness, a high CD4 count (>300 cells/µL) at diagnosis, and lesion enhancement on imaging.

Since the availability of the newer antiretroviral drugs, a number of studies have documented increased survival in patients with MRL—in one, median survival increased from 11 weeks in controls to 46.4 weeks in those on HAART (Clifford *et al.*, 1999). More recently, the combination of HAART with the anti-CMV drug Cidofovir, these figures have been further improved on.

The decline noted in opportunistic infections since HAART applies to the neurological complications. The Multicentre AIDS Cohort Study (MACS) found a significant decline in incidence in HIV dementia between 1990 and 1992, when most patients were on antiretroviral monotherapy from 21.1/1000 person-years to 10.5 in the period 1996–1998, when most patients were on HAART (Sacktor *et al.*, 2001). Similar reductions were noted for cryptococcal meningitis, toxoplasmosis, and CNS lymphoma. There was, however, no significant change in the incidence rates for PML.

## PRINCIPLES OF THERAPY

### Antiretroviral Therapy

In the two decades of the AIDS epidemic, improvement in morbidity and mortality was achieved in the first decade as a result of improved diagnosis and treatment regimens for opportunistic infections, including primary and secondary prophylaxis. During the second decade, dramatic strides have been made in the armamentarium of anti-HIV drug regimens. Whereas in 1987, there was only one antiretroviral drug, in 2001 there are more than 16, with more in developmental stages. As a result, the issues regarding when to start antiretroviral therapy, which drugs to use, assessment of resistance patterns, and treatment failures have become increasingly complex. These decisions should now be made by physicians who manage HIV patients full-time.

At present, antiretroviral drugs fall into three categories:

- Nucleoside analogue reverse transcriptase inhibitors (NRTIs; e.g., zidovudine (AZT), didanosine (ddI), zalcitabine (ddC), stavudine (d4T), lamivudine (3TC), abacavir (ABC), and combivir (a combination of zidovudine and lamivudine).
- Nonnucleoside reverse transcriptase inhibitors (NNRTIs; e.g., delaviridine, nevirapine, and efavirenz).
- Protease inhibitors (Pls; e.g., saquinavir, indinivir, nelfinavir, amprenavir, and ritonavir).

In essence, the rationale for combination drug therapy in HIV, as in other medical disciplines such as cancer chemotherapy, is to use drugs that act at different phases of the virus life cycle. Because the attrition of CD4 cells is related to high viral loads, reductions in the latter result in variable degrees of immune reconstitution as manifested by a rising CD4 count.

When HAART was introduced in 1996, it was postulated that a "hit early and hit hard" approach would result in complete viral eradication. It has become apparent that this is not the case in most patients. One placebo-controlled trial in primary HIV infection showed short-term benefit only (Kinloch-de Loës et al., 1995). Any putative benefits need to be balanced by the toxicity profile, such as lipodystrophy, and the associated risks of drug resistance.

The most recent guidelines from the British HIV Association (BHIVA) recommend deferring treatment in patients with a CD4 count >350 cells/µL whatever the viral load. Between 200 and 350 cell/µL, treatment should be considered in patients with a rapid decline in the CD4 count or a high viral load. Because compliance is a major issue in terms of efficacy, patients' wishes also need to be considered. Patients with CD4 counts <200 cells/µL should be offered antiretroviral therapy whatever the viral load. All patients with AIDS should be treated with antiretroviral therapy (BHIVA Guidelines, 2001). The United States Department of Health and Human Services' guidelines differ in that they recommend therapy for patients with CD4 counts in the range 200–350 cells/µL, and >350 cells/µL therapy should be considered if the viral load is high.

There is a variety of HAART regimens available, and there are no definitive controlled trials to demonstrate the clinical superiority of any one regimen. The BHIVA guidelines, for example, recommend two NRTIs plus an NNRTI. Other combinations to be considered include two NRTs plus one or two Pls.

These drugs have a significant toxicity profile, some of which are class specific and some of which are drug specific (Table VI).

The NRTIs ddC, ddI, and d4T have all been shown to cause a dose-dependent peripheral neuropathy. The association with 3TC is not well documented. Zidovudine causes a myopathy but has not been described to cause a neuropathy. Mitochondrial toxicity as a result of inhibition of DNA polymerase may be the underlying mechanism for these NRTI-related neuromuscular side effects. Nerve biopsy specimens from patients with nucleoside neuropathies demonstrate disrupted cristae similar to those seen in animal models (Lewis and Dalakas, 1995). The same mechanism may well explain the other effects encountered with this class of drug, which include pancreatitis, fulminant hepatic failure, and lactic acidosis. Abnormalities of fat distribution (lipodystrophy), in which, patients present with fat wasting from the buttocks, face, and limbs, with accumulation around the internal viscera, resulting in a distended abdomen, have been associated with both NRTIs and Pls.

The clinical presentation of these drug-related neuropathies is similar to that of the common HIV-related distal sensory peripheral neuropathy. However, these drug-induced neuropathies are more likely to be painful, have an abrupt onset, and progress rapidly. After cessation of the offending drug, there may be a paradoxical worsening of neuropathic symptoms for up to 4 weeks ("coasting"). On stopping the drug, an improvement in neuropathic symptoms can be expected in the majority of these drug-related neuropathies, unless there is a concomitant underlying HIV-associated DSPN.

The early monotherapy studies of ddC quoted an incidence of neuropathy in 12%–46% of patients. The reported rates for ddI ranged from 13%–34% and 12%–21% for d4T. More recent data such as the AIDS Clinical Trials Group (ACTG) protocol 175 found much lower incidence rates. One possible explanation is that the drugs are prescibed at higher CD4 counts, making it less likely that patients have had an underlying clinical or subclinical HIV-related neuropathy.

The risk of neuropathy is increased when combinations of neurotoxic drugs are used. This is especially apparent when hydroxyurea is used to potentiate the antiretroviral effects of ddI and d4T. Moore et al. found that the combined use of ddI, d4T, and hydroxyurea was associated with a 7.8-fold increase in the risk of sensory neuropathy compared with ddI alone. The use of d4T plus ddI without hydroxyurea was associated with a 3.5-fold increased risk compared with ddI and a 2.5-fold risk compared with d4T alone (Moore et al., 2000).

## Treatment of HIV-Related Disorders

### HIV Dementia

After the introduction of zidovudine in 1987, the incidence of HIV dementia declined significantly. One retrospective study reported a decline in incidence from 53% to 10% (Portegies et al., 1989). Further corroborative evidence for the beneficial effect of zidovudine on HIV dementia came from pathological studies showing a reduction in the numbers of multinucleated giant cells—the hallmark for HIV dementia—in those who had been prescribed the drug when compared with historical controls (Gray et al., 1991). Pizzo et al. demonstrated improvements, on psychometric testing and in some cases on CT scans, in 21 children with neurodevelopmental deficits after treatment with intravenous infusions of zidovudine (Pizzo et al., 1988). A randomized double-blind placebo-controlled trial of zidovudine at doses of 2000 mg/day and 1000 mg/day found evidence of benefit only at the higher dose (Sidtis et al., 1993).

With the introduction of combination therapies, in addition to the epidemiological evidence of decline in the incidence of HIV dementia, there is evidence of clinical benefit in terms of improved performance on tests of psychomotor speed—a sensitive index of HIV dementia. There is evidence that the plasma viral load may drive the HIV-related complications, both central and peripheral, as evidenced by the finding that individuals with a baseline plasma HIV RNA >30,000 copies/ml had a relative hazard for dementia 8.5 times greater than those with a load <3000 copies/ml. Thus, a reduction in the peripheral viral load may well have more of a central nervous system benefit (Childs et al., 1999).

An, as yet, unresolved issue is the importance of CSF penetrance of the antiretroviral drugs, with lipid-soluble drugs having better penetration than protein-bound drugs. A related concern is the possible development of protected sanctuary sites for the virus, such as within the brain, where the antiretrovirals have poor penetration. On the basis of CSF/serum distribution ratios, the NRTIs zidovudine, stavudine, and abacavir seem to have the best penetrance into the CSF (Enting et al., 1998). Nevirapine, an NNRTI, also has a favorable profile. The PIs, with high protein binding, have poor penetrance; indinivir, which has the lowest protein binding, has the highest CSF penetration.

As yet, there are insufficient data to recommend any particular combination treatment for HIV dementia, but it would seem reasonable to include at least one or two drugs that have good CSF penetrance. A recent study addressed this issue of whether using multiple drugs with good CSF penetrance would have a better outcome in terms of psychomotor slowing compared with the use of a single penetrating drug. Suprisingly, no significant differences were apparent (Sacktor et al., 2001).

Other mechanisms of neuronal injury, such as damage by cytokines and free radicals, have also been researched as potential avenues of therapy. Drugs investigated to date have included deprenyl (an antiapoptotic agent), nimodpine (a calcium channel blocker), and lexipafant (a platelet-activating factor antagonist) (Clifford, 2000).

### HIV-Associated Neuropathy

**Distal Sensory Peripheral Neuropathy/ NRTI-Related Neuropathy.** In patients with symptoms and signs of a peripheral neuropathy, if incriminating neurotoxic drugs are being prescribed (such as ddI, ddC, or d4T), it may be necessary to stop them, because it is difficult to discriminate, on clinical criteria or neurophysiologically, between these drug-induced and HIV-related neuropathies. However, if the drugs are proving effective in terms of controlling the HIV infection, this may be a difficult decision to make.

The drugs used in symptomatic treatment of neuropathic pain in AIDS have usually been extrapolated from those used in other conditions causing neuropathies such as diabetes. The antiepileptic drug gabapentin, which has been shown to be effective in diabetic neuropathy, is frequently used as first-line symptomatic therapy, even though to date there are no published data on HIV-related neuropathy (Backonja et al., 1998). The following have all been studied in randomized controlled trials with no evidence of benefit—acupuncture (Shlay et al., 1998), amitriptyline (Kieburtz et al., 1998; Shlay et al., 1998), peptide T (Simpson et al., 1996), and mexiletene (Kieburtz et al., 1998). In a small randomized, placebo-controlled trial, lamotrigine showed evidence of benefit in terms of pain scores (Simpson et al., 2000). The starting dose was 25 mg/day for 2 weeks, gradually titrating the dose to 150 mg twice daily. The most common side effect necessitating withdrawal was a rash. A phase II trial of recombinant human nerve growth factor injected subcutaneously twice weekly had a beneficial effect on neuropathic pain and pain sensitivity (McArthur et al., 2000). An extended open-label study confirmed this finding, but there was no improvement of the neuropathy as assessed by neurological examination, quantitative sensory testing, and epidermal nerve fiber density (Schifitto et al., 2001).

In a small pilot study, improved peripheral nerve function was demonstrated using quantitative assessments of thermal perception thresholds for warmth, cold, and heat pain in patients who responded to antiretroviral therapy compared with those who did not

(Martin *et al.*, 2000). This raises the possibility that reducing the peripheral viral load may reduce the "drive" resulting in peripheral nerve damage in a similar fashion to the central nervous system complications of dementia.

A small study described reduced levels of serum acetyl carnitine in patients who had a neuropathy develop on ddI, ddC, and/or d4T (Famularo *et al.*, 1997). This has been the impetus for a number of trials investigating acetyl carnitine as a treatment option. Because the carnitines facilitate beta-oxidation of fatty acids in mitochondria, a depletion results in accumulation of toxic fatty acids. They also enhance both the levels and the binding of nerve growth factor, thus facilitating nerve regeneration.

**Acute and Chronic Demyelinating Polyneuropathies; Vasculitic Neuropathy.** The acute demyelinating neuropathy (Guillain-Barré syndrome) responds in a similar manner as do non-HIV patients with beneficial results to intravenous immunoglobulin and plasma exchange. Similarly, the chronic demyelinating neuropathy (CIDP) responds to immunoglobulin and corticosteroids (Cornblath, 1988).

Mononeuritis multiplex related to vasculitis requires treatment with corticosteroids and antiretroviral drugs if it is HIV related (Manji *et al.*, 1997) or part of the diffuse infiltrative lymphocytic syndrome (DILS) (Moulignier *et al.*, 1997; Price, 1998). If the nerve biopsy shows evidence of CMV, then ganciclovir or foscarnet therapy is indicated (Said *et al.*, 1991).

### Vacuolar Myelopathy

Apart from symptomatic treatments, only recently a case report described significant improvement in one patient with an HIV-related myelopathy after treatment with HAART—d4T, 3TC, and nefinavir (Staudinger and Henry, 2000). There is a hypothesis that vacuolar myelopathy may be related to abnormal transmethylation metabolism. A pilot study showed improvements with oral L-methionine 6 g/day in two divided doses (DiRocco *et al.*, 1998). The results of a placebo-controlled trial are awaited.

## Opportunistic Infections

### Toxoplasmic Encephalitis

Empirical antibiotic therapy for toxoplasmosis is used in patients with focal acute or subacute cerebral lesions on CT or MRI. The diagnosis also needs to be considered in diffuse encephalopathies, because rare cases of a diffuse encephalitic illness with *T. gondii* are described in the literature (Gray *et al.*, 1989).

The treatment of choice is a combination of sulfadiazine with pyrimethamine. Pyrimethamine is an inhibitor of the hydrofolate reductase, an enzyme that converts hydrofolic acid to tetrahydrofolinic acid. Folinic acid, but not folic acid, is added to the regimen to prevent the myelosuppressive effects of pyrimethamine. Clindamycin and atovaquone are effective alternative drugs to sulfadiazine.

In patients with mass lesions when there is significant cerebral edema that may be life-threatening, corticosteroids need to be administered. Any response, clinically or radiologically, may be due to either the anti-toxoplasma therapy or the corticosteroids. After a period of stabilization, the corticosteroids can be reduced—any deterioration then warrants a consideration of a stereotactic brain biopsy, because the mass lesions may be due to lymphoma or other infectious diseases such as tuberculoma.

### Cryptoccocal Meningitis

Before the onset of the AIDS epidemic, the combination of amphotericin B and flucytosine (5FC) was shown to be superior to amphotericin alone (Bennett *et al.*, 1979). Amphotericin B alone has been claimed to be as effective as when combined with flucytosine in patients with AIDS (Chuck and Sande, 1989; Kovacs *et al.*, 1985; Zuger *et al.*, 1986). These last authors showed retrospectively that flucytosine had to be stopped because of toxic side effects (in particular bone marrow suppression) in 53% of patients receiving the combination. Other authors, however, found the combination to be more successful (Eng *et al.*, 1986; Larsen *et al.*, 1990). A later study randomly assigned patients to amphotericin 0.7 mg/kg/day IV with or without 5FC, 100 mg/kg/day orally for 2 weeks followed by either fluconazole or itraconazole, 400 mg/day orally for the next weeks. After 10 weeks there was no difference between the two groups in the percentage of patients with negative CSF cultures (van der Horst *et al.*, 1997).

Amphotericin B has a high binding rate to plasma proteins and exhibits poor CSF penetration. Because of the risk of thrombophlebitis, a central venous catheter should be used. During or after the infusion, fever, rigor, malaise, and vomiting may occur. Other side effects are anemia, thrombocytopenia, renal tubular acidosis, hypokalemia, and hypomagnesemia. A dose-related impairment of renal function is usually reversible after cessation of treatment. Liposomal amphotericin preparations, although more expensive, have fewer nephrotoxic side effects.

Fluconazole, a triazole compound, is an oral preparation with good CNS penetration, low toxicity, and a long half-life. Trials comparing amphotericin B at a dose of 0.3–0.5 mg/kg/day versus fluconazole 200 mg/day showed no significant differences in outcome. However, the higher rate of early deaths in the fluconazole group may have been related to the more rapid clearance of cryptococcus from the CSF in the amphotericin group (Saag *et al.*, 1992). However, it should be noted that the doses of both fluconazole and amphotericin were rather low.

Current opinion favors the use of amphotericin B with or without 5FC in patients with severe cryptococcal meningitis for at least 2 weeks. This would include patients with any of the following features that have been shown to be associated with poor outcome— abnormal mental status, CSF cryptococcal antigen titer >1 : 1024, and a CSF white cell count of <20 cells/μL. In milder cases fluconazole, 400 mg/day, would seem a reasonable option.

Maintenance therapy is necessary to prevent relapses. Lifelong maintenance therapy with fluconazole, 200 mg/day, is superior to weekly intravenous amphotericin in terms of fewer relapses and fewer side effects (Bozzette *et al.*, 1991; Powderly *et al.*, 1992).

A particular complication that needs vigilance is the development of elevated intracranial pressure in patients with cryptococcal meningitis, which is associated with a high mortality rate. This should be managed by repeated lumbar punctures, or if this becomes necessary too often, by the placement of a lumbar or ventricular drain. Acetazolamide may have a role in reducing production of CSF.

## Progressive Multifocal Leukoencephalopathy

Early anecdotal reports suggested that cytosine arabinoside (Ara-C) may play a role in the treatment of PML. A multicenter trial of 57 patients infected with HIV and with biopsy-proven PML found no significant difference in survival when the following three groups were compared: antiretroviral therapy alone, antiretroviral therapy plus intravenous Ara-C, and antiretroviral therapy plus intrathecal Ara-C (Hall *et al.*, 1998).

The use of alpha-interferon was prompted by its therapeutic benefit in anogenital warts induced by human papillomavirus. Alpha-interferon seems to have both an antiviral effect and an immune-enhancing effect. A retrospective open-label observational study found a significantly increased median survival time in the treatment arm. The dose used was either alpha-interferon (usually 2b) 3 million units subcutaneously daily or 5 million units three times a week (Huang *et al.*, 1998). The side effects of interferon treatment included leukopenia, pancytopenia, depression, and fatigue. A larger randomized study is required before specific recommendations can be made.

After the introduction of HAART, there were increasing reports of improvement in cases of HIV-related PML. This was not surprising, because in non-HIV cases of PML, improvements were described in patients once any immunosuppressive treatments had been stopped. Furthermore, it has been demonstrated that an interaction may occur with the HIV nuclear protein Tat, increasing the transcriptional activity of JC virus in glial cells (Chowdhury *et al.*, 1993). One study documented a fourfold increase in survival in patients on HAART (Clifford *et al.*, 1999).

The anti-CMV drug cidofovir has been shown to have in vitro activity against the JC virus. A multicenter observational study showed that patients on HAART and cidofovir had better clearance of JC virus from the CSF and also showed neurological improvement or stability compared with those treated with HAART only (De Luca *et al.*, 2000). Cidofovir has several serious side effects, including nephrotoxicity, which is a dose-dependent renal tubular acidosis, neutropenia, and ocular hypotonia.

## Opportunistic Tumors

In suspected cases of primary CNS brain lymphoma, cranial irradiation combined with corticosteroids led to clinical improvement in 76% of the patients, halted disease progression in 14%, and improved CT findings in 69% (Baumgartner *et al.*, 1990). The mean survival was also higher in the corticosteroid/radiotherapy (124 days) group compared with untreated patients (42 days). The most common causes of death in patients who received radiotherapy were opportunistic infections, whereas in the untreated group progression of the tumor was the most common. There is at present no clear role for chemotherapy.

Although with the introduction of HAART, there has been a decline in the incidence of PCNSL, it could be speculated that an improvement in immune status may have a beneficial effect on the tumor.

In clinical practice, patients with cerebral mass lesions, which are most frequently due to either toxoplasmosis or primary CNS lymphoma, empirical treatment is given for toxoplasmosis. If this fails, the issue of a stereotactic brain biopsy arises. Many patients refuse this procedure, and the decision depends on the, at best, modest benefits of radiotherapy in primary CNS lymphoma versus the potential side effects and also the quality of life caused by HIV disease.

The management of systemic non-Hodgkin's lymphoma includes imaging studies of the brain, chest, abdomen, and pelvis, as well as a bone marrow aspirate and biopsy. A lumbar puncture should also be performed to find evidence of meningeal infiltration. Treatment includes craniospinal irradiation together with systemic and intrathecal chemotherapy. High-dose chemotherapy may result in a higher rate of opportunistic infections, and low-dose chemotherapy has been advocated (Levine et al., 1991). Only about 50% of patients respond to treatment. In those who do, the median survival is about 15 months, although a few may live more than 2 years (Levine et al., 1991).

# REFERENCES

Adler, M. (2001). Development of the epidemic. In "ABC of AIDS", 5th ed. (M. Adler, Ed.), pp. 1–5. BMJ Publishing, London.

Andreula, C. F., Burdi, N., and Carella, A. (1993). CNS cryptococcosis in AIDS: spectrum of MR findings. J. Comput. Assist. Tomogr. 17(3), 438–441.

Arnaudo, E., Dalakas, M. C., Shanske, E. et al. (1991). Depletion of muscle mitochondrial DNA in AIDS patients with zidovudine induced myopathy. Lancet. 337, 508–510.

Arribas, J. R., Storch, G. A., Clifford, D. B. et al. (1996). Cytomegalovirus encephalitis. Ann. Intern. Med. 125, 577–587.

Backonja, M., Beydoun, A., Edwards, K. R. et al. (1998). Gabapentin for the symptomatic treatment of painful neuropathy in patients with diabetes mellitus. JAMA. 280, 1831–1836.

Barohn, R. J., Gronseth, G. S., Leforce, B. R. et al. (1993). Peripheral nervous system involvement in a large cohort of human immunodeficiency virus infected individuals. Arch. Neurol. 50, 167–171.

Batman, P. A., Miller, A. R., Sedgwick, P. M. et al. (1991). Autonomic denervation in jejunal mucosa of homosexual men infected with HIV. AIDS. 5, 1247–1252.

Baumgartner, J. E., Rachlin, J. R., Beckstead, J. H., Meeker, T. C., Levy, R. M., Wara, W. M., and Rosenblum, M. L. (1990). Primary central nervous system lymphomas: Natural history and response to radiation therapy in 55 patients with acquired immunodeficiency syndrome. J. Neurosurg. 73, 206–211.

Bennett, J. E., Dismukes, W. E., Duma, R. J., Medoff, G., Sande, M. A., Gallis, H., Leonard, J., Fields, B. T., Bradshaw, M., Haywood, H., McGee, Z. A., Cate, T. R., Berry, C. D., Hooton, T. M., Collier, A. C. et al. (1979). Neurologic relapse after benzathine penicillin therapy for secondary syphilis in a patient with HIV infection. N. Eng. J. Med. 301, 126–131.

Berengeur, J., Moreno, S., Laguna, F. et al. (1992). Tuberculous meningitis in patients with the human immunodeficiency virus. N. Engl. J. Med. 326, 668–672.

Berger, J. R. (1991). Neurosyphilis in human immunodeficiency virus type I seropositive individuals: A prospective study. Arch. Neurol. 48, 700–702.

Berger, J. R., Levy, R. M., Flomenhoft, D. et al. (1998). Predictive factors for prolonged survival in acquired immunodeficiency syndrome acquired progressive multifocal leucoencephalopathy. Ann. Neurol. 44, 341–349.

Berger, J. R., Pall, L., Lanska, D. et al. (1998). Progressive multifocal leucoencephalopathy in patients with HIV infection. J. Neurovirol. 4, 59–68.

BHIVA Writing Committee on behalf of the BHIVA Executive Committee (2001). BHIVA guidelines for the treatment of HIV infected adults with antiretroviral therapy. HIV Medicine 2, 276–313.

Bishburg, E., Sunderam, E., Reichman, L. B. et al. (1986). Central nervous system tuberculosis with the acquired immunodeficiency syndrome and its related complex. Ann. Intern. Med. 105, 210–213.

Bissuel, F., Berruyer, M., Causse, X. et al. (1992). Acquired protein S deficiency: correlation with advanced disease in HIV-1 infected patients. J. Acquir. Immun. Defic. Syndr. 5, 484–489.

Bossolasco, S., Nilsson, A., de Milito, A. et al. (2001). Soluble CD 23 in cerebrospinal fluid: a marker of AIDS related non-Hodgkin's lymphoma in the brain. AIDS. 15, 1109–1113.

Bowman, F. H., Skolsky, R., Hes, D. et al. (1998). Variable progression of HIV dementia. Neurology. 50, 1814–1820.

Bozzette, S. A., Larsen, R. A., Chiu, J. et al. (1991). A placebo controlled trial of maintenance therapy with fluconazole after treatment of cryptococcal meningitis in AIDS. N. Engl. J. Med. 324, 580–584.

Bradley, W. G., and Verma, A. (1996). Painful vasculitic neuropathy: relief of pain with prednisolone therapy. Neurology. 47, 1446–1451.

Brettle, R. (2001). HIV and AIDS: natural History of HIV/AIDS. In "Medicine." (A. Lever, Ed.), pp. 12–15. Medicine Publishing Co.

Brilla, R., Nabavi, D. G., Schulte-Altedorneburg, G. et al. (1999). Cerebral vasculopathy in HIV infection revealed by trans-cranial doppler. Stroke. 30, 811–813.

Brinkman, R., Schwinn, A., Narayan, O. et al. (1992). Human immunodeficiency virus infection in microglia: Correlation between cells infected in the brain and cells cultured from infectious brain tissue. Ann. Neurol. 31, 361–365.

Budka, H., Wiley, C. A., Kleihues, P. et al. (1991). HIV associated disease of the nervous system: review of the nomenclature and proposal for neuropathology based terminology. Brain Path. 1, 143–152.

Calabrese, L. H., Estes, M., Yen-Lieberman, B. et al. (1989). Systemic vasculitis in association with human immunodeficiency virus infection. Arthritis Rheum. 32, 569–576.

Carne, C. A., Tedder, R. S., Smith, A. et al. (1985). Acute encephalopathy coincident with seroconversion for anti-HTLV-III. Lancet. ii, 1206–1208.

Childs, E. A., Lyles, R. S., Selnes, O. A. et al. (1999). Plasma viral load and CD4 cell count predict HIV associated dementia and sensory neuropathy. Neurology. 52, 607–613.

Chimelli, L., and Scaravelli, F. (1991). Morphological changes in the autonomic nervous system of patients with AIDS. In "Proceedings of the Seventh International Conference on Neuroscience of HIV infection." Padova, Italy. p. 89.

Chowdhury, M. M., Kundu, M., Khalili, K. et al. (1993). GA/GC rich sequence confers Tat responsiveness to human neurotropic virus promoter, JCV 1, in cells derived from the central nervous system. Oncogene. 8, 887–892.

Chuck, S. L., and Sande, M. A. (1989). Infections with Cryptococcus neoformans in the acquired immunodeficiency syndrome. N. Engl. J. Med. 321, 794–799.

Cirillo, S. F., and Rosenblum, M. L. (1990). Use of Ct and MR imaging to distinguish intracranial lesions and to define the need for biopsy in AIDS patients. J. Neurosurg. 73, 720–724.

Cinque, P., Scarpellini, P., Vago, L. et al. (1996). Diagnosis of central nervous system complications in HIV infected patients: cerebrospinal fluid analysis by polymerase chain reaction. AIDS. 11, 1–17.

Clifford, D. (2000). Human immunodeficiency virus associated dementia. Arch. Neurol. 57, 321–324.

Clifford, D., Yiannoustsos C., Glickman, M. et al. (1999). HAART improves prognosis in HIV—associated progressive multifocal leucoencephalopathy. Neurology. 52, 623–625.

Cobbs, C. G., Warner, J. F., and Alling, D. W. (1979). A comparison of amphotericin B alone and combined with flucytosine in the treatment of cryptococcal meningitis. N. Engl. J. Med. 310, 126–131.

Connor, M. D., Lammie, G. A., Bell, J. E. et al. (2000). Cerebral infarction in adult AIDS patients. Stroke. 31, 2117–2126.

Cornblath, D. R., and McArthur, J. C. (1988). Predominantly sensory neuropathy in patients with AIDS and AIDS related complex. Neurology. 38, 794–796.

Cornblath, D. R., McArthur, J. C., Kennedy, P. G. E. *et al.* (1987). Inflammatory demyelinating peripheral neuropathies associated with human T-cell lymphotrophic virus type III infection. *Ann. Neurol.* 21, 32–40.

Crowe S. M., Carlin B., Stewark K. L. *et al.* (1991). Predictive value of CD4 lymphocyte numbers for the development of opportunistic infections and malignancies in HIV infected persons. *J. Infect Dis.* 4, 770–776.

Dalakas, M. C., Illa, I., Pezeshkpour, G. H. *et al.* (1990). Mitochondrial myopathy caused by long term zidovudine therapy. *N. Engl. J. Med.* 322, 1098–1105.

Dalakas, M. C., Pezeshkpour, G., Gravell, M. *et al.* (1986). Polymyositis associated with AIDS retrovirus. *JAMA.* 256, 2381–2383.

De Angelis, L. M., Wong, E., Rosenblum, M. *et al.* (1992). Epstein-Barr virus in Aids and non Aids primary central nervous system lymphoma. *Cancer.* 70, 1607–1611.

De la Monte, S. M., Gabuzda, D. H., Ho, D. D. *et al.* (1988). Peripheral neuropathy in the acquired immunodeficiency syndrome. *Ann. Neurol.* 23, 485–492.

De Luca, A., Fantoni, M., Tartaglione, T. *et al.* (2000). Response to cidofovir after failure of antiretroviral therapy alone in AIDS associated progressive multifocal leucoencephalopathy. *Neurology.* 52, 891–892.

Denning, D. W., Anderson, J., Rudge, P. *et al.* (1987). Acute myelopathy associated with primary infection with human immunodeficiency virus. *BMJ.* 294, 143–149.

Dirocco, A., Tagliati, M., Danisi, F. *et al.* (1998). A pilot study of I—methionine for the treatment of AIDS associated vacuolar myelopathy. *Neurology.* 51, 266–268.

Drabick, J. J., and Tramont, E. C. (1990). Utility of the VDRL test in HIV seropositive patients. *N. Engl. J. Med.* 322, 271.

Dunbar, N., and Brew, B. (1996). Neuropsychological dysfunction in AIDS: a review. *J. NeuroAIDS.* 1(3), 73–102.

Ellis, R. J., Hsia, K., Spector, S. A. *et al.* (1997). Cerebrospinal fluid human immunodeficiency virus type–1RNA levels in neurocognitively impaired individuals with acquired immunodeficiency syndrome. *Ann. Neurol.* 42, 679–688.

Eng, R. H. K., Bishburg, E., Smith, S. M., and Kapila, R. (1986). Cryptococcal infections in patients with acquired immune deficiency syndrome. *Am. J. Med.* 81, 19–23.

Enting, R. H., Hoetelmans, R. M., Lange, J. M. *et al.* (1998). Antiretroviral drugs and the central nervous system. *AIDS.* 12, 1941–1955.

Epstein, L. G., and Gendleman, H. E. (1993). Human immunodeficiency virus type-1 infection of the nervous system: pathogenic mechanisms. *Ann. Neurol.* 33, 429–436.

Famularo, G., Moretti, S., Marcellini, S. *et al.* (1997). Acetyl carnitine deficiency in AIDS patients with neurotoxicity on treatment with antiretroviral nucleoside analogues. *AIDS.* 11, 185–190.

Fuller, G. N., Jacobs, J. M., and Guiloff, R. J. (1993). Nature and incidence of peripheral nerve syndromes in HIV infection. *J. Neurol. Neurosurg. Psychiatr.* 56, 372–381.

Gherardi, R. K., Chretien, F., Deelfau-Larue, M.-H. *et al.* (1998). Neuropathy in diffuse infiltrative lymphocytosis syndrome. *Neurology.* 50, 1041–1044.

Gillespie, S. M., Chang, Y., Lemp, G. *et al.* (1991). Progressive multifocal leucoencephalopathy in persons infected with the human immunodeficiency virus. San Francisco, 1981–1989. *Ann. Neurol.* 30, 597–604.

Glass, J. D., Wesselingh, S. L., Selnes, O. A. *et al.* (1993). Clinical-neuropathological correlation in HIV associated dementia. *Neurology.* 43, 2230–2237.

Glesby, M. J., Moore, R. D., and Cjaisson, R. E. (1995). Clinical spectrum of herpes zoster in adults infected with human deficiency virus. *Clin. Infect. Dis.* 21, 370–375.

Gottlieb, M. S., Schroff, R., Schanker, H. M. *et al.* (1981). Pneumocystis carinii pneumonia and mucosal candidiasis in previously healthy homosexual men: evidence of a new acquired cellular immunodeficiency. *N. Engl. J. Med.* 305, 1425–1431.

Grant, I. H., Gold, J., Rosenblum, M. *et al.* Toxoplasma gondii serology in HIV infected patients: The development of CNS toxoplasmosis in AIDS. *AIDS* 4, 519–521.

Gray, F., Geny, C., Dournon, E., Fenelon, G., Lionnet, F., and Gherardi, R. (1991). Neuropathological evidence that zidovudine reduces incidence of HIV-infection of brain. *Lancet.* 337, 852–853.

Gray, F., Gherardi, R., Wingate, E., Singate, J., Fenelon, G., Gaston, A., and Sobel, A. (1989). Diffuse encephalitic cerebral toxoplasmosis in AIDS: Report of four cases. *J. Neurol.* 236, 273–277.

Graybill, J. R., Sobel, J., Saag, M. *et al.* (2000). Diagnosis and management of intracranial pressure in patients with AIDS and cryptococcal meningitis. *Clin. Infect. Dis.* 30, 47–54.

Guiloff, R. J. (1989). Neurological opportunistic infections in Central London. *J. R. Soc. Med.* 82, 278–280.

Guiloff, R. J., and Fuller, G. N. (1992). Other neurological diseases in HIV-1 infection: Clinical aspects. In "Clinical Neurology: Neurological Aspects of Human Retroviruses." (P. Rudge, Ed.), pp. 175–209. Balliere and Tindall, London.

Guiloff, R. J. and Tan, S. V. (1992). Central nervous system opportunistic infections in HIV disease: clinical aspects. In "Clinical Neurology: Neurological aspects of human retroviruses." (P. Rudge, Ed.), pp. 103–154. Balliere and Tindall London.

Haas, J., Bolan, G., Larsen, S. *et al.* (1990). Sensitivity of treponemal tests for detecting prior treated syphilis during human immunodeficiency virus infection. *J. Infect. Dis.* 162, 862–866.

Hall, C., Dafni, I., Simpson, D. *et al.* (1998). Failure of cytarabine in progressive multifocal leucoencephalopathy associated with human immunodeficiency virus infection. *N. Engl. J. Med.* 338, 1345–1351.

Hamed, L. M., Schatz, N. J., and Galetta, S. L. (1988). Brainstem ocular motility defects and AIDS. *Am. J. Ophthalmol.* 106, 437–442.

Ho, D. D., Pomerantz, R. J., and Kaplan, J. C. (1987). Pathogenesis of infection with the human immunodeficiency virus. *N. Engl. J. Med.* 278, 278–286.

Hoffmann, M., Berger, J., Nath A. *et al.* (2000). Cerebrovascular disease in young, HIV-infected black Africans in the KwaZulu Natal Province of South Africa. *J. Neurovirol.* 6, 229–236.

Holland, N. R., Power, C., Matthews, V. P. *et al.* (1994). Cytomegalovirus encephalitis in acquired immunodeficiency syndrome. *Neurology.* 44, 507–514.

Huang, P. P., McMeeking, A. A., Stempien, M. J. *et al.* (1997). Cytomegalovirus disease presenting as a focal brain mass: report of two cases. *Neurosurgery.* 40, 1074–1078.

Huang S. S., Skolansky, R. L., Dal Pan, G. J. *et al.* (1998). Survival prolongation in HIV—associated progressive multifocal leucoencephalopathy treated with alpha—interferon: an observational study. *J. Neurovirol.* 4, 324–332.

Illa, I., Nath, A., Dalakas, M. *et al.* (1991). Immunocytochemical and virological characteristics of HIV associated inflammatory myopathies. *Ann. Neurol.* 29, 474–481.

Kalayjian, R. C., Cohen, M. L., Bonomo, R. *et al.* (1993). Cytomegalovirus ventriculoencephalitis in AIDS. *Medicine.* 72, 67–77.

Kamin, S. S., and Petito, C. K. (1991). Idiopathic myelopathies with white matter vacuolation in non-acquired immunodeficiency patients. *Hum. Pathol.* 22, 816–824.

Kieburtz, K., Simpson, D., Yiannoustous, D. *et al.* (1998). A randomised trial of amitriptyline and mexiletine for painful neuropathy in HIV infection. *Neurology.* 51, 1682–1688.

Kinloch-de Loës, S., Hirschel, B., Hoen, B. *et al.* (1995). A controlled trial of zidovudine in primary HIV infection. *N. Engl. J. Med.* 333, 408–413.

Kovacs, J. A., Kovacs, A. A., Polis, M., Wright, W. C., Gill, V. J., Tuazon, C. U., Gelmann, E. P., Lane, H. C., Longfied, R., Overturf, G., Macher, A. M., Fauci, A. S., Parrillo, J. E., Bennett, J. E., and Masur, H. (1985). Cryptococcosis in the acquired immunodeficiency syndrome. *Ann. Intern. Med.* **103**, 533–538.

Kovacs, J. A., and Masur, H. (2000). Prophylaxis against opportunistic infections in patients with Human Immunodeficiency Virus Infection. *N. Engl. J. Med.* **342**, 1416–1429.

Lanska, M. J., Lanska, D. J., Schmidley, J. W. *et al.* (1988). Syphilitic polyradiculopathy in an HIV positive man. *Neurology.* **38**, 1297–1301.

Larsen, R. A., Leal, M. A. E., and Chan, L. S. (1990). Fluconazole compared with amphotericin B plus flucytosine for cryptococcal meningitis in AIDS: A randomized trial. *Ann. Intern. Med.* **113**, 183–187.

Levine, A. M., Sullivan-Halley, J., Pike, M. C. *et al.* (1991). Human immunodeficiency virus related lymphoma. Prognostic factors predictive of survival. *Cancer.* **6**, 2466–2472.

Levy, J. A. (1990). Features of HIV and the host response that influence progression to disease. *In* "The Medical Management of AIDS." (M. A. Sande and P. A. Volberding, Eds.), pp. 233–237. W.B. Saunders, Philadelphia.

Lewis, W., and Dalakas, M. C. (1995). Mitochondrial toxicity of antiviral drugs. *Nat. Med.* **1**, 417–422.

MacGowan, D. J., Scelasa, S. N., Waldron, M. *et al.* (2001). An ALS like syndrome with new HIV infection and complete response to antiretroviral therapy. *Neurology.* **57**, 1094–1097.

Mahe, A., Bruet, A., Chabin, E. *et al.* (1989). Acute rhabdomyolysis coincident with primary HIV-1 infection. *Lancet.* **ii**, 454–455.

Malessa, R. (1991). Veränderungun am Rückenmark. *In* "HIV Infektion und Nervensystem." (A. A. Moller and H. Backmund, Eds.), pp. 97–100.

Manji, H., LeCroix, C., and Said, G. (1997). The clinical and laboratory features of HIV-1 associated necrotizing arteritis peripheral neuropathy. *J. Neurol.* **244**(suppl 3), S67.

Manji, H., and Miller R. (2001). Progressive Multifocal Leucoencephalopathy—progress in the AIDS era (editorial). *J. Neurol. Neurosurg. Psychiatry* **69**, 569–571.

Martin, C., Solders, G., Sonnerberg, A. *et al.* (2000). Antiretroviral therapy may improve sensory function in HIV infected patients—a pilot study. *Neurology* **54**, 2120–2127.

Masur, H., Michelis, M. A., Greene, J. B. *et al.* (1981). An outbreak of community acquired pneumocystis carinii pneumonia: initial manifestation of cellular immune dysfunction. *N. Engl. J. Med.* **305**, 1431–1438.

McArthur J. C., Cohen B. A., Farzedegan, H. *et al.* (1988). Cerebrospinal fluid abnormalities in homosexual men with and without neuropsychiatric findings. *Ann. Neurol.* **23**(suppl), S34–S37.

McArthur, J. C., Hoover, D. R., Bacellar, M. A. *et al.* (1993). Dementia in AIDS patients: incidence and risk factors. *Neurology.* **43**, 2245–2252.

McArthur, J. C., McLernon, D. R., Cronin, M. F. *et al.* (1997). Relationship between human immunodeficiency virus associated dementia and viral load in cerebrospinal fluid and brain. *Ann. Neurol.* **42**, 689–698.

McArthur J. C., Yiatannoustsos, C., Simpson, D. M. *et al.* (2000). A phase 2 trial of nerve growth factor for sensory neuropathy associated with HIV infection. *Neurology.* **54**, 1080–1088.

Mhiri, C., Baudrimont, M., Bonne, G. *et al.* (1991). Zidovudine myopathy: a distinctive disorder associated with mitochondrial dysfunction. *Ann. Neurol.* **29**, 606–614.

Miller, R. F., Fox, J. D., Waite, J. C. *et al.* (1995). Herpes simplex virus type 2 encephalitis and concomitant CMV infection in a patient with AIDS. *Geritourin Med.* **71**, 262–264.

Miller, R. F., Hall-Craggs, M. A., Costa, D. C. *et al.* (1998). Magnetic resonance imaging, thallium-201 SPECT and laboratory analyses for discrimination of cerebral lymphoma and toxoplasmosis in AIDS. *Sex. Transm. Infect.* **74**(4), 258–264.

Miller, R. G., Storey, J. R., and Greco, C. M. (1990). Ganciclovir in the treatment of progressive AIDS—related polyradiculopathy. *Neurology.* **40**, 569–574.

Moore, R. D., Wong, W. M., Keruly, J. C. *et al.* (2000). Incidence of neuropathy in HIV-1 infected patients on monotherapy versus those on combination therapy with didanosine, stavudine and hydroxyurea. *AIDS.* **14**, 273–278.

Morgello, S., Wolfe, D., Godfrey, E., *et al.* (1995). Mitochondrial abnormalities in HIV associated myopathy. *Acta Neuropathol.* **90**, 366–374.

Moulignier, A., Authier, F.-J., Baudrimort, M. *et al.* (1997). Peripheral neuropathy in human immunodeficiency virus infected patients with the diffuse infiltrative lymphocytosis syndrome. *Ann. Neurol.* **41**, 438–445.

Moulignier, A., Moulonguet, A., Pialoux, G. *et al.* (2001). Reversible ALS—like disorder in HIV infection. *Neurology.* **57**, 995–1001.

Musher, D. M., Hamill, R. J., Baughn, R. E. *et al.* (1990). Effect of the human immunodeficiency virus (HIV) on the course of syphilis and on the response to treatment. *Ann. Intern. Med.* **113**, 872–881.

Narcisco, M., Koizumi, K., Koike, T. *et al.* (2001). Reversal of HIV associated motor neuron syndrome after highly active antiretroviral therapy. *J. Neurol.* **248**, 233–234.

Narciso, P., Galgani, M. D., Del Grosso, B. *et al.* (2001). Acute disseminated encephalitis as a manifestation of primary HIV infection. *Neurology.* **57**, 1493–1496.

Nishio, M., Koizumi, K., Moriwaka, F. *et al.* (2001). Reversal of HIV associated motor neuron syndrome after HAART (letter). *J. Neurol.* **248**, 233–234.

Palella, F. J., Delaney, K. M., Morrman, A. C. *et al.* (1998). Declining morbidity and mortality among patients with advanced human immunodeficiency virus infection. *N. Engl. J. Med.* **338**, 853–860.

Perriens, J., and Piot, P. (1993). Worldwide epidemiology of HIV infection. *In* "Focus on HIV." (H. C. Neu, J. A. Levy, and R. A. Weiss, Eds.), pp. 3–19. Churchill-Livingstone, Edinburgh.

Petito, C. K., Cho, E. S., Lemann, W. *et al.* (1986). Neuropathology of acquired immunodeficiency syndrome (AIDS): an autopsy review. *J. Neuropathol. Exp. Neurol.* **45**, 635–646.

Petito, C. K., Navia, B. A., Cho, E. S. *et al.* (1985). Vacuolar myelopathy pathologically resembling subacute combined degeneration in patients with the acquired immunodeficiency syndrome. *N. Engl. J. Med.* **312**, 874–879.

Pfister, H. W., Einhaupl, K. M., Baumlattner, U. *et al.* (1989). Dissociated nystagmus as a common symptom of oculomotor disorders in HIV infected patients. *Eur. Neurol.* **29**, 277–280.

Pinto, A. (1996). AIDS and cerebrovascular disease. *Stroke.* **27**, 538–543.

Pizzo, P. A., Eddy, J., Falloon, J. *et al.* (1988). Effect of continuous intravenous infusion of zidovudine (AZT) in children with symptomatic HIV infection. *N. Engl. J. Med.* **319**, 889–896.

Portegies, P., de Gans, J., Lange, J. M. A., Derix, M. M. A., Speelman, H., Bakker, M., Danner, S. A., and Goudsmit, J. (1989). Declining incidence of AIDS dementia complex after introduction of zidovudine treatment. *BMJ.* **299**, 819–820.

Portegies, P., Enting, R. H., deGaqns, J., Algra, P. R., Derix, M. M. A., Lange, J. M. A., and Goudsmit, J. (1993). Presentation and course of AIDS-dementia complex: 10 years of followup in Amsterdam, The Netherlands. *AIDS.* **7**, 669–675.

Porter, S. B., and Sande, M. A. (1992). Toxoplasmosis of the central nervous system in the acquired immunodeficiency sndrome. *N. Engl. J. Med.* **327**, 1643–1648.

Post, M. J., Hensley, G. T., Moskowowitz, L. B. *et al.* (1986). Cytomegalic inclusion virus encephalitis in patients with AIDS; CT, clinical and pathologic correlation. *AJN* **7**, 275–280.

Powderly, W. G. (1993). Cryptococcal meningitis and AIDS. *Clin. Infect. Dis.* **17**, 837–842.

Powderly, W. G., Saag, M. S., Cloud, G. A. *et al.* (1992). A controlled trial of fluconazole or amphotericin B to prevent relapse of cryptococcal meningitis in patients with the acquired immunodeficiency syndrome. *N. Engl. J. Med.* **326**, 793–798.

Price, R. W. (1998). Neuropathy complicating diffuse infiltrative lymphocytosis. *Lancet* **352**, 592–594.

Raffi, F., Aboulker, J.-P., Michelet, C. *et al.* (1997). A prospective study of criteria for the diagnosis of toxoplasmic encephalitis in 186 AIDS patients. *AIDS.* **11**, 177–184.

Renold, C., Sugar, A., Chave, J.-P. *et al.* (1992) Toxoplasmic encephalitis in patients with the acquired immunodeficiency syndrome. *Medicine.* **71**, 224–239.

Saag, M., Powderly, W., Cloud, G. *et al.* (1992). Comparison of amphotericin B with fluconazole in the treatment of acute AIDS associated cryptococcal meningitis. *N. Engl. J. Med.* **326**, 83–89.

Saag, M. S., Powderly, W. G., Cloud, G. A., Robinson, P., Grieco, M. H., Sacktor, N. C., Bacellar, H., Hoover, D. R. *et al.* (1996). Psychomotor slowing in HIV infection: a predictor of dementia, AIDS and death. *J. Neurovirol.* **2**, 404–410.

Sacktor, N., Lyles, R. H., Skolasky, M. A. *et al.* (2001). HIV associated neurologic disease incidence changes: Multicenter AIDS Cohort Study, 1990–1998. *Neurology.* **56**, 257–260.

Sacktor, N., Tarwater, P. M., Skolasky, M. A. *et al.* (2001). CSF antiretroviral drug penetrance and the treatment of HIV—associated psychomotor slowing. *Neurology.* **57**, 542–544.

Said, G., Lacroix, C., Chemoulli, P. *et al.* (1991). Cytomegalovirus neuropathy in acquired immunodeficiency syndrome. A clinical and pathological study. *Ann. Neurol.* **29**, 136–146.

Schifitto, G., Yiannoutsos, C., Simpson, D. M. *et al.* (2001). Long-term treatment with recombinant nerve growth factor for HIV associated sensory neuropathy. *Neurology.* **57**, 1313–1316.

Setinek, U., Wondousch, E., Jellinger, S. *et al.* (1995). CMV infection of the brain in AIDS. *Acta Neuropathol.* (Berlin) **90**, 511–515.

Sharkey, P. K., Thompson, S. E., Sugar, A. M., Tuazon, C. U., Fisher, J. F., Hyslop, N., Jacobsen, J. M., Hafner, R., Dismukes, W. E., and the NIAID Mycoses Study Group and the AIDS Clinical Trial Group. (1992). Comparison of amphotericin B with fluconazole in the treatment of acute AIDS-associated cryptococcal meningitis. *N. Engl. J. Med.* **326**, 83–89.

Shlay, J. C., Chaloner, K., Max, M. *et al.* (1998). Acupuncture and amitriptyline for pain due to HIV related neuropathy. *JAMA.* **280**, 1590–1595.

Sidtis, J. J., Gatsonis, C., Price, R. W., Singer, E. J., Collier, A. C., Richman, D. D., Hirsch, M. S., Schaerf, F. W., Fischl, M. A., Kieburtz, K., Simpson, D., Koch, M. A., Feinberg, J., Dafni, U., and the AIDS Clinical Trials Group. (1993). Zidovudine treatment of the AIDS dementia complex: Results of a placebo-controlled trial. *Ann. Neurol.* **33**, 343–349.

Silber, E., Sonnenberg, P., Koornhof, H. J. *et al.* (1998). Dual infective pathology in patients with cryptococcal meningitis. *Neurology.* **51**, 1213–1215.

Silva, N., O'Bryan, L., Medeiros, E. *et al.* (1999). Trypanosoma cruzi meningoencephalitis in HIV-1 infected patients. *J. Acqr. Immune. Def. Syndr.* **20**, 342–349.

Simpson, D. M., and Bender, A. N. (1988). Human immunodeficiency virus associated myopathy: analysis of 11 patients. *Ann. Neurol.* **24**, 79–84.

Simpson, D. M., Citak, K. A., Godfrey, E. *et al.* (1993). Myopathies associated with human immunodeficiency virus and zidovudine. Can their effects be distinguished? *Neurology.* **43**, 971–976.

Simpson, D. M., Dorfman, D., Olney, R. K. *et al.* (1996). Peptide T in the treatment of painful distal neuropathy associated with AIDS. *Neurology.* **47**, 1254–1259.

Simpson D. M., Katzenstein, D. A., Hughes, M. D. *et al.* (1998). Neuromuscular function in HIV infection: analysis of a placebo controlled combination antiretroviral trial. *AIDS.* **12**, 2425–2432.

Simpson, D., Olney, R., McArthur, J. C. *et al.* (2000). A placebo controlled trial of lamotrigine for painful HIV associated neuropathy. *Neurology.* **54**, 2115–2119.

Simpson, D. M., and Tagliati, M. (1995). Nucleoside analogue-associated peripheral neuropathy in human immunodeficiency virus infection. *J. Acquir. Immune. Defic. Syndr. Hum. Retrovirol.* **9**, 153–161.

Sindrup, J. H., Weismann, K., and Wantzin, G. L. (1986). Syphilis in HTLV-III infected male homosexuals. *AIDS Res.* **2**, 285–288.

So, Y. T., Holtzman, D. T., Abrams, D. J. *et al.* (1988). Peripheral neuropathy associated with acquired immune deficiency syndrome: Prevalence and clinical features from a population based survey. *Arch. Neurol.* **45**, 945–948.

So, Y. T., and Olney R. K. (1994). Acute lumbosacral polyradiculopathy in acquired immunodeficiency syndrome: experience in 23 patients. *Ann. Neurol.* **35**, 53–58.

Stahl, C., Wideman, C., Spira, T. *et al.* (1993). Protein s deficiency in men with long-term human immunodeficiency virus infection. *Blood.* **81**, 1801–1807.

Staudinger, R., and Henry, K. (2000). Remission of HIV myelopathy after highly active antiretroviral therapy. *Neurology.* **54**, 267–268.

Stimmler, M. M., Quismorio, F. P., McGhee, W. G. *et al.* (1989). Anticardiolipin antibodies in acquired immunodeficiency syndrome. *Arch. Intern. Med.* **149**, 1833–1835.

Sweeny, B. J., Manji, H., Mller, R. F. *et al.* (1994). Cortical and subcortical JC virus infection: two unusual cases of AIDS associated progressive multifocal leucoencephalopathy. *J. Neurol. Neurosurg. Psychiat.* **57**, 994–997.

Takahasi, K., Wesslingh, S. I., Griffin, D. E. *et al.* (1996). Localisation of HIV-1 in human brain using polymerase chain reaction/in situ hybridization and immunohistochemistry. *Ann. Neurol.* **39**, 705–711.

Tan, S. V., Guiloff, R. J., Scaravilli, F. *et al.* (1993). Herpes simplex type 1 encephalitis in acquired immunodeficiency syndrome. *Ann. Neurol.* **34**, 619–622.

Toggas, S., Masliah, E., Rockenstein, E. *et al.* (1993). Neurotoxicity of viral proteins – HIV-1 gp 120 effects on transgenic mice. *Clin. Neuropathol.* **12** (Suppl. 1), S4–5.

Toulon P., Lamine, M., Ledjev, I. *et al.* (1993). Heparin cofactor 11 deficiency in patients infected with the human immunodeficiency virus. *Thromb. Haemostat.* **70**, 730–735.

Tross, S., Price, R. W., Navia, B. A. *et al.* (1988). Neuropsychological characterisation of the AIDS dementia complex. *AIDS.* **2**, 81–88.

Vago, L., Nebuloni, M., Sala, E. *et al.* (1996). Coinfection of the central nervous system by CMV and herpes simplex type 1 and 2 in AIDS patients. *Acta Neuropathol.* **92**, 404–408.

Van der Hoorst, C. M., Gaag, M. S., Cloud, G. A. *et al.* (1997). Treatment of cryptococcal meningitis associated with the acquired immunodeficiency syndrome. *N. Engl. J. Med.* **337**, 15–21.

Veenstra, J., Krol, A., van Praag, R. M. *et al.* (1995). Herpes zoster, immunological deterioration and disease progression in HIV-1 infection. *AIDS.* **9**, 1153–1158.

Vinters, H. V., Kwok, M. K., Ho, H. W. *et al.* (1989). Cytomegalovirus in the nervous system of patients with the acquired immune deficiency syndrome. *Brain.* **112**, 245–268.

Von Einsiedel, R. W., Fife, T. D., Aksamit, A. J. *et al.* (1993). Progressive multifocal leucoencephalopathy in AIDS: a clinico-pathologic study and review of the literature. *J. Neurol.* **240**, 391–406.

Weber, T. (1999). Cerebrospinal fluid analysis for the diagnosis of Human Immunodeficiency Virus related neurologic disease. *Semin. Neurol.* **19**(2), 223–233.

Wesslingh, S. L., Power C., Glass, J. D. *et al.* (1993). Intracerebral cytokine messenger RNA expression in acquired immunodeficiency syndrome dementia. *Ann. Neurol.* **33**(6), 576–582.

Working Group of the American Academy of Neurology AIDS Task Force. (1991). Nomenclature and research case definitions for neurologic manifestations of human immunodeficiency virus type-1 infection. *Neurology.* **41**, 778–785.

Yechoor, V. K., Shandera, W. X., Rodriguez, P. *et al.* (1996). Tuberculous meningitis among adults with and without HIV infection. *Arch. Intern. Med.* **156**, 1710–1716.

Zeman, A., and Donaghy, M. (1991). Acute infection with human immunodeficiency virus presenting with neurogenic urinary retention. *Genitourin. Med.* **67**, 345–347.

Zuger, A., Louie, E., Holzman, R. S., Simberkoff, M. S., and Rahal, J. J. (1986). Maintenance amphotericin B for cryptococcal meningitis in the acquired immunodeficiency syndrome (AIDS). *Ann. Intern. Med.* **85**, 481–489.

## CHAPTER 52
# Tropical Neurology

J. S. Chopra and I. M. S. Sawhney

Parasitic infestations of the nervous system in true sense are diseases that were mostly confined to "tropical regions" of the world, thereby the nomenclature of tropical diseases or tropical disorders of the nervous system. These geographical areas of the world do have distinct historical backgrounds in location, culture, economy, and weather, although the world is rapidly becoming as one people in view of large migration of people from one continent to the other. Moreover, the ever-increasing rapid transportation of people in the world has broken the age-old boundaries between the nations, and we are witnessing these days the tropical disease in the nontropical areas and vice versa. The parasitic and tropical diseases of the nervous system are being dealt with in two chapters of this book, Chapters 48 and this chapter.

The following diseases are discussed in this chapter: neurosysticercosis, schistosomiasis, malaria, ameobiasis, trypanosomiasis, filariasis, and leishmaniasis.

## NEUROCYSTICERCOSIS (NCC)

### Clinical Aspects

#### Definition

NCC is a disease in which larvae of intestinal infestation, *Taenia solium*, lodge in the brain. There is marked tendency for the muscle and central nervous system to be affected by this parasite. There is variability in the clinical and pathological features, which depends on the inflammatory response surrounding the cysticerci, their location, size, and the quantum of infestation. The *Taenia solium* tapeworm is found only in the human intestine, and humans are the definite host. The head of the worm called the scolex is attached to the mucosa of the intestine. The proglottids are the distal segments of this tapeworm and are filled with eggs, which, when mature, are passed in human feces. Pigs, the intermediate hosts ingest these eggs, which mature in to larvae (*cysticeri*) in its intestinal tissue. These are oval shaped, vesicular, filled with fluid, and have an invaginated scolex. Ingestion of poorly cooked pork by man makes the human a definitive host. However, it is well known now that even a large number of vegetarians become intermediate hosts by ingesting ova in the contaminated food or by autoinfection through the fecal-oral route or by reverse peristalsis. Poor sanitation and unhygienic conditions prevalent in the underdeveloped and developing countries are the reasons for the widespread prevalence of NCC in such countries. The external coverings of the ovum are dissolved by the gastric juice, and a viable embryo is released, which passes through the capillaries and spreads throughout the body.

### Clinical Aspect

Variable clinical features have been recorded in NCC, which may at times be nonspecific. The symptoms depend on the area of the brain involved, the number of lesions, and the immune response of the patient to the parasites (Chopra and Sawhney, 2001; Del Brutto *et al.*, 1998). The most common presentation of NCC is one or another form of epilepsy. Partial motor seizures with or without generalization are the most common and seen in 50%–80% of patients with parenchymal lesions but observed less often in other forms of presentation. Cysticercosis on the other hand has been seen in 2.2%–9.6% of nonselected cases (Mahajan and Chopra, 1975). One thousand thirty-eight cases of epilepsy were screened by Chopra *et al.* (1981) for cysticercosis. Cysticercosis serological tests were positive in 26% of cases, and the positivity rate was 29% in patients with focal seizures as their initial presentation. The positivity was 2% in the general population. It is known that viable and living cysticerci in the brain may not cause any symptoms. However, degenerating cysts with inflammation and consequent surrounding edema, encephalitis, and vascular involvement cause severe symptoms.

Chronic meningitis-like presentation is another manifestation of NCC. A large number of cysticerci infestations of the meninges and parenchyma produce such presentation, which is known as the meningoencephalitic response. The patients have headache, vomiting, seizures, altered sensorium, and visual disturbances. Focal neurological deficits in the form of monoparesis or hemiparesis may occur. Cranial nerve palsies have been observed. The visual loss could be due to raised intracranial pressure or even a cyst located over the fundus, the only location where a living, moving parasite can be seen in the cyst on careful funduscopy. Single ring-enhancing CT lesions in all age groups are one of the common causes of epilepsy in the Indian subcontinent (Chandy *et al.*, 1991; Chopra *et al.*, 1992). These patients may have focal or generalized seizures, and clinical examination may not show any neurological deficit, although some may have postictal paresis of the affected limbs for a short period. Gait ataxia, abnormal movements, and brainstem dysfunctions have also been observed, depending on the location and number of cysts (Sotelo, 1988).

Massive cerebral edema has also been observed in NCC termed "pseudotumor cerebri." This is usually due to a large number of cysts scattered all over the brain (Figure 1) or a few cysts in the ventricular system. The clinical presentation in a large number of such patients is headache and vomiting or visual disturbances—a classical triad of symptoms. It is slowly progressive, and many patients were treated for migraine or intracranial space-occupying lesions before the neuroimaging era. These patients, in addition, may have a motor or sensory focal deficit or seizures. Cysticercus arachnoiditis or granular ependymitis may present with hydrocephalus. (Milenkovic *et al.*, 1982). Even a single cyst in the aquaduct may sometimes cause (Rangel-Guerra and Martinez, 1986) symptoms akin to a colloidal cyst when the cysticercus cyst is located in the third ventricle. The patient may sometimes lose consciousness, recover, and vice versa. Intense immune response from the host may further aggravate the brain edema. This type of presentation is more common in children and women, and the presenting symptoms may be clouding of consciousness, diminution of vision,

headache, vomiting, and bilateral papilledema. Some patients go to ophthalmologists for visual loss and are observed to have bilateral papilledema or even optic atrophy without any other symptoms.

Psychiatric manifestations as the presenting symptoms have also been observed in NCC. Such patients have visual or auditory hallucinations, lack of concentration and, with a mistaken diagnosis before the imaging era, have even been given electroconvulsive therapy before being referred to neurologists. However, these symptoms are due to cerebral edema or hydrocephalous from direct infestation with parasites (Chopra and Sawhney, 2001).

The authors have observed some of the rare manifestations of NCC such as a cyst located in the eye muscle or a cyst over the fundal head. One 11-year-old girl had with flashes of light, and a cyst was located over the optic radiation; another young boy was seen with symptoms of demmyelinating disease, with a cyst in the brainstem. Dementia is another rare type of presentation.

Spinal cysticercosis is also a rare presentation. Hematogenous spread usually causes intramedullary cysts, but extramedullary cysts are as a result of migration of larvae from brain by way of the subarachroid space. Isolated medullary forms are rare (Chopra and Sawhney, 2001). The cysts in the spinal cord may mimic root pains, gradually progressive weakness, and motor and sensory manifestations, which are usually seen in intramedullary or extramedullary tumors. Very rarely presentation may be like amyotrophic lateral sclerosis or a prolapsed intervertebral disc. Spinal cord manifestations are seen in 1%–5% of patients of NCC.

Involvement of skeletal muscle is a fairly common presentation in cysticercosis; however, it is invariably asymptomatic. Rarely, psuedohypertrophic changes in limb muscles and less commonly trunk muscles are observed (Figure 2). Such patients complain of motor weakness; muscles are swollen and may be tender in one third of patients. The exact pathogenesis of this condition is ill understood; however, it is believed that it is an allergic reaction by cysticercus infiltration into muscles that are edematous. Myopathic features in such muscles have been observed histopathologically, as well as on electromyography. Sawhney *et al.* (1976) and Wadia *et al.* (1988) have remarkably described the muscle infiltration in cysticercosis by imaging techniques.

### Differential Diagnosis

NCC should be differentiated from the following diseases.

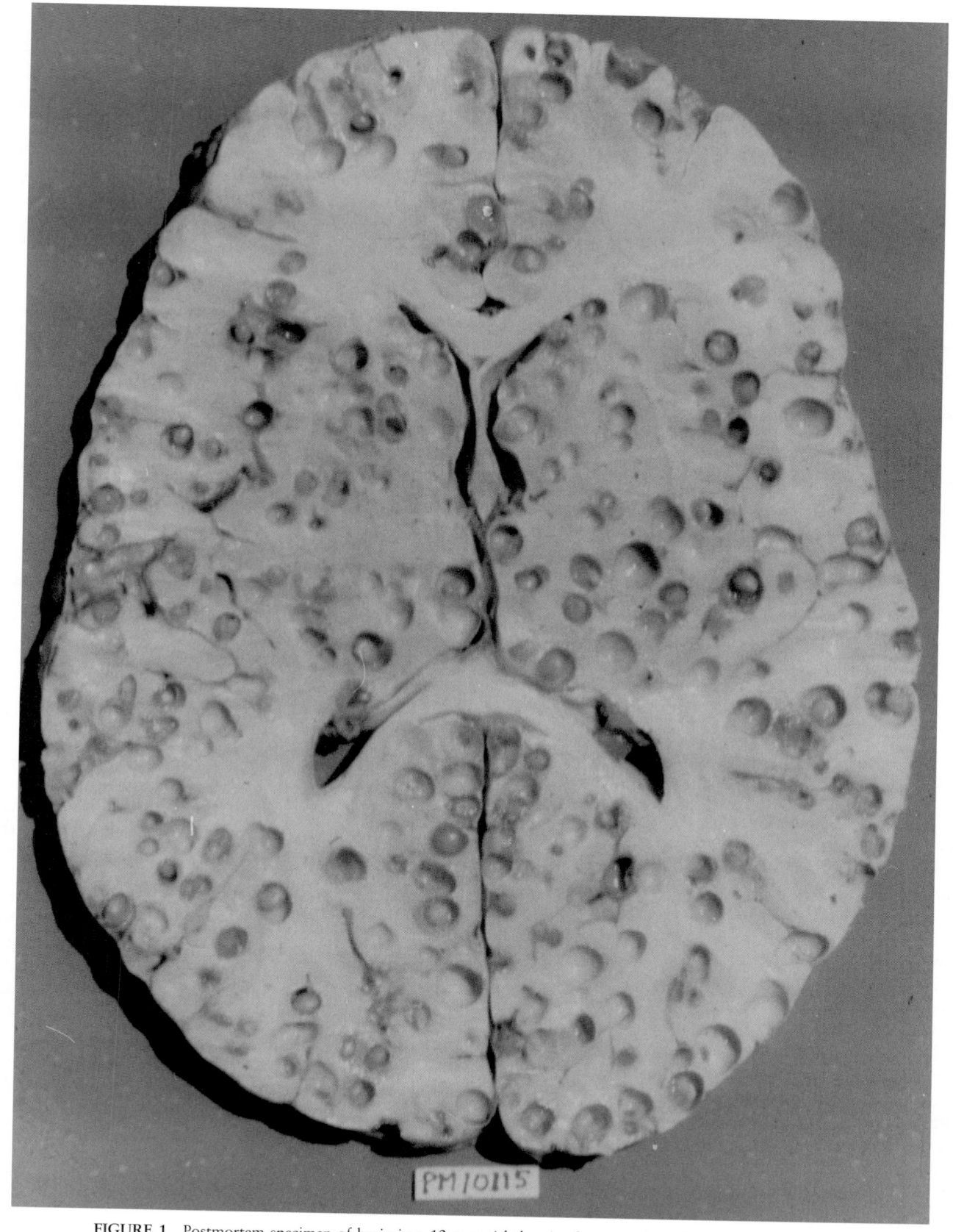

**FIGURE 1** Postmortem specimen of brain in a 12-year-girl showing large number of cysts in a case of cerebral cysticercosis. (Courtesy Dr. A. K. Banerjee, Professor of Histopathology, Postgraduate Institute of Medical Education and Research, Chandigarh, B. I. Churchill Livingstone, New Delhi, India).

| Tuberculosis | The prevalence of tuberculosis may be even more in areas where cysticercosis is endemic. A patient with intracranial tuberculoma can also have focal or generalized seizures and headache, with or without fever. Imaging may not always be helpful, because in both ring lesions with surrounding edema are seen. However, the "target" lesion is more common in tuberculoma compared with NCC, in which degenerating cyst in various evolutionary stagesis usually visible on MRI. <br><br> Tuberculosis in other parts of the body or immunological tests are helpful in differentiating this. The menigoencephalitic form of NCC may be confused with tubercular meningitis, but CSF examination could be diagnostic. |
| --- | --- |
| Brain abscesses, fungal abscesses, toxoplasma abscesses | The patient may be afebrile with such intracranial lesions; CT and MRI may show similar types of lesions, and the clinical presentation may be with headache and seizures. CSF analysis and immunological studies are of great help in differentiating these. |
| Primary or metastatic brain lesions | These could present with headache and seizures and need to be differentiated from NCC. Imaging techniques including angiography are helpful in differentiating these. |
| Other cystic lesions— colloidal cyst, arachnoidal cyst, and cholesterol cyst | Any of these cysts, including a cysticercus cyst, in the aqueduct of Sylvius or third ventricle could produce similar kinds of symptoms of headache, vomiting, hydrocephalus, and seizures. Appropriate tests, including imaging techniques, will be of great help in differentiation. Rarely, surgical removal would be necessary for diagnosis. |
| Pseudotumor cerebri | This should also be considered in the differential diagnosis, especially when the headache, vomiting, and seizures persist in a patient with bilateral papilledema, and the initial CT is not contributing. Other causes of pseudotumor cerebri need to be ruled out. A follow-up CT or MRI will ultimately be diagnostic. |

## Natural Coarse

Cysticercosis is the most common parasitic disease of the nervous system, and the involvement of the CNS could be as high as 50%–70%. It is known that more than 50 million people are infected by NCC throughout the world, and almost 50,000 die from it worldwide. It is probably one of the most frequent neurological disorders, particularly in the rural areas of Asia, Africa, and Latin America (Figure 3). The prevalence is maximum in South America, Southeast Asia, Mexico, Brazil, Chile, and some of the East European countries such as Poland, Germany, and Romania. NCC is often seen in the neurological wards in Spain and Portugal (Monteiro *et al.*, 1992). It has been reported that prevalence in rural areas of Mexico and Ecuador is between 1%–25% of the population. A review of 20,000 autopsies at "The General Hospital of Mexico" revealed that the prevalence of NCC in general population is 2.38% (Villagran and Rabiela, 1988). It has been noticed in India that disease is as common in vegetarians as in non-vegetarians (Ahuja *et al.*, 1983; Chopra *et al.*, 1981).

NCC is a common disease now in Southern United States because of an increase in immigration to the United States from endemic areas. Most cases are reported from California, inhabited by more than 5 million immigrants from Asia and Latin America. Household contact with *Taenia solium* carriers have shown a number of cases of NCC in United States, especially in people who have never traveled to endemic areas (Kruskal *et al.*, 1993; Schantz *et al.*, 1992).

Cysticercosis in African and Asian countries is also related to socioeconomic and cultural factors. Poor public health measurers, poverty, and unhygienic environments are the greatest adverse factors for spread of cysticercosis in these countries. Cysticercosis is almost unknown in some of the Muslim countries because of the religious ban on consumption of pork among Muslims. Pork from open fields is invariably and excessively consumed by people of low social strata in India and many other Saharan African countries (Chopra *et al.*, 1981; Mahajan *et al.*, 1982; Michel *et al.*, 1993), because it is the cheapest meat available.

There is no gender or age bar for NCC. The youngest child seen by the authors with from NCC was 3 years old, and it has also been seen in elderly, although it generally affects adults in the fourth and fifth decades (Chopra and Sawhney, 2001). A case of congenital NCC from transplacental transmission of cysticerci is also reported. The usual incubation period is 4 years, but it

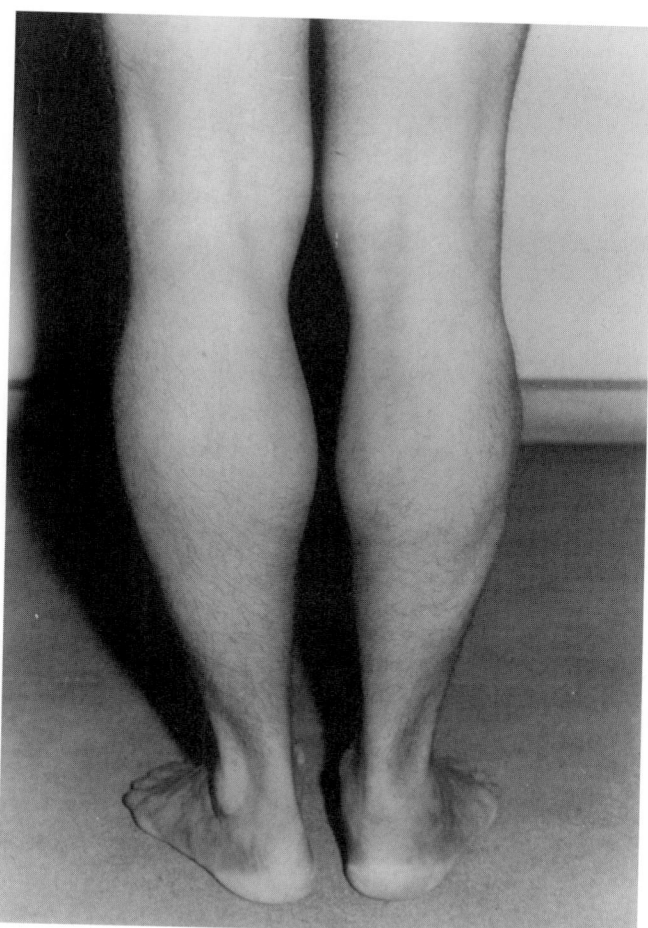

**FIGURE 2** Psuedohypertrophy of leg muscles in a male who presented with muscle pains and motor weakness along with seizures. Muscle biopsy confirmed cysticercus cysts. (Increased girth of left calf muscles compared with right.)

may range from less than 1 year to two or three decades. Humans are the definite host, however, both humans and pigs may behave as intermediate hosts. Each segment or proglottid has reproductive organs of both sexes and contains thousands of eggs. The eggs can survive in soil, water, and vegetation for a long time. The eggs swallowed by the pig enter the systemic circulation after the liberation of oncospheres (embryos) that cross the bowel wall and turn into cysticerci in various tissues and ultimately die. However, if poorly cooked infected pork is consumed by humans the cysticerci break down in the stomach, and the larvae evaginate in the small intestine, mature, and grow into the *Taenia solium*, and proglottids are formed. The eggs ingested by humans can also cross the bowel wall by a similar process as in pigs and lodge in various tissues, including the brain. Humans acquire cysticercosis from food contaminated with *Taenia solium* eggs and also by the anal-oral route in the individuals who are harboring adult parasites in the bowel. In India and many other countries, vegetarians are infected by this process. In fact, we have observed in India that almost 50% of patients of NCC are vegetarian and of the remaining 50% not more than 10% have consumed pork. Moreover, it should also be noted that the chances of pork remaining undercooked in India are remote, because the method of cooking in this country overcooks pork or any other meat preparation. Therefore unhygienic handling of food by carriers of the parasite is the major route of infection in the underdeveloped or developing countries.

The cysticerci lodged in tissues perish with time or may get calcified and remain in the tissues forever. This is the natural course of the disease in most; however, in certain human hosts, depending on the immunological reaction, various symptoms and sings are observed as already described. It is known that cysticerci have antigenic properties because of the many proteins, and this stimulates the production of specific anticysticercal antibodies (Grogl *et al.*, 1985). Several cysts in the brain can stimulate massive brain edema, which may even prove fatal. The host's reaction to cysticerci is variable in various individuals, and the antibodies so created may not give full protection against the disease (Fisser *et al.*, 1980).

The natural course of the cysticerci is their evolution into four stages. The first is the vesicular stage, which elicits very little inflammation, and the cysticerci can remain lodged in this form from years to decades, because the host develops immune tolerance to the parasites (Fisser, 1989). The cysticerci may perish in a few cases because of a severe immunological response from the host. In this stage the cysticerci are translucent and contain transparent fluid. Next is the colloidal stage; the wall become thick vesicular, the fluid turns turbid, and the scolex starts degenerating. There is a thick collagen capsule around the parasite, which also provokes a severe inflammatory reaction. Imaging shows a ringlike or nodular enhancement with contrast CT. The lesions are surrounded by edema. The third is the granular stage in which there is astrocystic gliosis around cysts; a thick vesicular wall, and hyperdense lesions, which enhance with contrast medium, are shown on imaging. The last is the calcified stage in which the parasite transforms into a small mineralized nodule; there is intense gliosis, and imaging shows small calcified specks.

The inflammatory reaction surrounding cysticerci in brain mostly consists of lymphocytes, plasma cells, and eosinophils, which initiate edema and gliosis. This also usually induces multiple changes in the brain parenchyma, the meninges, the cerebral ventricles, and the spinal cord. Meningeal cysticerci provoke edema, inflammatory reactions and meningeal signs, even initiating an exudative response (Sotelo and Del Brutto, 1999).

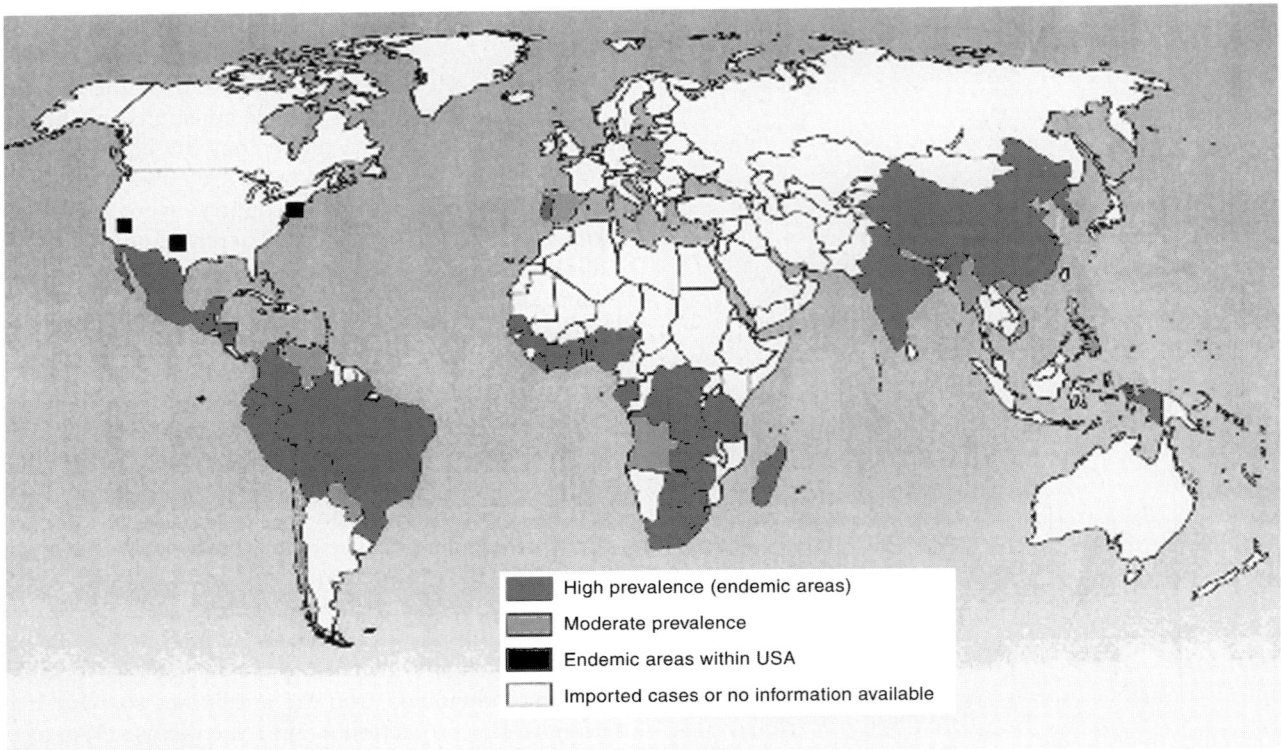

*World map showing endemic areas of neurocysticercosis.*

FIGURE 3  World map showing endemic areas of cysticercosis (with permission from the World Health Organization.)

## Principles of Therapy

In view of the pathogenesis, pathology, inflammatory reaction, and edema, the principles of medical therapy need to be aimed at these factors. In addition, it is also mandatory that treatment should be directed toward the *Taenia solium* infestation in the intestines. As a matter of fact, a patient with NCC requires symptomatic drugs and specific anticysticercal therapy. Only two specific drugs have stood the test of time, praziquantel and albendazole. It has been shown that praziquantel is effective against cysticerci, and 60%–70% of brain cysticerci may disappear within 2 weeks of treatment. It is an isoquinoline and has been in use since 1979 (Robles *et al.*, 1987). Recently, single-day therapy with praziquantel has been advocated, but larger studies would be required to prove that it is as effective as 2-week therapy. However, praziquantel is more expensive than albendazole, and it is a common knowledge that cysticercosis is more common in the low socioeconomic strata of population, who can ill afford treatment with praziquantel, although it is an effective medicine.

Albendazole is cheaper and as effective as praziquantel, if not more, in its anticysticercal properties. It is an imidazole, and in most cases requires 4 weeks of oral therapy, although a study has shown that its efficacy is also good with 8-day therapy (Sotelo and Del Brutto, 1999). It is known that albendazole destroys 75%–90% of brain cysts and has been considered superior to praziquantel in a number of trials (Sotelo *et al.*, 1990; Takayanagui and Jardim, 1992). It has more destructive power to eliminate meningial and ventricular cysts than only parenchymal cysts (Del Brutto and Sotelo, 1990). Cesticidal therapy also accelerates radiological disappearance of viable NCC, although the clinical outcome of the patient is better. Having treated innumerable patients of NCC, we have come to a definite conclusion that the control of seizures in patients with NCC is much better if the anticonvulsants are combined with cesticidal therapy, especially in patients who have multiple cysts in the brain. In one study it was observed that seizure control was 83% when these patients also received cesticidal therapy in addition to anticonvulsants, but seizure control was only 27% in those who did not receive cesticidal therapy with anticonvulsants (Del Brutto *et al.*, 1988). A single cyst in the brain should be treated with cesticidal drugs, which is controversial, because most forms of disease are self-limiting. It is also controversial that patients who present with headache or hydrocephalus and harbor parenchymal brain cysticerci should be treated with anticonvulsants, even when the patient has no seizures.

We believe from our observations that a short course of cesticidal therapy for a week will be beneficial for a single intracranial cyst, because it is expected that such a patient may harbor cysticerci in other tissues, and some of the very small cysticerci may even escape detection with imaging. Moreover, cesticidal therapy will also eliminate the parasitic infestation in the intestine, which is likely to exist with NCC. We do not believe from our experience that it is necessary to use anticonvulsants in patients who present with NCC without seizures.

Symptomatic medical therapy, apart from anticonvulsants, in patients with seizures is as essential as the cesticidal therapy. Corticosteroid use is often required in patients of brain edema from NCC, especially if the cysts are multiple and lodged in the ventricles, brain parenchyma, subarachnoid space, and spinal cord. In the absence of steroids a ruptured cyst may produce a severe meningeal response. Vascular involvement with cysticerci provoke angiitis, and a patient may have cerebral infarcts (Del Brutto, 1992). Intravenous corticosteroids may be essential in such circumstances to prevent angiitis. Prolonged use of steroids, even for months, would be required for basal cysticerci to provoke basal exudates and involvement of cranial nerves with hydrocephalus. Cysticidal therapy in some patients may also provoke acute destruction of parasites, causing intense brain inflammation and edema, which could be life-threatening and requires corticosteroid therapy for prevention. However, it must be understood that steroid therapy, along with praziquantel, may reduce the plasma level of this drug up to 50% (Vazquez et al., 1987). The reverse is true with combined therapy of albendazole and steroids, in which the plasma levels of albendazole have been reported to be enhanced (Jung et al., 1990).

Patients who complain of headache, nausea, and vomiting may get relief with analgesics and antiemetic drugs. In rare instances with massive brain edema, IV manitol, in addition to steroids, may be required. The treatment of hydrocephalus should follow the standard methods of managing obstructed hydrocephalus. The hydrocephalus is secondary to arachnoiditis from cysticerci. There seems to be no danger of dissemination of larva through the shunt. The removal of cisternal cysts or strictly intraventricular cysts remains surgical. These are removed only when they obstruct or reach the fourth ventricle. In rare circumstances bilateral surgical decompression may be required if the cerebral edema is life-threatening and there is poor response to medical therapy of decompression (Wani et al., 1981). Surgical procedures were common for removal of a tumorous mass before the imaging era, which on histopathological examination were observed as a collection of cysticerci. A constant-flow shunt without a valve mechanism has been introduced recently to avoid retrograde flow of spinal CSF into the ventricular system. Retrograde CSF flow from the spinal subarachnoid space carrying inflammatory cells, cytokines, and immunoglobins to the shunt site is an important reason for shunt malfunction. The preliminary studies are encouraging (Sotello and Del Brutto, 1999).

## Diagnosis

NCC should be suspected in young adults with persistent headaches with or without seizures and hailing from endemic areas. The diagnosis usually depends on the imaging techniques of CT or MR and immunological investigations. Subcutaneos or muscular palpable cysts if available can be of great help, although it is known that cysts are less common in America compared with African and Asian cysticercosis. However, a definite diagnosis is generally made by identification of the parasite either in nervous tissue or biopsy of the cyst elsewhere in the body. A parasite visible in the posterior chamber of eye or over the fundal head with ophthalmoscopy can also be diagnostic.

Immunodiagnostic techniques are of great value in diagnosis. A variety of methods are useful such as indirect hemagglutination, complement fixation, immunoelectrophoresis, indirect immunoabsorbance, enzyme-linked immunosorbent assay (ELISA), and immunoblot assays (Garcia and Sotelo, 1991; Mahajan and Chopra, 1975; Rosas et al., 1986). Such studies can be complementary to the neuroimaging techniques in diagnosing NCC but cannot be used alone to confirm the diagnosis. An immunoblot diagnostic test on serum has shown greater than 98% sensitivity and specificity. However, in patients with a single ring-enhancing lesion, it falls to 60%–80%. Sensitivity is also reduced if CSF rather than serum immunoblot is used (Tsang and Garcia, 1996). Skull skiagrams may show calcified cysticerci, and reported figures are as high as 40%, but reported incidence from India is less than 1% (Chopra et al., 1981).

The most important radiological tests are the imaging techniques of CT or MR, which have revolutionized our diagnostic accuracy of NCC. The CT may show massive bilateral edema and chincked ventricles, suggestive of a pseudotumer cerebri–type of presentation of NCC. CT and MRI can also recognize the state of development of cysticerci in the cerebral parenchyma. The lesions may show low attenuation with or without enhancement. Cystic lesions showing the scolex as a brilliant nodule or multiple calcific lesions can be characteristic or diagnostic of NCC (Suss et al., 1986). Degenerating cysticerci, which are more symptomatic, may be isodense or hyperdense, and edema around them usually gives the impression of a ring or a nodular enhancement by contrast medium (Kramer *et al.*, 1989). Hundreds or thousands of intraparenchymal cysts may be demonstrated by CT or MR in the severe encephalitic form of NCC, which may prove fatal. Subarachnoid NCC on CT or MR may show hydrocephalus, abnormal enhancement of the leptomeninges, arachnoid cysts, and cerebral infarcts. The racemose form with grapelike clusters of cysts or very large cysts with multiple septations occurs when parasites lodge within the ventricles or subarachnoid space. Such lesions may sometimes be difficult to detect with CT, but MR with superior resolution can detect these easily (Greenberg, 1995). Cysts not detected by CT can be picked up by MR, and the authors have seen many such patients who presented with seizures. MRI is also better than CT in detecting lesions at the base of brain, brainstem, and spinal cord. However, CT is a better for detecting calcifications.

Electroencephalography may show abnormalities in 30% of patients with NCC (Chopra *et al.*, 1981). There could be focal or diffuse slowing or unilateral/bilateral epileptic discharges. Even periodic lateralized epileptiform discharges have been recorded (Virmani *et al.*, 1977).

## Practical Management

The treatment of established NCC is controversial and depends on the associated inflammatory reactions, as well as clinical and pathological features. Symptomatic and specific anticysticercal drugs are the cornerstone of therapy. Symptomatic therapy with conventional anticonvulsant drugs is the definite therapy to control the seizures. Carbamazepine or phenytoin is best for control of seizures caused by NCC.

### Carbamazepine

Carbamazepine is most effective against partial seizures with or without secondary generalization. The dose of this drug must be adjusted in response to clinical effect, because there is great variability of response in different individuals. The starting daily dose in adults is 5 mg/kg body weight and in children 5–10 mg/kg divided in to two to three daily doses and gradually increased to 10–25 mg/kg for children to a maximum of 600–800 mg daily. In adults, the gradual increase may be up to 1000–2000 mg daily. The usual effective plasma concentration range is 7–14 μg/ml in adults (Sillanpaa, 1997).

**Adverse Effects.**   Neurotoxicity of drowsiness, dizziness, dyskinesia, ataxia, and visual disturbance are adverse effects of carbamazepine. There is less effect on cognition and behavior. Hematologic toxicity of thrombocytopenia, aplastic anemia, agranulocytosis, and pancytopenia are extremely rare. Skin rash has been observed in up to 17% of patients. Hepatotoxicity of enzyme elevation has been observed in 5%–22% Cardiac conduction defects may be observed in the elderly, and it should be used with great caution in those with a pacemaker. Rarely, proteinuria, hematuria, oliguria, and renal failure have been reported. Gastrointestinal side effects of diarrhea, nausea, vomiting, stomatitis, or glossitis have been observed rarely. Most of the side effects are reversible when the drug is discontinued.

### Phenytoin

Rapid concentration in plasma can be achieved with 15–20 mg/kg of body weight daily. The average effective dose in children is 5–10 mg/kg of body weight. Clinically, the effective phenytoin plasma levels range from 10–20 μg/ml. Higher doses may be used in children with status epilepticus caused by NCC. The initial doze of 300 mg/day results in concentrations between 10 and 20 μg/ml in approximately 45% of patients, less than 10 μg/ml in 35%, and greater than 20 μg/ml in 20% (Bauer and Blouin, 1982). Therefore it is important to make adjustments.

**Adverse Reactions.**   Nystagmus, ataxia, and incoordination are common side effects with this drug. Lateral-gaze nystgmus may be observed at plasma concentrations of 15–30 μg/ml, ataxia appears above 30 μg/ml, and at levels of 40 μg/ml there is lethargy or aggravation of seizures (Kutt and McDowell, 1968). Irreversible cerebellar Purkinji cell degeneration may occur with chronic intoxication or with severe acute intoxication. Gingival hyperplasia, hirustism, acne, and morbilliform skin rash are other side effects. Rare side effects are Stevens-Johnson syndrome; hypersensitivity reactions, including antinuclear antibodies, megaloblastic anaemia, and lowered serum folate levels. The authors have observed phenytoin encephalopathy in a 19-year-old man who had plasma levels of 97.4 μg/ml. He remained in a coma for a week. Self-medication of phenytoin in intractible epilepsy in this patient over many years also caused irreversible cerebellar degeneration as seen on imaging.

### Corticosteroids

The death of cerebral cysticerci, especially if multiple, can lead to an immense immunological reaction with brain edema, which can even prove fatal.

Intravenous dexamethasone, 4–8 mg every 6 hours, is recommended for the encephalitic form of NCC with massive brain edema in adults. Half of this dose can be given in children for similar problem. High doses of intravenous dexamethasone are also recommended in patients with angiitis caused by NCC. Oral corticosteroids are advised in less life-threatening brain edema at a daily dose of 1–2 mg/kg of body weight. Very rarely, a combination of corticosteroids and I.V. manitol is recommended in massive brain edema.

## Praziquantel

The recommended dose of praziquantel for NCC is 50 mg/kg/day for 14 days. Clinical improvement in up to 87.2% of patients has been reported (Groll, 1982); 60%–70% disappearance of brain parenchymal cysticerci has also been reported by various authors with a similar dosage (Sotello and Del Brutto, 1999). Recently, a single-day regimen of praziquantel and corticosteroids has been advocated. Praziquantel is given in three oral doses of 25 mg/kg body weight every 2 hours. It is followed 5 hours later by 10 mg of dexamethasone IM, and subsequently 10 mg of IM dexamethasone for the next 3 days. This is necessary to prevent the inflammatory reaction after the sudden destruction of cysticerci. Moreover, it has been noticed that simultaneous administration of praziquantel and corticosteroids lowers the plasma levels of praziquantel, which results in lowering the uptake of praziquantel into the cysticerci (Sotelo and Del Bruotto, 1999).

**Adverse Reactions.** The effects are transient and dose related. Abdominal discomfort, particularly nausea, vomiting, and pain, may access. Headache, dizziness, and drowsiness may also occur. Rarely, pruritus, urticaria, skin rash, muscle and joint pains may occur. Inflammatory reactions in NCC may produce meningismus, seizures, mental changes, and pleocytosis in the CSF. It is considered safe in children older than 4 years of age.

## Albendazole

Alberdazole has been shown to destroy 75%–90% of the cysts in the brain parenchyma and even considered better than praziquantel. It is administered orally with two divided doses of 15 mg/kg body weight/day given for 30 days (Escobedo *et al.*, 1987). Sotelo and colleagues have further shown that at similar doses, but given for 8 days, this drug is as effective as a monthly treatment. Dexamethasone, 10 mg IM, is given for first 4 days (Sotello *et al.*, 1988). However, in India from our own experience, albendazole given for 30 days is more effective in eradication of cerebral cysticerci compared with a shorter course. To avoid an inflammatory reaction, we invariably administer corticosteroids at a dose of 1 mg/kg body weight along

with albendazole at least for the first week. Treatment of a single brain parenchymal cyst in an earlier stage of evolution with cysticidal therapy is controversial. We advocate an 8-day course of albendazole, 15 mg/kg body weight, in two divided doses for eradication of not only the cerebral cyst but also eradication of *Taenia solium* in the gut. Earlier degeneration of the single cyst reduces the duration of anticonvulsive therapy in such patients.

**Adverse Reactions.** Adverse reactions are few and transient. Headache, nausea, dizziness, diarrhea, and abdominal pain may access. Rarely, the hepatic enzymes increase, and jaundice or chemical cholestasis may occur. There may be fever, fatigue, hair loss, leukopenia, and thrombocytopenia.

## Treatment No Longer Recommended

Anticysticercal drugs are not recommended for severe cysticercotic meningoencephalitis until the increased intracranial pressure caused by massive brain edema recedes with medical decompressive measures, including steroids. Anticysticercal drugs should also not be used in a patient with combined parenchymal cysts and hydrocephalus until a shunt has been installed to drain the ventricles. Anticysticercal medication is not recommended for calcific granulomas, which are sequelae of automatic destruction of cysts maybe months or years before the neuroimaging performed on a patient. Anticysticercal drugs should be used with caution for precariously located cysts in the brainstem, spinal cord, ventricles, aquaduct of Sylvius, or over the fundus or eye chambers. Albendazole should not be given to patients with hepatic cirrhosis. Both albendazole and praziquantel are not recommended in pregnancy and in children younger than 2 years of age.

# SCHISTOSOMIASIS (BILHARZIASIS)

## Definition

This infection is caused by trematode parasites of *S. mansoni*, *S. japonicum*, and *S. haematobium*. It is one of the most widespread parasitic diseases that affects the human race. It is also considered the most deadly parasitic disease after malaria. *S. mansoni* is most prevalent in Africa; the West Indies; South America, particularly Brazil; the Nile Valley in Africa; and in the Arabian peninsula. *S. japonicum* is seen mostly in the Asian countries of Japan, Taiwan, the Philippines, and China. *S. haematobium* is observed in Egypt, the Middle East, and some of the African countries. Schistosoma worms are also called blood flukes, which inhabit the circulatory system and thereby are capable of producing inflammatory lesions throughout the human body, including the central nervous system.

## Clinical Aspects

The clinical manifestations are based on three types of responses from the affected tissues. Soon after exposure, patients notice pruritus or what is also called cercarial dermatitis known as "*swimmer's itch.*" The exposed skin becomes erythematous. A massive infection may cause cough and fever—a picture of pnuemonitis and infection with *S. mansoni* at this stage causes meningoencephalitis or generalized vasculitis. Within 10–15 days, patients enter into the acute stage of schistosomiasis, called the stage of toxemia. There is systemic allergic reaction and patients get a high fever with shivering, known as Katayama's fever. The patient has headache, generalized urticaria, lymphadenopathy, and muscle pains. There is an increase of antigenic output from ova and worms. There is increased sweating and joint pains, and the liver and spleen may be enlarged along with lymphadenopathy.

The acute toxemia stage lasts 2–8 weeks, and in most the symptoms subside with or without treatment. Morbidity may be higher than mortality; the later is usually low.

Systemic involvement is the rule in chronic schistosomiasis. The granulomatous reaction over the retained ova in the intestinal submucosa may show polyp formation or ulcerate and is the cause of intermittent diarrhea containing mucus and blood. Chronic infections are also observed in other tissues once the worms complete intravascular migration and lodge in the venules of portal system, mesentry, urinary bladder, etc. The patient has hepatosplenomegaly, which may lead to portal hypertension and fibrotic changes in the liver. These manifestations are seen with the chronic infections of *S. mansoni* and *S. japonicum*. Ultimately, the patient may have with ascites, esophageal varices, and bleeding.

*S. haematobium* in the chronic stage causes a granulomatous reaction in the submucosa of urinary tract and bladder. These lesions fibrose and calcify, causing bladder neck obstruction, obstructive uropathy, stones in the ureters, and hematuria.

## Neuoroschistosomiasis

Ova of any of the three species may affect the central nervous system; however, *S. japonicum* is known mostly to affect the brain, and *S. mansoni* and *S. haematobium* often involve the spinal cord. Cerebral schistosomiasis is one of the major causes of epilepsy in the Far East (Olveda and Domingo, 1987). The seizures are mostly partial and noticed more often in men. There may or may not be any focal neurological deficit. The seizures are generally noticed 6 months–9 years after the infection. Congregation of ova around the cerebral blood vessels diminish cerebral blood flow and may cause cerebral stroke (Khalil, 1975). A granulomatous lesion from a fairly large collection of the worm ova may present like a space-occupying lesion with headache, vomiting, focal neurological deficit, speech disturbances, ataxia, and involvement of higher mental functions, depending on the location of the gramuloma. Hepatosplenic schistosomiasis may itself result in hepatic encephalopathy after hepatic failure. A patient with acute Katayama's fever may also have acute encephalopathy, with the neurological manifestations of confusion, delirium, behavioral changes, partial or generalized seizures, affected vision, raised intracranial signs including papilledema, pyramidal tract signs, and finally the patient may lapse in to coma (Hayashi *et al.*, 1984; Kirchhof and Nash, 1984). There may be spontaneous relief from these symptoms or the relief could be initiated with specific medication.

Spinal schistosomiasis is an important cause of paraparesis/paraplegia in some of the countries of the Middle East. The important sites of infestation are the upper sacral, lumbar, and lower thoracic areas. Patients have clinical manifestations of cauda equina and conus medullaris lesions. Gradual cord compression from a granulomatous lesion is another form of presentation. Transverse myelitis from the vascular cord infarct may be seen acutely. Men are more affected than women because of occupational hazards, such as farming and irrigating fields. Myelopathy may manifest 5 weeks after infection and up to 6 years (Bird, 1978). Children and tourists to the endemic area may have myelopathy in the shortest incubation period, because both are less immune to parasites than adults. The initial symptoms are pain in the lower back precipitated by coughing and sneezing if the granulomatous lesion is at cauda equina or conus medullaris region. Urinary symptoms may be another manifestation followed by weakness in the legs. Acute transverse myelitis as usual is seen with flaccid paraplegia with areflexia. This may happen with involvement of the anterior spinal artery or its branches. There is complete loss of superficial sensations, with preservation of posterior column sensations. The slowly progressive paraplegia is invariably spastic. The spinal cord lesions are mostly from *S. mansoni* and *S. haematobium*.

Peripheral neuropathy and myopathy have also been observed in schitosomiasis (El-Gengaihy *et al.*, 1991; Esam and El-Gengaihy, 1999). The neuropathy is mostly sensory motor and has been observed more often

| Other inflamatory granulomas Tuberculomas and neurocysticercosis | Both these diseases are commonly seen in many tropical countries in which schiztosomiasis is also endemic. These need to be differentiated, because they can also cause seizures, focal neurological deficits, fever, signs of meningeal irritation, and raised intracranial pressure. Spinal tuberculomas and neurocysticercosis can also be seen with back pain, radiculopathy, and slowly progressive myelopathy. Imaging techniques, involvement of the liver and urinary bladder, stool and urine examination could be helpful in the diagnosis. |
| Primary or metastatic lesions in the brain or spinal cord | Clinical presentation may be similar to schistosomiasis and needs to be differentiated, especially in the endemic areas. However, differentiation could be easy with appropriate investigations and the finding of schistosomiasis lesions in other tissues. |
| Inflammatory lesions of the brain or spinal cord abscess | These can also be seen with fever, focal or generalized seizures with focal neurological deficit or signs, of spinal cord involvement. However, imaging with CT or MR can clinch the diagnosis. |

with the hepatosplenic form of schistosomiasis. Muscle wasting may be present around the shoulder girdles. Recurrent brachial and lumbosacral plexus involvement have also been noted, mostly because of parainfection or allergic manifestations (Marra, 1983) of the antigen antibody reaction.

## Differential Diagnosis

Schistosomiasis should always be considered in endemic areas when the patient is seen with partial or generalized seizures with or without focal neurological deficit, acute or slowly progressive paraplegia, lower back pain, or sphincter disturbances. In most cases the diagnosis is confirmed with immunological tests, imaging techniques, or urine and stool examination. However, a number of conditions need to be specifically differentiated, because these can also cause similar neurological symptoms.

## Natural Course

Man is affected by swimming or using infected fresh water in which infectious larvae (cercariae) are released by snails—the intermediate hosts. The cercariae penetrate the skin or the mucous membrane, the developing worms enter the portal and hepatic venous system in the case of S. mansoni and S. japonicum and urinary venous system in the case of S. haematobicum. The larvae grow in to adult worms and after passing through portal vein enter the venules draining the pelvic viscera. Each worm may deposit hundreds to thousands of eggs each day. The ova pass in urine in the case of S. haematobium and

in stools in the case of S. mansoni and S. japonicum after penetrating the wall of the urinary bladder or the rectum. The liberated ova after entering fresh water hatch, and the embryos multiply in the intermediate host, the snail. The cercariae in thousands enter the fresh water after being liberated from snails and infect humans, thereby completing the life cycle. This is the natural course; however, some ova remain in the body and through various venous plexuses and channels, they lodge in the body tissues. The various clinical symptoms are due to the lodging of ova in various tissues. The central nervous system is involved by the transmission of ova through the vertebral venous plexus. The ova may lodge in the brain or spinal cord through the vertebral venous plexus or through the internal jugular vein in the brain. Rarely arterial embolization of ova occur through the arteriovenous fistula in the lungs (Ghaly and El-Banhawy, 1973). The ova penetrate or extravasate around the perivascular tissues by an enzymatic process (File, 1995). The main lesions are in the intestine or liver and the urinary bladder, but other tissues such as central nervous system, skin, muscles, lungs, myocardium, and uterus in women may also be involved.

Most of the symptoms described under clinical aspects may recede spontaneously or with specific therapy in 2–8 weeks. Morbidity is greater in this disease than mortality, which is low, unless there is profound infection with parasites. The granulomatous lesions usually consist of one or numerous ova that are encircled by mononuclear and plasma cells, epitheloid, cells,. and even giant cells. Eosinophil leukocytes may also be abundant in the lesions.

The prevalence of cerebral schistosomiasis is variable. Philippine General Hospital autopsy records have shown ova of *S. japonicum* in 6.5% of the brains of patients who had schiostosomiasis, and the brain was involved in 26% of the hepatosplemic form of *S. mansoni* (Pittella, 1985; Scrimgeour and Gajdusek, 1985). The brain may also be involved in 30%–50% of patients with urinary schistosomiasis from *S. haematobium*.

It is estimated that more than 200 million people are affected with schistosomiasis (Larotski and Davis, 1981), and there could be 600 million who are at risk of this infection (WHO Expert Committee, 1993). Those who live in the developing countries of South America, Asia, and Africa are at greater risk, and the disease is prevalent in almost 70 counties of the world (Mahmoud, 1977). The disease is likely to affect many tourists and migrants from nonendemic to endemic areas. A large number of American soldiers had schistosomiasis during World War II while stationed in the Philippines.

Men are more affected with schistosomiasis compared to women because of more exposure outdoors, with farming, irrigation, and postings in endemic areas. Children are more at risk with schistosomiasis because of poor immune protection.

There was an outbreak of schistosomiasis among Europeans and Americans who rafted in the Ethiopean river Omo (Istre *et al.*, 1984). Similar infections with *S. haematobium* occurred in peace corps volunteers from United States, who had central nervous system involvement while working in Africa. In the natural course, schistosomiasis can affect different organs of the body already enumerated, and fulminant infection can be fatal if untreated. Systemic complications can be profound, and therefore this disease must be treated with specific medication in its early stage.

## Diagnosis

Stool examination should be performed, because it can contain ova of *S. mansoni*, and urine examination may give the clue for infection with *S. haematobium*. In neuroschistosomiasis, the examination of stools, urine, and eiosinophilia in peripheral blood may not be helpful. However, eiosinophilia in the cerebrospinal fluid would be an indicator of conus medullaris and cauda equina granuloma, although CSF may be clear in most patients. There is a mild rise in CSF proteins.

Antibodies to eggs, cercarial, and the adult worm can be detected by a battery of serological tests. Antischistosomial antibodies are shown in the blood by the ELISA test, and specific antibodies in the patients' serum can be detected by an immunoblast test (Tsang and Wilkins, 1991). The ELISA test on CSF can be helpful, and a negative CSF ELISA nearly rules out myelopathy caused by schistosomiasis.

CT or MRI brain scans may show focal granulomas, which usually develop around eggs, could be multiple, enhance with contrast medium, and show perimesial edema. Rarely, hemorrhage can be detected with these techniques. CT or MRI of the spinal cord shows enhancement of the granuloma at the lower end or the enlargement of the cord. CT myelography would show a partial or complete block at the lower thoracic or upper lumber vertebra (Haribgai *et al.*, 1991). Ultrasonography of the abdomen may show hepatic granulomas or urinary tract abnormalities. Rectal, liver, or urinary bladder biopsy is usually helpful and may be performed if noninvasive tests are not diagnostic.

## Principles of Therapy

The principles of therapy need to be directed toward the following:

1. Relief of symptoms such as seizures, fever, headache, vomiting, and focal neurodeficits in the brain or the spinal cord. The seizures should be controlled with anticonvulsants, appropriate analgesics, and antiedema to relieve intracranial pressure from single or multiple granulomas in the brain. Steroids may also be required for relief of brain edema.
2. Cauda equina or spinal cord compression or a large brain granuloma needs to be removed with surgical intervention.
3. Eradication of the parasites with specific medicine— praziquantel.
4. Steroids should be used to avoid inflammatory and granulomatous reactions before there is destruction of tissues.

## Practical Management

### Praziquantel

**Praziquantel** is the specific drug of choice for all types of schistosomiasis (i.e., *S. mansoni, S. japonicum* and *S. haematobium*). The adult worms are eradicated with this drug, and further ova formation is restricted. A single oral dose of 40–60 mg/kg body weight is the standard treatment. It may be divided in to two or three doses given every 4–6 hours (Watt *et al.*, 1986); 20 mg/kg body weight at an interval of every 4 hours in a single day is also practiced by some. Recently, some authors (Haribgai *et al.*, 1991) have advocated praziquantel, 40 mg/kg body weight; given daily for a period of 2 weeks. However, this drug regimen is monitored with clinical improvement, repeated radiological assessment in the number and size of lesions, repeated serological, and parasitological observations.

An alternate therapy used now for *S. haematobium* infection is metrifonate. It is given orally, 7.5 mg/kg body weight, weekly for 3 weeks. Currently, a combination of praziquantel and oxamiquine even in low doses is considered effective in the treatment of *S. mansoni*. Oxamiquine can be administered orally as a single dose (15–20 mg/kg body weight) for eradication of *S. mansoni*.

Corticosteroids are usually given to counteract allergic phenomenon as a combination with antischistosomal therapy for meningoencephalitis, which is sometimes observed with schistosomiasis or inflammatory lesions of the CNS. Prednisolone orally, 1.5–2.0 mg/kg body weight daily, is administered for 3 weeks and then gradually decreased.

Surgical decompression for granulomas causing spinal cord compression is required urgently along with the specific antischistosomal therapy. Recovery may be complete with combined medical treatment and surgical decompression, except in cases in which acute necrotic myelitis has occurred. One third of patients usually may not show any changes even with these combined measures.

A combination of antischistosomal drugs and surgical resection may also be required for large schistosomal granulomas in the brain. Corticosteroids at doses already mentioned should also be given. Patients who have seizures are treated with conventional anticonvulsants.

### Anticonvulsants

Seizures in schistosomiasis should be controlled with conventional anticonvulsants. The drugs of choice are carbamazepine and phenytoin. Carbamazepine is advised in adults as 5 mg/kg of body weight and in children it may be given in a dosage of 5–10 mg/kg of body weight. The drug is given in two to three divided doses. The daily dosage is gradually increased according to control of seizures.

Phenytoin is given in a dosage of 15–20 mg/kg body weight, and effective rapid plasma concentration can be achieved with this medication in adults. In children the effective plasma concentration can be achieved with 5–10 mg/kg body weight, however, it may be increased if necessary in status epilepticus.

**Adverse reactions** with praziquantel and the anticonvulsants carbamazepine and phenytoin have already been listed in the section on "Neurocysticercosis," and readers are advised to look for the same in that section.

### Treatment No Longer Recommended

Radical removal of granulomas is not recommended, unless such lesions are causing features of increased pressure in the brain or neurological deficits in the spinal cord. Medical therapy can look for noncompressive granulomas. Laminectomy for biopsy to confirm the schistosomal granuloma is also not indicated, because satisfactory diagnosis can be established with noninvasive methods.

## CEREBRAL MALARIA

### Clinical Aspects

#### Definition

Cerebral malaria is defined as a symmetrical encephalopathy characterized by the presence of unarousable coma lasting for more than 30 minutes in a febrile patient with exclusion of other encephalopathies, especially bacterial meningitis and locally prevalent viral encephalitis, and confirmation of infection by the presence of the asexual form of *Plasmodium falciparum* in the blood (Warrell, 1989). *P. falciparum* infection is often associated with hypoglycemia, uremia, and hponatremia, which may result in alteration of conscious without direct cerebral involvement. In adults, coma is required for more than 6 hours after a generalized seizure to exclude a transient postictal state, although it has been reduced to 1 hour in children (Molyneux *et al.*, 1989).

#### Clinical Features

Cerebral malaria is seen with acute diffused febrile encephalopathy, and focal neurological deficit is relatively uncommon. Coma is often accompanied by multisystem dysfunction. The presence of anemia, jaundice, and acidotic breath should raise the clinical suspicion of cerebral malaria. Papilledema is unusual, and retinal haemorrhages are seen in 15% patients (Looareesuwan *et al.*, 1983). The incidence of seizures varies from 10%–50%. Dhamija *et al.* (1989) have given a detailed description of clinical manifestations, including various neurological syndromes (Table I).

African children in the endemic areas usually have cerebral malaria develop at the age of 40–45 months. It is rarely encountered after the age of 10 in those exposed to *P. falciparum* since birth. The incidence of seizures is high, and prolonged seizures are associated with poor outcome. African children rarely have renal failure or pulmonary edema develop. Most of the deaths occur within 24 hours of starting treatment, usually with brainstem signs, respiratory arrest, and overwhelming acidosis.

**Table I  Clinical Presentation in Cerebral Malaria**

1. Acute
   A. Aymptomatic
   B. Cerebral disorders
      Acute febrile encephalopathy
      Coma
      Cranial nerve palsies
      Hemiplegia
      Paraplegia
   C. Spinal disorders
      Myelitis
      ALS or tabes dorsalis–like picture
   D. Peripheral nerves
      Polyneuritis
      Guillian-Barre–like picture
   E. Acute psychiatric manifestations
      Neurasthenia
      Malarial psychosis
   F. Combined
      Disseminated encephalomyelitis
2. Late sequel
   Seizures
   Involuntary movements
      Dyskinesia
      Chorea
      Parkinsonism
   Cerebellar syndrome
   Psychiatric syndrome

## Diagnosis

Early diagnosis is the key to the prevention of complications and mortality. A high index of suspicion is essential in dealing with a patient with febrile encephalopathy from an endemic area. The examination of a Giemsa- or Field-stained blood smear by light microscopy is the most common and economical method used for confirmation of the diagnosis. A thick smear is used for the detection of malarial parasites and a thin smear for the identification of the species and intensity of parasitemia. The smears may be negative in more than 50% of cases because of lack of technical expertise or scanty parasitemia consequent to microvascular sequestration of parasitized red blood cells. Fluorescent microscopy and polymerase chain reaction–based methods are more sensitive. The recently introduced immunoassays, which detect malarial antigen in the blood, are promising. There are two parasite antigens in use—histidine-rich protein-2 (HRP-2), which is produced by falciparum only, and parasite lactate dehydrogenase antigen (pLDH) produced by all four species.

## Natural Course

Malaria continues to be widely prevalent in many tropical countries of the world, although its incidence has declined significantly in the developed countries. It is estimated that more than 40% of 2073 million people residing in 100 countries are exposed to the risk of malaria (WHO, 1990). There are four species of human malaria, but *P. falciparum* causes nearly all the neurological complications and deaths. It infects as many as 300–500 million people, causing up to 3 million deaths each year (Hoffman, 1996; WHO, 1996) About 1% of *P. falciparum* infections progress to cerebral malaria, a severe and potentially life-threatening encephalopathy.

With appropriate treatment, full recovery of consciousness takes a median of 2 days in a patient with a Glasgow Coma Scale score of <11. Occasionally, patients may take more than a week to recover. The overall mortality of cerebral malaria is approximately 20% (Warrell *et al.*, 1990). It is greatly enhanced by other vital organ dysfunction. Mortality is 8% in patients with isolated cerebral malaria, whereas it rises to 50% with associated renal failure and metabolic acidosis (Newton *et al.*, 2000). Respiratory irregularity, hypoxia, pulmonary edema, pneumonia, and shock are the usual terminal features. Mortality can be reduced with the availability of intensive care facilities.

## Principles of Therapy

Cerebral malaria is a medical emergency demanding urgent assessment and treatment. Fever, impaired consciousness, and convulsions associated with other clinical features in a patient coming from an endemic area should raise the possibility of cerebral malaria. Therapy should be initiated even without laboratory diagnosis if a clinical index of suspicion is high. The delay in initiating specific therapy leads to an increase in complications and mortality. The management of cerebral malaria is similar to any seriously ill unconscious patient and requires intensive care therapy with skilled nursing care and meticulous fluid–electrolyte balance. Adults with severe malaria are prone to respiratory distress syndrome and may need ventilation and monitoring of central venous and pulmonary wedge pressure. Blood transfusion is indicated if the packed cell volume falls below 20%. Hemofiltration should be started early in patients with renal failure and severe metabolic acidosis.

## Specific Antimalarial Therapy

Antimalarial therapy unequivocally affects the outcome of cerebral malaria and should be started as soon as possible. The following drugs have been in current use.

**Quinine.** Quinine is considered the mainstay of treatment because of widespread chloroquin resistance

(White, 1996). It must be given with an intravenous loading dose to achieve therapeutic plasma concentration of the drug as soon as possible. The loading dose can be safely given in patients with renal and hepatic dysfunction and pregnant women. Oral treatment should be substituted when the patient becomes conscious and can swallow normally. Hypotension is the most significant side effect of quinine, especially if the infusion is given too rapidly. It prolongs ventricular repolarization and the QT interval. It can cause hypoglycemia, thrombocytopenia, urticaria, and psychosis.

**Artemether and Artesunate.** These are the derivatives of traditional Chinese antimalarial Qinghaosu obtained from the plant *Artemisia annua*. They are simple to administer by the intramuscular route and are rapid acting. Intramuscular and rectal administration is as effective as intravenous therapy. There are no apparent local or systemic side effects. In experimental studies on primates artemether produced an unusual and selective pattern of damage auditory brainstem nuclei in a higher dose (Brewer *et al.*, 1994). It has been reported to show prolonged recovery time from coma (Hien *et al.*, 1996). Artemesinin is associated with a higher recrudescence rate, particularly when the treatment lasts less than 5 days (Tripathi *et al.*, 1999). These compounds clear parasitemia and fever faster than cinchona alkaloids, but there is no improvement in mortality. In a recent open randomized trial in African children, intramuscular artemether was compared with intravenous quinine. Artemether conferred no survival advantage over quinine in children with life-threatening cerebral malaria (Taylor *et al.*, 1998). A meta-analysis of randomized trials indicates that Artemether reduces the mortality by one fifth in adults, but there is no convincing difference in children (McIntosh and Olliaro, 1998). Nevertheless, because of their simplicity in administration, safety, and increasing drug resistance to quinine in some parts of the world, they may become the treatment of choice.

**Mefloquine.** Mefloquine is not used commonly in the treatment of cerebral malaria, because it is available only in tablet form. However, remarkable results have been seen with a single dose (25 mg/kg) administered through a nasogastric tube (Di Perri *et al.*, 1999). This study suggests that mefloquine deserves consideration if parenteral drugs are not available for treatment of cerebral malaria.

**Combination Therapy.** Combination therapy is advocated in those parts of the world where resistant strains are endemic. Triple-drug therapy with quinine, cotrimoxazole, and tetracycline has shown good results.

In a prospective study, Waiz *et al.* (1995) treated 254 cases of cerebral malaria with combination therapy. One hundred patients were treated with quinine and cotrimoxazole, and 154 patients were given triple therapy—quinine, cotrimoxazole, and tetracycline. The study reflected that administration of tetracycline along with cotrimoxazole and quinine reduced the mortality significantly.

### Ancillary Treatment

Seizures, hypoglycemia, and hyperpyrexia are common complications of cerebral malaria, which should be detected early and treated promptly. Seizures in malaria are often recurrent; 84% of the seizures are complex and most often focal in onset (Waruiru *et al.*, 1996). Intracranial sequestration of parasites or their toxins is the most likely cause of seizures. Some of the antimalarial drugs (e.g., chloroquin) may precipitate seizures. Recurrent seizures should be treated with intravenous loading of phenytoin. Phenobarbitone reduces the frequency of seizures in children, but higher doses are needed in children to prevent convulsions. In a recent double-blind controlled trial in children, phenobarbital (20 mg/kg) reduced seizures by 50% but doubled the mortality (Crawley *et al.*, 2000). Hypoglycemia is common in young patients with life-threatening disease and pregnant women. Those who are treated with quinine may become hypoglycemic, because it is a potent stimulus for islet cell insulin secretion.

There is no effective vaccine against falciparum (Targett, 1995). Creation of a successful vaccine has been hampered by the complexity of the parasite's cycle. Researchers from the US Centers for Disease Control and Prevention in Atlanta, Georgia, are evolving a recombinant, multivalent vaccine against falciparum. It is different from other vaccines, because it contains multiple antigens and targets the parasite at several stages in its life cycle.

### Practical Management

- Acute febrile encephalopathy associated with anemia, jaundice, acidotic breathing, or splenomegaly in a patient from an endemic area should raise the clinical suspicion of cerebral malaria.
- Start antimalarial treatment if the index of clinical suspicion is high pending confirmation of the diagnosis.
- Intravenous quinine hydrochloride is the drug of choice. The following regimen is recommended:
  — Loading dose: 7 mg/kg over 30 minutes by infusion pump.

— It should be immediately followed by 10 mg/kg over 4 hours, to be repeated every 8–12 hours.
— Start oral quinine as soon as possible and complete 7-day course.
— Add 1 g/day tetracycline in four divided doses for 7 days in nonpregnant adults in areas with resistant strains.

- Artemesinin derivatives serves as an alternate regimen. These are not approved in all countries as yet.
- Artemether is given intramuscularly. Artesunate is used intravenously and is reconstituted with bicarbonate solution immediately before use. The recommended dose for both the compounds is:
  — Loading dose: 3.2 mg/kg.
  — It is immediately followed by 1.6 mg/kg every 12–24 hours.
  — Change to oral mefloquine, single dose 15–25 mg/kg; maximum dose 1500 mg, as soon as possible.
- Monitor hypoglycemia.

### Treatments No Longer Recommended

Steroids are of no therapeutic benefit in cerebral malaria and lead to increased incidence of infection and gastrointestinal bleeding (Warrell *et al.*, 1982). Mannitol may be used in cases of cerebral edema (Newton *et al.*, 1997). Pentoxifylline (Looareesuwan *et al.*, 1998) and iron chelation therapy with deferoxamine (Gordeuk *et al.*, 1992) have been tried without any success. Other modes of therapy directed against pathogenetic mechanisms leading to cerebral malaria have been tried. Monoclonal antibodies against TNF, ICAM-1, knob proteins, and other cytokines are steps in this direction. The results are conflicting.

## CEREBRAL AMEBIASIS

Cerebral amebiasis is a rare parasitic manifestation of the brain caused by two different groups of amoebae—free-living amebae and pathogenic intestinal amebae. Amebae belonging to the genera *Naegleria*, *Acanthamoeba*, and *Balamuthia* are free-living, amphizoic, and opportunistic protozoa that are ubiquitous in nature. These amebae are found in soil, water, and air samples from all over the world. *Entamoeba histolytica* is the pathogenic agent for classic amebic dysentery. It may occasionally lead to cerebral involvement. Cerebral amebiasis is seen with the following clinical syndromes:

1. Primary amebic meningoencephalitis (PAM)
2. Granulomatous amebic encephalitis (GAE)
3. Amebic brain abscess

### Primary Amebic Meningoencephalitis

#### Clinical Aspects

**Definition.** Primary amebic meningoencephalitis is an acute, fulminant, rapidly fatal meningoencephalitis that usually affects previously healthy young adults and children who have been swimming and diving in hot summer months or in heated swimming pools. *Negleria fowleri* is the etiologic agent.

**Clinical Features.** Patients are seen with sudden onset of generalized headache, fever, nausea, and vomiting. Signs of meningitis and encephalitis are evident soon. There is rapid progression to coma. Seizures are seen in a few cases. Marked raised intracranial tension and decerebrate rigidity have been reported in terminal cases. The disease closely mimics acute purulent meningitis. Abnormalities of smell (parosmia), taste (ageusia), and early-onset ataxia may help to differentiate it from bacterial meningitis.

**Diagnosis.** The patient's age, season, and history of recent swimming may suggest the diagnosis of PAM. However, there are no distinctive features to confirm the diagnosis on clinical grounds. Neuroimaging with CT and MRI may be normal at the onset of the illness; the findings are nonspecific, with evidence of brain edema (Kidney and Kim, 1998; Schumaker *et al.*, 1995). Computed tomography shows obliteration of the cisterns surrounding the brainstem. Enhancement of the cisterns and subarachnoid space is seen on intravenous contrast. Serological tests are not useful, because most of the patients die too rapidly to produce antibodies.

The CSF pressure is almost always increased. It shows modest leukocytosis with a number of erythrocytes. The protein content is elevated, and the glucose concentration may be normal or low. The findings may be confused with those in *Herpesvirus hominis* or bacterial meningitis (Ma *et al.*, 1990). The definitive diagnosis of PAM is made on the visual detection of amebic trophozoites in the CSF or brain tissue. These may be easily mistaken for macrophages or epithelial cells. Motile amebic trophozoites can be observed in the CSF when one or two drops of unstained CSF is examined under the low-power microscope. Smears may be stained with Giemsa or Wright, demonstrating typical amebic trophozoites.

#### Natural Course

*Negleria* is an ameboflagellate that has three stages in the life cycle: trophozoite, cyst, and a temporary flagellate stage. *Negleria fowleri*, a human pathogen, is thermophilic, tolerating temperatures of 40–45°C. It has been isolated from thermally polluted water and sewage wastes. Most human infections have been associated with swimming in warm waters, but other reported sources of infection include tap water and hot baths. Although there has not been any epidemic, 16 cases occurred from the same lake in Virginia and 19 cases from the same stream-fed swimming pool in Czechoslovakia (Ma *et al.*, 1990). From 1966–1990, 63 cases were recorded at the Centers for Disease Control and Prevention (Visvesvara and Stehr-Green, 1990). All patients died, although amebae were identified in the CSF in four cases that received specific chemotherapy with amphotericin. The disease was rapidly fatal, with a mean interval between onset of symptoms and death of 6.4 days. So far, 179 cases have been reported, 81 in the United States alone (Martinez and Visvesvara, 1997).

### Principles of Therapy

Early diagnosis and treatment is essential for successful outcome of medical therapy, because PAM has a fulminant course and a uniformly poor prognosis. Patients need all the supportive care for a comatose patient. The specific therapy includes high-dose intravenous and intrathecal amphotericin. A combination of amphotericin B and miconazole has been recommended (Martinez and Visvesvara, 1997). No more than four cases have been documented to survive.

**Specific Therapy.** Treatment with antifungal agents is empirical. There are no guidelines or randomized trials available.

*Amphotericin B.* It is an amphoteric polyene antibiotic commonly used in systemic fungal infections. For CNS infections it is given by slow intravenous or intrathecal route. It normally crosses the blood–brain barrier poorly, but penetration may be improved if the meninges are inflamed. The most common and most serious side effect is renal toxicity. A degree of renal impairment occurs in 80% of cases and is reversible when the drug is stopped. Hypokalemia is seen in 25% of patients and requires potassium chloride supplementation. Injection frequently results initially in chills, fever, headache, and tinnitus.

*Miconazole.* It is a synthetic azole with a broad spectrum antimycotic activity. It is given by intravenous infusion for systemic infections. It crosses the blood–brain barrier poorly. Unwanted side effects are infrequent, and those commonly seen are gastrointestinal disturbances, but pruritus, blood dyscrasias, and hyponatremia are also reported. The intravenous injection may be associated with anaphylactic reaction, dysrhythmia, and fever.

**Prevention.** *N. fowleri* is susceptible to chlorine, and proper disinfection of water in swimming pools, whirlpools, and Jacuzzis with chlorine is a reasonable preventive measure. The National Health and Medical Research Council of Australia has recommended the use of chlorine in swimming pools at 1 mg/L if the water is less than 26°C, at least 2 mg/L if the temperature exceeds 26°C, and 3 mg/L for temperature over 28°C (Martinez and Visvesvara, 1997).

### Practical Treatment

- Start intravenous amphotericin only after confirmation of diagnosis.
- It is prudent to give 1-mg intravenous test dose, followed by rapidly escalating doses.
- Premedication with acetaminophen or aspirin or the addition of 25 mg hydrocortisone to the infusion decreases fever and chills.
- It may be given in a dose of 1 mg/kg/day. The duration of treatment depends on the patient's response.

## Granulomatous Amebic Encephalitis

### Clinical Aspects

**Definition.** Granulomatous amebic encephalitis is a subacute or chronic granulomatous encephalitis caused by opportunistic free-living amebae of genus *Acanthamoeba* and *Balamuthia mandrillaris*. It is usually seen in debilitated, malnourished individuals, in patients undergoing immunosuppressive therapy, and in acquired immunodeficiency syndrome (AIDS).

**Clinical Features.** The clinical manifestations are mainly those of focal or localized encephalopathy associated with severe meningeal irritation and encephalitis. It is a chronic, clinically protracted illness. Fever is sporadic and generally low grade. Headache is insidious, occurs early in the course of disease, and may be associated with seizures.

Several localizing signs in the form of hemiparesis, cranial nerve palsies (mainly III and VI), and focal deficit is seen, depending on the area of the brain involved. Stiff neck is seen in most of cases. Personality changes and drowsiness appear as the disease advances.

**Diagnosis.** The clinical profile is not specific. Chronic or subacute meningoencephalitis in the setting of immunosuppression should raise the clinicians' index of suspicion. CT scan and MRI are of limited diagnostic value. Neuroimaging shows areas of multiple infarcts or granulomas. Both enhancing and nonenhancing lesions have been observed (Lowichik *et al.*, 1995;

Schumacher *et al.*, 1995). CSF shows pleocytosis with an abundance of lymphocytes or polymophonuclear cells; glucose is usually low with a moderate increase in the protein content. Diagnosis is confirmed by brain biopsy. Examination of frozen sections with hematoxylin and eosin may establish the diagnosis in a few minutes.

### Natural Course

The life cycle of *Acanthameba* has only two stages: trophozoite and cyst. The organisms are widespread in nature. Normal human serum may contain inhibitors that prevent the growth of acanthamebae. Because the host defences are compromised, they multiply and disseminate in the body, including the CNS. The CNS lesions may be preceded by skin or lung lesions. Martinez and Visvesvara (1997) reviewed the world literature and found 166 cases of GAE, 103 caused by *Acanthamoba* and 63 caused by *Balamuthia*. Of the 103 cases caused by *Acanthameba*, 72 have been reported in the United States alone, more than 50 in persons with AIDS. The disease is invariably fatal, and death comes within 8 days to several months after the onset of symptoms (Martinez, 1985).

### Treatment

There is no effective treatment for GAE. Most of the cases have been diagnosed postmortem. Adjunctive therapy is usually not effective because of underlying disease processes such as AIDS. Most of the patients have other systemic complications, such as renal failure, anemia, respiratory distress syndrome, and hypoglycemia, which should be managed accordingly.

Although pentamidine, sulfadiazine, and ketoconazole seen to be effective in vitro, their efficacy is questionable in patients with a compromised immune system. Prophylaxis is equally ineffective, because no single drug is effective against both cysts and trophozoites, other opportunistic organisms may invade the host and increase the risk of multidrug toxicity (Slater *et al.*, 1994). Prevention is difficult, because the organisms are ubiquitous and most infections occur in patients with immunosuppression. Periodic inspection of water tanks, air filters, and plumbing systems is useful.

## Amebic Brain Abscess

### Clinical Aspects

**Definition.** Amebic brain abscess is a rare complication of infection caused by the pathogenic amoeba, *Entameba histolytica*, which usually causes intestinal amebiasis. Sometimes it can lead to serious extraintestinal disease in organs such as the liver, lung, pleural and pericardial spaces. Occasionally, brain abscess is produced during hematogenous spread of the organisms.

**Clinical Features.** The clinical manifestations depend on the site and size of the abscess. Abscesses may be small and multiple (Figure 4), or there may be a single large abscess. In 53 brain lesions, basal ganglia and frontal lobes were the most common sites of lesions (Bia and Barry, 1986). A large abscess is seen with raised intracranial tension and focal neurological deficit. Seizures are common. Most cases have associated hepatic amebic abscess.

**Diagnosis.** The diagnosis is suspected in case of extraintestinal amebiasis presenting with headache, seizures, and focal neurological deficits. CT and MRI scans show findings of an abscess. Serological tests for *E. histolytica* lend further support to the diagnosis. The diagnosis is confirmed on demonstration of *E. histolytica* on brain biopsy or abscess aspirate.

### Natural Course

*E. histolytica* is acquired by ingestion of viable cysts from fecally contaminated water or food. Motile trophozoites are released from the cysts in the small intestine. Amebic brain abscess is seen in patients with chronic amebiasis associated with hepatic amebiasis. Approximately 10% of the world population is infected with *E. histolytica*. There is a high incidence of the disease in developing countries, particularly in Mexico, India, Africa, and the nations of central America. Most of the patients remain asymptomatic carriers or have mild intestinal disease. CNS involvement is rare. In an autopsy series from Mexico, Lombardo *et al.* (1964) documented amebiasis in 210 (5.8%) cases. Cerebral involvement was seen in 17 (8.1%) cases. It is far higher than the reported incidence of 0.6%–4.7% in other series. Banerjee *et al.* (1983) from India reported four cases of cerebral amebiasis among 80 cases of hepatic amobiasis seen over 18 years. It carried a poor prognosis in the past. With early diagnosis and therapy, there are many cases of cure on the record (Ohnishi *et al.*, 1994).

### Principles of Therapy

The ancillary therapy includes treatment for seizures and raised intracranial tension. The anticonvulsants effective for partial seizures with secondary generalization are recommended. Patients may need mannitol or

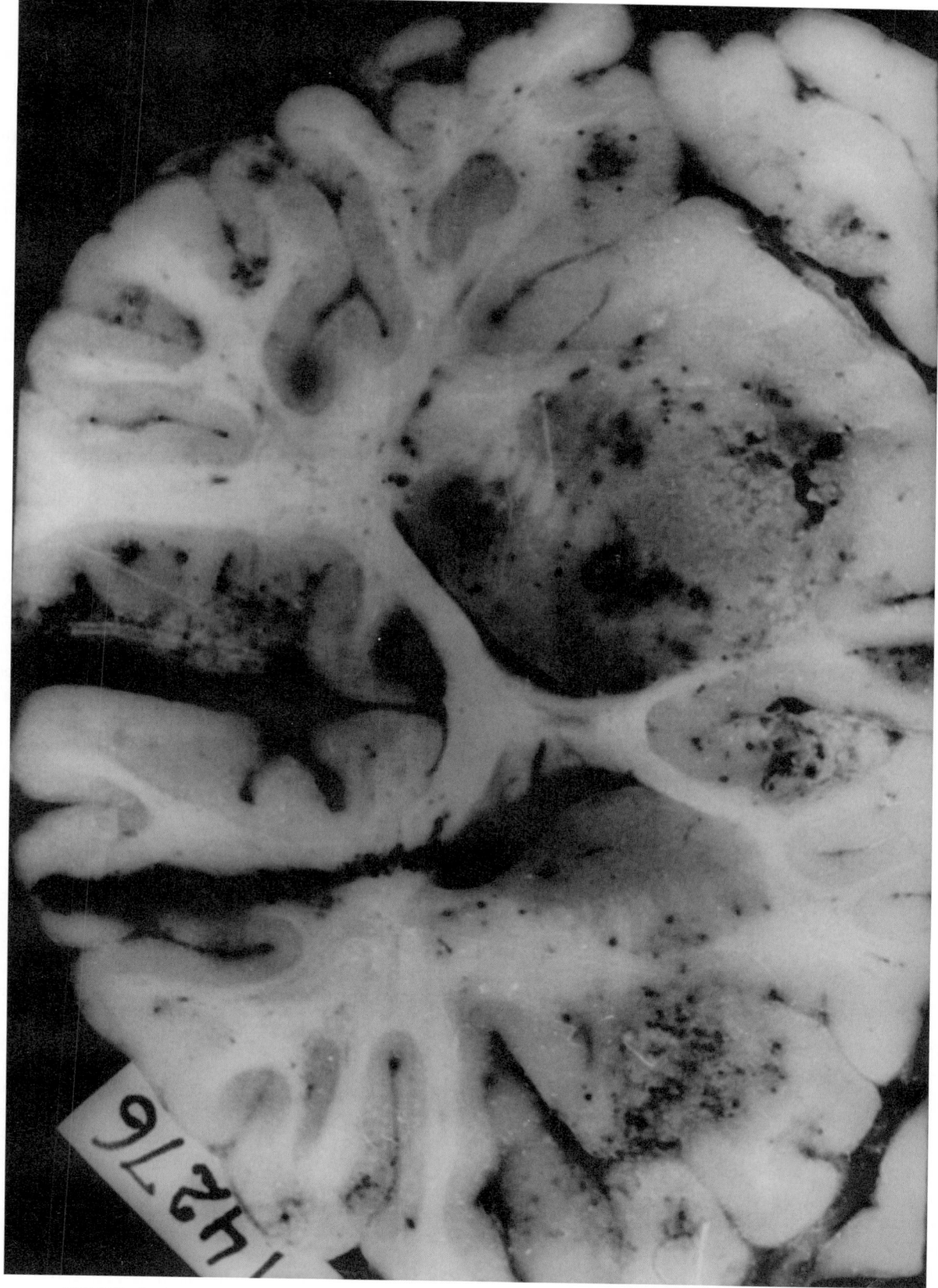

**FIGURE 4** Autopsy specimen of brain from a 3-year-old boy showing multiple necrotic lesions caused by secondary *Entamoeba histolytica* infection. The child also had colonic, hepatic, and pulmonary amebiasis. (Courtesy Dr. A. K. Banerjee, Professor Histopathology, Postgraduate Institute of Medical Education and Research, Chandigarh (India).

dexamethasone for raised intracranial tension. Most of the patients need drainage of cerebral and hepatic abscesses.

The specific medication for amebic cerebral abscess is the same as for hepatic abscess. The following drugs are commonly used:

**Metronidazole.** Metronidazole is 5-nitroimidazole, which kills the trophozoites of *E. histolytica*. It is one of the most effective drugs for invasive amebiasis. Both oral and intravenous preparations are available. It is distributed rapidly throughout the body tissues, reaching high concentrations in the body fluids, including CSF. The unwanted side effects include bitter or metallic taste in the mouth, minor gastrointestinal disturbances, headache, and dizziness. It produces a disulfiram-like reaction, and alcohol should be strictly avoided.

**Tinidazole.** Tinidazole is a nitroimidazole still not approved in some of the Western countries. It seems be more effective and better tolerated than mitronidazole.

### Practical Management

- Metronidazole is given at 750 mg orally or intravenously tid for 5–10 days.
- Most patients have associated liver abscess. Both cerebral and hepatic abscesses need surgical drainage if large.
- Use iodoquinolol, 650 mg tid for, 20 days to eradicate intestinal infection.

### Treatments No Longer Recommended

Emetine hydrochloride and chloroquin have been used in the past in the treatment of amebic abscess. These drugs should be avoided because of cardiovascular side effects.

## TRYPANOSOMIASIS

Trypanosomiasis is caused by protozoa of family Trypanosomatidae and is endemic in parts of Africa and Latin America. Although neurological involvement is seen in trypanosomiasis prevalent in both subcontinents, the incidence and clinical presentation differs considerably. It is described as its continent of origin—human African trypanosomiasis (sleeping sickness) or American trypanosomiasis (Chaga's disease). Human African trypanosomiasis (HAT) has significant neurological manifestations, whereas chronic Chaga's disease presents with cardiac and gastrointestinal lesions.

### Human African Trypanosomiasis

#### Clinical Aspects

**Definition.** HAT is a demyelinating encephalitis produced by flagellate protozoa *T. gambiense* and *T. rhodesiense*, resulting in sleeping sickness. The disease is transmitted to man by bite of African tsetse fly belonging to genus *Glossina*.

**Clinical Features.** A chancre or trypanome develops at the site of inoculation within hours of the bite of the infected tsetse fly. It may be associated with fever and localized lymphadenopathy. A lymphosanguine stage develops as the protozoan desseminates throughout the body. The patient has fever, generalized lymphadenopathy, pruritus, and skin eruptions. It may coexist with minor neurological and endocrine disturbances.

The meningoencephalitic stage develops insidiously over a period of months to years. Cognitive abnormalities and sleep disturbance is seen in most patients. Behavioral disorders and psychosis are present in some patients. Hypersomnolence is common during the daytime, and nocturnal insomnia is seen in 39% of the cases (Bouteille and Dumas, 1999). Motor dysfunction is frequent. Loss of muscle tone, especially of the neck muscles, and ptosis is seen on awakening. Extrapyramidal rigidity and gait disturbances are seen as the disease progresses. Disturbances of consciousness and dementia dominate the terminal stage of the disease.

**Diagnosis.** Anemia and increased sedimentation rate and IgG and IgM levels in a patient from an endemic area with the clinical features of the disease should evoke further search for trepanosomes. CSF shows lymphocytic pleocytosis, raised proteins, and elevated IgM levels. Card agglutination trypanosomiasis test, indirect immunofluorescence antibody test, and ELISA have been used to detect antibodies to variable and invariable antigens. The presence of circulating specific antigens is the best indicator of recent or ongoing infection. A first-generation antigen detection test already exists using polyclonal antibodies, but sensitivity and specificity are low. Further tests using ELISA are being developed. Polymerase chain reaction (PCR) is a significant advance in the detection of low parasitemia in the CSF (Bromidge *et al.*, 1993). EEG findings are nonspecific. It may show mild alteration of background rhythm with bilateral synchronous bursts of delta waves, particularly in the frontal region (Tapie *et al.*, 1996). Neuroimaging shows a diffuse zone of hypodensity, indicating demyelination of the centrum semiovale and periventricular regions.

## Natural Course

The parasites are transmitted by the blood-sucking tsetse fly. The insects acquire the infection by sucking blood from the mammalian host. After many cycles of multiplication in the midgut of the vector, the parasites migrate to the salivary glands. Further transmission takes place, when they are inoculated during the subsequent blood meal. HAT occurs in 36 countries, where about 200 foci are numbered. Cameroon, Chad, Congo, Republic of Central Africa, Sudan, Uganda, Zaire, and Angola are well-known foci. More than 50 million people living 15 degrees north and south are at the risk of contracting the disease. About 25,000 new cases are reported each year, but it is an underestimate (Kuzoe, 1993). The disease has a slow progression over months and years. Demyelinating encephalitis is the terminal stage. It leads to dementia, coma, progressive cachexia, and death.

## Principles of Therapy

The currently used drugs for the treatment of HAT are divided into two groups, depending on their ability to cross the blood–brain barrier. Pentamidine and suramin are effective in the first stage of the disease only, because they do not penetrate the blood–brain barrier. Patients with CNS involvement should be treated with melarsolal, eflornithine, or nifurtimox, because they reach the CNS at effective levels.

Therapy should be chosen on the basis of infecting organism (*T. gambiense* or *T. rhodesiense*), presence or absence of CNS disease, adverse reactions, and drug resistance.

**Pentamidine.** Pentamidine is used in stage I HAT. *T. rhodesiense* is resistant to it in some of the endemic areas. It is rapidly taken up in the parasites by a high-affinity energy-dependent carrier and thought to interact with DNA. It is given 4 mg/kg daily intramuscularly or intravenously for 10 days. It is eliminated slowly, and fairly high concentrations of the drug persist in kidney, liver, and spleen for several months. Frequent immediate side effects include nausea, vomiting, tachycardia, and hypotension. These reactions are short lasting, and there is no need to stop the therapy. Nephrotoxicity, abnormal liver function tests, neutropenia, hypoglycemia, and sterile abscesses are the other side effects.

**Suramin.** Suramin is highly effective in stage I disease. It has a selective action against trypanosomal enzymes. It should be administered under close supervision because of the side effects. It is given 1 g by slow intravenous infusion on days 1, 3, 7, 14, and 21. One patient in 20,000 has a immediate, potentially fatal reaction, resulting in shock and seizures. A 100–200 mg intravenous test dose should be administered to detect hypersensitivity. Renal toxicity is the most important adverse effect. Transient proteinuria often appears during treatment. Urinalysis should be carried out before each dose, treatment should be discontinued if porteinuria increases or casts and red cells appear in the sediment.

**Eflorinthine.** Eflorinthine is highly effective for both stages of *T. gambiense*. Its efficacy in *T. rhodesiense* infection is not determined. The recommended dose is 400 mg/kg/day intravenously in four divided doses for 2 weeks, followed by 300 mg/day orally for 3–4 weeks. Anemia, diarrhea, thrombocytopenia, seizures, and hearing loss are common side effects.

**Melarsoprol.** Melarsoprol is effective in both stages of the disease. However, it is not used as a first-line drug for stage I disease because of its high toxicity. In debilitated patients, suramin is administered for 2–4 days before starting this therapy. It is highly toxic and should be given with great care under close observation. The incidence of reactive encephalopathy is as high as 18% in some of the series. It manifests with high fever, headache, tremor, impaired speech, and drowsiness. Severe cases may progress to coma and death. Burri *et al.* (2000) have compared a 26-day standard Angolan schedule of three series of four daily injections of melarsoprol at doses increasing from 1.2–3.6 mg/kg within each series with a 7-day interval between series with the new treatment schedule. The new schedule was composed of 10 daily injections of 2.2 mg/kg. The outcome and serious side effects were comparable in both the groups. However, the skin reactions were more common with the new treatment schedule.

## Practical Treatment

- The arsenical melarsoprol is the drug of choice for treatment for HAT with CNS involvement. It is highly effective in East African trypanosomiasis.
- The drug should be given to adults in three courses of 3 days each.
- The dosage is 2–3.6 mg/kg/day intravenously in three divided doses for 3 days followed a week later by 3.6 mg/kg/day also in three divided doses and for 3 days. The latter course is repeated 10–21 days later.
- The total dose of melarsoprol should not exceed 1620 mg, and a single intravenous dose should not exceed 200 mg.

# NEUROFILARIASIS

Filariasis is a nematodal infection prevalent in tropical countries. Of the seven filarial nematodes infecting humans, four are highly pathogenic. *Onchocerca volvulus* is mainly responsible for neurofilariasis. *Loa loa* causes edematous skin swellings and neurological sequelae. *Wuchereria bancrofti* is known for obstruction of the lymphatic systems, and a few cases of CNS involvement are on the record. *Dracunculus medinensis* or Guinea worm causes severe skin manifestations but no neurological disease.

## Clinical Aspects

### Definition

Neurofilariasis is a filarial infection of the nervous system predominantly caused by *Onchocerca volvulus* and *Loa loa* resulting in blindness and epilepsy.

### Clinical Manifestations

Musculoskeletal pain is the first manifestation of onchocerciasis in the tropics. Patients may have crippling backache described as parasitic rheumatism. Muscle weakness, failure to grow, and weight loss have been described. Ocular lesions are by far the most serious consequence of *Onchocerca* infection, frequently resulting in visual impairment and blindness. Jilek-Aall (1999) has documented that that microfilariae of O. *volvulus* are responsible for epilepsy in a large population in Tanzania. Druet-Cabanac (1999) carried out a case-control study in the Central African Republic and found a positive correlation of onchocerciasis with epilepsy. A number of these patients have poor vision, edema of face and hands, dwarfism, ataxia, and psychomotor retardation. Encephalitis presenting with headache, somnolence, neck stiffness, finally deepening coma and death caused by pulmonary edema has been reported with *Loa loa* infection. Hypereosinophilic syndrome associated with encephalopathy, peripheral neuropathies, myocarditis, and thromboembolic lesions in the nervous system has been described (Sehadri *et al.*, 1991).

### Diagnosis

The clinical features associated with skin lesions and eosinophilia are suggestive of filarial infection. Monoclonal antibodies and DNA probes have been developed for detecting filarial antigens (Chandrashekar, 1997). Definitive diagnosis of onchocerciasis is made by excision of skin nodule or more commonly demonstration of microfilariae in a skin snip. Skin snips are made by lifting the skin with the tip of needle and excision of a 1–3 mm piece of skin. It is incubated for 2–4 hours in tissue culture medium or saline. Microfilariae can be visualized under a low-power microscope.

## Natural Course

Humans are infected with filarial larvae through an insect bite. The larvae transform into adults in 1–2 years. The female filaria periodically sheds millions of microfilariae. Microfilariae are responsible for the neurological manifestations. Direct contact between adult filaria and the brain has rarely been reported. The disease takes a subacute and chronic course. Acute exacerbations of allergic reaction often caused by death of microfilariae are well known. Ottenson (1993) estimates that more than 100 million people worldwide are infected with one of the filarial parasites. Onchocerciasis is common in West Africa. The Taraba River Valley in Nigeria has one of the densest foci in the world. *Loa loa* filariasis is common in Zaire.

## Principles of Therapy

The therapy is directed toward prevention of development of irreversible lesions and the treatment of symptoms. Surgical excision is recommended if the skin lesions are located on the head, it is because of the proximity of the microfilaria producing adult worms to eye. Chemotherapy remains the mainstay of treatment.

### Diethylcarbamazine (DEC)

Diethylcarbamazine is recommended for lymphatic filariasis and is also effective in the treatment of loiasis. It has been used in the past for onchocerciasis. It should be used under medical supervision, because it induces an inflammatory response (Mazzotti reaction) in up to 25% of patients. It may lead to encephalitis and death in a few cases. One of the most serious side effects is aggravation of existing eye disease.

### Suramin

Suramin is a potent, but potentially toxic, microfilarial agent. Mazzotti's reaction is common. It is highly nephrotoxic, and renal functions should be closely monitored during treatment. It is no longer a first-line drug in the treatment of neurofilariasis and is only recommended if total cure is necessary.

### Ivermectin

Ivermectin is a semisynthetic macrocyclic lactone active against microfilariae. Several large-scale trials have been conducted in heavily *O. volulus*–infested areas of West Africa with success (Jilek-Aall, 1999). The treatment leads to very mild reactions, consisting of myalgia, transient edema of face and extremities, pruritus, or papular rash in 1%–10% of cases. On the basis of collaborative research of eight countries with WHO, De Sole (1989) concluded that ivermectin is safe for large-scale treatment of onchocerciasis. However, because of possible side effects in patients with high microfilaria load, monitoring by resident nurses for 36 hours is recommended (Rothova *et al.*, 1989). Encephalopathy has been reported in cases of *Loa loa* with high microfilaria load treated with ivermectin (Boussinesq, 1998). It is the treatment of choice for onchocerciasis. Contraindications to treatment include pregnancy, breast feeding, any CNS disorder that may increase the CNS penetration and age less than 5 years.

### Albendazole

Albendazole may be useful in cases of loiasis resistant to DEC or where DEC cannot be used (Klion *et al.*, 1999).

## Practical Treatment

- Onchocerciasis should be treated with ivermectin. It is given orally as a single dose of 150 μg/kg either yearly or semiannually. The effect may last for 6 months.
- DEC may be used in loiasis. Heavy infections should be initially treated with a low dose of 0.5 mg/kg/day. The patient should be started on prednisolone, 40–60 mg/day. If there are no significant side effects from DEC, its dose is gradually increased to 8 mg/kg/day, and steroids are tapered off rapidly.

# LEISHMANIASIS

## Clinical Aspects

### Definition

Leishmaniasis is caused by the flagellates of the genus *Leishmania*. It is a parasitic disease transmitted by the bite of sand flies. Leishmaniasis is also known as "kala-azar," "black fever," or "black sickness." This disease commonly manifests either in cutaneous (skin) form or in visceral form, in which some of the internal organs are involved. Muscosal involvement may also often be seen along with the cutaneous form. It is a vector-borne zoonosis, and common reservoir hosts are cannines, small mammals, and rodents. Humans are incidental hosts. After infection, amastigotes disseminate to the reticuloendothelial system mostly involving the mononuclear phagocyte system in the dermis, nasal, oral, and pharyngeal areas. There is polyclonal humoral stimulation that results in hyperglobulinemia but with inadequate cell-mediated response and defective T cells to activate macrophages to destroy the parasite. The four most important species for humans are *L. donovani*, *L. tropica*, *L. mexicana*, and *L. brazilliensis*. Visceral leishmaniasis is typically caused by *L. donovani* but not always exclusively.

### Clinical Features

**Visceral Leishmaniasis.** Visceral leishmaniasis may present as an acute, subacute, or chronic course. In most it becomes symptomatic, but in a few it may remain subclinical. The main manifestations are fever and enlargement of the spleen out of proportion to enlargement of the liver, which can be massive in some patients. The parasites live in the liver, spleen, and bone marrow and reproduce rapidly. In addition to fever, the patient feels tired, the skin turns grey in "kala-azar," which is an Indian word for black fever. If untreated, the patient becomes malnourished and cachetic. Reticuloendothelial cell hyperplasia occurs in the liver and spleen, which may gradually cause complete destruction of these organs, and the patient may die in 6 months to 2 years (Herwaldt, 1998; Pearson and de Queiroz, 1996).

In geographical areas in which both HIV infection and leishmaniasis exist, visceral leishmaniasis is becoming an important opportunistic infection among persons with HIV infection. The patients are anemic and may have pancytopenia.

The cutaneous leishmaniasis is also called "oriental sore," "bouton d' orient," "Allepo boil," or "Baghdad sore." It has also been classified as new-world cutaneous leishmaniasis (American) or old world.

The neurological manifestations are relatively rare in leishmaniasis (Chunge *et al.*, 1985). Peripheral neuropathy, motor, sensory, or cranial nerve palsy may occur in this disease (Hashim *et al.*, 1995; Karak *et al.*, 1998; Mustafa, 1965). Very rarely a Guillain-Barre–type of acute presentation has been observed. The reason for affection of peripheral nerves in leishmaniasis is debatable. Malnutrition and poor nutrition may be responsible for sensory and, motor neuropathy; however, the acute Guillain-Barre–type of manifestation could also immunologically related. Occlusion of the vasa nervosum is not reported to explain isolated cranial nerve palsy, which leaves the autoimmune process for involvement of an isolated cranial nerve. Leishmaniasis as opportunistic affection in AIDS virus HIV may be an additional factor in the presentation of peripheral neuropathy in AIDS. The immune system is weakened in HIV infection, which results in explosive reproduction of leishmaniasis parasites. It is expected that leishmaniasis patients will increase with the expansion of AIDS, as has recently been noticed in Spain (Nothdurft, 2000). Footdrop, hearing loss, and burning feet are other neurological manifestations.

### Differential Diagnosis

The ulcerated form of leishmaniasis needs to be differentiated from the tropical ulcer, cutaneous diphtheria, mycobacterium ulcerans, amebic ulcer, tuberculosis, sarcoidosis, lepromatous leprosy, yaws, and syphilis. However, leishmaniasis can be confirmed by identification of the organisms by Giemsa smear on split-skin biopsy.

For visceral leishmaniasis the differential diagnosis is wide, and this includes typhoid fever, brucellosis, lymphoma, milliary tuberculosis, chronic malaria with tropical splenomegaly syndrome, hepatic schistosomiasis, and even leukemia with hepatosplenomegaly. The nonspecific features that may support visceral leishmaniasis are normocytic normochronic anemia, leukopenia, and thrombocytopenia. A raised erythrocyte sedimentation rate, 1gG, and total globulins may also help in the diagnosis. However, confirmation of the diagnosis can be made by the demonstration of amastigotes (Leishman-Donovan bodies) from lymph nodes, bone marrow or splenic aspirate (almost 60%–98% confirmation of diagnosis, respectively), and liver biopsy (60%).

Peripheral neuropathy caused by leishmaniasis in endemic areas needs to be differentiated from leprosy, nutritional neuropathy, alcoholic neuropathy, toxic neuropathy, peripheral neuropathy in AIDS, rarely peripheral neuropathy seen in chronic malaria, and cirrhosis. Appropriate investigations such as nerve biopsy and serological tests would help in the diagnosis.

### Natural Course

Leishmaniasis occurs in the tropics of America, African countries north of equator, some countries of Asia, and southern Europe. It is endemic both from tropics and subtropics and has been encountered from deserts to rain forests. Visceral leishmaniasis has been reported in 47 countries (Figure 5). It is endemic in Eastern India and Southern Sudan. All age groups are vulnerable. The population affected is in the rural and periurban areas. It has been estimated that more than 12 million people are affected with this disease in the world. There is no geographical limit for this disease, and leishmaniasis is on the rise in the temperate areas of southern Europe. Many people in Spain have this disease, and even several hundred patients annually are registered in Switzerland. The parasites usually lie in the bone marrow for very long periods and when the immune system becomes weak, the parasites reproduce and affect other organs. Visceral leishmaniasis is the most dangerous, and if it remains untreated, it is fatal.

Leishmaniasis infects a large number of vertebrates including man. The life cycle involves a vertebrate host (e.g., man) and a sand fly as a vector, which transmits parasites among the vertebrate hosts. The parasites in the vector are called promastigotes and reproduce asexually in the gut. The promastigotes are injected into the vertebrate host when the vector bites it. The promastigotes change to amastigotes after entering the host's cells. The amastigotes reproduce in the host's cell, and on the death of the cell the amastigotes are released and infect the other cells. It is the killing of host's cells that generates symptoms and the pathological condition associated with leishmaniasis. If the amastigotes collect at the site of the sand fly bite and do no spread, the result is cutaneous leishmaniasis, which often heals without any treatment. When the amastigotes spread to other organs, it causes visceral leishmaniasis, and if they spread to mucous membranes, then mucocutaneous leishmaniasis evolves (Nandy et al., 1997). If ingested by feeding sand flies, amastigotes are transformed back into promastigotes in about 7 days to be ready for reinfection.

The disease occurs in both sexes; however, is seen more often in men, who, because of their outdoor activities, are more prone to bites from sand flies.

### Principles of Therapy

Parasites not only can be transmitted in humans by the sand fly bite but also parenterally through blood transfusions or through contaminated needles. Congenital transmission is rare. Macrophages in the skin at the site of the bite are first infected, and dissemination occurs throughout the reticuloendothelial system. Highly effective therapy is essential, because visceral leishmaniasis is a fatal disease if it remains untreated.

The principles of therapy are eradication of the parasites from various organs (in visceral leishmaniasis) and from dermal and mucosal lesions. Pentavalent antimonial compounds are the most important agents as the first line of therapy in leishmaniasis both for eradication of the parasites and to hold further reproduction. Within the first few days of this therapy the patient

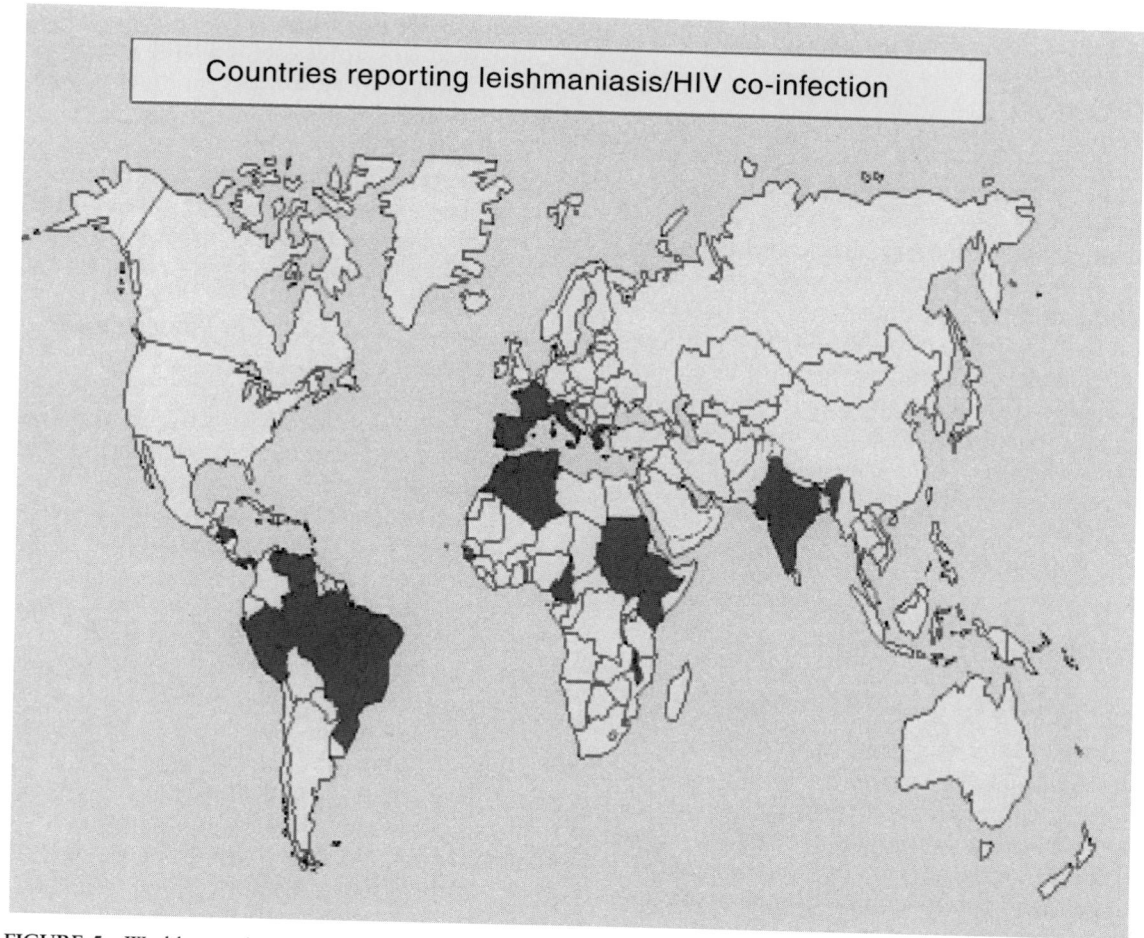

**FIGURE 5** World map showing endemic areas of leishmaniasis (with permission from the World Health Organization.)

becomes afebrile; however, organomegalies such as splenomegaly and hepatomegaly may recede to normal after weeks or months.

A permanent cure from this disease may not be possible in all patients; however, nonrecurrence of the disease 6 months after medication is considered a good sign of cure. A relapse may still be possible in such patients. A search for a concomitant HIV infection should be made in those with poor response to therapy or frequent relapses. Drugs resistant to antimonial compounds have been reported from India and Sudan, which may be as high as 40%. Alternative therapies are recommended in such patients. The scientists have not been able to evolve any vaccine for the prevention of leishmaniasis, even though the search has been going on for more than three decades. The success of any vaccine depends on the surface structure in the membrane. The leishmania pathogens are able to change their surface structure in the membrane any time and therefore eliminate the hope for a useful vaccine. Antimonial agents are very toxic, and it has been estimated that up to 15%

of patients die from the injections of this drug. Therefore epidemics of leishmaniasis require a reliable and effective drug that should also be inexpensive and could be given orally.

Only recently, miltefosine, has been given a trial in the Kala-Azar Research Centre in Brahmpura and Kala Azar Medical Research Centre, Banaras Hindu University in Varanasi, both centers are located in India. Oral treatment with miltefosine has been found to be successful in 98% of patients (Herwaldt, 1999; Jha *et al.*, 1999). Miltefosine is perhaps the only drug that can be given orally for treatment of leishmaniasis. Initially, the drug alleviates the pain and reduces the fever and generalized weakness, it cures the disease in about 4 weeks. The active ingredient in this drug is hexadecylphosphocholine, and it was discovered in Goettingen by Prof. Hansjorg Eibl from the Max Planck Institute of Biophysical Chemistry and Prof. Clemens Unger, then working at the University Hospital and presently at the clinic for tumor biology at the Albert Ludwig University in Freiburg, Germany.

## Diagnosis

Leishmaniasis can be confirmed by identification of the organism by Giemsa smear. The specimen is obtained from the infected site, which may be a histological section apart from this smear. The presumptive amastiogotes can be identified as a nucleus and rod-shaped kinetoplast under oil immersion by light microscopy. Splenic aspiration is the most sensitive and reliable method of diagnosis almost to 98%. Culture of the specimen in Novy-Mac Neal-Nicolle medium is also helpful in the diagnosis. Antibodies can also be detected by ELISA, countercurrent immunoelectrophoresis, and by direct agglutination tests. PCR can also be helpful in arriving at the diagnosis. Isoenzyme analysis of cultured promastigotes helps in identification of species, and these can also be identified by DNA probes or by use of monoclonal antibodies.

## Practical Management

### Visceral Leishmaniasis

Parenteral therapy is necessary, because kala-azar is a fatal disease if it remains untreated. Parenteral therapy is necessary since no oral preparation of antimonial agent is available. Pentavalent antimonial (Sb′) agent has been the drug of choice for decades; sodium stibogluconate is the active ingredient. It is given in a dosage of 20 mg of sb′/kg of body weight intravenously or intramuscularly daily for a period of 4 weeks. The response to therapy is usually seen in about 7 days, when the patient becomes afebrile and the general condition improves. Prolonged administration with sodium stibogluconate may be necessary when there is drug resistance or there is coexisting HIV infection. A combination of drugs may also become necessary under such circumstances. Single-dose therapy has recently been tried, but larger series may be required to prove its usefulness (Pirisi, 2000).

**Adverse Reactions.** Peripheral neuropathy is a side effect of sodium stibogluconate therapy. Other side effects are body and muscle pains, arthralgia, chemical pancreatitis, cardiac rhythm abnormalities, and hepatonephrotoxicity (Laguna del et al., 1994; Rai et al., 1994). These are reversible side effects that may not necessitate a stoppage of therapy.

**Alternative Therapies.** Amphotericin B is considered a potent alternative therapy, but it is potentially more toxic. It is given in a total dosage of 10–20 mg/kg body weight in divided doses of 0.5–1 mg/kg body weight intravenously daily, but some administer it on alternate days to avoid early adverse reactions such as nephrotoxicity, blood dyscrasias, hypotension, general malaise, and electrocardiographic changes mostly caused by hypokalemia. Liposomal amphotericin B is considered less toxic but equally effective, especially over the organs that are rich in macrophages and generate less nephrotoxicity.

Pentamidine is another alternative drug to be given alone or as an adjunct to other drugs. It is administered intravenously or intramuscularly in a daily or alternate-day regimen for a total of 15 days. The usual dosage is 2–4 mg/kg body weight. This drug causes pain at the injection site and may cause hypoglycemia, hypotension, nausea, and vomiting. Prolonged courses may be necessary with this drug.

Paromoxine or its equivalent aminoglycoside aminosidine is also considered effective in combination with an antimonial compound for treatment of leishmaniasis, especially in those resistant to an antimony agent. The aminosidine can be given intravenously or intramuscularly in a daily dosage of 10–15 mg/kg body weight for 2–4 weeks.

Miltefosine will perhaps be the future drug of choice that can be given orally. This drug is still in WHO trials and not available commercially; 98% success has been achieved (Jha et al., 1999) with the oral administration of 100 mg of miltefosine daily, whereas in the trials 50 mg, 100 mg, and 150 mg (0.8–2.5 mg/kg/day) has been tried. It is very likely that this new drug will be able to eradicate kala-azar in the near future.

Recently it has been observed that macrophages can be activated with cytokine immunotherapy with subcutaneous injections of recombinant interferon γ or gramlocyte macrophage colony stimulating factor as an adjunct with other therapies.

### Cutaneous Leishmaniasis

No treatment is usually needed; however, for large and multiple lesions or cosmetically unsightly lesions, local therapy with paromomycine or sodium stibogluconate injections may become necessary. The mucosal dissemination of cutaneous lesions also requires medical therapy. Pentavalent antimonial therapy with stibogluconate at a dosage of 20 mg/kg body weight either intravenously or intramuscularly daily for 3 weeks would show a clinical response. Depending on the clinical response, shorter courses may be necessary. The response to therapy is the flattening and later disappearance of skin lesions. Relapse is also the possibility.

Pentamidine intramuscularly at a dosage of 3 mg/kg body weight, a maximum of four doses given on alternate days, is considered effective as an alternate to the antimonial compound. Currently available oral agents for dermal leishmaniasis are imidazole, ketocanazole, and intraconazole, which are administrated for 4 weeks at an adult dosage of 600 mg daily and 200 mg twice daily, respectively. These are also considered effective.

Cryotherapy and heat therapy are alternative measures, which can be used for dermal leishmaniasis.

Sodium stibogluconate, 20 mg/kg body weight given daily intravenously or intramuscularly for 4 weeks, is the treatment of choice for mucosal leishmaniasis. Amphotericin B is the next alternative available for mucosal leishmaniasis.

**Prognosis.** Neurologically, sensory loss and hearing loss improve with medical therapy in about 2 weeks, but recovery of the motor component in peripheral neuropathy is slower. The burning sensation in the feet also improves in about 2 weeks. Because neurological complications are part of visceral leishmaniasis, the response to therapy of neurological complication go along with the visceral regression of the disease.

# REFERENCES

Ahuja, G. K., Roy, S., Kamla, J. G., and Virmani, V. (1983). Cerebral cysticercosis. *J. Neurol. Sci.* 35, 365–374.

Banerjee, A. K., Bhatnagar, R. K., and Bhusnurmath S. R. (1983). Secondary cerebral amebiasis. *Trop. Geogr. Med.* 35, 333–336.

Bauer, L. A., and Blouin, R. A. (1982). Age and phenytoin kinetics in adult epileptics. *Clin. Pharmacol. Ther.* 31, 301–304.

Bia, F. J., and Barry, M. (1986). Parasitic infections of the central nervous system. *Neurology Clin.* 4, 171–205.

Bird, A. V. (1978). Schistosomiasis of nervous system. *In* "Handbook of clinical Neurology." (P. J. Vinken and G. W. Brown, Eds.), vol. 35, pp. 231–241. Amsterdam Elsevier/North Holland.

Boussinesq, M., Gardon, J., Gardon-Wendel, N. *et al.* (1998). Three probable cases of Loa loa encephalopathy following ivermectin treatment for onchocerciasis. *Am. J. Trop. Med. Hyg.* 58, 461–469.

Bouteille, B., and Dumas, M. (1999). Human African trypanosomiasis. *In* "Neurology in the Tropics." (J. S. Chopra and I. M. S. Sawhney, Eds.), Vol. I, pp. 324–336. B. I. Churchill Livingstone, New Delhi.

Brewer, T. G., Peggins, J. O., Grate, S. J. *et al.* (1994). Neurotoxicity in animals due to artether and artemether. *Trans. R. Soc. Trop. Med. Hyg.* 88(suppl-1), 533–536.

Bromidge, T., Gibson, W. C., Hudson, K. *et al.* (1993). Identification of Trypanosoma brucei gambiense by PCR amplification of varient surface glycoprotein genes. *Acta Trop.* 53, 107–119.

Burri, C., Nkunku, S., Merolle, A. *et al.* (2000). Efficacy of new, concise schedule for melarsoprol in treatment of sleeping sickness caused by Trypanosoma brucei gambiense: arandomised trial. *Lancet* 355(9213), 1419–1425.

Chandrashekar, R. (1997). Recent advances in diagnosis of filarial infections. *Indian J. Exp. Biol.* 35, 18–26.

Chandy, M. J., Rajshekhar, V., and Ghosh, S. (1991). Single small enhancing lesions in Indian patients with epilepsy, clinical, radiological and pathological consideration. *J. Neurol. Neurosurg. Psychiat.* 54, 702–705.

Chopra, J. S., and Sawhney, I. M. S. (2001). Parasite infections. *In* "Text Book of Neurology." (J. S. Chopra, G. Arjundas, and S. Prabhakar, Eds.), pp. 242–263. Churchill Livingstone, New Delhi.

Chopra, J. S., Kaur, U., and Mahajan, R. C. (1981). Cysticercosis and Epilepsy; a clinical and serologic study. *Trans. R. Soc. Trop. Med. Hyg.* 75, 518–520.

Chopra, J. S., Sawhney, I. M. S., Suresh, N., Prabhakar, S., Dhand, U. K., and Suri, S. (1992). Vanishing CT lesions in epilepsy. *J. Neurol. Sci.* 107, 40–49.

Chunge, C. M., Gachihi, G., and Muigai, R. (1985). Is neurological involvement possible in visceral Leishmaniasis. *Trans. R. Soc. Trop. Med. Hyg.* 79(6), 872.

Crawley, J., Waruiru, C., Mithwani, S. *et al.* (2000). Phenobarbitone halves the seizure frequency, but doubles the mortality in childhood cerebral malaria: results of a double blind, randomised controlled trial. *Lancet.* 26, 701–706.

De Sole, G., Remme, J., Awadzi, K. *et al.* (1989). Adverse reactions after large scale treatment of onchocerciasis with ivermectin: combined results from eight community trials. *Bull. WHO* 67(6), 707–719.

Del Brutto, O. H. (1992). Cysticercosis and cerebrovascular disease: a review. *J. Neurol. Neurosurg. Psychiat.* 55, 252–254.

Del Brutto, O. H., and Sotelo, J. (1990). Albendazole therapy for subarachnoid and ventricular cysticercosis. *J. Neurosurg.* 72, 816–817.

Del Brutto, O. H., Garcia, E., Talamas, O., and Sotelo, J. (1988). Sex related severity of inflammation in parenchymal brain cysticercosis. *Arch. Intern. Med.* 42, 544–546.

Del Brutto, O. H., Santibanez, R., Noboa, C. A., Aguirre, R., Diaz, E., and Alarcon, T. A. (1992). Epilepsy due to neurocysticercosis: analysis of 203 patients. *Neurology* 42, 389–392.

Dhamija, R. M., Banerjee, A. K., and Venkataraman, S. (1989). Cerebral malaria. *Progr. Clin. Neurosci.* 5, 315–328.

Di Perri, G., Olliaro, P., Ward, S. *et al.* (1999). Rapid absorption and clinical effectiveness of intra gastric mefloquinine in the treatment of cerebral malaria in African children. *J. Antimicrob. Chemother.* 44, 573–576.

Druet-Cabanac, M., Preux, P. M., Bouteille, B. *et al.* (1999). Onchocerciasis and epilepsy: a matched case control study in Central African Republic. *Am. J. Epidemiol.* 149, 565–570.

El-Gengaihy, M. E., Soliman, H. M., Mangoud, A. M. *et al.* (1991). Myopathic changes in bilharzial hepatosplenomegaly, clinical and histopathological studies. *Zagazing Med. Assoc.* 4(2), 125–143.

Esam, M., and El-Gengaihy, M. E. (1999). Schistosomiasis (Bilharziasis) *In* "Neurology in Tropics." (J. S. Chopra and I. M. S. Sawhney, Eds.), pp. 288–302. B. I. Churchill Livingstone, New Delhi.

Escobedo, F., Penagos, P., Rodriguez, J., and Sotelo, J. (1987). Albendazole therapy for neurocysticercosis. *Arch. Intern. Med.* 147, 738–741.

File, S. (1995). Interaction of schistosome eggs with vascular endothelium. *J. Parasitol.* 8(2), 234–238.

Fisser, A. (1989). Taenia solium cysticercosis: some mechanism of parasite survival in immunocompetent hosts. *Acta Leidensia.* 57, 259–263.

Fisser, A., Woodhouse, E., and Larralde, C. (1980). Human cysticosis: antigen-antibodies and nonresponders. *Clin. Exp. Immunol.* 39, 27–37.

Garcia, E., and Sotelo, J. (1991). A new compliment fixation test for the diagnosis of neurocysticercosis in cerebrospinal fluid. *J. Neurol.* 238, 379–392.

Ghaly, A. F., and El-Banhawy, A. (1973). Schistosomiasis of spinal cord. *J. Path.* 111, 57–60.

Gordeuk, V., Thuma, P., Brittenham, G. *et al.* (1992). Effect of iron chelation therapy on recovery from deep coma in children with cerebral malaria. *N. Engl. J. Med.* 327, 1473–1477.

Greenberg, J. O. (1995). "Neuroimaging." Mc Graw-Hill, New York.

Grogl, M., Estrada, J. J., McDonald, G., and Kuhn, R. E. (1985). Antigen-antibody analyses in neurocystcercosis. *J. Parasitol.* 71, 433–442.

Groll, E. W. (1982). Chemotherapy of human cysticercosis with praziquantel. *In* "Cysticercosis: Present State of Knowledge and

Perspectives." (A. Fisser, K. Williams, J. P. Laelette, and C. Larralde, Eds.), p. 207. Academic Press, London.

Haribgai, H. C., Bahigjee, A. I., Bill, P. L. A. *et al.* (1991). Spinal cord schistosomiasis: a clinical, laboratory and radiological study with a note on the therapeutic aspect. *Brain* 114, 709–726.

Hashim, F. A., Ahmed, A. E., el Hassan, M., el Mubarak, M. H., Yagi, H., Ibraham, E. N., and Ali, M. S. (1995). Neurologic changes in visceral leishmaniasis. *Am. J. Trop. Med. Hyg.* 52, 149–154.

Hayashi, M., Matsuda, H., Tormis, J. S., and Blas, B. T. (1984). Clinical study on cerebral schistosomiasis japonicum on leyte island Philippines: follow up study 6 years after treatment with antischistosomal drugs. *Southeast Asian J. Trop. Med. Public Health* 15, 498–501.

Herwaldt, B. L. (1998). Leishmaniasis. *In* "Harrison's Principles of Internal Medicine." (A. S. Fauci, E. Braunwald, K. J. Isselbacher, J. D. Wilson, J. B. Martin, D. L. Kasper, S. L. Houser, and D. L. Longo, Eds.), pp. 1189–1193. Mc Graw-Hill, New York.

Herwaldt, B. L. (1999). Miltefosine—The long awaited therapy for Visceral Leishmaniasis? *N. Engl. J. Med.* 341(24), 1840–1842.

Hien, T. T., Day, N. P. J., Phu, N. H. *et al.* (1996). A controlled trial of artemether or quinine in Vietnamese adults with severe falciparum malaria. *N. Eng. J. Med.* 335, 76–83.

Hoffman, S. L. (1996). Artemether in severe malaria: still too many deaths. *N. Eng. J. Med.* 335, 124–126.

Istre, G. R., Fontaine, R. E., Tarr, J., and Hopkings, R. S. (1984). Acute schistosomiasis among Americans rafting the Omo River, Ethiopia. *JAMA* 251(4), 508–510.

Jha, T. K., Sunder, S., Thakur, C. P., Bachmann, P., Karbwang, J., Fischer, C., Voss, A., and Berman, J. (1999). Miltefosine, an oral agent for treatment of Indian Visceral Leishmaniasis. *N. Engl. J. Med.* 341(24), 1795–1800.

Jilek-All, L. (1999). Neurofilariasis. *In* "Neurology in Tropics." (J. S. Chopra and I. M. S. Sawhney, Eds.), pp. 259–272. B. I. Churchill Livingstone, New Delhi.

Jung, H., Hurtado, M., Medina, M. T., Sanchez, M., and Sotelo, J. (1990). Dexamethasone increases plasma levels of albendazole. *J. Neurol.* 237, 279–280.

Karak, B., Garg, R. K., Misra, S., and Sharma, A. M. (1998). Neurological manifestations in a patient with Visceral Leishmaniasis. *Postgrad. Med. J.* 74, 423–425.

Khalil, H. H. (1975). Multiple measures of altered cerebral blood flow in portopulmonary and portosplenic schistosomiasis. Proceedings of the International Conference on Schistosomiasis, Cairo. 137–147.

Kidney, D. D., and Kim, S. H. (1998). CNS infections with free-living ambeas: Neuroimaging findings. *A.J.R.* 171, 809–812.

Kirchhoff, L. V., and Nash, T. E. (1984). A case of schistosomiasis japonicum: resolution of CAT scan detected cerebral abnormalities without specific therapy. *Am. J. Trop. Med. Hyg.* 33(6), 1155–1158.

Klion, A. D., Horton, J., and Nutman, T. B. (1999). Albendazole therapy for loiasis refractory to diethylcarbamazepine treatment. *Clin. Infect. Dis.* 29, 680–682.

Kramer, L. D., Locke, G. E., Byrd, S. E., and Daryalagi, J. (1989). Cerebral cysticercosis: documentation of natural history with CT. *Radiology* 171, 459–462.

Kruskal, B. A., Maths, L., and Teele, D. W. (1993). Neurocysticercosis in a child with no history of travel outside the continental United States. *Clin. Infect. Dis.* 16, 290–292.

Kutt, H., and McDowell, F. (1968). Management of epilepsy with disphenylhydanatoin sodium. Dosage regulation for problem patients. *JAMA.* 203, 969–972.

Kuzoe, F. A. S. (1993). Current situation of African Trypanosomiasis. *Acta Trop.* 54, 153–162.

Laguna del Estal, P., Calabrease, S., Zabala, J. A., and Martin, T. (1994). Neurological toxicity from pentavalent antimonials during the treatment of visceral leishmaniasis. *Med. Clin. (Barc).* 102(7), 276–277.

Larotski, L., and Davis, A. (1981). The schistosomiasis problem in the world: result of WHO questionnaire survey. *Bull. World Health Organ.* 59, 115–127.

Lombardo, L., Alonso, P., Arroyo, L. S. *et al.* (1964). Cererbral amoebiasis. *J. Neurosurg.* 21, 704–708.

Looareesuwan, S., Warrell, D. A., White, N. J. *et al.* (1983). Retinal hemorrhage, a common sign of prognostic significance in cerebral malaria. *Am. J. Trop. Med. Hyg.* 32, 911–915.

Looareesuwan, S., Wilaiaratana, P., Vannaphan, S. *et al.* (1998). Pentoxifylline as an ancillary treatment for severe falciparum malaria in Thailand. *Am. J. Trop. Med. Hyg.* 58, 348–353.

Lowichik, A., Rollins, N., Delgado, R. *et al.* (1995). Leptomyxid amebic meningoencephalitis mimicking brain stem glioma. *Am. J. Neuroradiol.* 16, 926–929.

Ma, P., Visvesvara, G. S., Martinez, A. J. *et al.* (1990). Naegleria and Acanthamoeba infections: review. *Rev. Infect. Dis.* 12, 490–513.

Mahajan, R. C., and Chopra, J. S. (1975). Cysticercosis amongst cases of epilepsy and intracranial space occupying lesions. Proceedings of the National Seminar on Epilepsy, Bangalore (India) 95–98.

Mahajan, R. C., Chopra, J. S., and Ganguly, N. K. (1982). Human cysticercosis and epilepsy; a serologic study. *In* "Cysticercosis; Present State of Knowledge and Perspectives." (A. Fisser, K. Willms, J. P. Laclete, C. Larralde *et al.*, Eds.), pp. 171–178. Academic Press, New York.

Mahmoud, A. A. (1977). Schistosomiasis current concepts. *N. Engl. J. Med.* 297, 1329–1331.

Marra, T. M. (1983). Recurrent lumbosacral and brachial plexopathy associated with schistosomiasis. *Arch. Neurol. (Chicago)* 40, 586–588.

Martinez, A. J. (1985). Free-living Amebas: Natural History, Prevention, Pathology and Treatment of Disease. CRC Press, Inc., Boca Raton, Florida.

Martinez, A. J., and Visvesvara, G. S. (1997). Free-living, amphizoic and opportunistic amebas. *Brain Path.* 7, 583–598.

McIntosh, H. M., and Olliaro, P. (1998). Treatment of severe malaria with artemisinin derivatives. A systematic review of randomised controlled trials. *Med. Trop. (Mars).* 58, 61–62.

Michel, P., Callies, P., Raharison, H., Guyon, P., Holvoet, L., and Genin, C. (1993). Epidemiologie de la cysticercose a Madagascar. *Bull. Soc. Path. Ex.* 86, 62–67.

Milenkovic, Z., Penev, G., Stojanovic, D., and Jovicic Antovic, P. (1982). Cysticercosis cerebri involving the lateral ventricle. *Surg. Neurol.* 18, 94–96.

Molyneux, M. E., Taylor, T. E., Wirima, J. J. *et al.* (1989). Clinical features and prognostic indicators in paediatric cerebral malaria: a study of 131 comatose Malawian children. *Q. J. Med.* 71, 441–459.

Monteiro, L., Coelho, T., and Stocker, A. (1992). Neurocysticercosis. A review of 231 cases. *Infection* 20, 61–65.

Mustafa, D. (1965). Neurological disturbances in visceral leishmaniasis. *J. Trop. Med. Hyg.* 68(10), 240–250.

Nandy, A., Addy, M., Banerjee, D., Guha, S. K., Maji, A. K., and Saha, A. M. (1997). Laryngeal involvement during post kala-azar dermal leishmaniasis in India. *Trop. Med. Int. Health* 2(4), 371–373.

Newton, C. R., Crawley, J., Sowumni, A. *et al.* (1997). Intracranial hypertension in African children with cerebral malaria. *Arch. Dis. Child.* 76, 219–226.

Newton, C. R. J. C., Hien, T. T., and White, N. (2000). Cerebral malaria. *J. Neurol. Neurosurg. Psychiatry* 69, 433–441.

Nothdurft, C. (2000). Cure for fatal tropical disease-oral treatment of leishmaniasis. Press Release, Max Planck Institute for Biological Chemistry. Feb 18, 1–6.

Ohnishi, K., Murata, M., Kojima, H. *et al.* (1994). Brain abscess due to infection with Entamoeba histolytica. *Am. J. Trop. Med. Hyg.* **51**, 180–182.

Olveda, R. M., and Domingo, E. O. (1987). Schistosomiasis japonicum. *Bailleire's Clin. Trop. Med. Communicable Dis.* **2**(2), 397–417.

Ottenson, E. A. (1993). Filarial infections. *Infect. Dis. Clin. North Am.* **7**, 619–633.

Pearson, R. D., and De Queiroz Sausa, A. (1996). Clinical spectrum of leishmaniasis. *Clin. Infect. Dis.* **22**, 1–13.

Pirisi, A. (2000). Single-dose treatment shows effectiveness for Indian visceral leishmaniasis. *Lancet* **356**(9235), 1086.

Pittella, J. E. H. (1985). Vascular changes in cerebral schistosomiasis mansoni: A histopathological study of fifteen cases. *Am. J. Trop. Med. Hyg.* **34**(5), 898–902.

Rai, U. S., Kumar, H., Kumar, U., and Amitabh, V. (1994). Acute renal failure and 9th, 10th nerve palsy in a patient of kala-azar treated with stibogluconate. *J. Assoc. Physicians India* **42**(4), 338.

Rangel-Guerra, R., and Martinez, H. R. (1986). Diagnostico differential de la estenosis acueducto de Silvio en el adulto. *Rev. Invest. Clin. (Mex).* **38**, 21–27.

Robles, C., Sedano, A. M., Vargas, T. N., and Galindo, V. S. (1987). Long term results of praziquantel therapy in neurocysticercosis. *J. Neurosurg.* **66**, 359–363.

Rosas, N., Sotelo, J., and Nieto, D. (1986). ELISA in the diagnosis of neurocysticercosis. *Arch. Neurol.* **43**, 353–356.

Rothova, A., Vander Lelij, A., Stilma, J. S. *et al.* (1989). Side effects of ivermectin in treatment of onchocerciasis. *Lancet* **1**, 1439–1441.

Sawhney, B. B., Chopra, J. S., and Banerji, A. K. (1976). Psuedohypertrophic myopathy in cysticercosis. *Neurol. Minneap.* **26**, 270–272.

Schantz, P. M., Moore, A. C., Munoz, J. *et al.* (1992). Neurocysticercosis in orthodox Jewish Community in New York City. *N. Engl. J. Med.* **327**, 692–695.

Schumaker, D. J., Tien, R. D., and Lane, K. (1995). Neuroimaging findings in rare amebic infections of the central nervous system. *A.J.N.R.* **16**, 930–935.

Scrimgeour, E. M., and Gajdusek, D. C. (1985). Involvement of central nervous system in schistoma mansoni and S haematobium infection. A review. *Brain* **108**, 1023–1038.

Seshadri, S., Narula, J., and Chopra, P. (1991). Aymptomatic eosinophilic myocarditis: 2 + 2 = 4 or 5. *Int. J. Cardiol.* **31**, 348–349.

Sillanpaa, M. (1997). Specific antiepileptic medication. *In* "The Treatment of Epilepsy—Principles and Practice." (E. Wyllie, Ed.), pp. 808–823. Williams and Wilkins. Baltimore.

Slater, C. A., Sickel, J. Z., Visvesvara, G. S. *et al.* (1994). Successful treatment of disseminated Acanthamoeba infection in an immunocompromised patient. *N. Eng. J. Med.* **331**, 85–87.

Sotello, J., Penagos, P., Escobedo, F., and Del Brutto, O. H. (1988). Short course of albendazole therapy for neurocysticercosis. *Arch. Neurol.* **45**, 1130–1133.

Sotelo, J. (1988). Cysticercosis. *In* "Handbook of Clinical Neurology." (P. J. Vinken, W. Bruyn, and H. L. Klawans, Eds.). Vol. 52, pp. 529–534. Elsevier, Amsterdam.

Sotelo, J., and Del Brutoo, O. H. (1999). Neurocysticercosis. *In* "Neurology in Tropics." (J. S. Chopra and I. M. S. Sawhney, Eds.), pp. 227–243. B I Churchill Livingstone, New Delhi.

Sotelo, J., Del Brutto, O. H., Penagos, P. *et al.* (1990). Comparison of therapeutic regimen of anticysticercal drugs for parenchymal brain cysticercosis. *J. Neurol.* **237**, 69–72.

Suss, R. A., Maravilla, K. R., and Thompson, J. (1986). MR imaging of intracranial cysticercosis: comparison with CT and anatomopathologic features. *A.J.N.R.* **7**, 235–242.

Takayanagui, O. M., and Jardim, E. (1992). Therapy for neurocysticercosis: Comparison between albendazole and praziquantel. *Arch. Neurol.* **49**, 290–294.

Tapie, P., Buguet, A., Tabaraud, F. *et al.* (1996). Electroencephalographic and polygraphic features in sleeping sickness and healthy African subjects. *J. Clin. Neurophysiol.* **13**, 339–344.

Targett, G. A. T. (1995). Malaria vaccine—now and the future. *Trans. R. Soc. Trop. Med. Hyg.* **89**, 585–587.

Taylor, T. E., Wills, B. A., and Courval, J. M. (1998). Intrmuscular artemether vs intravenous quinine: an open, randomised trial in Malawian children with cerebral malaria. *Trop. Med. Int. Health* **3**, 3–8.

Tripathi, B. K., Agarwal, A. K., and Gupta, B. (1999). Artemisinin derivatives for falciparum malaria. *J. A. P. I.* **47**, 230–235.

Tsang, V. C. M., and Garcia, H. H. (1996). Immonublot diagnostic test (EITB) for *Taenia solium* cysticercosis and its contribution to definition of this under-recognized but serious public health problem. *In* "Teniasis/Cisticercosis por T. Solium." (H. H., Garcia and S. M., Martinez, Eds.), pp. 259–269. Ed Universo, Lima.

Tsang, V. C. W., and Wilkins, P. P. (1991). Immunodiagnosis of schistosomiasis screen with FAST-ELISA and confirm with immunoblot. *Clin. Lab. Med.* **11**, 1029–1039.

Vazquez, M. L., Jung, H., and Sotelo, J. (1987). Plasma levels of praziquantel decreases when dexamethasone is given simultaneously. *Neurology* **37**, 1561–1562.

Villagran, U. J., and Rabiela, O. J. E. (1988). Cisticercosis humana, studio clinico y patologico de 481 cases de autopsia. *Patologia (Mex).* **26**, 149–154.

Virmani, V., Roy, S., and Kamla, G. (1977). Periodic lateralized epileptiform discharges in a case of diffuse cysticercosis. *Neuropaediatrics* **8**, 196.

Visvesvara, G. S., and Stehr-Green, J. K. (1990). Epidemiology of free-living ameba infections. *J. Protozool.* **37**, 25S–33S.

W. H. O. (1990). Severe and complicated malaria. *Trans. R. Soc. Trop. Med. Hyg.* **84**(suppl-2), 1–65.

W. H. O. (1996). World malaria situation in 1993. Figs. 3–6.

Wadia, N., Desai, S., and Bhat, M. (1988). Disseminated cysticercosis. *Brain* **III** 597–614.

Waiz, A., Hossain, M. R., Chakraborty, B. *et al.* (1995). Triple drug therapy with quinine, cotrimoxazole and tetracycline in the management of cerebral malaria—A review of 254 cases. *Bangladesh Med. Res. Counc. Bull.* **21**, 77–80.

Wani, M. A., Banerji, A. K., Tandon, P. N., and Bhargava, S. (1981). Neurocysticercosis—some uncommon presentations. *Neurol. India.* **29**, 58–63.

Warrell, D. A. (1989). Cerebral malaria. *Q. J. M.* **7**, 369–371.

Warrell, D. A., Looareesuwan, S., Warrell, M. J. *et al.* (1982). Dexamethasone proves deleterious in cerebral malaria. A double-blind trial in 100 comatose patients. *N. Eng. J. Med.* **306**, 313–319.

Warrell, D. A., Molyneux, M. E., and Beales, P. F. (1990). Severe and complicated malaria. *Trans. R. Soc. Trop. Med. Hyg.* **84**(suppl-2), 1–65.

Waruiru, C., Newton, C. R., Forster, D. *et al.* (1996). Epileptic seizures and malaria in Kenyan children. *Trans. R. Soc. Trop. Med. Hyg.* **90**, 152–155.

Watt, G., Adapon, B., Long, G. W., Fernnando, M. T; Ranoa, C. P., and Cross, J. H. (1986). Praziquantel in treatment of cerebral schistosomiasis. *Lancet* **2**, 529–532.

White, N. J. (1996). The treatment of malaria. *N. Eng. J. Med.* **335**, 800–806.

WHO Expert Committee. (1993). Public Health Impact of schistosomiasis disease and mortality. *Bull. World Health Organ.* **71**, 657–662.

## CHAPTER 53
# Multiple Sclerosis

Roland Martin, Reinhard Hohlfeld, and H. F. McFarland

## INTRODUCTION

Multiple sclerosis (MS) is the most frequent demyelinating central nervous system (CNS) disease among Northern Europeans and Northern Americans. Although the etiology of MS is still unclear, it is now widely accepted that an autoimmune response directed at myelin antigens contributes to the pathogenesis. Findings supporting this hypothesis include the histopathology of the MS lesion, which reflects an inflammatory process consisting mainly of lymphocytes and macrophages, the association with certain HLA-class II antigens, and parallels to a well-described animal model for autoimmune demyelinating disease, experimental allergic encephalomyelitis (EAE). Pathoanatomically, MS is characterized by perivenular inflammatory lesions that are found throughout the white matter in the brain and spinal cord, are often periventricular, and show various degrees of demyelination but also axonal loss (Prineas, 1985; Trapp *et al.*, 1998). Because of the widespread location of lesions, almost any clinical sign of CNS damage may be observed. The clinical course is highly variable, but most often the disease begins with a relapsing remitting (RR-MS) course that later evolves into a secondary progressive (SP-MS) or progressive relapsing (PR-MS) one (Figure 1) (McFarlin and McFarland, 1982; Noseworthy *et al.*, 2000). Less commonly, the disease insiduously progresses from the onset (primary chronic progressive MS; PP-MS) (Figure 1). The variable and unpredictable course has made the assessment of therapeutic efficacy extremely difficult. This chapter will not only summarize the standard therapeutic regimens that have evolved in the last decades but also mention newer therapeutic strategies that aim at modulating the autoimmune response. The reader should note that regional differences exist with respect to therapeutic approaches. These differences may become more marked if approval for the use of a partic-ular therapy differs between countries, or if there are differences in availability of a particular drug. This chapter will provide a general overview of therapeutic approaches.

### Clinical Aspects

#### Clinical Course

MS is a disease of young adulthood and mostly manifests between the ages of 20 and 40 years. Only a few percent of patients have the disease develop before 15 and after 55 years of age. Females are affected 1.5–2 times more frequently than males. As mentioned previously the clinical course in MS is variable. Most often, MS begins as a relapsing remitting disease characterized by acute exacerbations or episodes of acute neurological dysfunction that improve, often completely, over a period of weeks without any treatment. The presenting symptoms and signs differ slightly between patients with onset at younger ages compared with those with later onset. In younger patients, RR-MS often starts mono-symptomatically with either optic neuritis (36%) or paresthesias (33%). More than half of previously healthy patients with optic neuritis later develop clinically definite MS. Recent studies have shown that the presence of MRI lesions (>3) at the time of optic neuritis increases the risk of later development of MS to greater than 50% (Tumani *et al.*, 1998), but the MRI abnormalities may also be viewed as reflecting existing, clinically silent disease. The presence of oligoclonal CSF Ig bands further increases the predictive value of a positive MRI. Pareses alone or in combination with sensory impairment are more often encountered in older patients (50%). New signs typically develop over several hours up to a few days and will often be reported to fluctuate at the beginning. Paroxysmal symptoms or

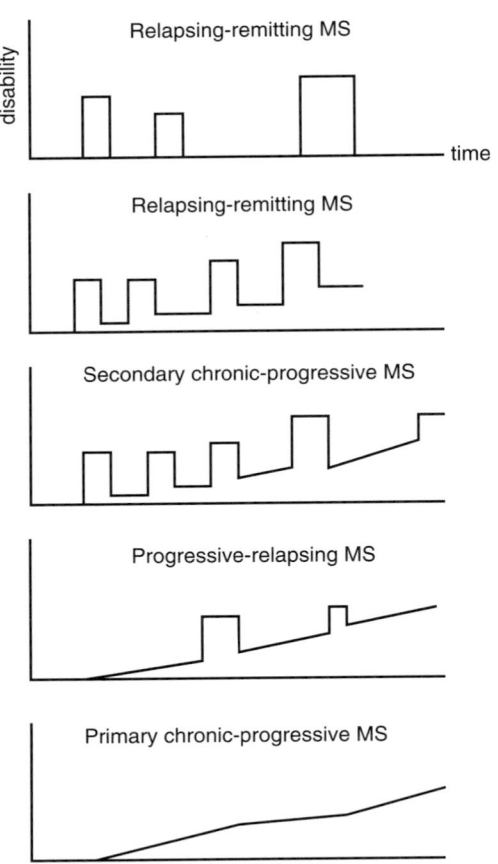

**FIGURE 1** Schematic diagram of the different clinical courses of multiple sclerosis.

signs such as neuralgic pain or dystonic movements that only last for seconds up to several minutes may occur but are not considered separate exacerbations. After an initial exacerbation, the development of new signs or reappearance of old symptoms that last more than 24 hours qualify as a relapse, unless elevated body temperature (e.g., during infection) accounts for transient worsening. Although some patients will continue to have a course characterized by relapses with complete recovery throughout their lives, most often, with time, the recovery from relapses becomes incomplete, and patients are left with increased disability after each attack. Most often exacerbations involve worsening of previous signs and symptoms, but in about 20% of cases new signs will develop. Short-lived worsening or reappearance of pre-existing symptoms, for example, during hot summer days or a hot bath, are due to impaired conduction in demyelinated axons (Uhthoff phenomenon). If MRI is used to monitor the inflammatory activity in MS longitudinally, up to 30-fold more bursts of new disease activity can be observed compared with the number of clinical exacerbations. After onset of RR MS in early adulthood, between 25–50% of the patients experience a first relapse within 1 year after the initial

clinical signs, up to 50% within 2 years, and two thirds within 3 years after onset. The relapse rates range between 0.2–1.3/year. Factors precipitating relapses include viral infections. An increased risk of exacerbation has also been described during the postpartum period (Confavreux *et al.*, 1998). Although bacterial infections do not seem to increase the risk of exacerbation, infections such as those of the urinary tract can cause significant symptomatic worsening and need to be considered when evaluating patients for clinical exacerbation. Various anecdotal reports have suggested a relationship between worsening or exacerbations and stress, trauma, and surgery. However, there is no convincing evidence for these associations.

### Prognosis

Most often, exacerbations do not persist for longer than 8 weeks and those that do often do not remit completely. The chances of recovery are, however, not only dependent on the duration of a relapse but also on the type of signs. Paresthesia, optic neuritis, and diplopia, at least in the initial phase of the disease, usually remit completely, whereas pareses, cerebellar signs, and autonomic dysfunction are associated with a poorer prognosis and tendency for incomplete remission. It is impossible to establish an accurate prognosis in individual MS patients, but some features seem to affect the general estimate for the future prognosis. In general, females, younger age of onset, and presentation with sensory symptoms or optic neuritis are associated with a better prognosis. In contrast, older onset, especially in males, and presentation with motor or cerebellar symptoms are associated with a poor prognosis. If the course of the disease over the initial few years is mild and restricted mainly to the optic nerve or sensory symptoms with complete recovery after each exacerbation, the long-term prognosis is better than for patients with early motor involvement or with evidence of progression early in the course of the disease. Substantial increases in CSF pleocytosis, intrathecal IgG synthesis, and CSF MBP may also point toward a poorer prognosis. The increasing use of MRI has added a new dimension to the prognostic evaluation of MS. Evidence of active disease on MRI and especially persistent activity on serial MRI studies point to a worse prognosis. Bursts of contrast-enhancing lesions correlate highly with disease exacerbations (Smith *et al.*, 1993). Brain parenchymal atrophy, marked reductions in magnetization transfer ratios, reduction of cord diameter at C2, and persistent T1 holes all indicate a poor long-term prognosis (McFarland *et al.*, 2002). The lesion load during the monosymptomatic stage indicates the risk for a second relapse during the following 2 years, as well as the extent of disability 2, 5, and 10 years after the initial manifestation (Morissey *et al.*, 1993). However, the correlations between individual MRI parameters and disease prognosis are often not strong and will need further investigation in long-term studies.

## Survival

MS is not considered a life-shortening disease. Studies of the effect of MS on survival generally indicate patients with MS have only a slightly shorter life span than healthy controls and that the mean survival after onset is 25 years or longer (Weinshenker, 1994). This represents a significant improvement from the first half of the century (Kesselring, 1992). Furthermore, MS does not invariably lead to severe disability. About one third or more of patients will not have permanent functional impairment of professional or nonprofessional activities, and about 50% of patients are disabled to the point to need a cane to walk half a block after 15 years of MS (Weinshenker, 1994). Malignant courses with progression over several months to a few years leading to death are rare. Life-threatening situations may occur when large lesions develop in the upper cervical segments or the brainstem. In severely impaired patients, immobility, dysfunction of the lower cranial nerves causing difficulties with swallowing or respiration, or severe bladder dysfunction may all contribute to mortality.

## Diagnosis

The diagnosis of MS is based primarily on clinical findings reflecting involvement of multiple sites in the CNS occurring at different times, involvement separated in both time and space. The diagnostic criteria proposed by Poser *et al.* (1983) are listed in Table I. There is no diagnostic test specific for MS, but findings from electrophysiological examinations of visual, acoustic, and somatosensory evoked potentials, cerebrospinal fluid (CSF) examination, and magnetic resonance imaging (MRI) can be used to support the diagnosis at a stage when the clinical findings are not sufficient to clinically diagnose definite MS. In patients who have clinical evidence of only a single lesion but a history of events separated in time or in patients with evidence for two lesions but a clinical history of only one event, CSF abnormalities can be used to establish the diagnosis of laboratory supported MS (Poser *et al.*, 1983). Similarly, evoked responses can be used to identify subclinical lesions establishing the existence of a second lesion in a patient with only one area of involvement on examination. Abnormalities on visual-evoked responses (VER) or somatosensory-evoked responses (SER) are most frequent and useful clinically. Finally, the evidence of multiple lesions can be aided by results from MRI, and

**TABLE I** Diagnostic Criteria for Relapsing Remitting Multiple Sclerosis*

Definite MS
  Postmortem histopathological diagnosis
Clinically definite MS
  Two attacks and clinical evidence of two separate lesions
  Two attacks; clinical evidence of one lesion, *and* paraclinical evidence of another lesion
Laboratory-supported definite MS
  Two attacks, clinical evidence of one lesion *or* paraclinical evidence of another, separate lesion, *and* CSF findings (oligoclonal bands/elevated IgG[†])
  One attack, clinical evidence of two separate lesions, *and* CSF findings (oligoclonal bands/elevated IgG)
  One attack, clinical evidence of one lesion, *and* paraclinical evidence of another, separate lesion, *and* CSF findings (oligoclonal bands/elevated IgG)
Clinically probable MS
  At least two relapses with monofocal clinical findings
  One exacerbation with multifocal clinical *or* monofocal clinical and additional mono- *or* multifocal neurophysiological or neuroradiological findings
Laboratory-supported, probable MS
  At least two relapses with monofocal clinical findings and elevated intrathecal Ig production with or without oligoclonal immunoglobulin bands
  One exacerbation with multifocal clinical or monofocal neurophysiological or neuroradiological findings and elevated intrathecal immunoglobulin production with or without oligoclonal bands
Clinically possible or questionable MS
  Clinical, neurophysiological, and neuroradiological findings with or without compatible CSF findings that are not sufficiently characteristic, but so far no other suspected disease

*Modified according to Poser *et al.* (1983).
[†]The oligoclonal bands are of much higher predictive value than the elevated IgG.

criteria relating to the number of abnormalities found on MRI and the likelihood of MS have been published (Barkhof *et al.*, 1997) (Tintore *et al.*, 2000). It is important to note that areas of increased signal on T2-weighted MRI are not diagnostic of MS, and when MRI findings are used to complement the diagnosis, the clinical findings and history of the patient must be consistent with MS; the diagnosis of MS continues to rest on the clinical picture; however, the role of MRI is steadily increasing. The newest diagnostic guidelines incorporate MRI parameters and allow us to diagnose MS, when a formal clinical diagnosis was not possible according to the preceding. Overall, these new diagnostic guidelines (see Table II) will simplify the diagnosis and help to guide early treatment decisions (McDonald *et al.*, 2001). The typical CSF findings are summarized in Table III.

Because of the psychological and social burden posed by the diagnosis of disease such as MS, the neurologist has to evaluate the evidence for the disease carefully, and a number of differential diagnoses must be considered (Table IV). Until recently, many neurologists have

**TABLE II   New Diagnostic Guidelines for Multiple Sclerosis**

**Multiple sclerosis** (dissemination of lesions in time and space; relying on clinical signs and MRI\*)
*Magnetic resonance imaging* (adapted from Barkhof *et al.* (1997) and Tintoré *et al.* (2000)):
  MRI criteria for dissemination in space
    One gadolinium enhancing lesion or nine T2 hyperintense lesions if no Gd-enhancing lesion
    At least one infratentorial lesion
    At least one juxtacortical lesion
    At least three periventricular lesions
    *Note*: Lesions will ordinarily be larger than 3 mm in diameter; one spinal cord lesion can substitute for one brain lesion
  MRI criteria for dissemination of lesions in time:
    If a first scan was 3 or more months after the onset of the clinical event, the presence of a Gd-enhancing lesion is sufficient to demonstrate
      dissemination in time; the lesion should not be implicated in the original clinical event. If there is no enhancing lesion, a follow-up scan
      is required (at >3 months). A new T2- or Gd-enhancing lesion at this time fulfills the criteria.
    If the first scan was performed less than 3 months after the onset of the clinical event, a second scan done 3 months or more after the
      clinical event showing a new Gd-enhancing lesion provides sufficient evidence for dissemination in time. If no enhancing lesion is
      present, a further scan at ≥3 months later showing a T2- or Gd-enhancing lesions is sufficient.

\*Modified from McDonald *et al.* (2001).
*Note*: the proposed new guidelines distinguish only the diagnosis of MS or possible MS. The latter applies if the outcome of the diagnostic evaluation is not conclusive according to the above criteria (i.e., the diagnosis is equivocal for those at risk of MS) and if other differential diagnoses have been excluded.

been reluctant to make the diagnosis early in the disease course. However, the results from a large optic neuritis study (Beck *et al.*, 1992, 1995)\*\*, as well as those from trials using interferon-beta during monosymptomatic stages (Controlled High-Risk Subjects Avonex Multiple Sclerosis Prevention Study, CHAMPS (Jacobs *et al.*, 2000)\*\* and Early Treatment Of Multiple Sclerosis (ETOMS)) indicate that the development of clinically definite MS can be delayed for more than 9 months (Comi *et al.*, 2001)\*\*. According to these data, MS should be treated as early as possible, perhaps already during monosymptomatic stages such as optic neuritis or transverse myelitis, when a firm clinical diagnosis is not possible. Larger MRI lesion load or the presence of enhancing lesions can be used as further aids to initiate treatment.

### Implications of MRI on the Natural History of MS

Until recently, MS was thought to be inactive in the early phase of the disease between exacerbations. Over the past several years, MRI studies have indicated that subclinical disease can occur even during periods of remission. Longitudinal studies have now shown that substantial disease activity as evidenced by active lesions seen on Gd-enhanced MRIs occurs in patients who are clinically stable, indicating that MS is an active disease even during periods of remission (Stone *et al.*, 1995). The occurrence of new or active lesions is not constant. Instead, lesions seem to occur in bursts of increased activity, and in some patients these bursts seem to occur with periodic regularity (Harris *et al.*, 1991). The non-constant frequency has important implications on the design of trials using MRI as an outcome measure. Although most MRI activity in the cerebrum is clinically silent, when clinical exacerbations do occur, they tend to be associated with bursts of MRI activity. This observation suggests that the likelihood of a lesion to occur in a clinically sensitive area of the CNS such as the spinal cord or brainstem is increased during periods of increased general disease activity in the cerebrum. When the extent or bulk of T2 abnormalities is determined using computer-assisted analysis, the area of abnormal T2 signal is also found to increase in the RR MS patients with evidence of disease on Gd-enhanced images. When several different MRI techniques and their findings are compared with clinical course and severity, the accumulation of T1 holes, the cord diameter at C2, the degree of brain parenchymal atrophy, and the reduction in the magnetization transfer ratio and the loss of N-acetyl aspartate by MR spectroscopy have the potential to serve as a better surrogate marker than Gd-enhancing lesions or T2 burden prognosis (McFarland *et al.*, 2002). Each of these parameters visualizes more specifically the characteristic tissue damage (e.g., T1-holes reflect demyelination, glial scarring, and axonal damage), and these findings provide strong

**TABLE III   Cerebrospinal Fluid Findings in Multiple Sclerosis\***

| | |
|---|---|
| Lymphomonocytic pleocytosis | Present in about 30%–70% of patients; usually <30 cells/μl |
| Total protein | Usually normal |
| Intrathecal IgG synthesis | Increased in >80% of patients |
| Oligoclonal immunoglobulin bands | Positive in up to 95% of patients |

**TABLE IV   Differential Diagnostic Aspects of Multiple Sclerosis**

Symptoms occurring at multiple times but attributable to a single focal lesion
1. Pathology in the midline, the brainstem, the base of the brain and skull, the atlanto-occipital border, the spinal cord (malignancies, granulomas, angiomas, meningoencephalitis, arachnopathy, vascular myelopathy, syringomyelia, skeletal malformations)

Symptoms attributable to multiple focal lesions of other causes
2. Infections and inflammatory diseases
   Para- and postinfectious, as well as postvaccinal encephalomyelitides (Lyme encephalomyelitis, syphilis, cysticercosis, echinococcosis, multiple abscesses, for example, caused by *Toxoplasma gondii*, HTLV-I–associated myelopathy/tropical spastic paraparesis [HAM/TSP], M. Behçet, systemic lupus, and other cerebral vasculitides, progressive multifocal leukoencephalopathy [PML]
3. Vascular diseases
   Vasculitides of various origins, multiple emboli, CADASIL
4. Neoplastic diseases
   Metastases, lymphoma, leukemia, histiocytosis
5. Hereditodegenerative diseases (e.g., mitochondrial encephalomyopathies, Leber's optical atrophy)
   Leukodystrophies, in particular adrenoleukodystrophy and adrenomyeloneuropathy, axonal dystrophies, cerebellar degenerations

Chronic intoxications
6. With bromium, barbiturates, diphenylhydantoin, subacute myelo-opticoneuropathy (SMON)

Vitamin deficiencies
7. Of vitamin $B_{12}$, vitamin E

Monofocal or monosymptomatic multiple sclerosis
8. Hemiplegic MS, purely spinal MS, optic neuritis without other symptoms, and signs or supportive laboratory/MRI findings

Symptoms that should make the diagnosis doubtful
9. Lack of ocular symptoms (no optic neuritis, no oculomotor deficits), lack of sensory symptoms, lack of autonomic symptoms (no bladder, bowel or sexual dysfunction), lack of multifocal symptoms/signs, lack of typical CSF findings, lack of delayed evoked potentials, lack of T2-weighted MRI lesions, lack of clinical remissions, in particular in young patients, significant involvement of the peripheral nervous system.

support for the use of MRI as a surrogate measure of disease activity in MS.

## Epidemiology and Genetics of MS

Incidence rates for MS range between 4–8 new cases per 100,000 in Northern European and North American countries. The prevalence in these areas ranges between 60–100/100,000. In general, prevalence rises with increasing latitudes on both hemispheres (south of 46° latitude in Europe and 38° latitude in North America the prevalence is 5–15/100,000; north of that it is 30–60/100,000), although exceptions are observed in certain areas and ethnic groups. Hungarian gypsies, for example, have a strikingly lower (2/100,000) prevalence than the general Hungarian population (30–50/100,000). Socioeconomic standard, hygiene conditions, nutrition, and environmental factors have been considered reasons for the differences in prevalence. In addition, the distribution of predisposing genes in certain populations and ethnic groups may contribute to this observation. Furthermore, the appearance of MS "epidemics" in formerly free populations like that followed up by Kurtzke on the Faroe islands suggest that environmental factors are probably involved in the induction of MS. The rise in prevalence rates in the last decades is most likely due to longer life expectancies and courses of the disease but also due to improved diagnostic tools, in particular MRI.

Considerable evidence supports an influence of genetic factors on disease susceptibility for MS, although susceptibility is clearly not determined by a single gene but rather a quantitative trait to which multiple genes contribute. A genetic influence is supported by population, family, and twin studies. The prevalence rates are about 20 times higher in first-degree relatives of MS patients. Among those, daughters of female MS patients carry the highest risk. Concordance rates for identical twins range between 25% and 35%. Population and family studies and whole genome screens have identified that genes of the HLA complex confer susceptibility for MS, particularly DRB1*1501, DRB5*0101, DQA1*0101, and DQB1*0602. In Northern European and Northern American Caucasians HLA-DRw15 Dw2 (formerly DR2 Dw2; gene designation see earlier) and DQw6 (formerly DQw1; gene designation see above), in Italian and Arab MS populations HLA-DR4, and in Mexican and Japanese MS populations HLA-DR6 have been found to be associated with disease. It is clear from these studies, however, that other genetic components are likely to contribute. T-cell receptor, myelin, cytokine, chemokine, immunoglobulin, and complement genes have been implicated, and certain restriction fragment length polymorphisms have been found overrepresented in MS patients in individual studies, but their contribution is still controversial. As far as the two major forms of MS are concerned, recent reports suggest that they might be genetically distinct entities.

## Assessment of Treatment Efficacy

Because of the variable and unpredictable course of MS, the assessment of therapeutic efficacy has been extremely difficult, and the reader is referred to an overview (Paty and McFarland, 1998). In evaluating therapeutic claims, the reader must remember to examine critically the clinical trials supporting the claims, and various aspects of clinical trials deserve mention, including the entry criteria, the outcome measures, whether the study was blinded and included a placebo group, and the statistical methods applied to the results. It is beyond the scope of this chapter to review trial design in detail. It is important to note, however, that assessment of clinical efficacy in MS has been difficult. Various outcome measures have been used, including exacerbation rate, impairment, disability, and quality of life. The Kurtzke scale or the Expanded Disability Status Scale (EDSS) and sometimes the Scripps Neurological Rating Scale (SNRS) as a measure of disability have been used in most recent studies and currently represent the standard. Recently, the Multiple Sclerosis Functional Composite (MSFC, consisting of timed 25-foot walk, 9-hole peg test, and paced serial auditory additions test) has been evaluated and may be used more often in the future. There is general agreement that problems exist with the EDSS. The recent evidence from MRI studies indicate that clinical disease does not fully reflect the level of disease activity (i.e., up to 30 times more increases in disease activity than clinical relapses are diagnosed by Gd-enhancing MRI), and that in chronic progressive patients considerable disability can exist in individuals with only minimal disease by MRI. These observations suggest that MS is heterogeneous (Lucchinetti et al., 2000) and that clinical outcome measures may often be a poor reflection of disease activity. Some therapies that will be examined later in this chapter have shown suggestive evidence of efficacy. Does this mean that the effect is slight in most patients, or does it mean that a few patients respond well to the treatment, but the trial design is inadequate to identify this effect? Furthermore, there is now considerable evidence that patients with mild RR-MS or even isolated clinical syndromes (e.g., optic neuritis), should be treated early (Comi et al., 2001; Jacobs et al., 2000). Although there is no question that elimination of disability is the goal of all treatments, the question as to how to assess whether a treatment works within a reasonable time span and using a reasonable sample size is currently being rethought. Increasing emphasis is being placed on MRI as a secondary outcome measure in most current trials in MS. Although there is debate as to the correlation between certain MRI parameters and clinical disease activity, MRI is evolving as a valu-able surrogate measure of disease activity in MS, and it is expected that further improvements such as the preceding MRI parameters will capture the heterogeneity and disease course even better in the near future (McFarland et al., 2002). The discrepant data from the studies of interferon-β-1b in secondary progressive MS in Europe (European Study Group et al., 1998**) and North America (results not reported yet) respectively (efficacious in the former, but not in the latter) indicate that inflammatory activity declines with progression of disease and that anti-inflammatory treatments are only effective during the inflammatory phase.

For practical purposes and outside of clinical studies, it is important to try to document objective measures of the patient's disability to capture disease progression as accurately as possible. Walking distances are simple measures and can be tested relatively easily. Recommendations as to which MRI parameters should be routinely documented, ideally before onset of treatment, 6 months to 1 year after treatment onset, and whenever treatment failure is suspected, are summarized in Table V.

## PRINCIPLES OF THERAPY

Although the cause of MS is unknown, considerable evidence including the pathology, the susceptibility genes, and the parallels between MS and EAE, have pointed to an autoimmune pathogenesis in MS. Because the immunological process producing disease has

---

TABLE V   Recommendation about Which MRI Parameters Should Be Documented before Onset of Treatment, after 6 Months to 1 Year, or when Treatment Failure Is Suspected

Standard MRI sequences for the evaluation of the status of MS
  Axial T1 scans (3-mm slices); careful positioning/repositioning using anatomical guidelines at each examination
  Axial T2 (fast spin echo) scans
  Gadolinium-contrasted axial T1 scans*
  Midsagittal scans to assess the corpus callosum
  Axial FLAIR (fluid retention inversion recovery) sequence
Optional MRI exams
  Magnetization transfer (MT) images to calculate the MT-ratio (indirect measure for demyelination and other structural abnormalities)
  Assessment of atrophy (third ventricle width; brain parenchymal fraction or other)
  Multivoxel magnetic resonance spectroscopy (MRS) (measure of N-acetyl-aspartate to determine axonal loss)
  Volumetric assessment of T1 holes, T2 lesion load

---

*The standard dose of gadolinium-DTPA is 0.1 mmol/kg. A higher sensitivity is achieved by delaying the contrast examination by 15 minutes or by administering triple-dose gadolinium-DTPA (0.3 mmol/kg).

only partially been identified, therapies have generally involved drugs or treatments with wide immunosuppressive properties. Recently, findings derived from the EAE model have kindled interest in targeting more specific components of the immune response. To understand these new directions, we will briefly describe the immunopathological events and the immunological background of EAE and MS.

## Pathogenesis of MS and Experimental Autoimmune Encephalomyelitis and Therapeutic Implications

Considerable information relating to possible mechanisms leading to immune-mediated demyelination has been obtained from studies of EAE, but it is important to note that EAE is not an exact model for MS, and care must be used in extrapolating from the animal model to the human disease (Martin et al., 1992). Nevertheless, the EAE model has been useful in examining the immunopathological responses that can produce demyelination and in examining potential treatments. Further, the models of relapsing EAE in the rodent are important for studying immunoregulatory processes.

EAE can be induced in a number of susceptible animal strains by the injection of myelin antigens such as myelin basic protein (MBP) and proteolipid protein (PLP) in appropriate adjuvants. T lymphocytes specific for the inducing antigens can adoptively transfer disease to naive animals showing that EAE is caused by autoimmune T cells. Antibodies against myelin components such as myelin oligodendroglia glycoprotein (MOG) or galactocerebrosides may contribute to the extent of demyelination but do not induce EAE themselves. Recently, it was also shown that immunization with glial proteins other than those derived from myelin, for example with S-100, may induce an inflammatory panencephalitis without demyelination (Kojima et al., 1994). On the basis of current knowledge, potentially autoreactive T cells are part of the physiological T-cell repertoire. Environmental or somatic events, for example viral infections, may lead to the activation of encephalitogenic T cells in the peripheral blood by molecular mimicry or bystander activation. Activated cells are capable of crossing the blood–brain barrier and entering the parenchyma. Adhesion molecules, matrix metalloproteinases, and chemokines are involved in the site-specific transmigration of T cells and monocytes. It is not clear whether the release of lymphokines (e.g., interferon [IFN]-γ and tumor necrosis factor [TNF]-α/β) and other factors such as oxygen radicals and nitric oxide alone or in combination with direct cell-mediated lysis by T cells are responsible for demyelination, or whether cells that are recruited during the inflammatory process into the lesion are important as well. Lymphokines up-regulate the local expression of class I and class II histocompatibility antigens on endothelial cells and glial components, thus facilitating antigen presentation and possibly also enhancing the tissue damage by cytolytic T cells. Encephalitogenic T cells dominate the early events of plaque development, whereas later other T cells are found as well. Along a chemokine gradient, macrophages/monocytes will migrate into the lesion where they are activated by T-cell–derived lymphokines and start to strip myelin from axons before they phagocytose and degrade fragmented myelin. Resident cells such as microglia and astroglia are probably able to fulfill many of the functions of macrophages such as cytokine release, phagocytosis, and antigen presentation. The pathogenic events that lead to the MS lesion and tissue damage are schematically depicted in Figure 2. Recent pathological studies of MS autopsy and biopsy tissue have provided strong support to the notion that MS is a heterogeneous disease (Lucchinetti et al., 2000). On the basis of the lesion composition of T cells, macrophages, antibody and complement deposition, and degree to which oligodendrocytes are primarily affected, four different lesion patterns have been described by Lucchinetti et al. (2000). Their data indicate that the relative role of each of these components differs between patients and, accordingly, is responsible for clinical heterogeneity and variable treatment responses.

In recent years, T cells that mediate EAE in different animal strains such as Lewis rats, SJL, and PL mice have been characterized in detail. Before we start to introduce individual therapies, the characteristics of encephalitogenic T cells will be summarized, because they also seem to play an important role in all disease subtypes, although to various extents. They are CD4+ and recognize antigen in the context of those MHC-class II antigens that confer susceptibility for EAE in the different strains. Because of their lymphokine profiles (i.e., secretion of IFN-γ and TNF-α/β), they belong to the subclass of T helper 1 (Th1) cells. For each susceptible animal strain, one or a few different peptide(s) of the disease-inducing autoantigens, MBP, PLP, myelin oligodendroglia glycoprotein (MOG), myelin oligodendroglia basic protein (MOBP), or oligodendrocyte-specific protein (OSP), have been identified as immunodominant. For example, in PL mice mice the N-terminal amino acids (AA) of MBP are encephalitogenic and the middle region of MBP in SJL mice and in Lewis rats. In addition to the peptide-MHC complex, the T-cell receptors (TCR) of encephalitogenic T cells have been analyzed in detail, and a striking restriction to a few TCR α- and β-chains has been described both in Lewis rats and in B10/PL and PL mice, with Vβ 8.2 and Vα 4 being

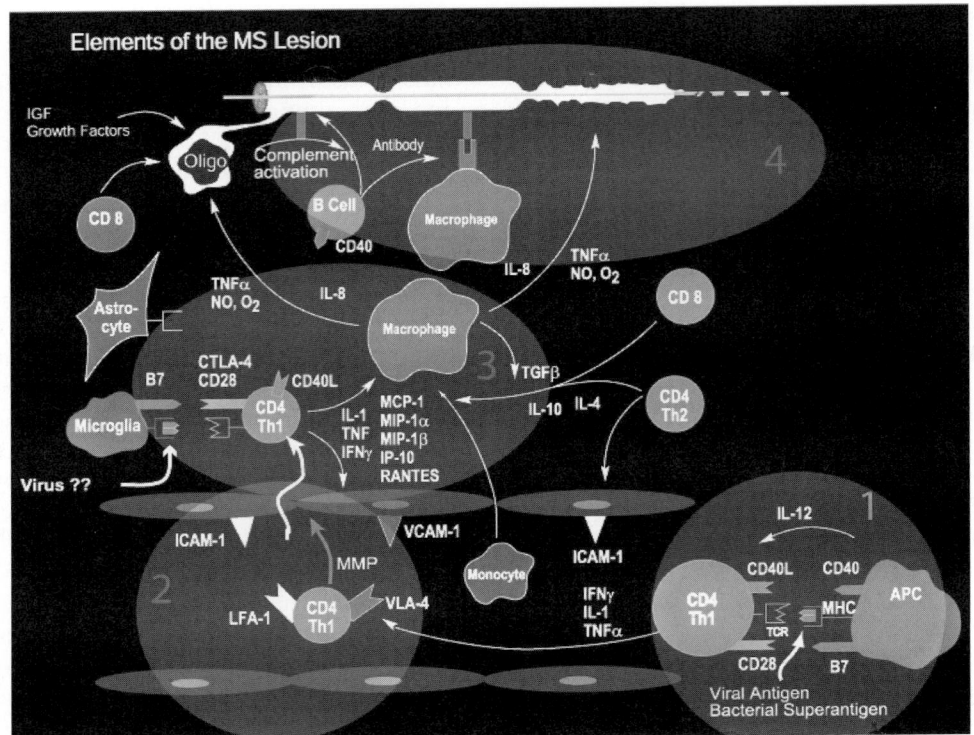

**FIGURE 2** Schematic depiction of the pathogenetic steps of immune activation and lesion formation in MS. (1) Peripheral activation of autoreactive CD4+ T cells. (2) Adhesion of activated autoreactive, myelin-specific T cells to the cerebrovascular endothelium, opening of the blood brain barrier, and transmigration into white matter tissue. (3) Local activation of microglia, astrocytes, further attraction of blood-derived monocytes/macrophages, T cells and B cells, and generation of a proinflammatory environment. (4) Damage of the myelin sheath, oligodendrocytes, and axons by a number of effector mechanisms including myelin-specific antibodies, macrophages, microglia, complement, oligodendrocyte apoptosis, and cytokines.

the predominant TCRs expressed. Based on the characterization of the trimolecular complex (Hohlfeld, 1989) consisting of autoantigenic peptide, MHC molecule, and TCR, immunotherapies have been developed that intervene in the disease process with high specificity. These therapies include peptides blocking the MHC molecule, anergy induction of encephalitogenic cells by means of inactivated antigen presenting cells (APC) with coupled autoantigen, monoclonal antibodies against the MHC-class II molecule, against the CD4 receptor or the TCR, altered peptide ligands (APL) of the autoantigen, and others (Hohlfeld *et al.*, 1997). In addition, therapeutic protocols have been developed to successfully vaccinate against encephalitogenic cells by using whole inactivated, encephalitogenic T cells themselves or peptides of their TCR. Another way to inactivate the autoantigen-specific T-cell response that has recently been used is the administration of the autoantigen, for example MBP, orally or by way of inhalation. This approach is called oral tolerization and, depending on the dose of antigen, acts by way of anergy or the

induction of regulatory CD8+ T cells that consecutively secrete transforming growth factor (TGF)-β, an immunosuppressive cytokine. Other therapies try to antagonize effector mechanisms of EAE-mediating T cells, such as antibodies against TNF-α/β or soluble TNF receptors, or the adhesion of those cells to brain endothelial cells by antibodies against adhesion molecules such as alpha 4 integrin or intracellular adhesion molecule 1 (ICAM1).

After the studies in EAE, similar immunological events have been examined in MS. In humans, the responses to MBP and PLP have been characterized in detail. As an example, MBP-specific T cells can be demonstrated in patients with MS, as well as in healthy controls. They are CD4+, largely HLA-DR restricted, often secrete TNF-α/β and IFN-γ, are cytotoxic, and recognize immunodominant peptide regions that overlap with encephalitogenic peptides in EAE. Controversy still exists as to whether human MBP-specific T cells express a restricted TCR profile. Because MBP-specific T cells can also easily be isolated from healthy individuals, their contribution to the pathogenesis is not yet clear; however, recent experiments in transgenic mice expressing human MBP-specific TCR and a clinical trial with an altered peptide ligand (APL) of MBP strongly indicate that MBP is encephalitogenic in humans (Bielekova et al., 2000; Madsen et al., 1999). However, rather than being due to a unique population of T cells in patients, disease may occur when exogenous factors or somatic events lead to abnormal regulation or to an increased state of activation of T cells, which are not deleted from the T-cell repertoire. This hypothesis would be compatible with evidence supporting a multigene influence in MS (Becker et al., 1998).

## CURRENTLY APPLIED THERAPEUTIC STRATEGIES

### Glucocorticoids

Glucocorticoids (GCS) and adrenocorticotropic hormone (ACTH) have long represented the cornerstone for the treatment of relapses in MS (Myers, 1992). They have potent anti-inflammatory and immunosuppressive effects, influence multiple functions, and have been used extensively in treating various immunologically mediated diseases. Because of both redistribution and deletion, there is a rapid decrease in lymphocytes and monocytes/macrophages. T lymphocytes, in particular T-helper cells, are more severely affected than B cells. At the same time, neutrophil numbers increase because of release from bone marrow and endothelial surfaces. Antibody-mediated responses, including allergic reactions, are suppressed as well as graft rejections. However, although total lympoid tissue shrinks during prolonged hypercorticism, both cellular and humoral immunity is still generally intact. Besides influencing the numbers of cells of the immune system, GCS downregulates the secretion of several lymphokines such as IL-1, IL-2, IL-4, IL-5, IL-6, IFN-$\gamma$, and TNF-$\alpha/\beta$. The anti-inflammatory effects include the suppression of the secretion of arachidonic acid metabolites such as prostaglandins and leukotriene, the decrease of capillary dilation, edema, fibrin deposition, and migration of inflammatory cells into the tissue, as well as the blockade of lysosomal enzyme degranulation.

Despite the effects of steroids on the immune system, the effectiveness of GCS in the treatment of MS is poorly understood. Although oral prednisone largely replaced IV ACTH, prednisone was never studied in a definitive clinical trial. Over the past several years administration of high doses (500–1000 mg/day) of IV methylprednisolone (MP) has become the most commonly used treatment for acute episodes. Again, little solid data exist supporting the effectiveness of high-dose MP. In a small study comparing MP to placebo, a significant improvement in EDSS scores was noted in patients treated with MP. Some benefit of MP was also noted in the recent study of optic neuritis comparing oral prednisone, placebo, and IV MP (Beck et al., 1992). Patients treated with MP had a quicker improvement in vision. Patients treated with oral prednisone, in distinction, did worse than those treated with placebo. The value of IV MP is supported by the demonstration that it can reverse the blood–brain barrier changes reflected by gadolinium enhancement on MRI. Gd enhancement is stopped within hours after administration (Miller et al., 1992). In addition, the blockade of lymphokine release from autoreactive cells may interrupt the cascade of events that leads to demyelination. Unfortunately, the reversal in gadolinium enhancement after a course of IV MP lasts only for a few weeks. Longer effects have been reported when MP is administered for 10 days. Long-term administration of either high- or low-dose GCS is generally thought to be contraindicated, except in unusual circumstances, but repeated single IV MP bolus doses with 3-day oral tapers were well tolerated and demonstrated a strong trend toward efficacy in a MRI-controlled trial (Kümpfel et al., 2001). Little evidence exists supporting a beneficial effect of GCS including IV MP on CP MS.

While GCS and especially IV MP may speed reversal in episodes of acute worsening, there is no evidence that GCS affect the long-term course of disease a month or slightly more beyond administration. Data from the multicenter optic neuritis study suggested that the occurrence of additional neurological events sufficient to establish the diagnosis of clinically definite MS in patients with monosymptomatic optic neuritis was longer in those patients treated with IV MP as compared to placebo or oral prednisone (Beck et al., 1993). These findings could suggest that IV MP alters the natural history of MS, but additional studies will be necessary to resolve this important question.

GCS will remain the treatment of choice for exacerbations even in patients receiving other drugs such as IFN-$\beta$. Future clinical studies focusing on optimal dose and duration of GCS treatment are needed. The effect of GCS treatment on the natural history of MS must also be resolved in properly designed clinical trials.

Intrathecal administration has been used at various MS centers over the past 20 years, but has been largely abandoned because of side effects consisting of arachnitis and because of the lack of documented clinical evidence of efficacy. However, one controlled trial has reported intrathecal GCS (triamcinolone-acetonide) to be beneficial in patients with predominantly spastic paraparesis (Rohrbach et al., 1988). Since the benefit consists primarily of a reduction in spasticity, a comparison with intrathecal baclofen is needed.

## Immunomodulators

Besides substances that have a broader effect on the immune system and the potential for more severe adverse reactions, interest during the last years has focused on the development of less harmful immunomodulators or highly specific immunotherapies. IFN-β (type I interferon) and glatiramer acetate (GA; formerly also known as copolymer-1) belong in the first group of immunomodulating substances. IFN-γ or type II interferon is secreted by lymphocytes and natural killer cells only and has many enhancing effects on T cell responses against foreign and autoantigens. Besides its antiviral effects, it activates cytolytic T lymphocytes, stimulates the secretion of other lymphokines and upregulates the expression of HLA molecules on many cells including astro- and microglial cells. As outlined before, it is currently believed that Th1 cells due to their ability to secrete IFN-γ and TNF-α/β contribute to the pathology in EAE and MS. It was therefore no surprise that a phase II trial with IFN-γ had to be stopped after several months due to an increase in exacerbations (Panitch *et al.*, 1987). IFN-β, a type I interferon, is believed to exert quite different effects on the immune response and has been approved for several years now for the treatment of RR MS. We will briefly mention the action of both immunomodulating substances already approved for RR MS, i.e., IFN-γ and GA, below before summarizing the most important therapeutic strategies that are currently being tested or are in the planning.

### Interferon-Beta (IFN-β)

On viral infection, IFN-β is secreted by lymphocytes, macrophages, and dendritic cells but also by a variety of other cells. It not only inhibits viral spread, but, more importantly, counteracts many of the immuno-stimulatory effects of IFN-γ. Administration of IFN-β results in the down-regulation of delayed-type hyper-sensitivity responses of the secretion of IFN-γ and TNF-α/β. The expression of MHC antigens decreases as does the proliferative response of T cells. In addition, a rise in the expression of the immunomodulatory cytokine IL-10 was noted not only in vitro but also during treatment of MS patients in vivo. As a consequence, many of the mechanisms that are involved in immune activation during an exacerbation, as well as those that are part of the effector phase of lesion development (secretion of IFN-γ and TNF-α/β), are counter-acted by IFN-β. Furthermore, IFN-β interferes with T-cell proliferation, induces the expression of soluble adhesion molecules (s-vascular cell adhesion molecule; s-VCAM), and inhibits the activity of matrix metallo-proteinases. Particularly, the two latter effects are probably critical for its action, because MRI-controlled

clinical trials have documented a profound inhibition of blood–brain barrier permeability and rapid cessation of Gd-enhancing MRI lesions (Stone *et al.*, 1995). Very recent data suggest that IFN-β, however, also enhances the expression of a number of molecules that are believed to be relevant for Th1 differentiation (e.g., IL-12 receptor β2 chain) and Th1 recruitment (e.g., the chemokine receptor CCR5 and the chemokines IP-10, Rantes) (Wandinger *et al.*, 2001). It is therefore difficult at the moment to pinpoint which effect of IFN-β is most important for its activity in MS in vivo, but a number of studies are ongoing to examine this question further.

After the initial observation of a clinical benefit from intrathecal IFN-β (Jacobs *et al.*, 1987), a number of large multicenter trials have been conducted during the last decade in various stages of MS, and these have firmly established IFN-β as the treatment of choice in RR-MS, SP-MS with inflammatory activity, and recently also in monosymptomatic stages of MS (Comi *et al.*, 2001; European Study Group, *et al.*, 1998; Interferon-B MS Study Group, 1993**; Jacobs *et al.*, 1996**, 2000). Three preparations of IFN-β are currently available: IFN-β 1b (recombinant unglycosylated IFN-β with a substitution of cysteine in position 17 by serine and removal of the aminoterminal methinonine; IFN-β-ser-17 = Betaseron®, or Betaferon®) (Interferon-B MS Study Group, 1993; Paty *et al.*, 1993**), and two preparations of IFN-β 1a (recombinant, glycosylated IFN-β produced in Chinese hamster ovary cells) i.e. Avonex® and Rebif®. Two doses of IFN-β 1b (1.6 and 8 million international units, IU) SC every other day were compared with placebo over a period of 3 years. The high dose reduced the attack frequency by one-third and cut serious attacks by 50% (Interferon-B MS Study Group, 1993). The decline in attacks started early (after 2 months of onset) and lasted throughout the trial. The number of MRI lesions dropped dramatically (80% for the medians of activity; 83% for the appearance of active, and 75% for the number of new lesions, whereas an increase of MRI-documented disease burden was observed in the placebo arm) (Paty *et al.*, 1993). The effects with the lower dose were smaller, but still notable. When the EDSS score was used as an outcome measure, the differences between the treatment and placebo arm were less dramatic, in part because of the mild disease of the patients and the short follow-up (for early RR-MS). On the basis of the reduction of attack and MRI lesion frequencies, it is expected that the clinical differences will become more pronounced over time. Similar results were observed with the two glycosylated IFN-β 1a preparations (Avonex® and Rebif®; the former injected as $6 \times 10^6$ IU once weekly IM; the latter injected three times a week at $6 \times 10^6$ or $12 \times 10^6$ IU SC). Sustained worsening of disability was chosen as a primary outcome measure in the IFN-β 1a, Avonex® trial, and a reduction of progression could be demonstrated in RR-MS (Jacobs *et al.*, 1996). Both IFN-β 1a (Rebif®) and IFN-β 1b (Betaseron®) were also tested in RR- and SP-MS, and after the demonstration in a large-scale, controlled, randomized, double-blind and multicenter phase III trial of IFN-β 1b (Betaseron®) in Europe (European Study Group, 1998) that this compound slows progression in SP-MS, IFN-β 1b (Betaseron®) was also approved by the European regulatory agencies (EMEA) for SP-MS. This effect on SP-MS was not confirmed by a similarly large US phase III in SP-MS, and after careful comparison of the study results and patient populations, it was concluded that the patients in the US study were slightly further advanced in their disease course and consequently that IFN-β 1b (Betaseron®) is probably effective during earlier stages of SP-MS, but less so at later stages, when inflammation declines and the disease course is mainly driven by degenerative aspects and oligodendrocyte death or scar formation (Molyneux *et al.*, 2000). The trial of Avonex® in SP-MS (IMPACT) has not concluded yet. Finally, both IFN-β 1a preparations have been tested in monosymptomatic demyelinating disease or early MS (CHAMPS and ETOMS) (Comi *et al.*, 2001; Jacobs *et al.*, 2000) and it was documented that the onset of clinically definite MS can be delayed for 10 and 9 months, respectively.

Because considerable controversy currently exists as to the use of the various IFN-β preparations, we will briefly summarize our own position and practice of applying them. The comparison of the large-scale clinical trials, as well as smaller studies, show the following: There is a dose-dependent effect of the various IFN-β preparations (i.e., higher doses up to a certain level are more effective than lower doses). The IFN-β preparation and its route of administration seem to influence the side effect profile and induction of neutralizing antibodies, whereas the route of administration does not seem to change bioavailability. Consequently, IFN-β 1b (Betaseron®; $8 \times 10^6$ IU SC every other day) and the high dose of Rebif® ($12 \times 10^6$ IU SC 3 times a week) are more effective than Avonex® at $6 \times 10^6$ IU once a week IM (Durelli *et al.*, 2001). However, because of the lack of glycosylation, the SC route of administration and possibly the change in amino acid sequence, IFN-β 1b (Betaseron®) seems to be most immunogenic (i.e., induces neutralizing antibodies more frequently than the glycosylated forms of IFN-β with the normal sequence and causes more side effects). On the basis of the preceding, we recommend IFN-β 1b (Betaseron®) or Rebif in cases of higher disease activity either clinically or by MRI, and IFN-β 1a (Avonex®) in cases with milder disease or when the dosing regimen (i.e., once weekly versus every other day) and potential local side effects are a particular concern, because these are clearly less frequent with IFN-β 1a (Avonex®). We do not believe that there are significant differences in the biological activity of the three preparations of IFN-β, but rather that the observed variations are due to the differences in dose and route of administration. Four-year follow-up in the PRISMS-4 trial (Rebif, low and high dose) showed that the outcomes are consistently better for the patients treated for the full period, that there is evidence for dose-dependency and that efficacy is reduced by neutralizing antibodies (PRISMS-4, 2001).

### Glatiramer Acetate (GA; Copolymer 1, Cop 1)

GA is a synthetic polypeptide containing the amino acids L-alanine, L-glutamic acid, L-lysine, and L-tyrosine at fixed molar ratios ($6.0 : 1.9 : 4.7 : 1.0$; molecular weight between 4.7 and 12 kDa; note that the molar ratio and the length varied slightly during the last decades in different studies), but with a random sequence (Arnon, 1996; Duda *et al.*, 2000). Immunological studies during the last few years have explained a number of different mechanisms of action of GA in vitro and in vivo. These include the inhibition of antigen presentation by competing with antigenic peptides for binding to HLA-DR binding grooves, the induction of anergy and TCR antagonism in certain T-cell lines/clones, and the induction of immunoregulatory Th2 cells that are capable of down-modulating Th1-mediated disease activity in MS via bystander suppression (Gran *et al.*, 2000; Neuhaus *et al.*, 2000). The latter effect has been shown by a number of laboratories, and it was also demonstrated that GA-specific T cells with a Th2 phenotype are often cross-reactive with MBP after long-term treatment of MS patients in vivo. Bystander suppression is thus the mode of action that probably plays the most important role in immune modulation in vivo. After the demonstration that GA is capable of suppressing EAE, an open trial and a randomized, double-blind placebo-controlled study showed beneficial effects (fewer relapses, higher percentages of clinically stable or improved and less patients with deterioration in the GA group) in RR-MS. In a similar trial in patients with CP-MS, divergent results were obtained in the two participating centers, with one obtaining beneficial effects in the GA-treated group and no difference between placebo- and GA-treated group in the other (Bornstein *et al.*, 1991**). In the meantime, a large-scale, controlled, multicenter, double-blind phase III clinical trial of 20 mg GA/day SC has confirmed its efficacy in RR-MS by showing a roughly 30% reduction in exacerbation rate (Johnson *et al.*, 1995**). These results held up during longer follow-up, and a recent MRI-controlled trial of GA in Europe also demonstrated that GA reduces contrast-enhanced MRI activity after approximately 6 months of administration (Johnson et al., 1998**; Comi et al., 2001**). This delayed effect on MRI is different from the profound and early inhibition of MRI activity by IFN-β. Although the exact reason is not clear at this point, it is assumed that the induction of bystander suppression by means of regulatory T cells requires more time than the quick closing of the blood–brain barrier by IFN-β. Furthermore, to be effective, Th2 immunomodulatory cells probably need access to the target tissue at least for some time before they can exert their suppressive effect. GA is administered subcutaneously and, besides local reactions at the injection site and rare hypersensitivity reactions (chest tightness, palpitations), was safe and well tolerated. In our experience and based on the review of the various trials, GA is probably slightly less active than the high dose interferon β preparations, but is overall tolerated best. We recommend its use primarily in patients with milder RR MS (clinically or by MRI) or those who have already experienced significant side effects or have developed neutralizing antibodies under treatment with one of the IFN-β preparations.

### Intravenous Immunoglobulins (iv-IG)

IV-Ig, a polyvalent pool of immunoglobulins from multiple donors, is a well-established treatment in other autoimmune disorders such as idiopathic thrombocytopenic purpura and Guillain-Barré syndrome and has recently also been tried in MS. The mechanism of action of IV-Ig is not clear yet but may include the capture of antigen, immunoregulation by network interactions between T and B cells, and acting by means of inhibitory Fc receptors. Recent trials with IV-Ig at monthly intervals (0.15–0.2 mg/kg/month; and higher doses) showed a mild, but clear, reduction of exacerbation rate to a similar magnitude as observed with IFN-β (Fazekas et al., 1997**; Achiron et al., 1998**). There was also a trend toward slowing disease progression. On the basis of experimental data that polyvalent immunoglobulins may foster remyelination, it was speculated that IV-Ig might also act by means of this mechanism and improve myelin repair. Data for this activity in MS are, however, still lacking. Because of its high cost, IV-Ig is currently considered a treatment alternative if other strategies have failed or cannot be administered safely (i.e., during pregnancy).

### Immunosuppressive Treatment

The various immunosuppressive treatments and their actions that have been adapted from the therapy of cancer or graft rejection will be summarized in the following. They can be separated into two general categories. The first are the relatively nonselective immunosuppressants or chemotherapeutics like azathioprine (AZA), cyclophosphamide (CTX), mitoxantrone (MIX), or methotrexate (MTX). The second category includes drugs that more specifically target immune cells such as cyclosporine A (CSA), FK506, rapamycin, or deoxyspergualine (DSG). Only those treatments that are currently used will be mentioned in the following.

### Mitoxantrone (MIX)

Mitoxantrone (MIX), a synthetic anthracycline derivative, has been shown to be of some benefit in small phase II studies in relapsing progressive and primary chronic progressive MS patients that have failed to respond to azathioprine or cyclosporine (Mauch et al., 1992*) but was proven effective in a phase III European multicenter, randomized, placebo-controlled study in active RR-MS and progressive MS (Edan et al., 1997**). On the basis of the latter trial, in which two doses ($12 \, mg/m^2$ and $5 \, mg/m^2$ IV every 3 months for 2 years) were given, MIX was recently approved for the treatment of severe RR-MS and SP-MS by the FDA. MIX showed a significant dose-related effect both on relapse rate (0.21 vs 0.6), time to first relapse (>24 months vs 15 months), measures of sustained disability as measured by EDSS and ambulation index, and the clinical benefit was paralleled by a reduction of MRI activity. The drug acts on both DNA and RNA synthesis, its molecular action is, however, not yet fully defined. The interference with nucleic acid synthesis explains its influence on rapidly dividing and metabolically active cells such as tumor cells and cells participating in physiological processes such as immune responses. MIX is given as a single infusion ($12 \, mg/m^2$ in 250 ml of 0.9% sodium chloride or 5% dextrose) over 30 minutes. Antiemetics (ondansetron or tropisetron) should be given to prevent nausea. To maintain immunosuppression over longer periods of time, MIX has to be administered every 3 months. Although the acute toxicity is lower than that seen in patients receiving induction therapy with cyclophosphamide, MIX along with other anthracycline derivatives (doxorubicin and daunorubicin) is cardiotoxic. Congestive heart failure and decreases in the left ventricular ejection fraction have been observed with cumulative doses of $140 \, mg/m^2$ and higher (cumulative probability for congestive heart failure 2.6%). Patients with preexisting cardiac problems should therefore not receive MIX. In addition, each patient receives an electrocardiogram and echocardiography before initiation of therapy and after every 6 months of therapy (Edan et al., 2001). If the left ventricular ejection fraction is reduced by ≥50%, treatment should be halted until it has normalized. Frequent analysis during the first 2 weeks after therapy and then weekly of laboratory values, including blood counts, liver enzymes, uric acid, and renal laboratory values, should be performed, because myelosuppression occurs within a few days after therapy. Patients should be informed that a bluish discoloration of urine and of the sclera may occur within 24 hours after administration. Besides cardiac problems, adverse reactions include bleeding, nausea and vomiting, jaundice, infections, renal failure, alopecia, seizures and headaches, cough and dyspnea, and allergic reactions. The risks of administration must therefore carefully be weighed against the desired effects, and the drug should be given only to patients with severe RR-MS or SP-MS.

### Azathioprine (AZA)

AZA is used extensively in organ transplantation and has been applied widely in the treatment of MS, particularly in Europe. In a number of European countries, AZA is still frequently used in RR-MS (Hughes, 1992; Yudkin *et al.*, 1991), and it has recently been approved for use in MS in Germany. AZA acts predominantly through its active metabolite 6-mercaptopurine. This metabolite competes with its analogue, hypoxanthine, a central component of nucleic acid synthesis, and therefore has widespread consequences, especially on DNA and RNA synthesis. In vitro studies have demonstrated effects on both T-cell and B-cell functions, in particular the inhibition of surface receptor expression (for example CD2), blocking of mitogen-driven responses, and of the induction of antibody responses. When given at the time of induction of an autoimmune disease, such as EAE or transplantion, it is extremely effective. This effect is, however, due to a broad immunosuppressive effect rather than a specific intervention. The efficacy of AZA in other autoimmune disease, including rheumatoid arthritis, myasthenia, and polymyositis, its use in transplantation, and its ability to modify EAE have all supported its use in MS. AZA has a good therapeutic ratio for an immunosuppressive drug; it is tolerated reasonably well and generally without serious side effects at the doses used clinically. A number of open and controlled trials have demonstrated a trend toward a beneficial effect in RR-MS. In 1988, a large and well-designed trial of AZA was performed and again demonstrated a trend toward a beneficial effect in the form of fewer exacerbations in patients on AZA, but the differences failed to reach statistical significance (Group, 1988**). The results of all the AZA trials have been re-examined in a meta-analysis, and a significant effect for AZA in RR-MS has been demonstrated (Weiner *et al.*, 1993). The results also raised the possibility that AZA treatment may affect the course of disease.

Despite its small, but consistent, effect and although it is currently the best-known immunosuppressive drug, patients may experience adverse reactions. With IFN-β and glatiramer-azetate as approved therapies of RR-MS, AZA remains an alternative when these drugs have failed or in countries where their cost prohibits widespread use. AZA is recommended for patients with RR-MS who show frequent relapses or more disabling exacerbations. Its combination with IFN-β is currently being evaluated in a number of small trials (Milanese *et al.*, 2001). The practical aspects of the administration of AZA and the other immunosuppressants will be given in the next section.

### Cyclophosphamide (CTX)

CTX is an alkylating agent because of its ability to intercalate into the DNA helix. It acts primarily on rapidly dividing cells from tissues and certain tumor cells. The desired actions (i.e., immunosupression and decrease of tumor growth) and the adverse effects (i.e., leukopenia, hemorrhagic cystitis, amenorrhea, oligospermia, and transient alopecia) are caused by DNA cross-linking. CTX affects the immune system by decreasing T cells of various subclasses, B-cell numbers and function, and induces a shift from Th1 to Th2 cytokine secretion. Its efficacy in autoimmune inflammatory diseases such as lupus nephritis, Wegener's granulomatosis, and various vasculitides has led to its use in MS. EAE could be blocked by the administration of CTX, but the exact mode by which this is achieved is not clear.

Over the past decades several studies have examined CTX in the chronic progressive phase of the disease. Although the results of these studies were inconclusive, they suggested that CTX might have a beneficial effect on reducing progression. Interest in CTX increased in 1983 after the report of a controlled, but unblinded study examining i.v. CTX and ACTH in CP MS that showed a significant benefit for CTX. The initial protocol incorporated an intense induction phase consisting of 400 to 500 mg daily for up to two weeks (Mackin *et al.*, 1992; Weiner *et al.*, 1993**). Small studies re-examining the usefulness of CTX provided conflicting results, but in 1991 a well-designed Canadian study failed to confirm a beneficial effect of CTX in CP MS (Noseworthy *et al.*, 1991**). There has been considerable controversy regarding the meaning of these divergent results. Because of the side effects including malignancies of the bladder and lymphopoietic system, the authors believe its use should be limited to cases with frequent and severe relapses or rapid progression and in which other treatments have been exhausted. Mackin *et al.* (1992) have pointed out that future studies should address the questions of which dosages of CTX should be used, whether induction therapy is required or whether concomitant administration of steroids is necessary. At present, we tend to prefer a protocol that was adopted from the treatment of lupus nephritis because of less adverse reactions and because it is well established for the treatment of other autoimmune diseases. A small, open-label and uncontrolled study of the above monthly dosing regimen in MS patients showed that i.v. CTX profoundly inhibits MRI activity and probably has a place in the therapy of patients with very active disease or those who have failed other treatments (Gobbini et al., 1999*).

### Cyclosporine A (CSA)

CSA, a cyclical undecapeptide isolated from two species of soil fungi, targets the immune system more specifically than azathioprine or cyclophosphamide. Since its immunosuppressive effects were first noted by Borel, it has revolutionized transplant surgery, leading to a dramatic increase in graft survival after cardiac, kidney, liver, or pancreas transplants. The effects on the immune system are numerous and mediated through the interaction with cyclophilin, one of two intracellular protein receptors to which cyclosporin binds. Its primary effect is the modification of lymphokine gene transcription, including the down-regulation of IL-2 expression, but also of IL-1, IL-3, migration inhibition factor, and IFN-γ. Most of the in vitro activities such as inhibition of mitogen-driven proliferative responses of T cells, responses to alloantigens in mixed lymphocyte reactions, the blockade of cytotoxic T-cell activity, and the inhibition of T-helper cells are explained by its molecular action. In addition to these effects on lymphocytes, CSA enhances the production of prostaglandins by monocytes. In acute and relapsing remitting EAE, both prevention and improvement of already established disease were observed. CD4+ cells that had been recovered from lymph nodes during CSA treatment and cultured in vitro were capable of suppressing the adoptive transfer of EAE by T cells. Thus, it has been speculated that the effectiveness of CSA in EAE originates from the in vivo expansion of CD4+ suppressor-inducer cells. The findings in EAE must, however, be viewed with caution, because exacerbations of disease have been observed with low doses of CSA.

CSA has been examined in MS in three clinical trials. In the first, CSA at a dose of 5 mg/kg was compared with AZA (Kappos et al., 1988**). The results indicated that there was little difference between CSA and AZA, and the latter had considerably less side effects. CSA at a higher dose of 7 mg/kg was compared with placebo in another study (Rudge et al., 1989**). Some inconsistency in results existed between the two centers, but overall there was a trend toward fewer episodes of worsening and less progression in the CSA-treated group. Finally, CSA has been examined in a large, controlled, double-blind study in North America (MSS Group, 1990**) at a dose of 6 mg/kg. CSA resulted in a small delay in progression, but significant side effects, especially nephrotoxicity, were noted. Because of the potential of CSA to produce kidney damage, AZA is a more reasonable choice if disease progression warrants immunosuppressive therapy. FK 506 and rapamycin are immunosuppressive drugs with a similar mechanism of action as CSA. Although initial reports indicated that toxicity of FK 506 would be less at equivalent immunosuppressive doses, there are currently no data on FK 506 in MS. If CSA is used, its plasma level should be followed, and renal function must be monitored (e.g., by periodical measures of glomerular function rate [GFR]), especially if serum creatinine levels increase by 20% or more.

### Experimental and Future Therapies

We will not mention each of the many experimental therapies. A recent review comprehensively summarized the entire spectrum of experimental therapies in MS and their immunological rationale (Hohlfeld, 1997). The following paragraphs summarize some of the therapies that have recently been tested, are currently being tested, or are potential candidates for future trials.

### Total Lymphoid Irradiation (TLI)

Preliminary and later more definitive results showed that irradiation of the lymphoid system (total lymphoid irradiation, TLI; with a dose of 1980 cGy in 11 fractions of 180 cGy each) had some beneficial effects in CP-MS. These earlier results could not be confirmed by a randomized, double-blind, placebo-controlled study of TLI in CP-MS (Wiles et al., 1994). TLI has a mild effect in slowing the accumulation of T2 lesion burden, and its efficacy on clinical and MRI parameters is improved by concomitant administration of oral prednisone. Because there are multiple other options for immunosuppression, TLI can currently not be recommended for the treatment of CP-MS.

### Plasmapheresis and Immunoadsorption

These treatments are well established in Guillain-Barré syndrome and myasthenia gravis; however, clear evidence that they are effective in MS is still lacking. Plasmapheresis has been applied in a number of trials and mostly in patients with chronic progressive and severe MS (Hauser et al., 1983; Weiner et al., 1990**; Khatri et al., 1985**). The results are controversial and difficult to interpret, because plasma exchange was usually combined with other treatments such as CTX, prednisone, ACTH, and immunoglobulins. In one study, plasmapheresis was beneficial (Khatri et al., 1985**) but generally failed to be effective in the treatment of MS. In life-threatening exacerbations, for example with brainstem manifestations, plasmapheresis may be used if high-dose steroids fail to improve the course but can otherwise not be recommended on the basis of available experiences. A theoretical advantage of plasmapheresis versus IV-Ig lies in its ability to remove other substances than antibodies (e.g., toxic metabolites, cytokines, or peptides).

### 2-Chlorodeoxyadenosine (Cladribine; 2-CdA)

The nucleoside 2-chlorodeoxyadenosine (cladribine; 2-CdA) has been used extensively to treat hematological malignancies. A recent randomized, double-blind, placebo-controlled trial with 51 patients with chronic progressive disease during which 2-CdA was given at 0.7 mg/kg monthly showed stabilization or improvement of clinical scores (EDSS and Scripps scores), as well as a reduction in demyelinated parenchymal volume and CSF oligoclonal bands (Sipe et al., 1994). The treatment was well tolerated except for one patient who had severe, but reversible, marrow depression develop. Besides general marrow suppression, 2-CdA caused a transient decrease of B cells and a long-standing reduction of CD4 lymphocytes that may account for its therapeutic effect in MS (Beutler et al., 1996). Despite its profound effect on gadolinium-enhancing MRI activity (>90% reduction), only a trend toward improvement of clinical measures was observed (Romine et al., 1999). Although little data on the long-term use of 2-CdA are available, the effect on MRI activity and information from small phase II studies suggest that 2-CdA is a reasonable alternative if other treatments have failed and patients are clinically very active.

### Methotrexate (MTX)

On the basis of findings in other autoimmune diseases, the folic acid analogue, methotrexate (MTX), another chemotherapeutic drug that has been applied in various malignancies and autoimmune diseases, has recently been used in chronic progressive MS. MTX inhibits the enzyme dihydrofolate reductase leading to a partial depletion of tetrahydrofolate cofactors and, as a consequence, to the inhibition of purine and thymidylate synthesis. Because of its mechanism of action, MTX acts on rapidly dividing cells. Low-dose (7.5 mg weekly) oral MTX was administered in a randomized, double-masked, placebo-controlled 2-year study to 60 patients with chronic progressive MS (Goodkin et al., 1995**). MTX was well tolerated in this phase II trial and significantly reduced the rate and frequency of progression compared with placebo, but its effect was particularly noteworthy on upper extremity function. This has been interpreted as indicating that lower extremity deficits pre-existed for too long to allow for any improvement, whereas the reduction of upper extremity function could still be slowed by MTX treatment. MTX is thus a relatively nontoxic treatment option for CPMS, but as mentioned for 2-CdA, the initial data are too preliminary and should be confirmed in larger trials before it can be recommended. Furthermore, it is our impression that MTX may be underdosed in the preceding regimen and that higher doses that are used in active rheumatoid arthritis (e.g., 12 mg weekly) should be examined as well.

### Autologous Transplants of Hemopoietic Stem Cells

Autologous transplants of hemopoietic stem cells (CD34+) have been tried in a number of fulminantly progressing and life-threatening autoimmune diseases such as systemic sclerosis, systemic lupus erythematosus, but also in MS patients in a number of small trials. This therapy is based on the assumption that an aberrant immune system can be purged by total body irradiation with methylprednisolone or chemotherapeutic conditioning regimens with either high-dose cyclophosphamide or the BEAM regimen (BCNU, etoposide, cytosine arabinoside, and melphalan). After this "purging" of the mature immune system, patients are infused with their own CD34+ hemopoietic stem cells after these have been purified from leukaphereses and depleted from T cells. Several trials are currently underway or in the planning, and these will probably clarify whether the preceding goal (i.e., reconstitution of a patient's dysregulated immune system with a "de novo" immune system) can be reached (Burt et al., 1995). At present, the existing experience from European and US trials and case reports shows that the procedure has substantial mortality (in the range of 10%) and thus should only be considered in rapidly progressing cases (Fassas et al., 2000). However, these figures vary greatly between different centers. The fact that it is at present difficult to give a firm prognosis in individual MS patients and because one only wants to subject a patient who is rapidly progressing, but will still benefit from the procedure, the identification of suitable patients is a formidable task. Furthermore, it is possible that the reconstitution of the immune system with autologous precursor cells will again lead to aberrant immune reactivity and that one therefore needs to consider allogeneic stem cell transplants in the future. A recent trial documented that inflammatory MRI activity in the CNS is suppressed for sustained periods (Fassas *et al.*, 2000).

### Other Treatments

A number of other therapeutic interventions, including antibodies against CD4, CD52 (Campath), intracellular adhesion molecule 1 (ICAM-1), very late activation antigen 4 (VLA-4), and other adhesion molecules, the vaccination against autoantigen-specific T cells by administration of inactivated MBP-specific T cells or TCR peptides, the blockade of antigen-presentation by MHC-peptide complexes and administration of TGF-β, the use of antibodies against TNF-α or application of soluble TNF receptor molecule, and the application of global or specific phosphodiesterase (PDE) inhibitors (e.g., the PDEIV inhibitors rolipram and mesopram) have recently been tested, are in clinical trials, or will be examined in the near future. Data from the study of an anti-VLA-4 antibody (Antegren®) will be released soon, and preliminary evidence indicates that blocking of adhesion molecules results in a reduction of blood–brain barrier opening and disease exacerbations (Tubridy *et al.*, 1999). The disease worsening by anti-TNF approaches (Coles *et al.*, 1999) and the marked side effects of e.g. TGF-β (Calabresi *et al.*, 1998) demonstrate not only the complex activities of some of these immune mediators but also that, in the case of TNF-α, the cytokine may not only be detrimental but could also have a role in myelin repair. Different from rheumatoid arthritis where TNF inhibitors have now a firm place in the therapeutic armamentarium, this cytokine seems to exert more complex functions in MS, and this is most likely caused by the differences in target tissue (i.e., joint vs CNS). Recent considerations include the transplantation of glial and neural progenitor cells and/or growth factors to foster remyelination, but it is too early to apply these to patients. This list is not complete, and it is premature to comment on the efficacy of these treatments. However, it has to be stressed that important lessons have been learned from a number of these trials. The administration of CAMPATH, a monoclonal antibody against CD52, for example, has initially worsened disease by means of cytokine release from activated lymphocytes, then, during the phase of T cell depletion, led to disease improvement and reduction of MRI activity, before atrophy and further disease progression resumed (Coles *et al.*, 1999). Linomide, an agent that stimulates NK cell activity, initially showed promise in EAE and MS, before two phase III trials had to be halted because of unexpected cardiotoxicity (Tan *et al.*, 2000). Finally, two recent phase II trials with an altered peptide ligand (APL) based on the immunodominant MBP peptide showed disease exacerbations and poor tolerability at the high dose (Bielekova *et al.*,

2000) and a trend toward disease improvement and induction of immunomodulatory Th2 cells at a lower dose (Kappos *et al.*, 2000). It became clear from these data that myelin-specific T cells are in fact potentially encephalitogenic in humans and that specific immunotherapies have to be approached with caution, particularly with respect to finding the best dose of the compound.

For further details on experimental approaches to immunomodulation in MS, we refer to a recent review (Hohlfeld, 1997).

## PRACTICAL MANAGEMENT

Because the cause of MS is unknown, treatment focuses on suppressing or modulating the supposed pathogenetic process at different levels. This paragraph will summarize the currently recommended treatment regimens and also address the therapy of choice for different courses and stages of the disease, as well as for special situations such as pregnancy. The authors are aware of the controversial issues of some of the treatments mentioned in the following. None is ideal for achieving the major goals such as preventing further relapses or halting progression. According to recent studies, IFN-β and GA are superior to other regimens in RR-MS. Furthermore, the disease heterogeneity and differences between patients, as well as experience from the many large scale trials and phase IV pharmacosurveillance, indicate that there are individual responders and non-responders to some of the treatments. Considering that each of the approved therapies shows some effect, but clearly only a partial response and not in every patient, we need to devise strategies that allow us to identify and apply combination therapies (e.g., of several immunomodulators, an immunomodulating substance with an immunosuppressant, or one of these combined with a remyelinating strategy).

### Therapy of Individual Exacerbations and Optic Neuritis

#### Glucocorticosteroids (GCS)

GCS are considered the standard therapy for acute exacerbations and monosymptomatic disease such as optic neuritis suggestive of MS. As mentioned previously, definitive evidence of efficacy is lacking. Also lacking is a demonstration that a particular form, dose, or route of administration of GCS is more effective in treating exacerbations of MS or that GCS are more effective than ACTH. However, the effectiveness of IV MP on Gd enhancement and the apparent effectiveness

**TABLE VI    Treatment of Relapses with High-Dose Intravenous Corticosteroids***

| | |
|---|---|
| Days 1–5 | 1000 mg methylprednisolone IV in 250 ml 5% glucose solution as a short infusion over 30 min–1 h, in the morning. |
| Days 6 and 7 | 100 mg oral prednisolone. |
| Days 8 and 9 | 75 mg oral prednisolone. |
| Days 10 and 11 | 50 mg oral prednisolone. |
| Days 12 and 13 | 25 mg oral prednisolone. |

Precautions during high-dose steroid treatment.

The drug should be administered as a short infusion over 30–60 min rather than as a single injection, outpatient treatment is planned, the first dose should be given in the presence of a physician, nurse practitioner, nurse, and with emergency treatment available for the case of anaphylaxis or cardiovascular problems.

Although gastric ulcers are rare, the routine administration of H₂-blockers (ranitidine-HCl 150 mg bid) or antacids or equivalent treatment is recommended until the oral taper is finished. Special caution (gastroscopy before onset of therapy) is needed in patients with a history of ulcers.

Patients at risk for thrombosis should receive low dose SC heparin.

Substitution with oral potassium, if necessary.

A history of tuberculosis is no contraindication, but chest x-ray examination is advisable before treatment is started, and isoniazid (300 mg daily as a single dose; check for hepato- and neurotoxicity, especially visual impairment) and pyridoxine should be given in parallel. If the chest x-ray examination is equivocal, full tuberculostatic combination therapy is necessary.

Fetal malformations may be caused by steroid administration during the first 3 months. Later, even high-dose steroids can be administered with relative safety under the supervision of an experienced obstetrician.

During the entire course of steroid therapy, patients should stay home and rest as much as possible. Admission to a hospital is, however, not required.

High-dose IV steroids should not be given when patients show signs of acute infections (such as fever >38.5°C or leukocytosis) before initiation of the treatment.

Any signs of psychosis, including depression, hallucinations, or paranoid ideation, require the appropriate professional intervention. Patients at risk or with a history of psychosis should not be treated on an outpatient basis. With patients compromised by restlessness and sleep disturbances, prophylactic benzodiazepines (diazepam, 5 mg, or a short-acting benzodiazepine in the evening) are helpful.

Latent or manifest diabetes require frequent monitoring of blood and urine glucose levels and accordingly adjustment of diet and SC insulin administration.

Hypertonia requires frequent monitoring of blood pressure. Adjust antihypertensive drugs accordingly. Thiazide diuretics may lead to pronounced loss of potassium, which should then be substituted.

Glaucoma requires frequent monitoring of intraocular tension and adjustment of therapy if necessary.

*The dosages of the oral tapering may be modified; see text.

of high-dose IV MP in the treatment of optic neuritis indicate that this treatment is preferable to oral GCS. Monthly single pulses of IV MP are currently being assessed as maintenance therapy (see earlier). First results are promising, but it is too early to assess whether such a regimen is recommendable in patients failing both IFN-β and GA.

Dose and administration (see Table VI). Most commonly, MP is administered at a dose of 500–1000 mg/day with a preference for the higher dose. Administration may be divided into four doses of 250 mg, but administration as a single dose in the morning is acceptable and in common use, because it is more practical. The drug is administered in 250 ml of 5% glucose in a brief infusion over 30 minutes to 1 hour. A minimum of three daily doses is given, but treatment for 5 days is becoming more common, particularly in severe episodes. At many centers, high-dose IV MP is followed by a tapering course of oral prednisone, whereas others prefer to administer high doses of IV MP for 3–5 days only. Again, the importance of the follow-up course of oral GCS is not documented but thought to further prevent reopening of the blood–brain barrier. Some MS centers routinely administer IV MP for 10 days without oral GCS. Most often, oral GCS after IV MP are administered daily for approximately 10–12 days beginning with a dose of 80–100 mg (approximately 1 mg/kg) and tapering to 10 mg before stopping. If a patient has a positive history of gastric problems, either sucralfate (4 × 1 g), aluminium- and magnesium-hydroxide, or H₂-blockers (cimetidine or ranitidine) are given to prevent the development of gastric ulcers. Likewise, patients with risk factors for deep vein thrombosis or a history of thrombosis are routinely given low-dose heparin during the course of corticosteroids. In patients who feel agitated or have sleeping difficulties during treatment, either diphenhydramin-HCl (50 mg) or short-acting diazepam derivatives may be used.

The effectiveness of GCS in the chronic progressive phase of MS has not been established. Although it is unlikely that GCS have a significant effect on the overall course of patients with CP-MS, they may have a place in the treatment of episodes of worsening when superimposed on a CP course. The authors have, however, frequently seen a transient improvement or halt of progression when a course of high-dose IV GCS was given in patients with CP-MS. Such a course may also be used to replace the induction protocol when CTX therapy is initiated, although controlled data addressing this issue are lacking. One report has suggested that intrathecal GCS in the form of triamcinolone-acetonide may benefit patents with spastic paraparesis. Forty to 80 mg of a crystal suspension of triamcinolone-acetonide are administered intrathecally under sterile conditions one to three times at 3-day intervals. Lumbar arachnopathy was observed with earlier steroid preparations, but not with triamcinolone-acetonide.

**Contraindications and Adverse Effects.** Intravenous MP has been widely used and found to cause minimal adverse reactions. The problems that arise from long-term oral treatment are well known and include osteoporosis, gastric ulcers, hyperglycemia, glycosuria, redistribution of body fat, myopathy, neuropathy, cataract, glaucoma, hypertension, and psychosis to name only the most important. These side effects are not usually observed with the high-dose IV steroid regimens. Before treatment, the medical history of the patient should be reviewed in detail and should include complications from previous GCS treatments, history of gastric ulcers, diabetes, thrombosis, and infectious processes, as well as concurrent medications. Many patients will complain of changes, in taste and flushing. Gastric ulcers are uncommon, but gastric complaints requiring symptomatic treatment are common. Emotional changes, including mild euphoria, anxiety, and difficulty sleeping, are common. Other transient side effects, including hypertension and glyosuria, may occur. GCS may unmask or accentuate various infectious processes. Before treatment, patients should be evaluated carefully for urinary tract infections and tuberculosis. Transient oral or vaginal candidiasis may occur. Although aseptic necrosis is more commonly associated with the use of long-term GCS, it may occur during courses of high-dose therapy, and complaints of hip pain must be evaluated, preferably by MRI. The most serious, and fortunately, very rare side effects of high-dose IV MP are acute cardiac arrhythmias and anaphylactic reactions, both of which can be life threatening. Although side effects probably increase with repeated administration, up to four courses per year have been used by the authors without serious side effects.

In patients with particularly severe or life-threatening exacerbations (e.g., brainstem involvement, paraplegia or tetraplegia, or severe cerebellar signs), and in those who show an incomplete or no response to high-dose IV steroids, plasmapheresis (see earlier) should be considered.

### Therapy of RR-MS

Corticosteroids remain the first step in the treatment of acute episodes of worsening during the relapsing remitting phase of the disease. As mentioned earlier, the preceding data from the CHAMPS and ETOMS trials ask for the initiation of treatment as early as possible (Jacobs *et al.*, 2000; Comi *et al.*, 2001). Both IFN-β and GA have been found to be effective in reducing the frequency of exacerbations and are generally free of serious side effects, although the experience remains limited to several years of use. The side effects of beta-interferon are bothersome, however, and the treatment by either drug is expensive if the patient must cover its cost. In patients who are free of exacerbations for 2 or more years, treatment with IFN-β needs to be weighed carefully. The authors consider the clinical disease activity (frequency and severity of relapses) and the activity by MRI (number of lesions, T2 lesion load, presence of Gd-enhancing lesions, T1 lesions, atrophy) in their decision regarding which of the approved drugs is preferable. The most active patients receive IFN-β 1b, because its higher dose ($8 \times 10^6$ IU/every other day SC; Betaseron®) is probably more effective than the $6 \times 10^6$ IU IM/weekly of IFN-β 1a (Avonex®). The other IFN-β 1a (Rebif®), which is available in two doses (6 and $12 \times 10^6$ IU/3 times weekly), is probably equivalent to IFN-β 1b at the higher dose. Patients with milder disease or when the flulike symptoms and injection-site reactions are a concern, receive either IFN-β 1a once weekly $6 \times 10^6$ IU (Avonex®) or GA. The latter is clearly the best tolerated but needs to be injected every day. Oral GA is currently in clinical testing, but no data are available yet. As mentioned earlier, the authors routinely use MRI in assessing the disease activity and treatment response, but because of its cost, this is not possible for all patients. The parameters that should routinely be addressed if MRI is used to monitor disease activity and treatment response are provided in Table V. In summary, treatment with IFN-β or GA should be considered in patients with mild neurological involvement or even those relatively stable if there is evidence of active disease on Gd-enhanced MRI. In locations where IFN-β is available, it should be the first choice in treating patients with active RR-MS; however, GA seems to be an equivalent alternative to IFN-β-1a and preferable in patients with milder disease, or it can be used to replace IFN-β if flulike symptoms persist or if local injection site reactions are severe. If IFN-β and GA are not effective in reducing attacks, are not tolerated, or are unavailable, other treatments such as immunosuppressive drugs should be considered. It is important to note that the preceding recommendations are the current preferences of the authors and based on the available data. Particularly, the question of the optimal dose is still the subject of investigation.

**TABLE VII** Therapeutic Considerations in the Various Stages and Courses of MS Based on the Currently Available and Approved Drugs

| Stage or Course of Disease | Type of Treatment or Drug of Choice |
| --- | --- |
| Optic neuritis and monosymptomatic disease | High-dose IV steroids with oral taper at exacerbations*; discuss early treatment with patient and recommend IFN-β or glatiramer acetate (GA) treatment if MRI indicates a high probability of transition to or already existing MS. |
| RR-MS with relapses <2/yr, primarily sensory symptoms, low MRI activity | High-dose IV steroids with oral taper at exacerbations*; IFN-β, or GA continuously (first choice), azathioprine (second choice). |
| RR-MS with frequent relapses >2/yr, severe symptoms, high MRI activity | IFN–β–1b (Betaseron® or high-dose, i.e., $12 \times 10^6/3$ times/wk, of Rebif®) continuously (first choice), IFN-β (Avonex®) (second choice), azathioprine (third choice), IV-Ig (fourth choice). |
| RR-MS with secondary chronic progression progressive-relapsing disease (severe disease, rapidly progressing disability)** | IFN-β (Betaseron® or high-dose, i.e., $12 \times 10^6/3$ times/wk Rebif®) continuously (first choice as long as exacerbations are still present), mitoxantrone (also first choice); cyclophosphamide monthly pulses with or without IV steroid induction therapy (methotrexate, cladribine are alternatives; see text). |
| Primary chronic progressive MS (PP-MS), secondary chronic progressive MS without exacerbations | Currently no proven efficacy for any treatment regimen, mitoxantrone and cyclophosphamide may be tried (methotrexate, cladribine are alternatives; see text). |

*High-dose IV steroids with oral taper is always the treatment of choice for exacerbations.
**In patients with clear relapse activity and at an early time point in the transmission from RR- to SP-MS, IFN-beta is preferable, if progression is substantial, relapses are not present anymore and at later points after the transition mitoxantrone is the drug of choice among the approved drugs for MS.

## Immunomodulation of Relapsing-Remitting MS with IFN-β 1b or -1a

On the basis of data from several multicenter, prospective, double-blind, and controlled trials (Interferon-B MS Group: 1993; Paty *et al.*, 1993) recombinant IFN-β-1b with a serine for cysteine substitution in position 17, no glycosylation (Betaseron®), and two fully glycosylated preparations of recombinant IFN-β 1a (Avonex®; Rebif®) (Jacobs *et al.*, 1996; PRISMS-4, 2001) have been approved for the treatment of RR-MS in the United States and Europe. These three IFN-β preparations reduce the exacerbation rate to similar extents of approximately 30%, although the higher-dosed IFN-β 1b and Rebif® (exact dosing and route of administration later) have a more profound effect on MRI activity, indicating that $6 \times 10^6$ IU IFN-β 1a IM (Avonex®) may be underdosed (Durelli *et al.*, 2001). It was recently demonstrated that the early administration of IFN-β in cases with monosymptomatic demyelinating disease may delay the progression to MS by several months (Comi *et al.*, 2001; Jacobs *et al.*, 2000), arguing for treatment initiation as early as possible. The use of IFN-β in secondary chronic progressive disease is currently a controversial issue, mainly because two large-scale trials of IFN-β 1b in patients with SP-MS have shown different results (European Study Group, 1998). The European trial demonstrated a significant reduction in disease progression and was therefore halted before reaching the end of the study (European Study Group, 1998), whereas the US study with the same compound and also in SP-MS failed to confirm these results (data not reported yet). A careful comparison of the patient demography, as well as their MRI and clinical activity,

indicates that slightly different patient populations were treated. The patients in the European study had more exacerbations and overall probably more inflammation, whereas the patients in the US study seemed to have entered a more advanced phase of the disease with less inflammation. IFN-β is now the treatment of choice for RR-MS, and IFN-β 1b or the higher dose of IFN-β 1a (Rebif®; where available) are preferable over the lower dose (Avonex®) in patients with higher disease activity either based on the number and severity of exacerbations or the disease activity signs by MRI. On the other hand, the once weekly IM administration of $6 \times 10^6$ IU of IFN-β 1a (Avonex®) has not only less local side effects at the injection site, but also fewer flulike symptoms, and its dosing schedule is preferred by many physicians and patients. The preferred use of these compounds and of glatiramer acetate and others in the various forms of MS is outlined in Table VII.

Patients should be taught self-administration by a member of the medical team knowledgeable about the use of IFN-β. For the first few days, only half the dose is administered accompanied by anti-inflammatory drugs (for example acetaminophen $3 \times 500$ mg–$3 \times 1000$ mg) to prevent flulike symptoms. Other nonsteroidal anti-inflammatory drugs or low doses of oral steroids may be used for the same purpose. The overall rate of adverse reactions is low and includes mild anemia, neutropenia and thrombopenia, transient flulike reactions (that respond well to indomethacin and acetaminophen), inflammation at the injection site, transient fever, nausea, headache, fatigue, chills, sweating and myalgia, and mild-to-moderate elevation of liver enzymes. Rare adverse reactions are cardiac arrhythmia and allergic reactions. One suicide occurred in the third treatment year of the Betaseron® multicenter trial in RR-MS, but depression was not a common problem during further therapeutical trials and did not seem to be increased based on postmarketing experience. Most of the benign side effects are observed during the first 3 months of therapy. Although neutralizing antibodies against IFN-β 1b, and to a lesser extent against IFN-β 1a (Avonex®), have been observed in a considerable percentage of patients at some point during treatment, they may disappear and do not correlate consistently with a reduction in the treatment response. If neutralizing antibodies at higher titers seem to reduce treatment efficacy (e.g., if an increase of exacerbations or MRI activity are observed, it is not advisable to switch from one IFN-β preparation to another, because the antibodies cross-react with all preparations).

Pregnancy during IFN-β therapy is no absolute indication for an interruption, because in more than 100 pregnancies for which conception occurred during treatment, the children were born healthy and without birth defects. IFN-β is, however, known to induce abortions. The patient should be informed about this potential hazard and treatment discontinued during pregnancy.

The major problems of the long-term use of IFN-β are the high cost, unpleasant route of administration, and the frequency of side effects, although these are rarely severe. IFN-β is not an alternative to high-dose steroids during acute exacerbations. If the clinical response disappears or patients have a very active MRI, particularly Gd-enhancing lesion, we recommend determination of whether the patient developed neutralizing antibodies (NAB). Higher titer NAB inhibit IFN-β, and it is not useful in those instances to switch to a different IFN-β preparation, because the NAB are usually reactive to all preparations. In individuals who have responded for some time and then show a reduction in treatment efficacy, combination therapy with azathioprine is currently under investigation in clinical trials.

## Immunomodulation of RR-MS with Glatiramer Acetate

Glatiramer acetate (GA; copolymer-1, Cop-1; Copaxone®) has evolved into a good alternative to IFN-β and has also been approved for the treatment of RR-MS in the United States and the United Kingdom after successful testing in a large-scale, double-blind, randomized and controlled trial that showed a reduction of the exacerbation rate of about 30% (Johnson *et al.*, 1995). Long-term follow-up for more than 6 years demonstrated sustained benefit of patients (Johnson *et al.*, 1998). Furthermore, GA, which is administered daily at 20 mg SC, is tolerated very well and has overall shown very few side effects. These range from local reactions,

pain at the injection site, to an unusual type of hypersensitivity that is accompanied by palpitations and chest tightness, very rarely true immediate-type hypersensitivity. The palpitation-type side effect does not build up over time but rather occurs in an erratic fashion, early after injection and overall rarely, and is thus different from systemic immediate-type hypersensitivity reactions. Although the different modes of action of IFN-β and GA would favor their use as a combination therapy, it is not yet clear whether their combined use is not only prohibitively expensive but also therapeutically useful. Initial data indicate that the combination is safe and that the two treatments do not block each other (Lublin *et al.*, 2001). A controlled, double-blind, multicenter phase III trial of oral GA is under way, after it was shown that oral GA inhibits EAE.

GA seems to be a good alternative to IFN-β preparations. In patients with active RR-MS with profound MRI activity, we prefer IFN-β-1b or Rebif, whereas GA is probably equivalent to IFN-β 1a (Avonex®) in individuals with mild disease courses and less severe exacerbations (see also Table VII).

## Immunosuppression of RR-MS with Azathioprine

Despite controversy about its efficacy, each study of AZA in MS has shown a trend toward a reduction in relapse frequency. No beneficial effects have been observed in chronic progressive MS. As long as IFN-β and GA are not generally available in some countries, AZA remains a reasonable alternative in patients with RR-MS.

AZA is given at 2–3 mg/day in 1–3 doses. We usually start with 50 mg once in the morning and increase to the full dose over 4 weeks or longer if gastrointestinal side effects occur. Subdividing the dose may also lower the likelihood of the latter side effects. Treatment effects cannot be expected before 3 months after onset of therapy and are judged by a 10%–15% increase in mean erythrocyte volume and raise in relative reticulocyte number to >15/1000. As an additional laboratory parameter, treatment aims at a white blood cell count of 3000–5000/μL. If leukocytes are continuously <3000/μL, the dose needs to be adjusted. A small percentage of patients experience gastrointestinal problems (malaise, nausea, vomiting) at the onset of therapy. This is rarely observed when the dose is increased slowly, and further gastrointestinal reactions may be prevented by symptomatic treatment with antiemetics and temporarily lowering the dose. In some patients, gastrointestinal side effects develop later and may be so severe that AZA must be discontinued. Although not documented by clinical studies, AZA is potentially teratogenic, and contraceptive measures are required in both male and female patients and should be continued for at least 6 months (i.e., approximately two cycles of spermatogenesis) after discontinuing AZA in male patients. The following blood studies should be performed during AZA treatment: Blood cell count and differential blood count including MCV and reticulocytes every week during the first month, every second week during months 2 and 3, and later every 4–8 weeks; liver enzymes (SGOT, SGPT), conjugated and unconjugated bilirubin, alkaline phosphatase every 4 weeks during the first 3 months, every 3 months thereafter; general physical examination every 6 months. In general, evidence does not support a correlation between the therapeutic effect and leukopenia. The dose should therefore not be increased intentionally to lower the white blood count if the other blood parameters (see earlier) and the clinical course indicate a therapeutic effect.

Contraindications for treatment with AZA include chronic liver disease, chronic infections, pregnancy, and intolerance to AZA. Besides the desired actions on immune responsiveness and bone marrow function, the following adverse reactions may be observed:

- Tendency toward intrahepatic cholestasis (alcohol consumption and any other potentially hepatotoxic drug should be discontinued).
- Malaise, nausea, vomiting at onset (transient reduction of dose required as mentioned earlier).
- Hyperuricemia: if treatment is required, use uricosuric drugs. Because allopurinol significantly enhances the toxicity of azathioprine, the dose of AZA should be reduced to 25%.
- During purulent infections with temperatures >38.5°C, transient discontinuation of treatment and initiation of specific antibiotic therapy.

Pregnancy during AZA therapy is no absolute indication for an interruption, because by far most of the children conceived during treatment were born healthy. The patient should be informed about the potential hazard, and treatment should be discontinued during pregnancy.

Regarding the length of AZA treatment, it should, in theory, be given indefinitely. In practice, therapy will be continued for a minimum of 2 years without exacerbations before tapering can be attempted. Ideally, lack of disease activity should be documented by MRI. Higher incidences for lymphatic tumors have been observed in patients receiving various transplants and AZA therapy but have not been confirmed for AZA-treated MS patients. Finally, AZA may be considered as a combination to add to IFN-β, particularly in individuals in which the clinical course and MRI indicate that the patients initially responded but later developed neutralizing antibodies (see earlier).

## Treatment of RR-MS with CSA

Because its comparison to azathioprine failed to demonstrate superior activity in the treatment of MS and because its immunosuppressive effect is accompanied by considerable side effects, CSA is only an option for patients who did not respond to AZA and who do not have access to IFN-β and GA. Young patients with mild RR-MS should not receive CSA.

CSA is available as an oily liquid and although much higher doses are used in transplant medicine, a starting dose of 5–6 mg/kg body weight divided in two doses is recommended by most studies that examined CSA for the treatment of MS. The desired concentration of CSA in whole blood (12–24 hours after the last administration) ranges between 150 and 400 ng/ml (ELISA with a polyclonal antibody). The whole blood levels and blood urea nitrogen (BUN) and creatinine levels should be tested every 2 weeks during the first 2 months and from then on every 4 weeks. Glomerular filtration rate (GFR) should be determined by radioisotope clearance before onset and once every year during therapy. If serum creatinine values change by 20% or more, the dose of CSA should be reduced in increments of 1 mg/kg until the serum creatinine returns to within 20% of the starting value. GFR values should be maintained within 25% of their starting value. The bad taste of CSA can be improved by admixing juices.

**Contraindications and Adverse Reactions.** Patients with hypertonus and kidney diseases should not receive CSA. Acute renal toxicity can occur with excessively high whole-blood levels (because of overdosage for example; single dialysis may then be required). Adverse reactions of CSA therapy include renal dysfunction, hypertension, hirsutism, anemia, paresthesia, gingival hyperplasia, headache, tremor, nausea, liver dysfunction, arthralgia, skin changes (coarse facies), mental disturbance, fatigue, and seizures. An increased incidence of cancer was described for long-term CSA therapy in transplant patients, but there is no experience so far with respect to the risk for MS patients to have tumors develop. Severely acute, but potentially reversible, CNS toxicity includes confusion, cortical blindness, tetraparesis, seizure, and coma and may have a pathoanatomical correlate in abnormal white matter signal intensities. Side effects usually developed early during the first 3–6 months of therapy and then remained stable.

## IV-Ig

IV-Ig is another alternative medication in cases with RR-MS if one of the approved therapies or AZA is not possible for a variety of reasons. The initial dose of IV-

Ig is 2 g/kg body weight administered over 2–5 days. Subsequently, maintenance doses of 0.15–0.2 mg/kg are given in monthly intervals.

**Contraindications and Adverse Reactions.** Patients with IgA deficiency may develop severe systemic reactions, and IgA deficiency has therefore to be excluded by determination of serum immunoglobulin isotypes. If IgA deficiency exists, immunoglobulin pools that are free of IgA should be administered. Fast infusion of IV-Ig may lead to systemic hypersensitivity, as well as renal failure or cardiac insufficiency, because of volume overload. It is therefore recommended to control the infusion rate and exclude pre-existing renal or cardiac insufficiency. Rarely, IV-Ig infusion leads to aseptic meningitis, whereas slight fever, malaise, and urticaria are more frequently observed but less worrisome. Patients should be alerted that IV-Ig may interfere with vaccinations with live vaccines.

The costs of IV-Ig are very high, and, although the aforementioned evidence indicates that this treatment is effective, it should be reserved for patients who have had the approved therapies fail or where these are not available. Whether IV-Ig is more effective than AZA is difficult to judge, because the latter has not been tested under modern trial conditions.

### Treatment of Severe RR-MS, SP-MS, and CP-MS

#### Mitoxantrone (MIX)

As already mentioned, MIX has recently been approved for active forms of RR-MS and SP-MS. It is administered as a single infusion of 12 mg/m² every 3 months until the maximum safe dosage of 140 mg/m² has been reached.

**Practical Considerations.** Before initiating therapy with MIX, an ECG and echocardiogram should be performed. These should be repeated after 3–6 months to exclude a reduction of the left ventricular ejection fraction below 50%. To prevent nausea and vomiting, antiemetics (promethazine HCl 25–50 mg orally or 12.5–25 mg IV, ondansetron HCl, or tropisetron HCl or equivalent) are given 15–30 minutes before and 4 hours after infusion. Tropisetron HCl may be given as a single 5-mg tablet or injection per day. Ondansetron HCl is given either as an 8-mg tablet three times per day or as a 32-mg short infusion 30 minutes before and 4 and 8 hours after MIX administration. Sufficient fluid intake of at least 2–3 L should be assured during the day of infusion, and patients should be alerted that a bluish discoloration of their urine will occur.

**Contraindications and Adverse Effects.** MIX should not be given to patients with pre-existing cardiac problems such as cardiomyopathy. MIX is closely related to doxorubicin and daunorubicin and, although clearly less toxic, shares their potential for cardiotoxicity. Caution is required in patients with pre-existing myelosuppression, liver, or renal problems.

It is currently not clear up to which dose MIX can be administered, and because of the preceding potential cardiotoxicity, MIX should not be given beyond the upper dose of 140 mg/m². This limits the long-term use of MIX.

#### Cyclophosphamide (CTX)

CTX is preferred in patients with frequent relapses that are unresponsive to other immunsoppressive or immunomodulating treatments or those who reached a critical step in their disease course where further deterioration would lead to severe functional impairment (Table VII). The previous experiences with CTX have been summarized earlier. Our treatment recommendations are based on the results of these trials and our own clinical experiences with a slightly modified protocol. Controversial issues and questions that should be addressed in the future will be mentioned at the end.

Induction therapy may still be considered in very severe cases and consists of 600 mg/m² IV CTX in 250 ml 0.9% NaCl solution as a short infusion over 30 minutes on days 1, 4, 7, 10, and 13. Depending on whether daily WBCs show lymphopenia (<5%) or leukocyte counts <4000/μL before that, the fifth dose may be omitted. Infusions should be given in the morning, because nausea and vomiting are more likely to appear if the therapy is started in the afternoon. We do not use the induction therapy any more but rather a modified treatment protocol (see later) that has been adapted from the therapy of lupus nephritis.

CTX infusions should be accompanied by the following steps:

- To lower bladder toxicity, one ampule mesna (200 mg) is administered at hours 1, 4, 8, and 12 after the infusion besides sufficient oral and IV fluids (at least 3 L/day) to provide high urinary fluid excretion.
- To prevent nausea and vomiting, antiemetics (promethazine HCl 25–50 mg orally or 12.5–25 mg IV, ondansetron HCl, or tropisetron HCl or equivalent) are given 15–30 minutes before and 4 hours after infusion. Tropisetron HCl may be given as a single 5-mg tablet or injection per day. Ondansetron HCl is given either as an 8-mg tablet three times per day or as a 32-mg short infusion 30 minutes before and 4 and 8 hours after CTX administration.
- Oral prednisone or methylprednisolone, 1 mg/kg body weight, are given in the mornings of the cyclophosphamide infusions to lower general side effects. Some centers give 1 g IV MP instead of the lower oral dose.

- A cooling cap (hat) may be provided for the first 4–6 hours during and after infusion to prevent hair loss (To our knowledge, there is no controlled data on the benefit of this measure).

If induction therapy has been administered, boosters with 700 mg/m² follow every 2 months. During the first few cycles, weekly WBC should be performed to determine the nadir of leukocytes and lymphocytes. If induction therapy is used, we extend it a longer period of time than in published trials (Macklin *et al.*, 1992), because we frequently observed severe leukopenia (WBC <700/μL) when CTX is administered in short intervals. As already widely practiced in nephrology and rheumatology, there is a trend toward using monthly single CTX infusions without induction therapy (see later). This protocol has already successfully been applied by one of the authors under frequent MRI monitoring (Gobbini *et al.*, 1999).

**Future Directions with Cyclophosphamide and Alternative Treatment Regimen.** The data from the aforementioned and other studies suggest that the boosters are more important than the induction therapy for maintaining the treatment effect. It should therefore be tested in the near future whether the induction therapy with cyclophosphamide can be replaced by a course of high-dose IV steroids or be left out completely. We administer cyclophosphamide monthly for 9 months (beginning with 600 mg/m² and then monthly increase the dose by 100 mg until a desired nadir of 2000/μL leukocytes is reached) and then every other month (six times) without giving the induction protocol (Gobbini *et al.*, 1999). It should be stressed, however, that evidence from large trials with this protocol is still lacking.

**Adverse Reactions and Contraindications.** These are similar, but more pronounced, than what has been mentioned for AZA. The drug has a considerably higher toxicity than AZA and affects all rapidly dividing cell systems and tissues. Bladder toxicity, nausea and vomiting, and hair loss (only observed after the induction therapy, usually reversible) are the most frequent acute adverse reactions, whereas infertility and the potential risk for neoplasms to develop are the most worrisome potential long-term side effects. Oral cyclophosphamide should not be given in MS, because the rate of adverse reactions is considerably higher. Prior treatment with allopurinol significantly prolongs the serum half-life of CTX. CTX doses should therefore be reduced in such cases.

A number of alternatives including 2-chlorodeoxyadenosine and methotrexate (see earlier) are available when MIX or CTX fail to slow the progression of the disease. We prefer, however, not to generally recommend them before larger trials have confirmed the efficacy that was demonstrated in initial studies.

## THERAPY OF MS IN SPECIAL SITUATIONS

### Pregnancy and Contraception during Immunosuppressive Therapy

Neither pregnancy nor the use of oral contraceptives seem to have an effect on the overall course of MS. In general, patients with MS have a reduction in the frequency of exacerbations during pregnancy, whereas exacerbations increase during the immediate (6 months) postpartum period (Confavreux *et al.*, 1998). During pregnancy, IV-Ig therapy is an option in patients with frequent and severe relapses before pregnancy. Some neurologists elect to treat patients prophylactically with steroids during the 6-month postpartum period, but the usefulness of this approach is not established and, if used, should probably be restricted to those patients with very active disease. Physicians caring for patients with MS should take time to carefully discuss the potential risks of pregnancy in relationship to a patient's particular course, because considerable misinformation exists both with respect to the effect of pregnancy on the disease and in regard to the genetic aspects of MS. Although the effects of pregnancy on the course of the disease are minimal except in patients with severe disease and the genetic risk to offspring of patients with MS is only slightly increased, the patient should have an opportunity to clearly understand these risks. In addition, patients with MS should seek out obstetricians knowledgable in dealing with MS patients. Many physicians are concerned about the use of anesthesia in MS. Although little experience is available concerning this issue, casuistic evidence indicates that spinal, but not epidural, anesthesia may increase the risk of an exacerbation.

Safe contraception is required during immunosuppressive therapy and for at least 6 months afterwards (approximately two cycles of spermatogenesis). Pregnancy during immunosuppressive therapy has generally not been associated with fetal malformations, although as newer drugs are used, often little information is available.

Despite potential risks by the administration of glucocorticosteroids in pregnant women, they can be administered relatively safely after the first trimester. In case of severe exacerbations (considerable impairment of motor, cerebellar or visual function, for example, but not with minor sensory problems) during pregnancy, high-dose steroids should be given as mentioned earlier.

## Seizures

Although seizures are not a common sequel of MS, the risk is two to three times greater than that in the general population. If seizures occur, they should be treated with standard therapy.

## Vaccinations

There are no controlled studies on the effect of vaccination on exacerbations in MS. Limited studies failed to demonstrate an increased risk of exacerbations associated with influenza vaccination. However, the risk of exacerbations is increased during and after acute viral infections. The induction of exacerbations by live virus vaccines is therefore theoretically possible but usually greatly outweighed by the benefits of vaccination and the much higher risk posed by natural infection. Casuistic reports mention, however, the possible relation to live virus vaccines and a few inactivated viral vaccines such as that against hepatitis B. Oral polio vaccination with live virus should be replaced in MS patients by the subcutaneously administered inactivated vaccine (Salk vaccine). If relatives in the household are orally vaccinated, special hygienic care should be taken by the patient to avoid infection with the vaccine virus. Influenza vaccination is recommended for patients with a reasonable risk of exposure and particularly those patients with an increased risk of complications for the infection (for example patients older than 60 years).

## Heat

A large percentage of MS patients do not tolerate heat (hot bath, hot summer days) and experience worsening of motor function or reappearance or worsening of sensory deficits (Uhthoff phenomenon). Because this worsening is caused by a transient decrease of nerve conduction in previously affected areas of the CNS, the patient should be informed that it does not indicate new inflammatory activity. Patients with pronounced Uhthoff phenomenon may benefit from 4-aminopyridine, slowly increasing the dose to 10–15 mg tid (see later).

## Surgery and Anesthesia

Surgery and general anesthesia probably do not induce exacerbations, but the reported data on that are sparse. Elective surgery should, however, be planned in a stable phase of the disease whenever possible. No special therapeutic steps are required.

# SUPPORTIVE MEASURES AND SYMPTOMATIC TREATMENT

Supportive measures such as physiotherapy or changes of professional activity, as well as symptomatic therapies including treatment of bladder problems, spasticity, and fatigue to name only a few, are to be adjusted to the individual needs of the MS patient. We will try to cover the most important aspects, although we are aware of the fact that many of these measures are not established in the sense that their effacy was tested in controlled trials.

## Physiotherapy

Regular exercise and physiotherapy are an important aspect of the overall care of the patient with MS. A therapy program designed to maintain muscle strength and range of motion should be considered in all patients. Even those patients with mild RR-MS should be encouraged to maintain a regular program of exercise (Petajan et al., 1996). During acute and severe relapses, passive exercises are preferred, whereas active physiotherapy should be started as soon as the exacerbation improves. The type of exercise that is preferred in MS patients that have motor deficits and spasticity will be covered elsewhere in this book (Chapter 87). Continuous physiotherapy is necessary in all patients with severe motor dysfunction. It should not only try to preserve remaining motor function but also help the patient regain physical strength after exacerbations. In addition, it is important to teach the patient alternative strategies if impaired motor function cannot be recovered or if fluctuating spasticity with certain movements are a problem. Finally and maybe most importantly, it helps to motivate the patient to overcome physical impairments.

## General and Daily Activity of MS Patients

Unless patients already have considerable physical impairment, the diagnosis of MS does not mean that a patient has to adjust her or his life in a special way to the disease. In contrast, no restrictions should be given, unless a patient wants to pursue extreme physical acitivities (for example, marathon running). The patient should be encouraged not to change the desired activities, in particular not to stay away from them, but rather increase physical activity and prefer a mixed diet that is rich in protein, vitamins, and fiber, but low in fat, particularly animal fat. The efficacy of special diets has not been proven yet, but it was reported that highly unsaturated fatty acids may exert a moderate immunosuppressive effect. If patients wish to follow a special diet, one should only advise against it, if the diet is unbalanced or devoid of necessary nutritional ingredients, because the placebo effect and the feeling to do something positive by themselves is not to be underestimated. In addition to these general recommendations, special problems like balance trouble, walking difficulties, and the appropriate measures that can make life easier should be discussed with the patient, physician, physical therapist, and social workers. These measures include the use of a cane or four-pronged cane, walker, or scooter, as well as the installation of grab bars, handrails, a shower chair at home, or of special devices in the car.

## Professional Activity and MS

Although it is not known that certain professions pose a special risk to MS patients, a patient should be advised not to choose a profession that constantly requires heavy physical activity. As mentioned before, because of the highly variable course, such advice can only be given if a treating neurologist oversees at least 2–3 years of an individual disease course. Before that, any prognostic predictions are highly speculative, and even after longer follow-up, the disease may always change the severity and course.

## Fatigue

Fatigue is a frequent complaint of MS patients and usually most severe in the afternoon hours. If regular physiotherapy, exercise programs, coffee or tea, and planning of the daily schedule have been exhausted, either pemoline (beginning with 37.5 mg/day up to 75 mg as single dose in the morning), amantadine HCl (100 mg twice a day), or low-dose fluoxetine HCl (10–20 mg in the morning) should be tried in these patients. Recently, modafinil (Provigil®), 100–200 mg in the morning, has shown promising results in MS patients with marked fatigue.

## Psychiatric Symptoms and Disorders

The emotional well-being of the patient with MS is an important part of their total medical care. Because MS is a chronic disease with serious implications for many aspects of an individual's life, emotional problems including depression are not uncommon. Although pharmacological treatment is often indicated, creating an environment for the patient to explore the impact of a disease such as MS on their personal life is a critical part of the effective treatment. If depression is severe, referral to a psychiatrist is needed. Treatment with antidepressants such as fluoxetine, sertraline, or desipramine should be considered. Drug therapy should be combined with the appropriate pyschotherapy. Physicians should keep in mind that these psychotropic drugs may effect urodynamics in a positive way in MS. Thus, the overall pharmacological approach should be evaluated in patients when adding or stopping individual drugs.

## Spasticity

Spasticity is one of the most frequent symptoms observed in MS patients. Several drugs are available for reducing spasticity, including baclofen, tizanidin, tetrazepam, and diazepam. The principles of antispastic therapy and the exact dosage and the question when to administer baclofen through an intrathecal port using a subcutaneously implanted pump will be outlined elsewhere in this book (Chapter 87).

## Acquired Pendular Nystagmus

Acquired forms of nystagmus frequently impair vision (oscillopsia). This can be a very distressing symptom in MS. Patients experience an illusory motion of the environment that interferes with their ability to watch television and read. Memantine (15–60 mg PO per day) is usually very effective for symptomatic treatment of this condition. Gabapentin (up to 900 mg/day) has also been shown to be effective in some patients in a double-blind, controlled study (Bandini *et al.*, 2001**).

## Pain

The therapy of the different types of pain is covered separately in this book (Chapters 7 and 11). Because a number of paroxysmal sensations, including pain (for example when the neck is bent forward), are due to ephaptic transmission of impulses, carbamazepine (beginning with $2 \times 100$ mg of retarded carbamazepine and then slowly increasing the dose to therapeutic levels; under control of the blood levels) or gabapentin should be tried. Desipramine may be given as an alternative. The prostaglandin $E_1$ analogue misoprostol has shown benefit in patients with trigeminal neuralgia (Reder and Arnason, 1995).

## Tremor

In some patients, intention tremor may be the single most debilitating sign. Ballistic movements may become so severe, for example, while eating, that they interfere with all aspects of daily life and may even throw a patient out of bed in case intention tremor interferes with leg movements. Carbamazepine seems to be superior to the following drugs and should be tried first, starting at a similar dose that was described for the treatment of pain. Isoniazid (800–1200 mg/day in three to four single doses) and beta-blockers (propranolol 3–4 × 10–20 mg/day) or clonazepam (starting with 0.5 mg and slowly increasing the dose to 2–4 mg/day) may be tried, but none has shown consistent effects in larger groups of patients. Stereotactic interventions with lesioning of the medial parts of the ventrolateral thalamic nuclei should only be considered in the most severe cases and if a center with special experience is accessible for the patient. Tremor-inhibiting strategies should also be assessed together with a physiotherapist (e.g., the use of wrist weights may alleviate some of the problems of hand tremor).

## Bladder Dysfunction

Attention and appropriate care of bladder disturbance is one of the most important aspects of caring for a patient with MS. Although MS cannot be cured, effective management of bladder disturbance can be achieved in most cases. More than half of the patients develop bladder problems, initially most often urgency and frequency, later more frequently urinary retention or incontinence, at some point in the disease course. The individual situation requires differential treatment that will be covered elsewhere in this book (Chapters 103 and 104). Oxybutnin chloride, an anticholinergic that relaxes the smooth muscles of the bladder (5 mg 2–3 times daily) or imipramine HCl, an antidepressant (75–150 mg as a single dose in the evening or twice daily) may be given, but a number of other drugs that differentially influence the various systems involved in continence and emptying the bladder (see elsewhere in this book) may also be considered. The use of intraurethral or a suprapubic permanent catheter, as well as surgical interventions, should be postponed as long as possible. If the residual bladder postmicturition contents repeatedly exceed 100 ml when determined sonographically or by single cathetherization, self-catheterization needs to be started. The patient should be instructed how to perform self-catheterization by a member of the nursing staff or a physician. Because self-catheterization is more difficult in men and may often lead to urethral traumatization and transient bleeding, it may become necessary to insert a suprapubic catheter. A suprapubic cathether may also be necessary in women, particularly when they have motor and coordination deficits or ataxia. Because of the greatly reduced incidence of infections with suprapubic catheters, they are always preferable to urethral catheterization for longer periods of time unless it is expected that bladder problems will resolve quickly. If spasticity is interfering with micturition, an antispastic drug (for example, baclofen beginning with 3 × 5 mg/day) should be administered. Because of the increased risk of urinary tract infections, oral fluid intake should not be restricted.

## Problems in Defecation

Stool incontinence is uncommon during the course of MS, but when it occurs, it has severe implications in social adjustments. Stool incontinence may be treated with probanthine. Frequently, patients have constipation rather than incontinence. Restricted fluid intake because of fear of urinary urgency, lack of physical exercise, and MS-associated disturbances of intestinal motility are the most frequent reasons. These problems related to constipation should be discussed with the patient. Treatment will initially focus on increasing physical activity and the fluid intake of the patient supplemented by a fiber-rich diet. If these measures fail, stool softener should be tried before giving stimulating laxatives (bisacodyl, 5-mg tablets 2–3 once a day) transiently.

## Sexual Dysfunction

Similar to symptoms of disturbed bladder and bowel function, sexual dysfunction is experienced by up to 70% at some point during the course of the disease. These problems may be organic in nature, for example, caused by demyelination in central autonomic tracts, or have a secondary psychogenic background. If the problems occurred during an exacerbation together with other symptoms, sexual dysfunction may be temporary, and the situation should be explained to the patient. If sexual dysfunction is a problem of longer duration, a number of approaches may be taken including sildenafil (Viagra®) prostheses, or vasoconstricting injections (for more detail see Chapter 104). Besides the latter treatment, therapies aimed at the psychological impact with focus on coping methods for the couple are recommended.

### 4-Aminopyridine and 3–4-Diaminopyridine

Both aminopyridine derivatives are potassium channel blockers. They may cause improvement of symptoms through prolongation of nerve action potentials and consequently restoration of conduction along previously blocked demyelinated axons. The experimental efficacy of IV 4-aminopyridine was confirmed in a randomized, double-blind, placebo-controlled, crossover study (van Diemen *et al.*, 1992**) in MS patients. Oral administration of 4-aminopyridine is also possible, but most formulations that are currently tested are impaired by the short half-life of the drug. Slow-release formulations will probably be available soon, when ongoing trials have been evaluated. Therapeutic doses are in the range of $3 \times 5$ mg–$3 \times 10$ mg (up to $3 \times 15$ mg). Maximal effective doses that are tolerated by an individual patient must be titrated carefully. Side effects include light-headedness, vertigo, nausea, dizziness, and rarely seizures. Treatment with aminopyridines may be especially useful in patients with temperature-sensitive symptoms, those with long-standing progressive disease, and those that are at risk to lose critical motor capabilities (for example, ability to transfer from a wheelchair to the toilet or bed without assistance).

### COMPLICATIONS

The number and severity of complications increases with the duration of the disease. Frequent complications include increasing immobility caused by spasticity, paresis, ataxia, and deficits in coordination and fine motor skills. Furthermore, impaired bladder and bowel function often lead to urinary tract infection and decubital ulcers. Blindness may result from relapsing bilateral optic neuritis. As in other bed-ridden patients, thromboembolism and pneumonia rarely develop. Osteoporosis may be the result of prolonged steroid therapy and immobility. It should occur less often when high-dose IV rather than long-term oral steroids are administered; continuous physiotherapy and exercise programs, as well as a balanced, calcium-supplemented diet or estrogen substitution in postmenopausal women are part of the therapeutic plan. Any severe complications like swallowing or respiratory difficulties during exacerbations with lesions in the brainstem and recurrent urinary tract infections with sepsis or pneumonia may finally lead to death in severely affected patients. Those complications have, however, decreased over the last decades.

### TREATMENTS NO LONGER RECOMMENDED

The treatments available to date are still far from ideal. Especially, patients that are severely affected or have rapidly progressive disease will therefore often try alternative treatments, including herbs, homeopathic treatment regimens, or special diets. These therapies have usually not been tested carefully and according to recommended guidelines. Their efficacy is therefore generally unknown. As long as they are not harmful or too costly, it is nevertheless not recommended to advise the patient against them, because such measures may still have a placebo effect and give the patient the feeling that she or he is actively fighting the disease. Diets that lack essential nutrients, injections of sera, animal proteins, or even fresh cells carry, however, a number of risks and should be avoided.

### SPECIAL FORMS OF MS

Special forms of MS or clinically closely related entities include neuromyelitis optica or Devic's disease; Schilder's disease, which probably represents a form of MS manifesting in childhood, and leukoencephalitis periaxialis concentrica, also known under the name concentric sclerosis or Balo's disease. Finally, an acute, non-purulent encephalomyelitis is known under the name of MS of Marburg type. Because of the rarity of these syndromes, no larger trials have been conducted, but therapy is mainly based on casuistic experiences. Because these varieties of MS usually run a worse course than RR-MS or CP-MS, more aggressive treatments have been recommended. They include cyclophosphamide together with high-dose IV steroids, as well as plasmapheresis.

### REFERENCES

Achiron, A., Gabbay, U., Gilad, R., Hassin-Baer, S., Barak, Y., Gornish, M. *et al.* (1998). Intravenous immunoglobulin treatment in multiple sclerosis. Effect on relapses. *Neurology* 50(2), 398–402.

Arnon, R. (1996). The development of Cop1 (Copaxone), an innovative drug for the treatment of multiple sclerosis: personal reflections. *Immunol. Lett.* 50, 1–15.

Bandini, F., Castello, E., Mazzella, L., Mancardi, G. L., and Solaro, C. (2001). Gabapentin but not vigabitrin is effective in the treatment of acquired nystagmus in multiple sclerosis: How valid is the GABAergic hypothesis? *J. Neurol. Neurosurg. Psychiatry* 71, 107–110.

Barkhof, F., Filippi, M., Miller, D. H., Scheltens, P., Campi, A., Polman, C. H. *et al.* (1997). Comparison of MRI criteria at first presentation to predict conversion to clinically definite multiple sclerosis. *Brain* 120(Pt 11), 2059–2069.

Beck, R. W., Cleary, P. A., Anderson, M. M. *et al.* (1992). A randomized, controlled trial of corticosteroids in the treatment of acute optic neuritis. *N. Engl. J. Med.* **326**, 581–588.

Beck, R. W., Cleary, P. A., Trobe, J. D., Kaufman, D. I., Kupersmith, M. J., Paty, D. W. *et al.* (1993). The effect of corticosteroids for acute optic neuritis on the subsequent development of multiple sclerosis. *N. Engl. J. Med.* **329**, 1764–1769.

Becker, K. G., Simon, R. M., Bailey-Wilson, J. E., Freidlin, B., Biddison, W. E., McFarland, H. F. *et al.* (1998). Clustering of non-major histocompatibility complex susceptibility candidate loci in human autoimmune diseases. *Proc. Natl. Acad. Sci. USA* **95**, 9979–9984.

Beutler, E., Sipe, J. C., Romine, J. S., Koziol, J. A., McMillan, R., and Zyroff, J. (1996). The treatment of chronic progressive multiple sclerosis with cladribine. *Proc. Natl. Acad. Sci. USA* **93**(4), 1716–1720.

Bielekova, B., Goodwin, B., Richert, N., Cortese, I., Kondo, T., Afshar, G. *et al.* (2000). Encephalitogenic potential of the myelin basic protein peptide (amino acids 83–99) in multiple sclerosis: Results of a phase II clinical trial with an altered peptide ligand. *Nat. Med.* **6**, 1167–1175.

Bornstein, M. B., Miller, A. I., Slagle, S., Weitzman, M., Drexler, E., Keilson, M. *et al.* (1991). A placebo-controlled, double-blind, randomized, two-center, pilot trial of Cop 1 in chronic progressive multiple sclerosis. *Neurology* **41**, 533–539.

Burt, R. K., Burns, W., Ruvolo, P., Fischer, A., Shiao, C., Guimaraes, A. *et al.* (1995). Syngeneic bone marrow transplantation eliminates V beta 8.2 T lymphocytes from the spinal cord of Lewis rats with experimental allergic encephalomyelitis. *J. Neurosci. Res.* **41**, 526–531.

Calabresi, P. A., Fields, N. S., Maloni, H. W., Hanham, A., Carlino, J., Moore, J. *et al.* (1998). Phase I trial of transforming growth factor beta 2 in chronic progressive, MS. *Neurology* **51**, 289–298.

Coles, A. J., Wing, M. G., Molyneux, P., Paolillo, A., Davie, C. M., Hale, G. *et al.* (1999). Monoclonal antibody treatment exposes three mechanisms underlying the clinical course of multiple sclerosis. *Ann. Neurol.* **46**(3), 296–304.

Comi, G., Filippi, M., Barkhof, F., Durelli, L., Edan, G., Fernandez, O. *et al.* (2001). Effect of early interferon treatment on conversion to definite multiple sclerosis: a randomised study. *Lancet* **357**(9268), 1576–1582.

Comi, G., Filippi, M., and Wolinsky, J. S. (2001). European/Canadian multicenter, double-blind, randomized, placebo-controlled study of the effects of glatiramer acetate on magnetic resonance imaging—measured disease activity and burden in patients with relapsing multiple sclerosis. European/Canadian Glatiramer Acetate Study Group. *Ann. Neurol.* **49**(3), 290–297.

Confavreux, C. *et al.* (1998). Rate of pregnancy-related in multiple sclerosis. *N. Engl. J. Med.* **339**, 285–291.

Duda, P. W., Schmied, M. C., Cook, S. L., Krieger, J. I., and Hafler, D. A. (2000). Glatiramer acetate (Copaxone(R)) induces degenerate, Th2-polarized immune responses in patients with multiple sclerosis. *J. Clin. Invest.* **105**, 967–976.

Durelli, L., Ghezzi, A., Montanari, E., Zaffaroni, M., Bergui, E., Verdun, E. *et al.* (2001). The independent comparison of interferon (INCOMIN) trial: a multicenter randomized trial comparing clinical and MRI efficacy of IFN beta-1a and beta-1b in multiple sclerosis. *Neurology* **56**, A148.

Edan, G., Le Page, E., Taurin, G., Le Duff, F., Kerdoncuff, V., De Marco, O. *et al.* (2001). Safety profile of mitoxantrone in a cohort of 293 multiple sclerosis patients. *Neurology* **56**, A149.

Edan, G., Miller, D. H., Clanet, M. *et al.* (1997). Therapeutic effect of mitoxantrone combined with methylprednisolone in multiple sclerosis: a randomized multicenter study of active disease using MRI and clinical criteria. *J. Neurol. Neurosurg. Psychiatry* **62**, 112–118.

European Study Group on Interferon-beta 1b in secondary progressive MS, Miller, D., Polman, C., Pozilli, C., and Thompson, A. (1998). Interferon beta-1b delays progression of disability in secondary progressive multiple sclerosis: results of a European multicentre randomised trial. *Lancet* **352**, 1491–1497.

Fassas, A., Anagnostopoulos, A., Kazis, A., Kapinas, K., Sakellari, I., Kimiskidis, V. *et al.* (2000). Autologous stem cell transplantation in progressive multiple sclerosis—an interim analysis of efficacy. *J. Clin. Immunol.* **20**(1), 24–30.

Fazekas, F., Deisenhammer, F., Strasser-Fuchs, S., Nahler, G., and Mamoli, B. (1997). Randomised placebo-controlled trial of monthly intravenous immunoglobulin therapy in relapsing-remitting multiple sclerosis. Austrian Immunoglobulin in Multiple Sclerosis Study Group. *Lancet* **349**(9052), 589–593.

Gobbini, M. I., Smith, M. E., Richert, N. D., Frank, J. A., and McFarland, H. F. (1999). Effect of open label pulse cyclophosphamide therapy on MRI measures of disease activity in five patients with refractory relapsing-remitting multiple sclerosis. *J. Neuroimmunol.* **99**, 142–149.

Goodkin, D. E., Rudick, R. A., VanderBrug Medendorp, S., Daughtry, M. M., Schwertz, K. M., Fischer, J. *et al.* (1995). Low-dose (7.5 mg) oral methotrexate reduces the rate of progression in chronic progressive multiple sclerosis. *Ann. Neurol.* **37**, 30–40.

Gran, B., Tranquill, L. R., Chen, M., Bielekova, B., Zhou, W., Dhib-Jalbut, S. *et al.* (2000). Mechanisms of immunomodulation by glatiramer acetate. *Neurology* **55**, 1704–1714.

Group. British and Dutch Multiple Sclerosis Azathioprine Trial Group. (1988). Double-masked trial of azathioprine in multiple sclerosis. British and Dutch Multiple Sclerosis Azathioprine Trial Group. *Lancet* **2**(8604), 179–183.

Group. The Lenercept Multiple Sclerosis Study Group and The University of British Columbia MS/MRI Analysis Group. (1999). TNF neutralization in MS: results of a randomized, placebo-controlled multicenter study. *Neurology* **53**, 457–465.

Harris, J. O., Frank, J. O., Patronas, N., McFarlin, D. E., and McFarland, H. F. (1991). Serial gadolinium-enhanced magnetic resonance imaging scans in patients with early, relapsing-remitting multiple sclerosis: Implication for clinical trials and natural history. *Ann. Neurol.* **29**, 548–555.

Hauser, S. L., Dawson, D. M., Lehrich, J. R., Beal, M. F., Kevy, S. V., and Weiner, H. L. (1983). Immunosuppression and plasmapheresis in chronic progressive multiple sclerosis. Design of a clinical trial. *Arch. Neurol.* **40**(11), 687–690.

Hohlfeld, R. (1989). Neurological autoimmune disease and the trimolecular complex of T lymphocytes. *Ann. Neurol.* **25**, 531–538.

Hohlfeld, R. (1997). Biotechnological agents of the immunotherapy of multiple sclerosis. Principles, problems and perspectives. *Brain* **120**, 865–916.

Hughes, R. A. C. (1992). Treatment of multiple sclerosis with azathioprine. *In* "Treatment of Multiple Sclerosis." (R. A. Rudick and D. E. Goodkin, Eds.) pp. 157–172. Springer, London.

Jacobs, L., Salazar, A. M., and Herndon, R. (1987). Intrathecally administered natural human fibroblast interferon reduces exacerbations of multiple sclerosis: results of a multicenter double-blind study. *Arch. Neurol.* **44**, 589–595.

Jacobs, L. D., Beck, R. W., Simon, J. H., Kinkel, R. F., Brownscheidle, C. M., Murray, T. J., *et al.* (2000). Intramuscular interferon beta-1a therapy initiated during a first demyelinating event in multiple sclerosis. CHAMPS Study Group. *N. Engl. J. Med.* **343**, 898–904.

Jacobs, L. D., Cookfair, D. L., Rudick, R. A., Herndon, R. M., Richert, J. R., Salazar, A. M. *et al.* (1996). Intramuscular interferon beta-1a for disease progression in multiple sclerosis. *Ann. Neurol.* **39**, 285–294.

Johnson, K. P., Brooks, B. R., Cohen, J. A., Ford, C. C., Goldstein, J., Lisak, R. P. *et al.* (1995). Copolymer 1 reduces relapse rate and improves disability in relapsing-remitting multiple sclerosis: Results of a phase-III multicenter, double-blind, placebo-controlled trial. *Neurology* 45, 1268–1276.

Johnson, K. P., Brooks, B. R., Cohen, J. A., Ford, C. C., Goldstein, J., Lisak, R. P. *et al.* (1998). Extended use of glatiramer acetate (Copaxone) is well tolerated and maintains its clinical effect on multiple sclerosis relapse rate and degree of disability. Copolymer 1 Multiple Sclerosis Study Group. *Neurology* 50, 701–708.

Kappos, L., Comi, G., Panitch, H., Oger, J., Antel, J., Conlon, P. *et al.* (2000). Induction of a non-encephalitogenic type 2 T helper-cell autoimmune response in multiple sclerosis after administration of an altered peptide ligand in a placebo-controlled, randomized phase II trial. *Nat. Med.* 6, 1176–1182.

Kappos, L., Patzold, U., Dommasch, D., Poser, S., Haas, J., Krauseneck, P. *et al.* (1988). Cyclosporine versus azathioprine in the long-term treatment of multiple sclerosis–results of the German multicenter study. *Ann. Neurol.* 23(1), 56–63.

Kesselring, J. (1992). "Multiple Sklerose." Kohlhammer, Stuttgart.

Khatri, B. O., McQuillen, M. P., Harrington, G. J., Schmoll, D., and Hoffmann, R. G. (1985). Chronic progressive multiple sclerosis: double-blind controlled study of plasmapheresis in patients taking immunosuppressive drugs. *Neurology* 35(3), 312–319.

Kojima, K., Berger, T., Lassmann, H., Hinze-Selch, D., Zhang, Y., Gehrmann, J. *et al.* (1994). Experimental autoimmune panencephalitis and uveoretinitis transferred to the Lewis rat by T lymphocytes specific for the S100b molecule, a calcium binding protein of astroglia. *J. Exp. Med.* 180, 817–829.

Kümpfel, T., Schumann, E., Then Bergh, F., Gottschalk, M., Auer, D., Holsboer, F. *et al.* (2001). Monthly high-dose methylprednisolone pulse therapy in patients with multiple sclerosis: an MRI controlled study with single crossover design. *Neurology* 56, A76.

Lublin, F., Cutter, G., Elfont, R., Khan, O., Lisak, R., McFarland, H. F. *et al.* (2001). A trial to assess the safety of combining therapy with interferon beta-1a and glatiramer acetate inpatients with relapsing, MS. *Neurology* 56, A148.

Lucchinetti, C., Brück, W., Parisi, J., Scheithauer, B., Rodriguez, M., and Lassmann, H. (2000). Heterogeneity of multiple sclerosis lesions: implications for the pathogenesis of demyelination. *Ann. Neurol.* 47, 707–717.

Mackin, G. A., Dawson, D. M., Hafler, D. A., and Weiner, H. L. (1992). Treatment of multiple sclerosis with cyclophosphamide. *In* "Treatment of Multiple Sclerosis." (R. A. Rudick and D. E. Goodkin, Eds.). pp. 199–216. Springer, London.

Madsen, L. S., Andersson, E. C., Jansson, L., Krogsgaard, M., Engberg, J. *et al.* (1999). A humanized model for multiple sclerosis using HLA-DR2 and a human T-cell receptor. *Nat. Genet.* 23, 343–347.

Martin, R., McFarland, H. F., and McFarlin, D. E. (1992). Immunological aspects of demyelinating diseases. *Annu. Rev. Immunol.* 10, 153–187.

Mauch, E., Kornhuber, H. H., Krapf, H., Fetzer, U., and Laufen, H. (1992). Treatment of multiple sclerosis with mitoxantrone. *Eur. Arch. Psychiatry Clin. Neurosci.* 242(2–3), 96–102.

McDonald, W. I., Compston, D. A. S., Edan, G., Goodkin, D. E., Hartung, H. P., Lublin, F. D. *et al.* (2001). Diagnostic criteria for MS: guidelines from the International Panel on the Diagnosis of MS. *Ann. Neurol.* 50, 121–127.

McFarland, H. F., Barkhof, F., Antel, J., and Miller, D. H. (2002). The role of MRI as a surrogate outcome measure in multiple sclerosis. *Mult. Scler.* 8, 40–51.

McFarlin, D. E., and McFarland, H. F. (1982). Multiple sclerosis. (Part 1). *N. Engl. J. Med.* 307, 1183–1188.

McFarlin, D. E., and McFarland, H. F. (1982). Multiple sclerosis. (Part 2). *N. Engl. J. Med.* 307, 1246–1251.

Milanese, C., La Mantia, L., Salmaggi, A., and Caputo, D. (2001). Azathioprine and interferon beta-1b treatment in relapsing-remitting multiple sclerosis. *J. Neurol. Neurosurg. Psychiatry* 70(3), 413–414.

Miller, D. H., Thompson, A. J., Morissey, S. P., MacManus, D. G., Moore, S. G., Kendall, B. E. *et al.* (1992). High dose steroids in acute relapses of multiple sclerosis: MRI evidence for a possible mechanisms of therapeutic effects. *J. Neurol. Neurosurg. Psychiatry* 55, 450–453.

Molyneux, P. D., Kappos, L., Polman, C., Pozzilli, C., Barkhof, F., Filippi, M. *et al.* (2000). The effect of interferon beta-1b treatment on MRI measures of cerebral atrophy in secondary progressive multiple sclerosis. European Study Group on Interferon beta-1b in secondary progressive multiple sclerosis. *Brain* 123(Pt 11), 2256–2263.

Morissey, S. P., Miller, D. H., Kendall, B. E., Kingsley, D. P. E., Kelly, M. A., Francis, D. A. *et al.* (1993). The significance of resonance imaging abnormalities at presentation with clinically isolated syndromes suggestive of multiple sclerosis. A 5-year follow-up study. *Brain* 116, 135–146.

MSS Group. (1990). Efficacy and toxicity of cyclosporine in chronic progressive multiple sclerosis. A randomized, double-blind, placebo-controlled clinical trial. *Ann. Neurol.* 27, 591–605.

Neuhaus, O., Farina, C., Yassouridis, A., Wiendl, H., Then Bergh, F., Dose, T. *et al.* (2000). Multiple sclerosis: Comparison of copolymer-1-reactive T cell lines from treated and untreated subjects reveals cytokine shift from T helper 1 to T helper 2 cells. *Proc. Natl. Acad. Sci. USA* 97, 7452–7457.

Noseworthy, J. H., Luccinetti, C., Rodriguez, M., and Weinshenker, B. G. (2000). Multiple sclerosis. *N. Engl. J. Med.* 343, 938–952.

Noseworthy, J. H., Vandervoort, M. K., Penman, M., Ebers, G., Shumak, K., Seland, T. P. *et al.* (1991). Cyclophosphamide and plasma exchange in multiple sclerosis. *Lancet* 337(8756), 1540–1541.

Panitch, H. S., Hirsch, R. L., Schindler, J., and Johnson, K. P. (1987). Treatment of multiple sclerosis with gamma interferon: Exacerbations associated with activation of the immune system. *Neurology* 37, 1097–1102.

Paty, D. W., Li, D. K. B., the UBC MS/MRI Study Group, the IFNB Multiple Sclerosis Study Group. (1993). Interferon beta-1b is effective in relapsing-remitting multiple sclerosis. II. MRI analysis results of a multicenter, randomized, double-blind, placebo-controlled trial. *Neurology* 43, 662–667.

Paty, D. W., and McFarland, H. F. (1998). Magnetic resonance techniques to monitor the long term evolution of multiple sclerosis and to monitor definitive clinical trials. *J. Neurol. Neurosurg. Psychiatry* 64, S47–S51.

Petajan, J. H., Gappmaier, E., White, A. T., Spencer, M. K., Mino, L., and Hicks, R. W. (1996). Impact of aerobic training on fitness and quality of life in multiple sclerosis. *Ann. Neurol.* 39, 432–441.

Poser, C. M., Paty, D. W., and Scheinberg, L. (1983). New diagnostic criteria for multiple sclerosis: guidelines for research protocols. *Ann. Neurol.* 13, 227–231.

PRISMS-4 (2001). Long-term efficacy of interferon-beta-1a in relapsing MS. *Neurology* 56, 1628–1636.

Prineas, J. W. (1985). The neuropathology of multiple sclerosis. *In* "Handbook of Clinical Neurology, Demyelinating Diseases." (P. J. Vinken, G. W. Bruyn, H. L. Klawans, and J. C. Koetsier, Eds.) pp. 213–257. Elsevier Scian, Amsterdam/New York.

Reder, A. T., and Arnason, B. G. W. (1995). Trigeminal neuralgia in multiple sclerosis relieved by a prostaglandin E analogue. *Neurology* 45, 1097–1100.

Rohrbach, E., Kappos, L., Städt, D., Kaiser, D., Hennes, A., Dommasch, D. *et al.* (1988). Intrathecal versus oral corticosteroid therapy of spinal symptoms in multiple sclerosis: a double-blind controlled trial. *Neurology* 38(Suppl. 1), 256.

Romine, J. S., Sipe, J. C., Koziol, J. A., Zyroff, J., and Beutler, E. (1999). A double-blind, placebo-controlled, randomized trial of cladribine in relapsing-remitting multiple sclerosis. *Proc. Assoc. Am. Physicians* **111**(1), 35–44.

Rudge, P., Koetsier, J. C., Mertin, J., Mispelblom Beyer, J. O., Van Walbeek, H. K., Clifford Jones, R. *et al.* (1989). Randomised double blind controlled trial of cyclosporin in multiple sclerosis. *J. Neurol. Neurosurg. Psychiatry* **52**(5), 559–565.

Sipe, J. C., Romine, J. S., Koziol, J. A., McMillan, R., Zyroff, J., and Beutler, E. (1994). Cladribine in treatment of chronic progressive multiple sclerosis. *Lancet* **344**, 9–13.

Smith, M. E., Stone, L. A., Albert, P. S. *et al.* (1993). Clinical worsening in multiple sclerosis is associated with increased frequency and area of gadolinium enhancing MRI lesions. *Ann. Neurol.* **33**, 480–489.

Stone, L. A., Frank, J. A., Albert, P. S., Bash, C., Smith, M. E., Maloni, H. *et al.* (1995). The effect of interferon-b on blood-brain-barrier disruptions demonstrated by contrast-enhanced magnetic resonance imaging in relapsing-remitting multiple sclerosis. *Ann. Neurol.* **37**, 611–619.

Stone, L. A., Smith, M. E., Albert, P. S., Bash, C. N., Maloni, H., Frank, J. A. *et al.* (1995). Blood-brain barrier disruption on contrast-enhanced MRI in patients with mild relapsing-remitting multiple sclerosis: relationship to course, gender, and age. *Neurology* **45**, 1122–1126.

Tan, I. L., Lycklama a Nijeholt, G. J., Polman, C. H., Ader, H. J., and Barkhof, F. (2000). Linomide in the treatment of multiple sclerosis: MRI results from prematurely terminated phase-III trials. *Mult. Scler.* **6**(2), 99–104.

Teitelbaum, D., Milo, R., Arnon, R., and Sela, M. (1992). Synthetic copolymer 1 inhibits human T-cell lines specific for myelin basic protein. *Proc. Natl. Acad. Sci. USA* **89**, 137–141.

The Interferon-B Multiple Sclerosis Study Group. (1993). Interferon beta-1b is effective in relapsing-remitting multiple sclerosis I. Clinical results of a multicenter, randomized, double-blind, placebo-controlled trial. *Neurology* **43**, 655–661.

Tintore, M., Rovira, A., Martinez, M. J., Rio, J., Diaz-Villoslada, P., Brieva, L. *et al.* (2000). Isolated demyelinating syndromes: comparison of different MR imaging criteria to predict conversion to clinically definite multiple sclerosis. *A.J.N.R. Am. J. Neuroradiol.* **21**(4), 702–706.

Trapp, B. D., Peterson, J., Ransohoff, R. M., Rudick, R. M., Moerk, S., and Boe, L. (1998). Axonal transection in the lesions of multiple sclerosis. *N. Engl. J. Med.* **338**, 278–285.

Tubridy, N., Behan, P. O., Capildeo, R., Chaudhuri, A., Forbes, R., Hawkins, C. P. *et al.* (1999). The effect of anti-alpha4 integrin antibody on brain lesion activity in, MS. The UK Antegren Study Group. *Neurology* **53**(3), 466–472.

Tumani, H., Tourtellotte, W. W., Peter, J. B., and Felgenhauer, K. (1998). Acute optic neuritis: combined immunological markers and magnetic resonance imaging predict subsequent development of multiple sclerosis. The Optic Neuritis Study Group. *J. Neurol. Sci.* **155**(1), 44–49.

van Diemen, H. A., Poman, C. H., van Dongen, T. M., van Loenen, A. C., Nauta, J. J., Taphoorn, M. J., van Walbeek, H. K., and Koetsier, J. C. (1992). The effect of 4-aminopyridine on clinical signs in multiple sclerosis: A randomized, placebo-controlled, double-blind, cross-over study. *Ann. Neurol.* **32**, 123–130.

van Diemen, H. A. M., Polman, C. H., van Dongen, M. M. M. M., Nauta, J. J. P., Strijers, R. L. M., van Loenen, A. C. *et al.* (1993). 4-aminopyridine induces functional improvement in multiple sclerosis: a neurophysiological study. *J. Neurol. Sci.* **116**, 220–226.

van Oosten, B. W., Barkhof, F., Truyen, L., Baringa, J. B., Bertelsmann, F. W., van Blomberg, B. M. *et al.* (1996). Increased MRI activity and immune activation in two multiple sclerosis patients treated with the monoclonal anti-tumor necrosis factor antibody cA2. *Neurology* **47**, 1531–1534.

Wandinger, K. P., Stürzebecher, C.-S., Bielekova, B., Detore, G., Rosenwald, A., Staudt, L. M., McFarland, H. F., and Martin, R. (2001). The complex immunomodulatory effects of interferon-b in multiple sclerosis include the upregulation of T helper 1-associated marker genes. *Ann. Neurol.* **50**, 349–357.

Weiner, H. L., Dau, P. C., Khatri, B. O., Petajan, J., Birnbaum, G., McQuillen, M. P. *et al.* (1990). Double-blind study of true, vs. sham plasmapheresis in patients being treated with immunosuppression for acute attacks of multiple sclerosis. *Prog. Clin. Biol. Res.* **337**(283), 283.

Weiner, H. L., Mackin, G. A., Orav, E. J., Hafler, D. A., Dawson, D. M., LaPierre, Y. *et al.* (1993). Intermittent cyclophosphamide pulse therapy in progressive multiple sclerosis: Final report of the Northeast Cooperative Multiple Sclerosis Treatment Group. *Neurology* **43**, 910–918.

Weinshenker, B. G. (1994). Natural history of multiple sclerosis. *Ann. Neurol.* **36**, S6–S11.

Wiles, C. M., Omar, L., Swan, A. V., Sawle, G., Frankel, J., Grunewald, R. *et al.* (1994). Total lymphoid irradiation in multiple sclerosis. *J. Neurol. Neurosurg. Psychiatry* **57**(2), 154–163.

Yudkin, P. L., Ellison, G. W., Ghezzi, A., Goodkin, D. E., Hughes, R. A. C., McPherson, K. *et al.* (1991). Overview of azathioprine in multiple sclerosis. *Lancet* **338**, 1051–1055.

*Infections and Inflammatory Diseases*

# CHAPTER 54
# Prion Diseases

Richard Knight and Bob Will

Currently, there is no effective treatment for Creutzfeldt-Jakob disease (CJD). However, an increasing understanding of the underlying disease mechanisms has suggested a number of therapeutic possibilities. The fundamental disease mechanisms in CJD are the conversion of $PrP^C$ to $PrP^{Sc}$ and the aggregation and deposition of $PrP^{Sc}$ related amyloid.

Therapeutic proposals can be grouped together under three main headings:

1. Reducing $PrP^C$ production
2. Prevention of the posttranslational alteration and accumulation of $PrP^{Sc}$
3. Prevention of neurotoxicity of $PrP^{Sc}$ and amyloid deposits

In general, present treatment suggestions are based on theory, in vitro experiment, and animal model data, with little, if any, definitive evidence of efficacy or safety in human disease.

## CLINICAL ASPECTS (DIFFERENTIAL DIAGNOSIS)

### Introduction

CJD is a human prion disease, one of the transmissible spongiform encephalopathies (TSEs) (Table I). Like all prion diseases, CJD is invariably progressive and fatal. Although rare, it is of intense current scientific and popular interest, largely because of the appearance of a clinicopathological variant (variant CJD, vCJD), which is considered to be due to the transmission of bovine spongiform encephalopathy (BSE) from cattle to man. There are concerns that this form of CJD may affect considerable numbers of people with the additional possibility of secondary human-to-human spread of disease.

CJD exists in the following four clinicopathological forms (Table II):

1. **Sporadic CJD (spCJD)** is the most common form, with a worldwide distribution and an annual incidence of around 1 per million of the population per year. It is typically a disease of middle to late life, and its cause is unknown. No consistent evidence supports any suggestion that it is environmentally acquired. It may result from spontaneous conversion of a normal cellular protein ($PrP^C$) to an abnormal form ($PrP^{Sc}$), or a somatic mutation of the *PRNP* gene responsible for $PrP^C$ production (Prusiner, 1999).

2. **Genetic CJD (gCJD)** is an autosomal dominantly inherited disease reflecting an underlying mutation of the *PRNP* gene. It is rare, accounting for only a few cases per year in many countries, although there are relatively high incidence areas (e.g., Israel (Gambetti *et al.*, 1999). The underlying mutation is thought to be directly causal, presumably by means of an inherently unstable structure of the mutant $PrP^C$ protein. Although conceivable that the mutations are susceptibility factors conferring particular sensitivity to some unrecognized environmental agent, there is no evidence to support this view.

3. **Iatrogenic CJD (iCJD)** is the result of accidental transmission of CJD (usually spCJD) during medical or surgical treatment. Reported instances include neurosurgical instruments, depth EEG electrodes, corneal transplants, the surgical use of human dura mater grafts, and human pituitary hormone treatments (Brown *et al.*, 1992, 2000; Palmer and Collinge, 1997). There has been much consideration of the possibility of transmission of CJD by means of blood and blood products, but no proven instance has been reported in humans (Will and Kimberlin, 1998). There is, as of yet, no proven instance of iatrogenic transmission of vCJD; however, the potential risks increase the need for a safe, effective, prophylactic treatment.

TABLE I  The Transmissible Spongiform Encephalopathies

| Disease | Host | Notes |
|---|---|---|
| Animal diseases | | |
| Scrapie | Sheep, goats | Naturally occurring animal diseases |
| Chronic wasting disease | Mule deer, elk | Naturally occurring animal diseases |
| Transmissible mink encephalopathy | Mink on mink farms | Cause uncertain |
| BSE | Cattle | Caused by contamination of feed |
| Feline spongiform encephalopathy | Domestic cats Cat species in zoos | Caused by BSE Contamination of feed |
| Human diseases | | |
| Kuru | Man | Confined to Papua New Guinea, now rare |
| CJD | Man | Different forms: see Table II |
| Gerstman Sträussler Schencker Syndrome | Man | Rare familial disease |
| Fatal familial insomnia | Man | Rare familial disease |

**4. Variant CJD (vCJD)**, formerly new variant CJD, was first reported in 1996 and is thought to be due to the transmission of BSE from cattle to man by means of food (Knight, 1999; Will *et al.*, 1996). Currently (July, 2002), 124 (definite or probable) cases have been confirmed in the United Kingdom, with six in France, one in Italy, one in the USA and one in the Republic of Ireland.

## Clinical Features

### Sporadic CJD

The characteristic feature of sporadic CJD is rapidly progressive dementia, and the mean total illness duration is about 4 months. There are a range of associated neurological signs, most frequently ataxia and myoclonus, which in combination with the rapidity of progression allow distinction from more common forms of dementia. Terminally, there is akinetic mutism and death is usually due to intercurrent infection. Atypical clinical presentations include a pure cerebellar syndrome or cortical blindness, and rarely the clinical course is protracted with survival extending to more than a year or longer (Will *et al.*, 1999).

### Variant CJD

The clinical illness in vCJD is relatively stereotypical, but is more protracted than sporadic CJD. There is an initial phase lasting on average 6 months, dominated by

TABLE II  Clinicopathological Forms of CJD

| Type | UK Frequency* 1996 | Notes |
|---|---|---|
| Classical sporadic | 66% | Random worldwide distribution |
| Genetic familial | 10% | Associated with *PRNP* mutations, autosomal dominant |
| Iatrogenic | 7% | Accidental transmission, surgical hormone treaments, etc |
| New variant | 17% | First reported 1996, Essentially in UK, Shown to be due to BSE |

*Note. Figures are approximate and for UK only. Based on deaths caused by definite or probable CJD.

psychiatric symptoms, including depression, anxiety, and withdrawal. This is followed by progressive ataxia and dementia with associated involuntary movements that may be myoclonic, choreiform, or dystonic. About half the cases describe painful sensory symptoms in the limbs or trunk, which can be present from the outset. The terminal phase is similar to sporadic CJD, and the mean survivial is 14 months (Will *et al.*, 1999, 2000).

### Genetic CJD

The clinical features in genetic forms of CJD are heterogeneous and to some extent correlate with the underlying mutation in the *PRNP* gene, although there can be great variability both within and between families. All genetic forms of CJD involve a progressive and fatal neurological disorder, but there is marked variation in clinical presentation, examples include ataxia, spastic paraparesis, autonomic dysfunction, or a neuropsychiatric disorder. The clinical phenotype in association with mutations at codon 200 of the *PRNP* gene is similar to sporadic CJD. Overall, genetic forms of CJD present about 10 years earlier than sporadic cases and have a more protracted clinical course, which can extend to many years (Will *et al.*, 1999).

### Iatrogenic CJD

CJD has been transmitted iatrogenically by a number of routes, but these can be divided into central (e.g., neurosurgical instruments, dura mater grafts) and peripheral (e.g., human growth hormone). The clinical features after central transmission are similar to sporadic CJD, whereas after peripheral infection, there is a progressive cerebellar syndrome, and dementia may not occur until late in the illness, if at all (Will *et al.*, 1999).

### Diagnosis and Differential Diagnosis

### General

In general, the diagnosis of CJD requires neurological expertise. Absolutely definitive confirmation of the diagnosis can be made only by neuropathological examination of the brain. A vital part of any clinical approach to a suspected case of CJD remains the exclusion of other possible neurological illnesses. However, clinical diagnostic criteria have been developed and validated allowing a premortem diagnosis, with a relatively high degree of certainty, in most of cases. However, as with all such diagnostic criteria, their reliability is critically dependent on their use by clinicians with expertise and experience.

The essentials are :

1. The exclusion of other diagnoses
2. A characteristic clinical course
3. Supportive investigations, in particular, the EEG, CSF analysis, and MR imaging

### spCJD

In typical cases of spCJD the diagnosis can be made clinically with some confidence, because the phenotype is relatively distinctive and the rapidity of progression very unusual in other causes of dementia. In any patient with rapid cognitive decline it is essential to exclude a metabolic disorder, an inflammatory disorder such as encephalitis, or a cerebral mass lesion. Once these diagnoses are excluded by investigation, the major differential diagnosis of spCJD is nonspecific encephalopathy, and in many such cases recovery or sustained improvement may exclude the diagnosis of CJD. In suspected cases of CJD that die the main differential diagnosis is Alzheimer's disease, which can be associated both with myoclonus and a subacute course. In the rare cases of spCJD with an extended course and only gradual deterioration, it may be impossible to reach a diagnosis of CJD in life.

The accepted clinical diagnostic criteria are given in Table III.

A diagnosis of "probable" spCJD, as defined by these criteria, carries approximately a 95%–97% certainty, which is significantly higher than for many "probable" diagnostic categories in medicine (WHO, 1998).

The EEG has been very useful in supporting a diagnosis of spCJD but is being used less often now that CSF 14-3-3 analysis is available. In most cases, the EEG shows, at some stage of the illness, the presence of periodic, biphasic or triphasic complexes (Steinhoff *et al.*, 1996). However, this characteristic pattern may appear late in the illness and may not develop at all. It is also not absolutely specific for spCJD, being seen in other conditions, including, occasionally, Alzheimer's disease.

Routine CSF analysis is normal, although there may be a mild-to-moderate total protein elevation. The CSF 14-3-3 test is probably the most useful supportive diagnostic test. The 14-3-3 test is simply a normal brain protein that has no known intrinsic role in CJD pathophysiology and is released into the CSF in a variety of conditions when there is neuronal destruction. However, in appropriate clinical contexts, most other causes of elevated 14-3-3 CSF levels can be distinguished from CJD on other grounds. Therefore, it has not only a high sensitivity for a diagnosis of spCJD but also a high specificity, providing the test is performed in appropriately selected patients. The 14-3-3 test is now

---

**TABLE III    Diagnostic Criteria for Sporadic CJD**

1. *Definite*
   Neuropathologically (histologically and/or immunocytochemically) confirmed
2. *Probable/possible*
   I  Rapidly progressive dementia
   II  Myoclonus
       Visual or cerebellar problems
       Pyramidal or extrapyramidal features
       Akinetic mutism
   Probable: I + at least 2 of II + Typical periodic EEG findings *or* Possible + Positive 14-3-3
   Possible: I + at least 2 of II + Duration of <2 y

incorporated into the accepted diagnostic criteria (Zerr *et al.*, 1998, 2000a). The principal indication for MR imaging of the brain in this context is the exclusion of alternative diagnoses. However, the cerebral MRI may show certain characteristics in a proportion of cases (Finkenstaedt, 1996). This requires further evaluation, and MR findings are not included in the currently accepted diagnostic criteria.

## vCJD

Currently, there are about three times as many suspect cases of vCJD as subsequently confirmed cases. The major differential diagnosis is nonspecific encephalopathy, and, as in spCJD, clinical improvement or recovery excludes the diagnosis of vCJD. Metabolic, structural, and inflammatory disorders must be excluded by investigation, and important differential diagnoses include cerebral vasculitis and Wilson's disease, both of which are potentially treatable. In patients who die the most frequent alternative diagnosis is spCJD, and it is important to recognize that spCJD does occur in younger age groups, albeit rarely (Will *et al.*, 2000). Applying established diagnostic criteria for vCJD and in particular the findings on MRI brain scan may allow relatively accurate distinction between vCJD and spCJD in life, but a definitive diagnosis depends on neuropathological examination. All cases of vCJD to date have been methionine homozygous at codon 129 of the *PRNP* gene, and the clinical and pathological phenotype of BSE infection in association with other codon 129 genotypes cannot be predicted.

The accepted clinical diagnostic criteria are given in Table IV.

The EEG, in vCJD, has never shown the periodic complexes characteristic of spCJD. It usually shows a progressive nonspecific abnormality but may be surprisingly normal for a time, even in the presence of a significant neuropsychiatric illness.

As with all forms of CJD, routine CSF analysis is normal, although there may be a mild-to-moderate total protein elevation. Although a positive CSF 14-3-3 test may provide diagnostic support, 14-3-3 analysis does not have the same degree of sensitivity in vCJD as it does in spCJD, and this is not included in the presently accepted clinical diagnostic criteria (Green *et al.*, 2001).

The most useful supportive diagnostic test for vCJD is cerebral MRI. Aside from its role in excluding other possible diagnoses, in approximately 80% of cases, a characteristic abnormality is seen in the posterior thalamus and has been termed the "pulvinar sign" (Zeidler *et al.*, 2000). Although similar radiological abnormalities have been seen in other illnesses, they should be readily distinguishable from vCJD on other grounds.

**TABLE IV**    **vCJD Diagnostic Criteria**

I
A—Progressive neuropsychiatric disorder
B—Duration of illness >6 mon
C—Routine investigations do not suggest an alternative diagnosis
D—No history of potential iatrogenic exposure
E—No evidence of a familial form of TSE

II
A—Early psychiatric symptoms*
B—Persistent painful sensory symptoms†
C—Ataxia
D—Myoclonus or chorea or dystonia
E—Dementia

III
A—EEG does not show the typical appearance of sporadic CJD‡ (or no EEG performed)
B—Bilateral pulvinar high signal on MRI scan

IV
A—Positive tonsil biopsy§
Definite: I A **and** neuropathological confirmation of vCJD‖
Probable: I **and** 4/5 of II **and** III A **and** III B        **OR**
                    I **and** IV A‖
Possible: I **and** 4/5 of II **and** III A

---

*Depression, anxiety, apathy, withdrawal, delusions.
†This includes both frank pain and/or dysesthesia.
‡Generalized triphasic periodic complexes at approximately 1/s.
§Tonsil biopsy is *not* recommended routinely or in cases with EEG appearances typical of sporadic CJD but may be useful in suspect cases in which the clinical features are compatible with vCJD and MRI does not show bilateral pulvinar high signal.
‖Spongiform change and extensive PrP deposition with florid plaques, throughout the cerebrum and cerebellum.

## gCJD

The autosomal dominant inheritance of gCJD means that, in most cases, there is a relevant family history to aid diagnosis. However, some cases of CJD related to an underlying *PRNP* gene mutation seem to have no such family history. The clinical features of gCJD vary somewhat according to the nature of the mutation and other factors, such as the *PRNP* gene codon 129 polymorphism. However, *PRNP* gene analysis is possible on a simple blood sample, and this allows clear diagnosis of genetic cases. The possibilty of gCJD should be considered in hereditary neurological disorders of uncertain etiology and perhaps also in apparently sporadic cases of atypical neurodegenerative diasease.

## iCJD

The identification of antecedent exposure to a recognized risk factor is essential to the diagnosis of iatrogenic CJD. However, links between cases may not be obvious, unless previous surgical records are consulted, for example, the use of a dura mater graft in a previous neurosurgical procedure. In addition, the clinical features of human growth hormone (hGH)–related CJD are relatively distinct and characteristic, being essentially a progressive cerebellar syndrome, with dementia appearing at a late stage of illness, if at all. The critical component in the diagnosis of iCJD is obtaining a detailed medical and surgical history in CJD suspects, particularly if there is an early cerebellar syndrome or the patient is young.

## NATURAL COURSE

### Incidence/Prevalence

#### spCJD

SpCJD is a rare disease, with an incidence of approximately 1 case per million population per year. Because the duration of illness is measured in months in most cases, the prevalence of spCJD is equivalent to the incidence. SpCJD occurs on a worldwide basis (Will *et al.*, 1999).

#### vCJD

To date there have been 124 cases of vCJD identified in the United Kingdom since 1995, and, although clearly a very rare disease, there has been a significant increase in the incidence year by year, with current estimates indicating a doubling every 3 years. Because BSE infection in humans may have a potentially prolonged, but unknown, incubation period, estimates of possible total numbers of future cases have a wide range. The possibility of a significant epidemic cannot be excluded. Six cases of vCJD have been identified in France, one in The Republic of Ireland, one in Italy and one in the USA.

#### gCJD

Overall, the incidence of gCJD is approximately one tenth the incidence of spCJD, although this varies from country to country, and there are some ethnic or geographical isolates with 60–100 times the incidence of gCJD, notably in Libyan-born Israelis and in Slovakia (Gambetti *et al.*, 1999). Although the survival in gCJD is generally longer than in sCJD, the prevalence of gCJD is very low in most countries.

#### iCJD

To date, worldwide, there have been more than 100 cases of both hGH-related CJD and human dura mater–

related CJD (Brown *et al.*, 2000). The other forms of iCJD are exceedingly rare.

## General Epidemiology

### spCJD

SpCJD is a disease predominantly affecting individuals in late middle age, with a mean age at death of about 65 years. However, a wide age range can be affected, and small numbers of cases have been described in individuals in their teens or in their 80s or 90s. SpCJD occurs worldwide, and most studies have shown no evidence of a geographical clustering of cases. Furthermore, case-control studies have provided no consistent evidence of an increased risk through occupation, diet, or animal exposures, and the cause of spCJD is unknown (Zerr *et al.*, 2000b). No good evidence of an environmental source of infection exists, and the current favored hypothesis is that spCJD is caused by a spontaneous somatic mutation in the *PRNP* gene or perhaps the spontaneous conversion of PrP to the disease-associated form (Prusiner, 1999).

### vCJD

One characteristic of vCJD is that it affects younger individuals than spCJD or gCJD. The average age at death is 29 years, with clinical onset occurring in individuals as young as 12 years. Until recently, the oldest affected patient was 53 years old at death, but the identification of a case aged 74 years has significantly extended the age range. Cases of vCJD have been identified throughout the United Kingdom, but the incidence is about double in the North of the United Kingdom compared with the South (Cousens *et al.*, 2001). The reason for this is unknown but may reflect differences in dietary exposures to the BSE agent. The favored hypothesis is that transmission of BSE to the human population was through past dietary exposure to infection from cattle, probably to bovine CNS tissues containing high titers of infectivity (Knight, 1999). The cases of vCJD in Ireland and the USA may have been infected in the United Kingdom, but the cases in France and Italy had never visited the United Kingdom, implying indigenous exposure to BSE.

### gCJD

Overall, the age at death in gCJD is about 10 years less than spCJD. Cases usually occur in familial aggregates, but in about a third of cases there is no relevant family history. gCJD occurs worldwide, but there is variation from country to country both in incidence and the distribution of specific *PRNP* mutations (Gambetti *et al.*, 1999).

## iCJD

Although cases of iCJD have occurred worldwide, there is marked variation in incidence form country to country. Most cases of dura mater–related CJD have been identified in Japan, probably related to a high local use of this material in neurosurgery (Brown *et al.*, 2000). A specific product, Lyodura, has been implicated in nearly all cases of dura mater–related CJD, regardless of country of use. hGH-related CJD has occurred predominantly in countries using locally produced rather than commercially sourced hormone, notably France, the United Kingdom, and the United States. Many countries have now banned the use of human dura mater grafts and recombinant growth hormone largely replaced hGH in 1985.

## Risk Factors

### spCJD

Case-control studies have failed to identify any consistent risk factor that increases the likelihood of spCJD developing. The only recognized factor that increases risk is the genotype at codon 129 of the *PRNP* gene, with methionine homozygosity in about 80% of cases of spCJD compared with about 30% of the general Caucasian population with this genotype (Windl *et al.*, 1996).

### vCJD

The risk factors for the development of vCJD are a young age, residence in the United Kingdom, and methionine homozygosity at codon 129 of *PRNP*. The reason for the young age of cases is unknown. Of 111 cases of vCJD to date, 108 have a history of residence in the United Kingdom for a period of years and most throughout life. This relates to the geographical risk of exposure to BSE, and there have been more than 180,000 cases of BSE in the United Kingdom compared with much smaller numbers of cases of BSE in other countries in Europe. The occurrence of six cases of vCJD, who had lived exclusively in France, implies local exposure to either BSE in French cattle or to BSE contaminated exports from the United Kingdom. All tested cases of vCJD have been methionine homozygous at codon 129 of *PRNP*. It is possible that cases of human BSE infection with an alternative codon 129 genotype may occur in the future, because variations at this locus may extend the incubation period.

## gCJD

The consensus is that gCJD is caused by the underlying mutation that leads to instability in PrP structure and eventually the self-replicating disease-associated form of PRP. There is no evidence of an environmental influence on susceptibility to gCJD.

## iCJD

Risk factors for iCJD relate to relevant antecedent medical or surgical exposures. In contrast to other forms of human prion disease, homozygosity, either valine or methionine, is a risk factor for hGH-related CJD. In iatrogenic CJD caused by mater grafts, methionine homozygosity is a risk factor.

## Natural Course

### General

All forms of CJD are invariably progressive and ultimately fatal. There is progressive neurological impairment, typically involving dementia, cerebellar, pyramidal and other dysfunction, often with involuntary movements. The final stage of illness is usually one of total dependence, often to the degree of an akinetic mute state. Death results from intercurrent illness such as pneumonia.

### spCJD

The clinical course in spCJD is typified by the rapid accumulation of multifocal and severe neurological deficits, leading to death within months. Neuropathologically, there is severe neuronal loss in association with the other pathognomonic histological features, spongiform change, and astrocytic gliosis. It is unlikely that the neurological deficits could be reversed, unless diagnosis was possible at a very early stage.

### vCJD

The median duration of illness in vCJD is 14 months, and the later stages are clinically similar to spCJD with progressive and devastating multifocal neurological impairment. The early clinical course in vCJD involves psychiatric symptoms, with a minority of cases having neurological symptoms and signs that might suggest an underlying neurodegenerative process. The distinction from a purely psychiatric illness may be impossible early in the clinical course, and an important objective is to identify markers for the true diagnosis before the development of neurological deficits.

It is likely that there was significant human exposure to the BSE agent, particularly in the United Kingdom, and that a proportion of the exposed population will have vCJD developing. Prion diseases are characterized by prolonged asymptomatic incubation periods extending to years or decades, during which it might be possible to use prophylactic therapy to prevent the development of subsequent neurological disease. There is currently no test to identify humans (or animals) that are infected with BSE, and the development of such a test will be essential to targeting prophylactic therapy.

### gCJD

In many cases of gCJD, the clinical picture and prognosis are similar to spCJD. In other forms the clinical course can be protracted with focal neurological deficits only gradually evolving, often over many years. With the advent of genetic testing, at-risk family members can now be screened, with consent, for mutations of *PRNP*, and mutation-positive individuals may be candidates for prophylactic therapy, should this be developed in the future.

### iCJD

The duration of illness in iCJD is in general similar to spCJD, although some cases of dura mater–related CJD and GH-related CJD have a more protracted course, with survival extending to well over 1 year. The population of individuals at risk of GH-related CJD can be, and in many countries has been, identified. The risk of CJD developing after GH-therapy is significant, and the importance of developing a prophylactic therapy apparent.

## PRINCIPLES OF THERAPY

### Introduction

No therapeutic intervention has yet been shown to slow, halt, or reverse disease progression in humans in any form of CJD. However, concern about the possibility of an epidemic has made treatment a greater priority, and advances in the understanding of the nature of prion diseases has perhaps made its development a more realistic possibility.

### Underlying Disease Mechanisms

The neuropathological characteristics are essentially neurodegenerative in nature typified by spongiform change, neuronal loss, astrocytic proliferation, and gliosis, in the absence of any inflammatory features. The essential finding is that of deposition of an abnormal isoform (PrP$^{Sc}$) of a normal cellular protein (PrP$^{C}$). Despite its degenerative pathological condition, CJD is transmissible under certain conditions. The precise nature of the transmissible agent is still not entirely clear, although the dominant view is that either PrP$^{Sc}$ itself is the agent or is the most important component of it (Palmer and Collinge, 1997; Prusiner, 1999). The importance of PrP$^{C}$ in prion disease is evident in the fact that PrP$^{0/0}$ mice are resistant to scrapie (Büeler *et al.*, 1993) and that overexpression of PrP$^{C}$ in mice results in spontaneous neurodegenerative disease (Westaway *et al.*, 1994).

PrP$^{C}$ is a normal host-encoded protein, transcribed from the *PRNP* gene, which is situated on chromosome 20 in humans. Its precise physiological function is unclear, although there are accumulating data on its properties and possible functions, which are discussed later. The gene is highly conserved across mammalian species, suggesting an important role maintained through evolution and PrP mRNA is particularly expressed in the brain. Ablation of the PrP gene in mice was suggested as an experimental means of trying to determine the role of this normal protein. Two PrP$^{0/0}$ mice lines were created, which proved to be viable and appeared phenotypically normal (Büeler *et al.*, 1992; Manson *et al.*, 1994). However, further reports described altered circadian activity rhythms and neurophysiological abnormalities (Colling *et al.*, 1996; Collinge *et al.*, 1994; Tobler *et al.*, 1996). The significance of these abnormalities was uncertain, and some other investigators could not reproduce them (Herms *et al.*, 1995; Lledo *et al.*, 1996). The PrP$^{0/0}$ mouse model is, of course, one of congenital deficiency rather than acquired deficiency in later life, conceivably allowing the development of alternative mechanisms not available to the previously normal animal. However, relatively recent work based on an adult deficiency model is interesting. Bigenic mice have been created, expressing inducible PrP transgenes, and when these animals were rendered PrP deficient as adults by the administration of doxycycline, they remained healthy for more than 1.5 years (Tremblay *et al.*, 1998). A third line of PrP$^{0/0}$ mice (*Ngsk* PrP$^{0/0}$) was created, and these animals developed ataxia and loss of cerebellar Purkinje cells (Sakaguchi *et al.*, 1996). A fourth line (*RcmO* PrP$^{0/0}$) also developed ataxia (Moore, 1997). These apparently conflicting results may be explained, at least in part, by another protein related to PrP [21]. A gene, desinated *prnd* (the doppel gene) is located 16-Kb downstream of *prnp*, encoding for a novel protein designated doppel (Dpl). This protein has an approximately 25% identity with known prion proteins. Dpl mRNA, like PrP mRNA, is expressed during embryogenesis, but, unlike PrP, it is expressed only at low levels in the adult central nervous system (CNS). In the two lines of $PrP^{0/0}$ mice that develop ataxia and Purkinje cell degeneration, Dpl is up-regulated in CNS cells, but this does not occur in a $PrP^{0/0}$ line that is ataxia-free (Moore *et al.*, 1999). This suggests that Dpl may provoke neurodegeneration in PrP-deficient mice, and the study of Dpl may lead to a deeper understanding of prion biology.

$PrP^C$ is a copper-binding protein, with copper probably being bound to the octapeptide-repeat region of the protein, the binding being co-ordinated by four histidine residues (Brown *et al.*, 1997). Interestingly, the octapeptide-repeat region is one of the best-conserved regions of mammalian $PrP^C$. This region has been found to aid the resistance of cultured cerebellar cells to copper toxicity and oxidative stress (Brown *et al.*, 1998). Copper has been reported to stimulate endocytosis of $PrP^C$ from the cell surface in cultured neuroblastoma cells, with the suggestion that $PrP^C$ could act as a means for uptake of copper ions from the extracellular space (Pauly and Harris, 1998). Copper has been found to facilitate the restoration of protease resistance and infectivity in a study of $PrP^{Sc}$ that had been initially disrupted by the use of guanidine hydrochloride (McKenzie *et al.*, 1998).

$PrP^C$ seems to influence the activity of Cu/Zn superoxide dismutase (Cu,Zn SOD) (Brown and Besinger, 1998), and thus deficiency of $PrP^C$ may lead to neurotoxicity by means of reduction of neuronal resistance to oxidative stress. $PrP^C$ itself seems to have SOD activity and that this activity is abolished by deletion of the copper-binding octapeptide-repeat region (Brown *et al.*, 1999). $PrP^C$ has been found to affect glutamate uptake by astrocytes, which is of potential importance, because elevated extracellular glutamate may be excitotoxic to neurons or may exacerbate neuronal degeneration (Brown and Mohn, 1999). There is therefore increasing evidence that $PrP^C$ has an important role in cellular resistance to oxidative stress and related mechanisms, and that its copper-binding properties are fundamental to this.

The toxicity of $PrP^{Sc}$ itself may be difficult to disentangle from the effects of $PrP^C$ function loss. There are reports concerning the toxicity of a peptide fragment of PrP (PrP106–126), with evidence suggesting that the toxicity requires the presence of $PrP^C$ and also that it may be mediated by astrocytes (Brown *et al.*, 1996, 1998, 1999). However, one study has failed to confirm neurotoxicity of PrP106–126 (Kunz *et al.*, 1999).

$PrP^{Sc}$, being transcribed from the same host gene, has the same amino acid sequence as the normal protein but undergoes a posttranslational conformational change, resulting in a much greater β-sheet structure and different physicochemical properties. In particular, $PrP^{Sc}$ is relatively protease resistant and accumulates forming amyloid structures and, in some instances, amyloid plaques in the affected tissue. Amyloid is a generic term referring to a group of diverse extracellular protein deposits that have certain common ultrastructural and physicochemical features. Amyloidosis is characterized by the extracellular deposition of fibrillar proteins with a β-pleated sheet conformation that are relatively insoluble and relatively resistant to protease digestion. In different amyloid diseases, different types of protein may be present in the tissue deposition and, in CJD, the deposition involves $PrP^{Sc}$. Amyloid deposition seems to involve the binding of a basement membrane glycoprotein or proteoglycan to an amyloidogenic protein. Sulfated glycosaminoglycans are known to be present in all types of amyloid. Generally, an "amyloidogenic protein" is a precursor protein, or fragment thereof, which can be converted into an insoluble amyloid fibril in vitro, and a number have been described, including prion protein (or fragments thereof). Understanding of the nature of amyloid and its formation has accumulated from the study of a number of diseases, including Alzheimer's disease, but it is not clear how much of this general knowledge applies equally to specific illnesses such as CJD.

When a protein is synthesized, the interaction of its amino acid side chains and other factors cause folding of the polypeptide backbone, resulting in a specific three-dimensional conformation that is critical in determining the biological properties of the protein. Some proteins seem to require interaction with other molecules, such as "molecular chaperones," to achieve their final biologically active conformation. The participation of one or more such chaperones in $PrP^C$/$PrP^{Sc}$ conversion is considered by some to be likely, and a particular, as of yet, uncharacterized macromolecule has been labeled "protein X." However, no specific protein has yet been identified (Prusiner, 1999). The essential mechanism that is thought to underlie the progression of prion disease is that of continuing, increasing $PrP^{Sc}$ production; the abnormal isoform interacting with $PrP^C$ and somehow inducing its conversion to $PrP^{Sc}$. It is not entirely clear whether the primary pathogenic event is the loss of $PrP^C$, the accumulation of $PrP^{Sc}$, or both, and there may also be some other related factor. This uncertainty has clear implications for the rational development of treatments.

## Therapy: General Considerations

Three general therapeutic circumstances might be considered:

1. The first concerns the actual treatment of established disease. In this situation, one would like to begin treatment as early on in the disease process as possible, and this requires early diagnosis, which can be difficult. However, given the inevitable and often rapid progression of CJD, the assessment of real therapeutic effect should not be too problematic.

2. The second concerns the possibility of prophylactic treatment in those who might be considered at particular risk of CJD developing at some future point, for example, presymptomatic treatment of individuals with a known *prnp* mutation or those who have been exposed to some potential source of infection. In this situation, some form of presymptomatic diagnostic test could be very helpful, and one would be particularly keen to select a treatment with minimal or no toxicity.

3. Finally, some therapeutic methods could conceivably have additional uses, in an in vitro setting, by reducing potential infectivity in body tissues or fluids used in medical procedures or in medical product manufacture. This does not represent direct therapeutic intervention but is a means of prevention of disease transmission.

In the development of therapy, one needs to consider the development of better and earlier diagnostic tests and also methods of safely and quickly evaluating the potential usefulness of any proposed treatment.

Increased understanding of the molecular biology of CJD has stimulated interest in the development of treatments. However, although detailed knowledge of the disease mechanism allows rational speculation and experimental design, it is noteworthy that many successful treatments in medicine have been discovered somewhat fortuitously, and rational treatment design has not always produced practically effective methods.

The essential requirement of any medical treatment is that it be shown to be effective, nontoxic, and relatively convenient to use in actual clinical practice. Theoretical predictions, in vitro experiments, and animal models are important initial steps but are no substitute for the ultimate demonstration of therapeutic usefulness. There is also some evidence that potential therapeutic agents may have differential effects in different forms of prion disease or with different prion strains (Adjou *et al.*, 1996; Xi *et al.*, 1992). This has important implications for the useful extrapolation of animal model data to human forms of CJD and might suggest that different treatments may be required for different types of CJD (for example, sporadic and variant). Progress is being made in the preliminary stages of CJD treatment development, but real clinical benefit still seems some way off.

## Therapy: Specific Possibilities

### Introduction

The present understanding of the molecular biology of CJD suggests a number of possible therapeutic interventions, based on the central role of *PRNP* and PrP in disease.

### Reducing PrP^C Production

$PrP^C$ is essential for the formation of $PrP^{Sc}$, and possibly necessary for $PrP^{Sc}$ to exert a neurotoxic effect (Brandner *et al.*, 1996; Brown *et al.*, 1996). Thus, reducing $PrP^C$ production is a therapeutic option. This possibility could be considered both as a treatment for established disease or as a prophylactic method, given the resistance of $PrP^{0/0}$ mice to scrapie (Büeler *et al.*, 1993).

However, the highly conserved nature of the *PRNP* gene suggests an important biological role, the evidence from $PrP^{0/0}$ mice experiments including the role of Dpl protein needs further consideration, and it is still uncertain as to the precise relative pathogenetic contributions of loss of $PrP^C$ and toxicity of $PrP^{Sc}$. It is also possible that the reported abnormalities in some $PrP^{0/0}$ mice might be more relevant in humans. Further understanding of the normal cellular role of $PrP^C$ is necessary before one can be confident that reducing its activity will not result in important side effects. The accumulating data concerning the role of copper in PrP function and prion disease might well open up additional therapeutic avenues.

### Prevention of the Posttranslational Alteration and Accumulation of PrP^C

The abnormal folding of $PrP^C$, or the binding of $PrP^C$ to $PrP^{Sc}$, might be a treatment target to prevent $PrP^{Sc}$ formation or further $PrP^C$ conversion. In addition, the deposition and accumulation of $PrP^{Sc}$ might somehow be inhibited. The prevention of $PrP^{Sc}$ production is an obvious, and perhaps the most reasonable, therapeutic target. The therapeutic use of stabilizing chaperones or interfering with putative chaperones involved in disease processes such as protein X requires a greater understanding of the role of such molecules and the identification of specific proteins, especially with regard to human disease. The proposals concerning the use of antibodies or nucleic acid molecules to disrupt disease processes are very speculative, and there is no evidence of actual therapeutic effect. Confirmatory evidence is required as to the true specificity of the 15B3 antibody for the disease-specific ($PrP^{Sc}$) form of PrP (Korth *et al.*, 1997).

## Preventing the Neurotoxic Effects of PrP$^{Sc}$ and Amyloid Deposits

Understanding the mechanisms of any intrinsic PrP$^{Sc}$ toxicity or the ways in which amyloid deposition disrupts cellular function might suggest therapeutic possibilities. The prevention of PrP$^{Sc}$ accumulation and amyloid formation is desirable if such intracellular deposition is intrinsically toxic. However, it is not clear that amyloid deposition per se is indeed toxic. Aside from the question as to the intrinsic toxicity of amyloid, the claims in relation to this are based on a number of assumptions concerning amyloid formation and deposition in general. The degree to which knowledge derived from one disease is applicable to another is uncertain and there is little, if any, direct evidence concerning CJD itself. The rational prevention of intrinsic PrP$^{Sc}$ toxicity requires more complete understanding of its nature and mechanisms. There is accumulating evidence that oxidative stress is important in prion disease pathogenesis, but the suggestion that anti-oxidant treatments may be helpful in humans with actual disease is entirely speculative.

## PRACTICAL MANAGEMENT

Many of the treatments discussed in the following are speculative suggestions or theoretical ideas. Some have supportive evidence in the form of in vitro or in vivo experimental results. There have been a few claims of improvement in individual patients, claims that, so far, have not received subsequent support. There is no organized treatment trial evidence to support the use of any compound. Some suggestions are based on patent registrations, without further published evidence. Often therapeutic claims are based on experimental strains of TSEs, including scrapie (Pocchiari *et al.*, 1991), and it is not clear whether one can extrapolate to human disease. The issue of treatment possibilities with a review of registered patents has been discussed in two recent articles (Knight, 2000; Prout and Larner, 1998).

### Treatments Aimed at the Disease Process

#### Introduction

A number of therapeutic claims have been made in the literature either based on anecdotal human case reports or on animal experimental work, along with discussions as to how therapy might be designed, but there are no proven effective therapies for human CJD.

## Reduction of PrP$^{C}$ Production

As the preceding discussion indicates, the background to this possible therapeutic option is complex and uncertain.

1. The use of antisense oligonucleotides to block PrP$^{C}$ production has been suggested.
2. The possible use of a specific oligonucleotide to down-regulate the expression of an aberrant endogenous prion protein gene has also been discussed.

## Inhibiting PrP$^{Sc}$ Production/ Deposition/Accumulation

There are several possible therapies aimed at preventing or limiting PrP$^{Sc}$ production and/or deposition. Some substances are thought to have a potential role by way of stabilizing the structure of PrP$^{C}$ or by preventing its binding with PrP$^{Sc}$ and thus inhibiting the autocatalytic process. A number of compounds are considered to have useful effects in inhibiting amyloid formation or deposition and, usually by extrapolation, are considered therefore to have potential use in CJD. The general underpinning being the notion that, whatever the particular disease process, amyloid formation has sufficient common pathogenesis to be approached by considerations not initially linked to specific diseases.

**Protein-Stabilizing Agents.** It might be possible to use chemical chaperones that may help to produce normal protein conformation or induce defective proteins to assume biologically correct structures. A fair number of such protein-stabilizing agents have been proposed, including dimethylsulfoxide (DMSO), trimethylamine N-oxide (TMAO), and various polyols and sugars. However, DMSO is known to be a superoxide generator, which would have theoretical, if not actual, disadvantages in prion disease.

**Pentosan.** Sulfated polyanions, including pentosan, can prolong the incubation period or prevent clinical illness in experimental studies in rodents. These effects are most pronounced if the treatment is given at or before the time of experimental infection, and the effects diminish and eventually disappear with increasing delay between infection and treatment. There is no evidence that these drugs have any effect on clinically affected animals, perhaps because these compounds do not usually cross the blood–brain barrier. Their mechanism of action is unknown, but presumably any action must be in peripheral tissues to influence agent replication or neuroinvasion. There is a theoretical possibility that this class of drugs could be used prophylactically in humans exposed to an environmental source of infection, but it is of note that the animal models involved injection, usually intraperitoneally, of large treatment doses, and there is the potential for side effects in this class of compounds, which have anticoagulant activity. Furthermore, on current evidence, for prophylactic treatment to be effective this would have to be given very shortly after any exposure (Brimacmbe *et al.*, 1999; Diringer and Ehlers, 1991; Ehlers and Diringer, 1984; Farquhar *et al.*, 1986).

**Tetrapyrroles.** A series of structurally related tetrapyrroles (porphyrins and phthalocyanines) have been shown to inhibit $PrP^{Sc}$ formation in cell-free systems and cell culture and to delay disease in TSE animal models. In cell-culture experiments, some of the tested compounds result in a reduction in the levels of $PrP^{Sc}$ in scrapie-infected neuroblastoma cells to 2% of control values. In the cell-free systems, complete inhibition of $PrP^{Sc}$ formation can be demonstrated. Some of these compounds have been used in animal experiments, with treatment being administered simultaneously with an infective challenge of hamster-adapted 263K scrapie strain to transgenic mice overexpressing hamster PrP. Significant increases in survival time have resulted.

**Acridine/Phenothiazines.** Recent studies in scrapie-infected tissue cultures have shown that acridine and phenothiazine derivatives inhibit the accummulation of $PrP^{Sc}$, leading to the proposition that these drugs, and in particular quinacrine or chlorpromazine, may be "immediate" candidates for treating human prion diseases (Korth *et al.*, 2001). Both drugs cross the blood–brain barrier and have been used in human medicine for many years. Quinacrine is more likely to be used in practice, because this was most efficient in the cell culture studies, and chlorpromazine may cause significant side effects, such as drowsiness, in patients who are already cognitively impaired. Although it is clear that the efficiency of a drug in a cell culture system may not necessarily be translated into clinical efficacy, the relentlessly progressive and fatal outcome in human prion diseases has led to understandable pressures to prescribe quinacrine in vCJD and other forms of CJD. It is essential that novel treatments are formally assessed, and it is to be hoped that uncertainties on the optimum dosage regimen, sideeffects, and efficacy of quinacrine in the treatment of human prion diseases are systematically examined.

**Antibodies.** The development of antibodies to prion protein has potential uses in diagnosis, disease treatment, and prophylaxis. One proposal concerns the provision of specific monoclonal antibodies able to detect PrP, including one antibody (15B3), which, it is claimed, can recognize disease-specific $PrP^{Sc}$ (Korth *et al.*, 1997).

The antibodies are produced by immunization of $PrP^{0/0}$ mice with a new recombinant fragment of PrP. In addition, anti-idiotype antibodies are described that bind with the binding region of the original monoclonal antibody, thereby mimicking features of the original antigen. Aside from the diagnostic uses of such agents, it is proposed that the formation of an antigen-antibody complex between the presented antibodies and the prior protein could be used to inhibit disease processes. The relevant proposed effects seem to be either by blocking sites on $PrP^{C}$ or by occupying distinct sites on the $PrP^{Sc}$ molecule. It is suggested that, by such means, established disease might be treated, or prophylactic protection may be afforded. In addition, the potential to identify $PrP^{Sc}$ in various tissues or fluids might have a valuable role in preventing the use of infectious material. The claims concerning the disease-specific antibody are predicated on its ability to distinguish between $PrP^{C}$ and $PrP^{Sc}$. The claims relating to antibodies against $PrP^{C}$ are predicated on the idea that loss of normal cellular prion protein is not harmful.

**Nucleic Acid Molecules.** Another speculative possibility is the identification and isolation of nucleic acid molecules capable of distinguishing the $PrP^{C}$ and $PrP^{Sc}$ isoforms, again not only providing diagnostic possibilities but also therapeutic ones by binding specifically with $PrP^{C}$ and thus preventing its conversion to $PrP^{Sc}$. Such active agents could perhaps be combined with carrier molecules to form pharmaceutically useful compounds.

**Anti-amyloidogenic Agents.** Certain antiamyloidogenic agents, including nonsteroidal anti-inflammatory agents, and certain novel compounds have been proposed as possible treatments. The essential principle is that of selecting agents that bind to one or more amyloidogenic proteins, thereby stabilizing its existing conformation or rendering the protein resistant to the degradation or denaturation, which results in amyloid formation.

### Limiting the Toxicity of $PrP^{Sc}$

On the assumption that $PrP^{Sc}$ is intrinsically toxic and thus directly responsible for pathogenesis, therapeutic suggestions have been made in relation to reducing or preventing this toxicity.

For example, identifying agents to protect cells from the cytotoxic effect of β-amyloid protein in various diseases including CJD in man.

As discussed earlier, there is evidence that oxidative stress has an important role in the neurodegenerative changes of CJD. Thus, antioxidant compounds might have a role in therapy.

## Symptomatic Treatment

### General

There is no proven treatment for the underlying disease process in any human prion disease. The critical management issue is making a clinical diagnosis and explaining this with sensitivity to the family and, if possible, to the patient. This type of communication is an integral part of neurological practice, but there may be specific concerns in CJD because of widespread public awareness and because of anxieties about the risk of onward transmission. Currently, there is no evidence of transmission of any form of human prion disease through social contact or from mother to child.

### Treatment of Mental and Behavioral Symptoms

In sporadic CJD a minority of cases develop extreme distress and fear, often in association with visual hallucinations. Treatment with antipsychotic drugs or sedation may be necessary. Depression is a prominent early feature in vCJD and may also occur in other human prion diseases (e.g., familial CJD and iatrogenic CJD). Treatment with antidepressants can temporarily alleviate these symptoms, and in the small minority of cases of vCJD with persistent first-rank symptoms antipsychotic drugs may be required. Behavioral disturbance can occur in any human prion disease but is not readily amenable to treatment and is usually rapidly superceded by progressive neurological deficits.

Terminally patients are usually unaware but can appear distressed, and grasp reflexes may be misinterpreted as voluntary indications of discomfort. Opiates have been used to minimize distress in the later stages.

### Treatment of Pain

Pain is not a feature of the illness itself in most cases of human prion disease, with the exception of vCJD in which painful sensory symptoms occur in about half the cases. The pain is presumed to be thalamic in origin and can be difficult to treat. Some patients have responded to gabapentin, sodium valproate, or carbamazepine, but in some cases stronger analgesics have been required.

### Treatment of Myoclonus

Myoclonus occurs in most cases of CJD and may respond to treatment with sodium valproate or clonazepam. In many cases the myoclonus is a relatively late feature, occurring in association with severe cognitive impairment, but treatment is often requested by the family who may find the involuntary movements distressing. In this context sedation may reduce or abolish the myoclonus.

### Management of Dysphagia

The ability to voluntarily swallow is lost in most cases of CJD or VCJD, usually as a preterminal event. Whether to proceed to some form of artificial feeding is a complex issue, which should be discussed in detail with the patient's family. In sporadic CJD many families take the view that ensuring that the patient is comfortable is the main priority and that artificial feeding may unnecessarily prolong the illness. In vCJD the clinical course is more protracted, and some families have requested artificial feeding, usually through a percutaneous endoscopic gastrostomy. Gastroscopes exposed to prion agents cannot be definitively sterilized, and the risk of secondary transmission from potentially contaminated instruments must be considered if the instrument is to be re-used. In the United Kingdom dedicated gastroscopes for use in CJD cases are held at the National CJD Surveillance Unit.

## TREATMENTS NO LONGER RECOMMENDED

A range of medications have been tried in CJD without success. This includes antivirals such as idoxuridine, vidarabine (Furlow et al., 1982) and acyclovir (David et al., 1984), interferon (Kovanen et al., 1980), dapsone (Mannuelidis, et al., 1998), and steroids. Antifungals, such as amphotericin B have been used with some experimental support but no reported success in human illness (Massulo et al., 1992; McKenzie et al., 1994), There were reports of improvement after treatment with amantidine, but observational data on subsequently treated cases have not confirmed any beneficial effect (Braham, 1971; Sanders and Dunn, 1973). An extensive range of other drugs, including antibiotics and antivirals, have failed to show discernible benefit in both experimental settings and in patients with CJD (Tateishi, 1981).

## REFERENCES

Adjou, K. T., Demaimay, R., Lasmézas, C. I., Seman, M., Deslys, J.-P., and Dormont, D. (1996). Differential effects of a new amphotericin B derivative, MS-8209, on mouse BSE and scrapie: implications for the mechanism of action of polyene antibiotics. *Res. Virol.* **147**, 213–218.

Braham, J. (1971). Jakob-Creutzfeldt disease: treatment by Amantadine. *BMJ* **4**, 212–213.

Brandner, S., Isenmann, S., Raeber, A., et al. (1996). Normal host prion protein necessary for scrapie-induced neurotoxicty. *Nature* **379**, 339–343.

Brimacombe, D. B., Bennett, A. D., Wusteman, F. S., Gill, A. C., Dann, J. C., and Bostock, C. J. (1999). Characterization and polyanion-binding properties of purified recombinat prion protein. *Biochem. J.* 342, 605–613.

Brown, D. R. (1998). Prion protein-overexpressing cells show altered response to a neurotoxic prion protein peptide. *J. Neurosci. Res.* 54, 331–340.

Brown, D. R. (1999). Prion protein peptide neurotoxicity can be mediated by astrocytes. *J. Neurochem.* 73, 1105–1113.

Brown, D. R., and Besinger, A. (1998). Prion protein expression and superoxide dismutase activity. *Biochem. J.* 334, 423–429.

Brown, D. R., Qin, K., Herms, J. W., *et al.* (1997). The cellular prion protein binds copper *in vivo*. *Nature* 390, 684–668.

Brown, D. R., and Mohn, C. M. (1999). Astrocytic glutamate uptake and prion protein expression. *Glia* 25, 282–292.

Brown, D. R., Schmidt, B., and Kretzschmar, H. A. (1996). Role of microglia and host prion protein in neurotoxicity of a prion protein fragment. *Nature* 380, 345–347.

Brown, D. R., Schmidt, B., and Kretzschmar, H. A. (1998). Effects of copper on survival of prion protein knockout neurons and glia. *J. Neurochem.* 70, 1686–1693.

Brown, D. R., Wong, B.-S., Hafiz, F., Clive, C., Haswell, S. J., and Jones, I. M. (1999). Normal prion protein has an activity like that of superoxide dismutase. *Biochem. J.* 344, 1–5.

Brown, P., Preece, M., Brandel, J.-P., Sato, T., McShane, L., Zerr, I., Fletcher, A., Will, R. G., Pocchiari, M., Cashman, N. R., d'Aignaux, J. H., Cervenakova, L., Fradkin, J., Schonberger, L. B., and Collins, S. J. (2000). Iatrogenic Creutzfeldt-Jakob disease at the millennium. *Neurology* 55, 1075–1081.

Brown, P., Preece, M. A., and Will, R. G. (1992). "Friendly fire" in medicine: hormones, homografts, and Creutzfeldt-Jakob disease. *Lancet* 340, 24–27.

Büeler, H., Aguzzi, A., and Sailer, A., *et al.* (1993). Mice devoid of PrP are resistant to scrapie. *Cell* 73, 1339–1347.

Büeler, H., Fischer, M., Lang, Y., Bluethmann, H., Lipp, H. P., DeArmond, S. J., *et al.* (1992). Normal development and behaviour of mice lacking the neuronal cell-surface PrP protein. *Nature* 356, 577–582.

Colling, S. B., Collinge, J., and Jefferys, J. G. (1996). Hippocampal slices from prion protein null mice: disrupted Ca(2+)-activated K+ currents. *Neurosci. Lett.* 209(1), 49–52.

Collinge, J., Whittington, M. A., Sidle, K. C. L., *et al.* (1994). Prion protein is necessary for normal Synaptic function. *Nature* 370, 295–297.

Cousens, S., Smith, P. G., Ward, H., Everington, D., Knight, R. S. G., Zeidler, M., Stewart, G., Smith-Bathgate, E. A. B., Macleod, M. A., Mackenzie, J., and Will, R. G. (2001). Geographical distribution of variant Creutzfeldt-Jakob disease in Great Britain, 1994–2000. *Lancet* 2001, 1002–1007.

David, A. S., Grant, R., and Ballantyne, J. P. (1984). Unsucscessful treatment of Creutzfeldt-Jakob disease with acylovir. *Lancet* i, 512–513.

Diringer, H., and Ehlers, B. (1991). Chemoprophyaxis of scrapie in mice *J. Gen. Virol.* 72, 457–460.

Ehlers, B., and Diringer, H. (1984). Dextran sulphate 500 delays and prevents mouse scrapie by impairment of agent replication in spleen. *J. Gen. Virol.* 65, 1325–1330.

Farquhar, C. F., and Dickinson, A. G. (1986). Prolongation of scrapie incubation period by an injection of dextran 500 within the month before or after infection. *J. Gen. Virol.* 67, 463–473.

Finkenstaedt, M., Szudra, A., Zerr, I., Poser, S., Hise, J. H., Stoebner, J. M., and Weber, T. (1996). MR imaging of Creutzfeldt-Jakob disease. *Radiology* 199, 793–798.

Furlow, T. W., Whitley, R. J., and Wilmes, F. J. (1982). Repeated suppression of Creutzfeldt-Jakob disease with vidarabine *Lancet* ii, 564–565.

Gambetti, P., Petersen, R. B., Parchi, P., Chen, S. G., Capellari, S., Goldfarb, L., Gabizon, R., Montagna, P., Lugaresi, E., Piccardo, P., and Ghetti, B. (1999). Inherited prion diseases. *In* "Prion Biology and Diseases" (S. B. Prusiner, Ed.), pp. 509–583. Cold Spring Harbour, New York.

Green, A. J. E., Thompson, E. J., Stewart, G. E., Zeidler, M., Mackenzie, J. M., Macleod, M. A., Ironside, J. W., Will, R. G., and Knight, R. S. G. (2001). Use of 14-3-3 and other brain-specific proteins in CSF in the diagnosis of variant Creutzfeldt-Jakob disease. *JNNP* 2001, 744–748.

Herms, J. W., Kretzschmar, H. A., Titz, S., and Kller, B. U. (1995). Patch-clamp analysis of synaptic transmission to cerebellar Purkinje cells of prion protein knockout mice. *Eur. J. Neurosci.* 7, 2508–2512.

Knight, R. (1999). The relationship between new variant Creutzfeldt-Jakob disease and bovine spongiform encephalopathy. *Vox Sanguinis* 76, 203–208.

Knight, R. (2000). Therapeutic possibilities in CJD: patents 1996–1999. *Exp. Opin. Ther. Patents.* 10(1), 49–57.

Korth, C., May, B. C. H., Cohen, F. E., and Prusiner, S. B. (2001). Acridine and phenothiazine derivatives as pharmacotherapeutics for prion disease. *Proc. Natl. Acad. Sci. USA* 98(17), 9836–9841.

Korth, C., Stierli, B., Streit, P., *et al.* (1997). Prion (PrP$^{Sc}$)-specific epitope defined by a monoclonal antibody. *Nature* 390, 74–77.

Kovanen, J., Haltia, M., and Cantell, K. (1980). Failure of interferon to modify Creutzfeldt-Jakob disease. *B.M.J.* 280, 902–903.

Kunz, B., Sandmeier, E., and Christe, N. P. (1999). Neurotoxicity of prion peptide 106–126 not confirmed. *FEBS. Lett.* 458, 65–68.

Lledo, P.-M., Tremblay, P., DeArmond, S. J., Prusiner, S. B., and Nicoll, R. A. (1996). Mice deficient for prion protein exhibit normal neuronal excitability and synaptic transmission in the hippocampus *Proc. Natl. Acad. Sci. USA* 93, 2403–2407.

Manson, J. C., Clarke, A. R., Hooper, M. L., Aitchison, L., McConnell, I., and Hope, J. (1994). 129/Ola mice carrying a null mutation in PrP that abolishes mRNA production are developmentally normal. *Mol. Biol.* 8, 121–127.

Manuelidis, L., Fritch, W., and Zaitsev, I. (1998). Dapsone to delay symptoms in Creutzfeldt-Jakob disease. *Lancet* 352, 456.

Masullo, C., Macchi, G., Xi, Y. G., and Pocchiari, M. (1992). Failure to ameliorate Creutzfeldt-Jakob disease with Amphotericin B therapy. *J. Infect. Dis.* 165, 784–785.

McKenzie, D., Bartz, J., Mirwald, J., Olader, D., Marsh, R., and Aiken, J. (1998). Reversibility of scrapie inactivation is enhanced by copper. *J. Biol. Chem.* 273, 255545–25547.

McKenzie, D., Kaczkowski, J., Marsh, R., and Aiken, J. (1994). Amphotericin B delays both Scrapie agent replication and PrP-res accumulation early in infection *J. Virol.* 68, 7534–7536.

Moore, R. (1997). Gene targeting studies at the mouse prion protein locus. *PhD Thesis, University of Edinburgh, Edinburgh.*

Moore, R. C., Lee, I. Y., Silverman, G. L., *et al.* (1999). Ataxia in prion protein (PrP)-deficient mice is associated with upregulation of the novel PrP-like protein doppel. *J. Mol. Biol.* 292, 797–817.

Palmer, M. S., and Collinge, J. (1997). Prion diseases: an introduction. *In* "Prion Diseases", (J. Collinge and M. S. Palmer, Eds.), pp. 1–17. Oxford University Press Inc, New York.

Pauly, P. C., and Harris, D. A. (1998). Copper stimulates endocytosis of the prion protein. *J. Biol. Chem.* 273, 33107–33110.

Pocchiari, M., Salvatore, M., Ladogana, A., *et al.* (1991). Experimental drug treatment of scrapie: a pathogenetic basis for rationale therapeutics. *Eur. J. Epidemiol.* 7, 556–561.

Prout, K. A., and Larner, A. J. (1998). Emerging therapeutic possibilities in prion diseases: patents 1993–1998. *Expert. Opin. Therap. Patents* 8(9), 1099–1108.

Prusiner, S. B. (1999). An introduction to Prion biology and diseases. *In* "Prion Biology and Diseases" (S. B. Prusiner, Ed.) pp. 1–66. Cold Spring Harbour Laboratory Press, New York.

Sakaguchi, S., Katamine, S., Nishida, N., *et al.* (1996). Loss of cerebellar Purkinje cells in aged mice homozygous for a disrupted *PrP* gene. *Nature* 380, 528–531.

Sanders, W. L., and Dunn, T. L. (1973). Creutfeldt-Jakob disease treated with amantidine. A report of two cases. *JNNP* 36, 581–584.

Steinhoff, B. J., Racker, S., Herrendorf, G., Poser, S., Grosche, S., Zerr, I., Kretzschmar, H., and Weber, T. (1996). Accuracy and reliability of periodic sharp wave complexes in Creutzfeldt-Jakob disease. *Arch. Neurol.* 53, 162–165.

Tateishi, J. (1981). Antibiotics and antivirals do not modify experimentally induced Creutzfeldt-Jakob disease in mice. *JNNP* 44, 723–724.

Tobler, I., Gauss, S. E., Deboer, T., *et al.* (1996). Altered circadian activity rhythms and sleep in mice devoid of prion protein. *Nature* 380(6575), 639–642.

Tremblay, P., Meiner, Z., Galou, M., *et al.* (1998). Doxycycline control of prion protein transgene expression modulates prion disease in mice. *Proc. Natl. Acad. Sci. USA* 13, 12580–12585.

Westaway, D., DeArmond, S. J., Cayetano-Canlas, J., *et al.* (1994). Degeneration of skeletal muscle, peripheral nerves, and the central nervous system in transgenic mice overexpressing wild-type proteins. *Cell* 76(1), 117–129.

Will, R. G., Alpers, M. P., Dormont, D., Schonberger, L. B., and Tateishi, J. (1999). Infectious and sporadic prion diseases. In "Prion Biology and Diseases," (S. B. Prusiner, Ed.), pp. 465–507. Cold Spring Harbour Laboratory Press, New York.

Will, R. G., Ironside, J. W., Zeidler, M., Cousens, S. N., Estibeiro, K., Alperovitch, A., Poser, S., Pocchiari, M., Hofman, A., and Smith, P. G. (1996). A new variant of Creutzfeldt-Jakob disease in the UK. *Lancet* 347, 921–925.

Will, R. G., and Kimberlin, R. H. (1998). Creutzfeldt-Jakob disease and the risk from blood or blood products. *Vox Sanguinis* 75, 178–180.

Will, R. G., Zeidler, M., Stewart, G. E., Macleod, M. A., Ironside, J. W., Cousens, S. N., Mackenzie, J., Estibeiro, K., Green, A. J. E., and Knight, R. S. G. (2000). Diagnosis of new variant Creutzfeldt-Jakob disease. *Ann. Neurol.* 47, 575–582.

Windl, O., Dempster, M., Estibeiro, J. P., Lathe, R., De Silva, R., Esmonde, T., Will, R., Springbett, A., Campbell, T. A., Sidle, K. C. L., Palmer, M. S., and Collinge, J. (1996). Genetic basis of Creutzfeldt-Jakob disease in the United Kingdom: a systematic analysis of predisposing mutations and allelic variation in the PRNP gene. *Hum. Gene.* 98, 259–264.

World Health Organization. (1998). "WHO Manual for Strengthening Diagnosis and Surveillance of Creutzfeldt-Jakob Disease," pp. 1–75. Geneva.

Xi, Y. G., Ingrosso, L., Lagonda, A., Masullo, C., and Pocchiari, M. (1992). Amphotericin B treatment dissociates *in vivo* replication of the scrapie agent from PrP accumulation. *Nature* 356, 598–601.

Zeidler, M., Sellar, R. J., Collie, D. A., Knight, R., Stewart, G., Macleod, M. A., Ironside, J. W., Cousens, S., Colchester, A. F. C., Hadley, D. M., and Will, R. G. (2000). The pulvinar sign on magnetic resonance imaging in variant Creutzfeldt-Jakob disease. *Lancet* 355, 1412–1418.

Zerr, I., Bodemer, M., Gefeller, O., Otto, M., Poser, S., Wiltfang, J., Windl, O., Kretzschmar, H. A., and Weber, T. (1998). Detection of 14-3-3 protein in the cerebrospinal fluid supports the diagnosis of Creutzfeldt-Jakob disease. *Ann. Neurol.* 43, 32–40.

Zerr, I., Brandel, J.-P., Masullo, C., Wientjens, D., De Silva, R., Zeidler, M., Granieri, E., Sampaolo, S., van Duijn, C., Delasnerie-Laupretre, N., Will, R., and Poser, S. (2000b). European surveillance on Creutzfeldt-Jakob disease: A case-control study for medical risk factors. *J. Clin. Epidimiol.* 53, 747–754.

Zerr, I., Pocchiari, M., Collins, S., Brandel, J.-P., de Pedro Cuesta, J., Knight, R. S., Bernheimer, H., Cardone, F., Delasnerie-Laupretre, N., Cuadrado Corrales, N., Ladogana, A., Bodemer, M., Fletcher, A., Awan, T., Ruiz Bremon, A., Budka, H., Laplanche, J.-L., Will, R. G., and Poser, S. (2000a). Analysis of EEG and CSF 14-3-3 proteins as aids to the diagnosis of Creutzfeldt-Jakob disease. *Neurology* 55(6), 811–815.

CHAPTER 55

# Critical Care Neurology

S. Schwab, R. Kollmar, D. F. Hanley, and Werner Hacke

## GENERAL PRINCIPLES IN THE MANAGEMENT OF NEUROLOGICAL CRITICAL CARE PATIENTS

### Airway Management and Pulmonary Function

Respiratory disorders in neurological patients usually do not occur from primary lung disease. Basically, neurological respiratory disorders can be subdivided into three different origins: (1) disturbances in breathing mechanics, protecting reflexes and central respiratory drive; (2) physiological and psychological stress, requiring artificial ventilation, (3) when sedation is necessary. In many neurological disorders the patients are awake while experiencing the respiratory distress. In neuromuscular disease a disturbance of breathing mechanics together with an absence of gag reflexes is an indication for intubation. Although respiratory problems are rare in the first few hours after an acute stroke, pulmonary complications constitute a major cause of morbidity and mortality in intensive care unit (ICU) patients with cerebrovascular disease. The maintenance of adequate ventilation and oxygenation is an important prerequisite for the preservation of metabolic turnover in the marginal zone of an ischemic stroke, the so-called penumbra.

### Respiratory Dysfunction in Neurocritical Care Patients

Respiratory dysfunction may develop as the result of a reduction in alveolar gas exchange caused by atelectasis or pneumonia in immobilized patients, hypoventilation caused by impaired central respiratory drive or critical care neuropathy, or upper airway obstruction caused by oropharyngeal muscular dysfunction. Patients with an extensive supratentorial infarction, vertebrobasilar infarction, seizure activity, or massive intracranial or subarachnoid hemorrhage are generally unconscious and at particular risk of aspiration. Respiratory failure, caused by fatigue of the respiratory muscles, is most frequently seen in Guillan-Barré syndrome and myasthenia gravis, in which the respiratory function is completely reversible after therapeutic intervention. The broad subject of mechanical ventilation cannot be captured in a few pages, so this chapter focuses on some basic principles for airway management in neurological patients.

### Intubation and Ventilation

Syndromes with a decreased level of consciousness or disruption of gas exchange (physiological or anatomical) generally indicate the need for airway protection. The indications for endotracheal intubation and artificial ventilation are summarized in Table I. Initially, orotracheal intubation is preferred to nasotracheal intubation because of the high incidence of paranasal sinusitis, common technical problems during intubation, tube contamination during nasal passage, and the smaller diameters of nasal tubes.

Anesthesia in a critically ill neurological patient presents a variety of specific problems. In general, it has to be assumed that all patients have a full stomach. In addition, laryngoscopy and intubation often affect hemodynamics, which can be deleterious in the case of high-grade carotid artery stenosis. Some drugs may worsen already increased intracranial pressure (ICP). Such situations may require the use of ICP-reducing but CNS-depressing drugs (fentanyl or thiopental). The effects of a difficult neurological examination have to be compared with desirable hemodynamic or ICP effects. We, therefore, recommend short-acting agents such as sufentanil (70 µg), followed by etomidate (0.3–0.5 mg/kg) or propofol (1.5–3 mg/kg) in conjunc-

TABLE I   Indications for Endotracheal Intubation and Ventilation

$pO_2$ <50–60 mmHg despite $O_2$ administration by nasal probe or mask
$pCO_2$ >50–60 mmHg
Vital capacity <500–800 ml
Risk of aspiration because of loss of protective airway reflexes
Tachypnea >35
Dyspnea with use of accessory muscles
Severe respiratory acidosis

tion with a depolarizing neuromuscular blocking agent such as succinylcholine (1.2 mg/kg) given in rapid sequence. Syndromes with depressed neuromuscular function require careful consideration to select the best neuromuscular blocking agents. There are many syndromes in which succinylcholine use can lead to life-threatening hyperkalemia. For further sedation, the use of analgosedative drugs is recommended (e.g., fentanyl, 0.05 mg/ml, and midazolam, 1.8 mg/ml, by infusion pump). Most intubated patients with acute cerebrovascular diseases require ventilatory support (initial ventilator setting: see Table II).

Ventilator settings should be adjusted to provide a $pO_2$ of ≥80 mmHg and a $pCO_2$ between 38 and 42 mmHg. Inspiration/expiration (I:E) ratio should initially be set at 1:2 and positive end-expiratory pressure (PEEP) at 4 mmHg. Pressure- or volume-controlled ventilation modes can be applied. Inspiratory flow should be set at 30 L/min and tidal volume at 12 ml/kg. Higher PEEP levels are often necessary for adequate oxygenation of patients with pulmonary disorders, because PEEP improves oxygenation by preventing or reducing atelectasis, increasing functional residual capacity, and reducing pulmonal shunting. The assumption that application of higher PEEP levels causes an increase in ICP could not be confirmed in a recent study. However, higher PEEP levels result in increased intrathoracic pressure and reduced venous return and could affect cerebral perfusion pressure (CPP) by lowering mean arterial blood pressure (MAP). Assuming a patent cerebral autoregulation, the MAP decrease would be compensated by a dilation of the cerebral arterioles, resulting in an ICP increase. This effect, though, can be counteracted by adjusting the central venous pressure

accordingly. Similary, it was shown that alterations of the I:E ratio from 1:2 to 1:1 do not influence ICP or CPP and could therefore be readily applied in patients with acute stroke (Georgiadis *et al.*, 2001). In general, early intubation and vigorous physical chest therapy can help prevent pulmonary complications, which improves the outcome of patients with all kinds of neurological diseases.

## Tracheotomy (Table III)

With the development of high-volume, low-pressure cuffs, the risk of tracheal stenosis has markedly decreased. Nevertheless, prolonged orotracheal or nasotracheal intubation may cause laryngeal damage and a phonation disability. Furthermore, tracheostomy improves the patient's ability to trigger the respirator and facilitates weaning. Hygienic care, including suctioning and pulmonary toilet, are less difficult in patients with a tracheostomy. When the benefits of better patient mobility and care outweigh the risks and complications, tracheotomy should be performed regardless of the length of orotracheal intubation. In general, tracheotomy has to be performed 2–3 weeks after orotracheal intubation because of increased probability of oral ulcers and/or laryngeal damage.

## Ventilation Mode

Mechanical ventilation provides alveolar ventilation and thereby removes carbon dioxide from the body. Mechanical ventilatory support also improves ventilation-to-perfusion relationships, ensuring adequate oxygenation. There are many ways to accomplish these goals. In general, one should choose a mode of ventilation that maintains oxygenation and is comfortable to the patient, who would receive preferably as few dugs for sedation as possible. Continuous positive airway pressure (CPAP) allows the patient to initiate spontaneous respiration at an airway pressure above baseline. CPAP usually is combined with mechanical ventilatory support. This ventilation mode is indicated in patients with some spontaneous respiration and can be used for noninvasive ventilatory support.

Controlled mechanical ventilation (CMV) involves delivery of a programmed selected tidal volume. This

TABLE II   Initial Ventilator Setting

IMV or CMV mode
Tidal volume 12 ml/kg
Respiratory rate $pCO_2$ <40 mmHg (8–12 breaths/min)
$FIO_2$ = 1,0
I:E = 1:2–3
Inspiratory flow 30 L/min

TABLE III   Indications for Tracheotomy

Coma and need for ventilatory support for >14 days
Bronchial cleansing/protection
Swallowing disturbances with risk of aspiration
Laryngeal obstruction
Prolonged weaning

mode does not provide or permit spontaneous breathing between the mechanical ventilator breaths. It is indicated for patients with acute neurotrauma such as severe stroke, SAH, basilar artery thrombosis, elevated ICP, and others. Obviously, the disadvantage of this ventilation mode is the dependence on sedation to eliminate respirator drive. Assist-control ventilation permits the patient to make spontaneous respiratory efforts. This triggers the ventilator to deliver a preprogrammed tidal volume to the patient. The trigger can vary in sensitivity, resulting in a greater or lesser inspiratory effort for the patient. Intermittent mandatory ventilation (IMV) is a combination of machine-initiated breathing at a fixed rate and spontaneous ventilation. The tidal volume is preselected for the mechanical breaths. In general, this mode is the appropriate one for weaning from the ventilator.

Pressure support ventilation (PSV) supplies a pressurized breath during spontaneous breathing. It allows the patient to maintain control over his or her own ventilatory process while decreasing the overall work of breathing. PSV can be used as the primary mode of ventilation and as a weaning mode. The patient must have an intact, stable respiratory drive to use PSV. It is important to recognize that patients with a changing airway resistance (bronchspasm) can receive varying amounts of tidal volume and be inadequately ventilated. Weaning from the ventilator should be initiated when several clinical and laboratory criteria are fulfilled. For patients who still need a Fio2 >0.45 and who have significant chest x-ray abnormalities, weaning should be postponed. Weaning should be done gradually, but in patients who were on the ventilator only for a short time, T-tube weaning can be achieved easily. For all other patients weaning should be done with effective rest at night and ventilatory support over this time. Establishment of a set plan for the weaning and the close monitoring of the patient's progress are essential for all efforts. This includes ensuring adequate sleep and nutrition, optimizing posture and beginning rehabilitation of the entire patient.

### Neurogenic Pulmonary Edema (NPE)

The NPE is defined as pulmonary edema that develops after an acute damage to the CNS with the exclusion of other causes (aspiration, myocardial insufficiency, etc.). It is observed after a variety of brain lesions such as head trauma, intracerebral hemorrhage, SAH, hemorrhage of the cerebellum, ischemic stroke, brain tumors, spinal infarctions, multiple sclerosis, brainstem lesions, and epileptic seizures. The incidence is not clear, because only retrospective studies were done. For example, in isolated brain trauma, the incidence is estimated to be 0.1%–0.8%. Other studies

suggest a highly increased incidence of NPE of 31% in SAH (Weir et al., 1978). Mortality is unclear at all, because there are no clinical studies for NPE alone. The pathophysiological mechanisms are not clear, but a combination of factors may contribute, such as left atrial hypertension, systemic hypertension, and increased sympathetic activity (Malik, 1985). An elevated ICP may lead to increased sympathic and parasymapthic excitation. Therefore, hemodynamic alterations increase the microvascular pressure, which leads to increased fluid evasion and reduced drainage capacity of the pulmonary capillaries, Subsequently, fluid shifts to the intersitium and the alveoli. Also, increased microvascular pressure leads to temporary damage of the endovascular tight junction followed by increased permeability of the endothelium. This results in a protein-rich alveolar fluid. Furthermore, lesions of the central nervous system may induce arrhythmias, and repolarization disturbances, with increased cardiac enzyme levels, may occur after SAH, head trauma, or epileptic seizures (Samuels, 1993). As in NPE, an increased sympathetic drive with elevated catecholamine levels leads to excitotoxicity with consequent cell death and leakage of cardiac enzymes into serum. Because the pathophysiology of NPE is not clear, symptomatic treatment is recommended. Four main goals arise: reduction of an elevated ICP, improvement of the oxygen delivery to the organism, reduction of cardiac preload and afterload, and improvement of myocardial inotropy. For prevention and treatment, the administration of benzodiazepines, barbiturates, β-blockers, calcium channel blockers, and free radical scavengers, which intervene in the afore described cascade, may be favorable, especially when given early in the clinical course, although there are no prospective studies supporting this hypothesis.

## TEMPERATURE

Temperature management can be critical in neurointensive care. In animal models hyperthermia has been shown to have deleterious effects on outcome after cerebral ischemia (Ginsberg et al., 1992), head trauma (Marion et al., 1993), and seizure. Conversely, mild hypothermia has been shown to be protective. The extent of protection is not adequately explained by reduction in cerebral metabolic rate (Nemoto et al., 1994), suggesting that hypothermia has additional beneficial effects such as decreased free radical production or reduction in neurotransmitter neurotoxicity. Such observations indicate, in the absence of conclusive trials for nontraumatic neurological conditions, that a minimal approach would be to aggressively treat and prevent the occurrence of fever. For routine care, maintenance of temperature at 37.5°C using cooling blankets and antipyretic drugs should be the overall goal.

## FLUIDS AND NUTRITION

### Fluid Balance

Fluid and electrolyte disturbances are frequently observed in patients treated in the NCCU. They may occur as a result of sympathetic responses to ischemic or hemorrhagic neuronal injury or may be the effect of fluid and electrolyte substitution, the nutritional regimen, or the administration of diuretics and other drugs. Sympathetic nervous system stimulation induces adrenal release of aldosterone, which causes sodium retention and kaliuresis and activates the renin-angiotensin system, affecting renal function. Antidiuretic hormone (ADH) secretion may also be affected by CNS lesions, which results in sodium and water retention and decreased urine output (SIADH). Diabetes insipidus occurs when ADH release is inhibited.

Fluid disturbances can be assessed by (1) clinical observation (signs of exsicosis, edema); (2) evaluation of fluid intake and output with respect to calculated fluid requirements (see Table IV); (3) measurement of the central venous pressure (2–12 cm $H_2O$) by means of a central venous line or measurement of the pulmonary capillary wedge pressure by means of pulmonary catheter to detect hypovolemic states; and (4) measurement of the serum osmolarity (see Table V), urine osmolarity, and sodium concentration.

Sodium salts are the main electrolytes of the extracellular fluid and account for more than 90% of its osmolarity There is a close interrelationship between sodium and water shifts. Sodium concentrations and the hydration state of the patient determine the main guidelines for the treatment of fluid disturbances. Table VI gives an overview of the common causes of isotonic volume depletion.

The treatment of choice is to immediately administer standard fluids by way of the enteral or parenteral route. When fluid and electrolyte losses continue, careful fluid balancing and monitoring of the central venous pressure (CVP) are necessary to determine the amount of fluids

needed, especially in nonresponsive patients. In patients with concomitant left ventricular failure, the chest x-ray, the echocardiographic findings, or pulmonary catheterization may help to avoid fluid overload.

### Hyponatremia

Hyponatremia can occur as a result of the syndrome of inappropriate secretion of antiduretic hormone (SIADH), the so-called syndrome of "cerebral salt wasting," or inappropriate free water administration. SIADH is generally associated with hypervolemia and cerebral salt wasting with hypovolemia. Both syndromes may be associated with elevated urinary sodium concentrations, making differentiation between the two syndromes difficult when relying on laboratory testing alone (Burcar et al., 1997; Solomon et al., 1984).

A rapidly increasing sodium concentration can produce permanent neurological damage because of central pontine myelinolysis. Thus, such emergent

**TABLE V   Serum Osmolarity Evaluation Approach**

Measurement of osmolarity, always including serum sodium, glucose, and urea
Calculation of osmolarity: mosmol/L $H_2O = 2 \times Na + Glucose + Urea$ (mmol/L)
Calculation of the osmotic gap (measured—calculated osmolarity)
Causes for increased osmotic gap >10 mosmoL/L $H_2O$
    Reduced serum water
        Hyperlipidemia
        Hyperproteinemia
        Additional low-molecular-weight substances: mannitol, ethanol, methanol, ethyleneglycol, and other toxins <150 kDa
Laboratory error

**TABLE IV   Assessment of the Daily Fluid Requirement**

| Basal requirement | 30 ml/kg body weight |
|---|---|
| Output | Urine output the day before |
| | Stool water: 100 ml (more in diarrhea) |
| | Loss through enteral probes or drainages |
| Insensible loss (skin and airway) | 800 ml (ventilated patients: 400 ml) |
| | 500 ml/°C > 37°C |

**TABLE VI   Common Causes of Isotonic Volume Depletion**

1. Inadequate fluid intake
   Impaired thirst mechanisms (hypothalamic lesions, old age)
   Impaired ability to maintain adequate fluid intake (altered consciousness, altered mentation, dysphagia, paresis)
   Inadequate fluid administration during critical care
2. Renal loss
   Polyuric kidney diseases (see also Figure 1)
   Diuretics
   Osmotic diuresis (mannitol, sorbitol, diabetic glucosuria)
   Endocrine (adrenal or anterior pituitary insufficiency)
3. Extrarenal loss
   Gastrointestinal (diarrhea, vomiting, gastric suction, fistulas, abdominal sequestration)
   Pulmonary (hyperventilation, mechanical ventilation without warming and humidification)
   Cutaneous (sweating, fever, high ambient temperature, wounds, burns)

therapy should be reserved for only the most critical sort of situation. If the hyponatremia is not associated with such serious neurological deterioration, and if the patient is judged to be hypervolemic, fluid restriction is an initial reasonable approach. If, however, hypovolemia is already present, sodium can be added in a titrated manner to the patient's infusion regimen. Up to 4 g of sodium can be added to the enteral diet. Alternatively, or in addition, the patient can have all intravenous fluids be 0.9% saline, or 10% saline can be infused, titrated with 10 ml/h.

## Hypernatremia

Hypernatremia can occur in NCCU patients from nonketotic diabetic coma, dehydration from lack of fluid intake or diuretic use, inappropriate hypertonic fluid administration, diabetes insipidus, or panhypopituitarism. It can be associated with thirst, irritability, seizures, intracranial hemorrhage, or coma. Once hypernatremia occurs, its treatment is compounded by effects of decreasing osmolarity on brain water and ICP.

Diabetes insipidus (DI) can occur when disease processes affect the pituitary gland or its vascular supply. It should be expected when urine output is inappropriately increased in combination with severe hypernatremia and hypovolemic hypotension. Diagnosis is made by evaluating the output of diluted urine in combination with hypertonic serum. Specific gravity of urine is near 1.001, with osmolarity less than 200 despite a serum osmolarity that may be greater then 320 mOsm.

In the hypovolemic patient, which is the usual scenario, therapy is directed at replacement of the total body water deficit:

$$\text{TBW deficit} = 0.6 \times \text{Weight (kg)} - (140/[\text{Na}^+]_p) \times 0.6 \times \text{Weight (kg)}$$

The water deficit is replaced over 24–48 hours at a rate no faster than 1–2 mEq/L/h. Hypernatremia should be treated slowly to avoid seizures and cerebral edema. In central DI, administration of 2–5 units of aqueous vasopressin or 1–5 µg of its analogue desmopressin (DDAVP) will effectively reduce water diuresis.

## NUTRITION

Most patients who are admitted to a NCCU for treatment of an acute neurological disease are unlikely to have any nutritional deficit and vitamin or mineral deficiencies, so that generally accepted guidelines can be applied.

The caloric requirements should be estimated individually to avoid malnutrition and, on the other hand, to avoid overfeeding. As a rule, a proportion of 1 part lipids, 1 part protein, and 3 parts carbohydrates should be chosen. Malnutrition predisposes to infections and weaning problems and may cause catabolism, whereas overfeeding may lead to hyperglycemia, hypertriglyceridemia, steatosis hepatis, cholestasis, and respiratory stress by increased $CO_2$ production. The basal energy expenditure is generally assumed to be about 25 kcal/kg body weight, which may be increased by 40% in severe illness. Calorie requirements can be determined more precisely by indirect calorimetry, by which the respiratory quotient (RQ) can be calculated, thus providing a measure of the resting energy expenditure. An RQ close to 0.7 may reflect inadequate feeding, whereas an RQ of more than 1.0 may indicate overfeeding.

Because this technique is not available for daily routine in most cases, the calorie requirement should be estimated using the Harris-Benedict equation, which provides a satisfactory method to develop an initial estimate of basal energy expenditure (BEE) based on weight and gender.

BEE (men) = 66.47 + 13.75W + 5.00H − 6.76A
BEE (women) = 655.10 + 9.56W + 1.85H − 4.68A
W = Weight (kg)
H = Height (cm)
A = Age (years)

Total caloric requirements can be estimated by multiplying the BEE with a stress factor, depending on the underlying conditions and illness.

Head trauma patients can have a 1.5–2-fold increase in metabolic needs associated with massive nitrogen loss. Protein intakes have to be adjusted as needed according to the urine urea nitrogen test results. If $N_2$ excretion exceeds the protein equivalent of 2.5 g/kg, higher protein intakes usually will not promote $N_2$ retention but will drive ureagenesis instead. Continued assessment of nutritional status as an indication of trends in a patient's nutritional status can be obtained from serial assessment of visceral protein levels, physical examination, and weight. Other more sophisticated measures of nutrition such as calorimetry or nitrogen balance can also be obtained for problematic cases.

## Feeding Routes and Strategies

Although an adequate fluid and electrolyte supply should be initiated at the beginning of ICU treatment, the institution of complete, optimized nutrition, including lipid administration, has to be established from day 3 of ICU treatment. Vitamin and mineral supplements may be added after 1 week but must be given sooner if pre-existent malnutrition is suspected. On admission to the NCCU all patients receive nothing by mouth until they have undergone neurological evaluation or have recovered from the initial event to adequately self-feed. If after 2 days, self-nutrition is not likely, enteral nutrition is started, usually by way of gastric or duodenal tube. Even in patients receiving high doses of CNS depressants or neuromuscular blockade, the enteral approach should be attempted, provided bowel movement is present. If enteral nutrition is not successful, as judged by high residual volume (>50 ml per 2 hours) with repeated attempts and with metoclopramide, then perenteral nutrition is implemented by means of a centrally placed venous catheter.

Enteral nutrition has the advantages of lower cost, simpler application, and a lower risk of sepsis than parenteral nutrition and also uses the normal physiological functions of digestion and absorption, which maintain the intestinal mucosa. Intestinal function and motility must be monitored regularly (bowel sounds, aspiration of gastric residuals) and may be supported by motility stimulants such as metoclopramide or cisapride. However, there are pharmacological interactions with risk of ventricular arrhythmias. Continuous pump-assisted infusion is better tolerated than bolus administration of food. Patients who will require long-term enteral nutritional support should be scheduled early for percutaneous endoscopic gastrostomy (PEG), which is better tolerated by the patient and will cause less local complications. A common complication of enteral feeding is diarrhea as a result of the hyperosmolar electrolyte solutions or because of quickly advanced enteral feeding after an extended period of fasting or parenteral nutrition. Another complication may be gastric retention as an effect of multiple medical conditions and their pharmacological treatments, with the risk of regurgitation and pulmonary aspiration. If motitlity stimulants do not work, postpyloric feeding (endoscopically placed nasoduodenal or nasojejunal probe) or parenteral nutrition will be required.

Parenteral nutrition is indicated in cases of imminent intubation or operation, gastrointestinal leakage, ileus, pancreatitis, or other conditions in which the patient's gastrointestinal tract is unable to tolerate oral or enteral feeding for at least 5–7 days. For short-term parenteral nutrition, a peripheral venous line is suggested, using formulas of not more than 1000 mOsmoL/kg, which may be combined with lipid solutions. Long-term parenteral nutrition, which meets the patient's full caloric and protein requirements without giving an excess fluid volume requires hyperosmolar formulas of up to 1800 mOsmoL/kg water. Because they may irritate the venous endothelium, they have to be administered through central venous lines (see later).

Complications of parenteral feeding include central, catheter-associated risks (arterial puncture, pneumothorax, central venous thrombosis, catheter sepsis) and metabolic problems. Hyperglycemia, a common side effect, often requires continuous insulin infusion; hypoglycemia after discontinuation of parenteral nutrition may be prevented by tapering the formula slowly. Liver function abnormalities, which are also very common, with mild to moderate elevation of serum liver enzyme activity and bilirubin are usually benign and self-limiting.

## CENTRAL VENOUS LINES

A central venous catheter provides direct information about the intravascular volume and is a safe route of administration for fluids, parenteral nutrition, and medication. Indications for central venous catheterization are summarized in Table VII.

The safest approach is through the antecubital vein; in patients who need multilumen catheters, the subclavian or internal jugular vein approach is recommended. In patients with elevated ICP, positioning of internal jugular vein catheters and the catheters themselves can impede central venous drainage and further increase ICP. Thus, subclavian catheters are preferred in these patients.

In some cases a pulmonary artery catheter may help to monitor and to guide fluid and cardiocirculatory therapy in patients with severe heart failure and to distinguish the type of cardiocirculatory insufficiency (e.g., cardiogenic shock: low output and high filling pressure; sepsis: high output and low filling pressure). However, the effectiveness of pulmonary catheterization is still a matter of controversy; its application should be limited to centers with special expertise (technical know-how, experience in interpreting the data).

## SEDATION

Patients treated on an ICU or NCCU are exposed to a number of environmental, medical, and psychological factors that can cause excessive stress and anxiety. Envi-

TABLE VII    Indications for Central Venous Catheterization

- Measurement of the central venous pressure
- Parenteral nutrition
- Administration of hyperoncotic or locally aggressive substances
- Dialysis, hemofiltration
- Guide for transvenous pacemakers, pulmonary catheters, and bulb catheters
- Lack of adequate veins for peripheral lines

ronmental factors include unpleasant and unfamiliar visual and auditory stimuli or low temperature. Psychologically caused stress may be due to the awareness of severe and disabling illness; sleep disturbances and deprivation; or anxiety, confusion, and other psychiatric disorders that may be pre-existent or associated with the treated neurological disorder or the NCCU therapy (drugs, mechanical ventilation, weaning). Medical factors include respiratory failure, pain, endocrine disorders, infection, sepsis, cardiovascular impairment, hypoxemia, and drug effects. It is also very important to treat agitation, because it may increase morbidity and mortality (Zander *et al.*, 1995). Inadequate sedation, as well as analgesia, may cause postaggression syndrome, thereby increasing the metabolic rate and oxygen consumption and promoting hemodynamic instability. In patients with pulmonary failure, the postaggression syndrome may make effective ventilation difficult or even impossible. Furthermore, agitation may lead to problems in nursing care and to risk of self-injury.

There is no clear-cut, ideal strategy for therapy of agitation. Pharmacological treatment includes administration of benzodiazepines, neuroleptics, barbiturates, and other drugs. However, sedation is particularly problematic in the NCCU patient; it impedes the ability to assess the very problem for which NCCU admission is required. The inability to assess the course of the illness can result in a prolonged stay and morbidity on the ICU related to undiagnosed, but potentially treatable, neurological deterioration. The effects of benzodiazepines can be antagonized, although reversal risks the occurrence of seizures, heart failure, and ICP increase. In some patients benzodiazepines can have paradoxical effects and cause exacerbation of agitation. Given for prolonged periods benzodiazepines can accumulate in tissues, making arousal time quite prolonged. As with most GABAergic drugs, there may be brain-protective side effects. However, they seem to have a more pronounced effect to decrease blood pressure and can produce disorientation and hypernatremia.

Propofol has a short half-life, which allows, even with high doses given over long periods, rather quick arousal times. It decreases blood pressure and may have proconvulsant properties. It should be used cautiously in patients with triglyceride intolerance, because it is delivered in a triglyceride emulsion vehicle.

Barbiturates such thiopental are occasionally used for sedation. Given over a short period of time, barbiturates are generally not problematic, even though they regularly cause arterial hypotension and respiratory depression. However, with prolonged use they accumulate in tissues (liver, fat), making arousal time quite prolonged. Barbiturates may have brain-protective properties and can be used to control elevated ICP.

## ANALGESIA

Anxiety, stress, and behavioral disturbances may be due to pain. Some form of pain will be experienced by almost every patient during some stage of the disease and may have cardiocirculatory and metabolic effects. Sedation is never an appropriate substitute for adequate analgesia. Prostaglandin inhibitors such as paracetamole or salicylates should primarily used for pain relief but often do not prove be sufficient. In this case opioids should be given, such as fentanyl (0.1 mg IV) or buprenorphine (0.2–0.4 mg SL, IM, or IV), which are widely used analgetic substances during ICU treatment. Alfentanil has the advantage of a short redistribution half-life (about 11 minutes) and a short elimination half-life (about 94 minutes); however, it is expensive and should therefore not be used for prolonged anesthesia. The principal advantages of opioids are their potent analgesic properties without associated hemodynamic side effects even in high-risk patients. The sedating side effect and suppression of cough reflexes are only desired in ventilated patients, thus, tramadol (50–100 mg PO, SC, rectally, IM, or IV) or tilidin (50–100 mg PO) should be preferred in awake, spontaneously breathing patients.

### Blood Pressure Management

Blood pressure management is an important issue in neurological critical care patients. Systemic hypertension may exacerbate cerebral edema or intracranial hemorrhage or have deleterious cardiopulmonary effects such as pulmonary edema or myocardial ischemia. On the other hand, blood pressure decreases can lead to insufficient cerebral perfusion. There are a variety of drugs for treatment of arterial hypertension. Calcium channel antagonist drugs, which may have brain-protective effects, are used primarily for antihypertensive therapy. These include verapamil, diltiazem, nifedipine, nimodipine, and nicardipine. Nimodipine and nicardipine, developed specifically for brain protection purposes, have been assessed in numerous studies with several reports, suggesting they lead to protection versus ischemic brain damage. All calcium antagonists are vasodilators and can increase ICP. Concern has been raised in North America that the use of sublingual antagonists in acute stroke may be associated with undesirable hypotension. Peripheral vasodilators such as nitroprusside, nitroglycerin, and hydralazine may cause hyperemic intracranial hypertension and should be restricted to therapy of severe, otherwise untreatable, hypertension. In cerebral ischemia, only extreme hypertensive blood pressure >200 mmHg should be treated with either sympatholytic drugs such as β-blockers, clonidine, or angiotensin-correcting enzyme (ACE) inhibitors or with urapidil, which is commonly used in Europe.

### Hypotension

Often, vasopressors are used inappropriately in the initial treatment of hypotension. When hypovolemia is thought to be a contributing factor, an initial fluid challenge should be administered. The choice of fluid remains uncertain, but both crystalloid or colloid infusion are generally associated with hemodilution. For the therapy of hypotension there is not a consensus regarding the most appropriate pressor to use. Dobutamine is primarily a $\beta_1$-stimulator and is used in the treatment of severe left ventricular failure and cardiogenic shock. At least in theory, increasing cardiac output will increase cerebral blood flow. An alternative catecholamine approach is to use primary vasoconstrictor drugs such as phenylephrine or norepinephrine to increase arterial blood pressure. For many patients with limited cardiac function, this approach is not helpful. In this situation the combination of dobutamine to enhance contractility and a constrictor is helpful.

## MANAGEMENT OF ELEVATED ICP

The assessment and treatment of ICP is fundamental to the management of any central nervous system trauma today. Intracranial hypertension affects an injured brain in several ways: Elevation of ICP may ultimately result in herniation of the brain through the incisura, the foramen magnum, or subfalcine openings, thus resulting in distortion and further injury to vital brain structures. A continuous rise of ICP may also cause secondary ischemic insults. Additional ischemia is a major factor producing further clinical deterioration after any neuronal injury. The main goal of ICP treatment is to minimize or, if possible, to eliminate secondary ischemic insults resulting from ischemia. As for critical care management of head-injured patients,

the focus of ICP therapy has shifted from purely pressure-oriented management to a cerebral perfusion pressure–oriented regimen.

### Definition of Intracranial Hypertension

Thresholds of intracranial hypertension have never been conclusively established, and the critical thresholds may vary among individuals and disease processes. However, ICP readings of 20 mmHg or more are usually considered the cut-off point of intracranial hypertension.

### ICP Monitoring

ICP monitoring provides the physician with additional data for rational ICP management. ICP devices can be positioned at epidural, subdural, intraparenchymal, or intraventricular sites. The ICP can be obtained using fluid-coupled (ventricular catheter, subarachnoid bolt) or non-fluid-coupled (fiberoptic, pneumatic devices) systems (see Table VIII). Although each has advantages, no system works satisfactorily to date. The main disadvantages are unreliable measurements and risk of infections. Currently, intraparenchymal ICP probes, which are easy to handle and give exact measurements, are used most widely. Invasive monitoring of ICP may be suggested if:

- The patient is suspected to be at risk for elevated ICP.
- The patient is comatose and has an intracranial pathological condition.
- The prognosis of the disease calls for aggressive therapy.

### ICP and Cerebral Perfusion Pressure

Cerebral perfusion pressure (CPP) is defined as MAP minus ICP. One of the primary goals of medical management is maintaining a sufficient CPP. This is of par-

**TABLE VIII  ICP-Monitoring Devices**

| ICP monitor | Advantage | Disadvantage |
|---|---|---|
| Ventricular catheter | Drainage and exact measures | Major surgery, high infection rate |
| Subarachnoid bolt | Low infection rate, not invasive | Unreliable, obstruction of device possible |
| Pneumatic device | Accurate measures | Expensive, dislocation possible |
| Fiberoptic device | Accurate measures, subdural, intraparenchymal, and intraventricular location | No further calibration possible, breakage risk |

ticular importance, because the cerebral autoregulation may be impaired in and around the ischemic lesion, and the blood flow changes passively in these regions as the perfusion pressure changes. In addition, the lower limit of autoregulation may be shifted toward a higher level in patients with long-standing arterial hypertension. Although it is likely that lowering blood pressure in the acute phase can be deleterious, it is less certain whether normal blood pressure should be raised. In analogy to patients with head trauma, it seems prudent to raise the CPP >70 mmHg in patients with arterial hypotension, although the effects of this strategy have not been determined. The blood pressure should only be lowered if marked hypertension is present (systolic BP >220 mmHg or diastolic BP >120 mmHg), and this should be done carefully to avoid a sudden drop in the blood pressure or hypotensive values.

All treatment modalities that lower ICP at the expense of blood pressure must be considered dangerous. Hypovolemia should be strictly avoided. Adequate volumes of crystalloid, colloid, and, if necessary, blood products should be given. A central venous pressure of 8 mmHg might reflect isovolemia. In patients with decreased systemic vascular resistance, pressors such as dopamine and epinephrine may be required. In patients with impaired cardiac (CO) output, dobutamine can be used to increase CO. Taking all this into account, it is important to be aware of the fact that the optimal CPP is not known. Recommendations from the literature differ, giving values between 50 and 70 mmHg.

### Treatment of Elevated ICP

Therapy of intracranial hypertension is primarily directed at removing the cause insofar as possible. This is not always possible. In such cases therapy is aimed at controlling ICP, in the hope that the primary cause of the intracranial hypertension will resolve. Controlling ICP is thus a supportive maneuver, intended to preserve viable neuronal tissue until the high ICP situation resolves. Therapeutic maneuvers involve one of five classes of therapy: decrease cerebral blood volume, decrease CSF volume, remove dead or injured brain tissue, remove masses or hematomas, and remove the calvarium to permit unopposed outward brain swelling.

Neurological deterioration subsequent to increasing brain edema is a common phenomenon in patients with large infarcts or intracerebral hemorrhage. Several treatment options for brain space-occupying edema are currently used. Traditionally, the mainstay of conventional therapy included ventilation, sedation, blood pressure control, hyperventilation, osmotic agents, and barbiturates. The recommendations for these therapies are based on small case series, evidence from animal experiments, or theoretical observations (Hacke et al., 1994). None of them has been evaluated in a randomized study.

Whether ICP monitoring should be included in the routine management of patients with large ischemic strokes is still a matter of controversy (Schwab et al., 1996). However, without knowledge of the ICP, CPP- or ICP-targeted therapies seem to be impossible.

As a basic procedure, the head should be positioned upright at approximately 20°. Euvolemia should be established before elevating the head because of the risk of MAP and CPP drop in hypovolemic patients. Fever has detrimental effects in cerebral ischemia and edema formation and should be rigorously treated. Electrolyte imbalance should be avoided, especially hyponatremia, which potentially aggravates the developing cerebral edema. Late reperfusion, extreme arterial hypertension, and the combination of both may increase brain edema significantly and should be prevented.

## OSMOTHERAPY

Hypertonic, low-molecular-weight solutions such as mannitol, sorbitol, glycerol, or hypertonic saline are used to reduce the brain water content by creating an osmotic gradient between brain and plasma, drawing water into the plasma. An intact blood–brain barrier is essential for establishing an osmotic gradient. It has been assumed that brain tissue dehydration is more pronounced on the side contralateral to the lesion, where the brain tissue is preserved. Hypertonic solutions decrease an elevated ICP and, therefore, may be beneficial in emergency situations in an acutely deteriorating patient before therapies such as hematoma evacuation or decompressive surgery can be initiated. However, their effect is not sustained. The long-term effects of repeated treatments with hypertonic solutions are still unknown. Repeated infusion of mannitol may aggravate cerebral edema by allowing the osmotic substances to migrate through a damaged blood–brain barrier into the brain tissue, thereby reversing the osmotic gradient. Furthermore, osmotic agents predominantly lead to dehydration and shrinkage of normal brain tissue and may facilitate displacement of brain tissue and even increase the risk of hemorrhage. However, these largely theoretical considerations have not been substantiated to date in clinical studies. Other complications of osmotic agents are electrolyte imbalances, hypervolemia with cardiac failure, and renal dysfunction. Serum osmolarity should be closely monitored.

In principle, the same dilemma as encountered with osmotic agents applies to all other treatment strategies: All these therapies can transiently decrease elevated ICP, whereas the long-term effects are less clear or even potentially noxious.

*Hyperventilation* causes serum and CSF alkalosis-induced cerebral vasoconstriction and reduces cerebral blood volume and ICP. The effects of hyperventilation are short-lived because of rapid compensation of CSF alkalosis and rebound vasodilation. Hyperventilation is performed in the intubated patient by increasing tidal volume or rate. A rebound increase in ICP occurs if normoventilation is resumed too rapidly. Deep hyperventilation may reduce the cerebral blood flow, leading to additional ischemic damage. Therefore, the $CO_2$ levels should not fall below 30 mmHg.

*Steroids* still continue to be administered in patients with ischemic stroke and intracerebral hemorrhage. Even so, several studies have demonstrated no benefit of this therapy; on the contrary, they show a higher risk of infection and gastrointestinal hemorrhage. From our viewpoint there is no indication for the routine administration of steroids in severe postischemic brain edema or after intracerebral hemorrhage.

*Drainage of CSF* is an effective method of lowering ICP, especially when the ventricular space is not collapsed. This is true, for instance, in patients with acute hydrocephalus after subarachnoid hemorrhage or space-occupying cerebellar infarction. The drawback of this method is a relatively high infection rate of 2%–10%, which increases when drainage is maintained over 10 days.

*Short-acting barbiturates* are frequently used in patients with elevated ICP. Barbiturates promptly and significantly reduce CBF and ICP. In addition to their action as a sedative, barbiturates reduce $CMRO_2$, induce vasoconstriction, and inhibit free radical–induced lipid peroxidation. Barbiturate therapy will usually decrease ICP. Barbiturates are mostly used when intracranial hypertension is refractory to other treatments. In addition, by blunting shivering, they can facilitate use of hypothermia therapy.

There are several adverse effects of prolonged barbiturate therapy for high ICP. Hypotension and respiratory depression may occur. Blood pressure may need to be pharmacologically supported, and dysfunction of the gastrointestinal tract may occur, making it difficult to sustain enteral nutrition. The use of barbiturates is limited because of various side effects, such as hypotension, cardiac depression, hepatotoxicity, and predisposition to infection, and two studies failed to demonstrate any long-term beneficial effects of prolonged barbiturate coma in patients with elevated ICP. Thiopental can be given 1–4 mg/kg, repeated as needed to control ICP. It is subsequently infused as needed to control ICP. Pentobarbital therapy is initiated with a loading dose of 10 mg/kg over 30 minutes. The maintenance dose is usually 3–5 mg/kg/h. Infusion can be

**TABLE IX    Conventional Therapy of Raised ICP**

Osmotherapy
- Glycerol 10%: 4 × 125 or 250 ml/24 h
- Mannitol 15%: 4–6 × 100 ml
- Hypertonic saline (NaCl 7.5% peak serum sodium of 165 mol)

Controlled hyperventilation
- Controlled mandatory ventilation (CMV)
- No higher PEEP (>10 mmHg if possible)
- $paco_2$: 30–35 mmHg

Barbiturates
- Pentobarbital administration of 3–5 mg/kg/h
- Continuous infusion
- EEG (burst-suppression pattern) monitoring
- Hemodynamic monitoring with Swan-Ganz catheter

THAM-buffer solution
Continuous infusion 1 mmol/kg/h by central venous line

increased until a burst-suppression pattern arises on the EEG. Increasing the barbiturate dose beyond that needed to produce burst suppression is unlikely to provide further decreases in ICP, because further decrements in CMR are not thought to occur at doses beyond that needed for burst suppression.

## Alternative Agents to Treat Elevated ICP

### Tromethamine

Intravenous administration of tris-hydroxy-methyl-aminomethane (THAM) or tromethamine induces vasoconstriction and thereby decreases ICP. The mode of action is not clearly understood. THAM increases the capacity of CSF to buffer pH changes and neutralizes acidosis-mediated vasodilation. THAM, 60 mmol in 100 ml 5% dextrose, may be infused over 45 minutes as a test dose. ICP should fall by 10–15 mmHg. A central venous line should be used for continuous IV infusion (1 mmoL/kg/h), because the substance may cause severe tissue necrosis. At high doses it can impair ventilation, which usually is not a problem, because patients requiring THAM are usually ventilated. However, THAM is also nephrotoxic, and therefore we use it only when all other conservative therapies have failed (Wolf *et al.*, 1993).

### Etomidate

Etomidate decreases ICP by decreasing CBF and CMR. It does not have as potent hemodynamic effects as the barbiturates. It is not suitable for prolonged infusion to control ICP because of inhibition of adrenal corticosteroid synthesis. Nonetheless, it can be used for brief periods as an adjunct in ICP control, especially if there is hemodynamic concern. The dose is 0.1–0.3 mg/kg IV.

## Propofol

Propofol is a hypnotic drug with a very rapid onset and a short duration of action. It depresses $CMRO_2$. In patients with head injuries, propofol reduced ICP. Side effects include a drop in systemic blood pressure, which may impair CPP, limiting its use for ICP control.

## Lidocaine

Lidocaine decreases synaptic transmission either directly or through blockade of the sodium channels. It also reduces $CMRO_2$ and acts as a vasoconstrictive. Lidocaine may also lower ICP, but its efficacy over time is not established yet. It is not typically used in prolonged therapy of intracranial hypertension. Rather it is most useful given 0.5–1.5 mg/kg for acute treatment of high ICP, particularly if associated with airway manipulation.

## Invasive Monitoring in Neurocritical Care Patients

Clear guidelines for instituting multimodal monitoring in severe stroke have not yet been established. ICP monitoring may provide data for sensible treatment of elevated ICP. However, ICP monitoring to evaluate CCP may not reflect the needs of the individual patient if it is the only method used (Bullock *et al.*, 1995; Rosner *et al.*, 1995; Unterberg *et al.*, 1997). There is evidence that the continuous measurement of the partial oxygen pressure of brain tissue (pbrO_2) by microprobes within brain tissue of the frontal white matter may represent an important alternative in the monitoring of comatose patients on neurological intensive care units (Dings *et al.*, 1998; Kiening *et al.*, 1996; van Santbrink *et al.*, 1996). Several experimental studies have shown that the continuous measurement of the pbrO_2 in animals significantly reflects changes in blood oxygenation, ventilation, ICP, and CPP (Dings *et al.*, 1998; Kiening *et al.*, 1996). Clinical experience with pbrO_2 monitoring has mainly been gained from patients with severe brain injury (van den Brink *et al.*, 1998; van Santbrink *et al.*, 1996; Yundt *et al.*, 1998).

New monitoring systems allow the introduction of several probes into the viable brain such as ICP, temperature, and PbrO_2 either in the injured hemisphere or bilaterally within the white matter of the frontal lobe. PbrO_2 is registered using a polarographic microproble, and temperature is measured continuously with microprobes. Data are processed in a computer to compensate for the temperature dependency of PbrO_2. Probes are inserted through a single screw.

Multimodal monitoring not only helps to control therapy but also helps to optimize the timing of therapies in two ways: (1) by avoiding a potentially dangerous therapy too early, and (2) by identifying those patients who will profit most from this therapy at an early stage in the course of the disease.

## External Ventricular Drain (EVD) Management

An EVD (ventriculostomy) may be placed to provide drainage for hydrocephalus or provide a means to measure ICP. Such drains require meticulous attention to their position relative to the patient and sterility. A drainage level relative to the midbrain is ordered if drainage is desired according to a certain ICP level. Alternatively, the EVD can be maintained clamped, with the pressure monitored. In such cases, the EVD will be initiated manually if ICP reaches a sustained level beyond a prescribed threshold. When the drain is open, there are several important considerations. If the drain is kept too high and the patient cannot drain CSF sufficiently, hydrocephalus or elevated ICP will result. If the drain is too low, excessive drainage can occur. If this allows gradients to arise, a herniation syndrome, usually across the midline (if the drain is contralateral from a mass lesion), can result. In addition, if there is an unruptured cerebral aneurysm, too low placement can produce an increase in transmural gradients across the aneurysm and encourage rupture. An addition, too low placement can encourage formation of a subdural hematoma as the brain retracts from the skull and tension is placed on bridging veins. It remains an unresolved problem what to do about abrupt brief and normal increases in ICP as may occur with coughing or other causes of a Valsalva maneuver. Such events result in a brief overflow of CSF, followed then by a period of low ICP, which might encourage aneurysmal bleeding or subdural hematoma.

When the EVD is clamped, good monitoring must be used to detect when the ICP is above the predetermined injury threshold in a given patient. Exceeding this limit indicates the need for drainage. When the drain is clamped to monitor, the usual arterial line flush system used for most pressure-monitoring purposes must never be attached. Infection is a major concern with an EVD. Published data indicate that the risk of EVD infection rises dramatically after 5 days of placement. Intrathecal antibiotic infusions have been used but not documented to decrease the infection rate and possibly encourage growth of resistant organisms. Subcutaneous tunnelling of the EVD catheter is one method, which may be effective to decrease the incidence of infection. Because of these concerns, most neurosurgeons change EVDs weekly or more often. Alternatively, some monitor WBC counts in the CSF to assess for occurrence of infection, although the necessary violation of the system to obtain such information may actually increase the infection rate.

## NEUROPHYSIOLOGICAL ASSESSMENT

The main goals of neurophysiological monitoring are to evaluate central nervous system functions in unresponsive patients and predict neurological deficit (Britt et al., 1981; Karnaze et al., 1982). In principle, electroencephalograms reflect cortical electrical activity deriving from synergy of cortical and subcortical structures, whereas evoked potentials comprise the sum of electrical responses at various levels of the neuroaxis within a particular sensory pathway after repetitive stimuli.

### Evoked Potential Monitoring

Although there are a number of useful features of evoked potentials in the assessment of unresponsive patients receiving narcotics and analgesics, there are several limitations as well: (1) data acquisition, signal processing, and analysis are difficult and require sophisticated machinery and software and well-trained staff; (2) only short latency-evoked potentials are resistant to external variables such as anesthetics, barbiturates, and hypothermia; and (3) the specificity of signal pattern abnormalities is low.

Several studies of nontraumatic coma revealed that the early bilateral absence of cortical somatosensory-evoked potentials is associated with death, persistent coma, locked-in-syndrome, or the development of a vegetative state (Krieger et al., 1993). The anatomical level of brainstem destruction, brainstem infarction by basilar artery occlusion, and brainstem impairment after massive cerebellar or hemispheric infarction have been correlated with abnormal patterns of BAEP (Nagao et al., 1979; Steiner et al., 1998).

Motor-evoked potentials (MEP) can be safely recorded in patients on a neurointensive care unit and yield important results. In patients with brainstem lesions, MEPs correlate with radiological findings, predict final motor function more accurately than clinical findings, and are a reliable diagnostic tool for assessing motor function in unresponsive patients (Schwarz et al., 2000).

### EEG Monitoring

Well-known benefits of EEG include detection of seizure activity (particulary when electrical-mechanical dissociation has occurred), close correlation with the cerebral metabolic rate, and sensitivity to hypoxia. EEG monitoring may also help to manage medically induced coma.

### Transcranial Doppler Ultrasonography

Transcranial Doppler ultrasonography (TCD) provides important information concerning the intracranial vasculature, the collateral flow, and signal characteristics of the various arteries (Aaslid et al., 1986). In the NCCU, the basal cranial arteries can be examined by TCD at the bedside. There is a strong correlation between TCD-blood flow velocity and angiographically confirmed stenosis and vasospasm. Therefore, TCD can reliably detect cerebral vasospasm in subarachnoidal hemorrhage and thus be used to guide specific therapy (Grosset et al., 1993). If the ICP is raised, a progressive reduction in end-diastolic velocity is seen, as reflected by the elevated pulsatility index. Intermittent TCD monitoring in acute cerebrovascular occlusion may accompany thrombolytic therapy to demonstrate recanalization.

With transcranial color-coded duplex sonography (TCCS) intracranial hemodynamics and parenchymal structures can be imaged at the bedside. This technique provides reliable information regarding midline shift in space-occupying hemispheric stroke (Bertram et al., 2000; Gerriets et al., 2001). Furthermore, TCCS has been shown to identify intracerebral hemorrhage, a possible complication of malignant hemispheric infarction (Hacke et al., 1996).

## FREQUENT CARDIOVASCULAR DISORDERS AND PROBLEMS

Cardiovascular diseases appear frequently on the NCCU, because neurological patients are commonly older, and acute stages of neurological disease may exacerbate former cardiac disease. In addition, cardiac disease itself may be the cause for neurological disease (stroke) or be part of the differential diagnosis (e.g., syncope). However, neurological diseases can cause cardiac complications themselves. Cerebral lesions, including thalamus and the insula as centers of vegetative regulation, may cause arrhythmias, hypertension and hypotension, and pulmonary edema. Acute polyradiculitis may also cause life-threatening arrhythmias.

### Cardiac Diagnostics and Differential Diagnosis

#### Electrocardiography and Echocardiography

Cardiac monitoring on the NCCU requires at least ECG and blood pressure assessment. Bipolar standard leads (Einthoven I, II, III) can be recorded continuously by means of a monitor. For exact cardiac analysis, ECG-diagnostic includes a 12-channel ECG. It shows cardiac rhythm, mean axis, hypertrophy, and intervals. A 24-hour ECG should be done if intermittent arrhythmia is part of the differential diagnosis.

Transthoracic echocardiography (TTE) allows the anatomical evaluation of the heart and surrounding larger vessels. The M mode is useful for motion analysis. Directions of blood flow and flow velocities can be measured, and the profile of the TCD shows hemodynamic parameters and pressure gradients. TTE as an emergency procedure should be done if pulmonary embolism or pericardial effusion is suspected. For the differential diagnosis of cardiac thomboembolism or endocarditis, TEE is required. Its optical resolution and sensitivity allow better detection of intracardiac thrombi, vegetations, and alterations of the atrial septum. In addition, intravenous injection of contrast agents with or without the Valsalva maneuver is of high sensitivity for detection of an open foramen ovale with right-left shunt.

### Chest X-ray

An antirapasticin chest x-ray is required for every newly admitted patient on the NCCU for follow-up, even if there is no actual obvious cardiac or pulmonary problem. Acute respiratory or cardiac insufficiency, suspected pneumonia, or central venous lines require follow-up chest x-ray.

Chest pain may come along with life-threatening conditions, mandating immediate diagnosis and treatment (see Table X). These include acute myocardial infarction (MI), unstable angina, aortic dissection, pulmonary embolism, esophageal rupture, and tension pneumothorax. Complicating the issue is the fact that the heart, aorta, lungs, esophagus, mediastinum, and upper abdominal viscera share interconnecting sensory fibers, making it difficult to distinguish MI from the more benign causes of chest pain. Any discomfort from the jaw to the upper abdomen should be viewed as a chest pain equivalent.

### Characteristics of Chest Pain of Cardiovascular Origin

Most frequent characteristics of acute chest pain of cardiovascular origin are shown in the following Table XI.

### Myocardial Infarction

Myocardial infarction (MI) is defined as an acute coronary thrombosis caused by a ruptured arteriosclerotic plaque and inflammatory altered plaque. The vascular occlusion leads to a decreased coronary perfusion and therefore a transmural or nontransmural infarction. The main goals in acute MI should be the early and long-lasting reperfusion of the occluded coronary vessel within 6 (up to 12) hours after MI and maintaining an adequate systemic perfusion.

Cardiac enzymes may be normal in early stages of MI (including the first 3 to 4 hours), and the ECG can be difficult to interpret (elevated T sign, but no ST elevation or depression). Only serial ECGs and cardiac enzyme diagnostic tests ensure the diagnosis. If the clinical situation and the ECG clearly (monophasic ST elevation over 1 mm in two limb leads or chest leads over 2 mm with ST depression in the other chest leads) indicate MI, immediate therapy should be started

**TABLE X** Differential Diagnosis of Acute Chest Pain

| Organ | Disease | Noninvasive diagnostics |
| --- | --- | --- |
| Heart | Myocardial infarction | ECG, Lab[a] |
| | Instable angina pectoris | ECG, Lab[a] |
| | Pericarditis | ECG, spiral CT |
| Aorta | Aortic dissection | TEE, spiral CT |
| Lungs/pleura | Pulmonary embolism | BGA, chest x-ray, ultrasonography, ventilation-perfusion scintigraphy |
| | Pleuritis | Chest x-ray |
| | Pneumothorax | Chest x-ray |
| Thoracic cage | Tietze's syndrome | X-ray of thoracic cage, CT/MRI |
| Gastrointestinal tract | Esophagitis | Endoscopy |
| | Ulcus ventriculi/duodeni | Endoscopy |
| | Pancreatitis | Lab[b], ultrasonography, CT |
| | Biliary cholic/cholecystitis | Lab[c], ultrasonography |

Lab[a]: CK, CK-MB, LDH, GOT, Troponin I or T.
Lab[b]: Amylases, lipase, calcium, hematology, CRP, sedimentation rate.
Lab[c]: Hepatic enzymes, bilirubin, ALP, hematology, CRP, sedimentation rate.

TABLE XI    Characteristics of Chest Pain of Cardiovascular Origin

| | |
|---|---|
| Instable angina pectoris | Changing chest pain, precipitated with or without exertion, radiating to the jaw or left arm; ECG alterations (e.g., negative T) not obligate, often positive troponin I or T when Ch or CK-MK is negative. Pain can relieved quickly (<5 min) by sublingual TNG. |
| Acute myocardial infarction | Persisting mostly retrosternal pain, no pain relief by TNG; monophasic ST elevation in ECG*; CK elevation with significant CK-MB portion (>6%) and positive troponin I or T[†]. |
| Aortic dissection | Sudden onset of severe anterior or posterior chest pain, with "ripping" quality; maximal pain may travel if dissection propagates. Additional symptoms relate to obstruction of aortic braches (stroke, MI), dyspnea (acute aortic regurgitation), or symptoms of low cardiac output caused by cardiac tamponade (dissection into pericardial sac). Hypertension or hypotension, blood pressure or pulse differences. TN-I/TN-T, CK negative. |
| Pulmonary embolism | Sudden onset of dyspnea, hemoptysis, sinus tachycardia, hyperventilation ($po_2$ and $pco_2$ decreased, lack of oxygen saturation when $O_2$ is given)[‡]; ECG (RV failure, right bundle branck block, $S_1Q_3$ type, p pulmonale), deep venous thrombosis, CK, or TN-I/TN-T elevation may appear. |

*Monophasic ST elevation >1 mm in two limb leads or chest leads >2 mm with ST depression in the other chest leads.
[†]TN-I/TN-T, CK, and CK-MB, as well as ECG may not clearly indicate MI during acute stages.
[‡]A normal BGA does not necessarily rule out pulmonary embolism. However, a ventilation-perfusion scintigraphy should be done if PE is likely.

(thrombolysis or PTCA). Negative cardiac enzymes do not exclude MI (Gershlick *et al.*, 2001). If the basic diagnostics do not lead to the diagnosis, additional procedures should be done: blood gas analyses, extended laboratory investigations (e.g., bilirubin, lactate), TEE, abdominal ultrasonography, ventilation-perfusion-scinitgraphy of the lungs, CT or MRI, cardiac diagnostic (coronary or pulmonary angiography), endoscopy.

### Basic Treatment

Oxygen delivery and an appropriate analgesia/sedation is essential for the patient. Morphine sulfate (2–10 mg IV) should be administered to reduce pain and cardiac preload and afterload. In the case of arterial hypertension or left cardiac failure, glyceroltrinitrate or isosorbide dinitrate should be given. Heparin (5000 Iu IV) and acetylsalicylate (500 mg IV) should prevent further occlusion of cardiac vessels. Antiarhythmics, diuretics, catecholamines, nitrate, PTCA, and coronary stenting is indicated in arrhythmias, heart insufficiency, pulmonary edema, or cardiac shock. Young patients with tachycardia and hypertension and no contraindications should additionally be treated with β-blockers within 12 hours after MI. The use of ACE inhibitors 12–25 hours after MI may support the remodeling of the injured heart. There are basically three methods for the causal treatment of MI.

- Thrombolysis: In acute MI within 6 hours in patients <75 years, ST alteration in two or more leads, and no contraindications. Thrombolytic agents are strptokinase (1.5 min IE IV over 1 h) or rt-PA (0.75 mg/kg body weight in 30 minutes followd by 0.5 mg/kg within 60 minutes).

- PTCA: PTCA is indicated in patients with an anterior MI, manifest heart insufficiency, age >75, after bypass surgery, diabetes mellitus. PTCA can be combined with stent implantation and consecutive administration of gycoprotein-IIb/IIIa receptor antagonists. PTCA is followed by full-dose heparin.
- Emergency bypass surgery.

### Cardiac Arrhythmias

Arrhythmias may be seen in the presence or absence of structural heart disease; they are more serious in the former. Conditions that provoke arrhythmias include (1) myocardial ischemia, (2) CHF, (3) hypoxemia, (4) hypercapnia, (5) hypotension, (6) electrolyte disturbances (e.g., K, Ca, Mg), (7) drug toxicity (digoxin), (8) caffeine, (9) alcohol, and CNS diseases (SAH, stroke, GBS).

Besides medical history and clinical examination, the ECG with 12 leads and rhythm analyses are most important. A 24-hour ECG should be done if arrhythmia is suspected to be transient. Certainly, the causes of arrhythmias should be narrowed by laboratory analyses and echocardiography. The following section gives a short, systematic overview of common arrhythmias and treatment. Additional information should be found in textbooks of internal medicine.

### Characteristics, Differentiation, and Therapy of Different Arrhythmias

Basically, cardiac arrhythmias can be divided into tachyarrhythmia and bradyarrhythmia. We only refer to tachycardias, because they are more common on the NCCU.

## Tachycardia and Tachyarrhythmia

Tachycardia is caused by increased pacemaker activity (sinus tachycardia), heterotopic pacing activity, or re-entry mechanisms. The later are the most clinically relevant tachycardias. Types of tachycardias are subdivided by the localization of the initial pacing activity. Supraventricular tachycardias (SVT) develop proximal to the AV conduction system and show a small QRS complex. Ventricular tachycardias (VT) arise distal to the AV conduction system, and a wide QRS complex is typical. Tachycardias can be subdivided based on the length of QRS complex and the rhythm as described in the following:

---

Heart rate >100/min:
QRS <0.12 s:     RR-Interval regulary spaced:
1. Sinas tachycardia
2. Atrial flutter or fibrillation with 2 : 1 or 3 : 1 block
3. AV reentry tachycardia
RR-Interval irregulary spaced:
1. Multifocal atrial tachycardia
2. Atrial fibrillation

QRS <0.12 s:     RR-Interval regulary spaced:
1. Ventricular tachycardia
2. Supraventricular tachycardia with prolonged AV-conduction
RR-Interval irregulary spaced:
1. Supraventricular tachycardias with aberrant ventricular conduction

---

### Sinus Tachycardia (ST)

ST has mostly noncardiac causes and is a physiological response of the organism. Consequently, treatment depends on the cause of ST (hypovolemia, hypoxia, sepsis, etc.). The main indication for pharmacological treatment is myocardial ischemia and infarction. Therefore, β-blockers should be used (e.g., metoprolol).

### Atrial Flutter and Fibrillation (AF)

Treatment of AF has two main reasons. First, AF shortens the diastolic interval and therefore decreases the diastolic filling of the left ventricle. A subsequent decrease of the ventricular compliance leads to a lower cardiac output. Second, AF can lead to formation of an atrial thrombus, which may be the cause for ischemic stroke.

The acute treatment of AF focuses more on slowing the heart rhythm than re-establishing the sinus rhythm.

Therefore, agents are listed in the following, and it has to be noted that only procainamide can convert AF to sinus rhythm (Gershlick et al., 2001; Kongsgaard et al., 2000).

**Verapamil:** HF can be controlled in 70% of patients, 0.075 mg/kg in 2 minutes, second dose of 0.15 mg/kg IV. Effects last for about 1 hour. If effective, continuous infusion (5–20 mg/h). Reduce by 50% in liver insufficiency. To diminish blood pressure reduction, infuse 10 ml 10% calcium gluconate over 5 minutes. Do not combine with β-blockers (AV blocks).

**Diltiazem:** Effects similar to verapamil. Less negative inotropic effects than verapamil; 0.25 mg/kg IV, over 2 minutes. If no effects, repat after 10 minutes in a dose of 0.35 mg/kg. If effective, continuos infusion (10–15 mg/h).

**β-receptor blockers:** In hyperadrenergic situations (e.g., MI). Use $\beta_1$-selective β-blockers (e.g., esmolol or metoprolol). Esmolol is given as a bolus of 0.5 mg/kg body weight over 30 seconds, and followed by continuos infusion of 50 μg/kg body weight over 4 minutes. Esmolol can be titrated. Metoprolol is given in 5 mg over 2 minutes. Repeat every 5 minutes if ineffective until the 15 mg are given in total.

**Magnesium:** 2 g $MgSO_4$ infusion over 15 minutes is effective for control of HF.

**Digoxin:** Although digoxin is used for control of HF, it is not useful for acute treatment.

## NEUROLOGICAL DISORDERS

### Cerebral Ischemia

Criteria for intensive care management of patients with acute cerebral ischemia include decreased vigilance state; respiratory insufficiency; signs of increased ICP; and severe cardiopulmonary, renal, septic, or metabolic complications. The clinical course and treatment of patients with basilar occlusion and cerebellar infarctions are desribed elsewhere. In the following, the course and treatment of space-occupying MCA infarction is described.

### *Clinical Course of Space-Occupying MCA Infarction*

The clinical course of severe MCA infarction follows a predictable pattern in most patients (Hacke et al., 1996; Krieger et al., 1999; Wijdiks and Diringer, 1989). At the time of initial clinical assessment within the first few hours after onset of symptoms, patients with large supratentorial infarcts are typically fully awake. Mild drowsiness may occasionally be present. Unresponsiveness caused by aphasia or hemineglect should not be misinterpreted as loss of consciousness. Fixed conjugate eye and head deviation toward the affected side are typical findings in patients with large supratentorial strokes, although this clinical sign is neither highly sensitive nor specific (Hacke *et al.*, 1996; Tijssen *et al.*, 1991). Bilateral motor signs, coma, posturing, or pupillary abnormalities are usually *not* present in the very early phase of large supratentorial infarcts.

Neurological deterioration occurs during the first 24 hours in most patients with large supratentorial infarcts, corresponding to the development of brain edema. The patient loses consciousness to varying degrees from drowsiness to coma, and the neurological deficit that already exists may worsen, if at all possible. Pupillary enlargement and areactivity to light, at first unilaterally on the side of the infarction and later bilaterally, nausea, vomiting, posturing, and abnormal breathing patterns are signs of secondary brainstem dysfunction caused by impending herniation.

In the uncal herniation syndrome, as extensively described by Plum and Posner (Plum and Posner, 1993), reaction of the pupil unilateral to the lesion first becomes sluggish, and then the pupil gradually enlarges, accompanied by progressive loss of consciousness. If the opposite cerebral peduncle is compressed, hemiparesis develops ipsilaterally to the lesion. The progression of the clinical signs reflects the brainstem damage in a craniocaudal direction from the midbrain to the medulla, manifesting as loss of oculomotor responses, contralateral pupillary enlargement, loss of corneal reflexes, posturing, and terminally leading to respiratory arrest and cardiac arrhythmias. Central herniation syndrome is characterized by loss of consciousness and bilaterally small, reactive pupils in the initial phase.

If ICP is being monitored, it is typically only moderately elevated (ca. 20 mmHg) at the first onset of deterioration. ICP values will subsequently rise over the next 24–48 hours. Elevated ICP is a reliable prognostic sign, and an ICP exceeding 30 mmHg is usually associated with a fatal course.

## Prognosis

With medical therapy only, the outcome in patients with large supratentorial infarcts is generally poor, and mortality rates from 55%–80% have been reported (Hacke *et al.*, 1996; Krieger *et al.*, 1999; Wijdiks *et al.*, 1989). Most patients who deteriorate neurologically after the first few hours will eventually die. Various predictors for deterioration and poor clinical outcome have been identified. Regarding vascular pathosis, distal ICA occlusion almost uniformly indicates fatal outcome. Proximal occlusion of the MCA stem is also an unfavorable radiological finding. Complete MCA plus ACA infarcts and panhemispheric infarcts are usually

lethal. With regard to clinical signs, progressive loss of consciousness during the first few hours after onset of symptoms is probably the most reliable predictor of a poor outcome (Hacke *et al.*, 1996). Rapid onset of neurological deterioration with loss of consciousness during the first 6 hours indicates an aggressive course of the disease and is associated with a high mortality. Once coma, pupillary abnormalities, or abnormal breathing patterns are observed, the further course will be fatal in most patients.

The extent of brain edema depends largely on the infarct size and location but also shows a substantial individual variability. As a general rule, young and middle-aged patients have less compensation capacity for space-occupying intracranial lesions than older patients with cerebral atrophy. As a consequence, younger patients tend to have elevated ICP develop more often, and an infarct of the same size may result in higher mortality in these patients compared with older stroke victims.

## General Management

Large hemispheric infarcts must be recognized in the emergency department as a life-threatening condition that requires prompt and massive intervention. After stabilization of the airway, breathing, and circulation, the initial diagnostic workup and transfer to a neuro-intensive care unit should not be delayed. If indicated, early reperfusion therapies can be initiated in the emergency department.

Venous access, continuous monitoring of blood pressure, ECG, and pulse oxygenation are part of routine intensive care measures. Continuous ECG monitoring is especially important, because neurogenic cardiac arrhythmias are frequently observed, particularly in patients with large right-hemispheric infarcts.

Although respiratory problems are uncommon on presentation, their frequency sharply increases within the first 24 hours, reflecting increasing brain edema and brainstem dysfunction. Most patients with large infarcts require ventilatory support. Indications for intubation and mechanical ventilation are hypoventilation with either hypercapnia or hypoxia or maintenance of stable airways in patients with reduced consciousness and a high risk of aspiration. Both hypercapnia and hypoxia lead to vasodilation, resulting in an increase of intracranial blood volume and thereby raising the ICP. Although about 60% of patients with large infarcts who have to be intubated eventually will die despite maximum conservative therapy. Some have a chance of a satisfactory outcome. If the decision for full medical management has been made, elective intubation should not be delayed to the point when additional complications from hypoxia or aspiration have already developed.

Patients with large strokes have an increased risk of secondary hemorrhage into the infarcted area. This is one of the reasons IV heparin, which has been advocated for many years despite the lack of evidence supporting its use, should be restricted to patients with a clear indication for anticoagulant therapy (e.g., patients with prosthetic heart valves). However, stroke patients in general are at high risk for venous thrombosis and pulmonary embolism and should be prophylactically treated with low-dose heparin.

Intravenous or intra-arterial thrombolysis with tissue plasminogen activator as a revascularizing therapy are treatment options that can only be used in the very early course of cerebral ischemia, before a large infarction has resulted. If baseline radiological findings demonstrate changes that are suggestive of a large infarction, thrombolytic therapy is contraindicated because of an overly high risk of intracranial hemorrhage (Hacke *et al.*, 1995). The use of thrombolytic therapy is also questionable in the presence of distal ICA occlusion, because revascularization rarely occurs in these patients.

### New Investigational Therapies of Severe Postischemic Brain Edema

Recognition of the high mortality in patients with large hemispheric infarcts and the obvious ineffectiveness of medical therapy alone has prompted innovative therapeutic approaches for these patients. Although the effects of these therapies have not yet been proven in randomized trials, the results of pilot studies are promising.

### Decompressive Surgery

Space-occupying hemispheric infarction, the so-called malignant MCA infarction, has a high mortality and morbidity even with optimal conservative treatment. In the past 15 years several studies have shown that decompressive surgery is a possible treatment strategy for otherwise uncontrollable, increased ICP after severe hemispheric stroke. Surgical decompression seems to be effective in lowering increased ICP and preventing transtentorial herniation and has been tested extensively in head trauma patients, with varying results, as well as in patients with space-occupying cerebellar infarctions, with a significant decrease in mortality and morbidity in comatose patients. Several cohorts of patients with large MCA infarction showed that decompressive surgery could reduce mortality to less than 50%. Our previously reported initial results revealed a mortality of 34%, with most survivors only mildly to moderately disabled. This is particularly important, because 75% of the patients in this series showed the clinical signs of uncal herniation. Because the clinical course of patients with severe MCA stroke is highly predictable, waiting for a pupillary dilation causes an unnecessary delay, because allowing mesencephalic ischemia to occur potentially worsens the prognosis. Therefore, we now recommend surgery within the first 24 hours after stroke onset on clinical and neuroradiological grounds. We advocate surgery within 24 hours when patients show the early neuroradiological criteria of complete MCA infarction together with a further clinical deterioration. Taking these two major points together, without surgical intervention the further clinical course in these patients is highly predictable. With these criteria for the early selection of our patients for decompressive surgery, we could reduce mortality to only 16% (Schwab *et al.*, 1998).

### Induced Hypothermia in Severe Ischemic Stroke

Normal body temperature is 37°C, although there is a significant diurnal variation of ±0.6°C. Body core temperature can be measured at varying sites; the shell temperatures are measured either sublingually, axillary, or on the skin. Core temperatures reflect tympanic membrane, esophageal, rectal, bladder, and pulmonary artery temperature measurements. Hypothermia is defined as mild hypothermia (−33°), moderate (−29°), and deep (<28°C).

## Clinical Application

Recently, the results of the first clinical trial on the use of moderate hypothermia in severe MCA infarction were published (Schwab *et al.*, 1998). Hypothermia was induced at a mean of 14 hours after the ischemic injury and maintained over 72 hours to overcome the maximum brain swelling, which is known to occur between days 2 and 5 after ischemia. In this trial all patients fulfilled the criteria for diagnosis of a "malignant" MCA infarction. However, the mortality was only 44%, and the survivors reached a favorable outcome, with a mean Barthel index of 70.

Hypothermia significantly reduced the ICP, a finding similar to the results of Marion *et al.* (1993) and Shiozaki *et al.* (1993), who used hypothermic therapy in traumatic brain injuries. However, rewarming the patients constantly led to a secondary rise of ICP, which required additional ICP therapy with mannitol. In some cases, it even exaggerated the initial ICP levels. It is known that the rewarming period is a high-risk time in brain injury, because metabolic needs may outstrip oxygen delivery at vari-ous temperatures. Hence, rewarming really is considered the "critical phase" of hypothermic therapy. This rebound after rewarming might be due to a proposed hypermetabolic response after induced hypothermia, because it was described after cardiopulmonary bypass surgery.

## Side Effects

Hypothermia affects virtually every organ system. Ventricular ectopy and fibrillation limit the extent of hypothermia, but this is only known to occur at temperatures <30°C. In the aforementioned study, pneumonia was the only severe side effect of moderate hypothermia. Other side effects of hypothermia, which have been shown in animal studies, are clotting abnormalities and coagulopathy. In baboons, systemic hypothermia led to an increase in the bleeding times. In men, the enzymatic reactions of the coagulation cascade were shown to be strongly inhibited by hypothermia. Pancreatitis with high serum amylase and lipase levels was also observed after hypothermic therapy. The association between hypothermia and pancreatitis is still only poorly understood.

Moderate hypothermia can decrease ICP and lowers mortality in patients with severe postischemic brain edema. Important side effects are reduction of platelet count, increased rate of pneumonia, and elevation of serum amylase and lipase levels. Our results suggest a beneficial effect of moderate hypothermia in the treatment of severe space-occupying MCA infarction.

## Polyradiculitis

Patients with the suspected diagnosis of acute polyradiculitis should be transferred to a medical center with a neurological intensive care unit, because severe forms can develop a variety of life-threatening complications. These include respiratory insufficiency, involvement of cranial nerves with insufficient swallowing or high risk of aspiration, impairment of the autonomic system including cardiac arrhythmias, hypertension and hypotension or paralytic ileus, and, in rare cases, electrolyte imbalance (Chalela *et al.*, 2001).

### Respiratory Insufficiency and Intubation

Respiratory insufficiency can occur from weakness of respiratory muscles and bulbar symptoms, including impending aspiration. Vital capacity is important for the decision of intubation and artificial ventilation (Chalela *et al.*, 2001). When the vital capacity is reduced to <25 ml/kg and pO$_2$ can only kept stable with increasing inspired oxygen concentration, intubation should be considered. Even though delayed intubation can reduce the risk for atelectasis and pneumonia, an emergency intubation should be avoided. Nasal intubation is preferred, because the time course of artificial ventilation

is hard to predict. Tracheostomy should be considered after 1 weak of nasal or oral intubation, because this makes the weaning easier (Lawn *et al.*, 2001).

### Autonomic Dysfunction

In 45% of patients who have polyradiculitis, autonomic dysfunction is observed (Asahina *et al.*, 2002). Often they are moderate or not recognized. However, one of the most dangerous complications is the occurrence of cardiac arrhythmias, which can be life-threatening (tachyarrhythmia and asystole caused by VT or vagal stimulation). Therefore, ECG monitoring is recommended. Because cardiac arrhythmias appear during bronchial toilet or other nursing procedures, a syringe of atropine (1 mg) should be present at the bedside of the patient. If cardiac arrhythmias appear, a transient cardiac pacemaker should be installed. β-Blockers should only be used for treatment of tachycardias, because they may lead to hypotension.

### Thromboembolic Complications

Severe paresis and mechanical ventilation increase the risk for deep vein thrombosis (DVT) and therefore life-threatening pulmonary embolism. Consequently, subcutaneous heparin (500 IU, SQ, 2 × 3/day) or low-molecular-weight heparin (30 mg, SQ, 2×/day), thromboembolic hose, and/or compression stockings are necessary for prevention of DVT.

## Encephalitis

Clinical features of encephalitis are described elsewhere in detail.

### Indications for Intensive Care Management

Most frequently, complications such as seizures, myoclonus and dystonia, catatonia, dysphagia, and electrolyte imbalance are the indications for intensive care unit admission. However, affection of the brainstem and cerebral edema may be life threatening.

### Therapy of Psychosis

The differentiation of extrapyramidal features caused by the underlying disease from side effects of neuroleptic therapy may be a problem during the treatment of psychotic or catatonic states. The malignant neuroleptic syndrome requires immediate cessation of neuroleptic therapy, but symptoms may occur for weeks. If neuroleptic therapeutics are stopped, benzodiazepines are recommended for further therapy. Overall, neuroleptics with anticholinergic effects (e.g., haloperidol) preferably should be used. However, haloperidol itself can cause severe rigidity and decreased responsiveness and therefore should be given in the lowest possible dose and duration. Because almost all drugs (benzodiazepines, barbiturates, neuroleptics) for sedation accumulate, and patients may require long sedation, a change of medication is recommended every 3–6 weeks.

### Myoclonus and Dyskinesia

The treatment of myoclonus and dyskinesia is difficult, because they do not respond well to drug therapy. If they affect the patient's respiratory muscles and stability, muscle relaxants should be used. Because most encephalitis patients are conscious, adequate sedation should be added. Benzodiazepines can be used, and the dosage has to be increased over 1 or 2 weeks; up to 100 mg/h of diazepam may be necessary.

### Intracranial Pressure and Cerebral Edema

Refer to the section above.

## Meningitis

### Intensive Therapy

All patients with purulent meningitis should be treated on an intensive care unit, because life-threatening complications appear frequently and may develop within minutes. Complications can be subdivided into intracranial and extracranial origin; 50% of patients with purulent meningitis have extracranial life-threatening complications develop, including septic shock, cardiac arrhythmias, adult respiratory diction syndrome (ARDS), and disseminating intervascular coagulation (DIC). Intracranial complications may be hard to detect. If clinical improvement with appropriate antibiotic therapy does not appear within 2–4 days, a cranial CT scan and, if necessary, an arterial angiography should be performed to detect arterial vasospasm, mycotic aneurysm, or septic sinus thrombosis. In addition, transcranial Doppler ultrasonographic examination is useful for the detection and monitoring of intracranial vasospasms. The treatments of the complications are listed in the relevant chapters. Corticosteroids and nimodipin in combination with increase of blood pressure to sufficient levels are required for therapy of vasospasms (Aronin et al., 2000; Ries et al., 1997).

Two features should also be focused on: if complications require intubation and mechanical ventilation of the patient, nasal intubation should be avoided, because possible infectious foci include otitis and sinusitis. If antibiotic therapy does not improve the patients status or temperature and infectious parameters are increasing, an unknown septic focus should be searched for. A cranial CT with a bone window of the sinus should be done. Every patient with meningitis should be seen by ENT specialists to detect cranial foci.

## Head Trauma

Basically, there are two kinds of accident-related brain damage. The primary damage includes injury immediately at the time of accident (lacerations, skull fractures, contusion, intracranial hemorrhage, diffuse axonal injury). Prevention of the impact itself is the only therapy for primary head injury. The severity of the head injury is the major determinant for the later outcome. Emergency care and intensive care unit treatment must prevent and/or minimize the secondary brain injury. Pragmatic therapy in the initial phase after traumatic brain injury consists of improving cardiorespiratory function (ameliorating hypoxia, hypercarbia, circulation instability, and hypoxemia) and reducing ICP. Therefore, space-occupying masses have to be removed, and pharmacological therapy has to be initiated for elevated ICP (Brain Trauma Foundation, 1995).

### Clinical Brain Trauma Classification

The most important factor determining the degree of severity of the posttraumatic brain damage are the duration of the posttraumatic disturbance of consciousness and the presence or absence of brainstem dysfunction. Open- and closed-head injury are important marks for further therapy.

A direct trauma of the skull leads most likely to skull fractures with dura damage and extracranial hematoma (epidural hematoma) or cerebral contusion hemorrhage. Acceleration or deceleration leads most likely to injury of the white matter, the corpus callosum, and the brainstem.

The Glasgow Coma Scale (GCS) (see Table XII) is the internationally established standard for primary assessment. A GCS of 3–8 points indicates severe head trauma, moderate is 9–12 points, and mild is 13–15 points. Limitation of the GCS includes the missing pupillomotoric reaction, brainstem reflexes, and side differences.

Assessment of the pupillomotoric function is one of the most important factors in the examination of head trauma. One-sided mydriasis is either the consequence of an ipsilateral intracranial mass by means of compression of the oculomotoric nerve at the posterior clinoid or the consequence of a direct trauma of the bulbus. The intracranial mass is ipsilateral in 85% of cases. Subdural hematoma especially can lead to contralateral mydriasis. Further on, the contralateral brain crus presses against the edge of the tentorium in the "Kernohan phenomenon," which leads to ipsilateral hemisymptomatic referral to the intracranial mass.

One-sided miosis may be seen in traumatic Horner's syndrome. If the pupillomotoric reaction is missing, an artificial eye bulb or an previous eye surgery should be considered. A bilateral wide, fixated pupil indicates central herniation. These patients have a worse outcome, because only 3.5% have a total recovery.

The rest of the brainstem reflexes, including the oculocephalic, ocluovestibular, corneal, and gag reflex, provide additional information about the function of the brainstem. The oculocephalic reflex should only been examined if injury of the cervical spine is excluded.

### Emergency Measures

Morbidity and mortality of brain trauma depend largely on how rapidly after the trauma the vital functions of breathing and circulation are made secure (Gildenberg and Makela, 1985). An accident casualty found unconscious must be regarded as polytraumatized and adequately protected (cave: cervical spine), until additional extracerebral injuries have been ruled out. Treatment priority is the securing the airway. Unconscious casualties and patients with a GCS <9 should be intubated in every case. The later should be controlled by mechanical ventilation to prevent hypoxia, respiratory acidosis, and hypercapnia. Sufficient sedation prevents increase of ICP during intubation. Extension of the cervical spine should be avoided, instead the Jackson position should be used. The neck should be stabilized.

TABLE XII  Glasgow Coma Scale

| Category | Reaction | Score |
|---|---|---|
| Eyes open | Spontaneously | 4 |
| | To command | 3 |
| | To pain stimuli | 2 |
| | Absent | 1 |
| Motor reaction | Follows commands | 6 |
| | Localizes stimulus | 5 |
| | Normal flexor withdrawl | 4 |
| | Abnormal flexion | 3 |
| | Extension posture | 2 |
| | No movement | 1 |
| Verbal reactions | Oriented | 5 |
| | Confused | 4 |
| | Isolated words | 3 |
| | Inarticulate sounds | 2 |
| | None | 1 |

Operative intervention in this ultra-early phase is strictly for the treatment of acute life-threatening secondary injuries. Normally, first priority is stabilization of vital functions. It is a serious treatment error to demand an unreasonably long transport route of the polytraumatized head-injured patient before thoracic and abdominal bleeding is cared for or reliably ruled out, and vital functions are stabilized.

### Mild Brain Trauma

Mild head trauma can lead to a short-lasting unconsciousness of up to 1 hour, followed by recovery, seldom to only volatile twilight state (Brain Trauma Foundation, 1995). However, the initial unconsciousness is deep. Up to 90% of the injured with mild brain trauma still complain 3 months after the trauma of occasional or lasting headache, feelings of giddiness, nausea, sickness, and sleep disorders (Levin *et al.*, 1987).

An x-ray examination should exclude cranial fractures (anteroposterior, lateral, or bilateral and frontonuchal views), and head wounds should be immediately treated. In all skull fractures, a CT should be carried out, even in the absence of alarming clinical symptoms, to exclude intracerebral hemorrhage. Any hint of cervical trauma demands an x-ray investigation of the cervical spine (minimal lateral view), including the C7 vertebra. Systemic sedation and analgesia are only used if necessary to examine the level of consciousness.

Inpatient observation for at least 24 hours is advisable, because severe posttraumatic intracranial complications, foremost hematomas, can appear even after mild craniocerebral trauma. In case of only brief loss of consciousness and favorable domestic circumstances, the patient can be admitted at home after excluding a skull fracture by x-ray investigation. Headache should be treated with mild and simple analgesics.

### Moderate and Severe Brain Trauma

Patients with moderate or severe brain trauma require continuous monitoring of the heart rate, arterial blood pressure, pulsoximetry, body temperature, as well as central venous pressure and urine excretion. Goals for treatment include an arterial oxygen saturation >95%, a $pO_2$ >100 mmHg, a $pCO_2$ between 35 and 40 mmHg, an MAP >90 mmHg, normovolemia, good electrolyte balance, normoglycemia, and normothermia.

**ICP.** Fifty to 70% of patients with a severe head trauma show an increase of ICP over 20 mmHg. Consequently, these patients (GCS <9) require invasive ICP monitoring (Brain Trauma Foundation, 2000). Also, patients operated for intracranial mass should receive an ICP probe. Advantages and disadvantages of different ICP probes are described earlier.

**Cerebral Oxygenation.** Recently, there have been different approaches for measuring cerebral oxygenation. Invasive monitoring of cerebral oxygenation is indicated in patients with severe head trauma. One possibility is a jugular-bulb catheter, which measures the jugular-venous $O_2$ saturation ($SJVO_2$). Combined with arterial $O_2$ concentration ($SaO_2$), the cerebral oxygenation consumption can be calculated:

$$O_2ER = (SaO_2 - SJVO_2)/SaO_2.$$

The $SJVO_2$ should be kept over 50% (normal, 60%–80%) by elevating the blood pressure or $pCO_2$. Desaturation has been shown to be associated with increased mortality. Another possibility for assessment of the cerebral $O_2$ content is the intraparenchnymal $O_2$ probe with a Clark probe. Normal values are between 15 and 30 mmHg, and a decrease <15 mmHg indicates low cerebral perfusion. Even though this method is highly reproducible, it is not possible to assess global oxygenation of the brain. Methods like [133]Xenon or [99m]Tc-HMPAO-SPECT are limited to certain neurological centers.

**Electrophysiological Monitoring and Diagnostics.** The conventional EEG is mostly used for adjusting the barbiturate dose in the case of elevated ICP ("burst suppression") and seizure activity. Certainly, EEG is frequently used for the diagnosis of brain death. Measurement of evoked potentials is an important method for prognosis in brain trauma. Bilaterally missing SEP indicates a poor prognosis, whereas unilateral nonresponse is difficult to interpret.

### Treatment of Elevated ICP

ICP has to be lowered if there are clinical signs, CCT findings, or ICP probe measures of elevated ICP. There two major concepts for treatment of elevated ICP after brain trauma: the hydrostatic-osmotic and the CCP-driven concept. Both approaches remain questionable.

The Lund concept is based on the assumption that posttraumatic edema formation results from gradients between hydrostatic and colloid osmotic pressure. The goal of this concept is to keep the colloid osmotic pressure high and the capillary pressure in the extracellular space low. Therefore, metoprolol and clonidine are used to induce hypotension, whereas dihydroergotamine is used for elevation of precapillary resistance. Fluid balance is driven negatively. Until recently,

there were no clinical studies for this. This management cannot be not recommended, because CPP is lowered.

The CPP-oriented concept assumes that the cerebral autoregulation curve is shifted to a higher cerebrovascular resistance. The goal of this treatment is to increase the arterial blood pressure by reactive vasoconstriction and therefore lowering the ICP by means of decreased intracranial blood volume. Catecholamines and volume administration are used to increase the blood pressure. Still, this concept is not proven to be superior to "conventional" management.

Until now, standard management for elevated ICP included a stepwise scheme that unites CPP and ICP as regulatory parameters (see Table XIII).

If elevated ICP is not responsive to conventional management, administration of barbiturates may be useful. The following preconditions have to be fullfilled if barbitruates will be used: normovolemia, normotonia, preserved $CO_2$ reactivity of the cerebral vessels, EEG activity. After administering a testing dose of 6–11 mg/kg body weight thiopental as a bolus resulting in a decrease of the ICP, further boluses can be administered (150–200 mg) to lower ICP to <20 mmHg. The subsequent continuos infusion of thiopental (3–5 mg/kg body weight) should be done under EEG control. Severe reduction of blood pressure should be treated by volume and catecholamine administration. Thiopental can be reduced if ICP remains at <20 mmHg over 12 hours.

Another experimental approach for the treatment of elevated ICP is the use of hypothermia, which has been shown to be effective in small cohorts. Recently, a clinical study has proven a lack of effect for induction of hypothermia after acute brain injury (Clifton *et al.*, 2001).

THAM is another possibility to reduce therapeutically resistant elevated ICP. Details are described earlier.

Besides this conservative management for elevated ICP, surgical decompression may be considered. The described four approaches should be considered as experimental, because the clinical benefit was not proven in any of them in controlled studies.

### Traumatic Hematoma

Traumatic hematoma complicates the consequences of the primary head injury. They may become life threatening by space-occuping mass effects leading to a sudden deterioration of the patient's status. The speed of development and anatomical site of the hemorrhages determine the clinical signs and the urgency of treatment (Marshall, 1990; Prasad *et al.*, 1997). Every secondary deterioration of the patient's status, including worsening of consciousness and new focal neurological signs, requires an immediate CT, even if the initial CT was normal.

TABLE XIII    Treatment of Elevated ICP in Brain Injury

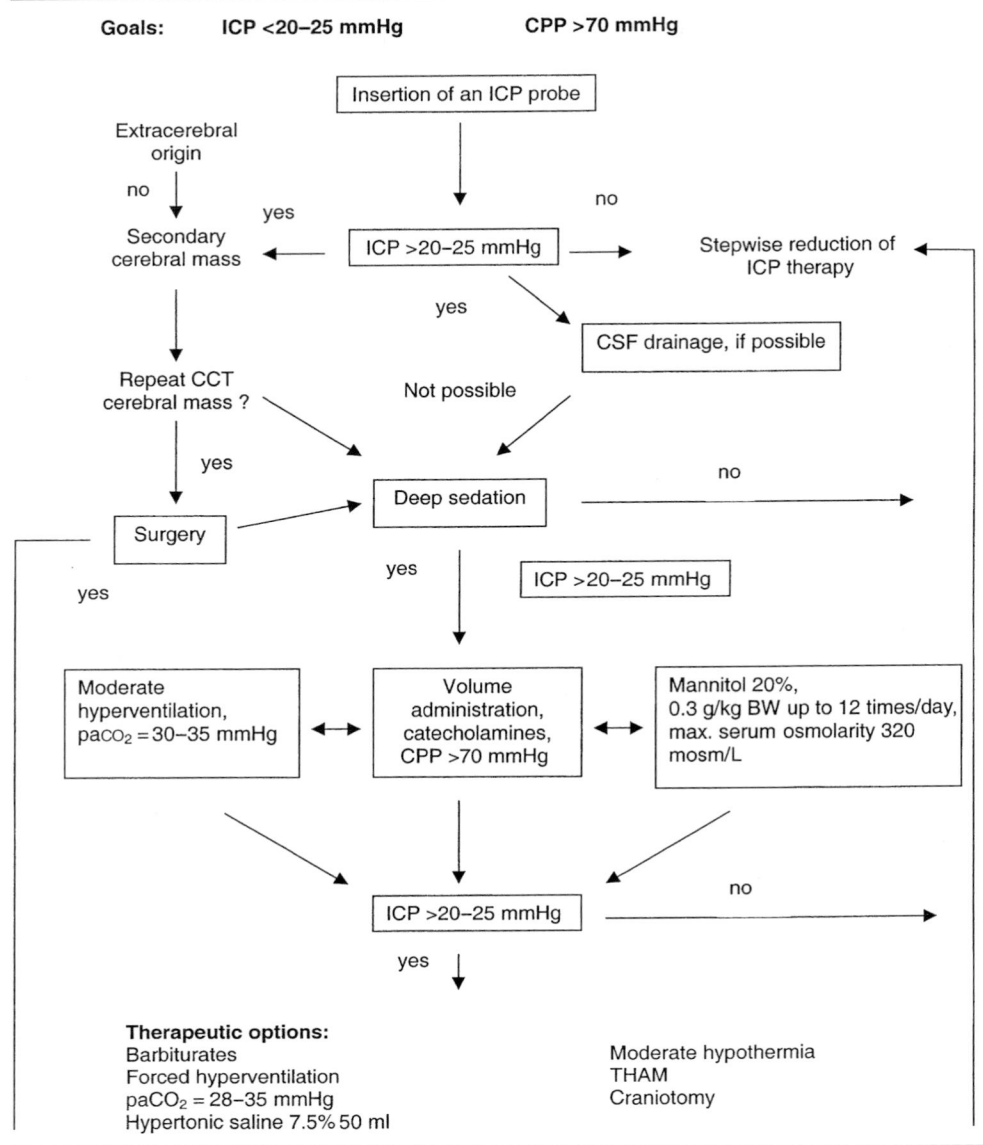

**Goals:    ICP <20–25 mmHg          CPP >70 mmHg**

*Epidural Hematoma*

Epidural hematoma (EDH) has an incidence of 1%–6% in all brain traumas. The middle meningeal artery and its branches are usually the source of bleeding. Otherwise, dural vessels, including the dural sinus and hemorrhage from fractures, may be causative.

Most frequently, EDH is localized in the temporolateral and frontal regions, in the posterior fossa, and at the vertex. Bilateral EDH has an incidence of 5%.

The most frequent reasons for a mortality of 15% from acute EDH are delayed diagnostic and surgical therapy. Mortality is directly associated with the level of consciousness at surgical decompression. Therefore,

mortality in a conscious patient is almost zero, whereas it increases in comatose patients up to 40%. Consequently, immediate diagnostic and surgical therapy is essential for a good outcome. Overall, there is not a uniform clinical picture indicating EDH. The sequence of initial unconsciousness followed by a lucid interval appears in only 20%. Overall, EDH should be checked for in patients with sudden unconsciousness in head trauma. Because skull fractures are seen at surgery in 80%–90% and conventional x-ray reveals only half of the skull fractures, CT should be the diagnostic approach of first choice. Even if the initial CT is normal, secondary worsening of level of consciousness should about one to secondary EDH.

Treatment of EDH is operative and should be performed early. Only in exceptional cases (hematoma with no midline shift or no tendency to enlarge) can one wait, provided CT is repeated. EDH of more than 25 cc should be operated in the unconscious patient. With threatening signs and symptoms of acute herniation, IV osmotherapy (e.g., 20% mannitol solution; 1 g/kg body weight) can be begun on the way to surgery.

### Acute Subdural Hematoma

Fifteen percent of all severe head traumas are associated with an acute subdural hematoma (aSDH). aSDH indicates a severe brain trauma, whereas venous bleeding is the most common cause of subacute or chronic SDH. Mortality is proportionately high. Mostly men between 45 and 60 years are affected. The space-occupying effect of aSDH is mostly amplified through additional regional or global brain swelling, which can lead to extensive midline displacement, disproportionate to the volume of the hematoma. In the case of a hyperacute, intensely space-occupying aSDH, the chance of survival is low, even when an immediate operation is performed.

The indication for surgical evacuation is given by the clinical state, the patient's medical history, and, most important, the CT findings. In acute trauma, the indication is a volume of hematoma of more than 25 cc and/or a midline shift of more than 5 mm. The therapy of choice is the immediate and radical removal of the hematoma and arrest of hemorrhage to reduce brain pressure and relieve brainstem deterioration. An osteoclastic craniotomy with enlargement of the dura is the surgical method of choice. Postoperative ICP monitoring is recommended for all injured patients with a severe preoperative disturbance of consciousness.

### Intracerebral Hematoma

More than 80% of intracerebral hematomas arise in the temporal poles and bases and in the orbital frontal brain surfaces. Intracerebral hemorrhage in the posterior fossa occurs rarely but is especially dangerous because of mass effects on the brainstem. The treatment of increased ICP is according to the general principles of NCCU management. There are no binding criteria for the decision of surgical intervention and its timing.

### Chronic Subdural Hematoma

Chronic subdural hematomas (cSDH) are mostly collections of venous origin that first became symptomatic after exceeding a critical hematoma size. Onset of bleeding and initial symptoms may be delayed more than 20 days (El-Kadi *et al.*, 2000). The initial trauma is mild or so trifling that it is not mentioned by the patient.

Predisposing factors, apart from nonspecific age-related changes of the dura, are alcoholism, epilepsy, anticoagulant treatment and other disorders of homeostasis, dialysis, and drastic reduction of ICP through a CSF shunt. The cSDH increases in size by afterbleeding that results from local hyperfibrinolysis on the hematoma surface (Sauter, 2000). The therapy of choice is operative hematoma evacuation. External drainage through burr hole trephination adequately reduces 80%–90% of all cSDH. In other cases a craniotomy with open hematoma evacuation is indicated.

## MALIGNANT HYPERTHERMIA

Malignant hyperthermia (MH) is a potentially life-threatening disease. The incidence of clinical apparent MH is estimated to 1:15,000 in children and 1:50,000 in adults. MH is seen in patients with an asymtomatic familial CK increase (30%–50% with increased risk) (Weglinski *et al.*, 1997) and a central or multicore disease (with some exceptions increased MH risk is obligate). In addition, MH may appear in Duchenne-Becker-Kiener dystrophia, myotonia, Schwartz-Jampel syndrome, periodic paresis, and congenital myodystrophia. Overall, patients with muscle disease have an increased risk for MH. Special substances like halothane or succinylcholine trigger an acute increase of intracellular calcium with activation of actin-myosin filaments. After exposure to these agents, clinical symptoms may appear even within minutes. Clinically, a massive rigor possibly mandating intubation, sinus tachycardia, and patchy skin is observed (Wappler *et al.*, 2001).

### Clinical Syndrome of MH

Trigger substances are as follows:

Potent volatile anesthetics: Halothane, methoxyflurane, isoflurane, cyclopropane, ether
Depolarizing muscle relaxants: succinylcholine, decamethonium

Obligatory predisposition is as follows:

Autosomal dominant MH (mapped to 19q, 13.1, 7q, 3q13.1)
Central core disease
King-Deborough syndrome

## Clinical Sequence

Beginning: tachycardia, tachyarrhythmias, unstable systolic blood pressure, tachypnea, cyanosis, increasing muscle rigidity

Following: metabolic and respiratory acidosis, increasing body temperature without shivering, increasing muscle rigidity, rhabdomyolysis

Complications: hyperkalemia, myoglobinuric renal failure, heart failure or arrest, consumption coagulopathy, brain edema.

## Principles of Therapy

If possible, survivors of MH should subsequently avoid exposure to triggering drugs. In susceptible subjects, MH is probably prevented by dantrolene. Dantrolene may be given orally several days before surgery or, with better control of plasma levels, it can be administered intravenously (2 mg/kg body weight) shortly before surgery.

When MH appears, early recognition almost certainly improves outcome. Monitoring of expiratory $CO_2$ concentration, pulse, temperature, central venous pressure, ECG, blood gases, and electrolytes should permit early recognition. The delivery of the trigger substances must be stopped as soon as possible. Anesthesia tubes should be changed if halothane or methoxyflurane was used. If the anesthesia cannot be stopped, one must switch to a nontriggering anesthetic. If metabolic and respiratory acidosis appear, the patient should be hyperventilated with 100% oxygen. According to the blood gases and acid-base values, sodium bicarbonate should be administered IV (2–4 mq/kg body weight). Diuresis should be maintained with $K^+$-free solutions and, if necessary, diuretics, in case of hyperkalemia, hemoconcentration, and myoglobinemia. The patient should be quickly cooled down with external and internal cooling. If the temperature exceeds 40°C or increases rapidly, the patient should be wetted with warm water; chilled IV fluids should be administered; and a cold lavage of the stomach, rectum, bladder, or peritoneum should be started. Icepacks can be placed in the axillae and groin. In addition, a cooling blanket is useful. Dantrolene should be nicked with sterile water for reconstitution. A reasonable initial dose is 3 mg/kg IV, which should be given rapidly. A maintenance infusion of 1 mg/kg/h can be continued until the crisis ends. The maximum daily dose is 10 mg/kg body weight (Denborough et al., 1998; Wappler et al., 2001).

Postoperative care includes continuous monitoring in an intensive care unit. Cardiac glycosides, calcium-containing substances, sympathomimetics, and parasympatholytics should be avoided.

## STATUS EPILEPTICUS

Status epilepticus (SE) is defined in earlier chapters. It has an incidence of 10–24/10,000/year, which is increased in known cases of epilepsia. The mortality of 11%–34% correlates with older (>60 years) age of the patient, an acute symptomatic genesis, and the duration of the SE. The mortality is >30% in duration over 1 hour, whereas shorter SE (<1 hour) leads to a mortality of 2.7% (Lockey, 2002).

The most frequent causes of SE include cerebrovascular lesions, drug withdrawl (alcohol, anticonvulsants, other hypnotizing agents), drug intoxication (e.g., cocaine), and CNS infections. Posthypoxic brain damage after prolonged resuscitation is an important cause for a myoclonic state, associated with a poor prognosis. Other causes include metabolic derangement and head trauma. Chronic diseases such as severe epilepsy, alcoholism, and intracerebral lesions (brain tumors, ischemia) can lead to SE even after years.

SE can be classified into generalized SE or partial SE as described in detail in other chapters.

### Therapy

Overall, it is accepted that SE should be treated rapidly and aggressively, because grand mal seizured lead to neuronal damage after 20–30 minutes. EEG monitoring may be helpful for therapeutic control.

### Grand Mal Seizures

Single, generalized epileptic tonic-clonic seizures do not require pharmacological treatment. In contrast, SE demands emergency treatment, because a correlation has been shown between increasing resistance to pharmacological treatment and duration of SE. Table XVII shows the pharmacological treatment of SE. Aggressive treatment of SE requires monitoring of the cardiopulmonary status of the patient. Oxygen should be given in every case (De Lorenzo et al., 1992; Lockey, 2002). A venous line should be inserted, and glucose-enriched infusions should be administered to treat hypoglycemic-mediated epileptic seizures.

Benzodiazepines are still the medication of first choice because of their effect within minutes (Henriksen, 1998). Lorazepam is administered in a dosage of 0.005–0.1 mg/kg body weight and is effective longer than diazepam, which as given as 0.1–0.3 mg/kg body weight. If SE is not interrupted by this medication within 10 minutes, phenytoin (20 mg/kg body weight) should be administered IV. The effects of phenytoin appear with a latency of 20–25 minutes. Alternately, valproate can be used starting with a bolus of 600–900 mg given over 30 minutes. If SE still is not interrupted, which is the case in 3%–15%, a barbiturate should be used. If all described therapeutic approaches fail, narcosis should be initiated.

## Treatment of Focal Status

It requires same medication as seen in a grand mal status, but the effective dosis is usually less.

# ALCOHOL WITHDRAWAL DELIRIUM

Alcohol delirium and Wernicke's encephalopathy are the most important and dangerous alcohol-related diseases. Basically, three scenarios are seen on the neuro-intensive care unit: (1) an alcoholic is admitted to the hospital because of an non-alcohol-related disease. The onset of alcohol withdrawal delirium must be avoided. (2) An alcoholic is admitted to the hospital with an alcohol withdrawal syndrome. The development of complete delirium has to be avoided. (3) An alcoholic is admitted because of alcohol withdrawal delirium.

Because even a mild reduction of alcohol intake or alcohol excess may initiate alcohol delirium, the blood alcohol level has to be measured initially. Increased alcohol levels should lead to a careful anti-delirient treatment. Only 3%–15% of alcoholics have delirium, but 12%–23% have repeated and increasing severity of delirium.

**TABLE XIV** Emergency Treatment of a Status Epilepticus

---

Initial emergency treatment:
  Lorazepam 2–4 mg IV (8 mg)
  Diazepam 10–20 mg IV (60 mg)
  Clonazepam 1–2 mg IV (6–12 mg)
  Midazolam 10–15 mg IM
Basic approaches:
  IV line; stabilization of cardiopulmonary function; laboratory
    (electrolytes, glucose); bolus of 50-mg glucose, thiamine IV;
    oxygen
Repeated benzodiazepine application
  Phenytoin
    20 mg/kg IV (50 mg/min)
    Bolus of 750 mg <12 h (max. daily dosage: 15–20 mg/kg)
  Sodium-valproate
    Bolus of 600–900 mg/(30 min)
    Infusion 1500 mg <24 h (max. daily dosage: 4–9 g)
Phenobarbital
  20 mg/kg IV; 50–75 mg/min
  Bolus 200 mg IV; 10–20 mg/kg (max. daily dose: 800–1000 mg)
Narcosis with artificial ventilation
  Thiopental IV 100–250 mg starting dose, followed by 3–5 mg/kg/h
  Alternatively propofol or midazolam

---

Chronic alcohol abuse leads to global depression of the brain function, and counterregulatory mechanisms outlast the withdrawal and lead to the clinical symptoms.

Chronic alcohol abuse leads to increased activity of the glutaminergic system and therefore to cerebral seizures during withdrawal. Consequently, carbamazepine is used as antiglutaminergic agent. The compensatory down-regulation of GABA inhibition leads to agitation and epileptic seizures during withdrawal. GABAergic therapeutics, such as clomethiazole or benzodiazepines, are useful in this case. The inhibitory alpha$_2$-receptors are reduced in alcoholics. During alcohol withdrawal, sympathic hyperactivation leads to tachycardia, sweating, hypertonus, etc. The reduction of dopaminergic receptors leads to a delayed, but overshooting, increase of dopamine receptors. This explains the productive psychotic symptoms, which can be treated by neuroleptics. In addition, cholinergic insufficiency causes cognitive deficits.

The alcohol withdrawal delirium has a variety of differential diagnoses, including alcoholic hallucination, Wernicke-Korsakoff syndrome, schizophrenia, manic episodes, drug withdrawal, intoxication with cholinesterase inhibitors, bacterial or viral meningitis, epilepsy, cerebrovascular diseases, hyperthyreosis, or metabolic encephalopathies.

## Therapy

Alcohol withdrawal delirium is a medical emergency and requires treatment in an intensive care unit. Basically, control and stabilization of the vital functions are required. Routine blood analyses should include alcohol levels. CT or MR imaging is indicated if seizures, focal neurological signs, Wernicke's encephalopathy appear. Examination of the CSF should rule out meningoencephalitis. Vitamin B$_1$ should be given in a dosage of 50–100 mg IM followed by glucose-enriched infusions. Common therapy includes volume substitution, often controlled by a central venous line, calcium substitution, and slow equalization of hyponatremia (Cave: central pontine myelinolysis). The specific therapy of alcohol withdrawal delirium should sedate without impairing vital reflexes, increase the epileptic seizure threshold, decrease autonomic hyperactivity, and be antipsychotic. The pharmacological therapy should not prolong the delirium and must be administered after the state of delirium. If an alcohol-addicted patient is treated on the ICU because of other diseases, there is commonly the chance of withdrawal syndrome. Therefore, these patients have to be observed carefully, and prophylactic vitamin B$_1$ (100 mg/day) should be given. If sedation is necessary, GABAergic agents (benzodiazepines, barbiturates, clomethiazole) should be used. The use of carbamazepine, benzodiazepines, or clomethiazole is indicated if seizures appear or former delirium is known. The use of alcohol itself is helpful but should be avoided. Signs of an incomplete delirium include vegetative symptoms like tremor, sweating, restlessness, sleep disorders, and hallucinations. In these cases, complete delirium can normally be prevented by the administration of carbamazepine, benzodiazepines, or clomethiazole. The full clinical picture of the delirium has to be treated with the same medication combined with haloperidol and clonidine (Mayo-Smith *et al.*, 1997).

# REFERENCES

Aaslid, R., Huber, P., and Nornes, H. (1986). A transcranial Doppler method in the evaluation of cerebrovascular spasm. *Neuroradiology* **28**, 11–16.

Aronin, S. I. (2000). Bacterial meningitis: principles and practical aspects of therapy. *Curr. Infect. Dis. Rep.* **2**, 337–344.

Asahina, M., Kuwabara, S., Suzuki, A., and Hattori, T. (2002). Autonomic function in demyelinating and axonal subtypes of Guillain-Barre syndrome. *Acta Neurol. Scand.* **105**, 44–50.

Bertram, M., Khoja, W., Ringleb, P., and Schwab, S. (2000). Transcranial colour-coded sonography for the bedside evaluation of mass effect after stroke. *Eur. J. Neurol.* **7**, 639–646.

Brain Trauma Foundation. (1995). "Guidelines for the Management of Severe Head Injury." BTF, New York.

Britt, C. W., Jr. (1981). Nontraumatic "spindle coma": clinical, EEG, and prognostic features. *Neurology* **31**, 393–397.

Bullock, R., Chesnut, R. M., Clifton, G., et al. (1995). Guidelines for the management of severe head injury. In: "The American Association of Neurological Surgeons & the Brain Trauma Foundation." Brain Trauma Foundation, New York.

Burcar, P. J., Norenberg, M. D., and Yarnell, P. R. (1977). Hyponatremia and central pontine myelinolysis. *Neurology* **27**, 223.

Chalela, J. A. (2001). Pearls and pitfalls in the intensive care management of Guillain-Barre syndrome [review]. *Semin. Neurol.* **21**, 399–405.

Clifton, G. L., Miller, E. R., Choi, S. C., Levin, H. S., McCauley, S., Smith, K. R., Jr, Muizelaar, J. P., Wagner, F. C. Jr, Marion, D. W., Luerssen, T. G., Chesnut, R. M., and Schwartz, M. (2001). Lack of effect of induction of hypothermia after acute brain injury. *N. Engl. J. Med.* **344**, 556–563.

Defining acute mild head injury in adults: a proposal based on prognostic factors, diagnosis, and management [review] (2001). *J. Neurotrauma* **18**(7), 657–664.

DeLorenzo, R. J., Towne, A. R., Pellock, J. M., and Ko, D. (1992). Status epilepticus in children, adults, and the elderly. *Epilepsia* **33 Suppl 4**, S15–S25.

Denborough, M. (1998). Malignant hyperthermia [review]. *Lancet* **352**(9134), 1131–1136.

Dings, J., Jager, A., Meixensberger, J., and Roosen, K. (1998). Brain tissue pO$_2$ and outcome after severe head injury. *Neurol. Res.* **20**, S71–S75.

El-Kadi, H., Miele, V. J., and Kaufman, H. H. (2000). Prognosis of chronic subdural hematomas [review]. *Neurosurg. Clin. North Am.* **11**, 553–567.

Georgiadis, D., Schwartz, S., Baumgartner, R. W., Veltkamp, R., and Schwab, S. (2001). Influence of positive end-expiratory pressure on intracranial pressure and cerebral perfusion pressure in patients with acute stroke. *Stroke* **32**, 2088–20922.

Gerriets, T., Stolz, E., Konig, S., et al. (2001). Sonographic monitoring of midline shift in space-occupying stroke: an early outcome predictor. *Stroke* **32**, 422–447.

Gershlick, A. H. (2001). The acute management of myocardial infarction [review]. *Br. Med. Bull.* **59**, 89–112.

Ginsberg, M. D., Sternau, L. L., Globus, M. Y., Dietrich, W. D., and Busto, R. (1992). Therapeutic modulation of brain temperature: relevance to ischemic brain injury. *Cerebrovasc. Brain Metab. Rev.* **4**, 189.

Grosset, D. G., Straiton, J., McDonald, I., Cockburn, M., and Bullock, R. (1993). Use of transcranial Doppler sonography to predict development of a delayed ischemic deficit after subarachnoid hemorrhage [see comments]. *J. Neurosurg.* **78**, 183–187.

Hacke, W., Kaste, M., Fieschi, C., Toni, D., Lesaffre, E., von Kummer, R., Boysen, G., Bluhmki, E., Höxter, G., Mahagne, M. H., and Hennerici, M. (1995). Intravenous thrombolysis with recombinant tissue plasminogen activator for acute hemispheric stroke. *JAMA* **274**, 1017–1025.

Hacke, W., Schwab, S., and De Georgia, M. (1994). Intensive care of acute ischemic stroke. *Cerebrovasc. Dis.* **4**, 385–392.

Hacke, W., Schwab, S., Horn, M., Spranger, M., De Georgia, M., and von Kummer, R. (1996). "Malignant" middle cerebral artery infarction. *Arch. Neurol.* **53**, 309–315.

Henriksen, O. (1998). An overview of benzodiazepines in seizure management. *Epilepsia* **39**, S2–S6.

Karnaze, D. S., Marshall, L. F., and Bickford, R. G. (1982). EEG monitoring of clinical coma: the compressed spectral array. *Neurology* **32**, 289–292.

Kiening, K. L., Unterberg, A. W., Bardt, T. F., Schneider, G. H., and Lanksch, W. R. (1996). Monitoring of cerebral oxygenation in patients with severe head injuries: brain tissue pO$_2$ vs. jugular vein oxygen saturation. *J. Neurosurg.* **85**, 751–757.

Kongsgaard, E., and Aass, H. (2000). Management of atrial flutter. *Curr. Cardiol. Rep.* **Jul;2**(4), 314–321.

Krieger, D., Adams, H., Rieke, K., Schwarz, S., Forsting, M., and Hacke, W. (1993). Prospective evaluation of the prognostic significance of evoked potentials in acute basilar occlusion. *Crit. Care Med.* **21**, 1169–1174.

Krieger, D. W., Demchuk, A. M., Kasner, S. E., Jauss, M., and Hantson, L. (1999). Early clinical and radiological predictors of fatal brain swelling in ischemic stroke. *Stroke* **30**, 287–292.

Lawn, N. D., Fletcher, D. D., Henderson, R. D., Wolter, T. D., and Wijdicks, E. F. (2001). Anticipating mechanical ventilation in Guillain-Barre syndrome. *Arch. Neurol.* **58**, 893–898.

Levin, H. S., Mattis, S., Ruff, R. M., Eisenberg, H. M., Marshall, L. F., Tabaddor, K., High, W. M. Jr, and Frankowski, R. F. (1987). Neurobehavioral outcome following minor head injury: a three-center study. *J. Neurosurg.* **66**, 234–243.

Lockey, A. S. (2002). Emergency department drug therapy for status epilepticus in adults. *Emerg. Med. J.* **19**, 96–100.

Malik, A. B. (1985). Mechanisms of neurogenic pulmonary edema. *Circ. Res.* **57**, 1–18.

Marion, D. W., Obrist, W. D., Carlier, P. M., Penrod, L. E., and Darby, J. M. (1993). The use of moderate therapeutic hypothermia for patients with severe head injuries: A preliminary report. *J. Neurosurg.* **79**, 354–362.

Marion, D. W., Penrod, L. E., Kelsey, S. F., Obrist, W. D., Kochanek, P. M., Palmer, A. M., Wisniewski, S. R., and DeKosky, S. T. (1997). Treatment of traumatic brain injury with moderate hypothermia. *N. Engl. J. Med.* **336**, 540–546.

Maroon, J. C., and Nelson, P. B. (1979). Hypovolemia in patients with subarachnoid hemorrhage: therapeutic implications. *Neurosurgery* **4**, 223.

Marshall, L. F. (1990). Surgical treatment of extracerebral lesions in head injury. In "Craniocerebral trauma" (L. H. Pitts and F. C. Wagner eds.), pp. 37–48. Thieme, New York.

Mayo-Smith, M. F. (1997). Pharmacological management of alcohol withdrawal. A meta-analysis and evidence-based practice guideline. American Society of Addiction Medicine Working Group on Pharmacological Management of Alcohol Withdrawal. JAMA 278, 144–151.

Nagao, S., and Moody, R. A. (1979). Acute intracranial hypertension and auditory brain-stem responses. Part 1: changes in the auditory brain-stem and somatosensory evoked responses in intracranial hypertension in cats. J. Neurosurg. 51, 669–676.

Nemoto, E. M., Klementavicius, R., Melick, J. A., and Yonas, H. (1994). Effect of mild hypothermia on active and basal cerebral oxygen metabolism and blood flow. Adv. Exp. Med. Biol. 361, 469.

Plum, F., and Posner, J. B. (1983). The diagnosis of stupor and coma. 3rd ed. FA Davis, Philadelphia.

Prasad, K., Browman, G., Srivastava, A., and Menon, G. (1997). Surgery in primary supratentorial intracerebral hematoma: a meta-analysis of randomized trials. Acta Neurol. Scand. 95, 103–110.

Ries, S., Schminke, U., Fassbender, K., Daffertshofer, M., Steinke, W., and Hennerici, M. (1997). Cerebrovascular involvement in the acute phase of bacterial meningitis. J. Neurol. 244, 51–55.

Rosner, M. J., Rosner, S. D., and Johnson, A. H., (1995). Cerebral perfusion pressure: management protocol and clinical results. J. Neurosurg. 83, 949–962.

Samuels, M. A. (1993). Cardiopulmonary aspects of acute neurologic disease. In "Neurological and Neurosurgical Intensive Care," 3rd ed. (A. H. Ropper, ed.), pp. 103–120. Raven, New York.

Sauter, K. L. (2000). Percutaneous subdural tapping and subdural peritoneal drainage for the treatment of subdural hematoma. Neurosurg. Clin. North Am. 11, 519–524.

Schwab, S., Aschoff, A., Spranger, M., Albert, F., and Hacke, W. (1996). The value of intracranial pressure monitoring in acute hemispheric stroke. Neurology 47, 393–398.

Schwab, S., Schwarz, S., Spranger, M., Keller, E., Bertram, M., and Hacke, W. (1998). Efficacy and safety of moderate hypothermia in the therapy of patients with acute MCA stroke. Stroke 29, 2641–2466.

Schwarz, S., Hacke, W., and Schwab, S. (2000). Magnetic evoked potentials in neurocritical care patients with acute brainstem lesions. J. Neurol. Sci. 172, 30–37.

Shiozaki, T., Sugimoto, H., Taneda, M., Yoshida, H., Iwai, A., Yoshioka, T., and Sugimoto, T. (1993). Effect of mild hypothermia on uncontrollable intracranial hypertension after severe head injury. J. Neurosurg. 79, 363–368.

Shiozaki, T., Kato, A., Taneda, M., Hayakata, T., Hashiguchi, N.,

Tanaka, H., Shimazu, T., and Sugimoto, H. (1999). Little benefit from mild hypothermia therapy for severely head injured patients with low intracranial pressure. J. Neurosurg. 91, 185–191.

Solomon, R. A., Post, K. D., and McMuirtry, J. G. III. (1984). Depression of circulating blood volume in patients after subarachnoid hemorrhage implications for the management of symptomatic vasospasm. Neurosurgery 15, 354.

Steiner, T., Jauss, M., and Krieger, D. (1998). Hemicraniectomy for massive cerebral infarction: evoked potentials as presurgical prognostic factors. J. Stroke Cerebrovasc. Dis. 7, 132–138.

The Brain Trauma Foundation. (2000). The American Association of Neurological Surgeons. The Joint Section on Neurotrauma and Critical Care. Indications for intracranial pressure monitoring [review]. J. Neurotrauma. 17, 479–491.

Tijssen, C. C., Schulte, B. P. M., and Leyten, A. C. M. (1991). Prognostic significance of conjugate eye deviation in stroke patients. Stroke 22, 200–202.

Unterberg, A. W., Kiening, K. L., Härtl, R., Bardt, T., Sarrafzadeh, A. S., and Lanksch, W. R. (1997). Multimodal monitoring in patients with severe head injury: Evaluation of the Effect of treatment on cerebral oxygenation. J. Trauma: Injury, Infection, Crit. Care 42, S32–S37.

van den Brink, W. A., van Santbrink, H., Avezaat, C. J., et al. (1998). Monitoring brain oxygen tension in severe head injury: the Rotterdam experience. Acta Neurochir. Suppl. 71, 190–194.

van Santbrink, H., Maas, A. I., and Avezaat, C. J. (1996). Continuous monitoring of partial pressure of brain tissue oxygen in patients with severe head injury. Neurosurgery 38, 21–31.

Wappler, F. (2001). Malignant hyperthermia [review]. Eur. J. Anaesthesiol. 18, 632–652.

Weglinski, M. R., Wedel, D. J., and Engel, A. G. (1997). Malignant hyperthermia testing in patients with persistently increased serum creatine kinase levels. Anesth. Analg. 84, 1038–1041.

Weir, B. K. (1978). Pulmonary edema following fatal aneurysm rupture. J. Neurosurg. 49, 502–507.

Wijdiks, E. F. M., and Diringer, M. N. (1989). Middle cerebral artery territory infarction and early brain swelling: progression of age on outcome. Mayo Clin. Proc. 73, 829–836.

Wolf, A., Levi, L., Marmorou, A., et al. (1993). Effect of THAM upon outcome in severe head insury: An randomized prospective clinical trial. J. Neurosurg. 78, 54–59.

Yundt, K. D., Grubb, R., Jr., Diringer, M. N., and Powers, W. J. (1998). Autoregulatory vasodilation of parenchymal vessels is impaired during cerebral vasospasm. J. Cereb. Blood Flow Metab. 18, 419–424.

Zander, J. F., and Bourke, D. B. (1995). Pain relief anf sedation. In "Neuro Critical Care" (W. Hacke, ed.), pp. 116–124. Springer, Berlin.

*Intensive Care in Neurology*

## CHAPTER 56

# Increased Intracranial Pressure

## Mark T. Keegan and Eelco F. M. Wijdicks

Intracranial pressure (ICP) is maintained within strict parameters by means of multiple compensatory mechanisms. Alterations in the level of ICP may occur in pathological states as diverse as traumatic head injury and fulminant hepatic failure. Whatever the cause of the rise in ICP, the endpoint, if untreated, can be compression of brain tissue, ischemic neuronal damage, and herniation of brain contents. Treatment of increased ICP requires both general and specific measures, and invasive monitoring of ICP may be used to guide therapy. This chapter discusses the pathophysiology of, and conditions associated with, elevated ICP. The indications for ICP monitoring, and the monitoring devices themselves, are discussed. Traditional and new methods of treating increased ICP are described in detail, backed up by the evidence to support their use. The pros and cons of novel therapies are argued, and recommendations for treatments to avoid are made.

## INTRODUCTION

The normal adult ICP is 5–15 mmHg when measured with the patient in the horizontal position. Physiological elevations occur with coughing, sneezing, Valsalva maneuver, and position changes, although these are distributed equally throughout the craniospinal axis and are tolerated well. ICP tends to be lower in infants and children. The interaction of raised ICP with intracranial pathosis produces neurological problems. A variety of cerebral insults may cause an elevated ICP (defined as ICP >20 mmHg for a sustained period of time). Cerebral ischemia and direct compression of brain tissue may occur, leading to neuronal injury and poor neurological outcome. Sustained elevations greater than 25–30 mmHg may be fatal. Untreated increased ICP leads to a shift of brain tissue and brain stem or global ischemia when arterial blood pressure cannot overcome

ICP. The final result is global hemispheric injury leading to severe disability, vegetative state, and, not uncommonly, brain death. Early detection and treatment of intracranial hypertension with the goal of maintaining adequate cerebral perfusion pressure (CPP) (defined as mean arterial pressure minus ICP) serves to minimize neurological damage. Knowledge of the determinants of ICP and the consequences of alterations of this pressure are important in the care of patients with acute cerebral pathosis. This chapter will discuss the pathophysiology, clinical manifestations, and consequences of elevated ICP. Methods of diagnosis and monitoring, in addition to both generally applicable and specific therapies, will be reviewed.

## ANATOMICAL CONSIDERATIONS

The volume within an adult skull is approximately 1600 ml. The components that comprise total intracranial volume include the brain parenchyma (87%, 1400 ml), the cerebrospinal fluid (9%, 140 ml), and the intracranial blood, divided into the arterial and venous portions (4%, 60 ml). Eighty percent of the brain volume is water, and 20% of this is sequestered in the extracellular space. These separate components are enclosed in the calvarium, which is rigid and nondistensible (except in the cases of craniotomy, skull fracture, and in infancy before closure of the fontanelles.) Because of the skull's rigidity, the total volume within it remains constant. The cerebrospinal fluid (CSF) is secreted by the choroid plexus at a rate of 0.3–0.4 ml/min, and total CSF volume is replaced, on average, every 8 hours. As a person ages, brain volume decreases, and CSF volume increases in a reciprocal fashion. Translocation of the CSF from the intracerebral space into the intrathecal space allows compensation for a condition that would tend to increase ICP, but its role is minimal. As will be discussed later, removal of CSF (e.g., using ventriculostomy) remains a simple method of reducing CSF pressure.

The intracranial contents are separated by the tentorium into superior and inferior compartments that communicate by means of the tentorial incisura. The cranial compartment is linked to the spinal compartment by way of the foramen magnum. In the horizontal position, the pressures within the craniospinal axis are equal. The increase in the size of any one component of the cerebral volume (e.g., an increase in the CSF compartment because of hydrocephalus or an increase in the blood compartment because of sagittal sinus thrombosis) must be compensated for by a reduction in the size of another. If inadequate compensation occurs, regional pressures will increase. Early in the process, "global" ICP may be normal and not reflective of ongoing cerebral compromise. Later, however, global ICP also increases, leading to tissue compression and shift. First described more than a century ago, the relationship between the individual volumes and the total volume of the brain is known as the Monro-Kellie hypothesis. Simply stated, it means that the volume of brain tissue plus the volume of CSF plus the volume of blood is equal to a constant.

Because of the blood–brain barrier (BBB), the brain's extracellular environment is closely regulated. However, pathological states may cause disruption of the BBB leading to an increase in extracellular water and consequent expansion of the brain. In addition, brain parenchyma itself may increase in volume caused by, for example, a growing brain tumor. Tumors may also cause disruption of the BBB as they increase in size.

## CLINICAL ASPECTS

### Causes of Elevated ICP

The pathological entities that lead to increased ICP may be divided into four broad groups: mass lesions, cerebral edema, CSF hypervolemia, and increased intracranial blood volume. An increase in the volume of the brain as a result of an increase in sodium and water content is known as cerebral edema, and this may be subdivided into vasogenic, cytotoxic, and interstitial. Vasogenic edema is typified by increased permeability of brain capillary endothelium, which allows entry of plasma proteins and other macromolecules. Clinical disorders that produce this form of edema tend to cause contrast enhancement on computed tomography (CT) scans and increased signal density on magnetic resonance imaging (MRI). They include tumor, abscess, hemorrhage, contusion, infarction, purulent meningitis, multiple sclerosis, and toxic encephalopathy. Swelling of the cellular elements of the brain characterizes cytotoxic edema. The swelling may be accompanied by a decrease in the volume of the brain extracellular fluid. The CT and MRI alterations seen in vasogenic edema are not seen in cytotoxic edema. Hypoxia, osmotic dysequilibrium, and acute plasma hypo-osmolality can cause cytotoxic edema. The ischemic brain edema seen after cerebral arterial occlusion initially shows evidence of cytotoxic edema, and this progresses later to vasogenic edema. The third type of cerebral edema is known as interstitial edema and is typified by obstructive hydrocephalus. Obstruction of the CSF circulation causes an increase in brain interstitial fluid.

The causes of increased ICP are multiple. They may involve trauma or arise de novo. Table I contains an abbreviated list. There is considerable overlap among the categories.

### Diagnosis of Increased ICP

#### Signs and Symptoms

Signs and symptoms of elevated ICP can be relatively nonspecific in the earlier stages and difficult to recognize clinically. The late-appearing classical triad of Cushing (hypertension, bradycardia, and Cheyne-Stokes respiration) is less often seen now because of advances in diagnosis and treatment but remains

TABLE I    Causes of Elevated ICP

| Intracranial mass | Cerebral edema | CSF hypervolemia | Increased intracranial blood volume |
|---|---|---|---|
| Tumor | Hypoxia | Obstructive hydrocephalus | Hypertensive encephalopathy |
| Subdural hematoma | Osmotic dysequilibrium | Nonobstructive hydrocephalus | Cerebral venous sinus thrombosis |
| Epidural hematoma | High-altitude cerebral edema | | |
| Subarachnoid hemorrhage | Eclampsia | | |
| | Fulminant hepatic failure | | |
| | Diabetic ketoacidosis | | |
| | Encephalitis | | |
| | Intoxications | | |

common in surgery of the posterior fossa. The duration of elevation of ICP and the rapidity of onset determine the clinical manifestations because of the opportunity (or lack thereof) for compensatory mechanisms to work. In addition, the location of the pathosis (global or regional and the specific regional location) can considerably influence the clinical presentation. For example, infratentorial lesions that raise the ICP may cause early brainstem dysfunction and will be seen with cranial nerve palsies, cerebellar signs, and cardiorespiratory disturbances, features that may be late and ominous signs with supratentorial pathosis. Distortion of brain tissue, whether it is direct or indirect, or the development of ischemia because of elevated ICP, causes symptoms. The expected signs may be altered in the presence of medications (e.g., atropine, neuromuscular blockers, or sedative agents) or comorbid conditions (e.g., ocular disorders).

Classically, raised ICP is seen with a bifrontal or occipital headache, which is worse in the morning or late evening, and which increases with stooping forward, Valsalva maneuver, or coughing. The headache is likely caused by traction on the vessels at the base of the brain or by dural stretching. Although the typical headache is highly suggestive of elevated ICP, it is not diagnostic and may even be absent. Vomiting may be present, and this often at least partially relieves the headache. The patient may report visual disturbances, including visual blurring and obscurations, defined as second-lasting blind spells. Visual disturbances are thought to be due to retinal ischemia and papilledema and indicate that vision may be in jeopardy. Papilledema is thought to be present within 48 hours of ICP elevation and occurs because of axoplasmic stasis caused by the transmission of ICP through the subarachnoid space to the optic disc. Visual field defects (enlargement of the blind spot, central or paracentral scotoma, peripheral field constriction) are also common. Abnormalities of the eye movements may occur, predominantly abducens palsy. Despite severe papilledema, however, visual changes may be absent or minimal. Diagnosis of increased ICP is, like the diagnosis of many medical conditions, more difficult in children. Irritability, altered feeding habits, vomiting, and somnolence may be the only features. In infants, clinical examination may reveal the presence of a tense fontanelle or the "setting sun" eye sign. Signs of intracranial hypertension are summarized in Table II.

### Imaging of Increased ICP

In addition to clinical examination, elevated ICP may be diagnosed by the use of additional tests, both invasive and noninvasive. Plain skull radiographs may indi-

**TABLE II  Signs of Intracranial Hypertension**

| Produced by increased ICP alone | Produced by brain distortion and herniation |
| --- | --- |
| Headache | Abnormal motor responses (decorticate or decerebrate posturing) |
| Vomiting | Third cranial nerve palsy |
| Papilledema | Sluggish light responses |
| Drowsiness | Internuclear ophthalmoplegia |
| | Pinpoint pupils |
| | Impaired brainstem reflexes |
| | Arterial hypertension |
| | Bradycardia |
| | Cheyne-Stokes, ataxic breathing |
| | Apnea |

cate a skull fracture, leading to suspicion of intracranial pathosis. Imaging of the brain by means of CT or MRI may reveal evidence of disease processes that give rise to elevations in ICP (for example, a tumor or intracerebral hemorrhage). CT evidence of raised ICP includes sulcal effacement, compression of the cisterns, midline shift, and distortion of nearby structures by a mass lesion. The degree of midline shift may be assessed by the lateral displacement of the pineal gland or septum pellucidum. The absence of the fourth ventricle or any new mass in the posterior fossa is a neurological emergency. CT imaging is more widely available than MRI, and the images can be obtained more quickly. MRI, however, is more sensitive and provides images in multiple planes. Radiologic imaging gives only an estimate of the severity of the elevation in ICP, and imaging studies must be reviewed in the context of the clinical neurological status of the patient. Lumbar puncture may be performed to quantify the ICP, but it should be used with caution because of the possibility of precipitating herniation by altering craniospinal pressure gradients. Evidence of cerebral herniation may be obtained from imaging studies.

### Differential Diagnosis

The differential diagnosis of elevated ICP is wide. Many conditions may cause signs and symptoms such as headache and vomiting. Knowledge of the patient's coexisting disease processes, however, may give rise to the suspicion of intracranial pathosis. Papilledema, when observed, points strongly toward raised ICP but is not diagnostic. Once a diagnosis of elevated ICP is made, further investigation must be carried out to explain the cause of the elevation in ICP. When CT and MRI are normal, an MR venogram should be performed to exclude venous occlusive disease.

## NATURAL COURSE

### Compensatory Mechanisms

Various physiological mechanisms exist to maintain ICP within a narrow range. These multiple compensatory mechanisms serve to provide a buffering action. Each mechanism accounts for a different degree of compensation, and the capacity to buffer influences the extent to which the ICP is altered as a result of the change. The plot of the relationship between ICP (on the *y* axis) and intracranial volume (on the *x* axis) is known as the intracranial elastance curve. (Elastance is defined as the change in pressure per unit change in volume.) Langfitt and colleagues (1965) formally defined the ICP–volume curve. There are three parts to the curve (Figure 1). Along Part I, the system shows spatial compensation, and ICP increases minimally even with relatively large changes in intracranial volume. There comes a point (II on the figure) at which the buffering capacity has been exhausted, and thereafter (III) any change in volume will result in a large increase in ICP. The slope of the pressure–volume curve (dP/dV) defines the elastance of the system. When the elastance of the neuraxis is high, for any given increase in volume, there will be a large increase in pressure. The elastance of the system increases exponentially at increasing intracranial mass volumes. The inverse of elastance (dV/dP) is known as compliance. Compliance is a measure of the distensibility of the craniospinal system. When the system has a high compliance (a desirable situation), a large increase in volume produces a relatively small increase in ICP. The pressure–volume curve

as shown is an idealized curve, and the relationships may not be the same if a mass lesion expands rapidly, allowing little, if any, time for the body's protective spatial buffering mechanisms to take effect. Thus, a rapidly expanding temporal artery hematoma will cause an elevation in ICP earlier than in the case of a slow-growing glioma. The pressure volume curve also markedly depends on age. Atrophy allows for more room with any new masses.

### Herniation Syndromes

If compensation for pathological states that increase ICP is inadequate, pressure gradients develop between various areas within the skull. These pressure gradients ultimately cause a shift in brain tissue, leading to herniation. The diagram (Figure 2) shows the four principal sites of potential herniation. Diencephalic structures may herniate centrally through the tentorium cerebelli (transtentorial), or the cingulate gyrus may be squeezed under the falx cerebri (subfalcine). The uncus of the temporal lobe may also be forced underneath the tentorium cerebelli (uncal). The cerebellar tonsils may herniate alone, or as part of a massive brain shift, through the foramen magnum (tonsillar herniation.) In addition, in the case of a skull fracture or after a craniectomy, portions of brain tissue may translocate through the opening in the skull (transcalvarial).

The herniation syndromes differ both in their clinical presentation and their mortality rates. Expanding mass lesions in the supratentorial space tend to herniate

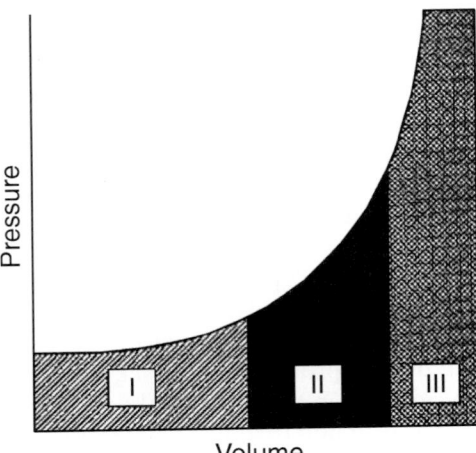

FIGURE 1    The ICP volume curve. Along Part I, the system shows spatial compensation, and ICP increases minimally, even with relatively large changes in intracranial volume. There comes a point (II) at which the buffering capacity has been exhausted, and thereafter (III) any change in volume will result in a large increase in ICP. (From Wijdicks, 1997, by permission of Mayo Foundation).

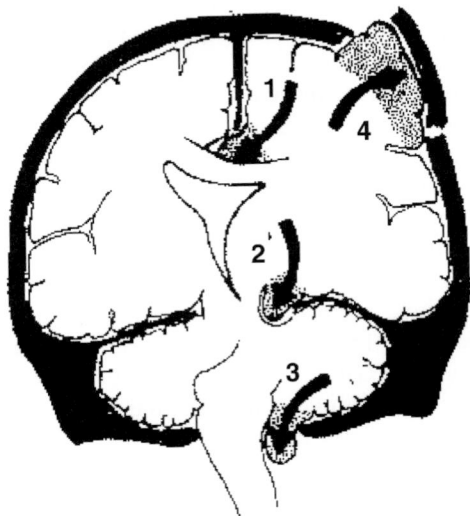

FIGURE 2    Potential sites of herniation. Sites through which brain tissue may herniate when compensatory mechanisms have been exhausted. 1, Cingulate; 2, temporal; 3, cerebellar; 4, transcalvarial. (From Fishman, 1975, with permission. *Copyright 1975, Massachusetts Medical Society. All rights reserved.*)

downward. Expanding infratentorial lesions may precipitate upward displacement of infratentorial structures, especially if the pressure above is lowered (e.g., by CSF drainage). Lesions in the posterior fossa can cause herniation of the cerebellar tonsils through the foramen magnum (causing respiratory or cardiac arrest) or push the superior cerebellar vermis through the temporal incisure, causing obstructive hydrocephalus. Mass lesions arising from the midline tend to cause central diencephalic transtentorial herniation, whereas laterally placed lesions tend to cause uncal herniation. The clinical manifestations of these conditions differ.

Central herniation is characterized by diencephalic dysfunction and tends to be associated with more slowly expanding lesions. Small, minimally reactive pupils are typical of central herniation, and its appearance is much less dramatic than a unilateral fixed pupil. Conjugate gaze and the oculocephalic and oculovestibular reflexes are preserved as long as the brainstem remains intact. Later, when the Edinger Westphal nucleus is damaged, the pupils become fixed and dilated in the midposition. Early in the course, the motor response to painful stimuli is preserved, but as the herniation syndrome progresses, decorticate or decerebrate posturing may be observed. Its gradual onset of brainstem dysfunction follows a rostrocaudal course.

Uncal herniation is more likely seen with a laterally placed, rapidly expanding mass lesion. The medial portion of the uncus may be forced through the tentorial incisura, or structures may be shifted horizontally to compromise the oculomotor nerve or the posterior cerebral artery. A decrease in responsiveness is seen in uncal herniation syndromes but could be a reflection of lateral midbrain displacement rather than tissue herniation. Unilateral dilation of the pupil is seen in the early stages of uncal herniation. When the contralateral cerebral peduncle is compressed, hemiparesis is on the same side as the lesion. Often, the early appearance of bilateral Babinski signs does indicate bilateral brainstem involvement. The patient eventually exhibits decerebrate rigidity.

Both central and uncal herniation can eventually result in midbrain compression, and this is associated with a poor prognosis. However, the point of no return is not well defined. Cerebellar tonsillar herniation can rapidly lead to compression of the brainstem. The oculovestibular and oculocephalic reflexes may remain intact with early midbrain compression, but later both are lost. Cheyne-Stokes breathing is eventually replaced by neurogenic hyperventilation. Further herniation leads to compression of the pons and medulla and rapidly evolves to fixed, dilated, midposition pupils, flaccid motor response and apneustic breathing. Medullary compression is a terminal condition and is associated with irregular respiration and hypotension.

## Outcome of Patients with Elevated ICP

Morbidity and mortality are closely related to increased ICP (Marmarou, 1991). Studies of patients with head injury have shown that worse outcomes correlate with persistent readings of ICP >20 mmHg. Expected mortality figures for patients with severe head injury and ICP values of <20 mmHg, >20 mmHg, >40 mmHg, and >60 mmHg have been reported as 18%, 45%, 74%, and 100%, respectively (Miller, 1983). The level of ICP has been correlated with outcome in other pathological states, including herpes simplex encephalitis (Barnett, 1988). However, it does not necessarily follow that therapies that decrease ICP could improve overall patient outcome.

## PRINCIPLES OF THERAPY

As described earlier, a reduction in the ICP depends on the decrease in the volume of one or more intracranial compartments. Therapy may be directed at each compartment, and there are significant differences in the extent and speed of the therapeutic effect. Reduction of CSF volume may be performed quickly using a ventriculostomy device. Manipulation of cerebral blood flow after alterations in lung ventilation also impacts quickly, but this effect is transient. Resection of brain tissue (normal or abnormal) causes long-lasting improvement in ICP levels, although this is less readily performed and may have significant adverse neurological consequences, because swollen tissue may not be distinguished from dead tissue (Table III).

There is a complex interdependence between ICP, cerebral blood flow, cerebral blood volume, cerebral perfusion pressure, and mean arterial blood pressure (MAP). The relationships are important only insofar as they relate to their influence on the volumes of the various intracranial compartments. Cerebral blood flow is related to cerebral perfusion pressure and inversely

**TABLE III** Intracranial Compartments and Techniques for Manipulation of their Volume

| Compartment | Volume control methods |
| --- | --- |
| Cells (including neurons, glia, tumors, and extravasated blood) | Surgical removal |
| Fluid (intracellular and extracellular) | Diuretics |
| | Steroids (principally tumors) |
| Cerebrospinal fluid | Drainage |
| Blood | |
|    Arterial side | Decrease cerebral blood flow |
|    Venous side | Improve cerebral venous drainage |

Adapted from Drummond and Patel, 2000, with permission.

related to cerebrovascular resistance. Cerebral autoregulation involves the maintenance of a constant cerebral blood flow. It operates when MAP is between 50 and 150 mmHg and is accomplished by alterations in cerebrovascular resistance, especially at the level of the cerebral arterioles. As MAP decreases, the resistance vessels dilate and allow blood to flow more easily into the brain. When the limits of autoregulation have been exceeded, cerebral blood flow follows MAP passively. There are circumstances under which cerebral autoregulation is regionally or globally impaired (e.g., after a subarachnoid hemorrhage, stroke, or other acute brain injury, in the presence of a tumor or vascular malformation, or in extreme metabolic derangements such as severe hypoxemia). In these cases, cerebral blood flow is related to cerebral perfusion pressure in a linear fashion. In patients who are chronically hypertensive, the cerebral autoregulation curve has been adjusted and is shifted to the right.

Cerebral blood flow is related to both $Pao_2$ and $Paco_2$. Carbon dioxide is the most potent regulator and is open to manipulation by mechanical ventilation. In the absence of a ventriculostomy, intracranial blood is the compartment most amenable to rapid alteration (by induced alterations in $Paco_2$). Between a $Paco_2$ of 24 and 100 mmHg there is an increase of $2 \, ml \cdot 100 \, g^{-1} \cdot min^{-1}$ in cerebral blood flow for each mmHg increase in $Paco_2$ (Sulek, 1998). In patients who have limited spatial buffering capacity, the acute increase in cerebral blood volume as a result of hypoventilation may be enough to cause a significant increase in ICP. By corollary, therapeutic hyperventilation may be used to compensate for increasing ICP.

When the $Pao_2$ level falls to critical levels (50 mmHg) cerebral blood flow will increase. Prolonged hypoxia also induces lactic acidosis, which itself causes vasodilation. Vasoconstriction occurs at oxygen partial pressures greater than 300 mmHg, probably to protect against cerebral oxygen toxicity, which can cause seizures. Cerebral blood flow (CBF) may also be influenced by temperature (CBF and cerebral metabolic rate increase 7%/°C increase in temperature, up to 42°C, after which CBF decreases) and by blood viscosity (polycythemia decreases CBF, anemia increases it).

## PRACTICAL MANAGEMENT

### Monitoring of ICP

Patients in whom the ICP is suspected to be elevated should have frequent monitoring of their neurological status by both nursing and medical personnel. Even in today's era of high-technology medicine, bedside clinical assessment remains vital. The Glasgow Coma score

is particularly useful in patients with elevated ICP. To more closely appreciate changes in ICP that may not manifest as clinical alterations, the use of ICP monitoring is advocated by many. In addition, ICP monitoring may allow a more rational use of therapies that could be harmful if used indiscriminately (e.g., mannitol, hyperventilation). Surprisingly, there have not been properly performed, prospective, randomized control trials to prove the usefulness of ICP monitoring in patients suspected of having elevated ICP. (However, the same can be said of many other frequently used medical devices such as pulmonary artery catheters). ICP monitoring may be indicated when there is a chance of preventing secondary brain injury. ICP monitoring would be helpful in early detection of intracranial mass lesions, may limit indiscriminate use of therapies to control ICP, and can reduce ICP using CSF drainage when a ventriculostomy is in place. The indications for ICP monitoring in specific disorders are summarized in Table IV.

A variety of techniques measure ICP. All require placement by means of a surgical hole in the skull. The devices may be fluid-coupled (a column of fluid transmits the pressure wave from the CSF to the pressure transducer) or they may use light (fiberoptics) or gas (pneumatics) to transmit the pressure wave. The "gold standard" is the measurement of intraventricular fluid pressure either directly or through a CSF reservoir. These ventriculostomy devices are especially useful in cases of acute or obstructive hydrocephalus, because decompression of a "tight brain" may be achieved by draining off CSF. The ventricular catheters are usually connected by means of a stopcock mechanism to an external transducer that allows intermittent ICP readings. The relative advantages and disadvantages of the various types of catheter are reviewed in Table V. The use of prophylactic antibiotics is controversial. The risk of ventriculitis is 1%–3%, and the cost of prophylaxis (e.g., using ceftazidime or vancomycin) is substantial.

Accurate, noninvasive ICP monitoring has long been sought after but remains a "holy grail." A number of techniques are promising but have not been proven to be satisfactory. *Transcranial Doppler (TCD)* is a noninvasive bedside test that gives some indication of CBF. The pulsatility index represents the relationship between systolic flow (influenced by blood pressure) and dias-

TABLE IV    Indications for Monitoring ICP

Traumatic brain injury (with Glasgow Coma Scale <7)
Fulminant hepatic failure with Grade 3 or 4 hepatic encephalopathy
Intracranial mass with perilesional edema
Barbiturate-induced coma
When using neuromuscular blocking agents in patients with
   multiple trauma
Encephalitis

**TABLE V**  ICP Monitoring Devices

| Device | Advantages | Disadvantages |
|---|---|---|
| Ventriculostomy | "Gold standard"<br>Simple<br>Most accurate<br>May be connected to external drainage system to withdraw CSF and decrease ICP<br>May administer medications directly to CSF<br>Can perform intracranial pressure volume measurements | Risk of infection (1%–3%) that increases after 5 days<br>Risk of hemorrhage<br>Needs to be rezeroed regularly<br>Needs to be flushed at intervals<br>May be difficult to place, especially if ventricles are compressed |
| Intraparenchymal fiberoptic transducer | Low incidence of infection<br>More accurate than subarachnoid bolts and epidural transducers | Does not allow CSF drainage by itself<br>Less accurate after 5 days because of "drift"<br>Fiberoptic catheter may leak |
| Subarachnoid bolt | Less chance of infection<br>Relatively easy to place | Unable to withdraw CSF<br>Unable to inject drugs<br>Prone to error<br>May become occluded<br>Prone to displacement |
| Epidural fiberoptic monitor | Significantly less risk of infection, because dura is not opened<br>Lower risk of seizures and hemorrhage compared with intraventricular device<br>Especially useful in patients prone to coagulopathy | Cannot inject drugs<br>Sensor must be closely applied to dura, parallel to plane of dura<br>Cannot perform intracranial pressure volume measurement<br>Prone to malfunction, baseline drift, and displacement |

tolic flow (influenced by cerebrovascular resistance, mainly ICP). It provides the best TCD correlate of ICP (Homberg *et al.*, 1993). TCDs are more useful as a trend marker in an individual patient as opposed to providing a true assessment of CPP. Furthermore, erroneous interpretations may be obtained at extreme values of CBF, and small changes in vessel diameter can markedly affect the calculation of flow. In extremes of increased ICP, only systolic peaks or reversed diastolic flow are seen. *Tympanic membrane displacement* (Reid *et al.*, 1990) is another indirect method of ICP assessment based on the fact that CSF and perilymph communicate through the cochlear aqueduct. By means of an impedance audiometer fitted into the external auditory canal, an estimate of ICP may be obtained from displacement of the tympanic membrane. More investigation of this technique is needed. The transmission of ultrasonic waves through the head has also been tested as a means to noninvasively measure ICP, but this technology is still at an early stage of development. Other promising new techniques are venous TCD of the Rosenthal vein (increased flow caused by displacement) (Schoser *et al.*, 1999) and pupillometry (in which a reduction in constriction velocity occurs causing so-called sluggish pupils before ICP increases.)

The ICP waveform is notable for three distinct peaks. The first peak (percussion wave) originates from the pulsation of the choroid plexus and is normally the most prominent. It tends to be relatively small in conditions of elevated ICP as the second and third waves increase in amplitude, and the waveform assumes a more rounded configuration. The second peak (dicrotic wave) is transmitted from pulsations of the major cerebral arteries. The third peak is known as the tidal wave. Although the waveforms are identifiable by looking at a short section of ICP tracing, ICP trend data are obtained over hours. The trend may show marked surges in ICP. For example, in the first 6–8 hours after head injury, in patients without a space-occupying lesion on CT scan, low ICP (mean, 15 mmHg) with a relatively stable mean is associated with a favorable outcome. However, in head-injured patients who manifest a stable elevated ICP (mean, >20 mmHg) with limited expression of the slow vasogenic waves, acute hydrocephalus or generalized brain swelling may be present.

Lundberg (1960) described three patterns of cyclic CSF pressure waves. Plateau waves (A waves) are sudden increases in ICP of 50–80 mmHg that may last for 5–20 minutes and often decrease rapidly. They may be the result of a "vasodilatory cascade" initiated by a reduction in CPP. Plateau waves are strong indicators of increasing intracranial elastance and reflect cerebral ischemia. They may be precipitated by suctioning, bladder catheterization, central line placement, and other activities required for ongoing care of the patient. Intravenous administration of lidocaine, barbiturates, or propofol may blunt these responses. B waves are pressure increases of less than 20 mmHg that occur once or twice per minute. They may be precursors of A waves. Lundberg C waves are of limited duration and amplitude and do not have major clinical significance (Czosnyka, 2000).

*Jugular bulb monitoring.* Measurement of the jugular venous oxygen saturation (SjvO2) using an oximeter-tipped catheter introduced retrogradely into the jugular bulb, may be of value in patients with elevated ICP (Cruz, 1998). As CBF decreases, SjvO2 decreases in a nonlinear fashion, because cerebral oxygen extraction is increased. SjvO2 also decreases if hyperventilation causes vasoconstriction or if fever or seizures occur. An increase in ICP related to cerebral hyperemia may lead to a decreased oxygen extraction and consequent rise in SjvO2. Normal SjvO2 is approximately 60%, and the ischemic threshold, while not clearly established, probably ranges from 24%–50%.

## Treatment of Increased ICP

The treatment of patients with elevated ICP consists of a combination of general medical and nursing practices in addition to therapy directed specifically at the underlying cause. Cerebral infections should be treated, tumors resected, and hematomas drained, because without treatment of the underlying cause of elevated ICP, general measures are ultimately doomed to failure. Patients should be admitted to an intensive care unit, preferably a unit specializing in the critical care of the brain-injured patient. When standard measures have failed and the patient has refractory intracranial hypertension (ICP, >20 mmHg for more than 10 minutes), more aggressive measures need to be instituted.

### General Management Principles

a. The patient with suspected or proven elevation in ICP should be admitted to a neurological and neurosurgical intensive care unit. Events that tend to increase ICP should be avoided or minimized. These can include frequent washing and changing of position. Neurological evaluation must be performed at frequent intervals and documented on a time-coded flow sheet. If a significant change occurs, prompt imaging using CT or MRI scans should be considered.

b. The patient's head should be maintained in a midline neutral position to avoid potential compression of jugular veins and consequent cerebral venous congestion. Endotracheal tube ties, if present, should not be tight around the patient's neck. Many intensive care neurologists advocate head elevation to 30 degrees, providing it does not jeopardize MAP and hence CPP. This position helps to improve venous outflow (which is the principal determinant of ICP in the normal person). Ideally, ICP monitoring should dictate the degree to which the head may be elevated. The Trendelenburg position should be avoided if at all possible. When placing central venous lines, the patient should be prepared and draped before assuming the Trendelenburg position to minimize the length of time spent with the head down. In some, an additional bolus of lidocaine or barbiturate may be needed.

c. Agitation and anxiety should be treated with reassurance with or without the use of benzodiazepines (e.g., midazolam 1–2 mg prn) or haloperidol (2.5–20 mg prn). Pain should be controlled adequately. Narcotic analgesics may be required for this purpose (e.g., codeine 15–60 mg prn, morphine 1–2 mg prn, fentanyl 50–100 μg prn). Mechanically ventilated patients may need sedation to minimize patient-ventilator dyssynchrony. Propofol is especially useful for this purpose, because stopping the infusion usually quickly terminates its effects, thus allowing neurological assessment. The drug should be infused and titrated to effect with doses of 25–100 μg/kg/min being typical. Bolus doses (10–50 mg) may be used to quickly control agitation, but the drug must be used cautiously, because it may cause significant hypotension by decreasing systemic vascular resistance.

d. In intubated patients, tracheobronchial suctioning should be sufficient to allow adequate oxygenation and ventilation, but frequent and multiple passes should not be allowed because of the danger of raising ICP. Intravenous lidocaine $(1 \, mg \cdot kg^{-1})$ may be used to blunt the response to suctioning.

e. The patient should be kept euvolemic. No evidence supports dehydration as a method to decrease ICP, and, in fact, it may cause harm by leading to hypotension, hemoconcentration, and hyperviscosity. Unless there are special circumstances, only isotonic fluids such as 0.9% saline or lactated Ringer's solution should be administered. The extra free water in 0.45% NaCl or D5W may potentially exacerbate cerebral edema. Systemic hypo-osmolality (<280 mOsm/L) should be corrected.

f. Fever should be investigated and treated promptly, because it will increase the metabolic rate and potentially increase ICP. Acetaminophen and cooling blankets are indicated, and the use of indomethacin may be considered (see later). Occasionally, morphine and bromocriptine need to be used in a patient with severe dysautonomia from diencephalic seizures (sudden tachycardia, tachypnea, shivering, and hyperthermia). Induction of moderate hypothermia (core temperature 32°–35°C) for 24 hours by use of cooling blankets may decrease intracranial pressure, but outcome has not been shown to be better in head-injured patients (Clifton *et al.*, 2001). Conversely, rewarming of a moderately hypothermic patient should be deferred in a patient with increased ICP.

g. In patients deemed to be at high risk for seizures, prophylaxis with intravenous fosphenytoin, 15–

20 mg/kg, loading dose should be instituted. However, the efficacy of prophylactic anticonvulsants in reducing the incidence of delayed development of epilepsy in head-injured patients remains unproven (Chesnut, 1996). Carbamazepine or sodium valproate are reasonable alternatives. If seizures do not occur, the prophylactic medications may be discontinued after 7 days. If seizures do occur, they should be treated promptly with barbiturates, benzodiazepines, or antiepileptic medications. Seizures may cause metabolic acidosis, hypoxemia, hypercarbia, and aspiration. Even in sedated and paralyzed patients, seizures can increase CBF, blood volume, and ICP.

h. Systemic blood pressure should be carefully controlled to optimize CPP. A goal CPP of 70–120 mmHg should be maintained. Reduction below this may lead to secondary hypoxic-ischemic damage, and increases above this range can lead to hyperperfusion and worsening of edema. This target may need to be modified in light of other clinical factors. Pressor medications (phenylephrine infusion beginning at $10 \mu g \cdot min^{-1}$ and titrated upward to effect, or dopamine beginning at $2 \mu g \cdot kg^{-1} \cdot min^{-1}$) may be used to raise MAP to a level sufficient to ensure an adequate CPP. Vasodilator antihypertensive agents (such as hydralazine and sodium nitroprusside) should be avoided, because they have the potential to increase ICP. If the cerebral perfusion pressure is excessively high (e.g., >120 mmHg), consideration should be given to its reduction by means of antihypertensive medications such as labetalol (5 mg IV initially, increasing until the desired effect is achieved.) This may lead to a concomitant decrease in ICP. Invasive monitoring (central venous pressure or pulmonary artery catheterization) may be required as a guide to fully optimize a patient's hemodynamics and improve CPP. Whether to use ICP or CPP as a target for treatment is controversial (see later).

i. Deep venous thrombosis prophylaxis: Patients with raised ICP will likely spend much of their time in bed and are thus at risk for deep venous thrombosis. Compression stockings and sequential compression devices should be used. Subcutaneous heparin may be appropriate but is contraindicated if there is a hemorrhagic cause to the raised ICP, or a high likelihood of hemorrhage exists.

j. Ulcer prophylaxis. With intracranial pathosis there is a risk of gastrointestinal ulceration (so-called Cushing's ulcer). The risk is increased if steroids are used to decrease intracerebral edema, in patients on mechanical ventilation, with underlying coagulopathies, or with prior nonsteroidal anti-inflammatory drug use. Consequently, $H_2$-receptor antagonists or proton pump inhibitors should be administered.

k. The use of antibiotics should be considered if there are signs of infection, and many institutions routinely administer prophylactic antibiotics when an ICP monitor or shunt is in situ (e.g., cefazolin 1 g tid) (see earlier).

l. As part of the patient's general care, adequate nutrition must be administered. The enteral route is preferable, and nutrition may be administered by way of nasogastric tube, gastrostomy, or jejunostomy. Specific formulations may be needed in the case of, for example, hepatic or renal failure to avoid precipitating further elevations in ICP. Parenteral nutrition, peripheral or central, may be necessary in cases of gastrointestinal dysfunction.

m. Systemic hyperglycemia has been shown to increase the likelihood of ischemic brain damage (Wass and Lanier, 1996) and is common, because of the catecholamine surge that may accompany intracranial hemorrhage or other pathosis. Steroid administration will also tend to increase the plasma glucose level. Insulin should be administered to maintain normoglycemia, and an intravenous infusion may be required. At the other end of the spectrum, even brief periods of hypoglycemia may be injurious to the brain. Normoglycemia is desirable.

n. Of all the methods used to decrease ICP, the use of a ventriculostomy device to drain CSF works most quickly. CSF drainage is especially useful in patients with acute hydrocephalus because of aneurysmal subarachnoid hemorrhage or expanding cerebellar masses. Devices can be automatically set to drain CSF passively when the ICP rises to a preset level, or active drainage may be undertaken. In the operating room, malleable needles may be placed into the intrathecal space in the lumbar region and used to drain CSF to allow the neurosurgeon to work on a "tight brain." The use of CSF drainage in patients with closed head injuries is controversial, although the practice is used by many. Drainage may be impeded by a large volume of intraventricular blood or because of compression of the frontal horns caused by surrounding edema.

o. A sudden increase of previously controlled ICP should lead the clinician to consider performing a repeat imaging study (such as a CT) to rule out the possibility of a new mass lesion (e.g., intracranial hemorrhage) that may be amenable to surgical intervention.

p. If ICP cannot be controlled by other methods, swollen brain tissue may be resected at surgery. Obviously, this procedure may lead to significant neurological defects, but it may be the only option to avoid herniation left available in the setting of refractory intracranial hypertension.

### Specific Aspects of Management

**Airway Management.** Maintenance of a patent airway is essential in patients with raised ICP. Endotracheal intubation should be performed if the patient's Glasgow Coma score is 8 or less or is rapidly declining. Intubation should also be considered in patients with head injuries and associated facial or airway trauma because of the potential for swelling and subsequent airway compromise. A physician experienced in airway management should perform endotracheal intubation. An adequate level of anesthesia must be assured before intubation is attempted, and this is usually achieved by means of an intravenous induction agent such as propofol, sodium thiopental, or etomidate. Ketamine is a poor choice for induction of the patient with raised ICP, because it tends to further elevate ICP. Despite adequate anesthesia, laryngoscopy can have significant hemodynamic consequences, and intravenous lidocaine or esmolol may decrease the potential for adverse events. The use of neuromuscular blockers for emergency intubation is controversial, but, assuming a satisfactory airway examination and a skilled operator, their use is encouraged in patients with raised ICP, because coughing or "bucking" will further raise ICP. Nondepolarizing neuromuscular blockers are preferred because of the theoretical risk of increasing the ICP by the use of succinylcholine. However, succinylcholine is indicated when a "rapid sequence induction and intubation" is necessary. Other drugs used in anesthetic practice that increase ICP include the volatile anesthetic gases, alfentanil, sodium nitroprusside, hydralazine, nitroglycerine, and nicardipine. The new methods of noninvasive ventilation are less applicable to patients with elevated ICP, because they do not provide a secured airway.

**Ventilatory Strategies.**    When possible, $PaO_2$ should be maintained at 80 mmHg or higher to ensure adequate cerebral oxygenation. Acute hyperventilation is an effective method to decrease ICP. Hyperventilation to a $PaCO_2$ of between 30 and 35 mmHg may be used to decrease CBF and thus cerebral blood volume. In refractory cases, hyperventilation to a $CO_2$ of 25 mmHg may be appropriate for a limited amount of time. Vasoconstriction is mediated by a change in the pH of the CSF. CBF decreases by 40% approximately 30 minutes after reduction of the $PaCO_2$ by 15–20 mmHg. The effects of hyperventilation are negated within several hours by bicarbonate buffering, which returns the CSF pH to normal. In addition, after several hours, CBF returns to 90% of baseline, and there is a potential for overshoot. There is no evidence that patients with severe head injury need prophylactic hyperventilation. Rather it should be reserved as a treatment modality for rapid treatment of acute rises in ICP unresponsive to other measures.*** Aggressive prophylactic hyperventilation has, in fact, been associated with adverse outcomes (Muizelaar *et al.*, 1991), presumably because of cerebral vasoconstriction and a decrease in CBF to below an ischemic threshold. In addition, the oxygen–hemoglobin dissociation curve may be shifted to the left, thus reducing the amount of oxygen delivered to the ischemic brain tissue. The patient's $CO_2$ should be returned to normocapnia gradually to avoid "rebound" increases in CBF.

Therapeutic hyperventilation may be deleterious to patients with emphysema and the obesity-hypoventilation syndrome, because a sudden reduction in $PaCO_2$ may cause hypotension. At the other end of the spectrum, permissive hypercapnia, a lung-protective strategy used in the modern-day management of acute respiratory distress syndrome (ARDS) is contraindicated in patients with elevated ICP. The use of pressure-controlled, inverse-ratio ventilation, also useful in patients with ARDS, does not seem to be associated with significant changes in ICP. ICP changes are influenced more by $PaCO_2$ than by airway pressures.

Positive end-expiratory pressure (PEEP) is used to recruit alveoli and improve oxygenation in mechanically ventilated patients. It may, however, increase ICP in two ways. Venous return may be impeded, thus increasing cerebral venous pressure and hence ICP. For this to occur, the cerebral venous pressure must at least equal the ICP. More importantly, PEEP may decrease systemic blood pressure, especially in hypovolemic patients. This may cause a reflex increase in cerebral blood volume and consequent elevation in ICP. In noncompliant lungs (in which PEEP is especially likely to be needed for oxygenation) the hemodynamic effects of PEEP are less, and hypotension-mediated increases in cerebral blood volume are less likely to occur. When using PEEP, one must weigh the potential improvement in systemic (and cerebral) oxygenation against the theoretical adverse effects PEEP may have on the brain.

**Osmotic Diuretics.***    Diuretics are one of the mainstays of treatment of raised ICP associated with brain edema. Available diuretics (both osmotic and nonosmotic) include mannitol, hypertonic saline, furosemide, urea, glycerol, and albumin. The drug most commonly used is 20% mannitol, which is given in a dose of 0.25–1.0 g/kg. The dose may be repeated every 2–6 hours as long as the patient's serum osmolality is less than 320 mOsm/kg. Intravenous mannitol causes an immediate plasma expansion that lowers the hematocrit and blood viscosity while increasing CBF and cerebral oxygenation. Mannitol favors the movement of brain water from the brain interstitium into the vasculature. In addition, it has a direct osmotic effect on neural cells. It reduces brain volume and elastance, provided there is an osmotic gradient between the blood and the brain. The effects last only a few hours, and a rebound phenomenon may occur. Mannitol is effective only in areas where there is an intact BBB, and therefore normal brain is most likely to shrink. An increase in plasma osmolality of 10 mOsm/kg may be effective therapeutically. Mannitol may be used in combination with furosemide for an additive effect on ICP reduction and duration of diuresis. Mannitol sets up the osmotic gradient, which draws fluid from the brain. The enhancement of water excretion by furosemide serves to maintain that gradient. Mannitol may also offer a degree of cerebral protection by its ability to scavenge free radicals, increase CBF because of transient hypervolemia, and decrease viscosity because of hemodilution. It may also increase CSF absorption. Urea and glycerol, although they are effective osmotic diuretics, have other less-desirable effects, and their use has largely been superseded by mannitol.

**THAM.** ** Tromethamine (dose, 1 ml·kg$^{-1}$·hr$^{-1}$) is a buffering agent that may be used to control increased ICP. THAM enters the CSF compartment and reduces cerebral acidosis. It alkalinizes without increasing plasma sodium or Paco$_2$. It is used in situations in which mannitol is contraindicated (e.g., renal failure) and may also help treat rebound from hyperventilation (Wolf et al., 1993). The drug has not been shown conclusively to improve outcome and so cannot be recommended as a first-line therapy. However, it may be useful in patients with renal failure.

**Corticosteroids.** Steroids have been used for many years in neurosurgical practice and are effective in reducing the edema associated with tumors.*** Clinical improvement after administration of steroids may be seen within 24 hours, but reduction of ICP may not occur for 48–72 hours. The mechanism of the beneficial effect of steroids is not well defined, but they may improve the "viscoelastic properties" of the intracranial space before edema reduction actually occurs. They may be helpful in restoring areas of vascular permeability that have been damaged, decreasing CSF production, and decreasing oxygen-mediated free radical production. Corticosteroids should not be routinely administered to the adult patient with raised ICP after head injury. They have not been shown to be beneficial and may be harmful (Allen and Ward, 1998). Fluorinated steroids may offer some new hope (Grumme et al., 1995).**

**Barbiturates.** Barbiturates have long been known to decrease ICP.*** They act by decreasing cerebral metabolic rate, decreasing CBF, and inhibiting free radical–mediated lipid peroxidation. Because of the associated manifestations of barbiturate therapy (including hypotension and apnea), barbiturates are reserved for situations in which ICP cannot be lowered by other means. A goal of barbiturate therapy for increased ICP

is the attainment of electroencephalogram (EEG) burst suppression, at which point neuronal metabolism serves only to preserve processes needed for cellular survival. Thus the cellular ischemic threshold is elevated. Barbiturates also decrease CBF and cerebral blood volume. The barbiturate most commonly used to reduce ICP is pentobarbital. It is administered as a loading dose of 5–12 mg·kg$^{-1}$ followed by a maintenance dose of 1–4 mg·kg$^{-1}$·hr$^{-1}$. An alternative is sodium thiopental, which in doses of 10–55 mg·kg$^{-1}$, given incrementally, will produce EEG burst suppression. Barbiturates administered at these doses may cause significant hemodynamic changes, necessitating pharmacological blood pressure support, and so the use of therapeutic "barbiturate coma" must be a well-considered decision. In addition, endotracheal intubation and mechanical ventilation are essential. High-dose barbiturates have been shown to be of use in lowering raised ICP associated with Reye's syndrome, head injury, fulminant hepatic failure, encephalitis, and focal cerebral ischemia (Bingaman and Frank, 1995). The mortality of head-injured patients with increased ICP refractory to pentobarbital exceeds 90%. However, although barbiturates do help to prevent patient death from refractory intracranial hypertension, there is little evidence that overall outcome is improved by such therapy. More recent data suggest that in a subset of patients who have an overreactive vasoconstrictive response, pentobarbital may worsen outcome because of its potential to induce cerebral hypoxemia (Cruz, 1996).

**Anesthetic Agents and ICP.** Intravenous anesthetic and sedative agents are associated with a parallel decrease in CBF and cerebral metabolic rate. Generally, they do not adversely affect intracranial compliance. They may be used as a treatment for elevations of ICP in patients who are intubated and ventilated. As mentioned earlier, ketamine is an exception. The volatile anesthetic agents cause a dose-dependent cerebral vasodilation, some of which may be offset by hyperventilation. They are not recommended as treatments for raised ICP but may be used in the operating room under controlled conditions when patients with raised ICP require general anesthesia to allow, for example, resection of a tumor.

## Management of Elevated ICP in Special Circumstances

**Fulminant Hepatic Failure.** Brain edema and intracranial hypertension are associated with liver failure and are one of the principal causes of death in fulminant hepatic failure (Blei, 1991). The major cause of elevated ICP in hepatic failure is an increase in brain water content. Brain edema is especially common in hyperacute liver failure, in which there is an interval of 7 days or less between the onset of jaundice and the development of hepatic encephalopathy. The presence of excess ammonia is a feature of liver failure, and there is a proven relationship between arterial ammonia concentrations and the development of cerebral herniation (Clemmesen et al., 1999). Brain astrocytes convert the ammonia to glutamine, which acts as an osmolyte and promotes cellular swelling. This leads to an increase in cerebral perfusion and consequent brain swelling. CT scans may be helpful to identify patients with cerebral edema, although their negative predictive value is not as good (Wijdicks et al., 1995). Lactulose is used in an attempt to decrease the production of ammonia in the gut. Ornithine aspartate and zinc have also been administered in an effort to increase the liver's capacity to clear the toxins responsible for hepatic encephalopathy, but the results are inconclusive. Two other mechanisms, namely the presence of inflammatory cytokines and the loss of cerebrovascular autoregulation caused by cerebral arterial vasodilatation, have also been implicated in elevations of ICP in fulminant hepatic failure (FHF). The benefits of invasive ICP monitoring must be balanced against the potential risks in a coagulopathic patient, although epidural catheters are frequently placed without complication. ICP monitoring in the context of FHF should be considered only for patients who are candidates for liver transplantation and should be part of a clinical protocol for the management of FHF. Treatment of FHF-related elevated ICP relies on conventional methods but will ultimately fail unless the underlying liver disease has been treated. Orthotopic liver transplantation is the treatment modality of choice in this circumstance, and survival rates of 60% or greater have been achieved in patients with FHF and cerebral edema. Bioartificial liver devices and auxiliary hepatic transplantation have been used as a "bridge" to either transplantation or spontaneous recovery and have been shown to be of use in the reduction of ICP and the prevention of neurological complications of FHF (Hughes and Williams, 1996). Recently, Jalan et al. (1999) have reported favorable results of the use of moderate levels of hypothermia in this setting.**

**Pseudotumor Cerebri.**    Pseudotumor cerebri (PTC) is a syndrome of raised ICP without evidence of a space-occupying lesion. The annual incidence is 0.9/100,000 in the general population, rising to 19.3/100,000 in women aged 20–44 years who are 20% or more above their ideal body weight (Friedman, 1999). The pathogenesis has not been fully established, but it is likely a disorder of CSF homeostasis, with the most widely accepted theory suggesting impaired CSF absorption at the level of the arachnoid granulations. The condition may be associated with endocrinopathies or the use of tetracyclines, vitamin A, or oral contraceptives. Almost all patients with PTC have headache, and the hallmark clinical sign is papilledema. Transient visual disturbances and pulsatile tinnitus may also be noted. Ventricular size is normal when visualized by imaging studies. A dilated optic nerve sheath, an empty sella, and enlargement of the subarachnoid space may be seen. CSF examination is essential for the diagnosis of PTC. In this condition, CSF pressure (measured at lumbar puncture in the lateral decubitus position with legs relaxed) is greater than 250 mm $H_2O$ in the obese patient and greater than 200 mm $H_2O$ in the nonobese patient. Medical management includes diet, weight loss, and diuretics (including the carbonic anhydrase inhibitor acetazolamide [1–4 g daily in divided doses] and furosemide). Corticosteroids (e.g., prednisone, 60–80 mg/day) rapidly decrease the ICP but are not suitable for chronic use because of side effects. Repeated lumbar punctures may also be used. Headaches associated with PTC may be managed with the same medications and techniques that are used in the management of migraine. Untreated, PTC leads to secondary optic atrophy and permanent visual loss. If there is visual loss or worsening of vision not attributable to macular edema, surgical intervention is indicated. The two major surgical treatments are optic nerve sheath fenestration and CSF shunting (usually lumboperitoneal). They are not recommended for the treatment of headache alone. In addition to the aforementioned measures, any specific cause of PTC requires treatment.

**ICP and Subarachnoid Hemorrhage, Head Trauma, and Hydrocephalus.**    The treatment of elevated ICP in the setting of these pathological conditions is discussed in Chapters 36, 57, and 60, respectively.

**High-Altitude Cerebral Edema.**    High-altitude cerebral edema is a feature of acute mountain sickness. It usually occurs at elevations above 2500 m and is more common in unacclimatized individuals. The clinical features include severe headaches, truncal ataxia, confusion, focal neurological signs, nausea, and vomiting. The syndrome may progress to seizures, obtundation, and coma. Retinal hemorrhages and papilledema may be seen in 50% of patients. The mainstay of treatment is immediate descent for at least 600 m, continuing until symptoms improve. Oxygen should be administered and dexamethasone 4–8 mg, given every 6 hours. A portable hyperbaric chamber may be used if immediate descent is impossible.

**Anoxia.**    After anoxia or an anoxic-ischemic event, cerebral edema may become evident on CT, but ICP does not rise to dangerous levels. There is no need for monitoring or aggressive medical management. Similar recommendations apply in patients with anoxic-ischemic injury caused by near-drowning.

### Controversial Points

**ICP or CPP.** There has been controversy as to the goal for therapies to treat raised ICP. Some workers believe that, in addition to, or instead of, attempting to decrease ICP, maintaining an adequate CPP is the key to successful outcome. CBF is maintained at a constant level across a wide range of CPPs (50–150 mmHg) when autoregulation is intact. When pathologic conditions impair cerebral autoregulation, CBF follows CPP in a straight-line relationship. The evidence for management of elevated ICP using CPP-directed therapy is not supported by a randomized, double-blind trial. Rosner and colleagues (1995) investigated the use of CPP as a target for managing patients with raised ICP after traumatic brain injury and found that there was an advantage over patients studied in the Traumatic Coma Data bank in whom ICP was used as the primary endpoint. However, Juul and colleagues (2000) suggested that ICP management should override CPP management. In their study, patients with an ICP of ≥20 mmHg had a significantly higher mortality rate and a poorer outcome than patients with an ICP of <20 mmHg, regardless of CPP.

**Therapy Based on Brain Volume Regulation and Preserved Microcirculation.** A new therapy for the treatment of posttraumatic brain edema and elevations in ICP is based on brain volume regulation and preservation of cerebral microcirculation ("the Lund protocol") (Eker et al., 1998). The proponents of this therapy believe that total brain edema is the major adverse factor in severe brain-injured patients and results from an impaired BBB. By lowering intracapillary hydrostatic pressure, preserving normal colloid osmotic pressure, and maintaining normovolemia, the therapy seeks to promote interstitial fluid resorption. Dihydroergotamine and thiopental are used to constrict precapillary resistance vessels, and dihydroergotamine also reduces intracranial volume by venoconstriction. Metoprolol and clonidine are administered to decrease intracapillary pressure by reducing systemic blood pressure and catecholamine release. The therapy potentially decreases CBF and risks cerebral ischemia. It remains controversial, because it contradicts the prevailing view of providing maximized CPP and has not gained widespread acceptance.

### Therapies That Do Not Work

**Glucocorticoids in Head Injury.** Glucocorticoids were used routinely in the 1970s and 1980s for the treatment of intracranial hypertension related to head injury. However, several controlled studies failed to show beneficial effects on outcome or control of raised ICP. The drugs are generally ineffective against cytotoxic edema. There is no role for the use of steroids in the setting of mass effect related to intracranial hemorrhage (Pourgavin et al., 1987) or cerebral infarction. Steroids are of use in the treatment of intracranial hypertension related to edema associated with intracranial tumors.

**Overzealous Hyperventilation.** Hyperventilation to $Paco_2$ levels of less than 25 mmHg and prolonged hyperventilation have not been shown to improve outcome in patients with elevated ICP. In fact, patients seem to do worse when these therapies are used. The use of "prophylactic hyperventilation" is also not recommended.

**Tirilazad.** The 21-aminosteroid derivative, tirilazad mesylate, was investigated to determine its efficacy in the treatment of head injury and associated elevations in ICP. A large prospective trial failed to show an improvement in outcome (Marshall et al., 1998).

**Hypothermia.** Hypothermia decreases neuronal metabolic requirements beyond the decrease caused by the achievement of EEG burst suppression. A recent multicenter randomized control trial (Clifton et al., 2001) showed that induced hypothermia (to 33°C) did decrease mean ICP in head-injured patients. However, outcome was not improved, and the therapy was associated with increased morbidity. Hypothermia does seem to have improved outcome in a group of patients with raised ICP in the setting of fulminant hepatic failure (Jalan et al., 1999).

### New Developments, The Future

**Decompressive Craniotomy, Unilateral Versus Bilateral.** Decompressive craniotomy, as a last resort for treatment of otherwise medically refractory increased ICP, has been performed but without much enthusiasm in the neurological community. However, a recent series of a well-defined population of 49 patients (Munch et al., 2000) found a decrease in ICP and brain shift but no impact on outcome. Prospective trials have not been performed, and data have been gathered by analysis of comparatively small heterogenous cohorts.

**Hypertonic Saline.** Hypertonic saline (NaCl solutions of 3% or greater) has been advocated as a treatment for cerebral edema and intracranial hypertension (Qureshi and Suarez, 2000). It is thought to act by means of an osmotic mechanism, although hemodynamic and anti-inflammatory effects may play a role. Data collection is still at the preliminary stage, and there have not been sufficient controlled clinical trials to recommend its routine use. However, hypertonic saline may be useful in the treatment of refractory intracranial hypertension, especially in the setting of traumatic brain injury.*

**Indomethacin.** Data from animal models and randomized control studies on preterm infants suggest that the nonsteroidal anti-inflammatory agent indomethacin can significantly reduce CBF (Harrigan *et al.*, 1997). Case series suggest that administration of 30–50 mg IV indomethacin may decrease CBF and ICP and improve CPP in patients with traumatic brain injury (Slavik and Rhoney, 1999). Use of the drug in this clinical setting has not progressed beyond the experimental stage.

## SUMMARY

Elevated ICP may be precipitated by a variety of pathological conditions. Irreversible neurological deterioration may be caused when the body's compensatory mechanisms are insufficient to stop the formation of intracerebral pressure gradients. The management involves the use of generally applicable interventions to decrease the ICP to normal levels, as well as treatment of the underlying cause.

## REFERENCES

Allen, C. H., and Ward, J. D. (1998). An evidence-based approach to management of increased intracranial pressure. *Crit. Care Clin.* 14, 485–495.

Barnett, G. H., Ropper, A. H., and Romeo, J. (1988). Intracranial pressure and outcome in adult encephalitis. *J. Neurosurg.* 68, 585–588.

Bingaman, W. E., Frank, J. I. (1995). Malignant cerebral edema and intracranial hypertension. *Neurol. Clin.* 13, 479–509.

Blei, A. T. (1991). Cerebral edema and intracranial hypertension in acute liver failure. Distinct aspects of the same problem. *Hepatology* 13, 376–379.

Bullock, R., Chesnut, R., Clifton, G., Ghajar, J., Marion, D. W., and Narayan, R. K. (1996). Guidelines for the management of severe head injury. New York: Brain Trauma Foundation.

Chesnut, R. M. (1996). Treating raised intracranial pressure in head injury. *In* "Neurotrauma" (R. K. Narayan, J. E. Wilberger, and J. T. Povlishock, Eds.). New York: McGraw-Hill.

Clemmesen, J. O., Larsen, P. S., Kondrup, J., Hansen, B. A., and Ott, P. (1999). Cerebral herniation in patients with acute liver failure is correlated with arterial ammonia concentration. *Hepatology* 29, 648–653.

Clifton, G. L., Miller, E. R., Choi, S. C., *et al.* (2001). Lack of effect of induction of hypothermia after acute brain injury. *N. Engl. J. Med.* 344, 556–563.

Cruz, J. (1996). Adverse effects of pentobarbital on cerebral venous oxygenation of comatose patients with acute traumatic brain swelling: Relationship to outcome. *J. Neurosurgery* 85, 758–761.

Cruz, J. (1998). The first decade of continuous monitoring of jugular bulb oxyhemoglobin saturation: Management strategies and clinical outcome. *Crit. Care Med.* 26, 344–351.

Czosnyka, M. (2000). Monitoring intracranial pressure. *In* "Textbook of Neuroanaesthesia and Critical Care" (B. F. Matta, D. K. Menon, and J. M. Turner, Eds.). London: Greenwich Medical Media.

Drummond, J. C., and Patel, P. M. (2000). Neurosurgical anesthesia. *In* "Anesthesia," 5th ed., (R. D. Miller, Ed.), Philadelphia: Churchill Livingstone.

Eker, C., Asgeirsson, B., Grande, P. O., Schalen, W., and Nordstrom, C. H. (1998). Improved outcome after severe head injury with a new therapy based on principles for brain volume regulation and preserved microcirculation. *Crit. Care Med.* 26, 1881–1886.

Friedman, D. I. (1999). Pseudotumor cerebri. *Neurosurg. Clin. North Am.* 10, 609–621.

Fishman, R. A. (1975). Brain edema. *N. Engl. J. Med.* 293, 706–711.

Grumme, T., Baethmann, A., Kolodziejczyk, D., *et al.* (1995). Treatment of patients with severe head injury by triamcinolone: A prospective, controlled multicenter clinical trial of 396 cases. *Res. Exp. Medx. (Berl)* 195, 217–229.

Harrigan, M. R., Tuteja, S., and Neudeck, B. L. (1997). Indomethacin in the management of elevated intracranial pressure: a review. *J. Neurotrauma* 14, 637–650.

Homberg, A. M., Jakobsen, M., and Enevoldsen, E. (1993). Transcranial Doppler recordings in raised intracranial pressure. *Acta. Neurol. Scand.* 87, 488–493.

Hughes, R. D., and Williams, R. (1996). Use of bioartificial and artificial liver support devices. *Semin. Liver Dis.* 16, 435–444.

Jalan, R., Damink, S. W. M. O., Deutz, N. E. P., Lee, A., and Hayes, P. C. (1999). Moderate hypothermia for uncontrolled intracranial hypertension in acute liver failure. *Lancet* 354, 1164–1168.

Juul, N., Morris, G. F., Marshall, S. B., and Marshall, L. F. (2000). Intracranial hypertension and cerebral perfusion pressure: influence on neurological deterioration and outcome in severe head injury. The Executive Committee of the International Selfotel Trial. *J. Neurosurg.* 92, 1–6.

Langfitt, T. W., Weinstein, J. D., and Kassell, N. F. (1965). Cerebral vasomotor paralysis produced by intracranial hypertension. *Neurology* 15, 622.

Lundberg, N. (1960). Continuous recording and monitoring of ventricular fluid pressure in neurosurgical practice. *Acta. Psychiat. Neurol. Scand.* 140(Suppl), 36.

Marmarou, A., Anderson, R. L., Ward, J. D., *et al.* (1991). Impact of ICP instability and hypotension on outcome in patients with severe head trauma. Report on The Traumatic Coma Bank Data. *J. Neurosurg.* 75(Suppl), 59–66.

Marshall, L. F., Maas, A. I. R., Marshall, S. B., *et al.* (1998). A multicenter trial on the efficacy of using tirilazad mesylate in cases of head injury. *J. Neurosurg.* 89, 519–525.

Miller, J. D. (1983). Significance and management of intracranial hypertension in head injury. *In* "Intracranial Pressure V." (S. Ishii, H. Nagai, and M. Brock, Eds.). Berlin: Springer-Verlag.

Muizelaar, J. P., Marmarou, A., Ward, J. D., Kontos, H. A., Choi, S. C., Becker, D. P., Gruemer, H., and Young, H. F. (1991). Adverse effects of prolonged hyperventilation in patients with severe head injury: a randomized clinical trial. *J. Neurosurg.* 175, 731–739.

Munch, E., Horn, P., Schurer, L., Piepgras, A., Paul, T., and Schmiedek, P. (2000). Management of severe traumatic brain injury by decompressive craniectomy. *Neurosurgery* 47, 315–322.

Pourgavin, H., Bhoopat, T. W., Viriyavejakul, A., *et al.* (1987). Effects of dexamethasone in primary supratentorial intracerebral hemorrhage. *N. Engl. J. Med.* 316, 1229–1233.

Qureshi, A. I., and Suarez, J. I. (2000). Use of hypertonic saline solutions in treatment of cerebral edema and intracranial hypertension. *Crit. Care Med.* 28, 3301–3313.

Reid, A., Marchbanks, R. J., Burge, D. M., Martin, A. M., Bateman, D. E., Pickard, J. D., and Brightwell, A. P. (1990). The relationship between intracranial pressure and tympanic membrane displacement. *Br. J. Audiol.* **24,** 123–129.

Rosner, M. J., Rosner, S. D., and Johnson, A. H., (1995). Cerebral perfusion pressure: Management protocol and clinical results. *J. Neurosurg.* **83,** 949–962.

Schoser, B. G., Riemenschneider, N., and Hansen, H. C. (1999). The impact of raised intracranial pressure on cerebral venous hemodynamics: a prospective venous transcranial Doppler ultrasonography study. *J. Neurosurg.* **91,** 744–749.

Slavik, R. S., and Rhoney, D. H. (1999). Indomethacin: a review of its cerebral blood flow effects and potential use for controlling intracranial pressure in traumatic injury patients. *Neurol. Res.* **21,** 491–499.

Sulek, C. A. Intracranial pressure. (1998). *In* "Clinical Neuroanesthesia," 2nd ed. (R. F. Cucchiara, S. Black, and J. D. Michenfelder, Eds.), New York: Churchill Livingstone.

Wass, C. T., and Lanier, W. L. (1996). Glucose modulation of ischemic brain injury—Review and clinical recommendations. *Mayo Clin Proc* **71,** 801–812.

Wijdicks, E. F. M., Plevak, D. J., Rakela, J., and Wiesner, R. H. (1995). Clinical features of cerebral edema in fulminant hepatic failure. *Mayo. Clin. Proc.* **70,** 119–124.

Wijdicks, E. F. M. (1997). The clinical practice of critical care neurology. Philadelphia: Lippincott-Raven.

Wolf, A. L., Levi, L., Marmarou, A., Ward, J. D., Muizelaar, P. J., Choi, S., Young, H., Rigamonti, D., and Robinson, W. L. (1993). Effect of THAM upon outcome in severe head injury: a randomized prospective clinical trial. *J. Neurosurg.* **78,** 54–59.

## CHAPTER 57

# Traumatic Brain Injury

Andres M. Salazar, Patrick Cooper, and James Ecklund

---

## CLINICAL ASPECTS

Traumatic brain injury (TBI) is the leading cause of death and disability in young adults in the United States. Brain injury may be caused by any of several types of head trauma, including the more typical closed head injury, in which rapid acceleration or deceleration produces shearing and other forces in the brain and impacts against the frontal and temporal fossae of the skull; direct impact to the head; and penetration by a bullet or other foreign object. Although some details of the pathosis of these types of trauma may differ, the principles of acute and especially long-term management are similar in most cases.

TBI is traditionally classified by its severity, although the current definitions are imperfect, and distinctions between mild, moderate, and severe head injury can often be blurred. For example, acute management of a comatose patient with a moderate head injury may be similar to that of one with a more severe injury, and a patient with little or no initial loss of consciousness may harbor more serious and even life-threatening pathosis, such as a delayed hematoma. Nevertheless, the distinctions are generally useful in guiding the approach to the patient.

Proper evaluation and diagnosis of the TBI patient is made especially challenging by the multifactorial nature and complexity of the pathologic conditions induced by the trauma and their evolving character. We present a practical approach to the diagnosis and management of the TBI patient, along with a brief discussion of the basic pathologic conditions that drive that management.

### Patient Evaluation and Diagnosis in Severe TBI

Moderate and severe head injuries are usually defined by a Glasgow Coma Scale (GCS) score of 8–12 and less than 8, respectively (Masters and McClean, 1987) A high index of suspicion for other injuries, including neck and spine fractures, is critical in the evaluation of these patients. There may be additional organ system injuries in up to 60% of patients (Saul and Ducker, 1982), with a significant incidence of associated spine injury. As initial resuscitation efforts are implemented, a quick and accurate diagnostic assessment of the head-injured patient must also be accomplished. The events of the injury, such as the speed of the vehicle, height of the fall, eyewitness reports at the scene, can be helpful in the accurate assessment of the patient. Past medical information, when available, is also essential for optimal patient management. For example, a critical underlying condition that may have led to the trauma can easily be overlooked.

The cornerstone tenets of basic resuscitation and life support must be instituted and adhered to promptly (Bullock *et al.*, 2000). Information gathered from the measurement of vital signs, oxygen saturation, and other standard laboratory studies are critical in triage and further management. Neurological evaluation must be focused yet thorough. Diagnosis of many injuries can be made before confirmation from radiographic imaging. For example, inspection for periorbital ecchymoses (raccoon's eyes), postauricular ecchymoses (Battle's sign), Cerebrospinal fluid (CSF) rhinorohea/otorrhea, and hemotympanum can all provide evidence of a basal skull fracture. Auscultation over the carotid arteries may reveal a bruit associated with traumatic dissection, or a bruit over the globe of the eye may indicate possible carotid-cavernous fistula. Palpation for cranial deformities, defects, entry or exit wounds, or crepitus can assist in diagnosing fractures or penetrating trauma. Assessment of the cranial nerves, pupillary responses, level of consciousness and mental state, motor examination, sensory examination, and evaluation for pathological reflexes will establish a

*Neurological Disorders: Course and Treatment, Second Edition*

neurological baseline for evaluation of progress or deterioration.

Pupillary responses can be a particularly useful part of the neurological examination (Chesnut *et al.*, 2000) Close inspection of pupillary diameter and light reflex is important for both diagnosing intracranial pathosis and for providing valuable prognostic information. For example, a unilateral fixed and dilated pupil is seen most often in uncal herniation, whereas bilateral midposition, pinpoint, and dilated pupils are usually present in various stages of central herniation (Fisher, 1995). Prognostically, a patient with bilaterally absent pupillary light reflex has at least a 70% positive predictive value for poor outcome (dead, vegetative, or severely disabled). However, it is key to note that when evaluating pupillary diameter and light reflex for prognostic significance, one must first ensure that hypotension and hypoxia are corrected, and effects of orbital trauma or drugs have been considered (e.g., risk of third nerve damage). Pupils should be also reassessed after surgical evacuation of intracranial mass lesions (Chesnut *et al.*, 2000). Finally, not all pupillary abnormalities are indicative of transtentorial herniation. Metabolic and pharmacological effects, as well as prior corneal surgery, can result in misleading examination results and possible false localizing signs.

The patient's level of consciousness is initially determined by response to voice, loud sounds, light, and pain. Level of consciousness is usually the first parameter to be evaluated and is often one of the most sensitive clinical indicators of significant brain injury. Level of consciousness is quantified and recorded for sequential evaluation using the GCS (Table I) (Chesnut *et al.*, 2000; Teasdale and Jennett, 1974). The initial GCS score is useful as a prognostic indictor in TBIs. There is an increasing probability of poor outcome with a decreasing GCS score in a continuous, stepwise manner. However, in using GCS score for prognosis, the following cautions are indicated: (1) the score should be measured in a standardized way; (2) it should be measured after pulmonary and hemodynamic resuscitation; (3) it should be measured after pharmacological sedation or paralytic agents have metabolized;

and (4) it should be measured by trained medical personnel. If the initial GCS is reliably obtained and not skewed by medications or prehospital intubation, only about 20% of patients with the worst initial GCS will survive, and fewer will have a functional survival (Chesnut *et al.*, 2000). Finally, frequent reassessment of the level of consciousness to detect deterioration is a critical element in the management of the head-injured patient.

Motor examination in an awake patient is recorded by grading strength in major muscle groups. If the patient is uncooperative, symmetrical or appropriate movement in all extremities is assessed in response to noxious central and peripheral stimuli. The presence of decerebrate or decorticate posturing indicates a more severe injury or central herniation, and unilateral weakness raises the suspicion of uncal herniation or lateralized intracranial mass. Weakness and sensory abnormalities also raise the suspicion of a spine or root injury.

The initial history and brief examination also provide a risk appraisal for significant intracranial mass lesions. Moderate-risk patients are those with a history of alteration or loss of consciousness, progressive headache, alcohol or drug intoxication, acute posttraumatic seizures, unreliable or inadequate history, age <2 years old, vomiting, posttraumatic amnesia, signs of basilar skull fractures, multiple trauma, serious facial injury, possible skull penetration or depressed skull fracture, or suspected child abuse. Patients at high risk of intracranial mass are those with depressed level of consciousness (not caused by drugs, alcohol, metabolic derangements, or seizures), focal neurological findings, decreasing levels of consciousness, penetrating injury, and depressed skull fracture (Duus *et al.*, 1994; Feuermanet *et al.*, 1988; Ingebrigtsen and Romner, 1996; Masters *et al.*, 1987; Stein, 1992). Those head-injured patients who do not have any of the moderate or high-risk findings are usually at low-risk for intracranial surgical lesions, although they may still have symptoms of headache, dizziness, or superficial scalp abnormalities such as hematomas, lacerations, contusions, or abrasions (Masters *et al.*, 1987).

TABLE I    Glasgow Coma Scale (Coma Score {E + V + M} = 3–15)

| Points: (3–15 possible) | Best eye opening (E) | Best verbal response (V) | Best motor response (M) |
|---|---|---|---|
| 6 | — | — | Obeys commands |
| 5 | — | Oriented | Localizes pain |
| 4 | Spontaneous | Confused | Withdraws to pain |
| 3 | To speech | Inappropriate | Abnormal flexion |
| 2 | To pain | Incomprehensible | Extensor response |
| 1 | None | None | None |

### Radiographic Evaluation

Radiographic evaluation and diagnosis in acute head trauma patients is focused mainly on the use of computed tomography (CT) scans (Chesnut et al., 2000). Noncontrast head CT of the brain is the first study of choice to evaluate the head trauma victim who has any moderate (Stein and Ross, 1990) or high-risk criteria for intracranial injury, specifically patients with a GCS less than 15, those who have focal neurological deficits, and those with posttraumatic amnesia. In addition, CT scan should be used before general anesthesia for other procedures, because delayed deterioration cannot be detected by serial examinations in this case. CT information combined with the patient's clinical presentation guides treatment and provides prognostic estimates. Hemorrhage can occur in the subarachnoid, subdural, epidural, and intraventricular spaces and in the brain parenchyma. Extra-axial hematomas should be evacuated when associated with significant mass effect. This procedure should not be delayed, because it has been shown that there is a significantly poorer outcome with surgical delays greater than 4 hours (Marshall et al., 1983; Seelig et al., 1981). Epidural hematomas that are rapidly evacuated have a much better prognosis than subdural hematomas, because the underlying brain injury is generally less severe (Marshall et al., 1983; Wilberger et al., 1991). Intraventricular hemorrhage will generally require ventricular drainage and has also been associated with a poor prognosis. The size, mass effect, and location of intraparenchymal hemorrhage, as well as neurological findings, will guide the need for surgical evacuation.

Obliteration of the basilar cisterns from mass effect portends a worse outcome as does diffuse hypodensity typical of cerebral hypoxia. Hydrocephalus can result from an obstructive lesion, can be easily diagnosed on CT, and should be rapidly treated. Pneumocephalus can indicate a basilar skull fracture or a penetrating injury. Normal findings on initial head CT should not lull the clinician into a false sense of security. Close observation is essential for even patients with mild TBI, because delayed hematomas may develop (Borovich et al., 1985; Bucci et al., 1986; Feuerman et al., 1988; Marshall et al., 1983). Patients with severe head injury usually undergo a second CT scan 3–5 days after the initial insult and again between days 10 and 14. A CT scan is indicated sooner for any previously stable head-injured patient with a new neurological deterioration or unexplained rise in intracranial pressure.

Certain other features of initial CT examination are useful for prognosis in severe head injury. Abnormalities on initial CT examination are present in approximately 90% of patients with severe head injury. In patients with severe head trauma, those with CT pathosis on initial scan have a less favorable prognosis than those severe head trauma patients with normal initial head CTs. In fact, the outcome in patients with a normal CT on admission is primarily related to concomitant extracranial injuries. However, as noted earlier, the absence of abnormalities on CT at admission does not preclude the occurrence of increased intracranial pressure (ICP), and significant new lesions may develop in up to 40% of patients (Chesnut et al., 2000). In addition, the use of CT classifications of head injury, CT status of the basal cisterns, CT presence and grading of traumatic subarachnoid hemorrhage (tSAH), CT presence of midline shift, and CT presence of intracranial mass lesions have all been found to have additional prognostic significance.

Skull films have essentially been replaced by CT scans. Although the presence of a skull fracture is associated with an increased probability of a hematoma, normal skull films are found in a significant number of patients with intracranial lesions and therefore are very insensitive in detecting possibly significant pathosis. In addition, the findings on skull films have been found to affect patient management in only 0.4%–2% of head-injured patients (Ingebrigtsen and Romner, 1996).

Although MRI is not usually appropriate for evaluation of acute head injuries, it can be very useful in the long-term management of moderate and severe TBI. The longer acquisition time, decreased access to the patient during the study, increased difficulty with patients requiring special equipment such as ventilators and IV pumps, and poor sensitivity for detecting acute blood (compared with CT) all make MRI impractical for acute radiological assessment (Snow et al., 1986). After the acute period, MRI can be valuable to better characterize questionable CT findings. MRI provides improved imaging of small white matter changes such as punctate hemorrhages in the corpus callosum indicative of diffuse axonal injury (Levin et al., 1989). MRI and MRA can also be valuable as tools to diagnose trauma-related vascular pathosis, such as carotid or vertebral dissections.

### "Mild" TBI

So-called mild TBI (MTBI) accounts for about 70% of TBI hospital admissions and for a much larger, uncertain group of patients who are not hospitalized or do not seek medical attention (Krauss and Nourjah, 1989). Because of the large number of patients involved and their accompanying morbidity, MTBI may have a greater overall economic impact than more severe TBI—it has been estimated that more than half of workdays lost after TBI result from "mild" injuries.

MTBI is variably defined as any TBI/concussion with loss of consciousness of 0–30 minutes, GCS of 13–15, posttraumatic amnesia/ confusion <24 hours, and no evidence of contusion or hematoma on CT. Nevertheless, MTBI, even without loss of consciousness, has been repeatedly associated with measurable abnormalities in cognition, attention, and behavior, as well as documented qEEG and neuropathological changes (Oppenheimer, 1968; Thatcher et al., 1989) (see below under "pathogenesis"). Abnormalities on cognitive tasks have been repeatedly documented after MTBI and typically include disturbances of attention, information processing, and memory (Gentilini et al., 1989; Gronwall, 1989; Levin et al., 1987; Ruff et al., 1989; Warden et al., 2001). As might be expected, MTBI also has a significant psychosocial impact (Dikmen et al., 1989).

More than 75% of MTBI patients report some somatic and/or cognitive symptoms over the first several weeks after injury; these can have important functional, social, and economic implications. Symptoms include headache, dizziness or vertigo, blurred vision, fatigue, sleep disturbance, irritability, depression, anxiety, and poor memory and concentration. Typically, these symptoms improve markedly over the first 3 months after injury, but some can persist longer. The term "postconcussion syndrome" is often applied when this complex of symptoms is persistent (Levin et al., 1987; Rutherford, 1989). More recently, there has been increasing attention to MTBI in sports (Thurman et al., 1998). Athletic populations offer several advantages in the study of MTBI, including the generally high preinjury health and motivation of athletes, the ability to conduct preinjury testing, and the relative accuracy of the time of the injury. Barth and colleagues studied a large cohort of college football players before and after injury and documented significant attention deficits persisting as long as 5 days after a minor "ding" without loss of consciousness (Barth et al., 1989; Macciocchi et al., 1996). Similar findings are reported in soccer players (Matser et al., 1999).

On the basis of available experimental data, it is reasonable to expect that the principal pathologic conditions occurring after a mild TBI or concussion may be a diffuse cortical neuronal dysfunction, selective vulnerability of certain neurons, and a modest amount of diffuse axonal injury (DAI). Microvascular injury with alterations in autoregulation and uncoupling of blood flow and metabolism has also been described in MTBI but may be more important in severe TBI and with repeated MTBI in pediatric and adolescent patients, as in the "second impact syndrome" (Bergsneider et al., 2000; Junger et al., 1997; McCrory and Berkovic, 1998). Finally, hematomas occur with some frequency in patients who might otherwise be classified as having MTBI (Eisenberg, 1989; Williams et al., 1990).

Specialized cognitive batteries have been developed that may be especially useful in this population,

including the "Sideline Assessment of Concussion (SAC) and the "Automated Neuropsychological Assessment Metric" (ANAM) (Bleiberg et al., 1998; McCrea et al., 1998).

## NATURAL COURSE AND PROGNOSIS

The average annual incidence of hospitalized closed head injury (CHI) in the United States was about 200/100,000 two decades ago (Frankowski et al., 1985); more recently those statistics have decreased to about 100/100,000, possibly as an artefact of changes in managed care and hospitalization but also because of improved prevention efforts (Thurman et al., 1999). It has been estimated that each year approximately 75,000 Americans die and another 75,000 suffer permanent disability because of TBI. The total cost of TBI has been estimated at about $38 billion per year in the United States, including $4.5 billion for direct care and $33.5 billion in injury-related work loss and lost income from death (Max et al., 1991).

### Acute Prognosis in Severe TBI

The projected outcome after TBI is important to consider. A relatively few features associated with TBI seem to contain the most significant prognostic information (Braakman et al., 1980; Chesnut et al., 2000; Jennett et al., 1979; Stablein et al., 1980). Features such as age of the patients, clinical indices indicating severity of brain injury (GCS), and results of CT scans and ICP monitoring have been found to be the most useful in assessing patient prognosis. A working group convened by the Brain Trauma Foundation, with endorsement of the American Association, of Neurological Surgeons, the Neuro-trauma Committee of the World Health Organization, and the Brain Injury Association, have evaluated the literature on prognostic indicators in head injury and developed recommendations adhering to the concepts of evidence-based practice (Chesnut et al., 2000). The study group evaluated GCS score, patient age, pupillary diameter and light reflex, presence of hypotension, and CT scan features in an effort to determine reliable prognostic indicators for outcome. The findings are discussed in more detail under "Diagnosis" and are summarized in Table II.

### Long-Term Outcome

Measurement of long-term outcome from TBI remains a challenge. Functional measurement instruments include the Glasgow Outcome Score; the Dis-

TABLE II Early Prognostic Indicators in Severe TBI

| Indicator evaluated | Comments |
| --- | --- |
| Glasgow Coma Scale score | • The probability of poor outcome increases with a low GCS. <br> • GCS should be measured in a standardized way after resuscitation. <br> • The GCS should be measured by trained medical personnel. |
| Patient age | • The probability of poor outcome increases with age. |
| Pupillary diameter/light reflex | • Bilaterally absent pupillary light reflex worsens prognosis. <br> • Pupillary measurement parameters: <br> —A measured pupil difference of 1 mm defines asymmetry. <br> —A fixed pupil has no response to bright light. <br> —A dilated pupil has a size >4 mm. <br> —Measure pupils after pulmonary and hemodynamic resuscitation. <br> —Exclude orbital trauma. <br> —Pupils should be measured by trained medical personnel. |
| Hypotension | • Systolic blood pressure <90 mmHg has a 67% positive predictive value for poor outcome and, when combined with hypoxia, a 79% positive predictive value. <br> • Accurate monitoring by arterial line is the method of choice. <br> • Blood pressure should be measured as frequently as possible, and hypotension duration should be documented. <br> • Blood pressure should be measured by trained medical personnel. |
| Computed tomography | • Abnormal CT scan worsens prognosis. <br> • CT classification of head injury impacts prognosis. <br> • Compression of basal cisterns worsens prognosis. <br> • Traumatic subarachnoid hemorrhage worsens prognosis. <br> • Midline shift >5 mm worsens prognosis over age 45. <br> • Intracranial mass lesions worsen prognosis <br> —Hematoma volume correlates with outcome. <br> —Prognosis is worse with acute subdural than with epidural hematoma. |

Modified from Chesnut et al., 2000.

ability Rating Scale (DRS); the Rancho los Amigos scale; the functional independence measure (FIM); and various neuropsychological, behavioral, and quality-of-life measures. However, return to gainful employment is probably the best overall measure of long-term outcome (Wehman et al., 1989). Approximately 50% of patients who survive severe TBI eventually return to work. In recent studies, return to work was also the single best correlate of perceived quality of life (Melamed et al., 1992).

Accurate predictors of outcome are also important to patients, to their families, and to caregivers in understanding recovery and planning for care. In addition to the acute predictors outlined previously, longer term outcome predictors also include preinjury intelligence and education, history of behavioral and psychiatric difficulties, age, and neurological residua. In a large multidisciplinary study of survivors of penetrating head injury, the presence of seven factors significantly predicted unemployment some 10–15 years after injury; these were hemiparesis, epilepsy, visual field loss, verbal memory loss, visual memory loss, psychological problems, and violent behavior. These factors represent different domains of brain function and were relatively equipotent in the model. In other words, it was the number of impaired domains, not impairment in any one particular domain, that predicted unemployment. The brain may thus compensate for injury by using whichever functional domains are still available to it (Schwab et al., 1993).

Finally, recent work also suggests that certain genotypes, such as the presence of the APOE4 allele, may affect recovery and outcome (Jordan et al., 1997; Mayeux et al., 1995). It is likely that additional genetic variants, such as those of neurotransmitter receptors or transporters, will also be linked to recovery and behavioral outcome.

## PATHOGENESIS AND PRINCIPLES OF THERAPY

TBI is now generally viewed as a multidimensional and dynamic process. Much of the pathosis evolves during the first few days after trauma, often with devastating secondary injury, and functional recovery can continue for years. The pathological changes described later may be due less to the injury itself than to an uncontrolled vicious circle of biochemical and physiological events set in motion by the trauma. The severe

TBI patient also often has multiple systemic abnormalities such as changes in nutrition (Bullock *et al.*, 2000), cardiopulmonary status, (Clifton *et al.*, 1985), circulating catecholamines, and coagulation (Hulka *et al.*, 1996; Selladurai *et al.*, 1997) that may be directly related to the brain injury and may have a profound impact on treatment. Thus, a rational approach to therapy should be predicated on an understanding of the multidimensional pathosis of TBI and its evolution.

## Pathology of TBI

Over the past two decades our understanding of the pathology of CHI has gradually evolved (Gennarelli *et al.*, 1982; Hume Adams *et al.*, 1985). At least six distinct parallel components can be identified in the pathology of TBI: (1) focal injury, including hematomas and contusions; (2) diffuse microvascular injury with loss of autoregulation, which has been implicated as playing an important role in acute brain swelling; (3) superimposed classical hypoxic-ischemic injury; (4) DAI, which may be responsible for much of the disability in more severe TBI but also occurs in patients with mild TBI (Povlishock, 1985); and (5) selective neuronal loss, especially in the hippocampus and reticular thalamus, possibly secondary to an excitotoxic mechanism (Ross *et al.*, 1993). (6) Finally, recent electrophysiological evidence now suggests a diffuse neuronal (gray matter) dysfunction that may be the most subtle and sensitive measure of mild TBI and concussion. The first five of these pathologic features have been reproduced in animal models of angular acceleration without impact; and possibly, except for DAI, all can also be features of penetrating head injury (Gennarelli and Thibault, 1985).

### Focal Injury

Focal injuries include intracerebral and extracerebral hematomas and focal contusions. Hematomas are most common after the rapid acceleration or deceleration that occurs in a fall or another form of impact, especially in the elderly. Delayed hematomas, which can occur in patients who initially seem to be at low risk but then deteriorate rapidly, are particularly important. Small hematomas can be treated conservatively, but delaying the surgical removal of large hematomas for longer than 4 hours after injury significantly increases mortality and morbidity.

Focal contusions may occur under the site of impact, but by far the most common locations after acceleration-deceleration injury are in the orbitofrontal and anterior temporal lobes, where the brain abuts the base of the skull. A relatively typical pathological and clinical picture can often be seen, and the most troubling clinical sequelae are usually behavioral and cognitive abnormalities referable to these areas of the brain. Both hematomas and contusions can undergo secondary delayed expansion. Patients with such injuries thus require particularly close observation in the acute period. Hematomas and contusions are also significant risk factors for the development of posttraumatic epilepsy.

### Diffuse Microvascular Injury

Diffuse microvascular damage also has been implicated as a major component of both closed and penetrating TBI. Depending on the severity of injury, early changes may include loss of cerebrovascular autoregulation, with decreased responses to changes in carbon dioxide and perfusion pressure and an initial, transient systemic hypertension (Proctor *et al.*, 1988). The loss of autoregulation makes the brain particularly susceptible to fluctuations in systemic blood pressure; otherwise tolerable hypotension can thus result in cerebral ischemic damage in the patient with TBI. In addition, altered vascular sensitivity to circulating catecholamines can lead to vasoconstriction and further focal ischemia or reperfusion injury, as well as brain swelling. The microvascular pathology includes an endothelial change probably involving oxygen free radical–induced decreases in endothelial nitric oxide, with a concomitant vasoconstriction. More recent studies have also shown an initial hyperglycolysis with a dissociation of cerebral blood flow and metabolism (Bergsneider *et al.*, 1997; Verma, 2000). Control of this pathogenetic mechanism with neuroprotectants remains a critical challenge in the acute management of the severely injured patient.

### Hypoxia-Ischemia

The classic pathosis of hypoxia-ischemia, which primarily involves the hippocampus and the vascular border zones of the brain, is often superimposed on the other, more specific pathology of TBI. The traumatized brain is particularly sensitive to hypoxia-ischemia, possibly because of the metabolic demands already placed on neurons by the trauma itself (Ishige *et al.*, 1987; Verma, 2000). The most significant improvements in the survival of patients with TBI have resulted from recognition of the importance of this component and its prevention with early resuscitation, largely through training of paramedics and the development of emergency transport systems.

### Diffuse Axonal Injury (DAI)

DAI is a major cause of persistent, severe neurologic deficits after CHI. It was originally described as a shearing injury in the hemispheric white matter, corpus callosum, and brainstem, usually accompanied by some hemorrhage (Strich, 1961). When severe, it is manifested clinically by immediate and prolonged loss of consciousness (Gennarelli *et al.*, 1982). In milder cases, petechial hemorrhages in the white matter or blurring of the gray matter–white matter junction are typically seen on MRI, especially in coronal slices.

More recent work with mild and moderate injury in animals, however, shows that the typical light microscopic histopathology of DAI may not emerge until 12–24 hours after injury. The principal early change is a cytoskeletal disruption with a disturbance of axonal flow (Povlishock, 1985). If severe enough, such axonal injury can lead to wallerian degeneration with neuronal loss and diffuse target deafferentation as an active process (Povlishock, 1992).

One implication of this work is that such axonal injury may occur even after mild brain injury and in the absence of pathological change in any other vascular, neural, or glial elements (Oppenheimer, 1968; Saatman *et al.*, 1998). Some of the cognitive changes seen after MTBI have thus been postulated to be related to DAI and deafferentation. Another equally important implication is that a potential therapeutic window exists during which neuroprotective treatments may prevent total axonal disruption or dysfunction (Povlishock and Jenkins, 1995).

### Neuronal Loss/Excitotoxic Injury

Selective vulnerability of gabergic thalamic reticular nucleus neurons receiving glutaminergic afferents from orbitofrontal cortex has been shown to occur after head injury (Ross *et al.*, 1993). This occurrence has also been described after "mild" head injury in animal models, along with hippocampal (CA-1, CA-3) and cerebellar (Purkinje cell) loss (Hicks *et al.*, 1993; Kotapka *et al.*, 1991; Ross *et al.*, 1993). Although the reasons for the vulnerability of these cells is unclear, it may be related to glutamate excitotoxicity and to their special metabolic needs and the central role they play in brain function (Verma, 2000). The reticular thalamus is an integral part of the reticular activating system, which maintains alertness. Thus, even subtle dysfunction of this region after an injury may well help explain the fatigue, reaction time, and other attention disturbances that are some of the most troubling components of the postconcussion syndrome (Van Zomeren and Deelman, 1976). Likewise, hippocampal vulnerability may account for the memory deficits seen after TBI.

### Diffuse Gray Matter Dysfunction

In addition to selective neuronal vulnerability, recent evidence from quantitative electroencephagram (EEG) and quantitative MRI studies suggests that in mild TBI or concussion the most common abnormality may be diffuse gray matter dysfunction that is manifested primarily by changes in brain electrical activity as measured by EEG coherence, phase, and power (Thatcher *et al.*, 1989; Thatcher *et al.*, 1998a; Thatcher *et al.*, 1998b; Thatcher *et al.*, 2001). These alterations may, in turn, reflect a relative loss of neuronal membrane electrical efficiency, probably as a consequence of failure of neuronal energy metabolism and the ATP-driven neuronal ion pumps. Such failure would not be unexpected in the face of a probable diffuse excitotoxic or other challenges in the early period after TBI. Although the pathoanatomical correlates of such dysfunction are likely to be extremely subtle, after TBI there is a strong correlation between these changes and changes in the MRI T2 signal in the gray matter, which in turn is thought to reflect the *functional* integrity of neuronal membranes (Thatcher *et al.*, 1998a; Thatcher *et al.*, 1998b). These EEG changes also correlate with neuropsychological performance, suggesting that they could also be responsible for the cognitive and reaction time changes after MTBI. Restoration of neuronal energy metabolism might be expected to ameliorate these changes.

### Secondary Injury and Neuroprotective Therapies

It is not unusual for a patient with TBI who initially is relatively stable and awake or in a light coma to deteriorate rapidly. Although delayed hematoma or expanding contusions that are amenable to surgery account for many of these cases, many are also related to uncontrolled brain swelling that may not respond to conventional management.

Over the past decade, delayed secondary injury at the cellular level has come to be recognized as a major contributor to the ultimate tissue loss after central nervous system (CNS) trauma, stroke, and other injury (McIntosh *et al.*, 1996; Siesjo and Siesjo, 1996). A cascade of biochemical events involving a multitude of systems has been shown to be set in motion in injured tissue. These include changes in excitatory amino acids, calcium and magnesium, arachidonic acid metabolites, and the formation of oxygen free radicals, excitotoxic neurotransmitters such as glutamate (Faden *et al.*, 1989), and various kinins and cytokines. These changes initiated by the primarily injured tissue result in progressive secondary injury to otherwise viable tissue through a number of mechanisms (e.g., by producing further ischemia [by means of vasospasm, clot formation, or secondary vascular occlusion]); by injuring neurons and glia directly; by activating macrophages and initiating inflammation; by producing brain swelling [edema or hyperemia]; or by establishing conditions favorable to secondary infection).

Some of these pathologic conditions, such as focal hematomas or microvascular injury with brain swelling, can result in death of the patient. Others, such as DAI and selective neuronal injury, will mainly result in death or dysfunction of certain neurons and thus have implications for long-term outcome. Evaluation of the efficacy of specific therapies should therefore be relevant to the particular pathosis targeted by the treatment. Likewise, specific outcome measures should be selected to reflect the expected disease in that particular injury. For example, patient survival after TBI may not be the optimal outcome measure for evaluating the efficacy of acute treatment with neuroprotective agents such as glutamate antagonists. On the other hand, such neuroprotective strategies might be expected to have the greatest impact on the function of those brain areas that are most susceptible to excitotoxic injury after TBI. Because of this multidimensional pathosis, it is likely that optimal management of severe TBI patients will ultimately consist of combination therapy with multiple pathological targets, applied sequentially or concomitantly.

Medical therapy of TBI has generally been limited to osmotic agents and/or corticosteroids; the latter have been shown to be not only unhelpful but possibly detrimental in head injury (Deutschman et al., 1987). Recent, large TBI trials of superoxide dismutase (PEG-SOD) (Young et al., 1996), the lipid peroxidation inhibitor Tirilazad (Marshall and Marshall, 1995), the calcium channel blocker Nimodipine, and several glutamate antagonists have failed to show a clear benefit.

Excitotoxic injury, most often caused by the neurotransmitter glutamate, may be one of the more important mechanisms of neuronal death after traumatic or ischemic injury. Excessive release of glutamate and other neurotransmitters unleashes a chain of cellular events that deplete neuronal energy stores and result in cell death or apoptosis. These findings suggest that the use of glutamate antagonists may play a role in the acute treatment of TBI (Bullock, 1995). However, because of the importance of glutamate in the brain, receptor blockade is usually accompanied by intolerable side effects. Another related neuroprotective strategy that reduces glutamate release and seems to protect neurons is moderate systemic hypothermia. A recent controlled pilot study showed long-term benefit for patients with severe TBI (Marion et al., 1997). However, a larger multicenter study failed to confirm these findings (Clifton et al., 2001).

A different and relatively novel approach to neuroprotection seeks to enhance neuronal energy metabolism and mitochondrial function after injury. Because it addresses a "final common path" in neuronal dysfunction and/or death, this approach is hypothesized to provide protection from a variety of the stressors resulting in secondary injury after brain trauma, including excitotoxic, oxidative, or calcium-induced mechanisms (Verma, 2000). Depletion of neuronal energy stores is a consequence not only of the increased demands placed on membrane pumps by excitotoxic and other stressors but also a failure of energy production. A number of studies have shown mitochondrial dysfunction after brain injury, including degradation of the pyruvate dehydrogenase complex and a failure of the mitochondrial membrane with release of cytochrome-$c$ into the cytosol and consequent activation of apoptotic pathways. Cyclosporin has been shown to have a neuroprotective effect based on stabilization of the mitochondrial membrane (Buki et al., 1999; Sullivan et al., 1999). Both creatine and the three-carbon sugar, pyruvate has recently been shown to have marked neuroprotective effects in animal models of TBI along with a stabilizing effect on the mitochondrial membrane (Sailor et al., 2000; Sullivan et al. 2000). These compounds are inexpensive and nontoxic and are both in early clinical trials for brain injury.

## PRACTICAL MANAGEMENT

### Management of MTBI

Perhaps the most important element in the management of patients with MTBI is the clinician's recognition that these patients' complaints have a structural, organic basis. This finding has repeatedly been confirmed by pathological studies of animal models and humans, by neuroimaging (with MRI, positron-emission tomography (PET), single-photon emission computed tomography (SPECT), and magnetoencephalography), and by computerized electroencephalography. A substantial range of severity of injury exists within the category of MTBI (Culotta et al., 1996). Findings on CT and MRI are usually normal, although skull fractures, focal cortical contusions, small petechial hemorrhages, and disruption of the gray matter–white matter junction often occur. Patients with an admission GCS score of 13 or 14 have a much higher incidence (up to 28%) of abnormal findings on CT than do patients with a GCS score of 15. Similarly, patients with clinically "mild" TBI complicated by cerebral contusion (about 20%) have a 6-month outcome that is more consistent with that of patients with moderate head injury (Williams et al., 1990).

Postconcussive symptoms, which include headache, dizziness, fatigue, and documented deficits in cognition, can be seen even after mild injuries without loss of consciousness. In general, the prognosis for recovery is very good, with most cognitive and somatic sequelae improving markedly by 3 months, and 85% of patients have no disabling symptoms at 1 year of follow-up (Alexander, 1995; Levin *et al.*, 1987). In the small percentage of patients who have postconcussive complaints and disability for longer than 1 year, psychogenic factors can often contribute to the persistence of symptoms. Early intervention that includes patient and family education, symptomatic treatment of headache and dizziness, and management of fatigue (with the goal of the patient's gradual return to full activity) is most important. Patients who have persistent symptoms of anxiety, depression, or both also need appropriate diagnosis and treatment, preferably by a psychiatrist who has experience with TBI.

## Management of Moderate and Severe TBI

The acute management of moderate to severe head injuries is being quietly revolutionized by the advent of "evidence-based" medicine, and in particular by the publication and dissemination of the *Guidelines for the Management and Prognosis of Severe Traumatic Brain Injury* by the Brain Trauma Foundation (BTF) (Bullock *et al.*, 1996; Bullock *et al.*, 2000). These have been recently updated, and the full text is now available on the Internet at *www.braintrauma.org*. The evidence-based guideline process is a scientific one in which experts in a field come together to address selected critical management questions through a full review the literature following strict pre-established rules of scientific evidence. This allows for a ranking of final recommendations in a hierarchy that reflects their level of certainty. For example, for therapeutic questions, the highest level of certainty is attained by prospective, controlled randomized trials, which often justify recommending a treatment "standard"; whereas in the lowest level are case reports or published "expert opinions," which may only allow for a treatment "option." In this section, we will refer to the BTF document liberally, but it is important to recognize that, for various reasons, randomized controlled trials may never become available to address certain questions and that recommendations below the level of "standard" will still usually provide the most valuable guidance available on a particular management issue. Another important role of the evidence-based medicine process is that it helps to focus clinical research efforts on medical questions for which there is the greatest information need.

### Resuscitation

The acutely traumatized brain is particularly sensitive to additional hypoxic-ischemic injury. Thus, resuscitation to prevent hypotension, hypoxia, hemorrhage, and cardiovascular collapse must be initiated immediately, especially in the face of possible severe head trauma. Cardiopulmonary resuscitation efforts that address hypotension and hypoxia are crucial prophylactic steps in preventing the negative effects of increased intracranial pressure (Bullock *et al.*, 2000; Chesnut *et al.*, 1993; Fearnside *et al.*, 1993; Pigula *et al.*, 1993). Airway and shock management must be the first priority in all trauma patients. The loss of cerebrovascular autoregulation places the brain at increased risk for cerebral ischemia from systemic hypotension, and levels of hypercarbia tolerated by the normal brain can lead to critical marginal increases in intracranial pressure in the patient with head injury (Vassar *et al.*, 1993). Over the past decades, there has been a marked improvement in outcome in head injury as a result of emergency care systems that include early prehospital intubation and resuscitation (Bullock *et al.*, 2000; Roy, 1987).

An organized team approach to the acute management of the patient with TBI is essential and includes prehospital, intensive care unit, and post-ICU care (Bullock *et al.*, 2000). Early management can be predicated on clinical evidence of intracranial hypertension, such as signs of herniation (unilateral/bilateral pupillary dilation, asymmetrical pupillary reactivity, motor posturing, and deterioration in neurological examination). An algorithm option for the initial management of severe head injury, based on evidence-based guidelines, is presented in Figure 1 (Bullock *et al.*, 2000).

If there are no clinical signs of transtentorial herniation, appropriate sedation and pharmacological relaxation may be used as needed for safe patient transport. It is clear that sedation and neuromuscular blockade may hinder the initial evaluation and treatment of the head trauma patient (Marion and Carlier, 1994). Pharmacological relaxation in the absence of evidence of herniation should be limited to those situations in which sedation alone is not adequate for safe transport and resuscitation. Because pharmacological relaxation limits the neurological examination of the pupils, short-acting agents are most appropriate. Prophylactic neuromuscular blockade in the head trauma patient without evidence of intracranial hypertension is not appropriate, because it has been found to be associated with a higher incidence of pneumonia and sepsis without an improvement in outcome (Hsiang *et al.*, 1994).

Early and aggressive blood pressure and oxygenation management is essential. Abundant evidence suggests that systolic blood pressures less than 90 mmHg or hypoxia (cyanosis or $PaO_2 < 60$ mmHg by arterial blood gas) is associated with increased morbidity and mortality (Chesnut *et al.*, 1993; Fearnside *et al.*, 1993; Pigula *et al.*, 1993). Evidence also exists that enhanced blood pressure may improve outcome from severe head injury (Vassar *et al.*, 1993). Attesting to the importance of adequate blood pressure in the head-injured patient is

Initial management

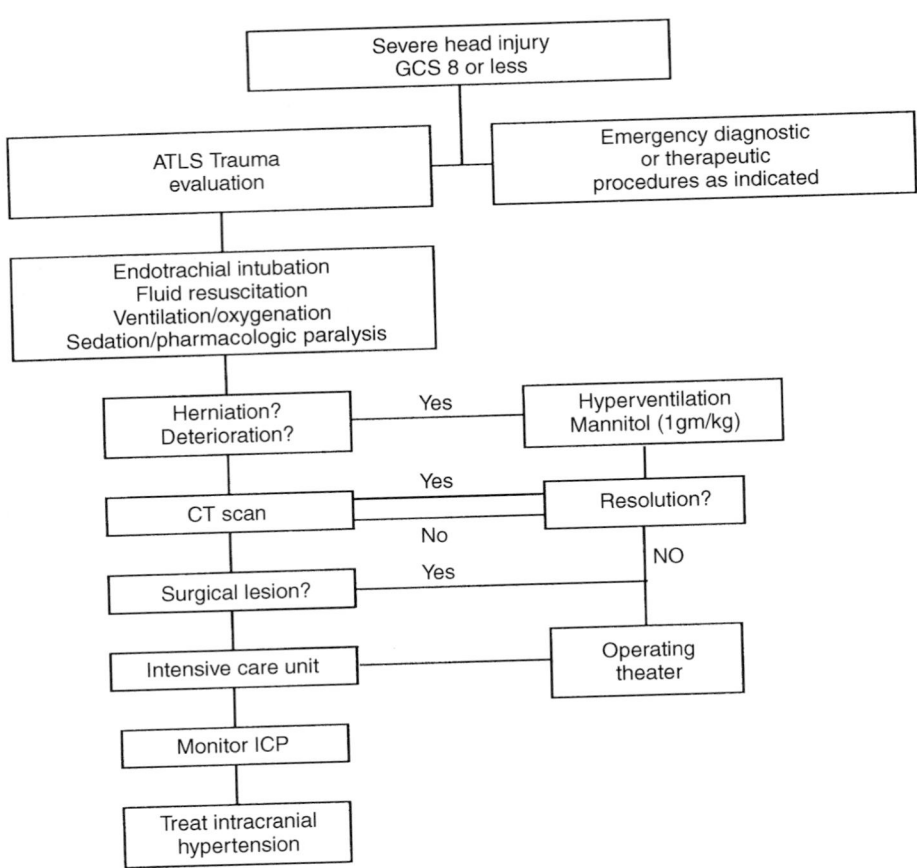

**FIGURE 1**    Initial management. Modified from Bullock *et al.*, 2000.

the finding that a systolic blood pressure <90 mmHg has a 67% positive predictive value (PPV) for poor outcome and, when combined with hypoxia, a 79% PPV (Chesnut *et al.*, 2000). A single recording of a hypotensive episode is generally associated with a doubling of mortality and a marked increase in morbidity from a given head injury (Bullock *et al.*, 2000). The importance of adequate blood pressure control is accentuated, because it is treatable.

The choice of resuscitation fluids varies widely, with most authors advocating Ringer's lactate or normal saline as the initial resuscitative crystalloid bolus (Bullock *et al.*, 2000). The end point for early fluid resuscitation is a value of 90 mmHg systolic pressure (Vassar *et al.*, 1993); however, once intracranial pressure monitoring has been established, the cerebral perfusion pressure should dictate ultimate blood pressure goals. Prehospital administration of mannitol versus placebo in traumatic brain injured patients has not shown a significant difference in mortality (Erhard *et al.*, 1996; Favre *et al.*, 1996). However, in the setting of obvious increased intracranial pressure or a herniation syndrome, mannitol administration is a well-established and effective treatment (Gaab *et al.*, 1990; Schwartz *et al.*, 1984; Smith *et al.*, 1986).

Hyperventilation of the head-injured patient in the early resuscitation period should be limited to avoid iatrogenic cerebral ischemia. Although brief periods of hyperventilation may be required in an acute herniation setting, it should be used sparingly, and prolonged hyperventilation should be avoided. It is known that hyperventilation reduces intracranial pressure by causing cerebral vasoconstriction and resultant decreased cerebral blood flow. In the head-injured patient, cerebral blood flow is often already impaired, and aggressive hyperventilation can exacerbate cerebral ischemia (Bouma and Muizelaar, 1992; Bouma *et al.*, 1991; Bouma *et al.*, 1992; Marion and Carlier, 1994). Hyperventilation to an arterial $PaCO_2$ less than 30 mmHg will continue to reduce cerebral blood flow, without a consistent continued decrease in intracranial pressure (Obrist *et al.*, 1984).

## Operative Management

The cornerstone of operative management of TBI focuses on increasing cerebral perfusion pressure by decreasing intracranial pressure, preventing infection, and decreasing secondary injury. The skull is a closed container occupied by brain, blood, and CSF. Transient hyperventilation and mild head elevation reduce blood volume while risking ischemia. CSF can be displaced through normal pathways if open, and brain volume can be decreased with mannitol and other diuretics. However, in the setting of severe brain injury these interventions can prove inadequate. Operative intervention may be required to monitor intracranial pressure, provide CSF diversion, evacuate hematomas, or remove contused or edematous brain. Surgery is also used for repair of any significant skull fractures or dural tears. Despite surgical procedures aimed at correcting or temporizing intracranial mass lesions, Class II data continue to demonstrated an unfavorable outcome in patients with traumatic intracranial mass lesions (PPV, 78%). The presence of mass lesions in patients older than 45 carries a PPV of 79% in predicting poor outcome (as defined by dead or vegetative). Hematoma volume is also correlated with patient outcome (Chesnut et al., 2000).

Perhaps the simplest operative intervention in the treatment of TBI is the placement of an intraventricular catheter. Catheter placement is advantageous in the severely head-injured patient for monitoring intracranial pressure and providing therapeutic CSF drainage. A Silastic catheter is generally placed into the frontal horn through a single burr hole. With an external ventricular drainage device in place, one can reduce intracranial pressure directly by means of CSF drainage, provide continuous monitoring of intracranial pressure, and promote drainage of blood products in the presence of an intraventricular hemorrhage (Ropper and King, 1984).

The use of exploratory burr holes has largely fallen out of favor as a diagnostic tool because of current general easy availability of the CT scanner. Situations in which exploratory burr holes may still be used include a hemodynamically unstable multitrauma patient who requires emergency surgery before a head CT and develops a dilated pupil, and a patient dying of rapid transtentorial herniation or brainstem compression in a remote location who does not improve with more conservative immediate stabilizing measures (mannitol or hyperventilation) (Mahoney et al., 1981). The use of burr holes alone for the drainage of acute hematomas is often unsatisfactory because of the high viscosity of acute blood products. If an exploratory burr hole reveals an extra-axial hemorrhage, the usual definitive management is a craniotomy. The usual sites of exploratory burr holes include bilateral temporal, frontal, and parietal, with the possibility of a posterior fossae burr hole in selected cases.

Epidural hematomas occur in approximately 1% of head trauma admissions. The usual teaching is that epidural hematomas are seen with typical brief loss of consciousness followed by a lucid period and progression to obtundation, contralateral hemiparesis, and ipsilateral pupil dilation. However, only <10%–27% of head trauma victims with epidural hematomas have this classical picture. The most common source of bleeding is the middle meningeal artery. The treatment for epidural hematomas is rapid operative decompression of any symptomatic lesion. Surgical intervention is needed for signs of local mass effect, signs of herniation (drowsiness, pupil changes, hemiparesis), or cardiorespiratory abnormalities. Surgery can also be indicated for acute asymptomatic epidural hematomas >1 cm at their thickest point or epidural hematomas in pediatric patients. Occasionally, smaller asymptomatic lesions can be managed nonoperatively if proven to be radiographically stable. The objectives of surgery for acute epidural hematomas are to remove the clot and provide hemostasis to prevent reaccumulation (Bucci et al., 1986; Kaye et al., 1985; Rivas et al., 1988; Roda et al., 1983; Ugrinovski et al., 1983).

Acute subdural hematomas are usually caused by the accumulation of blood around a parenchymal laceration or the more typical tearing of surface or bridging vessels. Subdural hematomas are often associated with severe underlying brain injury, resulting in a higher morbidity and mortality compared with acute epidural hematomas despite operative intervention (Chesnut et al., 2000). Treatment options depend on the patient's clinical condition and the size of the hematoma. A craniotomy is often required to remove the collection of acute subdural blood, whereas a subacute or chronic hematoma can often be removed through one or two burr holes. In one study delay in surgery for acute subdural hematoma longer than 4 hours after the traumatic event increased mortality from 59%–69%, and functional survival was also decreased (26%–16%) (Wilberger et al., 1991). Although the trend was not found to be significant, rapid evacuation of a symptomatic acute subdural hematoma and aggressive postoperative management provide the best chance for a good outcome from this disease.

Decompressive craniotomy is controversial but has been used in posttraumatic patients with severe cerebral edema to provide increased room for swelling. A recent small study shows that decompressive bifrontal craniectomy provides a statistical advantage over medical treatment and should be considered in the management of intractable posttraumatic cerebral swelling. If the operation can be accomplished before the intracranial pressure value exceeds 40 torr for a sustained period and within 48 hours of the time of injury, the potential to influence outcome is greatest (Polin *et al.*, 1997). Patients will require a cranioplasty 3–12 months after a decompressive craniotomy is performed.

Operative intervention is rarely required for basilar skull fractures or closed nondepressed skull fractures. Most CSF leaks associated with basilar skull fractures will heal spontaneously, but occasionally one will require temporary CSF diversion or an open operative repair. Fractures with significant depressed fragments will require surgical elevation. Open skull fractures and penetrating brain wounds usually require operative debridement and closure, although the extent of debridement and removal of retained bone or metal fragments in such cases remains controversial (Brandvold *et al.*, 1990).

## ICU Management

After a mass lesion has been surgically treated or excluded, the severe head-injured patient should be managed in an ICU. Preventing secondary insults to the brain remain the principal goal of therapy. The same principles of acute management such as preventing hypoxia, hypotension, and excessive intracranial pressure are applied to this stage of management.

Despite the lack of a prospective randomized clinical trial establishing the efficacy of intracranial pressure monitoring in improving outcome from severe head trauma, there is abundant published clinical experience that supports its use. The main benefits of intracranial pressure monitoring are derived from its help in earlier detection of intracranial mass lesions, help with guiding the use of other intracranial pressure control therapies, its use in reducing intracranial pressure by CSF drainage (and result in improved cerebral perfusion), and its help in determining prognosis. Current guidelines recommend that intracranial pressure be monitored in head-injured patients with GCS of 3–8, who have an abnormal head CT or who have a normal head CT and any two of the following factors: age older than 40 years, motor posturing, or a systolic blood pressure less than 90 mmHg. In practical terms, when a patient's GCS is 7 or less, the quality of the examination is reduced to where subtle clinical changes can be missed. Therefore, many neurosurgeons will elect to monitor intracranial pressure in any patient with a GCS score less than 8 (Bullock *et al.*, 2000). The particular monitoring technique used is determined by the neurosurgeon and the facilities available. An intraventricular catheter is recommended because of the ability to simultaneously drain CSF, although fiberoptic intraparenchymal, epidural, or subdural transducers can also be used (Bullock *et al.*, 2000).

The current standard measures for control of elevated intracranial pressure include sedation, paralysis, controlled hyperventilation, mannitol and other osmotics, ventricular drainage, and barbiturate coma (Figure 3 or Table IV) and are usually undertaken in that sequence to maintain an intracranial pressure of less than 20 mmHg. It is recommended as a guideline that mannitol be given in intermittent boluses of 0.25–1.0 gm/kg every 4 hours as needed, but serum osmolarity should be kept below 320 because of risks of renal compromise. Use of a Foley catheter is strongly recommended to monitor urine output and maintain euvolemia through adequate fluid replacement. Barbiturate coma has recently been shown to significantly improve outcome in patients younger than 45 years with otherwise uncontrolled elevated intracranial pressure. This measure is essentially the last step recommended in the nonsurgical control of intracranial hypertension. Barbiturate coma is induced with pentobarbital at an initial loading dose of 10 mg/kg intravenous over 30 minutes, and serum levels should then be maintained at 3–4 mg/dl, with dosages of about 1 mg/kg/h (Bullock *et al.*, 2000).

Intracranial pressure should not be managed in isolation from other hemodynamic parameters. Cerebral perfusion pressure should be maintained above 60–70 mmHg in the head-injured patient. Most patients seem to benefit from the higher value. This sometimes requires pressors for blood pressure support and remains especially critical during the use of barbiturate coma.

Any significant intracranial pressure elevation should prompt appropriate treatment and intervention (Table III), but simultaneously other treatable causes should be investigated. Delayed hematomas, progressive edema, enlarging contusions, or hydrocephalus must not be overlooked and are easily diagnosed by repeat CT scan. Similarly, other causes such as seizures, airway compromise, ischemia, fever, and other medical conditions should be considered.

## Posttraumatic Epilepsy

The risk of epilepsy in patients with closed-head injury is about 2%–5% overall and about 10%–20% in patients with severe closed-head injury (Annegers *et al.*, 1998). Some studies have shown a higher incidence of seizures in patients with depressed skull fractures (15%), hematomas (31%), and penetrating brain wounds (50%) (Jennett, 1975; Salazar *et al.*, 1985). In all cases, the risk decreases markedly with time after injury. In one study, high-risk patients with penetrating

**TABLE III**    Measures to Control Intracranial Hypertension

| Intervention | Reason |
| --- | --- |
| *Routine measures* | |
| Elevate head of bed to 30–45 degrees | Enhances venous outflow and reduces mean carotid pressure |
| Keep neck straight and free from compression | Enhances venous outflow |
| Avoid hypotension (systolic blood pressure < 90 mmHg) | Maintains cerebral perfusion pressure |
| Ventilate to normocarbia ($pCO_2$ = 35–45 mmHg) | Prophylactic hyperventilation reduces cerebral blood flow and is associated with worse outcome |
| Mild sedation with codeine 30–60 mg IM every 4 h as needed | Decreases sympathetic tone and hypertension caused by moving and muscle contractions |
| Noncontrast head CT | Rule out surgical lesion |
| *Specific measures for continued elevated ICP* | |
| Heavy sedation and/or paralysis | Decreases sympathetic tone and hypertension caused by moving and muscle contractions |
| Remove 3–5 ml of CSF if intraventricular catheter is present. | Reduced intracranial volume |
| Mannitol, 0.25–1.0 mg/kg q6h. Keep serum osmolarity <320 and avoid hypotension and hypovolemia | Decreases brain edema. May improve blood rheology. |
| Hyperventilation to $pCO_2$ = 25 mmHg *only* for brief periods, *if needed* for tentorial herniation | Decreases cerebral blood volume |
| Barbiturate coma (pentobarbital) Loading dose of 10 mg/kg, then 1 mg/kg/h. | Decreases metabolic demands, decreases intracranial pressure. |

From Bullock *et al.*, 2000.

head injury could be 95% certain of remaining seizure free if they had no seizures during the first 3 years after injury (Weiss *et al.*, 1986). On the other hand, some 10 years after injury, their relative risk of epilepsy developing was still significantly higher than for a normal age-matched population (Salazar *et al.*, 1985).

The debate over the use of prophylactic anticonvulsants in head-injured patients must be separated into two questions: (1) Are anticonvulsants indicated in a patient with posttraumatic epilepsy (PTE)?, and (2) Do prophylactic anticonvulsants or other drugs prevent the onset of PTE? In light of data suggesting that most patients with one late posttraumatic seizure will have recurrent seizures for some time, the answer of most clinicians to the first question is probably yes, and the drugs of choice for this situation have generally been the standard anticonvulsants phenytoin, phenobarbital, and carbamazepine.

The use prophylactic anticonvulsants to prevent the onset of PTE is the more controversial issue. Recent controlled, randomized studies have now shown that the use of phenytoin, phenobarbital, carbamazepine, or valproate does not prevent the development of PTE beyond the first week after injury (Glotzner *et al.*, 1983; Temkin *et al.*, 1990; Temkin *et al.*, 1999). Thus, it is now recommended as a standard of care that these medications not be used to prevent PTE in patients who have not had a seizure (Bullock, Chesnut *et al.*, 2000). However, in light of the sensitivity of the acutely traumatized brain to the secondary insult of a grand mal seizure, routine short-term use (for 1–2 weeks after

injury) of phenytoin or carbamazepine in high-risk patients with acute TBI is recommended as an option. Carbamazepine may be preferable, because it can help control agitation in some patients.

Evidence suggests that iron-catalyzed lipid peroxidation may partly mediate the development of PTE, and inhibitors of lipid peroxidation, such as methylprednisolone and alpha-tocopherol, can prevent iron-induced epilepsy in animals, but no clinical trials have yet explored this or any other neuroprotective avenues to PTE prevention (Willmore, 1990).

## TBI Rehabilitation

The young adult brain has a remarkable capacity to compensate for many aspects of injury naturally, and the principal objective of rehabilitation should be to stimulate and guide this natural process. Postacute TBI management differs from that of other neurological conditions. Although TBI survivors are generally young and healthy before injury and often achieve dramatic recoveries from TBI, residual cognitive and especially behavioral and executive function problems are a major source of persistent long-term disability. In contrast to stroke survivors, their youth also makes this a population of whom more long-term functional demands are generally made, particularly with respect to return to gainful employment. Rehabilitation goals, interventions, and outcome measures are thus often different for TBI survivors. The goals of therapy should be recovery of the patient's independence and his or her reintegration into the community within limits. The prevention of maladaptive behaviors is an important secondary goal.

During the past two decades, the field of TBI rehabilitation has blossomed. A profusion of therapies, including coma stimulation, cognitive rehabilitation, speech therapy, occupational therapy, and recreational therapy, have become available for TBI. Most of these use some combination of didactic and functional approaches to the patient. Although there is consensus about the benefits of some forms of rehabilitation for TBI patients, the type, intensity, and duration of rehabilitation that is best for a given patient remains hotly debated, and these at times expensive interventions have typically not been subjected to the degree of scientific scrutiny for efficacy and cost-effectiveness that is generally expected of new therapies in other fields of medicine. This scarcity of scientific evaluation may also make it more difficult to focus rehabilitation efforts on those therapeutic elements most likely to benefit the patient. In the absence of clear guidelines, limited resources can be consumed in the early phases of recovery for the evaluation and treatment of specific deficits that may improve spontaneously or that may be of limited importance to the ultimate goal of independence. Some intensive programs may even be counterproductive, particularly if they inadvertently exhaust the patient or foster continued dependence. Potentially more efficient and cost-effective interventions, such as training in community reintegration skills and certain forms of behavioral management, may ultimately be omitted for lack of resources.

Prior studies suggest that an interdisciplinary inpatient rehabilitation approach may improve outcome after TBI (Aronow, 1987; Cope *et al.*, 1991; Haffey and Abrams, 1991; Hall and Cope, 1995; Prigatano *et al.*, 1994; Ruff and Niemann, 1990), and that severe TBI patients benefit more from early versus late inpatient rehabilitation (Cope and Hall, 1982). In general, however, TBI rehabilitation studies have been poorly controlled, if at all. Criticisms include the lack of standardized interventions or settings, uneven patient inclusion criteria, lack of controls or of random assignment to treatment conditions, and lack of meaningful or standard outcome criteria (Cope *et al.*, 1991; Evans and Ruff, 1992; High *et al.*, 1995; Levin, 1990; Volpe and McDowell, 1990).

Given the difficulty of accurately predicting TBI outcome, many of these issues are unlikely to be satisfactorily resolved other than by prospective, randomized, controlled clinical studies. For example, one recent large prospective, randomized, controlled trial compared an intensive 8-week in-hospital program of cognitive rehabilitation with a limited home rehabilitation program in soldiers recovering from moderate to severe TBI (Braverman *et al.*, 1999; Salazar *et al.*, 2000; Warden *et al.*, 2000). At 1 and 2 years after injury there was no difference between the two groups in return to work, fitness for military duty, behavioral, neuropsychological, or quality-of-life measures. However, patients with more severe injuries (traumatic unconsciousness >1 hour) seemed to benefit more from the inpatient program, whereas less severely injured patients seemed to benefit more from home therapy.

Thus, a fundamental challenge for practitioners in the field of rehabilitation remains to properly tailor therapies that maximize the individual patient's natural potential for recovery. TBI rehabilitation can be expensive and extremely demanding of patients and therapists alike, but it can also be very rewarding. The scarcity of solid medical evidence to inform patient and physician choices in TBI rehabilitation should not be taken as a reason to deny patients the demonstrated benefits of rehabilitation but rather as a challenge to properly evaluate the costs and benefits of individual therapies.

# REFERENCES

Alexander, M. (1995). Mild traumatic brain injury: pathophysiology, natural history, and clinical management. *Neurology* 45, 1253–1260.

Annegers, J., Hauser, W., *et al.* (1998). A population based study of seizures after traumatic brain injuries. *N. Engl. J. Med.* 338, 20–24.

Aronow, H. U. (1987). Rehabilitation effectiveness with severe brain Injury: translating research into policy. *J. Head Trauma Rehabil.* 2, 24–36.

Barth, J., Alves, W., *et al.* (1989). Mild head injury in sports: neuropsychological sequelae and recovery of function. *In "Mild Head Injury"* (H. Levin, H. Eisenberg, and A. Benton, Eds.). New York: Oxford, Oxford University Press: 257–275.

Bergsneider, M., Hovda, D., *et al.* (2000). Dissociation of cerebral glucose metabolism and level of consciousness during the period of metabolic depression following human traumatic brain injury. *J. Neurotrauma* 17, 389–401.

Bergsneider, M., Hovda, D., *et al.* (1997). Cerebral hyperglycolysis following severe TBI in humans: a positron emission tomography study. *J. Neurosurg.* 86, 241–251.

Bleiberg, J., Halpern, E., *et al.* (1998). Future directions for the neuropsychological assessment of sports concussion. *J. Head Trauma Rehabil.* 13, 36–44.

Borovich, B., Braun, J., *et al.* (1985). Delayed onset of traumatic extradural hematoma. *J. Neurosurg.* 63, 30–34.

Bouma, G. J., and Muizelaar, J. P. (1992). Cerebral blood flow, cerebral blood volume, and cerebrovascular reactivity after severe head injury. *J. Neurotrauma* 9(Suppl 1), S333–348.

Bouma, G. J., Muizelaar, J. P., *et al.* (1992). Blood pressure and intracranial pressure-volume dynamics in severe head injury: relationship with cerebral blood flow. *J. Neurosurg.* 77, 15–19.

Bouma, G. J., Muizelaar, J. P., *et al.* (1991). Cerebral circulation and metabolism after severe traumatic brain injury: the elusive role of ischemia. *J. Neurosurg.* 75, 685–693.

Braakman, R., Gelpke, G. J., *et al.* (1980). Systematic selection of prognostic features in patients with severe head injury. *Neurosurgery* 6, 362–370.

Brandvold, B., Levi, L., *et al.* (1990). Penetrating craniocerebral injuries in the Israeli involvement in the Lebanese conflict. 1982–1985. *J. Neurosurg.* **72**, 15–21.

Braverman, S., Spector, J., *et al.* (1999). A multidisciplinary TBI inpatient rehabilitation program for active duty service members as part of a randomized clinical trial. *Brain Injury* **13**, 405–415.

Bucci, M. N., Phillips, T. W., *et al.* (1986). Delayed epidural hemorrhage in hypotensive multiple trauma patients. *Neurosurgery* **19**, 65–68.

Buki, A., Okonkwo, D., *et al.* (1999). Postinjury cyclosporin A administration limits axonal damage and disconnection in traumatic brain injury. *J. Neurotrauma* **16**, 511–521.

Bullock, R. (1995). Strategies for neuroprotection with glutamate antagonists. Extrapolating from evidence taken from the first stroke and head injury studies. *Ann. N. Y. Acad. Sci.* **765**, 272–278.

Bullock, R., Chesnut, R., *et al.* (2000). Guidelines for the management of severe traumatic brain injury. Nutrition. *J. Neurotrauma* **17**, 539–547.

Bullock, R., Chesnut, R., *et al.* (2000). Guidelines for the management of severe traumatic brain injury. Role of antiseizure prophylaxis following head injury. *J. Neurotrauma* **17**, 549–553.

Bullock, R., Chesnut, R., *et al.* (2000). Guidelines for the management of severe traumatic brain injury. Trauma systems. *J. Neurotrauma* **17**, 457–462.

Bullock, R., Chesnut, R., *et al.* (2000). Guidelines for the management of severe traumatic brain injury: Guidelines for cerebral perfusion pressure. *J. Neurotrauma* **17**, 507–511.

Bullock, R., Chesnut, R., *et al.* (2000). Guidelines for the management of severe traumatic brain injury: Indications for intracranial pressure monitoring. *J. Neurotrauma* **17**, 479–491.

Bullock, R., Chesnut, R., *et al.* (2000). Guidelines for the management of severe traumatic brain injury: Intracranial pressure treatment threshold. *J. Neurotrauma* **17**, 493–495.

Bullock, R., Chesnut, R., *et al.* (2000). Guidelines for the management of severe traumatic brain injury: Recommendations for intracranial pressure monitoring technology. *J. Neurotrauma* **17**, 497–506.

Bullock, R., Chesnut, R., *et al.* (2000). Guidelines for the management of severe traumatic brain injury: Initial management. *J. Neurotrauma* **17**, 463–469.

Bullock, R., Chesnut, R., *et al.* (2000). Guidelines for the management of severe traumatic brain injury: Resuscitation of blood pressure and oxygenation. *J. Neurotrauma* **17**, 471–478.

Bullock, R., Chesnut, R., *et al.* (1996). Guidelines for the management of severe head injury. *J. Neurotrauma* **13**, 639–731.

Chesnut, R., Ghajar, J., *et al.* (2000). Early indicators of prognosis in severe traumatic brain injury. *J. Neurotrauma* **17**, 557–627.

Chesnut, R. M., Marshall, L. F., *et al.* (1993). The role of secondary brain injury in determining outcome from severe head injury. *J. Trauma* **34**, 216–222.

Chesnut, R. M., Marshall, S. B., *et al.* (1993). Early and late systemic hypotension as a frequent and fundamental source of cerebral ischemia following severe brain injury in the Traumatic Coma Data Bank. *Acta. Neurochir. Suppl.* **59**, 121–125.

Clifton, G., Miller, E., *et al.* (2001). Lack of effect of induction of hypothermia after acute brain injury. *N. Engl. J. Med.* **344**, 556–563.

Clifton, G., Robertson, C., *et al.* (1985). Management of the cardiovascular and metabolic responses to severe head injury. *In* "Central Nervous System Trauma Status Report" (D. Becker and J. Povlishock, Eds.), Bethesda, MD: NINCDS, NIH: 139–159.

Cope, D. N., Cole, J. R., *et al.* (1991). Brain injury: Analysis of outcome in a post acute rehabilitaton system. Part 1: General analysis. *Brain Injury* **5**, 111–125.

Cope, D. N., Cole, J. R., *et al.* (1991). Brain injury: Analysis of outcome in a post-acute rehabilitation system. Part 2: Subanalyses. *Brain Injury* **5**, 127–139.

Cope, D. N., and Hall, K. (1982). Head injury rehabilitation: Benefit of early intervention. *Arch. Phys. Med. Rehabil.* **63**, 433–437.

Culotta, V., Sementilli, M., *et al.* (1996). Clinicopathological heterogeneity in the classification of mild head injury. *Neurosurgery* **38**, 245–250.

Deutschman, C., Konstantinides, F., *et al.* (1987). Physiological and metabolic response to isolated closed head injury: Part 2: Effects of steroids on metabolism. Potentiation of protein wasting and abnormalities of substrate utilization. *JNS* **66**, 388–395.

Dikmen, S., Temkin, N., *et al.* (1989). Neuropsychological recovery, relationship to psychosocial function and postconcussional complaints. *In* "Mild Head Injury" (H. Levin, H. Eisenberg, and A. Benton, Eds.). New York: Oxford University Press: 229–240.

Duus, B. R., Lind, B., *et al.* (1994). The role of neuroimaging in the initial management of patients with minor head injury. *Ann. Emerg. Med.* **23**, 1279–1283.

Eisenberg, H. (1989). Mild to moderate brain injury clinical diagnosis. *In* "Mild Head Injury" (H. Levin, H. Eisenberg, and A. Benton, Eds.). Cambridge: Blackwell: 95–105.

Erhard, J., Waydhas, C., *et al.* (1996). [Preclinical diagnosis and management in severe craniocerebral trauma]. *Unfallchirurg* **99**, 534–540.

Evans, R. W., and Ruff, R. M. (1992). Outcome and value: A perspective on rehabilitation outcomes achieved in acquired brain injury. *J. Head Trauma Rehabil.* **7**, 24–36.

Faden, A., Demediuk, P., *et al.* (1989). The role of excitatory amino acids and NMDA receptors in traumatic brain injury. *Science* **244**, 798–800.

Favre, J. B., Ravussin, P., *et al.* (1996). [Hypertonic solutions and intracranial pressure]. *Schweiz Med. Wochenschr* **126**, 1635–1643.

Fearnside, M. R., Cook, R. J., *et al.* (1993). The Westmead Head Injury Project outcome in severe head injury. A comparative analysis of pre-hospital, clinical and CT variables. *Br. J. Neurosurg.* **7**, 267–279.

Feuerman, T., Wackym, P. A., *et al.* (1988). Value of skull radiography, head computed tomographic scanning, and admission for observation in cases of minor head injury. *Neurosurgery* **22**, 449–453.

Fisher, C. M. (1975). Clinical syndromes in cerebral thrombosis, hypertensive hemorrhage, and ruptured saccular aneurysm. *Clin. Neurosurg.* **22**, 117–147.

Fisher, C. M. (1995). Brain herniation: a revision of classical concepts. *Can J. Neurol. Sci.* **22**, 83–91.

Frankowski, R., Annegers, J., *et al.* (1985). The descriptive epidemiology of head trauma in the United States. *CNS Trauma Status Report*. D. Becker and J. Povlishock. Bethesda, MD: NINDS: 33–84.

Gaab, M. R., Seegers, K., *et al.* (1990). A comparative analysis of THAM (Tris-buffer) in traumatic brain oedema. *Acta. Neurochir. Suppl.* **51**, 320–323.

Gennarelli, T., and Thibault, L. (1985). Biological models of head injury. *In* "Central Nervous System Trauma Status Report—1985" (D. Becker and J. Povlishock, Eds.). Bethesda, MD: NINDS, NIH: 391–404.

Gennarelli, T., Thibault, L., *et al.* (1982). Diffuse axonal injury and traumatic coma in the primate. *Ann. Neurol.* **12**, 564–574.

Gentilini, M., Nichelli, P., *et al.* (1989). Assessment of attention in mild head injury. *In* "Mild Head Injury" (H. Levin, H. Eisenberg, and A. Benton, Eds.). New York: Oxford University Press: 163–175.

Glotzner, F., Haubitz, I., *et al.* (1983). Anfallsprophylaxe mit carbamazepine nach schweren schadelhirnverletzungen. *Neurochirurgia* **26**, 66–79.

Gronwall, D. (1989). Cumulative and persisting effects of concussion on attention and cognition. *In* "Mild Head Injury" (H. Levin, H. Eisenberg, and A. Benton, Eds.). New York: Oxford University Press: 153–162.

Haffey, W. J., and Abrams, D. L. (1991). Employment outcomes for participants in a brain injury work re-entry program: Preliminary findings. *J. Head Trauma Rehabil.* **6,** 24–34.

Hall, K. M., and Cope, D. N. (1995). The benefit of rehabilitation in traumatic brain injury: A literature review. *J. Head Trauma Rehabil.* **10,** 1–13.

Hicks, R., Smith, D., *et al.* (1993). Mild experimental brain injury in the rat induces cognitive deficits associated with regional neuronal loss in the hippocampus. *J. Neurotrauma* **10,** 405–414.

High, W., Bloake, C., *et al.* (1995). Critical analysis of studies evaluating the effectiveness of rehabilitation after traumatic brain injury. *J. Head Injury Rehabil.* **10,** 14–26.

Hsiang, J. K., Chesnut, R. M., *et al.* (1994). Early, routine paralysis for intracranial pressure control in severe head injury: is it necessary? *Crit. Care Med.* **22,** 1471–1476.

Hulka, F., Mullins, R., *et al.* (1996). Blunt brain injury activates the coagulation process. *Arch. Surg.* **131,** 923–927.

Hume Adams, J., Graham, D. I., *et al.* (1985). Contempory neuropathological considerations regarding brain damage in head injury. *In* "Central Nervous System Trauma Status Report" (D. P. Becker and J. T. Povlishock, Eds.). Bethesda, MD: NINCDS, NIH: 65–77.

Ingebrigtsen, T., and Romner, B. (1996). Routine early CT-scan is cost saving after minor head injury. *Acta. Neurol. Scand* **93,** 207–210.

Ishige, N., Pitts, L., *et al.* (1987). Effect of hypoxia on traumatic brain injury in rats: Part 1. *NeuroSurg* **30,** 848–854.

Jennett, B., Teasdale, G., *et al.* (1979). Prognosis of patients with severe head injury. *Neurosurgery* **4,** 283–289.

Jennett, W. (1975). *Epilepsy after non-missile head injuries.* Chicago: Yearbook Medical Publishers.

Jordan, B., Relkin, N., *et al.* (1997). Apolipoprotein E epsilon 4 associated with chronic traumatic brain injury in boxing. *JAMA* **278,** 136–140.

Junger, E., Newell, D., *et al.* (1997). Cerebral autoregulation following minor head injury. *J. Neurosurg.* **86,** 425–432.

Kaye, E. M., Cass, P. R., *et al.* (1985). Chronic epidural hematomas in childhood: increased recognition and non-surgical management. *Pediatr. Neurol.* **1,** 255–259.

Kotapka, M., Gennarelli, T., *et al.* (1991). Selective vulnerability of hippocampal neurone in acceleration induced experimental head injury. *J. Neurotrauma* **8,** 247–258.

Krauss, J., and Nourjah, P. (1989). The epidemiology of mild head injury. *In* "Mild Head Injury" (H. Levin, H. Eisenberg, and A. Benton, Eds.). New York: Oxford U. Press: 8–22.

Levin, H., Mattis, S., *et al.* (1987). Neurobehavioral outcome following minor head injury: A three-center study. *J. Neurosurg.* **66,** 234–243.

Levin, H. S. (1990). Cognitive rehabilitation: Unproved but promising. *Arch. Neurol.* **47,** 223–224.

Levin, H. S., Amparo, E. G., *et al.* (1989). Magnetic resonance imaging after closed head injury in children. *Neurosurgery* **24,** 223–227.

Macciocchi, S., Jane, J., *et al.* (1996). Neuropsychological functioning and recovery after mild head injury in collegiate athletes. *Neurosurgery* **39,** 510–514.

Mahoney, B. D., Rockswold, G. L., *et al.* (1981). Emergency twist drill trephination. *Neurosurgery* **8,** 551–554.

Marion, D., Penrod, L., *et al.* (1997). Treatment of traumatic brain injury with moderate hypothermia. *N. Engl. J. Med.* **336,** 540–546.

Marion, D. W., and Carlier, P. M. (1994). Problems with initial Glasgow Coma Scale assessment caused by prehospital treatment of patients with head injuries: results of a national survey. *J. Trauma* **36,** 89–95.

Marshall, L., and Marshall, S. (1995). Pitfall and advances from the international tirilazad trial in moderate and severe head injury. *J. Neurotrauma* **12,** 929–932.

Marshall, L., Toole, B., *et al.* (1983). The National Traumatic Coma Data Bank, Part II: Patients who talk and deteriorate: Implications for treatment. *J. Neurosurg.* **59,** 285–288.

Masters, S. J., McClean, P. M., *et al.* (1987). Skull x-ray examinations after head trauma. Recommendations by a multidisciplinary panel and validation study. *N. Engl. J. Med.* **316,** 84–91.

Matser, E., Kessels, A., *et al.* (1999). Neuropsychological impairment in amateur soccer players. *JAMA* **282,** 971–973.

Max, W., MacKenzie, E., *et al.* (1991). Head injuries: Costs and consequences. *J. Head Trauma Rehabil.* **6,** 76.

Mayeux, R., Ottman, R., *et al.* (1995). Synergistic effect of traumatic head injury and apolipoprotein-e4 in patient with alzheimer's disease. *Neurology* **45,** 555–557.

McCrea, M., Kelly, J., *et al.* (1998). Standardized assessment of concussion (SAC) On-site mental status evaluation of the athlete. *J. Head Trauma Rehabil.* **13,** 27–35.

McCrory, P., and Berkovic, S. (1998). Second impact syndrome. *Neurology* **50,** 677–683.

McIntosh, T., Smith, D., *et al.* (1996). Therapeutic approaches for the prevention of secondary brain injury. *Eur. J. Anaesthesiol.* **13,** 291–309.

Melamed, S., Grosswasser, Z., *et al.* (1992). Acceptance of disability, work involvement and subjective rehabilitation status of traumatic brain-injured (TBI) patients. *Brain Injury* **6,** 233–243.

Obrist, W. D., Langfitt, T. W., *et al.* (1984). Cerebral blood flow and metabolism in comatose patients with acute head injury. Relationship to intracranial hypertension. *J. Neurosurg.* **61,** 241–253.

Oppenheimer, D. (1968). Microscopic lesions in the brain following head injury. *J. Neurol. Neurosurg. Psychiatry* **31,** 299.

Pigula, F. A., Wald, S. L., *et al.* (1993). The effect of hypotension and hypoxia on children with severe head injuries. *J. Pediatr Surg.* **28,** 310–314; discussion 315–316.

Polin, R. S., Shaffrey, M. E., *et al.* (1997). Decompressive bifrontal craniectomy in the treatment of severe refractory posttraumatic cerebral edema. *Neurosurgery* **41,** 84–92; discussion 92–94.

Povlishock, J. (1985). The morphopathologic responses to head injuries of varying severity. *In* "Central Nervous System Trauma Status report" (D. Becker and J. Povlishock, Eds.). Washington, DC: National Institute of Neurological and Communicative Disorders and Stroke (NINCDS), National Institutes of Health (NIH): 443–452.

Povlishock, J. (1992). Traumatically induced axonal injury: pathogenesis and pathobiological implications. *Brain Pathol.* **2,** 1–12.

Povlishock, J., and Jenkins, L. (1995). Are the pathobiological changes evoked by traumatic brain injury immediate and irreversible? *Brain Patholo.* **5,** 415–426.

Prigatano, G. P., Klonoff, P. S., *et al.* (1994). Productivity after neuropsychologically oriented millieu rehabilitation. *J. Head Trauma Rehabil.* **9,** 91–102.

Proctor, H., Palladino, G., *et al.* (1988). Failure of autoregulation after closed head injury: an experimental model. *J. Trauma* **28,** 347–352.

Rivas, J. J., Lobato, R. D., *et al.* (1988). Extradural hematoma: analysis of factors influencing the courses of 161 patients. *Neurosurgery* **23,** 44–51.

Roda, J. M., Gimenez, D., *et al.* (1983). Posterior fossa epidural hematomas: a review and synthesis. *Surg. Neurol.* **19,** 419–424.

Ropper, A. H., and King, R. B. (1984). Intracranial pressure monitoring in comatose patients with cerebral hemorrhage. *Arch. Neurol.* **41,** 725–728.

Ross, D. T., Graham, D. I., *et al.* (1993). Selective loss of neurones from the thalamic reticular nucleus following severe human head injury. *J. Neurotrauma* **10**, 151–165.

Roy, P. D. (1987). The value of trauma centres: a methodologic review. *Can J. Surg.* **30**, 17–22.

Ruff, R., Levin, H., *et al.* (1989). Recovery of memory after mild head injury: a three center study. *In* "Mild Head Injury" (H. Levin, H. Eisenberg, and A. Benton, Eds.). New York: Oxford University Press: 176–188.

Ruff, R. M., and Niemann, H. (1990). Cognitive rehabilitation versus day treatment in head-injured adults: Is there an impact on emotional and psychosocial adjustment. *Brain Injury* **4**, 339–347.

Rutherford, W. (1989). Postconcussion symptoms: relationship to acute neurological indices, individual differences, and circumstances of injury. *In* "Mild Head Injury" (H. Levin, H. Eisenberg, and A. Benton, Eds.). New York: Oxford University Press: 217–228.

Saatman, K., Graham, D., *et al.* (1998). The neuronal cytoskeleton is at risk after mild and moderate brain injury. *J. Neurotrauma* **15**, 1047–1058.

Sailor, K., Vemuganti, R., *et al.* (2000). *Pyruvate treatment is neuroprotective after traumatic brain injury.* Annual Meeting, Society for Neuroscience, New Orleans, Society for Neuroscience Abstracts.

Salazar, A., Jabbari, B., *et al.* (1985). Epilepsy after penetrating head injury, I: Clinical correlates. *Neurology* **35**, 1406–1414.

Salazar, A., Warden, D., *et al.* (2000). Cognitive rehabilitation for traumatic brain injury—a randomized trial. *JAMA* **283**, 3075–3081.

Saul, T. G., and Ducker, T. B. (1982). Effect of intracranial pressure monitoring and aggressive treatment on mortality in severe head injury. *J. Neurosurg.* **56**, 498–503.

Schwab, K., Grafman, J., *et al.* (1993). Residual impairments and work status 15 years after penetrating head injuries: report from the Vietnam Head Injury Study. *Neurology* **43**, 95–103.

Schwartz, M. L., Tator, C. H., *et al.* (1984). The University of Toronto head injury treatment study: a prospective, randomized comparison of pentobarbital and mannitol. *Can. J. Neurol. Sci.* **11**, 434–440.

Seelig, J. M., Becker, D. P., *et al.* (1981). Traumatic acute subdural hematoma: major mortality reduction in comatose patients treated within four hours. *N. Engl. J. Med.* **304**, 1511–1518.

Selladurai, B., Vickneswaran, M., *et al.* (1997). Coagulopathy in acute head injury-a study of its role as a prognostic indicator. *Br. J. Neurosurg.* **11**, 398–404.

Siesjo, B., and Siesjo, P. (1996). Mechanisms of secondary brain injury. *Eur. J. Anaesthesiol.* **13**, 247–268.

Smith, H. P., Kelly, Jr., D. L., *et al.* (1986). Comparison of mannitol regimens in patients with severe head injury undergoing intracranial monitoring. *J. Neurosurg.* **65**, 820–824.

Snow, R. B., Zimmerman, R. D., *et al.* (1986). Comparison of magnetic resonance imaging and computed tomography in the evaluation of head injury. *Neurosurgery* **18**, 45–52.

Stablein, D. M., Miller, J. D., *et al.* (1980). Statistical methods for determining prognosis in severe head injury. *Neurosurgery* **6**, 243–248.

Stein, S. C., and Ross, S. E. (1990). The value of computed tomographic scans in patients with low-risk head injuries. *Neurosurgery* **26**, 638–640.

Stein, S. C., and Ross, S. E. (1992). Moderate head injury: a guide to initial management. *J. Neurosurg.* **77**, 562–564.

Strich, S. (1961). Shearing of nerve fibres as a coause of brain damage due to head injury. A pathological study of twenty cases. *Lancet* **2**, 443.

Sullivan, P., Geiger, J., *et al.* (2000). Dietary supplement creatine protects against traumatic brain injury. *Ann. Neurol.* **48**, 723–729.

Sullivan, P., Thompson, M., *et al.* (1999). Cyclosporin A attenuates acute mitochondrial dysfunction following traumatic brain injury. *Exp. Neurol.* **160**, 226–234.

Teasdale, G., and Jennett, B. (1974). Assessment of coma and impaired consciousness. A practical scale. *Lancet* **2**(7872), 81–84.

Temkin, N., Dikmen, S., *et al.* (1999). Valproate therapy for prevention of posttraumatic seizures: a randomized trial. *J. Neurosurg.* **91**, 593–600.

Temkin, N., Dikmen, S., *et al.* (1990). A randomized, double blinded study of phenytoin for the prevention of post-traumatic seizures. *N. Engl. J. Med.* **323**, 497–502.

Thatcher, R., Biver, C., *et al.* (1998a). Biophysical linkage between MRI and EEG amplitude in closed head injury. *Neuroimage* **7**, 352–367.

Thatcher, R., Biver, C., *et al.* (1998b). Biophysical linkage between MRI and EEG coherence in closed head injury. *Neuroimage* **8**, 307–326.

Thatcher, R., North, D., *et al.* (2001). An EEG severity index of traumatic brain injury. *J. Neuropsychiatry Clin. Neurosci.* **13**, 77–87.

Thatcher, R., Walker, R., *et al.* (1989). EEG discriminant analysis of mild head trauma. *EEG Clin. Neurophysiol.* **73**, 93–106.

Thurman, D., Alverson, C., *et al.* (1999). Traumatic brain injury in the United States: a public health perspective. *J. Head Trauma Rehabil.* **14**, 602–615.

Thurman, D., Branche, C., *et al.* (1998). The epidemiology of sports related traumatic brain injuries in the United States: recent developments. *J. Head Trauma Rehabil.* **13**, 1–8.

Ugrinovski, J., Ruskov, P., *et al.* (1983). [Emergency management of intracerebral hematoma of nontraumatic origin]. *Acta. Chir. Iugosl.* **30**, 265–270.

Van Zomeren, A., and Deelman, B. (1976). Differential effects of simple and choice reaction time after closed head injury. *Clin. Neurol. Neurosurg.* **79**, 81–90.

Vassar, M. J., Fischer, R. P., *et al.* (1993). A multicenter trial for resuscitation of injured patients with 7.5% sodium chloride. The effect of added dextran 70. The Multicenter Group for the Study of Hypertonic Saline in Trauma Patients. *Arch. Surg.* **128**, 1003–1011; discussion 1011–1013.

Verma, A. (2000). Opportunities for neuroprotection in traumatic brain injury. *J. Head Trauma Rehabil.* **15**, 1149–1161.

Volpe, B. T., and McDowell, F. H. (1990). The efficacy of cognitive rehabilitation in patients with traumatic brain injury. *Arch. Neurol.* **47**, 220–222.

Warden, D., Bleiberg, J., *et al.* (2001). Persistent prolongation of simple reaction time in sports concussion. *Neurology.* **57**(3), 524–526.

Warden, D., Salazar, A., *et al.* (2000). A home program of rehabilitation for moderately severe traumatic brain injury patients. *J. Head Trauma Rehabil.* **15**, 1092–1102.

Wehman, P., Kreutzer, J., *et al.* (1989). Employment outcomes of persons following traumatic brain injury: Pre-injury, post-injury, and supported employment. *Brain Injury* **3**, 397–412.

Weiss, G., Salazar, A., *et al.* (1986). Predicting posttraumatic epilepsy in penetrating head injury. *Arch. Neurol.* **43**, 771–773.

Wilberger, J. E., Jr., Harris, M., *et al.* (1991). Acute subdural hematoma: morbidity, mortality, and operative timing. *J. Neurosurg.* **74**, 212–218.

Williams, D., Levin, H., *et al.* (1990). Mild head injury classification. *Neurosurgery* **27**, 422–428.

Willmore, L. (1990). Post-traumatic epilepsy: cellular mechanism and implications for treatment. *Epilepsia* **31**(suppl. 3), S-67–73.

Young, B., Runge, J., *et al.* (1996). Effects of pegorgotein on neurologic outcome of patients with severe head injury—a multicenter, randomized controlled trial. *JAMA* **276**, 538–543.

CHAPTER 58

# Malignant Hyperthermia and Neuroleptic Malignant Syndrome

Michael Weller and Frank Lehmann-Horn

## FEVER, FEVER OF CENTRAL ORIGIN, AND HYPERTHERMIA

The regulation of the body temperature is a function of the hypothalamus. *Fever* is defined as a state in which the setpoint for the body temperature is up-regulated as part of a physiological response pattern (e.g., to counteract a bacterial infection) (Saper and Breder, 1994). The fever response of the hypothalamus is mediated by humoral messenger molecules such as the cytokines, interleukin-1β, tumor necrosis factor-α, β- and γ-interferons, and prostaglandin metabolites. These effector mechanisms are thought to be identical to those that serve to maintain body temperature on environmental cold challenges. Physiological mechanisms to maintain body temperature include centralization of perfusion, resulting in reduced heat dissipation through the skin, reduced sweat production, and heat producton through shivering. Conversely, dissipation of heat involves enhanced peripheral blood flow and sweating. *Hyperthermia*, in contrast to fever, does not result from an altered setpoint in the hypothalamus but an imbalance between heat production and heat dissipation, associated with a failure of endogenous mechanisms of body temperature regulation. For instance, strong exogenous heat loads result in dangerous hyperthermia in individuals with deficient capacity to dissipate heat (e.g., because of regular intake of anticholinergic drugs). The concept of *fever-of-central-origin* is too often used to explain the cause of high body temperature in patients with various types of central nervous system disease and is insufficiently supported by experimental and clinical data. Hypothalamic dysfunction commonly results in hypothermia or poikilothermia, a state in which body temperature follows environmental temperature but does not result in hyper-

thermia. Any hyperthermic condition in a patient with central nervous system disease therefore requires careful analysis before the assumption of a fever-of-central-origin.

## MALIGNANT HYPERTHERMIA

### Clinical Aspects

Malignant hyperthermia is a pharmacogenetic disease, that is, an autosomal dominant trait predisposing to abnormal muscle responses to the administration of volatile anesthetics such as ether, halothane, isoflurane, sevoflurane, desflurane, or enflurane, and depolarizing muscle relaxants such as succinylcholine (Jurkat-Rott *et al.*, 2000). The classical syndrome of malignant hyperthermia is characterized by muscle spasms, preferentially of the masseter; tachyarrhythmia; metabolic and respiratory acidosis; rhabdomyolysis; myoglobinuria; and disseminated intravascular coagulation (Table I). Hyperthermia is the consequence of the hypermetabolic state and a late sign of the disease process. A pathological increase of myoplasmic calcium concentrations is the pathophysiological hallmark of malignant hyperthermia. The earliest clinical sign is often spasm of the masseter muscles in response to succinylcholine. This sign, however, is not specific for malignant hyperthermia but may also be observed in other muscle disorders in the absence of the full-blown syndrome of malignant hyperthermia. Thus, masseter spasms observed during the initiation of anesthesia require alertness to the possibility of malignant hyperthermia; however, only the development of additional symptoms or signs necessitates the cessation of anesthesia. Nevertheless, patients revealing a masseter spasm on exposure to volatile anesthetics or succinylcholine should be advised to undergo a neurological examination, possibly including an in vivo contracture test (see later) to determine susceptibility to malignant hyperthermia. Approximately half of the malignant hyperthermia trait carriers show elevated serum creatine kinase levels. Individuals with familial chronic creatine kinase elevation without corresponding history may also have an increased risk of complications from anesthesia. Consequently, both close relatives of malignant hyperthermia patients and individuals exhibiting unexplained familial elevated creatine kinase should undergo the *in vitro* contracture test (see later) before elective surgery. The clinical diagnosis of malignant hyperthermia is commonly made when the respective clinical syndrome develops during anesthesia. Because of the circumstances under which malignant hyperthermia becomes apparent, the differential diagnosis includes primarily intolerance of other agents, other causes of rhabdomyolysis, sepsis, toxic shock, or thyrotoxicosis. The combination of hyperthermia and rigidity is also characteristic of neuroleptic malignant syndrome, lethal catatonia, malignant dopamine deficiency syndrome, and akinetic crisis of Parkinson's disease. The muscle histology of malignant hyperthermia–susceptible individuals is normal except after rhabdomyolysis caused by an anesthesia-related event. Apparently, heart muscle is not a target of the disease. The peripheral and central nervous systems are only indirectly involved in malignant hyperthermia. Whether carriers of the trait may develop the syndrome of malignant hyperthermia in the absence of certain anesthetic agents has remained controversial.

More than 50% of all malignant hyperthermia trait carriers have mutations in the gene encoding the ryanodine receptor (RYR1), a channel protein that controls calcium release from the sarcoplasmic reticulum of skeletal muscle cells. Ryanodine is a plant alkaloid that acts as a ligand for this channel protein. More than 20 point mutations of the RYR1 gene associated with malignant hyperthermia susceptibility have been described (Brandt *et al.*, 1999; McCarthy *et al.*, 2000). Further loci believed to carry malignant hyperthermia susceptibility genes have been located to chromosomes 1q, 3q, 5, 7q, and 17q, including the main subunit of the voltage-gated dihydropyridine receptor of skeletal muscle (Jurkat-Rott *et al.*, 2000).

The heterogeneity of mutations affecting the RYR1 gene, as well as the putative non-RYR1 mutations accounting for a minority of the malignant hyperthermia cases, illustrates why routine genetic testing for malignant hyperthermia may not become available in the near future. An in vitro contracture test has therefore been devised that detects abnormal contractures of freshly biopsied muscle in response to halothane or caffeine (EMHG, 1984). This test classifies individuals as *malignant hyperthermia–susceptible* (MHS) with pathological responses to both agents, *malignant hyperthermia–equivocal* (MHE) with such a response to one of two agents, and *malignant hyperthermia–normal* (MHN) with normal responses to either agent (EMHG, 1984). The specificity of the in vitro contracture test may have been overestimated, because positive responses have been observed in patients with other muscle disorders, including progressive muscle dystrophies. However, a recent survey defined a sensitivity of 98.5% and a specificity of at least 81.8% for the in vitro contracture test (Brandt *et al.*, 1999). A novel test substitutes the ryanodine receptor–specific agonist, 4-chloro-m-cresol, for caffeine (Baur *et al.*, 2000).

## Natural Course

The incidence of malignant hyperthermia is approximately 1/20,000 anesthesias in children and 1 in 100,000 anesthesias in adults. Patients with central core disease, a morphologically defined myopathy, or King-Denborough syndrome are obligatory risk patients for malignant hyperthermia. Most patients with central core disease carry mutations in the RYR1 gene, which is also affected in more than half of the patients with malignant hyperthermia (Brandt *et al.*, 1999; McCarthy *et al.*, 2000). Even repetitive event-free anesthesias do not exclude a carrier state of malignant hyperthermia because of the low penetrance of this trait. Thus, exogenous factors such as coexposure to volatile anesthetic agents and succinylcholine, the doses of these agents, and strenuous physical exercise or mental stress preceding surgery may precipitate the first manifestation of malignant hyperthermia. Malignant hyperthermia–like symptoms may also occur in patients with neuromuscular disorders different from malignant hyperthermia susceptibility, but the pathogenesis is different in these patients.

Malignant hyperthermia is a dangerous condition with significant mortality. The mortality of approximately 75% in the sixties has been lowered to less than 10% because of the introduction of dantrolene. This drug inhibits the release of calcium from the sarcoplasmic reticulum. Causes of lethal outcome include cardiac arrhythmias caused by hyperkalemia, disseminated intravascular coagulation, malignant cerebral edema, and renal failure caused by myoglobin released from muscle. An episode of malignant hyperthermia may be followed by a prolonged, predominantly proximal myopathy with transient swelling caused by rhabdomyolysis. Abnormal fatigue may persist for months.

## Principles of Therapy

Principles of therapy include management of the acute episode of malignant hyperthermia and the prevention of further episodes in patients with a history of malignant hyperthermia or individuals at risk because of familial malignant hyperthermia and elevated serum creatine kinase levels. In the event of malignant hyperthermia, anesthesia should be halted and specific and symptomatic therapy instituted: application of dantrolene and balancing of acid base and fluid. Dantrolene must be available in the operation theater during any surgical procedure requiring general anesthesia. The administration of dantrolene should be continued until the syndrome remits. If patients susceptible to malignant hyperthermia need to undergo surgery, the possibility of regional anesthesia should be considered, or anesthetic agents with no intrinsic risk should be administered. Such agents include nitrous oxide, barbiturates, opiates, etomidate, ketamine, benzodiazepines, and nondepolarizing muscle relaxants such as pancuronium. Dantrolene prophylaxis for individuals at risk undergoing surgery is no longer recommended.

## Practical Management

Early recognition of the condition is the most important step to successful therapy. The combination of increased end-expiratory $CO_2$ and metabolic acidosis is a specific early sign of malignant hyperthermia. Persisting masseter spasm in response to succinylcholine should probably already alert the clinician not to further administer volatile anesthetic agents and succinylcholine. If surgery is absolutely necessary, anesthesia is continued using safe agents. Metabolic and respiratory acidosis are treated by hyperventilation with 100% $O_2$ and intravenous sodium bicarbonate (2–4 mEq/kg body weight). Further management depends on the state of acid base and fluid balances. Fluids have to be administered with constant monitoring of diuresis and potassium levels, because dangerous hyperkalemias may occur. Renal function is specifically endangered by myoglobinuria.

The instant administration of a bolus of dantrolene (2.5 mg/kg body weight) and subsequent infusion according to the clinical course is the most important therapeutic measure. The daily dose of dantrolene should not exceed 10 mg/kg body weight. Even high doses of dantrolene do not cause weakness in healthy individuals, but patients with myopathy may experience weakness when exposed to dantrolene. Except for a risk of cholestasis with prolonged use for weeks, dantrolene has no relevant side effects. After remission of the acute episode of malignant hyperthermia, intensive care monitoring is mandatory until all vital functions have normalized. It has remained controversial for how long dantrolene should be administered. Administration for several days would be rather unusual. Digitalis glycosides, α/β-adrenergic agonists, parasympatholytics, and calcium-containing agents should be avoided during the first days after malignant hyperthermia has resolved. Creatine kinase levels in serum and myoglobinuria are helpful parameters to monitor the remission of malignant hyperthermia.

Several pharmacological approaches to malignant hyperthermia have proven not to be successful. These include the administration of calcium channel blockers, procainamide, lidocaine, α- and β-adrenergic receptor agonists and antagonists, and corticosteroids. As indicated earlier, oral dantrolene prophylaxis is obsolete.

# NEUROLEPTIC MALIGNANT SYNDROME

## Clinical Aspects

Neuroleptic malignant syndrome was originally described as a life-threatening complication of the pharmacotherapy of schizophrenic psychoses with neuroleptic agents (Kornhuber and Weller, 1994). The key clinical features of neuroleptic malignant syndrome, extrapyramidal disturbances and hyperthermia, are attributed to an acute relative deficiency of dopaminergic neurotransmission in basal ganglia and hypothalamus (Table I). Dopamine deficiency may result in relative overactivity of glutamatergic systems, which are believed to mediate part of the clinical syndrome. An identical condition has been observed after administration of the antiemetic agent, metoclopramide, in children; and in patients with Huntington's disease treated with the dopamine-depleting agent, tetrabenazine. The entity of a "neuroleptic" malignant syndrome may be questioned, because the clinical syndrome has also been reported after the administration of various other agents that are neither neuroleptic drugs nor known to inhibit dopaminergic neurotransmission (e.g., lithium or tricyclic antidepressants). Moreover, there is significant syndromal overlap with the malignant dopamine deficiency syndrome and the akinetic crises of patients with Parkinson's disease. Thus, "neuroleptic malignant syndrome" may in fact be a misnomer, but nevertheless most authorities still agree on its salient features. Moreover, numerous case reports of "neuroleptic malignant syndrome" attributed to nonneuroleptic drugs would not meet stringent criteria for that diagnosis. Although neuroleptic malignant syndrome was originally considered an idiosyncratic reaction of sensitive patients, the definition of various risk factors indicates that almost any individual may have neuroleptic malignant syndrome develop under certain circumstances.

The typical clinical features, extrapyramidal disturbances and signs of autonomic dysregulation, notably hyperthermia, develop within hours, days, or even weeks of neuroleptic drug therapy. The extrapyramidal disturbances of neuroleptic malignant syndrome include rigidity, tremor, retrocollis, opisthotonus, trism, oculogyric crises, and choreoathetosis. Autonomical disturbances include, in addition to hyperthermia, pallor, diaphoresis, tachypnea, tachycardia, hypertension or hypotension, and urinary retention. The level of consciousness is fluctuating but commonly reduced. Laboratory abnormalities include an elevation of serum creatine kinase levels, often of several hundred units, elevation of liver enzyme levels, increased erythrocyte sedimentation rate, and leukocytosis. The elevation of serum creatine kinase levels is no independent diagnostic marker of neuroleptic malignant syndrome, because neuroleptic drug-induced hyperthermia is also associated with elevations of serum creatine kinase in the absence of extrapyramidal disturbances or other features of neuroleptic malignant syndrome (O'Dwyer and Sheppard, 1993).

In contrast to malignant hyperthermia, which is a primary disturbance of calcium metabolism in skeletal muscle, both central nervous system and muscular processes contribute to the pathogenesis of neuroleptic malignant syndrome. As in malignant hyperthermia, the source of elevated serum creatine kinase in neuroleptic malignant syndrome is muscle. However, the precise mechanism of creatine kinase liberation from muscle in neuroleptic malignant syndrome is unclear. Probably, rigidity and hyperthermia act in synergy to promote damage to muscle cell membranes. The elevation of creatine kinase is unlikely to be a direct toxic effect of neuroleptic drugs, because too many chemically unrelated agents cause neuroleptic malignant syndrome and because creatine kinase is elevated in the malignant dopamine deficiency syndrome (see later) in the absence of exposure to exogenous agents. Also, rigidity alone does not cause creatine kinase elevations in Parkinson's disease or other neurological disorders associated with rigidity. The pathogenetic importance and differential diagnostic relevance of lowered serum iron levels in neuroleptic malignant syndrome has remained controversial. Patients with a history of neuroleptic malignant syndrome do not have an increased risk of malignant hyperthermia.

Operational criteria for the diagnosis of neuroleptic malignant syndrome (Gurrera et al., 1992) have not met with relevant success, and poorly defined adverse reactions to novel drugs continue to be published as "neuroleptic malignant syndromes." The most important differential diagnoses of neuroleptic malignant syndrome are catatonic psychoses, notably lethal catatonia (Fleischhacker et al., 1990), and febrile disorders, including pneumonia and sepsis in neuroleptic drug–treated patients, misdiagnosed as neuroleptic malignant syndrome. The clinical history will easily allow one to distinguish neuroleptic malignant syndrome from malignant hyperthermia. Heat stroke is associated with strenuous physical exercise and heat exposure, and creatine kinase elevation and extrapyramidal signs are not found. Clinically relevant is the differential diagnosis of neuroleptic malignant syndrome and serotonin syndrome (Carbone, 2000; Sternbach, 1991). Serotonin syndrome has most often been diagnosed in patients receiving the combination of L-tryptophan, inhibitors of synaptic serotonin uptake, and inhibitors of monaminoxidase. Diagnostic criteria include hyperthermia, diaphoresis, tremor, diarrhea, and psychoorganic changes, although probably no creatine kinase elevation.

## Natural Course

The incidence of neuroleptic malignant syndrome in neuroleptic drug–treated patients is in the range of 0.1%–2.5%. Risk factors for the development of the syndrome include rapid increase of dose and depot application of neuroleptic drugs, dehydration, Parkinson's disease, pre-existing brain lesions, and oligophrenia. The natural course depends mainly on the time point at which the diagnosis is made. In the early years of the neuroleptic era, a lethal outcome was not uncommon. For instance, when the extrapyramidal signs were misinterpreted to signify progressive catatonia, drug doses were increased, and outcome was commonly poor. Causes of death were renal failure because of massive rhabdomyolysis, pulmonary embolism, or complicating infections. However, withdrawal of the culprit agent(s) and appropriate supportive care result in full recovery without adverse sequelae in almost all patients today, provided the diagnosis is made in time. Elderly patients are still at a higher risk of poor outcome or severe complications from neuroleptic malignant syndrome for several reasons. These patients are more often treated with high-potency neuroleptics to avoid the cardiovascular and sedative side effects of low-potency neuroleptic drugs; they are more prone to the erroneous intake of higher drug doses; they are less likely to exhibit prominent autonomic symptoms initially; and they are more often admitted to nonneurological or nonpsychiatric institutions, where the diagnosis of neuroleptic malignant syndrome may be more easily missed. Moreover, early symptoms such as confusion and desorientation are more readily attributed to other neurological diseases, such as Parkinson's disease or Alzheimer's disease in the elderly than in younger patients.

## Principles of Therapy

The first principle of management of neuroleptic malignant syndrome is to withdraw the agent(s) thought to be responsible for the condition. Supportive therapy should include hydration, lowering of body temperature, and prevention of complications such as pneumonia, deep vein thrombosis, and renal failure from rhabdomyolysis. In addition to these measures, which are beyond doubt efficient in the management of neuroleptic malignant syndrome, several more specific pharmacological approaches to neuroleptic malignant syndrome have been proposed based on current concepts of its pathophysiology. In analogy to malignant hyperthermia, it has been speculated that there is a primary abnormality in muscle of affected individuals and that dantrolene might be a suitable agent to interfere with the peripheral aspects of the condition. Moreover, several lines of evidence suggest that neuroleptic malignant syndrome results from a relative deficiency of dopaminergic neurotransmission in basal ganglia and hypothalamus (Kornhuber and Weller, 1994). Consequently, the literature contains a series of case reports that describe benefitial effects of dopaminomimetic agents such as L-dopa, bromocriptine, lisuride, or amantadine in neuroleptic malignant syndrome. However, independent reviews did not arrive at the conclusion that the administration of dantrolene or dopaminomimetics truely affects outcome once the offending agent has been withdrawn (Rosebush et al., 1991; Rosenberg and Green, 1989; Sakkas et al., 1991). In theory, dopaminomimetic agents would seem to be of benefit when the neuroleptic agent *cannot* be withdrawn (e.g., when depot neuroleptics have caused neuroleptic malignant syndrome). However, this has not been documented to be the case in the published series of neuroleptic malignant syndrome. Psychiatric patients with a history of neuroleptic malignant syndrome may not have an increased risk of recurrence in the absence of further risk factors even when challenged with slowly increasing doses of the agent that previously caused neuroleptic malignant syndrome. These observations question the notion that neuroleptic malignant syndrome is an idiosyncratic drug reaction. Theoretical considerations suggest that the atypical neuroleptic agent, clozapine, which has little probability of causing extrapyamidal side effects, is particularly useful for the treatment of psychosis in patients with a history of neuroleptic malignant syndrome (Weller and Kornhuber, 1992). Anecdotal evidence suggests that electroconvulsive therapy may be benefitial for patients with neuroleptic malignant syndrome (Troller and Sachdev, 1999). Presumably, it is difficult to determine whether the neuroleptic malignant syndrome or the underlying condition (e.g., psychotic depression or catatonia) responds to this mode of therapy. Electroconvulsive therapy is therefore probably a good choice for patients with neuroleptic malignant syndrome who require treatment of their underlying mental disorders within 2 weeks of diagnosis of the neuroleptic malignant syndrome. Interestingly, no single episode of malignant hyperthermia was found in a review of patients with neuroleptic malignant syndrome who subsequently received anesthesia for electroconvulsive therapy (Troller and Sachdev, 1999), confirming the independence of both disorders. Patients who have had an episode of neuroleptic malignant syndrome should carry with them a note indicating under what conditions and in response to what medication their neuroleptic malignant syndrome developed.

## Practical Management

Controlled trials addressing the management of neuroleptic malignant syndrome are not available. The most important therapeutic measure is the identification and withdrawal of the causative agent. Dehydration needs to be corrected with close monitoring of electrolytes. Body temperature should be lowered to 38°–38.5°C by physical measures and antipyretics such as paracetamol. Subcutaneous heparin should be administered for the prevention of deep vein thrombosis. Benzodiazepines, notably lorazepam ($3 \times 2.5$ mg), should be given for sedation. The specific pharmacological therapy of neuroleptic malignant syndrome has remained controversial. We advocate a trial with intravenous amantadine ($3 \times 200$ mg/day) for patients who do not show major improvement within 24 h of withdrawal of all neuroleptic agents. Amantadine ($3 \times 100$ mg) may be continued orally after 3 days if deemed helpful. This recommendation is based on personal experience and not better based than approaches to treat the condition with L-dopa, bromocriptine, or lisuride (Kornhuber and Weller, 1994). Furthermore, a trial of dantrolene may be appropriate if the clinical condition does not remit within 2–3 days. Prolonged severe variants of neuroleptic malignant syndrome, as well as possibly the underlying psychiatric condition, may respond to electroconvulsive therapy (Troller and Sachdev, 1999).

Anticholinergic agents should not be given to patients with neuroleptic malignant syndrome, because they impair heat dissipitation through the periphery by means of sweating and because their prodelirant effects may interfere with clinical monitoring. Classical neuroleptic drugs may not be continued if the differential diagnosis between lethal catatonia and neuroleptic malignant syndrome is uncertain.

## MALIGNANT DOPAMINE DEFICIENCY SYNDROME AND AKINETIC CRISIS OF PARKINSON'S DISEASE

### Clinical Aspects

Rapid withdrawal of dopaminomimetic agents may trigger a condition in patients with Parkinson's disease that is almost impossible to distinguish from neuroleptic malignant syndrome both clinically and presumably pathophysiologically. This syndrome, which may be labeled *malignant dopamine deficiency syndrome*, has been observed not only after withdrawal of L-dopa and dopamine receptor agonists such as bromocriptine but also after withdrawal of the N-methyl-D-aspartate type–glutamate receptor antagonist, amantadine. The withdrawal of all anti-Parkinsonian medication, previously advocated as a *drug holiday* and thought to be helpful for some patients, is discouraged because of the danger of precipitating the malignant dopamine deficiency syndrome (Table I).

The akinetic crisis of patients with Parkinson's disease is another condition that resembles neuroleptic malignant syndrome. These crises are also characterized by rigidity, hyperthermia, and even elevated serum creatine kinase. Risk factors for akinetic crises include intercurrent infections, exogenous heat loads, and fluid restriction. These may reflect poor general care of patients with Parkinson's disease and also predispose to irregular drug intake and thus a malignant dopamine deficiency syndrome. Akinetic crises have long been known to respond well to intravenous amantadine, confirming a role for interactions of dopaminergic and gluta-

matergic systems in the pathophysiology of all three conditions, neuroleptic malignant syndrome, malignant dopamine deficiency syndome, and akinetic crisis (Kornhuber and Weller, 1994).

### Natural Course

Malignant dopamine deficiency syndrome and akinetic crisis are serious conditions that may often be overlooked. This is because notably elderly patients are often admitted to and treated in nonneurological institutions and because complications such as pneumonia and urinary tract infections are common in this group of patients. Early recognition of the condition is associated with a good outcome.

### Principles of Therapy

Symptomatic management corresponds to that of neuroleptic malignant syndrome. Malignant dopamine deficiency syndrome and akinetic crisis are commonly iatrogenic disorders and illustrate the need for sufficient dopaminergic medication that needs to be (re)instituted. Prevention includes appropariate measures to ensure regular and proper intake of prescribed medications.

### Practical Management

There are no controlled trials, but clinical experience suggests that intravenous amantadine (3 × 200 mg/day) is effective in reversing the symptoms of akinetic crisis. By extrapolation, this should also hold true for malignant dopamine deficiency syndrome. Dopaminergic

TABLE I    Clinical Features and Management of Hyperthermia Syndromes

|  | Clinical features | Management |
| --- | --- | --- |
| Malignant hyperthermia | Hyperthermia, muscle spasms, tachyarrhythmia, cyanosis, myoglobinuria | Cessation of anesthesia or change of anesthetic agents and muscle relaxants; monitoring of body temperature, electrolytes (potassium), blood gases, and acid base and fluid balance; **dantrolene\*** |
| Neuroleptic malignant syndrome | Rigidity and other extrapyramidal signs, hyperthermia, reduced vigilance | *Necessary measures:* **Withdrawal of causative agent(s)\***; rehydration, subcutaneous heparin for prevention of deep vein thrombosis, lorazepam for agitation *Possible measures:* Amantadine, L-dopa, bromocriptine, lisuride, dantrolene |
| Malignant dopamine deficiency syndrome and akinetic crisis of Parkinson's disease | Signs of underlying disease, hyperthermia, reduced vigilance, dehydration | Rehydration, subcutaneous heparin for prevention of deep vein thrombosis, **reinstitution of anti-Parkinson medications\*** |

medication in patients with Parkinson's disease should be readjusted. Symptomatic therapy includes rehydration and prevention of internistic complications, in essence as outlined for neuroleptic malignant syndrome. Similar to neuroleptic malignant syndrome, anticholinergic agents should not be administered to patients with malignant dopamine deficiency syndrome or akinetic crisis.

# REFERENCES

Baur, C. P., Bellon, L., Felleiter, P., Fiege, M., Fricker, R., Glahn, K., Heffron, J. J., Herrmann-Frank, A., Jurkat-Rott, K., Klingler, W., Lehane, M., Ording, H., Tegazzin, V., Wappler, F., Georgieff, M., and Lehmann-Horn, F. (2000). A multicenter study of 4-chloro-m-cresol for diagnosing malignant hyperthermia susceptibility. *Anesth. Analg.* **90**, 200–205.

Brandt, A., Schleithoff, L., Jurkat-Rott, K., Klingler, W., Baur, C., and Lehmann-Horn, F. (1999). Screening of the ryanodine receptor gene in 105 malignant hyperthermia families: novel mutations and concordance with the in vitro contracture test. *Hum. Mol. Genet.* **8**, 2055–2062.

Carbone, J. R. (2000). The neuroleptic malignant and serotonin syndromes. *Emerg. Med. Clin. North. Am.* **18**, 317–325.

European Malignant Hyperthermia Group (EMHG). (1984). A protocol for the investigation of malignant hyperpyrexia (MH) susceptibility. *Br. J. Anaesth.* **56**, 1267–1269.

Fleischhacker, W. W., Unterweger, B., Kane, J. M., and Hinterhuber, H. (1990). The neuroleptic malignant syndrome and its differentiation from lethal catatonia. *Acta Psychiatr. Scand.* **81**, 3–5.

Gurrera, R. J., Chang, S. S., and Romero, J. A. (1992). A comparison of diagnostic criteria for neuroleptic malignant syndrome. *J. Clin. Psychiatry* **53**, 56–62.

Jurkat-Rott, K., McCarthy, T., and Lehmann-Horn, F. (2000). Genetics and pathogenesis of malignant hyperthermia. *Muscle Nerve* **23**, 4–17.

Kornhuber, J., and Weller, M. (1994). Neuroleptic malignant syndrome. *Curr. Opin. Neurol.* **7**, 353–357.

McCarthy, T. V., Quane, K. A., and Lynch, P. J. (2000). Ryanodine receptor mutations in malignant hyperthermia and central core disease. *Hum. Mutat.* **15**, 410–417.

O'Dwyer, A. M., and Sheppard, N. P. (1993). The role of creatine kinase in the diagnosis of neuroleptic malignant syndrome. *Psychol. Med.* **23**, 323–326.

Rosebush, P. I., Stewart, T., and Mazurek, M. F. (1991). The treatment of neuroleptic malignant syndrome. Are dantrolene and bromocriptine useful adjuncts to supportive care? *Br. J. Psychiatry* **159**, 709–712.

Rosenberg, M. R., and Green, M. (1989). Neuroleptic malignant syndrome: review of response to therapy. *Arch. Intern. Med.* **149**, 1927–1931.

Sakkas, P., Davis, J. M., Janicak, P. G., and Wang, Z. (1991). Drug treatment of the neuroleptic malignant syndrome. *Psychopharmacol. Bull.* **27**, 381–384.

Saper, C. B., and Breder, C. D. (1994). The neurologic basis of fever. *N. Engl. J. Med.* **330**, 1880–1886.

Sternbach, H. (1991). The serotonin syndrome. *Am. J. Psychiatry* **148**, 705–713.

Trollor, J. N., and Sachdev, P. S. (1999). Electroconvulsive treatment of neuroleptic malignant syndrome: a review and report of cases. *Aust. N. Z. J. Psychiatry* **33**, 650–659.

Weller, M., and Kornhuber, J. (1992). Clozapine rechallenge after an episode of "neuroleptic malignant syndrome." *Br. J. Psychiatry* **161**, 855–856.

*Intensive Care in Neurology*

# CHAPTER 59
# Acute Intoxications

Friedrich von Rosen and John C. M. Brust

## INCIDENCE AND SEVERITY OF INTOXICATIONS

Acute exposure to potentially toxic substances occurs in 1%–2% of the population per year. Only 10%–20% of exposures are symptomatic; they are minor in approximately 90%, moderate in 10%, and major in less than 1%. More than 50% of poison exposures concern children less than 6 years of age. There is no overall gender difference; however, there is a male preponderance in illicit drug intoxication and a female preponderance in intentional overdose of therapeutic drugs.

Whereas the usual cause of intoxication in children less than 10 years of age is accidental exposure, drug abuse or attempted suicide is common in older children and adults. Accidental or iatrogenic drug overdose is most common in aged persons. The most comprehensive data source on poison exposures is the annual report of the American Association of Poison Control Centres (AAPCC) Toxic Exposure Surveillance System (TESS). Unfortunately, most fatalities and probably a high proportion of severe intoxications with common substances are not reported to the TESS and therefore not included in the report. A comparison of 1994 data shows 16,527 poisoning deaths in the death certificate statistics of the National Center of Health Statistics and 766 deaths reported by TESS (Hoppe-Roberts *et al.*, 2000). According to the 1998 AAPCC report, 480,000 persons with intoxications were managed in health care facilities, and 61,000 were admitted to critical care in the United States. The most common substances involved are cleaning agents, analgesics, cosmetics, cold medications, and plants. The emergency departments of community hospitals treat intoxications with ethanol, sedatives, antidepressants, drugs of abuse, analgesics, and oral antidiabetics. Traditional medicines and pesticides are leading causes of severe intoxications in many developing countries (Nhachi and Kasilo, 1992).

More than 16,000 fatal intoxications per year are estimated to occur in the United States and several thousands in Germany. Thirty to 50% of lethal poisonings are suicidal (Hoppe-Roberts *et al.*, 2000). Carbon monoxide, opiates, the combination of ethanol and illicit drugs, tricyclic antidepressants, cocaine, and hypnotics cause most lethal intoxications (Hoppe-Roberts *et al.*, 2000). Eighty to 90% of the victims are either found dead or die before reaching a hospital (Chafee-Bahamon, 1983). Among those reaching a hospital, acetaminophen, cardiovascular drugs, cocaine and other stimulants, cyclic antidepressants, hypnotics, irreversible monoamine oxidase (MAO) inhibitors, and salicylates cause most fatalities.

## SYMPTOMS, COMPLICATIONS, AND DIAGNOSTIC STRATEGIES

Any patient with unexplained nonfocal neurological symptoms may suffer from an acute intoxication. A high proportion of head injury victims are intoxicated, usually with ethanol and/or illicit drugs. A history of drug abuse, a psychiatric disorder, or a recent adverse life event can precede intentional overdose. Circumstantial evidence (syringe, drug container, farewell letter) may point to drug overdose.

Most toxic substances lead to obtundation or coma when taken in high doses. Other neurological and computed tomography (CT) signs can narrow the differential diagnosis (Table I). Substance-specific clinical symptoms and signs are present in only 5% of patients. The syndromatic combination of symptoms and signs characterizes intoxications with certain drug groups:

1. Sympathomimetic syndrome. Tachycardia, arterial hypertension, diaphoresis, hyperventilation, pupillary

**TABLE I**  Differential Diagnosis of Neurological Symptoms and Signs in Acute Intoxications

| | |
|---|---|
| Mydriasis | Sympathomimetic drugs (amphetamines, "ecstasy," clonidine, cocaine, LSD); withdrawal state (alcohol, opiates, and opioids) |
| | Mydriasis with absent light reflex: anticholinergic drugs (atropine, scopolamine, cyclic antidepressants, clozapine, glutethimide, antihistamine drugs); combination of carbamazepine and venlafaxine |
| Miosis | Barbiturates; bromide; chloral hydrate; opiates and opioids; phencyclidine; benzodiazepines; ethanol; cholinergic drugs (pilocarpine, nicotine, physostigmine, organophosphates); propoxyphene |
| Gaze paralysis and VOR suppression | Awake patient: botulism; phenytoin; Wernicke's encephalopathy |
| | Comatose patient: the former or barbiturates; benzodiazepines (rare); other CNS depressants; carbamazepine; alcohols; narcotics |
| Dystonia | Neuroleptic drugs; atypical neuroleptic drugs; metoclopramide; cyclic antidepressants; nefazodone; cocaine; phencyclidine; L-dopa and dopa-agonists; phenytoin; carbamazepine; felbamat; chloroquine; thallium; manganese |
| Chorea | Phenytoin; amphetamines; cocaine; cimetidine; L-dopa and dopaminergic agonistis; lithium; ethanol; toluene; manganese; type II pyrethroid insecticides |
| Tremor | Sympathomimetic drugs; methylxanthines; cocaine; amphetamines; antiepileptic drugs; cyclic antidepressants; SSRIs; lithium, cyclosporine; tacrolimus; arsenic; mercury; manganese; trichloroethylene; organic solvents; organophosphates; carbamates; scorpion venom, withdrawal state |
| Ataxia | Ethanol, methanol, ethylene glycol; antiepileptics; sedative/hypnotic agents; phencyclidine; gamma-hydroxybutyrate; organic solvents; lead; manganese; methylmercury; arsenic; ciguatoxin; saxitoxin |
| Fasciculations | Organophosphates (early sign); lithium; scorpion venom |
| Myoclonus | Bismuth; cadmium; chloralose; clozapine; methylbromide; phencyclidine; cyclic antidepressants; MAO inhibitors; tetraethyl lead; organic mercury; methaqualone |
| Increased muscle tone, hyperreflexia | Phenothiazine neuroleptic drugs; cyclic antidepresants; phencyclidine; metaqualone; irreversible MAO inhibitors; serotonin syndrome (see text); strychnine; thyroid hormones; scorpion venom |
| Decreased muscle tone, hyporeflexia | Sedative/hypnotic drugs in high doses (exception metaqualone); muscle relaxants; organophosphates; carbamate; botulinus toxin |
| Seizures | Withdrawal (ethanol, benzodiazepines, barbiturates); carbon monoxide; cyclic antidepressants; clozapine; cocaine; isoniazide; theophylline and other methyxanthines; amphetamines; "ecstasy"; secondary to hypoglycemia; less common local anesthetics; MAO inhibitors; neuroleptic drugs; antiepileptic drugs; salicylate; antihistamines; antiarrhythmic drugs; baclofen; bupropione; chloroquine; cyclosporine; camphor; ergotamines; lead, lithium; LSD, methanol; phencyclidine; strychnine; tacrolimus; many other drugs |
| Brain swelling | Salicylate; methanol; ethylene glycol; lead (in children); zinc; vitamin A; valproate; "ecstasy"; secondary to acute hepatic failure (e.g., after poisoning with acetaminophen, paraquat, *Amanita phalloides*, potassium dichromate); secondary to hypoventilation and hypoxia (e.g., opioids) |
| Hypoventilation | Sedative and hypnotic drugs (barbiturates, benzodiazepines, opiates); alcohols; organophosphates; secondary to deep coma in many intoxications |
| Hyperventilation | Substances leading to metabolic acidosis (e.g., methanol, ethylene glycol; paraldehyde, salicylate, carbon monoxide, dinitrophenol); secondary to hypoxia in toxic fume inhalation |
| Hypothermia | Alcohol (ethanol, methanol); gamma-hydroxybutyrate; cholinergic drugs; secondary to hypotonia and coma in CNS depressant intoxications; NSAIDs (rare) |
| Hyperthermia | Anticholinergic drugs; salicylate; NSAIDs; amphetamine; LSD; cocaine; phencyclidine; mescaline; "ecstasy" (MDMA); *Datura*; MAO inhibitors; methaqualone; cyclic antidepressants; phenothiazine and butyrophenons (malignant neuroleptic syndrome, Chapter 58); diphenhydramine; thyroid hormones; withdrawal of dopaminergic medication; scorpion venom |

CNS, central nervous system; LSD, lysergieacid dietylonical. MDMA, 3,4-methylenedioxymethamphetamine; MAO, monoaminooxidase; NSAID, nonsteroidal anti-inflammatory drugs, SSRI, serotonin reuptake inhibitors.

dilation, confusion, and seizures: cocaine or amphetamines, methylxanthines, irreversible MAO inhibitors.

2. Cholinergic (muscarinic, nicotinic, or both) syndrome. Muscarinic symptoms are bradycardia, increased salivation and bronchial and intestinal secretions, diaphoresis, emesis, defecation, urination, and miosis. Nicotinic symptoms are tachycardia, arterial hypertension, muscle fasciculation, and muscle paralysis. Certain insecticides and tobacco ingestion (in children) cause a nicotinic syndrome. Nicotinic symptoms prevail over muscarinic symptoms in severe pyridostigmine intoxication. Causes of a muscarinic syndrome are acetylcholine, pilocarpine, carbachol, pyridostigmine, and some mushrooms. Organophosphate and carbamate insecticides produce a combination of muscarinic and nicotinic symptoms.

3. Anticholinergic syndrome. Tachycardia, dry skin, hyperthermia, mydriasis with loss of pupil response, decreased secretions, urinary retention, delirium, hallucinations: cyclic antidepressants, antihistamines, anticholinergic anti-Parkinson agents, neuroleptic drugs, scopolamine, belladonna alkaloids, *Datura stramonium* (jimson weed).

4. Opiate syndrome. Hypotension, hypoventilation, miotic but reactive pupils, central nervous septum (CNS) depression: morphine, heroin, and other opiates.

5. Sedative-hypnotic syndrome. Somnolence, stupor, or coma; decreased muscle tone combined with hypoventilation and hypotension: sedatives or hypnotics including ethanol.

Spontaneous hyperventilation is observed in many neurological emergencies. When combined with a pronounced metabolic acidosis (and an increased anion gap), diabetic ketoacidosis, uremia, lactic acidosis, and intoxications with ethanol, methanol, ethylene glycol, paraldehyde, or salicylic acid are likely causes.

Extensor and flexor posturing is most commonly seen in coma patients after structural brain lesions, but it may also occur in cases of sedative drug poisoning (Greenberg and Simon, 1982). Partial or complete suppression (sometimes asymmetric) of oculocephalic and vestibulo-ocular reflexes is usually interpreted as a sign of structural brainstem damage (e.g., after basilar artery thrombosis), but it is also described in certain intoxications (Table I). Only drugs with anticholinergic activity directly abolish reactivity of the pupils. Unreactive pupils after overdose of other drugs are secondary to anoxia-ischemia or to terminal CNS depression. Asymmetrical ocular signs (pupillary asymmetry, skew deviation) and ocular bobbing suggest structural brain damage.

## NONSPECIFIC THERAPEUTIC MEASURES

Acute death after intoxication is most often caused by apnea, pulmonary edema, heart failure, or severe cardiac arrhythmia. The initial treatment of the poison victim aims at the stabilization of cardiopulmonary function and on the prevention of irreversible organ failure. The vulnerability of the brain to even short interruptions of energy metabolism must always be considered.

Arterial hypotension with insufficient cerebral blood flow, severe hypoventilation or hypoxemia, blockade of either the oxygen transport to the organs (carbon monoxide, methemoglobin-generating substances) or the respiratory chain (cyanide, hydrogen sulfide), and fuel shortage (severe hypoglycemia) may cause irreversible brain damage within minutes. Selective neurotoxicity with structural damage to the central or peripheral nervous system is rare in acute intoxications. Methanol, ethylene glycol, the glutamatergic agonist domoic acid (a contaminant of shellfish, especially mussels), and 1-methyl-4-phenyl-1,2,3,6-tetrahydropyridine (MPTP, an unintended by-product in the manufacture of a meperidine-like opiate drug) are important examples of acute neurotoxicity (Krogsgaard-Larsen and Hansen, 1992; Tipton and Singer, 1993). The treatment of acute poisoning should include the following steps:

1. Stabilization of vital functions and CNS-oriented emergency therapy
2. Use of specific antidotes
3. Poison removal
4. Observation
5. Supportive intensive care

## STABILIZATION OF VITAL FUNCTIONS

Stabilization of respiration and circulation follows the established "ABC" resuscitation criteria: immediate life-threatening problems of airway, breathing, and circulation must be identified and treated. Vital signs must be monitored closely. An intravenous line should be placed. The neurological examination includes assessment of consciousness, respiratory pattern and depth, reactivity and equalitiy of the pupils, spontaneous and reflex eye movements, motor reactions, and tendon reflexes. The oculocephalic maneuver should not be performed unless cervical trauma has been excluded.

Initial therapy comprises glucose (1 ml/kg of 50% glucose in adults, 4 ml/kg of a 10% solution in infants and children), thiamine (100 mg) plus multivitamins IV, and oxygen (5 L/min by nasal mask or 100% oxygen in suspected carbon monoxide or cyanide intoxication). Naloxone, 0.4 mg, in repetitive doses up to 2 mg or more IV is given to comatose patients with suspected opiate poisoning.

Status epilepticus should initially be treated with slow intravenous administration of a benzodiazepine. Diazepam in 5- to 10-mg doses can be given up to 30 mg. Lorazepam (initial dose 2–4 mg and cumulative dose up to 10 mg) is an appropriate alternative. In intoxications that severely interfere with cerebral energy metabolism, specific therapy is most important. In severe isoniazid intoxication with status epilepticus, pyridoxine up to 5 g IV must be given over 5–30 minutes. Status epilepticus in severe theophylline overdose may be refractory to benzodiazepines but reactive to drug removal by hemoperfusion.

## SPECIFIC ANTIDOTES

A small, but important minority of intoxications can be treated with antidotes (26,000 annual cases in the 1998 AAPCC reports (Litovitz et al., 1999). Antidote effectiveness is proven for the following intoxications with level 1 evidence: acetaminophen (acetylcysteine solution), iron (deferoxamine mesylate), methanol or ethylene glycol (ethanol solution or fomepizole), acute opioid poisoning (naloxone), anticholinergic agents (physostigmine salicylate), and organophosphate insecticides (pralidoxime chloride). Guidelines for the stocking of emergency antidotes in the United States have recently been published (Dart *et al.*, 2000). High-urgency life-saving antidotes (e.g., cyanide kit) have to be used immediately, low-urgency life-saving antidotes (e.g., acetylcysteine solution) are given after hospital admission. The use of supportive antidotes (e.g., flumazenil) is optional. Antidote use must always be combined with adequate monitoring and supportive treatment. The half-life of the toxin may be much longer than the half-life of the antidote, making close monitoring and repetitive or continuous application necessary.

## POISON ELIMINATION

Ingestion is by far the most common route of poison intake. Induced emesis, gastric lavage, single- or multiple-dose activated charcoal, cathartics, and whole bowel irrigation are methods of gut decontamination. The American Academy of Clinical Toxicology in collaboration with the European Association of Poisons Centers and Clinical Toxicologists have recently reviewed the evidence and published guidelines for the use of each method of gut decontamination (American Academy of Clinical Toxicology, 1999; Barceloux *et al.*, 1997; Chyka and Seger, 1997; Krenzelok *et al.*, 1997; Tenenbein, 1997; Vale, 1997).

Gastric emptying using induced emesis or gastric lavage has traditionally been one of the first therapeutic measures in the emergency department. Although time-honored, data to support or exclude the benefit of these measures are insufficient. The position statements discourage the routine administration of syrup of ipecac in the emergency department (Krenzelok *et al.*, 1997) and recommend gastric emptying only after ingestion of a potentially life-threatening amount of a toxin if the poison intake was within 60 minutes (Vale, 1997). Contraindications to both methods are ingestion of a corrosive substance (acid or alkali) or a substance that causes severe lung damage when aspirated (e.g., some hydrocarbons) and patients at risk for gastric hemorrhage or perforation. Gastric lavage should only be used with protected airways.

The dose recommendations for syrup of ipecac are 5–10 ml in children between 6 and 12 months, followed by 120–240 ml water, 10 ml in children 1–12 years, followed by 120–240 ml water, 15 ml in persons 12 years or older, followed by 240 ml water. Emesis occurs in 90% of patients and may continue for 20–30 minutes. After pretreatment with atropine, 0.5 mg IV, gastric lavage is performed using a 36–40 French gastric tube in adults and a 28–32 French tube in children. After oral introduction of the tube, the stomach contents are aspirated and saved for analysis. The stomach is repeatedly filled with portions of 100–300 ml lukewarm tap water or isotonic saline (mandatory in children) until the lavage fluid is clear (at least 5 L in total).

Single-dose activated charcoal may be considered if a patient has ingested a potentially toxic amount of an absorbable toxin up to 1 hour previously (Chyka and Seger, 1997). Although many studies have demonstrated that multiple-dose activated charcoal increases the removal of many different substances significantly, an effect on morbidity or mortality of poisoned patients has not been proven. The current recommendation is that multiple-dose activated charcoal should be considered after the ingestion of a life-threatening amount of carbamazepine, dapsone, phenobarbital, quinine, or theophylline (American Academy of Clinical Toxicology, 1999). Activated charcoal does not bind to alcohols, iron, or lithium. An initial dose of 0.5–1.0 g/kg is given and can be repeated every 2–4 hours. The airway should be intact or protected. Intestinal obstruction is a contraindication.

There are no definite indications for the use of cathartics in the management of the poisoned patient. If a cathartic is used, it should be limited to a single dose (Barceloux *et al.*, 1997). Sorbitol (1–2 ml/kg of a 70% solution) is the preferred substance, whereas saline cathartics (magnesium citrate, magnesium sulfate, sodium sulfate) have more side effects. Whole bowel irrigation (WBI) with high-molecular-weight polyethylene glycol and isoosmolar electrolyte solutions (PEG-ELS), a procedure also used in preparation of a coloscopy, is a safe and efficacious method of gut decontamination. PEG-ELS is instilled over a nasogastric tube with the patient in a sitting position. The usual dose is 2 L/h in adults and 0.5 L/h in children for 2–6 hours until the rectal effluent is similar to the infusate. There are no controlled clinical studies establishing indications for the use of WBI. WBI may be considered for potentially toxic ingestions of sustained-release or enteric-coated drugs and is an option for potentially toxic ingestions of iron, lead, zinc, or packets of illicit drugs (Tenenbein, 1997).

Forced diuresis, hemodialysis, and hemoperfusion each enhance the elimination of certain toxins. Anecdotal reports or small case series describe the removal of many different drugs or chemicals with hemodialysis or hemoperfusion (Ellenhorn *et al.*, 1998). The effect of these treatment modalities on the outcome of specific intoxications is not proven by controlled studies.

Hemodialysis is used most often for severe lithium, theophylline, ethylene glycol, salicylate, and methanol intoxications. Hemoperfusion is used for poisoning with theophylline, long-acting barbiturates, and meprobamate.

## DRUG SCREENING AND TOXICOLOGICAL ANALYSIS

In patients with suspected poisoning or altered mental state of unknown etiology, emergency laboratory measurements should be obtained for complete blood count, arterial blood gases, carboxyhemoglobin (identifying CO intoxication), methemoglobin, serum electrolytes, osmolality, glucose, urea, creatinine, and prothrombin time. Anion gap and osmolar gap are calculated using serum electrolyte and blood gas results.

For toxicological analysis >10 ml ethylenediamine tetraceceticacid (EDTA) or heparinized blood, >10 ml native blood, >50 ml first urine, and 50–100 ml of the first gastric lavage portion are drawn. Rapid quantitative drug screening usually comprises barbiturates, benzodiazepines, amphetamines, cocaine, and opiates. Phencyclidine is part of screening sets in the United States, but not in Europe. A more comprehensive "general toxicology" screen should be obtained in seriously ill patients, looking for intoxications with specific treatment options (acetaminophen, ethanol, ethylene glycol, methanol, lithium, salicylate, and theophylline). Many laboratories routinely offer quantitative measurements of digoxin, carbamazepine, phenytoin, theophylline, and valproate. Overall, it is rare to find an unsuspected drug in a poisoned patient.

## MONITORING AND INTENSIVE CARE

Patients with the following symptoms or problems should be monitored and treated in an intensive care unit:

- Several seizures or status epilepticus
- Stupor or coma
- Respiratory insufficiency
- Pulmonary edema
- Prolongation of the QRS complex or severe cardiac arrhythmias
- Arterial hypotension
- Suspected intoxication with a critical amount of a substance causing delayed symptoms (acetaminophen, diquat, ethylene glycol, heavy metals, MAO inhibitors, methanol, paraquat)

After attempted suicide a psychiatric consultant should be involved, addressing the possible transfer to a psychiatric ward.

## PRACTICAL MANAGEMENT OF SPECIFIC INTOXICATIONS

### Analgesics

#### Acetaminophen

Acetaminophen is a widely used analgesic and antipyretic over-the-counter (OTC) medication. Intentional overdose is very common in Great Britain and the United States, accidental overdose occurs mainly in children. The case fatality rate is low. Poor nutrition, alcohol abuse, acquired immuno deficiency syndrome (AIDS), or concurrent use of drugs inducing the cytochrome P-450 enzymes increases the individual susceptibility to hepatic damage.

1. Uptake: Oral or rectal; toxic dose 5–15 g. Single acute toxic threshold dose for severe liver damage in adults 150–250 mg/kg.
2. Mechanism: Accumulation of toxic metabolites when normal metabolic pathways (glutathione dependent) are saturated.
3. Symptoms and signs: Initially gastrointestinal symptoms; after a latent period of 18–72 hours, abdominal pain, liver swelling, signs of hepatic failure; impairment of renal function. A minority of severe acetaminophen intoxications leads to fulminant hepatic failure after 3–6 days with a mortality of 50%. Hepatic encephalopathy with brain edema, diffuse bleeding, or septicemia may lead to death.
4. Treatment: Gastrointestinal decontamination (lavage, activated charcoal) within 1–4 hours. Plasma acetaminophen concentrations should be monitored during the first hours (Rumack et al., 1981). Antidote therapy with IV N-acetylcysteine in high-dose intoxications (plasma acetaminophen concentration above the lower treatment line corresponding to 1 mmol/L (150 mg/l) at 4 hours or an ingestion >100 mg/kg body weight) as early as possible (level I evidence for use within 8 hours, level III evidence for use later than 15 hours after exposure). The relative superiority of one of three different conventional dosing regimens (20-hour IV, 48-hour IV, and 72-hour oral dosing regimens) or a shorter 36-hour oral dosing regimen (Woo et al., 2000) has not been demonstrated. The 20-hour intravenous regimen starts with loading dose of 150 mg/kg in 200 ml 5% dextrose over 15 minutes, followed by a continuous infusion of 50 mg/kg in 500 ml 5% dextrose over 4 hours and 100 mg/kg in 1000 ml 5% dextrose over 16 hours. Diphenhydramine is given in cases of systemic anaphylactic reaction against intravenous N-acetylcysteine. Oral methionine, 2.5 g every 4 hours (four doses), is effective if started within 10 hours after

toxic exposure, but it has no advantage over N-acetyl-cysteine. The role of cimetidine as an adjunctive therapy is unproven. Emergency liver transplantation is indicated in patients with a poor prognosis under conservative treatment.

### Salicylates

Salicylates are widely used analgesic, antipyretic, anti-inflammatory, and anticoagulant substances. They are common causes of accidental (children) and intentional poisoning. Severe intoxications are frequently misdiagnosed.

1. Uptake: Oral, rarely dermal (children). Severe poisoning with >300 mg/kg.
2. Mechanism: Salicylates increase the sensitivity of respiratory centers, uncouple oxidative phosphorylation, and inhibit carbohydrate and lipid metabolism.
3. Symptoms and signs: Neurological symptoms are hypacusis (cochlear hearing loss), tinnitus, vertigo, headache, nausea and vomiting, confusion, agitation, delirium, hallucinations, and later lethargy, coma, and seizures. Brain swelling may occur. A metabolic acidosis compensated by hyperventilation and respiratory alkalosis progresses to a combined and often severe metabolic and respiratory acidosis. Coma, hyperpyrexia, cardiovascular and respiratory failure, pulmonary edema, coagulation disturbances, bleeding, hypoglycemia, and inappropriate secretion of antidiuretic hormone are common in life-threatening intoxications.
4. Treatment: There are no antidotes. The coingestion of acetaminophen should be excluded. Gastric lavage within 60 minutes; the use of multiple-dose activated charcoal is not supported by evidence. Urinary alkalinization using sodium bicarbonate and monitoring the urine pH between 7.5 and 8.5. Hemodialysis is recommended at a serum salicylate level >100–120 mg/dl. Hypoglycemia, hypokalemia, and metabolic acidosis should be corrected.

### Anticonvulsants

Intoxications with anticonvulsant drugs are seen in epileptic patients and their family members. Intentional overdose, chronic intoxication caused by pharmacokinetic interactions or nonlinear pharmakokinetics (phenytoin), and accidental ingestion (children) all occur. Sedation, cognitive decline, nausea, vertigo, and ataxia can be caused by most anticonvulsants. Phenobarbital, primidone, carbamazepine, phenytoin, and probably topiramate have a smaller therapeutic safety margin than valproate, benzodiazepines, and most newer antiepileptic substances. Anticonvulsant overdose can paradoxically cause seizures or status epilepticus.

### Carbamazepine

Carbamazepine is used as an anticonvulsant, mood-stabilizing, and antineuropathic drug. Overdose is not frequent but may be fatal.

1. Uptake: Oral (severe intoxications after single dose of 4–20 g, leading to plasma levels above 20 μg/ml).
2. Mechanism: Inhibition of membrane sodium conductance.
3. Symptoms and signs: Neurological symptoms progress from ataxia, nystagmus, and drowsiness to either tremor, agitation, dyskinesia, and hallucinations or seizures and coma. Nausea and vomiting are common. Severe poisoning causes deep coma, myocardial depression, and hypoventilation. Inadequate secretion of antidiuretic hormone can occur.
4. Treatment: No antidotes. Gastric lavage within 60 minutes. Repetitive activated charcoal. Elimination methods are ineffective. Complete heart block can be treated with atropine, 0.5–1.0 mg IV, or a temporary cardiac pacemaker. Diazepam or lorazepam for the treatment of seizures. Supportive care.

### Phenytoin

Phenytoin intoxications are common in epileptic patients; the case fatality rate is low. Intoxications often develop subacutely because of the nonlinear pharmakokinetics and interactions with other drugs.

1. Uptake: Oral, intravenous. Lethal dose >7500 mg.
2. Mechanism: Inhibition of sodium conductance.
3. Symptoms and signs: Nystagmus, ataxia, nausea, and drowsiness in mild intoxications; confusion, hyperreflexia, severe ataxia, hypermetabolism, and arrhythmias (bradycardia, conduction block, ventricular rhythm, ventricular fibrillation) occur in more severe intoxications. Seizures can occur, and hyperglycemia and nonketotic coma are described.
4. Treatment: No antidote. Gastric lavage within 1 hour after ingestion; the value of activated charcoal is unproven. Methods to enhance elimination are ineffective. Routine cardiologic monitoring is not needed in uncomplicated intoxications. Complete heart block should be treated with atropine, 0.5–1.0 mg IV, or placement of a temporary pacemaker. Diazepam or lorazepam for seizures. Fluids and dopamine to treat hypotension.

## Valproic Acid and Sodium Valproate

The acute toxicity of valproate is low. Valproate can unpredictably cause hepatic necrosis (especially in debilitated children), hyperammonemia, hepatic encephalopathy, and acute pancreatitis.

1. Uptake: oral, rarely IV (iatrogenic).
2. Mechanism: Increases the action of GABA, inhibits voltage-gated sodium and potassium channels.
3. Symptoms and signs: Nausea, miosis, tremor, somnolence, confusion, progressing to stupor or coma, respiratory depression, rarely seizures.
4. Treatment: No antidote. Gastric lavage and activated charcoal after ingestion of sustained-release valproate. Other elimination measures are not indicated. Seizure treatment with diazepam or lorazepam. Supportive treatment.

## Newer Antiepileptic Agents

Felbamate, gabapentine, lamotrigine, levotiracetam, tiagabine, topiramate, and vigabatrine were introduced between 1990 and 2000. Symptoms and signs of overdose are similar but usually less severe to those of traditional antiepileptic drugs. Treatment is supportive.

## Antidepressants

### Cyclic Antidepressants

Used for treatment of depression and chronic pain syndromes, cyclic antidepressants are frequently involved in suicide attempts. Eight to 15% of drug intoxications are caused by cyclic antidepressants. Safety margin and lethal doses vary widely between substances. The classic tricyclic antidepressants (TCAs) amitriptyline, desipramine, dothiepine, doxepine, and nortriptyline may cause severe intoxications. Most newer agents are less toxic (e.g., trazodone and mianserin).

1. Uptake: Oral.
2. Mechanism: Reuptake blockers of norepinephrine, serotonin, and dopamine in the CNS; central and peripheral anticholinergic activity; quinidine-like effect on the myocardium.
3. Symptoms and signs: Combinations of central nervous, autonomic, and cardiac symptoms developing within 30 minutes to 6 hours. Peripheral anticholinergic effects include mydriasis, dry mouth, urinary retention, tachycardia, and hyperthermia with dry skin. CNS symptoms range from confusion, agitation, and hallucinations to deep coma. Duration of coma is short. Seizures are common in severe poisoning, occur early after ingestion, are brief, and are usually self-terminating. Cardiovascular toxicity predominates in

life-threatening intoxications. Prolongation of the Q-T time, broad QRS complexes, severe arrhythmias, and decreased myocardial contractility may lead to heart failure and death, usually within 6 hours. Hypotension is the result of both peripheral vasodiltation and loss of cardiac inotropy.

4. Treatment: No antidote. Intravenous line and cardiac monitoring. Induced emesis should be avoided. Large-volume gastric lavage (>5 L) is recommended. Single-dose or multiple-dose activated charcoal decreases blood levels of TCAs and can be considered in patients with normal bowel sounds. Hemodialysis and hemoperfusion are not effective. Specific therapies for cardiac disturbances include lidocaine for tachyarrhythmias, temporary cardiac pacing for bradycardia, norepinephrine for hypotension, and extracorporeal circulation in refractory circulatory failure (Goodwin et al., 1993).

### Selective Serotonin Reuptake Inhibitors (SSRIs)

Fluoxetine, fluvoxamine, paroxetine, sertraline, and citalopram are SSRIs with little activity on other transmitter systems. They are used for the treatment of depression, anxiety, and obsessive-compulsive behavior. SSRIs are rarely fatal when taken alone. Almost all fatalities involving SSRIs have involved coingestion of other substances.

1. Uptake: Oral.
2. Mechanism: selective serotonin reuptake inhibition at the presynaptic membrane.
3. Symptoms and signs: No or minor symptoms in moderate overdose (less then 30 times the common daily dose), drowsiness, tremor, nausea and vomiting in more severe overdose, epileptic seizures, electrocordiogun (ECG) changes, and decreased consciousness after ingestion of very high amounts (Barbey and Roose, 1998). The coingestion of a SSRI and an MAO inhibitor, lithium, carbamazepine, or a dopaminergic agent may cause a serotonin syndrome characterized by shivering, tremor, myoclonus, rigidity, fever, tachycardia, sweating, confusion, agitation, seizures, and coma. The serotonin syndrome has less frequently followed use or overdose of a single serotonergic agent (Curry et al., 1998; Sternberg, 1991).
4. Treatment: Decontamination within 60 minutes, supportive treatment. Intensive care monitoring for severe serotonin syndrome.

### Novel Antidepressant Drugs

Venlafaxine and mirtazapine are noradrenaline and serotonin reuptake inhibitors (NARI) with antidepressant and antianxiety effect. A serotonin syndrome may develop after overdose of venlafaxine alone or in combination with an MAO inhibitor or an SSRI. Fatal intoxications have been described. Nefazodone is a phenylpiperazine antidepressant with pharmacological actions different from TCA and SSRI. It causes a down-regulation of 5-HT$_{2a}$ receptors. Intoxications are usually mild; the most common manifestations are drowsiness, nausea, and dizziness; hypotension may occur (Benson et al., 2000).

### MAO Inhibitors

Irreversible MAO A and B inhibitors (tranyl-cypromine, phenelzine, isocarboxazid) are second-line antidepressants. The antidepressant moclobemide is a reversible selective MAO A inhibitor. The selective MAO B inhibitor selegeline is used in the treatment of Parkinson's disease. Poisoning with this agent is rare.

1. Uptake: Oral, lethal dose for irreversible MAO inhibitors 4–8 mg/kg.
2. Mechanism: Accumulation of monoaminergic neurotransmitters.
3. Symptoms: Delayed and prolonged sympatho-mimetic syndrome with mydriasis, agitation, tremor, muscular rigidity, myoclonus, seizures, tachycardia, hypertension, and hyperthermia, in severe poisoning, coma and hypotension. Rhabdomyolysis, disseminated coagulation, and acute renal failure are serious compli-cations. Moclobemide overdose is usually benign. Fatal-ities have followed concomitant use of MAO inhibitors and other sympathomimetic agents such as ephedrine or phenylpropanolamine. Fatality can also occur in patients taking MAO inhibitors who ingest foods con-taining tyramine (e.g., aged cheese, smoked meat, or red wine). The combination of moclobemide and a TCA, an SSRI, or a NARI can cause life-threatening cardiac arrhythmia or a severe serotonin syndrome. Fatal sero-tonergic symptoms have also followed concomitant administration of an MAO inhibitor and the opiate analgesic meperidine (Brown and Linter, 1987).
4. Treatment: No antidote. Gastric lavage and acti-vated charcoal to be considered within 60 minutes. Cardiac monitoring, hospital observation for at least 24 hours. Sodium nitroprusside or phentolamine for severe hypertension; hypotension in severe poisoning should be treated with high-dose dopamine or norepinephrine. External cooling, sedation, and dantrolene, 1–2.5 mg/kg IV, in hyperthermic patients. Benzodiazepines and phenytoin for seizure treatment.

### Neuroleptic Drugs

Phenothiazine (e.g., chlorpromazine), thioxanthene, butyrophenone (e.g., haloperidol), benzamide (e.g., sul-prite), and benzisoxazole (risperidone) compounds are used primarily as antipsychotics. Intentional or acci-dental overdose is common in psychiatric patients and members of their households.

1. Uptake: Oral, rarely intramuscular or intravenous.
2. Mechanism: Blockade of postsynaptic dopamine receptors; central and peripheral anticholinergic effects of some phenothiazines; effects on other transmitter systems; quinidine-like effects on the heart (some phenothiazines).
3. Symptoms and signs: Confusion, somnolence, coma, hypotension, hypoventilation, and cardiac arrhythmias. Acute dystonic reactions (up to 23%). Akathisia (up to 20%), Parkinsonism, commonly observed in older patients, is reversible with removal of the drug. Tardive dyskinesia (choreiform or dystonic) is frequently seen after long-term treatment and may persist or even worsen after removal of the drug. Muscle rigidity and severe extrapyramidal symptoms, hyper-thermia, signs of autonomic dysregulation, depressed consciousness, and elevation of serum creatinine kinase characterize the neuroleptic malignant syndrome, which develops in 0.1%–2.5% of patients exposed to neu-roleptic drugs, usually within 24–72 hours (Kornhuber and Weller, 1994).
4. Treatment of acute intoxications: Early gastro-intestinal decontamination; benzodiazepines and pheny-toin for seizure control; lidocaine and phenytoin to treat tachyarrhythmias; intravenous fluid and pressor agents for hypotension. Hemodialysis and hemop-erfusion are ineffective. Neuroleptic malignant syn-drome is treated with cooling, benzodiazepines, the central dopamine agonist bromocriptine, and dantro-lene (which inhibits calcium release from muscle sar-coplasmic reticulum). Persistent tardive dyskinesia can be treated with agents that deplete CNS dopamine stores (e.g., reserpine).

### Clozapine

Atypical neuroleptic for patients with refractory schizophrenia. The use is limited by a risk of 1%–2% of severe neutropenia. Clozapine has no or very little extrapyramidal effects and is used in low doses for the treatment of psychosis in Parkinson's disease.

1. Uptake: Oral, rarely IV. A first exposure to as little as 400 mg in adults and 50–200 mg in children can cause deep coma and may be life-threatening. In patients taking clozapine, doses of >2000 mg are life-threatening; 12% of these patients die.
2. Mechanism: Clozapine has central and peripheral affinity to many receptor systems.
3. Symptoms and signs: Anticholinergic symptoms (mydriasis, urinary retention, dry skin, decreased

bowel movements, tachycardia), hypotension, cardiac arrhythmia, hypersalivation, CNS suppression or agitation, confusion and hypertension. Generalized seizures are common. Respiratory depression or cardiac arrhythmia leads to death.

4. Treatment: No specific antidote. Close monitoring of respiration and circulation. Physostigmine IV for anticholinergic symptoms. Cardiac arrhythmias can be treated with digoxin and potassium supplementation but not with quinidine or procainamide. Fluids and dopamine for hypotension; epinephrine can lead to worsening of hypotension. Benzodiazepines and phenytoin for seizure control.

### Olanzapine

Olanzapine overdose is much less dangerous than clozapine intoxications.

1. Uptake: Ingestion, usually suicidal.
2. Mechanism: Atypical antipsychotic drug with affinity to many dopamine, serotonin and other neurotransmitter receptor subtypes. Low affinity to $D_2$-receptors. Little extrapyramidal side effects in the usual dose range.
3. Symptoms and signs: Agitation, dysarthria, dystonia, akathisia, somnolence, coma, seizures, tachycardia, hypertension, or hypotension. A neuroleptic malignant syndrome may occur. Fatalities are rare but have been reported after doses of 450 mg or more. Death is caused by apnea or cardiac arrhythmia.
4. Treatment: No antidote available. Gastric lavage and activated charcoal. Cardiocirculatory monitoring. Supportive treatment.

## Sedative-Hypnotics

### Barbiturates

Barbiturates are used as hypnotics, anesthetics, and anticonvulsants. Abuse, dependency, intentional overdose, and fatalities were common until barbiturates were replaced by the safer benzodiazepines as first-line hypnotics. Short-acting barbiturates are still abused as "recreational" drugs.

1. Uptake: Oral, rarely intravenous, except in treating status epilepticus.
2. Mechanism: Agonistic effect on the GABA A-receptor, causing CNS depression.
3. Symptoms and signs: CNS depression ranging from somnolence with ataxia, nystagmus, and vertigo to deep coma with loss of all brainstem reflexes (except the pupillary light response). Hypoventilation and hypotension are the major problem in barbiturate poisoning.

4. Treatment: Gastric decontamination within 1 hour and repeated administration of activated charcoal are optional. Artificial ventilation and blood pressure support may be necessary. Fluid replacement is preferable to pressors, which reduce cardiac output; blood pressure is often restored with institution of assisted respiration. Forced diuresis and urinary alkalinization are used for long-acting barbiturates unless there is anuria secondary to shock. Hemodialysis is recommended for deeply comatose patients intoxicated with long-acting compounds (level 3 evidence, Lindberg *et al.*, 1992). Supportive intensive care treatment in severe poisoning.

### Benzodiazepines

Benzodiazepines are widely used as tranquilizers, hypnotics, anxiolytic agents, muscle relaxants, and anticonvulsants. Abuse and intentional or accidental intoxications, often in combination with other drugs, are common, including iatrogenic intoxication caused by accumulation of long-acting benzodiazepines. Although pure benzodiazepine intoxications are usually not fatal, mixed intoxications are more dangerous. The elimination of diazepam, chlordiazepoxide, and their active metabolites can be markedly prolonged in elderly or hepatopathic patients.

1. Uptake: Oral (rapid absorption), intravenous.
2. Mechanism: Binds to GABA-A receptors and potentiates inhibitory GABA action.
3. Symptoms and signs: Weakness, ataxia, and somnolence progressing to stupor and coma. Moderate hypoventilation (potentiated by opiates or ethanol).
4. Treatment: Monitoring and supportive treatment. Stuporous or comatose patients should by treated in critical care units. Gastrointestinal decontamination within 60 minutes after ingestion. The short-acting benzodiazepine antagonist flumazenil can be given starting with 0.1 mg and adding 0.1 mg every 30 seconds to a maximal dose of 2.5 mg (Weinbroum *et al.*, 1996) or a starting dose of 0.2 mg IV adding incremental doses of up to 0.5 mg every minute (Spivey *et al.*, 1993). If the patient does not react or rapidly falls back into coma, a nonbenzodiazepine or a mixed intoxication should be suspected. Responders remain fully awake for a mean of 45 minutes but >80% become sedated again. Further bolus doses of 0.5–1.0 mg when needed or a continuous infusion of 0.5–1 mg/h can be given to keep the patient awake. In a double-blind trial, continuous infusion of flumazenil, 0.5 mg/h, did not prevent complications in severe benzodiazepine intoxications (Chern *et al.*, 1998).

## Drugs of Abuse

### Opiates

These natural (morphine, codeine), semisynthetic (e.g., hydromorphone, heroin, oxycodone), and synthetic agents (e.g., meperidine, buprenorphine, methadone, pentazocine, fentanyl) are used in medicine as strong analgesics (morphine, synthetic agents) or antitussive drugs (codeine). Opiates produce euphoria and lead to physical and psychic dependence. Intentional or accidental overdose, often as multidrug poisoning and often fatal, is common in addicts. Human immunodeficiency virus (HIV) infection is prevalent in parenteral heroin users.

1. Uptake: Intravenous or subcutaneous (especially heroin), inhalation (opium, heroin), oral (methadone, codeine).
2. Mechanism: Binding to μ-, κ-, and δ-endorphin receptors, which mediate analgesic, respiratory depressant, antitussive, euphoric, and sedative effects. Tolerance develops to all effects. A withdrawal syndrome follows discontinuation of opiate consumption.
3. Symptoms and signs: Overdose leads to coma, respiratory depression, and miosis. Death is secondary to apnea. Brain edema, pulmonary edema, and rhabdomyolysis are common after initial resuscitation. Seizures are rare and warrant consideration of concomitant cocaine toxicity or ethanol withdrawal.
4. Treatment: Stabilization of vital signs (intubation, respirator use, pressor agents); decontamination measures after recent drug ingestion; use of the antidote naloxone IV starting with 0.4 mg and incremental doses up to 2–10 mg. Sublingual or endotracheal application of naloxone is also effective. Repetitive doses may be necessary because of the short half-life of naloxone. Its use can precipitate withdrawal symptoms. Drug screening for the exclusion of multidrug poisoning. Exclusion of head injury; supportive intensive care; monitoring for at least 24 hours after severe heroin and 72 hours after severe methadone poisoning.

### Cocaine

This psychostimulant drug is extracted from coca leaves. Illicit use has a high prevalence in the United States (0.9% of the adult population) and other countries. Multidrug abuse (e.g., in combination with heroin) is common.

1. Uptake: Nasal (snorting), inhalation (smoking), intravenous, ingestion (rare).
2. Mechanism: Blockade of central dopamine reuptake; central and peripheral norepinephrine effects; vasoconstriction, local anaesthetic secondary to blockade of sodium channels.
3. Symptoms and signs: Hypertension, tachycardia, cardiac arrhythmia, tachypnea, hyperthermia, tremor, dyskinesia, hyperreflexia, confusion, hallucinations, psychosis, delirium, mydriasis, tonic-clonic seizures, and coma. Acutely, cocaine can induce cerebral vasoconstriction, brain infarction, and intracerebral hemorrhage. Approximately half of the patients with hemorrhagic stroke have an underlying aneurysm or arteriovenous malformation. Myocardial infarction is even more common than stroke in regular cocaine users.
4. Treatment: Elimination measures are inefficient because of the rapid absorption and short half-life of cocaine. Supportive and symptomatic treatment, intensive care in severe intoxications. Sedation with benzodiazepines. Diazepam (up to 30 mg IV) and phenytoin for seizure control. Alpha-adrenergic blockers or nitroprusside for tachycardia and hypertension. Artificial ventilation and cardiac monitoring may be necessary; sodium bicarbonate to treat metabolic acidosis. Cooling, antipyretics, and, if necessary, neuromuscular blockers in severe hyperthermia. Multidrug poisoning should be suspected.

### Cannabis (Hashish and Marijuana)

The leaves (marijuana) and concentrated resin (hashish) of Cannabis sativa plants contain the psychotropic agent tetrahydrocannabinol. Although cannabis consumption is illicit in most countries, worldwide prevalence of use is second only to ethanol and tobacco.

1. Uptake: Inhalation, ingestion.
2. Mechanism: Binding to a brain cannabinoid receptor.
3. Symptoms and signs: Cannabis overdose is not directly lethal. Effects include mild euphoria and impairment of thinking, concentration, and motor abilities. Hallucinations, psychosis, and panic may follow high doses. Tachycardia occurs, and angina pectoris can be precipitated in persons with preexistent cardiovascular disease. Seizures are rare.
4. Treatment: Hospitalization and supportive treatment may be indicated in patients with cardiac or psychiatric symptoms. No other specific therapy is necessary.

### Phencyclidine (PCP) and Ketamine

The synthetic drug phencyclidine is used as an anesthetic in veterinary medicine. A related compound, ketamine, is used in human anesthesiology. PCP is a popular illicit drug in the United States; abuse is rare in

Europe. Ketamine abuse is a recent phenomenon in members of the rave culture.

1. Uptake: Inhalation (smoked), sniffing (ketamine), oral, intravenous.
2. Mechanism: Antagonism at the N-methyl-D-aspartate (NMDA) glutamate receptor.
3. Symptoms and signs: Agitation, psychosis, ataxia, nystagmus, miosis, hyperreflexia, tremor, and diaphoresis. In severe intoxications, myoclus, seizures, hyperthermia, rhabdomyolysis, hypersalivation, coma, and hypoventilation are common. The action of ketamine is short (usually 30–60 minutes), whereas PCP effects last many hours.
4. Treatment: Gastric lavage within 1 hour after ingestion; forced diuresis; supportive intensive care, including airway protection, clearing of bronchial secretions, respiratory support, treatment of hypertension, hyperthermia, and seizures. Benzodiazepines for anxiety or agitation. An acute psychosis after PCP use can be treated with haloperidol IM up to 5 mg every hour. Phenothiazine neuroleptics should be avoided. Physical restraint may be necessary. Cardiorespiratory status must be closely monitored.

### Ecstasy—MDMA and Related Substances

Methylendioxymetamphetamine (MDMA, "ecstasy") was developed in 1912 as an anorectic agent. MDMA and the related substances methylendioxyethylamphetamine (MDEA) and methylendioxyamphetamine (MDA) became popular psychotropic street drugs in the 1980s. MDMA causes amphetamine-like effects (increased vigilance, euphoria, excitement, decreased appetite) and hallucinogenic effects (visual and auditory hallucinations, illusions, altered perception of time). Besides stimulation similar to that caused by amphetamines, they usually induce a pleasant, easily controllable emotional state with relaxation, fearlessness, and increased empathy. A number of pharmacological studies support the hypothesis that these drugs make up a distinct class of psychoactive substances, which have been designated "entactogens."

1. Uptake: Oral.
2. Mechanism: Increase of serotonin and, to a lesser extent, dopamine in synaptic clefts. After prolonged use, MDMA and related compounds selectively damage serotonergic nerve endings in animals and probably in humans.
3. Symptoms and signs: Tremor, paresthesias, hyperthermia, seizures, coma, disseminated intravascular coagulation, rhabdomyolysis, renal and hepatic failure. Chronic neurotoxicity may lead to memory and learning impairment even in abstinent ex-users

(Gouzoulis-Mayfrank et al., 2000). Psychiatric complications of long-term use are depression, anxiety, and personality changes.
4. Treatment: Supportive.

### Gamma Hydroxybutyrate (GHB)

GHB, a colorless liquid, is a physiological active metabolite of GABA with medical use as an anesthetic drug. Abuse became popular during the 1990s because of its euphorigenic and sedative effects and its alleged anabolic action. GHB has acquired the reputation of a "date-rape" drug. Ingested alone or together with ethanol or psychostimulants, GHB is rapidly metabolized through the tricarboxylic acid cycle to succinic acid and $CO_2$.

1. Uptake: Oral, rarely IV.
2. Mechanism: GABA-B receptor agonist.
3. Symptoms and signs: Euphoria, disinhibition, anterograde amnesia, decreased consciousness, ataxia, analgesia, bradycardia, and hypothermia. Consciousness is usually regained within 5 hours.
4. Treatment: Supportive treatment. Withdrawal symptoms mimicking ethanol withdrawal may occur after chronic abuse.

## Inhalant Abuse

Recreationally abused volatile substances include organic solvents, nitrous oxide, and nitrites. They are euphorigenic, inexpensive, and available without restrictions. Users inhale plastic cement, glue, spray paint, lacquer thinner, gasoline, cleaning fluids, and industrial solvents. Xylene, acetone, n-butane (bottled fuel gas), benzene, trichlorethyl (used for dry cleaning), and gasoline have a high acute toxicity and can cause sudden deaths. Chronic neurotoxicity is well documented after regular consumption of toluene (encephalopathy), n-hexane (polyneuropathy), and nitrous oxide (myeloneuropathy). Acute or chronic intoxication caused by exposure during work is described in painters, parquet floorers, and factory workers.

### Toluene

A colorless, liquid, volatile, aromatic hydrocarbon, constituent of glues, paints, and thinners, toluene is probably the most widely abused inhalant.

1. Uptake: Inhalation.
2. Mechanism: Alcohol-like effect, exact mechanism unknown.

3. Symptoms and signs: Tiredness, dizziness, nausea; at higher doses euphoria, confusion; loss of consciousness, probably respiratory depression; muscle weakness secondary to hypokalemia, hypophosphatemia, or rhabdomyolysis; nausea, vomiting, abdominal pain, cardiocirculatory depression. Cerebral and cerebellar atrophy, pyramidal signs, ataxia, dysarthria, hearing loss, optic nerve atrophy, and dementia may follow prolonged abuse (Cassitto, 1994)
4. Treatment: Supportive treatment in acute intoxication. Withdrawal programs.

### N-Hexane and Methyl-N-Butyl Ketone

A colorless, volatile aliphatic hydrocarbon, N-hexane is a constituent of glues, thinners, cleaning fluids, and industrial solvents.

1. Uptake: Inhalation
2. Mechanism: Toxicity is caused by 2-5-hexanedione (2-5-HD), a metabolite of both substances. It reacts with ε-amine residues, finally leading to cross-linking of neurofilament proteins and disturbance of axonal transport.
3. Symptoms and signs: Subacute sensorimotor polyneuropathy in exposed industrial workers; subacute or acute, often severe polyneuropathy with autonomic dysfunction in "glue sniffers." Pyramidal signs suggest encephalopathy or myelopathy. Postexposure deterioration ("coasting") of up to 5 months is followed by slow recovery over 1–3 years (Bruyn and Yaqub, 1994).
4. Treatment: Cessation of exposure or abuse; no specific therapy available.

### Nitrous Oxide

Colorless, anaesthetic gas (laughing gas). Recreational abuse occurs especially in medical professionals.

1. Uptake: Inhalation.
2. Mechanism: Euphorigenic and sedating effect, the molecular mechanism of which is not well understood. A myeloneuropathy is caused by inactivation of vitamin $B_{12}$ by nitrous oxide. Patients with borderline vitamin $B_{12}$ deficiency who undergo anesthesia using nitrous oxide can also have a myeloneuropathy develop but more often megaloblastic anemia develops, whereas the chronic exposure of recreational users is more likely to cause myeloneuropathy without anemia.
3. Symptoms and signs: Nitrous oxide leads to a myeloneuropathy similar to the subacute combined degeneration of vitamin $B_{12}$ deficiency with sensory impairment and signs of pyramidal tract dysfunction.
4. Treatment: Discontinuation of nitrous oxide abuse, supplementation of vitamin $B_{12}$, and methionine (Stacy et al., 1992).

### Alcohols and Glycols

### Ethanol

Alcohol (ethanol) overdose with or without coingestion of other drugs is the most common acute intoxication and is responsible for a high number of fatalities. Neurological disorders caused by chronic alcohol abuse are discussed in Chapter 71.

1. Uptake: Oral; lethal dose 5–8 g/kg, higher in persons with alcohol tolerance.
2. Mechanism: CNS depressant, increases fluidity of neuronal membranes, but also directly (albeit nonspecifically) binds to numerous proteins, including domains of neurotransmitter receptors. Ethanol directly facilitates GABA transmission and inhibits glutamate transmission. With chronic use, glutamate receptors are up-regulated, probably contributing to withdrawal signs and long-term toxicity.
3. Symptoms and signs: Euphoria; psychomotor activation; impaired motor coordination, judgment, and cognition at low serum levels. With higher doses lethargy progresses to coma, and respiratory depression progresses to apnea.
4. Treatment: Gastrointestinal decontamination within 60 minutes only after ingestion of very high doses. Thiamine, 100 mg IV, plus multivitamins; monitoring of blood glucose and correction of hypoglycemia with glucose infusion; toxicology screen to identify multidrug intoxications; diagnosis or exclusion of other potential causes of coma (head injury, meningitis, liver failure); supportive treatment includes respiratory support in deeply comatose patients; although hemodialysis can rapidly lower the ethanol level, its therapeutic role has not been well studied.

### Ethylene Glycol

Ethylene glycol is a colorless, water-soluble liquid with sweet taste used as solvent and antifreeze. Use for wine adulteration has been reported. Ingestion of ethylene glycol is rare but dangerous. Most cases are suicidal; children and alcoholics may drink antifreeze accidentally. The TESS documented more than 2000 cases treated in health care facilities in 1998.

1. Uptake: Oral with rapid absorption.
2. Mechanism: Central nervous depression resembling the effects of ethanol. Severe metabolic acidosis

develops after 4–12 hours, caused by glycolic acid. Oxalic acid, a further metabolite, chelates calcium, leading to tetany, cardiac arrhythmia, and later deposition of oxalate stones in the kidney and renal failure.

3. Symptoms and signs: Dose-dependent signs of CNS depression begin shortly after ingestion and resemble ethanol intoxication: euphoria, ataxia, dysarthria, vertigo, nystagmus, somnolence progressing to coma, respiratory depression, and circulatory collapse (Karlson-Stiber and Persson, 1992). Symptoms and signs caused by the toxic metabolites are delayed: hyperventilation, hypotension, pulmonary edema, coma, and seizures. Laboratory features are a prominent metabolic acidosis with a high anion gap and hyperosmolarity.

4. Treatment: Gastric lavage is probably of little help. Activated charcoal early after consumption; ethanol infusion to a serum level of 100 mg/dl or higher for 3 to 6 days (competitive agonist of the enzyme alcohol dehydrogenase, blocking the production of the toxic metabolites glycolic acid and oxalic acid). Alternatively, infusion of fomepizole, a competitive alcohol dehydrogenase inhibitor. The recommended dosing scheme is a loading dose of 15 mg/kg of body weight, followed by bolus doses of 10 mg/kg every 12 hours. The bolus doses are increased to 15 mg/kg after 48 hours and continued every 12 hours until serum ethylene glycol concentration is low (Brent et al., 1999). Hemodialysis is recommended if the ethylene glycol concentration is >8 mmol/L or renal function is impaired. Correction of metabolic acidosis with sodium bicarbonate; supportive critical care; seizure treatment with phenytoin and benzodiazepines.

## Methanol

Used as fuel, denaturant of ethanol solutions, and a component of a variety of industrial solutions. Home-distilled alcoholic beverages have been responsible for epidemics of methanol poisoning.

1. Uptake: Oral, rapid adsorption, minimal lethal dose in adults >30 ml.

2. Mechanism: Methanol has a direct depressant effect on the CNS. Conversion to the toxic metabolites formaldehyde and formic acid by the enzyme alcohol dehydrogenase. Formic acid inhibits cytochrome-c oxidase and causes cellular hypoxia with metabolic acidosis, retinal damage, lesions of the subcortical white matter, basal ganglia, and cerebellar cortex, and cardiocirculatory depression.

3. Symptoms and signs: Early manifestations of methanol include nausea, vomiting, vertigo, ataxia, somnolence progressing to coma, and seizures. Hyperventilation, visual symptoms, and myocardial depression occur after a delay and are due to accumulation of the toxic metabolite formic acid. Laboratory abnormalities include hyperosmolarity with increased osmolar gap and metabolic acidosis with an increased anion gap. A serum formate exceeding 50 mg/dl is predictive of an unfavorable outcome with permanent sequelae. Blindness and parkinsonism are residues after severe intoxications (Bruyn et al., 1994).

4. Treatment: Gastrointestinal decontamination is probably of little help. Activated charcoal is not indicated. Ethanol infusion to maintain a serum ethanol level >100 mg/dl is indicated in symptomatic patients. As with ethylene glycol, fomepizole (Antizol) IV can be given in a loading dose of 15 mg/kg, followed by bolus doses of 10 mg/kg every 12 hours and after 48 hours, 15 mg/kg every 12 hours until the serum methanol concentration is less than 20 mg/dl (Brent et al., 2001). Haemodialysis if serum methanol >50 mg/dl, pH is low and inadequately responding to sodium bicarbonate, or visual symptoms are present (Brent et al., 2001). Sodium bicarbonate to maintain arterial pH >7.3.

## REFERENCES

American Academy of Clinical Toxicology and European Association of Poisons Centres and Clinical Toxicologists (1999). Position statement and practice guidelines on the use of multi-dose activated charcoal in the treatment of acute poisoning. J. Toxicol. Clin. Toxicol. 37, 731–751.

Barbey, J. T., and Roose, S. P. (1998). SSRI safety in overdose. J. Clin. Psychiatry 59(Suppl. 15), 42–48.

Barceloux, D., Mc.Guigan, M., and Hartigan-Go, K. (1997). Position statement: cathartics; American Academy of Clinical Toxicology; European Association of Poisons Centers and Clinical Toxicologists. J. Toxicol. Clin. Toxicol. 35, 743–752.

Benson, B. E., Mathiason, M., Dahl, B., Smith, K., Foley, M. M., Easom, L. A. J., and Butler, A. Y. (2000). Toxicities and outcomes associated with nefazodone poisoning: An analysis of 1338 exposures. Am. J. Emerg. Med. 18, 587–592.

Brent, J., McMartin, K., Phillips, S., Burkhart, K. K., Donovan, J. W., Wells, M., and Kulig, K., for the Methylpyrazole for Toxic Alcohols Study Group. (1999). Fomepizole for the treatment of ethylene glycol poisoning. N. Engl. J. Med. 340, 832–838.

Brent, J., McMartin, K., Phillips, S., Aaron, C., and Kulig, K., for the Methylpyrazole for Toxic Alcohols Study Group. (2001). Fomepizole for the treatment of methanol poisoning. N. Engl. J. Med. 344, 424–429.

Brown, B., and Linter, S. (1987). Monoamine oxidase inhibitors and narcotic analgesics: a critical review of the implications for treatment. Br. J. Psychiatry 151, 210–212.

Bruyn, G. W., and Yaqub, B. A. (1994). Neurotoxic effects of n-hexane and methyl-n-butyl ketone. In "Handbook of Clinical Neurology, Vol. 20 (64); Intoxications of the Nervous System, Part I" (F. A. Wolff, Ed.) Amsterdam: Elsevier, 81–94.

Bruyn, G. W., al-Deeb, S. M., Yaqub, B. A., and Vielvoye, G. J. (1994). Methanol intoxication. In "Handbook of Clinical Neurology, Vol. 20 (64): Intoxications of the Nervous System, Part I" (F. A. de Wolff, Ed.) Amsterdam: Elsevier, 95–106.

Cassitto, M. G. (1994). Organic solvents and the nervous system. In "Handbook of Clinical Neurology, Vol. 20 (64): Intoxications of the Nervous System, Part I" (F. A. de Wolff, Ed.). Amsterdam: Elsevier, 39–61.

Chafee-Bahamon, C. (1983). Epidemiology of serious poisoning. *Clin. Toxicol. Rev.* 5, 1–2.

Chern, C. H., Chern, T. L., Wang, L. M., Hu, S. C., Deng, J. F., and Lee, C. H. (1998). Continuous flumazenil infusion in preventing complications arising from severe benzodiazepine intoxication. *Am. J. Emerg. Med.* 16, 238–241.

Chyka, P. A., and Seger, D. (1997). Position statement: single-dose activated charcoal. American Academy of Clinical Toxicology; European Association of Poisons Centres and Clinical Toxicologists. *J. Toxicol. Clin. Toxicol.* 35, 721–741.

Curry, S. C., Mills, K. C., and Graeme, K. A. (1998). Neurotransmitters. *In* "Goldfrank's Toxicologic Emergencies," 6th Ed. (L. R. Goldfrank, N. E. Flomenbaum, N. A. Lewin, R. S. Weisman, M. A. Howland, and R.S. Hoffman Eds.). Stamford CT: Appleton & Lange, 137–171.

Dart, R. C., Goldfrank, L. R., Chyka, P. A., Lotzer D., Woolf, A. D., McNally, J, Snodgrass, W. R., Olson, K. R., Scharman, E., Geller, R. J., Spyker, D., Kraft, M., and Lipsy, R. (2000).Combined evidence-based literature analysis and consensus guidelines for stocking of emergency antidotes in the United States. *Ann. Emerg. Med.* 36, 126–132.

Ellenhorn, M. J., Schonwald, S., Ordog, G., and Wasserberger, J. (1997). Ellenhorn's medical toxicology, 2nd ed. Baltimore: Williams & Wilkins.

Goodwin, D. A., Lally, K. P., and Null, D. J., Jr. (1993). Extracorporeal membrane oxygenation support for cardiac dysfunction from tricyclic antidepressant overdose. *Crit. Care Med.* 21, 625–627.

Gouzoulis-Mayfrank, E., Daumann, J., Tuchtenhagen, F., Pelz, S., Becker, S., Kunert, H. J., Fimm, B., and Sass, H. (2000). Impaired cognitive performance in drug free users of recreational ecstasy (MDMA). *J. Neurol. Neurosurg. Psychiatry* 68, 719–725.

Greenberg, D. A., and Simon, R. P. (1982). Flexor and extensor postures in sedative drug-induced coma. *Neurology* 32, 448–451.

Hoppe-Roberts, J. M., Lloyd, L. M., and Chyka, P. A. (2000). Poisoning mortality in the United States: Comparison of national mortality statistics and poison control center reports. *Ann. Emerg. Med.* 35, 440–448.

Karlson-Stiber, C., and Persson, H. (1992). Ethylene glycol poisoning: Experiences from an epidemic in Sweden. *J. Toxicol. Clin. Toxicol.* 30, 565–574.

Kornhuber, J., and Weller, M. (1994). Neuroleptic malignant syndrome. *Curr. Opin. Neurol.* 7, 353–357.

Krenzelok, E. P., McGuigan, M., Lheur, P. (1997). Position statement: ipecac syrup. American Academy of Clinical Toxicology; European Association of Poisons Centres and Clinical Toxicologists. *J. Toxicol. Clin. Toxicol.* 35, 699–709.

Krogsgaard-Larsen, P., and Hansen, J. J. (1992). Naturally occurring excitatory amino acids as neurotoxins and leads in drug design. *Toxicol. Lett.* 64–65, 409–416.

Lindberg, M. C., Cunningham, A., and Lindberg, N. H. (1992). Acute phenobarbital intoxication. *South. Med. J.* 85, 803–807

Litovitz, T. L., Klein-Schwartz, W., Caravati, E. M., Youniss, J., Crouch, B., and Lee, S. (1999). 1998 Annual report of the American Association of Poison Control Centers toxic exposure surveillance system. *Am. J. Emerg. Med.* 17, 435–490.

Nhachi, C. F., and Kasilo, O. M. (1992). The pattern of poisoning in urban Zimbabwe. *J. Appl. Toxicol.* 12, 435–438.

Rumack, B. H., Peterson, R. C., Koch, G. G., and Amara, I. A. (1981). Acetaminophen overdose: 662 cases with evaluation of oral acetylcysteine treatment. *Arch. Intern. Med.* 141, 380–385.

Seeff, L. B., Cuccherini, B. A., Zimmerman, H. J., Adler, E., and Benjamin, S. B. (1986). Acetaminophen hepatotoxicity in alcoholics. *Ann. Intern. Med.* 104, 399–404.

Spivey, W. H., Roberts, J. R., and Derlet, R. W. (1993). A clinical trial of escalating doses of flumazenil for reversal of suspected benzodiazepine overdose in the emergency department. *Ann. Emerg. Med.* 22, 1813–1821.

Stacy, C. B., Di Rocco, A., and Gould, R. J. (1992). Methionine in the treatment of nitrous-oxide-induced neuropathy and myeloneuropathy. *J. Neurol.* 239, 401–403.

Sternberg, H. (1991). The serotonin syndrome. *Am. J. Psychiatry* 148, 705–713.

Tenenbein, M. (1997). Position statement: whole bowel irrigation. American Academy of Clinical Toxicology; European Association of Poisons Centres and Clinical Toxicologists. *J. Toxicol. Clin. Toxicol.* 35, 753–762.

Tipton, K. F., and Singer, T. P. (1993). Advances in our understanding of the mechanisms of the neurotoxicity of MPTP and related compounds. *J. Neurochem.* 61, 1191–1206.

Vale, J. A. (1997). Position statement: gastric lavage. American Academy of Clinical Toxicology and European Association of Poisons Centres and Clinical Toxicologists. *J. Toxicol. Clin. Toxicol.* 35, 711–719.

Weinbroum, A., Rudick, V., Sorkine, P., Nevo, Y., Halpern, P., Geller, E., and Niv, D. (1996). Use of flumazenil in the treatment of drug overdose: A double-blind and open clinical study in 110 patients. *Crit. Care Med.* 24, 199–206.

Woo, O. F., Mueller, P. D., Olson, K. R., Anderson, I. B., and Kim, S. Y. (2000). Shorter duration of oral N-acetylcysteine therapy for acute acetaminophen overdose. *Ann. Emerg. Med.* 35, 363–368.

## CHAPTER 60
# Hydrocephalus

Joseph R. Madsen, Rodolfo Hakim, and R. Michael Scott

The word "hydrocephalus" means water in the head, and it applies to a clinical condition of progressive neurological risk and injury resulting from unbalanced production and absorption of the cerebrospinal fluid (CSF). Hydrocephalus is a dynamic condition that can cause severe and sometimes acute brain injury and death, but it can be managed by rectifying the flow of CSF. Previous divisions of hydrocephalus into communicating and noncommunicating groups have become less useful, because newer diagnostic imaging modalities often allow clearer definition of the site of obstruction to CSF flow with certainty. Newer classification schemes are based on pathological processes leading to the dynamic abnormalities, including the actual point of obstruction. These advances have significant relevance to the treatment of hydrocephalus (Rekate, 1996).

## HISTORY

Hydrocephalus has been recognized since antiquity, and the cerebrospinal fluid was certainly mysterious to the early observers. Hippocrates (460–377 B.C.) recognized that the accumulation of water in the brain caused it to swell. This swelling occurred in children (from buildup of water either over the surface or within the brain) before the cranial sutures were fused. Hippocrates has been credited with removing CSF through the anterior fontanelle (Fisher, 1951). Galen (A.D. 130–200) recognized that the ventricles communicated with each other and thought that they possessed an animal spirit called "pituita," which exuded from the pituitary body. He described the fluid that bathed the brain as dynamic and that its absorption occurred along the base of the brain through the cribriform plate and the pituitary body (Fisher, 1975). Thomas Willis (1621–1675) identified the choroid plexus, and the

pineal gland elaborated the fluid. He was the first one who proposed that CSF drained into a venous system, but he thought this absorption occurred in the cribriform plate (Aronyk, 1993).

Around the 1550s, Andreas Vesalius (1514–1564) was the first to recognize, with the description of a hydrocephalic child, that the accumulation of CSF occurred within the ventricular system and that this excess of fluid was destructive to the brain (Russell, 1949). Franciscus Sylvius (1614–1672) described the "aqueduct," a small vulnerable tube that served as a pathway for CSF, and in 1701 Antonio Pacchioni (1665–1726) described the arachnoid "bodies" or granulations, which he thought secreted instead of absorbed CSF (Woolam, 1957). Alexander Monro II (1733–1817) described the interventricular foramen (Fisher, 1973). In 1842, Francois Magendie (1783–1855) showed that anatomically there was a communication between the ventricular and subarachnoid spaces and that fluid flowed through this aperture. He initially thought that the CSF was formed in the subarachnoid space and flowed into the ventricular system by way of the midline fourth ventricle foramen. In 1855 Hubert von Luschka (1820–1875) described lateral apertures in the fourth ventricle (Aronyk, 1993).

Giovanni Battista Morgagni (1682–1771) was the first to describe bulging fontanels and widening of the cranial sutures with skull enlargement in a child with hydrocephalus and that the head of an adult did not enlarge when this fluid accumulated. He also described two different forms of ventricular dilation: one of them had dilation of all the ventricles, whereas the second one showed dilated lateral ventricles with a normal fourth ventricle. Robert Whytt (1714–1766) described a series of clinical events in cases of closed-suture hydrocephalus, such as the consequences of raised intracranial pressure and rostrocaudal cerebral herniation (Aronyk, 1993; Fisher, 1975).

In 1854, about 200 years after Willis had suggested that the production of CSF occurred in the choroid plexus, Luschka and Faivre confirmed it. Afterwards, it was postulated by Faivre, Valentin, and, later, by Weed and Cushing, that histologically the choroid plexus was a secretory structure; CSF was actually seen to be released by it (Aronyk, 1993; Fisher, 1975).

Toward the later part of the 19th century, Key and Retzius finally established the correct CSF flow pattern in 1876. Also, Corning did the first puncture of the subarachnoid space in a living human in 1885, and, shortly after, Quincke perfected this procedure and in 1891 described its intermittent use to control hydrocephalus (Aronyk, 1993).

During the first half of the 20th century, investigators elaborated on the physiology and possible surgical concepts for CSF and hydrocephalus. Dandy, Weed, Blackfan, Katzenelbogen, Friedenwald, and several others clearly demonstrated the effects on the ventricular system by obstructive lesions, the role of the choroid plexus, and the probable important result after its surgical resection (Fisher, 1973).

Advances in CSF physiology have been provided during the second half of the 20th century with CSF perfusion techniques, the development of ultrafine blood studies, electron microscopic methods, MRI flow studies, and the development of different surgical methods to treat hydrocephalus (Bergstrand et al., 1988; Fisher, 1973; Pappenheimer et al., 1962).

## CSF PHYSIOLOGY

### CSF Formation and Absorption

The lack of noninvasive and reliable techniques to measure CSF formation and absorption rates in humans have been discouraging for the understanding of CSF physiology and pathophysiology (Fishman, 1992). In the 1930s, Masserman and later Sjoqvist calculated the rate of CSF formation in patients by measuring the time needed for the CSF pressure to return to its original value after removing a determined volume of fluid. It was concluded that CSF was secreted at a rate of 0.32–0.36 ml/min. Because this technique assumed that the rates of CSF formation and absorption were not altered after draining fluid or in pathological states, it was open to criticism. It was calculated with this technique, and later with the ventriculolumbar perfusion technique, that the normal total CSF production rate per day is about 500 ml (Fishman, 1992).

In the 1940s and 1950s, when radioisotopes became available, attempts were made to determine the CSF formation rate by calculating the rates of exchange of its individual components. Even though these studies gave some knowledge on the blood–CSF barrier, they did not provide information regarding bulk formation and absorption of CSF (CSF turnover), but the ventriculocisternal and ventriculolumbar methods in the 1960s and 1970s did. In 1962, by applying the principles underlying the renal clearance techniques, Pappenheimer et al. adapted the ventriculocisternal technique developed earlier by Royer and Leusen (Pappenheimer et al., 1962). The technique basically consists of inserting an inflow cannula into a lateral ventricle and an outflow cannula in the cisterna magna. An artificial CSF perfusion containing a nonmetabolized macromolecule marker that is nondiffusable through the ventricular walls (inulin or radioiodated serum albumin) was pumped through at a known and constant rate. It was later found that inulin (mol wt 1300–1800 daltons) definitely diffused into the adjacent tissue and therefore [131]I-labeled serum albumin (mol wt 69,000 daltons) and yet another macromolecule (blue-labeled dextran, with a mol wt of $2 \times 10^6$ daltons) were preferred instead. By knowing the concentration of the macromolecule and the volume of influx and eflux of the perfusate, a clearance of that particular substance could be calculated. The inflow rates can be measured and varied, whereas the outflow system allows one to modify the differential pressure between the ventricle and the cistern.

The CSF formation and absorption rates can be calculated as follows (Heisey et al., 1962; Lorenzo et al., 1970; Welch, 1975):

$$\text{Formation rate } (Vf) = Vi(Ci - Co)/Co$$

$Vi$ is the rate of perfusion and $Ci$ and $Co$ are the inflow and outflow concentrations of the marker, respectively, at steady state.

Absorption rate $(Va) = Vi + Vf - Vo$, where $Vo$ is the outflow rate at the steady state.

To prevent the absorption of the perfusate through the normal drainage channels (arachnoid granulations) and also to obtain a regular outflow, Pollay and Davson (1963) modified the setup in a rabbit by connecting a suction device to the outflow cannula. With this new setup, $Va$ can be ignored and therefore the new equation is:

$$Vf = Vo - Vi$$

The absolute rates of CSF formation were found to vary in different species from 0.325 μL/min in the mouse to 350 μL/min in man (Davson and Segal, 1996). These are obviously determined by the size of the CSF space.

The rate of CSF turnover, however, does not vary considerably in different species. If we consider the choroid plexus to be the main or only source of CSF production, then the rate of production per unit weight of choroid plexus is a constant. Therefore, the CSF production rate varies between 0.2 and 0.5 ml/min/g of choroid plexus (Davson, 1967) or, in relation to the size of the CSF space, between 0.2% and 0.5% of the total normal CSF volume per minute (Davson, 1967). This would mean that an adult human with a total CSF volume of 120 ml would have a rate somewhere between 14 ml/h and 36 ml/h.

### Radiological Studies

The first method that permitted noninvasive study of CSF motion was the MRI. Flow through the aqueduct was documented since the 1980s as a variation in the signal intensity within the aqueduct during the cardiac cycle (Bergstrand et al., 1985). Afterwards, the "flow void phenomenon" was described and later used to evaluate pulsatile flow within the aqueduct and other CSF spaces (Sherman and Citrin, 1986). Later, the direction of aqueductal CSF flow was illustrated as caudal in systole and cephalad in diastole; the reported approximate flow velocity within the aqueduct was 3–5 mm/s (Mascalchi et al., 1988).

Brain motion was initially studied noninvasively with ultrasonography. Medially directed brain motions were documented; the interpretation was that these compressed the lateral and third ventricles (Bergstrand et al., 1988). Afterwards, quantitative MRI was used to show brain motion. A caudal movement during the cardiac systole, especially of the midportion of the brain, including the corpus callosum, diencephalon, and the brainstem, was documented. First, these MR studies contradict Bering's theory that the pressure wave that initiates CSF flow is generated by the expansion of the choroid plexus. This is suggested because the corpus callosum is displaced downward during the cardiac systole instead of the expected upward or outward displacement. Second, they suggest that the CSF pulsations are a result of the transmission of the cardiac pulsations into the intracranial compartment through the intracranial arteries first and then the capillary bed (brain parenchyma) (Bergstrand et al., 1988).

### CSF Pressure

Lundberg, in 1960, found that the CSF pressure measured in the decubitus lateral position in a group of normal persons varied between 70 and 180 mm $H_2O$; this group included children and adults of different sizes (Lundberg, 1960). Welch and Friedman (1960) examined the hydraulic characteristics of the arachnoid villi by forcing liquid through samples of African green

monkey dura (Welch and Friedman, 1960), and later Davson et al. (1970) found the same by infusing additional fluid into the lateral ventricles of rabbits (Davson et al., 1970). There was a good correlation with their results and the ones found in humans: 147–190 cm $H_2O$ min/ml$^{-1}$ for the rabbit and 13 cm $H_2O$ min/ml$^{-1}$ in man (Cutler et al., 1968). The lower resistance in man is consistent with the much larger volume turnover of CSF compared with the rabbit. Therefore, because the CSF pressures in different species are within a similar range, the resistance to turnover or absorption will vary inversely with the volume flow of CSF.

The pressure of the CSF ($P_{csf}$) is induced by the rate (q) of CSF production, the hydraulic resistance (R) of the circuit, and the venous pressure ($P_v$) into which the fluid drains (SSS). Because the CSF has to flow through a series of hydraulic resistances before draining into the superior sagittal sinus, including the valvelike mechanism of the arachnoid villi, this assures us that normally the intraventricular pressure is higher than the venous pressure. The resistance produced by the arachnoid villi in normal subjects has been found to be 45 mm $H_2O$ (Hakim, 1985). The superior sagittal sinus pressure in adults, measured in a horizontal position, has been reported to be 75 mm $H_2O$ (Cutler et al., 1968; Davson, 1967, 1984; Pappenheimer et al., 1962). This pressure progressively drops within the venous system to approximately 0 mm $H_2O$ at the right atrium. One must superimpose the hydrostatic pressure distribution on these values once the individual leaves the horizontal position.

An important known fact is that the venous pressure in the body is referenced at the right atrium to external atmospheric conditions. It has been found that when tilting monitored subjects at different angles with respect to a horizontal plane, the venous pressures changed in greater values when the position was 90° (vertical position) and also the further away from the right atrium. Therefore the right atrium is considered the hydrostatic indifference point (HIP) for the venous system: it is the zero reference for venous pressures (Burch and Winston, 1943). This does not mean that the pressure necessarily has a value of zero with respect to the atmosphere at this level. It means that whatever its numerical value, this is the level where, when changing the position of the body, the reading of the pressure will change the least.

Interestingly and well-known, the blood pressure differentials between the arterial and venous systems are maintained at each level irrespective of the subject's angle of inclination (Rushmer, 1976). This means that the hydrostatic factor's value is the same in both systems, and therefore the pressure differentials across the capillary bed are kept unchanged whether the person is positioned horizontally or vertically.

Within the craniospinal cavity, because the CSF drains through the superior sagittal sinus into the venous system, the same rules when hydraulics are applied. The HIP is also at the level of the right atrium, and the pressure differentials between the CSF system and the other systems (venous and arterial) are maintained regardless of the body's angle of inclination.

## MECHANISMS FOR HYDROCEPHALUS

Clearly, for the ventricles to enlarge or decrease in size within the cranium, something else has to yield to be able to accommodate the changes in CSF volume.

Hydraulically, an interesting concept is to try to understand the brain both as a material and as a structure. As a material, it would not be compressible within the rigid skull, and therefore the ventricular size would not be modified by changing the intraventricular pressure. As a structure made of a "viscoelastoplastic" material, the ventricles could change their size, because the compartments of the brain capable of volumetric change would be the parenchymal extracellular fluid and the intraparenchymal venous system; these are considered the mechanical "give" of the brain, because they are exposed to the atmospheric pressure through the extracranial venous system. Therefore, the brain parenchyma is comparable to an open-cell sponge at a microscopic level (Hakim, 1985; Hakim et al., 1976).

Under normal conditions, the ventricular size and the brain parenchyma are unchanged, and therefore there is no tissue distortion, because an equilibrium exists between two opposing pressures; these are the intraparenchymal pressure ($P_p$) and the CSF pressure ($P_{csf}$). For the ventricular size to remain unchanged, the differential intraventricular CSF pressure ($P_{ei}$), which is equal to ($P_{csf}$) minus ($P_p$), must be equal to zero; this means ($P_{csf}$) is equal to ($P_p$). If ($P_{csf}$) > ($P_p$), then ($P_{ei}$) > 0, and the ventricular walls will be pushed toward the periphery of the brain, and fluid and intraparenchymal venous blood are squeezed out of the parenchyma. At this point hydrocephalus develops. Even if both pressures ($P_{csf}$ and $P_p$) were to become equal again, the brain parenchyma will have yielded bioplastically and will remain in this new steady-state condition with dilated ventricles, having reached the mechanics of normal pressure hydrocephalus. In theory, the treatment of hydrocephalus requires first the lowering of ($P_{csf}$) below ($P_p$) for the spongelike parenchyma to refill and therefore decrease the ventricular size. After the ventricles reach their "normal size", ($P_{csf}$) should be increased until it is equal to ($P_p$), thus avoiding slit ventricles (Hakim, 1985).

## CLINICAL ASPECTS

Hydrocephalus may occur at any age, and its clinical course is influenced by the time of onset, the rate at which the intracranial pressure increases, the duration of increased intracranial pressure, and the pre-existence of structural lesions (Menkes, 1990). Up to about 2 years of age, macrocrania is invariably the presenting complaint, but if hydrocephalus develops after this age, this sign is overshadowed by other neurological manifestations. The head circumference should alert the physician when it is 2 standard deviations above normal (98th percentile for the age) and/or when there is a discontinuity in the slope of the head circumference growth curve. In infants the skull can be thin, therefore accentuating the "cracked-pot" sound on percussion. The scalp can be shiny with sparse hair and dilated veins, and the sutures are diastatic. The anterior fontanelle, which is normally flat or slightly concave when the infant is held erect and not crying, can be bulging with excessive tension. Ocular signs may include the sunsetting sign caused by pressure of the third ventricle's suprapineal recess on the mesencephalic tectum impairing the vertical gaze centers; esotropia caused by unilateral or bilateral abducens nerve paresis; nystagmus, ptosis, and a decreased pupillary light response. Papilledema is not common because of the open sutures, but retinal venous engorgement and sometimes early signs of optical atrophy may be seen in the funduscopic examination.

When hydrocephalus is rapidly progressive, despite the open sutures, headache, somnolence, irritability, emesis, seizures, and signs of brainstem dysfunction (bradycardia and respiratory arrhythmia) caused by herniation can occur.

Hydrocephalus in children after the sutures have closed and in adults can manifest initially with sporadic headaches, which in time will progress and may be accompanied by nausea, vomiting, and sometimes early morning abdominal pain. The headaches, which at the beginning improve after emesis and/or upright posture, are usually bifrontal initially and more severe in the morning, because the intracranial pressure is further increased in the recumbent position. With the progression of the symptoms, the headache becomes generalized and continuous and usually awakens the patient at night. Occasionally, neck pain and stiffness may become evident and should caution the examiner, because this could be a sign of descending cerebellar tonsils through the foramen magnum. Visual disturbances such as a decrease in visual acuity, diplopia (from abducens paresis), and Parinaud's sign (from pressure on the quadrigeminal plate) can be found. If a patient complains of episodic "graying out" of vision, it should be considered an emergency, because this means that the optic nerve is severely affected by the high intracranial pressure. Papilledema is commonly found in patients with closed sutures and chronic hydrocephalus. Also a decline in school or work performance, memory loss, changes in personality, and lower limb spasticity (caused by stretched corticospinal tracts around the dilated lateral ventricles) can be seen. Although unusual, the head bobbing sign (apparently caused by pressure from the dilated third ventricle to the medial aspect of the dorsomedial nuleus of the thalamus) and endocrine disorders can occur.

## RADIOLOGICAL IMAGING

At present, three different imaging modalities are used to document hydrocephalus: these are ultrasonography, computed tomography (CT), and MRI. Each has its own advantages and disadvantages.

Ultrasonography of the head can be done as long as the anterior fontanelle is open. Axial, sagittal, and coronal images can be obtained. It is a simple and innocuous procedure that allows one to visualize the ventricular size and shape but generally insufficient anatomical detail to determine the etiological diagnosis in many cases. It has been very useful in the diagnosis and follow-up of intrauterine hydrocephalus.

A CT scan can be performed quickly, and it will show much more detailed anatomy than the ultrasonography, allowing one to visualize with greater resolution the ventricular and the subarachnoid spaces and the brain parenchyma. Many times it can help in localizing and revealing the nature of the CSF obstacle. Its disadvantages, however, include the poor brainstem and posterior fossa visualization, the fact that only axial images can be obtained, and, because it emits ionizing radiation, the risk of cataract formation, especially in infants requiring multiple scans.

MRI is usually the examination of first choice when better anatomical detail is needed for establishing the cause of hydrocephalus. It will allow for a better morphological definition of the various CNS structures, CSF spaces, periventricular edema, and lesions. Various planes can be obtained (axial, sagittal, and coronal); bony structures usually do not interfere in the identification of intrathecal structures, and, like ultrasonography, it is not a source of ionizing radiation. Among the disadvantages are an increased length of time to complete the examination and the fact that general anesthesia is often required in younger children or patients who cannot tolerate the length of the examination. Also, the presence of metallic implants may contraindicate the examination and/or create artifacts that will interfere in the interpretation of the images. Cine MRI can be useful for evaluating CSF flow and in localizing the obstruc-

tion site in cases of noncommunicating hydrocephalus (Kadowaki et al., 1995).

## PRINCIPLES OF THERAPY AND PRACTICAL MANAGEMENT (SEE ALSO CHAPTER 56)

The goals in the treatment of hydrocephalus consist of decreasing the intracranial pressure when elevated and the reversal of the brain parenchyma distortion caused by ventriculomegaly. This way the patient's cognitive function and neurological outcome can be maximized. White matter edema and demyelination followed by neuronal cell death can occur early in the course of hydrocephalus (Weller et al., 1972). An infarction in the territory of the posterior cerebral artery because of its occlusion against the tentorium can occur in cases of severe and/or acute untreated hydrocephalus, with subsequent cortical blindness, herniation, and death.

Once it is considered that the patient with hydrocephalus needs permanent treatment, two options can be offered at present. The first one, useful in cases of either communicating or in noncommunicating hydrocephalus, is to implant a shunt to divert the CSF into another cavity in the body. The second option, suitable only in noncommunicating hydrocephalus, is a third ventriculostomy. Both options bypass the site of CSF obstruction. However, before considering these two options, the cause of hydrocephalus should be determined, because many times by treating only the underlying cause the patient may be cured. Even though the resection of an evident causal lesion will not always re-establish the normal CSF flow, in as much as 80% of the posterior fossa tumor resections it will (Saint-Rose, 1996).

Temporary pharmacological treatment can sometimes be offered in cases in which the hydrocephalus could be considered transient, such as in the case of a premature infant with an intraventricular hemorrhage. By allowing the CSF to clear up, the normal fluid pathways could re-establish and the need for a shunt insertion might be obviated. Such medications will decrease the choroid plexus CSF secretion (acetazolamide, 100 mg/kg/day; furosemide, 1 mg/kg/day). However, the pharmacological treatment is not an option for long-term use, especially in neonates, because they can easily become metabolically decompensated.

Other means of controlling hydrocephalus temporarily are the ventricular tap, the lumbar puncture, and their continuous external drainage counterparts. The ventricular tap is useful, especially in neonates with an open fontanelle. One must be cautious and always keep in mind that the needle used is traversing the brain parenchyma to reach one of the lateral ventricles, and therefore one should not change its direction once the needle has entered the intracranial cavity. Lumbar puncture and ventricular tap are contraindicated in noncommunicating hydrocephalus. This is because after removing spinal CSF, a pressure gradient will be created between the two noncommunicated CSF compartments, and therefore a downward displacement (or upward when tapping a ventricle) of neural structures can occur. It is therefore wise to defer these examinations until after a CT or an MRI has been performed. In cases of acute hydrocephalus, a 1 g/k IU bolus of mannitol can prevent death from herniation during the minutes before performing an emergency CSF diversion procedure.

External drainage of CSF can be performed either through a ventricular catheter or through a lumbar catheter (only for communicating hydrocephalus) connected to an external drainage bag. CSF can drain continuously at an opening pressure set by placing the dripping chamber of the drainage bag at the desired height measured from the right atrium of the heart. Although these methods are very useful, they have a high risk of CSF contamination and require constant monitoring of the patient.

The implantation of a shunt to treat hydrocephalus is a standard procedure. There are several different kinds of shunts available. A basic shunt system consists of a ventricular catheter, which receives the CSF, a valve, which controls the opening pressure of the system, and a distal catheter, which drains the fluid to either the venous system or a cavity where it can be reabsorbed. The ventricular and the distal catheters are made out of silicone elastomer rubber.

In general terms, valves can have one of three different mechanisms: a slit mechanism, in which the CSF pressure deforms and opens the slit in the silicone elastomer tubing and thus allows it to flow; a diaphragm mechanism, also made of silicone elastomer, which also deforms to allow CSF to flow; and the ball, cone, and spring mechanism, which regulates the opening pressure with stainless-steel spring holding a sphere over a conical opening. These valves are available in different predetermined fixed opening pressures. There are also systems available with the ball, cone, and spring mechanism in which the opening pressure can be readjusted noninvasively as often as needed. It is important that the valve shunt system does not change its working characteristics with time.

As a foreign body, a shunt can easily get contaminated, especially during implantation. Appropriately 90% of the infections occur during the first 6 months after surgery, and 70% of these occur during the first month (Choux *et al.*, 1992). These are often indolent and manifest as an intermitent shunt malfunction, but

sometimes wound infection, erythematous shunt track, abdominal pain, or signs of meningitis may become evident. The manifestations of an infected ventriculoatrial shunt can include low fever, pulmonary embolism, congestive heart failure, and an immune complex nephritis. The treatment of choice in the case of shunt infection is to remove the entire system and give the appropriate antibiotic therapy. An external ventriculostomy might be necessary until the infection is controlled, after which a totally new shunt system is reimplanted.

## NORMAL PRESSURE HYDROCEPHALUS (SEE ALSO CHAPTER 77)

The concept of normal pressure hydrocephalus was introduced to the medical literature in 1964 (Hakim, 1964). This condition became evident after observing that a group of patients with symptomatic progressive hydrocephalus and normal CSF pressure improved after a lumbar puncture by decreasing the CSF pressure below the initial normal values. Although most of the cases occur in patients older than 65 years, it has been diagnosed in all ages (Black, 1980; Vanneste, 1994).

Etiologically, the patients with NPH are classified in one of two groups: the first group, NPH with a known cause, consists of the patients who have a medical history of subarachnoid hemorrhage, subdural hematomas, trauma, intracranial tumors, meningitis, intracranial surgery, etc., and the second group, idiopathic NPH consists of the patients in cobone no known cause can be identified.

Today, the most accurate diagnosis is still made by the clinical presentation, which consists of gait disturbance, urinary incontinence, and dementia in the presence of ventriculomegaly and normal CSF pressure. The syndrome can sometimes be confusing when it occurs simultaneously with other diseases that can course with these symptoms. Two entities that can be difficult to differentiate clinically from NPH and that are untreatable at present are Alzheimer's disease and Binswagner's disease.

The gait disturbance is usually the first symptom that appears in NPH. Initially, it may become manifest in different forms such as leg tiredness when standing or walking, poor balance, difficulty walking on the stairs, and diffuse weakness (Fisher, 1982). With time, the steps may become shorter and turning becomes slow to the point of the inability to stand or walk; patients sometimes describe this as "feeling glued to the ground" (clinically described as "magnetic gait"). This is in part due to an excess of activity of antigravity muscles. On examination one can also find a widened base and, when the patient is asked to turn around, there can be swaying of the feet followed by advancement of only one foot while rotating on the other.

Urinary incontinence may develop initially as increased frequency. When it appears before the mental impairment, some patients describe it as feeling that the bladder is full, but they are unable to reach the bathroom fast enough because of their gait difficulty. As the disease progresses, frontal lobe incontinence is added to the picture, and therefore the patient is unaware of the need to urinate. In advanced stages there can also be fecal incontinence.

The mental impairment is of the subcortical type, resembling that seen in frontal disorders and therefore inattention, forgetfulness, recent memory impairment, inability to analyze acquired knowledge, and decrease in interest, initiative, spontaneity and communication may be present. Nonverbal tasks (drawing, copying, digit symbols, and puzzle assembly) are more compromised than the verbal tasks. Voluntary movements become delayed and slow as the disease progresses, and afterward the patient may become abulic (Black, 1980; Fisher, 1982; Hakim, 1964; Vanneste, 1994; Vassilouthis, 1984).

The tests more frequently used to corroborate with the clinical diagnosis of NPH are the CT, MRI, and lumbar puncture. Phase-contrast cine-MRI is being used in some centers as an adjunct to the previous tests. Isotope cisternography has proven to be less accurate for evaluating NPH than the combined clinical and CT criteria (Vanneste *et al.*, 1992). The lumboventricular perfusion tests have been reliable to some physicians but unpredictable to others, whereas the measurement of cerebral metabolism and blood flow has been somewhat inconsistent at present (Vanneste, 1994).

## REFERENCES

Aronyk, K. E. (1993). The history and classification of hydrocephalus. *In* (A. Butler and D. G. McLone, Eds.), pp. 677–705. "Neurosurgery Clinics of North America," vol 4. Saunders, Philadelphia.

Bergstrand, G., Bergstrom, M., and Nordell, B., *et al.* (1988). Cardiac gated MR imaging of cerebrospinal fluid flow. *J. Comput. Assist. Tomogr.* 9, 1003–1006.

Black, P. M. (1980). Idiopathic normal-pressure hydrocephalus. Results of shunting in 62 patients. *J. Neurosurg.* 52, 371–377.

Burch, G. E., and Winston, T. (1943). The phlebometer. A new apparatus for direct measurement of venous pressure in large and small veins. *JAMA.* 123, 91–92.

Choux, M., Genitori, L., Lang, G., and Lena, G. (1992). Shunt implantation: reducing the incidence of shunt infection. *J. Neurosurg.* 77, 875–880.

Cutler, R. W. P., Page, L., Galicich, J., and Watters, G. V. (1968). Formation and absorption of cerebrospinal fluid in man. *Brain.* 91, 707–720.

Davson, H. (1967). Physiology of the Cerebrospinal Fluid." Boston, Little, Brown, and Co. Great Britain.

Davson, H. (1984). Formation and drainage of the cerebrospinal fluid. *In* "Hydrocephalus" (K. Shapiro, A. Marmarou, and H. Portnoy, Eds.), pp. 3–40. Raven Press, New York.

Davson, H., Hollingsworth, G., and Segal, M. B. (1970). The mechanism of drainage of the cerebrospinal fluid. *Brain.* 93, 665–678.

Davson, H., and Segal, M. B. (1996). "Physiology of the CSF and Blood-Brain Barriers." CRC Press, Boca Raton, 201.

Deck, M. D. F., and Potts, D. G. (1969). Movements of ventricular fluid levels due to cerebrospinal fluid formation. *Am. J. Roentgenol. Radium Ther. Nucl. Med.* 106, 354–368.

Fisher, C. M. (1982). Hydrocephalus as a cause of disturbances of gait in the elderly. *Neurology* 32, 1358–1363.

Fisher, R. G. (1951). Surgery of the congenital anomalies. *In* "A History of Neurological Surgery" (A. E. Walker Ed.), pp. 334–351. Williams & Wilkins, Baltimore.

Fisher, R. G. (1975). The cerebrospinal fluid. *Mayo Clinic Proc.* 50, 482–486.

Fishman, R. A. (1992). "Cerebrospinal Fluid in Diseases of the Nervous System", 2nd ed. pp. 25–26. W. B. Saunders Company, Philadelphia.

Hakim, C. (1985). The physics and physicopathology of the hydraulic complex of the cental nervous system. PhD thesis, Massachusetts Institute of Technology, Cambridge, Massachusetts.

Hakim, S. (1964). Some observations on CSF pressure. Hydrocephalic syndrome in adults with "normal" CSF (recognition of a new syndrome). Thesis No. 957, Javeriana School of Medicine, Bogotá, Colombia.

Hakim, S., Venegas J., and Burton, J. (1976). The physics of the cranial cavity, hydrocephalus and normal pressure hydrocephalus: mechanical interpretation and mathematical model. *Surg. Neurol.* 5, 187–210.

Heisey, S. R., Held, D., and Pappenheimer, J. R. (1962). Bulk flow and diffusion in the cerebrospinal fluid system of the goat. *Am. J. Physiol.* 203, 775–781.

Kadowaki, C., Hara, M., Numoto, M., *et al.* (1995). Cine magnetic resonance imaging of aqueductal stenosis. *Child. Nerv. Syst.* 11, 107.

Lorenzo, A. V., Page, K. K., and Watters, G. V. (1970). Relationship between cerebrospinal fluid formation, absorption and pressure in human hydrocephalus. *Brain.* 93, 679–692.

Lundberg, N. (1960). Continuous recording and control of ventricular fluid pressure in neurosurgical practice. *Acta. Psychiatrica. Neurol. Scand. Suppl.* 149.

Mascalchi, M., Ciraolo, L., Tananfani, N., *et al.* (1988). Cardiac-gated phase MR imaging of aqueductal CSF flow. *J. Comput. Assist. Tomogr.* 12, 923–926.

Menkes, J. H. (1990). Textbook of Child Neurology. p. 260. Lea & Febiger, Pennsylvania.

Pappenheimer, J. R., Neisey, S. R., Jordan, E. F., and Downer, J. de C. (1962). Perfusion of the cerebral ventricular system in unanesthetized goats. *Am. J. Physiol.* 203, 763–774.

Rekate, H. L. (1996). Cerebrospinal fluid circulation. *In* "Scientific Foundations of Neurology." (A. N. Guthkelch and K. E. Misulis, Eds.), pp. 323–336. Blackwell Science, Cambridge, MA.

Rushmer, R. F. (1976). "Cardiovascular Dynamics." W. B. Saunders Company, Philadelphia.

Russell, D. S. (1949). Observations on the pathology of hydrocephalus. Medical Resarch Council, Special Report Series No. 265, pp. 2–9. Her Majestty's Stationary Office, London.

Saint-Rose, C. (1996). Hydrocephalus in childhood. *In* "Youmans Neurological Surgery" (J. Youmans, Ed.). p. 890. WB Saunders, Philadelphia.

Sherman, J. L., and Citrin, C. M. (1986). Magnetic resonance demonstration of normal CSF flow. *A.J.N.R.* 7, 3–6.

Vanneste, J. (1994). Three decades of normal pressure hydrocephalus: are we wiser now? *J. Neurol. Neuorosurg. Psychiatry.* 57, 1021–1025.

Vanneste, J., Augustijn, P., Davies, G. A., Dirven, C., and Tan, W. F. (1992). Normal-pressure hydrocephalus: Is cisternography still useful in selecting patients for a shunt? *Arch. Neurol.* **49,** 366–370.

Vassilouthis, J. (1984). The syndrome of normal-pressure hydrocephalus. *J. Neurosurg.* **61,** 501–509.

Welch, K. (1975). The principles of physiology of the cerebrospinal fluid in relation to hydrocephalus including normal pressure hydrocephalus. *In* "Advances in Neurology. Current Reviews." (W. J. Friedlander, Ed.), pp. 247–332. Raven Press, New York.

Welch, K., the Friedman, V. (1960). The cerebrospinal fluid valves. *Brain.* **83,** 454.

Weller, R. O., and Shulman, K. (1972). Infantile hydrocephalus: clinical, histological, and ultrastructural study of brain damage. *J. Neurosurg.* **36,** 255.

Woolam, D. H. M. (1957). The historical significance of cerebrospinal fluid. *Med. Hist.* **1,** 91.

CHAPTER 61

# Palliative Care in the Terminal Stage of Neurological Diseases

Raymond Voltz, Gian Domenico Borasio, and Russell Portenoy

The increasing call for legalization of active euthanasia in many countries (Foley, 1997; Pollard, 2001) is relevant to many patients with neurological diseases. Currently, active euthanasia is legal in The Netherlands, and 5.4% of deaths caused by multiple sclerosis (MS) and 4.1% of deaths caused by amyotrophic lateral sclerosis (ALS) involve active euthanasia or physician-assisted suicide (van der Wal *et al.*, 1996). The rate of physician-assisted death in these two neurological diseases is about twice that reported for malignant tumors.

Most palliative care specialists believe that the wish for physician-assisted death is strongly associated with the inadequate availabilty of the clinical services and interventions subsumed by the term "palliative care" (Foley, 1997; Singer and Siegler, 1990). Indeed, many specialists believe that optimal palliative care can abolish the call for active euthanasia in almost all instances (Doyle *et al.*, 1998; Foley, 1997). It is the poor availability of palliative care in most countries that should and could be changed rather than legalization of physician-assisted suicide or active euthanasia. This way one buttresses the ethical, religious, and political concerns that are commonly invoked as reasons for caution in allowing access to these practices.

Traditionally, palliative care has focused on end-of-life issues in the cancer population. Specialist-level palliative care has been practiced in hospices and palliative care units, where most patients are close to death from malignant neoplasms. In recent years, however, palliative care has become viewed as a model approach for the management of the diverse problems that undermine quality of life in all types of patients with life-threatening illnesses. The principles that guide palliative care clearly apply to the management of patients with progressive neurological disorders (Voltz, 1998; Voltz *et al.*, 1997). Recent position statements from the American Academy of Neurology have concluded that neurologists have a duty to provide adequate palliative care to patients with incurable neurological disease, that support for assisted suicide or active euthanasia for these patients is not indicated, and that neurologists need improved education concerning palliative care (Bernat *et al.*, 1996; The Ethics and Humanities Subcommittee of the AAN, 1996; The Ethics and Humanities Subcommittee of the AAN, 1998).

## DEFINITION OF PALLIATIVE CARE

An expert committee of the World Health Organization (World Health Organization, 1986) defined palliative care as ". . . the active total care of patients whose disease is not responsive to curative treatment. Control of pain, of other symptoms, and of psychological, social and spiritual problems is paramount. The goal of palliative care is the achievement of the best possible quality of life for patients and their families." More recently, a committee of the United States' Institute of Medicine (Committee on Care at the End of Life 1997) observed that ". . . palliative care seeks to prevent, relieve, reduce, or sooth the symptoms of disease or disorder without effecting a cure . . . Palliative care in this broad sense is *not* restricted to those who are dying or those enrolled in hospice programs . . ." These definitions imply that palliative care is an interdisciplinary, therapeutic approach that focuses on the comprehensive management of the physical, psychological, social, and spiritual needs of patients with progressive incurable illnesses and their families. As such, it is a model that applies throughout the course of the illness and includes interventions primarily intended to maintain quality of life or attenuate suffering.

The need for palliative care typically intensifies during the period before and immediately after the death of a patient. As death approaches, the numerous interventions that constitute palliative care must begin to include those that specifically address end-of-life concerns. Communication, continuing assessment of the goals of care, and an understanding of the applicable ethics become important obligations of professional caregivers. At this time, patients and families must be reassured that values and decisions will be elicited and respected, comfort will become the major priority, practical needs at home will be addressed, and psychosocial and spiritual distress will be managed. Excellent palliative care at the end of life may enhance the opportunity for the patient and family to achieve a sense of growth, resolve differences, and find a comfortable closure. It can reduce the suffering and fear associated with the dying and prepare the family for bereavement.

Palliative care is *both* an approach to patient care that should be routinely integrated with life-prolonging therapies and a specialization in medicine and other disciplines. Palliative medicine is the medical specialty dedicated to excellence in palliative care. Indeed, in some countries, such as the United Kingdom, Canada, Australia, and New Zealand, palliative medicine is a recognized subspecialty. The evolution of palliative medicine is reflected in the publication of major medical texts devoted to the subject (Berger *et al.*, 1998; Doyle *et al.*, 1998).

Neurologists routinely care for patients with progressive incurable illnesses and should be trained to integrate palliative care practices into routine neurological management. When access to specialized palliative care programs exist, neurologists should understand their potential role and perhaps contribute directly to their success. Unfortunately, these outcomes are not realized in most settings. Palliative medicine has yet to be integrated into the basic or continuing education of neurologists, and most neurologists have limited skills in this discipline. The reasons for this are multiple and relate to a recognized gap between established guidelines for the care of patients with advanced illness and the beliefs and practices of many neurologists (Carver *et al.*, 1999). This gap may involve a traditional focus on diagnosis rather than therapeutics, a devotion to disease-related therapies in a manner similar to most medical practice, limited education in communication, persistent legal myths about end-of-life care (Meisel *et al.*, 2000), and the tendency of professionals to maintain unrealistic attitudes and conceptualisations related to hope (Carter *et al.*, 1998; Emanuel *et al.*, 2000).

## ISSUES IN COMMUNICATION

Effective palliative care starts with good communication skills. Breaking bad news, like the diagnosis of a progressive disease, is neither easy nor standardizable. It should be a multistep process involving the patient and his or her relatives. Typically, it begins by exploring the depth of understanding that the patient already possesses and the amount of information that the patient wants to know. All questions should be answered frankly, but it is not necessary to provide extensive information at one time. The goal is to allow the patient to attain an understanding that facilitates coping, preserves hope, and enhances planning and decision making. The physician must strive to be open, empathic, and compassionate. Humor is a sign of living and should therefore not be forgotten, even in a serious condition (Killeen, 1991). Most patients who have undergone long diagnostic odysseys from one hospital to another are at least partially relieved by having a label for their condition at last (Faulkner *et al.*, 1994; Johnston *et al.*, 1996).

Establishing the goals of care repeatedly throughout the course of the illness is one of the most important goals of communication. Cross-cultural differences must be taken into account, especially in difficult end-of-life decisions such as withdrawing ventilator support (Hayashi *et al.*, 1991; Sakakihara *et al.*, 2000) or stopping steroids (Hardy *et al.*, 2001).

Communicating prognosis is another challenge. Patients or relatives may inquire about the length of time remaining before death. Although it is important to answer honestly, the question should be explored for any subtexts (for example, should treatment be stopped?), and answers that communicate specific times should be avoided. Statements about a specific number of days or weeks will most probably be wrong, and ranges like "several days" or "days to weeks" would seem preferable. Methods to assist in prognostication have been developed and validated for cancer patients (Caraceni *et al.*, 2000) but cannot be generalized to neurological patients.

Relatives are an invaluable source of information about the patient's state, and for some types of information there is a high correlation with the patient's account (Field *et al.*, 1995). However, caution should be exercised when using proxy judgments to assess symptom distress, quality of life, or presumptive end-of-life wishes of an incompetent patient (Finlay and Dunlop, 1994; Tierney *et al.*, 1992).

**TABLE I** Risk Factors for Bereavement

| | |
|---|---|
| Patient | Young |
| Illness | Short, protracted, disfiguring, distressing |
| Death | Sudden, traumatic |
| Relationship | Ambivalent, hostile, dependent |
| Main carer | Young, own children, physical or mental illness, concurrent crisis, little or no support |

Palliative care views the family as the unit of care, and good communication with the family is essential. Near the end of the patient's life, this communication may help with anticipatory greaving and preparation for bereavement. After the death of the patient, relatives should be encouraged to see the body, but if there is reluctance to do so, this must be respected (Cathcart, 1988). Risk factors for pathological bereavement should be considered when addressing the needs of the family (see Table I; Adam, 1997).

## ADVANCE CARE PLANNING

Advance care planning is the process by which a person who has decisional capacity makes plans for health care that become active if he or she becomes incapacitated. This type of planning is considered to be an important aspect of palliative care for patients with life-threatening illnesses. The value of advance care planning derives from the observation that unnecessary and unwanted terminal hospital admission, intubation, or other forms of intensive care in a dying patient may prevent a peaceful death. Planning ahead, which should be initiated and guided by a physician, increases the likelihood that the care received at the end of life will be consistent with the patient's values and wishes and that the excessive "medicalization" of dying can be avoided. From this perspective, advance care planning should be seen as an essential element in patient-centered neurological management (Bernat, 2001; Miles et al., 1996).

Advance care planning can take several forms. The designation of an individual as the agent to make decisions in the event of incapacity is perhaps the most important. In some regions, this is an informal process, in which the designated person is simply identified to the physician. In other locales, the approach requires a legal document, which is known as a power of attorney for health care decisions or a health care proxy (Voltz et al., 1998).

Another form of advance care planning is the so-called "living will," which is a document that indicates the patient's preferences concerning resuscitation and the use of life-prolonging therapies, such as feeding tubes. "Do Not Resuscitate" orders, which focus on patients' preferences concerning cardiopulmonary resuscitation is a specific type of advance care planning that has become important in hospitals in the United States.

Advance care planning may therefore focus on any number of potential scenarios. Depending on the diagnosis, a variety of difficult situations can be foreseen. The typical symptoms in the terminal phase of neurological disease should be anticipated by the neurologist and explained to the patient and family. It should be stressed that intensive palliative care can prevent suffering and that open discussion about aggressive life-prolonging interventions, such as intubation, is helpful to avoid a crisis and ensure that treatment reflects the medical realities and the values and desires of the patient. Important decisions, such as a refusal of terminal intubation, should be incorporated into the advance directive, which should be as specific as possible. Oral advance directives should be noted in the medical records (Voltz et al., 1998).

The importance of advance directives again highlights the necessity of open and honest communication with the patient and family throughout the course of the illness. Patients can change their minds about the care that they want as they live with the disease. For this reason, it may be necessary to revisit the discussion about advance directives periodically. After a patient loses capacity and the advance directive becomes operative, there may yet be difficult challenges in discussing treatment options with the proxy or establishing clarity about the medical realities. Patients should be strongly encouraged to have an advance directive, but it should be understood that this set of pronouncements does not obviate the need for careful assessment, ongoing discussion, and ethical decision making on the part of the neurologist.

## PSYCHOSOCIAL CARE

In palliative care, as defined by the WHO, "the control of psychological, social and spiritual problems is paramount." Palliative care requires a multidisciplinary team approach (Table II), with close collaboration between the team members. At any one time, a different member of the team may be the most important person for the patient and his or her family. The psychosocial care of the relatives is as important as that of the patient (Goldblatt and Greenlaw, 1989). The needs and fears of the patients' children deserve special attention. Patients' associations may provide invaluable help and assistance and should be involved in patient care from the very beginning. When appropriate, referral to

TABLE II  Palliative Care: Who Is Involved?

Chaplain
Counselor
Dietitian
Hospice worker
Patients' associations
Nurse
Occupational therapist
Physical therapist
Physician
Psychologist
Relatives
Social worker

a tertiary care center with an interdisciplinary team may ease the burden on the practicing neurologist and may also be a means of providing hope.

## SPIRITUAL CARE AND BEREAVEMENT

Spiritual care is a very important, but often overlooked, part of palliative care. The word "spiritual" has several implications. Sykes (2000) defines it as "the need to find within present existence a sense of meaning," which may or may not involve a religious framework. A simple structured interview to assess the patients' spiritual needs has been recently developed (Puchalski and Romer, 2000). Spiritual care should encompass the whole family as a means of preventing problems during bereavement. It is important to acknowledge that the process of bereavement in terminal illness actually starts immediately after the diagnosis is communicated, in the form of so-called anticipatory grief, and that callous

delivery of the diagnosis may affect the psychological adjustment to bereavement.

## SYMPTOMS

Excellence in symptom control is part of all clinical practice and is fundamental to adequate palliative care. Efforts to improve comfort often use a starting point for improving palliative care generally. Neurologists should be familiar with the therapies that may be useful in managing symptoms that typically produce distress in populations with advanced neurological illness.

Few detailed studies of symptom prevalence have been conducted in populations with neurological disorders. A study from St Christopher's Hospice in the United Kingdom showed that 89% of ALS patients received opioids in the terminal phase, mostly for control of dyspnea; no patient choked to death, and 94% died peacefully (O'Brien et al., 1992). This experience of a peaceful death of ALS patients was recently confirmed in an outpatient clinic setting (Neudert et al., 2001). Similar studies of symptom prevalence in conditions such as MS, brain tumors, and dementia are available but incomplete (see Table III; Bausewein et al., 2000; Clanet and Brassat, 2000; Lloyd-Williams et al., 1996).

Systematic epidemiological symptom surveys are needed in populations with incurable neurological illnesses. This type of information is available for the cancer population and has helped define symptom prevalence and characteristics, the factors that contribute to symptom distress, and the outcomes that might be expected from the expert application of symptom management approaches in the clinical setting

TABLE III  Symptoms in the End Stage of ALS, MS, Dementia, and Oncological Patients

| ALS | MS | Dementia | Tumors |
|---|---|---|---|
| Weakness | Cognition deficit | Dyspnea | Fatigue |
| Immobility | Pain | Pain | Anorexia |
| Constipation | Spasticity | Pyrexia | Pain |
| Any pain | Ataxia | | Nausea |
| Dysarthria/dysphonia | Tremor | | Constipation |
| Musculoskeletal pain | Immobility | | Alteration of |
| Insomnia | Impaired sensation | | consciousness |
| Cramping pain | Pressure sores | | Death rattle |
| Ankle edema | Dysarthria | | Incontinence |
| Dyspnea | Dysphagia | | Restlessness |
| Drooling | Incontinence | | Dyspnea |
| Anxiety/agitation | Sexual dysfunction | | Sweating |
| Depression | Contractures | | |
| Weight loss/anorexia | | | |
| Vomiting | | | |
| Oral candidiasis | | | |
| Incontinence | | | |

(Boyd, 1993; Twycross and Lichter, 1998; Ventafridda *et al.*, 1990).

## Routes of Drug Administration

The oral route is generally preferred for the pharmacological management of symptoms (O'Neill, 1994). Unnecessary intravenous lines can reduce the mobility of the patients and may even be the limiting factor preventing a discharge. Some drugs, such as the opioid fentanyl and the anticholinergic scopolamine, can be delivered by means of transdermal patch, and this approach is a simple and effective alternative to oral drug delivery.

When dysphagia or gastrointestinal disease precludes the oral route, alternative delivery drug systems are needed. In addition to the transdermal route, the sublingual and rectal routes can be used with selected drugs and patients. There are few sublingual formulations; however, and the rectal route is usually a short-term solution, with particular usefulness at the very end of life.

Parenteral drug administration, if necessary, can be delivered as a continuous infusion by means of the intravenous or the subcutaneous route. Continuous subcutaneous infusion is a widely accepted technique among palliative care specialists. The needle can usually be kept in one site for days to a week or more. Several milliliters per hour can usually be delivered comfortably and, if volume becomes a problem, the addition of hyarluronidase to the infusate can often obviate it. Examples of important drugs and possibilities of their route of administration in the terminal phase are given in Table IV.

## Dyspnea

Dyspnea, if not appropriately treated, is a fearsome symptom. The vicious circle dyspnea-anxiety-dyspnea must be broken as soon as it occurs. There is extensive literature on cancer-associated dyspnea that is relevant to the problems encountered in neurological patients (Ahmedzai, 1998; Bruera *et al.*, 1990; Dudgeon, 1994).

The causes of dyspnea are multifactorial and may relate to primary cardiac or pulmonary dysfunction, muscle weakness, general debility, anxiety, and other factors. Unless the patient is imminently dying, a simple clinical evaluation can usually help clarify the potential causes and determine whether any are treatable. There is no correlation between the symptom of dyspnea and spirometry, and this diagnostic measure may be omitted (Heyse-Moore *et al.*, 2000).

When managing the dyspneic patient, communication with the patient and family is particularly important. It should emphasized that the patient will not choke to death and that nursing procedures, such as suction and chest physiotherapy, will keep the chest clear (Neudert *et al.*, 2001; O'Brien *et al.*, 1992). The patient and family may benefit from knowing that somnolence related to hypercapnea will probably occur if the dyspnea is associated with declining pulmonary function and that this process of "going to sleep" will further relieve the patient's suffering.

A variety of treatments may be considered for the management of dyspnea (Table V). Opioids are widely accepted as the mainstay approach. Most studies have been performed in patients with chronic obstructive pulmonary disease, and there are a few trials in cancer

**TABLE IV  Possible Routes of Drug Administration and Important Medications in Palliative Therapy**

| | |
|---|---|
| Oral | All drugs |
| Sublingual | Hyoscine, lorazepam |
| Transdermal | Fentanyl, hyoscine |
| Subcutaneous | Morphine, diclofenac, metoclopramide, haloperidol, midazolam, octreotide, dexamethasone |
| Rectal | Morphine, oxycodone, diclofenac, domperidone, diazepam |

Modified from Adam, 1997, with permission from the BMJ Publishing Group.

**TABLE V  Therapy of Dyspnea**

1. Treat reversible causes if present (e.g., bronchospasm, heart insufficiency, pneumonia, hyperhydration)
2. Nonmedical measures: explanation, reassurance, positioning patient, breathing exercises, distraction and relaxation techniques, calm presence of family, cool draught, physical therapy, fan, space
3. Intermittent dyspnea
   - Relieve anxiety: lorazepam (0.5–2 mg sublingually) or diazepam (2.5–5 mg PR); use opioids (e.g., starting 5–10 mg oral or 2.5–5 mg SC morphine in the opioid-naïve patient
   
     OR
   
     25%–50% of baseline dose in opioid-treated patients, and titrating as needed)
   - If severe, midazolam IV 2.5–5 mg slowly, or start at 10 mg/24 h constant infusion
4. Constant dyspnea
   - Opioids (e.g., morphine, starting 5–10 mg oral or 2.5–5 mg SC morphine every 4 hours in the opioid-naïve patient or 25%–50% of baseline dose in opioid-treated patients, and titrating as needed; may use controlled release oral opioids)
   - Regular inhaled opioids (not routinely recommended)
   - If cough is the main problem, use hydrocodone
   - Diazepam or midazolam 2.5–5 mg, mainly as add-on nocté
5. Oxygen
   - Only if clinically manifest hypoxia is present
   - Side effects: respiratory depression, restricts patient's mobility, psychological dependency, may prevent discharge home

patients and in normal individuals (Boyd and Kelly, 1997; Davis, 1997). Oral opioids are commonly used in the clinical situation. In one study of patients with far-advanced cancer, 5 mg of parenteral morphine reduced dyspnea and did not cause respiratory depression (Bruera *et al.*, 1990\*\*). This regimen also proved to be efficious in two recent controlled trials (Bruera *et al.*, 1993\*\*\*; Mazzocato *et al.*, 1999\*\*\*). In the opioid-naïve patient, a starting dose of morphine 5–10 mg orally every 4 hours, or the equivalent dose of another opioid, ensures safety and can be rapidly titrated to optimize effects. In those who are already receiving opioids for pain, the starting dose is typically 25%–50% higher than the baseline dose. In urgent situations, repeated boluses of morphine at short intervals, such as 1.5 mg every 10 min in the opioid-naïve patient, may be effective (Kumar *et al.*, 2000\*).

Nebulized morphine (Zeppetella, 1997\*) and fentanyl (Ahmedzai, 1998\*) have also been used for dyspnea, but controlled studies suggest that they are not better than nebulized saline (Noseda *et al.*, 1997\*\*) and may produce side effects (Lang and Jedeikin, 1998). Additional studies are needed before this approach can be recommended in routine practice.

Benzodiazepines also are frequently used in the management of dyspnea, despite the lack of well-controlled studies in neurological patients. A dose of 25 mg diazepam has been shown to have a negative effect on exercise tolerance and blood gases, but a single 5-mg dose in chronic obstructive airway disease patients improved sleep duration without worsening nocturnal hypoxemia (Woodcock *et al.*, 1981\*). Lorazepam and midazolam may both be useful in treatment of terminal dyspnoea (McNamara *et al.*, 1991\*).

Although oxygen should not be applied on a routine basis (Ahmedzai, 1998), it has been shown to significantly reduce dyspnea in a double-blind, crossover study, provided there is a proven hypoxia at the outset (Bruera *et al.*, 1993\*\*\*). There may be psychological value in providing oxygen even if hypoxemia cannot be demonstrated.

Other strategies have been used to manage dyspnea. Promethazine has been shown to reduce dyspnea in patients with chronic lung disease, but no studies in the neurological patient are available. Some surveys suggest that acupuncture may be effective against dyspnea (Filshie *et al.*, 1996\*; Pan *et al.*, 2000\*). No benefit has been shown with nebulized lignocaine (Wilcock *et al.*, 1994).

## Death Rattle

A disturbing noise called death rattle occurs in about half of all dying patients and may contribute significantly to the distress of family members (Morita *et al.*, 2000; Twycross and Lichter, 1998). Death rattle may be caused by salivary and/or bronchial secretions (Bennett, 1996). In patients with brain tumors, this symptom may be relatively prevalent and require aggressive therapy (Bennett, 1996; Morita *et al.*, 2000\*\*). The usual approach to treatment involves the administration of an anticholinergic drug combined with additional measures, like suction. Hyoscine (scopolamine), atropine, and glycopyrrolate are used. Atropine may be centrally excitatory, which can be controlled by benzodiazepines. In a recent prospective study, good symptom control was achieved in 71% of patients (see Table V; Morita *et al.*, 2000\*\*). Especially in comatose patients, *noisy tachypnea*, with a respiratory rate of 30–50/min, also may give the impression of severe distress. Morphine parenterally titrated to a rate of respiration of 10–15/min may be useful (Twycross and Lichter, 1998\*).

## Restlessness

Terminal "restlessness" is an imprecise label, and management depends on accurate assessment. In some cases, restlessness needs to be differentiated from myoclonus with minimal additional evidence of delirium; this is pure motor restlessness without mental disturbance (Adam, 1997). Other types of pure motor phenomena that may be termed restlessness include akathisia, resulting from the initiation of neuroleptics or opioids and restless legs syndrome (caused by, e.g., uremia). In the latter case, therapy with an opioid, a benzodiazepine, or a L-dopa agonist may be considered (Earley *et al.*, 2000).

In other cases, restlessness may be related to a specific reversible factor, such as pain, a distended bladder or rectum, dyspnea, the inability to move, a paradoxical reaction to benzodiazepines, or a drug or nicotine withdrawal (Back, 1992; Kranjnik and Zylic, 1995; Twycross and Lichter, 1998).

In most cases, however, restlessness is associated with a broader delirium in the context of advanced illness. Specific treatment of the delirium may relieve the problem (see later). If restlessness is severe, the use of a

---

**TABLE VI    Therapy of Death Rattle**

1. Explain to family
2. Stop hyperhydration
3. Gentle aspiration and lateral placement
4. 10–20 mg N-butyl-scopolamine SC (no central effect)
5. Alternatively, hyoscine (scopolamine base or hydrobromide) or atropine 0.5–1 mg SC or IM every 2–4 h; if atropine or hyoscine is used, add anxiolytic (e.g., lorazepam)
6. If necessary, add midazolam, 5 mg SC

**TABLE VII    Therapy of Restlessness**

1. Treat causes
2. Calm family
3. Benzodiazepines (e.g., lorazepam, 0.5–2 mg every 1–4 h [PO, SL, IV, IM]; midazolam, single dose 2.5–10 mg; 30–60 mg in 24 h [IV, IM, SC]; diazepam, 5–10 mg every 4–12 h [PO, PR, IV])
4. If necessary, sedating neuroleptic (e.g., levomepromazine, 10–20 mg every 4–8 h [PO, SC, IV, IM]; haloperidol, 0.5–2 mg [PO, SC, IV, IM] every hour until calm, then bid dose)

benzodiazepine or a neuroleptic for the purpose of inducing sedation often is appropriate (Table VII).

## Delirium

Delirium may be transient and potentially reversible, or it may be an indicator of the start of the dying process (Caraceni *et al.*, 2000; Lawlor *et al.*, 2000). Cardinal features are an acute onset with fluctuating course, presence of an underlying medical condition that could account for an encephalopathy, reduced or fluctuating sensorium, and attention deficit combined with cognitive and perceptual disturbance. Symptoms may worsen during the evening and night, or with sedating medication. Subtypes can be distinguished according to altered psychomotor activity as hypoactive, hyperactive, or mixed (Liptzin and Levkoff, 1992).

Delirium is present in 28%–44% of cancer patients on admission to a palliative care unit (Lawlor *et al.*, 2000; Pereira *et al.*, 1997). It may be present in more than 90% of dying patients (Caraceni *et al.*, 2000; Lawlor *et al.*, 2000). The differences in occurrence rates may be explained by nonstandardized diagnostic criteria and heterogenous study populations. There have been no studies to define the epidemiology of delirium in populations with progressive neurological disorders.

Delirium is most often multifactorial. Baseline vulnerability factors (e.g., male gender, advanced age, cachexia, hepatic impairment, CNS disease, impaired functional status) are in most cases not amenable to correction. However, there are a number of potentially reversible precipitants (Caraceni *et al.*, 2000; Lawlor *et al.*, 2000), and in a recent prospective series the syndrome was reversible in up to 49% of cases (Lawlor *et al.*, 2000). Potentially reversible factors include centrally acting medications, drug or alcohol withdrawal, dehydration, infection, hypercalcemia, and nonconvulsive status epilepticus.

The management of the delirious patient begins with a search for potentially reversible causes. The extent of the evaluation must be guided by the goals of care and judgments concerning the likelihood of imminent death.

The treatment of potentially reversible causes must similarly be guided by a careful assessment. Dehydration, electrolyte abnormalities, infection, and medication toxicity can all be addressed with simple interventions, but even these treatments may not be appropriate if death is imminent and the goals of care are limited to comfort in the dying process.

In a recent double-blind study, haloperidol, chlorpromazine, and lorazepam were compared in the treatment of delirious hospitalized patients with AIDS (Breitbart *et al.*, 1996***); the results suggest that haloperidol and chlorpromazine are more effective than lorazepam, equal to each other, and associated with few extrapyramidal side effects. In addition to the low risk of extrapyramidal movement disorder, these drugs can cause hypotension. They also can lower the seizure threshold, and antiepileptic therapy should be increased or initiated in patients at high risk for this complication. Newer neuroleptics, such as clozapine, risperidone, and olanzepine, have not yet been evaluated but are likely to be safer and are being used empirically. Combinations of a neuroleptic and other drugs, such as tricyclic antidepressants, or treatment with benzodiazepines to reduce restlessness, may occasionally worsen delirium.

Treatment of a hypoactive delirium using methylphenidate or another psychostimulant has been suggested but should be undertaken cautiously because of a risk of worsening perceptual disturbances (Morita *et al.*, 2000). Also, propofol has been advocated for the therapy of agitation associated with terminal delirium because of its rapid onset, short half-life, and lack of accumulation (Mercadante *et al.*, 1995*; Moyle, 1995*). A pragmatic approach to the management of delirium is shown in Table VIII.

In terminal agitated delirium, the management of the delirium may be very difficult and may pose the

**TABLE VIII    Therapy of Delirium**

1. Treat reversible causes (e.g., hyperhydration, drugs, infection, hypercalcemia, nonconvulsive status)
2. General supportive measures—light, reassurance, few and familiar faces
3. Calm presence of family
4. Neuroleptics (e.g., haloperidol, 0.5–2 mg [PO, SC, IV, IM] every hour until calm, then bid dose; SC infusion 5–30 mg over 24 h; indicated if drug toxicity, altered sensorium, metabolic disturbance): methotrimeprazine/levopromazine, SC bolus 25 mg, SC infusion up to 250 mg over 24 h; indicated if alternative or additional sedation necessary)
5. Benzodiazepines (*only in addition to neuroleptics*), if anxiety and distress, risk of seizure: lorazepam, 0.5–2 mg (PO, SL, IM, IV), or midazolam, 5–20 mg (IV, SC), or diazepam, 10–20 mg (PR)

*Avoid slippery slope of inappropriate sedation*

question of terminal sedation to manage the agitation, a benzodiazepine, such as lorazepam or midazolam, a barbiturate, or propofol, may be useful. A subcutaneous infusion of midazolam is a common approach in some palliative care units and may be initiated at 0.5–1 mg/h and be titrated higher as needed (McNamara *et al.*, 1991*). Because the main metabolite of midazolam, α-hydroxy-midazolam, accumulates in terminal renal failure, the dose should be constantly reviewed (Naritoku and Sinha, 2000).

## Drowsiness

There are very few studies on the level of consciousness of patients who are dying with neurological diseases. Depending on the underlying pathosis, some patients will lose consciousness long before death, whereas others will be lucid up to the very end. In studies of cancer patients, between 6% and 30% were fully conscious until death or less than 15 minutes before death (Twycross and Lichter, 1998). A recent study on the terminal phase in ALS patients showed that 27% of patients had been alert and communicating 5 minutes or less before death, whereas 62% were asleep and 11% were comatose at time of death (Neudert *et al.*, 2001).

Given the variability in consciousness at the end of life, family members should be informed that the patient may still hear them despite clouded consciousness. They should be encouraged to speak to the patient. Vegetative reactions such as tachycardia, tachypnea, and sweating should be closely monitored, because they may signal potentially treatable distressing symptoms.

The management of somnolence, like the management of delirium, is strongly influenced by the goals of care. If the patient is perceived to be imminently dying, there should be no effort to reverse somnolence. If, however, it is appropriate to treat, the first step involves an assessment of potentially reversible causes (see Table IX). If treatment of these factors is insufficient, a trial of a psychostimulant drug, such as

methylphenidate, dextroamphetamine, or modafinil, may be considered.

## Epileptic Seizures

Even in the terminal phase of disease, nonconvulsive status epilepticus should always be considered if a sudden change in mentation occurs. If the patient has tonic-clonic seizures, immediate therapy may consist of diazepam 10 mg PR, or midazolam, 5 mg parenterally. This can be followed by diazepam, 10 mg PR, every hour, until the seizures stop, then 20 mg PR at night. Alternatively, midazolam, 5–10 mg SC or IV, may be given every hour until the seizures stop, after which a subcutaneous infusion can be initiated at 30 mg/24 h; the dose of the infusion can then be adjusted as needed (MacNamara *et al.*, 1991). There is a prolongation of the half-life of midazolam after sustained infusion, and the dose often will need ongoing adjustment, particularly if the goal is to gradually awaken the patient (Naritoku and Sinha, 2000). If steroids are tapered in the terminal phase of brain tumor patients, antiepileptic (and analgesic) treatment should be increased prophylactically.

## Myoclonus

The causes of terminal myoclonus parallel those responsible for delirium (Twycross and Lichter, 1998). If the cause for myoclonus is not reversible, symptomatic therapy may be indicated if the patient experiences the movements as distressing (Table X).

## Pain

In contrast to cancer pain, relatively little is known about pain in populations with varied advanced neurological disorders. The limited survey data available suggest that pain is a significant problem in some populations (Bausewein *et al.*, 2000; Neudert *et al.*, 2001), but prevalence is probably less in neurological disorders than metastatic cancers (Lichter and Hunt, 1990).

---

**TABLE IX    Causes of Loss of Consciousness**

Raised intracranial pressure
Epileptic seizure
Hypoxia
Infection
Drug side effects (anticholinergics, benzodiazepines, opioids, H$_2$-blockers, phenothiazine, steroids)
Laboratory (calcium, sodium, glucose, ketones, uremia, hepatic encephalopathy, endocrine disturbance)
Intoxication (alcohol, drugs)

---

**TABLE X    Therapy of Myoclonus**

| | |
|---|---|
| • Hypoxic | Piracetam |
| • Metabolic or drug induced | Benzodiazepines (e.g., midazolam, 5–10 mg SC every hour; diazepam, 5–10 mg PR every hour; clonazepam; lorazepam, PO, SL) |

Despite this, neurological patients who experience pain should receive aggressive analgesic therapy using the principles of pain therapy developed for cancer pain (for details see Chapter 11). There is probably undertreatment of pain in populations with neurological diseases as there is in the cancer population (Zenz *et al.*, 1995).

An expert committee of the World Health Organization (WHO, 1986) promulgated the concept of an "analgesic ladder" approach to the management of cancer pain. This has been an important teaching tool, emphasizing as it does the need to select drugs with maximal analgesic efficacy sufficient to address the pain complaint. The approach encourages treatment by the least invasive route possible, individualization of the dose, and the treatment of side effects. This analgesic ladder approach should be viewed as a broad guideline, which may be shaped as needed by the evolving science of pain management, including the growing availability of new opioid delivery systems and new analgesic drugs.

Patients with acute or chronic, mild-to-moderate pain usually are first treated with acetaminophen or a nonsteroidal anti-inflammatory drug (NSAID). This therapy might be combined with drugs selected to treat a side effect of the analgesic (e.g., a proton pump inhibitor to reduce the risk of gastroduodenopathy from the NSAID). It also might be combined with, or supplanted by, a specific "adjuvant analgesic." The latter drugs comprise a very diverse group that are marketed for reasons other than pain but are analgesic in selected circumstances. Patients with mild-to-moderate neuropathic pain, for example, might be primarily treated with an analgesic anticonvulsant, such as gabapentin, or an antidepressant.

Patients with moderate-to-severe chronic pain usually are treated with an opioid. When the pain in moderate, a combination formulation that mixes an opioid and a nonopioid typically is tried first. Alternatively, a single-entity opioid, such as morphine, can be used at a relatively low dose. Again, the opioid can be combined with a drug to prevent or treat side effects or an adjuvant analgesic that can potentially provide additive analgesia. In some cases, such as the patient with "crescendo" neuropathic pain, an adjuvant analgesic, such as a corticosteroid, may be started before the opioid, or in tandem with it.

Individualization of the dose is most important in the effort to optimize an opioid regimen. There is no one correct dose or dose range. The absolute dose is immaterial as long as there is a favorable balance between analgesia and side effects. The optimal dose is found through a process of dose titration, in which the dose is increased (typically 25%–100%) at intervals until favorable effects occur or the patient reports intolerable and unmanageable side effects. Aggressive side effect management, such as laxative therapy, is essential to optimize this balance between analgesia and side effects.

### Nausea and Vomiting

In patients with neurological diseases, nausea and vomiting may be caused by diverse etiologies, including raised intracranial pressure or direct brainstem pathosis. Other causes (e.g., gastrointestinal disease or medications) should be identified and treated appropriately (Adam, 1997).

Treatment of nausea should attempt to target the cause, if possible, and also provide primary antiemetic treatment. For patients with space-occupying lesions, inflammatory foci in the brain, or bowel obstruction, corticosteroid therapy may be very useful (Vecht *et al.*, 1994**). Other primary antiemetic approaches include neuroleptics, anticholinergic drugs, cannabinoids, and serotonin antagonists (Table XI). The latter agents can be highly effective (Macleod, 2000*) but are very expensive and are typically tried after other drugs are shown to be ineffective.

**TABLE XI  Therapy of Terminal Nausea and Vomiting**

| Class | Mechanism | Drugs |
|---|---|---|
| 5-HT$_3$ antagonists | 5-HT$_3$ blockade | Ondansetron (8 mg slowly IV up to 8 hourly) |
| | | Granisetron (3 mg slowly IV up to 8 hourly) |
| Substituted benzamide | 5-HT$_3$/dopamine blockade | Metoclopramide (10–30 mg, 2–4 hourly, PO, SC) |
| Corticosteroids | Unknown | Dexamethasone (2–4 mg 8 hourly) |
| Butyrophenones | Dopamine blocker | Haloperidol (1.6–5.0 mg/day/PO, SC) |
| Phenothiazines | Dopamine blocker | Prochlorperazine (5–20 mg PO, 12.5–25 mg IM) |
| Benzodiazepines | Benzodiazepine receptor blocker | Lorazepam (0.5–2 mg PO) |

## Thirst

As long as the patient can express thirst, there typically is no need to provide fluids artificially (Micetich *et al.*, 1983; O'Neill, 1994). An excess of fluid replacement may lead to overhydration followed by exacerbation of symptoms (e.g., caused by increased brain edema). A slight terminal dehydration may actually be beneficial (Oliver, 1984), and there is no association between symptoms like thirst or dry mouth and the amount of fluid intake (Burge, 1993). On the other hand, a dry mouth can be a severely distressing symptom, which should be prevented with regular mouth care (see Table XII; Ventafridda *et al.*, 1998).

The use of fluid replacement in confused or comatose dying patients is controversial. Some experts believe that the administration of at least 700 ml/24 h in these patients, which can be achieved by subcutaneous infusion (hypodermoclysis) (MacDonald and Fainsinger, 1994), should be tried unless the patient is fluid overloaded. In some cases, this hydration may lessen delirium and apparently improve comfort. In contrast, other experts perceive that dehydration is a natural stage before death, and artificial rehydration does little but prolong the dying process. Clearly, the decision to provide fluids must be made on a case-by-case basis after a careful assessment of the goals of care.

## Depression

In terminal cancer patients, depression requiring treatment is often underdiagnosed and undertreated (Berney *et al.*, 2000; Lloyd-Williams *et al.*, 1999). This is probably also the case in those with advanced neurological disorders. The clinical diagnosis relies on identification of the affective state (e.g., pervasive sadness, hopelessness, and helplessness) and cognitions (e.g., suicidal thoughts) rather than somatic signs, because these are highly prevalent in the end stage of any disease. Even in the terminal phase of disease, pharmacological therapy may be instituted, and at least 50% of the patients may be expected to profit from it (Table XIII).

Fluoxetine, desipramine, and mianserin have been shown to be effective in cancer patients (Razavi *et al.*,

#### TABLE XII  Management of Terminal Dry Mouth

1. Good and regular oral hygiene (every 2 h, show to relatives)
2. 1–2 ml of water by pipette or syringe every 30–60 min
3. Smooth lips with vaseline or dexpanthenol
4. Moist air
5. If candidiasis, treat with nystatin
6. Pieces of pineapple, ice cream in small portions, crushed ice cubes in fabrics, etc.

#### TABLE XIII  Therapy of Depression

1. Rule out underlying potentially reversible causes (medication, metabolic, organic brain disorders)
2. Try to identify psychosocial factors
3. If onset of action should be rapid, psychostimulants (e.g., methylphenidate, 5–30 mg/day PO; dextroamphetamine, 5–30 mg/day PO; pemoline, 37.5–150 mg/day PO)
4. If onset of action may take weeks, tricyclics (e.g., amitryptiline, imipramine, doxepine, desipramine, nortriptyline, all 25–125 mg/day PO) or serotonin reuptake inhibitors (e.g., fluoxetine (20 mg/day), sertraline (50–200 mg/day)
5. For agitation and anxiety benzodiazepines (e.g., lorazepam [0.5–2 mg])
6. Phytotherapy (hypericum extract, 1050 mg/day)

1996\*\*; Van Heeringen and Zivkov, 1996\*\*). These may take weeks before the onset of effect, but there are reports that faster responses can occur in the medically ill. Even tricyclics may be effective within 1 week (Kugaya *et al.*, 1999\*). Psychostimulants also have been described to be useful and typically exert an effect very quickly (Lloyd-Williams *et al.*, 1999\*). Phytotherapy (e.g., St. John's wort) has been reported to be effective in some studies (Philipp *et al.*, 1999\*), but other trials fail to show benefit.

Some patients are at increased risk for suicide, including those with MS, spinal cord lesions, and some groups of patients with epilepsy (Stenager and Stenager, 1992). For MS patients, the cumulative lifetime risk is 1.95% and is highest for men with an onset of disease before age 30 (Stenager *et al.*, 1992). Nearly 30% of 145 deaths in a cohort of MS outpatients were due to suicide (Sadovnik *et al.*, 1991), but only 1 of 171 deths in an ALS cohort (Neudert *et al.*, 2001). In cancer patients, about one fifth has a high desire for hastened death, which is significantly associated with a clinical diagnosis of depression and hopelessness (Breitbart *et al.*, 2000). Optimal palliative therapy, better antidepressant therapy, and better family counseling may help to reduce these alarming figures.

## Fatigue

Fatigue is a very prevalent symptom in advanced disease, including neurological disorders such as MS (Krupp and Elkins, 2000). In a recent study in cancer patients, the prevalence of "severe subjective fatigue," defined as fatigue greater than that experienced by 95% of the control group, was found to be 75% (Stone *et al.*, 1999). Fatigue severity was significantly associated with pain and dyspnea scores in these patients. Causes of fatigue still remain obscure, but many potential causes can be indentified during a routine assessment. These include anemia, metabolic disturbances, medication toxicity, sleep disturbance, and major depression. The first step in the management of fatigue should include this assessment and judgments concerning the primary treatment of potential causes (Portenoy and Itri, 1999; Portenoy and Miaskowski, 1998). Some patients with fatigue warrant trials of pharmacotherapy. The drugs that have been used include the psychostimulants, corticosteroids, amantadine, and activating antidepressants (such as buproprion). In a recent controlled trial, methylphenidate (up to 60 mg) was tested vs pemoline (150 mg) or placebo in patients with AIDS. Both methylphenidate and pemoline were equally effective and led to significant improvement of the quality of life and a decreased level of depression and psychological distress (Breitbart *et al.*, 2001***). Similar outcomes have been reported during modafinil therapy of MS patients (Terzoudi *et al.*, 2000***).

## Palliative Sedation

Palliative sedation has been advocated as a means to address refractory suffering at the end of life (Hardy, 2000; Rousseau, 2000). Sedation means the deliberate induction of unconsciousness (Cherny and Portenoy, 1994; Mount and Hamilton, 1994). Refractory suffering is difficult to define, and its therapy must be done on an individual basis (Cherny and Portenoy, 1994; Mount and Hamilton, 1994). Although palliative sedation is usually not indicated unless death is perceived as imminent, there is no difference in survival time between sedated and nonsedated patients (Fainsinger *et al.*, 2000; Stone *et al.*, 1997). The ethical basis for sedation in the imminently dying is the "principle of double effect." This principle states that an action with a foreseen negative outcome, such as the possibility of accelerated death from the sedation, is acceptable if the intention is to produce a beneficent outcome, such as the relief of suffering, and the good outcome is more important to achieve than the negative outcome is to avoid (The Ethics and Humanities Subcommittee of the AAN, 1993). Given the ethical acceptability of sedation in these circumstances, clinicians should approach the prospect with openness. If intractable symptoms, for example, produce profound suffering, the possibility of sedation can be discussed with the patient, the family, and the medical team. These discussions and their outcome should be documented.

To induce sedation, benzodiazepines, barbiturates, opioids, and other drugs (e.g., propofol) can be considered. Barbiturates (Truog *et al.*, 1992) and benzodiazepines are usually preferred. Whatever drug is used, the dosing must be individually slowly titrated upward and stopped once satisfactory symptom control is acchieved.

## CONCLUSION

Providing adequate palliative care is a duty of every neurologist (The Ethics and Humanities Subcommittee of the AAN, 1996). There are good reasons to believe that appropriate skills in this area will reduce the call for active euthanasia and physician-assisted suicide. No matter how dire the diagnosis and prognosis, clinicians should never devastate patients and their families by saying "I'm sorry, but there's nothing more I can do for you." Rather, the words of Dame Cicely Saunders, the founder of the hospice movement, ring true: "You matter until the last moment of your life. We will do all we can not only to help you to die peacefully, but also to live until you die."

## REFERENCES

Adam, J. (1997). ABC of palliative care. The last 48 hours. *BMJ* **315**(7122), 1600–1603.

Ahmedzai, S. (1998). Palliation of respiratory symptoms. *In* "Oxford Textbook of Palliative Medicine," 2nd ed. (D. Doyle, G. W. C. Hanks, and N. Macdonald, Eds.) Oxford University Press.

Back, I. N. (1992). Terminal restlessness in patients with advanced malignant disease. *Palliat. Med.* **6**, 293–298.

Bausewein, C., Hau, P., Dudel, C., Hartenstein, R., Bogdahn, U., and Voltz, R. (2000). How do patients with brain tumors die? *Akt. Neurol.* **27**, S211–212.

Berger, A., Portenoy, R. K., Weismann, D. E., Eds. (1998). Supportive Oncology. Philadelphia: Lippincott.

Bennett, M. I. (1996). Death rattle: an audit of hyoscine (scopolamine) use and review of management. *J. Pain Sympt. Manage* **12**(4), 229–233.

Bernat, J. L., Goldstein, M. L., and Viste, K. M. (1996). The neurologist and the dying patient. *Neurology* **46**, 598–599.

Bernat, J. L. (2001). Plan ahead: How neurologists can enhance patient-centered medicine. *Neurology* **56**(2), 144–145.

Berney, A., Stiefel, F., Mazzocato, C., and Buclin, T. (2000). Psychopharmacology in supportive care of cancer: a review for the clinician. III. Antidepressants. *Support Care Cancer* **8**(4), 278–286.

Borasio, G. D., Voltz, R., and Miller, R. G. (2001). Palliative care in amyotrophic lateral sclerosis. *Neurol. Clin.* **19**, 829–847.

Boyd, K. J. (1993). Short terminal admissions to a hospice. *Palliat. Med.* **7**, 289–294.

Boyd, K. J., and Kelly, M. (1997). Oral morphine as symptomatic treatment of dyspnoea in patients with advanced cancer. *Palliat. Med.* **11**(4), 277–281.

Breitbart, W., Marotta, R., Platt, M. M., Weisman, H., Derevenco, M., Grau, C., Corbera, K., Raymond, S., Lund, S., and Jacobson, P. (1996). A double-blind trial of haloperidol, chlorpromazine, and lorazepam in the treatment of delirium in hospitalized AIDS patients. *Am. J. Psychiatry* **153**(2), 231–237.

Breitbart, W., Rosenfeld, B., Kaim, M., and Funesti-Esch, J. (2001). A randomized, double-blind, placebo-controlled trial of psychostimulants for the treatment of fatigue in ambulatory patients with human immunodeficiency virus disease. *Arch. Intern. Med.* **161**(3), 411–420.

Breitbart, W., Rosenfeld, B., Pessin, H., Kaim, M., Funesti-Esch, J., Galietta, M., Nelson, C. J., and Brescia, R. (2000). Depression, hopelessness, and desire for hastened death in terminally ill patients with cancer. *JAMA* **284**(22), 2907–2911.

Bruera, E., de Stoutz, N., Velasco-Leiva, A., Schoeller, T., and Hanson, J. (1993). Effects of oxygen on dyspnoea in hypoxaemic terminal cancer patients. *Lancet* **342**, 13–14.

Bruera, E., Legris, M. A., Kuehn, N., and Miller, M. J. (1998). Hypodermoclysis for the administration of fluids and narcotic analgesics in patients with advanced cancer. *J. Pain Sympt. Manage* **5**, 218–220.

Bruera, E., MacEachern, T., Ripamonti, C., and Hanson, J. (1993). Subcutaneous morphine for dyspnea in cancer patients. *Ann. Intern. Med.* **119**(9), 906–907.

Bruera, E., Macmillan, K., Pither, J., and MacDonald, R. N. (1990). Effects of morphine on dyspnoea of terminal cancer patients. *J. Pain Sympt. Man.* **5**, 341–344.

Burge, F. I. (1993). Dehydration symptoms of palliative care cancer patients. *J. Pain Sympt. Manage* **8**, 454–464.

Caraceni, A., Nanni, O., Maltoni, M., Piva, L., Indelli, M., Arnoldi, E., Monti, M., Montanari, L., Amadori, D., and De Conno, F. (2000). Impact of delirium on the short term prognosis of advanced cancer patients. Italian Multicenter Study Group on Palliative Care. *Cancer* **89**(5), 1145–1149.

Carter, H., McKenna, C., MacLeod, R., and Green, R. (1998). Health professionals' responses to multiple sclerosis and motor neurone disease. *Palliat. Med.* **12**(5), 383–394.

Carver, A. C., Vickrey, B. G., Bernat, J. L., Keran, C., Ringel, S. P., and Foley, K. M. (1999). End-of-life care: a survey of US neurologists' attitudes, behavior, and knowledge. *Neurology* **53**(2), 284–293.

Cathcart, F. (1988). Seeing the body after death. *BMJ* **297**, 997–998.

Cherny, N. I., and Portenoy, R. K. (1994). Sedation in the management of refractory symptoms: guidelines for evaluation and treatment. *J. Palliat. Care* **10**/2, 31–38.

Clanet, M. G., and Brassat, D. (2000). The management of multiple sclerosis patients. *Curr. Opin. Neurol.* **13**(3), 263–270.

Committee on Care at the End of Life, Division of Health Care Services, Institute of Medicine (1997). "Approaching Death." (M. J. Field and C. K. Cassel, Eds.) Washington DC: National Academy Press, p. 31.

Davis, C. L. (1997). ABC of palliative care. Breathlessness, cough, and other respiratory problems. *BMJ* **315**(7113), 931–934.

Doyle, D., Hanks, G. W. C., and MacDonald, N. (Eds). (1998). Oxford Textbook of Palliative Medicine, 2nd ed., Oxford University Press.

Dudgeon, D. (1994). Dyspnea: Ethical Concerns. *J. Palliat. Care* **10**/3, 48–51.

Earley, C. J., Allen, R. P., Beard, J. L., and Connor, J. R. (2000). Insight into the pathophysiology of restless legs syndrome. *J. Neurosci. Res.* **62**(5), 623–628.

Emanuel, E. J., Fairclough, D. L., and Emanuel, L. L. (2000). Attitudes and desires related to euthanasia and physician-assisted suicide among terminally ill patients and their caregivers. *JAMA* **284**(19), 2460–2468.

Fainsinger, R. L., Waller, A., Bercovici, M., Bengtson, K., Landman, W., Hosking, M., Nunez-Olarte, J. M., and deMoissac, D. (2000). A multicentre international study of sedation for uncontrolled symptoms in terminally ill patients. *Palliat. Med.* **14**(4), 257–265.

Faulkner, A., Maguire, P., and Regnard, C. (1994). Breaking bad news—a flow diagram. *Palliat. Med.* **8**, 145–155.

Field, D., Douglas, C., Jagger, C., and Dand, P. (1995). Terminal illness: views of patients and their lay carers. *Palliat. Med.* **9**, 45–54.

Filshie, J., Penn, K., Ashley, S., and Davis, C. L. (1996). Acupuncture for the relief of cancer-related breathlessness. *Palliat. Med.* **10**, 145–150.

Finlay, I. G., and Dunlop, R. (1994). Quality of life assessment in palliative care. *Ann. Oncol.* **5**(1), 13–18.

Foley, K. M. (1997). Competent care for the dying instead of physician-assisted suicide. *N. Engl. J. Med.* **336**(1), 54–58.

Goldblatt, D., and Greenlaw, J. (1989). Starting and stopping the ventilator for patients with amyotrophic lateral sclerosis. *Neurol. Clin.* **7**, 789–806.

Hardy, J. (2000). Sedation in terminally ill patients. *Lancet* **356**(9245), 1866–1867.

Hardy, J. R., Rees, E., Ling, J., Burman, R., Feuer, D., Broadley, K., and Stone, P. (2001). A prospective survey of the use of dexamethasone on a palliative care unit. *Palliat. Med.* **15**(1), 3–8.

Hayashi, H., Kato, S., and Kawada, A. (1991). Amyotrophic lateral sclerosis patients living beyond respiratory failure. *J. Neurol. Sci.* **105**(1), 73–78.

Heyse-Moore, L., Beynon, T., and Ross, V. (2000). Does spirometry predict dyspnoea in advanced cancer? *Palliat. Med.* **14**(3), 189–195.

Johnston, M., Earll, L., Mitchell, E., Morrison, V., and Wright, S. (1996). Communicating the diagnosis of motor neurone disease. *Palliat. Med.* **10**, 23–34.

Killeen, M. E. (1991). Clinical clowning: Humor in hospice care. *Am. J. Hospice Palliat. Care* **5**/6, 23–27.

Kranjnik, M., and Zylicz, Z. (1995). Terminal restlessness and nicotine withdrawal. *Lancet* **346**, 1044.

Krupp, L. B., and Elkins, L. E. (2000). Fatigue and declines in cognitive functioning in multiple sclerosis. *Neurology* **55**(7), 934–939.

Kugaya, A., Akechi, T., Nakano, T., Okamura, H., Shima, Y., and Uchitomi, Y. (1999). Successful antidepressant treatment for five terminally ill cancer patients with major depression, suicidal ideation and a desire for death. *Support Care Cancer* **7**(6), 432–436.

Kumar, K. S., Rajagopal, M. R., and Naseema, A. M. (2000). Intravenous morphine for emergency treatment of cancer pain. *Palliat. Med.* **14**(3), 183–188.

Lang, E., and Jedeikin, R. (1998). Acute respiratory depression as a complication of nebulised morphine. *Can. J. Anaesth.* **45**(1), 60–62.

Lawlor, P. G., Gagnon, B., Mancini, I. L., Pereira, J. L., Hanson, J., Suarez-Almazor, M. E., and Bruera, E. D. (2000). Occurrence, causes, and outcome of delirium in patients with advanced cancer: a prospective study. *Arch. Intern. Med.* **160**(6), 786–794.

Lichter, I., and Hunt, E. (1990). The last 48 hours of life. *J. Palliat. Care* **6**, 7–11.

Liptzin, B., and Levkoff, S. E. (1992). An empirical study of delirium subtypes. *Br. J. Psychiatry* **161**, 843–845.

Lloyd-Williams, M. (1996). A survey of palliative care given to patients with end-stage dementia. *Palliat. Med.* **10**, 63 (abstr).

Lloyd-Williams, M., Friedman, T., and Rudd, N. (1999). A survey of antidepressant prescribing in the terminally ill. *Palliat. Med.* **13**(3), 243–248.

MacDonald, S. M., and Fainsinger, R. L. (1994). Symptom control: the problem areas (letter). *Palliat. Med.* **8**, 167–169.

Macleod, A. D. (2000). Ondansetron in multiple sclerosis. *J. Pain Sympt. Manage* **20**(5), 388–391.

Mazzocato, C., Buclin, T., and Rapin, C. H. (1999). The effects of morphine on dyspnea and ventilatory function in elderly patients with advanced cancer: a randomized double-blind controlled trial. *Ann. Oncol.* **10**(12), 1511–1514.

McNamara, P., Minton, M., and Twycross, R. G. (1991). Use of midazolam in palliative care. *Palliat. Med.* **5**, 244–249.

Meisel, A., Snyder, L., and Quill, T. (2000). Seven legal barriers to end-of-life care: myths, realities, and grains of truth. *JAMA* **284**(19), 2495–2501.

Mercadante, S., DeConno, F., and Ripamonti, C. (1995). Propofol in terminal care. *J. Pain Sympt. Manage* **10**, 639–642.

Miles, S. H., Koepp, R., and Weber, E. P. (1996). Advance end-of-life treatment planning. A research reveiw. *Arch. Intern. Med.* **156**, 1062–1068.

Micetich, K. C., Steinecker, P. H., and Thomasma, D. C. (1983). Are intravenous fluids morally required for a dying patient? *Arch. Intern. Med.* **143**, 975–978.

Morita, T., Otani, H., Tsunoda, J., Inoue, S., and Chihara, S. (2000). Successful palliation of hypoactive delirium due to multi-organ failure by oral methylphenidate. *Support Care Cancer* **8**(2), 134–137.

Morita, T., Tsunoda, J., Inoue, S., and Chihara, S. (2000). Risk factors for death rattle in terminally ill cancer patients: a prospective exploratory study. *Palliat. Med.* **14**(1), 19–23.

Mount, B., and Hamilton, P. (1994). When palliative care fails to control suffering. *J. Palliat. Care* **10**(2), 24–26.

Moyle, J. (1995). The use of propofol in palliative medicine. *J. Pain Sympt. Manage* **10**, 643–646.

Naritoku, D. K., and Sinha, S. (2000). Prolongation of midazolam half-life after sustained infusion for status epilepticus. *Neurology* **54**(6), 1366–1368.

Neudert, C., Oliver, D., Wasner, M., and Borasio, G. D. (2001). The course of the terminal phase in patients with amyotrophic lateral sclerosis. *J. Neurol.* **248**(7), 612–616.

Noseda, A., Carpiaux, J. P., Markstein, C., Meyvaert, A., and de Maertelaer, V. (1997). Disabling dyspnoea in patients with advanced disease: lack of effect of nebulized morphine. *Eur. Respir. J.* **10**(5), 1079–1083.

O'Brien, T., Kelly, M., and Saunders, C. (1992). Motor neuron disease: a hospice perspective. *BMJ* **304**, 471–473.

O'Neill, W. M. (1994). Subcutaneous infusions—a medical last rite. *Palliat. Med.* **8**, 91–93 (editorial).

Oliver, D. (1984). Terminal dehydration. *Lancet* **2**(8403), 631 (letter).

Pan, C. X., Morrison, R. S., Ness, J., Fugh-Berman, A., and Leipzig, R. M. (2000). Complementary and alternative medicine in the management of pain, dyspnea, and nausea and vomiting near the end of life. A systematic review. *J. Pain Sympt. Manage* **20**(5), 374–387.

Pereira, J., Hanson, J., and Bruera, E. (1997). The frequency and clinical course of cognitive impairment in patients with terminal cancer. *Cancer* **79**(4), 835–842.

Philipp, M., Kohnen, R., and Hiller, K. O. (1999). Hypericum extract versus imipramine or placebo in patients with moderate depression: randomised multicentre study of treatment for eight weeks. *BMJ* **319**(7224), 1534–1538.

Pollard, B. J. (2001). Can euthanasia be safely legalized? *Palliat. Med.* **15**(1), 61–65.

Portenoy, R. K., and Miaskowski, C. (1998). Assessment and management of cancer-related fatigue. *In* "Principles and Practice of Supportive Oncology." (A. Berger, D. Weissman, and R. K. Portenoy, Eds.) Philadelphia: JB Lippincott, 109–118.

Portenoy, R. K., and Itri, L. M. (1999). Cancer-related fatigue: guidelines for evaluation and management. *Oncologist* **4**, 1–10.

Puchalski, C., and Romer, A. L. (2000). Taking a spiritual history allows clinicians to understand patients more fully. *J. Palliat. Med.* **3**, 129–137.

Razavi, D., Allilaire, J. F., Smith, M., Salimpour, A., Verra, M., Desclaux, B., Saltel, P., Piollet, I., Gauvain-Piquard, A., Trichard, C., Cordier, B., Fresco, R., Guillibert, E., Sechter, D., Orth, J. P., Bouhassira, M., Mesters, P., and Blin, P. (1996). The effect of fluoxetine on anxiety and depression symptoms in cancer patients. *Acta. Psychiatr. Scand.* **94**(3), 205–210.

Rousseau, P. (2000). The ethical validity and clinical experience of palliative sedation. *Mayo. Clin. Proc.* **75**(10), 1064–1069.

Sadovnick, A. D., Eisen, K., Ebers, G. C., and Paty, D. W. (1991). Cause of death in patients attending multiple sclerosis clinics. *Neurology* **41**, 1193–1196.

Sakakihara, Y., Kubota, M., Kim, S., and Oka, A. (2000). Long-term ventilator support in patients with Werdnig-Hoffmann disease. *Pediatr. Int.* **42**(4), 359–363.

Singer, P. A., and Siegler, M. (1990). Euthanasia- a critique. *N. Engl. J. Med.* **322**, 1881–1883.

Stenager, E. N., and Stenager, E. (1992). Suicide and patients with neurologic diseases—methodologic problems. *Arch. Neurol.* **49**, 1296–1303.

Stenager, E. N., Stenager, E., Koch-Henriksen, N., Bronnum-Hansen, H., Hyllested, K., Jensen, K., and Bille-Brahe, U. (1992). Suicide and multiple sclerosis: an epidemiological investigation. *JNNP* **55**, 542–545.

Stone, P., Hardy, J., Broadley, K., Tookman, A. J., Kurowska, A., and A'Hern, R. (1999). Fatigue in advanced cancer: a prospective controlled cross-sectional study. *Br. J. Cancer* **79**(9–10), 1479–1486.

Stone, P., Phillips, C., Spruyt, O., and Waight, C. (1997). A comparison of the use of sedatives in a hospital support team and in a hospice. *Palliat. Med.* **11**(2), 140–144.

Sykes, N. (2000). End-of-life care in ALS. *In* "Palliative Care in Amyotrophic Lateral Sclerosis." (D. Oliver, G. D. Borasio, and D. Walsh, Eds.) Oxford: Oxford University Press, 159–168.

Terzoudi, M., Gavrielidou, P., Heilakos, G., Visviki, K., and Karageogiou, C. (2000). Fatigue in multiple sclerosis: Evaluation and a new pharmacological approach. *Neurology* **54**, A61 (suppl. 3).

The Ethics and Humanities Subcommittee of the American Academy of Neurology. (1993). Position statement: Certain aspects of the care and management of profoundly and irreversibly paralyzed patients with retained consciousness and cognition. *Neurology* **43**, 222–223.

The Ethics and Humanities Subcommittee of the American Academy of Neurology. (1996). Palliative care in neurology. *Neurology* **46**, 870–872.

The Ethics and Humanities Subcommittee of the American Academy of Neurology. (1998). Assisted suicide, euthanasia, and the neurologist. *Neurology* **50**(3), 596–598.

Tierney, M. C., Bertonia, A., Nores, A., Fisher, R. H., and Senn, J. S. (1992). How reliable are advance directives for health care? A study of attitudes of the healthy and unwell to treatment of the terminally ill. *Ann. R. Coll. Phys. Surg. Can.* **25**, 267–270.

Truog, R. D., Berde, C. B., Mitchell, C., and Grier, H. E. (1992). Barbiturates in the care of the terminally ill. *N. Engl. J. Med.* **327**, 1678–1682.

Twycross, R. G., and Lichter, I. (1998). The terminal phase. *In* "Oxford Textbook of Palliative Medicine," 2nd ed. (D. Doyle, G. W. C. Hanks, and N. Macdonald, Eds.) Oxford: Oxford University Press.

van der Wal, G., and Onwuteaka-Philipsen, B. D. (1996). Cases of euthanasia and assisted suicide reported to the public prosecutor in North Holland over 10 years. *BMJ* **312**, 612–613.

van Heeringen, K., and Zivkov, M. (1996). Pharmacological treatment of depression in cancer patients. A placebo-controlled study of mianserin. *Br. J. Psychiatry* **169**(4), 440–443.

Vecht, C. J., Hovestadt, A., Verbiest, H. B., van Vliet, J. J., and van Putten, W. L. (1994). Dose-effect relationship of dexamethasone on Karnofsky performance in metastatic brain tumors: a randomized study of doses of 4, 8, and 16 mg per day. *Neurology* **44**(4), 675–680.

Ventafridda, V., Ripamonti, C., de Conno, F., Tamburini, M., and Cassileth, B. R. (1990). Symptom prevalence and control during cancer patients last days of life. *J. Palliat. Care* **6**, 7–11.

Voltz, R. (1998). Neurology and palliative care. *Progr. Palliat. Care* **6**, 151–152 (editorial).

Voltz, R., Akabayashi, A., Reese, C., Ohi, G., and Sass, H. M. (1998). End-of-life decisions and advance directives in palliative care: a cross-cultural survey of patients and health-care professionals. *J. Pain Sympt. Manage* **16**(3), 153–162.

Voltz, R., Akabayashi, A., Reese, C., Ohi, G., and Sass, H. M. (1997). Organization and patients' perception of palliative care: a cross-cultural comparison. *Palliat. Med.* **11**(5), 351–357.

Wilcock, A., Corcoran, R., and Tattersfield, A. E. (1994). Safety and efficacy of nebulized lignocaine in patients with cancer and breathlessness. *Palliat. Med.* **8**, 35–38.

Woodcock, A., Gross, E., and Geddes, D. (1981). Drug treatment of breathlessness: contrasting effects of diazepam and promethazine in pink puffers. *BMJ* **283**, 343–346.

World Health Organization: Cancer Pain Relief. (1986). Geneva, World Health Organization, (Technical Report Series No. 804).

Zenz, M., Zenz, T., Tryba, M., and Strumpf, M. (1995). Severe undertreatment of cancer pain: a 3-year survey of the German situation. *J. Pain Sympt. Manage* **10**(3), 187–191.

Zeppetella, G. (1997). Nebulized morphine in the palliation of dyspnoea. *Palliat. Med.* **11**(4), 267–275.

CHAPTER 62

# Primary Tumors of the Central and Peripheral Nervous System

Michael Weller and David G. T. Thomas

## INTRODUCTION

This chapter covers primary tumors of the central and peripheral nervous system. Primary cerebral lymphomas (Chapter 63), brain metastases (Chapter 64), and leptomeningeal metastases (Chapter 65) from nonnervous system tumors are dealt with elsewhere. From a clinical point of view, the common criteria for malignancy, local infiltration and metastasis, do not adequately reflect the impact of brain tumors on affected patients. Even the most malignant brain tumor rarely metastasizes outside the central nervous system (CNS). Conversely, a tumor with benign pathological characteristics may cause significant morbidity and may be lethal when causing obstructive hydrocephalus or interfering with the function of vital cerebral structures.

The new WHO classification of brain tumors is summarized in Table I (Kleihues and Cavenee, 2000). The cause of most brain tumors remains elusive. Some individuals are at increased risk for specific brain tumors because of an inherited disorder (Table II). For most brain tumors, genetic factors seem not to be predisposing. Similarly, few environmental factors related to brain tumor development have been identified. Irradiation may increase the incidence of meningiomas 10-fold, and that of gliomas threefold to sevenfold. No role for cellular phones, high-tension wires, head trauma, or dietary compounds has been demonstrated (DeAngelis, 2001). Extensive animal studies have shown that brain tumors can be induced virally or chemically, raising the possibility that similar factors might be operating in humans. Epidemiological data to estimate the incidence, age distribution, and disease-related survival for patients with brain tumors are available from the Central Brain Tumor Registry of the United States (CBTRUS) (Table III) and the

Surveillance, Epidemiology, and End Results (SEER) data (Table IV) (Davis *et al.*, 1999).

The clinical presentation of brain tumors (Table V) depends on the type and location of the tumor as outlined in more detail later. Appropriate neuroimaging is mandatory once the tentative diagnosis of a probable brain tumor has been made. The major available imaging modalities are computed tomography (CT) scans, magnetic resonance imaging (MRI), and angiography. CT reveals the presence of a tumor by abnormal density values or displacement and distortion of adjacent structures. Vasogenic edema arising as a consequence of a neoplasm follows the white matter extracellular fluid spaces, because they are more easily distended than those of the gray matter. This gives rise to a characteristic appearance on the scan. The diagnostic yield and information available from CT are improved by intravenous contrast enhancement. Extravasation of contrast agent occurs across a damaged blood–brain barrier, and the resulting pattern of enhancement often provides clues to the underlying pathosis. CT has the advantage of being able to demonstrate bony detail, which may aid surgical planning. Data are rapidly obtained, which is useful with restless uncooperative patients. CT is also a suitable modality for the monitoring of patients after therapy.

MRI is now the preferred diagnostic and monitoring imaging modality for patients with brain or spinal tumors. One advantage of MRI scans is the lack of exposure to radiation. The second advantage is the ability to demonstrate pathological processes, for example, demyelination or WHO grade I/II gliomas, which are not easily detected by CT. The third advantage is the ability to obtain similar resolution in any plane, thus providing a three-dimensional image. The final advantage is the lack of bony artifact, which is

**TABLE I    WHO Classification of Brain Tumors**

**Tumors of neuroepithelial tissue**

Astrocytic tumors

| | |
|---|---|
| 9400/3* | Diffuse astrocytoma |
| 9420/3 | Fibrillary astrocytoma |
| 9410/3 | Protoplasmic astrocytoma |
| 9411/3 | Gemistocytic astrocytoma |
| 9401/3 | Anaplastic astrocytoma |
| 9440/3 | Glioblastoma |
| 9441/3 | Giant cell glioblastoma |
| 9442/3 | Gliosarcoma |
| 9421/1 | Pilocytic astrocytoma |
| 9424/3 | Pleomorphic xanthoastrocytoma |
| 9384/1 | Subependymal giant cell astrocytoma |

Oligodendroglial tumors

| | |
|---|---|
| 9450/3 | Oligodendroglioma |
| 9451/3 | Anaplastic oligodendroglioma |

Mixed gliomas

| | |
|---|---|
| 9382/3 | Oligoastrocytoma |
| 9382/3† | Anaplastic oligoastrocytoma |

Ependymal tumors

| | |
|---|---|
| 9391/3 | Ependymoma |
| 9391/3† | Cellular |
| 9393/3 | Papillary |
| 9391/3† | Clear cell |
| 9391/3† | Tanycytic |
| 9392/3 | Anaplastic ependymoma |
| 9394/1 | Myxopapillary ependymoma |
| 9383/1 | Subependymoma |

Choroid plexus tumors

| | |
|---|---|
| 9390/0 | Choroid plexus papilloma |
| 9390/3 | Chorioid plexus carcinoma |

Glial tumors of uncertain origin

| | |
|---|---|
| 9430/3 | Astroblastoma |
| 9381/3 | Gliomatosis cerebri |
| 9444/1† | Chordoid glioma of the third ventricle |

Neuronal and mixed neuronal-glial tumors

| | |
|---|---|
| 9492/0 | Gangliocytoma |
| 9493/0† | Dysplastic gangliocytoma of the cerebellum (Lhermitte-Duclos) |
| 9412/1† | Desmoplastic infantile astrocytoma/ganglioglioma |
| 9413/0† | Dysembryoplastic neuroepithelial tumor |
| 9505/1 | Ganglioglioma |
| 9505/3† | Anaplastic ganglioglioma |
| 9506/1† | Central neurocytoma |
| 9506/1† | Cerebellar liponeurocytoma |
| 8680/1 | Paraganglioma of the filum terminale |

Neuroblastic tumors

| | |
|---|---|
| 9522/3 | Olfactory neuroblastoma (esthesioneuroblastoma) |
| 9523/3 | Olfactory neuroepithelioma |
| 9500/3 | Neuroblastoma of the adrenal gland and sympathetic nervous system |

Pineal parenchymal tumors

| | |
|---|---|
| 9361/1 | Pineocytoma |
| 9362/3 | Pinealoblastoma |
| 9362/3† | Pinealis parenchymal tumor of intermediate differentiation |

Embryonal tumors

| | |
|---|---|
| 9501/3 | Medulloepithelioma |
| 9392/3 | Ependymoblastoma |
| 9470/3 | Medulloblastoma |
| 9471/3 | Desmoplastic medulloblastoma |
| 9474/3† | Large-cell medulloblastoma |
| 9472/3 | Medullomyoblastoma |
| 9470/3† | Melanotic medulloblastoma |
| 9473/3 | Supratentorial primitive neuroectodermal tumor (PNET) |
| 9500/3 | Neuroblastoma |
| 9490/3 | Ganglioneuroblastoma |
| 9508/3 | Atypical teratoid/rhabdoid tumor |

**Tumors of peripheral nerves**

| | |
|---|---|
| 9560/0 | Schwannoma (neurinoma, neurilemmoma) |
| 9560/0† | Cellular |
| 9560/0† | Plexiform |
| 9560/0† | Melanotic |
| 9540/0 | Neurofibroma |
| 9550/0 | Plexiform |
| 9571/0† | Perineurioma |
| 9571/0† | Intraneural perineurioma |
| 9571/0† | Soft tissue perineurioma |
| 9540/3† | Malignant peripheral nerve sheath tumor (MPNST) |
| 9540/3† | Epitheloid |
| 9540/3† | MPNST with divergent mesenchymal and/or epithelial differentiation |
| 9540/3† | Melanotic |
| 9540/3† | Melanotic psammomatous |

**Tumors of the meninges**

Tumors of meningothelial cells

| | |
|---|---|
| 9530/0 | Meningioma |
| 9531/0 | Meningothelial |
| 9532/0 | Fibrous (fibroblastic) |
| 9537/0 | Transitional (mixed) |
| 9533/0 | Psammomatous |
| 9534/0 | Angiomatous |
| 9530/0† | Microcystic |
| 9530/0† | Secretory |
| 9530/0† | Lymphoplasmacyte-rich |
| 9530/0† | Metaplastic |
| 9538/1† | Clear cell |
| 9538/1† | Chordoid |
| 9539/1† | Atypical |
| 9538/3† | Papillary |
| 9538/3† | Rhabdoid |
| 9530/3 | Anaplastic meningioma |

Mesenchymal, nonmeningothelial tumors

| | |
|---|---|
| 8850/0 | Lipoma |
| 8861/0 | Angiolipoma |
| 8880/0 | Hibernoma |
| 8850/3 | Liposarcoma (intracranial) |
| 8815/0† | Solitary fibrous tumor |
| 8810/3 | Fibrosarcoma |
| 8830/3 | Malignant fibrous histiocytoma |
| 8890/0 | Leiomyoma |
| 8890/3 | Leiomyosarcoma |

*continues*

**TABLE I**    *Continued*

| | | | | |
|---|---|---|---|---|
| 8900/0 | Rhabdomyoma | | 9731/3 | Plasmacytoma |
| 8900/3 | Rhabdomyosarcoma | | 9930/3 | Granulocytic sarcoma |
| 9220/3 | Chondroma | | | |
| 9220/3 | Chondrosarcoma | | **Germ cell tumors** | |
| 9180/0 | Osteoma | | 9064/3 | Germinoma |
| 9180/3 | Osteosarcoma | | 9070/3 | Embryonal carcinoma |
| 9210/0 | Osteochondroma | | 9071/3 | Yolk sac tumor |
| 9120/0 | Hemangioma | | 9100/3 | Choriocarcinoma |
| 9133/1 | Epithelioid hemangioendothelioma | | 9080/1 | Teratoma |
| 9150/1 | Hemangiopericytoma | | 9080/0 | Mature |
| 9120/3 | Angiosarcoma | | 9080/3 | Immature |
| 9140/3 | Kaposi sarcoma | | 9084/3 | Teratoma with malignant transformation |
| **Primary melanocytic lesions** | | | 9085/3 | Mixed germ cell tumors |
| *8728/0*† | Diffuse melanocytosis | | | |
| *8728/1*† | Melanocytoma | | **Tumors of the sellar region** | |
| 8720/3 | Malignant melanoma | | 9350/1 | Craniopharyngeoma |
| *8728/3*† | Meningeal melanomatosis | | *9351/1*† | Adamantinous |
| **Tumors of uncertain histogenesis** | | | *9352/1*† | Papillary |
| 9161/1 | Hemangioblastoma | | 9582/0 | Granular cell tumor |
| | | | | |
| **Lymphomas and hematopoietic neoplasms** | | | **Metastatic tumors** | |
| 9590/3 | Malignant lymphoma | | | |

*Morphology code of the *International Classification of Diseases for Oncology* (ICD-O) and the *Systematized Nomenclature of Medicine* (SNOMED). Behavior is coded /0 for benign tumors, /1 for low or uncertain malignant potential or borderline malignancy (/2 for in situ lesions), and /3 for malignant tumors.

†The italicized numbers are provisional codes proposed for the third edition of ICD-O. They should be incorporated into the next edition of ICD-O but are subject to change (Kleihues and Cavenee, 2000).

From Kleinues, P., and Cavenee, W. K. (Eds.) (2000). WHO classification of tumours: Pathology and genetics of tumours of the nervous system. Lyon, IARC Press, with permission.

particularly useful for the imaging of the posterior fossa. Certain limitations of MRI have been outlined with respect to CT scans. These are slow scans, with which certain materials are not compatible, and a lack of bony detail, if this should be required for surgical planning. Gadolinium-mediated contrast enhancement plays an important role in the differential diagnosis and follow-up of patients with brain tumors. Gadolinium-enhanced MRI may also be useful for patients with an allergy to CT contrast media, because it causes no similar allergic reaction. CT and MRI are now the most important tools not only for the diagnosis of brain tumors but also to assess the efficacy of nonsurgical brain tumor therapies (Macdonald *et al.*, 1990) (Table VI). Cerebral angiography has lost most of its diagnostic role to CT or MRI. However, it is still a supplementary technique to aid in surgical planning or allow for preoperative embolization of vessels feeding the tumor. Although it cannot provide a similar degree of resolution as intra-arterial angiography, intravenous digital subtraction angiography provides sufficient information for surgical planning in most cases. Positron emission tomography (PET) requires a cyclotron and a team of specialists to run it. Thus, cost implications will probably ensure that PET remains a research technique for the near

future. PET may be a suitable technique to monitor metabolic activity in a tumor and to differentiate recurrent tumor from radiation necrosis. Ventriculography, pneumograms, and isotope scans are only of historical significance.

The definitive diagnosis of any tumor of the nervous system requires histological assessment of tumor tissue obtained by biopsy, partial, or complete resection. Freehand biopsy techniques have rapidly declined in frequency as surgeons have become more familiar with stereotactic methods. The latter technique is based on Cartesian mathematics, which states that any point in space can be referenced to a predetermined fixed point by three coordinates. In neurosurgical procedures these coordinates can be calculated from a CT or MRI scan and transcribed to a stereotactic apparatus attached to the patient's skull and a needle passed to the target by means of a burr hole (Vindlacheruvu *et al.*, 1999). A variety of targets within the same lesion should be biopsied to improve the diagnostic accuracy, which may approach 100%. Stereotactic biopsy is used routinely to biopsy lesions involving eloquent areas of cortex, the sensorimotor strip, visual areas, the diencephalon, and the brainstem. Stereotactic biopsies are associated with a mortality of 0%–3% and a permanent

**TABLE II** Genetic Predispositions for Brain Tumors

| | Incidence | Gene locus | Gene | Gene product | Type of nervous system tumors | Skin changes | Other manifestations |
|---|---|---|---|---|---|---|---|
| Neurofibromatosis I (von Recklinghausen) | 1:4,000 | 17q11 | NF1 | Neurofibromin | Neurofibroma, malignant peripheral nerve sheath tumor (MPNST), optic nerve glioma, astrocytoma | Café-au-lait spots, axillary freckling | Iris hamartomas, bone lesions, pheochromocytoma, leukemia |
| Neurofibromatosis II | 1:40,000 | 22q12 | NF2 | Merlin (schwannomin) | (Bilateral) vestibular schwannoma, schwannoma, meningioma, spinal ependymoma, astrocytoma, glial hamartias, cerebral calcifications | — | Posterior lens opacities, retinal hamartoma |
| von Hippel-Lindau syndrome | 1:40,000 | 3p25 | VHL | VHL protein | Hemangioblastoma | — | Retinal hemangioblastoma, renal cell carcinoma, phaeochromocytoma, visceral cysts |
| Tuberous sclerosis (Bourneville-Pringle) | 1:5,000 | 9q34 16p13 | TSC1 TSC2 | Hamartin Tuberin | Subependymal giant cell astrocytoma, cortical tubers | Cutaneous angiofibroma (*adenoma sebaceum*), peau chagrin, subungual fibromas | Cardiac rhabdomyomas, adenomatous polyps of duodenum and small intestine, lung and renal cysts, lymphangioleiomyomatosis, renal angiomyolipoma |
| Li-Fraumeni syndrome | 143 families | 17p13 | p53 (mostly) | TP53 | Astrocytoma, PNET | — | Breast carcinoma, bone and soft tissue sarcoma, adrenocortical carcinoma, leukemia |
| Cowden syndrome | rare | 10q23 | PTEN (MMAC1) (mostly?) | PTEN | Dysplastic gangliocytoma of the cerebellum (Lhermitte-Duclos), megalencephaly | Multiple trichilemmomas, fibromas | Hamartomatous polyps of the colon, thyroid neoplasms, breast carcinoma |
| Turcot syndrome | 160 cases | 5q21 3p21 7p22 | APC hMLH1 hPSM2 | APC hMLH1 hPSM2 | Medulloblastoma Glioblastoma | — Café-au-lait spots | Colorectal polyps Colorectal polyps |
| Nevoid basal cell carcinoma (Gorlin) syndrome | 1:57,000 | 9q31 | PTCH | Ptch | Medulloblastoma | Multiple basal cell carcinomas, palmar and plantar pits | Jaw cysts, ovarian fibromas, skeletal abnormalities |

Modified from Kleinues, P., and Cavenee, W. K. (Eds.) (2000). WHO classification of tumours: Pathology and genetics of tumours of the nervous system. Lyon, IARC Press, with permission.

**TABLE III  Epidemiology of Brain Tumors***

| Tumor | % of all brain tumors | Incidence per 100,000/year | Mean age at diagnosis |
|---|---|---|---|
| Tumors of neuroepithelial tissue | | | |
| Diffuse astrocytoma (protoplasmic, fibrillary) | 1.3 | 0.17 | 47 |
| Anaplastic astrocytoma | 4.3 | 0.54 | 50 |
| Glioblastoma | 22.6 | 2.94 | 62 |
| Pilocytic astrocytoma | 1.8 | 0.22 | 17 |
| Unique astrocytoma variants | 0.5 | 0.06 | 35 |
| Oligodendroglioma | 2.6 | 0.32 | 41 |
| Anaplastic oligodendroglioma | 0.6 | 0.07 | 46 |
| Ependymoma/anaplastic ependymoma | 2 | 0.25 | 36 |
| Ependymoma variants | 0.3 | 0.04 | 37 |
| Mixed glioma | 1 | 0.13 | 40 |
| Astrocytoma, not otherwise specified | 8.1 | 1.01 | 47 |
| Glioma malignant, not otherwise specified | 2.6 | 0.33 | 46 |
| Choroid plexus | 0.3 | 0.04 | 22 |
| Neuroepithelial | 0.2 | 0.02 | 42 |
| Benign and malignant neuronal/glial, neuronal and mixed | 1 | 0.12 | 24 |
| Pineal parenchymal | 0.2 | 0.02 | 28 |
| Embryonal/primitive/medulloblastoma | 1.8 | 0.22 | 14 |
| Tumors of cranial and spinal nerves | | | |
| Nerve sheath, benign and malignant | 6.5 | 0.86 | 51 |
| Other tumors of cranial and spinal nerves | 0 | 0 | 36 |
| Tumors of the meninges | | | |
| Meningioma | 24 | 3.15 | 62 |
| Other mesenchymal, benign and malignant | 0.4 | 0.04 | 39 |
| Hemangioblastoma | 0.9 | 0.12 | 46 |
| Lymphomas and hemopoietic neoplasms | | | |
| Lymphoma | 4.1 | 0.51 | 54 |
| Germ cell tumors and cysts | | | |
| Germ cell tumors, cysts and heterotopias | 0.6 | 0.08 | 22 |
| Tumors of the sellar region | | | |
| Pituitary | 8 | 1.02 | 50 |
| Craniopharyngioma | 0.9 | 0.11 | 36 |
| Local extensions from regional tumors | | | |
| Chordoma/chondrosarcoma | 0.2 | 0.03 | 60 |
| Unclassified tumors | | | |
| Hemangioma | 0.3 | 0.04 | 41 |
| Neoplasm, unspecified | 2.8 | 0.37 | 63 |
| All other | 0 | 0.1 | 49 |

*Data from www.cbtrus.org/2000/table2000-2.htm (incidence rate adjusted to year 2000 U.S. standard population).

**TABLE IV  Relative Survival at 2 and 5 Years from the Diagnosis of a Primary Brain Tumor**

| | Survival at 2 years (%) | Survival at 5 years (%) |
|---|---|---|
| Diffuse astrocytoma | 67 | 49 |
| Anaplastic astrocytoma | 46 | 31 |
| Glioblastoma | 9 | 3 |
| Pilocytic astrocytoma | 91 | 87 |
| Oligodendroglioma | 80 | 63 |
| Anaplastic oligodendroglioma | 61 | 38 |
| Ependymoma/anaplastic ependymoma | 80 | 67 |
| Mixed glioma | 74 | 59 |
| Astrocytoma, not otherwise specified | 45 | 35 |
| Malignant glioma, not otherwise specified | 34 | 27 |
| Neuroepithelial | 63 | 48 |
| Neuronal/glial, neuronal and mixed | 61 | 46 |
| Embryonal/medulloblastoma | 70 | 56 |
| All brain and other CNS | 36 | 28 |

From Davis *et al.*, 1999. Copyright © 1999 American Cancer Society. Reprinted by permission of Wiley-Liss, Inc., a subsidiary of John Wiley & Sons, Inc.

**TABLE V  Relative Frequency of Symptoms in 653 Patients with Gliomas**

| Symptom | At presentation (%) | At time of surgery (%) |
|---|---|---|
| Epilepsy | 38 | 54 |
| Headache | 35 | 72 |
| Mental changes | 17 | 52 |
| Hemiparesis | 10 | 43 |
| Vomiting | 8 | 32 |
| Dysphasia | 7 | 27 |
| Impaired consciousness | 5 | 25 |
| Visual failure | 4 | 18 |
| Hemianesthesia | 3 | 14 |
| Hemianopia | 2 | 8 |
| Cranial nerve palsy | 2 | 11 |
| Other | 2 | 7 |

Modified from Thomas and Graham, 1980.

TABLE VI    CT- or MRI-Based Response Criteria According to Macdonald *et al.* (1990)

| | |
|---|---|
| Complete response (CR) | Disappearance of all enhancing tumor on consecutive CT or MRI, determined by two observations not less than 4 weeks apart, steroid-free, and without progression of neurological symptoms or signs |
| Partial response (PR) | At least a 50% decrease in tumor size on consecutive CT or MRI, determined by two observations not less than 4 weeks apart, steroid-stable or reduced, and without progression of neurological symptoms or signs |
| Stable disease (SD) | A decrease in tumor size less than 50%, or increase in tumor size less than 25%, on consecutive CT or MRI, determined by two observations not less than 4 weeks apart, steroid-stable or reduced |
| Progressive disease (PD) | A greater than 25% increase in tumor size or de novo requirement for steroids as decided by the principal investigator |

neurological morbidity of 0%–6% (Thomas and Nouby, 1989).

## GENERAL PRINCIPLES OF THERAPY

### Symptomatic Treatment

*Vasogenic edema* is often a major reason for clinical deficits in patients with brain tumors. Edema is more common in malignant tumors such as glioblastoma and less frequent in histologically benign lesions. The pathogenesis of brain edema is incompletely understood but is likely to involve the release of soluble factors by tumor cells such as vascular endothelial–derived growth factor (VEGF). Brain tumor–associated edema is classified as vasogenic edema and is uniquely responsive to corticosteroids. Steroids such as dexamethasone are therefore of central importance for the treatment of patients with brain tumors. The benefit derived from steroids must be weighed against their potentially deleterious side effects on long-term use: immunosuppression, gastrointestinal ulceration, osteoporosis, depression, skin changes, myopathy, and enhanced risk of thromboembolism. Furthermore, the evidence for counterproductive effects of steroids during the chemotherapy of WHO grade III/IV gliomas is rather compelling (Weller *et al.*, 1997), including steroid-induced decreases in tumor perfusion, stabilization of the blood–tumoral barrier, and promotion of resistance to apoptosis. Similar antagonism of steroids and drug cytotoxicity seems to be operative in other tumors as well. Boswellic acids (sold as H15 tablets) are phytotherapeutic agents that are very popular in Western Europe and are hypothesized to inhibit edema formation and even tumor growth in patients with malignant gliomas. Although high concentrations of boswellic acids are cytotoxic to glioma cells in vitro, our own preliminary clinical data indicate that this agent has little activity against glioblastoma itself or glioblastoma-associated cerebral edema but some clinically relevant activity against edema associated with radiochemotherapy-induced neurotoxicity (Streffer *et al.*, 2001).

### Surgery

Surgery is not only required for the definitive diagnoses of brain tumors but represents the single most important therapeutic measure for many tumors of the nervous system. Advances in imaging, general surgical principles, microsurgical techniques, and anesthesia have considerably increased the survival benefit achieved by neurosurgical interventions for many tumor entities. Tumors that may be cured by surgery alone include meningiomas, schwannomas, and pituitary tumors. Conversely, surgery is not curative for diffuse astrocytomas or malignant brain tumors, and adjuvant radiotherapy and chemotherapy are required to improve the prognosis for these patients. Ventriculoperitoneal shunting to relieve hydrocephalus and placement of ventricular reservoirs for intrathecal chemotherapy are specific surgical procedures for patients with posterior fossa lesions and leptomeningeal metastases, respectively.

### Radiotherapy

Radiotherapy is an essential component of treatment for most tumors of the nervous system that cannot be resected in toto. Radiotherapy is delivered externally in a series of single fractions (fractionated radiotherapy) or as a single dose (radiosurgery) or interstitially by radioactive seed implantation (brachytherapy). *Fractionated radiotherapy* is a common approach to diffuse astrocytomas, notably anaplastic gliomas and glioblastomas, medulloblastomas, and ependymomas. *Radiosurgery* has a role in the management of some benign lesions such as schwannoma. *Brachytherapy* (interstitial radiotherapy) can in general be used for most nonresectable neoplasms but is now mainly used for WHO grade II astrocytomas (Kreth *et al.*, 1995). Common target volumes of radiotherapy include the lesion only, the involved field, whole brain radiotherapy, and craniospinal radiotherapy. The lesion only is a target mostly for the radiosurgery of benign lesions such as schwannoma. The involved field includes the lesion (e.g., defined by contrast-enhanced CT or MRI) plus a safety margin (e.g., of 20 mm). This target volume is commonly irradiated in patients with malignant gliomas. Whole brain radiotherapy is commonly used for primary cerebral lymphoma (Chapter 63) and metastatic brain disease (Chapter 64). Craniospinal radiotherapy has a role in the management of some intrinsic brain tumors, including germ cell tumors, ependymomas, and medulloblastomas.

The modes of action of radiotherapy are not clearly defined and may differ depending on the type of lesion. One major target of radiotherapy is the proliferative activity of tumor cells. In that regard, proliferation is a rather tumor-specific target in the brain, because neurons never, and astrocytes rarely, divide in the normal brain. The inhibition of tumor cell proliferation is probably an important effect of radiotherapy in glial tumors. In contrast, relevant cytolytic effects of radiotherapy may not be achieved in most glial tumors. Direct induction of cell death with virtual disappearance of tumor masses is seen primarily in primary cerebral lymphoma (Chapter 63) but also in germ cell tumors and medulloblastomas. Molecular factors determining tumor cell responses to radiotherapy resemble those for chemotherapy and are discussed later. Furthermore, radiotherapy is likely to compromise the vascular supply of brain tumors, notably of rapidly growing tumors with heavy neovascularization such as glioblastoma. Factors that influence tumor responses and the brain's tolerance of radiotherapy are the total dose administered, the size of the fractions, the volume of brain irradiated, the age of the patient, and the type of lesion.

## Chemotherapy

Chemotherapy may be administered as a second-line therapy after surgery and radiotherapy have failed, in the adjuvant setting after surgery, commonly combined with radiotherapy, or as a postsurgical or postbiopsy "neoadjuvant" treatment (preirradiation chemotherapy). Chemotherapy has assumed a major role in the treatment of oligodendroglial tumors. The role of chemotherapy in the treatment of medulloblastoma, germinoma, and glioblastoma has remained controversial, although some regimens have been documented to be active in these tumors as well. In principle, the efficacy of chemotherapy depends on the efficacy of drug delivery to tumor cells and on the intrinsic sensitivity of tumor cells to this drug. Delivery on systemic administration thus depends on tumor perfusion, specifically on the microvasculature in the tumor, the presence of arteriovenous shunting, and the distance of viable tumor cells from the blood vessels, which determines how far an agent must diffuse or be transported by bulk flow to

reach the tumor cell. Tight junctions between capillary endothelial cells of the normal brain form the blood–brain barrier. Choroidal epithelial cells provide an equivalent barrier between blood and the ventricular cerebrospinal fluid. Thus lipid-soluble, nonpolar drugs gain entry to the cerebral extracellular spaces and are able to diffuse through the brain far more readily than polar nonlipid-soluble drugs. Disruption of the blood–brain barrier within and in the vicinity of the tumor will therefore be another factor that modulates the efficacy of chemotherapy. The blood–brain barrier is not only a mechanical barrier but also a pharmacological barrier, because brain endothelial cells express drug transporters such as the P-glycoprotein, encoded by the multidrug resistance (mdr) gene, or the mdr-associated protein, MRP-1. These drug transporters, which may also be expressed by tumor cells, prevent the intracellular accumulation of various structurally unrelated drugs, including vincristine, doxorubicin, and etoposide or teniposide.

The molecular and cellular biological makeup of a tumor cell provides a second level, where resistance or sensitivity to chemotherapy are determined. Molecular genetic changes that have been postulated to confer resistance to radiotherapy or chemotherapy in brain tumors, notably gliomas, include loss of p53 wild-type activity, enhanced epidermal growth factor (EGF) receptor activity, and enhanced expression of antiapoptotic BCL-2 family members or inhibitor-of-apoptosis protein (IAP) family members. All of these changes are thought to render tumor cells resistant to spontaneous cell death by means of apoptosis and by extrapolation are expected to protect from induced modes of cell death as well. *In vitro* studies have failed to confirm that the determination of such molecular changes allows us to predict the chemosensitivity of human malignant glioma cell lines (Weller *et al.*, 1998). Similarly, assessing the expression of apoptosis-regulatory gene products in human gliomas in vivo did not lead to the identification of p53 or BCL-2 family protein levels as predictors of response to radiochemotherapy (Strik *et al.*, 1999). The first breakthrough possibly came in the field of predicting the response to therapy in oligodendroglial tumors where specific chromosomal losses have been associated with favorable responses to adjuvant chemotherapy (Cairncross *et al.*, 1998). The same molecular alteration may also predict a favorable response to therapy in patients with glioblastomas (Ino *et al.*, 2000). Moreover, inactivation of the DNA repair gene, $O^6$-methylguanine-DNA methyltransferase (MGMT) may predict the response to alkylating agents in gliomas (Esteller *et al.*, 2000).

The cytotoxic agents commonly used for the chemotherapy of brain tumors are summarized in Tables VII and VIII. Chemotherapy can be single agent

**TABLE VII**    Chemotherapeutic Agents Used for the Treatment of Brain Tumors

| Drug | Mode of action | Dose | Tumor | Major toxicity* | References |
|---|---|---|---|---|---|
| ACNU (nimustine) | Alkylating agent | 100 mg/m²† × 6 wk | Anaplastic glioma, glioblastoma | Pulmonary fibrosis | Takakura et al., 1986 |
| BCNU (carmustine) | Alkylating agent | 150–200 mg/m² × 6 wk | Anaplastic glioma, glioblastoma | Pulmonary fibrosis | Walker et al., 1978, 1980 |
| CCNU (lomustine) | Alkylating agent | 130 mg/m²† × 6 wk<br>110 mg/m²‡ × 6 wk | Anaplastic glioma, glioblastoma, medulloblastoma | Pulmonary fibrosis | See PCV and CCV, Table VIII |
| Carboplatin | DNA cross-linking and adduct formation | Various schedules, e.g., 360 mg/m² × 4 wk | Germinoma, oligodendroglioma | Peripheral neuropathy, nephrotoxicity | Brandes et al., 1998 |
| Cisplatin | DNA cross-linking and adduct formation | Various schedules, e.g., 100 mg/m² × 4 wk | Germinoma, oligodendroglioma | Peripheral neuropathy, nephrotoxicity, ototoxicity | Grossman et al., 1997 |
| Etoposide (VP16) | Topoisomerase II inhibitor | Various schedules | Germinoma, anaplastic glioma, glioblastoma | Diarrhea | Balmaceda et al., 1996<br>Baranzelli et al., 1997 |
| Hydroxyurea | Antiproliferative, inhibition of ribonucleoside disphosphate reductase | 20 mg/kg continuously | Meningioma | Overall moderate, gastrointestinal toxicity | Schrell et al., 1995 |
| Procarbazine | Alkylating agent | 130–150 mg/m²<br>D1–D28 × 4 wk | Anaplastic glioma, glioblastoma | Allergy (peripheral neuropathy) | Rodriguez et al., 1989<br>Brandes et al., 1999<br>Yung et al., 2000 |
| Tamoxifen | Antiestrogen, protein kinase C inhibitor | 20–200 mg daily | Anaplastic glioma, glioblastoma | Nausea, hepatotoxicity | Couldwell et al., 1996<br>Brandes et al., 1999 |
| Temozolomide | Alkylating agent | 200 mg/m² D1–D5 × 4 wk | Anaplastic glioma, glioblastoma | Diarrhea | Friedman et al., 1998<br>Yung et al., 1999, 2000 |
| Teniposide (VM26) | Topoisomerase II inhibitor | Various schedules | Anaplastic glioma, glioblastoma | Overall moderate | Brandes et al., 1998 |
| Topotecan | Topoisomerase I inhibitor | 1.5 mg/m² D1–5 × 3 wk | Anaplastic glioma, glioblastoma | Overall moderate | Macdonald et al., 1996 |
| Vincristine | Antimicrotubule agent | 1.4 mg/m² (max.: 2 mg)§ | Anaplastic glioma, glioblastoma, medulloblastoma | Peripheral neuropathy | Cairncross et al., 1994<br>Packer et al., 1997 |

†monotherapy.
‡combination chemotherapy.
*except myelosuppression.
§PCV protocol.

or combination therapy. Combination therapy is used with the rationales that two or more drugs act in synergy or act on different subclones within the same tumor. Nitrosoureas are the single agents most commonly used for the chemotherapy of brain tumors, and most combination schemes are nitrosourea-based, including the widely used PCV regimen (Table VIII). Because proliferation in the normal brain is low and limited to few astrocytes and endothelial cells, proliferation should be a rather tumor-specific target in the brain. Yet, the fraction of cycling cells even in glioblastomas at any given point does not exceed 40% and is often much lower, and sustained drug levels in the tumor bed are difficult to achieve. Accordingly, the focus for gliomas has shifted away from agents specifically acting on the cell cycle to cell cycle-independently acting agents, and to prolonged, preferably oral administration as opposed to high-dose pulse therapies, which are effective in chemosensitive tumors such as primary cerebral lymphomas.

**TABLE VIII    Combination Chemotherapy Protocols**

| Protocol | Design | Tumor | References |
|---|---|---|---|
| PCV | Procarbazine 60 mg/m² D8–D21<br>CCNU 110 mg/m² D1<br>Vincristine 1.4 mg/m² D8 + D29<br>× (6–)8 wk | (Anaplastic)<br>Oligodendroglioma<br>Anaplastic astrocytoma<br>Glioblastoma | Levin et al., 1990<br>Streffer et al., 2000 |
| ACNU/VM26 | ACNU 90 mg/m² D1<br>VM26 60 mg/m² D1–D3<br>× 6 wk | Anaplastic astrocytoma<br>Glioblastoma | German NOA-01 trial (unpublished) |
| CCV | CCNU 75 mg/m² D1<br>Cisplatin 70 mg/m² D1<br>Vincristine 1.4 mg/m² D1, D8, D15<br>× 6–7 wk | Medulloblastoma | Packer et al., 1994<br>Kortmann et al., 2000 |
| CV | Carboplatin 175 mg/m²<br>Vincristine 1.5 mg/m² | Grade I/II astrocytoma of childhood | Packer et al., 1997 |
| CE | Carboplatin 300 mg/m² D1<br>Etoposide (VP16) 150 mg/m² D2–D3<br>× 4 wk (and other schedules) | Recurrent oligodendroglioma<br>Germ cell tumors | Balmaceda et al., 1996<br>Baranzelli et al., 1997<br>Streffer et al., 2000 |

The concept of synergistic effects of radiochemotherapy in the treatment of brain tumors has not been supported by clinical data yet. Chemotherapeutic regimens that have activity against selected brain tumors have this activity on their own and not specifically in the context of radiochemotherapy. Although further efforts to develop potent combinations of radiotherapy and chemosensitizers (e.g., topotecan or temozolomide) are warranted, the combination of these treatment modalities may even result in apparent antagonism (e.g., as observed with the addition of cisplatin to radiotherapy of childhood brainstem astrocytomas) (Freeman et al., 2000).

*Preirradiation chemotherapy* has attracted a lot of interest, because higher drug levels in the tumor may be achieved when tumor vasculature has not been damaged by prior radiotherapy and because chemotherapy followed by radiotherapy is predicted to be tolerated better than the reverse sequence. This strategy has been highly successful in primary cerebral lymphoma (Chapter 63) and is beginning to be evaluated in oligodendroglial tumors. So far, preirradiation chemotherapy has not led to a significant progress in the treatment of medulloblastoma or germ cell tumors.

*High-dose chemotherapy* followed by autologous bone marrow transplantation has been evaluated in a series of small studies for the treatment of newly diagnosed or recurrent malignant brain tumors, preferentially in children but also in young adults (Finlay et al., 1996). Although some patients experienced durable responses with that treatment, high-dose chemotherapy is still an experimental treatment and has not gained a well-defined place in the treatment of any brain tumor. Medulloblastoma is probably the best candidate for a brain tumor responsive to high-dose chemotherapy followed by bone marrow rescue.

*Chemotherapy with blood–brain barrier disruption* is another approach aimed at increasing drug exposure of malignant brain tumor cells. RMP-7, a bradykinin agonist, has been used as a barrier-disrupting agent in phase I clinical trials in conjunction with carboplatin, both with intravenous (Ford et al., 1998) and intra-arterial application (Cloughesy et al., 1999). The clinical value of this approach remains uncertain.

Alternative routes of administering chemotherapy have also been explored. *Intra-arterial chemotherapy* has mostly been performed with BCNU and cisplatin and was found to be rather toxic to eye and brain. Other complications were catheter-related thromboembolic events. The most important objection to intra-arterial chemotherapy is on anatomical and pathological grounds, because the growth of brain tumors is not limited to the territory of single brain arteries. *Intrathecal chemotherapy* is ineffective against solid brain tumors but may have a role when intrinsic brain tumors seed to the subarachnoid space (Chapter 65). *Local (interstitial) chemotherapy* seeks to circumvent the problems of systemic toxicity and of inefficient drug delivery. The most advanced approach is to implant BCNU wafers into the resection cavity after surgery (Brem et al., 1995), but this strategy must still be considered experimental (Engelhard, 2000).

Controversy on the assessment of efficacy for new chemotherapeutic regimens, specifically in malignant gliomas (Perry et al., 1997), is ongoing. For instance, Grant et al. (1997) have demonstrated that the type of an initial response to chemotherapy other than progression, that is, partial response, minor response, or stable disease, does not predict the progression-free survival. Therefore, although the best response to treatment achieved may be a useful response parameter for phase II studies that try to define the biological activity of a new agent, phase III clinical trials that aim at changing the natural course of disease to a greater extent than a standard treatment should look primarily at the progression-free survival.

## Other Pharmacological Therapies

The disappointing results achieved with surgery, radiotherapy, and chemotherapy in the treatment of most malignant brain tumors of adulthood have led to the evaluation of numerous alternative approaches of medical therapy for these tumors. Differentiation therapies may in theory reverse the malignant phenotype of cancer cells that have escaped standard cytotoxic therapies. Candidate differentiation agents that have been used in the treatment of malignant gliomas with little success include α/β-interferons, retinoids, and lovastatin. A recent study of phenylacetate for recurrent malignant gliomas failed to reveal convincing activity for this agent as well (Chang *et al.*, 1999). Similarly, candidate inhibitors of angiogenesis such as thalidomide have not shown efficacy yet (Fine *et al.*, 2000). Many brain tumor cells show strong migratory and invasive properties that are responsible for progressive neurological morbidity and preclude a complete surgical resection. Accordingly, candidate antimigratory/antiinvasive agents such as marimastat are currently being explored for their possible role in the management of brain tumors. None of these agents has been shown to provide a tumor control or survival advantage over standard treatment.

## Somatic Gene Therapy

Glioblastoma has been the major target for somatic suicide gene therapy of human cancers. Suicide gene therapy is based on the delivery by means of mostly viral gene transfer of genes encoding enzymes not normally expressed in mammalian cells. These enzymes convert a systemically administered prodrug to a cytotoxic agent. In theory, the specificity of this approach relies on the systemic administration of a nontoxic prodrug that will be selectively converted into a toxic agent in those cells that have been transduced with the enzyme-encoding viral vector. The use of retroviral vectors would limit transduction to cycling cells. Clinical studies have focused on the activation of ganciclovir by thymidine kinase, but current preclinical studies examine alternative systems such as cytosine deaminase plus 5-

fluorouracil. A phase III clinical trial including 248 patients with newly diagnosed glioblastoma did not reveal an advantage of standard radiotherapy plus thymidine kinase/ganciclovir suicide gene therapy compared with radiotherapy alone (Rainov, 2000). Despite this disappointing first phase III trial, gene therapy is still a promising therapeutic approach to malignant brain tumors. The major obstacle at present seems to be the poor delivery of therapeutic transgenes to tumor cells.

## Immunotherapy

Immunotherapy has remained the most fascinating approach to cancer, because it is considered to provide the only option not only for palliation but actually for cures of cancer. Malignant brain tumors seem to be suitable candidates for such approaches, because they create an immunosuppressed environment in the brain and because their almost uniform failure to metastasize outside the brain is compatible with the hypothesis that extracerebral immune surveillance is sufficient to control tumor spread there. Although no approach of immunotherapy, including various vaccination strategies, has proven efficacy for any brain tumor, continuing efforts focus on the therapeutic neutralization of immunosuppressive molecules released by tumor cells, notably transforming growth factor-β (TGF-β) and more recently various different (e.g., dendritic cell-based) vaccination strategies.

## Side Effects from Tumor Therapy

The treatment of tumors of the nervous system is associated with significant side effects that are often difficult to differentiate from the consequences of tumor progression. Surgery-associated morbidity has been steadily declining, whereas significant morbidity and even mortality arises from the adjuvant therapies, mainly radiotherapy and chemotherapy, that are used more extensively today than 20 years ago (Keime-Guibert *et al.*, 1998). The side effects of whole-brain radiotherapy are discussed in Chapter 64. The side effects of *craniospinal radiotherapy* are briefly reviewed here, because this target volume is mostly used for the treatment of intrinsic brain tumors such as medulloblastoma or ependymoma. Affected individuals are often children or young adults, raising specific concerns with regard to long-term toxicity. Craniospinal radiotherapy delivered to children may cause a number of physical side effects, including short stature, bone marrow depression, alopecia, and hormonal problems in females. Irradiation to the growing spine was originally thought to be responsible for short stature, but the latter may also result from irradiation-induced pituitary dysfunction with subsequent growth hormone deficiency. If the condition is recognized at an early stage, growth hormone supplements can be prescribed and the side effect avoided. Bone marrow depression can lead to anemia, leukopenia, and thrombocytopenia. Usually this is a mild self-limiting side effect that does not interfere with treatment if high-dose chemotherapy is withheld until the bone marrow recovers. The alopecia induced by cranial irradiation is self-limiting, and hair may regrow normally. In females the field of irradiation places the ovaries at risk, with the subsequent possibility of hormonal problems in later life. The testes are not included within the field. The risk of radiation myelopathy is dose-dependent (Schultheiss *et al.*, 1995). This complication is rare with doses below less than Gy. Occasional patients have been reported to respond to anticoagulation (Chapter 64) or hyperbaric oxygen therapy, but the efficacy of these measures has not been adequately assessed.

*Chemotherapy* has the predicted side effect of mostly transient myelosuppression, which is managed as outlined later, as well as some drug-specific side effects that are summarized in Table VII. The lowest value for white blood cells and platelets is referred to as the nadir. This peak of myelosuppression is observed between 7 and 14 days after treatment for most cytotoxic agents but is delayed for 4–6 weeks for the nitrosoureas. The nadir determines the scheduling of the chemotherapy cycles. Guidelines for the management of myelosuppression and associated complications are as follows. Patients receiving chemotherapy usually need their white blood cell (WBC) and platelet counts monitored weekly. More frequent blood tests, twice per week, are required when neutrophils are <1500/μl or platelets <50,000/μl. Daily controls are mandatory with neutrophils <1000/μl or platelets <25,000/μl. When neutrophils are decreasing rapidly to fall below 500/μl or are <500/μl, granulocyte colony–stimulating factor (G-CSF, 300 μg daily) should be administered until the neutrophils have been >500/μl for 2 days. Patients with neutrophils <500/μl without fever or other signs of infection should receive prophylactic antibiotics (e.g., ofloxacin, 200 mg daily, and fluconazole, 100 mg daily) until the neutrophil count exceeds 500/μl. Patients receiving steroids concurrently should receive trimethoprim (80 mg)/sulfamethoxazole (400 mg) daily, in addition, as long as steroids are given and for 2 more weeks after steroids have been discontinued. This is because *Pneumocystis carinii* pneumonia is a common dangerous complication encountered in patients with brain tumors receiving chemotherapy, notably those receiving concurrent radiotherapy and steroids. Lymphopenia seems to be responsible for the enhanced sensitivity of brain tumor patients to *P. carinii* pneumonia.

Patients in myelosuppression who have fever or other signs of infection develop should be admitted to the hospital and should receive intravenous antibiotics (e.g., piperacillin/tazobactam, $3 \times 4.5$ g daily plus gentamycin, 240 mg daily). Patients with platelet counts <10,000/μl should be monitored closely for signs of hemorrhage, preferably as inpatients, and should receive platelet transfusions until counts have recovered to at least 25,000/μl without further transfusions.

Some drug-specific precautions must be taken as well. Patients receiving nitrosoureas should receive pulmonary function assessment before chemotherapy and after every three courses of therapy. Patients who have serious allergic reactions to procarbazine develop while receiving PCV chemotherapy should not be continued on procarbazine but switched, e.g., to lomustine [CCNU] monotherapy at a higher dose [130 mg/m$^2$ × 6–8 weeks]. Also, during PCV chemotherapy, vincristine will be omitted, should clinically relevant polyneuropathy develop.

### General Practical Management

*Steroids* are effective agents for the control of brain tumor–associated edema. All patients with newly diagnosed brain tumors who have significant edema on neuroimaging should receive steroids, commonly dexamethasone at 4 mg qid, when surgery is planned, with the important exception of a lesion suggestive of primary cerebral lymphoma (Chapter 63). When surgery is not planned (e.g., in patients with multiple brain lesions suggestive of metastasis from a known primary neoplasm), lower doses of steroids are sufficient for clinical palliation (Chapter 64). Large intravenous doses of steroids (dexamethasone at 40–100 mg) should only be given to patients vitally endangered by mass effect and impending herniation or rapidly progressive sensorimotor dysfunction from a spinal tumor. After surgery, steroids should be tapered as clinically feasible according to the rule: *as little as possible, as much as necessary*. Steroids should be withdrawn before chemotherapy whenever feasible and should no longer be used as antiemetics because of potential interference with the efficacy of chemotherapy (Weller *et al.*, 1997).

The role of *anticonvulsants* in the management of brain tumor patients is controversial. The following guidelines are based on consensus rather than controlled clinical trials. Anticonvulsants should be given to patients with brain tumors who are seen with a seizure because of the high likelihood of recurrence. Most neurosurgeons prefer to administer anticonvulsants to all brain tumor patients undergoing a surgical intervention. Valproic acid and carbamazepine are the agents of choice, valproic acid having the advantage of the availability of an intravenous formulation. Controversy regarding the risk of valproic acid–associated bleeding complications is ongoing. Neither larger published series (Anderson *et al.*, 1997) nor our own experience suggest that the perioperative use of valproic acid poses a significant risk to brain tumor patients undergoing surgery, provided that appropriate presurgical evaluation of coagulation parameters including bleeding time is performed. Patients who do not have seizures after surgery should be withdrawn from anticonvulsants at approximately 3 months after surgery. Patients who continue having seizures must be switched to another agent or to a combination of two or three anticonvulsants. In general, patients who achieve tumor control are much more likely to live seizure-free without anticonvulsants than patients who do not, and progressive tumor often causes seizures despite the administration of anticonvulsants. Importantly, anticonvulsants may have profound effects on the metabolism of cancer chemotherapeutic agents, as exemplified for paclitaxel (Chang *et al.*, 1998a), and may thereby mask the efficacy of such agents.

Patients with brain tumors, notably malignant gliomas, have an increased risk of *deep vein thrombosis* and pulmonary embolism (Marras *et al.*, 2000). The reasons for this include immobilization, limb pareses, administration of steroids, and the release of thrombogenic substances by tumor cells. Whether radiotherapy and chemotherapy contribute to the increased incidence of thrombosis has remained unclear. Although malignancies are commonly considered a contraindication for anticoagulation, numerous studies have documented that anticoagulation is safe in brain tumor patients as long as patients with large space-occupying lesions with mass effect are excluded. Therefore, most brain tumor patients who have deep vein thrombosis or pulmonary embolism should receive the same treatment as otherwise healthy individuals. Anticoagulation should not be given to patients undergoing chemotherapy. The introduction of low-molecular-weight heparins that have a favorable pharmacokinetic profile compared with warfarin offers a new alternative to anticoagulation in patients with brain tumors, but this approach needs to be evaluated in a prospective clinical trial.

## TUMORS OF NEUROEPITHELIAL TISSUE

### Astrocytic Tumors

#### Clinical Aspects

Astrocytic tumors include lesions of very different biological dignity, ranging from benign lesions such as pilocytic astrocytoma to the most malignant type of brain tumor, glioblastoma. The median ages at diagnoses and survival rates at 2 and 5 years for various types of astrocytic tumor are provided in Tables III and IV. Pilocytic astrocytoma (WHO grade I) is a tumor mostly encountered in childhood that is often cerebellar in location, may reveal large cysts on imaging, and may be cured by resection alone. Optic nerve gliomas are mostly grade I lesions and may be associated with neurofibromatosis type I in approximately a third of cases. Diffuse astrocytomas (WHO grade II), often referred to as low-grade gliomas, are infiltrative lesions typically localized in the white matter of the cerebral hemispheres and more rarely in the pons, hypothalamus, and spinal cord in adults. Epilepsy is the most common presenting feature of supratentorial grade II astrocytomas. Brainstem gliomas are seen with cranial nerve palsies, whereas pyramidal tract signs and ataxia are late features. Spinal astrocytomas are commonly slowly growing tumors often associated with cyst formation or a syrinx. Clinically, there is frequently a slow course of a relapsing, remitting nature not dissimilar to multiple sclerosis, which can lead to diagnostic errors. Midline spinal pain, dissociated sensory impairment, and subtle motor signs, initially affecting one limb but progressing sequentially to involve all four limbs in tumors localized in the cervical region or both legs in tumors localized in the thoracic region, suggest the diagnosis. On imaging, diffuse astrocytomas appear as an area of low or mixed density with little, if any, mass effect. Two thirds demonstrate moderate edema, but calcification is infrequent.

Anaplastic astrocytomas (WHO grade III) and glioblastomas (WHO grade IV) (both collectively referred to as *malignant gliomas*) may arise anywhere in the brain, but the frontal and temporal lobes are most commonly affected. Glioblastomas may develop apparently de novo by clinical history (primary glioblastoma) or by way of malignant progression from a grade II or III astrocytoma (secondary glioblastoma). Approximately 5% of the malignant gliomas are multifocal by neuroimaging at diagnosis, and 5%–10% seed in the subarachnoid space in the course of the disease. Features frequently noted at the time of diagnosis of glioblastoma are a short history of less than 6 months, a developing focal neurological deficit, either motor or sensory, headaches that may be intermittent or constant, focal or generalized epileptic seizures, or an alteration in personality. CT or MRI will demonstrate an intrinsic, poorly demarcated area of mixed density/intensity. Frequently there is surrounding vasogenic edema that may be excessive even with small tumors. The pattern of contrast enhancement tends to be an irregular ring, but a thick regular ring is not infrequent, reminiscent of metastasis. The center of the tumor is frequently a low-density region, reflecting a necrotic or cystic area. The percentages of CT contrast enhancement are 96% for glioblastoma, 57% for anaplastic glioma, and 21% for grade II glioma (Lote *et al.*, 1998). Conversely, 66 of 513 tumors with contrast enhancement were of "low-grade" histology in the same study.

## Natural Course

The WHO grade is a strong predictor of outcome among the astrocytic tumors (Table IV). Cures are probably very rare for all types of astrocytic tumors except for pilocytic astrocytomas. The prognosis for patients with pilocytic astrocytoma is excellent. Malignant transformation is a rare event that may occur spontaneously (0.9%) or after radiotherapy (1.8%) (Tomlinson *et al.*, 1994). Whereas pilocytic astrocytoma is thus a rather stable histological entity, the diffuse WHO grade II astrocytomas show an intrinsic tendency toward malignant progression to anaplastic astrocytoma or glioblastoma at the time of recurrence or progression. Gemistocytic astrocytomas seem to have the highest probability of malignant progression. Age at diagnosis ≤40, seizures at presentation, minimal residual tumor after surgery, Karnofsky status ≥70%, and a diagnosis of oligodendroglioma or oligoastrocytoma rather than astrocytoma are favorable prognostic factors in patients with WHO grade II gliomas (Leighton *et al.*, 1997; Philippon *et al.*, 1993). Diffuse astrocytomas are a common histological finding in the brainstem, specifically in children, and in the spinal cord in adults.

The overall prognosis for anaplastic astrocytomas and glioblastomas is poor. Malignant gliomas arising from a previously diagnosed grade II glioma do not have a better prognosis than the de novo (primary) malignant gliomas (Dropcho and Soong, 1996). The most important predictors of favorable outcome are young age and high Karnofsky score at diagnosis. Surgery, radiotherapy, and chemotherapy may prolong progression-free and overall survival for some months in glioblastoma to several years in some patients with anaplastic astrocytoma. The rate of long-term survival for glioblastoma defined by 36 months survival is 2.2% (Scott *et al.*, 1999). Compared with the general population of glioblastoma patients, such patients were younger, more often had macroscopic tumor resection and received adjuvant chemotherapy, and had lower proliferative activity assessed by Ki-67 labeling in their tumors. Conversely, the median survival of malignant glioma patients aged 70 or older may be less than 4 months, and no survival advantage is conferred by radiotherapy in patients older than 80 (Meckling *et al.*, 1996). The course of gliosarcoma corresponds to that of glioblastoma (Galanis *et al.*, 1998a). Whether giant cell glioblastomas carry a better prognosis than other glioblastomas, has remained unclear. There are no

differences in the treatment strategies for gliosarcoma or giant cell glioblastoma compared with other glioblastomas.

*Brainstem gliomas* present a subset of astrocytic tumors that poses specific clinical problems. These tumors are often diffuse grade II tumors but may also be of higher grade and are more common in children than in adults. They are often diagnosed by neuroimaging alone, especially in children. Survival figures in children are 37% at 1 year, 20% at 2 years, and 13% at 3 years (Kaplan *et al.*, 1996). Survival rates at 2 years and 5 years of 44% and 34% and a median survival of 19 months were observed in a series including many adult patients (median age, 29.5 years) (Schild *et al.*, 1998b). The course of brainstem gliomas has been hypothesized to be more favorable in adults than in children (Landolfi *et al.*, 1998), but there are limited data on the response to therapy and outcome in adult patients with histologically verified brainstem gliomas. Astrocytomas, mostly of grade II, represent 25% of all spinal tumors in adulthood. Among the *spinal astrocytomas*, survival rates at 10 years were high for pilocytic astrocytoma (81%) but low for fibrillary (grade II) astrocytoma (15%) (Minehan *et al.*, 1995). The prognosis of the rare WHO grade III/IV spinal astrocytomas is poor despite multimodality treatment, similar to grade III/IV astrocytomas originating elsewhere in the CNS (Linstadt *et al.*, 1989).

## Principles of Therapy

The major therapeutic approaches to astrocytic tumors at diagnosis and at recurrence are summarized in Table IX. *Surgery* is the principal treatment of pilocytic astrocytoma and has a definite role in the treatment of diffuse astrocytomas (WHO grade II), both at presentation and at recurrence. Although surgical interventions are necessary to establish a diagnosis in anaplastic astrocytomas and glioblastomas, no prospective controlled study has determined that efforts at macroscopic resection are superior to biopsy in WHO grade III/IV gliomas. Although most retrospective analyses support aggressive resection (Hess, 1999), other authors fail to note a prognostic impact of extent of resection (Kreth *et al.*, 1999). Because complete removal is never possible, prevention of neurological deficits has a higher priority than attempts at "complete" tumor resection. Yet, even second gross surgical resections should be considered in patients with favorable prognostic factors who are seen with recurrent anaplastic astrocytoma or glioblastoma. The interval from the preceding resection should be 6–12 months to make this intervention meaningful, and such patients should be candidates for further adjuvant treatment after the second resection.

**TABLE IX** Standard Treatments for Glial Tumors

| | Management at presentation | Management at recurrence or progression |
|---|---|---|
| Pilocytic astrocytoma (WHO grade I) | Resection* | Re-resection*<br>Radiotherapy if resection is not possible* |
| Diffuse astrocytoma (WHO grade II) | Resection<br>Biopsy and wait-and-see<br>Biopsy and radiotherapy | Resection<br>Radiotherapy* (chemotherapy) |
| Oligodendroglioma and oligoastrocytoma (WHO grade II) | Resection<br>Biopsy and wait-and-see<br>Biopsy and chemotherapy or radiotherapy | Resection and chemotherapy* or radiotherapy* |
| Anaplastic astrocytoma (WHO grade III) | Resection plus radiotherapy*<br>(chemotherapy) | Re-resection plus chemotherapy* or<br>stereotactic radiotherapy |
| Anaplastic oligodendroglioma and oligoastrocytoma (WHO grade III) | Resection plus chemotherapy*<br>or radiotherapy* | Resection and chemotherapy* or radiotherapy* |
| Glioblastoma (WHO grade IV) | Resection (or biopsy) plus<br>Radiotherapy*** (chemotherapy) | (Re-resection plus) chemotherapy* or<br>stereotactic radiotherapy |

*Involved-field radiotherapy* is a treatment option for the rare pilocytic astrocytomas that progress and are not accessible to surgery (e.g., in the optic chiasm in children) (Tao *et al.*, 1997). The role of radiotherapy in the management of diffuse astrocytomas is controversial. The EORTC trial 22844 disclosed no difference in terms of survival with the dose levels of 45 Gy administered in 5 weeks and 59.4 Gy administered in 6.6 weeks (Karim *et al.*, 1996). Although radiotherapy is an effective treatment in terms of delaying tumor growth, the EORTC trial 22845 has shown that radiotherapy (54 Gy) at diagnosis provides no advantage compared with radiotherapy at clinical progression (Karim *et al.*, 2002). Importantly, this trial also failed to support the notion that early radiotherapy promotes the malignant progression of low-grade gliomas. Interstitial radiotherapy (brachytherapy) is an alternative strategy of radiotherapy for WHO grade I/II gliomas that seems to be effective (Kreth *et al.*, 1995) but that has not been compared with conventional external fractionated radiotherapy in a randomized trial. On the basis of the role of age and extent of resection as prognostic factors (Bauman *et al.*, 1999), a current US study will pursue a wait-and-see strategy in patients younger than 40 who have had a macroscopically complete resection. Patients older than 40 or patients with an incomplete resection will be randomly assigned to radiotherapy alone or radiotherapy followed by PCV chemotherapy. As indicated previously, radiotherapy is the single most important therapeutic option for patients with astrocytomas of brainstem and spinal cord (Minehan *et al.*, 1995; Schild *et al.*, 1998b).

Involved-field radiotherapy is the standard treatment for patients with anaplastic astrocytoma and glioblastoma. Despite extensive efforts to define optimum doses and schedules, no major differences of various radiotherapy regimens have emerged, compared with conventional doses of 54–60 Gy delivered in 1.8–2 Gy fractions (Werner-Wasik *et al.*, 1996). Accelerated schedules using higher single fractions may be reasonable for patients with poor prognostic factors and thus short life expectancy, moreover, such schedules are cost-effective (Brada *et al.*, 1999). Because older patients (>70–75 years of age) may not benefit from radiotherapy (Meckling *et al.*, 1996), symptomatic treatment and supportive care only should be considered in these patients. Historical studies indicate that radiotherapy after surgery prolongs the median survival by approximately 6 months compared with surgery alone (Table X). Numerous approaches to enhance the efficacy of radiotherapy have failed, including efforts to overcome the hypoxia of glioma cells (Miralbell *et al.*, 1999) and various combinations with chemotherapy in which chemotherapy was primarily designed to act as a radiosensitizer. At recurrence, radiosurgery or a stereotactic fractionated boost of 15–20 Gy may be considered for some patients with circumscribed lesions (Shepherd *et al.*, 1997). The superior survival of patients with malignant glioma who receive radiosurgery may have been a consequence of selection bias, because patients eligible for radiosurgery represent a favorable prognostic subgroup (Curran *et al.*, 1993).

*Chemotherapy* plays no role in the management of pilocytic astrocytoma in adults but has been shown to be effective in grade I/II gliomas specifically of childhood, both at recurrence or in the preirradiation setting (Packer *et al.*, 1997). Specifically in young infants less than 3 years of age, postsurgical chemotherapy may allow one to delay radiotherapy. Although chemotherapy provides no cure, durable responses and stabilization for years may be achieved. The role of chemotherapy for anaplastic astrocytoma and glioblastoma as an add-on to radiotherapy (first-line therapy) and at recurrence are still under investigation. In the setting of adjuvant chemotherapy added to postsurgical radiotherapy, multiple chemotherapy studies of the recent two decades have failed to demonstrate a major advantage of radiotherapy plus chemotherapy over radiotherapy alone for WHO grade III/IV gliomas (Table X). No further chemotherapy trial has indicated a better effect of adjuvant chemotherapy in grade III/IV gliomas than the classical BCNU trials (Walker *et al.*, 1978; 1980). With the addition of BCNU monotherapy to surgery and radiotherapy, the proportion of patients surviving more than 18 months increased from 4%–19% and from 15%–27%, respectively. A meta-analysis of 16 randomized clinical trials revealed a relative increase in survival afforded by chemotherapy of 23.4% at 1 year and of 52.4% at 2 years (Fine *et al.*, 1993). Furthermore, a re-evaluation of two large malignant glioma trials (Green *et al.*, 1983; Walker *et al.*, 1980), confirmed that chemotherapy increased long-term survival specifically in the subset of glioblastoma patients and independently of major prognostic factors (DeAngelis *et al.*, 1998). Many authors believe that the combination of procarbazine, vincristine, and lomustine (CCNU), the PCV regimen, is the most active regimen for grade III/IV gliomas, although this has only been demonstrated for a small series of anaplastic gliomas in a trial that has been subject to much criticism (Levin *et al.*, 1990). Thus the superiority of PCV over BCNU in the RTOG trials has been contested (Prados *et al.*, 1999). Moreover, the negative British Medical Research Council (MRC) trial for adjuvant postirradiation PCV in grade III/IV gliomas (The Medical Research Council Brain Tumor Working Party, 2001) casts further doubts on the efficacy of adjuvant chemotherapy added to standard radiotherapy as part of the first-line therapy for malignant glioma. Admittedly, this trial used a PCV regimen that differed in dosages and scheduling from the classical PCV regimen used in most centers (Table VIII). The efficacy of temozolomide during and after radiotherapy compared with radiotherapy alone will be evaluated in the EORTC trial 26981. In general, older patients (>60–65 years) seem to benefit less from chemotherapy than younger patients but are more likely to experience severe side effects from chemotherapy. We use chemotherapy only exceptionally for patients older than 70 years, usually for recurrent disease and not as part of the first-line therapy, and these patients should be in good general condition.

On the basis of the assumption that the sequence of chemotherapy followed by radiotherapy would be superior to the reverse sequence because of higher drug levels in the tumor and lower neurotoxicity, several studies have readdressed the approach of "neoadjuvant" preir-radiation chemotherapy for anaplastic astrocytoma and glioblastoma (Table X). Most of these studies have been disappointing, with the exception of temozolomide, which achieved a rate of progression-free survival at 4 months of 55% (Friedman *et al.*, 1998), and a regimen of BCNU and cisplatin, which was probably even better

in terms of preventing tumor progression but was also rather toxic (Grossman *et al.*, 1997). Interestingly, the latter study confirms that high initial response rates do not translate into increased survival (Grant *et al.*, 1997) (Table X). Moreover, a subsequent randomized trial based on this regimen failed to demonstrate a survival advantage of BCNU plus cisplatin followed by radiotherapy over radiotherapy alone. Overall survival in these studies was in the range of conventional treatment of resection followed by radiotherapy, suggesting that preirradiation chemotherapy, even though not effective, did not negatively influence outcome in these trials.

The role of chemotherapy for WHO grade III/IV gliomas is better defined in the setting of recurrence or progression after radiotherapy. Several chemotherapy protocols may induce transient responses and prolong survival in such patients. A meta-analysis of single-arm phase II clinical trials of chemotherapy for recurrent WHO grade III/IV gliomas performed at a single center revealed that only 9% of the patients had CR or PR and that 21% of the patients were progression-free at 6 months (glioblastoma, 15%; anaplastic astrocytoma, 31%). The median progression-free survival was 9 weeks for glioblastoma and 13 weeks for anaplastic astrocytoma. The median overall survival was 25 weeks for glioblastoma and 47 weeks for anaplastic astrocytoma (Wong *et al.*, 1999). A recent randomized trial for recurrent glioblastoma reported a small, but statistically significant, advantage of temozolomide over procarbazine: the survival at 6 months was 21% with temozolomide compared with 8% for procarbazine (Yung *et al.*, 2000). This study suffers from the failure to exclude patients in the procarbazine arm that had previously received procarbazine as part of the PCV regimen. A less equivocal efficacy for temozolomide was demonstrated for recurrent anaplastic astrocytoma (Yung *et al.*, 1999). Further, local nitrosourea-based chemotherapy may prolong progression-free survival in some patients with recurrent glioma (Brem *et al.*, 1995).

### Practical Management

*Pilocytic astrocytomas* are best treated by surgical resection, which may be curative.* Recurrences should be treated surgically as well whenever feasible. Radiotherapy (54 Gy, 1.8–2 Gy fractions per day) to the involved field should be considered only when surgery can no longer be offered. Chemotherapy plays no role in the management of these tumors, except for optic nerve gliomas in childhood.

*WHO grade II astrocytomas* should be resected where feasible without risk of neurological deficits. Second surgery should be considered at recurrence. Radiotherapy at the time of histological diagnosis should be offered to patients with clinical deficits other

TABLE X   Results from Selected Clinical Trials in Malignant Gliomas

**Radiotherapy ± chemotherapy**

| Reference | Tumor | n | Management | Response | Progression-free survival (mo) | Median survival (mo) | Survival at 1 year (%) | 2 years (%) |
|---|---|---|---|---|---|---|---|---|
| Walker et al., 1978 | Malignant glioma* | 42 | S+ | | | 3.3 | 3 | 0 |
| | | 68 | S+ BCNU | | | 4.3 | 12 | 0 |
| | | 93 | S + RT | | | 8.4 | 24 | 1 |
| | | 100 | S + RT + BCNU | | | 8.1 | 32 | 5 |
| Walker et al., 1980 | Malignant glioma | 81 | S + MeCCNU | | | 5.6 | 15 | 8 |
| | | 94 | S + RT | | | 8.4 | 35 | 10 |
| | | 92 | S + RT + BCNU | | | 11.9 | 50 | 15 |
| | | 91 | S + RT + MeCCNU | | | 9.8 | 37 | 12 |
| Green et al., 1983 | Malignant glioma | 141 | S + RT + cortisone | | | 10 | | 6 |
| | | 124 | S + RT + BCNU | | | 12.5 | | 16 |
| | | 134 | S + RT + BCNU + cortisone | | | 10 | | 18 |
| | | 128 | S + RT + procarbazine | | | 12 | | 23 |
| Levin et al., 1990 | Glioblastoma | 29 | S + RT + hydroxyurea + BCNU | | | 14 | | 10 |
| | | 31 | S + RT + hydroxyurea + PCV | | | 12.5 | | 22 |
| | Anaplastic glioma | 37 | S + RT + hydroxyurea + BCNU | | | 22 | | 40 |
| | | 36 | S + RT + hydroxyurea + PCV | | | 36 | | 60 |
| Hildebrand et al., 1994 | Malignant glioma | 134 | S + RT | | 6.7 | 10.4 | | 12 |
| | | 135 | S + RT + BCNU + dibromodulcitol | | 8.1 | 13 | | 21 |
| Levin et al., 1995 | Glioblastoma, newly diagnosed | 83 | S + acc.RT/carboplatin + PCV | | 6.1 | 12.4 | | |
| Brandes et al., 1998 | Glioblastoma, newly diagnosed | 56 | S + RT/carboplatin/VM26 plus BCNU | | 7.5 | 12.5 | | 20 |
| Laperriere et al., 1998 | Malignant astrocytoma | 69 | S + RT | | | 13.2 | | |
| | | 71 | S + RT + brachytherapy | | | 13.8 | | |
| Prados et al., 1998 | Glioblastoma | 245 | S + RT/hydroxyurea + | | 7.7 | 13.1 | 56 | 23 |
| | Anaplastic glioma newly diagnosed | 110 | BCNU/6-thioguanine | | 65.8 | | 93 | 80 |
| Rainov, 2000 | Glioblastoma, newly diagnosed | 124 | S + RT | | 6.1 | 11.8 | 55 | |
| | | 124 | S + RT + HSV/TK | | 6 | 12.2 | 50 | |

| Reference | Histology | No. of patients | Treatment | Response | Median time to progression (months) | Median survival (months) | | |
|---|---|---|---|---|---|---|---|---|
| Medical Research Council, 2001 | Anaplastic astrocytoma | 60 | S + RT | | | 13 | | 37 |
| | Anaplastic astrocytoma | 53 | S + RT + PCV | | | 15 | | 43 |
| | Glioblastoma | 226 | S + RT | | 6 | 9 | | 8 |
| | Glioblastoma newly diagnosed | 223 | S + RT + PCV (modified PCV) | | | 9.5 | | 9 |
| **Preirradiation chemotherapy** | | | | | | | | |
| Fetell et al., 1997 | Glioblastoma, newly diagnosed | 33 | Paclitaxel | No CR or PR | <1[‡] | 11.8[§] | Nd | nd |
| Grossman et al., 1997 | Glioblastoma or anaplastic astrocytoma, newly diagnosed | 52 | BCNU + cisplatin | 17 PR | nd | 13[§] | 62 | 19 |
| Friedman et al., 1998 | Glioblastoma or anaplastic astrocytoma, newly diagnosed | 33 | Temozolomide | 3 CR, 15 PR | | | | |
| Weller et al., 2001 | Glioblastoma, newly diagnosed | 21 | Gemcitabine | No CR or PR | 2.3[‡] | 11[§] | 48[§] | nd |
| **Recurrent glioma** | | | | | | | | |
| Brem et al., 1995 | Recurrent glioma | 110 | S + interstitial BCNU | | | 7.2 | | |
| | | 112 | S + placebo polymer | | | 5.4 | | |
| Brandes et al., 1999a | Recurrent glioblastoma | 28 | Procarbazine plus tamoxifen | 1 CR, 8 PR | 3 | 6.3 | | |
| | Recurrent anaplastic astrocytoma | 23 | | 1 CR, 5 PR | 7.7 | 13.3 | | |
| Wong et al., 1999 | Anaplastic glioma | 150 | Various | 21 CR or PR | 3 | 11 | 47 | |
| | Glioblastoma | 225 | | 14 CR or PR | | | | |
| Yung et al., 1999 | Anaplastic astrocytoma at first relapse | 111 | Temozolomide | 8 CR, 31 PR | 2.1 | 5.8 | 21 | |
| | | | | | 5.4 | 14.5 | | |
| Yung et al., 2000 | Glioblastoma at first relapse | 112 | Temozolomide | 6 PR | 2.9 | | | |
| | | 113 | Procarbazine | 6 PR | 1.8 | | | |
| Kappelle et al., 2001 | Recurrent glioblastoma | 63 | PCV | 2 CR, 5 PR | 3 | 7.7 | | |

*malignant glioma trials commonly included 70–90% of glioblastomas.
†S surgery, RT radiotherapy.
‡with chemotherapy alone.
§with chemotherapy followed by radiotherapy.

than seizures from the tumor. This applies to almost all patients with grade II astrocytomas of the spinal cord. Radiotherapy may be delayed until clinical progression in patients who are asymptomatic except for seizures.* Most centers will administer fractionated external radiotherapy, but interstitial radiotherapy performed in specialized centers may be equally effective. WHO grade II astrocytomas that are not surgically accessible and progress after radiotherapy may be treated with chemotherapy on an individual basis. PCV or temozolomide seem to be the most suitable drug regimens for adults (Tables VII and VIII) in this setting. In the case of response or stable disease, chemotherapy should be held after four cycles of PCV or eight cycles of temozolomide. CT or MRI are performed at 6-month intervals or at any time of clinical progression. Ample evidence shows some activity of carboplatin plus vincristine in the pediatric age group of patients with grade I/II glioma (Packer *et al.*, 1997).

Surgery followed by radiotherapy (54 Gy, 1.8–2 Gy fractions per day) is the treatment of choice for *anaplastic astrocytoma*.* The role of chemotherapy as an add-on treatment to radiotherapy as the first-line therapy in anaplastic astrocytoma is not well-defined, but nitrosourea-based regimens or temozolomide are reasonable regimens in that setting. At recurrence after radiotherapy, second surgery should be considered. Irrespective of whether second surgery is feasible, patients with recurrent anaplastic astrocytoma who are in good general condition (Karnofsky performance >60%) should receive chemotherapy, either nitrosourea-based (PCV) (Levin *et al.*, 1990) or temozolomide (Yung *et al.*, 1999).* In the case of response or stable disease, chemotherapy should be held after four cycles of PCV or other nitrosourea-based chemotherapy or eight cycles of temozolomide. CT or MRI are performed in 4-month intervals or at any time of clinical progression.

*Glioblastoma* is treated by resection or biopsy followed by radiotherapy (54–60 Gy, 1.8–2 Gy fractions per day).*** A role for chemotherapy as an add-on to radiotherapy as the first-line treatment has not been clearly defined (Table X). Chemotherapy may be considered for patients with favorable prognostic factors and should be either nitrosourea monotherapy or nitrosourea-based (e.g., PCV) or temozolomide. In the case of response or stable disease, chemotherapy should be held after 6 cycles of PCV or 12 cycles of temozolomide. CT or MRI are performed in 3-month intervals or at any time of clinical progression. At progression or recurrence, second surgery, reirradiation, and chemotherapy may be considered. Second surgery is recommended for patients with an interval from diagnosis of 6–12 months and with an accessible location of the lesion. Reirradiation, probably better administered as hypofractionated stereotactic radiotherapy (e.g., four

fractions of 5 Gy) than radiosurgery, may be an option for patients with circumscribed lesions (Shepherd *et al.*, 1997). Chemotherapy, the most commonly used treatment at recurrence, should be given to patients with favorable prognostic factors, notably Karnofsky performance status >60%.* Suitable regimens are either nitrosourea-based (BCNU, PCV) or temozolomide in nitrosourea-naïve patients and temozolomide in nitrosourea-pretreated patients (Yung *et al.*, 2000). The management of WHO grade III/IV brainstem and spinal cord gliomas in adults follows the guidelines for supratentorial gliomas, except that efforts at resection cannot be undertaken.

### Other Rare Astrocytomas

*Pleomorphic xanthroastrocytoma* is commonly a supratentorial WHO grade II lesion of children and young adults (Giannini *et al.*, 1999). These tumors should be managed with an attempt at radical surgical resection. A postoperative wait-and-see strategy is recommended for most patients even after incomplete resections. Radiotherapy should be administered to progressive lesions that cannot be accessed surgically. *Subependymal giant cell astrocytoma* is a WHO grade I lesion strongly associated with tuberous sclerosis (Table II). These tumors can grow to large sizes and may impair CSF flow by obstructing the foramen of Monro. Symptomatic lesions should be resected. Radiotherapy is not indicated even after incomplete resection.*

### Oligodendrogliomas and Mixed Gliomas

#### Clinical Aspects

Oligodendroglioma and oligoastrocytoma (WHO grade II), as well as their anaplastic variants (WHO grade III), are considered together here, because their current management is similar. Moreover, although one might assume that the astrocytic component determined a less-favorable prognosis in patients with oligoastrocytoma, most studies have not confirmed differences in outcome between "pure" oligodendroglial and "mixed" oligoastrocytic tumors (Kyritsis *et al.*, 1993; Streffer *et al.*, 2000; Wallner *et al.*, 1988). Mixed tumors are of clonal origin, indicating that one transformed precursor cell may form two morphologically distinct tumor cell populations. Most oligodendrogliomas are supratentorial in location. More than 50% arise within the frontal lobes, and 20% are bifrontal and involve the white matter. The most common clinical presentation is epilepsy, and a common radiological feature is calcification (90%). Oligodendrogliomas are infiltrative tumors, frequently involving the basal ganglia or corpus callosum.

## Natural Course

The survival rates for oligodendrogliomas are much better than for astrocytic tumors (Donahue *et al.*, 1997) (Table IV). Almost all data on the natural course and response to therapy in oligodendroglioma are retrospective (Table XI). Because it remains unclear whether cures are possible and because some oligodendrogliomas seem to spontaneously cease growing for years, the available, noncontrolled data need to be interpreted with caution. Moreover, the histological criteria for grading oligodendrogliomas vary between the different studies. Predictors of favorable outcome resemble those for astrocytic tumors and include young age, frontal location, gross surgical resection, high Karnofsky score, and lack of contrast enhancement on neuroimaging.

## Principles of Therapy

There is no defined standard treatment for oligodendroglioma. Although surgery followed by radiotherapy is often considered standard, surgery has not proven to be superior to biopsy and a wait-and-see strategy. Also, surgery followed by radiotherapy has not been demonstrated to be superior to surgery and a wait-and-see strategy. Radiotherapy is commonly administered to the involved field. Although anaplastic oligodendrogliomas may seed in the subarachnoid space in more than 30% of the patients in the course of the disease, prophylactic craniospinal irradiation is not recommended. Among the retrospective studies, there is no agreement whether radiotherapy is beneficial (Celli *et al.*, 1994; Gannet *et al.*, 1994; Lindegaard *et al.*, 1987; Wallner *et al.*, 1988) or not (Bullard *et al.*, 1987; Nijjar *et al.*, 1993). Retrospective studies indicate that tumor doses of 60 Gy are required for adequate local control (Allison *et al.*, 1997; Wallner *et al.*, 1988). PCV chemotherapy has become an established treatment for oligodendroglioma (Cairncross *et al.*, 1994). Although PCV is considered the standard regimen of chemotherapy for these tumors, this has largely historical reasons in that PCV has been most extensively examined. It is reasonable to assume that other regimens, including temozolomide, will be active against oligodendroglioma as well. Specific molecular alterations, that is, losses of chromosomes 1p and 19q, have been proposed to signify a favorable response to PCV chemotherapy, whereas alternative molecular changes do not (Cairncross *et al.*, 1998; Smith *et al.*, 2000). With PCV being active at progression after radiotherapy, as well as in the "neoadjuvant" postsurgery preirradiation chemotherapy setting (Cairncross *et al.*, 1994; Paleologos *et al.*, 1999; Streffer *et al.*, 2000), the standard of therapy for oligodendroglioma and anaplastic oligo-

dendroglioma is currently a matter of debate. Macroscopic tumor resection, previously considered standard, has been questioned and might be no better than biopsy, because chemotherapy alone induces high rates of complete and partial remissions. Similarly, many centers now prefer to treat patients with up-front ("neoadjuvant") chemotherapy and to delay radiotherapy until progression or recurrence. Several studies trying to address these controversial issues are currently ongoing, including EORTC trial 26951, which compares radiotherapy followed by PCV with radiotherapy alone, and the German NOA-04 phase III trial, which compares radiotherapy followed by PCV or temozolomide with the reverse sequence.

## Practical Management

The following recommendations apply to patients who are treated outside clinical trials. Oligodendrogliomas and oligoastrocytomas should be biopsied or resected, with the concept that prevention of neurological deficits is of higher priority than extent of resection. Grade II tumors may be followed by neuroimaging without adjuvant therapy and will be treated with PCV at progression.* Grade III tumors will be treated with four cycles of PCV after surgery.* Patients progressing during PCV chemotherapy will receive either radiotherapy immediately or radiotherapy after a second resection (54 Gy, 1.8–2 Gy fractions daily).* Patients progressing after PCV and radiotherapy may respond to a second course of PCV (Streffer *et al.*, 2000), temozolomide, paclitaxel (Chamberlain and Kormanik, 1997), or carboplatin plus etoposide (Cairncross *et al.*, 1994; Streffer *et al.*, 2000).*

## Ependymal Tumors

### Clinical Aspects and Natural Course

The ependymal tumors include ependymoma (WHO grade II), anaplastic ependymoma (WHO grade III), myxopapillary ependymoma (WHO grade I), and subependymoma (WHO grade I). *Ependymomas* may occur in all age groups and at all sites along the ventricles and spinal canal. They are common in children and are typically found infratentorially in this age group (Nazar *et al.*, 1990). Although ependymal tumors altogether account for approximately 25% of all spinal tumors in adulthood, a spinal location is uncommon in children. Infratentorial and spinal ependymomas are equally common in adults and are diagnosed most often in the third and fourth decades of life. Ependymomas tend to metastasize within the subarachnoid space, regardless of whether there has been a surgical

TABLE XI  Outcome and Impact of Treatment in Oligodendroglioma

| Reference | Tumor | n | Management | Response | Progression-free survival (mo) | Median survival (mo) | Survival at 5 years (%) | 10 years (%) |
|---|---|---|---|---|---|---|---|---|
| Bullard et al., 1987 | Oligodendroglioma | 71 | S + RT (37) | | 28 | 62 | | |
| | | | S (34) | | 27 | 54 | | |
| Lindegaard et al., 1987 | Oligodendroglioma | 170 | S + RT (108) | | | 38 | 36 | |
| | | | S (62) | | | 26.5 | 27 | |
| Wallner et al., 1988 | Oligodendroglioma | 25 | S + RT (14) | | | 132 | 80 | 56 |
| | | | S (11) | | | 60 | 55 | 18 |
| Kyritsis et al., 1993 | Anapl. oligodendroglioma | 8 | S + RT + ChT (mostly PCV) | 1 CR, 3 PR, 4 SD | 8 | >15 | | |
| | Anapl. oligodendroglioma | 9 | S + RT, PCV at PD | 1 CR, 2 PR, 5 SD | 18 | >71 | | |
| | Anapl. oligoastrocytoma | 12 | S + RT + PCV | 2 CR, 3 PR, 7 SD | Not reached | >13 | | |
| | Anapl. oligoastrocytoma | 5 | S + RT ± ChT, ChT at PD | 1 PR, 4 SD | 6 | >22 | | |
| Gannett et al., 1994 | Oligodendroglioma | 41 | S + RT (27) | | 79 | 84 | 83 | 46 |
| | | | S (14) | | 42 | 47 | 51 | 36 |
| Cairncross et al., 1994 | Anapl. oligo., post RT | 15 | PCV | 5 CR, 6 PR | 15.4 | | | |
| | Anapl. oligo., newly dg. | 9 | PCV | 4 CR, 3 PR | >17.2 | | | |
| Celli et al., 1994 | Oligodendroglioma | 105 | S + RT (77) | | | 76 | 57 | |
| | | | S (28) | | | 37 | 36 | |
| Shaw et al., 1994 | Oligoastrocytoma | 71 | S ± RT | | | 70 | 55 | 29 |
| Kim et al., 1996 | Oligodendroglioma/ | 19 | S + PCV followed by RT | 6 CR, 12 PR | 15.4 | 49.8 | | |
| | Oligoastrocytoma WHO grades II, III, IV | 12 | S + RT followed by PCV | 4 CR, 7 PR | 25.5 | >65 | | |
| Chamberlain and Kormanik, 1997 | Recurrent oligodendroglioma | 20 | Paclitaxel | 3 PR, 7 SD | | 4–5 | | |
| Soffietti et al., 1998 | Oligodendroglioma, oligoastrocytoma | 26 | S ± RT followed by PCV | 3 CR, 13 PR | 24 | | | |
| Van den Bent et al., 1998 | Recurrent oligodendroglioma, oligoastrocytoma | 52 | PCV | 9 CR, 24 PR | 10 | 20 | | |
| Paleologos et al., 1999 | Anaplastic or aggressive oligodendroglioma | 30 | S + PCV | 12 CR, 9 PR | | | | |
| Streffer et al., 2000 | Oligodendroglioma, oligoastrocytoma, anaplastic oligodendroglioma, anaplastic oligoastrocytoma | 27 | S + RT, PCV at PD (11) | 1 CR, 5 PR, 5 SD | >18 | | | |
| | | | S + RT + PCV (5) | 3 CR, 2 SD | No PD at 12–54 | | | |
| | | | S + PCV (11) | 2 CR, 3 PR, 6 SD | >14 | | | |

procedure, and these metastases may remain clinically silent. This possibility must be assessed by craniospinal MRI and CSF cytologic examination at diagnosis and follow-up. Metastases outside the CNS do occur. Young age, incomplete resection, and cerebrospinal dissemination are unfavorable prognostic factors (Timmermann *et al.*, 2000). The 5-year survival rates are in the range of 50% or more for children and adults (Nazar *et al.*, 1990; Needle *et al.*, 1997; Vanuytsel *et al.*, 1992). The outcome for patients with spinal tumors seems to be much better than for the patients with intracranial tumors (Helseth and Mørk, 1989). *Anaplastic ependymomas* are more often localized intracranially than in the spinal cord. Most, but not all, retrospective analyses indicate that the anaplasia of ependymoma is associated with poor outcome (Rezai *et al.*, 1996; Schild *et al.*, 1998a). *Myxopapillary ependymomas* occur almost exclusively in the conus cauda region and are typically a tumor of young adults. Although histologically benign, these tumors have a high rate of local recurrence and less often distant recurrence when incompletely resected (Rezai *et al.*, 1996). *Subependymomas* are benign lesions that develop preferentially in the fourth ventricle (>50%) and the lateral ventricles (>30%).

## Principles of Therapy and Practical Management

Surgery and radiotherapy are the mainstay of treatment for ependymomas. Because the extent of resection is a prognostic factor (Nazar *et al.*, 1990), attempts at complete surgical removal are warranted for all types of ependymal tumor in all locations.* Excellent results have been reported with surgery alone even for intramedullary ependymomas, although with an insufficiently long follow-up to derive conclusions regarding the risk of recurrence when no postoperative radiotherapy is administered (Epstein *et al.*, 1993). In principle, all ependymomas are radiosensitive tumors (Linstadt *et al.*, 1989; Schild *et al.*, 1998a; Vanuytsel *et al.*, 1992; Whitaker *et al.*, 1991). Controversies regarding the role of radiotherapy include the timing of radiotherapy, the dose, and the target volume. Late side effects from radiotherapy are a particular concern in patients with ependymomas, because ependymomas are mainly tumors of children and early adulthood. These issues will need to be addressed in future trials, and no firm recommendations can be made at present (Table XII). Postoperative radiotherapy of the involved field is probably sufficient for supratentorial *ependymomas* and *anaplastic ependymomas*, because the initial tumor site is the most common site of failure (Timmermann *et al.*, 2000). Craniospinal radiotherapy is probably required for patients with leptomeningeal seeding from ependymoma or anaplastic ependymoma detected by neuroimaging or CSF cytology. We recommend that patients with infratentorial or spinal ependymoma do not receive prophylactic craniospinal irradiation in the absence of CSF seeding. There is no conclusive evidence that craniospinal irradiation reduces the risk of spinal seeding (Lyons and Kelly, 1991). Fractionated stereotactic radiotherapy or radiosurgery may be an option for some patients with a local recurrence. There is no proven role for chemotherapy in the management of ependymomas (Timmermann *et al.*, 2000), although current approaches to combine multiagent chemotherapy with radiotherapy show promising activity (Kühl *et al.*, 1998; Needle *et al.*, 1997). Our own unpublished experience with nitrosourea-based chemotherapy for adult ependymoma progressing after radiotherapy has been disappointing. As indicated earlier, *myxopapillary ependymomas* and *subependymomas* should be removed surgically whenever feasible. Patients with incompletely resected myxopapillary ependymomas should receive involved-field radiotherapy (Schild *et al.*, 1998a). Patients with incompletely resected subependymomas may be followed without adjuvant treatment until progression. In general, the possibility of second and third resections should be evaluated for all types of ependymal tumors at progression, even for symp-

**TABLE XII** Treatment Recommendations for Ependymoma

| | Frequency (%) | Surgery | Adjuvant therapy |
|---|---|---|---|
| Ependymoma or anaplastic ependymoma | 79 | Supratentorial, complete or incomplete resection | Involved-field radiotherapy* |
| | | Infratentorial, complete or incomplete resection | Involved-field radiotherapy (or craniospinal radiotherapy)* |
| | | Spinal | Involved-field radiotherapy* |
| | | Any, plus positive CSF cytology | Craniospinal radiotherapy* |
| Myxopapillary ependymoma | 13 | Complete resection | Observation* |
| | | Incomplete resection | Involved-field radiotherapy* |
| Subependymoma | 8 | Complete or incomplete resection | Observation* |

tomatic lesions in the setting of leptomeningeal seeding. A recommendation for the type of radiotherapy determined by the type and location of tumor and by the extent of surgery is provided in Table XII. The involved field should receive 54 Gy (5 × 1.8–2 Gy per week). The craniospinal axis should receive 36 Gy (5 × 1.8–2 Gy per week). Chemotherapy should be given in the context of clinical trials where feasible.

## Choroid Plexus Tumors

### Clinical Aspects and Natural Course

Choroid plexus papillomas (WHO grade I) are rare benign tumors, forming less than 1% of intracranial neoplasms, typically found in the trigone of the lateral ventricle of children and the fourth ventricle in adults, where they may extend into the cerebellopontine angle by way of the foramen of Luschka, or very rarely the third ventricle in adults. Plexus papillomas increase CSF production and hence present with the features of raised intracranial pressure. In young children, an increase in head circumference or failure to thrive may be the only clues. However, the more typical picture of headaches, nausea and vomiting, deteriorating consciousness level, and papilledema may supervene. Less frequent presentations are ataxia, hemiparesis, or seizures. Choroid plexus carcinomas (WHO grade III) are locally invasive tumors that may metastasize in the CSF space. Survival rates at 5 years are in the range of 100% for papillomas and 40% for plexus carcinomas (Pencalet *et al.*, 1998).

### Principles of Therapy and Practical Management

Surgical excision of choroid plexus tumors should be attempted and may be curative for papillomas.* Incompletely resected papillomas may be followed without further treatment. Postoperative radiotherapy may be given to patients with completely resected choroid carcinomas, although there is no proof of efficacy in this setting (Chow *et al.*, 1999; Pencalet *et al.*, 1998). Radiotherapy is recommended after an incomplete resection of choroid plexus carcinoma. Adjuvant chemotherapy (e.g., platin/etoposide-based) may be preferred to radiotherapy in children younger than 3 years of age. Craniospinal radiotherapy may be required for patients with symptomatic leptomeningeal metastases. The role of chemotherapy for recurrent or progressive choroid plexus tumors is poorly defined (Valencak *et al.*, 2000) and must therefore be individualized. Conversely, small series indicate a role for postbiopsy "neoadjuvant" platin-based chemotherapy to allow for a subsequent effort at a gross total resection (Greenberg, 1999).

## Glial Tumors of Uncertain Origin

*Astroblastoma* is a very rare glial tumor mainly of young adulthood that is of uncertain dignity (Brat *et al.*, 2000). These tumors are found most often in the hemispheres but may occur almost anywhere in the CNS. They may be cystic. Given the limited clinical data available, no firm treatment recommendations can be made. Astroblastomas lacking anaplastic features may be managed by gross surgical resection. Incompletely resected astroblastomas will recur. They should probably be treated with postsurgical adjuvant radiotherapy. The role of chemotherapy has not been defined. Astroblastomas with anaplastic features progressing after surgery and radiotherapy may be treated with chemotherapy (e.g., following the guidelines outlined for glioblastoma). The survival rate at 5 years is probably low in patients with astroblastoma with anaplastic features (Thiessen *et al.*, 1998). *Chordoid glioma of the third ventricle* is a rare tumor of adults attributed a dignity of WHO grade II. Surgical removal is often not possible, and a role for adjuvant therapy has not been defined. *Gliomatosis cerebri* defines the diffuse growth of neoplastic astrocytes affecting more than two lobes. The WHO classification attributes a grade of III to gliomatosis cerebri. Clinically, the condition may be impossible to differentiate from a progressive diffuse WHO grade II astrocytoma. Epilepsy and personality changes are common presenting features. Other clinical symptoms and signs are determined by the site of neoplastic growth. The condition may occur throughout childhood and adulthood. Surgical options are essentially limited to biopsy. The course of the disease is variable, but the median survival may not greatly exceed 1 year (Jennings *et al.*, 1995). The role of radiotherapy is not well-defined (Cozad *et al.*, 1996), but one larger series reported a median survival of 38.4 months after radiotherapy (Kim *et al.*, 1998). If radiotherapy is administered, large volume or even whole-brain radiotherapy, including involved brainstem or spinal cord, should be performed. Chemotherapy using barrier-penetrating agents (e.g., CCNU and procarbazine or temozolomide) may be tried either as a second-line therapy for patients progressing after radiotherapy or even before radiotherapy, notably in patients with histological detection of oligodendroglial tumor elements. In most patients, only transient prevention of progression (stable disease) has been observed by the authors using either treatment modality (Herrlinger *et al.*, 2002).

## Neuronal and Mixed Neuronal-Glial Tumors

These tumors share a neuronal differentiation and an overall favorable prognosis. *Gangliocytoma* (WHO grade I) and *ganglioglioma* (WHO grade II) may occur in any age group and at any site in the CNS. They are treated by gross surgical resection (Lang *et al.*, 1993). Gangliogliomas are a common cause of temporal lobe epilepsy. They rarely exhibit anaplastic features and even features of glioblastoma in their glial aspects. In that case, these tumors should be treated according to the guidelines for anaplastic astrocytoma and glioblastoma. *Desmoplastic infantile astrocytoma* and *ganglioglioma* are WHO grade I lesions of infants that are cured by resection. There is probably no potential for malignant progression. Similarly, dysembryoplastic neuroepithelial tumors that occur mainly in young adults are WHO grade I lesions that do not progress even after partial resection. *Central neurocytoma* (WHO grade II), a tumor of all age groups but predominantly diagnosed in early to middle adulthood, is commonly located in the lateral ventricles and presents with symptoms and signs of raised intracranial pressure. Surgical resection is the treatment of choice (Ashkan *et al.*, 2000). Incompletely resected neurocytomas may be observed and reoperated or irradiated at progression (Kim *et al.*, 1997; Schild *et al.*, 1997). Craniospinal dissemination has occurred in few patients. *Cerebellar neurolipocytoma* is a WHO grade I/II lesion of adults that should be treated by gross surgical resection. *Paragangliomas* may occur in association with autonomic ganglia outside the CNS. Within the CNS, they are almost always located in the region of the filum terminale and of WHO grade I. They are cured by resection in most instances, but local recurrence has been reported.

## Neuroblastic Tumors

### Olfactory Neuroblastoma (Esthesioneuroblastoma)

This is a rare neuroectodermal tumor thought to be derived from olfactory receptor cells. The tumor may exhibit features of neuroendocrine differentiation. Although olfactory neuroblastoma may occur in all age groups, there are peaks of incidence around the ages of 20 and 50. Gross surgical resection is the only chance of a cure, and the extent of resection is a predictor of favorable outcome (Koka *et al.*, 1998). Postoperative radiotherapy is indicated after complete and incomplete resection.* Currently both neoadjuvant preresection and preirradiation chemotherapy, commonly platinbased, as well as second-line chemotherapy for inoperable tumors progressing after radiotherapy seem to assume a role in the management of these tumors (McElroy *et al.*, 1998; Polin *et al.*, 1998).

### Neuroblastic Tumors of the Sympathetic Nervous System and the Adrenal Gland

These tumors include neuroblastoma, ganglioneuroblastoma, and ganglioneuroma. They are typically diagnosed in children, 96% in the first decade and 3.5% in the second decade of life, and thus play no major role in adult neuro-oncology. They originate from the adrenal glands or abdominal, thoracic, cervical, or pelvic sympathetic ganglia. The prognosis is better for young infants than for older children affected by these tumors and depends on the staging at diagnosis. Neuroblastomas are paradigmatic for malignancies that may regress spontaneously. The management includes surgery, radiotherapy, and chemotherapy (Berthold and Hero, 2000).

## Pineal Parenchymal Tumors

These tumors include pineocytoma (WHO grade II), pineoblastoma (WHO grade IV), and pineal parenchymal tumors of intermediate differentiation. They are seen in patients less than 30 years of age, with two thirds of the patients being between 10 and 21 years of age. Presenting features are either those of raised intracranial pressure caused by aqueduct stenosis, disturbed oculomotor function, or chiasmal compression. Endocrine disturbances are frequent and include diabetes insipidus, pubertas praecox, or anterior pituitary dysfunction. Pineoblastoma is a malignant tumor resembling medulloblastoma that commonly seeds in the CSF. Pineocytoma is a rather benign tumor, whereas pineal parenchymal tumors of intermediate differentiation carry an intermediate prognosis. The diagnostic workup must include craniospinal MRI and CSF analysis. The 1-year, 3-year, and 5-year survival rates were estimated at 100%, 100%, and 67% in patients with pineocytoma and 88%, 78%, and 58% for the other types of pineal parenchymal tumors (Schild *et al.*, 1993).

Although there are no sufficient data on these rare tumors, attempts at radical resection are probably indicated for all pineal parenchymal tumors. Incompletely resected pineocytomas should be observed and reoperated, or irradiated at progression, or both. All other pineal parenchymal tumors should be irradiated (e.g., locally at 54 Gy in 1.8–2 Gy fractions to the tumor region). Craniospinal irradiation (36 Gy) is indicated to treat or prevent spinal seeding (Schild *et al.*, 1993).* Cisplatin-based chemotherapy may be administered at progression or recurrence after radiotherapy (Galanis *et al.*, 1997).

## Embryonal Tumors

### Clinical Aspects and Natural Course

Medulloepitheliomas, ependymoblastomas, the different variants of medulloblastoma, supratentorial primitive neuroectodermal tumors (PNET), and the other rare embryonal tumors are all WHO grade IV malignant tumors and typically encountered in childhood or early adulthood. Only 25% of medulloblastomas, the most common embryonal tumor, are seen after 15 years of age. The roof of the fourth ventricle is the usual site of origin, and brainstem infiltration is common. A lateral origin within the cerebellar hemisphere is more common in teenagers and young adults. A supratentorial tumor with histological features of medulloblastoma is classified as a supratentorial PNET, but the genetic interrelations between medulloblastoma and supratentorial PNET remain controversial. Medulloblastoma frequently is seen with the features of raised intracranial pressure caused by obstructive hydrocephalus. Ataxia and cranial nerve palsies may also be found. Approximately 30% of medulloblastomas have spread to other sites within the CNS at diagnosis, and, unlike most other CNS neoplasms, metastases may also form outside the CNS, particularly in bone, lymph nodes, and lung, in 5%–25% of the patients in the course of disease (Rochkind et al., 1991). Age less than 3 years and CNS metastases at presentation predict poor outcome (Zeltzer et al., 1999). The prognostic impact of histological medulloblastoma subtype has remained controversial, specifically whether desmoplastic medulloblastoma has a better prognosis than other types of medulloblastoma. Surgery, radiotherapy, and chemotherapy have a defined role in the management of medulloblastoma (see later) and induce a 5-year survival of approximately 50% both in the pediatric and adult age group (Brandes et al., 1999b; Packer, 1999).

### Principles of Therapy

Because the extent of postoperative residual tumor is a prognostic factor (Bloom and Bessell, 1990), macroscopic removal of the tumor should be attempted. Other treatment-related prognostic factors in retrospective analyses were a radiation dose to the posterior fossa greater than 50 Gy and prophylactic radiotherapy to the craniospinal axis (Bloom and Bessell, 1990). Craniospinal radiotherapy is now the standard type of radiotherapy for all patients with medulloblastoma, with the possible exception of young infants less than 3 years of age. Most children who undergo this treatment and survive grow up to become normal, completely independent adults, only a minority demonstrating learning difficulties or behavioral problems. The role of adjuvant chemotherapy or neoadjuvant chemotherapy before radiotherapy in the treatment of medulloblastoma is not clearly defined (Kortmann et al., 2000; Krischer et al., 1991; Kühl et al., 1998). The addition of chemotherapy, using vincristine, CCNU, and pred-

nisone (VCP), during and after chemotherapy may prolong the disease-free survival in patients with medulloblastoma with unfavorable prognostic factors such as incomplete resection or metastatic disease in the CNS (Evans et al., 1990; Tait et al., 1990). This triple chemotherapy plus radiotherapy was superior to the *eight-drugs-in-one day* (8-in-1) schedule plus radiotherapy in a randomized study (Zeltzer et al., 1999). Promising results in poor-risk patients were also achieved with the triple combination of CCNU, vincristine, and cisplatin (Packer et al., 1994). Cisplatin-based chemotherapy also induced high response rates, albeit of moderate duration, in adult patients with progressive or recurrent embryonal tumors (Galanis et al., 1997). The management of supratentorial PNET corresponds to that of medulloblastoma, but response to treatment and overall prognosis seem to be worse than for medulloblastoma (Paulino and Melian, 1999). High-dose chemotherapy followed by autologous stem cell transplantation is currently being evaluated (Finlay et al., 1996).

### Practical Management

Adult patients with medulloblastoma should have surgical removal of the lesion* and receive craniospinal radiotherapy, 36 Gy, 5 × 1.8 Gy per week, to the neuraxis, followed by a 19.8-Gy boost in 1.8 Gy fractions to the tumor region and to any solid metastatic lesion within the CNS.* Children should be treated within ongoing multicenter trials that usually contain craniospinal radiotherapy (Kortmann et al., 2000). High-risk adult patients with macroscopic residual tumor or spinal lesions or positive CSF cytologic findings should receive adjuvant chemotherapy (e.g., CCNU, cisplatin, and vincristine) (Packer et al., 1994) (Table VIII).* Preirradiation chemotherapy should only be administered in the context of controlled clinical trials. Recurrent disease may be treated by stereotactic radiotherapy if the recurrence is circumscribed, or by CCNU, cisplatin, or vincristine chemotherapy in chemonaïve patients, or by cisplatin-based or other experimental chemotherapy protocols.

## TUMORS OF PERIPHERAL NERVES

### Schwannoma

#### Clinical Aspects and Natural Course

Schwannomas (neurilemmomas, neurinomas, neuromas) are benign tumors (WHO grade I) arising from the neuroectodermal Schwann cells of the peripheral (mostly vestibular) nerves at the transition of central and peripheral myelin. *Vestibular schwannomas* form 70%–80% of all lesions in the cerebellopontine angle, are slightly more common in women, and have a peak incidence in the fourth and fifth decades. The most common clinical presentation is progressive unilateral sensorineural hearing loss, often extending over months to years. Occurring initially at higher frequencies, progression will relentlessly continue to deafness, often in conjunction with tinnitus and an unsteady gait because of internal auditory artery occlusion as the tumor increases in size. Episodes of sudden deterioration of hearing are also not uncommon. Features commonly noted later are facial nerve paresis, trigeminal neuralgia, cerebellar signs, and hydrocephalus. Bilateral vestibular schwannomas are strongly associated with *neurofibromatosis type II* (Table II).

Schwannomas of spinal nerves represent approximately 25% of all spinal tumors. There is an equal gender incidence. Commonly, the tumors are intradural-extramedullary, although an intradural-extradural lesion is well recognized. They arise from a dorsal root and therefore lie posterolateral to the dentate ligament. Radicular pain is the most common presenting feature, with a subsequent pyramidal tract distribution of weakness caused by lateral cord compression with cervicothoracic locations. Sensory symptoms are present only later in the presentation. Multiple, typically peripheral schwannomas are indicative of a rare genetic disorder distinct from neurofibromatosis type II, now referred to as *schwannomatosis* (MacCollin *et al.*, 1996). Schwannomatosis is often a sporadic disease. Its course is probably more favorable than that of neurofibromatosis type II (Seppälä *et al.*, 1998).

## Principles of Therapy and Practical Management

Older patients (>65 years) with asymptomatic or oligosymptomatic vestibular schwannomas may be observed and monitored for tumor progression or clinical deterioration. Younger patients should receive treatment as outlined later. The younger the patient, the more likely the tumor is to grow and produce neurological symptoms. A number of surgical approaches, middle cranial fossa, suboccipital, and translabyrinthine, have been described for vestibular schwannomas. The translabyrinthine approach is used for intracanalicular tumors in patients who have lost hearing at presentation. Complete resection is curative. Using either surgical approach, intraoperative facial nerve monitoring should be used to preserve facial nerve function. A large survey of outcome after surgery performed at various centers across the United States revealed a 44% rate of facial weakness, a 11% rate of CSF leakage, a 9% rate of persistent imbalance, and a

8% rate of persistent or progressive tumor on imaging (Wiegand *et al.*, 1996). In contrast, the best surgical series approached a rate of complete resection of 98% and reported a recurrence rate <1% in patients with vestibular schwannoma who did not have neurofibromatosis type II (Samii and Mathies, 1997). The recurrence rate is probably lowest after a total translabyrinthine removal (Shelton, 1995). Whether hearing is preserved after surgery depends on the size of the tumor and on the surgical approach (Post *et al.*, 1995). Few clinical series provide adequate information on the preinterventional and postinterventional hearing status.

Alternative nonsurgical treatment options for vestibular schwannomas include radiosurgery (Kondziolka *et al.*, 1998) and fractionated stereotactic radiotherapy (Kagei *et al.*, 1999). The success of radiosurgery depends on the quality of planning and will probably be further improved with the more widespread use of high-resolution MRI to define the target volume of irradiation. The largest series of radiosurgery with an adequate follow-up of at least 5 years reported a local tumor control rate of 98% (Kondziolka *et al.*, 1998). Facial nerve function was preserved in 79% of the patients at 5 years. Approximately, 50% of the patients maintained their preinterventional hearing. There was a rate of tumor shrinkage of 62%, and 33% remained stable. No neurological deterioration occurred later than 28 months after radiosurgery. These data compare favorably with the data from most surgical series. Accordingly, radiosurgery would seem to be at least an equieffective, but presumably safer, strategy of treatment for patients with tumors less than 30 mm of diameter (Pollock *et al.*, 1995).* Larger tumors, as well as those tumors that produce progressive neurological symptoms (e.g., from brainstem compression at diagnosis) should be treated by resection whenever feasible.* Neither chemotherapy nor hormonal treatment have a role in the management of these tumors. Patients with vestibular schwannomas should be examined by audiometry to assess the degree of preinterventional hearing impairment.

Solitary spinal schwannomas, which are commonly diagnosed because they are symptomatic, should be resected. The involved nerve root must be sacrificed, although the resultant sensory deficit may pass unnoticed by the patient because of sensory cutaneous nerve overlap. Complete excision provides a cure. Surgery should be restricted to symptomatic lesions in patients with multiple schwannomas, notably in schwannomatosis.

## Other Tumors of Peripheral Nerves

Solitary *neurofibromas* are slowly growing tumors (WHO grade I) that are composed of Schwann cells and connective tissue. They are more often found peripherally than centrally and can be rather painful. Multiple neurofibromas are the leading feature of neurofibromatosis type I. A high rate of malignant progression (5%) is seen with plexiform neurofibromas. Neurofibromas are cured by complete resection. *Perineurioma* is a rare benign tumor (WHO grade I) composed of perineurial cells that occurs intraneurally or in soft tissues. Intraneural perineurioma may be diagnosed by biopsy and observed, because complete resection requires sacrifice of peripheral nerve function. Soft tissue perineuriomas can be removed and will not recur. *Malignant peripheral nerve sheath tumor* (MPNST), previously referred to as neurogenic sarcoma, malignant schwannoma, or anaplastic neurofibroma, is a WHO grade III or IV lesion. More than 50% of the patients with MPNST have a mutation in the NF1 gene. Another 10% of MPNST arise in previously irradiated fields. MPNST is commonly treated with gross total resection and focal adjuvant radiotherapy. The survival at 5 years is in the range of 34% (Ducatman *et al.*, 1986) to 39% (DeCou *et al.*, 1995) to 52% (Wong *et al.*, 1998).

## TUMORS OF THE MENINGES

### Tumors of Meningothelial Cells

#### Clinical Aspects and Natural Course

*Meningiomas* are usually benign, slowly growing tumors, most commonly located at the sites of arachnoid granulations and are among the most common intracranial and spinal tumors (Table III). They are rare in children and become more frequent in middle and late adulthood. Women are more often affected than men by WHO grade I tumors. Up to 10% of the patients harbor more than one meningioma, among them patients with neurofibromatosis type II (Table II). Losses on chromosome 22q, which includes the NF2 locus, are also common (>50%) in patients with sporadic meningiomas. Most meningiomas are grade I tumors. The proportion of grade II and grade III are approximately 5% and 1%, respectively. Men are more often affected by WHO grade II/III meningiomas than women. A predisposing factor, which seems to contribute to the formation of these tumors, is therapeutic irradiation to the scalp or the pituitary. The female preponderance and an association with breast cancer have raised the question of an endocrine modulation of the growth of meningiomas. This clinical evidence has been supported by laboratory studies confirming the presence of receptors for progesterone and glucocorticoids on meningioma cells. The expression of progesterone receptors seems to be a favorable prognostic factor and is lost in the higher grade (II/III) meningiomas (Hsu *et al.*, 1997). Accordingly, tamoxifen or other endocrine treatments, including somatostatin analogs, have not assumed a role in the adjuvant treatment of meningiomas.

On imaging, meningiomas have the general characteristics of an extra-axial mass in that they exhibit well-defined margins, occupy a peripheral position, and cause displacement of brain or spinal cord. Contrast enhancement tends to be uniform. A supratentorial meningioma will usually produce epilepsy, headaches, or a focal neurological deficit. An infratentorial or intraventricular meningioma can lead to headache, deteriorating consciousness, or papilledema because of raised intracranial pressure caused by obstructive hydrocephalus, or to a focal neurological deficit, usually ataxia or a cranial nerve palsy. Meningiomas also account for 25% of all spinal tumors. Most of these tumors are intradural-extramedullary in position, but intradural-extradural, pure extradural, and intramedullary tumors have all been described. Midline back pain is the most common early feature, frequently accompanied by vague sensory symptoms. Motor and bladder disturbances occur later in the presentation. Multiple sclerosis is frequently considered in the differential diagnosis.

Meningiomas of all WHO grades may infiltrate the brain locally. Metastatic disease is very rare; is most often detected in lung, bone, and liver; and is strongly associated with the WHO grade. Anaplastic meningiomas are most likely to seed outside the brain, but rare instances of seeding from WHO grade I after surgery have been reported. The survival of patients with WHO grade I meningiomas may exceed 90% at 10 years. In contrast, median survival rates less than 2 years have been reported for patients with anaplastic meningiomas (Perry *et al.*, 1999). Another series reported a median recurrence-free survival of 11.9 and 2 years, and 5-year and 10-year survival rates of 95% and 79%, and 64.3% and 34.5%, for patients with atypical (grade II) and anaplastic (grade III) meningiomas (Palma *et al.*, 1997).

#### Principles of Therapy

Most meningiomas are benign and never change this biological behavior. Calcified lesions often fail to demonstrate growth over years. The case for intervention in elderly patients, those with no clinical deficit, those with tumors without evidence of radiological progression, and those with unfavorably placed tumors must therefore be carefully considered. Conversely, raised intracranial pressure, tumor-associated edema, seizures, and radiologically documented progression call for intervention. If treatment is indicated, surgery remains the mainstay of therapeutic intervention (Giombini *et al.*, 1984; Mirimanoff *et al.*, 1985). Interventional radiologists may offer the occlusion of the vascular supply simultaneously with the angiogram. There are several reasons why curative surgery is often not possible with meningiomas: an anatomically difficult tumor location, adhesion to surrounding vital structures, diffuse infiltration of sinuses or veins, and osteoplastic growth. Complete surgical removal is almost always possible in patients with meningiomas of the convexity but is often impossible with meningiomas of the sphenoid wing or the olfactory groove. Hence, recurrence becomes a major problem. Overall, the recurrence rate for meningiomas is 20% within 10 years after complete resection, and 80% of the tumors progress after partial resection (Table XIII). Gross surgical resection is also the treatment of choice for spinal meningiomas (Solero *et al.*, 1989).

Even WHO grade I meningiomas are moderately radiosensitive, such that radiotherapy has a distinct role in their management (Glaholm *et al.*, 1990). Radiotherapy is most often used for the local control of inoperable meningiomas (e.g., of the cavernous sinus or the sphenoid ridge). In these locations, the local control rates at 10 years are 90% and 69%, respectively (Nutting *et al.*, 1999). Conventional external beam radiotherapy has been the most common form of treatment, but radiosurgery has been studied as well, with local control rates of 96% at 2 years and of 93% after 5–10 years (Kondziolka *et al.*, 1991, 1999). Radiosurgery may be the treatment of choice for small inoperable skull base meningiomas that are increasing in size on serial scans. Given the often irregular shape of meningiomas and their location in the vicinity of critical structures, stereotactic conformal radiotherapy would seem to be a promising strategy of radiotherapy for meningiomas as well, but long-term follow-up for this approach is not available (Alheit *et al.*, 1999).

Aggressive chemotherapy plays no role in the treatment of WHO grade I tumors. Approximately 20% of progressive meningiomas respond to continuous chemotherapy using hydroxyurea (Schrell *et al.*, 1997). Stabilization in response to interferon-α has been reported as well (Kaba *et al.*, 1997).

The adequate treatment of WHO grade II and III meningiomas has remained controversial. Aggressive surgery is probably useful, but the prevention of neurological deficits should be the highest priority. Some series failed to identify an impact of radiotherapy or chemotherapy on the disease (Younis *et al.*, 1995). Other authors suggest that aggressive initial surgery followed by adjuvant radiotherapy is required for long-term tumor control (Dziuk *et al.*, 1996). A prospective study combining surgery, adjuvant radiotherapy, and chemotherapy (cyclophosphamide, doxorubicin, vincristine) resulted in a progression-free survival of 4.6 years and a median survival of 5.3 years (Chamberlain, 1996). Conservative medical approaches for these tumors, as for progressive WHO grade I lesions, include hydroxyurea and interferon-α (Kaba *et al.*, 1997; Schrell *et al.*, 1997). The role of more aggressive types of chemotherapy in the management of grade II/III meningiomas is still poorly defined.

### Practical Management

If an intervention is necessary according to the guidelines outlined earlier, surgery with an attempt at complete resection is the strategy of choice for most WHO grade I meningiomas.* Incompletely resected asymptomatic lesions may be observed. CT is recommended at 1 year and 3 years and in longer intervals thereafter. Incomplete excision of a symptomatic meningioma, inoperable or recurrent symptomatic lesions, or a critical site of an asymptomatic tumor (e.g., cavernous sinus) are valid reasons to administer radiotherapy either up-front or as postoperative adjuvant therapy.*

**TABLE XIII**  Progression-Free Survival in Patients with WHO Grade I Meningiomas

|  | Patients (*n*) | Therapy | Progression-free survival at 5 years (%) | 10 years (%) | 15 years (%) |
|---|---|---|---|---|---|
| Mirimanoff *et al.*, 1985 | 145 | Complete resection | 93 | 80 | 68 |
|  | 80 | Incomplete resection | 63 | 45 | 9 |
| Taylor *et al.*, 1988 | 86 | Complete resection | 86 | 77 | 65 |
|  | 29 | Incomplete resection | 42 | 18 | 18 |
|  | 13 | Incomplete resection plus RT | 82 | 82 |  |
| Goldsmith *et al.*, 1994 | 117 | Incomplete resection plus RT | 89 | 77 | 77 |
| Nutting *et al.*, 1999 | 33 (sphenoid) | Incomplete resection plus RT | 81 | 69 |  |
|  | 37 (parasellar) |  | 100 | 90 |  |
|  | 12 (other) |  | 100 | 100 |  |

Radiotherapy to meningiomas usually entails a dose of 60 Gy administered in 1.8–2.0 Gy daily fractions. In the absence of surgical or radiotherapeutic treatment options, a trial of hydroxyurea (20 mg/kg daily) or interferon-α may be warranted for progressive WHO grade I meningiomas (Table VII).

Asymptomatic grade II meningiomas may be observed as grade I lesions, even after incomplete resection but should be monitored more closely (e.g., in 6-month intervals for the first 2 years). The adjuvant therapy of completely resected grade III meningiomas has remained controversial. We believe that observation with imaging controls in 6-month intervals for the first 2 years is justified. Incompletely resected grade III lesions should be irradiated (60 Gy, 1.8–2 Gy fractions) after surgery. Local recurrences should be treated surgically when feasible. Stereotactic reirradiation or radiosurgery may be suitable second-line options for some patients. Experimental chemotherapy (e.g., using hydroxyurea, interferon-α, or carboplatin plus etoposide), may be evaluated for the treatment of malignant meningiomas progressing after radiotherapy, notably in the presence of metastatic disease.

### Mesenchymal, Nonmeningothelial Tumors

These include a whole variety of tumors ranging from WHO grade I to WHO grade IV (Table I). Systemic metastases are common in all primary cerebral sarcomas. Primary *cerebral fibrosarcoma* is an aggressive tumor with a poor prognosis even after attempts at complete surgical resection followed by radiotherapy (Gaspar *et al.*, 1983). *Hemangiopericytoma* is a lesion of intermediate dignity (WHO grade II/III) previously classified among the meningiomas. These tumors may occur in almost any age group and are preferentially localized in the vicinity of the large venous sinuses and in the occipital region. The rate of local recurrence even after complete resection is very high and approaches 85%–95% at 15 years for the survivors (Guthrie *et al.*, 1989; Vuorinen *et al.*, 1996). Histological predictors of unfavorable outcome include higher proliferation indices, high cellularity, nuclear polymorphism, hemorrhage, and necrosis. Postoperative radiotherapy is recommended for all patients with hemangiopericytomas.* At recurrence, surgical options should be evaluated, previously not irradiated lesions should be treated with radiotherapy, and radiosurgery may induce prolonged local control for small lesions (Galanis *et al.*, 1998b). Chemotherapy has not shown efficacy against this tumor. Hemangiopericytoma has one of the highest rates of extracranial metastases among intracranial tumors: the likelihood of metastasis was 13% at 5 years,

33% at 10 years, and 64% at 15 years in one series (Guthrie *et al.*, 1989).

### Primary Melanocytic Lesions

These are rare conditions and include diffuse leptomeningeal melanocytosis and neurocutaneous melanosis, melanocytoma, and primary meningeal malignant melanoma. The latter may spread within the leptomeninges (malignant meningeal melanomatosis). Diffuse melanocytosis is seen with mental disturbances, seizures, and symptoms and signs of hydrocephalus. Neurocutaneous melanosis is a phakomatosis with multiple associated abnormalities. A shunting procedure may alleviate some symptoms in diffuse melanocytosis, but the prognosis is commonly poor. Malignant meningeal melanoma is an aggressive, rather radioresistant tumor that metastasizes systemically similar to primary extracerebral malignant melanomas. These tumors should be treated as outlined for parenchymal and leptomeningeal metastases from malignant melanomas originating outside the CNS (Chapters 64 and 65).

### Tumors of Uncertain Histogenesis (Hemangioblastoma)

*Hemangioblastomas* form approximately 2% of both primary intracranial and spinal lesions. Up to one quarter of cases are associated with Von Hippel-Lindau syndrome, an autosomal dominant neuroectodermal syndrome (Table II). The associated features include retinal angioma, pheochromocytomas, pancreatic, hepatic, renal, or adrenal adenomas or angiomas, and cerebellar, supratentorial, and spinal hemangiomas, which may be multiple. Hemangioblastomas are benign, slow-growing lesions whose most common site of origin is the cerebellum. Spinal cord lesions usually arise at the cervical or thoracic cord level. They can be intramedullary or extramedullary. If intramedullary, they are frequently associated with a syrinx. A supratentorial origin is unusual, and here the tumor may be solid rather than cystic. A posterior fossa lesion may manifest with raised intracranial pressure or focal neurological signs, nystagmus or ataxia being common. Spinal lesions may be seen as cord compression, spinal subarachnoid hemorrhage, or syringomyelia (Murota and Symon, 1989).

Surgery should aim at completely excising the lesion, because this is the only chance of a cure.* Recurrent tumor should also be managed surgically. A complete excision is not always feasible in the brainstem or with a spinal or supratentorial location. However, even in cases of solid, highly vascular lesions, an en-bloc resection should be considered (Richard *et al.*, 2000). Hemangioblastomas are relatively radioresistant. External beam irradiation is an option for the local control of inoperable or multiple lesions (Smalley *et al.*, 1990). The role of stereotactic radiosurgery remains unproven (Chang *et al.*, 1998b).

# GERM CELL TUMORS

## Clinical Aspects and Natural Course

Intracranial germ cell tumors are rare (<1%, Table III) and commonly affect children or young adults. The possibility that they are secondary tumors arising from a gonadal primary tumor must be considered. Histologically, the primary intracranial germ cell tumors are impossible to distinguish from primary extracranial germ cell tumors that, for example, arise in the gonads. Germ cell tumors are histologically varied and are composed of germinoma, embryonal carcinoma, yolk sac tumor, choriocarcinoma, teratoma, and mixed germ cell tumor, each with varying characteristics and hence prognosis (Jennings *et al.*, 1985; Sawamura *et al.*, 1998a). Although several types of intracranial germ cell tumors are distinguished, the diagnosis of a mixed germ cell tumor is common. Germinomas are the most frequently encountered variant and are highly invasive, solid neoplasms. Germ cell tumors are typical midline tumors preferentially found within the pineal gland or the suprasellar cistern. Accordingly, Parinaud syndrome and progressive hydrocephalus are typical clinical presentations. Germ cell tumors seed by means of the CSF compartment to establish metastases. This makes cytological examination of the CSF mandatory before intervention. Extraneural metastases have rarely been described as well. The diagnosis of a germ cell tumor may also be supported by serum and CSF tumor markers. α-Fetoprotein (AFP) suggests a diagnosis of yolk sac tumor or teratoma, β-human chorionic gonadotrophin (βHCG) suggests choriocarcinoma or a germ cell tumor containing βHCG-secreting cells, and placental alkaline phosphatase (PLAP) suggests mostly germinoma but may be elevated in other germ cell tumors as well. The differential diagnosis of germ cell tumors is clinically relevant, because the radiosensitivity of pure germinomas seems to determine their overall favorable prognosis compared with the other germ cell tumors. The interpretation of survival data from earlier studies is difficult because of the tendency to report various tumors of the pineal and suprasellar region together and to treat them without

histological verification of the diagnosis and because of the small number of patients involved in many of the series.

## Principles of Therapy

Unlike other intracranial tumors, serum and CSF analysis may be diagnostic for certain types of germ cell tumor. Accordingly, histological confirmation may not be necessary in all cases, for instance, in the setting of diagnostic CSF findings and typical features on neuroimaging. Biopsy alone is probably sufficient for all pure germinomas, and no primary attempt at complete resection should be made. However, patients with symptomatic tumors that are unresponsive to radiotherapy and chemotherapy are candidates for resection. In general, the aims of surgery have been to establish the diagnosis of the type of germ cell tumor and to alleviate hydrocephalus if present. The exception was the mature teratoma, which may be removed by means of an open operation. With increasingly more sophisticated techniques of neurosurgery, the role of surgery especially for the secreting germ cell tumors with their less-favorable prognosis may need to be reassessed. The histology, degree of dissemination, whether CSF cytology or tumor markers are positive on analysis, and the condition of the patient will all influence the further management. Germinomas are thought to be more radiosensitive than the other germ cell tumors, although excellent survival data with radiotherapy alone have also been reported for mixed germ cell tumors (Shibamato *et al.*, 1997). A high proportion of germinomas can be cured by radiotherapy alone, and the rate of complete responses approaches 100% (Bamberg *et al.*, 1999; Hardenbergh *et al.*, 1997).* Given the high complete response rates at 30 Gy craniospinal radiotherapy plus 15-Gy tumor boost administered in 1.5-Gy fractions, further dose reductions of 24-Gy craniospinal irradiation plus 16-Gy tumor boost are currently being evaluated in the SIOP-CNS-GCT-96 study in Europe.

Germinomas are also responsive to chemotherapy, and chemotherapy followed by radiotherapy, at reduced dose or as involved-field rather than craniospinal radiotherapy, may achieve tumor control rates similar to radiotherapy alone at higher doses (Sawamura *et al.*, 1998b). As indicated earlier, nongerminoma germ cell tumors are less responsive to radiotherapy, but chemotherapy followed by radiotherapy and further adjuvant chemotherapy may result in 4-year survival rates of a least 60% in these tumors (Robertson *et al.*, 1997). The chemotherapeutic agents most commonly used in the treatment of germ cell tumors are cisplatin plus etoposide. Altogether, chemotherapy alone as currently administered (Balmaceda *et al.*, 1996; Baranzelli *et al.*, 1997) seems to be less effective than radiotherapy alone (Bamberg *et al.*, 1999).

### Practical Management

With the possible exception of mature teratoma, surgery alone provides no adequate therapy for germ cell tumors. Attempts at complete removal are discouraged in germinomas, because these tumors are highly radiosensitive. Surgery is obsolete if a typical radiological appearance is supported by diagnostic CSF findings. Ideally, patients with intracranial germ cell tumors should be managed within ongoing trials. Because both craniospinal radiotherapy alone and chemotherapy followed by focal radiotherapy are effective treatment regimens that have not been compared in a controlled clinical trial, no firm recommendation for the management of germ cell tumors outside clinical trials can be made.

## TUMORS OF THE SELLAR REGION

Pituitary adenomas are covered in Chapter 102. For all tumors of the sellar region, a thorough visual assessment and endocrinological profile must be performed before any intervention and hormonal deficiences replaced as required. *Craniopharyngeomas* are histologically benign tumors (WHO grade I) that arise from the epithelial remnants of the embryonic craniopharyngeal duct. They represent 2.5%–4% of all cerebral neoplasms and commonly arise in the suprasellar region where they form 20% of all suprasellar tumors. Macroscopic examination reveals a solid or cystic tumor. They can arise at any age but are most frequent in childhood. The tumors are in intimate contact with the third ventricle and associated structures. This localization gives rise to the clinical picture and makes aggressive surgical intervention often difficult. In children these lesions frequently are seen with symptoms of raised intracranial pressure caused by blockage of the foramen of Monro. Adults usually have a visual deficit, most commonly bitemporal hemianopia, or alternatively an endocrine disturbance, manifestations of pituitary insufficiency or diabetes insipidus being most common. Most the tumors are cystic and partially calcified. The recurrence-free survival at 10 years after surgery, with or without radiotherapy, is in the range of 60%–95% (Crotty *et al.*, 1995; Rajan *et al.*, 1993; Yasargil *et al.*, 1990). If surgery has left residual tumor volume, then radiotherapy is recommended to reduce the risk of recurrence (Rajan *et al.*, 1993). Yet, the functional outcome for children treated with surgery and radiotherapy is often poor (Habrand *et al.*, 1999). A cystic lesion can be treated with intracavity brachytherapy under stereotactic control (Voges *et al.*, 1997). Craniopharyngeomas that recur and progress after surgery and radiotherapy and can no more be treated surgically may respond for years to chemotherapy with doxorubicin and lomustine (Lippens *et al.*, 1998). Tumor responses (25%) have also been observed with continuous interferon-α therapy (Jakacki *et al.*, 2000). Intracavitary chemotherapy is not an established therapy for recurrent craniopharyngioma and is potentially neurotoxic.

*Granular cell tumors of the neurohypophysis* are rare WHO grade I tumors of the sellar region (Schaller *et al.*, 1998). They occur in adults and are more common in women. Granular cell tumors should be resected. Progressive or recurrent tumors that cannot be resected may respond to radiotherapy.

## OTHER TUMORS

### Chordoma

Chordomas are not included in the WHO classification of brain tumors (Table I), but they may cause significant neurological morbidity. Chordomas are rare, slow-growing neoplasms thought to derive from primitive notochordal remnants. They arise in the sacrococcygeal region (50%), clivus (35%), or vertebra (15%). Histologically, they demonstrate benign features, but they are locally invasive and highly destructive lesions with a marked propensity to recur after surgical intervention. The locally destructive nature of these lesions causes bone pain to be a dominant clinical symptom. Associated clinical features will depend on the site of the lesion. A sacral lesion can give rise to rectal dysfunction or a palpable mass, a vertebral tumor may cause paraparesis, and cranial localization may lead to headaches or cranial nerve palsies. Tumors arising within the skull base are usually in the sphenoid or clivus regions, although an eccentric origin in the cerebellopontine angle has been described in up to one third of the cases. Extension into the sphenoid sinus or nasopharynx does occur, leading to a submucosal swelling. The mean survival for patients with clivus chordomas is probably >5 years and with sacrococcygeal chordomas up to 20 years (Bergh *et al.*, 2000). Metastases are a rare and late feature in this condition. Spread usually occurs to the lungs, lymph nodes, or liver.

The aim of surgery is macroscopic tumor resection. Unfortunately, this is seldom achieved because of the unfavorable location and infiltration of vital structures that occur with these tumors. Because of the high local recurrence rate, postoperative radiotherapy, notably fractionated proton radiotherapy, is now considered to be the treatment of choice for skull base chordomas (Austin-Seymour *et al.*, 1989).* This mode of radiotherapy is able to provide high tumor doses while sparing critical neighboring structures such as the brainstem or optic chiasm. Because chordomas are not very radiosensitive, doses of up to 70 Gy are delivered, resulting in a disease-free survival at 5 years of 76% (Austin-Seymour *et al.*, 1989). Brainstem toxicity developed in 17 of 367 patients receiving radiotherapy for skull base tumors. Risk factors for toxicity included the radiation dose to the brainstem, number of surgical procedures, and diabetes (Debus *et al.*, 1997). Recurrent chordomas may also be treated with radiosurgery, but the experience with this approach is limited (Miller *et al.*, 1997). Chemotherapy has no defined role in the treatment of chordomas.

## Epidermoid and Dermoid Tumors

These tumors are not included in the WHO classification of brain tumors (Table I). They are rare, benign, cystic lesions, malformations rather than neoplasms, which have frequently grown to large sizes at the time of diagnosis. Epidermoids may contain cholesterol, whereas dermoids contain cutaneous appendages such as hair follicles and sebaceous and sweat glands. Epidermoids most frequently arise in the cerebellopontine angle or suprasellar cistern and are typically diagnosed in early to middle adulthood. Dermoids may occur in the same locations but are diagnosed more often in childhood. Clinical features vary with the location. Suprasellar tumors tend to cause visual disturbance, whereas trigeminal neuralgia is the most common symptom of the cerebellopontine angle epidermoid. The content of these lesions may rupture and cause chemical meningitis and vasospasm. Gross total resection is the treatment of choice for symptomatic lesions. This is frequently impossible because of tumor adherence to local vital structures, such that only subtotal or partial resections may be performed (Rubin *et al.*, 1989). Whether asymptomatic lesions, which are radiologically stable, should be resected, is debatable. Radiotherapy and chemotherapy have no role in the treatment of these lesions.

## REFERENCES

Alheit, H., Saran, F. H., Warrington, A. P., Rosenberg, I., Perks, J., Jalali, R., Shepherd, S., Beardmore, C., Baumert, B., and Brada, M. (1999). Stereotactically guided conformal radiotherapy for meningiomas. *Radiother. Oncol.* 50, 145–150.

Allison, R. R., Schulsinger, A., Vongtama, V., Barry, T., and Shin, K. H. (1997). Radiation and chemotherapy improve outcome in oligodendroglioma. *Int. J. Radiat. Oncol. Biol. Phys.* 37, 399–403.

Anderson, G. D., Lin, Y. X., Berge, C., and Ojemann, G. A. (1997). Absence of bleeding complications in patients undergoing cortical surgery while receiving valproate treatment. *J. Neurosurg.* 87, 252–256.

Ashkan, K., Casey, A. T. H., D'Arrigo, C., Harkness, W. F., and Thomas, D. G. T. (2000). Benign central neurocytoma. A double misnomer? *Cancer* 89, 1111–1120.

Austin-Seymour, M., Munzenrider, J., Goitein, M., Verhey, L., Urie, M., Gentry, R., Birnbaum, S., Ruotolo, D., McManus, P., Skates, S., Ojemann, R. G., Rosenberg, A., Schiller, A., Koehler, A., and Suit, H. (1989). Fractionated proton radiation therapy of chordoma and low-grade chondrosarcoma of the base of the skull. *J. Neurosurg.* 70, 13–17.

Balmaceda, C., Heller, G., Rosenblum, M., Diez, B., Villablanca, J. G., Kellie, S., Maher, P., Vlamis, V., Walker, R. W., Leibel, S., and Finlay, J. L., for the First International Central Nervous System Germ Cell Tumor Study. (1996). Chemotherapy without irradiation—a novel approach for newly diagnosed CNS germ cell tumors: results of an international cooperative trial. *J. Clin. Oncol.* 14, 2908–2915.

Bamberg, M., Kortmann, R. D., Calaminus, G., Becker, G., Meisner, C., Harms, D., and Göbel, U. (1999). Radiation therapy for intracranial germinoma: results from the German cooperative prospective trials MAKEI 83/86/89. *J. Clin. Oncol.* 17, 2585–2592.

Bauman, G., Lote, K., Larson, D., Stalpers, L., Leighton, C., Fisher, B., Wara, W., Macdonald, D., Stitt, L., and Cairncross, J. G. (1999). Pretreatment factors predict overall survival for patients with low-grade glioma: a recursive partitioning analysis. *Int. J. Radiat. Oncol. Biol. Phys.* 45, 923–929.

Baranzelli, M. C., Patte, C., Bouffet, E., Couanet, D., Habrand, J. L., Portas, M., Lejars, O., Lutz, P., Le Gall, E., and Kalifa, C. (1997). Nonmetastatic intracranial germinoma. The experience of the French Society of Pediatric Oncology. *Cancer* 80, 1792–1797.

Berthold, F., and Hero, B. (2000). Neuroblastoma: current drug therapy recommendations as part of the total treatment approach. *Drugs* 59, 1261–1277.

Bloom, H. J. G., and Bessell, E. M. (1990). Medulloblastoma in adults: A review of 47 patients treated between 1952 and 1981. *Int. J. Radiat. Oncol. Biol. Phys.* 18, 763–773.

Brada, M., Sharpe, G., Rajan, B., Britton, J., Wilkins, P. R., Guerroro, J., Hines, F., Traish, D., and Ashley, S. (1999). Modifying radical radiotherapy in high grade gliomas; shortening the treatment time through acceleration. *Int. J. Radiat. Oncol. Biol. Phys.* 43, 287–292.

Brandes, A. A., Rigon, A., Zampieri, P., Ermani, M., Carollo, C., Altavilla, G., Turazzi, S., Chierichetti, F., and Fiorentino, M. V. (1998). Carboplatin and teniposide concurrent with radiotherapy in patients with glioblastoma multiforme. A phase II study. *Cancer* 82, 355–361.

Brandes, A. A., Ermani, M., Turazzi, S., Scelzi, E., Berti, F., Amistà, P., Rotilio, A., Licata, C., and Fiorentino, M. V. (1999a). Procarbazine and high-dose tamoxifen as a second-line regimen in recurrent high-grade gliomas: a phase II study. *J. Clin. Oncol.* 17, 645–650.

Brandes, A. A., Palmisano, V., and Monfardini, S. (1999b). Medulloblastoma in adults: clinical characteristics and treatment. *Cancer Treat. Rev.* 25, 3–12.

Brat, D. J., Hirose, Y., Cohen, K. J., Feuerstein, B. G., and Burger, P. C. (2000). Astroblastoma: clinicopathologic features and chromosomal abnormalities defined by comparative genomic hybridization. *Brain Pathol.* 10, 342–352.

Brem, H., Piantadosi, S., Burger, P. C., Walker, M., Selker, R., Vick, N. A., Black, K., Sisti, M., Brem, S., Mohr, G., Muller, P., Morawetz, R., Schold, S. C., for the Polymer Brain Tumor Treatment Group. (1995). Placebo-controlled trial of safety and efficacy of intraoperative controlled delivery by biodegradable polymers of chemotherapy for recurrent gliomas. *Lancet* 345, 1008–1012.

Buckner, J. C., Peethambaram, P. P., Smithson, W. A., Groover, R. V., Schomberg, P. J., Kimmel, D. W., Raffel, C., O'Fallon, J. R., Neglia, J., and Shaw, E. G. (1999). Phase II trial of primary chemotherapy followed by reduced-dose radiation for CNS germ cell tumors. *J. Clin. Oncol.* 17, 933–940.

Bullard, D. E., Rawlings, C. E., Phillips, B., Cox, E. B., Schold, S. C., Burger, P., and Halperin, E. C. (1987). Oligodendroglioma: an analysis of the value of radiation therapy. *Cancer* 60, 2179–2188.

Cairncross, J. G., Macdonald, D., Ludwin, S., Lee, D., Cascino, T., Buckner, J., Fulton, D., Dropcho, E., Stewart, D., Schold, C., Wainman, N., and Eisenhauer, E., for the National Cancer Institute of Canada Clinical Trials Group. (1994). Chemotherapy for anaplastic oligodendroglioma. *J. Clin. Oncol.* 12, 2013–2021.

Cairncross, J. G., Ueki, K., Zlatescu, M. C., Lisle, D. K., Finkelstein, D. M., Hammond, R. R., Silver, J. S., Stark, P. C., Macdonald, D. R., Ino, Y., Ramsay, D. A., and Louis, D. N. (1998). Specific genetic predictors of chemotherapeutic response and survival in patients with anaplastic oligodendrogliomas. *J. Natl. Cancer Inst.* 90, 1473–1479.

Celli, P., Nofrone, I., Palma, L., Cantore, G., and Fortuna, A. (1994). Cerebral oligodendroglioma: prognostic factors and life history. *Neurosurgery* 35, 1018–1034.

Chamberlain, M. C. (1996). Adjuvant combined modality therapy for malignant meningiomas. *J. Neurosurg.* 84, 733–736.

Chamberlain, M. C., and Kormanik, P. A. (1997). Salvage chemotherapy with paclitaxel for recurrent oligodendrogliomas. *J. Clin. Oncol.* 15, 3427–3432.

Chang, S. M., Kuhn, J. G., Rizzo, J., Robins, H. I., Schold, S. C., Spence, A. M., Berger, M. S., Mehta, M. P., Bozik, M. E., Pollack, I., Gilbert, M., Fulton, D., Rankin, C., Malec, M., and Prados, M. D. (1998a). Phase I study of paclitaxel in patients with recurrent malignant glioma: a North American Brain Tumor Consortium report. *J. Clin. Oncol.* 16, 2188–2194.

Chang, S. D., Meisel, J. A., Hancock, S. L., Martin, D. P., McManus, M., and Adler, J. R., Jr. (1998b). Treatment of hemangioblastomas in von Hippel-Lindau disease with linear accelerator-based radiosurgery. *Neurosurgery* 43, 28–34.

Chang, S. M., Kuhn, J. G., Robins, H. I., Schold, S. C., Spence, A. M., Berger, M. S., Mehta, M. P., Bozik, M. E., Pollack, I., Schiff, D., Gilbert, M., Rankin, C., and Prados, M. D. (1999). Phase II study of phenylacetate in patients with recurrent malignant glioma: a North American Brain Tumor Consortium report. *J. Clin. Oncol.* 17, 984–990.

Chow, E., Reardon, D. A., Shah, A. B., Jenkins, J. J., Langston, J., Heideman, R. L., Sanford, R. A., Kun, L. E., and Merchant, T. E. (1999). Pediatric choroid plexus neoplasms. *Int. J. Radiat. Oncol. Biol. Phys.* 44, 249–254.

Cloughesy, T. F., Black, K. L., Gobin, Y. P., Farahani, K., Nelson, G., Villablanca, P., Kabbinavar, F., Viñeula, P., and Wortel, C. H. (1999). Intra-arterial cereport (RMP-7) and carboplatin: a dose escalation study for recurrent malignant gliomas. *Neurosurgery* 44, 270–279.

Couldwell, W. T., Hinton, D. R., Surnock, A. A., DeGiorgio, C. M., Weiner, L. P., Apuzzo, M. L., Masri, L., Law, R. E., and Weiss, M. H. (1996). Treatment of recurrent malignant gliomas with chronic oral high-dose tamoxifen. *Clin. Cancer Res.* 2, 619–622.

Cozad, S. C., Townsend, P., Morantz, R. A., Jenny, A. B., Kepes, J. J., and Smalley, S. R. (1996). Gliomatosis cerebri. Results with radiation therapy. *Cancer* 78, 789–793.

Crotty, T. B., Scheithauer, B. W., Young, W. F., Jr., Davis, D. H., Shaw, E. G., Miller, G. M., and Burger, P. C. (1995). Papillary craniopharyngioma: a clinicopathological study of 48 cases. *J. Neurosurg.* 83, 206–214.

Curran, W. J., Scott, C. B., Weinstein, A. S., Martin, L. A., Nelson, J. S., Phillips, T. L., Murray, K., Fischbach, A. J., Yakar, D., Schwade, J. G., Corn, B., and Nelson, D. F. (1993). Survival comparison of radiosurgery-eligible and -ineligible malignant glioma patients treated with hyperfractionated radiation therapy and carmustine: a report of radiation therapy oncology group 83–02. *J. Clin. Oncol.* 11, 857–862.

Davis, F. G., McCarthy, B. J., Freels, S., Kupelian, V., and Bondy, M. L. (1999). The conditional probability of survival of patients with primary malignant brain tumors. Surveillance, epidemiology, and end results (SEER) data. *Cancer* 85, 485–491.

DeAngelis, L. M. (2001). Brain tumors. *N. Engl. J. Med.* 344, 114–122.

DeAngelis, L. M., Burger, P. C., Green, S. B., and Cairncross, J. G. (1998). Malignant glioma: who benefits from adjuvant chemotherapy? *Ann. Neurol.* 44, 691–695.

Debus, J., Hug, E. B., Liebsch, N. J., O'Farrel, D., Finkelstein, D., Efird, J., and Munzenrider, J. E. (1997). Brainstem tolerance to conformal radiotherapy of skull base tumors. *Int. J. Radiat. Oncol. Biol. Phys.* 39, 967–975.

DeCou, J. M., Rao, B. N., Parham, D. M., Lobe, T. M., Bowman, L., Pappo, A. S., and Fontanesi, J. (1995). Malignant peripheral nerve sheath tumors: the St. Jude Children's Research Hospital experience. *Ann. Surg. Oncol.* 2, 524–529.

Donahue, B., Scott, C. B., Nelson, J. S., Rotman, M., Murray, K. J., Nelson, D. F., Banker, F. L., Earle, J. D., Fischbach, J. A., Asbell, S. O., Gaspar, L. E., Markoe, A. M., and Curran, W. (1997). Influence of an oligodendroglial component on the survival of patients with anaplastic astrocytomas: a report of radiation therapy oncology group 83-02. *Int. J. Radiat. Oncol. Biol. Phys.* 38, 911–914.

Dropcho, E. J., and Soong, S. J. (1996). The prognostic impact of prior low grade histology in patients with anaplastic gliomas; a case-control study. *Neurology* 47, 684–690.

Ducatman, B. S., Scheithauer, B. W., Piepgras, D. G., Reiman, H. M., and Ilstrup, D. M. (1986). Malignant peripheral nerve sheath tumors. A clinicopathologic study of 120 cases. *Cancer* 57, 2006–2021.

Dziuk, T. W., Woo, S., Butler, E. B., Thornby, J., Grossman, R., Dennis, W. S., Lu, H., Carpenter, L. S., and Chiu, J. K. (1998). Malignant meningioma: an indication for initial aggressive surgery and adjuvant radiotherapy. *J. Neuro-Oncol.* 37, 177–188.

Engelhard, H. H. (2000). The role of interstitial BCNU chemotherapy in the treatment of malignant glioma. *Surg. Neurol.* 53, 458–464.

Epstein, F. J., Farmer, J. P., and Freed, D. (1993). Adult intramedullary spinal cord ependymomas: the result of surgery in 38 patients. *J. Neurosurg.* 79, 204–209.

Esteller, M., Garcia-Foncillas, J., Andion, E., Goodman, S. N., Hidalgo, O. F., Vanaclocha, V., Baylin, S. B., and Herman, J. G. (2000). Inactivation of the DNA-repair gene MGMT and the clinical response of gliomas to alkylating agents. *N. Engl. J. Med.* 343, 1350–1354.

Evans, A. E., Jenkin, R. D. T., Sposto, R., Ortega, J. A., Wilson, C. B., Wara, W., Ertel, I. J., Kramer, S., Chang, C. H., Leikin, S. L., and Hammond, G. D. (1990). The treatment of medulloblastoma: results of a prospective randomized trial of radiation therapy with and without CCNU, vincristine and prednisone. *J. Neurosurg.* 72, 572–582.

Fetell, M. R., Grossman, S. A., Fisher, J. D., Erlanger, B., Rowinsky, E., Stockel, J., Piantodosi, S., for the New Approaches to Brain Tumor Therapy Central Nervous System Consortium. (1997). Preirradiation paclitaxel in glioblastoma multiforme: efficacy, pharmacology, and drug interactions. *J. Clin. Oncol.* **15**, 3121–3128.

Fine, H. A., Dear, K. B., Loeffler, J. S., Black, P. M., and Canellos, G. P. (1993). Meta-analysis of radiation therapy with and without adjuvant chemotherapy for malignant gliomas in adults. *Cancer* **71**, 2585–2597.

Fine, H. A., Figg, W. D., Jaeckle, K., Wen, P. Y., Kyritsis, A. P., Loeffler, J. S., Levin, V. A., Black, P. M., Kaplan, R., Pluda, J. M., and Yung, W. K. (2000). Phase II trial of the antiangiogenic agent thalidomide in patients with recurrent high-grade gliomas. *J. Clin. Oncol.* **18**, 708–715.

Finlay, J. L., Goldman, S., Wong, M. C., Cairo, M., Garvin, J., August, C., Cohen, B. H., Stanley, P., Zimmerman, R. A., Bostrom, B., Geyer, J. R., Harris, R. E., Sanders, J., Yates, A. J., Boyett, J. M., and Packer, R. J., for the Children's Cancer Group. (1996). Pilot study of high-dose thiotepa and etoposide with autologous bone marrow rescue in children and young adults with recurrent CNS tumors. The Children's Cancer Group. *J. Clin. Oncol.* **14**, 2495–2503.

Ford, J., Osborn, C., Barton, T., and Bleehen, N. M. (1998). A phase I study of intravenous RMP-7 with carboplatin in patients with progression of malignant glioma. *Eur. J. Cancer* **34**, 1807–1811.

Freeman, C. R., Kepner, J., Kun, L. E., Sanford, R. A., Kadota, R., Mandell, L., and Friedman, H. (2000). A detrimental effect of a combined chemotherapy-radiotherapy approach in children with diffuse intrinsic brain stem gliomas. *Int. J. Radiat. Oncol. Biol. Phys.* **47**, 561–564.

Friedman, H. S., McLendon, R. E., Kerby, T., Dugan, M., Bigner, S. H., Henry, A. J., Ashley, D. M., Krischer, J., Lovell, S., Rasheed, K., Marchev, F., Seman, A. J., Cokgor, I., Rich, J., Stewart, E., Colvin, O. M., Provenzale, J. M., Bigner, D. D., Haglund, M. M., Friedman, A. H., and Modrich, P. L. (1998). DNA mismatch repair and O6-alkylguanine-DNA alkyltransferase analysis and response to Temodal in newly diagnosed malignant glioma. *J. Clin. Oncol.* **16**, 3851–3857.

Fuller, D., and Bloom, J. G. (1988). Radiotherapy for chordoma. *Int. J. Radiat. Oncol. Biol. Phys.* **15**, 331–339.

Galanis, E., Buckner, J. C., Schomberg, P. J., Hammack, J. E., Raffel, C., and Scheithauer, B. W. (1997). Effective chemotherapy for advanced CNS embryonal tumors in adults. *J. Clin. Oncol.* **15**, 2939–2944.

Galanis, E., Buckner, J. C., Dinapoli, R. P., Scheithauer, B. W., Jenkins, R. B., Wang, C. H., O'Fallon, J. R., and Farr, G. (1998a). Clinical outcome of gliosarcoma compared with glioblastoma multiforme: North Central Cancer Treatment Group results. *J. Neurosurg.* **89**, 425–430.

Galanis, E., Buckner, J. C., Scheithauer, B. W., Kimmel, D. W., Schomberg, P. J., and Piepgras, D. G. (1998b). Management of recurrent meningeal hemangiopericytoma. *Cancer* **82**, 1915–1920.

Gannett, D. E., Wisbeck, W. M., Silbergeld, D. L., and Berger, M. S. (1994). The role of postoperative irradiation in the treatment of oligodendroglioma. *Int. J. Radiat. Oncol. Biol. Phys.* **30**, 567–573.

Gaspar, L. E., Mackenzie, I. R. A., Gilbert, J. J., Kaufmann, J. C. E., Fisher, B. F., Macdonald, D. R., and Cairncross, J. G. (1993). Primary cerebral fibrosarcomas. Clinicopathologic study and review of the literature. *Cancer* **72**, 3277–3281.

Giannini, C., Scheithauer, B. W., Burger, P. C., Brat, D. J., Wollan, P. C., Lach, B., and O'Neill, B. P. (1999). Pleomorphic xanthroas-trocytoma: what do we really know about it? *Cancer* **85**, 2033–2045.

Giombini, S., Solero, C. L., and Morello, G. (1984). Late outcome of operations for supratentorial convexity meningiomas. Report on 207 cases. *Surg. Neurol.* **22**, 588–594.

Glaholm, J., Bloom, H. J. G., and Crow, J. H. (1990). The role of radiotherapy in the management of intracranial meningiomas: The Royal Marsden Hospital experience with 186 patients. *Int. J. Radiat. Oncol. Biol. Phys.* **18**, 755–761.

Goldsmith, B. J., Wara, W. M., Wilson, C. B., and Larson, D. A. (1994). Postoperative irradiation for subtotally resected meningiomas. A retrospective analysis of 140 patients treated from 1967–1990. *J. Neurosurg.* **80**, 195–201.

Grant, R., Liang, B. C., Slattery, J., Greenberg, H. S., and Junck, L. (1997). Chemotherapy response criteria in malignant glioma. *Neurology* **48**, 1336–1340.

Green, S. B., Byar, D. P., Walker, M. D., Pistenmaa, D. A., Alexander, E., Jr., Batzdorf, U., Brooks, W. H., Hunt, W. E., Mealey, J., Jr., Odom, G. L., Paoletti, P., Ransohoff, J., 2nd., Robertson, J. T., Selker, R. G., Shapiro, W. R., Smith, K. R., Jr., Wilson, C. B., and Strike, T. A. (1983). Comparisons of carmustine, procarbazine, and high-dose methylprednisolone as additions to surgery and radiotherapy for the treatment of malignant glioma. *Cancer Treat. Rep.* **67**, 121–132.

Greenberg, M. L. (1999). Chemotherapy of choroid plexus carcinoma. *Child's Nerv. Syst.* **15**, 571–577.

Grossman, S. A., Wharam, M., Sheidler, V., Kleinberg, L., Zeltzman, M., Yue, N., and Piantadosi, S. (1997). Phase II study of continuous infusion carmustine and cisplatin followed by cranial irradiation in adults with newly diagnosed high-grade astrocytoma. *J. Clin. Oncol.* **15**, 2596–2603.

Guthrie, B. L., Ebersold, M. J., Scheithauer, B. W., and Shaw, E. G. (1989). Meningeal hemangiopericytoma: histopathological features, treatment, and long-term follow-up of 44 cases. *Neurosurgery* **25**, 514–522.

Habrand, J. L., Ganry, O., Couanet, D., Rouxel, V., Levy-Piedbois, C., Pierre-Kahn, A., and Khalifa, C. (1999). The role of radiation therapy in the management of craniopharyngioma: a 25-year experience and review of the literature. *Int. J. Radiat. Oncol. Biol. Phys.* **44**, 255–263.

Hardenbergh, P. H., Golden, J., Billet, A., Scott, R. M., Shrieve, D. C., Silver, B., Loeffler, J. S., and Tarbell, N. J. (1997). Intracranial germinoma: the case for lower dose radiation therapy. *Int. J. Radiat. Oncol. Biol. Phys.* **39**, 419–426.

Helseth, A., and Mørk, S. J. (1989). Primary intraspinal neoplasms in Norway, 1955 to 1986. A population-based survey of 467 patients. *J. Neurosurg.* **71**, 842–845.

Herrlinger, V., Felsberg, J., Küker, W., Bornemann, A., Plasswilm, L., Knobbe, C. B., Strik, H., Wick, W., Meyermann, R., Dichgans, J., Bamberg, M., Reifenberger, G., and Weller, M. (2002). Gliomatosis cerebri. Molecular pathology and clinical course. *Ann. Neurol.* **52**, 390–399.

Hess, K. R. (1999). Extent of resection as a prognostic variable in the treatment of gliomas. *J. Neuro-Oncol.* **42**, 227–231.

Hildebrand, J., Sahmoud, T., Mignolet, F., Brucher, J. M., Afta, D., and the EORTC Brain Tumor Group. (1994). Adjuvant therapy with dibromodulcitol and BCNU increases survival in adults with malignant gliomas. *Neurology* **44**, 1479–1483.

Hsu, D. W., Efird, J. T., and Hedley-Whyte, E. T. (1997). Progesterone and estrogen receptors in meningiomas: prognostic implications. *J. Neurosurg.* **86**, 113–120.

Ino, Y., Zlatescu, M. C., Sasaki, H., Macdonald, D. R., Stemmer-Rachamimov, A. O., Jhung, S., Ramsay, D. A., Von Deimling, A., Louis, D. N., and Cairncross, J. G. (2000). Long survival and therapeutic responses in patients with histologically disparate high-grade gliomas demonstrating chromosome 1p loss. *J. Neurosurg.* **92,** 983–990.

Jakacki, R. I., Cohen, B. E., Jamison, C., Mathews, V. P., Arenson, E., Longee, D. C., Hilden, J., Cornelius, A., Needle, M., Heilman, D., Boaz, J. C., and Luerssen, T. G. (2000). Phase II evaluation of interferon-α-2a for progressive or recurrent craniopharyngiomas. *J. Neurosurg.* **92,** 255–260.

Jennings, M. T., Gelman, R., and Hochberg, F. (1985). Intracranial germ-cell tumors: natural history and pathogenesis. *J. Neurosurg.* **63,** 155–167.

Jennings, M. T., Frenchman, M., Shehab, T., Johnson, M. D., Creasy, J., LaPorte, K., and Dettbarn, W. D. (1995). Gliomatosis cerebri presenting as intractable epilepsy during early childhood. *J. Child Neurol.* **10,** 37–45.

Kaba, S. E., DeMonte, F., Bruner, J. M., Kyritsis, A. P., Jaeckle, K. A., Levin, V., and Yung, W. K. A. (1997). The treatment of recurrent unresectable and malignant meningiomas with interferon alpha-2B. *Neurosurgery* **40,** 271–275.

Kagei, K., Shirato, H., Suzuki, K., Isu, T., Sawamura, Y., Sakamoto, T., Fukuda, S., Nishioka, T., Hashimoto, S., and Miyasaka, K. (1999). Small-field fractionated radiotherapy with or without stereotactic boost for vestibular schwannoma. *Radiother. Oncol.* **50,** 341–347.

Kappelle, A. C., Postma, T. J., Taphoorn, M. J., Groeneveld, G. J., Van den Bent, M. J., van Groeningen, C. J., Zonnenberg, B. A., Sneeuw, K. C., and Heimanns, J. J. (2001). PCV chemotherapy for recurrent glioblastoma multiforme. *Neurology* **56,** 118–120.

Kaplan, A. M., Albright, A. L., Zimmerman, R. A., Rorke, L. B., Li, H., Boyett, J. M., Finlay, J. L., Wara, W. M., and Packer, R. J. (1996). Brainstem gliomas in children. *Pediatr. Neurosurg.* **24,** 185–192.

Karim, A. B. M. F., Maat, B., Hatlevoll, R., Menten, J., Rutten, E. H. J. M., Thomas, D. G. T., Mascarenhas, F., Horiot, J. C., Parvinen, L. M., Van Reijn, M., Jager, J. J., Fabrini, M. G., Van Alphen, A. M., Hamers, H. P., Gaspar, L., Noordman, E., Pierart, M., and Van Glabbeke, M. (1996). A randomized trial on dose-response in radiation therapy of low-grade cerebral glioma: European Organization for Research and Treatment of Cancer (EORTC) study 22844. *Int. J. Radiat. Oncol. Biol. Phys.* **36,** 549–556.

Karim, A. B. M. F., Afra, D., Cornu, P., Bleehan, N., Schraub, S., De Witte, O., Darcel, F., Stenning, S., Pierart, M., and Van Glabbeke, M. (2002). Randomised trial on the efficacy of radiotherapy for cerebral low-grade glioma in the adult: European Organization for Research and Treatment of Cancer Study 22845 with the Medical Research Council Study BRO4: An interim analysis. *Int. J. Radiat. Biol. Phys.* **52,** 316–324.

Keime-Guibert, F., Napolitano, M., and Delattre, J. Y. (1998). Neurological complications of radiotherapy and chemotherapy. *J. Neurol.* **245,** 695–708.

Kim, L., Hochberg, F. H., Thornton, A. F., Harsh, IV, G. R., Patel, H., Finkelstein, D., and Louis, D. N. (1996). Procarbazine, lomustine, and vincristine (PCV) chemotherapy for grade III and grade IV oligoastrocytomas. *J. Neurosurg.* **85,** 602–607.

Kim, D. G., Paek, S. H., Kim, I. H., Chi, J. G., Jung, H. W., Han, D. H., Choi, K. S., and Cho, B. K. (1997). Central neurocytoma: the role of radiation therapy and long-term outcome. *Cancer* **79,** 1995–2002.

Kim, D. G., Yang, H. J., Park, I. A., Chi, J. G., Jung, H. W., Han, D. H., Choi, K. S., and Cho, B. K. (1998). Gliomatosis cerebri: clinical features, treatment, and prognosis. *Acta Neurochir.* **140,** 755–762.

Kleihues, P., and Cavenee, W. K. (2000). World Health Organization classification of tumours. Pathology & genetics. Tumours of the nervous system. Lyon: IARC Press.

Koka, V. N., Julieron, M., Bourhis, J., Janot, F., Le Ridant, A. M.,

Marandas, P., Luboinski, B., and Schwaab, G. (1998). Aesthesioneuroblastoma. *J. Laryngol. Otol.* **112,** 628–633.

Kondziolka, D., Lunsford, L. D., Coffey, R. J., and Flickinger, J. C. (1991). Stereotactic radiosurgery of meningiomas. *J. Neurosurg.* **74,** 552–559.

Kondziolka, D., Lunsford, D., McLaughlin, M. R., and Flickinger, J. C. (1998). Long-term outcomes after radiosurgery for acoustic neuromas. *N. Engl. J. Med.* **339,** 1426–1433.

Kondziolka, D., Levy, E. I., Niranjan, A., Flickinger, J. C., and Lunsford, L. D. (1999). Long-term outcomes after meningioma radiosurgery: physician and patient perspectives. *J. Neurosurg.* **91,** 44–50.

Kortmann, R. D., Kühl, J., Timmermann, B., Mittler, U., Urban, C., Budach, V., Richter, E., Willich, N., Flentje, M., Berthold, F., Slavc, I., Wolff, J., Meisner, C., Wiestler, O., Sörensen, N., Warmuth-Metz, M., and Bamberg, M. (2000). Postoperative neoadjuvant chemotherapy before radiotherapy as compared to immediate radiotherapy followed by maintenance chemotherapy in the treatment of medulloblastoma in childhood: results of the German prospective randomized trial HIT '91. *Int. J. Radiat. Oncol. Biol. Phys.* **46,** 269–279.

Kreth, F. W., Faist, M., Warnke, P. C., Rossner, R., Volk, B., and Ostertag, C. B. (1995). Interstitial radiosurgery of low-grade gliomas. *J. Neurosurg.* **82,** 418–429.

Kreth, F. W., Berlis, A., Spiropoulou, V., Faist, M., Scheremet, R., Rossner, R., Volk, B., and Ostertag, C. B. (1999). The role of tumor resection in the treatment of glioblastoma multiforme in adults. *Cancer* **86,** 2117–2123.

Krischer, J. P., Ragab, A. H., Kun, L., Kim, T. H., Laurent, J. P., Boyett, J. M., Cornell, C. J., Link, M., Luthy, A. R., and Camitta, B. (1991). Nitrogen mustard, vincristine, procarbazine, and prednisone as adjuvant chemotherapy in the treatment of medulloblastoma: A Pediatric Oncology Group study. *J. Neurosurg.* **74,** 905–909.

Kühl, J., Müller, H. L., Berthold, F., Kortmann, R. D., Deinlein, F., Maaß, E., Graf, N., Gnekow, A., Scheurlen, W., Göbel, U., Wolff, J. E. A., Bamberg, M., Kaatsch, P., Kleihues, P., Rating, D., Sörensen, N., and Wiestler, O. D. (1998). Preradiation chemotherapy of children and young adults with malignant brain tumors: results of the German pilot trial HIT '88/'89. *Klin. Pädiatr.* **210,** 227–233.

Kyritsis, A. P., Yung, W. K. A., Bruner, J., Gleason, M. J., and Levin, V. A. (1993). The treatment of anaplastic oligodendrogliomas and mixed gliomas. *Neurosurgery* **32,** 365–370.

Landolfi, J. C., Thaler, H. T., and DeAngelis, L. M. (1998). Adult brainstem gliomas. *Neurology* **51,** 1136–1139.

Lang, F. F., Epstein, F. J., Ransohoff, J., Allen, J. C., Wisoff, J., Abbott, I. R., and Miller, D. C. (1993). Central nervous system gangliogliomas. Part 2: clinical outcome. *J. Neurosurg.* **79,** 867–873.

Laperriere, N. J., Leung, P. M. K., McKenzie, S., Milosevic, M., Wong, S., Glen, J., Pintilie, M., and Bernstein, M. (1998). Randomized study of brachytherapy in the initial management of patients with malignant astrocytoma. *Int. J. Radiat. Oncol. Biol. Phys.* **41,** 1005–1011.

Leighton, C., Fisher, B., Bauman, G., Depiero, S., Stitt, L., Macdonald, D., and Cairncross, G. (1997). Supratentorial low-grade glioma in adults: an analysis of prognostic factors and timing of radiation. *J. Clin. Oncol.* **15,** 1294–1301.

Levin, V. A., Silver, P., Hannigan, J., Wara, W. M., Gutin, P. H., Davis, R. L., and Wilson, C. B. (1990). Superiority of post-radiotherapy adjuvant chemotherapy with CCNU, procarbazine, and vincristine (PCV) over BCNU for anaplastic gliomas: NCOG 6G61 final report. *Int. J. Radiat. Oncol. Biol. Phys.* **18,** 321–324.

Levin, V. A., Maor, M. H., Thall, P. F., Yung, W. K. A., Bruner, J., Sawaya, R., Kyritsis, A. P., Leeds, N., Woo, S., Rodriguez, L., and

Gleason, M. J. (1995). Phase II study of accelerated fractionation radiation therapy with carboplatin followed by vincristine chemotherapy for the treatment of glioblastoma multiforme. *Int. J. Radiat. Oncol. Biol. Phys.* **33**, 357–364.

Lindegaard, K. F., Mørk, S. J., Eide, G. E., Halvorsen, T. B., Hatlevoll, R., Solgaard, T., Dahl, O., and Ganz, J. (1987). Statistical analysis of clinicopathological features, radiotherapy, and survival in 170 cases of oligodendroglioma. *J. Neurosurg.* **67**, 224–230.

Linstadt, D. E., Wara, W. M., Leibel, S. A., Gutin, P. H., Wilson, C. B., and Sheline, G. E. (1989). Postoperative radiotherapy of primary spinal cord tumors. *Int. J. Radiat. Oncol. Biol. Phys.* **16**, 1397–1403.

Lippens, R. J. J., Rotteveel, J. J., Otten, B. J., and Merx, H. (1998). Chemotherapy with adriamycin (doxorubicin) and CCNU (lomustine) in four children with recurrent craniopharyngioma. *Eur. J. Paed. Neurol.* **2**, 263–268.

Lote, K., Egeland, T., Hager, B., Skullerud, K., and Hirschberg, H. (1998). Prognostic significance of CT contrast enhancement within histological subgroups of intracranial glioma. *J. Neuro-Oncol.* **40**, 161–170.

MacCollin, M., Woodfin, W., Kronn, D., and Short, M. P. (1996). Schwannomatosis: a clinical and pathologic study. *Neurology* **46**, 1072–1079.

Macdonald, D. R., Cascino, T. L., Schold, S. C., and Cairncross, J. G. (1990). Response criteria for phase II studies of supratentorial malignant glioma. *J. Clin. Oncol.* **8**, 1277–1280.

Macdonald, D., Cairncross, G., Stewart, D., Forsyth, P., Sawa, C., Wainman, N., and Eisenhauer, E., for the National Clinical Institute of Canada Clinical Trials Group (NCIC CTG). (1996). Phase II study of topotecan in patients with recurrent malignant glioma. *Ann. Oncol.* **7**, 205–207.

Marras, L. C., Geerts, W. H., and Perry, J. R. (2000). The risk of venous thromboembolism is increased throughout the course of malignant glioma. An evidence-based review. *Cancer* **89**, 640–646.

McElroy, E. A., Buckner, J. C., and Lewis, J. E. (1998). Chemotherapy for advanced esthesioneuroblastoma: the Mayo clinic experience. *Neurosurgery* **42**, 1023–1028.

Meckling, S., Dold, O., Forsyth, P. A. J., Brasher, P., and Hagen, N. A. (1996). Malignant supratentorial glioma in the elderly: is radiotherapy useful? *Neurology* **47**, 901–905.

The Medical Research Council Brain Tumor Working Party. (2001). Randomized trial of procarbazine, lomustine, and vincristine in the adjuvant treatment of high-grade astrocytoma: A Medical Research Council trial. *J. Clin. Oncol.* **19**, 509–518.

Miller, R. C., Foote, R. L., Coffey, R. J., Gorman, D. A., Earle, J. D., Schomberg, P. J., and Kline, R. W. (1997). The role of stereotactic radiosurgery in the treatment of malignant skull base tumors. *Int. J. Radiat. Oncol. Biol. Phys.* **39**, 977–981.

Minehan, K. J., Shaw, E. G., Scheithauer, B. W., Davis, D. L., and Onofrio, B. M. (1995). Spinal cord astrocytoma: pathological and treatment considerations. *J. Neurosurg.* **83**, 590–595.

Miralbell, R., Mornex, F., Greiner, R., Bolla, M., Storme, G., Hulshof, M., Bernier, J., Denekamp, J., Rojas, A. M., Pierart, M., Van Glabbeke, M., and Mirimanoff, R. O. (1999). Accelerated radiotherapy, carbogen, and nicotinamide in glioblastoma multiforme: report of European Organization for Research and Treatment of Cancer Trial 22933. *J. Clin. Oncol.* **17**, 3143–3149.

Mirimanoff, R. O., Dosoretz, D. E., Linggood, R. M., Ojemann, R. G., and Martuza, R. L. (1985). Meningioma: analysis of recurrence and progression following neurosurgical resection. *J. Neurosurg.* **62**, 18–24.

Murota, T., and Symon, L. (1989). Surgical management of hemangioblastoma of the spinal cord: a report of 18 cases. *Neurosurgery* **25**, 699–708.

Nazar, G. B., Hoffman, H. J., Becker, L. E., Jenkin, D., Humphreys,
R. P., and Hendrick, E. B. (1990). Infratentorial ependymomas in childhood: prognostic factors and treatment. *J. Neurosurg.* **72**, 408–417.

Needle, M. N., Goldwein, J. W., Grass, J., Cnaan, A., Bergman, I., Molloy, P., Sutton, L., Zhao, H., Garvin, J. H., and Phillips, P. C. (1997). Adjuvant chemotherapy for the treatment of intracranial ependymoma of childhood. *Cancer* **80**, 341–347.

Nijjar, T. S., Simpson, W. J., Gadalla, T., and McCartney, M. (1993). Oligodendroglioma. The Princess Margaret Hospital experience. (1958–1964). *Cancer* **71**, 4002–4006.

Nutting, C., Brada, M., Brazil, L., Sibtain, A., Saran, F., Westbury, C., Moore, A., Thomas, D. G. T., Traish, D., and Ashley, S. (1999). Radiotherapy in the treatment of benign meningioma of the skull base. *J. Neurosurg.* **90**, 823–827.

Packer, R. J. (1999). Childhood medulloblastoma: progress and future challenges. *Brain Dev.* **21**, 75–81.

Packer, R. J., Sutton, L. N., Elterman, R., Lange, B., Goldwein, J., Nicholson, H. S., Mulne, L., Boyett, J., D'Angio, G., Wechsler-Jentzsch, K., Reaman, G., Cohen, B. H., Bruce, D. A., Rorke, L. B., Molloy, P., Ryan, J., LaFond, D., Evans, A. E., and Schut, L. (1994). Outcome for children with medulloblastoma treated with radiation and cisplatin, CCNU, and vincristine chemotherapy. *J. Neurosurg.* **81**, 690–698.

Packer, R. J., Ater, J., Allen, J., Phillips, P., Geyer, R., Nicholson, H. S., Jakacki, R., Kurczynski, E., Needle, M., Finlay, J., Reaman, G., and Boyett, J. M. (1997). Carboplatin and vincristine chemotherapy for children with newly diagnosed progressive low-grade gliomas. *J. Neurosurg.* **86**, 747–754.

Paleologos, N. A., Macdonald, D. R., Vick, N. A., and Cairncross, J. G. (1999). Neoadjuvant procarbazine, CCNU, and vincristine for anaplastic and aggressive oligodendroglioma. *Neurology* **53**, 1141–1143.

Palma, L., Celli, P., Franco, C., Cervoni, L., and Cantore, G. (1997). Long-term prognosis for atypical and malignant meningiomas: a study of 71 surgical cases. *J. Neurosurg.* **86**, 793–800.

Paulino, A. C., and Melian, E. (1999). Medulloblastoma and supratentorial primitive neuroectodermal tumors. An institutional experience. *Cancer* **86**, 142–148.

Pencalet, P., Sainte-Rose, C., Lellouch-Tubiana, A., Kalifa, C., Brunelle, F., Sgouros, S., Meyer, P., Cinalli, G., Zerah, M., Pierre-Kahn, A., and Renier, D. (1998). Papillomas and carcinomas of the choroid plexus in children. *J. Neurosurg.* **88**, 521–528.

Perry, J. R., DeAngelis, L. M., Schold, S. C., Burger, P. C., Brem, H., Brown, M. T., Curran, W. J., Scott, C. B., Prados, M. D., Kaplan, R., and Cairncross, J. G. (1997). Challenges in the design and conduct of phase III brain tumor therapy trials. *Neurology* **49**, 912–917.

Perry, A., Scheithauer, B. W., Stafford, S. L., Lohse, C. M., and Wollan, P. C. (1999). "Malignancy" in meningiomas: a clinicopathologic study of 116 patients, with grading implications. *Cancer* **85**, 2046–2056.

Philippon, J. H., Clemenceau, S. H., Fauchon, F. H., and Foncin, J. F. (1993). Supratentorial low-grade astrocytomas in adults. *Neurosurgery* **32**, 554–559.

Polin, R. S., Sheehan, J. P., Chenelle, A. G., Munoz, E., Larner, J., Phillips, C. D., Cantrell, R. W., Laws, E. R., Newman, S. A., Levine, P. A., and Jane, J. A. (1998). The role of preoperative adjuvant treatment in the management of esthesioneuroblastoma: the University of Virginia experience. *Neurosurgery* **42**, 1029–1037.

Pollock, B. E., Lunsford, L. D., Kondziolka, D., Flickinger, J. C., Bissonette, D. J., Kelsey, S. F., and Jannetta, P. J. (1995). Outcome analysis of acoustic neuroma management: a comparison of microsurgery and stereotactic radiosurgery. *Neurosurgery* **36**, 215–229.

Post, K. D., Eisenberg, M. B., and Catalano, P. J. (1995). Hearing preservation in vestibular schwannoma surgery: what factors influence outcome? *J. Neurosurg.* **83**, 191–196.

Prados, M., Larson, D. A., Lamborn, K., McDermott, M. W., Sneed, P. K., Wara, W. M., Chang, S. M., Mack, E. E., Krouwer, H. G. J., Chandler, K. L., Warnick, R. E., Davis, R. L., Rabbitt, J. E., Malec, M., Levin, V. A., Gutin, P. H., Phillips, T. L., and Wilson, C. B. (1998). Radiation therapy and hydroxyurea followed by the combination of 6-thioguanine and BCNU for the treatment of primary malignant brain tumors. *Int. J. Radiat. Oncol. Biol. Phys.* 40, 57–63.

Prados, M. D., Scott, C., Curran, W. J., Jr., Nelson, D. F., Leibel, S., and Kramer, S. (1999). Procarbazine, lomustine, and vincristine (PCV) chemotherapy for anaplastic astrocytoma: a retrospective review of Radiation Therapy Oncology Group protocols comparing survival with carmustine or PCV adjuvant chemotherapy. *J. Clin. Oncol.* 17, 3389–3395.

Rainov, N. G., on behalf of the GL1328 International Study Group. (2000). A phase III clinical evaluation of herpes simplex virus type 1 thymidine kinase and ganciclovir gene therapy as an adjuvant to surgical resection and radiation in adults with previously untreated glioblastoma multiforme. *Human Gene Therapy* 11, 2389–2401.

Rajan, B., Ashley, S., Gorman, C., Jose, C. C., Horwich, A., Bloom, H. J. G., Marsh, H., and Brada, M. (1993). Craniopharyngeoma—long-term results followig limited surgery and radiotherapy. *Radiother. Oncol.* 26, 1–10.

Rezai, A. R., Woo, H. W., Lee, M., Cohen, H., Zagzag, D., and Epstein, F. J. (1996). Disseminated ependymomas of the central nervous system. *J. Neurosurg.* 85, 618–624.

Richard, S., David, P., Marsot-Dupuch, K., Giraud, S., Beroud, C., and Resche, F. (2000). Central nervous system hemangioblastomas, endolymphatic sac tumors, and von Hippel-Lindau disease. *Neurosurg. Rev.* 23, 1–22.

Robertson, P. L., DaRosso, R. C., and Allen, J. C. (1997). Improved prognosis of intracranial non-germinoma germ cell tumors with multimodality therapy. *J. Neuro-Oncol.* 32, 71–80.

Rochkind, S., Blatt, I., Sadeh, M., and Goldhammer, Y. (1991). Extracranial metastases of medulloblastoma in adults: literature review. *J. Neurol. Neurosurg. Psychiatry* 54, 80–86.

Rodriguez, L. A., Prados, M., Silver, P., and Levin, V. A. (1989). Reevaluation of procarbazine for the treatment of recurrent malignant central nervous system tumors. *Cancer* 64, 2420–2423.

Rubin, G., Scienza, R., Pasqualin, A., Rosta, L., and Da Pian, R. (1989). Craniocerebral epidermoids and dermoids: a review of 44 cases. *Acta Neurochir.* 97, 1–16.

Samii, M., and Matthies, C. (1997). Management of 1000 vestibular schwannomas (acoustic neuromas): surgical management and results with an emphasis on complications and how to avoid them. *Neurosurgery* 40, 11–23.

Sawamura, Y., Ikeda, J., Shirato, H., Tada, M., and Abe, H. (1998a). Germ cell tumours of the central nervous system: treatment consideration based on 111 cases and their long-term clinical outcomes. *Eur. J. Cancer* 34, 104–110.

Sawamura, Y., Shirato, H., Ikeda, J., Tada, M., Ishii, N., Kato, T., Abe, H., and Fujeda, K. (1998b). Induction chemotherapy followed by reduced-volume radiation therapy for newly diagnosed central nervous system germinoma. *J. Neurosurg.* 88, 66–72.

Schaller, B., Kirsch, E., Tolnay, M., and Mindermann, T. (1998). Symptomatic granular cell tumor of the pituitary gland: case report and review of the literature. *Neurosurgery* 42, 166–170.

Schild, S. E., Scheithauer, B. W., Schomberg, P. J., Hook, C. C., Kelly, P. J., Frick, L., Robinow, J. S., and Buskirk, S. J. (1993). Pineal parenchymal tumors. Clinical, pathologic, and therapeutic aspects. *Cancer* 72, 870–880.

Schild, S. E., Scheithauer, B. W., Haddock, M. G., Schiff, D., Burger, P. C., Wong, W. W., and Lyons, M. K. (1997). Central neurocytoma. *Cancer* 79, 790–795.

Schild, S. E., Nisi, K., Scheithauer, B. W., Wong, W. W., Lyons, M. K., Schomberg, P. J., and Shaw, E. G. (1998a). The results of radio-

therapy for ependymomas: the Mayo clinic experience. *Int. J. Radiat. Oncol. Biol. Phys.* 42, 953–958.

Schild, S. E., Stafford, S. L., Brown, P. D., Wood, C. P., Scheithauer, B. W., Schomberg, P. J., Wong, W. W., Lyons, M. K., and Shaw, E. G. (1998b). The results of radiotherapy for brainstem tumors. *J. Neuro-Oncol.* 40, 171–177.

Schrell, U. M. H., Rittig, M. G., Anders, M., Koch, U. H., Marschalek, R., Kiesewetter, F., and Fahlbusch, R. (1997). Hydroxyurea for treatment of unresectable and recurrent meningiomas. II. Decrease in the size of meningiomas in patients treated with hydroxyurea. *J. Neurosurg.* 86, 840–844.

Schultheiss, T. E., Kun, L. E., Ang, K. K., and Stephens, L. C. (1995). Radiation response of the central nervous system. *Int. J. Radiat. Oncol. Biol. Phys.* 31, 1093–1112.

Scott, J. N., Newcastle, N. B., Brasher, P. M. A., Fulton, D., MacKinnon, J. A., Hamilton, M., Cairncross, J. G., and Forsyth, P. (1999). Which glioblastoma multiforme patient will become a long-term survivor? A population-based study. *Ann. Neurol.* 46, 183–188.

Seppälä, M. T., Sainio, M. A., Haltia, M. J. J., Kinnunen, J. J., Setälä, K. H., and Jääskelainen, J. E. (1998). Multiple schwannomas: schwannomatosis or neurofibromatosis type 2? *J. Neurosurg.* 89, 36–41.

Shaw, E. G., Scheithauer, B. W., O'Fallon, J. R., and Davis, D. H. (1994). Mixed oligoastrocytomas: a survival and prognostic factor analysis. *Neurosurgery* 34, 577–582.

Shelton, C. (1995). Unilateral acoustic tumors: how often do they recur after translabyrinthine removal? *Laryngoscope* 105, 958–966.

Shepherd, S. F., Laing, R. W., Cosgrove, V. P., Warrington, A. P., Hines, F., Ashley, S. E., and Brada, M. (1997). Hypofractionated stereotactic radiotherapy in the management of recurrent glioma. *Int. J. Radiat. Oncol. Biol. Phys.* 37, 393–398.

Shibamato, Y., Takahashi, M., and Sasai, L. (1997). Prognosis of intracranial germinoma with syncytiotrophoblastic giant cells treated by radiation therapy. *Int. J. Radiat. Oncol. Biol. Phys.* 37, 505–510.

Smalley, S. R., Schomberg, P. J., Earle, J. D., Laws, E. R., Scheithauer, B. W., and O'Fallon, J. R. (1990). Radiotherapeutic considerations in the treatment of hemangioblastomas of the central nervous system. *Int. J. Radiat. Oncol. Biol. Phys.* 18, 1165–1171.

Smith, J. S., Perry, A., Borell, T. J., Lee, H. K., O'Fallon, J., Hosek, S. M., Kimmel, D., Yates, A., Burger, P. C., Scheithauer, B. W., and Jenkins, R. B. (2000). Alterations of chromosome arms 1p and 19q as predictors of survival in oligodendrogliomas, astrocytomas, and mixed oligoastrocytomas. *J. Clin. Oncol.* 18, 636–645.

Soffietti, R., Rudà, R., Bradac, G. B., and Schiffer, D. (1998). PCV chemotherapy for recurrent oligodendrogliomas and oligoastrocytomas. *Neurosurgery* 43, 1066–1073.

Solero, C. L., Fornari, M., Giombini, S., Lasio, G., Oliveri, G., Cimino, C., and Pluchino, F. (1989). Spinal meningiomas: review of 174 operated cases. *Neurosurgery* 125, 153–160.

Streffer, J., Schabet, M., Bamberg, M., Grote, E. H., Meyermann, R., Voigt, K., Dichgans, J., and Weller, M. (2000). A role for preirradiation PCV chemotherapy for oligodendroglial brain tumors. *J. Neurol.* 247, 297–302.

Streffer, J., Bitzer, M., Schabet, M., Dichgans, J., and Weller, M. (2001). Response of radiochemotherapy-associated cerebral edema to a phytotherapeutic agent, H15. *Neurology* 561, 1219–1221.

Strik, H., Deininger, M., Streffer, J., Grote, E., Wickboldt, J., Dichgans, J., Weller, M., and Meyermann, R. (1999). Bcl-2 family protein expression in initial and recurrent glioblastomas: modulation by radiochemotherapy. *J. Neurol. Neurosurg. Psychiatry* 67, 763–768.

Tait, D. M., Thornton-Jones, H., Bloom, H. J. G., Lemerle, J., and Morris-Jones, P. (1990). Adjuvant chemotherapy for medulloblastoma: the first multi-centre control trial of the International Society of Pediatric Oncology (SIOP I). *Eur. J. Cancer* 26, 464–469.

Takakura, K., Abe, H., Tanaka, R., Kitamura, K., Miwa, T., Takeuchi, K., Yamamoto, S., Kageyama, N., Handa, H., Mogami, H., Nishimoto, A., Uozumi, T., Matsutani, M., and Nomura, K. (1986). Effects of ACNU and radiotherapy on malignant glioma. *J. Neurosurg.* **64**, 53–57.

Tao, M. L., Barnes, P. D., Billett, A. L., Leong, T., Shrieve, D. C., Scott, R. M., and Tarbell, N. J. (1997). Childhood optic chiasm gliomas: radiographic response following radiotherapy and long-term clinical outcome. *Int. J. Radiat. Oncol. Phys.* **39**, 579–587.

Taylor, B. W., Marcus, R. B., Friedman, W. A., Ballinger, W. E., and Million, R. R. (1988). The meningioma controversy: postoperative radiation therapy. *Int. J. Radiat. Oncol. Biol. Phys.* **15**, 299–304.

Thiessen, B., Finlay, J., Kulkarni, R., and Rosenblum, M. K. (1998). Astroblastoma: does histology predict biologcal behavior? *J. Neuro-Oncol.* **40**, 59–65.

Thomas, D. G. T., and Graham, D. I. eds. (1980). Brain tumours: Scientific basis, clinical investigation and current therapy, Butterworth; London: 48.

Thomas, D. G. T., and Nouby, R. M. (1989). Experience in 300 cases of CT-directed stereotactic surgery for lesion biopsy and aspiration of haematoma. *Br. J. Neurosurg.* **3**, 321–325.

Timmermann, B., Kortmann, R. D., Kühl, J., Meisner, C., Slavc, I., Pietsch, T., and Bamberg, M. (2000). Combined postoperative irradiation and chemotherapy for anaplastic ependymomas in childhood: results of the German prospective trials HIT 88/89 and HIT 91. *Int. J. Radiat. Oncol. Biol. Phys.* **46**, 287–295.

Tomlinson, F. H., Scheithauer, B. W., Hayostek, C. J., Parisi, J. E., Meyer, F. B., Shaw, E. G., Weiland, T. L., Katzmann, J. A., and Jack, C. R. (1994). The significance of atypia and histologic malignancy in pilocytic astrocytoma of the cerebellum: a clinicopathologic and flow cytometric study. *J. Child Neurol.* **9**, 301–310.

Valencak, J., Dietrich, W., Raderer, M., Dieckmann, K., Prayer, D., Hainfellner, J. A., and Marosi, C. (2000). Evidence of therapeutic efficacy of CCNU in recurrent choroid plexus papilloma. *J. Neuro-Oncol.* **49**, 263–268.

Van den Bent, M. J., Kros, J. M., Heimans, J. J., Pronk, L. C., Van Groeningen, C. J., Krouwer, H. G. J., Taphoorn, M. J. B., Zonnenberg, B. A., Tijssen, C. C., Twijnstra, A., Punt, C. J. A., and Boogerd, W., for the Dutch Neuro-oncology Group. (1998). Response rate and prognostic factors of recurrent oligodendroglioma treated with procarbazine, CCNU, and vincristine chemotherapy. *Neurology* **51**, 1140–1145.

Vanuytsel, L. J., Bessel, E. M., Ashley, S. E., Bloom, H. J. G., and Brada, M. (1992). Intracranial ependymoma: long-term results of a policy of surgery and radiotherapy. *Int. J. Radiat. Oncol. Biol. Phys.* **23**, 313–319.

Vindlacheruvu, R. R., Casey, A. T. H., and Thomas, D. G. T. (1999). MRI-guided stereotactic brain biopsy: a review of 33 cases. *Br. J. Neurosurgery* **13**, 143–147.

Voges, J., Sturm, V., Lehrke, R., Treuer, H., Gauss, C., and Berthold, F. (1997). Cystic craniopharyngeoma: long-term results after intracavitary irradiation with stereotactically applied colloidal beta-emitting radioactive sources. *Neurosurgery* **35**, 1001–1010.

Vuorinen, V., Sallinen, P, Haapasalo, H., Visakorpi, T., Kallio, M., and Jääskelainen, J. (1996). Outcome of 31 intracranial haemangiopericytomas: poor predictive value of cell proliferation indices. *Acta Neurochir.* **138**, 1399–1408.

Walker, M. D., Alexander, E., Hunt, W. E., MacCarty, C. S., Mahaley, M. S., Mealey, J., Norrell, H. A., Owens, G., Ransohoff, J., Wilson, C. B., Gehan, E. A., and Strike, T. A. (1978). Evaluation of BCNU and/or radiotherapy in the treatment of anaplastic gliomas. A cooperative clinical trial. *J. Neurosurg.* **49**, 333–343.

Walker, M. D., Green, S. B., Byar, D. P., Alexander, E., Batzdorf, U., Brooks, W. H., Hunt, W. E., MacCarty, C. S., Mahaley, M. S., Mealey, J., Owens, G., Ransohoff, J., Robertson, J. T., Shapiro, W. R., Smith, K. R., Wilson, C. B., and Strike, T. A. (1980). Randomized comparisons of radiotherapy and nitrosoureas for the treatment of malignant glioma after surgery. *N. Engl. J. Med.* **303**, 1323–1329.

Wallner, K. E., Gonzales, M., and Sheline, G. E. (1988). Treatment of oligodendrogliomas with or without post-operative irradiation. *J. Neurosurg.* **68**, 684–688.

Weller, M., Schmidt, C., Roth, W., and Dichgans, J. (1997). Chemotherapy of human malignant glioma: prevention of efficacy by dexamethasone? *Neurology* **48**, 1704–1709.

Weller, M., Rieger, J., Grimmel, C., Van Meir, E. G., De Tribolet, N., Krajewski, S., Reed, J. C., Von Deimling, A., and Dichgans, J. (1998). Predicting chemoresistance in human malignant glioma cells: the role of molecular genetic analyses. *Int. J. Cancer* **79**, 640–644.

Weller, M., Streffer, J., Wick, W., Kortmann, R. D., Heiss, E., Küker, W., Meyermann, R., Dichgans, J., and Bamberg, M. (2001). Pre-irradiation gemcitabine chemotherapy for newly diagnosed glioblastoma: a phase II study. *Cancer* **91**, 423–427.

Werner-Wasik, M., Scott, C. B., Nelson, D. F., Gaspar, L. E., Murray, K. J., Fischbach, J. A., Nelson, J. S., Weinstein, A. S., and Curran, W. J., Jr. (1996). Final report of a phase I/II trial of hyperfractionated and accelerated hyperfractionated radiation therapy with carmustine for adults with supratentorial malignant gliomas. Radiation Therapy Oncology Group Study 83-02. *Cancer* **77**, 1535–1543.

Whitaker, S. J., Bessell, E. M., Ashley, S. E., Bloom, H. J. G., Bell, B. A., and Brada, M. (1991). Postoperative radiotherapy in the management of spinal cord ependymoma. *J. Neurosurg.* **74**, 720–728.

Wiegand, D. A., Ojemann, R. G., and Fickel, V. (1996). Surgical treatment of acoustic neuroma (vestibular schwannoma) in the United States: report from the Acoustic Neuroma Registry. *Laryngoscope* **106**, 58–66.

Wong, W. W., Hirose, T., Scheithauer, B. W., Schild, S. E., and Gunderson, L. L. (1998). Malignant peripheral nerve sheath tumor: analysis of treatment outcome. *Int. J. Radiat. Oncol. Biol. Phys.* **42**, 351–360.

Wong, E. T., Hess, K. R., Gleason, M. J., Jaeckle, K. A., Kyritsis, A. P., Prados, M. D., Levin, V. A., and Yung, W. K. (1999). Outcomes and prognostic factors in recurrent glioma patients enrolled onto phase II clinical trials. *J. Clin. Oncol.* **17**, 2572–2578.

Yasargil, M. G., Curcic, M., Kis, M., Siegenthaler, G., Teddy, P. J., and Roth, P. (1990). Total removal of craniopharyngiomas. Approaches and long-term results in 144 patients. *J. Neurosurg.* **73**, 3–11.

Younis, G., Sawaya, R., DeMonte, F., Hess, K. R., Albrecht, S., and Bruner, J. M. (1995). Aggressive meningeal tumors: review of a series. *J. Neurosurg.* **82**, 17–27.

Yung, W. K. A., Prados, M. D., Yaga-Tur, R., Rosenfeld, S. S., Brada, M., Friedman, H. S., Albright, R., Olson, J., Chang, S. M., O'Neill, A. M., Friedman, A. H., Bruner, J., Yue, N., Dugan, M., Zaknoen, S., and Levin, V. A., for the Temodal Brain Tumor Group. (1999). Multicenter phase II trial of temozolomide in patients with anaplastic astrocytoma or anaplastic oligoastrocytoma at first relapse. *J. Clin. Oncol.* **17**, 2762–2771.

Yung, W. K. A., Albright, R. E., Olson, J., Fredericks, R., Fink, K., Prados, M. D., Brada, M., Spence, A., Hohl, R. J., Shapiro, W., Glantz, M., Greenberg, H., Selker, R. G., Vick, N. A., Rampling, R., Friedman, H., Phillipps, P., Bruner, J., Yue, N., Osoba, D., Zaknoen, S., and Levin, V. A. (2000). A phase II study of temozolomide vs. procarbazine in patients with glioblastoma multiforme at first relapse. *Br. J. Cancer* **83**, 588–593.

Zeltzer, P. M., Boyett, J. M., Finlay, J. L., Albright, A. L., Rorke, L. B., Milstein, J. M., Allen, J. C., Stevens, K. R., Stanley, P., Li, H., Wisoff, J. H., Geyer, J. R., McGuire-Cullen, P., Stehbens, J. A., Shurin, S. B., and Packer, R. J. (1999). Metastasis stage, adjuvant treatment, and residual tumor are prognostic factors for medulloblastoma in children: conclusions from the Children's Cancer Group 921 randomized phase III study. *J. Clin. Oncol.* **17**, 832–845.

CHAPTER 63

# Primary and Secondary Lymphoma of the Central Nervous System

Martin Begemann, Martin Schabet, and Lisa M. DeAngelis

---

Patients with malignant lymphoma frequently have nervous system involvement. This can take the form of primary central nervous system (CNS) lymphoma or metastatic disease from systemic lymphoma. Both primary and metastatic lymphomas are almost always non-Hodgkin's lymphomas (NHL); Hodgkin's disease is rare, either metastatic or primary. Other neurological complications in patients with systemic NHL are common, including paraneoplastic syndromes and opportunistic infections, but they are not the focus of this chapter.

## PRIMARY CENTRAL NERVOUS SYSTEM LYMPHOMA

### Pathology, Cytogenetics, and Molecular Biology

Originally described by Bailey in 1929 as perithelial sarcoma, denoting its close proximity to the vascular endothelium, primary CNS lymphoma (PCNSL) also appears in the older literature as perivascular sarcoma, malignant reticuloendotheliosis, microglioma, and reticulum cell sarcoma. Since 1994, the Revised European-American Lymphoma (REAL) Classification system identifies lymphomas as being B-cell, T-cell, and Hodgkin's disease (for review see Morgello 1995). PCNSL is a B-cell tumor in almost all cases, with primary T-cell neoplasms occurring in <2% of patients. According to the working formulation, most PCNSLs are large cell or large cell immunoblastic subtypes. PCNSL is usually located in the basal ganglia, corpus callosum or cerebellum and is commonly in contact with either the ependyma of the ventricles or the subarachnoid space (Nishiura *et al.*, 1980; Shibata, 1989). Periventricular growth of PCNSL in the absence of focal masses has also been described at autopsy (Miller *et al.*, 1994).

In immunosuppressed patients with PCNSL, Epstein-Barr virus (EBV) can be detected in the CSF or in parenchymal tumor tissue in all patients (Castagna *et al.*, 1997; Cinque *et al.*, 1993). In these patients, PCNSL develops from an EBV-driven B-cell clone that comes from latently infected immortalized B-cells whose proliferation is normally controlled by T-cell–mediated immunity. However, no viral origin for nonimmunosuppressed patients has been detected. Cytogenetic and molecular analyses have been carried out on PCNSL samples. Cytogenetic studies (Itoyama *et al.*, 1994) showed abnormalities of chromosomes 1 (1q2l), 6 (6q15; 6q2l), 7 (7q15), and 14 (14q24; 14q32). Comparative genomic hybridization (CGH) confirmed genomic imbalances in 95% of PCNSLs; most common were losses of 6q (47%; with a common deletion of 6q21–q22) and gains of 12q (63%), 18q (37%), and 22q (37%). Other gains involved 1q, 9q, 11q, 12p, 16p, and 17p (each 26%; Weber *et al.*, 2000).

Studies on p53 yielded conflicting results: Overexpression (Nozaki *et al.*, 1998) and mutations (Koga *et al.*, 1994) of p53 were identified in some cases of PCNSL. Other studies report that p53 was undetectable (Deckert-Schlüter *et al.*, 1998) or rarely mutated (Cobbers *et al.*, 1998). Furthermore, bcl-2 and bcl-6 were overexpressed in 20%–50% of PCNSL (Cobbers *et al.*, 1998; Deckert-Schlüter *et al.*, 1998; Nozaki *et al.*, 1998), but overexpression was not seen in p53-expressing PCNSL tumors (Nozaki *et al.*, 1998). This indicates that one of several central pathways that govern apoptosis is affected in most PCNSLs. In fact, in a recent study 66% of PCNSLs contained ≤10% apoptotic cells detectable by TUNEL (terminal deoxynucleotidyl transferase-mediated UTP-nick-end labeling assay, Deckert-Schlüter *et al.*, 1998). In addition to p53, other evidence points to abnormal control of the $G_1$-S cell cycle checkpoint in PCNSL. About 50% of PCNSL samples from nonimmunocompromised patients show deletions of p15 and p16 (Kumanishi *et al.*, 1996) or deletions or methylations of 5' CpG island methylation of p16 (Cobbers *et al.*, 1998; Zhang *et al.*, 1998). The expression of p16 was examined by reverse transcriptase–mediated polymerase chain reaction (RT-PCR) and immunocytochemistry: Immunoreactivity for p16 was either absent or restricted to single tumor cells intermingled among negative tumor cells; mRNA expression was absent in 7 of 11 PCNSL and weakly present in 4 of 11 PCNSL (Cobbers *et al.*, 1998). P16 is known to inhibit and bind the cyclin D–regulated CDK4 and 6 kinases that control the transition into the late $G_1$ phase of the cell cycle. Inactivation of p16 by deletion would release that block, and the cell would progress into the S phase of the cell cycle. In a recent study, all PCNSLs were MIB-1 positive, and 53% of them showed a high proliferative activity, with more than 20% MIB-1–positive cells (Deckert-Schlüter *et al.*, 1998). What exactly regulates apoptosis and cell division in PCNSL is not yet known, but once identified, specific molecular abnormalities might be reasonable targets for future chemotherapy agents.

There was no evidence for expression of c-myc, mdm2, or EBV in the immunocompetent host (Itoyama *et al.*, 1994; Nozaki *et al.*, 1998). Bcl-x, bax were present rarely and bak was never detected in a recent series (Deckert-Schlüter *et al.*, 1998).

### Clinical Aspects

Most PCNSLs arise as brain tumors, and patients are initially seen with focal neurological deficits and symptoms and signs of increased intracranial pressure. About 10% of patients have diffuse periventricular growth, which mainly leads to personality changes. Rarely patients have symptoms and signs of diffuse leptomeningeal lymphoma such as psycho-organic syndrome, headache, ataxia, cranial nerve palsies, back pain, radicular pain, and sensory and motor deficits (for further details on leptomeningeal metastasis see Chapter 65). About 10% of patients have visual disturbances caused by infiltration of the vitreous body and the uvea. On slit-lamp examination yet another 10% show asymptomatic intraocular growth (DeAngelis *et al.*, 1990; Herrlinger *et al.*, 1999; Hochberg *et al.*, 1988). Spinal cord involvement is very rare, and is usually manifested by paraparesis at a sensory level. Table I summarizes our recommendations for evaluation of presumptive PCNSL.

**TABLE I    Evaluation For Diagnosis of Presumed Primary CNS Lymphoma**

Immunocompetent patients
  Required
    MRI with and without contrast of the brain
    Lumbar puncture with cytology for malignant cells and
      $\beta_2$-microglobulin
    Serum LDH and $\beta_2$-microglobulin
    Slit-lamp examination of the eyes
    HIV test
  Consider but not required
    CT chest, abdomen, and pelvis
    Bone marrow aspiration
Additionally in immunocompromised host
  SPECT/PET scan
  CSF for EBV DNA by PCR

On plain CT scan cerebral lymphomas are usually isodense or hyperdense lesions. They occur in the subcortical white matter, often in proximity to the ventricles. Lesions are solitary in 60%–70% of cases, and multifocal, usually bilateral, in 30%–40%. Contrast enhancement, which is seen in more than 90% of cases, is mostly homogeneous with slightly irregular borders. Frequently, there is little surrounding edema compared with comparably sized metastases or malignant gliomas. Cases with diffuse periventricular growth show subependymal contrast enhancement. As usual, MRI gives better delineation of the lesions and edema than CT (for review Herrlinger *et al.*, 1999). Figure 1 shows a typical MRI of a patient with PCNSL at presentation and after treatment. The radiological signs of leptomeningeal tumor growth are discussed in Chapters 62 and 65.

In cases with diffuse periventricular or leptomeningeal tumor growth, CSF protein content and lactate levels are almost always increased. However, mild elevation of CSF protein concentration (<100 mg/dl) is present in 80%–90% of all patients with PCNSL. The identification of malignant cells may be difficult, especially when cell counts are low or a reactive lymphocytosis is also present. Because most PCNSLs represent B-cell lymphomas, the demonstration of a monoclonal B-cell population in the CSF by immunocytochemistry or immunoglobulin gene rearrangement analysis may facilitate the diagnosis. At presentation, malignant cells are found in 20%–30% and suspicious cells in 30% of cases. After complete remission under therapy, leptomeningeal recurrence with subarachnoid seeding of tumor cells occurs in 40% of cases (Hochberg *et al.*, 1988). Leptomeningeal relapse may accompany brain or ocular recurrence or be isolated (DeAngelis *et al.*, 1990; Hochberg *et al.*, 1988; Chapter 65). In immunosuppressed patients, identification of EBV DNA in the CSF by PCR can establish the diagnosis (Antinori

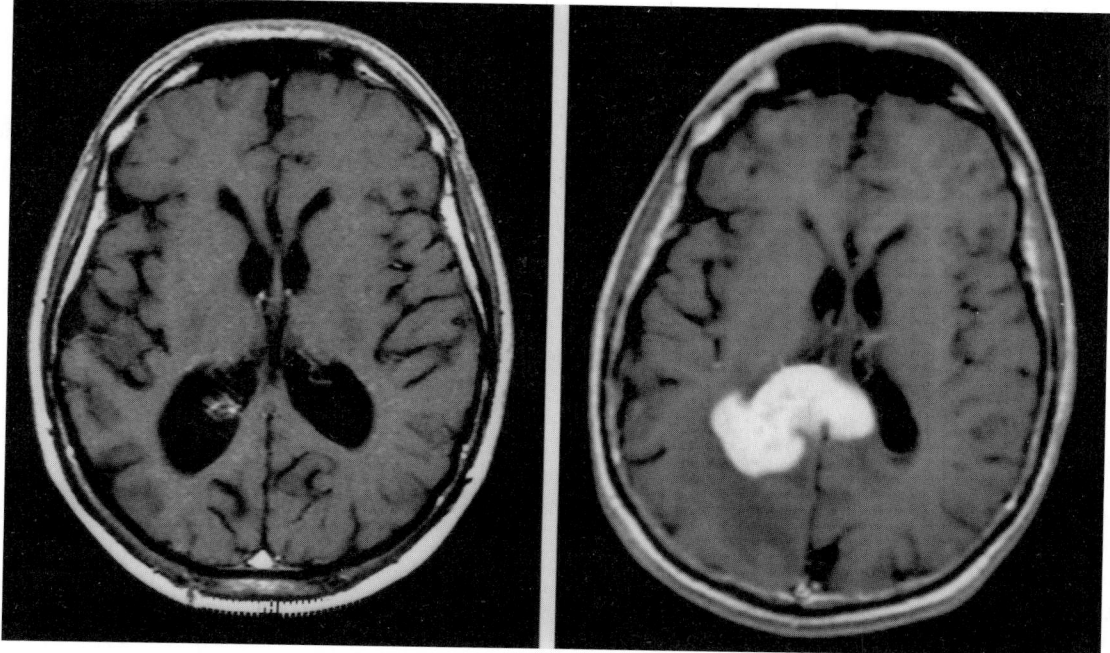

**FIGURE 1** Gadolinium-enhanced MRI scans demonstrating complete response of PCNSL to high-dose methotrexate, lomustine, procarbazine, and vincristine. The diffuse enhancement and periventricular location of the mass is characteristic for PCNSL.

et al., 1999ab; Castagna et al., 1997; Cingolani et al., 2000; Cinque et al., 1993).

The differential diagnosis of cerebral lymphoma includes malignant glioma and metastases, although these lesions rarely enhance diffusely and usually have a necrotic center. Focal inflammatory lesions (e.g., toxoplasmosis) also have to be considered, particularly in the AIDS population (Baumgartner et al., 1990; Chamberlain and Kormanik, 1999; Chapter 51). Other differential diagnoses that may be considered are sarcoidosis, multiple sclerosis, CNS involvement of systemic lupus erythematosus, various vasculitides, and leukoencephalopathies (DeAngelis, 1990). Radiological findings atypical for lymphoma include calcification or cyst formation. Hematoma may be seen in immunosuppressed patients but is rare in immunocompetent individuals. PCNSL shows a reduction of tumor volume after corticosteroids in 20%–60% of patients (Hochberg et al., 1988, 1991; Macdonald et al., 1995). Even long-lasting remissions of PCNSL have been seen after corticosteroids (Herrlinger et al., 1996). The diagnosis of PCNSL should rely on a stereotactic brain biopsy or the demonstration of malignant lymphocytes in the CSF or vitreous in cases of ocular involvement. When patients have initially been treated with corticosteroids, biopsy often proves negative. In these cases diagnosis may be secured after discontinuation of therapy and biopsy of a recurrent lesion. In some cases

diagnosis of lymphoma has to rely on the exclusion of other causes and the corticosteroid sensitivity of the lesion itself as evidenced on imaging (DeAngelis et al., 1999; Herrlinger et al., 1999; Hochberg et al., 1988).

## Natural Course

CNS lymphomas account for about 0.9%–4.2% of all extranodal non-Hodgkin's lymphomas and for 0.85%–6.6% of all primary CNS tumors in 1999 autopsy studies, with the higher percentage seen in the more recent series (Hao et al., 1999; Jellinger et al., 1975). However, the incidence of PCNSL has tripled over the past 20 years, and presently it accounts for 4% of all intracranial neoplasms. Review of the surveillance, epidemiology, and end results (SEER) showed a continued increase in PCNSL after introduction of MRI even in low HIV risk registries, suggesting a true increase in the immunocompetent population (O'Neill et al., 2001).

CNS lymphomas may arise at any age. In immunocompetent patients, the median age at diagnosis is 55 years, and the median interval from first symptom(s) to diagnosis is 2–3 months. Three quarters of these patients are initially seen between 45 and 70 years of age. Even though rare in childhood, PCNSL has also been described in unimmunosuppressed children

(Baleydier *et al.*, 2001; Dahlborg *et al.*, 1996, 1998; Kai *et al.*, 1998; Silfen *et al.*, 2001).

Patients with AIDS have a risk of 2%–6% and are initially seen at a median age of 37–39 years. In addition to AIDS, predisposing conditions for PCNSL include immunosuppressive therapies (transplant patients), congenital immunodeficiencies (e.g., ataxia-teleangiectasia), or acquired immunodeficiencies (e.g., systemic lupus erythematosus, tuberculosis; for summary see Table II). The risk of PCNSL amounts to 1%–2% for patients with renal transplants, representing about 50% of lymphomas in this patient group. CNS lymphomas arise at a median of 9 months and a range of 5–46 months after kidney or heart transplantation, depending on the intensity of immunosuppressive therapy (Penn and Porat, 1995). PCNSL has also been described as a secondary malignancy (DeAngelis, 1991). The risk from less intensive immunosuppression, as usually applied in autoimmune disease, is unknown. In AIDS patients, PCNSL may occur at any stage of the disease but tends to occur late, and patients usually have CD4 counts well below 100/mm$^3$. With the increase of AIDS, primary CNS lymphoma was estimated to become the most frequent CNS tumor (Eby *et al.*, 1988; Hochberg *et al.*, 1988; Schabet, 1999); however, the availability of highly active antiretroviral therapy (HAART) has dramatically reduced the incidence of PCNSL in this population (Ammassari *et al.*, 2000; Grulich, 1999; Jacobson *et al.*, 1999; Jones *et al.*, 1999; Rabkin *et al.*, 1999; Sparano *et al.*, 1999).

Immunocompetent patients have a median survival of 3 months without therapy, which increases to 5 months with surgery alone and to about 1 year with the addition of radiotherapy. Whole-brain radiotherapy gives a 2-year survival of 15%–30% and a 5-year survival of 3%. Prognosis is best in younger patients with solitary lymphoma and without initial leptomeningeal dissemination. In a prospective radiation therapy oncology group (RTOG) study using 40 Gy to the whole brain and a boost of 20 Gy to the primary lesion, age and Karnofsky performance score (KPS) were the most important prognostic factors. Patients older than 60 years or with a KPS of less than 60 had a median survival of about 6 months compared with 22 months for younger patients in better condition (Nelson *et al.*, 1992; Schultz *et al.*, 1996; Table III). Survival is further increased by adding aggressive chemotherapy; 2-year survival has been reported for 54%–70% and 5-year survival for up to 45% of patients (Table III). However, as many as 50% of patients have severe dementia develop (DeAngelis, 1999). Recently, chemotherapy alone has been shown to yield comparable survival to radiotherapy alone (Table III). Several trials are under way to clarify the role of chemotherapy and radiotherapy and the combination of both regarding disease control and late neurotoxicity (Hoang-Xuan and Delattre, 1999).

Recurrence of neurological signs and symptoms usually heralds regrowth of tumor but may indicate leukoencephalopathy or radiation necrosis (Chapters 64 and 65). Fewer than 10% of patients have systemic metastasis at autopsy, and most have only microscopic disease that is asymptomatic. Most patients, however, eventually die from recurrence of tumor in the CNS (Table III; Chapters 64 and 65).

Patients with AIDS have a median survival of 1–2 months when treated symptomatically. With whole-brain irradiation, 50% of patients survive 2–5 months, and 10% patients survive 9 months. The few patients eligible for radiochemotherapy survive 10–12 months (Table IV). Two thirds of patients with AIDS-related PCNSL eventually die from opportunistic infections, including CNS infections, and not from recurrent PCNSL. Recent data suggest that the introduction of highly active antiretroviral therapy (HAART) with antiviral therapy directed against EBV can induce sustained remissions from AIDS-related PCNSL.

There are several case reports of spontaneous remission of PCNSL (Al-Yamany *et al.*, 1999; Heinzlef *et al.*, 1996; Sugita *et al.*, 1988) some of which have occurred in the context of AIDS (Daniels *et al.*, 1992). These spontaneous regressions are different than the regressions observed after steroids, which are due to the oncolytic effects of steroid on lymphoma cells (Homo-Delarche *et al.*, 1984). Spontaneous regressions of systemic lymphomas have been ascribed to natural killing activity in nonimmunocompetent patients (Ono *et al.*, 1996). In fact, natural killing activity was greater in patients with spontaneous regression of lymphoma than in healthy controls (Ono *et al.*, 1996). Presumably, the mechanisms of spontaneous regressions in PCNSL are similar but have not been studied. Long-term remission of AIDS-related PCNSL while on HAART in the

**TABLE II    PCNSL in Immunocompromised Host**

Congenital immunodeficiency syndromes
  Wiskott-Aldrich syndrome
  Ataxia-teleagiectasia
  Severe combined immunodeficiency
  Dysglobulinemia with T-lymphocyte depression
  X-linked immunoproliferative disorders
AIDS
Transplantation (e.g., organ transplantation)
Miscellaneous pre-existing conditions
  Sjögren's syndrome
  Rheumatoid arthritis
  Systemic lupus erythematosus
  Sarcoidosis
  Vasculitis
  Tuberculosis

TABLE III    Treatment Results in Immunocompetent Patients with Primary Non-Hodgkin's Lymphoma of the CNS (Number of Treated Patients In Brackets)

| Reference | No. pats. | WBRT (Gy) | Chemotherapy | Survival Median (months) | 2-year (%) |
|---|---|---|---|---|---|
| Hochberg and M., 1988 | 61 | 30–60* | HD MTX | 14 | 40 |
| Pollak et al., 1989 | 27 | 30–60 | HD CHOP (5) MTX + IT MTX | 24 | 54 |
| Nelson et al., 1992 | 41 | 60 | | 12 | 28 |
| Chamberlain & Levin, 1992 | 16 | 55–62 | HU, PCV | 41 | 70 |
| DeAngelis et al., 1992 | 31 | 40–54 | HD + IT MTX | 43 | 76 |
| | 16 | | HD Ara-C | 22 | 43 |
| Ferracini et al., 1993 | 147 | 45–50 | CVP (80) | 10 | 18 |
| | | | | 12 (RT) | 30 |
| | | | | 14 (RT + CVP) | 34 |
| Glass et al., 1994 | 25 | 30–44 | HD MTX | 33 | 60 |
| Schultz et al., 1996 | 52 | 41–59 | | 16 | 42 |
| | | | | | 30 |
| Dahlberg et al., 1996 | 39 | | MTX IA + CPM + PCZ | 41 | NA |
| Guha-Thakurta et al., 1999 | 31 | | HD MTX | 30 | 63 |
| Ng et al., 2000 | 10 | | HD MTX | 36 | NA |
| O'Brien et al., 2000 | 46 | 45–54 / 36 | HD MTX | 33 | 62 |
| Herrlinger et al., 2000 | 28 | 50 | HD MTX, IT MTX HD Ara-C (5) | 11 (all) / 41 (RT + chemo.) | |

**Abbreviations:** WBRT = whole brain radiotherapy, NA = not available, IT = intrathecal/intraventricular, IA = intraarterial, HD MTX = high dose methotrexate, Ara-C = cytarabine, CHOP = cyclophosphamide, doxorubicin, vincristine, prednisone, HU = hydroxyurea, PCV = procarbazine, CCNU, vincristine, CVP = cyclophosphamide, vincristine, prednisone, CPM = cyclophosphamide, PCZ = procarbazine.
*48 patients had 40 Gy neuraxis RT.

absence of chemotherapy or radiation therapy might be related to activation of the immune system (McGowan and Shah, 1998; Raez et al., 1999). In fact, reconstitution of NK cell activity in HIV-1–infected individuals while receiving HAART has been reported (Weber et al., 2000). Since the introduction of HAART, a decrease in the incidence in AIDS-PCNSL was seen in several studies (Ammassari et al., 2000; Grulich, 1999;

Jacobson et al., 1999; Jones et al., 1999; Rabkin et al., 1999; Sparano et al., 1999).

## Principles of Therapy

Important clinical studies regarding therapy of PCNSL in immunocompetent patients are summarized

TABLE IV    Treatment Results in HIV Positive Patients with Primary Non-Hodgkin's Lymphoma of the CNS (Number of Treated Patients in Brackets)

| | No. pats. | WBRT (Gy) | Chemotherapy | Survival median in mo. |
|---|---|---|---|---|
| Formenti et al., 1989 | 10 | 20–50 (10) | MTX iv | 2, 5 |
| Baumgartner et al., 1990 | 55 | 40 (29) | | 1 (without RT) |
| | | | | 4 (with RT) |
| Ling et al., 1994 | 41 | 38–62 (35) | | 3 |
| Forsyth et al., 1994 | 10 | 40 (9) | MTX it/iv (9) | 3, 5 |
| | | 54 (9) | PCZ, TTP (8) | |
| | | | MTX, PCZ, TTP (1) | |
| Chamberlain and K., 1999 | 67 | 30 (52) | PCZ, CCNU, VCR (7) | 1, 5 (without RT) |
| | | | | 4 (with RT) |
| | | | | 13 (with RT and chemotherapy) |

**Abbreviations:** B = biopsy, WBRT = whole brain radiotherapy, RT = radiotherapy, it = intrathecal/- ventricular, iv = intravenous, MTX = methotrexate, PCZ = procarbazine, TTP = thiotepa, VCR = vincristine, CCNU = lomustine.

in Table III (based on retrospective or nonrandomized studies). Surgical resection alone increases median survival by only 2 months. Radical surgery of the usually deep-seated CNS lymphomas carries a risk of severe neurological deficits. It is generally agreed that stereotactic biopsy is adequate to secure diagnosis before the start of definitive treatment.

Radiotherapy is delivered to the whole brain because of the multifocal growth and spread of PCNSL. It significantly increases median survival to 12–18 months (Table III). Different authors observed a threshold dose of 30–50 Gy but no clear dose dependency of survival. A tumor boost does not improve local control (Bataille et al., 2000; Nelson, 1999), because relapse is as frequent within the boosted field as outside of it. Some authors have recommended radiation of the neuraxis with 24–40 Gy to treat the frequently encountered leptomeningeal seeding; however, there is no evidence that neuraxis RT improves survival, and it can compromise subsequent chemotherapy administration. Recently, systemic chemotherapy with drugs or regimens that cross the blood–brain barrier such as procarbazine, lomustine, vincristine (PCV), high-dose methotrexate, and/or high-dose cytarabine has improved the outcome of PCNSL patients. Combined radiochemotherapy using preirradiation intravenous high-dose methotrexate ($>1\,g/m^2$) with or without postirradiation high-dose cytarabine ($>3\,g/m^2$) has improved survival compared with radiotherapy alone but has about a 50% incidence of severe late neurotoxicity (DeAngelis, 1999; Schlegel et al., 1999; Table III). Most authors believe that intravenous methotrexate with doses exceeding $3\,g/m^2$ is sufficient to treat the CSF compartment, obviating the need for supplemental intrathecal treatment, but this has not been established. High-dose intravenous methotrexate has been shown to be well tolerated in the elderly, resulting in high response rates (Abrey et al., 2000; Ng et al., 2000) and sparing patients from cognitive adverse effects seen often after radiation therapy (DeAngelis, 1999). Recently, a methotrexate-based chemotherapy using osmotic blood–brain barrier disruption was developed showing longer survival and improved neuropsychological performance compared with the group that underwent cranial irradiation (Dahlborg et al., 1996). Similarly, high-dose methotrexate with procarbazine and vincristine yielded similar survival times with or without radiation therapy in patients older then 60 years (Abrey et al., 2000).

In patients with recurrent PCNSL after intravenous high-dose methotrexate treatment and radiotherapy, long-lasting remission has been observed with PCV treatment alone (Herrlinger et al., 2000).

There are few studies regarding therapy of PCNSL in patients with AIDS (Table IV). Radiotherapy improves median survival by 2–3 months. A few selected patients can tolerate chemotherapy and survive longer. As reviewed previously, a few reports suggest that HAART may lead to long-lasting remission in AIDS-related PCNSL (McGowan and Shah, 1998; Raez et al., 1999). Furthermore, hydroxyurea ($400–700\,mg/m^2$/day) has been shown to lead to remission in AIDS-related PCNSL (Slobod et al., 2000). Hydroxyurea supposedly eradicates episomal EBV from tumor cells. However, prospective studies to confirm the use of HAART or hydroxyurea for the treatment of PCNSL are not yet available. On the other hand, high-dose methotrexate regimens with or without cranial irradiation are effective in a subset of patients with PCNSL related to AIDS (Chamberlain and Kormanik, 1999; Forsyth et al., 1994; Jacomet et al., 1997).

## Practical Management

In patients who have CNS lesion(s) suspicious for lymphoma, corticosteroids should be withheld until histological confirmation. If edema warrants treatment, osmotic drugs such as mannitol should be used. The diagnostic procedure of choice is a stereotactic biopsy. However, diagnosis may also rely on the demonstration of a monoclonal B-cell population in the CSF by immunocytochemistry or immunoglobulin gene rearrangement analysis. Patients with AIDS should undergo lumbar puncture with screening of CSF for EBV by means of PCR, as well as either Thallium-201 SPECT or FDG-PET scanning (Table I). Combined positive PCR for EBV on CSF *and* Thallium-201 SPECT or FDG-PET scans have 100% specificity and positive predictive value for PCNSL in AIDS (Antinori et al., 1999ab).

After confirmation of the diagnosis, corticosteroid treatment should be started with dexamethasone (e.g., 16 mg/day). Clinical stabilization with or without tumor regression is usually achieved with corticosteroids within a few days. Dexamethasone treatment may be maintained and then slowly tapered.

Patients without immunodeficiency or renal insufficiency should receive high-dose methotrexate-containing chemotherapy before radiotherapy. We currently give five cycles of systemic methotrexate $3.5\,g/m^2$ every 14 days in combination with vincristine ($1.4\,mg/m^2$ intravenously with a maximum of 2.8 mg) and procarbazine ($100\,mg/m^2$/day orally for 7 days during cycles 1, 3, and 5). Others use the high-dose methotrexate as a single agent (Ng et al., 2000; Sandor et al., 1998). High-dose cytarabine may follow the methotrexate. In complete responders no further therapy may be justified, especially in older patients. However, incomplete or partial responders who are young, as well as in stable disease, and any patient with progressive disease should receive whole-brain irradiation to 45 Gy. Patients with positive CSF cytological findings should receive intrathecal chemotherapy concurrently with systemic chemotherapy.

At relapse, PCV, high-dose cytarabine, temozolomide, or thiotepa are options (Herrlinger *et al.*, 2000). Another regimen for recurrent or refractory PCNSL contained dexamethasone, VP 16, ifosfamide, and carboplatin (Takasu *et al.*, 2000) but had a high relapse rate. Intrathecal therapy is indicated in patients with leptomeningeal relapse. Details of the intrathecal chemotherapy, including recommended single and maximal cumulative doses of cytostatics and their side effects and the side effects of radiotherapy, are discussed in Chapter 65.

Patients in poor general condition may receive cranial irradiation before or without systemic chemotherapy for palliative care. These patients should not receive high-dose methotrexate after radiotherapy because of the potential for enhanced neurotoxicity, but PCV as adjuvant therapy can prolong survival (Chamberlain and Levin, 1992; Herrlinger *et al.*, 2000). Treatment for patients with AIDS is decided on an individual basis. Whole-brain irradiation is used for palliative treatment. High-dose methotrexate-based chemotherapy, or a combination of antiviral and antiretroviral therapy, is appropriate for patients in good clinical condition.

Recently, several authors reported the use of high-dose chemotherapy with hematopoietic stem cell transplantation. In one study 22 patients with refractory or recurrent PCNSL were treated with intensive chemotherapy with thiotepa, busulfan, and cyclophosphamide followed by hematopoietic stem cell rescue (Soussain *et al.*, 2001). Patients underwent intensive chemotherapy only if they had responded to salvage treatment with cytarabine and etoposide. After completion of treatment, the probability of survival at 3 years was 60%. This regimen proved feasible and effective but was toxic in patients older than 60 years of age. Another study reported two patients with PCNSL (one leptomeningeal, one with cauda equina and brain parenchymal involvement) who received carmustine (15 mg/kg to a maximum of 550 mg/m²) or 12 Gy fractionated total-body irradiation, followed by etoposide (60 mg/kg) and cyclophosphamide (100 mg/kg) with transplantation of G-CSF–mobilized and purged peripheral blood progenitor cells (Alvarnas *et al.*, 2000). These patients also received intrathecal treatment with methotrexate and cytarabine or both. Both patients remain alive and in remission 1085 and 3704 days after transplantation (Alvarnas *et al.*, 2000). Intense chemotherapy followed by hematopoietic stem-cell rescue seems feasible and effective in a subset of patients with primary CNS lymphoma. Finally, a recent case report described remission by a graft-versus-lymphoma effect after allogeneic peripheral blood stem cell transplantation for recurrent PCNSL (Varadi *et al.*, 1999). Whether this graft-versus-lymphoma effect can be used for the treatment of PCNSL has not yet been studied systematically.

Because the optimal treatment for PCNSL has not been established, patients should be considered for clinical trials. Patients and caregivers can find information about clinical studies through the following websites: *http://clinicaltrials.gov/ct/gui/c/b*, *http://cnetdb.nci.nih.gov/trialsrch.shtml*, as well as *http://cancertrials.nci.nih.gov* or by 1-800-4-CANCER.

## SECONDARY CNS MANIFESTATION OF SYSTEMIC NON-HODGKIN'S LYMPHOMA

### Clinical Aspects

Secondary CNS involvement from systemic NHL occurs in the leptomeninges 75% of the time. Symptoms and signs include psycho-organic syndrome, headache, ataxia, cranial nerve palsies, back pain, radicular pain, sensory and motor deficits (for further details on leptomeningeal metastasis see Chapter 65). The remaining 25% of patients have epidural metastases, and intracerebral metastases occur in <1%. In 50% of patients with epidural metastasis, it is the first manifestation of the systemic disease. The clinical aspects of spinal and cerebral lesions are discussed in Chapters 62 and 64.

### Natural Course

The incidence of systemic NHL is 5/100,000, and the peak age of onset lies between 45 and 50 years old. Approximately 5%–15% of all patients, especially younger ones with lymphoblastic or diffuse undifferentiated histological findings, have CNS manifestations develop during their disease. In a series of 462 children with noncleaved-cell lymphoma, 49 had CNS disease at diagnosis (Gururangan *et al.*, 2000). In a series of 152 patients with NHL, excluding lymphoblastic lymphoma and small noncleaved-cell lymphoma, 12 patients had CNS involvement, mostly patients with high serum LDH concentration (Tomita *et al.*, 2000). In 50% of patients with CNS involvement neurological symptoms and signs develop at the time of systemic progression with seeding of the bone marrow, the testes, and the peripheral blood. The median interval between the diagnosis of systemic lymphoma and CNS disease ranges from 6–9 months. In the remaining half, they are present at diagnosis (Liang *et al.*, 1989, 1990;

Mackintosh *et al.*, 1982; Mead *et al.*, 1986; Recht *et al.*, 1988).

In general, patients with leptomeningeal metastasis have a median survival of only 3 months (see Table V and Chapter 65); however, patients with an isolated CNS relapse who are younger than 30 years of age and are treated aggressively have a median survival of 22 months (Mackintosh *et al.*, 1982). Also, children with small noncleaved-cell lymphoma who had CNS disease at diagnosis have a 45% 3-year event-free survival with appropriate therapy (Gururangan *et al.*, 2000). Median survival of patients with epidural metastasis has been reported to range from 12–18 months (Haddad *et al.*, 1976; Mead *et al.*, 1986; Raz *et al.*, 1984). Most patients with nervous system metastases from NHL die from progressive systemic disease, whereas progressive CNS lymphoma is the cause of death in only about 20% of patients (Hoemi-Simon *et al.*, 1987; Liang *et al.*, 1989; Recht *et al.*, 1988).

## Principles of Therapy

Results from representative clinical studies are detailed in Table V (also see Chapter 65). Patients with parenchymal metastases are considered in the clinical series of patients with brain metastases from solid tumors (see Chapter 64). The basic therapy of leptomeningeal metastasis of lymphoma is intrathecal chemotherapy with methotrexate or cytarabine and irradiation of the clinically involved level of the neuraxis. A standard treatment plan for secondary CNS complications in NHL is summarized in Table VI. Radiation doses between 20 and 40 Gy have an equivalent clinical outcome (Mackintosh *et al.*, 1982). Mackintosh *et al.* (1982) have shown an increase of median survival from 8.5 to 22 months by combined whole-brain irradiation and intrathecal chemotherapy compared with radiotherapy alone in patients who are less than 30 years old and have no evidence of progressive systemic

**TABLE V** Treatment Results in Patients with Secondary CNS Manifestations of Systemic Non-Hodgkin's Lymphomas of the CNS (Number of Treated Patients in Brackets)

| Author | No. pat. | Therapy radiotherapy (Gy) | Chemotherapy | Survival median (mo.) | 2-Year survival (%) |
|---|---|---|---|---|---|
| **Leptomeningeal** | | | | | |
| Griffin *et al.*, 1971 | 21 | W 30 (8) T 15–30 (?) | MTX IT | 3 | 5 |
| Young *et al.*, 1979 | 38 | ?? | MTX/Ara-C IT | 3 | 8 |
| Herman et al., 1979 | 49 | ?? | MTX/Ara-C IT | 2 | 4 |
| Mackintosh *et al.*, 1982 | 97 | W 20–40 (66) | MTX/Ara-C IT (64) MTX IV (9) | 2 22[a] | 8 40 |
| Ersboll *et al.*, 1985 | 38 | W 20–? (26) | MTX/Ara-C IT | 2.5 | |
| Hoerni-Simon *et al.*, 1987 | 30 | W 17–36 (9) | MTX IT (20) ? IV (24) | 3.5[b] | |
| Recht *et al.*, 1988 | 90 | W 24–30 (68) | MTX/ARA-C IT | 0.5[c] 2[d] 4[e] | 13 |
| Liang *et al.*, 1989 | 41 | W ? (34) S ? (2) | MTX/Ara-C IT (25) | 3 | 7 |
| Tomita *et al.*, 2000 | 12 | ?? | Systemic Chemo. | 4.5 | |
| Alvarnas *et al.*, 2000 | 13 | ?? | Systemic Chemo. | NA | 46% 5-year event-free survival |
| Gururangan *et al.*, 2000 | 49* | W 8-26 (36) | MTX/Ara-C IT | NA | 45% 3-year event-free survival |
| **Epidural Spinal Metastasis** | | | | | |
| Haddad *et al.*, 1976 | 72 | 35–40 | | 14 | 42 |
| Raz *et al.*, 1984 | 7 | 40 | | 12 | 2/7 |
| Mead *et al.*, 1986 | 9 | 30–40 | Systemic Chemo. (3) | 18 | 4/9 |

**Abbreviations:** W = whole brain, S = spinal axis, = tumor region, ? = not indicated, it = intrathecal/-ventricular, iv = intravenous, MTX = methotrexate. Systemic chemo. = combination regimen for systemic disease.
[a] patients under 30 years of age without progressive systemic disease.
[b] mean survival time.
[c] untreated.
[d] with radiotherapy.
[e] with intrathecal/-ventricular MTX therapy with/without radiotherapy.
*CNS disease at first diagnosis.

TABLE VI   Standard Treatment Plan of Metastatic NHL to the Nervous System

Leptomeningeal metastasis
  Intraventricular/intrathecal chemotherapy
  Irradiation of neurological focus with 30 Gy
  Systemic chemotherapy
Intracerebral secondary lymphoma
  Corticosteroids
  Systemic chemotherapy
  Whole-brain irradiation with 30 Gy (2–3 Gy single fraction, respectively)
Epidural secondary lymphoma
  Corticosteroids
  Systemic chemotherapy
  Irradiation of spinal focus with 30 Gy

disease. Intradural spinal lesions have been effectively irradiated with focal doses of 15–30 Gy. Recht et al. (1988) advocate intrathecal chemotherapy alone (without radiotherapy) for patients with minor neurological symptoms and signs. Several authors have reported a better outcome for patients who have been treated by means of an Ommaya reservoir compared with patients who had lumbar intrathecal chemotherapy (Mackintosh et al., 1982; Raz et al., 1984; Recht et al., 1988).

In most cases intrathecal chemotherapy leads to remission, with at least temporary clearance of malignant cells from the CSF. However, intrathecal chemotherapy must be avoided in patients with obstruction of CSF flow. This may be evident on cranial CT or MRI in patients with hydrocephalus. Hydrocephalus can be relieved by ventriculoperitoneal shunt with resolution of symptoms of increased intracranial pressure. However, intrathecal chemotherapy cannot be effectively administered after placement of a shunt, and therapy must be delivered by alternative means such as RT, systemic chemotherapy, or occasionally intrathecal chemotherapy by means of lumbar puncture. Some patients have symptoms of increased intracranial pressure, but enlarged ventricles are not seen on neuroimaging. These patients may also benefit from ventriculoperitoneal shunt. Apart from hydrocephalus, significant spinal lesions can also impair CSF flow. An assessment of CSF dynamics can be made with a CSF indium study, which should be obtained before the start of intrathecal chemotherapy in any patient where there is a concern that CSF flow is impaired. This is essential, because decreased efficacy and enhanced neurotoxicity are related to poor CSF flow dynamics. Because 80% of patients die from progressive systemic disease, some authors recommend additional systemic chemotherapy at the time of CNS treatment (Liang et al., 1989; Recht et al., 1988).

Some authors have treated leptomeningeal metastasis with systemic methotrexate in doses of several grams per day followed by leucovorin rescue. This approach has the theoretic advantage of penetrating areas of bulky tumor, reaching all areas of the subarachnoid space, and being independent of CSF flow dynamics. This therapy seems as effective as intrathecal chemotherapy, but has a higher rate of side effects, mainly myelosuppression. (Mackintosh et al., 1982; Skarin et al., 1977). The same holds true for systemic high-dose cytarabine (Amadori et al., 1984; Morra et al., 1993). Occasionally, nitrosoureas such as ACNU, BCNU, or CCNU have been used without impressive results (Freund et al., 1983; Mackintosh et al., 1982). Alvarnas et al. (2000) using high-dose therapy with hematopoietic cell transplantation achieved complete responses in 8 of 11 patients with secondary CNS disease. Actuarial 5-year event-free survival was 46%.

Spinal epidural metastasis with compression of the cord may be treated with surgery or radiotherapy using 30–40 Gy. Surgery may be important to confirm the diagnosis in those who have epidural spinal cord compression as the first manifestation of their disease. However, surgery followed by radiotherapy does not improve functional outcome or survival compared with radiotherapy alone. Therefore, we usually treat patients with radiotherapy only unless the diagnosis is unknown (Haddad et al., 1976; Mead et al., 1986; Raz et al., 1984). Alternately, some patients with epidural lymphoma can be successfully treated with systemic chemotherapy, thus minimizing the myelotoxic effects of spinal radiotherapy (Wong et al., 1996).

The early treatment of microscopic CNS disease—so-called CNS prophylaxis—has reduced the risk of CNS relapse from 50%–70% to 5%–15% in children with acute lymphocytic leukemia (Coleman et al., 1981). The effectiveness of "prophylactic" radiotherapy and chemotherapy has also been shown in children with NHL (Duque-Hammershaimb et al., 1983; Murphy and Hustu, 1980). CNS prophylaxis has not been evaluated in adults with NHL. Some authors recommend prophylactic CNS treatment only for patients with advanced diffuse undifferentiated lymphoma who have bone marrow involvement, epidural metastases, Waldeyer ring involvement, or otherwise extensive extranodal disease (Recht et al., 1988; Young et al., 1979). Others recommend this type of therapy for patients with lymphoblastic, diffuse undifferentiated, diffuse histiocytic lymphoma or small noncleaved-cell lymphoma (Haddy et al., 1991; Mackintosh et al., 1982). Usually, intrathecal lumbar or intraventricular chemotherapy is applied simultaneously with systemic chemotherapy (Recht et al., 1988).

## Practical Management

The diagnosis of leptomeningeal metastasis should be confirmed by the demonstration of malignant cells in the CSF or tumor nodules in the subarachnoid space on enhanced MR scan. If the anatomy of the patient permits, an Ommaya reservoir is inserted for intraventricular chemotherapy. We give 12 mg methotrexate or 50 mg cytarabine every 3–4 days for 3 weeks and then continue with decreasing frequency. If intraventricular chemotherapy is not possible, it can be administered by lumbar puncture, although this is a less effective route. Radiotherapy should be administered to the symptomatic site (e.g., the cauda equina for leg pain and weakness), usually for a total dose approximating 30 Gy in 3 Gy fractions (Table V; based on retrospective or nonrandomized studies). Cranial radiotherapy is not required for all patients with leptomeningeal lymphoma and should be reserved for those with cranial nerve or cerebral symptoms. To avoid side effects intrathecal chemotherapy should be held during or delayed until completion of radiotherapy. The details of intrathecal chemotherapy, including recommended dosages of cytostatics and their neurotoxicity, are outlined in Chapter 65 and the side effects of radiotherapy in Chapters 64 and 65.

Patients with intracerebral solid lesions are initially treated with high-dose corticosteroids (see earlier) unless there is a diagnostic question, and then steroids should be withheld until a biopsy has been obtained. Depending on the stage of the patient's systemic disease, these patients may be treated in a comparable fashion to those with PCNSL (see earlier) or according to high-dose treatment protocols with hematopoietic cell transplantation as used in systemic NHL (Alvarnas *et al.*, 2000). For those in poor clinical condition with advanced systemic disease, palliative whole-brain irradiation to a total dose of 30 Gy in 2–3 Gy fractions is more appropriate. Most of these patients have concurrent leptomeningeal involvement and should be treated with intrathecal chemotherapy (see Chapter 65).

Patients with known NHL who have a spinal epidural lesion leading to compression of the cord receive a single high dose of corticosteroid such as 100 mg dexamethasone and then 8 mg dexamethasone four times daily for the first week of radiotherapy. Radiotherapy is given to the tumor area for a total dose of 30 Gy in 3 Gy fractions. Corticosteroids are tapered off by the end of radiotherapy. Surgical decompression is indicated in patients with an unknown primary tumor and may occasionally be appropriate for a patient who has epidural relapse after prior radiotherapy (also see earlier and Chapter 68).

## CNS MANIFESTATION OF SYSTEMIC HODGKIN'S LYMPHOMA

### Clinical Aspects

Primary CNS Hodgkin's disease (HD) is very rare and reviewed elsewhere (Biagi *et al.*, 2000; Herrlinger *et al.*, 2000). The differentiation of HD from NHL is based on the demonstration of Reed-Sternberg cells, as well as the presence of CD30 antigen in conjunction with negative staining for leukocyte antigen (LCA, CD45) and negative staining for B and T cell markers. Remissions with median follow-up of 12 months (range, 6–18 months) were obtained with complete resection, whole-brain radiation, and chemotherapy (e.g., MTX, or procarbazine, lomustine, vincristine).

The primary neurological complication of HD is epidural spinal metastases, accounting for 90% of nervous system metastatic disease. Epidural spinal metastases are located in the thoracic spine in two thirds of cases and more frequently in the lumbar than in the cervical spine in the remaining patients. Patients have back pain and progressive paraplegia. Spinal epidural metastases usually are detected by MRI. Alternately, myelogram will demonstrate unilateral or circumferential extradural lesions that sometime lead to complete obstruction of CSF flow. CT may reveal adjacent paraspinal lesions. Plain x-ray films show vertebral body destruction in only 30% of cases and, therefore, are a poor diagnostic test. MRI should be done in any patient suspected of having an epidural tumor. Metastases within CNS parenchyma are rare. Usually, they can be attributed to cortical invasion from leptomeningeal metastases leading to cranial nerve palsies, focal neurological deficits, and signs and symptoms of increased intracranial pressure. Intracranial metastases are visualized by CT or MRI. Malignant cells are rarely found in CSF, even in cases with proven meningeal tumor growth (Sapozink and Kaplan, 1983; Shenoy *et al.*, 1987). The differential diagnosis for spinal paraplegia in Hodgkin patients, including radiation myelopathy, is discussed later and in Chapter 68.

### Natural Course

The overall incidence of HD is 2–6/100,000, with peak incidences in the third and after the fourth decade. Men are affected twice as often as women. Two thirds of patients with secondary CNS manifestations, which typically occur 2–5 years after initial diagnosis, are between 20 and 40 years of age. Four to 8% of patients with nodular sclerosing or mixed histological findings (classification of Lukes and Butler, 1966) have spinal

TABLE VII   Treatment Results in Patients with Secondary CNS Manifestations of Systemic Hodgkin's Lymphomas of the CNS (Number of Treated Patients in Brackets)

| Author | No. pat. | Localization | Therapy (no. pat.) | | | Survival | |
|---|---|---|---|---|---|---|---|
| | | | Surgery | Radiotherapy | Chemo. | Median | 2-year (%) |
| Haddad *et al.*, 1976 | 22 | epidural/spinal | S (22) | T 35–40 | | 5 yrs. | 80 |
| Schertel *et al.*, 1985 | 18 | epidural/spinal | S (5) | T 36 | iv[a] (3) | 2 yrs. | 50 |
| | 2 | intramedullary | | | iv[a] (3) | | |
| | 9 | cerebral/mening. | S (7) | W 45 | iv[a] (6) | 10 mo. | 15 |
| | 6 | intracerebral | | | | | |
| Sapozink and Kaplan, 1983 | 12 | cerebral | B (4) | W 20–40 | iv[b] (6) | 10 mo. | 25 |
| | | | | T 20–50 | MTX it (1) | | |

**Abbreviations:** Pat. = patients, B = biopsy, S = subtotal or complete tumor resection, W = whole brain, T = tumor region, it = intrathecal/-ventricular, iv = intravenous.
[a] cyclophosphamide, CCNU, BCNU, COPP- or ABVD-protocol.
[b] cyclophosphamide, CCNU, vinblastine, vincristine, chlorambucil, TOPP (Thiotepa)-, MOPP- or ABVD-protocol.

epidural metastases, but <0.5% of all patients have intracranial metastases (Antonio *et al.*, 2000; Sapozink and Kaplan, 1983). Median survival of patients with spinal epidural tumor ranges from 2–5 years for patients who are treated with radiotherapy and chemotherapy. Patients with intracranial metastases survive 10 months after radiochemotherapy and about 2 months without specific treatment (Aabo *et al.*, 1986; Sapozink and Kaplan, 1983; Schertel *et al.*, 1985; Table VII).

## Principles of Therapy

Results of the available retrospective clinical studies are summarized in Table VII, and a standard treatment plan for CNS complications of Hodgkin's disease is presented in Table VIII (based on retrospective or nonrandomized studies). In patients with spinal epidural Hodgkin's lymphoma, the tumor area is irradiated with approximately 30 Gy. This treatment usually results in

TABLE VIII   Standard Treatment Plan of Metastatic Hodgkin's Disease to the Nervous System

Epidural secondary Hodgkin's-lymphoma
  Corticosteroids
  Systemic chemotherapy
  Irradiation of spinal focus (30 Gy)
Intracerebral secondary Hodgkin's-lymphoma
  Corticosteroids
  Systemic chemotherapy
  Whole-brain irradiation (30 Gy)
  ± Boost to tumor region (50 Gy)
Leptomeningeal metastasis
  Intraventricular/intrathecal chemotherapy
  Irradiation of neurological focus (30 Gy)
  Systemic chemotherapy

good clinical improvement and a 5-year-survival rate of 30% (Schertel *et al.*, 1985). With surgical tumor resection before radiotherapy Haddad *et al.* (1976) observed a 5-year-survival rate of 50%. The authors themselves, however, have attributed this improved outcome in their surgical series of patients to a selection of patients with less-advanced disease. Functional improvement is not superior using surgery plus radiotherapy compared with radiotherapy alone. Systemic chemotherapy is indicated in patients with additional metastases elsewhere in the body and may be beneficial in the occasional patient with epidural metastasis alone, when the patient has no or minimal neurological deficits. Patients with cerebral metastases have usually been treated with whole-brain irradiation with 30–36 Gy and a boost of 14 Gy to the tumor region. This treatment improves clinical signs and symptoms in three quarters of patients and leads to a median survival of 10 months (Sapozink and Kaplan, 1983; Schertel *et al.*, 1985). The therapeutic value of surgery before radiotherapy has not been evaluated. Some patients who received additional chemotherapy after whole-brain radiation therapy because of concomitant systemic metastasis and had a complete remission had no intracranial relapse for more than 2 years (Sapozink and Kaplan, 1983). The role of systemic chemotherapy in the treatment of brain metastasis from HD is unknown but may be beneficial in light of its efficacy in the treatment of cerebral NHL.

## Practical Management

Patients with known HD who have a spinal epidural lesion leading to compression of the cord receive a single high dose of corticosteroid such as 100 mg dexamethasone and then 8 mg dexamethasone four times daily for the first week of radiotherapy. Corticosteroids are tapered by the end of radiotherapy. Radiotherapy is given to the tumor area for approximately 30 Gy in 2–3 Gy fractions. Surgical decompression is indicated in patients with an unknown primary tumor and occasionally for patients who relapse in a previously irradiated region (also see earlier and Chapter 68). In the rare patient with intracranial lesions, biopsy should be considered to exclude other processes but is unnecessary when tumor can be demonstrated on CSF cytological examination. Patients are initially treated with high-dose corticosteroids. Systemic chemotherapy using high-dose methotrexate may be considered and followed by whole-brain irradiation again with 36 Gy in 2 Gy fractions and a boost of 14 Gy to the tumor region. It is uncertain whether a boost is really helpful in this situation, given the experience with PCNSL. In cases with diffuse leptomeningeal metastasis, intrathecal chemotherapy and radiotherapy are indicated as detailed in Chapter 65. The side effects of radiotherapy and intrathecal chemotherapy are discussed in Chapters 64 and 65. In all cases with CNS manifestations of HD, additional systemic therapy should be considered using conventional protocols.

# REFERENCES

Aabo, K., and Walbom-Joergensen, S. (1986). Central nervous system complications by malignant lymphomas: Radiation schedule and treatment results. *Int. J. Radiat. Oncol. Biol. Phys.* **12**, 187–202.

Aboulafia, D. M. (1998). Regression of acquired immunodeficiency syndrome-related pulmonary Kaposi's sarcoma after highly active antiretroviral therapy. *Mayo Clin. Proc.* **73**, 439–443.

Abrey, L. E., Yahalom J., and DeAngelis, L. M. (2000). Treatment for primary CNS lymphoma: the next step. *J. Clin. Oncol.* **18**, 3144–3150.

Al-Yamany, M., Lozano, A., Nag, S., Laperriere, N., and Bernstein, M. (1999). Spontaneous remission of primary central nervous system lymphoma: report of 3 cases and discussion of pathophysiology. *J. Neuro-Oncol.* **42**, 151–159.

Alvarnas, J. C., Negrin, R. S., Horning, S. J., Long, G. D., Schriber, J. R., Stockerl-Goldstein, K., Tierney, K., Wong, R., Blume, K. G., and Chao, N. J. (2000). High-dose therapy with hematopoetic cell transplantation for patients with central nervous system involvement by non-Hodgkin's lymphomas. *Biol. Blood Marrow Transplant.* **6**, 352–358.

Amadori, S., Papa, G., Avvisati, G., Petti, M. C., Motta, M., Salvagnini, M., Meloni, G., Martelli, M., Monarca, B., and Mandelli, F. (1984). Sequential combination of systemic high-dose ara-c and asparaginase for the treatment of central nervous system leukemia and lymphoma. *J. Clin. Oncol.* **2**, 98–101.

Ammassari, A., Cingolani, A., Pezzotti, P., De Luca, D. A., Murri, R., Giancola, M. L., Larocca, L. M., and Antinori, A. (2000). AIDS-related focal brain lesions in the era of highly active antiretroviral therapy. Neurology 55, 1194–1200.

Antinori, A., Cingolani, A., De Luca, A., Gaidano, G., Ammassari, A., Larocca, L. M., and Ortona, L. (1999b). Epstein-Barr virus in monitoring the response to therapy of acquired immunodeficiency syndrome-related primary central nervous system lymphoma. *Ann. Neurol.* **45**, 259–261.

Antinori, A., De Rossi, G., Ammassari, A., Cingolani, A., Murri, R., Di Giuda, D., De Luca, A., Pierconti, F., Tartaglione, T., Scerrati, M., Larocca, L. M., and Ortona, L. (1999a). Value of combined approach with thallium-201 single-photon emission computed tomography and Epstein-Barr virus DNA polymerase chain reaction in CSF for the diagnosis of AIDS-related primary CNS lymphoma. *J. Clin. Oncol.* **17**, 554–560.

Antonio, G., Dahlstrom, J., Chandran, K. N., and O'Neil, R. (2000). Cerebellopontine angle Hodgkin's disease. *Australas. Radiol.* **44**, 115–117.

Baleydier, F., Galambrun, C., Manel, A. M., Guibaud, L., Nicolino, M., and Bertrand, Y. (2001). Primary lymphoma of the pituitary stalk in an immunocompetent 9-year-old child. *Med. Pediatr. Oncol.* **36**, 392–395.

Bataille, B., Delwail, V., Menet, E., Vandermarcq, P., Ingrand, P., Wager, M., Guy, G., and Lapierre, F. (2000). Primary intracerebral malignant lymphoma: report of 248 cases. *J. Neurosurg.* **92**, 261–266.

Baumgartner, J. E., Rachlin, J. R., Beckstead, J. H., Meeker, T. C., Levy, R. M., Wara, W. M., and Rosenblum, M. L. (1990). Primary central nervous system lymphomas: natural history and response to radiation therapy in 55 patients with acquired immunodeficiency syndrome. *J. Neurosurg.* **73**, 206–211.

Biagi, J., MacKenzie, R. G., Lim, M. S., Sapp, M., and Berinstein, N. (2000). Primary Hodgkin's disease of the CNS in an immunocompetent patient: a case study and review of the literature. *Neuro-Oncol.* **2**, 239–243.

Castagna, A., Cinque, P., d'Amico, A., Messa, C., Fazio, F., and Lazzarin, A. (1997). Evaluation of contrast-enhancing brain lesions in AIDS patients by means of Epstein-Barr virus detection in cerebrospinal fluid and 201thallium single photon emission tomography. *AIDS* **11**, 1522–1523.

Chamberlain, M. C., and Kormanik, P. A. (1999). AIDS-related central nervous system lymphomas. *J. Neuro-Oncol.* **43**, 269–276.

Chamberlain, M. C., and Levin, V. A. (1992). Primary central nervous system lymphoma: a role for adjuvant chemotherapy. *J. Neuro-Oncol.* **14**, 271–275.

Cingolani, A., Gastaldi, R., Fassone, L., Pierconti, F., Giancola, M. L., Martini, M., De Luca, A., Ammassari, A., Mazzone, C., Pescarmona, E., Gaidano, G., Larocca, L. M., and Antinori, A. (2000). Epstein-Barr virus infection is predictive of CNS involvement in systemic AIDS-related non-Hodgkin's lymphomas. *J. Clin. Oncol.* **18**, 3325–3330.

Cinque, P., Brytting, M., Vago, L., Castagna, A., Parravicini, C., Zanchetta, N., D'Arminio Monforte, A., Wahren, B., Lazzarin, A., and Linde, A. (1993). Epstein-Barr virus DNA in cerebrospinal fluid from patients with AIDS-related primary lymphoma of the central nervous system. *Lancet* **342**, 398–401.

Cobbers, J. M., Wolter, M., Reifenberger, J., Ring, G. U., Jessen, F., An, H. X., Niederacher, D., Schmidt, E. E., Ichimura, K., Floeth, F., Kirsch, L., Borchard, F., Louis, D. N., Collins, V. P., and Reifenberger, G. (1998). Frequent inactivation of CDKN2A and rare mutation of TP53 in PCNSL. *Brain Pathol.* **8**, 263–276.

Coleman, C. N., Cohen, J. R., Burke, J. S., and Rosenberg, S. A. (1981). Lymphoblastic lymphoma in adults: Results of a pilot protocol. *Blood* **57**, 679–684.

Dahlborg, S. A., Henner, W. D., Crossen, J. R., Tableman, M., Petrillo, A., Braziel, R., and Neuwelt, E. A. (1996). Non-AIDS primary CNS lymphoma: First example of a durable response in a primary brain tumor using enhanced chemotherapy delivery without cognitive loss and without radiotherapy. *Cancer J. Sci. Am.* **2**, 166.

Dahlborg, S. A., Petrillo, A., Crossen, J. R., Roman-Goldstein, S., Doolittle, N. D., Fuller, K. H., and Neuwelt, E. A. (1998). The potential for complete and durable response in nonglial primary brain tumors in children and young adults with enhanced chemotherapy delivery. *Cancer J. Sci. Am.* **4**, 110–124.

Daniels, D., Lowdell, C. P., and Glaser, M. G. (1992). The spontaneous regression of lymphoma in AIDS. *Clin Oncol (R Coll Radiol)* **4**, 196–197.

DeAngelis, L. M. (1990). Primary central nervous system lymphoma imitates multiple sclerosis. *J. Neuro-Oncol* **9**, 177–181.

DeAngelis, L. M. (1991). Primary central nervous system lymphoma as a secondary malignancy. *Cancer* **67**, 1431–1435.

DeAngelis, L. M. (1999). Primary CNS lymphoma: treatment with combined chemotherapy and radiotherapy. *J. Neuro-Oncol.* **43**, 249–257.

DeAngelis, L. M., Yahalom, J., Heinemann, J-H, Cirrincione, C., Thaler, H. T., and Krol, G. (1990). Primary CNS lymphoma: Combined treatment with chemotherapy and radiotherapy. *Neurology* **40**, 80–86.

DeAngelis, L. M., Yahalom, J., Thaler, H. T., and Kher, U. (1992). Combined modality therapy for primary CNS lymphoma. *J. Clin. Oncol.* **10**, 635–643.

Deckert-Schlüter M., Rang, A., and Wiestler, O. D. (1998). Apoptosis and apoptosis-related gene products in primary non-Hodgkin's lymphoma of the central nervous system. *Acta. Neuropathol.* **96**, 157–162.

Duque-Hammershaimb, L., Wollner, N., and Miller, D. R. (1983). LSA2-L2 Protocol treatment of stage IV non-Hodgkin's lymphoma in children with partial and extensive bone marrow involvement. *Cancer* **52**, 39–43.

Eby, N. L., Gruffermann, S., Flannelly, C. M., Schold, S. C., Vogel, F. S., and Burger, P. C. (1988). Increasing incidence of primary brain lymphoma in the US. *Cancer* **62**, 2461–2465.

Ersbøll, J., Schultz, H. B., Thomsen, B. L., Keiding, N., and Nissen, N. I. (1985). Meningeal involvement in non-Hodgkin's lymphoma: Symptoms, incidence, risk factors and treatment. *Scand. J. Haematol.* **35**, 487–496.

Ferracini, R., Pileri, S., Bergmann, M., Sabattini, E., Rigobello, L., Gambacorta, M., Galli, C., Manetto, V., Frank, G., Godano, U., Spagnoli, F., Casadei, G., Azzolini, U., Falini, B., and Gulotta, F. (1993). Non-Hodgkin lymphomas of the central nervous system. Clinico-pathologic and immunohistochemical study of 147 cases. *Pathol. Res. Pract.* **189**, 249–260.

Ferreri, A. J., Reni, M., Dell'Oro, S., Ciceri, F., Bernardi, M., Camba, L., Ponzoni, M., Terreni, M. R., Tomirotti, M., Spina, M., and Villa, E. (2001). Combined treatment with high-dose methotrexate, vincristine and procarbazine, without intrathecal chemotherapy, followed by consolidation radiotherapy for primary central nervous system lymphoma in immunocompetent patients. *Oncology* **60**, 134–140.

Formenti, S. C., Gill, P. S., Lean, E., Rarick, M., Meyer, P. R., Boswell, W., Petrovich, Z., Chak, L., and Levine, A. M. (1989). Primary central nervous system lymphoma in AIDS. *Cancer* **63**, 1101–1107.

Forsyth, P. A., Yahalom, J., and DeAngelis, L. M. (1994). Combined-modality therapy in the treatment of primary central nervous system lymphoma in AIDS. *Neurology* **44**, 1473–1479.

Freund, M., Ostendorf, P., Gärtner, V. H., and Waller, H. D. (1983). CNS manifestations in non-Hodgkin lymphomas (NHL). *Klin. Wschr.* **61**, 903–909.

Glass, J. P., Gruber, M. L., Cher, L., and Hochberg, F. H. (1994). Preirradiation methotrexate chemotherapy of primary central nervous system lymphomas: long-term outcome. *J. Neurosurg.* **81**, 188–195.

Griffin, J. W., Thompson, R. W., Mitchinson, M. J., De Kiewiet, J. C., and Welland, F. H. (1971). Lymphomatous leptomeningitis. *Am. J. Med.* **51**, 200–208.

Grulich, A. E. (1999). AIDS-associated non-Hodgkin's lymphoma in the era of highly active antiretroviral therapy. *J. Acquir. Immune Defic. Syndr.* **21** (Suppl 1), S27–S30.

Guha-Thakurta, N., Darnek, D., Pollack, C., and Hochberg, F. H. (1999). Intravenous methotrexate as initial treatment for primary CNS lymphoma: Response to treatment and quality of life of patients. *J. Neuro-Oncol.* **43**, 259–268.

Gururangan, S., Sposto, R., Cairo, M. S., Meadows. A. T., and Finlay, J. L. (2000). Outcome of CNS disease at diagnosis in disseminated small noncleaved-cell lymphoma and B-cell leukemia: a children's cancer group study. *J. Clin. Oncol.* **18**, 2017–2025.

Haddad, P., Thaell, J. F., Kiely, J. M., Harrison, Jr E. G., and Miller, R. H. (1976). Lymphoma of the spinal extradural space. *Cancer* **38**, 1862–1866.

Haddy, T. B., Adde, M. A., and Magrath, I. T. (1991). CNS involvement in small noncleaved-cell lymphoma: is CNS disease per se a poor diagnostic sign? *J. Clin. Oncol.* **9**, 1973–1982.

Hao, D., DiFrancesco, L. M., Brasher, P. M., deMetz, C., Fulton, D. S., DeAngelis, L. M., and Forsyth, P. A. (1999). Is primary CNS lymphoma really becoming more common? A population-based study of incidence, clinicopathological features and outcomes in Alberta from 1975 to 1996. *Ann. Oncol.* **10**, 65–70.

Heinzlef, O., Poisson, M., and Delattre, J. Y. (1996). Spontaneous regression of primary cerebral lymphoma. *Rev. Neurol. (Paris)* **152**, 135–138.

Herman, T. S., Hammond, N., Jones, S. E., Butler, J. J., Byrne, Jr G. E., and McKelvey, E. M. (1979). Involvement of the central nervous system by non-Hodgkin's lymphoma. *Cancer* **43**, 390–397.

Herrlinger, U., Brugger, W., Bamber, M., Küker, W., Dichgans, J., and Weller, M. (2000). PCV salvage chemotherapy for recurrent primary CNS lymphoma. *Neurology* **54**, 1707–1708.

Herrlinger, U., Klingel, K., Meyermann, R., Kandolf, R., Kaiserling, E., Kortmann, R. D., Melms, A., Skalej, M., Dichgans, J., and Weller, M. (2000). Central nervous system Hodgkin's lymphoma without systemic manifestation: case report and review of the literature. *Acta Neuropathol.* **99**, 709–714.

Herrlinger, U., Schabet, M., Brugger, W., Kortmann. R. D., Kanz, L., Bamberg, M., Dichgans, J., and Weller, M. (2000). Primary CNS lymphoma 1991–1997: outcome and late adverse effects after combined modality treatment. *Cancer* **91**, 130–135.

Herrlinger, U., Schabet, M., Eichhorn, M., Petersen, D., Grote, E. H., Meyermann, R., and Dichgans, J. (1996). Prolonged corticosteroid-induced remission in primary central nervous system lymphoma: report of a case and review of the literature. *Eur. Neurol.* **36**, 241–243.

Herrlinger, U., Schabet, M., Petersen, D., and Krauseneck, P. (1999). Primary CNS lymphoma: from clinical presentation to diagnosis. *J. Neuro-Oncol.* **43**, 219–226.

Hoang-Xuan, K., and Delattre, J.-Y. (1999). Ongoing protocols for non-AIDS primary central nervous system lymphoma. *J Neuro-Oncol.* **43**, 287–291.

Hochberg, F. H., Loeffler, J. S., and Prados, M. (1991). The therapy of primary brain lymphoma. *J. Neuro-Oncol.* **10**, 191–201.

Hochberg, F. H., and Miller, D. C. (1988). Primary central nervous system lymphoma. *J. Neurosurg.* **68**, 835–853.

Hoerni-Simon, G., Suchaud, J. P., Eghbali, H., Coindre, J. M., and Hoerni, B. (1987). Secondary involvement of the central nervous system in malignant non-Hodgkin's lymphoma: A study of 30 cases in a series of 498 patients. *Oncology* **44**, 98–101.

Homo-Delarche, F. (1984). Glucocorticoid receptors and steroid sensitivity in normal and neoplastic human lymphoid tissues: a review. *Cancer Res.* **44**, 431–437.

Itoyama, T., Sadamori, N., Tsutsumi, K., Tokunaga, Y., Soda, H., Tomonaga, M., Yamamori, S., Masuda, Y., Oshima, K., and Kikuchi, M. (1994). Primary central nervous system lymphomas. Immunophenotypic, virologic, and cytogenetic findings of three patients without immune defects. *Cancer* **73**, 455–463.

Jacobson, L. P., Yamashita, T. E., Detels, R., Margolick, J. B., Chmiel, J. S., Kingsley, L. A., Melnick, S., and Munoz, A. (1999). Impact of potent antiretroviral therapy on the incidence of Kaposi's sarcoma and non-Hodgkin's lymphomas among HIV-1-infected individuals. Multicenter AIDS Cohort Study. *J. Acquir. Immune Defic. Syndr.* **21**, S34–S41.

Jacomet, C., Girard, P. M., Lebrette, M. G., Farese, V. L., Monfort, L., and Rozenbaum, W. (1997). Intravenous methotrexate for primary central nervous system non-Hodgkin's lymphoma in AIDS. *AIDS* **11**, 1725–1730.

Jellinger, K., Radaskiewicz, T. H., and Slowik, F. (1975). Primary malignant lymphomas of the central nervous system in man. *Acta Neuropathol.* **6**, 95–102.

Jones, J. L., Hanson, D. L., Dworkin, M. S., Ward, J. W., and Jaffe, H. W. (1999). Effect of antiretroviral therapy on recent trends in selected cancers among HIV-infected persons. Adult/Adolescent Spectrum of HIV Disease Project Group. *J. Acquir. Immune Defic. Syndr.* **21** (Suppl 1), S11–S17.

Kai, Y., Kuratsu, J., and Ushio, Y. (1998). Primary malignant lymphoma of the brain in childhood. *Neurol. Med. Chir. (Tokyo)* **38**, 232–237.

Koga, H., Zhang, S., Ichikawa, T., Washiyama, K., Kuroiwa, T., Tanaka, R., and Kumanishi, T. (1994). Primary malignant lymphoma of the brain: demonstration of the p53 gene mutations by PCR-SSCP analysis and immunohistochemistry. *Noshuyo Byori* **11**, 151–155.

Kumanishi, T., Zhang, S., Ichikawa, T., Endo, S., and Washiyama, K. (1996). Primary malignant lymphoma of the brain: demonstration of frequent p16 and p15 gene deletions. *Jpn. J. Cancer Res.* **87**, 691–695.

Liang, R., Chiu, E., and Loke, S. L. (1990). Secondary central nervous system involvement by non-Hodgkin's lymphoma: the risk factors. *Hematol. Oncol.* **8**, 141–145.

Liang, R. H., Woo, EK., Yu, Y. L., Todd, D., Chan, T. K., Ho, F. C., Tso, S. C., and Chum, J. S. (1989). Central nervous system involvement in non-Hodgkin's lymphoma. *Eur. J. Cancer Clin. Oncol.* **25**, 703–710.

Ling SM, Roach M, Larson DA, and Wara W. M. (1994). Radiotherapy of primary central nervous system lymphoma in patients with and without human immunodeficiency virus. *Cancer* **73**, 2570–2582.

Lukes, R. J., and Butler, J. J. (1966). The pathology and nomenclature of Hodgkin's disease. *Cancer Res.* **26**, 1063–1081.

Macdonald, D. R. (1995). New therapies of primary CNS lymphomas and oligodendrogliomas. *J. Neuro-Oncol.* **24**, 97–101.

Mackintosh, F. R., Colby, T. V., Podolsky, W. J., Burke, J. S., Hoppe, R. T., Rosenfelt, F. P., Rosenberg, S. A., and Kaplan, H. S. (1982). Central nervous system involvement in non-Hodgkin's lymphoma: An analysis of 105 cases. *Cancer* **49**, 586–595.

McGowan, J. P., and Shah, S. (1998). Long-term remission of AIDS-related primary central nervous system lymphoma associated with highly active antiretroviral therapy. *AIDS* **12**, 952–954.

Mead, G. M., Kennedy, P., Smith, J. L., Thompson, J., Macbeth, F. R., Ryall, R. D., Williams, C. J., and Whitehouse, J. M. (1986). Involvement of the central nervous system by non-Hodgkin's lymphoma in adults: A review of 36 cases. *Q. J. Med.* **60**, 699–714.

Miller, D. C., Hochberg, F. H., Harris, N. L., Gruber, M. L., Louis, D. N., and Cohen, H. (1994). Pathology with clinical correlations of primary central nervous system non-Hodgkin's lymphoma. The Massachusetts General Hospital experience 1958–1989. *Cancer* **74**, 1383–1397.

Morgello, S. (1995). Pathogenesis and classification of primary central nervous system lymphoma: an update. *Brain Pathol.* **5**, 383–393.

Morra, E., Lazzarino, M., Brusamolino, E., Pagnucco, G., Castagnola, C., Bernasconi, P., Orlandi, E., Corso, A., Santagostino, A., and Bernasconi, C. (1993). The role of systemic high-dose cytarabine in the treatment of central nervous system leukemia. *Cancer* **72**, 439–445.

Murphy, S. B., and Hustu, H. O. (1980). A randomized trial of com-

bined modality therapy of childhood non-Hodgkin's lymphoma. *Cancer* **45**, 630–637.

Nelson, D. F. (1999). Radiotherapy in the treatment of primary central nervous system lymphoma (PCNSL). *J. Neuro-Oncol.* **43**, 241–247.

Nelson, D. F., Martz, K. L., Bonner, H., Nelson, J. S., Newall, J., Kerman, H. D., Thomson, J. W., and Murray, K. J. (1992). Non-Hodgkin's lymphoma of the brain: can high dose, large volume radiation therapy improve survival? Report on a prospective trial by the radiation therapy oncology group (RTOG). *Int. J. Radiat. Oncol. Biol. Phys.* **23**, 9–17.

Ng, S., Rosenthal, M. A., Ashley, D., and Cher, L. (2000). High-dose methotrexate for primary CNS lymphoma in the elderly. *J. Neuro-Oncol* **2**, 40–44.

Nishiura, I., Takeuchi, J., Handa, H., and Kang, S. S. (1980). Intracranial malignant lymphoma. *No Shinkei Geka* **8**, 839–844.

Nozaki, M., Tada, M., Mizugaki, Y., Takada, K., Nagashima, K., Sawamura, Y., and Abe, H. (1998). Expression of oncogenic molecules in primary central nervous system lymphomas in immunocompetent patients. *Acta Neuropathol.* **95**, 505–510.

O'Brien, P., Roos, D., Pratt, G., Liew, K., Barton, M., Poulsen, M., Oliver, I., and Trotter, G. (2000). Phase II multicenter study of brief single-agent methotrexate followed by irradiation in primary CNS lymphoma. *J. Clin. Oncol.* **18**, 519–526.

O'Neill, B. P., Janney, C. A., Olson, J. E., Cerhan, J. R., Kurtin, P. J., Schiff, D., and Kaplan, R. S. (2001). The continuing increase in primary central nervous sytem non-hodgkin's lymphoma (PCNSL): a surveillance, epidemiology, and end results (SEER) analysis. Proceedings of the American Society of Clinical Oncology, Abstract# 208. *J. Clin. Oncol.* **20** (1), 53a.

Ono, K., Kikuchi, M., Funai, N., Matsuzaki, M., and Shimamoto, Y. (1996). Natural killing activity in patients with spontaneous regression of malignant lymphoma. *J.Clin. Immunol.* **16**, 334–339.

Paulus, W. (1999). Classification, pathogenesis and molecular pathology of primary CNS lymphomas. *J. Neuro-Oncol.* **43**, 203–208.

Penn, I., and Porat, G. (1995). Central nervous system lymphomas in organ allograft recipients. *Transplantation* **59**, 240–244.

Rabkin, C. S., Testa, M. A., Huang, J., and Von Roenn, J. H. (1999). Kaposi's sarcoma and non-Hodgkin's lymphoma incidence trends in AIDS Clinical Trial Group study participants. *J. Acquir. Immune Defic. Syndr.* **1**, S31–S33.

Raez, L., Cabral, L., Cai, J. P., Landy, H., Sfakianakis, G., Byrne, G. E., Hurley, J., Scerpella, E., Jayaweera, D., and Harrington, W. J. (1999). Treatment of AIDS-related primary central nervous system lymphoma with zidovudine, ganciclovir, and interleukin 2. *AIDS Res. Hum. Retroviruses* **15**, 713–719.

Raz, I., Siegal, T., Siegal, T., and Polliack, A. (1984). CNS involvement by non-Hodgkin's lymphoma. *Arch. Neurol.* **41**, 1167–1171.

Recht, L., Straus, D. J., Cirrincione, C., Thaler, H. T., and Posner, J. B. (1988). Central nervous system metastases from non-Hodgkin's lymphoma: Treatment and prophylaxis. *Am. J. Med.* **84**, 425–435.

Sandor, V., Stark-Vancs, V., Pearson, D., Nussenblat, R., Whitcup, S. M., Brouwers, P., Patronas, N., Heiss, J., Jaffe, E., deSmet, M., Kohler, D., Simon, R., and Wittes, R. (1998). Phase II trial of chemotherapy alone for primary CNS and intraocular lymphoma. *J. Clin. Oncol.* **16**, 3000–3006.

Sapozink, M. D., and Kaplan, H. S. (1983). Intracranial Hodgkin's disease. *Cancer* **52**, 1301–1307.

Schabet, M. (1999). Epidemiology of primary CNS lymphoma. *J. Neuro-Oncol.* **43**, 199–201.

Schertel, L., Fischer, G., Mohring, R., and Ritter, S. (1985). Zur Strahlentherapie zerebralen und spinalen Hodgkin-Befalls. *Röntgenpraxis* **38**, 400–407.

Schlegel, U., Pils, H., Oehring, R., and Blümcke, I. (1999). Neurologic sequelae of treatment of primary CNS lymphomas. *J. Neuro-Oncol.* **43**, 277–286.

Schultz, C., Scott, C., Sherman, W., Donahue, B., Fields, J., Murray, K., Fisher, B., Abrams, R., and Meis-Kindblom, J. (1996). Preirradiation chemotherapy with cytoxan, adriamycin, vincristine, and decadron (CHOD) for primary central nervous system lymphomas (PCNSL). *J. Clin. Oncol.* **14**, 556–564.

Shenoy, U. A., Kushner, J. P., and Schumann, G. B. (1987) Cytologic diagnosis and monitoring of Hodgkin's disease in cerebrospinal fluid: A case report. *Diagn. Cytopathol.* **3**, 323–325.

Shibata, S. (1989). Sites of origin of primary intracerebral malignant lymphoma. *Neurosurgery* **25**, 14–19.

Silfen, M. E., Garvin, J. H. Jr, Hays, A. P., Starkman, H. S., Aranoff, G. S., Levine, L. S., Feldstein, N. A., Wong, B., and Oberfield, S. E. (2001). Primary central nervous system lymphoma in childhood presenting as progressive panhypopituitarism. *J. Pediatr. Hematol. Oncol.* **23**, 130–133.

Skarin, A. T., Zuckermann, K. S., Pitman, S. W., Rosenthal, D. S., Moloney, W., Frei, E., and Canellos, G. P. (1977). High-dose methotrexate with folinic acid in the treatment of advanced non-Hodgkin lymphoma including CNS involvement. *Blood* **50**, 1039–1047.

Slobod, K. S., Taylor, G. H., Sandlund, J. T., Furth, P., Helton, K. J., and Sixbey, J. W. (2000). Epstein-Barr virus-targeted therapy for AIDS-related primary lymphoma of the central nervous system. *Lancet* **356**, 1493–1494.

Soussain, C., Suzan, F., Hoang-Xuan, K., Cassoux, N., Levy, V., Azar, N., Belanger, C., Achour, E., Ribrag, V., Gerber, S., Delattre, J.-Y., and Leblond, V. (2001). Results of intensive chemotherapy followed by hematopoietic stem-cell rescue in 22 patients with refractory or recurrent primary CNS lymphoma or intraocular lymphoma. *J. Clin. Oncol.* **19**, 742–749.

Sparano, J. A., Anand, K., Desai, J., Mitnick, R. J., Kalkut, G. E., and Hanau, L. H. (1999). Effect of highly active antiretroviral therapy on the incidence of HIV-associated malignancies at an urban medical center. *J. Acquir. Immune Defic. Syndr.* **1**, S18–S22.

Sugita, Y., Shigemori, M., Yuge, T., Iryo, O., Kuramoto, S., Nakamura, Y., and Morimatsu, M. (1988). Spontaneous regression of primary malignant intracranial lymphoma. *Surg. Neurol.* **30**, 148–152.

Takasu, S., Wakabayashi, T., Kajita, Y., Hatano, N., Hatano, H., Usui, T., Kinoshita, T., and Yoshida, J. (2000). Effectiveness of DeVIC chemotherapy for recurrent primary central nervous system lymphoma. *No Shinkei Geka* **28**, 789–794.

Tomita, N., Kodama, F., Sakai, R., Koharasawa, H., Hattori, M., Taguchi, J., Fujita, H., Tanabe, J., Fujisawa, S., Fukawa, H., Harano, H., Kanamori, H., Miyashita, H., Matsuzaki, M., Ogawa, K., Motomura, S., Maruta, A., and Ishigatsubo, Y. (2000). Predictive factors for central nervous system involvement in non-Hodgkin's lymphoma: significance of very high serum LDH concentrations. *Leuk. Lymphoma* **38**, 335–343.

Varadi, G., Or, R., Kapelushnik, J., Naparstek, E., Nagler, A., Brautbar, C., Amar, A., Kirschbaum, M., Samuel, S., Slavin, S., and Siegal, T. (1999). Graft-versus-lymphoma effect after allogeneic peripheral blood stem cell transplantation for primary central nervous system lymphoma. *Leuk. Lymphoma* **34**, 185–190.

Weber, T., Weber, R. G., Kaulich, K., Actor, B., Meyer-Puttlitz, B., Lampel, S., Buschges, R., Weigel, R., Deckert-Schlüter, M., Schmiedek, P., Reifenberger, G., and Lichter, P. (2000). Characteristic chromosomal imbalances in primary central nervous system lymphomas of the diffuse large B-cell type. *Brain Pathol.* **10**, 73–84.

Wong, E. T., Portlock, C. S., O'Brien, J. P., and DeAngelis, L. M. (1996). Chemosensitive epidural spinal cord disease in non-Hodgkins lymphoma. *Neurology* **46**, 1543–1547.

Young, R. C., Howser, D. M., Anderson, T., Fisher, R. I., Jaffe, E., and DeVita, Jr, V. T. (1979). Central nervous system complications of non-Hodgkin's lymphoma. The potential role for prophylactic therapy. *Am. J. Med.* **66**, 435–443.

Zhang, S. J., Endo, S., Ichikawa, T., Washiyama, K., and Kumanishi, T. (1998). Frequent deletion and 5' CpG island methylation of the p16 gene in primary malignant lymphoma of the brain. *Cancer Res.* **58**, 1231–1237.

*Tumors and Developmental Disorders*

CHAPTER 64

# Brain Metastases from Systemic Solid Tumors

Michael Weller

## CLINICAL ASPECTS

This chapter covers the management of patients with brain metastases from systemic solid tumors. Central nervous system (CNS) metastases from systemic lymphomas, leptomeningeal metastases, and spinal metastases are dealt with in Chapters 63, 65, and 68. Brain metastases are a major clinical problem in adult patients with lung cancer, breast cancer, melanoma, and renal cell carcinoma (Table I). Brain metastases in younger patients and children originate more often from sarcomas, rhabdomyosarcomas, and germ cell tumors. Tumor cells reach the CNS mainly by way of the bloodstream and tend to seed in small arterial vessels or capillaries in the border zone of white matter and cortex, especially in the watershed areas. Although the different areas of the brain are essentially affected as predicted by their volume, metastases are relatively more frequent in deep and midline structures, including the basal ganglia, and the cerebellum. For unknown reasons, uterine and gastrointestinal tumors preferentially metastasize to the posterior fossa. Lymphatically and venously drained tumor cells from systemic primary tumors need to pass the lung before they reach the brain. Most patients with brain metastases therefore also have lung metastases, which may only be detected at autopsy. Approximately 50% of brain metastases are seen clinically as single lesions (Delattre *et al.*, 1988). Autopsy, however, reveals multiple brain metastases in approximately 75% of patients affected by brain metastases. Neurological symptoms and signs usually develop within weeks. Patients have headache (50%), hemiparesis (50%), mental changes (30%), seizures (15%–20%), cranial nerve palsies, or symptoms and signs of raised intracranial pressure (Cairncross and Posner, 1983). Approximately 5%–

10% of the patients, notably those with metastatic melanoma or choriocarcinoma, have acute deterioration caused by tumoral hemorrhage or noncommunicating hydrocephalus.

Computed tomography (CT) or magnetic resonance imaging (MRI) are the most important tools for the diagnosis of brain metastases. On CT scan, brain metastases usually appear as round hypodense or isodense lesions with a variable degree of perifocal edema. Metastases from melanomas, colon carcinomas, or choriocarcinomas may be hyperdense. Contrast enhancement is usually homogeneous but may be ringlike in metastases with central necrosis. MRI with gadolinium enhancement is more sensitive than CT in the detection of small lesions and is the preferred technique to assess metastasis to the brain whenever available. The differential diagnoses of brain metastases include primary benign and malignant brain tumors, abscesses, hemorrhagic or ischemic infarction, and parasitoses.

There is a history of cancer in approximately 70% of the patients with brain metastases. In a third of the patients, however, brain metastasis is the first manifestation of malignancy, most commonly in patients with lung cancer. Patients with cerebral metastases of unknown origin require a thorough clinical examination notably of the skin, as well as chest x-ray, mammography, ultrasonography of the abdomen, including the kidneys, and stool examination. If chest x-ray and ultrasonography of the abdomen do not disclose the primary lesion, the patients will have a CT scan of the thorax, abdomen, and pelvis. Blood tests, notably tumor markers, are of limited value in detecting a primary lesion. In the context of brain metastases, elevations of serum neuron-specific enolase suggest small cell lung cancer, whereas the detection of α-fetoprotein (AFP) or human β-chorionic gonadotropin (βhCG) in serum or

TABLE I   Frequency of Single or Multiple Brain Metastases Diagnosed Clinically or at Autopsy, and Interval from Diagnosis of Primary Tumor to Diagnosis of Brain Metastases*

| Primary tumor | Frequency (%) | | Interval (mo) | |
|---|---|---|---|---|
| | Clinical | Autopsy | Median | Range |
| Lung cancer | | | | |
|   Small cell | 30–45 | 30–70 | 2.6 | 0–15 |
|   Adenocarcinoma | 24–30 | 50 | 2 | 0–66 |
|   Squamous carcinoma | 30 | 40 | 0.2 | 0–31 |
| Breast cancer | 10–20 | 20–40 | 23 | 0–121 |
| Melanoma | 20–45 | 40–90 | 36 | 3–83 |
| Kidney or bladder cancer | 20 | 20 | 39 | 19–119 |
| Colorectal cancer | 4 | 6–10 | 22 | 0–48 |

*Data from: Cascino *et al.*, 1983; Doyle, 1982; Reddy *et al.*, 1983; Retsas and Gershuny, 1988; Robin *et al.*, 1982; Rosner *et al.*, 1986; Sørensen *et al.*, 1988; Zimm *et al.*, 1981.

cerebrospinal fluid (CSF) may indicate metastases from germ cell tumors. Because no primary tumor will be found in up to half of the patients with brain metastases in the absence of a known cancer (Salvati *et al.*, 1995b; Van de Pol *et al.*, 1996), it spending too much time and resources to detect a primary lesion, should be avoided especially in patients with rapidly progressing neurological deficits. Such patients are commonly candidates for surgical resection or for biopsy in the case of multiple lesions or if the location prohibits complete tumor removal. Stereotactic biopsy may be performed with the patient under local anaesthesia and is therefore feasible even in patients who are not eligible for surgery. Such biopsies are diagnostic in more than 90% of the patients. The procedure has a morbidity rate of 3%–4%, and a mortality rate <1% (Hall, 1998; Sawin *et al.*, 1998). The surgical procedure will often, but not always, provide essential clues to the site of origin of the primary lesion. Cerebral metastases in the absence of a known malignancy are more likely to have a strong impact on quality of life and survival than the primary lesion and should therefore receive appropriate treatment without delay. Some authors have advocated routine biopsy in all cases of presumptive brain metastases, because 11% (6 of 54) of suspected single brain metastases in patients with known malignancies revealed a different histological finding in a prospective trial (Patchell *et al.*, 1990). For practical purposes, however, patients with a known malignancy who have multiple as opposed to single lesions suggestive of brain metastases develop are commonly treated according to the histological findings of their primary tumor without confirmation of the histological findings by biopsy or resection.

## NATURAL COURSE

More than 20% of all patients with systemic malignancies have brain metastases develop in the course of their disease. The frequency is highest in malignant melanoma and small cell lung cancer (45%) and breast and renal cell cancer (20%). Lung cancer as a very common cancer thus accounts for 50% of all brain metastases, breast cancer for 15%–20%, gastrointestinal tumors, melanomas and urogenital tumors for approximately 5–10% each, and unknown primary tumors for 10% of all brain metastases (Delattre *et al.*, 1988). The median survival after diagnosis of brain metastases is 1 month without treatment and 2 months with symptomatic treatment, including corticosteroids. Patients usually die from increasing intracranial pressure with cerebral herniation and brainstem compression. Radiotherapy alone results in an improvement of the neurological status in approximately 70% of the patients and increases the median survival to 3–6 months (Table II). The median survival may increase to 2–21 months with surgery followed by whole-brain radiotherapy in patients with single brain metastases from less radiosensitive tumors (Table III). Median survival is not the most important endpoint to assess the benefit of therapy for patients with brain metastases, because most of these patients die from systemic tumor progression and not from their brain metastases (Cairncross and Posner, 1983). Instead, clinical response rates and the duration of these responses are critical parameters to assess the benefit of therapy.

Predictors of long-term survival after the manifestation of brain metastases include absent or stable extracranial disease, a long interval between the diag-

TABLE II   Outcome in Patients with Multiple or Single Brain Metastases Treated with Whole Brain Radiotherapy*

| Reference | Primary tumor | Patients (n) | Radiotherapy (total Gy/n wk) | Clinical or radiographic response (%) | Median survival (mo) | Survival at 1 year (%) |
|---|---|---|---|---|---|---|
| Hendrickson, 1977 | All | 1001 | 10/1–40/3 | 50 | 3.7 | 15 |
| Chassard et al., 1988 | All | 196 | 24/4–50/4–5 | 65 | 4.2 | 11 |
| Hoskin et al., 1990 | All | 164 | 35/3 | 86 | 3.7 | 15 |
| Mandell et al., 1986 | Lung, NSCLC[+] | 69 | 25/1–39/4 | 83 | 4.0 | 18 |
| Patchell et al., 1986 | Lung, NSCLC | 43 | 30–55/? | 83 | 9.0 | 42 |
| Carmichael et al., 1988 | Lung, SCLC | 23[‡] | 40/4 | 73 | 7 | ? |
|  |  | 36[§] |  | 56 | 3 |  |
| Ryan et al., 1995 | Lung, all | 416 | 20/1–30/2 | ? | 3.3 | 8 |
| Postmus et al., 1998 | Lung, SCLC | 20 | 30/2 | 50 | 4.7 | ? |
| Kamby and Soerensen, 1988 | Breast | 106 | 30/2–40/3 |  | 3.3 | 17 |
| Boogerd et al., 1993 | Breast | 137 | 30/3 | 60 | 4 | 19 |
| Corn et al., 1995 | Ovarian | 32 | 20/1–52/4 | 71 | 4 | 19 |
| Carella et al., 1980 | Melanoma | 60 | 10/1–40/4 | 76 | 2.5–3.5 |  |
| Choi et al., 1985 | Melanoma | 194 | 30/1–48/2 | 40 | 3 | 12 |
| Ziegler and Cooper, 1986 | Melanoma | 72 | 30/2 | 63 | 5 |  |
| Reddy et al., 1983 | Urogenital[‖] | 68 | 20/1–40/4 | 30–80 |  |  |
| Maor et al., 1988 | Kidney | 39 | 30/2 | 30 | 2 | 15 |
| Wronski et al., 1997 | Kidney | 119 | Various |  | 4.4 | 16.8 |

*Some patients in these different series were also treated surgically or with chemotherapy.
[+]NSCLC, non-small cell lung cancer.
[‡]At presentation.
[§]Delayed metastases.
[‖]Kidney, 43; testis, 11; bladder, 7; prostate, 7.

nosis of the primary tumor and of brain metastases, supratentorial location of the metastases, a single brain metastasis, good neurological function, and younger age. Poor prognostic factors include a diagnosis of non-small cell lung cancer, old age, poor neurological function, multiple brain metastases, infratentorial location of the metastases, advanced systemic disease, or acute neurological deterioration (Borgelt et al., 1980; Diener-West et al., 1989; Zimm et al., 1981). The histological findings of the primary tumor are not relevant predictors of survival. Some patients with metachronous brain metastases, that is, metastases developing years after the manifestation of the primary tumor, from breast or non-small cell lung cancer may survive for several years. Single patients with brain metastases are apparently cured by radiotherapy (Cairncross et al., 1979).

## PRINCIPLES OF THERAPY

### Symptomatic Treatment

The most important symptomatic treatment measures in patients with brain metastases are the control of increased intracranial pressure by corticosteroids and the control of seizures by anticonvulsants. Corticos-teroids are the cornerstone of pharmacological management of brain edema in patients with brain metastases. Symptomatic improvement is achieved in 70% of the patients for about 1 month with corticosteroid treatment alone (Weissman, 1988). A controlled prospective clinical trial demonstrated that dexamethasone at 1 mg qid was less toxic but as effective as 4 mg qid in patients with brain metastases (Vecht et al., 1994). Furthermore, improvement of neurological function is faster and the outcome in patients with poor neurological function better when corticosteroids are added to radiotherapy compared with radiotherapy alone (Borgelt et al., 1980). Hyperosmolar agents such as mannitol or forced hyperventilation, in addition to high-dose steroids, are rarely necessary as life-saving measures in acutely deteriorating patients with brain metastases.

The role of anticonvulsants for patients with brain metastases is not clearly defined (Cohen et al., 1988). We recommend that patients who have seizures receive anticonvulsants for no more than 3 months when the brain metastases respond to treatment. Progressive disease in these patients commonly provokes further seizures, making anticonvulsive treatment necessary until death. Prophylactic anticonvulsants are given perioperatively when resection of metastases is planned. Because seizures are potentially life-threatening in

**TABLE III** Outcome in Patients with (Mostly) Single Brain Metastases Treated with Surgery Alone (S), with Surgery and Subsequent (Mostly) Whole-Brain Radiotherapy (S + RT), with Radiotherapy Alone (RT), or with Radiosurgery (RS)

| Reference | Tumor | n | Therapy | Mortality (%)* | Median survival (mos) | Survival at 1 year (%) | Survival at 2 years (%) | Survival at 4 years (%) |
|---|---|---|---|---|---|---|---|---|
| Störtebecker, 1954 | All | 125 | S | 25 | 3.6 | 21 | | 3 |
| Richards and McKissock, 1963 | All | 108 | S ± RT | 32 | 5 | 17 | 8 | |
| Lang and Slater, 1964 | All | 208 | S | 22 | 4 | 20 | 13 | |
| Vieth and Odom, 1965 | All | 155 | S | 15 | 6 | 14 | 8 | |
| Haar and Patterson, 1972 | All | 167 | S | 11 | 6 | 22 | | 4 |
| Ransohoff, 1975 | All | 100 | S ± RT | 10 | | 38 | 13 | |
| Winston et al., 1980 | All | 79 | S ± RT | 10 | 5 | 22 | 10 | |
| White et al., 1981 | All | 122 | S + RT | 6 | 7 | 30 | 15 | 5 |
| Gagliardi and Mercuri, 1983 | All | 325 | S + RT | 10 | >6 (mean) | 22 | 3 | |
| Yardeni et al., 1984 | All | 74 | S + RT | 15 | 6.6 | 30 | 15 | |
| Sundaresan and Galicich, 1985 | All | 125 | S ± RT | 6 | 12 | 50 | 25 | 12 |
| Smalley et al., 1987 | All | 34 | S + RT | | 21 | 68 | 46 | 23 |
| | All | 51 | S | | 11.5 | 46 | 23 | 14 |
| DeAngelis et al., 1989b | All | 79 | S + RT | 7 | 20.6 | 48 | | |
| | All | 19 | S | 7 | 14.4 | 47 | | |
| Patchell et al., 1990 | All | 25 | S + RT | 4 | 9 | 45 | 0 | |
| | All | 23 | RT | 4 | 3.5 | 5 | 4 | |
| Vecht et al., 1993 | All | 32 | S + RT | 9 | 10 | 41 | 19 | |
| | All | 31 | RT | | 6 | 23 | 10 | |
| Bindal et al., 1993 | All | 30 | S† + RT | 3 | 6 | 23 | 0 | 0 |
| | All | 26 | S‡ + RT | 3 | 14 | 55 | 32 | 11 |
| | All | 26 | S§ + RT | 0 | 14 | 50 | 30 | 15 |
| Coffey et al., 1991 | All | 24 | RS‖ | | 10 | 33 | | |
| Fuller et al., 1992 | All | 27 | RS | | 5 | | | |
| Mehta et al., 1992 | All | 40 | RS | | 6.5 | 30 | | |
| Engenhardt et al., 1993 | All | 69 | RS | | 6 | 28 | | |
| Jokura et al., 1994 | All | 25 | RS | | 8.5 | 8 | | |
| Flickinger et al., 1994 | All | 116 | RS | | 11 | 45 | 20 | |
| Voges et al., 1994 | All | 46 | RS | | 11.5 | 45 | | |
| Alexander et al., 1995 | All | 248 | RS | | 9.4 | 44 | 20 | |
| Bindal et al., 1996 | All | 62 | S ± RT | | 16.4 | 58 | 36 | |
| | All | 31 | RS ± RT | | 7.5 | 27 | 10 | |
| Sneed et al., 1999 | All | 62 | RS | | 11.3 | 48 | | |
| | All | 43 | RS + RT | | 11.1 | 46 | | |
| Magilligan et al., 1976 | Lung, NSCLC | 22 | S ± RT | 0 | 14 | 45 | 23 | |
| Gagliardi and Mercuri, 1983 | Lung, all | 150 | S + RT | 11 | 5.8 (mean) | 16 | 4 | |
| Sundaresan and Galicich, 1985 | Lung, all | 50 | S ± RT | 6 | 18 | | 38 | |
| Mandell et al., 1986 | Lung, NSCLC | 35 | S + RT | 3 | 16 | 66 | 37 | 20 |
| Patchell et al., 1986 | Lung, NSCLC | 43 | S + RT | 2 | 19 | 65 | 32 | |
| Armstrong et al., 1994 | Lung, NSCLC | 32 | S | | 14 | 55 | 30 | 12 |
| | | 32 | S + RT | | 10 | 30 | 19 | 8 |
| Wronski et al., 1995a | Lung, NSCLC | 231 | S ± RT | 3 | 11 | 46 | 24 | 14 |
| Gagliardi and Mercuri, 1983 | Breast | 18 | S + RT | 6 | 7 (mean) | 24 | 0 | |
| Sundaresan and Galicich, 1985 | Breast | 8 | S ± RT | 6 | 12 | | 25 | |
| Gagliardi and Mercuri, 1983 | Melanoma | 7 | S + RT | | 7.2 (mean) | 0 | | |
| Sundaresan and Galicich, 1985 | Melanoma | 14 | S ± RT | 6 | 6 | | 14 | |
| Choi et al., 1985 | Melanoma | 32 | S + RT | ? | 7 | ? | | |
| Wornom et al., 1986 | Melanoma | 18 | S | 11 | 8 | 25 | 13 | |
| Hagen et al., 1990 | Melanoma | 19 | S + RT | ? | 6.5 | 41 | 30 | |
| | | 16 | S | ? | 8.5 | 36 | 15 | |
| Gagliardi and Mercuri, 1983 | Kidney | 14 | S + RT | 7 | 10.5 (mean) | 67 | 0 | |
| Sundaresan and Galicich, 1985 | Kidney | 14 | S ± RT | 6 | 6 | | 31 | |
| Wronski et al., 1996 | Kidney | 50 | S ± RT | 10 | 12.6 | 51 | 24 | |
| Culine et al., 1998 | Kidney | 68 | (S) ± RT | | 7 | | | |
| Mori et al., 1998 | Kidney | 35 | RS ± RT | | 11 | 43 | 22 | |
| Wronski et al., 1995b | Sarcoma | 25 | S ± RT | | 6.6 | 40 | 16 | |
| Salvati et al., 1998 | Sarcoma | 15 | S ± RT | 0 | 9.3 | 33 | | |
| Gagliardi and Mercuri, 1983 | Unknown | 127 | S + RT | 11 | 6.1 (mean) | 30 | 3 | |
| Sundaresan and Galicich, 1985 | Unknown | 6 | S ± RT | 6 | 5 | | 17 | |
| Eapen et al., 1988 | Unknown | 27 | S + RT | | 7 | 25 | | |

*Perioperative mortality within 30 days.
†Resection of 1 of 2 or more lesions.
‡Resection of 2 or more lesions.
§Resection of single lesions.
‖The radiosurgery (RS) studies included patients with heterogenous pretreatment and posttreatment regimens.

patients with increased intracranial pressure, prophylactic treatment may also be considered for patients with multiple large metastatic lesions during the first weeks of radiotherapy. We do not recommend routine administration of anticonvulsants for all patients with brain metastases. Phenytoin, valproate, and carbamazepine are probably equally effective in controlling seizures in brain tumor patients. We advocate the preferential use of valproate, because phenytoin may exhibit undesirable drug interactions with systemic chemotherapy agents (e.g., paclitaxel) and because carbamazepine is not available for intravenous administration. Terminally ill patients may be treated with benzodiazepines, which are easily administered orally, intravenously, or rectally.

## Surgery

Specific modes of cancer therapy for patients with brain metastases include surgery, radiotherapy, and chemotherapy. Surgery is most commonly used as part of a combined therapeutic approach involving macroscopic tumor resection followed by whole-brain radiotherapy. In general, surgery is a useful therapeutic measure for: (1) patients with single rather than multiple metastases; (2) patients in good general condition and neurological performance status; (3) patients with no or stable extracranial disease (e.g., no change in tumor status during the previous 3 months); (4) patients with a radioresistant tumor; (5) patients with an unknown primary tumor or a lesion of diagnostic uncertainty; and (6) patients with an accessible lesion, without a high risk of a disabling neurological deficit on resection. There are exceptions to these guidelines. Even patients with two or three metastases are candidates for surgery if the other criteria are met. This situation is not uncommon in patients with metachronous metastases from gastrointestinal or renal cell cancer. No attempt at resection should be undertaken in patients with small-cell lung cancer or lymphoma, because these cancers are usually radiosensitive and chemosensitive.

## Radiotherapy

Radiotherapy is the single most important therapeutic measure in the management of brain metastases. The radiosensitivity of brain metastases depends on the histological findings of the tumor (Table II). Parallel to the rates of clinical improvement of up to 90% that are achieved with radiotherapy, CT reveals major—commonly defined as more than 50% regression of contrast-enhancing lesions—or minor responses in 60%–80% of patients with metastases from non-small cell or small cell lung cancer and approximately 75% of patients with metastases from breast cancer. In contrast, brain metastases from melanoma respond in only 10% of cases, notwithstanding the clinical improvement in 40%–70% of the patients, which may be largely due to the use of steroids (Cairncross and Posner, 1983; Katz, 1981). Similarly, radiographic response rates in the range of 10% are obtained in patients with urogenital tumors originating from kidney, bladder, prostate, or testes, although these patients show clinical response rates of 30%–50% (Maor et al., 1988; Reddy et al., 1983). The same holds true for gastrointestinal tumors (Cascino et al., 1983). In contrast, metastases from germ cell tumors are radiosensitive (Lester et al., 1984; Fossa et al., 1999).

Because the detection even of a *single brain metastasis* signifies the capacity of the primary tumor to seed in the brain, the presence of further microscopic lesions in the brain must be suspected. Accordingly, whole-brain radiotherapy, including optic nerve sheaths and posterior fossa, has assumed a major role in the management of patients with brain metastases. There may be improved local control of metastatic brain disease for patients receiving postsurgical whole-brain radiotherapy compared with surgery alone (Armstrong et al., 1994; Smalley et al., 1987; Patchell et al., 1998). However, the median survival is not prolonged, and severe side effects, notably radiation-induced dementia, were observed in 10% of the patients surviving for 1 year or more in earlier studies (DeAngelis et al., 1989b; Hagen et al., 1990).

The approach of whole-brain radiotherapy is particularly common in patients with *multiple brain metastases*. Because MRI is more sensitive for the detection of small metastases than CT and because MRI is now the preferred technique to assess patients with brain metastases, fewer patients may be diagnosed with single brain metastases than 10–20 years ago. The results of representative therapeutic studies on whole-brain radiotherapy for patients with brain metastases are summarized in Table II.

Retrospective studies that included patients with different primary tumors and treated with different regimens of whole-brain radiotherapy reported median survival times from 2–9 months. The Radiation Therapy Oncology Group (RTOG) conducted several prospective randomized trials that included more than 2000 patients and that yielded essentially similar results. Fractionation schedules of 20 Gy in 1 week, 30 Gy in 2 weeks, 30 Gy in 3 weeks, 40 Gy in 3 weeks, 40 Gy in 4 weeks, or 50 Gy in 4 weeks did not differ with respect to neurological function, duration of remission or median survival (Borgelt et al., 1980, 1981; Hendrickson *et al.*, 1983; Kurtz *et al.*, 1981). No advantage of one or the other fractionation schedule became apparent in 561 patients with favorable prognostic factors either (Gelber *et al.*, 1981). Accelerated schedules of 20 Gy in 1 week or 30 Gy in 2 weeks were most efficient in inducing rapid neurological improvement in approximately 50% of the patients by the end of 2 weeks. A boost to the tumor region in addition to whole-brain radiotherapy did not improve outcome (Hoskin *et al.*, 1990).

As indicated previously, surgical resection of one or more metastases followed by whole-brain radiotherapy is considered the treatment of choice in most centers. Given improved surgical techniques and a decrease in perioperative and postoperative mortality, the median survival has increased from 3–7 months to 5–20 months, and 1-year survival rates have exceeded 50% in some series. A synopsis of studies broken down by tumor entity where feasible is provided in Table III. The results of such studies need to be interpreted with caution. Few studies were randomized. Not only the techniques of treatment have changed over time but also the selection criteria of patients. Modern data on control groups receiving best supportive care only are not available.

The advantage of surgery followed by radiotherapy over radiotherapy alone has been well documented in patients with single brain metastases. The RTOG studies included 218 patients who had surgical resection of their brain metastases followed by whole-brain radiotherapy (Hendrickson *et al.*, 1983). The improvement of neurological function lasted longer in patients who had complete resection of a single metastasis before radiotherapy than in patients who had radiotherapy alone. Resection was a prognostic factor, especially in patients with good neurological function and with melanoma or renal cell cancer as primary tumors. In contrast, resection was less effective in patients with non-small cell lung cancer and seemed to be without benefit in patients with breast or colorectal cancer. Ensuing studies including two prospective, randomized trials (Patchell *et al.*, 1990; Vecht *et al.*, 1993) showed significantly increased survival of patients with single brain metastases from non-small cell lung cancer and non-lung cancer after surgical resection followed by whole-brain radiotherapy compared with radiotherapy alone (Mandell *et al.*, 1986; Patchell *et al.*, 1986, 1990; Vecht *et al.*, 1993). A third randomized trial did not show this advantage of surgery (Mintz *et al.*, 1996).

Several approaches to improve the efficacy of radiotherapy for brain metastases have focused on the use of radiosensitizing agents. Such agents will overcome the intrinsic radioresistance of some solid tumors or confer radiosensitization to tumor cells in hypoxic tumor regions. Negative studies include whole-brain radio-

therapy (30 Gy) plus the candidate radiosensitizing agents metronidazole (Aiken *et al.*, 1984) or lonidamine (DeAngelis *et al.*, 1989a).

## Prophylactic Cranial Irradiation (PCI)

Whole-brain radiotherapy at 20–30 Gy in 2 Gy fractions is used as a prophylactic measure in patients with small cell lung cancer who receive a complete response after systemic chemotherapy. PCI results in a reduced incidence of subsequent brain metastases but does not significantly prolong survival (Arriagada *et al.*, 1995; Gregor *et al.*, 1997; Lucas *et al.*, 1986; Rusch *et al.*, 1989). Similar results have been reported for non-small cell lung cancer (Stuschke *et al.*, 1999). PCI should not be given concurrently with chemotherapy or to patients with extensive disease at presentation (Glantz *et al.*, 1997).

## Radiosurgery

Single fraction percutaneous stereotactic radiotherapy, also known as *radiosurgery*, is increasingly used in the management of brain metastases and is often a suitable alternative to conventional surgery. Radiosurgery can be applied as the primary treatment of single or multiple lesions smaller than 3 cm in diameter or as a secondary treatment in patients with a recurrent lesion in a previously irradiated field. Its advantages compared with surgery are the short hospitalization time and the lack of operative morbidity and mortality. Radiosurgery is effective for the local control of metastases from radiosensitive and radioresistant tumors. The local control rates are in the range of 73%–94% (Alexander *et al.*, 1996). Maximal tolerated doses were defined to be 24 Gy, 18 Gy, and 15 Gy for lesions ≤20 mm, 21–30 mm, and 31–40 mm in diameter, respectively (Shaw *et al.*, 2000). Similar to conventional surgery, radiosurgery is usually combined with adjuvant whole-brain radiotherapy (Fuller *et al.*, 1992). A recent randomized comparative study of radiosurgery followed by whole-brain radiotherapy vs whole-brain radiotherapy alone confirmed that the responses to whole-brain radiotherapy were only transient and that local control was much better with radiosurgery plus whole-brain radiotherapy (Kondziolka *et al.*, 1999). Conversely, recent retrospective data indicate that whole-brain radiotherapy may be safely delayed in patients receiving radiosurgery (Sneed *et al.*, 1999), with the possible benefit of less neurotoxicity from radiotherapy. Eventually, radiosurgery may also be a treatment option for patients with metastatic disease of the brainstem that is commonly not surgically accessible (Huang *et al.*, 1999).

Overall, the results of radiosurgery plus radiotherapy seem not to be as good as those obtained with surgery followed by radiotherapy (Table III). Some authors concluded, based on retrospective analysis, that surgery is superior to radiosurgery in comparable populations of patients (Bindal *et al.*, 1996). However, this view may be biased, because radiosurgery has often been used in patients with inoperable single or multiple metastatic lesions or with advanced systemic disease. These patients may have had a worse prognosis than the common surgical patients from the onset. The efficacy of radiosurgery may therefore be underestimated at present.

## Pharmacotherapy

Pharmacotherapy including conventional cancer chemotherapy plays a limited role in the treatment of brain metastases. The major reason why brain metastases respond poorly to chemotherapy is probably the chemoresistance of the primary tumors. Most of the tumors that metastasize to the brain, with the exception of small cell lung cancer and breast cancer, are chemoresistant cancers (e.g., non-small cell lung cancer, malignant melanoma, or renal cell carcinoma). Moreover, most patients receiving chemotherapy for brain metastases have previously received whole-brain radiotherapy, raising the possibility that brain metastases derive from cells selected already for resistance to irradiation and therefore to genotoxic stress in general.

The role of the blood–brain barrier in protecting brain metastases from the cytotoxic effects of cancer chemotherapy has probably been overestimated (Lesser, 1996). Brain metastases exhibit fenestrated endothelial linings that do not resemble the blood–brain barrier. The latter, in contrast, is characterized by intercellular tight junctions bridging nonfenestrated endothelial cells. Because this barrier may still be intact at the interface of blood vessels and the CSF space, measuring CSF drug levels is not a suitable parameter to assess whether a drug will reach brain metastases or not. Accordingly, numerous studies have confirmed adequate concentrations of cytotoxic drugs in resected metastases but not in the CSF (Lesser, 1996). Moreover, the blood vessels supplying brain metastases are not characterized by the expression of the multidrug resistance (mdr-1) gene encoding the P-glycoprotein. This protein is highly expressed in blood vessels of the normal brain and prevents the intracellular accumulation of structurally unrelated drugs such as vincristine, doxorubicin, or etoposide (VP16). Protection from systemic chemotherapy by the blood–brain barrier can thus only be postu-

lated for the infiltrative edge of growing metastases. These considerations explain why brain metastases may occasionally develop or grow during chemotherapy-induced regression of extracerebral metastases—if the cells grow in a microenvironment protected by the blood–brain barrier—whereas drugs that are known to be mdr substrates, such as the topoisomerase II inhibitor etoposide (VP16) or the topoisomerase I inhibitor topotecan, exhibit activity against brain metastases from lung cancer—because the metastatic lesions do not contain barrier-expressing mdr genes. The blood–brain barrier and mdr-type drug transport mechanisms at the endothelial border will assume an important role for future efforts to supplement or replace whole-brain radiotherapy by chemotherapy for the sterilization of cerebral micrometastases.

Patients with brain metastases commonly receive concomitant chemotherapy during radiotherapy or postradiotherapy chemotherapy for the control of extracerebral metastases. The efficacy of combined systemic chemotherapy and whole-brain radiotherapy compared with radiotherapy alone, or of surgery followed by chemotherapy alone compared with surgery followed by radiotherapy alone, in the control of brain metastases has not been investigated systematically. One possible role of chemotherapy could be the adjuvant setting where patients would receive chemotherapy to target micrometastases and to prolong brain-specific progression-free survival (Nakagawa *et al.*, 1993). In principle, this type of adjuvant chemotherapy could be given after surgery to delay the need for whole-brain radiotherapy, or after radiotherapy to delay the time of progression or recurrence.

Other studies evaluating chemotherapy for patients with brain metastases have included patients who progressed after surgery and radiotherapy. These observations allow us to reflect more realistically the efficacy of chemotherapy but usually represent data obtained in a population of heavily pretreated patients, specifically patients harboring brain metastases likely to be formed by radioresistant tumor cell clones. In principle, patients with brain metastases should receive the same chemotherapy that would be used to treat extracerebral metastases from the same tumor.

Of all cancers, *lung cancer* has been most extensively assessed for the response of brain metastases to systemic chemotherapy (Postmus and Smit, 1999). A response rate, including complete and partial responses, of 43% was achieved with high-dose etoposide in patients with small cell lung cancer (Postmus *et al.*, 1989). Ensuing studies showed that responses were also achieved with conventional dose regimens, using etoposide or teniposide (Postmus *et al.*, 1995). Encouraging results mostly in patients with non-small cell lung cancer were obtained with cisplatin (Nakagawa *et al.*, 1993) and cis-

**TABLE IV    Outcome in Patients with Brain Metastases Treated with Chemotherapy in Addition to Surgery (S) or Radiotherapy (RT)**

| Reference | Tumor | n | Chemotherapy | Response | Progression-free survival (mos) | Median survival (mos) | Survival at 1 year (%) | Survival at 2 years (%) |
|---|---|---|---|---|---|---|---|---|
| Rosner et al., 1986 | Breast | 100 | CFP or CFPMV or MVP or CA* | 10 CR, 40 PR | | 5.5 | 31 | |
| Cocconi et al., 1990 | Breast | 22 | Cp + VP16 (±RT) | 5 CR, 7 PR | 5.8 | 13.5 | 55 | |
| Boogerd et al., 1992 | Breast | 22 | CMF or CAF | 2 CR, 10 PR | | 5.8 | | |
| Jacquillat et al., 1990 | Melanoma | 39 | Fotemustine | 2 CR, 9 PR | | 6.1 | 21 | |
| Chang et al., 1994 | Melanoma | 34 | Dacarbazine + fotemustine | 2 CR, 2 PR | | 4.5 | 20 | 5 |
| Nakagawa et al., 1993 | Lung | 25 | S ± RT + Cp | | 7.6 (mean) | 15 (mean) | 44 | 20 |
| | Lung | 25 | S ± RT + other chemotherapy | | 6.1 (mean) | 11 (mean) | 32 | 4 |
| | Lung | 39 | S ± RT | | 4.2 (mean) | 7 (mean) | 5 | 3 |
| Twelves et al. 1990 | Lung, SCLC | 14 | C + V + VP16 | 1 CR, 8/14 PR | | 7 | | |
| Kristjansen et al., 1993 | Lung, SCLC | 21 | Cp + VM26 + V | 4 CR, 7 PR | | 3.5 | | |
| Pujol et al., 1994 | Lung, NSCLC | 12 | Fotemustine | 1 CR, 1 PR | | | | |
| Postmus et al., 1995 | Lung, SCLC | 80 | VM26 | 6 CR, 20 PR | | 2.9 | | |
| Cotto et al., 1996 | Lung, NSCLC | 31 | Cp + fotemustine | 2 CR | | 4 | | |
| Malacarne et al., 1996 | Lung, all | 30 | Carboplatin + VP16 | 3 CR, 7 PR | | 5.7 | | |
| Ardizzoni et al., 1997 | Lung, SCLC | 7 | Topotecan | 3 CR, 1 CR | | | | |
| Kaba et al., 1997 | Rec. breast | 28 | TPcDCc-FHu | 7 PR | 4 | 8 | | |
| | Rec. lung, NSCLC | 39 | | 1 CR + 4 PR | 3 | 6 | | |
| | Rec. lung, SCLC | 9 | | 3 CR + 3 PR | 6.5 | 8.3 | | |
| | Rec. melanoma | 9 | | 0 CR, 0 PR | 1.5 | 3 | | |
| Minotti et al., 1998 | Lung, NSCLC | 23 | Cp + VM26 | 3 CR + 5 CR | | 5 | | |
| Franciosi et al., 1999 | Lung, NSCLC | 43 | Cp + VP16 | 3 PR + 10 PR | 4 | 7.5 | 25 | 10 |
| | Breast | 56 | | 7 CR, 14 PR | 4 | 7.2 | 32 | 9 |
| | Melanoma | 8 | | no CR or PR | 2.6 | 4 | 0 | |

*A, adriamycin/doxorubicin; C, cyclophosphamide; Cc, CCNU; Cp, cisplatin; D, dibromodulcitol; F, 5-fluorouracil; Hu, hydroxyurea; M, methotrexate; P, prednisone; Pc, procarbazine; T, thioguanine; V, vincristine; VM26, teniposide; VP16, etoposide; RT, radiotherapy.

platin plus etoposide (Franciosi et al., 1999). The overall response rates are in the range of 15%–40%, better in small cell lung cancer than in non-small cell lung cancer, and of only a few months duration (Table IV). Brain metastases from *ovarian cancer* may well respond to chemotherapy (Cooper et al., 1994) and should be treated as other metastases from this type of cancer outside the brain. The best results of systemic chemotherapy with response rates of about 50% and median survival times ranging from 3–8 months have been reported for patients with brain metastases from *breast cancer* (Boogerd et al., 1992; Rosner et al., 1986). Only 1 in 10 patients showed no response of the brain metastases when systemic metastases responded (Cocconi et al., 1990). On the other hand, the brain as the first site of relapse is more common in patients who received adjuvant chemotherapy for the management of the primary lesion than in those who did not (Boogerd et al., 1993). This observation suggests that brain metastases may find a protected site in the brain during systemic chemotherapy. Last, there are case

reports on responses of brain metastases from breast cancer to treatment with antiestrogens, mostly tamoxifen, but also megestrol acetate (Lesser, 1996). Approximately 10%–30% of *malignant melanomas* metastatic to the brain respond to systemic chemotherapy. Brain metastases from malignant melanoma that progress after radiotherapy may show transient responses or stabilization to chemotherapy using dacarbazine and fotemustine (Chang et al., 1994), fotemustine (Jacquillat et al., 1990; Kleeberg et al., 1995), the PCV (procarbazine, CCNU, vincristine) regimen, or temozolomide (Chapter 62). There is commonly a close correlation between the response of extracerebral and intracerebral metastases from melanoma (Jacquillat et al., 1990). Chemotherapy is a rather effective treatment for patients with brain metastases from *germ cell tumors* and *choriocarcinomas* (Athanassiou et al., 1983; Rustin et al., 1986, 1989; Spears et al., 1991). Particularly good response rates were observed with alternating courses of etoposide/methotrexate, actinomycin, and vincristine/cyclophosphamide in choriocarcinoma metastatic to the brain (Rustin *et al.*, 1989). Cisplatin-based chemotherapy resulted in a 45% 5-year survival rate in patients with brain metastases from malignant germ cell tumors at diagnosis (Fossa *et al.*, 1999). Rare brain metastases from *thyroid cancer* are rather refractory to treatment, including radiotherapy or radioactive iodine (Salvati *et al.*, 1995a; Samuel and Shah, 1997; Chiu *et al.*, 1997). Brain metastases from *renal cell carcinoma* and *colorectal cancer* usually do not respond to chemotherapy, consistent with the biological behavior of the primary tumors.

The *intra-arterial application* of cytotoxic drugs has been pursued with the aim to increase the local drug concentration in the metastatic tissue. The early studies have used cisplatin or nitrosoureas and did not result in major responses. Yet, intra-arterial chemotherapy was associated with severe side effects, including blindness and deafness (Williams *et al.*, 1995). Other types of local therapy (e.g., intralesional or intrathecal chemotherapy) play no role in the treatment of solid brain metastases. Furthermore, neither immunotherapeutic strategies nor somatic gene therapy has gained a place in the management of patients with brain metastases.

### Recurrent Brain Metastases

The prognosis for patients with recurrent brain metastases is generally poor. The recurrence of brain metastases may not always be easily distinguished from focal radiation necrosis by CT or MRI alone. Positron emission tomography (PET) may reveal metabolically active tumor, whereas necrotic tissue is supposed to be inactive. If PET is not available or not informative, biopsy may be necessary to clarify the nature of the lesion. The treatment options for patients with recurrent brain metastases depend on the prior treatment. Surgical removal of relapsed brain metastases is an option that may be of benefit in patients with favorable prognostic factors and notably cancers that are resistant to cytotoxic therapies, such as non-small cell lung cancer (Arbit *et al.*, 1995; Bindal *et al.*, 1993, 1995; Sundaresan *et al.*, 1988). Whole-brain radiotherapy may be given to patients who have not had whole-brain radiotherapy before. Radiosurgery is a suitable approach for patients who have had either surgery or whole-brain radiotherapy before (Chen *et al.*, 2000). Interstitial radiotherapy (brachytherapy) using radioactive iodine or iridium may be an option but is less frequently used than a few years ago (Heros *et al.*, 1988). Aggressive chemotherapy protocols may induce transient responses in a significant proportion of patients with recurrent brain metastases (Kaba *et al.*, 1997) (Table IV).

### Concomitant Leptomeningeal Metastasis

In the absence of increased intracranial pressure, which might lead to neurological deterioration after a lumbar puncture, CSF analysis should be part of the diagnostic workup of patients with brain metastases to identify tumor cell dissemination in the subarachnoid space. The diagnosis and treatment of leptomeningeal metastases are outlined in Chapter 65.

## PRACTICAL MANAGEMENT

### Symptomatic Treatment

Symptomatic treatment measures are initiated as soon as the diagnosis of brain metastases is made. A minority of patients with brain metastases are seen as emergencies with progressive neurological deficits and loss of consciousness. Patients with potentially controlable disease, who are unconscious and are likely to have cerebral herniation develop, should be intensively treated using high-dose steroids, hyperosmolar agents, and even forced hyperventilation. After clinical stabilization, the further treatment of these patients follows the same principles as in all other patients. Patients with occlusion hydrocephalus caused by infratentorial metastases may need immediate external ventricular drainage.

Most patients with mild or moderate symptoms and signs of increased intracranial pressure or focal neurological symptoms caused by local perifocal edema receive steroids after the tentative diagnosis of brain metastases has been made. In patients with no known primary tumor, steroids should not be given with lesions radiographically consistent with primary CNS lymphoma. Only patients with severe neurological deficits (e.g., prominent aphasia or hemiparesis) should receive an intravenous bolus, e.g., of 40 mg dexamethasone. Before surgical therapy, except for a ventricular draining procedure, dexamethasone at 4 mg qid should be administered for 2–3 days to prevent exacerbation of edema during and after surgery. Patients who are not candidates for surgery are treated with 1 mg qid initially (Vecht *et al.*, 1994), especially before the initiation of radiotherapy. The dose of steroids is increased as needed and tapered or discontinued near the end or after the completion of radiotherapy. Steroid therapy requires continuous prophylactic treatment to prevent gastrointestinal ulceration with antacids, $H_2$-receptor antagonists, or protein pump inhibitors. Individualized measures to prevent osteoporosis and imbalances of salts and fluids need to be considered as well.

The principles of *anticonvulsant treatment* in patients with brain metastases have been outlined earlier.

We do not advocate the routine use of anticonvulsants in these patients. Daily doses, therapeutic plasma levels, and side effects of anticonvulsants are detailed in Chapter 20. Hospitalized patients with advanced systemic disease who seem to have a life expectancy of less than 2–3 months despite treatment, who are unable to communicate, or who get somnolent or unconscious during radiotherapy should only receive symptomatic therapy involving steroids, anticonvulsants, and analgesics.

## Surgery, Radiotherapy, and Chemotherapy for Single or Multiple Brain Metastases

With the exception of small cell lung cancer, germinoma, or lymphoid or hematological cancers, single brain metastases should be resected,** following the guidelines provided earlier. Alternately, or in patients who are not eligible for surgery, radiosurgery should be performed if the lesion measures 3 cm or less in diameter.* Patients with advanced systemic disease may be treated with whole-brain radiotherapy upfront. Surgical resection or radiosurgery of single brain metastases is commonly followed by whole-brain radiotherapy with 36 Gy, 4 × 3 Gy fractions per week. A boost of 9 Gy may be given to sites of incompletely resected metastases. In the presence of several favorable prognostic factors, we recommend a regimen of 36–45 Gy in 5 × 2 Gy fractions per week, with the aim to increase the duration of neurological remissions and to reduce the likelihood of late side effects of radiotherapy. As indicated previously, there is little evidence from controlled clinical trials to support this widespread use of whole-brain radiotherapy. There is therefore justification to delay whole-brain radiotherapy in patients with completely resected brain metastases, notably from radioresistant cancers. To avoid the late side effects of radiotherapy, we withhold whole-brain radiotherapy in patients with completely resected single brain metastases from renal or gastrointestinal cancer, notably if these metastases are metachronous. Such patients have often experienced long-term remissions after surgery alone. Brain metastases from radiosensitive tumors such as small cell lung cancer do not require resection and are treated with whole-brain radiotherapy using 36 Gy in 4 × 3 Gy fractions per week and possibly a boost of 9 Gy to the tumor region. As indicated earlier, 2 Gy single fractions are preferred in patients with several favorable prognostic factors.

Most patients with multiple brain metastases are treated with whole-brain radiotherapy.** Exceptions include patients with severe neurological deficits such as progressive aphasia or hemiparesis attributable to one specific lesion who may also be candidates for surgery or patients meeting the criteria favoring resection of multiple brain metastases outlined earlier. Whole-brain radiotherapy for multiple brain metastases is applied as indicated previously.

Selected regimens of chemotherapy per tumor type are summarized in Table V. These regimens may be administered to patients who are not candidates for radiotherapy but who are in adequate physical and mental condition for such an approach (e.g., have a Karnofsky index of at least 50%). Preirradiation chemotherapy is an attractive approach that should be pursued in the context of clinical trials.

## Side Effects from Radiotherapy and Chemotherapy

Whole-brain radiotherapy carries significant risks of acute and chronic toxicity, which depend on the total dose and on the fractionation schedule (Cairncross and Posner, 1983; Crossen et al., 1994; Marks et al., 1981). Alopecia may be expected in all patients and is incompletely reversible. Early neurological side effects occur in about 20% of patients and include nausea and vomiting, fever, and neurological deterioration. These early side effects are attributed to radiation-induced vascular leakage, which may exacerbate pre-existing edema. They respond well to increased doses of corticosteroids (Cairncross and Posner, 1983; Weissman, 1988).

TABLE V    Selected Chemotherapy Regimens for Patients with Brain Metastases

| Drug | Schedule | Major toxicity | Tumor type | Reference |
|---|---|---|---|---|
| Carboplatin (Cb) + etoposide (VP16) | Cb 300 mg/m² D1 + VP16 120 mg D1–3 × 4 weeks | Myelosuppression | SCLC, NSCLC | Malacarne et al., 1996 |
| Cisplatin (Cp) + etoposide (VP16) | Cp 100 mg/m² D1 + VP16 100 mg/m²D1,3,5 × 3 wk | Myelosuppression | Breast cancer, NSCLC | Franciosi et al., 1999 |
| Fotemustine | 100 mg/m², e.g., D1,8 × 6 wk | Myelosuppression | NSCLC, melanoma | Pujol et al., 1994 |
| Temozolomide | 200 mg/m² D1–5 × 4 wk | Myelosuppression | Melanoma | Unpublished |
| Teniposide (VM26) | 150 mg/m² D1,3,5 × 3 wk | Myelosuppression | SCLC | Postmus et al., 1995 |
| Topotecan | 1.5 mg/m² D1–5 × 3 wk | Myelosuppression | SCLC, NSCLC | Ardizzoni et al., 1997 |

Vertigo, nausea, and fatigue developing at 1–3 months after radiotherapy affect 20%–25% of the patients. This condition is attributed to transient demyelination, tends not to respond to corticosteroids, is usually mild, and remits spontaneously in most patients (Bleyer and Byrne, 1988). The clinical syndrome of dementia, ataxia, and urinary incontinence associated radiologically with leukencephalopathy and global brain atrophy, or radiation necrosis that often is seen as a space-occupying lesion, are observed in approximately 10% of the patients who have been irradiated for brain metastases and survive for 1 year or longer (DeAngelis et al., 1989b; Sundaresan and Galicich, 1985). As indicated earlier, the rate of leukencephalopathy depends on the fractionation and the cumulative dose of radiation (DeAngelis et al., 1989c; Marks et al., 1981). Patients with space-occupying radiation necroses and symptomatic edema may be managed with high doses of corticosteroids or anticoagulation (Glantz et al., 1994), but surgical resection of the necrotic tissue is sometimes necessary in neurologically deteriorating patients. Side effects from radiosurgery include acute mild headache and nausea and delayed steroid-sensitive transient edema with temporary worsening of neurological symptoms in about 10% of the patients. Delayed radionecrosis of neural tissue has been reported for about 3% of the patients but may become a critical issue with increasing use of this technique and longer patient follow-up (Engenhardt et al., 1993; Fuller et al., 1992).

## Treatments no Longer Recommended

A second course of whole-brain radiotherapy in patients with relapsing or progressive brain metastases is highly neurotoxic. The response rate is in the range of 30%, whereas survival may be prolonged for some weeks (Hazuka and Kinzie, 1988). Radiosurgery has largely replaced the more invasive approach of interstitial radiotherapy (brachytherapy) for recurrent brain metastases. Intraarterial chemotherapy of brain metastases with or without hyperosmotic opening of the blood–brain barrier may have severe side effects and has not proven to be superior to conventional chemotherapy (Williams et al., 1995).

## REFERENCES

Aiken, R., Leavengood, J. M., Kim, J. H., Deck, M. D. F., Thaler, H. T., and Posner, J. B. (1984). Metronidazole in the treatment of metastatic brain tumors. Results of a controlled clinical trial. J. Neuro-Oncol. 2, 105–111.

Alexander III, E., Moriarty, T. M., Davis, R. B., Wen, P. Y., Fine, H. A., Black, P. M., Kooy, H. M., and Loeffler, J. S. (1995). Stereotactic radiosurgery for the definitive, noninvasive treatment of brain metastases. J. Natl. Cancer Inst. 87, 34–40.

Alexander III, E., Moriarty, T. M., and Loeffler, J. S. (1996). Radiosurgery for metastases. J. Neuro-Oncol. 27, 279–285.

Arbit, E., Wronski, M., Burt, M., and Galicich, J. H. (1995). The treatment of patients with recurrent brain metastases. A retrospective analysis of 109 patients with nonsmall cell lung cancer. Cancer 76, 765–773.

Ardizzoni, A., Hansen, H., Dombernowsky, P., Gamucci, T., Kaplan, S., Postmus, P., Giaccone, G., Schaefer, B., Wanders, J., and Verweij, J., for the EORTC Early Clinical Studies Group and New Drug Development Office, and the Lung Cancer Cooperative Group. (1997). Topotecan, a new active drug in the second-line treatment of small-cell lung cancer: a phase II study in patients with refractory and sensitive disease. J. Clin. Oncol. 15, 2090–2096.

Armstrong, J. G., Wronski, M., Galicich, J., Arbit, E., Leibel, S. A., and Burt, M. (1994). Postoperative radiation for lung cancer metastatic to the brain. J. Clin. Oncol. 12, 2340–2344.

Arriagada, R., Le Chevalier, T., Borie, F., Rivière, A., Chomy, P., Monnet, I., Tardivon, A., Viader, F., Tarayre, M., and Benhamou, S. (1995). Prophylactic cranial irradiation for patients with small-cell lung cancer in complete remission. J. Natl. Cancer Inst. 87, 183–190.

Athanassiou, A., Begent, R. H. J., Newlands, E. S., Parker, D., Rustin, G. J. S., and Bagshawe, K. D. (1983). Central nervous system metastases of choriocarcinoma. 23 years experience at Charing Cross Hospital. Cancer 52, 1728–1735.

Bindal, R. K., Sawaya, R., Leavens, M. E., and Lee, J. J. (1993). Surgical treatment of multiple brain metastases. J. Neurosurg. 79, 210–216.

Bindal, R. K., Sawaya, R., Leavens, M. E., Hess, K. R., and Taylor, S. H. (1995). Reoperation for recurrent metastatic brain tumors. J. Neurosurg. 83, 600–604.

Bindal, A. K., Bindal, R. K., Hess, K. R., Shiu, A., Hassenbusch, S. J., Shi, W. M., and Sawaya, R. (1996). Surgery versus radiosurgery in the treatment of brain metastasis. J. Neurosurg. 84, 748–754.

Bleyer, W. A., and Byrne, T. N. (1988). Leptomeningeal cancer in leukemia and solid tumors. Curr. Probl. Cancer 12, 181–238.

Boogerd, W., Dalesio, O., Bais, E. M., and van der Sande, J. J. (1992). Response of brain metastases from breast cancer to systemic chemotherapy. Cancer 69, 972–980.

Boogerd, W., Vos, V. W., Hart, A. A. M., and Baris, G. (1993). Brain metastases in breast cancer; natural history, prognostic factors and outcome. J. Neuro-Oncol. 15, 165–174.

Borgelt, B., Gelber, R., Kramer, S., Brady, L. W., Chang, C. H., Davis, L. W., Perez, C. A., and Hendrickson, F. R. (1980). The palliation of brain metastases: Final results of the first two studies by the Radiation Therapy Oncology Group. Int. J. Radiat. Oncol. Biol. Phys. 6, 1–9.

Borgelt, B., Gelber, R., Larson, M., Hendrickson, F., Griffin, T., and Roth, R. (1981). Ultra-rapid high dose irradiation schedules for the palliation of brain metastases: Final results of the first two studies by the Radiation Therapy Oncology Group. Int. J. Radiat. Oncol. Biol. Phys. 7, 1633–1638.

Cairncross, J. G., Chernik, N. L., Kim, J. K., and Posner, J. B. (1979). Sterilization of cerebral metastases by radiation therapy. Neurology 29, 1195–1202.

Cairncross, J. G., and Posner, J. B. (1983). The management of brain metastases. In Martinus Nijhoff, (M. D. Walker, Ed.) Oncology of the Nervous System Boston, The Hague, Dordrecht, Lancaster: 341–377.

Carella, R. J., Gelber, R., Hendrickson, F., Berry, H. C., and Cooper, J. S. (1980). Value of radiation therapy in the management of patients with cerebral metastases from malignant melanoma:

Radiation Therapy Oncology Group brain metastases study I and II. *Cancer* 45, 679–683.

Carmichael, J., Crane, J. M., Bunn, P. A., Glatstein, E., and Ihde, D. C. (1988). Results of therapeutic cranial irradiation in small cell lung cancer. *Int. J. Radiat. Oncol. Biol. Phys.* 14, 455–459.

Cascino, T. L., Leavengood, J. M., Kemeny, N., and Posner, J. B. (1983). Brain metastases from colon cancer. *J. Neuro-Oncol.* 1, 203–209.

Chang, J., Atkinson, H., A'Hern, R., Lorentzos, A., and Gore, M. E. (1994). A phase II study of the sequential administration of dacarbazine and fotemustine in the treatment of cerebral metastases from malignant melanoma. *Eur. J. Cancer* 30A, 2093–2095.

Chassard, J. L., Zouai, M. E., Gérard, J. P., Dutou, L., Mornex, F., Lacroze, M., Ardiet, J. M., and Chauvin, F. (1988). La radiothérapie des métastases cérébrales. 196 cas traités de 1973 à 1981. *Rev. Neurol. Paris* 144, 489–493.

Chen, J. C. T., Petrovich, Z., Giannotta, S. L., Yu, C., and Apuzzo, M. L. J. (2000). Radiosurgical salvage therapy for patients presenting with recurrence of metastatic disease to the brain. *Neurosurgery* 46, 860–866.

Chiu, A. C., Delpassand, E. S., and Sherman, S. I. (1997). Prognosis and treatment of brain metastases in thyroid carcinoma. *J. Clin. Endocrinol. Metab.* 82, 3637–3642.

Choi, K. N., Withers, H. R., and Rotman, M. (1985). Intracranial metastases from melanoma. Clinical features and treatment by accelerated fractionation. *Cancer* 56, 1–9.

Cocconi, G., Lottici, R., Bisagni, G., Bacchi, M., Tonato, M., Passalacqua, R., Boni, C., Belsanti, V., and Bassi, P. (1990). Combination therapy with platinum and etoposide of brain metastases from breast carcinoma. *Cancer Invest.* 8, 327–334.

Coffey, R. J., Flickinger, J. C., Bissonette, D. J., and Lunsford, L. D. (1991). Radiosurgery for solitary brain metastases using the cobalt-60 gamma unit: methods and results in 24 patients. *Int. J. Radiat. Oncol. Biol. Phys.* 20, 1287–1295.

Cohen, N., Strauss, G., Lew, R., Silver, D., and Recht, L. (1988). Should prophylactic anticonvulsants be administered to patients with newly-diagnosed cerebral metastases? A retrospective analysis. *J. Clin. Oncol.* 6, 1621–1624.

Cooper, K. G., Kitchener, H. C., and Parkin, D. E. (1994). Cerebral metastases from epithelial ovarian carcinoma treated with carboplatin. *Gynecol. Oncol.* 44, 318–323.

Corn, B. J., Greven, K. M., Randall, M. E., Wolfson, A. H., Kim, R. Y., and Lanciano, R. M. (1995). The efficacy of cranial irradiation in ovarian cancer metastatic to the brain: analysis of 32 cases. *Obstet. Gynecol.* 86, 955–959.

Cotto, C., Berille, J., Souquet, P. J., Riou, R., Croisile, B., Turjman, F., Giroux, B., Brune, J., and Trillet-Lenoir, V. (1996). A phase II trial of fotemustine and cisplatin in central nervous system metastases from non-small cell lung cancer. *Eur. J. Cancer* 32A, 69–71.

Crossen, J. R., Garwood, D., Glatstein, E., and Neuwelt, E. A. (1994). Neurobehavioral sequelae of cranial irradiation in adults: a review of radiation-induced encephalopathy. *J. Clin. Oncol.* 12, 627–642.

Culine, S., Bekradda, M., Kramar, A., Rey, A., Escudier, B., and Droz, J. P. (1998). Prognostic factors for survival in patients with brain metastases from renal cell carcinoma. *Cancer* 83, 2548–2553.

DeAngelis, L. M., Currie, V. E., Kim, J. H., Krol, G., O'Hehir, M. A., Farag, F. M., Young, C. W., and Posner, J. B. (1989a). The combined use of radiation therapy and lonidamine in the treatment of brain metastases. *J. Neuro-Oncol.* 7, 241–247.

DeAngelis, L. M., Mandell, L. R., Thaler, H. T., Kimmel, D. W., Galicich, J. H., Fuks, Z., and Posner, J. B. (1989b). The role of postoperative radiotherapy after resection of single brain metastases. *Neurosurgery* 24, 798–805.

DeAngelis, L. M., Delattre, J. Y., and Posner, J. B. (1989c). Radiation-induced dementia in patients cured of brain metastases. *Neurology* 39, 789–796.

Delattre, J. Y., Krol, G., Thaler, H. T., and Posner, J. B. (1988). Distribution of brain metastases. *Arch. Neurol.* 45, 741–744.

Diener-West, M., Dobbins, T. W., Phillips, T. L., and Nelson, D. F. (1989). Identification of an optimal subgroup for treatment evaluation of patients with brain metastases using RTOG study 7916. *Int. J. Radiat. Oncol. Biol. Phys.* 16, 669–673.

Doyle, T. J. (1982). Brain metastasis in the natural history of small-cell lung cancer 1972–1979. *Cancer* 50, 752–754.

Eapen, L., Vachet, M., Catton, G., Danjoux, C., McDermot, R., Nair, B., Girard, A., Genest, P., Stewart, D., and Gerig, L. (1988). Brain metastases with an unknown primary: a clinical perspective. *J. Neuro-Oncol.* 6, 31–35.

Engenhardt, R., Kimmig, B. N., Höver, K. H., Wowra, B., Romahn, J., Lorenz, W. J., van Kaick, G., and Wannenmacher, M. (1993). Long-term follow-up for brain metastases treated by percutaneous stereotactic single high-dose irradiation. *Cancer* 71, 1353–1361.

Flickinger, J. C., Kondziolka, D., Lunsford, L. D., Coffey, R. J., Goodman, M. L., Shaw, E. G., Hudgins, W. R., Weiner, R., Harsh IV, G. R., Sneed, P. K., and Larson, D. A. (1994). A multi-institutional experience with stereotactic radiosurgery for solitary brain metastasis. *Int. J. Radiat. Oncol. Biol. Phys.* 28, 797–802.

Fossa, S. D., Bokemeyer, C., Gerl, A., Culine, S., Jones, W. G., Mead, G. M., Germa-Luch, J. R., Pont, J., Schmoll, H. J., and Tjulandin, S. (1999). Treatment outcome of patients with brain metastases from malignant germ cell tumors. *Cancer* 85, 988–997.

Franciosi, V., Cocconi, G., Michiara, M., Di Constanzo, F., Fosser, V., Tonato, M., Carlini, P., Boni, C., and Di Sarra, S. (1999). Front-line chemotherapy with cisplatin and etoposide for patients with brain metastases from breast carcinoma, nonsmall cell lung carcinoma, or malignant melanoma. A prospective study. *Cancer* 85, 1599–1605.

Fuller, B. G., Kaplan, I. D., Adler, J., Cox, R. S., and Bagshaw, M. A. (1992). Stereotaxic radiosurgery for brain metastases: the importance of adjuvant whole brain irradiation. *Int. J. Radiat. Oncol. Biol. Phys.* 23, 413–418.

Gagliardi, F. M., and Mercuri, S. (1983). Single metastases in the brain: late results in 325 cases. *Acta Neurochir.* 68, 253–262.

Gelber, R. D., Larson, M., Borgelt, B. B., and Kramer, S. (1981). Equivalence of radiation schedules for the palliative treatment of brain metastases in patients with favorable prognosis. *Cancer* 48, 1749–1753.

Glantz, M. J., Burger, P. C., Friedmann, A. H., Radtke, R. A., Massey, E. W., and Schold, S. C. (1994). Treatment of radiation-induced nervous system injury with heparin and warfarin. *Neurology* 44, 2020–2027.

Glantz, M. J., Choy, H., and Yee, L. (1997). Prophylactic cranial irradiation in small cell lung cancer: rationale, results, and recommendations. *Semin. Oncol.* 24, 477–483.

Gregor, A., Cull, A., Stephens, R. J., Kirkpatrick, J. A., Yarnold, J. R., Girling, D. J., Macbeth, F. R., Stout, R., and Machin, D., on behalf of the United Kingdom Coordinating Committee for Cancer Research (UKCCCR) and the European Organization for Research and Treatment of Cancer (EORTC). (1997). Prophylactic cranial irradiation is indicated following complete response to induction therapy in small cell lung cancer: results of a multicentre randomised trial. *Eur. J. Cancer* 33, 1752–1758.

Haar, F., and Patterson, R. H. (1972). Surgery for metastatic intracranial neoplasm. *Cancer* 30, 1241–1245.

Hagen, N. A., Cirrincione, C., Thaler, H. T., and DeAngelis, L. M. (1990). The role of radiation therapy following resection of single brain metastasis from melanoma. *Neurology* 40, 158–160.

Hall, W. A. (1998). The safety and efficacy of stereotactic biopsy for intracranial lesions. *Cancer* 82, 1749–1755.

Hazuka, M. B., and Kinzie, J. J. (1988). Brain metastases: results and effects of re-irradiation. *Int. J. Radiat. Oncol. Biol. Phys.* 15, 433–437.

Hendrickson, F. R. (1977). The optimum schedule for palliative radiotherapy for metastatic brain cancer. *Int. J. Radiat. Oncol. Biol. Phys.* **2**, 165–168.

Hendrickson, F. R., Lee, M. S., Larson, M., and Gelber, R. D. (1983). The influence of surgery and radiation therapy on patients with brain metastases. *Int. J. Radiat. Oncol. Biol. Phys.* **9**, 623–627.

Heros, D. O., Kasdon, D. L., and Chun, M. (1988). Brachytherapy in the treatment of recurrent solitary brain metastases. *Neurosurgery* **23**, 733–737.

Hoskin, P. J., Crow, J., and Ford, H. T. (1990). The influence of extent and local management on the outcome of radiotherapy for brain metastases. *Int. J. Radiat. Oncol. Biol. Phys.* **19**, 111–115.

Huang, C. F., Kondziolka, D., Flickinger, J. C., and Lunsford, L. D. (1999). Stereotactic radiosurgery for brainstem metastases. *J. Neurosurg.* **91**, 563–568.

Jacquillat, C., Khayat, D., Banzet, P., Weil, M., Avril, M. F., Fumoleau, P., Namer, M., Bonneterre, J., Kerbrat, P., Bonerandi, J. J., Bugat, R., Montcuquet, P., Audhuy, B., Cupissol, D., Lauvin, R., Grosshans, E., Vilmer, C., Prache, C., and Bozzarri, J. P. (1990). Chemotherapy by fotemustine in cerebral metastases of disseminated malignant melanoma. *Cancer Chemother. Pharmacol.* **25**, 263–266.

Jokura, H., Takahashi, K., Kayama, T., and Yoshimoto, T. (1994). Gamma knife radiosurgery of a series of only minimally selected metastatic brain tumors. *Acta Neurochir. (Suppl.)* **62**, 77–82.

Kaba, S. E., Kyritsis, A. P., Hess, K., Yung, W. K. A., Mercier, R., Dakhil, S., Jaeckle, K. A., and Levin, V. A. (1997). TPDC-FuHu chemotherapy for the treatment of recurrent metastatic brain tumors. *J. Clin. Oncol.* **15**, 1063–1070.

Kamby, C., and Soerensen, P. S. (1988). Characteristics of patients with short and long survivals after detection of intracranial metastases from breast cancer. *J. Neuro-Oncol.* **6**, 37–45.

Katz, H. R. (1981). The relative effectiveness of radiation therapy, corticosteroids, and surgery in the management of melanoma metastatic to the central nervous system. *Int. J. Radiat. Oncol. Biol. Phys.* **7**, 897–906.

Kleeberg, U. R., Engel, E., Israels, P., Bröcker, E. B., Tilgen, W., Kennes, C., Gérard, B., Lejeune, F., Glabbeke, M. V., and Lentz, M. A. (1995). Palliative therapy of melanoma patients with fotemustine. Inverse relationship between tumor load and treatment effectiveness. A multicentre phase II trial of the EORTC-Melanoma Cooperative Group (MCG). *Melanoma Res.* **5**, 195–200.

Kondziolka, D., Patel, A., Lunsford, L. D., Kassam, A., and Flickinger, J. C. (1999). Stereotactic radiosurgery plus whole brain radiotherapy versus radiotherapy alone for patients with multiple brain metastases. *Int. J. Radiat. Oncol. Biol. Phys.* **45**, 427–434.

Kristjansen, P. E. G., Sørensen, P. S., Hansen, M. S., and Hansen, H. H. (1993). Prospective evaluation of the effect on initial brain metastases from small cell lung cancer of platinum-etoposide based induction chemotherapy followed by an alternating multidrug regimen. *Ann. Oncol.* **4**, 579–593.

Kurtz, J. M., Gelber, R., Brady, L. W., Carella, R. J., and Cooper, J. S. (1981). The palliation of brain metastases in a favorable patient population: a randomized clinical trial by the Radiation Therapy Oncology Group. *Int. J. Radiat. Oncol. Biol. Phys.* **7**, 891–895.

Lang, E. F., and Slater, J. (1964). Metastatic brain tumors: Results of surgical and non-surgical treatment. *Surg. Clin. North Am.* **44**, 865–872.

Lesser, G. L. (1996). Chemotherapy of cerebral metastases from solid tumors. *Neurosurg. Clin. North Am.* **7**, 527–536.

Lester, S. G., Morphis II, J. G., Hornback, N. B., Williams, S. D., and Einhorn, L. H. (1984). Brain metastases and testicular tumors: need for aggressive therapy. *J. Clin. Oncol.* **2**, 1397–1403.

Lucas, C. F., Robinson, B., Hoskin, P. J., Yarnold, J. R., Smith, I. E., and Ford, H. T. (1986). Morbidity of cranial relapse in small cell lung cancer and the impact of radiation therapy. *Cancer Treat. Rep.* **70**, 565–570.

Magilligan, D. J., Rogers, J. S., Knighton, R. S., and Davila, J. C. (1976). Pulmonary neoplasm with solitary cerebral metastasis. Results of combined excision. *J. Thorac. Cardiovasc. Surg.* **72**, 690–696.

Malacarne, P., Santini, A., and Maestri, A. (1996). Response of brain metastases from lung cancer to systemic chemotherapy with carboplatin and etoposide. *Oncology* **53**, 210–213.

Mandell, L., Hilaris, B., Sullivan, M., Sundaresan, N., Nori, D., Kim, J. H., Martini, N., and Fuks, Z. (1986). The treatment of single brain metastasis from non-oat cell lung carcinoma. Surgery and radiation versus radiation therapy alone. *Cancer* **58**, 641–649.

Maor, M. H., Frias, A. E., and Oswald, M. J. (1988). Palliative radiotherapy for brain metastases in renal carcinoma. *Cancer* **62**, 1912–1917.

Marks, J. E., Baglan, R. J., Prassad, S. C., and Blank, W. F. (1981). Cerebral radionecrosis: incidence and risk in relation to dose, time, fractionation and volume. *Int. J. Radiat. Oncol. Biol. Phys.* **7**, 243–252.

Mehta, M. P., Rozental, J. M., Levin, A. B., Mackie, T. R., Kubsad, S. S., Gehring, M. A., and Kinsella, T. J. (1992). Defining the role of radiosurgery in the management of brain metastases. *Int. J. Radiat. Oncol. Biol. Phys.* **24**, 619–625.

Minotti, V., Crinó, L., Meacci, M. L., Corgna, E., Darwish, S., Palladino, M. A., Betti, M., and Tonato, M. (1998). Chemotherapy with cisplatin and teniposide for cerebral metastases in non-small cell lung cancer. *Lung Cancer* **20**, 93–98.

Mintz, A. H., Kestle, J., Rathbone, N. P., Gaspar, L., Hugenhoetz, H., Fisher, B., Duncan, G., Shingley, P., Foster, G., and Levine, M. (1996). A randomized trial to assess the efficacy of surgery in addition to radiotherapy in patients with a single brain metastasis. *Cancer* **78**, 1470–1476.

Mori, Y., Kondziolka, D., Flickinger, J. C., Logan, T., and Lunsford, L. D. (1998). Stereotactic radiosurgery for brain metastasis from renal cell carcinoma. *Cancer* **83**, 344–353.

Nakagawa, H., Fujita, T., Izumimoto, S., Miyawaki, Y., Kubo, S., Nakajima, Y., Tsuruzono, K., Kodama, K., Higashiyama, M., Doi, O., and Hayakawa, T. (1993). Cis-diamminedichloroplatinum (CDDP) therapy for brain metastasis of lung cancer. II. Clinical effects. *J. Neuro-Oncol.* **16**, 69–76.

Patchell, R. A., Cirrincione, C., Thaler, H. T., Galicich, J. H., Kim, J. H., and Posner, J. B. (1986). Single brain metastases: surgery plus radiation or radiation alone. *Neurology* **36**, 447–453.

Patchell, R. A., Tibbs, P. A., Walsh, J. W., Dempsey, R. J., Maruyama, Y., Kryscio, R. J., Markesbery, W. R., Macdonald, J. S., and Young, B. (1990). A randomized trial of surgery in the treatment of single metastases to the brain. *N. Engl. J. Med.* **322**, 494–500.

Patchell, R. A., Tibbs, P. A., Regine, W. F., Dempsey, R. J., Mohivddin, M., Kryscio, R. J., Markesbery, W. R., Foom, K. A., and Young, B. (1998). Postoperative radiotherapy in the treatment of single metastases to the brain. A randomized trial. *J. Am. Med. Assoc.* **280**, 1485–1489.

Postmus, P. E., and Smit, E. F. (1999). Chemotherapy for brain metastases of lung cancer: a review. *Ann. Oncol.* **10**, 753–759.

Postmus, P. E., Haaxma-Reiche, H., Sleijfer, D. T., Kirkpatrick, A., McVie, J. G., Kleisbauer, J. P., and the EORTC Lung Cancer Cooperative Group. (1989). High dose etoposide for brain metastases of small cell lung cancer. A phase II study. *Br. J. Cancer* **59**, 254–256.

Postmus, P. E., Smit, E. F., Haaxma-Reiche, H., van Zandwijk, N., Ardizzoni, A., Quoix, E., Kirkpatrick, A., Sahmoud, T., and Giaconne, G., for the EORTC Lung Cancer Cooperative Group. (1995). Teniposide for brain metastases of small-cell lung cancer: a phase II study. *J. Clin. Oncol.* **13**, 660–665.

Postmus, P. E., Haaxma-Reiche, H., Gregor, A., Groen, H. J. M., Lewinski, T., Scolard, T., Kirkpatrick, A., Curran, D., Sahmoud, T., and Giaccone, G. (1998). Brain-only metastases of small cell lung cancer; efficacy of whole brain radiotherapy. An EORTC phase II study. *Radiother. Oncol.* **46**, 29–32.

Pujol, J. L., Monnier, A., Berille, J., Cerrina, M. L., Douillard, J. Y., Rivière, A., Grandgirard, A., Gouva, S., Bizzari, J. P., and Le Chevalier, T. (1994). Phase II study of nitrosourea fotemustine as single-drug chemotherapy in poor-prognosis non-small-cell lung cancer. *Br. J. Cancer* **69**, 1136–1140.

Ransohoff, J. (1975). Surgical management of metastatic tumors. *Semin. Oncol.* **2**, 21–27.

Reddy, S., Hendrickson, F. R., Hoeksema, J., and Gelber, R. (1983). The role of radiation therapy in the palliation of metastatic genitourinary tract carcinomas. A study of the Radiation Therapy Oncology Group. *Cancer* **52**, 25–29.

Retsas, S., and Gershuny, A. R. (1988). Central nervous system involvement in malignant melanoma. *Cancer* **61**, 1926–1934.

Richards, P., and McKissock, W. (1963). Intracranial metastases. *BMJ* **1**, 15–18.

Robin, E., Bitran, J. D., Golomb, H. M., Newman, S., Hoffman, P. C., Desser, R. K., and DeMeester, T. R. (1982). Prognostic factors in patients with non-small cell bronchogenic carcinoma and brain metastases. *Cancer* **49**, 1916–1919.

Rosner, D., Nemoto, T., and Lane, W. W. (1986). Chemotherapy induces regression of brain metastases in breast carcinoma. *Cancer* **58**, 832–839.

Rusch, V. W., Griffin, B. R., and Livingston, R. B. (1989). The role of prophylactic cranial irradiation in regionally advanced non-small cell lung cancer. A Southwest Oncology Group Study. *J. Thorac. Cardiovasc. Surg.* **98**, 535–539.

Rustin, G. J. S., Newlands, E. S., Bagshawe, K. D., Begent, R. H. J., and Crawford, S. M. (1986). Successful management of metastatic and primary germ cell tumors in the brain. *Cancer* **57**, 2108–2113.

Rustin, G. J. S., Newlands, E. S., Begent, R. H. J., Dent, J., and Bagshawe, K. D. (1989). Weekly alternating etoposide, methotrexate, and actinomycin D/vincristine and cyclophosphamide chemotherapy for the treatment of CNS metastases of choriocarcinoma. *J. Clin. Oncol.* **7**, 900–903.

Ryan, G. F., Ball, D. L., and Smith, J. G. (1995). Treatment of brain metastases from primary lung cancer. *Int. J. Radiat. Oncol. Biol. Phys.* **31**, 273–278.

Salvati, M., Cervoni, L., and Celli, P. (1995a). Solitary brain metastases from thyroid carcinoma: study of 6 cases. *Tumori* **81**, 142–143.

Salvati, M., Cervoni, L., and Raco, A. (1995b). Single brain metastases from unknown primary malignancies in CT-era. *J. Neuro-Oncol.* **23**, 75–80.

Salvati, M., Cervoni, L., Caruso, R., Gagliardi, F. M., and Delfini, R. (1998). Sarcoma metastatic to the brain: a series of 15 cases. *Surg. Neurol.* **49**, 441–444.

Samuel, A. M., and Shah, D. H. (1997). Brain metastases in well-differentiated carcinoma of the thyroid. *Tumori* **83**, 608–610.

Sawin, P. D., Hitchon, P. W., Follett, K. A., and Torner, J. C. (1998). Computed imaging-assisted stereotactic brain biopsy: a risk analysis of 225 consecutive cases. *Surg. Neurol.* **49**, 640–649.

Shaw, E. G., Scott, C., Souhami, L., Dinapoli, R., Kline, R., Loeffler, J., and Farnan, N. (2000). Single dose radiosurgical treatment of recurrent previously irradiated primary brain tumors and brain metastases: final report of RTOG protocol 90-05. *Int. J. Radiat. Oncol. Biol. Phys.* **47**, 291–298.

Smalley, R., Schray, M. F., Laws, E. R., and O'Fallon, J. R. (1987). Adjuvant radiation therapy after surgical resection of solitary brain metastases: association with pattern of failure and survival. *Int. J. Radiat. Oncol. Biol. Phys.* **13**, 1611–1616.

Sneed, P. K., Lamborn, K. R., Forstner, J. M., McDermott, M. W., Chang, S., Park, E., Gutin, P. H., Phillips, T. L., Wara, W. M., and Larson, D. A. (1999). Radiosurgery for brain metastases: is whole brain radiotherapy necessary? *Int. J. Radiat. Oncol. Biol. Phys.* **43**, 549–558.

Sørensen, J. B., Hansen, H. H., Hansen, M., and Dombernowsky, P. (1988). Brain metastases in adenocarcinoma of the lung: frequency, risk groups, and prognosis. *J. Clin. Oncol.* **6**, 1474–1480.

Spears, W. T., Morphis, J. G., Lester, S. G., Williams, S. D., and Einhorn, L. H. (1991). Brain metastases and testicular tumors: long-term survival. *Int. J. Radiat. Oncol. Biol. Phys.* **22**, 17–22.

Störtebecker, T. P. (1954). Metastatic tumors of the brain from a neurosurgical point of view. A follow-up study of 158 cases. *J. Neurosurg.* **1**, 84–111.

Stuschke, M., Eberhardt, W., Pöttgen, C., Stamatis, G., Wilke, H., Stüben, G., Stäblen, F., Wilhelm, H. H., Menker, H., Teschler, H., Müller, R. D., Budach, V., Seeber, S., and Sack, H. (1999). Prophylactic cranial irradiation in locally advanced non-small-cell lung cancer after multimodality treatment: long-term follow-up and investigations of late neuropsychological effects. *J. Clin. Oncol.* **17**, 2700–2709.

Sundaresan, N., and Galicich, J. H. (1985). Surgical treatment of brain metastases. Clinical and computerized tomography evaluation of the results of treatment. *Cancer* **55**, 1382–1388.

Sundaresan, N., Sachdev, V. P., DiGiacinto, G. V., and Hughes, J. E. O. (1988). Reoperation for brain metastases. *J. Clin. Oncol.* **6**, 1625–1629.

Twelves, C. J., Souhami, R. L., Harper, P. G., Ash, C. M., Spiro, S. G., Earl, H. M., Tobias, J. S., Quinn, H., and Geddes, D. M. (1990). The response of cerebral metastases in small cell lung cancer to systemic chemotherapy. *Br. J. Cancer* **61**, 147–150.

Van den Pol, M., van Aalst, V. C., Wilmink, J. T., and Twijnstra, A. (1996). Brain metastases from an unknown primary tumour: which diagnostic procedures are indicated? *J. Neurol. Neurosurg. Pychiatry* **61**, 321–323.

Vecht, C. J., Haaxma-Reiche, H., Noordijk, E. M., Padberg, G. W., Voormolen, J. H. C., Hoekstra, F. H., Tans, J. T. J., Lambooij, N., Metsaars, J. A. L., Wattendorff, A. R., Brand, R., and Hermans, J. (1993). Treatment of single brain metastasis: radiotherapy alone or combined with neurosurgery? *Ann. Neurol.* **33**, 583–590.

Vecht, C. J., Hovestadt, A., Verbiest, H. B. C., van Vliet, J. J., and van Putten, W. L. J. (1994). Dose-effect relationship of dexamethasone on Karnofsky performance in metastatic brain tumors: a randomized study of doses of 4, 8, and 16 mg per day. *Neurology* **44**, 675–680.

Vieth, R. G., and Odom, G. L. (1965). Intracranial metastases and their neurosurgical treatment. *J. Neurosurg.* **23**, 375–383.

Voges, J., Treuer, H., Erdmann, J., Schlegel, W., Pastyr, O., Müller, R. P., and Sturm, V. (1994). Linac radiosurgery in brain metastases. *Acta Neurochir (Suppl.)* **62**, 72–76.

Weissman, D. E. (1988). Glucocorticoid treatment for brain metastases and epidural spinal cord compression: a review. *J. Clin. Oncol.* **6**, 543–551.

White, K. T., Fleming, T. R., and Laws, E. R. (1981). Single metastasis to the brain. *Mayo Clin. Proc.* **56**, 424–428.

Williams, P. C., Henner, W. D., Roman-Goldstein, S., Dahlborg, S. A., Brummett, R. E., Tableman, M., Dana, B. W., and Neuwelt, E. A. (1995). Toxicity and efficacy of carboplatin and etoposide in conjunction with disruption of the blood-brain tumor barrier in the treatment of intracranial neoplasms. *Neurosurgery* **37**, 17–28.

Winston, K. R., Walsh, J. W., and Fischer, E. G. (1980). Results of operative treatment of intracranial metastatic tumors. *Cancer* **45**, 2639–2645.

Wornom, I. L., Smith, J. W., Soong, S. J., McElvein, R., Urist, M. M., Balch, C. M. (1986). Surgery as palliative treatment for distant metastases of melanoma. *Ann. Surg.* **204,** 181–185.

Wronski, M., Arbit, E., Burt, M., and Galicich, J. H. (1995a). Survival after surgical treatment of brain metastases from lung cancer: a follow-up study of 231 patients treated between 1976 and 1991. *J. Neurosurg.* **83,** 605–616.

Wronski, M., Arbit, E., Burt, M., Perino, G., Galicich, J. H., and Brennan, M. F. (1995b). Resection of brain metastases from sarcoma. *Ann. Surg. Oncol.* **2,** 392–399.

Wronski, M., Arbit, E., Russo, P., and Galicich, J. H. (1996). Surgical resection of brain metastases from renal cell carcinoma in 50 patients. *Urology* **47,** 187–193.

Wronski, M., Maor, M. H., Davis, B. J., Sawaya, R., and Levin, V. A. (1997). External radiation of brain metastases from renal carcinoma: a retrospective study of 119 patients from the M.D. Anderson Cancer Center. *Int. J. Radiat. Oncol. Biol. Phys.* **37,** 753–759.

Yardeni, D., Reichenthal, E., Zucker, G., Rubeinstein, A., Cohen, M., Israeli, J., and Shalit, M. N. (1984). Neurosurgical management of single brain metastasis. *Surg. Neurol.* **21,** 377–384.

Ziegler, J. C., and Cooper, J. S. (1986). Brain metastases from malignant melanoma: conventional vs. high-dose-per-fraction radiotherapy. *Int. J. Radiat. Oncol. Biol. Phys.* **12,** 1839–1842.

Zimm, S., Wampler, G. L., Stablein, D., Hazra, T., and Young, H. F. (1981). Intracerebral metastases in solid-tumor patients: natural history and results of treatment. *Cancer* **48,** 384–394.

*Tumors and Developmental Disorders*

CHAPTER 65

# Leptomeningeal Metastasis

Michael Weller

## CLINICAL ASPECTS

Leptomeningeal metastasis signifies the systemic dissemination of malignancy. This complication of cancer may become increasingly common with improved systemic therapy of various types of cancer (Bleyer and Byrne, 1988; Chamberlain, 1994). Tumor cells may reach the subarachnoid space (1) by means of hematogenous seeding to the leptomeninges, (2) by means of migration from solid parenchymal brain or medullary metastases or from metastases to the choroid plexus, or (3) *per continuitatem* from bony skull or vertebral or epidural metastases or along cranial and spinal nerves. Tumor cell spreading occurs within the leptomeninges and in the subarachnoid space, notably in the basal cisterns, the sylvian fissure, and the lumbosacral dural sac. Some patients have solid leptomeningeal lesions; other patients have a significant load of floating cancer cells in the cerebrospinal fluid (CSF) compartment; many patients have a combination of both. Depending on the underlying malignancy, leptomeningeal metastasis is also referred to as carcinomatous, sarcomatous, melanomatous, lymphomatous, gliomatous, or simply neoplastic meningitis. Among solid tumors, leptomeningeal metastasis is common in breast cancer, lung cancer, and malignant melanoma, roughly correlating with the frequency of solid brain metastases in solid tumors (Chapter 64). Renal cell carcinoma is an exception in that solid parenchymal brain metastases are much more common than leptomeningeal metastasis. Furthermore, tumor cell seeding in the subarachnoid space is rather common in some hematological and lymphoid neoplasias. Thus, up to 30% of patients with primarily extracerebral non-Hodgkin's lymphoma have leptomeningeal spreading in the course of disease (Chamberlain, 1994). Central nervous system (CNS) seeding is also very common in childhood acute lymphoblastic leukemia (Schrappe

*et al.*, 1998), whereas the frequency is less than 1% in patients with chronic lymphocytic leukemia (Bower *et al.*, 1997; Cramer *et al.*, 1996). The distribution of primary tumors in large series of patients with leptomeningeal metastasis is summarized in Table I. The frequency of leptomeningeal metastasis per tumor type is difficult to estimate. A postmortem estimate in a large series of 2375 unselected cancer patients was 8% (Posner and Chernik, 1978). The incidence was 11% in a series of 526 patients with small cell lung cancer (Rosen *et al.*, 1982). Leptomeningeal metastasis also arises as a complication of primary brain tumors, notably in patients with germinomas, medulloblastomas, and primitive neuroectodermal tumors (PNET). Staging in the latter neoplasms includes CSF analysis and craniospinal magnetic resonance imaging (MRI) (Chapter 62). Subarachnoid spread may also evolve in the course of disease in 10% of malignant glioma and 20% of ependymoma patients (Onda *et al.*, 1989).

Leptomeningeal metastasis is commonly diagnosed in patients with advanced cancer but may occasionally be the first distant metastasis or even the primary manifestation of cancer, the latter usually in patients with lung cancer. Approximately half of the patients with leptomeningeal metastasis have solid parenchymal brain metastases as well. Two thirds of the patients with leptomeningeal metastasis have other systemic metastases. The typical presentation of leptomeningeal metastasis is that of a patient with known malignancy who has neurological symptoms and signs develop that are not attributable to a single anatomical lesion site. The clinical features include nausea and vomiting, symptoms and signs of raised intracranial pressure; cranial nerve palsies, most commonly affecting oculomotor, facial, and vestibulocochlear nerves; head, neck, and back pain; and symptoms and signs of spinal cord disease such as radicular pain or sensory loss, focal

TABLE I    Distribution of Primary Malignancies in Published Series of Patients with Leptomeningeal Metastasis

|  | All | Breast | Lung | Melanoma | Lymphoma | Others |
|---|---|---|---|---|---|---|
| Theodore and Gendelmann, 1981 | 33 | 21 | 5 | 5 |  | 2 |
| Trump et al., 1982 | 25 | 16 | 2 |  | 4 | 3 |
| Wasserstrom et al., 1982 | 90 | 46 | 23 | 11 |  | 10 |
| Giannone et al., 1986 | 22 | 10 | 7 | 1 | 2 | 2 |
| Hitchins et al., 1987 | 44 | 11 | 16 | 1 | 3 | 13 |
| Stewart et al., 1987 | 23 | 3 | 7 |  | 7 | 6 |
| Pfeffer et al., 1988 | 98 | 33 | 8 |  | 36 | 21 |
| Grossman et al., 1993 | 52 | 25 | 12 |  | 10 | 5 |
| Siegal et al., 1994 | 31 | 10 |  |  | 13 | 8 |
| Tübingen (1980–1991, unpublished) | 90 | 24 | 17 | 8 | 12 | 29 |

weakness, and bladder disturbances. Personality and cognitive changes, hemipareses, and seizures suggest that solid brain metastases have developed as well. The differential diagnosis of leptomeningeal metastasis includes subacute and chronic meningitis, tuberculosis, borreliosis, neurosarcoid, fungal infections, and angiitis of the CNS.

Neuroradiology plays an important role in the diagnosis of leptomeningeal metastasis (Chamberlain, 1994; Freilich et al., 1995). Cranial computed tomography (CCT) may reveal enlarged ventricles, and contrast enhancement may disclose small tumor nodules or diffuse tumor growth along the ventricle walls in 25%–50% of the patients with leptomeningeal metastasis. MRI is more sensitive than CCT and is also suitable to screen the spinal canal for possible involvement. Although spinal manifestations may also be detected using myelography and postmyelography CT, MRI is the preferred technique whenever available. [111]Indium or [99]technetium ventriculography may be a useful adjunct technique for the diagnosis and for treatment planning (Chamberlain, 1994). In one study, CSF circulation was abnormal in 19 of 31 patients (61%) with leptomeningeal metastasis and was normalized by focal radiotherapy in 11 of 19 patients (Glantz et al., 1995). Moreover, patients with normal CSF circulation initially or after radiotherapy had a better outcome than patients with abnormal CSF flow that was not corrected by radiotherapy, and patients with abnormal CSF flow were at an increased risk of developing neurotoxic side effects from intrathecal chemotherapy. In fact, CSF compartments that are not reached by intrathecally administered drug because of abnormal CSF circulation are likely to be responsible for failure to respond to therapy or may be the site of, or origin of, early relapse (protected site effect).

The detection of neoplastic cells in the CSF is the unequivocal proof of tumor cell seeding in the sub-arachnoid space. However, repeated lumbar punctures are often necessary to detect neoplastic cells, and routine CSF cytological examination may need to be complemented by specific tests, such as immunochemical analysis for distinct tumor-associated antigens (e.g., markers for epithelial cells). The diagnostic value of a third puncture after two negative CSF analyses with adequate workup is low (Chamberlain, 1994). Thus, appropriate chemoradiotherapy for leptomeningeal metastasis has occasionally to be instituted in the absence of a diagnostic CSF cytological examination.

Several other CSF findings support the diagnosis of leptomeningeal metastasis. CSF pressure is greater than 150 mmH$_2$O in more than half of the patients. CSF lactate exceeds 3–4 μmol/L in many patients, and thus, in the absence of massive pleocytosis suggestive of meningitis, is probably the most helpful additional CSF parameter for the detection of tumor cell seeding in the subarachnoid space. The CSF cell count is elevated in more than 70% and total CSF protein in more than 80%, of the patients. CSF glucose has been reported to be decreased in 40%–80% of the patients. Thus, the typical CSF pattern for newly diagnosed leptomeningeal metastasis is moderate pleocytosis with increased lactate, increased protein, and elevated CSF pressure. CSF cytological examination, cell counts, and lactate are suitable monitoring parameters for patients with leptomeningeal metastasis who receive chemoradiotherapy. Recently, the assessment of numerical chromosomal aberrations by fluorescent in situ hybridization (FISH) has been introduced as a novel approach to monitor response to treatment in patients with leptomeningeal metastasis (Van Oostenbrugge et al., 2000).

An elevated IgG index in 40% and oligoclonal bands on isoelectric focusing of CSF proteins in 30% of the patients indicate an inflammatory and immunological response to tumor cell invasion in the subarachnoid space (Weller et al., 1992), but these changes are not of diagnostic value. The same applies to a series of "tumor markers" that have been examined for their possible diagnostic use in establishing a diagnosis of leptomeningeal metastasis when cytological examination has failed. These candidate tumor markers include carcinoembryonic antigen (CEA), human milk fat globulin (HMFG)-1, β-glucuronidase, lactate dehydrogenase (LDH) and fibronectin for carcinomas, and $\beta_2$-microglobulin for hematological and lymphoid malignancies. The determination of these parameters is of little clinical value because most cases are diagnosed safely by CSF cytological examination and craniospinal MRI and because the specificity of these parameters is too low. Only the determination of α-fetoprotein (AFP) and human beta chorionic gonadotropin (βhCG) is useful in the diagnosis and follow-up of patients with germ cell tumors (Chapter 62).

## NATURAL COURSE

Leptomeningeal metastasis is often a sign of widespread dissemination of malignancy. The prognosis for untreated patients with leptomeningeal metastasis from solid tumors is very poor, with a median survival of 6–8 weeks. The prognosis for patients with CSF seeding from hematological and lymphoid malignancies is somewhat better. Focal radiotherapy of symptomatic lesions and intrathecal chemotherapy may prolong the median survival to 3–6 months and results in a survival at 1 year of 5%–25% (Chamberlain and Kormanik, 1998; Grant et al., 1994; Grossman et al., 1993; Pfeffer et al., 1988; Schabet et al., 1986; Theodore and Gendelman, 1981; Wasserstrom et al., 1982; Weller et al., 1993) (Table II). Single long-term survivors maintain a good quality of life. Two thirds of the patients who receive specific therapy for their leptomeningeal metastasis die from systemic disease progression and associated complications but not from CNS metastases. Negative prognostic factors include low Karnofsky score, cranial nerve palsies, elder age, low CSF glucose, and high CSF protein (Boogerd et al., 1991; Grossman et al., 1993). In a more recent study, female gender, longer duration of neurological symptoms, absence of clinical cerebral leptomeningeal involvement, and absence of elevated CSF protein were independent predictors of longer survival (Balm and Hammack, 1996). Leptomeningeal metastasis from breast cancer and hematological and lymphoid malignancies responds better to chemoradiotherapy than metastasis from lung cancer and malignant melanoma in most series (Table II). However, even some patients with malignant melanomatosis may survive for many months with intrathecal methotrexate (MTX) plus radiotherapy (Weller et al., 1993). Failure to respond to systemic or locoregional chemoradiotherapy may result from intrinsic tumor cell resistance to radiotherapy and chemotherapy or insufficient drug distribution in the subarachnoid space. Some patients have intolerable neurological or systemic toxicity from treatment, precluding further treatment. Intrathecal chemotherapy alone, even if local control is achieved, is of little value in patients with progressive solid parenchymatous brain metastases or progressive systemic extra-CNS metastases.

## PRINCIPLES OF THERAPY

### General Principles

The treatment of leptomeningeal metastasis is always palliative in patients with solid tumors and also in most patients with leptomeningeal seeding from hematological or lymphoid malignancies. This needs to be considered when planning the treatment for these patients, with specific consideration of quality of life issues. Because the detection of leptomeningeal metastasis by neuroradiological methods or by CSF analysis signifies widespread cancer cell dissemination in the CSF, therapeutic strategies need to target the whole CSF compartment. The major modes of treatment include intrathecal chemotherapy, systemic chemotherapy, and radiotherapy.

Because of the physiological flow patterns of CSF, *intrathecal chemotherapy* should be administered by means of reservoirs connected to a lateral ventricle (Ommaya, Rickham) rather than lumbar puncture. Although a randomized trial has never been performed, intraventricular application is considered to be more efficient than lumbar application, because higher and more homogenous concentrations of drug are reached within the CSF (Bleyer and Byrne, 1988). Moreover, once the reservoir has been placed, this route of administration is less painful and not associated with postlumbar puncture headaches. The injection into a ventricular reservoir is also safer than the lumbar route, because epidural leakage of CSF and erroneous injection of drugs into newly formed spaces, resulting in local toxicity without efficient drug delivery to the CSF, may occur on repeated lumbar applications. Lumbar port systems have high local complication rates and do not circumvent the problem that the lumbar route does not follow the natural flow of CSF. Therefore, lumbar injections of cytotoxic agents should be given only for the palliation of intractable lumbar pain in terminally ill patients or if the placement of a ventricular reservoir is not possible.

Solid leptomeningeal metastases that have their own blood supply are accessible for systemic chemotherapy (see later). However, most chemotherapy drugs, except

**TABLE II**  Response to Chemoradiotherapy in Patients with Leptomeningeal Metastasis

| Reference | Primary tumor | n | Median survival (mos) | Range of survival (mos) | Chemotherapy | Radiotherapy |
|---|---|---|---|---|---|---|
| Trump et al., 1982 | All | 25 | 5.7 | | V*: MTX, 10 mg, day 1, and thiotepa, 10 mg, day 4, 8–12 wk, then monthly, systemic chemotherapy (n = 19) | Focal (n = 25) 25–30 Gy |
| | Breast | 16 | | | | |
| | Lung | 2 | | | | |
| | Lymphoma | 4 | | | | |
| | Others | 3 | | | | |
| Wasserstrom et al., 1982 | All | 90 | 5.8 | 1–29 | V: MTX, 7 mg/m$^2$, or Ara-C 30 mg/m$^2$, 2x/wk; systemic chemotherapy (?) | Focal (n = 86) 24 Gy |
| | Breast | 46 | 7.2 | 1–29 | | |
| | Lung | 23 | 4.0 | 1–10 | | |
| | Melanoma | 11 | 3.6 | 1–12 | | |
| | Others | 10 | 6.3 | 2–12 | | |
| Giannone et al., 1986 | All | 22 | 2.5 | 1–6+ | V: MTX, 12 mg, + Ara-C, 40 mg + thiotepa 15 mg, 2x/wk, systemic chemotherapy (n = 7) | Focal (n = 11) dose? |
| | Breast | 10 | | | | |
| | Lung | 7 | | | | |
| | Others | 5 | | | | |
| Hitchins et al., 1987 | All | 44 | A: 2.8 | | A. V/L: MTX, 15 mg, 3x/wk | Whole brain (n = 17) Dose? |
| | Breast | 11 | B: 1.6 | | B. V/L: MTX, 15 mg + Ara-C, 50 mg/m$^2$, 3x/wk | Spinal (n = 4) |
| | Lung | 16 | | | | Neuraxis (n = 1) |
| | Others | 17 | | | Systemic chemotherapy or radiotherapy (n = 30) | |
| Stewart et al., 1987 | All | 23 | 2.2 | 1–34+ | V: MTX, 15 mg, + cortisone, 15 mg, days 1 and 8, Ara-C, 75 mg + thiotepa, 7.5 mg, days 3 and 10, × 2 wk, systemic chemotherapy? | Whole brain (n = 13) 30 Gy |
| | Breast | 3 | 8.5 | | | |
| | Lung | 7 | 2.0 | | | |
| | Lymphoma | 7 | 3.3 | | | |
| | Others | 6 | | | | |
| Pfeffer et al., 1988 | All | 98 | | | V/L: MTX, 12.5 mg, or Ara-C/ thiotepa (doses ?), systemic (?) | Focal (?) 30–50 Gy |
| | Breast | 33 | 2.5 | 1–40+ | | |
| | Lung | 8 | 7.0 | 2–10 | | |
| | Lymphoma | 36 | 6.0 | 1–86+ | | |
| | Brain | 8 | 24 | 1–53+ | | |
| | Others | 13 | | | | |
| Boogerd et al., 1991 | All breast | A: 44 | 3 | ?–41 | A: MTX, 5 mg, whenever CSF MTX levels were <1 μmol/L, until clearing of CSF from tumor cells, additional systemic chemotherapy? B: systemic only | Whole brain (n = 3) Focal (n = 7) Unspecified (n = 10) |
| | | B: 14 | 3 | | | |
| Grossman et al., 1993 | All | 52 | A: 4 | 0–14 | A. V: MTX, 10 mg, 2x/wk | Whole brain (?) |
| | Breast | 25 | B: 3.5 | | B. V: thiotepa, 10 mg 2x/wk | Spinal focus (?) |
| | Lung | 12 | | | Systemic (?) | 30 Gy |
| | Lymphoma | 10 | | | | |
| | Others | 5 | | | | |
| Chamberlain and Kormanik, 1997 | Lymphoma | 22 | 10 | 3–24 | V: MTX, Ara-C, Thiotepa, plus systemic (n = 15) | Brain (n = 4) Spine (n = 4) |
| Glantz et al., 1998 | All | 16 | 13.8 | ?–52+ | High-dose MTX (8 g/m$^2$) IV × 2–4 wk | None? |
| | Breast | 1 | | | | |
| | Lung | 2 | | | | |
| | Lymphoma | 8 | | | | |
| | Brain | 5 | | | | |
| Glantz et al., 1999a | All | 61 | A: 3.5 | ? | A: V/L: Cytarabine (DepoCyt) 50 mg × 2 wk | Focal brain or spine (n = 12) |
| | Breast | 22 | B: 2.6 | | B: V/L: MTX, 10 mg 2x/wk | |
| | Lung | 10 | | | Systemic (n = 13) | |
| | Melanoma | 5 | | | | |
| | Brain | 14 | | | | |
| | Others | 10 | | | | |

*V, ventricular; L, lumbar.

MTX and cytosine arabinoside (cytarabine, Ara-C) at high doses, and probably topotecan and temozolomide, do not achieve cytotoxic concentrations in the CSF when administered systemically. Floating cells in the CSF, which are not directly reached by the blood circulation, cannot be eliminated by most cytotoxic agents when administered systemically. Conversely, intrathecal therapy does not cause severe myelosuppression or other drug-specific side effects associated with systemic chemotherapy. MTX has been most widely used for the intrathecal chemotherapy of cancer. MTX is commonly administered twice weekly.

There are sufficient clinical data to support the clinical use outside controlled clinical trials for Ara-C (Fulton et al., 1982) and thiotriethylenephosphoramide (thio-TEPA), now commonly labeled thiotepa (Gutin et al., 1977). Further drugs, including topotecan and temozolomide, are currently under investigation but are still experimental agents at present. Other cytotoxic agents (e.g., vincristine) are highly neurotoxic and must never be applied intrathecally.

The role of *systemic chemotherapy* for the management of leptomeningeal metastasis has remained controversial. Intrathecal chemotherapy has commonly been favored by neurosurgeons and neurologists, because they tend to see patients with neurological problems who may even have the CNS as the only site of metastatic disease. Conversely, patients with disseminated metastases, who *also* have leptomeningeal metastasis, among many others, will be treated by medical oncologists in most institutions. It is a matter of debate whether patients receiving systemic chemotherapy benefit from additional intrathecal chemotherapy. There is no reason to assume that leptomeningeal metastasis would respond poorer to systemic chemotherapy than metastases at other sites as long as they can be accessed by the systemic circulation. Moreover, systemic chemotherapy, in contrast to intrathecal chemotherapy, may not have the risk of rapid recolonization of the CSF compartment from the periphery after an initial CSF clearing from malignant cells. Numerous studies have suggested a role for systemic chemotherapy in the treatment of leptomeningeal metastasis. For instance, some responses have been observed with high-dose etoposide in patients with small cell lung cancer (Postmus et al., 1989). Systemic chemotherapy plus focal radiotherapy was as effective as intraventricular MTX in a retrospective analysis of patients with breast cancer (Boogerd et al., 1991). Accordingly, intrathecal chemotherapy has been suggested to be dispensable in breast cancer patients receiving appropriate systemic treatment (Joyson et al., 1994). Siegal et al. (1994) also documented partial and complete treatment responses in patients with leptomeningeal metastasis from different primary tumors who received systemic chemotherapy

only. Furthermore, systemic high-dose Ara-C may be effective in leptomeningeal spreading of hematological and lymphoid neoplasias (Morra et al., 1993), and systemic high-dose MTX ($8\,g/m^2$ in 4 hours) resulted in peak CSF levels of $3.7$–$55\,\mu M$ and clearing of the CSF from malignant cells in 13 of 16 patients with leptomeningeal metastasis and induced an impressive median survival of 13.8 months (Glantz et al., 1998). Two of the three nonresponders had malignant glioma. Thus, some authors believe that intrathecal chemotherapy plays no role in the management of leptomeningeal metastasis and should be discouraged because of significant acute and delayed toxicity (Bokstein et al., 1998).

Because intrathecal chemotherapy is not effective against solid leptomeningeal lesions, intrathecal chemotherapy is commonly combined with *radiotherapy* if solid lesions are present. Furthermore, radiotherapy is a useful option when neurological deficits are evolving and progressing rapidly, because radiotherapy is a rather reliably effective palliative treatment. The most common strategies include whole-brain radiotherapy plus focal radiotherapy to spinal lesions or focal radiotherapy to cranial and spinal lesions. Because of the myelosuppression associated with craniospinal irradiation in patients who are often heavily pretreated, this large target volume is only rarely irradiated. However, there are exceptions (e.g., intrinsic brain tumors with leptomeningeal seeding such as medulloblastoma or ependymoma) (see Chapter 62).

The response rates to combined modality treatment defined by neurological stabilization or remission of neurological deficits are in the range of 40%–80% for breast and lung cancer and even higher for hematological and lymphoid malignancies but only 10%–20% for malignant melanoma. Even single patients with leptomeningeal spread from malignant gliomas may respond (Grant et al., 1992). Responses defined by CSF clearing from malignant cells may be achieved in up to 80% of the patients with hematological or lymphoid neoplasias and up to 50% of the patients with solid tumors (Boogerd et al., 1991; Chamberlain and Kormanik, 1997; Glantz et al., 1998; Wasserstrom et al., 1982). The data on median survival and range of survival by primary tumor are summarized in Table II.

### Intrathecal Chemotherapy

#### MTX

MTX is most commonly used for intrathecal chemotherapy. MTX is an inhibitor of dihydrofolate reductase, the enzyme that generates reduced folates for purin synthesis, and thus inhibits DNA synthesis. The therapeutic concentration of MTX in the CSF is in the range of 1 μM. MTX is a classical chemotherapeutic agent for the systemic treatment of hematological and lymphoid neoplasms. However, MTX is also used for the intrathecal therapy of leptomeningeal metastasis from many solid neoplasms that are not treated with MTX systemically. The efficacy of MTX against leptomeningeal tumor cell spread is established by consensus, although a randomized clinical trial has not been performed.

## Ara-C

Ara-C is a cytidine analog antimetabolite that inhibits DNA synthesis. The therapeutic concentrations for leukemia and lymphoma cells are in the range of 0.1–40 μM. These levels may be achieved in the CSF with systemic high-dose Ara-C chemotherapy, because CSF levels reach 40% of the plasma levels on continuous infusion. The intrathecal application has the advantage of fewer systemic side effects. Ara-C is thought to be less active against leptomeningeal metastasis from solid tumors and has a shorter half-life in the CSF than MTX. Ara-C is therefore preferentially used for the treatment of hematological and lymphoid neoplasias. DepoCyt is a depot application of Ara-C designed to achieve cytotoxic levels of Ara-C for prolonged periods (Chamberlain et al., 1993, 1995). Compared with free Ara-C, 50 mg twice a week, DepoCyt at 50 mg biweekly achieved higher response rates and better quality of life (Glantz et al., 1999b). In a second randomized trial, depot Ara-C was not superior to MTX in terms of responses and median survival but produced a longer neurological progression-free survival (Glantz et al., 1999a).

## Thiotepa

Thiotepa is a DNA-alkylating agent that damages cancer cells in a cell cycle–independent manner. It is sufficiently lipophilic to reach the CSF after systemic administration. As with Ara-C, prevention of systemic side effects is the rationale for intrathecal thiotepa therapy. Intraventricular injection of 10 mg thiotepa results in CSF concentrations of 0.1–1 mM, which are cytotoxic to leukemia, breast and lung cancer, melanoma, and glioma cells. Thiotepa has been shown to be equally effective as MTX in the management of leptomeningeal metastasis from solid tumors (Grossman et al., 1993).

## Other Agents

In addition to MTX, cytarabine, and thiotepa, several other cytotoxic agents have been evaluated in phase I/II studies for the intrathecal chemotherapy of lep-

tomeningeal metastasis: dacarbacine for malignant melanoma (Champagne and Silver, 1992), ACNU for solid and intrinsic brain tumors (Kochi et al., 1993; Levin et al., 1989; Ushio et al., 1998), diaziquone (Berg et al., 1992), 4-hydroperoxycyclophosphamide and mafosfamide (Slavc et al., 1998). None of these drugs has gained a defined role in the clinical management of leptomeningeal metastasis. They should be administered only within clinical trials. Two promising agents, topotecan and temozolomide, that have shown activity against solid brain metastasis from small cell lung cancer (Chapter 64) and against malignant glioma (Chapter 62), respectively, are currently being evaluated for toxicity and efficacy on intrathecal administration.

### Intrathecal Combination Chemotherapy

Intrathecal combination chemotherapy has been evaluated with the rationale that two or three drugs in combination might exert antitumor effects in synergy. The combination of MTX and Ara-C was not superior to MTX alone (Hitchins et al., 1987). Further phase II studies not only confirmed a lack of superiority of combination therapy but showed enhanced toxicity of combined MTX and Ara-C (Nakagawa et al., 1992), MTX, Ara-C, and thiotepa (Giannone et al., 1986) or this triple combination plus steroids (Stewart et al., 1987).

### Radiotherapy

Because intrathecal chemotherapy is not an effective therapy for solid lesions with a diameter of several millimetres, it is often combined with whole-brain radiotherapy or focal cerebral and spinal radiotherapy. We recommend radiotherapy as part of the regimen for leptomeningeal metastasis from solid tumors with few exceptions. For instance, patients with no MRI evidence of solid cerebral or spinal lesions who have primarily CSF seeding by floating cancer cells and presumably only thin layers of tumor cells in the ventricle walls and surrounding the CSF compartment may be treated with intrathecal chemotherapy alone. Furthermore, patients with leptomeningeal seeding from hematological and lymphoid malignancies, who commonly receive systemic therapy in addition to intrathecal therapy, may not benefit from additional radiotherapy.

Although radiotherapy is an effective palliative treatment for leptomeningeal metastasis, timing, fractionation, and target volumes of radiotherapy have remained controversial. Craniospinal irradiation is, in principle, a rational approach, because this targets the total presumptive space of tumor cell dissemination. However, craniospinal irradiation is usually only administered to patients with medulloblastoma or germinoma and leptomeningeal metastasis but rarely to patients with leptomeningeal metastasis from primary tumors originating outside the CNS. This is because of significant myelosuppression associated with craniospinal irradiation in patients who are often heavily pretreated. The sequence of chemotherapy and radiotherapy and the choice of agents determine the risk of toxicity from combined modality treatment. Radiotherapy followed by chemotherapy is more likely to cause neurotoxicity of the delayed leukencephalopathy type than the reverse sequence, and certain drugs such as topotecan or gemcitabine may act as radiosensitizers in vivo when administered during radiotherapy. Of note, leukencephalopathy may also develop after systemic and intrathecal chemotherapy alone, in the absence of adjuvant radiotherapy (Siegal *et al.*, 1994). Whole-brain radiotherapy should be considered in patients with parenchymal brain or cerebellar metastases and in patients with cranial nerve palsies. Whole-brain radiotherapy should also be given to patients who cannot be treated with a ventricular reservoir and receive lumbar chemotherapy only instead. Circumscribed solid spinal lesions should be treated with focal radiotherapy. Focal cerebral radiotherapy may also be appropriate to restore proper CSF circulation before intraventricular chemotherapy (Glantz *et al.*, 1995).

## Immunotherapy

Immunotherapeutic approaches do not play an important role in the management of leptomeningeal metastasis. Immunotherapy includes immunotoxin therapy and immunoradiotherapy, administration of cytokines into the CSF, and active cellular immunotherapy. Immunotoxin therapy and immunoradiotherapy are types of passive immunotherapy that rely on the conjugation of an antibody directed to a target protein of interest (e.g., epidermal growth factor receptor [EGFR] or interleukin 4 receptor), either to a toxin (e.g., bacterial pseudomonas or diphtheria toxin) or on the radioactive labeling of such antibodies. Radiolabeled HMFG1 antibodies have been studied in a larger series of human patients with leptomeningeal metastasis (Moseley *et al.*, 1991). The response rate was low. There were significant side effects, including arachnoid adhesions, aseptic meningitis, and seizures. In a retrospective survey of 52 patients, PNET patients treated with antibodies to the fetal L1 antigen showed the best responses to this mode of treatment (Coakham and Kemshead, 1998).

The intrathecal administration of cytokines such as α- or β-interferon or interleukin-2 seeks to enhance immune cell homing to the subarachnoid space and immune-mediated clearing of the CSF from cancer cells. These strategies have not assumed clinical relevance so far but have induced severe neurotoxic side effects. Similarly, active cellular immunotherapy (e.g., using autologous or heterologous lymphokine-activated killer (LAK) cells) currently plays no role in the management of leptomeningeal metastasis.

## Somatic Gene Therapy

The most advanced approach of somatic gene therapy for cancer is based on the transduction of cancer cells with the virally encoded enzyme, thymidine kinase (TK), using (retro)viral vectors followed by systemic treatment with the nontoxic prodrug ganciclovir that is activated to a toxic metabolite by the action of TK. The retroviral approach is suitable for the locoregional therapy of cancers in compartments where chiefly cancer cells proliferate, whereas host cells are quiescent, including brain but also subarachnoid space. Feasibility and some efficacy of this approach have been confirmed in an animal model of leptomeningeal metastasis from gliosarcoma in the rat (Ram *et al.*, 1994). Reports of TK/ganciclovir gene therapy for human patients with leptomeningeal metastasis have not been published.

## Tumor-Specific Considerations

Given the limited number of agents that are available for intrathecal chemotherapy and the overall dismal prognosis, the type of primary lesion is often given little attention when devising the treatment strategy for patients with leptomeningeal metastasis. In fact, the systemic manifestations of cancer and concurrent solid brain lesions are the most important factors determining the choice of therapy. However, some tumor-specific considerations may also be helpful in arriving at an optimal treatment strategy.

Hormone-dependent cancers (e.g., breast cancer) may be responsive to hormonal ablation, even at the stage of cerebrospinal metastasis (Boogerd *et al.*, 2000). Similarly, remission of neoplastic meningitis from prostate cancer in response to hormonal ablation has been documented (Mencel *et al.*, 1994).

CSF relapse is a common site of failure after systemic chemotherapy for leukemia and systemic non-Hodgkin's lymphoma. Although these disorders require systemic chemotherapy, additional intrathecal chemotherapy is probably a useful adjunct (Ribeiro *et al.*, 1995) and may contribute to median survival times of 10 months in non-Hodgkin's lymphoma (Chamberlain and Kormanik, 1997). CNS seeding and CNS recurrence are also very common in childhood acute lymphoblastic leukemia and can often be managed using multimodality treatment consisting of systemic and intrathecal chemotherapy and radiotherapy (Schrappe *et al.*, 1998).

Although the prognosis for patients with subarachnoid seeding from malignant gliomas is considered very poor, some patients experience transient stabilization in response to multimodality treatment (Grant *et al.*, 1992; Pradat *et al.*, 1999). Leptomeningeal metastasis from malignant gliomas usually requires not only intrathecal therapy but also systemic chemotherapy because of solid lesions, which in most patients have previously been irradiated. We believe that such patients may well be treated with systemic chemotherapy alone (e.g., the combination of procarbacine, CCNU, and vincristine [PCV]), which contains two agents with relevant CSF penetration, procarbazine and CCNU, or with temozolomide (Chapter 62). Our own unpublished experience with intrathecal chemotherapy for leptomeningeal metastasis from glioma using MTX or cytarabine has been disappointing. Encouraging results have been obtained with intrathecal thiotepa, although several patients received concomitant systemic chemotherapy in that series (Witham *et al.*, 1999). Ventriculolumbar perfusion using ACNU is an invasive approach of intrathecal chemotherapy that has been reported to induce responses in a significant proportion of patients (Kochi *et al.*, 1993; Ushio *et al.*, 1998). However, concomitant systemic chemotherapy and radiotherapy in most patients precludes a conclusion as to the contribution of ventriculolumbar ACNU perfusion to the responses observed in this cohort of patients.

### Individual Choice of Treatment

The treatment plan for the individual patient with leptomeningeal metastasis must take into consideration several aspects of the patient's current condition and medical history. It is reasonable to distinguish whether a patient has preferentially solid leptomeningeal metastasis, or floating cancer cells in the CSF compartment, or both. Floating tumor cells or thin tumor cell layers covering cord and brainstem are predicted to respond better to intrathecal chemotherapy, whereas solid lesions are likely to be reached by the systemic route. Furthermore, the presence of concurrent solid brain metastases and systemic metastases determines the choice of treatment. Some guidelines for decision making that may prove helpful in shaping individual treatment strategies are provided in Table III. If there is both solid and floating leptomeningeal tumor cell burden, the approaches need to be combined.

If leptomeningeal metastasis is mainly solid and if there are no solid brain or systemic metastases, the patient should receive focal spinal radiotherapy which may be combined with systemic chemotherapy. If solid brain metastases are present, the patient should also receive whole-brain radiotherapy, possibly combined with systemic chemotherapy. If systemic lesions are present, but no solid cerebral lesions, the patient should receive systemic chemotherapy and in most instances also focal spinal radiotherapy. Patients with solid cerebral, spinal, and systemic lesions may require systemic chemotherapy plus whole-brain radiotherapy plus focal spinal radiotherapy.

If tumor staging reveals no relevant extra-CNS disease and if tumor cell burden is primarily of the floating type, the patient should have an Ommaya reservoir placed and should receive intrathecal chemotherapy. If solid brain metastases are also present, the patient should receive whole-brain radiotherapy, and if spinal solid lesions are present, the patient should also receive focal radiotherapy to these lesions. In that case, six intrathecal injections, twice a week, should be administered before radiotherapy, and intrathecal chemotherapy may be resumed after radiotherapy has been completed. If there are widespread systemic manifesta-

**TABLE III    Guidelines for the Management of Leptomeningeal Metastasis***

| Type | Solid brain metastases | Systemic metastases | Therapeutic strategy |
|------|------------------------|---------------------|----------------------|
| Solid | No | No | (Systemic chemotherapy plus) focal spinal radiotherapy |
| | Yes | No | (Systemic chemotherapy plus) whole brain radiotherapy plus focal spinal radiotherapy |
| | No | Yes | Systemic chemotherapy (plus focal spinal radiotherapy) |
| | Yes | Yes | Systemic chemotherapy plus whole brain radiotherapy plus focal spinal radiotherapy |
| Floating | No | No | Intrathecal chemotherapy |
| | Yes | No | Intrathecal chemotherapy plus whole brain radiotherapy |
| | No | Yes | Systemic chemotherapy (plus intrathecal chemotherapy) |
| | Yes | Yes | Systemic chemotherapy plus whole brain radiotherapy (plus intrathecal chemotherapy) |

tions of cancer, systemic chemotherapy should be considered. Depending on the type of cancer and the type of systemic chemotherapy, intrathecal chemotherapy may not be necessary (e.g., in regimens containing high-dose MTX or high-dose Ara-C or other drugs with good CSF penetration, such as nitrosoureas, procarbazine, temozolomide, or topotecan). The recommendations in Table III are guidelines only, which may help to find a reasonable treatment regimen for most patients with leptomeningeal metastasis.

## PRACTICAL MANAGEMENT

### Intrathecal Chemotherapy

Before the initiation of intrathecal chemotherapy, white blood cell counts should be greater than 3000/μl, platelets greater than 100,000/μl, and serum creatinine before MTX therapy less than 1.5 mg/dl. When these parameters are not met, careful monitoring is mandatory. As outlined earlier, intraventricular therapy through a reservoir is preferable to repeated lumbar punctures. Correct placement of the ventricular catheter tip in the anterior horn of the lateral ventricle has to be verified by CT before the first ventricular drug injection. The first application through the reservoir can be administered immediately after surgery. When instant treatment seems to be necessary for clinical reasons, intrathecal therapy should be started by the lumbar route, and a reservoir should be put in place as soon as possible. Some patients have intracranial pressure such that external shunting or placement of a reservoir are initial therapeutic measures.

The surgical placement of a ventricular catheter has a mortality of 0.5% and a rate of perioperative morbidity, including hemorrhage, infection and reversible neurological deficits, of 2%–10%. Tube dislocation and CSF leakage require surgical revisions in 5% of the patients. Approximately 5% of the patients have focal leukencephalopathy develop around the ventricular catheter tip. This is attributed to CSF dissection into the white matter, probably as a result of CSF pulsations or chronic increases in CSF pressure, occasionally necessitating removal of the catheter (Leman et al., 1988).

Absolute sterility during drug injections into the reservoir is of utmost importance: 5%–10% of the patients with a ventricular reservoir have meningitis or ventriculitis. The most common infectious agents are *Staphylococcus epidermidis* in adults and *Propionibacterium acnes* in children (Boogerd et al., 1991; Obbens et al., 1985; Pfeffer et al., 1988). Patients with leptomeningeal metastasis may be at higher risk of infection because of disseminated malignancy and previous

radiochemotherapy. Although many centers treat reservoir infections with systemic or intraventricular antibiotics, definitive clearing of bacteria may in our experience only be achieved by removing reservoir and connecting tubes.

Before treatment, the skin covering the access to the reservoir has to be carefully desinfected. The accessible chamber of the reservoir is pressed with the thumb several times and monitored for spontaneous refilling. The injection may only be performed if free circulation of CSF in the reservoir system can be demonstrated. A 22-gauge needle is used to withdraw an amount of CSF equivalent to the volume that will be administered during intrathecal chemotherapy. The drug is then injected during 10–15 minutes by repetitive injection and reaspiration to ensure proper mixing with the CSF. After the injection, the reservoir should be pumped several times to support distribution of the drug in the CSF. Lumbar applications should be performed using an atraumatic needle (18–22 gauge). After repetitive injection and reaspiration to support mixing of drug and CSF, the patient should remain recumbent for 4 hours with the bed tilted head-down approximately 10 degrees to promote diffusion of the agent toward the cranial CSF spaces.

Cytotoxic agents are prepared for injection strictly according to the instructions of the manufacturers. A predilution in artificial CSF is not necessary. Solutions hyperosmolar to CSF must not be used. We do not advocate the use of intrathecal steroid coinjection for the prevention of side effects from intrathecal chemotherapy such as chemical ventriculitis or meningitis. White blood cells, platelets, and hemoglobin should be monitored weekly during intrathecal chemotherapy, especially when combined with (focal) radiotherapy. Intrathecal chemotherapy plus whole-brain radiotherapy induce myelosuppression in 5%–10% of the patients. Patients with seizures should receive anticonvulsive treatment (e.g., carbamazepine or valproic acid). Phenytoin should not be given, because phenytoin is more likely to cause undesirable drug interactions.

MTX is the drug of choice for intrathecal chemotherapy irrespective of the histological findings of the primary tumor in most centers.* The drug is administered twice weekly. Single doses are 12 mg for intraventricular and 15 mg for lumbar MTX therapy. To prevent systemic side effects, the patients receive a *leucovorin rescue*, that is, 15 mg folinic acid orally starting at 6 hours after the MTX injection, every 6 hours for 48 hours. Dose intensification of MTX seems to be feasible, although convincing evidence of enhanced efficacy has not been presented (Fizazi et al., 1996).

The side effects of intrathecal MTX include acute disturbances that are mostly reversible, as well as delayed severe sequelae, specifically affecting the white matter, which are probably irreversible in most patients. A reversible aseptic chemical meningitis, often associated with significant pleocytosis, may develop within a few hours of intrathecal MTX application in 5%–40% of the patients. This may be confused with meningitis or ventriculitis in patients receiving serial lumbar punctures or patients with a ventricular reservoir. Subacute side effects during the first 2 weeks of intrathecal MTX therapy (10%) include headache, nausea, dizziness, fever, and mild CSF pleocytosis. The immediate cause of neurological events such as seizures or of progressive neurological deficits is often difficult to determine in patients with metastatic brain disease who receive multiple pharmacological agents including intrathecal chemotherapy. Chronic progressive irreversible leukencephalopathy developing with a delay of several weeks up to years and resulting in brain atrophy, dementia, seizures, and focal neurological deficits is the most severe complication of intrathecal MTX chemotherapy. The most extensive data on the neurotoxicity of MTX stem from older leukemia trials (Bleyer and Byrne, 1988). Severe leukencephalopathy is particularly common when MTX is administered after rather than before radiotherapy. For this reason four to six MTX applications should be given before radiotherapy during combined modality treatment of leptomeningeal metastasis.

If MTX therapy fails, we treat leptomeningeal seeding from leukemia and lymphoma with Ara-C (50 mg) and leptomeningeal metastasis from solid tumors with thiotepa (10 mg). In terms of side effects, leukencephalopathies and also myelopathies with ascending sensorimotor deficits have been observed after intrathecal Ara-C injection, but these neurological side effects are probably less common with Ara-C than with MTX. Similarly, the risk of leukencephalopathy developing from combined modality treatment using thiotepa and radiotherapy is probably lower than for the combination of MTX and radiotherapy.

Individual dose adjustments (e.g., square meter body surface area or kilogram body weight) are not required, because the CSF has a rather constant volume of 150 ml from the age of 3. Like MTX, Ara-C and thiotepa are commonly administered twice weekly. Cumulative doses exceeding 150 mg MTX or 700 mg Ara-C should be avoided. A cumulative dose for thiotepa has not been defined. In patients scheduled to receive sandwich chemoradiotherapy, we commonly administer six injections over 3 weeks, followed by whole-brain radiotherapy plus focal radiotherapy of spinal lesions.

Although we continue intrathecal chemotherapy with once-weekly injections during radiotherapy, other centers prefer not to administer cytotoxic agents intrathecally during radiotherapy for safety considerations in terms of long-term toxicity. However, there are no data to support that MTX during radiotherapy is more neurotoxic than MTX after radiotherapy. After radiotherapy, the injections are commonly repeated at weekly intervals. The longer intrathecal therapy is administered, the more there is a need for individual management, especially after radiotherapy. The life-threatening nature of the condition has to be weighed against the increasing risk of neurotoxicity, notably leukencephalopathy, with time and doses of drugs and irradiation.

The goals of treatment are stabilization or regression of neurological deficits and a normalization of CSF findings, notably clearing of the CSF from malignant cells. Abnormal CSF cytological findings, CSF cell counts if initially abnormal, and CSF lactate are suitable parameters to monitor the efficacy of therapy and to arrive at further management decisions in the course of the disease.

Some specific *questions that may arise* are as follows:

*When is intrathecal chemotherapy considered a failure and stopped or changed?* In the case of primary intrathecal chemotherapy without irradiation, we would aim at clearing the CSF of malignant cells within 2 weeks, that is, after the effects of the third injection can be assessed. However, we would only stop treatment or change the agent if the CSF findings are clearly progressive or if clinical deterioration attributable to leptomeningeal disease and not to single solid metastases becomes apparent. Most patients will immediately go on to receive whole-brain radiotherapy plus focal radiotherapy when MTX has failed and will continue to receive intrathecal chemotherapy using Ara-C or thiotepa if malignant cells are still found in the CSF after completion of radiotherapy.

*How long will an effective regimen of intrathecal chemotherapy be continued after CSF clearing from malignant cells?* It is reasonable practice to hold intrathecal chemotherapy after two to three consecutive CSF analyses negative for malignant cells. Alternatively, intrathecal therapy may be modified to maintenance therapy of one injection per 1–3 months, but this strategy has not been evaluated in a clinical trial. Thus, at the beginning of therapy, it is our practice to complete 3 weeks of intrathecal chemotherapy even if CSF findings normalize after the first or second injection. Later in the course of treatment, commonly after completion of radiotherapy, we stop intrathecal chemotherapy after two normal CSF analyses, and we do not administer routine maintenance therapy. On relapse, we will use the same drug that induced remission previously. Occasional patients show clinical progression suggestive of progressive leptomeningeal metastasis in the absence of progressive CSF findings, and these patients tend to benefit from reinstitution of therapy (e.g., using weekly injections for a month).

*How long shall a clinically effective regimen be continued in the absence of CSF clearing from malignant cells?* CSF clearing from malignant cells is theoretically a much better parameter to assess the efficacy of intrathecal chemotherapy than clinical status, because clinical improvement may also result from concurrent radiotherapy or changes in steroid medication or changes in anticonvulsant medication. Yet, CSF findings are prone to artifact, and the detection of malignant cells is often time-consuming and expensive (e.g., when malignant cell populations can only be adequately assessed using immunochemistry). Thus, because the treatment of leptomeningeal metastasis is palliative, patients should sometimes be treated according to clinical criteria only (e.g., symptoms or signs of intracranial pressure) even with unremarkable or stable CSF findings. Conversely, deteriorating CSF findings in the absence of clinical symptoms or signs may not necessitate immediate reinstitution of chemotherapy.

## Systemic Chemotherapy

The role of systemic chemotherapy is different for leptomeningeal metastasis from hematological or lymphoid malignancies compared with solid tumors. Patients with leukemia or disseminated non-Hodgkin's lymphoma almost always receive systemic chemotherapy in addition to intrathecal chemotherapy. Some regimens of systemic chemotherapy (e.g., those containing high-dose MTX or high-dose Ara-C) are likely to have significant activity against CSF seeding in the absence of intrathecal chemotherapy.* Such patients are commonly treated by medical oncologists and not by neurologists.

The most common reason to treat patients with leptomeningeal metastasis from solid tumors with systemic chemotherapy is the presence of symptomatic extra-CNS metastases or the presence of solid brain metastases that have recurred or progressed after whole-brain radiotherapy. However, systemic chemotherapy is also a useful option if multiple solid lesions would otherwise require craniospinal irradiation. The choice of systemic chemotherapy should be based on current recommendations for the primary tumor. PCV polychemotherapy or temozolomide may be recommended for leptomeningeal metastasis from malignant gliomas and melanomas. Details on these regimens are presented in Chapter 62.

## Radiotherapy

Radiotherapy is part of the treatment strategy for most patients with leptomeningeal metastasis, notably from solid tumors. Focal radiotherapy to spinal lesions is given in 2 Gy fractions up to a total dose of 30–36 Gy, usually enclosing one vertebra above and one vertebra below the lesion in the target volume.* Focal cerebral lesions are irradiated in 3 Gy fractions to 36 Gy total dose. Whole-brain radiotherapy is also commonly administered in 3 Gy fractions to a total dose of 36 Gy.* Focal cerebral lesions may be treated with a fractionated or radiosurgical boost as outlined in detail in Chapter 64. For sandwich chemoradiotherapy, whole-brain radiotherapy is commonly delayed for up to 3 weeks (see earlier). Whole-brain radiotherapy is always part of the regimen upfront when the patient will receive full palliation but cannot be treated with intraventricular, but only lumbar, intrathecal chemotherapy. Few patients have extensive solid leptomeningeal lesions that have not responded to, but rather developed, during systemic chemotherapy. These patients may receive craniospinal irradiation initially, and no intrathecal chemotherapy is administered during craniospinal irradiation. Failure to respond to intrathecal chemotherapy may also be a reason to administer craniospinal irradiation. During radiotherapy patients will receive corticosteroids as clinically required, but a standard dose (e.g., of dexamethasone [4 × 1 mg/day]) should be given to all patients. In addition to myelosuppression, spinal irradiation may cause radiation myelopathy, radiation colitis, and radiation esophagitis. The side effects of whole-brain radiotherapy are discussed in more detail in Chapter 64.

## Supportive Care

Most patients with leptomeningeal metastasis benefit from low doses of corticosteroids (e.g., dexamethasone at 4 × 1 mg/day) even when no radiotherapy is given. Therefore, symptomatic patients with leptomeningeal metastasis should receive corticosteroids. Given the presumptive dose dependency of steroid-induced side effects, we recommend starting patients at 4 × 1 mg/day and increasing the dose slowly as deemed necessary. Efforts to withdraw steroids should be undertaken in patients who seem to have responded to chemotherapy or radiotherapy on an individual basis.

Guidelines for the treatment and prevention of seizures in brain tumor patients are provided in Chapters 62 and 64. Anticonvulsants do not need to be given to patients with leptomeningeal metastasis prophylactically, even after a reservoir has been placed. Many centers prescribe anticonvulsants prophylactically when patients have solid brain metastases, but there are no data to support this strategy. We do not recommend prophylactic anticonvulsive treatment of these patients, but the first seizure will be an indication to treat for at least 3 months. Decisions to withdraw anticonvulsant medication after an event-free interval may be made on an individual basis, response to cancer therapy probably being the main factor influencing this decision.

# REFERENCES

Balm, M., and Hammack, J. (1996). Leptomeningeal carcinomatosis. Presenting features and prognostic factors. *Arch. Neurol.* 53, 626–632.

Berg, S., Balis, F., Zimm, S., Murphy, R. F., Holcenberg, J., Sato, J., Reaman, G., Steinherz, P., Gillespie, A., Doherty, K., and Poplack, D. G. (1992). Phase I/II trial and pharmacokinetics of intrathecal diaziquone in refractory meningeal malignancies. *J. Clin. Oncol.* 10, 143–148.

Bleyer, W. A., and Byrne, T. N. (1988). Leptomeningeal cancer in leukemia and solid tumors. *Curr. Probl. Cancer* 12, 185–238.

Bokstein, F., Lossos, A., and Siegal, T. (1998). Leptomeningeal metastases from solid tumors: a comparison of two prospective series treated with and without intra-cerebrospinal fluid chemotherapy. *Cancer* 82, 1756–1763.

Boogerd, W., Dorresteijn, L. D., van der Sande, J. J., de Gast, G. C., and Bruning, P. F. (2000). Response of leptomeningeal metastases from breast cancer to hormonal therapy. *Neurology* 55, 117–119.

Boogerd, W., Hart, A. A. M., Sande, J. J., and Engelsman, E. (1991). Meningeal carcinomatosis in breast cancer: prognostic factors and influence of treatment. *Cancer* 67, 1685–1695.

Bower, J. H., Hammack, J. E., McDonnell, S. K., and Tefferi, A. (1997). The neurologic complications of B-cell chronic lymphocytic leukemia. *Neurology* 48, 407–412.

Chamberlain, M. C. (1994). New approaches to and current treatment of leptomeningeal metastases. *Curr. Opin. Neurol.* 7, 492–500.

Chamberlain, M. C., Khatibi, S., Kim, J. C., Howell, S. B., Chatelut, E., and Kim, S. (1993). Treatment of leptomeningeal metastasis with intraventricular administration of depot cytarabine (DTC 101): a phase I study. *Arch. Neurol.* 50, 261–264.

Chamberlain, M. C., and Kormanik, P. A. (1997). Non-AIDS-related lymphomatous meningitis: combined modality therapy. *Neurology* 49, 1728–1731.

Chamberlain, M. C., and Kormanik, P. A. (1998). Carcinoma meningitis secondary to non-small cell lung cancer: combined modality therapy. *Arch. Neurol.* 55, 506–512.

Chamberlain, M. C., Kormanik, P., Howell, S. B., and Kim, S. (1995). Pharmacokinetics of intralumbar DTC-101 for the treatment of leptomeningeal metastases. *Arch. Neurol.* 52, 912–917.

Champagne, M. A., and Silver, H. K. B. (1992). Intrathecal dacarbazine treatment of leptomeningeal malignant melanoma. *J. Natl. Cancer Inst.* 84, 1203–1204.

Cramer, S. C., Glaspy, J. A., Efird, J. T., and Louis, D. N. (1996). Chronic lymphocytic leukemia and the central nervous system: a clinical and pathological study. *Neurology* 46, 19–25.

Coakham, H. B., and Kemshead, J. T. (1998). Treatment of neoplastic meningitis by targeted radiation using [131]I-radiolabelled monoclonal antibodies. *J. Neuro-Oncol.* 38, 225–232.

Fizazi, K., Asselain, B., Vincent-Salomon, A., Jouve, M., Dieras, V., Palangie, T., Beuzeboc, P., Dorval, T., and Pouillart, P. (1996). Meningeal carcinomatosis in patients with breast carcinoma. Clinical features, prognostic factors, and results of a high-dose intrathecal methotrexate regimen. *Cancer* 77, 1315–1323.

Freilich, R. J., Krol, G., and DeAngelis, L. M. (1995). Neuroimaging and cerebrospinal fluid cytology in the diagnosis of leptomeningeal metastasis. *Ann. Neurol.* 38, 51–57.

Fulton, D. S., Levin, V. A., Gutin, P. H., Edwards, M. S. B., Seager, M. L., Stewart, J., and Wilson, C. B. (1982). Intrathecal cytosine arabinoside for the treatment of meningeal metastases from malignant brain tumors and systemic tumors. *Cancer Chemother. Pharmacol.* 8, 285–291.

Giannone, L., Greco, F. A., and Hainsworth, J. D. (1986). Combination intraventricular chemotherapy for meningeal neoplasia. *J. Clin. Oncol.* 4, 68–73.

Glantz, M. J., Cole, B. F., Recht, L., Akerley, W., Mills, P., Saris, S., Hochberg, F., Calabresi, P., and Egorin, M. J. (1998). High-dose intravenous methotrexate for patients with nonleukemic cancer: is intrathecal chemotherapy necessary? *J. Clin. Oncol.* 16, 1561–1567.

Glantz, M. J., Hall, W. A., Cole, B. F., Chozik, B. S., Shannon, C. M., Wahlberg, L., Akerley, W., Marin, L., and Choy, H. (1995). Diagnosis, management, and survival of patients with leptomeningeal cancer based on cerebrospinal fluid-flow status. *Cancer* 75, 2919–2931.

Glantz, M. J., Jaeckle, K. A., Chamberlain, M. C., Phuphanich, S., Recht, L., Swinnen, L. J., Maria, B., LaFollette, S., Schumann, G. B., Cole, B. F., and Howell, S. B. (1999a). A randomized controlled trial comparing intrathecal sustained-release cytarabine (DepoCyt) to intrathecal methotrexate in patients with neoplastic meningitis from solid tumors. *Clin. Cancer Res.* 5, 3394–3402.

Glantz, M. J., Jaeckle, K. A., Chamberlain, M. C., Phuphanich, S., Recht, L., Swinnen, L. J., Maria, B., LaFollette, S., Schumann, G. B., Cole, B. F., and Howell, S. B. (1999b). Randomized trial of a slow-release versus a standard formulation of cytarabine for the intrathecal treatment of lymphomatous meningitis. *J. Clin. Oncol.* 17, 3110–3116.

Grant, R., Naylor, B., Greenberg, H. S., and Junck, L. (1994). Clinical outcome in aggressively treated meningeal carcinomatosis. *Arch. Neurol.* 51, 457–461.

Grant, R., Naylor, B., Junck, L., and Greenberg, H. S. (1992). Clinical outcome in aggressively treated meningeal gliomatosis. *Neurology* 42, 252–254.

Grossman, S. A., Finkelstein, D. M., Ruckdeschel, J. C., Trump, D. L., Moynihan, T., and Ettinger, D. S., for the Eastern Cooperative Oncology Group. (1993). Randomized prospective comparison of intraventricular methotrexate and thiotepa in patients with previously untreated neoplastic meningitis. *J. Clin. Oncol.* 11, 561–569.

Gutin, P. H., Levi, J. A., Wiernik, P. H., and Walker, M. D. (1977). Treatment of malignant meningeal disease with intrathecal thioTEPA: a phase II study. *Cancer Treat. Rep.* 61, 885–887.

Hitchins, R. N., Bell, D. R., Woods, R. L., and Levi, J. A. (1987). A prospective randomized trial of single-agent versus combination chemotherapy in meningeal carcinomatosis. *J. Clin. Oncol.* 5, 1655–1662.

Joyson, G. C., Howell, A., Harris, M., Morgenstern, G., Chang, J., and Ryder, W. D. (1994). Carcinomatous meningitis in patients with breast cancer. An aggressive disease variant. *Cancer* 74, 3135–3141.

Kochi, M., Kuratsu, J., Mihara, Y., Takaki, S., Seto, H., Uemura, S., and Ushio, Y. (1993). Ventriculolumbar perfusion of 3-[(4-amino-2-methyl-5-pyrimidinyl)methyl]-1-(2-chloroethyl)-1-nitrosourea hydrochloride. *Neurosurgery* 33, 817–823.

Leman, W., Wiley, R. G., and Posner, J. B. (1988). Leukoencephalopathy complicating intraventricular catheters: clinical, radiographic and pathologic study of 10 cases. *J. Neuro-Oncol.* 5, 67–74.

Levin, V. A., Chamberlain, M., Silver, P., Rodriguez, L., and Prados, M. (1989). Phase I/II study of intraventricular and intrathecal ACNU for leptomeningeal neoplasia. *Cancer Chemother. Pharmacol.* 23, 301–307.

Mencel, P. J., De Angelis, L. M., and Motzer, R. J. (1994). Hormonal ablation as effective therapy for carcinomatous meningitis from prostatic carcinoma. *Cancer* 73, 1892–1894.

Morra, E., Lazzarino, M., Brusamolino, E., Pagnucco, G., Castagnola, C., Bernasconi, P., Orlandi, E., Corso, A., Santagostino, A., and Bernasconi, C. (1993). The role of systemic high-dose cytarabine in the treatment of central nervous system leukemia. *Cancer* **72**, 439–445.

Moseley, R. P., Benjamin, J. C., Ashpole, R. D., Sullivan, N. M., Bullimore, J. A., Coakham, H. B., and Kemshead, J. T. (1991). Carcinomatous meningitis: antibody-guided therapy with I-131 HMFG1. *J. Neurol. Neurosurg. Psychiatry* **54**, 260–265.

Nakagawa, H., Murasawa, A., Kubo, S., Nakajima, S., Nakajima, Y., Izumoto, S., and Hayakawa, T. (1992). Diagnosis and treatment of patients with meningeal carcinomatosis. *J. Neuro-Oncol.* **13**, 81–89.

Obbens, E. A. M. T., Leavens, M. E., Beal, J. W., and Lee, Y. (1985). Ommaya reservoir in 387 cancer patients: a 15-year experience. *Neurology* **35**, 1274–1278.

Onda, K., Tanaka, R., Takahashi, H., Takeda, N., and Ikuta, F. (1989). Cerebral glioblastoma with cerebrospinal fluid dissemination. A clinicopathological study of 14 cases examined by complete autopsy. *Neurosurgery* **25**, 533–540.

Pfeffer, M. R., Wygoda, M., and Siegal, T. (1988). Leptomeningeal metastases. Treatment results in 98 consecutive patients. *Isr. J. Med. Sci.* **24**, 611–618.

Posner, J., and Chernik, N. (1978). Intracranial metastases from systemic cancer. *Adv. Neurol.* **19**, 579–591.

Postmus, P. E., Haaxma-Reiche, H., Berendsen, H. H., and Sleijfer, D. T. (1989). High-dose etoposide for meningeal carcinomatosis in patients with small cell lung cancer. *Eur. J. Cancer Clin. Oncol.* **25**, 377–378.

Pradat, P. F., Hoang-Xuan, K., Cornu, P., Mokhtari, K., Martin-Duverneuil, N., Poisson, M., and Delattre, J. Y. (1999). Treatment of meningeal gliomatosis. *J. Neuro-Oncol.* **44**, 163–168.

Ram, Z., Walbridge, S., Oshiro, E. M., Viola, J. J., Chiang, Y., Mueller, S. N., Blaese, R. M., and Oldfield, E. H. (1994). Intrathecal gene therapy for malignant leptomeningeal neoplasia. *Cancer Res.* **54**, 2141–2145.

Ribeiro, R. C., Rivera, G. K., Hudson, M., Mulhern, R. K., Hancock, M. L., Kun, L., Mahmoud, H., Sandlund, J. T., Crist, W. M., and Pui, C. H. (1995). An intensive re-treatment protocol for children with an isolated CNS relapse of acute lymphoblastic leukemia. *J. Clin. Oncol.* **13**, 333–338.

Rosen, S. T., Aisner, J., Makuch, R. W., Mathews, M. J., Ihde, D. C., Whitacre, M., Glatstein, E. J., Wiernik, P. H., Lichter, A. S., and Bunn, P. A. (1982). Carcinomatous leptomeningitis in small cell lung cancer. A clinicopathologic review of the National Cancer Institute experience. *Medicine* **61**, 45–53.

Schabet, M., Kloeter, I., Adam, T., Heidemann, E., and Wiethölter, H. (1986). Diagnosis and treatment of meningeal carcinomatosis in 10 patients with breast cancer. *Eur. Neurol.* **25**, 403–411.

Schrappe, M., Reiter, A., and Riehm, H. (1998). Prophylaxis and treatment of neoplastic meningeosis in childhood acute lymphoblastic leukemia. *J. Neuro-Oncol.* **38**, 159–165.

Siegal, T., Lossos, A., and Pfeffer, M. R. (1994). Leptomeningeal metastases: analysis of 31 patients with sustained off-therapy response following combined-modality therapy. *Neurology* **44**, 1463–1469.

Slavc, I., Schuller, E., Czech, T., Hainfellner, J. A., Seidl, R., and Dieckmann, K. (1998). Intrathecal mafosfamide therapy for pediatric brain tumors with meningeal dissemination. *J. Neuro-Oncol.* **38**, 213–218.

Stewart, D. J., Maroun, J. A., Hugenholtz, H., Benoit, B., Girard, A., Richard, M., Russell, N., Huebsch, L., and Drouin, J. (1987). Combined intraommaya methotrexate, cytosine arabinoside, hydrocortisone and thiotepa for meningeal involvement by malignancies. *J. Neuro-Oncol.* **5**, 315–322.

Theodore, W. H., and Gendelmann, S. (1981). Meningeal carcinomatosis. *Arch. Neurol.* **38**, 696–699.

Trump, D. L., Grossman, S. A., Thompson, G., Murray, K., and Wharam, M. (1982). Treatment of neoplastic meningitis with intraventricular thiotepa and methotrexate. *Cancer Treat. Rep.* **66**, 1549–1551.

Ushio, Y., Kochi, M., Kitamura, I., and Kuratsu, J. (1998). Ventriculolumbar perfusion of 3-[(4-amino-2-methyl-5-pyrimidinyl)methyl]-1-(2-chloroethyl)-1-nitrosourea hydrochloride for subarachnoid dissemination of gliomas. *J. Neuro-Oncol.* **38**, 207–212.

Van Oostenbrugge, R. J., Hopman, A. H., Arends, J. W., Ramaekers, F. C., and Twijnstra, A. (2000). Treatment of leptomeningeal metastases evaluated by interphase cytogenetics. *J. Clin. Oncol.* **18**, 2053–2058.

Wasserstrom, W. R., Glass, J. P., and Posner, J. B. (1982). Diagnosis and treatment of leptomeningeal metastases from solid tumors. *Cancer* **49**, 759–772.

Weller, M., Stevens, A., Sommer, N., Schabet, M., and Wiethölter, H. (1992). Tumor cell dissemination triggers an intrathecal immune response in neoplastic meningitis. *Cancer* **69**, 1475–1480.

Weller, M., Stevens, A., Sommer, N., and Wiethölter, H. (1993). Intrathecal IgM response in disseminated cerebrospinal metastasis from malignant melanoma. *J. Neuro-Oncol.* **16**, 55–59.

Witham, T. F., Fukui, M. B., Meltzer, C. C., Burns, R., Kondziolka, D., and Bozik, M. E. (1999). Survival of patients with high grade glioma treated with intrathecal thiotriethylenephosphoramide for ependymal or leptomeningeal gliomatosis. *Cancer* **86**, 1347–1353.

*Tumors and Developmental Disorders*

## CHAPTER 66
# Paraneoplastic Syndromes

Jerome B. Posner

Paraneoplastic syndromes affecting the nervous system are rare but devastating complications of systemic cancer. The neurological disorder usually precedes identification of the cancer and can affect any portion of the nervous system, including brain, spinal cord, peripheral nerves, and muscle. A single area or cell type of the nervous system may be affected, or the entire neuraxis may be involved. The pathogenesis of paraneoplastic syndromes involving the nervous system is thought to be immune mediated: the current hypothesis is that antigens usually expressed only in neurons are expressed in a cancer; the immune system recognizes the antigen in the cancer as foreign and mounts an immune response that slows the growth of the tumor but damages the nervous system. The diagnosis of a paraneoplastic syndrome is made either by identifying a small cancer in a patient with a neurological disorder or by identifying paraneoplastic autoantibodies in the serum of patients. The treatment involves identification and treatment of the causal cancer and immunosuppression to suppress both the humoral and cellular immune response.

## INTRODUCTION

Paraneoplastic syndromes are defined as dysfunction of an organ or tissue caused by a malignant neoplasm but not directly related to invasion of the affected organ or tissue by the primary tumor or its metastases. Paraneoplastic syndromes can affect almost any organ or tissue, including the nervous system. Examples of non-nervous system paraneoplastic syndromes include severe weight loss (cachexia) associated with small and occult lung cancers, anemia associated with cancers that do not invade the bone marrow, and a variety of endocrine disorders including the syndrome of inappropriate antidiuretic hormone secretion, hypercalcemia, and Cushing's syndrome.

The nervous system can also be affected by tumors that do not directly invade neural structures (Posner, 1995). Examples of such nonmetastatic complications include vascular disorders (e.g., brain hemorrhage associated with thrombocytopenia or disseminated intravascular coagulation or infarction associated with nonbacterial thrombotic endocarditis, and/or the cancer-induced hypercoagulable state), opportunistic infections in patients immunosuppressed by the tumor (e.g., lymphoma) or its treatment (e.g., neutropenia), and direct side effects of radiation (e.g., dementia, myelopathy) or chemotherapy (e.g., cognitive dysfunction, seizures). All of these disorders can be considered paraneoplastic. However, for the purposes of this chapter, paraneoplastic syndromes affecting the nervous system are defined as disorders of nervous system function that are caused by cancer but cannot be ascribed to metastases or to the destruction of vital systemic organs (e.g., liver, kidney) by the tumor or its treatment. Clinically and pathologically identical disorders occur in patients without cancer but at a lesser incidence.

## CLASSIFICATION

Paraneoplastic syndromes can affect any portion of the nervous system (Table I). Only a single area (e.g., limbic system, peripheral nerves) or a single cell type of the nervous system may be involved (e.g., Purkinje cell, in paraneoplastic cerebellar degeneration), or multiple levels of the nervous system may be involved (e.g., encephalomyeloradiculitis). Furthermore, although the presence of paraneoplastic antibodies (see later) is an important diagnostic tool, identical clinical syndromes may be associated with different antibodies, or a single autoantibody may be associated with several different clinical syndromes, and not all patients with a paraneoplastic syndrome harbor identifiable paraneoplastic antibodies in their serum. Thus, a bewildering variety

**TABLE I  Some Paraneoplastic Syndromes of the Nervous System**

Brain and cranial nerves
  Limbic encephalitis (Gultekin *et al.*, 2000)
  Brainstem encephalitis (Barnett *et al.*, 2001)
  Cerebellar degeneration (Cao *et al.*, 1999)
  Opsoclonus-myoclonus (Bataller *et al.*, 2001)
  Visual syndromes
    Cancer-associated retinopathy (CAR) (Goldstein *et al.*, 1999)
    Optic neuritis (Lieberman *et al.*, 1999)
  Chorea (Croteau *et al.*, 2001)
  Parkinsonism (Golbe *et al.*, 1989)
Spinal cord
  Necrotizing myelopathy (Rudnicki and Dalmau, 2000)
  Inflammatory myelitis (Babikian *et al.*, 1985; Scully *et al.*, 1988)
  Motor neuron disease (ALS) (Younger, 2000)
  Subacute motor neuronopathy (Schold *et al.*, 1979)
  Stiff person syndrome (Brown and Marsden, 1999; Silverman, 1999)
Dorsal root ganglia
  Sensory neuronopathy (Graus *et al.*, 2001)
Peripheral nerves: (Antoine *et al.*, 1999; Rudnicki and Dalmau, 2000)
  Autonomic neuropathy (Lee *et al.*, 2001)
  Acute sensorimotor neuropathy
    Polyradiculoneuropathy (Guillain-Barré) (Lisak *et al.*, 1977)
    Brachial neuritis (Lachance *et al.*, 1991)
  Chronic sensorimotor neuropathy (Antoine *et al.*, 1999)
  Vasculitic neuropathy (Blumenthal *et al.*, 1998)
  Neuromyotonia (Lahrmann *et al.*, 2001; Vincent, 2000)
Neuromuscular junction
  Lambert-Eaton myasthenic syndrome (Carpentier and Delattre, 2001)
  Myasthenia gravis (Vernino *et al.*, 1999)
Muscle
  Polymyositis/dermatomyositis (Stockton *et al.*, 2001)
  Necrotizing myopathy (Levin *et al.*, 1998)
  Myotonia (Pascual *et al.*, 1994)

of antibodies and clinical syndromes have been described, too many to discuss each of them here. Instead, the following paragraphs describe general considerations that apply to most or all paraneoplastic syndromes. Discussion of individual disorders can be found in the references (Table I).

# INCIDENCE

Neurological paraneoplastic syndromes are not common, but their exact incidence is unknown. Incidence figures vary, depending on how one defines the illness. For example, if one only considers those syndromes that cause sufficient symptoms to prompt referral to a neurologist, paraneoplastic syndromes probably affect fewer than 1% of patients with cancer (Camerlingo *et al.*, 1998; Chio *et al.*, 1988). Perhaps the most common syndrome, Lambert-Eaton myasthenic syndrome (LEMS), may affect as many as 3% of patients with small-cell lung cancer, and peripheral neuropathy may be even more common in patients with the osteosclerotic form of multiple myeloma, but cerebellar degeneration, encephalomyelitis, opsoclonus, and photoreceptor degeneration are much less common. However, if one assesses minor neurological abnormalities not likely to prompt neurological consultation in patients with cancer, the figures are substantially higher. Erlington and colleagues (1991) found that about 50% of patients with small-cell lung cancer who were examined neurologically demonstrated findings including weakness (44%), abnormal deep tendon reflexes (32%), and ataxia or nystagmus (5%). We have found minor neurological abnormalities in about 70% of patients with small-cell lung cancer at their first visit to an oncologist at Memorial Sloan-Kettering Cancer Center. Electrophysiological studies indicate peripheral nerve abnormalities in a substantial portion of cancer patients (Lipton *et al.*, 1987). Whether these electrophysiological disorders are metabolic, nutritional, or paraneoplastic is unclear.

The importance of paraneoplastic syndromes lies not in their frequency but in the fact that the neurological disorder, if appropriately diagnosed, may lead to the identification of a small and potentially curable cancer. In addition, paraneoplastic syndromes may cause severe disability at a time when the cancer is causing little or no symptoms. Some patients with lung cancer may have paraneoplastic syndromes develop that mimic metastatic or other nonmetastatic neurological complications of the cancer. Paraneoplastic syndromes are also biologically important, because they are sometimes (but not always) associated with indolent tumor growth (Maddison *et al.*, 1999), suggesting a better prognosis with respect to the tumor in patients with paraneoplastic syndromes than in patients with a similar cancer but without a paraneoplastic syndrome. Autoantibodies found in the serum of some patients with paraneoplastic syndromes (see later) have served as probes to identify genes usually expressed only in neurons (Szabo *et al.*, 1991).

The rarity of paraneoplastic syndromes means that a busy oncologist may see only a few patients with a paraneoplastic syndrome in a lifetime of practice. However, the situation is different for the neurologist. Even though paraneoplastic syndromes are rare, certain neurological disorders strongly suggest that cancer is the cause (i.e., the neurological disorder is paraneoplastic). For example, in about two thirds of patients with the LEMS, the cause is paraneoplastic, and the offending cancer is usually small-cell lung cancer. In about 10%–15% of patients with myasthenia gravis, thymoma (rarely Hodgkin's disease involving the thymus) is the underlying cause. Perhaps as many as 20% of patients with a subacutely developing sensory neuronopathy (i.e., dorsal root ganglionopathy) are associated with cancer. About 20% of adults with opsoclonus/myoclonus have cancer as the underlying cause. Thus, for the neurologist, paraneoplastic syndromes seem a more common problem than they do to the oncologist.

## PATHOGENESIS

A central problem in the understanding of paraneoplastic syndromes is how a small tumor in the lung or ovary, often too small to be detected by standard imaging techniques and occasionally only microscopic in size, can cause a disorder in which all the Purkinje cells in the cerebellum disappear (Verschuuren *et al.*, 1996). In the past investigators have suggested that substances such as cytokines or other toxins elaborated by the tumor or by cells reacting to the tumor might poison the nervous system, that the tumor and the nervous system may compete for an essential substrate, or that the neurological disorder may represent an opportunistic infection in an immunosuppressed patient. Although there are specific examples of each of these pathogenetic mechanisms causing nervous system dysfunction (Table II), the current hypothesis is that most paraneoplastic syndromes are immune mediated. The hypothesis is as follows:

1. Antigens normally restricted to the nervous system (occasionally restricted to nervous system and testis (Rosenfeld *et al.*, 2001), also an immunologically privileged site) are ectopically expressed in a cancer.
2. The immune system recognizes the antigen in the cancer as foreign, although current evidence indicates that the antigen is not mutated.
3. An immune response is generated. The immune response can be humoral (B-cell mediated), T-cell mediated, or both.

4. The immune response attacks some or all of the areas of the nervous system that express the same antigen expressed in the tumor, resulting in nervous system damage.
5. The immune response also attacks the tumor, often keeping it small or even leading to its destruction (Byrne *et al.*, 1997).

The evidence for immune mediation of paraneoplastic syndromes is best seen in the LEMS associated with small-cell lung cancer. P/Q-type voltage-gated calcium channels are expressed at the presynaptic neuromuscular junction and in all small-cell lung cancers. The antibodies recognizing the antigen circulate in the serum of patients with the clinical syndrome. If the antibodies are removed by plasma exchange, the patient's clinical symptoms improve. If the antibodies are given to an experimental animal, the electrophysiological abnormality found in patients with LEMS can be reproduced.

In other paraneoplastic syndromes the evidence is less compelling. Several pieces of evidence do suggest an autoimmune pathogenesis:

1. In some patients with paraneoplastic syndromes, high titers of antibodies that react only with the tumors and the nervous system are found in serum (Table III).
2. In patients with central nervous system paraneoplastic syndromes, there is a relatively higher titer of antibody in the CSF than in the serum, suggesting that the antibody is being synthesized intrathecally (Furneaux *et al.*, 1990).
3. In some patients with paraneoplastic syndromes, antigen-specific T-cells are found in the blood, CSF, and brain (Albert *et al.*, 2000).
4. Some investigators report the presence of antibody in neurons in the brains of patients who die with antibody-positive paraneoplastic syndromes (Dalmau *et al.*, 1991).
5. In experimental animals, immunization with a paraneoplastic antigen may retard the growth of a tumor expressing that antigen (Carpentier *et al.*, 1998).

In some instances, paraneoplastic antigens are expressed in all tumors of a specific type, even though only a small minority of patients mount an immune reaction. Examples are the anti-Hu antigen and the P/Q-type voltage-gated calcium channels in small-cell lung cancers. In other instances, the antigens seem to be expressed in only a subset of tumors, for example, the Yo antigens in ovarian cancer. Even in this instance the antigens are expressed in many patients who do not develop a paraneoplastic syndrome.

That paraneoplastic syndromes were immune mediated was first suggested by Dorothy Russell, an English

**TABLE II**  Some Possible Pathogenic Mechanisms of Paraneoplastic Syndromes

| Hypothesis | Example |
|---|---|
| "Toxin" secreted by tumor | ACTH = Cushing's syndrome |
| | PTHRP* = hypercalcemia |
| Competition for essential substrate | Use of glucose by sarcomas = hypoglycemia |
| | Carcinoid tumors compete with brain for tryptophan = pellagra-like syndrome |
| Opportunistic infection | Papovavirus = progressive multifocal leukoencephalopathy |
| Autoimmune process | Lambert-Eaton myasthenic syndrome |

*Parathyroid hormone–related protein.

**TABLE III    Antineuronal Antibody–Associated Paraneoplastic Disorders**

| Antibody | Neuronal reactivity | Protein antigens | Cloned genes | Tumor | Paraneoplastic symptoms | References |
|---|---|---|---|---|---|---|
| Anti-Hu | Nucleus> cytoplasm (all neurons) | 35–40 Kd | HuD, HuC Hel-Nl | SCLC, neuroblastoma, sarcoma, prostate | PEM, PSN, PCD, autonomic dysfunction | (Graus et al., 2000) |
| Anti-Yo | Cytoplasm Purkinje cells | 34, 62 Kd | CDR34, CDR62 | Ovary, breast, lung | PCD | (Okano et al., 1999) |
| Anti-Ri | Nucleus> cytoplasm (CNS neurons) | 55, 80 | Nova | Breast, Gyn lung, bladder | Ataxia/ opsoclonus | (Jensen et al., 2000) |
| Anti-Tr | Cytoplasm Purkinje cells | ? | — | Hodgkin's | PCD | (Peltola et al., 1998) |
| Anti-VGCC | Presynaptic NMJ | VGCC 64 Kd | P/Q type MysB | VGCC, SCLC | LEMS | (Carpentier and Delattre, 2001) |
| Antiretinal | Photoreceptor, ganglion cells | 23, 65, 145, 205 Kd | Recoverin | SCLC, melanoma, Gyn | CAR, MAR | (Maeda et al., 2001) |
| Antiamphiphysin | Presynaptic | 128 Kd | Amphiphysin | Breast, SCLC | Stiff-person syndrome, PEM | (Saiz et al., 1999b) |
| Anti-CRMP5 (Anti-CV2) | Oligodendrocytes, neurons, cytoplasm | 66 Kd | CRMP5 (POP66) | SCLC, thymoma | PEM PCD, chorea, sensory neuropathy | (Yu et al., 2001) |
| Anti-PCA-2 | Purkinje cytoplasm and other neurons | 280 Kd | | SCLC | PEM, PCD, LEMS, others | (Vernino and Lennon, 2000) |
| Anti-Mal | Neurons (subnucleus) | 40 Kd | Mal | Lung, others | Brainstem, PCD | (Rosenfeld et al., 2001) |
| Anti-Ma2 | Neurons (subnucleus) | 41.5 Kd | Ma2 | Testis | Limbic brainstem encephalitis | (Rosenfeld et al., 2001) |
| ANNA 3 | Nuclei, Purkinje cells | 170 Kd | | Lung cancer | Sensory neuropathy, PEM | (Chan et al., 2001) |

Gyn, gynecological cancer; CAR, cancer-associated retinopathy; MAR, melanoma-associated retinopathy; PEM, paraneoplastic encephalomyelitis; PSN, paraneoplastic sensory neuronopathy; PCD, paraneoplastic cerebellar degeneration; VGCC, voltage-gated calcium channel; SCLC, small-cell lung cancer; NJM, neuromuscular junction.

neuropathologist, who suggested that the inflammatory infiltrates found in the nervous system might be immune rather than infection-mediated. How the immune response damages the nervous system is not entirely known. The Lambert-Eaton syndrome and myasthenia gravis are largely antibody mediated. Cellular immunity through cytotoxic T-cells may play an important role in paraneoplastic cerebellar degeneration (Greenlee, 2000) and paraneoplastic encephalomyelitis (Tanaka et al., 1999).

How the immune response arises is unclear. One study suggested that patients with small-cell lung cancer who had paraneoplastic syndromes develop were more likely to express HLA antigens on the surface of tumor cells than were those who did not have paraneoplastic syndromes develop (Dalmau et al., 1995). Others have suggested that an antibody response is not mounted until the tumor has spread to lymph nodes. In fact, most patients with a paraneoplastic syndrome have lymph node involvement. In some instances, the tumor is found only in the lymph nodes, and the primary tumor is never found. This occurs in some patients with anti-Yo positive breast cancer and in some with anti-Hu positive small-cell lung cancer.

## DIAGNOSIS

Several aspects of the diagnosis of paraneoplastic syndromes deserve consideration. Between two thirds and three fourths of patients with paraneoplastic syndromes have their neurologic symptoms develop before the cancer is identified. Thus, the neurologist is likely to see the patient before the oncologist. For the neurologist encountering a patient with a suspected paraneoplastic disorder who is not known to have cancer, the first task is to identify the disorder as paraneoplastic. Certain clinical clues help:

1. Paraneoplastic syndromes usually begin acutely or subacutely and evolve rapidly to cause severe disability. The rapid evolution of symptoms helps differentiate them from similar appearing degenerative diseases of the nervous system, which usually evolve much more slowly. Creutzfeldt-Jacob disease (CJD), which can affect the cerebellum and cerebral cortex, is an exception; the disorder, especially when atypical, may mimic a paraneoplastic syndrome. Distinguishing a paraneoplastic syndrome from CJD may be difficult. CSF pleocytosis indicates a paraneoplastic or other inflammatory disorder. CSF 14-3-3 protein, however, may be positive in some paraneoplastic syndromes as well as CJD (Saiz et al., 1999a).

2. When paraneoplastic syndromes affect the central nervous system, the spinal fluid often reveals pleocytosis, elevated immunoglobulins, and oligoclonal bands. The pleocytosis often disappears after several weeks; the immunoglobulins remain elevated.

3. The clinical disorder often stabilizes. Many patients evolve rapidly over several weeks to several months but then stabilize, neither regressing nor improving over time.

4. As indicated earlier, certain neurological disorders such as LEMS strongly suggest the presence of an underlying cancer.

5. Finally, measurement of specific paraneoplastic antibodies in the serum may aid in the diagnosis. If the physician suspects that a patient has a paraneoplastic syndrome, serum should be studied for appropriate autoantibodies (Table III). At times the physician may order a specific antibody test based on the patient's clinical state. For example, for a patient who either is or was in the past a heavy smoker and who has evidence of encephalomyelitis, measurement of the anti-Hu antibody is likely to be fruitful. Such a measurement would not be fruitful in a nonsmoking woman who has subacutely developing cerebellar degeneration. In that instance, the anti-Yo antibody is likely to be positive. However, there is sufficient overlap among antibody-specific clinical syndromes that measurement of a single antibody may miss the diagnosis (Table III). For example, a patient with cerebellar symptoms and signs may be anti-Hu, anti-Yo, anti-Ma2, or other antibody positive. Some patients will have no identifiable paraneoplastic antibody in their serum. Thus, the best approach when a paraneoplastic syndrome is suspected is to screen serum for antineuronal antibodies using both immunohistochemistry and Western blotting (Dalmau and Posner, 1994).

Once a neurologist concludes that the neurological disorder may be paraneoplastic, the next step is a search for the underlying cancer. If paraneoplastic antibodies have been found, the search can often be narrowed to one or a few causal neoplasms (Table III). For example, the presence of the anti-Yo antibody strongly suggests either a breast or a gynecological cancer, although on occasion other cancers have been encountered (Meglic et al., 2001). In a like manner the presence of the Ta (Ma2 antibody) strongly suggests testicular cancer as the cause. If the antibody-directed search identifies an early cancer, treatment may not only cure the cancer but may also stabilize or at times reverse the neurological symptoms. In many instances, either no well-characterized paraneoplastic antibody is present in the serum or the directed search does not identify a cancer in the usual causal organ. In that instance, a total body search is necessary. This initially includes a CT scan of the chest, abdomen, and pelvis; measurement of the blood for tumor antigens (e.g., CEA, breast cancer antigen, prostate specific antigen); and a careful general physical examination. Recently, whole-body positron emission tomography (PET) scans using 18-fluorodeoxyglucose have been used to search for small cancers (Delbeke and Martin, 2001; Rees et al., 2001). Whole-body PET scans promise to play an increasingly important role in the search for cancer underlying paraneoplastic syndromes, because very small lesions can often be identified.

When a patient known to have cancer has neurologic symptoms or signs develop, the neurologist is often asked to determine whether the illness is paraneoplastic; usually it is not, and the neurologist must carefully consider and rule out other neurological complications of cancer before concluding that the illness might be paraneoplastic (Posner, 1995). Thus, appropriate MR or PET scans looking for metastases, consideration of side effects of therapy, and a search for metabolic disorders must proceed along with evaluation for a paraneoplastic syndrome. Only when a paraneoplastic antibody has been found in the patient's serum and the antigen recognized by that antibody is found in the underlying tumor (Graus et al., 2001) can the physician conclude unequivocally that the patient's disorder is paraneoplastic. If a typical paraneoplastic antibody is present but the antigen is not found in the underlying tumor, physicians should search for a second primary tumor.

In patients with known cancer, paraneoplastic antibodies can arise either shortly after the cancer is identified or after the cancer has been successfully treated, at a time of relapse. Thus, even in a patient with a remote history of cancer and the appropriate clinical setting, a search should be made for recurrence of the lesion.

## TREATMENT

Paraneoplastic syndromes are considered to be immune mediated. Accordingly, two treatment approaches have been used:

1. *Remove the source of antigenic challenge.* This is done by treating the underlying tumor. For many paraneoplastic syndromes this is the most and indeed the only approach that seems to be effective (Levy *et al.*, 2001; Vigliani *et al.*, 2001). For example, in anti-Hu–associated paraneoplastic encephalomyelitis an inverse correlation exists between severity of the neurological disease at diagnosis and delay between the onset of the neurological disease and the diagnosis. Thus, the earlier the diagnosis is made, the better the patient's neurological condition. Furthermore, in multivariant analysis, treatment of the tumor with or without concomitant immunotherapy is an independent predictor of stabilization of neurological dysfunction for 6 months or more (Graus, *et al.*, 2001).

2. *Suppress the immune reaction.* If the disease is antibody mediated, as it is in LEMS and myasthenia gravis, plasma exchange or intravenous immunoglobulin (IVIg) may prove effective (Bain *et al.*, 1996). If the disease is T-cell–mediated, as is suspected in many of the CNS disorders such as anti-Hu–positive encephalomyelitis, drugs such as tacrolimus or more recently mycophenolate mofetil (Schneider *et al.*, 2001) may be tried. In many paraneoplastic disorders the exact pathogenesis is unknown, but humoral and cell-mediated immunity may both play a role, thus requiring the physician to suppress both arms of the immune system.

There are no established protocols for the treatment of most of the paraneoplastic syndromes, but the worried physician encountering a deteriorating patient usually uses a combination of either plasma exchange or IVIg and an immunosuppressant agent such as corticosteroids, cyclophosphamide, or tacrolimus. There is no established protocol for immunosuppressive treatment. Keime-Guibert and colleagues (2000) administered IVIg at a dose of 0.5 gm/kg/day for 5 days, intravenous methylprednisolone 1 g/day for 3 days, and intravenous cyclophosphamide 600 mg/m² for 1 day on day 4. If there was evidence of improvement or stability, the treatment was repeated three times at 3-week intervals. If the patient improved after the third treatment, maintenance treatment with 0.5 gm/kg of IVIg,

**TABLE IV    Treatment of Paraneoplastic Neurological Syndromes**

| | |
|---|---|
| Paraneoplastic syndromes that usually respond to treatment | |
| LEMS | Tumor, plasma exchange, IVIg, 3,4 diaminopyridine |
| Myasthenia gravis | Tumor, plasma exchange, IVIg, immunosuppressants |
| Dermatomyositis | Steroids, immunosuppressants, IVIg |
| Opsoclonus/myoclonus (pediatrics) | Steroids, ACTH, tumor |
| Carcinoid myopathy | Tumor, cyproheptadine |
| Neuropathy (osteosclerotic myeloma) | Tumor resection, radiation |
| Paraneoplastic syndromes that may respond to treatment | |
| Vasculitis (nerve/muscle) | Steroids, cyclophosphamide |
| Opsoclonus/myoclonus (adults) | Steroids, tumor, protein A column, clonazepam, diazepam, baclofen |
| PCD (Hodgkin's) | Tumor |
| Opsoclonus/ataxia (anti-Ri) | Steroids, cyclophosphamide |
| Guillain-Barré (Hodgkin's) | Tumor, plasma exchange, IVIg |
| Stiff person syndrome | Tumor, steroids, diazepam, baclofen, IVIg |
| Neuromyotonia | Plasma exchange |
| Paraneoplastic syndromes that usually do not respond to treatment | |
| Paraneoplastic cerebellar degeneration (PCD) | |
| SCLC (irrespective of anti-Hu) | |
| Anti-Yo antibodies (cancer of ovary, breast) | |
| Paraneoplastic encephalomyelitis/sensory neuronopathy | |
| Limbic encephalopathy | |
| Cerebellar degeneration | |
| Brainstem encephalopathy | |
| Myelopathy | |
| Sensory neuronopathy | |
| Autonomic dysfunction (central or peripheral) | |
| Cancer-associated retinopathy | |
| Paraneoplastic syndromes that may improve spontaneously | |
| Acute motor neuronopathy and lymphoma | |
| PCD associated with Hodgkin's | |
| Acute polyradiculopathy associated with Hodgkin's | |
| Limbic encephalopathy | |
| Opsoclonus/myoclonus (pediatric and adult population) | |

intravenous methylprednisolone 1 g, and intravenous cyclophosphamide 600 mg/m², was delivered in 1 day monthly for 6 months (Keime-Guibert et al., 2000). There is less experience with tacrolimus. One investigator gave 0.15 mg/kg a day for 14 days followed by 0.3 mg/kg a day for 7 days (Albert et al., 2000). This course decreased T-cells in the spinal fluid but added no substantial effect on the clinical course. For most paraneoplastic syndromes immunotherapy is not effective (Croteau et al., 2001; Keime-Guibert et al., 2000). However, isolated case reports describing responses to various immunotherapeutic endeavors encourage the physician to combine immunotherapy with cancer treatment in an otherwise desperate situation.

## PROGNOSIS

The prognosis for the paraneoplastic syndrome varies with the nature of the syndrome (Table IV). Some disorders such as LEMS and myasthenia gravis respond quite well to immunosuppression and subsequently to treatment of the underlying tumor. The peripheral neuropathy associated with osteosclerotic myeloma generally resolves when the tumor is treated by radiotherapy. A few disorders, such as opsoclonus/myoclonus in adults, either respond to treatment of the underlying tumor and/or immunosuppression or resolve spontaneously. In many instances, it is not clear whether the paraneoplastic syndrome resolved on its own or whether it was a response to treatment. Many of the disorders involving the central nervous system, such as encephalomyelitis associated with cancer or paraneoplastic cerebellar degeneration, respond poorly to treatment, although they may stabilize when the underlying tumor is treated. These disorders do not seem to respond to plasma exchange or immunosuppressants. However, even in those disorders that generally do not respond, an occasional patient is reported who either improves or recovers, sometimes spontaneously but usually associated with some kind of treatment.

The reason for the different prognosis in different paraneoplastic syndromes probably has to do with the underlying disease. LEMS and myasthenia gravis are diseases of the neuromuscular junction, areas that, because there is no loss of the parent neuron, can recover function once the causal insult has resolved. Disorders such as paraneoplastic cerebellar degeneration are usually associated with neuronal loss and, because they evolve subacutely and by the time treatment is undertaken the neurons are dead, recovery is impossible. Some CNS disorders such as opsoclonus/myoclonus do not involve cellular loss and, in fact, may be associated with no identifiable pathology. Thus, these disorders, like LEMS, have the potential for recovery.

One of the questions that often arises is whether immunosuppression for treatment of the paraneoplastic syndrome worsens the tumor. No evidence for such a phenomenon has been reported. Most reports, when describing no response of the paraneoplastic syndrome to immunosuppression, do not demonstrate exacerbation of the tumor.

## REFERENCES

Albert, M. L., Austin, L. M., and Darnell, R. B. (2000). Detection and treatment of activated T cells in the cerebrospinal fluid of patients with paraneoplastic cerebellar degeneration. Ann. Neurol. 47, 9–17.

Antoine, J. C., Mosnier, J. F., Absi, L., Convers, P., Honnorat, J., and Michel, D. (1999). Carcinoma associated paraneoplastic peripheral neuropathies in patients with and without anti-onconeural antibodies. J. Neurol. Neurosurg. Psychiatry 67, 7–14.

Babikian, V. L., Stefansson, K, Dieperink, M. E., et al. (1985). Paraneoplastic myelopathy: antibodies against protein in normal spinal cord and underlying neoplasm [letter to the editor]. Lancet 2, 49–50.

Bain, P. G., Motomura, M., Newsom-Davis, J., Misbah, S. A., Chapel, H. M., Lee, M. L., Vincent, A., and Lang, B. (1996). Effects of intravenous immunoglobulin on muscle weakness and calcium-channel autoantibodies in the Lambert-Eaton myasthenic syndrome. Neurology 47, 678–683.

Barnett, M., Prosser, J., Sutton, I., Halmagyi, G. M., Davies, L., Harper, C., and Dalmau, J. (2001). Paraneoplastic brain stem encephalitis in a woman with anti-Ma2 antibody. J. Neurol. Neurosurg. Psychiatry 70, 222–225.

Bataller, L., Graus, F., Saiz, A., Vilchez, J. J., and Spanish Opsoclonus MS. (2001). Clinical outcome in adult onset idiopathic or paraneoplastic opsoclonus-myoclonus. Brain 124, 437–443.

Blumenthal, D., Schochet, S., Jr., Gutmann, L., Ellis, B., Jaynes, M., and Dalmau, J. (1998). Small-cell carcinoma of the lung presenting with paraneoplastic peripheral nerve microvasculitis and optic neuropathy. Muscle Nerve 21, 1358–1359.

Brown, P., and Marsden, C. D. (1999). The stiff man and stiff man plus syndromes. J. Neurol. 246, 648–652.

Byrne, T., Mason, W. P., Posner, J. B., and Dalmau, J. (1997). Spontaneous neurological improvement in anti-Hu associated encephalomyelitis. J. Neurol. Neurosurg. Psychiatry 62, 276–278.

Camerlingo, M., Nemni, R., Ferraro, B., Casto, L., Partziguian, T., Censori, B., and Mamoli, A. (1998). Malignancy and sensory neuropathy of unexplained cause. Arch. Neurol. 55, 981–984.

Cao, Y., Abbas, J., Wu, X., Dooley, J., and van Amburg, A. L. (1999). Anti-Yo positive paraneoplastic cerebellar degeneration associated with ovarian carcinoma: case report and review of the literature. Gynecol. Oncol. 75, 178–183.

Carpentier, A. F., and Delattre, J. Y. (2001). The Lambert-Eaton myasthenic syndrome. Clin. Rev. Allergy Immunol. 20, 155–158.

Carpentier, A. F., Rosenfeld, M. R., Delattre, J.-Y., Whalen, R. G., Posner, J. B., and Dalmau, J. (1998). DNA vaccination with HuD inhibits growth of a neuroblastoma in mice. Clin. Cancer Res. 4, 2819–2824.

Chan, K. H., Vernino, S., and Lennon, V. A. (2001). ANNA-3 anti-neuronal nuclear antibody: Marker of lung cancer-related autoimmunity. Ann. Neurol. 50, 301–311.

Chio, A., Brignolio, F., Meineri, P., Rosso, M. G., Tribolo, A., and Schiffer, D. (1988). Motor neuron disease and malignancies: results of a population-based study. J. Neurol. 235, 374–375.

Croteau, D., Owainati, A., Dalmau, J., and Rogers, L. R. (2001). Response to cancer therapy in a patient with a paraneoplastic choreiform disorder. *Neurology* 57, 719–722.

Dalmau, J., Furneaux, H. M., Rosenblum, M. K., Graus, F., and Posner, J. B. (1991). Detection of the anti-Hu antibody in specific regions of the nervous system and tumor from patients with paraneoplastic encephalomyelitis/sensory neuronopathy. *Neurology* 41, 1757–1764.

Dalmau, J., Graus, F., Cheung, N.-K. V., Rosenblum, M. K., Ho, A., Canete, A., Delattre, J.-Y., Thompson, S. J., and Posner, J. B. (1995). Major histocompatibility proteins, anti-Hu antibodies, and paraneoplastic encephalomyelitis in neuroblastoma and small cell lung cancer. *Cancer* 75, 99–109.

Dalmau, J., and Posner, J. B. (1994). Neurologic paraneoplastic antibodies (anti-Yo; anti-Hu; anti-Ri): The case for a nomenclature based on antibody and antigen specificity. *Neurology* 44, 2241–2246.

Delbeke, D., and Martin, W. H. (2001). Positron emission tomography imaging in oncology. *Radiol. Clin North Am.* 39, 883–917.

Erlington, G. M., Murray, N. M., Spiro, S. G., and Newsom-Davis, J. (1991). Neurological paraneoplastic syndromes in patients with small cell lung cancer. A prospective survey of 150 patients. *J. Neurol. Neurosurg. Psychiatry* 54, 764–767.

Furneaux, H. M., Reich, L., and Posner, J. B. (1990). Autoantibody synthesis in the central nervous system of patients with paraneoplastic syndromes. *Neurology* 40, 1085–1091.

Golbe, L. I., Miller, D. C., and Duvoisin, R. C. (1989). Paraneoplastic degeneration of the substantia nigra with dystonia and Parkinsonism. *Movement Dis.* 4, 147–152.

Goldstein, S. M., Syed, N. A., Milam, A. H., Maguire, A. M., Lawton, T. J., and Nichols, C. W. (1999). Cancer-associated retinopathy. *Arch. Ophthalmol.* 117, 1641–1645.

Graus, F., Keime-Guibert, F., Reñe, R., Benyahia, B., Ribalta, T., Ascaso, C., Escaramis, G., and Delattre, J. Y. (2001). Anti-Hu-associated paraneoplastic encephalomyelitis: analysis of 200 patients. *Brain* 124, 1138–1148.

Greenlee, J. E. (2000). Cytotoxic T cells in paraneoplastic cerebellar degeneration. *Ann. Neurol.* 47, 4–5.

Gultekin, S. H., Rosenfeld, M. R., Voltz, R., Eichen, J., Posner, J. B., and Dalmau, J. (2000). Paraneoplastic limbic encephalitis: neurological symptoms, immunological findings and tumour association in 50 patients. *Brain* 123, 1481–1494.

Jensen, K. B., Dredge, B. K., Stefani, G., Zhong, R., Buckanovich, R. J., Okano, H. J., Yang, Y. Y., and Darnell, R. B. (2000). Nova-1 regulates neuron-specific alternative splicing and is essential for neuronal viability. *Neuron* 25, 359–371.

Keime-Guibert, F., Graus, F., Fleury, A., René, R., Honnorat, J., Broet, P., and Delattre, J. Y. (2000). Treatment of paraneoplastic neurological syndromes with antineuronal antibodies (Anti-Hu, Anti-Yo) with a combination of immunoglobulins, cyclophosphamide, and methylprednisolone. *J. Neurol. Neurosurg. Psychiatry* 68, 479–482.

Lachance, D. H., O'Neill, B. P., Harper, C. M., Jr., Banks, P. M., and Cascino, T. L. (1991). Paraneoplastic brachial plexopathy in a patient with Hodgkin's disease. *Mayo Clin. Proc.* 66, 97–101.

Lahrmann, H., Albrecht, G., Drlicek, M., Oberndorfer, S., Urbanits, S., Wanschitz, J., Zifko, U, A., and Grisold, W. (2001). Acquired neuromyotonia and peripheral neuropathy in a patient with Hodgkin's disease. *Muscle Nerve* 24, 834–838.

Lee, H. R., Lennon, V. A., Camilleri, M., and Prather, C. M. (2001). Paraneoplastic gastrointestinal motor dysfunction: Clinical and laboratory characteristics. *Am. J. Gastroenterol.* 96, 373–379.

Levin, M. I., Mozaffar, T., Al-Lozi, M. T., and Pestronk, A. (1998). Paraneoplastic necrotizing myopathy—Clinical and pathologic features. *Neurology* 50, 764–767.

Levy, E. I., Harris, A. E., Omalu, B. I., Hamilton, R. L., Branstetter, B. F., and Pollack, I. F. (2001). Sudden death from fulminant acute cerebellitis. *Pediatr. Neurosurg.* 35, 24–28.

Lieberman, F. S., Odel, J., Hirsch, J., Heinemann, M, Michaeli, J., and Posner, J. (1999). Bilateral optic neuropathy with IgGkappa multiple myeloma improved after myeloablative chemotherapy. *Neurology* 52, 414–416.

Lipton, R. B., Galer, B. S., Dutcher, J. P., Portenoy, R. K., Berger, A., Arezzo, J. C., Mizruchi, M., Wiernik, P. H., and Schaumberg, H. H. (1987). Quantitative sensory testing demonstrates that subclinical sensory neuropathy is prevalent in patients with cancer. *Arch. Neurol.* 44, 944–946.

Lisak, R. P., Mitchell, M., Zweiman, B., Orrechio, E., and Asbury, A. K. (1977). Guillain-Barre syndrome and Hodgkin's disease: Three cases with immunological studies. *Ann. Neurol.* 1, 72–78.

Maddison, P., Newsom-Davis, J., Mills, K. R., and Souhami, R. L. (1999). Favourable prognosis in Lambert-Eaton myasthenic syndrome and small-cell lung carcinoma [letter]. *Lancet* 353, 117–118.

Maeda, T., Maeda, A., Maruyama, I., Ogawa, K., Kuroki, Y., Sahara, H., Sato, N., and Ohguro, H. (2001). Mechanisms of photoreceptor cell death in cancer-associated retinopathy. *Invest. Ophthalmol. Vis. Sci.* 42, 705–712.

Meglic, B., Graus, F., and Grad, A. (2001). Anti-Yo-associated paraneoplastic cerebellar degeneration in a man with gastric adenocarcinoma. *J. Neurol. Sci.* 185, 135–138.

Okano, H. J., Park, W. Y., Corradi, J. P., and Darnell, R. B. (1999). The cytoplasmic Purkinje onconeural antigen cdr2 down-regulates c-Myc function: implications for neuronal and tumor cell survival. *Genes Dev.* 13, 2087–2097.

Pascual, J., Sanchez-Pernaute, R., Berciano, J., and Calleja, J. (1994). Paraneoplastic myotonia. *Muscle Nerve* 17, 694–695.

Peltola, J., Hietaharju, A., Rantala, I., Lehtinen, T., and Haapasalo, H. (1998). A reversible neuronal antibody (anti-Tr) associated paraneoplastic cerebellar degeneration in Hodgkin's disease. *Acta Neurol. Scand.* 98, 360–363.

Posner, J. B. (1995). "Neurologic Complications of Cancer." F.A. Davis, Philadelphia.

Rees, J. H., Hain, S. F., Johnson, M. R., Hughes, R. A., Costa, D. C., Ell, P. J., Keir, G., and Rudge, P. (2001). The role of [(18)F]fluoro-2-deoxyglucose-PET scanning in the diagnosis of paraneoplastic neurological disorders. *Brain* 124, 2223–2231.

Rosenfeld, M. R., Eichen, J. G., Wade, D. F., Posner, J. B., and Dalmau, J. (2001). Molecular and clinical diversity in paraneoplastic immunity to Ma proteins. *Ann. Neurol.* 50, 339–348.

Rudnicki, S. A., and Dalmau, J. (2000). Paraneoplastic syndromes of the spinal cord, nerve, and muscle. *Muscle Nerve* 23, 1800–1818.

Saiz, A., Dalmau, J., Butler, M. H., Chen, Q., Delattre, J.-Y., De Camilli, P., and Graus, F. (1999b). Anti-amphiphysin I antibodies in patients with paraneoplastic neurological disorders associated with small cell lung carcinoma. *J. Neurol. Neurosurg. Psychiatry* 66, 214–217.

Saiz, A., Graus, F., Dalmau, J., Pifarré, A., Marín, C., and Tolosa, E. (1999a). Detection of 14-3-3 brain protein in the cerebrospinal fluid of patients with paraneoplastic neurological disorders. *Ann. Neurol.* 46, 774–777.

Schneider, C., Gold, R., Reiners, K., and Toyka, K. V. (2001). Mycophenolate mofetil in the therapy of severe myasthenia gravis. *Eur. Neurol.* 46, 79–82.

Schold, S. C., Cho, E. S., Somasundaram, M., and Posner, J. B. (1979). Subacute motor neuronopathy: a remote effect of lymphoma. *Ann. Neurol.* 5, 271–287.

Scully, R. E., Mark, E. J., McNeely, W. F., *et al.* (Eds). (1988). Case records of the Massachusetts General Hospital. Case 14-1988. *N. Engl. J. Med.* 318, 903–915.

Silverman. I. E. (1999). Paraneoplastic stiff limb syndrome. *J. Neurol. Neurosurg. Psychiatry* **67**, 126–127.

Stockton, D., Doherty, V. R., and Brewster, D. H. (2001). Risk of cancer in patients with dermatomyositis or polymyositis, and follow-up implications: a Scottish population-based cohort study. *Br. J. Cancer* **85**, 41–45.

Szabo, A., Dalmau, J., Manley, G., Rosenfeld, M., Wong, E., Henson, J., Posner, J., and Furneaux, H. (1991). HuD, a paraneoplastic encephalomyelitis antigen, contains RNA-binding domains and is homologous to Elav and Sex-lethal. *Cell* **67**, 325–333.

Tanaka, K., Tanaka, M., Inuzuka, T., Nakano, R., and Tsuji, S. (1999). Cytotoxic T lymphocyte-mediated cell death in paraneoplastic sensory neuronopathy with anti-Hu antibody. *J. Neurol. Sci.* **163**, 159–162.

Vernino, S., Auger, R. G., Emslie-Smith, A. M., Harper, C. M., and Lennon, V. A. (1999). Myasthenia, thymoma, presynaptic antibodies, and a continuum of neuromuscular hyperexcitability. *Neurology* **53**, 1233–1239.

Vernino, S., and Lennon, V. A. (2000). New Purkinje cell antibody (PCA-2): Marker of lung cancer-related neurological autoimmunity. *Ann. Neurol.* **47**, 297–305.

Verschuuren, J., Chuang, L., Rosenblum, M. K., Lieberman, F., Pryor, A., Posner, J. B., and Dalmau, J. (1996). Inflammatory infiltrates and complete absence of Purkinje cells in anti-Yo associated paraneoplastic cerebellar degeneration. *Acta Neuropathol.* **91**, 519–525.

Vigliani, M. C., Palmucci, L., Polo, P., Mutani, R., Schiffer, D., De Luca, S., and De Zan, A. (2001). Paraneoplastic opsoclonus-myoclonus associated with renal cell carcinoma and responsive to tumour ablation. *J. Neurol. Neurosurg. Psychiatry* **70**, 814–815.

Vincent, A. (2000). Understanding neuromyotonia. *Muscle Nerve* **23**, 655–657.

Younger, D. S. (2000). Motor neuron disease and malignancy. *Muscle Nerve* **23**, 658–660.

Yu, Z. Y., Kryzer, T. J., Griesmann, G. E., Kim, K. K., Benarroch, E. E., and Lennon, V. A. (2001). CRMP-5 neuronal autoantibody: Marker of lung cancer and thymoma-related autoimmunity. *Ann. Neurol.* **49**, 146–154.

CHAPTER 67

# Pseudotumor Cerebri (Idiopathic Intracranial Hypertension)

Ulrich Wüllner and James J. Corbett

Pseudotumor cerebri (PTC) is a disorder of increased intracranial pressure (ICP) of unknown cause. Headache and bilateral papilledema with its associated loss of vision are the leading findings. The correct diagnosis is made following the modified Dandy criteria (Table I); the treatment is directed at reducing ICP to prevent visual loss.

## CLINICAL ASPECTS

Unilateral or bilateral headache, often pulsatile in character and occasionally accompanied by nausea and vomiting, is the most common (90%–100%) and usually also the first symptom, followed by transient visual obscurations (60%–70%), self-audible bruits (60%), photopsias (50%), diplopia with paralysis of the VIth nerve (20%–40%), loss of visual acuity (20%), and loss of vision (5%) (Giuseffi *et al.*, 1991; Radhakrishnan *et al.*, 1993b). Paralysis of the VIIth nerve, facial pain, and numbness of the face and the upper extremities are much less frequent findings, although somewhat more common in children. Infants and young children may have irritability and apathy rather than headache. Visual field testing yields early enlargement of the blind spot and peripheral constriction, especially nasal inferior (30%–60%). It is the most sensitive and meaningful way to monitor the course of the disease (Wall and George, 1991). Bilateral papilledema is a hallmark of PTC; however, a few patients present with unilateral papilledema or even normal optic discs (Chari and Rao, 1991; Johnson *et al.*, 1991; Marcelis and Silberstein, 1991).

Other causes of headache and papilledema must be excluded. In particular an intracranial mass, intracranial venous thrombosis, deformity or obstruction of the ventricular system, chronic infectious and tumorous meningeal affections, and disturbances of the CSF circulation caused by increased spinal protein synthesis such as in Guillain–Barré syndrome or a spinal tumor should be sought (Table II). To this end imaging of the brain (CT, MRI) and subsequent lumbar puncture (LP) must be performed. Optic disc abnormalities like optic papillitis and a central retinal venous thrombosis should also be considered, but these are almost always unilateral. Drusen of the optic disc in combination with migraine may be a problem, but a CT scan of the orbits will show small mineralized spots at the level of the disc.

In PTC, the composition of the CSF is chemically and cytologically normal. The diagnosis should never be made otherwise, unless CSF pressure is increased and all other diagnostic tests are normal. The CSF pressure at the initial LP is usually increased to 200–550 mm $H_2O$. In almost half the patients the increased ICP has led to an empty sella on CT scan; typically no ventricular enlargement is found (Weisberg, 1985). A thickening of the optic nerve sheath is sometimes visible on MRI. CSF pressure monitoring reveals a wavelike pattern with periods of pressure build up and shorter periods of near normal pressure, suggestive of an imbalance between CSF production and resorption (Johnston and Paterson, 1974).

## NATURAL COURSE

The incidence of PTC in the general population is 1/100,000, whereas in obese women in the United States, aged 15–44, it is increased to 10–20/100,000. The female/male ratio is approximately 8:1. A different situation is encountered in children; obaesity does

**TABLE I    Modified Dandy Criteria for PTC***

Awake and alert patient with signs and symptoms of increased ICP
No localizing neurological signs
Normal neurodiagnostic studies except for increased CSF pressure

not seem to be as important a factor and no gender predilection has been found (Radhakrishnan *et al.*, 1993a,b).

No randomized prospective studies of the different treatment regimens or the spontaneous course of PTC have been performed to date. PTC seems to be self-limited clinically, and in most patients medical treatment seems to lower ICP enough to allow papilledema to cease. However, ICP may remain increased despite treatment for years, and relapses occur in as many as 30% of patients (Durcan *et al.*, 1988).

Permanent loss of vision is the major risk, occurring in 5%–10%, although 20%–70% of patients may have some sort of visual field loss. Neither the duration of symptoms nor the degree of the papilledema is clearly predictive of visual loss, but concomitant hypertension seems to be a risk factor (Wall, 1991). A large number of retrospective studies reported the association of PTC with various conditions, including endocrine and metabolic disorders (Addison's disease, Cushing's disease, hypoparathyroidism) and medication, for example, tetracycline, vitamin A, and withdrawal from steroid therapy (for comprehensive review, see Digre and Corbett, 2001). These associations were not confirmed in two recent case-controlled studies, in which only obesity and especially recent weight gain occurred with statistically greater frequency in patients with PTC than in controls (Giuseffi *et al.*, 1991; Ireland *et al.*, 1990). However, the relatively small sample size (40 and 81, respectively) precluded rigorous testing. If the factors mentioned previously or others play a causative role, it remains to be determined why most of the patients who have PTC develop are obese women of childbearing age and whether there is a common underlying abnormality. Although the high incidence among

**TABLE II    Differential Diagnosis of Papilledema and Headache**

Intracranial mass
Intracranial venous sinus thrombosis
Deformity or obstruction of the ventricular system
Chronic infectious and tumorous meningeal affections
Increased spinal protein synthesis
Optic disc drusen and migraine
Optic papillitis
Central retinal venous thrombosis

young obese women is suggestive of an endocrine dysfunction, so far no endocrine hypothesis has been substantiated, and the etiopathogenesis of PTC remains unknown.

## PRINCIPLES OF THERAPY

The treatment is directed at preventing visual loss (Corbett and Thompson, 1989). Because the precise pathophysiological mechanisms leading to PTC are not known, all measures at reducing CSF pressure derive from personal experience or open studies. As a rule, only patients with symptoms (loss of vision, transient visual obscurations, diplopia, or persistent headache) need to be treated. Asymptomatic PTC requires no specific therapy other than close (initially every 2–4 weeks) observation of the visual fields, visual acuity, and contrast sensitivity to detect pathological changes in optic nerve function.

Sustained weight loss is effective for treating papilledema caused by PTC and most likely prevents the reoccurrence of PTC but usually is not sufficient to lower CSF pressure acutely. Weight loss is an effective, if difficult, treatment of idiopathic intracranial hypertension (IIH) and has been done with diet (Johnson *et al.*, 1998; Kupersmith *et al.*, 1991; Newborg, 1958), as well as by the more draconian method of bariatric surgery (Sugerman *et al.*, 1995) in morbidly obese patients. Twenty percent to 30% of patients benefit from a single (the first, "diagnostic") LP. Repeated LP and in some cases lumboperitoneal shunts have been used, although some patients "will risk visual loss rather than return for serial lumbar punctures" (Wall and George, 1991). For medical treatment, the most frequently used drugs are acetazolamide, furosemide, and steroids. Acetazolamide, a carbonic anhydrase inhibitor, can effectively decrease CSF production; furosemide is also effective. Steroids are widely used to treat increased ICP of various causes; however, the side effects, especially weight gain and the reccurrence of papilledema after withdrawal from chronic steroid therapy, makes this an undesirable treatment. Hyperosmotic agents (glycerol) also have their advocates; the potential side effects, however, limit their use. In cases in which association of PTC with specific medication use is suspected, elimination of that therapy is warranted.

Surgical optic nerve sheath fenestration is the current method of choice to prevent a rapid progression of visual loss if the conservative approaches fail (Anderson and Flaharty, 1992; Kelman *et al.*, 1991; Pearson *et al.*, 1991). The optic nerve sheath fenestration is a drainage site produced by slits or windows cut in the optic nerve dural sheath which permit drainage of CSF into the orbit where it is absorbed (Hamed *et al.*, 1992).

## PRACTICAL MANAGEMENT

To establish the diagnosis of PTC, a CT scan or MRI and MRV is performed in all suspected patients before the LP. MRI is superior for imaging the optic nerve, and MRV allows more accurate exclusion of venous thrombosis (Biousse et al., 1999). The visual field should be measured by kinetic Goldmann or static Humphrey perimetry. The progression of visual loss determines the type of intervention; surgical decompression (optic nerve sheath fenestration) should be performed in the first place, if loss of vision is a presenting sign or visual loss occurs rapidly (Wall and George, 1991). Most patients, however, will initially have LP and medical treatment using acetazolamide (500–1000 mg twice daily) or (Furosemide Lasix) (40–160 mg/day). Serum potassium must be monitored to prevent hypokalemia. Metabolic acidosis is a constant finding with acetazolamide. All overweight patients are started on a controlled high-fiber 1200 kcal diet (Kupersmith et al., 1991). Corticosteroid use is discouraged because of the multiple serious side effects and the tendency for steroid use to drag on for months to years. One of us (UW) uses repeated LP to lower CSF pressure. To this end, CSF is drained daily (25 ml) until the pressure remains below 180 mm H$_2$O; subsequently, LP is performed at weekly or monthly intervals to monitor CSF pressure. Lumboperitoneal shunt is advocated by some, and it is effective in the short term for relief of headache and papilledema (Burgett et al., 1999). More than half are singly or repeatedly revised, primarily for nerve root irritation or low pressure headache. Complications are many, and virtually all develop a Chiari I and/or a cervical syrinx (Chumas et al., 1993; Digre and Corbett, 2001). Although lumboperitoneal shunt is described as an effective means to treat PTC, a recent multiple center review revealed a success rate of only 27% and many serious side effects (Rosenberg et al., 1993). In our opinion, if the vision (acuity or field) of an affected eye continues to decline or loss of vision progresses despite medical treatment or repeated LP, optic nerve sheath fenestration should be performed, preferably before visual acuity has been compromised.

## REFERENCES

Anderson, R. L., and Flaharty, P. M. (1992). Treatment of pseudotumor cerebri by primary and secondary optic nerve sheath decompression. Am. J. Ophthalmol. 113, 599–601.

Biousse, V., et al. (1999). Isolated intracranial hypertension as the only sign of cerebral venous thrombosis. Neurology 53, 1537–1542.

Burgett, R. A., Purvin, V. A., and Kawasaki, A. (1999). Lumboperitoneal shunting for pseudotumor cerebri. Neurology 49, 734–739.

Chari, C., and Rao, N. S. (1991). Benign intracranial hypertension—its unusual manifestations. Headache 31, 599–600.

Chumas, P. D., Armstrong, D. C., Drake, J. M., et al. (1993). Tonsillar herniation: the rule rather than the exception after LP shunting in the pediatric population. J. Neurosurg. 78, 568–573.

Corbett, J. J., and Thompson, H. S. (1989). The rational management of idiopathic intracranial hypertension. Arch. Neurol. 46, 1049–1051.

Digre, K. D., and Corbett, J. J. (2001). Idiopathic intracranial hypertension (pseudotumor cerebri): A reappraisal. Neurologist 7, 2–69.

Durcan, F. J., Corbett, J. J., and Wall, M. (1988). The incidence of pseudotumor cerebri: Population studies in Iowa and Louisiana. Arch. Neurol. 45, 875–877.

Fishman, R. A. (1984). Brain edema and disorders of intracranial pressure. In "Merritt's Textbook of Neurology" (M. P. Rowland, ed.), pp. 206–215. Lea & Febiger, Philadelphia.

Giuseffi, V., Wall, M., Siegel, P. Z., and Rojas, P. B. (1991). Symptoms and disease associations in idiopathic intracranial hypertension (pseudotumor cerebri): A case-control study. Neurology 41, 239–244.

Hamed, L. M., Tse, D. T., Glaser, J. S., Byrne, S. F., and Schatz, N. J. (1992). Neuroimaging of the optic nerve after fenestration for management of pseudotumor cerebri. Arch. Ophthalmol. 110, 636–639.

Ireland, B., Corbett, J. J., and Wallace, R. B. (1990). The search for the cause of idiopathic intracranial hypertension. Arch. Neurol. 47, 315–320.

Jacobson, D. M., et al. (1999). Serum vitamin A concentration is elevated in idiopathic intracranial hypertension. Ophthalmology 105, 2313–2317.

Johnston, I., Hawke, S., Halmagyi, M., and Teo, C. (1991). The pseudotumor syndrome. Disorders of cerebrospinal fluid circulation causing intracranial hypertension without ventriculomegaly. Arch. Neurol. 48, 740–747.

Johnston, I. H., and Paterson, A. (1974). Benign intracranial hypertension, II: cerebrospinal fluid pressure and circulation. Brain 97, 301–312.

Kelman, S. E., Sergott, R. C., Cioffi, G. A., Savino, P. J., Bosley, T. M., and Elman, M. J. (1991). Modified optic nerve decompression in patients with functioning lumboperitoneal shunts and progressive visual loss. Ophthalmology 98, 1449–1453.

Kupersmith, M. J., et al. (1991). Effects of weight loss on the course of idiopathic intracranial hypertension in women. Neurology 50, 1094–1098.

Marcelis, J., and Silberstein, S. D. (1991). Idiopathic intracranial hypertension without papilledema. Arch. Neurol. 48, 392–399.

Pearson, P. A., Baker, R. S., Khorram, D., and Smith, T. J. (1991). Evaluation of optic nerve sheath fenestration in pseudotumor cerebri using automated perimetry. Ophthalmology 98, 99–105.

Radhakrishnan, K., Ahlskog, J. E., Cross, S. A., Kurland, L. T., and O'Fallon, W. M. (1993a). Idiopathic intracranial hypertension (pseudotumor cerebri). Descriptive epidemiology in Rochester, Minn, 1976 to 1990. Arch. Neurol. 50, 78–80.

Radhakrishnan, K., Thacker, A. K., Bohlaga, N. H., Maloo, J. C., and Gerryo, S. E. (1993b). Epidemiology of idiopathic intracranial hypertension: A prospective and case-control study. J. Neurol. Sci. 18–28.

Rosenberg, M. L., Corbett, J. J., Smith, C., Goodwin, J., Sergott, R., Savino, P., and Schatz, N. (1993). Cerebrospinal fluid diversion procedures in pseudotumor cerebri. Neurology 43, 1071–1072.

Selhorst, J. B., Kulkentraharn, K., Cynthil, R., et al. Vitamin A hypothesis in idiopathic pseudotumor cerebri. Ann. Neurol. 42, 457.

Smith, J. L. (1985). Whence pseudotumor cerebri? *J. Clin. Neuro-Ophtalmol.* **5**, 55–56.

Spoor, T., and McHenry, J. G. (1993). Long-term effectiveness of optic nerve sheath decompression for pseudotumor cerebri. *Arch. Ophthalmol.* **111**, 632–635.

Sugerman, H. J., Felton, W. L. III, Salvant, J. B., *et al.* (1995). Effects of surgically-induced weight loss on idiopathic intracranial hypertension in morbid obesity. *Neurology* **45**, 1655–1659.

Wall, M. (1991). Idiopathic intracranial hypertension. *Neurol. Clin.* **9**, 73–95.

Wall, M., and George, D. (1991). Idiopathic intracranial hypertension. A prospective study of 50 patients. *Brain* **114**, 155–180.

Weisberg, L. A. (1985). Computed tomography in benign intracranial hypertension. *Neurology* **35**, 1075–1078.

CHAPTER 68

# The Syndromes of Spinal Cord Dysfunction

Volker Dietz and Robert R. Young

When one makes diagnostic and therapeutic decisions concerning spinal cord syndromes, the time course of their development plays an important role regardless of the origin. With acute spinal cord lesions, the diagnosis must be made and appropriate treatment begun within a few hours. With chronically developing spinal cord syndromes, there is less urgency, so diagnostic and therapeutic measures should not be undertaken except when conditions are optimal. Radiological diagnosis should include magnetic resonance (MR) imaging. Transfer to a specialized center with appropriate expertise should not only be undertaken for patients with traumatic spinal cord syndromes but also for those whose spinal cord dysfunctions are of unclear origin. Table I is a summary outline of the most important spinal cord syndromes (see also Woolsey and Young, 1995). Etiologically speaking, a spinal cord syndrome can be due to trauma, ischemia, inflammation, chronic compression, degenerative myelopathy, or tumor (Hughes, 1989).

Rehabilitation of a paraplegic or tetraplegic patient starts immediately after the lesion occurs. With an acute lesion, various studies must be undertaken quickly in a rational, well-coordinated sequence. To optimize the prognosis and restrict the progression of an acute traumatic spinal lesion, the following is of essential importance (cf. Braakman and Penning, 1976; Guttmann, 1976): quick rescue by professionally trained paramedical personnel who accompany the patient to a center that has specialized teams who can assess the exact diagnosis and institute appropriate treatment within the first few hours (Table II).

## TRAUMATIC SPINAL CORD LESIONS

### Clinical Aspects

#### Acute Stage

The prevalence of acute spinal cord injury (SCI) is estimated to be 30–50 cases per million population per year (Sett and Crockard, 1991). Paralysis of the upper and lower extremities, so-called quadriplegia or tetraplegia, accounts for 54% of the injured patients; the remaining 46% are paraplegic. Traumatic SCI predominantly occurs in young men (61% are between 16 and 30 years old); 20% or less of the patients are women. About half the patients have neurologically complete injuries when they are first seen.

After a traumatic spinal cord injury, it is essential immediately to differentiate with certainty between paraplegic and quadriplegia. If the level of injury within the cervical spinal cord ascends even a few millimeters, it can lead to a change in functional status from relative self-independency to complete dependency. History/anamnesis concerning the trauma is important to get information about the forces that acted on the patient. The patient must be asked about the *localization of any pain*. In an unconscious patient, a cervical cord lesion can be recognized by breathing, which is exclusively diaphragmatic. More caudal spinal cord lesions may eventually be detected only at a later stage. Therefore, every unconscious trauma patient should be treated as having a SCI until it has been ruled out.

During the acute stage, the principal symptoms include loss of voluntary movements and sensory perceptions, as well as dysfunction of bladder and bowel. When paresis of muscles in all four limbs is present, associated with absent sensation to pinprick within

**TABLE I    Spinal Cord Syndromes**

| Syndrome | Motor/reflexes | | | Sensory | | Vegetative | |
|---|---|---|---|---|---|---|---|
| | Paresis | Tendon reflexes | Babinski sign | Position sense | Pain/ temperature | Bladder dysfunction | Blood pressure dysfunction |
| Spinal cord lesion | | | | | | | |
| C3/C4 | Spastic tetraplegia | ++ | ++ | -- | -- | Detrusor/sphincter-dyssynergia | +/Autonomic dysreflexia |
| Th8 | Spastic paraplegia | ++ | ++ | -- | -- | Detrusor/sphincter-dyssynergia | + |
| Conus/cauda | Flaccid paraplegia | -- | -- | -- | -- | Detrusor areflexia | -- |
| Brown Séquard's syndrome | Spastic monoparesis Ipsilateral | ++ Ipsilateral | + Ipsilateral | -- Ipsilateral | -- Contralateral | No | -- |
| Central cord syndrome | Spastic tetraparesis (arms > legs proximal > distal) | + (+) | + | +/-- | -- | (Detrusor-dysreflexia) | (+) |
| Anterior spinal artery syndrome | Para- (tetra)— paresis | + | + | + | +/- | Detrusor/ dysreflexia/ areflexia | (+) |
| Dorsal column syndrome (uncommon) | Motor function preserved | + | --/+ | -- | -- | No | -- |
| Spinal shock | Flaccid plegia | -- | +/-- | -- | -- | Detrusor areflexia | + |

+, normal, present; ++, exaggerated; – reduced, lacking.

**TABLE II    Early-Stage Treatment of a Paraplegic or Tetraplegic Patient after Injury**

| Accident site | Rescue helicopter | Medical center | | |
|---|---|---|---|---|
| Medical management ↓ | Transportation | Examinations ↓ | Treatment ↓ | |
| Anamnesis | | General | Central intravenous drug therapy | → Compensation of volume deficit → High dosage of prednisolone → Blood replacement |
| ↓ | | ↓ | ↓ | → Heparin (thrombosis) → Oxygenation → Anticholinergic drug (bradycardia) |
| Clinical examination | | Neurological | Care of associated injuries | → General surgery → Hematopneumothorax |
| ↓ | | | ↓ | → Bone fractures |
| Rescue and evacuation | | | Bladder and bowel therapy | → Suprapubic catheter with closed system (or indwelling catheter) |
| | | ↓ | ↓ | → Activation of bowel function (paraffin oil, prostigmine) → Intravenous or oral feeding |
| | | Radiological ↓ | Body positioning ↓ | |
| | | Laboratory | Ventilatory assistance | |

dermatomes C6 to Th1 [thumb (C6), little finger (C8), elbow (Th1)] and from the thorax downward, the lesion is at the level of the cervical cord causing quadriplegia. When only voluntary leg movements are impaired and sensation is lost in the legs, paraplegia can be assumed.

### Early Course

The neurological examination must be repeated several times during the first few days to clearly document the level and severity of a spinal cord lesion. In 1992, the American Spinal Cord Injury Association

**TABLE III** ASIA Impairment Scale

A = *Complete*: No motor or sensory function in the sacral segments S4–S5.

B = *Incomplete*: Sensory, but not motor function is preserved below the neurological level and extends through the sacral sements S4–S5.

C = *Incomplete*: Motor function is preserved below the neurological level, and most key muscles below the neurological level have a muscle grade less than 3.

C = *Incomplete*: Motor function is preserved below the neurological level, and most key muscles below the neurological level have a muscle grade greater than or equal to 3.

E = *Normal*: Motor and sensory function are normal.

From Ditunno (1992). Reprinted with permission.

(ASIA) published standards for neurological and functional classification of SCIs that have been revised twice (Ditunno, 1992; Maynard *et al.*, 1997). Depending on the motor and sensory deficits, a patient can be categorized according to the ASIA impairment scale (Table III) if the deficits have been rated according to the standard ASIA chart. Reflex examination during the early stage after trauma provides information about the presence of spinal cord shock (which may last for weeks) during which tendon reflexes are absent.

The initial traumatic lesion is in most cases associated with hemorrhage and edema within the spinal cord. Soon, at the site of the lesion, cellular infiltrates and ischemia occur, associated with tissue necrosis. Later, degenerative changes of bones and vertebral disks may occur, which can lead to secondary compression of the cord.

Early after the trauma, *spinal concussion* should be distinguished from a *spinal cord lesion*. Spinal concussion represents an impairment of spinal cord functions with clinical symptoms that are completely reversible within 72 hours (Zimpfer and Bernstein, 1990). A spinal cord lesion is associated with morphological changes and persistent clinical symptoms—incomplete or complete paraplegia or quadriplegia. Nevertheless, neurological deficits may recover during the first 3 months after a traumatic lesion.

Interestingly, total unilateral transection of descending motor tracts combined with 85%–90% transection of the contralateral tracts in human beings leads to total paralysis of the lower limbs that can, however, recover, so that walking, albeit with spastic paraparesis, may return after 2 months (Nathan, 1994). The similarity of this recovery to effects of partial lesions in rats and cats suggests a common mechanism; that is, plasticity of local spinal circuits with adaptation to inputs by a small number of remaining descending axons.

## Dysfunction of the Autonomic Nervous System

SCIs, particularly lesions above T6, compromise sympathetic nervous system activity with consequences for *cardiovascular function*. Because sympathetic outflow from the cord occurs between T1 and L2, sympathetic tone throughout most of the body in quadriplegics and high-level paraplegics is severely or completely impaired (depending on the completeness and level of the injury). This results acutely in hypotension and bradycardia or, in the extreme, with increases in parasympathetic tone, in asystole. The latter tends to occur particularly during manipulations of the patient's oropharynx, which produce increased vagal activity.

Regulation of *blood pressure* is impaired, first as *orthostatic hypotension* (because of loss of vasoconstriction and absent pumping function of leg muscles) and later as *dysreflexia* accompanied by hypertension. The latter is a consequence of nociceptive input to the cord from the bladder (e.g., overdistention or inflammation), bowel (distention), or other structures caudal to the cord lesion. Symptoms of dysreflexia include profuse perspiration, pounding headache, and pain; signs include extreme hypertension, which may result in intracerebral hemorrhage.

Regulation of *body temperature* is usually impaired with lesions above T10 and, when the lesion is above C8, affects the whole body. This leads, for example, to hyperthermia when the patient is in a hot environment, because equilibration of body temperature by peripheral vasodilation and perspiration is impossible. Similarly, hypothermia results when the patient's environment is cold, because vasoconstriction and shivering are no longer possible.

*Respiration* is impaired with a lesion above T6 when most intercostal muscles are paralyzed. With a lesion above C5, diaphragmatic breathing also is impaired, so these patients usually require a respirator for several weeks to avoid prolonged atelectasis and pneumonia. Respiratory complications can develop at any time after SCI and are particularly likely in quadriplegics and high-level paraplegics during the acute stage, when aspiration, fractured ribs, and lung contusions may compound existing problems. Vital capacity is an extremely useful index of ventilatory sufficiency and should be determined daily for the first month. Quadriplegics with high-level lesions (C5 and above) require ventilatory assistance when the vital capacity falls below 1.0–1.2 L or if pneumonia develops. Because such assistance is often required for more than 10 days, tracheostomy should not be needlessly delayed. Conditions causing respiratory problems include decreased diaphragmatic excursion and impaired cough effectiveness caused by paralysis of abdominal and intercostal muscles with decreased expiratory force. In certain patients with lesions rostral to the diaphragmatic motoneurons (i.e., above C3), functional electrical stimulation of the phrenic nerves can provide adequate respiration for many years (see Chapter 84).

Contractility of the *bladder* is lost after SCI, and the bladder may remain areflexic for weeks to months, even in patients with upper motoneuron lesions in whom return of such reflexes is anticipated. Advances in the management of the neurogenic bladder have constituted some of the most important developments in the care of patients with SCI. Overflow of urine or reflux into the ureters and kidneys occurs when drainage (suprapubic or intermittent urethral catheterization) is not instituted early after the trauma. After spinal shock, the form bladder dysfunction takes depends on the level of the lesion. Usually, reflex bladder function develops, which in some cases can be used to empty it sufficiently.

After SCI, *peristalsis* ceases within 24 hours as a result of spinal shock, and this produces gastrointestinal problems that usually persist for 3–4 days. The combination of absent bowel sounds with hypotension (in quadriplegics) or decreased hematocrit may mislead one to suspect an acute condition of the abdomen. Severe spinal cord lesions are always associated with impaired *sexual functions*. With lesions above the level of T9/10, reflex erections and ejaculations can eventually develop.

## The Evolution of Signs and Symptoms

A severe spinal cord lesion is initially associated with spinal shock below the level of lesion. Spinal shock is characterized by flaccid paresis, absent tendon reflexes, and failure of the autonomic nervous system (paralytic ileus and loss of bladder function). Some conus reflexes are preserved (e.g., the bulbocavernosus reflex).

Spinal shock can last several (3–6) weeks. As it disappears, exaggerated tendon reflexes develop. Persistent areflexia may result from myelomalacia below the level of lesion. Nonneuronal factors (e.g., transformation of muscle and connective tissue), contribute to the development of spastic hypertonia and the occurrence of muscle spasms (Hiersemenzel *et al.*, 2000). This might also result in nociceptive input to the cord.

Partial recovery of spinal cord function can take place during the first 2–3 months after trauma. In addition to the clinical examination, electrophysiological recordings improve predictions about the prognosis of motor and sensory deficits even at an early stage after the trauma and even in unconscious patients. By using nerve conduction studies, any contribution of peripheral

lower motoneuron impairment can be assessed within the first few weeks after injury (Curt and Dietz, 1996a). With these recordings, the outcome in terms of a flaccid versus spastic hand can be predicted in patients with cervical injuries and, consequently, appropriate physical and occupational therapy can be applied at an early stage. With a combination of neurographic and somatosensory-evoked potential (SSEP) recordings from ulnar and median nerves, eventual hand function (i.e., passive versus active hand function) can be predicted (Curt and Dietz, 1996b). With early SSEP studies stimulating nerves in the legs combined with recordings of leg muscle potentials evoked by transcranial magnetic stimulation, prognosis for locomotor capability can be made (Curt and Dietz, 1997; Curt *et al.*, 1998).

Spastic muscle tone develops over weeks and increases for several months; these changes in muscle tone are not reversible (for further information about spastic symptoms see Chapter 87). When tendon reflexes reappear after spinal shock, a reflex bladder usually develops, which makes special urologic management necessary.

## Principles of Therapy

### General Concepts

For many years, a feeling of hopelessness dominated this field, but that is becoming less prevalent. Experimental evidence for regeneration of central axons after injury (Cheng *et al.*, 1996; Schnell and Schwab, 1990; for review see Schwab and Bartholdi, 1995) is currently modifying this view to a more optimistic one.

Treatment concepts for SCI fall into three separate time frames. Treatment of *acute* SCI is directed at interrupting the cascade of secondary injury processes, thus limiting tissue damage. This approach seeks to arrest or even reverse sensorimotor dysfunction. The second therapeutic phase deals with *consequences* of SCI. Rehabilitative efforts aim to stabilize the current status and train reflexes and residual circuits to achieve optimal living conditions for the patient who has a given deficit. In the future, a third treatment approach will be directed toward enhancement of axonal *regeneration*. Many substances have been tested in experimental neurotrauma models, but few have been evaluated in extensive dose-response studies. Lack of such data has made it difficult for certain promising drugs to be taken into clinical trials, so this exciting approach remains experimental at this time.

*Steroids* have been used extensively in animal studies and in clinical trials (Bracken *et al.*, 1985, 1997). Treatment of SCI with steroids was initially based on their anti-inflammatory actions and their effectiveness in treating cerebral edema. Results from experimental studies on SCI with steroids are best reviewed in Faden and Salzman (1992). Biochemical studies showed that very high doses of the synthetic steroid methylprednisolone also exert nonspecific free radical scavenging effects (Braughler and Hall, 1982). A series of animal experiments, conducted while the first U.S. national acute spinal cord injury study (NASCIS 1) was in progress, suggested that higher doses of methylprednisolone than those used in that study might be needed to achieve a therapeutic response. Therefore, NASCIS 2 began in 1985 using high-dose methylprednisolone, naloxone, or placebo (Bracken *et al.*, 1990). Those patients who received methylprednisolone within 8 hours of injury eventually achieved significantly better sensory and motor function than those who received placebo. Delayed treatment with methylprednisolone was, however, associated with decreased neurological recovery (Bracken and Holford, 1993). Because of its systemic effects, methylprednisolone may not be suitable for multiply traumatized patients.

According to the NASCIS 3 trial (Bracken *et al.*, 1997), steroid therapy should be maintained for 48 hours if it is initiated 3–8 hours after injury. In contrast to the results reported in NASCIS 2 and 3, recent studies in the rat failed to show a protective effect of methylprednisolone during the early phase of ischemic necrosis (Bartholdi and Schwab, 1995). Instead, a pronounced anti-inflammatory effect with massive reduction of the recruitment of neutrophils and macrophages was observed. Recently, the use of high-dose methylprednisolone has been questioned, because the initial evidence that it improves neurological recovery seems not to be supported by a systematic meta-analysis (Hurlbert, 2000; Short *et al.*, 2000).

*Gangliosides* are supposed to play important roles in normal neuronal development and differentiation (Gorio, 1988). A study testing the efficacy of combinations of GM-1 with high-dose methylprednisolone (Geisler, 1993) did not provide convincing evidence of a beneficial effect of this treatment (Constantini and Young, 1994).

The opiate receptor antagonist *naloxone* has been shown to facilitate sensorimotor recovery in cats after spinal cord contusion and to prevent the development of posttraumatic ischemia (Faden *et al.*, 1981; Young *et al.*, 1981). However, a randomized clinical trial in spinal cord–injured patients failed to show any significant improvement with naloxone compared with placebo (Bracken *et al.*, 1990).

The noncompetitive *NMDA antagonist dizocilpine* was shown to improve multiple pathophysiological parameters after SCI and ischemia (Hao *et al.*, 1992). Serious side effects of competitive and noncompetitive NMDA blockers have hindered clinical trials so far.

### General Measures Used during the Acute Stage

While rescuing an injured person, every effort must be made to prevent all active and passive movement of the spine. After ensuring that ventilation and circulation are adequate, the patient should be log rolled onto his or her back, with extremities in the anatomical position if no resistance is met. The natural hollows of the cervical and lumbar lordoses should be supported, using towels or clothes, without hyperextending or flexing the spine. Any brisk movement can lead to worsening of the spinal cord lesion. In quadriplegic patients, head movements must be minimized by a collar. Because of vasomotor paralysis and absence of shivering, the possibility of hypothermia must be recognized, and the patient should be protected by a blanket. The acute paraplegic or quadriplegic patient should not take food or fluid orally because of the danger of aspiration. Opiates (or their derivatives) should not be used for pain relief, so as to avoid worsening of ventilation and circulation.

The injured person should be transferred immediately to a hospital with 24-hour emergency and intensive care facilities capable of handling trauma victims, preferably one that is closely affiliated with a spinal cord injury center, where the patient also can be rehabilitated. Medical personnel should ensure the adequacy of the person's airway, administer oxygen, start an intravenous infusion, and perform any other life-saving measures, such as inserting an intrathoracic tube if pneumothorax or hemothorax is present.

### Treatment during the Acute Stage

When the patient arrives at the medical center, immediate assessment of his or her respiratory and circulatory status and level of consciousness is necessary to determine the need for supportive treatments (Table IV), such as ventilatory assistance after intubation, cardiac monitoring, vasoactive drugs, or use of a transvenous pacemaker.

A *central venous catheter* should be inserted, because a spinal cord–injured patient usually requires infusion therapy for several days. Access through arm veins should be avoided, because the position of arms and hands will continuously be changed. To prevent shock caused by loss of blood volume and to secure circulation and tissue oxygenation, infusions of human albumin and electrolytes should be given. Infusion volumes should be adjusted by monitoring central venous pressure to avoid pulmonary edema.

To reduce secondary worsening of spinal cord lesions by *edema* of the cord, monotraumatic SCIs should be treated with high doses of methylprednisolone (NASCIS 2 and 3; Bracken *et al.*, 1990, 1997). However, this

**TABLE IV** Pragmatic Therapy of an Acute Spinal Cord Lesion

*Optimal positioning* of the body with support of cervical and lumbar lordoses

*"En bloc" transportation*; over longer distances by means of helicopter to a medical center

Infusions of human albumin and electrolytes over a *central venous catheter*

*Methylprednisolone i.v.* application (monotraumatic cases) within 8 hr:*** Bolus injection with 30 mg/kg body weight within 15 minutes; followed by an infusion of 0.9% NaCl over 45 minutes; followed by 5.4 mg/kg/h methylprednisolone over (23) 48 hours

*Radiological diagnosis* and *decision* about conservative or operative treatment

In the case of *bradycardia*: atropine 0.25–0.5 mg sc 8 h

*Thrombosis prophylaxis* with low-molecular-weight heparin (Nadroparin): 0.3 ml (≤50 kg body weight) to 0.8 ml (≥90 kg) sc/die

Provision of sufficient *blood oxygenation*

*Pain* relief by (e.g., metamizol (2–5 ml in 250 ml NaCl 0.9%, up to 4×/day)

Suprapubic *bladder drainage* by catheter

*Bowel motility* absent: Prostigmine 0.5 mg sc, 3–4×/day

Physical therapy for *respiration* and passive limb movements

Regular (every 3 hours) *"en bloc" turning* of the body

Occupational therapy of *hand function*

sc: subcutaneous.

acute therapy was recently questioned (Short *et al.*, 2000) and further investigations are needed.

To avoid *stress ulceration* of the stomach or duodenum with potential bleeding, prophylaxis with antacids and histamine-blocking agents should be undertaken in patients with a history of stomach problems. Cimetidine has been advocated for prophylaxis against acid peptic disease. Prophylactic antibiotics are not necessary to prevent pneumonia. Application of mucolytics may improve expectoration of bronchial secretions.

*Bradycardia* can be treated with anticholinergic drugs. When bradycardia occurs frequently, a transvenous pacemaker can be inserted for 3–5 days until the cardiovascular system accommodates. Because bradycardia is caused by unopposed vagal tone on the sinoatrial node, vagal stimulation should be avoided. Improper tracheal suctioning may lead to sinus arrest, so the heart rate should be closely monitored during this procedure. During the first few days, administration of an anticholinergic drug may be necessary.

Because paraplegic or quadriplegic patients have an increased risk of *deep vein thrombosis* during the first few weeks after trauma, prophylaxis with low-molecular-weight heparin (without dihydroergotamine) should be provided. Such therapy is contraindicated when there is hemothorax, brain damage, or severe associated injuries.

For *pain relief*, metamicol can be used, but opiates should be avoided.

Except for a few contraindications (anticoagulation, pregnancy), suprapubic drainage of the *bladder* represents standard European therapy during the early phase after trauma. Such drainage is thought to represent the best prophylaxis against infection. It must be performed under ultrasonic control when the bladder is filled with more than 400 ml. Urine draining through a suprapubic or an indwelling urethral catheter must be connected to a closed system. This continuous drainage should be used during the polyuric phase (1–2 weeks) or as long as intravenous infusion therapy is needed. Alternately, particularly in the United States, intermittent urethral catheterization is performed every 3–4 hours.

Even before *bowel motility* has normalized, the patient is allowed to take light food or fluids perorally to avoid mucosal atrophy and other complications. Most patients will have lost both the urge to defecate and voluntary control of the anal sphincter, so a bowel program to train the bowel to empty regularly should begin as soon as bowel sounds return. Bowel motility may be supported, for example, by peroral use of paraffin oil or stool softeners. Prostigmine (sc) may be used when bowel motility does not recover spontaneously.

Quadriplegics with high-level lesions (C5 and above) have *respiratory problems* and may require ventilatory assistance. In these patients the thorax is in a permanent inspiratory position causing respiratory problems, including impaired cough effectiveness leading to accumulation of secretions in the tracheobronchial tree with a danger of atelectasis. These patients should receive intensive respiratory therapy, including intermittent positive pressure breathing, chest percussion, and assisted coughing every 6 hours. This therapy is supplemented by regular turning of the body.

Proper *positioning* of the patient is important to prevent pressure sores and contractures of the peripheral joints. The shoulders require special attention in patients with quadriplegia, because the shoulder capsule may contract, leading to painful capsulitis, which makes full range of motion difficult and painful. Positioning the patient with the legs slightly elevated facilitates venous drainage and reduces stasis. Patients must be turned safely and properly. They can be lifted and then log rolled, or log rolled using a drawsheet. The patient must be turned "en bloc" every 2 or 3 hours day and night. In addition, the feet must be positioned so that a plantar-flexed position is avoided to prevent contractures.

For *operative interventions* on the vertebral column, the following indications exist: (1) failure to reduce a fracture/dislocation or to restore acceptable alignment by closed methods; (2) failure to carry out closed reduction early enough; (3) instability of a fracture after laminectomy, because this increases the chance of a late-developing deformity; (4) progressive deficit ascending beyond two segments above the initial level of injury or progressive transverse neurologic deterioration at the level of bony injury in a patient initially without or with an incomplete neurological deficit; (5) acute decompression of the cord by removal of a hematoma or fragments of bone or disk from the neural canal to avoid progression of the spinal cord lesion; (6) stabilization of the vertebral column to allow mobilization earlier than would be possible with spontaneous healing of the fracture. A detailed discussion of indications for surgery and operative procedures can be found in the review by Bohlman and Ducker (1992).

### Pragmatic Therapy (see Table IV)

Pragmatic therapies for an acute spinal cord lesion are outlined in Table IV. The various measures listed describe adequate treatment for a quadriplegic patient and must be selected and adapted for an individual paraplegic or quadriplegic patient according to the level and severity of the lesion.

## Transition from Acute Care to Rehabilitation: Principles of Therapy and Pragmatic Therapy

Rehabilitation of a patient with SCI should initially be conducted together with acute care. Management must flow continuously from emphasis on acute care through the steps leading up to the ultimate rehabilitative efforts. The length of time a patient should be kept immobilized varies with the type of injury, its location, and whether surgery has been performed. Spine surgery is usually performed to facilitate early mobilization. After cervical injuries, early mobilization should be undertaken with caution. Patients who are ambulatory should use halo devices for 6–8 weeks. A body jacket is used by all patients with thoracolumbar or lumbar injuries. The degree of functional independence achieved after mobilization depends on the level and the severity of the lesion and the adequacy of training. Recovery of functions lost because neurapraxic lesions of spinal nerve fibers can occur during the first 3–4 months after trauma.

The *cardiovascular system* accommodates rather quickly to the effects of sympathetic insufficiency, and treatment of bradycardia can usually be discontinued after about 2 weeks. Syncopal episodes may occur in all patients with complete or nearly complete quadriplegia when they first sit up. *Pragmatic therapy* to facilitate accommodation of the cardiovascular system during the early phases of mobilization includes use of a tilt table, the angle of tilt being increased as the patient is able to sustain a systolic blood pressure of 80 to 90 mmHg for about an hour at any given angle. To facilitate venous return, quadriplegic (and paraplegic) patients should wear elastic stockings for the first 4 months while sitting and during training on the tilt table.

Patients with lesions above T6 usually experience *autonomic dysreflexia*, which is characterized by extreme hypertension leading to pounding headache and flushing (Curt *et al.*, 1997). This complication, also known as sympathetic spasticity, develops after reflex activity, especially detrusor reflex activity, returns. Dysreflexia results from an uncontrolled increase of spinal sympathetic activity released because of disconnection of the spinal cord from the brain centers. Associated bradycardia is triggered by the baroreceptors, because vagal innervation remains intact. Autonomic hyperreflexia is most often caused by noxious stimuli arising in the bladder or bowel. *Pragmatic therapy* is directed to relieving the underlying cause, for example, removing an obstruction (changing the catheter) and using topical anesthetic in the bowel or bladder to decrease noxious afferent stimuli.

Patients with spinal cord injury are at special risk of *thromboembolic disease* because of tissue factors released at the time of injury, immobilization, and reduced leg muscle pump function. The risk is greatest during the first 2 months after injury. Pulmonary embolization, as a complication of venous thrombosis in the legs, is associated with a relatively high morbidity and mortality in quadriplegics. Therefore, important routine diagnostic procedures include daily measurements of leg and thigh circumferences and, in the case of suspected thrombosis, Doppler sonography. Phlebography is most reliable for showing the location and extent of a clot. Manifestations of pulmonary embolism may be obvious or nonspecific, but the diagnosis is supported by laboratory findings of a fall in $pO_2$, electrocardiographic evidence of right heart strain, and a pattern of diminished perfusion on lung scan. Prophylactic administration of low-molecular-weight heparin reduces but does not eliminate the risk of deep venous thrombosis. *Pragmatic therapy* includes preventive measures, such as heparin (depending on body weight, 0.3–0.8 ml sc/day low-molecular-weight heparin), having the patient wear elastic stockings, chronic electrical stimulation of leg muscles, and passive range of motion exercises. When a diagnosis of deep vein thrombosis is made, intravenous heparin must be started at once and continued for 10 to 14 days while the patient is being immobilized.

The *respiratory system* usually accommodates well to the altered breathing mechanics. Nevertheless, chest physiotherapy must be continued, and vital capacity must be routinely followed. Patients with lesions above C5 may require ventilator support for prolonged periods. Because quadriplegics rely completely on the diaphragm for breathing and have no additional respiratory muscle reserve, they are liable to fatigue easily. Therefore, if a ventilator is required, it should be used until the patient can oxygenate adequately on room air. As *pragmatic therapy* some quadriplegics with high lesions require ventilator support for several weeks to months and others require it permanently. In some of these latter patients, functional electrical stimulation of the phrenic nerves has been used successfully for years. In less severely affected patients, intermittent positive pressure breathing and mucolytics are useful.

*Defecation* can be regulated in patients with upper motoneuron lesions who have active conus reflexes. The gastrocolic reflex can be used after the patient has been fed and a suppository containing glycerin is inserted against the rectal wall to stimulate peristalsis. In patients with lower motoneuron lesions who have a flaccid anus, one can try the same procedure, but their stool often must be removed manually. As *pragmatic therapy*, the bowel program should be performed at a regular time every day or every other day, and the stool should be kept soft by attention to the diet and use of stool softeners.

For management of the *neurogenic bladder*, intermittent catheterization is the preferred method of training the bladder after the suprapubic or urethral catheter is removed (i.e., after the polyuric phase; see the section on treatment during the acute stage). Fluid balance should be checked, and bladder overdistension should be avoided. The quantity of fluid taken by patients should be limited in the early phase of rehabilitation to 150 ml to 200 ml every 2 hours, and patients must be catheterized every 4 to 6 hours. The program should be continued until adequate voiding occurs. Patients who must have a suprapubic or indwelling Foley catheter for longer periods are certain to develop chronic bacteriuria. A balance between bacteria and host defenses can be achieved by maintaining a high fluid intake and a high urine output. This requires maintenance of an unobstructed conduit from the bladder and keeping the perineal area as clean as possible at all times. Reflex voiding can be facilitated by bladder tapping or Crede maneuvers (suprapubic manual pressure) every 2 to 4 hours. Residual volumes must be regularly assessed by ultrasonographic examination or postvoiding catheterization. If the residual urine volume is consistently less than 80 to 100 ml, intermittent catheterization can be reduced or discontinued. Reducing the frequency of catheterization while residual volumes remain high predisposes to infection. Patients with areflexic or hyporeflexic bladders are taught to empty their bladders by straining and by suprapubic pressure. Drug therapy for improvement of bladder function depends on the type of bladder dysfunction, which itself depends on the level of spinal cord lesion (Table I). Drug therapy is meant to achieve adequate voiding and to avoid incontinence. By intravesical injections of botulinum-toxin A, exaggerated bladder contractions can be suppressed, which consequently leads to continence in hyperreflexive bladder dysfunction (Schurch *et al.*, 2000). Drugs that are used with different types of bladder dysfunctions are listed in Table V. Local injection of botulinum toxin into the external urethral sphincter (Schurch *et al.*, 1996b) or, if all else fails, sphincterotomy improves flow rates, leads to lower voiding pressure and residual urine volume, and reduces autonomic dysreflexia. For *pragmatic therapy* see Table V.

Fundamental to the management of *skin problems* in patients with spinal cord injuries is prevention of pressure ulcers. This includes intermittent relief of pressure from weight-bearing surfaces and frequent visual inspection of the skin by the patient himself or by a nurse. Conservative treatment of pressure ulcers consists of complete relief of weight bearing and friction, adequate protein intake, and if necessary, restoration of hemoglobin to normal values. If the ulcers do not heal with these measures, surgical treatment is usually required after infection has been controlled. *Pragmatic therapy* consists of cleaning, disinfection, debridement, stimulation of granulation, and induction of epithelial tissue.

For the treatment of *spastic symptoms*, physiotherapeutic measures are preferred in mobile patients. Recent studies (Dietz *et al.*, 1995, 1998) have shown that a locomotor pattern can be both induced and trained in complete and incomplete paraplegic patients with a positive effect on spastic symptoms, but only incomplete patients profit directly in terms of their walking capa-

**TABLE V    Drugs Used for Treatment of Paralyzed Bladder Function**

| Hypoactive bladder | Hyperactive bladder | Neck of bladder | Spastic sphincter |
|---|---|---|---|
| Parasympaticomimetic drugs (e.g., distigmibromide or neostigmine, 5–15 mg/day) | Anticholinergic drugs (e.g., oxybutinin, 5 mg 3×/day) | *High tone* α-Blocking drugs (e.g., prazinosin/prazosin 1 mg 1–3×/day) | Local Injection of botulinum-toxin A; (250 mμ Dysport, repet. 2×/2 months) |
|  | Intravesical injections of botulinum-toxin A | *Low tone* α-Stimulating drugs (e.g., midodrine, 2.5 mg 3×/day) | Spasmolytic drugs (e.g., Baclofen) |

bilities. For further details, including *pragmatic therapy*, see Chapter 87.

One of the serious posttraumatic complications, not infrequently seen in quadriplegic patients, is *heterotopic ossification* (HO). Why bone forms in periarticular connective tissue (primarily around the hips and less often the knees) in some patients after SCI is not known. HO should be suspected when muscles about the susceptible joints become swollen and warm and when serum alkaline phosphatase levels are elevated. This bone formation can be documented by radionuclide scanning (3 phases). Early recognition is required to begin early treatment. *Pragmatic therapy* is difficult. The early treatment includes radiation therapy (2 Gy on five successive days) in combination with indomethacin (50 mg three times a day) for 3 months and immobilization of the affected joint. Surgery of the bone formation should only be performed when the process is inactive (i.e., 1–2 years after onset).

Secondary development of *syringomyelia* can occur above and below the level of the lesion several years after trauma. This usually is associated with increasing pain and spastic symptoms. After MRI diagnosis (Schurch *et al.*, 1996a) is made, operative decompression should be considered to prevent further deterioration; surgery rarely leads to improved function.

With injuries at any level, *genital functions* are impaired. Spinal cord injury does not generally affect the ability of female patients to conceive, because viable ova are produced once the menstrual cycle returns about 3–8 months after injury. Although vaginal delivery is usually possible in paraplegic patients, cesarean section is required in most quadriplegic patients. Male patients often become infertile because of a reduction in the number and quality of sperm, which usually occurs several months after injury. Furthermore, the synergistic smooth muscle activity required to ejaculate is compromised, particularly in quadriplegics, so sperm may be washed into the bladder and become inactive. Alterations in potency caused by spinal cord injury vary, depending on the extent of the lesion. In general, reflex erections are not difficult to produce in quadriplegics, but ejaculation occurs rather rarely. *Pragmatic therapy* can improve impaired sexual function (i.e., erections can be restored) by injections of prostaglandin E$_1$ (1 ml or 10 µg and, if needed, repeated higher doses such as 20 µg) into the erectile organ (Lue and Tanagho 1987). The peroral use of soldenafil (50 mg) (Giuliano *et al.*, 1999) is the first choice for improving sexual function. If this therapy is not successful, intracavernous injections of papaverine (12 mg) in combination with phentolamine (2 mg) can be used (Kapoor *et al.*, 1993). In these circumstances, frequent follow-up of the patient is recommended.

## SPINAL CORD LESIONS OF NONTRAUMATIC ORIGIN

All of the different spinal cord syndromes listed in Table I can be of traumatic or nontraumatic origin. For nontraumatic spinal cord lesions of vascular, inflammatory, degenerative, and neoplastic origin, with few exceptions (such as the use of methylprednisolone, which is applicable only during the first few hours after injury), the general pragmatic therapies, as outlined in the preceding paragraphs, can be adapted and used. Beyond these, additional measures may need to be undertaken and will be listed in the furthering (McKinley *et al.*, 1999).

### Vascular Spinal Cord Lesions

#### Clinical Aspects

Vascular spinal cord lesions most frequently occur suddenly, associated with segmental radiating pain, or during operations on the aorta. The latter case mainly concerns abdominal aortic aneurysms (aneurysm dissecans), which are important to diagnose at an early stage because of their therapeutic consequences to prevent unwelcome sequelae. Spontaneous recovery of function is similar in patients with spinal cord lesions of traumatic or vascular origin (Iseli *et al.*, 1999). The main differential diagnoses include vascular malformations (Hughes, 1989) and thrombosis caused by a tumor. Diagnosis involves the use of imaging procedures (plain x-rays, myelography, and MRI). Only if these examinations are suspicious for a vascular malformation, is spinal angiography indicated. The most frequently occurring vascular lesion of the cord is the anterior spinal artery syndrome (see Table I).

#### Specific Therapy

Primary therapy should be directed to the underlying cause of the thrombosis. If this is not possible, general supportive measures and control of circulation should be instituted. Rheological and hemodilution therapy have not been proven useful. The same holds true for use of steroids. No clear beneficial effects have been documented from substances that inhibit aggregation of thrombocytes (acetylsalicylic acid).

If a *vascular malformation* has led to spinal ischemia, the procedure to be used depends on the kind of malformation, its extent, and localization (see Chapter 34). The two most frequent approaches are embolization and operative extirpation of the malformation, both of which usually lead to favorable results (Behrens and

Thron, 1999). The goal must be to eliminate the shunt, which most frequently is an arteriovenous fistula (Criscuolo *et al.*, 1989; Koenig *et al.*, 1989) to reduce the possibility of this venocongestive myelopathy (Kim *et al.*, 1995). Epidural hemorrhages must quickly be decompressed as would be the case with a tumor (see later).

## Acute Myelitis

### Clinical Aspects

Acute transverse myelitis usually occurs within 24 hours, independent of an infectious agent. The extent and severity of deficit depend on the localization and completeness of the inflammatory lesion. The upper limbs are affected less than 25% of the time. Dorsal columns are significantly less affected than motor tracts. The level of lesion is frequently marked by a symmetrical zone of allodynia and hyperpathia. Development of symptoms is associated in 50% of cases with fever. The course is unpredictable with about one third of patients having no remission, one third with a partial remission, and one third having a good recovery. Persistence of deficits over more than 3 months indicates poor prognosis. The peak occurrence of myelitis is between 15 and 40 years of age without a gender preference.

Acute transverse myelitis is associated with various viral and bacterial infectious diseases, but in most patients (around 60%) no infectious agent is found on routine blood and cerebrospinal fluid examinations, and no association with a generalized infection is obvious. People with an impaired immune system are at greater risk for myelitis. In about 60% of cases, transverse myelitis represents the first manifestation of multiple sclerosis (Scott *et al.*, 1998). For diagnosis, examination of cerebrospinal fluid, including cytology and gamma globulins, and MRI (Imai *et al.*, 1997) should be performed. In addition, a serological search for neurotropic viruses or bacteria should be undertaken. Elevation of protein and cells is found in the cerebrospinal fluid.

### Specific Therapy

Treatment during the acute stage of myelitis consists of antiviral therapy (acyclovir or ganciclovir), if appropriate, or if one is suspicious of a bacterial infection, antibiotics (Klekamp and Samii, 1999). Combination therapy with steroids is recommended (1 mg/kg prednisone/day for 14 days with tapering over 10 days [Lisak, 1985] or, alternatively, 1 g methylprednisolone/day iv for 7 days with tapering over 2 days [(Kincaid and Dyken, 1985; Stüve and Zamvil, 1999)].

## Chronic Cervical Myelopathy

### Clinical Aspects

Cervical spondylotic myelopathy is the most frequent myelopathy of the elderly. It is caused by progressive degenerative changes of the cervical spine superimposed on a developmentally narrow spinal canal. Degenerative changes of the cervical column such as osteophytes are present in most people older than 40 years of age, but they become symptomatic in only a small proportion of subjects. This happens when there is narrowing of the spinal canal of less than 13 mm (normal anterior/posterior diameter = 17 mm), so the spinal cord (diameter about 10 mm) becomes compressed if the spine is bent. Most harmful are marked hyperextensions of the spine. In addition to bone formation and ligamentous fixations, soft disks and vascular factors play an important role in the development of chronic cervical myelopathy (Bradley, 1984).

Chronic cervical myelopathy is characterized by a spastic-ataxic gait disorder that slowly worsens with time. Fasciculations and paresthesias in the arms occur quite frequently. Impairment of bladder and bowel function and radiating radicular pain are less common. The combination of spastic paraparesis and atrophic paresis in cervical segments is pathognomonic.

Recordings should be made of somatosensory-evoked potentials from stimulation of leg and arm nerves and recordings of hand and leg muscle EMG potentials evoked by transcranial magnetic stimulation of the motor cortex (Curt and Dietz, 1997; Curt *et al.*, 1998; Kiers and Chiappa, 1995). Delayed and reduced amplitudes of these potentials are characteristic findings. By stimulating both median and ulnar nerves, the level of cord compression within the cervical region can frequently be assessed (Curt and Dietz, 1995). A diagnosis requires confirmation by MRI.

The differential diagnosis includes soft disks, syringomyelia (see Chapter 69), spinal tumors, and amyotrophic lateral sclerosis (see Chapter 81). Less often, multiple sclerosis or funicular myelopathy (see later) also must be considered.

### Specific Therapy

The variable course of the disease and the lack of controlled clinical trials make it difficult to decide on the appropriate form of therapy. Only mild functional disturbances, lack of progression, and old age justify an attempt to adopt conservative therapy based on immobilization, provided that close clinical follow-up is used (Thier and Dichgans, 1992). In the other cases, when a correct diagnosis is established, decompression of the cord is the therapy of choice. Surgery is mainly prophylactic to prevent the usually progressive gait deterioration. A decision to operate early becomes easier if compression exists at only one or two levels (Jeffreys, 1986). If surgery is considered necessary, stenosis confined to less than three segments should be treated by use of an anterior procedure (Samii *et al.*, 1989) and a ventral fusion of the adjacent vertebrae may be performed for stabilization of the segments. A posterior approach is more promising for more extended stenoses (Thier *et al.*, 1992). An operation should be followed by physical training, strengthening of neck muscles, and a cervical collar for about 6 weeks.

## Tumors Producing Spinal Cord Lesions

### Clinical Aspects

Spinal lesions caused by tumors usually become symptomatic in one of two ways. Rapid progression of deficits occurs when a malignant spinal tumor or metastasis is present. Radicular pain frequently precedes the deficits by weeks or months. Slow progression, sometimes over years, is characteristic of benign or semibenign tumors. It is most important to recognize the latter at an early stage, because complete operative removal of the tumor is frequently possible and can be associated with recovery of spinal cord function (Hoshimaru *et al.*, 1999). To make an early diagnosis, one must be aware that sensory and motor symptoms usually begin much more caudally than the lesion; for example in the feet, even though the tumor is located at a cervical level. Even later, with more severe sensory and motor deficits, the tumor is usually several segments higher than the upper limit of the sensory disturbance. Recordings of somatosensory potentials, obtained by stimulation of leg and arm nerves, usually help diagnose the level of lesion. Further confirmation should be obtained by MRI. Cerebrospinal fluid examination should be performed with caution, and neurosurgeons should be standing by for emergency operative decompression if sudden deterioration occurs because of herniation of the tumor.

*Neurinomas* are seen predominantly in the cervical spinal canal. They are characterized by a radicular lesion usually preceding compression of the spinal cord and by elevated protein in the cerebrospinal fluid. *Meningiomas* are the typical benign tumor affecting the thoracic spinal cord with a predominant occurrence in women older than 40. *Chordomas, ependymomas,* and *lipomas* are rather rare benign tumors that are prevalent in the lumbosacral region (McCormick *et al.*, 1990). The most frequently occurring malignant or semimalignant intraspinal tumors are gliomas and glioblastomas (Cohen *et al.*, 1989) (see also Chapter 62).

Other tumors that are benign 85% of the time may arise from the spine, (e.g., hemangiomas, osteoblastomas, eosinophilic granulomas, and giant cell tumors); less frequent are malignant tumors, such as plasmocytomas, lymphoma and osteosarcomas, chondrosarcomas, or Ewing sarcomas (Greenberg *et al.*, 1980). *Metastases* usually appear as extradural tumors. Most metastases arise from bronchial and breast carcinoma, followed in frequency by prostate and renal cell carcinomas (Mulder and Dale, 1985; Stark *et al.*, 1982; Wong *et al.*, 1990).

### Specific Therapy

For benign tumors, the therapy of choice is early decompression of the spinal cord, extirpation of the tumor, and, if necessary, stabilization of the spine. For hemangiomas of the vertebrae (50% of benign tumors of the spine), an operation is usually not possible. In these cases embolization can possibly reduce the epidural extent of the vascular malformation. For intraspinal tumors (e.g., glioma, glioblastoma) and for extraspinal malignant tumors, curative therapy is not available. If a spinal neoplasm is diagnosed by imaging and the course suggests a malignant tumor, steroid therapy is indicated using high-dose methylprednisolone (initially 200 mg IV, with tapering over weeks). If possible, operative decompression of the spinal cord should follow, which also permits a biopsy of the tumor. Subsequently, depending on the malignancy of the tumor, radiation with 36 Gy, in fractions between 1.5 and 1.8 Gy, can be performed (Loblaw and Laperriere, 1998; Mulder and Dale 1985). The same procedure should also be applied when metastases compressing the spinal cord cause significant symptoms for the patient. Radiation of the tumor without operative decompression can lead to worsening symptoms because of the edema induced by radiotherapy. In the case of a complete spinal cord lesion lasting more than 24 hours, remission of paraplegia cannot be expected regardless of the tumor. In this case, an operation is only indicated if stabilization of the spine or a biopsy is needed.

## REFERENCES

Bartholdi, D., and Schwab, M. E. (1995). Methylprednisolone inhibits early inflammation processes but not ischemic cell death after experimental spinal cord lesion in the rat. *Brain Res.* **672**, 172–186.

Behrens, S., and Thron, A. (1999). Long-term follow-up and outcome in patients treated for spinal dural artero-venous fistula. *J. Neurol.* **246**, 181–185.

Bohlman, H. H., and Ducker, T. B. (1992). Spine trauma in adults. *In* "The Spine" (R. H. Rothman and F. A. Simeone, Eds.), pp. 973–1104. WB Saunders Company, Philadelphia.

Braakman, R., and Penning, L. (1976). Injuries of the cervical spine. *In* (P. J. Vinken and G. W. Bruyn, Eds.), "Handbook of Clinical Neurology". North Holland Publishing Company, Amsterdam.

Bracken, M. B., and Holford, T. R. (1993). Effects of timing of methylprednisolone or naloxone administration on recovery of segmental and long-tract neurological function in NASCIS 2. *J. Neurosurg.* 79, 500–507.

Bracken, M. B., Shepard, M. J., Collins, W. F., Holford, T. R., Young, W., Baskin, D. S., Eisenberg, H. M., Flamm, E., Leo-Summers, L., Maroon, J., Marshall, L. F., Perot, P. L., Piepmeier, J., Sonntag, V. K. H., Wagner, F. C., Wilberger, J. E., and Winn, H. R. (1990). A randomized, controlled trial of methylprednisolone or naloxone in the treatment of acute spinal cord injury. *N. Engl. J. Med.* 322, 1405–1411.

Bracken, M. B., Shepard, M. J., and Hellenbrand, K. G. (1985). Methylprednisolone and neurological function one year after spinal cord injury. *J. Neurosurg.* 63, 704–713.

Bracken, M. D., Shepard, M. J., Holford, T. R. *et al.* (1997). Administration of methylprednisolone for 24 or 48 hours or tirilazad mesylate for 48 hours in the treatment of acute spinal cord injury. Results of the third national acute spinal cord injury randomized controlled trial *JAMA* 277, 1597–1604.

Bradley, W. G. (1984). Myelopathies affecting anterior horn cells. *In* (P. J. Dyck, P. K. Thomas, E. H. Lambert, and M. Bunge, Eds.), "Peripheral Neuropathy", 2nd ed., pp. 1351–1367. WB Saunders, Philadelphia.

Braughler, J. M., and Hall, E. D. (1982). Correlation of methylprednisolone levels in cat with its effects on (Na2+K+)-ATPase, lipid peroxidation, and alpha motor neuron function. *J. Neurosurg.* 56, 838–844.

Cheng, H., Cao, Y., and Olson, L. (1996). Spinal cord repair in adult paraplegic rats: partial restoration of hind limb function. *Science* 273, 510–513.

Cohen, A. R., Wisoff, J. H., Allen, J. C., and Epstein, F. (1989). Malignant astrocytomas of the spinal cord. *J. Neurosurg.* 70, 50–54.

Constantini, S., and Young, W. (1994). The effects of methylprednisolone and the ganglioside GM1 on acute spinal injury in rats. *J. Neurosurg.* 80, 97–111.

Criscuolo, G. R., Oldfield, E. H., and Doppmann, J. L. (1989). Reversible acute and subacute myelopathy in patients with dural arteriovenous fistulas. Foix-Alajouanine syndrome reconsidered. *J. Neurosurg.* 70, 354–359.

Curt, A., and Dietz, V. (1996a). Neurographic assessment of intramedullar motoneuron lesions in cervical spinal cord injury: consequences for hand function. *Spinal Cord* 34, 326–332.

Curt, A., and Dietz, V. (1996b). Traumatic cervical spinal cord injury. Relation between somato-sensory evoked potentials, neurological deficit, and hand function. *Arch. Phys. Med. Rehabil.* 77, 48–53.

Curt, A., and Dietz, V. (1997). Ambulatory capacity in spinal cord injury: significance of somatosensory evoked potentials and ASIA protocols in predicting outcome. *Arch. Phys. Med. Rehabil.* 78, 39–43.

Curt, A., Keck, M. E., and Dietz, V. (1998). Functional outcome following spinal cord injury: significance of motor evoked potentials. *Arch. Phys. Med. Rehabil.* 79, 81–86.

Curt, A., Nitsche, B., Rodic, B., Schurch, B., and Dietz, V. (1997). Assessment of autonomic dysreflexia in patients with spinal cord injury. *J. Neurol. Neurosurg. Psychiat.* 62, 473–477.

Dietz, V., Colombo, G., Jensen, L., and Baumgartner, L. (1995). Locomotor capacity of spinal cord in paraplegic patients. *Ann. Neurol.* 37, 574–582.

Dietz, V., Wirz, M., Curt, A., and Colombo, G. (1998). Locomotor pattern in paraplegic patients. Training effects and recovery of spinal cord function. *Spinal Cord* 36, 380–390.

Ditunno, J. F. (1992). New spinal cord injury standards. *Paraplegia* 30, 90–91.

Faden, A. I., Jacobs, T. P., Mougey, E., and Holaday, J. W. (1981). Endorphins in experimental spinal injury: Therapeutic effect of naloxone. *Ann. Neurol.* 10, 326–332.

Faden, A. I., and Salzman, S. (1992). Pharmacological strategies in CNS trauma. *TINS* 13, 29–35.

Geisler, F. H. (1993). GM-1 ganglioside and motor recovery following human spinal cord injury. *J. Emerg. Med.* 11, 49–55.

Giuliano, F., Hulting, C., El Masry, W. S., Smith, M. D., Osterloh, I. A., Orr, M., and Maytom, A. (1999). Randomized trial of sildenafil for the treatment of erectile dysfunction in spinal cord injury. Sildenafil Study Group. *Ann. Neurol.* 46, 15–21.

Gorio, A. (1988). Gangliosides as a possible treatment affecting neuronal repair processes. *In* "Advances in Neurology. Functional Recovery in Neurological Disease". (S. G. Waxman, Ed.), pp. 523–530. Raven Press. New York.

Greenberg, H. S., Kim, J. H., and Posner, J. B. (1980). Epidural spinal card compression from metastatic tumor: Results with a new treatment protocol. *Ann. Neurol.* 8, 361–366.

Guttmann, Sir L. (1976). "Spinal cord injuries—Comprehensive Management and Research", ed. 2. Blackwell Scientific Publications. Oxford.

Hao, J. X., Watson, B. D., Xu, X. J., Wiesenfeld-Hallin, Z., Seiger, A., and Sundstrom, E. (1992). Protective effect of the NMDA antagonist MK-801 on photochemically induced spinal lesions in the rat. *Exp. Neurol.* 118, 143–152.

Hiersemenzel, L. P., Curt, A., and Dietz, V. (2000). From spinal shock to spasticity. Neuronal adaptation to a spinal cord injury. *Neurology* 54, 1574–1582.

Hoshimaru, M., Koyama, T., Hashimoto, N., and Kikuchi, H. (1999). Results of microsurgical treatment for intramedullary spinal cord ependymomas, analysis of 36 cases. *J. Neurosurg.* 44, 264–269.

Hughes, J. T. (1989). Vascular disorders of the spinal cord. *In* (J. F. Toole, ed.) "Vascular Diseases, Part III, Handbook of Clinical Neurology," Vol. 55. pp. 95–106. Elsevier Amsterdam.

Hurlbert, R. J. (2000). Methylprednisolon for acute spinal cord injury: an inappropriate standard of care. *J. Neurosurg.* 93 (Suppl. 1), 1–7.

Imai, T., Matsumoto, H., Ohkubo, Y., Shizukawa, H., Chiba, S., and Kobayashi, N. (1997). A case of acute transverse myelitis affecting the entire length of the spinal cord. *Eur. Neurol.* 37, 247–248.

Iseli, E., Cavigelli, A., Dietz, V., and Curt, A. (1999). Prognosis and recovery in ischaemic and traumatic spinal cord injury, clinical and electrophysiological evaluation. *J. Neurol. Neurosurg. Psychiat.* 66, 1–4.

Jeffreys, R. V. (1986). The surgical treatment of cervical myelopathy due to spondylosis and disc degeneration. *J. Neurol. Neurosurg. Psychiat.* 49, 353–361.

Kapoor, V. K., Chakal, A. S., Jyoti, S. P., Mundkur, Y. S., Kotwal, S. V., and Metita, V. K. (1993). Intracavernous papaverine for impotence in spinal cord injured patients. *Paraplegia* 31, 675–677.

Kiers, L., and Chiappa, K. H. (1995). Motor and somatosensory evoked potentials in spinal cord disorders. *In* (R. R. Young and R. M. Woolsey, Eds.), "Diagnosis and Management of Disorders of the Spinal Cord". pp. 153–169. W.B. Sauders, Philadelphia.

Kim, R. C., Gutierrez, P. A., Vulpe, M., Young, R. R., and Choi, B. H. (1995). Foix-Alajouanine syndrome: a venocongestive myelopathy. *J. Spinal. Cord Medicine* 18, 280.

Kincaid, J, C., and Dyken, M. L. (1985). Myelitis and myelopathy. *In* (A. B. Baker and R. J. Joint, Eds.), "Clinical Neurology" (revised series), Vol. 3: 48, Harper & Row, Philadelphia. 1–32.

Klekamp, J., and Samii, M. (1996). Surgical results of spinal meningeomas. *Acta Neurochir.* (Suppl) 65, 77–81.

Koenig, E., Thron, A., Schrader, V., and Dichgans, J. (1989). Spinal arteriovenous malformations and fistulae: clinical, neuroradiological and neurophysiological findings. *J. Neurol.* 236, 260–266.

Lisak, R. P. (1985). Acute transverse myelitis. In "Current Therapy in Neurologic Disease" (R. T. Johnson, Ed.), pp. 398–400. Mosby Year Book, St. Louis, MO.

Loblaw, D. A., and Laperriere, N. L. (1998). Emergency treatment of malignant extradural spinal cord compression, an evidence-based guideline. *J. Clin. Oncol.* **16**, 1613–1624.

Lue, T. F., and Tanagho, E. A. (1987). Physiology of erection and pharmacological management of impotence. *J. Urol.* **134**, 829.

Maynard Jr, M. F., Bracken, M. B., Creasy, G., Ditunno, J. F., Donovan, W.H., Ducker, T.B., Garber, S.L., Marino, R. J., Stover, S. L., Tator, C. H., Waters, R. L., Wildberger, J. E., and Young, W. (1997). International standards for neurological and functionional classification of spinal cord injury. *Spinal Cord* **35**, 266–274.

McKinley, W. O., Seel, R. T., and Hardmann, J. T. (1999). Nontraumatic spinal cord injury, incidence, epidemiology and functional outcome. *Arch. Phys. Med. Rehabil.* **80**, 619–623.

McCormick, P. C., Torres, R., Post, K. D., and Stein, B. M. (1990). Intramedullary ependymoma of the spinal cord. *J. Neurosurg.* **72**, 523–532.

Mulder, D. W., and Dale, A. J. D. (1985). Spinal cord tumors and discs. *In* (A. B. Baker and R. J. Joint, Eds.), "Clinical Neurology" (revised series) Vol. 3. pp 1–28. Harper & Row, Philadelphia.

Nathan, P. W. (1994). Effects on movement of surgical incisions into the human spinal cord. *Brain* **117**, 337–346.

Samii, M., Volkening, D., Sepherina, A., Penkert, G., and Baumann, H. (1989). Surgical treatment of myeloradiculopathy in cervical spondylosis: a report on 438 operations. *Neurosurg. Rev.* **12**, 283–290.

Schnell, L., and Schwab, M. E. (1990). Axonal regeneration in the rat spinal cord produced by an antibody against myelin-associated neurite growth inhibitors. *Nature* **343**, 269–272.

Schurch, B., Hauri, D., Rodic, B., Curt, A., Meyer, M., and Rossier, A. B. (1996). Botulinum toxin as a treatment of detrusor sphincter dyssynergia: a prospective study in 24 spinal cord injured patients. *J. Urol.* **155**, 1023–1029.

Schurch, B., Schmid, D. M., and Stöhrer, M. (2000). Botulinum-A toxin injections to treat neurogenic incontinence in spinal cord injured patients. *N. Engl. J. Med.* **342**, 665.

Schurch, B., Wichmann, W., and Rossier, A. B. (1996). Post-traumatic syringomyelia (cystic myelopathy): a prospective study of 449 patients with spinal cord injury. *J. Neurol. Neurosurg. Psychiat.* **60**, 61–67.

Schwab, M. E., and Bartholdi, D. (1995). Degeneration and regeneration of axons in the lesioned spinal cord. *Physiol. Rev.* **76**, 310–370.

Scott, T. F., Bhagavatula, K., Snyder, P. J., and Chieffe, C. (1998). Transverse myelitis. Comparison with spinal cord presentations of multiple sclerosis. *Neurology* **50**, 429–433.

Sett, P., and Crockard, H. A. (1991). The value of magnetic resonance imaging (MRI) in the follow up management of spinal injury. *Paraplegia* **29**, 396–410.

Short, D.J., El Masry, W. S., and Jones, P. W. (2000). High dose methylprednisolone in the management of acute spinal cord injury—a systematic review from a clinical perspective. *Spinal Cord* **38**, 273–286.

Stark, R. J., Henson, R. A., and Evans, S. J. W. (1982). Spinal metastases. *Brain* **105**, 189–213.

Stüve, O., and Zamvil, S. S. (1999). Pathogenesis, diagnosis, and treatment of acute disseminated encephalomyelitis. *Curr. Opin. Neurol.* **12**, 395–401.

Thier, P., Dichgans, J., and Grote, E. H. (1992). Die zerivkale spondylotische myelopathie. *Akt. Neurol.* **19**, 119–131.

Wong, D. A., Fornasier, V. L., and MacNab, I. (1990). Spinal metastases: The obvious, the occult, and the imposters. *Spine* **15**, 1–4.

Woolsey, R. M., and Young, R. R. (1995). The clinical diagnosis of disorders of the spinal cord. *In* (R. R. Young and R. M. Woolsey, Eds.), "Diagnosis and Management of Disorders of the Spinal Cord." pp. 135–144. W.B. Saunders, Philadelphia.

Young, W., Flamm, E. S., Demopoulos, H. V., Tomasula, J. J., and Decrescito, V. (1981). Effect of naloxone on posttraumatic ischemia in experimental spinal contusion. *J. Neurosurg.* **55**, 209–219.

Zwimpfer, T. J., and Bernstein, M. (1990). Spinal cord concussion. *J. Neurosurg.* **72**, 894–900.

## CHAPTER 69

# Syringomyelia and Syringobulbia

Louis R. Caplan and Henry J. M. Barnett

## CLINICAL ASPECTS

### Definition and Pathology

Syringomyelia is a term used to describe conditions characterized by eccentric spinal fluid cavities within the spinal cord. The term *syrinx* is derived from the Greek word that means *pipe* or *tube*; it has come to be a name given to a hollow, fluid-filled cavity. Syringomyelic cavities usually include the central gray matter of the spinal cord but extend asymetrically into the anterior or posterior horns, at least on one side, and often affect the root entry zone white matter on that side. The cavities often have irregular outpouchings that extend into the other side of the spinal cord.

Syringomyelia rarely occurs as an isolated disorder but is almost always associated with other conditions such as developmental anomalies, inflammatory conditions, trauma, and spinal cord tumors. Different mechanisms account for the development of syringomyelic cavities. The cavities are usually filled with clear or slightly yellow fluid that contains variable amounts of proteinaceous material. Only the syringeal cavities associated with tumors contain fluid, which is bright to dark yellow and viscous because of a high protein content. Syringomyelic cavities often incorporate the residue of the central canal, in which cases they are partly lined by ependyma. Other areas of the cystic cavities are lined by glia and vascular connective tissue elements. Syrinxes are most common in the cervical and upper thoracic spinal cord and often extend rostrally up to the second cervical segment. Less common is extension caudally into the caudal thoracic and lumbar spinal cord and extention cephalad into the brainstem, although these extentions do occur. The central gray commissure, which contains pain fibers that cross to ascend in the ventral spinothalamic tracts, is invariably interrupted, accounting for the classic sign of dissociated sensory loss; that is, regions of loss of pain and temperature that have retained touch perception. The anterior horn that contains motor neurons is nearly always involved, at least on one side, accounting for atrophy and weakness often in one upper limb. The centrally placed lesions also interrupt the reflex arc of afferent fibers that enter in the dorsal roots that will synapse in the anterior horn nerve cells, accounting for loss of reflexes in the upper limbs.

The term *hydromyelia* should be reserved for a cavity containing fluid within a dilatation of the persisting central canal of the spinal cord. Hydromyelia has a free communication with the cerebrospinal fluid from the ventricles by a continued patency at the obex of the fourth ventricle (Foster, 1987). The distinction between hydromyelia and syringomyelia is only of clinical consequence in making decisions regarding surgical treatment.

Cavities, slits, and clefts may also be present in the caudal portion of the brainstem, especially in the medulla, and are referred to as syringobulbia. The clefts can extend into the pons but rarely involve the midbrain. These clefts occur most frequently in the regions of the floor of the fourth ventricle, where normal indentations are found. The most common anatomical structures included in the clefts are the vestibular nuclei, nuclei within the lateral medullary tegmentum and pontine reticular formation, the descending tract of nerve V, and hypoglossal nerve fibers as they exit from the base of the medulla just lateral to the pyramid.

### Frequency

The advent of magnetic resonance imaging (MRI) has now allowed recognition of many syringomyelic lesions that were previously undetected. Before MRI, the prevalence of syringomyelia wa estimated to be about

8.4/100,000 individuals, and about 0.4% of neurologic patients were thought to have syringomyelia (Schliep, 1978).

## Etiology and Classifications

Although the pathophysiological mechanisms of syringomyelia remain controversial, there are four main groups of patients: those with spinal tumors, developmental anomalies, trauma, and inflammatory conditions. Any process that interferes with spinal fluid circulation or leads to abnormalities in the blood supply of the central portion of the spinal cord can cause syringomyelic cavities to form.

### Neoplastic

Intramedullary tumors, especially gliomas and ependymomas, often affect the same region of the cord as syrinxes caused by other etiologies. Some spinal cord tumors are predominantly cystic, and others have prominent cystic components. Ependymomas and gliomas are the most common intramedulalry tumors associated with syrinx development. Hemangioblastomas and tumors in patients with neurofibromatosis often have prominent cystic components. Cystic regions overlap because of tumor cavitation and secretion of fluid (Peerless and Durward, 1983). The protein content of the cyst fluid is usually high. The clinical course may be more acute and progressive than in other causes. MRI can now in most instances differentiate neoplastic cavities from those that are developmental.

### Associated with Posterior Fossa Lesions

The most common abnormality is the Chiari malformation, but some patients have the Dandy-Walker syndrome, and some have early life posterior fossa arachnoiditis. These disorders interupt normal circulation of cerebrospinal fluid, leading to persistence of the central canal of the spinal cord. The open obex of the fourth ventricle leads into the residually patent central canal. This patent communication with the fourth ventricle leads to the designation of these spinal cavities as a communicating form of syringomyelia. These are the largest group of syringomyelic patients by far (about 80%).

Most investigators and clinicians consider syrinxes to be a developmental anomaly that enlarges slowly during life. Gardner posited that syringomyelia was caused by a pressure imbalance during uterine life and a crowding in the posterior fossa (Barnett *et al.*, 1973; Gardner, 1965). He theorized that the lateral foramina of Luschka of the fourth ventricle were late in opening, producing a temporary increase in posterior fossa pressure. The increased pressure led to herniation of the cerebellar tonsils into the upper cervical region, causing a Chiari-type malformation. If the pressure buildup was severe, then a full Arnold-Chiari–type malformation would occur. Gardner posited that pressure was conveyed from the fourth ventricle to the spinal cord through the central canal and that the increased pressure in the canal could lead to clefts and cystic outpouchings (Barnett *et al.*, 1973; Gardner, 1965). The mechanism of subsequent syrinx formation and growth is posited to be related to abnormal CSF flow dynamics usually with communication of the central spinal cord canal and the syrinx with the ventricular system.

Evidence to support the Gardner theory includes the following: (1) patients with Arnold-Chiari malformations often have syringomyelia; (2) posterior fossa anomalies such as herniated cerebellar tonsils, congenital bands, posterior fossa arachnoiditis, and membranous scarring at the foramen of Magendie (Matsumoto and Symon, 1989) are often found in patients with syringomyelia; (3) after surgical decompression of the fourth ventricle, symptoms and signs often improve or stabilize; and (4) medullary clefts are present at weak points in the fourth ventricle.

The Chiari malformation is thought to favor the inflow of spinal fluid into the central canal by a valve mechanism (Aboulker, 1979; Du Boulay *et al.*, 1974; Oldfield *et al.*, 1994; Terae *et al.*, 1994; Williams, 1969, 1979). After a Valsalva maneuver or other activities that increase intracranial pressure, the pressure inside the head is higher than intraspinal pressure, and a water-hammer pulse effect develops. This allows the intracranial pressure to dissipate into the spinal cord cyst (Oakes, 1985). The mechanism for so-called hydromyelia may be persistence of the central canal of the spinal cord (in humans it is generally obliterated) (Milhorat *et al.*, 1994). The high frequency of coexisting spina bifida, meningoceles, myelomeningoceles, and myelomeningodysplasia also favors maldevelopment as the cause of syringomyelic cavitation.

### Inflammatory Conditions with Arachnoiditis

Tuberculous meningitis and other forms of chronic inflammation of the spinal pia and arachnoid membranes have been associated with syringomyelia and syringomyelic syndromes. Scarring of the pia and arachnoid can also follow pyogenic infctions and bleeding from vascular malformations. In patients who have syringomyelia develop, the arachnoiditis tends to be very constrictive. Arachnoidal scarring affects the arteries and veins within the meninges. The constrictive arachnoidal scarring affects microvascular flow in small blood vessels within the spinal cord. One mechanism of syrinx formation is chronic ischemia to the central portions of the spinal cord with subsequent cavitation (Abbleby *et al.*, 1969; Caplan *et al.*, 1990). Arachnoiditis and meningeal scarring in the posterior fossa could also produce a block of spinal fluid circulation, thus altering normal spinal fluid dynamics and interupting drainage of fluid from the central canal region (Caplan *et al.*, 1990). Extramedullary spinal cord tumors have also been associated with syrinx cavities usually below the level of the tumor (Caplan *et al.*, 1990). The mechanism of cavity development in these patients may also relate to spinal block.

### Posttraumatic Cavities

Spinal trauma is followed by development of syringomyelic cavities in 1%–3% of patients. The trauma is usually major, causing papaplegia or tetraplegia. Syringomyelic cavities develop after a delay of months to years (10 years on average) (Barnett and Jouse, 1976; Rossier *et al.*, 1985). Higher incidence rates have recently been reported with the advent of MRI (up to about 20%) (Squirer and Lehr, 1994). Trauma may produce necrosis and microcyst formation as contributory factors (Barnett and Jousse, 1976; McDonald *et al.*, 1988; McLean *et al.*, 1973). Usually there is a central region of myelomalacic necrotic tissue that is produced during the acute trauma (Barnett and Jousse, 1976). Some patients have hematomyelia develop after trauma, with hemorrhage into the spinal cord. Absorption of blood and cavitation followed by an increased fluid content within the central canal can then lead to delayed enlargement and extention of the posttraumatic syringomyelic cavities (Barnett and Jousse, 1976; Anwer and Fisher, 1996).

Spinal block may occasionally be an important factor in the formation of syringomyelic cavities after trauma. Cavities usually extend rostrally and can reach the medulla; sometimes posttraumatic syringomyelic cavities extend caudally from the region of original trauma. Extension of the cavities is most likely explained by transmission of venous back pressure, which can be related to Valsalva maneuvers and other factors that even temporarily increase intraspinal pressure. Patients sometimes report that coughing, sneezing, and straining can be followed by a quick ascent or descent of sensory symptoms or pain. Trauma can also lead to arachnoidal scarring and spinal block (Caplan *et al.*, 1990). Arachnoidal adhesions can also promote extension of cavities by fixing and tethering the spinal cord (Barnett and Jousse, 1976). The fluid contained in posttraumatic

**TABLE I**

Type 1.  Syringomyelia with obstruction of the foramen magnum and dilation of the central canal
    A.  With type 1 Chiari malformation
    B.  With other obstructive foramen magnum lesions
Type II.  Syringomyelia without obstruction of the foramen magnum (idiopathic type)
Type III.  Syringomyelia with other spinal cord or meningeal diseases
    A.  Spinal cord tumors usually intramedullary
    B.  Traumatic myelopathy
    C.  Spinal arachnoiditis and pachymeningitis
    D.  Secondary spinal cord softening from cord compression (extramedullary tumor, spondylosis) or infarction (e.g., vascular malformation)
Type IV.  Pure hydromyelia (developmental dilation of the central canal usually with hydrocephalus

syringomelic cavities is identical to cerebrospinal fluid (CSF). Cavities fill from the CSF by means of the Virchow-Robin spaces around blood vessels.

### Idiopathic

At times and increasingly often since the advent of more frequent MRI imaging of the spine, syringomyelic cavities are found that do not fit into any of the preceding categories. These cavities have varying length and are usually not associated with severe symptoms or signs.

A classification of syringomyelic cavities suggested by Barnett and colleagues (Barnett *et al.*, 1973) is shown in Table I.

### Clinical Findings

Symptoms usually begin during the third to fifth decades of life but can be noted in adolescence. The onset is usually insidious. In occasional patients there is an abrupt acute onset of symptoms (Anwer and Fisher, 1996). In many patients a painless burn or hand atrophy calls attention to the disease, and the patient cannot say when the difficulty began (Victor and Ropper, 2001). In some patients numbness (i.e., loss of feeling in an upper limb) is the first reported symptom. Early symptoms often relate to lateralized root entry zone and gray matter involvement. Weakness and atrophy in one arm and hand, and unwitting burns and other injuries of the fingers are common. Weakness most often develops in the hand, forearms, and shoulder girdle, first on one side. Patients may notice fasciculations and loss of muscle bulk in these regions. Scoliosis caused by unequal strength in the paraspinal muscles is present in about 50% of patients with the congenital form of syringomyelia.

Pain is often an early and prominent symptom and is present in about half of patients (Van den Bergh et al., 1990; Victor and Ropper, 2001). Expansion of cavities can manifest as increasing pain. The pain is deep and aching in quality and can be quite severe. Pain is most often felt in the neck and shoulders but may follow a dermatomal distribution, radiating especially in the arms or upper trunk. Pain may be precipitated by coughing, sneezing, and straining. Paraesthesias are often present, especially in the upper limbs. Some patients become aware of decreased pain and temperature sensibility especially in the hands. Later in the course leg spasticity with altered gait and loss of control of urine and bowels may develop.

Examination usually shows loss of pain and temperature sensation in areas of the upper limbs and in a cape-like distribution across the shoulders and the lower neck. Dissociated sensory loss, meaning the loss of pain and temperature sensation with preservation of touch in some areas, is the hallmark of syringomyelia and is due to involvement of crossing pain and temperature fibers in the region near the central spinal canal. Extension of the dissociated sensory loss more caudally than expected from the location of the syrinx is related to a lesion of the spinothalamic tract. Deep tendon reflexes are nearly always lost in the involved arm.

Scoliosis and a smaller breast and chest development are usually found on the side of the involved arm. That arm also shows considerable muscle atrophy especially in the hand. As the syrinx enlarges, the upper extremity becomes weaker. Later the cavity creates pressure on the long white matter tracts and loss of position and vibration sense in the lower extremities, and spasticity and extensor plantar reflexes are found. Sphincter dysfunction and impotence develop with the long tract abnormalities.

Analgesic joints may be exposed to chronic recurrent trauma, and gross joint deformities (Charcot's joints) may occur similar to those found in luetic tabes dorsalis. Patients with syringomyelia may have sharp lancinating, stabbing limb pains similar to those also found in tabes. Autonomic dysfunction in the form of Horner's syndrome and postural hypotension is also relatively common and is due to involvement of the intermediolateral spinal gray column.

The most common symptoms of syringobulbia are diplopia, episodic vertigo, trigeminal distribution pain or analgesia, dysarthria, and hoarseness (Victor and Ropper, 2001). On examination, the findings are most often attributable to clefts on one side of the brainstem. The most common signs are unilateral tongue atrophy, analgesia and loss of temperature sensation on one side of the face, palatal and vocal cord paralysis, absent gag reflex, and nystagmus.

## Imaging and Laboratory Diagnosis

Plain x-ray films may show abnormalities of the skull and bony skeleton. Scoliosis, widening of the spinal canal, basilar invagination or impression or platybasia, atlanto-occipital fusion, and Klippel-Feil anomaly may be found (Oakes, 1985). MRI is the diagnostic procedure of choice. A cyst containing fluid with a signal intensity similar to CSF is usually found in the cervical and thoracic regions and produces widening of the cord. Cyst margins are often irregular, and the cyst may contain folds or septations (Mancall and McCormick, 1995). The C1 and C2 areas are often of normal size and do not show a syrinx. Evaluation of the craniocervical junction by MRI sagittal images that include the upper cervical region and the posterior fossa are important. Hydrocephalus may also be present. Chiari-type malformations, platybasia, and other craniocervical abnormalities are common. In patients with spinal cord tumors, irregular heterogeneous signal abnormalities may be shown on MRI in addition to the cystic cavity. Gadolinium enhancement may help define tumors and arachnoiditis. The use of different relaxation times and gadolinium as a contrast medium usually allows an answer to the question of whether there is a communicating syringomyelia or a tumor. New techniques (cine-MRI with presaturation bolus tracking) may be helpful in analyzing CSF dynamics. CSF flow measurements with velocity encoding can be performed using MRI, and changes in systolic and diastolic velocities can determine the effectiveness of surgery and other treatments (Brugieres et al., 2000).

CT myelography is also used in diagnosis. The myelo-CT in some patients may show an inflow of contrast medium into the syringomyelic cavity. Instillation of water-soluble contrast media into the subarachnoid space can determine whether the cyst communicated with the ventricular system and the CSF around the spinal cord. The cyst can be punctured using a lateral cervical approach under fluoroscopic control. It thereby can be drained, and the fluid content can be characterized.

The CSF is uaually normal but can contain elevated protein (especially if there is a tumor or spinal block) and a few lymphocytes.

Electrophysiology (electromyography and somatosensory-evoked potentials) may be helpful in determining the extent and level of the functional deficit, but findings of conduction slowing and denervation of muscles are not specific.

Differential diagnosis has to consider (depending on the clinical presentation) motor neuron disease, tumors of the lower brainstem and medulla, cervical spondylotic myelopathy, multiple sclerosis, and various causes of neuropathy and myelopathy.

## NATURAL COURSE

The onset is usually insidious, and progression is variable, depending on the cause of the syringomyelia. Rarely, the onset is acute and symptoms rapidly progress (Anwer and Fisher, 1996). Patients with spinal cord tumors tend to develop symptoms at an older age than those with congenital anomalies, and they may progress more rapidly. Some patients may have plateaus during which their symptoms and physical findings do not change appreciably for years. The course of the disease depends mainly on the pathophysiological mechanism of development of the syrinx. Investigations of the natural course are inconclusive, because all studies report on selected patient groups. Most authors agree that most patients will not die from the disease. About 60% have a chronic progressive course, 25% have a changing course with stationary and progressive episodes, whereas 15% have no signs of progression of symptoms (Anderson *et al.*, 1985; McIllroy and Richardson, 1965; Schliep, 1978). Fits of coughing, sneezing, and other maneuvers with sudden changes of intraspinal and intracerebral pressure may lead to unexpected deteriorations in the course of the disease. The most common symptoms are pain syndromes, paresis with nuclear atrophy of the upper limbs and cranial nerves, and abnormalities of long tract functions with spasticity and ambulatory restriction. After a 10-year course of the disease, about two thirds of the patients report severe pain (Anderson *et al.*, 1985). Aspiration is a risk in patients with syringobulbia. Infections of the lungs and bladder and autonomic disturbances with skin lesions and trauma influence the long-term prognosis.

## PRINCIPLES OF THERAPY

Choice of treatment will depend heavily on the cause and classification of the syrinx, the severity of clinical signs, and the presence of evidence of continuing progression toward functional disability caused by worsening of neurological symptoms and signs. Therapy attempts to limit the enlargement of the syrinx by removal of the cause when possible (e.g., by removing an intramedullary or extramedullary spinal cord tumor), or correction of abnormal CSF dynamics by surgically repairing the causative congenital anomaly or

aquired disease, or by surgically decompressing the spinal cord and draining or shunting the contents of the syrinx either into the subarachnoid space or the peritoneum. Surgery cannot be promised to reverse already existing functional loss but is aimed at attempting to halt any progression of disability.

In patients with spinal cord tumors that are associated with syringomyelic cavities, removal of the tumor is often possible. Patients with Von Hippel-Lindau disease (usually with spinal hemangioblastomas) and those with von Recklinghausen disease more often have cystic lesions that are amenable to surgical extirpation (Oakes, 1985).

In patients who have Chiari malformations and other craniocervical anomalies that alter CSF flow, craniocervical decompression, that is, foramen magnum decompression or a ventriculoatrial or ventriculoperitoneal shunt to drain the accompanying hydrocephalus may improve symptoms and halt progression of the syrinx, but there are risks of these procedures. The choice of the best neurosurgical procedure depends on the anatomical situation around the craniocervical junction and brainstem, as well as the type of spinal cavitation. Surgical intervention should alleviate pressure on the nervous structures and impede the buildup of new cavitations. In addition to surgical therapy, other measures to treat symptoms may be useful. Physiotherapy for the training of remaining muscular strength, treatment of spasticity (see Chapter 87), and therapy of acute and chronic pain (see Chapter 11) are often useful for the patient. The sequelae of lesions of the autonomic nervous system of insensitivity to pain and temperature may also require attention and symptomatic treatment. In patients with unfavorable progression of symptoms, continuous medical and psychological support is necessary.

## PRACTICAL MANAGEMENT

### Surgery

The decision for invasive therapy is difficult. The choice of the best neurosurgical procedure depends primarily on the origin of the syrinx and whether the cause of the syrinx is due to obstructive phenomena in the craniocervical subarachnoid space or another mechanism (Peerless and Durward, 1983). Rapid progression of symptoms should prompt consideration of surgery, whereas patients with long-standing courses and mild-to-moderate relatively stable deficits should be treated preferentially by nonsurgical methods. An especially unfavorable predictor is a long course of the disease (more than 2 years). Pain is the symptom that may be relieved effectively by surgery, whereas weakness and atrophies will hardly be changed. The few studies, including recent reports on surgical therapy, do not allow for comparison of the different applied techniques with respect to the differential indications for each treatment (Hankinson, 1970; Isu et al., 1989; Logue and Rice-Edwards, 1981; Love and Olafson, 1966; Matsumoto and Symon, 1989; Oakes, 1985; Padovani et al., 1989; Park et al., 1989; Peerless and Durward, 1983; Sgouros and Williams, 1995; Stevens et al., 1993; Wiedemayer et al., 1989, 1994). Surgery is used to alleviate pressure on nervous system structures, restore the extramedullary circulation of cerebrospinal fluid, and thereby prevent enlargement and extension of the syrinx.

Hydrocephalus and evidence of increased intracranial pressure are found in an important number of patients with Chiari malformations and syrinxes. In that circumstance, the recommendation is that a valve-regulated ventriculoperitoneal shunt be placed as the preferred first treatment modality (Oakes, 1985). Weakness, sensory abnormalities, spasticity, and cranial nerve abnormalities may be reversed by the shunting.

Most patients with syrinxes and Chiari malformations do not have increased intracranial pressure. In these patients, the surgical procedure of choice is a posterior fossa decompression with a suboccipital craniectomy and cervical laminectomy. The laminectomy is carried down to the level of the displaced posterior fossa components, the caudally displaced cerebellum and brainstem. It is not necessary to expose the upper portion of the syrinx cavity (Oakes, 1985). A follow-up of 388 patients who had surgical decompression reported improvement in almost 50%, in 25% deterioration, in 2.5% death, and the rest remained unchanged (Nogues, 1987). Several other studies suggest that, especially in symptomatic syringomyelia with Arnold-Chiari malformation, surgical decompression of the foramen magnum sometimes with additional shunts leads to clinical improvement or stabilization in about 80% of patients (Dyste et al., 1989; Hankinson, 1970; Logue and Rice-Edwards, 1981; Pillay et al., 1991). In the series of 56 cases of Logue and Rice-Edwards, occipital and cervical pain was usually relieved by surgical decompression; upper motor neuron type weakness was usually also improved, but shoulder and arm pain and segmental sensory and motor signs were not relieved by surgery (Logue and Rice-Edwards, 1981). In a review of 141 patients by Stevens and colleagues, good surgical outcome was achieved in half of those patients who had minor descent of the cerebellar tonsils below the foramen magnum, but only 12% of those patients who had a major cerebellar ectopia had a satisfactory result after surgical decompression (Stevens et al., 1993). When there is arachnoiditis and obstruction of CSF drainage around the fourth ventricle, reconstruc-

tive neurosurgery within the possterior fossa may be successful.

Drainage of syringomyelic cavities is more controversial. A number of different procedures have been used, but the success rate has been variable. Especially in syringomyelia caused by tumors and in syringomyelia after trauma, some authors used to recommend needle aspiration of cyst contents and myelotomy with syringostomia (opening the cavity into the subarachnoid space) (Love and Olafson, 1966; Powers et al., 1984; Sgouros and Williams, 1995; Suzuki et al., 1985), but our experience with this approach convinces us that it is not very helpful and should not be recomended. This procedure can occasionally result in stabilization, but success is often only temporary. Love and Olafson found that 30% of 40 patients who had drainage of a syrinx had excellent outcomes (Love and Olafson, 1966). Among 73 patients who had drainage of a syringomyelic cavity about half remained stable during a 10-year period after surgery, whereas 15% had serious complications of the surgery (Sgouros and Williams, 1995). Syringoarachnoid shunts (Tator and Briceno, 1988; Tator et al., 1982), and syringoperitoneal (Van Calenbergh et al., 1990; Wester et al., 1989), as well as syringopleural shunts are recommended by some surgeons in those patients in whom the cause of syringomyelia is not a tumor or a treatable malformation of the craniocervical junction. Syringoperitoneal shunts may be superior to syringoarachnoid shunts, because the drainage into the peritoneum is clearly determined and may be checked, whereas drainage into the subarachnoid space may be functionally insufficient (Firsching and Sanker, 1993; Peerless and Durward, 1983). Draining of a syrinx into the subarachnoid space often becomes obstructed, and consequently reflux into the syrinx may occur (Suzuki et al., 1985). Drainage into the subarachnoid space has usually been ineffective in our experience. In our opinion syringoperitoneal shunts are the treatment of choice in patients with syringomyelic cavities that are not caused by posterior fossa obstructive lesions. Sometimes repeated shunting is necessary.

Occasional patients with cord-destroying lesions who have permanent total paralysis below syringomyelic cavities may benefit from cordectomy if a catheter for drainage is placed into the rostral end of the severed cord cavity and connected to the peritoneal space. This procedure should not be done in patients with arachnoiditis.

The worst outlook is for patients with syringomyelia caused by chronic adhesive arachnoiditis. Patients with this variety of syringomyelia often fail to improve after syringoperitoneal or other surgical treatment, because the arachnoiditis continues to progress and continues to impede circulation of CSF.

Symptomatic treatment of various symptoms is also important. Tricyclic antidepressents such as amitryptiline, carbamazepine, or gabapentin can be used in an attempt to mitigate chronic pain. In patients with severe scoliosis, consultation and possibly therapy by an orthopedic surgeon may be necessary. Physiotherapy is helpful to maintain muscle strength and function. Baclofen or tizanidine can be used to alleviate spasticity.

## TREATMENTS NO LONGER RECOMMENDED

Radiotherapy in syringomyelia is not supported by any controlled study. Plugging the communication between the fourth ventricle and the central canal (a procedure recommended by Gardner) during posterior fossa decompression in patients with Chiari malformations has been abandoned. There have been complications with this procedure, and the results are no better than with simple decompression (Victor and Ropper, 2001).

## REFERENCES

Aboulker, J. (1979). La syringomyelie et les liquides intraarachidiens. *Neurochirurgie* 25 (Suppl. 1), 1–44.

Anderson, N. E., Willoughby, E. W., and Wrightson, P. (1985). The natural history of the influence of surgical treatment in syringomyelia. *Acta Neurol. Scand.* 71, 472–479.

Anwer, U. E., and Fisher, M. (1996). Acute and atypical presentations of syringomyelia. *Eur. Neurol.* 36, 215–218.

Appleby, A., Bradley, W. G., Foster, J. B., Hankinson, J., and Hudgson, P. (1969). Syringomyelia due to chronic arachnoiditis at the foramen magnum. *J. Neurol. Sci.* 8, 451–464.

Barnett, H. J. M., Foster, J. B., and Hudgson, P. (1973). "Syringomyelia." WB Saunders, Philadelphia.

Barnett, H. J. M., and Jousse, A. T. (1976). Posttraumatic syringomyelia (cystic myelopathy). *In* "Handbook of Clinical Neurology" (P. J. Vinken and G. W. Bruyn, Eds.), Vol 26. Injuries of the Spine and Spinal Cord part II. (Braakman Ed.). pp. 113–157. North-Holland Publ Co. Amsterdam.

Barnett, H. J. M., Jousse, A. T., and Ball, M. J. (1973). Pathology and pathogenesis of progressive cystic myelopathy as a late sequel of a spinal cord injury. *In* "Syringomyelia" (H. J. M. Barnett, J. B. Foster, and P. Hudgson, eds.), pp. 179–219. W. B. Saunders, Philadelphia.

Brugieres, P., Idy-Peretti, I., Iffenecker, C. *et al.* (2000). CSF flow measurement in syringomyelia. *Am J. Neuroradiol.* 21, 1785–1792.

Caplan, L. R., Norohna, A. B., and Amico, L. L. (1990). Syringomyelia and arachnoiditis. *J. Neurol. Neurosurg. Psychiatr.* 53, 106–113.

DuBoulay, G., Shah, S. H., Currie, J. C., and Logue, V. (1974). The mechanism of hydromyelia in Chiari type I malformations. *Br. J. Radiol.* 47, 579–587.

Dyste, G. N., Menezes, A. H., and VanGilder, J. C. (1989). Symptomatic Chiari malformations. *J. Neurosurg.* 71, 159–168.

Firsching, R., and Sanker, P. (1993). MRI follow-up in syringomyelia: Observation from twelve cases. *Acta Neurochir.* 123, 206–207.

Foster, J. B. (1987). Hydromyelia. *In* "Malformations, Handbook of Clinical Neurology" (N. C. Myrianthopoulos, Ed.), Vol. 6(50). pp. 425–433. Elsevier, Amsterdam.

Gardner, W. J. (1965). Hydrodynamic mechanisms of syringomyelia: Its relationship to myelocele. *J. Neurol. Neurosurg. Psychiat.* 28, 247–259.

Hankinson, J. (1970). Syringomyelia and the surgeon. *In* "Modern trends in Neurology," (D. Williams, Ed.), Vol 5. pp. 127–148. Butterworth, London.

Isu, T., Iwasaki, Y., Akino, M., and Abe, H. (1989). Surgical treatment of syringomyelia: Selection of surgical procedures. *Neurol. Med. Chir. Tokyo* 29, 728–734.

Logue, V., and Rice-Edwards, M. (1981). Syringomyelia and its surgical treatment—An analysis of 75 patients. *J. Neurol. Neurosurg. Psychiat.* 44, 273–284.

Love, J. G., and Olafson, R. A. (1966). Syringomyelia: a look at surgical therapy. *J. Neurosurg.* 24, 714.

Mancall, E. L., and McCormick, P. C. (1995). Syringomyelia. *In* "Merritt's Textbook of Neurology" (L. P. Rowland, Ed.), pp. 750–753. Williams & Wilkins, Baltimore.

Matsumoto, T., and Symon, L. (1989). Surgical management of syringomyelia—Current results. *Surg. Neurol.* 32, 258–265.

McDonald, R. L., Findlay, J. M., and Tator, C. H. (1988). Microcystic spinal cord degeneration causing posttraumatic myelopathy: Report of two cases. *J. Neurosurg.* 68, 466–471.

McLean, D. R., Miller, J. D. R., Allen, P. B. R., and Ezzeddin, S. R. (1973). Posttraumatic syringomyelia. *J. Neurosurg.* 39, 485–492.

McIllroy, W. J., and Richardson, J. C. (1965). Syringomyelia: A clinical review of 75 cases. *Can. Med. Assoc. J.* 93, 731–734.

Milhorat, T. H., Kotzen, R. M., and Anzil, A. P. (1994). Stenosis of central canal of spinal cord in man: Incidence and pathological findings in 232 autopsy cases. *J. Neurosurg.* 80, 716–722.

Nogues, M. A. (1987). Syringomyelia and syringobulbia. *In* "Malformations: Handbook of Clinical Neurology" (N. C. Myrianthopoulos, Ed.), Vol. 6(50). pp. 443–464. Elsevier, Amsterdam.

Oakes, W. J. (1985). Chiari malformations, hydromyelia, syringomyelia. *In* "Neurosurgery" (R. H. Wilkins and S. S. Rengadriry, Eds.), pp. 2113–2123. New York. McGraw Hill.

Oldfield, E. H., Muraszko, K., Shawker, T. H., and Patronas, N. J. (1994). Pathophysiology of syringomyelia associated with Chiari I malformation of the cerebellar tonsils. Implications for diagnosis and treatment. *J. Neurosurg.* 80, 3–15.

Padovani, R., Cavallo, M., and Gaist, G. (1989). Surgical treatment of syringomyelia: Favourable results with syringosubarachnoid shunting. *Surg. Neurol.* 32, 173–180.

Park, T. S., Cail, W. S., Broaddus, W. C., and Walker, M. G. (1989). Lumboperitoneal shunt combined with myelotomy for treatment of syringohydromyelia. *J. Neurosurg.* 70, 721–727.

Peerless, S. J., and Durward, Q. J. (1983). Management of syringomyelia: A pathophysiological approach. *Clin. Neurosurg.* 30, 531–576.

Pillay, P. K., Awad, I. A., Little, J. R., and Hahn, J. F. (1991). Surgical management of syringomyelia: A five year experience in the era of magnetic resonance imaging. *Neurol. Res.* 13, 3–9.

Powers, S. K., Edwards, M. S. B., Boggan, J. E., Pitts, L. H., Gutin, P. H., Hosobuchi, Y., Adams, J. E., and Wilson, C. B. (1984). Use of argon surgical laser in neurosurgery. *J. Neurosurg.* 60, 523–530.

Rossier, A. B., Foo, D., Shillito, J., and Dyro, F. M. (1985). Posttraumatic cervical syringomyelia. Incidence, clinical presentation, electrophysiological studies, syrinx protein and results of conservative and operative treatment. *Brain* 108, 439–461.

Schliep, G. (1978). Syringomyelia and syringobulbia. *In* "Handbook of Clinical Neurology" (P. J. Vinken and G. W. Bruyn, Eds.), Vol. 32, pp. 255–327. Elsevier, Amsterdam.

Sgouros, S., and Williams, B. (1995). A critical appraisal of drainage in syringomyelia. *J. Neurosurg.* 82, 1–10.

Squier, M. V., and Lehr, R. P. (1994). Post-traumatic syringomyelia. *J. Neurol. Neurosurg. Psychiatr.* **57**, 1095–1098.

Stevens, J. M., Serva, W. A., Kendall, B. *et al.* (1993). Chiari malformation in adults: relation of morphologic aspects to clinical features and operative outcome. *J. Neurol. Neurosurg. Psychiat.* **56**, 1072–1077.

Suzuki, M., Davis, C., Symon, L., and Gentili, F. (1985). Syringoperitoneal shunt for treatment of cord cavitation. *J. Neurol. Neurosurg. Psychiat.* **48**, 620–627.

Tator, C. H., and Briceno, C. (1988). Treatment for syringomyelia with a syringosubarachnoid shunt. *Can. J. Neurol. Sci.* **15**, 48–57.

Tator, C. H., Meguro, K., and Rowed, D. W. (1982). Favorable results with syringosubarachnoid shunts for treatment of syringomyelia. *J. Neurosurg.* **56**, 517–523.

Terae, S., Miyasaka, K., Abe, S., Abe, H., and Tashiro, K. (1994). Increased pulsatile movement of the hindbrain in syringomyelia associated with the Chiari malformation: cine-MRI with presaturation bolus tracking. *Neuroradiology* **36**, 125–129.

Van Calenbergh, F., Hoorens, G., and Van den Bergh, R. (1990). Syringomyelia: A retrospective study. Part II. Diagnostic and therapeutic approach. *Acta Neurol. Belg.* **90**, 100–110.

Van den Bergh, R., Hoorens, G., and Van Calenbergh, F. (1990). Syringomyelia: A retrospective study. Part I: Clinical features. *Acta Neurol. Belg.* **90**, 93–99.

Victor, M., and Ropper, A. H. (2001). "Adams and Victor's principles of Neurology," 7th ed. pp. 1337–1345. McGraw Hill, New York.

Wester, K., Kjosavik, I. F., and Midgard, R. (1989). Multicystic syringomyelia treated with a single, non-valved syringoperitoneal shunt: Fast and near-complete MRI normalization. *Acta Neurochir. Wien* **98**, 148–152.

Wiedemayer, H., Nau, H. E., Rauhut, F., Grote, W., and Gerhard, L. (1994). Operative treatment and prognosis of syringomyelia. *Neurosurg. Rev.* **17**, 37–41.

Wiedemayer, H., Nau, H. E., Reinhardt, V., Gerhard, L., and Grote, W. (1989). Syringomyelie aus neurochirurgischer Sicht. *Nervenarzt* **60**, 17–25.

Williams, B. (1969). The distending force in the production of communicating syringomyelia. *Lancet* **2**, 189–193.

Williams, B. (1980). On the pathogenesis of syringomyelia: A review. *J. Roy. Soc. Med.* **73**, 798–806.

CHAPTER 70

# Malformations and Neurocutaneous Disorders

M. Bähr and B. L. Schlaggar

Malformations of the central nervous system (CNS) are irreversible structural defects caused by disturbances of normal prenatal or postnatal development. Normal maturation of the CNS involves a predictable sequence of stages, each precisely timed in its evolution. With an insult to one or more of those stages, a brain malformation can result. The specific types of malformations seen are determined by the timing of the insult to the brain, the duration of that insult, and its severity (Barth, 1992; Volpe, 2001a;b). Some insults are brief (e.g., a one time exposure to a toxin), whereas others may occur over many weeks or throughout an entire gestation (e.g., recurrent uterine bleeding, gestational diabetes mellitus, maternal alcohol abuse, genetic defects). The clinical consequences of these events are determined by the regions of brain affected and the extent of any cerebral malformation that results.

The neurological deficits caused by some cerebral malformations are rarely apparent immediately after birth, because subtle neurological abnormalities usually cannot be appreciated at that time, even by the most skilled examiner. Thus, these cerebral dysplasias are often not detected until later in life, not uncommonly by chance when neuroradiological studies are done for unrelated reasons (e.g., in a school child who is having difficulty learning to read). The rate of developmental abnormalities of the brain may be as high as 3% (Kalter and Warkany, 1983).

The most definitive way to confirm the presence of a CNS malformation and to delineate its pathology is by direct histopathological examination (Friede, 1989). In the absence of such examination, high-resolution imaging techniques—computed cranial tomography (CCT) and especially magnetic resonance imaging (MRI)—will almost always demonstrate those developmental CNS disorders that are of greatest clinical importance (Barkovich, 1995).

## CLASSIFICATION

An understanding of the sequence of events in neuroembryology provides a framework for an appropriate classification scheme for the array of malformations seen clinically (for excellent and detailed textbook reviews see Sarnat and Menkes, 2000 and Volpe, 2001a,b). The neurons and glia of the CNS are derived from a specialized region of ectoderm called the neural plate. Adjacent mesoderm induces the neural plate to form the neural tube during the first 4 weeks of gestational life, a process called neurulation. Between the third and fourth gestational weeks, the lateral margins of the neural plate close over dorsally to form the neural tube. During this closure, the neural crest cells are formed, and these give rise to dorsal root ganglia, sensory ganglia of cranial nerves, autonomic ganglia, Schwann cells, and cells of the pia and arachnoid. The neural tube gives rise to the CNS and spinal cord. Closure of the neural tube proceeds rostrally and caudally; the rostal end closes at approximately 23–25 days, the caudal end at about 25–27 days (McLone and Dias, 1994). With an insult or a genetic defect acting during this time, a disorder of neurulation can result (Table I). The rostral end of the neural tube later gives rise to antecedents of the telencephalon. Lesions that interfere with CNS development when the rostral neural tube is closing are expressed mainly as midline defects and malformations of the prosencephalon (telencephalon and diencephalon) (Table I). Accordingly, interference with closure of the caudal neural tube manifests as meningomyelocele, for example.

The cavity of the neural tube gives rise to the ventricular system, while dividing neuroepithelial cells located at the ventricular surface in the wall of the neural tube generate progenitors for neurons and glia. Most of these progenitors migrate outward from the

TABLE I   Gestational Timing of Some CNS Malformations

| Type of malformation | Gestational stage |
|---|---|
| I.   Disorders of neurulation | 3–4 weeks of gestation |
|       Craniorachischisis totalis | 3 weeks |
|       Anencephaly | 4 weeks |
|       *Encephaloceles* | 4 weeks |
|       *Meningomyeloceles* | 4 weeks |
| II.  Malformations of the prosencephalon | 5–10 weeks of gestation |
|       Atelencephaly, aprosencephaly | 5 weeks |
|       *Holoprosencephaly* | 5–6 weeks |
|       *Septooptic dysplasia* | 6–7 weeks |
| III. Disorders of proliferation and/or migration | 2–5 months of gestation |
|       *Schizencephalies* | 2 months |
|       *Lissencephalies (agyria)* | 3 months |
|       *Pachygyria, polymicrogyria* | 3–5 months |
| IV.  Disturbances of differentation | 2–6 months of gestation |
|       Microcephaly, megalencephaly | 2–4 months |
|       *Neurocutaneous disorders* | 2–4 months or later |
|       *Corpus callosum hypo-/aplasia* | 3–5 months |
|       *Aicardi syndrome* | |
|       Colpocephaly | 2–3 months |
|       *Congenital vascular malformations and CNS tumors* | 2–3 months |
|       Aqueduct stenosis | 4 months |
|       *Encephaloclastic* | 3 months–perinatal |
|           Porencephaly | |
|           Multicystic encephalopathies, | |
|           Hydranencephaly | |
| V.   Cerebellar malformations | 4 weeks–1 year postnatal |
|       *Chiari malformations* | 4 weeks |
|       Cerebellar hemisphere hypoplasia and aplasia | 6 weeks |
|       Cerebellar vermis hypoplasia and aplasia | 6–10 weeks |
|       | 7–10 weeks |
|       *Dandy-Walker malformations* | |
| VI.  Disorders of myelination | 7 months–first years postnatal |
|       Hypomyelination, dysmyelination, delayed myelination | |

Disorders described in the text appear in italics.

neutral tube between the 6th and 24th gestational week. When normal proliferation or migration of these progenitors is disrupted, disturbances of normal organization (i.e., gyral formation, lamination) of the cerebral cortex are seen (Table I). On reaching their final destination, neurons and glia differentiate to their characteristic mature forms; if this process is disturbed, errors of differentiation result (Table I). The cerebellum is a portion of brain with a relatively protracted development (between the first gestational month and the first postnatal year), making it vulnerable to malformation over a long developmental period (Table I). Malformations of the cerebellum are frequently associated with cerebral malformations. One of the latest stages in maturation of the CNS is myelination of the fiber tracts. CNS myelination can be interfered with between the late stages of pregnancy and through the first postpartum years (Table I). (Disorders of myelination are not dis-

cussed in this chapter. Interested readers are referred to Lyon *et al.* [1996]. This is a brief but excellent textbook on the neurology of hereditary neurological diseases of childhood, including leukodystrophies.) Abnormalities of the ventricular system and disorders of the cerebrovascular system are described elsewhere (Chapters 41 and 60). Developmental malformations of the leptomeninges can result in CSF-filled arachnoid cysts, which are usually classified according to their location along the neuraxis (Table I).

## PRINCIPLES OF THERAPY

### General Principles

The therapeutic options for treatment of CNS malformations are presently limited to prevention and symptomatic management. Because disorders of neurulation can be caused by folic acid–deficient states, prevention of folic acid deficiency is of utmost importance (Fishman, 2000). Hence, periconceptional folic acid (0.4 mg/day) use is recommended*** (Thompson *et al.*, 2000). In addition, avoidance of folic acid antagonists (such as some anticonvulsants and antibiotics) by women of child-bearing age is recommended** (Hernandez-Diaz *et al.*, 2000).

Most instances of brain malformations are sporadic, but specific genetic disorders are increasingly recognized (see later). Hence another form of prevention is genetic counseling. Most cerebral malformations, particularly the migrational disorders and many of the neurocutaneous disorders (discussed later), are accompanied by seizures that require treatment with anticonvulsants and/or neurosurgery (see Chapter 20 and 21). Interferences with normal CSF flow typically manifest as hydrocephalus and may require medical treatment to slow the formation of CSF (e.g., acetazolamide or furosemide for the treatment of communicating hydrocephalus) or surgical shunting of CSF, usually from a lateral ventricle to the peritoneum (e.g., V-P shunting for treatment of noncommunicating hydrocephalus caused by Dandy-Walker, Chiari, or other malformation or from an aqueductal stenosis or interventricular foraminal blockade as in tuberous sclerosis) (for treatment of hydrocephalus see Chapter 60). Other disorders that are associated with intracranial tumors (such as acoustic neuromas in neurofibromatosis 2) or with arachnoid cysts (as in Aicardi syndrome) may call for surgical treatment. On occasion, direct surgical intervention may result in a cure (as with a Chiari I malformation).

## DISORDERS OF NEURULATION

Among the most frequent CNS malformations are disorders caused by incomplete or defective formation of the neural tube. Incomplete closure of the cranial end of the neural tube results in *anencephaly* (which is lethal) or in *meningoceles* or *meningoencephaloceles* (also known as *encephalomeningoceles*). Defective closure of the caudal neural tube results in spinal *meningoceles* or more frequently, spinal *meningomyeloceles* (also known as *myelomeningoceles*). Complete failure of neural tube closure results in *craniorachisis totalis*, which is fatal. The prenatal diagnosis of these disorders is usually quite easily accomplished with an ultrasonographic examination, although serum alphafetoprotein screening seems to yield a greater diagnostic sensitivity (Pilu *et al.*, 2000).

### Encephaloceles

*Meningoencephaloceles* (usually shortened in name to *encephaloceles*) are less common than meningomyeloceles, with a prevalence of about 1 in 10,000. Roughly two dozen inherited syndromes associated with encephalocele are listed on the Online Medelian Inheritance in Man (OMIM) (*www3.ncbi.nlm.gov/Omim*). In Western countries, 85% of encephaloceles are posterior in location, whereas anterior encephaloceles are more common in Asia (Aicardi, 1992a). With the posterior lesions, tissue from the occipital lobe, cerebellum, or brainstem can be contained within the sac, and not uncommonly there are also accompanying malformations of brain (often agenesis of the corpus callosum but also other malformations of cerebrum, cerebellum, or mesencephalon). In half of the cases, there is associated hydrocephalus. Somatic malformations are frequent as well. Hamartomatous cerebral tissue is frequently found within the wall of the encephalocele. When the walls of the sac are thinned, CSF may escape, and meningitis can develop. The treatment of choice for smaller encephaloceles that do not contain viable brain tissue (10%–20% of the occipital lesions) (i.e., cranial meningoceles) is surgical excision and closure of the underlying cutaneous defect. With encephaloceles that contain brain tissue, and particularly those where the sac is large, the prognosis even with surgery is generally poor (Date *et al.*, 1993). In general, outcome is more favorable for patients with anterior than with posterior encephaloceles (Brown and Sheridan-Pereira, 1992). For a discussion of surgical treatment strategies, see Mori (1985) and Humphreys (1994).

### Meningomyeloceles

Incomplete closure of the caudal neural tube occurs most often in the lumbar region (80%) and results in the formation of *meningoceles* or *meningomyeloceles*. The treatment of these lesions is determined to some extent by their severity, following therapeutic guidelines similar to those applied to meningoencephaloceles earlier. Most physicians advocate early surgery, because the likelihood of reversing any associated deficits tends to decrease with age (May, 1992). With rare exceptions, meningoceles and meningomyeloceles should be surgically excised. Lumbar (including thoracolumbar and lumbosacral) meningomyeloceles are often associated with a paraplegia and usually a neurogenic bladder disorder as well. By contrast, most patients with lesions that extend below S1 are ultimately able to walk unaided (Liptak *et al.*, 1992). About 90% of patients with lumbar meningomyeloceles have hydrocephalus develop from an associated Chiari malformation (accompanied by an aqueductal stenosis in 40%–75% of cases); by contrast, with occipital, cervical, thoracic, or sacral lesions, the incidence of hydrocephalus is about 60% (this occurs only with meningomyeloceles, not with uncomplicated meningoceles [Aicardi, 1992a]) (see Chapter 60 for treatment of hydrocephalus). Long-term neuropediatric, neurosurgical, urological, and orthopedic follow-up is required for all patients in whom a myelocele and especially a meningomyelocele is repaired. It is important to watch for late complications, (such as retethering of the cord or hind brain herniation), so these can be dealt with promptly.

The question of optimal mode of obstetrical delivery for fetal meningomyelocele has been raised. Many centers act on the belief that the long-term outcome is bolstered by cesarean section and avoid vaginal delivery (Luthy *et al.*, 1991). A recent retrospective study suggests that in isolated fetal meningomyelocele mode of delivery has no clear impact on outcome (Merrill *et al.*, 1998).

Tulipan and colleagues have argued that in utero repair of meningomyelocele may reduce the degree of hindbrain herniation normally seen in patients with myelomeningocele, thereby decreasing the morbidity associated with the Chiari type II malformation (see later), including brainstem dysfunction, hydrocephalus, and syringomyelia (Tulipan *et al.*, 1998).

## MIDLINE MALFORMATIONS, DEVELOPMENTAL MALFORMATIONS OF THE PROSENCEPHALON

### Holoprosencephaly (Disorder of Prosencephalon Cleavage)

Several variants of *holoprosencephaly* (HPE; an undivided forebrain) have been described, based on the appearance of the observed defects. These include alobar, semilobar, lobar, and abortive lobar forms and aplasia of the olfactory bulbs and tracts. In alobar, the most severe form, one finds absence of the interhemispheric fissure with a single-sphered cerebrum and a common ventricle, a membranous roof over the third ventricle, undivided basal ganglia, absent olfactory bulbs and tracts, and hypoplasia of the optic nerves, with the cerebral cortex surrounding the single ventricle exhibiting the cytoarchitecture of the hippocampus and other limbic structures; the cortex often also shows disordered neuronal migration. The diencephalon is also undivided. In semilobar, the interhemispheric fissure is evident posteriorly. In lobar, there is absence of the anterior corpus callosum and continuity of the frontal lobes and basal ganglia. There are often accompanying cranial malformations (e.g., *microcephaly*), and facial malformations (e.g., hypotelorism, flattened nose, bilateral cleft lip and palate, single central incisor), as well as frequent developmental defects of the cardiac, gastrointestinal, genitourinary, and musculoskeletal systems (Volpe, 2001a). The severity of the facial feature dysmorphism correlates fairly well with the degree of cerebral malformation leading to the axiom that "the face predicts the brain." For example, cyclopia, predicts alobar pathosis, whereas mild hypotelorism is more consistent with semilobar or lobar pathosis.

The frequency of HPE is 1 in 15,000 live births, with a 60-fold greater incidence (~1 in 250) in aborted embryos (Cohen, 1989; Kinsman *et al.*, 2000). Chromosome abnormalities are increasingly recognized. At least 12 different loci have been associated with HPE (Wallis and Muenke, 2000). Familial instances have shown varied patterns of inheritance, including autosomal dominant, autosomal recessive, X-linked recessive. Being the infant of a diabetic mother seems to increase the risk of HPE 10-fold (Volpe, 2001b). Disturbances of several factors/genes have recently been implicated as potential causes of holoprosencephaly, such as mutations of telencephalin; a telencephalon-specific glycoprotein, SIX3; a homeobox gene of the sine oculis family of transcription factors, sonic hedgehog; ZIC2 or TGIF, a homeodomain protein that represses transcription of certain proteins (Kinsman *et al.*, 2000; Wallis and Muenke, 2000).

The clinical pictures seen are extremely variable and include a wide spectrum of abnormalities of the cranium, brain, and extracranial structures, with the lobar forms typically relatively less affected. The patients are usually severely mentally retarded and frequently have seizures. Infants with alobar HPE ususally die within the first few months. The prenatal diagnosis of HPE is usually relatively easy and can be made from the 16th week of pregnancy by ultrasonography (Chervenak *et al.*, 1985).

Management of HPE is directed at the attendant neurological problems of seizures, nonprogressive encephalopathy, pulmonary aspiration, blindness, and spastic para/tetraparesis and hydrocephalus.

Hypothalamic/pituitary axis abnormalities such as deficient thyroid-, growth-, adrenocorticotropic-, and antidiuretic-hormones must be suspected, and, if present, specific interventions are needed (Cameron *et al.*, 1999).

### Septo-Optic Dysplasia

Septo-optic dysplasia (SOD) represents a disorder of development of the midline prosencephalic structures. These structures include the commisural, chiasmatic, and hypothalamic plates. Differential involvement of these plates produces the array of malformations seen clinically. For example, in SOD, which typically includes underdevelopment or absence of the septum pellucidum, optic nerves, and optic chiasm, the commisural and chiasmatic plates are involved. Accordingly, agenesis of the corpus callosum can be seen. With involvement of the hypothalamic plate there can be hypothalmic/pituitary dysfunction. The constellation of absent septum pellucidum, optic nerve hypoplasia, hypothalamic/pituitary insufficiency is called *DeMorsier syndrome*. Minor *forme frustes* have been recognized such as isolated agenesis of the corpus callosum and septum pellucidum cyst. There may be accompanying malformations of the olfactory system, other parts of the rhinencephalon, and other midline structures. Midline prosencephalon malformations can also be seen with disorders of cerebral cortex proliferation/migration such as *SOD and schizencephaly* and *Aicardi syndrome*.

Clinical features of SOD include varying degrees of visual disturbances, developmental delay, and endocrine abnormalities (especially growth hormone deficiency and central diabetes insipidus) (Aicardi, 1992a) (for treatment, see Chapters 102–104). Hypotensive and or hypoglycemic crisis accounts for two out of three deaths in these patients (Cameron *et al.*, 1999). The etiology is unknown. Recently, a novel homeobox gene called Hesx1 on chromosome 3p21.1–p21.2 has been implicated in the development of septo-optic dysplasia (Dattani *et al.*, 2000). Nonetheless, familial SOD is decidedly uncommon, suggesting that most cases are sporadic, with risk of recurrence less than 5%.

## DISORDERS OF PROLIFERATION AND/OR MIGRATION

### Schizencephaly

This deficit, also referred to in the literature as *agenetic porencephaly* and *familial porencepahly*, occurs when there is complete agenesis of a portion of the neuronal germinative zone, resulting in unilateral or bilateral clefts extending through the entire cerebral hemisphere from the ependymal lining of the lateral ventricles to the pial covering of cortex (Barkovich, 1995). Frequently classified as a defect in migration, schizencephaly also likely represents a disorder of proliferation taking into account the apparent failure of portion(s) of the neuronal germinative zone. A vascular insult has been implicated (Barkovich and Kjos, 1992). The clefts are lined by thickened pachygyric or polymicrogyric cortex; their lips can be fused (closed-lip *schizencephaly*) or separated (open-lip schizencephaly), with hydrocephalus a frequent accompaniment of the latter variety (Byrd *et al.*, 1989; Yakovlev and Wadsworth, 1946). Motor disabilities and seizures are present in most cases, whether bilateral or unilateral (Barkovich and Kjos, 1992). With a unilateral cleft with fused lips, intelligence is often normal; with a unilateral cleft with open lips, there is usually mild-to-moderate developmental retardation; with bilateral clefts, there is usually severe retardation, early onset of seizures, major motor handicaps, and frequent blindness (often secondary to optic nerve hypoplasia) (Barkovich, 1995).

Recent findings support the role of germline mutations in the homeobox gene EMX2 in patients with severe schizencephaly (Brunelli *et al.*, 1996). The role of EMX2 in the development and differentiation is not yet clear, but a recent study suggests its contribution in the differentiation of neocortical areas (Bishop *et al.*, 2000).

## Lissencephaly (Agyria) and Pachygyria

The most severe migrational abnormality is lissencephaly (smooth brain), which can be diagnosed by CCT scan and particularly well by MRI (Barkovich, 1995). In lissencephaly the brain has few if any gyri (hence *agyria* in the extreme). Two general types of *lissencephaly* have been distinguished. In type I, or classical lissencephaly, the cerebral wall is like that of a 12-week old fetus, with an outermost cell-poor layer, a diffuse cellular layer (containing neurons characteristic of lower layers of cortex), a zone of columns of heterotopic neurons, and an innermost band of white matter. Additional findings can include enlarged ventricles and hypoplasia of the corpus callosum. In type II lissencephaly, the cortex consists of clusters and circular arrays of neurons, with no recognizable lamination or other organization, separated by glial and vascular septa, with prominent neuronal heterotopias (thus the term "cobblestone lissencephaly"). A striking feature of cobblestone lissencephaly is the apparent loss of integrity of the pial surface that typically defines the outermost surface of the cortical mantle. It seems that migrating neurons stream through the compromised pia and then travel haphazardly and tangentially once beyond the compromised pial boundary. Recognized

syndromes consistently involving type II lissencephaly commonly have an associated muscular dystrophy and can have retinal abnormalities (discussed later) (Volpe, 2001b).

In *pachygyria*, the features are similar to those of lissencephaly but less marked, with a paucity of gyri that are unusually broad and with an abnormally thickened and poorly organized cerebral cortex. However, the number of neurons is not clearly fewer (Volpe, 2001b).

Most commonly, type I lissencephaly occurs either alone (*isolated lissencephaly sequence; ILS*) or with accompanying craniofacial and extracranial abnormalities (*Miller-Dieker syndrome; MDS*). Children with isolated type I lissencephaly show bitemporal hollowing, a small jaw, and early hypotonia, with later spastic quadriparesis, seizures beginning in the first year, and severe mental retardation. These children characteristically become microcephalic during their first year. In isolated pachygyria, similar but milder clinical findings are seen. The radiological findings in type I lissencephaly are best seen with MRI, which shows a smooth cortical surface with disorganization of the cortical architecture and accompanying colpocephaly (discussed later) in all cases. Type I lissencephaly is usually sporadic, although occasionally it is of autosomal recessive inheritance (Dobyns, 1989). In MDS, in addition to bitemporal hollowing and a small jaw, there is a characteristic facies, with a short, upturned nose; a long, thin, and protuberant upper lip; and a flattened midface. Accompanying genital, cardiac, and limb abnormalities are seen. The neurology of MDS is similar to that of ILS, although in MDS the neurological abnormalities tend to be even more severe. These abnormalities consist of severe mental retardation, seizures, and spastic quadriparesis. Death before 1 year of age is common.

ILS seems to be a microdeletion of 17p13.3 involving the LIS1 locus (Ledbetter *et al.*, 1992), whereas MDS is due to a larger deletion/translocation involving LIS1 and encompassing surrounding gene(s) in the sub-band p13.3 (Dobyns *et al.*, 1993). The gene product of LIS1, the regulatory gamma subunit of platelet activating factor (PAF) acetylhydrolase, is directly implicated in neuronal migration in the neopallium and cerebellum (Sweeney *et al.*, 2000). The diagnosis of MDS can be made by commercially available fluorescent in situ hybridization.

In addition to somatic line mutations in LIS1, type I lissencephaly can also be seen in men with mutation of the doublecortex (DCX) gene located at Xq22.3–q23. The name "double cortex" comes from the cortical appearance in affected women. Whereas men have a typically lissencephalic brain, affected women can have subcortical band heterotopia where there is a concentric but subcortical "band" of neurons that have failed to migrate to the cortical plate proper, thus a "double cortex." The overlying cortex might be thin and simplified, but it can also appear normal. The explanation for the gender difference stems from the fact that women, by virtue of random inactivation of one X chromosome in each somatic cell (lyonization; Barr body formation), are mosaic for the mutation. Cortical neurons whose inactivated X chromosome contains the wild-type allele will express only mutant gene product and consequently have abnormal migration. Those cortical neurons whose inactivated X chromosome contains the mutant allele will make normal gene product and migrate normally. The gene product of DCX is doublecortin, a microtubule-associated protein required for neuronal migration to the cerebral cortex (Gleeson, 2000). Mutation analysis for LIS1 and DCX is essential in determining the etiology of the disease in patients and can be helpful in determining the recurrence risk in families since LIS-1 is autosomal and doublecortin is X-linked (Gleeson, 2000).

An interesting neuroimaging clue that can help distinguish LIS-1 from DCX involvment is that the agyria appears more profound occipitally in LIS-1 cases, whereas it is more impressive frontally in DCX cases. The neurobiological significance of this observation has not been explained but is likely to relate to molecular gradients in the developing telencephalon already demonstrated to function in the differentiation of the neocortex (e.g., Bishop *et al.*, 2000).

The three disorders most strongly associated with type II lissencephaly are the Walker-Warburg syndrome (WWS), Fukuyama congenital muscular dystrophy (FCMD), and muscle-eye-brain disease (MEB). Each is inherited by autosomal recessive patterns (Dobyns, 1987). As noted earlier, these conditions each involve type II lissencephaly and congenital muscular dystrophy. In *Walker-Warburg syndrome*, there is *h*ydrocephalus, *a*gyria, *r*etinal *d*ysplasia, ± encephalocele (thus the acronym HARD ±E syndrome) with accompanying macrocrania (congenital or less often, developing in the first year), hypoplasia, or absence of the cerebellar vermis with or without Dandy-Walker malformation, and muscular dystrophy. Neurologically, severe seizures and mental retardation are seen, as in type I lissencephaly, to which marked hypotonia and weakness are added, because of the accompanying muscle disease. The putative locus for WWS is 9q31 (Toda *et al.*, 1995). In contrast to WWS, *FCMD* has less severe lissencephaly and no retinal dysplasia. Myocardial fibrosis can be seen. FCMD is seen more commonly in Japanese people. The gene locus for FCMD is also known to be at 9q31, raising the possiblity that FCMD and WWS are "genetically identical" (Toda *et al.*, 1995). The predicted protein of the FCMD gene, called fukutin, is thought to be a secreted extracellular matrix component (Kobayashi *et al.*, 1998). This putative function could account for both loss of pial integrity and muscular dystrophy when fukutin is deficient. Although MEB is phenotypically very similar to WWS, a genetic locus has recently been identified at 1p34–p32 (Cormand *et al.*, 1999). In principle, WWS may be allelic to both FCMD and MEB. Most children with type II lissencephaly die in early infancy (Aicardi, 1992a).

### Polymicrogyria

In polymicrogyria the cerebral cortex is subdivided by a large number of very small folds causing its external surface to look like that of a wrinkled chestnut. The polymicrogyria can be of four distinct layers (as seen in destructive encephalopathies of ischemic or infectious type [e.g., in hydranencephaly or cytomegalovirus disease]), perhaps occuring during cortical migration. The polymicrogyria can be nonlayered as with disorders of peroxisomes such as *Zellweger syndrome and neonatal adrenoleukodystrophy*. In patients with *Zellweger syndrome*, a lethal panperoxisomal disorder, and in mice lacking the Pxr1 import receptor for peroxisomal matrix proteins, the absence of peroxisomes leads to abnormal neuronal migration (Gressens *et al.*, 2000). Children with polymicrogyria typically have frequent seizures and severe developmental delay.

Symmetrical forms of polymicrogyria have been noted in both sporadic and familial forms. In these patients one might find bilaterial perisylvian, frontal, or occipital polymicrogyria. Deficits can range from extremely mild and localized to severe mental retardation and seizures. The causes of symmetrical polymicrogyrias are not known but are presumed to be genetic.

Seizures accompanying the neuronal migratory disorders are often refractory to anticonvulsants. When hydrocephalus coexists, for example, with lissencephaly, shunting may be needed, although the clinical outcome is unlikely to be influenced substantially. Because neuronal migration that gives rise to formation of the cerebral cortex is associated temporally and causally with development of the corpus callosum, the migrational disorders discussed previously are frequently accompanied by underdevelopment or agenesis of the corpus callosum.

## DISTURBANCES OF DIFFERENTIATION

The disorders classically listed as malformations of differentiation (Table I) (Friede, 1989) could be included in other categories as well, but for the sake of simplicity they will be addressed here. Note that the neurocutaneous disorders are discussed in a separate section later. *Agenesis of the corpus callosum* is a common malformation with a prevalence determined by CCT scanning, ranging from 0.74%–2.3% (Sarnat, 1992). From a developmental standpoint, it can be the result of the aberrant development of the prosencephalic commisural plate, as discussed earlier. The agenesis can be either complete or can occur in partial form, with the missing portion usually the posterior part, because the corpus callosum develops in an anteroposterior direction. Although cases may occur in isolation, other brain abnormalities, such as Chiari II malformation or neuronal migrational disorders, frequently coexist (Barkovich and Norman, 1988). Dysmorphic-appearing cingulate cortex, commonly with everted gyri and misdirected Probst bundles, is frequently seen in patients with complete agenesis of the corpus callosum (Barkovich and Norman, 1988). This observation is consistent with the notion that the cingulate callosal fibers pioneer the corpus callosum (Koester and O'Leary, 1994). When accompanied by chorioretinal lacunae, vertebral abnormalities, infantile spasms, and developmental delay in a girl, the combination is called *Aicardi syndrome*. There are many other causes of callosal agenesis, including a variety of chromosome defects (e.g., trisomy 8), toxins (e.g., maternal alcohol abuse), and metabolic disorders (e.g., nonketotic hyperglycinemia) (Aicardi, 1992a).

When agenesis of the corpus callosum occurs alone, accompanying symptoms may be absent or mild. In such cases, the agenesis is often found by chance when a CCT or MRI scan is done, often for unrelated reasons. When there are accompanying cerebral abnormalities, mental retardation and seizures are frequent findings. Extracranial malformations can coexist and can affect the face as well as the cardiovascular, genitourinary, gastrointestinal, and musculoskeletal systems (Kozlowski and Ouvrier, 1993). Agenesis of the corpus callosum is accompanied by elevation of the third ventricle, a radial rather than horizontal orientation of cerebral gyri (producing a "sunburst" appearance), and very often a malformed lateral ventricular system, which maintains its fetal morphology with enlargement of the occipital horns, an appearance called *colpocephaly* (Yakovlev and Wadsworth, 1946). Colpocephaly also occurs with other brain pathosis, including destructive encephalomalacias, with particular involvement of periventricular white matter (e.g., periventricular leukomalacia) and a number of developmental disorders of brain in which myelination is impaired. (Aicardi, 1992a; Herskowitz *et al.*, 1985).

Isolated cysts of the septum pellucidum or a cavum vergae are usually unaccompanied by any clinical symptoms.

## ENCEPHALOCLASTIC LESIONS: HYDRANENCEPHALY AND PORENCEPHALY

The cerebral ventricles can dilate because of loss of brain parenchyma (hydrocephalus *ex vacuo*) or can be enlarged on a maldevelopmental basis, as in some cases of colpocephaly (discussed earlier). The term *porencephaly* refers to a cavity or cavities in cerebrum (Gr. porus = hole), acquired either in utero or in early postnatal life. Although such lesions can be consequences of trauma, infection, and hemorrhage, ischemia is the most common cause. Porencephalic cysts of ischemic causation are usually found in the territory of middle cerebral artery supply (when unilateral, left-sided in 80%). Such ischemia has many possible causes, including vascular maldevelopment, vasospasm (as with cocaine exposure), embolic vascular occlusion (as from placental fragments with placental infarction), or vascular thrombosis (as with hypernatremic dehydration, disseminated intravascular coagulation, or a variety of hypercoagulable states) (Volpe, 2001c). True porencephalies always communicate with the cerebral subarachnoid space, and they often communicate with the ventricular system as well. The porencephalic defects are sometimes surrounded by microgyria, the presence or absence of which probably depends on the time of the insult (Aicardi, 1992a). Most porencephalic cysts do not need treatment, but occasionally they enlarge and cause an increase in intracranial pressure because of an accompanying ball valve that interferes with CSF outflow from the cyst. In that circumstance, surgical extirpation or shunting is needed. Furthermore, these cysts can be intimately associated with epileptogenic foci in medically intractable epilepsy requiring surgical removal.

Bilateral porencephaly that is extreme in degree, with most of the cerebral hemispheres reduced to CSF-filled sacs, is called *hydranencephaly*. When caused by ischemia, hydranencephaly is usually a sequel to bilateral cerebral infarction in the distribution of both internal carotid arteries (Halsey, 1987). Hydrocephalus often coexists with hydranencephaly and is amenable to treatment, but the cerebral defects of hydranencephaly are not. On occasion, severe hydrocephalus alone may mimic hydranencephaly; in such instances, even with extreme hydrocephalus, the outlook is much better than with hydranencephaly.

When ischemic stroke in the perinate is suspected, a comprehensive screen for genetic and acquired risk factors is recommended (Gunther *et al.*, 2000). The risk of recurrence of ischemic stroke or other manifestations of thrombophilia after perinatal ischemic stroke is presently under investigation but is likely to be low. Accordingly, the role of antiplatelet and anticoagulation therapy in these patients is not known.

# CEREBELLAR MALFORMATIONS

## Chiari Malformations

Chiari malformations are complex disorders of early embryonic development. These are the most common dysplasias of the cerebellum and are associated with a wide variety of malformations of the rhombencephalon, mesencephalon, diencephalon, and telencephalon in different combinations. The Chiari malformations are classified into types I, II, III, and IV.

### Chiari I Malformation

**Clinical Aspects and Natural Course.** Chiari I malformations are unilateral or bilateral herniations of the cerebellar tonsils down through the foramen magnum, with or without accompanying caudal displacement of the medulla oblongata (Figure 1). Hydrocephalus, syringomyelia/bulbia, and malformation of the skull base and upper cervical spine (i.e., atlantoaxial dislocation, platybasia) are frequently associated. Clinical symptoms are often delayed until late in the third or fourth decade of life. Occipital headaches are frequent, and lower cranial nerve signs can be seen, with horizontal gaze evoked and downbeat nystagmus particularly characteristic. Ten percent of the patients have cerebellar signs. Paroxysmal bulbar signs and or headache brought on with cough or the Valsava moreover should prompt consideration of Chiari I malformation. Chiari I can be seen dramatically in infancy, with bulbar dysfunction such as dysphagia, stridor, apnea, and aspiration. A particularly concerning complication is development of sleep apnea, which can be the only clinical appearance of the syndrome (Zolty et al., 2000). A reversible acquired Chiari I malforma-

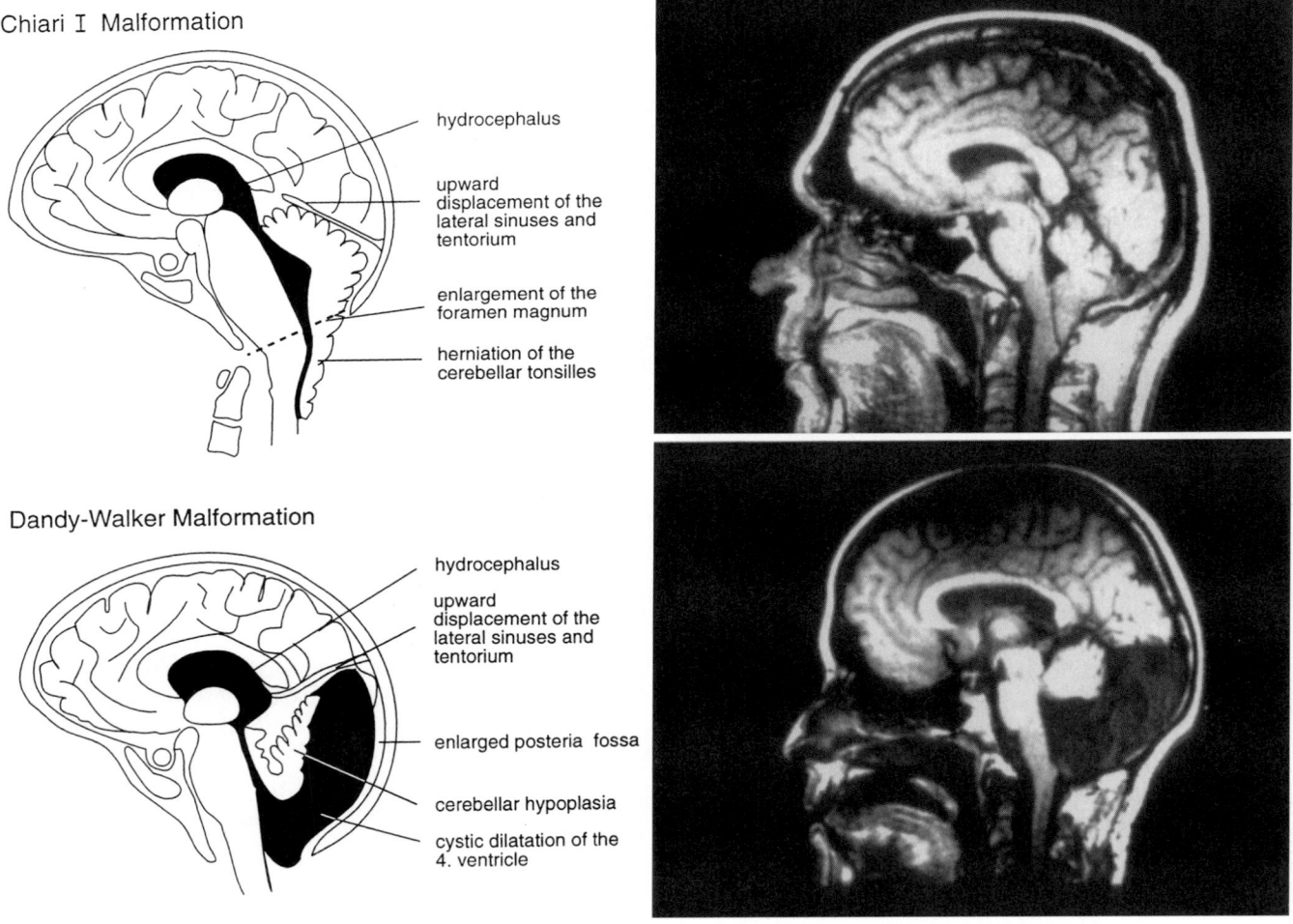

Chiari I Malformation

- hydrocephalus
- upward displacement of the lateral sinuses and tentorium
- enlargement of the foramen magnum
- herniation of the cerebellar tonsilles

Dandy-Walker Malformation

- hydrocephalus
- upward displacement of the lateral sinuses and tentorium
- enlarged posteria fossa
- cerebellar hypoplasia
- cystic dilatation of the 4. ventricle

**FIGURE 1** Typical features of a Chiari I and a Dandy-Walker malformation. On the left side, a schematic drawing of the main neuroradiological findings in the sagittal plane and on the right side, original sagittal MR pictures are shown.

tion has been described with CSF leakage after lumbar puncture (Samii *et al.*, 1999).

Most centers with an MRI facility are equipped to perform CSF flow studies to demonstrate expected flow diminution at the foramen magnum; a finding consistent with symptomatic Chiari I malformations (Bhadelia *et al.*, 1995).

**Principles of Therapy.**   If the malformation is symptomatic, suboccipital decompression is recommended with exploration of the fourth ventricle and its outlet foramina and duraplasty (Munshi *et al.*, 2000; see also treatment recommendations for Chiari II malformation). This intervention will reverse the symptoms in more than two thirds of patients, although up to one third of those treated will relapse within the next 3 years (Levy *et al.*, 1983). The benefit from a surgical decompression in general seems to be greatest in patients who have only cerebellar symptoms and when the cerebellar dislocation is not below the C1 level (Stevens *et al.*, 1993), although evidence for relief of pediatric headache exists (Weinberg *et al.*, 1998).

### Chiari II Malformation

**Clinical Aspects and Natural Course.**   The Chiari II malformation consists of inferior displacement of the medulla, fourth ventricle, and cerebellar vermis through the foramen magnum into the upper cervical canal. This displacement is accompanied by elongation and thinning of the upper medulla and lower pons and persistence of the embryonic flexure of these structures, along with bony defects of the foramen magnum, occiput, and upper cervical vertebrae. An accompanying myelomeningocele, most often in the lumbosacral region, with hydrocephalus is almost invariably present. Symptoms of brainstem dysfunction are of particular concern. These include vocal cord paralysis with laryngeal stridor, obstructive and central apnea, and dysphagia (with the last sometimes resulting aspiration pneumonia). There may be associated torticollis or retrocollis, scoliosis, and hydromyelia (Paul *et al.*, 1983). Maldevelopment of the skull base and upper cervical spine, as with Chiari I, can be seen. The symptoms are usually present at birth, but in some cases they may be first diagnosed later in childhood. In most cases of myelomeningocele with Chiari malformation, there are also accompanying abnormalities of cerebral cortical development, most often polymicrogyria, with or without disordered lamination, and neuronal heterotopias. Migrational anomalies of the cerebellum and brainstem are also noted (Gilbert *et al.*, 1986). Mental retardation is often a consequence, caused either by the accompanying cerebral dysgenesis or by the hydrocephalus, especially if the latter has been shunted,

and complicating infections have necessitated shunt revisions (Aicardi, 1992a).

**Principles of Therapy.**   Accompanying meningomyeloceles should be treated surgically during the first days of life, and associated hydrocephalus should be shunted. If there is evidence for brainstem compression, with paraparesis, tetraparesis, dysphagia, torticollis, or retrocollis, suboccipital decompression should be carried out (Carmel, 1983; Haines and Berger, 1991). This decompressive surgery can result in complete reversal of the neurological signs. Most patients with type II Chiari malformations need placement of a V-P shunt, particularly if there is an associated aqueductal stenosis (Mori, 1985). These shunts not uncommonly obstruct, and the resulting increase in intracranial pressure—if unrelieved—can be damaging to the brain and life-threatening. Thus, blocked shunts should be revised promptly. All patients with surgically treated Chiari II malformations should be followed carefully long term, because delayed complications (such as retethering of the spinal cord or hindbrain herniation) are not infrequent and call for reoperation. In general, indications for surgical decompression of Chiari II malformations are similar to those for type I (Venes *et al.*, 1986). The outcome of patients with a Chiari II malformation who undergo surgical decompression, however, seems to be less favorable compared with patients with Chiari I malformations, depending on the grade of cerebellar ectopia (Stevens *et al.*, 1993). Thus, patients with a level of the cerebellar ectopia below the C2 level have a 69% risk of deterioration by 18 months compared with an improvement of symptoms in about 50% of the patients with a cerebellar ectopia not below the C1 level.

### Chiari III Malformation

**Clinical Aspects and Natural Course.**   Encephaloceles located in the low occipital or high cervical regions, when combined with deformities of the lower brainstem, skull base, and upper cervical vertebrae characteristic of the Chiari type II malformation, comprise the Chiari III malformation. This type of encephalocele contains cerebellum in virtually all cases and occipital lobe in about one half. Partial or complete agenesis of the corpus callosum occurs in two thirds (Volpe, 20001a). The treatment strategies of encephaloceles have already been delineated.

### Chiari IV Malformation

This type of malformation consists only of cerebellar hypoplasia and is not part of the Chiari deformity as it is now understood (Friede, 1989).

## Dandy-Walker Malformation

**Clinical Aspects and Natural Course.** The *Dandy-Walker malformation* includes partial or complete absence of the cerebellar vermis, cystic dilation of the fourth ventricle, enlargement of the posterior fossa, upward displacement of the intracranial torcular and lateral venous sinuses and tentorium cerebelli, and accompanying hydrocephalus; findings are best seen with MRI (Figure 1). The disturbance in Dandy-Walker malformation seems to be primarily a delay or total failure of the foramen of Magendie (exiting from the fourth ventricle) to open, leading to a build-up of CSF pressure and cystic dilation of the fourth ventricle. Despite subsequent opening of the foramina of Lushka (usually patent in Dandy-Walker malformation), cystic dilation of the fourth ventricle and impairment of CSF flow persist (Volpe, 2001a). Prenatal diagnosis is relatively easy from the 18th to 20th week (Newman, 1982). The prevalence of Dandy-Walker malformation is between 1 in 25,000 and 1 in 35,000, with girls more often affected than boys. Most cases are sporadic (Aicardi, 1992a). The dominant clinical feature in Dandy-Walker malformation is development of hydrocephalus early in life, usually with striking enlargement of the occipital skull. The neurological signs are those of increased intracranial pressure and others more specifically related to posterior fossa abnormalities, including lower cranial nerve palsies, nystagmus, and ataxia. Associated abnormalities of the cerebrum occur in up to 70% of cases, the most important being agenesis of the corpus callosum and disordered neuronal migration. As a result, mental retardation is frequent. Systemic abnormalities (e.g., cardiac, urinary) are present in 20%–80% of cases (Volpe, 2001a). In the absence of developmental abnormalities of the cerebrum and with successful treatment of the hydrocephalus, the outlook is generally good. The mortality in such cases has varied from 10% (Hirsch *et al.*, 1984) to 40% (Pascual-Castroviejo *et al.*, 1991), with the presence or absence of accompanying cerebral and extracranial malformations a major determinant of these outcome figures. For those cases detected in utero or in the neonatal period, the outcome is usually unfavorable, with about a 40% mortality and cognitive deficits in 75% of the survivors (Volpe, 2001a). Even with treatment, only 50% of children with Dandy-Walker malformation achieve an IQ of 80 or greater. With early surgery, the outlook for cognitive development is usually better.

**Principles of Therapy.** Surgical treatment of the Dandy-Walker malformation is complicated by the presence of the fourth ventricular cyst. Unroofing of the cyst is usually not effective (Edwards and Raffel, 1987). Placement of a cystoperitoneal shunt is also usually not effective, because the aqueduct in Dandy-Walker malformations typically does not allow adequate CSF flow. Thus, the preferred approach is to shunt both the cyst and the lateral ventricle at the same time, connecting the two catheters by a Y-connector to the peritoneal catheter (Osenbach and Menezes, 1992).

## Arachnoid Cysts

### Clinical Aspects and Natural Course

*Arachnoid cysts* are a common finding on CCT or MRI, with a frequency of 4%; they are seen more often in males. These cysts are fluid-filled cavities that develop within a duplication or anomolous splitting of the arachnoid membrane or between the arachnoid and the pia (Sarnat and Menkes, 2000) They may or may not communicate with the subarachnoid space (Aicardi, 1992c), which may be differentiated by uptake of contrast-enhancing substances from the CSF. Symptoms occur in only 10%–30% of cases (Harsh *et al.*, 1986). These cysts are usually classified according to their location on the neuraxis. They are more often supratentorial than infratentorial, with 50%–60% of the total found in the middle cranial fossa (Sarnat and Menkes, 2000) (Figure 2). They vary greatly in size. Location and size determine the presence or absence of clinical symptoms. Even large cysts can be asymptomatic. When located in the midline, they can compress the cerebral aqueduct or a foramen of Monro and cause obstructive hydrocephalus (Raimondi *et al.*, 1980). If located in the posterior fossa, they may simulate a tumor and can cause hydrocephalus by obstructing CSF flow. A minority of arachnoid cysts, similar to porencephalic cysts, have a ball–valve type opening and can accumulate CSF, causing progressive enlargement. Clinical symptoms associated with arachnoid cysts are extremely variable. In addition to those of intracranial hypertension from secondary hydrocephalus, there may be visual impairment, cranial nerve palsies, gait ataxia, seizures, endocrine abnormalities, and deformation of the skull (De Meyer, 1985). Focal symptoms can be caused by local pressure effects, hemorrhage into a cyst, or rupture of a cyst wall.

### Principles of Therapy

Symptomatic arachnoid cysts should be treated. When there is accompanying obstructive hydrocephalus, the cyst and the ventricular system should both be shunted with a common valve adapted to both systems to circumvent pressure differences between the two (Raimondi *et al.*, 1980). Complications include shunt failure, infection, and subdural hematoma (from

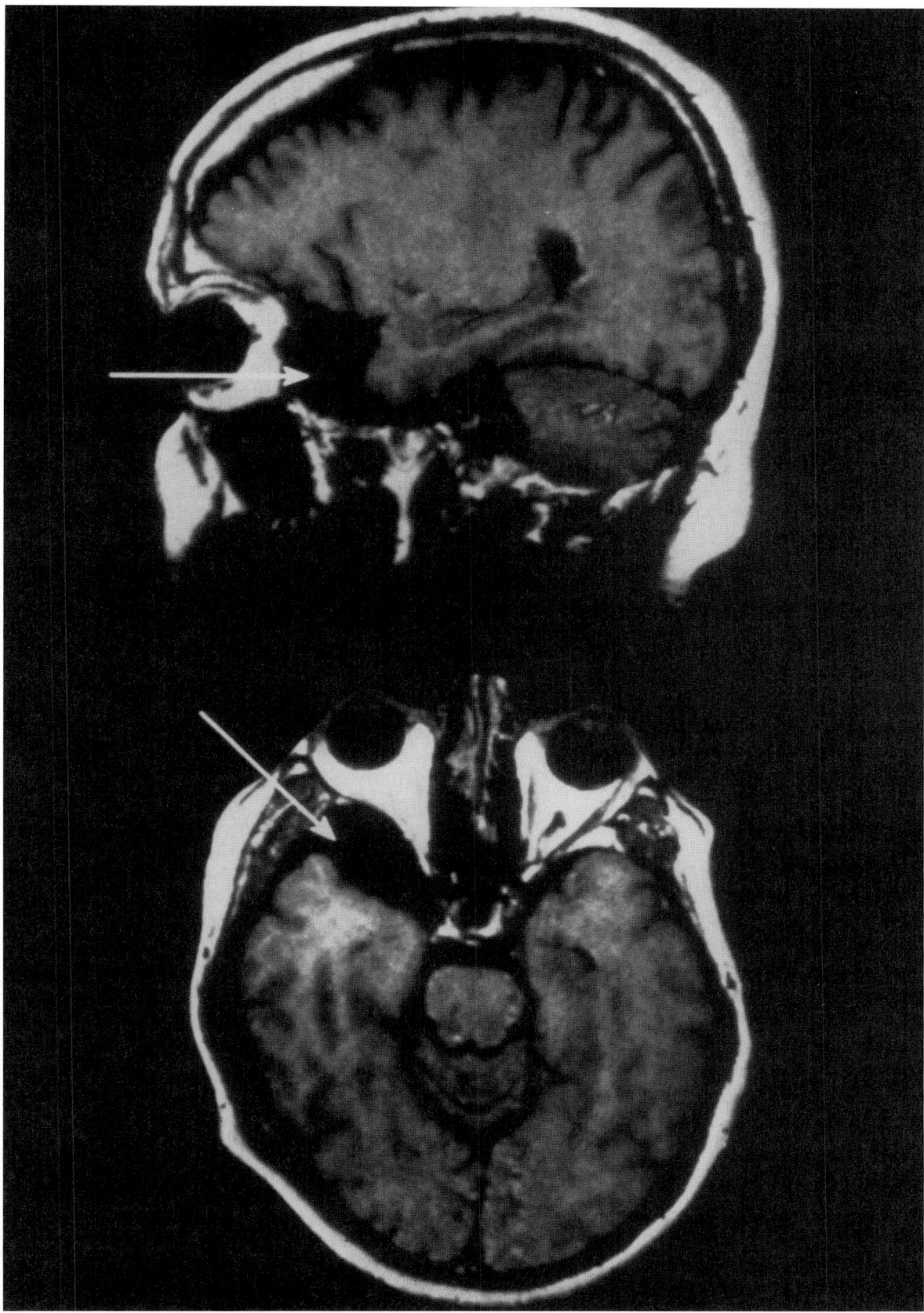

**FIGURE 2**  Typical location of an arachnoid cyst (arrows indicate the position of the cyst on the MR pictures).

excessive or rapid decompression of the cerebral hemispheres). As an alternative to shunting, these cysts can sometimes be treated effectively by fenestration of their walls into the surrounding normal CSF spaces (Baskin and Wilson, 1984; Raffel and McComb, 1994). Some of the cysts can be surgically excised, but the location of others results in a high rate of complication with attempted removal (Rappaport, 1993).

## NEUROCUTANEOUS DISORDERS

The skin, eye, and central nervous system share a common neuroectodermal origin. The neurocutaneous disorders or *phakomatoses* (Gr. phakoma = birth spot or mother spot) are a complex group of conditions affecting two or three of these neuroectodermal derivatives simultaneously and other organs as well. Most are genetic. Systemic manifestations of these disorders commonly include skeletal overgrowth, hamartomatous tumors (such as fibromas and myomas), and vascular malformations (particularly angiomas) that can involve the skin, eye, or CNS. The degree to which different tissues are involved is extremely variable. Therapies differ depending on the symptoms that are present or the ones that can be anticipated and may include excision of tumors or angiomas that are causing seizures and shunting of CSF for treatment of accompanying hydrocephalus. Genetic linkage studies have been useful in diagnosing these disorders in more mildly affected patients in whom the diagnosis might otherwise not have been appreciated. Table II summarizes the known patterns of inheritance in different neurocutaneous disorders.

### Neurofibromatosis (NF)

*Neurofibromatosis* refers mainly to two different disorders, both of autosomal dominant inheritance, NF 1 and NF 2. Each has a broad range of clinical expression. Following the recommendations of the National Institutes of Health Consensus Development Conference (1988), classical neurofibromatosis (von Recklinghausen disease) has been referred to as NF 1, whereas neurofibromatosis restricted to the CNS, with accompanying bilateral VIII nerve tumors, is referred to as NF 2. Other variants described, accounting for less than 1% of all cases of NF (Pont and Elster, 1992), are referred to as NF 3–8 (Mackool and Fitzpatrick, 1992). NF 1 is the most frequent form, with a prevalence of 1 in 3000 to 1 in 4000 (Braffman and Naidich, 1994a; Gomez, 1991; Listernick and Charrow, 1990), accounting for 90% of all NF cases. NF1 has a penetrance of almost

**TABLE II**  Neurocutaneous Disorders

Autosomal dominant inheritance
  *Neurofibromatosis*
  *Tuberous sclerosis*
  *v. Hippel-Lindau disease*
  *Nevoid basal cell carcinoma*
  *syndrome (Gorlin-Goltz syndrome)*
  [Hypomelanosis Ito]
Autosomal recessive inheritance
  Ataxia teleangiectatica
  Louis-Bar*
  [Xeroderma pigmentosum]
  Cockayne syndrome*
  [Fucosidose]
  [Phenylketonuria]
  [Homozystinuria]
  [Argininosuccinaciduria and Citrullämia]
  [Carboxylase-deficency]
  [Neuroichthyosis]
  Sjogren-Larsson syndrome*
  Refsum's disease*
  Giant-Axon neuropathy*
  [Werner Syndrome]
  [Progeria]
  Familiar dysautonomia*
  [Chediak-Higashi disease]
X-chromosomal inheritance
  Fabry's disease*
  [Kinky hair disease]
  [Incontinentia pigmenti]
Sporadic
  [Neurocutaneous melanosis]
  *Sturge-Weber syndrome*
  Klippel-Trenaunay syndrome*

Syndromes that are indicated with an asterisk (*) are described in other chapters, rare disorders, indicated with [ ] should be further studied in the specialized literature of neuropediatrics. Disorders described in the text appear in italics.
  Modified according to Gomez (1991).

100% and variable expressivity; its mutation rate of 50% (Aoki *et al.*, 1989) is one of the highest among all autosomal dominant diseases. The responsible gene has been mapped to band 11.2 of the long arm of chromosome 17 (Barker *et al.*, 1987), with the protein normally encoded by this gene called neurofibromin. Neurofibromin has sequence similarities to a family of proteins called guanidine triphosphatase–activating proteins (GAPs), which are involved in the inactivation of the proto-oncogene *ras*. *Ras* proteins are important for a variety of cellular processes, including cell proliferation and differentiation, and tumor formation. Neurofibromin acts as a tumor-suppressor gene by inhibiting *Ras* activity. Biochemical analysis of tumors in NF 1 gave evidence of hyperactive Ras, which may be a target for future therapies (Weiss *et al.*, 1999). Neurofibromin is expressed at different levels in several organs, which may explain the heterogeneity of the pathologic condition in NF 1. The clinical variability may also be contributed to by secondary somatic mutations (Gutmann and Collins, 1993).

NF 2 is an autosomal dominant disorder that has been mapped to the 11.2 band of the long arm of chromosome 22 (Aicardi, 1992b). It affects about 1 in 50,000 individuals and is typically manifested at a later age than NF 1 (Mulvihill *et al.*, 1990), usually during or soon after puberty (Elster, 1992). NF 2 patients typically have bilateral vestibular schwannomas develop, which are sometimes associated with schwannomas at other locations, meningeomas, ependymomas, or juvenile subcapsular lenticular opacities. The disease is caused by inactivating mutations in a tumor suppressor gene on 22q12, which encodes Merlin, a member of the protein 4.1 superfamily (Rouleau *et al.*, 1993), which regulates differentiation and proliferation in various cell types (McCartney *et al.*, 2000).

## Clinical Aspects

The initial diagnosis of NF 1 is usually made by detection of multiple light brown, typically oval, skin lesions or café-au-lait spots (present in 99% of patients, but often not seen until infancy or later childhood), axillary

---

**TABLE III     Diagnostic Criteria of NF 1 and 2**

NF-1

Two or more of the following:
1. Six or more café-au-lait macules more than 5 mm in greatest diameter in prepubertal individuals and more than 15 mm in greatest diameter in postpubertal individuals
2. Two or more neurofibromas of any type or one plexiforme neurifibroma
3. Freckling in the axillary or inguinal regions
4. Optic glioma
5. Two or more Lisch nodules (iris-hamartomas)
6. A distinctive osseous lesion such as sphenoid dysplasia or thinning of long bone cortex with or without pseudoarthrosis
7. A first-degree relative (parent, sibling, or offspring) with NF-1 by the above criteria

NF-2

Either of the following
1. Bilateral eight nerve masses seen with appropriate imaging techniques (e.g., CT or MRI)
   or
2. A first-degree relative with NF-2 and either a unilateral eighth nerve mass,
   or
   two of the following:
   — Neurofibroma
   — Meningeoma
   — Glioma
   — Schwannoma
   — Juvenile posterior subcapsular lenticular opacity

---

or inguinal freckles (usually very numerous), and cutaneous or subcutaneous neurofibromas. The 1988 NIH criteria for the diagnosis of NF 1 are listed in Table III.

### Intracranial and Intraspinal Tumors

About 5%–10% of NF 1 patients eventually develop intracranial tumors. The most common of these are optic pathway gliomas (OPT; optic pathway tumor). OPTs occur in 5%–15% of children with NF 1 (Braffman and Naidich, 1994a; Gomez, 1991) and are one of the most frequent tumors in the first decade of life. Of all patients presenting with an OPT, about 25% have NF 1 (Pont and Elster, 1992). Gliomas are also found in NF 1 in the brainstem and hypothalamus (Braffman and Naidich, 1994a). Neurofibromas, and less often, schwannomas, can involve spinal nerve roots in NF 1. When spinal nerve roots are involved it is typically the dorsal more often than the ventral root. Optic gliomas are not found in NF 2, but other intracranial tumors (meningiomas and schwannomas) and spinal tumors (schwannomas and ependymomas) are seen. The most characteristic intracranial tumor in NF 2, however, is the acoustic schwannoma, present in at least 90% of cases, almost always bilaterally. These tumors are rarely found in NF 1. Cutaneous manifestations are uncommon in NF 2, with few or no café-au-lait spots found in most patients (Aicardi, 1992b).

### Other Tumor Manifestations in NF

Adolescents with NF also have an increased frequency of pheochromocytomas (0.5%–1%), ganglioneuromas, carotid glomus tumors, rhabdomyosarcomas, Wilms' tumors and leukemia. In about 5% of patients with NF 1, a neurofibroma will undergo malignant transformation to a fibrosarcoma. This usually occurs after the age of 10 years and occurs most often with plexiform neurofibromas, the most common extracranial tumor in NF 1 (Pont and Elster, 1992). Plexiform neuromas are locally aggressive tumors with a tendency to centripetal growth (Braffman and Naidich, 1994a). Plexiform neurofibromas can undergo malignant transformation to one of the clinically most aggressive cancers associated with neurofibromatosis 1 (NF1), the malignant peripheral nerve sheath tumor (MPNST) (King *et al.*, 2000). The development of MPNST occurs in about 5% of identified plexiform neurofibromas. Thus when a patient with a plexiform neurofibroma notes an acute change in that tumor or new pain associated with it, a concerted effort to identify a malignant transformation is recommended. This includes MRI of the plexiform tumor, CT of the chest, and referral to a surgical oncologist (Gutmann, 1998). By contrast, cutaneous neurofibromas are not thought to undergo malignant transformation.

### Other Neurological Symptoms

Frequent brain-based neurological findings in NF 1 include mental retardation (8%) and learning problems (40%). There is an increased frequency of macrocrania (15%–45%) in NF 1 (Aicardi, 1992b), usually secondary to macrencephaly (large brain) or less often from hydrocephalia (usually from aqueductal stenosis). The brains of such children, in addition to being large, often show abnormalities in gyral formation and cortical architecture, which can explain the cognitive and learning deficits that are frequently found (Rosman and Pearce, 1975). Glial nodules are often found in the cerebral cortex of NF 1 patients and may be important contributors to the increased frequency of seizures seen in NF 1. MRI scans of patients with NF 1 show small punctate or confluent foci of increased signal intensity within brain parenchyma on T2-weighted images, especially in cerebellum, brainstem, globus pallidus, thalamus, and supratentorial white matter. These foci, probable hamartomas, occur in 60%–80% of patients (Duffner et al., 1989; Elster, 1992). They are more frequent in younger NF 1 patients, resolving with increasing age, and are rarely seen after age 20 (Aoki et al., 1989). Their precise significance is unknown, but the degree of their distribution (i.e., widespread versus focal) correlates with measured intelligence in children with NF 1. A more widespread distribution correlates with lower measured intelligence (Denckla et al., 1996).

### Anomalies of the Skeleton

About 0.5%–1% of all NF 1 patients have congenital pseudoarthroses (usually of the tibia or radius). In about 2% of NF 1 patients, scoliosis or kyphoscoliosis develops, usually in the second decade of life. This occurs most often in the lower cervical and upper thoracic regions. Cardiorespiratory compromise may result. Underdevelopment of the greater wing of the sphenoid bone, which forms the posterior wall of the orbit, is not uncommonly seen in NF 1 and causes a pulsatile exophthalmos.

### Malformations of the Vasculature

Vaso-occlusive disease can be seen in NF, with the abdominal aorta, and the iliac, mesenteric, renal, or cerebral arteries beeing common sites for such involvement. A Moya-Moya vascular syndrome can develop with or without a history of radiation therapy. Arterial hypertension is found in about 1% of NF patients, most often in adolescence; this is usually caused by renal artery stenosis or by a pheochromocytoma. Cardiovascular malformations are found in 2.3% of NF patients (Lin et al., 2000).

### Natural Course

Except for optic glioma, which usually is seen in the first decade of life, most of the clinical signs of NF develop after puberty. As indicated earlier, these signs are extremely variable, because they depend on the nature and location of the causative pathosis. Overall, life expectancy in NF is somewhat shortened.

### Principles of Therapy

The treatment of NF is entirely symptomatic. Anticipation of the wide array of potential problems is a necessity. Surgical treatment may be cosmetic or may be needed to remove tumors from one or more body sites. Radiation of the tumors is usually without benefit in NF 1 (optic glioma is a possible exception) and should generally be avoided, because the consequences of encephalopathies and other complications of radiation can be more severe than those of the underlying lesions (Gomez, 1991). The biological behavior of OPTs in NF 1 varies with their location and age of presentation. Hence the management of children with OPT and NF 1 depends on the location of the tumor and age of the child at presentation. Tumors anterior to the chiasm frequently remain stable for many years, whereas more posteriorly located gliomas tend to be invasive. OPT, which is seen before the age 6, is more likely to grow rapidly and progress (Schroder et al., 1999). Several authors have cautioned, however, that spontaneous improvement/regression of OPTs occurs, also more commonly in younger patients (Perilongo et al., 1999; Rossi et al., 1999). In general, surgery has played a limited role in the management of OPT, reserved for the most aggressive or poorly located. Until recently, the most experience has been with the use of radiation therapy localized to the tumor. However, the use of radiation therapy in younger children (<6 years old) is highly associated with endocrinological, cerebrovascular (e.g., Moya-Moya syndrome) and neuropsychological adverse sequelae (Cappeli et al., 1998). Recently, the usefulness of chemotherapeutics, carboplatinum and vincristine, has been advocated, particularly in younger children (Listernick et al., 1999; Silva et al., 2000). At this time, a reasonable approach to OPT in children with NF 1 is to document progression before intervention. If intervention is warranted, chemotherapy should be considered, particularly for younger children. Surgery and radiation therapy should be limited to selected cases. Therefore, depending on a variety of circumstances, treatment of OPT in NF 1 should include regular ophthalmological and neuroradiological (MRI) follow-up with the possiblity of no intervention, chemotherapy, radiation therapy, surgery, or some combination of these modalities (Aicardi, 1992a; Gutmann, 1998).

Shunting is frequently needed for hydrocephalus caused by third ventricular obstruction (from an optic glioma) or from aqueductal stenosis. Acoustic neuromas should be surgically excised, which is accomplishable in most cases. When excision is not possible, radiation therapy can be considered, particularly in elderly patients, in others at high surgical risk, and in those who have been operated on one side with development of a contralateral tumor with hearing that is still satisfactory on that side (Flickinger *et al.*, 1993; Maire *et al.*, 1992). Table IV provides an overview of the major craniocerebral manifestations of NF and therapeutic options to be considered. Special educational assistance and behavior modification should be provided to children with NF who have learning and behavioral problems.

Recently, application of vitamin D$_3$ analogues was found effective in improving the pigmentation of the café au lait spots in NF patients (Nakayama *et al.*, 1999).

## Tuberous Sclerosis (TS)

*Tuberous sclerosis* (eponym: *Bourneville-Pringle disease*) is another neurocutaneous disorder of autosomal dominant inheritance. Multiple organ systems are affected with hamartomas seen in brain, retina, skin,

heart, kidneys, and lungs. Other organs can be involved as well, although less often. The disease is of dominant inheritance, but 60% of cases seem to arise by spontaneous mutation (Smirniotopoulos and Murphy, 1992). The prevalence varies from 1 in 10,000 to 1 in 40,000, figures that are probable underestimates because of the variability of symptoms, which are quite minor at times, and the frequency of entirely asymptomatic patients (Ahlsen *et al.*, 1994). Gene loci have been identified on chromosome 9 (Gomez, 1991) and chromosome 16 (Nellist *et al.*, 1993), which encode the tuberous sclerosis complex (TSC) genes TSC1 (hamartin) and TSC2 (tuberin), respectively (Crino and Henske, 1999). Both genes are expressed widely in the brain and interact in signalling pathways that regulate cellular differentiation and tumor suppression (Miloloza *et al.*, 2000). About 50% of the TSC families show linkage to TSC1 and 50% to TSC2.

### Clinical Aspects

Tuberous sclerosis is classically characterized by the clinical triad of mental retardation, epilepsy, and so-called adenoma sebaceum (facial angiofibroma), with onset before the age of 10. Many patients, rather than showing this classical triad, have incomplete forms. The most common skin lesions in TS are hypopigmented macules (present in 90%; seen best under ultraviolet light), facial angiofibromas (seen in about 50%, usually after 5 years of age), and fibrous plaques (usually found on the forehead or scalp, and sometimes at birth as the earliest diagnostic sign of TS) (Gomez, 1991). CNS

TABLE IV    Craniocerebral Manifestations of NF, Clinical Relevance, and Therapeutic Options

| Type of manifestation | Preferred location | Clinical aspects | Principles of therapy |
|---|---|---|---|
| 1. Tumors | | | |
| Neurofibromas, neurilemmomas | Cranial nerves VIII, V, IX–XII | Often bilateral, VIIIth nerve symptoms | Excision, stereotactic radiation |
| Meningiomas | Convexity, falx cerebri | Multiple tumors occur, recurrent manifestation | Surgical excision |
| Vascular tumors of brain | Cerebellum | | Surgical excision |
| Astrocytomas | Optic pathways | Early manifestation of the disease | Chemotherapy radiation therapy (surgical excision) |
| Diffuse gliomatosis | Cerebrum, brainstem | Rare manifestation, no clinical signs | No treatment required, radiation without influence on natural course |
| Hamartomas | Hypothalamus | Endocrine dysfunction | Surgical excision, if possible |
| 2. Malformations | | | |
| Migrational disorders | Cerebrum | Epilepsy, mental retardation | Treatment of seizures |
| Stenosis of the aqueductus cerebri | Aqueductus cerebri | Occlusive hydrocephalus | V-P shunting |
| Sceletal features | Pseudoarthrosis, absence of the greater sphenoidal wing | Proptosis and pulsating exophthalmus | Plastic surgery |
| Vascular malformations | Aorta, iliac, mesenteric, renal and cerebral arteries | Secondary hypertension | Symptomatic |

involvement includes three types of pathoses, cortical tubers, subependymal nodules, and cerebral white matter migration lines. Cortical tubers are hard nodules with increased astrocytes, decreased neurons, giant cells, and disturbed lamination. By radiological and histopathological methods, they may be indistinguishable from focal cortical dysplasia. Tubers are most commonly located in the cerebrum but can be found throughout the neuraxis. Subependymal astroglial nodules are frequently calcified, contrast-enhancing, and are found on MRI or CCT of the brain. They can transform to become subependymal giant cell astrocytomas. When located at the foramina of Monro (seen in 2%–26% of cases) or the third ventricle, they can obstruct CSF flow and cause hydrocephalus; rarely, they undergo malignant transformation. Cerebral white matter migration lines are seen on MRI as white matter lesions containing clusters of heterotopic giant cells that produce an abnormal signal nearly identical to that produced by the cortical tubers (Braffman and Naidich, 1994a). They are likely indicative of aberrant neuronal migration contributing to the pathogenesis of TSC. One or more of the preceding neuroradiological findings is found in at least 90% of patients with TS.

Almost half the patients have retinal or optic nerve hamartomas. One quarter have periungual or subungual fibromas develop (more frequent on the toes and in females, usually appearing after puberty). Grouped papules or shagreen patches (connective tissue naevus) usually occur on the dorsal trunk (seen in 20%) (Gomez, 1991).

Despite the advances of molecular genetics of TSC, the diagnosis remains a clinical one.

A National Institutes of Health Consensus Conference has recently updated the diagnostic criteria for tuberous sclerosis complex (Table V; adapted from Roach et al., 1998). Note that these new diagnostic criteria do not include genetic or familial criteria. The NIH consensus paper also includes recommendations for diagnostic and surveillance screening for TSC in both children and adults (Hyman and Whittemore, 2000) (Table VI). MRI is the brain imaging modality of choice for a suspected case or the initial diagnostic evaluation, although it is less sensitive for demonstrating subependymal nodules compared with CCT. CCT is recommended for screening an "asymptomatic" parent, child, or first-degree relative at the time of diagnosis of an affected individual as long as the physical examination is negative. Follow-up imaging in children should be on the order of every 1–3 years, but modified per clinical scenario (e.g., known nodule at foramen of Monro). In adults surveillance neuroimaging can be less frequent than in children. Neurodevelopmental testing should be performed in children as part of the initial

**TABLE V    Current Recommended Diagnostic Criteria for Tuberous Sclerosis Complex**

Major features
  Facial angiofibroma ("adenoma sebaceum") or forehead plaque
  Nontraumatic ungual or periungual fibroma
  Hypomelanotic macules (>3)
  Shagreen patch (nevus)
  Multiple retinal nodule hamartomas
  Cortical tuber*
  Subependymal nodule
  Subependymal giant cell astrocytoma
  Cardiac rhabdoyoma
  Lymphangiomyomatosis†
  Renal angiomyolipoma†
Minor features
  Multiple randomly distributed dental enamel pits
  Hamartomatous rectal polyps (histologic confirmation suggested)
  Bone cysts (radiographic confirmation sufficient)
  Cerebral white matter migration lines*
  Gingival fibroma
  Nonrenal hamartoma (histological confirmation suggested)
  Multiple renal cysts (histological confirmation suggested)
  Retinal achromic patch
  "Confetti" skin lesions

Definite TSC: either 2 major or 1 major and 2 minor features.
Probable TSC: 1 major and 1 minor feature.
Possible TSC: either 1 major or >1 minor features.
*When cerebral cortical dysplasia and cerebral white matter migration tracts occur together, they should be counted as one rather than two features.
†When both lymphangioleiomyomatosis and renal angiomyolipomas are present, other features should be present before definite TSC can be diagnosed.
Based on NIH Consensus, adapted from Roach et al., [1998], and published with permission.

evaluation and before primary school matriculation. An EEG should be performed when clinically driven and not as part of the initial evaluation. Funduscopic examination should be performed as part of the initial evaluation and then as indicated clinically (Hyman and Whittemore, 2000) (Table VI).

Seizures develop in 80%–90% of patients with TS and are the presenting symptom in as many as 75%. They usually begin in the first year of life and may be partial or generalized; common patterns are tonic/clonic seizures, atypical absences, and especially infantile spasms. Mental retardation is seen in about 50% of patients with TS and is closely linked with the occurrence of seizures, particularly when seizures are of early onset. TS patients without seizures, or those in whom seizures develop late, usually have normal intelligence (Wiss, 1992). Most patients with TS have behavioral/psychiatric problems, and some become frankly autistic (Ahlsen et al., 1994; Curatolo et al., 1991).

TABLE VI  Diagnostic and Surveillance Screening in Tuberous Sclerosis Complex

| | "Asymptomatic" parent, child, or first-degree relative at time of diagnosis of affected individual | Suspected case or initial diagnostic evaluation | Known case and no symptoms in referable organ | | Known case and symptoms or findings previously documented | |
|---|---|---|---|---|---|---|
| | | | Child | Adult | Child | Adult |
| Fundoscopic examination | + | + | − | − | + | +[e] |
| Brain MRI | +[a] | + | +[b] | +[c] | + | +[d] |
| Brain EEG | − | −[e] | − | − | + | +[d] |
| Cardiac ECG, Echo | −[f] | − | − | − | +[d] | +[d] |
| Renal CT, MRI, US | +[b] | + | +[i] | +[b] | +[g] | +[d] |
| Dermatological screen | + | + | − | − | + | +[g] |
| Neurodevelopment | − | +[j] | +[k] | − | +[d] | +[d] |
| Pulmonary CT | − | − | − | +[l] | +[d] | +[d] |

[a]With negative physical examination results, CT is recommended.
[b]Every 1 to 3 years.
[c]Probably less frequently than in children.
[d]As clinically indicated.
[e]Unless seizures are suspected, this is generally not useful for diagnosis.
[f]Unless needed for diagnosis.
[g]Every 6 months to 1 year until involution or size stabilization occurs.
[h]Ultrasound is generally recommended because of cost, although local imaging expertise may vary.
[i]Every 3 years until adolescence.
[j]Generally for children only.
[k]Recommended for children at the time of beginning first grade.
[l]For women at age 18 years.
Modified from, and published with permission, Hyman, M.H., Whittemore, V.H., Archives of Neurology. Volume 57, page 664. Copyright [2000], American Medical Association.

Systemic involvement in TS beyond that listed in the diagnostic criteria include periosteal new bone formation, pulmonary cysts and fibrosis, and splenic cysts (Gomez, 1991). Cardiac rhabdomyomas and renal angiomyolipomas are quite common, occuring in nearly half of TSC patients (Gomez, 1991).

## Natural Course

The neurocutaneous manifestations of TS are usually recognized early in childhood. Although the life expectancy in TS is shortened, patients with no or minimal symptoms may have a normal life span. The clinical manifestations are protean, reflecting the multisystem involvement.

## Principles of Therapy

Cognitive development is best in patients with TS who are treated early and effectively (Gomez, 1985). Infantile spasms (IS) are typically well managed with adrenocorticotrapic homone (ACTH), with sodium valproate, nitrazepam, and clonazepam as acceptable second-line drugs. Recently, vigabatrin has been successfully used in several studies with TSC patients with complete cessation of the seizures in 95% and complete cessation of the infantile spasms in 54% (Hancock and Osborne, 1999), suggesting that vigabatrin can be considered as an alternate first-line monotherapy of infantile spasms. Of course, the risk of irreversible peripheral retinal injury must be considered before initiating vigabitrin. At present, a prospective head-to-head study of vigabatrin and ACTH for IS in TSC is not available. However, such studies do exist for IS per se (reviewed in Wong and Trevathan, 2001). In general, vigabatrin and ACTH likely have comparable efficacy for the treatement of IS, with better tolerance reported for vigabatrin (Wong and Trevathan, 2001).

Other seizure types are treated as outlined in Chapter 20. Excision of one or more cortical tubers may be indicated if they serve as seizure foci in patients in whom medical treatment has been ineffective (Koh et al., 2000). Obstructive hydrocephalus is usually treated with shunting. Occasionally, a giant cell astrocytoma needs to be surgically resected from the region of a foramen of Monro. Facial angiofibromata can be treated with laser surgery.

A point of controversy at present is the putative risk of whole cell pertussis vaccine facilitating onset of infantile spasms in susceptible TSC patients. It is probably reasonable to withold the whole cell pertussis vaccine in susceptible patients at this time. Future work should be directed at evaluating the safety of the acellular pertussis vaccine.

Referral to a genetic counsellor is recommended for parents who wish to have more children and for individuals with TSC who are of child-bearing age.

### Sturge-Weber Syndrome (SWS, Encephalo-Trigeminal-Angiomatosis)

#### Clinical Aspects

Sturge-Weber syndrome (SWS) is a congenital malformation of the cranial vasculature. It is of sporadic occurrence. The prevalence is unknown but seems to be less common than NF or TS. The causative gene has not been found. In its complete form, there is a capillary/venous angioma of the face (nevus flammeus), most characteristically seen over the forehead and upper eyelid (the area supplied by first division of 5th cranial nerve, V1), with venous angiomatous involvement of the underlying leptomeninges (usually overlying the ipsilateral occipital lobe [Alexander and Norman, 1972]) and of the choroid in the ipsilateral eye. These findings can be unilateral or bilateral. Ischemic/hypoxic injury of the underlying brain (caused by stasis of blood in the overlying angioma and from impaired venous drainage from the cerebral cortex) frequently results in secondary calcifications, often paralleling pial blood vessels ("railroad track" appearance on cranial x-ray film). Also, abnormalities in gyral formation and cortical architecture can be seen (Wohlwill and Yakovlev, 1957). Reduced size of the underlying cerebral hemisphere can sometimes be seen on CCT or MRI soon after birth. Seizures, present in up to 90% of patients, are usually the first symptom of SWS and typically begin in early infancy. A hemiparesis contralateral to the cutaneous angioma is seen in up to two thirds of patients, a homonymous hemianopia is seen in almost all (Braffman and Naidich, 1994b). Mental retardation occurs in 60% (Pascual-Castroviejo et al., 1993), particularly in those with seizures and bihemispheric disease (Gomez, 1991). By contrast, normal intelligence occurs when seizures are absent, even with bihemispheric lesions (Gomez and Bebin, 1987). The clinical diagnosis is usually not difficult, especially if typical signs such as a unilateral facial angioma, seizures, developmental delay, and a contralateral hemiparesis are found. Radiologically, the intracranial angiomas are best seen with contrast-enhanced MRI (which shows enhancing thickened leptomeninges) or by MRI angiography (which shows a pial blush overlying the affected hemisphere[s]), but CCT is better than MRI for demon-strating underlying cerebral calcifications (Marti-Bonmati et al., 1992). Angiographic study shows impaired superficial venous drainage and shunting of blood to the deep venous system, with development of abnormally dilated deep veins (Benedikt et al., 1993; Braffman and Naidich, 1994b; Vogl et al., 1993).

#### Differential Diagnosis

The primary differential diagnosis of the facial angioma seen in SWS is the syndrome of Klippel-Trenaunay, which includes facial/somatic hemihypertrophy, facial/somatic cutaneous nevi, seizures, and intracranial calcifications. In addition, celiac disease, Dandy-Walker syndrome, and hereditary neurocutaneous angiomatosis should be considered.

#### Natural Course

The prognosis of Sturge-Weber syndrome depends on the location and extent of the cerebral and ocular lesions. The neurological deterioration in SWS has been described as "saltatory," indicating abrupt changes in function with intervening periods of less-dramatic change. Patients at risk for neuro-ocular symptoms are only those in whom the cutaneous angioma involves the area supplied by V1 (Enjolras et al., 1985). The seizures are often partial, sometimes with secondary generalization, with a postictal paralysis (Todd) frequently found. The intracranial vascular malformations rarely bleed. With angiomatous involvement of the choroid, glaucoma is a common accompaniment. When this occurs in prenatal life, the child is typically born with congenital glaucoma with an accompanying large eye (buphthalmos). Angiomas can involve other organs as well. A unilateral facial angioma does not rule out bilateral cerebral involvement. Ten to 15% percent of patients with SWS do not have a facial angioma (Gomez and Bebin, 1987).

#### Principles of Therapy

Cerebral lesions, particularly when unilateral and accompanied by medically intractable seizures, can be treated surgically early in life (Gomez and Bebin, 1987). Corticectomy, lobectomy, or even hemispherectomy has been recommended. The greatest success is seen with operation on young patients in whom there has been no neurological deterioration. The cutaneous nevi can be treated with laser therapy with a good cosmetic result. Glaucoma can be treated medically and surgically. Enteric-coated aspirin at 3–5 mg/kg/day might be useful in patients with symptoms consistent with transient ischemic events. (Practitioners should follow guidelines comparable to those proposed for the treatment of children with Kawasaki disease.) Genetic counseling should address the sporadic nature of the syndrome suggesting <5% chance of recurrence.

## Von Hippel-Lindau Disease (VHL)

### Clinical Aspects and Natural Course

*Von Hippel-Lindau* (VHL) disease is a disorder of autosomal dominant inheritance in which several organs of mesenchymal origin are affected. VHL patients typically have hemangioblastomas (vascular lesions with tumor cells) in the brain, particularly in the cerebellum (found in 35%–60% of patients [Hubschmann et al., 1981]); less often, hemangioblastomas are found in the medulla and in the retina. These patients also have tumor cysts (Lindau cysts) in other organs (particularly kidney, pancreas, and epididymis). They may also have renal cell carcinoma develop. Because 75% of the CNS hemangioblastomas are located in the posterior fossa, such patients usually are seen with headache, vomiting, and gait and limb ataxia, usually in the second to fourth decade (Braffman and Naidich, 1994b; Gomez, 1991) and rarely in the pediatric setting. Primary spinal involvement is rare, but spinal hemangioblastomas can be seen in as many as one quarter of the patients with brain involvement (Neumann et al., 1992). Once a cerebellar hemangioblastoma is found, it is important to search for these tumors outside the CNS, particularly in the retina, where hemangioblastomas are present in one half to two thirds of patients with VHL disease (Hardwig and Robertson, 1984). They are multiple in one third to two thirds of patients and bilateral in 20%–50% (Huson et al., 1986). The diagnosis can be made under any one of the following three circumstances: (1) CNS and retinal hemangioblastomas; (2) CNS or retinal hemangioblastoma plus one of the following: renal, pancreatic, hepatic, or epididymal cysts; pheochromocytoma; renal cancer; (3) a definite family history of VHL plus one of the following: CNS or retinal hemangioblastoma; renal, pancreatic, hepatic, or epididymal cysts; pheochromocytoma; renal cancer (Michels, 1987). The earliest clinical evidence of VHL disease is often the finding of a retinal hemangioma, most frequently first seen in the third decade but sometimes as early as the first. The retinal lesions are often multiple. As they grow, the retina may detach. Ten to 20% of patients with retinal hemangioblastomas also have an intracranial tumor, and thus by definition have VHL disease (Gomez, 1991). Renal cell carcinoma (hypernephroma) occurs in 25%–40% of patients with VHL (Horton et al., 1976), pheochromocytoma in about 10% (Hubschmann et al., 1981). Polycythemia is often found with cerebellar hemangioblastoma, and hematuria occurs in VHL patients with renal tumors. Polycythemia is likely due to excessive renal production of erythropoetin.

The diagnosis of VHL disease is further supported by a positive family history. The incidence of the disorder is 1 in 50,000 (Neumann et al., 1992). The mean life expectancy is about 50 years, with deaths from renal cell carcinoma or from cerebellar hemangioblastoma resulting in this reduced life expectancy. The gene locus for VHL has been mapped to chromosome 3p25–p26. The protein product is thought to have tumor suppressor function (Latif et al., 1993). The VHL tumor suppressor gene has been demonstrated to be required for cell cycle exit (Pause et al., 1998). Mutations in the gene result in constitutive up-regulation/expression of hypoxia-inducible genes like HIF1α, which in turn may be rseponsible for vascular tumors (Ohh et al., 2000).

### Principles of Treatment

Surgical removal of cerebellar hemangioblastoma is the treatment of choice in VHL patients. Radiotherapy is indicated in only those patients who have inoperable posterior fossa tumors (e.g., of the medulla) with progressing neurological signs. Stereotactic radiation may have a role to play. In 20% of patients, hemangioblastomas recur after initial surgery. Retinal angiomas are usually treated by cryosurgery or photocoagulation (Huson et al., 1986). VHL patients and others at risk for that disorder should be examined regularly, looking for (recurrent) hemangioblastomas and other evidence of the disease. When operable spinal tumors are found, they should be promptly excised, particularly if clinical symptoms are present. Therapeutic phlebotomy might be necessary in some cases of polycythemia.

## Nevoid Basal Cell Carcinoma Syndrome (Gorlin-Goltz Syndrome)

### Clinical Aspects and Natural Course

The nevoid basal cell carcinoma syndrome, or *Gorlin-Goltz or Gorlin syndrome*, is of autosomal dominant inheritance. It is characterized by multiple basal cell carcinomas of the face, neck, and upper trunk that develop between puberty and age 35. The name nevoid basal cell carcinoma syndrome is misleading, however, because only 50% of these patients 20 years or older manifest the basal cell carcinomas, and it is only the rare lesion that becomes aggressive. Odontogenic keratocysts are characteristic of the disorder; they are seen late in the first decade and peak during the second or third decade (Gorlin, 1987). They are rarely symptomatic unless they become infected, although they can cause pathological fractures of the mandible or maxilla. Migrational abnormalities in the cerebrum occur, and seizures are frequent. Associated findings include skeletal abnormalities (such as frontal and parietal bossing, platybasia, and kyphoscoliosis); brain tumors (such as cerebellar medulloblastomas); other brain abnormalities (such as hydrocephalus); extraneural tumors (such as ovarian sarcomas); and endocrine abnormalities (such as hypogonadotrophic hypogonadism) (Gomez, 1991). Life expectancy is shortened. The pathogenesis of the disorder is unknown.

### Principles of Treatment

The clinical course in nevoid basal cell carcinoma is extremely variable. Complications, such as hydrocephalus, should be treated actively. If the cysts found in this disorder are infected, antibiotics should be given. Larger cysts may require surgical removal. If there is an accompanying medulloblastoma, it should be treated surgically rather than with radiation, because the latter predisposes to development of new skin lesions or malignant transformation of those already present (Gomez, 1991).

## REFERENCES

Ahlsen, G., Gillberg, I. C., Lindblom, R., and Gillberg, C. (1994). Tuberous sclerosis in Western Sweden. *Arch. Neurol.* **51**, 76–81.

Aicardi, J. (1992a). Malformations of the CNS. *In* "Diseases of the Nervous System in Childhood, Clinics in Developmental Medicine," No. 115/118, pp. 108–202. MacKeith Press, London.

Aicardi, J. (1992b). Neurocutaneous diseases and syndromes. *In* "Diseases of the Nervous System in Childhood, Clinics in Developmental Medicine," No. 115/118, pp. 203–39. MacKeith Press, London.

Aicardi, J. (1992c). Tumours of the CNS and other space-occupying lesions. *In* "Diseases of the Nervous System in Childhood, Clinics in Developmental Medicine," No. 115/118, pp. 780–849. MacKeith Press, London.

Alexander, G. L., and Norman, R. M. (1972). Sturge-Weber syndrome. *In* "Handbook of Clinical Neurology," (P. J. Vinken and G. W. Bruyn, Eds.), Vol. 14, pp. 223–240. North-Holland Publishing Company, Amsterdam.

Aoki, S., Barkovich, A. J., Nishimura, K., Kjos, B. O., Machida, T., Cogan, P., Edwards, M., and Norman, G.(1989). Neurofibromatosis types 1 and 2: Cranial MR findings. *Radiology* **172**, 527–534.

Barker, D., Wright, E., Nguyen, K., Cannon, L., Fain, P., Goldgar, D., Bishop, D. T., Carey, J., Baty, B., Kivlin, J., Willard, H., Waye, J. S., Creig, G., Leinwand, L., Nakamura, Y., O'Connell, P., Leppert, M., Lalouel, J.-M., White, R., and Scolnick, M. (1987). Gene for von Recklinghausen neurofibromatosis in pericentrometric region of chromosome 17. *Science* **236**, 1100–1102.

Barkovich, A. J. (1995). "Pediatric Neuroimaging," Second edition. Raven Press, New York.

Barkovich, A. J., and Kjos, B. O. (1992). Schizencephaly: Correlation of clinical findings with MR characteristics. *AJNR* **13**, 85–94.

Barkovich, A. J., and Norman, D. (1988). Absence of the septum pellucidum: A useful sign in the diagnosis of congenital brain malformations. *AJNR* **9**, 1107–1114.

Barth, P. G. (1992). Migrational disorders of the brain. *Curr. Opin. Neurol. Neurosurg.* **5**, 339–343.

Baskin, D. S., and Wilson, C. B. (1984). Transsphenoidal treatment of non-neoplastic intrasellar cysts. A report of 39 cases: *J. Neurosurg.* **60**, 8–13.

Benedikt, R. A., Brown, D. C., Walker, R., Ghaed, V. N., Mitchell, M., and Geyer, C. A. (1993). Sturge-Weber syndrome: Cranial MR imaging with Gd-DTPA. *AJNR* **14**, 409–415.

Bhadelia, R. A., Bogdan, A. R., Wolpert, S. M., Lev, S., Appignani, B. A., and Heilman, C. B. (1995). Cerebrospinal fluid flow waveforms: analysis in patients with Chiari I malformation by means of gated phase-contrast MR imaging velocity measurements. *Radiology* **196**, 195–202.

Bishop, K. M., Goudreau, G., and O'Leary, D. D. (2000). Regulation of area identity in the mammalian neocortex by Emx2 and Pax6. *Science* **288**, 344–349.

Braffman, B., and Naidich, T. P. (1994a). The phakomatoses: Part I. Neurofibromatosis and tuberous sclerosis. *Neuroimaging Clin. North Am.* **4**, 299–324.

Braffman, B., and Naidich, T. P. (1994b). The phakomatoses: Part II. von Hippel-Lindau disease, Sturge-Weber syndrome, and less common conditions. *Neuroimaging Clin. North Am.* **4**, 325–348.

Brown, M. S., and Sheridan-Pereira, M. (1992). Outlook for the child with a cephalocele. *Pediatrics* **90**, 914–919.

Brunelli, S., Faiella, A., Capra, V., Nigro, V., Simeone, A., Cama, A., and Boncinelli, E. (1996). Germline mutations in the homeobox gene EMX2 in patients with severe schizencephaly. *Nat Genet.* **12**, 94–96.

Byrd, S. E., Osborn, R. E., Bohan, T. P., and Naidich, T. P. (1989). The CT and MR evaluation of migrational disorders of the brain, Part II. Schizencephaly, heterotopia and polymicrogyria. *Pediatr. Radiol.* **19**, 219–222.

Cameron, F. J., Khadilkar, V. V., and Stanhope, R. (1999). Pituitary dysfunction, morbidity and mortality with congenital midline malformation of the cerebrum. *Eur. J. Pediatr.* **158**, 97–102.

Cappeli, C., Grill, J., Raquin, M., Pierre-Kahn, A., Lellouch-Tubiana, A., Terrier-Lacombe, M. J., Habrand, J. L., Couanet, D., Brauner, R., Rodriquez, D., Hartmann, O., and Kalifa, C. (1998). Long-term follow up of 69 patients treated for optic pathway tumours before the chemotherapy era. *Arch. Dis. Child.* **79**, 334–338.

Carmel, P. W. (1983). Management of the Chiari malformations in childhood. *Clin. Neurosurg.* **30**, 385–406.

Chervenak, F. A., Isaacson, G., Hobbins, J. C., Chitkara, U., Tortora, M., and Berkowitz, R. L. (1985). Diagnosis and management of fetal holoprosencephaly. *Obstet. Gynecol.* **66**, 322–326.

Cohen, M. M. (1989). Perspectives on holoprosencephaly: Part I, Epidemiology, genetics, and syndromology. *Teratology* **40**, 211–235.

Cormand, B., Avela, K., Pihko, H., Santavuori, P., Talim, B., Topaloglu, H., de la Chapelle, A., and Lehesjoki, A.-E. (1999). Assignment of the muscle-eye-brain disease gene to 1p32-p34 by linkage analysis and homozygosity mapping. *Am. J. Hum. Genet.* **64**, 126–135.

Crino, P. B., and Henske, E. P. (1999). New developments in the neurobiology of the tuberous sclerosis complex. *Neurology* **53**, 1384–1390.

Curatolo, R., Cusmai, R., Cortesi, F., Chiron, C., Jambaque, I., and Dulac, O. (1991). Neuropsychiatric aspects of tuberous sclerosis. *In* Proceedings of Annals of the New York Academy of Sciences: Tuberous Sclerosis and Allied Disorders. *Clin. Cell. Mol. Stud.* **615**, 8–16.

Date, I., Yagyu, Y., Asari, S., and Ohmoto, T. (1993). Long-term outcome in surgically treated encephalocele. *Surg. Neurol.* **40**, 125–130.

Dattani, M. L., Martinez-Barbera, J., Thomas, P. Q., Brickman, J. M., Gupta, R., Wales, J. K., Hindmarsh, P. C., Beddington, R. S., and Robinson, I. C. (2000). Molecular genetics of septo-optic dysplasia. *Horm. Res.* 53, 26–33.

De Meyer, W. (1985). Arachnoid and porencephalic cysts. *In* "Current Therapy in Neurologic Disease 1985–1986." (R. T. Johnson, Ed.) pp. 98–101. BC Decker Inc, Philadelphia.

Denckla, M. B., Hofman, K., Mazzocco, M. M., Melhem, E., Reiss, A. L., Bryan, R. N., Harris, E. L., Lee, J., Cox, C. S., and Schuerholz, L. J. (1996). Relationship between T2-weighted hyperintensities (unidentified bright objects) and lower IQs in children with neurofibromatosis-1. *Am. J. Med. Genet.* 67, 98–102.

Dobyns, W. B. (1987). Developmental aspects of lissencephaly and the lissencephaly syndromes. *Birth Defects Original Article Series* 23, 225–241.

Dobyns, W. B. (1989). The neurogenetics of lissencephaly. *Neurol. Clin.* 7, 89–105.

Dobyns, W. B., Reiner, O., Carrozzo, R., and Ledbetter, D. H. (1993). Lissencephaly. A human brain malformation associated with deletion of the LIS1 gene located at chromosome 17p13. *JAMA* 15, 270(23), 2838–2842.

Duffner, P. K., Cohen, M. E., Seidel, F. G., and Shucard, D. W. (1989). The significance of MRI abnormalities in children with neurofibromatosis. *Neurology* 39, 373–378.

Edwards, M. S., and Raffel, C. (1987). Discussion of articles by Maria *et al.*, and Golden *et al. Pediatr. Neurosci.* 13, 45–51.

Elster, A. D. (1992). Radiologic screening in the neurocutaneous syndromes: Strategies and controversies. *AJNR* 13, 1078–1082.

Enjolras, O., Riche, M. C., and Merland, J. J. (1985). Facial portwine stains and Sturge-Weber syndrome. *Pediatrics* 76, 48–51.

Fishman, M. A. (2000). Birth defects and supplemental vitamins. *Curr. Treat. Options. Neurol.* 2, 117–122.

Flickinger, J. C., Lunsford, L. D., Linskey, M. E., Duma, C. M., and Kondziolka, D. (1993). Gamma knife radiosurgery for acoustic tumors: multivariate analysis of four year results. *Radiother. Oncol.* 27, 91–98.

Friede, R. L. (1989). "Developmental Neuropathology," 2nd Ed., Springer-Verlag, Berlin, New York.

Gilbert, J. N., Jones, K. L., Rorke, L. B., Chernoff, G. I., and James, H. E. (1986). Central nervous system anomalies associated with meningomyelocele, hydrocephalus, and the Arnold-Chiari malformation: Reappraisal of theories regarding the pathogenesis of posterior neural tube closure defects. *Neurosurgery* 18, 559–564.

Gleeson, J. G. (2000). Classical lissencephaly and double cortex (subcortical band heterotopia): LIS1 and doublecortin. *Curr. Opin. Neurol.* 13, 121–125.

Gomez, M. R. (1985). Phakomatoses. *In* "Current Therapy in Neurologic Disease 1985–1986." (R. T. Johnson, Ed.) pp 119–122. BC Decker Inc, Philadelphia.

Gomez, M. R. (1991). Neurocutaneous diseases. *In* "Neurology in Clinical Practice," Vol. II, "The Neurological Disorders," (W. G. Bradley, R. B. Daroff, G. M. Fenichel, and C. D. Marsden, Eds.), Chapter 67, pp. 1323–1342. Butterworth-Heinemann, Boston-London.

Gomez, M. R., and Bebin, E. M. (1987). Sturge-Weber syndrome. *In* "Neurocutaneous diseases: A Practical Approach," (M. R. Gomez, Ed.), 40, pp. 356–367. Butterworths, Boston.

Gorlin, R. J. (1987). Nevoid basal cell carcinoma syndrome. *In* "Neurocutaneous Diseases: A Practical Approach," (M. R. Gomez, Ed.), Chapter 5, pp. 67–79. Butterworths, Boston.

Gressens, P., Baes, M., Leroux, P., Lombet, A., Van Veldhoven, P., Janssen, A., Vamecq, J., Marret, S., and Evrard, P. (2000). Neuronal migration disorder in Zellweger mice is secondary to glutamate receptor dysfunction. *Ann. Neurol.* 48, 336–343.

Gunther, G., Junker, R., Strater, R., Schobess, R., Kurnik, K., Kosch, A., Nowak-Gottl, U., and Group, F. T. (2000). Symptomatic ischemic stroke in full-term neonates: role of acquired and genetic prothrombotic risk factors. *Stroke* 31, 2437–2441.

Gutmann, D. H. (1998). The diagnosis and management of neurofibromatosis 1. *Neurologist* 8, 313–326.

Gutmann, D. H., and Collins, F. S. (1993). Neurofibromatosis Type I. Beyond positional cloning. *Arch. Neurol.* 50, 1185–1193.

Haines, S. J., and Berger, M. (1991). Current treatment of Chiari malformations types I and II: A survey of the pediatric section of the American Association of Neurological Surgeons. *Neurosurgery* 28, 353–357.

Halsey, J. H. (1987). Hydranencephaly. *In* "Handbook of Clinical Neurology," (P. J. Vinken, G. W. Bruyn, H. L. Klawans, and N. C. Myrianthopoulos, Eds.), Vol. 50, Revised Series 6, Chapter 19, pp. 337–353. Elsevier Science Publishers, Amsterdam.

Hancock, E., and Osborne, J. P. (1999). Vigabatrin in the treatment of infantile spasms in tuberous sclerosis: literature review. *J. Child. Neurol.* 14, 71–74.

Hardwig, P., and Robertson, D. M. (1984). Von Hippel-Lindau disease: A familial, often lethal, multi-system phakomatosis. *Ophthalmology* 91, 263–270.

Harsh, G. R., Edwards, M. S. B., and Wilson, C. B. (1986). Intracranial arachnoid cysts in children. *J. Neurosurg.* 64, 835–842.

Hernandez-Diaz, S., Werler, M. M., Walker, A. M., and Mitchell, A. A. (2000). Folic acid antagonists during pregnancy and the risk of birth defects. *N. Engl. J. Med.* 343, 1608–1614.

Herskowitz, J., Rosman, N. P., and Wheeler, C. B. (1985). Colpocephaly: Clinical, radiologic, and pathogenetic aspects. *Neurology* 35, 1594–1598.

Hirsch, J-F., Pierre-Kahn, A., Renier, D., Sainte-Rose, C., and Hoppe-Hirsch, E. (1984). The Dandy-Walker malformation. A review of 40 cases. *J. Neurosurg.* 61, 515–522.

Horton, W. A., Wong, V., and Eldridge, R. (1976). Von Hippel-Lindau disease: Clinical and pathological manifestations in nine families with 50 affected members. *Arch. Intern. Med.* 136, 769–777.

Hubschmann, O. R., Vijayanathan, T., and Countee, R. W. (1981). Von Hippel-Lindau disease with multiple manifestations: Diagnosis and management. *Neurosurgery* 8, 92–95.

Humphreys, R. P. (1994). Encephalocele and dermal sinuses *In* "Pediatric Neurosurgery: Surgery of the Developing Nervous System." (W. R. Cheek, A. E. Marlin, D. G. McLone, D. H. Reigel, and M. L. Walker, Eds.), 3rd ed., Chapter 5, pp. 96–103. WB Saunders, Philadelphia.

Huson, S. M., Harper, P. S., Hourihan, M. D., Cole, G., Weeks, R. D., and Compston, D. A. S. (1986). Cerebellar haemangioblastoma and von Hippel-Lindau disease. *Brain* 109, 1297–1310.

Hyman, M. H., and Whittemore, V. H. (2000). National Institutes of Health consensus conference: tuberous sclerosis complex. *Arch. Neurol.* 57 (5), 662–665.

Kalter, H., and Warkany, J. N. (1983). Congenital malformations (second of two parts). *N. Engl. J. Med.* 308, 491–497.

King, A. A., Debaun, M. R., Riccardi, V. M., and Gutmann, D. H. (2000). Malignant peripheral nerve sheath tumors in neurofibromatosis 1. *Am. J. Med. Genet.* 93, 388–392.

Kinsman, S. L., Plawner, L. L., and Hahn, J. S. (2000). Holoprosencephaly: recent advances and new insights. *Curr. Opin. Neurol.* 13, 127–132.

Kobayashi, K., Nakahori, Y., Miyake, M., Matsumura, K., Kondo-Iida, E., Nomura, Y., Segawa, M., Yoshioka, M., Saito, K., Osawa, M., Hamano, K., Sakakihara, Y., Nonaka, I., Nakagome, Y., Kanazawa, I., Nakamura, Y., Tokunaga, K., and Toda, T. (1998). An ancient retrotransposal insertion causes Fukuyama-type congenital muscular dystrophy. *Nature* 394, 388–392.

Koester, S. E., and O'Leary, D. D. M. (1994). Axons of early generated neurons in cingulate cortex pioneer the corpus callosum. *J. Neurosci.* 14, 6608–6620.

Koh, S., Jayakar, P., Dunoyer, C., Whiting, S. E., Resnick, T. J.,

Alvarez, L. A., Morrison, G., Ragheb, J., Prats, A., Dean, P., Gilman, J., and Duchowny, M. S. (2000). Epilepsy surgery in children with tuberous sclerosis complex: presurgical evaluation and outcome. *Epilepsa* **41**, 1206–1213.

Kozlowski, K., and Ouvrier, R. A. (1993). Agenesis of the corpus callosum with mental retardation and osseous lesions. *Am. J. Med. Genet. (Neuropsychiatric Genetics)* **48**, 6–9.

Latif, F., Tory, K., Gnarra, J., Yao, M., Duh, F.-M., Orcutt, M. L., Stackhouse, T., Kuzmin, I., Modi, W., Geil, L., Schmidt, L., Zhou, F., Li, H., Chen, M. H. W. F., Glenn, G., Choyke, P. Walther, M. M., Weng, Y. Duran, D.-S. R., Dean, M., Glavac, D., Richards, F. M. Crossey, P. A., Ferguson-Smith, M. A., Paslier, D. L., Chumakov, I., Cohen, D. C., Chinault, A. C. Maher, E. R., Linehan, W. M., Zbar, B., and Lerman, M. I. (1993). Identification of the Hippel-Lindau disease tumor suppressor gene. *Science* **260**, 1317–1320.

Ledbetter, S. A., Kuwano, A., Dobyns, W. B., and Ledbetter, D. H. (1992). Microdeletions of chromosome 17p13 as a cause of isolated lissencephaly. *Am. J. Hum. Genet.* **50**(1), 182–189.

Levy, W. J., Mason, L., and Hahn, J. F. (1983). Chiari Malformation presenting in adults: a surgical experience in 127 cases. *Neurosurgery* **12**, 377–390.

Lin, A. E., Birch, P. H., Korf, B. R., Tenconi, R., Niimura, M., Poyhonen, M., Armfield, Uhas, K., Sigorini, M., Virdis, R., Romano, C., Bonioli, E., Wolkenstein, P., Pivnick, E. K., Lawrence, M., and Friedmann, J. M. (2000). Cardiovascular malformations and other cardiovascular abnormalities in neurofibromatosis 1. *Am. J. Med. Genet.* **95**, 108–117.

Liptak, G. S., Shurtleff, D. B., Bloss, J. W., Baltus-Hebert, E., and Manitta, P. (1992). Mobility aids for children with high-level myelomeningocele: Parapodium versus wheelchair. *Dev. Med. Child Neurol.* **34**, 787–796.

Listernick, R., and Charrow, J. (1990). Neurofibromatosis type 1 in childhood. *J. Pediatr.* **116**, 845–853.

Listernick, R., Charrow, J., Tomita, T., and Goldman, S. (1999). Carboplatin therapy for optic pathway rumors in children with neurofibromatosis type-1. *J. Neurooncol.* **45**, 185–190.

Luthy, D. A., Wardinsky, T., Shurtleff, D. B., Hollenbach, K. A., Hickok, D. E., Nyberg, D. A., and Benedetti, T. J. (1991). Cesarean section before the onset of labor and subsequent motor function in infants with meningomyelocele diagnosed antenatally. *N. Engl. J. Med.* **324**, 662–666.

Lyon, G., Adams, R. A., and Kolodny, E. H. (1996). "Neurology of Hereditary Metabolic Diseases of Children". McGraw-Hill, New York.

Mackool, B. T., and Fitzpatrick, T. B. (1992). Diagnosis of Neurofibromatosis by cutaneous examination. *Semin. Neurol.* **12**, 358–363.

Maire, J. P., Floquet, A., Darrouzet, V., Guerin, J., Bebear, J. P., and Caudry, M. (1992). Fractionated radiation therapy in the treatment of stage III and IV cerebello-pontine angle neurinomas: Preliminary results in 20 cases. *Int. J. Radiat. Oncol. Biol. Phys.* **23**, 147–152.

Marti-Bonmati, L., Menor, F., Poyatos, C., and Cortina, H. (1992). Diagnosis of Sturge-Weber syndrome: Comparison of the efficacy of CT and MR imaging in 14 cases. *AJR* **158**, 867–871.

May, P. (1992). Pediatric neurosurgery. *Curr. Opin. Neurol. Neurosurg.* **5**, 25–29.

McCartney, B. M., Kulikauskas, R. M., LaJeunesse, D. R., and Fehon, R. G. (2000). The neurofibromatoisis-2 homologue, Merlin, and the tumor suppressor expanded function together in Drosophila to regulate cell proliferation and differentiation. *Development* **127**, 1315–1324.

McLone, D. G., and Dias, M. S. (1994). Normal and abnormal early development of the nervous system. *In* "Pediatric Neurosurgery: Surgery of the Developing Nervous System," (W. R. Cheek, A. E. Marlin, D. G. McLone, D. H. Reigel, and M. L. Walker, Eds.), 3rd ed., Chapter 1, pp. 3–39. WB Saunders, Philadelphia.

Merrill, D. C., Goodwin, P., Burson, J. M., Sato, Y., Williamson, R., and Weiner, C. P. (1998). The optimal route of delivery for fetal meningomyelocele. *Am. J. Obstet. Gynecol.* **179**, 235–240.

Michels, V. V. (1987). Von Hippel-Lindau disease, *In* "Neurocutaneous Diseases: A Practical Approach," (M. R. Gomez, Ed.), Chapter 4, pp. 53–66. Butterworths, Boston.

Miloloza, A., Rosner, M., Nellist, M., Halley, D., Bernaschek, G., and Hengstschlager, M. (2000). The TSC1 gene product, hamartin, negatively regulates cell proliferation. *Hum. Mol. Genet.* **9**, 1721–1727.

Mori, K. (1985). "Anomalies of the Central Nervous System. Neuroradiology and Neurosurgery." Thieme-Stratton Inc, New York.

Mulvihill, J. J., Parry, D. M., Sherman, J. L., Pikus, A., Kaiser-Kupfer, M. I., and Eldridge, R. (1990). Neurofibromatosis 1 (Recklinghausen disease) and neurofibromatosis 2 (bilateral acoustic neurofibromatosis). *Ann. Intern. Med.* **113**, 39–52.

Munshi, I., Frim, D., Stine-Reyes, R., Weir, B. K., Hekmatpanah, J., and Brown, F. (2000). Effects of posterior fossa decompression with and without duraplasty on Chiari malformation-associated hydromyelia. *Neurosurgery* **46**, 1384–1389.

Nakayama, J., Kiryu, H., Urabe, K., Matsuo, S., Shibata, S., Koga, T., and Furue, M. (1999). Vitamin D3 analogues improve cafe au lait spots in patients with von Recklinghausen's disease: experimental and clinical studies. *Eur. J. Dermatol.* **9**, 202–206.

National Institutes of Health Consensus Development Conference. (1988). Neurofibromatosis. Conference statement. *Arch. Neurol.* **45**, 575–578.

Nellist, M., Ward, C. J., and Roelfsema, J. H. (1993). Identification and characterization of the tuberous sclerosis gene on chromosome 16. *Cell* **75**, 1305–1315.

Neumann, H. P. H., Eggert, H. R., Scheremet, R., Schumacher, M., Mohadjer, M., Wakhloo, A. K., Volk, B., Hettmannsperger, U., Riegler, P., Schollmeyer, P., and Wiestler, O. (1992). Central nervous system lesions in von Hippel-Lindau syndrome. *J. Neurol. Neurosurg. Psychiat.* **55**, 898–901.

Newman, G. C., Buschi, A. I., Sugg, N. K., Kelly, T. E., and Miller, J. Q. (1982). Dandy-Walker syndrome diagnosed in utero by ultrasonography. *Neurology* **32**, 180–184.

Ohh, M., Park, C. W., Ivan, M., Hoffman, M. A., Kim, T. Y., Huang, L. E., Pavletich, N., Chau, V., and Kaelin, W. G. (2000). Ubiquitination of hypoxia-inducible factor requires direct binding to the beta-domain of the von Hippel-Lindau protein. *Nat. Cell. Biol.* **2**, 423–427.

Osenbach, R. K., and Menezes, A. H. (1992). Diagnosis and management of the Dandy-Walker malformation: 30 years of experience. *Pediatr. Neurosurg.* **18**, 179–189.

Pascual-Castroviejo, I., Diaz-Gonzalez, C., Garcia-Melian, R. M., Gonzalez-Casado, I., and Munoz-Hiraldo, E. (1993). Sturge-Weber syndrome: Study of 40 patients. *Pediatr. Neurol.* **9**, 283–288.

Pascual-Castroviejo, I., Velez, A., Pascual-Pascual, S. I., Roche, M. C., and Villarejo, F. (1991). Dandy-Walker malformation: analysis of 38 cases. *Child. Nerv. Syst.* **7**, 88–97.

Paul, K. S., Lye, R. H., Strang, F. A., and Dutton, J. (1983). Arnold-Chiari malformation. Review of 71 cases. *J. Neurosurg* **58**, 183–187.

Pause, A., Lee, S., Lonergan, K. M., and Klausner, R. D. (1998). The von Hippel-Lindau tumor suppressor gene is required for cell cycle exit upon serum withdrawal. *Proc. Natl. Acad. Sci. U S A* **95**, 993–998.

Perilongo, G., Moras, P., Carollo, C., Battistella, A., Clementi, M., Laverda, A., and Murgia, A. (1999). Spontaneous partial regression of low-grade glioma in children with neurofibromatosis-1: a real possibility. *J. Child. Neurol.* **14**, 352–356.

Pilu, G., Perolo, A., Falco, P., Visentin, A., Gabrielli, S., and Bovicelli, L. (2000). Ultrasound of the fetal central nervous system. *Curr. Opin. Obstet. Gynecol.* **12**, 93–103.

Pont, M. S., and Elster, A. D. (1992). Lesions of skin and brain: modern imaging of the neurocutaneous syndromes. *AJR* **158**, 1193–1203.

Raffel, C., and McComb, J. G. (1994). Arachnoid cysts, *In* "Pediatric Neurosurgery: Surgery of the Developing Nervous System," (W. R. Cheek, A. E. Marlin, D. G. McLone, D. H. Reigel, and M. L. Walker, Eds.), 3rd ed., Chapter 6, pp. 104–110. WB Saunders, Philadelphia.

Raimondi, A. J., Skimoji, T., and Gutierrez, F. A. (1980). Suprasellar cysts: surgical treatment and results. *Child Brain* **7**, 57–72.

Rappaport, Z. H. (1993). Suprasellar arachnoid cysts: Options in operative management. *Acta Neurochir.* **122**, 71–75.

Roach, E. S., Gomez, M. R., and Northrup, H. (1998). Tuberous sclerosis complex conference: revised clinical diagnostic criteria. *J. Child. Neurol.* **13**, 624–628.

Rosman, N. P., and Pearce, J. (1967). The brain in multiple neurofibromatosis (von Recklinghausen's disease): A suggested neuropathological basis for the associated mental defect. *Brain* **90**, 829–838.

Rossi, L. N., Truilzi, F., Parazzini, C., and Maninetti, M. M. (1999). Spontaneous improvement of optic pathway lesions in children with neurofibromatosis type 1. *Neuropediatrics* **30**, 205–209.

Rouleau, G. A., Merel, P., Lutchman, M., Sanson, M., Zucman, J., Marineau, C., Hoang-Xuan, K., Demczuk, S., Desmaze, C., Plougastel, B., *et al.* (1993). Alteration in a new gene encoding a putative membrane-organizing protein causes neurofibromatosis type 2. *Nature* **363**, 515–521.

Samii, C., Mobius, E., Weber, W., Heienbrok, H. W., and Berlit, P. (1999). Pseudo Chiari type I malformationsecondary to cerebrospinal fluid leakage. *J. Neurol.* **246**, 162–164.

Sarnat, H. B. (1992). Disorders of the lamina terminalis: midline malformations of the forebrain. *In* "Cerebral Dysgenesis: Embryology and Clinical Expression," (H. B. Sarnat, Ed.), Chapter 4, pp. 167–244. Oxford University Press, New York.

Sarnat, H. B., and Menkes, J. H. (2000). Neuroembryology, genetic programming, and malformations of the nervous system. *In* "Child Neurology," 6th ed (J. H. Menkes and H. B. Sarnat, Eds.), Chapter 4, pp. 277–400. Lippincott, Williams & Wilkins, Phildadelphia.

Schroder, S., Bauman-Schroder, U., Hazim, W., Haase, W., and Mautner, V. F. (1999). Long-term outcome of gliomas of the visual pathway in type 1 neurofibromatosis. *Klin. Monatsbl. Augenheilkd.* **215**, 349–354.

Silva, M. M., Golmand, S., Keating, G., Marymount, M. A., Kalapurakal, J., and Tomita, T. (2000). Optic pathway hypothalamic gliomas in children under three years of age: the role of chemotherapy. *Pediatr. Neurosurg.* **33**, 151–158.

Smirniotopoulos, J. G., and Murphy, F. M. (1992). The phakomatoses. *AJNR* **13**, 725–746.

Stevens, J. M., Serva, W. A. D., Kendall, B. E., Valentine, A. R., and Ponsford, J. R. (1993). Chiari malformation in adults: relation of morphological aspects to clinical features and operative outcome. *J. Neurol. Neurosurg. Psychiatr.* **56**, 1072–1077.

Sweeney, K. J., Clark, G. D., Prokscha, A., Dobyns, W. B., and Eichele, G. (2000). Lissencephaly associated mutations suggest a requirement for the PAFAH1B heterotrimeric complex in brain development. *Mech. Dev.* **92**, 263–271.

Thompson, S., Torres, M., Stevenson, R., Dean, J., and Best, R. (2000). Periconceptional vitamin use, dietary folate and occurrent neural tube defected pregnancies in a high risk population. *Ann. Epidemiol.* **10**, 476.

Toda, T., Yoshioka, M., Nakahori, Y., Kanazawa, I., Nakamura, Y., and Nakagome, Y. (1995). Genetic identity of Fukuyama-type congenital muscular dystrophy and Walker-Warburg syndrome. *Ann. Neurol.* **37**, 99–101.

Tulipan, N., Hernanz-Schulman, M., and Bruner, J. P. (1998). Reduced hindbrain herniation after intrauterine myelomeningocele repair: A report of four cases. *Pediatr. Neurosurg.* **29**, 274–278.

Venes, J. L., Black, K. L., and Latack, J. L. (1986). Preoperative evaluation and surgical management of the Arnold-Chiari II malformation. *J. Neurosurg.* **64**, 363–370.

Vogl, J., Stemmler, J., Bergman, C., Pfluger, T., Egger, E., and Lissner, J. (1993). MR and MR angiography of Sturge-Weber syndrome. *AJNR* **14**, 417–425.

Volpe, J. J. (2001a). Neural tube formation and prosencephalic development. *In* "Neurology of the Newborn" 4th ed., Chapter 1, pp. 1–44. WB Saunders, Philadelphia.

Volpe, J. J. (2001b). Neuronal proliferation, migration, organization, and myelination. In "Neurology of the Newborn" 4th ed., Chapter 2, pp. 45–99. WB Saunders, Philadelphia.

Volpe, J. J. (2001c). Hypoxic-ischemic encephalopathy: neuropathology and pathogenesis. *In* "Neurology of the Newborn" 4th ed., Chapter 8, pp. 296–330. WB Saunders, Philadelphia.

Wallis, D., and Muenke, M. (2000). Mutations in holoprosencephaly. *Hum. Mutat.* **16**, 99–108.

Walsh, C. A. (2000). Genetics of neuronal migration in the cerebral cortex. *Ment. Retard Dev. Disabil. Res. Rev.* **6**, 34–40.

Weinberg, J. S., Freed, D. L., Sadock, J., Handler, M., Wisoff, J. H., and Epstein, F. J. (1998). Headache and Chiari I in the pediatric population. *Pediatr. Neurosurg.* **29**, 14–18.

Weiss, B., Bollag, G., and Shannon, K. (1999). Hyperactive Ras as a therapeutic target in neurofibromatosis type 1. *Am. J. Med. Genet.* **26**, 89(1), 14–22.

Wiss, K. (1992). Neurocutaneous disorders: Tuberous sclerosis, incontinentia pigmenti and hypomelanosis of Ito. *Semin. Neurol.* **12**, 364–373.

Wohlwill, F. J., and Yakovlev, P. I. (1957). Histopathology of meningofacial angiomatosis (Sturge-Weber's disease) *J. Neuropath. Exp. Neurol.* **16**, 341–364.

Wong, M., and Trevathan, E. (2001). Infantile spasms. *Pediatr. Neurol.* **24**, 89–98.

Yakovlev, P. L., and Wadsworth, R. C. (1946). Schizencephalies. *J. Neuropathal. Exp. Neurol.* **5**, 116–206.

Zolty, P., Sanders, M. H., and Pollack, I. F. (2000). Chiari malformation and sleep-disordered breathing: a review of diagnostic and management issues. *Sleep* **23**, 637–643.

*Metabolic and Progressive Disorders*

CHAPTER 71

# Acute and Chronic Alcohol-Related Disorders

Peter Thier

---

Alcohol abuse damages various parts of the central and the peripheral nervous system. Which part of the nervous system is affected depends on the extent and the duration of abuse, nutrition, and a variety of poorly understood individual factors. Alcohol is oxidized by the hepatic alcohol dehydrogenase to acetaldehyde, a metabolite that is highly cytotoxic. It is therefore most probably of major importance in the pathogenesis of alcohol-related damage to both internal organs and the nervous system. Moreover, it has been suggested that acetaldehyde might also play a role in the formation of alcohol dependence by forming condensation products with monoamines such as serotonin, adrenalin, or noradrenalin with properties similar to those of morphinelike alkaloids (Davis and Walsh, 1970). However, it is rather unlikely that the necessary concentrations for the synthesis of the condensation products prevail under in vivo conditions, given that the degradation of acetaldehyde to acetic acid by the mitochondrial enzyme acetaldehyde dehydrogenase exceeds its formation by alcohol oxidation many times over, and therefore, alternative concepts on the origin of alcohol dependence have come to the fore (e.g., Merikangas, 1990; Meyer, 2000). In addition to the cytotoxicity of the alcohol metabolite acetaldehyde, probably the most important factor contributing to the formation of alcohol-related cell damage is malnutrition. Chronic abuse of alcohol is almost always accompanied by malnutrition, and alcoholism is by far the most important cause of malnutrition in the Western world. The alcoholic provides the larger part of his or her need for calories by the intake of alcoholic beverages that contain few vitamins and minerals. About 40l of beer or 200l of wine is needed to cover the daily requirements for B vitamins. However, malnutrition is not the only factor responsible for vitamin defi-

ciency. Another factor is the reduced intestinal uptake of vitamins and the impairment of hepatic storage and activation. Although the supply of vitamins is reduced, the metabolism of alcohol increases the demand beyond normal. Substitution of vitamins in alcoholics, therefore, usually requires amounts that exceed the demand of healthy subjects.

In addition to the direct toxic effects of alcohol and its metabolites and the consequences of the alcohol-related vitamin deficiency, the interaction of alcohol with the GABA-benzodiazepine receptor also has to be taken into account in an attempt to understand alcohol-related disorders of the nervous system. The interaction between alcohol and the GABA-benzodiazepine receptor is probably the major pathogenetic basis responsible for the symptoms of alcohol withdrawal.

## ALCOHOL INTOXICATION

### Clinical Aspects

The so-called normal alcohol intoxication is a transient exogenous psychosis resulting from acute intake of sufficiently large amounts of alcohol. However, the clinical picture does not simply reflect the blood alcohol level. Additional factors such as the general physical shape of the subject, personality, and environmental influences also contribute to the clinical picture. For practical purposes the differentiation of three stages of intoxication (after Feuerlein, 1989) has proven to be applicable.

1. Mild intoxication (blood alcohol between 0.5‰ and 1.5‰): Psychomotor capacities are reduced, loss of inhibitions, increased sociability, increased volubility,

increased thirst for action, reduced control, positional nystagmus.

2. Medium level intoxication (blood alcohol between 1.5‰ and 2.5‰): Euphoria or aggressive irritability, reduced self-criticism, daze, behavior strongly dependent on external cues, primitive, explosive reactions.

3. Severe intoxication (blood alcohol greater than 2.5‰): Disturbance of consciousness, disorientation, free-floating anxiety, and agitation. As a consequence of alcohol-induced dysfunction of the vestibulocerebellar system—ataxia, dizziness, dysarthria, nystagmus. Especially in cases of progressive disturbance of consciousness care has to be taken to exclude additional factors such as concomitant craniocerebral injury, possibly with intracranial hemorrhage, intoxication with hypnotics, or other kinds of psychotropic drugs, or metabolic coma.

4. Pathological intoxication: In predisposed subjects, for instance, those with preceding brain trauma, even small amounts of alcohol may cause states of agitation or drowsy states with disorientation, misjudgment of location and persons, illusions, anxiety, fury, and proneness to acts of violence. There is usually complete amnesia for these conditions.

### Therapy

Neither the mild nor the medium-level intoxication usually requires therapy. In cases of agitation, however, it is important to remember that not only alcohol but also tranquilizers taken in addition to alcohol or even alone may be responsible. It follows that tranquilizers should never be given in an attempt to sedate the agitated drunken patient. Rather the drug of choice is haloperidol, which is highly effective in the treatment of agitation and unlike less potent neuroleptics has a low risk of cardiovascular side effects. Initially 5–10 mg is administered, preferably intravenously. Another 1–2 doses can be given at intervals of 30 minutes. The maximum dose should not exceed 60 mg in 24 hours. Although clearly a second choice, sedating neuroleptics such as levomepromazine (25 mg IM) or chlorprothixene (30–50 mg IM) may be considered. The treatment of severe alcohol intoxication follows standard protocols also valid for intoxications of other cause. There is no specific antidote to alcohol. Guidelines are provided in Chapter 59.

### Ineffective or Obsolete Therapies

The hypothesis that the effects of alcohol on the central nervous system are partially mediated by central opioid mechanisms (see introduction) has been responsible for attempts to use the opioid antagonist naloxone in the treatment of alcohol intoxication. However, controlled studies have failed to demonstrate that naloxone, given in a dose of 0.4 mg, is able to counteract heavy alcohol intoxication (Nuotto et al., 1983). Experiments based on animal models of intoxication have basically come to the same conclusion: ethanol-induced respiratory depression and motor disturbances are not counteracted by naloxone (Lignian et al., 1983).

Ro 15-4513 is an imidazodiazepine that acts as a so-called inverse benzodiazepine agonist, having effects opposite to that of benzodiazepines and alcohol. Although this profile seemed to suggest initially that this compound might be useful as a sort of sobering-up pill, later work has shown that Ro 15-4513 has substantial proconvulsive and anxiety-promoting effects, rendering it useless and even dangerous for the treatment of alcohol intoxication (Lister and Nutt, 1987). Also the hope that the partial alcohol antagonistic effects of the $\alpha_2$-antagonists atipamezol and idazoxane might be used to modulate the intoxicating effects of alcohol has not materialized (Lister et al., 1989).

## ALCOHOLIC HALLUCINOSIS

### Clinical Aspects

This rare condition (Glass, 1989; Selzer, 1980; Soyka, 1989) is characterized by auditory hallucinations, usually without clouding of senses and consciousness. Typically, the person affected will have hallucinations in which members of his or her family, neighbors, and acquaintances, speak. These conversations will typically center around the person's socially unacceptable behavior and will contain reproaches, accusations, and threats. The patient, who is unaware of the unreality of these conversations, will react comprehensively by either trying to defend himself or herself or by showing signs of retreat with depression, anxiety, or even panic. Hallucinations usually occur during the night. Visual hallucinations are rare exceptions. Signs of autonomic or motor disturbances are not part of the clinical picture. The mechanisms causing alcoholic hallucinosis remain unknown.

### Natural Course

Alcoholic hallucinosis has an acute or gradual onset, typically during a phase of abstinence 1–2 weeks after a bout of drinking. Alcohol hallucinosis is, however, not a part of alcohol withdrawal (see later), although mixtures of the two may occur. Without treatment, it usually subsides within a period of time ranging from a few days up to 6 months. Although the prognosis is usually benign, in about 10%–20% of the patients there is a chronic course. In these cases, the disease may end up in a clinical picture not unlike chronic schizophrenia. Apart from such rare cases, the tight temporal relation of alcoholic hallucinosis to previous bouts of drinking, its comparatively short duration, and its reversibility should make it easy to discriminate this alcohol-related disorder from schizophrenia.

## Therapy

The major alternative in the treatment of alcoholic hallucinosis is the choice between anxiolytic drugs such as chlordiazepoxide and the use of neuroleptics such as haloperidol. Whereas the proponents of anxiolytic drugs fear the proconvulsive effect of neuroleptics (Selzer, 1980), the proponents of the latter hold that the proconvulsive potential of neuroleptics is usually overestimated (Soyka, 1989, 1992). They, on the other hand, consider the antipsychotic potential of neuroleptics to be indispensable for the effective treatment of hallucinations. Because results based on a controlled study comparing the respective efficacies of these two groups of drugs are not available, we will present treatment regimens for both. However, in view of the comparatively lower risk of anxiolytic drugs, we would suggest starting with them and resorting to neuroleptics only if anxiolytic medication is ineffective. At any rate, treatment should be carried out in hospital in a calm and well-illuminated room with sufficient emotional support, thus helping the patient to feel secure and relaxed.

### Anxiolytics

Treatment is initiated with 25–100 mg of chlordiazepoxide. If necessary, additional doses may be delivered every 4 hours. Intramuscular application is possible, but painful. Moreover, the absorption of intramuscularly delivered chlordiazepoxide is slow and not very reliable. Active metabolites of chlordiazepoxide will accumulate in the presence of $H_2$-antagonist cimetidine but not ranitidine in cases of liver cirrhosis. The diet should be well-balanced, and fluid intake should be sufficient. To prevent the development of Wernicke's disease, thiamine should be supplemented orally as part of a multivitamin preparation.

### Neuroleptics

Neuroleptics are given if the stabilization achieved with anxiolytics turns out to be insufficient. Depending on age, weight, and state of health, initially 2–10 mg of haloperidol is administered orally or, if necessary, intramuscularly or intravenously. If needed, up to 10 mg/h up to a daily maximum of 60 mg may be added.

## ALCOHOL WITHDRAWAL

### Clinical Aspects

Probably every alcoholic will have some degree of symptoms of uncomplicated alcohol withdrawal after the cessation or reduction of prolonged heavy ingestion of alcohol. The more severe variety of alcohol withdrawal, however, alcohol withdrawal delirium, will be found in only a minority (between 1% and 15% according to different sources) of those having ended drinking. It has been suggested that increased concentrations of toxic metabolites such as ammonia resulting from alcohol-related liver disease are necessary for the development of the full-fledged alcoholic delirium. The main symptom of all varieties of alcohol withdrawal is a 6–8 Hz tremor of the hands, which occasionally may also involve the tongue and the eyelids. The tremor is increased by emotional disturbance but also by motor activities. Less frequently, alcohol withdrawal may cause nausea and vomiting, a feeling of weakness, anxiety, depression, insomnia, brisk tendon reflexes, myoclonus, hyperkinesia, grand mal seizures, and signs of vegetative hyperactivity such as tachycardia, hypertension, or hyperhydrosis. In addition, the patient may complain of dryness of the mouth and headache. Although vivid hallucinations are a characteristic feature of alcoholic delirium, more transient and less vivid hallucinations (visual, auditory, or tactile) may also occur as part of uncomplicated acute alcohol withdrawal, without justifying the diagnosis of alcoholic delirium. Alcoholic delirium is clearly differentiated from uncomplicated alcohol withdrawal by disturbed orientation and clouded consciousness. Hallucinations during alcoholic delirium are predominantly visual, typically consisting of sequences of scenes with small moving objects such as animals. The person affected is usually extremely restless. The alcoholic delirium is always life-threatening, however, not as a consequence of the psychotic decompensation but because of the severe vegetative dysregulation, which typically includes hyperthermia, respiratory disturbances, and cardiovascular decompensation.

### Natural Course

Uncomplicated alcohol withdrawal typically starts 8–12 hours after cessation or reduction of chronic alcohol consumption, peaks after 24–36 hours, and wears off after 5–7 days. However, more subtle residual symptoms such as an alteration of the sleep pattern may be detectable for a couple of months (Meyer, 1989). The full-fledged alcoholic delirium usually occurs on the second or third day. It terminates after rarely more than 3–5 days in a deep sleep. Alcohol withdrawal seizures of the grand mal type are usually observed before the full-fledged alcoholic delirium has developed. Until 20 years ago, when the only means of treatment available were alcohol and paraldehyde, the mortality of alcoholic delirium was on the order of 10%–20%. More recent measures of treatment have been able to lower this rate to 1%–5%.

## Principles of Therapy

The major purpose of the treatment of alcohol withdrawal is to avoid the development of alcoholic delirium or, in the case of an existing alcoholic delirium, to shorten its duration and thereby to shorten the life-threatening dysregulation of vital functions.

Alcohol withdrawal reflects increased excitability of various parts of the central nervous system. Similar to benzodiazepines, alcohol acts as a modulator at the GABA-benzodiazepine receptor complex increasing GABA-mediated inhibition (Mhatre et al., 1993; Sanna and Harris, 1993). The enhancement of GABA-mediated inhibition is lost in alcohol withdrawal. The fact that benzodiazepines are effective in the treatment of alcohol withdrawal can therefore be easily explained by assuming that one substance enhancing GABA-mediated inhibition is replaced by a second one. Those components of alcohol withdrawal such as tremor or tachycardia, which reflect sympathetic overactivity, result from loss of GABA-ergic inhibition of noradrenergic cell groups in the locus coeruleus. This sympathetic overactivity can be influenced quite directly, without the need to interfere with GABA-ergic neurotransmission, by using catecholamine agonists or antagonists. Although the loss of suppression of GABA-ergic inhibition is by far the most important factor contributing to increased excitability in alcohol withdrawal, it is not the only one. Other factors such as hypomagnesia, frequently found during chronic alcohol consumption and during alcohol withdrawal, has been suggested as contributing to increased excitability (Flink, 1981; McIntyre, 1984).

With a few exceptions, all the drugs that have proven to be effective in the treatment of alcohol withdrawal and alcoholic delirium are able, albeit with varying potency, to replace alcohol as the addictive drug. Their application, therefore, should be strictly controlled and confined to the acute phase of alcohol withdrawal.

More than 150 compounds have been tried in the treatment of alcohol withdrawal (Nutt et al., 1989), and there is still no consensus as to the drug of choice. This is at least partially because the necessary distinction between uncomplicated alcohol withdrawal and alcoholic delirium is not always made. The consequence of this is that drugs that are able to suppress the mild symptoms of alcohol withdrawal sufficiently are recommended for the treatment of alcoholic delirium (Schied and Mann, 1989). The suggested management that follows tries to avoid this lack of differentiation. It must be remembered that, although drugs are of considerable importance, they are by no means the only option in the treatment of alcohol withdrawal. Nursing and psychotherapeutic measures may contribute considerably to the stabilization of the patient with alcohol withdrawal and should be part of any therapeutic regimen independent of the specific drug chosen (Castaneda and Cushman, 1989).

## Practical Management

It is only when mild alcohol withdrawal symptoms are present that drugs may be unnecessary. In all other cases drug treatment is indispensable. The major group of drugs that have proven to be effective in the treatment of alcohol withdrawal including alcoholic delirium are benzodiazepines. In Europe clomethiazole is considered a major alternative to benzodiazepines. Although clomethiazole is not available in the United States, there are good reasons to consider it in the treatment of alcohol withdrawal.

### Benzodiazepines

This group of drugs allows for reliable treatment of uncomplicated alcohol withdrawal syndromes. Although less efficient than clomethiazole in the treatment of the full-fledged alcoholic delirium (McGrath, 1975), they offer the advantage of smaller risks (larger therapeutic window, lower potential for the development of addiction) compared with clomethiazole. Benzodiazepines are usually well tolerated, and furthermore they have the advantage of increasing the seizure threshold. A large number of different benzodiazepines have been shown to be effective, among them chlordiazepoxide, diazepam, oxacepam, clobazepam, midazolam, flunitrazepam, phenyzepam, and alprazolam. The dosage of all benzodiazepines is chosen such as to suppress all withdrawal symptoms or, alternatively, to sedate the patient. The treatment of severe alcoholic delirium may require the supplementary application of beta-blockers or clonidine (see Chapters 28 and 55 for dosage and side effects) to alleviate tremor and high blood pressure, which are usually insufficiently treated by benzodiazepines.

Despite the ongoing controversy in the literature as to the question of the benzodiazepine of choice, our admittedly subjective estimation is that diazepam is the preferred benzodiazepine in the treatment of alcohol withdrawal. Diazepam has a large therapeutic window, is absorbed rapidly after oral delivery, and subsequently, is distributed rapidly in the brain (see, e.g., Nutt *et al.*, 1989). Another advantage of diazepam is that its active metabolite has a comparatively long half-life, thus helping to produce a more sustained response on the symptoms of alcohol withdrawal. Table I summarizes guidelines for dosage.

Chlordiazepoxide may be considered as one of the alternatives to diazepam. However, it has the disadvantage that it is absorbed and distributed more slowly than diazepam. The dosage of chlordiazepoxide is analogous to that of diazepam (see Table I): 25 mg of chlordiazepoxide replaces 10 mg of diazepam. The elimination of both diazepam and chlordiazepoxide is slowed down if severe liver disease is present. In such cases, some authors give preference to the benzodiazepines lorazepam or oxacepam, whose elimination is relatively independent of oxidative metabolism by the liver. Both benzodiazepines are administered orally every 6 hours in doses of 1–3 mg in the case of lorazepam and 15–60 mg in the case of oxacepam. Because of the short half-life, the interval between subsequent doses should not be increased.

### Clomethiazole

Clomethiazole is a substance that chemically is a fragment of the thiamine molecule. It facilitates both GABA-ergic and glycinergic inhibition. Clomethiazole has profound anticonvulsive, sedative, and anxiolytic effects and has proven superior to any other drug in the treatment of full-fledged alcoholic delirium. This is the reason that despite its risks this compound, at least in Europe, is still considered to be the drug of choice in the treatment of delirium. The marked decline of delirium mortality in Europe (from 10%–15% down to approximately 0.5%, the most favorable figure) is usually attributed to the introduction of clomethiazole. Clomethiazole combines a high degree of effectiveness in the suppression of all the symptoms of alcoholic delirium with very favorable pharmacokinetic properties,

**TABLE I    Drug Treatment of Alcohol Withdrawal Syndrome**

| Pre-existing cardiopulmonary disease? | |
| --- | --- |
| Yes: Diazepam | No: Clomethiazole |
| **Initial treatment** <br> 10 mg/h until resolution of symptoms (diazepam loading). <br>   For treatment of severe delirium (only in intensive care unit!) up to 1 g of diazepam may be required. A supplementary dose of clonidine or beta-blockers is recommended in cases of insufficient control of blood pressure and tremor. | 2–4 capsules (192 mg each); 30 min afterwards, up to 6 more capsules during the first 2 hours if symptoms have not resolved. |
| **Maintenance dose** <br> 20 mg/6 h. From day 2 on, gradual reduction of dose. | 2 capsules every 1–2 h; maximal dose, 24 capsules within 24 h. From day 2 on, gradual reduction of dose. |
| **Caution** <br> Patient must remain capable of responding to external stimuli. There is danger of aspiration in patients with alcoholic delirium; therefore, IV or IM (deltoid muscle). Alternative, 25 mg chlordiazepoxide substitute for 10 mg diazepam. | Exceeding the maximal dose as well as parenteral application in the treatment of severe delirium requires intensive care conditions! |
| **Interaction** <br> Ranitidine should be preferred to cimetidine in the prophylaxis of peptic ulcers, because the latter inhibits diazepam and chlordiazepoxide elimination by the liver. | |
| **Parenteral application** | If oral medication turns out to be insufficient and intubation, artificial respiration, and control of cardiovascular functions are guaranteed (intensive care unit!) <br> Initially, 60–150 drops/min of a 0.8% solution until a superficial sleep is setting in, then maintenance by 10–20 drops/min (maximum 20 g/24 h). <br> From day 1–3 on, transition to oral medication and rapid reduction of dose. |

allowing for rapid onset and easy control of its effects even after oral medication. Clomethiazole is not, however, without problems. Addiction is induced in a considerable percentage of cases after only 2–3 weeks of use. This is why the use of chlomethiazole has to be restricted to a hospital setting and must be limited in time. After delivery of high doses, especially when applied parenterally (see later), there is a risk of respiratory problems caused by respiratory center depression and obstruction of the airways from increased bronchial secretion. Furthermore, severe and difficult-to-control hypotension may occur. The risk of these side effects is increased in patients who already have pre-existing cardiovascular or respiratory disease. Because even comparatively small amounts of clomethiazole may cause severe complications in this group of patients, benzodiazepines should be given preference. If, however, clomethiazole is chosen irrespective of the increased risk, continuous monitoring of cardiovascular and respiratory functions and the means for prompt intervention are essential. On the other hand, in patients without pre-existing cardiopulmonary disease, the treatment with low-to-moderate amounts of oral clomethiazole (up to 24 capsules in 24 hours) may be considered safe. It is usually sufficient if started early enough during alcohol withdrawal. Because clomethiazole tablets may cause esophageal ulceration, capsules are preferred. If treatment of a severe alcoholic delirium requires the application of more than 24 capsules in 24 hours, higher dosage clomethiazole or parenteral clomethiazole may be administered, provided control of cardiopulmonary functions is guaranteed and rapid intervention, including intubation and ventilation, is possible. Table I summarizes guidelines for the dosage of clomethiazole.

## Further Options in the Treatment of Alcohol Withdrawal

Although benzodiazepines and clomethiazole are still the drugs of choice in the treatment of alcohol withdrawal, the search for even more efficient and less dangerous alternatives keeps going. We mention three alternatives. The first one, gamma-hydroxy-butyric acid, seems to hold the greatest promise.

### Gamma-Hydroxy-Butyric Acid

Although discovered nearly 40 years ago, the full potential of gamma-hydroxy-butyric acid (GHBA), a precursor and metabolite of GABA, in the treatment of alcohol withdrawal symptoms and alcohol dependency has only recently become clear. GHBA, which mimics many of the central nervous effects of alcohol, seems to be as effective as benzodiazepines in the management of

withdrawal (Addolorato *et al.*, 1999; Gallimberti *et al.*, 1989; Lenzenhuber *et al.*, 1999). When administered orally at a dosage of 50 mg/kg/day, it reduces the withdrawal symptoms of tremor, sweating, nausea, depression, anxiety, and agitation without causing major sedation. When given together with haloperidol or clonidine, GHBA allows substantial savings of the supplementary drug (Lenzenhuber *et al.*, 1999). The onset of the effects is rapid, and major side effects were absent in these studies. GHBA is currently in clinical use only in a few countries such as Italy. Its proven efficacy and lack of major risks when used in a well-defined clinical setting suggests that it might become a major choice among the drugs used in the treatment of alcohol withdrawal. As discussed later, GHBA seems also to be useful in the treatment of alcohol dependency (see this paragraph also for more information on GHBA side effects).

### Clonidine

Clonidine is a compound that reduces the level of sympathetic activity by binding presynaptically to central alpha$_2$-adrenoreceptors. There is no doubt that clonidine is as effective as benzodiazepines and clomethiazole with respect to symptoms such as tremor or sweating and probably even more effective regarding the necessary stabilization of blood pressure, heart rate, and respiration (Baumgartner, 1988). However, its effect on sleep disturbance (Björkvist, 1975) and especially its ability to prevent hallucinations and seizures is insufficient (Robinson *et al.*, 1989). Because of these considerations, we do not recommend clonidine monotherapy as an alternative to benzodiazepines or clomethizole. However, because the profiles of clonidine and benzodiazepines cover different aspects of alcohol withdrawal, the combined use of clonidine with benzodiazepines is a well-established option in the treatment of severe alcoholic delirium (see Chapter 28 on intensive care for dosage and side effects).

### Combination Therapy with Clomethiazole and Haloperidol

Some authors recommend the combination of clomethiazole and haloperidol (up to 60 mg/day) as an alternative to clomethiazole monotherapy for alcoholic delirium (e.g., Pfitzer *et al.*, 1988). The rationale underlying this suggestion is that the psychotic decompensation contributes to the life-threatening autonomic dysregulation. However, we are not aware of any controlled studies that prove the superiority of combinations of haloperidol and clomethiazole or alternatively haloperidol and benzodiazepines over the respective monotherapy without haloperidol. We will come back to a discussion of haloperidol monotherapy later in this chapter.

## Supportive Measures

Independent of the drug chosen for the treatment of alcohol withdrawal, it is necessary to ensure a well-balanced diet including an adequate supply of B vitamins (see the section on Wernicke's disease). Care should be given to correcting any electrolyte imbalance (see the section on central pontine myelinolysis for details on hyponatremia). Magnesium, given orally, may be able to suppress mild alcohol withdrawal symptoms. It may be given as a supplement to clomethiazole or benzodiazepines, even without regular control of serum levels, provided renal function is unimpaired. Dosage is $3 \times 100$–$150$ mg magnesium aspartathydrochloride or magnesium citrate per day before meals (daily magnesium requirements of adults, 400–600 mg). Parenteral application of magnesium, balanced according to the deficit measured, may also be used. Because both benzodiazepines and clomethiazole give sufficient protection against seizures, the application of additional anticonvulsive medication is usually not required (discussed later).

## Less Effective, Ineffective, or Unnecessary Treatments

Endogeneous levels of corticosteroids are elevated during alcohol withdrawal, reflecting activation of the hypothalamic-pituitary-adrenal axis. Suppressing this system by exogenous corticosteroids might ameliorate the symptoms of alcohol withdrawal. Actually, according to some studies, comparatively small amounts of corticosteroids (3 mg of dexamethasone every 12 hours) may be effective in cases of severe alcoholic delirium unresponsive to benzodiazepine or clomethiazole (Tormey and Chambers, 1988). However, a recent study, in which 4 mg dexamethasone was administered intravenously in addition to standard treatment with the benzodiazepine lorazepam, could not find any difference as to the amount of lorazepam needed (Adinoff and Pols, 1997). Beta-blockers influence only some of the symptoms. For instance, propanolol reduces tremor and tachycardia, whereas the remainder of the symptoms respond similarly to placebo (Ladewig et al., 1977). Unlike propanolol, another beta-blocker, atenolol, also reduces agitation and restlessness (Gottlieb, 1988), but like all other beta-blockers, it does not prevent seizures. Favorable effects, at least, on the symptoms of comparatively mild withdrawal have been reported for the anticonvulsive drugs carbamazepine and valproic acid (Wilbur and Kulik, 1981). For instance, about 800 mg of carbamazepine per day has been found to be as effective as benzodiazepines (Malcolm et al., 1989). However, carbamazepine does not match first-choice drugs such as clomethiazole in its ability to prevent the development of a full-fledged alcoholic delirium (Palsson, 1981). As to valproic acid, no convincing evidence has been presented so far that it is of any use in the treatment of alcoholic delirium. Diphenylhydantoin does not have a place in the treatment of withdrawal-induced seizures. Its effect on withdrawal-induced seizures does not surpass the effects of placebo (Alldredge et al., 1989). Phenothiazines are not only useless but also dangerous! They increase the probability of withdrawal seizures, because they lower the thresholds for seizures, independent of alcohol withdrawal. Furthermore, there is evidence that they increase the frequency of severe forms of alcoholic delirium. This is most probably a consequence of their anticholinergic side effects. Neuroleptics of the butyrophenone type, such as haloperidol or benperidol, have much lower anticholinergic side effects and lower seizure thresholds than phenothiazines. However, their effects on the life-threatening autonomic dysregulation is clearly insufficient, and it therefore comes as no surprise that the incidence of severe and also lethal forms of alcoholic delirium is higher when butyrophenones compared with clomethiazole are used (Athen et al., 1977; Schied et al., 1986). Application of butyrophenones is justified only in those rare cases of alcohol withdrawal turning into alcoholic hallucinosis (see earlier). Piracetam is a GABA derivative whose effectiveness has not been proven. The application of barbiturates or paraldehyde, the only therapeutic options in former times, has become unnecessary in view of the current availability of far less toxic alternatives. Alcohol no longer has a place in the treatment of alcohol withdrawal. Although it suppresses mild withdrawal symptoms, it is inadequate to treat the full-fledged alcoholic delirium.

## WERNICKE'S DISEASE

### Clinical Aspects

Wernicke's disease (Wernicke's encephalopathy) is characterized by eye muscle paresis, gaze paresis and nystagmus, pupillary dysfunction, autonomic dysregulation including hypotension and hypothermia, and seizures. Accompanying mental disturbances may include confusion, disorientation, apathy, and varying degrees of clouding of consciousness from drowsiness to coma. In addition, cerebellar symptoms, similar to those following atrophy of the anterior cerebellar vermis (see later), may be observed. Symptoms may occur in various combinations, and they may be accompanied by the signs of alcoholic polyneuropathy. The onset of symptoms is usually quite acute. The cause of this syndrome is always thiamine deficiency, which may be especially deleterious in genetically predisposed persons. Although alcoholism-related malnutrition is by far the most frequent cause of thiamine deficiency, there are also other causes. Thiamine deficiency may result from malnutrition of other kinds, including excessive fasting, inadequate parenteral nutrition (excessive intake of carbohydrates), hemodialysis, uremia, repeated vomiting (for instance, during pregnancy), disseminated tuberculosis, carcinomas of the upper gastrointestinal tract, and disseminated tumors of the hemopoietic system. Thiamine contributes to the metabolism of carbohydrates. Its pyrophosphate is a coenzyme involved in glycolysis, in the tricarbon cycle, and in the hexose-monophosphate shunt. It is unclear whether thiamine deficiency affects brain function by means of a disturbed carbohydrate metabolism or indirectly by means of alterations of the metabolism of various CNS transmitter(s) candidates such as serotonin, glutamate, or aspartate.

Korsakoff's psychosis (amnesic or amnestic-confabulatory psychosis) is characterized by a profound loss of short-term and long-term memory (retrograde amnesia) and an impaired ability to acquire new information (anterograde amnesia). Other cognitive functions are spared or only mildly impaired. Confabulations, a term referring to the fabrication of stories and accounts of events, are typically observed in the early phase of the psychosis and during recovery. Like Wernicke's disease, Korsakoff's psychosis is in most cases a consequence of alcoholism-related thiamine deficiency. In such cases it represents the common denominator for the cognitive symptoms of Wernicke's disease. However, typical isolated Korsakoff's psychosis may also be observed after lesions in the diencephalon or the temporal lobes of other origin; for instance, as a result of herpes simplex encephalitis or tumors.

The structural changes characteristic of Wernicke's disease are symmetrical, hemorrhagic, spongiform lesions, located in the thalamus and hypothalamus in the vicinity of the ventricles, close to the aqueduct, and in the floor of the fourth ventricle. In addition, cerebellar disease may occur. The characteristic lesions are found in 0.8%–4.7% of unselected autopsy cases. In only 20% of the neuropathological cases with evidence of Wernicke's disease had this diagnosis been made on clinical grounds during life (Harper, 1983; Harper et al., 1989). According to Harding et al. (2000), loss of neurons in the anterior thalamic nuclei is confined to patients with alcoholic Korsakoff's psychosis, whereas other brain regions known to be involved in the encoding and retrieval of memory may be affected in

Wernicke patients both with and without Korsakoff's psychosis.

## Natural Course

The acute ocular symptoms of Wernicke's disease and also the milder mental symptoms such as drowsiness or impaired concentration will improve rapidly (within 2–24 hours) if thiamine supplementation is introduced. Nystagmus and ataxia, however, may last much longer, albeit with reduced intensity. The good response to thiamine suggests that, at this stage of the disease, these symptoms result mainly from functional and not from structural alterations. On the other hand, the amnestic-confabulatory psychosis, if present, usually does not respond to thiamine. Only about 20% of all patients with Korsakoff's psychosis show satisfactory recovery. This suggests that the amnestic-confabulatory psychosis is mainly due to irreversible structural alterations. In other words, it is obviously of utmost importance to substitute thiamine as early as possible, before the symptoms of Korsakoff's psychosis have developed.

## Therapeutic Principles

There is consensus in the literature that parenteral application of high-dose thiamine is necessary. The doses usually given by far exceed the amounts of thiamine needed to fill up the tissue stores and cover the daily requirements of healthy subjects. Depending on the study, daily doses ranging from 50–300 mg have been recommended. In most cases the dose exceeds the actual requirements, but this does not pose a problem, because even higher doses of thiamine have to be given to provoke toxic side effects. Even long-term application of 150 mg/day will be tolerated without signs of toxicity.

The real problem with thiamine substitution therapy is not toxicity but intolerance, independent of dose. Both the frequency and the severity of these allergic reactions, which may end lethally, are usually underestimated. Reliable numbers on the incidence of thiamine intolerance are not available. However, a clue is given by the fact that in the first 15 years of thiamine substitution therapy more than 200 incidents were observed after parenteral application of thiamine, at least 6 of which were lethal. It is possible that the thiamine treatment and not the underlying disease itself may be blamed for some fatal outcomes of Wernicke's disease. This is suggested by the fact that Wernicke's disease and thiamine intolerance share symptoms such as seizures, clouded consciousness including coma, severe autonomic dysregulation with tachycardia, hypotension, difficulty in breathing, and finally severe shock. Other symptoms include erythema, urticaria, precordial and epigastric pain, and vomiting. An unidentified compound contained in commercial preparations of thiamine seems to be responsible for the anaphylactoid-type reaction (Blum *et al.*, 1974). Some uncontrolled observations suggest that the risk of parenterally given thiamine is reduced by the simultaneous application of other B-complex vitamins. It is also claimed occasionally that the risk of parenteral thiamine may be smaller if thiamine is given intramuscularly rather than intravenously. On the other hand, at least one of the fatal outcomes reported in the literature resulted from intramuscular injection of thiamine. Compared with parenteral thiamine, oral thiamine seems to be relatively safe. There are only a few reports of mild intolerance reactions after oral thiamine (Acharya *et al.*, 1969).

An effective treatment of the memory deficits of patients with persisting Korsakoff's psychosis is not yet available. Attempts to improve memory functions by boosting the noradrenergic or serotonergic neuromodulatory systems have yielded contradictory results. Assuming that damage of noradrenergic projections surrounding the third and the fourth ventricles and the aqueduct in between is the major cause of the memory deficits in Korsakoff's psychosis, the central noradrenalin agonist clonidine has been tried in several studies. Whereas McEntee and Mair (1980) reported a significant improvement of some memory-related functions after oral application of clonidine, later studies (O'Carroll *et al.*, 1993; Martin *et al.*, 1983) failed to reveal a significant improvement. DL-threo-3,4-dihydrohyphenylserine (DOPS), a nonphysiological precursor of noradrenalin has been shown to yield a significant improvement in one of several measures of memory-related functions in a double-blind study comparing DOPS with placebo (Langlais *et al.*, 1988). However, the other measures of memory functions remained unaffected, suggesting that the benefit to be expected from DOPS treatment will be very limited at best. Suggesting that, alternatively, damage of the serotonergic raphe system might underlie Korsakoff's psychosis, Martin *et al.* (1988) tested the serotonin uptake inhibitor fluvoxamin. They found that application of this compound led to an improvement of short-term memory functions in a small group of patients with Korsakoff's psychosis. Unfortunately, the beneficial effect of this drug has not been corroborated by later work (O' Carroll *et al.*, 1994). Recently, anticholinesterase drugs have been suggested for the treatment of alcoholic Korsakoff syndrome (Angunawela and Barker, 2001). However, as yet, relevant observations on the potential of this approach are still lacking.

## Practical Management

Wernicke's disease is the only neurological disease unequivocally requiring parenteral application of thiamine. We recommend the following procedure: initially, 50 mg of thiamine is given intravenously and 50 mg intramuscularly. On each following day an additional intramuscular injection of 50 mg of thiamine is given until the patient is able to eat normal food. In addition to thiamine a multivitamin preparation should be given. Some authors have suggested the prophylactic application of steroids (e.g., 0.5 mg synacthen) to prevent thiamine intolerance reactions. In our view, however, the frequency of this complication, although significant, is too small to justify the routine application of steroids, which by themselves introduce considerable risks. First of all, alcoholics are particularly prone to ulcers of the gastrointestinal tract. Second, the steroid-induced glucose mobilization will increase the thiamine requirements.

Despite early and correct treatment, the mortality of acute Wernicke's disease remains high at 17%. Fatal outcome is usually due to autonomic dysregulation resulting from diencephalic lesions. Patients at risk should, therefore, be treated prophylactically. In this case, the thiamine derivative benfotiamine (300 mg/day, given orally) should be preferred. Unlike thiamine, it is well absorbed after oral administration. However, note that benfotiamine application cannot be considered an established alternative to parenteral application of thiamine in Wernicke's disease.

A proven pharmacological therapy for Korsakoff's psychosis is not available.

## ALCOHOLIC CEREBELLAR ATROPHY

### Clinical Aspects

This alcohol-related disease is about twice as frequent as typical Wernicke's disease. Men are affected much more frequently (11:1) than women. The incidence of alcoholic cerebellar atrophy peaks in the fifth decade. The clinical picture is characterized by severe ataxia of stance and gait with conspicuous intersegmental instability. These disturbances are accentuated by eye closure. Although the result of the knee-heel test is ataxic, coordination of hands and arms is unaffected or is only mildly disturbed. Oculomotor abnormalities may include saccadic pursuit, impaired suppression of the vestibulo-ocular reflex, and gaze-evoked nystagmus. The structural substrate of the disease is a degeneration of the cerebellum that emphasizes the anterior and middle parts of the cerebellar vermis (Adams, 1976). Purkinje cells are more affected than are other cerebellar neurons. There is only a loose correlation between the degree of cerebellar atrophy and the degree of functional disturbance. This might indicate that processes that are not understood may allow for varying degrees of functional compensation. Mild clinical signs of cerebellar dysfunction are found in about a third of all chronic alcoholics (Scholz et al., 1986).

### Natural Course

In most cases the symptoms develop subacutely over the course of several weeks. If alcohol consumption has ceased, a further progression of the deficits can be avoided, and in some cases an impressive improvement may even be observed. Conversely, if alcohol consumption should continue, a progressive increase in the deficits must be expected. Furthermore, the deficits may be accentuated by intercurrent disease such as infections or alcoholic delirium.

### Therapeutical Principles

The bulk of the data available suggests that lack of thiamine is the major factor causing alcoholic cerebellar atrophy in man (Adams and Victor, 1989). Although rats chronically fed with alcohol show a loss of cerebellar granule cells and interneurons as well as alterations of Purkinje cells (Tavares et al., 1987; Wenisch et al., 1998), in humans chronic alcohol consumption without concomitant thiamine deficiency does not lead to cell loss. On the other hand, thiamine-deficient alcoholics show profound neuronal loss involving the Purkinje cell population, as well as atrophy of the molecular layer, changes that emphasize the vermis and the flocculi (Baker et al., 1999). These findings strongly suggest that alcoholic cerebellar atrophy is a variant of Wernicke's disease. The patients affected regularly have malnutrition, and occasionally rapid and excessive loss of weight precedes the exacerbation of symptoms. The view that alcohol by itself is not the primary pathogenetic factor is also supported by reports describing the development of symptoms during phases of abstention from alcohol. Furthermore, malnutrition not accompanied by alcohol consumption may cause cerebellar ataxia indistinguishable from the deficits caused by alcoholic cerebellar atrophy. It remains unclear why a lack of thiamine may cause Wernicke's disease in some patients and a selective lesion of parts of the cerebellum in others.

### Practical Management

The most important therapeutic step is a cessation of alcohol consumption. Assuming that thiamine defi-

ciency is the major pathogenetic factor, thiamine and other B vitamins are supplemented according to the protocol suggested for the treatment of Wernicke's disease. These measures should be accompanied by physiotherapy.

## ALCOHOL-RELATED CEREBRAL ATROPHY

Chronic alcohol consumption causes cerebral atrophy, which involves both white and gray matter and shows regional specificity, with the frontal lobes being particularly affected (Kril and Halliday, 1999). Even alcoholics with no overt functional deficits show on average significantly wider ventricles and cortical sulci on CT or MRI scans than healthy controls. Neuropathological studies using unbiased stereological methods have provided evidence that a loss of neurons contributes to the atrophy delineated by neuroradiological techniques (Kril et al., 1997). On the other hand, a number of investigations has demonstrated that the neuroradiological signs of cerebral atrophy are partially, although by no means completely, reversible (Carlen et al., 1978; Carlen and Wilkinson, 1987; Muuronen et al., 1989). The reduction of cerebral atrophy seems to be accompanied by an improvement of cognitive function (Muuronen et al., 1989). The extent of this improvement, though, is still a matter of debate, and the anatomical basis of the mechanisms responsible for the reduction of the cerebral atrophy after cessation of alcohol consumption delineated by neuroradiological examinations remains unclear. The reduction of cerebral atrophy takes some weeks to months. MRI studies, which have failed to show a significant uptake of water by white matter (Schroth et al., 1988), are at odds with the view that the reduction of atrophy reflects reunfolding of the brain by movement of water and electrolytes into brain tissue. These findings rather suggest that the reduction might be a consequence of regeneration, taking place at cellular levels. One attractive possibility would be the regrowth and reshaping of dendritic trees, including the regeneration of dendritic spines (Carlen and Wilkinson, 1987; Lescaudron et al., 1989). These considerations have so far not led to any suggestions as to how to promote the reduction of alcohol-induced cerebral atrophy.

## CENTRAL PONTINE MYELINOLYSIS

### Clinical Aspects

Central pontine myelinolysis is an osmotic demyelination disorder that typically affects the central parts of the pons, without being confined to this part of the brain. Accompanying extrapontine lesions may be found in about 10% of all cases, in the thalamus, cerebellum, internal capsule, basal ganglia, and other parts of the brain (Adams *et al.*, 1959; Weissman and Weissman, 1989; Wright *et al.*, 1979). The spectrum of clinical symptoms ranges from mild signs of pontine dysfunction to a locked-in syndrome. Typically there are signs of pseudobulbar paralysis with impaired articulation and swallowing, the symptoms of pyramidal tract dysfunction with hyperreflexia, extensor plantar responses, quadriparesis, and impaired consciousness (Goebel and Herman-Ben Yur, 1976; Pfister *et al.*, 1985).

Most studies available suggest that central pontine myelinolysis is due to too rapid correction of hyponatremia (e.g., Illowsky and Laureno, 1987; Norenberg *et al.*, 1982), causing transient extracellular hyperosmolarity. Chronic alcoholism is only one cause of an imbalance of electrolytes leading to hyponatremia. It is probably the most important cause of central pontine myelinolysis, because it is frequently accompanied by liver damage. Several recent observations suggest that liver dysfunction might increase the susceptibility of brain tissue to fluctuations in serum sodium levels (e.g., Estol *et al.*, 1989).

### Natural Course

In former years, central pontine myelinolysis was looked on as a disease that usually ended fatally and was usually diagnosed postmortem. With the advent of modern brain imaging techniques (CT scan, MRI) the number of reports on patients, in whom the diagnosis had been made in life, has increased. These cases are often characterized by comparatively mild clinical symptoms and comparatively good recovery.

### Therapy

In response to hyonatremia, organic osmolytes move into the extracellular space to make the cells more iso-osmolar with respect to serum. Rapid correction of chronic hyponatremia leads to an overshoot of extracellular brain sodium and chloride, while at the same time, the concentrations of organic molecules remain low. The consequence is net extracellular hyperosmolarity and therefore shrinkage of the brain, imposing physical stress on brain cells and blood vessel endothelium, which in turn release myelinolytic factors (Norenberg, 1983; Norenberg and Papendick, 1984). Animal experiments have shown that steroids may interfere with this mechanism by protecting endothelial cells. Such experiments furthermore suggest that

colchicine may be beneficial, by blocking secondary macrophage invasion into myelinoclastic lesions (Oh, 1990; Rojiani *et al.*, 1988).

Plasmapheresis has been considered in an attempt to remove myelinolytic factors and immunoglobulins to block their action. In selected patients, plasmapheresis (Bibl *et al.*, 1999), high-dose corticosteroids (Nakano *et al.*, 1996) or intravenous immunoglobulins (Finsterer *et al.*, 2000) have been found effective. The same holds for treatment with daily applications of thyroid-releasing hormone (Chemaly *et al.*, 1998). However, in the absence of well-controlled studies, these approaches remain fully experimental, and no general recommendation to use either of them can be made. In any case, symptomatic measures such as those necessary to prevent pneumonia, embolism, or decubitus have to be taken. In the case of locked-in syndrome ventilation of the patient may be necessary.

Despite its considerable risk it would be wrong to abstain from correcting hyponatremia. The reason is that especially acute hyponatremia leads to a number of potentially life-threatening CNS complications (brain edema, seizures, respiratory arrest, impaired consciousness). However, correction of hyponatremia has to be carried out very slowly under tight control. Serum sodium levels should not increase more than 0.6 mmol/l/h (Brunner *et al.*, 1990). Moreover, complete normalization or even overshoot into hypernatremia must be avoided. Treatment should be stopped when serum sodium levels corresponding to only mild hyponatremia (between 121 and 134 mmol/l) have been reached (Ayus *et al.*, 1987). In case of an excessive correction rate, hypotonic fluids should be given to re-lower the sodium rapidly to pretreatment levels. Unlike acute hyponatremia, chronic hyponatremia is usually comparatively well tolerated and treated sufficiently by fluid restriction.

## MARCHIAFAVA-BIGNAMI DISEASE

### Clinical Aspects

Marchiafava-Bignami disease is characterized by a necrosis of the central parts of the corpus callosum, accompanied by glial sclerosis of cortical layer III, the origin and target of callosal fibers, and occasionally also necrosis of the anterior and posterior commissures. It is usually observed after excessive and long-term (greater than 20 years) consumption of alcoholic beverages, mainly red wine. The cause of the callosal necrosis is unknown. It has been suggested that Marchiafava-Bignami disease bears a certain relation to cyanide-induced encephalopathies observed in experimental animals (Brion, 1976). This has prompted the speculation that the callosal necrosis might actually be mediated by demyelinating endogenous cyanides, released by an alcoholism-related disturbance of vitamin $B_{12}$ metabolism. However, no compelling evidence in support of this speculation has so far been presented, and the pathogenesis of this rare disease remains poorly understood.

The clinical picture of Marchiafava-Bignami disease is comparatively ill defined. There may be progressive mental decline, seizures, dysarthria, disturbances of stance and gait, increased muscle tone, signs of pyramidal tract dysfunction, grasping, and disturbances of consciousness of varying degrees sometimes leading to coma. Ocular motor gaze paresis is unusual in Marchiafava-Bignami disease.

### Natural Course

The onset of the disease is typically acute. The disease has been thought to be progressive and in most cases to end fatally after only a few days to a few months. However, with the advent of modern brain imaging techniques, the number of reports of more benign courses, characterized by long-lasting remissions and outcome with comparatively mild residual deficits has increased (Baron *et al.*, 1989; Canaple *et al.*, 1992; Tomasini *et al.*, 1992). No specific measure is known that influences the natural course of the disease reliably. However, recently a case of successful high-dose intravenous corticoid treatment has been reported (Kikkawa *et al.*, 2000), suggesting that corticosteroids might be considered as an option in rapidly progressive cases. Otherwise, treatment is confined to symptomatic measures.

## ALCOHOLIC PELLAGRA ENCEPHALOPATHY

Alcoholic pellagra encephalopathy is characterized by morphological changes and clinical signs that are very similar to those of endemic pellagra encephalopathy. We mention this disease here because there is evidence that pellagra encephalopathy in alcoholics may be due to correction of thiamine or pyridoxine deficiencies without concomitant application of nicotinic acid (Hauw *et al.*, 1988; Serdaru *et al.*, 1988).

## ALCOHOLIC MYELOPATHY

### Clinical Aspects

In rare instances chronic alcoholics may have the signs of a progressive myelopathy develop caused by dysfunction of the more lateral and dorsal parts of the spinal cord. These patients are seen with spastic paraparesis, sensory disturbances, and bladder dysfunction. The older literature has favored the notion that alcoholic myelopathy is actually a hepatic myelopathy, a metabolic intoxication of the spinal cord, secondary to liver dysfunction comparable to hepatic encephalopathy (see Chapter 72). However, this view has been questioned, because several cases of typical alcoholic myelopathy without significant liver dysfunction have been reported.

### Natural Course

The symptoms develop progressively. Abstention from alcohol stops the progression but is not usually followed by a significant improvement (Sage *et al.*, 1984).

### Practical Management

In the case of chronic hepatic insufficiency with portocaval shunting, the treatment follows the recommendations given for the treatment of hepatic encephalopathy (see Chapter 72). Abstention from alcohol is, of course, mandatory. Vitamin deficiencies should be corrected, and the necessary symptomatic measures taken.

## ALCOHOLIC POLYNEUROPATHY

### Clinical Aspects

Alcoholic polyneuropathy is predominantly axonal, affecting both afferent and efferent fibers. The legs are usually affected earlier and more severely, and the progression is from distal to proximal in both legs and arms. The clinical picture is characterized by pain, paresthesias, and paresis. The character of the pain is variable. It may be dull and constant or else sharp and lancinating. Other complaints may be cramping sensations in the muscles of the feet and calves, coldness of the feet, or conversely feelings of heat and burning. Pain and paresthesias are usually worsened by contact. Examination reveals pain in muscles and more superficially located nerves with pressure, loss of tendon reflexes in the legs, less frequently in the arms, paresis, atrophy, and sensory disturbances emphasizing the legs. The involvement of autonomic fibers is indicated by excessive sweating of the soles and palmar surfaces of the hand, thinning of the skin, alteration of nails, and erectile dysfunction. Involvement of cranial nerves is rare and, if present, is usually confined to the oculomotor and lower cranial nerves. Occasionally sensory and motor disturbances are minor, but nevertheless, subjective complaints may be considerable. Electromyography of the muscles involved reveals signs of denervation. Nerve conduction velocities are reduced to a varying degree. A study by Scholz and co-workers (1986) indicates that these electrophysiological tests are quite sensitive. About half of their sample of chronic alcoholics showed clinical signs of polyneuropathy, and electrophysiological tests were positive in two thirds.

## Natural Course

The course is variable. Alcoholic polyneuropathy may not be apparent clinically, despite continued alcohol consumption. Alternatively, it may lead to a more or less stable clinical picture or cause a steady increase of the deficits. Rarely, alcoholic polyneuropathy may be seen as acute-onset sensorimotor polyneuropathy, clinically mimicking Guillain-Barré syndrome (Vandenbulcke and Janssens 1999; Wöhrle et al., 1998). Patients may recover within a couple of months of complete alcohol abstention. Even patients who initially have been wheelchair bound may achieve improvement. The speed of recovery is determined largely by the time required for regeneration of the most markedly afflicted nerve. Less than a couple of months may be adequate for a substantial improvement of the clinical picture in mild cases. Prognosis does not depend on age.

## Therapeutic Principles

As discussed earlier, vitamin deficiencies including thiamine, other B-complex vitamins, or folic acid are regularly found in alcoholics. Probably the most important factor to be considered apart from the vitamin deficiencies mentioned is alcohol and its major toxic metabolite acetaldehyde. It is known that deficiencies of any of the vitamins mentioned may cause peripheral nerve damage. The importance of folic acid in the pathogenesis of alcoholic polyneuropathy has been shown by the correlation between biochemical markers of folic acid deficiency and electrophysiological measures of the degree of polyneuropathy (Gimsing et al., 1989). On the other hand, the contribution of thiamine deficiency, previously thought to be well established, has been questioned more recently. The reason is that no correlation between the extent of thiamine deficiency and the occurrence and extent of the polyneuropathy could be found (e.g., Meyer et al., 1981). On the other hand, benfotiamine, a thiamine derivative, which (unlike thiamine) is excellently absorbed after oral administration, has been shown to be effective in the treatment of alcoholic polyneuropathy. A medication of 320 mg/day

benfotiamine over 4 weeks followed by 120 mg/day in the following 4 weeks led to a significant improvement of sensory and motor functions and a tendency toward improvement of pain and coordination, provided other B vitamins were not added (Woelk et al., 1998). No adverse side effects have been noted. In view of the established efficacy of benfotiamine, the application of thiamine in its risky parenteral form is not justified in the treatment of alcoholic polyneuropathy.

It has been hoped that treatment with gangliosides might promote nerve regeneration and reinnervation. However, despite some early optimistic reports (Bassi et al., 1982; Mamoli and coworkers, 1980), convincing evidence for the efficacy of this approach is lacking. On the other hand, the intramuscular application of gangliosides bears the risk of inducing acute polyneuritis of the Guillain-Barré type.

## Practical Management

Management consists of complete abstention from alcohol. In addition, a well-balanced diet containing all the necessary vitamins should be guaranteed, supplemented by oral benfotiamine (320 mg/day for 4 weeks, followed by 120 mg/day for at least 4 further weeks). The nursing and physical therapy measures required are described in Chapters 88 and 89 (polyneuropathies). There and in Chapter 11 (pain) recommendations for the symptomatic treatment of the painful dysesthesias and hyperpathia will be found.

# NUTRITIONAL OPTIC NEUROPATHY

## Clinical Aspects

The patient affected by nutritional optic neuropathy (tobacco-alcohol amblyopia) reports blurred or loss of vision, evolving gradually within several days up to a few weeks. Examination discloses bilateral, symmetrical scotomas that emphasize the central or centrocaecal parts of the visual field. The scotomas are larger for colored than for white stimuli. Pallor of the temporal parts of the optic disc may be observed. The substrate for these deficits is a bilateral, symmetrical demyelination of the papillomacular bundle of the optic nerve, which may lead to fiber loss. The lesion starts in the retrobulbar region and progresses retinofugally. The retina typically shows a loss of parafoveal ganglion cells. Tobacco-alcohol neuropathy is observed in about 1 of 200 hospitalized alcoholics. However, mild optic nerve damage may be much more frequent in alcoholics. This is suggested by a study presented by Meinck and Adler (1982), who found alterations of VEP parameters in

40% of a sample of alcoholics, most of whom had a completely normal ophthalmolgical examination.

## Natural Course

Without treatment, nutritional optic neuropathy leads to an irreversible optic nerve atrophy. Even with optimal therapy (see later), full recovery can no longer be expected if the symptoms have persisted for more than a couple of weeks (Aulhorn, 1989).

## Therapeutic Principles

A disease resembling tobacco-alcohol amblyopia in every respect may be evoked by deficiencies in any of the B vitamins, as well as by exposure to cyanides or methanol or combinations of these risk factors. It has been suggested that the final common pathway may be an acquired mitochondrial impairment (Sadun, 1998), which for unknown reasons would emphasize specific parts of the retina while sparing other metabolically active tissues. However, as yet, no attempts to stop the disease by trying to booster mitochondrial functions have become known. At any rate, even if alcohol consumption continues, recovery from nutritional optic neuropathy is possible, provided B vitamins are supplemented (Dreyfus, 1977).

## Practical Management

Therapy consists mainly of an oral substitution of B vitamins and a well-balanced diet. As mentioned before, the extent and speed of recovery critically depend on the extent and duration of the deficits before onset of therapy.

## ALCOHOLIC MYOPATHIES

At least three different types of myopathy may be observed in association with alcohol abuse. The first type is acute and characterized by rhabdomyolysis and myoglobinuria, the second type is also acute. It is differentiated from the first one by the occurrence of hypokalemia and the absence of rhabdomyolysis. Finally, the third form is chronic.

## Acute Alcoholic Rhabdomyolysis

### Clinical Aspects

Alcohol is probably the most frequent cause of acute rhabdomyolysis with myoglobinuria. However, compa-rable syndromes may be evoked by intoxications of other kinds, including carbon monoxide, opiates, barbiturates, snake venom, and various industrial toxins (for a review, see Görtz et al., 1978). The clinical picture is defined by extreme pain, swelling, and inability to move the affected muscles. Every muscle of the limbs and the trunk may be involved, although eye and facial muscles are usually spared. The decomposition of muscle fibers causes hyperkalemia, excessive rise of serum creatine kinase levels, and acute kidney failure. Observations based on animals suggest that acute alcoholic rhabdomyolysis may result from acute food withdrawal after chronic exposure to alcohol (Haller and Drachman, 1980). However, details of the pathogenesis still have to be worked out. It should, however, be emphasized that shifts of serum potassium levels, which are essential in the pathogenesis of hypokalemic alcohol myopathy (discussed next), do not seem to be involved in the development of acute alcoholic rhabdomyolysis (Haller et al., 1984; Hallgren et al., 1980; Perkoff, 1972).

### Natural Course and Therapy

A specific therapy for acute alcoholic rhabdomyolysis is not known, and treatment is therefore confined to symptomatic measures among which the management of acute kidney failure is prominent. The necessary steps include enforced diuresis, strict control of water and electrolytes, and possibly hemodialysis. Mortality from toxic rhabdomyolysis, including acute alcoholic rhabdomyolysis, is high. For instance, three of the four patients presented by Görtz and coworkers (1978) died despite early hemodialysis. Observations like this have prompted others to discuss early and repeated plasmapheresis as a therapeutic option. Observations that would allow one to assess the usefulness of this suggestion are not available. If the patient survives the acute kidney failure, a significant recovery of the motor deficits is possible, provided strict abstention from alcohol is maintained. Otherwise further bouts of acute rhabdomyolysis may occur.

## Hypokalemic Alcoholic Myopathy

### Clinical Aspects

This second form of acute alcohol-related myopathy is distinguished from the previous one by a different clinical picture, pathogenesis, and therapy (Finsterer et al., 1998; Sharief et al., 1997). It is caused by hypokalemia, which in alcoholics results from frequent vomiting and increased loss of potassium through the gastrointestinal tract, whereas renal excretion of potassium is not increased. Within several days up to a few weeks painless paresis develops, mainly affecting the proximal parts of the limbs, and serum potassium levels drop below 2 mmol/l. Serum levels of liver and muscle enzymes are significantly increased. Biopsy of affected muscles reveals necrosis of single muscle fibers and development of vacuoles.

### Therapy

Treatment consists of slow infusion of up to 20 mmol/h of potassium chloride or potassium lactate with continuous monitoring of the electrocardiogram. The overall amount of potassium given is determined by the size of the deficit. Special care has to be taken in cases of renal dysfunction. The paresis will usually recede within 7–14 days of treatment. Serum enzyme levels will return to normal within the same period of time.

## Chronic Alcoholic Myopathy

### Clinical Aspects

This is a painless type of myopathy that typically develops subacutely or chronically in two thirds of all alcohol misusers. It may lead to considerable atrophy and paresis of the proximal limb muscles. Usually the pelvic girdle musculature is affected much more than the shoulder girdle musculature. Histological investigation shows that damage mainly involves type II muscle fibers (Langohr *et al.*, 1989). The condition is often accompanied by cardiomyopathy. Furthermore, it may be accompanied by alcoholic polyneuropathy, which usually involves the more distal parts of the limbs (see the section on alcoholic neuropathy). The extent of the myopathy (and also the cardiomyopathy) linearly correlates with the overall amount of alcohol consumed during the patient's lifetime (Urbano-Marquez *et al.*, 1989). Blood tests are usually characterized by signs of alcoholic liver damage with increased levels of γ-glutamyl transpeptidase, glutamic oxaloacetic transaminase, and glutamic pyruvic transaminase, although levels of creatine phosphokinase are usually normal. According to Oh (1976) full-blown chronic alcoholic myopathy is found in about 1% of all hospitalized alcoholics. However, milder manifestations with little or even absent clinical symptoms may be demonstrated in up to two thirds of all alcohol misusers by electromyographical or histological examination. The cause of chronic alcoholic myopathy remains a puzzle. Possible pathogenetic factors that have been suggested involve damage to muscle fiber membranes or mitochondria by toxic alcohol metabolites or free radicals, intracellular edema caused by concomitant liver disease, decrease of glycolytic enzymes brought about by the demands of alcohol metabolism, reduced calcium transients and immunological disturbances.

### Course and Therapy

Once again, abstention from alcohol accompanied by a well-balanced diet and physiotherapy are the essential measures. According to Fernandez-Sola *et al.* (2000), recovery is possible, even in cases in which abstention is not complete and low-dose (≤60 g ethanol/day) alcohol consumption is maintained. Recovery from alcoholic myopathy is usually faster and more complete than recovery from alcoholic polyneuropathy.

## TREATMENT OF ALCOHOL DEPENDENCY

The preceding paragraphs have emphasized that strict abstention from alcohol is indispensable if recovery from alcohol-related diseases of the nervous system is to be achieved or at least further progression avoided. Provided that strict lasting abstention is ensured, a normal life expectancy may be achieved even after many years of alcohol dependency (Bullock *et al.*, 1992). In other words, not only the devastating psychological and social consequences of alcoholism but also its somatic consequences require an uncompromising fight against alcohol dependency. Although other therapeutic measures such as drugs may be important or even essential in some particular forms of alcohol-related disease, they will never make the treatment of alcohol dependency unnecessary. It is still controversial (Feuerlein, 1989) as to whether there are alcoholics who are able to return to a controlled intake of alcoholic beverages. At any rate, even if some alcoholic patients are able to achieve this, their number would still be negligibly small compared with those who will never be able to return to controlled drinking. In other words, there is no reason to give up complete and continuing abstention from alcohol as the main therapeutic objective.

A major prerequisite for any promising treatment of alcohol dependency is that the patient is able to see that he or she has a severe loss of self-control. The alcoholic must be able to realize the somatic and social consequences of this deficit and decide voluntarily in favor of abstention. The patient has to develop a sense of self-responsibility. This does not mean to say that the offer of treatment should depend on the patient's initiative. Rather the physician, social workers, and people close to the patient should try to promote the necessary motivation and understanding. With regard to the psychotherapeutic and sociotherapeutic instruments available, the reader is referred to more detailed discussions on the subject in the literature (e.g., Feuerlein, 1989; Schied, 1989). In the remainder of this section we will discuss old and new pharmacological approaches that show promise for the treatment of alcohol dependency and may become part of an integrated treatment plan involving education, psychological, and social support. The drugs considered in the treatment of alcoholism fall in three groups. The agents of the first and oldest group induce aversion to alcohol. The substances of the second, probably the most promising group, block craving for alcohol and drugs. Finally, the members of the third group comprise serotonergic substances also ameliorating depression and associated psychiatric disorders.

Disulfiram is the only relevant representative of the group of substances inducing aversion to alcohol. It is an irreversible blocker of acetaldehyde dehydrogenase, leading to accumulation of acetaldehyde. Under disulfiram medication, about 10–30 minutes after the intake of alcohol a very unpleasant reaction occurs, characterized by nausea, vomiting, sleepiness, hypotension, tachycardia, and flush. The fact that this reaction is only partially reproduced by infusion of acetaldehyde suggests that the alcohol-disulfiram reaction cannot be fully attributed to accumulation of acetaldehyde. Rather the disulfiram-induced blockade of additional enzymes such as dopamine-beta-hydroxylase probably contributes to the reaction. The disulfiram-alcohol reaction can be stopped by the intravenous injection of antihistamines or alternatively by high oral doses of ascorbic acid. Disulfiram is given in oral doses of 0.2–0.5 g/day, depending on body weight, if possible under supervision. The disulfiram effect will last for 3 up to 10 days after the last dose of the drug. Disulfiram implants (one-time dose of 800–1000 mg) are an attractive alternative, promising to guarantee compliance. However, in view of the fact that the bioavailability of disulfiram from implants has not been demonstrated, at present, implants cannot be recommended.

There have been rare reports of absent disulfiram-alcohol reactions despite intake of disulfiram. Conversely, mild symptoms reminiscent of a full-fledged disulfiram-alcohol reaction including sleepiness or dizziness may occur quite frequently even without intake of alcohol. Other side effects of disulfiram that may be observed in the absence of alcohol intake are seizures (Price and Silberfarb, 1976), liver damage (Ranek and Buch Andreasen, 1977), teratogenic damage (if applied during the first 3 month of pregnancy; Nora, 1977), psychosis (Reisner et al., 1968), polyneuropathy, disturbances of taste, optic neuropathy, ataxia (Kwentus and Major, 1979), and bone marrow suppression (Casagrande and Michot, 1989).

Despite its use for 50 years, the evidence for a beneficial effect of disulfiram is far from being convincing.

For instance, a large multicenter study carried out by Fuller et al. (1986) could not find a difference in the number of abstinent patients treated with the alcohol-sensitizing drug disulfiram compared with those treated with placebo. The major predictor for the success of the treatment turned out to be compliance with the drug regimen, emphasizing once more the importance of motivation and self-control. In this study, the amount of alcohol consumed was smaller in those nonabstinent patients who were on disulfiram compared with those taking placebo. Several other studies also show that the drinking frequencies, the number of drinking days, and the amount of alcohol consumed is reduced. However, disulfiram does not enhance abstinence rates (Garbutt et al., 1999; Hughes and Cook, 1997). Although the benefit from disulfiram monotherapy seems to be limited, recent work suggests that it might be useful as supplement to the anticraving drugs, discussed below, improving their effectiveness (Besson et al., 1998).

In conclusion, disulfiram may be considered an option in those patients who have proved to be unresponsive to psychotherapeutic and sociotherapeutic measures after careful consideration of the possible risks and side effects (see e.g., Banys, 1988, for further discussion). Disulfiram should not be given in cases of liver disease, epilepsy, psychosis, and disturbances of hematopoiesis. Further contraindications are fresh peptic or duodenal ulcers and coronary artery disease.

Although the potential of disulfiram seems to be limited, the substances reducing craving, the urge to drink, hold considerable promise for the future (see Garbutt et al. for an analysis of the studies available). Their common mode of action seems to be a facilitation of central satiation signals. Three of them, acamprosate, naltrexone, and GHBA, are currently in clinical use in a number of countries. Acamprosate (calcium-acetyl-homotaurinate) is an analogue of GABA known to block calcium channels and glutamate receptors. Acamprosate, administered in amounts of 1300–2000 mg/day orally, has been shown to reduce the drinking frequency and the number of drinking days. Several studies also found that acamprosate increased abstinence rates (Sass et al., 1996). The side effects are diarrhea, dizziness, increased sexual desire, and itching. Naltrexone is an antagonist of endogenous opioids such as β-endorphin and enkephalins. When given in amounts of 50 mg/day, it has been shown to reduce both drinking frequency and relapse rate. Nalmefene, another antagonist of endogenous opioids, has a similar efficacy when applied in amounts of 10–40 mg/day. The side effects of the endogenous opioids are nausea and headache. GHBA (gamma-hydroxy-butyric acid), a precursor and metabolite of GABA, may not only reduce withdrawal symptoms but also help to maintain abstinence after withdrawal by effectively attenuating alcohol craving. Treatment with 50 mg/kg/day reduces the number of daily drinks by approximately 50% and increases the days of abstinence by a factor of three (Gallimberti *et al.*, 2000; Moncini *et al.*, 2000). During the 1980s, GHBA was widely available over-the-counter in health food stores, purchased by bodybuilders for its ability to stimulate growth hormone release. Later it became popular as a drug, offering a pleasant, alcohol-like, hangover-free "high" with potent prosexual effects. However, when used in a standardized clinical setting, such as in Italy since 1991, abuse of GHBA seems to be very rare and serious side effects absent (Beghe and Carpanini, 2000). The known side effects can be largely understood as reflecting the alcohol-mimicking action of GHBA and include disinhibition and prosexual effects, sedation and sleepiness, dizziness, headache, hypotension, seizures, and at very high doses cardiac and respiratory depression. GHBA may lead to addiction and, actually, to GHBA withdrawal, when applied outside a clinical setting.

The third group of drugs for the treatment of alcohol dependency is composed of serotonergic compounds such as the serotonin-uptake inhibitors zimelidine, citalopram, fluoxetine, fluvoxamine, the serotonin agonist (5-HT$_{1a}$) buspirone hydrochloride or the serotonin antagonist (5-HT$_3$) ondastron hydrochloride, and lithium. The hope that these and related drugs might be useful in fighting alcohol dependency was originally prompted by experiments that showed that an increase of serotonergic activity in animals led to a reduction in alcohol consumption. However, the literature on the effect of these drugs in humans is less convincing, with only few controlled studies yielding mixed results. Garbutt and coworkers (1999), who recently reanalyzed the observations available, conclude that the literature does not support the use of any of the compounds of this group.

# REFERENCES

Acharya, V., Store, S. D., and Golwalla, A. F. (1969). Anaphylaxis following ingestion of aneurine hydrochloride. *J. Indian Med. Assoc.* **52**, 84–85.

Adams, A. (1976). Nutritional cerebellar degeneration. *In* "Handbook of Clinical Neurology" (P. J. Vinken and G. W. Bruyn, Eds.), Vol. 28, Metabolic and Deficiency Disease of the Nervous System. pp. 271–283. Elsevier, Amsterdam.

Adams, A., and Victor, M. (1989). "Principles of Neurology", 4th ed. McGraw-Hill, New York.

Adams, R. D., Victor, M., and Mancall, E. L. (1959). Central pontine myelinolysis: A hitherto undescribed disease occurring in alcoholic and malnourished patients. *Arch. Neurol. Psychiatry* **81**, 154–172.

Adinoff, B., and Pols, B. (1997). Dexamethasone in the treatment of the alcohol withdrawal syndrome. *Am. J. Drug Alcohol Abuse* **23**, 615–622.

Addolorato, G., Balducci, G., Capristo, E., Attilia, M. L., Taggi, F., Gasbarrini, G., and Ceccanti, M. (1999). Gamma-hydroxybutyric acid (GHB) in the treatment of alcohol withdrawal syndrome: a randomized comparative study versus benzodiazepine. *Alcohol Clin. Exp. Res.* **23**, 1596–604.

Alldredge, B. K., Lowenstein, D. H., and Simon, R. P. (1989). Placebocontrolled trial of intravenous diphenyhydantoin for short-term treatment of alcohol withdrawal seizures. *Am. J. Med.* **87**, 645–648.

Angunawela, I. I., and Barker, A. (2001). Anticholinesterase drugs for alcoholic Korsakoff syndrome. *Int. J. Geriatr. Psychiatry* **16**, 338–339.

Athen, D., Hippius, H., Meyendorf, R., Riemer, C. H., and Steiner, R. H. (1977). Ein Vergleich der Wirksamkeit von Neuroleptika und Clomethiazol bei der Behandlung des Alkoholdelirs. *Nervenarzt* **48**, 528–532.

Aulhorn, E. (1989). Die Tabak-Alkohol-Amblyopie. *In* "Der chronische Alkoholismus Grundlagen, Diagnositik, Therapie" (H. W. Schied, H. Heimann, and K. Mayer, Eds.), pp. 163–174. Gustav Fischer Verlag, Stuttgart and New York.

Ayus, J. C., Radha, K., Krothapalli, K., and Arieff, A. I. (1987). Treatment of symptomatic hyponatremia and its relation to brain damage. *N. Engl. J. Med.* **317**, 1190–1195.

Banys, P. (1988). The clinical use of disulfiram (antabuse): A review. *J. Psychoact. Drugs* **20**, 243–261.

Baker, K. G., Harding, A. J., Halliday, G. M., Kril, J. J., and Harper, C. G. (1999). Neuronal loss in functional zones of the cerebellum of chronic alcoholics with and without Wernicke's encephalopathy. *Neuroscience* **91**, 429–438.

Baron, R., Heuser, K., and Moarioth, G. (1989). Marchiafava-Bignami disease with recovery diagnosed by CT and MRI: Demyelination affects several CNS structures. *J. Neurol.* **236**, 364–366.

Bassi, S., Albizzati, M. G., Calloni, E., and Frattola, L. (1982). Electromyographic study of diabetic and alcoholic polyneuropathic patients treated with gangliosides. *Muscle Nerve* **5**, 352–356.

Baumgartner, G. R. (1988). Clonidine versus chlordiazepoxide in acute alcohol withdrawal: A preliminary report. *South. Med. J.* **81**, 56–60.

Beghe, F., and Carpanini, M. T. (2000). Safety and tolerability of gamma-hydroxybutyric acid in the treatment of alcohol-dependent patients. *Alcohol* **20**, 223–225.

Besson, J., Aeby, F., Kasas, A., Lehert, P., and Potgieter, A. (1998). Combined efficay of acamprosate and disulfiram in the treatment of alcoholism: a controlled study. *Alcohol Clin. Exp. Res.* **22**, 573–579.

Bibl, D., Lampl, C., Gabriel, C., Jungling, G., Brock, H., and Kostler, G. (1999). Treatment of central pontine myelinolysis with therapeutic plasmapheresis. *Lancet* **353**, 1155.

Björkvist, S. E. (1975). Clonidine in alcohol withdrawal. *Acta Psychiatr. Scand.* **52**, 256–263.

Blum, K. U., Kasemir, H., and Scharfe, W. (1974). Untersuchungen zur Pathogenese der anaphylaktischen Reaktionen nach Thiaminapplikation. *Verh. Dtsch. Ges. Inn. Med.* **80**, 1569–1571.

Brion, S. (1976). Marchiafava-Bignami syndrome. *In* "Handbook of Clinical Neurology" (P. H. Vinken and G. W. Bruyn, Eds.), Vol. 28. pp. 317–332. Elsevier, Amsterdam.

Brunner, J. E., Redmond, J. M., Haggar, A. M., Kruger, D. F., and Elias, S. B. (1990). Central pontine myelinolysis and pontine lesions after rapid correction of hyponatremia: A prospective magnetic resonance imaging study. *Ann. Neurol.* **27**, 61–66.

Bullock, K. D., Reed, R. J., and Grant, I. (1992). Reduced mortality risk in alcoholics who achieved longterm abstinence. *JAMA* **257**, 668–672.

Canaple, S., Rosa, A., and Mizon, J. P. (1992). Maladie de Marchiafava-Bignami: Disconnexion interhemispherique, evolution favorable. Aspect neuroradiologique. *Rev. Neurol.* **148**, 638–640.

Carlen, P. L., and Wilkinson, D. A. (1987). Reversibility of alcohol-related brain damage: Clinical and experimental observations. *Acta Med. Scand. Suppl.* **717**, 19–26.

Carlen, P. L., Wortzmann, G., Holgate, R. C., Wilkinson, D. A., and Rankin, J. G. (1978). Reversible cerebral atrophy in recently abstinent chronic alcoholics measured by computed tomography scans. *Science* **200**, 1076–1078.

Casagrande, G., and Michot, F. (1989). Alcohol-induced bone marrow damage: Status before and after a 4-week period of abstinence from alcohol with or without disulfiram. A randomized bone marrow study in alcohol-dependent individuals. *Blut* **59**, 231–236.

Castaneda, R., and Cushman, P. (1989). Alcohol withdrawal: A review of clinical management. *J. Clin. Psychiatry* **50**, 278–284.

Chemaly, R., Halaby, G., Mohasseb, G., Medlej, R., Tamraz, J., and el-Koussa, S. (1998). Myelinolyse extra-pontine: traitement par T.R.H. [Extrapontine myelinolysis: treatment with TRH]. *Rev. Neurol. (Paris)* **154**, 163–165.

Davis, V. E., and Walsh, M. J. (1970). Alcohol, amines and alkaloids: A possible biochemical basis for alcohol addiction. *Science* **167**, 1005–1006.

Dreyfus, P. M. (1977). Diseases of the nervous system in chronic alcoholics. *In* "The Biology of Alcoholism", (B. Kissin and H. Begleiter, Eds.), Plenum, New York and London.

Estol, C. J., Faris, A. A., Martinez, A. J., and Ahdab-Barmada, M. (1989). Central pontine myelinolysis after liver transplantation. *Neurology* **39**, 493–498.

Fernandez-Sola, J., Nicolas, J. M., Sacanella, E., Robert, J., Cofan, M., Estruch, R., and Urbano-Marquez, A. (2000). Low-dose ethanol consumption allows strength recovery in chronic alcoholic myopathy. *QJM* **93**, 35–40.

Feuerlein, W. (1989). "Alkoholismus—Mißbrauch and Abhängigkeit," 4th Ed. Thieme, Stuttgart and New York.

Finsterer, J., Engelmayer, E., Trnka, E., and Stiska, M. (2000). Immunoglubulins are effective in pontine myelinolysis. *Clin. Neuropharmacol.* **23**, 110–113.

Finsterer, J., Hess, B., Jarius, C., Stollberger, C., Budka, H., and Mamoli, B. (1998). Malnutrition-induced hypokalemic myopathy in chronic alcoholism. *J. Toxicol. Clin. Toxicol.* **36**, 369–373.

Flink, E. B. (1981). Magnesium deficiency. Etiology and clinical spectrum. *Acta Med. Scand. Suppl.* **647**, 125–137.

Fuller, R. K., Branchey, L., Brightwell, D. R., Derman, R. M., Emrick, C. D., Iber, F. L., James, K. E., Lacoursiere, R. B., Lee, K. K., Lowenstam, I., Manny, I., Neiderhiser, D., Nocks, J. J., and Shaw, S. (1986). Disulfiram treatment of alcoholism. A veterans administration cooperative study. *JAMA* **256**, 1149–1155.

Gallimberti, L., Canton, G., Gentile, N., Ferri, M., Cibin, M., Ferrara, S. D., Fadda, F., and Gessa, G. L. (1989). Gamma-hydroxybutyric acid for treatment of alcohol withdrawal syndrome. *Lancet* **2**, 787–789.

Gallimberti, L., Spella, M. R., Soncini, C. A., and Gessa, G. L. (2000). Gamma-hyroxybutyric acid in the treatment of alcohol and heroin dependence. *Alcohol* **20**, 257–262.

Garbutt, J. C., West, S. L., Carey, T. S., Lohr, K. N., and Crews, F. T. (1999). Pharmacological treatment of alcohol dependence. A review of the evidence. *JAMA* **281**, 1218–1325.

Gimsing, P., Melgaard, B., Andersen, K., Vilsstrup, H., and Hippe, E. (1989). Vitamin B$_{12}$ and folate function in chronic alcoholic men with peripheral neuropathy and encephalopathy. *J. Nutr.* **119**, 416–424.

Glass, I. B. (1989). Alcoholic hallucinosis: A psychiatric enigma. 1. The development of an idea. *Br. J. Addict.* **84**, 29–41.

Goebel, H. H., and Herman-Ben Yur, P. (1976). Central pontine myelinolysis. *In* "Handbook of Clinical Neurology" (P. J. Vinken and G. W. Bruyn, Eds.), Vol. 28, pp. 285–316. Elsevier, Amsterdam.

Görtz, B., Kunst, H., and Heitman, R. (1978). Toxische Rhabdomyolyse nach Alkohol- und Medikamentenvergiftungen. *Dtsch. Med. Wochenschr.* **103**, 121–123.

Gottlieb, L. D. (1988). The role of beta blockers in alcohol withdrawal syndrome. *Postgrad. Med.* **29**, 169–174.

Haller, R. G., Carter, N. W., Ferguson, E., and Knochel, J. P. (1984). Serum and muscle potassium in experimental alcoholic myopathy. *Neurology* **34**, 529–532.

Haller, R. G., and Drachman, D. B. (1980). Alcoholic rhabdomyolysis: An experimental model in the rat. *Science* **208**, 412–414.

Hallgren, R., Lundin, L., Roxin, L. E., and Venge, P. (1980). Serum and urinary myoglobin in alcoholics. *Acta Med. Scand.* **208**, 33–39.

Harding, A., Halliday, G., Caine, D., and Kril, J. (2000). Degeneration of anterior thalamic nuclei differentiates alcoholics with amnesia. *Brain* **123**, 141–154.

Harper, C. (1983). The incidence of Wernicke's encephalopathy in Australia-a neuropathological study of 131 cases. *J. Neurol. Neurosurg. Psychiatry* **46**, 593–598.

Harper, C., Gold, J., Rodriguez, N., and Perdices, N. (1989). The prevalence of the Wernicke-Korsakoff syndrome in Sydney, Australia: A prospective necropsy study. *J. Neurol. Neurosurg. Psychiatry* **52**, 282–285.

Hauw, J. J., De-Baecque, C., Hausser-Hauw, C., and Serdaru, M. (1988). Chromatolysis in alcoholic encephalopathies. Pellagra-like changes in 22 cases. *Brain* **111**, 843–857.

Hughes, J. C., and Cook, C. C. (1997). The efficacy of disulfiram: a review of outcome studies. *Addiction* **92**, 381–395.

Illowsky, B. P., and Laureno, R. (1987). Encephalopathy and myelinolysis after rapid correction of hyponatremia. *Brain* **110**, 855–867.

Kikkawa, Y., Takaya, Y., and Niwa, N. (2000). A case of Marchiafava-Bignami disease that responded to high-dose intravenous corticosteroid administration. *Rinsho Shinkeigaku* **40**(11), 1122–1125.

Kril, J. J., and Halliday, G. M. (1999). Brain shrinkage in alcoholics: a decade on and what have we learned? *Prog-Neurobiol.* **58**, 381–387.

Kril, J. J., Halliday, G. M., Svoboda, M. D., and Cartwright, H. (1997). The cerebral cortex is damaged in chronic alcoholics. *Neuroscience* **79**, 983–998.

Kwentus, J., and Major, L. F. (1979). Disulfiram in the treatment of alcoholism. *J. Stud. Alcohol* **40**, 428–446.

Ladewig, D., Levin, P., Gastpar, M., Gerking, P., and Roth, E. (1977). "Betablocker und Zentralnervensystem," (H. Kielholz, ed.), pp. 154–162. Huber, Bern.

Langlais, P. J., Mair, R. G., Whalen, P. J., McCourt, W., and McEntee, W. J. (1988). Memory effect of DL-threo-3,4-dihydroxyphenylserine (DOPS) in human Korsakoff's disease. *Psychopharmacology (Berlin)* **95**, 250–254.

Langohr, H. D., Tröster, H., and Zimmermann, C. W. (1989). Alkoholenzymopathie der Muskulatur. *In* "Der chronische Alkoholismus. Grundlagen, Diagnositik, Therapie." (H. W. Schied, H. Heimann, and K. Mayer, Eds.), pp. 229–238. Gustav Fischer Verlag, Stuttgart and New York.

Lenzenhuber, E., Muller, C., Rommelspacher, H., and Spies, C. (1999). Gamma-hydroxybutyrate for treatment of alcohol withdrawal syndrome in intensive care patients. A comparison between two symptom-oriented concepts. *Anaesthesist* **48**, 89–96.

Lescaudron, L., Jaffard, R., and Verna, A. (1989). Modifications in number and morphology of dendritic spines resulting from chronic ethanol consumption and withdrawal: A Golgi study in the mouse anterior and posterior hippocampus. *Exp. Neurol.* **106**, 156–163.

Lignian, H., Fontaine, J., and Askenasi, R. (1983). Naloxone and alcohol intoxication in the dog. *Hum. Toxicol.* **2**, 221–225.

Lister, R. G., Durcan, M. J., Nutt, D. J., and Linnoila, M. (1989). Attenuation of ethanol intoxication by alpha-2 adrenoceptor antagonists. *Life Sci.* 44, 111–119.

Lister, R. G., and Nutt, D. J. (1987). Is Ro 15-4513 a specific alcohol antagonist? *Trends Neurosci.* 10, 223–225.

McEntee, W. J., and Mair, R. G. (1980). Memory enhancement in Korsakoff's psychosis by clonidine: Further evidence for a noradrenergic deficit. *Ann. Neurol.* 7, 466–470.

McGrath, S. D. (1975). A controlled trial of clomethiazole and chlordiazepoxide in the treatment of the acute withdrawal phase of alcoholism. Conference on Alcoholism, pp. 81–90. Longmans, London.

McIntyre, N. (1984). The effects of alcohol on water, electrolytes and minerals. *Contemp. Issues Clin. Biochem.* 1, 117–134.

Malcolm, R., Ballenger, J. C., Sturgis, E. T., and Anton, R. (1989). Double-blind controlled trial comparing carbamazepine to oxacepam treatment of alcohol withdrawal. *Am. J. Psychiatry* 146, 617–621.

Mamoli, B., Brunner, G., Mader, R., and Schanda, H. (1980). Effects of cerebral gangliosides in the alcoholic polyneuropathies. *Eur. Neurol.* 19, 324–326.

Martin, P. R., Adinoff, B., Eckardt, M. J., Stapleton, J. M., Bone, G. A., Rubinow, D. A., Lane, E. A., and Linnoila, M. (1988). Effective pharmacotherapy of alcoholic amnestic disorder with fluvoxamine. Preliminary findings. *Arch. Gen. Psychiatry* 46, 617–621.

Martin, P. R., Ebert, M. H., Gordon, E. K., and Kopin, I. J. (1983). Central and peripheral catecholamine metabolism during clonidine treatment of alcohol amnestic disorder. *Clin. Pharmacol. Ther.* 33, 19–27.

Meinck, H. M., and Adler, L. (1982). Optikusaffektion bei Alkoholabhängigkeit-Früherkennung durch das visuell evozierte Potential. *Nervenarzt* 53, 644–646.

Merikangas, K. R. (1990). The genetic epidemiology of alcoholism. *Psychol. Med.* 20, 11–22.

Meyer, J. G., Neundörfer, B., Rether, R., Walker, G., and Bayerl, J. (1981). Über die Beziehung zwischen alkoholischer Polyneuropathie und Vitamin B1, B12 und Folsäure. *Nervenarzt* 52, 329–332.

Meyer, R. E. (1989). Prospects for a rational pharmacotherapy of alcoholism. *J. Clin. Psychiatry* 50, 403–412.

Meyer, R. E. (2000). Craving: what can be done to bring the insights of neuroscience, behavioral science and clinical science into synchrony. *Addiction Suppl.* 2, S219–27.

Mhatre, M. C., Pena, G., Sieghart, W., and Ticku, M. K. (1993). Antibodies specific for GABA-A receptor alpha subunits reveal that chronic alcohol treatment down-regulates alpha-subunit expression in rat brain regions. *J. Neurochem.* 61, 1620–1625.

Moncini, M., Masini, E., Gambassi, F., and Mannaioni, P. F. (2000). Gamma-hyroxybutyric acid and alcohol-related syndromes. *Alcohol* 20, 285–291.

Muuronen, A., Bergman, H., Hindmarsh, T., and Telakivi, T. (1989). Influence of improved drinking habits on brain atrophy and cognitive performance in alcoholic patients: A 5-year follow-up study. *Alcohol (N.Y.)* 13, 137–141.

Nakano, H., Ohara, Y., Bandoh, K., and Miyaoka, M. (1996). A case of central pontine myelinolysis after surgical removal of a pituitary tumor. *Surg. Neurol.* 46, 32–36.

Nora, A. H. (1977). Limb-reduction anomalies in infants born to disulfiram-treated alcoholic mothers (letter). *Lancet* 2, 664.

Norenberg, M. D. (1983). A hypothesis of osmotic endothelial injury: A pathogenetic mechanism in central pontine myelinolysis. *Arch. Neurol.* 40, 66–69.

Norenberg, M. D., Leslie, K. O., and Robertson, A. S. (1982). Association between rise in serum sodium and central pontine myelinolysis. *Ann. Neurol.* 11, 128–135.

Norenberg, M. D., and Papendick, R. E. (1984). Pathogenetic factors in electrolyte-induced myelinolysis. *Ann. Neurol.* 16, 140.

Nuotto, E., Palva, E. S., and Lahdenranta, U. (1983). Naloxone fails to counteract heavy alcohol intoxication. *Lancet* 2, 167.

Nutt, D., Adinoff, B., and Linnoila, M. (1989). Benzodiazepines in the treatment of alcoholism. *Recent Dev. Alcohol.* 7, 283–313.

O'Carroll, R. E., Moffoot, A. P., Ebmeier, K. P., and Goodwin, G. M. (1994). Effects of fluvoxamine treatment on cognitive functioning in the alcoholic Korsakoff syndrome. *Psychopharmacology (Berl).* 116, 85–88.

O'Carroll, R. E., Moffoot, A., Ebmeier, K. P., Murray, C., and Goodwin, G. M. (1993). Korsakoff's syndrome, cognition and clonidine. *Psychol. Med.* 23, 341–347.

Oh, M. S. (1990). Prevention of myelinolysis in rats by dexamethasone or colchicine. *Am. J. Nephrol.* 10, 158–161.

Oh, S. J. (1976). Alcoholic myopathy, an electrophysiologic study. *Electromyogr. Clin. Neurophysiol.* 16, 205–218.

Palsson, A. (1981). Die Wirksamkeit frühzeitiger Clomethiazol-Medikation zur Prävention des Delirium tremens. Eine retrospektive Studie über das Ergebnis verschiedener medikamentöser Behandlungsstrategien in den Psychiatrischen Kliniken Helsingborgs 1975–1980. In "Clomethiazol" (J. G. Evans, et al., Eds.), pp. 114–119. Verlag für angewandte Wissenschaften, München.

Perkoff, G. T. (1972). Alcoholic myopathy. *Annu. Rev. Med.* 22, 125–132.

Pfister, H. W., Einhäupl, K. M., and Brandt, T. (1985). Mild central pontine myelinolysis: A frequently undetected syndrome. *Eur. Arch. Psychiatry Neurol. Sci.* 235, 134–139.

Pfitzer, F., Schuchardt, V., and Heitmann, R. (1988). Die Behandlung schwerer Alkoholdelirien. *Nervenarzt* 59, 229–236.

Price, T. R., and Silberfarb, P. M. (1976). Disulfiram-induced convulsions without challenge by alcohol. *J. Stud. Alcohol* 37, 980–982.

Ranek, L., and Buch Andreasen, P. (1977). Disulfiram hepatotoxicity. *Br. J. Med.* 2, 94.

Reisner, H., Krypsin-Exner, K., Mader, R., and Schnaberth, G. (1968). Zur Frage der Antabuspsychosen. *Z. Nervenheilkunde* 26, 331.

Robinson, B. J., Robinson, G. M., Maling, T. J., and Johnson, R. H. (1989). Is clonidine useful in the treatment of alcohol withdrawal? *Alcohol (N.Y.)* 13, 95–98.

Rojiani, A. M., Prineas, J. W., and Chow, E. S. (1988). Alteration in myelin degradation in electrolyte induced demyelination (EID) by steroid/colchicine (abstract). *J. Neuropathol. Exp. Neurol.* 47, 307.

Sadun, A. (1998). Acquired mitochondrial impairment as a cause of optic nerve disease. *Trans. Am. Ophthalmol. Soc.* 96, 881–923.

Sage, J. I., Van Itert, R. L., and Lepore, F. E. (1984). Alcoholic myelopathy without substantial liver disease. A syndrome of progressive dorsal and lateral column dysfunction. *Arch. Neurol.* 41, 999–1001.

Sanna, E., and Harris, R. A. (1993). Recent developments in alcoholism: neuronal ion channels. *Recent Dev. Alcohol.* 11, 169–186.

Sass, H., Soyka, M., Mann, K., and Zieglgänsberger, W. (1996). Relapse prevention by acamprosate: results from a placebo controlled study in alcohol dependence. *Arch. Gen. Psychiatry* 53, 673–680.

Schied, H. W. (1989). Konzepte der Alkoholismus-Behandlung. In "Der chronische Alkoholismus. Grundlagen, Diagnostik, Therapie" (H. W. Schied, H. Heinmann, and K. Mayer, Eds.), pp. 253–266. Gustav Fischer Verlag, Stuttgart and New York.

Schied, H. W., Braunschweiger, M., and Schupmann, A. (1986). Die Behandlung des Delirium tremens in den Psychiatrischen Krankenhäusern der Bundesrepublik Deutschland. In "Clomethiazol" (J. G. Evans, et al., Eds.), pp. 121–125. Verlag für angewandte Wissenschaften, München.

Schied, H. W., and Mann, K. (1989). Die Behandlung des Delirium tremens und des Alkoholentzugssyndroms. *In* "Der chronische Alkholismus. Grundlagen, Diagnostik, Therapie." (H. W. Schied, H. Heimann, and K. Mayer, Eds.), pp. 285–300. Gustav Fischer Verlag, Stuttgart and New York.

Scholz, E., Diener, H. C., Dichgans, J., Langohr, H. D., Schied, W., and Schupmann, A. (1986). Incidence of peripheral neuropathy and cerebellar ataxia in chronic alcoholics. *J. Neurol.* **233**, 212–217.

Schroth, G., Naegele, T., Kloe, U., Mann, K., and Petersen, D. (1988). Reversible brain shrinkage in abstinent alcoholics, measured by MRI. *Neuroradiology* **30**, 385–389.

Selzer, M. L. (1980). Alcoholism and alcoholic psychoses. *In* "Comprehensive Textbook of Psychiatry" (H. I. Kaplan, A. M. Freedman, and B. J. Sadock, Eds.), Vol. 1–3, 3rd ed. pp. 1629–1644. Williams & Wilkins, Baltimore, Maryland and London.

Serdaru, M., Hausser-Hauw, C., Laplane, D., Buge, A., Castaigne, P., Goulon, M., Lhermitte, F., and Hauw, J. J. (1988). The clinical spectrum of alcoholic pellagra encephalopathy. A retrospective analysis of 22 cases studied pathologically. *Brain* **111**, 829–842.

Sharief, M. K., Robinson, S. F., and Swash, M. (1997). Hypokalaemic myopathy in alcoholism. Neuromuscul. Disord. 7: 533–535.

Soyka, M. (1989). Die Alkoholhalluzinose-einige Überlegungen zu Ätiologie, Verlauf und Therapie. *Nervenheilkunde* **8**, 128–133.

Soyka, M. (1992). Neuroleptic treatment in alcoholic hallucinosis—no evidence for increased seizure risk. *J. Clin Psychopharmacol.* **12**, 66–67.

Tavares, M. A., Paula-Barbossa, M. M., and Verwer, R. W. (1987). Alcohol-induced granule cell loss in the cerebellar cortex of the adult rat. *Exp. Neurol.* **78**, 574–582.

Tomasini, P., Guillot, D., Sabbah, P., Brosset, C., Salamand, P., and Briant, J. F. (1992). Marchiafava-Bignami's disease with a favourable course. Apropos of a case. *Ann. Radiol.* **36**, 319–322.

Tormey, W. P., and Chambers, J. P. M. (1988). Efficacy of dexamethasone in benzodiazepine-resistant delirium tremens *Lancet* **11**, 1340f.

Vandenbulcke, M., and Janssens, J. (1999). Acute axonal polyneuropathy in chronic alcoholism and malnutrition. *Acta Neurol. Belg.* **99**, 198–201.

Urbano-Marquez, A., Estruch, R., Navarro-Lopez, F., Grau, J. M., Mont, L., and Rubin, E. (1989). The effects of alcoholism on skeletal and cardiac muscle. *N. Engl. J. Med.* **320**, 409–415.

Weissman, J. D., and Weissman, B. M. (1989). Pontine myelinolysis and delayed encephalopathy following the rapid correction of acute hyponatremia. *Arch. Neurol.* **46**, 926–927.

Wenisch, S., Fortmann, B., Steinmetz, T., Kriete, A., Leiser, R., and Bitsch, I. (1998). 3-D confocal laser scanning microscopy used in morphometric analysis of rat Purkinje cell dendritic spines after chronic ethanol consumption. *Anat. Histol. Embryol.* **27**, 393–397.

Wilbur, R., and Kulik, F. A. (1981). Anticonvulsant drugs in alcohol withdrawal: Use of phenytoin, primidone, carbamazepine, valproic acid, and the sedative convulsants. *Am. J. Hosp. Pharm.* **38**, 1138–1143.

Woelk, H., Lehrl, S., Bitsch, R., and Kopcke, W. (1998) Benfotiamine in treatment of alcoholic polyneuropathy: an 8-week randomized controlled study (BAP I Study). *Alcohol-Alcohol.* **33**, 631–638.

Wöhrle, J. C., Spengos, K., Steinke, W., Goebel, H. H., and Hennerici, M. (1998). Alcohol-related acute axonal polyneuropathy: a differential diagnosis of Guillain-Barre syndrome. *Arch. Neurol.* **55**, 1329–1334.

Wright, D. G., Laureno, R., and Victor, M. (1979). Pontine and extrapontine myelinolysis. *Brain* **102**, 361–385.

*Metabolic and Progressive Disorders*

CHAPTER 72

# Metabolic and Toxic Encephalopathies

Jörg B. Schulz and Allen I. Arieff

---

Metabolic encephalopathies are conditions of the central nervous system that are not due to structural abnormalities. Selected metabolic encephalopathies, however, can result in secondary structural damage to the central nervous system. Encephalopathies are often characterized clinically by the occurrence of symptoms of organic psychosis. Depending on the type and severity of encephalopathy, these symptoms range from subtle personality changes, deterioration of memory, and cognitive functions to confusion and disturbances of consciousness. Neurological symptoms, including myoclonus, asterixis, weakness, and seizures, may also be present. Metabolic encephalopathies include essentially any disorders associated with a nonpsychiatric cause of alteration in consciousness that does not have a macroscopic structural basis. Thus, the list of metabolic encephalopathies is extensive (Table I).

Two common themes emerge with regard to presumptive causes of metabolic encephalopathies in systemic disease: impaired substrate delivery (glucose or oxygen) to the brain or release of circulating substances by the systemic disease, which cross the blood–brain barrier and causes diffuse neuronal and cellular dysfunction. The former, implicated in hypoxic-ischemic encephalopathy and hypoglycemia, may result in irreversible brain injury. The latter, implicated in most metabolic encephalopathies associated with organ system dysfunction or with systemic infection, may be largely reversible if the underlying disorder is treated. Although these delineations do not always hold true, this mechanistic differentiation can be of great importance in determining treatment and prognosis.

This chapter will deal with those metabolic encephalopathies that are secondary to hepatic and renal disease, hypercalcemia, and a number of toxic encephalopathies caused by chronic exposure to heavy metals. Other types of metabolic and toxic encephalopathies not covered in this chapter include alcoholism (Chapter 71), acute drug intoxications (Chapter 59), and disturbances of fluids, electrolytes, and acid–base metabolism (Chapter 55).

## HEPATIC ENCEPHALOPATHY

### Clinical Aspects

The clinical presentation of hepatic encephalopathy is highly variable. In its latent form, there are often no clinically overt symptoms. However, performance evaluation with the number connection test and other tests of attention and motor skills gives pathological results (Groeneweg *et al.*, 1998; Schomerus and Schreiegg, 1993; Tarter *et al.*, 1984), showing that there are groups of cirrhotic patients whose lives are impacted by their condition. Clinically, hepatic encephalopathy can be divided into four stages, depending on the severity of the encephalopathy (Table II [Conn, 1994]). Psychic symptoms range from impaired ability to concentrate, dysphoria, and agitation in stage 1 to personality changes, lethargy, disorientation in stage 2. Stages 3 and 4 are characterized by progressive clouding of consciousness. Neurologic symptoms consist of flapping tremor (asterixis), ataxia (stage 2), muscular rigidity, and hyperreflexia (stage 3). Seizures are uncommon in patients with hepatic encephalopathy. These changes are paralleled by progressive slowing of the EEG. Clinically, hepatic encephalopathy is difficult to distinguish from other metabolic encephalopathies. A history of liver disease, laboratory tests indicative of liver disease, and increased arterial ammonia levels are the basis for making the diagnosis. However, the blood ammonia

TABLE I    Mechanisms of Metabolic Encephalopathies

Impaired substrate delivery
  Hypoxia-ischemia
  Hypoglycemia
Organ failure
  Hepatic encephalopathy
  Renal failure
  Pancretic encephalopathy
  Carbon dioxide narcosis
Circulating cytokines (putative)
  Sepsis
  Multisystem organ failure
Electrolyte abnormalities
  Hyponatremia
  Hypernatremia
  Hypercalcemia
Autoantibodies
  Paraneoplastic syndromes
  Collagen vascular diseases

may be in the normal range, although this finding is infrequent (Fraser and Arieff, 1985). In diagnostically difficult cases, the observation of paroxysmal triphasic EEG waves may help to verify the diagnosis.

## Natural Course

*Fulminant hepatic encephalopathy* develops in the course of acute liver failure caused by viral hepatitis or intoxication with agents that are toxic to the liver (mushroom toxins, acetaminophen, ethylene glycol) (Worthley, 1994). In the United States and Western Europe, acetaminophen or paracetamol ingestion as a suicide attempt or gesture, and accidental ingestion by children, are particularly common causes of the condition. The prognosis of this condition is poor. Of the patients, 80% die unless liver transplantation is performed (Smith and Ciferni, 1993). Death is often caused by cerebral edema and may occur within days of the onset of the disorder. The prognosis tends to be better in patients with acute liver failure caused by hepatitis A and acetaminophen toxicity. Because the initial signs and symptoms of fulminant hepatic failure are frequently referable to the brain, neurologists are often among the first to be involved in the care of these patients. Nausea, vomiting, and abdominal pain associated with delirium or mania are common in the initial stages of the disorder. These symptoms evolve rapidly to a comatose state associated with generalized seizures.

*Portal systemic encephalopathy* is due to chronic liver failure, and in the United States and Western Europe, the most common cause is cirrhosis, often caused by alcoholism. The course and prognosis of portal systemic encephalopathy depend on the severity of the underlying liver disease. Patients who have latent hepatic encephalopathy may suddenly decompensate because of gastrointestinal bleeding, an increase in dietary protein; constipation; infection; excessive consumption of alcohol; and administration of tranquilizers, analgetics, or diuretics. Patients with portal systemic encephalopathy may also deteriorate without obvious reasons. The prognosis of coma caused by hepatic failure is poor: only 35% of comatose patients make a good recovery within 1 year and regain the ability to live independently, whereas most others either do not

TABLE II    Encephalopathy Index

| Findings | Weighing factor | Grade | | | | |
|---|---|---|---|---|---|---|
| | | 0 | 1 | 2 | 3 | 4 |
| Psychic symptoms | 3 | Normal | Impaired ability to concentrate, short attention span, dysphoria, agitation, anxiety | Personality changes, inappropriate behavior, lethargy, disorientation, impaired cognition | Somnolence, stupor, gross confusion, able to respond appropriately to noxious stimuli | Coma |
| Number connection test | 1 | <30 s | 31–50 s | 51–80 s | 81–120 s | >120 s |
| Asterixis | 1 | 0 | Rarely (2–4/min) | Occasionally, irregularly (6–8/min) | Frequently (10–60/min) | Permanent |
| EEG | 1 | Normal | 7–8/s | 5–7/s | 3–5/s | <3/s |
| Arterial ammonium concentration | 1 | <150 μg/dl | 151–200 μg/dl | 201–250 μg/dl | 251–300 μg/dl | >300 μg/dl |

Maximal Index: 28 points.
Modified from Conn (1994).

recover or survive with severe disabilities (Levy *et al.*, 1981).

## Principles of Therapy

Both spontaneous portal systemic shunt and hepatic parenchymal dysfunction contribute to the pathogenesis of hepatic encephalopathy. As a consequence, the concentration of a number of gut-derived neurotoxic compounds, such as ammonia, mercaptans, and phenol rises in blood and brain (Hazell and Butterworth, 1999). These compounds are formed in the gastrointestinal tract by bacterial degradation of proteins and lipids. Furthermore, the concentration of manganese, which is normally removed by the hepatobiliary route, increases in plamsa and accumulates in the brain (Butterworth *et al.*, 1995; Krieger *et al.*, 1995). Manganese deposition in the globus pallidus in chronic liver failure results in signal hyperintensity on T1-weighted magnetic resonance imaging and may be responsible for the extrapyramidal symptoms characteristic of portal systemic encephalopathy (Hazell and Butterworth, 1999). In addition, changes of central neurotransmission are involved in the pathogenesis of hepatic encephalopathy. These manifestations of hepatic encephalopathy are caused by multiple factors, including disturbances of the blood–brain barrier that may allow excess amounts of GABA and benzodiazepine-like compounds, which are formed in the gastrointestinal tract, to enter the brain (Mullen *et al.*, 1990). Because of a reduced rate of oxidation of aromatic amino acids, the balance between aromatic and branched-chain amino acids is shifted toward the side of aromatic amino acids. Higher concentrations of aromatic amino acids in the brain lead to inhibition of the synthesis of noradrenaline and dopamine and the formation of "false" transmitters, such as octopamine, with reduced intrinsic activity (Fraser and Arieff, 1985). The cerebral metabolic rate for both oxygen and glucose is reduced in portal systemic encephalopathy. Such reductions parallel the onset of overt clinical symptoms and are directly proportional to the deterioration of neurologcial status. However, descreased brain glucose use in early portal systemic encephalopathy is most probably the result of decreased energy demand after a reduction of energy needs (Hindfelt *et al.*, 1977).

The neural cell most vulnerable to liver failure is the astrocyte. In *acute liver failure* the astrocyte undergoes swelling, resulting in increased intracranial pressure. The expression of the astrocyte glutamate transporter GLT-1 is reduced, leading to increased extracellular concentrations of glutamate. In *chronic liver failure*, the astrocyte undergoes characteristic changes known as Alzheimer type II astrocytosis. In portal encephalopathy resulting from chronic liver failure, astrocytes manifest altered expression of several key proteins and enzymes including monoamine oxidase A and neuronal nitric oxide synthase.

## Practical Management

### Fulminant Hepatic Failure

Emergency liver transplantation is the only known effective therapy of acute hepatic encephalopathy. Liver transplantation reduces the mortality from 80% to less than 30% (Rivera-Penera *et al.*, 1997; Smith and Ciferni, 1993)*. The goal of therapy is to support patients so that they remain candidates for orthopic liver transplantation. This requires specialized care, preferably in a transplant center, and admission to an intensive care unit. However, currently only 10% of patients with acute liver failure receive a donor liver. High-volume plasma exchange with fresh donor plasma transiently improves the clinical state in patients with acute liver failure but is not curative. Plasma exchange may be useful to temporarily maintain the patient to obtain a donor liver for emergency transplantation (Riviello *et al.*, 1990). Brain edema occurs in many patients with stage 3 and 4 encephalopathy (Ware *et al.*, 1971). These patients require treatment for their cerebral edema, but there are a limited number of therapeutic maneuvers (Davies *et al.*, 1994; Worthley, 1994). Such therapy includes mannitol (rapid IV infusion of 100–200 ml of a 20% solution), hyperventilation to induce hypocapnia, and thiopentone (Davies *et al.*, 1994; Strauss *et al.*, 1999)*. Dexamethasone is of little value. The head should be elevated to 20 degrees. Head-up tilting of more than 20 degrees may be deleterious, because it reduces cerebral perfusion in hepatic encephalopathy (Lee, 1993)*.

### Portal Systemic Encephalopathy

Management of chronic hepatic encephalopathy is usually conservative, because there is no definite cure. Most of the therapies that are clearly effective in the treatment of portal systemic encephalopathy are believed to work by influencing ammonia metabolism. A number of clearly defined conditions, such as gastrointestinal bleeding, increase in dietary protein, constipation, electrolyte imbalances, or infection, may precipitate encephalopathy in susceptible patients. These factors must be identified and rapidly corrected. Specific therapeutic measures are aimed at detoxifying the gastrointestinal tract. Toward this end, protein intake is restricted, and disaccharides and nonabsorbable antibiotics are administered (Conn *et al.*, 1977; Worthley, 1994). Although lactulose and neomycin are mainstays of therapy, the clinical trials supporting their use would not meet current standards for drug approval required by the U. S. Food and Drug Administration (FDA). However, clinical experience with these compounds is extensive, and they should be used. When hepatic encephalopathy has been precipitated by constipation, gastrointestinal bleeding, or heavy protein intake, gut cleaning by orally given cathartics and rectal enemas is important. In addition, it has often been attempted to correct the imbalances of amino acid and neurotransmitter metabolism by administration of branched chain amino acids or benzodiazepine antagonists. The duration of treatment is exclusively determined by the clinical picture. Chronic hepatic encephalopathy often requires lifelong treatment (Morgan, 1991).

Table III gives an overview about the practical management of decompensated chronic hepatic encephalopathy.

In patients with acute exacerbation of chronic encephalopathy, daily oral protein intake is limited to 20–30 g/day. After 2 to 3 days, protein intake is gradually increased by 10 g/day until a daily protein intake of 1 g/kg body weight is reached and can be tolerated. In some patients, protein tolerance is reduced so much that a permanent protein restriction of 0.5 g/kg body weight is necessary. Vegetable protein is better tolerated than animal protein (Uribe *et al.*, 1982). During protein restriction sufficient caloric supply of at least 1600 kcal/day must be maintained by intravenous infusion of glucose solutions or other carbohydrates.

Lactulose and lactitol are nonabsorbable disaccharides that produce an acid intracolic environment that results in increased binding and excretion of ammonia. In addition, lactulose and lactitol have laxating actions. Currently, lactulose is the mainstay in the treatment of hepatic encephalopathy. However, a number of recent studies show that lactitol is as efficacious as lactulose

and has fewer side effects (flatulence). The required dosages of both compounds are identical (Blanc *et al.*, 1992). In acute decompensation of hepatic encephalopathy, lactulose or lactitol given at a dose of 20–30 g/h until gut cleaning is achieved. For long-term treatment an individual dose is to be chosen to produce two or three semiformed stools per day (Conn *et al.*, 1977).

Nonabsorbable antibiotics inhibit formation of toxins by reducing the growth of intestinal bacteria and the production of ammonia. In severe cases, neomycin or paromomycin are given in doses of 6–8 g/day through a nasogastric tube. For long-term treatment maximal daily doses are 2 g. Because 1%–3% of neomycin or paromomycin is absorbed, it may exert ototoxic and nephrotoxic actions. Therefore, regular tests of otological and renal function are mandatory during treatment. In infected patients who receive systemic antibiotic treatment additional treatment with nonabsorbable antibiotics is unnecessary. Metronidazol (2–3 × 200–400 mg/day orally) and vancomycin may serve as reserve antibiotics.

If conventional treatment with protein restriction, disaccharides, and antibiotics fails, a trial with intravenous infusion of branched-chain amino acids (1 g/kg body weight per day) can be attempted (Alexander *et al.*, 1989). In chronic hepatic encephalopathy long-term oral supplementation with branched-chain amino acids (0.3 g/kg body weight/day) (Marchesini *et al.*, 1990)** or 3 × 3–6 g/day L-ornithine-L-aspartate (Stauch *et al.*, 1998)** may be given as an adjunct to conventional therapy. A number of studies report favorable effects of the benzodiazepine antagonist flumazenil in acute or decompensated hepatic encephalopathy (Gyr and Meier, 1991); however, data from double-blind, placebo-controlled trials are not convincing. Flumazenil was only beneficial in a selected subset of cirrhotic patients, and the observed functional improvement was only temporary (Gyr *et al.*, 1996; Barbaro *et al.*, 1998)**. Flumazenil should be used if an intoxication with benzodiazepines is suspected.

TABLE III  Practical Management of Decompensated Chronic Hepatic Encephalopathy*

| | Compound | Dosage (acute phase) | Dosage (chronic phase) |
|---|---|---|---|
| Protein intake | | 20–30 g/day | 0.5–1 g/kg/day |
| Nonabsorbable disaccharides | Lactulose | 20–30 g/h PO | Individual dose PO |
| | Lactitol | 20–30 g/h PO | Individual dose PO |
| Antibiotics | Neomycin | 6–8 g/day PO | Max. 2 g/day PO |
| | Paromomycin | 6–8 g/day PO | Max. 2 g/day PO |
| | Metronidazol | 0.4–1.2 g/day PO | |
| Branched-chained amino acids | | 1 g/kg/day IV | 0.3 g/kg/day PO |
| L-ornithine-L-aspartate | | | 3 × 3–6 g/day |

# CENTRAL PONTINE MYELINOLYSIS

Central pontine myelinolysis (CPM) is a cerebral demyelinating lesion located in the center of the pons, which was initially described by Adams and associates (Adams *et al.*, 1959) as "...a single, sharply outlined focus of myelin destruction which indiscriminately affected all fiber tracts." Although some patients with CPM may also have occasional extrapontine demyelinating lesions, these are generally symmetrical and are infrequently present. When strict diagnostic criteria (either pathological or radiological) are used, virtually all patients with central pontine myelinolysis have had severe associated medical conditions, which have included alcoholism, advanced liver disease, extensive burns, sepsis, Hodgkin's disease, or other malignancies (Cole and Richardson, 1964; McKee *et al.*, 1988; Messert *et al.*, 1979; Singh *et al.*, 1994; Sullivan and Pfefferbaum, 2001; Wszolek and McComb, 1989). The condition is often asymptomatic (Chason *et al.*, 1964; Sullivan and Pfefferbaum, 2001).

## Natural Course

Hyponatremia is a common finding in patients with liver disease (Arieff and Papadakis, 1988). Hyponatremia is a major risk factor for survival among patients with hepatic insufficiency, although it is unclear whether the hyponatremia actually increases the mortality or is merely a marker for more severe degrees of hepatic insufficiency. Recent studies suggest that the major determinants of the occurrence of permanent brain damage in patients with liver disease and hyponatremia include at least six factors (Papadakis and Arieff, 1990): (1) the occurrence of a hypoxic episode, usually respiratory arrest from hyponatremia; (2) correction of serum sodium above 128–132 mmol/L within the initial 48 hours of therapy; (3) elevation of serum sodium by more than 25 mmol within the initial 24–48 hours of therapy; (4) the presence of hypernatremia (serum sodium above 145 mmol/L) in patients with liver disease receiving lactulose; (5) the presence of associated medical conditions known to result in abnormalities of the blood–brain barrier, such as hepatic cirrhosis or metastatic cancer; (6) delay in institution of therapy for symptomatic hyponatremia. The mortality in patients who have both liver disease and serum sodium below 120 mmol/L accompanied by central nervous symptoms is probably greater than 40%. Both extrapontine cerebral demyelinating lesions and CPM are also frequent complications of liver transplantation (Estol *et al.*, 1989). Among 11 patients who had either CPM or extrapontine cerebral demyelinating lesions develop after liver transplantation, all had a "slow" but sub-

stantial rise in the serum sodium, and most developed hypernatremia, but none were hyponatremic.

In general, patients with advanced liver failure who have been treated with liver transplantaton are at major risk for the development of cerebral demyelinating lesions. In contrast to patients without hepatic insufficiency, this risk seems to be exacerbated by upward changes in plasma sodium. Thus, the presence of liver failure in patients in whom there is an increase of serum sodium is a major risk factor for the development of cerebral demyelinating lesions (Ayus *et al.*, 1987). Currently (after 1988), the diagnosis of CPM is frequently established by radiological criteria using either CT or MRI of the brain (Arieff, 1994). However, the diagnosis of CPM has often been suggested when radiological examination (CT and/or MRI) was either not carried out or disclosed only extrapontine cerebral demyelinating lesions. Because CPM is a distinct pathological entity and not a clinical syndrome or diagnosis, if the distinct pathological findings are absent, CPM is not present.

# UREMIC ENCEPHALOPATHY

## Clinical Aspects

The natural history of renal failure and its clinical manifestations have changed since the advent of dialysis programs and kidney transplantation. New neurological syndromes have been defined as a consequence of both increased longevity for hemodialysis patients and the complications of therapy. Those with chronic renal failure who have not yet received dialytic therapy may have signs of neural dysfunction develop composed of sensorial clouding, dysarthria, gait ataxia, asterixis, action tremor, multifocal myoclonus, seizures, delirium, and coma. One or more of the signs may predominate, and their fluctuation from day to day, or sometimes from hour to hour, is characteristic of metabolic encephalopathy. The clinical features of uremic encephalopathy do not correlate well with any single laboratory abnormality but seem to be related in many patients to the rate of development of renal failure. Thus, stupor and coma are not uncommon in acute renal failure, whereas the CNS may be deranged to a lesser degree in chronic renal failure, despite greater degrees of azotemia. Among patients with chronic renal failure, depression and anxiety, maladaptive and uncooperative behavior, and frank psychosis are common (Levy and Cohen, 2001). Even after the institution of otherwise adequate maintenance dialysis therapy, patients may continue to manifest more subtle nervous system dysfunction such as impaired mentation, generalized weakness, and peripheral neuropathy. The central nervous system disorders of both untreated renal failure and that persisting despite dialysis are referred to as uremic encephalopathy (Fraser and Arieff, 1999). The dialytic treatment of end-stage renal disease has itself been associated with the emergence of at least two distinct disorders of the central nervous system: dialysis disequilibrium syndrome (discussed later) and dialysis dementia (also discussed later) (Fraser and Arieff, 2001). The dialysis disequilibrium syndrome is a consequence of the initiation of dialysis therapy in a minority of patients with end-stage renal failure (Arieff et al., 1979). Dialysis dementia is a progressive, generally fatal encephalopathy that can affect patients on chronic hemodialysis and children with chronic renal failure who have not been treated with dialysis (Andreoli et al., 1984).

## Natural Course

In addition to the preceding manifestations of neurological dysfunction specifically related to uremia, dialysis, or both, a number of other neurological disorders occur with increased frequency in patients who have end-stage renal disease and are being treated with chronic hemodialysis. These include subdural hematoma, certain electrolyte disorders, vitamin deficiencies, drug intoxication, hypertensive encephalopathy, and acute trace element intoxication. All of these must be considered in patients with chronic renal failure who manifest an altered mental state (Fraser and Arieff, 2001). Patients with end-stage renal disease are also at risk for other organic brain disease and metabolic encephalopathies developing, which can affect the general population (Levy and Cohen, 2001). Therefore, when a patient with end-stage kidney disease is seen with altered mental status, a thorough and complete evaluation is necessary (Fraser and Arieff, 2001).

The early symptoms of uremic encephalopathy may include disturbances of concentration and attention; with progression of the encephalopathy, a deliriant state may occur with agitation, misperceptions, illusions, and visual hallucinations (Teschan and Arieff, 1985). The most severe stage of uremic encephalopathy is characterized by coma. Possible accompanying neurological symptoms include myoclonus, asterixis, ataxia, hyperreflexia, and increased muscular tone. Generalized tonic-clonic seizures used to be common among patients with acute renal failure but were rarely observed in the 1990s. Dialysis rapidly improves the clinical situation. The long-term prognosis depends to a large extent on the underlying renal disease (Fraser and Arieff, 2001), although both dialysis and renal transplantation carry a mortality (Held et al., 1994).

## Therapy

In uremia, there is a rise in the concentration of a number of small, water-soluble molecules that may have neurotoxic effects on the nervous system (Fraser and Arieff, 2001). The severity of the encephalopathy is only loosely correlated with any single laboratory abnormality but seems to be related to the rate of development of renal failure. Hemodialysis or peritoneal dialysis is the only effective therapeutic measure to remove these compounds from the blood. A definite treatment may be achieved by renal transplantation. Although seizures associated with renal failure are very rare since the 1990s, they must occasionally be treated. Diazepam is probably the treatment of choice for interruption of seizures related to renal failure, although dialysis is virtually always indicated. Because the pharmacokinetics of anticonvulsant drugs are altered in uremic patients, the following rules have to be observed. Phenobarbital and primidone may accumulate in renal failure. Dialysis may cause a rapid fall of the plasma concentration of such compounds. Therefore, neither compound is well suited for anticonvulsant therapy in uremia. The half-time of decay of phenytoin is shortened in uremia. Plasma levels of phenytoin are approximately 25% of those shown by control subjects for identical dosages. The volume of distribution of drugs is larger in uremia, which is the major reason that plasma levels are lower. Therefore, phenytoin must be given three times per day. However, these factors are offset by higher plasma levels of free, unbound phenytoin because of the decreased protein binding of drugs that attends renal failure. Because the rate of diffusion of anticonvulsant drugs into the brain is proportional to the free drug concentration, the latter is the determinant of anticonvulsant benefit. These considerations also apply for valproic acid, whereas carbamazepine is given in the same way as in nonuremic patients.

## DIALYSIS DISEQUILIBRIUM SYNDROME

### Clinical Aspects

Dialysis disequilibrium syndrome is a clinical syndrome that occurs in patients being treated with hemodialysis. The syndrome was first described in 1962 (Kennedy et al., 1962). A host of neurological problems may arise during or after hemodialysis or peritoneal dialysis, including headache, nausea, emesis, blurring of vision, muscular twitching, disorientation, hypertension, tremors, and seizures. Dialysis disequilibrium sydrome has been expanded to include milder symptoms, such as muscle cramps, anorexia, restlessness, and dizziness. The more serious neurological sequelae were seen frequently during the 1960s, when patients with advanced uremia were dialyzed aggressively; today, with patients entering dialysis programs early in the course of renal failure and with shortened dialysis times, they are far less common but continue to occur. Children and elderly people have a higher risk for development of disequilibrium than people in other age groups. Many patients manifest exophthalmos and increased intra-ocular pressure at the height of the syndrome, conditions that, if present, are helpful in the differential diagnosis. The other clinical correlates of the syndrome include intracranial pressure, papilledema, and generalized slowing of the EEG. The incidence is very high among patients with pre-existing neurological disease, such as head trauma, recent stroke, or malignant hypertension.

## Pathophysiology

Substantial evidence points to shifts of water into the brain as the cause of the dialysis disequilibrium syndrome. It was originally postulated that the rapid reduction in blood urea could not be paralleled by brain urea because of the influence of the blood–brain barrier. This effect was believed to produce an osmotic gradient between blood and brain, causing movement of water into brain and resulting in encephalopathy, increased intracranial pressure, and cerebral edema and was termed "reverse urea effect." Experimental models of disequilibrium have shown that an osmotic gradient between brain and blood is not created by the movement of urea alone; however, the principle of the hypothesis has been supported by the demonstration that unidentified osmotically active substances are present in the brain in the dialyzed uremic animal, creating an osmotic gradient between brain and blood that results in shifts of water into the brain. Arieff and collegues have shown that a decrease in the intracellular pH of the cerebral cortex attends experimental disequilibrium, reflecting an increased production of organic acids (Arieff et al., 1976); the latter are osmotically active solutes that may account for the osmotic gradient.

## Natural Course

The symptoms of dialysis disequilibrium syndrome are usually self-limited, but recovery may take several days. It seems that present methods of dialysis have altered the clinical picture. Most reports of seizures, coma, and death were published before 1970. The symptoms of dialysis disequilibrium syndrome as reported in the last two decades (1980–2000) have generally been mild, consisting of nausea, weakness, headache, fatigue, and muscle cramps. It is also unclear

whether any patient ever actually died from dialysis disequilibrium syndrome or in fact from other associated neurological complications of dialysis, such as acute stroke, subdural hematoma, subarachnoid hemorrhage, or hyponatremia. Recently, the diagnosis of dialysis disequilibrium syndrome has become a "wastebasket" for a number of disorders that can occur in patients with renal failure and may affect the central nervous system. It is important to recognize that the diagnosis of dialysis disequilibrium syndrome should be one of exclusion (Arieff, 1994).

## Therapy

Prevention of dialysis disequilibrium syndrome is largely achieved by "slow" dialysis, that is, low blood flow rates, at frequent intervals (every 1 to 2 days). Furthermore, dialysis disequilibrium syndrome has been treated by the addition of osmotically active solute (glucose, glycerol, albumin, urea, fructose, NaCl, mannitol) to the dialysate or by substitution of sodium bicarbonate for sodium lactate (or acetate) in the dialysate. With the technique of pure ultrafiltration the patient is subjected to ultrafiltration without dialysis. The net result is loss of fluid without the patient undergoing dialysis. The role of ultrafiltration followed by dialysis in prevention of dialysis disequilibrium syndrome is currently being evaluated.

The technique of chronic ambulatory peritoneal dialysis is currently in use worldwide. With this technique, patients undergo continuous low-volume peritoneal dialysis for 24 h/day. Symptoms of dialysis disequilibrium syndrome have not been presently reported in patients using this mode of dialysis.

## DIALYSIS DEMENTIA

### Clinical Aspects

Dialysis dementia (also known as dialysis encephalopathy, progressive myoclonic dialysis encephalopathy, and hemodialysis encephalopathy) was first clearly documented by several groups from 1972–1976 (Alfrey et al., 1972; 1976; Chokroverty et al., 1976; Flendrig et al., 1976; Mahurkar et al., 1973). A mixed dysarthria and dysphasia with dysgraphia has been reported as one of the earliest signs of dialysis dementia in up to 95% of cases. The patient may initially have a stuttering, hesitant speech that only occurs during and immediately after dialysis. Initially, the patient may also be more apathetic and become depressed. As the disorder progresses, language function becomes more severely and persistently involved. Myoclonic jerks occur in up to 80% of cases, and patients may become both ataxic and dyspraxic (Fraser and Arieff, 1994; Raskin, 2001). Convulsions develop in up to 60% in the later stages, and psychosis with hallucinations and paranoid delusions may be prominent. Frank dementia is obvious in more than 95% of patients. Preterminally, the patient becomes immobile and mute (Chokroverty et al., 1976). Death usually occurs as a consequence of aspiration pneumonia (Arieff et al., 1979).

Abnormalities in the EEG may precede clinically overt symptoms by up to 6 months. Intermittent bursts of high-voltage slowing and spike and wave activity are noted, particularly in the frontal leads (Hughes and Schreeder, 1980). As the disease progresses, the normal background activity also deteriorates to slow frequencies. The EEG has been said to be pathognomonic, but a similar pattern may also be seen in patients with other metabolic encephalopathies. The EEG may show an initial deterioration after treatment with desferrioxamine has begun (see below). Neuroimaging studies and analysis of CSF are of no positive help in making the diagnosis of dialysis dementia but are of use in excluding other diagnoses if the clinical picture is atypical. The role of serum aluminium concentrations and the desferrioxamine infusion test are discussed briefly in the following.

## Natural Course

The disorder is progressive and invariably fatal unless treated. Dialysis dementia may be part of a multisystem disorder that includes vitamin D–resistant osteomalacia, proximal myopathy, and non-iron-deficient, microcytic, hypochromic anaemia. In Europe, between 1976 and 1977, the prevalence of dialysis dementia was 600/100,000 dialysis patients, although there was a wide variation between centers (see later). The mean age of those affected in a large series was 50 years, with an age range of 21–68 (Jack et al., 1983). Mean onset of symptoms after hemodialysis had begun was 35 months in the same series (range, 0.5–112 months). Death occurred 6–9 months after the onset of symptoms in most untreated cases. The current prevalence of dialysis dementia has been estimated at around 0.6% to 1.0% of dialysis patients.

## Pathophysiology of Dialysis Dementia

An early finding was the marked geographical variation in the incidence of the dementia, suggesting the involvement of an environmental toxin. High concentrations of tin and decreased rubidium concentrations in the brains of patients with dialysis dementia were noted first. Subsequent work confirmed an 11-fold increased concentration of aluminium in the cerebral cortex of patients with dialysis dementia compared with a threefold increase of nondemented dialysed patients (Arieff et al., 1979). These findings were rapidly linked to the aluminium concentration in the dialysate water supply (Jack et al., 1983; Raskin, 2001). The European Dialysis and Transplant Association determined that 92% of cases of dialysis dementia were linked with untreated or "soft" water compared with only 6% of cases who had received deionized water. It is now recognized that reducing concentrations of aluminium in water to less than 20 µg/L by reverse osmosis seems to prevent the onset of the disease in patients who have just started dialysis. Sporadic cases of dementia still occur, however, and may relate in part because of the use of phosphate binding gels such as aluminium hydroxide. Even absorption of aluminum by oral ingestion of these aluminium-containing agents can lead to considerable accumulation of aluminium (Kaehny et al., 1977). However, because the use of these binders is so widespread, other, as yet unrecognized, factors must be involved, given the rarity of sporadic cases (Fraser and Arieff, 1994; Raskin, 2001).

How aluminium interferes with neuronal function to cause the dementia and why the transition between reversible and irreversible brain dysfunction occurs are still unknown. Potential mechanisms include complexing with high-energy phosphates, impaired enzymatic function, deoxyribonucleic acid binding, impaired hydrolysis of phosphoinositides, impaired microtubular function, reduced calmodulin activity through binding, and reduced neurotransmitter uptake. Several of these mechanisms have only been demonstrated in in vitro models, and they are probably not mutually exclusive. Neurofibrillary material has been found in cortical neurons of patients dying from dialysis dementia. There are, however, considerable differences both in the composition of the tangle material and its distribution compared with Alzheimer's disease.

Concentrations of aluminium in CSF are of no help in making the diagnosis of dialysis dementia (Raskin, 2001). Serum aluminium concentrations are of only limited assistance. Dialysis dementia has been reported in patients with serum aluminum concentrations ranging from 15 to in excess of 1000 µg/L (normal range, <15 µg/L). Although the dementia is uncommon with serum concentrations <50 µg/L, such concentrations by no means exclude the diagnosis (Raskin, 2001).

## Principles of Therapy

The use of aluminium-free dialysate may arrest, or even improve, the established case, but because aluminium is so avidly bound to plasma protein, very little is actually removed at subsequent dialyses. Desferrioxamine infusions are the mainstay of treatment of dialysis dementia, improving up to 70% of patients, sometimes to normal. Desferrioxamine binds aluminium with greater avidity than plasma protein and tissue binding sites. The chelated complex has a molecular weight of 600 and so is removed by dialysis. The usual dosage is 4–6g/wk, 1–2g given intravenously during the last 2 hours of each dialysis session. The clinical improvement is slow; a year of treatment may be required. How long treatment must be continued has not yet been determined (Raskin, 2001). There is a similarity to chelation treatments used for other neurological illness (for example, D-penicillamine therapy for Wilson's disease) in that there may be a period of paradoxical clinical and EEG worsening after treatment is begun. The mechanism for this is uncertain, but the deterioration may be profound and occasionally fatal.

## OTHER NEUROLOGIC COMPLICATIONS OF RENAL FAILURE

### Stroke in Patients Treated with Chronic Hemodialysis

Cerebrovascular disease is a common cause of death in chronic hemodialysis patients (Iseki et al., 2002), and stroke represents the second most frequent cause of death (Iseki and Fukiyama, 1996; Iseki et al., 1993) (the three most frequent are heart attack, stroke, and infection (Mazzuchi et al., 2000). In the United States and Western Europe (including Israel), cardiovascular disease is far more common in dialysis patients than in the rest of the population (Parfrey, 1993; Rostand et al., 1991). Part of the reason may be that the major cause of end-stage renal disease in the United States is diabetes mellitus (27%), far less than in Europe (19%) and Japan (10%) (Breyer, 1995; Held et al., 1990). Among the factors that which doubtless contribute to the high incidence of stroke in patients with end-stage renal disease treated with hemodialysis are the high incidence of hypertension, the large number of such patients who have diabetes mellitus, and the accelerated arteriosclerosis in such patients (Foley and Harnett, 1997). In addition, uremic patients tend to have high cholesterol and obesity and tend to smoke cigarettes (Bronner et al., 1995). There is a high incidence of chronic infection in dialysis patients, which leads to elevated blood levels of atherogenic risk factors, such as cytokines, which seem to contribute to the increased incidence of stroke in such patients (Ayus and Sheikh-Hamad, 1998; Zimmerman et al., 1999).

There is substantial recent evidence that chronic inflamation plays a role in the pathogenesis of cardiovascular disease. Cytokines released from involved tissues stimulate the liver to synthesize acute phase proteins, including C-reactive protein (CRP). Elevated levels of CRP and other cytokines constitute an independent risk factor for cardiovascular disease, including stroke (Ballou and Kushner, 1992). The effects of dialysis reuse on cytokine production has not been evaluated but may be important, because reuse could theoretically lead to more contamination of dialysate (Wratten et al., 2000).

Recent knowledge of the pathogenesis of stroke has led to a major expansion in the opportunities for prevention of stroke. Common preventive measures include treatment of hypertension, cessation of smoking, lowering of plasma cholesterol, control of plasma glucose (in diabetic pateints), weight loss, increased exercise, and decreased alcohol consumption (Bronner et al., 1995). Other possible preventive measures include dietary antioxidants, low-dose aspirin, and a decrease of intake of saturated fatty acids (Bronner et al., 1995). Although treatment of hypertension is known to decrease the incidence of stroke, not all antihypertensive agents are of equal efficacy. In general, only beta-blockers, thiazide diuretics, and angiotensin-converting enzyme (ACE) inhibitors have been shown to reduce the incidence of stroke, whereas alpha-adrenergic blocking agents and calcium channel blockers may not.

### Sexual Dysfunction in Uremia

Disturbances in sexual function are a common complication of chronic renal failure (Droller and Anderson, 1993; Palmer, 1999). These complications include erectile dysfunction, decreased libido, and decreased frequency of intercourse (Campese and Liu, 1990). Studies in uremic rats showed that erectile impairment was associated with a disturbance in nitric oxide synthetase gene expression (Abdel-Gawad et al., 1999). Sexual dysfunction in men with end-stage renal disease maintained on hemodialysis is common, and previously, impotence was observed in at least 50% of such patients (Massry et al., 1997). A number of abnormalities associated with renal failure seem to be important in the genesis of impotence. There are abnormalities in autonomic nervous system function (Campese and Liu, 1990), impairment in arterial and venous systems of the penis (along with vascular disease in other vascular beds), hypertension (many drugs used to treat hypertension cause secondary impotence), and other associated endocrine abnormalities. There are also the associated effects of aging, with impotence observed in more than 50% of men older than 60 years who do not have renal failure (Lamberts *et al.*, 1997). Patients with end-stage renal disease have a high incidence of cardiovascular disease (Meeus *et al.*, 2000), which impairs penile vessels along with those of the rest of the body (Rostand *et al.*, 1991). The incidence of hypertension is also higher in end-stage renal disease patients than the rest of the population, and hypertension is a major contributor to vascular disease. Many drugs used to treat hypertension can lead to impotence (calcium channel blockers, thiazides, guanethidine). The incidence of depression is high in patients with end-stage renal disease, and many drugs used to treat depression can lead to impotence (phenothiazines, tricyclics, fluxetine) (Levy and Cohen, 2001). Although appreciaton of the aforementioned abnormalities may increase our understanding, until very recently, there was little that could be done other than to discontinue certain drugs used to treat hypertension. There are now a number of drugs that can successfully treat impotence (Goldstein *et al.*, 1998).** Alprostadil was successful but had to be delivered transurethrally. In particular, sildenafil can be administered orally and is highly effective, even in men who have cardiovascular disease (Herrmann *et al.*, 2000)** or uremia (Utiger, 1998).**

## ENCEPHALOPATHY ASSOCIATED WITH RENAL TRANSPLANTATION

### Rejection Encephalopathy

This may be more common in young recipients of transplants. More than 80% of cases occur within 3 months of transplantation, but cases have been reported up to 2 years after surgery. The syndrome most commonly is seen with convulsions, confusion, and headache, combined with systemic features of graft rejection. The EEG, neuroimaging, and CSF findings are nonspecific. The release of cytokines in the rejection process may be important in the pathophysiology of this condition. Symptomatic treatment of the seizures is usually necessary, but the prognosis overall is good for complete recovery (Patchell, 1994).

### Posterior Leukoencephalopathy Syndrome

Recently, a reversible posterior leukoencephalopathy syndrome has been described in a heterogeneous group of patients, including those undergoing renal, liver, and bone marrow transplantation and immunosuppressive treatment with either tacrolimus (FK506) or cyclosporine (Hinchey *et al.*, 1996). Abrupt increases in blood pressure are probably central in the pathophysiology of the condition, which presents with headaches,

vomiting, confusion, seizures, cortical blindness, and other visual abnormalities. Brain MRI confirms extensive bilateral white matter abnormalities suggestive of edema in the posterior regions of the cerebral hemispheres. Providing the syndrome is recognized and appropriate antihypertensive treatment is instituted in combination with a reduction or withdrawal of the immunosuppressive agent, the outcome is excellent. Others have criticized the term "reversible posterior leukoencephalopathy syndrome" and have pointed out that the condition is clinically and radiographically similar to the previously well-characterized disorders of hypertensive encephalopathy and cyclosporin-induced neurotoxicity (Schwartz, 1996).

## HYPERCALCEMIC ENCEPHALOPATHY

### Clinical Aspects

Hypercalcemia is a rare cause of a metabolic encephalopathy. In particular, patients with malignant tumors with or without bone metastases, plasmacytoma, sarcoidosis, and primary hyperparathyroidism, and patients treated with dihydrotachysterol are at risk of hypercalcemic encephalopathy developing. In addition, hypercalcemia may be induced by intoxication with vitamin D. Acute hypercalcemia is characterized by a combination of renal (polyuria, exsiccosis, nephrocalcinosis), gastrointestinal (nausea, vomiting, obstipation, rarely acute pancreatitis), cardiovascular (shortened Q-T interval on the electrocardiogram, enhanced sensitivity to digitalis), and neuropsychological symptoms (confusion, hallucinations, disturbance of consciousness, affective disorders, muscular hypotonia).

### Principles of Therapy

The decision to treat mainly depends on the magnitude of the hypercalcemia. Levels of 14 mg/100 ml and more require immediate aggressive therapy, whereas less-aggressive treatment is often sufficient at levels less than 14 mg/100 ml. However, the calcium level cannot serve as the only guide to therapy, because the severity of symptoms also depends on the velocity with which calcium levels rise. It must be further considered that the fraction of ionized calcium varies with blood pH and serum albumin concentrations. Therefore, the clinical situation must be taken into account when deciding about the type of therapy.

The major goals of the therapy of hypercalcemia are to enhance renal excretion of calcium and inhibit bone resorption. The first therapeutic step is to correct the dehydration present in almost all patients with hypercalcemia. Restoration of the intravascular volume leads to a decline of calcium levels by dilution and increased glomerular filtration rate. The increased glomerular filtration rate and excretion of sodium enhance the renal excretion of calcium. Renal excretion of calcium is further accelerated by loop diuretics and calcitonin (Stevenson, 1988). Thiazides should not be used, because they enhance reabsorption of calcium. Mobilization of calcium from the bones is reduced by diphosphonates, calcitonin, and mithramycin (Bilezikian, 1992). A flowchart indicating the general rules in the management of hypercalcemia is given in Figure 1. Table IV summarizes the dosages of the compounds used for treatment of hypercalcemia.

## Practical Management

Hydration with isotonic NaCl infusions is the first therapeutic step in the management of the hypercalcemic patient. A widely used regimen is to administer 3–5 L of isotonic (154 mmol/L) NaCl daily by intravenous infusion. When symptoms of fluid overload appear, furosemide is administered at doses of 20–60 mg every 2 hours. In older patients or patients with reduced fluid tolerance, furosemide should be given at lower doses of 10–20 mg every 6–12 hours from the beginning of the infusion therapy. Therapy with volume expansion and diuretics requires close monitoring of fluid and electrolyte balance.

In cases of mild hypercalcemia without manifest symptoms fluid restoration alone is adequate. With high calcium levels (>3.5 mmol/L; 14 mg/100 ml), more aggressive treatment should be initiated. If the situation is not life threatening and symptoms are moderate, biphosphonates are recommended. The following biphosphonates are available worldwide: clodronate, etidronate, and pamidronate. The recommended dosages of each compound and the maximal duration of treatment are given in Table IV. Pamidronate is the most effective compound. Treatment should be interrupted if serum calcium concentration is in the upper normal range or if it drops by more than 2–3 mg/100 ml within the first 2 or 3 days of treatment. Prolonged treatment is associated with some risk of hypocalcemia. Oral long-term treatment with pamidronate (2 × 150 mg) or clodronate (4 × 4–800 mg/day for a maximum of 6 months) has the potential to inhibit growth of bone metastases and prevent recurrent hypercalcemia.

If hypercalcemia is life threatening (>4 mmol/L; 16 mg/100 ml) or if the clinical symptoms require rapid lowering of the calcium levels, there are other effective

**TABLE IV**  Compounds Used in the Management of Decompensated Hypercalcemia*

| Compound | Dosage (per day) | Duration of treatment |
|---|---|---|
| Biphosphonates | | |
| Clodronate | 4–6 mg/kg IV over 2–5 h | 5 days |
| Etidronate | 7.5 mg/kg IV over 4 h | 7 days |
| Pamidronate | 15–45 mg/kg IV over 4 h | 6 days |
| | or 90 mg IV over 24 h | 1 days |
| | or 1200 mg PO | 5 days |
| Calcitonin | 2 × 4 units/kg SC or IM | 3 days |
| Mithramycin | 25 µg/kg IV over 6 h | 10 days (every second day) |

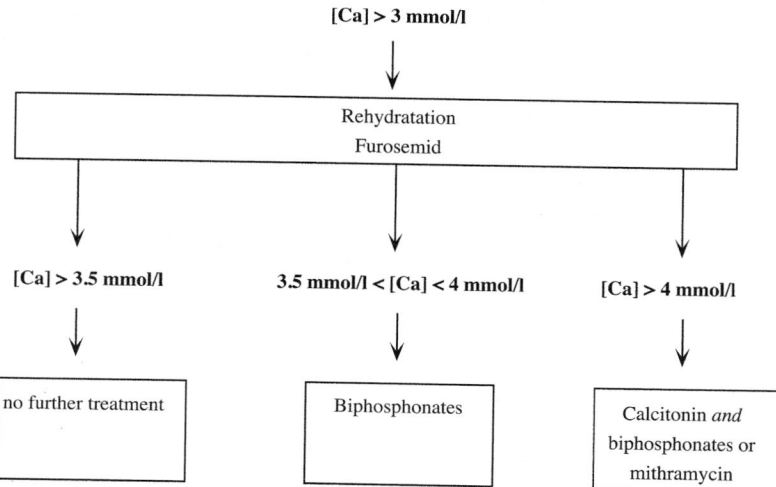

**FIGURE 1**  Practical management of hypercalcemic encephalopathy. Modified from Bilezikian (1992)*.

therapeutic regimens (Bilezekian, 1992). Calcitonin therapy is recommended in addition to rehydration, because calcitonin is the most rapidly acting inhibitor of calcium mobilization from the bones. Calcitonin is given at a dose of 4 U/kg body weight subcutaneously every 12 hours. Because the action of calcitonin is weak and short lived, it must be given in combination with biphosphonates or mithramycin. Mithramycin is given intravenously at a dose of 25 μg/kg body weight over a period of 6 hours. The dose can be repeated several times at intervals of at least 48 hours. Renal or hepatic failure, thrombocytopenia, and coagulopathy are relative contraindications to the use of mithramycin.

After stabilization of calcium levels below 3 mmol/L (12 mg/100 ml), appropriate measures should be taken to treat the underlying disease (malignancy, adenoma of the parathyroid glands).

## SEPTIC ENCEPHALOPATHY

### Clinical Aspects

Encephalopathy is a common complication of sepsis. It has been reported to occur in 8%–70% of septic patients and is the most common form of encephalopathy among patients in intensive care units. The large variation in the reported incidence of septic encephalopathy probably results from the different definitions of sepsis and encephalopathy used (Papadopoulos et al., 2000). It is often the first manifestation of sepsis. Septic patients with encephalopathy have a higher mortality than those without encephalopathy. Septic encephalopathy is probably underdiagnosed, because many critically ill patients receive treatments such as sedation, mechanical ventilation, or neuromuscular junction blockade that mask the signs of neural dysfunction. Septic patients may also have renal or liver failure, acute respiration distress syndrome, electrolyte disturbances, acid-base alterations, hypo/hyperthermia, or endocrine abnormalities. These associated conditions make it difficult for the effects of sepsis on the brain to be studied in isolation. Nevertheless, the onset of encephalopathy often precedes these abnormalities, suggesting that septic encephalopathy is not caused by them. EEGs may be more sensitive than bedside testing in identifying septic encephalopathy and may be valuable in intensive care units where clinical assessment is difficult.

The pathophysiology of septic encephalopathy is likely to be multifactorial. The cause involves reduced cerebral blood flow and oxygen extraction by the brain, cerebral edema, and disruption of the blood–brain barrier that may arise from the action of inflammatory mediators on the cerebrovascular endothelium, abnor-mal neurotransmitter composition of the reticular-activating system, impaired astrocyte function, and neuronal degeneration. A failure to detect disseminated cerebral microabscesses postmortem; a similar proportion of septic patients with gram-negative bacteremia, gram-positive bacteremia, fungemia, or with no identified causative organism to develop an encephalopathy; and the fact that encephalopathy occurs in noninfectious conditions such as pancreatitis suggest that infecting organisms and/or their toxins are less likely to directly cause an encephalopathy.

### Principles of Therapy

Currently there is no effective treatment.

## HEAVY METAL ENCEPHALOPATHIES

### Clinical Aspects

#### Acute Encephalopathy

The most common clinical neurological syndrome that results from intoxication is an acute but nonspecific encephalopathy. In its mildest form, the main symptoms are headache and an associated sense of light-headedness; these resolve within minutes to hours after the exposure is terminated. Neurological examination is usually normal even during the period of maximal symptoms. In a more severe form, there may also be confusion and irritability, impaired judgment, alteration in the level of consciousness, tinnitus, numbness and paresthesias, ataxia, a sense of weakness, and nausea and vomiting; recovery may take 24 hours or more but is usually complete. If exposure continues, alteration in the level of conssciousness supervenes and may progress to coma; seizures may also occur. If this stage is reached, morbidity and mortality increase sharply.

#### Chronic Encephalopathy

The encephalopathy that results from chronic exposure to neurotoxins is not well characterized. In fact, there is no incontrovertible evidence of a chronic neurotoxic encephalopathy. The most commonly cited symptoms are mild and nonspecific. They include headache, dizziness or light-headedness, difficulty concentrating, memory impairment, irritability, sleep disturbances (both insomnia and hypersomnolescence), loss of libido, depression, parasthesias, and weakness. Such symptoms may be caused by any number of medical diseases unrelated to toxin exposure or by exposure to neurotoxins unrelated to the workplace (e.g., alcoholism). Furthermore, identical symptoms may occur with primary psychological diseases and may easily claimed by malingerers. The routine neurological examination and EEG and EPs are generally normal.

### Heavy Metal Exposure

#### Lead

Many cases of chronic lead intoxication are due to industrial exposure, usually by inhalation of lead oxide. Domestic intoxications occur only rarely and are due mostly to lead contamination of drinking water. Bullets lodged in the body may be the cause of chronic lead intoxication even after years. Organic lead compounds (tetraethylated lead, tetramethylated lead) have been widely used as an adjuvant in gasoline.

Bones contain about 90% of the lead body burden, with a half-time of decay of 20 years. Blood and soft tissue compartments of lead have a half-time of 30–40 days. The toxic action of lead is due to blockade of the synthesis of porphyrins.

After acute exposure, blood lead levels, urinary excretion of lead, and δ-aminolevulinic acid are increased. For diagnosis of chronic lead intoxication, determination of free erythrocyte protoporphyrin is more helpful, because the levels of this compound remain elevated even after termination of exposure. After an exposure that dates back for several years, mobilization with 0.5 g calcium–disodium–ethylenediamine tetraacetate (EDTA) given intravenously is helpful for making the diagnosis. Urinary excretion of more than 1 g of lead in 24 hours is assumed to be pathological (Bleecker, 1987).

#### Mercury

Inorganic mercury is taken up by inhalation; exposure occurs when technical equipment (thermometers) breaks. Outbreaks of organic mercury poisoning occurred after the consumption of grain treated with methyl mercurial fungicides and of fish poisoned with mercury (Minamata disease). The chronic toxic action of mercury is due to enzyme inhibition.

Whereas after acute exposure to inorganic mercury, gastrointestinal and renal symptoms predominate, organic mercury compounds cause sensory and visual disturbances (numbness of fingers and tongue, blindness). Chronic intoxication may lead to an encephalopathy with tremor, ataxia, dysarthria, and irritability (Marsh, 1985). For diagnosis, mercury levels are determined in blood and urine.

#### Thallium

The wide use of thallium (I) sulfate as rat poison led to numerous intoxications. After intake of thallium, the clinical symptoms usually start after a latency of several days. Alopecia is the hallmark of thallium intoxication. In addition, thallium causes a painful sensory neuropathy that may resemble acute polyradiculitis. Thallium encephalopathy is characterized by restlessness, confusion, and ataxia.

In acute intoxications, thallium levels in urine are increased. For diagnosis of chronic thallium intoxication, thallium content of hair should be determined.

#### Manganese

Manganese intoxication occurs in an industrial setting, usually caused up uptake of manganese (IV) oxide. Chronic exposure may lead to an encephalopathy with a symptom complex resembling Parkinson's disease. For diagnosis, excretion of manganese in urine is measured after mobilization with EDTA.

#### Selenium

Selenium toxicity has been described in humans living in areas where the soil selenium content is relatively high, contributing to high selenium levels in vegetation and the food chain. Early symptoms of selenium toxicity include impaired vision, depressed appetite, and a tendency to wander in circles, which may progress to various degrees of paralysis and death from respiratory failure (Hogberg and Alexander, 1986). Other symptoms have been described and include peripheral anesthesia, pain, and hyperreflexia, sometimes progressing to numbness, convulsions, paralysis, and altered motor function (Hogberg and Alexander, 1986). Selenium is an essential metal, and the NRC Council's Food and Nutrition Board recommends 200 μg selenium per day as the maximum safe upper limit for adult human intake, whereas about 70 μg selenium per day is required to maintain Se balance.

#### Aluminum

Aluminum is among the most abundant elements in the earth's crust. Among patients with chronic renal failure, the presence of an increased body aluminum burden has an adverse effect on overall mortality (Chazan et al., 1988). More specifically, an increased body aluminum burden (estimated by the desferrioxamine infusion test) has been associated with memory impairment and increased severity of myoclonus and decreased motor strength (Sprague et al., 1988).

Aluminum toxicity is almost exclusively associated with metabolic encephalopathy. In individuals other than those with renal failure, aluminum enters the body through oral ingestion of aluminum-containing antacids, after which some aluminum apparently is absorbed. The retention of aluminum after oral administration of aluminum salts is greater in patients with renal failure than in normal subjects (Kaehny et al., 1977). However, the typical daily dietary aluminum intake is 10–100 mg. This quantity of dietary aluminum is more than enough to account for the entire increase of brain aluminum observed in patients with either Alzheimer's disease or dialysis dementia. Other forms of dementia besides Alzheimer's disease are associated with increased aluminum levels in the brain. These include amyotrophic lateral sclerosis and parkinsonism-dementia on Guam (Garruto et al., 1986). In general, when aluminum is given orally to humans or laboratory animals, most investigators have found aluminum to have no apparent harmful sequelae, although it has resulted in increased brain aluminum content. Aluminum with or without calcium and silicone, in the presence of an abnormal blood–brain barrier may exert toxic effects on the CNS and is involved in the pathogenesis of several different forms of dementia. In particular, aluminum has been implicated in the pathogenesis of dialysis dementia (see earlier section on dialysis dementia in this chapter).

## Principles of Therapy

The basic principles of the treatment of heavy metal–induced encephalopathies are (1) to avoid further exposure, (2) to prevent gastrointestinal absorption and reabsorption (colestyramine in mercury intoxication, ferric hexacynaoferrate (II) in thallium intoxication), and (3) to remove heavy metal ions from the body by formation of excretable complexes with chelating agents.

Table V gives an overview about specific antidotes, dosage, and mode of application. The duration of therapy depends on the clinical situation and the amount of metal excretion in urine. The use of chelating agents is associated with a considerable risk of toxic and allergic side effects (bone marrow depression, renal damage). Therefore, frequent blood cell counts and laboratory tests of renal function are required. Mobilization of heavy metals by chelating agents may lead to temporary deterioration of the clinical situation. In these cases, forced diuresis or hemodialysis is necessary. Because chelating agents also increase the excretion of essential metals, such as magnesium and zinc, substitution of these metals is recommended during long-term therapy.

Chelation is the formation of a metal ion complex in which the metal ion is associated with a charged or uncharged electron donor, referred to as a ligand. The ligand may be monodentate, bidentate, or multidentate; that is, it may attach or coordinate using one or two or more donor atoms. Bidentate ligands form ring structures that include the metal ion and the two ligand atoms attached to the metal. Metals may react with O—, S—, and N— containing ligands present in the form of —OH, —COOH, >C=O, —S—S—, —NH2, and >NH. A resultant metal complex is formed by a coordinate bond (coordination compound), in which both electrons are contributed by the ligand (Klaassen, 1990).

**TABLE V** Antidotes in Heavy Metal Intoxications*

| Compound | Metals | Mode of application | Dosage (per day) |
|---|---|---|---|
| Calcium-disodium-ethylenediamine tetraacetate (EDTA) | Anorganic Pb, Mn | IV infusion | 20 mg/kg continuously or in 3 fractions |
| 2,3-dimercaptosuccinic acid (DMSA) | Pb | PO | 700 mg/m²/day for 4 wk in children with chronic intoxication and blood levels >44 µg/dl (course may be repeated) |
| Calcium-trisodium-ethylenetriamine (DTPA) | Mn, anorganic Pb | IV infusion | 1 g |
| Dimercaprol (BAL) | Anorganic Hg, Pb | IM injection | Day 1–2: 6 × 25 mg/kg<br>Day 3–4: 4 × 2.5 mg/kg<br>Day 5–6: 2 × 2.5 mg/kg |
| 2,3-Dimercapto-propan-1-sulfonate (DMPS) | Hg | PO, IV injection | 3 × 100–200 mg<br>Day 1: 6 × 250 mg<br>Day 2: 4 × 250 mg<br>Day 3: 3 × 250 mg |
| D-Penicillin | Hg, Pb | PO | 600 mg (several fractions) |
| Colestyramine | Organic Hg | PO | 16–24 g (several fractions) |
| Ferric hexacyano-ferrate (II) (Prussian blue) | Th | PO | Initially 3 g, followed by 3–20 g continuously |

Chelating agents (drugs) vary in their specificity for toxic metals. Ideal chelating agents should be water soluble, resistant to biotransformation, able to reach sites of metal storage, capable of forming nontoxic complexes with toxic metals and of being excreted from the body, and should have a low affinity for essential metals, particularly calcium and zinc (Klaassen, 1990). The general properties of chelating agents that are of current interest will be briefly described.

1. BAL (2,3-dimercaptopropanol) was the first clinically usefu chelating agent. It was developed during World War II as a specific antagonist to vesicant arsenical war gases based on the observation that arsenic has an affinity for sulfhydryl-containing substances. BAL, a dithiol compound with two sulfur atoms on adjacent carbon atoms, competes with the critical binding sites responsible for the toxic effects. BAL forms stable chelates in vivo with many toxic metals, including inorganic mercury, antimony, bismuth, cadmium, chromium, cobalt, gold, and nickel. However, it is not necessarily the treatment of choice for toxicity to these metals. BAL is a potentially toxic drug, and its use may be accompanied by multiple side effects. Although treatment with BAL will increase the excretion of cadmium, there is a concomitant increase in renal cadmium concentration, so that its use in case of cadmium toxicity is usually avoided. It does, however, remove inorganic mercury from the kidneys, but it is not useful in the treatment of alkyl mercury or phenyl mercury toxicity. BAL also enhances the toxicity of selenium and tellurium, so it is not to be used to remove these metals from the body.

2. DMPS (2,3-dimercapto-1-propanesulfonic acid) is a water-soluble derivative of BAL developed in response to BAL's toxicity and unpleasant side effects. DMPS has been shown to reduce blood lead levels in children and has the advantage over EDTA in that it is administered orally and does not seem to have toxic side effects.

3. Calcium EDTA is the calcium disodium salt of EDTA. The calcium salt must be used for clinical purposes, because the sodium salt has greater affinity for calcium and will produce hypocalcemic tetany. However, the calcium salt will bind lead, with displacement of calcium from the chelate. EDTA is poorly absorbed from the gastrointestinal tract, so it must be given parenterally, which distributes it rapidly throughout the body. It has long been the method of choice for the treatment of lead toxicity. Calcium EDTA is potentially nephrotoxic, so it should be administered only when indicated clinically.

4. DMSA (meso-2,3,-dimercatosuccinic acid; Succimer), like DMPS, is a chemical analogue of BAL. More than 90% of DMSA is in the form of a mixed disulfide in which each of the sulfur atoms is in disulfide linkage with a cysteine molecule. The drug is of current interest clinically because of its ability to lower blood lead levels. It has advantages over EDTA, because it is given orally and has greater specificity for lead. It may be safer than EDTA in that it does not enhance excretion of calcium and zinc to the same degree.

5. Penicillamine (p,B-dimethylcysteine), a hydrolytic product of penicillin, has been used for the removal of copper in persons with Wilson's disease and for the removal of lead, mercury, and iron. It is also important to note that penicillamine removes other physiologically essential metals, including zinc, cobalt, and manganese. Attached to its use is the risk of inducing a hypersensitivity reaction with a wide spectrum of undesirable immunological effects, including skin rash, blood dyscrasia, and possibly proteinuria and nephrotic syndrome.

6. DTPA or diethylenetriaminepentaacetic acid has chelating properties similar to those of EDTA. The calcium salt ($CaNa_2$ DPTA) must be used clinically because of DPTA's high affinity for calcium. It has been used for chelation of plutonium and other actinide elements but with mixed success.

7. Deferroxamine is a hydroxylamine isolated as the iron chelate of *Streptomyces pilosus* and is used clinically in the metal-free form. It has a remarkable affinity for ferric iron and a low affinity for calcium. It is poorly absorbed from the gastrointestinal tract so it must be given parenterally. It has been widely used for the treatment of aluminum toxicity (Altmann *et al.*, 1989). There are a variety of toxic effects including hypotension, skin rashes, and, possibly, cataract formation.

## Practical Management

### Lead

Acute lead intoxication is classically treated with EDTA, which is no longer available in Germany. Instead, DMPS is used. DMPS is given slowly intravenously at an initial dose of $6 \times 250$ mg on the first day of treatment, $4 \times 250$ mg on the second day, and $3 \times 250$ mg on the third day. After that day, doses ($1-3 \times 250$ mg) are chosen according to the clinical state. Alternately, DMPS can be given orally ($3 \times 100-200$ mg). In chronic intoxications, DMPS is given orally from the beginning. Intoxications with organic lead compounds are treated subsequently with D-penicillamine at a dose of 600 mg/day over a period of several months.

Lead encephalopathy is more frequent in children than in adults. The encephalopathy is characterized by lethargy that may progress to somnolence, personality changes, and ataxia. In addition, chronic lead intoxication may be the cause of motor neuropathy, anemia, and lead colics (Marsh, 1985).

However, thousands of children, especially poor children living in deteriorated urban housing, are exposed to enough lead to produce cognitive impairment. In 1991, the Food and Drug Administration licensed succimer (dimercaptosuccinic acid), the first approved oral lead chelator for use in children with blood lead levels of at least 45 μg/100 ml (2.2 μmol/L) (Nightingale, 1991). Succimer reduced blood lead levels at least as well as parenteral treatment with edetate calcium disodium in children with blood levels of 30 μg/100 ml (1.4 μmol/L) or greater (Graziano et al., 1988), a level associated with cognitive impairment but not the other symptoms of lead poisoning. A recent randomized, placebo-controlled, double-blind study asked whether the cognitive impaiment of children with blood lead levels of 20–44 μg/100 ml (1.0–2.1 μmol/L) would improve from up to three 26-day courses of treatment with succimer (Rogan et al., 2001)**. Although treatment with succimer lowered blood lead levels, it did not improve scores on tests of cognition, behavior, or neuropsychological function in children with blood lead levels less than 45 μg/100 ml. Because the regimen was expensive and a significant burden on the family, treatment was not recommended for children with blood lead levels less than 45 μg/100 ml.

### Mercury

In acute intoxication with inorganic mercury, the basic aim is to remove unabsorbed mercury from the gastrointestinal tract (gastric lavage, rinse with proteinaceous solution). In severe intoxication, forced diuresis or dialysis should follow. The classical chelating agent for mercury intoxication is dimercaprol, which is no longer available in Germany. Both in intoxications with organic and inorganic mercury, treatment with intravenous or oral DMPS is recommended (Clarkson et al., 1981). Dosages correspond to those in lead intoxication. To increase excretion of mercury with feces, colestyramine is given at doses of 16–24 g/day orally.

### Thallium

Treatment of acute thallium intoxication consists of gastric lavage and application of ferric hexacyanoferrate (II) (Prussian blue) at a dose of initially 3 g by means of gastric tubing. Application of Prussian blue is then continued at a dose of 3–20 g/day (Stevens et al., 1974). Duration of treatment depends on the amount of thallium excretion with the feces. Simultaneously, laxatives are given. To increase renal excretion, forced diuresis is performed. Severe intoxications require hemodialysis (De Groot and van Heijst, 1988).

### Manganese

In earlier states of manganese intoxication, a trial of calcium-trisodium-ethylenetriamine-pentacetate (DPTA) may be successful: 1 g of DPTA is infused within 6 hours in 250 ml saline. This dose (1 g/day) is given for 6 days, after which a break of at least 3 days should be made. Manganese-induced parkinsonism is not sensitive to treatment with chelating agents. Instead, symptomatic treatment with dopaminergic drugs is required.

## REFERENCES

Abdel-Gawad, M., Huynh, H., and Brock, G. B. (1999). Experimental chronic renal failure-associated erectile dysfunction: Molecular alterations in nitric oxide synthase pathway and IGF-I system. *Mol. Urol.* **3**, 117–125.

Adams, R. D., Victor, M., and Mancall, E. L. (1959). Central pontine myelinolysis: A hitherto undescribed disease occurring in alcoholic and malnourished patients. *Arch. Neurol. Psychiat.* **81**, 154–172.

Alexander, W. F., Spindel, E., Harty, R. F., and Cerda, J. J. (1989). The usefulness of branched chain amino acids in patients with acute or chronic hepatic encephalopathy. *Am. J. Gastroenterol.* **84**, 91–96.

Alfrey, A. C., LeGendre, G. R., and Kaehny, W. D. (1976). The dialysis encephalopathy syndrome: possible aluminium intoxication. *N. Engl. J. Med.* **294**, 184–188.

Alfrey, A. C., Mishell, J. M., Burks, J., Contiguglia, S. R., Rudolph, H., Lewin, E., and Holmes, J. H. (1972). Syndrome of dyspraxia and multifocal seizures associated with chronic hemodialysis. *Trans. Am. Soc. Artif. Intern. Organs.* **18**, 257–261, 266–257.

Altmann, P., Dhanesha, U., Hamon, C., Cunningham, J., Blair, J., and Marsh, F. (1989). Disturbance of cerebral function by aluminum in haemodialysis patients without overt aluminum toxicity. *Lancet* **ii**(8653), 7–12.

Andreoli, S. P., Beegstein, J. M., and Sherrard, D. J. (1984). Aluminium intoxication from aluminium-containing phosphate binders in children with azotemia not undergoing dialysis. *N. Engl. J. Med.* **310**, 1079.

Arieff, A. I. (1994). Dialysis disequilibrium syndrome: current concepts on pathogenesis and prevention. *Kidney Int.* **45**, 629–635.

Arieff, A. I., Cooper, J. D., Armstrong, D., and Lazarowitz, V. C. (1979). Dementia, renal failure, and brain aluminum. *Ann. Intern. Med.* **90**, 741–747.

Arieff, A. I., Guisado, R., Massry, S. G., and Lazarowitz, V. C. (1976). Central nervous system pH in uremia and the effects of hemodialysis. *J. Clin. Invest.* **58**, 306–311.

Arieff, A. I., and Papadakis, M. A. (1988). Hyponatremia and hypernatremia in liver disease. *In* "The Kidney in Liver Disease" (M. Epstein, ed.), pp. 73–88, Williams & Wilkins, Baltimore.

Ayus, J. C., Krothapalli, R. K., and Arieff, A. I. (1987). Treatment of symptomatic hyponatremia and its relation to brain damage. A prospective study. *N. Engl. J. Med.* **317**, 1190–1195.

Ayus, J. C., and Sheikh-Hamad, D (1998). Silent infection in clotted hemodialysis access grafts. *J. Am. Soc. Nephrol.* **9**, 1314–1317.

Ballou, S. P., and Kushner, I. (1992). C-reactive protein and the acute phase response. *Adv. Intern. Med.* 37, 313–336.

Barbaro, G., Di Lorenzo, G., Soldini, M., Giancaspro, G., Bellomo, G., Belloni, G., Grisorio, B., Annese, M., Bacca, D., Francavilla, R., and Barbarini, G. (1998). Flumazenil for hepatic encephalopathy grade III and IVa in patients with cirrhosis: an Italian multicenter double-blind, placebo-controlled, cross-over study. *Hepatology* 28, 374–378.

Bilezikian, J. P. (1992). Mangement of acute hypercalcemia. *N. Engl. J. Med.* 326, 1196–1203.

Blanc, P., Daures, J. P., Rouillon, J. M., Peray, P., Pierrugues, R., Larrey, D., Gremy, F., and Michel, H. (1992). Lactitol or lactulose in the treatment of chronic hepatic encephalopathy: results of a meta-analysis. *Hepatology* 15, 222–228.

Bleecker, M. I. (1987). Heavy metals and industrial toxins. *In* "Current therapy in Neurological Disease" (R. T. Johnson, ed.), pp. 329–333, B. C. Decker, Toronto.

Breyer, J. A. (1995). Medical management of nephropathy in type I diabetes mellitus: Current recommendations. *J. Am. Soc. Nephrol.* 6, 1523–1529.

Bronner, I. L., Kanter, D. S., and Manson, J. E. (1995). Primary prevention of stroke. *N. Engl. J. Med.* 333, 1392–1400.

Butterworth, R. F., Spahr, L., Fontaine, S., and Layrargues, G. P. (1995). Manganese toxicity, dopaminergic dysfunction, and hepatic encephalopathy. *Metab. Brain. Res.* 10, 259–267.

Campese, V. M., and Liu, C. L. (1990). Sexual dysfunction in uremia. Endocrine and neurological alterations. *Contrib. Nephrol.* 77, 1–14.

Chason, J. L., Landers, J. W., *et al.* (1964). Central pontine myelinolysis. *J. Neurol. Neurosurg. Psychiatry* 27, 317–321.

Chazan, J. A., Blonsky, S. L., Abuelo, J. G., and Pezzullo, J. C. (1988). Increased body aluminum: An independent risk factor in patients undergoing long-term hemodialysis? *Arch. Intern. Med.* 148, 1817–1820.

Chokroverty, S., Bruetman, M. E., Berger, V., and Reyes, M. G. (1976). Progressive dialytic encephalopathy. *J. Neurol. Neurosurg. Psychiatr.* 39, 411–419.

Clarkson, T. W., Magos, I., Cox, C., Greenwood, M., Amin-Zaki, L., Majeed, M. A., and Al-Damluji, S. F. (1981). Tests of efficacy of antidotes for removal of methylmercury in human poisoning during the Iraq outbreak. *J. Pharmacol. Exp. Ther.* 218, 1–9.

Cole, M., Richardson, E. P., *et al.* (1964). Central pontine myelinolysis: Further evidence relating the lesion to malnutrition. *Neurology* 14, 165–170.

Conn, H. E., Leevy, C. M., Vlahcevic, Z. R., Rodgers, J. B., Maddrey, W. C., Seeff, L., and Levy, L. L. (1977). Comparison of lactulose and neomycin in the treatment of portal systemic encephalopathy. *Gastroenterology* 72, 573–583.

Conn, H. O. (1994). Quantifying the severity of hepatic encephalopathy. *In* "Hepatic Encephalopathy: Syndromes and Therapies" (H. O. Conn and J. Bircher, Eds.), pp. 13–26, Medi-Ed Press, Bloomington.

Davies, M. H., Mutimer, D., Lowes, J., Elias, E., and Neuberger, J. (1994). Recovery despite impaired cerebral perfusion in fulminant hepatic failure. *Lancet* 343, 1329–1330.

De Groot, G., and van Heijst, A. N. (1988). Toxicokinetic aspects of thallium poisoning. Methods of treatment by toxin elimination. *Sci. Total Environ.* 71, 411–418.

Droller, M. J. and Anderson, J. R. (1993). NIH consensus conference. Impotence. *JAMA* 270, 83–90.

Estol, C. J., Faris, A. A., Martinez, A. J., and Ahdab-Barmada, M. (1989). Central pontine myelinolysis after liver transplantation. *Neurology* 39, 493–498.

Flendrig, J. A., Kruis, H., and Das, H. A. (1976). Aluminum and dialysis dementia. *Lancet* 1, 235.

Foley, R. N., and Harnett, J. D. (1997). Cardiovascular complications of end-stage renal disease. *In* "Diseases of the Kidney" (R. W.

Schrier and C. W. Gottschalk, Eds.), Vol. III, pp. 2647–2660. Little, Brown & Co., Boston.

Fraser, C. L., and Arieff, A. I. (1985). Hepatic encephalopathy. *N. Engl. J. Med.* 313, 865–873.

Fraser, C. L., and Arieff, A. I. (1994). Metabolic encephalopathy as a complication of renal failure: mechanisms and mediators. *New Horizons* 2, 518–526.

Fraser, C. L., and Arieff, A. I. (2001). Nervous system manifestations of renal failure. *In* "Diseases of the Kidney" (R. W. Schrier, C. W. Gottschalk, Eds.), Vol. III, pp. 2769–2794, Lippincott Williams & Wilkins, Philadelphia.

Garruto, R. M., Swyt, C., Yanagihara, R., Fiori, C. E., and Gajdusek, D. C. (1986). Intraneuronal co-localization of silicon with calcium and aluminum in amyotrophic lateral sclerosis and parkinsonism with dementia of Guam. *N. Engl. J. Med.* 315, 711–712.

Goldstein, I., Lue, T. F., Padma-Nathan, H., Rosen, R. C., Steers, W. D., and Wicker, P. A. (1998). Oral sildenafil in the treatment of erectile dysfunction. Sildenafil Study Group. *N. Engl. J. Med.* 338, 1397–1404.

Graziano, J. H., Lolacono, N. J., and Meyer, P. (1988). Dose-response study of oral 2,3-dimercaptosuccinic acid in children with elevated blood lead concentrations. *J. Pediatr.* 113, 751–757.

Groeneweg, M., Quero, J. C., De Bruijn, I., Hartmann, I. J., Essink-bot, M. L., Hop, W. C., and Schalm, S. W. (1998). Subclinical hepatic encephalopathy impairs daily functioning. *Hepatology* 28, 45–49.

Gyr, K., and Meier, R. (1991). Flumazenil in the treatment of portal systemic encephalopathy—an overview. *Intensive Care Med.* 17, Suppl. 1, S39–S42.

Gyr, K., Meier, R., Haussler, J., Bouletreau, P., Fleig, W. E., Gatta, A., Holstege, A., Pomier-Layrargues, G., Schalm, S. W., Groeneweg, M., Scollo-Lavizzari, G., Ventura, E., Zeneroli, M. L., Williams, R., Yoo, Y., and Amrein, R. (1996). Evaluation of the efficacy and safety of flumazenil in the treatment of portal systemic encephalopathy: a double blind, randomised, placebo controlled multicentre study. *Gut* 39, 319–324.

Hazell, A. S., and Butterworth, R. F. (1999). Hepatic encephalopathy: An update of pathophysiologic mechanisms. *Proc. Soc. Exp. Biol. Med.* 222, 99–112.

Held, P. J., Brunner, F., Odaka, M., Garcia, J. R., Port, F. K., and Gaylin, D. S. (1990). Five year survival for end stage renal disease patients in the USA, Europe and Japan, 1982–1987. *J. Am. Soc. Nephrol.* 15, 451–457.

Held, P. J., Port, F. K., and Webb, R. L. (1994). United States Renal Data System Annual Data Report. VI. Patient survival. *Am. J. Kidney Dis.* 24 Suppl. 2, S76–S87.

Herrmann, H. C., Chang, G., Klugherz, B. D., and Mahoney, P. D. (2000). Hemodynamic effects of sildenafil in men with severe coronary artery disease. *N. Engl. J. Med.* 342, 1622–1626.

Hinchey, J., Chaves, C., Appignani, B., Breen, J., Pao, L., Wang, A., Pessin, M. S., Lamy, C., Mas, J. L., and Caplan, L. R. (1996). A reversible posterior leukoencephalopathy syndrome. *N. Engl. J. Med.* 334, 494–500.

Hindfelt, B., Plum, F., and Duffy, T. E. (1977). Effects of acute ammonia intoxication on cerebral metabolism in rats with portocaval shunts. *J. Clin. Invest.* 59, 386–396.

Hogberg, J., and Alexander, J. (1986). Specific metals. *In* "Handbook on the Toxicology of Metals" (L. Friberg, G. Nordberg, and V. B. Vouk, Eds.), pp. 482–512, Elsevier, Amsterdam.

Hughes, J. R., and Schreeder, M. T. (1980). EEG in dialysis encephalopathy. *Neurology* 30, 1148–1154.

Iseki, K., and Fukiyama, K. (1996). Predictors of stroke in patients receiving chronic hemodialysis. *Kidney Int.* 50, 1672–1675.

Iseki, K., Kawazoe, N., Osawa, A., and Fukiyama, K. (1993). Survival analysis of dialysis patients in Okinawa, Japan (1971–1990). *Kidney Int.* 43, 404–409.

Iseki, K., Tozawa, M., Iseki, C., Takishita, S., and Ogawa, Y. (2002). Demographic trends in the Okinawa Dialysis Study (OKIDS) registry (1971–2000). *Kidney Int.* **61**, 668–675.

Jack, R., Rabin, P. L., and McKinney, T. D. (1983). Dialysis encephalopathy: a review. *Int. J. Psychiatry Med.* **13**, 309–326.

Kaehny, W. D., Hegg, A. P., and Alfrey, A. C. (1977). Gastrointestinal absorption of aluminum from aluminum-containing antacids. *N. Engl. J. Med.* **296**, 1389–1390.

Klaassen, C. D. (1990). Heavy metals and heavy metal antagonists. *In* "Goodman and Gillman's The Pharmacological Basis of Therapeutics" (A. G. Gillman, T. W. Rall, A. S. Nies, and P. Taylor. Eds.), pp. 1592–1614, Pergamon Press, New York.

Kennedy, A. C., Lintonm, A. L., and Eaton, J. C. (1962). Urea levels in cerebrospinal fluid after hemodialysis. *Lancet* **1**, 410.

Krieger, D., Krieger, S., Jansen, O., Gass, P., Theilmann, L., and Lichtnecker, H. (1995). Manganese and chronic hepatic encephalopathy. *Lancet* **346**, 270–274.

Lamberts, S. W. J., van den Beld, A. W., and van der Lely, A. J. (1997). The endocrinology of aging. *Science* **278**, 419–424.

Lee, W. M. (1993). Acute liver failure. *N. Engl. J. Med.* **329**, 1862–1872.

Levy, N. B., and Cohen, L. M. (2001). Central and peripheral nervous system in uremia: Psychiatric and psychosocial considerations. *In* "Massry and Glassock's Textbook of Nephrology" (S. G. Massry and R. J. Glassock, Eds), Vol. 1, pp. 1279–1282, Lippincott Williams & Wilkins, Philadelphia.

Levy, D. E., Bates, D., Caronna, J. J., Cartlidge, N. E. F., Knill-Jones, R. P., Lapinski, R. H., Singer, B. H., Shaw, D. A., and Plum, F. (1981). Prognosis in nontraumatic coma. *Ann. Intern. Med.* **94**, 293–301.

Mahurkar, S. D., Salta, R., Smith, E. C., Dkar, S. K., Meyers Jr., L., and Dunea, G. (1973). Dialysis dementia. *Lancet* **1**(7817), 1412–1415.

Marchesini, G., Dioguarda, F. S., Bianchi, G. P., Zoli, M., Bellati, G., Roffi, L., Martines, D., and Abbiati, R. (1990). Long-term oral branched amino acid treatment in chronic hepatic encephalopathy. A randomized double-blind casein-controlledtrial. The Italian Multicenter Study Group. *J. Hepatol.* **11**, 92–101.

Marsh, D. O. (1985). The neurotoxicity of mercury and lead. *In* "Neurotoxicity of Industrial and Commercial Chemicals" (J. L. O'Donoghue, ed.), pp. 159–169, CRC Press, Boca Raton.

Massry, S. G., Smogorzewski, M. J., *et al.* (1997). Metabolic and endocrine dysfunctions in uremia. *In* "Diseases of the Kidney" (R. W. Schrier and C. W. Gottschalk, Eds.), Vol. III, pp. 2661–2698, Little, Brown & Co., Boston.

Mazzuchi, N., Carbonell, E., and Fernandez-Cean, J. (2000). Importance of blood pressure control in hemodialysis patient survival. *Kidney Int.* **58**, 2147–2154.

McKee, A., Winkelman, M. D., and Banker, B. Q. (1988). Central pontine myelinolysis in severely burned patients: Relationship to serum hyperosmolality. *Neurology* **38**, 1211–1217.

Meeus, F., Kourilsky, O., Guerin, A. P., Gaudry, C., Marchais, S. J., and London, G. M. (2000). Pathophysiology of cardiovascular disease in hemodialysis patients. *Kidney Int.* **58**(Suppl. 76), S140–S147.

Messert, B., Orrison, W. W., Hawkins, M., and Quaglieri, C. E. (1979). Central pontine myelinolysis. Considerations on etiology, diagnosis, and treatment. *Neurology* **29**, 147–160.

Morgan, M. Y. (1991). The treatment of chronic hepatic encephalopathy. *Hepatogastroenterology* **38**, 377–387.

Mullen, K. D., Szauter, K. M., and Kaminsky-Russ, K. (1990). "Endogenous" benzodiazepine activity in body fluids of patients with hepatic encephalopathy. *Lancet* **336**, 81–83.

Nightingale, S. L. (1991). From the Food and Drug Administration. *JAMA* **265**, 1802.

Padma-Nathan, H. P., Hellstrom, W. J. G., Kaiser, F. E., Labasky, R. F., Lue, T. F., Nolten, W. E., Norwood, P. C., Peterson, C. A., Shabisgh, R., Tam, P. Y., *et al.* (1997). Treatment of men with erectile dysfunction with transurethral alprostadil. *N. Engl. J. Med.* **336**, 1–7.

Palmer, B. F. (1999). Sexual dysfunction in uremia. *J. Am. Soc. Nephrol.* **10**, 1381–1388.

Papadakis, M. A., Fraser, C. L., and Arieff, A. I. (1990). Hyponatraemia in patients with cirrhosis. *Quart. J. Med.* **76**, 675–688.

Papadopoulos, M. C., Davies, D. C., Moss, R. F., Tighe, D., and Bennett, E. D. (2000). Pathophysiology of septic encephalopathy: a review. *Crit. Care Med.* **28**, 3019–3024.

Parfrey, P. S. (1993). Cardiac and cerebrovascular disease in chronic uremia. *Am. J. Kidney Dis.* **21**, 77–80.

Patchell, R. A. (1994). Neurological complications of organ transplantation. *Ann. Neurol.* **36**, 688–703.

Raskin, N. H. (2001). Neurological complications of renal failure. *In* "Neurology and General Medicine" (M. J. Aminoff, ed.), pp. 293–306, Churchill Livingstone, New York.

Rivera-Penera, T., Moreno, J., Skaff, C., McDiarmid, S., Vargas, J., and Ament, M. E. (1997). Delayed encephalopathy in fulminant hepatic failure in the pediatric population and the role of liver transplantation. *J. Pediatr. Gastroenterol. Nutr.* **24**, 128–134.

Riviello, J. J., Haligan, G. E., Dunn, S. P., Widzer, S. J., Foley, C. M., Breningstall, G. N., and Grover, W. D. (1990). Value of plasmapheresis in hepatic encephalopathy. *Pediatr. Neurol.* **6**, 388–390.

Rogan, W. J., Dietrich, K. N., Ware, J. H., Dockery, D. W., Salganik, M., Radcliffe, J., Jones, R. L., Ragan, N. B., Chisolm, J. J., Jr., and Rhoads, G. G. (2001). The effect of chelation therapy with succimer on neuropsychological development in children exposed to lead. *N. Engl. J. Med.* **344**, 1421–1426.

Rostand, S. G., Brunzell, J. D., Cannon, R. D., and Victor, R. G. (1991). Cardiovascular complications in renal failure. *J. Am. Soc. Nephrol.* **2**, 1053–1062.

Schomerus, H., and Schreiegg, J. (1993). Prevalence of latent portasystemic encephalopathy in an unselected population of patients with liver cirrhosis in general practice. *Z. Gastroenterol.* **31**, 231–234.

Schwartz, R. B. (1996). A reversible posterior leukoencephalopathy syndrome. *N. Engl. J. Med.* **334**, 1743; discussion 1746.

Singh, N., Yu, V. L., and Gayowski, T. (1994). Central nervous system lesions in adult liver transplant recipients. *Medicine (Baltimore)* **73**, 110–118.

Smith, S. L., and Ciferni, M. (1993). Liver transplantation for acute hepatic failure: a review of clinical experience and management. *Am. J. Crit. Care* **2**, 137–144.

Sprague, S. M., Corwin, H. L., Tanner, C. M., Wilson, R. S., Green, B. J., and Goetz, C. G. (1988). Relationship of aluminum to neurocognitive dysfunction in chronic dialysis patients. *Arch. Intern. Med.* **148**, 2169–2172.

Stauch, S., Kircheis, G., Adler, G., Beckh, K., Ditschuneit, H., Gortelmeyer, R., Hendricks, R., Heuser, A., Karoff, C., Malfertheiner, P., Mayer, D., Rosch, W., and Steffens, J. (1998). Oral L-ornithine-L-aspartate therapy of chronic hepatic encephalopathy: results of a placebo-controlled double-blind study. *J. Hepatol.* **28**, 856–864.

Stevens, W., van Peteghem, C., Heyndrickx, A., and Barbier, R. (1974). Eleven cases of thallium intoxication treated with prussian blue. *Int. J. Clin. Pharmacol. Ther. Toxicol.* **10**, 1–22.

Stevenson, J. C. (1988). Current management of malignant hypercalcaemia. *Drugs* **36**, 229–238.

Strauss, G. I., Hogh, P., Moller, K., Knudsen, G. M., Hansen, B. A., and Larsen, F. S. (1999). Regional cerebral blood flow during mechanical hyperventilation in patients with fulminant hepatic failure. *Hepatology* 30, 1368–1373.

Sullivan, E. V., and Pfefferbaum, A. (2001). Magnetic resonance relaxometry reveals central pontine myelinolysis in clinically asymptomatic alcoholic men. *Alcohol Clin. Exp. Res.* 25, 1206–1212.

Tarter, R. E., Hegedus, A. M., Van Thiel, D. H., Schade, R. R., Gavaler, J. S., and Starzl, T. E. (1984). Nonalcoholic cirrhosis associated with neuropsychological dysfunction in the absence of overt evidence of hepatic encephalopathy. *Gastroenterology* 86, 1421–1427.

Teschan, P. E., and Arieff, A. I. (1985). Uremic and dialysis encephalopathies. *In* "Cerebral Energy Metabolism and Metabolic Encephalopathy" (D. W. McCandless, ed.), pp. 263–285, Plenum, New York.

Uribe, M., Marquez, M. A., Ramos, G. G., Ramos-Uribe, M. H., Vargas, F., Villalobos, A., and Ramon, C. (1982). Treatment of chronic portal-systemic encephalopathy with vegetable and animal protein diets: a controlled crossover study. *Dig. Dis. Sci.* 27, 1109–1115.

Utiger, R. D. (1998). A pill for impotence. *N. Engl. J. Med.* 338, 1458–1459.

Ware, A. J., D'Agostino, A. N., and Combes, B. (1971). Cerebral edema: A major complication of massive hepatic necrosis. *Gastroenterology* 61, 877–884.

Worthley, L. I. G. (1994). Hepatic failure. *In* "Synopsis of Intensive Care Medicine" (L. I. G. Worthley and N. Mathews, Eds.), pp. 553–564, Churchill Livingstone, Edinburgh.

Wratten, M. L., Tetta, C., Ursini, F., and Sevanian, A. (2000). Oxidant stress in hemodialysis: Prevention and treatment strategies. *Kidney Int.* 58 (Suppl. 76), S126–S132.

Wszolek, Z. K., McComb, R. D., Pfeiffer, R. F., Steg, R. E., Wood, R. P., Shaw Jr., B. W., and Markin, R. S. (1989). Pontine and extrapontine myelinolysis following liver transplantation. *Transplantation* 48, 1006–1012.

Zimmerman, J., Herrlinger, S., Pruy, A., Metzger, T., and Wanner, C. (1999). Inflamation enhances cardiovascular risk and mortality in hemodialysis patients. *Kidney Int.* 55, 648–658.

CHAPTER 73

# Inherited and Noninherited Ataxias

Thomas Klockgether and Johannes Dichgans

The inherited and noninherited ataxias comprise a wide spectrum of disorders, with ataxia as the leading symptom. In most of these disorders, ataxia is due to degeneration of the cerebellar cortex and its afferent or efferent fiber connections. An etiological classification of the inherited and noninherited ataxias is given in Table I.

## FRIEDREICH'S ATAXIA (FRDA)

### Clinical Aspects

FRDA is an autosomal recessively inherited ataxia with early disease onset. The major clinical signs of FRDA are progressive gait and limb ataxia, dysarthria, lower limb areflexia, loss of proprioception and hypertrophic cardiomyopathy. Further symptoms include muscle weakness, distal muscle wasting, extensor plantar responses, oculomotor disturbances with square wave jerks and reduced vestibulo-ocular reflex, optic atrophy, skeletal deformities (scoliosis, pes cavus), and diabetes mellitus. MRI typically shows cervical spinal cord atrophy without major cerebellar atrophy. Nerve conduction studies reveal an axonal form of sensory neuropathy (Dürr *et al.*, 1996).

In 1996, a homozygous, intronic GAA repeat expansion in a novel gene named *X25* was identified as the most frequent mutation causing FRDA. Less than 4% of FRDA patients are compound heterozygotes with one allele carrying the GAA repeat expansion and the other a point mutation (Campuzano *et al.*, 1996; Cossee *et al.*, 1999). Because of the mutation, tissue levels of the *X25* gene product, frataxin, are severely reduced. Frataxin is a mitochondrial protein. Loss of frataxin leads to mitochondrial iron overload. As a consequence, mitochondrial respiratory activity declines, resulting in increased production of free radicals (Babcock *et al.*, 1997).

A genetic test demonstrating the GAA repeat expansion is widely available and can be used to confirm a clinical diagnosis of FRDA. Genetic testing is particularly useful in atypical cases with preserved muscle reflexes and late disease onset. If the genetic test is negative, serum levels of vitamin E should be determined, because the clinical phenotype of FRDA can be mimicked by ataxia with isolated vitamin E deficiency (AVED). To distinguish FRDA from hereditary motor and sensory neuropathies (HMSN) nerve conduction studies are useful.

### Natural Course

The prevalence of FRDA is estimated at 2–3/100,000. Mean age at onset is 15 years, ranging from 2–51 years (Dürr *et al.*, 1996). FRDA is a progressive disease leading to disability and premature death. Median latency to become wheelchair-bound after disease onset is 11 years. Life expectancy after disease onset is estimated at 35–40 years (Klockgether *et al.*, 1998). Age of onset and progression rate are partly determined by the GAA repeat length of the shorter allele: in patients with longer expansions, disease onset is earlier and progression faster (Dürr *et al.*, 1996).

### Principles of Therapy

Because frataxin deficiency leads to increased production of free radicals; free radical scavengers are currently investigated in FRDA. Rustin *et al.* recently reported that idebenone (5 mg/kg/day), a short-chain quinone analogue acting as a free radical-scavenger given over 4–9 months decreased the left ventricular mass index in three FRDA patients (Rustin *et al.*, 1999)*. Oral treatment with idebenone for 8 weeks

**TABLE I    Classification of Ataxias**

Inherited ataxias
  Autosomal recessive ataxias
    Friedreich's ataxia (FRDA)
    Ataxia telangiectasia (AT)
    Autosomal recessive spastic ataxia of Charlevoix-Saguenay
      (ARSACS)
    Abetalipoproteinemia
    Ataxia with isolated vitamin E deficiency (AVED)
    Refsum's disease
    Cerebrotendinous xanthomatosis
    Early-onset cerebellar ataxia (EOCA)
  Autosomal dominant ataxias
    Spinocerebellar ataxias (SCA)
    Episodic ataxias (EA)
Noninherited ataxias
  Multiple system atrophy, cerebellar type (MSA-C)
  Sporadic adult-onset ataxia of unknown origin
Symptomatic ataxias
  Alcoholic cerebellar degeneration
  Ataxia caused by other toxic reasons
  Ataxia caused by acquired vitamin deficiency or metabolic
    disorders
  Paraneoplastic cerebellar degeneration
  Immune-mediated ataxias

significantly decreased urinary concentrations of 8-hydroxy-2'-deoxyguanosine, a marker of oxidative stress (Schulz et al., 2000). Randomized, controlled trials with this compound have not yet been completed.

Antiataxic drugs such as 5-hydroxytryptophan, buspirone, and amantadine are ineffective or only marginally effective in FRDA (Botez et al., 1996; Trouillas et al., 1995).

### Practical Manangement

FRDA patients should receive physiotherapy and speech therapy. With progression of the disease most patients will require walking aids and a wheelchair. If necessary, FRDA patients should receive standard medical treatment for cardiomyopathy and diabetes mellitus. A trial of idebenone (5 mg/kg/day)* can be made. However, efficacy of this treatment has not yet been proven. Other centrally acting compounds such as 5-hydroxytryptophan, buspirone, and amantadine are not recommended.

## ATAXIA TELANGIECTASIA (AT)

### Clinical Aspects

AT is an autosomal recessively inherited multisystem disorder characterized by cerebellar ataxia with an onset in early childhood, choreoathetosis, an inability to initiate saccadic eye movements (oculomotor apraxia), oculocutaneous telangiectasias, a high incidence of neoplasia, increased radiosensitivity, and recurrent sinopulmonary infections. The gene affected in AT, *ATM*, encodes a member of the phosphoinositol-3 kinase family involved in cell cycle checkpoint control and DNA repair (Savitsky et al., 1995). More than 200 distinct mutations distributed over the entire gene have been reported, making routine genetic testing difficult. A probable diagnosis can be based on the characteristic clinical presentation, elevated serum levels of α-fetoprotein (AFP), and in vitro demonstration of radiosensitivity.

Very rarely, a disorder similar to AT is caused by mutations in the double-strand break repair gene hMRE11 (Stewart et al., 1999). Ataxia with oculomotor apraxia resembles AT with respect to the neurological symptoms but lacks the nonneurological symptoms of AT. This disorder has recently been mapped to a locus on chromosome 9q (Nemeth et al., 2000).

### Natural Course

The incidence of AT has been estimated at 0.3/100,000 live births (Swift et al., 1986). AT usually begins at 2–4 years after the child has learned to walk. Adult-onset cases of AT are extremely rare. Most patients need wheelchairs at the age of 10 years. Life expectancy is severely reduced because of recurrent infections and neoplasia. Most patients die in their third decade.

### Principles of Therapy

Although the gene defect causing AT has been found and the cellular pathogenesis is partly understood, effective therapies are not available. In particular, there is no way to improve ataxia. Infections and neoplasias require medical management. Prophylactic measures to avoid infections are of great importance.

### Practical Management

AT patients should receive physiotherapy and speech therapy. Physiotherapy is of particular importance, because it may help to prevent pulmonary infections. With progression of the disease most patients will require walking aids and a wheelchair.

In principle, children with AT should receive standard vaccinations. However, there is concern that live vaccinations can cause severe illness in AT patients. Therefore, AT patients should receive live vaccinations only after a thorough immunological evaluation has been performed.

Treatment of infections should be initiated early and maintained over a prolonged time. Usually, infections require intravenous or oral application of broad-spectrum antibiotics. Administration of immunoglobulins can be considered in patients with repeated infections. However, standard immunoglobulin preparations are often poorly tolerated by AT patients, owing to IgA deficiency. In these patients, a switch to preparations with low or absent IgA levels is required.

Treatment of malignant neoplasias is a particular problem, because AT patients have increased sensitivity to radiation and chemotherapy. Therefore, conventional radiotherapy should be avoided, and chemotherapy should be administered only on an individual basis (Eyre *et al.*, 1988; Sandoval and Swift, 1998).

## AUTOSOMAL RECESSIVE SPASTIC ATAXIA OF CHARLEVOIX-SAGUENAY (ARSACS)

ARSACS is an autosomal recessive disorder clinically characterized by progressive ataxia and marked spasticity. In affected children, muscle reflexes are always exaggerated. With progression of the disease, a neuropathy may develop, leading to loss of ankle jerks. Molecular genetic studies in an isolated population in Quebec, Canada, identified causative mutations in a novel gene encoding a large protein with a heat-shock domain (Engert *et al.*, 2000). Linkage to the same locus was established in a large Tunisian family with a similar phenotype (Mrissa *et al.*, 2000).

There is no effective therapy for ARSACS. A minority of patients with pronounced spasticity may benefit from antispastic drugs.

## ABETALIPOPROTEINEMIA

Abetalipoproteinemia is an autosomal recessively inherited disorder characterized by onset of diarrhea soon after birth and slow development of a neurological syndrome thereafter. The neurological syndrome consists of ataxia, weakness of the limbs with loss of tendon reflexes, disturbed sensation, and retinal degeneration. Abetalipoproteinemia is caused by mutations in the gene encoding a subunit of a microsomal triglyceride transfer protein (Sharp *et al.*, 1993). As a consequence, circulating apoprotein B–containing lipoproteins are almost completely missing, and the patients are unable to absorb and transport fat and fat-soluble vitamins. The neurological symptoms are due to vitamin E deficiency.

Management of abetalipoproteinemia consists of a diet with reduced fat intake and vitamin supplementation. Intake of dietary fat should be restricted to 25% of the total daily calories. One third of daily fat should stem from food sources, two thirds should be given as medium-chain triglycerides. Patients receive an adaequate supply of essential fatty acids (Kohlschuetter, 2000).

Despite the principal absorption defect, vitamin E is supplemented orally, because patients are able to secrete very small amounts of apoprotein B–containing lipoproteins. Recommended doses are 50–100 mg/kg/day. In addition, vitamin A (200–400 IU/kg/day) and vitamin K (5 mg every 2 weeks) are given. Levels of vitamin E and A should be closely monitored. Vitamin supplementation should be started as early as possible. Restoration of vitamin E levels will lead to clinical improvement or arrest of further deterioration (Kohlschuetter, 2000)*.

## ATAXIA WITH ISOLATED VITAMIN E DEFICIENCY (AVED)

AVED is an autosomal recessively inherited disorder with a phenotype resembling FRDA. AVED patients carry homozygous mutations of the gene encoding the α-tocopherol transport protein, a liver-specific protein that incorporates vitamin E into very low-density lipoproteins (Ouahchi *et al.*, 1995). As a consequence, vitamin E is rapidly eliminated.

Because there is no absorption deficit, oral supplementation of vitamin E at a dose of 800–2000 mg/day is recommended (Martinello *et al.*, 1998)*. Levels of vitamin E should be closely monitored.

## REFSUM'S DISEASE

Refsum's disease is a rare autosomal recessively inherited disorder caused by mutations in the gene encoding phytanoyl-CoA hydroxylase that is involved in the α-oxidation of phytanic acid (Jansen *et al.*, 1997). The clinical phenotype of Refsum's disease is caused by accumulation of phytanic acid in body tissues. Clinically, Refsum's disease is characterized by ataxia, demyelinating sensorimotor neuropathy, pigmentary retinal degeneration, deafness, cardiac arrhythmias, and ichthyosis-like skin changes. Whereas ocular and hearing problems are usually slowly progressive, there may be acute exacerbations that are precipitated by low caloric intake and mobilization of phytanic acid from adipose tissue.

Refsum's disease is treated by dietary restriction of phytanic acid from 50–100 mg contained in a normal

Western diet to less than 10 mg/day. The diet should provide adaequate caloric intake to prevent mobilization of phytanic acid from adipose stores. Details of the dietary management are given by Gibberd et al. (Gibberd et al., 1985)*. With good dietary supervision ataxia and neuropathy may improve. In contrast, the progressive loss of vision and hearing cannot be prevented.

In acute exacerbations, plasma exchange (four sessions over a period of 7–21 days) is effective in lowering phytanic acid levels and improving neurological and cardiac function. Plasmapheresis may be also considered in patients in whom dietary control is insufficient (Harari et al., 1991).

# CEREBROTENDINOUS XANTHOMATOSIS

Cerebrotendinous xanthomatosis is an autosomal recessively inherited lipid storage disorder with accumulation of cholestanol and cholesterin in various tissues. The disorder is due to mutations of the gene encoding 27-hydroxylase (Leitersdorf et al., 1993). The clinical syndrome includes xanthomatous swelling of the tendons, cataracts, and slowly progressive neurological symptoms, including ataxia, pyramidal signs, and cognitive decline.

Cerebrotendinous xanthomatosis is treated by oral administration of chenodeoxycholate (750 mg/day) (Berginer et al., 1984)*. This treatment results in a marked drop of plasma cholestanol levels and prevents further progression of the neurological syndrome. In early stages of the disease, clinical improvement may be achieved. Cataracts and xanthomatous swelling of the tendons are not affected by this treatment. Treatment can be further improved by addition of HMG CoA reductase inhibitors such as simvastatin or lovastatin (Peynet et al., 1991)*.

# EARLY-ONSET CEREBELLAR ATAXIA (EOCA)

EOCA is used to denote those ataxias with an onset before the age of 20 years in which the etiology is unknown. It is assumed that most EOCAs are autosomal recessive disorders. Apart from cerebellar ataxia, patients may have a variety of additional symptoms, including retinal degeneration (Hallgren syndrome), hypogonadism (Holmes syndrome), optic atrophy (Behr syndrome), cataracts and mental retardation (Marinesco-Sjögren syndrome), and myoclonus (Ramsay Hunt syndrome). Because the etiology of none of these disorders is known, rational treatment approaches are not available.

# SPINOCEREBELLAR ATAXIAS (SCA)

## Clinical Aspects

The SCAs are a genetically heterogeneous group of autosomal dominantly inherited progressive ataxia disorders. In most affected families ataxia is not an isolated symptom but occurs in combination with a variety of other neurological symptoms. These symptoms include supranuclear ophthalmoplegia, optic atrophy, basal ganglia symptoms, dementia, sensory loss, and amyotrophy. Rarely, ataxia is associated with progressive visual loss caused by retinal degeneration (SCA7) or with epilepsy (SCA10). In about one third of the families, all affected members have an almost pure cerebellar syndrome (SCA5,6,8,11,14) (Klockgether et al., 2000).

Until now, 13 different gene loci (SCA1–8,10–14) have been found to be associated with SCA. In eight of these disorders, the affected genes have been cloned, and the mutations have been identified. In five (SCA1–3,6,7), the mutation is an expanded CAG repeat within the coding region of the respective genes. SCA8 is caused by a CTG repeat expansion in the 3'-untranslated region, SCA10 by an intronic pentanucleotide repeat expansion, and SCA12 by a CAG repeat expansion in the 5'-untranslated region of the respective genes. In all other SCAs, the mutations remain unknown (Klockgether et al., 2000).

Genetic tests for SCA1–3, 6, and 7 are widely available. These tests can be used to assign a patient with a dominantly inherited ataxia to one of these mutations. With some regional variation, these mutations account for 50%–70% of all families with dominant ataxia.

## Natural Course

The prevalence of the dominant ataxias is estimated to be 0.9–1.3/100,000 (Polo et al., 1991). Mean age of onset of most SCAs is 30–40 years, with considerable variation between and within families. SCA6 has a later disease onset, with an average of 50 years (Schöls et al., 1998). Median latency to become wheelchair-bound is 17 years in families with accompanying noncerebellar symptoms and 26 years in families with a pure cerebellar syndrome. Patients who have additional noncerebellar symptoms die prematurely 20–25 years after disease onset. Life expectancy is not significantly reduced in patients with a pure cerebellar syndrome (Klockgether et al., 1998)

## Principles of Therapy

Studies of the molecular pathogenesis of the SCAs have not yet resulted in development of therapies that are available for use in humans.

It has been repeatedly claimed that drugs that increase neurotransmission at central 5-HT receptors improve cerebellar ataxia (Trouillas *et al.*, 1995). Three recent studies investigated the antiataxic effect of the anxiolytic 5-HT$_{1A}$ receptor agonist buspirone. Results of an open-label study of 20 patients with different forms of degenerative cerebellar ataxia suggested an antiataxic action of buspirone at a dose of 30–60 mg/day (Lou *et al.*, 1995)*. The efficacy of buspirone was confirmed in a randomized, placebo-controlled study of 19 patients with ataxia caused by cerebellar cortical atrophy (Trouillas *et al.*, 1997)**. In contrast, another study did not report a favorable effect of buspirone in ataxia (Hassin-Baer *et al.*, 2000)*.

## Practical Management

All patients should receive physiotherapy and speech therapy if necessary. Many patients will require walking aids and a wheelchair. Although efficacy has not been definitely proven, a trial of buspirone can be made. Buspirone is given in increasing doses up to 40 mg/day**.

Accompanying symptoms can often be improved by appropriate treatment. These symptoms include parkinsonsism (Chapter 74), restless legs syndrome (Chapter 82), spasticity (Chapter 87), dystonia (Chapter 78), epilepsy (Chapter 20), bladder dysfunction (Chapter 104), and muscle cramps (Chapter 100). For management of these symptoms the reader is referred to the respective chapters.

## EPISODIC ATAXIAS (EA)

### Clinical Aspects

EAs are autosomal dominant disorders characterized by intermittent attacks of ataxia. To date, two different genetic and clinical variants are known. Missense mutations in a brain potassium channel gene, KCNA1, result in EA-1 (Browne *et al.*, 1994). Clinically, EA-1 is characterized by brief attacks of ataxia and dysarthria. The attacks last for seconds to minutes and may occur several times per day. They are often provoked by movements and startle. Apart from ataxia, the attacks may have dystonic or choreic features. EA-1 is associated with interictal myokymia (i.e., twitching of small muscles around the eyes or in the hands). Ataxia and gaze-evoked nystagmus are absent between attacks.

Truncating mutations of the CACNA1A gene encoding the $\alpha_{1A}$ voltage-dependent calcium channel subunit have been found in families with EA-2. Missense mutations of the same gene are associated with familial hemiplegic migraine, whereas a CAG repeat expansion in the 3′ end of the gene causes SCA6 (Ophoff *et al.*, 1996;

Zhuchenko *et al.*, 1997). Compared with EA-1, attacks in EA-2 last longer and are precipitated by emotional stress and exercise but not by startle. The attacks are often associated with vertigo, nausea, and vomiting. Some individuals who may or may not have episodic ataxia have slowly progressive ataxia and cerebellar atrophy. About half of the patients report headaches that meet criteria for migraine.

EAs are often misdiagnosed as psychogenic disorders, epilepsy, or migraine. Because genetic tests are not routinely available, the diagnosis is based on a carefully taken history and clinical examination. In EA-1, interictal EMG of muscles displaying myokymia shows spontaneous repetitive discharges that subside after nerve blockade. Imaging studies give normal results. In contrast, EA-2 patients may have mild cerebellar atrophy. In addition, MR spectroscopy shows an elevated cerebellar pH in these patients (SappeyMarinier *et al.*, 1999).

### Natural Course

EAs are rare disorders. Reliable information on the incidence and prevalence of EAs is lacking. EA-1 starts in early childhood. It has a favorable prognosis in that this disorder does not result in permanent disability. In most patients attacks become milder with increasing age. The age of onset in EA-2 varies from 2–30 years. Some patients have persistent or slowly progressive ataxia.

### Principles of Therapy

EA-1 and EA-2 are caused by mutations of genes encoding a potassium channel (KCNA1) and a calcium channel (CACNA1A) subunit, respectively. The mechanisms leading to neuronal dysfunction have recently been explored in electrophysiological studies performed in frog oocytes or neuronal cell lines wherein the respective mutant genes were expressed artificially. Acetazolamide is a drug known to prevent attacks in a number of paroxysmal disorders caused by defective ion channel. In muscle disorders, its mode of action seems to involve an enhancement of currents through calcium-activated potassium channels (Tricarico *et al.*, 2000). It remains to be shown whether this effect is direct or mediated by changes of pH. In general, acetazolamide is effective in EA-2. Some patients with EA-1 also benefit from acetazolamide, others respond to carbamazepine or phenytoin, a number of cases remain refractory to therapy. Systematic clinical studies exploring the efficacy of acetazolamide or other compounds in EA have not been performed (Griggs and Nutt, 1995).

## Practical Management

Some patients learn to prevent attacks by avoiding provoking stimuli. In particular, EA-1 patients avoid sudden abrupt movements.

If medical treatment is required, acetazolamide is used to prevent attacks. The effect of acetazolamide is more reliable in EA-2 than in EA-1. If treatment is necessary, patients are typically started on a low dose (125 mg/day), which is then gradually increased (500–700 mg/day) until a satisfactory suppression of attacks is achieved (Griggs et al., 1978)*. Paraesthesias that frequently occur under acetazolamide may be reduced by oral potassium supplementation. As a second-line treatment carbamazepine and phenytoin can be tried. However, the efficacy of these drugs is questionable.

## MULTIPLE SYSTEM ATROPHY (MSA)

### Clinical Aspects

MSA is a sporadic, adult-onset disease characterized by neurodegeneration in the basal ganglia, brainstem, cerebellum, and intermediolateral cell columns of the spinal cord. MSA encompasses the former disease categories striatonigral degeneration, sporadic olivopontocerebellar atrophy, and Shy-Drager syndrome. The ultrastructural hallmark of MSA is the presence of oligodendroglial cytoplasmic inclusions. MSA patients have various combinations of parkinsonism, cerebellar ataxia, and autonomic failure (orthostatic hypotension, urinary incontinence) (Gilman et al., 1999). Cerebellar ataxia is present in more than half of MSA patients (Wenning et al., 1997).

The diagnosis of MSA is made clinically. Observation of autonomic symptoms (orthostatic fall in blood pressure by 30 mmHg systolic or 15 mmHg diastolic, urinary incontinence) is of particular importance when establishing a clinical diagnosis of MSA. MRI typically shows cerebellar and brainstem atrophy and signal abnormailies in the basal ganglia and brainstem, but these changes may occur rather late in the course of the disease (Schulz et al., 1994).

Patients with sporadic late-onset ataxia without autonomic failure require a careful diagnostic workup. A number of these patients may develop autonomic symptoms with further progression of their disease, suggesting a diagnosis of MSA. Others may have a hereditary ataxia, either late-onset FRDA or SCA. The probability of finding an SCA mutation is particularly high in sporadic late-onset ataxia patients with incomplete or uninformative family history. Most importantly, sporadic late-onset ataxia patients may have an acquired, nongenetic disease. The most frequent causes are alcoholism, other toxic causes, malignant disease (paraneoplastic cerebellar degeneration), vitamin deficiency, and immune-mediated cerebellar damage. Even with careful clinical workup, the cause remains unknown in a considerable portion of patients with adult-onset sporadic ataxia. Compared with MSA these patients have a more favorable prognosis.

### Natural Course

The prevalence of MSA is 4.4/100,000 (Schrag et al., 1999). The mean age at disease onset is 55 years. MSA takes a unrelentlessly progressive course. After a median latency of 6 years, MSA patients become wheelchairbound. The median life expectancy after disease onset is 9 years (Klockgether et al., 1998).

### Principles of Treatment

There is no curative or preventive treatment for MSA. Among the cardinal symptoms of MSA, cerebellar ataxia, parkinsonism, and autonomic failure, ataxia is most difficult to treat. Botez et al. reported that a 3- to 4-month treatment with amantadine (200 mg/day) improved reaction time and movement time in a group of 30 patients with olivopontocerebellar atrophy, most of which may have had MSA (Botez et al., 1996)*.

### Practical Management

MSA patients should receive physiotherapy and speech therapy. With progression of the disease, all patients will require a wheelchair. For symptomatic treatment of ataxia, amantadine in increasing doses up to 2–3 × 100 mg/day can be tried*. For management of parkinsonism (Chapter 75) and autonomic failure (Chapters 103 and 104) the reader is referred to the respective chapters.

## SPORADIC ADULT-ONSET ATAXIA OF UNKNOWN ORIGIN

In many patients with sporadic adult-onset progressive ataxia, the underlying cause remains unknown. It is estimated that these patients are twice as frequent as patients with the cerebellar type of MSA. Age of onset is around 55 years, and life expectancy is almost normal. Most of the patients have isolated cerebellar atrophy with little or no involvement of the brainstem.

In most patients of this group, cerebellar ataxia is the prominent symptom. However, pyramidal signs and sensory disturbances may occur. Sporadic adult-onset ataxia can be differentiated from MSA by the lasting absence of severe autonomic failure. In contrast, the clinical phenotype of sporadic adult-onset ataxia may resemble that of SCA patients. There are no specific treatment approaches for sporadic adult-onset ataxia.

## ALCOHOLIC CEREBELLAR DEGENERATION

Alcoholic cerebellar degeneration is probably the most common form of chronic cerebellar ataxia, although reliable estimates of prevalence are not available. Clinically, alcoholic cerebellar degeneration is characterized by ataxic gait and stance without major involvement of the upper extremities. A 3-Hz postural tremor while standing with closed eyes is highly characteristic. Ataxia occurs subacutely in heavy drinkers and may then stabilize for years. Symptoms may progress, particularly in those who continue to drink. The pathological changes consist of a loss of the Purkinje cell layer of the upper vermis and the anterior parts of the cerebellar hemispheres.

It is not entirely clear whether alcoholic cerebellar degeneration is due to nutritional deficiency of vitamin $B_1$ (thiamine), as in Wernicke's encephalopathy, or whether it is due to the toxic actions of alcohol, or both (Butterworth, 1995). Strict abstinence improves ataxia, whereas ataxia progresses in patients who continue to drink (Diener et al., 1984).

It is essential that patients undergo a cure for alcoholism and completely stop drinking. In addition, vitamin $B_1$ is supplemented. Initially, 50 mg is given intravenously and intramuscularly. Intramuscular injections are repeated for several days until supplementation is continued with an oral vitamin $B_1$ preparation. In addition to vitamin $B_1$ a multivitamin preparation is recommended (Chapter 71)*.

## ATAXIA CAUSED BY OTHER TOXIC REASONS

A number of compounds may lead to cerebellar degeneration and persistent ataxia after chronic intake. These compounds include anticonvulsant drugs, in particular phenytoin, anticancer drugs (5-fluorouracil, cytosine arabinoside) and solvents. Intoxication with lithium salts may also lead to irreversible ataxia.

It is essential to stop further exposition. In epileptic patients, phenytoin should be replaced by an anti-convulsant drug that is less prone to damage the cerebellum.

## ATAXIA CAUSED BY ACQUIRED VITAMIN DEFICIENCY OR METABOLIC DISORDERS

### Vitamin $B_1$

Wernicke's encephalopathy is an acute or subacute encephalopathy caused by deficiency of vitamin $B_1$ (thiamine) typically occurring in chronic alcoholics. Wernicke's encephalopathy may also result from excessive fasting, repeated vomiting, and prolonged parenteral nutrition without adaequate vitamin supplementation. In addition to ataxia, the clinical syndrome includes eye muscle paresis, peripheral neuropathy, seizures, and mental confusion. If not treated adaequately, Wernicke's encephalopathy may result in a chronic amnesic state, Korsakoff's psychosis. There is close relationship of Wernicke's encephalopathy and alcoholic cerebellar degeneration, because vitamin $B_1$ is thought to play a prominent role in both disorders. Immediate parenteral application of high doses of vitamin $B_1$ is necessary. Initially, 50 mg is given intravenously and intramuscularly. Intramuscular injections are repeated for several days until supplementation is continued with an oral vitamin $B_1$ preparation. In addition to vitamin $B_1$ a multivitamin preparation is recommended (Chapter 71)*.

### Vitamin $B_{12}$

Vitamin $B_{12}$ deficiency causes macrocytic anemia, an axonal form of sensorimotor polyneuropathy and subacute combined degeneration of the spinal cord. Ataxia is often a prominent symptom of vitamin $B_{12}$ deficiency. The most frequent cause is lack of intrinsic factor caused by gastric disease. Vitamin $B_{12}$ (hydroxy-cobalamin) is given intramuscularly at a dose of 1000 µg/day until neurological symptoms improve. Subsequently, the interval between applications is expanded to 3–4 days for a year, followed by lifelong application of 1000 µg/mo*.

### Vitamin E

Acquired vitamin E deficiency may occur as a consequence of malabsorption in gastrointestinal diseases such as celiac disease, cystic fibrosis, short-bowel syndrome, biliary atresia, and intrahepatic cholestasis. Patients have ataxia of gait and stance, dysarthria, and sensory neuropathy with loss of tendon reflexes

(Harding *et al.*, 1982). To stop further progression, intramuscular application of vitamin E at a dose of 100–200 mg/day should be initiated as early as possible*. Most patients are also deficient in other vitamins, which should be supplemented together with vitamin E.

### Hypothyroidism

Cerebellar ataxia is a rare neurological complication of hypothyroidism. The pathogenesis of this syndrome is unclear. Ataxia is completely relieved after adaequate substitution of thyroid hormone.

## PARANEOPLASTIC CEREBELLAR DEGENERATION (PCD)

PCD is an immune-mediated disorder that may occur in association with almost every tumor. Most frequently, however, small-cell lung cancer, cancer of the breast and ovary, and lymphoma are involved. Paraneoplastic encephalomyelitis/sensory neuropathy (PEM/SN) is another paraneoplastic syndrome. In contrast to PCD, PEM/SN affects multiple areas of the central nervous system, dorsal root ganglia, and autonomic nerves. In 20% of the patients with PEM/SN, however, cerebellar ataxia is the presenting symptom.

In many, but not all patients with PCD, antibodies are found in serum and CSF that react with antigens expressed by the nervous system and the tumor. These antibodies do not cause cerebellar degeneration. Rather, they are disease markers. Anti-Hu antibodies are found in association with small-cell lung cancer, anti-Yo antibodies mainly with ovarian cancer, anti-Tr with lymphoma, and anti-Ri antibodies with various malignancies.

Ataxia in PCD has a subacute onset and rapidly progresses to severe disability. In most cases, ataxia precedes the detection of the underlying tumor. Ataxia involves upper and lower extremities and is accompanied by dysarthria and variable degrees of dysphagia. The clinical syndrome is similar in all types of PCD, with the exception of PCD associated with anti-Ri antibodies. A highly characteristic feature of this disorder is the presence of opsoclonus leading to oscillopsia.

At disease onset, CT or MRI do not show major cerebellar atrophy. However, cerebellar atrophy usually develops in the further disease course. A suspected diagnosis of PCD is confirmed by demonstration of specific antibodies. However, absence of antibodies does not rule out a diagnosis of PCD. In cases with suspected or proven PCD, a careful search for the underlying tumor is required. If this search is negative, it has to be repeated every 6 months for 3 years.

In general, PCD neither responds to treatment of the underlying tumor nor to immunosuppressive therapy. However, exceptions to this rule have been observed in single cases (David *et al.*, 1996; Paone and Jeyasingham, 1980).

## IMMUNE-MEDIATED ATAXIAS

Recently, it has been recognized that there are immune-mediated cerebellar degenerations other than PCD. Ataxia may occur in patients with cryptic gluten sensitivity and circulating antigliadin IgG antibodies. These patients do not necessarily have manifest celiac disease with diarrhea, although half of them have mucosal changes in distal duodenal biopsy specimens compatible with celiac disease (Hadjivassiliou *et al.*, 1996). It has been proposed that this disorder is labeled "gluten ataxia" (Hadjivassiliou *et al.*, 1998). Clinically, patients have cerebellar ataxia and signs of sensory neuropathy. A gluten-free diet is recommended for patients with gluten ataxia, although randomized controlled trials have not yet been completed (Hadjivassiliou *et al.*, 1996)*.

Very rarely, ataxia is part of a polyglandular endocrine autoimmune syndrome in patients with circulating antibodies directed against glutamic acid decarboxylase (GAD). Improvement of ataxia after intravenous application of immunoglobulins has been reported in a single case (Abele *et al.*, 1999)*.

## REFERENCES

Abele, M., Weller, M., Mescheriakov, S., Bürk, K., Dichgans, J., and Klockgether, T. (1999). Cerebellar ataxia with glutamic acid decarboxylase autoantibodies. *Neurology* 52, 857–859.

Babcock, M., De Silva, D., Oaks, R., Davis-Kaplan, S., Jiralerspong, S., Montermini, L., Pandolfo, M., and Kaplan, J. (1997). Regulation of mitochondrial iron accumulation by Yfh1p, a putative homolog of frataxin. *Science* 276, 1709–1712.

Berginer, V. M., Salen, G., and Shefer, S. (1984). Long-term treatment of cerebrotendinous xanthomatosis with chenodeoxycholic acid. *N. Engl. J. Med.* 311, 1649–1652.

Botez, M. I., Botez-Marquard, T., Elie, R., Pedraza, O. L., Goyette, K., and Lalonde, R. (1996). Amantadine hydrochloride treatment in heredodegenerative ataxias: A double blind study. *J. Neurol. Neurosurg. Psychiatry* 61, 259–264.

Browne, D. L., Gancher, S. T., Nutt, J. G., Brunt, E. R. P., Smith, E. A., Kramer, P., and Litt, M. (1994). Episodic ataxia/myokymia syndrome is associated with point mutations in the human potassium channel gene, *KCNA1*. *Nat. Genet.* 8, 136–140.

Butterworth, R. F. (1995). Pathophysiology of alcoholic brain damage: synergistic effects of ethanol, thiamine deficiency and alcoholic liver disease. *Metab. Brain Dis.* 10, 1–8.

Campuzano, V., Montermini, L., Moltò, M. D., Pianese, L., Cossée, M., Cavalcanti, F., Monros, E., Rodius, F., Duclos, F., Monticelli, A., Zara, F., Cañizares, J., Koutnikova, H., Bidichandani, S. I., Gellera, C., Brice, A., Trouillas, P., De Michele,

G., Filla, A., De Frutos, R., Palau, F., Patel, P. I., Di Donato, S., and Mandel, J. L. (1996). Friedreich's ataxia: Autosomal recessive disease caused by an intronic GAA triplet repeat expansion. *Science* **271**, 1423–1427.

Cossee, M., Dürr, A., Schmitt, M., Dahl, N., Trouillas, P., Allinson, P., Kostrzewa, M., Nivelon, C. A., Gustavson, K. H., Kohlschutter, A., Muller, U., Mandel, J. L., Brice, A., Koenig, M., Cavalcanti, F., Tammaro, A., De, M. G., Filla, A., Cocozza, S., Labuda, M., Montermini, L., Poirier, J., and Pandolfo, M. (1999). Friedreich's ataxia: point mutations and clinical presentation of compound heterozygotes. *Ann. Neurol.* **45**, 200–206.

David, Y. B., Warner, E., Levitan, M., Sutton, D. M., Malkin, M. G., and Dalmau, J. O. (1996). Autoimmune paraneoplastic cerebellar degeneration in ovarian carcinoma patients treated with plasmapheresis and immunoglobulin. A case report. *Cancer* **78**, 2153–2156.

Diener, H. C., Dichgans, J., Bacher, M., and Guschlbauer, B. (1984). Improvement of ataxia in alcoholic cerebellar atrophy through alcohol abstinence. *J. Neurol.* **231**, 258–262.

Dürr, A., Cossee, M., Agid, Y., Campuzano, V., Mignard, C., Penet, C., Mandel, J. L., Brice, A., and Koenig, M. (1996). Clinical and genetic abnormalities in patients with Friedreich's ataxia. *N. Engl. J. Med.* **335**, 1169–1175.

Engert, J. C., Berube, P., Mercier, J., Dore, C., Lepage, P., Ge, B., Bouchard, J. P., Mathieu, J., Melancon, S. B., Schalling, M., Lander, E. S., Morgan, K., Hudson, T. J., and Richter, A. (2000). ARSACS, a spastic ataxia common in northeastern Quebec, is caused by mutations in a new gene encoding an 11.5-kb ORF. *Nat. Genet.* **24**, 120–125.

Eyre, J. A., Gardner-Medwin, D., and Summerfield, G. P. (1988). Leukoencephalopathy after prophylactic radiation for leukaemia in ataxia telangiectasia. *Arch. Dis. Child* **63**, 1079–1080.

Gibberd, F. B., Billimoria, J. D., Goldman, J. M., Clemens, M. E., Evans, R., Whitelaw, M. N., Retsas, S., and Sherratt, R. M. (1985). Heredopathia atactica polyneuritiformis: Refsum's disease. *Acta Neurol. Scand.* **72**, 1–17.

Gilman, S., Low, P. A., Quinn, N., Albanese, A., Ben Shlomo, Y., Fowler, C. J., Kaufmann, H., Klockgether, T., Lang, A. E., Lantos, P. L., Litvan, I., Mathias, C. J., Oliver, E., Robertson, D., Schatz, I., and Wenning, G. K. (1999). Consensus statement on the diagnosis of multiple system atrophy. *J. Neurol Sci.* **163**, 94–98.

Griggs, R. C., Moxley, R. T., Lafrance, R. A., and McQuillen, J. (1978). Hereditary paroxysmal ataxia: response to acetazolamide. *Neurology* **28**, 1259–1264.

Griggs, R. C., and Nutt, J. G. (1995). Episodic ataxias as channelopathies. *Ann. Neurol.* **37**, 285–287.

Hadjivassiliou, M., Gibson, A., Davies Jones, G. A., Lobo, A. J., Stephenson, T. J., and Milford Ward, A. (1996). Does cryptic gluten sensitivity play a part in neurological illness? *Lancet* **347**, 369–371.

Hadjivassiliou, M., Grunewald, R. A., Chattopadhyay, A. K., Davies, J. G., Gibson, A., Jarratt, J. A., Kandler, R. H., Lobo, A., Powell, T., and Smith, C. M. (1998). Clinical, radiological, neurophysiological, and neuropathological characteristics of gluten ataxia. *Lancet* **352**, 1582–1585.

Harari, D., Gibberd, F. B., Dick, J. P., and Sidey, M. C. (1991). Plasma exchange in the treatment of Refsum's disease (heredopathia atactica polyneuritiformis). *J. Neurol. Neurosurg. Psychiatry* **54**, 614–617.

Harding, A. E., Muller, D. P., Thomas, P. K., and Willison, H. J. (1982). Spinocerebellar degeneration secondary to chronic intestinal malabsorption: a vitamin E deficiency syndrome. *Ann. Neurol.* **12**, 419–424.

Hassin-Baer, S., Korczyn, A. D., and Giladi, N. (2000). An open trial of amantadine and buspirone for cerebellar ataxia: a disappointment. *J. Neural Transm.* **107**, 1187–1189.

Jansen, G. A., Ofman, R., Ferdinandusse, S., Ijlst, L., Muijsers, A. O., Skjeldal, O. H., Stokke, O., Jakobs, C., Besley, G. T., Wraith, J. E., and Wanders, R. J. (1997). Refsum disease is caused by mutations in the phytanoyl-CoA hydroxylase gene. *Nat. Genet.* **17**, 190–193.

Klockgether, T., Lüdtke, R., Kramer, B., Abele, M., Bürk, K., Schöls, L., Riess, O., Laccone, F., Boesch, S., Lopes Cendes, I., Brice, A., Inzelberg, R., Zilber, N., and Dichgans, J. (1998). The natural history of degenerative ataxia: a retrospective study in 466 patients. *Brain* **121**, 589–600.

Klockgether, T., Wüllner, U., Spauschus, A., and Evert, B. (2000). The molecular biology of the autosomal-dominant cerebellar ataxias. *Mov. Disord.* **15**, 604–612.

Kohlschuetter, A. (2000). Abetalipoproteinemia. In "Handbook of Ataxia Disorders." (T. Klockgether, ed.), M. Dekker, New York.

Leitersdorf, E., Reshef, A., Meiner, V., Levitzki, R., Schwartz, S. P., Dann, E. J., Berkman, N., Cali, J. J., Klapholz, L., and Berginer, V. M. (1993). Frameshift and splice-junction mutations in the sterol 27-hydroxylase gene cause cerebrotendinous xanthomatosis in Jews or Moroccan origin. *J. Clin. Invest.* **91**, 2488–2496.

Lou, J. S., Goldfarb, L., McShane, L., Gatev, P., and Hallett, M. (1995). Use of buspirone for treatment of cerebellar ataxia—An open-label study. *Arch. Neurol.* **52**, 982–988.

Martinello, F., Fardin, P., Ottina, M., Ricchieri, G. L., Koenig, M., Cavalier, L., and Trevisan, C. P. (1998). Supplemental therapy in isolated vitamin E deficiency improves the peripheral neuropathy and prevents the progression of ataxia. *J. Neurol. Sci.* **156**, 177–179.

Mrissa, N., Belal, S., Hamida, C. B., Amouri, R., Turki, I., Mrissa, R., Hamida, M. B., and Hentati, F. (2000). Linkage to chromosome 13q11–12 of an autosomal recessive cerebellar ataxia in a tunisian family. *Neurology* **54**, 1408–1414.

Nemeth, A. H., Bochukova, E., Dunne, E., Huson, S. M., Elston, J., Hannan, M. A., Jackson, M., Chapman, C. J., and Taylor, A. M. R. (2000). Autosomal recessive cerebellar ataxia with oculomotor apraxia (Ataxia-telangiectasia-like syndrome) is linked to chromosome 9q34. *Am. J. Hum. Genet.* **67**, 1320–1326.

Ophoff, R. A., Terwindt, G. M., Vergouwe, M. N., Van Eijk, R., Oefner, P. J., Hoffman, S. M. G., Lamerdin, J. E., Mohrenweiser, H. W., Bulman, D. E., Ferrari, M., Haan, J., Lindhout, D., Van Ommen, G. J. B., Hofker, M. H., Ferrari, M. D., and Frants, R. R. (1996). Familial hemiplegic migraine and episodic ataxia type-2 are caused by mutations in the Ca²⁺ channel gene CACNL1A4. *Cell* **87**, 543–552.

Ouahchi, K., Arita, M., Kayden, H., Hentati, F., Ben Hamida, M., Sokol, R., Arai, H., Inoue, K., Mandel, J. L., and Koenig, M. (1995). Ataxia with isolated vitamin E deficiency is caused by mutations in the α-tocopherol transfer protein. *Nat. Genet.* **9**, 141–145.

Paone, J. F., and Jeyasingham, K. (1980). Remission of cerebellar dysfunction after pneumonectomy for bronchogenic carcinoma. *N. Engl. J. Med.* **302**, 156.

Peynet, J., Laurent, A., De Liege, P., Lecoz, P., Gambert, P., Legrand, A., Mikol, J., and Warnet, A. (1991). Cerebrotendinous xanthomatosis: treatments with simvastatin, lovastatin, and chenodeoxycholic acid in 3 siblings. *Neurology* **41**, 434–436.

Polo, J. M., Calleja, J., Combarros, O., and Berciano, J. (1991). Hereditary ataxias and paraplegias in Cantabria, Spain. An epidemiological and clinical study. *Brain* **114**, 855–866.

Rustin, P., vonKleistRetzow, J. C., ChantrelGroussard, K., Sidi, D., Munnich, A., and Rotig, A. (1999). Effect of idebenone on cardiomyopathy in Friedreich's ataxia: a preliminary study. *Lancet* **354**, 477–479.

Sandoval, C., and Swift, M. (1998). Treatment of lymphoid malignancies in patients with ataxia-telangiectasia. *Med. Pediatr. Oncol.* **31**, 491–497.

SappeyMarinier, D., Vighetto, A., Peyron, R., Broussolle, E., and Bonmartin, A. (1999). Phosphorus and proton magnetic resonance spectroscopy in episodic ataxia type 2. *Ann. Neurol.* **46**, 256–259.

Savitsky, K., Bar-Shira, A., Gilad, S., Rotman, G., Ziv, Y., Vanagaite, L., Tagle, D. A., Smith, S., Uziel, T., Sfez, S., Ashkenazi, M., Pecker, I., Frydman, M., Harnik, R., Patanjali, S. R., Simmons, A., Clines, G. A., Sartiel, A., Gatti, R. A., Chessa, L., Sanal, O., Lavin, M. F., Jaspers, N. G. J., and Shiloh, Y. (1995). A single ataxia telangiectasia gene with a product similar to PI-3 kinase. *Science* **268**, 1749–1753.

Schöls, L., Krüger, R., Amoiridis, G., Przuntek, H., Epplen, J. T., and Riess, O. (1998). Spinocerebellar ataxia type 6: genotype and phenotype in German kindreds. *J. Neurol. Neurosurg. Psychiatry* **64**, 67–73.

Schrag, A., Ben Shlomo, Y., and Quinn, N. P. (1999). Prevalence of progressive supranuclear palsy and multiple system atrophy: a cross-sectional study. *Lancet* **354**, 1771–1775.

Schulz, J. B., Dehmer, T., Schols, L., Mende, H., Hardt, C., Vorgerd, M., Burk, K., Matson, W., Dichgan, J., Beal, M. F., and Bogdanov, M. B. (2000). Oxidative stress in patients with Friedreich ataxia. *Neurology* **55**, 1719–1721.

Schulz, J. B., Klockgether, T., Petersen, D., Jauch, M., Muller-Schauenburg, W., Spieker, S., Voigt, K., and Dichgans, J. (1994). Multiple system atrophy: natural history, MRI morphology, and dopamine receptor imaging with 123IBZM-SPECT. *J. Neurol. Neurosurg. Psychiatry* **57**, 1047–1056.

Sharp, D., Blinderman, L., Combs, K. A., Kienzle, B., Ricci, B., Wager Smith, K., Gil, C. M., Turck, C. W., Bouma, M. E., Rader, D. J., and *et al.* (1993). Cloning and gene defects in microsomal triglyceride transfer protein associated with abetalipoproteinaemia. *Nature* **365**, 65–69.

Stewart, G. S., Maser, R. S., Stankovic, T., Bressan, D. A., Kaplan, M. I., Jaspers, N. G., Raams, A., Byrd, P. J., Petrini, J. H., and Taylor, A. M. (1999). The DNA double-strand break repair gene hMRE11 is mutated in individuals with an ataxia-telangiectasia-like disorder. *Cell* **99**, 577–587.

Swift, M., Morrell, D., Cromartie, E., Chamberlin, A. R., Skolnick, M. H., and Bishop, D. T. (1986). The incidence and gene frequency of ataxia-telangiectasia in the United States. *Am. J. Hum. Genet.* **39**, 573–583.

Tricarico, D., Barbieri, M., and Camerino, D. C. (2000). Acetazolamide opens the muscular KCa2+ channel: a novel mechanism of action that may explain the therapeutic effect of the drug in hypokalemic periodic paralysis. *Ann. Neurol.* **48**, 304–312.

Trouillas, P., Serratrice, G., Laplane, D., Rascol, A., Augustin, P., Barroche, G., Clanet, M., Degos, C. F., Desnuelle, C., Dumas, R., Michel, D., Viallet, F., Warter, J. M., and Adeleine, P. (1995). Levorotatory form of 5-hydroxytryptophan in Friedreich's ataxia: Results of a double-blind drug-placebo cooperative study. *Arch. Neurol.* **52**, 456–460.

Trouillas, P., Xie, J., Adeleine, P., Michel, D., Vighetto, A., Honnorat, J., Dumas, R., Nighoghossian, N., and Laurent, B. (1997). Buspirone, a 5-hydroxytryptamine$_{1A}$ agonist, is active in cerebellar ataxia—Results of a double-blind drug placebo study in patients with cerebellar cortical atrophy. *Arch. Neurol.* **54**, 749–752.

Wenning, G. K., Tison, F., Ben Shlomo, Y., Daniel, S. E., and Quinn, N. P. (1997). Multiple system atrophy: a review of 203 pathologically proven cases. *Mov. Disord.* **12**, 133–147.

Zhuchenko, O., Bailey, J., Bonnen, P., Ashizawa, T., Stockton, D. W., Amos, C., Dobyns, W. B., Subramony, S. H., Zoghbi, H. Y., and Lee, C. C. (1997). Autosomal dominant cerebellar ataxia (SCA6) associated with small polyglutamine expansions in the $\alpha_{1A}$-voltage-dependent calcium channel. *Nat. Genet.* **15**, 62–69.

## CHAPTER 74
# Parkinsonism

Wolfgang H. Oertel and Stanley Fahn

## DEFINITION, CLASSIFICATION, AND DIFFERENTIAL DIAGNOSIS

### Clinical Definition—Signs and Symptoms

Parkinsonism (Parkinson syndrome—PS) is characterized by akinesia, rigidity, and/or tremor at rest with, in general, an asymmetric onset. This chapter deals with Parkinson's disease(s) (PD), the most common form(s) of parkinsonism, and with secondary parkinsonism. For atypical parkinsonism, see Chapter 75, for dementias, including dementia of the Lewy body type, see Chapters 29 and 75.

In PD, in addition to the three cardinal motor features, alterations in posture and in postural reflexes, autonomic symptoms, pain, neuropsychological and psychiatric changes such as depression and dementia are often observed.

*Akinesia* includes slowness (bradykinesia), reduction (hypokinesia), and lack (akinesia) of spontaneous automatic and voluntary movements. It may first manifest itself as disturbed fine motor control of a hand, on buttoning up buttons or writing (micrographia), as a slightly masked face (hypomimia), as monotonous low volume, sometimes hoarse speech, as reduced swinging of one arm during walking, as a tendency to drag one leg, or as a small-stepped gait. For possible unspecific (i.e., nonneurological symptoms before manifestation of cardinal motor symptoms) see Gonera *et al.*, 1997. Performing two movements at the same time becomes difficult. Later on, start hesitation, freezing or propulsion, rarely retropulsion, or lateropulsion can appear and lead to falls. The reasons for these instabilities of posture—often resistant to therapy—are unclear. Possibly a combination of bradykinesia, rigidity, and reduced preparatory *postural reflexes* is responsible.

*Axial akinesia* leads to disturbed movement of the trunk and proximal muscles. Patients complain of difficulty turning in bed or turning while standing or walking, rising from a low chair, or sitting down again. Clinically, "en bloc" turning of the body is observed. Axial akinesia is usually combined with rigidity.

*Rigidity* is due to a tonic cocontraction of agonists and antagonists and is clinically felt as an increased resistance to passive extension and flexion of a joint that, unlike spasticity, is relatively equal between flexors and extensors ("lead pipe") but slightly greater in the former. This may lead to a characteristic flexed "simian" posture. Rigidity may also underlie the common occurrence of "rheumatic"-type pains in muscles and joints, mimicking a frozen shoulder or lumbago, which often initially cause a misdiagnosis (discussed later, see Table IV). Cogwheeling can be explained by the presence of tremor underlying the rigidity.

*Rest tremor* is due to an alternating agonist–antagonist activity, predominantly in the distal muscles of the arm (pill rolling, pronation–supination), less often in the legs, and sometimes in the chin. Neck, tongue, or lip tremor is unusual. Rest tremor classically exhibits a frequency of 4–7 Hz (Lance *et al.*, 1963), is enhanced by either emotional or mental stress or during relaxation, and lessens on voluntary movement and decreases or disappears during sleep. Only during early stages of the disease higher resting tremor frequencies up to 9 Hz may occur. Both rest tremor and a faster postural tremor (as observed in essential tremor) can occur together in up to 30%–60% of subjects with PD and may require a combination of therapies (discussed later, see Table XVI). The frequency of this postural tremor lies between 5 (7) and 12 Hz (Deuschl *et al.*, 2000; Findley *et al.*, 1981). Some patients have only a postural tremor, and some have no tremor at all.

*Complex disorders of movement* can occur as part of PD itself. These include paradoxic kinesis (sudden and short lived, nearly normal mobility in situations of danger and extreme stress) and freezing (sudden immobility, in which the patient's feet stick to the ground). Complex dyskinetic movements such as chorea or dystonia of arms and legs (mostly distally) or of the head or trunk are frequently associated with chronic L-dopa treatment (discussed later).

*Autonomic symptoms* include sialorrhea, seborrhea, and dandruff, disturbed sweating with hyperhidrosis or hypohidrosis (leading to altered heat tolerance), male impotence late in the course of the disease (see in contrast, multiple system atrophy, see Chapter 75), impaired micturition with increased urinary frequency and urgency and sometimes incontinence or retention, constipation, and disturbances of heart rate and blood pressure regulation. In many patients weight loss can occur (Levi *et al.*, 1990) despite normal appetite and sufficient caloric intake.

*Neuropsychiatric symptoms* include depression and bradyphrenia (slowness of thinking with unimpaired intellectual capacity). In about 30% dementia will develop at sometime during the course of the disease. Whether dementia is due to PD alone or to the additional presence (comorbidity) of varying combinations of Alzheimer's disease, of disseminated vascular lesions ("multiinfarct dementia"), of normal pressure hydrocephalus, or due to dementia of the Lewy body type (DLB) (see Chapter 75) can often be clarified only at postmortem examination and sometimes not even then.

### Nosological Classification

Parkinsonism may be classified into three groups:

1. PD of either:
   a. Hereditary etiology
   b. Sporadic etiology
2. Symptomatic (secondary) parkinsonism with known etiology
3. Parkinsonism as part of other neurodegenerative disorders, such as DLB, progressive supranuclear palsy (PSP), multiple system atrophy (MSA), or corticobasal ganglionic degeneration (CBGD) (see Table I, see Chapter 75)

These neurodegenerative disorders (previously incorrectly termed "parkinsonism plus") exhibit, in addition to parkinsonism, autonomic, cerebellar, pyramidal, oculomotor and/or cortical signs, spontaneous hallucinations or dementia (in case of DLB).

Table I contains a classification of parkinsonism.

### Differential Diagnosis, Clinical History, and Red Flags Casting Doubts on the Diagnosis of PD

The clinical history and the neurological examination should concentrate especially on particular symptoms and signs (see Tables II and III) to avoid the most frequent misdiagnoses (see Table IV). Symptomatic parkinsonism may be suggested by a history of strokes or stepwise progression or by a history of encephalitis. Previous bacterial meningitis, subarachnoid hemorrhage, or brain trauma can lead to normal-pressure hydrocephalus. Urinary incontinence or impaired cognitive function, often observed by relatives or friends, is also common in this condition. The drug history should focus on classical and atypical neuroleptics, antiemetics (metoclopramide, alizapride), reserpine-containing antihypertensive drugs, and the calcium-antagonists flunarizine or cinnarizine. The clinical history should also include questions related to exposure to potential dopaminergic neurotoxins or human immunodeficiency virus (see Table I and II). The importance of hereditary factors for PD is increasingly recognized. Seven genes or genetic loci causing PD have so far been identified (see Table I) and are in part associated with a different time of disease onset and progression rate than the "typical" sporadic PD type. In patients with parkinsonism starting before the age of 50, Wilson's disease must be excluded. One should seek evidence of autonomic failure, pyramidal and cerebellar signs in case the patient has multiple system atrophy of the striatonigral degeneration (parkinsonian) type (MSA–SND; MSA-P) when parkinsonism predominates, or of the sporadic olivopontocerebellar atrophy (cerebellar) type (MSA–sOPCA; MSA-C) when cerebellar signs prevail. Diurnal fluctuations of cognitive functions and spontaneous fluctuating hallucinations combined with parkinsonism are suggestive of DLB. Isolated postural or action tremor with no evidence of akinesia or rigidity, except cogwheeling, suggests essential tremor.

At the level of general practice, essential tremor in the elderly, subcortical vascular encephalopathy (SVE), or normal pressure hydrocephalus is still frequently misdiagnosed as PD (Trenkwalder *et al.*, 1995). In general practice 25% of patients receiving antiparkinsonian therapy and/or carrying the clinical diagnosis of parkinsonism do not have parkinsonism, but rather have essential tremor, subcortical vascular encephalopathy, dementia, and other causes, 50% have probable PD (hereditary or sporadic type) and 25% have secondary parkinsonism (Meara *et al.*, 1999) (see Table I).

Table II gives a practical guide for the differential diagnosis of parkinsonism. The most frequent misdiagnoses, especially in the early stage of PD, are listed in Table IV. Signs and symptoms ("red flags") that cast doubt on the diagnosis of PD and help especially in the clinical differentiation between PD on one hand and MSA, PSP, and DLB on the other hand are found in Table III.

**TABLE I**  Classification of Parkinsonism (Parkinsonian Syndromes; PS)

Primary
    Parkinson's disease (PD)

Hereditary PD

| Gen/mode of inheritance | Chromosomal location | Gene product | Protein function |
|---|---|---|---|
| Park1 AD | 4q21–q23 | Alpha-synuclein | |
| Park2 AR | 6q25.2–q27 | Parkin | Ubiquitin-ligase |
| Park3 AD | 2p13 | Unknown | |
| Park4 AD | | | |
| Park5 AD | 4p14 | Ubiquitin-C-terminal Hydrolase L1 | |
| Park6 AR | 1p35–p36 | Unknown | |
| Park7 AR | | Unknown | |

Sporadic PD
  Secondary (symptomatic)
  Infectious or postinfectious
    Postencephalitic (encephalitis lethargica)
    AIDS encephalopathy
    Encephalitis of other causes
    Lyme borreliosis (?)
  Toxic
    CO, $CS_2$, cyanide, manganese, methanol, $n$-hexane (?)
  Drug induced
    Presynaptic dopamine depletors: reserpine, tetrabenazine
    Alpha-methyl-dopa (?)[a]
    Neuroleptics: butyrophenones, phenothiazines, thioxantines, substituted benzamides
      including antiemetics (metoclopramide, alizapride)
    Calcium antagonists: cinnarizine, flunarizine, amlodipine (?), diltiazem (?)
    NSAIDs (indomethacin)
    Cyclosporin
    Valproate
    Lithium (rare)
  Metabolic
    Hypoparathyroidism
  Posttraumatic
    Pugilist's encephalopathy
  Vascular
    Subcortical vascular encephalopathy (SVE; lower body parkinsonism/arteriosclerotic
      pseudoparkinsonism)
  Caused by tumor

Neurodegenerative disorders
  Corticobasal ganglionic degeneration (CBGD)[b]
  Multiple system atrophy (MSA)[c]
    Striatonigral degeneration/MSA-Parkinson type (MSA-SND/MSA-P = MSA with
      predominant parkinsonian symptoms, atypical features, and autonomic dysfunction)
    Olivopontocerebellar atrophy (sporadic type)/MSA-cerebellar type (MSA-sOPCA/MSA-C =
      MSA with predominant cerebellar symptoms and autonomic dysfunction or
      parkinsonian symptoms)
  Progressive supranuclear palsy (PSP) (Steele–Richardson–Olszewski syndrome, SRO)[d]
  Dementia
  Creutzfeldt–Jakob disease (CJD)
  Dementia of the Lewy body type (DLB)
  Normal pressure hydrocephalus (NPH)
  Parkinson dementia–ALS complex (Guam)
  Senile dementia of the Alzheimer type (SDAT)
Hereditary disorders
  Hallervorden–Spatz disease
  Huntington's chorea (Westphal variant)
  Wilson's disease

[a]Duvoisin and Marsden (1975).
[b]Rinne *et al.* (1994), Watts *et al.* (1994).
[c]Quinn (1989, 1994), Wenning *et al.* (1994, 1995).
[d]Jackson *et al.* (1983), Maher and Lees (1986).

**TABLE II   Approach to the Differential Diagnosis of PD and Other Parkinsonian Syndromes**

| | |
|---|---|
| Clinical history (patient, partner): | Time of first symptoms, previous course of the disease, degree of disability, psychological and psychiatric changes, changes in cognitive functions, sleep disorders, disorders of temperature regulation, syncope, frequent falls, constipation, urinary incontinence or retention, early male impotence, early female reduced genital sensitivity (*check for MSA–SND [P-type]*), diurnal fluctuations in cognition and hallucinations (the later independent of dopamimetics—check for DLB) |
| | Drug history, drug addiction |
| Medical history: | Arterial hypertension, cardiac arrhythmia, diabetes mellitus, stroke, encephalitis, meningitis, HIV infection, brain trauma |
| Occupation: | Exposure to toxins, pesticides |
| Family history: | Hereditary neurological disorders |
| Neurological status: | Mimic, speech (volume, melody), body posture, arm swing on walking, fluidity of movement, fine motor control, muscle tone, tremor (rest, action, postural, intention tremor—*check for MSA–sOPCA*) (C-type), asymmetry or symmetry of motor deficits, oculomotor disturbancies (vertical paresis, up or down (*check for PSP*), saccades, vestibulo–ocular reflex, suppression of VOR by fixation, eye–head coordination, corneal Kayser–Fleischer ring (*check for Wilson's disease*), pareses, reflex differences, pyramidal signs |
| | Abnormal involuntary movements: chorea, athetosis, ballism, painful or painless dystonia, myoclonus, cerebellar ataxia, abasia |
| | Autonomic dysfunction: (*check for MSA–SND [P-type], MSA–sOPCA [C-type]*) bradycardia, orthostatic hypotension (blood pressure test, supine vs. upright), impaired sweat production, seborrhea, dandruff, sialorrhea, urinary incontinence or retention |
| Excluding Wilson's disease: | If patient is younger than 50 years, measure serum copper and ceruloplasmin concentrations and perform slit-lamp examination for Kayser–Fleischer rings |
| Imaging—general rules: | Typical clinical PD and convincing response to L-dopa or dopamine agonist—*no brain imaging* |
| | Doubtful parkinsonism, any hint of atypical clinical symptoms or signs or poor or unusual response to L-dopa or dopamine agonist—*brain imaging (CT, MRI, in exceptional cases dopamine transporter SPECT, IBZM-SPECT)* |
| Imaging procedure: | |
| Cranial CT: | Identify normal pressure hydrocephalus, tumor, periventricular lucencies (leukoaraiosis), lacunar infarcts (*check for SVE*), familial calcification of basal ganglia, cerebellar nuclei, and frontal cortex |
| Magnetic resonance imaging (MRI-T$_2$): | Exclusion of hypointensive areas in the (dorsolateral) posterior putamen (1.5 T), slitlike hyperintensity at the lateral edge of the putamen (*check for MSA–SND [P-type], MSA–sOPCA [C-type]*), hyperintensities in globus pallidus and substantia nigra (vascular changes?), marked hypointensity in globus pallidus and substantia nigra (combined with eye of the tiger sign) (*check for Hallervorden–Spatz disease, neuroacantocytosis*) |
| Pharmacological test: | L-dopa test, apomorphine |
| Further diagnostic procedures: | According to clinical examination and in case of comorbidity, to exclude rare causes, which may complicate or enhance parkinsonian signs, the following screening tests may be carried out: continuous wave, transcranial and duplex doppler sonography, ECG, ESR, differential blood cell count, liver, kidney and thyroid function, electrolytes |
| Summary: | 1. The diagnosis of PD is clinical; in the absence of other distinguishing features, the chance of a parkinsonian patient with classical pill-rolling rest tremor having PD vs. MSA is 100:1 |
| | 2. In uncomplicated PD, the results of all the preceding investigations are normal—except dopamine transporter SPECT |
| | 3. In a clinically typical case with onset >50 years, *no* additional investigation is necessary |
| | 4. Investigations (generally available in 2001) help only insofar as they may reveal or exclude other causes of parkinsonism |
| | 5. In practice, about 10%–25% of patients still carrying a diagnosis of PD at death, have not had PD |

## Technical Diagnostic Procedures

In patients with "true" parkinsonism, who respond well to L-dopa or another dopamimetic treatment and with no clinical features atypical for PD, no investigations are needed, apart from excluding Wilson's disease in individuals with onset younger than 50 years. However, if any atypical feature (symptom or sign) is present, structural and functional imaging may be helpful.

Cranial computed tomography (cCT), as a screening method, may show evidence of normal pressure hydrocephalus, vascular disease, or space-occupying lesions. These may (1) be the cause of secondary parkinsonism,

(2) the cause of atypical features in a patient who already has PD (comorbidity), or (3) represent incidental findings. Magnetic resonance imaging (MRI) has been found to be superior to cCT, because T$_2$-weighted imaging (1.5 T) in patients with atypical parkinsonism may show (1) hypointense areas in the dorsolateral and posterior putamen or (2) a slitlike hyperintensity at the lateral border of the putamen, adjacent to the external capsule (Konagaya *et al.*, 1995; Lang *et al.*, 1994; Schrag *et al.*, 2000; Schwarz *et al.*, 1996; Testa *et al.*, 1993). Both signs are very suggestive of MSA (see Chapter 75). In PSP a reduction of the brainstem (mickey mouse sign) has been reported mainly in

TABLE III   Red Flags: Clinical Signs and Symptoms That May Cast Doubt on a Diagnosis of Parkinson's Disease (PD)[a]

Rapid progression
Early postural instability and falls, especially backwards
Falls in the first year
Permanently wheelchair bound despite therapy
Irregular jerky tremor or myoclonus
Autonomic disturbances (orthostatic hypotension, syncope, male impotence early in the course of the disease, female reduced genital sensitivity early in the course of the disease, urinary incontinence or retention, anhydrosis)
Akinetic symptoms mainly in the lower extremities (prevailing gait disorder)
Partial or absent response to a sufficient dose of L-dopa (800–1000 mg/day, if tolerated)
Absent or atypical L-dopa–induced abnormal involuntary movements (AIMs), especially in the face and neck, often elicited by a very small L-dopa dose
Abnormal eye movements, beyond the degree observed in Parkinson's disease
Marked dysarthria
Marked dysphagia
Disproportionate antecollis
Contractures
Cold dusky hands, Raynaud's phenomenon induced by ergot preparations
Respiratory stridor, sighs, increased snoring or groaning
Fluctuations in cognitive function
Spontaneous fluctuating hallucinations (i.e., independent of antiparkinsonian therapy)

[a]Modified from Quinn (1994); *Movement Disorders* 3, Butterworth-Heinemann, with permission.

TABLE IV   Diseases Frequently Misdiagnosed as Parkinsonism or Parkinson's Disease and Vice Versa

Essential tremor
Normal pressure hydrocephalus
Arteriosclerotic pseudoparkinsonism (subcortical vascular encephalopathy [SVE], chronic vascular encephalopathy [CVE] or Binswanger's disease)
Depression
Progressive supranuclear palsy
Multiple system atrophy
Cortico-basal ganglion degeneration
Dementia of the Lewy body type
Senile dementia of the Alzheimer type (SDAT)
Hemiparesis
Hemisensory syndrome (without sensory loss)
Rheumatism
Polymyalgia rheumatica
Frozen shoulder
Lumbago or sciatica

advanced stages. MRI also allows identification of vascular and nonvascular lesions in other basal ganglia regions and elsewhere in the CNS.

Positron emission tomography (PET) is an expensive research tool and will not be discussed here (Brooks DJ, 2000a,b; Leenders *et al.*, 1990). Single-photon emission CT (SPECT) with β-CIT (Brücke *et al.*, 1994; Laruelle *et al.*, 1993; Marek *et al.*, 1994) or FP-CIT (Benamer *et al.*, 2000) images the dopamine transporter (DAT) located at the presynaptic nerve terminal of the nigrostriatal tract. Because of the preclinical phase of nigrostriatal degeneration in PD and other parkinsonian syndromes, density of presynaptic ligands is reduced in PD and in parkinsonism as part of another neurodegenerative disorder except L-dopa–responsive dystonia, but not or rarely in subjects with neuroleptic-induced parkinsonism. Dopamine transporter imaging is then useful, if a clinical diagnosis cannot be made because of subtle clinical signs and/or in the presence of isolated rest, action, or atypical tremor and subjective complains of bradykinesia, so that even a movement disorder specialist hesitates to make the diagnosis of possible PD (see Table V).

SPECT with [123]I-iodobenzamide (IBZM) demonstrates the postsynaptic density of dopamine D$_2$-receptors on striatal neurons. Postsynaptic ligand binding is normal or slightly increased in untreated PD but may be slightly reduced after chronic L-dopa therapy and is definitely decreased during therapy with dopamine agonists. It is—at least at the advanced stage—usually decreased in MSA and PSP (Schulz *et al.*, 1994; Schwarz *et al.*, 1992).

## PARKINSON'S DISEASE(S)

### Clinical Features and Subtypes of PD

The diagnosis of established PD requires the presence of at least two of the cardinal features (discussed earlier), one of which has to be akinesia. Other focal neurological signs must be absent or alternatively explained by the presence of a second independent disease of the CNS. After exclusion of other potential causes for parkinsonism, an asymmetric (unilateral) onset with a classical rest tremor, a clear-cut response to L-dopa or apomorphine (discussed later) or both, and the appearance of L-dopa–associated response fluctuations and dyskinesias during the course of the disease makes the clinical diagnosis of PD probable (see Table V).

Within this classical presentation, PD can present as one of three main clinical syndromes:

1. Mixed type: Akinesia, rigidity, and rest tremor are fairly equally present.
2. Akinetic–rigid type: Rest tremor is minimal or absent.
3. Tremor-dominant type: Rest tremor predominates, and akinesia or rigidity is minimal or absent.

TABLE V  Diagnostic Criteria for PD[a]

| Level of diagnostic certainty for PD | Clinical symptoms and signs, pharmacological response |
| --- | --- |
| Possible | Progressive disorder |
| | Presence of at least two of the three cardinal features of parkinsonism: akinesia, rigidity, rest tremor |
| | No atypical features |
| Probable | Criteria as described for possible PD *and* at least two of the following: |
| | Marked response to L-dopa |
| | L-dopa–related response fluctuations or L-dopa–induced dyskinesia |
| | Asymmetry of signs |
| Definite | Clinical criteria as described for probable PD |
| | At postmortem examination, degeneration of pigmented neurons in the SNc, predominantly in the ventrolateral tier; Lewy bodies present in SNc; no oligodendroglial inclusion bodies |

[a]Modified from Gibb and Lees (1990a); Ward and Gibb (1990).

If age of onset is taken as the leading criterion, an early-onset type (<40 years) is differentiated from the classical type (onset >40 years; mean onset time, 55 years). Because of advances in genetic research, these clinical subtypes will likely be redefined. For example, the early-onset type is often due to homozygous mutations in the Parkin gene (Park2) (Lücking *et al.*, 1998, 2000). Compound heterozygosity of the Parkin alleles (each of the two Parkin alleles carries a different mutation) can lead to different age of onset and different clinical phenomenology (Klein *et al.*, 2000). Nevertheless, so far these different genetic etiologies have not influenced treatment strategies. They still rely on phenomenology, the results of clinical trials, and clinical experience.

## Epidemiology, Course, and Life Expectancy in PD

PD is a common basal ganglia disorder. Prevalence of parkinsonian syndromes overall ranges from 100–200/100,000 inhabitants, the (median) annual incidence ranges between 12 and 20/100,000 (Rajput *et al.*, 1984; Tanner and Aston, 2000). The age distribution shows a continuing steep rise in incidence with advancing age from 5:100,000 less than 54 years of age, through 32:100,000 at 55–64 years, 113:100,000 at 65–74 years, to 254:100,000 at 75–84 years (Rajput *et al.*, 1984). A recent longitudinal population-based study of a cohort of 4341 individuals initially free of PS and aged 65–84 years found the average annual incidence per 100,000 of PS to be 530 and of clinically diagnosed PD to be 326 (Baldereschi *et al.*, 2000).

According to Mutch *et al.* (1986) the prevalence of PD is 12.5 per 100,000 in the age group 40–44, markedly increases to 240 per 100,000 in the age group between 60 and 64, and to 707 per 100,000 in the age group between 70 and 74. The reported median prevalence of PD is 110 per 100,000 (Tanner and Aston, 2000). Another study based on a random sample of 467 community residents aged 65 years or older reported a prevalence of any type of a parkinsonian syndrome to be even 15% for those aged 65–74, 29% for those 75–84, and 52% for those 85 and older (Bennett *et al.*, 1996). The cumulative incidence to suffer from a Parkinson syndrome at the age of 90 years is 7.5% (Bower *et al.*, 1999). In most studies, parkinsonism is slightly more common in men than in women.

Parkinson's disease is insidious in onset and progressive. Onset is usually in middle to late life. The underlying degenerative process in PD is progressive, despite the use of symptomatic drug therapy. Some long-term retrospective studies (Diamond *et al.*, 1987; Fischer, 1987) suggest that the extent of pretreatment disability will be again reached by the patient after 7–10 years under therapy. Because of the development and approval of several new antiparkinsonian agents in the last 10 years, these figures may be more favorable for the year 2002.

The mortality of untreated patients with parkinsonism was reported to be 1.5–3 times higher than for age-matched controls and to rise with increasing duration of the disease. Before the levodopa era, about 30% of untreated patients with parkinsonism were disabled or had died after 5 years, increasing to 60% after 5–9 years and to 80% after 10–14 years (Hoehn and Yahr, 1967). These data may be too pessimistic, because during the period covered, other akinetic-rigid neurodegenerative syndromes such as MSA or PSP, which have a more rapid course, went largely unrecognized as separate disease entities. With L-dopa treatment (discussed later), the mortality fell to approach that of the normal population (Markham and Diamond, 1986), and the average age of patients with PD increased from 67–69 years to 72–73 years, and the duration of the disease from 9–10 years to 13–14 years (Marttila, 1987; Uitti *et al.*, 1993). However, in recent analyses, mortality in Parkinson's disease was reported to remain increased despite low-dose levodopa/carbidopa therapy, and no additional benefit was gained from early use of bromocriptine (Hely *et al.*, 1999; Montastruc *et al.*, 2001). The relative risk of death associated with PD was estimated at 2.3 (1.8–3.0). The risk for death in men with PD was higher (3.1) than in women (1.8) (Berger *et al.*, 2000).

The various clinical subtypes of PD show differences in the course of the disease. Patients with an early onset and of the mixed or akinetic–rigid type respond well to L-dopa therapy but experience L-dopa–associated response fluctuations and dyskinesias earlier and to a more severe degree than patients of the tremor-dominant type. Part of these patients belong to the juvenile akinetic rigid type, homozygous for the Park 2 mutation. Patients with late onset of the tremor-dominant type may experience less benefit from L-dopa. In particular, larger doses may be required to improve rest tremor than akinesia and rigidity. Akinesia and rigidity may initially be hardly present and progress more slowly over the years, so that on average tremor-dominant patients have a better prognosis than patients of the akinetic–rigid or mixed type (Jankovic *et al.*, 1995). Sometimes the most extreme example of this is called *benign tremulous PD* (BTPD), which is often mistaken for benign essential tremor (BET) or BET preceding the development of PD.

### Cause of Death in PD

Parkinsonian patients share with the normal population the four most common causes of death. These are cardiovascular disorders, malignant tumors, cerebrovascular disease, and bronchopneumonia. In contrast, intercurrent infections (e.g., urinary infection), infected pressure sores, and the sequelae of accidents and operations—considered to be secondary to disease-related immobility—lead significantly more frequently to death than in the age-matched general population (Hoehn and Yahr, 1967).

### Etiology and Pathogenesis of PD

**Hereditary PD.** Linkage analysis has so far failed to demonstrate a role for candidate genes, such as those for catalase, superoxide dismutase 1, brain-derived neurotrophic factor, tyrosine hydroxylase, and the dopamine transporter (Gasser *et al.*, 1994). Under the hypothesis of an interplay of genetic and environmental factors, further linkage analysis in PD has focused on the largely hepatic enzyme complex cytochrome P450-debrisoquine-4-hydroxylase, monoamine oxidase B, or changes in *N*-methylation and sulfation (see for review Gasser *et al.*, 2001; Oertel *et al.*, 1996).

Until 2001, seven genes or genetic loci have been identified to cause PD. These genes represent about 2% of the cause of all PD. Of the individuals with an onset <20 years about 75% are considered to have a hereditary type of PD. In a further 10%–15% of PD patients, another affected family member can be found. Of the seven different hereditary forms, four follow an autosomal-dominant and three an autosomal-recessive trait. Of the genes identified (see Table I), at least three are related to the ubiquitin-proteasome pathway, an important mechanism for protein degradation. In the center of interest lie the proteins alpha-synuclein and

parkin. Mutated alpha-synuclein is the gene product of the Park1 mutation. Alpha-synuclein is not only present in the Lewy bodies of the autosomal-dominant Park1 form of PD, but also of sporadic PD. In addition, it influences membrane and synaptic vesicle function and thus may interfere with dopamine transporter function at the presynaptic nerve terminal. Mutated parkin is the gene product of the Park2 mutation. Parkin possesses ubiquitin-ligase activity and interacts with at least a glycosylated special form (alphaSpo22) of alpha-synuclein (Shimura *et al.*, 2001). Thus, because of these mutations, protein degradation is impaired, leading to protein aggregation and, for example, likely formation of inclusion bodies such as Lewy bodies, with the exception of most Park2 mutations. For further details see Chung *et al.*, 2001.

**Sporadic PD.** The cause(s) of sporadic PD is (are) still unknown. As individuals age, their number of dopaminergic nerve cells decreases, but this reduction does not lead to clinically manifest parkinsonism. Increasing age, further unidentified genetic factors, and endogenous or exogenous substances that are toxic to dopaminergic cells (or to the glial cells on which they depend) have been proposed as possible causes.

Several such dopaminergic neurotoxins are known. For example human dopaminergic neurons can selectively be damaged by the substance MPTP (*N*-methyl-4-phenyl-1,2,3,6,-tetrahydropyridine) leading to a clinical picture extremely similar to PD.

In experimental animals MPTP or the naturally occurring substance rotenone (Betarbet *et al.*, 2000) can induce a phenotype similar to parkinsonism. In transgenic mice, the ratio between the synaptic membrane-located dopamine transporter (DAT) and the vesicular monoamine transporter 2 (VMAT 2) (Fumagalli *et al.*, 1999; Gainetdinov *et al.*, 1997) determines the susceptibility to the dopaminergic neurotoxin MPTP. Respective association studies on the candidate genes VMAT and DAT for PD are, however, negative (Plante-Bordeneuve *et al.*, 1997).

These data have promoted speculation that exposure to a so far unknown exogenous or endogenous factor may contribute to the lesion of the substantia nigra (Calne and Langston, 1983; Golbe *et al.*, 1990). On the basis of epidemiological studies, toxins and pesticides have been proposed as possible environmental etiological factors for sporadic PD. Herewith in line are toxicological studies in experimental animals suggesting that exposure not to one but to several toxins may promote the development of nigrostriatal degeneration (Priyadarshi *et al.*, 2000, 2001).

In the substantia nigra of patients with pathologically proven PD, postmortem studies have shown a reduction of complex I activity in the mitochondrial respiratory chain. The relevance of this finding to the etiology or pathogenesis of PD is presently unclear. Unlike classical mitochondrial diseases, maternal (X-linked) transmission of PD has not been universally demonstrated (see, for review, Orth and Schapira, 2001; Schapira, 1994).

## Neuropathology

The neuropathological diagnosis of PD is based on (1) depigmentation and cell loss from substantia nigra pars compacta, especially its ventrolateral tier (Gibb *et al.*, 1990; Fearnley and Lees, 1991), whose dopaminergic neurons project to the striatum; together with (2) the presence of the intracytoplasmic eosinophilic inclusions (Lewy bodies) in some of the surviving nigral neurons. These Lewy bodies are also present in numerous other brain regions in PD, for example, in the cortex, the (noradrenergic) locus ceruleus, the dorsal vagal motor nucleus, the (serotoninergic) raphe nuclei of the mesencephalon and pons, in the hypothalamic area, and in sympathetic ganglia. Furthermore, Lewy bodies and cell loss also occur in the cholinergic nucleus basalis of Meynert (Jellinger, 1990). The presence of Lewy bodies in brain is not specific to PD but is a necessary precondition for its neuropathological diagnosis (see Table V). Because of genetic research, the Lewy body has again become a major research focus, because it contains alpha-synuclein, parkin, synphilin-1, ubiquitin, and other proteins as part of the protein aggregopathy.

## Changes in Neurotransmitters

Changes in neurotransmitter concentrations in PD relate predominantly to the dopaminergic nigrostriatal system. The dopamine deficit in the putamen (70%–95% decrease) predominates both in the early and late stages of disease (Kish *et al.*, 1988). In the late stages, a more widespread but mild to moderate dopamine deficit is found in numerous other areas of the central nervous system.

In parallel with this loss of dopamine one finds reduced activity of the enzyme L-tyrosine hydroxylase (TH), the rate-limiting enzyme of dopamine synthesis and of the enzyme L-dopa-decarboxylase (AADC) (see Figure 1). PD also affects multiple other neurotransmitter systems. Reduced concentrations of serotonin (5-hydroxytryptamine) and of the $\gamma$-aminobutyric acid (GABA) synthesizing enzyme glutamic acid decarboxylase (GAD) in the striatum and the substantia nigra are likely sequelae of a down-regulation of both systems, because 5-hydroxytryptophan and GABA predominantly inhibit the nigrostriatal system (Hornykiewicz,

1982). The decrease in serotonin concentration may be related to the known degeneration of serotoninergic neurons in raphe nuclei or the effect that L-dopa therapy has in reducing serotonin production, because L-dopa competes with 5-hydroxytryptophan, the precursor of serotonin, at the enzyme AADC.

In the nucleus basalis of Meynert and in the cortex, the activity of the acetylcholine-synthesizing enzyme choline acetyltransferase (CAT) is reduced. In parkinsonian patients with dementia, this decrease of cholinergic activity is particularly prominent. In contrast, cholinergic interneurons in the striatum are overactive, because they are no longer sufficiently inhibited by dopamine (see Figure 2).

## The Basis of Therapeutic Principles

### The Dopaminergic Synapse and Dopaminergic Receptors

Figure 1 depicts a dopaminergic synapse in the striatum with its essential features. Five distinct dopamine receptors ($D_1$ to $D_5$) so far have been identified. They can be divided into $D_1$-like ($D_1$, $D_5$) and $D_2$-like ($D_2$, $D_3$, $D_4$) receptors. Thus far, all but the $D_5$-receptor have been shown to have relevance for the clinician (Creese *et al.*, 1983; Kebabian and Calne, 1979; Shiosaki *et al.*, 1994; Strange, 1993).

In the nigrostriatal dopaminergic system, $D_1$- and $D_2$-receptors are postsynaptically localized, the former mainly on the indirect and the latter mainly on the direct striatopallidal output pathway (see Figure 2). In contrast, the $D_3$-receptor is considered to be located in part postsynaptically and in part on the presynaptic nigrostriatal nerve terminal. It may also correspond to a subpopulation of the so-called dopamine autoreceptor.

Dopamine is removed from the synaptic cleft into the presynaptic nigrostriatal terminal by means of a reuptake mechanism, mediated by a dopamine transporter protein (DAT). Dopamine can be metabolized by autooxidation, by the intracytoplasmic enzyme monoamine oxidase B (MAO B), located mostly in glia, and by the enzyme catechol-O-methyltransferase (COMT) (discussed later; also see Figure 1).

### Cholinergic Neurotransmission

Centrally active anticholinergic antiparkinsonian drugs act by means of muscarinic receptors. Anticholinergics reduce the activity of striatal cholinergic interneurons and therefore ameliorate the functional dopaminergic–cholinergic imbalance in the striatum. Secondarily, they impair the already reduced activity of cholinergic neurons in the area of the substantia innom-

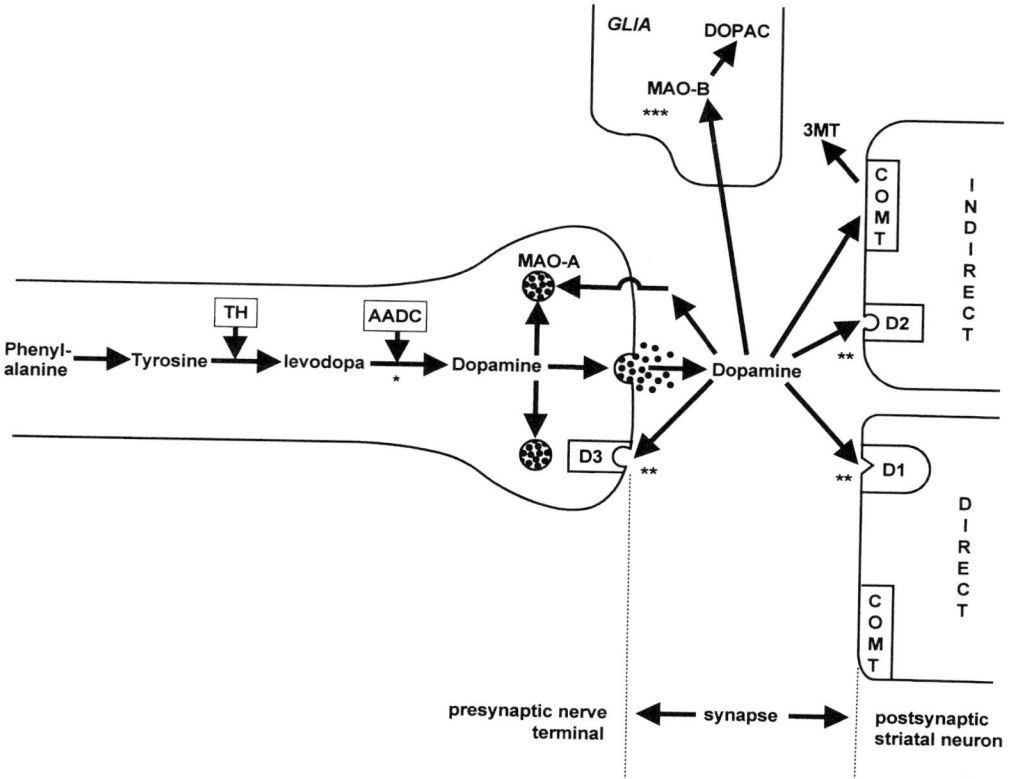

**FIGURE 1** Schematic diagram of a nigrostriatal synapse. In the presynaptic nerve terminal, phenylalanine is converted to tyrosine, tyrosine is converted by tyrosine hydroxylase to L-dopa, and L-dopa is metabolized by the enzyme L-dopa-decarboxylase (aromatic L-amino acid decarboxylase, AADC) to the transmitter dopamine. Dopamine is released and binds postsynaptically to dopamine $D_1$- and $D_2$-receptors. Presynaptically, it binds to the dopamine $D_3$-receptor (autoreceptor). Dopamine is removed from the synaptic cleft by a reuptake mechanism. It is also metabolized by means of the enzymes monoamine oxidase B (MAO B) and catechol-O-methyltransferase (COMT). Dopamine can also be synthesized from L-dopa in nondopaminergic neurons in the striatum. COMT, catechol-O-methyltransferase; D1, dopamine $D_1$-receptor; D2, dopamine $D_2$-receptor; D3, dopamine $D_3$-receptor; AADC, aromatic L-amino acid decarboxylase (levodopa decarboxylase); DOPAC, dihydroxyphenylacetic acid; L-dopa, L-dihydroxyphenylalanine; MAO A, monoamine oxidase A; MAO B, monoamine oxidase B; TH, tyrosine hydroxylase; 3 MT, 3-methoxytyramine; O, vesicular storage of dopamine; ◄〜, dopamine reuptake; *, action of L-dopa; **, action of dopamine agonist; ***, action of MAO B inhibitor.

inata–nucleus basalis of Meynert, which project to hippocampus and cortex. Anticholinergics, therefore, have the potential to pharmacologically impair cognitive function (discussed later).

### Excitatory Amino Acids and Glutamate Receptors

Glutamate, the most important excitatory amino acid transmitter in the central nervous system, acts by means of at least four classes of glutamate receptors: the *N*-methyl-D-aspartate (NMDA) receptor, the α-amino-3-hydroxy-5-methyl-4-isoxazol-propionic acid (AMPA) receptor, the kainate receptor, and the metabotrophic receptor (Young and Fagg, 1990). These receptors are present in varying densities in the different nuclei of the

basal ganglia, where glutamate is known to be the neurotransmitter of the corticostriatal and the subthalamic–pallidal afferents. Degeneration of midbrain dopaminergic neurons in PD is accompanied by an increased level of excitatory amino acid transmission in the striatum and through the subthalamic afferents to the internal segment of the globus pallidus–substantia nigra pars reticulata.

In the circuitry of the basal ganglia, one differentiates between a *direct* and an *indirect* striatal output pathway (see legend of Figure 2). In PD patients the striatal neurons of the indirect pathway are thought to be overstimulated by an increase in corticostriatal glutamate release, which is normally under inhibitory control through a dopamine $D_2$-receptor–mediated dopaminer-

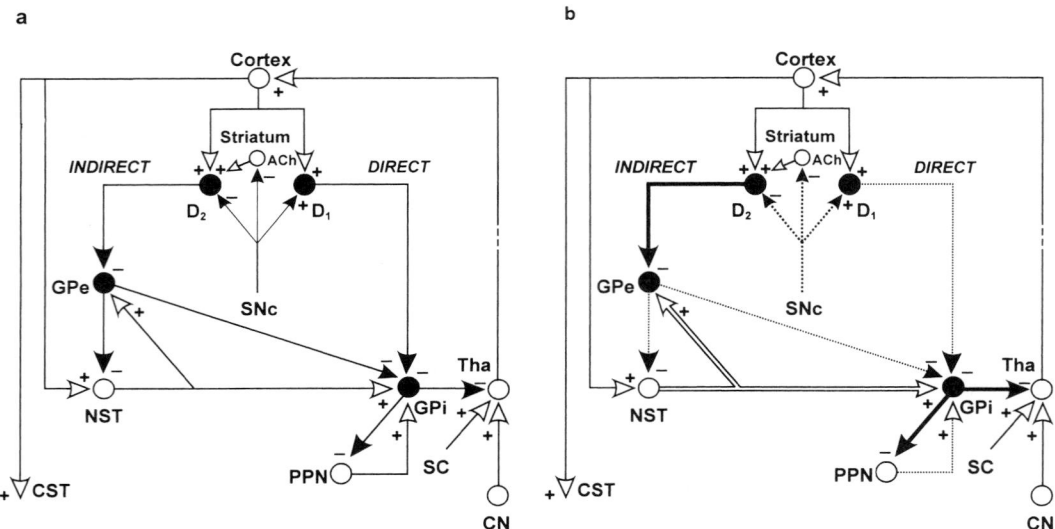

a    b

**FIGURE 2** Neurochemical anatomy of the basal ganglia: basal ganglia nuclei, circuitry, and neurotransmitters. (a) The circuitry of basal ganglia in normal people. Open circles and open arrows represent the excitatory neurotransmitter glutamate. PPN is peripeduncular nucleus, SC is the spinal cord, CST is the corticospinal tract, and CN is the cerebellar nuclei. For example, the connections from subthalamic nucleus (NST) to external globus pallidus (GPe) and internal globus pallidus (GPi) are glutamatergic and therefore excitatory. The dopaminergic nigrostriatal projection inhibits striatal cells bearing dopamine $D_2$-receptors and activates striatal neurons bearing dopamine $D_1$-receptors. Filled circles and filled arrows represent the inhibitory neurotransmitter γ-aminobutyric acid (GABA). For example, the connection (direct pathway) from the striatum (putamen and caudate nucleus) to internal globus pallidus (GPi) contains both GABA and the neuropeptides substance P (SP) and dynorphin. The projection (indirect pathway) from the striatum to external globus pallidus (GPe) contains both GABA and the neuropeptide metenkephalin. Part of the connection from the substantia nigra pars compacta (SNc) to the striatum contains, in addition to dopamine, the neuropeptide cholecystokinin. (Modified from Penney and Young [1986]; modified from Oertel [1994].) (b) Parkinson's disease (PD). In the *indirect pathway*, degeneration of the dopaminergic nigrostriatal projection (SNc to striatum) leads to a reduced inhibition of striatal neurons bearing dopamine $D_2$-receptors. The inhibitory GABA-ergic connection from striatum to external globus pallidus (GPe) is overactive (connection from striatum to external globus pallidus is indicated by a thick arrow). Consequently, GABA-ergic neurons in the GPe are overinhibited; therefore, activity in the GABA-ergic pathway from external globus pallidus (GPe), which normally inhibits the subthalamic nucleus (NST), is reduced (broken line), causing overactivity of the NST (connection with double open lines). This increased activity in turn results in enhanced activation of the inhibitory connection from the internal globus pallidus (GPi) to the thalamic nuclei (THA). In the *direct pathway*, degeneration of the dopaminergic nigrostriatal connection (SNc to striatum) leads to a reduced activation of dopamine $D_1$-receptor–bearing neurons in striatum. As a result activity is reduced in the direct GABA-ergic projection from striatum to internal globus pallidus (GPi) (broken line). Hence, the GPi receives a reduced inhibitory input, so again is overactive. In summary, the internal globus pallidus (GPi) receives enhanced activation by means of the indirect pathway and decreased inhibition by means of the direct pathway. Both changes result in a markedly increased activity of the neurons in the internal globus pallidus, whose overactive GABA-ergic projection then overinhibits thalamic neurons (indicated by thick arrow from GPi to THA).

gic mechanism (see Figure 2). In PD, the overactive *indirect* striatal projection to the external part of the globus pallidus provides an increased level of inhibition to the neurons of the external part of the globus pallidus. Hence, the external globus pallidus is underactive (inhibited) and provides a reduced level of inhibition to the neurons of the subthalamic nucleus. The subthalamic nucleus, in turn, is overactive (disinhibited) and overstimulates the internal part of the globus pallidus and its connecting pathway to the thalamus. This pallidothalamic projection is of inhibitory nature and over-

active in PD, thus overinhibiting thalamic neurons. Features of parkinsonism therefore can be improved by therapy with an NMDA receptor antagonist (Klockgether and Turski, 1993) or by surgical lesioning or stimulation of the subthalamic nucleus (Bergman *et al.*, 1990; Limousin *et al.*, 1995) or the internal pallidum (Dogali *et al.*, 1995; Krack *et al.*, 2000; Laitinen *et al.*, 1992; Obeso *et al.*, 2000; Siegfried and Lippitz, 1994) (see Chapter 76; see also the legend to Figure 2). In contrast to the indirect striatopallidal neurons, the inhibitory neurons of the *direct* striatopallidal pathway (striatum to the internal part of the globus pallidus) are excited by dopamine through the dopamine $D_1$-receptors. Hence, in PD, the direct pathway is underactive (i.e., less inhibitory).

According to this simplified scheme, in PD the internal globus pallidus (GPi) receives enhanced activation through the indirect pathway and decreased inhibition through the direct pathway. Both changes result in a markedly increased activity of the neurons in the internal globus pallidus, whose overactive GABAergic projection then overinhibits thalamic neurons.

## Therapeutic Strategies

This chapter covers pharmacotherapy and other nonsurgical therapy of PD. For surgical therapy see Chapter 76.

### Drug Therapy

L-dopa. Levodopa, the direct precursor of dopamine, was the first dopaminomimetic substance to be introduced as symptomatic therapy of PD. Oral L-dopa is taken up by a neutral amino-acid uptake mechanism from the duodenum to the blood. Its pharmakokinetics with a plasma half-life of $1\frac{1}{2}$ hours provides a pulsatile mode of L-dopa supply to the brain. In contrast to dopamine, levodopa passes the blood–brain barrier (BBB) by the uptake mechanism. It is taken up by remaining nigrostriatal nerve endings and possibly also by glia and serotoninergic neurons (Melamed et al., 1980). It is then (intraneuronally) decarboxylated by the enzyme levodopa decarboxylase (DDC = aromatic L-amino-acid decarboxylase [AADC]). Dopamine is released from the presynaptic terminal and binds to the various presynaptic and postsynaptic dopamine receptors (see Figure 1).

The dosage of plain levodopa (2–8 g/day) needed to treat PD and its peripheral side effects are usually reduced by a factor of 4 or 5 when it is combined with a peripheral decarboxylase inhibitor (AADC-I). Nowadays, administration of plain levodopa therefore is not recommended, if not obsolete.

L-dopa (henceforth defined as the combination of levodopa plus peripheral decarboxylase inhibitor—AADC-I) is nearly available worldwide. It possesses the best short-term ratio of motor efficacy to neuropsychiatric side effects of all antiparkinsonian agents and is still the "gold standard" of drug therapy for most patients in all stages of PD.

Fixed combinations of levodopa with the respective peripheral decarboxylase inhibitors benserazide (in Madopar) and carbidopa (in Sinemet and others) are available. As a rule of thumb, L-dopa should be prescribed in the lowest possible dosage that will give adequate but not necessarily maximal benefit. About 15% of patients with parkinsonism are considered to be primarily unresponsive to L-dopa therapy. Most of these patients do not have PD (see Tables I, III, and IV).

Theoretically, exogenous L-dopa treatment could lead, by means of auto-oxidation and deamination by the enzyme monoamine oxidase B, to the production of toxic metabolites (formation of hydroxyl radicals with subsequent lipid peroxidation) and thus to an additional therapy-induced lesion of the dopaminergic system. Long-term studies and practical experience speak against this hypothesis, and patients with PD receiving long-term L-dopa treatment have, compared with untreated patients, a prolonged to nearly normal life expectancy (Agid et al., 1999). However, it is not yet clear whether long-term survival may be even longer with a combination therapy of L-dopa with a dopamine agonist, a COMT-inhibitor, amantadine and/or monoamine oxidase B inhibitor (or both) or, for example, after several years of initial monotherapy with, for example, a dopamine agonist followed by combination therapy (dopamine agonist plus L-dopa or another non-L-dopa antiparkinsonian agent) than under L-dopa monotherapy (discussed later; Montastruc et al., 1994, Rascol et al., 2000; Oertel et al., submitted; The Parkinson Study Group, 1993; Parkinson's Disease Research Group, UK, 1993; Parkinson Study Group, 2000).

Standard preparations of L-dopa (levodopa plus AADC-I) should be started with 50 mg of L-dopa in the morning and gradually increased, according to the degree of parkinsonism, benefit, and side effects, by 50–100 mg every 3–7 days normally to reach 100–200 mg tid in the first 3–6 months. Benefit is usually noticed after a few days, but some patients may need several weeks or rarely even months to begin improving. The maximal effect of L-dopa therapy often is reached only 3–6 weeks after the last increase of dosage. Therefore, in patients with mild parkinsonian symptoms, a stepwise and slow increase of L-dopa dosage is justified.

Only two internationally marketed products, (i.e., Madopar and Sinemet) are mentioned in the text. For further different brand names in individual countries please consult Oertel and Dodel (1995) and newer national listings of drugs. For most patients, there are no known therapeutic differences between benserazide and carbidopa. Both maximally, but not necessarily completely, inhibit peripheral AADC in a dosage of about 75 mg/day (i.e., Madopar 125 tid or Sinemet plus tid). However, when levodopa is combined with carbidopa in a ratio of 10:1 (as in Sinemet 250/25 or Sinemet 100/10), some patients on low dosages, particularly when beginning treatment or receiving substan-

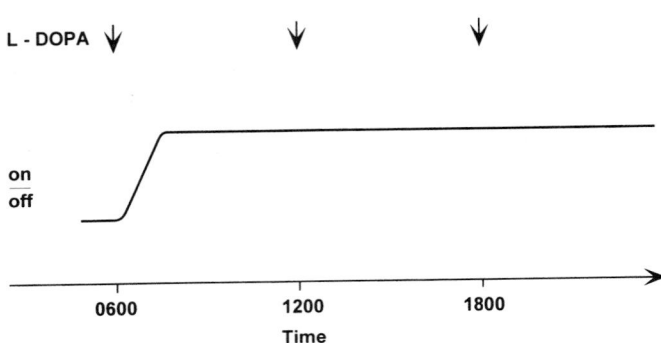

FIGURE 3    Smooth response during the day despite L-dopa tid. Occurrence of early morning akinesia may be an early sign of impending end-of-dose akinesia during the day. (From Marsden *et al.* (1982) with permission.)

tially less than 75 mg/day of carbidopa, may experience peripheral dopaminergic side effects or be mistakenly considered as L-dopa nonresponders. Traditionally, L-dopa is initially given three times daily, although some patients' symptoms initially are well controlled with bid dosage. A constant clinical effect, the "L-dopa honeymoon" (see Figure 3), can usually be observed on such a regimen, despite rising and falling plasma levels of levodopa (see Figure 7, see Marsden *et al.*, 1982). When L-dopa preparations are stopped because of doubts about the diagnosis, it may take up to 7 days or longer for the signs of parkinsonism to re-emerge (long-duration response; see Figure 4). On average, after about 2 years of treatment, the duration of action of the individual L-dopa dose will seem to gradually shorten. Although there is some real shortening of duration of

action of single dose, a more important factor is the progressive increase in the amplitude of the L-dopa response, making end-of-dose effects noticeable for the first time (Lees and Stern, 1983) (see Figure 5). In more advanced disease, a single standard L-dopa dose may have a mean latency of 40 minutes to a therapeutic effect (latency to "on") and a mean duration of action of 160 minutes (medium-duration response; Esteguy *et al.*, 1985). However, smaller doses given in attempts to fractionate the intake of L-dopa will last for a shorter period, with a more variable response and more frequent instances of "dose failure" (see Figure 6; also discussed later). In a minority of patients a much longer latency to on (delayed on) or even dose failure is observed in the morning or afternoon (after lunch). This phenomenon is either due to impaired gastric mobility, to dilution of L-dopa in the contents of the stomach (Djaldetti *et al.*, 1996), or to L-dopa-protein interactions (see later).

In most PD patients monotherapy with standard L-dopa is eventually associated with motor fluctuations (change from the immobil phase to the mobil and back to the immobil phase) and dyskinesias—together termed motor complications. In high dosages L-dopa may lead earlier to L-dopa–associated motor fluctuations (see Figure 7) and dyskinesias than low-dose L-dopa therapy (Poewe *et al.*, 1986; Rajput *et al.*, 1984).

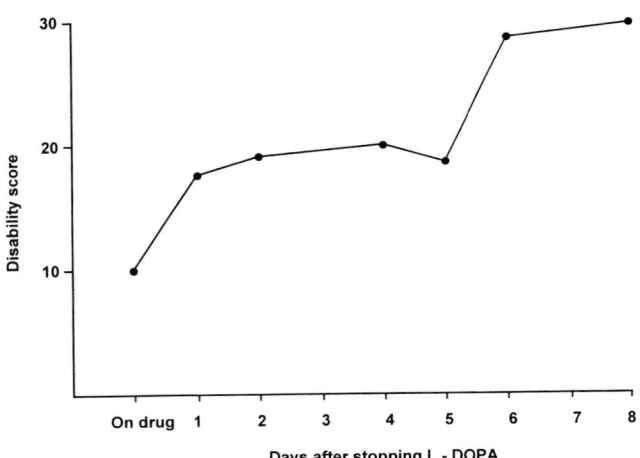

FIGURE 4    Late deterioration after withdrawal of L-dopa. The patient shows a moderate deterioration during the first two days of L-dopa withdrawal. After six days, a more severe, late deterioration is observed: long-duration response. (From Marsden *et al.* (1982) with permission.)

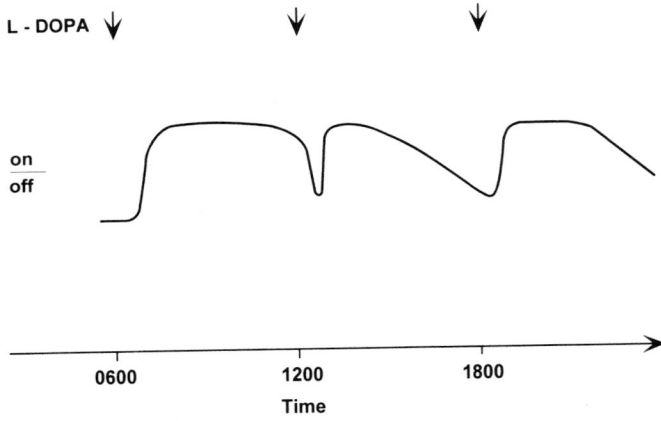

FIGURE 5    End-of-dose deterioration. The patient experiences early morning akinesia but is turned "on" by the first L-dopa dose. At the end of the first cycle and shortly after the second dose of L-dopa is taken, the patient turns "off." A number of patients are led to erroneously believe that taking the second dose is responsible for this phenomenon, whereas it likely relates to the "wearing off" ("end-of-dose akinesia") of the first dose before the second has had time to act. However, "beginning-of-dose worsening" of parkinsonian signs, particularly rest tremor, can also be the first sign of a dose beginning to have an effect (Lees and Stern, 1983). The second dose taken at mealtime shows a more pronounced end-of-dose akinesia. At 1800, the deterioration has occurred already before the third dose of L-dopa. Compare with Figure 3. (From Marsden *et al.* (1982) with permission.)

A 5-year trial with de novo PD patients comparing standard L-dopa (mean daily dose 426 mg/day) with controlled release L-dopa (mean daily bioavailable dose of controlled release 510 mg/day) resulted in about 21% of motor fluctuations or dyskinesia (Koller *et al.*, 1999).

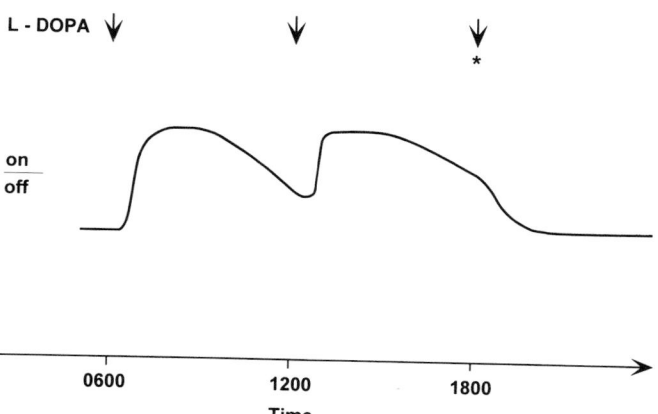

**FIGURE 6** End-of-dose deterioration after the first and second L-dopa doses plus lack of effect of the third oral L-dopa dose (*) ("dose failure" to oral L-dopa). Several reasons, for example, a large meal or a high-protein diet taken around 1800, can be responsible for this lack of efficacy. If required, an increase in the oral L-dopa dose or a subcutaneous injection of apomorphine may overcome this "off" period. (From Marsden *et al.* (1982) with permission.)

From active comparative trials in de novo PD patients comparing dopamine agonist therapy and L-dopa standard monotherapy, the L-dopa treatment arm provided the following data: 2-year treatment with L-dopa (mean daily dose 550 mg/day, Parkinson Study Group, 2000) resulted in development of dyskinesia or motor fluctuations in 51%. After 3-year treatment with L-dopa (mean daily dose 504 mg/day at endpoint, Oertel *et al.*, submitted) dyskinesia was observed in 26% and motor complications in 43% of patients. A 5-year treatment with L-dopa (mean daily dose 753 mg/day, Rascol *et al.*, 2000) resulted in a cumulative incidence of dyskinesia of 45%. On one hand, the general trend is similar. On the other hand, the reported percentages of observed motor complications relate either to motor fluctuations **and/or** dyskinesia (Koller *et al.*, 1999) or only to dyskinesias (Oertel *et al.*, submitted; Rascol *et al.*, 2000). The differences may in part be explained by the different L-dopa dosages administered. In addition, different definitions were used for the assessment and scoring of motor complications in the individual studies. Thus the data illustrate the limitations of study comparisons.

Abrupt withdrawal of L-dopa in advanced patients risks the development of an akinetic crisis or a neuroleptic malignant-like syndrome (discussed later). The side effects of L-dopa are listed in Table VI.

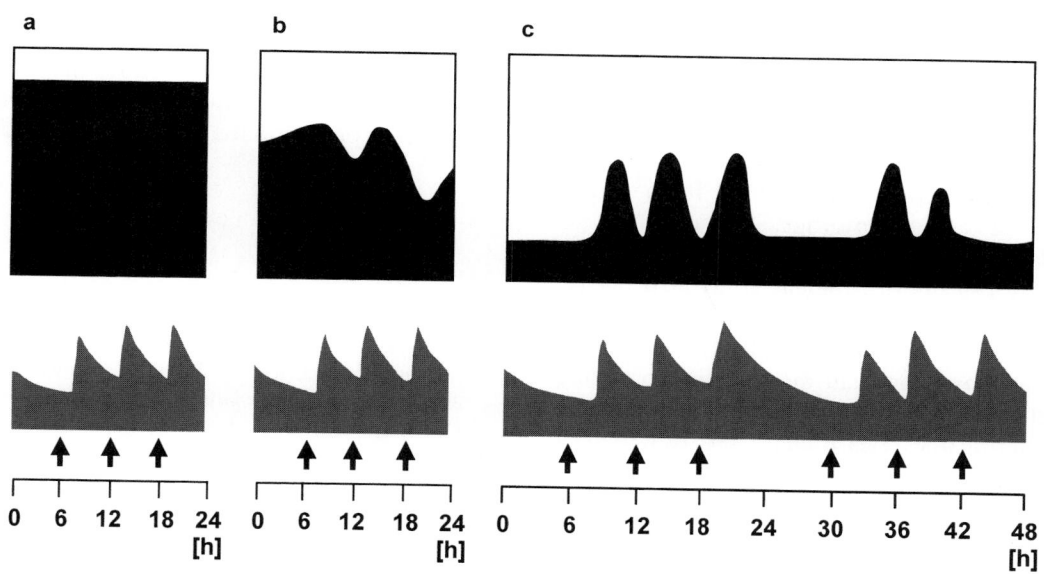

**FIGURE 7** Synopsis of development of L-dopa–related response fluctuations with time. The extent of the clinical benefit (black area) is dependent on the plasma level (lower half of the figure) of levodopa (areas with plasma peaks), which in turn is dependent on the timing of drug intake (vertical arrows). (a): Initial constant L-dopa response (compare Figures 3 and 4)—long duration response. (b): Daily response fluctuations with waning of the effect of each dose just before the intake of the subsequent one (wearing-off; compare Figure 5). (c): Rapid onset and offset of drug effect (on–off fluctuation). The overall efficacy is reduced on day 1 (0–24 hr), followed by marked nocturnal akinesia (24–30 hr), then lack of effect of the third dose of medication on day 2 (42–48 hr; compare Figure 6). Modified from Scholz and Oertel (1993).

TABLE VI    Side Effects of L-dopa

| | |
|---|---|
| **Early** | |
| Gastrointestinal: | Nausea or vomiting, usually at the beginning of treatment, because tolerance develops over a week or two in most patients. Management: take tablets with food; add domperidone 20 mg t.i.d. |
| Autonomic: | Orthostatic hypotension, arrhythmia, tachycardia, tachyarrhythmia (usually in people who already have cardiac disorders—the role of L-dopa in causing rhythm disturbances is controversial). Under anaesthesia sudden unexpected fluctuations in arterial blood pressure. Increased sweating. |
| Others: | Rarely dysgeusia. |
| **Late** | L-dopa-associated response fluctuations, painless dyskinesia (peak dose chorea, on-phase dyskinesia), painful dystonia (off-phase dystonia). Hypo- or hypersomnia, confusion, agitation, vivid dreams, illusions of presence, visual illusions, pseudo (insight retained) and true (insight lost) hallucinations (nearly always visual), paranoid psychosis, hypomania, hypersexuality. |

TABLE VII    L-Dopa Preparations: Latency to "On," Absorption, and Duration of Clinical Response

| L-dopa | Latency to "on" (min) | Duodenal absorption (%)[a] | Duration of clinical response[b] (hr) |
|---|---|---|---|
| Standard | 40 | 90–100 | 1–3 |
| Controlled release | | | |
| Madopar CR/HBS[c]/ Sinemet CR[d] | 75 | ca. 70 | 2–4 |
| Rapid (dispersible) | 25–30 | 90–100 | 1–3 |

[a]Absorption may be impaired by a protein (neutral amino acid)-rich diet or certain antibiotics (discussed later in text).

[b]As assessed in PD patients with (marked) end-of-dose fluctuations.

[c]Opening the capsule of Madopar CR/HBS eliminates the controlled-release mechanism.

[d]Breaking Sinemet CR in half reduces the controlled-release mechanism, thus altering the release characteristics toward the standard preparation. Sinemet CR is better absorbed with food, at least as a single dose in healthy young volunteers (Yeh *et al.*, 1989).

Two *controlled-release* (CR) L-*dopa preparations* are available: Madopar CR (Madopar HBS; hydrodynamically balanced system) and Sinemet CR. Both have about 70% bioavailability relative to standard L-dopa preparations (90%–100%). When given alone, the latency to "on" is longer (for standard L-dopa about 40 minutes; for controlled-release L-dopa about 75 minutes; for rapid acting L-dopa about 25–30 minutes, see below), but the duration of the response to CR preparations is usually also longer, provided enough of the drug is administered (see Table VII). In the intermediate stage of PD on average, the duration of response of the standard L-dopa or the rapid-acting L-dopa formula is 1–3 hours, whereas controlled-release L-dopa acts for about 2–4 hours.

Opening the capsule of Madopar CR/HBS eliminates the controlled-release mechanism; likewise breaking Sinemet CR in half reduces the controlled-release mechanism, thus altering the release characteristics toward the standard preparation.

Transport and absorption of controlled-release L-dopa (CR L-dopa), as for standard L-dopa discussed previously, are prone to changes in gastric motility and competition with neutral amino acids at the L-dopa uptake site in the duodenum and through the BBB (see discussion of a low-protein diet). Whether and when a given dose of CR L-dopa will be effective (that is, the occurrence of latency to "on") can be less predictable with CR L-dopa than with standard L-dopa. However, paradoxically, Sinemet CR, at least as a single

dose in healthy young volunteers, is absorbed better if taken with food than in the fasting state (Yeh *et al.*, 1989).

The optimal adjustment to CR L-dopa therapy can be very time consuming. An increase in (peak dose or on-phase) dyskinesia in the afternoon can occur because of progressive accumulation of the drug. End-of-dose or off-period dystonia often improves, but sometimes it can instead dramatically increase in intensity and duration (abnormal involuntary movements are discussed later). Table VIII contains practical guidelines on how to switch from standard L-dopa to CR L-dopa.

TABLE VIII    Steps in Changing from Standard L-dopa to Controlled Release L-dopa

1. Calculate total daily standard intake (as mg levodopa).
2. If changing completely to CR L-dopa, you will need to give 30%–40% more of the latter and space the doses somewhat further apart, initially every 4–5 hr. This spacing will probably soon need to be shortened to every 3–4 hr, but starting less frequently avoids early overdosage.
3. A patient who already has fluctuations will need a "kick-start" of 50–100 mg standard L-dopa with the first and often with some of the subsequent doses of CR L-dopa.
4. Changeover can be done with either one dose at a time or, sometimes, completely from one day to the next. In either case (particularly in the latter), careful *follow-up with feedback* from the patient or spouse and *further adjustments* are essential.
5. Be on the lookout for possible increased and unpredictable "latency to on," dose failure, and dosage-build-up phenomena, which may cause *increased dyskinesia* or confusion as the day progresses (time-bomb effect).

TABLE IX  Dissociation Constants of Dopaminergic and Antidopaminergic Drugs for Dopamine Receptor Subtypes ($K_i$ in nM)[a]

| Agonist | D$_1$-like group D$_1$-clones | D$_5$-clones | D$_2$-like group D$_2$-clones | D$_3$-clones | D$_4$-clones |
|---|---|---|---|---|---|
| Dopamine | | | | | |
| High | 250 | **18** | **33** | **4** | **30** |
| Low | 2000 | 228 | 924 | **31** | 50–150 |
| Apomorphine | | | | | |
| High | — | — | **39** | — | **4** |
| Low | 417 | 252 | 86 | **32** | — |
| Bromocriptine[b] | 672 | 454 | **12** | **4.8** | 285 |
| Pergolide | | | | | |
| High | **0.8** | — | **0.8** | — | — |
| Low | 1400 | 900 | **2.0** | **2** | — |
| Clozapine | 140 | 250 | 70 | 500 | **9** |
| Haloperidol | **30** | **40** | **0.6** | **3** | **5** |

*Note.* The data for pharmacological affinity of the various agonists and antagonists are adapted from Seeman and van Tol (1994) and Strange (1993) and are obtained in tissue from pig, rabbit (Seeman and van Tol, 1993), and largely for human receptors expressed in recombinant cell systems (Strange, 1993). High = high-affinity state of the receptors. Low = low-affinity state of the receptors.

Values printed bold are <40.

Also note the ratio of the affinity of haloperidol to the D$_1$-like and D$_2$-like group (see the discussion of neuroleptic malignant syndrome).

[a]The smaller the value for $K_i$ in nM, the higher is the affinity of the dopamine agonist or antagonist for the receptor investigated.

[b]High- and low-affinity states for bromocriptine are identical.

Three main indications for the use of CR L-dopa have been proposed. The first two are supported by empirical evidence, but the third must remain theoretical until the results of controlled trials are available:

1. *Nocturnal akinesia.* Sleep disturbances in PD patients can be due to an increase of motor symptoms during the second part of the night, when the effect of a late-evening administration of a standard L-dopa preparation wanes. In this situation administration of Madopar CR/Madopar HBS 125 (100 mg levodopa/ 25 mg benserazide) 1–2 (in rare cases 3) capsules or 1 tablet of Sinemet CR 250 (200 mg levodopa/50 mg carbidopa) to 1 tablet and/or a progressively increasing dose of a dopamine agonist at bedtime can improve nocturnal akinesia substantially. However, as a precaution, the patient has to be warned about the possibility of nocturnal dyskinesias, restlessness, insomnia, or vivid dreams. Sometimes bedtime CR L-dopa improves early morning dystonia the next day, but also paradoxically, in rare instances, the latency to "on" or the maximal effect of the first L-dopa dose can instead be less predictable and/or less marked the following morning. Bedtime CR-L-dopa should then be re-substituted with standard L-dopa at bedtime and, if necessary, an additional standard L-dopa given during the night.

2. *Mild to intermediate predictable L-dopa–related wearing-off (end-of-dose deterioration).* These response fluctuations can also improve on CR L-dopa. However, the administration of a standard L-dopa preparation shortly before or with the first CR dose of the day is normally necessary to provide a "kick-start"; that is, to turn the patient "on" within 30–60 minutes (Table VII). Subsequently, overlapping therapy with a CR L-dopa preparation can be continued over the rest of the day, although sometimes additional doses of standard L-dopa may again be needed. The timing and size of the doses during the day has to be individually tailored. Although it was originally hoped that the clinical benefit from a single dose of CR L-dopa might last 5–6 hours, in practice, these preparations usually need to be given every 3–4 hours. The more severe and advanced response fluctuations (wearing-off) become, the more a standard preparation is required with each dose of CR L-dopa. In the case of a very severe response fluctuation (on–off, discussed later), even the combination of standard with CR L-dopa may give disappointing results,

and these patients may worsen instead. This is particularly the case in patients with diphasic ("beginning" and "end-of-dose") dyskinesias (see Figure 8). If absorption of the CR preparation is unpredictable or delayed, a patient may experience a long akinetic, rigid (off) period in the morning only to become dyskinetic for hours in the afternoon ("time-bomb effect", A. Lees, personal communication, 1995).

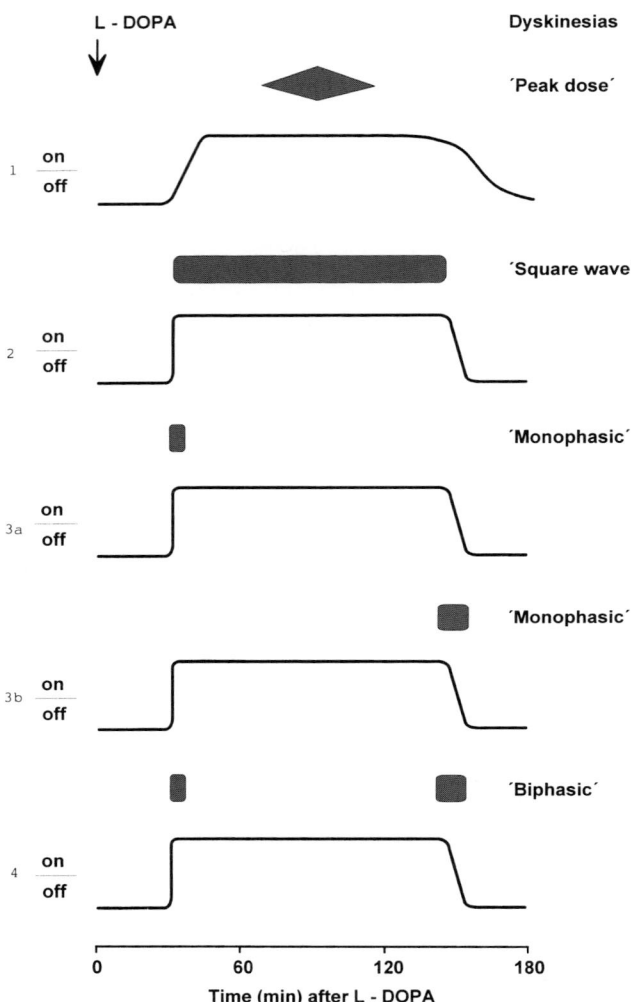

**FIGURE 8** Relation of dyskinesias to drug intake. (1 and 2) At effective L-dopa levels (patient fully "on"). On period dyskinesias, (1) is peak-dose dyskinesia; (2) is on-phase dyskinesia ("square wave" or "plateau dyskinesia"), where the dyskinesia is usually mobile, painless, choreic, rarely dystonic, or choreodystonic. (3 and 4) At intermediate L-dopa levels (patient not yet fully "on"). Biphasic dyskinesias, which include *higher threshold dyskinesias* (mobile, usually painless dystonic, choreic, ballistic, or stereotypic dyskinesias) and *lower threshold dyskinesias* (slow or fixed, often painful dystonia): (3) monophasic dyskinesia (a) when effect of L-dopa is waxing and (b) when effect of L-dopa is waning; (4) biphasic ("beginning" and "end-of-dose") dyskinesia. Modified from Marsden *et al.* (1982) with permission.

3. *Initial L-dopa therapy* of de novo *PD patients.* When L-dopa therapy is needed, some physicians start with a CR preparation, in the *hope* that its use will, in the long term, result in a lower or delayed prevalence of motor response fluctuations and other side effects than does the administration of a standard L-dopa preparation from the onset. Various experimental data in animal models favor the use of CR preparations (Chase *et al.*, 1993), but so far it has not been proven that this treatment policy conveys a definite advantage in patients (Koller *et al.*, 1999). In a 5-year double-blind active comparator trial comparing standard levodopa/carbidopa with controlled-released levodopa/carbidopa about 21% in both groups had motor fluctuations or dyskinesia. This relatively low occurrence of motor complications may be explained by the slower progression of PD patients who remained in the trial until the end point (Koller *et al.*, 2000; van Horn and Schiess, 2000) and the definition of how motor complications were scored.

In a few countries, a *rapid-acting L-dopa preparation* (Madopar dispersible) is available. It is provided as a tablet that rapidly disperses in fluid. This preparation is either swallowed and then dissolves in the gastric fluid, or it is first dispersed in fluid (water) and then drunk. After drinking a solution of Madopar dispersible (50 mg or 100 mg levodopa/12.5 mg or 25 mg benserazide), the latency to "on" is often about 10 minutes shorter than with a standard L-dopa preparation. However, the on-period duration with dispersible L-dopa is about equal to, or even slightly shorter than, that after standard L-dopa formulations (Fornadi *et al.*, 1994; Stocchi *et al.*, 1994). This dispersible L-dopa preparation is useful (1) in patients who take a long time to turn on with their first morning dose, (2) in those who experience afternoon akinesia on standard L-dopa therapy, (3) in patients who have difficulty swallowing, and (4) to finely titrate L-dopa dosage in rare patients who are extremely sensitive to the drug. As an alternative, the patient can chew standard L-dopa or take it together with cola, because this acidic sparkling beverage disintegrates the L-dopa tablet rapidly.

In akinetic crisis (discussed later) swallowing difficulties can make the oral administration of L-dopa and other antiparkinsonian agents extremely difficult and inadequate if a nasogastric tube cannot be swallowed or endoscopically placed.

In this instance levodopa solution can also be given by IV infusion for a short period but has to be specially ordered from the producer. Five levodopa ampoules (25 mg/5 ml) are mixed with 225 ml of 5% glucose up to a total volume of 250 ml and infused for 25–80 minutes at a rate of 3–10 ml/min. (Ideally, benserazide or carbidopa should be also given through a nasogastric tube but can be omitted if swallowing is not possible or either substance is not available, in which case the levodopa infusion rate needs to be higher (10–20 ml/min; watch out for volume overload, hyponatraemia, and venous thrombosis).

Because IV levodopa is not readily available, subcutaneous apomorphine is preferred (see later).

The therapeutic benefit of interruptions in L-dopa therapy for several days to 14 days (drug holiday) is low (Kofmann, 1984). After 6–12 months, the problems that initiated drug withdrawal (on–off symptoms, dyskinesia, dystonia, psychosis) mostly reappear. Drug holidays also carry a high risk of an akinetic crisis (discussed later) with a bedridden state and the danger of secondary complications such as aspiration pneumonia or deep venous thrombosis. Therefore, drug holidays are today obsolete in most clinics (Nutt *et al.*, 1994a). However, L-dopa may still sometimes need to be briefly withdrawn in a patient who is toxically overdosed with medication.

**Dopamine Agonists.** Dopamine agonists (DAs) act directly on postsynaptic or presynaptic dopamine receptors. Therefore, unlike L-dopa, they do not require a synthetic step to become effective. Available DAs (Table Xa,b) are classified as ergot and nonergot derivatives.

Five oral ergot DAs are presently marketed with various availability throughout the world: bromocriptine, cabergoline, alpha-dihydroergocriptine, lisuride, and pergolide, with bromocriptine and pergolide the most widely used ergot dopamine agonists. Two nonergot dopamine agonists, pramipexole and ropinirole, are available.

All dopamine agonists are predominantly dopamine $D_2$-receptor agonists. Bromocriptine has a slight dopamine $D_1$-receptor antagonist (partial agonist) effect, and there may be a mild $D_1$-receptor agonist effect of lisuride. Binding of bromocriptine to the $D_2$-receptor is 10-fold and of lisuride is 100-fold stronger than that of dopamine. Pergolide is an agonist at both $D_1$- and $D_{2/3}$-receptors, with similar affinity to the $D_2$- and $D_3$-receptor subtypes.

Cabergoline is predominantly a $D_2/D_3$-receptor agonist, with a preponderance for the $D_2$-receptor subtype; its $D_1$-receptor activity is low. Pramipexole and ropinirole are $D_2$-/$D_3$-receptor agonists and are virtually devoid of $D_1$-receptor activity. Pramipexole has a pre-

**TABLE X** Duration of Effect for Various Antiparkinsonian Agents[a]

| Substance (dose) | Medium duration response (hr) | Plasma half-life (hr) | Peak plasma level (hr) | Peak clinical response (hr) | Long-term response (days) |
|---|---|---|---|---|---|
| Amantadine (100 mg PO) | 1–8 | — | 2–8 | 2–4 | 2–4 |
| Benzhexol (trihexyphenidyl) | | | | | |
| (2 mg PO) | 1–6 | — | — | 2–4 | 2–4 |
| (5–10 mg PO) | — | 3.7 (SD 4.4) | 1.5 | | |
| Levodopa[b] | | | | | |
| (250 mg) + PDI | 0.5 (late[f])–24 (early[f]) | 1.5 (0.5–2.0) | 1.1 (0.5–3) | | Up to 7–14 days |
| (25–50 mg) | | | | | |
| MAO B inhibitor (selegiline) | >24 | Irreversible enzyme inhibitor | — | | Enzyme activity returns to normal after 3–8 weeks |
| *Dopamine agonists* | | | | | |
| Bromocriptine (2.5 mg PO) | 1–6 | 6 | 0.5–3 | 1.5–3.5 | Uncertain |
| Cabergoline (PO) | 24 | 65 (–110) | — | — | Uncertain |
| alpha-DHEC (PO) | — | 15 | — | — | Uncertain |
| Lisuride (1 mg PO) | 1–3 | >2.5–3 | 0.5–2.5 | 1–2 | Uncertain |
| Pergolide (4 mg PO) | 1.5–8 | 24 ? (8–16–42) | — | 2–6 | Uncertain |
| Pramipexole (PO) | — | 8–12 | — | — | Uncertain |
| Ropinirole (PO) | — | 6–8 | — | — | Uncertain |
| Apomorphine[c,d] (3 mg = 50 µg/kg) | 20–60 min | | 8–20 min[e] | 10–25 min | Uncertain |

[a]Modified from Marsden *et al.* (1982); Liebermann and Goldstein (1982).
[b]Response assessed on drug withdrawal from long-term treatment with levodopa/AADC-I.
[c]From Przedborski *et al.* (1995).
[d]From Gancher *et al.* (1995).
[e]Possibly depending on apomorphine concentration of solution.
[f]Stage of disease.

ponderance for the $D_3$-receptor subtype, ropinirole, a $D_2$-receptor subtype preponderance (see Table Xa).

Table IX shows the different affinities to the $D_1$-, $D_2$- and $D_3$-receptors, based on *in vitro* assays. Pharmacokinetic properties of the different dopamine agonists such as bioavailability, plasma half-life, duration of response, peak plasma levels, clinical effect, long-term response, interaction with food, and means of metabolism are listed in Table X and Xb.

Therapeutic doses, side effects, contraindications, and relative dosage equivalents of the different DAs are listed in Table XII.

### General Therapeutical Aspects

The principal hypothesis of DA therapy is to provide a somehow continuous dopamimetic stimulation of dopamine receptors. This continuous dopaminergic stimulation is thought to be associated with a lower incidence and less severity of motor fluctuations and dyskinesia than seen with the more pulsatile mode of L-dopa monotherapy (Blanchet *et al.*, 2001; Olanow *et al.*, 2000).

The use of dopamine agonists is indicated in any stage of PD. Thus, DAs are registered for initial monotherapy in *de novo* (previously untreated) PD patients and for add-on therapy in combination with L-dopa and/or any other antiparkinsonian drug in nonfluctuating and fluctuating patients without or with dyskinesia.

The therapeutic effect on akinesia, rigidity, and rest tremor in the recommended dose ranges is considered to be somewhat less than that obtained with L-dopa.

For *de novo* patients this statement is based on results of long-term open-label and double-blind active comparator trials: in about 40%–50% *de novo* (previously untreated) PD patients with symptoms in Hoehn and Yahr stage I and II were sufficiently controlled with DA monotherapy (without or under coverage with the peripheral dopamine $D_2$-receptor antagonist domperidone) in the first 2–3 years of therapy. Thus, monotherapy with a dopamine agonist delayed the use of L-dopa therapy in these patients by 2–3 years. In the other half of patients, symptoms usually reached a degree of severity after 6 months –3 years that required additional L-dopa (L-dopa rescue). In addition to this symptomatic effect, initial dopamine agonist monotherapy has been shown to delay and reduce the occurrence and severity mainly of dyskinesia but not necessarily of motor fluctuation (Oertel *et al.*, submitted; Parkinson's Disease Research Group in the United Kingdom, 1993; Parkinson Study Group, 2000; Rascol *et al.*, 2001; Rinne, 1989).

In L-dopa–treated nonfluctuating patients, the additional use of DAs allows reduction of the L-dopa dose,

**TABLE Xa**  Receptor Specificity of Dopamine Receptor Agonists

| Drug | $D_1$ | $D_2$ | $D_3$ | $D_2/D_3$ |
|------|-------|-------|-------|-----------|
| Apomorphine | + | +++ | NA | NA |
| Bromocriptine | − | ++ | + | 9:1 |
| Cabergoline | + | +++ | ++ | NA |
| α-Dihydroergocryptine | +/− | +++ | NA | NA |
| Lisuride | ± | +++ | +++ | NA |
| Pergolide | + | +++ | +++ | 1:1 |
| Pramipexole | 0 | ++ | +++ | 1:7 |
| Ropinirole | 0 | +++ | ++ | 3:1 |

Based on Sautel *et al.* (1995), Piercey *et al.* (1996), Watts (1997), and Gerlach *et al.* (2001).

+ = Agonist (low affinity), ++/+++ = agonist (high affinity), ± = partial agonist, − = no affinity, NA = Information not available.

**TABLE Xb**  Pharmacokinetic Properties of Orally Active Dopamine Agonists

| Dopamine agonist unchanged in urine (%) | Oral bioavailability (%) | $t_{max}$(min) | Protein binding (%) | $t_{1/2}$ (hr) | Dosing schedule | Metabolism |
|---|---|---|---|---|---|---|
| Bromocriptine 2.5–5.5 | 8 | 70–100 | 90–96 | 3–8 | tid | Complete |
| Cabergoline <4 | 50–80 | 120–240 | 40–42 | 65–105 | Once or twice daily | Extensive |
| Pergolide not available | 20 | 60–120 | 90 | 24 | tid | Extensive |
| Pramipexole ≈90% | >90 | 60–180 | <20 | 8–12 | tid | ≈10% |
| Ropinirole 5–10 | 55 | 90 | ≤40 | 6–8 | tid | Extensive |

tid = three times daily; $t_{max}$ = time to peak plasma concentration; $t_{1/2}$ = elimination half-life.

**TABLE XI  Changing from One Dopamine Agonist to Another**

This exchange depends on the clinical duration of action of each of the two drugs. In patients whose fluctuations are not controlled by their existing dopamine agonist, such a change is more likely to be from a shorter acting drug such as bromocriptine or lisuride to a longer acting one such as cabergoline, α-DHEC, pergolide, pramipexole, or ropinirole.

Dosage equivalence is given in Table XII and duration of action in Table X. We recommend a gradual switch over, dose by dose, over a period of several weeks or a rapid switchover in 1 or 2 days.

An example might be a gradual change over a week from bromocriptine, 5 mg 3-hourly 5× daily to pergolide 0.5 mg 4 hourly 4× daily.

If a complete switchover from one day to the next is performed, inpatient conditions or daily monitoring (i.e., per telephone) is generally advisable.

Examples for complete changeover in 1 day—clinical experience, no evidence-based medicine: from bromocriptine 5 × 5 mg. Lower evening dose of the newly introduced agonist initially on purpose to reduce change of newly developing visual hallucinations.

| Bromocriptine | 5—5—5—5—5 |
| Cabergoline | 2—2 |
| | 4—0 |
| | 3—1 |
| Pergolide | 0.5—0.5—0.5—0.25 |
| Pramipexole | 0.7—0.7—0.35 |
| Ropinirole | 5—5—3 |

again under the assumption to delay and reduce the onset, prevalence, and severity of motor complications.

In L-dopa–treated fluctuating patients, the addition of DAs allows one to ameliorate, if not abolish, motor fluctuations. In patients with mild to marked dyskinesia during on, dopamine agonists allow one to substantially reduce L-dopa and thus often to reduce the extent and severity of dyskinesia, provided the herefore necessary intermediate to high dose of the dopamine agonist is tolerated.

DAs should initially be increased gradually (see Table XII; Table XV). This recommendation is valid in *de novo* (so far drug naive) patients and in advanced patients when dopamine agonists are added to another therapy. Unlike the case with L-dopa, often only after many weeks of slow upward titration of the dose is the full therapeutic benefit from DAs observed.

In about 10% of patients, low doses of DAs may lead to an increase of parkinsonian symptoms, possibly caused by the predominant stimulation of the presynaptic dopamine receptor (autoreceptor), which leads to a reduction of dopamine release from the presynaptic terminal. At higher doses, the direct stimulation of postsynaptic $D_1$- and $D_2$-/$D_3$-receptors prevails.

High-dose dopamine agonist therapy has been advocated by few groups with the goal to minimize or even entirely withdraw L-dopa treatment and in the hope of reducing dyskinesia and motor fluctuations. It requires a careful selection of patients, a long-term tedious uptitration, and time-consuming and frequent medical attention. Its long-term effects were not studied.

Switching from one agonist to another is possible, either abruptly or with an overlapping regimen by reducing the first and simultaneously increasing the new dopamine agonist (Goetz *et al.*, 1999). On the basis of clinical experience, a range of equivalence doses can be recommended for the different agonists and compared with L-dopa. Table XI discusses the changeover from one dopamine agonist to another.

In rare occasions, a double agonist treatment, although expensive, can be tried either in *de novo* PD patients, if a patient wishes to avoid the addition of L-dopa as long as possible, or in advanced patients, when the addition of one dopamine agonist is not sufficient to control for fluctuations and/or dyskinesia, and a further increase of the dopamine agonist is not tolerated because of side effects. Then the addition of a second agonist, for example, in combinaton with an ergot or nonergot agonist will sometimes lead to further improvement of motor symptoms without enhancing the side effect profile of the first agonist.

Additional administration of the peripheral dopamine $D_2$-receptor antagonist domperidone (20–30 mg tid half an hour before each intake of DA) can markedly reduce the occurrence of gastrointestinal side effects and may help attenuate postural hypotension (see discussion of antiemetics). In fact, such domperidone pretreatment together with DA monotherapy in *de novo* parkinsonian patients can allow one to reach doses averaging 50–60 mg/day of bromocriptine, 6–14 mg/day of cabergoline, 2–6 mg/day of lisuride, 5–15 mg/day of pergolide, 4.5 mg of pramipexole, and higher and up to 40 mg/day of ropinirole, usually without major peripheral side effects. When such doses of dopamine agonists are used, clinical improvement of parkinsonian symptoms can be similar in a subgroup of patients in Hoehn and Yahr stage 1 and 2 to that obtained with L-dopa therapy.

In summary, on the basis of several open-label trials (Hely *et al.*, 1994; Montastruc *et al.*, 1994; Parkinson's Disease Study Group in the United Kingdom, 1993; Rinne, 1983, 1989; Stern and Lees, 1983) and more recently in randomized multicenter double-blind active comparator long-term studies, the advantages of (initial) DA monotherapy seem to be twofold: first the low incidence and severity of drug-induced involuntary movements (i.e., dyskinesia and dystonia [discussed later]) at least in the first 3–5 years and second a low incidence of response fluctuations (wearing-off or on–off phenomena) (Oertel *et al.*, submitted; Parkinson Study Group (PSG), 2000; Rascol *et al.*, 2000; Rinne *et al.*, 1998).

The reduced incidence and lower severity of dyskinesia is also retained in those patients who initially received dopamine agonist monotherapy and subse-

TABLE XII    Therapeutic Doses, Dosage Equivalents, and Side Effects of Oral Dopamine Agonists

| Dopamine agonist | Usual dosage range |
|---|---|
| Apomorphine SC | 2–5 (–8) mg (40–50 µgmg/kg) per injection |
| Bromocriptine | 5–45 (–120) mg/day (tablet at 2.5 mg, caps. at 5, 10 mg) |
| Cabergoline | 2–6 (–15) mg/day (tablet at 1, 2, 4 mg) |
| α-Dihydroergocriptine | 15–60 (–180) mg/day (tablet at 15, 30, 60 mg) |
| Lisuride | 0.4–3 (–6) mg/day (tablet at 0.2 mg) |
| Pergolide | 0.75–5 (–15) mg/day (tablet at 0.05, 0.25, 1 mg) |
| Piribedil | 10–30 mg/day (tablet at 10 mg) |
| Pramipexole | 0.1–3–4.5 mg/day (salt) (tablet at 0.125, 0.25, 1 mg) |
|  | 0.7–2.1–3 mg/day (free base) (tablet at 0.088, 0.18, 0.7 mg) |
| Ropinirole | 3–9–24 (–40) (tablet at 1, 2, 5, 10 mg) |

| Dosage equivalence | Approximately |
|---|---|
| L-dopa | 100 mg |
| Apomorphine | 3–5 mg (40–50 µgmg/kg) |
| Bromocriptine | 10–15 mg |
| Cabergoline | 2 mg |
| α-Dihydroergocriptine | 30 mg ? |
| Lisuride | 1 mg |
| Pergolide | 1 mg |
| Pramipexole | 1–1.5 mg (salt) |
|  | 0.7–1 mg (free base) |
| Ropinirole | 3–5 mg |

Side effects of dopamine agonists

| Frequent | Rare | Contraindications |
|---|---|---|
| Orthostatic hypotension | Vasospasm | Recent myocardial infarct |
| Nausea | Gastric bleeding | Gastric or duodenal ulcer |
| Vomiting | Erythromelalgia | |
| Constipation | Sleep disorders (insomnia) | |
| Psychosis | Sudden-onset somnolence ("sleep attacks") | |
| Dyskinesia[a] | Leg edema | |
|  | Pleuropulmonary and retroperitoneal fibrosis | |

*Note.* For plasma half-life see Table X.

[a]Rarely with dopamine agonist monotherapy. If dyskinesia is present under L-dopa therapy, addition of a dopamine agonist may initially augment dyskinesia.

quently required L-dopa rescue (Parkinson Study Group, 2000; Rascol *et al.*, 2000; Rinne *et al.*, 1998).

### Individual Compounds: Ergot Agonists

For the ergot agonists, their similarity of the ergoline-ring structure to the endogenous monoamines explains their action on dopaminergic, adrenergic, and serotoninergic receptors (dopamine receptor subtypes were discussed previously). Absorption of ergot DAs in the gastrointestinal tract, unlike L-dopa, is hardly influenced by food intake. However, the bioavailability of most ergot derivatives is reduced to 10%–20% because of metabolism in the liver (first-pass effect).

The side effect profiles of available ergot agonists generally consist of more orthostatic hypotension and gastrointestinal and psychiatric symptoms than observed with L-dopa (see Tables VI and XII).

Extensive experience exists with *bromocriptine* as add-on therapy to L-dopa in early, intermediate, and advanced PD and as monotherapy in *de novo* PD patients (Hely *et al.*, 1994; Lieberman *et al.*, 1983; Montastruc *et al.*, 1994; Parkinson's Disease Research Group in the United Kingdom, 1993; Rascol *et al.*, 1984; Rinne, 1985, 1989; Stern and Lees, 1983). Most of our clinical knowledge of the efficacy and side effects of ergot DAs is derived from studies with this compound.

*Bromocriptine* is a dopamine $D_2$-receptor agonist with a weak dopamine $D_1$-antagonist effect. Its motor effects last for about 2–6 hours. Treatment with bromocriptine in monotherapy or combination therapy should be started with 1.25 mg (in some cases even with

0.625 mg) at night, preferably with domperidone cover. Under outpatient conditions, total daily dosage should be increased by 1.25–2.5 mg/week. Average doses are 3–5 × 2.5 to 5–10 mg/day, but many patients may need up to 3 × 20 mg/day to derive significant benefit.

*Lisuride* is mainly a dopamine $D_2$-receptor agonist with a duration of action of 2–4 hours. Lisuride also binds to the 5-HT$_{1a}$ serotonin receptor type. Oral therapy with lisuride starts with 0.1 mg at night. Under outpatient conditions, the total daily dose should be increased by 0.2 mg/week, rising up to 3 × 0.4–1 mg/day. In patients with marked response fluctuations, the dose may be distributed to 6 × 0.2–0.6 mg/day (Lieberman *et al.*, 1983). Lisuride is water soluble and therefore can be injected as a bolus IV or SC or even infused (Obeso *et al.*, 1985; Quinn *et al.*, 1983). In highly selected patients with severe response fluctuations, subcutaneous continuous infusions by means of mini-pumps have been used (see discussion of apomorphine) but seem to cause more psychiatric side effects than apomorphine. However, the injectable form is available only on special order from the company.

*Pergolide* activates both $D_1$- and $D_{2/3}$-receptors. It possesses a duration of action of 4–6 hours. Compared with apomorphine and bromocriptine, it has higher affinities for the $D_2$- and $D_3$-receptors. Oral therapy with pergolide should be started with 0.05 mg/day and initially increased gradually over 2–7 weeks to a dose of 0.25 mg tid. Maximal registered dose is 5 mg/day. Pergolide is effective as monotherapy (Barone *et al.*, 1999) (preferentially together with domperidone for the first couple of weeks). In a 3-year active comparator trial with *de novo* PD patients, the effect of strict monotherapy of pergolide (3.23 mg/day at endpoint) and a strict L-dopa monotherapy (504 mg/day at endpoint) were studied. With pergolide, more than 50% had their motor symptoms sufficiently controlled in the pergolide arm. Monotherapy with pergolide delayed the onset and severity of dyskinesia and motor complications significantly compared with L-dopa. At the end of the study 8% of the pergolide-treated patients and 26% of the L-dopa–treated patients had dyskinesia (highly significant difference); 30.6% of the pergolide group and 43.8% of the L-dopa group had motor complications (Oertel *et al.*, submitted). Pergolide is also effective as add-on therapy to L-dopa (Goetz *et al.*, 1983; Olanow *et al.*, 1994; Pezzoli *et al.*, 1994; Wolters *et al.*, 1994). Because of its longer plasma half-life, pergolide can improve marked response fluctuations in some patients in whom bromocriptine has failed to do so. Side effects are similar to those of bromocriptine and lisuride.

*Cabergoline* is an ergoline derivative with predominantly $D_2$-receptor agonist properties with a plasma half-life of about 65–110 hours. Clinical improvement after a single dose of 1.5–2.5 mg was observed after 4 and 24 hours and in some patients even after 48 hours. This feature allows once-a-day administration, usually in the morning. Clinical experience shows, that in a subgroup, twice-a-day administration (i.e., in the morning and at noon) gives a more stable response than using the daily dose at once in the morning. In both single and bid dosing, when titration has reached the fully efficacious dose, careful follow-up observation of (increasing) adverse effects for 4–8 weeks is required, because the steady-state efficacy (i.e., a stable dose of cabergoline) is reached after 4 weeks. In patients with marked nocturnal akinesia with or without off-dystonia at night, even a late-evening dose is recommended, provided the patient does not have insomnia or develop visual hallucination or even psychosis.

Cabergoline reduces parkinsonian features in *de novo* patients and should be slowly increased, starting with 0.5 mg/daily in the morning to be increased to 1 mg/daily after 3 days. At 1 week a further increase by 1 mg/week is recommended to reach at 3 mg/daily, 3 weeks, and if tolerated, a further slow increase is possible in the following months. As with other dopamine agonists, cabergoline monotherapy allows one to sufficiently control motor symptoms in a subgroup of *de novo* PD patients and reduces dyskinesia in the monotherapy group and in the group receiving the combination with L-dopa (L-dopa rescue therapy) as shown in an active comparator trial against L-dopa (Rinne *et al.*, 1998a).

In advanced PD patients, cabergoline (4–5 mg/day in combination with L-dopa) decreased the degree of motor fluctuations and also dyskinesia, and this beneficial response was maintained over more than 2 years follow-up. The side-effect profile seems, apart from a low incidence of gastrointestinal signs and symptoms, similar to that of bromocriptine (Ahlskog *et al.*, 1994; Inzelberg *et al.*, 1996; Lera *et al.*, 1993; Rabey *et al.*, 1994). In even higher doses (6–15 mg) cabergoline is able to substantially reduce off time and also dyskinesia, if side effects are low and tolerable and if L-dopa can be (in exchange with the high dose of cabergoline) substantially decreased.

*α-Dihydroergocriptine* is available in few European countries. Its half-life is reported to be 10–15 hours. Initial treatment is 5 mg bid. Dosing ranges between 15–60 mg/tid. The compound has been studied as monotherapy in *de novo* PD patients against placebo and as adjunct therapy to L-dopa in advanced fluctuating patients. Publications are limited, and it is not clear whether this substance is similarly efficacious as or whether it provides an advantage over bromocriptine or any other available dopamine agonist in respect to efficacy, safety, or adverse event profile.

A further DA, the piperazine derivative *piribedil*, is mainly a $D_2$-receptor agonist; its pharmacological features are less well known. The compound Piribedil is registered for treatment of PD in only a few countries. We are not aware of any study that compares it with other agents. On the basis of limited personal experience (WHO), piribedil is helpful in the control of rest tremor in some patients.

### Nonergot Derivatives

*Pramipexole* (Mierau *et al.*, 1995) and *ropinirole* are nonergot dopamine agonists. Both substances are $D_2$-$D_3$-receptor agonists, with pramipexole having the relatively higher affinity to the $D_3$-receptor than ropinirole. The dose of *pramipexole* is either given as the salt or as the free base; 0.7 mg of the free base (as marketed for example in some European countries) is equivalent to 1 mg of the salt (as marketed for example in the United States). Pramipexole has a half-life of 8–12 hours. It shows a relative low incidence of peripheral side effects such as nausea or orthostatic hypotension (Wermuth *et al.*, 1994). At least some patients who experience marked peripheral side effects under ergot agonists, even with additional domperidone treatment, tolerate pramipexole. An initally effective dose of pramipexole of 0.25–0.5 mg/tid can be achieved in 3 weeks.

Ropinirole has a half-life of 6–8 hours. To reach an effective dose requires in general more than a month. According to an active comparator trial (ropinirole versus bromocriptine), its side effect profile is more similar to that of ergot agonists, but some patients also tolerate it better than ergot DAs.

Both compounds are clinically effective for the motor symptoms of PD in *de novo* PD patients (Parkinson Study Group, 2000; Rascol *et al.*, 2000) and as add-on therapy to L-dopa (Liebermann *et al.*, 1997; Weiner and Factor, 2000), compared with placebo.

*Ropinirole* has been tested in a multicenter double-blind controlled 5-year long-term active comparator trial against L-dopa to show that the incidence of dyskinesia is substantially reduced after 5 years: 16% of patients randomly assigned at baseline completed the 5-year trial with ropinirole monotherapy (i.e., true monotherapy arm), and 5% of this group showed dyskinesia. Twenty percent of patients who had been started on ropinirole monotherapy initially, but required additional L-dopa (L-dopa rescue) to gain adequate symptomatic relief (i.e., ropinirole and L-dopa arm), developed in dyskinesia. In contrast, of the patient group who had been started on L-dopa and remained throughout the 5-year trial on L-dopa monotherapy, 46% had dyskinesia develop (Rascol *et al.*, 2000). The two groups (ropinirole without or with L-dopa versus L-dopa), however, showed no significant difference in respect to motor fluctuations. Similar results were obtained in a 2-year

report on a double-blind multicenter randomized trial comparing pramipexole and L-dopa monotherapy. Patients either on pramipexole monotherapy or a combination of pramipexole/L-dopa (L-dopa rescue) showed significantly less motor complications than the patients treated with L-dopa monotherapy at endpoint (Parkinson Study Group, 2000).

Pramipexole also is particularly effective in the therapy of rest tremor in mild to advanced PD, either without or in conjunction with L-dopa (Pogarell *et al.*, 2002).

Both nonergot DAs have been reported to induce excessive daytime sleepiness and sudden onset of sleep (so called sleep attacks) in a minority of patients without or with previous daytime sleepiness (Frucht *et al.*, 1999). This phenomenon is more likely to occur during up-titration than in the dose-maintenance phase. It is dose related, for example, more often observed above a dose of 2.1 mg free base/3 mg salt of pramipexole or 16 mg of ropinirole. It is now clear that sudden onset of sleep is a feature of any dopamimetic compound (see sleep impairment).

Thus far, one comparative trial between the ergot DA bromocriptine and the nonergot DA pramipexole has been published in advanced fluctuating PD patients. This study was powered for safety and not for efficacy. Both DAs were well tolerated and significantly more efficacious than placebo, with a slight tendency for pramipexole to be more effective (Guttman, 1997).

In respect to *de novo* patients, neither a direct comparison between ropinirole and pramipexole nor an active comparator trial between any ergot DA and a nonergot DA, powered for efficacy, has been performed so far. Therefore, it is difficult to state which of the DAs is superior to any other one in respect to long-term efficacy, adverse effect profile, safety or delay in, occurrence of or reduction in severity of motor complications.

*Apomorphine* is presently the most potent DA available for the treatment of PD. It is a dopamine $D_1$-, $D_2$-, and $D_3$-receptor agonist (see Tables IX, X, and XII) and is registered in a number of European countries and Canada for the therapy of PD. Subcutaneous bolus injections or subcutaneous continuous infusion with a pump improve akinesia, rigidity, and rest tremor. Because of rapid subcutaneous absorption, response to a bolus occurs after 10–15 minutes (latency to "on"), and the effect lasts 20–60 (120) minutes, depending on the dose given. The effect of oral or rectal apomorphine is less reliable than, and inferior to, its subcutaneous administration. Common side effects include yawning (this sign in fact indicates that apomorphine has reached the CNS), nausea, emesis, or postural hypotension. The concomitant use of domperidone (20–30 mg tid) started 1–3 days before apomorphine administration reduces or even abolishes peripheral side effects. A single dose of 50 mg domperidone half an hour before apomorphine can also reduce side effects. Psychiatric side effects occur less frequently with apomorphine than ergot or non-ergot derivatives.

In specialized centers, apomorphine is offered to parkinsonian patients for four indications:

1. *Apomorphine test.* This is done to test the response of the CNS to dopaminomimetics (Hughes *et al.*, 1990; Oertel *et al.*, 1989). After pretreatment with domperidone, usual dopaminomimetic therapy is stopped overnight. A baseline score (UPDRS-scale III, motor part, and/or measurement of finger tapping or walking speed) in defined "off" stage is obtained. Injection of 2–5 mg (40–50 µg/kg) apomorphine subcutaneously usually leads to marked improvement of parkinsonian symptoms and signs. Of those with a positive test (reduction of UPDRS III by at least 20%), 88% will respond to chronic oral L-dopa treatment after 1 year. Evaluation of apomorphine tests in *de novo* patients requires experience and time because of their low initial UPDRS score (Gasser *et al.*, 1992).

2. *Severe off-periods.* Subcutaneous injection of apomorphine (2–5 mg, in rare cases up to 10 mg) offers the most reliable escape from marked off periods, which may include not only severe akinesia, rigidity, or rest tremor but also depression, anxiety, panic attacks, painful (biphasic) dystonia (discussed later), and speech and swallowing difficulties. Some patients with marked response fluctuations can benefit from the continuous subcutaneous delivery of apomorphine by means of an infusion mini-pump. The required dose varies but is usually 3–6 mg apomorphine/h (range, 2–20 mg/h) (Colzi *et al.*, 1998; Frankel *et al.*, 1990; Gervason *et al.*, 1993; Poewe *et al.*, 1988a; Stibe *et al.*, 1988). Special needles, preferably made out of plastic, should be used, because apomorphine has a low pH. Particular attention to the subcutaneous infusion sites is needed, because allergic reactions (in this case the topic application of a steroid containing cream may be helpful), aseptic necrosis, or infection may occur. Ultrasound may speed reduction of painful nodules. Rarely, hemolytic anemia or eosinophilia have been observed.

3. *Akinetic crises.* This is discussed as a separate section.

4. *Perioperative and postoperative treatment.* Surgery is discussed in a separate section.

**Catechol-O-Methyl-Transferase Inhibitors.** Catechol-O-methyl-transferase (COMT) is a major metabolizing enzyme both for dopamine and L-dopa. The respective metabolites are 3-O-methyl-dopamine and 3-O-methyl-dopa. 3-O-methyl dopa is a nontoxic metabolite that cannot be used in the synthesis of dopamine and does not interfere with the effect of L-dopa (at least not at blood concentrations derived from conventional therapy). Inhibition of COMT is an important approach for extending the duration of L-dopa plasma concentration. Thus, the main indication for COMT inhibition is to lessen the abruptness of predictable wearing-off of the L-dopa effect and subsequently to extend daily "on" time in patients with motor fluctuations.

COMT inhibition occurs the first day of administration; thus, the effect is already observed on the first day(s). Titration is not required, in contrast to treatment with dopamine agonists.

Most COMT is present in peripheral tissues, particularly the liver, intestinal tract, and kidneys. Comparatively low COMT activity is found in the brain, and dopaminergic nigrostriatal neurons are devoid of COMT. In fact, the main levodopa-catabolizing enzyme in the brain is monoamine oxidase (MAO), not COMT. In addition, the effect of dopamine is mainly terminated by re-uptake rather than by catabolism. On the basis of these considerations, a peripheral COMT inhibitor should be sufficient to extend and enhance the action of L-dopa and reduce the potential for central side effects and interactions. On the other hand, it is evident that monotherapy with a COMT-inhibitor is without any rationale.

The COMT inhibitor entacapone is registered for therapy in PD in many countries, whereas the COMT inhibitor tolcapone is available with restrictions in a few countries such as Switzerland, Norway, and the United States (see later). Both act peripherally. When given with oral L-dopa and to patients with PD, the area under the curve for L-dopa plasma concentration is markedly increased. This is due principally to a prolongation of the plasma elimination half-life of L-dopa and, at the same time, the peripheral production and concentration of the metabolite 3-O-methyl-dopa is markedly decreased. However, COMT inhibitors do not tend to increase the peak plasma concentration of the drug (Dingemanse *et al.*, 1995; Nutt *et al.*, 1994b). On the basis of the latter pharmacokinetic property, COMT inhibition can be used to extend the L-dopa effect (i.e., "on" time up to 30%–60% without increasing dyskinesias or other peak-action adverse effects in most patients (Kaakkola *et al.*, 1994; Limousin *et al.*, 1993; Merello *et al.*, 1994; Nutt *et al.*, 1994b; Roberts *et al.*, 1993). In a minority of patients the effect of L-dopa is enhanced, and likewise motor complications can be enhanced or appear for the first time. In patients already with dyskinesia in "on-time," this "on-time" with dyskinesia will be prolonged. On initiating a COMT-inhibitor treatment, it is possible to reduce each dose of L-dopa. Alternatively or concomitantly, one can space the latency between the individual L-dopa dosages. Both strategies reduce or eliminate newly occurring peak dose dyskinesia and may allow to even reduce previously present dyskinesia. This decrease in L-dopa may range between 10% and 30% compared with prior daily L-dopa intake. However, COMT inhibitors do not possess a central "L-dopa to sparing" role, because the prolongation of the peripheral availability of levodopa leads rather to an increase in the amount of levodopa that finally enters the brain and thus to an increase in the amount of levopamine produced in the brain. In summary, COMT inhibition prolongs and thus optimizes L-dopa therapy without introducing, apart from the rare side effect of diarrhea, a new side effect profile.

*Entacapone.* Entacapone, (E)-2-cyano-N,N-diethyl-3-(3,4-dihydroxy-5-nitrophenyl) propenamide) inhibits dose-related peripheral COMT, which is near-maximal with the recommended dose of 200 mg taken with each L-dopa dose. Orally administered, it is rapidly and almost completely absorbed. Its distribution half-life is around 20 minutes, and its elimination half-life is between 1.4 and 3.4 hours. Entacapone is catabolized by the liver to mostly glucuronide conjugates or metabolites of entacapone. A small fraction is excreted unchanged. Several studies have shown that, despite a prolongation in the duration of the L-dopa action, there is no or hardly an increase in peak plasma levodopa concentrations (Kaakkola et al., 1994; Myllyla et al., 1993). However, multiple dosing may lead to an increase in peak plasma L-dopa concentration throughout the day, because the second dose will meet a higher baseline L-dopa level than the first, and the third dose may meet an even higher baseline level than the second (etc.). Two large multicenter controlled trials have been conducted with Parkinson patients experiencing motor fluctuations. In both studies (Parkinson Study Group, 1997; Rinne et al., 1998b), a 200-mg dose of entacapone or placebo was added to each levodopa dose in a previously optimized drug regimen. The primary measure of clinical efficacy was the change in "on-time." After 6 months, "on-time" assessed at baseline increased by approximately 1–1.3 hours. In all trials conducted so far, the addition of entacapone to L-dopa in fluctuating PD patients was associated with a significant reduction in "off" time an increase in "on" time and an improvement in overall clinical performance.

More recently, two further studies were conducted in PD patients, including patients without (stable) as well as patients with mild to intermediate wearing off (Poewe et al., 2002). In the nonfluctuating group of patients there was an improvement of the UPDRS II (activity of daily living) score in the entacapone patients, whereas patients receiving placebo showed an impairment in ADL scores (Poewe et al., unpublished).

Entacapone is marketed under the name Comtess (Orion Pharma Pharmaceuticals) or Comtan (Novartis, Switzerland).

*Tolcapone (limited availability).* Tolcapone (3,4-dihydroxy-methylnitrobenzophenone) is rapidly absorbed with a $T_{max}$ of 1.5 hours; its elimination half-life is 2.3 hours. Tolcapone's maximal effect is achieved with a 200-mg dose. This dose yields an approximately doubled area under curve and elimination half-life of L-dopa. The 200-mg dose of tolcapone inhibits more than 80% of systemic COMT activity. The recommended dose is 100 mg tolcapone tid (i.e., every 6 hours).

Tolcapone, in high doses in experimental animals, also has a central effect. In addition, in experiments on rodents, tolcapone enhances CNS levels of S-adenosyl methionine (Da Prada et al., 1994), a substance claimed to have an antidepressant effect.

Two multicenter placebo-controlled trials with patients with marked wearing off under previously optimized therapy showed that a tolcapone dose of 200 mg tid (at 6-hour intervals) for 3–12 months significantly reduces mean "off-time" by 50%. Compared with placebo, 100 mg tid reduced mean "off-time" by 31%, although this finding did not reach statistical significance. COMT inhibition allowed the clinicians to decrease mean L-dopa intake by 21% and 24% with 100 and 200 mg tid tolcapone dose, respectively (Adler et al., 1998; Baas et al., 1997; Kurth et al., 1997; Myllylä et al., 1997; Rajput et al., 1997).

Tolcapone has also been studied in patients without fluctuations. These stable responders improved in clinical ratings over 6 months when receiving tolcapone in their UPDRS II (activities of daily living). At the same time in the tolcapone 100 and 200 mg tid group, a mean reduction in daily L-dopa intake was possible by 21 mg/day and 32 mg/day, respectively. In contrast, the placebo-treated group had to increase L-dopa by a mean of 47 mg/day (Waters et al., 1997). It is unclear whether early combination of L-dopa plus a COMT inhibitor would delay and reduce the occurrence and severity of motor complications (wearing off, dyskinesia, dystonia) or of psychosis in the long term.

An indication of COMT inhibitors in the treatment of nonfluctuators or even *de novo* patients is not presently available.

*Adverse effects of COMT-inhibitors.* As expected with a compound that enhances dopaminergic effects, adverse effects included dyskinesia and nausea. Because tolcapone blocks peripheral COMT activity to a higher extent than entacapone, it is administered three times a day, whereas entacapone is recommended to be given in conjunction with each L-dopa dose. Tolcapone, when initiated, tends to increase dyskinesia in patients with already existing dyskinesia in "on" to a larger extent than entacapone; on the other hand, in general, it allows a larger reduction of the daily L-dopa dose than entacapone.

A significant nondopamimetic adverse reaction is the occurrence of diarrhea, mostly in the first 1–2 months of therapy. This diarrhea is so severe that in half of the patients, the diarrhea requires discontinuation of the COMT inhibitor. In the other half, diarrhea disappears over a couple of days. Comparing entacapone and tolcapone, diarrhea has been less frequently and less severely encountered with entacapone than tolcapone. This diarrhea can be successfully treated by the addition of cumin (Eichhorn, Selzer, Oertel, unpublished observation, 1998).

No adverse effects occur from combining tolcapone with the MAO-B inhibitor selegeline in L-dopa–treated patients (Davis *et al.*, 1995).

In the first 9 months after the marketing of tolcapone, three cases of fatal hepatotoxicity were reported in patients receiving tolcapone. Although a cause-and-effect relationship was not established, suspicion of a rare idiosyncratic toxicity has led to the cancellation of marketing of tolcapone in the European community and Canada. Limitations are placed on the indication of the drug in the United States, Switzerland, Norway, and several other countries worldwide. These restrictions include recommendations by the Food and Drug Administration in the United States and authorities in the European Union that tolcapone should only be used if other drugs fail in the treatment of motor fluctuations. In addition, liver enzyme function tests (transaminases) are to be monitored every 2 weeks under tolcapone therapy for the first 6 months and every 4 weeks under tolcapone treatment thereafter. The use of tolcapone and the relation of benefit, efficacy, and safety is currently being re-evaluated in the European Union.

**Monoamine Oxidase B Inhibitor—Selegiline.** Selegiline selectively and irreversibly blocks intraneuronal and extraneuronal monoamine oxidase B (MAO B) (see Figure 1). It therefore reduces or retards the metabolism of dopamine. It has also been reported to decrease dopamine re-uptake (Riederer and Jellinger, 1983). The standard dosage ranges from 5–10 mg/day (see Table XV). Positron emission tomography data have demonstrated that inhibition of MAO B enzyme activity is irreversibly blocked for at least 3 and possibly up to 8 weeks (Leenders *et al.*, 1987). The addition of selegiline to L-dopa may allow the L-dopa dose to be reduced by 10%–15%, maximally 30%, in about 50% of patients without losing clinical benefit. Predictably, early mild L-dopa response fluctuations often can be reduced by the addition of selegiline (Heinonen and Rinne, 1989). All the side effects of L-dopa are potentially enhanced by selegiline. Thus, peak dose dyskine-

sias can appear or increase when selegiline is added. Also, the addition of selegiline may, for the first time, precipitate vivid dreams or hallucinations, so that this agent should either not be given or used with extreme caution in patients with a history of psychosis. According to present knowledge, selegiline should also not be combined with selective serotonin reuptake inhibitors (SSRIs). Moreover, because of its prolonged depression of MAO B activity, selegiline should probably be discontinued for a full 6 weeks before an SSRI is started (see discussion of depression). In respect to other antidepressants, a combination of selegiline with amitriptyline or mianserin has been used by the authors without problems of drug interaction.

The issue of whether selegiline is neuroprotective is controversial. In animals, selegiline, by inhibiting MAO B, blocks the conversion of MPTP, an exogenous proneurotoxin, to $MPP^+$, which is toxic to dopaminergic neurons (Heikkila *et al.*, 1984). Pretreatment with selegiline therefore prevents MPTP-induced parkinsonism in primates. It has been hypothesized that an MAO B inhibitor might be able to retard the progression of human PD, even though this is clearly different from MPTP-induced parkinsonism in animals (in which true Lewy bodies are not found). Second, inhibition of MAO B diminishes the production of hydrogen peroxide, which in turn gives rise to the reactive hydroxyl radical (see the discussion of standard L-dopa preparations); however, it is not known whether MAO B inhibition will in turn enhance auto-oxidation of dopamine.

Double-blind prospective placebo-controlled studies (Allain *et al.*, 1991; Myllyla *et al.*, 1992; Tetrud and Langston, 1989; The Parkinson Study Group, 1989a,b) designed to test this hypothesis have been undertaken. In one of those, the DATATOP study (The Parkinson Study Group, 1993) 400 *de novo* PD patients received selegiline, and another 400 received the placebo with the following results.

Selegiline very significantly delayed the need for L-dopa therapy by about 9 months. The endpoint was when L-dopa was necessary to maintain functional independence. This observation has been confirmed by two other studies (Palhagen *et al.*, 1998; Przuntek *et al.*, 1999). It could be explained either by a symptomatic effect of selegiline, as has already been described by European clinicians in the 1970s and 1980s (Birkmayer *et al.*, 1983) or by a combination of a symptomatic effect and a neuroprotective effect. However, in the 3-year follow-up of the DATATOP study (Parkinson Study Group, 1996a,b), patients subsequently receiving L-dopa after initial therapy with either placebo or selegiline showed no difference with respect to clinical symptoms, the L-dopa amount taken, or to the rate of development of wearing off, dyskinesias, "on/off," or freezing when data were analyzed from the time of entry into the original study. Thus it remains open whether the policy of initial selegiline treatment to delay L-dopa therapy conveys a long-term advantage.

A subanalysis of patients who had received selegiline but had failed to show clinical improvement after initiation of therapy (i.e., did not exhibit an early symptomatic effect of selegiline) showed that their delay until needing L-dopa was longer than the delay of the group who improved clinically with placebo, suggesting a small neuroprotective effect of selegiline. However, in view of the PET studies and the long central action of selegiline, the washout time of 1 (later 2) month used in this study may have been too short to eliminate a symptomatic effect of selegiline in this group. In another study (Olanow et al., 1995), four groups of 30 patients each were treated with different combinations. Two groups received a placebo either with bromocriptine or L-dopa, and two groups received selegiline in combination either with bromocriptine or L-dopa. After 12 months, selegiline and placebo were stopped for 7 weeks, and then bromocriptine and L-dopa were withdrawn and the endpoint analysis carried out 1 week later. Thus, as in the later stages of the DATATOP study, there was a total selegiline washout time of 2 months. The group with selegiline therapy showed significantly less worsening of "off treatment" UPDRS score than placebo controls. However, the washout time for bromocriptine or L-dopa may likewise have not been long enough to exclude a confounding symptomatic therapy factor.

Another randomized but open study has failed to demonstrate that an initial combination of selegiline and L-dopa from the beginning is superior to L-dopa monotherapy with respect to clinical efficacy and frequency of side effects, such as motor fluctuations, dyskinesia, or psychosis, after 3 years (Parkinson's Disease Research Group in the United Kingdom, 1993). In a second interim analysis of the same study allocating subjects according to intention-to-treat, a significantly higher death rate after a mean of 5.6 years of treatment was found in the L-dopa–selegiline arm than in the L-dopa–alone arm. The hazard ratio of 1.57:1 ($p = 0.0152$) was calculated, which is equivalent to a 50%–60% excess mortality among patients in the L-dopa–selegiline arm. The "on" treatment analysis showed a similar, although nonsignificant, trend. With respect to the clinical outcome, both the L-dopa–selegiline and the L-dopa–alone groups had a similar rate of dyskinesias developing and end-of-dose fluctuations, despite a clear L-dopa–sparing effect of additional selegiline (Parkinson's Disease Research Group of the UK, 1995). The mortality results in this randomized but open study were unexpected, and it remains undetermined whether and how they are causally related to selegiline treatment. Furthermore, the design and statis-

tical analysis of the study has been subject to numerous criticisms. The article stimulated the discussion on whether initial therapy with selegiline, followed by L-dopa (Ben-Shlomo et al., 1999; Parkinson Study Group, 1996b), the immediate combination therapy of selegeline and L-dopa (Myllylä et al., 1992), or initial L-dopa-therapy followed eventually by adjunctive therapy with selegiline (Gerlach et al., 1996; Olanow et al., 1998) reduces or even increases mortality (Lees on behalf of the PDRG-UK, 1995; Thorogood et al., 1998) in subjects with PD. All in all, this controversy has not been finally resolved, because the data analysed stem from (1) studies with different designs, (2) different populations of patients ranging from the typical study patient (nearly healthy apart from PD) to patients with a certain degree of major co-morbidity and/or at the old age range, and (3) different medical specialities, because study patients were taken care of by neurologists, general practitioners, or geriatricians. If all evidence provided is taken into consideration (Ben-Shlomo and Lees on behalf of the PDRG-UK, 2001; Donnan et al., 2000; Van Gerpen and Ahlskog, 2001), it seems that in the average PD patients the administration of selegiline, together with L-dopa, does not increase long-term mortality in PD patients; however, PD patients with marked cardiac disease may be a group at risk, because selegiline has been reported to impair cardiac conductivity in such a subgroup (Churchyard et al., 1997).

Only one study compared L-dopa monotherapy, early combination therapy of L-dopa with a dopamine agonist, and triple combination therapy with L-dopa, a dopamine agonist, and selegiline. This study demonstrated the superiority of early combination therapy (L-dopa and dopamine agonist) over monotherapy, but not of triple therapy over early combination therapy. However, it was not of a triple-blind, placebo-controlled design (Rinne, 1989).

In summary, at this time the putative neuroprotective efficacy of selegiline is considered unlikely in that current evidence suggests, but is insufficient to show, a true (robust) neuroprotective action. Nonetheless, because neuroprotection is such an important concept, some patients may elect to take selegiline in the possibility that it may be neuroprotective, and because selegiline carries a low risk (at least) in early stages of otherwise healthy PD patients for serious adverse reactions, there seems to be no serious consequences associated with taking the drug. With respect to symptomatic treatment, monotherapy with selegeline in PD is mildly efficacious, the effect is not robust and may be short lived. In patients with a stable nonfluctuating response on L-dopa or dopamine agonists, the efficacy of MAO B inhibitors has not been shown convincingly, but may possibly be useful in some patients. In the treatment of motor fluctuations in PD MAO B inhibitor–adjuvant therapy to L-dopa has a positive (mild) effect; however, it may not be long lasting in some patients (Rascol *et al*., 2002).

**Amantadine.** The antiparkinsonian effect of the antiviral agent amantadine (Table XIII) was discovered by chance in 1969 (Schwab *et al*., 1969). Several possible mechanisms of action have been discussed: It may increase dopamine synthesis; it promotes release of catecholamines from presynaptic stores by an amphetamine-like action; it may be a dopamine and noradrenaline presynaptic re-uptake blocker; and it also has a mild anticholinergic action. Its most important pharmacological mechanism seems to be its antagonistic action at the N-methyl-D-aspartate (NMDA) receptor. Memantine, chemically closely related to amantadine, has also been demonstrated to be an antagonist at the NMDA receptor (Bormann, 1989; Kornhuber *et al*., 1989; see previous discussion of glutamate receptors). Thus, amantadines can influence glutamatergic neurotransmission at the corticostriatal synapse or at the subthalamic–internal pallidal synaptic level (see Figure 2).

Amantadine has several clinically established effects: (1) it influences akinesia and rigidity, (2) it has a mild effect on rest tremor, and more recently (3) its antidyskinetic effect was described. With an initial infusion of amantadine sulfate, the severity of L-dopa–induced dyskinesia was reduced, and at the same time "on"-time increased (Ruzicka *et al*., 1999; Verhagen Metman *et al*., 1998). Follow-up oral treatment with 100–200 mg tid allowed maintenance of the antiakinetic and antidyskinetic effects and in addition a reduction in L-dopa (Del Dotto *et al*., 2001—see Table XIII). Therapy for all three indications is started with 1 to 2 × 50 mg, to be increased to an average dosage of 100 mg tid with a maximal dosage of 200 mg tid.

Because of the advent of more effective dopaminomimetic agents, amantadine is at present not a first-choice treatment but can help delay the need for L-dopa in young parkinsonian subjects. Placebo or active comparator controlled trials in *de novo* PD patients are lacking. In some patients, the effect of amantadine wanes after several months, but many continue to derive benefit and can worsen considerably when the drug is stopped. Amantadine can also be beneficial as an adjunctive antiakinetic therapy in intermediate or even advanced PD. The three actions together make amantadine a type of stabilizing add-on therapeutic option for L-dopa–treated patients with motor fluctuations **and** peak dose dyskinesia during "on"-time.

The drug, mainly in high doses, can cause visual hallucinations and confusion, especially in patients with impaired renal function, in patients taking additional anticholinergic therapy, or in subjects with impaired intellectual function. When amantadine and anticholinergics are withdrawn, this process should be as gradual as possible to avoid rebound phenomena or confusional states (see discussion of cognitive impairment). Intake of amantadine later than 4 PM may cause insomnia or disturbed sleep pattern. A rare side effect is epileptic seizures, potentially occurring in patients with additional cerebral pathology or renal insufficiency, in whom the dose of amantadine should be dramatically reduced. In Austria, Germany, and Switzerland, amantadine is available as a ready-to-use infusion and is used for the treatment of akinetic crises (discussed later). The clinical effect of this infusion can be approximated by 3 × 200 mg amantadine orally every 2 hours.

**Budipine.** Budipine, a diphenyl-dipiperidine derivate, has been reported to possess *in vitro* a weak anticholinergic and antihistaminic effect and to weakly inhibit MAO B. Further studies have shown that budipine is also, like the amantadines, an NMDA receptor antagonist (Klockgether *et al*., 1993). Its effect on akinesia, rigidity, and rest tremor are comparable with that of amantadine, but it may particularly help a subpopulation of patients with rest tremor (Spieker *et al*., 1999).

**TABLE XIII    Amantadine: Dosage and Side Effects**

| Route of administration | Indication | Dosage |
|---|---|---|
| Oral | Mild parkinsonism | 2–3 × 100 mg |
| | Intermediate and advanced parkinsonism | 2–3 × 100 mg + L-dopa; maximal dose 3 × 200 mg |
| | Antiakinetic effect | |
| | Antidyskinetic effect | Reduce dose, if renal function is impaired |
| Intravenous[a,b] | For akinetic crisis | 200 mg amantadine sulfate in 500 ml NaCl |

Side effects

Sleep disturbance, restlessness, agitation, anorexia, nausea, livedo reticularis, distal edema of lower extremities

Psychiatric symptoms (predominantly elementary visual hallucinations), prevalence high with high doses and in patients with cerebral lesions of other cause

Very rare: epileptic seizure in patients with pre-existing cerebral lesion or impaired renal function, supraventricular tachycardia

Can be combined with

L-dopa, with dopamine agonists, MAO-B-inhibitor and/or COMT-inhibitor. If combined with anticholinergics, enhancement of anticholinergic effect possible, then reduce dose; frequency of psychosis can increase

[a]Ready-to-use solution; registered only in Austria, Germany, and Switzerland.
[b]3 × 200 mg PO every 2 hours results in a clinical effect similar to IV infusion.

The recommended daily dose is 10–30 mg tid; however, its therapeutic range is small, and in most patients its effect wanes in a few months. In patients with rest tremor that is refractory to other medications, a trial of budipine may be justified (LeWitt and Truong, 1990). Side effects include sedation, restlessness, insomnia, dry mouth, and dizziness. Prolongation of QTc time has limited its use, which requires regular ECG monitoring. The drug is only available in Germany with the preceding restrictions.

**Anticholinergics.** The use of anticholinergics dates back to Charcot (1880; belladonna extract). These compounds were traditionally thought to predominantly influence rigidity and rest tremor. Comparative efficacy data on the various synthetic registered substances are not available. Anticholinergic doses should be increased slowly. They can cause a number of unwanted side effects, including urinary retention, constipation, blurred vision caused by impaired accommodation, or precipitation of latent narrow-angle glaucoma. They can also cause a dry mouth, which may help sialorrhea (discussed later), a frequent and unpleasant feature of PD.

Anticholinergics are contraindicated in patients with incipient or manifest dementia and should be used with caution, if at all, in elderly subjects. High doses do not seem to be more effective than low to intermediate doses. The higher the dose of anticholinergics and the age of the patient, the greater is the risk of precipitating an organic confusional state, difficulty with memory or concentration, or worsening frontal lobe symptoms (Sadeh *et al.*, 1982; Syndulko *et al.*, 1981). These neuropsychiatric side effects call for withdrawal of anticholinergics, but their abrupt termination should be avoided, if possible, because parkinsonism can increase or may even become more marked than before initiation of therapy (rebound phenomenon). In addition, rapid reduction or withdrawal of anticholinergics can also cause disorientation or a confusional state (anticholinergic withdrawal syndrome; see discussion of dopaminergic psychosis). Doses of anticholinergics are listed in Table XVI (treatment of rest tremor).

**Clozapine, an Atypical Neuroleptic.** Clozapine is an atypical neuroleptic. It possesses a relatively high affinity for the dopamine $D_4$-receptor and a comparably low affinity for $D_2$- and $D_1$-receptors (see Table IX). It also exhibits anticholinergic and antihistaminic components, and especially acts as a $5\text{-HT}_2$ serotonin receptor antagonist. In contrast to classical neuroleptics, akinesia and abnormal involuntary movements (AIMs) are rarely induced by clozapine in patients with psychosis. In patients with PD, drug-induced worsening of parkinsonism has usually only been observed in doses >100 mg/day. Clozapine is especially effective for the treatment of dopaminomimetic psychosis (Friedman, 1991; Scholz and Dichgans, 1985; Wolters *et al.*, 1990) (discussed later). It may also be helpful in the treatment of drug-resistant rest tremor (Friedman and Lannon, 1990; Pakkenberg and Pakkenberg, 1986) (discussed later) and may reduce or suppress L-dopa–induced dyskinesias (Arevalo and Gershanik, 1993; Bennett *et al.*, 1993). However, in several countries clozapine is at present only licensed for use in psychosis in patients with schizophrenia and/or who are resistant to classical neuroleptics.

Side effects of clozapine can include induction of arterial hypotension, hypersalivation (despite an anticholinergic component), sedation, promotion of epileptic seizures, and in about 1% of patients, the induction of agranulocytosis (Alvir *et al.*, 1993). This potentially fatal side effect requires mandatory registration with the manufacturer and regular blood counts, initially weekly and later biweekly. The dose required for the treatment of L-dopa/dopamine agonist–induced psychosis ranges from 6.25–100 mg, and averages between 12.5 and 50 mg nightly. Rarely, an additional dose (6.25–25 mg) is required during the day (discussed later).

Quetiapine, another atypical neuroleptic, may be a useful alternative to clozapine (see later for PD and psychosis).

**Antiemetics.** Metoclopramide, a commonly employed antiemetic, is a $D_2$-receptor antagonist of the substituted benzamide class and is in the strict sense a classic neuroleptic (Table I). It can pass the blood–brain barrier (BBB) to induce acute dystonic reactions, tardive dyskinesia or dystonia, or drug induced parkinsonism (discussed later).

In contrast, domperidone, another antiemetic drug, is a peripherally acting $D_2$-receptor antagonist that does not usually cross the BBB. By blocking dopamine $D_2$-receptors in the chemoreceptor trigger zone in the area postrema, situated in the floor of the fourth

**TABLE XIV   Hoehn and Yahr Scale for Parkinson's Disease**

| | |
|---|---|
| Stage 1: | Unilateral disease |
| Stage 2: | Bilateral disease, without impairment of balance |
| Stage 3: | Mild to moderate bilateral disease, some postural instability, physically independent |
| Stage 4: | Severe disability, still able to walk or stand unassisted |
| Stage 5: | Wheelchair bound or bedridden unless aided |

Optional stages 1.5 and 2.5

| | |
|---|---|
| Stage 1.5: | Unilateral plus axial involvement |
| Stage 2.5: | Mild bilateral disease, with recovery on pull test |

Reprinted with permission from Lippincott Williams & Wilkins (1967).

ventricle in the medulla oblongata and functionally outside the BBB, domperidone can act as an antiemetic without causing or worsening parkinsonism. In addition, domperidone helps to reduce or prevent L-dopa and dopamine agonist–induced nausea and vomiting (Pollak *et al.*, 1981; Quinn *et al.*, 1981; also discussed later and see Tables VII and XIV). It may also ameliorate dopaminomimetic-induced constipation or arterial hypotension. In extremely rare cases, domperidone has presumably crossed the BBB to induce dyskinesia or dystonia (Franck and Noel, 1984; Pellegrino *et al.*, 1990).

## Nonmedical Conservative Therapy

**Low-Protein Diet.** L-Dopa is actively transported from the duodenum into the blood, as well as across the BBB. Large neutral amino acids compete with L-dopa for the same transport mechanism (Woodward *et al.*, 1993). Therefore, protein intake can lead to reduced L-dopa plasma levels and reduced cerebral availability of L-dopa (Carter *et al.*, 1989; Pincus and Barry, 1987; Riley and Lang, 1988). Patients in whom food intake clearly interferes with the efficacy of L-dopa should take their L-dopa 1 hour before or 1–2 hours after meals (but watch for gastrointestinal side effects of L-dopa). To minimize the interference of L-dopa from diet, a rapidly dispersible L-dopa preparation may be used. Alternatively, the individual L-dopa dose taken with a meal can be increased. Patients with L-dopa response fluctuations, usually those who already notice a definite "protein" effect, can sometimes experience a marked reduction in duration and severity of the off-phases when taking a reduced and repartitioned protein diet (10–20 mg protein during the day, 50–60 mg protein taken in the evening). Nevertheless, adequate protein intake is essential or else the patients will lose weight (discussed previously).

Oral L-dopa medication in the perioperative phase (altered absorption, gastrointestinal motility, comedication with antibiotics) is discussed later.

**Physiotherapy.** Physical therapy is an important element in the management of PD. The patient with parkinsonism has difficulty initiating simple motor programs necessary for an initial movement in a sequence. The amplitude and speed of simple repetitive movements progressively decreases (fatigues) or is reduced. Simultaneous (standing up and reaching out with the hand) or sequential performance of simple motor tasks is impaired. The goal of physical therapy is to help patients recall and initiate movements and to retrain sequences of movements that used to be performed in an automatic way. Music and learned "internal" rhythms can improve rhythmical patterns of movement.

The consequences of rigidity can be reduced by isotonic movements, which help to keep joints mobile and prevent contractures. Group physical therapy also helps compensate for reduced social contacts. According to some uncontrolled observations, early initiation of physical therapy may help to minimize the drug dosage required (Comella *et al.*, 1994; Dam *et al.*, 1996). Initial supervised training should lead to self-conducted sessions at home. Regular physical activities (if possible, 1 hour of walking per day) are recommended, because it can be considered efficacious for inducing improvement for the duration of the therapy provided. It is not established whether there is any long-term benefit of such therapy once the physical therapy program has concluded. Physical exercise should not be taken to excess, because consequent stress and fatigue can be counterproductive, and in some patients the duration of the action of an individual L-dopa dose may shorten.

## Stereotactic Neurosurgery (for Details See Chapter 76)

In patients who in principle are responding to dopamimetic therapy, but symptoms cannot be controlled sufficiently any more, even by the most sophisticated pharmacotherapy or would require a drug regimen with an untolerable degree of adverse effects, functional neurosurgery should be discussed as possible therapy.

**Rest Tremor.** Stereotactic (functional) neurosurgery is an effective treatment for drug-resistant rest tremor. Two techniques are available: (1) coagulation (thalamotomy) either in the nucleus ventralis oralis posterior or intermedialis thalami (VOP/VIM) or in the zona incerta and (2) high-frequency stimulation either in the nucleus ventralis intermedialis thalami or in the subthalamic nucleus. Preconditions are:

- Marked rest tremor (preferably unilateral)
- Drug resistance to long-term and combined pharmacotherapy or unacceptable side effects of drug therapy despite effectiveness
- No other cerebral disorders such as cerebrovascular disease or dementia or major depression

**Motor Fluctuations and/or Dyskinesia/Dystonia.** Functional neurosurgery for marked motor fluctuations with or without severe dyskinesia/dystonia include pallidotomy in the internal globus pallidus and high-frequency stimulation of the internal globus pallidus or the subthalamic nucleus (Krack *et al.*, 2000). For details see Chapter 76.

## Psychosocial Support

Patients with parkinsonism tend to become socially withdrawn. Nonverbal (hypomimia, reduced body language) and verbal (monotonous, low volume) communication is impaired. Most patients feel insecure and stigmatized in public and at social activities, because either parkinsonian symptoms (especially rest tremor and dysarthrophonia) or L-dopa–induced dyskinesia can increase under psychological stress. Detailed information on the nature of PD, its course and prognosis, and the availability of many different effective therapies should be offered to both patient and caregiver. Partners and the rest of the family are all affected by the patient's PD and should be involved as much as possible. Among other benefits, this can increase understanding of the patient's symptoms, signs, moods, and their changes in response to therapy (fluctuations). Explaining the desired effects and possible side effects of drugs will increase compliance. All of this information helps keep the patient independent for as long as possible. Spouses learn to keep or regain "time for themselves," which is necessary to recuperate from the burden of caring. Depending on the stage of the disease, topics such as employment, health care (help at home, social services, moving to a home for the elderly, or home care) should be addressed (Ellgring *et al.*, 1993). PD societies or associations, many with local branches, exist in numerous countries worldwide and should be indicated to patient and spouse.

## General Principles and Recommendations for Treatment of PD

The management of PD is a multidisciplinary approach. It relies mainly on a wide range of effective drugs. In addition, surgical approaches are enjoying a renaissance. The education and support of patients and spouses (in conjunction with PD societies) and the timely provision of physiotherapy, speech and occupational therapy, dietetic, and social work inputs are also important (Oertel and Ellgring, 1995; Quinn, 1995).

### Stages of PD

The clinical signs or symptoms can be repeatedly assessed during the course of the disease by means of a variety of clinical rating scales (Columbia Rating Scale, Hoehn and Yahr Scale, Kings College Hospital Rating Scale, Northwestern University Disability Scale (NUDS), Webster Rating Scale, Unified Parkinson's Disease Rating Scale (UPDRS); for overview see Martinez-Martin and Bermejo-Pareja, 1988). In practice, the simplest is the Hoehn and Yahr scale (see Table XIV), which was designed to describe different stages in PD evolution and monitor progression and not to detect improvement caused by therapy. Despite its limitations,

it is used here as a guideline for therapeutic recommendations according to the different stages of PD. In contrast, the UPDRS is sensitive to improvements caused by therapy and has been extensively used in clinical pharmacological studies.

### Initiation of Symptomatic Drug Therapy

The decision whether to treat the symptoms of a patient with medication in the early, relatively mild disease is taken jointly by the patient and physician. The doctor's role is to explain the various available therapeutic options and their advantages and drawbacks. The patient's individual situation (age, employment, lifestyle, retirement) and expectations need to be considered. A range of effective symptomatic therapies for early PD exist, although none of the options discussed here is perfect. Our own choice of treatment depends greatly on the age of the subject. Younger patients develop fluctuations and dyskinesia to a more severe degree, and sooner, after starting L-dopa therapy. They will likely require pharmacotherapy for decades, tolerate polypharmacy better, and dementia is uncommon. In contrast, older patients develop motor complications to a milder degree and later after starting L-dopa. They may have other diseases and succumb to them or to old age before developing motor complications. They are less tolerant to polypharmacy and may have clinical or subclinical cognitive impairment (Quinn, 1995). It should be pointed out that over the past years, the choice of how to start drug treatment in early PD has become less controversial. The therapeutic options seem clearest at the extremes of age and less certain in subjects with onset 5 years either side of the 65-year median. Mainly because of the results of several long-term (2–5 years) multicenter double-blind active comparator trials on dopamine agonist versus L-dopa therapy, PD patients, if <65–70 years of age and otherwise healthy, will initially receive a dopamine agonist if available and affordable.

### Initiation of Symptomatic Drug Therapy for Akinetic–Rigid and Mixed Types of PD

Symptomatic treatment of the akinetic–rigid and the mixed type PD (see Table XV) is usually initiated in Hoehn and Yahr stage 1 or 2 with—depending on the age of the patient—one of the following options, listed in a sequence of personal preference.

**Elderly (>70 Years) Patients—The Pragmatic Approach.** When symptomatic treatment for PD is needed in otherwise healthy patients, we recommend

**TABLE XV**   Practical Guidelines for the Initiation of Symptomatic Drug Therapy for the Akinetic–Rigid and Mixed Types of PD[a]

A—Patient suffering from PD, otherwise healthy:
A-1. Elderly (≥70 years) patients*
  a. Initial treatment
    Start with L-dopa standard (levodopa plus benserazide or carbidopa) 50 mg in the morning, increase by a total of 50 mg every 1–3 days up to 100–200 mg tid, as required, in the first 3–6 months.
    A combination of standard L-dopa and controlled-release L-dopa is possible (see Table VII).
  b. Subsequent treatment
    If wearing-off develops, particularly if L-dopa daily dosage has already reached 600 mg, cautiously add a dopamine agonist,* initially covered by domperidone 20 mg tid.
  Start with either
    bromocriptine 1.25 mg (in rare cases 0.625 mg) at night, increasing by 1.25–2.5 mg/week, average dose is 2.5 (range up to 10 mg) tid
    or cabergoline 0.5–1 mg in the morning, increasing by 1 mg/week, average dose is 3 (range up to 9 mg) once a day
    or lisuride 0.1 mg at night, increasing by 0.1–0.2 mg/week, average dose is 0.4–1 mg tid
    or pergolide 0.05 mg at night, increasing 0.05 mg/day over 2 weeks to a dose of 3 × 0.25 mg, average dose is 0.5–1.5 mg tid
    or pramipexole 0.125 mg (salt) in the morning, increasing by 0.125 mg (salt) every day until 0.25 mg (salt) tid is reached, average dose is 0.25–1 mg (salt) tid
    or ropinirole 1 mg in the morning, increasing by 1 mg every day until 2 mg tid is reached, average dose is 2–8 mg tid
    If addition of dopamine agonist is not tolerated, despite domperidone, try adding COMT inhibitor (entacapone 200 mg) to each L-dopa dose or amantadine (100 mg bid) or selegeline, and avoid anticholinergics.
A-2. Young to intermediate/average age (≤70 years) patients***
  a. Initial treatment
  a-a.
    Begin with dopamine agonist with domperidone (20 mg tid) cover, as outlined under A-1b.
    Daily dosage increments and higher total dosages can often be achieved in young patients (e.g., bromocriptine up to 60 mg/day; lisuride or pergolide up to 5 mg/day, cabergoline 6–10 mg, pramipexole 3–4.5 mg/day (salt); ropinirole 24–30 mg/day)—see also **Table XI and XII.**
  a-b.
    If a more rapid, reliable effect is needed, start as described under A-1a or
  a-c.
    —although without convincing evidence of superiority to A-1a, start with capsule of Madopar CR (100/25 mg) bid or a tablet of Sinemet CR (100/25 mg) (half a tablet of Sinemet CR (200/50 mg) bid, rising to tid and, if necessary, up to 200/50 mg (2 × 100/25 mg) L-dopa CR preparation tid
  b. Subsequent treatment***
    If began with a dopamine agonist, add L-dopa standard (as in A-1a) or L-dopa slow release (as in A-2a-c).
    If began with L-dopa standard or slow release, add a dopamine agonist (as in A-1b) and try to keep the L-dopa dose as low as possible.
A-3. Before symptomatic therapy is needed—potentially in all groups but particularly in young patients (see A-2) you may.
  consider selegiline 10 mg in the morning or 5 mg bid to delay need for therapy with dopaminomimetic drugs (L-dopa, dopamine agonists) (see previous discussion on the use of monoamine oxidase inhibitors (selegiline)).***
A-4. Alternative *de novo* symptomatic therapies
  a. Amantadine 100 mg in the morning or bid*
  b. Anticholinergic, such as benzhexol (trihexyphenidyl) 2 mg bid or tid*
A-5. If initial dopaminomimetic therapy is not tolerated.
  See A-3 or A-4, followed by A-1 or A-2 after 6–12 months.
B—Multimorbid Parkinson patients:
B-1. All age groups*
  a. Initial treatment
    as outlined under A-1a.
  b. Subsequent treatment
  b-a.
    If wearing-off develops, particularly if L-dopa daily dosage has already reached 600 mg, add the COMT inhibitor entacapone 200 mg with each dose of L-dopa. If dyskinesia newly occurs, reduce individual L-dopa dose and/or extend latency to next L-dopa/entacapone intake.
  b-b. The younger the patient, the more a dopamine agonist is preferred to a COMT inhibitor. Cautiously add dopamine agonist as outlined under A-1b.

*After reaching first likely effective maintenance dosage, wait at least 4 weeks to observe development of full efficacy and adverse events; this is of particular importance for cabergoline.
[a]See also Olanow *et al.* (2001).

starting with a standard L-dopa preparation given thrice daily. This regimen is simple, cheap, effective, and widely available. It also has the best short-term therapeutic index (benefit vs side effect ratio) in this population. If and when these patients develop some "wearing off" on this regimen, we would either further increase the dosage frequency of L-dopa while trying to keep the L-dopa dose less than about 600 mg/day (see Tables VII and XV) or cautiously add a dopamine agonist, building up to a thrice or four times daily regimen (discussed in the next section), depending on the half-life of the agonist. If available, use additional domperidone, if required. If a dopamine agonist is not tolerated, a COMT inhibitor is added to each L-dopa dose. Alternatively, we would either further increase the dosage frequency of L-dopa while still trying to keep the L-dopa dose less than about 600 mg/day or would try to change to a combination of L-dopa standard and controlled-release preparation (see Tables VII and XV).

Elderly patients may have multiple pathological conditions (cardiac disease, metabolic disorders such as diabetes mellitus, gastritis, peptic ulcer) or may have intellectual impairment. In this group of multimorbid PD patients we again start the symptomatic therapy with L-dopa therapy first. If wearing off becomes manifest, we recommend a COMT inhibitor (entacapone) or an increase in dosage frequency of L-dopa as the second or third step, respectively. Later on, if tolerated, a dopamine agonist may be added. Use amantadine with caution in multimorbid patients. Avoid anticholinergics in this age group.

**Young to Intermediate Age (<70 Years) Patients.** When symptomatic treatment is needed, we would usually start with a dopamine agonist (with domperidone cover) (discussed later). However, if rapid and predictable benefit is required (e.g., job is at risk), we would instead begin with an L-dopa standard preparation. There is as yet no evidence of an advantage of controlled-release L-dopa preparations over standard L-dopa preparations in *de novo* therapy. If starting with an agonist, we would delay adding L-dopa for as long as possible. In contrast, if starting with L-dopa, we would introduce an agonist as soon (1–2 months) as possible after a sustained response to L-dopa has been achieved in an attempt to keep the dosage of L-dopa as low as possible. In multimorbid patients between 55 and 70 years of age in principle we use the same strategy as previously outlined for multimorbid patients >70 years of age.

In multimorbid PD patients among the less than 55 years (arbitrary age limit), we attempt to use, first, a dopamine agonist with domperidone cover followed by addition of L-dopa, if and when necessary. However, if marked side effects are observed, we do not hesitate to switch to L-dopa to keep the patient independent and capable of working as long as possible.

**Before Symptomatic Treatment Is Needed.** In the early disease, at a stage when symptomatic therapy may not yet be required, it is at present uncertain whether giving selegiline conveys any long-term advantage to the patient. However, although it is still not clear whether this drug has a neuroprotective effect, we nevertheless know that it can be used to delay the need for L-dopa by a mean of about 9 months and consider this therapy optional. The standard dosage is 10 mg, given as one dose in the morning or 5 mg bid.

**Alternative *de Novo* Symptomatic Therapies.** The use of either amantadine or an anticholinergic (in younger patients) can also probably delay the need for dopaminergic medication, provided these mild antiparkinsonian agents result in acceptable relief of symptoms with no or few side effects. We, however, do not often use them as initial therapy.

When stronger treatment is subsequently required, start or add L-dopa or a dopamine agonist (with domperidone cover) and proceed as previously described.

**If Initial Dopaminomimetic Therapy Is Not Tolerated.** If, despite domperidone, a patient does not tolerate initial dopaminomimetic therapy (L-dopa, dopamine agonist), try selegiline or amantadine monotherapy or (if young) an anticholinergic. After 6–12 months try to add or start a dopaminomimetic drug again (with domperidone cover).

Recommended therapy for PD of the tremor-dominant type is described later in the text and in Table XVI.

*L-Dopa—Early (Initial Monotherapy) or Late?*

Despite the shortcomings of L-dopa and the advent of dopamine agonists, levodopa plus AADC-I (L-dopa) remains the "gold standard," the single most effective symptomatic treatment for PD. What controversy there is centers on *when* it should be introduced in an individual patient (Fahn and Bressman, 1984; Lees, 1994; Markham and Diamond, 1981, 1986; Quinn, 1995). Early initiation of L-dopa therapy takes place as soon as the patient complains of mild symptoms, which are not necessarily causing functional disability. Late initiation of L-dopa therapy takes place when functional disability persists, despite the use of dopamine agonists, amantadine, anticholinergics, or MAO B inhibitors. Physicians who favor early therapy argue that L-dopa does not adversely influence the course of the disease and that late initiation of L-dopa therapy will deprive the patient of months or years of early, good L-dopa response. Early L-dopa therapy may also demonstrate to the patient the best possible effect that drug therapy can provide, and a positive result helps to support the clinical diagnosis of probable PD. On the other hand, delaying L-dopa therapy for months or even years allows the patient time to live with and accept the diagnosis before a commitment to chronic replacement therapy and allows the physician to assess the tempo of the disease process. Most patients can manage satisfactorily for at least a year after diagnosis without symptomatic therapy. A minority may even prefer to cope without medication for up to 3 years (Lees, 1994).

As discussed later, L-dopa–associated response fluctuations and dyskinesias usually occur early and severely in patients with young-onset PD or with high-dose L-dopa monotherapy (Fahn and Bressman, 1984; Kostic et al., 1991; Quinn et al., 1987), and these aspects sway most physicians toward delaying levodopa therapy in young-onset patients as long as possible.

In practice, the decision of when to start L-dopa therapy is made jointly by physician, patient, and spouse in light of the information provided by the physician. The authors recommend delaying L-dopa therapy in young-onset (<40 years of age) and intermediate-aged patients (<70 years) but have no hesitation in introducing it early in older-onset patients (>70 years). After use of non-L-dopa strategies or a dopamine agonist monotherapy in the former, one passes logically to combination therapy of the dopamine agonist with L-dopa (discussed later; see Table XV), allowing the daily L-dopa dose to be kept reasonably low. The following section discusses a number of possibilities, the choice of which depends on the individual needs of the patient (optimal vs suboptimal benefit) and the desire to spare as much L-dopa as possible. As a rule, the younger the patient, the less and the later L-dopa should be used. This rule holds also for multimorbid PD patients.

### Dopamine Agonist—De Novo Monotherapy

Patients with young-onset PD and up to the intermediate age range (70 years)(Hoehn and Yahr stages 1 and 2) should initially be treated with dopamine agonist monotherapy. This advice stems from theoretical considerations (continuous dopaminergic stimulation; Olanow et al., 2000) and is supported by a number of randomized open-label trials and more recent double-blind randomized multicenter active comparator trials on de novo patient groups. In one of the earlier open-label studies, 31 de novo PD patients received 3-year initial monotherapy of high doses of bromocriptine (average of 52 mg/day) alone to which L-dopa rescue (471 mg/day) was added later. They were compared with 29 de novo PD patients receiving L-dopa (average 569 mg/day) alone from the onset. Domperidone (60 mg/day) was used when needed to prevent side effects. Motor complications were fewer and appeared later in the group with initial bromocriptine therapy (in 56% after a mean of 4.9 years of treatment) than in the L-dopa monotherapy group (in 90% after a mean of 2.7 years of therapy). "Wearing off" was observed on average after 4.5 years in the bromocriptine/L-dopa group, whereas it occurred, on average, after only 2.9 years in the L-dopa monotherapy group. Peak dose dyskinesia occurred in 3 patients treated with bromocriptine/L-dopa compared with 14 L dopa–treated patients (Montastruc et al., 1994). Similar data have been presented by Hely et al. (1994).

Four double-blind randomized multicenter long-term active comparator studies on dopamine agonist versus L-dopa have confirmed these data for cabergoline (Rinne et al., 1998; 2–5 years), pergolide (Oertel et al., submitted; 3 years), pramipexole (Parkinson Study Group, 2000; 2 years), and ropinirole (Rascol et al., 2000; 5 years). For details, see previous sections on individual dopamine agonists.

In summary, de novo agonist monotherapy is effective and well tolerated in about half of the de novo PD patients for 2 to 3 years and thus allows a delay in the initiation of L-dopa therapy in this subgroup. It enables postponement of the onset and reductive of the occurrence and severity of L-dopa–associated side effects (wearing off, dyskinesias) (see also the following section).

### Early Combination Therapy (Dopamine Agonist Followed by L-Dopa)

De novo otherwise healthy PD patients who have successfully been initiated on dopamine agonist monotherapy will eventually need the addition of L-dopa (about 50% in the first 3 years). According to the preceding active comparator trials, these patients still benefit from the initial dopamine agonist treatment, because they show, although to a lesser degree than patients who can remain on monotherapy with a dopamine agonist, a delay in onset and a reduced degree of severity of L-dopa–associated motor complications, especially of dyskinesia compared with the group on L-dopa monotherapy (see also previous section on dopamine agonists).

Initial monotherapy with a dopamine agonist is recommended particularly for parkinsonian patients with disease onset before age 50 years, because they tend to develop marked response fluctuations and dyskinesias early in the course of L-dopa treatment. Because these patients need optimal therapy to continue working, every effort should be made to ensure that they are given an effective dosage of the dopamine agonist (under domperidone cover). To reach this dosage, however, a long time for up-titration, sometimes several months, is required. If dopamine agonist treatment is not tolerated, the patient may be switched to a low but effective dose of L-dopa. Alternatively, if high agonist doses are not tolerated, L-dopa may be added to a low dose of agonist.

### Early or Late Combination Therapy (L-Dopa Followed by Dopamine Agonist)

The addition of a dopamine agonist to L-dopa early in the course of treatment may allow the dose of L-dopa to be (1) reduced, (2) kept lower than otherwise would have been needed, or (3) kept constant for a longer period of time. The concept of early combination therapy is based on the hypothesis that the two drugs at a lower dose will be as effective as L-dopa given at high dosage, but at the same time be associated with a reduced frequency and delayed occurrence of response fluctuations and dyskinesias. The results of several studies are in keeping with this concept (Fischer *et al.*, 1984; Montastruc *et al.*, 1994; Rinne, 1985, 1989). However, definite proof is still lacking, because none of them has been carried out under strict randomized double-blind criteria. In addition, no one has ever tested whether a group treated with such early combination therapy will do better when switched back to L-dopa monotherapy than a group treated with L-dopa monotherapy from the very beginning. Nevertheless, theoretical arguments and clinical experience lend support to the early combination of L-dopa and a dopamine agonist.

The later addition of a dopamine agonist in a patient who is already fluctuating on L-dopa often will improve akinesia and rigidity. It will reduce off-time and prolong on-time, and thus the degree and frequency of response fluctuations (Hoehn, 1985). On the other hand, addition of dopamine agonists can increase mobile dyskinesias (discussed later) unless L-dopa dosage is reduced accordingly, but in general helps to reduce fixed, painful dystonia (also discussed later). In patients with very severe, long-standing response fluctuations, however, even the addition of dopamine agonists may fail to diminish the severity and duration of the "off" phase.

### MAO B Inhibitor—De Novo Monotherapy

Monotherapy with MAO B inhibitors in early PD (Hoehn and Yahr stages 1 and 2) allows L-dopa to be delayed by 9 months on average. Some, but not all, of these patients will notice a mild symptomatic effect of the drug. Side effects are very rare, apart from occasional insomnia (Allain *et al.*, 1991; Myllyla *et al.*, 1992; The Parkinson Study Group, 1993).

Selegiline may be a useful drug in the add-on treatment of L-dopa–associated motor fluctuations in PD. Many other categories of drugs exist, including dopamine agonists, COMT inhibitors, and amantadine for the treatment of motor fluctuations. The effect of selegiline may not be dramatic or long lasting. Nonetheless this drug, because of its safety profile, should be considered for patients with mild motor fluctuations on levodopa therapy if potential adverse effects are impor-tant in the selection of other adjunct therapy (Rascol *et al.*, 2002).

### Drug Therapy for Tremor-Dominant Type of PD

Drug therapy for prominent rest tremor is often un-satisfactory, and the patient should be informed about this dilemma. On the other hand, tremor-dominant parkinsonism tends to have a better motor prognosis than the mixed or akinetic–rigid type. The symptom is visible, fluctuates during the day, and is enhanced by stress. Rest tremor draws attention to the patient in public but, in contrast to akinesia and gait disturbance, it rarely causes significant functional disability, because it classically lessens or disappears on movement. Conveying these general remarks to the patient often improves compliance and increases acceptance of the symptom.

Treatment for predominant tremor at rest can be ini-tiated either with L-dopa, dopamine agonists, or alter-natively with anticholinergics if the patient is cognitively intact and not too old (see Table XVI).

Antihistamines with additional anticholinergic prop-erties may also be useful. If anticholinergic monother-apy is not sufficient, it can be combined with an L-dopa preparation and vice versa. It is important to realize that L-dopa *is* effective against rest tremor, provided a high enough dose is given. However, it also has disadvan-tages in that a low starting dose may make tremor worse, and a high effective dose may cause intolerable dyskinesias or lead to worsening of tremor, when the patient is about to turn on, or to rebound worsening of tremor when the patient turns off.

Of the different dopamine agonists, pramipexole has recently been tested in a double-blind placebo-controlled trial in patients who experienced rest tremor despite optimized pharmacotherapy. It showed a marked tremorlytic effect at standard doses (mean, 3 mg/day) (Pogarell *et al.*, 2002). Whether other dopamine agonists have a similar property remains to be studied, preferentially in active comparator monotherapy trials against pramipexole or L-dopa, because so far no controlled study has indicated whether any of the dopamine agonists influences rest tremor better than L-dopa.

Amantadine possesses some antitremor effects. Tri-cyclic antidepressants can improve rest tremor by means of their anticholinergic or sedating effect and can also be combined with L-dopa or anticholinergics. Budipine (discussed later), an NMDA receptor antagonist, can also reduce rest tremor. In individual cases and for special occasions, the administration of a short-acting benzodiazepine is justified.

Several groups also recommend a trial of low-dose clozapine (6.25–37.5 mg/day; Friedman and Lannon,

**TABLE XVI** Pharmacotherapy of Rest Tremor

| Rest tremor | Substance (generic) | Dose |
|---|---|---|
| **Nearly exclusively rest tremor** | | |
| Anticholinergics | Biperiden | 3 × 2–4 mg |
| | Bornaprine | 3 × 2–4 mg |
| | Methixine | 3 × 2.5–5 (–10) mg |
| | Trihexyphenidyl | 3 × 2–5 mg |
| L-dopa | See Table VII, XV | |
| Dopamine agonist** | See Table XII, XV | |
| Amantadine* | | 2–3 × 100 mg |
| Antihistamine* | Diphenhydramine | 2 × 25–50 mg |
| | Orphenadrine | 1–2 × 200 mg |
| **Rest tremor and mild akinetic rigid syndrome** | | |
| L-dopa** | See Table VII, XV | |
| Dopamine agonist** | See Table XII, XV | |
| Amantadine* | | 2–3 × 100 mg |
| **If emotion or psychological stress markedly enhances rest tremor** | | |
| Addition of β-blockers (see later) | | |
| Sedating antidepressants | | |
| **Resistant to preceding therapy** | | |
| Budipine | | 3 × 10–30 mg |
| Clozapine | | 12.5–37.5 mg |
| **Postural tremor component** | | |
| Addition of β-blockers | Propranolol | 3 × (20)–40 mg, if not effective |
| | Metoprolol | 1–2 × 50–200 mg |
| Try | Primidone | 25–100 mg/day |

1990) as a last resort for rest tremor before turning to stereotactic surgery.

If a marked action or postural tremor component is present, a trial of a nonselective β-receptor blocker, such as propranolol 20–40 mg tid, or primidone, 50–200 mg/day (best given at night), can be helpful; sometimes this can also alleviate rest tremor (Marsden and Owen, 1970, personal observation). Drug therapy for rest tremor is summarized in Table XVI.

## Experimental Therapy

### Experimental Drug Therapy

L-**threo DOPS.** L-threo-DOPS is a precursor of noradrenaline. Peripheral metabolism of this drug may attenuate postural hypotension. It has been proposed that parkinsonian freezing may be due to the central noradrenergic deficit arising from pathology in the locus ceruleus. It has been claimed that L-threo-DOPS, by correcting this deficit, improves freezing in PD (Kondo,

1993). However, whether peripherally administered L-threo-DOPS can reliably raise central noradrenergic transmission and whether it has any proven significant effect on freezing remain controversial issues.

### Cerebral Transplantation of Dopamine-Synthesizing Cells

See Chapter 76.

### Pallidotomy and High-Frequency Stimulation of the Pallidum or Subthalamic Nucleus

See Chapter 76.

## Medication of Unproven Efficacy

N-adenosine-dinucleotide (NADH) therapy and oxy-ferriscorbone ("iron") infusion have not been found to be effective.

**TABLE XVII**  Classification of Response Fluctuations and Involuntary Abnormal Movements

| | |
|---|---|
| Hypokinetic phenomena | |
| Related to individual doses (predictable) | Medium duration |
| | End-of-dose akinesia |
| | Wearing-off |
| | Early morning akinesia |
| | Afternoon akinesia |
| | Nocturnal akinesia |
| Unrelated to individual doses (unpredictable) | Short duration |
| | Freezing[a] |
| | Paradoxic akinesia[a] |
| | Paradoxic kinesis[a] |
| | Medium duration |
| | Paroxysmal on–off |
| | Rapid oscillations |
| | Sudden switching-off |
| | Yo-yoing |
| Hyperkinetic phenomena | |
| Therapy related | Mobil dyskinesias, usually painless—Choreic, dystonic, or choreodystonic |
| | Peak-dose dyskinesias (dyskinesias at maximal mobility) |
| | Plateau or square-wave dyskinesias (on-phase dyskinesias) |
| | Biphasic "beginning" and "end-of-dose" dyskinesias |
| | Slow or fixed dyskinesia, dystonia, usually painful |
| | End-of-dose or off-period dystonia |
| | Early morning dystonia |
| | Biphasic dystonia |
| Disease related | Dystonia[a] |

*Note.* See also Figure 8.
[a]Disease-related phenomenon.

## Special Aspects of Therapy

### Fluctuations of Parkinsonian Symptoms and Response Fluctuations Caused by Therapy

Fluctuations of parkinsonian symptoms can be either disease or therapy related. A classification of these fluctuations is found in Table XVII. Two major states are differentiated: "on" and "off." When "on," the patient is mobile and—later in the course of the disease—often dyskinetic, and symptoms such as akinesia, rigidity, or rest tremor are mild or absent. The "off" state sees a return of parkinsonism, often accompanied by other unpleasant nonmotor symptoms.

*Disease-related fluctuations* are paradoxic kinesis (sudden good mobility in extremely emotional situa-tions—anxiety, excitement) and freezing (paradoxic akinesia, "start hesitation"). *Freezing* describes a sudden short-term immobility; for example, when passing through a door. Freezing can also be stress induced.

*Drug therapy–related response fluctuations* include wearing-off, a predictible motor phenomenon related to individual drug doses, and unpredictable motor phenomena with no obvious relation to individual doses such as paroxysmal on–off phenomena. The latter are characterized by rapid switches between marked akinesia and good (usually excessive) mobility. Response fluctuations related to L-dopa pharmacokinetics (wearing-off, end of dose akinesia) are due partly to the progressive degeneration of nigral neurons leading to a reduced presynaptic storage and buffering capacity (see Figure 5); evidence suggests that postsynaptic changes may also be important (Bravi *et al.*, 1994).

Any additional factor influencing cerebral availability of L-dopa, such as impaired gastrointestinal absorption (retarded emptying of the stomach, either caused by disease or enhanced by L-dopa, dopamine agonists, or anticholinergics) or impaired transport of L-dopa by means of the duodenal mucosa into the blood or by way of the BBB into the CNS may contribute to "end of dose akinesia." Furthermore, neutral amino acids interfere with the membrane-bound carrier system of L-dopa. Thus, intake of L-dopa with a protein-rich meal results in its diminished absorption into the bloodstream and impaired transport into the CNS (see the discussion of a low-protein diet). Wearing-off can eventually be replaced by the paroxysmal on–off phenomenon. These random oscillations are partly due to pharmaco-dynamic postsynaptic changes (Bravi *et al.*, 1994; Fabbrini *et al.*, 1988; Mouradian *et al.*, 1987, 1988, 1989a,b, 1990) but are also partly iatrogenic, as a consequence of giving progressively smaller individual doses of L-dopa.

During long-term L-dopa therapy, the thresholds for its antiakinetic and dyskinetic effects approach each other. Therefore, the therapeutic window for L-dopa to induce good mobility without dyskinesias becomes progressively narrower, and eventually mobility is achieved only at the expense of concomitant dyskinesia (on-phase, plateau, or square-wave dyskinesias; see Figure 8). All response fluctuations (see Figures 5–9; Table XVII) increase in intensity and frequency during the course of the disease under L-dopa therapy.

**End-of-Dose Akinesia (Wearing-Off).** The first signs of the transition from a constant drug response (long duration response of the "levodopa honeymoon"; see Figures 3 and 5) to drug-related response fluctuations are increases in nocturnal akinesia, early morning akinesia before the first drug intake, or afternoon akinesia. Later on, gradual waning of the benefit from each L-dopa dose ("end-of-dose" akinesia or wearing-off) will occur just before the next L-dopa dose is due or even just after the L-dopa has been taken. Frequently, phases of marked immobility are observed in the early afternoon as part of physiological diurnal fluctuations or because of impaired absorption of L-dopa after meals.

Suggested therapeutic options include the following:

- Take L-dopa well (30–60 minutes) before meals.
- Add dopamine agonist.
- Add COMT inhibitor.
- Add selegiline.
- Change to a combination of L-dopa standard and L-dopa controlled-release preparation.
- Shorten the intervals between L-dopa administration and reduce size of individual doses to avoid overdosage (NOTE: this may result in a less-predictable response, because smaller doses have a lesser chance of producing the threshold L-dopa level required to "turn on").
- Take the first L-dopa dose immediately on rising.
- Eat a low-protein diet during the day.
- Take L-dopa as a dispersible or liquid formulation for early morning or afternoon akinesia.
- Add amantadine.

**Paroxysmal On–Off Phenomenon.** This term refers to sudden loss of efficacy over seconds or minutes and a similarly rapid reappearance of benefit.

Often there is no obvious relationship to the time of drug intake, but sequential on–off charting over a week or two may reveal such a pattern. This change between "on" and "off" can occur several times a day. Synonyms for the paroxysmal on–off phenomenon are *random oscillations*, *yo-yoing*, and *sudden switching-off*. Because this phenomenon occurs in advanced stages of the disease, on-phases are usually accompanied by dyskinesia; therefore the patient experiences a rapid transition from dyskinesia in "on" to akinesia in the "off" phase.

Suggested therapeutic options (they are limited) include the following:

- See the preceding discussion of wearing-off for different therapeutic options.
- Fewer, higher doses of L-dopa may be preferable to more frequent, lower doses, because at this stage of the disease the effect of higher individual doses of L-dopa is much more predictable for the patient.
- Administer apomorphine by subcutaneous intermittent injections or continuous infusion (mini-pump).
- Administer duodenal infusion of levodopa or

levodopa-methylester in selected cases (Sage *et al.*, 1988), although this is not practical in most patients.

**Freezing.** Freezing describes very short-lived, rapid-onset gait disturbances with start hesitation. The patient is suddenly unable to walk "as if the feet are nailed to the ground." Freezing quite often leads to falls, especially forward onto the knees. Freezing is briefer than the paroxysmal on–off phenomenon and more likely in confined spaces, in doorways, on turning, if suddenly interrupted, or if trying to do two movements at the same time. Freezing is rarely observed in the early years of disease during L-dopa treatment or agonist treatment (Giladi, 2001). It can often be overcome by visual or acoustic stimuli, such as marks on the floor or a horizontal attachment to the bottom of a walking stick. It is important to recognize two distinct forms of freezing. In some patients, freezing occurs exclusively in "off" periods or at times of wearing-off and can therefore be improved by the usual measures to increase time "on" (discussed earlier in this section and in the next section). The second and, unfortunately, more common type of freezing occurs randomly (not related to doses or drug intake) and is rarely improved by adjustments in therapy (see also the discussion of L-threo-DOPS previously).

### Abnormal Involuntary Movements (Dyskinesias)

Abnormal involuntary movements (other than tremor) in PD can occasionally be related to the disease itself but are much more often drug induced. The occurrence of dyskinesia is presently explained by the concept of pulsatile versus continuous (tonic) stimulation of postsynaptic dopamine receptors in the striatum (Olanow *et al.*, 2000). Two main categories are seen: mobile and fixed dyskinesias.

**Mobile Dyskinesias.** Mobile dyskinesias (hyperkinesias) may be choreic, dystonic, or choreodystonic (choreoathetoid) and are painless. They predominantly affect the extremities, less frequently the neck and trunk, and can sometimes arise even in the first year of L-dopa therapy, but even rarely under dopamine agonist monotherapy. The longer the duration of the (pulsatile) L-dopa treatment, the higher is their frequency and severity. Treatment with high L-dopa doses seems to promote their occurrence. Dyskinesias tend to appear first at the time of the highest L-dopa plasma level (peak dose dyskinesia), when they can often be reduced or abolished by reducing the L-dopa dose. Drugs to treat dyskinesias (such as classical neuroleptics) are not useful, because they also exacerbate parkinsonism.

In the advanced stages, dyskinesias occur during the whole time of mobility (on-phase dyskinesias, square-wave dyskinesias, plateau dyskinesias) and disappear only when the off-phase appears. Rarely painless choreic, dystonic, ballistic, or stereotypic (mobile) dyskinesias occur at the beginning or end of the on-phase or both (biphasic dyskinesias; see Figure 8). Biphasic dyskinesias may also be fixed (painful dystonias; see the next section). One might expect biphasic dyskinesias to be reduced by overlapping oral doses of L-dopa, but usually they are made worse. Peak-dose (and occasionally biphasic) dyskinesias have been reported to decrease under high-dose dopamine agonist therapy (continuous dopamimetic stimulation), allowing concomitant marked reduction of L-dopa or with continuous parenteral infusion of L-dopa, lisuride, or apomorphine (Chase *et al.*, 1993; Colzi *et al.*, 1998; Stocchi *et al.*, 1988) and with duodenal infusions of L-dopa (Sage *et al.*, 1988). Treatment with CR L-dopa can sometimes reduce peak-dose dyskinesias (Goetz *et al.*, 1989) but often markedly increases biphasic dyskinesias (see controlled release L-dopa preparations). If modifications in pharmacotherapy cannot control dyskinesia without reducing total on-time, functional neurosurgery (see Chapter 76; thermocoagulation, high-frequency deep brain stimulation) is recommended in selected cases.

Suggested therapeutic options include the following:

- Discuss with the patient whether the dyskinesias are an acceptable "price" to pay for mobility, a mild to intermediate degree of dyskinesias often bothers the spouse more than the patient.
- Discuss whether the patient prefers increased time "on" with dyskinesia or less time "on" with less dyskinesia.
- Suggest intake of drugs with meals (may help peak-dose, or precipitate biphasic dyskinesias).
- Add a dopamine agonist, if not already prescribed, to reduce the amount of L-dopa. If tolerated, increase dopamine agonist as high as possible to reduce the amount of L-dopa to a minimum.
- Add amantadine, 100–200 mg bid or tid, to reduce severity of dyskinesia and at the same time to prolong on-time.
- COMT inhibitor. This may initially increase dyskinesia. Reduce or/and adapt individual L-dopa dose or latency to next intake.
- Try CR L-dopa preparations (see the earlier discussion of CR L-dopa. Caution: peak-dose, and especially biphasic, dyskinesia and dystonia can increase).
- MAO B inhibitors may increase peak-dose dyskinesias; reduce or withdraw such agents.

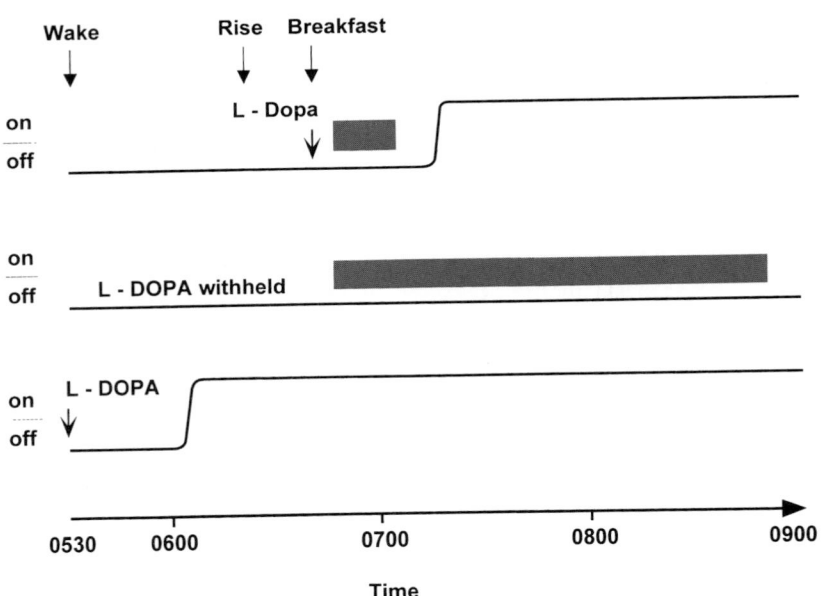

**FIGURE 9**   Early morning dystonia (indicated by dark areas). This patient's early morning dystonia regularly occurred just after taking his first L-dopa dose at breakfast, lasted 20 minute, and resolved before he turned "on" (top). When he omitted his L-dopa medication, the dystonia still occurred but was more severe and lasted 3 hours (center). When he took his first L-dopa very early, in bed, he was free of dystonia on rising. (From Marsden *et al.* (1982), with permission.)

- In emergency cases (extremely prolonged dyskinesia), use liquid haloperidol (5–30 mg) in increasing doses until dyskinesia is blocked.

**Dystonia—Fixed or Slow Dyskinesia.** Although dystonia is very rarely seen in untreated PD, some instances of spasmodic torticollis or a dystonic foot have been reported 0.5–25 years before parkinsonism became manifest. In contrast, dystonia, sometimes painful and fixed, frequently occurs under chronic L-dopa therapy. Dystonic "spasms or cramps" tend to be biphasic, occurring at the beginning and the end of the dose. They also can be monophasic at either the beginning or end of L-dopa/dopamine agonist action (see Figure 8).

A frequent example is the early morning dystonia in toes or foot after rising (see Figure 9) or dystonia occurring just before the next L-dopa intake as akinesia and rigidity are returning. Dystonia presents typically as focal dystonia in the form of involuntary flexion, extension, or fanning of the toes (striatal toe, not to be confused with a Babinski response), often with plantar flexion and inversion of one or both ankles (foot dystonia). Any other focal dystonia (e.g., arm dystonia, torticollis) is possible. In patients with hemiparkinsonism or marked asymmetry, dystonia in general is more pronounced on the more affected side. Pure dystonia in the "on" period is less common, and it should always be clarified whether it is not instead a sign of biphasic dystonia.

Fixed dystonia subsides as soon as the patient turns on, and therefore in general responds promptly and well to a subcutaneous injection of apomorphine (2–5 mg) (see the earlier discussion of apomorphine). If apomorphine is not available, a dispersible or liquid form of L-dopa may be an acceptable (sometimes less satisfactory) alternative. Treatment with classical neuroleptics or tetrabenazine may abolish dystonia but turns the patient fully "off." Good results have been described with the addition of long-acting dopamine agonists to L-dopa medication. Amantadine may also help to reduce dystonia.Some patients may respond favorably to baclofen (5–40 mg/day). Local intramuscular injection of botulinum toxin A can help in some cases of marked drug-related focal dystonias such as spasmodic torticollis (Gasser et al., 1991) and the striatal toe. Anticholinergics sometimes help (Poewe et al., 1988b). If all else fails, lithium carbonate can be effective for off-period painful dystonia (Quinn and Marsden, 1986).

Suggested therapeutic options include the following:

- Take the first L-dopa morning dose as early as possible.
- For early morning dystonia, try dispersible or liquid L-dopa, if available.

- Take a nocturnal CR L-dopa preparation.
- Take a dopamine agonist at bedtime, preferably a long-acting one.
- Take a combination of L-dopa with a long-acting dopamine agonist.
- Add amantadine.
- Administer a subcutaneous apomorphine injection.
- Take a daytime CR L-dopa preparation (Caution: dystonia may actually be enhanced or its duration prolonged).
- Take anticholinergics.
- In individual cases, administer a local IM injection of botulinum toxin A.
- If all else fails, administer lithium carbonate.

### Akinetic Crisis

Akinetic crisis is a life-threatening state. In contrast to the short-term off-phases in advanced Parkinson's disease, lasting minutes to hours, akinetic crises last for days. They usually build up over days to weeks but can develop rapidly over the course of 24 hours. The patient may be completely immobilized, unable to walk or stand, with rigid limbs fixed in a flexed position. Continuous resting tremor may be observed. Speech may be unintelligible, and patients may become dehydrated, because they are often unable to eat or drink. The risks of deep venous thromboses, pressure sores, urinary tract infections, or pneumonia are high in patients with akinetic crisis. Hyperthermia can also occur (discussed in the next section) analogous to the neuroleptic malignant syndrome. Marked weight loss and disturbance of fluid and electrolyte balance and secondary cardiovascular and respiratory changes may be present. This acute exacerbation of parkinsonism may be either disease or therapy related. The latter includes underdosage of dopaminergic drugs, withdrawal of antiparkinsonian medication (drug holidays, perioperative phase, administration of antibiotics), or secondary absorption failure (e.g., caused by ileus, swallowing difficulties, gastroenteritis). Any other severe infection such as pneumonia or appendicitis should be searched for. The most frequent misdiagnoses are malignant neuroleptic syndrome, catatonia, major depression, or akinetic mutism.

Suggested therapeutic options include the following:

- If akinetic crisis is the result of withdrawal of L-dopa, treatment with levodopa should be restarted (orally or by dissolved dispersible or crushed tablet or opened capsule through a nasogastric tube) at an identical or even higher dose than before. Doses should be gradually increased to an effective dose over 1–2 days.
- If akinetic crisis is due to underdosing of levodopa, its dose should be increased daily by 100–300 mg until a response is seen.

- Intensive care conditions with facilities for artificial ventilation should be available.
- Apomorphine in a single subcutaneous bolus injection of 2–10 mg is effective within 10–15 minutes, and its action lasts 30–60 min (maximally 120 min), depending on dosage. If this approach is effective, a continuous subcutaneous infusion can be used (discussed previously). Apomorphine given by a SC route should be infused at an initial rate of 1–2 mg/h, with increases of 1–3 mg/h every 12 hours. Some groups consider a 8- to 12-hours break at night, if the condition of the patient allows. Maximal daily infused amounts have been up to 170–360 mg apomorphine. Potential side effects include nausea, vomiting, orthostatic hypotension, bradycardia, and other dysrhythmias. Concomitant treatment with domperidone is recommended to prevent peripheral side effects but is possible only if a nasogastric tube is in place. Ideally, domperidone, 20 mg tid, should be given orally for 24 hours before apomorphine, but in an emergency, one can give 50–60 mg 30–60 minutes before apomorphine. Domperidone may not be necessary in patients already established on L-dopa, dopamine agonist long-term therapy, or both.
- Intravenous amantadine infusions are available and used only in Austria, Germany, and Switzerland (as discussed previously).
- For severe akinesia, levodopa can be administered IV at a dosage of 1–2 mg/kg/h. Concomitant therapy with AADC-I such as benserazide or carbidopa is suggested for optimal efficacy, although this is not necessary or possible in emergencies, because they have to be given through a nasogastric tube.
- Levodopa methylester (LDME) is an alternative and can be given in a smaller, more concentrated volume (equivalent to 200 mg L-dopa/ml (see discussion of IV administration of L-dopa). These intravenous preparations can also be given by nasogastric tube, but are not commercially available and have to be ordered specially from the pharmaceutical companies (levodopa from Hoffmann-LaRoche, Basel, Switzerland; LDME from Chiesi Pharmaceuticals, Milan, Italy, or its local distributors).
- Ancillary measures include parenteral fluids with electrolyte and caloric substitution, antithrombotic prophylaxis (low-dose heparin SC and elastic stockings), pneumonia prophylaxis (physiotherapy), and skin care (including regular turning).

**Malignant Levodopa Withdrawal Syndrome—Hyperthermia.** The rare malignant levodopa withdrawal syndrome, a variant of akinetic crisis, occurs when the drug is abruptly stopped or its dose quickly

**TABLE XVIII**    Management of the Akinetic Crisis and Malignant Levodopa Withdrawal Syndrome[a]

1. Provide dopamine substitution
   Apomorphine infusion SC (discussed in text)
   Levodopa infusion IV (discussed in text)
   Amantadine infusion (in Austria, Germany, and Switzerland only)
   Restart oral L-dopa (100–200 mg every 2–4 hr) or dopamine agonists (see Table XV, initially bromocriptine 2.5–10 mg, lisuride 0.1–0.4 mg, pergolide 0.25–0.5 mg, pramipexole 0.25–0.5 mg)
2. General supportive measures
   IV fluids, electrolytes, and calories
   Low-dose heparin
   Antibiotics (after cultures), if evidence of infection
3. Whole-body cooling (fever >40°C)
4. Dantrolene sodium in cases with marked creatine kinase elevations (effectiveness not proven, optional)
   Start orally (4 × 25–50 mg/day) or parenterally with 2.5 mg/kg IV infusion followed by 5–10 mg/kg/day IV as infusions or 4–5 bolus injections

[a]Modified from Poewe and Oertel (1994).

reduced. It has also been reported in patients with impaired gastrointestinal absorption or during drug holidays (discussed previously). Physical findings are as for akinetic crisis plus marked hyperthermia, tachypnea, tachycardia, and mental status changes such as confusion, disorientation, hallucinations, or somnolence. Serum creatine kinase (CK) concentrations may be markedly raised, often *without* increase of the erythrocyte sedimentation rate. The condition can be misdiagnosed as a combination of akinetic crisis plus infection or as neuroleptic malignant syndrome (NMS).[2]

Suggested therapeutic options (see Table XVIII and the preceding section) include the following:

- Neurological intensive care is essential.
- Physical measures for whole body cooling are required.
- Other causes of fever or tachycardia should be excluded, such as infection, pulmonary embolism, or myocardial infarction.
- The acute worsening of PD should be treated with L-dopa or levodopa methylester by means of nasogastric tube (or with IV levodopa or levodopa-methylester) or by SC apomorphine.
- In Austria, Germany, and Switzerland, amantadine infusions may be used.
- As soon as possible, patients should be restarted on their previous daily oral dose of L-dopa or dopamine agonist.
- Dantrolene sodium may be considered for patients with markedly increased CK. Its efficacy, however, has not been evaluated in controlled trials.

**TABLE XIX** Management of Dopaminergic Drug-Induced Psychosis in PD

Consider whether the patient has dementia of the Lewy body type
1. Assess hydration of patient, administer fluids orally or parenterally
2. Reduce antiparkinsonian agents
   Reduce or, if possible, gradually withdraw dopamine agonists, selegiline, anticholinergics, amantadines, and tricyclic antidepressants, usually in that order
   Finally, if necessary, reduce L-dopa to minimum effective dose
   One should be cautious, because abrupt discontinuation of these medications (especially of anticholinergics, amantadines, or tricyclic antidepressants with an anticholinergic component) can result in a withdrawal syndrome and worsening confusion
3. Add antipsychotic agents
   Try quetiapine, 25 mg at night, in mild cases or with 50 mg in more severe cases, increase by 25 mg every second or third day, depending on psychosis control, up to a maximum of 100–125 mg bid
   If psychosis is not controlled after about 2–4 weeks, substitute quetiapine with clozapine in a ratio of 2–3 : 1;
   if clozapine is chosen as first-line drug, start with 6.25–12.5 mg at night in mild cases; increase up to 100 mg/day (3–4 × 25 mg/day to 2 × 50 mg/day) in severe cases; weekly blood counts are obligatory to detect the development of agranulocytosis
   if neither drug is tolerated or fails to be effective, try olanzapine, risperidone, later sulpiride, thioridazine
   if the psychosis is refractory, add classical neuroleptics
   For example, haloperidol 3–10 mg/day, even higher, if necessary
4. Take general measures to prevent complications
   Low-dose heparin (immobilized patients)
   Consider broad-spectrum antibiotics in immobilized febrile patients

tive confusional state. Dopamine agonists should be withdrawn and reintroduced as outlined previously for L-dopa; however, respective systematic studies of any kind, especially in respect to the different half-life of the available ergot and nonergot dopamine agonists, are lacking. For simplification of the postoperative regimen, dopamine agonists can be substituted with a mild increase in oral L-dopa dosage. Anticholinergics should not be withdrawn abruptly before an operation, because an anticholinergic withdrawal syndrome with marked deterioration of parkinsonism and confusional state may occur, even with a latency of 2–3 days postoperatively. Therefore, if possible, anticholinergics should be gradually reduced or withdrawn some weeks before surgery. Rapid withdrawal of amantadine can also lead to a marked deterioration of parkinsonian symptoms and a confusional state.

No interference with anesthesia has been ascribed to MAO B inhibitors, but because they may influence the metabolism of peripheral monoamines, they may, for theoretical reasons, be withdrawn about 6–8 weeks before surgery and reintroduced postoperatively.

**Parenteral Medication.** Parenteral therapy with antiparkinsonian agents is sometimes necessary; for example, after an extensive abdominal operation. The same therapeutic measures are recommended as listed in the discussion of akinetic crisis (see Table XVIII). If the patient has a nasogastric tube, it is possible to administer all antiparkinsonian agents through it, apart from apomorphine.

## Surgery

Major surgery and general anesthesia can both precipitate a dopaminergic psychosis, which in turn is more likely if the patient is elderly or on polytherapy, particularly including dopamine agonists, amantadine, or anticholinergics. These psychotic or confusional episodes usually settle spontaneously with sufficient hydration (if not, see the discussion of dopaminergic psychosis).

**Oral Medication.** Oral L-dopa medication should be continued preoperatively up to the evening before the planned surgery (if surgery is in the afternoon, an early morning dose is justified). Postoperatively L-dopa therapy should be initiated as soon as possible, but often the preoperative dose is less effective. This may be due to altered absorption of L-dopa or gastrointestinal motility, to infection, or to comedication with an antibiotic; for example, a macrolide (Brion *et al.*, 1992). Any compensatory increase in L-dopa dosage must be cautious because of patients' susceptibility to a postopera-

## Anesthesia

Classical neuroleptics (dopamine $D_2$-receptor blockers) are contraindicated. In principle the choice of general, regional (epidural), or local anesthesia should be made according to surgical requirements and the experience of the anesthetist. For general anesthesia, a combination of opiates and benzodiazepines is possible. To introduce anaesthesia good, although limited, experience exists with the hypnotic substance propofol. Inhalation anesthetics can be combined with nitric oxide. Among the halogenated inhalation anesthetics, enflurane or isoflurane sensitizes the myocardium to catecholamines less than halothane.

Regional anesthesia carries the risk of intraoperative and anesthesia-related hypotension, which should be avoided by careful monitoring.

Local anesthesia should ideally not include the addition of adrenaline, a recommendation that dates back to the time when levodopa preparations did not contain AADC-I. Octapressin can be used as an alternative addition to local anesthesia for example in dental surgery.

Patients often report a deterioration of parkinsonism after surgery under general or regional anesthesia that occurs independently of the use of classical neuroleptics, the cause of which is unknown.

### Difficulties in Swallowing

Impaired swallowing is admitted by up to 50% of PD patients, if directly asked. Technically (barium swallow, esophageal pressure assessment), 95% of patients show some type of impairment in swallowing (Fischer, 1984). This complaint is probably due largely to degenerative changes in the dorsal vagal nucleus, but Lewy body pathology in the esophageal plexus may contribute as well.

Suggested therapeutic options in severe cases are the following:

- Change to dispersible or liquid L-dopa.
- Provide percutaneous endoscopic gastrectomy.
- Open capsules of L-dopa or grind tablets and administer the L-dopa powder or the suspension through a nasogastric tube.
- Administer apomorphine by injection or pump.

### Autonomic Nervous System–Related Symptoms

**Orthostatic Hypotension.** Impaired regulation of cardiovascular function, especially orthostatic hypotension, frequently occurs in the late stages of PD. However, cardiovascular dysfunction early in the disease should be a "red flag" to consider the possibility of MSA, in which autonomic failure occurs earlier and more commonly than in PD (see Table III). Supine arterial hypertension may be seen in about 25% of patients who have orthostatic hypotension on rising.

Orthostatic hypotension is often first noticed when L-dopa or dopamine agonist therapy is introduced. Domperidone (10–20 mg tid) can sometimes reduce or abolish this early side effect, and after several weeks,

**TABLE XX    Therapeutic Measures for Orthostatic Hypotension**

Head-up tilt of bed at night
Elastic stockings or tights
Increased salt intake
Ephedrine 3 × 15–45 mg/day
midodrine 3 × 2, 5–10 mg/day
octreotide (?)
yohimbine (?)
L-threo-DOPS (?),
Fludrocortisone 0.05–0.3 mg/day

domperidone can be reduced and finally withdrawn as tolerance develops. The postural hypotensive symptoms can be helped by using elastic support stockings or tights, a high-salt diet, a head-up tilt of the bed at night, and by rising slowly from a sitting to a standing position. Therapy with ephedrine, midodrine, octreotide, and yohimbine has given inconsistent results. If these measures fail, the mineralocorticoid fludrocortisone, 0.05–0.3 mg nightly, may be given. L-threo-DOPS, a precursor of noradrenaline, currently registered only in Japan, has been claimed to substantially improve orthostatic hypotension (see Table XX and discussion of MSA therapy in Chapter 75). In extremely rare cases, arterial hypertension has been induced by L-dopa or a dopamine agonist (for example by bromocriptine, cabergoline, lisuride, or pramipexole). However, L-dopa and bromocriptine usually cause a moderate reduction in mean blood pressure. If hypertension is present in a patient with untreated PD, it will often come under control when PD is treated with dopamimetics, obviating the need for any specific antihypertensive treatment.

**Bladder Function.**    About two thirds of PD patients have urinary disturbances. Age- and gender-related disturbances, such as consequences of traumatic childbirth in women and hypertrophy of the prostate gland in men, also need to be considered. Symptoms include urgency, frequency, and nocturia and sometimes hesitancy. Incontinence of urine is much less common in PD and usually a late feature, and it is often secondary to akinesia, slowing the necessary preparations for passing water. Cystometric and urodynamic investigations are recommended before medical therapy is introduced. Two types of detrusor dysfunction, hyperactivity and hypoactivity, can be observed.

Imipramine, 50–100 mg/day, or oxybutynine, 2.5–5 mg bid to tid, reduce the much more frequent detrusor hyperactivity (increased frequency and urge incontinence). However, too high a dose of the peripherally acting anticholinergic may instead precipitate urinary retention caused by detrusor atony. Detrusor hypoactivity is rare in PD and, when seen, in fact may be caused by anticholinergics or tricyclic antidepressants. Whether cholinomimetic substances, such as carbachol, or choline esterase-inhibitors, such as distigmine bromide, donezepil, or rivastigmine, can worsen parkinsonism has not been carefully studied, so these substances should be administered with caution, if at all. Finally, urinary incontinence or retention always calls for careful reconsideration of a diagnosis of PD (see Chapter 75 on multiple system atrophy).

A substantial postmicturition residue of >150 ml is an indication for clean intermittent self-catheterization (or catheterization by the spouse). In the very advanced stages, a urethral or suprapubic catheter may become necessary.

**Male Sexual Function.** Disturbance of male sexual function may be drug induced (anticholinergics, tricyclic antidepressants, β-blockers). However it can also occur spontaneously in some young PD patients < age 50. The most frequent complaint is of impaired erection rather than difficulties with ejaculation. However, impotence (erectile dysfunction) is often a very early feature of MSA, which should be considered as a possible differential diagnosis to PD (see Chapter 75). Endocrine parameters are unchanged in PD. Beneficial effects of L-dopa or dopamine agonists are sometimes reported by patients. Occasionally, an aphrodisiac effect with increased libido (but often impaired performance) is observed with L-dopa and dopamine agonists.

Erectile failure can be improved by oral yohimbine (2 mg, 1 hour before intercourse), by intracavernosal injection of papaverine, a penis implant, or more recently by sildafenil (Zesiewicz *et al.*, 2000), or sublingual or SC apomorphine. Rarely, dopamine agonists can cause priapism (for further therapeutic options, the discussion of MSA therapy, Chapter 75; Koller *et al.*, 1990).

**Gastrointestinal Function.** In addition to impaired motility of the esophagus, emptying of the stomach is often retarded. Therapy with L-dopa, dopamine agonists, and compounds with an anticholinergic component (anticholinergics, tricyclic antidepressants, amantadine, antihistamines) can augment this symptom, leading to reduced bioavailability of drugs.

Disorders of intestinal function, especially constipation, increase with age. About one third of PD patients less than and two thirds greater than age 60 complain about this symptom, the severity of which also increases with age. Anticholinergics and amantadine increase the frequency and degree of constipation. Physical measures (high fluid and fiber intake, exercise) are recommended. Domperidone (see discussion of antiemetics) should be tried. Cisapride, a substance apparently acting independently of the dopamine system, seems to be of (short term) help (Jost and Schimrigk, 1993), but the compound occasionally worsens parkinsonism. Polyethyleneglycol (Macrogol) is an inert substance. It increases water content of the stool, thus softens it and thereby improves defecation, and reduces pain associated with defecation. It can improve gut function, even in patients who have defecation only once every 2 weeks. Initial treatment is started with one half to three bags dissolved in water and drunk before the meal. After several weeks, dosage can be adjusted to one half to one bag a day or even one half a bag every other day (Eichhorn and Oertel, 2001). Controlled long-term data are lacking; however, clinical experience shows maintained efficacy over more than a year in most of the patients.

**Sialorrhea.** Sialorrhea is a frequent complaint caused by akinesia of swallowing. However, basal and reflex production of saliva is reduced about 50% in PD patients compared with controls (Fischer, 1984). Anticholinergics, including peripherally acting anticholinergics such as atropine derivatives, can reduce hypersalivation in many patients. Like L-dopa, dopamine agonists can improve this symptom by reducing akinesia. When severe and resistant to changes in medication, this distressing symptom can sometimes be improved by irradiating the salivary glands or by local injection of botulinum toxin.

**Thermoregulation, Seborrhea, and Melanoma.** Impaired thermoregulation can lead to hyperthermia on very hot days in patients with advanced PD, in part caused by impaired sweating, sometimes combined with marked L-dopa–induced dyskinesia (see also the discussion of dyskinesia under CR L-dopa preparation). Administration of anticholinergics or amantadine is then contraindicated, because they reduce sweat production and may lead to an increase in body temperature. In such a rare case, immediate admission to the hospital is mandatory to introduce physical measures to reduce temperature, if possible under intensive care conditions (discussed previously).

Oily skin and seborrhea may be due to lack of dopaminergic inhibitory control of melanin-stimulating hormone (MSH) excretion and are sometimes improved by L-dopa or dopaminomimetic therapy.

Some patients report a change in taste (either an abnormal sensation or reduced taste) on L-dopa therapy. Olfactory impairment, a common and early feature of PD, may also play a role.

The *Physicians' Desk Reference* cautions against the use of L-dopa in PD patients with a history or presence of melanoma. According to a review of the literature, however, the use of L-dopa seems to be safe in PD patients with melanoma (Weiner *et al.*, 1993b).

**Weight Loss.** Many patients with PD gradually lose weight, until they plateau at a new, lower weight. Hypothalamic dysfunction has been postulated as a possible cause for this.

Clearly patients with PD are not immune to developing other more sinister causes of weight loss, but most commonly investigations for other possible causes are negative and are probably best restricted to those patients with prolonged and progressive weight loss or positive clues to another etiology.

PD patients on a low-protein diet are also in danger of losing weight and should receive dietetic advice (discussed earlier).

### Depression

Depression has a great impact on quality of life in PD. There is general agreement that the frequency of episodes of sustained depression in PD is about 40% (literature values range from 10%–90% of patients, Dooneief *et al.*, 1992; Mayeux, 1990; for review see Poewe and Luginger, 1999). Depression in PD is characterized by loss of self-esteem, hopelessness, worthlessness, pessimism about the future, and sadness with a high prevalence of anxiety and panic attacks. Although depressed parkinsonian patients often have suicidal ideation, the suicide rate is low (Cummings *et al.*, 1999).

Longitudinal studies showed that the prevalence of depression is not related to the stage of the illness (Starkstein *et al.*, 1992). Depression was most common in stages I, III, and IV. Furthermore, there is growing evidence that different subgroups of patients may be more vulnerable to depression than others. In a significant proportion of patients depression may precede motor signs and symptoms for years (Santamaria *et al.*, 1986). It seems to be more frequent in patients with the akinetic–rigid than the tremor-dominant type (Seiler *et al.*, 1992). Transient short-lived, in part marked off-period, depression often occurs in fluctuating patients and disappears as soon as the patient turns "on" (Nissenbaum *et al.*, 1987). Nevertheless, in a substantial proportion of patients, the depression does not improve with dopaminomimetics, and the degree of depression does not seem to correlate with the degree of motor impairment.

It is still unknown to what extent depression in PD is a symptom of an underlying organic change, a psychological reaction to the disease, or a mixture of both mechanisms. With respect to the organic hypothesis, an imbalance of the transmitter systems dopamine/noradrenaline/serotonin is assumed. Mayeux found decreased levels of the serotonin metabolite 5-hydroxyindoleacetic acid (5-HIAA) in the CSF of depressed parkinsonian patients compared with parkinsonian patients without depression and normal controls (Mayeux, 1990; Mayeux *et al.*, 1984). Data from functional imaging studies using PET showed relative hypometabolism involving the caudate and orbital-inferior area of the frontal lobe in PD patients with depression (Mayberg *et al.*, 1990). Several risk factors have been proposed, including early onset of parkinsonian symptoms, a positive family history, and female gender. Recently, a genetic variation of the serotonin transporter gene has been proposed to be a risk factor for the development of anxiety and depression in PD (Menza *et al.*, 1999).

Different strategies for therapeutic approach are available (Oertel *et al.*, 2001). Concerning reactive depression, mostly associated with the diagnosis of PD, psychosocial counseling should be initiated. Especially for treatment of depression related to "off" periods, dopamine-replenishing strategies with L-dopa, selegiline, and dopamine agonists proved to be effective.

The MAO B inhibitor selegiline has been reported to possess mild antidepressant properties. Likewise recent open-label reports on pramipexole in a small group of psychiatric depressed patients suggest an antidepressive component besides its efficacy for motor symptoms (Perugi *et al.*, 2001). Tricylic and tetracyclic antidepressants (TCAs) represent a traditionally available group of medication for the treatment of depression in PD, and many studies, although of limited quality, have suggested the efficacy of TCAs. However, the anticholinergic side effects of the TCAs can cause urinary retention; acute glaucoma and decrease of cognitive function, on the other hand, may add to improvement of motor symptoms. Our experience is best and most extensive with amitriptyline, dothiepin, doxepine, and prothiaden. In the morning, stimulating drugs such as imipramine (25 mg) should be administered, whereas in the evening sedating antidepressants are advantageous: amitriptyline (25–150 mg; normal or slow-release preparation), dothiepin (25–150 mg), doxepine (10–100 mg, normal or slow-release preparation), or maprotiline (10–30 mg). These drugs in low dosage also have a good hypnotic effect and are non-addictive.

The SSRIs have equal antidepressant efficacy to TCAs but may have a better side effect profile, particularly in the elderly. Because of the serotonin syndrome, they should not be given with, or shortly after withdrawing, selegiline. Besides described efficacy of SSRIs in the treatment of depressed PD patients, increased motor disability has rarely been observed after the use of fluoxetine (Steur, 1993), fluvoxamine (Wils, 1992), and paroxetine (Jimenez-Jimenez, 1994). Two open-label studies on sertraline for the treatment of depression in PD found no decline in motor function (Hauser *et al.*, 1997; Meara *et al.*, 1998). A recently published study with paroxetine suggested that paroxetine at the dosage of 20 mg/day does not worsen parkinsonian symptoms, despite observing increased tremor in one patient (Ceravolo *et al.*, 2000).

Data concerning novel antidepressant drugs in PD like reboxetine (selective noradrenaline reuptake inhibitor), venlafaxine (norepinephrine and serotonin reuptake inhibitor), and mirtazapine (noradrenergic and specific serotonergic antidepressant) is rare. Our experience is favorable with mirtazapine, which is thought to have an early antidepressant effect. In addition, we use the sedative, sleep-inducing side effect of low dose mirtazapine for the treatment of agitated PD patients during the night.

Lithium can also sometimes worsen parkinsonism, even at therapeutic doses.

Finally, the use of electroconvulsive therapy (ECT) has been reported to be effective for temporary alleviation of severe depression in PD with immediate effects and further benefits on motor symptoms (Moellentine et al., 1998).

### Impairment of Cognitive Function and Dementia

Cumulatively, 30%–40% of PD patients will develop dementia (Mayeux et al., 1992; Rajput et al., 1984). However, patients with young-onset PD only very rarely develop dementia before the age of 60, despite severe disease of long duration. This suggests that PD per se is not usually sufficient to cause dementia, and some additional factor related to age, in many cases involving Alzheimer's disease changes, may be necessary for clinical dementia to appear. A second cause of dementia in PD is the so-called DLB type (see Chapters 29 and 75) related to the cortical density and distribution of Lewy bodies. As a third possibility, in some PD patients dementia may be related to an impairment of multiple transmitter systems.

Some patients, particularly older ones, may develop impairment of short-term memory on anticholinergic drugs (De Smet et al., 1982; Sadeh et al., 1982). Therefore, anticholinergics and other substances with an anticholinergic component (amantadine, budipine, tricyclic antidepressants) are contraindicated, or at least should be used with great caution, in elderly patients and in those with any evidence of cognitive impairment. In patients with manifest dementia but also in apparently nondemented patients, anticholinergics can cause an acute confusional state (discussed previously).

Suggested therapeutic options include withdrawal of anticholinergics and related substances, gradually if possible. Caution: the patient may experience anticholinergic withdrawal syndrome.

For treatment of DLB type see Chapters 29 and 75.

### Sleep Impairment

Disease-related sleep impairment is present in up to 90% of PD patients and is usually characterized by a prolonged latency to fall asleep, reduced sleep efficiency with reduced proportions of slow wave and rapid eye movement (REM) sleep phases. On one hand sleep impairment is partly caused by the degeneration of specific brainstem neurons in PD. On the other hand, these sleep disorders are due in part to nocturnal akinesia, which can be an early sign of response fluctuations. Difficulty turning in bed correlates with the degree of nocturnal akinesia and occurs most often in patients who also have response fluctuations during the day. Other reasons for disturbed sleep include frequent nycturia, comorbidity with REM sleep behavior disorder, obstructive sleep apnea syndrome, restless legs syndrome, or period limb movements during sleep (PLMS).

Therapy of nocturnal akinesia consists of the administration of L-dopa or dopamine agonists just before going to bed. CR L-dopa or long-acting dopamine agonists at bedtime are presently the treatment of choice. If the patient wakes up during the night with marked rigidity, akinesia, or rest tremor, an additional dose of L-dopa or dopamine agonist is justified. Sleep quality is improved at a similar degree as motor symptoms are reduced: latency to sleep onset is shorter, and an increase in REM and non-REM sleep is observed.

Drug-induced sleep impairment is observed predominantly in the later stages of the disease. Amantadine given late in the afternoon or dopamine agonists may increase latency to fall asleep or induce insomnia (waking up earlier). The influence of L-dopa on sleep depends on the time of intake. Administration of a standard preparation 2 hours before falling asleep does not show a marked influence, but L-dopa intake 1 hour before going to sleep leads to reduction of REM activity in the early hours of the night. Increased dream activity or even visual hallucinations on waking up during the night can be a precursor of exogenous dopaminomimetic-induced psychosis (for its therapy see next section). Impaired serotonin metabolism caused by chronic L-dopa treatment has also been suggested as a possible cause of sleep impairment.

Antidepressants with a sedating component often help to induce sleep. Nocturnal confusion or disorientation can, in exceptional cases, be treated for a short time with chlormethiazole (1–2 capsules nightly). Benzodiazepines can sometimes help sleep disorders in PD but often are not beneficial in the long term, particularly when the disturbed sleep is due to re-emergence of parkinsonism during the night. Clonazepam (0.5–2.0 mg at bedtime), however, can be used for the treatment of REM sleep behavior disorder. Panic attacks during nocturnal akinesia (off period) are treated by eliminating the off period and/or respond to lorazepam (0.5–2.0 mg). When sleep is disturbed because of psychosis, clozapine (25–50 mg) or quetiapine (25–100 mg) nightly is recommended (see the next section).

Daytime sleepiness is another frequently observed feature in PD and is often caused by sleep impairment. Dopaminergic medication by itself is, however, also known to have a sedating effect, for example half an hour after intake. Sudden onset of sleep ("sleep attacks") leading to car accidents have been reported under therapy with the nonergot dopamine agonists pramipexole and ropinirole (Frucht *et al.*, 1999). Patients taking these dopamine agonists should be informed not to drive a vehicle until they have gained sufficient experience to gauge whether their mental performance is adversely affected. The current legal regulatios concerning driving ability however, are not identical in the United States and the European Union countries. Furthermore, several studies have now provided convincing evidence that sudden onset of sleep is an adverse effect of any dopaminergic medication in PD (Ferreira *et al.*, 2000; Möller *et al.*, 2002; Sanjiv *et al.*, 2001).

### Dopaminergic Psychosis

Exogenous drug-induced psychosis can occur with all types of antiparkinsonian medication. Its management represents a major, if not the therapy-limiting, problem in patients with advanced PD. Psychotic episodes also tend to occur after surgery under general and sometimes regional anesthesia (discussed previously). These patients are often dehydrated and may improve after rehydration (see Table XIX). Drug-induced psychosis should not be confused with confusional states with disorientation, as observed predominantly in patients with dementia or those who are taking anticholinergics (for drug-independent psychosis see also Chapters 29 and 75, Dementia of the Lewy body type).

Prevalence rates of drug-induced psychosis range from 3%–85% (Doraiswamy *et al.*, 1995). It is generally accepted that about 20%–30% of patients with PD develop confusion, hallucinations, or paranoid delusions during long-term therapy with L-dopa or dopamine agonists. Confusion is dose related and occurs more frequently with ergot and nonergot agonists than with L-dopa. The frequency and severity of these exogenous psychoses increase with duration of disease and duration of therapy. Further main risk factors include advanced age, comorbidity with cerebrovascular disease, and cognitive impairment, for example, caused by comorbidity with Alzheimer-type changes (Giladi *et al.*, 2000). The occurrence of sleep disruption, vivid dreams, or nightmares is believed to be an early sign of psychosis in PD. These "precursors" often have to be asked about specifically, in particular in patients receiving dopamimetic therapy for nocturnal akinesia. This phenomenon can usually be controlled by reducing the dose. If medication is not adjusted (see later), some patients will develop visual illusions or pseudo-hallucinations (with retained insight). Later, true hallucinations (with loss of insight) or paranoid psychosis may appear. Patients may be agitated and disoriented. The most common features of psychosis in PD are visual hallucinations (Fenelon *et al.*, 2000), whereas hallucinations in other modalities are rare. Other features are delusions that are characterized by unrealistic beliefs of spousal infidelity, persecution, and jealousy. The severest level is reached when patients become aggressive to others and themselves. Additional features may include tachycardia, sweating, and sometimes hypertension and worsening of L-dopa–induced dyskinesias.

The etiology of drug-induced psychosis in PD remains poorly understood. Overstimulation of mesocorticolimbic dopamine receptors and abnormalities in dopamimetic and serotonergic systems are discussed (Wolters, 1999).

Rarely, fluctuating patients may experience short-lived psychotic episodes exclusively in "off" periods or at night, which terminate abruptly when the patient turns "on" (Nissenbaum *et al.*, 1987). These episodes may improve with treatment modifications designed to increase time "on," including paradoxically an increase in dopaminomimetic medication.

The treatment of psychosis in PD consists of different steps. First, several general factors triggering psychosis in PD (e.g., infection or dehydration) should be eliminated. In a further step, antiparkinsonian drugs (e.g., anticholinergics, amantadine, and dopamine agonists) should be slowly withdrawn or, if necessary and preferentially under in-patient conditions, taken off. Later, L-dopa should be reduced. If motor symptoms get worse after reduction of antiparkinsonian medications or psychotic symptoms cannot be controlled, therapy with an atypical neuroleptic should be initiated (Juncos, 1999).

Clozapine represents the best studied and effective atypical neuroleptic for treatment of drug-induced psychosis in PD, with a low incidence of extrapyramidal side effects. It probably blocks $D_4$-receptors located in the projection areas of the mesocorticolimbic system and possesses a relatively low $D_1$- and $D_2$-dopamine receptor–blocking activity (see Table IX). However, clozapine use is rarely associated with leukopenia and agranulocytosis (about 1%). Patients treated with clozapine therefore must have a weekly blood cell count for the first 18 weeks and then monthly (Alvir *et al.*, 1993). For this reason, in psychiatry, in most countries, clozapine is licensed as a second-line drug, only to be used if other neuroleptics fail. On the basis of numerous open-label studies (first by Scholz and Dichgans, 1985) and two recent multicenter, double-blind placebo-controlled trials, low-dose clozapine (range, 6.25–50, maximally 100 mg/day) is very effective in treating dopaminomimetic psychosis in PD without worsening parkinsonism (French Clozapine Parkinson Study Group, 1999, Parkinson Study Group, 1999).

Because of side effects like severe orthostatic hypotension, tachycardia, and sedation, clozapine should be increased slowly.

It may also be used in the treatment of amantadine-induced psychosis. However, it may not be effective for psychosis in patients with concomitant cerebrovascular disorders or marked cerebral atrophy, in whom it may even induce aggressive behavior (for other side effects, see previous discussion of clozapine).

The combination of clozapine with benzodiazepines should be used with caution, because marked arterial hypotension has been encountered with this combination.

Olanzapine is an atypical neuroleptic drug with greater affinity blocking 5-HT2A receptors compared with $D_2$-receptors. Several open-label studies have been published on psychosis in PD (Friedman and Factor, 2000). Some studies noted a beneficial effect of olanzapine (mean range, 5–7.5 mg/day) in the treatment of psychotic symptoms in PD without extrapyramidal side effects; however, most of the studies indicated a worsening of motor symptoms. A recent double-blind comparison trial of olanzapine and clozapine in hallucinating patients with PD showed significant declines in motor function in the olanzapine group compared with baseline function and to the patients receiving clozapine (Goetz et al., 2000).

Risperidone is another antipsychotic with a low incidence of extrapyramidal side effects. There have been a few open-label studies with risperidone in PD; with mean daily dosages from 0.5–1.9 mg/day (Ford et al., 1994). These studies proposed risperidone to be an effective treatment for psychosis in PD; however, risperidone worsened parkinsonian symptoms in several patients and can therefore not be generally recommended.

Quetiapine is a new atypical antipsychotic drug without inducing agranulocytosis. Quetiapine has close pharmacological resemblance to clozapine. Fernandez et al. (1999) performed one of the first open-label studies using 50 mg quetiapine per day in parkinsonian patients for sufficient treatment of psychosis. According to our experience (Brandstädter et al., 2002), for sufficient control of psychosis under inpatient conditions, higher dosages of quetiapine (between 50–150 mg/day, rarely 225 mg/day) are often required, and compared with clozapine, the equivalent dosage of quetiapine seems to be approximately two to three times greater. So far, there have been no placebo-controlled multicenter studies with quetiapine or active comparator trials to confirm these encouraging data.

Classic neuroleptic drugs (for example, haloperidol should) not be used, because they worsen motor symptoms and diminish response to dopaminergic drugs, effects that may persist for several days after the neuroleptic is stopped. If clozapine or quetiapine is not available, the previously listed "atypical" neuroleptics or classical neuroleptics with a low incidence of extrapyramidal side effects such as sulpiride, thioridazine, but not zotepine (Arnold et al., 1994), may be tried as a transient compromise.

### Pregnancy—Parkinson's Disease and Therapy

No systematic study has been conducted on this topic. According to the few reports available, five women delivered six normal infants while taking levodopa and carbidopa (Ball and Sagar, 1995; Cook and Klawans, 1985; Golbe, 1987). Eight women were reported to receive levodopa and benserazide during pregnancy (Allain et al., 1989; Bauherz, 1994; Graevenitz et al., 1996). Four women delivered a healthy child; in two, abortion was induced; one had a spontaneous abortion, and one woman was lost to follow-up. A healthy child was born after pregnancy under therapy with a combination of levodopa/benserazide and selegiline (Kupsch and Oertel, 1998). No report of human teratogenicity in association with levodopa therapy is known. In addition, benserazide and carbidopa (Merchant et al., 1994) do not pass the placenta readily. So, on the basis of present knowledge, levodopa should not be automatically withheld in PD patients during pregnancy. On the other hand, estrogens have a predominantly antidopaminergic effect. Accordingly, worsening of parkinsonian symptoms has been reported in 10 of 17 pregnancies followed (Golbe, 1987).

## SECONDARY (SYMPTOMATIC) PARKINSONISM

The term secondary parkinsonism subsumes all parkinsonian syndromes with a known etiology.

### Postencephalitic Parkinsonism

Postencephalitic parkinsonism may be diagnosed if the patient has had encephalitis lethargica (von Economo) with its typical features (disorders of consciousness, sleep disorders). Persistent parkinsonism rarely, if ever, occurs after other encephalitides of identified origin (coxsackie B, spotted fever [epidemic typhus], measles). The diagnosis may be supported by a previous inflammatory response in cerebrospinal fluid during the acute illness, which sometimes is followed by persisting oligoclonal bands. Nowadays, this type of parkinsonism is extremely rare. Its therapy corresponds broadly to that of PD (discussed earlier).

## Arteriosclerotic or Vascular Pseudoparkinsonism (Subcortical Vascular Encephalopathy)

Arteriosclerotic pseudoparkinsonism (Critchley, 1929) is clinically distinct from PD. It may arise from multiple infarcts, basal ganglia lacunes, or subcortical white matter (vascular) encephalopathy (SVE; Binswanger's disease, leukoaraiosis). Clinically, it exhibits locomotor difficulty mainly of the lower limbs. The gait disturbance that it causes therefore has also been called *lower body parkinsonism* (Jankovic, 1990), *parkinsonian ataxia* (Thompson and Marsden, 1987), or *gait apraxia*. It rarely, if ever, mimics classical parkinsonism with akinesia and rest tremor of the arms or legs. Tendon reflexes frequently are increased. A typical symptom is the so-called magnetic foot response: The patient has extreme difficulties in starting ("gait ignition failure") but when in motion may walk nearly normally, or with a "màrche petits pas." Rare cases respond partially to L-dopa or dopamine agonists. In general, this syndrome responds minimally or not at all to dopaminomimetic medication. In some late cases, it is difficult to differentiate arteriosclerotic pseudoparkinsonism from early normal-pressure hydrocephalus (discussed in the next section). A similar "senile gait disorder" may occur in the absence of clinical or imaging evidence of vascular disease.

A history of or the presence of vascular risk factors—such as arterial hypertension, hypercholesterolaemia, diabetes mellitus, arterial emboli either caused by cardiac arrhythmia, a patent atrial septal defect, or intimal irregularities of the extracranial and intracranial major arteries—support the suspected diagnosis. These vascular risk factors can lead to multiple small infarctions in the white matter. A vascular etiology of this pseudoparkinsonism is possible when it begins acutely or develops in a stepwise fashion, when it occurs after cerebral ischemia or generalized hypoxia, or when additional pyramidal tract signs or dementia are present. The white matter lesions of Binswanger's disease commonly coexist with one or more small lacunar infarcts, some of which may be visible on CT and most of which are detectable on MRI scanning.

Only preliminary data on the prevalence of arteriosclerotic pseudoparkinsonism exist (Trenkwalder et al., 1995). In postmortem studies 1%–7% of all cases diagnosed with "parkinsonism" were found to be of vascular etiology (Jellinger, 1986; Sutcliffe et al., 1985). This syndrome is distinct from the excessively rare instance of vascular true parkinsonism caused by a dopamine deficit in the striatum as a result of an isolated infarct in the substantia nigra or in the course of the nigrostriatal projection, which responds to L-dopa.

In arteriosclerotic pseudoparkinsonism, a trial with L-dopa and other dopaminomimetics is justified but usually unsuccessful. Amantadine, anticholinergics, or MAO B inhibitor can also be tried but rarely lead to useful improvement. Instead, the best (but still modest) results are obtained from training focused on walking and balance skills. Therapy should aim at controlling cardiac and vascular risk factors, particularly arterial hypertension, and by giving an anticoagulant or a platelet adhesion or aggregation inhibitor when indicated.

## Hydrocephalus-Associated Parkinsonism

Parkinsonism may be associated with communicating or noncommunicating hydrocephalus. It may occur before, be relieved by, or develop after shunt insertion and may occasionally improve with L-dopa (Curran and Lang, 1994). Early normal pressure hydrocephalus can mimic SVE. In such cases, imaging (cCT, MRI) is not always diagnostic, and the best guide to whether shunting will help is marked improvement in gait after removal of 50 ml CSF by lumbar puncture (see Chapter 77).

## Toxin-Induced Parkinsonism

### Manganese

Workers in manganese mines or the manganese refining industry may develop a parkinsonian syndrome after 1–2 years' (rarely after only 1–2 months) exposure. After a manganese psychosis (insomnia, somnolence, emotional lability with inappropriate laughing and crying, and rarely visual and auditory hallucinations), parkinsonian features such as akinesia and rigidity, rarely rest tremor, and cerebellar signs or impotence occur. L-Dopa has been claimed to reduce the neurological symptoms in some patients (Fahn, 1977), but a report (Huang et al., 1993) has not confirmed this observation.

### Carbon Monoxide

Survivors of severe carbon monoxide intoxication may often, after a delay of some weeks or months, develop parkinsonism, dementia, emotional lability with inappropriate laughing and crying, or occasionally hyperkinesias (athetosis, ballism). The parkinsonism is either static or partially regressive. A progressive course has not been described (Lee and Marsden, 1994). L-Dopa has been reported to help some of these patients (Fahn, 1977). However, in our own experience, dopaminomimetics, including L-dopa, have had no beneficial effect.

### N-*Methyl-3-Phenyl-1,2,3,6,Tetrahydropyridine (MPTP)*

MPTP, a side product in the synthesis of meperidine, a synthetic heroin substitute, has induced acute parkinsonism in drug addicts. The severity of this MPTP-induced parkinsonism syndrome ranges from mild to extremely severe. These victims respond very well to L-dopa and dopamine agonists. However, they also develop the complications of L-dopa therapy early after its introduction and to a severe degree. This may be a special feature of MPTP-induced parkinsonism or may simply reflect the young age at disease onset of the affected addicts (Langston and Ballard, 1984). The treatment of MPTP-induced parkinsonism corresponds broadly to that of PD (discussed earlier under drug therapy).

### Other Causes

Single cases of toxic parkinsonism have been reported after methanol or cyanide intoxication, exposure to *n*-hexane (Pezzoli *et al.*, 1995), or petroleum waste ingestion (Tetrud *et al.*, 1994). Reversible parkinsonism has been described during alcohol withdrawal.

### Tumor-Associated Parkinsonism

Akinetic parkinsonian syndromes are rarely observed in the presence of intracranial tumors such as medially or centrally located meningiomas or gliomas close to the anterior corpus callosum. The clinical picture should not be confused with "gait apraxia," as observed with arteriosclerotic pseudoparkinsonism or normal-pressure hydrocephalus or obstructive hydrocephalus (both discussed previously). However, the latter disorder is occasionally seen when a tumor in the posterior fossa obstructs cerebrospinal fluid circulation.

### Parkinsonism Related to Metabolic Disorders

#### Hypoparathyroidism

Idiopathic, nonarteriosclerotic intracerebral calcifications of the basal ganglia do not lead to neurological symptoms. In contrast, autosomal-dominant or autosomal-recessive familial diseases or postoperative hypoparathyroidism can lead to calcification of basal ganglia (usually together with other structures) and the clinical picture of parkinsonism, dementia, chorea, or athetosis. The diagnosis is established by measuring calcium, phosphate, and parathyroid hormone levels in serum. With appropriate treatment, both the basal ganglia calcification and clinical symptoms may decrease or even reverse in the case of hypoparathyroidism but are irreversible if hypoparathyroidism is not present (Fahn, 1977).

#### Wilson's Disease

Wilson's disease must be excluded by biochemical and ophthalmological (slit-lamp examination for Kayser–Fleischer rings) tests and, if necessary, by liver biopsy with subsequent biochemical assessment of copper concentration in all patients with parkinsonism beginning less than 50 years of age (see Chapter 80).

#### Disorders of Folate Metabolism

A parkinsonian syndrome can be observed in the course of the hereditary disease "dihydrobiopterin reductase deficiency," which causes reduced activity of tyrosine hydroxylase in the brain (Woody *et al.*, 1989). Treatment with folic acid improves symptoms within a few days.

### Drug-Induced Parkinsonism

*This issue is still of great practical importance.*
According to Rajput *et al.* (1984), 7% of all parkinsonism is drug induced. In patients residing in geriatric institutions, up to 50% of newly incident diagnoses of parkinsonism were found to be drug induced (Stephen and Williamson, 1984). The clinical signs and symptoms can be identical to those of PD, including the presence of rest tremor and asymmetry. *Table I lists the various classes of substances that most often induce parkinsonism.*
Lithium in therapeutic doses rarely causes basal ganglia symptoms. However, occasionally akinesia, rigidity, and cerebellar symptoms such as intention tremor, saccadic pursuit, and dysarthria can occur and are reversible after reduction or withdrawal of lithium. Flunarizine and cinnarizine, and possibly amlodipine and diltiazem, can induce a reversible parkinsonian syndrome, in general appearing after weeks or months of therapy. They may also unmask subclinical PD (Chouza *et al.*, 1986). Flunarizine may also induce a concomitant dyskinesia. After withdrawal of the neuroleptic drugs, in the most patients resolution of the symptoms and signs of parkinsonism occurs over 4–8 weeks, although complete remission may take months and even up to 1.5 years.

## REFERENCES

Adler, C. H., Singer, C., O'Brien, C., Hauser, R. A., Lew, M. F., Marek, K. L., Dorflinger, E., Pedder, S., Deptula, D., and Yoo, K. (1998).

Randomized, placebo-controlled study of tolcapone in patients with fluctuating Parkinson disease treated with levodopa-carbidopa. Tolcapone Fluctuator Study Group III. *Arch. Neurol.* **55**(8), 1089–1095.

Agid, Y., Ahlskog, E., Albanese, A., Calne, D., Chase, T., De Yebenes, J., Factor, S., Fahn, S., Gershanik, O., Goetz, C., Koller, W., Kurth, M., Lang, A., Lees, A., Lewitt, P., Marsden, D., Melamed, E., Michel, P.P., Mizuno, Y., Obeso, J., Oertel, W., Olanow, W., Poewe, W., Pollak, P., Tolosa, E., *et al.* (1999). Levodopa in the treatment of Parkinson's disease: a consensus meeting. *Mov. Disord.* **14**(6), 911–913.

Ahlskog, J. E., Muenter, M. D., Maraganore, D. M., Matsumoto, J. Y., Lieberman, A., Wright, K. F., and Wheeler, K. (1994). Fluctuating Parkinson's disease: Treatment with the long-acting dopamine agonist cabergoline. *Arch. Neurol.* **51**, 1236–1241.

Allain, H., Bentue-Ferrer, D., Milon, D., Moran, P., Jacquemard, F., and Defawe, G. (1989). Pregnancy and parkinsonism: A case report without problem. *Clin. Neuropharmacol.* **12**, 217–219.

Allain, H., Coughard, J., Neukirch, H.-C., and the FSMT members (1991). Selegiline in *de novo* parkinsonian patients: The French selegiline multicenter trial (FSMT). *Acta Neurol. Scand.* **84**, 73–78.

Alvir, J. M. J., Lieberman, J. A., Saffermann, A. Z., Schwimmer, J. L., and Schaaf, J. A. (1993). Clozapine induced agranulocytosis: Incidence and risk factors in the United States. *N. Engl. J. Med.* **329**, 162–167.

Arevalo, G. J., and Gershanik, O. S. (1993). Modulatory effect of clozapine in Parkinson's disease: A preliminary study. *Mov. Disord.* **8**, 349–354.

Arnold, G., Trenkwalder, C., Schwarz, J., and Oertel, W. H. (1994). Zotepine reversibly induces akinesia and rigidity in Parkinson's disease patients with resting tremor or drug-induced psychosis. *Mov. Disord.* **9**, 238–240.

Baas, H., Beiske, A. G., Ghika, J., Jackson, M., Oertel, W. H., Poewe, W., and Ransmayr, G. (1997). COMT inhibition with tolcapone reduces the "wearing-off" phenomenon and levodopa reuqirements in fluctuating parkinsonian patients. *J. Neurol. Neurosurg. Psychiat.* **63**, 421–428.

Ball, M. C., and Sagar, H. J. (1995). Levodopa in pregnancy. *Mov. Disord.* **10**, 115.

Baldereschi, M., Di Carlo, A., Rocca, W. A., *et al.* (2000). Parkinson's disease and parkinsonism in a longitudinal study; two-fold higher incidence in men. *Neurology.* **55**(9), 1358–1363.

Barone, P., Bravi, D., Bermejo-Pareja, F., Marconi, R., Kulisevsky, J., Malagu, S., Weiser, R., and Rost, N. (1999). Pergolide monotherapy in the treatment of early PD: a randomized, controlled study. Pergolide Monotherapy Study Group. *Neurology.* **53**(3), 573–579.

Bauherz, G. (1994). Pregnancy and Parkinson's disease. *New Trends Clin. Neuropharmacol.* **8**, 142.

Benamer, T. S., Patterson, J., Grosset, D. G., Booij, J., de Bruin, K., van Royen, E., Speelman, J. D., Horstink, M. H., Sips, H. J., Dierckx, R. A., Versijpt, J., Decoo, D., Van Der Linden, C., Hadley, D. M., Doder, M., Lees, A. J., Costa, D. C., Gacinovic, S., Oertel, W. H., Pogarell, O., Hoeffken, H., Joseph, K., Tatsch, K., Schwarz, J., and Ries, V. (2000). Accurate differentiation of parkinsonism and essential tremor using visual assessment of $^{123}$I-FP-CIT SPECT imaging. *Mov. Disord.* **15**, 503–510.

Bennett, D. A., Beckett, L. A., Murray, A. M., *et al.* (1996). Prevalence of parkinsonian signs and associated mortality in a community population of older people. *N. Engl. Med.* **334**(2), 71–76.

Bennett, J. P., Landow, E. R., and Schuh, L. A. (1993). Suppression of dyskinesias in advanced Parkinson's disease. II. Increasing daily clozapine doses suppress dyskinesias and improve parkinsonian symptoms. *Neurology.* **43**, 1551–1555.

Ben-Shlomo, Y., Head, J., and Lees, A. J. (1999). Mortality in DATATOP. *Ann. Neurol.* **45**(1), 138–139.

Ben-Shlomo, Y., and Lees, A. (2001). Selegiline and mortality in subjects with Parkinson's disease: a longitudinal community study. *Neurology.* **57**(2), 369; discussion 369–370.

Berger, K., Breteler, M. M., Helmer, C., Inzitari, D., Fratiglioni, L., Trenkwalder, C., Hofman, A., and Launer, L. J. (2000). Prognosis with Parkinson's disease in europe: A collaborative study of population-based cohorts. Neurologic Diseases in the Elderly Research Group. *Neurology.* **54**(11 Suppl 5), S24–S27.

Bergman, H., Wichmann, T., and DeLong, M. R. (1990). Reversal of experimental parkinsonism by lesions of the subthalamic nucleus. *Science.* **249**, 1436–1438.

Betarbet, R., Sherer, T. B., MacKenzie, G., Garcia-Osuna, M., Panov, A. V., and Greenamyre, J. T. (2000). Chronic systemic pesticide exposure reproduces features of Parkinson's disease. *Nat. Neurosci.* **3**(12), 1301–1306.

Birkmayer, W., Knoll, J., Riederer, P., and Youndim, M. B. H. (1983). Deprenyl leads to prolongation of L-DOPA efficacy in Parkinson's disease. *Mod. Probl. Pharmacopsychiat.* **19**, 170–176.

Blanchet, P. J., Calon, F., Morissette, M., Goulet, M., Grondin, R., Levesque, D., Bedard, P. J., and Di Paolo, T. (2001). Regulation of dopamine receptors and motor behavior following pulsatile and continuous dopaminergic replacement strategies in the MPTP primate model. *Adv. Neurol.* **86**, 337–344.

Bormann, J. (1989). Memantine is a potent blocker of N-methyl-D-aspartate (NMDA) receptor channels. *Eur. J. Pharmacol.* **166**, 591–592.

Bower, J. H., Maraganore, D. M., McDonnell, S. K., and Rocca, W. A. (1999). Incidence and distribution of parkinsonism in Olmsted Country, Minnesota. *Neurology* **52**, 1214–1220.

Brandstadter, D., and Oertel, W. H. (2002). Treatment of drug-induced psychosis with quetiapine and clozapine in Parkinson's disease. *Neurology.* **8**;**58**(1), 156–162.

Bravi, D., Mouradian, M. M., Roberts, J. W., Davis, T. L., Sohn, Y. H., and Chase, T. N. (1994). Wearing-off fluctuations in Parkinson's disease: Contribution of postsynaptic mechanisms. *Ann. Neurol.* **36**, 27–31.

Brion, N., Kollenbach, K., Marion, M. H., Grègoire, A., Advenier, C., and Pays, M. (1992). Effect of a macrolide (Spiramycin) on the pharmacokinetics of L-DOPA and carbidopa in healthy volunteers. *Clin. Neuropharmacol.* **15**, 229–235.

Brooks, D. J. (2000a). Morphological and functional imaging studies on the diagnosis and progression of Parkinson's disease. *J. Neurol.* **247** Suppl. 2II, 11–18.

Brooks, D. J. (2000b). PET studies and motor complications in Parkinson's disease. *Trends. Neurosci.* **23**(10 Suppl), S101–S108.

Brücke, T., Asenbaum, S., Pozzera, A., Hornykiewicz, S., Harasko-van der Mee Wenger, C., Koch, G., Pirker, W., Wüber, C. H., Müller, C. H., Kornhuber, J., Angelberge, P., and Podreka, I. (1994). Dopaminergic nerve cell loss in Parkinson's disease quantified with [$^{123}$I]-β-CIT and SPECT correlates with clinical findings. *Mov. Disord.* **9**(Suppl. 1), 120.

Calne, D. B., and Langston, J. W. (1983). Aetiology of Parkinson's disease. *Lancet.* **II**, 1457–1459.

Carter, J. H., Nutt, J. G., Woodward, W. R., Hatcher, L. F., and Trotman, T. L. (1989). Amount and distribution of dietary protein affects clinical response to levodopa in Parkinson's disease. *Neurology.* **39**, 552–556.

Ceravolo, R., Nuti, A., Piccinni, A., Dell'Agnello, G., Bellini, G., Gambaccini, G., Dell'Osso, L., Murri, L., and Bonuccelli, U. (2000). Paroxetine in Parkinson's disease: effects on motor and depressive symptoms. *Neurology.* **55**(8), 1216–1218.

Charcot, J. M. (1880). "Lecons sur les maladies du systeme nerveux faites a la Salpetriere," 4th ed. Bourneville, Delahaye & Lecrosnier, Paris.

Chase, T. N., Engber, T. M., and Mouradian, M. M. (1993). Striatal dopaminoceptive system changes and motor response complications in L-dopa-treated patients with advanced Parkinson's disease. In "Parkinson's Disease—From Basic Research to Treatment" (H. Narobayashi, T. Nagatsu, N. Yanagisawa, and Y. Mizuno, Eds.), Advances in Neurology, Vol. 60, pp. 181–185. Raven, New York.

Chouza, C., Caamano, J. L., Aljanati, R., Scaramelli, A., De Median, O., and Romero, S. (1986). Parkinsonism, tardive dyskinesia, akathisia, and depression induced by flunarizine. Lancet. I, 1303–1304.

Chung, K. K., Zhang, Y., Lim, K. L., Tanaka, Y., Huang, H., Gao, J., Ross, C. A., Dawson, V. L., and Dawson, T. M. (2001). Parkin ubiquitinates the alpha-synuclein-interacting protein, synphilin-1: Implications for Lewy-body formation in Parkinson's disease. Nat. Med. 7, 1144–1150.

Churchyard, A., Mathias, C. J., Boonkongchuen, P., and Lees, A. J. (1997). Autonomic effects of selegiline: possible cardiovascular toxicity in Parkinson's disease. J Neurol Neurosurg Psychiatry. 63(2), 228–234.

Colzi, A., Turner, K., and Lees, A. J. (1998). Continuous subcutaneous waking day apomorphine in the long term treatment of levodopa induced interdose dyskinesias in Parkinson's disease. J. Neurol. Neurosurg. Psychiatry. 64(5), 573–576.

Comella, C. L., Stebbins, G. T., Brown-Toms, N., and Goetz, C. G. (1994). Physical therapy and Parkinson's disease: A controlled clinical trial. Neurology. 44, 376–378.

Cook, D. G., and Klawans, H. L. (1985). Levodopa during pregnancy. Clin. Neuropharmacol. 8, 93–95.

Creese, I., Sibley, O. R., Hamblin, M. W., and Leff, S. E. (1983). Dopamine receptors in the central nervous system. In "Molecular Pharmacology of Neurotransmitter Receptors" (I. Segawa, Ed.), pp. 125–134. Raven, New York.

Critchley, M. (1929). Arteriosclerotic parkinsonism. Brain. 52, 23–83.

Cummings, J. L., and Masterman, D. L. (1999). Depression in patients with Parkinson's disease. Int. J. Geriatr. Psychiatry. 14(9), 711–718. Review.

Curran, T., and Lang, A. E. (1994). Parkinsonian syndromes associated with hydrocephalus: Case reports, a review of the literature, and pathophysiological hypotheses. Mov. Disord. 9, 508–520.

Dam, M., Tonin, P., and Casson, S. (1996). Effects of conventional and sensory-enhanced physiotherapy on disability of Parkinson's disease patients. Adv. Neurol. 69, 551–555.

Da Prada, M., Borgulya, J., Napolitano, A., and Zürcher, G. (1994). Improved therapy of Parkinson's disease with tolcapone, a central and peripheral COMT inhibitor with an S-adenosyl-L-methionine–sparing effect. Clin. Neuropharmacol. 17(Suppl. 3), S26–S37.

Davis, T. L., Roznoski, M., and Burns, R. S. (1995). Effects of tolcapone in Parkinson's patients taking L-dihydroxyphenylalanine/carbidopa and selegiline. Mov. Disord. 10(3), 349–351.

De Smet, Y., Ruberg, M., Serdaru, M., Dubois, B., Lhermitte, F., and Agid, Y. (1982). Confusion, dementia and anticholinergics in Parkinson's disease. J. Neurol. Neurosurg. Psychiat. 45, 1161–1164.

Del Dotto, P., Pavese, N., Gambaccini, G., Bernardini, S., Metman, L. V., Chase, T. N., and Bonuccelli, U. (2001). Intravenous amantadine improves levodopa-induced dyskinesias: an acute double-blind placebo-controlled study. Mov. Disord. 16(3), 515–520.

Deuschl, G., Raethjen, J., Baron, R., Lindemann, M., Wilms, H., and Krack, P. (2000). The pathophysiology of parkinsonian tremor. A review. J. Neurol. 247 Suppl 5, V 33–48.

Diamond, S. G., Markham, C. H., Hoehn, M. M., McDowell, F. H., and Muenter, M. D. (1987). Multicenter study of Parkinson

mortality with early versus later dopa treatment. Ann. Neurol. 22, 8–12.

Dingemanse, J., Jorga, K. M., Schmitt, M., et al. (1995). Integrated pharmacokinetics and pharmacodynamics of the novel catechol-O-methyltransferase inhibitor tolcapone during first administration to humans. Clin. Pharmacol. Ther. 57, 508–517.

Djaldetti, R., Baron, J., Ziv, I., and Melamed, E. (1996). Gastric emptying in Parkinson's disease: patients with and without response fluctuations. Neurology. 46(4), 1051–1054.

Dogali, M., Fazzini, D. O., Kolodny, E., Eidelberg, D., Sterio, D., Devinsky, O., and Beri'c, A. (1995). Stereotactic ventral pallidotomy for Parkinson's disease. Neurology. 45, 753–761.

Donnan, P. T., Steinke, D. T., Stubbings, C., Davey, P. G., and MacDonald, T. M. (2000). Selegiline and mortality in subjects with Parkinson's disease: a longitudinal community study. Neurology. 55(12), 1785–1789.

Dooneief, G., Mirabello, E., Bell, K., Marder, K., Stern, Y., and Mayeux, R. (1992). An estimate of the incidence of depression in idiopathic Parkinson's disease. Arch. Neurol. 49, 305–307.

Doraiswamy, M., Martin, W., Metz, A., and Deveaugh-Geiss, J. (1995). Psychosis in Parkinson's disease: diagnosis and treatment. Prog. Neuropsychopharmacol. Biol. Psychiatry. 19(5), 835–846. Review.

Duvoisin, R. C., and Marsden, C. D. (1975). Alpha-methyldopamine as a dopaminergic agonist. In (D. B. Calne, T. N. Chase, and A. Barbeau, Eds.), "Advances in Neurology" Vol. 9, pp. 243–248. Raven Press, New York.

Eichhorn, T. E., and Oertel, W. H. (2001). Macrogol 3350/electrolyte improves constipation in Parkinson's disease and multiple system atrophy. Mov. Disord. 16(6), 1176–1177.

Ellgring, H., Seiler, S., Perleth, B., Frings, W., Gasser, T., and Oertel, W. H. (1993). Psychosocial aspects of Parkinson's disease. Neurology. 43(Suppl.), 41–44.

Esteguy, M., Bonnet, A. M., Kefalos, J., Lhermitte, F., and Agid, Y. (1985). Le test a la L-DOPA dans la maladie de Parkinson. Rev. Neurol. Paris. 141, 413–415.

Fabbrini, G., Mouradian, M. M., Juncos, J. L., Schlegel, J., Mohr, E., and Chase, T. N. (1988). Motor fluctuations in Parkinson's disease: Central pathophysiological mechanisms, Part I. Ann. Neurol. 24, 366–371.

Fahn, S. (1977). Secondary parkinsonism. In "Scientific Approaches to Clinical Neurology" (E. S. Goldensohn and S. H. Appel, eds.), pp. 1159–1189. Lea & Febiger, Philadelphia.

Fahn, S., and Bressman, S. B. (1984). Should levodopa therapy for Parkinsonism be started early or late? Evidence against early treatment. Can. J. Neurol. Sci. 11, 200–206.

Fearnley, J. M., and Lees, A. (1991). Ageing and Parkinson's disease: Substantia nigra regional selectivity. Brain 114, 2283–2301.

Fenelon, G., Mahieux, F., Huon, R., and Ziegler, M. (2000). Hallucinations in Parkinson's disease: prevalence, phenomenology and risk factors. Brain. 123 (Pt 4), 733–745.

Ferreira, L. J., Galitzky, M., Montastruc, J. L., Rascol, O. (2000). Sleep attacks and Parkinson's disease treatment. Lancet 355, 1333–1334.

Fernandez, H. H., Friedman, J. H., Jacques, C., and Rosenfeld, M. (1999). Quetiapine for the treatment of drug-induced psychosis in Parkinson's disease. Mov. Disord. 14(3), 484–487.

Findley, L. J., Gresty, M. A., and Halmagyi, G. M. (1981). Tremor, the cogwheel phenomenon and clonus in Parkinson's disease. J. Neurol. Neurosurg. Psychiat. 44, 534–546.

Fischer, P. A. (1984). "Vegetative Störungen beim Parkinson-Syndrom." Editiones Roche, Basel.

Fischer, P. A. (1987). Long-term course in Parkinson's syndrome and cerebral polypathy (Parkinson plus). Adv. Neurol. 45, 235–238.

Fischer, P. A., Przuntek, H., Maier, M., and Welzel, D. (1984). Kombinationsbehandlung früher Stadien des Parkinson-Syndroms mit Bromocriptin und Levodopa. *Dtsch. Med. Wschr.* **109**, 1279–1283.

Ford, B., Lynch, T., and Greene, P. (1994). Risperidone in Parkinson's disease (letter). *Lancet.* **344**, 681.

Fornadi, F., Milani, F., and Werner, M. (1994). Madopar dispersible in the treatment of advanced Parkinson's disease. *Clin. Neuropharmacol.* **17**(Suppl. 3), S7–S15.

Franck, X., and Noel, J. (1984). Acute extrapyramidal dysfunction after domperidone administration. Report of a case. *Helv. Paediatr. Acta.* **39**, 285–288.

Frankel, J. P., Lees, A. J., Kempster, P. A., and Stern, G. M. (1990). Subcutaneous apomorphine in the treatment of Parkinson's disease. *J. Neurol. Neurosurg. Psychiat.* **53**, 96–101.

French Clozapine Parkinson Study Group. (1999). Clozapine in drug-induced psychosis in Parkinson's disease. *Lancet.* **353**(9169), 2041–2042.

Friedman, J. H. (1991). The management of levodopa psychoses. *Clin. Neuropharmacol.* **14**, 283–295.

Friedman, J. H., and Factor, S. A. (2000). Atypical antipsychotics in the treatment of drug-induced psychosis in Parkinson's disease. *Mov. Disord.* **15**(2), 201–211. Review.

Friedman, J. H., and Lannon, M. C. (1990). Clozapine-responsive tremor in Parkinson's disease. *Mov. Disord.* **5**, 225–229.

Frucht, S., Rogers, J. D., Greene, P. E., Gordon, M. F., and Fahn, S. (1999). Falling asleep at the wheel: motor vehicle mishaps in persons taking pramipexole and ropinirole. *Neurology.* **52**(9), 1908–1910.

Fumagalli, F., Gainetdinov, R. R., Wang, Y. M., Valenzano, K. J., Miller, G. W., and Caron, M. G. (1999). Increased methamphetamine neurotoxicity in heterozygous vesicular monoamine transporter 2 knock-out mice. *J. Neurosci.* **19**(7), 2424–2431.

Gainetdinov, R. R., Fumagalli, F., Jones, S. R., and Caron, M. G. (1997). Dopamine transporter is required for in vivo MPTP neurotoxicity: evidence from mice lacking the transporter. *J. Neurochem.* **69**(3), 1322–1325.

Gancher, S. T., Nutt, J. G., and Woodward, W. R. (1995). Apomorphine infusional therapy in Parkinson's disease: Clinical utility and lack of tolerance. *Mov. Disord.* **10**, 37–43.

Gasser T. (2001). Genetics of Parkinson's disease. *J. Neurol.* **248**(10), 833–840.

Gasser, T., Fritsch, G., Arnold, G., and Oertel, W. H. (1991). Botulinum-toxin A in orthopedic surgery (Letter). *Lancet* **338**, 761.

Gasser, T., Schwarz, J., Arnold, G., Trenkwalder, C., and Oertel, W. H. (1992). Apomorphine test for dopaminergic responsiveness in patients with previously untreated Parkinson's disease. *Arch. Neurol.* **49**, 1131–1134.

Gasser, T., Wszoled, Z. K., Trofatter, J., Ozelius, L., Uitti, R. J., Lee, C. S., Gusella, J., Pfeiffer, R. F., Calne, D. B., and Breakefield, X. O. (1994). Genetic linkage studies in autosomal dominant parkinsonism: Evaluation of seven candidate genes. *Ann. Neurol.* **36**, 387–396.

Gerlach, M., Riederer, P., and Vogt, H. (1996). Effect of adding selegeline to levodopa in early, mild Parkinson's disease. "On treatment" rather than intention to treat analysis should have been used. *BMJ.* **312**(7032):704; discussion 704–705.

Gerlach, M., Reichmann, H., Riederer, P. (2001). "Die Parkinson Krankheit," 2nd edition. Springer, Wien.

Gervason, C. L., Pollak, P. R., Limousin, P., and Perret, J. E. (1993). Reproducibility of motor effects induced by successive subcutaneous apomorphine injections in Parkinson's disease. *Clin. Neuropharmacol.* **16**, 113–119.

Gibb, W. R. G., Fearnley, J. M., and Lees, A. J. (1990). The anatomy and pigmentation of the human substantia nigra in relation to selective neuronal vulnerability. *In* "Parkinson's Disease—Anatomy, Pathology, and Therapy" (M. B. Streifler, A. D. Korczyn, E. Melamed, and M. B. H. Youdim, Eds.), Advances in Neurology, Vol. 53, pp. 31–40. Raven, New York.

Giladi, N. (2001). Freezing of gait. Clinical overview. *Adv. Neurol.* **87**, 191–197.

Giladi, N., Treves, T. A., Paleacu, D., Shabtai, H., Orlov, Y., Kandinov, B., Simon, E. S., and Korczyn, A. D. (2000). Risk factors for dementia, depression and psychosis in long-standing Parkinson's disease. *J. Neural. Transm.* **107**(1), 59–71.

Goetz, C. G., Blasucci, L. M., Leurgans, S., and Pappert, E. J. (2000). Olanzapine and clozapine: comparative effects on motor function in hallucinating PD patients. *Neurology.* **55**(6), 789–794.

Goetz, C. G., Blasucci, L., and Stebbins, G. T. (1999). Switching dopamine agonists in advanced Parkinson's disease: is rapid titration preferable to slow? *Neurology.* **52**(6), 1227–1229.

Goetz, C. G., Tanner, C. M., Gilley, D. W., and Klawans, H. L. (1989). Development and progression of motor fluctuations and side effects in Parkinson's disease: Comparison of Sinemet CR versus carbidopa/levodopa. *Neurology* **39**(Suppl. 2), 63–66.

Goetz, C. G., Tanner, C. M., Glantz, R., and Klawans, H. L. (1983). Pergolide in Parkinson's disease. *Arch. Neurol.* **40**, 785–787.

Golbe, L. I. (1987). Parkinson's disease and pregnancy. *Neurology.* **37**, 1245–1249.

Golbe, L. I., Di Iorio, G., Bonavita, V., Miller, D. C., and Duvoisin, R. C. (1990). A large kindred with autosomal dominant Parkinson's disease. *Ann. Neurol.* **27**, 276–282.

Gonera, E. G., van't Hof, M., Berger, H. J., van Weel, C., and Horstink, M. W. (1997). Symptoms and duration of the prodromal phase in Parkinson's disease. *Mov. Disord.* **12**, 871–876.

Graevenitz, K. S., Schulmann, L. M., and Rewel, R. S. (1996). Levodopa in pregnancy. *Mov. Disord.* **11**, 115–116.

Guttman, M. (1997). Double-blind comparison of pramipexole and bromocriptine treatment with placebo in advanced Parkinson's disease. International Pramipexole-Bromocriptine Study Group. *Neurology.* **49**(4), 1060–1065.

Hauser, R. A., and Zesiewicz, T. A. (1997). Sertraline for the treatment of depression in Parkinson's disease. *Mov. Disord.* **12**(5), 756–759.

Heikkila, R. E., Manzino, L., Cabbat, F. S., and Duvoisin, R. C. (1984). Protection against the dopaminergic neurotoxicity of 1-methyl-1,2,5,6-tetrahydropyridine by monoamine oxidase inhibitors. *Nature* **311**, 467–469.

Heinonen, E. H., and Rinne, U. K. (1989). Selegiline in the treatment of Parkinson's disease. *Acta Neurol. Scand.* **126**, 103–111.

Hely, M. A., Morris, J. G. L., Reid, W. G., O'Sullivan, D. J., Williamson, P. M., Rail, D., Broe, G. A., and Margrie, S. (1994). The Sydney Multicentre Study of Parkinson's disease: A randomised, prospective five year study comparing low dose bromocriptine with low dose levodopa–carbidopa. *J. Neurol. Neurosurg. Psychiat.* **57**, 903–910.

Hely, M. A., Morris, J. G., Traficante, R., Reid, W. G., O'Sullivan, D. J., Williamson, P. M. (1999). The sydney multicentre study of parkinson's disease : progression and mortality at 10 years. *J. Neurol. Neurosurg. Psychiatry.* **67**(3), 300–307.

Hoehn, M. M. (1985). Results of chronic levodopa therapy and its modification by bromocriptine in Parkinson's disease. *Acta Neurol. Scand.* **71**, 97–106.

Hoehn, M. M., and Yahr, D. M. (1967). Parkinsonism: Onset, progression and mortality. *Neurology* **17**, 427–442.

Hornykiewicz, O. (1982). Brain neurotransmitter changes in Parkinson's disease. *In* "Movement Disorders" (C. D. Marsden and S. Fahn, Eds.), pp. 41–48. Butterworth, London.

Huang, C. C., Lu, C. S., Chu, N. S., Hochberg, F., Lilienfeld, D., Olanow, W., and Calne, D. B. (1993). Progression after chronic manganese exposure. *Neurology* **43**, 1479–1483.

Hughes, A. J., Lees, A. J., and Stern, G. M. (1990). Apomorphine test to predict dopaminergic responsiveness in parkinsonian syndromes. *Lancet*. **II**, 32–34.

Inzelberg, R., Nisipeanu, P., Rabey, J. M., Orlov, E., Catz, T., Kippervasser, S., Schechtmann, E., and Korczyn, A. D. (1996). Double blind comparison of cabergoline and bromocriptine in Parkinson's disease patients with motor fluctuations. *Neurology*. **47**, 785–788.

Jackson, J. A., Jankovic, J., and Ford, J. (1983). Progressive supranuclear palsy: Clinical features and response to treatment in 16 patients. *Ann. Neurol.* **13**, 273–278.

Jankovic, J. (1990). Lower body (vascular) parkinsonism [letter]. *Arch. Neurol.* **47**, 728.

Jankovic, J., Beach, J., Schwartz, K., and Contant, C. (1995). Tremor and longevity in relatives of patients with Parkinson's disease, essential tremor, and control subjects. *Neurology*. **45**, 645–648.

Jellinger, K. (1986). Pathology of parkinsonism. *In* "Recent Developments in Parkinson's Disease" (S. Fahn, C. D. Marsden, and P. Teychenne, Eds.), pp. 33–66.

Jellinger, K. (1990). New developments in the pathology of Parkinson's disease. *In* "Parkinson's Disease—Anatomy, Pathology, and Therapy" (M. B. Streifler, A. D. Korczyn, E. Melamed, and M. B. H. Youdim, Eds.), Advances in Neurology, Vol. 53, pp. 1–16. Raven, New York.

Jimenez-Jimenez, F. J., Tejeiro, J., Martinez-Junquera, G., Cabrera-Valdivia, F., Alarcon, J., and Garcia-Albea, E. (1994). Parkinsonism exacerbated by paroxetine. *Neurology*. **44**(12), 2406.

Jost, W. H., and Schimrigk, K. (1993). Cisapride treatment of constipation in Parkinson's disease. *Mov. Disord.* **8**, 339–343.

Juncos, J. L. (1999). Management of psychotic aspects of Parkinson's disease. *J. Clin. Psychiatry*. **60** Suppl 8, 42–53.

Kaakkola, S., Teravainen, H., Ahtila, S., Rita, H., and Gordin, A. (1994). Effect of entacapone, a COMT inhibitor, on clinical disability and levodopa metabolism in parkinsonian patients. *Neurology*. **44**, 77–80.

Kebabian, J. W., and Calne, D. B. (1979). Multiple receptors for dopamine. *Nature*. **277**, 93–96.

Kish, S. J., Shannak, K., and Hornykiewicz, O. (1988). Uneven pattern of dopamine loss in the striatum of patients with idiopathic Parkinson's disease: Pathophysiologic and clinical implications. *N. Engl. J. Med.* **318**, 876–881.

Klein, C., Pramstaller, P. P., Kis, B., Page, C. C., Kann, M., Leung, J., Woodward, H., Castellan, C. C., Scherer, M., Vieregge, P., Breakefield, X. O., Kramer, P. L., and Ozelius, L. J. (2000). Parkin deletions in a family with adult-onset, tremor-dominant parkinsonism: expanding the phenotype. *Ann. Neurol.* **48**(1), 65–71.

Klockgether, T., Jacobsen, P., Lüschmann, P. A., and Turski, L. (1993). The antiparkinsonian agent budipine is an *N*-methyl-D-aspartate antagonist. *J. Neural. Transm. Park. Dis. Dement. Sect.* **5**, 101–106.

Klockgether, T., and Turski, L. (1993). Toward an understanding of the role of glutamate in experimental parkinsonism: Agonist sensitive sites. *Ann. Neurol.* **34**, 585–593.

Kofman, O. S. (1984). Are levodopa "drug-holidays" justified? *Can. J. Neurol. Sci.* **11**, 206–209.

Koller, W. C., Hutton, J. T., Tolosa, E., and Capilldeo, R. (1999). Immediate-release and controlled-release carbidopa/levodopa in PD: a 5-year randomized multicenter study. Carbidopa/Levodopa Study Group. *Neurology*. **22**;53(5), 1012–1019.

Koller, W. C., Hutton, J. T., Tolosa, E., and Capilldeo, R. (2000). Immediate-release and controlled-release carbidopa/levodopa in PD: a 5-year randomized multicenter study. Carbidopa/Levodopa Study Group (Comment). *Neurology*. **12**;55(1), 156–157.

Koller, W., Lees, A., Doder, M., and Hely, M. (2001). Randomized trial of tolcapone versus pergolide as add-on to levodopa therapy in Parkinson's disease patients with motor fluctuations. *Mov. Disord.* **16**(5), 858–866.

Koller, W. C., Verteren-Overfield, B., Williamsen, A., Busenbork, K., Naoh, J., and Parrish, D. (1990). Sexual dysfunction in Parkinson's disease. *Clin. Neuropharmacol.* **13**, 461–463.

Konagaya, M., Konagaya, Y., and Iida, M. (1995). Clinical and magnetic resonance imaging study of extrapyramidal symptoms in multiple system atrophy. *J. Neurol. Neurosurg. Psychiat.* **57**, 1528–1531.

Kondo, T. (1993). L-threo-DOPS in advanced parkinsonism. *In* "Parkinson's Disease—From Basic Research to Treatment" (T. Nagatsu, N. Yanagisawa, and Y. Mizuno, Eds.), Advances in Neurology, Vol. 60, pp. 660–665. Raven, New York.

Kornhuber, J., Bormann, J., Retz, W., Hübers, M., and Riederer, P. (1989). Memantine displaces [$^3$H]MK 801 at therapeutic concentrations in postmortem human frontal cortex. *Eur. J. Pharmacol.* **166**, 589–590.

Kostic, V., Przedborski, S., Flaster, E., and Sternic, N. (1991). Early development of levodopa—induced dyskinesias and response fluctuations in young-onset Parkinson's disease. *Neurology*. **41**, 202–205.

Krack, P., Poepping, M., Weinert, D., Schrader, B., and Deuschl, G. (2000). Thalamic, pallidal, or subthalamic surgery for Parkinson's disease (Review)? *J. Neurol.* **247** Suppl 2, II122–134.

Kupsch, A., and Oertel, W. H. (1998). Selegiline, pregnancy, and Parkinson's disease. *Mov. Disord.* **13**(1), 175–176.

Kurth, M. C., Adler, C. H., Hilaire, M. S., *et al.* (1997). Tolcapone improves motor function and reduces levodopa requirement in patients with Parkinson's disease experiencing motor fluctuations: a multicenter, double-blind, randomized, placebo-controlled trial. Tolcapone Fluctuator Study Group I. *Neurology*. **48**, 81–87.

Laitinen, L. V., Bergenheim, A. T., and Hariz, M. I. (1992). Leksell's posteroventral pallidotomy in the treatment of Parkinson's disease. *J. Neurosurg.* **77**, 487–488.

Lance, J. W., Schwab, R. S., and Peterson, E. A. (1963). Action tremor and the cogwheel phenomenon in Parkinson's disease. *Brain* **86**, 95–110.

Lang, A. E., Curran, T., Provias, J., and Bergeron, C. (1994). Striatonigral degeneration: Iron deposition in putamen correlates with the slitlike void signal of magnetic resonance imaging. *Can. J. Neurol. Sci.* **21**, 311–318.

Langston, J. W., and Ballard, P. (1984). Parkinsonism induced by 1-methyl-4-phenyl-1,2,3,6-tetrahydropyridine (MPTP): Implications for treatment and the pathogenesis of Parkinson's disease. *Can. J. Neurol. Sci.* **11**, 160–165.

Laruelle, M., Baldwin, R. M., and Malison, R. T. (1993). SPECT Imaging of dopamine and serotonin transporters with [$^{123}$I]βCIT: Pharmacological characterization of brain uptake in nonhuman primates. *Synapse*. **13**, 295–309.

Lee, M. S., and Marsden, C. D. (1994). Neurological sequelae following carbon monoxide poisoning clinical course and outcome according to the clinical types and brain computed tomography scan finding. *Mov. Disord.* **9**, 550–558.

Leenders, K. L., Aquilonius, S. M., Fowler, J. S., *et al.* (1987). Localization and quantification of monoamine oxidase (MAO-B) in brain in healthy volunteers and parkinsonian patients using PET. *J. Cereb. Flow Metab.* 7, S370.

Leenders, K. L., Salmon, E. P., Tyrrell, P., Perani, D., Brooks, D. J., Sager, H., Jones, T., Marsden, C. D., and Frackowiak, R. S. J. (1990). The nigrostriatal dopaminergic system assessed *in vivo* by positron emission tomography in healthy volunteer subjects and patients with Parkinson's disease. *Arch. Neurol.* 47, 1290–1298.

Lees, A. J. (1994). Levodopa substitution: The gold standard. *Clin. Neuropharmacol.* 17, S1–S6.

Lees, A. J. (1995). Comparison of therapeutic effects and mortality data of levodopa and levodopa combined with selegiline in patients with early, mild Parkinson's disease. Parkinson's Disease Research Group of the United Kingdom. *BMJ.* 311(7020), 1602–1607.

Lees, A. J., and Stern, G. M. (1983). Sustained low dose levodopa therapy in Parkinson's disease. A 3-year follow-up. *In* "Experimental Therapeutics of Movement Disorders" (S. Fahn, D. B. Calne, and I. Shoulson, Eds.), Advances in Neurology, Vol. 37, pp. 9–15. Raven Press, New York.

Lera, G., Vaamonde, J., Rodriguez, M., and Obeso, J. A. (1993). Cabergoline in Parkinson's disease: Long-term follow-up. *Neurology.* 43, 2587–2590.

Levi, S., Cox, M., Lugon, M., Hodkinson, M., and Tomkins, A. (1990). Increased energy expenditure in Parkinson's disease. *B.M.J.* 301, 1256–1257.

LeWitt, P. A., and Truong, D. D. (1990). Budipine in Parkinson's disease. *In* "Therapy of Parkinson's Disease" (W. C. Koller and G. Paulson, Eds.), pp. 399–404. Marcel Dekker, New York.

Lieberman, A. N., and Goldstein, M. (1982). Treatment of advanced Parkinson's disease with dopamine agonists. *In* "Movement Disorders" (C. D. Marsden and S. Fahn, Eds.), pp. 146–165. Butterworth Scientific, London.

Lieberman, A. N., Goldstein, M., Gopinathan, G., Neophytides, A., Leibowitz, M., Walker, R., and Hiesinger, E. (1983). Lisuride in Parkinson's disease and related disorders. *In* "Lisuride and Other Dopamine Agonist" (D. B. Calne, R. Horowski, R. McDonald, and W. Wuttke, eds.), pp. 419–429.

Lieberman, A., Ranhosky, A., and Korts, D. (1997). Clinical evaluation of pramipexole in advanced Parkinson's disease: results of a double-blind, placebo-controlled, parallel-group study. *Neurology.* 49(1), 162–168.

Limousin, P., Pollak, P., Benazzouz, A., Hoffmann, D., Le-Bas, J. F., Broussolle, E., Perret, J. E., and Benabid, A. L. (1995). Effect of parkinsonian signs and symptoms of bilateral subthalamic nucleus stimulation. *Lancet* 345 (8942), 91–95.

Limousin, P., Pollak, P., Gervason-Tournier, C. L., Hommel, M., and Perret, J. E. (1993). COMT inhibitor, plus levodopa in Parkinson's disease. *Lancet.* 341, 1605.

Lucking, C. B., Abbas, N., Durr, A., Bonifati, V., Bonnet, A. M., de Broucker, T., De Michele, G., Wood, N. W., Agid, Y., and Brice, A. (1998). Homozygous deletions in parkin gene in European and North African families with autosomal recessive juvenile parkinsonism. The European Consortium on Genetic Susceptibility in Parkinson's Disease and the French Parkinson's Disease Genetics Study Group. *Lancet.* 352(9137), 1355–1356.

Lucking, C. B., Durr, A., Bonifati, V., Vaughan, J., De Michele, G., Gasser, T., Harhangi, B. S., Meco, G., Denefle, P., Wood, N. W., Agid, Y., and Brice, A. (2000). Association between early-onset Parkinson's disease and mutations in the parkin gene. French Parkinson's Disease Genetics Study Group. *N. Engl. J. Med.* 342(21), 1560–1567.

Maher, E. R., and Lees, A. J. (1986). The clinical features and natural history of the Steele-Richardson-Olszewski syndrome (progressive supranuclear palsy). *Neurology.* 36, 1005–1008.

Marek, K. L., Seibyl, J. P., Sandridge, B., Fussell, B., Smith, E. O., Baldwin, R. M., Zoghbi, S., Hoffer, P. B., and Innis, R. B. (1994). SPECT imaging with [I-$^{123}$]β-CIT demonstrates striatal dopamine transporter loss in hemi-Parkinsonism. *Mov. Disord.* 9(Suppl. 1), 103.

Markham, C. G., and Diamond, S. G. (1981). Evidence to support early levodopa therapy in Parkinson's disease. *Neurology.* 31, 125–131.

Markham, C. G., and Diamond, S. G. (1986). Modification of Parkinson's disease by long-term levodopa treatment. *Arch. Neurol.* 43, 405–407.

Marsden, C. D., Parkes, J. D., and Quinn, N. (1982). Fluctuations of disability in Parkinson's disease—Clinical aspects. *In* "Movement Disorders" (C. D. Marsden and S. Fahn, Eds.), pp. 96–122. Butterworth Scientific, London.

Martinez-Martin, P., and Bermejo-Pareja, F. (1988). Rating scales in Parkinson's disease. *In* "Parkinson's Disease and Movement Disorders" (J. Jankovic and E. Tolosa, Eds.), pp. 235–242. Urban & Schwarzenberg, Munich.

Marttila, R. J. (1987). Epidemiology. *In* "Handbook of Parkinson's Disease" (W. C. Koller, Ed.), pp. 35–50. Marcel Dekker, New York.

Mayberg, H. S., Starkstein, S. E., Sadzot, B., Preziosi, T., Andrezejewski, P. L., Dannals, R. F., Wagner, H. N. Jr., and Robinson, R. G. (1990). Selective hypometabolism in the inferior frontal lobe in depressed patients with Parkinson's disease. *Ann. Neurol.* 28(1), 57–64.

Mayeux, R. (1990). The "Serotonin Hypothesis" for depression in Parkinson's disease. *In* "Parkinson's Disease—Anatomy, Pathology, and Therapy" (M. B. Streifler, A. D. Korczyn, E. Melamed, and M. B. H. Youdin, Eds.), Advances in Neurology, Vol. 53, pp. 163–166. Raven, New York.

Mayeux, R., Denaro, J., Hemenegildo, N., Marder, K., Tang, M. X., Cote, L. J., and Stern, Y. (1992). A population-based investigation of Parkinson's disease with and without dementia: Relationship to age and gender. *Arch. Neurol.* 49, 492–497.

Mayeux, R., Stern, Y., Cote, L., and Williams, J. B. (1984). Altered serotonin metabolism in depressed patients with parkinson's disease. *Neurology.* 34(5), 642–646.

Meara, J., and Hobson, P. (1998). Sertraline for the treatment of depression in Parkinson's disease. *Mov. Disord.* 13(3), 622.

Meara, J., Bhowmick, B. K., and Hobson, P. (1999). Accuracy of diagnosis in patients with presumed Parkinson's disease. *Age Ageing* 28, 335–336.

Melamed, E., Hefti, F., and Wurtman, R. J. (1980). Nonaminergic striatal neurons convert exogenous L-DOPA to dopamine in Parkinsonism. *Ann. Neurol.* 8, 558–563.

Menza, M. A., Palermo, B., DiPaola, R., Sage, J. I., and Ricketts, M. H. (1999). Depression and anxiety in Parkinson's disease: possible effect of genetic variation in the serotonin transporter. *J. Geriatr. Psychiatry Neurol.* 12(2), 49–52.

Merchant, C. A., Cohen, G., Mytilineou, C., DiRocco, A., Moros, D., Molinari, S., and Yahr, M. D. (1995). Human transplacental transfer of carbidopa/levodopa. *J. Neural Transm. Park. Dis. Dement. Sect.* 9, 239–242.

Merello, M., Lees, A. J., Webster, R., Bovingdon, M., and Gordin, A. (1994). Effect of entacapone, a peripherally acting catechol-*O*-methyltransferase inhibitor, on the motor response to acute treatment with levodopa in patients with Parkinson's disease. *J. Neurol. Neurosurg. Psychiat.* 57, 186–189.

Mierau, J., Schneider, F. J., Ensinger, H. A., Chio, C. L., Lajiness, M. E., and Huff, R. M. (1995). Pramipexole binding and activation of cloned and expressed dopamine $D_2$, $D_3$ and $D_4$ receptors. *Eur. J. Pharmacol.* 290, 29–36.

Moellentine, C., Rummans, T., Ahlskog, J. E., Harmsen, W. S., Suman, V. J., O'Connor, M. K., Black, J. L., and Pileggi, T. (1998). Effectiveness of ECT in patients with parkinsonism. *J. Neuropsychiatry Clin. Neurosci.* **10**(2), 187–193.

Möller, C., Stiasny, K., Krüger, P. H., and Oertel, W. H. (2002). Sudden onset somnolence under dopamine agonist treatment—driving simulation and sleep tests. *Mov. Disord.* **17**, 474–481.

Montastruc, J. L., Rascol, O., Senard, J. M., and Rascol, A. (1994). A randomized controlled study comparing bromocriptine to which levodopa was later added, with levodopa alone in previously untreated patients with Parkinson's disease: A five year follow up. *J. Neurol. Neurosurg. Psychiat.* **57**, 1034–1038.

Montastruc, J. L., Desboeuf, K., Lapeyre-Mestre, M., Senard, J. M., Pascol, O., and Brefel-Courbon, C. (2001). Long-term mortality results of the randomized controlled study comparing bromocriptine to which levodopa was later added with levodopa alone in previously untreated patients with Parkinson's disease. *Mov. Disord.* **16**(3), 511–514.

Mouradian, M. M., Heuser, I. J. E., Baronti, F., and Chase, T. N. (1990). Modification of central dopaminergic mechanisms with continuous levodopa infusion therapy for advanced Parkinson's disease. *Ann. Neurol.* **27**, 18–23.

Mouradian, M. M., Heuser, I. J. E., Baronti, F., Fabbrini, G., Juncos, I. L., and Chase, T. N. (1989a). Pathogenesis of dyskinesias in Parkinson's disease. *Ann. Neurol.* **25**, 523–526.

Mouradian, M. M., Juncos, J. L., Fabbrini, G., and Chase, T. N. (1987). Motor fluctuations in Parkinson's disease: Pathogenetic and therapeutic studies. *Ann. Neurol.* **22**, 475–479.

Mouradian, M. M., Juncos, J. L., Fabbrini, G., Schlegel, J., Bartko, J. J., and Chase, T. N. (1988). Motor fluctuations in Parkinson's disease: central pathophysiological mechanisms. Part II. *Ann. Neurol.* **24**, 372–378.

Mouradian, M. M., Juncos, J. L., Fabbrini, G., and Chase, T. N. (1989b). Motor fluctuations in Parkinson's disease. *Ann. Neurol.* **25**, 633–634.

Mutch, W. J., Dingwall-Fordyce, I., Downie, A. W., Paterson, J. G., and Roy, S. K. (1986). Parkinson's disease in a Scottish city. *BMJ. (Clin. Res. Ed.)* **292**(6519), 534–536.

Myllyla, V. V., Jackson, M., Larsen, J. P., and Baas, H. (1997). Efficacy and safety of tolcapone in levodopa-treated Parkinson's disease patients with "wearing-off" phenomenon: a multicentre, double-blind, randomized, placebo-controlled trial. *Eur. J. Neuro.* **4**(4), 333–341.

Myllyla, V. V., Sotaniemi, K. A., Vu-orinen, J. A., and Heinonen, R. H. (1992). Selegiline as initial treatment in *de novo* parkinsonian patients. *Neurology* **42**, 339–343.

Myllyla, V. V., Sotaniemi, K. A., Illi, A., *et al.* (1993). Effects of entacapone, a COMT inhibitor, on the pharmacokinetics of levodopa and on cardiovascular responses on patients with Parkinson's disease. *Eur. J. Clin. Pharmacol.* **45**, 419–423.

Nissenbaum, H., Quinn, N. P., Brown, R. G., Toone, B., Gotham, A.-M., and Marsden, C. D. (1987). Mood swing associated with the "on-off" phenomenon in Parkinson's disease. *Psychol. Med.* **17**, 899–904.

Nutt, J. G., Carter, J. H., and Woodward, W. R. (1994a). Effect of brief levodopa holidays on the short-duration response to levodopa: Evidence for tolerance to the antiparkinsonian effects. *Neurology* **44**, 1617–1622.

Nutt, J. G., Woodward, W. R., Beckner, R. N., Stone, C. K., Berggren, K., Carter, J. H., and Gancher, S. T. (1994b). Effect of peripheral catechol-O-methyltransferase inhibition on the pharmacokinetics and pharmacodynamics of levodopa in parkinsonian patients. *Neurology* **44**, 913–919.

Obeso, J. A., Martinez-Lage, J. M., Luquin, M. R., and Bolio, N. (1985). Intravenous lisuride infusion for Parkinson's disease. *Ann. Neurol.* **18**, 252.

Obeso, J. A., Benabid, A. L., and Koller, W. C. (2000). Deep brain stimulation for Parkinson's disease and tremor. *Neurology* **55**(Suppl 6), S1–S66.

Oertel, W. H. (1994). Basalganglien-Erkrankungen. *In* "Klinische Pathophysiologie" (W. Siegenthaler, Hrsg.), 7. Aufl., pp. 865–871. Thieme-Verlag, Stuttgart.

Oertel, W. H., Bandmann, O., Eichhorn, T., and Gasser, T. (1996). Peripheral Markers in Parkinson's Disease—An Overview. *In* "Parkinson's disease" (L. Battistin, G. Scarlato, T. Caraceni, and S. Ruggieri, Eds.), Advances of Neurology, Vol. 69, pp. 283–291. Lippincott-Raven, New York.

Oertel, W. H., and Dodel, R. C. (1995). Special Article: International guide to drugs for Parkinson's disease. *Mov. Disord.* **10**, 121–131.

Oertel, W. H., and Ellgring, H. (1995). Education of Parkinson patients and their partners: Understanding the biomedical basis and interventions for psychosocial problems. *Patient Edu. Counsel.* **26**, 71–79.

Oertel, W. H., Gasser, T., Ippisch, R., Trenkwalder, C., and Poewe, W. (1989). Apomorphine test for dopaminergic responsiveness. *Lancet* **I**, 1262–1263.

Oertel, W. H., Hoglinger, G. U., Caraceni, T., Girotti, F., Eichhorn, T., Spottke, A. E., Krieg, J. C., and Poewe, W. (2001). Depression in Parkinson's disease. An update (Review). *Advances of Neurology.* **86**, 373–383.

Olanow, C. W., Fahn, S., Muenter, M., Klawans, H., Hurtig, H., Stern, M., Shoulson, I., Kurlan, R., Grimes, J. D., Jankovic, J., Hoehn, M., Markham, C. H., Duvoisin, R., Reinmuth, O., Leonard, H. A., Ahlskog, E., Feldman, R., Hershey, L., and Yahr, M. D. (1994). A multicenter double-blind placebo-controlled trial of pergolide as an adjunct to Sinemetþ in Parkinson's disease. *Mov. Disord.* **9**, 40–47.

Olanow, C. W., Hauser, R. A., Gauger, L., Malapira, T., Koller, W., Hubble, J., Buschenbark, K., Lilienfeld, D., and Esterlitz, J. (1995). The effect of deprenyl and levodopa on the progression of Parkinson's disease. *Ann. Neurol.* **38**, 771–777.

Olanow, C. W., Myllyla, V. V., Sotaniemi, K. A., Larsen, J. P., Palhagen, S., Przuntek, H., Heinonen, E. H., Kilkku, O., Lammintausta, R., Maki-Ikola, O., and Rinne, U. K. (1998). Effect of selegiline on mortality in patients with Parkinson's disease: a meta-analysis. *Neurology* **51**(3), 825–830.

Olanow, W., Schapira, A. H., and Rascol, O. (2000). Continuous dopamine-receptor stimulation in early Parkinson's disease (Review). *Trends Neurosci.* **23**(10 Suppl), S117–126.

Olanow, C. W., Watts, R. L., and Koller, W. C. (2001). An algorithm (decision tree) for the management of Parkinson's disease (2001): Treatment guidelines. *Neurology* **54**(Suppl. 5), S1–S88.

Orth, M., and Schapira, A. H. (2001). Mitochondria and degenerative disorders. *Am. J. Med. Genet.* **106**(1), 27–36.

Pakkenberg, H., and Pakkenberg, B. (1986). Clozapine in the treatment of tremor. *Acta Neurol. Scand.* **73**, 295–297.

Palhagen, S., Heinonen, E. H., Hagglund, J., Kaugesaar, T., Kontants, H., Maki-Ikola, O., Palm, R., and Turunen, J. (1998). Selegiline delays the onset of disability in de novo parkinsonian patients. Swedish Parkinson Study Group. *Neurology* **51**(2), 520–525.

Parkinson Study Group. (1996a). Impact of deprenyl and tocopherol treatment on Parkinson's disease in DATATOP subjects not requiring levodopa. *Ann. Neurol.* **39**, 29–36.

Parkinson Study Group. (1996b). Impact of deprenyl and tocopherol treatment on Parkinson's disease in DATATOP patients requiring levodopa. *Ann. Neurol.* **39**, 37–45.

Parkinson Study Group. (1997). Entacapone improves motor fluctuations in levodopa-treated Parkinson's disease patients. *Ann. Neurol.* **42**, 747–755.

Parkinson Study Group. (1999). Low-dose clozapine for the treatment of drug-induced psychosis in Parkinson's disease. *N. Engl. J. Med.* **340**(10), 757–763.

Parkinson Study Group. (2000). Pramipexole vs levodopa as initial treatment for Parkinson disease: A randomized controlled trial. Parkinson Study Group. *JAMA.* **284**(15), 1931–1938.

Parkinson's Disease Research Group in the United Kingdom. (1993). Comparisons of therapeutic effects of levodopa, levodopa and selegiline, and bromocriptine in patients with early, mild Parkinson's disease: three year interim report. *BMJ.* **307**, 469–472.

Parkinson's Disease Research Group of the United Kingdom. (1995). Comparison of therapeutic effects and mortality data of levodopa and levodopa combined with selegiline in patients with early, mild Parkinson's disease. *BMJ.* **311**, 1602–1607.

Pellegrino, M., Sacco, M., and Lotti, A. (1990). Extrapyramidal syndrome caused by moderate overdosage of domperidone: Description of a case. *Pediatr. Med. Chir.* **12**, 205–206.

Penney, J. B., and Young, A. B. (1986). Striatal inhomogeneities and basal ganglia function. *Mov. Disord.* **1**, 3–15.

Perugi, G., Toni, C., Ruffolo, G., Frare, F., and Akiskal, H. (2001). Adjunctive dopamine agonists in treatment of resistant bipolar II depression: an open case series. *Pharmacopsychiatry* **34**(4), 137–141.

Pezzoli, G., Antonini, A., Barbieri, S., Canesi, M., Perbellini, L., Zecchinelli, A., Mariani, C. B., Bonetti, A., and Leenders, K. L. (1995). *n*-Hexane induced parkinsonism: Pathogenic hypotheses. *Mov. Disord.* **10**, 279–282.

Pezzoli, G., Martignoni, E., Pacchetti, C., Angeleri, V. A., Lamberti, P., Muratorio, A., Bonuccelli, U., De Mari, M., Foschi, N., Cossutta, E., Nicoletti, F., Giammona, F., Canesi, M., Scarlato, G., Caraceni, T., and Moscarelli, E. (1994). Pergolide compared with bromocriptine in Parkinson's disease: A multicenter, crossover, controlled study. *Mov. Disord.* **9**, 431–436.

Piercey, M. F., Hoffmann, W. E., Smith, M. W., and Hyslop, D. K. (1996). Inhibition of dopamine neuron firing by pramipexole, a dopamine $D^3$ receptor-preferring agonist: comparison to other dopamine receptor agonists. *Eur. J. Pharmacol.* **312**, 35–44.

Pincus, J. H., and Barry, K. M. (1987). Influence of dietary protein on motor fluctuations in Parkinson's disease. *Arch. Neurol.* **44**, 270–272.

Plante-Bordeneuve, V., Taussig, D., Thomas, F., Said, G., Wood, N. W., Marsden, C. D., and Harding, A. E. (1997). Evaluation of four candidate genes encoding proteins of the dopamine pathway in familial and sporadic Parkinson's disease: evidence for association of a DRD2 allele. *Neurology* **48**(6), 1589–1593.

Poewe, W. H., Deuschl, G., Gordin, A., Kultalahti, E. R., Leinonen, M. (2002). Efficacy and safety of entacapone in Parkinson's disease patients with suboptimal levodopa respanse: A 6-month randomized placebo controlled double-blind study in Germany and Austria (celomen study). *Acta. Neurol. Scand.* **105**, 245–255.

Poewe, W., Kleedorfer, B., Gerstenbrand, F., and Oertel, W. H. (1988a). Subcutaneous apomorphine in Parkinson's disease. *Lancet* **I**, 943.

Poewe, W. H., Lees, A. J., and Stern, G. M. (1986). Low-dose L-dopa therapy in Parkinson's disease: a 6-year follow-up study. *Neurology* **36**, 1528–1530.

Poewe, W. H., Lees, A. J., and Stern, G. M. (1988b). Dystonia in Parkinson's disease: Clinical and pharmacological features. *Ann. Neurol.* **23**, 73–78.

Poewe, W., and Luginger, E. (1999). Depression in Parkinson's disease: impediments to recognition and treatment options (Review). *Neurology* **52**(7 Suppl 3), S2–6.

Poewe, W., and Oertel, W. H. (1994). Parkinson's disease: akinetic crisis and dopamimetica-induced psychosis. In "Neurocritical Care" (W. Hacke, D. F. Hanley, K. M. Einhaupl, T. P. Bleck, and M. N. Diringer, Eds.), pp. 883–887. Springer-Verlag, Berlin.

Pogarell, O., Gasser, T., Spieker, S., van Hilten, J. J., Meier, D., Pollentier, S., and Oertel, W. H. (2002). Pramipexole is effective in early and advanced Parkinson's disease with predominant tremor—results of a multicenter double-blind randomized placebo controlled study. *J. Neurol. Neurosurg. Psychiatry* **72**, 713–720.

Pollack, P., Gaio, J. M., Hommel, M., Pellat, J., and Chateau, R. (1981). Etude aigue de l'association bromocriptine et dompéridone dans le syndrom parkinsonien. *Therapie* **36**, 671–676.

Priyadarshi, A., Khuder, S. A., Schaub, E. A., and Shrivastava, S. (2000). A meta-analysis of Parkinson's disease and exposure to pesticides. *Neurotoxicology* **21**(4), 435–440.

Priyadarshi, A., Khuder, S. A., Schaub, E. A., and Priyadarshi, S. S. (2001). Environmental risk factors and Parkinson's disease: a metaanalysis. *Environ. Res.* **86**(2), 122–127.

Przedborski, S., Levivier, M., Raftopoulos, C. H., Naini, A. B., and Hildebrand, J. (1995). Peripheral and central pharmacokinetics of apomorphine and its effect on dopamine metabolism in humans. *Mov. Disord.* **10**, 28–36.

Przuntek, H., Conrad, B., Dichgans, J., Kraus, P. H., Krauseneck, P., Pergande, G., Rinne, U., Schimrigk, K., Schnitker, J., and Vogel, H. P. (1999). SELEDO: a 5-year long-term trial on the effect of selegiline in early Parkinsonian patients treated with levodopa. *Eur. J. Neurol.* **6**(2), 141–150.

Quinn, N. (1989). Multiple system atrophy—the nature of the beast. *J. Neurol. Neurosurg. Psychiat.* (Suppl.), 78–89, Review.

Quinn, N. (1994). Multiple system atrophy. In "Movement Disorders" (C. D. Marsden and S. Fahn, Eds.), Vol. 3, chapter 13, pp. 262–281. Butterworth-Heinemann, London.

Quinn, N. (1995). Drug treatment of Parkinson's disease. *BMJ.* **310**, 575–579.

Quinn, N., Critchley, P., and Marsden, C. D. (1987). Young onset Parkinson's disease. *Mov. Disord.* **2**, 73–91.

Quinn, N., Illas, A., Lhermitte, F., and Agid, Y. (1981). Bromocriptine and domperidone in the treatment of Parkinson disease. *Neurology* **31**, 662–667.

Quinn, N., and Marsden, C. D. (1986). Lithium for painful dystonia in Parkinson's disease. *Lancet* **i**, 1377.

Quinn, N., Marsden, C. D., Schachter, M., Thompson, C., Lang, A. E., and Parkes, J. D. (1983). Intravenous Lisuride in extrapyramidal disorders. In "Lisuride and other Dopamine Agonists" (D. B. Calne, R. Horowski, R. J. Mcdonald, and W. Wuttke, Eds.), pp. 383–393. Raven, New York.

Rabey, J. M., Nissipeanu, P., Inzelberg, R., and Korczyn, A. D. (1994). Beneficial effect of cabergoline, new long-lasting $D_2$ agonist in the treatment of Parkinson's disease. *Clin. Neuropharmacol.* **17**, 286–293.

Rajput, A. H., Offort, K., Beard, C. M., and Kurland, L. T. (1984). Epidemiological survey of dementia in Parkinsonism and control population. In "Parkinson-Specific Motor and Mental Disorders" (R. G. Hassler and J. F. Christ, Eds.), *Adv. Neurol.*, Vol. 40, pp. 229–234. Raven, New York.

Rajput, A. H., Martin, W., Saint, Hilaire, M. H., Dorflinger, E., and Pedder, S. (1997). Tolcapone improves motor function in parkinsonian patients with the "wearing-off" phenomenon: a double-blind, placebo-controlled, multicenter trial. *Neurology* **49**:1066–1071.

Rajput, A. H., Stern, W., and Laverty, W. H. (1984). Chronic low dose levodopa therapy in Parkinson's disease. An argument for delaying levodopa therapy. *Neurology* **34**, 991–997.

Rascol, O., Brooks, D. J., Korczyn, A. D., De Deyn, P. P., Clarke, C. E., and Lang, A. E. (2000). A five-year study of the incidence of dyskinesia in patients with early Parkinson's disease who were treated with ropinirole or levodopa. 056 Study Group. *N. Engl. J. Med.* 342(20), 1484–1491.

Rascol, O., Ferreira, J. J., Thalamas, C., Galitsky, M., and Montastruc, J. L. (2001). Dopamine agonists. Their role in the management of Parkinson's disease (Review). *Advances of Neurology* 86, 301–309.

Rascol, O., Goetz, C., Koller, W., Poewe, W., and Sampaio, C. (2002). Treatment interventions for Parkinson's disease: An evidence based assessment. *Lancet.* 359, 1589–1598.

Riederer, P., and Jellinger, K. (1983). Neurochemical insights into monoamine oxidase inhibitors, with special reference to deprenyl (selegiline). *Acta Neurol. Scand.* 95(Suppl.), 43–55.

Riley, D., and Lang, A. E. (1988). Practical application of a low-protein diet for Parkinson's disease. *Neurology* 38, 1026–1031.

Rinne, U. K. (1983). Dopamine agonists in the treatment of Parkinson's disease. *In* "Experimental Therapeutics of Movement Disorders" (S. Fahn, D. B. Calne, and J. Shoulson, Eds.), Advances in Neurology, Vol. 37, pp. 141–150. Raven, New York.

Rinne, U. K. (1985). Combined bromocriptine-levodopa therapy early in Parkinson's disease. *Neurology* 35, 1196–1198.

Rinne, U. K. (1989). Combination of a dopamine agonist: MAO-B-inhibitor and levodopa-a new strategy in the treatment of early Parkinson's disease. *Acta Neurol. Scand.* 126, 165–169.

Rinne, U. K., Bracco, F., Chouza, C., Dupont, E., Gershanik, O., Marti, Masso, I. F., Montastruc, J. L., Marsden, C. D., and PKDS009 study group. (1998a). Early treatment of Parkinson's disease with cabergoline delays the onset of motor complictions: results of a double-blind L-dopa controlled trial. *Drugs* 55(Suppl 1), 23–30.

Rinne, J. D., Le, M. S., Thompson, P. D., and Marsden, C. D. (1994). Corticobasal degeneration: A clinical study of 86 cases. *Brain* 117, 1183–1196.

Rinne, U. K., Larsen, J. P., Siden, A., Worm, Petersen, J., and the Nomecomt Study Group. (1998b). Entacapone enhances the response to levodopa in parkinsonian patients with motor fluctuations. *Neurology* 51, 1309–1314.

Roberts, J. W., Cora-Locatelli, G., Bravi, D., Amantea, M. A., Mouradian, M. M., and Chase, T. N. (1993). Catechol-O-methyl transferase inhibitor tolcapone prolongs levodopa/carbidopa action in parkinsonian patients. *Neurology* 43, 2685–2688.

Ruzicka, E., Streitova, H., Jech, R., Kanovsky, P., Roth, J., Rektorova, I., Mecir, P., Hortova, H., Bares, M., Hejdukova, B., and Rektor, I. (2000). Amantadine infusion in treatment of motor fluctuations and dyskinesias in Parkinson's disease. *J. Neural Transm.* 107, 1297–1306.

Sadeh, M., Braham, J., and Modan, M. (1982). Effects of anticholinergic drugs on memory in Parkinson's disease. *Arch. Neurol.* 39, 666–667.

Sage, J. I., Trooskin, S., Sonsalla, P. K., Heikkila, R., and Duvoisin, R. C. (1988). Long-term duodenal infusion of levodopa for motor fluctuations in parkinsonism. *Ann. Neurol.* 24, 87–89.

Sanjiv, C. C., Schulzer, M., Mak, E., Fleming, J., Martin, W. R., Brown, T., Calne, S. M., Tsui, J., Stoessl, A. J., Lee, C. S., and Calne, D. B. (2001). Daytime somnolence in patients with Parkinson's disease. *Parkinsonism Relat. Disord.* 7(4), 283–286.

Santamaria, J., Tolosa, E., and Valles, A. (1986). Parkinson's disease with depression: a possible subgroup of idiopathic parkinsonism. *Neurology* 36(8), 1130–1133.

Sautel, F., Griffon, N., Levesque, D., *et al.* (1995). A functional test identifies dopamine agonists selective for D3 versus D2 receptors. *Neuroreport.* 6, 329–332.

Schapira, A. H. (1994). Evidence for mitochondrial dysfunction in Parkinson's disease: A critical appraisal. *Mov. Disord.* 9, 125–138.

Scholz, E., and Dichgans, J. (1985). Treatment of drug-induced exogenous psychosis in parkinsonism with clozapine and fluperlapine. *Eur. Arch. Psychiat. Neurol. Sci.* 235, 60–64.

Scholz, E., and Oertel, W. H. (1993). Parkinson-syndrome. *In* "Therapie und Verlauf neurologischer Erkrankungen," 2nd ed. (T. Brandt, J. Dichgans, and H. C. Diener, Eds.), pp. 927–968. Verlag W. Kohlhammer, Stuttgart.

Schulz, J. B., Klockgether, T., Petersen, D., Jauch, M., Müller-Schauenburg, W., Spieker, S., Voigt, K., and Dichgans, J. (1994). Multiple system atrophy: natural history, MRI morphology, and dopamine receptor imaging with $^{123}$IBZM-SPECT. *J. Neurol. Neurosurg. Psychiat.* 57, 1047–1056.

Schwab, R. S., England, A. C., Jr., Poskanzer, D. C., and Young, R. R. (1969). Amantadine in the treatment of Parkinson's disease. *JAMA.* 208, 1168–1170.

Schwarz, J., Tatsch, K., Arnold, G., Gasser, T., Trenkwalder, C., Kirsch, C. M., and Oertel, W. H. (1992). $^{123}$I-iodobenzamide SPECT predicts dopaminergic responsiveness in patients with "de novo" parkinsonism. *Neurology* 42, 556–561.

Schwarz, J., Weis, S., Kraft, E., Tatsch, K., Mehraein, P., Vogl, T., and Oertel, W. H. (1996). Signal changes on MRI and increase of reactive microastrogliosis and iron in the putamen of two patients with multiple system atrophy. *J. Neurol. Neurosurg. Psychiat.* 60, 98–101.

Seeman, P., and van Tol, H. H. (1993). Dopamine receptor pharmacology. *Trends Pharmacol. Sci.* 15, 264–270.

Seiler, S., Perleth, B., Gasser, T., Ulm, G., Oertel, W. H., and Ellgring, H. (1992). Partnership and depression in Parkinson's disease. *Behav. Neurol.* 5, 75–81.

Shimura, H., Schlossmacher, M. G., Hattori, N., Frosch, M. P., Trockenbacher, A., Schneider, R., Mizuno, Y., Kosik, K. S., and Selkoe, D. J. (2001). Ubiquitination of a new form of alpha-synuclein by parkin from human brain: implications for Parkinson's disease. *Science* 293, 263–269.

Shiosaki, K., Jenner, P., Asin, K. E., Bianchi, B., Britton, S., Domenico, D. I., Hodges, L., Hong, Y., Lin, C. W., Michaelides, M. R., Mikusa, Y., Miller, T., Nikkel, A., Shmit, L., Stashko, M., Williams, M., and Witte, D. (1994). A potent and selective dopamine (DA) D1-receptor agonist with long-term anti-parkinsonian efficacy in MPTP-lesioned marmosets. *Mov. Disord.* 9(Suppl. 1), S431.

Siegfried, J., and Lippitz, B. (1994). Bilateral chronic electrostimulation of ventroposterolateral pallidum: A new therapeutic approach for alleviating all parkinsonian symptoms. *Neurosurgery* 35, 1126–1129.

Spieker, S., Eisebitt, R., Breit, S., Przuntek, H., Muller, D., Klockgether, T., and Dichgans, J. (1999). Tremorlytic activity of budipine in Parkinson's disease. *Clin. Neuropharmacol.* 22(2), 115–119.

Starkstein, S. E., Mayberg, H. S., Leiguarda, R., Preziosi, T. J., and Robinson, R. G. (1992). A prospective longitudinal study of depression, cognitive decline, and physical impairments in patients with Parkinson's disease. *J. Neurol. Neurosurg. Psychiatry.* 55(5), 377–382.

Stephen, P. J., and Williamson, J. (1984). Drug-induced parkinsonism in the elderly. *Lancet* II, 1082–1083.

Stern, G. M., and Lees, A. J. (1983). Sustained bromocriptine therapy in 50 previously untreated patients with Parkinson's disease. *In* "Experimental Therapeutics in Movement Disorders" (S. Fahn,

D. B. Calne, and J. Showson, Eds.), Advances in Neurology, Vol. 37, pp. 17–21. Raven, New York.

Steur, E. N. (1993). Increase of Parkinson disability after fluoxetine medication. *Neurology* 43, 211–3.

Stibe, C. M. H., Lees, A. J., Kempster, P. A., and Stern, G. M. (1988). Subcutaneous apomorphine in parkinsonian on–off oscillations. *Lancet* I, 403–406.

Stocchi, F., Ruggieri, S., Antonini, A., Baronti, F., Brughitta, G., Bellantuono, P., Bravi, D., and Agnoli, A. (1988). Subcutaneous lisuride infusion in Parkinson's disease: Clinical results using different modes of administration. *J. Neural. Transm.* 27(Suppl.), 27–33.

Stocchi, F., Ruggieri, S., Monge, A., Nordera, G., Bolner, P., Viselli, F., Bradmante, L., Quinn, N. P., de Pandis, F., and Manfredi, M. (1994). Clinical efficacy of single morning doses of different levodopa formulations. *Clin. Neuropharmacol.* 17(Suppl. 3), S16–S20.

Strange, P. G. (1993). Review: Dopamine receptors in the basal ganglia: Relevance to Parkinson's disease. *Mov. Disord.* 8, 263–270.

Sutcliffe, R. L. G., Prior, R., Mawby, B., and McQuillan, W. J. (1985). Parkinson's disease in the district of the Northampton Health Authority, United Kingdom: A study of prevalence and disability. *Acta Neurol. Scand.* 72, 363–379.

Syndulko, K., Gilden, E. R., Hansch, E. C., Potvin, A. R., Tourtelotte, W. W., and Potvin, J. H. (1981). Decreased verbal memory associated with anticholinergic treatment in Parkinson's disease. *Int. J. Neurosci.* 14, 61–66.

Tanner, C. M., and Aston, D. A. (2000). Epidemiology of Parkinson's disease and akinetic syndromes. *Curr. Opini. Neurol.* 13, 417–430.

Testa, D., Savoiardo, M., Fetoni, V., Strada, L., Palazzini, E., Bertulezzi, E., and Girotti, F. (1993). Multiple system atrophy: Clinical and MR observations in 42 cases. *Ital. J. Neurol. Sci.* 14, 211–216.

Tetrud, J. W., and Langston, J. W. (1989). The effect of deprenyl (selegiline) on the natural history of Parkinson's disease. *Science* 245, 519–522.

Tetrud, J. W., Langston, J. W., Irwin, I., and Snow, B. (1994). Parkinsonism caused by petroleum waste ingestion. *Neurology* 44, 1051–1053.

The Parkinson Study Group. (1989a). Effect of deprenyl on the progression of disability in early Parkinson's disease. *N. Engl. J. Med.* 321, 1364–1371.

The Parkinson Study Group. (1989b). DATATOP: A multicenter controlled clinical trial in early Parkinson's disease. *Arch. Neurol.* 46, 1052–1060.

The Parkinson Study Group. (1993). Effects of tocopherol and deprenyl on the progression of disability in early Parkinson's disease. *N. Engl. J. Med.* 328, 176–183.

Thompson, P. D., and Marsden, C. D. (1987). Gait disorder of subcortical arteriosclerotic encephalopathy: Binswanger's disease. *Mov. Disord.* 2, 1–8.

Thorogood, M., Armstrong, B., Nichols, T., and Hollowell, J. (1998). Mortality in people taking selegiline: observational study. *BMJ.* 317(7153), 252–254.

Trenkwalder, C., Schwarz, J., Gebhard, J., Ruland, D., Trenkwalder, P., Hense, H.-W., and Oertel, W. H. (1995). Starnberg trial on epidemiology of parkinsonism and hypertension: Prevalence of Parkinson's disease and related disorders assessed by a door-to-door survey of inhabitants older than 65 years. *Arch. Neurol.* 52, 1017–1022.

Uitti, R. J., Ahlskog, J. E., Maraganore, D. M., Muenter, M. D., Atkinson, E. J., Cha, R. H., and O'Brien, P. C. (1993). Levodopa therapy and survival in idiopathic Parkinson's disease: Olmsted County project. *Neurology* 43, 1918–1926.

Van Gerpen, J. A., and Ahlskog, J. E. (2001). Selegiline and mortality in subjects with Parkinson's disease: a longitudinal community study. *Neurology* 57(2), 368–369; discussion 369–370.

Van Horn, G., and Schiess, M. C. (2000). Immediate-release and controlled-release carbidopa/levodopa in PD: A 5-year randomized multicenter study. *Neurology.* 55(1), 156–157.

Verhagen Metman, L., Del Dotto, P., van den Munckhof, P., Fang, J., Mouradian, M. M., and Chase, T. N. (1998). Amantadine as treatment for dyskinesias and motor fluctuations in Parkinson's disease. *Neurology* 50(5), 1323–1326.

Ward, C. D., and Gibb, W. R. (1990). Research diagnostic criteria for Parkinson's disease. *In* "Parkinson's Disease—Anatomy, Pathology, and Therapy" (M. B. Streifler, A. D. Korczyn, E. Melamed, and M. B. H. Youdim, Eds.), *Advances of Neurology*, Vol. 53, pp. 245–249.

Waters, C. H., Kurth, M., and Bailey, P., *et al.* (1997). Tolcapone in stable Parkinson's disease: efficacy and safety of long-term treatment. The Tolcapone Stable Study Group. *Neurology* 49:665–671.

Watts, R. L. (1997). The role of dopamine agonists in early Parkinson's disease. *Neurology* 49(Suppl.), S34–S48.

Watts, R. L., Mirra, S. S., and Richardson, E. P., Jr. (1994). Corticobasal degeneration. *In* "Movement Disorders" (C. D. Marsden, and S. Fahn, Eds.), Vol. 3, pp. 282–299. Butterworth-Heinemann, London.

Weiner, W. J., Singer, C., Sanchez-Ramos, J. R., and Goldenberg, J. N. (1993b). Levodopa, melanoma, and Parkinson's disease. *Neurology* 43, 674–677.

Weiner, W. J., and Factor, S. A. (2000). Ropinirole as compared with levodopa in Parkinson's disease. *N. Engl. J. Med.* 343(12), 885.

Wenning, G. K., Ben Shlomo, Y., Magalhaes, M., Daniel, S. E., and Quinn, N. P. (1994). Clinical features and natural history of multiple system atrophy: An analysis of 100 cases. *Brain* 117, 835–845.

Wenning, G. K., Ben Shlomo, Y., Magalhaes, M., Daniel, S. E., and Quinn, N. P. (1995). Clinicopathological study of 35 cases of multiple system atrophy. *J. Neurol. Neurosurg. Psychiat.* 58, 160–166.

Wermuth, L., Boas, J., Gyring, J., Bisgaard, C., Boesen, F., Boisen, E., Clausen, J., Dupont, E., Erichsen, P., Hansen, E., Jensen, N. B., Karlsborg, M., Magnussen, I. *et al.* (1994). Pramipexol—A new dopamine agonist for treatment of Parkinson's disease: 11th International Symposium on Parkinson's disease, Rome. *New Trends Clin. Neuropharmacol.* 8, 280.

Wils, V. (1992). Extrapyramidal symptoms in a patient treated with fluvoxamine. *J. Neurol. Neurosurg. Psychiatry.* 55(4), 330–331.

Wolters, E. C. (1999). Dopaminomimetic psychosis in Parkinson's disease patients: diagnosis and treatment (Review). *Neurology* 52(7 Suppl 3), S10–3.

Wolters, E. C., Hurwitz, T. A., Mak, E., Teal, P., Peppard, F. R., Remick, R., Calne, S., and Calne, D. B. (1990). Clozapine in the treatment of parkinsonian patients with dopaminemimetic psychosis. *Neurology* 40, 832–834.

Wolters, E. C. H., Vermeulen, R. J., Kuiper, M. A., and Stoof, J. C. (1994). Dopamine agonist monotherapy in Parkinson's disease. *In* "Parkinson's Disease: Symptomatic versus Preventive Therapy 1994" (E. C. H. Wolters, Ed.), pp. 55–71. ICG Publications Dordrecht, The Netherlands.

Woody, R. C., Brewster, M. A., and Glasier, C. (1989). Progressive intracranial calcification in dihydrobiopterin deficiency prior to folinic acid therapy. *Neurology* 39, 673–675.

Woodward, W. R., Olanow, C. W., Beckner, R. M., Hauser, R. A., Gauger, L. L., Cedarbaum, J. M., and Nutt, J. G. (1993). The effect of L-dopa infusion with and without phenylalanine challenges in parkinsonian patients: Plasma and ventricular CSF

L-dopa levels and clinical responses. *Neurology* **43**, 1704–1708.

Yeh, K. C., August, T. F., Bush, D. F., Lasseter, K. C., Musson, D. G., Schwartz, S., Smith, M. E., and Titus, D. S. (1989). Pharmacokinetics and bioavailability of Sinemet CR: A summary of human studies. *Neurology* **39**(Suppl. 2), 25–38.

Young, A. B., and Fagg, G. (1990). Excitatory amino acid receptors in the brain: Membrane binding and receptor autoradiographic approaches. *Trends Pharmacol. Sci.* **11**, 126–133.

Zesiewicz, T. A., Helal, M., and Hauser, R. A. (2000). Sildenafil citrate (Viagra) for treatment of erectile dysfunction in men with Parkinson's disease. *Mov. Disord.* **15**, 305–308.

# CHAPTER 75
# Atypical Parkinsonism

Werner Poewe and Gregor K. Wenning

## MULTIPLE SYSTEM ATROPHY (MSA)

### Clinical Aspects and Differential Diagnosis

Multiple system atrophy (MSA) is a degenerative disorder of the central and autonomic nervous systems characterized by abnormal alpha-synuclein aggregation in oligodendroglia and neurons. Clinically, cardinal features include autonomic failure, parkinsonism, cerebellar ataxia, and pyramidal signs in any combination. Two major motor presentations can be distinguished clinically. Parkinsonian features predominate in 80% of patients (MSA-P subtype), cerebellar ataxia is the major motor feature in 20% of patients (MSA-C subtype) (Wenning *et al.*, 1994). Neuropathologically, the motor disorder of MSA-P is associated with striatonigral degeneration, and the motor disorder of MSA-C is associated with olivopontocerebellar atrophy. MSA-associated parkinsonism is dominated by progressive akinesia and rigidity. Jerky postural tremor and, less commonly, tremor at rest may be superimposed. Frequently, patients exhibit orofacial dystonia associated with a characteristic quivering high-pitched dysarthria. Postural stability is compromised early on; however, recurrent falls at disease onset are unusual in contrast to progressive supranuclear palsy (PSP). Differential diagnosis of MSA-P and Parkinson's disease (PD) may be exceedingly difficult in the early stages because of a number of overlapping features, such as rest tremor or asymmetrical akinesia and rigidity. Furthermore, L–dopa induced improvement of parkinsonism may be seen in 30% of MSA-P patients. However, the benefit is transient in most of these subjects, leaving 90% of the MSA-P patients L-dopa unresponsive in the long term. L-Dopa–induced dyskinesias affecting orofacial and neck muscles occur in 50% of MSA-P patients, sometimes in the absence of motor benefit (Bösch *et al.*, 2002). In most instances a fully developed clinical picture of MSA-P evolves within 5 years of disease onset, allowing a clinical diagnosis during follow-up (Wenning *et al.*, 2000). The cerebellar disorder of MSA is composed of gait ataxia, limb kinetic ataxia, and scanning dysarthria, as well as cerebellar oculomotor disturbances. Patients with MSA-C usually have additional noncerebellar symptoms and signs develop, but before doing so, they may be indistinguishable from other patients with idiopathic late-onset cerebellar ataxia, many of whom have a disease restricted clinically to cerebellar signs and pathologically to degeneration of the cerebellum and olives (see Chapter 73 for idiopathic late-onset cerebellar ataxia). Dysautonomia develops in virtually all patients with MSA. Clinically most important is urogenital dysfunction. Early impotence (erectile dysfunction) is virtually universal in men, and urinary incontinence (71%) or retention (30%), often early in the course or as presenting symptoms, are frequent (Wenning *et al.*, 1994). Disorders of micturition in MSA are due to changes in the complex peripheral and central innervation of the bladder (Beck *et al.*, 1994) and generally occur more commonly, earlier, and to a more severe degree than in PD. Urinary retention can be caused or exacerbated by benign prostatic hypertrophy in men or by perineal laxity secondary to difficult childbirth or uterine descent in women. In contrast, constipation occurs equally in PD and MSA. Symptomatic orthostatic hypotension is present in 68% but causes recurrent syncope in only 15% of MSA patients (Wenning *et al.*, 1994). L-Dopa or dopamine agonists may provoke or worsen orthostatic hypotension.

The clinical diagnosis of MSA rests largely on history and physical examination. Consensus criteria are now widely used for a clinical diagnosis of MSA (Gilman *et al.*, 1998). The MSA criteria specify three diagnostic categories of increasing certainty: possible, probable, and definite (Table II). A definite diagnosis requires a typical neuropathological lesion pattern and deposition

TABLE I    Clinical Domains, Features, and Criteria Used in the Diagnosis of MSA

I.   Autonomic and urinary dysfunction
     A.  Autonomic and urinary features
         1.  Orthostatic hypotension (by 20 mmHg systolic or
             10 mmHg diastolic)
         2.  Urinary incontinence or incomplete bladder emptying
     B.  Criterion for autonomic failure or urinary dysfunction in
         MSA
         Orthostatic falls in blood pressure (by 30 mmHg systolic or
         15 mmHg diastolic) or urinary incontinence (persistent,
         involuntary partial or total bladder emptying, accompanied
         by erectile dysfunction in men) or both
II.  Parkinsonism
     A.  Parkinsonian features
         1.  Bradykinesia (slowness of voluntary movement with
             progressive reduction in speed and amplitude during
             repetitive actions)
         2.  Rigidity
         3.  Postural instability (not caused by primary visual,
             vestibular, cerebellar, or proprioceptive dysfunction)
         4.  Tremor (postural, resting, or both)
     B.  Criterion for parkinsonism in MSA
         Bradykinesia plus at least one of items 2 to 4
III. Cerebellar dysfunction
     A.  Cerebellar features
         1.  Gait ataxia (wide-based stance with steps of irregular
             length and direction)
         2.  Ataxic dysarthria
         3.  Limb ataxia
         4.  Sustained gaze-evoked nystagmus
         Criterion for cerebellar dysfunction in MSA
         Gait ataxia plus at least one of items 2 to 4
IV.  Corticospinal tract dysfunction
     A.  Corticospinal tract features
         1.  Extensor plantar responses with hyperreflexia
     B.  Corticospinal tract dysfunction in MSA: no corticospinal
         tract features are used in defining the diagnosis of MSA

A feature (A) is characteristic of the disease, and a criterion (B) is a defining feature or composite of features required for diagnosis. Modified with permission from Gilman et al., 1999.

TABLE II    Diagnostic Categories of MSA

I.   Possible MSA
     One criterion plus two features from separate other domains.
     When the criterion is parkinsonism, a poor levodopa response
     qualifies as one feature (hence only one additional feature is
     required).
II.  Probable MSA
     Criterion for autonomic failure/urinary dysfunction plus poorly
     levodopa responsive pakinsonism or cerebellar dysfunction.
III. Definite MSA
     Pathologically confirmed by the presence of a high density of
     glial cytoplasmic inclusions in association with a combination
     of degenerative changes in the nigrostriatal and
     olivopontocerebellar pathways.

Modified with permission from Gilman et al., 1999.

sentations of MSA are associated with similar survival times (Ben Shlomo et al., 1997). However, MSA-P patients have a more rapid functional deterioration than MSA-C patients (Watanabe et al., 2002). Most MSA patients ultimately die from bronchopneumonia.

## Principles of Therapy

### Autonomic Failure

Unfortunately, there is no causal therapy of autonomic dysfunction available. Therefore the therapeutic strategy is defined by clinical symptoms and impairment of quality of life in these patients. Because of the progressive course of MSA, a regular review of the treatment is mandatory to adjust measures according to clinical needs. The concept to treat symptoms of orthostatic hypotension is based on the increase of intravasal volume and the reduction of volume shift to lower body parts when changing into an upright position. The selection and combination of the following options depends on the severity of symptoms and their practicability in

TABLE III    Exclusion Criteria for the Diagnosis of MSA

I.   History
     Symptomatic onset <30 years of age
     Family history of a similar disorder
     Systemic disease or other identifiable causes for features listed in
         Table I
     Hallucinations unrelated to medication
II.  Physical examination
     DSM IV criteria for dementia
     Prominent slowing of vertical saccades or vertical supranuclear
         gaze palsy
     Evidence of focal cortical dysfunction such as aphasia, alien
         limb syndrome, and parietal dysfunction
III. Laboratory investigation
     Metabolic, molecular genetic, and imaging evidence of an
         alternative cause of features listed in Table I

Modified with permission from Gilman et al., 1999.

of alpha-synuclein–positive glial cytoplasmic inclusions. The diagnosis of possible and probable MSA is based on the presence of clinical features listed in Table I. In addition, exclusion criteria have to be considered (Table III).

## Natural Course

One recent epidemiological survey has found a prevalence rate of 4.4/100,000, and another an incidence rate of 3/100,000/year (Bower et al., 1997; Schrag et al., 1999). The disease affects both men and women; it usually starts in the 6th decade and relentlessly progresses, with death occuring after an average of 9 years (Wenning et al., 1994). There is considerable variation of disease progression, with survival of more than 15 years in some instances. Importantly, both motor pre-the single patient but not on the extent of blood pressure drop during tilt test. Nonpharmacological options include sufficient fluid intake, high-salt diet, more frequent, but smaller, meals per day to reduce postprandial hypotension by spreading the total carbohydrate intake, and custom-made elastic body garments. During the night, head-up tilt increases intravasal volume up to 1 L within a week, which is particularly helpful to improve hypotension early in the morning. This is achieved by an increased secretion of renin caused by reduced renal perfusion pressure and reduced atrial natriuretic hormone because of lower atrial filling pressure. This approach is successful in particular in combination with fludrocortisone, which further supports natrium retention. The next group of drugs are the sympathomimetics. They include ephedrine (with both direct and indirect effects), which is often valuable in central autonomic disorders such as MSA. With higher doses, side effects include tremulousness, loss of appetite, and urinary retention in men. Among the large number of vasoactive agents that have been evaluated in MSA, only one, the directly acting α-agonist midodrine, meets the criteria of evidence-based medicine (Jankovic et al., 1993; Low et al., 1997). Side effects are usually mild and only rarely lead to discontinuation of treatment because of urinary retention or pruritus, predominantly on the scalp. Another promising drug seems to be the norepinephrine precursor L-threo-dihydroxy-phenylserine (L-threo-DOPS), which has been used for this indication in Japan for years and whose efficacy has now been shown by a recent open, dose-finding trial (Mathias et al., 2001).

If the preceding drugs do not produce the desired effect, then selective targeting is needed. The somatostatin analogue, octreotide, often is beneficial in postprandial hypotension (Alam et al., 1995), presumably because it inhibits release of vasodilatory gastrointestinal peptides (Raimbach et al., 1989); importantly, it does not enhance nocturnal hypertension (Alam et al., 1995). The vasopressin analogue, desmopressin, which acts on renal tubular vasopressin-2 receptors, reduces nocturnal polyuria and improves morning postural hypotension (Mathias et al., 1986). The peptide erythropoietin may be beneficial in some patients by raising red cell mass, secondarily improving cerebral oxygenation (Perera et al., 1995; Winkler et al., 2001). A broad range of drugs have been used in the treatment of postural hypotension (Mathias and Kimber, 1999). The value and side effects of many of these drugs have not been adequately determined in MSA patients using appropriate endpoints. Problems encountered in therapy include serious side effects (such as gastric ulceration with indomethacin and cardiac failure with pindolol), low bioavailability (with dihydroergotamine), and erratic blood pressure control and excessive supine hypertension (when combining monoamine oxidase inhibitors with tyramine).

In neurogenic bladder dysfunction clean intermittent catheterization three to four times per day is a widely accepted approach to prevent secondary consequences from failure to micturate. It can become necessary to provide the patient with a permanent transcutaneous suprapubic catheter if mechanical obstruction in the urethra or motor symptoms of MSA prevent uncomplicated catheterization. Pharmacological options with anticholinergic or procholinergic or alpha-adrenergic substances are usually not successful to adequately reduce postvoid residual volume in MSA, but anticholinergic agents like oxybutynin can improve symptoms of detrusor hyperreflexia or sphincter-detrusor dyssynergy in the early course of the disease (Beck et al., 1994). Recently, alpha-adrenergic receptor antagonists (prazosin and moxisylyte) have been shown to improve voiding with reduction of residual volumes in MSA patients (Sakakibara et al., 2000). Urological surgery must be avoided in these patients, because worsening of bladder control postoperatively is most likely (Beck et al., 1994).

The necessity of a specific treatment of sexual dysfunction needs to be evaluated individually in each MSA patient. Preliminary evidence in PD patients (Zesiewicz et al., 2000) suggests that sildenafil may also be successful in treating erectile failure in MSA patients; however, controlled trials are unavailable at present. Erectile failure in MSA may also be improved by oral yohimbine or by intracavernosal injection of papaverine or a penis implant (Beck et al., 1994).

Constipation is relieved by an increase in intraluminal fluid, which may be achieved by macrogol-water solution.

Inspiratory stridor develops in about 30% of patients. Continuous positive airway pressure (CPAP) may be helpful in some of these patients (Iranzo et al., 2000). In only about 4% is a tracheostomy needed and performed.

### Motor Disorder

**Parkinsonism.** Parkinsonism is the predominating motor disorder of MSA and therefore represents a major target for therapeutic intervention. L-Dopa replacement has been the mainstay of antiparkinsonian therapy in MSA; however, a sufficiently powered double-blind controlled trial has never been performed. Open-label studies suggest that, in contrast to PD patients, most MSA patients fail to benefit from treatment with L-dopa, except for a transient response in up to 30% of the patients (Parati et al., 1993; Wenning et al., 1994). Occasionally, a beneficial effect is evident only when seemingly unresponsive patients deteriorate after L-dopa withdrawal (Hughes *et al.*, 1992). L-dopa–induced dyskinesias may occur even in the absence of antiparkinsonian efficacy (Bösch *et al.*, 2002). Pre-existing orthostatic hypotension is often unmasked or exacerbated in L-dopa–treated MSA patients. In contrast, psychiatric or toxic confusional states seem to be uncommon (Wenning *et al.*, 1994). The overall poor response to L-dopa reflects both the loss of striatal dopamine receptors and disease downstream of the striatopallidal projections (Tison *et al.*, 1995).

Results with dopamine agonists have been even more disappointing (Lees, 1999). Severe psychiatric side effects occurred in a double-blind crossover trial of six patients on lisuride, with nightmares, visual hallucinations, and toxic confusional states (Lees and Bannister, 1981). Wenning *et al.*, (1994) reported a response to oral dopamine agonists in 4 of 41 patients. None of 30 patients receiving bromocriptine improved, but 3 of 10 who received pergolide had some benefit. Twenty-two percent of the L-dopa responders had good or excellent response to at least one orally active dopamine agonist in addition. MSA patients frequently report the appearance or worsening of postural hypotension after initiation of dopaminomimetic therapy, which may limit further increase in dosage. Antiparkinsonian effects were noted in 4 of 26 MSA patients treated with open-label amantadine (Wenning *et al.*, 1994); however, there was no significant improvement in a double-blind crossover study of nine patients with atypical parkinsonism, including five subjects with MSA (Colosimo *et al.*, 1996).

**Cerebellar Ataxia.** There is no effective therapy for the progressive ataxia of MSA-C. Occasional successes have been reported with cholinergic drugs, amantadine, 5-hydroxytryptophan, isoniazid, baclofen, and propanolol; for most patients these drugs proved to be ineffective. One intriguing observation is the apparent temporary exacerbation of ataxia by cigarette smoking (Graham and Oppenheimer, 1969; Johnsen and Miller, 1986). Nicotine is known to increase the release of acetylcholine in many areas of the brain and probably also releases noradrenaline, dopamine, 5-hydroxytryptophan, and other neurotransmitters. Nicotinic systems may therefore play a role in cerebellar function. Trials of nicotinic antagonists such as dihydrobeta-erythroidine might be worthwhile in cerebellar degenerations.

## Practical Management

Because of the small number of randomized controlled trials, the practical management of MSA is largely based on empirical evidence, retrospective and non-randomized studies respectively (*), or single randomized studies (**), except for 2 randomized controlled studies of midodrine (**, ***). The present recommendations are summarized in Table IV.

### Autonomic Disorder

Orthostatic hypotension is a cardinal feature of MSA that frequently excacerbates the disability arising from progressive motor disturbance. A number of simple nonpharmacological strategies such as elastic support stockings or tights, a high-salt diet, frequent small meals, head-up tilt of the bed at night, and rising slowly from a sitting to a standing position may all improve orthostatic symptoms and should be tried before resorting to drug therapy. If these measures fail, the mineralocorticoid fludrocortisone may be given at night (0.1–0.3 mg). If orthostatic blood pressure drop persists, sympathomimetics such as ephedrine (15–45 mg tid) or midodrine (2.5–10 mg tid) should be added to fludrocortison. Alternately, L-threo-DOPS (300 mg bid), a precursor of noradrenaline, may improve orthostatic hypotension; however, the drug is not available in many

TABLE IV  **Practical Management of MSA**

Pharmacotherapy
  For akinesia-rigidity, trial with
    L-dopa up to 1000 mg/day, if tolerated (*)
    Dopamine agonists as second-line antiparkinsonian drugs
      (dosing as for PD patients—see Chapter 74) (*)
    Amantadine as third-line drug, 100 mg up to three times daily*
  For focal dystonia
    Botulinum toxin A, doses and injection sites for blepharospasm
      and other focal dystonias are listed in Chapter 78 (*)
  For orthostatic hypotension
    Head-up tilt of bed at night (*)
    Elastic stockings or tights (*)
    Increased salt intake (*)
    Fludrocortisone 0.1–0.3 mg/day*
    Ephedrine 15–45 mg tid
    L-threo DOPS 300 mg bid (*)
    Midodrine 2.5–10 mg tid. (**, ***)
  For postprandial hypotension
    Octreotide 25–50 µg SC 30 min before a meal (*)
  For nocturnal polyuria
    Desmopressin (spray: 10–40 mcg/night or tablet: 100–
      400 mcg/night) (*)
  For bladder symptoms
    Oxybutynin for detrusor hyperreflexia (2.5–5 mg bid–tid)*
    Intermittent self-catheterization for retention or residual volume
      ~100 ml (*)
Other therapies
  Physiotherapy (*)
  Speech therapy (*)
  Occupational therapy (*)
  Percutaneous endoscopic gastrostomy (PEG; rarely needed in late
    stage) (*)
  Provision of wheelchair (*)
  CPAP (rarely tracheostomy) for inspiratory stridor (*)

countries. For most patients with MSA, the bladder symptoms pose a much greater problem. Frequency and urge incontinence are often helped by oxybutynin (2.5–5 mg bid to tid), but this peripherally acting anticholinergic may precipitate urinary retention. A substantial postmicturition residue of >100 ml is an indication for intermittent self-catheterization (or catheterization by the spouse). In the advanced stages of MSA, a urethral or suprapubic catheter may become necessary. Erectile failure can be improved by oral yohimbine (2.5–5 mg tid) or sildenafil (50–100 mg) or by intracavernosal injection of papaverine or a penis implant. Inspiratory stridor develops in about 30% of patients. CPAP may be beneficial and should be tried in patients with prominent stridor. A tracheostomy is rarely needed and performed.

### Motor Disorder

L-Dopa (up to 1000 mg/day, if tolerated) should be administered as first-line antiparkinsonian therapy. If high-dose L-dopa administration fails to induce a beneficial response, the chances of success with dopamine agonists are clearly limited. However, occasionally L-dopa–unresponsive MSA patients improve with dopamine agonists, and a trial is therefore warranted using titration schemes (including domperidone cover with 10 mg tds) that have been established for PD patients. If dopamine agonists are ineffective, amantadine (100 mg tds) should be administered, because beneficial antiparkinsonian effects may sometimes occur in MSA patients not responding to dopaminergic therapy. There is no effective drug therapy for cerebellar ataxia in MSA.

Blepharospasm, as well as limb dystonia, but not antecollis, may respond well to local injections of botulinum toxin A (see Chapter 78).

Because the results of drug treatment for MSA are generally poor, other therapies are all the more important. Physiotherapy helps maintain mobility and prevent contractures, and speech therapy can improve speech and swallowing and provide communication aids. Dysphagia may require feeding by means of a nasogastric tube or even percutaneous endoscopic gastrostomy (PEG). Occupational therapy helps to limit the handicap resulting from the patient's disabilities and should include a home visit. Provision of a wheelchair is usually dictated by the liability to falls because of postural instability and gait ataxia but not by akinesia and rigidity per se. Psychological support for patients and partners needs to be stressed.

TABLE V  Initial Clinical Features in 239 Cases with Postmortem Confirmed DLB*

| Feature | Cases | |
|---|---|---|
| | N | % |
| Dementia | 137 | 57 |
| Parkinsonism | 85 | 36 |
| Hallucinations | 31 | 13 |
| Depression | 21 | 9 |
| Fluctuating cognition | 16 | 7 |
| Falls | 14 | 6 |
| Paranoia | 12 | 5 |

*Wenning et al., unpublished data

## DEMENTIA WITH LEWY BODIES (DLB)

### Clinical Aspects and Differential Diagnosis

Dementia with Lewy bodies represents a still somewhat controversial entity defined by coexistent parkinsonism and progressive cognitive decline accompanied by spontaneous recurrent visual hallucinations and conspicuous fluctuations in alertness and cognitive performance (McKeith et al., 1996). Affected patients at postmortem examination show numerous Lewy bodies in many parts of the cerebral cortex, particularly neocortical and limbic areas in addition to the nigral Lewy body degeneration characteristic for PD. Progressive cognitive decline with particular deficits of visuospatial ability and frontal executive function is accompanied by usually only mild to moderately severe parkinsonism, which is often akineto-rigid without the classical parkinsonian rest-tremor. Recurrent visual hallucinations may occur without exposure to dopaminergic antiparkinsonian agents. Marked diurnal fluctuations in cognitive performance have been the most difficult-to-define feature of the disease but are often conspicuous to the environment. Table V summarizes initial clinical features described in a total of 239 cases with postmortem-confirmed DLB that have been compiled from one brain bank center and the published literature (Wenning et al., unpublished data).

The two main differential diagnoses include Alzheimer's disease (AD) and PD. AD is accompanied by extrapyramidal motor features in up to 30% of cases, and most patients diagnosed with DLB also show some Alzheimer change on postmortem brain examination, such that terms such as "Lewy body variant of Alzheimer's disease" have been proposed. Such terminologies illustrate the differential diagnostic difficulties that may arise when trying to distinguish between these two types of degenerative dementia. Another diagnostic problem is related to the fact that between 20% and 40% of patients with PD will become demented in the course of their illness (Mayeux *et al.*, 1992). Current consensus therefore is to *restrict a diagnosis of DLB only to patients with parkinsonism who develop dementia within 12 months of the onset of motor symptoms.* To improve the differential diagnosis of DLB, consensus criteria have been developed that establish possible and probable levels of diagnostic accuracy (McKeith *et al.*, 1996; Table VI). The specificity of a clinical diagnosis of probable DLB, made by using consensus criteria, is generally high (>85%), but sensitivity of case detection is lower and more variable (McKeith *et al.*, 1999). Though in a recent prospective validation study these criteria have shown a sensitivity of 83% and a specificity of 95% (McKeith *et al.*, 2000a).

## Natural History

Many neuropathologists claim that DLB represents the second most common form of dementia after AD, and autopsy-based studies suggest that DLB may account for up to 30% of dementia cases (Crystal *et al.*, 1990). Disability in DLB progresses at a significantly faster rate compared with the progression of PD. Mean survival in postmortem-confirmed DLB series has been less than 10 years. Risk factors for increased mortality in DLB and present at disease onset include older age, dementia, fluctuating cognition, and hallucinations (Wenning *et al.*, 2001). To date, there is no specific evidence for a genetic basis to DLB, and none of the mutations causing a parkinsonian phenotype has so far been shown to produce a phenotype similar to DLB and this is also true for common AD mutations.

## Principles of Therapy

### Treatment of Cognitive Dysfunction and Dementia

Progressive dementia and visual hallucinations are diagnostic hallmarks of DLB and together with the

**TABLE VI**  Consensus Criteria for the Clinical Diagnosis of Probable and Possible DLB (Modified from McKeith *et al.*, 1996)

1. Progressive cognitive decline (deficits on tests of attention and of frontal-subcortical skills and visuospatial ability may be especially prominent)
2. Presence of one ("possible DLB") or two ("probable DLB") core features
   Fluctuating cognition
   Recurrent visual hallucinations
   Spontaneous parkinsonism
3. Supportive features
   Repeated falls
   Syncope
   Transient loss of consciousness
   Neuroleptic sensitivity
   Systematized delusions
   Hallucinations in other modalities

behavioral disturbances accompanying them are at the center of therapeutic needs in this disorder. Although there is no satisfying treatment yet available, a number of important advances have been made in recent years.

**Cholinesterase Inhibitors.**  The pathophysiological basis of dementia in DLB is not fully understood and likely multifactorial. The pathological hallmark is diffuse Lewy Body deposition in limbic and neocortical areas, commonly associated with Alzheimer-type pathology. There is a variable and inconsistent correlation between the degree of Lewy Body pathology and the severity of dementia, but cortical neuronal dysfunction and cell loss are likely to play a role in the process of cognitive decline (Cummings, 1995). In addition, several studies have found marked loss of ChAT staining and up-regulation of cortical M1-muscarinic receptors in DLB brains (Perry *et al.*, 1999), indicating cholinergic denervation similar to but—according to some—more pronounced than in AD.

Consequently, a number of studies have assessed the impact of treatment with cholinesterase inhibitors on the cognitive and behavioral performance in patients with DLB. Only one of these, however, conforms to the standards of an adequately powered randomized, controlled trial and has tested the effects of rivastigmine against placebo in a prospective double-blind design (McKeith *et al.*, 2000b). All other available studies to date regarding the use of CEI's in DLB are uncontrolled open-label observations, generally in the form of case reports or small series with less than 15 patients.

*Rivastigmine.* Rivastigmine is a reversible inhibitor of cholinesterase and has been shown to improve cognitive function in AD. McKeith *et al.*, (2000b) included 120 patients with probable DLB and mild to moderately severe dementia into a prospective placebo-controlled double-blind trial of rivastigmine (up to total daily doses of 12 mg). Study duration was 20 weeks, and assessments were performed at baseline and weeks 12, 20, and again after a 3-week washout period at week 23, with subscores of four items of the neuropsychological inventory (NPI-4) thought most relevant to DLB—delusions, hallucinations, apathy, and depression—being the primary efficacy criterion. Ninety-two patients completed the trial, and there was a statistically significantly greater percentage of patients on verum (mean daily dose of 9.4 mg) vs placebo, showing at least a 30% improvement at week 20 over baseline (63% vs 30%) both in the ITT- and OC-analyses. Further symptom domains of the NPI that showed improvement with rivastigmine included indifference, anxiety, and aberrant motor behavior. Secondary outcome measures like clinical global change or MMSE scores were not statistically significantly different between the two groups, but there was a significant advantage for rivastigmine on a computerized cognitive assessment battery. More patients on rivastigmine (92%) had adverse events than on placebo (75%), with nausea, vomiting, anorexia, and somnolence being significantly more frequent with rivastigmine. Parkinsonian symptoms as assessed by the UPDRS did not change.

McKeith et al., (2000c) subsequently reported on a 3-month open-label extension study in 11 patients from their center originally included in the multicenter RCT. Patients received a mean dose of 9.6 mg/day and NPI scores for delusions, apathy, agitation, and hallucinations fell by 73%, 63%, 45%, and 27% over baseline. In 5 of 11 patients clinical improvement was judged very significant. In addition, parkinsonism as assessed by the UPDRS also showed some improvement.

*Donepezil.* Donepezil is another cholinesterase inhibitor with proven efficacy in AD. Contrary to rivastigmine, it has not been tested in a controlled fashion in DLB. Several open-label observational studies and case reports in DLB patients report improvements in cognitive function, including memory, attention, alertness with reduced fluctuations as well as behavioral benefits with reduced agitation, and aggressive behavior (Lanctot and Herrmann, 2000; Rojas-Fernandez, 2001; Shea et al., 1998; Skjerve and Nygaard, 2000). Several reports emphasize improvements in pre-existing hallucinations and delusions (Fergusson and Howard, 2000; Samuel et al., 2000; Aarsland et al., 1999); but there are also observations on worsening of parkinsonism (Shea et al., 1998).

*Tacrine.* A few small-scale open-label uncontrolled studies have been conducted on the use of tacrine in DLB (Lebert et al., 1998; Levy et al., 1994; Querfurth et al., 2000). Doses were between 80 and 120 mg/day and results varied. Querfurth and colleagues failed to detect significant differences over baseline using the Mattis Dementia Rating Scale in six DLB and six AD patients, although individual patients were classified as responders. Somewhat more than half of the patients in an uncontrolled trial by Lebert et al. including 20 AD and 19 DLB patients showed improvements in the DRS, whereas the remainder did not. DLB responders mainly showed improvements in verbal initiation and digit span.

**Antipsychotics.** Psychotic symptoms, in particular hallucinations and visual delusions, are among the cardinal symptoms of DLB and have been found in between 25% and 80% of patients in various series (Morris et al., 1998). Despite the differences between studies, it is probably correct to estimate that two thirds of DLB patients will have prominent psychotic behavior in the course of their illness. Hallucinatory psychosis often goes along with confusion, agitation, and sleep disorders and constitutes a major therapeutic challenge.

On the other hand, the use of neuroleptic drugs can induce severe complications regarding marked increases in parkinsonism with rigidity, postural imbalance, and falls, sometimes accompanied by mental status changes and other features of the neuroleptic malignant syndrome (McKeith et al., 1992a, 1992b). This was associated with increased mortality in both series published by McKeith and colleagues. Attention has therefore focussed on atypical neuroleptics to minimize extrapyramidal side effects.

*Risperidone.* Risperidone is a benzisoxazole derivative, and its antipsychotic activity is believed to be mediated through both D2 and 5HT2-receptor antagonism with somewhat greater 5HT2 compared with D2 affinity. There are no controlled trial data available on the use of risperidone to treat psychosis in DLB, but data from open label retrospective observations in small series suggest that it can induce marked worsening of parkinsonism and even aggravate confusion and agitation (McKeith et al., 1995; Rich et al., 1995). This was noted even in patients receiving low doses of 0.5–1.0 mg daily (Ballard et al., 1998; Rich et al., 1995). The potential of risperidone to aggravate parkinsonism has also been reported in patients with PD and drug-induced psychosis (Ford et al., 1994; Rich et al., 1995) so that its use in DLB is not recommended.

*Ondansetron.* Ondansetron is a selective 5-HT3 antagonist that has been tested in patients with PD based on a hypothesis that drug-induced psychosis could be mediated through excessive stimulation of 5HT-receptors by means of dopamine-mediated serotonine release (Zoldan et al., 1995). Open-label studies have yielded conflicting results in this indication (Eichhorn et al., 1996; Friedberg et al., 1998; Zoldan et al., 1995) and there are no reports on its use to treat psychosis in DLB.

*Olanzapine.* Olanzapine is a thienobenzodiazepine similar to clozapine with preferential affinity to 5HT2- over D2-receptors and selective affinity to mesolimbic over striatal dopamine receptors. Its potential to induce extrapyramidal side effects is therefore considered low and, contrary to clozapine, its use does not require blood count monitoring. In a small open-label study in eight DLB patients with psychosis and behavioral disturbances, low doses of 2.5–7.5 mg of olanzapine led to unacceptable worsening of parkinsonism or postural instability, whereas there was no or only marginal improvement of psychosis in the others (Walker et al., 1999). This is in line with the results from a recent double-blind prospective clozapine-controlled study in Parkinson's disease patients with drug-induced psychosis (Goetz et al., 2000) and argues against the use of this drug in DLB.

*Clozapine.* Clozapine is a dibenzodiazepine derivative with potent antipsychotic properties and is virtually free of extrapyramidal side effects when used in schizophrenic patients (Kane *et al.*, 1988). Its affinity within the dopamine system is preferentially for mesolimbic D1-receptors with relative sparing of striatal dopamine receptors, and there is also prominent binding to the D4-receptor subtype (Baldessarini and Frankenburg, 1991). The first report on the efficacy of clozapine to treat psychotic symptoms in a parkinsonian disorder was by Scholz and Dichgans (1985), who described marked efficacy in drug-induced psychosis in four patients with PD with low doses of around 60 mg of clozapine daily. This efficacy has only recently been confirmed in the setting of randomized placebo-controlled trials in PD patients with drug-induced psychosis (PSG, 1999; The French Clozapine Parkinson Study Group, 1999). Both these randomized controlled trials demonstrate marked efficacy in ameliorating or clearing drug-induced psychosis with clozapine doses as low as 6.25 mg/day and usually less than 50 mg/day. They also confirm that clozapine does not induce worsening of parkinsonian symptoms. One trial even showed statistically significant improvement in tremor (PSG, 1999). Another recent randomized controlled trial compared clozapine with olanzapine in the same type of patients and was prematurely stopped because of inacceptable worsening of parkinsonism in the olanzapine arm but nevertheless confirmed the antipsychotic efficacy of clozapine (Goetz *et al.*, 2000). With appropriate weekly blood count monitoring, there was only one withdrawal because of leukopenia less than 3000/mm$^3$; other side effects included sedation, weight loss, and small statistical increases in heart rate.

There are no controlled trials of clozapine in the treatment of psychosis in DLB. Rich and co-workers (1995) anecdotally report on two patients who had worsening of parkinsonism and confusion after treatment of risperidone, who were successfully switched to clozapine with subsequent resolution of psychotic behavior. On the other hand, Burke *et al.*, (1998) observed an increase in confusion and psychotic behavior but no worsening of parkinsonism in two DLB patients treated with clozapine with doses of 6.25–12.5 mg. These authors cite their experience as another example of "neuroleptic sensitivity," this time to clozapine, in DLB.

*Quetiapine.* Quetiapine is another dibenzodiazepine with close pharmacological resemblance to clozapine. A limited number of small uncontrolled clinical studies suggest antipsychotic efficacy in drug-induced psychosis in patients with PD (Fernandez *et al.*, 1999; Matheson and Lamb, 2000). One small uncontrolled study also assessed quetiapine given in a total daily dose of 25–300 mg, specifically in 10 patients with DLB-associated psychosis. According to a published abstract (Parsa *et al.*, 2000), psychosis as assessed by the brief psychiatric rating scale (BPRS) score improved in all patients after

6 months of follow-up without negative impact of motor function.

### Treatment of Parkinsonism and Other Noncognitive Features

Although parkinsonism, depression, autonomic dysfunction, and sleep disorders all contribute significantly to the clinical spectrum of DLB, no therapeutic intervention has been specifically assessed for any of these problems within an appropriately designed controlled trial specifically in DLB patients. Treatment decisions are therefore generally based on experience in other diseases like PD, depression, or MSA, although the underlying pathophysiological mechanisms may not always be the same.

## Practical Management

Management of patients with DLB has to use a multidimensional approach taking into account the cognitive decline and dementia that form the core clinical syndrome, the characteristic hallucinations, and visual delusions present in most cases, as well as dementia-associated behavioral symptoms and depression. Furthermore, parkinsonism is a therapeutic issue in these patients as are symptoms and signs of autonomic dysfunction and—not infrequently—sleep disorders like REM-associated behaviour disorder (RBD). The practical management is summarized in Table VII.

### Parkinsonism

L-Dopa is the "gold standard" of symptomatic efficacy in PD and is also the drug of first choice to treat parkinsonism in DLB. Doses are usually in the low-middle range between 300 and 500 mg of L-dopa plus decarboxylase inhibitor per day but may be increased if clinically required. Visual hallucinations and psychotic behavior can be dose-limiting. The best compromise between need for improvement of akinesia and rigidity and the risk of increasing psychotic behavior has to be sought. Dopamine agonists do not offer significant advantages over L-dopa in DLB but have a greater risk of inducing psychotic side effects. Although comparative trials between L-dopa and dopamine agonists are lacking and will probably never be performed in DLB, it is generally not recommended to use DA agonists in this condition. The same is true for anticholinergics, which may improve rigidity but at the cost of a significant potential to increase cognitive dysfunction. In DLB patients with a fluctuating motor response L-dopa can be combined with the COMT inhibitor entacapone given with every levodopa dose.

**TABLE VII  Practical Management of DLB**

Pharmacotherapy
  For parkinsonism[†]
    L-dopa 300–500 mg/day, if tolerated*
    Add entacapone (200 mg with each L-dopa dose) if wearing off
      is present(**, ***)
    Add amantadine in case of L-dopa–induced dyskinesias(**, ***)
    Avoid dopamine agonists and anticholinergics(*)
  For dementia
    Rivastigmine 3–12 mg/day (first-line drug)(**)
    Donepezil 5–10 mg/day (second-line drug)*
  For hallucinations and psychosis
    Clozapine if required (starting with 6.25 mg at night-time,
      occasionally titrating up to 50 mg; weekly monitoring of
      blood counts)*
    Try Quetiapine if weekly monitoring of blood counts proves to
      be impossible (starting with 25 mg, titrating up to 50–
      150 mg/day)*
    Avoid classical neuroleptics, risperidone, or olanzapine*
  For REM sleep behavior disorder
    Low-dose clonazepam (0.5 mg/day) for RBD*
  For depression
    SSRIs (sertraline 50 mg/day; fluoxetine 20 mg/day, paroxetine
      20 mg/day)(*)
    Mirtazapine (15–30 mg/day)(*)
    Avoid antimuscarinics
  For autonomic dysfunction
    See Table IV
Other therapies
  Physiotherapy(*)
  Speech therapy(*)
  Occupational therapy(*)
  Percutaneous endoscopic gastrostomy (PEG; rarely needed in late
    stage)(*)
  Provision of wheelchair(*)

[†]Recommendations for entacapone and amantadine are based on randomized, controlled trials in PD.

Amantadine has weaker antiparkinsonian efficacy compared with levodopa but, like the former, can also induce or increase psychosis in DLB. DLB patients are not candidates for antiparkinsonian deep brain surgery. Patients with significant dysarthria and dysphagia should receive speech therapy. Postural imbalance and falls are also symptoms generally not responsive to drug therapy and require physiotherapy.

## Dementia

On the basis of one randomized placebo-controlled trial rivastigmine can be considered efficacious in improving cognitive function and psychotic behavior in DLB. Doses range between 3 and 12 mg/day, with a usual mean target dose close to 10 mg. Main side effects are gastrointestinal with nausea and vomiting, and there is generally no negative impact on motor symptoms. Donepezil, although not tested in a randomized controlled fashion, is also likely efficacious in a similar

way as rivastigmine. Donepezil is started as a dose of 5 mg/day and can be increased to 10 mg/day. Side effects are similar to rivastigmine. Tacrine has been very poorly studied in DLB and, because of its worse safety profile compared with rivastigmine and donepezil, is not generally recommended.

## Hallucinations and Psychosis

Visual hallucinations, delusions, and psychotic behavior may improve when DLB patients are put on cholinesterase inhibitors like rivastigmine or donepezil. Still, add-on treatment with antipsychotics is frequently needed. As in PD patients, classic neuroleptics should be avoided because of their potential to significantly worsen motor symptoms. Unfortunately, this is also true for the atypical neuroleptics risperidone and olanzapine. Clozapine with starting bedtime doses of 6.25 mg/day and incomense in a range between 6.25 and 50 mg/day (rarely, 75–150 mg/day) is probably the current best option, although it may be less well tolerated in DLB patients compared with psychotic PD patients. Weekly blood count monitoring is cumbersome but inevitable for the first 6 months (to be followed by biweekly blood count controls). Quetiapine may therefore prove to be an easier-to-use option with a starting dose of 25 mg, which generally has to be increased in a range between 50 and 150 mg/day. The clinical study data on the use of quetiapine in DLB, however, are very limited so far.

## Sleep Disorder

Next to nocturnal hallucinosis and confusional states REM-sleep–associated behavior disorder (RBD) is another common cause for disturbed night sleep in patients with DLB. When reasonable suspicion for the presence of RBD emerges from interview with spouses or sleep laboratory studies, a trial of clonazepam starting with 0.5 mg/day may be tried. As with other benzodiazepines, patients should be closely monitored for possible paradoxical reactions and increased anxiety, agitation, or confusion.

## Autonomic Dysfunction

Symptomatic orthostatic hypotension may at times cause significant problems in DLB and should be treated with the same approaches as outlined for MSA.

## Depression

Treatment of depression in patients with DLB is entirely empirically based without any controlled trial data. Selective serotonin reuptake inhibitors (SSRIs) are frequently used with sufficient success (e.g., sertraline, 50 mg/day; fluoxetine, 20 mg/day). Antimuscarinic agents should be avoided because of their anticholinergic properties and potential to aggravate cognitive dysfunction. There are no data on the use of newer antidepressants like mirtazepine or reboxetine in patients with DLB, but a trial is warranted in patients not responding to SSRIs.

## PROGRESSIVE SUPRANUCLEAR PALSY (STEELE-RICHARDSON-OLSZEWSKI DISEASE)

### Clinical Aspects and Differential Diagnosis

Progressive supranuclear palsy (PSP), also known as Steele-Richardson-Olszewski (SRO) disease, is a sporadic neurodegenerative disorder defined neuropathologically as tauopathy with prominent subcortical neurofibrillary degeneration. Clinical hallmarks include parkinsonism that responds poorly to levodopa, supranuclear ophthalmoplegia (upward and downward before horizontal gaze palsy), recurrent falls, and dementia. Other features are slowing of vertical saccades, a staring facial expression reflecting both a markedly reduced blink rate and frontalis muscle overactivity, neck rigidity, erect posture, pseudobulbar palsy with dysphagia and dysarthria, palilalia or palilogia, and pseudobulbar crying, or laughing. The NINDS diagnostic criteria for PSP have been established as the "gold standard" during the last 5 years (Litvan et al., 1996a) and are shown in Table VIII. Although the NINDS criteria define supranuclear vertical gaze palsy as an obligatory feature for a diagnosis of probable PSP, this clinical sign is rather nonspecific, occurring in a range of other CNS disorders including vascular encephalopathy, DLB, corticobasal degeneration (CBD), Whipple's disease, and Huntington's disease. In some instances it can be difficult to distinguish PSP from PD patients during the first 2–3 years from symptom onset if the former do not yet clearly exhibit postural instability or ophthalmoplegia and when akinesia and rigidity may still show a response to levodopa. However, such a presentation is uncommon, and in most instances PSP can be readily discriminated from PD on clinical grounds (Table IX).

**TABLE VIII** NINDS-SPSP Clinical Criteria for the Diagnosis of PSP

| PSP | Mandatory inclusion criteria | Mandatory exclusion criteria | Supportive criteria |
|---|---|---|---|
| Possible | Gradually progressive disorder<br>Onset at age 40 or later | Recent history of encephalitis, alien limb syndrome, cortical deficits, focal frontal or temporoparietal atrophy | Symmetric akinesia or rigidity, proximal more than distal<br>Abnormal neck posture, especially retrocollis |
| | *Either* vertical (upward or downward gaze) supranuclear palsy palsy or both slowing of vertical saccades and prominent postural instability with falls in the first year of disease onset<br>No evidence of other diseases that could explain the foregoing features, as indicated by mandatory exclusion criteria | Hallucinations or delusions unrelated to dopaminergic therapy<br>Cortical dementia of Alzheimer's (severe amnesia and aphasia or agnosia, according to NINCDS-ADRA criteria)<br>Prominent, early cerebellar symptoms or prominent, early unexplained dysautonomia (marked hypotension and urinary disturbances) | Poor or absent response of parkinsonism to levodopa therapy<br>Early dysphagia and dysarthria<br><br>Early onset of cognitive impairment including at least two of the following: apathy, impairment in abstract thought, decreased verbal fluency, use or imitation behavior, or frontal release signs |
| Probable | Gradually progressive disorder<br>Onset at age 40 or later<br>Vertical (upward or downward gaze) supranuclear palsy *and* prominent postural instability with falls in the first year of disease onset<br>No evidence of other diseases that could explain the foregoing features, as indicated by mandatory exclusion criteria | Severe, asymmetrical parkinsonian signs (i.e., bradykinesia)<br>Neuroradiological evidence of relevant structural abnormality (i.e., basal gangliam or brainstem infarcts, lobar atrophy)<br>Whipple's disease, confirmed by polymerase chain reaction, if indicated | |
| Definite* | Clinically probable or possible PSP *and* histopathological evidence of typical PSP | | |

Litvan et al., 1996a, with permission from Lippincott Williams & Wilkins.
*Definite PSP is a clinicopathological diagnosis.

TABLE IX    Comparison of PSP and Parkinson's Disease

| Symptoms or sign | PSP | Parkinson's disease |
|---|---|---|
| Parkinsonian signs (akinesia, rigidity) | Symmetric | Asymmetric |
| Rest tremor | Very rare | Frequent |
| Rigidity | More pronounced axially (neck and trunk) than in the extremities | More pronounced in extremities than in neck and trunk |
| L-dopa response | Little or none; rarely moderate but transient | Excellent |
| L dopa–induced dyskinesias | None | Common |
| Dystonia | Neck > extremities | Extremities > neck (L-dopa induced) |
| Postural instability | Early | Late |
| Falls | Early | Late |
| Axial (trunk) position during walking | Erect | Flexed |
| Facial expression | Surprised, eyes widely opened, hypomimia, overactive frontalis muscle (skin folded on forehead, eyebrows raised) | Hypomimia |
| Blink rate | Very reduced | Reduced |
| Frontal-lobe behavior | Early, frequent | Late, uncommon |

## Natural Course

Recent epidemiological studies have established the disorder PSP as a frequent cause of atypical parkinsonism with an incidence rate of 5.3/100,000 (in the population aged >50 years) and age-adjusted prevalence of 6.4/100.000 (Bower *et al.*, 1997; Schrag *et al.*, 1999).

The mean age at disease onset is around 60 years, and mean survival is 6 years (Litvan *et al.*, 1996b). The disease is progressive despite any therapy. In the late stage PSP patients are wheelchair or bed bound. Their speech shows a characteristic growling and groaning. Dressing and feeding themselves are impossible. Marked difficulty in swallowing bears the danger of aspiration and consequent pneumonia. The patient ends in a state of immobility with akinesia, rigidity, and dystonia. The cause is unknown; however, in PSP (and CBD) pathological tau, composed of aggregated 4-repeat (E10+) tau isoforms, accumulates in cells and glia in subcortical and cortical areas. Interestingly, recent genetic studies have indicated that a specific haplotype of the tau gene is overrepresented in PSP and CBD, indicating a common genetic background of these tauopathies (Di Maria *et al.*, 2000).

## Principles of Therapy

### Parkinsonism

To date, there have been no randomized double-blind controlled trials of dopaminergic replacement therapies (L-dopa and dopamine agonists) in PSP patients except for one short-term trial of pergolide limited by small sample size (Jankovic, 1983). The literature evidence is based on anecdotal case reports; small-scale, open-label trials; or retrospective chart reviews using clinical, at times not validated, criteria for diagnosing PSP. Taken together, the results suggest that dopaminergic agents are usually ineffective in PSP, reflecting striatal dopamine receptor loss and lesions in nondopaminergic neurotransmitter systems, including cholinergic brainstem and basal forebrain nuclei. Overall, only around 20%–40% of the patients show transient mild benefit to L-dopa (Jackson *et al.*, 1983; Kompoliti *et al.*, 1998; Litvan and Chase, 1992; Nieforth and Golbe, 1993). L-Dopa–induced motor or psychiatric complications seem to be rare. In a review of 82 consecutive patients treated with open-label L-dopa (mean maximum daily dose, 1.015 mg) only 3 had mild dyskinesias, and 1 had an acute psychotic reaction when amitriptyline was added to the L-dopa regimen (Nieforth and Golbe, 1993). In theory, postsynaptic dopamine receptor agonists should be more effective in ameliorating PSP symptoms than L-dopa, because the former agents do not require metabolic conversion by degenerating dopaminergic neurons to become pharmacologically effective. In practice, however, dopamine agonists have proved no less ineffective. Williams and colleagues (1979) failed to demonstrate any significant improvement in most of nine PSP patients treated with bromocriptine in a double-blind placebo-controlled trial. However, transient antiparkinsonian efficacy has been observed in a retrospective chart review of 12 autopsy-proven PSP patients treated with bromocriptine (Kompoliti *et al.*, 1998). A modest 20% improvement of motor disability has also been reported in a double-blind controlled trial of pergolide conducted in three PSP patients (Jankovic, 1983). In contrast, lack of antiparkinsonian efficacy was reported for lisuride (Neophytides *et al.*, 1982) and pramipexole (Weiner *et al.*, 1999) in two small-scale open-label studies. Nevertheless, in the absence of more effective antiparkinsonian treatment these therapies should be tried when parkinsonism is present in PSP because of the possibility of a sometimes moderate, but useful, effect and because the lack of a sustained or marked benefit from L-dopa effectively rules out PD and may therefore support the diagnosis of PSP or other atypical parkinsonian disorders.

A double-blind crossover study of amantadine in atypical parkinsonism including a small number of PSP patients failed to demonstrate significant motor benefit (Colosimo et al., 1996).

A mild (rarely dramatic) symptomatic improvement of parkinsonism may be seen with tricyclic antidepressants in a minority of patients (Engel, 1996; Newman, 1985; Tamai and Almeida, 1997). Amitriptyline and desipramine were both shown to ameliorate parkinsonian features in a small-scale double-blind crossover trial of four PSP patients (Newman, 1985). In addition, gaze palsy was improved in one patient and apraxia of eyelid opening in two patients. Side effects were mild and included reversible urinary retention and dry mouth. A recent open-label trial in two patients with PSP reported antiparkinsonian efficacy and improved tolerability of low-dose amitriptyline (Engel, 1996). Nortyptyline may also be used in patients with PSP; however, future multicenter controlled trials are necessary to definitively establish the role of antidepressants.

Idazoxan is a potent and selective alpha$_2$-presynaptic inhibitor drug, whose overall effect is to increase noradrenaline neurotransmission. A double-blind crossover trial in nine PSP patients revealed significant improvement of mobility, balance, gait, and finger dexterity (Ghika et al., 1991). In contrast, efaroxan, a more potent enhancer of noradrenaline neurotransmission compared with Idazoxan, failed to confer any benefit to 14 PSP patients in a double-blind crossover trial (Rascol et al., 1998).

Methysergide, a serotonin antagonist, has been reported to improve swallowing, speech, parkinsonian features, and oculomotor disturbances in an open-label study of 12 patients (Rafal and Grimm, 1981), but this observation could not be confirmed by others (Duncombe and Lees, 1985; Paulson et al., 1981). Serotonin antagonists such as methysergide may cause severe side effects such as retroperitoneal fibrosis or pleuropulmonary fibrosis. Aniracetam, a metabolic activator with acetylcholine-like properties, was reported to improve motor and cognitive function in two PSP patients; however, there is no further evidence to substantiate these findings (Nagasaka et al., 1997). Other drugs with minimal or absent efficacy in PSP include amantadine, baclofen, bupropion, fluoxetine, selegeline, and valproate (Colosimo et al., 1996; Golbe et al., 1990; Nieforth and Golbe, 1993). Electroconvulsive therapy was reported to ameliorate motor dysfunction in some PSP patients; however, hospitalization was prolonged, and patients experienced treatment-induced confusion, thus limiting the usefulness of this technique (Barclay et al., 1996; Hauser and Trehan, 1994). Implantation of adrenal medullary tissue into the caudate nucleus has been performed in a few PSP patients; however, the procedure proved to be ineffective and hazardous, with substantial perioperative morbidity and mortality (Koller et al., 1989; Waxman et al., 1991).

### Oculomotor and Related Disturbances

Zolpidem, a short-acting hypnotic drug and selective agonist of the benzodiazepine receptor subtype BZ1, was shown to improve saccadic eye movements and parkinsonism in a double-blind crossover study of 10 patients with probable PSP; however, the benefit was limited by drowsiness, particularly at dosages higher than 5 mg/day (Daniele et al., 1999). Involuntary eye closure may be treated with botulinum toxin injections (Barclay and Lang, 1997; Müller et al., 2002; Polo and Jabbari, 1994; Piccione et al., 1997). Visual prisms are rarely of help; patients may instead resort to books on tape. Artificial tears are useful to avoid exposure keratitis secondary to decreased eye blink rate.

### Cognitive Disturbance

Administration of cholinergic agents is not beneficial; indeed, mental status and gait of patients may worsen with these drugs (Fabbrini et al., 2001; Foster et al., 1989, Litvan et al., 1994). To date, only three cholinergic agents have been used to treat PSP. Foster et al. (1989) evaluated the effects of RS-86, a M1—M2 muscarinic agonist, on motor and cognitive function in 10 PSP patients during a 9-week double-blind randomized controlled trial. No effects were found with regard to either cognitive tasks or motor functions. Litvan and colleagues reported lack of therapeutic efficacy for physostigmine and scopolamine in a double-blind placebo-controlled study of nine PSP patients (Litvan et al., 1994). Fabbrini and colleagues (2001) demonstrated lack of efficay for donepezil, a centrally acting cholinesterase inhibitor, in an open-label trial of six PSP patients.

### Other Features

Because gait instability shows only minimal or no response to drug therapy, weighted walkers should be considered. Swallowing disturbances should be regularly evaluated by speech therapists. Dysphagia can be managed by the use of straws, food thickeners, or soft processed food. Percutaneous endoscopic gastrostomy (PEG) appears to be more efficacious to avoid aspiration pneumonia in the long term than nasogastric tubes, which should be avoided. Patients can also be helped by a variety of communication aids. Although drooling can be managed with anticholinergics, these drugs should be used cautiously, because they can worsen patient's symptoms. Emotional incontinence has been reported to improve with amitriptyline (Newman, 1985).

## Practical Management

Because of the small number of randomized controlled trials, the practical management of PSP is largely based on empirical evidence, retrospective and non-randomized studies respectively (*), or a single randomized study (**). Although improvement is rarely seen, L-dopa (*) up to 1000 mg/day, if tolerated) should be administered as first-line therapy. Second-line drugs include amitriptyline (*) (orally up to 25 mg tid, occasionally higher doses up to 150 mg/d needed) and zolpidem (**) (5–10 mg od), both of which may improve motor impairment transiently for weeks or months. At present, there is no role for cholinergic, serotonergic, or noradrenergic drug therapy in PSP. In view of the limited pharmacological options, physiotherapy(*), occupational therapy(*), and speech therapy(*) are recommended. Dysphagia may require feeding by means of percutaneous endoscopic gastrostomy. Tarsal and pretarsal blepharospasm and limb dystonia can respond well to local injections of botulinum toxin A(*) (see Chapter 78).

# CORTICOBASAL DEGENERATION (CBD)

## Clinical Features and Differential Diagnosis

First described by Rebeiz and colleagues in 1967 (Rebeiz et al., 1967) as "corticodentatonigral degeneration with neuronal achromasia," almost 20 years elapsed before this rare neurodegenerative disorder has received renewed attention (Gibb et al., 1989; Riley et al., 1990; Watts et al., 1989).

Similar to the other conditions discussed in this chapter, corticobasal degeneration (CBD) is a multisystem disorder affecting the nigrostriatal motor system plus a variety of other subcortical structures, including variable cell loss in the thalamus, subthalamic nucleus, the pallidum, red nucleus, dentate nucleus, and scattered changes in other brainstem nuclei. In addition, there is prominent and usually asymmetrical cortical degeneration involving frontoparietal areas. The resulting clinical picture is one of a strikingly asymmetrical akineto-rigid parkinsonian syndrome associated with other movement disorders—most often dystonia and myoclonus—in combination with cortical signs, including apraxia and "alien limb" phenomena, cortical

sensory loss, and variable degrees of dysphasia. Between 30% and 50% of patients eventually show signs of depression and frontal lobe–type behavioral changes, including apathy or disinhibition, impulsiveness, and irritability (Cummings and Litvan, 2000; Rinne et al., 1994; Wenning et al., 1998) and some series suggest that 25% may become demented (Kompoliti et al., 1998).

The clinical picture of CBD in its full expression, including limb apraxia, alien limb behavior, and strictly asymmetrical parkinsonism, and jerky dystonia of the limbs is so characteristic that there is little room for clinical diagnostic error. On the other hand, early CBD with unilateral limb rigidity and clumsiness can be confused with idiopathic Parkinson's disease; patients with significant postural instability and falls may be confused with PSP when there is some additional limited vertical gaze or frontal executive dysfunction. However, the supranuclear gaze palsy in CBD usually affects horizontal and vertical gaze equally, whereas vertical gaze is more severely affected in PSP. CBD presenting with cognitive impairment is also difficult to diagnose, and there is some controversy that frontotemporal dementia and CBD may represent different ends of a single disease spectrum (Kertesz et al., 2000). Overall, diagnostic sensitivity to CBD is suboptimal even among expert neurologists, who only detected 30% of postmortem-confirmed CBD cases on the basis of clinical presentation at first visit (Litvan et al., 1997).

## Clinical Course

Although publications on CBD have multiplied because the late 1980s, there are still no available data on its incidence and prevalence. There seems to be no gender predominance (Wenning et al., 1998). Classically, patients with CBD are seen in the sixth or seventh decade of life with a unilateral jerky tremulous akinetic-rigid and apraxic extremity held in a fixed posture and displaying the alien limb syndrome (Rinne et al., 1994). However, presentations can vary widely; they may relate to difficulty in walking, speech, or, less commonly, limb sensation. Symptoms usually remain clearly asymmetrical, eventually spreading from the affected arm to the ipsilateral leg, and progress steadily until death, which usually occurs 4 to 8 years after disease onset. Bilateral parkinsonian features at first neurological evaluation predict shorter survival, particularly in the presence of frontal lobe dysfunction (Wenning et al., 1998). The course and pathogenesis of CBD remain to be resolved. There is abnormal aggregation of tau affecting both basal ganglia and motor cortex. Recent evidence suggests that there may be a genetic predisposition toward tau accumulation that is shared by PSP patients (Di Maria et al., 2000).

## Principles of Therapy

Overall, CBD is often considered an untreatable condition because of its relentless progression and the, at best, modest response to various symptomatic interventions. Nonetheless, at least temporary improvement can be achieved for several of the clinical problems highlighted in up to two thirds of patients, depending on the target symptom (Kompoliti et al., 1998). It has to be pointed out, however, that because of the rarity of CBD, where even tertiary referral centers usually follow less than 20 patients (Kompoliti et al., 1998), to date there has been no single controlled or even uncontrolled prospective clinical trial of any intervention, and all recommendations given here are based on retrospective uncontrolled case series.

### Treatment of Parkinsonism in CBD

Parkinsonism in CBD is dominated by rigidity and bradykinesia, whereas tremor is present in only 30%–50% of cases and is often irregular and jerky (Kompoliti et al., 1998; Wenning et al., 1998). In addition, parkinsonism contributes to the gait disorder of CBD, which becomes a major source of disability in the course of disease with marked postural instability and falls. Because of the impact of cerebellar dysfunction and apraxia, antiparkinsonian medications have a variable effect on the gait problems of CBD patients.

The drugs that have been used to improve parkinsonian features in CBD include levodopa, dopamine agonists, selegeline, amantadine, and anticholinergics. The largest case series with uncontrolled retrospective data on drug responses was published by Kompoliti and colleagues (1998), who had included 147 patients followed at eight movement disorder centers.

**L-dopa.** Eighty-seven percent of cases in the series by Kompoliti and coworkers had been exposed to therapeutic trials of L-dopa, with a mean daily dose of 300 mg (range, 100–2000 mg). Some clinical improvement of parkinsonism was noted in 26% of patients, and bradykinesia and rigidity responded the most, but there are no data on the magnitude or duration of this response. One patient each of a total of 33 responding to L-dopa was noted to improve regarding dystonic or alien limb features, whereas 5% had some degree of worsening either of parkinsonism or gait dysfunction or dystonia and myoclonus.

L-dopa–induced dyskinesias were not observed in this series, even with high-dose treatment, and there are also no other reports in the literature noting the occurrence of dyskinesias in response to L-dopa in CBD (Kompoliti et al., 2000). Gastrointestinal complaints were present in 15%, followed by confusion, dizziness, somnolence (4% each), and hallucinosis (2%).

*Dopamine agonists.* Dopamine agonists are less commonly used than levodopa to treat parkinsonian features of CBD. Twenty-five percent of the 147 patients collected by Kompoliti and associates (1998) had been treated with either pergolide or bromocriptine, and doses were analyzed as "pergolide equivalents" with a conversion ratio of 1:10. Mean agonist dose was thus reported as 0.82 mg of pergolide equivalent (range, 0.15–3.0 mg), but it is not clear from the report whether this was in monotherapy or as adjunct to levodopa. Not surprisingly, given the low mean doses, only 6% of patients showed some improvement—again no further details on the type and magnitude or duration of response are given.

Agonist-induced confusion was relatively common, affecting 14% of patients, whereas gastrointestinal side effects or dizziness was noted in 11% each.

**Other Antiparkinsonian Drugs.** Data on the efficacy and safety of other antiparkinsonian agents in CBD are even more fragmentary and unsystematic as for levodopa and dopamine agonists. Selegeline was used in 20% of cases of the preceeding series, improving parkinsonism in 3 of 30 patients, whereas amantadine was given to 24 patients (16% of that series), producing some benefit in 3 cases (13%). Other than tremor and rigidity, gait was reported to improve with amantadine. Thirty-eight patients (27%) had been exposed to anticholinergics—either trihexiphenidyl or benztropine—and four of these (10%) had improvements in rigidity or tremor. Side effects of such treatment, specifically cognitive dysfunction, are not commented on in this report or, indeed, any report in the literature.

### Treatment of Dystonia in CBD

Dystonia is one of the cardinal motor features of CBD, affecting between 50% and 80% of patients (Vanek and Jankovic, 2000), most often as asymmetrical limb dystonia producing jerky or fixed postural deformities that may the render the affected extremity functionally useless and may also be painful. It is therefore an area of great therapeutic need, but again there are no controlled prospective or controlled trials of antidystonic interventions on which to base treatment decisions. Retrospective uncontrolled observations point to the possible efficacy of drugs like anticholinergics, benzodiazepines, baclofen and—most strongly—of local botulinum toxin injections.

**Systemic Drug Therapy.** Anticholinergics and baclofen were reported to have improved dystonia in individual cases of the series of Goetz and co-workers and Kompoliti et al., and these authors, as well as Vanek and Jankovic, remark on the efficacy of clonazepam to improve myoclonic dystonia. Details on doses used, effect size, or side effects are absent from all these reports.

**Botulinum Toxin Injections.** Because asymmetrical focal limb dystonia is a typical presentation of dystonia in CBD (Rinne *et al.*, 1994; Vanek *et al.*, 2000; Wenning *et al.*, 1998), localized injections of botulinum toxin have been tried in this situation—in analogy to their successful use in adult-onset focal idiopathic dystonia. Again, no systematic or controlled trials of this form of treatment for dystonia in CBD are available. Of nine patients in the large series reported by Kompoliti *et al.* (1998), who received local botulinum toxin injections to treat focal limb dystonia, six (67%) reportedly showed some improvement. All 6 patients in the series of 66 cases reported by Jankovic and his colleague (Vanek and Jankovic, 2000) had some response of their focal dystonic symptoms to botulinum toxin injections—two of these experienced marked degrees of improvement of both dystonia and pain with this form of therapy. Unfortunately, none of these reports gives any detailed dose of botulinum toxin or muscle selection, so that it is difficult to derive practical treatment recommendations from them. Müller and colleagues (2002) recently reported on two patients with CBD who received successful treatment with botulinum toxin injections for focal limb dystonia. Dosages used were 40–120 units Dysport® for finger flexor muscles, 80–120 units Dysport® for wrist flexor and extensor muscles, and 160–240 units Dysport® for elbow flexor muscles.

### Treatment of Myoclonus in CBD

The use of benzodiazepines—most often clonazepam—and valproate, mysoline, and piracetam to treat myoclonic jerking in CBD has been reported in a number of studies in anecdotal fashion (Kompoliti *et al.*, 1998; Litvan, 1997; Vanek and Jankovic, 2000). Results are inconsistent, and the drug most often reported as beneficial is clonazepam, ameliorating myoclonic jerking in 23% of patients in the series of Kompoliti and coworkers (1998). Sedation is the most common side effect, affecting 26% of patients exposed to clonazepam in that report.

### Treatment of Cortical Dysfunction in CBD

Apraxia, dysphasia, cortical sensory loss, and alien limb behavior are hallmarks of the cortical disease in this condition. There is no evidence available that various methods of physiotherapy, occupational therapy, or speech therapy significantly reduce the disability caused by the symptoms or, indeed, affect their progression.

### Practical Management

Intervention should be based on the relative impact of various clinical features found in a given patient on his or her global disability. For the treatment of parkinsonism, levodopa given in maximum tolerated doses to achieve the best possible motor response (300–2000 mg) is the drug of first choice(*). Dopamine agonists are of no added value and definitely less well tolerated and also more expensive(*). The expected response rate to levodopa is in the order of 25% of patients. Anticholinergics can be useful in cases in which both rigidity and dystonia cause limb postural deformities(*). Local botulinum toxin injections are an option for focal limb dystonia and may also reduce associated pain and jerking(*). Depending on target muscles, doses between 40 and 300 units of Dysport® or 10 and 100 units of Botox per muscle may be used. Alternatively, myoclonus can be treated with clonazepam in doses between 0.5 and 2 mg/day(*).

Speech and swallowing problems are common as CBD progresses and should be treated with speech therapy(*); nasogastric or gastrostomy tube feeding for severe dysphagia are rarely necessary but should be considered in case of silent aspirations(*). Limb apraxia may show a limited response for a limited period in time to physiotherapy(*), and occupational therapy(*) may increase functional hand use, particularly when combined with antiparkinsonian and antidystonic pharmacotherapy.

## REFERENCES

### MSA/PSP

Alam, M., Smith, G., Bleasdale-Barr, K., Pavitt, D. V., and Mathias, C. J. (1995). Effects of the peptide release inhibitor, octreotide, on daytime hypotension and on nocturnal hypertension in primary autonomic failure. *J. Hypertens.* 13, 1664–1669.

Barclay, C. L., Duff, J., Sandor, P., and Lang, A. E. (1996). Limited usefulness of electroconvulsive therapy in progressive supranuclear palsy. *Neurology* 46, 1284–1286.

Barclay, C. L., and Lang, A. E. (1997). Dystonia in progressive supranuclear palsy. *J. Neurol. Neurosurg. Psychiat.* 62, 352–356.

Beck, R. O., Betts, C. D., and Fowler, C. J. (1994). Genitourinary dysfunction in multiple system atrophy: clinical features and treatment in 62 cases. *J. Urol.* 151, 1336–1341.

Ben-Shlomo, Y., Wenning, G. K., Tison, F., and Quinn, N. P. (1997). Survival of patients with pathologically proven multiple system atrophy: a meta-analysis. *Neurology* 48, 384–393.

Boesch, S. M., Wenning, G. K., Ransmayr, G., Poewe, W. (2002). Dystonia in multiple system atrophy. *J. Neurol. Neurosurg. Psychiatry* 72(3), 300–303.

Bower, J. H., Maraganore, D. M., McDonnell, S. K., and Rocca, W. A. (1997). Incidence of progressive supranuclear palsy and multiple system atrophy in Olmsted County, Minnesota, 1976 to 1990. *Neurology* 49, 1284–1288.

Colosimo, C., Merello, M., and Pontieri, F. E. (1996). Amantadine in parkinsonian patients unresponsive to levodopa: a pilot study. *J. Neurol.* 243, 422–425.

Daniele, A., Moro, E., and Bentivoglio, A. R. (1999). Zolpidem in progressive supranuclear palsy. *N. Engl. J. Med.* 341(7), 543–544.

Di Maria, E., Tabaton, M., Vigo, T., Abbruzzese, G., Bellone, E., Donati, C., Frasson, E., Marchese, R., Montagna, P., Munoz, D. G., Pramstaller, P. P., Zanusso, G., Ajmar, F., and Mandich, P. (2000). Coritcobasal degeneration shares a common genetic background with progressive supranuclear palsy. *Ann. Neurol.* 47, 374–377.

Duncombe, A. S., and Lees, A. J. (1985). Methysergide in progressive supranuclear palsy. *Neurology* 35, 936–937.

Engel, P. A. (1996). Treatment of progressive supranuclear palsy with amitriptyline: therapeutic and toxic effects. *J. Am. Geriatr. Soc.* 44, 1072–1074.

Fabbrini, G., Barbanti, P., Bonifati, V., Colosimo, C., Gasparini, M., Vanacore, N., and Meco, G. (2001). Donepezil in the treatment of progressive supranuclear palsy. *Acta Neurol. Scand.* 103, 123–125.

Foster, N. L., Aldrich, M. S., Bluemlein, L., White, R. F., and Berent, S. (1989). Failure of cholinergic agonist RS-86 to improve cognition and movement in PSP despite effects on sleep. *Neurology* 39, 257–261.

Ghika, J., Tennis, M., Hoffman, E., Schoenfeld, D., and Growdon, J. (1991). Idazoxan treatment in progressive supranuclear palsy. *Neurology* 41, 986–991.

Gilman, S., Low, P. A., Quinn, N., Albanese, A., Ben-Shlomo, Y., Fowler, C. J., Kaufmann, H., Klockgether, T., Lang, A. E., Lantos, P. L., Litvan, I., Mathias, C. J., Oliver, E., Robertson, D., Schatz, I., and Wenning, G. K. (1999). Consensus statement on the diagnosis of multiple system atrophy. *J. Neurol. Sci.* 163(1), 94–98.

Golbe, L. I., Langston, J. W., and Shoulson, I. (1990). Selegiline and Parkinson's disease. Protective and symptomatic considerations. *Drugs* 39, 646–651.

Graham, J. G., and Oppenheimer, D. R. (1969). Orthostatic hypotension and nicotine sensitivity in a case of multiple system atrophy. *J. Neurol. Neurosurg. Psychiat.* 32, 28–34.

Hauser, R. A., and Trehan, R. (1994). Initial experience with electroconvulsive therapy for progressive supranuclear palsy. *Mov. Disord.* 9, 467–469.

Hughes, A. J., Colosimo, C., Kleedorfer, B., Daniel, S. E., and Lees, A. J. (1992). The dopaminergic response in multiple system atrophy. *J. Neurol. Neurosurg. Psychiat.* 55, 1009–1013.

Iranzo, A., Santamaria, J., and Tolosa, E. (2000). Continuous positive air pressure eliminates nocturnal stridor in multiple system atrophy. Barcelona Multiple System Atrophy Study Group. *Lancet* 356, 1329–1330.

Jackson, J. A., Jankovic, J., and Ford, J. ( 1983). Progressive supranuclear palsy: clinical features and response to treatment in 16 patients. *Ann. Neurol.* 13, 273–278.

Jankovic, J. (1983). Controlled trial of pergolide mesylate in Parkinson's disease and progressive supranuclear palsy. *Neurology* 33, 505–507.

Jankovic, J., Gilden, J. L., Hiner, B. C., Kaufmann, H., Brown, D. C., Coghlan, C. H., Rubin, M., and Fouad-Tarazi, F. M. (1993). Neurogenic orthostatic hypotension: a double-blind, placebo-controlled study with midodrine. *Am. J. Med.* 95, 38–48.

Johnsen, J. A., and Miller, V. T. (1986). Tobacco intolerance in multiple system atrophy. *Neurology* 36, 986–988.

Koller, W. C., Morantz, R., Vetere-Overfield, B., and Waxman, M. (1989). Autologous adrenal medullary transplant in progressive supranuclear palsy. *Neurology* 39, 1066–1068.

Kompoliti, K., Goetz, C. G., Litvan, I., Jellinger, K., and Verny, M. (1998). Pharmacological therapy in progressive supranuclear palsy. *Arch. Neurol.* 55, 1099–1102.

Lees, A. J. (1999). The treatment of the motor disorders of multiple system atrophy. *In* "Autonomic Failure." (C. J. Mathias and R. Bannister, Eds.), pp. 357–363. Oxford University Press.

Lees, A. J., and Bannister, R. (1981). The use of lisuride in the treatment of multiple system atrophy with autonomic failure (Shy-Drager syndrome). *J. Neurol. Neurosurg. Psychiat.* 44, 347–351.

Litvan, I., Blesa, R., Clark, K., Nichelli, P., Atack, J. R., Mouradian, M. M., Grafman, J., and Chase, T. N. (1994). Pharmacological evaluation of the cholinergic system in progressive supranuclear palsy. *Ann. Neurol.* 36, 55–61.

Litvan, I., Chase, T. N. (1992). Traditional and experimental therapeutic approaches. *In* "Progressive Supranuclear Palsy." (I. Litvan and Y. Agid, Eds.), pp. 254–269. Oxford University Press.

Litvan, I., Agid, Y., Calne, D., Campbell, G., Dubois, B., Duvoisin, R. C., Goetz, C. G., Golbe, L. I., Grafman, J., Growdon, J. H., Hallett, M., Jankovic, J., Quinn, N. P., Tolosa, E., and Zee, D. S. (1996a). Clinical research criteria for the diagnosis of progressive supranuclear palsy (Steele-Richardson-Olszewski syndrome): report of the NINDS-SPSP international workshop. *Neurology* 47, 1–9.

Litvan, I., Mangone, C. A., McKee, A., Verny, M., Parsa, A., Jellinger, K., D'Olhaberriague, L., Chaudhuri, K. R., and Pearce, R. K. (1996b). Natural history of progressive supranuclear palsy (Steele-Richardson-Olszewski syndrome) and clinical predictors of survival: a clinicopathological study. *J. Neurol. Neurosurg. Psychiat.* 60, 615–620.

Low, P. A., Gilden, J. L., Freeman, R., Sheng, K. N., and McElligott, M. A. (1997). Efficacy of midodrine vs placebo in neurogenic orthostatic hypotension. A randomized, double-blind multicenter study. Midodrine Study Group. *JAMA* 277, 1046–1051.

Mathias, C. J., Fosbraey, P., da Costa, D. F., Thornley, A., and Bannister, R. (1986). The effect of desmopressin on nocturnal polyuria, overnight weight loss, and morning postural hypotension in patients with autonomic failure. *Br. Med. J.* (Clin Res Ed) 293, 353–354.

Mathias, C. J., and Kimber, J. R. (1999). Postural hypotension: Causes, clinical features, investigation, and management. *Annu. Rev. Med.* 50, 317–336.

Mathias, C. J., Senard, J. M., Braune, S., Watson, L., Aragishi, A., Keeling, J. E., and Taylor, M. D. (2001). L-threo-dihydroxyphenylserine (L-threo-DOPS; droxidopa) in the management of neurogenic orthostatic hypotension: A multi-national, multi-center, dose-ranging study in multiple system atrophy and pure autonomic failure. *Clin. Auton. Res.* 11(4), 235–242.

Mueller, J., Wenning, G. K., Wissel, J., Seppi, K., and Poewe, W. (2002). Botulinum toxin treatment in atypical parkinsonian disorders associated with disabling focal dystonia. *J. Neurol.* 249(3), 300–304.

Nagasaka, T., Togashi, S., Amino, A., Nitta, K., Shindo, K., and Shiozawa, Z. (1997). Aniracetam for treatment of patients with progressive supranuclear palsy. *Eur. Neurol.* 37, 195–198.

Neophytides, A., Lieberman, A. N., Goldstein, M., Gopinathan, G., Leibowitz, M., Bock, J., and Walker, R. (1982). The use of lisuride, a potent dopamine and serotonin agonist, in the treatment of progressive supranuclear palsy. *J. Neurol. Neurosurg. Psychiat.* 45, 261–263.

Newman, G. C. (1985). Treatment of progressive supranuclear palsy with tricyclic antidepressants. *Neurology* 35, 1189–1193.

Nieforth, K. A., and Golbe, L. I. (1993). Retrospective study of drug response in 87 patients with progressive supranuclear palsy. *Clin. Neuropharmacol.* 16, 338–346.

Parati, E. A., Fetoni, V., Geminiani, G. C., Soliveri, P., Giovannini, P., Testa, D., Genitrini, S., Caraceni, T., and Girotti, F. (1993). Response to L-DOPA in multiple system atrophy. *Clin. Neuropharmacol.* **16**, 139–144.

Paulson, G. W., Lowery, H. W., and Taylor, G. C. (1981). Progressive supranuclear palsy: pneumoencephalography, electronystagmography and treatment with methysergide. *Eur. Neurol.* **20**, 13–16.

Perera, R., Isola, L., and Kaufmann, H. (1995). Effect of recombinant erythropoietin on anemia and orthostatic hypotension in primary autonomic failure. *Clin. Auton. Res.* **5**, 211–213.

Piccione, F., Mancini, E., Tonin, P., and Bizzarini, M. (1997). Botulinum toxin treatment of apraxia of eyelid opening in progressive supranuclear palsy: report of two cases. *Arch. Phys. Med. Rehabil.* **78**, 525–529.

Polo, K. B., and Jabbari, B. (1994). Botulinum toxin-A improves the rigidity of progressive supranuclear palsy. *Ann. Neurol.* **35**, 237–239.

Rafal, R. D., and Grimm, R. J. (1981). Progressive supranuclear palsy: functional analysis of the response to methysergide and antiparkinsonian agents. *Neurology* **31**, 1507–1518.

Raimbach, S. J., Cortelli, P., Kooner, J. S., Bannister, R., Bloom, S. R., and Mathias, C. J. (1989). Prevention of glucose-induced hypotension by the somatostatin analogue octreotide (SMS 201–995) in chronic autonomic failure: haemodynamic and hormonal changes. *Clin. Sci. (Lond.)* **77**, 623–628.

Rascol, O., Sieradzan, K., Peyro-Saint-Paul, H., Thalamas, C., Brefel-Courbon, C., Senard, J. M., Ladure, P., Montastruc, J. L., and Lees, A. (1998). Efaroxan, an alpha-2 antagonist, in the treatment of progressive supranuclear palsy. *Mov. Disord.* **13**, 673–676.

Sakakibara, R., Hattori, T., Uchiyama, T., Suenaga, T., Takahashi, H., Yamanishi, T., Egoshi, K., and Sekita, N. (2000). Are alpha-blockers involved in lower urinary tract dysfunction in multiple system atrophy? A comparison of prazosin and moxisylyte. *J. Auton. Nerv. Syst.* **79**, 191–195.

Schrag, A., Ben-Shlomo, Y., and Quinn, N. P. (1999). Prevalence of progressive supranuclear palsy and multiple system atrophy: A cross-sectional study. *Lancet* **354**, 1771–1775.

Tamai, S., and Almeida, O. P. (1997). Nortriptyline for the treatment of depression in progressive supranuclear palsy. *J. Am. Geriatr. Soc.* **45**, 1033–1034.

Tison, F., Wenning, G. K., Daniel, S. E., and Quinn, N. P. (1995). The pathophysiology of parkinsonism in multiple system atrophy. *Eur. J. Neurol.* **2**, 435–444.

Watanabe, H., Saito, Y., Terao, S., Ando, T., Kachi, T., Mukai, E. *et al.* (2002). Progression and prognosis in multiple system atrophy: An analysis of 230 Japanese patients. *Brain* **125**(Pt 5): 1070–1083.

Waxman, M. J., Morantz, R. A., Koller, W. C., Paone, D. B., and Nelson, P. W. (1991). High incidence of cardiopulmonary complications associated with implantation of adrenal medullary tissue into the caudate nucleus in patients with advanced neurologic disease. *Crit. Care Med.* **19**, 181–186.

Weiner, W. J., Minagar, A., and Shulman, L. M. (1999). Pramipexole in progressive supranuclear palsy. *Neurology* **52**, 873–874.

Wenning, G. K., Ben-Shlomo, Y., Hughes, A., Daniel, S. E., Lees, A., and Quinn, N. P. (2000). What clinical features are most useful to distinguish definite multiple system atrophy from Parkinson's disease? *J. Neurol. Neurosurg. Psychiat.* **68**, 434–440.

Wenning, G. K., Ben Shlomo, Y., Magalhaes, M., Daniel, S. E., and Quinn, N. P. (1994). Clinical features and natural history of multiple system atrophy. An analysis of 100 cases. *Brain* **117**, 835–845.

Williams, A. C., Nutt, J., Lakes, C. R. *et al.* (1979). Actions of bromocriptine in the Shy-Drager and Steele-Richardson-Olszewski syndrome. *In* "Dopaminergic Ergots and Motor Control." (K. Fixe and D. B. Calne, Eds.), pp. 271–283. Pergamon Press, Oxford.

Winkler, A. S., Marsden, J., Parton, M., Watkins, P. J., and Chaudhuri, K. R. (2001). Erythropoietin deficiency and anaemia in multiple system atrophy. *Mov. Disord.* **16**, 233–239.

Zesiewicz, T. A., Helal, M., and Hauser, R. A. (2000). Sildenafil citrate (Viagra) for the treatment of erectile dysfunction in men with Parinson's disease. *Mov. Disord.* **15**, 305–308.

## DLB

Aarsland, D., Larsen, J. P., Lim, N. G., and Tandberg, E. (1999). Olanzapine for psychosis in patients with Parkinson's disease with and without dementia. *J. Neuropsychiatry Clin. Neurosci.* **11**, 392–394.

Baldessarini, R. J., and Frankenburg, F. R. (1991). Clozapine. A novel antipsychotic agent. *N. Engl. J. Med.* **324**, 746–754.

Ballard, C., Grace, J., McKeith, I. G. *et al.* (1998). Neuroleptic sensitivity in dementia with Lewy bodies and Alzheimer's disease. *Lancet* **351**, 1032–1033.

Burke, W. J., Pfeiffer, R. F., and McComb, R. D., (1998). Neuroleptic sensitivity to clozapine in dementia with Lewy bodies. *J. Neuropsychiatry Clin. Neurosci.* **10**(2), 227–229.

Crystal, H. A., Dickson, D. W., Lizardi, J. E., Davies, P. *et al.* (1990). Antemortem diagnosis of diffuse Lewy body disease. *Neurology* **40**, 1523–1528.

Cummings, J. L. (1995). Lewy body diseases with dementia: pathophysiology and treatment. *Brain Cognition* **28**, 266–280.

Eichhorn, T. E., Brunt, E., and Oertel, W. H. (1996). Ondansetron treatment of L-dopa induced psychosis. *Neurology* **47**, 1608–1609.

Fergusson, E., and Howard R. (2000). Donepezil for the treatment of psychosis in dementia with Lewy bodies. *Int. J. Geriatr. Psychiat.* **15**, 280–281.

Fernandez, H. H., Friedman, J. H., Jacques, C., and Rosenfeld, M. (1999). Quetiapine for the treatment of drug-induced psychosis in Parkinson's disease. *Mov. Disord.* **14**, 484–487.

Ford, B., Lynch, T., and Greene, P. (1994). Risperidone in Parkinson's disease. *Lancet* **434**, 1370–1371.

Friedberg, G., Zoldan, J., Weizman, A., and Melamed, E. (1998). Parkinson Psychosis Rating Scale: a practical instrument of grading psychosis in Parkinson's disease. *Clin Neuropharmacol.* **21**, 280–294.

Goetz, C. G., Blasucci, L. M., Leurgans, S., and Pappert, E. J. (2000). Olanzapine and clozapine: comparative effects on motor function in hallucinating PD patients. *Neurology* **55**, 748–749.

Kane, J., Honigfeld, G., Singer, J., and Meltzer, H. (1988). Clozapine for the treatment-resistant schizophrenic. A double-blind comparison with chlorpromazine. *Arch. Gen. Psychiatry* **45**, 789–796.

Lanctot, K. L., and Herrmann, N. (2000). Donepezil for behavioral disorders associated with Lewy bodies: a case series. *Int. J. Geriatr. Psychiatry* **15**, 338–345.

Lebert, F., Pasquier, F., Souliez, L., and Petit, H. (1998). Tacrine efficacy in Lewy body dementia. *Int. J. Geriatr. Psychiatry* **13**, 516–519.

Levy, R., Eagger, S., Griffiths, M., Perry, E. *et al.* (1994). Lewy bodies and response to tacrine in Alzheimer's disease. *Lancet* **343**, 176.

Mayeux, R., Denaro, J., Hemenegildo, N., Marder, K. *et al.* (1992). A population-based invstigation of Parkinson's disease with and without dementia. Relationship to age and gender. *Arch. Neurol.* **49**, 492–497.

Matheson, A. J., and Lamb, H. M. (2000). Quetiapine. A review of its clinical potential in the management of psychotic symptoms in Parkinson's disease. *CNS Drugs* **14**, 157–172.

McKeith, I. G., Ballard, C. G., Perry, R. H., Ince, P. G., O'Brien, J. T., Neill, D. *et al.* (2002a). Prospective validation of consensus criteria for the diagnosis of dementia with Lewy bodies. *Neurology* 54(5), 1050–1058.

McKeith, I., Del Ser, T., Spano, P., Emre, M. *et al.* (2000b). Efficacy of rivastigmine in dementia with Lewy bodies: a randomised, double-blind, placebo-controlled international study. *Lancet* 356, 2031–2036.

McKeith, I. G., Fairbairn, A., Perry, R. *et al.* (1992a). Neuroleptic sensitivity in patients with senile dementia of Lewy body type. *BMJ* 305, 673–678.

McKeith, I. G., Fairbairn A., Perry, E. K. *et al.* (1992b). Operational criteria for senile dementia of Lewy body type (SDLT). *Psychol. Med.* 22, 911–922.

McKeith, I. G., Galasko, D., Kosaka, K., Perry, E. K. *et al.* (1996). Consensus guidelines fort he clinical and pathologic diagnosis of dementia with Lewy bodies (DLB): report of the Consortium on DLB International Workshop. *Neurology* 47, 1113–1124.

McKeith, I. G., Galasko, D., Wilcock, G. K., and Byrne, E. (1995). Lewy body dementia—diagnosis and treatment. *Br. J. Psychia.* 167, 709–717.

McKeith, I. G., Grace, J. B., Walker, Z., Byrne, E. J. *et al.* (2000c). Rivastigmine in the treatment of dementia with Lewy bodies: preliminary findings from an open trial. *Int. J. Geriatr. Psychiatry* 15, 387–392.

McKeith, I. G., Perry, E. K., and Perry, R. H. (1999). Report of the second dementia with Lewy body international workshop: diagnosis and treatment consortium on dementia with Lewy bodies. *Neurology* 53, 902–905.

Morris, S. K., Olichney, J. M., and Corey-Bloom, J. (1998). Psychosis in dementia with lewy bodies. *Semin. Clin. Neuropsychiatry* 3(1), 51–60.

Parkinson Study Group. (1999). Low-dose clozapine for the treatment of drug-induced psychosis in Parkinson's disease. *N. Engl. J. Med.* 340, 757–763.

Parsa, M., Greenway, H., and Bastani, B. (2000). Quetiapine in the treatment of psychosis in patients with Parkinson's disease and dementia (Lewy body disease variant). Annual Meeting of the American Association for Geriatric Psychiatry, Mar 12–15, Miami, Florida.

Perry, E., Walker, M., Grace, J., and Perry, R. (1999). Acetylcholine in mind: a neurotransmitter correlate of consciousness? *Trends Neurosci.* 22, 273–280.

Querfurth, H. W., Allam, G. J., Geffroy, M. A., Schiff, H. B., and Kaplan, R. F. (2000). Acetylcholinesterase inhibition in dementia with Lewy bodies: results of a prospective pilot trial. *Dement. Geriatr. Cogn. Disord.* 11, 314–321.

Rich, S. S., Friedman, J. H., and Ott, B. R. (1995). Risperidone versus clozapine in the treatment of psychosis in six patients with Parkinson's disease and other akinetic-rigid syndromes. *J. Clin. Psychiatry* 56, 556–559.

Rojas-Fernandez, C. H., (2001). Successful use of donazepil for the treatment of dementia with Lewy bodies. *Ann. Pharmacother.* 35, 202–205.

Samuel, W., Caligiuri, M., Galasko, D. *et al.* (2000). Better cognitive and psychopathologic response to donepezil in patients prospectively diagnosed as dementia with lewy bodies: a preliminary study. *Int. J. Geriatr. Psychiatry* 15, 794–802.

Shea, C., MacKnight, C., Rockwood, K. (1998). Aspects of dementia. Donepezil for treatment of dementia with Lewy bodies: a case series of nine patients. *Int. Psychogeriatr.* 10, 229–238.

Skjerve, A., and Nygaard, H. A. (2000). Improvement in sundowning in dementia with Lewy bodies after treatment with donepezil. *Int. J. Geriatr. Psychiatry* 15, 1147–1151.

Scholz, E., and Dichgans, J. (1985). Treatment of drug-induced exogenous psychosis in parkinsonism with clozapine and fluperrlapine. *Eur. Arch. Psychiatry Neurol. Sci.* 235, 60–64.

The French Clozapine Parkinson Study Group. (1999). Clozapine in drug-induced psychosis in Parkinson's disease. *Lancet* 353, 2041–2041.

Walker, Z., Grace, J., Overshot, R., Satarsinghe, S. *et al.* (1999). Olanzapine in dementia with Lewy bodies: a clinical study. *Int. J. Geriatr. Psychiatry* 14, 459–466.

Zoldan, J., Friedberg, G., Livneh, M., and Melamed, E. (1995). Psychosis in advanced Parkinson's disease: treatment with ondansetron, a 5-HT3 receptor antagonist. *Neurology* 45, 1305–1308.

## CBD

Cummings, J. L., and Litvan, I. (2000). Neuropsychiatric aspects of corticobasal degeneration. *In* "Corticobasal Degeneration" (I. Litvan, C. G. Goetz, and A. E. Lang, Eds.), Advances in Neurology (Vol. 82), pp. 147–152. Lippincott Williams & Wilkins, Philadelphia.

Gibb, W. R. G., Luthert, P. J., Marsden, C. D. (1989). Corticobasal degeneration. *Brain* 112, 1171–1192.

Kertesz, A., Martinez-Lage, P., Davidson, W., and Munoz, D. G. (2000). The corticobasal degeneration syndrome overlaps progressive aphasia and frontotemporal dementia. *Neurology* 55, 1368–1375.

Kompoliti, K., and Goetz, C. G. (2000). Therapeutic approaches. *In* "Corticobasal Degeneration" (I. Litvan, C. G. Goetz, and A. E. Lang, Eds.). Advances in Neurology (Vol. 82), pp. 217–221. Lippincott Williams & Wilkins, Philadelphia.

Kompoliti, K., Goetz, C. G., Boeve, B. F., Maraganore, D. M., *et al.* (1998). Clinical presentation and pharmacological therapy in corticobasal degeneration. *Arch. Neurol.* 55, 957–961.

Litvan, I. (1997). Progressive supranuclear palsy and corticobasal degeneration. *Bailliere's Clin. Neurol.* 6, 167–185.

Litvan, I., Agid, Y., Goetz, C., *et al.* (1997). Accuracy of the clinical diagnosis of corticobasal degeneration: a clinicopathological study. *Neurology* 57, 1184–1189.

Mueller, J., Wenning, G. K., Wissel, J., Seppi, K., and Poewe, W. (2002). Botulinum toxin treatment in atypical parkinsonian disorders associated with disabling focal dystonia. *J. Neurol.* 249(3), 300–304.

Rebeiz, J. J., Kolodny, E. H., and Richardson, E. P. (1967). Cortico-dentationigral degeneration with neuronal achromasia: a progressive disorder of late adult life. *Trans. Am. Neurol. Assoc.* 92, 23–26.

Riley, D. E., Lang, A. E., Lewis, A., Resch, L., Ashby, P., Hornykiewicz, O., *et al.* (1990). Cortical-basal ganglionic degeneration. *Neurology* 40, 1203–1212.

Rinne, J. O., Lee, M. S., Thompson, P. D., and Marsden, C. D. (1994). Corticobasal degeneration: a clinical study of 36 cases. *Brain* 117, 1183–1196.

Vanek, Z. F., and Jankovic, J. (2000). Dystonia in corticobasal degeneration. *In* "Corticobasal degeneration" (I. Litvan, C. G. Goetz, and A. E. Lang, Eds.), Advances in Neurology (Vol. 82), pp. 61–67. Lippincott Williams & Wilkins, Philadelphia.

Watts, R. L., Mirra, S. S., and Young, R. R. (1989) Corticobasal ganglionic degeneration. (CBDG) with neuronal achromasia: clinical-pathological study of two cases. *Neurology* 39, 140.

Wenning, G. K., Litvan, I., Jankovic, J., Granata, R., *et al.* (1998). Natural history and survival of 14 patients with corticobasal degeneration confirmed at post-mortem examination. *J. Neurol. Neurosurg. Psychiatry* 64, 184–189.

CHAPTER 76

# Deep Brain Stimulation for Movement Disorders

Kai Bötzel and Paul Krack

## GENERAL REMARKS

### Historical Notes

Surgery for movement disorders has a long tradition. Meyers was the first to show that basal ganglia lesions could alleviate parkinsonian symptoms without inducing a paresis (Meyers, 1942). However, open surgery had unacceptably high risks. The advent of stereotactic surgery (Spiegel *et al.*, 1947) made precisely aimed intracerebral manipulations possible. Although the pallidum had been aimed at by some, Hassler's thalamotomy was the most frequent operation for parkinsonian tremor before the introduction of L-dopa (Hassler and Riechert, 1954; Riechert, 1962). When this effective medical treatment became available, functional neurosurgery for movement disorders disappeared almost completely. Only when it became clear that long-term L-dopa treatment can be complicated by motor fluctuations and dyskinesias, and with technical advances mainly in imaging, did interest in surgical therapies re-emerge.

Unilateral pallidotomies were reinvestigated by Laitinen and shown to improve not only all contralateral parkinsonian motor signs but also levodopa-induced dyskinesias (LID) (Laitinen, 1995). Bilateral pallidotomies, however, are considered to be prone to disabling side effects such as psychiatric and corticobulbar symptoms. This limits the value of pallidotomy in advanced bilateral disease.

Deep brain stimulation (DBS) of the Vim nucleus of the thalamus had been successfully applied to patients with a thermocoagulation in the opposite thalamus and was then found to be safely applicable as a simultaneous bilateral procedure (Benabid *et al.*, 1991). The emerging concept of basal ganglia circuitry (Alexander *et al.*, 1990) promoted exploration of the subthalamic

nucleus (STN) (Pollak *et al.*, 1993) and the globus pallidus internus (GPi) (Siegfried and Lippiz, 1994) as a DBS targets in Parkinson's disease. All parkinsonian symptoms can be improved with DBS of these two targets. Tremor of other than parkinsonian origin can still be best helped by thalamic stimulation. The GPi has attracted much interest recently as a DBS target for otherwise intractable dystonia.

### Pathophysiology

It is still not well understood which and how information is processed in the basal ganglia. The motor circuit of the basal ganglia originates in the frontal cortex, passes through striatum and globus pallidus, and enters the thalamus to be relayed back to the cortex. In experimental Parkinson's disease in monkeys, STN and GPi have been found to be overactive and, perhaps more important, to have abnormal patterns of activity (Bergman *et al.*, 1994). It is therefore reasonable to assume that the reduction of this abnormal activity by DBS (similar to thermocoagulation) is responsible for the relief of parkinsonian symptoms. In thalamic stimulation for the treatment of tremor, it is assumed that oscillatory circuits are interrupted by the stimulation. However, the exact mechanism of these local effects of DBS are not yet well understood. It is unclear whether it causes inhibition or excitation and whether axons or neurons are the crucial elements (Ashby, 2000). A direct blockade of voltage-gated currents of STN neurons by high-frequency stimulation has been shown in vitro (Beurrier *et al.*, 2001). The immediate cessation of tremor caused by DBS and the delayed response of dystonic symptoms are indicative of the notion that different mechanisms may be involved.

## Surgery

Surgery (whether ablative or for DBS) is usually performed with the patient under local anesthesia, because general anesthesia suppresses the target symptoms and does not allow for cooperation of the patient during intraoperative neurological and neurophysiological testing. Stereotactic coordinates can be defined by ventriculography, computed tomography (CT), magnetic resonance imaging (MRI) or a combination of these (Holtzheimer *et al.*, 1999; Schuurman *et al.*, 1999; Starr *et al.*, 1999). Visualization of the target structure is possible for STN and GPi, thus allowing for direct targeting based on the patient's individual anatomy. Microelectrode recordings of extracellular activity can be used to identify the anatomical target structure (Hariz and Fodstad, 1999; Vitek *et al.*, 1998). The target is also verified when intraoperative electrical stimulation causes immediate relief of symptoms (thalamus: suppression of tremor; STN: decrease of rigidity) and no stimulation-induced side effects (muscle twitches caused by stimulation of the pyramidal tract fibers, eye movements, or others).

The standard DBS lead (Medtronic Inc., Minneapolis; model 3387) has a row of four cylindrical contacts, each with a length of 1.5 mm and a diameter of 1.3 mm separated by 1.5 mm. For STN stimulation, a lead with smaller interelectrode distances is available (model 3389) to allow for higher spatial precision. Current can be passed between one or more of these contacts (cathode) and the metal stimulator case (anode), which is implanted underneath the skin near the clavicle. Also, bipolar stimulation can be performed between different electrodes. The stimulation consists of brief pulses (thalamus and STN: 60 μs; Gpi: 60–120 μs) with a frequency from 130–180 Hz. The stimulator applies a constant voltage, which can be adjusted telemetrically like the other parameters. Current consumption is proportional to the product of amplitude (voltage) * pulse width * frequency. This product is on average two to three times higher in GPi compared with STN stimulation, whereas it is comparable in STN and Vim stimulation. Battery longevity is approximately 6 years in Vim or STN DBS and 2 to 3 years in GPi DBS.

In many patients the placement of the electrode alone causes a significant relief of tremor or parkinsonian symptoms that may last for a few days (stun effect). In dystonia, stimulation effects seem to appear only with a latency of several hours, and sustained improvement has even been reported after some weeks. Detailed anatomical knowledge provides clues for correct repositioning when particular side effects are encountered (Hariz and DeSalles, 1997). Postoperatively, the electrode positions are documented by MRI. Although stimulators have been placed into the MRI magnet without problems (Tronnier *et al.*, 1999), the manufacturer advises against this. Most neurosurgeons implant the stimulators in a second operation and perform a postoperative MRI between these two operations. If an MRI needs to be performed in a case of a medical emergency, the stimulator must be switched off, and in addition, all parameters must be set to their lowest values, because the stimulator may accidentally be switched on in the MRI magnet.

## Advantages and Disadvantages of DBS versus Lesioning

The principle advantage of DBS over thermocoagulation is that all the effects induced are reversible by reducing the intensity of stimulation. This has an influence on the occurrence and severity of persistent side effects, which are generally less frequently seen with DBS*. In unilateral thermocoagulations of the *thalamus* the incidence of persistent side effects was 16% in five mostly retrospective studies (Goldman *et al.*, 1992; Jankovic *et al.*, 1995; Nagaseki *et al.*, 1986; Schuurman *et al.*, 2000; Wester and Hauglie Hanssen *et al.*, 1990). In three recent DBS studies the comparable incidence was only 5% (Benabid *et al.*, 1996; Koller *et al.*, 1997; Schuurman *et al.*, 2000). Bilateral thermocoagulations of the thalamus are generally not performed because of the high incidence of severe cognitive and behavioral side effects (Krayenbühl *et al.*, 1961).

Considering surgery in the GPi, unilateral thermocoagulation can be performed relatively safely with only moderate incidence of persistent side effects (Baron *et al.*, 1996). Bilateral GPi lesions are rarely performed, although when the posteroventral parts are lesioned, sustained side effects can be absent in Parkinson's disease (Laitinen, 1995) and dystonia (Ondo *et al.*, 1998). However, cognitive, psychiatric, and disabling corticobulbar dysfunction have been described, even when modern surgical techniques were applied (Ghika *et al.*, 1999; Trepanier *et al.*, 2000). In contrast, there is no relevant worsening of cognition with bilateral GPi or STN stimulation in young, nondemented patients (Ardouin *et al.*, 1999). STN operations are generally performed as bilateral electrode implantations, whereas lesions of STN have only rarely been performed (Alvarez *et al.*, 2001). Therefore, conclusions about the safety of unilateral or bilateral STN lesions are not yet possible. When bilateral surgery is necessary, as in most patients with Parkinson's disease, DBS is generally recommended*. Costs of DBS are higher than thermocoagulation, because the value of the implanted materials amounts to $9500 (US) for a unilateral operation. Additional costs are for consultations by a specialized neurologist and replacement of the stimulators every 3 to 6 years.

## Side Effects of DBS

*Complications of implantation* of DBS electrodes have included subdural and cerebral *hematomas*, which were reported in 2 of 218 patients (partly bilateral implantations) from six recent studies. Hemorrhage-related death occurred in one patient. Postoperative confusion is relatively frequent in the elderly (>70 years) and is usually reversible within a few days. The risk of *infection* of the implanted material is estimated to range between 3% and 4% in experienced hands (Pollak *et al.*, 1998). Infection most often occurs during the postoperative period, when the connector cable is temporarily connected to an external test stimulator. This is a reason for implanting electrodes and stimulators during a single operation. Skin erosions over the connector below the scalp may occur and constitute a source for delayed infections. In cases with clear signs of local infection, meningitis, or brain abscess, explantation of the material and antibiotic treatment are mandatory. The subcutaneous material can then be reimplanted after a delay of about 3 months.

*Disconnection* of one or more leads is suspected when the effect of stimulation decreases abruptly, possibly related to trauma that may have put strain on the cables. When the measured current is reduced to values near 7 μA, the stimulator is no longer connected, and plain x-ray films of the shoulder, neck, and head are to be initiated to demonstrate the site of disconnection. A *sudden decrease* in stimulation effects can sometimes be seen in patients with the Itrel II type stimulator. This stimulator can be deactivated and activated by the patient using a small magnet. Electromagnetic fields from other sources (electric tools, antitheft devices in department stores, metal detectors in airports) can accidentally deactivate the stimulator by the same mechanism. For some patients it is difficult to distinguish the effect of these episodes from normal fluctuations of mobility, and the accidental deactivation is then detected only some time later when the stimulators are checked by a specialist. The Kinetra stimulator allows the patient to check its function and to set some parameters by means of an external control device. Turning the stimulator on or off may not only be necessary in instances of suspected accidental failure but also when an ECG or other electrophysiological investigations are needed. These are normally not possible with the stimulators switched on.

Stimulation itself can cause transient and sustained side effects (Table I). *Speech problems* can be seen as

TABLE I  Probability of Stimulation-Induced Reversible Side Effects of DBS in Different Targets

| Side effect | STN | GPi | Vim |
|---|---|---|---|
| Dysarthria | + | + | +++ |
| Dyskinesia | +++ | + | (+) |
| Pyramidal tract | + | + | + |
| Cognitive | (+) | (+) | (+) |
| Affective/behavioral | + | + | − |
| Apraxia of lid opening | + | − | − |
| Ocular deviation | (+) | − | − |
| Weight gain | +++ | +++ | − |
| Worsening of akinesia | (+) | + | − |
| Paresthesia | (+) | − | + |
| Ataxia/dysmetria | + | − | + |

+++ = relative probability high; + = low probability; (+) = infrequent; − = not observed.

sustained problems (i.e., they persist until stimulation amplitude is reduced). They are rarely seen in unilateral stimulation, but are frequent in bilateral stimulation and are seen as dysarthria or reduced speech volume. In some patients mild sustained speech problems are tolerated and balanced against the benefit the patient gains by the stimulation. Other amplitude-dependent side effects can be due to electric diffusion into neighboring tissue and include, among others, tonic muscular contractions in cases of pyramidal tract stimulation, visual phosphenes in GPi stimulation, and eye movements in STN stimulation. In thalamic stimulation paresthesias may occur as transient phenomena (i.e., they vanish within seconds without reduction of stimulation amplitude). Side effects may also be caused by stimulation of the target structure. STN stimulation can trigger choreiform, ballistic, or dystonic *dyskinesias* that may have some clinical resemblance to L-dopa–induced dyskinesias. In the long term, dyskinesias induced by STN stimulation tend to disappear (Krack *et al.*, 1999). To avoid these symptoms, a gradual postoperative adjustment of stimulation parameters and medication is necessary. Dyskinesias may appear even several hours after new stimulation parameters are programmed. For this reason, it is not recommended to change stimulation parameters just before the weekend or in the afternoon.

*Neuropsychological complaints* after DBS surgery may concentrate on postoperative confusion and memory impairments. In detailed tests, no change in the overall cognitive performance in Parkinson's disease was seen several months after the surgery, but a decreased verbal fluency was repeatedly reported (Ardouin *et al.*, 1999; Trepanier *et al.*, 2000). Long-lasting side effects include the weight gain frequently seen after STN and pallidal stimulation and cases of eyelid dyspraxia (Limousin *et al.*, 1998) with STN DBS. The latter may require botulinum toxin injections into the pars palpebralis of the orbicularis oculi muscles in rare instances.

## Patient Selection and Exclusion in General

It is important to consider the expectations of the patient and his or her family when explaining the possible benefits. Postoperative patient satisfaction critically depends on a realistic explanation of the potential benefits and risks of the therapy. Patients therefore have to be able to understand the intended procedures and to cooperate during the operation and during tedious testing sessions. Dementia and psychosis must be excluded by history and specific tests. After DBS surgery, borderline dementia can be exacerbated (Saint-Cyr *et al.*, 2000; Trepanier *et al.*, 2000). Benign hallucinations caused by L-dopa medication are frequent in Parkinson's disease and are not generally taken as an argument to reject the patient from the operation. Candidates should be younger than 70 years. Whenever possible, it is recommended that the patient be admitted preoperatively for a few days to determine the off-drug symptoms and the levodopa-sensitivity (see later). A CT scan or MRI of the brain should be obtained to detect severe brain atrophy or indications of multiinfarct syndromes, which are contraindications. Documentation of preoperative and postoperative status should be performed along existing guidelines for Parkinson's disease (Defer *et al.*, 1999; Fahn *et al.*, 1987), tremor (Fahn *et al.*, 1993), and dystonia (Burke *et al.*, 1985).

## Anatomical Targets for DBS

For the symptoms of Parkinson's disease, stimulation of the STN achieves slightly better relief than GPi stimulation* (DBS-for Parkinson's-disease-study-group, 2001; Krack *et al.*, 1998d), especially when long-term observations are considered (Ghika *et al.*, 1999; Houeto *et al.*, 2000a; 2000b) (Table II). STN stimulation also has the advantage of less current consumption and, thus, extended battery life of the stimulators. Also, medication can on average be cut in half after STN stimulation but not after GPi stimulation. Tremor of causes other than Parkinson's disease is mainly an indication for stimulation of the Nc ventrointermedius (Vim) of the thalamus. Tremor in Parkinson's disease may also be alleviated by STN or GPi stimulation, especially when akinesia and rigidity, which are not improved by Vim stimulation, are present. For the treatment of dystonia, electrodes are placed in the GPi, because stimulation of the thalamus has been less successful to date.

### TABLE II    Symptoms and the Effect of DBS at Different Targets

| | Vim | GPi | STN |
|---|---|---|---|
| Akinesia | 0 | ++ | +++ |
| Rigidity | 0 | +++ | +++ |
| Off-period dystonia | 0 | ++ | +++ |
| On-period dyskinesias | (+) | +++ | − short term |
| | | | +++ long term |
| PD tremor | +++ | ++ | +++ |
| ET | ++ | ? | ? |
| Cerebellar tremor | + | ? | ? |
| Primary dystonia | ++ | +++ | ? |
| Secondary dystonia | + | ++ | ? |

+++ = excellent; ++ = good; + = moderate; (+) = mild symptom control; 0 = no effect; − = worsening; ? = unknown.

## DBS FOR PARKINSON'S DISEASE (PD)

### Principles of Therapy

#### Results of STN and GPi DBS

Both STN and GPi DBS induce major improvements in off-period symptoms of parkinsonism and in on-period dyskinesias. At 3–6 months after surgery, UPDRS motor scores in the off-drug condition are improved by 30%–60% on average in bilateral GPi DBS (DBS-for-Parkinson's-disease-study-group, 2001; Volkmann *et al.*, 1998), and by 40%–70% on average in bilateral STN DBS (DBS-for Parkinson's-disease-study-group, 2001; Houeto *et al.*, 2000b; Kumar *et al.*, 1998; Limousin *et al.*, 1998; Molinuevo *et al.*, 2000), whereas dyskinesias are improved by 80% with stimulation of either target. STN DBS can improve akinesia to the same extent as levodopa, whereas GPi DBS seems to be slightly less effective for this cardinal symptom (DBS-for Parkinson's-disease-study-group, 2001; Krack *et al.*, 1998d) Some patients with long-term failure of pallidal DBS had to be reoperated for electrode placement in the STN (Houeto *et al.*, 2000a). Patients who are severely akinetic generally require bilateral surgery. A patient who is unable to walk will not be sufficiently improved with a unilateral intervention, because the effects are mainly contralateral (Kumar *et al.*, 1999). Levodopa-induced dyskinesias (LID) are improved with stimulation of either target, but the mechanisms and time course are different. STN stimulation can transiently induce dyskinesias in the short term, but LID improve with time on chronic STN DBS, because patients are relieved from their motor fluctuations (Krack *et al.*, 1997a; 1999). GPi-DBS has a direct and immediate effect on LID (Krack *et al.*, 1998c). In the long term, the degree of improvement in LID is similar to DBS of either target. Tremor and rigidity are also improved to a great extent. Antitremor effects of DBS in STN are similar to those of DBS in Vim (Krack *et al.*, 1998a; 1997c; Rodriguez *et al.*, 1998). The frontal cognitive symptoms of PD do not improve with surgery. Patients benefit most from improvements in off-drug condition. The Schwab and England score reflecting changes in activities of daily living is mainly improved in off-drug condition. The main benefit in the on-drug condition is related to the decrease in LID. Although dyskinesias are socially disabling, they often do not lead to much functional deficit, and therefore the Schwab and England score in the on-drug condition shows only small improvements (Limousin *et al.*, 1998). Dopaminergic treatment can be reduced by about 50% on average in bilateral STN DBS, whereas it remains unchanged in bilateral GPi DBS (Krack *et al.*, 1998d). Although there are no randomized studies with sufficiently large numbers of patients comparing STN to GPi DBS, the arguments in favor of bilateral STN DBS presently are overwhelming.

### When to Consider Surgery for PD?

Surgery is not without side effects and therefore is not a treatment for early PD. Once patients are disabled despite optimal medical treatment, however, the question arises, whether surgery may be considered (Table III). The main prerequisite for improvement with STN or GPi DBS is good *levodopa sensitivity*. Indeed, if akinesia is not improved with levodopa, it will not improve with surgery of any target. Levodopa sensitivity is generally expressed as the percentage improvement of the motor score of the Unified Parkinson's Disease Rating Scale (UPDRS) (Fahn and Elton, 1987). The levodopa test is performed using a suprathreshold dose of levodopa in the morning, after overnight withdrawal of antiparkinsonian drugs (Krack *et al.*, 1998d). An improvement of at least 30% is a prerequisite for the diagnosis of PD. Good surgical candidates will have more than 50% improvement, and ideal candidates have more than 80% improvement (Table III). Patients with atypical parkinsonism (Chapter 75; Poewe and Wenning), who show only little response to levodopa, will not benefit from surgery. Because patients with

**TABLE III** DBS for PD: Inclusion Criteria

---

Idiopathic PD, evolution >5 y

Severely disabled despite optimal medical treatment, either by off-period signs (off-dystonia, akinesia, tremor) or by on-period dyskinesias.

Good levodopa response (>50% decrease in UPDRS motor score)

Biological age ≤ 70

Good general health, no anticoagulation

Cranial MRI within normal range

No cognitive deterioration, no severe depression, no psychosis

---

atypical parkinsonism may initially have a good response to levodopa, it is not recommended to operate on patients with suspected PD with a duration of less than 5 years.

Levodopa-induced dyskinesias (LID) per se can be so severely disabling as to warrant surgery. Moreover LID may induce *severe pain*, especially off-drug dystonia. Patients with disabling motor fluctuations and LID are the best surgical candidates, because they can expect improvement in off-period parkinsonism and LID. Tremor also improves with stimulation of either STN or GPi DBS. If tremor is the main symptom, thalamic stimulation is also effective (see later), and the effects of surgery on tremor can be better than those of levodopa. If tremor is the main symptom, therefore, levodopa sensitivity is less critical.

### Practical Management

After surgery, the neurologist has to adapt both antiparkinsonian medication and the electrical stimulation parameters to the needs of the patient. Possible therapeutic problems with deep brain stimulation result from the complex interactions of medical therapy and electrical stimulation. One of the challenges in the follow-up is differentiating between "genuine" stimulation-induced side effects versus symptoms of the disease that are uncovered by a combination of reduced dopaminergic therapy and inadequate stimulation effects. In the following, guidelines for postoperative management will be detailed.

### Setting of Stimulation Parameters in the Off-Drug Condition

The most important step during postoperative programming is the examination of the effectiveness and side effects of each of the four individual electrodes of the quadripolar lead of one side to determine the optimal electrode(s) for chronic stimulation. This examination takes place in the off-phase. The most important target symptom to assess is rigidity, because even small changes in rigidity can be assessed reliably. A change in rigidity always occurs within seconds (usually 20–30 seconds) in a fixed time relationship to the stimulation. The maximum effect is reached within 1 minute. Rigidity, in contrast to tremor and akinesia, is a more stable symptom, and its examination is much less dependent on the cooperation of the patient. Tremor can also be used as a target symptom, and the stimulation effect usually appears within seconds as well. Because of the variability of tremor, however, it is harder to determine a threshold. The evaluation of changes in bradykinesia is even more difficult. The evaluation of simultaneous bilateral diadochokinesis can be very useful when considering lateralized changes. However, the latency of effects on bradykinesia is quite variable. Moreover, bradykinesia is dependent on the motivation and fatigue of the patient. In addition to rigidity, akinesia, and tremor, off-phase dystonia is also directly improved by stimulation. For the STN, it has been found that the electrode with the best stimulation effect on rigidity also induces the best antiakinetic effect. This is different in GPi stimulation, in which stimulation of the most ventral electrodes has the greatest impact on rigidity (and on dyskinesias in the on-drug condition), whereas electrodes that are localized more dorsally have the best effect on akinesia (Bejjani *et al.*, 1997; Krack *et al.*, 1997b; 1998b).

The induction of dyskinesias with STN DBS predicts a favorable long-term outcome (Krack *et al.*, 1999). If the stimulation-induced dyskinesias (which can slowly develop over a period of several hours) are disabling to the patient, the amplitude must be reduced. The threshold for the induction of dyskinesias continually increases, and stimulation amplitude can gradually be increased accordingly. DBS of the dorsal GPi may also induce dyskinesias, along with an improvement in akinesia, although this happens more rarely and is less predictive of a long-term improvement in akinesia.

## Initial Parameter Selection

As a rule, rigidity and akinesia are much improved by activating one or two particular electrodes in a unipolar setting, whereas there is little effect from the two remaining ones. The electrode with the best clinical effect in the off-drug condition is selected for chronic stimulation. The initial setting is unipolar stimulation of a single electrode with 60 µs, 130 Hz. In STN DBS, the amplitude is very gradually increased until parkinsonian signs improve to the same level as with levodopa or until side effects occur. It may take 1 or 2 weeks until there is a good and stable effect. In GPi-DBS stimulation parameters can be increased more rapidly, because stimulation seldom induces disabling dyskinesias. The pulse width may also be increased. However, increasing the pulse width requires more energy to obtain the same effect than increasing the voltage (Moro *et al.*, 2000). In rare cases, further increase in the frequency may lead to further improvement, particularly in tremor. Here again, improvements may be obtained more economically by increasing the voltage. If a single electrode does not produce an optimal effect, unipolar stimulation of the two adjacent electrodes with the best clinical effects may be an alternative. If an increase in the stimulation parameters leads to a better effect, but only at the cost of side effects from current diffusion into adjacent struc-

tures, the more focal bipolar stimulation allows a better ratio between desired effect and side effect.

### Adjustment of the Stimulation Parameters in the On-Phase

The antiparkinsonian effect of stimulation and the effect of levodopa are roughly equivalent. An additive effect may be observed, but it rarely exceeds the "best-on" (Limousin *et al.*, 1998). Rigidity and action tremor associated with rigidity, especially, are improved significantly more by levodopa and stimulation than by levodopa alone. However, this does not result in a relevant functional improvement.

The examination in the on-phase is important to evaluate dyskinesias. It is necessary to determine how much levodopa the patient needs or tolerates postoperatively at a given stimulation intensity. Levodopa and STN stimulation may have additive effects on dyskinesias, particularly postoperatively. However, with chronic STN stimulation, the initial dyskinesia-inducing effect declines, and the threshold for dyskinesias increases (Krack *et al.*, 1999). If disabling dyskinesias occur, a reduction of either the stimulation parameters or the medication (see later) is necessary. With chronic STN stimulation, an average reduction of 50% in dopaminergic treatment is possible (Limousin *et al.*, 1998). DBS of the ventral GPi leads to a direct reduction in dyskinesias, as soon as the stimulation is switched on. If the voltage is increased too much, dyskinesias completely disappear, but the levodopa effects on akinesia may also disappear, and the patient may have a pure akinesia develop, without rigidity (Krack *et al.*, 1998b; 1998c). In this case, the stimulus intensity needs to be reduced.

### Adjustment of Stimulus Parameters in the Long Term

In the first weeks after surgery the effects of stimulation often decrease, and the threshold to induce side effects may also increase. This is related to changes in impedance that occur with resolution of edema and also to the reduction of the stun or microlesioning effect after stereotaxic lead implantation. In addition, there may be a loss of the long-term effects of drugs after reduction of dopaminergic medication. Patients can then be seen in the outpatient unit to restore the effect by carefully increasing the stimulus amplitude in increments of 0.2–0.5 V. At 3 months follow-up, the effects are generally stable, and only minor changes may be necessary over the following years related to progression of the disease.

If stimulation effects are unsatisfactory despite maximal tolerated increase in voltage (i.e., if they do not match the levodopa response), the following strategies should be pursued:

- Verify, that the best electrode of the quadripolar lead is being used
- Try two adjacent electrodes with monopolar stimulation
- Try increases in frequency up to 185 Hz
- Try to increase pulse width if a voltage >3.6 V is needed (only if Itrel II or Soletra are used as neurostimulator, not with Kinetra)
- If the results are unsatisfactory related to side effects from current diffusion, try bipolar stimulation

If these strategies are not successful, electrode position is probably not ideal, and symptom relief should be optimized through the adjustment of medication.

### Adjustment of Medication

In STN DBS, dopaminergic therapy can be reduced postoperatively as a result of an improvement in all (dopa-sensitive) cardinal symptoms. This is necessary because of the additive dyskinesia-inducing effect of STN stimulation and dopaminergic therapy. Levodopa not only affects the motor symptoms but also improves nonmotor symptoms. If dopaminergic therapy is reduced too much, this may lead to anhedonia, abulia, or even depression (Krack et al., 1998d). Depression generally develops only very gradually with the loss of the long-term effects of dopaminergic treatment. In these instances, an increase in dopaminergic medication is more effective than antidepressant therapy, but in severe depression, both strategies can be combined. Conversely, the psychotropic-stimulating effects of L-dopa may be intensified by STN stimulation, and a hypomanic condition may develop in the postoperative period. In this case, a reduction in dopaminergic therapy is appropriate (Krack et al., 2001b). After 12 months, dopaminergic therapy is typically reduced by approximately 50% compared with the preoperative regimen (Limousin et al., 1998). In GPi DBS, it is generally not possible to reduce dopaminergic medication (DBS-for-Parkinson's-disease-study-group, 2001; Krack et al., 1998d; Volkmann et al., 1998). Because GPi DBS has a direct antidyskinetic effect, patients tolerate higher doses of dopaminergic drugs without dyskinesias developing.

## DBS FOR TREMOR

Tremor occurs in different presentations as a symptom of various neurological diseases. Parkinsonian and essential tremor (ET) are normally well treatable with medication. If tremor becomes disabling despite optimal medical treatment, DBS for tremor may be considered. For cerebellar tremor there is no effective medical treatment, whereas DBS may have a dramatic effect in selected patients. In enhanced physiological tremor and drug-induced tremors, DBS is not indicated, because the underlying causes can be treated. Despite different pathomechanisms, surgery for different tremors concentrates on the Vim of the thalamus. In contrast to the almost equally effective thermocoagulation of the Vim, DBS causes fewer side effects ** and can be applied bilaterally without the danger of causing irreversible psychiatric symptoms (Schuurman et al., 2000). Recently, the subthalamic region (zona incerta) has been advocated as a target for DBS in cases of cerebellar tremor (Kitagawa et al., 2000). Systematic studies of this target have not yet been undertaken, however.

### Essential Tremor

#### Principles of Therapy

Essential tremor of the extremities can be improved significantly by Vim DBS**. Several single and multi-center studies included more than 100 patients and agree that tremor can be almost completely suppressed in 60%–100% of patients (Limousin et al., 1999). Bilateral implantations are frequently performed. In these cases, patients must be informed that a mild stimulation-induced dysarthria is possible. Severe voice tremor and head tremor can improve but often does not completely subside after bilateral DBS (Taha et al., 1999). Postoperative functional results are better in ET than with other forms of tremor, because patients are disabled by tremor only. This is also reflected by a better improvement in activities of daily living compared with PD patients, even if the effect on tremor intensity is generally less than in PD (Limousin et al., 1999).

#### Practical Management

When a good tremor suppression is achieved with the permanent electrode during the operation, there is no need for prolonged testing with external stimulators. Therefore, most groups implant the stimulator in the same session. During the first weeks after the operation, stimulus amplitude needs to be increased, as the micro-talamotomy effect wanes. Thereafter, slight increments of stimulus amplitude are sometimes necessary to maintain a complete tremor suppression, but otherwise tremor control is usually maintained for as long as the stimulation continues. Assessing the effect of stimulation in the outpatient setting should always include writing and drawing a spiral, because these tests are easily performed and can detect subtle action tremor, which may be ameliorated by increased stimulus amplitude. Switching off the stimulator causes a vigorous tremor in some patients, sometimes referred to as rebound tremor. If this is not the case, the stimulator may be switched off at night to prolong battery life span. Medication is no longer necessary in most patients.

## Tremor in Parkinson's Disease

### Principles of Therapy

In some patients with Parkinson's disease, the tremor can remain the only obvious symptom of disease for many years. DBS of the Vim can reduce the parkinsonian tremor completely or almost completely in more than 80% of the patients (Benabid et al., 1996; Koller et al., 1997; Limousin et al., 1999). Most patients with tremor-dominant PD will have akinesia and rigidity develop in the long term. Then motor fluctuations and dyskinesias become the main problems, whereas tremor is generally well controlled. Because the latter symptoms are not alleviated by Vim stimulation, STN stimulation has been suggested for parkinsonian tremor-(Krack et al., 1997c). STN stimulation reduces the tremor significantly, but the maximum effect may be delayed up to several weeks after the operation. The value of identification of tremor-related cells by microelectrode recordings in the sensorimotor area of STN is stressed (Krack et al., 1998a; Rodriguez et al., 1998). However, the patients included in the latter studies were mostly not tremor dominant or monosymptomatic patients, so that a direct comparison of the effectiveness of Vim and STN stimulation for monosymptomatic tremor in Parkinson's disease is still lacking.

### Practical Management

Patients are assessed in the on-drug and off-drug condition. If akinesia and rigidity are manifest when off-drug, STN stimulation should be considered. When these symptoms are absent, Vim stimulation may be considered, because it has an immediate effect on tremor and because postoperative management is easy in uncomplicated cases. To allow for effective intraoperative testing, dopaminergic medication must be tapered before the operation to ensure the presence of tremor. Postoperatively, medication is given according to the needs of the patients.

## Cerebellar Tremor

### Principles of Therapy

DBS for cerebellar tremor is not as straightforward as for essential tremor or PD. Generally, results are more variable, and frequently the effect vanishes with time. This is most relevant to patients with multiple sclerosis (MS), who constitute the largest group of cerebellar tremor patients treated with DBS. There are no data that allow for a safe conclusion about which patients will benefit from an operation. Most authors agree that candidates for an operation should have a disabling action (intention) and/or postural tremor of one or both upper limbs (Table IV). Patients with distal tremor are more likely to improve in the long term than are patients with proximal tremor, who frequently experience loss of the effectiveness of stimulation. It may be possible that these patients would benefit more from an implantation in the zona incerta, as recently proposed (Kitagawa et al., 2000). Dysmetria of the limbs must be differentiated from tremor, because only the latter can be influenced by DBS. Axial postural tremor, which may affect the head, can only be ameliorated with bilateral stimulation. Some groups do not recommend bilateral operations in MS, however, because the incidence of side effects seems to be higher than in other patients. In severely affected patients, an irregular, low-frequency tremor may be seen that also occurs during rest. For these patients the efficacy of DBS is very limited (Pollak et al., 1998).

For all candidates, a careful neurological examination has to be carried out to detect central palsy, which may be seen as spasticity, reduced individual finger movements, or both. Severe deficits will interfere with the postoperative function of the hand or arm and should exclude the patient from the operation. The same is true for massive sensory deficits. Absent stereognosia or absent limb position sense should take the patient from the list of DBS candidates. In rapidly progressive disease, functional improvement after DBS may only last for several months, and therefore only patients in

TABLE IV  Suggestions for Inclusion and Exclusion Criteria for DBS in Multiple Sclerosis

| Inclusion | Exclusion |
| --- | --- |
| Disabling action/intention tremor of upper limb | Significant mental alteration or dementia |
| Disabling postural tremor of upper limb | Significant sensory or motor deficits of the affected limb |
| No exacerbations of MS during 6 months | Regular postoperative consultation (2 months) not possible |

whom the disease has been stable for at least 6 months should be operated on. Only a few patients experience a complete cessation of the tremor after the operation (Geny *et al.*, 1996; Montgomery *et al.*, 1999). About 25%–50% may be able to drink from a cup postoperatively. Most patients regain the ability to grasp and manipulate objects without the danger of injuring themselves. However, this figure stems from short observation periods, and a realistic estimate is that only 30% of patients benefit from the operation after 1 year (Pollak *et al.*, 1998). In contrast to other patients, MS patients with DBS require more frequent adjustments of stimulus parameters during the postoperative course. Patients with immunosupressive therapy have an increased risk of infection and therefore should be excluded from DBS.

### Practical Management

Thorough explanation of the chances of the procedure to the patient and his relatives is necessary, because expectations tend to be high in this group. Correspondingly, patients frequently are not satisfied with the postoperative results and often react briskly. When the effectiveness of the stimulation wanes, an increase of stimulus amplitude or switching to other electrodes may be helpful. However, eventually stimulation may fail to suppress the tremor, even after many attempts to change stimulation parameters.

## DYSTONIA

### Principles of Therapy

DBS was applied only recently for patients with dystonia, with very encouraging results. These operations were performed after it had been shown that pallidotomy has a marked effect on L-dopa–induced dystonia in PD and is also highly effective in patients with primary dystonia (Bötzel *et al.*, 2000; Lozano *et al.*, 1997; Ondo *et al.*, 1998). Experience with DBS in dystonia is limited, and this treatment should therefore be considered experimental. At the moment there is clearly a need for studies comparing bilateral pallidotomy with pallidal DBS. Indications for DBS were generalized dystonias, segmental dystonias, and torticollis, the only focal type of dystonia for which surgery was done. The origin of dystonia was not a criterion for including or excluding a patient for an operation. Primary dystonias show better postoperative results, however, compared with secondary dystonias (Coubes *et al.*, 2000; Roubertie *et al.*, 2000). Of the children with generalized dystonia, those who were positive for the DYT1 gene

achieved a dramatic postoperative reduction of dystonia scores in the range of 90% (Coubes *et al.*, 2000). Most of these children were postoperatively able to walk without aid, and many could return to normal living. Adults with primary generalized or segmental dystonia can expect a 60%–70% postoperative improvement on the BFM dystonia scale. Individual patients may be free of all symptoms (Bötzel *et al.*, 2000). In contrast to PD, the effects of pallidal DBS are often delayed and progressive. Stimulation effects start during the first days of stimulation, but maximum effects may only be achieved after several months. Stimulation parameters tend to be higher than in PD, but the few observations reported do not allow for definite conclusions.

### Practical Management

When examining patients with generalized dystonia, special emphasis should be directed to detection of contractures of joints and muscles. Implantations should be bilateral except for cases with clear hemidystonia. Operations usually require general anesthesia. Propofol anesthesia, without relaxation, allows for monitoring of pyramidal tract side effects from macrostimulation, as well as for extracellular microrecordings and recordings of visually evoked potentials. It also is possible to interrupt propofol anesthesia for a short period of intraoperative testing. Postoperative stimulation should concentrate on the lower electrodes (just above the optic tract), and amplitudes and pulse width should be increased regularly as long as a clear effect appears, and side effects do not become intolerable. Children and adults with severe generalized dystonia may require a special postoperative setting and a prolonged postoperative hospitalization. In these cases physiotherapists and ergotherapists are required to help the patient learn new movement strategies. Medication for dystonia and pain should be reduced slowly when stimulation is effective.

## ALTERNATIVE SURGICAL TREATMENTS

### Thermocoagulation

#### Pallidotomy

Laitinen reinvestigated the pallidotomy performed by his teacher Leksell before the advent of levodopa. Not only did he confirm the antiakinetic effect of pallidotomy, but he also described its marked effect on levodopa-induced dyskinesias (Laitinen *et al.*, 1992). These results have been confirmed in many studies, thus firmly establishing the value of unilateral posteroventral pallidotomy (PVP) in PD (Baron *et al.*, 1996; Samuel *et al.*, 1998). On the average, values in the motor part of the Unified Parkinson's Disease Rating Scale (UPDRS III) are improved after 3–6 months by 26% in off-drug condition, whereas they do not significantly improve in the on-drug condition. Because PD is a bilateral disease, the effects of unilateral PVP can only be limited. Because of its high morbidity in the recent literature, bilateral pallidotomy was not recommended by a task force on surgery for PD (Hallett and Litvan, 1999). However, encouraging improvement was reported after bilateral pallidotomy in children with different dystonic syndromes (Ondo *et al.*, 1998). Detailed long-term follow-up and neuropsychological evaluation are still lacking.

### Subthalamotomy

On the basis of experimental work in the parkinsonian monkey, unilateral (Alvarez *et al.*, 2001), and bilateral (Gill and Heywood, 1997) subthalamotomies that directly targeted the sensorimotor part of the STN have recently been performed. The outcome was remarkable: there was 50% improvement in motor UPDRS values in a series of 11 patients with unilateral subthalamotomy (Alvarez *et al.*, 2001). The first two patients with bilateral STN lesions showed a short-term improvement of UPDRS III values by 72%, a figure that can only be compared with improvements after bilateral STN DBS, thus confirming the paramount role played by the STN in parkinsonism. However, one of the patients with a unilateral STN lesion required a pallidotomy after 12 months for persisting hemiballismus. In contrast to STN lesions, postoperative DBS-related dyskinesias can be managed by individually tailoring stimulation parameters and levodopa dosage (Krack *et al.*, 1999). Bilateral lesioning of the STN, which is bordered by the corticobulbar and corticospinal tracts, theoretically has a high potential for inducing a pseudobulbar syndrome. Other concerns are behavioral changes that have been shown to appear after STN lesions in the rat (Baunez *et al.*, 1995) or to occur as a reversible and manageable problem with STN DBS (Bejjani *et al.*, 1998; Krack *et al.*, 2001a). The number of patients who have undergone the operation is too small so far to allow any conclusions. Subthalamotomy must presently be considered an experimental therapy.

### Thalamotomy

Thalamotomy of the Vim is a very effective treatment for parkinsonian tremor. Small lesions with a volume of 40–60 mm³ can be sufficient to completely suppress it (Hirai *et al.*, 1983). Tasker found that 82% of his patients had complete tremor suppression 2 years after Vim thalamotomy, whereas 8% had significant persistent complications (Tasker *et al.*, 1983). The antitremor effects are longlasting.

Bilateral thalamic surgery causes deterioration of speech, dysphagia, or balance problems in more than 20% of patients (Speelman, 1991). Therefore thalamotomy is only performed unilaterally in patients with levodopa-resistant tremor-dominant PD. A recent randomized prospective study has demonstrated that thalamic lesions and thalamic DBS have the same effect on tremor amplitude. DBS, however, had less side effects, and the functional improvement was also greater in the DBS group than in the thalamotomy group (Schuurman *et al.*, 2000).

**Gamma Knife Thalamotomy.** For patients with parkinsonian or essential tremor gamma-knife thalamotomy may be an alternative to DBS and thermocoagulation (Ohye *et al.*, 2000; Young *et al.*, 2000). Tremor subsides approximately 1 year after irradiation. Most reported cases become tremor free, and side effects seem to be very rare. The availability of this technique is limited to only a few centers.

### Neurotransplantation in PD

The intrastriatal transplantation of embryonic mesencephalic tissue in patients with PD has been reported to cause therapeutic improvement (Freed *et al.*, 2001; Lindvall *et al.*, 1989). A subsequent increase in fluorodopa-uptake was shown in positron emission tomography (Sawle *et al.*, 1992), and graft survival and striatal reinnervation could also be demonstrated in postmortem histopathological studies (Kordower *et al.*, 1995). In the most successful cases, patients have been able to completely stop levodopa treatment after transplantation. Approximately two thirds of grafted patients have shown clinically useful, partial recovery of motor function: increased percent time in the on phase and reduced rigidity and hypokinesia during off phases bilaterally but predominantly on the side contralateral to the graft. Some patients also exhibited improved gait, speech, and balance and less dyskinesias after transplantations, but in most cases these symptoms have not signified any major, consistent changes (Freed *et al.*, 2001; Lindvall, 1999). Dyskinesias may even worsen after surgery (Defer *et al.*, 1996; Peschanski *et al.*, 1994), and delayed dyskinesias have been reported to occur independently of dopaminergic medication and as a direct result of unphysiological graft function (Fahn, 2000; Freed *et al.*, 2001). For the time being, however, many problems must be solved before neural transplantation can develop into a clinically useful and available therapeutic strategy.

# REFERENCES

Alexander, G. E., Crutcher, M. D., and DeLong, M. R. (1990). Basal ganglia-thalamocortical circuits: parallel substrates for motor, oculomotor, "prefrontal" and "limbic" functions. *Prog. Brain Res.* 85, 119–146.

Alvarez, L., Macias, R., Guridi, J., Lopez, G., Alvarez, E., Maragoto, C., Teijeiro, J., Torres, A., Pavon, N., Rodriguez-Oroz, M. C., Ochoa, L., Hetherington, H., Juncos, J., DeLong, M. R., and Obeso, J. A. (2001). Dorsal subthalamotomy for Parkinson's disease. *Mov. Disord.* 16, 72–78.

Ardouin, C., Pillon, B., Peiffer, E., Bejjani, P., Limousin, P., Damier, P., Arnulf, I., Benabid, A. L., Agid, Y., and Pollak, P. (1999). Bilateral subthalamic or pallidal stimulation for Parkinson's disease affects neither memory nor executive functions: a consecutive series of 62 patients. *Ann. Neurol.* 46, 217–223.

Ashby, P. (2000). What does stimulation in the brain actually do? *In* "Movement disorder surgery" Progress in neurological surgery (A. M. Lozano, Ed.) Vol. 15, pp. 236–245. Kagerer, Basel.

Baron, M. S., Vitek, J. L., Bakay, R. A., Green, J., Kaneoke, Y., Hashimoto, T., Turner, R. S., Woodard, J. L., Cole, S. A., McDonald, W. M., and DeLong, M. R. (1996). Treatment of advanced Parkinson's disease by posterior GPi pallidotomy: 1-year results of a pilot study. *Ann. Neurol.* 40, 355–366.

Baunez, C., Nieoullon, A., and Amalric, M. (1995). In a rat model of parkinsonism, lesions of the subthalamic nucleus reverse increases of reaction time but induce a dramatic premature responding deficit. *J. Neurosci.* 15, 6531–6541.

Bejjani, B. P., Damier, P., Arnulf, I., Bonnet, A. M., and Agid, Y. (1998). Acute major depression induced by subthalamic deep brain stimulation. *Mov. Disord.* 13, 123.

Bejjani, B., Damier, P., Arnulf, I., Bonnet, A. M., Vidailhet, M., Dormont, D., Pidoux, B., Cornu, P., Marsault, C., and Agid, Y. (1997). Pallidal stimulation for Parkinson's disease. Two targets? *Neurology* 49, 1564–1569.

Benabid, A. L., Pollak, P., Gao, D., Hoffmann, D., Limousin, P., Gay, E., Payen, I., and Benazzouz, A. (1996). Chronic electrical stimulation of the ventralis intermedius nucleus of the thalamus as a treatment of movement disorders. *J. Neurosurg.* 84, 203–214.

Benabid, A. L., Pollak, P., Gervason, C., Hoffmann, D., Gao, D. M., Hommel, M., Perret, J. E., and de Rougemont, J. (1991). Long-term suppression of tremor by chronic stimulation of the ventral intermediate thalamic nucleus. *Lancet* 337, 403–406.

Bergman, H., Wichmann, T., Karmon, B., and DeLong, M. R. (1994). The primate subthalamic nucleus. II. Neuronal activity in the MPTP model of parkinsonism. *J. Neurophysiol.* 72, 507–520.

Beurrier, C., Bioulac, B., Audin, J., and Hammond, C. (2001). High-frequency stimulation produces a transient blockade of voltage-gated currents in subthalamic neurons. *J. Neurophysiol.* 85, 1351–1356.

Bötzel, K., Bereznai, B., Steude, U., Jäger, M., and Gasser, T. (2000). Chronic high-frequency stimulation of the globus pallidus internus in different types of dystonia. *Mov. Disord.* 15, 168.

Brundin, P., Pogarell, O., Hagell, P., Piccini, P., Widner, H., Schrag, A., Kupsch, A., Crabb, L., Odin, P., Gustavii, B., Bjorklund, A., Brooks, D. J., Marsden, C. D., Oertel, W. H., Quinn, N. P., Rehncrona, S., and Lindvall, O. (2000). Bilateral caudate and putamen grafts of embryonic mesencephalic tissue treated with lazaroids in Parkinson's disease. *Brain* 123, 1380–1390.

Burke, R. E., Fahn, S., Marsden, C. D., Bressman, S. B., Moskowitz, C., and Friedman, J. (1985). Validity and reliability of a rating scale for the primary torsion dystonias. *Neurology* 35, 73–77.

Coubes, P., Roubertie, A., Vayssiere, N., Hemm, S., and Echenne, B. (2000). Treatment of DYT1-generalised dystonia by stimulation of the internal globus pallidus. *Lancet* 355, 2220–2221.

DBS-for-Parkinson's-disease-study-group (2001). Deep brain stimulation (DBS) of the subthalamic nucleus or globus pallidus pars interna in Parkinson's disease. *N. Engl. J. Med.* 345, 956–963.

Defer, G. L., Geny, C., Ricolfi, F., Fenelon, G., Monfort, J. C., Remy, P., Villafane, G., Jeny, R., Samson, Y., Keravel, Y., Gaston, A., Degos, J. D., Peschanski, M., Cesaro, P., and Nguyen, J. P. (1996). Long-term outcome of unilaterally transplanted parkinsonian patients. I. Clinical approach. *Brain* 119, 41–50.

Defer, G. L., Widner, H., Marie, R. M., Remy, P., and Levivier, M. (1999). Core assessment program for surgical interventional therapies in Parkinson's disease (CAPSIT-PD). *Mov. Disord.* 14, 572–584.

Fahn, S., and Elton, R. L. (1987). Unified Parkinson's Disease Rating Scale *In* "Recent Developments in Parkinson's Disease" (S. Fahn, C. D. Marsden, D. Calne, and M. Goldstein, Eds.), pp. 153–163. MacMillan Health Care Information, Florham Park, NJ.

Fahn, S., Tolosa, E., and Marin, C. (1993). Clinical rating scale for tremor *In* "Parkinson's Disease and Movement Disorders. 2nd Edition" (J. Jancovic and E. Tolosa, Eds.), pp. 271–280. Williams & Wilkins, Baltimore.

Fine, J., Duff, J., Chen, R., Chir, B., Hutchison, W., Lozano, A. M., and Lang, A. E. (2000). Long-term follow-up of unilateral pallidotomy in advanced Parkinson's disease. *N. Engl. J. Med.* 342, 1708–1714.

Freed, C. R., Greene, P. E., Breeze, R. E., Tsai, W., DuMouchel, W., Kao, R., Dillon, S., Winfield, H., Culver, S., Trojanowski, J. Q., Eidelberg, E., and Fahn, S. (2001). Transplantation of embryonic dopamine neurons for severe parkinson's disease. *N. Engl. J. Med.* 334, 710–719.

Geny, C., Nguyen, J. P., Pollin, B., Feve, A., Ricolfi, F., Cesaro, P., and Degos, J. D. (1996). Improvement of severe postural cerebellar tremor in multiple sclerosis by chronic thalamic stimulation. *Mov. Disord.* 11, 489–494.

Ghika, J., Ghika Schmid, F., Fankhauser, H., Assal, G., Vingerhoets, F., Albanese, A., Bogousslavsky, J., and Favre, J. (1999). Bilateral contemporaneous posteroventral pallidotomy for the treatment of Parkinson's disease: neuropsychological and neurological side effects. Report of four cases and review of the literature. *J. Neurosurg.* 91, 313–321.

Gill, S. S., and Heywood, P. (1997). Bilateral dorsolateral subthalamotomy for advanced Parkinson's disease. *Lancet* 350, 1224.

Goldman, M. S., Ahlskog, J. E., and Kelly, P. J. (1992). The symptomatic and functional outcome of stereotactic thalamotomy for medically intractable essential tremor. *J. Neurosurg.* 76, 924–928.

Hallett, M., and Litvan, I. (1999). Evaluation of surgery for Parkinson's disease: a report of the Therapeutics and Technology Assessment Subcommittee of the American Academy of Neurology. The Task Force on Surgery for Parkinson's Disease. *Neurology* 53, 1910–1921.

Hariz, M. I., and DeSalles, A. A. (1997). The side-effects and complications of posteroventral pallidotomy. *Acta Neurochir Suppl (Wien)* 68, 42–48.

Hariz, M. I., and Fodstad, H. (1999). Do microelectrode techniques increase accuracy or decrease risks in pallidotomy and deep brain stimulation? A critical review of the literature. *Stereotact. Funct. Neurosurg.* 72, 157–169.

Hassler, T., and Riechert, T. (1954). Indikationen und Lokalisationsmethode der gezielten Hirnoperationen. *Der Nervenarzt.* 25, 441–447.

Hirai, T., Miyazaki, M., Nakajima, H., Shibazaki, T., and Ohye, C. (1983). The correlation between tremor characteristics and the predicted volume of effective lesions in stereotaxic nucleus ventralis intermedius thalamotomy. *Brain* 106, 1001–1018.

Holtzheimer, P. E., 3rd, Roberts, D. W., and Darcey, T. M. (1999). Magnetic resonance imaging versus computed tomography for target localization in functional stereotactic neurosurgery. *Neurosurgery* 45, 290–297.

Houeto, J. L., Bejjani, P. B., Damier, P., Staedler, C., Bonnet, A. M., Pidoux, B., Dormont, D., Cornu, P., and Agid, Y. (2000a). Failure of long-term pallidal stimulation corrected by subthalamic stimulation in PD. *Neurology* 55, 728–730.

Houeto, J. L., Damier, P., Bejjani, P. B., Staedler, C., Bonnet, A. M., Arnulf, I., Pidoux, B., Dormont, D., Cornu, P., and Agid, Y. (2000b). Subthalamic stimulation in Parkinson disease: a multidisciplinary approach. *Arch. Neurol.* 57, 461–465.

Jankovic, J., Cardoso, F., Grossman, R. G., and Hamilton, W. J. (1995). Outcome after stereotactic thalamotomy for parkinsonian, essential, and other types of tremor. *Neurosurgery* 37, 263–270.

Kitagawa, M., Murata, J., Kikuchi, S., Sawamura, Y., Saito, H., Sasaki, H., and Tashiro, K. (2000). Deep brain stimulation of subthalamic area for severe proximal tremor. *Neurology* 55, 114–116.

Koller, W., Pahwa, R., Busenbark, K., Hubble, J., Wilkinson, S., Lang, A., Tuite, P., Sime, E., Lazano, A., Hauser, R., Malapira, T., Smith, D., Tarsy, D., Miyawaki, E., Norregaard, T., Kormos, T., and Olanow, C. W. (1997). High-frequency unilateral thalamic stimulation in the treatment of essential and parkinsonian tremor. *Ann. Neurol.* 42, 292–299.

Kordower, J. H., Freeman, T. B., Snow, B. J., Vingerhoets, F. J., Mufson, E. J., Sanberg, P. R., Hauser, R. A., Smith, D. A., Nauert, G. M., Perl, D. P. *et al.* (1995). Neuropathological evidence of graft survival and striatal reinnervation after the transplantation of fetal mesencephalic tissue in a patient with Parkinson's disease. *N. Engl. J. Med.* 332, 1118–1124.

Krack, P., Ardouin, C., Funkiewiez, A., Caputo, E., Benazzouz, A., Benabid, A. L., and Pollak, P. (2001a). What is the influence of STN stimulation on the limbic loop? *In* "Basal Ganglia and Thalamus in Health and Movement Disorders" (K. Kultas-Ilinsky and I. Ilinsky, Eds.), Kluwer Academic/Plenum Press, New York.

Krack, P., Benazzouz, A., Pollak, P., Limousin, P., Piallat, B., Hoffmann, D., Xie, J., and Benabid, A. L. (1998a). Treatment of tremor in Parkinson's disease by subthalamic nucleus stimulation. *Mov. Disord.* 13, 907–914.

Krack, P., Kumar, R., Ardouin, C., Limousin Dowsey, P., McVicker, J. M., Benabid, A. L., and Pollak, P. (2001b). Mirthful laughter induced by subthalamic nucleus stimulation. *Mov. Disord* 16, 867–875.

Krack, P., Limousin, P., Benabid, A. L., and Pollak, P. (1997a). Chronic stimulation of subthalamic nucleus improves levodopa-induced dyskinesias in Parkinson's disease. *Lancet* 350, 1676.

Krack, P., Pollak, P., Limousin, P., and Benabid, A. L. (1997b). Levodopa-inhibiting effect of pallidal surgery. *Ann. Neurol.* 42, 129–129.

Krack, P., Pollak, P., Limousin, P., Benazzouz, A., and Benabid, A. L. (1997c). Stimulation of subthalamic nucleus alleviates tremor in Parkinson's disease. *Lancet* 350, 1675.

Krack, P., Pollak, P., Limousin, P., Benazzouz, A., Deuschl, G., and Benabid, A. L. (1999). From off-period dystonia to peak-dose chorea: the clinical spectrum of varying subthalamic nucleus activity. *Brain* 122, 1133–1146.

Krack, P., Pollak, P., Limousin, P. *et al.* (1998b). Opposite motor effects of pallidal stimulation in Parkinson's disease. *Ann. Neurol.* 43, 180–192.

Krack, P., Pollak, P., Limousin, P., Hoffmann, D., Benazzouz, A., and Benabid, A. L. (1998c). Inhibition of levodopa-effects by internal pallidal stimulation. *Mov. Disord.* 13, 648–652.

Krack, P., Pollak, P., Limousin, P., Hoffmann, D., Xie, J., Benazzouz, A., and Benabid, A. L. (1998d). Subthalamic nucleus or internal pallidal stimulation in young onset Parkinson's disease. *Brain* 121, 451–457.

Krayenbühl, H., Wyss, O. A. M., and Yasargil, M. G. (1961). Bilateral thalamotomy and pallidotomy as treatment for bilateral parkinsonism. *J. Neurosurg.* 18, 429.

Kumar, R., Lozano, A. M., Kim, Y. J. *et al.* (1998). Double-blind evaluation of subthalamic nucleus deep brain stimulation in advanced Parkinson's disease. *Neurology* 51, 850–855.

Kumar, R., Lozano, A. M., Sime, E., Halket, E., and Lang, A. E. (1999). Comparative effects of unilateral and bilateral subthalamic nucleus deep brain stimulation. *Neurology* 53, 561–566.

Laitinen, L. V. (1995). Pallidotomy for Parkinson's disease. *Neurosurg. Clin. North. Am.* 6, 105–112.

Laitinen, L. V., Bergenheim, A. T., and Hariz, M. I. (1992). Leksell's posteroventral pallidotomy in the treatment of Parkinson's disease. *J. Neurosurg.* 76, 53–61.

Limousin, P., Krack, P., Pollak, P., Benazzouz, A., Ardouin, C., Hoffmann, D., and Benabid, A. L. (1998). Electrical stimulation of the subthalamic nucleus in advanced Parkinson's disease. *N. Engl. J. Med.* 339, 1105–1111.

Limousin, P., Speelman, J. D., Gielen, F., and Janssens, M. (1999). Multicenter European study of thalamic stimulation in parkinsonian and essential tremor. *J. Neurol. Neurosurg. Psychiat.* 66, 289–296.

Lindvall, O. (1999). Cerebral implantation in movement disorders: state of the art. *Mov. Disord.* 14, 201–205.

Lindvall, O., Rehncrona, S., Brundin, P., Gustavii, B., Astedt, B., Widner, H., Lindholm, T., Bjorklund, A., Leenders, K. L., Rothwell, J. C. *et al.* (1989). Human fetal dopamine neurons grafted into the striatum in two patients with severe Parkinson's disease. A detailed account of methodology and a 6-month follow-up. *Arch. Neurol.* 46, 615–631.

Lozano, A. M., Kumar, R., Gross, R. E., Giladi, N., Hutchison, W. D., Dostrovsky, J. O., and Lang, A. E. (1997). Globus pallidus internus pallidotomy for generalized dystonia. *Mov. Disord.* 12, 865–870.

Meyers, R. (1942). The modification of alternating tremors, rigidity and festination by surgery of the basal ganglia. *Res. Publ. Assoc. Res. Nerv. Ment. Dis.* 21, 602–665.

Molinuevo, J. L., Valldeoriola, F., Tolosa, E., and Rumià, J. (2000). Levodopa withdrawal after bilateral subthalamic nucleus stimulation in advanced Parkinson disease. *Arch. Neurol.* 57, 9883–9988.

Montgomery, E. B., Jr., Baker, K. B., Kinkel, R. P., and Barnett, G. (1999). Chronic thalamic stimulation for the tremor of multiple sclerosis. *Neurology* 53, 625–628.

Moro, E., Esselink, R., Xie, J., Fraix, V., Benabid, A. L., and Pollak, P. (2000). Role of electrical variables on subthalamic nucleus stimulation-induced antiparkinsonian effects. *Neurology* 54 **Suppl** 3, A282.

Nagaseki, Y., Shibazaki, T., Hirai, T., Kawashima, Y., Hirato, M., Wada, H., Miyazaki, M., and Ohye, C. (1986). Long-term follow-up results of selective VIM-thalamotomy. *J. Neurosurg.* 65, 296–302.

Ohye, C., Shibazaki, T., Ishihara, J., and Zhang, J. (2000). Evaluation of gamma thalamotomy for parkinsonian and other tremors: survival of neurons adjacent to the thalamic lesion after gamma thalamotomy. *J. Neurosurg.* 93 **Suppl** 3, 120–127.

Ondo, W. G., Desaloms, J. M., Jankovic, J., and Grossman, R. G. (1998). Pallidotomy for generalized dystonia. *Mov. Disord.* 13, 693–698.

Peschanski, M., Defer, G., N'Guyen, J. P., Ricolfi, F., Monfort, J. C., Remy, P., Geny, C., Samson, Y., Hantraye, P., Jeny, R. *et al.* (1994). Bilateral motor improvement and alteration of L-dopa effect in two patients with Parkinson's disease following intrastriatal transplantation of foetal ventral mesencephalon. *Brain* 117, 487–499.

Pollak, P., Benabid, A. L., Gross, C., Gao, D. M., Laurent, A., Benazzouz, A., Hoffmann, D., Gentil, M., and Perret, J. (1993). Effets de la stimulation du noyau sousthalamique dans la maladie de Parkinson. *Rev. Neurol. Paris.* 149, 175–176.

Pollak, P., Benabid, A. L., Krack, P., Limousin, P., and Benazzouz, A. (1998). Deep brain stimulation *In* "Parkinson's Disease and

Movement Disorders" (J. Jankovic and E. Tolosa, Eds.), pp. 1085–1101. Williams & Wilkins, Baltimore.

Riechert, T. (1962). Long-term follow-up of results of stereotaxic treatment of extrapyramidal disorders. *Confin. Neurol.* **22,** 356–363.

Rodriguez, M. C., Guridi, O. J., Alvarez, L., Mewes, K., Macias, R., Vitek, J., DeLong, M. R., and Obeso, J. A. (1998). The subthalamic nucleus and tremor in Parkinson's disease. *Mov. Disord.* **13 Suppl 3,** 111–118.

Roubertie, A., Cif, L., Vayssière, N., Tuffery, S., Hemm, S., Claustres, M., Bonafe, A., Frerebeau, P., Echenne, B., and Coubes, P. (2000). Symptomatic generalized dystonia: neurourgical treatment by continuous bilateral stimulation of the internal globus pallidus in eight patients. *Mov. Disord.* **15 Suppl 3,** 154.

Saint-Cyr, J. A., Trépanier, L. L., Kumar, R., and Lozano, A. (2000). Neuropsychological consequences of chronic bilateral stimulation of the subthalamic nucleus in Parkinson's disease. *Brain* **123,** 2091–2108.

Samuel, M., Caputo, E., Brooks, D. J., Schrag, A., Scaravilli, T., Branston, N. M., Rothwell, J. C., Marsden, C. D., Thomas, D. G., Lees, A. J., and Quinn, N. P. (1998). A study of medial pallidotomy for Parkinson's disease: clinical outcome, MRI location and complications. *Brain* **121,** 59–75.

Sawle, G. V., Bloomfield, P. M., Bjorklund, A., Brooks, D. J., Brundin, P., Leenders, K. L., Lindvall, O., Marsden, C. D., Rehncrona, S., Widner, H. *et al.* (1992). Transplantation of fetal dopamine neurons in Parkinson's disease: PET [18F]6-L-fluorodopa studies in two patients with putaminal implants. *Ann. Neurol.* **31,** 166–173.

Schuurman, P. R., Bosch, D. A., Bossuyt, P. M., Bonsel, G. J., van Someren, E. J., de Bie, R. M., Merkus, M. P., and Speelman, J. D. (2000). A comparison of continuous thalamic stimulation and thalamotomy for suppression of severe tremor. *N. Engl. J. Med.* **342,** 461–468.

Schuurman, P. R., de Bie, R. M., Majoie, C. B., Speelman, J. D., and Bosch, D. A. (1999). A prospective comparison between three-dimensional magnetic resonance imaging and ventriculography for target-coordinate determination in frame-based functional stereotactic neurosurgery. *J. Neurosurg.* **91,** 911–914.

Siegfried, J., and Lippitz, B. (1994). Chronic electrical stimulation of the VL-VPL complex and of the pallidum in the treatment of movement disorders: personal experience since 1982. *Stereotact. Funct. Neurosurg.* **62,** 71–75.

Speelman, J. D. (1991). "Parkinson's Disease and Stereotaxic Neurosurgery." Rodopi, Amsterdam.

Spiegel, E. A., Wyciss, H. T., Marks, M., and Lee, A. S. (1947). Stereotaxic apparatus for operations on the human brain. *Science* **106,** 349–350.

Starr, P. A., Vitek, J. L., DeLong, M., and Bakay, R. A. (1999). Magnetic resonance imaging-based stereotactic localization of the globus pallidus and subthalamic nucleus. *Neurosurgery* **44,** 303–313.

Taha, J. M., Janszen, M. A., and Favre, J. (1999). Thalamic deep brain stimulation for the treatment of head, voice, and bilateral limb tremor. *J. Neurosurg.* **91,** 68–72.

Tasker, R. R., Siqueira, J., Hawrylyshyn, P., and Organ, L. W. (1983). What happened to VIM thalamotomy for Parkinson's disease? *Appl. Neurophysiol.* **46,** 68–83.

Trepanier, L. L., Kumar, R., Lozano, A. M., Lang, A. E., and Saint Cyr, J. A. (2000). Neuropsychological outcome of GPi pallidotomy and GPi or STN deep brain stimulation in Parkinson's disease. *Brain Cogn.* **42,** 324–347.

Tronnier, V. M., Staubert, A., Hahnel, S., and Sarem Aslani, A. (1999). Magnetic resonance imaging with implanted neurostimulators: an in vitro and in vivo study. *Neurosurgery* **44,** 118–125.

Vitek, J. L., Bakay, R. A., Hashimoto, T., Kaneoke, Y., Mewes, K., Zhang, J. Y., Rye, D., Starr, P., Baron, M., Turner, R., and DeLong, M. R. (1998). Microelectrode-guided pallidotomy: technical approach and its application in medically intractable Parkinson's disease. *J. Neurosurg.* **88,** 1027–1043.

Volkmann, J., Sturm, V., Weiss, P., Kappler, J., Voges, J., Koulousakis, A., Lehrke, R., Hefter, H., and Freund, H. J. (1998). Bilateral high-frequency stimulation of the internal globus pallidus in advanced Parkinson's disease. *Ann. Neurol.* **44,** 953–961.

Wester, K., and Hauglie Hanssen, E. (1990). Stereotaxic thalamotomy–experiences from the levodopa era. *J. Neurol. Neurosurg. Psychiatry.* **53,** 427–430.

Young, R. F., Jacques, S., Mark, R., Kopyov, O., Copcutt, B., Posewitz, A., and Li, F. (2000). Gamma knife thalamotomy for treatment of tremor. *J. Neurosurg.* **93 Suppl 3,** 128–134.

CHAPTER 77

# Normal Pressure Hydrocephalus

Joachim K. Krauss and Michael Strupp

## CLINICAL FEATURES

Hakim and Adams introduced the concept of normal pressure hydrocephalus (NPH) in 1965, when they described the clinical triad of gait disturbance, cognitive impairment, and urinary incontinence in patients with communicating hydrocephalus and normal mean intracranial pressure (Adams *et al.*, 1965; Hakim and Adams, 1965). Almost 40 years later, a thorough understanding of the pathophysiology of NPH still remains illusive. Over the past few years, there have been conceptual changes regarding various aspects of NPH (Bradley, 2000; Krauss, 1999). Nowadays, it is generally agreed that the most frequent and important symptom of NPH is disturbance of gait (Krauss *et al.*, 2001). There is little evidence to support the earlier assumption that deficient cerebrospinal fluid (CSF) absorption is the priming event in the pathogenesis of NPH. Furthermore, it is generally accepted that the pathophysiological mechanisms of this disease involve reduced cerebral blood flow (CBF), a reduction in intracranial compliance, and a disturbance of CSF homeostasis. NPH most likely represents a final common pathway for a number of different entities. The only known effective and available treatment is shunt surgery. Because the benefit of CSF shunting is contingent primarily on the preoperative selection of appropriate candidates, the differential diagnosis of NPH continues to be a major challenge in clinical neurology and neurosurgery.

### Clinical Symptoms

The most frequent and characteristic symptom of NPH is disturbance of gait. It is usually the first sign of NPH and also the most likely to improve after shunting (Graff-Radford and Godersky, 1986; Graff-Radford *et al.*, 1989). The classic clinical triad becomes evident with progression of the disease and also includes urinary problems and cognitive deficits (Fisher, 1977). In addition to these three cardinal symptoms, a variety of other signs and symptoms may be present, in particular akinetic and bradykinetic movement disorders.

The phenomenology of gait disturbance varies widely, depending on the progression of the disease and coexistent features (Krauss *et al.*, 2001). In the early course of the disease, gait disturbance may be subtle. Patients may describe themselves as feeling dizzy or unstable. Objectively, their gait seems to be more careful, and they have difficulties with tandem gait. The number of steps may be increased when they turn. Later, gait becomes obviously unstable and may have a shuffling appearance similar to parkinsonian gait disorders. Patients may then complain of leg weakness, although formal neurological assessment does not usually show motor deficits. There is a decrease of walking speed, stride length, stride frequency, step height, and foot-floor clearance. The characteristic features of gait disturbance in NPH during this second phase of clinical deterioriation have been described as *magnetic gait* or as *walking as if the feet were glued to the floor* (Adams *et al.*, 1965). With further progression, hydrocephalic gait is characterized by increasing slowness. Patients are often barely able to walk without support. In late stages, the inability to walk can be embedded in a clinical picture resembling akinetic mutism. Some have difficulties standing up or turning around in bed. Postural instability is present in most NPH patients, and the tendency to fall backwards is increased (Blomsterwall *et al.*, 2000). Gait ignition/initiation failure has been described in 30% of patients with NPH, and freezing has been noted to occur in 56% (Petzinger *et al.*, 1994). Festination is rare.

The frequently used terms *gait apraxia* and *gait ataxia* are inappropriate. Little justifies their continued use, because there is no evidence of ideokinetic apraxia of the lower limbs. Even patients who are not able to stand may lift their legs when they are supine and perform cycling movements. Usually, there are no signs of cerebellar ataxia or dysmetria on tests such as the heel-knee test. It has to be noted that although gait disturbance is the most prominent and important symptom of NPH, it is neither pathognomonic nor unique (Elble *et al.*, 1992). Similar gait disturbances are seen in patients with frontal lobe pathology. Various gait disturbances in the elderly can be similar to each other regardless of their cause. The gait disturbance in the early phase may be indistinguishable from careful senile gait. The unstable gait seen later is also observed in other processes that affect the subcortical white matter such as subcortical vascular encephalopathy (SVE).

Cognitive deficits are almost always present; however, they are usually mild in the early phase of the disease and may be detected only by neuropsychological examination (Hütter and Krauss, 2001). The cognitive deficits are characterized mainly by impaired memory, decreased speed of information processing, and impaired executive function (Iddon *et al.*, 1999; Merten T, 1999). When cognitive deficits are more severe, the clinical picture is characterized by *subcortical dementia*. Symptoms such as psychomotor slowness, apathy, anhedonia, and bradyphrenia prevail. Some patients seem to be depressed. Signs of cortical dementia such as apraxia, aphasia, and agnosia are usually not evident. The clinical manifestation of dementia in a single patient may originate from coexistent diseases such as SVE or occasionally Alzheimer's disease. Severe isolated dementia makes a diagnosis of NPH rather unlikely.

Urinary problems are frequent. Initially, patients may complain only of sensations of urge without manifest incontinence. Later on, many patients become rapidly incontinent after urge. Even the most severely affected patients, however, still have some control over their micturition. *Incontinence sans gene* is seen only in patients with severe mental deficits and marked inability to walk.

Apart from gait, other motor activities are also decelerated in NPH patients (Blomsterwall *et al.*, 1995). Involvement of the upper extremities, however, is much less pronounced. Characteristic features are a reduction of spontaneous movements, decreased associated movements, and difficulties in performing fine motor tasks. About half of the patients exhibit hypomimia, hypokinesia, and bradykinesia (Krauss *et al.*, 1997a). Occasionally, tremor at rest, rigidity, and dystonia may be seen. In some patients, the clinical picture is dominated by a parkinsonian appearance (Curran and Lang, 1994; Jacobs and Kinkel, 1976). Sometimes, sleep apnea syndrome may be present. It is unclear, however, whether this is a genuine symptom of NPH or an increased co-occurrence (Kristensen *et al.*, 1998).

The frequency of clinical symptoms in patients with NPH has varied considerably among different series. These differences owe mainly to methodological issues. Remarkably, gait disturbance was found in almost all instances. After excluding other possible causes for movement disorders, parkinsonian symptoms included hypomimia in 55%, hypokinesia in 42%, bradykinesia in 55%, tremor at rest in 6%, rigidity in 8%, and dystonia in 3% (Krauss *et al.*, 1997a). Behavioral and psychiatric manifestations of NPH are rare.

### Differential Diagnosis

The idiopathic form of NPH usually becomes manifest in the sixth or seventh decade of life. Secondary chronic NPH may follow subarachnoid hemorrhage, meningitis, severe craniocerebral trauma, and intracranial surgery. It has been observed with schwannomas of the posterior fossa, megadolichobasilaris, spinal tumors, and Paget's disease. It may occur at any age. The most important differential diagnosis of idiopathic NPH of the elderly is SVE (Fisher, 1989; Kinkel *et al.*, 1985). Occasional co-occurrence of the two conditions was already noted in early reports (Earnest *et al.*, 1974; Koto *et al.*, 1977). Since the introduction of MR imaging, correlates of SVE have been frequently observed in patients with idiopathic NPH (Bradley *et al.*, 1991a; Jack *et al.*, 1987). The common link for this frequent co-occurrence might be arterial hypertension or diabetes mellitus. The prevalence of arterial hypertension (74%–83%) and diabetes mellitus (49%–52%) is significantly greater in patients with idiopathic NPH of the elderly than in age-matched controls (Graff-Radford and Godersky, 1987; Jacobs, 1977; Krauss *et al.*, 1996a). There is no difference in the prevalence of other vascular risk factors. Symptoms of idiopathic NPH of the elderly and SVE may be very similar, and in some patients it may not be possible to decide which of the two entities contributes most to a clinical symptom. Usually, subcortical dementia is more pronounced in advanced SVE, whereas the early appearance of gait disturbance and its dominance during the later course support a diagnosis of idiopathic NPH. Differential diagnosis in elderly patients with idiopathic or secondary NPH mainly includes other neurodegenerative diseases. Alzheimer's disease causes cortical dementia (see Chapter 29) not subcortical. In patients with parkinsonian symptoms and hydrocephalus, it has to be determined whether the movement disorders respond to L-dopa/dopamine agonists (i.e., whether there is a causal or a coincidental association). Parkinsonian symptoms may be part of the clinical picture of idiopathic NPH. NPH coexisting with Parkinson's disease or progressive supranuclear palsy has been noted repeatedly (Curran and Lang, 1994).

The differential diagnosis in patients with secondary NPH is usually uncomplicated, and the decision making for shunting is straightforward. This is particularly the case in patients with progressive clinical symptoms paralleled by increasing width of the ventricular system on imaging studies (Yasargil *et al.*, 1973), for example, a decompensated aqueduct stenosis, which is often characterized by headache and a faster progression of symptoms. In some patients, however, it may be difficult to determine whether the clinical picture is mainly due to the underlying disease that prompted the development of NPH or to progression of the hydrocephalus. Such difficulties mainly arise in patients after severe head injury with concomitant diffuse axonal injury or in patients who recover only gradually from a massive subarachnoid hemorrhage (Marmarou *et al.*, 1996).

### Diagnostic Studies

The confirmation of the diagnosis and the selection of appropriate candidates for surgery are of utmost importance. Over the years, a variety of diagnostic studies have been introduced. Because NPH probably represents the final common pathway for a number of different priming events and the issues of the co-occurrence with other diseases are complex, there is obviously no single diagnostic test to confirm or reject the diagnosis. It is feasible to use a combined diagnostic approach, taking advantage of morphological imaging, diagnostic CSF removal, and hydrodynamic tests.

#### Computed Tomography

Computed tomography usually demonstrates symmetrical widening of the lateral and third ventricles. The fourth ventricle may or may not be dilated. Often there is flattening of the sulcal relief over the parasagittal convexity. Generalized cortical atrophy argues against a diagnosis of NPH, whereas focal sulcal atrophy and widening of the sylvian fissures may indicate localized CSF accumulation caused by accompanying external hydrocephalus (Holodny *et al.*, 1998a); this is supported by the fact that these accumulations decrease after successful CSF shunting. Periventricular caps and rims may be clearly visible on CT scans, whereas diffuse white matter lesions indicate SVE.

#### Magnetic Resonance Imaging

The extent and severity of periventricular and deep white matter lesions can be delineated with magnetic resonance imaging. Sagittal images are useful for evaluating the patency of the aqueductus Sylvii. The correlation between the degree of the flow void and postoperative improvement is low or negative (Hakim and Black, 1998; Krauss *et al.*, 1997c). Phase-contrast MR

imaging may be suitable for quantifying the aqueductal CSF stroke volume and for differentiating between true hyperdynamic CSF flow and increased CSF flow velocity (Naidich *et al.*, 1993).

#### CSF Removal

Transient improvement of the clinical symptoms after diagnostic CSF removal (50 ml) has been increasingly used to confirm diagnosis and to predict therapeutic outcome (Larsson *et al.*, 1991). Standardized documentation including videotaping of the gait disturbance before and after CSF removal has proven useful for better evaluating its effect. Some investigators have doubted the accuracy of predicting outcome by a single lumbar puncture (Malm *et al.*, 1995), because it has been repeatedly demonstrated that the predictive accuracy of a negative CSF tap test is questionable. Therefore, if the patient does not respond to a single CSF removal, this may be repeated up to three times on following days or preferably continuous external lumbar CSF drainage (about 100–200 ml CSF drained daily for a period of 3–5 days) should be performed. This is highly accurate when predicting outcome after shunting (Chen *et al.*, 1994; Haan and Thomeer, 1988). However, patients have to be monitored more closely, and the complication rate including the occurrence of meningitis is higher. Two recent studies quantified the improvement of gait disturbance before and after the CSF tap test in NPH patients (Krauss *et al.*, 2001; Stolze *et al.*, 2000). Interestingly, there was no increase in cadence, although the velocity and stride length increased.

#### Invasive Tests of CSF Dynamics

Several tests allow investigation of the hydrodynamics of NPH and calculation of the compliance, resistance, or conductance (Kosteljanetz, 1986). Experience with these tests has differed from center to center, and their value in predicting outcome after shunt surgery has not been considered useful by all investigators. In clinical routine they are not generally recommended for making the diagnosis.

#### Cerebral Hemodynamics, Autoregulation, and Metabolism

SPECT and PET investigations repeatedly demonstrated subcortical hypoperfusion in NPH patients, in particular in the frontal and temporal white matter (Kristensen *et al.*, 1996; Vostrup *et al.*, 1987). Some studies have described postoperative improvement of the perfusion of the deep white matter. A pathological response to the acetazolamide challenge test, which usually increases cerebral blood flow (CBF), has been found in NPH patients, in particular, in the periventricular white matter (Tanaka *et al.*, 1997). This finding has been thought to indicate that the arterioles are already maximally dilated. Interestingly, it has been demonstrated that those NPH patients who show subcortical hypoperfusion and disturbed autoregulation improved postoperatively, in contrast to those patients with subcortical hypoperfusion but intact autoregulation. The latter constellation has been considered indicative of irreversible ischemia of the deep white matter.

### Biochemical CSF Markers

There have been several attempts to find specific diagnostic biochemical markers to confirm the diagnosis of NPH. This approach has been limited by methodological issues including the heterogeneity of the patients investigated, the differences in clinical symptoms, and the selection of control groups. In a recent study, NPH patients had higher CSF levels of neurofilament triplet protein (NLF) and of glial fibrillary acidic protein (GFAP) (Tullberg *et al.*, 1998). NLF levels were higher in patients with more severe clinical symptoms. On the other hand, there was a positive correlation between the NLF level and the degree of postoperative improvement after shunting. NLF thus could be a marker of progressive, but still reversible axonal damage in NPH. Choleocystokinin (CCK) has been found to be significantly decreased in patients with NPH; there is a statistically significant correlation between abnormal intracranial pressure recordings and the CCK level (Galard *et al.*, 1997). CSF sulfatide was described to be useful for distinguishing between NPH and SVE (Tullberg *et al.*, 2000). In another recent study, the level of tau protein was shown to be significantly higher in CSF from NPH patients than in control persons (Kudo *et al.*, 2000).

## NATURAL COURSE

### Epidemiology

Although no systematic studies have been conducted on the frequency of idiopathic NPH, it has been considered a rather rare disease with an estimated incidence of 1.3–2.2/million/year (Vanneste, 1994). Sudarsky and Ronthal (1983), however, found hydrocephalus to be responsible for the clinical picture in 2 of 50 patients (4%) with previously undiagnosed gait disorders. Fisher demonstrated hydrocephalic dilation of the ventricular system in 46 of 50 patients (92%) older than the age of 60 with otherwise unexplained disturbance of gait, whereas hydrocephalus was seen in only 6.3% of a group of similar age (Fisher, 1982). The Starnberg trial,

which was performed in the early 1990s in Germany, found an unexpectedly high prevalence of NPH by a door-to-door survey in a rural Bavarian population of individuals older than 65 years (0.4%, i.e., 4 of 982 inhabitants) (Trenkwalder *et al.*, 1995). The mean age at diagnosis of idiopathic NPH was 71 years in a group of 65 patients, aged 51–88 years. The gender ratio seems to be about equal, with a slightly higher, but not significant, preponderance of women.

There are no studies on the natural course of idiopathic NPH so far. Its spontaneous course seems to be characterized by insidious onset and variable progression over months and years. The time course of the progression also depends on the severity of coexistent SVE. Spontaneous regression of symptoms has not been described. Some patients show long-lasting improvement after a single lumbar puncture. Occasionally, the further course may be altered by repeated lumbar puncture on reappearance of symptoms.

The incidence and prevalence of secondary hydrocephalus are also largely unknown. Overall, about 50% of newly diagnosed NPH patients have idiopathic hydrocephalus and about 50% have secondary hydrocephalus. The occurrence of NPH in subarachnoid hemorrhage depends on the amount of blood in the cisterns at the time of bleeding and on the Hunt and Hess grading (Gruber *et al.*, 1999; Yasargil *et al.*, 1973). About 5%–20% of patients have NPH develop after higher grade subarachnoidal hemorrhage. After severe head injury, about 6% of patients will have NPH in the long term.

The natural course of secondary NPH differs from that of primary NPH. Onset of symptoms of NPH after subarachnoid hemorrhage is evident after a delay of days or weeks, and progression is much faster.

## PRINCIPLES OF THERAPY

### Pathomechanisms of Idiopathic NPH

CSF pressure in NPH is not entirely normal. First, baseline pressure is often slightly elevated compared with that of healthy persons. Second, long-term CSF monitoring studies have shown intermittent oscillations of 0.5–2/min, known as B-waves, which seem to be more frequent in patients with NPH (Black *et al.*, 1985; Symon and Dorsch, 1975). Despite these semantic difficulties, however, the term NPH has been commonly accepted as the diagnostic label for hydrodynamic active communicating hydrocephalus. Other terms such as *adult hydrocephalus syndrome* or *adult symptomatic hydrocephalus* are used more interchangeably and also include patients with ventriculomegaly caused by atrophy, although they might be more appropriate terms from a pathophysiological point of view (Vanneste, 1994; Symon and Hinzpeter, 1977).

Initially, idiopathic NPH was thought to be primarily a disorder of CSF absorption at the level of the arachnoid villi; however, this was not supported by leptomeningeal biopsies. Autopsy studies and biopsies have shown demyelinization, axonal lesions, and gliosis (Del Bigio, 1993; Del Bigio et al., 1997). Therefore, changes in the cerebral parenchyma seem to initiate the pathomechanisms, and altered fluid dynamics seen in patients with idiopathic NPH are secondary (Bradley, 2000; Bradley et al., 1991a; Krauss, 1998).

Ventriculomegaly may occur when the transmantle pressure (i.e., the difference in pressure between the ventricles and the subarachnoid space) is increased (Conner et al., 1984). In secondary hydrocephalus, especially when caused by subarachnoid hemorrhage, decreased CSF resorption may be the priming event. It must, however, be considered that a substantial amount of CSF resorption occurs also at the transcapillary or transvenular level. In patients with idiopathic NPH, both periventricular and deep white matter lesions were frequently found after the introduction of MR imaging (Bradley et al., 1991a; Jack et al., 1987; Krauss et al., 1997b). White matter lesions were thought to reduce periventricular tissue strength and alter the elastic properties, thus predisposing the ventricles to dilate under the CSF pulse pressure (Bradley et al., 1991a). Thus, defective CSF absorption could also be a secondary phenomenon in patients with idiopathic NPH. Ongoing clinical deterioration most probably results from decreased periventricular blood flow. This would be in accordance with the findings that periventricular CBF is increased in some patients who improve after shunting. One factor that could underly both SVE and idiopathic NPH is arterial hypertension (Bateman, 2000; Graff-Radford and Godersky, 1987; Krauss et al., 1996a). Arterial hypertension might also be linked to increased CSF pulsatility, which, particularly in combination with altered compliance, acts on the walls of the dilated ventricles and could contribute to ventriculomegaly and the progression of symptoms. Because the neuronal tissue in the white matter is ischemic, small arteries in the white matter are supposed to be maximally dilated, which would explain the loss of autoregulation and also the lack of response to acetazolamide. In a recent study, vascular compliance in the superior sagittal sinus in patients with NPH was shown to be lower than that of healthy subjects (Bateman, 2000). It was suggested that failure of craniospinal compliance to absorb vascular pulsations could be the initiating cause of NPH. CSF that is normally drained by the parenchymal veins could back up, cause an elevation of intracranial pressure, and subsequently generate pressure waves. This series of events would be consistent with the known fact that

the apparent diffusion coefficient is increased, indicating an increased interstitial edema in patients with NPH (Gideon et al., 1994; Tamaki et al., 1990). It would also be compatible with the high correlation of NPH with hypertension and arteriosclerosis, because the latter conditions exert stress on the available compliance because of the high input pulse pressure. Interestingly, after CSF shunting, vascular compliance in NPH patients became more like that in patients with ischemia instead of returning to normal (Bateman, 2000). However, it remains unclear why the previously normal venous resistance becomes elevated in elderly patients.

The occurrence of the specific symptoms of NPH; the classic triad of gait disturbance, urinary problems, and cognitive impairment; and also the manifestation of parkinsonian movement disorders are attributable to subcortical ischemia, which primarily affects the periventricular and deep white matter. Recent findings from morphological and dynamic imaging studies point to the unlikeliness of the older hypothesis that these symptoms are secondary only to mechanical stretching of periventricular and deep white matter fibers. The fact that the classic triad is usually only partially reversible with shunting indicates that both ischemic demyelination and hypoxia are relevant for the occurrence of these symptoms. A study on motor-evoked potentials in the preoperative and postoperative assessment of NPH patients showed that the gait disturbance is not associated with dysfunction of the pyramidal tract but that the sensorimotor integration underlying normal gait is disturbed (Zaaroor et al., 1997). Simple mechanical stretching of motor fibers also seems to be unlikely, given that the gait disturbance and other symptoms improve after shunting, although little or no change of ventriculomegaly is seen on postoperative imaging studies in most patients (Jacobs et al., 1976). Both functional disturbances and structural lesions most likely contribute to the development of clinical symptoms. Clinical improvement after CSF removal or shunting may be attributed to several factors, including increased periventricular blood flow along with improvement of metabolism subsequent to reduction of intracerebral pressure, decrease of CSF pulse pressure, and normalization of compliance. The coexistence of morphological white matter damage, on the other hand, may explain why and to what extent recovery of clinical symptoms is incomplete in most of these elderly patients. We have shown earlier that the extent and severity of both deep and periventricular white matter lesions correlate with the preoperative severity of symptoms (Krauss et al., 1996b). Also, the degree of overall clinical improvement was negatively correlated with the extension of both periventricular lesions and deep white matter lesions. This negative correlation was also noted when the analysis was conducted separately for each of the cardinal symptoms.

## TREATMENT OF NPH

The only available treatment for NPH at this time is CSF diversion. The most problematic aspect is the preoperative selection of patients. Other aspects are the surgical technique and the choice of different shunt systems. Medical therapies have not proven useful in the treatment of NPH. From a theoretical point of view, a reduction of arterial hypertension might be beneficial, preventing NPH or slowing further progression. However, no study has investigated this issue so far. Endoscopic ventriculocisternostomy is now the treatment of choice in aqueductal stenosis. Its use has also been explored in NPH, and some positive results have been reported (Mitchell and Mathew, 1999). There is not enough evidence, however, to support its routine use in clinical practice. Also, it is difficult to explain why and how patients with NPH should improve with ventriculocisternostomy.

## PRAGMATIC THERAPY

The differential diagnosis and the confirmation of the suspected diagnosis of NPH is still a problematic issue. It cannot be overstated that the most important aspect is the clinical diagnosis. In particular, gait disturbance plays the most important role in the diagnosis of idiopathic NPH of the elderly. In patients with secondary hydrocephalus, the diagnosis and the decision making for shunting is straightforward when the history is short and there is a clear progression of hydrocephalus with flattening of the sulcal relief in subsequent imagings. Shunting is clearly indicated in patients with idiopathic NPH who improve after one or several CSF lumbar punctures or continuous external lumbar CSF drainage. We do not consider patients with severe dementia, in particular those with cortical dementia associated with cortical atrophy, who do not have gait disturbance for further diagnostic workup. The most problematic group consists of patients with coexisting idiopathic NPH and SVE. Given the complexity of NPH and the frequency of other co-occurrent diseases, it is difficult to advise which diagnostic tests should be used in general. It is important for each center to develop and assess feasible algorithms, depending on the specific diagnostic instruments available.

The minimal diagnostic criteria for a diagnosis of NPH are the early appearance of gait disturbance and its predominance in the clinical picture, chronic hydrocephalus without cortical atrophy, and improvement of the gait disorder after CSF removal. The diagnosis is strongly supported by the results of hydrodynamic studies including a pressure volume index of <13 ml and a resistance of outflow >12 mmHg/ml/min, the occurrence of B waves during 70% or more of the recording time during overnight intracranial pressure monitoring, and the presence of high-amplitude or ramp-type B waves. The demonstration of white matter lesions in patients with suspected idiopathic NPH is not helpful for ultimately confirming or rejecting the diagnosis in a clinical setting. As described earlier, periventricular and deep white matter lesions of varying degrees are common findings on MR scans of patients with idiopathic NPH of the elderly. Although the occurrence of such lesions should not give rise to therapeutic nihilism, in general one cannot expect excellent improvement after CSF shunting in patients with severe lesions. Nevertheless, some patients still benefit from CSF diversion and show moderate or good improvement.

Treatment guidelines also differ depending on the stage of the disease. Patients at an early clinical stage with mild gait disturbance may be monitored initially with repeated CSF removal by means of lumbar puncture. If the disease progresses, patients can undergo CSF shunting without the necessity of extensive hydrodynamic tests. It is advisable to shunt patients with more marked symptoms and a longer history earlier, soon after the diagnosis has been confirmed by CSF removal and ancillary hydrodynamic tests. Those patients who are seen at a late stage of disease should be selected carefully; we recommend extensive counseling with the patients' relatives to weigh the expected benefits from shunting and the possibilities of long-term shunt dysfunction. If there is secondary deterioration of the clinical symptoms after initially successful improvement after CSF shunting, shunt dysfunction should be ruled out aggressively before further measures are taken.

New principles of shunt technology have evolved slowly over the past two decades (Czosnyka et al., 1998). For CSF shunting in patients with NPH, both ventriculoperitoneal (VP) and ventriculoatrial (VA) CSF diversion is being performed. In view of the possibly long-term systemic complications, it is advisable to use VP shunts in younger patients. There is no clear answer to the question, however, which form of CSF diversion is more advantageous in elderly patients with idiopathic NPH. From a theoretical point of view, the siphoning effect should be more pronounced the higher the hydrostatic column. Nevertheless, it seems that VP shunts are not associated with symptoms of overdrainage and subdural hematomas more often than VA shunts. Accurate positioning of the tip of the ventricular catheter during the operation is of utmost importance for the long-term function of the system. Ingrowth of tissue into the lumen of the catheter may occur if the tip of the catheter is located at the choroid plexus or at the wall of the ventricle (Del Bigio, 1998). Most frequently, differential pressure valves have been used in patients with NPH (Boon *et al.*, 1998). The flow-regulated–valve type is an alternative thought to act in a more physiological way. Programmable valve systems that allow percutaneous determination of the opening pressure of the valve have found widespread acceptance over the past few years (Reinprecht *et al.*, 1995; Zemack and Romner, 1997). These systems permit slow titration of the appropriate opening pressure in the individual patient and its adjustment in the later course. Different systems with up to 18 different pressure levels that can be programmed are commercially available. The option of adjusting the opening pressure is also useful in conservative treatment of overdrainage with subdural hematoma.

Despite these new developments, there is still the risk of overdrainage caused by the siphoning effect when the patient assumes an upright position. Antisiphoning devices have been used in the past to prevent

this effect. Their use has been limited, however, by an increased tendency for shunt dysfunction, in particular with constructions that were compressible by subcutaneous tissue pressure. New devices working on ball-valve hydrostatic mechanisms embedded in metallic capsules have recently become available (Sprung *et al.*, 1997).

## Outcome of CSF Shunting

Following the reports of impressive postoperative improvement of NPH patients after shunting in the mid-1960s, this treatment modality was soon used worldwide. Its widespread use and the relative simplicity of the surgical method, however, also resulted in its being used rather uncritically at times. Some considered shunting to be the best proof of whether a patient with severe dementia had NPH or not. Needless to say, this strategy resulted in many useless or even harmful surgical

**TABLE I**  Outcome and Complications of CSF Shunting in Recent Studies on Patients with Idiopathic Normal Pressure Hydrocephalus (NPH)

| Series (first author) | Number of patients | Postoperative improvement (% of patients) | Follow-up | Operative mortality (%) | Transient/reversible complications | Persistent complications |
|---|---|---|---|---|---|---|
| Graff-Radford, 1989 | 30 (5 patients witht sec. NPH) | 77 | 6 mo | 0 | * | * |
| Cardoso, 1989 | 19 | 100 | 3 mo | 0 | 5 shunt infection | 0% |
| Benzel, 1990 | 37 | 70 | * | 0 | 5% seizures | 3% (secondary to bilateral subdural hematoma) |
| Larsson, 1991 | 26 | 73 | 2.1 y | 0 | * | * |
| Vanneste, 1992 | 127 | 31 | 2 mo–8 y | 2 | 17% "severe complications" | 9% "severe residual deficits" |
| Raftopoulos, 1994/1996 | 23 | 96 / 91 | 1 y / 20 mo | 0 | 30% shunt dysfunction | 0% |
| Krauss, 1996/1997 | 50 | 90 | 19.4 mo | 0 | 8% (asymptomatic hemorrhage, infection) 22% shunt dysfunction | 0% |
| Boon, 1997/1998 | 95 | 76 | 6 mo | 0 | * | * |
| Kristensen, 1998 | 17 | 78 | 3 mo | 0 | * | * |

*Not detailed.

**TABLE II**  Summary of Predictors of Probable Individual Benefits from Shunt Operation in NPH

| Predictors | Predicted shunt responsiveness | |
|---|---|---|
| | Good | Poor |
| History | Short (<2 ys) | Long (>2 ys) |
| Clinical picture | Classic triad: gait disturbance >> dementia gait disturbance: initial symptom | Dementia > gait disturbance |
| CT scan | At least moderate ventriculomegaly | Severe vascular lesions |
| ICP monitoring (>12 h) | Mean intracranial pressure normal or slightly raised, increased relative frequency of B-waves (>10%) | Low frequency of B-waves (<5%) |
| Resistance to CSF flow (CSF infusion test) | $R_{out}$ > 12.5 mmHg/ml/min | $R_{out}$ < 12.5 mmHg/ml/min |

Modified from Gerloff and Pickard (1996).

interventions. There is a wide range of postoperative improvements in different series (Stein and Langfitt, 1974; Vassilouthis, 1984). The results of larger studies on patients with idiopathic NPH published in the past decade are summarized in Table I. With appropriate selection criteria, the relative proportion of patients who experience postoperative improvement of the symptoms has increased compared with that 20 years earlier. In general, postoperative improvement will be seen in 70%–90% of patients. The predictors of probable individual benefits are summarized in Table II. Patients with secondary hydrocephalus tend to improve more than patients with primary hydrocephalus. Complete normalization of clinical symptoms is rare in idiopathic NPH. The gait disorder usually improves much better than the cognitive deficits. There are very few studies on the long-term outcome of shunted NPH patients. In one study, it was noted that 57% of shunted patients died within 5 years postoperatively (Raftopoulos et al., 1996). The most frequent cause of death was ischemic cerebral apoplexia.

Adverse operative events during shunt surgery are rare and generally transient. The most important complications are intracerebral bleeding and meningitis. Adverse operative events usually occur in 5%–10% of cases. Even with new generations of shunt systems, shunt dysfunction is a problem in long-term follow-up. In prospective studies, shunt dysfunction is observed in up to 30% of patients and may manifest either as overdrainage or underdrainage. Shunt dysfunction should be promptly corrected. Persistent complications range between 0%–3% in most studies.

# REFERENCES

Adams, R. D., Fisher, C. M., Hakim, S. et al. (1965). Symptomatic occult hydrocephalus with "normal" cerebrospinal fluid pressure. A treatable syndrome. N. Engl. J. Med. 273, 117–126.

Bateman, G. A. (2000). Vascular compliance in normal pressure hydrocephalus. AJNR 21, 1574–1585.

Benzel, E. C., Pelletier, A. L., and Levy, P. G. (1990). Communicating hydrocephalus in adults: prediction of outcome after ventricular shunting procedures. Neurosurgery 26, 655–660.

Black, P. M., Ojemann, R. G., and Tzouras, A. (1985). CSF shunts for dementia, incontinence, and gait disturbance. Clin. Neurosurg. 32, 632–651.

Blomsterwall, E., Bilting, M., Stephensen, H. et al. (1995). Gait abnormality is not the only motor disturbance in normal pressure hydrocephalus. Scand. J. Rehabil. Med. 27, 205–209.

Blomsterwall, E., Svantesson, U., Carlsson, U., et al. (2000). Postural disturbance in patients with normal pressure hydrocephalus. Acta Neurol. Scand. 102, 284–291.

Boon, A. J., Tans, J. T., Delwel, E. J. et al. (1997). Dutch normal-pressure hydrocephalus study: prediction of outcome after shunting by resistance to outflow of cerebrospinal fluid. J. Neurosurg. 87, 687–693.

Boon, A. J., Tans, J. T., Delwel, E. J. et al. (1998). Dutch normal-pressure hydrocephalus study: randomized comparison of low- and medium-pressure shunts. J. Neurosurg. 88, 490–495.

Borgesen, S. E., Gjerris, F., and Soerensen, S. C. (1979). Intracranial pressure and conductance to outflow of cerebrospinal fluid in normal-pressure hydrocephalus. J. Neurosurg. 50, 489–493.

Bradley, W. G. (2000). Normal pressure hydrocephalus: new concepts on etiology and diagnosis. AJNR 21, 1586–1590.

Bradley, W. G., Whittemore, A. R., Watanabe, A. S. et al. (1991a). Association of deep white matter infarction with chronic communicating hydrocephalus: implications regarding the possible origin of normal-pressure hydrocephalus. AJNR 12, 31–39.

Bradley, W. G., Whittemore, A. R., Kortman, K. E. et al. (1991b). Marked cerebrospinal fluid void: indicator of successful shunt in patients with suspected normal-pressure hydrocephalus. Radiology 178, 459–466.

Bradley, W. G., Scalzo, D., Queralt, J. et al. (1996). Normal-pressure hydrocephalus: evaluation with cerebrospinal fluid flow measurements at MR imaging. Radiology 198, 523–529.

Cardoso, E. R., Piatek, D., Del Bigio, M. R. et al. (1989). Quantification of abnormal intracranial pressure waves and isotope cisternography for diagnosis of occult communicating hydrocephalus. Surg. Neurol. 31, 20–27.

Chen, I. H., Huang, C. I., Liu, H. C. et al. (1994). Effectiveness of shunting in patients with normal pressure hydrocephalus predicted by temporary, controlled-resistance, continuous lumbar drainage: a pilot study. J. Neurol. Neurosurg. Psychiatry 57, 1430–1432.

Conner, E. S., Foley, L., and Black, P. M. (1984). Experimental normal-pressure hydrocephalus is accompanied by increased transmantle pressure. J. Neurosurg. 61, 322–327.

Curran, T., and Lang, A. E. (1994). Parkinsonian syndromes associated with hydrocephalus: case reports, a review of the literature, and pathophysiological hypotheses. Mov. Disord. 9, 508–520.

Czosnyka, Z., Czosnyka, M., Richards, H. K. et al. (1998). Posture-related overdrainage: comparison of the performance of 10 hydrocephalus shunts in vitro. Neurosurgery 42, 327–334.

Del Bigio, M. R. (1993). Neuropathological changes caused by hydrocephalus. Acta Neuropathol. 85, 573–585.

Del Bigio, M. R. (1998). Biological reactions to cerebrospinal fluid shunt devices: a review of the cellular pathology. Neurosurgery 42, 319–326.

Del Bigio, M. R., Cardoso, E. R., and Halliday, W. C. (1997). Neuropathological changes in chronic adult hydrocephalus: cortical biopsies and autopsy findings. Can. J. Neurol. Sci. 24, 121–126.

Droste, D. W., and Krauss, J. K. (1993). Simultaneous recording of cerebrospinal fluid pressure and middle cerebral artery blood flow velocity in patients with suspected symptomatic normal pressure hydrocephalus. J. Neurol. Neurosurg. Psychiatry 56, 75–79.

Droste, D. W., and Krauss, J. K. (1997). Oscillations of cerebrospinal fluid pressure in nonhydrocephalic persons. Neurol. Res. 19, 135–138.

Droste, D. W., and Krauss, J. K. (1999). Intracranial pressure B-waves precede corresponding arterial blood pressure oscillations in patients with suspected normal pressure hydrocephalus. Neurol. Res. 21, 627–630.

Droste, D. W., Krauss, J. K., Berger, W. et al. (1994). Rhythmic oscillations with a wavelength of 0.5–2 min in transcranial Doppler recordings. Acta Neurol. Scand. 90, 99–104.

Earnest, M. P., Fahn, S., Karp, J. H. et al. (1974). Normal pressure hydrocephalus and hypertensive cerebrovascular disease. Arch. Neurol. 31, 262–266.

Elble, R. J., Hughes, L., and Higgins, C. (1992). The syndrome of senile gait. J. Neurol. 239, 71–75.

Fisher, C. M. (1977). The clinical picture in occult hydrocephalus. Clin. Neurosurg. 24, 270–284.

Fisher, C. M. (1982). Hydrocephalus as a cause of disturbances of gait in the elderly. *Neurology* **32**, 1358–1363.

Fisher, C. M. (1989). Binswanger's encephalopathy: a review. *J. Neurol.* **236**, 65–79.

Galard, R., Poca, M. A., Catalan, R. *et al.* (1997). Decreased cholecystokinin levels in cerebrospinal fluid of patients with adult chronic hydrocephalus syndrome. *Biol. Psychiatry* **41**, 804–809.

Gerloff, C., and Pickard, J. D. (1996). Normal pressure hydrocephalus. *In* "Neurological Disorders: Course and Treatment." (Brandt, *et al.*, Eds.) pp. 773–778. Academic Press, New York.

Gideon, P., Thomsen, C., Gjerris, F. *et al.* (1994). Increased self-diffusion of brain water in hydrocephalus measured by MR imaging. *Acta Radiol.* **35**, 514–519.

Graff-Radford, N. R., and Godersky, J. C. (1986). Normal pressure hydrocephalus: onset of gait abnormality before dementia predicts a good surgical outcome. *Arch Neurol.* **43**, 940–942.

Graff-Radford, N. R., and Godersky, J. C. (1987). Idiopathic normal pressure hydrocephalus and systemic hypertension. *Neurology* **37**, 868–871.

Graff-Radford, N. R., Godersky, J. C., and Jones, M. P. (1989). Variables predicting surgical outcome in symptomatic hydrocephalus in the elderly. *Neurology* **39**, 1601–1604.

Gruber, A., Reinprecht, A., Bavinzski, G. *et al.* (1999). Chronic shunt-dependent hydrocephalus after early surgical and early endovascular treatment of ruptured intracranial aneurysms. *Neurosurgery* **44**, 503–512.

Haan, J., and Thomeer, R. T. W. M. (1988). Predictive value of temporary external lumbar drainage in normal pressure hydrocephalus. *Neurosurgery* **22**, 388–391.

Hakim, S., and Adams, R. D. (1965). The special clinical problem of symptomatic hydrocephalus with normal cerebrospinal fluid pressure. *J. Neurol. Sci.* **2**, 307–327.

Hakim, R., and Black, P. M. (1998). Correlation between lumbo-ventricular perfusion and MRI-CSF flow studies in idiopathic normal pressure hydrocephalus. *Surg. Neurol.* **49**, 14–19.

Holodny, A. I., George, A. E., De Leon, M. J. *et al.* (1998a). Focal dilation and paradoxical collapse of cortical fissures and sulci in patients with normal-pressure hydrocephalus. *J. Neurosurg.* **89**, 742–747.

Holodny, A. I., Waxman, R., George, A. E. *et al.* (1998b). MR differential diagnosis of normal-pressure hydrocephalus and Alzheimer disease: significance of perihippocampal features. *AJNR* **19**, 813–819.

Hütter, B. O., and Krauss, J. K. (2002). The diagnostic value of the cognitive functional level in patients with suspected normal pressure hydrocephalus. *Zentr. Neurochirur.* in press.

Iddon, J. L., Pickard, J. D., Cross, J. J. *et al.* (1999). Specific patterns of cognitive impairment in patients with idiopathic normal pressure hydrocephalus and Alzheimer's disease: a pilot study. *J. Neurol. Neurosurg. Psychiatry* **67**, 723–732.

Jack, C. R., Mokri, B., Laws, E. R. *et al.* (1987). MR findings in normal-pressure hydrocephalus: significance and comparison with other forms of dementia. *J. Comput. Assist. Tomogr.* **19**, 923–931.

Jacobs, L. (1977). Diabetes mellitus in normal pressure hydrocephalus. *J. Neurol. Neurosurg. Psychiatry* **40**, 331–335.

Jacobs, L., and Kinkel, W. (1976). Computerized axial transverse tomography in normal pressure hydrocephalus. *Neurology* **26**, 501–507.

Jacobs, L., Conti, D., Kinkel, W. R. *et al.* (1976). "Normal-pressure" hydrocephalus. Relationship of clinical and radiographic findings to improvement following shunt surgery. *JAMA* **235**, 510–512.

Kinkel, W. R., Jacobs, R., Polachini, I. *et al.* (1985). Subcortical arteriosclerotic encephalopathy (Binswanger's disease): computed tomographic, magnetic resonance, and clinical correlations. *Arch. Neurol.* **42**, 951–959.

Kosteljanetz, M. (1986). CSF dynamics and pressure-volume relationships in communicating hydrocephalus. *J. Neurosurg.* **64**, 45–52.

Koto, A., Rosenberg, G., Zingesser, L. H. *et al.* (1977). Syndrome of normal pressure hydrocephalus: possible relation to hypertensive and arteriosclerotic vasculopathy. *J. Neurol. Neurosurg. Psychiatry* **40**, 73–79.

Krauss, J. K. (1998). Idiopathic normal pressure hydrocephalus: primary disorder of CSF absorption or parenchymal cerebral disease? The relevance of vascular risk factors and cerebral white matter lesions. *In* "Intracranial and Inner Ear Physiology and Pathophysiology." (A. Reid, R. Marchbanks, and A. Ernst, Eds.) pp. 67–78. Whurr Publishers, London.

Krauss, J. K. (1999). Der idiopathische Normaldruckhydrozephalus des älteren Menschen: Ueberblick und Perspektiven. *Akt Neurologie* **26**, 250–259.

Krauss, J. K., and Droste, D. W. (1994). Predictability of intracranial pressure oscillations in patients with suspected normal pressure hydrocephalus by transcranial Doppler ultrasound. *Neurol. Res.* **16**, 398–402.

Krauss, J. K., Droste, D. W., Bohus, M. *et al.* (1995). The relation of intracranial pressure B-waves to different sleep stages in patients with suspected normal pressure hydrocephalus. *Acta Neurochir* **136**, 195–203.

Krauss, J. K., and Regel, J. P. (1997). The predictive value of ventricular CSF removal in normal pressure hydrocephalus. *Neurol. Res.* **19**, 357–360.

Krauss, J. K., Droste, D. W., Vach, W. *et al.* (1996b). Cerebrospinal fluid shunting in idiopathic normal-pressure hydrocephalus of the elderly: effect of periventricular and deep white matter lesions. *Neurosurgery* **39**, 292–300.

Krauss, J. K., Regel, J. P., Vach, W. *et al.* (1996a). Vascular risk factors and arteriosclerotic disease in idiopathic normal pressure hydrocephalus of the elderly. *Stroke* **27**, 24–29.

Krauss, J. K., Regel, J. P., Droste, D. W. *et al.* (1997a). Movement disorders in adult hydrocephalus. *Mov. Disord.* **12**, 53–60.

Krauss, J. K., Regel, J. P., Vach, W. *et al.* (1997b). White matter lesions in patients with idiopathic normal pressure hydrocephalus and in an age-matched control group: a comparative study. *Neurosurgery* **40**, 491–496.

Krauss, J. K., Regel, J. P., Vach, W. *et al.* (1997c). Flow void of cerebrospinal fluid in idiopathic normal pressure hydrocephalus of the elderly: can it predict outcome after shunting? *Neurosurgery* **40**, 67–74.

Krauss, J. K., Faist, M., Schubert, M. *et al.* (2001). Evaluation of gait in normal pressure hydrocephalus before and after shunting. *In* "Gait Disorders, Advances in Neurology." (E. Ruzicka, M. Hallett, and J. Jankovic, Eds.)

Kristensen, B., Malm, J., Fagerlund, M. *et al.* (1996). Regional cerebral blood flow, white matter abnormalities, and cerebrospinal fluid hydrodynamics in patients with idiopathic adult hydrocephalus syndrome. *J. Neurol. Neurosurg. Psychiatry* **60**, 282–288.

Kristensen, B., Malm, J., and Rabben, T. (1998). Effects of transient and persistent cerebrospinal fluid drainage on sleep disordered breathing in patients with idiopathic adult hydrocephalus syndrome. *J. Neurol. Neurosurg. Psychiatry* **65**, 497–501.

Kudo, T., Mima, T., Hashimoto, R. *et al.* (2000). Tau protein is a potential biological marker for normal pressure hydrocephalus. *Psychiatry Clin. Neurosci.* **54**, 199–202.

Lamas, E., and Lobato, R. D. (1979). Intraventricular pressure and CSF dynamics in chronic adult hydrocephalus. *Surg. Neurol.* **12**, 287–295.

Larsson, A., Wikkelsö, C., Bilting, M. *et al.* (1991). Clinical parameters in 74 consecutive patients shunt operated for normal pressure hydrocephalus. *Acta Neurol. Scand.* **84**, 475–482.

Malm, J., Kristensen, B., Karlsson, T. *et al.* (1995). The predictive value of cerebrospinal fluid dynamic tests in patients with the idiopathic adult hydrocephalus syndrome. *Arch. Neurol.* 52, 783–789.

Marmarou, A., Foda, M. A., Bandoh, K. *et al.* (1996). Posttraumatic ventriculomegaly: hydrocephalus or atrophy? A new approach for diagnosis using CSF dynamics. *J. Neurosurg.* 85, 1026–1035.

McCullough, D. C., Harbert, J. C., Di Chiro, G. *et al.* (1970). Prognostic criteria for cerebrospinal fluid shunting from isotope cisternography in communicating hydrocephalus. *Neurology* 20, 594–598.

Meier, U., Kunzel, B., Zeilinger, F. S. *et al.* (1997). Zur Diagnostik des Normaldruckhydrocephalus. Ein Berechnungsmodell zur Ermittlung der ICP-abhängigen Resistance und Compliance. *Nervenarzt* 68, 496–502.

Merten, T. (1999). Neuropsychology of normal pressure hydrocephalus. *Nervenarzt* 70, 496–503.

Meyer, J. S., Tachibana, H., Hardenberg, J. P. *et al.* (1984). Normal pressure hydrocephalus: influences on cerebral hemodynamic and cerebrospinal fluid pressure—chemical autoregulation. *Surg. Neurol.* 21, 195–203.

Mitchell, P., and Mathew, B. (1999). Third ventriculostomy in normal pressure hydrocephalus. *Br. J. Neurosurg.* 13, 382–385.

Naidich, T. P., Altman, N. R., and Gonzalez-Arias, S. M. (1993). Phase contrast cine magnetic resonance imaging: normal cerebrospinal fluid oscillation and applications to hydrocephalus. *Neurosurg. Clin. North. Am.* 4, 677–705.

Penar, P. L., Lakin, W. D., and Yu, J. (1995). Normal pressure hydrocephalus: an analysis of aetiology and response to shunting based on mathematical modeling. *Neurol. Res.* 17, 83–88.

Petzinger, G., Perez, E., and Fahn, S. (1994). Motor features of normal pressure hydrocephalus. *Mov. Disord.* 9, 126.

Raftopoulos, C., Chaskis, C., Delecluse, F. *et al.* (1992). Morphological quantitative analysis of intracranial pressure waves in normal pressure hydrocephalus. *Neurol. Res.* 14, 389–396.

Raftopoulos, C., Deleval, J., Chaskis, C. *et al.* (1994). Cognitive recovery in idiopathic normal pressure hydrocephalus: a prospective study. *Neurosurgery* 35, 397–405.

Raftopoulos, C., Massager, N., Baleriaux, D. *et al.* (1996). Prospective analysis by computed tomography and long-term outcome of 23 adult patients with chronic idiopathic hydrocephalus. *Neurosurgery* 38, 51–59.

Reinprecht, A., Czech, T., and Dietrich, W. (1995). Clinical experience with a new pressure-adjustable shunt valve. *Acta Neurochir.* 134, 119–124.

Sahuquillo, J., Rubio, E., Codina, A. *et al.* (1991). Reappraisal of the intracranial pressure and cerebrospinal fluid dynamics in patients with the so-called "normal pressure hydrocephalus" syndrome. *Acta Neurochir.* 112, 50–61.

Sprung, C., Miethke, C., Shaken, K. *et al.* (1997). The importance of the dual-switch valve for the treatment of adult normotensive or hypertensive hydrocephalus. *Eur. J. Pediatr. Surg.* 7 **Suppl 1**, 38–40.

Stein, S. C., and Langfitt, T. W. (1974). Normal-pressure hydrocephalus: predicting the results of cerebrospinal fluid shunting. *J. Neurosurg.* 41, 463–474.

Stolze, H., Kuhtz-Buschbeck, J. P., Drücke, H. *et al.* (2000). Gait analysis in idiopathic normal pressure hydrocephalus—which parameters respond to the CSF tap test? *Clin. Neurophysiol.* 111, 1678–1686.

Sudarsky, L., and Ronthal, M. (1983). Gait disorders among elderly patients. A survey study of 50 patients. *Arch. Neurol.* 40, 740–743.

Symon, L., and Dorsch, N. W. C. (1975). Use of long-term intracranial pressure measurement to assess hydrocephalic patients prior to shunt surgery. *J. Neurosurg.* 42, 258–273.

Symon, L., and Hinzpeter, T. (1977). The enigma of normal pressure hydrocephalus: tests to select patients for surgery and to predict shunt function. *Clin. Neurosurg.* 24, 285–315.

Tamaki, N., Shirakuni, T., Ehara, K. *et al.* (1990). Characterization of periventricular edema in normal-pressure hydrocephalus by measurement of water proton relaxation times. *J. Neurosurg.* 73, 864–870.

Tanaka, A., Kimura, M., Nakayama, Y. *et al.* (1997). Cerebral blood flow and autoregulation in normal pressure hydrocephalus. *Neurosurgery* 40, 1161–1167.

Tedeschi, E., Hasselbalch, S. G., Waldemar, G. *et al.* (1995). Heterogeneous cerebral glucose metabolism in normal pressure hydrocephalus. *J. Neurol. Neurosurg. Psychiatry* 59, 608–615.

Trenkwalder, C., Schwarz, J., Gebhard, J. *et al.* (1995). Starnberg trial on epidemiology of parkinsonism and hypertension in the elderly. Prevalence of Parkinson's disease and related disorders assessed by a door-to-door survey of inhabitants older than 65 years. *Arch. Neurol.* 52, 1017–1022.

Tullberg, M., Mansson, J. E., Fredman, P. *et al.* (2000). CSF sulfatide distinguishes between normal pressure hydrocephalus and subcortical arteriosclerotic encephalopathy. *J. Neurol. Neurosurg. Psychiatry* 69, 74–81.

Tullberg, M., Rosengren, L., Blomsterwall, E. *et al.* (1998). CSF neurofilament and glial fibrillary acidic protein in normal pressure hydrocephalus. *Neurology* 50, 1122–1127.

Vanneste, J. A. L. (1994). Three decades of normal pressure hydrocephalus: are we wiser now? *J. Neurol. Neurosurg. Psychiatry* 57, 1021–1025.

Vanneste, J., Augustijn, P., Davies, G. A. *et al.* (1992a). Normal-pressure hydrocephalus. Is cisternography still useful in selecting patients for a shunt? *Arch. Neurol.* 49, 366–370.

Vanneste, J., Augustijn, P., Dirven, C. *et al.* (1992b). Shunting normal-pressure hydrocephalus: do the benefits outweigh the risks? *Neurology* 42, 54–59.

Vassilouthis, J. (1984). The syndrome of normal-pressure hydrocephalus. *J. Neurosurg.* 61, 501–509.

Vorstrup, S., Christensen, J., Gjerris, F. *et al.* (1987). Cerebral blood flow in patients with normal-pressure hydrocephalus before and after shunting. *J. Neurosurg.* 66, 379–387.

Wikkelsö, C., Andersson, H., Blomstrand, C. *et al.* (1986). Normal pressure hydrocephalus: predictive value of the cerebrospinal fluid tap-test. *Acta Neurol. Scand.* 73, 566–573.

Yasargil, M. G., Yonekawa, Y., Zumstein, B. *et al.* (1973). Hydrocephalus following spontaneous subarachnoid hemorrhage. Clinical features and treatment. *J. Neurosurg.* 39, 474–479.

Yokota, A., Matsoka, S., Ishikawa, T. *et al.* (1989). Overnight recordings of intracranial pressure and electroencephalography in neurosurgical patients. Part II: changes in intracranial pressure during sleep. *Sangyo Ika Daigaku Zasshi* 11, 383–391.

Zaaroor, M., Bleich, N., Chistyakov, A. *et al.* (1997). Motor evoked potentials in the preoperative and postoperative assessment of normal pressure hydrocephalus. *J. Neurol. Neurosurg. Psychiatry* 62, 517–521.

Zemack, G., and Romner, B. (1997). 5 years of clinical experience with the programmable Medos valve, 428 implanted valves in 388 patients. *Neurosurgery* 41, 737.

*Movement Disorders*

CHAPTER 78
# Dyskinesias

Helge Topka, J. Jankovic, and Johannes Dichgans

The disorders described in this chapter include various forms of involuntary activation of muscles from tonic dystonia to more phasic involuntary movements like tardive dyskinesias, ballism, or tics. Although these movement disorders are generally considered of central origin because of similarities in clinical features and therapeutical approach, hemifacial spasm, a peripherally induced movement disorder, will be included in this chapter. Tremors and the different types of choreic movement disorders are discussed in other chapters (86, 79).

## DYSTONIAS

The concept of dystonia was introduced by Oppenheim in 1911 (dystonia musculorum deformans). *Dystonia* is defined as a syndrome of sustained and patterned muscle contractions frequently causing twisting and repetitive movements or abnormal postures. Brief muscle contractions may additionally lead to jerklike (myoclonic dystonia) or abnormal postures such as torticollis. Different forms of dystonia are classified depending on the age at onset (see Table I), their cause (see Table II), and the distribution of the body parts involved (see Table III). Recent advances in the understanding of the genetics of dystonia have led to modifications of previous schemes used to classify the different causes of dystonia. As for a number of dystonic syndromes abnormal genes have been identified, it is now recommended to use the term *primary* for idiopathic (familial and sporadic) and *secondary* instead of symptomatic dystonias (Table III).

According to this classification, the term primary dystonia refers to syndromes in which signs of dystonia represent the only clinical manifestation of the disorder except for tremor that frequently accompanies dystonias (Münchau *et al.*, 2001). The most common form of childhood-onset primary dystonia is DYT1, an autosomal dominant disorder whose clinical features had first been described by Oppenheim (1911) and in which an abnormal mutation is located on chromosome 9q34.1. Primary forms of dystonia are also characterized by the fact that they are not associated with neurodegeneration (i.e., progressive loss of central nervous system (CNS) neurons). Secondary dystonias are defined as a group of syndromes in which dystonia is often accompanied by other neurological deficits indicating a specific cause for the dystonia, such as lesions involving the basal ganglia, certain drugs or toxins, metabolic causes, or heredito-degenerative disorders. These secondary dystonias are often referred to as *dystonia-plus syndromes* because in addition to dystonic symptoms concurrent parkinsonian signs, myoclonus, ataxia, and other neurological deficits are present. Examples of secondary dystonias (dystonia-plus syndromes) include DYT3, an X-linked disorder, and DYT5, an autosomal dominant dopa-responsive dystonia, both of which are associated with parkinsonism (Fahn *et al.*, 1998; Klein *et al.*, 1999).

Other clinical classification schemes such as classification by age at onset are also useful. In general, onset during childhood or adolescence and dystonia that initially affects the lower limbs is suggestive of a progressive disease that tends to generalize and most frequently is hereditary (DYT1). In contrast, adult-onset dystonia usually is characterized by focal involvement of neck or upper extremity muscles, does not show significant or very little progression, and in most cases is sporadic.

Despite recent advances in understanding the genetics and the molecular pathology of dystonia, treatment in most cases is still symptomatic at this point. Because treatment options of dystonia are the main focus of this manual, this particular chapter is organized to reflect the specifics of symptomatic treatment rather than dealing extensively with the molecular or genetic basis of the various types of dystonia. From a treatment point of

**TABLE I**   Classification of Dystonia According to Age of Onset

Childhood (1–12 y)
Adolescence (13–20 y)
Adults (>20 y)

view, it is most relevant to differentiate between generalized and focal forms of dystonia and to identify secondary and some of the dystonia-plus syndromes.

## Generalized Dystonia

### Clinical Aspects

Primary generalized dystonias show no consistent structural abnormality at autopsy nor has a metabolic defect common to all generalized dystonias been identified. Nevertheless, it is very likely that many or even all forms of primary dystonia are due to some neurochemical or neurophysiological abnormality. Even though genetic abnormalities have been found in a large proportion of primary dystonia (see Table III), its clinical presentation may be variable with early-onset generalized dystonia in some patients, whereas other family members have focal dystonia (Opal et al., 2001). From a genetic point of view, low penetrance and variable expressivity are hallmarks of primary hereditary dystonia. These features have also prompted the notion that additional environmental factors and/or modifier genes may play a role in the pathogenesis of these disorders (Opal et al., 2001).

At this point, it is not known whether current models of dystonia pathogenesis and pathophysiology are valid for all types of dystonia, sporadic or hereditary, primary or secondary. Most physiological studies have been performed in patients with what seemed to be sporadic and focal dystonia. Several neurophysiological studies

suggest defective control of interneuronal networks in the brainstem, probably caused by altered descending control from the basal ganglia. PET studies of dopa uptake in the striatum in patients with dystonia have produced conflicting results so far, negating a primary deficit in dopaminergic function (Spinella and Sheridan, 1994). Patients with torsion dystonia show overactivity of frontal cortical areas during hand movements, whereas activation of executive motor cortical areas is reduced compared with healthy subjects (Ceballos-Baumann et al., 1995); however, the significance of this observation has to yet to be clarified. Recent studies suggest that dystonia may be associated with reorganization of somatosensory and perhaps also motor cortical areas both in animal experiments (Byl et al., 1996) and in patients with writer's cramp (Bara-Jimenez et al., 1998; Elbert et al., 1998). Nevertheless, several lines of evidence seem to suggest that changes of cortical organization represent an epiphenomenon rather than a primary cause of dystonia. Intraoperative recordings in patients with focal dystonia demonstrated reorganization also in structures upstream of somatosensory and motor cortical areas such as the thalamus (Lenz et al., 1999). Along these lines, earlier studies indicated that involvement of the basal ganglia, in particular of the putamen and thalamus, is crucial, because these structures are involved in some forms of secondary dystonia (Lee and Marsden, 1994; Marsden et al., 1985). Further evidence implicating the basal ganglia in dystonia is provided by acute dystonic reactions to neuroleptic drugs, which bind to receptors in the striatum. In addition, drug-related dystonia is a frequent complication of levodopa treatment of Parkinson's disease (Kidron and Melamed, 1987). Indirect support for the notion that the basal ganglia may play an important role in the pathogenesis of dystonia stems from recent molecular genetics studies, demonstrating that the main neurochemical defect in dopa-responsive dystonia (DYT5),

**TABLE II**   Classification of Dystonia According to Its Topology

| Focal | Affecting a single body part | Eyelids (blepharospasm) |
| | | Mouth (oromandibular dystonia) |
| | | Larynx (spasmodic dysphonia) |
| | | Neck (torticollis) |
| | | Arm/hand (writer's cramp) |
| Segmental | Contiguous parts of the body | Cranial, at least two parts of cranial and neck musculature |
| | | Axial, neck, and trunk |
| | | Brachial, one arm and axial or both arms |
| | | Crural, one leg and trunk or both legs (with or without trunk) |
| Generalized | At least one leg plus another segment of the body | |
| Hemidystonia | Ipsilateral arm and leg (generally secondary) | |

**TABLE III**  Etiologic Classification of Dystonia

**Primary dystonia**
*Sporadic*
Idiopathic torsion dystonia (adult or childhood-onset, generalized, segmental, focal)
*Hereditary*
Autosomal-dominant
Classic "Oppenheims" dystonia (DYT1; chr. 9q34.1)
Adolescent-onset, mostly segmental (DYT6; chr. 8p21–8p22)
Adult-onset torticollis (DYT7; in some families chr. 18p)
"Non-DYT1" (DYT4; dysphonia, variable)
Autosomal-recessive
Early-onset (DYT2, segmental or generalized)
Dystonia-plus
*Sporadic*
Paroxysmal dystonia
Myoclonus-dystonia
Dystonia-parkinsonism
*Hereditary*
Autosomal-dominant
Paroxysmal dystonic choreoathetosis (attacks of dystonia/choreoathetosis, precipitation by stress, alcohol, fatigue; DYT8; chr. 2q33–q25)
Paroxysmal choreoathetosis with ataxia (attacks of dystonia/choreoathetosis, double vision, paresthesias, precipitation by exercise, alcohol, fatigue; DYT9; 1p21–p13.3)
Paroxysmal kinesigenic dystonia/choreoathetosis (attacks of dystonia/choreoathetosis, precipitation by sudden movement; DYT10)
Myoclonus-dystonia (responsive to alcohol; DYT11; chr. 11q23)
Dystonia-parkinsonism (acute or subacute onset, generalized dystonia; DYT12; chr. 19q)
Dopa-responsive dystonia (Segawa syndrome, dystonia with parkinsonian signs, marked diurnal variation, responsive to L-dopa treatment; DYT5a; chr. 14q22.1–q22.2)
Autosomal-recessive
Dopa-responsive dystonia (dystonia with parkinsonian signs, DYT5b; chr. 11p15.5)
X-linked recessive
X-linked dystonia-parkinsonism ("lubag," segmental or generalized; parkinsonian signs in 50% of patients, DYT3; chr. Xq13.1)

**Secondary dystonia**
Associated with neurodegenerative disorders
*Sporadic*
Parkinson's disease
Progressive supranuclear palsy
Multiple system atrophy
*Hereditary*
Wilson's disease
Huntington' disease
Hallervorden-Spatz' disease
Progressive pallidal degeneration
Ataxia telangiectasia
Neuroacanthocytosis (action-induced orofacial dystonia)

Rett's syndrome
Intraneuronal inclusion disease
Infantile bilateral striatal necrosis
Familial basal ganglia calcification
Spinocerebellar degeneration
Associated with metabolic disorders
Amino acid disorders
Homocystenuria
Glutaric acidemia
Methylmalonic acidemia
Hartnup's disease
Tyrosinosis
Lipid disorders
Metachromatic leukodystrophy
GM-1 gangliosidosis
GM-2 gangliosidosis
Hexosaminidase A and B deficiency
Ceroid lipofuscinosis
Dystonic lipidosis ("sea blue" histiocytosis)
Miscellaneous metabolic disorders
Mitochondrial encephalopathies (Leigh's disease, Leber's disease)
Lesch-Nyhan syndrome
Triosephosphate isomerase deficiency
Vitamin E deficiency
Caused by a known specific cause
Perinatal cerebral injury and kernicterus (athetoid cerebral palsy, delayed-onset dystonia)
Infection (viral encephalitis, encephalitis lethargica, Reye's syndrome, subacute sclerosing panencephalitis, AIDS, Creutzfeldt-Jakob disease; tuberculosis, syphilis)
Toxins (MN, CO, CS2, methanol, disulfiram, wasp sting)
Drugs (L-dopa, bromocriptine, neuroleptics, metoclopramid, fenfluramin, flecainid, anticonvulsants, Ca channel blockers, ergotamine)
Paraneoplastic syndromes
Focal CNS lesions (arteriovenos malformation, multiple sclerosis, central pontine myelinolysis, brain tumor, traumatic or vascular injury, thalamotomy, syringomyelia)
Focal PNS lesions (traumatic injury, sympathetic maintained pain syndromes)

**Psychogenic**

**Pseudodystonia (frequently torticollis)**
Atlanto-axial subluxation
Arnold-Chiari malformation
Focal seizures
Trochlear nerve palsy
Vestibular torticollis
Posterior fossa mass
Soft tissue mass
Congenital Klippel-Feil syndrome
Sandiffer's syndrome
Congenital muscular lesions
Stiff person syndrome

Adapted from Jankovic and Fahn (1993), Fahn *et al.* (1998), and Müller *et al.* (1998).

now categorized as dystonia-plus syndrome, is a marked reduction in dopamine synthesis affecting dopaminergic neurons in the substantia nigra and the striatum (Ichinose and Nagatsu, 1999).

Secondary dystonias (Table III) generally have a poor prognosis, unless the disease is caused by an identified metabolic defect that can be treated. For that reason, Wilson's disease has to be excluded in young patients with dystonia. As opposed to primary dystonia, secondary dystonias generally are seen clinically with additional nondystonic features.

### Natural Course

According to an epidemiological study from Rochester, Minnesota (Nutt *et al.*, 1988), the prevalence of generalized dystonia was estimated at 3.4/100,000 with an incidence of 0.2/100,000 per year. When all forms of dystonia were combined, the total prevalence was estimated to reach 39/100,000, approaching three quarters of the prevalence of multiple sclerosis, outnumbering by far other well-known neurological disorders, such as Huntington's disease. Of all patients with idiopathic generalized dystonia, 60% experienced disease onset between age 6 and 10. In 50% of those patients, dystonia progressed from focal, sometimes only occupationally induced dystonia, to generalized dystonia within 1 year (Inzelberg *et al.*, 1988). If dystonia starts before the age of 8 years, it usually progresses rapidly and becomes generalized, whereas late-onset dystonia generally has a better prognosis.

About 90% of the patients with generalized dystonia and childhood onset seem to be hereditary according to a survey from the United Kingdom (Fletcher *et al.*, 1990). Several linkage studies demonstrated an autosomal-dominant inheritance with a penetrance of 30–40% and variable expression. A heterogeneous genome could be localized on chromosome 9 q32–34 (called DYT1) in a non-Jewish kindred and in Jews originating from eastern Europe, who seem to be affected 10 times more often (Bressman *et al.*, 1994a; Warner *et al.*, 1993). DNA diagnostic testing detects GAG deletion on chromosome 9 (DYT1). Classic hereditary "Oppenheim's" dystonia (DYT1) usually starts during childhood, affecting the lower extremities and progressively generalizing. The presentation and manifestation, however, varies from patient to patient and even within families. Although some members of the family typically have foot dystonia or writer's cramp, some may evolve more rapidly and may even be complicated by a dystonic storm (Opal *et al.*, 2001). Genetic testing for DYT1 is recommended for patients with progressive torsion dystonia with onset before age 26. Testing may also be worthwhile after the age of 26 in patients having an affected relative with early-onset dystonia (Bressman *et al.*, 2000). Possible progression from focal onset to segmental or generalized dystonia and increased probability of dystonia in relatives of patients with cervical dystonia have suggested a common mode of inheritance also in types of dystonia in which no genetic abnormalities have yet been identified. At this point, linkage analyses have localized chromosomal abnormalities to chromosome 8p21–8p22 in primary adult-onset dystonia DYT6, to chromosome 18p in DYT7, and to chromosome 1p36.13–36.22 (Valente *et al.*, 2001); however, routine tests in individual patients are not yet available.

With respect to treatment options, it is particularly relevant to identify one type of dystonia that is categorized as dystonia-plus syndrome. Less than 10% of patients with childhood-onset generalized dystonia have a disorder, first described by Segawa (Segawa *et al.*, 1976), that is now called dopa-responsive dystonia (DRD; DYT5). In addition to dystonia, gait is affected, and parkinsonian bradykinesia and rigidity may develop. Diurnal variation of symptoms with worsening during the day is common, but not mandatory. As opposed to other forms of dystonia, dramatic improvement with levodopa therapy in a large proportion but not in all patients is the most characteristic feature of this disorder (Nutt and Nygaard, 2001; Steinberger *et al.*, 2000). Most cases maintain striking improvement while on continued levodopa therapy with doses between 50 and 1000 mg/day (Nygaard *et al.*, 1991). The DRD gene has been localized to chromosome 14 (Nygaard *et al.*, 1993). Routine diagnostic testing is difficult because of multiple mutations identified in the GTP-cyclohydrolase gene. Dystonia may accompany parkinsonism in patients with Parkinson's disease (Jankovic and Tintner, 2001), and this combination is particularly common in patients with mutations in the parkin gene on chromosome 6 (Tassin *et al.*, 2000).

### Principles of Therapy

Treatment of dystonia in most cases of primary or secondary cases is symptomatic. In generalized dystonia, drug treatment is recommended, whereas in most cases of focal dystonia, irrespective of any genetic background, local injections of botulinum toxin represent the treatment of choice***.

Animal models of dystonia so far have contributed only little to the understanding of human idiopathic dystonia or to pharmacological therapy of dystonia. Drugs affecting mainly dopaminergic neurotransmission within basal ganglia systems have been shown to be modestly effective in some patients with generalized dystonia (Jankovic and Fahn, 1998; Jankovic, 1998). Because of the possibility of DRD even in patients with an unusual clinical appearance for DRD (Steinberger *et al.*, 1999) and because of negligible toxicity, all therapeutic trials in childhood or young adult–onset generalized dystonia should start with levodopa therapy in a gradually increasing dosage (Table IV). If doses of 100–125 mg levodopa with decarboxylase-inhibitor three times a day do not show an obvious improvement, the patient will probably not benefit, but occasionally higher doses may be required. If a patient desires pregnancy, the decarboxylase-inhibitor carbidopa is pre-

**TABLE IV** Drug Treatment of Dystonia (in Succession of Recommendation)

| Drug | Initial dose per day | Increase | Maximal dose per day |
|---|---|---|---|
| L-dopa (+decarboxylase inhibitor) | 150 mg | 150 mg/wk | 300 mg tid |
| Trihexyphenidyl | 2 mg | 1–2 mg/wk | 15(–25) mg tid |
| Baclofen | 15 mg | 15 mg/wk | 3 × 40 mg |
| Carbamazepine | 200 mg | 400 mg/wk | 400 mg tid |
| Clonazepam | 1 mg | 1 mg/2–5 days | 4 mg tid |
| Tetrabenazine | 25 mg | 25 mg/3 days | 25–50 mg tid |
| Pimozide | 1 mg | 1–2 mg/wk | 6(–16) mg |
| Haloperidol | 1 mg | 0.5 mg/wk | 3 mg tid |
| Trihexiphenidyl |  |  | 6–30 mg |
| + Pimozide |  |  | 6–25 mg |
| + Tetrabenazine |  |  | 75 mg |

ferred to benserazide, because growth of fetal bones may be affected by benserazide, although uncomplicated pregnancies have been reported on medication with both drugs. Also, dopamine agonists (bromocriptine, pergolide, pramipexole, ropinirole, cabergoline) and anticholinergcs (see later) are effective in the treatment of DRD* (Nygaard et al., 1991).

If dopaminergic therapy fails, as it does in most childhood-onset dystonia, anticholinergics should be tried next. In a double-blind, prospective trial of high-dose trihexyphenidyl in 31 patients, 71% had a clinically significant response, although patients rarely reported dramatic improvement. After a mean follow-up of 2.4 years, 42% of patients continued to show a considerable or dramatic benefit from 30 mg, which was generally well tolerated** (Burke et al., 1986). To prevent potentially troublesome side effects of anticholinergic medication, the dosage should be increased slowly, until patients show enough benefit or experience intolerable side effects. Subjective complaints by patients treated with anticholinergic drugs include loss of concentration, memory disturbances, blurring of vision, and urinary retention. Memory encoding and speed of information processing have been shown to be affected after 2–4 months of high-dose trihexyphenidyl (15–74 mg), with older patients being more susceptible. However, after withdrawal, mental functions returned to baseline levels (Taylor et al., 1991). If central side effects become intolerable, dosage has to be reduced. Against peripherally induced side effects such as dry mouth or blurred vision, cholinesterase inhibitors like pyridostigmine or pilocarpine (3%) eyedrops can be tried.

If anticholinergics prove inefficient, other medication such as baclofen, benzodiazepines like diazepam or clonazepam, pimozide, and tetrabenazine may be tried, although rates of success are somewhat less. For otherwise intractable patients, Marsden (1984) proposed a combination of anticholinergics, tetrabenazine, and baclofen or pimozide*. Intrathecal baclofen infusions

may be helpful in some patients with disabling generalized dystonia (Greene, 1992). However, long-term experience has been rather disappointing. Of 14 patients treated with intrathecal baclofen by infusion pump for up to 6 years, only two patients showed clear clinical benefit (Walker et al., 2000).

Hemidystonias are generally symptomatic and may respond to thalamotomy (stereotaxic lesions in the thalamic nucleus ventralis lateralis) (Pettigrew and Jankovic, 1985). Dysarthria occurs in 6% of patients undergoing unilateral thalamotomy. That rate increases to 18% for bilateral surgery in generalized dystonia, with a mortality reaching 2% (Tasker et al., 1988). For patients with generalized dystonia and with little or no improvement on drug treatment, pallidotomy or implantation of pallidal stimulation electrodes may be considered (Kulisevsky et al., 2000; Lin et al., 1999; Tronnier and Fogel, 2000; Yoshor et al., 2001). For a more detailed discussion of surgical options in dystonia see Chapter 76.

*Paroxysmal kinesiogenic choreoathetosis/dystonia (PKC)* is characterized by brief and frequent dyskinetic attacks that most frequently are provoked by sudden movement. PKC is more common in men. Most cases are idiopathic, and careful taking of the family history frequently is suggestive of a familial disorder. The cause and pathophysiology of paroxysmal kinesiogenic choreoathetosis are unknown; an ion-channel disorder is discussed. In favor of this hypothesis is also the fact that treatment with antiepileptic drugs, in particular, carbamazepine, results in dramatic improvement in most cases* (Houser et al., 1999).

*Paroxysmal nonkinesiogenic dystonia (DYT 8)* is a familial disorder in which attacks are precipitated by stress, alcohol, or fatigue. The responsible gene has been located to chr. 2q33–q25 (Fouad et al., 1996). In some patients with paroxysmal nonkinesiogenic dystonia, treatment with acetazolamide may be helpful* (Bressman et al., 1988). Because of the low frequency of the syndrome, no controlled trials have been performed.

## Focal Dystonia

### Clinical Aspects

A variety of different dystonic disorders in which only isolated groups of muscles are affected are categorized as focal dystonias. The cause of focal dystonias is similar to generalized dystonias, and some patients with focal dystonia evolve into segmental or generalized dystonia. As opposed to generalized forms of dystonia, in only up to 25% of patients may a genetic basis of focal dystonia be suspected (Stojanovic et al., 1995). Several reports suggest that at least 10% of patients with focal dystonia and multifocal or segmental dystonia experienced peripheral trauma that precipitated or exacerbated the dystonic syndrome (Defazio et al., 1998; Fletcher et al., 1991; Samii et al., 2000). Onset of dystonia may be induced by basal ganglia dysfunction caused by peripheral injury in dystonia gene carriers (Jankovic, 1994b). The genetic basis of adult-onset focal dystonia has not been identified yet, but it is distinct from the DYT1 locus (Bressman et al., 1994b). Among the candidates are loci on chromosome 8 (DYT6) and chromosome 18 (DYT7).

### Natural Course

The prevalence of focal dystonia in the United States is estimated at 29.5/100,000 (Nutt et al., 1988) and in Japan at 6.1/100,000 (Nakashima et al., 1995). Clinical signs of focal dystonia usually develop within a few weeks with some day-to-day fluctuations with respect to the severity but do show only little or no progress over the years. Spontaneous remissions occur. Typically, signs of focal dystonia disappear spontaneously for days or even years in some 20% of patients. In most of these patients, however, symptoms recur after some time. As opposed to generalized dystonia, focal dystonia usually affects the face or neck muscles or muscles of the upper extremities. Only rarely, focal dystonia manifests itself in the lower extremities. With exception of X-linked recessive dystonia, which usually progresses after focal onset (Lee et al., 1976), secondary generalization from focal onset in the upper half of the body is rare. Some progress with involvement of few adjacent muscles may be observed in about 8% of patients (Marsden and Harrison, 1974). In the future, epidemiological data may provide a more precise picture of focal dystonia, as more knowledge on the genetic basis of focal dystonia accumulates.

### Principles of Therapy

The introduction of botulinum toxin (BoNTx) A for use by experienced physicians has completely changed

**TABLE V** Recommendations for Initiation of BoNTx Therapy in Focal Dystonia and Hemifacial Spasm (as Judged by the Report of the Therapeutic and Technology Assessment Subcommittee of the American Academy of Neurology; 1990)

| | |
|---|---|
| Blepharospasm | Effective, safe, suitable for primary therapy |
| Oromandibular dystonia | Effective and safe |
| Cervical dystonia | Effective, safe, suitable for primary therapy |
| Writer's cramp | Effective and safe |
| Spasmodic dysphonia | |
|    Adductor type | Effective and safe |
|    Abductor type | Experimental |
| Hemifacial spasm | Effective, safe, suitable for primary therapy |

the therapeutic approach to focal dystonias (Jankovic, 1994a). To date, local injections with botulinum toxin A, and more recently also botulinum toxin type B represent the most effective, albeit symptomatic, treatment of focal dystonia (see Table V). Injections of BoNTx serotypes A, B, and in a limited number of subjects also serotype F have been used in humans and allow dystonic muscles to denervate. A high selectivity, relatively long duration of action, lack of severe side effects when the injections are performed by experienced users, and little antigenicity contribute to the high therapeutic value of BoNTx. Double-blind, placebo-controlled studies have proven efficacy of BoNTx therapy for most of the important focal dystonias with serotype A*** and for cervical dystonia with serotype B***.

Although therapeutic use of BoNTx B in a clinical setting has just begun, BoNTx A is still the serotype most widely used. BoNTx A is one of seven serologically distinct neurotoxins produced by the anaerobic bacteria Clostridium botulinum. All major aspects of the cellular mechanisms of action of BoNTx (Blasi et al., 1993; Schiavo et al., 1993) have been identified. BoNTx is a protein with the molecular weight of 150,000 Da, consisting of two chains that are connected by a disulfide bridge. The heavy chain is responsible for rapid highly specific binding at acceptors located at the presynaptic cholinergic nerve terminals. After internalization and cleavage of the disulfide bridge, the light chain acts as a zinc-protease that specifically cleaves certain membrane proteins of the docking complex of synaptic vesicles. Although BoNTx A cleaves SNAP 25 (synaptosome associated protein), types B, D, and F cleave VAMP (vesicle associated protein), also called Synaptobrevin, each of which is required for exocytosis of acetylcholine transmitters (Schiavo et al., 1993). Injections in selected hyperactive muscles (see Figure 1) thus accomplish a graded selective chemical denervation, leading to atrophy of the paralyzed muscle. After the endplate degenerates, new collaterals sprout from

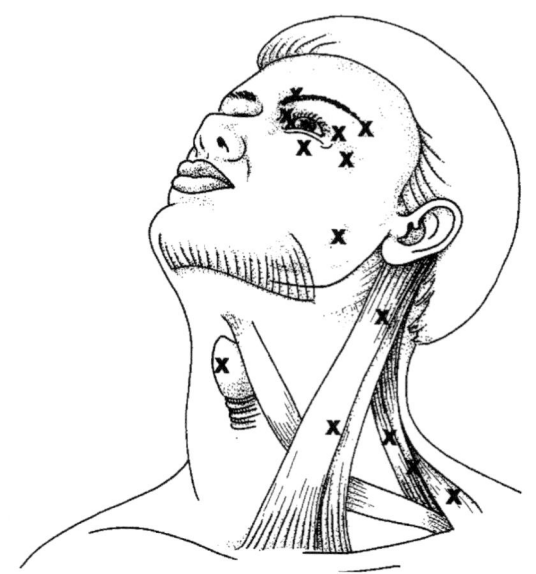

**FIGURE 1** Typical sites for injection of *Botulinum toxin* in the treatment of focal dystonias and hemifacial spasm.

the distal axon and finally connect with the muscle membrane and form new neuromuscular endplates (Alderson *et al.*, 1991; De Paiva *et al.*, 1999). On average, original muscle size and power as determined in the clinical examination will recover within 3–4 months after an injection. Electromyographic recordings from laryngeal muscles, however, have demonstrated subtle changes up to 1 year after injection of BoNTx (Davidson and Ludlow, 1996). Because of the high specificity of binding to acceptors at the neuromuscular endplate, no adverse systemic effects have been observed. Hematogenous spread of BoNTx is suggested by an increased jitter detected within single-fiber EMG of remote muscles, but a distal effect of BoNTx cannot be detected clinically. In addition to improvement of dystonia by inducing partial paralysis of injected muscles, additional mechanisms such as effects on muscle spindle innervation (Rosales *et al.*, 1996) may also play a role in the overall benefit derived from BoNTx treatment. Other mechanisms such as the induction of reorganization within the CNS after toxin injections (Giladi, 1997) remain highly speculative. Short-term side effects of BoNTx treatment are minor and localized. They result from local spread of the toxin, leading to paralysis of neighboring muscles, or from an excessive weakening of the injected muscles. No long-term unacceptable side effects have been observed so far. In patients who had repeated injections with BoNTx for blepharospasm, histological analysis of orbicularis oculi muscle after myectomy showed permanent morphological changes: the number of axon collaterals was increased, new endplates of variable size, and an ele-

vated number of endplates at a single muscle fiber could be detected (Alderson *et al.*, 1991; De Paiva *et al.*, 1999). Muscle biopsy from vastus lateralis muscles in patients who had received multiple injections of BoNTx A for treatment of cervical dystonia reported an increased frequency of angular atrophic type IIB fibers and a reduction in the mean size of IIA and IIB muscle fibers (Ansved *et al.*, 1997) without clinical signs of muscle weakness. Remote changes after botulinum toxin into facial muscles such as an enhanced immunoreactivity of various astroglial markers in motor cortex neurons has been reported in animal experiments (Laskawi *et al.*, 1997); however, at this point it remains unclear whether similar changes in molecular patterns occur in humans and whether these observation are at all relevant for the therapeutic use of the toxin in man.

Several studies have investigated the long-term effects of BoNTx therapy in various types of focal dystonia. Although there are some hints that subjective clinical efficacy of BoNTx A as reported by patients may decrease to a minor degree in patients with spasmodic torticollis (Brashear *et al.*, 2000), this does not seem to be generally the case (Jankovic and Fahn, 1998). Similarly, no changes in the efficacy or long-term side effects were observed in blepharospasm (Ainsworth and Kraft, 1995; Mauriello *et al.*, 1996; Nüssgens and Roggenkämper, 1995), writer's cramp (Karp *et al.*, 1994), or oromandibular dystonia (Tan and Jankovic, 1999b). Analyses of long-term treatment with BoNTx also confirmed the symptomatic nature of this type of therapy; however, in few patients, remission of focal dystonia after one or more BoNTx injections has been reported (Giladi *et al.*, 2000).

To maintain clinical improvement, injections in most cases have to be repeated every 3–6 months. A minimum of 12 weeks between injections has been recommended to avoid formation of antibodies against BoNTx A (Hatheway and Dang, 1994; Jankovic, 2002); however, systematic studies that investigate optimal treatment intervals are still lacking. In most centers, injections for treatment of cervical dystonia are repeated every 3 months. Shorter intervals are considered safe in conditions that require only small dosages such as blepharospasm; however, in rare instances secondary resistance to BoNTx A has been reported also in this situation (Smith and Ford, 2000). Some 3%–5% of patients have been reported to stop responding after several treatment cycles. A mouse assay, which is now marketed commercially by Northview Laboratories in the United States and by Speywood in the United Kingdom, has been used to detect circulating blocking antibodies in up to 18% of patients after repeat injections. These patients would not be expected to respond to subsequent injections with the same serotype but should respond to alternative serotypes (Cullis *et al.*, 1998; Lew *et al.*, 2000; Moyer and Settler, 1994; Tsui *et al.*, 1995). BoNTx B is now commercially available both in the United States and in Europe, and this serotype is particularly useful in patients who have developed blocking antibodies to BoNTxA. Clinical application of BoNTx F may be limited because of reduced and shortened efficacy (Greene and Fahn, 1993) and rates of secondary nonresponders because of antibody formation may be higher than those with the other serotypes (Chen *et al.*, 1998).

In most countries, BoNTx A is currently supplied by American (Allergan, trademark: Botox) and British (Ipsen, trademark: Dysport), and BoNTx B (Elan Pharmaceuticals; trademark: Neurobloc or Myobloc) (see Table VI). Protein concentrations per unit of activity vary between the different preparations, which may influence diffusion of the toxin, as well as the risk of blocking antibodies developing. According to clinical experience, about three (Odergren *et al.*, 1998) to fourfold (Sampaio *et al.*, 1997) more units of Dysport are required to achieve clinical effects similar to those achieved with Botox (Pickett and Hambleton, 1994). Although potency of Dysport has been recently enhanced by improvements in biological availability (Bigalke *et al.*, 2001; Rollnik *et al.*, 2000), more data, in particular randomized double-blind studies comparing the two available BoNTx A preparations, are needed. Because of the differences in potencies and in formulations of the different preparations, clinicians must be aware how the number of units in published investigations corresponds to the amount of toxin in use and what procedures of dilution were used. For practical purposes, we will report currently used dosages of both products, recognizing that as future studies and data become available, these recommendations will have to be adjusted.

If access to BoNTx therapy is not possible, or the patient does not respond to BoNTx, a variety of drugs can be used. Limited objective data are available concerning a differential effect of drugs on the various forms of focal dystonia. Anticholinergics generally

produce the highest rate of response. Other drugs used for generalized dystonia may be also useful in some patients with focal dystonia (Jankovic, 1998; Jankovic and Fahn, 1998) (see Table IV).

### Specific Entities

**Blepharospasm.**    *Clinical Aspects.* Blepharospasm is characterized by intermittent or sustained closure of both eyes due to involuntary contraction of the orbicularis oculi muscles. Symptoms usually start with bilaterally increased blinking. External stimuli, like bright light, driving a car, or experiencing stress, may worsen the symptoms. Some patients experience relief by sensory tricks such as placing a finger on the lateral margin of the orbit or by vocalizing when singing or humming. In principal, the cause and pathophysiology of blepharospasm are thought to be similar to other focal dystonias (Jankovic, 1988). Recent animal experiments suggest that adaptive changes in the excitability of the blink reflex similar to what has been reported in patients may result from the combination of subclinical lesions of the facial nerve and a (subclinical) reduction of dopaminergic neurons in the substantia nigra (Schicatano *et al.*, 1997). Observations in patients seem to provide some support for this hypothesis, indicating maladaptive changes in brainstem reflexes after facial nerve lesions (Baker *et al.*, 1997; Syed *et al.*, 1999). Apraxia of eyelid opening without spasms of the orbicularis oculi muscle has been differentiated from typical blepharospasm (Lepore, 1988). This eyelid motor disorder is analogous to *freezing* or *motor blocks*, and it is more appropriately referred to as *lid freezing*. It may accompany supranuclear palsy, multiple system atrophies, and other forms of atypical parkinsonism (Barclay and Lang, 1997) and is manifested by hyperactivity of the frontalis muscle accompanied by the inability to raise the eyelids without contraction of orbicularis oculi. Occasionally, bilateral ptosis in myasthenia gravis has to be differentiated from blepharospasm.

*Natural Course.* Blepharospasm affects patients most often between 45 and 65 years of age, with women outnumbering men 2 : 1. Many patients start with blepharospasm, but within a few months or years other cranial and later cervical muscles may become involved. In a survey of 264 cases, blepharospasm was the only dystonic feature (Grandas *et al.*, 1988). A genetic disposition is suggested by the fact that 20% of patients report movement disorders (including tremor) in first- and second-degree family members. Temporary partial or complete remissions occurred in 11.4% of patients affected.

*Principles of Therapy.* As in generalized dystonia, various drugs can be tried orally. Double-blind studies

**TABLE VI**    Recommendations of Initial Dosages of BoNTX A in the Treatment of Cervical Dystonia

|  | Dysport® | Botox® | MyoBloc® |
|---|---|---|---|
| Sternocleidomastoideus m. | 2.5–5 ng | 25–50 MU | 1250–2500 |
| Splenius capitis m. | 2.5–5 ng | 25–50 MU | 1250–2500 |
| Trapezius m. | 2.5–5 ng | 25–50 MU | 1250–2500 |
| Semispinalis m. | 1.25–2.5 ng | 10–25 MU | 500–1250 |
| Levator scapulae m. | 1.25–2.5 ng | 25–50 MU | 1250–2500 |
| Scalenii m. | 1.25–2.5 ng | 10–25 MU | 500–250 |

*Note.* For patients with a thin neck, such as women, doses should be reduced by 20–30%.

have been performed with trihexyphenidyl (Nutt *et al.*, 1984), tetrabenazine (Jankovic, 1982), and lisuride (Nutt *et al.*, 1985), each shown to be effective in some patients. However, in most patients, the degree of improvement is much more dramatic after BoNTx treatment. Efficacy of BoNTx injections has been proven in one small double-blind study and a large number of open-label series** (Jankovic, 1994a). Reported benefits vary within a range of 70%–90% of injected patients showing a moderate-to-marked improvement. Latency of improvement generally ranges between 2 and 5 days, with a duration of benefit lasting on the average 3 months. Repeated BoNTx injections have been performed over 10 years without a decrease in efficacy. No relevant permanent clinical side effects have been observed. Temporary side effects like local bruising, dry-eye symptoms, tearing, lid edema, diplopia, or ptosis are dependent on injection techniques, the injected dose, and the experience of the injecting physician. In one series (Elston, 1992), the rate of minor side effects did not exceed 11%; 60% minor side effects and 10% major side effects (including ptosis with covering of the pupil and double vision) were reported in an earlier series (Grandas *et al.*, 1988), but the frequency of these side effects has been substantially lowered as a result of refinements in the injection technique. Patients with apraxia of eyelid opening (eyelid freezing) can also be successfully treated with BoNTx if blepharospasm triggers the eyelid freezing. Although injections into the levator palpebrae are to be avoided, this muscle may be targeted to induce a temporary ptosis to protect the cornea in various eyelid disorders.

*Practical Management.* Because of clinical efficacy and safety in a large number of patients, local injections of BoNTx have become first-line treatment of blepharospasm. If BoNTx treatment is not available or if serious contraindications to BoNTx therapy exist, drug treatment may be offered as an alternative form of therapy. In that case, the anticholinergic trihexyphenidyl may be used in slowly increasing dosage (1–2 mg/wk) up to three times 10 mg/day (as tolerated). The pattern of injections sites has some effect on the rate and severity of side effects (Price *et al.*, 1997) and differs slightly among users. In most cases, BoNTx is injected into the extreme medial and lateral parts of the pretarsal orbicularis oculi muscle of the upper eyelids and into the lateral portions of the lower eyelids. Injections at the medial and lateral edge of the eyebrows may be used in addition to those mentioned in the upper eyelid. To reduce the risk of ptosis, injection into the middle of the upper eyelid where the levator palpebrae muscle is located must be avoided. Also injections at the inner orbital region of the orbicularis oculi muscles are associated with a larger rate of ptosis. We prefer to start with smaller dosages (80 Units Dysport/20 Units Botox

per eye), which can be increased in subsequent visits if the result is not satisfactory. Artificial tears are helpful in some patients to prevent exposure keratitis.

In a small and open-labeled study, local injection of doxorubicin was used to induce chemomyectomy in patients with blepharospasm and hemifacial spasm (Wirtschafter and McLoon, 1998). Initial responses and follow-up studies up to 1 year after the last injection suggest that long-term amelioration of symptoms may be achieved; however, large-scale and double-blind studies, in particular those comparing the effects with botulinum toxin treatment, are presently lacking.

*Treatments No Longer Recommended.* Surgical approaches such as facial nerve sectioning have now been abandoned. Only in rare patients who do not respond to either oral medication or to BoNTx injections, myectomy of the orbicularis oculi muscle and brow lift may be considered.

**Cranial Dystonia.** *Clinical Aspects.* Some patients with blepharospasm also exhibit oromandibular dystonia affecting masticatory, lower facial, or tongue muscles, often presenting as jaw closure or jaw opening dystonia (Jankovic, 1988). The combination of blepharospasm and oromandibular dystonia is categorized as cranial dystonia, previously also referred to as *Meige's syndrome* and *Brueghel's syndrome*. Similar to other forms of dystonia, oromandibular dystonia may be precipitated by trauma such as dental procedures or jaw injury (Sankhla *et al.*, 1998). Occasionally, cranial dystonia may be seen in a patient with Parkinson's disease, and Lewy bodies have been found in some autopsy cases, supporting heterogeneous pathological substrates for this disorder (Mark *et al.*, 1994).

*Natural Course.* Several studies confirmed a considerable female preponderance in cranial dystonia between 4:1 and 6:1. In cranial dystonia, blepharospasm is the initial symptom in 58%, with a mean age at diagnosis of 51 years. Spontaneous remissions are rare and only temporary.

*Practical Management.* According to several retrospective surveys, oral medication with trihexyphenidyl results in meaningful improvements in only one fifth of all patients, and the initial benefit is frequently not sustained*. A large number of other drugs have been tried in the treatment of oromandibular dystonia; however, these drugs have not yielded significant treatment success rates. In patients with *jaw closure dystonia* BoNTx injections into masseter, temporalis, and internal pterygoid muscles result in an improvement in more than 70% of patients, whereas 10% do not respond. EMG guidance is helpful in pterygoid injections. Dosage should be cautiously titrated, starting with 25 U Botox or 100 U Dysport into masseter and comparable muscles and less than half for the smaller muscles. In some patients injections cause transient weakness, impeding jaw closure, and possibly interfere with chewing and swallowing. *Jaw opening dystonia* is more complicated to treat. Injections into the external pterygoid muscles and the digastric or myohyoid muscles (submentalis complex) achieve satisfactory improvement in most patients. Long-term BoNTx treatment remains efficient in these cases (Tan and Jankovic, 1999b). Mild dysphagia may occur in up to a third of the patients treated with BoNTx injections for oromandibular or lingual dystonia, although rarely a liquid diet or tube feeding may be required.

**Cervical Dystonia.** *Clinical Aspects.* Cervical dystonia affects the neck muscles and causes the head to deviate in any direction, either as rotatory torticollis or laterocollis, retrocollis, or anterocollis. Initially, phasic or tremulous movements of the head eventually lead to fixed abnormal head posture. Pain in the contracting muscle is a common symptom, occurring in more than two thirds of patients. Involved muscles tend to develop hypertrophy. The severity of symptoms varies among patients, and some are unable to perform work or participate in activities of daily living. Sensory stimuli (geste antagonistique), such as lightly touching the lateral chin, can improve control of head posture. Electromyography shows cocontraction of various combinations of agonist and antagonist muscles. In most patients superimposed tremulous activity in addition to dystonic features is present, providing a possible link to essential tremor (Münchau, 2001; Pal *et al.*, 2000). Although rare symptomatic cases may show gross morphological abnormalities in the basal ganglia, no specific changes were found in pathologically studied idiopathic cases.

*Natural Course.* Most surveys of cervical dystonia found a female preponderance approaching 2:1. Onset of disease occurs in 70% of patients between ages 30 and 60, with different studies finding a median onset around 42 years (Jankovic *et al.*, 1991). Between 11% and 20% of patients experience a spontaneous remission, usually during the first 5 years of the disease. Remissions may last for years but are only infrequently permanent. One third of patients with cervical dystonia progress to dystonia of the face, upper limbs, or truncal muscles, and 23% display postural hand tremor resembling essential tremor (Chan *et al.*, 1991; Jankovic *et al.*, 1991). One third of untreated patients have been shown to have some abnormalities of swallowing. Some 10% of patients reported a head trauma within 3 months of the onset of dystonia. This supports the hypothesis that in addition to genetic factors, environmental factors, in particular minor changes of peripheral input, may help to trigger focal dystonias in subjects susceptible to the disease (Defazio *et al.*, 1998).

*Principles of Therapy.* A positive effect from oral drug treatment can be expected from anticholinergics (e.g., trihexyphenidyl) in up to 40% of patients according to a retrospective study by Greene* (1988). Drug trials with baclofen, benzodiazepines, or carbamazepine are considerably less efficient (in 11%–20%)*. Haloperidol has been tested in a double-blind trial, but the risk of potentially serious side effects, including tardive dyskinesia, outweighs the therapeutic potential. Retrospectively, haloperidol showed a rate of 46% responders, by far surpassing other dopamine antagonists (20%) or the dopamine-depletor tetrabenazine (11%) (Jankovic and Beach, 1997; Lang, 1988). The potential of using atypical neuroleptics such as clozapine (Burbaud *et al.*, 1998) can not be judged on the basis of the available data. Open-label studies suggest that clozapine may be effective to some degree (Karp *et al.*, 1999); however, its therapeutic value is limited by side effects and the need for hematological surveillance. Lisuride has been shown to improve symptoms statistically significantly in a double-blind drug trial; however, the effect is clinically not sufficient (Nutt *et al.*, 1985). The antiarrhythmic mexiletine, an oral formulation of lidocaine, has been reported to show some benefit in small open-label studies both in patients with focal (Ohara *et al.*, 1998) and generalized dystonia (Lucetti *et al.*, 2000). However, clinical effects are small, and the therapeutic value of the drug is substantially limited by severe gastrointestinal side effects that frequently require dose reduction or stopping of the medication. Although there is some debate as to which are optimal variables to assess clinical efficacy of BoNTx injections in cervical dystonia (Ceballos-Baumann, 2001), there is consensus that BoNTx can be considered a safe and highly effective therapy of cervical dystonia*** (Comella *et al.*, 2000). Double-blind studies not only confirm the superiority of BoNTx A injections compared with placebo (e.g., Blackie and Lees (1990); Brashear *et al.* (1999); Brin *et al.* (1999); but also to drug treatment such as trihexyphenidyl (Brans *et al.*, 1996). During recent years, the widespread availability of BoNTx has altered the management of cervical dystonias. BoNTx treatment rarely is able to normalize cervical dystonia completely but provides dramatic relief in most patients. Selection of injection sites and doses has to be tailored to the individual symptoms. The number of injections and dosages applied to individual muscles is still somewhat open to controversy. Currently, it is recommended to initiate therapy with a total dosage of 500 units Dysport (Poewe *et al.*, 1998). Larger initial doses yield more severe effects and a longer duration, however, are associated with a larger rate of side effects. According to various open and blinded studies, between 70% and 90% of patients experience a substantial improvement in head control (Jankovic and Brin, 1991). Pain relief follows 90% of treatments, with moderate to excellent pain reduction in 66%. Up to 6% of patients show no response, even after repeated injections, whereas up to 16% experienced a failure to respond after one or more treatments (Jankovic and Schwartz, 1991). Patients who are treated early during the course of the disease tend to respond better, probably because no contractures have evolved. The most frequent side effect is mild dysphagia, reported after up to 44% of treatments in a large British series (Anderson *et al.*, 1992) and 17% in a large American series (Jankovic and Schwartz, 1991). This difference might be due to the higher potency of the British product and the use of relatively higher doses when applying the British toxin. However, severe dysphagia requiring change to liquid diet is rare (2%) with both BoNTx preparations. Because women run a higher risk of dysphagia, probably because of their thinner necks, doses per muscle should be reduced by 20%–30%. Bilateral injections into the sternomastoid muscles for anterocollis considerably augment the risk of dysphagia. Localized neck weakness is the second most common complaint in up to 10% of treatments. Other less important adverse effects include general fatigue and malaise. In rare cases, a brachial plexopathy may result because of local trauma or as a result of immunological reaction (Glanzman *et al.*, 1990). A hands-on introduction to injections in cervical dystonia is provided by Anderson (Anderson *et al.*, 1992) and Jankovic (Jankovic, 1994a; Jankovic *et al.*, 1994); however, guided introduction by an experienced user is strongly recommended.

Only in the few nonresponders to BoNTx may surgical approaches be considered. Selective peripheral denervation, that is, dissecting the branches of the accessory nerve to the sternomastoid muscle and of the motor roots of C1–C6, has been reported to benefit up to 88% of patients* (Gauthier *et al.*, 1988) and may be offered to patients who have become nonresponsive after initial success with BoNTx. On the other hand, when considering surgery, one has to keep in mind that the results of surgery may be much less promising in patients who are primary nonresponders to BoNTx (Braun *et al.*, 1995).

*Practical Management.* With several studies available that confirm the safety and efficacy of BoNTx treatment in cervical dystonia, most would agree that BoNTx in most cases should be offered as primary form of therapy. In particular, if anticholinergics in doses up to 15 mg trihexyphenidyl per day have been ineffective, treatment with local injections of BoNTx should be initiated. Appropriate muscles for injections are identified by localization of pain and tenderness, by inspection and palpation for hyperactive muscles, and by the abnormal head posture. Although in most cases, elec-

tromyographic guidance is not required, additional use of electromyography has been advocated to reduce necessary BoNTx A dosages and to optimally identify muscles relevant for injection and therapeutic efficacy (Brans *et al.*, 1995; Dressler, 2000; Finsterer *et al.*, 1997). In complicated cases or for inexperienced neurologists, polymyographic EMG recordings may help to identify involved muscles by verification of hyperactivity or a lack of inhibition. Hollow EMG needles allow injection under EMG control. Multiple injection sites per muscle seem to contain the biological activity in the target muscle and prevent diffusion to adjacent muscles (Borodic *et al.*, 1994). For typical rotatory torticollis (more than 50% of patients) the sternomastoid muscle contralateral and the splenius capitis ipsilateral to the direction of chin rotation generally are injected; occasionally, injection into the trapezius or other neck muscles may be helpful (Table VI). Pure laterocollis can be controlled by injecting the ipsilateral sternomastoid and splenius muscles; in retrocollis both splenius capitis muscles and sometimes semispinalis capitis muscles are involved. Because dysphagia is generally a consequence of bilateral sternomastoid injection, these muscles should be injected using relatively low doses.

In patients who have clinical signs of secondary resistance to BoNTx, a test for detection of antibodies should be performed. Several attempts have been made in the past to restore sensitivity to BoNTx A in secondary nonresponders, including plasmapheresis. However, at this point, no simple regimen exists that allows for safe and reliable restoration of BoNTx A sensitivity. In these patients, injections of botulinum toxin B (Elan Pharmaceuticals, San Francisco, California) may be helpful (Brashear *et al.*, 1999; Brin *et al.*, 1999; Lew *et al.*, 1997, 2000). Several studies have shown that BoNTx B is effective in the treatment of cervical dystonia both in patients responsive to BoNTx A (Brashear *et al.*, 1999) and in secondary nonresponders (Brin *et al.*, 1999). The spectrum of side effects of BoNTx B is similar to type A; however, neurologists performing injections should be aware that the average dosage is much greater when measured in mouse units (MU) compared with the two commercially available preparations of BoNTx A. Total doses ranging between 2500 and 10,000 MU may be necessary to achieve clinical efficacy. In these dosages, treatment effects of BoNTx B usually last 12–16 weeks.

*Treatments No Longer Recommended.* Cervical cord stimulation has been shown to be ineffective in a controlled double-blind study (Goetz *et al.*, 1988). Surgical approaches with cervical rhizotomy, sectioning of the sternomastoid muscle, or thalamotomy have been given up for a selective denervation by sectioning of accessory branches to the sternomastoid and of the upper cervical motor roots (Bertrand and Molina-Negro, 1988). After therapy with BoNTx has become the mainstay of therapy for cervical dystonia, selective denervation should be recommended only for patients who have become secondary nonresponders to BoNTx injections (Braun and Richter, 1994; Krauss *et al.*, 1997). Because the organic origin of dystonia has become accepted, psychotherapy is no longer recommended. Biofeedback therapy may be offered for relaxation of dystonic muscles; however, sustained improvement has rarely been achieved.

**Spasmodic Dysphonia.** *Clinical Aspects.* Spasmodic dysphonia is the result of focal laryngeal dystonia leading to hoarseness, pitch breaks, limited intensity range, and poor intensity control of phonation. Up to 85% of patients (Blitzer and Brin, 1991) experience spasms of the adductor muscles with involuntary hyperadduction of the vocal fold, which leads to a strained, strangled voice quality. Abductor spasms are characterized by a breathy phonation. Voice tremors, abnormal breathing patterns, and other focal dystonias (in 20%–70% of patients) accompany the disorder.

*Natural Course.* Age of patients ranges between 20 and 70 years, with a female preponderance. The average age at onset of the large series described by Blitzer and Brin (1991) was 38 years. Speaking professionals (e.g., preachers, teachers) are overrepresented. Family histories show significant affections of relatives with essential tremor (27%), spasmodic dysphonia (11%), or other related movement disorders like Parkinson's disease and other focal dystonias (Pool *et al.*, 1991).

*Practical Management.* Because oral medication of anticholinergics or sedative drugs has not shown a uniform success, BoNTx therapy of adductor spasms may be offered as primary therapy. Its efficacy and safety have been shown by several large open series (Whurr *et al.*, 1998)* and a double-blind study** (Troung *et al.*, 1991). The injection technique is identical to EMG of the vocal folds. A hollow Teflon-coated EMG needle is directed percutaneously through the cricothyroid membrane into the thyro-arytenoid vocalis muscle complex. The position of the needle is controlled by EMG on phonation. Each side of the laryngeal muscle complex is injected with low doses (2.5 U Botox in 0.1 ml), or one side is injected with a higher dose (15 U Botox). Although both approaches are effective, the unilateral BoNTx injections may provide both superior and longer lasting benefits than bilateral BoNTx injections (Adams *et al.*, 1993). All reports agree that with either technique a dramatic improvement in most patients can be achieved improving speech function from 60% of normal function to an average of 90% of normal function with a duration of benefit from 3–4 months. Quantitative studies analyzing objective and subjective voice

characteristics before and after BoNTx A injections confirm improvement; however, optimal voice after treatment never quite seems to match normal voice (Langeveld *et al.*, 2001). Side effects of injections are frequent but not severe. Transient breathy hypophonia occurs in up to 50% of patients during the first 2 weeks, and up to 25% experience a clinically insignificant aspiration of fluids (Blitzer and Brin, 1991). Patients who had undergone recurrent nerve sectioning without success in earlier years may be treated on the contralateral side with careful adaptation of dose.

Because of the danger of laryngeal stridor and the necessity of temporary tracheostomy, treatment of abductor spasms is still experimental and should be performed only by experienced physicians. In contrast to the failure in most dysphonia patients, oral medication (e.g., with trihexyphenidyl) may yield success in one third of patients with abductor spasms (Blitzer and Brin, 1991) (see also Table V).

*Treatments No Longer Recommended.* Before BoNTx treatments proved so successful, sectioning of the recurrent laryngeal nerve was performed. Because of the hyperfunctioning of the contralateral dystonic vocal muscle, long-term success could be achieved only in about 36% of patients treated surgically (Aronson and DeSanto, 1983).

**Hand Cramps.** *Clinical Aspects.* Writer's cramp and other occupational dystonias like musician's cramp of various locations are triggered only when the patient performs a specific well-learned fine motor task. In writer's cramp after writing a few words the pen is gripped with excessive force, leading to abnormal postures of the fingers and wrist. Kinesiological EMG recordings show that the normal alternating short-duration burst pattern of agonist–antagonist muscles is replaced by co-contraction with overflow in remote muscles. Tension and discomfort in the fingers and forearm muscles are common. About one third of patients with dystonic writer's cramp display postural or writing tremor.

*Natural Course.* In the series of patients described by Marsden and Sheehy (1990), 5% experienced a spontaneous remission, which is more likely during the first 5 years. Relapses are frequent. Most patients manage to use their dominant hand, writing for limited purposes; only 15% have to give up completely. Roughly 5% report similarly affected first-degree relatives, but several families with dystonic writer's cramp have been reported. Contrary to other focal dystonias, the male/female ratio is 2 : 1, with a mean onset around the age of 39.

*Principles of Therapy.* Pharmacological therapy is even less effective than in other focal dystonias with only 10%–20% of patients experiencing some relief from anticholinergics. Writing with a typewriter or a personal computer may help some patients. Ranawaya and Lang (Ranawaya and Lang, 1991) proposed a writing device that helped 75% of patients to improve writing; however, it was considered bothersome by most patients. About one third of patients attempt to learn writing with the nondominant hand, which takes about 6 months. Unfortunately, 25% bone similar writer's cramp develop in the nondominant hand after an indefinite period of time (Sheehy *et al.*, 1988). BoNTx has been used in several small open-label or double-blind studies. Toxin injections were superior to placebo; however, they proved to be less efficient than in other focal dystonias. The muscles involved are more difficult to localize, and the dose of the toxin has to be titrated to weaken the muscles sufficiently without incapacitating them functionally. Also, frequently there is a tendency for the toxin to spread to adjacent muscles (Ross *et al.*, 1997). Several double-blind studies have failed to show improvement in objective measures of pen control or writing, but subjective improvements have been reported repeatedly in about 80%*** (Jankovic and Schwartz, 1993). Long-term success of BoNTx A therapy is somewhat smaller compared with cervical dystonia and blepharospasm (Karp *et al.*, 1994).

## SPONTANEOUS ORAL DYSKINESIAS

### Clinical Aspects

Spontaneous oral dyskinesias (by some authors called *senile dyskinesias*) are characterized by involuntary, irregular movements of oral, lingual, masticatory, and pharyngeal muscles. Grunting vocalizations may be a part of the syndrome. At this point, it is unclear whether spontaneous oral dyskinesias represent a disease entity or rather a minor variant of oromandibular dystonia. Frequent and severe involvement of the tongue and the lack of involvement of upper facial muscles may distinguish spontaneous oral dyskinesias from other facial dyskinesias. The diagnosis implies that there is no history of treatment with neuroleptics, levodopa, or comparable drugs. Contrary to drug-induced dyskinesias, spontaneous dyskinesias rarely affect the limbs. The frequency of this syndrome is high but not exactly known.

### Natural Course

It has been estimated that up to 10% of the elderly older than the age of 60 years are affected (Brion *et al.*, 1988). Most likely, the true incidence is somewhat smaller, because in a significant proportion of patients

careful taking of the drug history reveals that patients had been taking neuroleptics previously (Ticehurst, 1990). Approximately one third of patients improve over time, whereas the same proportion deteriorates. Complete spontaneous remissions have not been observed.

### Principles of Therapy

The cause is unclear; however, a shift toward a dopaminergic dominance in the striatal dopaminergic–cholinergic transmitter systems caused by degenerative processes has been hypothesized. In clinical practice dopamine depletors or dopamine antagonists proved to be most useful. No double-blind, placebo-controlled studies are available.

### Practical Management

*Tetrabenazine* is the treatment most frequently recommended. Starting with a tablet of 25 mg, dosage is increased every 3 days up to a maximum of 100 mg. In many patients 50 mg is sufficient. Effects are usually evident within a week. In some patients parkinsonism may develop after weeks, requiring a reduction of the daily dose. Alternatively, 2–3 mg pimozide may be tried. If individual drug treatment fails, a combination of both drugs might be helpful*.

*Tiapride* is a substituted benzamide derivative with $D_2$-receptor blocking activities*. It possesses good clinical antidyskinetic properties with only minor side effects, such as parkinsonism. Although it is frequently used in clinical practice, no controlled studies have been carried out in spontaneous orofacial dyskinesias. The initial dosage of 300 mg/day may be increased to 600 mg/day in one step when the effect is insufficient. When a decision to stop Tiapride medication has been made, tapering of the drug is recommended, because abrupt withdrawal frequently is associated with severe rebound effects.

## DRUG-INDUCED DYSKINESIA

### Acute Dyskinesias

#### Clinical Aspects

Acute dyskinesias may occur within hours or up to a few days after initiating treatment with classic (typical) neuroleptics such as antipsychotics or antiemetics (e.g., metoclopramide). They consist of intermittent or sustained muscle spasms and abnormal deviation of eyes, face, neck, and throat. Acute dyskinesias are considerably more frequent in children and young adults. In younger patients severe trunk and extremity dystonia is more common. Minor but uncomfortable dystonia tends to be overlooked. Intravenous or intramuscular anticholinergics are highly effective; in minor forms oral anticholinergics may suffice. Diphenhydramine and diazepam are equally effective. Acute dyskinesias disappear after drug withdrawal; however, anticholinergic treatment may be necessary for 1–2 days.

### Practical Management

When symptoms are severe, intravenous injection of anticholinergics (2 mg benztropine or 5 mg biperiden), diazepam (5–10 mg), or diphenhydramine (50 mg) may be necessary to stop the abnormal movement.

## Tardive Dyskinesias

### Clinical Aspects

Tardive dyskinesias (TD) represent a group of iatrogenic movement disorders caused by exposure to neuroleptics, drugs that block dopamine receptors. The most typical manifestation of TD are orofacial and lingual stereotypical movements characterized by repetitive, coordinated movements of the mouth such as pouting, sucking, licking, smacking of the lips, chewing, and facial grimacing. Less frequent are involuntary stereotypic movements of the trunk and extremities (Stacy et al., 1993). Another form of TD is tardive dystonia manifested usually by cranial dystonia, phasic retrocollis, opisthotonic posturing, and arm extension (Kang et al., 1988). Tardive tremor, myoclonus, and other movement disorders have also been reported in patients with TD (Miller and Jankovic, 1992). High-dose treatment with nearly all neuroleptics may induce TD as early as a few days or weeks after initiation of treatment. The role of the atypical neuroleptic clozapine, with a different profile of receptor blockade (relatively high affinity to 5-HT$_2$, cholinergic, histaminergic, and noradrenergic receptors, D4 rather than D1 and D2 receptors) and sulpiride is somewhat controversial. TD has been reported as a result of treatment with clozapine (Bruneau and Stip, 1998; Elliott et al., 2000) and olanzapine (Raja et al., 1999). However, these side effects seem to occur only rarely with the atypical neuroleptics compared with the typical neuroleptics. Risperidone, initially promoted as an atypical antipsychotic drug, has been associated with a relatively large number of tardive syndromes, suggesting a profile more similar to the typical rather than atypical neuroleptics (Krebs and Olie, 1999; Simpson and Lindenmayer,

1997; Tachikawa et al., 2000). As with other antipsychotic drugs, tardive reactions usually are prominent during rapid withdrawal of the drug (Ahmed et al., 1998).

Kane and colleagues (1988) found a cumulative incidence rate of TD to be 18.5% after 4 years and 40% after 8 years of neuroleptic treatment. In the same cohort of patients a duration of dyskinesia exceeding 6 months was observed in 11% and 22%, respectively. Several risk factors for the development of TD and tardive dystonia have been identified. In addition to exposure to neuroleptics, old age, female gender, affective disorder, the presence of other extrapyramidal signs such as drug-induced parkinsonism, intellectual impairment, and diabetes mellitus type II have been suggested to increase the risk of TD (Marsalek, 2000; Muscettola et al., 1999). Tardive dystonia seems to have a different risk profile than typical TD: it is more common in young male patients, particularly those with mood disorders. The Pisa syndrome, a rare dystonic syndrome involving a tonic deviation of the trunk to one side, in the absence of other dystonic symptoms, may be a forme fruste of tardive dystonia. Contrary to acute dyskinesias, onset occurs within 1 week to months of neuroleptic therapy. The Pisa syndrome frequently affects women with organic brain dysfunctions treated in gerontopsychiatric institutions. At this point, no accurate epidemiological data are available, but prevalence rates in such institutions have been estimated as high as 8.3% (Yassa et al., 1991).

### Natural Course

TD tends to first emerge or deteriorate after withdrawal of antipsychotic drugs. This exacerbation is often reversible. In younger outpatient populations, remission rates of 50%–90% have been observed, usually several months to 2 years after drug withdrawal (Kane et al., 1986). The natural history of tardive dystonia differs from that of TD to some extent (Kiriakakis et al., 1998). Symptoms may develop at any time, ranging from 4 days–23 years after initiation of antipsychotic treatment. At outset, the dystonia is focal, typically involving the craniocervical region; however, frequently, there is some progress to segmental or generalized forms if dystonia over months or years. Clinically, tardive dystonia is indistinguishable from idiopathic dystonia. Only a small proportion of patients experiences spontaneous remission over the period of observation; although discontinuation of neuroleptics seems to increase the chances of remission fourfold. Interestingly, in a series of 107 cases (Kiriakakis et al., 1998), a significant negative correlation between the occurrence of remission and treatment with benzodiazepines was observed. In another longitudinal study, TD improved after stopping neuroleptics in most patients over the years; however, the severity of parkinsonism seemed to worsen (Fernandez *et al.*, 2001).

### Principles of Therapy

Despite numerous reports of drug treatments for TD, no reliably effective therapy for TD has been established. Anticholinergics are helpful only to control acute dystonia, but this therapy is not helpful or actually may worsen TD. Prevention and early detection of reversible cases are the most important principles. Early extrapyramidal symptoms seem to correlate with subsequent development of TD. Therefore, drug-induced parkinsonism and akathisia should be managed by reducing or discontinuing the offending drug. Although anticholinergics have not consistently been associated with increased risk for TD, they may mask early warning signs (Tarsy, 1989). If possible, TD should be managed by withdrawing the offending drug as soon as possible. Sulpiride has been found to be effective to some extent in double-blind and placebo-controlled studies (Quinn and Marsden, 1984; Schwartz *et al.*, 1990), whereas the efficacy of tiapride is even less impressive (Auberger *et al.*, 1985). Atypical neuroleptics such as clozapine and risperidone have been found effective in reducing TD in some patients (Casey, 1998; Van Harten *et al.*, 1996). However, in a review of 25 studies, only 26% of participants experienced an improvement of 50% or more by any of the drugs tested (Feltner and Hertzman, 1993). Furthermore, these drugs may eventually cause or exacerbate TD. Tetrabenazine was found useful in at least three quarters of patients in a double-blind study and open-label trials (Jankovic and Beach, 1997; Jankovic and Orman, 1988). The advantage of tetrabenazine over all other neuroleptics is that this drug has never been documented to cause TD. TBZ may cause some dose-related side effects, such as parkinsonism, drowsiness, and depression, but these rarely require complete discontinuation of the drug. Because its application has virtually no side effects, vitamin E has attracted considerable interest. In several double-blind studies it proved effective, if dosages exceeded 1200 IU/day and trials lasted longer than 8 weeks (Adler *et al.*, 1993), although more recent studies have produced equivocal results. The basis for vitamin E is the theoretical possibility that neuroleptics cause oxidant stress.

In contrast to typical TD, tardive dystonia has been reported to respond to anticholinergics about 46% of cases; (Kang *et al.*, 1988). Other open-label reports suggested response to tetrabenazine, clonazepam, or even bromocriptine (Wojcik *et al.*, 1989).

### Practical Management

Treatment of typical TD has to be individualized but the following sequence may serve as a guide: (1) Carefully examine necessity of treatment and withdraw the offending drug as soon as the primary disorder is under control; (2) vitamin E, 1600 IU/day; (3) tetrabenazine starting with half of a 25-mg tablet, increasing dosage gradually up to a maximum of 200 mg. If continuing antipsychotic therapy is required, atypical neuroleptics should be preferred.

### Treatments No Longer Recommended

In controlled studies, valproate (Fisk and York, 1987), diltiazem (Loonen *et al.*, 1992), CDP-choline (Gelenberg *et al.*, 1989), essential fatty acid supplementation (Vaddadi *et al.*, 1989), buspirone (Moss *et al.*, 1993), and L-deprenyl (Goff *et al.*, 1993) among other drugs were ineffective in TD. Ambiguous results have been reported for clonazepam; it might be superior to a placebo in tardive dystonias more than in TD, but effects have been shown to diminish in continuing treatment (Thaker *et al.*, 1990).

## BALLISM

### Clinical Aspects

The term ballism refers to hyperkinetic, large-amplitude, flinging, involuntary movements about proximal joints. Usually, ballistic involuntary movements are restricted to one side of the body (hemiballism). With respect to the clinical appearance, hemiballism and hemichorea are synonyms. From the beginning of the last century, it is recognized that focal lesions involving the subthalamic nucleus may result in hemiballism (Jakob, 1923). Recent observations with deep brain stimulation confirm that inactivation of the subthalamic nucleus is associated with contralateral hemiballism in man (Limousin *et al.*, 1996). Other pertinent studies, however, revealed that the syndrome is not only caused by lesions of the subthalamic nucleus but also may originate from a number of lesions within efferent and afferent projections of the subthalamic nucleus (caudate, putamen, thalamus). Animal experiments confirmed that in addition to a lesion within the subthalamic nucleus itself, lesioning of afferent or efferent pathways connecting the subthalamic nucleus and the pallidum is associated with hemiballism (Martin, 1957). The cause of hemiballism is remarkably heterogeneous (Table VII). In man, ischemia is the most frequent cause, and, therefore, hemiballism usually evolves rather abruptly (Dewey and Jankovic, 1989; Vidakovic *et al.*, 1994).

TABLE VII   Etiology of Ballism/Hemiballism
_____

Vascular (hemorrhage or ischemia)
Arteriovenous malformations
Subarachnoidal hemorrhage
Infectious/parainfektious (toxoplasmosis and AIDS, *Cryptococcus*
    meningitis, tuberculous meningitis)
Drugs (oral contraceptives, estrogen, phenytoin, L-dopa)
Tumors
Hyperglycemia, diabetes mellitus
Traumatic
Multiple sclerosis
Other focal lesions within subthalamic nucleus
_____

## Natural Course

The course of the disorder is highly variable and depends on the cause of hemiballism. No data as to the incidence and prevalence of the condition are available. If hemiballism results from vascular lesions, abrupt onset is typical. In most patients, spontaneous remissions occurs within days or a few months. However, even if no single cause of the condition is identified, spontaneous remissions are thought to occur in some 75% of patients. In particular, lesions outside of the subthalamic nucleus are thought to be associated with a poor prognosis and are postulated to result from deficient adaptive reorganization of CNS pathways (Lang, 1985).

## Principles of Therapy and Practical Management

If the cause of hemiballism is identifiable, treatment should be directed to the underlying condition. Care should be taken to avoid injury because of the uncontrollable and violent involuntary movements. To ameliorate symptoms until spontaneous remission occurs, short-term symptomatic treatment with dopamine antagonists (phenothiazine, butyrophenone, tetrabenazine) may be helpful. Large-scale clinical studies investigating the efficacy of different dopamine antagonists or any other therapeutic interventions are currently lacking. Dopamine antagonistic therapy usually results in meaningful improvement in most patients irrespective of the cause. A small series of patients was treated with promethazine at dosages between 75 and 200 mg, tetrabenazine (50–150 mg) or haloperidol (1–3 mg) daily (Gilbert, 1975; Johnson and Fahn, 1977; Pearce, 1972)*. In rare instances, botulinum toxin injections in proximal limb muscles have been performed (Dressler *et al.*, 2000). In medically intractable patients, functional stereotaxic surgery in the region of the zona incerta and the base of the ventrolateral thalamus of the thalamus may be considered an alternative therapy (Krauss and Mundinger, 1996).

# TICS AND GILLES DE LA TOURETTE SYNDROME

## Clinical Aspects

Tics represent repetitive sudden movements involving different parts of the body. Movements involve either only few muscles (simple motor tics), such as frequent blinking, or may involve larger regions of the body, generating complex movements. Tics may also involve laryngeal muscles and other nasal and oral passages producing vocal (phonic) tics. Characteristically, patients report a sensation of inner stress immediately before the onset of the movement (premonitory sensations) that improves transiently after the execution of the movement. Some patients report that the tic movements are initiated deliberately to achieve the sensation of relief associated with the execution of the movement. Tics may occur temporarily (transient tic disorder), chronically, or as part of Gilles de la Tourette Syndrome (TS) (Table VIII). The cause of tic disorders is very heterogeneous; however, most cases are considered idiopathic (Table IX).

TS, characterized by a combination of various motor and phonic tics, represents the most common tic disorder (Jankovic, 2001). The syndrome often starts in childhood between ages of 2 and 18 (mean, 7) years as a simple tic such as frequent blinking, grimaces, head jerking, shoulder shrugging, and other simple and coordinated movements (Jankovic, 1993). Phonic tics include sniffing, snorting, throat clearing, barking, and repeating one's own words (palilalia), repeating the words of phrases of others (echolalia), and utterances of obscenities (coprolalia). Attention deficit with hyperactivity, obsessive-compulsive disorder, and other behavior comorbidities frequently accompany TS. Several imaging studies have attempted to identify a morphological substrate of TS; however, results are still somewhat controversial. In one MRI study, patients with TS had larger volumes in dorsal prefrontal and in parieto-occipital regions and smaller inferior occipital volumes (Peterson *et al.*, 2001). In contrast, earlier studies reported reduced size of left basal ganglia volumes with a reduction of the normal asymmetry (left greater than right) in basal ganglia size (Peterson *et al.*, 1993). No specific anatomical region has been associated with the development of tic disorders. Similarly, functional imaging studies are not conclusive at this point with respect to the neuroanatomy of the disorder. In one study using single photon emission tomography, significantly lower perfusion in the left lateral temporal area and asymmetrical perfusion in the dorsolateral frontal, lateral, and medial temporal areas in TS patients was reported (Chiu *et al.*, 2001). Positron emission tomography ($^{15}$O-H$_2$O) demonstrated an aberrant activ-

**TABLE VIII    Classification of Tic Disorders**

| | |
|---|---|
| Gilles de la Tourette syndrome | • Onset usually between 2 and 15, always before 21 y<br>Multiple, simple, or complex motor tics<br>• Multiple vocal tics<br>• Voluntary suppression of tics for minutes to hours<br>• Fluctuating severity in the course of weeks or months<br>Duration >1 y |
| Chronic tic disorder | • Onset at any age<br>• Multiple, simple or complex motor tics involving more than three muscle groups<br>• Voluntary suppression of tics for minutes to hours<br>No fluctuations<br>• Duration >1 y |
| Transient tic disorder | • Onset during childhood or adolescence<br>• Mostly simple, rarely complex motor tics<br>• Voluntary suppression of tics for minutes to hours<br>Fluctuating severity of tics over the course of weeks or months, then spontaneous remission<br>• Duration <1 y |

Modified after DSM III, American Psychiatric Association.

ity in the sensorimotor, language, executive, and paralimbic regions that was associated with the expression of motor and vocal tics (Stern *et al.*, 2000). In addition to alterations of dopaminergic receptor activity (Comings *et al.*, 2000; Ernst *et al.*, 1999; Müller-Vahl *et al.*, 2000; Noble, 2000; Thompson *et al.*, 1998), alterations of serotonergic and noradrenergic neurotransmission (Comings *et al.*, 2000; Heinz *et al.*, 1998; Lam *et al.*, 1996) are likely to play a role in the pathogenesis of the disorder.

There is general agreement that TS has a neurogenetic disorder, although no single gene has yet been identified. Earlier twin and family studies were compatible with autosomal-dominant transmission; however, recent studies suggest that the mode of inheritance is more complex. Candidate genes that are either involved as major genes or act as modifier genes are thought to be present on chromosome 2, 7, 8, and most likely also on chromosome 11 (Merette *et al.*, 2000; Petek *et al.*, 2001; Simonic *et al.*, 2001).

**Table IX    Etiologic Classification of Tic Disorders**

Primary
   Simple transient tics (duration <1 y)
   Persistent multiple motor tics during childhood with remission
   Chronic tic disorder with simple or multiple motor tics
   Senile tics (age >50 y)
   Gilles de la Tourette syndrome
Secondary
   In association with postrheumatic chorea
   Postencephalitic (encephalitis lethargica)
   CO intoxication
   Neuroacanthocytosis
   Posttraumatic
   Drug-induced (L-dopa, neuroleptics, carbamazepine, phenytoin,
     amphetamine, cocaine)
   Mental retardation, developmental disorders

### Natural Course

Boys with TS are affected three times more often than girls, but the overall prevalence may be as high as 3% of the general population. The symptoms seem to peak just before puberty, but most patients achieve partial or complete remission after the age 20. A prospective longitudinal cohort study by Burd *et al.* (2001) showed declining tic severity in 39 of 54 eligible patients who were diagnosed with TS in 1984 and 1985. Global functioning improved by 50%, and the average number of comorbidities decreased by 42%. At follow-up, 44% of patients were symptom-free, whereas only 22% were taking medication as adults. Predictors of remissions or persistence have yet to be identified (Coffey *et al.*, 2000). Only rarely, do tic disorders develop in adults. Of 411 patients with tic disorders seen in a US center, only 13 patients could be identified who were seen for the first time with a tic disorder (either idiopathic or secondary) after the age of 21 and did not have a previous history of childhood tics. (Chouinard and Ford, 2000). In our experience, however, most patients with "adult-onset" tics actually have recurrence of childhood-onset tics after a period of remission.

### Principles of Therapy

Given the favorable natural course of tic disorders, particularly after adolescence and in early adulthood, the decision to treat should be carefully considered and weighed against the possibility of adverse effects. A wide variety of different drugs have been used both in children and in adults. None of the drugs investigated so far has proven superior. Most commonly, dopamine-receptor blocking drugs have been used traditionally to control tics. However, the therapeutic value of these drugs is limited, because long-term therapy may be associated with the risk of inducing drug-related dyskinesias. In our experience, however, fluphenazine, pimozide, and risperidone seem to be the most powerful anti-tic medications. Clonidine, an alpha-adrenergic receptor agonist, was found to suppress tics in milder cases, but the effect could not be confirmed in a double-blind study (Goetz et al., 1987)*. The drug, however, may be effective in patients who have problems with their impulse control. Tetrabenazine has been found effective in about 50% of patients with different tic disorders in an open-label study* (Jankovic and Beach, 1997). Similar to clonidine, the drug does not induce TD. Modest clinical efficacy has also been attributed to clonazepam (Jankovic and Rohaidy, 1987) and flunarizine (Micheli et al., 1990). Tiapride was able to reduce frequency of tics by 50% in a double-blind crossover study** (Eggers et al., 1988). Pimozide showed less frequent adverse effects; however, haloperidol proved slightly more efficient (Shapiro et al., 1989). On the other hand, direct comparisons revealed somewhat favorable effects of risperidone and better tolerability compared with pimozide (Bruggeman et al., 2001). Nevertheless, despite proven efficacy in clinical studies, classic neuroleptics should not be used as a primary drug to avoid development of tardive syndromes. Atypical neuroleptics such as olanzapine have also been reported beneficial both in an open-label study* (Budman et al., 2001) and when comparing clinical efficacy with pimozide (Onofrj et al., 2000). In children, quetiapine (Parraga et al., 2001) was able to reduce frequency and severity of tics. In general, tapering of drugs avoids a rebound effect after withdrawal. Other substances that have been investigated in small double-blind or open-label studies include baclofen in doses up to 60 mg daily (Singer et al., 2001), the alpha$_2$-adrenergic agonist guanfacine, frequently used in the management of drug withdrawal (Scahill et al., 2001), and also the opioid antagonist naloxone (Van Wattum et al., 2000). Botulinum toxin injections do seem to reduce the frequency of tics to some extent (Kwak et al., 2000; Marras et al., 2001).

### Practical Management

To reduce dose-limiting side effects, dosages have to be titrated slowly until a drug effect can be verified (Table X). It may take up to several weeks to reach the effective dose. If a therapy is warranted in less severe cases, trials should start with clonidine, otherwise with olanzapine. In severely affected patients, individually tailored interventions like stress management training or creation of a favorable environment may be useful as an addition to drug treatment.

**Table X   Drug Treatment of Tic Disorders (in Succession of Recommendation)**

| Drug | Initial daily dose | Increase | Maximal daily dose |
|---|---|---|---|
| Clonidine | 2 × 0.075 mg | 0.075 mg/wk | 3 × 0.3 mg |
| Olanzapine | 2 × 5 mg | 5 mg/wk | 2 × 10 mg |
| Risperidone | 2 × 1 mg | 1 mg/wk | 4 mg |
| Clonazepam | 1 × 0.5 mg | 0 mg/2–5 days | 3 × 2 mg |
| Tiapride | 1 × 100 mg | 100 mg/2–5 days | 3 × 200 mg |
| Pimozide | 1 × 1 mg | 2 mg/wk | 3 × 6 mg |
| Haloperidol | 1 × 1 mg | 0.5 mg/wk | 3 × 4 mg |
| Fluphenazine | 1 × 1 mg | 0.5 g/wk | 3 × 4 mg |

### Treatments No Longer Recommended

Traditional behavior therapy techniques did not show a significant improvement after contingency management and massed practice (Sand and Carlson, 1973). Frontal or cingular leukotomy is only rarely helpful in severe cases manifested by disabling obsessive–compulsive disorders, although this surgical procedure has no effect on tics.

## HEMIFACIAL SPASM

### Clinical Aspects

Because hemifacial spasm (HFS) has to be considered a hyperactive dysfunction syndrome of the VIIth cranial nerve, it differs from focal dystonias and other dyskinesias by its peripheral origin. It is included in this chapter because of its similarities in symptoms and symptomatic therapy.

HFS is characterized by unilateral involuntary twitching and spasms sustained for seconds in muscles that are innervated by the VIIth cranial nerve (Wang and Jankovic, 1998). Only in less than 5% of patients is there evidence of bilateral manifestation; then the syndrome has to be differentiated from blepharospasm. In typical cases HFS starts with involvement of the orbicularis oculi muscle, slowly progressing to involve the lower facial muscles, occasionally including the platysma. About 7% of patients display atypical HFS with onset in buccal muscles and upward progression to the eye. The spasms often persist during sleep. Mild weakness of the orbicularis oris muscle frequently accompanies advanced cases.

At this point, there is agreement that HFS is caused by a compression of the VIIth nerve at the root exit zone, in most cases by a blood vessel. Other sources of nerve compression include hemangioma, schwannoma, or other tumors in the vicinity of the root exit zone. Electrophysiological studies have demonstrated ephaptic transmission between nerve fibers and additionally hyperactivity of the facial motornucleus as a result of the nerve irritation. When performing microvascular decompression as a therapy, surgeons have identified the posterior inferior cerebellar artery or the anterior inferior cerebellar artery in varying proportions and less often the vertebral artery as responsible for about 90% of the cases (Wilkins, 1991). Occasionally, a vein is identified (in less than 10%), whereas up to 25% of patients exhibit more than one vessel as a possible cause. With the advent of high-resolution MR imaging, identification of abnormal nerve-vessel contacts have become possible (Du et al., 1995; Hosoya et al., 1995). Currently, 3D-FT T2-weighted (CISS) and contrast-enhanced 3D-FT T1-weighted (turbo- FLASH) sequences and MR angiography using 3D-MT FISP images are thought to provide almost complete sensitivity (Girard et al., 1997). Although most cases of HFS can be clearly diagnosed clinically, topographic imaging using either computed tomography or MRI is warranted to exclude tumors and other nonvascular causes. EMG shows synchronous arrhythmic discharge in different ipsilateral facial muscles, and, frequently, ephaptic transmission can be confirmed (Nielsen, 1984). Differential diagnoses include facial synkinesis after VIIth nerve palsy (increased latencies of R1 and R2 of the blink reflex on the affected side), facial myokymia (constant rapid undulation of facial muscles, such as in multiple sclerosis), and facial tic (brief stereotypic complex movement).

## Natural Course

Auger and Whisnant (1990) reported an average prevalence rate of 14.5/100,000 in women and 7.4/100,000 in men, but a much higher prevalence is present in the Asian population. Age at onset ranged between 17 and 70 years, with a mean age of 45 years in a series of 106 patients studied by Ehni and Woltman (1945). In rare cases of bilateral HFS the latency of onset between the two sides varied between 1 and 15 years (Tan and Jankovic, 1999a). Generally, symptoms progress slowly, until all facial muscles are involved. Spontaneous remissions may occur but are always temporary, lasting less than a few months. Because of a shared compression of various adjacent nerves, HFS may be associated with acoustic or trigeminal nerve dysfunction. Subclinically impaired hearing accompanies HFS in up to 17% of patients and an abnormal acoustic middle ear reflex in 41% (Moller and Moller, 1985).

## Principles of Therapy

*Microvascular decompression* of the facial nerve is a highly effective therapy introduced by Gardner and pioneered by Jannetta (Wilkins, 1991). It removes the aberrant vessel from the nerve near its exit from the brainstem. After craniotomy, dissection is carried anteriorly across the skull base to visualize the IXth, Xth, and XIth cranial nerves and later the VIIth nerve anteromedially positioned to the auditory nerve. The surgeon has to inspect the area closely to decide how to decompress the VIIth nerve. Usually, the vessel will be placed away from the nerve and some material, like a Teflon or Ivalon sponge, will be positioned between the facial nerve and the offending vessel. Results depend considerably on the experience of the surgeon (Wilkins, 1991). The most experienced neurosurgeon, Jannetta, reported after more than 700 microvascular decompressions a complete response in 84%, a partial response in 7%, and failure in 9% after 10 years of follow-up. This includes 10% of patients undergoing surgery a second time after initial failure. In less-specialized centers, the rate of successful surgery may be somewhat smaller; however, overall, the outcome of microvascular decompression determined in other recent surveys in smaller series of patients seems to be comparable (Acevedo et al., 1997; Kondo, 1997; Zhang and Shun, 1995). Not all patients are symptom-free immediately after surgery. Successful improvement was observed in about 60% of patients within the first week after surgery, whereas rates of improvement may increase up 83% when re-evaluated at 6 months (Shin et al., 1997). Rates of success can be improved if intraoperative facial EMG monitoring is used to assist in identifying the offending vessel. Recurrence of HFS after microvascular decompression occurs in up to 10% of patients after an initially successful operation, most frequently, during the first 2 years after surgery (Payner and Tew, 1996). Complications center around inadvertent injury to the neighboring nerves and vessels. Before the advent of intraoperative auditory-evoked potential monitoring, profound ipsilateral hearing loss was encountered by experienced surgeons in 7%, with such monitoring in less than 2%, and with considerably augmented risk during a second microvascular decompression operation. In up to 4% of patients, facial weakness, often permanent, develops. A lethal outcome occurred in less than 1% by experienced surgeons, and serious complications like intracerebellar hematomas, brainstem infarction, or cerebral infarct or hemorrhage were encountered in up to 4% of patients (Wilkins, 1991). At this point, most surgeons agree in that brainstem auditory-evoked potential (BAEP) intraoperative monitoring should be performed during surgery to avoid postoperative auditory complications. Currently, the value of neuroendoscopy in surgical treatment of HFS is being evaluated (Abdeen *et al.*, 2000).

Although HFS is a chronic socially embarrassing disorder, and microvascular decompression has proven an effective surgical therapy of the condition, it is rarely disabling to a degree that urgently calls for a surgical procedure with considerable risks. Sufficient *symptomatic control* of HFS may be achieved by local injections of BoNTx or, if BoNTx injections are not possible, by anticonvulsant drugs. Electrophysiological findings suggest that an antidromic activation of motoneurons in the facial nuclei leads to an increased firing and excitability of the nucleus ("kindling"). This mechanism could explain the partial effectiveness of pharmacotherapy with membrane-stabilizing anticonvulsants (carbamazepine, phenytoin, gabapentine). However, in most cases, tolerable doses of anticonvulsants are not able to provide for complete control of spasms.

During the last few years, *BoNTx* injections have been widely used to manage HFS and now represent the treatment of choice*** (Mauriello *et al.*, 1996). Because patients with hemifacial spasm seem to be more sensitive to the toxin than patients with blepharospasm, relatively smaller doses can be used and effects last longer (Jankovic and Brin, 1991). A blind placebo-controlled study (Yoshimura *et al.*, 1992) and many open series with large samples of patients have proven the efficacy and safety of the treatment. Between 90% and 100% of patients in various studies improved after BoNTx injections. Although often abnormal movements do not disappear completely, a 75% reduction of movements can be achieved on average and maintained over many years by repeated injections. The effects last between 15 and 19 weeks; occasionally patients experience a relief of symptoms for more than a year. Side effects are less frequent than in blepharospasm and always temporary. Because of the desired weakening of orbicularis oculi muscles, minor side effects like lid edema, midfacial muscle weakness, and lip droop may be seen. In up to 8% a cosmetically observable paralysis of mouth angle was found, with discrete paralysis in 3%. So far no evidence of antibody formation in HFS patients or other serious systemic side effects have been reported.

### Practical Management

If symptoms are moderate or a neurological center with experience in therapy with botulinum toxin is not available, symptomatic therapy should start with carbamazepine. Alternatively, phenytoin or gabapentine may be tried. Dosage, precautions, and side effects follow the rules for the use of carbamazepine as an anticonvulsant or for trigeminal neuralgia. Chronicity of the disease allows for a slower dosage increase in HFS. Clearly, a major reduction of spasms can be expected only in early stages and mild cases.

Symptomatic therapy with BoNTx is highly effective, safe, and shows minimal complications. After reconstitution of the toxin and cleansing of the skin, injections are given into the extreme medial and lateral portions of the pretarsal orbicularis muscle of the upper eyelid and into the lateral portion of the lower eyelid. Additional injection at the transition between pars orbitalis and pars palpebralis of the orbicularis oculi muscle or at the lateral canthus may improve the rate of success. Differences in location do not cause significantly different effects, but the levator muscle should always be avoided, and the tip of the needle should be directed into the opposite direction. Often treatment of the orbicularis oculi muscle is sufficient to reduce muscle spasms in the lower face. If not, in subsequent visits small doses of BoNTx may be injected into muscles of the lower face, that is, zygomaticus, depending on the clinical picture. The risk of cosmetically or even functionally notable paralysis by injections into perioral muscles should serve as a warning not to inject near the lips. We generally start with 60 u Dysport or 15 u Botox. For follow-up injections after initially successful treatment, as well as for the first treatment in patients older than 70 years, the dosage may be reduced. If initial treatment is unsuccessful, dosage is increased at the subsequent visit.

*Microvascular decompression* of the facial nerve is only rarely necessary. Surgical treatment may be considered in young patients and in cases in which if BoNTx injections have not reduced muscle spasms sufficiently and the patient is disabled by the symptoms. An experienced surgeon should be recommended. Preoperatively, EMG may prove a pre-existing hidden paralysis of facial muscles. MR-imaging, including MR-angiography of the intracranial vessels, will provide an overview of the vascular situation and exclude a malformation. With high-resolution MRI, the offending vessel can be identified in advance in most subjects. During the surgical procedure, specific monitoring techniques, including brainstem auditory-evoked potentials, direct auditory nerve compound action potentials, and EMG of the facial nerve assist in identifying the offending vessel and reducing the risk of profound hearing loss.

### REFERENCES

Abdeen, K., Kato, Y., Kiya, N., Yoshida, K., and Kanno, T. (2000). Neuroendoscopy in microvascular decompression for trigeminal neuralgia and hemifacial spasm: technical note. *Neurol. Res.* **22**, 522–526.

Acevedo, J. C., Sindou, M., Fischer, C., and Vial, C. (1997). Microvascular decompression for the treatment of hemifacial spasm. Retrospective study of a consecutive series of 75 operated patients—electrophysiologic and anatomical surgical analysis. *Stereotact. Funct. Neurosurg.* 68(1–4), 260–265.

Adams, S. G., Hunt, E. J., Charles, D. A., and Lang, A. E. (1993). Unilateral versus bilateral botulinum toxin injections in spasmodic dysphonia: Acoustic and perceptual results. *J. Otolaryngol.* 22, 171–175.

Adler, L. A., Peselow, E., Rotrosen, J., Duncan, E., Lee, M., Rosenthal, M. et al. (1993). Vitamin E treatment of tardive dyskinesia. *Am. J. Psychiat.* 150, 1405–1407.

Ahmed, S., Chengappa, K. N., Naidu, V. R., Baker, R. W., Parepally, H., and Schooler, N. R. (1998). Clozapine withdrawal-emergent dystonias and dyskinesias: a case series. *J. Clin. Psychiatry* 59, 472–477.

Ainsworth, J. R., and Kraft, S. P. (1995). Long-term changes in duration of relief with botulinum toxin treatment of essential blepharospasm and hemifacial spasm. *Ophthalmology* 102, 2036–2040.

Alderson, K., Holds, J. B., and Anderson, R. L. (1991). Botulinum-induced alteration of nerve-muscle interactions in the human orbicularis oculi following treatment for blepharospasm. *Neurology* 41, 1800–1805.

Anderson, T. J., Rivest, J., Stell, R., Steiger, M. J., Cohen, H., Thompson, P. D. et al. (1992). Botulinum toxin treatment of spasmodic torticolli. *J. R. Soc. Med.* 85, 524–529.

Ansved, T., Odergren, T., and Borg, K. (1997). Muscle fiber atrophy in leg muscles after botulinum toxin type A treatment of cervical dystonia. *Neurology* 48, 1440–1442.

Aronson, A. E., and DeSanto, L. W. (1983). Adductor spastic dysphonia: Three years after recurrent nerve section. *Laryngoscope* 93, 1–8.

Auberger, S., Greil, W., and Rüther, E. (1985). Tiapride in the treatment of tardive dyskinesia: A double-blind study. *Pharmacopsychiatry* 18, 61–62.

Auger, R. G., and Whisnant, J. P. (1990). Hemifacial spasm in Rochester and Olmsted County, Minnesota. *Arch. Neurol.* 47, 1233–1234.

Baker, R. S., Sun, W. S., Hasan, S. A., Rouholiman, B. R., Chuke, J. C., Cowen, D. E. et al. (1997). Maladaptive neural compensatory mechanisms in Bell's palsy-induced blepharospasm. *Neurology* 49, 223–229.

Bara-Jimenez, W., Catalan, M. J., Hallett, M., and Gerloff, C. (1998) Abnormal somatosensory homunculus in dystonia of the hand. *Ann. Neurol.* 44, 828–831.

Barclay, C. L., and Lang, A. E. (1997). Dystonia in progressive supranuclear palsy. *J. Neurol. Neurosurg. Psychiatry* 62, 352–356.

Bertrand, C. M., and Molina-Negro, P. (1988). Selective peripheral denervation in 111 cases of spasmodic torticollis: Rationale and results. *Adv. Neurol.* 50, 637–643.

Bigalke, H., Wohlfarth, K., Irmer, A., and Dengler, R. (2001). Botulinum A toxin: Dysport improvement of biological availability. *Exp. Neurol.* 168, 162–170.

Blackie, J. D., and Lees, A. J. (1990). Botulinum toxin treatment in spasmodic torticollis. *J. Neurol. Neurosurg. Psychiat.* 53, 640–643.

Blasi, J., Chapman, E. R., Link, E., Binz, T., Yamasaki, S., De Camilli, P. et al. (1993). Botulinum neurotoxin A selectively cleaves the synaptic protein SNAP-25. *Nature* 365, 160–163.

Blitzer, A., and Brin, M. F. (1991). Laryngeal dystonia: A series with botulinum toxin therapy. *Ann. Otol. Rhinol. Laryngol.* 100, 85–89.

Borodic, G. E., Ferrante, R., Pearce, L. B., and Smith, K. (1994). Histologic assessment of dose-related diffusion and muscle fiber response after therapeutic botulinum A toxin injections. *Mov. Disord.* 9, 31–39.

Brans, J. W., de Boer, I. P., Aramideh, M., Ongerboer de Visser, B. W., and Speelman, J. D. (1995). Botulinum toxin in cervical dystonia: low dosage with electromyographic guidance. *J. Neurol.* 242, 529–534.

Brans, J. W., Lindeboom, R., Snoek, J. W., Zwarts, M. J., van Weerden, T. W., Brunt, E. R. et al. (1996). Botulinum toxin versus trihexyphenidyl in cervical dystonia: a prospective, randomized, double-blind controlled trial. *Neurology* 46, 1066–1072.

Brashear, A., Bergan, K., Wojcieszek, J., Siemers, E. R., and Ambrosius, W. (2000). Patients' perception of stopping or continuing treatment of cervical dystonia with botulinum toxin type A. *Mov. Disord.* 15, 150–153.

Brashear, A., Lew, M. F., Dykstra, D. D., Comella, C. L., Factor, S. A., Rodnitzky, R. L. et al. (1999). Safety and efficacy of NeuroBloc (botulinum toxin type B) in type A- responsive cervical dystonia. *Neurology* 53, 1439–1446.

Braun, V., and Richter, H. P. (1994). Selective peripheral denervation for the treatment of spasmodic torticollis. *Neurosurgery* 35, 58–62.

Braun, V., Richter, H. P., and Schroder, J. M. (1995). Selective peripheral denervation for spasmodic torticollis: is the outcome predictable? *J. Neurol.* 242, 504–507.

Bressman, S. B., Fahn, S., and Burke, R. E. (1988). Paroxysmal nonkinesigenic dystonia. *Adv. Neurol.* 50, 403–413.

Bressman, S. B., Heiman, G. A., Nygaard, T. G., Ozelius, L. J., Hunt, A. L., Brin, M. F. et al. (1994a). A study of idiopathic torsion dystonia in a non-Jewish family: Evidence for genetic heterogeneity. *Neurology* 44, 283–287.

Bressman, S. B., Hunt, A. L., Heiman, G. A., Brin, M. F., Burke, R. E., Fahn, S. et al. (1994b). Exclusion of the DYT1 locus in a non-Jewish family with early-onset dystonia. *Mov. Disord.* 9, 626–632.

Bressman, S. B., Sabatti, C., Raymond, D., de Leon, D., Klein, C., Kramer, P. L. et al. (2000). The DYT1 phenotype and guidelines for diagnostic testing. *Neurology* 54, 1746–1752.

Brin, M. F., Lew, M. F., Adler, C. H., Comella, C. L., Factor, S. A., Jankovic, J. et al. (1999). Safety and efficacy of NeuroBloc (botulinum toxin type B) in type A- resistant cervical dystonia. *Neurology* 53, 1431–1438.

Brion, S., Plas, J., Chevalier, J. F., and Dussaux, P. (1988). Dyskinesies acute et dyskinesies tardives. *Encephale* 14, 215–219.

Bruggeman, R., van der Linden, C., Buitelaar, J. K., Gericke, G. S., Hawkridge, S. M., and Temlett, J. A. (2001). Risperidone versus pimozide in Tourette's disorder: a comparative double-blind parallel-group study. *J. Clin. Psychiatry* 62, 50–56.

Bruneau, M. A., and Stip, E. (1998) Metronome or alternating Pisa syndrome: a form of tardive dystonia under clozapine treatment. *Int. Clin. Psychopharmacol.* 13, 229–232.

Budman, C. L., Gayer, A., Lesser, M., Shi, Q., and Bruun, R. D. (2001). An open-label study of the treatment efficacy of olanzapine for Tourette's disorder. *J. Clin. Psychiatry* 62, 290–294.

Burbaud, P., Guehl, D., Lagueny, A., Petiteau, F., and Bioulac, B. (1998). A pilot trial of clozapine in the treatment of cervical dystonia. *J. Neurol.* 245, 329–331.

Burd, L., Kerbeshian, P. J., Barth, A., Klug, M. G., Avery, P. K., and Benz, B. (2001). Long-term follow-up of an epidemiologically defined cohort of patients with Tourette syndrome. *J. Child. Neurol.* 16, 431–437.

Burke, R. E., Fahn, S., and Marsden, C. D. (1986). Torsion dystonia: A double-blind, prospective trial of high-dosage trihexyphenidyl. *Neurology* 36, 160–164.

Byl, N. N., Merzenich, M. M., and Jenkins, W. M. (1996). A primate genesis model of focal dystonia and repetitive strain injury: I. Learning-induced dedifferentiation of the representation of the hand in the primary somatosensory cortex in adult monkeys. *Neurology* 47, 508–520.

Casey, D. E. (1998). Effects of clozapine therapy in schizophrenic individuals at risk for tardive dyskinesia. *J. Clin. Psychiatry* **59**, 31–37.

Ceballos-Baumann, A. O. (2001). Evidence-based medicine in botulinum toxin therapy for cervical dystonia. *J. Neurol.* **248**(Suppl 1), 14–20.

Ceballos-Baumann, A. O., Passingham, R. E., Warner, T., Playford, E. D., Marsden, C. D., and Brooks, D. J. (1995). Overactive prefrontal and underactive motor cortical areas in idiopathic dystonia. *Ann. Neurol.* **37**, 363–372.

Chan, J., Brin, M. F., and Fahn, S. (1991). Idiopathic cervical dystonia: Clinical characteristics. *Mov. Disord.* **6**, 119–126.

Chen, R., Karp, B. I., and Hallett, M. (1998). Botulinum toxin type F for treatment of dystonia: long-term experience. *Neurology* **51**, 1494–1496.

Chiu, N. T., Chang, Y. C., Lee, B. F., Huang, C. C., and Wang, S. T. (2001). Differences in 99mTc-HMPAO brain SPET perfusion imaging between Tourette's syndrome and chronic tic disorder in children. *Eur. J. Nucl. Med.* **28**, 183–190.

Chouinard, S., and Ford, B. (2000). Adult onset tic disorders. *J. Neurol. Neurosurg. Psychiatry* **68**, 738–743.

Coffey, B. J., Biederman, J., Geller, D. A., Spencer, T., Park, K. S., Shapiro, S. J. *et al.* (2000). The course of Tourette's disorder: a literature review. *Harv. Rev. Psychiatry* **8**, 192–198.

Comella, C. L., Jankovic, J., and Brin, M. F. (2000). Use of botulinum toxin type A in the treatment of cervical dystonia. *Neurology* **55**(Suppl 5), S15–S21.

Comings, D. E., Gade-Andavolu, R., Gonzalez, N., Wu, S., Muhleman, D., Blake, H. *et al.* (2000). Comparison of the role of dopamine, serotonin, and noradrenaline genes in ADHD, ODD and conduct disorder: multivariate regression analysis of 20 genes. *Clin. Genet.* **57**, 178–196.

Cullis, P. A., O'Brien, C. F., Truong, D. D., Koller, M., Villegas, T. P., and Wallace, J. D. (1998). Botulinum toxin type B: an open-label, dose-escalation, safety and preliminary efficacy study in cervical dystonia patients. *Adv. Neurol.* **78**, 227–230.

Davidson, B. J., and Ludlow, C. L. (1996). Long term effects of botulinum toxin injections in spasmodic dysphonia. *Ann. Otol. Rhinol. Laryngol.* **105**, 33–42.

De Paiva, A., Meunier, F. A., Molgó, J., Aoki, K. R., and Dolly, J. O. (1999). Functional repair of motor endplates after botulinum neurotoxin A poisoning: Bi-phasic switch of synaptic activity between nerve sprouts and their parent terminals. *Proc. Natl. Acad. Sci. USA* **96**, 3200–3205.

Defazio, G., Berardelli, A., Abbruzzese, G., Lepore, V., Coviello, V., Acquistapace, D. *et al.* (1998). Possible risk factors for primary adult onset dystonia: A case-control investigation by the Italian Movement Disorders Study Group. *J. Neurol. Neurosurg. Psychiat.* **64**(1), 25–32.

Dewey, R. B., and Jankovic, J. (1989). Hemiballism-hemichorea: Clinical and pharmacologic findings in 21 patients. *Arch. Neurol.* **46**, 862–867.

Dressler, D. (2000). Electromyographic evaluation of cervical dystonia for planning of botulinum toxin therapy. *Eur. J. Neurol.* **7**, 713–718.

Dressler, D., Wittstock, M., and Benecke, R. (2000). Treatment of persistent hemiballism with botulinum toxin type A. *Mov. Disord.* **15**, 1281–1282.

Du, C., Korogi, Y., Nagahiro, S., Sakamoto, Y., Takada, A., Ushio, Y. *et al.* (1995). Hemifacial spasm: three-dimensional MR images in the evaluation of neurovascular compression. *Radiology.* **197**, 227–231.

Eggers, C., Rothenberger, A., and Berghaus, U. (1988). Clinical and neurobiological findings in children suffering from tic disease following treatment with tiapride. *Eur. Arch. Psychiat. Neurol. Sci.*, **237**, 223–229.

Ehni G, and Woltman, H. W. (1945). Hemifacial spasm. *Arch. Neurol.* **53**, 205–211.

Elbert, T., Candia, V., Altenmüller, E., Rau, H., Sterr, A., Rockstroh, B. *et al.* (1998). Alteration of digital representations in somatosensory cortex in focal hand dystonia. *Neuroreport* **9**, 3571–3575.

Elliott, E. S., Marken, P. A., and Ruehter, V. L. (2000). Clozapine-associated extrapyramidal reaction. *Ann. Pharmacother.* **34**, 615–618.

Elston, J. S. (1992). The management of blepharospasm and hemifacial spasm. *J. Neurol.* **239**, 5–8.

Ernst, M., Zametkin, A. J., Jons, P. H., Matochik, J. A., Pascualvaca, D., and Cohen, R. M. (1999). High presynaptic dopaminergic activity in children with Tourette's disorder. *J. Am. Acad. Child. Adolesc. Psychiatry* **38**, 86–94.

Fahn, S., Bressman, S. B., and Marsden, C. D. (1998). Classification of dystonia. *Adv. Neurol.* **78**, 1–10.

Feltner, D. E., and Hertzman, M. (1993). Progress in the treatment of tardive dyskinesia: Theory and practice. *Hosp. Commun. Psychiat.* **44**, 25–34.

Fernandez, H. H., Krupp, B., and Friedman, J. H. (2001). The course of tardive dyskinesia and parkinsonism in psychiatric patients: 14 years follow-up. *Neurology* **56**, 805–807.

Finsterer, J., Fuchs, I., and Mamoli, B. (1997). Quantitative electromyography-guided botulinum toxin treatment of cervical dystonia. *Clin. Neuropharmacol.* **20**, 42–48.

Fisk, G. G., and York, S. M. (1987). The effect of sodium valproate on tardive dyskinesia—Revisited. *Br. J. Psychiat.* **150**, 542–546.

Fletcher, N. A., Harding, A. E., and Marsden, C. D. (1990). A genetic study of idiopathic torsion dystonia in the United Kingdom. *Brain* **113**, 379–395.

Fletcher, N. A., Harding, A. E., and Marsden, C. D. (1991). The relationship between trauma and idiopathic torsion dystonia. *J. Neurol. Neurosurg. Psychiat.* **54**, 713–717.

Fouad, G. T., Servidei, S., Durcan, S., Bertini, E., and Ptacek, L. J. (1996). A gene for familial paroxysmal dyskinesia (FPD1) maps to chromosome 2q. *Am. J. Hum. Genet.* **59**, 135–139.

Gauthier, S., Perot, P., and Bertrand, G. (1988). Role of surgical anterior rhizotomies in the management of spasmodic torticollis. *Adv. Neurol.* **50**, 633–635.

Gelenberg, A. J., Wojcik, J., Falk, W. E., Bellinghausen, B., and Joseph, A. B. (1989). CDP-choline for the treatment of tardive dyskinesia. A small negative series. *Compr. Psychiat.* **30**, 1–4.

Giladi, N. (1997). The mechanism of action of botulinum toxin type A in focal dystonia is most probably through its dual effect on efferent (motor) and afferent pathways at the injected site. *J. Neurol. Sci.* **152**, 132–135.

Giladi, N., Meer, J., Kidan, H., and Honigman, S. (2000). Long-term remission of idiopathic cervical dystonia after treatment with botulinum toxin. *Eur. Neurol.* **44**, 144–146.

Gilbert, G. J. (1975). Response of hemiballismus to haloperidol. *JAMA* **233**, 535–536.

Girard, N., Poncet, M., Caces, F., Tallon, Y., Chays, A., and Martin-Bouyer, P. *et al.* (1997) Three-dimensional MRI of hemifacial spasm with surgical correlation. *Neuroradiology* **39**, 46–51.

Glanzman, R. L., Gelb, D. J., Drury, I., Bromberg, M. B., and Truong, D. D. (1990). Brachial plexopathy after botulinum toxin injections. *Neurology* **40**, 1143.

Goetz, C. G., Penn, R. D., and Tanner, C. M. (1988). Efficacy of cervical cord stimulation in dystonia. *Adv. Neurol.* **50**, 645–649.

Goetz, C. G., Tanner, C. M., Wilson, R. S., Carroll, V. S., Como, P. G., and Shannon, K. M. (1987). Clonidine and Gilles de la Tourette's syndrome: Double-blind study using objective rating methods. *Ann. Neurol.* **21**, 307–310.

Goff, D. C., Renshaw, P. F., Sarid, Segal, O., Dreyfuss, D. A., Amico, E. T., and Ciraulo, D. A. (1993). A placebo-controlled trial of selegiline (L-deprenyl) in the treatment of tardive dyskinesia. *Biol. Psychiat.* **33**, 700–706.

Grandas, F., Elston, J., Quinn, N., and Marsden, C. D. (1988). Blepharospasm: A review of 264 patients. *J. Neurol. Neurosurg. Psychiat.* **51**, 761–772.

Greene, P. (1992). Baclofen in the treatment of dystonia. *Clin. Neuropharmacol.* **15**, 276–288.

Greene, P., Shale, H., and Fahn, S. (1988). Experience with high dosages of anticholinergic and other drugs in the treatment of torsion dystonia. *Adv. Neurol.* **50**, 547–556.

Greene, P. E., and Fahn, S. (1993). Use of botulinum toxin type F injections to treat torticollis in patients with immunity to botulinum toxin type A. *Mov. Disord.* **8**, 479–483.

Hatheway, C. L., and Dang, C. (1994). Immunogenicity of the neurotoxins of Clostridium Botulinum. *In "Therapy with Botulinum Toxin."* (J. Jankovic and M. Hallett, Eds.), pp. 93–108. Dekker, New York.

Heinz, A., Knable, M. B., Wolf, S. S., Jones, D. W., Gorey, J. G., Hyde, T. M. *et al.* (1998). Tourette's syndrome: [I-123]beta-CIT SPECT correlates of vocal tic severity. *Neurology* **51**, 1069–1074.

Hosoya, T., Watanabe, N., Yamaguchi, K., Saito, S., and Nakai, O. (1995). Three-dimensional-MRI of neurovascular compression in patients with hemifacial spasm. *Neuroradiology.* **37**, 350–352.

Houser, M. K., Soland, V. L., Bhatia, K. P., Quinn, N. P., and Marsden, C. D. (1999). Paroxysmal kinesigenic choreoathetosis: a report of 26 patients. *J. Neurol.* **246**, 120–126.

Ichinose, H., and Nagatsu, T. (1999). Molecular genetics of DOPA-responsive dystonia. *Adv. Neurol.* **80**, 195–198.

Inzelberg, R., Kahana, E., and Korczyn, A. D. (1988). Clinical course of idiopathic torsion dystonia among Jews in Israel. *Adv. Neurol.* **50**, 93–100.

Jakob, A. (1923). *Die extrapyramidalen Erkrankungen.* Springer, Berlin.

Jankovic, J. (1982). Treatment of hyperkinetic movement disorders with tetrabenazine: A double-blind crossover study. *Adv. Neurol.* **11**, 41–47.

Jankovic, J. (1988). Etiology and differential diagnosis of blepharospasm and oromandibular dystonia. *In "Facial Dyskinesias: Advances of Neurology,"* Volume 49. (J. Jankovic and E. Tolosa, Eds.), pp. 103–116. Raven, New York.

Jankovic, J. (1993). Tourette's syndrome: Phenomenology, pathophysiology, genetics, epidemiology, and treatment. *In "Current Neurology,"* Volume 13. (S. H. Appel, Ed.), pp. 209–227. Mosby-Year Book, Chicago.

Jankovic, J. (1994a). Botulinum toxin in the treatment of dystonia and other disorders. *In "Current Neurology,"* Volume 14. (S. H. Appel, Ed.), pp. 207–229. Mosby-Year Book, Chicago.

Jankovic, J. (1994b). Posttraumatic movement disorders: Central and peripheral mechanisms. *Neurology* **44.** 2006–2014.

Jankovic, J., and Beach, J. (1997). Long-term effects of tetrabenazine in hyperkinetic movement disorders. *Neurology* **48**, 358–362.

Jankovic, J., and Brin, M. F. (1991). Therapeutic uses of botulinum toxin. *N. Engl. J. Med.* **324**, 1186–1194.

Jankovic, J., Brin, M. F., and Comella, C. (1994). *"Handbook of Botulinum Toxin Treatment for Cervical Dystonia."* Churchill-Livingstone, New York.

Jankovic, J. (1998). Dystonia: Medical therapy and botulinum toxin in dystonia. *In "Dystonia 3, Advances in Neurology"*, Vol 78, (S. Fahn, C. D. Marsden, and D. R. DeLong, Eds.), pp. 169–184. Lippincott-Raven, Philadelphia.

Jankovic, J., and Fahn, S. (1998). Dystonic disorders. *In "Parkinson's Disease and Movement Disorders,"* 3rd edition (J. Jankovic and E. Tolosa, Eds.), pp. 513–551. Williams and Wilkins, Baltimore.

Jankovic, J., Leder, S., Warner, D., and Schwartz, K. (1991). Cervical dystonia: Clinical findings and associated movement disorders. *Neurology* **41**, 1088–1091.

Jankovic, J., and Orman, J. (1988). Tetrabenazine therapy of dystonia, chorea, tics, and other dyskinesias. *Neurology* **38**, 391–394.

Jankovic, J., and Rohaidy, H. (1987). Motor, behavioral and pharmacologic findings in Tourette's syndrome. *Can. J. Neurol. Sci.* **14**, 541–546.

Jankovic, J., and Schwartz, K. (1993). The use of botulinum toxin in the treatment of hand dystonias. *J. Hand Surg.* **30**, 295–296.

Jankovic, J., and Schwartz, K. S. (1991). Clinical correlates of response to botulinum toxin injections. *Arch. Neurol.* **48**, 1253–1256.

Jankovic, J. (2001). Medical progress: Tourette's syndrome. *N. Engl. J. Med.* **345**, 1184–1192.

Jankovic, J. (2002). Botulinum toxin: Clinical implications of antigenicity and immunoresistance. *In "Scientific and Therapeutic Aspects of Botulinum Toxin."* (M. F. Brin, M. Hallett, and J. Jankovic, Eds.), Lippincott Williams & Wilkins, Philadelphia.

Jankovic, J., and Tintner, R. (2001). Dystonia and parkinsonism. *Parkinsonism Relat. Disord.* **8**, 109–121.

Johnson, W. G., and Fahn, S. (1977). Treatment of vascular hemiballism and hemichorea. *Neurology* **27**, 634–636.

Kane, J., Woerner, M. G., Borenstein, M., Wegner, J. M., and Lieberman, J. (1986). Integrating incidence and prevalence of tardive dyskinesia. *Psychopharmacol. Bull.* **22**, 254–258.

Kane, J. M., Woerner, M., and Lieberman, J. (1988). Tardive dyskinesia: Prevalence, incidence, and risk factors. *J. Clin. Psychopharmacol.* **8**, 52–56.

Kang, U. J., Burke, R. E., and Fahn, S. (1988). Tardive dystonia. *Adv. Neurol.* **50**, 415–429.

Karp, B. I., Cole, R. A., Cohen, L. G., Grill, S., Lou, J. S., and Hallett, M. (1994). Long-term botulinum toxin treatment of focal hand dystonia. *Neurology* **44**, 70–76.

Karp, B. I., Goldstein, S. R., Chen, R., Samii, A., Bara-Jimenez, W., and Hallett, M. (1999). An open trial of clozapine for dystonia. *Mov. Disord.* **14**, 652–657.

Kidron, D., and Melamed, E. (1987). Forms of dystonia in patients with Parkinson's disease. *Neurology* **37**, 1009–1011.

Kiriakakis, V., Bhatia, K. P., Quinn, N. P., and Marsden, C. D. (1998). The natural history of tardive dystonia. A long-term follow-up study of 107 cases. *Brain* **121**, 2053–2066.

Klein, C., Breakefield, X. O., and Ozelius, L. J. (1999). Genetics of primary dystonia. *Semin. Neurol.* **19**, 271–280.

Kondo, A. (1997). Follow-up results of microvascular decompression in trigeminal neuralgia and hemifacial spasm. *Neurosurgery* **40**, 46–51; discussion 51–52.

Krauss, J. K., Toops, E. G., Jankovic, J., and Grossman, R. G. (1997). Symptomatic and functional outcome of surgical treatment of cervical dystonia. *J. Neurol. Neurosurg. Psychiatry* **63**, 642–648.

Krauss, J. K., and Mundinger, F. (1996). Functional stereotactic surgery for hemiballism. *J. Neurosurg.* **85**, 278–286.

Krebs, M. O., and Olie, J. P. (1999). Tardive dystonia induced by risperidone. *Can. J. Psychiatry* **44**, 507–508.

Kulisevsky, J., Lleo, A., Gironell, A., Molet, J., Pascual-Sedano, B., and Pares, P. (2000). Bilateral pallidal stimulation for cervical dystonia: dissociated pain and motor improvement. *Neurology* **55**, 1754–1755.

Kwak, C. H., Hanna, P. A., and Jankovic, J. (2000). Botulinum toxin in the treatment of tics. *Arch. Neurol.* **57**, 1190–1193.

Lam, S., Shen, Y., Nguyen, T., Messier, T. L., Brann, M., Comings, D. *et al.* (1996). A serotonin receptor gene (5HT1A) variant found in a Tourette's syndrome patient. *Biochem. Biophys. Res. Commun.* **219**, 853–858.

Lang, A. E. (1985). Persistent hemiballismus with lesions outside the subthalamic nucleus. *Can. J. Neurol. Sci.* **12**, 125–128.

Lang, A. E. (1988). Dopamine agonists and antagonists in the treatment of idiopathic dystonia. *Adv. Neurol.* **50**, 561–570.

Langeveld, T. P., van, Rossum, M., Houtman, E. H., Zwinderman, A. H., Briaire, J. J., and Baatenburg de Jong, R. J. (2001). Evaluation of voice quality in adductor spasmodic dysphonia before and after botulinum toxin treatment. *Ann. Otol. Rhinol. Laryngol.* **110**, 627–634.

Laskawi, R., Rohlmann, A., Landgrebe, M., and Wolff, J. R. (1997). Rapid astroglial reactions in the motor cortex of adult rats following peripheral facial nerve lesions. *Eur. Arch. Oto. Rhino. Laryngol.* **254**, 81–85.

Lee, L. V., Pascasio, F. M., Fuentes, F. D., and Viterbo, G. H. (1976). Torsion dystonia in Panay, Philipines. *Adv. Neurol.* **14**, 137–151.

Lee, M. S., and Marsden, C. D. (1994). Movement disorders following lesions of the thalamus or subthalamic region. *Mov. Disord.* **9**, 493–507.

Lenz, F. A., Jaeger, C. J., Seike, M. S., Lin, Y. C., Reich, S. G., DeLong, M. R. *et al.* (1999). Thalamic single neuron activity in patients with dystonia: dystonia-related activity and somatic sensory reorganization. *J. Neurophysiol.* **82**, 2372–2392.

Lepore, F. E. (1988). So-called apraxias of lid movement. *Adv. Neurol.* **49**, 85–90.

Lew, M. F., Adornato, B. T., Duane, D. D., Dykstra, D. D., Factor, S. A., Massey, J. M. *et al.* (1997). Botulinum toxin type B: a double-blind, placebo-controlled, safety and efficacy study in cervical dystonia. *Neurology* **49**, 701–707.

Lew, M. F., Brashear, A., and Factor, S. (2000). The safety and efficacy of botulinum toxin type B in the treatment of patients with cervical dystonia: summary of three controlled clinical trials. *Neurology* **55**(12), S29–35.

Limousin, P., Pollak, P., Hoffmann, D., Benazzouz, A., Perret, J. E., and Benabid, A. L. (1996). Abnormal involuntary movements induced by subthalamic nucleus stimulation in parkinsonian patients. *Mov. Disord.* **11**, 231–235.

Lin, J. J., Lin, G. Y., Shih, C., Lin, S. Z., Chang, D. C., and Lee, C. C. (1999). Benefit of bilateral pallidotomy in the treatment of generalized dystonia. Case report. *J. Neurosurg.* **90**, 974–976.

Loonen, A. J., Verwey, H. A., Roels, P. R., van Bavel, L. P., and Doorschot, C. H. (1992). Is diltiazem effective in treating the symptoms of (tardive) dyskinesia in chronic psychiatric inpatients? A negative, double-blind, placebo-controlled trial. *J. Clin. Psychopharmacol.* **12**, 39–42.

Lucetti, C., Nuti, A., Gambaccini, G., Bernardini, S., Brotini, S., Manca, M. L. *et al.* (2000). Mexiletine in the treatment of torticollis and generalized dystonia. *Clin. Neuropharmacol.* **23**, 186–189.

Mark, M. H., Sage, J. I., Dickson, D. W., Heikkila, R. E., Manzino, L., Schwartz, K. O. *et al.* (1994). Meige syndrome in the spectrum of Lewy body disease. *Neurology* **44**, 1432–1436.

Marras, C., Andrews, D., Sime, E., and Lang, A. E. (2001). Botulinum toxin for simple motor tics: a randomized, double-blind, controlled clinical trial. *Neurology* **56**, 605–610.

Marsalek, M. (2000). Tardive drug-induced extrapyramidal syndromes. *Pharmacopsychiatry* **33 Suppl 1**, 14–33.

Marsden, C. D., and Harrison, M. J. G. (1974). Idiopathic torsion dystonia (dystonia musculorum deformans). A review of forty-two patients. *Brain* **97**, 793–810.

Marsden, C. D., Marion, M. H., and Quinn, N. (1984). The treatment of severe dystonia in children and adults. *J. Neurol. Neurosurg. Psychiat.* **47**, 1166–1173.

Marsden, C. D., and Sheehy, M. P. (1990). Writer's cramp. *Trends Neurosci.* **13**, 148–153.

Marsden, D., Obeso, J. A., and Zarranz, J. J. (1985). The anatomical basis of symptomatic hemidystonia. *Brain* **108**, 463–483.

Martin, J. P. (1957). Hemichorea (hemiballismus) without lesions in the corpus Luysii. *Brain* **80**, 1–10.

Mauriello, J. A., Jr., Leone, T., Dhillon, S., Pakeman, B., Mostafavi, R., and Yepez, M. C. (1996). Treatment choices of 119 patients with hemifacial spasm over 11 years. *Clin. Neurol. Neurosurg.* **98**, 213–216.

Merette, C., Brassard, A., Potvin, A., Bouvier, H., Rousseau, F., Emond, C. *et al.* (2000). Significant linkage for Tourette syndrome in a large French Canadian family. *Am. J. Hum. Genet.* **67**, 1008–1013.

Micheli, F., Gatto, M., Lekhuniec, E., Mangone, C., Fernandez Pardal, M., Pikielny, R. *et al.* (1990). Treatment of Tourette's syndrome with calcium antagonist. *Clin. Neuropharmacol.* **13**, 77–83.

Miller, L. G., and Jankovic, J. (1992). Drug-induced dyskinesias: An overview. In "Disorders of Movement in Psychiatry and Neurology." (J. B. Anthony and R. B. Young, Eds.), pp. 5–32. Blackwell Scientific Publications, Cambridge.

Moller, M. B., and Moller, A. R. (1985). Loss of auditory function in microvascular decompression for hemifacial spasm. Results in 143 consecutive cases. *J. Neurosurg.* **63**, 17–20.

Moss, L. E., Neppe, V. M., and Drevets, W. C. (1993). Buspirone in the treatment of tardive dyskinesia. *J. Clin. Psychopharmacol.* **13**, 204–209.

Moyer, E., and Settler, P. E. (1994). Botulinum toxin type B: Experimental and clinical experience. In "Therapy with Botulinum Toxin." (J. Jankovic and M. Hallet, Eds.). Dekker, New York.

Müller, U., Steinberger, D., and Nemeth, A. H. (1998). Clinical and molecular genetics of primary dystonias. *Neurogenetics* **1**, 165–177.

Müller-Vahl, K. R., Berding, G., Brucke, T., Kolbe, H., Meyer, G. J., Hundeshagen, H. *et al.* (2000). Dopamine transporter binding in Gilles de la Tourette syndrome. *J. Neurol.* **247**, 514–520.

Münchau, A., Schrag, A., Chuang, C., MacKinnon, C. D., Bhatia, K. P., Quinn, N. P., and Rothwell, J. C. (2001). Arm tremor in cervical dystonia differs from essential tremor and can be classified by onset age and spread of symptoms. *Brain* **124**, 1765–1776.

Muscettola, G., Barbato, G., Pampallona, S., Casiello, M., and Bollini, P. (1999). Extrapyramidal syndromes in neuroleptic-treated patients: prevalence, risk factors, and association with tardive dyskinesia. *J. Clin. Psychopharmacol.* **19**, 203–208.

Nakashima, K., Kusumi, M., Inoue, Y., and Takahashi, K. (1995). Prevalence of focal dystonias in the western area of Tottori Prefecture in Japan. *Mov. Disord.* **10**, 440–443.

Nielsen, V. K. (1984). Pathophysiology of hemifacial spasm: I. Ephaptic transmission and ectopic excitation. *Neurology* **34**, 418–426.

Noble, E. P. (2000). The DRD2 gene in psychiatric and neurological disorders and its phenotypes. *Pharmacogenomics* **1**, 309–333.

Nüssgens, Z., and Roggenkämper, P. (1995). Long-term treatment of blepharospasm with botulinum toxin type A. *Geriatr. J. Ophthalmol.* **4**, 363–367.

Nutt, J. G., Hammerstad, J. P., and Carter, J. (1984). Cranial dystonia: Double-blind crossover study of anticholinergics. *Neurology* **34**, 215–217.

Nutt, J. G., Hammerstad, J. P., Carter, J. H., and deGarmo, P. L. (1985). Lisuride treatment of focal dystonia. *Neurology* **35**, 1242–1243.

Nutt, J. G., Muenter, M. D., Aronson, A., Kurland, L. T., and Melton, L. J. (1988). Epidemiology of focal and generalized dystonia in Rochester, Minnesota. *Mov. Disord.* **3**, 188–194.

Nutt, J. G., and Nygaard, T. G. (2001). Response to levodopa treatment in dopa-responsive dystonia. *Arch. Neurol.* **58**, 905–910.

Nygaard, T. G., Marsden, C. D., and Fahn, S. (1991). Dopa-responsive dystonia: Long-term treatment response and prognosis. *Neurology* **41**, 174–181.

Nygaard, T. G., Wilhelmsen, K. C., Risch, N. J. *et al.* (1993). Linkage mapping of dopa-responsive dystonia (DRD) to chromosome 14q. *Nature Genet.* **5**, 386–391.

Odergren, T., Hjaltason, H., Kaakkola, S., Solders, G., Hanko, J., Fehling, C. *et al.* (1998). A double blind, randomised, parallel group study to investigate the dose equivalence of Dysport and Botox in the treatment of cervical dystonia. *J. Neurol. Neurosurg. Psychiatry* **64**, 6–12.

Ohara, S., Hayashi, R., Momoi, H., Miki, J., and Yanagisawa, N. (1998). Mexiletine in the treatment of spasmodic torticollis. *Mov. Disord.* **13**, 934–940.

Onofrj, M., Paci, C., D'Andreamatteo, G., and Toma, L. (2000). Olanzapine in severe Gilles de la Tourette syndrome: a 52-week double-blind cross-over study vs. low-dose pimozide. *J. Neurol.* **247**, 443–446.

Opal, P., Tintner, R., Jankovic, J., Leung, J., Breakfield, X. O., Friedman, J., and Ozelius, L. (2001). Intrafamilial phenotypic variability of the DYT1 dystonia: From asymptomatic TOR1A gene carrier status to dystonic storm. *Mov. Disord.* **17**, 339–345.

Oppenheim, H. (1911). Über eine eigenartige Krampfkrankheit des kindlichen und jugendlichen Alters (Dysbasia lordotica progressiva, Dystonia musculorum deformans). *Neurologisches Centralblatt* **19**.

Pal, P. K., Samii, A., Schulzer, M., Mak, E., and Tsui, J. K. (2000). Head tremor in cervical dystonia. *Can. J. Neurol. Sci.* **27**, 137–142.

Parraga, H. C., Parraga, M. I., Woodward, R. L., and Fenning, P. A. (2001). Quetiapine treatment of children with Tourette's syndrome: report of two cases. *J. Child Adolesc. Psychopharmacol.* **11**, 187–191.

Payner, T. D., and Tew, J. M., Jr. (1996). Recurrence of hemifacial spasm after microvascular decompression. *Neurosurgery* **38**, 686–690; discussion 690–691.

Pearce, J. (1972). Reversal of hemiballismus by tetrabenazine. *JAMA* **219**, 1345.

Petek, E., Windpassinger, C., Vincent, J. B., Cheung, J., Boright, A. P., Scherer, S. W. *et al.* (2001). Disruption of a novel gene (IMMP2L) by a breakpoint in 7q31 associated with Tourette syndrome. *Am. J. Hum. Genet.* **68**, 848–858.

Peterson, B., Riddle, M. A., Cohen, D. J., Katz, L. D., Smith, J. C., Hardin, M. T. *et al.* (1993). Reduced basal ganglia volumes in Tourette's syndrome using three-dimensional reconstruction techniques from magnetic resonance images [see comments]. *Neurology* **43**, 941–949.

Peterson, B. S., Staib, L., Scahill, L., Zhang, H., Anderson, C., Leckman, J. F. *et al.* (2001). Regional brain and ventricular volumes in Tourette syndrome. *Arch. Gen. Psychiatry* **58**, 427–440.

Pettigrew, L. C., and Jankovic, J. (1985). Hemidystonia: A report of 22 patients and a review of the literature. *J. Neurol. Neurosurg. Psychiat.* **48**, 650–657.

Pickett, A. M., and Hambleton, P. (1994). Dose standardisation of botulinum toxin [letter]. *Lancet* **344**, 474–475.

Poewe, W., Deuschl, G., Nebe, A., Feifel, E., Wissel, J., Benecke, R. *et al.* (1998). What is the optimal dose of botulinum toxin A in the treatment of cervical dystonia? Results of a double blind, placebo controlled, dose ranging study using Dysport. German Dystonia Study Group. *J. Neurol. Neurosurg. Psychiatry* **64**, 13–17.

Pool, K. D., Freeman, F. J., Finitzo, T., Hayashi, M. M., Chapman, S. B., Devous, M. D. *et al.* (1991). Heterogeneity in spasmodic dysphonia. *Arch. Neurol.* **48**, 305–309.

Price, J., Farish, S., Taylor, H., and O'Day, J. (1997). Blepharospasm and hemifacial spasm. Randomized trial to determine the most appropriate location for botulinum toxin injections. *Ophthalmology* **104**, 865–868.

Quinn, N., and Marsden, C. D. (1984). A double blind trial of sulpiride in Huntington's disease and tardive dyskindesia. *J. Neurol. Neurosurg. Psychiat.* **47**, 844–847.

Raja, M., Azzoni, A., and Maisto, G. (1999). Three cases of improvement of tardive dyskinesia following olanzapine treatment. *Int. J. Neuropsychopharmcol.* **2**, 333–334.

Ranawaya, R., and Lang, A. (1991). Usefulness of a writing device in writer's cramp. *Neurology* **41**, 1136–1138.

Rollnik, J. D., Matzke, M., Wohlfarth, K., Dengler, R., and Bigalke, H. (2000). Low-dose treatment of cervical dystonia, blepharospasm and facial hemispasm with albumin-diluted botulinum toxin type A under EMG guidance. An open label study. *Eur. Neurol.* **43**, 9–12.

Rosales, R. L., Arimura, K., Takenaga, S., and Osame, M. (1996). Extrafusal and intrafusal muscle effects in experimental botulinum toxin-A injection. *Muscle Nerve* **19**, 488–496.

Ross, M. H., Charness, M. E., Sudarsky, L., and Logigian, E. L. (1997). Treatment of occupational cramp with botulinum toxin: diffusion of toxin to adjacent noninjected muscles. *Muscle Nerve* **20**, 593–598.

Samii, A., Pal, P. K., Schulzer, M., Mak, E., and Tsui, J. K. (2000). Post-traumatic cervical dystonia: a distinct entity? *Can. J. Neurol. Sci.* **27**, 55–59.

Sampaio, C., Ferreira, J. J., Simoes, F., Rosas, M. J., Magalhaes, M., Correia, A. P. *et al.* (1997). DYSBOT: a single-blind, randomized parallel study to determine whether any differences can be detected in the efficacy and tolerability of two formulations of botulinum toxin type A—Dysport and Botox—assuming a ratio of 4 : 1. *Mov. Disord.* **12**, 1013–1018.

Sand, P. L., and Carlson, C. (1973). Failure to establish control over tics in the Gilles de la Tourette syndrome with behaviour therapy techniques. *Br. J. Psychiat.* **122**, 665–670.

Sankhla, C., Lai, E. C., and Jankovic, J. (1998). Peripherally induced oromandibular dystonia. *J. Neurol. Neurosurg. Psychiatry* **65**, 722–728.

Scahill, L., Chappell, P. B., Kim, Y. S., Schultz, R. T., Katsovich, L., Shepherd, E. *et al.* (2001). A placebo-controlled study of guanfacine in the treatment of children with tic disorders and attention deficit hyperactivity disorder. *Am. J. Psychiatry* **158**, 1067–1074.

Schiavo, G., Rossetto, O., Catsicas, S., Polverino, de, Laureto, P., DasGupta, B. R., Benfenati, F. *et al.* (1993). Identification of the nerve terminal targets of botulinum neurotoxin serotypes A, D, and E. *J. Biol. Chem.* **268**, 23784–23787.

Schicatano, E. J., Basso, M. A., and Evinger, C. (1997). Animal model explains the origins of the cranial dystonia benign essential blepharospasm. *J. Neurophysiol.* **77**, 2842–2846.

Schwartz, M., Moguillansky, L., Lanyi, G., and Sharf, B. (1990). Sulpiride in tardive dyskinesia. *J. Neurol. Neurosurg. Psychiat.* **53**, 800–802.

Segawa, M., Hosaka, A., Miyagawa, F., Nomura, Y., and Imai, H. (1976). Hereditary progressive dystonia with marked diurnal fluctuation. *Adv. Neurol.* **14**, 215–233.

Shapiro, E., Shapiro, A. K., Fulop, G., Hubbard, M., Mandeli, J., Nordlie, J. *et al.* (1989). Controlled study of haloperidol, pimozide and placebo for the treatment of Gilles de la Tourette's syndrome. *Arch. Gen. Psychiat.* **46**, 722–730.

Sheehy, M. P., Rothwell, J. C., and Marsden, C. D. (1988). Writer's cramp. *Adv. Neurol.* **50**, 457–472.

Shin, J. C., Chung, U. H., Kim, Y. C., and Park, C. I. (1997). Prospective study of microvascular decompression in hemifacial spasm. *Neurosurgery* **40**, 730–734; discussion 734–735.

Simonic, I., Nyholt, D. R., Gericke, G. S., Gordon, D., Matsumoto, N., Ledbetter, D. H. *et al.* (2001). Further evidence for linkage of Gilles de la Tourette syndrome (GTS) susceptibility loci on chromosomes 2p11, 8q22 and 11q23–24 in South African Afrikaners. *Am. J. Med. Genet.* **105**, 163–167.

Simpson, G. M., and Lindenmayer, J. P. (1997). Extrapyramidal symptoms in patients treated with risperidone. *J. Clin. Psychopharmacol.* **17**, 194–201.

Singer, H. S., Wendlandt, J., Krieger, M., and Giuliano, J. (2001). Baclofen treatment in Tourette syndrome: a double-blind, placebo-controlled, crossover trial. *Neurology* **56**, 599–604.

Smith, M. E., and Ford, C. N. (2000). Resistance to botulinum toxin injections for spasmodic dysphonia. *Arch. Otolaryngol. Head Neck Surg.* **126**, 533–535.

Spinella, G., and Sheridan, P. H. (1994). Research opportunities in dystonia: National Institute of Neurological Disorders and Stroke workshop summary. *Neurology* **44**, 1177–1179.

Stacy, M., Cardoso, F., and Jankovic, J. (1993). Tardive sterotypy and other movement disorders in tardive dyskinesias. *Neurology* **43**, 937–941.

Steinberger, D., Korinthenberg, R., Topka, H., Berghauser, M., Wedde, R., and Muller, U. (2000). Dopa-responsive dystonia: mutation analysis of GCH1 and analysis of therapeutic doses of L-dopa. German Dystonia Study Group. *Neurology* **55**, 1735–1737.

Steinberger, D., Topka, H., Fischer, D., and Muller, U. (1999). GCH1 mutation in a patient with adult-onset oromandibular dystonia. *Neurology* **52**, 877–879.

Stern, E., Silbersweig, D. A., Chee, K. Y., Holmes, A., Robertson, M. M., Trimble, M. *et al.* (2000). A functional neuroanatomy of tics in Tourette syndrome. *Arch. Gen. Psychiatry* **57**, 741–748.

Stojanovic, M., Cvetkovic, D., and Kostic, V. S. (1995). A genetic study of idiopathic focal dystonias. *J. Neurol.* **242**, 508–511.

Syed, N. A., Delgado, A., Sandbrink, F., Schulman, A. E., Hallett, M., and Floeter, M. K. (1999). Blink reflex recovery in facial weakness: an electrophysiologic study of adaptive changes. *Neurology* **52**, 834–838.

Tachikawa, H., Suzuki, T., Kawanishi, Y., Hori, M., Hori, T., and Shiraishi, H. (2000). Tardive dystonia provoked by concomitantly administered risperidone. *Psychiatry Clin. Neurosci.* **54**, 503–505.

Tan, E. K., and Jankovic, J. (1999a). Bilateral hemifacial spasm: a report of five cases and a literature review. *Mov. Disord.* **14**, 345–349.

Tan, E. K., and Jankovic, J. (1999b). Botulinum toxin A in patients with oromandibular dystonia: long-term follow-up. *Neurology* **53**, 2102–2107.

Tarsy, D. (1989). Neuroleptic-induced movement disorders. *In* "Disorders of Movement." (N. P. Quinn and P. C. Jenner, Eds.), pp. 361–393. Academic Press, London.

Tasker, R. R., Doorly, T., and Yamashiro, K. (1988). Thalamotomy in generalized dystonia. *Adv. Neurol.* **50**, 615–631.

Tassin, J., Durr, A., Bonnet, A. M., Gil, R., Vidailhet, M., Lucking, C. B. *et al.* (2000). Levodopa-responsive dystonia. GTP cyclohydrolase I or parkin mutations? *Brain* **123**, 1112–1121.

Taylor, A. E., Lang, A. E., Saint, Cyr, J. A., Riley, D. E., and Ranawaya, R. (1991). Cognitive processes in idiopathic dystonia treated with high-dose anticholinergic therapy: Implications for treatment strategies. *Clin. Neuropharmacol.* **14**, 62–77.

Thaker, G. K., Nguyen, J. A., Strauss, M. E., Jacobson, R., Kaup, B. A., and Tamminga, C. A. (1990). Clonazepam treatment of tardive dyskinesia: A practical GABAmimetic strategy. *Am. J. Psychiat.* **147**, 445–451.

Thompson, M., Comings, D. E., Feder, L., George, S. R., and O'Dowd, B. F. (1998). Mutation screening of the dopamine D1 receptor gene in Tourette's syndrome and alcohol dependent patients. *Am. J. Med. Genet.* **81**, 241–244.

Ticehurst, S. B. (1990). Is spontaneous orofacial dyskinesia an artefact due to incomplete drug history? *J. Geriatr. Psychiatry. Neurol.* **3**, 208–211.

Tronnier, V. M., and Fogel, W. (2000). Pallidal stimulation for generalized dystonia. Report of three cases. *J. Neurosurg.* **92**, 453–456.

Troung, D. D., Rontal, M., Rolnick, M., Aronson, A. E., and Mistura, K. (1991). Double-blind controlled study of botulinum toxin in adductor spasmodic dysphonia. *Laryngoscope* **101**, 630–634.

Tsui, J. K., Hayward, M., Mak, E. K., and Schulzer, M. (1995). Botulinum toxin type B in the treatment of cervical dystonia: a pilot study. *Neurology* **45**, 2109–2110.

Vaddadi, K. S., Courtney, P., Gilleard, C. J., Manku, M. S., and Horrobin, D. F. (1989). A double-blind trial of essential fatty acid supplementation in patients with tardive dyskinesia. *Psychiat. Res.* **27**, 313–323.

Valente, E. M., Bentivoglio, A. R., Cassetta, E., Dixon, P. H., Davis, M. B., Ferraris, A., Ialongo, T., Frontali, M., Wood, N. W., and Albanese, A. (2001). DYT13, a novel primary torsion dystonia locus, maps to chromosome 1p36.13–36.32 in an Italian family with cranial-cervical or upper limb onset. *Ann. Neurol.* **49**, 362–366.

Van Harten, P. N., Kampuis, D. J., and Matroos, G. E. (1996). Use of clozapine in tardive dystonia. *Progr. Neuropsychopharmacol. Biol. Psychiatry* **20**, 263–274.

Van Wattum, P. J., Chappell, P. B., Zelterman, D., Scahill, L. D., and Leckman, J. F. (2000). Patterns of response to acute naloxone infusion in Tourette's syndrome. *Mov. Disord.* **15**, 1252–1254.

Vidakovic, A., Dragasevic, N., and Kostic, V. S. (1994). Hemiballism: report of 25 cases. *J. Neurol. Neurosurg. Psychiatry* **57**, 945–949.

Walker, R. H., Danisi, F. O., Swope, D. M., Goodman, R. R., Germano, I. M., and Brin, M. F. (2000). Intrathecal baclofen for dystonia: benefits and complications during six years of experience. *Mov. Disord.* **15**, 1242–1247.

Wang, A., and Jankovic, J. (1998). Hemifacial spasm: Clinical correlates and treatments. *Muscle Nerve* **21**, 1740–1747.

Warner, T. T., Fletcher, N. A., Davis, M. B., Ahmad, F., Conway, D., Feve, A. *et al.* (1993). Linkage analysis in British and French families with idiopathic torsion dystonia. *Brain* **116**, 739–744.

Whurr, R., Nye, C., and Lorch, M. (1998). Meta-analysis of botulinum toxin treatment of spasmodic dysphonia: a review of 22 studies. *Int. J. Lang. Commun. Disord.* **33**, 327–339.

Wilkins, R. H. (1991). Hemifacial spasm: A review. *Surg. Neurol.* **36**, 251–277.

Wirtschafter, J. D., and McLoon, L. K. (1998). Long-term efficacy of local doxorubicin chemomyectomy in patients with blepharospasm and hemifacial spasm. *Ophthalmology* **105**, 342–346.

Wojcik, J., Falk, W. E., Fink, J. S., Cole, J. O., and Gelenberg, A. J. (1989). A review of 32 cases of tardive dystonia. *Am. J. Psychiat.* **148**, 1055–1059.

Yassa, R., Nastase, C., Cvejic, J., and Laberge, G. (1991). The Pisa syndrome (or pleurothotonus): prevalence in a psychogeriatric population. *Biol. Psychiat.* **29**, 942–945.

Yoshimura, D. M., Aminoff, M. J., Tami, T. A., and Scott, A. B. (1992). Treatment of hemifacial spasm with botulinum toxin. *Muscle Nerve* **15**, 1045–1049.

Yoshor, D., Hamilton, W. J., Ondo, W., Jankovic, J., and Grossman, R. G. (2001). Comparison of thalamotomy and pallidotomy for the treatment of dystonia. *Neurosurgery* **48**, 818–824; discussion 824–826.

Zhang, K. W., and Shun, Z. T. (1995). Microvascular decompression by the retrosigmoid approach for idiopathic hemifacial spasm: experience with 300 cases. *Ann. Otol. Rhinol. Laryngol.* **104**, 610–612.

CHAPTER 79

# Huntington's Disease and Sydenham's Chorea

Thomas Gasser and Karl Kieburtz

## HUNTINGTON'S DISEASE

### Clinical Aspects

Huntington's disease (HD) is an autosomal dominantly inherited disease caused by the expansion of a variable CAG-repeat sequence in a gene located on the short arm of chromosome 4, encoding a cytoplasmic protein that has been called "huntingtin" (Huntington's Disease Collaborative Research Group, 1993). Pathologically, HD is characterized by a progressive neuronal degeneration affecting predominantly small and medium-sized striatal GABAergic projection neurons with reactive gliosis (Reiner *et al.*, 1988), whereas neurons that contain somatostatin and neuropeptide Y as their transmitter are selectively spared.

Clinically, the classical triad of symptoms is comprised of (1) a movement disorder, usually characterized by chorea, but also by dystonia and (often less conspicuously) akinesia and rigidity; (2) emotional disturbances and changes in personality; and (3) cognitive impairment. Neurological examination usually reveals generalized chorea and a variable degree of cognitive impairment. Impersistence of sustained movement, particularly of gaze or of tongue protrusion and hand grip (milkmaid grip) may be a prominent feature. However, the movement disorder of HD is usually more complex and includes some degree of akinesia (Angelini *et al.*, 1998; Berardelli *et al.*, 1999) and dystonia (Louis *et al.*, 1999). In fact, the widespread use of molecular diagnosis for HD has shown that the phenotypic spectrum may even include late-onset levodopa-responsive parkinsonism (Racette and Perlmutter, 1998) and parkinson-plus syndromes (Reuter *et al.*, 2000), tics (Angelini *et al.*, 1998), or the symptoms of paroxysmal choreoathetosis (Scheidtmann *et al.*, 1997).

Apraxia for orolingual movements or manual dyspraxia can also interfere with motor function. Oculomotor functions are disturbed in most patients early in the course of the disease (impaired initiation and decreased velocity of volitional saccades), which can be confirmed by electronystagmography.

*Cognitive impairment* can be demonstrated in different areas, including memory, executive functions, and visuospatial abilities. The common denominator may be a reduction in mental speed and processing capability resulting from pathological changes in the frontostriatal circuitry (Brown and Marsden, 1988). Because the neurotransmitter dopamine plays a crucial role in these connections, dopaminergic PET markers have been found to correlate well with the degree of mental changes (Backman *et al.*, 1997).

In addition to dementia, psychiatric disturbances include major depression or, less frequently, bipolar manic-depressive illness and schizophreniform or atypical psychoses. Behavioral disturbances, particularly aggressiveness, may be a major burden for caregivers. Especially during the early phases of the disease, suicide and suicide attempts are common (Robins Wahlin *et al.*, 2000). If the *diagnosis* of HD is suspected, on the basis of the clinical picture, a positive family history, and the progressive nature of the disease, it may be confirmed by molecular genetic testing. A concealed or overlooked family history, nonpaternity, and *de novo* mutations are possible explanations for apparently sporadic cases of HD.

The neurodegenerative process can be assessed by neuroimaging studies. Caudate atrophy with an increase in bicaudate-cranial index is often visible on CT scans (Culjkovic *et al.*, 1999) or MRI. Hypometabolism of the putamen and caudate nucleus, as demonstrated by 18F-fluorodeoxyglucose PET is very sensitive and may precede the onset of symptoms (Antonini *et al.*, 1996). Binding of the radioligand raclopride, which labels dopamine receptors on GABAergic striatal neurons most severely affected in the disease, is reduced and can be used to monitor the progression of the disease (Antonini *et al.*, 1998). The degree of neuropsychological change also correlates significantly with the degree of cerebral atrophy and changes in dopamine innervation (Backman *et al.*, 1997) as demonstrated by PET.

Electrophysiological investigations are not usually helpful in the diagnosis. A reduction of the amplitude of evoked potentials (somatosensory-evoked potentials, auditory-evoked potentials) can be found, but latencies are within normal limits. The early response to mechanically or electrically evoked long latency reflexes may be absent. The differential diagnosis of HD is summarized in Table I.

### Etiology and Pathogenesis of HD

HD is inherited as an autosomal-dominant trait, with age-dependent penetrance. The molecular basis for HD is the expansion of a repetitive trinucleotide sequence within the coding region of a gene on chromosome 4p31 (called IT15, or huntingtin). In normal individuals, the triplet CAG, coding for the amino acid glutamine, is repeated between 5 and about 36 times. Expansion of this sequence to more than 39 triplets (and therefore more than 39 glutamine residues in the protein) is associated with HD (Duyao *et al.*, 1993; Huntington's Disease Collaborative Research Group, 1993). In the small number of cases that have been identified with repeat numbers between 36 and 39, some, but not all, individuals had the disease develop during a normal lifespan (incomplete penetrance [Rubinsztein *et al.*, 1996]). The expanded gene fragment can be detected in gene carriers by a polymerase chain reaction (PCR)–based assay, thus obviating the family studies that were previously necessary for predictive diagnosis. This test is also useful for confirming the diagnosis, particularly in sporadic or atypical cases. It should not be forgotten, however, that a diagnosis of HD usually affects the entire family, so appropriate counseling must be given.

The guidelines originally established by the World Federation of Neurology and the HD Association for the indirect molecular diagnosis using family studies should still be regarded essential in molecular diagnosis, particularly in presymptomatic testing. These guidelines include extensive counseling before and after the test, informed consent of the proband, the confidentiality of the test result, and the exclusion of testing of those younger than the age of 18 (International Huntington Association [IHA] and the World Federation of

**TABLE I**   Differential Diagnosis of Chorea in the Possible Context of Huntington's Disease

*Hereditary disorders associated with chorea*
  Inborn errors of metabolism (lysosomal storage diseases, phenylketonuria)
  Wilson disease
  Hallervorden-Spatz disease
  Benign familial chorea
  Lesch-Nyhan syndrome
  Gilles de la Tourette syndrome
  Paroxysmal choreoathetosis
  Dentatorubropallidoluysian atrophy
  Neuroacanthocytosis
  Spinocerebellar degenerations
*Metabolic disorders*
  Hypo- and hypernatremia
  Hypocalcemia
  Hypo- and hyperglycemia
  Hepatic and renal encephalopathy
  Hyperthyroidism
  Hypoparathyroidism
*Cerebrovascular disorders*
  Basal ganglia infarction
  Arteriovenous malformations
Pregnancy
Sydenham's chorea
Systemic lupus erythematosus
*Drug- or toxin-induced chorea*
  Antidopaminergic drugs
  Dopaminergic drugs
  Anticonvulsants
    Phenytoin
    Carbamazepine
  Steroids
    Contraceptives
    Anabolic steroids
  Opiates
  Antihistamines
  Diazoxide
  Digoxin
  Flunarizine
  INH
  Lithium
  Triazolam
  Tricyclic antidepressants
*Toxins*
  Carbon monoxide
  Manganese
  Mercury
  Thallium
  Toluene

Neurology [WFN] Research Group on Huntington's Chorea, 1994). Predicitive testing should always be performed within the setting of an experienced genetic counseling center.

The molecular mechanisms underlying this selective cell death remain unknown. It is likely that neurodegenerative disorders, which are caused by trinucleotide repeat expansions, such as HD or the spinocerebellar ataxias, are caused by a common pathogenic mechanism, which may involve specific interactions of the elongated polyglutamine domain with other cellular proteins, as well as the abnormal cleavage of the gene products and their aggregation as intranuclear inclusions (Brice, 1998; Gutekunst *et al.*, 2000).

Families with a clinical and/or neuropathological picture resembling HD, but without the trinucleotide expansion in the huntingtin gene, have been described, and two other gene loci, one on chromosome 4 in cases with recessive inheritance (Kambouris *et al.*, 2000), and one on chromosome 20 with dominant transmission (Xiang *et al.*, 1998), have been described. Therefore, a negative result of the molecular test rules out HD but not an inherited HD-like disorder caused by a mutation in another gene.

## Natural Course

HD occurs in all races with a prevalence of 2–7/100,000 with regional clusters. New ("de novo") mutations are rare but have been found to arise usually from a pool of intermediate-sized alleles. The age at onset is variable, most patients have symptoms develop in the fourth and fifth decade of life (mean, 38 years) (Conneally, 1984). The average duration of the disease is 19 years, with a range from 10–25 years. Approximately 10% of patients have an onset before 20 years of age (juvenile HD). In about 10%–15% of patients the disease becomes manifest only after the age of 55. However, once symptoms appear, the rate of progression and functional decline seems to be independent of age at onset.

Age of onset is inversely correlated with the length of trinucleotide repeat expansion (Duyao *et al.*, 1993). However, age of onset varies by up to 20–30 years for a given repeat number. Therefore, it is at present not possible to predict age of onset in an individual case. The triplet repeat expansion is unstable (i.e., its length tends to increase from one generation to the next) providing the molecular basis for the phenomenon of anticipation (earlier manifestation of the disease in later generations), which has been described in HD as in a number of other inherited neurologic disorders, which are also based on triplet repeat expansions. The observation that patients with juvenile onset of the disease have inherited the disease gene predominantly from their fathers is explained by the fact that expansion of the repeat seems to be especially likely during spermatogenesis.

Typically, the movement disorder of HD starts as subtle "restlessness" of the fingers, toes, and face, progressing to generalized chorea. Later in the course of the disease, the movement disorder assumes more dystonic features and eventually consists largely of an akinetic rigid syndrome. If the disease manifests during childhood or adolescence, ataxia, dystonia, myoclonus, akinesia, rigidity, and seizures may be predominant neurological features, rather than chorea (van Dijk *et al.*, 1986). This "akinetic rigid variant" has been called the "Westphal form" of the disease and differs from the classical choreic form neuropathologically; in akinetic rigid individuals, all striatal neurons are affected early in the course of the disease (Albin *et al.*, 1990), whereas in classical cases, neurons projecting to the lateral globus pallidus seem to degenerate predominantly. However, it is now believed that these cases do not represent a distinct subtype of the disease but rather that there is a continuum of signs and symptoms, primarily depending on the size of the triplet repeats but possibly also on other genetic modifying factors.

Subtle personality changes can preceed the manifestation of the motor disorder by several years. Patients may be irritable and suspicious, or display irresponsible, antisocial, or uncritical behavior. As the disease progresses, cognitive decline becomes obvious. Cognitive changes may also preceed frank motor or psychiatric signs (Hahn-Barma *et al.*, 1998).

During the late stages, the patient becomes bedridden, slowly deteriorating to a vegetative state. Because the motor disorder leads to immobility and swallowing difficulties, the patient finally dies as a consequence of complications such as aspiration pneumonia or sepsis (Lanska *et al.*, 1988).

## Principles of Therapy

Current drug therapy has no effect on the progression of disability, and the need for any pharmacological treatment should be carefully considered. Hyperkinesias and psychiatric symptoms may respond well to pharmacotherapy, but neuropsychological deficits and dementia remain untreatable. Pharmacological intervention in the treatment of the movement disorder of HD is aimed at restoring the balance of neurotransmitters in the basal ganglia. Neuroprotective or neurorestorative (i.e., neurotransplantation) treatment is still experimental.

Hyperkinesias are often treated with antidopaminergic drugs (see Practical Management and Table II). However, although these agents may reduce chorea, there is no evidence that they produce true functional benefit for the patient. In addition, they often cause side effects, including parkinsonism, sedation, apathy, and tardive dyskinesia superimposed on the primary movement disorder. Therefore, only very disabling chorea should be treated with the lowest possible doses of antidopaminergic drugs.

TABLE II    Drug Therapy of Chorea

| Generic name | Initial dosage (mg/day) | Therapeutic dosage (mg/day) |
| --- | --- | --- |
| Sulpiride | 100–200 | 200–1200 |
| Tetrabenazine | 25 | 50–200 |
| Haloperidol | 2 | 5–10 |
| Perphenazine | 4 | 8–12 |
| Pimozide | 2 | 12–16 |

Low doses of dopamine agonists, such as bromocriptine (10–15 mg/day) may also improve hyperkinesias, possibly by reducing dopaminergic transmission by virtue of their action on the presynaptic autoreceptor (Frattola et al., 1977). Empirically, a combination of low-dose antidopaminergic drugs with benzodiazepines is frequently used (Shoulson, 1986), or, preferably, nothing at all.

Psychiatric disturbances are treated according to general psychopharmacological principles with antidepressants or neuroleptics (Leroi and Michalon, 1998). Depression is common and should be vigorously treated, because it frequently responds very well. It has to be kept in mind that the anticholinergic action of antidepressants may exacerbate the movement disorder, so that newer antidepressants, such as the selective serotonin reuptake inhibitors (SSRIs) are often preferable. In addition, recent studies show that behavioral disturbances, particularly aggressive behavior, is effectively treated by SSRIs, such as fluoxetine (De Marchi et al., 2001) or sertraline (Ranen et al., 1996).

## Practical Management

### General Measures

The treatment of HD has to take into account not only the pharmacological, physical, and psychosocial treatment of the affected individual but also the support and genetic counseling of the family. Education of the patient and the family as to the nature and course of the disease is an important and difficult task for the physician caring for patients with HD. Nondirective genetic counseling should be offered, and, if appropriate, the options of predictive and prenatal testing may be discussed.

Psychological and psychosocial support of the patient and of the family should be provided and securing social services and assisting in long-term planning of care are important therapeutic interventions. HD lay organizations have provided valuable assistance in this task. Attention to nutrition is particularly important, because patients with chorea often lose weight and require high-caloric food supplements. In those with dysphagia, feeding by means of a gastrostomy should be considered.

### Treatment of Hyperkinesias

Different options for pharmacological treatment are summarized in Table II. Sulpiride is probably the safest and least problematic drug in terms of side effects (Quinn and Marsden, 1984). If this drug is ineffective, other antidopaminergic agents such as tetrabenazine (Asher and Aminoff, 1981), perphenazine (Fahn, 1973), haloperidol, or pimozide (Girotti et al., 1984) can be tried. Tetrabenazine seems to be effective, but drug-induced parkinsonism and depression may be a side effect with higher doses. Response to different antidopaminergic drugs may vary considerably. As stated previously, dosage should be kept as low as possible, and reduction or discontinuation should be considered frequently, because hyperkinesias tend to decrease with advancing disease. A considerable number of patients will report functional improvement after dose reduction or discontinuation of neuroleptics.

Use of the atypical neuroleptic clozapine has been considered, because it is not associated with tardive dyskinesias; however, one small double-blind study has shown only little benefit (van Vugt et al., 1997).

### Treatment of Akinetic-Rigid Symptoms

In akinetic-rigid forms of HD, antiparkinsonian drugs (see Chapter 74) have been tried, but their benefit is short lived and usually not pronounced. Comparative trials of levodopa, dopamine agonists, amantadine, or anticholinergics have not been carried out, so the choice of drug is purely empirical.

### Treatment of Psychiatric Symptoms

If depression is prominent, a trial with tricyclic antidepressants may be initiated (e.g., with doxepine [50–75 mg tid] or amitryptiline [25–150 mg/day]. Alternately, particularly when obsessive-compulsive or aggressive behavior is prominent, clomipramine or an SSRI (fluoxetine, 20–80 mg/day; sertraline, 25–200 mg/day; paroxitene, 20–50 mg/day) may be indicated.

### Experimental Treatments

At present, there is no neuroprotective or neurorestorative treatment of proven efficacy in HD. However, several approaches have good theoretical support (Kieburtz, 1999) and have appeared promising enough to warrant further clinical trials.

Several pharmacologic agents that have shown neuroprotective properties in animal and tissue culture systems have been investigated in relatively small clinical trials, among them Coenzyme Q10 (Feigin *et al.*, 1996) and the antiglutamatergic agents remacemide (Kieburtz *et al.*, 1996) and riluzole (Rosas *et al.*, 1999), but none of them so far has proven very effective.

Neuroprotective and restorative cell therapy approaches have also been explored. The neurotrophic factor CNTF has been applied directly to the brain using encapsulated cells genetically engineered to secrete this protein (Bachoud-Levi *et al.*, 2000a). Finally, the transplantation of fetal tissue to the damaged striatum has received wide attention. The rationale is based on the argument that the primary pathological process involves basically only striatal projection neurons and that other pathological condition, particularly in the cortex, are secondary phenomena, which can be ameliorated by striatal cell replacement (Peschanski *et al.*, 1995). The first clinical trials show only modest improvement (Bachoud-Levi *et al.*, 2000b). There is reason for optimism, however, because in a genetic mouse model of HD, the pathological condition and behavioral deficits seem to be at least partially reversible by blockade of the expression of the mutated huntingtin fragment (Yamamoto *et al.*, 2000).

### Ineffective or Obsolete

Trials with drugs that enhance GABAergic transmission such as INH (Stober *et al.*, 1983), valproic acid, or gamma-vinyl-GABA have been unsuccessful. Likewise, treatment with baclofen, which was based on the excitotoxic hypothesis of neural degeneration in HD, failed to improve symptoms or to slow the progression of the disease (Shoulson *et al.*, 1989).

### Huntington's Disease Lay Organizations

International Huntington's Association
Mr. Gerrit Dommerholt
President (ret) & Int Development Officer IHA
Callunahof 8
7217 ST Harfsen
The Netherlands
tel: 31-573-431 595
fax: 31-573-431 719

### United States
Huntington's Disease Society
158 West 29th Street, 7th Floor
New York NY 10001-5300, USA
tel: 1-212-242 1968
fax: 1-212-239 3430
E-mail: hdsainfo@hdsa.org
Website: http://www.hdsa.org/

The Hereditary Disease Foundation
11400 West Olympic Boulevard
Suite 855
Los Angeles, CA 90064-1560
USA
tel: 310-575-9656
fax: 310-575-9156
E-mail: cures@hdfoundation.org
Website: www.hdfoundation.org

### Canada
Huntington Society of Canada
151 Frederick Street, Suite 400
N2H 2M2 KITCHENER, ONTARIO, CANADA
tel: 1-519-749 7063
fax: 1-519-749 8965
E-mail: info@hsc-ca.org
Website: http://www.hsc-ca.org/

### UK
Huntington's Disease Association
108 Battersea High Street
London SW 11 3HP
tel: 071-223-7000

### Germany
Deutsche Huntington-Hilfe e.V.
Postfach 281251
47241 Duisburg
tel: 0203-788777

## SYDENHAM'S CHOREA

### Clinical Aspects

Sydenham's chorea (SC) is a disease of childhood and adolescence; most cases occur between 7 and 12 years of age. The clinical picture is characterized by the gradual evolution of chorea, subsiding between 5 and 15 weeks after onset. The movement disorder is most often generalized, but hemichorea may occur. Dysarthria and behavioral abnormalities are other common features of the disease, but other neurological abnormalities are rare (Nausieda *et al.*, 1980).

The disease is strongly associated with streptococcal infections, although their precise pathogenetic role remains obscure. A history and laboratory evidence of group A streptococcal infection is found in most, but not all, cases. An immunological cause has been proposed, because cross-reacting antibodies to neuronal structures have been demonstrated in streptococcal infection (Bronze and Dale, 1993). About one third of patients with SC have other manifestations of rheumatic disease, such as valvular heart disease.

## Natural Course

SC formerly accompanied one half of all cases with acute rheumatic fever (Aita, 1973), but since the 1960s, the incidence in developed countries has been declining steadily (Nausieda *et al.*, 1980). The streptococcal infection usually preceeds the onset of the movement disorder by 1–6 months. Recurrence occurs in less than 20% of cases, usually between 1 and 2 years after the initial manifestation, but more than two attacks are rare. Persistence of a mild, functionally insignificant, choreic movement disorder, action tremor or mild neuropsychologic deficit years after the initial episode is common (Nausieda *et al.*, 1983). Patients with a history of SC also seem to be more susceptible to chorea induced by drugs, such as oral contraceptives or phenytoin, and pregnancy. Accumulating evidence suggests that individuals with a history of SC in childhood may have a recrudescence or a decompensation of mild persistent chorea in late life, possibly caused by the effects of drugs, ischemia, and ageing on a damaged striatum (Gibb *et al.*, 1985).

## Principles of Therapy

Antibiotic treatment of a streptococcal infection for 10 days and secondary prophylaxis for at least 5 years should be initiated to minimize sequelae of rheumatic fever such as valvular heart disease (Table III). For the treatment of hyperkinesias, valproate has been used sucessfully in several trials (Daoud *et al.*, 1990). Use of neuroleptic drugs should be brief and dosage as low as possible to avoid tardive dyskinesia.

A beneficial effect of corticosteroids has been suggested by one retrospective study (Green, 1978).

**TABLE III    Drug Therapy of Sydenham's Chorea**

| Drug | Dosage | Duration |
| --- | --- | --- |
| Antibiotics | | |
| Primary prevention | | |
| Penicillin V | 1,000,000 IU/kg/day orally | 10 days |
| Penicillin G | 1,200,000 IU/day IM | Once |
| Erythromycin | 40 mg/kg/day | 10 days |
| Secondary prevention | | |
| Penicillin G | 1 × 1,200,000 IU/day IM | 5 y |
|  | Every 3–4 wk | (until 20 y of age) |
| Anticonvulsants | | |
| Valproic acid | 15–40 mg/kg/day | 4–5 wk |
| Carbamazepine | 10–20 mg/kg/day | 4–5 wk |
| Antidopaminergic drugs | | |
| Pimozide | 1–6 mg/day | 4–5 wk |
| Tetrabenazine | 5–10 mg/kg/day | 4–5 wk |
| Haloperidol | 1–3 mg/day | 4–5 wk |

## Practical Management

Suggestions for treatment are listed in Table III. If sodium valproate is ineffective, sulpiride or, if necessary, other antidopaminergic agents should be tried. Treatment should be discontinued after several weeks, because chorea usually subsides spontaneously.

## Symptomatic Chorea in Adulthood

Choreic syndromes may be secondary to a large number of structural and metabolic disturbances (see Table I). In the elderly population, ischemic lesions within the basal ganglia, particularly the subthalamic nucleus, may underlie "senile" hemiballismus or hemichorea. In these cases, the movement disorder usually subsides spontaneously over several weeks or months.

For all cases of symptomatic chorea, treatment, if necessary, should follow the lines described for Huntington's disease.

## REFERENCES

Aita, J. A. (1973). Neurologic manifestations of rheumatic fever. *Postgrad. Med.* 54, 82–86.

Albin, R. L., Reiner, A., Anderson, K. D., Penney, J. B., and Young, A. B. (1990). Striatal and nigral neuron subpopulations in rigid Huntington's disease: implications for the functional anatomy of chorea and rigidity-akinesia. *Ann. Neurol.* 27, 357–365.

Angelini, L., Sgro, V., Erba, A., Merello, S., Lanzi, G., and Nardocci, N. (1998). Tourettism as clinical presentation of Huntington's disease with onset in childhood. *Ital. J. Neurol. Sci.* 19, 383–385.

Antonini, A., Leenders, K. L., and Eidelberg, D. (1998). [11C]raclopride-PET studies of the Huntington's disease rate of progression: relevance of the trinucleotide repeat length. *Ann. Neurol.* 43, 253–255.

Antonini, A., Leenders, K. L., Spiegel, R., Meier, D., Vontobel, P., Weigell-Weber, M., Sanchez-Pernaute, R., de Yebenez, J. G., Boesiger, P., Weindl, A., and Maguire, R. P. (1996). Striatal glucose metabolism and dopamine D2 receptor binding in asymptomatic gene carriers and patients with Huntington's disease. *Brain* 119, 2085–2095.

Asher, S. W., and Aminoff, M. J. (1981). Tetrabenazine and movement disorders. *Neurology* 31, 1051–1054.

Bachoud-Levi, A. C., Deglon, N., Nguyen, J. P., Bloch, J., Bourdet, C., Winkel, L., Remy, P., Goddard, M., Lefaucheur, J. P., Brugieres, P., Baudic, S., Cesaro, P., Peschanski, M., and Aebischer, P. (2000a). Neuroprotective gene therapy for Huntington's disease using a polymer encapsulated BHK cell line engineered to secrete human CNTF. *Hum. Gene. Ther.* 11, 1723–1729.

Bachoud-Levi, A. C., Remy, P., Nguyen, J. P., Brugieres, P., Lefaucheur, J. P., Bourdet, C., Baudic, S., Gaura, V., Maison, P., Haddad, B., Boisse, M. F., Grandmougin, T., Jeny, R., Bartolomeo, P., Dalla, B. G., Degos, J. D., Lisovoski, F., Ergis, A. M., Pailhous, E., Cesaro, P., Hantraye, P., and Peschanski, M. (2000b). Motor and cognitive improvements in patients with Huntington's disease after neural transplantation. *Lancet* 356, 1975–1979.

Backman, L., Robins-Wahlin, T. B., Lundin, A., Ginovart, N., and Farde, L. (1997). Cognitive deficits in Huntington's disease are predicted by dopaminergic PET markers and brain volumes. *Brain* **120**, 2207–2217.

Berardelli, A., Noth, J., Thompson, P. D., Bollen, E. L., Curra, A., Deuschl, G., van Dijk, J. G., Topper, R., Schwarz, M., and Roos, R. A. (1999). Pathophysiology of chorea and bradykinesia in Huntington's disease. *Mov. Disord.* **14**, 398–403.

Brice, A. (1998). Unstable mutations and neurodegenerative disorders. *J. Neurol.* **245**, 505–510.

Bronze, M. S., and Dale, J. B. (1993). Epitopes of streptococcal M proteins that evoke antibodies that cross-react with human brain. *J. Immunol.* **151**, 2820–2828.

Brown, R. G., and Marsden, C. D. (1988). "Subcortical dementia": the neuropsychological evidence. *Neuroscience* **25**, 363–387.

Conneally, P. M. (1984). Huntington disease: genetics and epidemiology. *Am. J. Hum. Genet.* **36**, 506–526.

Culjkovic, B., Stojkovic, O., Vojvodic, N., Svetel, M., Rakic, L., Romac, S., and Kostic, V. (1999). Correlation between triplet repeat expansion and computed tomography measures of caudate nuclei atrophy in Huntington's disease. *J. Neurol.* **246**, 1090–1093.

Daoud, A. S., Zaki, M., Shakir, R., and al Saleh, Q. (1990). Effectiveness of sodium valproate in the treatment of Sydenham's chorea. *Neurology* **40**, 1140–1141.

De Marchi, N., Daniele, F., and Ragone, M. A. (2001). Fluoxetine in the treatment of Huntington's disease. *Psychopharmacology (Berl).* **153**, 264–266.

Duyao, M., Ambrose, C., Myers, R., Novelletto, A., Persichetti, F., Frontali, M., Folstein, S., Ross, C., Franz, M., and Abbott, M. (1993). Trinucleotide repeat length instability and age of onset in Huntington's disease. *Nat. Genet.* **4**, 387–392.

Fahn, S. (1973). Treatment of choreic movements with Perphenazin. *Adv. Neurol.* **1**, 281–289.

Feigin, A., Kieburtz, K., Como, P., Hickey, C., Claude, K., Abwender, D., Zimmerman, C., Steinberg, K., and Shoulson, I. (1996). Assessment of coenzyme Q10 tolerability in Huntington's disease. *Mov. Disord.* **11**, 321–323.

Frattola, L., Albiazzati, M. G., Spano, P. F., and Trabucchi, M. (1977). Treatment of Huntington's chorea with bromocriptine. *Acta. Neurol. Scand.* **56**, 37–45.

Gibb, W. R., Lees, A. J., and Scadding, J. W. (1985). Persistent rheumatic chorea. *Neurology* **35**, 101–102.

Girotti, F., Carella, F., Scigliano, G., Grassi, M. P., Soliveri, P., Giovannini, P., Parati, E., and Caraceni, T. (1984). Effect of neuroleptic treatment on involuntary movements and motor performances in Huntington's disease. *J. Neurol. Neurosurg. Psychiatry* **47**, 848–852.

Green, L. N. (1978). Corticosteroids in the treatment of Sydenham's chorea. *Arch. Neurol.* **35**, 53–54.

Gutekunst, C. A., Norflus, F., and Hersch, S. M. (2000). Recent advances in Huntington's disease. *Curr. Opin. Neurol.* **13**, 445–450.

Hahn-Barma, V., Deweer, B., Durr, A., Dode, C., Feingold, J., Pillon, B., Agid, Y., Brice, A., and Dubois, B. (1998). Are cognitive changes the first symptoms of Huntington's disease? A study of gene carriers. *J. Neurol. Neurosurg. Psychiatry* **64**, 172–177.

Huntington's Disease Collaborative Research Group. (1993). A novel gene containing a trinucleotide repeat that is expanded and unstable on Huntington's disease chromosomes. *Cell* **72**, 971–983.

International Huntington Association (IHA) and the World Federation of Neurology (WFN) Research Group on Huntington's Chorea. (1994). Guidelines for the molecular genetics predictive test in Huntington's disease. *Neurology* **44**, 1533–1536.

Kambouris, M., Bohlega, S., Al Tahan, A., and Meyer, B. F. (2000). Localization of the gene for a novel autosomal recessive neuro-degenerative Huntington-like disorder to 4p15.3. *Am. J. Hum. Genet.* **66**, 445–452.

Kieburtz, K. (1999). Antiglutamate therapies in Huntington's disease. *J. Neural Transm. Suppl.* **55**, 97–102.

Kieburtz, K., Feigin, A., McDermott, M., Como, P., Abwender, D., Zimmerman, C., Hickey, C., Orme, C., Claude, K., Sotack, J., Greenamyre, J. T., Dunn, C., and Shoulson, I. (1996). A controlled trial of remacemide hydrochloride in Huntington's disease. *Mov. Disord.* **11**, 273–277.

Lanska, D. J., Lavine, L., Lanska, M. J., and Schoenberg, B. S. (1988). Huntington's disease mortality in the United States. *Neurology* **38**, 769–772.

Leroi, I., and Michalon, M. (1998). Treatment of the psychiatric manifestations of Huntington's disease: a review of the literature. *Can. J. Psychiatry* **43**, 933–940.

Louis, E. D., Lee, P., Quinn, L., and Marder, K. (1999). Dystonia in Huntington's disease: prevalence and clinical characteristics. *Mov. Disord.* **14**, 95–101.

Nausieda, P. A., Bieliauskas, L. A., Bacon, L. D., Hagerty, M., Koller, W. C., and Glantz, R. N. (1983). Chronic dopaminergic sensitivity after Sydenham's chorea. *Neurology* **33**, 750–754.

Nausieda, P. A., Grossman, B. J., Koller, W. C., Weiner, W. J., and Klawans, H. L. (1980). Sydenham chorea: an update. *Neurology* **30**, 331–334.

Peschanski, M., Cesaro, P., and Hantraye, P. (1995). Rationale for intrastriatal grafting of striatal neuroblasts in patients with Huntington's disease. *Neuroscience* **68**, 273–285.

Quinn, N., and Marsden, C. D. (1984). A double blind trial of sulpiride in Huntington's disease and tardive dyskinesia. *J. Neurol. Neurosurg. Psychiatry* **47**, 844–847.

Racette, B. A., and Perlmutter, J. S. (1998). Levodopa responsive parkinsonism in an adult with Huntington's disease. *J. Neurol. Neurosurg. Psychiatry* **65**, 577–579.

Ranen, N. G., Lipsey, J. R., Treisman, G., and Ross, C. A. (1996). Sertraline in the treatment of severe aggressiveness in Huntington's disease. *J. Neuropsychiatry Clin. Neurosci.* **8**, 338–340.

Reiner, A., Albin, R. L., Anderson, K. D., D'Amato, C. J., Penney, J. B., and Young, A. B. (1988). Differential loss of striatal projection neurons in Huntington disease. *Proc. Natl. Acad. Sci. USA* **85**, 5733–5737.

Reuter, I., Hu, M. T., Andrews, T. C., Brooks, D. J., Clough, C., and Chaudhuri, K. R. (2000). Late onset levodopa responsive Huntington's disease with minimal chorea masquerading as Parkinson plus syndrome. *J. Neurol. Neurosurg. Psychiatry* **68**, 238–241.

Robins Wahlin, T. B., Backman, L., Lundin, A., Haegermark, A., Winblad, B., and Anvret, M. (2000). High suicidal ideation in persons testing for Huntington's disease. *Acta. Neurol. Scand.* **102**, 150–161.

Rosas, H. D., Koroshetz, W. J., Jenkins, B. G., Chen, Y. I., Hayden, D. L., Beal, M. F., and Cudkowicz, M. E. (1999). Riluzole therapy in Huntington's disease (HD). *Mov. Disord.* **14**, 326–330.

Rubinsztein, D. C., Leggo, J., and Coles, R. (1996). Phenotypic characterization of individuals with 30–40 CAG repeats in the Huntington disease (HD) gene reveals HD cases with 36 repeats and apparently normal elderly individuals with 36–39 repeats. *Am. J. Hum. Genet.* **59**, 16–22.

Scheidtmann, K., Schwarz, J., Holinski, E., Gasser, T., and Trenkwalder, C. (1997). Paroxysmal choreoathetosis—a disorder related to Huntington's disease? *J. Neurol.* **244**, 395–398.

Shoulson, I. (1986). On chorea. *Clin. Neuropharmacol.* **9 Suppl 2:**S85–99, S85–S99.

Shoulson, I., Odoroff, C., Oakes, D., Behr, J., Goldblatt, D., Caine, E., Kennedy, J., Miller, C., Bamford, K., and Rubin, A. (1989). A controlled clinical trial of baclofen as protective therapy in early Huntington's disease. *Ann. Neurol.* **25**, 252–259.

Stober, T., Schimrigk, K., Holzer, G., and Ziegler, B. (1983). Quantitative evaluation of functional capacity during isoniazid therapy in Huntington's disease. *J. Neurol.* **229,** 237–245.

van Dijk, J. G., van der Velde, E. A., Roos, R. A., and Bruyn, G. W. (1986). Juvenile Huntington disease. *Hum. Genet.* **73,** 235–239.

van Vugt, J. P., Siesling, S., Vergeer, M., van der Velde, E. A., and Roos, R. A. (1997). Clozapine versus placebo in Huntington's disease: a double blind randomised comparative study. *J. Neurol. Neurosurg. Psychiatry* **63,** 35–39.

Xiang, F., Almqvist, E. W., Huq, M., Lundin, A., Hayden, M. R., Edstrom, L., Anvret, M., and Zhang, Z. (1998). A Huntington disease-like neurodegenerative disorder maps to chromosome 20p. *Am. J. Hum. Genet.* **63,** 1431–1438.

Yamamoto, A., Lucas, J. J., and Hen, R. (2000). Reversal of neuropathology and motor dysfunction in a conditional model of Huntington's disease. *Cell* **101,** 57–66.

# CHAPTER 80
# Wilson's Disease

Andreas Straube and Phillip D. Swanson

## CLINICAL ASPECTS

Wilson's disease (hepatolenticular degeneration) is a genetic disorder of autosomal recessive inheritance. The estimated worldwide prevalence is 1 in 30,000 live births, with a carrier rate of 1 in 90 (Walshe and Yealland, 1992). Disturbance in hepatic copper metabolism leads to an increase in intracellular copper. It is now known that the abnormal gene in Wilson's disease is located at q14.3 on chromosome 13 (Bull *et al.*, 1993; Tanzi *et al.*, 1993; Yamaguchi *et al.*, 1993; Yang *et al.*, 1986). This gene (ATP7Bgene) codes for a copper-transporting P-type ATPase (Wilson ATPase). The structure of this ATPase resembles that of the enzyme that is defective in Menkes' steely hair disease, whose gene is on the X chromosome. The Wilson ATPase is expressed in liver and kidney and seems to be essential for hepatic copper excretion into bile and incorporation into ceruloplasmin. This ATPase is predominately present in a trans-Golgi network in the hepatocytes. Increasing concentrations of copper result in redistribution of the Wilson ATPase to a cytoplasmic vesicular compartment close to the bile canalicular membrane (Loudianos and Gitlin, 2000; Schaefer *et al.*, 1999). Copper accumulates in these vesicles and then is excreted into the bile.

More than 170 different mutations of the ATP7B gene have been described in Wilson's disease, as well as 40 normal allelic variants. The H1069Q mutation accounts for 30%–90% of mutations in Northern, Central, and Eastern European populations. The A778L mutation is found in 30% of Asian patients. Compound heterozygotes are more common than homozygotes (Maier-Dobersberger *et al.*, 1997). Because of the disturbed excretion of copper into the bile, accumulation of copper in cells induces a defect of mitochondrial energy metabolism, which then causes increased oxidative stress by free-radical formation (Gu *et al.*, 2000). Failure to incorporate copper into ceruloplasmin results in

secretion of a ceruloplasmin apoprotein that is rapidly degraded in the plasma (Loudianis and Gitlin, 2000), which is the reason for decreased plasma ceruloplasmin levels in Wilson's disease. The pathogenetic role of reduced synthesis or impaired function of plasma protein ceruloplasmin is not clear (Czaja *et al.*, 1987). Although a large portion of plasma copper is bound to ceruloplasmin, aceruloplasminemia is not associated with tissue copper deposition (Floris *et al.*, 2000). The gene encoding ceruloplasmin is located on the long arm of chromosome 3 (Yang *et al.*, 1986).

In the course of Wilson's disease, increased storage of copper occurs in liver, brain, cornea (Kayser-Fleischer ring in Descemet's membrane), and kidneys. Neurological symptoms can occur (1) secondary to deposition of copper in the brain and (2) as a result of hepatic encephalopathy caused by copper-induced liver damage. Neuropathological changes are especially prominent in the corpus striatum, which may appear grossly shrunken with cavitation. White matter sponginess may also be noted. Histologically, astrocytes proliferate and show large nuclei with vesiculation, especially in the putamen. Neuronal loss occurs in the striatum, globus pallidus, and other deep structures. Pericapillary concretions staining for copper and hemorrhages may be found. Less severe changes may occur in cerebral and cerebellar cortex (Duchen and Jacobs, 1992).

### Symptoms and Signs

The initial symptoms in patients with Wilson's disease are usually due to either cerebral involvement or liver failure. Patients homozygous for the H1069Q mutation are more likely to be seen with neurological rather than with hepatic symptoms (Maier-Dobersberger *et al.*, 1997). Forty percent–50% of patients initially demonstrate neurological or psychological symptoms. These symptoms are quite variable and include tremor, akinesia, dysarthria, dystonia, chorea, limb and gait ataxia, spasms, and personality changes. The tremor can resemble that of benign essential tremor or the resting tremor of Parkinson's disease at an early stage. It may later become coarse with a wing-beating quality. In the large series of patients seen by Walshe and Yealland (1992), about 20% had a history of earlier episodes of liver damage that did not establish the diagnosis. The most common initial symptoms or signs were dysarthria, hand tremor, and personality change. Some patients also exhibit a slight intellectual decline (Walshe, 1986). There are some observations that the apolipoprotein E genotype ε3/3 is associated with a delayed onset of symptoms, although there is no influence on the clinical phenotype (Schiefermeier et al., 2000).

Half of Wilson's disease patients have symptoms of liver disease develop either alone or before the onset of cerebral symptoms (Dobyns et al., 1979). Indications of liver involvement include poor growth, jaundice, hypersplenism, ascites, and hematemesis (Danks and Stevens, 1969). Rarely are other organs involved early on. Patients occcasionally have symptoms caused by involvement of the kidney (nephrotic syndrome), pancreas, or heart.

The most important clinical sign for the diagnosis of Wilson's disease is the Kayser-Fleischer ring. These brownish rings are localized in Descemet's membrane on the innermost layer of the limbus of the cornea and initially may be present in only a crescentic area. In suspected cases, if not obvious, they must be looked for carefully with a slit lamp. They are present in nearly all cases that have cerebral involvement but only in about one third of patients with exclusively hepatic involvement (Stremmel et al., 1991). After initiation of a succesful copper-depleting therapy, the Kayser-Fleischer ring may disappear. The diagnosis of Wilson's disease can and should be made before the onset of symptoms in close relatives of affected patients (Walshe, 1988b). Of 30 presymptomatic cases evaluated by Walshe, 11 had abnormal physical signs on examination, including 7 with Kayser-Fleischer rings. The remaining patients were diagnosed on the basis of biochemical abnormalities.

### Diagnosis

The diagnosis of Wilson's disease is confirmed with the assistance of the following laboratory findings (Walshe, 1988a):

- Increased excretion of copper in the urine (normal is less than $30\,\mu g/24\,h$; Wilson's disease, 100–$1000\,\mu g/24\,h$).

- Decreased serum level of total copper (normal is $85$–$145\,\mu g/dl$).
- Decreased serum level of ceruloplasmin (normal range is $25$–$45\,mg/dl$; about 5% of cases will have normal ceruloplasmin levels [Gibbs and Walshe, 1979]).
- Increased serum level of free (nonceruloplasmin) copper (normal range is $10$–$15\,\mu g/dl$; Wilson's disease, $20$–$100\,\mu g/dl$). The free copper can be calculated by subtracting from the total copper ($\mu g/dl$) that amount bound to ceruloplasmin (multiply by 3 the ceruloplasmin level [in $mg/dl$] to obtain the bound copper in $\mu g/dl$).
- Increased copper content in the liver (normal is less than $50\,\mu g/g$ dry weight (or $10\,\mu g/g$ wet weight); Livers from Wilson's disease heterozygotes contain from 39 to $213\,\mu g/g$ dry weight; Wilson's disease patients have greater than $250\,\mu g/g$ dry weight).

An abnormal radioactive copper test ($^{64}CuCl_2$), can also be used to find heterozygotes. This test measures incorporation of radioactive copper into newly formed ceruloplasmin, which is reduced in patients with Wilson's disease (Scheinberg and Sternlieb, 1984).

If the clinical picture is characteristic and the diagnosis of Wilson's disease is confirmed by serum and urine studies, a liver biopsy to determine copper content is not necessary. A liver biopsy, however, is important in differentiating asymptomatic homozygotes from heterozygotes. The radioactive copper test may not be available in some medical centers. Predictive testing for Wilson's disease using DNA markers is available on a research basis. Linkage analysis is possible for early diagnosis of at-risk sibs and for couples who have a child with Wilson's disease (Farrer et al., 1991).

### Imaging Techniques

Imaging techniques such as CT scan (Williams and Walshe, 1981) and MR scan (Starosta-Rubinstein et al., 1987) can also be helpful for diagnosis. These studies usually show lesions in the area of the putamen and caudate nucleus, as well as slight general atrophy. The MRI is more sensitive than the CT and may show increased signal intensity in the thalamus or basal ganglia on T2-weighted images. These lesions may help to document the effects of chelation therapy (Thuomas et al., 1993). The $^{123}I$-iodobenzamide (IBZM) single-photon emission computed tomography (SPECT) scan shows an almost linear correlation between the reduction of IBZM binding (measured as a ganglia to frontal cortex ratio) and the severity of neurological signs (Oertel et al., 1992).

## Differential Diagnosis

Wilson's disease must be distinguished from other chronic liver diseases, including the acquired type of hepatocerebral degeneration (Victor *et al.*, 1965), and from neurological conditions such as idiopathic or symptomatic parkinsonian syndromes, essential tremor, focal or segmental dystonias, Huntington's disease, Sydenham's chorea, the different cerebellar degenerations, and in cases with primary psychiatric symptoms, from schizophrenic psychosis and personality disorders.

If the diagnosis of Wilson's disease is not established during the first clinical examination (in about 25%–33% of patients), the mean delay of diagnosis thereafter is 13 months (Walshe and Yealland, 1992).

## NATURAL COURSE

The onset of clinical symptoms can be slowly progressive or acute. In general, two different types of course can be distinguished (Martin, 1968):

1. Onset between 5 and 20 years of age with a predominance of symptoms caused by liver damage, as well as intravascular hemolysis with renal failure. Without treatment, death will occur after 2–7 years.
2. Onset of the clinical signs between 20 and 40 years of age with a chronic course over 10–40 years. In these patients, neurological and psychiatric symptoms dominate. Rare cases have been diagnosed as late as 56 years of age (Bellary and Van Thiel, 1993). Oder *et al.* (1993) identified three subgroups with different neurological and neuropsychiatric symptoms: one group exhibits bradykinesia, rigidity, cognitive impairment, and organic mood syndrome, MRI shows dilatation of the third ventricle; the second group has ataxia and tremor develop, MRI shows focal thalamic lesions; the third group has dyskinesias and an organic personality syndrome develop. Imaging studies show focal lesions in the putamen and pallidum.

If treatment starts early enough, especially when the patient is still asymptomatic, the occurrence of clinical signs can be prevented, and patients will have a normal life expectancy (Arima *et al.*, 1977; Walshe, 1988b). After the occurrence of neurological signs, complete remission can be expected only in 20% of patients, partial remission of the neurological signs in 60% and of the psychiatric signs in 70% (Stremmel *et al.*, 1989). Walshe and Yealland (1993) reported the

results of chelation therapy in 137 patients. Fifty-seven patients achieved a complete remission and 36 a good response. There was a poor response in 24 and no response in 20 of the 137 patients after adequate chelation treatment. Nine patients who received little or no treatment died.

In a small number of patients with Wilson's disease, the clinical symptoms may actually progress, despite improvement of the copper balance (Brewer *et al.*, 1987a; Walshe, 1988a). This phenomenon usually occurs during the first 2–4 weeks of therapy and may be irreversible. A possible reason for clinical worsening is the toxic effect of the increased quantity of copper released from the liver by the initial treatment. Symptoms that may have prognostic value for the effect of treatment are not known (Walshe, 1988a). Denny-Brown (1964) stressed the poorer prognosis of patients with prominent dystonic signs compared with patients with predominant tremor.

## PRINCIPLES OF THERAPY

The main pathophysiological disturbances are due to the increased intracellular copper concentration and, less important, to the lack or the production of a functionally ineffective ceruloplasmin in the liver. Therefore, the main therapeutic approaches are to reduce (1) intracellular storage and (2) intestinal resorption of copper.

### Reduction of Copper Storage

The following agents are used to reduce copper storage:

1. D-Penicillamine (Cuprimine), a sulfhydryl-containing amino acid that forms chelates with copper, has been the standard drug since its introduction by Walshe in the 1950s.
2. Trientine hydrochloride (triethylene-tetramine dihydrochloride, Syprine) forms compounds with copper in blood and kidneys, but not in liver. The complexes are eliminated by the kidneys. This agent is recommended for patients who do not tolerate the side effects of penicillamine.
3. Tetrathiomolybdate is a new agent that has been shown in experimental studies to be an effective chelating agent in Wilson's disease (Brewer *et al.*, 1991, 1994).
4. Dimercaprol (British anti-Lewisite) or the water-soluble analogue Dimval can be used to chelate copper rapidly.

## Reduction of Intestinal Copper Resorption

The following agents have been used:

- Zinc acetate (Galzin) or zinc sulfate
- Potassium sulfate
- Diethyl dithio-carbamate

Reduction of copper in food with a maximum daily copper resorption of less than 1.5 mg has been advocated by some investigators. This can be attained by abstention from liver (5000–7000 mg/100 g), crabs (1500 mg/100 g), brewer's yeast (3.3 mg/100 g), mushrooms (1–2 mg/100 g), nuts (0.5–1.3 mg/100 g), and cheese (0.1–0.8 mg/100 g).

The goal of these therapies is to achieve and sustain a negative copper balance. Copper balance can be estimated by the method of Hill *et al.* (1987). The apparent copper balance is the difference between copper intake in the food (about 1 mg/day in the average American diet) and copper excreted in feces and urine. Because copper loss in the skin averages about 0.25–0.34 mg/day, a positive balance is defined as a difference between intake and excretion (in feces and urine) of greater than 0.25–0.34 mg/24 h.

## PRACTICAL MANAGEMENT

### Increase of Copper Excretion

#### *D-penicillamine

D-Penicillamine (Cuprimine) is considered by many neurologists to be the initial chelating agent of choice (Walshe, 1999). However, at present there is considerable controversy about the initial treatment choice (Brewer, 1999, 2000; Hoogenraad, 1998; LeWitt, 1999). At the beginning of treatment, penicillamine should be started at 250 mg daily and increased by 250 mg a day every 4–7 days to prevent too rapid mobilization of intracellular copper, which can be monitored by measurement of the urine copper excretion and which can lead to irreversible aggravation of the clinical symptoms (Brewer *et al.*, 1987a). The drug is normally administered 3–4 times a day, half an hour before meals, or at least 2 hours after a meal, to avoid reduced absorption caused by food. The capsules should not be opened or chewed but should be swallowed intact. In some cases, the daily penicillamine dose must be slowly increased to attain a daily urine copper excretion of more than 500 μg/24 h. In addition to the copper excretion in the urine, the serum free copper level can serve as a scale for the adjustment of the dose. The goal is a normal serum free copper level of 10–15 μg/dl and a 24-hour urine copper of less than 0.1 mg. If the free copper rises above 20 μg/dl, the penicillamine dose should again be increased. If the free copper is less than 10 μg/dl, the penicillamine dose can be decreased. The average dose for an adult is 1000 mg/day. However, this may vary significantly, depending on the individual, from between 750 and 1600 mg/day. Most patients show symptomatic improvement after 3–6 months, and then the serum copper levels and the copper excretion fall, which indicates that the copper stores are depleted. The maintenance level of the therapy is then usually 500–750 mg penicillamine daily. Supplementation with oral pyridoxine (25 mg/day in adults and 12.5 mg/day in children) is recommended for the prevention of D-penicillamine–induced optic neuropathy. Initial side effects of penicillamine include hypersensitivity reactions in the form of urticarial rash, wheezing and dyspnea, thrombocytopenia, leukopenia, pancytopenia, and nephropathy. These can be minimized by temporary reduction of the D-penicillamine dose or additional treatment with corticosteroids (20–30 mg/day) (Tanakow, 1991). Five percent to 10% of patients exhibit long-lasting intolerance to penicillamine (Brewer *et al.*, 1983), which calls for a change of drug. Other possible side effects of chronic D-penicillamine therapy include drug-induced lupus erythematosus, immune complex nephritis, myasthenia gravis, myositis, intestinal ulcers, pancytopenia, lymphoreticular neoplasia (caused by immunosuppressive effect of D-penicillamine), dermatological problems caused by changes in the connective tissue (caused by inhibition of linkage of collagen fibers), and macular or papular skin rashes. Some of these side effects are dose dependent.

For the first 4–6 months of therapy, follow-up every 2–4 weeks is recommended, with complete blood count, platelet count, liver function tests, and urinalysis, as well as Cu studies carried out. Thereafter, patients can be followed less frequently, approximately every 3–6 months, depending on the patient's clinical status.

#### *Trientine Hydrochloride

Trientine hydrochloride (triethylene-tetramine dihydrochloride, Syprine) is approved by the Food and Drug Administration for use by patients who cannot tolerate penicillamine. Brewer (2000) believes that this agent should be chosen in preference to penicillamine because of a lower likelihood of side effects. It forms a compound with copper in blood and the kidney but does not enter the liver. The complex is eliminated in the urine. The recommended daily dose is 750–1000 mg taken 3–4 times a day on an empty stomach. Adverse effects have included reactivation of drug-induced lupus, anorexia, abdominal pain, skin rash, rhabdomyolysis, and iron-deficiency anemia. It can also induce a sideroblastic anemia, especially when used in combination with zinc salts (Scheinberg *et al.*, 1987; Walshe and Yealland, 1993). Pyridoxine supplementation is not needed.

### *Tetrathiomolybdate

Tetrathiomolybdate has been used to treat copper poisoning in sheep. It has also been used in patients who were intolerant to both penicillamine and trientine. It is not yet commercially available but is presently under study as an initial chelating agent in patients with Wilson's disease (Brewer, 2000). The substance immediately blocks copper absorption and also combines with albumin and ceruloplasmin-bound copper in the blood, which makes the copper nontoxic. In preliminary studies with 6 and 17 patients, none showed worsening of Wilson's disease symptoms after initiating tetrathiomolybdate therapy (Brewer *et al.*, 1991, 1994). The doses used in these studies were 1.6–2.9 mg/kg (120–240 mg/daily). The drug was administered six times a day (single dose of 20–40 mg); three doses were given with the meals and the remaining doses between the meals (Brewer *et al.*, 1994). Tetrathiomolybdate can induce reversible bone marrow depression and in animals can interfere with epiphyseal growth, so it should probably be avoided in children (Spence and Suttle, 1980).

### *Dimercaprol (BAL in Oil)

Historically, the first really useful drug for copper chelation, dimercaprol is used relatively infrequently for rapid copper chelation in patients who are not doing well on conventional regimens. A recommended approach (Scheinberg and Sternlieb, 1995) is as follows: A 1.0-ml (100 mg) test dose is injected intramuscularly into one buttock the day before the first course of treatment is to begin. Then, a 3.0-ml injection is given each day for 5 days into alternate buttocks. The 5-day regimen is repeated three times (five injections per week). Each injection site should be marked, with subsequent injections placed 3–6 cm from previous sites. The injections are quite painful and can cause sterile abscesses. A water-soluble analogue, dimival, may be available in some countries (Walshe and Yealland, 1993). A second course would be only done if the patient deteriorated again.

### Reduction of Intestinal Copper Resorption

#### *Zinc Acetate or Zinc Sulfate

For patients in whom chelating agents cannot be used or in asymptomatic patients, zinc acetate or zinc sulfate is recommended at an average dosage of 50–100 mg PO tid given 30–60 minutes before meals. Some investigators advocate zinc therapy as the initial treatment of choice for Wilson's disease (Hoogenraad, 1998) or as the preferred maintenance therapy after initial chelation (Brewer, 1999, 2000). Zinc acetate is usually less irritating to the stomach than zinc sulfate (Hill *et al.*, 1987). The physiological effect of zinc is to compete with copper for binding sites on the carrier protein metallothionine, which is found in the intestinal mucous membrane and in the liver (Nartey *et al.*, 1987). Because the intestinal cells slough into the the lumen of the bowl with a turnover time of 6 days, copper absorption is prevented. (Brewer *et al.*, 2000). Over the long term, a negative copper balance can be attained by monotherapy with zinc (Brewer *et al.*, 1987b). Positive reports on long-term treatment exist (Hoogenraad *et al.*, 1979, Brewer *et al.*, 1998).

Although serious side effects are rare, 30% of patients initially complain of temporary nausea and lack of appetite. Stremmel *et al.* (1989) reported 3 patients of 11 who again had symptoms of Wilson's disease develop under zinc sulfate monotherapy. In general, the recommended dosage is $3 \times 50$ mg. If the urine 24-hour secretion of copper is less than 0.1 mg, the dosage can be lowered to $3 \times 25$ mg daily.

Little is known about the combination of D-penicillamine and zinc, but it is probable that $Zn^{2+}$ and D-penicillamine interact in the intestinal tract, thereby reducing the effectiveness of both substances. A comparison of Cu balance after treatment with zinc alone versus zinc plus a chelator showed little difference, suggesting little advantage to combined treatment (Brewer *et al.*, 1993). Because the combination of zinc salts with triethylene-tetramin dihydrochloride (Trientine) increases the risk of sideroblastic anemia, a combination is not recomended.

### Potassium Sulfate

Potassium sulfate forms a compound with copper in the intestinal tract and thus prevents copper resorption. The average dosage is 20–40 mg PO tid before meals. Patients are generally uncooperative because of the unpleasant taste of potassium sulfate, but no serious side effects are known.

### Low-copper Diet

The popularity of low-copper diets has generally lost out to the other therapeutic options. However, the benefits of moderate copper reduction (using food tables) may be underestimated. Foods with high copper content include shellfish, mushrooms, liver, chocolate, nuts, and broccoli.

## Liver Transplant

In cases with fulminant hepatitis or advanced cirrhosis of the liver, a liver transplant is recommended, which leads to persistent normalization of the serum copper and ceruloplasmin levels (Hefter *et al.*, 1991; Starzl *et al.*, 1971; Sternlieb, 1990). Dramatic improvement in severe neurological symptoms that had not responded to chelation therapy has also been reported after liver transplant (Polson *et al.*, 1987; Robles *et al.*, 1999).

## Symptomatic Treatment

In addition to these specific therapeutic steps, symptomatic treatment of symptoms can be tried. However, drugs used in Parkinson's disease such as L-dopa are disappointing. Some patients benefit from anticholinergic drugs such as benzhexol. Dantrolene and baclofen have not helped dystonia and spasms (Walshe and Yealland, 1993).

## Asymptomatic Patients

In these patients, zinc acetate monotherapy (3 × 50–100 mg/day) should be initiated with regular clinical follow-up. As in symptomatic patients, the goal of therapy is to achieve and maintain a normal serum free copper level of 10–15 µg/dl.

## Pregnancy

Because D-penicillamine and trientine are teratogenic in animal experiments, it is recommended that the dose should be not more than 750–1000 mg/day in the first two trimesters in patients whose Wilson's disease was well controlled and be reduced to 500 mg/day during the last trimester (Sternlieb, 2000). Concerns regarding possible teratogenic effects of both substances are not supported by the available data concerning pregnancies in patients with Wilson's disease (Sternlieb, 2000). As an alternative treatment, zinc (3 × 50–100 mg/day) may be used in pregnant patients with Wilson's disease. Brewer *et al.* report on a rate of 7.7% congenital defects in 26 live births compared with about 4% in the healthy population (Brewer *et al.*, 2000).

## REFERENCES

Arima, M., Takeshita, K., Yoshino, K., Kitahara, T., Suzuki, Y. (1977). Prognosis of Wilson's disease in childhood. *Eur. J. Pediatr.* **126**, 147–154.

Bellary, S. V., and Van Thiel, D. H. (1993). Wilson's disease: A diagnosis made in two individuals greater than 40 years of age. *J. Okla. State Med. Assoc.* **86**, 441–444.

Brewer, G. J. (1999a). Penicillamine should not be used as initial therapy in Wilson's disease. *Mov. Disord.* **14**, 551–554.

Brewer, G. J. (1999b). The treatment of Wilson's Disease. *In* "Copper Transport and its Disorders" (A. Leone and J. F. B. Mercer, Eds.), *Adv. Exp. Med. Biology*, **448**, 115–126.

Brewer, G. J. (2000). Recognition, diagnosis, and management of Wilson's disease. *Proc. Soc. Exp. Biol. Med.* **223**, 39–46.

Brewer, G. J., Dick, R. D., Johnson, V. C., Brunberg, J. A., Kluin, J., and Fink, J. K. (1998). Treatment of Wilson's disease with zinc: XV. Long-term follow-up studies. *J. Lab. Clin. Med.* **132**, 264–278.

Brewer, G. J., Dick, R. D., Johnson, V., Wang, Y., Yuzbasiyan-Gurkan, V., Kluin, K., and Fink, J. F. (1994). Treatment of Wilson's disease with ammonium tetrathiomolybdate. *Arch. Neurol.* **51**, 545–554.

Brewer, G. J., Dick, R. D., Yuzbasiyan-Gurkan, V., Tankanow, R., Young, A. B., and Kluin, K. J. (1991). Initial therapy of patients with Wilson's disease with tetrathiomolybdate. *Arch. Neurol.* **48**, 42–47.

Brewer, G. J., Hill, G. M., Prasea, A. S., Cossack, Z. T., and Rabbani, P. (1983). Oral zinc therapy for Wilson's disease. *Ann. Intern. Med.* **99**, 314–320.

Brewer, G. J., Hill, G. M., Dick, R. D., Nostrant, T. T., Sams, J. S., Wells, J. J., and Prasad, A. S. (1987b). Treatment of Wilson's disease with zinc: III. Prevention of reaccumulation of hepatic copper. *J. Lab. Clin. Med.* **109**, 526–531.

Brewer, G. J., Johnson, V. D., Dick, R. D., Hedera, P., Fink, J. K., Kluin, K. (2000). Treatment of Wilson´s disease with zinc. XVII: Treatment during pregnancy. *Hepatology* **31**, 364–370.

Brewer, G. J., Terry, C. A., AIsen, A. M., and Hill, G. M. (1987a). Worsening of neurologic syndrome in patients with Wilson's disease with initial penicillamine therapy. *Arch. Neurol.* **44**, 490–493.

Brewer, G. J., Yuzbasiyan-Gurkan, V., Johnson, C., Dick, R. D., and Wang, Y. (1993). Treatment of Wilson's disease with zinc: XI. Interaction with other anticopper agents. *J. Am. Coll. Nutr.* **12**, 26–30.

Bull, P. C., Thomas, G. R., Rommens, J. M., Forbes, J. R., and Cox, D. W. (1993). The Wilson's disease gene is a putative copper transporting P-type ATPase similar to the Menkes gene. *Nature Genet.* **5**, 327–337.

Czaja, M. J., Weiner, F. R., Schwarzenberg, F. J., Sternlieb, I., Scheinberg, I. H., Van Thiel, D. H., LaRusso, N. F., Giambrone, M. A., Kirschner, R., Koschinsky, M. L., *et al.* (1987). Molecular studies or ceruloplasmin deficiency in Wilson's disease. *J. Clin. Invest.* **80**, 1200–1204.

Danks, D. M., and Stevens, B. J. (1969). Diagnosis of Wilson's disease in children with liver disease. A report of two families. *Lancet* **1**, 22–25.

Denny-Brown, D. (1964). Hepatolenticular degeneration (Wilson's disease). Two different components. *N. Engl. J. Med.* **270**, 1149–1156.

Dobyns, W. B., Goldstein, N. P., and Gordon, H. (1979). Clinical spectrum of Wilson's disease. *Mayo Clin. Proc.* **54**, 35–42.

Duchen, L. W., and Jacobs, J. M. (1992). Nutritional deficiencies and metabolic disorders. *In* "Greenfield's Neuropathology" (J. H. Adams and L. W. Duchen, Eds.), Ed. 5, Chapter 13, pp. 838–841. Oxford University Press, New York.

Farrer, L. A., Bowcock, A. M., Hebert, J. M., Bonné;-Tamir, B., Sternlieb, I., Giagheddu, M., St. George-Hyslop, P., Frydman, M., Lössner, J., Demelia, L., Carcassi, C., Lee, R., Beker, R., Bale, A. E., Donis-Keller, H., Scheinberg, I. H., and Cavalli-Sforza, L. L. (1991). Predictive testing for Wilson's disease using tightly linked and flanking DNA markers. *Neurology* **41**, 992–999.

Floris, G., Medda, R., Padiglia, A., Musci, G. (2000). The physiopathological significance of ceruloplasmin. *Biochem. Physiol.* **60**, 1735–1741.

Gibbs, K., and Walshe, J. M. (1979). A study of the coeruloplasmin concentrations found in 75 patients with Wilson's disease, their kinships and various control groups. *Q. J. Med.* **48**, 447–463.

Gu, M., Cooper, J. M., Butler, P., Walker, A. P., Mistry, P. K., Dooley, J. S., Schapira, A. H. V. (2000). Oxidative-phosphorylation defects in liver of patients with Wilson's disease. *Lancet* **356**, 469–474.

Hefter, H., Rautenberg, W., Kreuzpainter, G., Arendt, G., Freund, H.-J., Pichlmayr, R., and Strohmeyer, G. (1991). Does orthotopic liver transplantation heal Wilson's disease? Clinical follow-up of two liver-transplanted patients. *Acta Neurol. Scand.* **84**, 192–196.

Hill, G. M., Brewer, G. J., Prasad, A. S., Hydrick, C. R., and Hartmann, D. E. (1987). Treatment of Wilson's disease with zinc. I. Oral zinc therapy regimens. *Hepatology* **7**, 522–528.

Hoogenraad, T. U. (1998). Zinc treatment of Wilson's disease. *J. Lab. Clin. Med.* **132**, 240–241.

Hoogenraad, T. U., Koevoet, R., and deRuyter Korver, E. G. W. M. (1979). Oral zinc sulphate as long-term treatment in Wilson's disease (hepatolenticular degeneration). *Eur. Neurol.* **18**, 205–211.

LeWitt, P. A. (1999). Penicillamine as a controversial treatment for Wilson's disease. *Mov. Disord.* **24**, 555–556.

Loudianos, G. and Gitlin, J. D. (2000). Wilson's disease. *Semin. Liver Dis.* **20**, 353–364.

Maier-Dobersberger, T., Ferenci, P., Polli, C., Balaac, P., Dienes, H. P., Kaserer, K., *et al.* (1997). Detection of the His1069Gln mutation in Wilson disease by rapid polymerase chain reaction. *Ann. Intern. Med.* **127**, 21–26.

Martin, P. (1968). Wilson's disease. In "Handbook of Clinical Neurology" (P. J. Vinken and G. W. Bruyn, Eds.), Vol. 6, pp. 267–278. North-Holland, American Elsevier, New York.

Nartey, N. O., Frei, J. V., and Cherian, M. G. (1987). Hepatic copper and metallothionine distribution in Wilson's disease (hepatolenticular degeneration). *Lab. Invest.* **57**, 397–401.

Oder, W., Prayer, L., Grimm, G., Spatt, J., Ferenci, P., Kollegger, H., Schneider, B., Gangl, A., and Deecke, L. (1993). Wilson's disease: Evidence of subgroups derived from clinical findings and brain lesion. *Neurology* **43**, 120–124.

Oertel, W. H., Tatsch, K., Schwarz, J., Kraft, E., Trenkwalder, C., Scherer, J., Weinzierl, M., Vogel, T., and Kirsch, C. M. (1992). Decrease of D2 receptors indicated by 123I-idobenzamide single-photon emission computed tomography relates to neurological deficit in treated Wilson's disease. *Ann. Neurol.* **32**, 743–748.

Polson, R. J., Rolles, K., Calne, R. Y., Williams, R., and Marsden, D. (1987). Reversal of severe neurological manifestations of Wilson's disease following orthotopic liver transplantation. *Q. J. Med.* **64**, 684–691.

Robles, R., Parrilla, P., Scillia, J., Ramírez, P., Bueno, F. S., Rodríguez, J. M., Luján, J. A., Fernandez, J. A., López, J. (1999). Indications and results of liver transplants in Wilson's diesease. *Transplant. Proc.* **31**, 2453–2454.

Schaefer, M., Roelofsen, H., Wolters, H., Hofmann, W. J., Müller, M., Kuipers, F., Stremmel, W., Vonk, R. J. (1999). Localization of the Wilson's diesease protein in human liver. *Gastroenterology* **117**, 1380–1385.

Scheinberg, I. H., Jaffe, M. E., and Sternlieb, I. (1987). The use of trientine in preventing the effects of interrupting penicillamine therapy in Wilson's disease. *N. Engl. J. Med.* **317**, 209–213.

Scheinberg, I. H. and Sternlieb, I. (1984). Wilson's disease. *Major Probl. Intern. Med.* **23**, 171.

Scheinberg, I. H. and Sternlieb, I. (1995). Treatment of the neurologic manifestations of Wilson's disease. *Arch. Neurol.* **52**, 339.

Schiefermeier, M., Kollegger, H., Madl, C., Polli, C., Oder, W., Kühn, H.-J., Berr, F., Ferenci, P. (2000). The impact of apolipoprotein E genotypes on age at onset of symptoms and phenotypic expression in Wilson´s disease. *Brain* **123**, 585–590.

Spence, J. A., and Suttle, N. F. (1980). A sequential study of the skeletal abnormalities which develop in rats given a small dietary supplement of ammonium tetrathiomolybdate. *J. Comp. Pathol.* **90**, 139–153.

Starosta-Rubinstein, S., Young, A. B., Kluin, K., Hill, G., Aisen, A. M., Gabrielsen, T., and Brewer, G. J. (1987). Clinical assessment of 31 patients with Wilson's disease. Correlations with structural changes on magnetic resonance imaging. *Arch. Neurol.* **44**, 365–370.

Starzl, T. E., Giles, G., Lilly, J. R., Takagi, H., Martineau, G., Hatgrimson, C. G., Penn, J., and Potnam, C. W. (1971). Indications for orthotopic liver transplantation: With particular reference to hepatomas, bilary atresia, cirrhosis, Wilson's disease and serum hepatitis. *Transplant. Proc.* **3**(No. 1), 308–312.

Sternlieb, I. (1990). Perspectives on Wilson's disease. *Hepatology* **12**, 1234–1239.

Sternlieb, I. (2000). Wilson's disease and pregnancy. *Hepatology* **31**, 531–532.

Stremmel, W., Meyerrose, K.-W., Niederau, C., Hefter, C., Kreuzpaintner, G., and Strohmeyer, G. (1991). Wilson's disease: Clinical presentation, treatment, and survival. *Ann. Intern. Med.* **115**, 720–726.

Stremmel, W., Niederau, C., and Strohmeyer, G. (1989). Diagnostik und Therapie stoffwechselbedingter Lebererkrankungen. *Muench. Med. Wochenschr.* **131**, 257–261.

Tankanow, R. M. (1991). Pathophysiology and treatment of Wilson's disease. *Clin. Pharm.* **10**, 839–849.

Tanzi, R. E., Petrukhin, K., Chernov, I. *et al.* (1993). The Wilson's disease gene is a copper transporting ATPase with homology to the Menkes disease gene. *Nature Genet.* **5**, 344–350.

Thuomas, K. A., Aquilonius, S. M., Bergstrröm, K., and Westermark, K. (1993). Magnetic resonance imaging of the brain in Wilson's disease. *Neuroradiology* **35**, 134–141.

Van Den Hamer, C. J. A., and Hoogenraad, T. U. (1984). 64Cu loading tests for monitoring zinc therapy in Wilson's disease. *Trace Elem. Med.* **1**, 84–87.

Victor, M., Adams, R. D., and Cole, M. (1965). The acquired (non-Wilsonian) type of chronic hepatocerebral degeneration. *Medicine (Baltimore)* **44**, 345–396.

Walshe, J. M. (1986). Wilson's disease. In "Handbook of Clinical Neurology" (P. J. Vinken, G. W. Bruyn, and H. I. Klawans, Eds.), Vol. 5(49), pp. 223–238. North-Holland, American Elsevier, New York.

Walshe, J. M. (1988a). Wilson's disease: Yesterday, today, and tomorrow. *Mov. Disord.* **3**, 10–29.

Walshe, J. M. (1988b). Diagnosis and treatment of presymptomatic Wilson's disease. *Lancet* **2**, 435–437.

Walshe, J. M. (1999). Penicillamine: The treatment of first choice for patients with Wilson's disease. *Mov. Disord.* **14**, 545–550.

Walshe, J. M., and Yealland, M. (1992). Wilson's disease: The problem of delayed diagnosis. *J. Neurol. Neurosurg. Psychiatry* **55**, 692–696.

Walshe, J. M., and Yealland, M. (1993). Chelation treatment of neurological Wilson's disease. *Q. J. Med.* **86**, 197–204.

Williams, F. J. B. and Walshe, J. M. (1981). Wilson's disease: Analysis of the cranial computerized tomographic appearances found in patients and the changes in responses to treatment with chelating agents. *Brain* **104**, 735–752.

Yamaguchi, Y., Heiny, M. E., and Gitlin, J. D. (1993). Isolation and characterization of a human cDNA as a candidate gene for Wilson's disease. *Biochem. Biophys. Res. Commun.* **197**, 271–277.

Yang, F., Naylor, S. L., Lum, J. B., Cutshaw, S., McCombs, J. L., Naberhaus, K. H., McGill, J. D. R., Adrian, G. S., Moore, C. M., Barnett, D. R., and Bowman, B. H. (1986). Characterization, mapping, and expression of the human ceruloplasmin gene. *Proc. Natl. Acad. Sci. USA* **83**, 3257–3261.

CHAPTER 81

# Upper and Lower Motor Neuron Disorders

Gian Domenico Borasio and Stanley H. Appel

Motor neuron disorders can be classified according to the structure involved (central motor neurons originating from the cortex, peripheral motoneurons from the brainstem nuclei or the ventral horn of the spinal gray matter), the age at onset, the course of the disease, the primary site of involvement (bulbar muscles vs extremities, proximal vs distal), and the genetic evidence (hereditary vs sporadic). In the following chapter we will primarily discuss degenerative motor neuron disorders with onset in juvenile and adult age.

## AMYOTROPHIC LATERAL SCLEROSIS (ALS)

### Clinical Aspects

In ALS there is clinical and pathological involvement of both upper and lower motor neurons, usually with sparing of the oculomotor nuclei and the spinal Onuf nucleus, which supplies the anal sphincter muscles (Hirano and Iwata, 1978). Mild pathological changes can also be found in the spinocerebellar tracts and the posterior columns, especially in familial cases, usually without accompanying symptoms or signs (Williams, 1991).

Most patients have asymmetrical, distal weakness of the arm or leg. Bulbar onset occurs in 20%–30% of all cases, but in older women more than 50% have bulbar symptoms (Li *et al.*, 1990). The clinical hallmark of ALS is the coexistence of neurogenic atrophy, weakness, and fasciculations caused by lower motor neuron degeneration together with hyperactive or incongruously present deep tendon reflexes, pyramidal tract signs, and increased muscle tone caused by corticospinal tract involvement. Muscle cramps are often

already present before diagnosis. Pseudobulbar affect with uncontrolled crying or laughter is common. Sensation, mentation, sphincter control, and extraocular motility are usually spared, with only 2%–3% of patients showing an associated dementia. The long-known absence of decubitus sores even in bedridden ALS patients is thought to be related to biochemical skin changes (Kolde *et al.*, 1996).

Although the classic clinical presentation leaves little doubt about the diagnosis, certain ALS cases can be difficult to diagnose, particularly at an early stage. The differential diagnosis of ALS includes (for a review, see Louwerse *et al.*, 1991):

- Physical causes (e.g., spondylotic myelopathy)
- Immune disorders (multifocal motor neuropathy with or without anti-GM1-antibodies, myasthenic syndromes, inclusion body myositis, paraneoplastic syndromes)
- Toxins (e.g., lead, organophosphates)
- Infections (syphilis, borreliosis, Creutzfeldt-Jakob disease)
- Metabolic disorders (e.g., diabetes, hyperthyroidism, hyperparathyroidism, porphyria)
- Enzyme deficiencies (hexosaminidase)
- Other neurological disorders (motor neuropathy, myopathic syndromes, multiple sclerosis)

The diagnosis of ALS is essentially a clinical one. Additional investigations may be required at early stages to exclude treatable diseases. Investigations that are required at presentation include a thorough neurophysiological workup (Brooks, 1999), thyroid hormone levels, and serum immune electrophoresis to check for monoclonal gammopathy (Gordon *et al.*, 1997). Anti-GM1-antibodies should be measured only if there is neurographic evidence for conduction block and/or a pure lower motor neuron syndrome (see Chapter 88). Magnetic resonance imaging of the spine may be necessary to exclude spondylotic myelopathy when signs and symptoms are restricted to one extremity. A spinal tap is often performed but is rarely contributory. A muscle biopsy should be considered when the presentation is atypical, and an alternative diagnosis such as inclusion body myositis is suspected. A tumor search is usually not warranted, except for female patients presenting with primary lateral sclerosis (PLS, see "ALS Variants" [Forsyth *et al.*, 1997]). Diagnostic criteria for ALS (the so-called "El Escorial" criteria) have been established on the basis of international expert consensus and have been recently revised (Brooks *et al.*, 1998). These criteria are primarily useful as inclusion criteria for therapeutic trials, but they also provide a guideline for the practicing neurologist (Figure 1).

### ALS Variants

The term primary lateral sclerosis (PLS) denotes a progressive disease of the corticospinal tracts, with or without corticobulbar involvement and with a generally slower course (Le Forestier *et al.*, 2001). Progressive bulbar palsy (PBP) indicates a progressive disease of the bulbar motor nuclei and has a worse prognosis (Bruyn, 1991). Most, but not all, patients presenting as PLS or

PBP ultimately have signs and symptoms of classical ALS develop. Progressive muscular atrophy (PMA) shows almost exclusive involvement of the lower motoneuron. The rapid clinical progression is the main distinguishing feature from SMA type IV. Some families with SOD1-linked familial ALS may present as PMA (Cervenakova *et al.*, 2000).

### Familial ALS

Around 5%–10% of all ALS cases are familial; most show an autosomal dominant inheritance pattern with a balanced gender ratio. In approximately 20% of patients with familial ALS (FALS), the genetic defect has been shown to reside in the gene for the enzyme superoxide dismutase on chromosome 21 (Rosen *et al.*, 1993), resulting in a toxic gain of function of the mutated enzyme (Julien, 2001). Apart from a slightly earlier age at onset (fourth to fifth decade), clinical presentation and course of the familial cases are indistinguishable from the sporadic ones (Figlewicz *et al.*, 1989). A rare form of recessive FALS has been linked to chromosome 2q33–q35 (Hentati *et al.*, 1994), and a form of FALS with frontotemporal dementia maps to chromosome 9q21–22 (Hosler *et al.*, 2000). Further loci and modifying genetic factors (e.g., deletion of the SMN2 gene [Veldink *et al.*, 2001]) are under study (for review, see Robberecht, 2000).

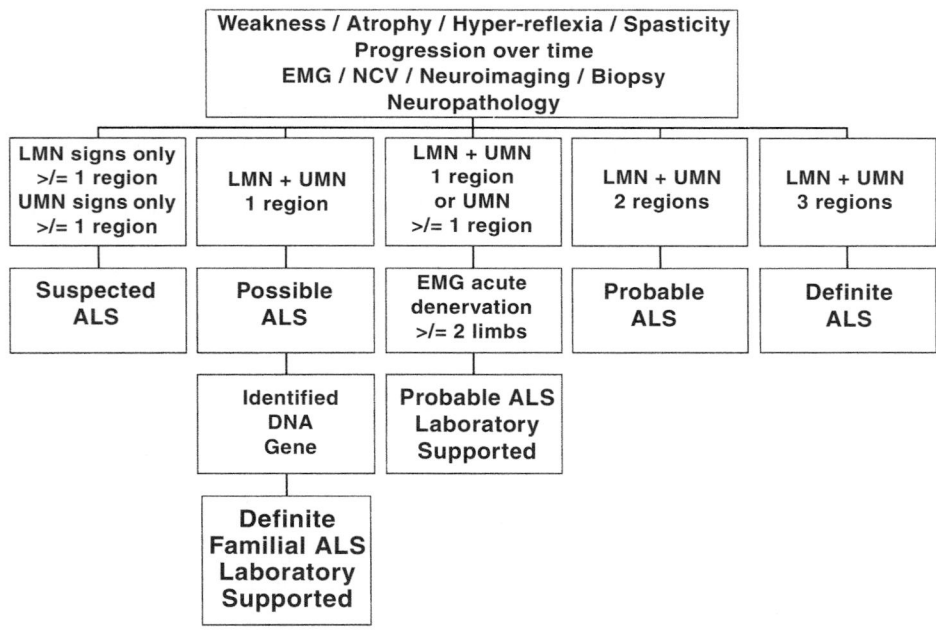

**FIGURE 1**    Revised El Escorial Criteria. UMN, upper motor neurons; LMN, lower motor neurons; NCV, nerve conduction velocities; EMG, electromyography. Regions are defined as bulbar, cervical, thoracic and lumbosacral. From Brooks *et al.* (1998) with permission.

## Natural Course

The annual incidence of ALS is about 1.5–2/100,000 population, with a prevalence of approximately 6–8/100,000 (Brooks, 1996). It accounts for about 0.1% of all adult deaths. A tendency towards an increased incidence of ALS has been reported (Lilienfeld *et al.*, 1989). Although rare cases may begin before the age of 20 years, most cases begin after age 40; the mean age at onset is around 58 (Gubbay *et al.*, 1985). The incidence of ALS, long thought to decline after age 70, seems instead to continue to rise with advancing age. Men are more commonly affected (1.5–2:1). The clustering of ALS cases in Guam and the Japanese Kii Peninsula has been attributed to a foodborne neuro-toxin—cycasin—(Spencer *et al.*, 1989), but the variable interval between exposure and onset of disease (often as long as several decades) has been difficult to explain by a toxin.

In ALS there is remarkable variability in the rate of disease progression (Haverkamp *et al.*, 1995). It is not uncommon to see patients with long phases of very slow progression, but bona fide remissions are very rare. Sudden worsening or relapses such as are common in multiple sclerosis do not usually occur in ALS. Patients with bulbar onset or PBP have the worst prognosis (median survival, 2–2.5 years), which is probably explained by the older age of onset in such patients. Average disease duration is in the range of 3–5 years, 10% of patients survive >10 years, and disease courses of >40 years have been reported (Grohme *et al.*, 2001). Survival is generally prolonged in younger patients (Caroscio *et al.*, 1984) and in patients with PLS (Le Forestier *et al.*, 2001). The terminal stage is character-ized by respiratory muscle weakness. The most common cause of death in ALS is respiratory failure, often in conjunction with aspiration pneumonia (Neudert *et al.*, 2001) (see also "Practical Management").

## Principles of Therapy

The cause of sporadic ALS is unknown. Conse-quently, every therapeutic attempt is experimental. Various hypotheses have been proposed, including immunologic abnormalities, metabolic disturbances, neurotransmitter imbalance, chronic intoxication (heavy metals, plant toxins), lack of neurotrophic support, and impaired DNA repair (for review, see Al-Chalabi and Leigh, 2000). A report of a high fre-quency of enteroviral DNA in postmortem tissue of ALS patients (Berger *et al.*, 2000) was not confirmed in a subsequent study (Walker *et al.*, 2001). The role of impaired mitochondrial energy metabolism has received increasing attention (Beal, 2000).

The finding of increased glutamate in the CSF and decreased glutamate uptake in spinal cord nerve termi-nals of ALS patients led to the hypothesis of a chronic excitotoxic mechanism underlying ALS pathogenesis. This hypothesis has been strengthened by the finding of a selective loss of a glial glutamate transporter in ALS patients (Rothstein *et al.*, 1995), but the exact signifi-cance of this finding is controversial (Meyer *et al.*, 1999). Several trials with substances that impinge on glutamate activity or metabolism (e.g., branched-chain aminoacids, dextromethorphan, and lamotrigine) showed no effect.

The antiglutamatergic agent Riluzole has been shown to alter the natural history of the disease and has been estimated to prolong life by about 3 months in patients with ALS (Lacomblez *et al.*, 1996)**. The drug is approved in the United States and Europe but not in Canada and Australia. A number of concerns about the therapeutic effect persist: the lack of benefit observed on most secondary measures of efficacy, the quite modest prolongation of survival, and the high costs (approxi-mately $10,000 per year in the United States, and £3000 per year in the United Kingdom). The pros and cons of Riluzole should be discussed with patient and family (AAN, 1997). Around 10%–15% of patients will dis-continue treatment because of side effects (asthenia, dizziness, gastrointestinal problems), and a few will have to stop because of significant liver enzyme eleva-tion, which requires monitoring. It is important to tell the patients in advance that there is no subjectively per-ceived benefit, such as increase in muscle strength, to be expected from Riluzole treatment and that the drug will not stop the disease from progressing.

The role of the immune system has been implicated in the pathogenesis of ALS (Appel *et al.*, 1995) based on clinical and basic science observations (Smith *et al.*, 1992, 1994). However, immunosuppressant therapy has been as relatively ineffective as has therapy targeted at other implicated processes, including free radical mechanisms, mitochondrial dysfunction, or altered calcium homeostasis. As noted earlier, even the effects of therapy targeted to inhibit glutamate excitotoxicity have been quite modest. It is likely that therapy aimed at any single pathogenic mechanism of disease might be marginally effective. Furthermore, it is unlikely that therapies will be successful if only targeted at the neuron. Because neurons in ALS spinal cord are sur-rounded by activated microglia that can release injuri-ous cytokines and free radical toxins, future therapies targeted at these inflammatory cells may well provide significant benefit.

Hopes for a new therapeutic approach have been raised by reports about trophic factors that can prevent motor neuron degeneration under experimental condi-tions. However, clinical trials with ciliary neurotrophic factor and brain-derived neurotrophic factor showed no clear clinical benefit (Anonymous Study Group, 1996, 1999). The results of a clinical trial with insulin-like growth factor I (IGF-I) showed a reduction of disease progression by about 25% (Lai *et al.*, 1997), which, however, was not confirmed in a parallel study (Borasio *et al.*, 1998a). Xaliproden, a compound with trophic activities in experimental models, was tested in two randomized studies. In patients already taking Riluzole, there was no clear evidence for an additional effect. Patients without Riluzole showed a small reduction in decline of respiratory function without a significant effect on survival (Meininger, 2000).

The establishment of transgenic mouse models for familial ALS (Gurney *et al.*, 1996) has allowed for in vivo screening of potential therapeutics. Of particular interest is the neuroprotective effect reported for the nonprescription dietary supplement creatine in this model system (Klivenyi *et al.*, 1999). Given the low toxicity of the substance, many patients choose to take creatine despite the absence of human studies. The usual dose is 5 g/day with 1 day pause/week.

Future avenues for curative treatment in ALS may include new glutamate antagonists, small molecular weight trophic substances (Borasio *et al.*, 1998b), neuroimmunophilin ligands, and possibly gene and stem cell therapy (Hottinger *et al.*, 2000). Physicians should encourage patient participation in clinical trials, which is often beneficial per se.

### Preventive Therapy

Clinically, the availability of genetic testing for relatives of SOD1-linked FALS patients poses a dilemma. On theoretical grounds, substances that are effective in the SOD1 transgenic mouse model (such as creatine, Riluzole, and possibly even red wine [Esposito *et al.*, 2000]) may be regarded as potentially preventive drugs for persons at risk for FALS. However, no data from human studies are available so far. Thus, the uncertain benefits must be weighed against the burden of knowledge of the high-risk status as carriers of the mutation. A decision can only be reached on a case-by-case basis through careful and empathic genetic counseling and discussion of the available options with FALS patients and their families. The guidelines for genetic testing in Huntington's disease provide a good model (see Chapter 79).

### Alternative Therapies

Many ALS patients turn to alternative treatments because of dissatisfaction with the effectiveness of available drugs. Often, this is not discussed with the physician because of fear of "condemnation" (Wasner *et al.*,

2001). Therefore, it is best to approach this subject proactively, offering to discuss whatever therapeutic options the patient might wish to try. This will enable the physician to protect the patient from treatments that might entail serious financial and/or medical risks while preserving hope and maintaining trust in the patient–physician relationship.

### Practical Management

Even in the absence of a satisfactory causative therapy, much can be done to alleviate the suffering of the patients and their families. The available evidence for the care of patients with ALS has been recently reviewed (Miller *et al.*, 1999). Almost all signs and symptoms of ALS can be ameliorated by palliative measures (Borasio *et al.*, 2001). Palliative care in ALS starts at the time of diagnosis and requires a coordinated interdisciplinary and multiprofessional approach (Table I; Oliver *et al.*, 2000), with special attention to psychosocial and spiritual issues (Robbins *et al.*, 2001).

*Breaking the news.* Once diagnosis is established, all aspects of the disease should be openly discussed with the patient and relatives. This requires considerable empathy on the physician's port and may often be best performed in a stepwise fashion as the disease progresses, underscoring positive aspects such as absence of sudden deteriorations; sparing of cognition, sensation, and continence; and variability of the disease course (Borasio *et al.*, 1998c). Emphasis should be placed on the available symptomatic treatment options.

*Muscle weakness* should be managed by regular exercise, never to the point of fatigue, and by the progressive installment of the proper assistive devices (cane, ankle-foot orthosis, wheelchair, aids for clothing and eating, etc.) to maintain as high as possible a degree of independence and mobility. Occasionally, a short-term increase of muscle strength may be achieved by pyridostigmine 40–60 mg tid.

*Dysphagia* should first be treated by an adjustment in diet consistency (a recipe book for ALS patients is available from the Muscular Dystrophy Association, see "Patients' Associations"). Specific swallowing techniques (such as supraglottic swallowing) can help to

---

TABLE I  Palliative Care in ALS: Who Is Involved?

| | |
|---|---|
| • Chaplain | • Physical therapist |
| • Counselor | • Physician |
| • Dietitian | • Psychologist |
| • Hospice worker | • Relatives |
| • Lay associations | • Social worker |
| • Nurse | • Speech therapist |
| • Occupational therapist | • Swallowing therapist |

prevent aspiration. When the oral caloric intake is insufficient (weight loss >10%), it is best to perform a percutaneous endoscopic gastrostomy (PEG), which is usually very well tolerated if the vital capacity is >50% at time of introduction. A radiologically inserted gastrostomy (RIG) may be an alternative in more advanced stages.

*Dysarthria* can lead to a complete loss of oral communicative abilities. Logopedic treatment is helpful at the beginning. Electronic communication devices can be a blessing in advanced cases.

*Dyspnea* is the most severe symptom in ALS. Fears of death by choking arise in almost all patients at the onset of dyspneic symptoms and should be met by a frank discussion of the terminal stage of the illness. The latter is characterized by a mostly nocturnal, hypercapnia-induced light coma (i.e., most patients die peacefully in their sleep) (Neudert *et al.*, 2001).

At the beginning of dyspnea, chest physiotherapy is helpful. Dyspneic attacks usually have a pronounced anxiety component and are best managed by short-acting benzodiazepines (lorazepam SL 0.5–1 mg). In more advanced stages, chronic nocturnal hypoventilation ensues, which may considerably hamper the patient's quality of life (the symptoms are shown in Table II).

Noninvasive intermittent ventilation by means of a mask (NIV) is an efficient and cost-effective means of alleviating these symptoms (Cazzolli and Oppenheimer, 1996), which may even prolong the patient's life span considerably (Aboussouan *et al.*, 1997). It should be discussed with the patient and family at the onset of symptoms of chronic hypoventilation or when the vital capacity drops below 50% (Miller *et al.*, 1999). They should be informed about the temporary nature of the measure, which is primarily directed toward improving quality of life rather than prolonging it (as opposed to tracheostomy). The problem with mechanical ventilation is usually not related to cost or technical difficulties but to the increasing care needs of ventilated patients. A slow progression, good communication skills, mild bulbar involvement, a strong motivation on the patient's side, and a supportive family environment argue in favor of the initiation of NIV. To be effective, NIV needs to be administered for at least 4 h/day, most preferably at night (Kleopa *et al.*, 1999). It is very important to reassure the patients that, whenever they may decide to stop NIV, all necessary care and appropriate medication will be available to ensure a peaceful death (Borasio and Voltz, 1998). The physician has a legal and ethical duty to honor a patient's request for discontinuation of such treatment (AAN, 1998). End-of-life decisions should be discussed in advance with the patient and family, and the physician should encourage and facilitate the establishment of appropriate advance directives (Borasio and Voltz, 2000).

*Psychological symptoms* usually consist of reactive depression, starting shortly after establishment of the diagnosis. In severe cases, antidepressants such as amitriptyline may be indicated (see Table I). Counseling should always be extended to the family members. Despite a reportedly high interest of ALS patients in physician-assisted suicide (Ganzini *et al.*, 1998), suicide attempts are uncommon (Neudert *et al.*, 2001).

*Sleep disorders* are usually a consequence of the inability to change position during sleep. Psychic problems, muscle cramps, fasciculations, dysphagia, and dyspnea can also impair sleep. Sedatives (see Table I) should be used sparingly, because they may impair residual muscle force.

*Thick mucous secretions*: This symptom, which results from a combination of diminished fluid intake and reduced coughing pressure, is difficult to treat. N-acetylcysteine is helpful only in a minority of cases. Suction is usually not fully effective unless performed by means of a tracheostomy. Both manually assisted coughing techniques and mechanical insufflation-exsufflation can assist in extracting excess mucus from the airway (Hanayama *et al.*, 1997). Intermittent positive vibration devices are specialized inhalators that deliver a pressurized, intermittent flow of nebulized saline with or without expectorants. They are used for 10–15 minutes at a time and can assist in clearing pulmonary and bronchial secretions (Gelinas and Miller, 2000). Physical therapy with vibration massage may also be helpful, especially in the initial stages.

Other symptoms of ALS, which can be relieved by appropriate medication, include *muscle cramps, fasciculations, spasticity, drooling*, and *pathological laughing/crying*. Treatment options for these symptoms are shown in Table III. For cases with therapy-refractory drooling, botulinum toxin injections in the salivary glands may be considered (Giess *et al.*, 2000).

*Psychosocial care.* In ALS, the burden of the relatives often exceeds that of the patients and deserves particu-

**TABLE II  Symptoms of Chronic Hypoventilation**

- Daytime fatigue and sleepiness, concentration problems
- Difficulty falling asleep, disturbed sleep, nightmares
- Morning headache
- Nervousness, tremor, increased sweating, tachycardia
- Depression, anxiety
- Tachypnea, dyspnea, phonation difficulties
- Visible efforts of auxiliary respiratory muscles
- Recurrent or chronic upper respiratory tract infections
- Cyanosis, edema
- Vision disturbances, dizziness, syncope
- Reduced appetite, weight loss, recurrent gastritis
- Diffuse pain in head, neck, and extremities

TABLE III  Symptomatic Medication in ALS (in Order of Recommendation)

| | Dosage* |
|---|---|
| **Fasciculations and muscle cramps** | |
| If mild: | |
|    Magnesium | 5 mmol qd–tid |
|    Vitamin E | 400 IE bid |
| If severe | |
|    Quinine sulfate | 200 mg bid |
|    Carbamazepine | 200 mg bid |
|    Phenytoin | 100 mg qd–tid |
| **Spasticity** | |
| Baclofen | 10–80 mg |
| Tizanidine | 6–24 mg |
| Memantine | 10–60 mg |
| Tetrazepam | 100–200 mg |
| **Drooling** | |
| Glycopyrrolate | 0.1–0.2 mg SC/IM tid |
| Transdermal hyoscine patches | 1–2 patches |
| Amitriptyline | 10–150 mg |
| Atropine/benztropine | 0.25–0.75 mg/1–2 mg |
| Clonidine | 0.15–0.3 mg |
| **Pathologic laughing/crying** | |
| Amitriptyline | 10–150 mg |
| Fluvoxamine | 100–200 mg |
| Lithium carbonate | 400–800 mg |
| L-Dopa | 500–600 mg |
| **Sedatives** | Dosage nocté |
| Chloral hydrate | 250–1000 mg |
| Diphenhydramine | 50–100 mg |
| Flurazepam (beware of respiratory depression) | 15–30 mg |

*Usual range of adult daily dosage; some patients may require higher dosages (e.g., of antispastic medication).

lar attention (Rabkin *et al.*, 2000). The physician must be particularly sensitive to the needs and fears of the patients' children and the importance of helping patients in their role as parents. Patients' associations (see list at end of chapter) may provide invaluable help and assistance and should be involved in patient care from the very beginning. Referral to a tertiary care center with an interdisciplinary team may ease the burden for the practicing neurologist and may also be a means of providing hope (a list of ALS centers worldwide can be found at www.wfnals.org).

*Spiritual care and bereavement.* The role of spiritual care is often underestimated. A recent study indicated that spirituality or religiousness may affect the use of PEG and NIV in ALS and may be a source of comfort to the patients (Murphy *et al.*, 2000). Cases of patients whose spiritual practice greatly enhanced their ability to cope with ALS have been reported (Borasio, 2001). A simple structured interview to assess the patients' spiritual needs has been recently developed (Puchalski and Romer, 2000). Spiritual care is not limited to patients but should encompass the whole family as a means of preventing problems during bereavement (McMurray, 2000), which may be particularly severe in ALS families (Martin and Turnbull, 2000). It is important to acknowledge that the process of bereavement in ALS actually starts immediately after the diagnosis is communicated, in the form of so-called "anticipatory grief," and that callous delivery of the diagnosis may affect the psychological adjustment to bereavement (Ackerman and Oliver, 1997).

*Terminal phase.* Approximately 90% of patients with ALS will die peacefully, mostly in their sleep. "Choking to death" is exceedingly rare (Neudert *et al.*, 2001). In the absence of artificial ventilation, the death process usually begins with the patients slipping from sleep into coma because of increasing hypercapnia. If restlessness or signs of dyspnea develop, morphine should be administered beginning with 2.5–5 mg PO, SL, SC, or IV every 4 hours (if necessary in combination with chlorpromazine as an antiemetic). Because morphine is not an anxiolytic drug, anxiety should be treated with lorazepam SL (beginning with 1–2.5 mg) or midazolam PO/SC (beginning with 1–2 mg). The dosage of morphine and anxiolytics should be increased until satisfactory symptom control is achieved. The potential of these drugs to induce respiratory depression is usually overestimated and is irrelevant in the terminal phase according to the doctrine of double effect (Borasio and Voltz, 1998).

Most patients with ALS wish to die at home. This can often best be achieved through enrollment of the patient in a hospice program (Oliver, 1996). It is advisable for the physician to initiate contact with the hospice institution well in advance of the terminal phase. If death at home is not possible, inpatient hospices or palliative care units should be considered. Hospice teams can also assist the relatives' bereavement after the patient's death.

### Treatments No Longer Recommended

Because of the poor prognosis of the disease, a great number of substances have been the subject of clinical testing. So far, none of them, although partly recommended for symptomatic relief, has been shown to influence the course of the disease.

The substances tested include vitamins ($B_{12}$, D, and E), idoxuridine, isoprinosine, transfer factor, tilerone, levamisole, lipoic acid, pancreas and liver extracts, proteinase inhibitors, lecithine, testosterone, L-dopa, threonine, amitriptyline, baclofen, naloxone, penicillamine, modified snake neurotoxin, magnesium pemoline, gangliosides, amantadine, guanidine, flunarizine, thyeotropin-releasing hormone (TRH), fresh-frozen plasma, interferon, steroids, cyclophosphamide, cyclosporin A, plasmapheresis, *n*-acetylcysteine, dextromethorphane, branched-chain aminoacids, lamotrigine, total lymphoid irradiation (for review see Mitsumoto, 1995). In recent years, further negative trials were performed with nimodipine, verapamil, IV immunoglobulins, and gabapentin (Miller *et al.*, 2001).

## SPINAL MUSCULAR ATROPHY (SMA)

In contrast to ALS, most of the cases of SMA are hereditary. The degenerative process is usually restricted to the spinal ventral horn motoneurons but may sometimes involve the motoneurons of the caudal cranial nerve brainstem nuclei. Because of the accumulation of molecular genetic evidence, the classification of the SMAs is undergoing constant changes, a process that will continue until all genes responsible for the hereditary SMAs have been cloned. For practical purposes, the classification shown in Table IV will be used, which reflects genetic cosiderations, age at onset, and the site of primary clinical involvement.

### Proximal SMA

#### SMA Type I–III

The clinical distinction between SMA I (acute infantile form, Werdnig-Hoffmann), SMA II (intermediate form), and SMA III (juvenile form, Kugelberg-Welander) has been overturned by genetic evidence showing that all three forms are due to mutations of the same locus on chromosome 5q. The inheritance mode is autosomal-recessive; dominant transmission is extremely rare. The disease is defined genetically by the presence of homozygous deletions/mutations in exon 7 and 8 of the telomeric survival motor neuron (SMN1) gene (Wirth, 2000). The number of copies of the centromeric SMN gene (SMN2) modifies the SMA phenotype (Brahe, 2000).

Clinically, the disease represents a continuum between an acute, congenital form with fatal outcome

**TABLE IV** Spinal Muscular Atrophies

| |
|---|
| Proximal SMA |
|   SMA type I–III (Werdnig-Hoffmann, intermediate, and Kugelberg-Welander type) |
|   SMA type IV |
|   Bulbospinal muscular atrophy (Kennedy's disease) |
| Distal SMA |
| Scapuloperoneal form |
| Monomelic amyotrophy |
|   Juvenile distal and segmental SMA of the upper extremities (Hirayama's disease) |

before 3 years of age (SMA I) and a chronic, juvenile-onset form with a more benign course (SMA III)(Zerres and Rudnik-Schöneborn, 1995). At the latter end of the disease spectrum, onset is usually insidious, with bilateral symmetrical atrophy and weakness of the pelvic girdle and proximal leg muscles, followed by the shoulder girdle and upper arm muscles. Fasciculations are present in about half the cases; tendon reflexes are weak to absent. EMG and muscle biopsy show neurogenic changes, and the CPK may be raised. The ability to walk can be retained for 10 years and longer; the life expectancy is almost normal.

#### SMA Type IV

The onset of this form is usually in the fourth decade of life. Both autosomal-recessive and autosomal-dominant transmission have been reported. The latter shows no linkage to the gene locus for SMA I–III (Kausch *et al.*, 1991). The autosomal-recessive form has a similar, but more prolonged, course, as described earlier for SMA III. Some patients may show SMN deletions (Zerres *et al.*, 1995). Patients usually need a walking aid 20 years from onset and have a normal life expectancy. The autosomal-dominant form carries a worse prognosis, the life expectancy being about 20 years from disease onset (Pearn *et al.*, 1978). Differentiation from muscular dystrophy or ALS may be difficult at early stages.

#### *Bulbospinal Muscular Atrophy (Kennedy's Disease)*

In this rare syndrome, the inheritance mode is X-linked recessive. The onset is between 20 and 40 years. Proximal weakness develops first in the lower limbs, then spreads to the shoulder girdle and bulbar muscles. Cranial motor nerve involvement results in tongue atrophy and fasciculations of temporal, facial, and tongue muscles with dysarthria and dysphagia. The clinical observations of concomitant gynecomastia, testicular atrophy, and reduced fertility has led to the identification of the androgen receptor gene as the site of mutation in this disorder (La Spada *et al.*, 1991). Sensory nerve action potentials are reduced or absent in 95% of patients (Ferrante and Wilbourn, 1997). Clinical presentation is variable; life expectancy is usually normal. Heterozygous female carriers may develop mild bulbar symptoms in later life (Mariotti *et al.*, 2000). High-dose testosterone treatment does not alter disease progression (Goldenberg and Bradley, 1996).

### Distal SMA

Chronic distal SMAs, which comprise about 10% of all SMAs, represent a heterogenous group with both autosomal-recessive and autosomal-dominant inheritance (Harding and Thomas, 1980). The onset ranges from early childhood to late adulthood. Distal atrophy and weakness mainly involve the legs; the upper limbs are affected in about one quarter of patients. The course is usually slowly progressive. Normal sensory action potentials are the main mean of differentiation from HMSN type II.

### Scapulo-Peroneal Form

Juvenile and adult onset, with various modes of inheritance, have been reported. This form usually starts in the peroneal muscles and follows a progressive course (Padberg, 1991). There is a remarkable incidence of foot deformities (pes cavus or equinovarus). EMG and muscle biopsy often show both myopathic and neurogenic changes. The differentiation from facio-scapulo-humeral muscle dystrophy may be difficult.

### Monomelic Amyotrophy

This disorder is characterized by insidious onset of wasting and weakness confined to one limb, initial slow progression, and subsequent stationary course. It occurs sporadically, mostly in males between 15 and 25 years of age. Most cases have been reported from far eastern countries (Gourie-Devi et al., 1984). About two thirds of these patients show a uniform clinical pattern of unilateral involvement of the hand and forearm, which has been termed juvenile distal and segmental SMA of the upper extremities (Hirayama's disease). The EMG usually shows neurogenic changes also in the homonymous contralateral muscles. The disease progression is arrested as a rule within 1–3 years. Compression of the lower cervical spinal cord during neck flexion leading to local ischemia has been postulated as a pathogenetic mechanism (Hirayama, 2000), although data from imaging studies argue in favor of an intrinsic motoneuronal degeneration (Schröder et al., 1999). Monomelic amyotrophy has also been describeed with a scapulo-humeral distribution of the quadriceps muscle, of the anterior crural muscles and/or the calf, and of the entire upper or lower limb (for review, see De Visser et al., 1991).

### Practical Management of the Spinal Muscular Atrophies

No therapy for the disease process is so far available. Prospective parents with a family history of SMA should receive genetic counseling and may consider prenatal testing. Young patients should be discouraged from engaging in professional activities involving hard physical labor. Symptomatic treatment consists mainly in continuous physiotherapy, which also helps to prevent joint contractures. Orthopedic surgery is often required to treat severe kyphoscoliosis resulting from long-standing weakness of the paraspinal muscles. In principle, all treatment options previously described for ALS patients may apply during the course of the disease, including non-invasive ventilation (Bach et al., 2000).

## HEREDITARY SPASTIC PARAPARESIS

Most "pure" cases of hereditary spastic paraparesis (HSP) are autosomal-dominant; autosomal-recessive and X-linked inheritance have also been described. The genetic basis is heterogenous, the first genes causing autosomal-dominant and X-linked HSP have recently been cloned (for review, see McDermott et al., 2001). Neuropathologically, there is caudally increasing degeneration of the corticospinal tracts. Motor delay is often found in early-onset patients (before age 35). In the late-onset type (age 40–65), leg weakness tends to be more severe. The clinical course is slowly progressive; some patients may eventually need crutches or become chair-bound. Life expectancy is normal. In contrast to the nonhereditary primary lateral sclerosis, mild sensory abnormalities and sphincter disturbances are not infrequent and are compatible with the diagnosis of a "pure" form of the disease. A wide range of associated clinical features in so-called complicated forms have been reported, for which the most common inheritance mode is autosomal-recessive (for review, see Reid 1999). Treatment consists of antispastic medication (see Table I) and physiotherapy. Occasionally, intrathecal baclofen application or intramuscular botulinum toxin may become necessary. Surgery may be required for severe pes cavus or Achilles tendon contractures. Genetic counseling should take age at onset and variation in severity into account (see also Chapter 107).

## POSTPOLIO SYNDROME

Increasing weakness and fatigue develop in about 20%–50% of patients who had paralytic poliomyelitis as they age. Most cases are due to aging or secondary diseases (e.g., arthritis). A small subset of patients shows progressive muscular atrophy and weakness, often with fasciculations, usually in muscles that had been most severely affected in the acute illness (for review, see Jubelt and Agre, 2000). The onset is gradual, starting on average 30–40 years after the acute illness. The new weakness is usually asymmetrical, progression is very slow (average of 1%–2% per year based on strength measurements [Grimby *et al.*, 1998]). Phases of little or no progression may last for 3–10 years (Dalakas, 1995). Although some patients show oligoclonal bands on CSF electrophoresis, there is no evidence for a reactivation of the poliovirus. The disease is thought to be due to excessive metabolic stress on remaining motor neurons over many years, leading to late degeneration. Treatment involves symptomatic measures, including mild physiotherapy, pacing physical activity, and reassuring patients about the benign prognosis. Amantadine, prednisone, and more recently pyridostigmine (Trojan *et al.*, 1999) have shown no benefit in controlled studies. Importantly, mild-to-moderate weakness in postpolio syndrome can be improved with nonfatiguing exercise (Agre *et al.*, 1997), whereas overuse of weakened muscles should be avoided. Bulbar and diaphragmatic muscle weakness may necessitate the institution of noninvasive ventilation.

## PATIENTS' ASSOCIATIONS

Patients with motor neuron disorders can obtain help and information from the following associations: in the United States, the Muscular Dystrophy Association (MDA, 3300 E. Sunrise Drive, Tucson, AZ 85718, www.mdausa.org) and the ALS Association (ALSA, 27001 Agoura Road, Suite 150, Calabasas Hills, CA 91301-5104, www.alsa.org); in Great Britain, the Motor Neuron Disease Association (MNDA, PO Box 246, Northampton NN1 2PR, UK, www.mndassociation.org); in Germany, the Deutsche Gesellschaft für Muskelkranke (DGM, im Moos 4, D-79112 Freiburg, www.dgm.org).

## REFERENCES

Aboussouan, L. S., Khan, S. U., Meeker, D. P., Stelmach, K., and Mitsumoto, H. (1997). Effect of noninvasive positive-pressure ventilation on survival in amyotrophic lateral sclerosis. *Ann. Intern. Med.* **127**, 450–453.

Ackerman, G. M., and Oliver, D. (1997). Psychosocial support in an outpatient clinic. *Palliat. Med.* **11**, 167–168.

Agre, J. C., Rodriquez, A. A., and Franke, T. M. (1997). Strength, endurance, and work capacity after muscle strengthening exercise in postpolio subjects. *Arch. Phys. Med. Rehabil.* **78**, 681–686.

Al-Chalabi, A., and Leigh, P. N. (2000). Recent advances in amyotrophic lateral sclerosis. *Curr. Opin. Neurol.* **13**, 397–405.

American Academy of Neurology. (1997). Practice advisory on the treatment of amyotrophic lateral sclerosis with riluzole: report of the Quality Standards Subcommittee of the American Academy of Neurology. *Neurology* **49**, 657–659.

American Academy of Neurology. (1998). Assisted suicide, euthanasia, and the neurologist. The Ethics and Humanities Subcommittee of the American Academy of Neurology. *Neurology* **50**, 596–598.

Anonymous. (1996). A double-blind placebo-controlled clinical trial of subcutaneous recombinant human ciliary neurotrophic factor (rHCNTF) in amyotrophic lateral sclerosis. ALS CNTF Treatment Study Group. *Neurology* **46**, 1244–1249.

Anonymous. (1999). A controlled trial of recombinant methionyl human BDNF in ALS: The BDNF Study Group (Phase III). *Neurology* **52**, 1427–1433.

Appel, S. H., Smith, R. G., Alexianu, M. F., Engelhardt, J. I., and Stefani, E. (1995). Autoimmunity as an etiological factor in sporadic amyotrophic lateral sclerosis. *Adv. Neurol.* **68**, 47–57.

Bach, J. R., Niranjan, V., and Weaver, B. (2000). Spinal muscular atrophy type 1: A noninvasive respiratory management approach. *Chest* **117**, 1100–1105.

Beal, M. F. (2000). Mitochondria and the pathogenesis of ALS. *Brain* **123**, 1291–1292.

Berger, M. M., Kopp, N., Vital, C., Redl, B., Aymard, M., and Lina, B. (2000). Detection and cellular localization of enterovirus RNA sequences in spinal cord of patients with ALS. *Neurology* **54**, 20–25.

Borasio, G. D. (2001). Meditation and ALS. *In* "Amyotrophic Lateral Sclerosis: A Comprehensive Guide to Management" (H. Mitsumoto and T. Munsat, Eds.), Demos Medical Publ., New York.

Borasio, G. D., Horstmann, S., Anneser, J. M. H., Neff, N. T., and Glicksman, M. A. (1998b). CEP-1347/KT7515, a JNK pathway inhibitor, supports the in vitro survival of chick embryonic neurons. *NeuroReport* **9**, 1435–1439.

Borasio, G. D., Robberecht, W., Leigh, P. N., Emile, J., Guiloff, R. J., Jerusalem, F., Silani, V., Vos, P. E., Wokke, J. H. J., Dobbins, T., and the European ALS/IGF-I study group (1998a). A placebo-controlled trial of insulin-like growth factor-I in amyotrophic lateral sclerosis. *Neurology* **51**, 583–586.

Borasio, G. D., and Voltz, R. (1998). Discontinuation of life support in patients with amyotrophic lateral sclerosis. *J. Neurol.* **245**, 717–722.

Borasio, G. D., Sloan, R., and Pongratz, D. E. (1998c). Breaking the news in amyotrophic lateral sclerosis. *J. Neurol. Sci.* **160** (**Suppl. 1**), 127–133.

Borasio, G. D., and Voltz, R. (2000). Advance directives. *In* "Palliative Care in Amyotrophic Lateral Sclerosis (Motor Neurone Disease)" (D. Oliver, G. D. Borasio, and D. Walsh, Eds.), pp. 36–41. Oxford University Press, Oxford.

Borasio, G. D., Voltz, R., and Miller, R. G. (2001). Palliative care in amyotrophic lateral sclerosis. *In* "Palliative Care" (A. Carver and K. Foley, Eds.), *Neurol. Clin.* **19**, 829–847.

Brahe, C. (2000). Copies of the survival motor neuron gene in spinal muscular atrophy: the more, the better. *Neuromuscul. Disord.* **10**, 274–275.

Brooks, B. R. (1996). Clinical epidemiology of amyotrophic lateral sclerosis. *Neurol. Clin.* **14**, 399–420.

Brooks, B. R. (1999). Diagnostic dilemmas in amyotrophic lateral sclerosis. *J. Neurol. Sci.* **165** (**Suppl. 1**), 1–9.

Brooks, B. R., Miller, R. G., Swash, M., and Munsat, T. (1998). El Escorial Revisited: Revised Criteria for the Diagnosis of Amyotrophic Lateral Sclerosis *In* "World Federation of Neurology Subcommittee of Motor Neuron Disease." Website: www.wfnals.org/Articles/elescorial1998.htm.

Bruyn, G. W. (1991). Progressive bulbar palsy in adults. *In* "Handbook of Clinical Neurology" (H. L. Klawans, P. J. Vinken, G. W. Bruyn, and J. M. B. V. de Jong, eds.), Vol. 59 (Diseases of the Motor System), pp. 217–229. North Holland, Amsterdam.

Caroscio, J. T., Calhoun, W. F., and Yahr, M. D. (1984). Prognostic factors in motor neuron disease: A prospective study of survival. *In* "Research Progress in Motor Neuron Diseases" (F. C. Rose, ed.), pp. 34–43. Bath, Pitman.

Cazzolli, P. A., and Oppenheimer, E. A. (1996). Home mechanical ventilation for amyotrophic lateral sclerosis: Nasal compared to tracheostomy-intermittent positive pressure ventilation. *J. Neurol. Sci.* **139 (Suppl)**, 123–128.

Cervenakova, L., Protas, I. I., Hirano, A., Votiakov, V. I., Nedzved, M. K., Kolomiets, N. D., Taller, I., Park, K. Y., Sambuughin, N., Gajdusek, D. C., Brown, P., and Goldfarb, L. G. (2000). Progressive muscular atrophy variant of familial amyotrophic lateral sclerosis (PMA/ALS). *J. Neurol. Sci.* **177**, 124–130.

Dalakas, M. C. (1995). The post-polio syndrome as an evolved clinical entity—Definition and clinical description. *Ann. NY Acad. Sci.* **753**, 68–80.

De Visser, M., Bolhui, P. A., and Barth, P. G. (1991). Differential diagnosis of spinal muscular atrophies and other disorders of motor neurons with infantile or juvenile onset. *In* "Handbook of Clinical Neurology", op. cit., pp. 367–382.

Esposito, E., Rossi, C., Amodio, R., Di Castelnuovo, A., Bendotti, C., Rotondo, T., Algeri, S., and Rotilio, D. (2000). Lyophilized red wine administration prolongs survival in an animal model of amyotrophic lateral sclerosis. *Ann. Neurol.* **48**, 686–687.

Ferrante, M. A., and Wilbourn, A. (1997). The characteristic electrodiagnostic features of Kennedy's disease. *Muscle Nerve* **20**, 323–329.

Figlewicz, D. A., McKenna-Yasek, D., Horvitz, R., Rouleau, G. A., and Brown, R. H. (1989). Epidemiological analysis of 90 families with hereditary amyotrophic lateral sclerosis. *Int. ALS-MND Update*, 9–10.

Forsyth, P. A., Dalmau, J., Graus, F., Cwik, V., Rosenblum, M. K., and Posner, J. B. (1997). Motor neuron syndromes in cancer patients. *Ann. Neurol.* **41**, 722–730.

Ganzini, L., Johnston, W. S., McFarland, B. H., Tolle, S. W., and Lee, M. A. (1998). Attitudes of patients with amyotrophic lateral sclerosis and their care givers toward assisted suicide. *N. Engl. J. Med.* **339**, 967–973.

Gelinas, D., and Miller, R. G. (2000). A treatable disease: a guide to the management of amyotrophic lateral sclerosis. *In* "Amyotrophic Lateral Sclerosis" (R. H. Brown Jr., V. Meininger, and M. Swash, Eds.), pp. 405–421. Martin Dunitz, London.

Giess, R., Naumann, M., Werner, E., Riemann, R., Beck, M., Puls, I., Reiners, C., and Toyka, K. V. (2000). Injections of botulinum toxin A into the salivary glands improve sialorrhoea in amyotrophic lateral sclerosis. *J. Neurol. Neurosurg. Psychiat.* **69**, 121–123.

Goldenberg, J. N., and Bradley, W. G. (1996). Testosterone therapy and the pathogenesis of Kennedy's disease (X-linked bulbospinal muscular atrophy). *J. Neurol. Sci.* **135**, 158–161.

Gordon, P. H., Rowland, L. P., Younger, D. S., Sherman, W. H., Hays, A. P., Louis, E. D., Lange, D. J., Trojaborg, W., Lovelace, R. E., Murphy, P. L., and Latov, N. (1997). Lymphoproliferative disorders and motor neuron disease: an update. *Neurology* **48**, 1671–1678.

Gourie-Devi, M., Suresh, T. G., and Shankar, S. K. (1984). Monomelic amyotrophy. *Arch. Neurol.* **41**, 388–394.

Grimby, G., Stalberg, E., Sandberg, A., and Sunnerhagen, K. S. (1998). An 8-year longitudinal study of muscle strength, muscle fiber size, and dynamic electromyogram in individuals with late polio. *Muscle Nerve* **21**, 1428–1437.

Grohme, K., v. Maravic, M., Gasser, T., and Borasio, G. D. (2001). A case of amyotrophic lateral sclerosis with a very slow progression over 44 years. *Neuromuscul. Disord.* **11**, 414–416.

Gubbay, S. S., Kahana, E., Zilber, N., Cooper, G., Pintov, S., and Leibowitz, Y. (1985). Amyotrophic lateral sclerosis: a study of its presentation and prognosis. *J. Neurol.* **232**, 295–300.

Gurney, M. E., Cutting, F. B., Zhai, P., Doble, A., Taylor, C. P., Andrus, P. K., and Hall, E. D. (1996). Benefit of vitamin E, riluzole, and gabapentin in a transgenic model of familiar amyotrophic lateral sclerosis. *Ann. Neurol.* **39**, 147–157.

Hanayama, K., Ishikawa, Y., and Bach, J. R. (1997). Amyotrophic lateral sclerosis. Successful treatment of mucous plugging by mechanical insufflation-exsufflation. *Am. J. Phys. Med. Rehabil.* **76**, 338–339.

Harding, A. E., and Thomas, P. K. (1980). Hereditary distal spinal muscular atrophy. *J. Neurol. Sci.* **454**, 337–348.

Haverkamp, L. J., Appel, V., and Appel, S. H. (1995). Natural history of amyotrophic lateral sclerosis in a database population. *Brain* **118**, 707–719.

Hentati, A., Bejaoui, K., Pericak-Vance, M. A., Hentati, F., Speer, M.C., Hung, W-Y., Figlewicz, D. A., Haines, J., Rimmler, J., Hamida, C. B., Hamida, M. B., Brown, R. H. Jr., and Siddique, T. (1994). Linkage of recessive familial amyotrophic lateral sclerosis to chromosome 2q33–q35. *Nat. Genet.* **7**, 425–428.

Hirano, A., and Iwata, M. (1978). Pathology of motor neurons with special reference to amyotrophic lateral sclerosis and related diseases. *In* "Amyotrophic Lateral Sclerosis" (T. Tsubaki and Y. Toyokura, eds.), pp. 107–133. University Park Press, Baltimore.

Hirayama, K. (2000). Juvenile muscular atrophy of distal upper extremity (Hirayama disease): focal cervical ischemic poliomyelopathy. *Neuropath.* **20 (Suppl.)**, 91–94.

Hosler, B. A., Siddique, T., Sapp, P. C., Sailor, W., Huang, M. C., Hossain, A., Daube, J. R., Nance, M., Fan, C., Kaplan, J., Hung, W. Y., McKenna-Yasek, D., Haines, J. L., Pericak-Vance, M. A., Horvitz, H. R., and Brown, R. H. (2000). Linkage of familial amyotrophic lateral sclerosis with frontotemporal dementia to chromosome 9q21–q22. *JAMA* **284**, 1664–1669.

Hottinger, A. F., Azzouz, M., Deglon, N., Aebischer, P., and Zurn, A. D. (2000). Complete and long-term rescue of lesioned adult motoneurons by lentiviral-mediated expression of glial cell line-derived neurotrophic factor in the facial nucleus. *J. Neurosci.* **20**, 5587–5593.

Jubelt, B., and Agre, J. C. (2000). Characteristics and management of postpolio syndrome. *JAMA* **284**, 412–414.

Julien, J. (2001). Amyotrophic lateral sclerosis. Unfolding the toxicity of the misfolded. *Cell* **104**, 581–591.

Kausch, K., Muller, C. R., Grimm, T., Ricker, K., Rietschel, M., Rudnick-Schöneborn, S., and Zerres, K. (1991). No evidence for linkage of autosomal dominant proximal spinal muscular atrophies to chromosome 5q markers. *Hum. Genet.* **86**, 317–318.

Kleopa, K. A., Sherman, M., Neal, B., Romano, G. J., and Heiman-Patterson, T. (1999). Bipap improves survival and rate of pulmonary function decline in patients with ALS. *J. Neurol. Sci.* **164**, 82–88.

Klivenyi, P., Ferrante, R. J., Matthews, R. T., Bogdanov, M. B., Klein, A. M., Andreassen, O. A., Mueller, G., Wermer, M., Kaddurah-Daouk, R., and Beal, M. F. (1999). Neuroprotective effects of creatine in a transgenic animal model of amyotrophic lateral sclerosis. *Nat. Med.* **5**, 347–350.

Kolde, G., Bachus, R., and Ludolph, A. C. (1996) Skin involvement in amyotrophic lateral sclerosis. *Lancet* **347**, 1226–1227.

Lacomblez, L., Bensimon, G., Leigh, P. N., Guillet, P., and Meininger, V. (1996). Dose-ranging study of riluzole in amyotrophic lateral sclerosis. *Lancet* **347**, 1425–1431.

Lai, E. C., Felice, K. J., Festoff, B. W., Gawel, M. J., Gelinas, D. F., Kratz, R., Murphy, M. F., Natter, H. M., Norris, F. H., and Rudnicki, S. A. (1997). Effect of recombinant human insulin-like growth factor-I on progression of ALS. A placebo-controlled study. The North America ALS/IGF-I Study Group. *Neurology* **49**, 1621–1630.

LaSpada, A. R., Wilson, E. M., Lubahn, D. B., Harding, A. E., and Fischbeck, K. H. (1991). Androgen receptor gene mutations in X-linked bulbar and spinal muscular atrophy. *Nature* **352**, 77–79.

Le Forestier, N., Maisonobe, T., Spelle, L., Lesort, A., Salachas, F., Lacomblez, L., Samson, Y., Bouche, P., and Meininger, V. (2001). Primary lateral sclerosis: further clarification. *J. Neurol. Sci.* **185**, 95–100.

Li, T. M., Alberman, E., and Swash, M. (1990). Clinical associations of 560 cases of motor neuron disease. *J. Neurol. Neurosurg. Psychiat.* **51**, 778–784.

Lilienfeld, D. E., Chan, E., Ehland, J., Godbold, J., Landrigan, P. J., Marsh, G., and Perl, D. P. (1989). Rising mortality from motoneuron disease in the USA 1962–1984. *Lancet* **1**, 710–713.

Louwerse, E. S., Sillevis Smitt, P. A. E., and de Jong, J. M. B. V. (1991). Differential diagnosis of sporadic amyotrophic lateral sclerosis, progressive spinal muscular atrophy and progressive bulbar palsy in adults. *In* "Handbook of Clinical Neurology", op. cit., pp. 383–424.

Mariotti, C., Castellotti, B., Pareyson, D., Testa, D., Eoli, M., Antozzi, C., Silani, V., Marconi, R., Tezzon, F., Siciliano, G., Marchini, C., Gellera, C., and Donato, S. D. (2000). Phenotypic manifestations associated with CAG-repeat expansion in the androgen receptor gene in male patients and heterozygous females: a clinical and molecular study of 30 families. *Neuromuscul. Disord.* **10**, 391–397.

Martin, J., and Turnbull, J. (2000). Lasting impact, and ongoing needs, in families months to years after death from ALS. *Amyotr. Lat. Scler.* **1 (Suppl. 3)**, S14–S15.

McDermott, C. J., Dayaratne, R. K., Tomkins, J., Lusher, M. E., Lindsey, J. C., Johnson, M. A., Casari G., Turnbull, D. M., Bushby, K., and Shaw, P. J. (2001). Paraplegin gene analysis in hereditary spastic paraparesis (HSP) pedigrees in northeast England. *Neurology* **56**, 467–471.

McMurray, A. (2000). Bereavement. *In:* "Palliative Care in Amyotrophic Lateral Sclerosis" (D. Oliver, G. D. Borasio, D. Walsh, Eds.), pp. 169–181. Oxford University Press, Oxford.

Meininger, V. (2000). "A Phase 3 Placebo-Controlled Trial of SR57746A in ALS." Presentation at the 11th International Symposium on ALS/MND, Aarhus, Danmark.

Meyer, T., Fromm, A., Munch, C., Schwalenstocker, B., Fray, A. E., Ince, P. G., Stamm, S., Gron, G., Ludolph, A. C., and Shaw, P. J. (1999). The RNA of the glutamate transporter EAAT2 is variably spliced in amyotrophic lateral sclerosis and normal individuals. *J. Neurol. Sci.* **170**, 45–50.

Miller, R. G., Moore, D. H. 2nd, Gelinas, D. F., Dronsky, V., Mendoza, M., Barohn, R. J., Bryan, W., Ravits, J., Yuen, E., Neville, H., Ringel, S., Bromberg, M., Petajan, J., Amato, A. A., Jackson, C., Johnson, W., Mandler, R., Bosch, P., Smith, B., Graves, M., Ross, M., Sorenson, E. J., Kelkar, P., Parry, G., and Olney, R. (2001). Phase III randomized trial of gabapentin in patients with amyotrophic lateral sclerosis. *Neurology* **56**, 843–848.

Miller, R. G., Rosenberg, J. A., Gelinas, D. F., Mitsumoto, H., Newman, D., Sufit, R., Borasio, G. D., Bradley, W. G., Bromberg, M. B., Brooks, B. R., Kasarskis, E. J., Munsat, T. L., Oppenheimer, E. A., and the ALS Practice Parameters Task Force (1999). Practice Parameter: The care of the patient with amyotrophic lateral sclerosis (an evidence-based review): report of the Quality Standards Subcommittee of the American Academy of Neurology: ALS Practice Parameters Task Force. *Neurology* **52**, 1311–1323.

Mitsumoto, H. (1995). New therapeutic approaches: rationale and results. *In* " Motor Neuron Disease" (P. N. Leigh and M. Swash, Eds.), pp. 419–441. Springer, London.

Murphy, P. L., Albert, S. M., Weber, C., Del Bene, M. L., and Rowland, L. P. (2000). Impact of spirituality and religiousness on outcomes in patients with ALS. *Neurology* **55**, 1581–1584.

Neudert, C., Oliver, D., Wasner, M., and Borasio, G. D. (2001). The course of the terminal phase in patients with amyotrophic lateral sclerosis. *J. Neurol.* **248**, 612–616.

Oliver, D. (1996). The quality of care and symptom control—The effects on the terminal phase of ALS/MND. *J. Neurol. Sci.* **139 (Suppl.)**, 134–136.

Oliver, D., Borasio, G. D., and Walsh, D., eds. (2000). "Palliative Care in Amyotrophic Lateral Sclerosis (Motor Neurone Disease)." Oxford University Press, Oxford.

Padberg, G. W. (1991). Special forms of spinal muscular atrophy. *In* "Handbook of Clinical Neurology", op. cit., 41–50.

Pearn, J. H., Hudgson, P., and Walton, J. N. (1978). A clinical and genetic study of spinal muscular atrophy of adult onset. *Brain* **101**, 591–606.

Puchalski, C., and Romer, A. L. (2000). Taking a spiritual history allows clinicians to understand patients more fully. *J. Palliat. Med.* **3**, 129–137.

Rabkin, J. G., Wagner, G. J., and Del Bene, M. (2000). Resilience and distress among amyotrophic lateral sclerosis patients and caregivers. *Psychosom. Med.* **62**, 271–279.

Reid, E. (1999). The hereditary spastic paraplegias. *J. Neurol.* **246**, 995–1003.

Robberecht, W. (2000). Oxidative stress in amyotrophic lateral sclerosis. *J. Neurol.* **247 (Suppl. 1)**, 1–6.

Robbins, R. A., Simmons, Z., Bremer, B. A., Walsh, S. M., and Fischer, S. (2001). Quality of life in ALS is maintained as physical function declines. *Neurology* **56**, 442–444.

Rosen, D. R., Siddique, T., Patterson, D., Figlewicz, D. A., Sapp, P., Hentati, A., Donaldson, D., Goto, J., O'Regan, J. P., Deng, H.-X., Rahmani, Z., Krizus, A., McKenna-Yasek, D., Cayabyab, A., Gaston, S. M., Berger, R., Tanzi, R. E., Halperin, J. J., Herzfeldt, B., Van den Bergh, R., Hung, W.-Y., Bird, T., Deng, G., Mulder, D. W., Smyth, C., Laing, N. G., Soriano, E., Pericak-Vance, M., Haines, J., Rouleau, G. A., Gusella, J. S., Horvitz, H. R., and Brown, R. H. (1993). Mutations in Cu/Zn superoxide dismutase gene are associated with familial amyotrophic lateral sclerosis. *Nature* **362**, 59–62.

Rothstein, J. D., Van Kammen, M., Levey, A. I., Martin, L. J., and Kuncl, R. W. (1995). Selective loss of glial glutamate transporter GLT-1 in amyotrophic lateral sclerosis. *Ann. Neurol.* **38**, 73–84.

Schröder, R., Keller, E., Flacke, S., Schmidt, S., Pohl, C., Klockgether, T., and Schlegel, U. (1999). MRI findings in Hirayama's disease: flexion induced cervical myelopathy or intrinsic motor neuron disease? *J. Neurol.* **246**, 1069–1074.

Smith, R. G., Alexianu, M. E., Crawford, G., Nyormoi, O., Stefani, E., and Appel, S. (1994). Cytotoxicity of immunoglobulins from amyotrophic lateral sclerosis patients on a hybrid motoneuron cell line. *Proc. Natl. Acad. Sci. USA* **91**, 3393–3397.

Smith, R. G., Hamilton, F., Hofmann, F., Schneider, T., Nastainczyk, W., Birnbaumer, L., Stefani, E., and Appel, S. H. (1992). Serum antibodies to L-type calcium channels in patients with amyotrophic lateral sclerosis. *N. Engl. J. Med.* **327**, 1721–1728.

Spencer, P. S., Palmer, V., and Kisby, G. (1989). Western Pacific amyotrophic lateral sclerosis and exposure to untreated Cycad seed. *Int. ALS-MND Update*, 30–31.

Trojan, D. A., Collet, J. P., Shapiro, S., Jubelt, B., Miller, R. G., Agre, J. C., Munsat, T. L., Hollander, D., Tandan, R., Granger, C., Robinson, A., Finch, L., Ducruet, T., and Cashman, N. R. (1999). A multicenter, randomized, double-blinded trial of pyridostigmine in postpolio syndrome. *Neurology* **53**, 1225–1233.

Veldink, J. H., van Den Berg, L. H., Cobben, J. M., Stulp, R. P., De Jong, J. M., Vogels, O. J., Baas, F., Wokke, J. H., and Scheffer, H. (2001). Homozygous deletion of the survival motor neuron 2 gene is a prognostic factor in sporadic ALS. *Neurology* **56**, 749–752.

Walker, M. P., Schlaberg, R., Hays, A. P., Bowser, R., and Lipkin, W. I. (2001). Absence of echovirus sequences in brain and spinal cord of amyotrophic lateral sclerosis patients. *Ann. Neurol.* **49**, 249–253.

Wasner, M., Klier, H., and Borasio, G. D. (2001). The use of alternative medicine by patients with amyotrophic lateral sclerosis. *J. Neurol. Sci.* **191**, 151–154.

Williams, D. B. (1991). Familial amyotrophic lateral sclerosis. *In* "Handbook of Clinical Neurology," op. cit., pp. 241–251.

Wirth, B. (2000). An update of the mutation spectrum of the survival motor neuron gene (SMN1) in autosomal recessive spinal muscular atrophy. *Hum. Mutat.* **15**, 228–237.

Zerres, K., and Rudnik-Schöneborn, S. (1995). Natural history in proximal spinal muscular atrophy. *Arch. Neurol.* **52**, 518–523.

Zerres, K., Rudnik-Schöneborn, S., Forkert, R., and Wirth, B. (1995). Genetic basis of adult-onset autosomal recessive spinal muscular atrophy. *Lancet* **346**, 741–742.

*Movement Disorders*

## CHAPTER 82
# Restless Legs Syndrome

Claudia Trenkwalder and Arthur S. Walters

Most patients with restless legs syndrome (RLS) complain about sleep disturbances associated with motor restlessness and therefore do not primarily seek help from a neurologist. The diagnosis of RLS should be considered when abnormal sensations of the legs occur at rest and are associated with motor restlessness and sleep disturbance but usually not with signs of local disease (see Table I). RLS may be either a primary or a secondary disorder. The primary condition is idiopathic and often familial with autosomal-dominant inheritance (Trenkwalder *et al.*, 1996; Walters *et al.*, 1986). A gene has not yet been identified for RLS, although a single gene disorder is assumed in large families. A putative chromosome (Number 12) has been identified in RLS patients with a presumed pseudo-dominant mode of inheritance (Desautels *et al.*, 2001). Secondary causes include iron deficiency, uremia, pregnancy, rheumatoid arthritis, spinal disease, and others. Recent epidemiological studies point to a high age-dependent prevalence of 5%–10% RLS in the general population.

## CLINICAL ASPECTS

The typical symptom is a "creepy," "crawly," "tense," "unbearable" (yet "not really painful") sensation in the calves, sometimes the feet, thighs, or even the arms. It may alternate from one limb to another. An unpleasant, often indescribable aching builds up in settings in which the patient is at rest (e.g., in theaters, on planes) but is most common when patients go to bed (Walters *et al.*, 1995). There is a concomitant urge to move the legs that only temporarily relieves the sensory disturbance. The RLS is a clinical diagnosis and determined by the presence of the four Minimal Diagnostic Criteria of the International Restless Legs Syndrome Study Group (Walters *et al.*, 1995). Addi-

tional features are often present but not obligatory for the diagnosis of RLS (see Table I). "Resting dyskinesias while awake" may also occur (Hening *et al.*, 1986) and may be occasionally recorded during daytime rest (Trenkwalder *et al.*, 1993). They follow a circadian pattern with worsening of symptoms in the evening or at night (Trenkwalder *et al.*, 1999). In many patients, the unfortunate outcome is insomnia and daytime sleepiness. The neurological examination in idiopathic RLS is usually normal without signs of polyneuropathy.

The RLS is accompanied by so-called periodic limb movements of sleep (PLMS) in about 80% of RLS patients (Montplaisier *et al.*, 1997; Walters *et al.*, 1990). The clue to this manifestation is usually provided by bed partners who report that they are aroused by the patients' regular, forceful leg movements. Surface electromyography (EMG) recordings from the tibialis anterior muscle can be used to document PLMS during sleep studies (Atlas task force of the American Sleep Disorders Association, 1993; Coleman, 1982). Although earlier referred to as nocturnal myoclonus, PLMS are produced by long-lasting EMG bursts (up to 5 seconds). The finding of a PLMS syndrome (>5 PLMS/h sleep) during polysomnography supports the clinical diagnosis of RLS (Lugaresi *et al.*, 1986; Montplaisir *et al.*, 1998). However, PLMS occur in a variety of sleep disorders and thus are not specific for RLS.

The differential diagnosis of RLS includes polyneuropathy with paresthesia or "burning feet," muscle cramps, and vascular disorders such as venous stasis, erythromelalgia, and arterial occlusion (Danek and Pollmächer, 1990; Walters *et al.*, 1995).

A positive therapeutic response to levodopa may further confirm the diagnosis of RLS, a standardized L-dopa test for RLS will be provided (K. Stiasny and C. Trenkwalder, personal communication, 2002).

**TABLE I**    Diagnostic Criteria for Restless Legs Syndrome

Obligatory criteria
  Desire to move the limbs usually associated with paresthesias or
    dysesthesias *and* motor restlessness *and*
  Symptoms are worse or exclusively present at rest (i.e., lying,
    sitting) with at least partial and temporary relief by activity *and*
  Symptoms are worse in evening or night
Additional features
Sleep disturbances and their consequences
Involuntary movements (periodic limb movements in sleep,
  involuntary limb movements while awake and at rest)
Chronic condition (any age, but more severe after middle age)
Family history
No neurological abnormalities in primary RLS
In secondary RLS, peripheral neuropathy, radiculopathy and
  other neurological disorders may be present

From Walters *et al.* (1995). This material is used by permission
of Wiley-Liss, Inc., a subsidiary of John Wiley & Sons, Inc.

## NATURAL HISTORY

RLS occurs in both genders and in all age groups but may show a predominant affection of postmenopausal women (Rothdach *et al.*, 2000). Recent epidemiological studies show even a higher prevalence of about 10% RLS affecteds in the general population (Lavigne and Montplaisir, 1994; Philips *et al.*, 2000; Rothdach *et al.*, 2000) than previously estimated from Gibb and Lees (1986).

Idiopathic RLS (approximately half of the cases) often begins in the second decade (Lugaresi *et al.*, 1986; Walters *et al.*, 1996; Winkelmann *et al.*, 2000). Familial cases may even manifest in childhood and are often misdiagnosed as "growing pains" or as attention deficit hyperactivity disorder (ADHD) (Picchietti and Walters, 1996). Characteristically, symptoms fluctuate spontaneously with asymptomatic periods of several weeks or only a few days. RLS affects quality of life and may lead to depression (Rothdach *et al.*, 2000) and suicidal ideation (Lauerma, 1991). Exacerbations have been reported in association with various drugs, especially with neuroleptic treatment of the dopamine $D_2$ receptor antagonist type (Ekbom, 1970; Kraus *et al.*, 1999; Vahedi *et al.*, 1994).

RLS occurs in 15%–20% of patients on hemodialysis (Collado-Seidel *et al.*, 1998; Winkelman *et al.*, 1996) and often resolves after renal transplantation. Of pregnant women, 19% are reported to have RLS that resolves after delivery (Goodman *et al.*, 1988); in familial RLS pregnancy may be the condition of first manifestation of symptoms (Winkelmann *et al.*, 2000).

## PRINCIPLES OF THERAPY

The etiology of RLS is still unknown, although central nervous system, especially spinal disinhibition phenom-

ena (Bara-Jimenez *et al.*, 2000; Yokota *et al.*, 1991) or brainstem structures (Bucher *et al.*, 1997), may be implicated in the pathophysiology. Subtle alterations of the dopaminergic nigrostriatal system of RLS patients are known from recent neuroimaging studies (Routtinen *et al.*, 2000; Turjanski *et al.*, 1999). In addition, the efficiency of dopaminergic agents in the treatment of RLS points to an involvement of the dopaminergic system in RLS (Walters *et al.*, 1986, 1988). There is no cure for idiopathic RLS, although medical therapy often relieves symptoms. In secondary RLS, treatment of the underlying disorder may alleviate or even cure RLS and is recommended as the first line of therapy. To temporarily suppress symptoms during a period of rest (e.g., a visit to the theater or during a plane flight) single doses of medicine, especially of levodopa, are helpful. Long-term medication is needed in chronic RLS with sleep disturbances and/or daytime sleepiness. Drug holidays may be tried, but in most cases continuous treatment is necessary, especially in older patients.

In the last years an increasing number of controlled studies of dopaminergic agents have been performed in RLS. Trials of secondary forms of RLS are rare. The studies that have been performed use outcome measures such as subjective patients' reports, the restless legs syndrome severity scale (Collado-Seidel *et al.*, 1999b), or objective data of polysomnography. Double-blind, placebo-controlled studies have shown L-dopa and various dopamine agonists to be effective in idiopathic RLS for relief of symptoms and improvement of subjective and objective sleep parameters (see Table II). Opioids such as oxycodone (Walters *et al.*, 1993), methadone, and propoxyphene and tilidine/naloxone also reduced both subjective complaints and periodic limb movements (see Table II).

Clonazepam significantly reduced sensory symptoms in one of two studies (Boghen *et al.*, 1986; Montagna *et al.*, 1984) although it has become a drug of second choice after dopaminergic agents and opioids. Carbamazepine was found to be effective for the parasthesias but not for reducing periodic limb movements (Telstadt *et al.*, 1984; Zucchoni *et al.*, 1989). Recent studies proved the efficacy of other anticonvulsants such as valproic acid (Ehrenberg *et al.*, 2000) and gabapentine (Mellick and Mellick, 1996). Also melatonin and magnesium (Hornyak *et al.*, 1998) have been reported to be efficient in RLS.

In uremic RLS, clonazepam and L-dopa were investigated in open and controlled studies (Read *et al.*, 1981; Trenkwalder *et al.*, 1995; Walker *et al.*, 1996). A placebo-controlled trial also found clonidine to be effective in uremic (Ausserwinkler and Schmitt, 1989) and idiopathic RLS (Wagner *et al.*, 1996). In an open study the patients with RLS secondary to iron deficiency benefited from iron substitution (O'Keeffe *et al.*, 1994), but

**TABLE II    Practical Drug Management in Restless Legs Syndrome**

In secondary RLS: Therapy of associated disorders (e.g., iron substitution [iron sulfate PO])

Dopaminergic agents (dosages taken from published studies; in single cases, depending on efficacy, higher dosages have been used)
L-dopa
 L-dopa/benserazide* (Restex), L-dopa/carbidopa[†] (Sinemet, Nacom, Striaton), and sustained-release preparations* (Restex retard, Sinemet CR)
  *Complaint: prolongation of sleep latency*
   50–200 mg standard L-dopa, 1 h before bedtime and, if necessary, additional dose at bedtime
  *Complaint: prolonged sleep latency and symptoms during the night, nocturnal awakenings*
   100–200 mg standard L-dopa, 1 h before bedtime combined with an additional dose of 100–200 mg sustained-release L-dopa
Dopamine agonists
 *Ergot derivats:*
  Bromocriptine* (Parlodel, Pravidel) 2.5–7.5 mg at night
  Pergolide* (Permax, Parkotil) 0.15–0.5 mg at night, beginning with 0.05 mg dose splitting recommended or in combination with domperidone (Motilium) 10–30 mg
  Cabergoline[†] (Cabaseril) 0.5–2 mg, taken 2 h before bedtime
  Lisurid[‡] (Dopergin) 0.2–0.6 mg, 1 h before bedtime
  Dihydroergocryptin[‡] (DHEC, Almirid) 5–40 mg 1 h before bedtime
 *Nonergot derivates:*
  Pramipexole* (Mirapex, Sifrol) 0.125–1.5 mg at night
  Ropinirole[‡] (Requip) 0.5–3 mg at night
 *Complaint: daytime symptoms*
  All dopamine agonists can be given also during daytime as bid or tid application.
Opioids
 Smaller doses preferable; oxycodone* (Roxycodone) 5 mg, 2 h before bedtime, at bedtime, in the middle of the night
 Tilidine[‡] (Valoron N) 25–100 mg at night, or sustained release[‡] (Valoron retard N) 50–100 mg at night, or bid
 Propoxyphene[†] 50–200 mg at night (Darvon)
 Tramadole[†] (Tramal) 50–150 mg at night
 Morphin sulfate[‡] (in severe cases; dosage titrated according to severity of symptoms; epidural application in single cases to decrease systemic effects)
Benzodiazepines (not for long-term treatment)
 Clonazepam* (Klonopin, Rivotril) 0.5–3 mg at night
 Alprazolam[‡] (Xanax, Tafil) 0.5–1.5 mg at night
Benzodiazepine receptor ligands*:
 Zolpidem 5–10 mg at night
Anticonvulsants
 Carbamazepine* (Tegretol, Tegretal, Timonil) 100–300 mg at night, rarely up to 1000 mg
 Valproic acid* (Ergenyl, Orfiril) 150–600 mg at night or bid
 Gabapentine[†] (Neurontin) 300–1500 mg tid or at night in single dosages[†]

*Randomized and placebo-controlled prospective study available.
[†]Randomized study or comparative study available.
[‡]Open study available or only case reports and clinical experience.
For further information on dopaminergic agents see the chapter "Parkinsonism."

those with elevated pre-treatment transferrin saturation or ferritin levels should be re-measured during iron therapy (Barton *et al.*, 2001). Patients without iron deficiency did not improve with iron substitution (Davis *et al.*, 2000).

## PRACTICAL MANAGEMENT (Table II)

According to international guidelines of the American Sleep Disorders Association, the drug of first choice is L-dopa followed by dopamine agonists like pergolide, pramipexole, and bromocriptine (Chesson *et al.*, 1999; Hening *et al.*, 1999). Opioids are estimated as the medication of second choice, and, in some patients who cannot tolerate dopaminergic agents as first choice med-

ication. Carbamazepine and clonidine are only given in rare cases. Valproic acid and especially gabapentine are preferred because of their efficacy and better tolerability. Benzodiazepines are helpful additional drugs, particularly in uremic RLS, or may be substituted by benzodiazepine receptor ligands (i.e., zolpidem) but alone usually do not abolish symptoms completely. Like opioids, benzodiazepines and benzodiazepine receptor ligands should be prescribed in a controlled manner.

PLMS in isolation, also known as nocturnal myoclonus and now called periodic limb movement disorder (PLMD, ASDA, 1993), does not necessarily require treatment unless sleep is compromised. In that case, patients mostly complain about daytime fatigue, and the bed partners report increased motor activity of the patients during sleep. Dopaminergic agents are the first choice of therapy, and L-dopa has been shown to reduce PLMS (Guilleminault *et al.*, 1987; Hening *et al.*, 1999). In summary, all medications that are helpful for RLS seem to be effective in PLMD, although only few have been studied in controlled trials. RLS in hemodialyzed uremic patients is often relieved by treatment with erythropoetin when this agent is used for primary therapy of accompanying anemia (Roger *et al.*, 1991), although RLS patients do not exhibit anemia more frequently than non-RLS-affected dialysis patients (Collado-Seidel *et al.*, 1998).

### Dopaminergic Agents

L-dopa in conjunction with benserazide or carbidopa is the treatment of first choice in idiopathic RLS (Chesson *et al.*, 1999; Hening *et al.*, 1999). If patients complain about increased sleep latency, 50–100 mg L-dopa should be taken 1 hour before bedtime. A slow escalation in dose with increments of 50 mg is recommended, with a maximum of 200 mg as a single dose for the most severe cases. A dose of 100–150 mg has been found to be effective in most patients (Trenkwalder *et al.*, 1995). Because L-dopa plasma levels are usually subtherapeutic by about 4 hours after intake of the standard preparation, RLS symptoms and PLMS may recur in the middle of the night. Therefore either a second dose of the standard preparation may be required during the night (Brodeur *et al.*, 1988), or a combination treatment with a single dose of standard and a sustained-release preparation of L-dopa may be superior. This combination treatment has the advantage of reduced nocturnal awakenings, thus improving sleep quality (Collado-Seidel *et al.*, 1999a). Patients with RLS treated with L-dopa have been followed systematically for 2 years (Becker *et al.*, 1993; von Scheele and Kempi, 1990), single cases now for up to 10 years. Thus far, long-term treatment with L-dopa has not resulted in serious side effects, especially not in the occurrence of dyskinesias that we see as a long-term side effect in Parkinson's disease. Although L-dopa is still effective, some patients complain about the development of either a morning rebound (Guilleminault *et al.*, 1993) or a phenomenon called augmentation (Allen and Earley, 1996). Augmentation means that patients experience RLS symptoms either earlier during the day or in other parts of their body in comparison to the time before dopaminergic treatment. This phenomenon is the main side effect in treatment with levodopa and is especially pronounced in patients with severe symptoms and daytime symptoms. Because of loss of efficacy over time, dosage increase or use of an additional medication may be necessary. These patients should be switched to a dopamine agonist either as monotherapy or in combi-

nation therapy. Unfortunately, no clinical studies on any trials for combination treatment in RLS are available. Patients with refractory RLS should be cautious about the potential negative effect of food, particularly protein, on absorption and efficacy of L-dopa. Recent studies have shown that all dopamine agonists that have been tested in RLS are an effective treatment appropriate for moderate and severe cases. Dopamine agonists should be used in chronic treatment, and a slow titration period is recommended for all agents, especially for the ergot derivates like pergolide, bromocriptine and cabergoline because of nausea. Bromocriptine, up to 7.5 mg, can be taken 1–3 hours before retiring (Walters *et al.*, 1988) beginning with 2.5 mg. Pergolide (0.15–0.75 mg, start with 0.05 mg) has been used successfully in a double-blind study and for the first time in multicenter studies and was efficient in respect to improvement of subjective and objective sleep quality and reduction of PLMS (Earley *et al.*, 1998; Silber *et al.*, 1997; Wetter *et al.*, 1999). Long-term results of more than 12 months treatment show a continuous efficacy of pergolide on RLS symptoms (Stiasny *et al.*, 2001). Augmentation seems to be only a minor problem in the treatment with pergolide and other dopamine agonists, but long-term data are still missing. For better tolerability, the combination treatment with domperidone, if available (not in the United States), is recommended at the beginning. Other dopamine agonists proved their efficacy in the treatment of RLS in open studies. Cabergoline reduced subjective symptoms and the frequency of the PLMS and improved sleep (Stiasny *et al.*, 2000) and subjective symptoms. Lisuride has shown to be efficient in an open trial (Benes *et al.*, 2000) as has dihydroergocriptine.

The nonergot dopamine agonists pramipexole and ropinirole improved the restless legs symptoms during the day and sleep, as well as periodic limb movements in idiopathic RLS. Controlled studies with pramipexole (Montplaisir *et al.*, 1999) and ropinirole (Saletu *et al.*, 2000) showed a more appropriate spectrum of side effects. Recent publications about sleep attacks in Parkinson's disease (Frucht *et al.*, 1999) occurring during the treatment with pramipexol suggest that these may also occur in RLS. For contraindications and other side effects of dopaminergic therapy, see Chapter 64.

### Opioids

If dopaminergic agents are not sufficiently efficient or if contraindications do not allow dopaminergic treatment, opioids should be given as the medication of second choice. Recent publications confirm not only the efficacy (Lauerma and Markkula, 1999; Walters *et al.*, 1993) but also the long-term beneficial effects

(Walters *et al.*, 2001) of opioids in RLS. In general, opioids are indicated in severe cases of RLS, as well as in about 15% of patients who do not benefit from dopaminergic treatment (von Scheele and Kempi, 1990). In a controlled study of oxycodone for idiopathic RLS, there was improvement of sleep efficiency and motor restlessness using a drug regimen of 5 mg oxycodone 2 hours before bedtime, at bedtime, and in the middle of the night (Walters *et al.*, 1993). Sustained-released preparations of opioids seem to improve the therapeutic effect, especially in patients who have daytime symptoms (Hening *et al.*, 1999; C. Trenkwalder and T. C. Wetter, personal observation, 2002). Opioids are also helpful in acute exacerbations of RLS (e.g., by neuroleptic agents or during immobilization or after surgery). They can be delivered epidurally for rapid relief of symptoms (Vahedi *et al.*, 1994). Patients with a history of addiction to either alcohol or any drugs should not be treated with opioids.

### Clonazepam

Because of possible substance dependence, clonazepam (Klonopin or Rivotril 0.5 mg at bedtime) should be prescribed cautiously. As with other benzodiazepines, tolerance may develop. Increase in dose may temporarily re-establish effectiveness but also may lead to significant side effects, such as ataxia and drowsiness. Episodes of nocturnal wandering, presumably related to the persisting urge to move, may occur during benzodiazepine treatment, particularly with triazolam and midazolam (Lauerma, 1991). Clonazepam reduced RLS symptoms and improved sleep but did not significantly decrease the frequency of periodic limb movements (Montagna *et al.*, 1984).

In the past years benzodiazepine receptor ligands such as zolpidem were prescribed more frequently instead of benzodiazepines, if problems falling asleep predominate the clinical picture, and may be combined with any dopaminergic drug.

### Anticonvulsants

Carbamazepine in a dose of 200–500 mg could improve RLS symptoms and sleep in 60% of RLS patients (Telstadt *et al.*, 1984). Because recent studies of valproic acid and especially gabapentine showed highly beneficial effects on sleep in RLS patients, one may prefer the latter two medications instead of carbamazepine. Valproate reduced periodic limb movements and improved sleep at a dose of 300–600 mg in patients with sleep disorders and PLMS (Ehrenberg *et al.*, 2000); gabapentine may be especially beneficial for patients

with painful sensations in RLS and can be prescribed at a dosage of 300 up to 1500 mg/day also for patients with severe RLS (Adler, 1997; Mellick and Mellick, 1996).

### Clonidine

In uremic RLS, clonidine 75 µg (bid) was effective without significant hypotensive side effects (Ausserwinkler and Schmidt, 1989); in idiopathic RLS a higher dose of clonidine has been given (0.1–1.0 mg; mean dose, 0.5 mg) and was moderately effective (Wagner *et al.*, 1996).

## TREATMENTS NO LONGER RECOMMENDED

In the past, many therapies were based on anecdotal observations, few of which have been tested in controlled trials. Many of these therapies can no longer be recommended. They include vibration of calves (Montagna *et al.*, 1984); avoidance of nicotine; avoidance of consumption of caffeine; avoidance of red wine at bedtime; vitamins; zinc; dextran; and chloroquine. Note that amitriptyline, a drug commonly used for control of neuropathic paresthesia, can in fact exacerbate RLS symptoms (Collado-Seidel *et al.*, 1999b), but there are also observations that it may help in painful forms of RLS (A. S. Walters, personal communication, 2002). Other antidepressants (trazodone, paroxetine, imipramine, trimipramine) may improve RLS symptoms in selected cases (Fleming *et al.*, 1988; Sewitch and Liebmann, 1988), although there are controversial observations (Sanz-Fuentenebro *et al.*, 1996).

## PLMS SYNDROME

PLMS syndrome in isolation, also earlier known as "nocturnal myoclonus" and renamed periodic limb movement disorder (PLMD; ASDA, 1993), does not necessarily require treatment unless sleep is compromised. In that case, dopaminergic agents are the first choice treatment, and L-dopa has been shown to reduce PLMS (Guilleminault *et al.*, 1987). Indeed, a medical regimen similar to RLS seems justified, because PLMS are a highly frequent phenomenon in patients with familial and sporadic RLS (Walters *et al.*, 1990). The beneficial treatment effect also applies for clonazepam (Peled and Lavie, 1987), pergolide (Staedt *et al.*, 1997), and valproic acid (Ehrenberg *et al.*, 2000), whereas 5-hydroxytryptophan and L-tryptophan (Guilleminault *et al.*, 1987) are not effective. Baclofen has only limited efficiency in reducing PLMS (Guilleminault and Flagg, 1984). In summary one may first try dopaminergic agents followed by clonazepam and anticonvulsants. Opioids are helpful in reducing the PLMS in patients with RLS and thus by analogy may be helpful in reducing the PLMS in patients without RLS (Walters *et al.*, 1993; 2001).

# REFERENCES

Adler, C. H. (1997). Treatment of restless legs syndrome with gabapentin. *Clin. Neuropharmacol.* **20**, 148–151.

Allen, R. P., and Earley, C. J. (1996). Augmentation of the restless legs syndrome with carbidopa/levodopa. *Sleep* **19**, 205–213.

Atlas Task Force of the American Sleep Disorders Association, Guilleminault, C., chairman (1993). Recording and scoring leg movements. *Sleep* **16**, 748–759.

Ausserwinkler, M., and Schmidt, P. (1989). Erfolgreiche Behandlung des "Restless leg"—Syndroms bei chronischer Niereninsuffizienz mit Clonidin. *Schweiz. Med. Wschr.* **119**, 184–186.

Bara-Jimenez, W., Aksu, M., Graham, B., Sato, S., and Hallet, M. (2000). Periodic limb movements in sleep: state dependent excitability of the spinal flexor reflex. *Neurology* **54**, 1609–1616.

Barton, J. C., Wooten, V. D., Acton, R. T. (2001). Hemochromatosis and iron therapy of restless legs syndrome. *Sleep Med.* **2**, 249–251.

Becker, P. M., Jamieson, A. O., and Brown, W. D. (1993). Dopaminergic agents in restless legs syndrome and periodic limb movements of sleep: Response and complications of extended treatment in 49 cases. *Sleep* **16**, 713–716.

Benes, H., Deißler, A., Clarenbach, P., Rodenbeck, A., and Hajak, G. (2000). Lisurid in the management of restless legs syndrome—an extended polysomnographic study. *Mov. Disord.* **15**(Suppl.3), 134.

Boghen, D., Lamothe, I., Elie, R., Godbout, R., and Montplaisir, J. (1986). The treatment of the restless legs syndrome with clonazepam: A prospective controlled study. *Can. J. Neurol. Sci.* **13**, 245–247.

Brodeur, C., Montplaisir, J., Godbout, R., and Marinier, R. (1988). Treatment of restless legs syndrome and periodic movements during sleep with L-dopa: a double-blind, controlled study. *Neurology* **38**, 1845–1848.

Bucher, S. F., Seelos, K. C., Oertel, W. H., Reiser, M., and Trenkwalder, C. (1997). Cerebral generators involved in the pathogenesis of the restless legs syndrome. *Ann. Neurol.* **41**, 639–645.

Chesson, A. L., Wise, M., Davila, D., Johnson, S., Littner, M., Anderson, M., Hartse, K., and Rafecas, J. (1999). Practice parameters for the treatment of restless legs syndrome and periodic limb movement disorder. An American Academy of Sleep Medicine Report. Standards of Practice Committee of the Academy of Sleep Medicine. *Sleep* **22**, 961–968.

Coleman, R. (1982). Periodic movements in sleep (nocturnal myoclonus) and restless legs sindrome. *In* "Sleeping and Waking Disorders: Indications and Techniques" (C. Guilleminault, Ed.), pp. 265–295. Addison-Wesley, Menlo Park, CA.

Collado-Seidel, V., Kazenwadel, J., Wetter, T. C., Kohnen, R., Winkelmann, J., Selzer, R., Oertel, W. H., and Trenkwalder, C. (1999a). A controlled study of additional sr-L-dopa in L-dopa-responsive restless legs syndrome with late-night symptoms. *Neurology* **52**, 285–290.

Collado-Seidel, V., Kohnen, R., Samtleben, W., Hillebrand, G. F., Oertel, W. H., and Trenkwalder, C. (1998). Clinical and biochem-

ical findings in uremic patients with and without restless legs syndrome. *Am. J. Kidney Dis.* **31**, 324–328.

Collado-Seidel, V., Winkelmann, J., and Trenkwalder, C. (1999b). Aetiology and treatment of restless legs syndrome. *CNS Drugs* **12**, 9–20.

Danek, A., and Pollmächer, T. (1990). Restless-legs-Syndrom: Klinik, Differentialdiagnose, Therapieansätze. *Nervenarzt* **61**, 69–76.

Davis, B. J., Rajput, A., Rajput, M. L., Aul, E. A., and Eichhorn, G. R. (2000). A randomized, double-blind placebo-controlled trial of iron in restless legs syndrome. *Eur. Neurol.* **43**, 70–75.

Desautels, A., Turecki, G., Montplaisir, J., Sequeira, A., Verner, A., and Rouleau, G. A. (2001). Identification of a major susceptibility locus for restless legs syndrome on chromosome 12q. *Am. J. Hum. Genet.* **69**, 1266–1270.

Earley, C. J., Yaffee, J. B., and Allen, R. P. (1998). Randomized, double-blind, placebo-controlled trial of pergolide in restless legs syndrome. *Neurology* **51**, 1599–1602.

Ehrenberg, B. L., Eisensehr, I., Corbett, K. E., Crowley, P. F., and Walters, A. S. (2000). Valproate for sleep consolidation in periodic limb movement disorder. *J. Clin. Psychopharmacol.* **20**, 574–578.

Ekbom, K. A. (1970). Restless legs. *In* "Handbook of Clinical Neurology" (P. J. Vinken and G. W. Bruyn, Eds.), Vol. 8, pp. 311–320. North-Holland, Amsterdam.

Fleming, J. A. E., Isomura, T., and Rungta, K. N. (1988). The effects of trazodone hydrochloride on periodic leg movements. *Sleep Res.* 39.

Frucht, S, Rogers, J. D., Greene, P. D., Gordon, M. F., and Fahn, S. (1999). Falling asleep at the wheel: motor vehicle mishaps in persons taking pramipexole and ropinirole. *Neurology* **52**, 1908–1910.

Gibb, W. R. G., and Lees, A. J. (1986). The restless legs syndrome. *Postgrad. Med. J.* **62**, 329–333.

Goodman, J. D. S., Brodie, C., and Ayida, G. A. (1988). Restless leg syndrome in pregnancy. *BMJ* **297**, 1101–1102.

Guilleminault, C., Cubel, M., and Philip, P. (1993). Dopaminergic treatment of restless legs and rebound phenomenon. *Neurology* **43**, 445.

Guilleminault, C., and Flagg, W. (1984). Effect of baclofen on sleep-related periodic leg movements. *Ann. Neurol.* **15**, 234–239.

Guilleminault, C., Mondini, S., Montplaisir, J., Mancuso, J., Cobasko, D., and Dement, W. C. (1987). Periodic leg movement, L-dopa, 5-Hydroxytryptophan, and L-Tryptophan. *Sleep* **10**, 393–397.

Hening, W. A., Allen, R., Earley, C., Kushida, C., Picchietti, D., and Silber, M. (1999). The treatment of restless legs syndrome and periodic limb movement disorder—An American Academy of Sleep Medicine Review. *Sleep* **22**, 970–998.

Hening, W. A., Walters, A. S., Kavey, N., Gidro-Frank, S., Côté, L., and Fahn, S. (1986). Dyskenisias while awake and periodic movements in sleep in restless legs syndrome: Treatment with opioids. *Neurology* **36**, 1363–1366.

Hornyak, M., Voderholzer, U., Hohagen, F., Berger, M., and Riemann, D. (1998). Magnesium therapy for periodic leg movements-related insomnia and restless legs syndrome: An open pilot study. *Sleep* **21**, 501–505.

Kraus, T., Schuld, A., and Pollmächer, T. (1999). Periodic leg movements in sleep and restless legs syndrome probably caused by olanzapine. *J. Clin. Psychopharmacol.* **19**, 478–479 (letter).

Lauerma, H. (1991). Nocturnal wandering caused by restless legs and short-acting benzodiazepines. *Acta. Psychiatr. Scand.* **83**, 492–493.

Lauerma, H., and Markkula, J. (1999). Treatment of restless legs syndrome with tramadol: an open study. *J. Clin. Psychiatry* **60**, 241–244.

Lavigne, G. J., and Montplaisier, J. Y. (1994). Restless legs syndrome and sleep bruxism: prevalence and association among Canadians. *Sleep* 17, 739–743.

Lugaresi, E., Cirignotta, F., Cuccagna, G., and Montagna, P. (1986). Nocturnal myoclonus and restless legs sindrome. *Adv. Neurol.* 43, 295–307.

Mellick, G. A., and Mellick, L. B. (1996). Management of restless legs syndrome with gabapentin. *Sleep* 19, 224–226.

Montagna, P., Sassoli de Bianchi, L., Zucconi, M., Cirignotta, F., and Lugaresi, E. (1984). Clonazepam and vibration in restless legs sindrome. *Acta. Neurol. Scand.* 69, 428–430.

Montplaisir, J., Boucher, S., Poirier, G., Lavigne, G., Lapierre, O., Lesperance, R. (1997). Clinical, polysomnographic, and genetic characteristics of restless legs syndrome: A study of 133 patients diagnosed with new standard criteria. *Mov. Disord.* 12, 61–65.

Montplaisir, J., Boucher, S., Nicolas, A., Lesperance, P., Gosselin, A., Rompre, P., and Lavigne, G. (1998). Immobilization tests and periodic leg movements in sleep for the diagnosis of restless leg syndrome. *Mov. Disord.* 13, 324–329.

Montplaisir, J., Nicolas, A., Denesle, R., and Gomez-Mancilla, B. (1999). Restless legs syndrome improved by pramipexole: a double-blind randomized trial. *Neurology* 52, 938–943.

O'Keeffe, S. T., Gavin, K., Lavan, J. N. (1994). Iron status and restless legs syndrome in the elderly. *Age Ageing* 23, 200–203.

Peled, R., and Lavie, P. (1987). Double-blind evaluation of clonazepam on periodic leg movements in sleep. *J. Neurol. Neurosurg. Psychiat.* 50, 1679–1681.

Philips, B., Young, T., Finn, L., Asher, K., Hening, W. A., and Purvis, C. (2000). Epidemiology of restless legs symptoms in adults. *Arch. Intern. Med.* 160, 2137–2141.

Picchietti, D. L., and Walters, A. S. (1996). Restless legs syndrome and periodic limb movement disorder in children and adolescents. Comorbidity with attention-deficit hyperactivity disorder. *Neurol. Clin.* 5, 729–740.

Read, D. J., Feest, T. G., and Nassim, M. A. (1981). Clonazepam: Effective treatment for restless legs syndrome in uraemia. *BMJ* 283, 885–886.

Roger, S. D., Harris, D. C., and Stewart, J. H. (1991). Possible relation between restless legs and anaemia in renal dialysis patients. *Lancet* 337, 1551.

Rothdach, A., Trenkwalder, C., Haberstock, J., Keil, U., and Berger, K. (2000). Prevalance and risk factors of RLS in an elderly population: The MEMO study. *Neurology* 54, 1064–1068.

Routtinen, H. M., Partinen, M., Hublin, C., Bergman, J., Haaparanta, M., Solin, O., and Rinne, J. O. (2000). An FDOPA PET study in patients with periodic limb movement disorder and restless legs syndrome. *Neurology* 54, 502–504.

Saletu, M., Anderer, P., Saletu, B., Hauer, C., Mandl, M., Oberndorfer, S., Zoghlami, A., and Saletu-Zyhlarz, G. (2000). Sleep laboratory studies in restless legs syndrome patients as compared with normals and acute effects of ropinirole. *Neuropsychobiology* 41, 190–199.

Sanz-Fuentenebro, F. J., Huidobro, A., and Tejadas-Rivas, A. (1996). Restless legs syndrome and paroxetine. *Acta Psychiatr. Scand.* 94, 482–484.

Sewitch, D. E., and Liebmann, K. O. (1988). Treatment of periodic movements in sleep "nocturnal myoclonus" with a low dosage of the tricyclic antidepressant imipramine. *Sleep Res.* 17, 256.

Silber, M. H., Shepard, J. W., and Wisbey, J. A. (1997). Pergolide in the management of restless legs syndrome: an extended study. *Sleep* 20, 878–882.

Staedt, J., Waßmuth, F., Ziemann, U., Hajak, G., Rüther, E., and Stoppe, G. (1997). Pergolide: Treatment of choice in restless legs syndrome (RLS) and nocturnal myoclonus syndrome (NMS). A double-blind randomized cross-over trial of pergolide versus L-Dopa. *J. Neural. Transm.* 104, 461–468.

Stiasny, K., Roebbecke, J., Schüler, P., and Oertel, W. H. (2000). The treatment of idiopathic restless legs syndrome (RLS) with the D2-agonist cabergoline—An open clinical trial. *Sleep* 23, 349–354.

Stiasny, K., Wetter, T. C., Winkelmann, J., Brandenburg, U., Penzel, T., Rubin, M., Hundemer, H. P., Oertel, W. H., and Trenkwalder, C. (2001). Long-term effects of pergolide in the treatment of restless legs syndrome. *Neurology* 56, 1399–1402.

Telstadt, W., Sørensen, O., Larsen, S., Lillevold, P. E., Stensrud, P., and Nyberg-Hansen, R. (1984). Treatment of the restless legs syndrome with carbamazepine: A double-blind study. *BMJ* 288, 444–446.

Trenkwalder, C., Bucher, S., Pröckl, D., Paulus, W., and Oertel, W. H. (1993). Bereitschaftspotential in idiopathic and symptomatic restless legs syndrome. *Electroencephic. Clin. Neurophysiol.* 89, 95–103.

Trenkwalder, C., Collado-Seidel, V., Gasser, T., and Oertel, W. H. (1996). Clinical symptoms and possible anticipation in a large kindred of familial restless legs syndrome. *Mov. Disord.* 11, 389–394.

Trenkwalder, C., Hening, W. A., Walters, A. S., Campbell, S. S., Rahman, K., and Chokroverty, S. (1999). Circadian rhythm of periodic limb movements and sensory symptoms of restless legs syndrome. *Mov. Disord.* 14, 102–110.

Trenkwalder, C., Stiasny, K., Pollmächer, T., Wetter, T. C., Schwarz, J., Kohnen, R., Kazenwadel, J., Krüger, H. P., Ramm, S., Künzel, R., and Oertel, W. H. (1995). L-DOPA therapy of uremic and idiopathic restless legs syndrome: a double-blind, crossover trial. *Sleep* 18, 681–688.

Turjanski, N., Lees, A. J., and Brooks, D. J. (1999). Striatal dopaminergic function in restless legs syndrome. 18F-dopa and 11C-raclopride PET studies. *Neurology* 52, 932–937.

Vahedi, H., Küchle, M., Trenkwalder, C., and Krenz, C.-J. (1994). Peridurale Morphiumanwendung bei Restless-legs Status. *Anästhesiol. Intensivmed. Notfallmed. Schmerzth.* 29, 368–370.

von Scheele, C., and Kempi, V. (1990). Long-term effect of dopaminergic drugs in restless legs. *Arch. Neurol.* 47, 1223–1224.

Wagner, M. L., Walters, A. S., Coleman, R. G., Hening, W. A., Grasing, K., and Chokroverty, S. (1996). Randomized, double-blind, placebo-controlled study of clonidine in restless legs syndrome. *Sleep* 19, 52–58.

Walker, S. L., Fine, A., and Kryger, M. H. (1996). L-DOPA/Carbidopa for nocturnal movement disorders in uraemia. *Sleep* 19, 214–218.

Walters, A. S., Hening, W. A., Côté, L., and Fahn, S. (1986). Dominantly inherited restless legs with myoclonus and periodic movements of sleep: A syndrome related to the endogenous opiates? *Adv. Neurol.* 43, 309–319.

Walters, A. S., Hening, W. A., Kavey, N., Chokroverty, S., and Gidro-Frank, S. (1988). A double-blind crossover trial of bromocriptine and placebo in restless legs syndrome. *Ann. Neurol.* 24, 455–458.

Walters, A. S., Hickey, K., Maltzman, J., Joseph, D., Hening, W., Wilson, V., and Chokroverty, S. (1996). A questionnaire study of 138 patients with restless legs syndrome: the night-walkers-survey. *Neurology* 46, 92–95.

Walters, A. S., Picchietti, D., Hening, W., and Lazzarini, A. (1990). Variable expressivity in familial restless legs syndrome. *Arch. Neurol.* 47, 1219–1220.

Walters, A. S., Wagner, M. L., Hening, W. A., Grasing, K., Mills, R., Chokroverty, S., and Kavey, N. (1993). Successful treatment of the idiopathic restless legs syndrome in a randomized double-blind trial of oxycodone versus placebo. *Sleep* 16, 327–332.

Walters, A. S., and The International Restless Legs Syndrome Study Group. (1995). Towards a better definition of the restless legs syndrome. *Mov. Disord.* 10, 634–642.

Walters, A. S., Winkelmann, J., Trenkwalder, C., Fry, J., Kataria, V., Wagner, M., Sharma, R., and Hening, W. (2001). Long-term opioid monotherapy in patients with restless legs syndrome. *Mov. Disord.* **16,** 1105–1109.

Wetter, T. C., Stiasny, K., Winkelmann, J., Buhlinger, A., Brandenburg, U., Penzel, T., Medori, R., Rubin, M., Oertel, W. H., and Trenkwalder, C. (1999). A randomized controlled study of pergolide in patients with restless legs syndrome. *Neurology* **52,** 944–950.

Winkelman, J. W., Chertow, G. M., and Lazarus, J. M. (1996). Restless legs syndrome in end-stage renal disease. *Am. J. Kidney Dis.* **28,** 372–378.

Winkelmann, J., Wetter, T. C., Collado-Seidel, V., Gasser, T., Dichgans, M., Yassouridis, A., and Trenkwalder, C. (2000). Frequency and characteristics of the hereditary restless legs syndrome in a population of 300 patients. *Sleep* **23,** 597–602.

Yokota, T., Hirose, K., Tanabe, H., and Tsukagoshi, H. (1991). Sleep-related periodic leg movements (nocturnal myoclonus) due to spinal cord lesion. *J. Neurol. Sci.* **104,** 13–18.

Zucconi, M., Coccagna, G., Petronelli, R., Gerardi, R., Mondini, S., and Cirignotta, F. (1989). Nocturnal myoclonus in restless legs syndrome: Effect of carbamazepine treatment. *Funct. Neurol.* **4,** 263–271.

*Movement Disorders*

CHAPTER 83

# Rehabilitation of Motor Function

A. J. Thompson and S. Hesse

## CLINICAL ASPECTS AND NATURAL COURSE

Stroke is a leading cause of disability and handicap in the industrialized world. Each year 750,000 subjects suffer a stroke in the United States, the prevalence is 200–300 patients/100,000 inhabitants (Williams *et al.*, 1999). Approximately 90% of these subjects have persisting neurological motor deficits, leading to disability and handicap, namely incompetence in their daily activities, impaired arm and hand function, and walking ability. Large outcome studies reported that only 5% of stroke survivors regain full arm function, and 20% of them cannot use their arms at all (Nakayama *et al.*, 1994). With respect to walking ability, approximately 75% of stroke sufferers regain limited walking ability within a period of 12 weeks, and 25% remain wheelchair-bound (Jorgensen *et al.*, 1995). Established prognostic factors with respect to recovery of motor function after stroke are the initial competence in daily activities, the initial motor deficit, and the presence of urinary incontinence. Furthermore the functional outcome after stroke depends on the extent and localization of the brain lesion. A lesion including the premotor cortex, the basal ganglia, and thalamic regions (Miyai *et al.*, 1999, 2000) diminished the mobility outcome.

A large group of patients, usually early in life, develop motor deficits after spinal cord injury (SCI), the prevalence in the United States is 200,000 persons, the estimated national economic impact of SCI is approximately $9.73 billion per year (Weaver *et al.*, 2000). In addition, progressive motor disorders can result from a range of chronic neurological conditions that include multiple sclerosis (MS), Parkinson's disease, nontraumatic spinal cord disease, and neuromuscular disorders. MS with a prevalence of 1/1000 affects around 350,000 people in Europe. Although considered a variable and

unpredictable condition, most patients move from a relapsing/remitting phase into a secondary progressive phase over time, and it has been estimated that 50% of those affected will require assistance to walk within 15 years of onset (Weinshenker, 1994). This is usually the result of weakness and spasticity of spinal cord origin but may also be complicated by cerebellar deficits (Alusi *et al.*, 2001) and sensory disturbance. Although Parkinson's disease which has a prevalence of up to 200/100,000, is regarded as a treatable condition, it is progressive in nature in most patients and it can result in considerable disability, which may be compounded by the comorbidities of the older patient.

## PRINCIPLES OF THERAPY

### Motor Rehabilitation Promotes Brain Plasticity

Nudo and coworkers studied the neural substrates of rehabilitative training on motor recovery after ischemic infarct in primates (Nudo *et al.*, 1996). A subtotal lesion confined to a small portion of the representation of one hand resulted in a further loss of hand territory in the adjacent, undamaged cortex of squirrel monkeys if no rehabilitation was applied. On the other hand, retraining of skilled hand use with the affected extremity resulted in prevention of loss of territory. In some instances, the hand representation even expanded after retraining, indicating a reshaping of the cortical organization of the thumb. Potential mechanisms of brain plasticity are the modulation of existing pathways, the growth of new axonal processes, and suppression of diaschisis. In healthy subjects, even rapid changes of the cortical network representing the thumb could be induced by practicing simple, repetitive movements of the thumb over a short period of time (Classen *et al.*, 1998).

Furthermore, Friel and Nudo (1998) analyzed the movements of the fingers of the squirrel monkeys in form and number when retrieving food pellets. After a small cerebral injury in the hand motor area, the animals showed a deficit in skilled finger use with the number and shape of finger flexions increased. After 1 month of rehabilitative training, the monkeys in most trials again retrieved the pellets using stereotypic movement patterns different from those used before the injury. This finding is not congruent with the traditional concept of so-called neurofacilitatory treatment techniques (e.g., Bobath, PNF), which aim at the restoration of a normal function rather than promoting compensatory movements.

The activation of ipsilateral sensorimotor pathways seems to contribute to the recovery of motor function. Functional imaging studies after stroke revealed an enhanced activation or blood flow of the ipsilateral sensorimotor area after stroke and subsequent motor recovery of the affected extremity (Cramer et al., 1997). Bilateral practice is said to promote the activation of ipsilateral sensorimotor pathways.

### A Task-Specific Repetitive Approach Is Most Promising (***)

With respect to the content of rehabilitative training, modern concepts of motor learning favor a task-specific, repetitive approach, with the patients voluntarily activating the affected extremities (Asanuma and Keller, 1991). Skillful learning, for instance of sports or violin playing, requires several thousand repetitions. Correspondingly, Bütefisch et al. (1995;*) showed that the repetitive training of isolated movements of the paretic wrist and fingers effected better arm and hand recovery than a classic therapy regimen focusing on reduction of muscle spasticity. The same principles can be applied to gait rehabilitation, where goal-setting treadmill therapy, enabling the repetitive practice of complex gait cycles, proved superior with respect to gait restoration in chronic paraparetic and hemiparetic patients (Dietz et al., 1995;* Hesse et al., 1995**; Wernig and Müller, 1992*). The theoretical background of locomotor therapy is the activation of spinal and supraspinal central pattern generators by practicing complex gait cycles instead of preparatory maneuvers while lying or sitting (Lovely et al., 1986).

Conventional physiotherapy either followed a traditional functional approach or applied various neurofacilitation techniques, such as the Brunnstroem technique with synergistic movements, propioceptive neuromuscular facilitation with spiral and diagonal movements, and neurodevelopmental therapy (Bobath) with reflex inhibitory movements. Albeit different, comparative outcome studies failed to prove any superiority, so far (for overview, see Ernst, 1990**). Recently, a Norwegian study compared a motor relearning program (Carr and Shepherd, 1987) based on the principles of a task-specific repetitive approach (see earlier) with a Bobath program in 61 acute stroke survivors. The group following the motor relearning program was discharged earlier (21 vs 34 days) and scored significantly better in general motor function. Competence in activities of daily living, however, did not differ at discharge (Langhammer and Stanghelle, 2000**). This result further promotes the concept of a task-specific repetitive approach.

### The Concept of Learned Nonuse (***)

The concept of "learned nonuse" means that repeated disappointment in attempts to use the affected arm in the acute and subacute phase could lead to negative reinforcement of using the affected arm. Consequently, the patients do not use the affected extremity, despite an existing motor potential. To overcome this learned nonuse, Taub et al. (1993***) introduced the constrained-induced movement therapy in arm rehabilitation after stroke after successful animal experiments. The patients wear a sling to immobilize the unaffected arm and are thus enforced to use the affected extremity repetitively and task specifically.

## PRACTICAL MANAGEMENT

### General Rehabilitation

#### Stroke Survivors Should Be Transferred to Stroke Units Offering a Comprehensive Rehabilitation Program as Soon as Possible (***)

The Copenhagen stroke study group (Jorgensen et al., 2000***) compared the outcome of 1241 stroke survivors who were either treated in general neurological and medical wards (GW) or in a specialized single stroke unit (SU). The stroke unit offered besides acute medical care a comprehensive rehabilitation program including rehabilitative nursing, physiotherapy, occupational and speech therapy, and neuropsychology. The relative risks of initial death, poor outcome, and 1-year and 5-year mortality rates were reduced by 40% on average in patients treated in the SU compared with the GW (Figure 1). Those who benefited most seemed to be the patients with the most severe stroke. The beneficial effect of stroke unit treatment was observed regardless of the patients' age, sex, comorbidity, and initial stroke severity. A chain of care supporting patients after dis-

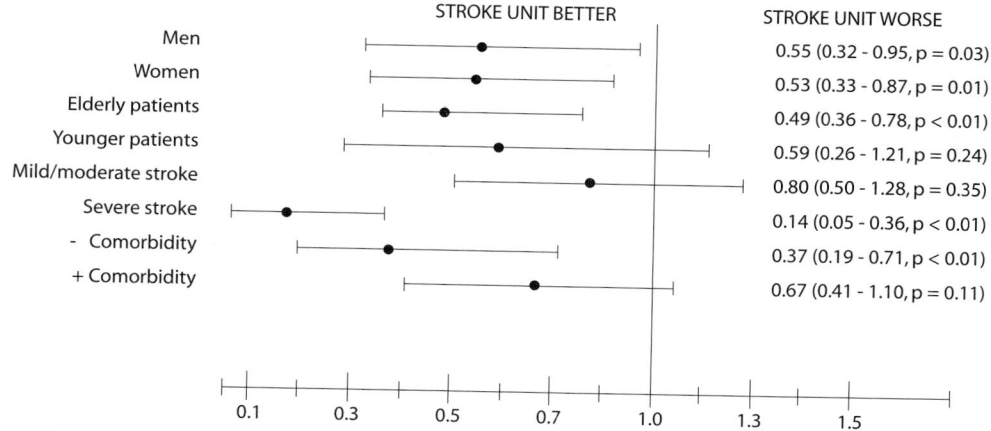

**FIGURE 1** Presentation of results for subgroups of the multivariate logistic regression analysis of death during hospital stay in patients treated in the stroke unit compared with in general neurological and medical wards. ORs and their 95% CIs and probability values are given (From Jorgensen *et al. Stroke* 2000; **31**, 436, with permission.).

charge could even optimize the effects of a stroke unit (Indredavik *et al.*, 2000**). For moderately affected stroke patients, a Swedish study compared early supported discharge (after 14 days) and continued rehabilitation at home by a specialized team vs inpatient routine rehabilitation with a median length of stay of 29 days. Both groups scored comparable over a 6-month observation period after stroke (von Koch *et al.*, 2000**), speaking in favor of the more cost-efficient concept of an early supported discharge after stroke in moderately affected patients.

### Greater Therapy Intensity Pays Off (***)

Kwakkel and coworkers 1 (1999***) studied 101 acute stroke victims, who were randomly assigned to a rehabilitation program with emphasis on arm training, a rehabilitation program with emphasis on leg training, or a control program in which the arm and leg were immobilized with an inflatable pressure splint. Each treatment regimen was applied for 30 minutes 5 days a week during the first 20 weeks after stroke. At week 20, the leg training group ($n = 31$) had higher scores than the control group ($n = 37$) for activities of daily living (ADL) ability, walking ability, and dexterity. The arm training group ($n = 33$) differed significantly from the control group only in dexterity. The authors concluded that greater intensity of leg rehabilitation improves functional recovery and that exercise therapy primarily induces treatment effects on the abilities at which training is specifically aimed. This finding is in line with previous studies that also reported a positive correlation between therapy intensity and motor function, particularly when the therapy content did not remain constant but covered different aspects of dis-

ability (Parry *et al.*, 1999;*** Sunderland *et al.*, 1992, 1994***).

### Detrimental Drugs Should Be Avoided and Drugs Promoting Brain Plasticity Should Be Prescribed (***)

Studies in laboratory animals indicate that certain centrally acting drugs (e.g., the antihypertensive clonidine and prazosine, neuroleptics and other dopamine receptor antagonists, benzodiazepines, and the anticonvulsants phenytoin and phenobabital) impair *behavioral* recovery after focal brain injury (Feeney *et al.*, 1992). A retrospective study compared a "detrimental" drug group ($n = 37$) (those who received at least one of the aforementioned drugs) with a "neutral" drug group ($n = 59$) for the outcome of upper and lower motor function and ADL ability (Goldstein *et al.*, 1995**). The detrimental drug group scored worse in all dependant variables 56 and 84 days after stroke. Correspondingly, the clinician should avoid the prescription of these "detrimental" drugs in stroke rehabilitation, if possible. On the other hand, amphetamine and methylphenidate could promote early poststroke recovery. In a small placebo-controlled study Walker-Batson and coworkers compared doses of 10 mg dextroamphetamine every fourth day for 10 sessions paired with physical therapy and placebo in 10 acute stroke victims (Walker-Batson *et al.*, 1995**). The administration of dextroamphetamine paired with physical therapy increased the rate and extent of motor recovery at the end of the study and at follow-up 12 months later. A transient side effect was sleeping difficulties in one patient each while taking the medication; later on no negative related side effects were observed. Grade and colleagues (1998**) compared a daily and gradually increased intake of 5–30 mg methylphenidate for 3 weeks with placebo in 21 acute stroke survivors. The experimental group was less depressed and scored better on motor and ADL scores at the end of the study. Side effects were not observed.

## Special Aspects of Upper Limb Motor Rehabilitation

### Repetitive Training of Isolated Movements (***)

The potential of repetitive training of isolated wrist movements was shown by Bütefisch and coworkers (1995**) in 27 hemiparetic patients within a multiple baseline approach (see preceding). The specific training consisted of repetitive hand and finger flexions and extensions against various loads and was carried out twice daily during 15 minutes in addition to conventional therapy. In contrast to the baseline of conventional NDT therapy, grip strength, peak force of isometric hand extensions, and peak accelerations of isotonic wrist extension improved significantly during the training period. The value of repetitive sensorimotor stimulation of the paretic upper limb was further confirmed in a large ($n = 100$) prospective study. The experimental group pushed a rocking chair with their paretic arm, fixed in an inflatable splint, repetitively whereas while in the control group the arm was rested on a cushion in front of the patient (Feys et al., 1998**). Patients in the experimental group performed better on the Brunnstroem-Fugl-Meyer test than those in the control group throughout the study period, but differences were only significant at follow-up. The treatment was most effective in patients with severe motor deficits and hemi-inattention. The authors attributed the effect of therapy to the repetitive stimulation of muscle activity.

### Constrained Induced Movement Therapy (***)

Constrained-induced movement therapy is gaining more and more acceptance (Figure 2). In their original article, Taub et al. recommended constraining the non-affected arm more than 90% of the working day (1993***). The restraint devices were worn for 14 days. On each weekday during this period patients spent 7 hours at the rehabilitation center and were given a variety of tasks to be carried by the paretic upper extremity for 6 hours. Before treatment, patients should be able to extend at least 10° at the metacarpophalangeal and interphalangeal joints and 20° at the wrists. A first clinical study compared two groups of chronic hemiparetic subjects; they either received the forced-use approach for 14 days or were assigned to an attention

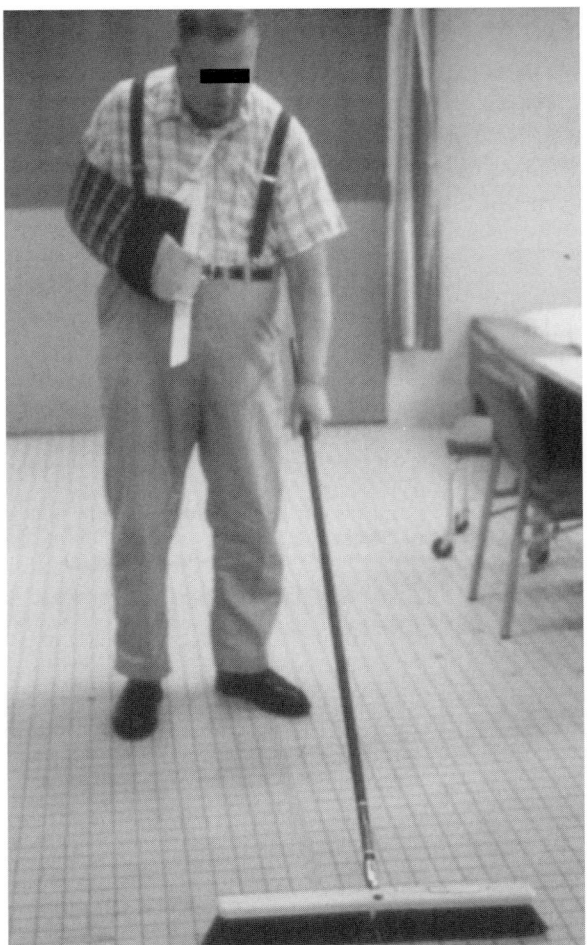

FIGURE 2   Constrained-induced movement therapy of a left hemiparetic subject. Immobilizing the right upper extremity enforces the use of the affected left arm.

comparison group. It received several procedures to focus attention on use of the impaired upper extremity. The constrained group scored significantly better on upper limb motor function and related ADL at the end of the study. These gains were maintained during a 2-year period of follow-up. Another controlled study with 66 chronic hemiparetic patients (either allocated to forced use or reference therapy of 14 days) partly confirmed these results with a small but lasting effect of forced use therapy on the dexterity of the affected arm (van der Lee et al., 1999**). For 23 acute stroke survivors, a recent controlled study showed less arm impairment at the end of treatment compared with a control group (Dromerick et al., 2000**).

### Bilateral Practice (*)

Instead of constraining the nonaffected arm, bilateral practice intends to use the facilitatory drive of the nonaffected onto the affected extremity. This phenomenon can be seen with many hemiparetic patients performing mirror movements when asked to activate their paretic muscles. Mudie and Matyas studied the effects of bilateral practice in eight subchronic s troke patients with mild impairment (1996*). Within a multiple baseline design, patients were instructed to place a block, drink from a glass, or target a peg either unilaterally with the affected arm or bilaterally. The bilateral practice proved superior with regard to upper limb motor function improvement. A more recent open study investigated bilateral arm training with rhythmic auditory cueing (Whitall et al., 2000*). The authors determined the effects of 6 weeks of bilateral practice on 14 patients with chronic hemiparetic stroke (median time after stroke, 30 months) immediately after stroke and at 2 months after training. Four 5-minute periods per session were performed with the use of a custom-designed machine. The patients showed significant and durable increase of their arm function, isometric strength, and range of motion of the affected side.

### Neurolytic Treatment of Spasticity with Botulinum Toxin A (***)

Botulinum toxin A (BTX) is nowadays accepted as the first treatment choice in the case of disabling focal spasticity. Before injection, realistic functional goals (i.e., ease of nursing, improvement of motor function, pain management) should be defined within the therapeutic team. In the case of upper limb spasticity, Simpson et al. were the first to report a randomized, double-blind placebo-controlled trial using BTX in 39 individuals with spastic upper extremities (1996**). The authors injected placebo or a total dosage of 75, 150, and 300 units BOTOX into three muscles: biceps brachii (65%), flexor carpi radialis (25%), and ulnaris (15% of total dosage). Only the treatment with the highest dose resulted in a statistically significant mean decrease of muscle tone 2, 4, and 6 weeks after injection. There were no adverse reactions. No significant differences were found between placebo and treatment groups for motor functions of the affected upper extremity, pain, caregiver dependency, and competence in daily activities. For the British product, Dysport, a large dose-finding randomized, placebo-controlled study in upper limb spasticity ($n = 83$) reported a relevant muscle tone reduction at doses of 500, 1000, and 1500 units (Bakheit et al., 2000**). Again, functional disability did not differ between treatment groups. Bhakta and coworkers reported in their randomized study on 40 patients with spasticity in a functionally useless arm a temporary reduction in carer burden at doses of 1000 units Dysport (2000**).

To increase the effectiveness of the toxin, the reader should pay attention to the correlation of terminal nerve end activity and the toxin uptake. Because in clinical practice spasticity is almost always accompanied by central paresis, the affected muscles are usually less active, on average, than normal or dystonic muscles. Correspondingly, the patients should use the injected extremities after injection. When not possible, electrical stimulation of the treated muscles can be used to artificially activate the terminal nerve ends. A viable protocol applies 3-second trains of charge-balanced constant current pulses (20 Hz, 0.2 ms, 50–90 mA) for 30 minutes five times/day 3 days after the injection. Hesse and coworkers conducted a placebo-controlled study on upper limb spasticity with four treatment arms. Only the group receiving verum (1000 units Dysport) plus electrical stimulation showed a relevant muscle tone reduction and improvement for such tasks as putting the arm through a sleeve (1998**). Other discussed means of increasing the effectiveness of the costly toxin are a larger dilution and the combination with serial casting after injection to address the accompanying changes of muscle properties by applying a tonic stretch (Reiter et al., 1998**).

### Electrical Stimulation (***)

Electrical stimulation has a long and varied history in motor rehabilitation. A recent controlled study included 60 patients with acute poststroke hemiplegia (Powell et al., 1999**). The patients were assigned randomly to one of two groups either receiving standard therapy or electrical stimulation (20 Hz, pulse width 0.2 ms, above motor threshold) of the wrist and finger extensors (3 times 30 minutes for 8 weeks) in addition to standard therapy. The experimental group significantly scored better with respect to isometric wrist strength and grasping function after 8 weeks. Benefits of electrical stimulation were most apparent in those patients with some residual motor function at the wrist. The study did not include any follow-up, also a sham therapy in the control group would have added to the significance of the study.

A novel approach is electromyography-triggered electrical stimulation of the wrist and finger extensors. The patient is instructed to voluntarily activate the paretic extensor muscle and the corresponding EMG signal is sensed. Once the muscle activity has reached a preset threshold, an external electrical stimulus elicits full wrist and finger extension. A randomized clinical study with 12 chronic (at least 1 year stroke interval) hemiparetic patients with a residual motor function followed a modified crossover treatment with one block of 12 sessions of electrical stimulation and another block of 12 treatment sessions in which the patients were encouraged to attempt wrist and finger extension without external assistance (Caraugh *et al.*, 2000**). During the experimental phases patients could significantly move more wooden blocks with their hand and displayed a higher isometric force of the wrist and finger extensors.

For tetraplegic patients, multichannel Functional Electrical Stimulation (e.g., the commercially availabe Freehand System®) offers grasping and releasing objects with the hand. Electrodes are attached to relevant muscles of the forearm and a pacemaker-type stimulator is surgically implanted into the chest. The patient controls the system by simple shoulder movements that are monitored by a shoulder position sensor. Most subjects used the system frequently for self-care, including eating, grooming, and brushing teeth. Lack of physical assistance during morning care to put on the system and incompatibility with multiple transfers to and from the wheelchair were perceived as two of the more common reasons for nonuse. After implantation, it took 6–8 months before the tetraplegic individuals could master the system (Carroll *et al.*, 2000; Davis *et al.*, 1998).

### Robot-Aided Simulation (**)

Hogan and coworkers introduced the MIT-Manus as a robotic device (1992). The patient holding an to a robotic arm can perform shoulder and elbow movements in multiple degrees of freedom, the device guides the limb and provides a sensorimotor experience that responds quickly, just like a "hand-over-hand" therapy. For evaluation, Volpe *et al.* (2000**) included 56 stroke patients who were either randomly assigned to the robot training for at least 25 hours or were exposed to the robotic device without training. Clinical data at the begin of the study were comparable in both groups; at the end of treatment the robot-trained group demonstrated improvement in motor outcome for the trained shoulder and elbow that did not generalize to the untrained wrist and hand.

### Special Aspects of Lower Limb Motor Rehabilitation

#### Locomotor Therapy (**)

In line with the task-specific repetitive approach, treadmill therapy with partial body weight support (BWS) enables nonambulatory hemiparetic subjects the repetitive practice of complex gait cycles at a very early stage (Figure 3). The harness substitutes for deficient equilibrium reflexes, the moving belt enforces locomotion, and part of the body weight is supported according to the extent of lower limb paresis. Initially, two therapists assist the movement, setting the paretic limbs

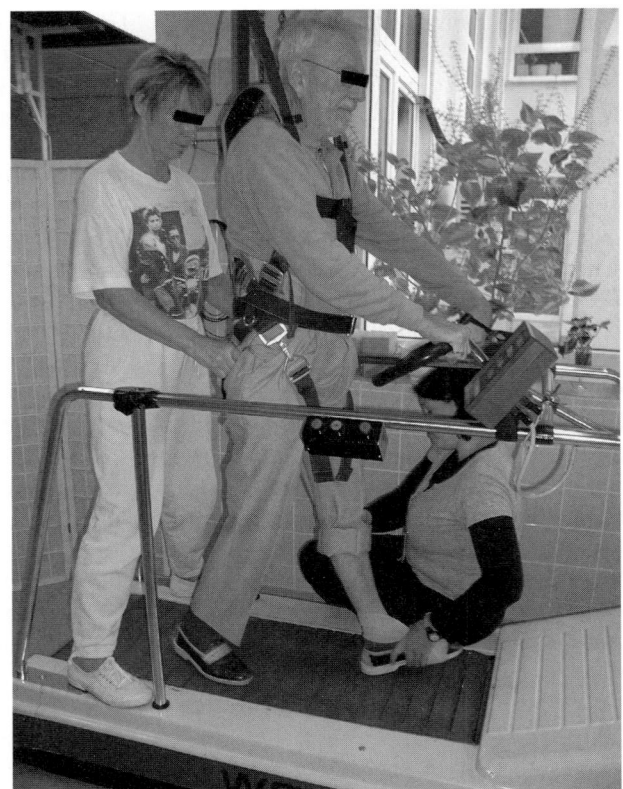

**FIGURE 3** Treadmill training with partial body weight support of a wheelchair-bound left hemiparetic patient.

and controlling the trunk movements, to practice not only repetitively but also in a correct manner. Within one session of 20 minutes, patients can perform up to 1000 steps compared with 50–100 steps during a conventional physiotherapy session.

For clinical evaluation in stroke patients, Hesse and coworkers conducted two single case design studies following an A-B-A design with 14 chronic, nonambulatory hemiparetic patients (1995*). During the two A-phases, the patients exclusively received the treadmill therapy for 3 weeks, and during the B-phase conventional physiotherapy following a classic NDT approach was administered for another 3 weeks. Gait ability and walking velocity improved substantially during the first A-phase, then leveled during the B phase, to increase again during the second A-phase. At the end of the study all patients could walk independently at least with verbal support; the statistical analysis revealed that the task-specific approach was superior with regard to gait restoration and improvement of walking velocity.

Visintin and coworkers compared treadmill therapy with and without BWS in 100 acute stroke survivors. The BWS group scored significantly better in mobility outcomes and walking velocity, endurance, and balance after a 6-week training period. Three months later, the BWS group continued to have significantly higher scores for walking velocity and motor recovery (Visintin *et al.*, 1998**).

Biomechanical studies revealed that hemiparetic patients walked more dynamic, symmetrical and less spastic on the treadmill compared with floor walking. Furthermore, the amount of body weight relief negatively correlated with the activity of relevant weight-bearing muscles; accordingly one should reduce the support as soon as possible. Clinical criterion was the patients' ability to carry their load adequately during the single stance phase of the paretic limb (i.e., without excessive knee flexion or swinging in the harness) (Hesse *et al.*, 1999). Several authors reported on the beneficial effect of locomotor therapy in wheelchair-bound paraparetic subjects, who could regain limited walking ability (100–200 m) after a training period of 1–6 months. The largest series was on 25 wheelchair-bound chronic patients with an interval of at least 6 months after trauma. Twenty-one regained limited walking ability with the help of a walker for a distance of up to 300 m. At follow-up several months to years after therapy, 19 of them could still ambulate independently (Wernig *et al.*, 1998). Dietz and colleagues even reported on the effects of treadmill therapy in paraplegic patients. Electromyographic recordings on the treadmill revealed a modulated activity of the calf muscles similar to healthy subjects after several months of locomotor therapy. The subjects, however, did not regain any walking ability on the floor. Future controlled studies will explain the potential of treadmill therapy with partial body weight support in spinal cord–injured patients (Dietz *et al.*, 1994).

The major inconvenience of treadmill therapy (and physiotherapy) is the great physical effort required by two therapists to assist the patients' gait. Therefore, one group designed an electromechanical gait trainer (Hesse *et al.*, 2000) and another group a computer-assisted version with the patients wearing a powered lower limb exoskeleton as most recent developments (Locomat, 2001). In the electromechanical gait trainer, the harness-secured patient is positioned on two footplates whose movements simulate natural walking, a servo-controlled motor supports the patients according to their abilities, and the vertical and horizontal trunk movements are controlled in a phase-dependent manner by ropes attached to the harness. The other system consists of a treadmill with body weight support, in addition, the patients wear a powered exoskeleton in which the programmable drives flex the hip and knee actively during the swing phase. A passive orthosis controls for the ankle movement. Future clinical outcome studies will detect their clinical value.

### Botulinum Toxin (**)

Botulinum toxin A (BTX) is the preferred option in the treatment of lower limb spasticity after stroke, traumatic brain (TBI), and spinal cord injuries (SCI). For the spastic dropfoot, several open studies showed that the intramuscular injection of BTX into the plantar flexors reduced muscle tone, improved the mode of initial contact, ankle range of motion with better advancement of the body, gait symmetry, and walking velocity up to 4 months each (Figure 4). Dynamic EMG recordings revealed a preferential diminution of the so-called premature activity of the plantar flexors already originat-

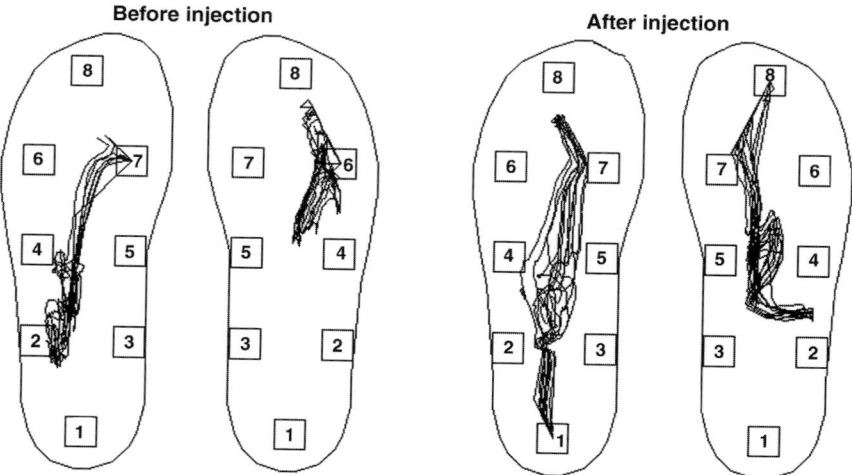

**FIGURE 4** Trajectories of the force point of action under the nonaffected and affected foot in a right hemiparetic patient before and 2 weeks after the injection of botulinum toxin A. Note a better mode of initial contact and better body advancement.

ing in the terminal swing (Hesse *et al.*, 1996). Burabaud and coworkers conducted a double-blind, placebo-controlled study in 24 individuals with spastic dropfoot (1996**). The authors injected EMG-guided placebo or a total dosage of 1000 units Dysport into the soleus, gastrocnemius, tibialis posterior, and flexor digitorum longus muscles. Patients reported a clear subjective improvement in foot spasticity after BTX but not after placebo administration. Significant changes were noted in Ashworth scale values for ankle extensors and invertors and for active ankle dorsiflexion up to 3 months after injection. Gait velocity was slightly but not significantly improved after the injection of BTX.

To increase the effectiveness of the costly toxin, patients should walk vigorously after injection referring to the positive correlation between terminal nerve end activity and toxin uptake (see earlier). When the patients are unable to walk, an electrical stimulation of the injected muscles (trains of 3 s, 25 Hz, 200 ìs, above motor threshold) has also been recommended five times for 30 minutes each day for 3 days after injection. Furthermore, the exclusive injection of the gastrocnemius muscle bellies is not sufficient in most cases; the soleus, tibialis posterior, long toe flexor, and in case of a severe ankle inversion, the tibialis anterior muscle should be additionally treated. To address the accompanying mechanical changes of spastic muscles, Reiter and coworkers studied the effects of a tonic stretch of the injected muscles. The authors compared two groups of hemiparetic patients with a spastic drop foot. One group received an injection of high-dose BTX into the plantar flexors, whereas the second group received a smaller dose of 50% plus a tonic stretch with the help of a tape. Both groups showed similar effects on muscle tone and gait function (Reiter *et al.*, 1998**). In summary, functional activity, electrical stimulation, a proper muscle selection, and a tonic stretch should help to promote the effectiveness of the treatment. Also, do not shake the vials after dilution, otherwise the protein may not dissolve but sticks to the wall of the vial.

### Electrical Stimulation of SCI Patients

Functional electrical stimulation (FES) for restoration of standing and walking in paraplegic subjects has been adopted since the 1980s (Kralj *et al.*, 1980). A four channel surface electrical stimulation method included both quadriceps muscles to provide standing, whereas stepping was induced by the stimulation of combined ankle dorsal flexion and flexion withdrawal response of the peroneal nerve. Alternating stimulation between the quadriceps muscle and peroneal nerve of both lower limbs was controlled with manual switches by the patients themselves. After a 6-week inpatient training program, 16 paraplegic patients had a mean standing duration of 22.6 minutes. The gait velocity of 12 patients ranged from 2.9–24.2 m/min, the maximum walking distance ranged from 4–335 m. Further positive effects of FES-assisted standing and walking were strengthening of lower limb muscles, prophylaxis of thrombosis, osteoporosis and joint contracture, improvement of skin and muscle blood perfusion, and improved thermoregulation. Nevertheless, because of flexor spasm and unrealistic expectations, 8 of the 16 patients had stopped the program completely, and 7 patients only practiced FES-assisted standing after 1 year. One patient only continued FES walking. In incomplete paraparctic and tetraparetic patients, FES could help to restore voluntary motor function of the lower limbs. Four patients with an incomplete cervical lesion who had been able to walk a short distance before treatment improved their gait velocity/walking distance without FES for a mean of +33.3%/+163.8% after 6 weeks of FES treatment (Hesse, 1998).

### Technical Aids

Walking sticks, canes, ankle-foot orthoses, and wheelchairs are an integral part of lower limb motor rehabilitation. A cane is enlarging the area of support, thereby promoting walking security. Nevertheless, some therapists are hesitant to prescribe canes; they fear that the patient will walk in a nonphysiological manner, particularly when using short canes. Comparative studies, however, did not find any difference between walking with and without a cane with respect to cycle parameters, gait symmetry, trunk kinematics, and lower limb muscle activity (Tyson, 1998). Also, the height of the walking stick did not influence any of the aforementioned gait parameters, the force supported by the cane was approximately 15% of the body weight, irrespective of the height of the stick. In summary, patients should use walking sticks or canes if necessary to minimize the risk of falls.

Ankle-foot orthoses (AFOs) help to clear the foot during the swing phase, stabilize the ankle, and can prevent ankle spraining caused by inversion. With AFOs, patients walk faster, more dynamic, balanced, and efficient, and the vastus medialis muscle is facilitated irrespective of the change of walking velocity (Hesse *et al.*, 1999). Irrespective of any particular model, an individually tailored prescription should consider the necessary plantar flexor stop during the swing (to clear the ground), the needed dorsiflexor stop during the stance (to control the body advancement), and the required amount of ankle stabilization in case of inversion. These parameters need to be checked manually before prescription. Cosmetic considerations also play a major role with many patients. Again, some therapists do not recommend the early and/or prolonged use of AFOs after stroke or traumatic brain injury. Clinical data do not support this notion (see previously), patients definitely profit from a well-adjusted AFO. A firm elastic bandage (stabilizing the ankle and preventing inversion) can be used as an intermediate solution within a rehabilitation setting.

Proper wheelchair fitting should consider at least the following points:

*Breadth*: the patient should sit comfortably so that one can put a hand between the pelvis and the edge of the armrest. A simple equation is maximum breadth of the shanks plus 5 cm, or plus 8 cm if the wheelchair is used outside extensively to allow for warm clothes.
*Depth*: should be 6 cm less than the length of the shanks, adjustable with a cushion.
*Height and foot rest*: if the feet are put on the foot rests, there should be a space of 2–3 cm between the shanks and the edge of the seat to prevent pressure sores. The adjustable footrests should be at least 8 cm above the ground to pass obstacles. If the patient propels the chair with his nonaffected foot, the height of the chair should be adjusted in such a way that the patient can put his complete soles on the ground to propel effectively.
*Back of the chair*: should end below the scapulae, or height of the back = distance of the seat to the armpit minus 10 cm.

The reader should keep in mind that these hints are minimal requirements, and that proper fitting is teamwork with physio-occupational therapists and technicians participating.

## Neurological Rehabilitation in MS

Patients with MS require comprehensive management that incorporates the expertise appropriate to their symptoms and is sufficiently flexible to respond rapidly to their changing pattern of need. They require information and support to facilitate their involvement in the management of their own condition, retain a sense of control, and maximize their independence (Hatch, 1997). The philosophy of rehabilitation, which emphasises patient education and self-management, is ideally suited to meet the needs of people with this fluctuating, unpredictable condition. This philosophy must be based on a thorough understanding of the mechanisms underlying disability and recovery in MS (Thompson,

2000). It is translated into practice by carrying out goal-orientated rehabilitation program that are based on expert multidisciplinary assessment and evaluated through appropriate clinical outcome measures (Thompson and Hobart, 1998) and monitoring of goal achievement (Rossiter *et al.*, 1998).

Tentative evidence now exists to support the use of components of the rehabilitation process including physiotherapy (Wiles *et al.*, 2001) and aerobic exercise (Petajan *et al.*, 1996). In the former randomized, controlled, crossover trial, 42 patients were allocated to one of six permutations of three 8-week treatment periods separated by an 8-week interval. The three treatment arms consisted of (1) physiotherapy at home, (2) outpatient therapy, and (3) no physiotherapy. A wide range of outcome measures was used. The primary outcome was The Rivermead Mobility Index. The study showed a highly significant result favoring therapy, either hospital or home based ($p < 0.001$) with hospital therapy being considerably cheaper than home based. Evaluation of the potential benefit of the full rehabilitation package has focused primarily on the inpatient setting and has attempted to identify which areas are likely to improve and whether that improvement is maintained after discharge into the community. Although recent studies have methodological limitations, they are an improvement on earlier work and, perhaps most importantly, demonstrate that such studies are feasible (Thompson, 2000). Despite small numbers, inpatient rehabilitation resulted in a significant benefit in disability and handicap in a UK study of patients with progressive MS and benefit in disability and aspects of quality of life in a less-disabled Italian cohort (Freeman *et al.*, 1997; Solari *et al.*, 1998). The first of these studies was waitlist controlled and included 66 patients. The treatment group underwent a short period (mean, 25 days) of inpatient rehabilitation. At 6 weeks assessment the treated group showed a significant benefit in both disability (measured on the Functional Independence Measure [FIM]) ($p < 0.001$) and handicap (London Handicap Scale) ($p < 0.01$). The Italian study included 50 mobile patients who recived either inpatient physiotherapy or a home exercise program. Benefit was again seen on the FIM at 15 weeks ($p = 0.004$), whereas in relation to quality of life significant improvement was only seen in the mental composite score at 3 and 9 weeks and not at the end of the study. The duration of benefit has not been looked at in a randomized controlled trial, but a single group study suggested that benefits in disability and handicap persist for approximately 6 months, whereas the positive effect on quality of life and emotional well-being continued for longer (Freeman *et al.*, 1999). This study emphasized the important link between hospital and community in attempting to provide a continuum of care for patients with MS and their families throughout the course of their condition (Thompson and Freeman, 2000). More recently, this link has been provided by MS nurse specialists (Johnson, 1997). Few studies have attempted to evaluate the different models of care available, but a recent study from Rome has suggested that coordinated multidisciplinary care may be more beneficial compared with medical care alone (Pozzilli *et al.*, 1999). The development of standards of care for the management of MS is an important first step and should serve to emphasize both patient needs and the scarcity of evidence available to guide the way in which they are met (Hatch *et al.*, 1999).

### Parkinson's Disease

Although a range of therapy disciplines, involved either singly or in combination, has been advocated in Parkinson's disease, it is acknowledged that less than 30% of patients receive any input. Although there is some evidence that attention to posture and stretching may improve muscle tone and decrease rigidity (Comella *et al.*, 1994), there is overall a scarcity of rigorous evidence showing clear benefit from therapy interventions targeting mobility, dexterity, and speech (Thompson and Playford, 2001; Pohl *et al.*, 1999). A recent small study (involving only 20 patients) suggested a potential benefit from the use of sensory cues that seemed to persist after input (Marchese *et al.*, 2000). Larger, adequately powered studies are required in this and other areas. A series of Cochrane reviews have recently been undertaken into the benefits of physiotherapy, occupational therapy, and speech and language therapy, and they have all arrived at the same conclusion, namely, that more evidence is required to justify and target therapy input in this chronic, disabling condition (Deane *et al.*, 2001).

### Treatments No Longer Recommended

Hemiparetic patients should no longer be treated on general wards that do not offer a comprehensive rehabilitation program. During motor rehabilitation, the exclusive or predominant application of tone-inhibiting maneuvers within neurofacilitatory techniques can no longer be recommended. The treatment of spasticity should be goal oriented with restoration of motor function as the primary target. Generally speaking, motor rehabilitation should aim at realistic goals by applying evidence-based methods. On the other hand, dogmatic approaches that exclude the use of other techniques can no longer be accepted. In the case of chronic progressive neurological disorders, it is no longer acceptable to carry out rehabilitation in a piecemeal fashion. Input needs to be coordinated, accessible, flexible, and responsive with good communication between acute hospital and community services.

## REFERENCES

### Natural Course

Alusi, S. H., Worthington, J., Glickman, S., and Bain, P. G. (2001). A study of tremor in multiple sclerosis. *Brain* **124**, 720–730.

Colcher, A., and Stren, M. (1999). Therapeutics in the neurorehabilitation of Parkinson's disease. *Neurorehabil. Neural. Repair* **13**, 205–218.

Jorgensen, H. S., Nakayma, H., Raaschou, H. O., and Olsen, T. S. (1995). Recovery of walking function in stroke patients: the Copenhagen stroke study. *Arch. Phys. Med. Rehabil.* **76**, 27–32.

Miyai, I., Suzuki, T., Kang, J., Kubota, K., and Volpe, B. T. (1999). Middle cerebral artery stroke that includes the premotor cortex reduces mobility outcome. *Stroke* **30**, 1380–1383.

Miyai, I., Suzuki, T., Kang, J., and Volpe, B. T. (2000). Improved functional outcome in patients with hemorrhagic stroke in putamen and thalamus compared with those with stroke restricted to the putamen or thalamus. *Stroke* **31**, 1365–1369.

Nakayma, H., Jorgensen, H. S., Raaschou, H. O., and Olsen, T. S. (1994). Recovery of upper extremity function in stroke patients: the Copenhagen study. *Arch. Phys. Med. Rehabil.* **75**, 852–857.

Weaver, F. M., Hammond, M. C., Guihan, M., and Hendricks, R. D. (2000). Department of Veterans Affairs Qality Enhancement Research Initiative for spinal cord injury. *Med. Care* **38**, 182–191.

Weinshenker, B. G. (1994). Natural history of multiple sclerosis. *Ann. Neurol.* **36, Suppl**, 6–11.

Williams, G. R., Jiang, J. G., Matchar, D. B., and Samsa, G. P. (1999). Incidence and occurrence of total (first-ever and recurrent) stroke. *Stroke* **30**, 2523–2528.

### Principles

Asanuma, H., and Keller, A. (1991). Neurobiological basis of motor learning and memory. *Concepts Neurosci.* **2**, 1–30.

Bütefisch, C., Hummelsheim, H., Denzler, P., and Mauritz, K. H. (1995). Repetitive training of isolated movements improves the outcome of the centrally paretic hand. *J. Neurol. Sci.* **130**, 59–68.

Carr, J. H., and Shepherd, R. B. (1987). A Motor Relearning Programme. pp. 1–25. William Heinemann, London.

Classen, J., Liepert, L., Wise, S. P., Hallett, M., and Cohen, L. G. (1998). Rapid plasticity of human cortical movement representation induced by practice. *J. Neurophysiol.* **79**, 1117–1123.

Cramer, C. C., Nelles, G., Benson, R. R., Kaplan, J. D., Parker, R. A., Kwong, K. K., Kennedy, D. N., Finklestein, S. P., and Rosen, B. R. (1997). A Functional MRI study of subjects recovered from hemiparetic stroke. *Stroke* **28**, 2518–2527.

Dietz, V., Colombo, G., Jensen, L., and Baumgartner, L. (1995). Locomotor capacity of spinal cord in paraplegic patients. *Ann. Neurol.* **37**, 574–582.

Ernst, E. (1990). A review of stroke rehabilitation and physiotherapy. *Stroke* **21**, 1081–1085.

Friel, M. F., and Nudo, R. J. (1998). Recovery of motor function after focal cortical injury in primates: compensatory movement patterns used during rehabilitative training. *Somatosens. Motor Res.* 15(3), 173–189.

Hesse, S., Bertelt, C., Jahnke, M. T., Schaffrin, A., Baake, P., Malezic, M., and Mauritz, K. H. (1995). Treadmill training with partial body weight support as compared to physiotherapy in non-ambulatory hemiparetic patients. *Stroke* 26, 976–981.

Langhammer, B., and Stanghelle, J. K. (2000). Bobath or motor relearning programme? A comparison of two different approaches of pyhsiotherapy in stroke rehabilitation: a randomized controlled study. *Clin. Rehabil.* 14, 361–369.

Lovely, R. G., Gregor, R. J., Roy, R. R., and Edgerton, V. R. (1986). Effects of training on the recovery of full-weight bearing stepping in the adult spinal cat. *Exp. Neurol.* 92, 421–435.

Nudo, R. J., Wise, B. M., SiFuentes, F., and Milliken, G. W. (1996). Neural substrates for the effects of rehabilitative training on motor recovery after ischemic infarct. *Science* 272, 1791–1794.

Taub, E., Miller, N. E., Novak, T. A., Cook, E. W., Fleming, W. C., Nepomuceno, C. S., Connell, J. S., and Crago, J. E. (1993). Technique to improve chronic motor deficit after stroke. *Arch. Phys. Med. Rehabil.* 74, 347–354.

Wernig, A., and Müller, S. (1992). Laufband locomotion with body weight support in persons with severe spinal cord injuries. *Paraplegia* 30, 229–238.

## Practical Management

### General Rehabilitation

Feeney, D. M., Gonzales, A., and Law, W. (1982). Amphetamine, haloperidol and experience interact to affect rate of recovery after motor cortex injury. *Science* 217, 855–857.

Goldstein, L. B., and the Sygen in Acute Stroke Study Investigators. (1995). Common drugs may influence motor recovery after stroke. *Neurology* 45, 865–871.

Grade, C., Redford, B., Chrostowski, J., Toussaint, L., and Blackwell, B. (1998). Methylpehnidate in early poststroke recovery: a double-blind, placebo-controlled study. *Arch. Phys. Med. Rehabil.* 79, 1047–1050.

Indredavik, B., Fjaertoft, H., Ekeberg, G., Loge, A. D., and Morch, B. (2000). Benefit of an extended stroke unit service with early supported dicharge. A randomized, controlled trial. *Stroke* 31, 2989–2994.

Jorgensen, H. S., Kammersgaard, L. P., Houth, J., Nakayama, H., Raaschou, H. O., Larsen, K., Hübbe, P., and Olsen, T. S. (2000). Who benefits from treatment and rehabilitation in stroke unit? A community-based study. *Stroke* 31, 434–439.

Kwakkel, G., Wagenaar, R. C., Twisk, J. W. R., Lankhorst, G. J., and Koetsier, J. C. (1999). Intensity of leg and arm training after primary middle-cerebral-artery stroke: a randomised trial. *Lancet* 354, 191–196.

Parry, R. H., Lincoln, N. B., and Vass, C. D. (1999). Effect of severity of arm impairment on response to additional physiotherapy early after stroke. *Clin. Rehabil.* 13, 187–198.

Sunderland, A., Fletcher, D., Bradley, E. L., Tinson, D. J., Langton Hewer, R., and Wade, D. T. (1994). Enhanced physical therapy for arm function after stroke: a one year follow up study. *J. Neurol. Neurosurg. Psychiatry* 57, 856–858.

Sunderland, A., Tinson, D. J., Bradley, E. L., Fletcher, D., Langton Hewer, R., and Wade, D. T. (1992). Enhanced physical therapy improves recovery of arm function after stroke. A randomised controlled trial. *J. Neurol. Neurosurg. Psychiatry* 55, 530–535.

von Koch, L., Holmquist, L. W., Kostulas, V., Almazan, J., and de Pedro-Cuesta, J. (2000). A randomized controlled trial of rehabil-itation at home after stroke in Southwest Stockholm: Outcome at six months. *Scand. J. Rehab. Med.* 32, 80–86.

Walker-Batson, D., Smith, P., Curtis, S., Unwin, H., and Greenlee, R. (1995). Amphetamine paired with phyical therapy accelerates motor recovery after stroke. Further evidence. *Stroke* 26, 2254–2259.

### Special Aspects of Upper Limb Motor Rehabilitation

Bakheit, A. M., Thilmann, A. F., Ward, A. B., Poewe, W., Wissel, J., Muller, J., Benecke, R., Collin, C., Muller, F., Ward, C. D., and Neumann, C. (2000). A randomized, double-blind, placebo-controlled, dose-ranging study to compare the efficacy and safety of three doses of Botulinum toxin type A (Dysport) with placebo in upper limb spasticity after stroke. *Stroke* 31, 2402–2406.

Bhakta, B. B., Cozens, J. A., Chamberlain, M. A., and Bamford, J. M. (2000). Impact of botulinum toxin type A on disability and carer burden due to arm spasticity after stroke: a randomised double blind placebo controlled trial. *J. Neurol. Neurosurg. Psychiatry* 69, 217–221.

Caraugh, J., Light, K., Kim, S., Thigpen, M., and Behrman, A. (2000). Chronic motor dysfunction after stroke. Recovering wrist and finger extension by electromyography-triggered neuromuscular stimulation. *Stroke* 31, 1360–1364.

Carroll, S., Cooper, C., Brown, D., Sorman, G., and Flood, S. (2000). Australian experience with the Freehand System® for restoring grasp in quadriplegia. *Aust. N. Z. J. Surg.* 70, 563–568.

Davis, S. E., Mulcahey, M. J., Smith, B. T., and Betz, R. R. (1998). Self-reported use of an implanted FES hand system by adolescents with tetraplegia. *J. Spinal Cord Med.* 21, 220–226.

Dromerick, A. W., Edwards, D. F., and Hahn, M. (2000). Does the application of constrained-induced movement therapy during acute rehabilitation reduce arm impairment after ischemic stroke?. *Stroke* 31, 2984–2996.

Feys, H. M., De Weerdt, W. J., Selz, B. E., Cox Steck, G. A., Spichiger, R., Vereeck, L. E., Putman, K. D., and Van Hoydonck, G. A. (1998). Effect of a therapeutic intervention for the hemiplegic upper limb in the acute phase after stroke. A single-blind, randomized, controlled multicenter trial. *Stroke* 29, 785–792.

Hesse, S., Reiter, F., Konrad, M., Uhlenbrock, D., and Jahnke, M. T. (1998). Botulinum toxin type A and short-term electrical stimulation in the treatment of upper limb flexor spasticity after stroke: a randomised, double-blind, placebo-controlled trial. *Clin. Rehabil.* 12, 381–388.

Hogan, N., Krebs, J., Charnnarong, J., Srikrishna, P., and Sharon, A. (1992). MIT-MANUS: a workstation for manual therapy and Training II. In *Proceedings of the SPIE Conference on Telemanipulator Technology* 833, 28.

Mudie, M. H., and Matayas, T. A. (1996). Upper extremity retraining following stroke: effects of bilateral practice. *J. Neuro. Rehabil.* 10, 167–184.

Powell, J., Pandyan, D., Granat, M., Cameron, M., and Stott, D. J. (1999). Electrical stimulation of wrist extensors in poststroke hemiplegia. *Stroke* 30, 1384–1389.

Reiter, F., Danni, M., Lagalla, G., Ceravolo, M. G., and Provinciali, L. (1998). Low-dose botulinum toxin with ankle taping for the treatment of spastic equinovarus foot after stroke. *Arch. Phys. Med. Rehabil.* 79, 532–535.

Simpson, D. M., Alexander, D. N., O'Brian, C. F., Tagliati, M., Aswad, A. S., Leon, J. M., Gibson, J., Mordaund, J. M., and Monaghan, E. P. (1996). Botulinum toxin type A in the treatment of upper extremity spasticity: a randomized, double-blind, placebo-controlled trial. *Neurology* 46, 1306–1310.

Smith, S. J., Ellis, E., White, S., and Moore, A. P. (2000). A doube-blind placebo-controlled study of botulinum toxin in upper limb spasticity after stroke and head injury. *Clin. Rehabil.* **14**, 5–13.

Van der Lee, J. H., Wagenaar, R. C., Lankhorst, G. J., Vogelaar, T. W., Deville, W. L., and Bouter, L. M. (1999). Forced use of the upper extremity in chronic stroke patients. *Stroke* **30**, 2369–2375.

Volpe, B. T., Krebs, H. I., Hogan, N., Edelstein, L., Diels, C., and Aisen, M. (2000). A novel approach to stroke rehabilitation. *Neurology* **54**, 1938–1944.

Whitall, J., McCombe, S., Waller, S., Silver, K. H. C., and Macko, R. F. (2000). Repetitive bilateral arm training with rhythmic auditory cueing improves motor function in chronic hemiparetic patients. *Stroke* **31**, 2390–2398.

### Special Aspects of Lower Limb Motor Rehabilitation

Bland, S. T., Schallert, T., Strong, R., Aronowski, J., and Grotta, J. C. (2000). Early exclusive use of the affected forelimb after moderate transient focal ischemia in rats. Functional and anatomic outcome. *Stroke* **31**, 1142–1152.

Burbaud, P., Wiart, L., Dubos, J. L., Gaujard, E., Debeillex, X., Joseph, P. A., Mazaux, J. M., Bioulac, B., Barat, M., and Lagueny, A. (1996). A randomised, double blind, placebo controlled trial of botulinum toxin in the treatment of spastic foot in hemiparetic patients. *J. Neurol. Neurosurg. Psychiatry* **61**, 265–269.

Chen, R., Cohen, L. G., and Hallet, M. (1997). Role of the ipsilateral motor cortex involuntary movement. *Can. J. Neurol. Sci.* **24**, 284–291.

Cramer, S. C., Nelles, G., Besnon, R. R., Kaplan, J. D., Parker, R. A., Kwong, K. K., Kennedy, D. M., Finklestein, S. P., and Rosen, B. R. (1997). A Functional MRI study of subjects recovered from hemiparetic stroke. *Stroke* **28**, 2518–2527.

Dietz, V., Colombo, O., and Jenson, L. (1994). Locomotor activity in spinal man. *Lancet* **344**, 1260–1263.

Hesse, S., Bertelt, C., Jahnke, M. T., Schaffrin, A., Baake, P., Malezic, M., and Mauritz, K. H. (1995). Treadmill training with partial body weight support as compared to physiotherapy in non-ambulatory hemiparetic patients. *Stroke* **26**, 976–981.

Hesse, S., Konrad, M., and Uhlenbrock, D. (1999). Treadmill walking with partial body weight support versus floor walking in hemiparetic subjects. *Arch. Phys. Med. Rehabil.* **80**, 421–427.

Hesse, S., Krajnik, J., Luecke, D., Jahnke, M. T., Gregoric, M., and Mauritz, K. H. (1996). Ankle muscle activity before and after botulinum toxin therapy for lower limb extensor spasticity in chronic hemiparetic patients. *Stroke* **27**, 455–460.

Hesse, S., Malezic, M., Lücke, D., and Mauritz, K.-H. (1998). Value of functional electrostimulation in paraplegic patients. *Nervenarzt* **69**, 300–305.

Hesse, S., Uhlenbrock, D., Werner, C., and Bardeleben, A. (2000). A mechanized gait trainer for restoring gait in non-ambulatory subjects. *Arch. Phys. Med. Rehabil.* **81**, 1158–1161.

Hesse, S., Werner, C., Konrad, M., Kirker, S., and Berteanu, M. (1999). Non-velocity related effects of a rigid double-stopped ankle-foot orthosis on gait and lower limb muscle activity of hemiparetic patients with an equinovarus deformity. *Stroke* **30**, 1855–1861.

Kralj, A., Bajd, T., and Turk, R. (1980). Electrical stimulation providing functional use of paraplegic patient muscles. *Med. Progr. Technol.* **15**, 1–10.

Locomat (2001). www.Hacoma.ch.

Reiter, F., Danni, M., Lagalla, G., Ceravolo, G., and Provinciali, L. (1998). Low-dose botulinum toxin with ankle taping for the treatment of spastic equinvarus foot after stroke. *Arch. Phys. Med. Rehabil.* **79**, 532–535.

Tyson, S. (1998). The influence of different walking aids on hemiplegic gait. *Clin. Rehabil.* **12**, 395–401.

Visintin, M., Barbeau, H., Korner-Bitensky, N., and Mayo, N. E. (1998). A new approach to retrain gait in stroke patients through body weight support and treadmill stimulation. *Stroke* **29**, 1122–1128.

Wernig, A., Nanassy, A., and Müller, S. (1998). Maintenance of locomotor abbilities following Laufband (treadmill) therapy in para- and tetraplegic persons: follow-up studies. *Spinal Cord* **36**, 744–749.

## Neurological Rehabilitation in Chronic Neurological Conditions

Comella, C. L., Stebbins, G T., Brown-Toms, N., and Goetz, C. G. (1994). Physical therapy and Parkinson's Disease: A controlled clinical trial. *Neurology* **44**, 376–378.

Deane, K. H. O., Ellis-Hill, C., Jones, D., Whurr, R., Ben-Shlomo, Y., Playford, E. D., and Clarke, C. (2001). Systemic review of rehabilitation therapies in Parkinson's disease. *Cochrane Database Syst. Rev.* **1**, CD002815.

Freeman, J. A., Langdon, D. W., Hobart, J. C., and Thompson, A. J. (1997). The impact of inpatient rehabilitation on progressive multiple sclerosis. *Ann. Neurol.* **42**(2), 236–244.

Freeman, J. A., Langdon, D. W., Hobart, J. C., and Thompson, A. J. (1999). Inpatient rehabilitation in multiple sclerosis: do the benefits carry over into the community? *Neurology* **52**(1), 50–56.

Hatch, J. (1997). Building partnerships. In "Multiple Sclerosis: Clinical Challenges and Controversies", (A. J. Thompson, C. H. Polman, and R. Hohlfeld, eds.), pp. 345–351. London, Martin Dunitz Ltd.

Hatch, J., Johnson, J., and Thompson, A. J. (1999). Standards of health care for people with multiple sclerosis: improving health care in the UK. *MS Management* **5**(1), 16–23.

Johnson, J. (1997). What can specialist nurses offer in caring for people with multiple sclerosis? In "Multiple Sclerosis: Clinical Challenges and Controversies", (A. J. Thompson, C. H. Polman, and R. Hohlfeld, Eds.), pp. 335–343. London, Martin Dunitz Ltd.

Marchese, R., Diverio, M., Zucchi, F., Lentino, C., and Abbruzzese, G. (2000). The role of sensory cues in the rehabilitation of parkinsonian patients: a comparision of two physical therapy protocols. *Mov. Disord.* **15**, 879–883.

Petajan, J. H., Gappmaier, E., White, A. T., Spencer, M. K., Mino, L., and Hicks, L. (1996). Aerobic training on fitness and quality of life in multiple sclerosis. *Ann. Neurol.* **39**(4), 432–441.

Pohl, M., Mrass, G. J., and Oertel, W. H. (1999). "Rehabilitation in Parkinson's Disease: the Treatment Options". (P. Le Witt and W. Oertel, Eds.), pp. 215–228. London, Martin Dunitz.

Pozzilli, C., Pisani, A., Palmisano, L., Battaglia, M. A., Fieschi, C., and the Roman Home Care Multiple Sclerosis Group. (1999). Service location in multiple sclerosis: home or hospital. In "Advances in Multiple Sclerosis: Clinical Research and Therapy". (S. Fredrikson and H. Link, Eds.), pp. 173–180. London, Martin Dunitz.

Rossiter, D. A., Edmondson, A., Al-Shahi, R., and Thompson, A. J. (1998). Integrated care pathways in multiple sclerosis: completing the audit cycle. *Multiple Sclerosis* **4**, 85–89.

Solari, A., Filippini, G., Gasco, P., Colla, L., Salmaggi, A., La Mantia, L., Farinotti, M., Eoli, M., and Mendozzi, L. (1999). Physical rehabilitation has a positive effect on disability in multiple sclerosis patients. *Neurology* **52**(1), 57–62.

Thompson, A. J. (2000). The effectiveness of neurological rehabilitation in multiple sclerosis. *J. Rehab. Res. Dev.* **37**(4), 455–461.

Thompson, A. J. and Freeman, J. A. (2000). Community services in multiple sclerosis: still a matter of chance. *J. Neurol. Neurosurg. Psychiatry* **69**, 728–732.

Thompson, A. J., and Hobart, J. C. (1998). Multiple sclerosis: assessment of disability and disability scales. *J. Neurol.* **245**, 189–196.

Thompson, A. J., and Playford, E. D. (2001). Rehabilitation for patients with Parkinson's disease *Lancet* **357**, 410–411.

Wiles, C. M., Newcombe, R. G., Fuller, K. J., Shaw, S., Furnival-Doran, J., Pickersgill, T. P., and Morgan, A. (2001). Controlled randomised crossover trial of the effects of physiotherapy on mobility in chronic multiple sclerosis. *J. Neurol. Neurosurg. Psychiatry* **70**, 174–179.

# CHAPTER 84
# Neural Prostheses

J. Quintern, R. J. Jaeger, and U. Baumann

Neural prostheses use electrical stimulation of neural tissue to restore sensory, motor, or autonomic functions in patients with lesions in the central nervous system (CNS) or in sense organs. In motor neural prostheses functionally useful movements of skeletal or smooth muscles are induced by sequences of electrically triggered action potentials in motor neurons or by the release of spinal reflexes. In neural prostheses for the restoration of sensory functions, afferent nerve fibers are electrically stimulated in such a sequence that the brain receives sensory information. Neural prostheses cannot effect a cure of lesions in the CNS, as the word "prosthesis" implies, but they are aids for a better use of the unaffected or uninjured parts of the nervous system. Today in many applications of neural prostheses, the prevention of secondary complications resulting from a lesion plays an almost equally important role as the restoration of the lost function itself (Yarkony *et al.*, 1992).

Although numerous laboratory prototypes of different neural prostheses for a variety of applications have been developed over the past 40 years, up to now only the cardiac pacemaker and the cochlear implant are in routine clinical use on a widespread basis. Some of the reasons for the slow propagation of this technology have been technical difficulties, unsatisfactory functional gain, poor electromechanical reliability, and the cumbersome handling of the existing systems, particularly with respect to external components. Since the first edition of this chapter in 1996, several new functional electrical stimulation (FES) systems (including implants) have emerged from the experimental stage to clinical trials or commercial availability. Because of the rapid progress in the fields of microelectronics, biocompatible materials, neurocomputing, and development of sensors, it can reasonably be expected that neural prostheses will become more reliable and widespread aids in the future.

The indication for the prescription of existing neural prostheses should be given by a clinician, who is familiar with this field. Not only must the neurological state of the patient be taken into account, but the cognitive state, the expected compliance, the social situation of the patient, and the possibilities for the clinical follow-up must also be clearly determined. After the patient has been provided with a neural prosthesis, a period of controlled training for the patient and of further adaptations of the system must follow. Specialized centers for fitting specific neural prostheses are listed in the FES Resource Guide (Teeter *et al.*, 1995; available through feswww.fes.cwru.edu).

In the United States, neural prosthetic systems used outside specific research laboratories have to be approved for marketing by the Food and Drug Administration (FDA; URL: www.fda.gov). In many other countries a somewhat similar body of laws has to be respected (e.g., in Canada, the Medical Devices Regulations of the Health and Welfare Bureau, in the European Union the Medical Devices Directive 93/42 EEC is translated into country-specific laws; URL: europa.eu.int/comm/). *This chapter is based primarily on the scientific research literature devoted to neural prostheses and therefore does not consider all laws and regulations in the various countries. Some research applications described in this chapter may not be covered by a government permission for a specific approved system (e.g., crutches may not be allowed as a mechanical balance aid for a specific neural prosthesis for stance). Before a neural prostheses system is used, the clinician and the user have to be sure that the system is approved for sale by the responsible state agency, and they have to follow the instructions in the manual for this system, so that the system is used for an approved application.*

## SOMATIC MOTOR NEURAL PROSTHESES

### General Considerations

The goal of motor neural prostheses through FES or functional neuromuscular stimulation (FNS) is the restoration of those motor functions that have been lost by upper motor neuron lesions. By low-frequency stimulation (approximately 12–30 pulses/s) with short impulses (pulse width <500 µs) action potentials in peripheral nerves or their endings are triggered, which promulgate the contraction of the innervated muscle.

### *Muscle Fatigue and Training for Muscle Restrengthening*

Rapid fatigue of the muscles subjected to FES is a major problem in neural prostheses. The problem is caused by fundamental differences between FES and the natural physiology of muscle activation. When using FES, the recruitment order obtained is inverted, because the largest diameter motor neurons have the lowest stimulation threshold. Therefore with low stimulation intensities these large diameter, rapidly fatiguing motor units are activated first. In addition to this recruitment problem, most current FES systems trigger the action potentials in all stimulated motor neurons of the respective nerve simultaneously. This is opposed to the naturally occurring asynchrony typically seen in normal motor neurons. Therefore the stimulation frequency must be above the range of 12–16 pulses/s to achieve a relatively smooth muscular contraction or tetanus. Finally, with the same electrode position and the same stimulation intensity, the same motor units are always activated, which defeats the goal of spatial summation. Some implanted systems evade this problem by the "carousel" stimulation, in which four or five electrodes around the same nerve are activated successively in different combinations (Happak et al., 1989*).

To improve fatigue resistance of muscles subjected to FES it is often necessary to restrengthen the paralyzed muscles with low-frequency electrical stimulation before FES can be applied for functional use. This muscle-restrengthening training causes an increase of muscle fiber diameter and force development (Belanger et al., 2000*); in some studies an increased fatigue resistance of the stimulated muscles was also found (Ragnarsson, 1988*; Stein et al., 1992*). Not only the pattern of neuronal discharge but also the physical conditions of contraction are important in determining the metabolic profile of human muscle fibers (Munsat et al., 1976*). In a recent controlled three-arm study in patients with acute spinal cord injury, an increase of lower limb and gluteal lean body mass could only be observed after FES-cycle ergometry but not after FES-induced unloaded isometric contractions (Baldi et al., 1998**). However, it is difficult to compare the different studies in the literature and to derive an optimized training strategy for patients with spastic paralysis from these data, because different muscles, different stimulation parameters, different duty cycles, different weekly training times, and different types of contraction have been used. As a pragmatic guideline, we recommend the use of stimulation frequencies around 16–20 pulses/s for training purposes and high stimulus intensities to recruit most of the muscle fibers of the target muscles. For the lower extremities an alternating pattern of stimulation of the respective muscle groups bilaterally is most common. The duration of the daily training can be slowly increased from 10 minutes to 1 hour within several months. Concomitantly, the duration of the duty cycle can be increased from 5–10 or 20 seconds. Initially, a single treatment per day can be increased to twice daily. Unloaded isotonic contractions seem to be less effective than isometric or loaded contractions; however, training protocols with a high amount of eccentric muscle contractions should be avoided, because this type of contraction has a high risk of damage to the muscle fibers. Training on a cycle or modified cycle ergometer powered with FES seems to be an effective training method, and, whenever possible, at least part of the training should be performed while the patient is standing and walking with FES.

### *Medical Advantages of Motor Neural Prostheses*

The use of FES not only prevents the atrophy of the stimulated muscles, but positive effects of FES on important cardiopulmonary parameters such as the tidal volume, cardiac output, and blood pressure regulation have been shown in several studies with different training modalities (Hooker et al., 1990*; Jacobs et al., 1997*; Petrofsky and Stacy, 1992*). FES of the lower extremities can also prevent venous blood pooling, which affects cardiac return, in patients with spinal cord injuries (Phillips et al., 1995*). These findings are especially relevant for patients with higher spinal lesions, because the chronic preload deficits, the reduction of sympathetic tone, and physical immobilization are leading to dramatic changes in cardiac structure and function (Kessler et al., 1986). Voluntary arm crank exercising, which is one of the standard methods for cardiovascular conditioning in paraplegic patients, is restricted by venous blood pooling in the lower limbs and by the low mass of muscles involved. Therefore, combined training with voluntary arm crank exercising and lower limb FES seems to be the most effective means to enhance cardiopulmonary fitness in patients with spinal cord injury (Phillips and Burkett, 1995*). The correct application of FES can help improve joint abnormalities secondary to spinal cord injury (Betz *et al.*, 1996*). The positive trophic effects and the increased skin and muscle perfusion after FES training (Kern, 1997*) may help to prevent pressure sores (Levine *et al.*, 1990). Lower limb FES has also been proposed as a means to prevent deep venous thrombosis in paralyzed or immobilized persons, because FES not only increases venous blood flow but also enhances plasma fibrinolytic activity (Katz *et al.*, 1987*). Many patients using lower extremity motor neural prostheses anecdotally report a facilitation of micturition and defecation. For several hours after the application of the stimulation there is a significant reduction of spasticity (Robinson *et al.*, 1988a*). Together with the increase of muscle force development after a few weeks of stimulation, the intensity or strength of spasticity is often increased. However, after some months of regular stimulation, in most cases this will decrease (Robinson *et al.*, 1988b*). Positive effects of FES on spasticity have also be shown in patients with hemiparesis after stroke (Mokrusch, 1997*). However, the stimulation parameters used for FES may not be the most appropriate for spasticity reduction, and some authors have reported increased spasticity when surface stimulation above motor threshold is used (Daly *et al.*, 1996*), whereas low-intensity skin stimulation leads to a marked reduction of spasticity. In neuroprosthetic applications of FES it is an advantage for patients to use their own muscles for the movement. Several authors reported positive carry-over effects of training with FES on voluntary motor control without FES in patients with incomplete spinal cord injury (Bajd *et al.*, 1999*) and stroke (Bogataj *et al.*, 1995**; Taylor *et al.*, 1999*); however, the findings were not consistent for all groups of patients. Osteoporosis is another severe problem in patients with spinal cord injuries. It is thought that isokinetic muscle training, standing, walking, and stationary bicycle ergometry with FES will increase the bone mineral density in the long tubular bones of the lower extremities. Although a significant increase in bone mineral content after an FES exercise period was found in a few studies (e.g., Belanger *et al.*, 2000*; Mohr *et al.*, 1997*), other investigators could not find such a significant increase (e.g., Needham-Shropshire *et al.*, 1997*).

### Requirements and Contraindications

An important precondition for the production of sufficient muscle force by FES is that a plurality of lower motor neurons innervating the target muscle are intact. Although depolarization of denervated muscle fibers can be induced by very long stimulation pulses (pulse width in the range of 100 ms), motor neural prostheses are not currently feasible in flaccid paralysis. However, direct stimulation of denervated muscles is a therapeu-

tic option to preserve the denervated muscle in cases of peripheral nerve damage when axonal regeneration can be expected or to prevent muscular atrophy and hyposomia of paralyzed extremities in growing children with Erb's palsy or meningomyelocele (Eichhorn *et al.*, 1984*).

Additional contraindications for the application of FES are severe contractures or very strong spasticity (Kralj and Bajd, 1989), sores in the proximity of the surface stimulation electrodes, severe additional neurological impairment (e.g., ataxia, apraxia, or dementia), decrepitude or a rapidly progressing disease as the reason for the paralysis. Intact proprioception (e.g., in patients with arteria-spinalis-anterior syndrome) may be advantageous for the application of motor neural prostheses in the sense that balance may be improved. However, many patients with partly intact sensation experience pain associated with the stimulation if surface electrodes are used. It must further be considered that transcutaneous electrical stimulation may disturb the function of cardiac pacemakers (Chen *et al.*, 1990) and other implanted medical devices. Furthermore, all kinds of metal implants may influence the electric field between the FES electrodes in such a way that the current density exceeds the limits of safety for the tissue. In patients with a spinal lesion at a Th5 or higher level the application of FES requires monitoring of the heart rate and blood pressure, because autonomic dysreflexia may lead to an excessive rise of blood pressure.

The use of motor neural prostheses requires vast cooperative efforts from the patient over a long period of time. In a psychological study (Bradley, 1994*), increases in depression and hostility were found in subjects who had unrealistic expectations for an FES exercise program. Therefore lack of motivation, insufficient intellectual abilities, psychological instability, and unrealistically high expectations are absolute contraindications for FES.

In view of the possibility of unrealistic expectations on the part of the patient, the application of neural prostheses for the restitution of motor functions (particularly walking) is questionable during the first few months after a spinal or cerebral lesion (this is not true for FES used for training of motor abilities).

### Stimulators and Electrodes

Surface electrodes or percutaneous wire electrodes are typically used together with external battery-powered devices that produce the electrical stimulation waveform. For all kinds of electrodes, biphasic impulses are recommended (Mortimer, 1981) to prevent ionic migration in the tissue. For stimulation by means of surface electrodes, suitable stimulation parameters are frequency of 16–25 pulses per second, current amplitude of 20–120 mA, and pulse width of 0–500 μs. The current and pulse width should be selected so that the charge delivered is less than 75 microcoulomb (μC) per pulse. When the conductivity of the electrodes is poor, a high-voltage drop at the transition between the electrodes and the skin may cause damage to the skin. Therefore, the stimulator should have a built-in impedance control. Self-adhesive electrodes made from conductive fabrics covered with a kind of conductive polymer (e.g., PALS™ electrodes, www.axelgaard.com) avoid the disadvantages of older surface electrode designs. Some researchers continue to explore percutaneous systems (in which only the electrodes are inside the body) (Shimada *et al.*, 1996*). However, infections and broken electrodes that hardly can be removed from the body can become serious problems.

Implanted FES systems normally consist of an external control box that transmits signals and power to one or two implanted receiver/stimulator units by means of a radio frequency link. For these systems electrodes such as epimysial, epineural, or cuff electrodes around the nerve have been used. The main advantages of implanted systems compared with external systems are less time needed for donning/doffing the system, more selective and more reproducible stimulation, and the possibility to stimulate muscles deep below the surface. Some patients have benefited from simple implanted motor neural prostheses with one to three channels for periods up to 15 years (Brindley, 1994*; Dobelle *et al.*, 1994*; Waters *et al.*, 1985*). However, implanted multichannel systems for the lower extremities are just at the stage of first clinical studies (Davis *et al.*, 1999; Kobetic *et al.*, 1999), and there is not yet an FDA- or CE-approved implantable system for stance or gait on the market. One of the main problems with conventional multichannel implanted systems is the numerous long wire leads that have to be fed through the tissue to different remote implantation sites. Intrafascicular electrodes (Yoshida and Horch, 1993), multipolar nerve cuff electrodes (Veraart *et al.*, 1993), and intramuscularly implanted single-channel microstimulators (Cameron *et al.*, 1997) have been developed to provide more selective stimulation without the need for numerous long cables. Up to now these advanced technologies have only been extensively tested in animal experiments.

### Control Strategies

Today, commercially available neural prostheses are open-loop systems without sensory feedback. However, the functional gain achieved with these neural prostheses is limited by the numerous nonlinearities in space and time of the human nervous and musculoskeletal systems and the difficult coordination between the neural prostheses and the preserved voluntary motor control. Future closed-loop systems with biomechanical sensors and advanced control strategies have the potential to allow better coordination between voluntary and stimulated movements (Donaldson and Yu, 2000) to compensate for the nonlinearities of musculotendon and body segmental dynamics (Quintern *et al.*, 1997) and to reduce muscle fatigue (Davis *et al.*, 1999).

### Stance and Gait in Paraplegia

#### Clinical Aspects

The main indication for stance and gait with neural prostheses (Kralj and Bajd, 1989) is a complete or incomplete but severe lesion of the spinal cord at a thoracic level. If the lesion is below T12, the lower motor neurons of the quadriceps are damaged in most cases; therefore these patients are not normally candidates for this type of neural prostheses. If the lesion is between C8 and T5, stance and gait with FES is possible in principle, but insufficient stability of the trunk and autonomic dysreflexia (blood pressure and heart rate) may cause problems. Because the patient must develop sufficient force in the upper extremities during stance and gait with FES for balance control and reasons of safety, stance in patients with a complete spinal cord lesion above C7 can only be achieved with mechanical aids (tilt table, etc.). Obesity and severe osteoporosis are also contraindications for FES-enabled stance and gait.

There are considerable differences in application of FES for stance and gait between complete and incomplete spinal cord injuries. Most of the references in this section refer primarily to individuals with complete injuries. Patients with incomplete paraparesis or tetraparesis, who still can produce enough voluntary muscle force for stance and/or gait, may not always profit from FES. However, if the gait of these patients is disturbed by a specific problem (e.g., a lack of control of the knee extension in the support phase of gait or insufficient dorsiflexion of the foot during the swing phase), a selective stimulation of the affected muscles during the respective phase of the gait cycle may considerably improve the gait pattern. Also, in a considerable subgroup of patients with incomplete spinal cord injuries, positive carryover effects of FES on voluntary motor control can be expected (Bajd *et al.*, 1999*; Hesse *et al.*, 1998), and some patients may switch from nonambulatory to ambulatory status or from therapeutic walking to functional walking.

#### Practical Management

**Stance with FES.** Although first demonstrated in 1960 (Kantrowitz, 1960), standing and ambulation by

FES has still not come to the point of regular and widespread use among individuals with spinal cord injury. Studies about patient selection for FES standing programs indicated that the actual number of individuals with paraplegia who can use FES to stand and who want to use FES is relatively small, perhaps on the order of 5%–10% (Jaeger *et al.*, 1990*; Rushton *et al.*, 1998*).

The simplest way to achieve standing for up to 20 minutes in paraplegic patients is by surface stimulation of both quadriceps femoris muscles (Kralj and Bajd, 1989). Additional stimulation of the gluteal muscles and the hamstrings with additional stimulation channels stabilizes the pelvis and allows the patient to assume a more correct posture of the trunk (Vossius *et al.*, 1987). Initially, standing up, quiet stance, and sitting down with FES should be practiced between parallel bars. If the patient feels safe, a walker or even crutches may be used for stance. At present, free stance with neural prostheses without requiring arm support is not yet possible, but research on closed-loop systems for free stance is going on in several laboratories (Hunt *et al.*, 1999).

**Ambulation with FES.** Up to now in the United States only one FES system has been approved by the FDA: the Parastep® I System (www.sigmedics.com). This 4–6 channel surface stimulation system allows muscle restrengthening training, stance, and walking over short distances for paraplegic patients (Graupe and Kohn, 1994). A walker with control switches is an integral part of the system. The Parastep system has also been approved for use in Canada. In the European Union the Krauth + Timmermann Microstim 8 (URL: www.krauth-timmermann.de), a versatile 8-channel surface stimulation system for muscle restrengthening, stance, gait, and cycling, has received CE approval.

In the last few years, follow-up studies of 15–17 patients with spinal cord injury each have been reported after training programs with the Parastep system (Brissot *et al.*, 2000*; Klose *et al.*, 1997*) or a similar system (Hesse *et al.*, 1998*). In all of these studies, positive training effects, but also a considerable variability of stance and gait performance, was observed between individuals. Four tetraplegic patients reached a mean standing duration of 6.8 min after 6 weeks, however, after 1 year three patients had stopped standing because of orthostatic problems. In the paraplegic patients mean standing durations of more than 20 minutes were reported and between 70% and 100% of the paraplegic patients achieved independent ambulation over 4–335 m; with a gait velocity from 2.9–24.2 m/min. Several psychological and physiological benefits of the training have been reported. On the other hand, the modest performance, the high metabolic cost, and

the cardiovascular strain during ambulation with these multichannel FES systems limited their long-term use, especially in patients with unrealistic expectations. Only 10%–30% of the paraplegic patients continued to use FES for therapeutic ambulation.

It should be noted that the most popular method of achieving standing (long leg braces) also has been reported to have a high rejection rate (Hussey and Stauffer, 1973*). However, it is the experience of the authors, that most properly selected paraplegic patients continue to use FES for muscle restrengthening, stance, and gait for several years, if regular training and follow-up sessions are offered in a special outpatient facility. Most complete paraplegic patients in an FES program will only achieve therapeutic standing and ambulation. Patients with incomplete spinal cord injuries may profit more from FES ambulation programs, and a considerable number of these patients may achieve functional walking (Bajd *et al.*, 1999*). For the latter patients, treadmill training (see Chapter "Rehabilitation of Motor Function") is an alternative treatment, which may also be combined with FES.

**Hybrid Systems for Stance and Ambulation.** Combinations of mechanical orthoses and FES are called hybrid systems. These are supposed to combine advantages of both systems. In the simplest case the ankle is stabilized during the previously described FES-assisted locomotion by an ankle-foot orthosis. On the other side of the spectrum are complex orthoses such as the ORLAU ParaWalker (Nene and Patrick, 1990) and the Reciprocation Gait Orthoses or "RGO" (Hirokawa *et al.*, 1996). The latter orthosis braces the ankle, knee, and hip joints and extends to provide mechanical support up to the middle of the trunk. Some of these orthoses transmit force (and energy) between different joints by cables during gait. Paraplegic patients achieve the best gait performance with respect to velocity and stability with these complex orthoses in combination with electrical stimulation of proximal muscles. Nevertheless, the free cadence walking speed with orthosis is much lower than that of able-bodied people or wheelchair users. Also, the metabolic cost at a given speed is much greater even if the cardiovascular stress is reduced by means of FES (Beillot *et al.*, 1996*). The overall use and long-term compliance with complex orthosis like the RGO, whether with or without stimulation, is rather poor (Sykes *et al.*, 1996). For locomotion on flat surfaces hybrid systems are still inferior to the wheelchair. Therefore, hybrid systems can rather be considered as a means for training and prevention of secondary complications than as a primary means of locomotion. A medical evaluation of 70 individuals with paraplegia showed that after an average of 14 weeks of training during which patients walked for

3 hours per week, significant reductions in spasticity, total cholesterol and low-density lipids, hydroxyproline/creatinine ratio, and increased knee extensor torque were evident. The limited but reasonable level function regained with the RGO II was associated with general improvement in the paraplegic's physiological condition if used for a minimum of 3–4 h/week (Solomonow et al., 1997*).

## Improvement of Gait in Hemiparesis

The gait of patients with hemiparesis is characterized mainly by insufficient flexion at the hip and knee and a lack of dorsiflexion at the ankle during the swing phase. Stimulation of the common peroneal nerve was initially introduced in the 1960s (Liberson et al., 1961) to evoke ankle dorsiflexion during the swing phase in patients with hemiparesis. Subsequently, multichannel stimulation systems were developed that allowed selection of stimulation patterns individually according to the motor deficiencies of the respective patients (Malezic et al., 1984*; Marsolais et al., 1990*). This makes it possible to mobilize hemiparetic patients faster than with conventional therapy. In a randomized, controlled study in 20 nonambulatory patients after stroke, multichannel FES combined with conventional physiotherapy proved superior to conventional physiotherapy alone (Bogataj et al., 1995**).

Numerous problems complicate the application of FES in hemiplegic patients. Because of the paretic arm and coincidently existing cognitive and neuropsychological deficits, the patients often have difficulties in donning, doffing, and operating these systems. Therefore the timing and control of the stimulation must be done automatically. The synchronization with the gait cycle is achieved in most cases by a switch under the foot. The application of complex multichannel systems is restricted to the clinical environment; at home, the patients must get along with only simple systems. Incorrect positioning of surface electrodes for stimulation of the peroneal nerve and triggering of the flexion reflex may result in a varus or valgus position of the foot. To guarantee constant stimulation effects and simple operation of the system, implantable one-channel stimulators have been developed and tested in clinical studies (Strojnik et al., 1987*; Waters et al., 1975, 1985*). The use of so-called natural sensors (e.g., recording activity from sensory nerves) is currently being studied (Haugland and Sinkjaer, 1999) as are tilt sensors on the stimulator to replace the footswitch as a trigger for the stimulation (Dai et al., 1996*).

Several studies have involved the Odstock Dropped Foot Stimulator (ODFS), a common peroneal stimulator. In a randomized controlled trial of 32 hemiplegic patients who had a single stroke, the mean increase in walking speed between the beginning and end of the trial was 20.5% in the FES group (when the stimulator was used) and 5.2% in the control group (physiotherapy alone). Also, the effort of walking measured by the physiological cost index (PCI) was statistically significantly improved in the FES group when the stimulator was used. No improvement in these parameters was measured in the FES group when the stimulator was not used (Burridge et al., 1997**). In a retrospective study of 151 patients who had used the ODFS for 4 1/2 months, short-term carryover effects to walking without the stimulator have also been found (Taylor et al., 1999). Problems with the ODFS encountered during these studies in some patients were skin irritation, increased spasticity, difficult electrode positioning, and insufficient reliability of the device.

FES has not yet emerged as a standard technique in the rehabilitation of these patients, except in a very few centers. Often the improvement of gait through simple mechanical orthoses is sufficient and more practical than FES. However, positive carryover effects of FES therapy to unaided gait can sometimes be observed. After walking with FES, some gait parameters, such as the dorsiflexion of the foot during the swing phase, remain improved for several hours after the stimulation has been switched off (Marsolais et al., 1990*). After several weeks of FES application, a consistent improvement of gait without stimulation has been reported; however, several months after the end of the FES therapy, this effect disappears, and the gait parameters of these patients have been shown not to differ from a control group that had not been treated with FES (Malezic et al., 1984*).

One illustration of the controversy still surrounding the use of FES in this application are the conflicting guidelines in the United Kingdom and the United States. The UK's National Clinical Guidelines for Stroke (Royal College of Physicians, 1999) recommends the use of FES for correction of dropped foot and reduction of the pain associated with a subluxed shoulder. In contrast, the Post-Stroke Rehabilitation Clinical Guideline Number 16 in the United States (Agency for Health Care Policy Research, 1995a) notes that although FES has been shown to produce short-term increases in muscle strength and motor control in selected patients, the effects of FES on function have not been adequately studied.

## Stroke—Upper Extremity

Use of FES to improve upper extremity function in individuals with stroke have focused mainly on shoulder subluxation and hand function.

The effectiveness of FES in the management of acute and chronic shoulder subluxation in hemiplegia was studied in a group less than 21 days after stroke and a group with onset of stroke more than 1 year ago. Subjects in each group were further assigned randomly to either a control subgroup or an experimental subgroup. The experimental subgroups received FES therapy in which the supraspinatus and posterior deltoid were induced to contract repetitively up to 6 h/day for 6 weeks. The early-onset group showed significant improvements in reducing subluxation as indicated by x-ray findings. The same effect was not shown for the late-onset group (Wang et al., 2000*).

The influence of FES of lower arm and hand muscles on biomechanical and functional movement parameters has been evaluated in studies with different design and between 10 and 60 akute or chronic hemiparetic patients. In a multiple baseline study, the influence of FES of the extensor and flexor carpi radialis muscles on biomechanical and functional movement parameters was compared with the effect of a standardized active repetitive training of hand and fingers. With the exception of spasticity in hand and finger flexors, repetitive FES did not improve biomechanical or functional motor parameters of the centrally paretic hand and arm. The repetitive motor training, however, seemed to improve biomechanical and functional movement parameters significantly (Hummelsheim et al., 1997*). Nevertheless, a randomized study of the effect of FES during rehabilitation of wrist extensors in 60 subjects showed improvements in isometric wrist extensor strength and grasping for the FES group compared with the control group after 8 weeks of therapy (Powell et al., 1999**). In another randomized controlled trial parametric analyses revealed significantly greater gains in the upper extremity component of the Fugl-Meyer Motor Assessment for the treatment group. However, no significant effect of neuromuscular stimulation on self-care function was found in the 28 patients who completed the study (Chae et al., 1998*). In a pilot study a new hybrid FES orthosis system for the upper limb, the Handmaster (URL: www.nessltd.com), was used by 10 patients with chronic stable hemiparesis. A statistically significant improvement was noted in all muscle tone/spasticity parameters measured, but functional tests have not yet been reported (Weingarden et al., 1998*).

In some FES applications the surface electromyogram (EMG) signals of centrally paretic muscles (especially wrist and finger extensors) trigger bursts of electrical stimulation that augment the voluntary muscle contraction. It is thought that this kind of therapy combines the advantages of conventional FES therapy with a kind of positive feedback that supports motor relearning. Two randomized controlled studies with small sample size ($n = 9$; $n = 11$) showed significant functional improvements of hand function in the FES groups compared with the control groups (Cauraugh et al., 2000**; Francisco et al., 1998*). However, no study has yet been published that compares EMG triggered FES with conventional FES.

### Restoration of Grasp in Tetraplegia

#### Indications and Contraindications

The indication for neural prostheses for the restoration of grasp functions is a cervical spinal cord injury between the C5 (possibly C4) and C6 level. Only patients with lost grasp functions in both hands may really profit from these neural prostheses. Additional prerequisites are preserved voluntary shoulder movements in two planes and the excitability of the most important muscles for the grip function by FES. Although therapeutic effects of FES on upper extremity spasticity have been described in the literature, uncontrolled upper extremity spasticity is considered a contraindication for upper extremity neural prostheses. FES application is especially complicated in the upper extremity by the presence of segmental denervation. In cases of lower motor neuron damage of relevant muscles for stabilization or grasping, tendon transfer surgery from innervated to noninnervated muscles may help to achieve the desired function (Keith et al., 1988*). Otherwise, the aforementioned general contraindications for FES are valid. For implanted systems contraindications also include active or recurrent sepsis.

#### Technology and Clinical Applications

The FREEHAND System®, an implantable FES system designed to restore hand function to individuals with quadriplegia received FDA clearance in 1997 in the United States (URL: http://www.neurocontrol.com/). More than 200 people have received the device worldwide. This system is intended to improve the ability to grasp, hold, and release objects in select patients with tetraplegia caused by C5 or C6 spinal cord injury. Potential adverse events from using the system include device malfunction, fibrosis and scarring, infection, rejection (immunological), skin irritation, surgical revision, and tissue breakdown. There are approximately 25 centers in the United States where individuals can determine whether this system is suitable for their needs.

Some follow-up studies of C5 and C6 tetraplegics after implantation of the freehand system with small sample size ($n = 2$–$6$) have been published. The most common activity performed with the system was self-care. It was also used for productivity activities such as writing, socialization, and manipulation of school and household objects. Reasons given for use of the system included perceived need and importance to perform an activity of daily living (ADL) in an independent fashion, physical ease of using the system, and availability of social supports that facilitated use. There were several reported barriers for use, such as incompatibility with multiple transfers to and from the wheelchair and lack of physical assistance during morning care to don the system (Davis *et al.*, 1998\*). In another study of six C5 individuals, all subjects were able to grasp, move, and release more objects with the neuroprosthesis than without it. Collective results for the eight ADL tests for all six subjects show that, in 73% of occasions, less physical assistance and/or adaptive equipment was required when the Freehand system was used compared with when it was not used. In 85% of occasions, subjects expressed a preference for using the neuroprosthesis. Twelve months after rehabilitation, five of the six subjects still used the neuroprosthesis daily or every second day (Carroll *et al.*, 2000\*). After implantation, it may take 6–8 months before individuals can use the system completely in daily life (Biering-Sorensen *et al.*, 2000\*).

FES for the restoration of grasp function can be realized with surface electrodes, even if the selective stimulation of atrophied muscles at the forearm may be difficult. One disadvantage of surface electrodes for this application is that an additional helper must mount the electrodes and connectors to the stimulator. No FES systems for grasp restoration by surface electrodes are currently on the American market; however, a company in Israel sells the Handmaster System (URL: www.nessltd.com). In Japan microcomputer-based FES systems with up to 64 stimulation outputs and percutaneous wire electrodes are used in a clinical study (Handa *et al.*, 1992).

In all systems the stimulation pattern during grasping, holding, and releasing an object has to be individually fitted to the patient. The control of the system by the patient is achieved by shoulder-position transducers (Keith *et al.*, 1988), chin switches, or voice control (Nathan, 1989).

## SENSORY NEURAL PROSTHESES

### Visual Prostheses

The first neural prosthesis that elicited worldwide scientific interest was the visual prosthesis (Brindley and Lewin, 1968). The visual cortex of a blind patient was directly stimulated by a matrix of 80 implanted electrodes. Although some patients could identify projected Braille writing, the clinical application of direct stimulation of the visual cortex was questionable. Because of high current density and electrochemical reactions, the implants remained functioning only several months to a few years (Donaldson, 1983). However, contemporary research along this line has continued and even advanced to the stage of implantation in human volunteers (Dobelle, 2000; Schmidt *et al.*, 1996). A new high-technology approach to a visual prosthesis is the implantation of a silicone matrix into the retina, which is in the stage of early animal experiments (Stett *et al.*, 2000).

### Auditory Prostheses (Cochlear Implants)

#### Technology

In contrast to the visual prostheses, the auditory prostheses (cochlear implants) have reached a widespread and successful clinical use by reason of advantageous anatomical and physiological conditions. Cochlear implants are electronic devices made up of two parts: an implanted component (receiver with stimulator) that is inserted during an operation usually done by an ENT surgeon and external battery-powered components worn on the head or body like a hearing aid. The stimulator case is placed in a bed drilled out in the mastoid region behind the ear. The most commonly used operative technique opens the middle ear cavity, and the electrode (typically on a strip of silicone) is inserted through a cochleostomy into the cochlea. The electric pulses emitted from the electrode bypass damaged hair cells in the inner ear and stimulate the auditory nerve directly. Cochlear implants are appropriate for both adults and children with severe to profound hearing impairment in both ears.

The success of the cochlear implant in restoring auditory communication in deaf and severely hearing-impaired patients (approximately 35,000 implantations were performed worldwide up to year 2000) led to a rapid development of this type of sensory prosthesis.

**Electrode Design.** Regular clinical application started with single-channel devices with electrodes in the middle ear in the proximity of the round window or in the scala tympani of the cochlea (House and Urban, 1973). Single-channel devices now have lost their importance because of poorer speech representation compared with multichannel devices (Cohen *et al.*, 1993\*\*). Multichannel devices use several point-shaped (MED-EL Combi-40, Clarion preformed electrode) or ring-shaped electrodes (Nucleus CI22m); newer electrode designs apply halfband electrodes (Clarion HiFocus, Nucleus 24 Contour). Current implants offer external reference electrodes on separate leads or attached to the housing of the implant to allow monopolar stimulation. For the ossified or malformed cochlea, special electrode designs are available (see later).

**Stimulator Design.** There are two basic approaches for the design of the stimulator: (1) hermetically sealed titanium case with attached platinum receiver coil/magnet embedded in silicone, (2) implant housings in compact ceramic cases comprising coil and magnet. Concerning the compliance with magnetic resonance imaging (MRI), some manufactures have approved the compatibility of their devices under certain circumstances up to 1.5 Tesla. Because children younger than the age of 2 are recipients of cochlear implants (see later), the urge to minimize the stimulator size led to a reduced implant thickness of as low as 3.9 mm (MED-EL Combi-40+).

**Signal Transmission.** All current cochlear implant designs use transcutaneous signal and energy transmission with the help of high-frequency electromagnetic waves (forward telemetry). A transmitting coil is positioned over the implant and fixated by use of a magnet, which finds its counterpart inside the implanted stimulator centered in a receiving coil. Previous electrode designs with a carbon percutaneous pedestal (Ineraid, Symbion) showed poor acceptability because of complications caused by frequent infections.

By means of backward telemetry (option incorporated in all current implant designs), information about the impedance of the electrodes can be derived for diagnostic purposes or to test the integrity of the implanted device. Some devices allow the measurement of neural responses to stimulation pulses, a feature that might improve speech processor fitting in small children or can give diagnostic evidence in case of auditory nerve malfunction (Shallop et al., 1999*).

**Stimulation Strategies.** Because improvements in electrode design, the advancement of speech-processing algorithms implemented in the external unit (speech processor) led to a continuous enhancement of speech recognition performance (Dillier et al., 1995**; Laizou, 1999). Whereas first implementations applied some sort of speech feature extraction, current developments use high-rate pulsatile stimulation (continuous interleaved sampler, CIS, [Wilson et al., 1993*]) or a compressed analog multichannel stimulation (simultaneous analog stimulation, SAS, [Kessler, 1999]). Although the speech recognition performance with up-to-date strate-

gies and modern cochlear implants is high (most patients are able to communicate on the telephone), some restrictions apply concerning speech understanding under adverse conditions and the reproduction of music. Future developments will tackle these constraints with advanced knowledge of the function of the auditory nerve and further improved signal processing (Morse and Roper, 2000).

### Patient Selection

The assessment process before cochlear implantation is very detailed; the aim is to establish whether a cochlear implant is the best possible treatment for an individual patient. Many evaluations are required, including medical, radiological, audiological, speech and language, and educational (in the case of children). The cochlear implant is therefore a prosthesis requiring a multidisciplinary team of specialists.

**Criteria.** A cochlear implant is indicated if the patient has bilateral profound to severe hearing loss and the integrity of the auditory nerve and the central auditory system is given. A brainstem implant is indicated only if the auditory nerve is bilaterally destroyed and the function of the central auditory system is not impaired (see later). General rules are outlined in a Consensus Statement published by the National Institutes of Health (NIH, 1995).

***Shifting Boarders: Residual Hearing, Age at Implantation.*** Over the years, a broadening of the selection criteria for a cochlear implant can be recognized (Lenarz, 1998). Criteria changed from bilateral deafness to a severe impairment with residual hearing that cannot be improved adequately by conventional therapy (hearing aids, surgery in the middle ear, etc.). A speech recognition score of 30% (aided condition) on open set word materials is presently considered an appropriate upper limit for preoperative performance in determining cochlear implant candidacy (Fraysse et al., 1998*; Müller-Deile et al., 1998). Children younger than 2 years of age were initially excluded from cochlear implant candidacy for a variety of reasons. Reasons ranged from concerns about the reliability of the diagnosis of a profound hearing loss in very young children to concerns about surgical safety and long-term durability of the device in a growing child. However, results from several recent studies have shown that children younger than 2 years of age can safely and successfully be implanted (Lenarz et al., 1999*; Rizer and Burkey, 1999). Moreover, the literature indicates that stimulation of the auditory system by a cochlear implant is sufficient to restore at least some aspects of central auditory pathway maturation, as reflected by age-related changes in the auditory-evoked potentials (Ponton and Don, 1995). It is nowadays widely accepted that cochlear implantation should be performed within the first 2 years of life if the child has congenital hearing loss, but in any case before the age of 4. A mandatory prerequisite is still the fitting of hearing aids and intensive hearing and speech training.

Even the geriatric population can benefit from cochlear implantation; although a reduction in the processing of sensory stimulation might exist, the elderly can process a new auditory code delivered by means of a cochlear implant (Waltzman et al., 1993*).

*Inner Ear Malformation, Syndromes.* In case of malformation of the inner ear, an implantation is feasible with an acceptable risk of complications (Cullington and Brown, 1998; Graham et al., 2000; Turrini et al., 1997; Weber et al., 1998*). Severe hearing disorder caused by syndromes requires the detection of other organic damage. Particularly the affection of visual, motoric, or cognitive areas is significant in respect to cochlear implantation.

*Onset of Hearing Loss.* If the onset of deafness occurs after the age of 7 years, a completely matured central auditory system with lasting synaptic connections between the neurons of the auditory cortex including the cortical association areas can be assumed. After cochlear implant surgery and activation of the implant, auditory sensations induced by electrical stimulation are compared with the patient's auditory memories, and a learning process accounts for an increased understanding of the artificially induced auditory sensations. Deafness leads to a degeneration of ganglion cells. However, the minimum number of cells required for good speech perception is quite low, and most implant users exceed this minimum requirement (Blamey, 1997). If a degradation of the central auditory system or the auditory associative memory exists, only very limited expectation for the development of speech understanding should be raised.

If the hearing loss is prenatal, perinatal, or postnatal or congenital, only incomplete development of the central auditory system takes place because of deprivation of the auditory system. If cochlear implantation is performed after the critical phase for language acquisition has already been completed (after the age of 7 or 10) this process cannot be made up completely (Snik et al., 1997*). It has been shown that probably because of disorganization of central auditory system pathways some prelingually deafened adults experienced somatosensory phenomena. Earlier auditory deprivation seems to produce greater central auditory alterations, and perceptible crossover between somatosensory and auditory signals may be the end result (McFeely et al., 1998*).

*Postmeningitic Deafness.* The development of postmeningitic deafness is independent of the patient's age but is in approximately 22% the cause of acquired deafness in children (Aso and Gibson, 1995). The infection can cause a reformation of connective tissue with successive ossification (obliteration) that prevents the insertion of the electrode in the usual way. For this reason the detection of postmeningitic hearing loss as soon as possible is mandatory to facilitate the implantation before the cochlea is obliterated (Dodds et al., 1997). High-resolution CTs (HRCT) can give radiographic evidence of ossification, but the absence of ossification on HRCT is no guarantee of cochlear patency at the time of implantation (Young et al., 2000*). Special surgical methods like drilling (Balkany et al., 1998), insertion into the scala vestibuli, or the media fossa approach (Colletti et al., 1998) and the use of special implants (shortened electrode, split array or double array) are required if the obliteration process already has begun (Lenarz et al., 1997*).

*Traumatic Deafness.* Laterobasal fractures commonly lead to traumatic hearing loss by traverse or longitudinal fracture of the temporal bone. In addition, labyrinthine hemorrhages; stretching or rupture of the auditory nerve; and damage of the central auditory system in the area of the brainstem, midbrain, and auditory cortex may be involved. It is essential to perform the diagnostic procedure as soon as possible, because the labyrinthine lesion threatens to lead to obliteration. The verification of a conductive auditory nerve and the integrity of the central auditory system is decisive for the indication of a cochlear implantation. As shown by Lenarz, only a minority of patients (11 from 47 in total in his study) with traumatic deafness will not benefit from cochlear implantation (Lenarz, 1998*).

### Preoperative Procedures

The preoperative diagnostic procedures, the implantation, and after treatments are all typically done at ENT departments. Before the implantation of an auditory prosthesis, most centers perform a test of the sound perception during electrical stimulation at the promontory or at the membrane of the round window (Ito et al., 1994*). This test is done to predict whether enough excitable ganglion cells remain and whether the central pathways are intact. However, there is still discussion about whether these preoperative measures correlate with the postoperative result (vanDijk et al., 1993). Although there is evidence that in the case of absent hearing sensations on preoperative electrostimulation the patient will have only limited benefit from the implant, this is not a necessary implication (Silverstein et al., 1994*). In small children, who cannot cooperate with the investigator, electrically evoked auditory brainstem responses (EABR) under general anesthesia have been proposed for preimplantation assessment (Frohne *et al.*, 1997).

### Possible Complications

The complication rate of cochlear implants reported in the literature is relatively low. If the operation is done by a well-trained surgeon, the risk for major complications such as facial nerve injury or incorrect electrode position can be minimized (Kempf *et al.*, 1999a\*). In rare cases of severe infection, removal of the cochlear implant might be necessary. The electrode should remain in the cochlea, thus keeping open the intracochlear space. Skin flap complication was reported between 1.57% and 3.66% of pediatric patients. A recent retrospective analysis of 697 cases reported an incidence of 5.4% for the development of cholesteatoma (Kempf *et al.*, 1999b\*). Other complications such as increased tinnitus, facial nerve stimulation, dizziness, or pain are reported sporadically.

The failure rate of the implant device itself is comparably low. Cochlear Corporation reports cumulative survival percentages (CSP) of 97.7% and 94.3% (adults/children) for the CI22M implant after 10 years of implant use. After a modification of the receiving antenna, the CSP raised to 98.3% 6 years after modification. It is still unclear whether there are adverse effects to electrode insertion trauma or to chronic electrical stimulation; no reports of new bone growth in the cochlea or deterioration of the nerve cells have been given.

### Outcomes

Cochlear implantation can greatly improve the quality of life; however, the signal received by means of the implant is not normal hearing, despite further development of stimulation strategies and electrode design. Therefore patients require an individual amount of time and training to become accustomed to it. Rehabilitation is an essential part to promote optimal benefit from the device, especially in the case of young children, who may never have heard before. When this prerequisite is fulfilled, early implanted (before the age of 7 years) congenital deaf children can acquire not only a high amount of speech understanding but also an improvement of speech intelligibility (Allen *et al.*, 1998\*).

Factors such as etiology, age, duration of deafness, and duration of implant use have been suggested as possible predictors of performance in postlinguistically deafened adults. A review of 13 studies that investigated these potential factors has shown a statistically significant effect of each factor in at least one of the cited studies and a composite analysis of the data derived from 1033 patients lead to the following conclusions: (1) Duration of deafness has a strong negative effect on auditory performance. (2) Age at implantation has a slight negative effect that increases after age 60. (3) Age at onset of deafness has little effect on auditory performance up to the age of 60 years. (4) Duration of experience has a strong positive effect on auditory performance. The effect of etiology is relatively weak and inconsistent with ganglion cell counts in the histological literature (Blamey *et al.*, 1996).

It has been demonstrated that patients supplied with current cochlear implant designs can reach a mean monosyllabic word score as high as 50% and a mean of 89% sentence understanding after 1 year of experience (Helms *et al.*, 1997\*). Most patients are able to communicate over the telephone.

### Auditory Brainstem Implant

The auditory brainstem implant has been used effectively to provide hearing sensations to individuals deafened by bilateral auditory nerve tumors (Shannon *et al.*, 1993\*). During tumor removal, the auditory brainstem implant is implanted into the lateral recess of the fourth ventricle, preferably by a translabyrinthine approach, and is intended to stimulate auditory neurons of the cochlear nucleus complex. Mild nonauditory sensations (primarily tingling in the head or torso) were encountered in some instances but could be managed by changing the stimulus characteristics or excluding electrodes. Testing of perceptual performance indicated significant benefit from the device for communication purposes, including sound-only sentence recognition scores in some patients ranging from 49%–58% and ability to converse on the telephone (Laszig *et al.*, 1999\*; Otto *et al.*, 1998\*). These results indicate that significant auditory benefit can be derived from direct multichannel electrical stimulation of the auditory portion of the human brainstem, although the outcomes are not comparable with a cochlear implant.

## AUTONOMIC MOTOR NEURAL PROSTHESES

Practicable neural prostheses are available today for the restoration of lost autonomic functions (especially neurogenic bladder-voiding dysfunctions and central respiratory dysfunctions). These devices offer the patient not only enhanced functionality and the prevention of secondary complications, but unlike motor neural prostheses, they help also to save time.

## Dynamic Cardiomyoplasty

A relatively new branch of autonomic motor neural prostheses is dynamic myoplasty. In this kind of application an autologous, transposed, conditioned, and electrically stimulated skeletal muscle is used as an artifical urethral or anal sphincter or to assist the heart in selected cases of end-stage heart disease. For the latter field of the so-called dynamic cardiomyoplasty (Chachques et al., 1992), the reader is refered to the literature on thoracic surgery (e.g., Jessup, 2000*).

## Phrenic (Diaphragm) Pacemaker

Electrical stimulation of the phrenic nerve is an alternative to mechanical ventilation in patients with congenital or acquired dysfunctions of the respiration center in the brainstem or with lesions of the upper cervical spinal cord. Phrenic pacemakers deliver bursts of electrical pulses to the phrenic nerves, which cause each hemidiaphragm to contract and the lungs to expand (inspiration). In contrast to mechanical ventilation, breathing by phrenic pacing is achieved by generation of negative pressure in a manner similar to physiological breathing. The present experience with phrenic pacemakers covers a total of several thousand patient-years of implantation in all age groups (2 months–89 years). Several patients have successfully undergone phrenic pacing for more than 25 years.

### Indications and Contraindications

The indications include cervical spinal cord injury, idiopathic central alveolar hypoventilation (Ondine's syndrome), and brainstem lesions of different origins (Dobelle et al., 1994*). Also, phrenic pacing has been applied as an ultimate therapy in a small number of patients to suppress intractable hiccups (Dobelle, 1999*).

Phrenic pacing typically fails in patients with chronic obstructive pulmonary disease (COPD) and in patients with severe kyphoscoliosis and deformations of the rib cage. The most important precondition for the use of phrenic pacemakers is that the phrenic nerve and the diaphragm are still intact, although phrenic pacing is still be possible if only a portion of the nerve fibers is damaged. In spinal lesions between C3 and C5, a considerable amount of the anterior root cells to the phrenic nerve are damaged; therefore, phrenic pacing in patients with spinal lesions at or below C3 may fail. Damage to the phrenic nerve induced by poliomyelitis, neuromuscular disorders, and mitochondrial myopathy may also be considered contraindications for the implantation of a phrenic pacemaker. The standard preopera-

tive test for an intact function of phrenic nerve and diaphragm is percutaneous electrical stimulation of the phrenic nerve at the neck, together with fluoroscopy of the thorax, surface electromyography of the diaphragm (Markand et al., 1984), and recording of the transdiaphragmatic pressure. As an alternative test for the decision-making process of phrenic pacing, the assessment of the central and peripheral motor pathway to the diaphragm using cortical and cervical magnetic stimulation has been proposed (Similowski, et al., 1996*). However, all noninvasive tests may produce false-negative results. The most unequivocal test is to surgically expose the nerve and stimulate it directly. Patients with irreparable damage to their phrenic nerves may be candidates for a nerve graft from one or more intercostal nerves to the phrenic nerve (Krieger and Krieger, 2000*). Although the risk of lasting damage to the phrenic nerve during the implantation of a pacemaker is relatively low (incidence ca. 5%) (Chervin and Guilleminault, 1994; Weese-Mayer et al., 1996*), the indication for implantation should be considered carefully in all patients with preserved spontaneous breathing during most times of the day and in children. Especially in patients with central sleep apnea syndrome, timed mechanical ventilation (e.g., nasal BiPAP or CPAP) should be the treatment of first choice (Thalhofer and Dorow, 1997).

### Clinical Considerations

Compared with positive pressure ventilation phrenic pacing has beneficial effects to the cardiovascular system, especially to forestall a cor pulmonale caused by long-term positive-pressure mechanical ventilation (Ishii et al., 1990*). Other advantages that have been attributed to phrenic pacing are less formation of secretions, a lower rate of pulmonary infections, reduced barotrauma of the lungs, and prevention of atrophy of the diaphragm. When it is possible to close the tracheostomy or to use tracheostomy tubes with speaking valves, tetraplegic patients with insufficient voluntary respiration may easily speak during phrenic pacing. In patients who need ventilatory assistance not only at night but also during the day, phrenic pacing allows much increased mobility and therefore a better quality of life (Esclarin et al., 1994*). Disadvantages of today's phrenic pacemakers include a lack of synchronization with the muscles of the upper airways and the accessory respiratory muscles. Therefore a tracheostomy is necessary. Closure can be considered in some patients with a lesion confined to the upper spinal cord with preservation of the brainstem, in patients with some preserved spontaneous ventilation, and in patients who are able to perform glossopharyngeal (frog) breathing for more than 10 minutes. Atelectasis with subsequent disturbances of gas exchange are not a serious problem in long-term phrenic pacing. Respiratory or other infections can reduce the response to diaphragm pacing, so that mechanical ventilation may become temporarily necessary. The implant itself can also be subject to direct or hematogenous infection, even years after implantation. MRI, lithotripsy, and diathermy are contraindicated to avoid damage to implanted components. This may restrict the use of phrenic pacemakers in patients in whom MRIs may be a necessary diagnostic option in the future. The initial costs for phrenic pacing are high compared with a portable ventilator and humidifier; however, relative savings on health, supervisory, and institutional care costs, as well as increased productivity and satisfaction of the patient may make up for the difference in costs (Esclarin *et al.*, 1994*).

One important factor for the success of diaphragm pacing is rehabilitation after surgery (Glenn *et al.*, 1988*; Glenn and Phelps, 1985) with reconditioning of the diaphragm, especially in tetraplegic patients. Although some tetraplegic patients may achieve 24-hour pacing within a few days, with unipolar stimulation it normally takes 3–6 months in adults, and longer in children, to condition the diaphragm (Chervin and Guilleminault, 1994). Pulse oximetry with an alarm and memory capacity should be used in all applications of phrenic pacemakers. In most case reports in the literature a backup ventilator has been demanded for all patients using phrenic pacemakers. The need for a backup ventilator has been questioned by some authors who believe that newer pacing systems are reliable enough to abandon the backup source (Dobelle *et al.*, 1994*).

### Technology and Clinical Applications

The most frequently used phrenic pacemaker has been developed by the group of W. Glenn at Yale University together with the manufacturer, Avery Laboratories, New York (Dobelle Group, URL: www.dobelle.com). This system uses one or two receivers implanted under the skin; each receiver delivers the electrical pulses to the phrenic nerve by means of a platinum cuff electrode. An external unit (stimulator/transmitter) sends power and control signals by radio frequency to the receiver(s). The ring antennae are taped to the skin over the receivers. Since the first implantation in 1966, it has undergone several improvements. In March of 1998, the latest version, the Mark IV breathing pacemaker obtained premarket approval from the FDA; it also received a CE Mark under the European Active Implantable Medical Device Directive.

In Europe, two different types of multichannel phrenic pacemakers have been developed. The ATROSTIM® phrenic nerve stimulator (ATROTECH

OY, P.O.box 28, SF-33721 Tampere, Finland; URL: www.hermia.fi/atrotech/) has one receiver for each side and a four- pole electrode for each nerve. Four-pole sequential stimulation is used by selecting four different combinations of electrodes for subsequent stimulation pulses. In this case it is assumed that the firing frequency of the single motor units is lower than the overall stimulation frequency (one fourth of the total stimulation frequency in the ideal case), but because of the intermingling of axons supplying different muscle fibers, a smooth fused contraction of the diaphragm will occur (Baer *et al.*, 1990).

In Austria, the group of H. Thoma in Vienna developed a system with one multichannel receiver for two pairs of four electrodes (manufactured by Medimplant GmbH, Laudongasse 10, A-1080 Vienna, Austria). Four wire loop electrodes, which are attached by microsurgical techniques to each phrenic nerve, are used for "carousel" stimulation (see previously) to prevent fatigue of the nerve and the diaphragm. As of 1992, 23 patients with complete ventilatory insufficiency of differing causes have been treated with this system (Mayr *et al.*, 1993*).

The advantage of the multichannel systems with "carousel" or four-pole sequential stimulation is less fatigue of the diaphragm compared with single-channel devices. The conditioning period after surgery (see later) may be shorter with multichannel systems, and full-time bilateral stimulation is more likely to be achieved. However, there is no comparative clinical study between the different phrenic pacemaker designs.

### Related Technology: FES for Cough

A related potential application of FES to the respiratory system is the stimulation of abdominal muscles to produce or enhance cough in persons with tetraplegia (Jaeger *et al.*, 1993; Linder, 1993). Secondary complications of the respiratory system caused by spinal cord injury are a major cause of death and rehospitalization in this population. The classic approach to the problem of absent or impaired cough has been the so-called "manually assisted" cough. In this maneuver, a therapist or other caregiver uses their hands to manually compress the abdomen, while the patient controls his upper airway. In this way, coughs can be obtained; this has been found to be clinically efficacious in clearing secretions. The disadvantage of the manual assist technique is that it requires a caregiver each time the patient wants to cough. Electrical stimulation of abdominal muscles for cough has been found to generate peak expiratory air flows at the mouth that are comparable to the manual assist (Jaeger *et al.*, 1993) and generate maximum expiratory pressures at the mouth that are significantly higher than spontaneous cough with no manual assist (Linder, 1993). The technique has been tested worldwide (Jaeger et al, 1994*) in approximately 100 patients. A commercially available stimulator is available for sale outside of the United States (Quik-Coff®; URL: www.bandb-medical.com). This technique has not yet been widely adopted and will require further study to document safety and efficacy.

## Voiding of Bladder and Bowels

### Indications and Contraindications

Neural prostheses for bladder voiding induce a contraction of the detrusor muscle by either stimulation of the bladder wall, its motor nerves, or the sacral spinal cord. These neural prostheses are indicated in patients with loss of voluntary bladder control because of spinal cord lesions ("neurogenic bladder"). This is especially true if urinary incontinence or increased post-void residuals are accompanied by repeated urinary tract infections, which occur despite conventional therapy (bladder training, intermittent catheterization, pharmacological therapy; see also Chapter 104). Contraindications to neural prostheses stimulating the motor nerves of the detrusor muscle include lesions of the efferent parasympathetical peripheral neurons (S2–S4) with insufficient intravesical pressure (<50 cm $H_2O$) during cystometry or intrarectal electrical stimulation (Brindley et al., 1986*). Other contraindications that have been mentioned in the literature are motor or sensory incomplete spinal lesions, low bladder capacity <200 ml, anatomical abnormalities of the lumbosacral spine, decubitus ulcers, and refusal or impossibility for intermittent catherterization (van Kerrebroeck et al., 1997*).

### Technology and Clinical Applications

The most successful clinical technique today is the stimulation of the sacral anterior nerve roots (Brindley, 1994*, 1995*; Creasey, 1993), typically preceded by dorsal rhizotomy. These neural prostheses have emerged to a commercially available system that is CE approved for sale in the European Union and approved by the FDA for sale in the United States. The so-called Finetech-Brindley (also known as the VOCARE) Bladder System is manufactured by Finetech Medical Limited (URL: www.finetech-medical.co.uk) and distributed by Neurocontrol Corporation (URL: www.neurocontrol.com). The system consists of a set of intradurally or extradurally implanted platinum electrodes for stimulation of the sacral anterior roots S2–S4 on both sides and a receiver/stimulator with three independent units for every pair of nerve roots. The receiver

is implanted into the abdominal wall. The stimulation parameters, which can be adjusted separately for the three pairs of nerve roots by the external electronics, are determined after surgery by intravesical manometry. Because the sacral anterior roots (especially S2) innervate not only the detrusor (autonomic branch) but also the external sphincter (pudendal branch), repetitive bursts of stimulation pulses are induced for micturition. After each burst the external sphincter relaxes faster than the detrusor, leading to a staccato voiding pattern.

Autonomic bladder hyperreflexia is a main problem in patients with spinal cord injuries, because it leads to a low-volume, high-pressure bladder reservoir, vesicouretral reflux in combination with urinary tract infections, and finally it may cause decompensation of the upper urinary tract (Barton et al., 1984). Therefore, in the last decade, the combination of sacral anterior root stimulation together with complete posterior sacral root rhizotomies has become the standard operation (Sauerwein, 1990*), leading to improved clinical results compared with sacral anterior root stimulation without sacral deafferentation. In a prospective clinical Dutch study with 52 spinal cord–injured patients treated with a Finetech-Brindley stimulator and sacral rhizotomies (van Kerrebroeck et al., 1997*), the number of fully continent patients rose from 10% before surgery to 70%–80% afterwards. In a French study with 96 patients, 83 (86%) were fully continent after implantation (Egon et al., 1998*). In both studies there was a significant increase in bladder capacity and bladder compliance and a significant decrease of residual volume and of the incidence of urinary tract infections. Together with the Dutch clinical study, a cost-effectiveness analysis of sacral anterior root stimulation with dorsal rhizotomies compared with conventional care of neurogenic bladder dysfunction was performed (Wielink et al., 1997*). The authors concluded that sacral anterior root stimulation with rhizotomy is a cost-effective method and that the costs for implantation, training, and follow-up are earned back in about 8 years after the implantation.

In male patients one important side effect of the sacral deafferentation is the loss of reflex penile erections. Although implant-driven erections are possible in about 65%–90% of the patients (Egon et al., 1998*; van der Aa et al., 1999*; see also later), the desire to preserve reflex erections is considered as a contraindication for the implantation with rhizotomy by some authors (van Kerrebroeck et al., 1997*). Stress incontinence is another possible side effect of total posterior rhizotomy from S2–S4. Less frequent, but serious, complications of the implantation of a sacral anterior root stimulator included CSF leaks (especially after intradural implantations), postoperative denervations (Colombel et al., 1992*), infection of the implant (about 1%, Brindley,

1994*), sepsis, and implant failure (Brindley, 1995*). From the first 500 implanted patients, 9% stopped using the implant, mainly because of inadequate implant-driven micturitions ($n = 28$), pain during stimulation ($n = 6$), or autonomic dysreflexia ($n = 3$) with uncontrolled hypertensive crisis, which was not prevented by sacral posterior rhizotomy (Schurch et al., 1998*). In another series, some patients required specific therapies (parasphincteral infiltrations, alpha-blockers, external sphincterotomy, etc.) after implantation to promote continence between periods of stimulation and complete bladder voiding (Isambert et al., 1993).

### Positive and Negative Side Effects

Penile erection that is suitable for sexual intercourse is achieved in about one third of the male patients using the sacral anterior root stimulator (Brindley et al., 1986*). With the Brindley-Finetech stimulator, a controlled voiding of the bowel is possible; however, the number of patients in the literature who use the implant for controlled defecation ranges from less than 3% (13 of 500 patients; Brindley, 1994*) to 71% (27 of 38 patients; van der Aa et al., 1999*).

### FES for Fecal Continence

A transposition of the gracilis muscle with chronic electrical stimulation of the transposed muscle ("dynamic graciloplasty") can be used as an anal neosphincter. A variety of case reports and small clinical studies in the literature indicate that this application of dynamic graciloplasty is a promising option for patients with end-stage fecal incontinence, especially for patients with resection of the anal sphincter during tumor surgery (Mander et al., 1999*). Improvement of continence can be expected in up to 90% of the patients; however, impaired evacuation of stool and implant-related infections are the main drawbacks (Niriella and Deen, 2000*).

### Neuromodulation for Treatment of Urinary Incontinence

Besides neural prostheses for bladder voiding, other therapeutic applications of electrical stimulation for improvement of bladder functions have been studied. S3 or S4 sacral nerve stimulation with an implantable stimulator (InterStim System, Medtronic Inc., URL: www.medtronic.com) has successfully been tested in several studies in patients with refractory urinary urge incontinence (Hassouna et al., 2000*). Series of daily intravesical transurethral bladder stimulation combined with biofeedback have been applied to children with myelodysplasia to gain conscious urinary

control and increase bladder capacity. The positive effects of earlier studies (Kaplan et al., 1989*) with respect to bladder capacity, development of detrusor contractions, and acquisition of bladder sensation with this time-consuming, labor-intensive program could not be verified in a prospective, randomized clinical trial (Boone et al., 1992**). Electrical stimulation of the pelvic floor muscles by surface electrodes (Kralj, 1999*) or percutaneous electrodes (Ishigooka et al., 1993*) and bladder stimulation by an intravaginal electrode (Primus, 1992*) have been tested with partial success in patients with female stress incontinence and neurogenic detrusor hyperactivity. In several studies electrical stimulation was combined with biofeedback. However, there are only a few controlled studies with rather small numbers of patients, and some authors could not find positive therapeutic effects of electrical stimulation in patients with stress incontinence (Luber and Wolde-Tsadik, 1997**).

### Restoration of Sexual Function

In most cases, women with spinal cord injury remain procreative, whereas the erectile and ejaculatory capabilities in men with spinal cord injury are seriously affected (Smith and Bodner, 1993). In about 70% of male patients with a complete spinal cord injury above the T12 level, a reflexogenic penile erection during mechanical stimulation (masturbation) is possible without technical aids, whereas pyschogenic erections are lacking.

Ejaculatory function is only preserved in about 10% of men with spinal cord injury (Bennett et al., 1988). Because penile erections occur in response to neural transmission from the pelvic parasympathetic nerves (which originate in S2–S4), a penile erection can be evoked by sacral anterior root stimulation. A significant number of patients who have a sacral anterior root stimulator implanted for bladder voiding (see earlier) use the stimulation also for sexual intercourse (Brindley et al., 1986*). However, a neurogenic disorder of penile erection alone is no indication for neural prostheses because of the potential risk of a lesion to the nerve roots and the uncertain effects over a longer period of time. Other less invasive forms of therapy should be preferred for achieving penile erections (see Chapter 104).

Vibratory stimulation of the penis is the preferred treatment for disorders of ejaculation (impotentia generandi) in men with spinal cord injury (Bennet et al., 1988). If this method fails, an ejaculation can still be achieved in about two thirds of all patients by electrical stimulation of sympathetic fibers of the hypogastric plexus (Brindley, 1984*). Electroejaculation has also successfully been applied in patients with psychogenic anejaculation and after retroperitoneal surgery. With both methods, subsequent artificial insemination of the spouse is necessary (intrauterine insemination, in vitro fertilization, gamete intrafallopian tube transfer, intracytoplasmic sperm injection). Usually electroejaculation is performed with a rectal probe, which carries the electrodes (e.g., the Seager LifeSTIM™ System, provided by Neurocontrol Corporation [URL: *www.neurocontrol*.com/products.htm]). The stimulation frequency is between 10 and 30 Hz (Perkash *et al.*, 1990*). In about one third of the patients using electroejaculation, a retrograde ejaculation into the urinary bladder takes place. In this case, the bladder has to be emptied before the electrical stimulation. A poststimulation catheter flush with Ringer's solution permits collection of sperm. In patients with intact sensation, electroejaculation may be performed under either general or spinal anesthesia. In patients with autonomic dysreflexia, rectal probe stimulation may cause noticeable increase in blood pressure. Therefore short-term control of blood pressure and eventually long-term medication to reduce blood pressure may be necessary. Because of its risks, vasaspiration, the third practical method to obtain semen from men who cannot ejaculate, should only be used if vibratory stimulation and electroejaculation have failed (Seager and Halstead, 1993). In chronic spinal cord injury, the quality of semen obtained with either of the forementioned methods is reduced from that seen in neurologically intact individuals with respect to mobility and viability, especially in patients with urinary infections, high pressure reflex voiding, and in patients using permanent catheters (Ohl *et al.*, 1989*). Several authors have reported that sperm quality is lower after electroejaculation compared with vibratory stimulation of the penis (Ohl *et al.*, 1997**), although the reasons are not quite well understood. Despite the reduced quality of sperm, pregnancy rates of 40% or more have been reported (Brindley, 1984*; Bennet *et al.*, 1988; Seager and Halstead, 1993) if repeated assisted ejaculation and intrauterine insemination was performed. Methods that have been proposed to improve the success rate of electroejaculation are consecutive-day electroejaculation (Mallidis *et al.*, 2000*), balloon catheters to tamponade the bladder neck for preventing retrograde ejaculation, chemical exposure of the spermatozoa (e.g., to pentoxifylline) (Sikka and Hellstrom, 1991*), and gamete micromanipulation.

## REFERENCES

Agency for Health Care Policy Research (1995). Post-Stroke Rehabilitation Clinical Guideline Number 16. AHCPR Publication No. 95–0062: May 1995 (available at http://hstat.nlm.nih.gov).

Allen, M. C., Nikolopoulos, T. P., and O'Donoghue, G. M. (1998). Speech intelligibility in children after cochlear implantation. *Am. J. Otol.* **19**, 742–746.

Aso, S., and Gibson, W. P. (1995). Surgical techniques for insertion of a multi-electrode implant into a postmeningitic ossified cochlea. *Am. J. Otol.* **16**, 231–234.

Baer, G. A., Talonen, P. P., Shneerson, J. M., Markkula, H., Exner, G., and Wells, F. C. (1990). Phrenic nerve stimulation for central ventilatory failure with bipolar and four-pole electrode systems. *PACE* **13**, 1061–1072.

Bajd, T., Kralj, A., Stefancic, M., and Lavrac, N. (1999). Use of electrical stimulation in the lower extremities of incomplete spinal cord injured patients. *Artif. Organs* **23**, 403–409.

Baldi, J. C., Jackson, R. D., Moraille, R., and Mysiw, W. J. (1998). Muscle atrophy is prevented in patients with acute spinal cord injury using functional electrical stimulation. *Spinal Cord* **36**, 463–469.

Balkany, T., Bird, P. A., Hodges, A. V., Luntz, M., Telischi, F. F., and Buchman, C. (1998). Surgical technique for implantation of the totally ossified cochlea. *Laryngoscope* **108**, 988–992.

Barton, C. H., Vaziri, N. D., Gordon, S., Tilles, S. (1984). Renal pathology in end-stage renal disease associated with paraplegia. *Paraplegia* **22**, 31–41.

Beillot, J., Carre, F., Le Claire, G., Thoumie, P., Perruoin-Verbe, B., Cormerais, A., Courtillon, A., Tanguy, E. O., Nadeau, G., Rochcongar, P., Dassonville, J. (1996). Energy consumption of paraplegic locomotion using reciprocating gait orthosis. *Eur. J. Appl. Physiol.* **73**, 376–381.

Belanger, M., Stein, R. B., Wheeler, G. D., Gordon, T., and Leduc, B. (2000). Electrical stimulation: can it increase muscle strength and reverse osteopenia in spinal cord injured individuals? *Arch. Phys. Med. Rehabil.* **81**, 1090–1098.

Bennet, C. J., Seager, S. W., Vasher, E. A., and McGuire, E. J. (1988). Sexual dysfunction and electroejaculation in men with spinal cord injury. *J. Urol.* **139**, 453–457.

Betz, R., Boden, B., Triolo, R., Mesgarzadeh, M., Gardner, E., and Fife, R. (1996). Effects of functional electrical stimulation on the joints of adolescents with spinal cord injury. *Paraplegia* **34**, 127–136.

Biering-Sorensen, F., Gregersen, H., Hagen, E., Haugland, M., Keith, M., Larsen, C. F., Leicht, B. P., Nielsen, F. H., Rabischong, E., and Sinkjaer, T. (2000). Improved function of the hand in persons with tetraplegia using electric stimulation via implanted electrodes VERNACULAR TITLE: Forbedret handfunktion hos tetraplegikere ved elektrisk stimulering via implanterede elektroder. *Ugeskr. Laeger.* **162**, 2195–2198.

Blamey, P. (1997). Are spiral ganglion cell numbers important for speech perception with a cochlear implant? *Am. J. Otol.* **18**, 6 **Suppl**, S11–S12.

Blamey, P., Arndt, P., Bergeron, F., Bredberg, G., Brimacombe, J., Facer, G., Larky, J., Lindstrom, B., Nedzelski, J., Peterson, A., Shipp, D., Staller, S., and Whitford, L. (1996). Factors affecting auditory performance of postlinguistically deaf adults using cochlear implants. *Audiol. Neurootol.* **1**, 293–306.

Bogataj, U., Gros, N., Kljajic, M., Acimovic, R., and Malezic, M. (1995). The rehabilitation of gait in patients with hemiplegia: a comparison between conventional therapy and multichannel functional electrical stimulation therapy. *Phys. Ther.* **75**, 490–502.

Boone, T. B., Roehrborn, C. G., and Hurt, G. (1992). Transurethral intravesical electrotherapy for neurogenic bladder dysfunction in children with myelodysplasia: a prospective, randomized clinical trial. *J. Urol.* **148**, 550–554.

Bradley, M. B. (1994). The effect of participating in a functional electrical stimulation exercise program on affect in people with spinal cord injuries. *Arch. Phys. Med. Rehabil.* **75**, 676–679.

Brindley, G. S. (1984). The fertility of men with spinal injuries. *Paraplegia* **22**, 337–348.

Brindley, G. S. (1994). The first 500 patients with sacral anterior root stimulator implants: general description. *Paraplegia* **32**, 795–805.

Brindley, G. S. (1995). The first 500 sacral anterior root stimulators: implant failures and their repair. *Paraplegia* **33**, 5–9.

Brindley, G. S., and Lewin, W. S. (1968). The visual sensations produced by electrical stimulation of the medial occipital cortex. *J. Physiol.* **196**, 479–493.

Brindley, G. S., Polkey, C. E., Rushton, D. N., and Cardozo, L. (1986). Sacral anterior root stimulators for bladder control in paraplegia: the first 50 cases. *J. Neurol. Neurosurg. Psychiat.* **49**, 1104–1114.

Brissot, R., Gallien, P., Le Bot, M. P., Beaubras, A., Laisne, D., Beillot, J., and Dassonville, J. (2000). Clinical experience with functional electrical stimulation-assisted gait with Parastep in spinal cord-injured patients. *Spine* **25**, 501–508.

Burridge, J. H., Taylor, P. N., Hagan, S. A., Wood, D. E., and Swain, I. D. (1997). The effects of common peroneal stimulation on the effort and speed of walking: a randomized controlled trial with chronic hemiplegic patients. *Clin. Rehabil.* **11**, 201–210.

Cameron, T., Loeb, G. E., Peck, R. A., Schulman, J. H., Strojnik, P., and Troyk, P. (1997). Micromodular implants to provide electrical stimulation of paralyzed muscles and limbs. *IEEE. Trans. Biomed. Eng.* **44**, 781–790.

Carroll, S., Cooper, C., Brown, D., Sormann, G., Flood, S., and Denison, M. (2000). Australian experience with the Freehand System for restoring grasp in quadriplegia. *Aust. N. Z. J. Surg.* **70**, 563–568.

Cauraugh, J., Light, K., Kim, S., Thigpen, M., and Behrman, A. (2000). Chronic motor dysfunction after stroke: recovering wrist and finger extension by electromyography-triggered neuromuscular stimulation. *Stroke* **31**, 1360–1364.

Chachques, J. C., Acar, C., Portoghese, M., Bensasson, D., Guibourt, P., Grare, P., Jebara, V. A., Grandjean, P. A., and Carpentier, A. (1992). Dynamic cardiomyoplasty for long-term cardiac assist. *Eur. J. Cardiothorac. Surg.* **6**, 642–647.

Chae, J., Bethoux, F., Bohine, T., Dobos, L., Davis, T., and Friedl, A. (1998). Neuromuscular stimulation for upper extremity motor and functional recovery in acute hemiplegia. *Stroke* **29**, 975–979.

Chen, D., Philip, M., Philip, P. A., and Monga, T. N. (1990). Cardiac pacemaker inhibition by transcutaneous electrical nerve stimulation. *Arch. Phys. Med. Rehabil.* **71**, 27–30.

Chervin, R. D., and Guilleminault, C. (1994). Diaphragm pacing: review and reassessment. *Sleep* **17**, 176–187.

Cohen, N. L., Waltzman, S. B., and Fisher, S. G. (1993). A prospective, randomized study of cochlear implants. The Department of Veterans Affairs Cochlear Implant Study Group. *N. Engl. J. Med.* **328**, 233–237.

Colletti, V., Fiorino, F. G., Carner, M., and Pacini, L. (1998). Basal turn cochleostomy via the middle fossa route for cochlear implant insertion. *Am. J. Otol.* **19**, 778–784.

Colombel, P., Egon, G., and Isambert, J. L. (1992). Electrostimulation des racines sacrees anterieures chez le blesse medullaire (bilan des 25 premiers cas). [Electrostimulation of anterior sacral nerve roots in spinal cord injury patients (evaluation of the 1st 25 cases)]. *Prog. Urol. (France)* **2**, 41–49.

Creasey, G. H. (1993). Electrical stimulation of sacral roots for micturition after spinal cord injury. *Urol. Clin. North. Am.* **20**, 505–515.

Cullington, H. E., and Brown, E. J. (1998). Bilateral otoacoustic emissions pass in a baby with Mondini deformity and subsequently confirmed profound bilateral hearing loss. *Br. J. Audiol.* **32**, 249–253.

Dai, R., Stein, R. B., Andrews, B. J., James, K. B., and Weiler, M. (1996). Application of Tilt Sensors in Functional Electrical Stimulation. *IEEE. Trans. Rehab. Engin.* **4**, 63–72.

Daly, J. J., Marsolais, E. B., Mendell, L. M., Rymer, W. Z., Stefanovska, A., Wolpaw, J. R., and Kantor, C. (1996). Therapeutic neural effects of electrical stimulation. *IEEE. Trans. Rehabil. Eng.* **4**, 218–230.

Davis, R., Houdayer, T., Andrews, B., and Barriskill, A. (1999). Paraplegia: prolonged standing using closed-loop functional electrical stimulation and Andrews ankle-foot orthosis. *Artif. Organs.* **23**, 418–420.

Davis, S. E., Mulcahey, M. J., Smith, B. T., and Betz, R. R. (1998). Self-reported use of an implanted FES hand system by adolescents with tetraplegia. *J. Spinal. Cord Med.* **21**, 220–226.

Dillier, N., Battmer, R. D., Döring, W. H., and Müller-Deile, J. (1995). Multicentric field evaluation of a new speech coding strategy for cochlear implants. *Audiology* **34**, 145–159.

Dobelle, W. M. H., D'Angelo, M. S., Goetz, B. F., Kiefer, D. G., Lallier, T. J., Lamb, J. I., and Yazwinsky, J. S. (1994). 200 cases with a new breathing pacemaker dispel myths about diaphragm pacing. *ASAIO J.* M244–M252.

Dobelle, W. H. (1999). Use of breathing pacemakers to suppress intractable hiccups of up to thirteen years duration. *ASAIO J.* **45**, 524–525.

Dobelle, W. H. (2000). Artificial vision for the blind by connecting a television camera to the visual cortex. *ASAIO J.* **46**, 3–9.

Dodds, A., Tyszkiewicz, E., and Ramsden, R. (1997). Cochlear implantation after bacterial meningitis: the dangers of delay. *Arch. Dis. Child.* **76**, 139–140.

Donaldson, N., and Yu., C. H. (2000). Experiments with CHRELMS patient-driven stimulator controllers for the restoration of function to paralysed legs. *Proc. Inst. Mech. Eng.* **214**, 1–20.

Donaldson, P. E. K. (1983). Engineering visual prostheses. *Eng. Med. Biol.* **21**, 14–18.

Eichhorn, K. F., Schubert, W., and David, E. (1984). Maintenance, training and functional use of denervated muscles. *J. Biomed. Eng.* **6**, 205–211.

Egon, G., Barat, M., Colombel, P., Visentin, C., and Isambert, J. L. (1998). Implantation of anterior sacral root stimulators combined with posterior sacral rhizotomy in spinal injury patients. *World J. Urol.* **16**, 342–349.

Esclarin, A., Bravo, P., Arroyo, O., Mazaira, J., Garrido, H., and Alcaraz, M. A. (1994). Tracheostomy ventilation versus diaphragmatic pacemaker ventilation in high spinal cord injury. *Paraplegia* **32**, 687–693.

Francisco, G., Chae, J., Chawla, H., Kirshblum, S., Zorowitz, R., Lewis, G., and Pang, S. (1998). Electromyogram-triggered neuromuscular stimulation for improving the arm function of acute stroke survivors: a randomized pilot study. *Arch. Phys. Med. Rehabil.* **79**, 570–575.

Fraysse, B., Dillier, N., Klenzner, T., Laszig, R., Manrique, M., Morera, P. C., Morgon, A. H., Müller-Deile, J., and Ramos Macias, A. (1998). Cochlear implants for adults obtaining marginal benefit from acoustic amplification: a European study. *Am. J. Otol.* **19**, 591–597.

Frohne, C., Lesinski, A., Battmer, R. D., and Lenarz, T. (1997). Intraoperative test of auditory nerve function. *Am. J. Otol.* **18**, 6 **Suppl**, S93–S94.

Glenn, W. W. L., Brouillette, R. T., Dentz, B., Fodstad, H., Hunt, C. E., Keens, T. G., Marsh, H. M., Pande, S., Piepgras, D. G., and Vanderlinden, R. G. (1988). Fundamental considerations in pacing of the diaphragm for chronic ventilatory insufficiency: A multicenter study. *PACE* **11**, 2121–2127.

Glenn, W. W. L., and Phelps, M. L. (1985). Diaphragm pacing by electrode stimulation of the phrenic nerve. *Neurosurgery* **17**, 974–984.

Graham, J. M., Phelps, P. D., and Michaels, L. (2000). Congenital malformations of the ear and cochlear implantation in children: review and temporal bone report of common cavity. *J. Laryngol. Otol. Suppl.* **25**, 1–14.

Graupe, D., and Kohn, K. H. (1994). "Functional Electrical Stimulation for Ambulation by Paraplegics—Twelve Years of Clinical Observations and Systems Studies." Krieger Publishing Co., Malabar.

Haugland, M., and Sinkjaer, T. (1999). Interfacing the body's own sensing receptors into neural prostheses devices. *Technol. Health Care* 7, 393–399.

Handa, Y., Handa, T., Ichie, M., Murakami, H., Hoshimiya, N., Ishikawa, S., and Ohkubo, K. (1992). Functional electrical stimulation (FES) systems for restoration of motor function of paralyzed muscles—versatile systems and a portable system. *Front. Med. Biol. Eng. (Netherlands)* 4, 241–255.

Happak, W., Gruber, H., Holle, J., Mayr, W., Schmutterer, CH., Windberger, U., Losert, U., and Thoma, H. (1989). Multi-channel indirect stimulation reduces muscle fatigue. Proc., 3rd Vienna Int Workshop Functional Electrostimulation. *Austrian Soc. Artifi. Organs, Vienna*, 163–165.

Hassouna, M. M., Siegel, S. W., à Nyeholt, A. L., Elhilali, M. M., van Kerrebroeck, P. E., Das, A. K., Gajewski, J. B., Janknegt, R. A., Rivas, D. A., Dijekma, H., Milam, D. F., Oleson, K. A., and Schmidt, R. A. (2000). Sacral neuromodulation in the treatment of urgency—frequency symptoms: a multicenter study on efficacy and safety. *J. Urol.* 163, 1849–1854.

Helms, J., Muller, J., Schon, F., *et al.* (1997). Evaluation of performance with the COMBI 40 cochlear implant in adults: A multicentric clinical study. *ORL J. Otorhinolaryngol. Relat. Spec.* 59, 23–35.

Hesse, S., Malezic, M., Lücke, D., and Mauritz, K. H. (1998). Stellenwert der funktionellen Elektrostimulation bei Patienten mit Querschnittlähmung. [Value of functional electrostimulation in patients with paraplegia]. *Nervenarzt* 69, 300–305.

Hirokawa, S., Solomonow, M., Baratta, R., and D'Ambrosia, R. (1996). Energy expenditure and fatiguability in paraplegic ambulation using reciprocating gait orthosis and electric stimulation. *Disabil. Rehabil.* 18, 115–122.

Hooker, S. P., Figoni, S. F., Glaser, R. M., Rodgers, M. M., Ezenwa, B. N., and Faghri, P. D. (1990). Physiologic responses to prolonged electrically stimulated leg-cycle exercise in the spinal cord injured. *Arch. Phys. Med. Rehabil.* 71, 863–869.

House, W. F., and Urban, J. (1973). Longterm result of electrode implantation and electronic stimulation of the cochlea in man. *Ann. Otol. Rhinol. Laryngol.* 82, 504–517.

Hummelsheim, H., Maier-Loth, M. L., and Eickhof, C. (1997). The functional value of electrical muscle stimulation for the rehabilitation of the hand in stroke patients. *Scand. J. Rehabil. Med.* 29, 3–10.

Hunt, K. J., Gollee, H., Jaime, R., and Donaldson, N. (1999). Feedback control of unsupported standing. *Technol. Health Care* 7, 443–447.

Hussey, R. W., and Stauffer, E. S. (1973). Spinal cord injury: Requirements for ambulation. *Arch. Phys. Med. Rehabil.* 50, 544–547.

Isambert, J. L., Egon, G., and Colombel, P. (1993). Adjuvant drug therapy: a review of 30 cases of sacral anterior root stimulator. *Neurourol. Urodyn.* 12, 513–515.

Ishigooka, M., Hashimoto, T., Izumiya, K., Katoh, T., Yaguchi, H., Nakada, T., Handa, Y., and Hoshimiya, N. (1993). Electrical pelvic floor stimulation in the management of urinary incontinence due to neuropathic overactive bladder. *Front. Med. Biol. Eng.* 5, 1–10.

Ishii, K., Kurosawa, H., Koyanagi, H., Nakano, K., Sakakibara, N., Sato, I., Noshiro, M., and Ohsawa, M. (1990). Effects of bilateral transvenous diaphragm pacing on hemodynamic function in patients after cardiac operations. *J. Thoracic. Cardiovasc. Surg.* 100, 108–114.

Ito, J., Tsuji, J., and Sakakihara, J. (1994). Reliability of the promontory stimulation test for the preoperative evaluation of cochlear implants: a comparison with the round window stimulation test. *Auris Nasus Larynx* 21, 13–16.

Jacobs, P. L., Nash, M. S., Klose, K. J., Guest, R. S., Needham-Shropshire, B. M., and Green, B. A. (1997). Evaluation of a training program for persons with SCI paraplegia using the Parastep® 1 ambulation system: part 2. Effects on physiological responses to peak arm ergometry. *Arch. Phys. Med. Rehabil.* 78, 794–798.

Jaeger, R. J., Langbein, E. W., and Kralj, A. R. (1994). Augmentation of cough by FES in tetraplegia: a comparison of results at three clinical centers. *Basic Appl. Myology* 4, 195–200.

Jaeger, R. J., Turba, R. M., Yarkony, G. Y., and Roth, E. J. (1993). Cough in spinal cord injured patients: comparison of three methods of cough production. *Arch. Phys. Med. Rehabil.* 74, 1358–1361.

Jaeger, R. J., Yarkony, G. M., Roth, E. J., and Lovell, L. (1990). Estimating the user population of a simple electrical stimulation system for standing. *Paraplegia* 28, 505–511.

Jessup, M. (2000). Dynamic cardiomyoplasty: expectations and results. *J. Heart Lung Transplant.* 19 (Suppl 8), 68–72.

Kantrowitz, A. (1960). Electronic physiologic aids. *In* "Report of the Maimonides Hospital." Brooklyn, NY.

Kaplan, W. E., Richards, T. W., and Richards, I. (1989). Intravesical transurethral bladder stimulation to increase bladder capacity. *J. Urol.* 142, 600–602.

Katz, R. T., Green, D., Sullivan, T., and Yarkony, G. (1987). Functional electric stimulation to enhance systemic fibrinolytic activity in spinal cord injury patients. *Arch. Phys. Med. Rehabil.* 68, 423–426.

Keith, M. W., Peckham, P. H., Thrope, G. B., Buckett, J. R., Stroh, K. C., and Menger, V. (1988). Functional neuromuscular stimulation neuroprostheses for the tetraplegic hand. *Clin. Orthopaed. Rel. Res.* 233, 25–33.

Kempf, H. G., Johann, K., and Lenarz, T. (1999a). Complications in pediatric cochlear implant surgery. *Eur. Arch. Otorhinolaryngol.* 256, 128–132.

Kempf, H. G., Tempel, S., Johann, K., and Lenarz, T. (1999b). [Complications of cochlear implant surgery in children and adults] Komplikationen der Cochlear Implant-Chirurgie bei Kindern und Erwachsenen. *Laryngorhinootologie* 78, 529–537.

Kern, H. (1997). Functional electrical stimulation in paraplegic spastic patients. *Artif. Organs* 21, 195–196.

Kessler, D. K. (1999). The CLARION Multi-Strategy Cochlear Implant. *Ann. Otol. Rhinol. Laryngol. Suppl.* 177, 8–16.

Kessler, K. M., Pina, I., Green, B., Burnett, B., Laighold, M., Bilsker, M., Palomo, A. R., and Myerburg, R. J. (1986). Cardiovascular findings in quadriplegic and paraplegic patients and in normal subjects. *Am. J. Cardiol.* 58, 525–530.

Klose, K. J., Jacobs, P. L., Broton, J. G., Guest, R. S., Needham-Shropshire, B. M., Lebwohl, N., Nash, M. S., and Green, B. A. (1997). Evaluation of a training program for persons with SCI paraplegia using the Parastep 1 ambulation system: part 1. Ambulation performance and anthropometric measures. *Arch. Phys. Med. Rehabil.* 78, 789–793.

Kobetic, R., Triolo, R. J., Uhlir, J. P., Bieri, C., Wibowo, M., Polando, G., Marsolais, E. B., Davis, J. A. Jr., and Ferguson, K. A. (1999). Implanted functional electrical stimulation system for mobility in paraplegia: a follow-up case report. *IEEE. Trans. Rehabil. Eng.* 7, 390–398.

Kralj, A., and Bajd, T. (1989). "Functional Electrical Stimulation, Standing and Walking after Spinal Cord Injury." CRC Press, Boca Raton, Florida.

Kralj, B. (1999). Conservative treatment of female stress incontinence with functional electrical stimulation. *Eur. J. Obstet. Gynecol. Reprod. Biol.* 85, 53–56.

Krieger, L. M., and Krieger, A. J. (2000). The intercostal to phrenic nerve transfer: An effective means of reanimating the diaphragm in patients with high cervical spine injury. *Plast. Reconstr. Surg.* **105**, 1255–1261.

Laizou, P. C. (1999). Signal-processing techniques for cochlear implants. *IEEE Eng. Med. Biol. Magazine* **18**, 34–46.

Laszig, R., Marangos, N., Sollmann, W. P., and Ramsden, R. T. (1999). Central electrical stimulation of the auditory pathway in neurofibromatosis type 2. *Ear Nose Throat J.* **78**, 110–117.

Lenarz, T. (1998). Cochlear Implants: selection criteria and shifting borders. *Acta Oto-Rhino-Laryngologica. Belg.* **52**, 183–199.

Lenarz, T., Battmer, R. D., Lesinski, A., and Parker, J. (1997). Nucleus double electrode array: a new approach for ossified cochleae. *Am. J. Otol.* **18, 6 Suppl**, 39–41.

Lenarz, T., Lesinski-Schiedat, A., von-der-Haar, H., Illg, A., Bertram, B., and Battmer, R. D. (1999). Cochlear implantation in children under the age of two: the MHH experience with the CLARION cochlear implant. *Ann. Otol. Rhinol. Laryngol. Suppl.* **177**, 44–49.

Levine, S. P., Kett, R. L., Cederna, P. S., and Brooks, S. V. (1990). Electric muscle stimulation for pressure sore prevention: tissue shape variation. *Arch. Phys. Med. Rehabil.* **71**, 210–215.

Liberson, W. T., Holmquest, H. J., Scot, D., and Dow, M. (1961). Functional electrotherapy: stimulation of the peroneal nerve synchronized with the swing phase of the gait in hemiplegic patients. *Arch. Phys. Med. Rehabil.* **42**, 101–105.

Linder, S. H. (1993). Functional electrical stimulation to enhance cough in quadriplegia. *Chest* **103**, 166–169.

Luber, K. M., and Wolde-Tsadik, G. (1997). Efficacy of functional electrical stimulation in treating genuine stress incontinence: a randomized clinical trial. *Neurourol. Urodyn.* **16**, 543–551.

Malezic, M., Stanic, U., Kljajic, M., Acimovic, R., Krajnik, J., Gros, N., and Stopar, M. (1984). Multichannel electrical stimulation of gait in motor disabled patients. *Orthopedics* **7**, 1187–1195.

Mallidis, C., Lim, T. C., Hill, S. T., Skinner, D. J., Brown, D. J., Johnston, W. I. H., and Gordon-Baker, H. W. (2000). Necrospermia and chronic spinal cord injury. *Fertil. Steril.* **74**, 221–227.

Mander, B. J., Wexner, S. D., Williams, N. S., Bartolo, D. C., Lubowski, D. Z., Oresland, T., Romano, G., and Keighley, M. R. (1999). Preliminary results of a multicenter trial of the electrically stimulated gracilis neoanal sphincter. *Br. J. Surg.* **86**, 1543–1548.

Markand, O. N., Kincaid, J. C., Pourmand, R. A., Morthy, S. S., King, R. D., Mahomed, Y., and Brown, J. V. (1984). Electrophysiologic evaluation of diaphragm by transcutaneous phrenic nerve stimulation. *Neurology* **34**, 604–614.

Marsolais, E. B., Kobetic, R., and Jacobs, J. (1990). Comparison of FES treatment in the stroke and spinal cord injury patient. *In* "Advances in External Control of Human Extremities" (D. Popovic, Ed.), Vol. X, pp. 210–224. Nauka, Belgrad.

Mayr, W., Bijak, M., Girsch, W., Holle, J., Lanmuller, H., Thoma, H., and Zrunek, M. (1993). Multichannel stimulation of phrenic nerves by epineural electrodes. Clinical experience and future developments. *ASAIO J.* **39**, M729–M735.

McFeely, W., Antonelli, P. J., Rodriguez, F. J., and Holmes, A. E. (1998). Somatosensory phenomena after multichannel cochlear implantation in prelingually deaf adults. *Am. J. Otol.* **19**, 467–471.

Mohr, T., Podenphant, J., Biering-Sœrensen, F., Galbo, H., Thamsborg, G., and Kjaer, M. (1997). Increased bone mineral density after prolonged electrically induced cycle training of paralyzed limbs in spinal cord injured man. *Calcif. Tissue Int.* **61**, 22–25.

Mokrusch, T. (1997). Behandlung der hirninfarktbedingten spastischen Hemiparese mit EMG-getriggerter Elektrostimulation [Treatment of stroke-induced hemiparesis with EMG-triggered electrostimulation]. *Neurol. Rehabil.* **3**, 82–86.

Morse, R. P., and Roper, P. (2000). Enhanced coding in a cochlear-implant model using additive noise: Aperiodic stochastic resonance with tuning. *Phys. Rev. E (Statistical Physics, Plasmas, Fluids, and Related Interdisciplinary Topics)* **61**, 5683–5692.

Mortimer, J. T. (1981). Motor prostheses. *In* "Handbook of Physiology" (J. M. Brookhart and V. B. Mountcastle, Eds.), Sect 1: The Nervous System, Pt II: Motor Control, pp. 155–187. American Physiol Soc, Bethesda, Maryland.

Müller-Deile, J., Rudert, H., Brademann, G., and Frese, K. (1998). [Cochlear implant for non-deaf patients?] Cochlear-Implant-Versorgung bei nicht tauben Patienten? *Laryngorhinootologie* **77**, 136–143.

Munsat, T. L., McNeal, D., and Waters, R. (1976). Effects of nerve stimulation on human muscle. *Arch. Neurol.* **33**, 608–617.

Nathan, R. H. (1989). Maximization of arm function in the C4 quadriplegic. *In* Proceedings of the 3rd Vienna International Workshop Functional Electrostimulation. pp. 179–182. Austrian Society of Artificial Organs, Vienna.

Needham-Shropshire, B. M., Broton, J. G., Klose, K. J., Lebwohl, N., Guest, R. S., and Jacobs, P. L. (1997). Evaluation of a training program for persons with SCI paraplegia using the Parastep 1 ambulation system: part 3. Lack of effect on bone mineral density. *Arch. Phys. Med. Rehabil.* **78**, 799–803.

Nene, A. V., and Patrick, J. H. (1990). Energy cost of paraplegic locomotion using the ParaWalker-electrical stimulation "hybrid" orthosis. *Arch. Phys. Med. Rehabil.* **71**, 116–120.

NIH. (1995). Cochlear implants in adults and children. NIH. Consensus. Statement. **13**, 1–30.

Niriella, D. A., and Deen, K. I. (2000). Neosphincters in the management of fecal incontinence. *Br. J. Surg.* **87**, 1617–1628.

Ohl, D. A., Bennett, C. J., McCabe, M., Menge, A. C., and McGuire, E. J. (1989). Predictors of success in electroejaculation of spinal cord injured men. *J. Urol.* **142**, 1483–1486.

Ohl, D. A., Sonksen, J., Menge, A. C., McCabe, M., and Keller, L. M. (1997). Electroejaculation versus vibratory stimulation in spinal cord injured men: Sperm quality and patient preference. *J. Urol.* **157**, 2147–2149.

Otto, S. R., Shannon, R. V., Brackmann, D. E., Hitselberger, W. E., Staller, S., and Menapace, C. (1998). The multichannel auditory brain stem implant: performance in twenty patients. *Otolaryngol. Head. Neck. Surg.* **118, 3 Pt 1**, 291–303.

Perkash, I., Martin, D. E., Warner, H., and Speck, V. (1990). Electroejaculation in spinal cord injured patients: simplified new equipment and technique. *J. Urol.* **143**, 305–307.

Petrofsky, J. S., Stacy, R. (1992). The effect of training on endurance and the cardiovascular responses of individuals with paraplegia during dynamic exercise induced by functional electrical stimulation. *Eur. J. Appl. Physiol.* **64**, 487–492.

Phillips, W., and Burkett, L. N. (1995). Arm crank exercise with static leg FNS in persons with spinal cord injury. *Med. Sci. Sports Exerc.* **27**, 530–535.

Phillips, W., Burkett, L. N., Munro, R., Davis, M., and Pomeroy, K. (1995). Relative changes in blood flow with functional electrical stimulation during exercise of the paralyzed lower limbs. *Paraplegia* **33**, 90–93.

Ponton, C. W., and Don, M. (1995). The mismatch negativity in cochlear implant users. *Ear Hear* **16**, 131–146.

Powell, J., Pandyan, A. D., Granat, M., Cameron, M., and Stott, D. J. (1999). Electrical stimulation of wrist extensors in poststroke hemiplegia. *Stroke* **30**, 1384–1389.

Primus, G. (1992). Maximal electrical stimulation in neurogenic detrusor hyperactivity: experiences in multiple sclerosis. *Eur. J. Med.* **1**, 80–82.

Quintern, J., Riener, R., and Rupprecht, S. (1997). Comparison of simulation and experiments of different closed-loop strategies for functional electrical stimulation: experiments in paraplegics. *Artif. Organs* **21**, 232–235.

Ragnarsson, K. T. (1988). Physiologic effects of functional electrical stimulation-induced exercises in spinal cord-injured individuals. *Clin. Orthop. Rel. Res.* **233**, 53–63.

Rizer, F. M., and Burkey, J. M. (1999). Cochlear implantation in the very young child. *Otolaryngol. Clin. North Am.* **32**, 1117–1125.

Robinson, C. J., Kett, N. A., and Bolam, J. M. (1988a). Spasticity in spinal cord injured patients: 1. Short-term effects of surface electrical stimulation. *Arch. Phys. Med. Rehabil.* **69**, 598–604.

Robinson, C. J., Kett, N. A., and Bolam, J. M. (1988b). Spasticity in spinal cord injured patients: 2. Inital measures and long-term effects of surface electrical stimulation. *Arch. Phys. Med. Rehabil.* **69**, 862–868.

Royal College of Physicians, National Clinical Guidelines (Stroke), National Electronic Health Library. www.rcplondon.ac.uk/pubs/ceeu_stroke_workingparty.htm, 1999.

Rushton, D. N., Barr, F. M., Donaldson, N. de N., Harper, V. J., Perkins, T. A., Taylor, P. N., and Tromans, A. M. (1998). Selecting candidates for a lower limb stimulator implant programme: a patient-centred method. *Spinal Cord* **36**, 303–309.

Sauerwein, D. (1990). Die operative Behandlung der spastischen Blasenlähmung bei Querschnittlähmung. Sakrale Deafferentation mit der Implantation eines sakralen Vorderwurzelstimulators. [Surgical treatment of spastic bladder paralysis in paraplegic patients. Sacral deafferentation with implantation of a sacral anterior root stimulator]. *Urologe A* **29**, 196–203.

Schmidt, E. M., Bak, M. J., Hambrecht, F. T., Kufta, C. V., O'Rourke, D. K., and Vallabhanath, P. (1996). Feasibility of a visual prosthesis for the blind based on intracortical microstimulation of the visual cortex. *Brain* **119**, 507–522.

Schurch, B., Knapp, P. A., Jaenmonod, D., Rodic, B., and Rossier, A. B. (1998). Does sacral posterior rhizotomy suppress autonomic hyper-reflexia in patients with spinal cord injury? *Br. J. Urol.* **81**, 73–82.

Seager, S. W. J., and Halstead, L. S. (1993). Fertility options and success after spinal cord injury. *Urol. Clin. North. Am.* **20**, 543–548.

Sikka, S. C., and Hellstrom, W. J. G. (1991). The application of pentoxifylline in the stimulation of sperm motion in men undergoing electroejaculation. *J. Androl.* **12**, 165–170.

Shallop, J. K., Facer, G. W., and Peterson, A. (1999). Neural response telemetry with the nucleus CI24M cochlear implant. *Laryngoscope* **109**, 1755–1759.

Shannon, R. V., Fayad, J., Moore, J., Lo, W. W., Otto, S., Nelson, R. A., and O'Leary, M. (1993). Auditory brainstem implant: II. Postsurgical issues and performance. *Otolaryngol. Head Neck Surg.* **108**, 634–642.

Shimada, Y., Sato, K., Abe, E., Kagaya, H., Ebata, K., Oba, M., and Sato, M. (1996). Clinical experience of functional electrical stimulation in complete paraplegia. *Spinal Cord* **Oct;34(10)**, 615–619.

Silverstein, H., Wanamaker, H. H., Rosenberg, S. I., Crosby, N., and Flanzer, J. M. (1994). Promontory testing in neurotologic diagnosis. *Am. J. Otol.* **15**, 101–107.

Similowski, T., Straus, C., Attali, V., Duguet, A., Jourdain, B., and Derenne, J. P. (1996). Assessment of the motor pathway to the diaphragm using cortical and cervical magnetic stimulation in the decision-making process of phrenic pacing. *Chest* **110**, 1551–1557.

Smith, E. M., and Bodner, D. R. (1993). Sexual dysfunction after spinal cord injury. *Urol. Clin. North Am.* **20**, 535–542.

Snik, A. F., Makhdoum, M. J., Vermeulen, A. M., Brokx, J. P., and van-den-Broek, P. (1997). The relation between age at the time of cochlear implantation and long-term speech perception abilities in congenitally deaf subjects. *Int. J. Pediatr. Otorhinolaryngol.* **41**, 121–131.

Solomonow, M., Reisin, E., Aguilar, E., Baratta, R. V., Best, R., and D'Ambrosia, R. (1997). Reciprocating gait orthosis powered with electrical muscle stimulation (RGO II). Part II: Medical evaluation of 70 paraplegic patients. *Orthopedics* **20**, 411–418.

Stein, R. B., Gordon, T., Jefferson, J., Sharfenberger, A., Yang, J. F., de Zepetnek, J. T., and Belanger, M. (1992). Optimal stimulation of paralyzed muscle after human spinal cord injury. *J. Appl. Physiol.* **72**, 1393–1400.

Stett, A., Barth, W., Weiss, S., Haemmerle, H., and Zrenner, E. (2000). Electrical multisite stimulation of the isolated chicken retina. *Vis. Res.* **40**, 1785–1795.

Strojnik, P., Acimovic, R., Vavken, E., Simic, V., and Stanic, U. (1987). Treatment of drop foot using an implantable peroneal underknee stimulator. *Scand. J. Rehabil. Med.* **19**, 37–43.

Sykes, L., Ross, E. R., Powell, E. S., and Edwards, J. (1996). Objective measurement of use of the reciprocating gait orthosis (RGO) and the electrically augmented RGO in adult patients with spinal cord lesions. *Prosthet. Orthot. Int.* **20**, 182–190.

Taylor, P. N., Burridge, J. H., Dunkerley, A. L., Wood, D. E., Norton, J. A., Singleton, C., and Swain, I. D. (1999). Clinical use of the Odstock dropped foot stimulator: its effect on the speed and effort of walking. *Arch. Phys. Med. Rehabil.* **80**, 1577–1583.

Teeter, J. O., Kantor, C., and Brown, D. L. (1995). FES Resource Guide. Published by FES Information Center, Cleveland, Olio.

Thalhofer, S., and Dorow, P. (1997). Central sleep apnea. *Respiration* **64**, 2–9.

Turrini, M., Orzan, E., Gabana, M., Genovese, E., Arslan, E., and Fisch, U. (1997). Cochlear implantation in a bilateral Mondini dysplasia. *Scand. Audiol. Suppl.* **46**, 78–81.

van der Aa, H. E., Alleman, E., Nene, A., and Snoek, G. (1999). Sacral anterior root stimulation for bladder control: clinical results. *Arch. Physiol. Biochem.* **197**, 248–256.

van Dijk, J. E., van Olpen, A. F., and Smoorenburg, G. F. (1993). Preoperative electrical nerve stimulation as one of the criteria for selection. *In* "Cochlear Implants: New Perspectives" (B. Fraysse and O. Deguine, Eds.), Adv Otorhinolaryngol., Vol. 48, pp. 103–107. Karger, Basel.

van Kerrebroeck, Ph. E. V., van der Aa, H. E., Bosch, J. L. H. R., Koldewijn, E. L., Vorsteveld, J. H. C., Debruyne, F. M. J., and the Dutch Study Group on Sacral Anterior Root Stimulation. (1997). Sacral rhizotomies and electrical bladder stimulation in spinal cord injury. Part I: Clinical and urodynamic analysis. *Eur. Urol.* **31**, 263–271.

van Savage, J. G., Perez-Abadia, G. P., Palanca, L. G., Bardoel, J. W., Harralson, T., Slaughenhoupt, B. L., Palacio, M. M., Tobin, G. R., Maldanado, C., and Barker, J. H. (2000). Electrically stimulated detrusor myoplasty. *J. Urol.* **164**, 969–972.

Veraart, C., Grill, W. M., and Mortimer, J. T. (1993). Selective control of muscle activation with a multipolar nerve cuff electrode. *IEEE. Trans. Biomed. Eng.* **40**, 640–653.

Vossius, G., Müschen, U., and Holländer, H. J. (1987). Multichannel stimulation of the lower extremities with surface electrodes. *In* "Advances in External Control of Human Extremities" (D. Popovic, Ed.), Vol. IX, pp. 193–203. Tanjug, Belgrad.

Waltzman, S. B., Cohen, N. L., and Shapiro, W. H. (1993). The benefits of cochlear implantation in the geriatric population. *Otolaryngol. Head Neck Surg.* **108**, 329–333.

Wang, R. Y., Chan, R. C., and Tsai, M. W. (2000). Functional electrical stimulation on chronic and acute hemiplegic shoulder subluxation. *Am. J. Phys. Med. Rehabil.* **79**, 385–390.

Waters, R. L., McNeal, D., and Perry, J. (1975). Experimental correction of footdrop by electrical stimulation of the peroneal nerve. *J. Bone. Joint Surg.* **57A**, 1074–1054.

Waters, R. L., McNeal, D. R., and Faloon, W. (1985). Functional electrical stimulation of the peroneal nerve for hemiplegia. Long-term clinical follow-up. *J. Bone. Joint. Surg.* **67A**, 792–793.

Weber, B. P., Dillo, W., Dietrich, B., Maneke, I., Bertram, B., and Lenarz, T. (1998). Pediatric cochlear implantation in cochlear malformations. *Am. J. Otol.* **19**, 747–753.

Weese-Mayer, D. E., Silvestri, J. M., Kenny, A. S., Ilbawi, M. N., Hauptman, S. A., Lipton, J. W., Talonen, P. P., Garcia, H. G., Watt, J. W., Exner, G., Baer, G. A., Elefteriades, J. A., Peruzzi, W.

T., Alex, C. G., Harlid, R., Vincken, W., Davis, G. M., Decramer, M., Kuenzle, C., Saeterhaug, A., and Schober, J. G. (1996). Diaphragm pacing with a quadripolar phrenic nerve electrode: an international study. *Pacing Clin. Electrophysiol.* **19**, 1311–1309.

Weingarden, H. P., Zeilig, G., Heruti, R., Shemesh, Y., Ohry, A., Dar, A., Katz, D., Nathan, R., and Smith, A. (1998). Hybrid functional electrical stimulation orthosis system for the upper limb: effects on spasticity in chronic stable hemiplegia. *Am. J. Phys. Med. Rehabil.* **77**, 276–281.

Wielink, G., Essink-Bot, M. L., van Kerrebroeck, Ph. E. V., Rutten, F. F. H. and the Dutch Study Group on Sacral Anterior Root Stimulation. (1997). Sacral rhizotomies and electrical bladder stimulation in spinal cord injury. 2. Cost-effectiveness and quality of life analysis. *Eur. Urol.* **31**, 441–446.

Yarkony, G. M., Roth, E. J., Cybulski, G. R., and Jaeger, R. J. (1992). Neuromuscular stimulation in spinal cord injury. II: Prevention of secondary complications. *Arch. Phys. Med. Rehabil.* **73**, 195–200.

Yoshida, K., and Horch, K. (1993). Selective stimulation of peripheral nerve fibres using dual intrafascicular electrodes. *IEEE. Trans. Biomed. Eng.* **40**, 492–494.

Young, N. M., Hughes, C. A., Byrd, S. E., and Darling, C. (2000). Postmeningitic ossification in pediatric cochlear implantation. *Otolaryngol. Head Neck Surg.* **122**, 183–188.

*Movement Disorders*

## CHAPTER 85
# Myoclonus

Mark Hallett and Helge Topka

## CLINICAL ASPECTS (DIFFERENTIAL DIAGNOSIS)

Myoclonus is a type of involuntary movement characterized by quick muscle jerks, either irregular or rhythmic (Hallett, 1997; Hallett *et al.*, 1987; Toro and Hallett, 1997). The term myoclonus may refer to a symptom that accompanies some other disorder of the nervous system, to a syndrome, or to a specific disease entity such as progressive myoclonic epilepsy (EPM1). Myoclonic movements are always simple in nature, and this is often a critical feature separating myoclonus from other types of involuntary movements. Myoclonus can be focal, involving only a few adjacent muscles; generalized, involving many or most of the muscles in the body; or multifocal, involving many muscles but in different jerks. Myoclonus can be spontaneous, can be activated or accentuated by voluntary movement (action myoclonus), and can be activated or accentuated by sensory stimulation (reflex myoclonus).

Myoclonic movements may be present by themselves or be seen in association with other neurological manifestations. The movements may be very slight and not bother the patient much or can be severe and so disruptive as to make voluntary tasks impossible. Myoclonus can affect virtually any muscle of the body and produce difficulty with many activities of daily living. With severe generalized myoclonus, patients may well not be able to feed themselves or walk. Standing and walking is often characterized by marked myoclonus leading to a "bouncy gait."

Myoclonus can be rhythmic and look like tremor. One such situation is with frequent, small myoclonic jerks of the fingers called minipolymyoclonus. Myoclonus can be negative, a paroxymal lapse of muscle innervation, as well as positive. Negative myoclonus is often called asterixis (Toro *et al.*, 1995).

Spontaneous myoclonus can be observed while the patient is sitting talking with the examiner. Because the myoclonus can be infrequent, the examiner must be patient. Action myoclonus is brought out by movement of the body part. Facial myoclonus is produced by closing the eyes or showing the teeth. Upper extremity myoclonus can be produced by postural movement but even more with kinetic movements such as the finger-nose-finger maneuver. Sometimes there can be a little difficulty in differentiating action myoclonus and dysmetria. Somatosensory stimulation may produce reflex myoclonus. Stretch of the fingers is a useful stimulus or taps on any part of the limb. Tendon taps can also be used, but then it is necessary to separate the short-latency tendon jerk from the longer latency reflex myoclonus. This can often be done but requires careful attention to see a double jerk of the body part. The latency of organic myoclonus is very brief and invariant. In psychogenic myoclonus, often the latency is long and variable. Sometimes a jerk may even occur if the tendon hammer is directed to the tendon, but just stops short of the skin.

Because myoclonus can reflect an underlying neurological condition, a full standard evaluation is often required including blood studies, neuroimaging, and cerebrospinal fluid evaluation.

Electrophysiological assessment can be a significant help in deciding whether the movement is myoclonus, and, if so, which type (Hallett, 1997; Hallett *et al.*, 1987; Shibasaki, 1988, 2000). More details about these methods will be given subsequently.

Some simple tics look identical to myoclonus and cannot be visually distinguished. A point in favor of tic is the ability to suppress the movements voluntarily, with a frequent concomitant rise in psychic tension dispelled when the movements resume. Some movements of chorea are quick, but slower movements and sustained postures are also present. The major differential diagnosis for rhythmic myoclonus is tremor, and the distinction here is often just convention. For example, palatal myoclonus is also referred to as palatal tremor.

Some disorders of the peripheral nervous system can be confused with myoclonus. Fasciculation is the spontaneous firing of a single motor unit. Myokymia typically looks like an irregular oscillation of a muscle, but it can have the appearance of small jerks. Hemifacial spasm is characterized by jerks of the facial muscles and occasional tonic spasms. Electrodiagnosis can help make the diagnosis, because they all show characteristic physiological findings with needle electromyography.

## PRINCIPLES OF THERAPY

There are many types of myoclonus, and there are no common etiological, physiological, or therapeutic features. For this reason, *recognizing that an involuntary movement is myoclonic in nature is only the beginning of the investigation.*

There have been many schemes proposed for classifying the large number of myoclonias, and there are at least two useful approaches, etiological and physiological (Hallett, 1994, 1997; Hallett *et al.*, 1987). The first consideration for therapy should be the etiological classification, because the cause should be dealt with first, if possible. This is clearly not always possible. Table I summarizes the etiology of myoclonus syndromes.

Second, the physiological classification becomes helpful, because it guides symptomatic treatment. Here, we will first discuss the physiological classification and its implications for treatment, because these principles will apply to all of the myoclonias.

The starting point for a physiological classification is to decide whether the myoclonus is a fragment of epilepsy. On this basis, the myoclonus is said to be epileptic or nonepileptic (Hallett, 1985). There are a number of subtypes in each category (Table II).

### Epileptic Myoclonus

A significant feature in favor of epileptic myoclonus is if whether the patient has epileptic seizures. This would be more definitive, of course, if the myoclonus was clearly a fragment of the seizure, including a part of the aura. The association of the myoclonus with paroxysmal activity in the EEG is also indicative, and more will be said about the electrophysiological evaluation subsequently.

**TABLE I    Etiology of Myoclonus**

**Physiological myoclonus**
Hiccup (singultus)
Physiological startle reaction, startle reflex
Myoclonus induced by exercise, orgasm, anxiety
Hypnic jerks (sleep starts)
Benign neonatal sleep myoclonus
Benign infantile myoclonus with feeding

**Primary myoclonus syndromes**
Essential myoclonus
Segmental myoclonus
Propriospinal myoclonus
Palatal myoclonus
Exaggerated startle disease
Hereditary hyperekplexia
Nocturnal myoclonus
Periodic movements of sleep

**Neurodegenerative disorders**
Unverricht-Lundborg (Ramsay-Hunt)
SCA 2
DRPLA
Parkinson's disease
Multiple system atrophy
Corticobasal degeneration
Huntington's disease
Wilson's disease
Neuroaxonal dystrophy
Hallervorden-Spatz disease
Progressive supranuclear palsy

**Storage diseases**
Lafora body disease
GM2 gangliosidosis (Tay-Sachs disease)
Ceroid-lipofuscinosis (Batten's disease)
Sialidosis (cherry-red-spot-myoclonus syndrome)
Noninfantile neuronopathic Gaucher's disease (type 3)
Biotin deficiency

**Infectious and parainfectious diseases**
Herpes-simplex encephalitis
Coxsackie encephalitis
Subacute sclerosing panencephalilitis
HIV encephalopathy
Opsoclonus-myoclonus syndrome (also paraneoplastic)
Progressive encephalitis with rigidity and myoclonus (PERM)

**Intoxications**
Bismuth
Heavy metals
Methyl bromide
Tricyclic antidepressants
Opioids
Lithium
Elective serotonin reuptake inhibitors
Drugs

**Metabolic disorders**
Hepatic failure
Hyponatremia
Hypoglycemia
Nonketotic hyperglycemia
Renal failure

**Dementias**
Creutzfeld-Jakob disease
Gerstmann-Sträussler-Scheinker syndrome
Alzheimer's disease
Rett syndrome
Angelman syndrome

**Physical encephalopathies**
Trauma
Heat stroke
Electric shock
Decompression injury

**Others**
Mitochondrial disorders
Dystonia
Focal CNS damage
Restless legs syndrome
Celiac disease
Hypoxia

**TABLE II** Classification and Differential Diagnosis of Myoclonus Syndromes

| | Epileptic myoclonus | | | Nonepileptic myoclonus |
| --- | --- | --- | --- | --- |
| | Cortical reflex myoclonus | Reticular reflex myoclonus | Primary generalized myoclonus | Heterogenous group of disorders including hiccup, sneeze, hypnic jerks, dystonia |
| Clinical appearance | Few adjacent muscles, multifocal, spontaneous or triggered by sensory stimuli or action | Generalized, proximal more than distal muscles, spontaneous or facilitated by sensory stimuli or action | Small, focal jerks, frequently only fingers, *or* generalized, whole-body movements, spontaneous | |
| Electrophysiology | Burst length 10–50 ms synchronous antagonist activity, EEG correlate (back-averaging, giant SEP) | Burst length 10–50 ms synchronous antagonist activity, EEG correlate (back-averaging) possible | Burst length 10–50 ms synchronous antagonist activity, EEG correlate (back-averaging) | Burst length 50–300 ms synchronous or asynchronous antagonist activity no EEG correlate |

Three types of epileptic myoclonus are now recognized, cortical reflex myoclonus, reticular reflex myoclonus, and primary generalized epileptic myoclonus (Hallett, 1997; Hallett *et al.*, 1987; Toro and Hallett, 1997).

Cortical reflex myoclonus is a fragment of focal or partial epilepsy. Each myoclonic jerk involves only a few adjacent muscles, but larger jerks with more muscles involved can be seen. The disorder is commonly multifocal and accentuated by action and sensory stimulation. Reticular reflex myoclonus is a fragment of a type of generalized epilepsy. These jerks are usually generalized with predominance, which is proximal more than distal and flexor more than extensor. Voluntary action and sensory stimulation increase the jerking. Primary generalized epileptic myoclonus is a fragment of primary generalized epilepsy. The most common clinical manifestation is small, focal jerks, often involving only the fingers; thus the myoclonus is sometimes called minipolymyoclonus. The term "minipolymyoclonus" was originally coined to refer to small jerks seen in patients with motor neuron disease. Minipolymyoclonus of central origin and minipolymyoclonus of peripheral origin have a similar clinical appearance and are probably most easily separated by the company they keep: epilepsy and muscle denervation, respectively. A second clinical presentation of primary generalized epileptic myoclonus is generalized, synchronized whole-body jerks, not unlike those seen with reticular reflex myoclonus. Electrodiagnosis can help define the type of epileptic myoclonus.

Negative myoclonus can be isolated, but it usually occurs together with positive myoclonus (Guerrini *et al.*, 1993; Tassinari *et al.*, 1995, 1998; Toro and Hallett, 1997). Clinically, the appearance is called asterixis, and there can be large movements, small multifocal movements, and such frequent movements that the

appearance is of an irregular tremor or tremulousness. To evoke negative myoclonus, the patient is asked to generate steady muscle force against gravity, for example during extension of the wrist and elevating the arms. In this situation, negative myoclonus causes sudden lapses in continuous muscle activity. Negative myoclonus can have a similar physiology to cortical myoclonus with electrophysiological correlates and production by sensory stimulation. There may be associated seizures.

## Nonepileptic Myoclonus

The types of nonepileptic myoclonus are particularly heterogeneous. It is important to recognize them, because each has its own treatment, and, in general, use of anticonvulsants is not valuable. Some physiological phenomena are included in this group, including hiccough, sneeze, and the hypnic jerk.

### Electrophysiological Separation of Epileptic and Nonepileptic Myoclonus

Electrophysiological assessment can be a significant help in deciding whether the movement is myoclonus, and, if so, which type (Hallett *et al.*, 1987, 1997; Shibasaki, 1988, 2000). Techniques to be used include electromyography to evaluate the EMG activity associated with the movement, EEG to determine whether there is an EEG event related to the movement (averaging of activity may be needed), analysis of whether the involuntary movement can be produced reflexly, and evaluation of the evoked response in the EEG to stimulation that produces reflex involuntary movements. The physiological characteristics of epileptic myoclonus are EMG burst length of 10–50 ms, synchronous antagonist activity, and an EEG correlate. Nonepileptic myoclonus shows EMG burst lengths of 50–300 ms, synchronous or asynchronous antagonist activity, and no EEG correlate.

With cortical reflex myoclonus, the EEG reveals a focal positive-negative event over the sensorimotor cortex contralateral to the jerk preceding both spontaneous and reflex-induced myoclonic jerks. With stimulus sensitivity, C-reflexes are seen and are correlated with giant somatosensory-evoked potentials. The EEG event associated with reflex jerks is a giant P1–N2 component of the somatosensory-evoked potential. Often the P1–N2 has exactly the same topography as the positive-negative event preceding the spontaneous myoclonus, but at times there are some differences. A final feature is that if the cranial nerve muscles are active, then the timing of onset of activation is from above downward; that is, the masseter (fifth cranial nerve) is active before the orbicularis oculi (seventh cranial nerve), which is itself active before the sternocleidomastoid (eleventh cranial nerve). With reticular reflex myoclonus, there are brief generalized EMG bursts lasting 10–30 ms triggered by sensory stimulation such as touch or muscle stretch or by action; the EEG correlates, if present, are not time-locked to the muscle activation, and the pattern of EMG activation in cranial nerve muscles is with the sternocleidomastoid muscle activated first and the other cranial nerve muscles activated in reverse numerical order. In primary generalized epileptic myoclonus, the EEG correlate is a slow, bilateral frontocentrally predominant negativity similar to the wave of a primary generalized paroxysm.

### Physiological Myoclonus

Hiccough and hypnic jerks are common forms of physiological myoclonus and are experienced by most normal individuals. Usually physiological myoclonus is transient, imposes little inconvenience, and requires no pharmacological treatment. If it occurs frequently, symptomatic forms should be excluded (especially in persistent hiccough). Hypnic jerks or sleep starts are brief, sudden jerks usually of one or both legs that occur while falling asleep. They are considered part of an arousal response mediated by the reticular activating system and often are associated with an arousal reaction in the EEG. They must be distinguished from other hypnagogic and hypnopompic phenomena such as periodic movements of sleep, nocturnal myoclonus, and early morning myoclonus of juvenile myoclonic epilepsy. Occasionally, hypnic jerks are frequent enough for individuals to seek medical help. In such cases it is often sufficient to explain the benign nature of the condition. Other forms of physiological myoclonus during sleep in adults and neonates are listed in Table II. Repetitive jerks of the extremities in healthy newborns during slow-wave sleep without concomitant EEG changes are referred to as benign neonatal sleep myoclonus. These jerks disappear spontaneously within 3–4 months and are often misdiagnosed as epileptic seizures (Di Capua et al., 1993).

## PRACTICAL MANAGEMENT OF EPILEPTIC MYOCLONUS

Unless a definite and treatable etiology has been identified, treatment of myoclonus is symptomatic. Studies investigating therapeutic approaches to myoclonus usually are limited in that only small numbers of patients are recruited, and the selection of patients frequently is rather heterogeneous. Thus, therapeutic recommendations are based on empirical observations rather than double-blinded, randomized controlled trials. Generally, the first approach is anticonvulsants. The most useful agents are clonazepam and valproate, both of which work by promoting GABA action in brain. Primidone may also play a role. Particularly in childhood, ethosuximide and lamotrigine should be considered (Wallace, 1998). Serotonergic agents such as 5-hydroxytryptophan (5-HTP) may also be effective. Piracetam has been demonstrated to be useful for epileptic myoclonus, particularly of the cortical type (Brown et al., 1993; Ikeda et al., 1996; Koskiniemi et al., 1998; Obeso et al., 1988; Van Vleymen and Van Zandijcke, 1996). Obeso and colleagues (Obeso et al., 1989) point out that two or three drugs in combination may well be better than a single drug. One report has suggested that acetazolamide might be a useful adjunct (Vaamonde et al., 1992).

Epileptic negative myoclonus is often treated just like positive myoclonus. In some refractory circumstances, however, ethosuximide can be particularly effective (Oguni et al., 1998). In the setting of childhood partial epilepsy, for example, negative myoclonus can be treated with ethosuximide (Capovilla et al., 1999). On the other hand, one must be aware that toxic levels of anticonvulsants (e.g., carbamazepine) may cause negative myoclonus.

## SPECIFIC ENTITIES

### The Epileptic Myoclonias

#### Degenerative Diseases

Progressive myoclonic epilepsy with the principal features of slowly progressive myoclonus and epilepsy is most frequently secondary to a degenerative disorder with some involvement of the cerebellum or cerebellar pathways. The disorder can be sporadic but is often familial. Typically the symptoms begin between 7 and 15 years of age and include action and reflex myoclonus, with particular photic sensitivity and grand mal seizures. Cerebellar ataxia is said to be common, but differentiation of ataxia and intention myoclonus is often very difficult. Occasional features include dementia, spasticity, myopathy, neuropathy, and deafness. The syndrome is clearly heterogeneous. Cases have been described under the names Unverricht-Lundborg syndrome, Ramsay Hunt syndrome, and "Baltic myoclonus epilepsy."

The most common form of progressive myoclonic epilepsy worldwide (EPM1) is now known to be an autosomal-recessive condition with mutation in the gene encoding cystatin B (chromosome 21q22.3), a cysteine protease inhibitor (Lehesjoki and Koskiniemi, 1998; Pennacchio et al., 1996). The most common mutation is an unstable expansion of a dodecamer minisatellite repeat unit in the promoter region of the cystatin B gene. Cystatin B mutations are now known to account for both "Mediterranean myoclonus" and for "Baltic" myoclonus, described mainly from Finland, thus solving a long-term controversy and proving that these two disorders represent one single disease entity. Myoclonus can be a feature of the autosomal-dominant cerebellar degenerations, such as SCA2 (Cancel et al., 1997). It also is a feature of dentatorubral-pallidoluysian atrophy (DRPLA), which manifests various combinations of chorea, myoclonus, seizures, ataxia, and dementia (Becher et al., 1997).

There are no current treatments for degenerative conditions, but there are some important therapeutic implications. It is of critical importance in these cases to recognize that phenytoin treatment may be associated with worsening of the condition (Eldridge et al., 1983). Patients can be treated successfully with other anticonvulsants as described later. The antioxidant N-acetylcysteine has been claimed to have considerable beneficial effect on the myoclonus in Unverricht-Lundborg disease*, but this is only an open-label observation (Ben-Menachem et al., 2000).

### Storage Diseases

Storage diseases may give rise to the syndrome of progressive myoclonic epilepsy, and for this reason the patients may be thought to have a degenerative condition. The most well-known entity is Lafora body disease. Other entities include lipidoses such as GM2 gangliosidosis (Tay-Sachs disease), ceroid-lipofuscinosis (Batten's disease), and sialidosis (cherry-red-spot-myoclonus syndrome).

There are at least two clinical differences between Lafora body disease and Baltic myoclonus epilepsy (Eldridge et al., 1983). Age of onset in Lafora body disease is 11–18 years, whereas the age of onset in Baltic myoclonus epilepsy is 6–16 years. For Lafora body disease dementia is present in two thirds of cases by 2 years after onset, and in all by 5 years; in contrast in Baltic myoclonic epilepsy, only a rare patient shows dementia in the first 5 years (provided they do not receive phenytoin). Occipital seizures are frequent in Lafora body disease. The diagnostic feature of Lafora body disease is the periodicacid–Schiff–positive inclusion body found in neurons throughout the gray matter of the brain, including the dentate nucleus of the cerebellum. These inclusions can sometimes be found in liver, skeletal muscle, and skin. For clinical purposes, the method of first choice for confirming the diagnosis is skin biopsy, particularly of axillary skin. The gene defect for Lafora body disease has been identified as a protein tyrosine phosphatase (PTPase; chromosome 6q23–25), and it can be referred to as EPM2 (Minassian et al., 1998, 1999; Serratosa et al., 1999).

With these storage diseases, there is no treatment, but screening and prenatal detection can be used for prevention.

Noninfantile neuronopathic Gaucher's disease (type 3; chromsome 1q21) can cause myoclonus. There are rapid advances in Gaucher's disease, with enzyme replacement therapy (macrophage-targeted glucocerebrosidase) already available for the nonneuronopathic forms and gene replacement treatment on the horizon.

Biotin deficiency is especially important to keep in mind, because replacement with biotin can effect a cure (Bressman et al., 1983). Diagnosis of biotin deficiency is established using a quantitative assay to determine biotinidase activitiy.

Mitochondrial disorders are increasingly recognized as common causes of myoclonus and the "Ramsay Hunt syndrome" in particular. Indeed, in some series most patients with progressive myoclonic epilepsy may have a mitochrondrial abnormality (Berkovic et al., 1986). One well-defined syndrome is myoclonus epilepsy and ragged red fiber syndrome (MERRF). A muscle biopsy looking for ragged red fibers might be helpful. Ekbom's syndrome of photomyoclonus, cerebellar ataxia, and cervical lipoma is associated with the tRNA(Lys) A8344G mutation in mitochondrial DNA (Traff et al., 1995). There is no good treatment for mitochondrial diseases at present.

### Dementias

Creutzfeldt-Jacob disease (subacute spongioform encephalopathy) frequently exhibits myoclonus as a relatively early feature. The myoclonus can be produced by external stimuli, such as noise, or can be spontaneous and rhythmic, associated with a periodic EEG. Patients with Alzheimer's disease can also exhibit myoclonus, although this feature typically occurs late in the illness. The myoclonus is multifocal and can be stimulus induced. Electrophysiological investigations in these two disorders show distinctive results, but both are similar to cortical reflex myoclonus.

Myoclonus is also a feature of some childhood retardation syndromes such as Rett syndrome (Guerrini *et al.*, 1998) and Angelman syndrome (Guerrini *et al.*, 1996). In both situations, the myoclonus is cortical in origin.

### Viral Encephalopathies

A variety of viruses and postviral syndromes cause myoclonus. Herpes simplex encephalitis is probably the most common example. Subacute sclerosing panencephalitis (SSPE) is frequently characterized by a slow periodic movement often called myoclonus, but its duration is on the order of 1 second, and it does not fit the definition formally. AIDS encephalopathy may be seen with a myoclonic dementia (Maher *et al.*, 1997).

### Metabolic Encephalopathies Including Endocrine Disorders

Disorders such as hepatic failure, renal failure, hyponatremia, hypoglycemia, and nonketotic hyperglycemia can give rise to myoclonus. Treatment should be directed to the underlying condition. For example, myoclonus in patients with chronic renal failure on hemodialysis can be due to aluminum toxicity and can be successfully treated with chelation therapy with deferoxamine mesylate (Sprague *et al.*, 1986).

Bhatia *et al.* have reported four patients with a progressive myoclonic ataxic syndrome and associated celiac disease (Bhatia *et al.*, 1995). The onset of the neurological syndrome followed the gastrointestinal and other manifestations of celiac disease while on a gluten-free diet, in the absence of overt features of malabsorption or nutritional deficiency. The neurological syndrome was dominated by action- and stimulus-sensitive myoclonus of cortical origin with mild ataxia and infrequent seizures. The condition progressed despite strict adherence to diet. No treatment is known.

The opsoclonus-myoclonus syndrome is easy to diagnose because of the dramatic clinical feature of opsoclonus (Pranzatelli, 1992). It may arise in a variety of settings, including infections, toxins, and a paraneoplastic syndrome. In childhood the syndrome is often associated with neuroblastoma. The syndrome is associated with a distinctive pattern of serum IgM and IgG binding to neural tissues and antigens (Connolly

*et al.*, 1997). If there is a tumor, this should be treated, and there may be symptomatic relief, but the disorder might be unaffected or even worsened. Symptomatic therapy that should be considered includes steroids or adrenocortilotropic hormone*. Trazadone may also be helpful.

### Toxic Encephalopathies Including Drug Side Effects

Toxic causes include bismuth, heavy metals, methyl bromide, and drugs. Offending drugs include tricyclic antidepressants, opioids, and lithium. A good deal of attention has been paid to the serotonin syndrome, which may be a consequence of use of the selective serotonin reuptake inhibitors (SSRIs). The serotonin syndrome is characterized by alterations in cognition (disorientation, confusion), behavior (agitation, restlessness), autonomic nervous system function (fever, shivering, diaphoresis, diarrhea), and neuromuscular disorders including ataxia, hyperreflexia, and myoclonus (Lane and Baldwin, 1997). Patients with the serotonin syndrome usually respond to discontinuation of drug therapy and supportive care alone. Only in rare cases treatment with an antiserotonergic agent such as cyproheptadine, methysergide, or propranolol may be required.

### Basal Ganglia Disorders

Parkinson's disease rarely manifests myoclonus. In two well-studied patients, the disorder was found to be cortical myoclonus (Caviness *et al.*, 1998). Myoclonus is a well-described feature of dopa-dyskinesias (Luquin *et al.*, 1992), and this would be its most frequent cause in patients with Parkinson's disease. Myoclonus is more common in multiple system atrophy. It may also be a feature of Huntington's disease, and in one case it was shown to be cortical myoclonus (Caviness and Kurth, 1997). Myoclonus is a prominent, characteristic feature of corticobasal degeneration (Kompoliti *et al.*, 1998). The myoclonus is stimulus sensitive with a short latency that may be helpful in differential diagnosis. Its pathophysiology is still debated as to whether it is cortical or subcortical in origin (Carella *et al.*, 1997; Strafella *et al.*, 1997).

### Physical Encephalopathies Including Hypoxia

These conditions include trauma, heat stroke, electric shock, decompression injury, and hypoxia. Posthypoxic myoclonus, the Lance-Adams syndrome (Lance and Adams, 1963), has received a great deal of attention since the demonstration that the myoclonus could be successfully treated with 5-hydroxytryptophan. The clinical syndrome as reported by Lance and Adams noted the precipitating feature of action and the association with cerebellar ataxia, postural lapses, gait disturbance, and grand mal seizures. Werhahn *et al.* reported the clinical and neurophysiological features of 14 patients with chronic posthypoxic myoclonus (Werhahn *et al.*, 1997). All patients had had a cardiorespiratory arrest, most caused by an acute asthmatic attack. All patients had multifocal action myoclonus, and 11 had additional stimulus-sensitive myoclonus. There was late improvement in the myoclonic syndrome and the level of disability in all but one patient. Cognitive deficits were found in seven patients and were usually mild. Electrophysiological investigation confirmed cortical action myoclonus in every case, although this could be combined with cortical reflex myoclonus, an exaggerated startle response, or brainstem reticular reflex myoclonus. The site of the responsible lesion in the brain is not certain, but there does seems to be a disorder of serotonin metabolism supported not only by the therapeutic response to 5-hydroxytryptophan but also by the reduction in CSF levels of 5-HIAA, which improves with successful therapy.

There are some special considerations in the treatment of posthypoxic myoclonus because of the specific deficiency of serotonin. 5-HTP, either alone or with carbidopa, is often beneficial in this condition. On the other hand, however, it is not often used as a drug of first choice because of adverse effects. Most patients will respond just as well to clonazepam, valproate, or clonazepam plus valproate*. Another drug that has been described to be useful in posthypoxic myoclonus is estrogen (Marsden *et al.*, 1982), although the mechanism of action is not clear.

### Focal CNS Damage

The cause of focal cortical myoclonus can be almost any type of focal cortical lesion; tumors, angiomas, and encephalitis should be suspected. Curiously, particularly in patients with epilepsia partialis continua, the cortex can appear normal in pathological examination. In a number of patients surgical excision of the excitable tissue has cured the myoclonus (Obeso *et al.*, 1985), and this approach should may be considered if symptomatic treatment with anticonvulsants fails.

## The Nonepileptic Myoclonias

### Dystonic Myoclonus and Fragments of Other Involuntary Movement Disorders

Patients with dystonia or chorea, for example, may have quick jerks, as well as more prolonged involuntary movements. Other "basal ganglia" disorders such as Wilson's disease, neuroaxonal dystrophy, Hallervorden-Spatz disease, and progressive supranuclear palsy may manifest myoclonus. The diagnosis is made on the basis of other clinical features. Epileptic myoclonus has been recognized in some circumstances in Parkinson's disease, multiple system atrophy, Huntington's disease, and corticobasal degeneration, as noted earlier. Treatments are available for Parkinson's disease, of course. Myoclonus in Huntington's disease can be improved with valproic acid*, perhaps because of the involvement of the GABA system in this disorder (Carella *et al.*, 1993). Focal myoclonus in the setting of dystonia can often be treated with focal injections with botulinum toxin.

### Essential Myoclonus

This term can be used for those patients whose sole neurological abnormality is myoclonus and who specifically do not have seizures, dementia, or ataxia. The EEG and other laboratory investigations should be normal. Familial cases, as well as sporadic cases, are seen. The most common features of the familial cases are autosomal-dominant inheritance with variable severity, equal involvement of males and females, onset in the first or second decade of life, and benign course compatible with normal life span. Essential myoclonus can be generalized or multifocal. The myoclonus is variable in amplitude, and in some cases, the jerks are so small that the disability can be minimal. Jerks can be present at rest and may be improved or worsened by action. Reflex myoclonus has not been described in this group.

In some families with essential myoclonus some involved patients also have essential tremor, and some family members had essential tremor without myoclonus. Some of these patients also may exhibit elements of dystonia. The essential tremor, myoclonus, and dystonia may all be sensitive to alcohol in these patients (Quinn, 1996).

Essential myoclonus is a heterogeneous disorder, and therapy is largely empirical. Alcohol is most likely to be beneficial but cannot be recommended for regular use. There is one report of the usefulness of propranolol*. There is some benefit with 5-hydroxytryptophan and clonazepam. Some patients have responded dramatically to benztropine* (Duvoisin, 1984).

### Exaggerated Startle

Startle is a normal phenomenon that can be said to be exaggerated if an excessive response occurs to a startling stimulus or if a startle response occurs to a stimulus that ordinarily would not be startling (Matsumoto and Hallett, 1994). There are a number of startle syndromes, including hereditary hyperekplexia, symptomatic hyperekplexia, startle epilepsy, and the Latah syndrome. Startle epilepsy is principally characterized by epileptic seizures triggered by sudden, unexpected stimuli and initiated by a startle.

Hereditary hyperekplexia is an autosomal-dominant disorder caused by a mutation in the alpha-1 subunit of the glycine receptor (Floeter and Hallett, 1993). This may lead to altered ligand binding or disturbance of the chloride ion–channel part of the receptor. Glycine is an inhibitory neurotransmitter in several spinal interneurons, including Renshaw cells, Ia inhibitory interneurons, and some Ib inhibitory interneurons. Physiological studies suggest normal recurrent inhibition but abnormal Ia reciprocal inhibition in hyperekplexia (Floeter et al., 1998). Affected babies are hypertonic and stiff when handled and have difficulty subsequently with walking. They develop excessive startles, which take two forms; the minor form is a simple brief startle jerk; the major form is a more prolonged tonic startle spasm. In these situations, the normal startle reflex is exaggerated, both in amplitude and in its failure to rapidly habituate. The afferent and efferent systems of the startle reflex in hyperekplexia are identical to those of the normal startle response, involving a similar or the same generator in the lower brainstem, probably in the medial bulbopontine reticular formation (Brown et al., 1991a; Matsumoto and Hallett, 1994; Matsumoto et al., 1992).

Clonazepam is the therapy of choice* (Matsumoto and Hallett, 1994*). Other benzodiazepines, such as diazepam and chlordiazepoxide, have been used with similar results.

### Noctural Myoclonus

There are a variety of types of myoclonus that occur during drowsiness or sleep. There are two physiological forms, the hypnic jerk and "physiological fragmentary myoclonus" characterized by small, multifocal jerks maximal in hands and face, but present diffusely. Pathological types of myoclonus include isolated periodic movements in sleep, restless legs syndrome with periodic movements in sleep, and excessive fragmentary myoclonus in NREM sleep. Myoclonus associated with epilepsy, intention myoclonus associated with semivolitional movements, and segmental myoclonus also occur in sleep but are not primarily nocturnal.

Periodic movements of sleep (PMS), or periodic limb movement disorder (PLMD), occurs in virtually all groups of patients referred to a sleep disorders laboratory, and the clinical correlation is not always clear. Patients with the restless legs syndrome often have PMS (Trenkwalder et al., 1996; Walters, 1995). Certainly, PMS can be asymptomatic for the patient, although, as with all types of nocturnal myoclonus, the disorder may cause distress to the patient's spouse. On some occasions, however, PMS can induce sleep fragmentation and excessive daytime sleepiness. The cause of PMS seems to be increased excitability of the flexor reflex mechanism of the spinal cord (Bara-Jimenez et al., 2000). The treatment of choice is dopaminergic therapy, but opiate therapy may also be used (Becker et al., 1993*; Prinz, 1995). Trials of dopamine agonists are particularly successful** (Earley et al., 1998; Pieta et al., 1998). For some patients clonazepam may be helpful*.

### Segmental Myoclonus Including Spinal Myoclonus and Palatal Myoclonus

In this disorder a segment of the spinal cord or brainstem produces spontaneous, persistent rhythmic repetitive discharges usually unaffected by sleep. A number of contiguous muscles produce synchronous contractions at a rate of 0.5–3 Hz and seems to be due to heightened spinal excitability (Di Lazzaro et al., 1996). Involved regions can be one limb, one limb and adjacent trunk, or both legs. Lesions of the spinal cord giving rise to focal movements include infection, degenerative disease, tumor, cervical myelopathy, and demyelinating disease, and it may follow spinal anesthesia or the introduction of contrast media into the CSF. Unlike palatal myoclonus, spinal myoclonus would only rarely be idiopathic. For spinal myoclonus, treatment should be directed to the underlying cause if possible. For example, surgery can ameliorate myoclonus caused by cervical cord compression. Antiviral therapy may improve segmental myoclonus associated with herpes zoster. On the other hand, sometimes drug treatment can be helpful such as valproate, L-5 HTP, clonazepam, or trihexyphenidyl*.

Propriospinal myoclonus is a special type of spinal myoclonus (Brown et al., 1991b; Chokroverty et al., 1992). It is clinically characterized by axial jerks that are nonrhythmic and that lead to symmetrical flexion of neck, trunk, hips, and knees. Jerks can be spontaneous or stimulus induced. The myoclonus can be induced by drowsiness and relaxation (Montagna et al., 1997). The myoclonus is identified with electrodiagnostic studies that show the myoclonus starting in midthoracic region and propagating slowly, about 5 m/sec, both rostrally and caudally. It has been seen in the setting of tetraplegia (Fouillet et al., 1995), a spinal inflammatory condition (Schulze-Bonhage et al., 1996), and Lyme disease (De la Sayette et al., 1996).

Palatal myoclonus, now preferably called palatal tremor, is most common in this group and has now been shown to consist of two separate disorders, essential palatal tremor, which manifests an ear click, and symptomatic palatal tremor, which is associated with cerebellar disturbances (Deuschl *et al.*, 1994). In essential palatal tremor, the movement is of the tensor veli palatini muscle, no cause is evident, and there are no other neurological manifestations. The movements stop during sleep. In symptomatic palatal tremor, the movement is of the levator veli palatini muscle, there is typically a brainstrem lesion, and the myoclonus can appear many months after the acute episode. Patients may have associated movements of other cranial nerve muscles, particularly the orbicular oris, and even have synchronous eye movements (oculopalatal myoclonus). The disease is typically in the dentato-olivary pathway leading to hypertrophy of the inferior olivary nucleus.

The ear click in essential palatal tremor is the symptom that requires therapy. For the ear click of essential palatal myoclonus, a number of drugs may be useful in individual cases such as clonazepam, tryptophan, carbamazepine, trihexyphenidyl, or ceruletide, but most seem refractory. Particular success recently has been reported with sumatriptan* (Jankovic *et al.*, 1997; Scott *et al.*, 1996). The ear click can be successfully treated with focal injection of botulinum toxin* (Deuschl *et al.*, 1991).

### Asterixis

Negative myoclonus may well have a subcortical origin, as well as a cortical origin (epileptic), although there are no clinical rules for separating them. The distinction is made on the basis of electrophysiological studies that may show EEG correlates with the cortical form, but this may not be a definite qualitative difference. Certainly, subcortical lesions may cause asterixis, but that does not necessarily imply where the movement is generated. It is difficult to treat asterixis whether cortical or subcortical, and a search for a metabolic or toxic cause should be the first course of action.

### Psychogenic Myoclonus

Myoclonus can also be psychogenic. Monday and Jankovic (Monday and Jankovic, 1993) reported the clinical features of 18 such patients. There were 13 women and 5 men with an age range of 22–75 years. The myoclonus was present for 1–110 months, and it was segmental in 10, generalized in 7, and focal in 1. Stress precipitated or exacerbated the myoclonic movements in 15 patients; 14 had a definite increase in myoclonic activity during periods of anxiety. The following findings helped to establish the psychogenic nature of the myoclonus: clinical features incongruous with "organic" myoclonus, including characteristic electrophysiological findings, evidence of underlying psychopathology, an improvement with distraction or placebo, and the presence of incongruous sensory loss

or false weakness. Over half of all patients with adequate follow-up improved after gaining insight into the psychogenic mechanisms of their movement disorder.

## REFERENCES

Bara-Jimenez, W., Aksu, M., Graham, B., Sato, S., and Hallett, M. (2000). Periodic limb movements in sleep: state-dependent excitability of the spinal flexor reflex. *Neurology* 54, 1609–1616.

Becher, M. W., Rubinsztein, D. C., Leggo, J., Wagster, M. V., Stine, O. C., Ranen, N. G. et al. (1997). Dentatorubral and pallidoluysian atrophy (DRPLA). Clinical and neuropathological findings in genetically confirmed North American and European pedigrees. *Mov. Disord.* 12, 519–530.

Becker, P. M., Jamieson, A. O., and Brown, W. D. (1993). Dopaminergic agents in restless legs syndrome and periodic limb movements of sleep: response and complications of extended treatment in 49 cases. *Sleep* 16, 713–716.

Ben-Menachem, E., Kyllerman, M., and Marklund, S. (2000). Superoxide dismutase and glutathione peroxidase function in progressive myoclonus epilepsies. *Epilepsy Res.* 40, 33–39.

Berkovic, S. F., Andermann, F., Carpenter, S., and Wolfe, L. D. (1986). Progressive myoclonus epilepsies: specific causes and diagnosis. *N. Engl. J. Med.* 315, 296–305.

Bhatia, K. P., Brown, P., Gregory, R., Lennox, G. G., Manji, H., Thompson, P. D. et al. (1995). Progressive myoclonic ataxia associated with coeliac disease. The myoclonus is of cortical origin, but the pathology is in the cerebellum. *Brain* 118, 1087–1093.

Bressman, S., Fahn, S., Eisenberg, M., and Maltese, W. (1983). Adult onset encephalopathy with myoclonus, ataxia, deafness, hemianopsia and hemiparesis responsive to biotin. *Ann. Neurol.* 14, 109–110.

Brown, P., Rothwell, J. C., Thompson, P. D., Britton, T. C., Day, B. L., and Marsden, C. D. (1991a). The hyperekplexias and their relationship to the normal startle reflex. *Brain* 114, 1903–1928.

Brown, P., Steiger, M. J., Thompson, P. D., Rothwell, J. C., Day, B. L., Salama, M. et al. (1993). Effectiveness of piracetam in cortical myoclonus. *Mov. Disord.* 8, 63–68.

Brown, P., Thompson, P. D., Rothwell, J. C., Day, B. L., and Marsden, C. D. (1991b). Axial myoclonus of propriospinal origin. *Brain* 114, 197–214.

Cancel, G., Durr, A., Didierjean, O., Imbert, G., Burk, K., Lezin, A. et al. (1997). Molecular and clinical correlations in spinocerebellar ataxia 2: a study of 32 families. *Hum. Mol. Genet.* 6, 709–715.

Capovilla, G., Beccaria, F., Veggiotti, P., Rubboli, G., Meletti, S., and Tassinari, C. A. (1999). Ethosuximide is effective in the treatment of epileptic negative myoclonus in childhood partial epilepsy. *J. Child. Neurol.* 14, 395–400.

Carella, F., Ciano, C., Panzica, F., and Scaioli, V. (1997). Myoclonus in corticobasal degeneration. *Mov. Disord.* 12, 598–603.

Carella, F., Scaioli, V., Ciano, C., Binelli, S., Oliva, D., and Girotti, F. (1993). Adult onset myoclonic Huntington's disease. *Mov. Disord.* 8, 201–205.

Caviness, J. N., Adler, C. H., Newman, S., Caselli, R. J., and Muenter, M. D. (1998). Cortical myoclonus in levodopa-responsive parkinsonism. *Mov. Disord.* 13, 540–544.

Caviness, J. N., and Kurth, M. (1997). Cortical myoclonus in Huntington's disease associated with an enlarged somatosensory evoked potential. *Mov. Disord.* 12, 1046–1051.

Chokroverty, S., Walters, A., Zimmerman, T., and Picone, M. (1992). Propriospinal myoclonus: a neurophysiologic analysis. *Neurology* 42, 1591–1595.

Connolly, A. M., Pestronk, A., Mehta, S., Pranzatelli, M. R., 3rd, and Noetzel, M. J. (1997). Serum autoantibodies in childhood opsoclonus-myoclonus syndrome: an analysis of antigenic targets in neural tissues. *J. Pediatr.* 130, 878–884.

De la Sayette, V., Schaeffer, S., Queruel, C., Bertran, F., Defer, G., Hazera, P. *et al.* (1996). Lyme neuroborreliosis presenting with propriospinal myoclonus. *J. Neurol. Neurosurg. Psychiatry* 61, 420.

Deuschl, G., Löhle, E., Heinen, F., and Lücking, C. (1991). Ear click in palatal tremor: its origin and treatment with botulinum toxin. *Neurology* 41, 1677–1679.

Deuschl, G., Toro, C., Valls-Solé, J., Zeffiro, T., Zee, D. S., and Hallett, M. (1994). Symptomatic and essential palatal tremor. 1. Clinical, physiological, and MRI analysis. *Brain* 117, 775–788.

Di Capua, M., Fusco, L., Ricci, S., and Vigevano, F. (1993). Benign neonatal sleep myoclonus: clinical features and video-polygraphic recordings. *Mov. Disord.* 8, 191–194.

Di Lazzaro, V., Restuccia, D., Nardone, R., Oliviero, A., Profice, P., Insola, A. *et al.* (1996). Changes in spinal cord excitability in a patient with rhythmic segmental myoclonus. *J. Neurol. Neurosurg. Psychiatry* 61, 641–644.

Duvoisin, R. C. (1984). Essential myoclonus: response to anticholinergic therapy. *Clin. Neuropharm.* 7, 141–147.

Earley, C. J., Yaffee, J. B., and Allen, R. P. (1998). Randomized, double-blind, placebo-controlled trial of pergolide in restless legs syndrome. *Neurology* 51, 1599–1602.

Eldridge, R., Iivanainen, M., Stern, R., Koerber, T., and Wilder, B. J. (1983). Baltic myoclonus epilepsy: hereditary disorder of childhood made worse by phenytoin. *Lancet* 2, 838–842.

Floeter, M. K., and Hallett, M. (1993). Glycine receptors: a startling connection. *Nature Genet.* 5, 319–320.

Floeter, M. K., Valls-Sole, J., Toro, C., Jacobowitz, D., and Hallett, M. (1998). Physiologic studies of spinal inhibitory circuits in patients with stiff-person syndrome. *Neurology* 51, 85–93.

Fouillet, N., Wiart, L., Arne, P., Alaoui, P., Petit, H., and Barat, M. (1995). Propriospinal myoclonus in tetraplegic patients: clinical, electrophysiological and therapeutic aspects. *Paraplegia* 33, 678–681.

Guerrini, R., Bonanni, P., Parmeggiani, L., Santucci, M., Parmeggiani, A., and Sartucci, F. (1998). Cortical reflex myoclonus in Rett syndrome. *Ann. Neurol.* 43, 472–479.

Guerrini, R., De Lorey, T. M., Bonanni, P., Moncla, A., Dravet, C., Suisse, G. *et al.* (1996). Cortical myoclonus in Angelman syndrome. *Ann. Neurol.* 40, 39–48.

Guerrini, R., Dravet, C., Genton, P., Bureau, M., Roger, J., Rubboli, G. *et al.* (1993). Epileptic negative myoclonus. *Neurology* 43, 1078–1083.

Hallett, M. (1985). Myoclonus: Relation to epilepsy. *Epilepsia* 26 (Suppl 1), S67–S77.

Hallett, M. (1994). Diagnosis and treatment of myoclonus. *In* "Mouvements Anormaux et Douleurs D'Origine Neurologique. Thérapeutique et Neurologie" (G. Defer, Ed.), 3, pp. 55–68. DERN, Paris.

Hallett, M. (1997). Myoclonus and myoclonic syndromes. *In* "Epilepsy: A Comprehensive Textbook" (J. J. Engel and T. A. Pedley, Eds.), 3, pp. 2717–2723. Lippincott-Raven, Philadelphia.

Hallett, M., Marsden, C. D., and Fahn, S. (1987). Myoclonus (Chapter 37). *In* "Handbook of Clinical Neurology" (P. J. Vinken, G. W. Bruyn, and H. L. Klawans, Eds.), 5 (49), pp. 609–625. Elsevier Science Publishers, Amsterdam.

Ikeda, A., Shibasaki, H., Tashiro, K., Mizuno, Y., and Kimura, J. (1996). Clinical trial of piracetam in patients with myoclonus: nationwide multiinstitution study in Japan. The Myoclonus/Piracetam Study Group. *Mov. Disord.* 11, 691–700.

Jankovic, J., Scott, B. L., and Evans, R. W. (1997). Treatment of palatal myoclonus with sumatriptan [letter]. *Mov. Disord.* 12, 818.

Kompoliti, K., Goetz, C. G., Boeve, B. F., Maraganore, D. M., Ahlskog, J. E., Marsden, C. D. *et al.* (1998). Clinical presentation and pharmacological therapy in corticobasal degeneration. *Arch. Neurol.* 55, 957–961.

Koskiniemi, M., Van Vleymen, B., Hakamies, L., Lamusuo, S., and Taalas, J. (1998). Piracetam relieves symptoms in progressive myoclonus epilepsy: a multicentre, randomised, double blind, crossover study comparing the efficacy and safety of three dosages of oral piracetam with placebo. *J. Neurol. Neurosurg. Psychiatry* 64, 344–348.

Lance, J. W., and Adams, R. D. (1963). The syndrome of intention or action myoclonus as a sequel to hypoxic encephalopathy. *Brain* 86, 111–136.

Lane, R., and Baldwin, D. (1997). Selective serotonin reuptake inhibitor-induced serotonin syndrome: review. *J. Clin. Psychopharmacol.* 17, 208–221.

Lehesjoki, A. E., and Koskiniemi, M. (1998). Clinical features and genetics of progressive myoclonus epilepsy of the Univerricht-Lundborg type. *Ann. Med.* 30, 474–480.

Luquin, M. R., Scipioni, O., Vaamonde, J., Gershanik, O., and Obeso, J. A. (1992). Levodopa-induced dyskinesias in Parkinson's disease: clinical and pharmacological classification. *Mov. Disord.* 7, 117–124.

Maher, J., Choudhri, S., Halliday, W., Power, C., and Nath, A. (1997). AIDS dementia complex with generalized myoclonus. *Mov. Disord.* 12, 593–597.

Marsden, C. D., Hallett, M., and Fahn, S. (1982). The nosology and pathophysiology of myoclonus. *In* "Movement Disorders" (C. D. Marsden and S. Fahn, Eds.), Neurology 2, pp. 196–248. Butterworth Scientific, London.

Matsumoto, J., Fuhr, P., Nigro, M., and Hallett, M. (1992). Physiological abnormalities in hereditary hyperekplexia. *Ann. Neurol.* 32, 41–50.

Matsumoto, J., and Hallett, M. (1994). Startle syndromes. *In* "Movement Disorders 3" (C. D. Marsden and S. Fahn, Eds.), pp. 418–433. Butterworth-Heinemann, Oxford.

Minassian, B. A., Lee, J. R., Herbrick, J. A., Huizenga, J., Soder, S., Mungall, A. J. *et al.* (1998). Mutations in a gene encoding a novel protein tyrosine phosphatase cause progressive myoclonus epilepsy. *Nat. Genet.* 20, 171–174.

Minassian, B. A., Sainz, J., Serratosa, J. M., Gee, M., Sakamoto, L. M., Bohlega, S. *et al.* (1999). Genetic locus heterogeneity in Lafora's progressive myoclonus epilepsy. *Ann. Neurol.* 45, 262–265.

Monday, K., and Jankovic, J. (1993). Psychogenic myoclonus. *Neurology* 43, 349–352.

Montagna, P., Provini, F., Plazzi, G., Liguori, R., and Lugaresi, E. (1997). Propriospinal myoclonus upon relaxation and drowsiness: a cause of severe insomnia. *Mov. Disord.* 12, 66–72.

Obeso, J. A., Artieda, J., Quinn, N., Rothwell, J. C., Luquin, M. R., Vaamonde, J. *et al.* (1988). Piracetam in the treatment of different types of myoclonus. *Clin. Neuropharm.* 11, 529–536.

Obeso, J. A., Artieda, J., Rothwell, J. C., Day, B., Thompson, P., and Marsden, C. D. (1989). The treatment of severe action myoclonus. *Brain* 112, 765–777.

Obeso, J. A., Rothwell, J. C., and Marsden, C. D. (1985). The spectrum of cortical myoclonus: from focal reflex jerks to spontaneous motor epilepsy. *Brain* 108, 193–224.

Oguni, H., Uehara, T., Tanaka, T., Sunahara, M., Hara, M., and Osawa, M. (1998). Dramatic effect of ethosuximide on epileptic negative myoclonus: implications for the neurophysiological mechanism. *Neuropediatrics* 29, 29–34.

Pennacchio, L. A., Lehesjoki, A. E., Stone, N. E., Willour, V. L., Virtaneva, K., Miao, J. *et al.* (1996). Mutations in the gene encoding cystatin B in progressive myoclonus epilepsy (EPM1). *Science* 271, 1731–1734.

Pieta, J., Millar, T., Zacharias, J., Fine, A., and Kryger, M. (1998). Effect of pergolide on restless legs and leg movements in sleep in uremic patients. *Sleep* **21**, 617–622.

Pranzatelli, M. R. (1992). The neurobiology of the opsoclonus-myoclonus syndrome. *Clin. Neuropharm.* **15**, 186–228.

Prinz, P. N. (1995). Sleep and sleep disorders in older adults. *J. Clin. Neurophysiol.* **12**, 139–146.

Quinn, N. P. (1996). Essential myoclonus and myoclonic dystonia. *Mov. Disord.* **11**, 119–124.

Schulze-Bonhage, A., Knott, H., and Ferbert, A. (1996). Pure stimulus-sensitive truncal myoclonus of propriospinal origin. *Mov. Disord.* **11**, 87–90.

Scott, B. L., Evans, R. W., and Jankovic, J. (1996). Treatment of palatal myoclonus with sumatriptan. *Mov. Disord.* **11**, 748–751.

Serratosa, J. M., Gomez-Garre, P., Gallardo, M. E., Anta, B., de Bernabe, D. B., Lindhout, D. *et al.* (1999). A novel protein tyrosine phosphatase gene is mutated in progressive myoclonus epilepsy of the Lafora type (EPM2). *Hum. Mol. Genet.* **8**, 345–352.

Shibasaki, H. (1988). AAEE minimonograph #30: electrophysiologic studies of myoclonus. *Muscle Nerve* **11**, 899–907.

Shibasaki, H. (2000). Electrophysiological studies of myoclonus. *Muscle Nerve* **23**, 321–335.

Sprague, S. M., Corwin, H. L., Wilson, R. S., Mayor, G. H., and Tanner, C. M. (1986). Encephalopathy in chronic renal failure responsive to deferoxamine therapy. Another manifestation of aluminum neurotoxicity. *Arch. Intern. Med.* **146**, 2063–2064.

Strafella, A., Ashby, P., and Lang, A. E. (1997). Reflex myoclonus in cortical-basal ganglionic degeneration involves a transcortical pathway. *Mov. Disord.* **12**, 360–369.

Tassinari, C. A., Rubboli, G., Parmeggiani, L., Valzania, F., Plasmati, R., Riguzzi, P. *et al.* (1995). Epileptic negative myoclonus. *In* "Negative Motor Phenomena" (S. Fahn, M. Hallett, H. O. Lüders, and C. D. Marsden, Eds.), 67, pp. 181–197. Lippincott-Raven Publishers, Philadelphia.

Tassinari, C. A., Rubboli, G., and Shibasaki, H. (1998). Neurophysiology of positive and negative myoclonus. *Electroencephalogr. Clin. Neurophysiol.* **107**, 181–195.

Toro, C., and Hallett, M. (1997). Pathophysiology of myoclonic disorders (Chapter 41). *In* "Movement Disorders. Neurologic Principles and Practice" (R. L. Watts and W. C. Koller, Eds.), pp. 551–560. McGraw-Hill, New York.

Toro, C., Hallett, M., Rothwell, J. C., Asselman, P. T., and Marsden, C. D. (1995). Physiology of negative myoclonus. *In* "Negative Motor Phenomena, Advances in Neurology, Volume 67" (S. Fahn, M. Hallett, H. O. Lüders, and C. D. Marsden, Eds.), pp. 211–217. Lippincott-Raven Publishers, Philadelphia.

Traff, J., Holme, E., Ekbom, K., and Nilsson, B. Y. (1995). Ekbom's syndrome of photomyoclonus, cerebellar ataxia and cervical lipoma is associated with the tRNA(Lys) A8344G mutation in mitochondrial DNA. *Acta Neurol. Scand.* **92**, 394–397.

Trenkwalder, C., Walters, A. S., and Hening, W. (1996). Periodic limb movements and restless legs syndrome. *Neurol. Clin.* **14**, 629–650.

Vaamonde, J., Legarda, I., Jimenez-Jimenez, J., and Obeso, J. A. (1992). Acetazolamide improves action myoclonus in Ramsay Hunt syndrome. *Clin. Neuropharm.* **15**, 392–396.

Van Vleymen, B., and Van Zandijcke, M. (1996). Piracetam in the treatment of myoclonus: an overview. *Acta Neurol. Belg.* **96**, 270–280.

Wallace, S. J. (1998). Myoclonus and epilepsy in childhood: a review of treatment with valproate, ethosuximide, lamotrigine and zonisamide. *Epilepsy Res.* **29**, 147–154.

Walters, A. S. (1995). Toward a better definition of the restless legs syndrome. The International Restless Legs Syndrome Study Group. *Mov. Disord.* **10**, 634–642.

Werhahn, K. J., Brown, P., Thompson, P. D., and Marsden, C. D. (1997). The clinical features and prognosis of chronic posthypoxic myoclonus. *Mov. Disord.* **12**, 216–220.

*Movement Disorders*

# CHAPTER 86
# Tremor

Peter Bain, Prithiva Navan, and Tipu Aziz

Tremors can be classified according to the state of activation of the tremulous body part or by the cause of the underlying disease process. This is important not only for correct treatment selection but also for the success of epidemiological and genetic studies. All healthy people have physiological (normal) tremor that can be enhanced in certain circumstances. In clinical practice essential tremor, parkinsonian tremor, the dystonic tremor syndromes, drug-induced tremor, and tremor associated with multiple sclerosis are the most commonly encountered pathological tremors. Tremor severity can be measured directly using a variety of clinical and physiological techniques or indirectly by assessing the impact of tremor on patients' lives by use of disability, handicap, or quality-of-life questionnaires. Treatment of tremor includes a variety of medications, depending on the diagnosis, and stereotactic surgery for disabling medically refractory tremor irrespective of origin.

*Tremor* was defined in a consensus statement from the Movement Disorder Society as a *rhythmic, involuntary oscillatory movement of a body part*. It is notable that amplitude is not a critical part of this definition, and small amplitude tremors might only be detected by sensitive electronic recording devices. In some patients with psychogenic tremor, tremor may be produced voluntarily, although this may not be a conscious act (Deuschl *et al.*, 1998).

## TREMOR ASSESSMENT AND MEASUREMENT

The severity of tremor can be measured with a variety of physiological and clinical techniques, including accelerometry, electromyography, computerized tracking tasks, graphic digitizing tablets, clinical rating scales, or spirography, as well as indirectly by functional per-

formance tests, for example the nine-hole pegboard test, and disability, handicap, and quality-of-life questionnaires (Bain, 1998). The frequency of a tremor can also be measured, but with the exception of primary orthostatic tremor, is not diagnostic, because the frequencies of many different tremor types overlap (Bain, 1993; Deuschl *et al.*, 1998).

## TREMOR CLASSIFICATION

Tremors are classified by the state of activity of the affected part of the body and/or the cause of the underlying disease (Bain 1993; Deuschl *et al.*, 1998), which is important, because assessment of prognosis and treatment depend on tremor type. Furthermore, accurate classification is vital to the success of therapeutic trials, as well as genetic and epidemiological studies.

### Phenomenological Classification of Tremor by State of Activation

*Rest tremor* is defined as tremor that occurs in a body part that is not voluntarily activated and is fully supported against gravity (ideally the patient should be relaxed and resting on a couch). Typically, the amplitude of rest tremor can be enhanced by subjecting the patient to mental exercise, for example counting down, or asking the patient to perform simple repetitive movements with another part of the body, for instance opening and closing the fist.

*Action tremor* is any tremor that is produced by voluntary muscle contraction and includes postural, kinetic, intention, isometric, and task specific tremors.

*Postural tremor* is present while voluntarily maintaining a position against the acceleration of gravity. A *position-specific postural tremor* is a variant of postural tremor in which tremor develops or becomes exacerbated during specific, usually visually guided, postures.

*Kinetic tremor* is tremor that occurs during any voluntary movement, regardless of whether the movement is goal directed (e.g., the finger-nose-finger test) or not (e.g., opening and closing the fist). The latter is termed *simple kinetic tremor*.

*Tremor during target directed movements—classic intention tremor* is present when tremor amplitude increases during a visually guided movement, and the possibility of position-specific tremor or a postural tremor released at the beginning or end of movement has been excluded. Typically tremor amplitude fluctuates significantly, sometimes even from beat to beat, as the target is approached. Intention tremor can be confused clinically with action myoclonus but can be segregated from it by physiological techniques and may be difficult to distinguish from other elements of ataxia, such as serial dysmetria (Alusi *et al.*, 1999).

*Task-specific kinetic tremor* may appear or become exacerbated during specific activities (e.g., primary writing tremor and occupational tremors).

*Isometric tremor* occurs as a result of muscle contraction against a rigid stationary object (e.g., pulling against a rigid bar).

*The cogwheel phenomenon* is a rhythmic brief increase in resistance during passive movement about a joint (Findley *et al.*, 1981).

*Froment's (muscle tone) sign* is present when there is an increased resistance to passive movements of a limb about a joint that can be detected, specifically when there is a voluntary action of another body part. This definition differs from Froment's original description (Froment and Gardere, 1926) but has become the most common and practical way of eliciting this sign.

These consensus definitions provide a useful framework for discussing the phenomenology of tremor (Deuschl *et al.*, 1998) but are only proposals and require constant refinement. In particular, the definition of *intention tremor* is unsatisfactory and needs further clarification. Similarly, the definition of a *position-specific postural tremor* could be ammended to a "tremor that develops or becomes exacerbated in a specific position" (rather than tasks). Although such semantic preoccupations seem to be rather abtruse, the difficulties involved in providing adequate terminology reflect the conceptual difficulties inherent to the study and analysis of tremor.

## Tremor Classification by Etiology

### Physiological and Enhanced Physiological Tremor

Healthy people have an asymptomatic (physiological) tremor that involves every part of the body, typically having a frequency of 8–12 Hz. Physiological tremor has multiple components that are reflected in the fast Fourier transform in the frequency domain but is principally a mechanical tremor that occurs even at rest as energy from perturbations (e.g., the cardiac impulse) are transmitted to the rest of the body. The amplitude of physiological tremor can be modulated by temperature and supraspinal influences, such as vision, by drugs that interact with β-receptors, and situations that increase sympathetic drive, such as anxiety, exercise, or hypoglycaemia, which may make it symptomatic (Arblaster *et al.*, 1990; Isokawa-Akesson and Komisaruk, 1985; Marsden 1984; Young *et al.*, 1975). β-Receptor agonists such as adrenaline and salbutamol enhance physiological tremor, whereas nonselective β-blockers (e.g., propranolol) and $\beta_2$-antagonists are more effective at preventing this than $\beta_1$-antagonists (e.g., atenolol), suggesting that these actions are mediated by means of peripheral $\beta_2$-receptors.

### Essential Tremor

At present, there is no laboratory test for definitively confirming a clinical diagnosis of essential tremor. Thus the presence of essential tremor is based on phenomenological criteria that were proposed by an ad hoc committee of the International Tremor Foundation *Tremor Investigation Group* (TRIG) (Deuschl *et al.*, 1995; Findley *et al.*, 1993). These diagnostic criteria were based on phenomenological evidence, because current knowledge does not permit definitions based on disability, genetic, or pathophysiological functions. The criteria developed by TRIG subclassify essential tremor into *definite*, *probable*, and *possible* essential tremor according to various inclusion and exclusion factors.

The inclusion criteria for *definite essential tremor* are the presence of bilateral visible and persistent postural tremor involving the hands or forearms, which may or may not be accompanied by kinetic tremor. The postural upper limb tremor can be asymmetrical, may fluctuate in amplitude, and tremor may affect other parts of the body. The tremor may or may not produce disability but must have been present for at least 5 years.

The exclusion criteria are: (1) the presence of other abnormal neurological signs with the exception of "cogwheeling" (palpable tremor on passive movement of the limbs) and Froment's sign; (2) the presence of known causes of enhanced physiological tremor (e.g., hyperthyroidism); (3) concurrent or recent exposure to tremorgenic drugs or the presence of a drug withdrawl state; (4) a history of direct or indirect trauma to the nervous system in the 3 months preceding the onset of tremor; (5) historical or clinical evidence for a psychogenic origin of tremor; and (6) tremor of sudden onset or evidence for a stepwise deterioration of the tremor.

The inclusion criteria for *probable* essential tremor are the same as those for *definite* essential tremor, except that the minimum duration of tremor was reduced to 3 years for probable cases. In addition, tremor that was confined to body parts other than the hands, including tremor of the head and postural tremor of the legs, was considered to be probable essential tremor. However, a head tremor associated with an abnormal head posture would suggest a diagnosis of dystonic tremor. The exclusion criteria for *probable* essential tremor are the same as for *definite* essential tremor, except that in addition primary orthostatic tremor, isolated voice tremor (because of the clinical difficulty of separating essential vocal tremor from dystonic vocal tremor), isolated position-specific or task-specific tremor, and isolated tongue or chin tremor are specifically excluded.

*Possible* essential tremor cases were subdivided into two types: type I, in which subjects satisfy the criteria for *definite* or *probable* essential tremor but exhibit either other recognizable neurological disorders, such as parkinsonism, dystonia, myoclonus, peripheral neuropathy, or restless leg syndrome, or other neurological signs of uncertain significance (i.e., which are insufficient to make the diagnosis of a recognizable neurological disorder). Such signs include mild extrapyramidal features, for example hypomimia, reduced arm swing, or mild bradykinesia; and type II, in which monosymptomatic and isolated tremors of uncertain relationship to essential tremor are present. The latter include position-specific tremor and task-specific tremors (e.g., occupational tremors, primary writing tremor, primary orthostatic tremor, isolated voice tremor, isolated postural leg tremor, and unilateral upper limb tremor). The exclusion criteria for *possible* essential tremor are the same as items 2–4 of *definite* essential tremor. Finally TRIG proposed a further subclassification of essential tremor into hereditary and presumed sporadic subtypes.

The TRIG criteria contributed toward standardizing the diagnosis of essential tremor, although the criteria for *definite* and *probable* essential tremor have subsequently been amalgamated and streamlined by Deuschl *et al.* (1998) under the term *classic* essential tremor, which is defined as a bilateral, largely symmetrical postural or kinetic tremor involving hands and forearms that is visible and persistent. Additional or isolated tremor of the head may occur but in the absence of abnormal posturing.

The exclusion criteria for *classic* essential tremor are the presence of the following:

1. Other abnormal neurological signs, especially dystonia
2. Known causes of enhanced physiological tremor, including current or recent exposure to tremorgenic drugs or the presence of a drug withdrawl state

3. Historical or clinical evidence of stepwise deterioration
4. Primary orthostatic tremor
5. Isolated voice tremor
6. Isolated position-specific or task-specific tremors, including occupational tremors and primary writing tremor
7. Isolated tongue or chin tremor
8. Isolated leg tremor

The TRIG criteria for *possible* essential tremor were controversial, and thus those patients with TRIG type I *possible* essential tremor were reclassified under their appropriate recognizable neurological disorder (e.g., parkinsonian tremor, dystonic tremor, or neuropathic tremor syndromes) (Deuschl *et al.*, 1998). Similarly, the conditions listed by TRIG as type II *possible* essential tremor have now been classified as separate disorders (Deuschl *et al.*, 1998), because a study performed by Bain *et al.* (1994) cast considerable doubt on the view that task-specific tremor and isolated tremors of the voice, tongue, facial muscles, head, and legs were part of the clinical spectrum of essential tremor (Biary and Koller, 1987; Critchley, 1949, 1972; Massey and Poulson 1985; Rosenbaum and Jankovic, 1988; Rothwell *et al.*, 1979). Thus the criteria for the diagnosis of essential tremor are still evolving and may need further refinement. For example, the inclusion of isolated head tremor as part of the definition of *classic* essential tremor by Deuschl *et al.* (1998) is highly contentious and probably wrong.

The age of onset of hereditary essential tremor has a bimodal distribution, with peaks in the second and fifth/sixth decades. Furthermore, segregation analysis indicates an autosomal-dominant inheritance with virtually complete penetrance by the age of 65 (Bain *et al.*, 1994). These data can be used to provide counseling to "at-risk" individuals about the probability of inheriting essential tremor, as well as the likelihood of such tremor causing disability or social handicap. Furthermore, Bain *et al.* (1994) did not detect any examples of the disease skipping a generation and thus producing an obligate gene carrier, which suggests that patients with late-onset ("senile tremor") or sporadic-action tremors may have different entities to essential tremor. Essential tremor begins to cause disability in the second decade, which progressively increases, so that between 12% and 25% of cases are compelled to change jobs or take early retirement (Bain *et al.*, 1994).

The prevalance of essential tremor varies according to the method and the diagnostic criteria used (Findley, 2000), with overall prevalence figures varying from 0.31%–4.0% (Louis *et al.*, 1995; Rajput *et al.*, 1984), although it is possible that this reflects genuine heterogeneity in different populations. The gene or genes responsible for hereditary essential tremor have not as yet been cloned. Analysis of the DNA from 15 families studied by Bain *et al.* (1994) provided substantial evidence against linkage between a gene causing essential tremor and the chromosome 9q32–34 markers argininosuccinate synthetase (ASS) and Abelson locus (ABL), a result that has subsequently been confirmed (Conway *et al.*, 1993; Durr *et al.*, 1993). However, two groups have established linkage of essential tremor to different chromosome loci: Gulcher *et al.* (1997) and Higgins *et al.* (1997) linking essential tremor to chromosome 3q13 and 2p22–25 with LOD scores of 3.71 and 5.92, respectively.

The pathophysiological mechanisms producing essential tremor are still poorly understood, although it is considered to involve CNS circuitry (Deuschl and Elble, 2000). This view is supported by the observation that focal lesions in the cerebellum, pons, or thalamus abolish or reduce essential tremor (Dupuis *et al.*, 1989; Duncan *et al.*, 1988; Hirai *et al.*, 1983). Current evidence suggests that abnormal cerebellar function has an important role in the genesis of essential tremor, because patients with advanced essential tremor may exhibit intention tremor and impaired tandem gait (Deuschl *et al.*, 2000; Singer *et al.*, 1994). In addition, a study of EMG recordings involving rapid wrist movements by patients with essential tremor revealed abnormalities in the timing of the second agonist burst, which may also be found in patients with cerebellar disease (Britton *et al.*, 1994); the authors speculated that delayed timing of the second agonist burst might induce tremor. However, Elble *et al.* (1994) were unable to corroborate this finding, which may have a simpler interpretation, namely, that tremor was entraining the patients' movements.

Techniques using positron emission tomography (PET) to measure the rate of $F^{18}$-dopa uptake by the basal ganglia have shown that the rate of $F^{18}$-dopa uptake by the putamen is normal in hereditary essential tremor (unlike Parkinson's disease [PD]). Patients with either "isolated" leg tremor or prominant rest tremor of the upper or lower limbs tend to show striatal $F^{18}$-dopa influx constants more typical of PD than hereditary essential tremor. In clinical studies no association with Parkinson's disease has been found, and thus patients with hereditary essential tremor should normally be told that they do not have an increased risk of PD (Cleeves *et al.*, 1988; Pahwa and Koller 1993; Roy *et al.*, 1983). However, some cases of sporadic essential tremor had a reduced rate of putaminal $F^{18}$-dopa uptake (Brooks *et al.*, 1992). The results of PET cerebral activation studies on essential tremor subjects are conflicting. Hallet and Dubinsky (1993) have shown that patients with essential tremor have glucose hypermetabolism in the medulla and thalami but not the cerebellum, supporting the view that the inferior olive and thalami are involved

in the circuits generating tremor. However, Wills *et al.* (1994) using $H_2 O^{15}$ failed to demonstrate olivary activation and found that the red nucleus and cerebellum were hypermetabolic, confirming the results of previous studies using $O^{15}$-labeled $CO_2$ of bilateral cerebellar activation in essential tremor that occurs both while the patients' tested arm is resting or exhibiting postural tremor (Colebatch *et al.*, 1990; Jenkins *et al.*, 1993).

**Treatment.** Approximately 50% of patients with hereditary essential tremor are alcohol responsive, and within an individual family there may also be a heterogeneous response to alcohol (Bain *et al.*, 1994). Propranolol** (80–320 mg/day) or primidone** (125–750 mg/day) are established treatments for essential tremor, although the impact of these drugs on patients' disability or quality of life have not been documented in sufficient detail (Bain, 1997). In cases of severe essential tremor in which the response to these medications, either alone or in combination, is unsatisfactory, stereotactic surgery is the main alternative, although a wide variety of "second-line" medications have their advocates (Koller *et al.*, 2000).

The choice of stereotactic surgery (see Chapter 76) is between either nucleus ventralis intermedius thalamotomy** or thalamic stimulation**. Both techniques can suppress or ablate contralateral limb tremor, but for essential tremor the latter is usually preferable, because essential tremor typically affects both hands, and bilateral thalmotomy is associated with significant morbidity, particularly with respect to speech (Bain, 1997; Pahwa *et al.*, 2000; Schurmann *et al.*, 2000). However, it is notable that after *Vim* stimulation, there is a greater recurrence rate in patients with essential tremor compared with those with parkinsonian tremor (Pahwa *et al.*, 2000).

### Parkinsonian Tremor

Tremor is usually ascribed to PD if the patient has any form of pathological tremor and fulfils the UK brain bank criteria for PD (Hughes *et al.*, 1993). Although "pill rolling" tremor is characteristic of the disease, it is uncommon. Pure rest tremor is also infrequent, and a combination of rest and postural tremor is more usual (Deuschl *et al.*, 1998). In 10%–20% of cases a rest tremor component never appears, but postural tremor occurs in most cases and is symptomatic in about 60% of patients (Elble and Koller, 1990). The tremors present in Parkinson's disease have been divided into three subtypes: type I consists of rest tremor or a rest and postural tremor that share the same frequency; type II rest tremor that is associated with an action tremor of a different frequency; and type III an isolated action tremor, although the latter is rare (Deuschl *et al.*, 1998). The tremor of PD is usually asymmetrical, typically starting in the fingers of one hand before spreading proximally to the wrist and forearm. Subsequently, over an average period of 2 years, it spreads to the ipsilateral foot. This hemitremulous state characteristically remains for several years before involvement of the contralateral limbs becomes apparent (Bain, 1993). However, occasionally this sequence of spread is delayed, or tremor first starts in a leg. The disease may cause tremor of the lips, tongue, or jaw, but it rarely causes significant vocal or head tremor (Elble and Koller, 1990). The term "benign tremulous PD" is sometimes used to describe patients in whom a parkinsonian tremor is present for many years, but the accompanying signs of parkinsonism remain very mild. Cogwheeling is palpable tremor, whereas cogwheel rigidity is palpable tremor associated with underlying extrapyramidal rigidity (Findley et al., 1981; Lance et al., 1963).

**Treatment.** PD is aggravated by anticholinesterase, which penetrates the blood–brain barrier, and this effect can be reversed by anticholinergics such as benztropine. Thus the mechanism by which anticholinergics diminish the severity of PD tremor is through a central effect exerted in the striatum. Anticholinergics are thought to modify the state of cholinergic excess that arises in response to a primary striatal dopamine deficiency (Barbeau, 1962). They have been used for the treatment of tremor dominant PD, particularly in young patients and in the early stages of the disease, in an attempt to delay the introduction of levodopa and to keep the daily dose of dopaminergic drugs to a minimum (Tolosa and Marin, 1995). Koller (1986) compared the effects of an anticholinergic, trihexiphenidyl** (benzhexol), with amantadine** and carbidopa-levodopa** on PD tremor. Nine untreated patients were given a 2-week supply of each of the three drugs in a double-blind trial. Five of nine patients preferred the anticholinergic. Objectively a 59% reduction in tremor amplitude was achieved with trihexiphendyl compared with 55% with carbidopa-levodopa and 23% with amantadine. The use of higher doses of anticholinergic did not yield better results. At present, the main role for amantadine (100–600 mg/day) is in the treatment of drug-induced dyskinesias rather than tremor (Tolosa and Marin, 1995; Koller and Herbster, 1987).

Anxiety can worsen parkinsonian rest tremor, and this is thought to be through beta-adrenergic receptor stimulation. This increase in PD tremor is preventable by beta-adrenergic antagonists. Propranolol** (80–240 mg/day) has been shown to reduce parkinsonian rest tremor by 70% and postural tremor by 50%, respectively (Tolosa and Marin, 1995). Despite these results, it is not widely prescribed (Bain and Findley, 1992). Propranolol is considered to have both a periph-

eral and central effect on PD tremor. However nadolol* (80–240 mg/day), a peripherally acting beta-adrenergic blocker, has also been shown to reduce parkinsonian tremor (Tolosa and Marin, 1995).

The effect of levodopa** (at doses of 100–1500 mg/day) on PD tremor was widely studied in the 1970s, and both bradykinesia and rigidity were thought to be more responsive to it than tremor. Furthermore, it was noted that tremor required higher doses of levodopa than akinesia or rigidity to decrease it. What is more, occasionally tremor re-emerged after rigidity had improved only to respond as the levodopa dosage was increased (Tolosa and Marin, 1995). A transient worsening of tremor by levodopa might be the result of activation of peripheral adrenergic receptors by catecholamines produced during the systemic metabolism the drug (Barbeau, 1962), because it was not found by Koller (1986), who investigated the effect of levodopa in combination with a peripheral dopa-decarboxylase inhibitor and found a 55% reduction of tremor.

Over the long term PD tremor remains sensitive to levodopa, with percentage improvements of 73% and 85% after 2 and 21 years, respectively (Tolosa and Marin, 1995). In contrast, speech, gait, and postural stability worsened and were less responsive to levodopa (Idanpaan-Heikilla et al., 1975). Both postural and rest tremor severity are reduced by levodopa. Koller (1986) demonstrated reductions of 46% and 58%, respectively (Tolosa and Marin, 1995).

Apomorphine** (3–30 mg/day by subcutaneous injection in divided doses) is a non-ergot dopaminergic compound first synthesised by Mathiesen and Wright in 1869, and the drug's use as a potential treatment for PD was first suggested by Weil in 1884. It is the most potent direct-acting dopamine agonist with affinity for both D1 and D2 dopamine receptors (Poewe and Wenning, 2000). Its effects on PD are thought to be due to its dopaminergic properties. Strian et al. (1972) found that apomorphine inhibited rest tremor but aggravated action tremor and was less effective on rigidity and bradykinesia. More recently subcutaneous apomorphine** infusion (beginning with 1 mg/h, during waking hours (maximum 100 mg/day), has found a niche as a treatment for fluctuating patients with severe breakthrough "off" period tremor or disability (Tolosa and Marin, 1995). A 50% reduction in the number of tremor-filled hours per day can be achieved, and its antiparkinsonian effects are similar to levodopa (Hughes et al., 1990).

Pramipexole* (1.5–4.5 mg/day) is an orally administered non-ergot dopaminergic compound that is an aminobenthiazole derivative. It is highly potent at the D2 dopamine receptor and has preferential affinity for the dopamine D3 receptor subtype (Piercey et al., 1996).

Pogarell *et al.* (1997) demonstrated in a combined analysis of three studies that the incidence of tremor was 7.2% for the pramipexole-treated group vs 14.1% for the placebo group. However, tremor was a secondary endpoint in these studies, and the results did not achieve statistical significance. Further studies are presently underway assessing pramipexole's tremorlytic properties (Bain, personal communication). The ergot alkaloids bromocriptine** (7.5–40 mg/day) and lisuride** (0.6–5 mg/day) both with predominant D2-receptor affinity, and pergolide, with both D2 and D1 receptor affinity, have all been shown to reduce tremor in de novo PD patients (Agnoli *et al.*, 1983; Rinne, 1983; Riopelle, 1987). Lewitt *et al.* (1982) compared bromocriptine and pergolide** and found that their antitremor effects were similar. Oral lisuride was also found to have a similar effect as bromocriptine in another study (Lewitt, 1999), but when administered intravenously, reduced rest tremor by 80% (Agnoli *et al.*, 1983). Consequently, it would seem that the tremorlytic effect of these various ergot derivative direct-acting dopamine agonists on PD tremor are equivalent, and the same probably applies to cabergoline* (1–4 mg/day) a newer ergot derivative that only needs to be ingested once daily and has predominant D2 receptor affinity (Lewitt, 1999).

Benzodiazepines may relieve the sensation of "internal tremor." Clonazepam* (0.5–2.5 mg/day) diminishes tremor of various different origins, PD (Lewitt, 1999). Loeb and Priano (1977) reported that clonazepam reduced PD tremor by 50% when administered either alone or in association with standard anti-parkinsonian drugs, but this observation was not obtained corroborated by Koller and Herbster (1987). Primidone, which is a first-line treatment for essential tremor (Bain, 1997), showed no effect on PD tremor (Heilman, 1984). The atypical neuroleptic clozapine* (50–300 mg/day) looks more promising, as it not only does not induce parkinsonian side effects but is also reported to reduce PD tremor (Gerlach et al., 1974). The mechanism by which it exerts its antitremor effect is unknown. However, clozapine causes agranulocytosis, which may be fatal, and thus routine monitoring of blood indices is required. It also commonly produces sedation (Idanpaan-Heikilla et al., 1975; Pakkenberg and Pakkenberg, 1986).

The surgical treatment of parkinsonian tremor (see Chapter 76) is currently undergoing a small renaissance. It is an option for patients who are refractory or intolerant of anti-parkinsonian medication, and the residual disability merits the risks involved in surgery. This type of surgery should only be conducted in the setting of a multidisciplinary team, as success depends on the skill, experience, technical availability, and habits of the surgical team (Jankovic, 1999). The patient should not have surgery if there is any general contraindication to it, a significant alteration of mental function, or a psychiatric state, such as a severe depressive state (Benabid *et al.*, 1998).

### Primary Orthostatic Tremor

The only tremor with a diagnostic frequency is primary orthostatic tremor. It has a characteristic frequency of between 14 and 18 Hz, although episodes of period doubling have been documented (Britton *et al.*, 1992). This is a rare condition in which the patient's principle complaint is of unsteadiness or discomfort on standing rather than walking, resulting from a high-frequency tremor of the lower limbs (Pazzaglia *et al.*, 1970). In many instances the tremor is more easily palpated than seen. It is associated with increased postural sway and can be diagnosed from force platform recordings (Yarrow *et al.*, 2001). Clonazepam* (0.5–2.5 mg/day) is the treatment of choice for orthostatic tremor, although some patients have been reported to respond to levodopa* (Wills *et al.*, 1999).

### Primary Writing Tremor

Primary writing tremor was first described by Rothwell *et al.*, in 1979, when they described a young man who had difficulty writing caused by bursts of tremor that occurred on pronation of the right forearm. Subsequently a number of similar cases have been described (Bain *et al.*, 1995; Cohen *et al.*, 1987; Elble *et al.*, 1990; Kachi *et al.*, 1985; Klawans *et al.*, 1982; Koller and Martyn, 1986; Ohye *et al.*, 1982; Ravits *et al.*, 1985; Rosenbaum and Jankovic, 1988). The issue of whether primary writing tremor is a variant of essential tremor, a type of dystonia, or a separate entity is controversial (Bain *et al.*, 1995; Elble and Koller, 1990; Fahn, 1984; Sheehy *et al.*, 1988). The evidence supporting the view that PWT is a variant of essential tremor is that both types of tremor have similar frequencies (4–8 Hz) and can be relieved (in about 30% and 50% of cases, respectively) by moderate amounts of alcohol (Bain *et al.*, 1994, 1995; Kachi *et al.*, 1985; Koller and Martyn, 1986; Ohye *et al.*, 1982). The opposing view, namely, that PWT is a type of writer's cramp, is supported by the following observations: (1) writer's cramp is task specific and in some cases both tremor and abnormal posturing are apparent (Sheehy and Marsden, 1982; Sheehy *et al.*, 1988); (2) tremor without any other movement disorder is one of the manifestations of idiopathic torsion dystonia (Bundey *et al.*, 1975; Fletcher *et al.*, 1990, 1991); and (3) one family has been described in which cases of writer's cramp, writing tremor, and nontask-specific tremor were all noted (Cohen *et al.*, 1987).

Sheehy *et al.* (1988) and Rosenbaum and Jankovic (1988) suggested a scheme in which some cases of primary writing tremor were related to essential tremor and others to torsion dystonia, but it is also conceivable that PWT is neither a form of essential tremor nor a type of dystonia (Bain *et al.*, 1995).

Primary writing tremor (PWT) has been subclassified into type A or B depending on whether tremor appeared during writing (type A, task-induced tremor) or while adopting the hand posture used for writing as well as during writing (type B, positionally sensitive tremor) (Bain *et al.*, 1995). The mean age of onset of PWT is about 50 years old, and a family history of primary writing tremor can be obtained from one third of patients with this condition (Bain *et al.*, 1995). However, there are patients in whom both involuntary abnormal posturing (dystonia) and tremor appeared during writing (Elble *et al.*, 1990). This last group of patients should be classified as having tremulous writer's cramp. However, difficulties arise because some tremulous patients voluntarily use unusual postures and excessive force to control the pen and thus seem to be dystonic. In PWT dominant hand writing speed (measured in letters per minute) is significantly impaired compared with that of age-matched controls (Bain *et al.*, 1995).

The pathophysiology of primary writing tremor is poorly understood. Tremor frequencies of between 5 and 7 Hz have consistently been found (Elble *et al.*, 1990; Kachi *et al.*, 1985; Ohye *et al.*, 1982; Ravits *et al.*, 1985), which is similar to the frequency of oscillation of the writing hand while normal subjects are writing (Bain *et al.*, 1995). The only exception is the original patient described by Rothwell *et al.* in 1979 in whom a 4–6 Hz tremor occurred in biceps, supinator, and pronator teres, whereas triceps fired at about 10 Hz. Surface polymyography performed during writing in patients with PWT has revealed a variety of EMG patterns in the flexor and extensor muscles of the forearm, including alternating, extensor activation alone, skipping from alternating to extensor activation, and cocontracting patterns (Bain *et al.*, 1995; Elble *et al.*, 1990; Kachi *et al.*, 1985; Ravits *et al.*, 1985). In a group of nine patients with PWT described by Kachi *et al.* (1985) the amplitude and latencies of the forearm stretch reflexes were normal in five of these patients, but the authors had difficulty reconciling this observation with the fact that tendon taps and muscle stretches caused tremor to appear in two and six of their nine patients, respectively. Ravits *et al.* (1985) recorded 5–20 mV cerebral potentials, which could be elicited by stretching pronator teres, and also noted that C-reflexes were absent. It is notable that two specific physiological differences between patients with PWT and writer's cramp were found by Bain *et al.* (1995), namely, that in patients with PWT no evidence for excessive "overflow" of the rhythmic EMG activity was recorded, whereas overflow is typical of dystonia. Furthermore, forearm reciprocal inhibition was normal in PWT, whereas in writer's cramp decreased presynaptic inhibition is present (Nakashima *et al.*, 1989).

About 30% of PWT patients respond to alcohol,* and some benefit may be obtained from treatment with propranalol* (80–240 mg/day), primidone* (125–750 mg/day), or botulinum toxin treatment (Bain *et al.*, 1995). In 1982 Ohye *et al.* recorded a very high incidence of irregular burst discharges within the thalamus during penetration and showed that contralateral stereotactic ventralis intermedius thalamotomy could successfully abolish PWT. There have been no postmortem studies of this condition.

### Neuropathic Tremor

Tremor is one of the manifestations of peripheral neuropathy (Bain *et al.*, 1996). It is observed in some patients with acute and chronic idiopathic demyelinating, hereditary motor and sensory and IgM paraproteinaemic neuropathies, and less often in the neuropathies associated with diabetes mellitus, diseases of the anterior horn cell, uremia, and porphyria (Dalakas *et al.*, 1984; Ridley, 1969; Said *et al.*, 1982; Smith *et al.*, 1984). The reason for the occurrence of tremor in some patients with these conditions but not others is not understood. Characteristically an action tremor is produced that resembles essential tremor, although a "pill-rolling" rest tremor, identical to that seen in PD, has been reported in one patient with demyelinating neuropathy (Mathews *et al.*, 1970). Studies of 6 patients with action tremor of the upper limbs associated with IgM paraproteinemic neuropathy showed that tremor frequency in the thumb correlated with ulnar nerve motor conduction velocity. The frequency in abductor pollicis brevis ranged from 3.7–5.5 Hz, although short episodes at lower frequencies (2.5–2.8 Hz) were recorded in two cases (Smith *et al.*, 1984). Bain *et al.* (1996) demonstrated, using magnetic brain stimulation, somatosensory-evoked potentials, and stretch reflex studies, that central conduction was not delayed. However, the short (M1) and long (M2) latency components of the forearm stretch reflexes were present, but the latencies of both components were delayed, and their sizes decreased. The duration of the M2 component was prolonged, probably as a result of dispersion. The cortical responses to somatosensory-evoked potentials were obtained in most patients but were delayed. Peripheral stimuli reset tremor phase in every case. These findings support the hypothesis that distorted mistimed peripheral inputs reach a central processor, which although intact is misled into producing tremor in certain parts of the body. The frequency of the resulting tremor is influenced by peripheral nerve conduction velocity. However, the absence of tremor from many patients with similar neuropathies suggests that an additional central fault may be present. It was suggested that delayed and diminished peripheral feedback, weakness, or both might induce the cerebellum to compensate by altering the gain or timing involved in one or more of the central motor control parameters in an attempt to maintain a normal movement profile. The capacity to do this may vary between patients, but in some cases a compensatory change in the gain of one or more motor control parameters (perhaps within the cerebellum) could alter the dynamics of the system beyond a threshold to a point at which the neural circuitry oscillates. Symptomatic treatment with primidone* (250–750 mg/day), propranolol* (80–320 mg/day), benzodiazepines or baclofen* (15–30 mg/day) is worth attempting, but the beneficial effect of these drugs is usually rather unimpressive (Bain *et al.*, 1996). Thus treatment should be focused on the underlying neuropathy.

### The Dystonic Tremor Syndromes

The dystonic tremor syndromes can be segregated into three types: (1) *dystonic tremor*: tremor that occurs in a body part that is also affected by abnormal dystonic posturing; (2) *tremor associated with dystonia*: tremor that occurs in a body part that is not affected by dystonia, but the patient has dystonia elsewhere (3) *dystonia gene-associated tremor*: tremor that is present as an isolated feature in an individual who has dystonia on genetic grounds (Deuschl *et al.*, 1998). The tremors evident in these dystonic tremor syndromes typically occur on action, examples are (1) a postural tremor that is apparent in the outstretched arms and is clinically indistinguishable from enhanced physiological or essential tremor. This type of tremor is often associated with spasmodic torticollis but may occur as the sole manifestation of hereditary torsion dystonia (Fletcher *et al.*, 1990); (2) a jerky irregular action tremor intermingled with sustained muscular spasms that can last several seconds (Yanagisawa and Goto, 1971). This type of tremor is often very disabling and can affect the muscles of the neck (tremulous spasmodic torticollis), face, trunk, and limbs; (3) a task-specific movement disorder in which tremulousness and jerky spasms develop concurrently during the performance of highly skilled acts (examples are tremulous writer's and typist's cramps and musician's incoordination syndrome) (Newmark and Hochberg, 1987; Sheehy and Marsden, 1982).

A combination of tremor and dystonia should prompt the clinician to investigate whether the dystonia is primary (idiopathic or hereditary) or secondary (symptomatic), because these are managed differently (Fahn *et al.*, 1987; Marsden and Quinn, 1990). Focal dystonic tremor, particularly tremulous spasmodic torticollis, can be treated by botulinum toxin injections into the rele-

vant musculature, whereas more widespread tremor is more appropriately treated with propranolol* (80–320 mg/day), primidone* (250–750 mg/day), or anticholinergic** medication (Moore, 1995). More recently, bilateral globus pallidus stimulation has been advocated for the treatment of spasmodic torticollis (Parkin *et al.*, 2001).

### Holmes ("Rubral") Tremor Syndrome

This syndrome is characterized by the presence of rest and intention tremors, typically at a frequency of less than 4.5 Hz. It was first described by Benedikt, although the first detailed description of this tremor was by Holmes (1904). There is usually a delay of about 4 weeks–2 years after an identifiable CNS lesion has developed before the tremor appears. In practice, this type of tremor is most commonly seen after brainstem/midbrain vascular disease but may occur after other disease. It has previously been termed "rubral" tremor, midbrain tremor, thalamic tremor, myorhythmia, or Benedikt's syndrome (Deuschl *et al.*, 1998). In one patient a Holmes ("rubral") tremor, which resulted from an arteriovenous malformation arising from the top of the basilar artery and draining through the midbrain through an anomalous vein could be suppressed by treatment with levodopa* (Findley and Gresty, 1981), which is now the main treatment for this condition (Deuschl *et al.*, 1998). Positron emission tomography (PET) studies have demonstrated dopaminergic denervation in some cases, but this finding is not ubiquitous (Remy *et al.*, 1995).

### Multiple Sclerosis and Cerebellar Tremors

The intention component of kinetic tremor is considered to be characteristic of cerebellar disease, although it is probable that lesions of the superior cerebellar peduncle, rather than the cerebellum itself, are responsible for this symptom (Holmes, 1904). Various types of postural tremor have also been described, including slow oscillations of the arms about the shoulders or legs about the hips (Holmes, 1917 and 1924). This type of tremor is referred to as titubation when it affects the head and trunk, and it can be particularly striking when a patient is standing. Cerebellar tremor is normally accompanied by disorders of ocular motility, especially nystagmus and dysmetria, and often by other signs, namely incoordination, dysdiadokinesis, pendular reflexes, and an unsteady gait.

Alusi *et al.* (2001a) studied the prevalence, subtypes, clinical features, and associated disability of tremor associated with multiple sclerosis (ms) in 100 randomly selected patients obtained from a clinic-based population. Tremor was detected in 58% of this population.

The median latency from the onset of MS to the appearance of tremor was found to be 11 years. Tremor affected the arms and less frequently the legs, head, and trunk but not the face, tongue, or jaw. All the patients had action, postural, or kinetic (including intention) tremor, but rest, Holmes' ("rubral") and primary orthostatic tremors were not detected. Tremor severity ranged from minimal in 27% to mild in 16% and moderate or severe in 15% of cases and correlated with the degree of dysarthria, dysmetria, and dysdiadochokinesia. Twenty-seven percent of the overall study population had tremor-related disability and 10% had incapacitating tremor.

Medical treatment of MS tremor is generally unrewarding, although carbamazepine*, clonazepam*, glutethamide*, hyoscine*, isoniazid*, ondansetron*, primidone* and tetrahydrocannibol* have been reported to produce some beneficial effect (Aisen et al., 1991; Clifford, 1983; Findley and Gresty 1981; Francis et al., 1986; Hallet et al., 1985; Henkin and Herishanu, 1991; Koller, 1984; Morrow et al., 1985; Rice et al., 1997; Sabra et al., 1982; Sechi et al., 1989). In our (uncontrolled) clinical experience, propranolol, at doses of 160–240 mg/day, can be useful, but its efficacy has not been demonstrated in a clinical trial (Koller, 1984), and the benefit seems to be temporary.

Stereotactic thalamotomy has been performed for the alleviation of disabling MS tremor since 1960 (Cooper, 1960), and more recently thalamic stimulation has been effectively used, but the results are critically dependent on careful patient selection (Alusi et al., 1999, 2001b; Geny et al., 1996; Haddow et al., 1997; Hooper and Whittle 1998; Nguyen et al., 1996, 1998). However, neurosurgeons have begun to target the nucleus ventralis oralis posterior (VOP), zona incerta, and subthalamic nucleus rather than nucleus ventralis intermedius (VIM) of the thalamus to alleviate MS tremor (Alusi et al., 1999 and 2001b), which is surprising, because there is evidence to suggest that VOP is the basal ganglia output nucleus and VIM the cerebellar input nucleus of the thalamus (Alusi et al., 2001b; Stein and Aziz, 2000). One explanation for this paradox may be that cerebellar tremors, like those seen in MS, are actually generated by the basal ganglia (Deuschl et al., 1999). Thus, when the cerebellum is malfunctioning, it may be better to eliminate the basal ganglia output to the motor cortex (Stein and Aziz, 2000).

### Drug-Induced Tremor

Numerous drugs are known to produce tremor in man. The most common miscreant is probably alcohol (withdrawal or prolonged heavy ingestion), which can cause an action tremor. An action tremor resembling enhanced physiological tremor can also be produced by a variety of substances, including salbutamol, adrenaline, amphetamine, lithium, caffeine, and steroids. In neurological and psychiatric practice, iatrogenic tremor is frequently seen, and both drug-induced parkinsonism and tremulous dyskinesias are common as are action tremors resulting from the use of tricyclic antidepressants. The antidopaminergic drugs used in the treatment of vertigo, nausea, vomiting, and schizophrenia can all incite a rest tremor indistinguishable from that seen in Parkinson's disease, but more typically these drugs cause a symmetrical action tremor. The anticonvulsants phenytoin, phenobarbitone, carbamazepine, and sodium valproate can all result in an action tremor when administered in sufficiently high doses for prolonged periods (Bain, 1993; Deuschl et al., 1998; Elble and Koller, 1990).

### Psychogenic Tremor

Psychogenic tremor is used to describe tremor that is produced or exacerbated voluntarily or "subconsciously" by a patient. The following phenomena suggest a psychogenic origin: (1) Sudden onset of tremor, which may be an emergency, go into remission, or both; (2) unusual clinical combinations of rest and postural/intention tremors; (3) decrease in tremor amplitude on distraction or changes in tremor frequency during voluntary movements of the contralateral hand; (4) the presence of the coactivation sign of psychogenic tremor, in which resistance to passive movement about a joint causes the appearance and disappearance of tremor to mirror changes in tone; (5) a clear medical history of a somatization disorder and the appearance of additional and unrelated neurological signs with tremor. Furthermore, loading may increase the amplitude of psychogenic tremor, unlike essential or parkinsonisn tremors that decrease with external loading (Deuschl et al., 1998; McAuley et al., 1998). However, it should be noted that organic tremors often present at times of great stress and may be paroxysmal, for example, in porphyria.

## THE PRACTICAL MANAGEMENT OF TREMOR

The appropriate management of patients with tremor depends on its severity and underlying origin. Propranolol (80–320 mg/day) and primidone (250–750 mg/day) are the main treatments for essential tremor but can be useful for patients with dystonic tremor syndromes. Propranolol (80–240 mg/day) has also been shown to reduce parkinsonian rest and postural tremors. Isolated head tremor or head tremors associated with essential or dystonic tremors respond to intramuscular botulinum toxin treatment. Neuropathic tremors and tremor associated with multiple sclerosis are difficult to treat effectively, but propranolol (80–240 mg/day) and clonazepam (0.5–2.5 mg/day) are worth trying.

Parkinsonian tremors respond variably to anticholinergic, levodopa, or direct-acting dopamine agonist therapy. Stereotactic surgery can alleviate contalateral limb tremor in patients with Parkinson's disease, essential tremor, dystonic tremor syndromes, writing tremor, Holmes' tremor, and tremulous multiple sclerosis. The preferred intracerebral target for parkinsonian tremor is the subthalamic nucleus rather than the nucleus ventralis intermedius of the thalamus, because the former also ameliorates rigidity and bradykinesia. The thalamus is still the target of choice for Holmes' tremor, as well as essential and dystonic tremor syndromes, although ventralis oralis posterior may be as effective as ventralis intermedius, whereas in MS the zona incerta may be a better target, particularly when there is a proximal tremor component. At present, deep brain stimulation is considered to be safer than lesional surgery, although the former requires considerably more postoperative servicing. Thus the treatment of patients with disabling tremor suffers from a conundrum: should patients tolerate their tremors, which are frequently inadequately controlled by medical therapy, or accept the risks involved in undergoing a stereotactic surgical procedure? At present most neurologists are conservative and rarely refer patients for surgery. What is perfectly clear is that if this situation is to alter the benefit/risk ratio, surgery must be maximized. This can only be achieved with a dedicated multidisciplinary team approach. The disabled tremulous patient deserves no less.

# REFERENCES

Agnoli, A., Ruggieri, S., Baidassarrw, M. *et al.* (1983). Dopaminergic ergots in parkinsonism. *In* "Lisuride and other Dopamine Agonists" (D. B. Calne, R. Horowski, R. S. Mcdonald, and W. Wuttke, Eds.), pp. 407–417. Raven Press, New York.

Aisen, M. L., Holzer, M., Rosen, M. *et al.* (1991). Glutethimide treatment of disabling action tremor in patients with multiple sclerosis and traumatic brain injury. *Arch. Neurol.* 48, 513–515.

Alusi, S. H., Glickman, S., Aziz, T. Z., and Bain, P. G. (1999). Tremor in multiple sclerosis. *J. Neurol. Neurosurg. Psychiatry* 66, 131–134.

Alusi, S. H., Worthington, J., Glickman, S., and Bain, P. G. (2001a). A study of tremor in multiple sclerosis. *Brain* 124, 720–730.

Alusi, S. H., Aziz, T. Z., Glickman, S., Stein, J., and Bain, P. G. (2001b). Stereotactic lesional surgery for the treatment of tremor in multiple sclerosis—a prospective case-controlled study. *Brain*, in press.

Arblaster, L. A., Elton, R. J., Lakie, M. *et al.* (1990). Human physiological tremor—a relationship with limb temperature. *J. Physiol.* 423, 71P.

Aziz, T. Z., and Bain, P. G. (1996). A multidisciplinary approach to tremor. *Br. J. Neurosurgery* 10, 435–437.

Bain, P. (1993). A combined clinical and neurophysiological approach to the study of patients with tremor. *J. Neurol. Neurosurg. Psychiatry* 56, 839–844.

Bain, P. G. (1997). The effectiveness of treatments for essential tremor. *Neurologist* 3, 305–321.

Bain, P. G. (1998). Clinical measurement of tremor. *Mov. Disord.* 13 (suppl 3), 77–80.

Bain, P. G., and Findley, L. J. (1992). Clinical aspects of parkinsonian tremor. "The Parkinson Papers." Vol. 1, pp. 1–4. Franklin Scientific Projects, London.

Bain, P. G., Findley, L. J., Britton, T. C. *et al.* (1995). Primary writing tremor. *Brain* 118, 1461–1472.

Bain, P. G., Britton, T. C., Jenkins, I. H. *et al.* (1996). Tremor associated with benign IgM paraproteinaemic neuropathy. *Brain* 119, 789–799.

Bain, P. G., Findley, L. J., Thompson, P. D. *et al.* (1994). A study of hereditary essential tremor. *Brain* 117, 805–824.

Barbeau, A. (1962). the pathogenesis of Parkinson's disease: a new hypothesis. *Can. Med. Assoc. J.* 87, 802–807.

Benabid, A. L., Benazzouz, A., Hoffmann, D. *et al.* (1998). Long-term electrical inhibition of deep brain targets in movement disorders. *Mov. Disord.* 13 (suppl 5), 119–125.

Biary, N., and Koller, W. C. (1987a). Essential tongue tremor. *Mov. Disord.* 2, 25–29.

Biary, N., and Koller, W. C. (1987b). Kinetic predominant tremor: effect of clonazepam. *Neurology* 37, 471–474.

Britton, T. C., Thompson, P. D., Day, B. L. *et al.* (1994). Rapid wrist movements in patients with essential tremor. The critical role of the second agonist burst. *Brain* 117, 39–47.

Britton, T. C., Thompson, P. D., Van der Kamp, W. *et al.* (1992). Primary orthostatic tremor: further observations in five cases. *J. Neurol.* 239, 209–217.

Brooks, D. J., Playford, E. D., Ibanez, V. *et al.* (1992). Isolated tremor and disruption of the nigrostriatal dopaminergic system: an F18-dopa PET study. *Neurology* 42, 1554–1560.

Bundey, S., Harrison, M. J. G., and Marsden, C. D. (1975). A genetic study of torsion dystonia. *J. Med. Genet.* 12, 12–19.

Cleeves, L., Findley, L. J., and Koller, W. C. (1988). Lack of association between essential tremor and Parkinson's disease. *Ann. Neurol.* 24, 23–26.

Clifford, D. B. (1983). Tetrahydrocannibol for tremor in multiple sclerosis. *Ann. Neurol.* 13, 669–671.

Cohen, L. G., Hallett, M., and Sudarsky, L. (1987). A single family with writer's cramp, essential tremor and primary writing tremor. *Mov. Disord.* 2, 109–116.

Colebatch, J. G., Findley, L. J., Frackowiak, R. S. J., Marsden, C. D., and Brooks, D. J. (1990). Preliminary report: activation of the cerebellum in essential tremor. *Lancet* 336, 1028–1030.

Conway, D., Bain, P. G., Warner, T. T. *et al.* (1993). Linkage analysis with chromosome 9 markers in hereditary essential tremor. *Mov. Disord.* 8, 374–376.

Cooper, I. S. (1960). Neurosurgical alleviation of intention tremor of multiple sclerosis and cerebellar disease. *N. Engl. J. Med.* 263, 441–444.

Critchley, M. (1949). Observations on essential (heredofamilial) tremor. *Brain* 72, 113–139.

Critchley, E. (1972). Clinical manifestations of essential tremor. *J. Neurol. Neurosurg. Psychiatry* 35, 365–372.

Dalakas, M. C., Teravainen, H., and Engel, W. K. (1984). Tremor as a feature of chronic relapsing and dysgammaglobulinaemic polyneuropathy. Incidence and management. *Arch. Neurol.* 47, 711–714.

Deuschl, G., Bain, P., Mitchell, B., and an Ad Hoc scientific committee. (1998). Consensus statement of the movement disorder society on tremor. *Mov. Disord.* 13, 2–23.

Deuschl, G., and Elble, R. J. (2000). The pathophysiology of essential tremor. *Neurology* **54 (Suppl 4)**, 14–20.

Deuschl, G., Wenzelburger, K., Loffler, K., Raethjen, J., and Stolze, H. (2000). Essential tremor and cerebellar dysfunction. Clinical and kinematic analysis of intention tremor. *Brain* **123**, 1568–1580.

Deuschl, G., Wilms, H., Krack, P. *et al.* (1999). Function of the cerebellum in Parkinsonian rest and Holmes' tremor. *Ann. Neurol.* **46**, 126–128.

Deuschl, G., Zimmermann, R., Genger, H., and Lucking, C. H. (1995). Physiologic classification of essential tremor. *In* "Handbook of Tremor Disorders" (L. J. Findley and W. C. Koller, Eds.), pp. 195–208. Marcel Decker, New York.

Duncan, R., Bone, I., and Melville, I. D. (1988). Essential tremor cured by infarction adjacent to the thalamus. *J. Neurol. Neurosurg. Psychiatry* **51**, 591–592.

Dupuis, M. J., Delwaide, P. J., Boucquey, D., and Gonsette. R. E. (1989). Homolateral disappearance of essential tremor after cerebellar stroke. *Mov. Disord.* **4**, 183–187.

Durr, A., Stevanin, G., Jedynak, C. P. *et al.* (1993). Familial essential tremor and idiopathic torsion dystonia are different genetic entities. *Neurology* **43**, 2212–2214.

Elble, R. J., Higgins, C., and Hughes, L. (1994). Essential tremor entrains rapid voluntary movements. *Exp. Neurol.* **126**, 138–143.

Elble, R. J., and Koller, W. C. (1990). "Tremor." John Hopkins University Press, Baltimore/London.

Elble, R. J., Moody, C., and Higgins, C. (1990). Primary writing tremor. A form of focal dystonia. *Mov. Disord.* **5**, 118–126.

Fahn, S. (1984). Rare tremors. *In* "Movement Disorders: Tremor" (L. J. Findley and R. Capildeo, Eds.), pp. 436–437. Macmillan Press, London and Basingstoke.

Fahn, S., Marsden, C. D., and Calne, D. B. (1987). Classification and investigation of dystonia. *In* "Movement Disorders 2" (C. D. Marsden and S. Fahn, Eds.), pp. 332–358. Butterworth, London.

Fletcher, N. A., Harding, A. E., and Marsden, C. D. (1990). A genetic study of idiopathic torsion dystonia in the United Kingdom. *Brain* **113**, 379–395.

Fletcher, N. A., Harding, A. E., and Marsden, C. D. (1991). A case control study of idiopathic torsion dystonia. *Mov. Disord.* **4**, 304–309.

Findley, L. J. (2000). Epidemiology and genetics of essential tremor. *Neurology* **54 (Suppl 4)**, 8–13.

Findley, L. J., and Gresty, M. A. (1981a). Tremor. *Br. J. Hosp. Med.* **26**, 16–32.

Findley, L. J., and Gresty, M. A. (1981b). Suppression of "rubral" tremor with levodopa. *BMJ* **281**, 1043.

Findley, L. J., Gresty, M. A., and Halmagyi, G. M. (1981). Tremor, the cogwheel phenomenon and clonus in Parkinson's disease. *J. Neurol. Neurosurg. Psychiatry* **44**, 534–546.

Findley, L. J., Koller, W. C., LeWitt, P. *et al.* (1993). Tremor Investigation Group (TRIG) classification and definition of tremor. *In* "Indications for and Clinical Implications of Botulinum Toxin Therapy" (Lord Walton of Detchant, Ed.), pp. 22–23. Royal Society of Medicine Services Ltd, London.

Francis, D. A., Grundy, D., and Heron, J. R. (1986). The response to isoniazid of action tremor in multiple sclerosis and its assessment using polarised light goniometry. *J. Neurol. Neurosurg. Psychiatry* **49**, 87–89.

Froment, J., and Gardere, H. (1926). La rigidite et la rue dentee Parkinsonione s'effacent an repos. *Rev. Neurol.* **1**, 52–53.

Geny, C., Nguyen, J. P., Pollin, B. *et al.* (1996). Improvement of severe postural cerebellar tremor in multiple sclerosis by chronic thalamic stimulation. *Mov. Disord.* **5**, 489–494.

Gerlach, J., Kappelhus, P., Helweg, E. *et al.* (1974). Clozapine and haloperidol in a single crossover trial. *Acta Psychiatr. Scand.* **50**, 410–424.

Gulcher, J. R., Jonsson, P., Kong, A. *et al.* (1997). Mapping of a familial essential tremor gene, FET1, to chromosome 3q13. *Nat. Genet.* **17**, 84–87.

Haddow, L. J., Mumford, C., and Whittle, I. R. (1997). Stereotactic treatment of tremor due to Multiple Sclerosis. *Neurosurg. Q.* **7**, 23–34.

Hallett, M., and Dubinsky, R. M. (1993). Glucose metabolism in the brain of patients with essential tremor. *J. Neurol. Sci.* **114**, 45–48.

Hallet, M., Lindsey, J. W., Allelstein, B. D., and Reiley, P. O. (1985). Controlled trial of isoniazid therapy for severe postural cerebellar tremor in multiple sclerosis. *Neurology* **32**, 1374–1377.

Heilman, K. M. (1984). Orthostatictrunkal tremor. *Arch. Neurol.* **41**, 880–881.

Henkin, Y, and Herishanu, Y. O. (1989). Primidone as a treatment for cerebellar tremor in multiple sclerosis. *Isr. J. Med. Sci.* **25**, 720–721.

Higgins, J. J., Pho, L. T., and Nee, L. E. (1997). A gene for essential tremor maps to chromosome 2p22–p25. *Mov. Disord.* **12**, 859–864.

Hirai, T., Miyazaki, M., Nakajima, H., Shibazaki, T., and Ohye, C. (1983). The correlation between tremor characteristics and the predicted volume of effective lesions in stereotaxic nucleus ventralis intermedius thalamotomy. *Brain* **106**, 1001–1018.

Holmes, G. (1904). On certain tremors in organic cerebral lesions. *Brain* **27**, 360–375.

Holmes, G. (1917). The symptoms of acute cerebellar injuries from gunshot wounds. *Brain* **40**, 461–535.

Holmes, G. (1922). The Croonian lectures on the clinical symptoms of cerebellar disease and their interpretation. *Lancet* **1**, 1177–1182.

Hooper, J., and Whittle, I. R. (1998). Long-term outcome after thalamotomy for movement disorders in multiple sclerosis. *Lancet* **352**, 1984.

Hughes, A. J., Daniel, S. E., Blankson, S. *et al.* (1993). A clinicopathologic study of 100 cases of Parkinson's disease. *Arch. Neurol.* **50**, 140–148.

Hughes, A. J., Lees, A. J., and Stern, G. M. (1990). Apomorphine in the diagnosis and treatment of parkinsonian tremor. *Clin. Neuropharmacol.* **13**, 312–317.

Idanpaan-Heikkila, S., Alhara, E., Olkinuora, M., and Palva, I. (1975). Clozapine and agranulocytosis. *Lancet* **2**, 611.

Isokawa-Akesson, M., and Komisaruk, B. R. (1985). Tuning the power spectrum of physiological finger tremor frequency with flickering light. *J. Neurosci. Res.* **14**, 373–380.

Jankovic, J. (1999). New and emerging therapies for Parkinson disease. *Arch. Neurol.* **56**, 785–790.

Jankovic, J., Cardoso, F., Grossman, R. G., and Hamilton, W. J. (1995). Outcome after stereotactic thalmotomy for parkinsonian, essential and other types of tremor. *Neurosurgery* **45**, 1743–1746.

Jenkins, I. H., Bain, P. G., Colebatch, J. *et al.* (1993). A positron emission tomography study of essential tremor: evidence for overactivity of cerebellar connections. *Ann. Neurol.* **34**, 82–90.

Kachi, T., Rothwell, J. C., Cowan, J. M. A., and Marsden, C. D. (1985). Writing tremor: its relationship to benign essential tremor. *J. Neurol. Neurosurg. Psychiatry* **48**, 545–550.

Klawans, H. L., Glantz, R., Tanner, C. M., and Goetz, C. G. (1982). Primary writing tremor: a selective action tremor. *Neurology* **32**, 203–206.

Koller, W. C. (1984). Pharmacologic trials in the treatment of cerebellar tremor. *Arch. Neurol.* **41**, 280–281.

Koller, W. C. (1986). Pharmacological treatment of parkinsonian tremor. *Arch. Neurol.* **43**, 126–127.

Koller, W. C., and Herbster, G. (1987). Adjuvent therapy of parkinsonian tremor. *Arch. Neurol.* **44**, 921–923.

Koller, W. C., Hristova, A., and Brin, M. (2000). Pharmacological treatment of essential tremor. *Neurology* **54 (Suppl 4)**, 30–38.

Koller, W. C., and Martyn, B. (1986). Writing tremor: its relationship to essential tremor. *J. Neurol. Neurosurg. Psychiatry* **49**, 220.

Lance, J. W., Scwab, R. S., and Peterson, E. A. (1963). Action tremor and the cogwheel phenomenon in Parkinson's disease. *Brain* **86**, 95–110.

Lewitt, P. A. (1999). Pharmacology of dopaminergic agonists for Parkinson's disease. *In* "Parkinson's Disease the Treatment Options" (P. A. Lewitt and W. Oertel, Eds.), pp. 159–186. Martin Dunitz, London.

Lewitt, P. A., Gopinathan, G., Ward, C. D. *et al.* (1982). Lisuride versus bromocriptine treatment in parkinson's disease: a double blind study. *Neurology* **32**, 69–72.

Lewitt, P. A., Ward, C. D., Larsen, T. A. *et al.* (1983). Comparison of pergolide and bromocriptine therapy in parkinsonism. *Neurology* **33**, 1009–1014.

Loeb, C., and Priano, A. (1977). Preliminary evaluation of the effects of clonazepam on parkinsonian tremor. *Eur. Neurol.* **15**, 143–145.

Louis, E. D., Marder, K., Cote, L. *et al.* (1995). Differences in the prevalence of essential tremor among elderly African Americans, whites and Hispanics in northern Manhattan, NY. *Arch. Neurol.* **52**, 1201–1205.

Marsden, C. D. (1984). Origins of normal and pathological tremor. *In* "Movement Disorders: Tremor" (L. J. Findley and R. Capildeo, Eds.), pp. 37–84. Macmillan Press, London.

Marsden, C. D., and Quinn, N. (1990). The dystonias (neurological disorders affecting 20,000 people in Britain). *BMJ* **300**, 139–144.

Massey, E. W., and Paulsen, G. W. (1985). Essential vocal tremor: clinical aharacteristics and response to therapy. *South. Med. J.* **78**, 316–317.

Mathews, W. B., Howell, D. A., and Hughes, R. C. (1970). Relapsing corticosteroid dependent polyneuritis. *J. Neurol. Neurosurg. Psychiatry* **33**, 330–337.

McAuley, J. H., Rothwell, J. C., Marsden, C. D., and Findley, L. J. (1998). Elecrophysiology aids in distinguishing organic from psychological tremor. *Neurology* **50**, 1882–1884.

Moore, P. (1995). "Handbook of Botulinum Toxin Treatment." Blackwell Science, Oxford.

Morrow, J., McDowel, H. *et al.* (1985). Isoniazid and action tremor in multiple sclerosis. *J. Neurol. Neurosurg. Psychiatry* **48**, 282–283.

Nakashima, K., Rothwell, J. C., Day, B. L. *et al.* (1989). Reciprocal inhibition between forearm muscles in patients with writer's cramp and other occupational cramps, symptomatic hemidystonia and hemiparesis due to stroke. *Brain* **112**, 681–697.

Newmark, J., and Hochberg, F. H. (1987). Isolated painless manual incoordination in 57 musicians. *J. Neurol. Neurosurg. Psychiatry* **50**, 291–295.

Nguyen, J. P., Feve, A., Cesaro, P., and Keravel, Y. (1998). Long term follow-up of patients with multiple sclerosis and action tremor treated by thalamic stimulation. *Mov. Disord.* **13 (Suppl)**, 132.

Nguyen, J. P., Feve, A, and Keravel, Y. (1996). Is electrostimulation preferable to surgery for upper limb ataxia. *Curr. Opin. Neurol.* **9**, 445–450.

Ohye, C., Miyazaki, M., Hirai, T. *et al.* (1982). Primary writing tremor treated by stereotactic selective thalmotomy. *J. Neurol. Neurosurg. Psychiatry* **45**, 988–997.

Pahwa, R., and Koller, W. C. (1993). Is there a relationship between Parkinson's disease and essential tremor. *Clin. Neuropharmacol.* **16**, 30–35.

Pahwa, R., Lyons, K., and Koller, W. C. (2000). Surgical treatment of essential tremor. *Neurologist* **54 (Suppl 4)**, 39–44.

Pakkenberg, H., and Pakkenberg, B. (1986). Clozapine in the treatment of tremor. *Acta Neurol. Scand.* **73**, 295–297.

Parkin, S., Aziz, T. Z., Gregory, R., and Bain, P. G. (2001). Bilateral globus pallidus stimulation for the treatment of spasmodic torticollis. *Mov. Disord.*, **16**, 489–493.

Pazzaglia, P., Sabattini, L., and Lugarese, E. (1970). Su di singulare disturbo della stazione evetta (osservazione di tre casi). *Ric. Freniatri* **96**, 450–459.

Piercey, M. F., William, E. H., Smith, M. W., and Hyslop, D. K. (1996). Inhibition of dopamine neuron firing by pramipexole, a dopamine D3 receptor-preferring agonist: comparison to other dopamine receptor agonists. *Eur. J. Pharmacol.* **312**, 35–44.

Poewe, W., and Wenning, G. K. (2000). Apomorphine: An under-utilised therapy for Parkinson's disease. *Mov. Disord.* **15**, 789–794.

Pogarell, O., Kunig, G., and Oertel, W. H. (1997). A non-ergot dopamine agonist, pramipexole, in the therapy of advanced Parkinson's disease: Improvement of Parkinsonian symptoms and treatment-associated complications. A review of three studies. *Clin. Neuropharmacol.* **1**, 28–35.

Rajput, A. H., Offord, K. P., Beard, C. M., and Kurland, L. T. (1984). Essential tremor in Rochester, Minnesota: a 45 year study. *J. Neurol. Neurosurg. Psychiatry* **47**, 466–470.

Ravits, J., Hallett, M., Baker, M., and Wilkins, D. (1985). Primary writing tremor and myoclonic writer's cramp. *Neurology* **35**, 1387–1391.

Remy, P., de Recondo, A., Defer, G. *et al.* (1995). Peduncular "rubral" tremor and dopaminergic denervation: a PET study. *Neurology* **45**, 472–477.

Ridley, A. (1969). The neuropathy of acute intermittent porphyria. *Q. J. Med.* **38**, 307–333.

Rinne, U. K. (1983). New ergot derivates in the treatment of Parkinson's disease. *In* "Lisuride and Other Dopamine Agonists" (D. B. Calne, R. Horowski, R. S. Mcdonald, and W. Wuttke, Eds.), pp. 431–432. Raven Press, New York.

Riopelle, R. J. (1987). Bromocriptine and the clinical spectrum of Parkinson's disease. *Can. J. Neurol. Sci.* **14**, 455–459.

Rosenbaum, F., and Jankovic, J. (1988). Focal task-specific tremor and dystonia: categorization of occupational movement disorders. *Neurology* **38**, 522–527.

Rothwell, J. C., Traub, M. M., and Marsden, C. D. (1979). Primary writing tremor. *J. Neurol. Neurosurg. Psychiatry* **42**, 1106–1114.

Roy, M., Boyer, L., and Barbeau, A. (1983). A prospective study of 50 cases of familial Parkinson's disease. *Can. J. Neurol. Sci.* **10**, 34–42.

Sabra, A. F., Hallet, M., Sudarsky, L., and Mullally, W. (1982). Treatment of action tremor in multiple sclerosis with isoniazid. *Neurology* **32**, 912–913.

Sechi, G. P., Zuddas, M., Piredda, M. *et al.* (1989). Treatment of cerebellar tremors with carbamazepine: A controlled trial with long term follow up. *Neurology* **39**, 1113–1115.

Said, G., Bathien, N., and Cesaro, P. (1982). Peripheral neuropathies and tremor. *Neurology* **32**, 480–485.

Schurmann, P. R., Bosch, A., Bossuyt, P. M. M. *et al.* (2000). A comparison of continuous thalamic stimulation and thalamotomy for suppression of severe tremor. *N. Engl. J. Med.* **342**, 461–468.

Sheehy, M., and Marsden, C. D. (1982). Writers cramp: a focal dystonia. *Brain* **105**, 461–480.

Sheehy, M. P., Rothwell, J. C., and Marsden, C. D. (1988). Writer's cramp. *In* "Advances in Neurology, Vol. 50, Dystonia 2" (S. Fahn and C. D. Marsden, Eds.), pp. 457–472. Raven Press, New York.

Singer, C., Sanchez-Ramos, J., and Weiner, W. J. (1994). Gait abnormality in essential tremor. *Mov. Disord.* **9**, 193–196.

Smith, I. S., Furness, P., and Thomas, P. K. (1984). Tremor in peripheral neuropathy. *In* "Movement Disorders: Tremor" (L. J. Findley and R. Capildeo, Eds.), pp. 399–406. Macmillan press, London.

Stein, J. F., and Aziz, T. Z. (1999). Does imbalance between basal ganglia and cerebellar outputs cause movement disorders? *Curr. Opin. Neurol.* **12**, 667–669.

Strian, F., Micheler, E., and Benkert, O. (1972). Tremor inhibition in Parkinson syndrome after apomorphine administration under

L-dopa and decarboxylase-inhibitor basic therapy. *Pharmacopsychiatr. Neuro-Psychopharmacol.* **5**, 198–205.

Tolosa, S., and Marin, C. (1995). Medical management of parkinsonian tremor. *In* "Handbook of Tremor Disorders" (L. J. Findley and W. C. Koller, Eds.), pp. 333–350. Marcel Dekker Inc, New York.

Wills, A. J., Brusa, L., Wang, H. C. *et al.* (1999). Levodopa may improve orthostatic tremor: case report and trial of treatment. *J. Neurol. Neurosurg. Psychiatry* **66**, 681–684.

Wills, A. J., Jenkins, I. H., Thompson, P. D. *et al.* (1994). Red nuclear and cerebellar but no olivary activation associated with essential tremor: a positron tomographic study. *Ann. Neurol.* **36**, 636–642.

Yanagisawa, N., and Goto, A. (1971). Dystonia musculorum deformans. Analysis with electromyography. *J. Neurol. Sci.* **13**, 39–65.

Yarrow, K., Brown, P., Gresty, M. A., and Bronstein, A. M. (2001). Force platform recordings in the diagnosis of primary orthostatic tremor. *Gait Posture* **13**, 27–34.

Young, R. R., Growdon, J. H., and Shahani, B. T. (1975). Beta adrenergic mechanisms in action tremor. *N. Engl. J. Med.* **293**, 950–953.

CHAPTER 87

# The Syndrome of Spastic Paresis

Volker Dietz and Robert R. Young

## CLINICAL ASPECTS

Spasticity is associated with numerous physical signs such as muscle hypertonia and exaggerated tendon reflexes including clonus. Lance (1980) defined *spasticity* as a velocity-dependent resistance of muscle to stretch caused by activation of tonic stretch reflexes. In contrast *rigidity* is defined as a continuous, plastic resistance throughout the range of passive movement not related to the velocity of stretch. In addition, in spasticity, antigravity muscles (arm flexors or leg extensors) are predominantly affected. Spastic signs are almost always accompanied by a variable degree of paresis that, together, constitute a syndrome known as *spastic paresis*.

Treatments for spasticity have been influenced by the assumption, which followed observation of the clinical signs in the chronic state, that exaggerated reflexes are responsible for muscle hypertonia. Drug and other forms of therapy, therefore, are usually directed toward producing reduction in activity of stretch reflexes. The function of these reflexes during natural movements and any causal relationships between exaggerated reflexes and the spastic movement disorder are frequently not considered. In reality, the physical signs of spasticity bear little relationship to a patient's disability, which is due to impairment of functional movement (for reviews see Chiara *et al.*, 1998; Dietz, 1992a, b; 1997; O'Dwyer and Ada, 1996; Young, 2000).

Clinical observations raised doubts about a simple relationship between reflex excitability, muscle tone, and disability. After an acute stroke, tendon reflexes may be exaggerated early, whereas spastic muscle tone develops over weeks. However, only after the development of spastic muscle tone can the patient support his or her body during stepping.

Neuronal regulation of functional movements, such as locomotion, is achieved by a complex interaction of spinal and supraspinal mechanisms. Rhythmic activation of leg muscles by spinal interneuronal circuits is modulated and adapted to the body's actual needs by multisensory afferent inputs. Electrical activity of leg muscles, which results from a close interaction between these different mechanisms, is translated into functionally modulated muscle tension by the mechanical properties of muscle fibers. Spinal programming and reflex activity are under supraspinal control. Disturbances of this supraspinal control lead to characteristic gait impairments seen with cerebellar and extrapyramidal disorders, as well as in spastic paresis.

## THE EVOLUTION OF SIGNS AND SYMPTOMS

After an acute lesion of pyramidal and extrapyramidal fibers, changes in physical signs occur over months and are only partially understood pathophysiologically. Initially, flaccid paresis is present, and tendon reflexes are absent, although acute parasagittal Rolandic lesions can produce an immediate but transient spastic dystonia (Russell and Young, 1969). Also, rather *rigid* muscle tone can suddenly develop after acute brainstem lesions. Flaccid paresis may last for weeks after a traumatic spinal cord lesion or acute stroke. Tendon reflexes may be exaggerated after a few days, whereas spastic muscle tone develops over weeks and increases over several months (Hiersemenzel *et al.*, 2000). These changes in muscle tone are not reversible and are usually more pronounced with spinal than cerebral lesions (Faist *et al.*, 1999).

It has been suggested that neuronal reorganization occurs after a central lesion in the cat (see Mendell, 1984) and in humans (see Carr *et al.*, 1993); this includes changes in cerebral areas after a spinal cord injury (Brühlmeier *et al.*, 1998; Lacourse *et al.*, 1999). These plastic changes may involve (1) novel connections (e.g., sprouting, functional strengthening of already available connections), (2) changes in strength of inhibition, and (3) denervation supersensitivity. Reduction of spinal presynaptic inhibition of group Ia fibers occurs (Burke and Ashby, 1972; Delwaide, 1973; Faist *et al.*, 1994) and seems to correlate with the excitability of tendon reflexes. In addition, after a few weeks, changes in mechanical properties occur in the leg extensor and arm flexor muscles, which may contribute to spastic muscle tone (for review see Dietz *et al.*, 1992b; 1997; 1999; O'Dwyer and Ada, 1996). Structural changes within the spastic muscle and of connective tissue become most prominent 1 year or more after an acute lesion (Hufschmidt and Mauritz, 1985; Sinkjaer *et al.*, 1993) and are described as muscle contractures (Skold *et al.*, 1999). Little is known about the time course of spastic symptoms after 1 year.

## PRINCIPLES OF THERAPY

### Pathophysiological Basis

At present, no direct therapy is available for improvement of central paresis. That is to say, disconnection of lower from higher motor centers cannot yet be remedied. Functional electrical stimulation (FES) of paralyzed muscles may compensate for certain aspects of the paresis but is still in a largely experimental stage (Popovic *et al.*, 1999; Yarkony *et al.*, 1992; see also Chapter 84), although a system providing FES of the distal upper limb in quadriplegic patients is now commercially available (NeuroControl Corporation, Cleveland, OH) and has been documented to increase users' ability to perform activities of daily living, increase their independence, and improve their quality of life (Bhadra and Peckham, 1997; Kilgore *et al.*, 1997; Wuolle *et al.*, 1999).

Treatment of spasticity is usually directed toward reduction of stretch reflex activity because of the assumption that exaggerated reflexes are responsible for increased muscle tone, which somehow accounts for the spastic movement disorder. Studies of muscle tone and reflex activity usually take place under passive motor conditions with the patient resting (cf., Thilmann *et al.*, 1991). On the other hand, extensive investigations of functional movements of leg (Berger *et al.*, 1984) and arm (Dietz *et al.*, 1991; Ibrahim *et al.*, 1993; Powers *et al.*, 1989) muscles did not reveal any causal relationship between exaggerated reflexes and disorders of movement. A reciprocal mode of leg muscle activation during gait is preserved in spasticity, but exaggerated stretch reflexes and spastic paresis are associated with an absence or reduction of the functionally essential polysynaptic (or long latency) reflexes. Tension development

during functional movements (Berger *et al.*, 1984) does not depend on exaggerated stretch reflexes. In fact, overall leg muscle activity is reduced during functional movements in patients with spasticity of spinal or cerebral origin. Electrophysiological and histological evidence shows that transformation of motor units takes place after a supraspinal lesion and regulation of muscle tone is achieved by a lower level of neuronal organization (for review see Dietz, 1997, 1999; O'Dwyer and Ada, 1996).

This default to a simpler mode of regulation of muscle tension after paresis caused by a spinal or a supraspinal lesion is basically advantageous for a patient; it enables the body to be supported during gait and, consequently, the patient to achieve mobility in almost every case after a unilateral lesion. Rapid movements are, however, no longer possible, because modulation of muscle activity is absent. After a severe spinal or supraspinal lesion, these transformed processes can become excessive with unwelcome sequelae such as painful spasms (Hiersemenzel *et al.*, 2000).

Therapeutic outcomes of patients with spastic paresis of either spinal or cerebral origin must also be evaluated after *physiotherapeutic* approaches. These should be used to train and activate residual motor functions and to prevent secondary complications such as muscle contractures and spasms. *Antispastic drug therapy* is the second therapeutic tool; it reduces muscle tone and spasms but may increase weakness (e.g., Hoogstraten *et al.*, 1988), which interferes with the performance of functional movements, although there is some evidence that intrathecal baclofen can reduce hyperactive reflexes and hemiparetic dystonia without producing weakness (Meythaler *et al.*, 1999). Antispastic drug therapy is, therefore, predominantly of benefit for immobilized patients in whom it reduces muscle tone and relieves muscle spasms, both of which may also improve nursing care for these patients. It may also be an essential precursor to effective FES.

### Action of Drugs

The aim of antispastic drug therapy is to reduce spastic muscle tone without reducing the voluntary force. Different sites of action are attributed to the various drugs (see Figure 1): (1) increased presynaptic inhibition of group I afferents, which leads to reduction of monosynaptic and oligosynaptic reflex activity (baclofen, clonazepam, diazepam); (2) inhibition of excitatory interneurons, which are interconnected in spinal reflex pathways (tizanidine, glycine); (3) reduced activation of peripheral intramuscular receptors (dantrolene, phenothiazine) and reduced muscle con-

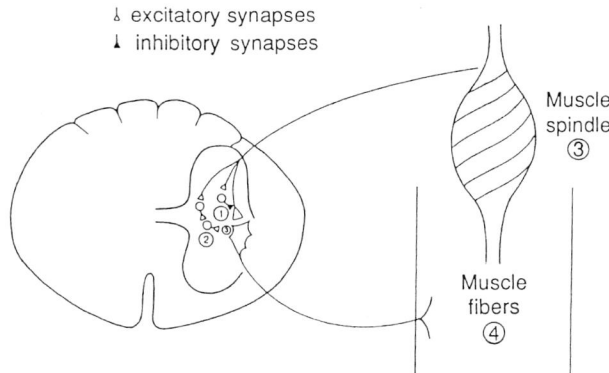

↓ excitatory synapses
⊥ inhibitory synapses

Muscle spindle ③

Muscle fibers ④

FIGURE 1  Presumed site of action of drugs with antispastic effects: (1) clonazepam/diazepam facilitate GABA-A mediated presynaptic inhibition; (2) baclofen inhibits activity of polysynaptic reflexes by GABA-B-receptor activation; (3) tizanidine acts on alpha₂-adrenergic receptors; (4) dantrolene reduces the sensitivity of peripheral intramuscular receptors and reduces release of calcium ions from the sarcoplasmic reticulum, which thus weakens muscle contraction.

traction force (dantrolene) (for review see Dietz, 1992b; Young and Delwaide, 1981).

The beneficial effects of antispastic drugs listed previously under (1) and (2) cannot necessarily be expected to improve functional movements considering the behavior of monosynaptic and polysynaptic reflexes during such movements. Clinical studies have shown that reduced activity of stretch reflexes does not produce an improvement in functional movements (tizanidine, Lapierre *et al.*, 1987; diazepam, Bes *et al.*, 1988; baclofen, Duncan *et al.*, 1976; Bass *et al.*, 1988; Corston *et al.*, 1981; Hoogstraten *et al.*, 1988; Stien *et al.*, 1987). These findings correspond to the experimental observation that abolishing hyperactive stretch reflexes of patients with spasticity does not result in an improvement of motor function (Landau, 1980; Thach and Montgomery, 1990). Recently, it was reported that gabapentin reduces the impairment induced by spasticity (Cutter *et al.*, 2000).

Similar considerations also apply to the peripherally acting drug, dantrolene. Its negative influence on contraction produces muscle paresis, which may hamper physiotherapy. Such failures to improve function, together with other side effects of these drugs, frequently impose limitations on the use of such drug therapy (Anderson, 1982; Meyler *et al.*, 1981).

In children with cerebral palsy, selective dorsal rhizotomy (McLaughlin *et al.*, 1998; Steinbock *et al.*, 1997; Wright *et al.*, 1998) or the application of botulinum toxin A (Corry *et al.*, 1997) improves clinical signs of spastic hemiplegia or diplegia. However, this was regarded as a "cosmetic" benefit (Corry *et al.*,

1997) because functional movements of the upper limbs were little changed. However, reduction in spastic dystonia of the lower limbs in such children is functionally beneficial.

These observations are not surprising, given the conditions under which antispastic drugs are usually tested: (1) effects of the antispastic drugs are not often tested during functional movements with recordings of biomechanical and electrophysiological parameters and (2) depending on the conditions, drug effects are observed not only on monosynaptic and polysynaptic reflexes but also on other neuronal mechanisms (for example, on Renshaw cell or fusimotor activity), so the interaction of these effects on motor function can hardly be controlled (Davidoff, 1985).

These warnings, which recommend cautious use of antispastic drugs, refer primarily to *mobile* patients who may require spastic muscle tone to be able to walk. Such warnings, however, are not valid for *immobilized* patients in whom increased paresis might be helpful to permit physiotherapy and nursing care. Such patients also benefit from improvement of painful spasms and clonus.

In *conclusion*, as shown by clinical and electrophysiological investigations, (1) drug-induced reduction of muscle tone by the diminution of exaggerated reflexes is not followed by an improvement in movement performance, but (2) in immobile patients antispastic drugs are beneficial, because they diminish the frequency and magnitude of painful sequelae of spasticity.

## PRACTICAL MANAGEMENT

Pharmacological management of spasticity is to a large extent empirically determined. Most studies in the literature have focused on reflex activity under artificial conditions. In fact, the few reports of the effects of antispastic drugs on functional movement failed to show any significant change. Similar conclusions can also be drawn for other nondrug treatments of spasticity. Adequately controlled trials have rarely been done, and several studies were empirically, not objectively, conducted. For an overview of methods for treating spasticity, see the following reviews: (Davidoff, 1985; Dietz, 1992b; Glenn and Whyte, 1990).

### Nonspecific Procedures

Painful flexor spasms and increased muscle tone frequently result from increased cutaneous reflexes induced by noxious or potentially painful afferent activity such as is associated with infections of the urinary tract, other infections combined with fever, and skin ulcerations, as well as by clothes irritating the skin. Consequently, worsening of spastic symptoms can frequently be alleviated by appropriate treatment of bladder function and skin care in paraplegic patients, as well as by early detection and management of the responsible factors (e.g., appropriate shoes or clothes).

## Physiotherapy

Although this statement is not based on hard data, physiotherapy represents a most definitive mode of treatment for mobile and immobilized spastic patients. Active and passive manipulative forms of physiotherapeutic treatment are of great importance for both groups of patients. On the one hand, residual motor functions can be improved by training. On the other hand, contractures of muscles and joints that are difficult to treat when established must be prevented at an early stage by frequent muscle stretching. Physiotherapy within a water-filled pool (i.e., underwater therapy) seems to be promising, because most movements are easier to perform during immersion. Exercise therapy should be directed toward treatment of those defined functions for which training is specifically indicated and for which benefits have been shown to depend on the intensity of training (Kwakkel et al., 1999).

Based on divergent empirical evidence, different physiotherapeutic procedures are being applied. Proprioceptive neuromuscular facilitation (PNF) and myofeedback techniques are meant to activate spinal motoneurons reflexly. The techniques of Bobath and Vojta are primarily used to treat children with cerebral palsy. Stereotyped movements become activated by such stimulation techniques when they are applied to specific dermatomes and joints. The Vojta method tries to activate complex movements that are believed to be programmed in the central nervous system. In contrast, the Bobath method tries to inhibit spastic symptoms in flexor muscles of the upper extremity and extensors of the lower extremity.

All these techniques hope to achieve the following benefits and goals:

1. Avoidance of secondary complications (i.e., pneumonia, skin ulcerations and deep vein thrombosis)
2. Prevention and treatment of muscle contractures
3. Reduction of muscle hypertonia
4. Training of posture and automatically performed movements by the induction of voluntarily initiated and controlled complex movements
5. Learning and training of coordinated movements by the involvement of tactile, auditory, vestibular and visual cues

6. Appropriate application of supportive aids, such as rollator, wheelchair, crutches, orthoses, and technical equipment (e.g., special shoes).

Each of these techniques is based on questionable theories. Controlled studies documenting positive effects of the treatment exist for none of them. Therefore, it is not yet possible to perform an appropriate evaluation and arrive at a recommendation based on the objective superiority of one of these techniques compared with another in the treatment of a given spastic patient. Nevertheless, physiotherapy must be part of a multidisciplinary integrated approach to patients. It also includes ergotherapeutic and nursing assistance. These all are meant to achieve greater mobility and, as far as possible, independence for the patient.

## Locomotor Training

Investigations (Visintin and Barbeau, 1989) have shown that in patients with incomplete paraplegia and those with spastic hemiplegia, a locomotor pattern can be elicited and trained under certain conditions. This training is based on observations made in cats with complete spinal lesions (Barbeau and Fung, 1992). Interactive locomotor training is performed on a treadmill with various percentages of the subjects' body weight (about 20%–50%) mechanically supported by an overhead harness using a strain-gauge transducer. As the subjects walk on the treadmill with a reduced load on their lower extremities, coordinated stepping movements and proper muscle activation can be facilitated by the moving treadmill. Optimal body weight support is provided and progressively reduced until the subjects can walk with full weight bearing. In most cases, leg movements must be assisted externally at the beginning of training, which limits the duration and success of the training. In the future, a driven gait orthosis will compensate for this drawback.

During the course of training, a progressively "normal" locomotor pattern is developed (Dietz et al., 1994; Visintin and Barbeau, 1989) and patients profit functionally from such training. Compared with the locomotor pattern found in healthy subjects, EMG activity in these patients is smaller in amplitude and less well modulated.

Although locomotor patterns can also be induced and trained in subjects with complete paraplegia (Dietz et al., 1994), therapeutic application of such locomotor training is presently restricted to patients with incomplete lesions, because only these patients profit directly in terms of mobility. Improvement of locomotor function by the activation of spinal locomotor centers may also be influenced by the repetitive elements of the ther-

TABLE I    Dosage in Drug Therapy for Spasticity

| Substance | Tablets | Beginning | Increase | Maximal |
|---|---|---|---|---|
| Primary choice | | | | |
| Baclofen*** | 5, 10, 25 mg | 2 × 5 mg/day | 2 × 5 mg/week | 4 × 20 mg up to 400 mg/day |
| Clonazepam*** | 0.5, 2 mg | 2 × 0.5 mg/day | 3 × 0.5 mg/week | 3 × 2 mg/day |
| Tizanidine** | 2, 4, 6 mg | 3 × 2 mg/day | 4–8 mg/wk | 36 mg/day |
| Secondary choice | | | | |
| Clonidine** | 0.075, 0.15, 0.3 mg | 2 × 0.075 mg/day | 0.075 mg/wk | 3 × 0.15 mg/day |
| Diazepam** | 2, 5, 10 mg | 2 × 2 mg/day | 2 × 4 mg/wk | ca. 3 × 20 mg/day |
| Tetrazepam* | 50 mg | 1 × 25 mg/day | 25 mg/day | 4–8 × 50 mg/day |
| Memantine* | 10 mg | 1 × 10 mg/day | 2 × 10 mg/wk | 3 × 20 mg/day |
| Dantrolene* | 25, 50 mg | 2 × 25 mg/day | 2 × 25 mg/wk | 4 × 50–100 mg/day |

apeutic program; animal experiments have shown that repetitive afferent input is essential for motor learning (Sakamato *et al.*, 1989). Even in chronic incomplete paraplegic patients, this training can successfully be applied (Wernig *et al.*, 1995).

## Drug Therapy

As a rule, the use of only one substance at a time is recommended, at least to begin with. There are patients who do best with modest doses of two medications that have different modes of action (baclofen and tizanidine, for example), so combination therapy may eventually be necessary. Because relief of spasms and muscle hypertonia may only be achieved at the cost of reduced muscle power, doses (see Table I) should be kept to minimum effective levels, especially in mobile patients. Almost all antispastic drugs may induce side effects, often consisting of drowsiness and nausea (for synopsis, see Table II).

### Drugs of First Choice

Best antispastic effects are reported for baclofen, tizanidine, and benzodiazepines (e.g., clonazepam). Therefore, these are the drugs of first choice for spastic patients. They are most effective in spasticity of spinal origin such as with multiple sclerosis and traumatic or neoplastic spinal cord lesions (see Glenn and Whyte, 1990).

*Baclofen* acts as a gamma-aminobutyric acid (GABA)-B agonist on a spinal level presynaptically and (less) postsynaptically. Monosynaptic stretch reflexes are depressed more effectively than polysynaptic reflexes, but flexor spasms are particularly reduced. Baclofen can alleviate spasms and muscle hypertonia in patients with spasticity (Duncan *et al.*, 1976; Hattab, 1980). Baclofen should be introduced slowly because of possible induction of drowsiness and hallucinations. It

should be withdrawn slowly so as not to produce a rebound increase in spasms (see Table I). Gabapentin, a GABA-related drug, is effective particularly for the treatment of painful muscle spasms (Cutter *et al.*, 2000).

*Tizanidine* is an imidazoline derivative closely related to *clonidine*. Both are thought to act on alpha-$_2$-adrenergic receptors in spasticity of supraspinal origin. It is suggested that these substances reduce the activity of polysynaptic reflexes, in many ways similar to the action of baclofen (Bes *et al.*, 1988; Stien *et al.*, 1987). Clonidine and tizanidine also have effects on the cord that are generally inhibitory; in part at least, they reduce the release of glutamate (and perhaps other excitatory amino acids). Clonidine and presumably tizanidine produce marked inhibition of short latency responses in alpha motoneurons to group II activity in the spinal cat (Schomburg and Steffens, 1988). Tizanidine also results in nonopiate analgesia by actions on alpha-$_2$-receptors in the spinal dorsal horn, which inhibit release of sub-

Table II    Side Effects of Drug Therapy for Spasticity

| | |
|---|---|
| Common effects: | Sedation, drowsiness, nausea, muscle weakness |
| Baclofen: | Nausea, vomiting, diarrhea, psychosis, and confusion (also after sudden withdrawal), ataxia, depression of respiratory and cardiovascular systems (intrathecal), headache (especially when kidneys are malfunctioning) |
| Tizanidine or clonidine: | Arterial hypotension (especially when hypotensive therapy is simultaneously being used), mouth dryness, indigestion |
| Clonazepam or diazepam: | Increased appetite, loss of libido, impaired menstruation, ataxia potentiated with alcohol; long-term effects: dependency, development of tolerance, sleeplessness, timidity, hallucinations |
| Memantine: | Agitation, pressure in the head, mouth dryness; contraindication: patients with liver dysfunction, confusion, pregnancy |
| Dantrolene: | Nausea, vomiting, anorexia, diarrhea, hepatic failure (especially in women older than 35 years or on estrogen therapy) |

stance P. This would diminish flexor reflex afferent (FRA)—mediated actions. Thus it comes as no surprise that clonidine alleviates spastic dystonia and reduces the frequency and severity of spasms in patients with spinal cord injury (Shefner *et al.*, 1992).

*Benzodiazepines* (e.g., clonazepam) amplify the inhibitory action of GABA-A at a presynaptic and post-synaptic level. Thereby, excitatory actions become dampened with a negative rebound. It is believed that increasing presynaptic inhibition in the spinal cord of patients with spasticity should reduce the release of exci-tatory transmitters from afferent fibers and thereby reduce the gain of monosynaptic and polysynaptic stretch reflexes and flexor reflexes. One can assume that these compounds work directly on the spinal cord (Davidoff, 1985). For diazepam, serious side effects such as development of tolerance, dependency, and drowsiness are reported (Glenn and Whyte, 1990). These side effects are somewhat less pronounced with clonazepam. Again, withdrawal must occur slowly (see Table I). Cumulative effects were observed with diazepam and other benzodiazepines, which have longer half-lives.

### Drugs of Second Choice

*Memantine* represents an amantadine derivative that seems to act primarily as an NMDA-receptor antago-nist (Seif el Nasr *et al.*, 1990). A few studies report effi-cacy on spasticity after brain lesions.

*Glycine* reduces experimentally induced hypertonia in animals because of its inhibitory action on neurons in the central nervous system. Oral administration of the simple amino acid glycine can alleviate the symptoms of spasticity. Similarly, an antispastic effect was described for L-threonine (Lee and Patterson, 1993), which is thought to modify spinal glycinergic transmission.

*Cannabinoids* may have significant beneficial effects on spasticity in a dosage (5 mg) that is well below the amount that induces altered states of consciousness (Maurer *et al.*, 1990; Meinck *et al.*, 1989). On the basis of animal experiments, they are thought to attenuate monosynaptic and polysynaptic reflexes.

*Dantrolene* acts primarily in the muscle itself, pro-ducing peripheral paresis because it reduces release of calcium ions from the sarcoplasmic reticulum, thereby preventing activation of the contractile apparatus. It is effective in all forms of spasticity because of its periph-eral target. Use of dantrolene is restricted, because its generalized paretic effects are frequently not well toler-ated by patients (Anderson, 1982). In addition, severe side effects such as toxic liver necrosis are reported, especially in women older than 35 and when used in combination with estrogen.

### Intrathecal Infusion of Baclofen

Even in immobilized patients with severe spastic symptoms, oral antispastic drugs are frequently not well tolerated in the long term because of their adverse effects but chronic intrathecal baclofen can efficiently reduce painful symptoms and has tolerable side effects (Latash *et al.*, 1989; Ochs *et al.*, 1989; Penn *et al.*, 1989). There are great advantages to this approach compared with the use of oral baclofen, also because intrathecal baclofen effectively combats spasticity caused by supraspinal lesions that respond poorly to oral baclofen. The intrathecal dose is minute (100–400 µg/day), but the antispastic effects, especially on muscle tone and spasms, are powerful, and systemic side effects are usually avoided. Severe spasticity can be trans-formed into flaccid paresis, which usually makes nursing easier. During the first month, some tolerance develops, which often makes an increase in dosage necessary (Coffey *et al.*, 1993). Most of this increase (mean from 182–528 µg/day) occurs within the initial 12 months after infusion pump implantation, and tolerance tends to plateau thereafter, with little need for dose increases. Only a few (6%) poor long-term responders have been reported (Penn, 1992). It is now clear that, in patients with severe spasticity caused by lesions at any level of the CNS, continuous intrathecal baclofen infusion is a safe and effective adjunct to physical therapy (Stewart-Wynne *et al.*, 1991). After termination of chronic treat-ment with intrathecal baclofen, lasting reduction of spasticity has been reported (Dressnandt and Conrad, 1996).

Effects of intrathecal baclofen can be tested, once or several times, by incremental lumbar injections of single doses (50, 75, or 100 µg), depending on the patient's response at each dose level. Continuous monitoring of blood pressure and respiration is necessary over about 6 hours. If a clear effect (e.g., reduced muscle tone) can be detected that lasts more than 4 hours, continuous drug application can be achieved by the subcutaneous implantation of a programmed electronic drug delivery system (e.g., the Medtronic system, see Ochs *et al.*, 1989). The intrathecal portion of the catheter leading from the pump should be placed in the thoracolumbar region. After 4 to 5 years, the pump and the battery it contains will need to be replaced. Several pump systems are now on the market. Extensive discussions with the patient about this therapeutic approach and the possi-ble side effects are necessary; close follow-up is impor-tant. Intrathecal baclofen is indicated primarily for patients with chronic spasticity in the legs, especially, but not necessarily, of spinal origin, to improve nursing and to alleviate painful muscle spasms and automatic movements of the legs.

The main side effects of intrathecal baclofen (for synopsis, see Table III) consist of drowsiness and somnolence, perhaps associated with depression of respiration. These side effects are usually due to an overdose during the bolus injection or overinfusion by the pump system, with excessive baclofen reaching the lower brainstem. Intensive care supervision is necessary in these situations; at present, no safe antidote is available. When complications occur, the pump system must be stopped and baclofen removed by insertion of a needle into the reservoir of the pump. By means of a lumbar puncture, cerebrospinal fluid (about 30 ml) should also be removed. The catheter system must eventually be repaired in as many as half the patients; its failure is the main cause of interruption of drug delivery (Schurch, 1993). Sudden cessation of intrathecal baclofen delivery constitutes an emergency, because it results in severe rebound spasticity.

## OTHER APPROACHES TO TREATMENT

### Local Antispastic Therapy

For the treatment of circumscribed muscle hypertonia, local injection of *botulinum toxin*, which acts to reduce release of acetylcholine from motor nerve endings, has become an established therapy (Davis and Jabbari, 1993; Al-Khodairy *et al.*, 1998). Local muscle hypertonia can be reduced by this reversible induction of peripheral paresis (chemical denervation), which usually lasts 3 to 4 months (Corry *et al.*, 1997). This also represents an elegant technique for the improvement of bladder function in patients with incomplete voiding caused by hypertonia of the sphincter externus muscle. In the latter case, botulinum toxin is injected three or four times within 3 months (i.e., 25 U Botox) through a special needle that, because it also allows simultaneous electromyographic recording of sphincter

**TABLE III**  Side Effects and Complications of Intrathecal Baclofen Therapy

- Pump dysfunction: Recurrence of spasticity
- Pump overfunction: Somnolence-coma-respiratory insufficiency; flaccid tetraparesis; monosynaptic- and polysynaptic reflexes abolished
- Dislocation or occlusion of catheter: recurrence of spasticity
- Hematoma and infection within the area of implantation
- Postlumbar puncture syndrome
- Cerebrospinal fluid leakage
- Meningitis
- Deep vein thrombosis in the legs after induction of flaccid paraparesis

muscle activity, helps locate the appropriate injection site (Schurch *et al.*, 1996).

### Treatments Less Frequently Recommended

There is a long history of neurosurgical alleviation of spasticity, specifically concerning localized treatment of spastic symptoms by interruption of the peripheral reflex arc. Selective *dorsal rhizotomy* (Laitinen *et al.*, 1983; Peacock and Staudt, 1991) or *dorsal longitudinal myelotomy* (Putty and Shapiro, 1991) is most commonly used in children with cerebral palsy. These procedures reduce afferent input responsible for increased muscle tone. Abnormal movement patterns, however, persist after spasticity is reduced (Giuliani, 1991; McLaughlin *et al.*, 1998; Wright *et al.*, 1998). Furthermore, although clinical signs are improved, impairment of functional movements is little changed (Corry *et al.*, 1997). Similarly, *infiltration of ventral roots or muscle nerves by phenol or alcohol* can transform a spastic into a flaccid paresis (Scott *et al.*, 1985). These treatments should rarely be used, because spasticity usually reappears after some months, and unwelcome sequelae, such as skin ulcerations caused by sensory loss in the corresponding dermatomes, are not uncommon.

Beneficial effects on spasticity are reported with *functional electrical stimulation* (FES) (Pease, 1998; Weingarden *et al.*, 1998) and by *transcutaneous electrical stimulation* of several muscles (Levin and Hui-Chan, 1992; Seib *et al.*, 1994). It is assumed that this effect, which is reported to be connected with a more phasic mode of muscle activation, is due to inhibition of EMG activity in the spastic extensor muscles. For most patients with moderate spasticity, this treatment is too awkward to be used regularly and may induce increased flexor spasms; negative results have been reported (Sonde *et al.*, 2000).

Improvement of spastic symptoms is also reported after chronic *stimulation of lobus anterior of the cerebellum* (Penn *et al.*, 1978) and of the *dorsal columns of the spinal cord* (Wiesendanger *et al.*, 1985). These effects were attributed to a reduction in stretch reflexes. However, negative results are also reported (Wiesendanger *et al.*, 1985). Because the outcomes are uncertain and damage to tissue within the central nervous system cannot be excluded using these approaches, such treatment at present is not recommended.

*Orthopaedic surgery* for spasticity is usually restricted to deformities of the feet and lower legs in children with cerebral palsy (Harryman, 1992). The most common operation consists of lengthening the Achilles tendon to compensate for shortening of the leg extensors with pes equinus deformity. This is done in an attempt to improve musculoskeletal alignment and, if the patient's feet then lie flat on the floor, balance is improved. However, such operations are also performed with reservations because of the danger of iatrogenically induced deformities (Baumann, 1986).

## STIFF-MAN SYNDROME

### Clinical Aspects

The stiff-man syndrome (Moersch-Woltman syndrome) is a rare disorder of motor function characterized by progressive rigidity, stiffness, and spasms of axial and extremity muscles associated with continuous EMG activity and superimposed painful muscle spasms that are often induced by startle or emotional stimuli (for synopsis of diagnostic criteria see Table IV). An association has been reported between stiff-man syndrome and epilepsy, insulin-dependent diabetes, and a variety of organ-specific autoimmune disorders (McEvoy, 1991). Antibodies directed against glutamic acid decarboxylase and against pancreatic islet cells have been detected in the serum and cerebrospinal fluid of about 60% of patients with stiff-man syndrome (Brown and Marsden, 1999; Dinkel et al., 1998). These findings suggest that stiff-man syndrome may be an autoimmune disease. Autoantibodies directed against GABA-ergic neurons are another useful diagnostic feature of the disease (Solimena and De Camilli, 1991).

A subgroup of patients with the stiff-man syndrome is likely to have a paraneoplastic autoimmune disease. Detection of antibodies against glutamic acid decarboxylase should be considered an indication to search for an occult breast cancer or small cell lung cancer (Folli, et al., 1993).

### Natural Course

Gradual worsening of symptoms is usually seen. A distinction was made between those with an acute illness leading to death within 1 year and those with a chronic course. The latter were divided into those with

TABLE IV    Diagnostic Criteria for Stiff Man Syndrome

1. Begins with stiffness and spasms of axial muscles
2. Successive spread of stiffness in proximal limb muscles with impairment of functional movements
3. Induction of painful muscle contractions by startle or emotional stimuli
4. Normal sensory and motor function during neurological examination (except for increased muscle tone)
5. Normal intellectual function
6. Alleviation of symptoms after treatment with benzodiazepines
7. No involvement of peripheral motor system

localized stiff limbs (Saiz et al., 1998) and those with general rigidity (i.e., stiff-man syndrome [Barker et al., 1998]). With progressive rigidity, patients usually become unable to walk or perform functional movements. With drug therapy, a favorable response is usually obtained in most patients.

### Practical Management

The efficacy of drugs that enhance GABA neurotransmission, such as diazepam (20–100 mg/day) and baclofen, represents a diagnostic criterion for this disease (Jog et al., 1992; McEvoy, 1991). Development of tolerance was reported only in 1 of 13 patients after 6 years of treatment (Lorish et al., 1989). Substantial improvement was also described for immunomodulatory agents (steroids, plasmapheresis, and intravenous immunoglobulin) (Khanlou and Eiger, 1999; Levy et al., 1999). An improvement of spasms was reported after paraspinal injection of botulinum toxin A (Davis and Jabbari, 1993). Clomipramine induces spasms and may be used as a provocative test in uncertain cases (Stöhr and Heckl, 1977).

## CONTINUOUS MUSCLE ACTIVITY OF PERIPHERAL NERVE ORIGIN (NEUROMYOTONIA)

### Clinical Aspects

The syndrome of continuous muscle activity of peripheral origin presents a relatively stereotyped clinical picture of muscle stiffness. Symptoms begin gradually with muscle stiffness, myokymia, and twitching at rest. Stiffness and "cramps" are more pronounced during and after muscle contraction. Distal, proximal, and cranial muscles are involved. Symptoms persist during sleep. Abnormal postures of the feet and hands are evident. Inspection of muscles reveals myokymia and fasciculations. Tendon reflexes are usually absent. The continuous motor unit and muscle fiber activity are caused by peripheral nerve hyperexcitability and are abolished by curare. The precise etiological mechanism is not known, but electrophysiological studies suggest an underlying neuropathy (Thompson, 1993). Recent observations point to an antibody-mediated ion channel dysfunction in the peripheral nervous system (Vincent, 1999; Vincent et al., 1998).

### Natural History

A typical feature of patients without an underlying peripheral neuropathy (or other disease) has been a benign course with sustained improvement over long periods. A hereditary form of sustained muscle activity of peripheral nerve origin is known to exist (Auger *et al.*, 1984).

## Practical Management

Plasma exchange seems to be superior to high-dose intravenous human immunoglobulin in the treatment of this condition (van den Berg, 1999). Carbamazepine and phenytoin, in doses determined by clinical symptoms, usually alleviate the muscle stiffness and result in tendon reflexes returning to normal.

## REFERENCES

Al-Khodairy, A. T., Gobelet, C., and Rossier, A. B. (1998). Has botulinum toxin type A a place in the treatment of spasticity in spinal cord injury patients? *Spinal Cord* 36, 954–958.

Anderson, T. P. (1982). Rehabilitation of patients with completed stroke. In "Krusen's Handbook of Physical Medicine and Rehabilitation," 3rd ed. (F. J. Kottke, G. K. Stillwell, and J. F. Lehmann, Eds.), pp. 583–603. Saunders, Philadelphia.

Auger, R. G., Daube, J. R., Gomez, M. R., and Lambert, E. H. (1984). Hereditary form of sustained muscle activity of peripheral nerve origin causing generalized myokymia and muscle stiffness. *Ann. Neurol.* 15, 13–21.

Barbeau, H., and Fung, J. (1992). New experimental approaches in the treatment of spastic gait disorders. In "Movement Disorders in Children." vol. 36. (H. Forssberg and H. Hirschfeld, Eds.), pp. 234–246. Karger, Basel.

Barker, R. A., Revesz, T., Thom, M., Marsden, C. D., and Brown, P. (1998). Review of 23 patients affected by the stiff man syndrome: clinical subdivision into stiff trunk (man) syndrome, stiff limb syndrome, and progressive encephalomyelitis with rigidity. *J. Neurol. Neurosurg. Psychiatry* 65, 633–640.

Bass, B., Weinshenker, B., Rice, G. P., Noseworthy, J. H., Cameron, M. G., Hader, W., Bouchard, S., and Ebers, G. C. (1988). Tizanidine versus baclofen in the treatment of spasticity in patients with multiple sclerosis. *Can. J. Neurol. Sci.* 15, 15–19.

Baumann, J. U. (1986). Behandlung kindlicher spastischer Fuss-Deformitäten. *Orthopäde* 15, 191–198.

Berger, W., Horstmann, G. A., and Dietz, V. (1984). Tension development and muscle activation in the leg during gait in spastic hemiparesis: The independence of muscle hypertonia and exaggerated stretch reflexes. *J. Neurol. Neurosurg. Psychiatry* 47, 1029–1033.

Bes, A., Eyssette, M., Pierrot-Deseilligny, E., Rohmer, F., and Warter, J. M. (1988). A multi-centre, double-blind trial of tizanidine, a new antispastic agent, in spasticity associated with hemiplegia. *Curr. Med. Res. Opin.* 10, 709–718.

Bhadra, N., and Peckham, P. H. (1997). Peripheral nerve stimulation for restoration of motor function. *J. Clin. Neurophysiol.* 14, 378–393.

Brown, P., and Marsden, C. D. (1999). The stiff man and stiff man plus syndromes. *J. Neurol.* 246, 648–652.

Brühlmeier, M., Dietz, V., Leenders, K. L., Roelcke, U., Missimer, Y., and Curt, A. (1998). How does the brain deal with a spinal cord injury? *Eur. J. Neurosci.* 10, 3918–3922.

Burke, D., and Ashby, P. (1972). Are spinal "presynaptic" inhibitory mechanisms suppressed in spasticity? *J. Neurol. Sci.* 15, 321–326.

Carr, L. J., Harrison, L. M., Evans, A. L., and Stephens, J. A. (1993). Patterns of central motor reorganization in hemiplegic cerebral palsy. *Brain* 116, 1223–1247.

Chiara, T., Carlos, J., Martin, D., Miller, R., and Nadeau, S. (1998). Cold effect on oxygen uptake, perceived exertion, and spasticity in patients with multiple sclerosis. *Arch. Phys. Med. Rehabil.* 79, 523–528.

Coffey, J. R., Cahill, D., Steers, W., Park, T. S., Ordia, J., Meythaler, J., Herman, R., Shetter, A. G., Levy, R., Gill, B. et al. (1993). Intrathecal baclofen for intractable spasticity of spinal origin: Results of a long-term multicenter study. *J. Neurosurg.* 78, 226–232.

Corry, I. S., Cosgrove, A. P., Walsh, E. G., McClean, D., and Graham, H. K. (1997). Botulinum toxin A in the hemiplegic upper limb: a double blind trial. *Dev. Med. Child. Neurol.* 39, 185–193.

Corston, R. N., Johnson, F., and Godwin-Austen, R. B. (1981). The assessment of drug treatment of spastic gait. *J. Neurol. Neurosurg. Psychiatry* 44, 1035–1039.

Cutter, N. C., Scott, D. D., Johnson, J. C., and Whiteneck, G. (2000). Gabapentin effect on spasticity in multiple sclerosis: a placebo-controlled, randomized, trial. *Arch. Phys. Med. Rehabil.* 81, 164–169.

Davidoff, R. A. (1985). Antispasticity drugs: mechanisms of action. *Ann. Neurol.* 17, 107–116.

Davis, D., and Jabbari, B. (1993). Significant improvement of stiff-person syndrome after paraspinal injection of botulinum toxin A. *Mov. Disord.* 8, 371–376.

Delwaide, P. J. (1973). Human monosynaptic reflexes and presynaptic inhibition. In "New Developments in Electromyography and Clinical Neurophysiology," vol. 3. (J. E. Desmedt, Ed.), pp. 508–522. Karger, Basel.

Dietz, V. (1992a). Human neuronal control of functional movements. Interaction between central programs and afferent input. *Physiol. Rev.* 72, 33–69.

Dietz, V. (1992b). Spasticity: exaggerated reflexes or movement disorder? In "Movement Disorders in Children," vol. 36. (H. Forssberg and H. Hirschfeld, Eds.), pp. 225–233. Karger, Basel.

Dietz, V. (1997). Neurophysiology of gait disorders: present and future applications. *Electroenceph. Clin. Neurophysiol.* 103, 333–355.

Dietz, V. (1999). Supraspinal pathways and the development of muscle tone dysregulation. *Dev. Med. Child. Neurol.* 41, 708–715.

Dietz, V., Colombo, G., and Jensen, L. (1994). Locomotor activity in spinal man. *Lancet* 344, 1260–1263.

Dietz, V., Trippel, M., and Berger, W. (1991). Reflex activity and muscle tone during elbow movements in patients with spastic paresis. *Ann. Neurol.* 30, 767–779.

Dinkel, K., Meinck, H. M., Jury, K. M., Karges, W., and Richter, W. (1998). Inhibition of gamma-aminobutyric acid synthesis by glutamic acid decarboxylase autoantibodies in stiff-man syndrome. *Ann. Neurol.* 44, 194–201.

Dressnandt, J., and Conrad, B. (1996). Lasting reduction of severe spasticity after ending chronic treatment with intrathecal baclofen. *J. Neurol. Neurosurg. Psychiatry* 60, 168–173.

Duncan, G. W., Shahani, B. T., and Young, R. R. (1976). An evaluation of baclofen treatment for certain symptoms in patients with spinal cord lesions: a double-blind, cross-over study. *Neurology* 26, 441–446.

Faist, M., Ertel, M., Berger, W., and Dietz, V. (1999). Impaired modulation of quadriceps tendon jerk reflex during spastic gait. Differences between spinal and cerebral lesions. *Brain* 122, 567–579.

Faist, M., Mazevet, D., Dietz, V., and Pierrot-Deseilligny, E. (1994). A quantitative assessment of presynaptic inhibition of Ia afferents in spastics: Differences in hemiplegics and paraplegics. *Brain* 117, 1449–1455.

Folli, F., Solimena, M., Cofiell, R., Austoni, M., Tallini, G., Fassetta, G., Bates, D., Cartilidge, N., Bottazzo, G. F., Piccolo, G., *et al.* (1991). Autoantibodies to a 128-kd synaptic protein in three women with the stiff-man syndrome and breast cancer. *N. Engl. J. Med.* 328, 546–551.

Glenn, M. B., and Whyte, J. (1990). "The Practical Management of Spasticity in Children and Adults." Lea & Febiger, Philadelphia, London.

Giuliani, C. A. (1991). Dorsal rhizotomy for children with cerebral palsy: Support for concepts of motor control. *Phys. Ther.* 72, 248–259.

Harryman, S. E. (1991). Lower-extremity surgery for children with cerebral palsy: Physical therapy management. *Phys. Ther.* 72, 16–24.

Hattab, J. R. (1980). Review of European clinical trials with baclofen. *In* "Spasticity: Disordered Motor Control." (R. G. Feldman, R. R. Young, and W. P. Koella, Eds.), pp. 71–85. Year Book, Chicago.

Hiersemenzel, L. P., Curt, A., and Dietz, V. (2000). From spinal shock to spasticity. Neuronal adaptation to spinal cord injury. *Neurology* 54, 1574–1582.

Hoogstraten, M. C., van der Ploeg, R. J., van der Burg, W., Vreeling, A., van Marle, S., and Minderhoud, J. M. (1988). Tizanidine versus baclofen in the treatment of spasticity in multiple sclerosis patients. *Acta Neurol. Scand.* 77, 224–230.

Hufschmidt, A., and Mauritz, K. H. (1985). Chronic transformation of muscle in spasticity: A peripheral contribution to increased tone. *J. Neurol. Neurosurg. Psychiatry* 48, 676–685.

Ibrahim, I. K., Berger, W., Trippel, M., and Dietz, V. (1993). Stretch-induced electromyographic activity and torque in spastic elbow muscles. *Brain* 116, 971–989.

Jog, M. S., Lambert, C. D., and Lang, A. E. (1992). Stiff-person syndrome. *Can. J. Neurol. Sci.* 19, 383–388.

Khanlou, H., and Eiger, G. (1999). Long-term remission of refractory stiff-man syndrome after treatment with intravenous immunoglobulin. *Mayo Clin. Proc.* 74, 1231–1232.

Kilgore, K. L., Peckham, P. H., Keith, M. W., Thrope, G. B., Wuolle, K. S., Bryden, A. M., and Hart, R. L. (1997). An implanted upper-extremity neuroprosthesis. *J. Bone. Joint. Surg.* 79A, 533–541.

Kwakkel, G., Wagenaar, R. C., Twisk, J. W. R., Lankhorst, G. J., and Koetsier, J. C. (1999). Intensity of leg and arm training after primary middle-cerebral-artery stroke: a randomized trial. *Lancet* 354, 191–196.

Lacourse, M. G., Lawrence, K. E., Cohen, M. J., and Young, R. R. (1999). Spinal cord injury. *In* "Exercise in Rehabilitation Medicine." (W. R. Frontera, D. M. Dawson, and D. M. Slovik, Eds.), pp. 267–292. Human Kinetics, Champaign, IL.

Laitinen, L. V., Nilsson, S., and Fugl-Meyer, A. R. (1983). Selective posterior rhizotomy for treatment of spasticity. *J. Neurosurg.* 58, 895–899.

Lance, J. W. (1980). Symposium synopsis. *In* "Spasticity: Disordered Motor Control." (R. G. Feldman, R. R. Young, and W. P. Koella, Eds.), pp. 485–495. Year Book, Chicago.

Landau, W. M. (1980). Spasticity: What is it? What is it not? *In* "Spasticity: Disordered Motor Control." (R. G. Feldman, R. R. Young, and W. P. Koella, Eds.), pp. 17–24. Year Book, Chicago.

Lapierre, Y., Bouchard, S., Tansey, C., Gendron, D., Barkas, W. J., and Francis, G. S. (1987). Treatment of spasticity with tizanidine in multiple sclerosis. *Can. J. Neurol. Sci.* 14, 513–517.

Latash, M. L., Penn, R. D., Carcos, D. M., and Gottlieb, G. L. (1989). Short-term effects of intrathecal baclofen in spasticity. *Exp. Neurol.* 103, 165–172.

Lee, A., and Patterson, V. (1993). Double-blind study of L-threonine in patients with spinal spasticity. *Acta Neurol. Scand.* 88, 334–338.

Levin, M. F., and Hui-Chan, C. W. (1992). Relief of hemiparetic spasticity by TENS is associated with improvement in reflex and voluntary motor functions. *Electroenceph. Clin. Neurophysiol.* 85, 131–142.

Levy, L. M., Dalakas, M. C., and Floeter, M. K. (1999). The stiff-person syndrome: an autoimmune disorder affecting neurotransmission of gamma-aminobutyric acid. *Ann. Intern. Med.* 131, 522–530.

Lorish, T. R., Thorstensson, G., and Howard, F. M. (1989). Stiff man syndrome updated. *Mayo Clin. Proc.* 64, 629–636.

Maurer, M., Henn, V., Dittrich, A., and Hofmann, A. (1990). Delta-9-tetrahydrocannabinol shows antispastic and analgesic effects in a single case double-blind trial. *Eur. Psychiat. Clin. Neurosci.* 240, 1–4.

McEvoy, K. M. (1991). Stiff-man syndrome. *Mayo Clin. Proc.* 66, 300–304.

McLaughlin, J. F., Bjornson, K. F., Astley, S. J., Graubert, C., Hays, R. M., Roberts, T. S., Price, R., and Temkin, N. (1998). Selective dorsal rhizotomy: efficacy and safety in an investigator-masked randomized clinical trial. *Dev. Med. Child. Neurol.* 40, 220–232.

Meinck, H.-M., Schönle, P. W., and Conrad, B. (1989). Cannabinoids on spasticity and ataxia in multiple sclerosis. *J. Neurol.* 236, 120–122.

Mendell, L. M. (1984). Modifiability of spinal synapses. *Physiol. Rev.* 64, 260–324.

Meyler, W. J., Bakker, H., Kok, J. J., Agoston, S., and Wesseling, H. (1981). The effect of dantrolene sodium in relation to blood vessels in spastic patients after prolonged administration. *J. Neurol. Neurosurg. Psychiatry* 44, 334–339.

Meythaler, J. M., Guin-Renfroe, S., and Hadley, M. N. (1999). Continuously infused intrathecal baclofen for spastic/dystonic hemiplegia. *Am. J. Phys. Med. Rehabil.* 78, 247–254.

Ochs, G., Struppler, A., Meyerson, B. A., Linderoth, B., Gybels, J., Gardner, B. P., Teddy, P., Jamous, A., and Weinmann, P. (1989). Intrathecal baclofen for long-term treatment of spasticity: A multicentre study. *J. Neurol. Neurosurg. Psychiatry* 52, 933–939.

O'Dwyer, N. J., and Ada, L. (1996). Reflex hyperexcitability and muscle contracture in relation to spastic hypertonia. *Curr. Opin. Neurol.* 9, 451–455.

Peacock, W. J., and Staudt, L. A. (1991). Functional outcomes following selective posterior rhizotomy in children with cerebral palsy. *J. Neurosurg.* 74, 380–385.

Pease, W. S. (1998). Therapeutic electrical stimulation for spasticity: quantitative gait analysis. *Am. J. Phys. Med. Rehab.* 77, 351–355.

Pedersen, E. (1974). Clinical assessment and pharmacologic therapy of spasticity. *Arch. Phys. Med. Rehabil.* 55, 344–356.

Penn, R. D. (1992). Intrathecal baclofen for spasticity of spinal origin: Seven years of experience. *J. Neurosurg.* 77, 236–240.

Penn, R. D., Gottlieb, G. L., and Agarwal, G. C. (1978). Cerebellar stimulation in man. Quantitative changes in spasticity. *J. Neurosurg.* 48, 779–786.

Penn, R. D., Savoy, S. M., Corcos, D., Latash, M., Gottlieb, G., Parke, B., and Kroin, J. S. (1989). Intrathecal baclofen for severe spinal spasticity. *N. Engl. J. Med.* 320, 1517–1521.

Popovic, M. R., Keller, T., Pappas, I., Morari, M., and Dietz, V. (1999). Grasping and walking protheses for stroke and spinal cord injured subjects. *Proc. Am. Control Conf.* 2, 1243–1247.

Powers, R. K., Campbell, D. L., and Rymer, W. Z. (1989). Stretch reflex dynamics in spastic elbow flexor muscles. *Ann. Neurol.* 25, 32–42.

Putty, T. K., and Shapiro, S. A. (1991). Efficacy of dorsal longitudinal myelotomy in treating spinal spasticity: A review of 20 cases. *J. Neurosurg.* 75, 397–401.

Russell, W. R., and Young, R. R. (1969). Missile wounds of the parasagittal rolandic area. *In* "Modern Neurology." (S. Locke, Ed.), pp. 289–302. Little, Brown and Company, Boston.

Saiz, A., Graus, F., Valldeoriola, F., Valls-Sole, J., and Tolosa, E. (1998). Stiff-leg syndrome: a focal form of stiff-man syndrome. *Ann. Neurol.* 43, 400–403.

Sakamato, T., Porter, L. L., and Asanuma, H. (1989). Functional role of the sensory cortex in learning motor skills in cats. *Brain Res.* **503**, 258–264.

Schomburg, E. D., and Steffens, H. (1988). The effect of DOPA and clonidine on reflex pathways from group II afferents to alpha-motoneurons in the cat. *Exp. Brain Res.* **71**, 442–446.

Schurch, B. (1993). Errors and limitations of the multimodality checking methods of defective spinal intrathecal pump system. Case report. *Paraplegia* **31**, 611–615.

Schurch, B., Hauri, D., Rodic, B., Curt, A., Meyer, M., and Rossier, A. B. (1996). Botulinum-A toxin as a treatment of detrusor-sphincter dyssynergia: a prospective study in 24 spinal cord injury patients. *J. Urol.* **155**, 1023–1029.

Scott, B. A., Weinstein, Z., Chiteman, R., and Pulliam, M. W. (1985). Intrathecal phenol and glycerin in metrizamide for treatment of intractable spasms in paraplegia. *J. Neurosurg.* **63**, 125–127.

Seib, T. P., Price, R., Reyes, M. R., and Lehmann, J. F. (1994). The quantitative measurement of spasticity: Effect of cutaneous electrical stimulation. *Arch. Phys. Med. Rehabil.* **75**, 746–750.

Seif el Nasr, M., Peruche, B., Rossberg, C., Mennel, H. D., and Krieglstein, J. (1990). Neuroprotective effect of memantine demonstrated in vivo and in vitro. *Eur. J. Pharmacol.* **185**, 19–24.

Shefner, J. M., Berman, S. A., Sarkarati, M., and Young, R. R. (1992). Recurrent inhibition is increased in patients with spinal cord injury. *Neurology* **42**, 2162–2168.

Sinkjaer, T., Toft, E., Larsen, K., Andreassen, S., and Hansen, H. (1993). Non-reflex and reflex mediated ankle joint stiffness in multiple sclerosis patients with spasticity. *Muscle Nerve* **16**, 69–76.

Skold, C., Levi, R., and Seiger, A. (1999). Spasticity after traumatic spinal cord injury: nature, severity and location. *Arch. Phys. Med. Rehabil.* **80**, 1548–1557.

Solimena, M., and De Camilli, P. (1991). Autoimmunity to glutamic acid decarboxylase (GAD) in stiff-man syndrome and insulin-dependent diabetes mellitus. *Trends Neurosci.* **14**, 452–457.

Sonde, L., Kalimo, H., Fernaeus, S. E., and Viitanen, M. (2000). Low TENS treatment on post-stroke paretic arm: a three-year follow-up. *Clin. Rehabil.* **14**, 14–19.

Steinbock, P., Reiner, A. M., Beauchamp, R., Armstrong, R. W., Cochrane, D. D., and Kestle, J. (1997). A randomized clinical trial to compare selective posterior rhizotomy plus physiotherapy with physiotherapy alone in children with spastic diplegic cerebral palsy. *Dev. Med. Child. Neurol.* **39**, 178–184.

Stewart-Wynne, E. G., Silbert, P. L., Buffery, S., Perlman, D., and Tan, E. (1991). Intrathecal baclofen for severe spasticity: Five years experience. *Clin. Exp. Neurol.* **28**, 244–255.

Stien, R., Nordal, H. J., Oftedal, S. I., and Slettebo, M. (1987). The treatment of spasticity in multiple sclerosis: A double-blind clinical trial of a new antispastic drug tizanidine compared with baclofen. *Acta. Neurol. Scand.* **75**, 190–194.

Stöhr, M., and Heckl, R. (1977). Das Stiff-man-syndrome. *Arch. Psychiat. Nervenkr.* **223**, 171–180.

Thach, W. T., and Montgomery, E. B. (1990). Motor systems. *In* "Neurobiology of Disease." (A. L. Pearlman and R. C. Collins, Eds.), pp. 168–196. Oxford Univ. Press, Oxford.

Thilmann, A. F., Fellows, S. J., and Garms, E. (1991). The mechanism of spastic muscle hypertonus: Variation in reflex gain over the time course of spasticity. *Brain* **114**, 233–244.

Thompson, P. D. (1993). Stiff muscles. *J. Neurol. Neurosurg. Psychiatry* **56**, 121–124.

Tsang, K. L., Fong, K. Y., and Ho, S. L. (1999). Localized neuromyotonia of neck muscles after radiotherapy for nasopharyngeal carcinoma. *Mov. Disord.* **14**, 1047–1949.

van den Berg, J. S., van Engelen, B. G., Boerman, R. H., and de Baets, M. H. (1999). Acquired neuromyotonia: superiority of plasma exchange over high-dose intravenous human immunoglobulin. *J. Neurol.* **246**, 623–625.

Vincent, A. (1999). Immunology of the neuromuscular junction and presynaptic nerve terminal. *Curr. Opin. Neurol.* **12**, 545–551.

Vincent, A., Jacobson, L., Plested, P., Polizzi, A., Tang, T., Riemersma, S., Newland, C., Ghorazian, S., Farrar, J., MacLennan, C., Willcox, N., Beeson, D., and Newsom-Davis, J. (1998). Antibodies affecting ion channel function in acquired neuromyotonia, in seropositive and seronegative myasthenia gravis, and in antibody-mediated arthrogryposis multiplex congenita. *Ann. NY. Acad. Sci.* **841**, 482–496.

Visintin, M., and Barbeau, H. (1989). The effects of body weight support on the locomotor pattern of spastic paretic patients. *Can. J. Neurol. Sci.* **16**, 315–325.

Weingarden, H. P., Zeilig, G., Heruti, R., Shemesh, Y., Ohry, A., Dar, A., Katz, D., Nathan, R., and Smith, A. (1998). Hybrid functional electrical stimulation orthosis system for the upper limb: effects on spasticity in chronic stable hemiplegia. *Am. J. Phys. Med. Rehabil.* **77**, 276–281.

Wernig, A., Müller, S., Nanassy, A., and Cagol, E (1995). Laufband therapy on rules of spinal locomotion is effective in spinal cord injured persons. *Eur. J. Neurosci.* **7**, 823–829.

Wiesendanger, M., Chapman, C. E., Marini, G., and Schorderet, D. (1985). Experimental studies of dorsal cord stimulation in animal models of spasticity. *In* "Clinical Neurophysiology in Spasticity." (P. J. Delwaide and R. R. Young, Eds.), pp. 205–219. Elsevier, Amsterdam.

Wright, F. V., Sheil, E. M., Drake, J. M., Wedge, J. H., and Naumann, S. (1998). Evaluation of selective dorsal rhizotomy for the reduction of spasticity in cerebral palsy: a randomized controlled trial. *Dev. Med. Child. Neurol.* **40**, 239–247.

Wuolle, K. S., Van Doren, C. L., Bryden, A. M., Peckham, P. H., Keith, M. W., Kilgore, K. L., and Grill, J. H. (1999). Satisfaction with and usage of a hand neuroprosthesis. *Arch. Phys. Med. Rehabil.* **80**, 206–213.

Yarkony, G. M., Roth, E. J., Cybulski, G. R., and Jaeger, R. J. (1992). Neuromuscular stimulation in spinal cord injury. II: Prevention of secondary complications. *Arch. Phys. Med. Rehabil.* **73**, 195–200.

Young, R. R., and Delwaide, P. J. (1981). Drug therapy: Spasticity. *N. Engl. J. Med.* **304**, 28–33, 96–99.

Young, R. R. (2000). Spastic paresis. *In* "Multiple Sclerosis: Diagnosis, Medical Management, and Rehabilitation." (J. S. Burks and K. P. Johnson, Eds.), pp. 299–306. Demos, New York.

CHAPTER 88

# Inflammatory and Infectious Polyneuropathy

Ian Sutton and John B. Winer

## GUILLAIN-BARRÉ SYNDROME

### Clinical Aspects

The typical patient with Guillain-Barré syndrome (GBS) is seen after a nonspecific antecedent infection with tingling in the extremities and ascending symmetrical weakness progressing to flaccid tetraparesis over a week or two. Of such patients, one in four will develop respiratory insufficiency requiring assisted ventilation develop. Then, after a plateau, the patient gradually begins to recover over several weeks, eventually being able to return to work many months after admission to the hospital. This is a rough outline of a typical GBS patient's history and is meant to emphasise two important points in the management of GBS. First, the outcome is often good, but it takes a long time to recover from severe neurological deficits and possible complications. Second, if a specific treatment is to be effective, it must be initiated very early in the course of the illness, at a stage when it is still possible to turn off the immunological processes leading to demyelination and axonal damage.

GBS has an annual incidence of 1–2/100,000 worldwide and occurs at all ages. An upper respiratory tract or gastrointestinal infection precedes the onset of neurological symptoms by 1–3 weeks in 70% of patients. A considerable number of viruses or bacteria have been reported to trigger GBS. Of these the most frequent is *Campylobacter jejuni*, which is reported to precede GBS in 15%–46% of cases. Serological studies have detected IgM antibodies against cytomegalovirus in 15% of patients who do not necessarily have clinical symptoms of infection. IgM antibodies against Epstein-Barr virus have also been detected in up to 8% of patients. An association with human immunodeficiency virus (HIV) infection is also recognized, although there are no controlled epidemiological studies. HIV-seropositive patients often have CSF pleocytosis, and GBS may be the first clinical manifestation of HIV. Serum antibodies against GM1 or other glycolipids occur in 9%–30%. It has been suspected that glycolipid antibodies are correlated with preceding *C. jejuni* infection and poorer recovery; however, this has remained controversial (Enders *et al.*, 1993; Vriesendorp *et al.*, 1993). Approximately 6% of GBS patients have been vaccinated recently, but this may not be significantly different from the frequency of immunization in a control population (Winer *et al.*, 1988). A recent study suggested a possible slight excess of about 1 case per million from influenza vacination (Lasky *et al.*, 1998). An association of GBS with rabies vaccine (if prepared from CNS tissue, which is now widely abandoned) and a certain swine influenza vaccine in the United States (in the 1970s) is established; an association with poliovirus, *Hemophilia influenzae*, typhoid vaccine, and tetanus toxoid is possible.

The exact pathogenesis of GBS is not fully understood. It is now recognized to be a clinical syndrome encompassing a number of pathological entities. About 75% of cases of GBS are demyelinating, with multifocal demyelination with mononuclear inflammatory cell infiltrates in the peripheral nerves. Cases of primarily axonal neuropathy involving either motor (acute motor axonal neuropathy; AMAN) or motor and sensory (acute motor and sensory neuropathy; AMSAN) make up the remaining 25% of cases. An autoimmune origin is generally assumed on the basis of the close association with infection and the histological similarities with experimental autoimmune neuritis, an animal model of GBS. It is suggested that activation of T cells against peripheral nerve antigens initiates the autodestructive process, and subsequently macrophages, cytokines, and antibodies lead to the full-blown inflammatory picture with demyelination.

## Diagnosis

Diagnosis is mainly clinical and based on rapidly evolving, often ascending, symmetrical paralysis and areflexia (Table I). Sensory symptoms such as paresthesia and pain are found in more than half of the patients, whereas sensory signs are much less common. Cranial nerve involvement is common. Facial weakness is present in approximately 50% at some time during the disease, and ophthalmoplegia is present in 10%. Dysautonomic signs such as sinus tachycardia, labile hypertension, or postural hypotension have been reported in two thirds of the patients. They are usually mild, but their recognition is crucial for effective management.

GBS is the most common cause of acute neuromuscular paralysis, and diagnosis is frequently straightforward, but because of clinical variations in presentation and the wide differential diagnosis, exact diagnostic criteria (Table I) have been formulated (Asbury and Cornblath, 1990). Helpful diagnostic aids in GBS are

TABLE I   Outline of Diagnostic Criteria for Guillain-Barré Syndrome*

Required are
    Progressive motor weakness of more than one limb
    Areflexia: distal areflexia with definite hyporeflexia of the biceps
        and knee jerks will suffice
Strongly supportive are
    Clinical features
    Progression for less than 4 weeks
    Relative symmetry
    Mild sensory symptoms or signs
    Cranial nerve involvement (particularly facial weakness)
    Slow but continuous recovery, usually beginning 2–4 wk after
        progression stops
    Autonomic dysfunction (tachycardia, labile hypertension)
    Absence of fever
    Typical CSF findings with albuminocytological dissociation after
        the first week
    Typical electrodiagnostic findings with slowing of motor
        conduction velocity, conduction block or abnormal temporal
        dispersion, prolonged distal latencies, slowed F-wave responses;
        conduction studies may be abnormal only after a few weeks
Features that cast doubt on the diagnosis of GBS are
    Marked, persistent asymmetry
    Initial or persistent sphincter dysfunction
    More than 50 cells/ml in CSF
    Polymorphonuclear cells in CSF
    Sharp sensory level

*Note.* These criteria do not include up to 15% of variant
syndromes.
    *Based on Asbury and Cornblath (1990).

both CSF analysis and electrophysiological investigations. In the CSF, there is an increased protein level with a normal or only slightly increased cell count (albuminocytological dissociation). Protein concentrations may not increase for 1 week and are occasionally normal throughout the illness. Electrophysiological investigations carried out at an early stage of the disease may show only delayed F-wave and H-reflex latencies or minor increases in distal motor latencies. Slowing of nerve conduction velocities, conduction block, and temporal dispersion of compound action potentials are the predominant findings at later stages.

The differential diagnosis includes other causes of acute inflammatory or toxic neuropathy outlined in this and Chapters 76–79 and 84–88. Nonneuropathic causes of acute tetraparesis such as poliomyelitis (asymmetrical), rabies (history of animal bite), spinal cord compression, or myelitis (sharp sensory level), myasthenia gravis, botulism (dilated pupils), acute myopathy, electrolyte disturbances, hysteria, or brainstem infarction may occasionally cause diagnostic difficulty.

## Clinical Variants

Clinical variants of GBS do not strictly meet the criteria in Table I but seem to overlap with typical GBS in some respect. Most clinicians believe that atypical variants of GBS are more common than typical presentations of the rarer diseases they mimic. *Fisher's syndrome* occurs in approximately 5% of cases of GBS in several series and consists of external ophthalmoplegia, ataxia, and areflexia. It usually begins with double vision followed a few days later by ataxia. The mechanism of the ataxia is uncertain, because CNS lesions have never been convincingly demonstrated, and proprioception is either normal or only mildly impaired. A mismatch of information from muscle spindles and other proprioceptive afferents within the cerebellum has been postulated. About one third of cases of Fisher's syndrome eventually progress to profound weakness, which together with the raised CSF protein, link the syndrome to GBS itself. IgG anti-GQ1b ganglioside antibodies have been associated with Fisher's syndrome and GBS with ophthalmoplegia (Chiba *et al.*, 1993). These antibodies have been shown to affect neuromuscular transmission in the mouse hemidiaphragm preparation (Roberts *et al.*, 1994) *Sensory polyneuritis* with areflexia accompanied by dysautonomia but little or no weakness is rare. It may develop more slowly and may be difficult to differentiate from other forms of sensory polyneuropathy. *Pure pandysautonomia* is characterized by a relatively selective involvement of sympathetic and parasympathetic functions more or less sparing the other peripheral nerve functions.

## Natural Course

Weakness progresses continuously and reaches a maximum after 2 weeks in approximately two thirds of the patients. Maximum severity varies considerably. More than 50% are bedbound, and at least half of those will require temporary artificial ventilation. Thanks to modern intensive care facilities the mortality rate has fallen to 3%–6% in most larger modern series. Causes of death are pneumonia, pulmonary embolus, respiratory failure, cardiac arrest, sepsis, or profound dysautonomia. By contrast, an estimated 20% of GBS patients are able to walk unassisted throughout their illness.

Recovery begins after a plateau phase of 1–4 weeks; that is, between 2 and 6 weeks after onset of symptoms. Full remission with no or very minor sequelae is seen in more than 70%. Severe residual weakness is nowadays found in less than 15%. Poor outcome correlates with older age, rapid progression, requirement for respiratory assistance, and electrophysiological findings (especially small compound muscle action potentials).

Recurrent GBS with up to seven episodes has been described in 2%–5% of the patients. Relapses are commonly associated with infections of various types. Rapid onset of paresis with subsequent complete or near complete recovery is the rule (Grand'Maison et al., 1992).

GBS in childhood is generally similar to the disease in adults but is usually less severe with a better prognosis.

## Principles of Therapy

*General medical and nursing care* are most essential in the treatment of a GBS patient (Hughes, 1990). Severely ill patients are best cared for in an intensive care unit, where frequent monitoring of vital functions is possible. Deteriorating respiratory capacity should prompt intubation and assisted ventilation before serious dyspnea or hypoxemia occurs. A small portable respirometer is useful in monitoring expiratory vital capacity. If it decreases to 15 ml/kg, there is immediate danger of respiratory muscle fatigue (Ropper and Kehne, 1985). Also, severe oropharyngeal weakness with the danger of aspiration and difficulty clearing secretions should initiate mechanical ventilation. If respiratory assistance is required for more than 10 days, tracheostomy should be performed. In tetraparetic patients correct positioning, frequent turning, and subcutaneous heparin are mandatory. Dysphagia may require feeding by a nasogastric tube. Decrease in blood pressure is best treated with volume replacement. Hypertension may require nifedipine. If episodes of profound bradycardia occur, especially during tracheal suction, a temporary pacemaker should be inserted.

Systolic hypertension and reduced R-R interval variation were significantly more common in patients who subsequently had serious arrhythmias develop (Winer and Hughes, 1988), and these simple autonomic function tests can help to predict the patients at risk. Pain has often been underestimated in GBS. It sometimes responds to conventional analgesics. Anticonvulsants, antidepressants, local capsaicin, corticosteroids, quinine, and morphine have all been recommended with variable success.

Psychological support and continuous but realistic reassurance are necessary to help the patient to overcome the long and distressing cause of uncertainty and helplessness. Rehabilitation must start as soon as possible and should include a coordinated team of physicians, nurses, physiotherapists, and occupational and speech therapists, as well as psychologists and social workers. Early physiotherapy is important to facilitate pulmonary toilet, prevent tendon shortening, and retain joint mobility. When muscle power returns, an active training and mobilization program should help to improve stamina. Utensil-holding devices help if ataxia is prominent. Orthotic devices, such as ankle-foot orthoses, assist in persistent muscle weakness. Most patients will eventually return to work; however, depression and even suicide have been reported after GBS and should be prevented by professional help.

## Specific Therapies

Among the specific therapies investigated only plasmapheresis and intravenous immunoglobulins (IVIg) have been shown to be effective in GBS. Corticosteroids alone are of no benefit (Hughes, 1991). Why plasmapheresis or immunoglobulin therapy help is not clearly understood. An inflammatory myelinotoxic serum factor may be removed by plasmapheresis. This idea is supported by the clinical impression that treatment response in some patients is extremely rapid and obviously not dependent on remyelination. On the other hand, high-dose immunoglobulins may be effective by blocking surface Fc receptors on macrophages, which are known to play an effective role in demyelination or by providing a source of anti-idiotypes that interfere with putative pathogenetic antimyelin antibodies or by swamping mechanisms designed to prevent the breakdown of natural immunoglobulin and shortening the half-life of putative antimyelin antibodies (Yu and Lennon, 1999).

*Plasmapheresis* was shown to be effective in GBS in two large randomized trials including more than 400 patients (French Cooperative Group on Plasma Exchange in Guillain-Barré Syndrome, 1987; The Guillain-Barré Syndrome Study Group, 1985). Plasmapheresis shortened the time to independent walking in both studies (by 4.5 and 6 weeks, respectively) and improved outcome after 1 year. The incidence of lasting severe motor disability, however, was not affected. Patients with disease duration less than 7 days had greater benefit.

Usually four to six treatments are performed on alternate days, exchanging 40–50 ml/kg plasma each time. Albumin was as efficient as fresh frozen plasma as replacement fluid. Treatment should be initiated in patients who cannot walk independently or those in whom worsening to such a level can be expected. In mild disease significant improvement by plasmapheresis has not been documented. Complications of plasmapheresis are allergic reactions, thrombosis, bleeding, electrolyte and protein depletion, hypocalcemia (caused by citrate), and bacterial and viral infection. Severe side effects are rare. Mortality rate is 3/10,000 procedures. Relative contraindications are hepatic failure, severe electrolyte disturbances, increased risk for cardiovascular complications, active infection, or bleeding disorders.

IVIg treatment seems to be as effective as plasmapheresis (Anonymous 1997; Van der Meché et al., 1992). In the Dutch study of more than 150 patients those that were treated with IVIg required less assisted ventilation and had a slightly better clinical outcome after 4 weeks than those receiving plasma exchange. However, patients treated with plasmapheresis responded rather poorly compared with the treatment arms of the French and American trials, and on follow-up assessment at 3 months there was no significant difference between the two treatment groups. IVIg was also started on average 1 day before plasmapheresis, which may have biased the results in favor of that treatment arm. In 1997 a large international multicenter trial compared plasma exchange, immunoglobulin, and a combination of both treatments. This study confirmed that plasma exchange and immunoglobulin are roughly equivalent in shortening the time taken for patients to regain the ability to walk unaided. A small but insignificant advantage was seen with both treatments combined. The mean change after 4 weeks on a 7-point disability scale was 0.9 for PE, 0.8 for IVIg, and 1.1 for both treatments. In light of this trial most neurologists favour IVIg over plasma exchange as the treatment of choice because of the lower morbidity. The recommended dose is 0.4 g/kg/day for 5 days, which should be given at 4.5 g/h. Side effects such as allergic reactions, tachycardia, and hypotension are infrequent. Contraindications are allergies against immunoglobulins, selective IgA deficiency (which is very rare), and severe cardiovascular problems. A clear recommendation for one of the several available immunoglobulin formulations has not been made.

Plasma exchange is a suitable treatment for patients who fail to respond to IVIg. Doubt still exits about whether a second course of IVIg is worthwhile, and plans for a multicenter trial to address this issue are well advanced. Pilot studies on the possible use of interferon-beta are also in progress.

### Practical Management

Supportive medical care is the mainstay of GBS therapy. ICU treatment is necessary in case of arrhythmias or incipient respiratory failure. Artificial ventilation should be initiated, when vital capacity drops to 15 ml/kg, or in case of severe oropharyngeal weakness. Tracheostomy is recommended when artificial ventilation is required for more than 10 days.

Rehabilitation must start as soon as possible. Physiotherapy should begin with entry to a hospital. Regular communication should be present between doctors, nurses, physiotherapists, speech and occupational therapists, psychologists, and social workers involved in the patient's care. This should help to coordinate the various therapies, avoid complications, assess progress, and eventually ease the patient's way back into a normal life.

IVIg should be administered as soon as possible after onset of the neuropathy if the patient cannot walk independently. In milder cases the decision to consider IVIg depends on the clinical course and the degree of disability. If the patient deteriorates after IVIg, the authors would recommend plasma exchange if easily available or if not a second course of IVIg.

## CHRONIC INFLAMMATORY DEMYELINATING POLYNEUROPATHY

### Clinical Aspects

Chronic inflammatory demyelinating polyneuropathy (CIDP) is now considered a disease entity separate from GBS, although common clinical and probably pathogenetic features exist. It is characterized by a more insidious onset and, importantly, by its response to immunosuppressive therapy. CIDP is rarer than GBS, and its annual incidence has been estimated to be 0.25–0.5/100,000 (Hughes, 1990). It may occur at any age, with men being slightly more often affected than women. Preceding infections or vaccinations are found in less than one third of the cases.

The prominent clinical feature is weakness, usually symmetrical and global. Areflexia is considered a mandatory diagnostic feature, although 10% of patients may only lose ankle reflexes. In slowly progressive disease, reflexes may be present for weeks. Sensory deficit is common and may be the major feature in 10%. Typically loss of large fiber function is prominent. Pain is a symptom in 20%, paresthesias in more than half of the patients. Cranial nerve involvement (facial weakness, ophthalmoplegia, bulbar weakness) occurs in 10%–20%. In a few patients with high CSF protein levels, papilledema has been described. Tremor (mostly action tremor) has been reported in a few patients, but the mechanism is uncertain. CNS demyelination detected by magnetic resonance imaging, sometimes associated with clinical features of multiple sclerosis, has been reported in up to 38% of CIDP patients in smaller series (Feasby *et al.*, 1990; Mendell *et al.*, 1987). However, this high incidence may be due to a selection bias. Exact diagnostic criteria (Table II) have been proposed but are less useful than in GBS because of greater variability. Variants may begin with a rapid onset (in one study in 16% of the patients) or with a purely sensory deficit (6% in McCombe *et al.*, 1987).

Clinical diagnosis is supported by typical findings in CSF, nerve conduction studies, and if diagnostic doubt still exists, nerve biopsy (Table II). A raised CSF protein is probably the most consistent laboratory finding and is detected in 95% of the patients. The differential diagnosis includes all chronic demyelinating neuropathies of both acquired and hereditary types. For therapeutic reasons it is important to distinguish CIDP from GBS, multifocal motor neuropathy, and neuropathy with monoclonal gammopathy of undetermined significance.

## Natural Course

A relapsing course is found in approximately half the patients, a progressive or monophasic course in the remainder. Older series report a mortality of about 10%, usually caused by infections, but now the natural course is nearly always modified by immunosuppressive therapy. Life-threatening disease requiring artificial ventilation is rare. Less than 10% of patients eventually have a severe disabling neurological deficit develop, whereas most make a good recovery.

## Principles of Therapy

The effectiveness of corticosteroids, plasmapheresis, and IVIg in CIDP has been established in controlled trials (Dyck *et al.*, 1982, 1986; Hahn *et al.*, 1996a, 1996b; Vermeulen *et al.*, 1993). Azathioprine in addition to corticosteroids does not result in a better outcome but may help to reduce the steroid dose (Hughes, 1990).

The preference and sequence of these treatment regimens is a matter of debate. We suggest that in moderately affected patients initial steroid treatment should suffice and can be slowly tapered after clinical improvement. Steroid withdrawal is a common cause of relapse, and the requirement for high doses of steroids is an indication for adding azathioprine. In patients with severe neurological deficit, IVIg is our preferred initial treatment followed by steroids and azathioprine. We usually reserve plasmapheresis for patients that remain very weak after IVIg.

Occasional patients fail to tolerate or deteriorate despite immunosuppressive treatment with steroids and azathioprine. Anecdotal reports support the use of cyclophosphamide (Good *et al.*, 1998) or cyclosporin (Barnet *et al.*, 1998) in such patients. There are also reports of successful use of beta-interferon (Hadden *et al.*, 1999) and alpha-interferon (Gorson *et al.*, 1998). Prophylaxis for osteoporosis in patients requiring prolonged steroids remains controversial, but our practice is to instigate prophylaxis in all patients with CIDP on steroid treatment, because it is unusual for treatment to last less than 6 months.

The general therapeutic guidelines outlined for GBS also apply to severely ill CIDP patients.

## Practical Management

IVIg infusions should be administered in patients with severe neurological deficit (difficulty walking or disabling sensory loss) or with a rapidly progressive course. If there is no improvement, plasmapheresis with six exchanges can be tried. Further details of plasmapheresis and IVIg treatment are described in the section on GBS.

---

**TABLE II  Outline of Diagnostic Criteria for CIDP**

Clinial criteria
   Progression of peripheral motor deficit for 2 mo (usually much longer)
   Hyporeflexia or areflexia
   Symmetrical proximal and distal weakness
   Large fiber sensory loss predominates over small fiber loss
Laboratory criteria
   Elevated CSF protein with normal or slightly raised cell count
   Nerve conduction studies with features of demyelination (reduction of conduction velocities; partial conduction block; prolonged distal latencies; absent or prolonged F-wave latencies)
   Nerve biopsy with features of demyelination and remyelination

*Note.* Based on Barohn *et al.* (1989) and Report from an Ad Hoc Subcommittee of the American Academy of Neurology AIDS Task Force (1991). A complete list of inclusion and exclusion criteria for research purposes is given there. The first two clinical criteria are considered mandatory.

Corticosteroids may be the only treatment required in milder cases. In patients with severe neurological deficit, it should be accompanied by IVIg from the beginning. Prednisone (or prednisolone) at 1–1.5 mg/kg should be given for at least 4 weeks. This is slowly tapered and changed to an alternate-day regimen. Dose reduction must depend on the clinical response but should be no faster than 5 mg every 2 weeks. Antacids and potassium supplements are necessary to avoid side effects. Prophylaxis with a biphosphonate, hormone replacement therapy, or vitamin D with calcium should be instigated if the average steroid dose is likely to be greater than 7.5 mg for more than 6 months.

Azathioprine (2–2.5 mg/kg) should be added if steroid reduction leads to neurological deterioration. It should be given for approximately 2 years, because the drug needs at least 6 months and possibly as long as 18 months to show its full potency. Severe side effects are leukopenia and impairment of liver function. White blood counts and liver enzymes should be monitored weekly for 8 weeks, then monthly. A small percentage (probably 5%) of patients have acute gastrointestinal symptoms develop and do not tolerate the drug.

## MULTIFOCAL MOTOR NEUROPATHY

### Clinical Aspects

Multifocal motor neuropathy (MMN) is a distinct nosological entity that is related to CIDP. MMN responds to immunosupression with IVIg and cyclophosphamide. However, plasma exchange and steroids are ineffective and may worsen the condition.

MMN typically produces distal and asymmetrical weakness usually beginning in the upper limbs. Nerve conduction studies reveal evidence of selective motor nerve demyelination, but although motor conduction block (MCB) has been used as the main electrodiagnostic criterion, there is no universally agreed definition of MCB. MCB may vary over time, and a >50% CMAP amplitude reduction may well be preceded by a smaller decremental response that is indicative of focal demyelination in the appropriate clinical setting. Furthermore, even though conduction block may be missed unless specifically sought, occasional patients with a typical MMN phenotype may not show any evidence of overt conduction block. Although 50%–85% of patients have high titers of IgM antibody reactive with the GM1 ganglioside, anti-GM1 sera do not seem to mediate conduction block (Hirota *et al.*, 1997).

### Natural Course

The natural course in most patients is a slow progression of symptoms leading to amyotrophy.

### Principles of Therapy

Intravenous immunoglobulin therapy is now a well-established treatment for MMN (Leger *et al.*, 2001), and 70% of patients will show signs of improvement after initial treatment. Patients with high titers of anti-GM1 antibodies are more likely to respond to IVIg, although anti-GM1 antibody titers do not fall with treatment (Leger *et al.*, 2001). Patients with no demonstrable conduction block are less likely to respond (Azulay *et al.*, 1994). However, the improved muscle strength seen in responders is rarely maintained, generally lasting for less than 12 weeks, and regular maintenance immunogloulin therapy only maintains this improvement in 60%. Cyclophosphamide has been successfully applied as a primary therapy with a 70% response rate but is less likely to be used than IVIg because of potential toxic side effects. Use of cyclophosphamide as an adjunct therapy to IVIg increases the interval between IVIg infusions and may lead to temporary cessation of both treatments (Meucci *et al.*, 1997).

Favorable results have been observed in an open study using Rituximab, an anti-CD20 monoclonal antibody that depletes B cells, in four patients with MMN that had become resistant to other forms of immunotherapy (Levine and Pestronk, 1999). Interferon-beta1a (IFN-beta1a) has been used to successfully treat three patients with MMN who no longer responded to treatment with IVIg and cyclophosphamide (Martina *et al.*, 1999). The finding in a second unblinded study that three of nine patients showed an improvement with IFN-beta1a that was more pronounced than with IVIg requires further investigation (Van den Berg-Vos *et al.*, 2000).

### Practical Management

Most patients with MMN and conduction block will improve after a 5-day course of IVIg at a dose of 0.4 g/kg/day. Improvement is rarely sustained, and regular maintenance therapy is usually required. Our practice is to administer one dose of 0.8 g/kg 2 months after the initial course and review maintenance dose and frequency of administration according to clinical response

In patients who fail to respond to IVIg, cyclophosphamide is probably the most frequently used drug, although no controlled data support its use in MMN.

# MULTIFOCAL ACQUIRED DEMYELINATING SENSORY AND MOTOR NEUROPATHY (MADSAM)—LEWIS-SUMNER SYNDROME

In 1982 five patients with a chronic asymmetric sensorimotor neuropathy that was most pronounced in the upper limbs and was associated with electrophysiological evidence of persistent multifocal conduction block were reported (Lewis *et al.*, 1982). Two recent studies (Saperstein *et al.*, 1999; Van den Berg-Vos *et al.*, 2000) have described similar patient groups, suggesting the existence of a distinct clinical entity that probably represents an asymmetrical form of CIDP but clearly differs form MMN. Patients are seen with a multifocal and asymmetrical acquired sensorimotor demyelinating neuropathy, with nerve conduction studies showing conduction block and other features of demyelination. Patients with MADSAM are more likely have an elevated CSF protein (82% vs 9% with MMN) and do not have elevated titers of anti-GM1 antibodies. Sural nerve biopsies in MADSAM show prominent evidence of demyelination, whereas only subtle abnormalities are seen in similar biopsies from patients with MMN. MADSAM neuropathy patients improve with IVIg, but in contrast to MMN, MADSAM responds to treatment with prednisolone.

# POLYNEUROPATHY WITH MONOCLONAL GAMMOPATHY

## Clinical Aspects

Monoclonal paraproteins are found in 10% of patients with peripheral neuropathy, and immunoelectophoresis or immunofixation of blood and urine is an essential investigation in patients with unexplained neuropathy. Most patients with detectable paraproteins have monoclonal gammopathy of undetermined significance (MGUS) (Table III).

MGUS is characterized by low levels of serum M protein (less than 3 g/dl), low levels or absent Bence Jones proteins, less than 5% bone marrow plasma cells and the absence of anemia, renal insufficiency, bone lesions, and hypercalcaemia at presentation. Seventeen percent of cases of MGUS will transform into a malignant condition after 10 years, and by 20 years 33% of cases will have become malignant (Kyle and Lust, 1990). In most cases the neuropathy associated with MGUS is a demyelinating neuropathy that resembles CIDP. Although neuropathies associated with MGUS are a heterogeneous group, comparisons between idiopathic CIDP and MGUS neuropathy reveal that patients with MGUS neuropathy usually have less weakness, more sensory impairment, and more abnormal sensory conduction studies than those with idiopathic CIDP (Simmons *et al.*, 1993a). However, in our experience patients with IgG and IgA paraproteins are usually clinically indistinguishable from idiopathic CIDP and respond in a similar fashion to immunotherapy.

Relatively homogeneous groups of patients with MGUS neuropathy can be identified in patients with IgM paraprotein–associated neuropathies. One group is formed by 50% of patients with an IgM paraproteinemia, in whom the paraprotein reacts against myelin-associated glycoprotein (MAG). Evidence suggests that anti-MAG antibodies are involved in the pathogenesis of the associated neuropathy that is characterized clinically by distal sensory loss, gait ataxia, and mild distal weakness. Electrodiagnostic studies show slow conduction velocities without evidence of conduction block and prolonged distal motor latencies. IgM paraproteins that react with gangliosides, notably GD1b, are found in some patients with autoimmune ataxic sensory neu-

**TABLE III  Clinical Diagnosis in Patients with a Paraprotein**

|  | A: Incidence of disorders in 856 patients with a para-protein | B: Incidence of disorders in 28 patients with a para-protein and neuropathy* |
|---|---|---|
| MGUS | 63% (541) | 57% (16) |
| Multiple myeloma | 12% (102) | 11% (3) |
| Amyloidosis | 9% (75) | 25% (7) |
| Lymphoma | 5% (43) | 0% |
| Plasmacytoma | 4% (33) | 0% |
| CLL | 3% (23) | 0% |
| Indolent myeloma | 2% (19) | 0% |
| γ-Heavy chain disease | 0% | 3.5% (1) |
| Waldenstrom's macroglobulinaemia | 2% (20) | 3.5% (1) |

Column A shows the clinical diagnosis in 856 patients with a serum monoclonal protein in whom the neuropathy status is unspecified (Kyle, 1992). Column B shows the clinical diagnosis in 28 of 279 with a paraprotein who also had a neuropathy (Kelly *et al.*, 1981).

ropathies. Another well-defined clinical phenotype is CANOMAD (Chronic Ataxic Neuropathy with Ophthalmoplegia, M-protein, Agglutination and Disialosyl antibodies) in which the IgM paraprotein reacts with gangliosides containing the disialosyl moiety (GD3, GD1b, GT1b, and GQ1b) (Quarles and Dalakas, 1996).

Paraproteinemic neuropathies arise infrequently in patients with osteolytic myeloma and Waldenstrom's macroglobulinaemia, occurring in only 5% of patients with these disorders. However, although osteosclerotic myeloma accounts for only 3% of all myelomas, it is complicated by neuropathy in half the cases and overlaps with the POEMS syndrome (polyneuropathy, organomegaly, endocrinopathy, M-protein, and skin changes). POEMS is characterized by an elevated CSF protein and associated with elevated levels of vascular endothelial growth factor (VEGF) that fall with treatment (Watanabe et al., 1998).

Seventeen percent of patients with primary AL amyloidosis have a peripheral neuropathy at presentation. Electrodiagnostic studies reveal an axonal neuropathy that clinically is seen with symptoms of autonomic and small fiber sensory dysfunction.

A generalized axonal sensorimotor polyneuropathy or less frequently a multifocal neuropathy can arise in patients with cryoglobulinemia. Cryoglobulins are monoclonal and polyclonal serum immunoglobulins that precipitate in the cold. Two thirds of patients with mixed cryoglobulinemic neuropathy have evidence of previous hepatitis C infection (Apartis et al., 1996).

### Natural Course

In paraproteinemic neuropathy with MGUS slow chronic progression is the rule. Most patients with anti-MAG IgM have a favorable prognosis even after several years (Smith, 1994). Osteosclerotic myeloma has a better prognosis than osteolytic myeloma, with some patients surviving more than 10 years. Neuropathy in the former improves with treatment of the osteosclerotic lesion, whereas neuropathy in the latter runs an independent course. The prognosis in amyloidosis is poor, and the median survival in patients with neuropathy is 25 months (Rajkumar et al., 1998).

### Principles of Treatment

Response to immunotherapy is variable in patients with paraproteinemic neuropathy. Patients with IgG or IgA paraproteins that are otherwise indistinguishable to CIDP tend to respond to immunotherapies in a similar manner to idiopathic CIDP. Patients with IgM parapro-

teins, especially with anti-MAG antibodies, show a poor response to both steroids (Nobile-Orazio et al., 2000) and IVIg therapy—18% (Dalakas et al., 1996) and 24% (Ellie et al., 1996). Plasma exchange results in improvement in paraproteinemic neuropathy with MGUS, but the effect is more marked in patients with IgA or IgG paraproteins than those with IgM paraproteins (Dyck et al., 1991). Furthermore, although 50% of patients with neuropathy and an IgM paraprotein are reported to improve with chlorambucil treatment, concomitant use of plasma exchange with chlorambucil is of no additional benefit (Oksenhendler et al., 1995). Two small uncontrolled studies of monthly pulsed cyclophosphamide for 6 months in paraproteinemic neuropathy–MGUS have been reported. In one trial patients also received concomitant plasma exchange, and improvement was observed in all four patients (IgM-associated paraproteinaemic neuropathy) (Blume et al., 1995), and in the second trial 16 patients (11 IgM, 5 IgG) also received pulsed methlyprednisolone for 6 months leading to improvement in 50% and stabilization in 38% (Notermans et al., 1996).

Recent interesting anecdotal reports of response to treatment with fludarabine (Wilson et al., 1999) and reduction in IgM titers after treatment with Rituximab, an anti-CD20 monoclonal antibody, require further investigation (Latov and Sherman, 1999; Levine and Pestronk, 1999).

The recognition of osteosclerotic myeloma–POEMS syndrome is important, because the neuropathy improves with radiation treatment of the associated bone lesion. Local treatment with 40–50 Gy leads to slow improvement that continues for years after the radiation therapy. In multiple widespread lesions, melphalan plus prednisolone therapy is indicated (Kyle and Dyck, 1993b). Tamoxifen use has also shown beneficial effects (Enevoldson and Harding, 1992). Coticosteroids, IVIg, azathioprine, and plasmapheresis are usually ineffective.

Waldenstrom's macroglobulinaemia is treated with an alkylating agent and a glucocorticoid, inducing a response in approximately 50% of patients. Accompanying polyneuropathy might improve, but this has been poorly documented (Dalakas et al., 1983). Multiple myeloma may respond to chemotherapy but usually without influence on the neuropathy.

Treatment of amyloidosis with melphalan and prednisolone is associated with a median survival of 18 months (Kyle et al., 1997) but does not result in improvement of the associated neuropathy.

Cryoglobulinemic neuropathy can be treated with prednisolone, plasma exchange, and other immunosupressive drugs. Those cases associated with hepatitis C infection respond to treatment with IFN-α (Misiani *et al.*, 1994). However, it should be noted that initial IFN-α treatment has been associated with precipitation of severe weakness before improvement after treatment with prednisolone and reintroduction of IFN-α (Scelsa *et al.*, 1998).

### Practical Management

1. A search for monoclonal gammopathy in serum and urine is mandatory in polyneuropathy of unknown cause.

2. In the case of gammopathy one has to exclude a malignant condition by bone marrow studies, radioisotope, and x-ray skeletal surveys. Regular monitoring of paraprotein levels in MGUS is required, because malignant transformation will occur in one third of cases followed up for 20 years.

3. Those patients with paraproteinemic neuropathies and MGUS with stable or slowly progressive disease should not be subjected to toxic immunotherapies.

4. Sural nerve biopsy may result in a diagnosis of amyloidosis in up to 86% of cases of amyloid neuropathy (Kyle and Dyck, 1993a); however, other studies show a much lower rate of diagnosis (Simmons *et al.*, 1993b). Therefore, a negative sural nerve biopsy does not exclude amyloidosis, and other tissues should be biopsied if amyloidosis is suspected.

5. In primary systemic amyloidosis therapy with melphalan (0.15 mg/kg daily for 7 days every 6 weeks) and prednisolone (0.8 mg/kg daily for 7 days every 6 weeks). Treatment should be continued for 2 years assuming there is no toxicity and the melphalan dose increased by 2 mg every 6 weeks until midcycle leukopaenia or thrombocytopenia develops.

6. Solitary lesions in osteosclerotic myeloma–POEMS syndrome should be treated with radiation (40–50 Gy). In cases with widespread lesions melphalan (0.15 mg/kg daily) plus prednisolone (20 mg three times daily) for 7 days every 6 weeks is recommended. This should be continued for approximately 2 years.

7. Waldenstrom's macroglobulinemia or multiple myeloma are treated according to oncological guidelines.

## SARCOID NEUROPATHY

### Clinical Aspects

Neurological involvement is found in 5% of patients with sarcoidosis. In two thirds of those the peripheral nervous system is affected, and most commonly there are multiple fluctuating cranial nerve palsies. Facial nerve palsy is by far the most common, and if solitary, it is indistinguishable from idiopathic Bell's palsy. Atypical neuropathies like mononeuritis multiplex or Guillain-Barré syndrome have also been reported and may be the first manifestation of sarcoidosis. Subacute or chronic sensorimotor neuropathy may also be present and shows prominent axonal degeneration (Gainsborough *et al.*, 1991). Granulomatous infiltration is thought to underlie pathological change in single-nerve lesions, but verification of sarcoidosis by sural biopsy may not be possible without clinical involvement (Matthews, 1993). In the more acute forms of demyelinating polyneuritis an immunological mechanism has also been postulated. Diagnosis is based on the findings of systemic sarcoidosis. The CSF shows raised levels of protein and angiotensin-converting enzyme (ACE) in most patients. Serum ACE levels are usually not diagnostic in neurosarcoidosis.

### Practical Management

All forms of peripheral nerve involvement may resolve spontaneously, as described in older series. Nowadays, corticosteroid therapy (prednisolone up to 1 mg/kg for several weeks) is usually recommended with good results (Gainsborough *et al.*, 1991) (see also Chapter 38). Our practice is to add azathioprine if high steroid doses are required for maintenance of treatment.

## VASCULITIC NEUROPATHY

### Clinical Aspects

Peripheral nerve involvement is found in approximately 50% of patients with polyarteritis nodosa (PAN), its Churg-Strauss variant, and essential mixed cryoglobulinemia. Neuropathy is also found in roughly 10% of patients with rheumatoid arthritis, Sjögrens syndrome, systemic lupus erythematosus (SLE), Wegener's granulomatosis, and giant cell arteritis. Rarely, neuropathy has been described in systemic sclerosis, mixed connective tissue disease, and in Behçet's syndrome. According to the low overall incidence of some of these syndromes, the neurologist is most often confronted with neuropathy in PAN and rheumatoid arthritis. Vasculitis with purely neuropathic symptoms makes up 30% of this group. This syndrome of vasculitic neuropathy without clinical evidence of other organ involvement shares the common clinical and pathological features.

Clinically, all forms of vasculitic neuropathy may appear as multifocal neuropathy or as distal symmetrical sensorimotor neuropathy. Multifocal neuropathy may be the first manifestation of PAN. By contrast, in rheumatoid arthritis a distal symmetrical sensorimotor neuropathy typically develops after a duration of 10 years. In vasculitic neuropathy pain is a common symptom, and cranial nerve involvement may be present. Cases of trigeminal neuropathy have repeatedly been described in Sjögren's syndrome and systemic sclerosis. A subacute neuropathy resembling Guillain-Barré syndrome may sometimes develop in SLE.

The pathogenesis of vasculitic neuropathy probably involves T-cell–mediated damage, as well as antibody-mediated mechanisms and immune complex deposits. The common denominator is a necrotizing vasculitis leading to ischemic nerve damage with predominantly axonal degeneration and mild secondary segmental demyelination. It is not possible to diagnose a specific vasculitic syndrome from nerve biopsy. Diagnosis rests on the biopsy findings frequently accompanied by lymphocytic infiltrate in active lesions together with features of a systemic disease. Electrophysiological studies show evidence for axonal neuropathy with relatively normal conduction velocities. CSF is usually normal.

### Natural Course

Peripheral neuropathy usually improves with treatment, but axonal damage is slow to repair, and recovery is often limited by the prognosis of the underlying systemic disease (Hawke *et al.*, 1991).

### Principles of Therapy and Practical Management

Controlled studies for vasculitic neuropathy are rare. From uncontrolled trials it seems that a combination of corticosteroids and cyclophosphamide is probably the best currently available therapy. Treatment must also be guided by the nature and extent of the systemic disease.

In PAN a combination of corticosteroids (prednisolone up to 1 mg/kg/day at onset, with gradual tapering) and oral cyclophosphamide (2 mg/kg/day; up to a maximum dose of 150 mg/day) should be administered until 3 months after remission (minimum 6 months and maximum 12 months). After induction of remission, azathioprine should be commenced at a dose of 1.5 mg/kg to prevent relapse. A similar approach seems sensible in most other syndromes. In rheumatoid vasculitis penicillamine and plasmapheresis may also be beneficial. Methotrexate is gaining support as an effective immunosuppressant among rheumatologists, although controlled supportive data are lacking. For treatment of temporal arteritis, see Chapter 32.

## NEURALGIC AMYOTROPHY

### Clinical Aspects

Acute brachial plexus neuropathy or neuralgic amyotrophy is believed to have an inflammatory or immune-mediated origin in most cases. This conclusion is based on the observation that neuralgic amyotrophy is often associated with antecedent infections or vaccinations. However, more than half the patients have no such event. The estimated annual incidence is 1.6/100,000 with a slight male predominance.

Abrupt onset of shoulder pain is the typical initial symptom. It lasts for days to a few weeks and spontaneously subsides. Weakness and often atrophy of shoulder girdle muscles then appears. The most commonly affected muscles are deltoid, supraspinatus, and infraspinatus followed by serratus anterior, biceps, and triceps. Rarely, phrenic nerve involvement results in diaphragmatic weakness. Sensory loss is less severe and less frequent than the observed motor deficits. CSF and brachial plexus MRI with gadolininium are normal. EMG sampling is invariably abnormal in affected muscles showing evidence of denervation. Nerve conduction studies usually reveal normal median and ulnar motor function, and on occasions reduced sensory nerve action potentials help distinguish a brachial plexus lesion from a more proximal lesion.

Differential diagnosis includes a wide variety of syndromes, namely, shoulder injuries, rotator cuff tears, traumatic mononeuropathies, cervical root syndromes, Lyme radiculitis, thoracic outlet syndromes, polymyalgia rheumatica, and plexus lesions caused by tumor, metastases, or after radiation. It should be noted that inflammatory neuralgic amyotrophy is usually clinically indistinguishable from the autosomal dominantly inherited form of this disorder, hereditary neuralgic amyotrophy (HNA), that has currently been linked to the 17q24–q25 locus. A second inherited form, as yet unlinked, is seen with a more chronic course (van Alfen *et al.*, 2000). It should be noted that although HNA is usually considered to be a separate entity from hereditary neuropathy with liability to pressure palsies (HNPP), brachial neuropathy can occasionally be the sole manifestation of the latter condition (Gil-Neciga *et al.*, 2000). Although recurrence of inflammatory neuralgic amyotrophy occurs in 5% of cases, recurrent attacks and a positive family history point toward HNA.

### Natural Course

Prognosis is good; rapid recovery is the rule in mild cases. In severely affected patients full recovery may take up to 3 years. Upper plexus lesions recover more quickly than lower plexus lesions.

## Principles of Therapy and Practical Management

High-dose prednisolone (40–60 mg) is effective in the management of pain and should be tapered after resolution of the pain, because there is no evidence that this treatment affects speed of recovery or leads to a reduction in the final disability. Immobilization with an arm sling will reduce pain and may prevent humeral subluxation, but daily physiotherapy is very important to prevent complications such as painful arc syndrome and contractures.

# LEPROSY

## Clinical Aspects

Leprosy is the world's most common treatable neuropathy. It is estimated that there are more than 5 million cases throughout the world. The highest disease rate is in the Far East, with a prevalence rate of 28 cases/10,000; 9000 cases are estimated in Europe (Noordeen, 1993). Infection is mainly transmitted by inhalation during close, prolonged, or repeated contact with infected untreated patients. Only a small proportion of the population is susceptible. Incubation time lasts from months to many years.

Infection with *Mycobacterium leprae*, an intracellular acid-fast rod, affects mainly the superficial nerves of the skin. The clinical manifestation varies considerably, depending on the state of cell-mediated immunity of an individual. The three major forms are *tuberculoid*, in which resistance is high; *borderline*, in which immune response is moderate; and *lepromatous* leprosy with low cellular immunity. Further subdivisions have been made clinically and histologically. Indeterminate leprosy is an early disease stage without determined outcome. For therapeutic reasons patients are grouped into two categories as *multibacillary* (consisting of the lepromatous and most borderline cases) or *paucibacillary* (tuberculoid, indeterminate, and the remaining borderline cases).

Diagnosis rests on the clinical signs of skin lesions, sensory loss, thickening of nerve trunks, and the demonstration of *M. leprae* in skin lesions. Differential diagnosis includes various forms of skin disorders, including psoriasis, seborrheic dermatitis, mycosis fungoides, syphilis, and tropical infections such as yaws, and cutaneous leishmaniasis. In multibacillary leprosy skin lesions are more numerous in the form of macules, papules, and nodules with variable sensory loss. In paucibacillary leprosy sensory loss is definite but limited to a few patches. Pain and temperature sensation are most severely affected. Painless burns or injuries should arouse suspicion of leprosy in patients from endemic areas. Thickened nerves may be felt on palpation, with common involvement of the ulnar, peroneal, and great auricular nerves. Tendon reflexes are typically normal even in advanced cases. Anhidrosis of tuberculoid macules is characteristic of chronic cases. CSF is normal. Conduction studies show segmental slowing in nerves at sites of enlargement. Skin smears show acid-fast bacilli in multibacillary patients. Skin biopsy specimens from lesions show the typical histological involvement of peripheral nerves, even in the absence of bacilli.

## Natural Course

Leprosy is rarely fatal. However, it leads to physical deformity in approximately one third of the patients, resulting in rejection and isolation by communities. Absorption of the hands and feet occurs in insensitive but actively used limbs and is probably due to repeated traumas and painless infection. In completely paralyzed extremities absorption does not occur.

Only a small proportion of infected individuals have signs develop, and mild forms are self-healing. The prognosis after treatment is good in the tuberculoid type, and even in lepromatous leprosy consequent treatment can prevent permanent sequelae.

The otherwise indolent course of leprosy may be interrupted by two types of reactions, both of which occur spontaneously or as complications of treatment. Reversal (type I) reaction may complicate borderline patients because of an increase in cellular reactivity, resulting in erythema, swelling of existing lesions, and appearance of new lesions. Erythema nodosum leprosum (type II reaction) may occur in lepromatous patients, is caused by an immune complex mechanism, and results in skin nodules, fever, lymphadenopathy, and arthralgias.

## Principles of Therapy

A number of effective drugs are available today. Dapsone (diamino diphenyl sulfone, DDS), clofazimine (CLO), and rifampicin (RMP) are most often used. Because of increasing resistance to DDS, multidrug therapy is now recommended by the World Health Organization (Noordeen, 1993). DDS and CLO are weakly bactericidal, whereas RMP is highly bactericidal. The main side effect of DDS is hemolysis, but anemia is usually mild, except in individuals with glucose-6 phosphate dehydrogenase deficiency in whom serious and life-threatening hemolysis can occur. Agranulocytosis and skin lesions are rare side effects. CLO may cause red coloration of the involved skin. RMP may be hepatoxic. Chemotherapy will render the patient noncontagious within days. Isolation is not required.

The treatment of complications includes analgesics and steroids. Thalidomide is very effective in erythema nodosum leprosum (400 mg daily) but should never be given to women of childbearing age. Because nerve compression may also occur, nerve release surgery should be considered if a progressive mononeuropathy is not explained by untreated disease or leprosy reaction. Because of insensitivity, wounds must be treated efficiently to avoid recurrent injury, which may eventually lead to traumatic or surgical amputation. Social rehabilitation may require a change of occupation to prevent limb trauma, information to relatives and employers on the nature of the disease, and sometimes cosmetic surgery for deforming lesions of the face and extremities.

### Practical Management

Multibacillary patients should be treated with RMP, 600 mg once a month (supervised), DDS, 100 mg daily (self-administered), and CLO, 50 mg daily (self-administered) and 300 mg once a month supervised. This should be continued for at least 2 years and if possible until smears become negative.

Paucibacillary patients should be treated with RMP, 600 mg once a month (supervised), and DDS, 100 mg (self-administered) for 6 months (Noordeen, 1993).

## HERPES ZOSTER

### Clinical Aspects

Varicella zoster virus is a member of the herpes virus family and causes two distinct diseases. Varicella (chickenpox) usually occurs in childhood and takes a benign course. Reactivation of latent virus in dorsal root ganglia leads to zoster (shingles), a painful sensory neuritis with vesicular cutaneous lesions within a dermatome. Clinical reactivation of zoster is associated with reduced cellular immunity that may occur with increasing age or immunocompromised states. This is reflected in the increased incidence of zoster in the elderly and immunocompromised. The annual incidence of zoster ranges from 0.4–1.6/1000 individuals <20 years of age to 4.5–11 cases/1000 individuals >80 years of age, and individuals with leukemia and HIV have a significantly higher incidence of zoster than age-matched controls (Kost and Straus, 1996). However, it should be noted that a population-based study on 590 patients showed that the relative risk for the incidence of newly diagnosed cancer within the first 5 years after herpes zoster was the same as in the general population (Ragozzino *et al.*, 1982).

Patients often report local pain, fever, and malaise 1–4 days before the zoster appears. The typical exanthema appears in the form of grouped vesicles on reddened bases that become pustules by 3–4 days, form crusts by 7–10 days, and normally resolve within 2–3 weeks. Diagnosis is normally easy, although herpes simplex infection should be considered in atypical lesions. CSF may show mild pleocytosis and increased protein in acute infection without signs of meningeal irritation. Virus isolation is rarely feasible. Serological investigations are sensitive. A Tzanck smear from the base of a lesion demonstrates multinucleated giant cells and can be used as rapid diagnostic sign.

Zoster most commonly affects thoracic dermatomes, followed by cranial nerves, lumbar, cervical, and sacral nerve lesions. In approximately 5% of patients the inflammation spreads to the adjacent motor roots and ventral horns with subsequent segmental motor weakness. Infection of the trigeminal nerve almost always involves the ophthalmic branch and may be accompanied by keratitis and also by ophthalmolplegia. Ramsay-Hunt syndrome is rare and consists of ear pain, radiation of pain to the ipsilateral tonsillar region, and vesicles in the external auditory canal. Facial palsy is common in this syndrome, and cranial nerves VIII and IX may also be affected. Other relatively rare neurological complications are encephalitis, myelitis, and angiitis with contralateral hemiparesis after ophthalmic zoster. They occur mostly within the first weeks after the onset of the exanthema, but the latent period sometimes lasts up to several months. Cases of polyneuritis indistinguishable from Guillain-Barré syndrome have been reported in association with zoster.

### Natural Course

An uncomplicated zoster infection is self-limiting, but 2% of nonimmunocompromised and five times as many immunocompromised individuals have a second episode of zoster develop. Recurrences commonly occur at the site of the primary zoster.

Virus dissemination occurs in less than 10% of all patients, but it can affect up to 50% in immunocompromised individuals with potentially life-threatening organ involvement. The mortality rate from encephalitis is 10%–20%, but full recovery is the rule in the remaining patients. Cerebrovascular complications from angiitis have a mortality rate of 25%, with only a third of the survivors recovering without deficit. Segmental motor neuropathy may cause atrophy but is fully reversible in 55%–75%. Outcome does not correlate with the degree of paresis.

Postherpetic neuralgia (PHN) is pain that persists in the area affected by herpes zoster and is a debilitating complication of zoster reactivation. The postherpetic pain is believed to be due to the lower activation thresholds and unprovoked discharges occurring in injured and regenerating peripheral neurons. This results in hyperexcitability of the dorsal horn, causing exaggerated CNS signal processing of peripheral inputs (Kost and Straus, 1996). Postherpetic pain may be either continuous, burning, tearing, or of paroxysmal jabbing and lancinating quality. Review of the literature is confusing, because PHN is variously defined at time points from pain persisting after crusting of the cutaneous lesion to residual pain at 6 months. Nearly all patients with reactivation of zoster experience acute pain, and the number reporting persistent pain decreases with time, but it should be noted that PHN can develop after a pain-free interval. The intractability and duration of pain tends to increase with age, but the prevalence of PHN is not increased in immunocompromised patients. Twelve months after the acute zoster infection, PHN persists in 4% of those less than 20 years of age and 48% of those older than 70 (Kost and Straus, 1996). Treatment with acyclovir (McKendrick et al., 1986), valacyclovir (Beutner et al., 1995), and famciclovir (Tyring et al., 1995) within the 72 hours after the appearance of the rash markedly reduces the median time to complete resolution of acute zoster-associated pain. However, studies looking for a reduction in the prevalence of PHN 2–6 months after acyclovir, valacyclovir, and famciclovir reveal conflicting results.

### Principles of Therapy

The nucleoside analog acyclovir is an efficacious agent in the management of cutaneous lesions, ophthalmic, and visceral complications of zoster infection if given within 72 hours after appearance of the rash. Low oral bioavailability of acyclovir prompted the development of the prodrug valacyclovir, which has proved more efficacious in the management of zoster-associated pain than acyclovir (Beutner et al., 1995). Another nucleoside analog famciclovir has also proved as efficacous as acyclovir in the treatment of zoster. Famciclovir can be administered less frequently than acyclovir, because the active metabolite of the former compound, penciclovir triphosphate, has a longer half-life than acyclovir.

Ophthalmic zoster should be promptly referred to an ophthalmologist to avoid local complications. In addition to oral therapy, topical (3.3%) acyclovir has been shown to be effective.

Concomitant administration of prednisolone (40 mg, tapered over 21 days) in addition to acyclovir significantly reduced the pain of acute zoster (Wood et al., 1994). However, because prednisolone had no influence on the incidence or severity of PHN at 6 months and was associated with adverse side effects, steroid use in acute zoster is not recommended.

Despite the accepted association of zoster with malignancies, there is as yet no clear evidence that otherwise healthy patients should be screened for such a condition (see earlier). However, HIV testing is to be considered in patients with zoster who belong to an HIV risk group.

Postherpetic neuralgia is difficult to treat. Mild analgesic drugs are rarely of benefit, and narcotic drugs are often of limited efficacy in neuropathic pain. For example, although intravenous morphine has proven benefit in the management of PHN (Rowbotham et al., 1991), another trial showed that codeine was no more efficacious than placebo (Max et al., 1988). Topical application of capsacin (Watson et al., 1993) and lidocaine (Rowbotham et al., 1995) has been shown to be of benefit in controlled trials. However, burning sensation resulting from application of capsacin makes this treatment intolerable in up to one third of patients, and some authorities argue that this side effect makes blinded studies with capsacin impossible. Amitriptyline is of proven benefit in providing relief from PHN (Watson et al., 1982) but is only effective in half the patients treated, and its use is limited by side effects. More recently, gabapentin has proved effective in the treatment of pain and sleep interference associated with PHN (Rowbotham et al., 1998). Carbamazepine is useful for paroxysmal pain but is ineffective for continuous pain.

Transcutaneous electrical nerve stimulation has also been beneficial in uncontrolled trials. Sympathetic block with lidocaine is not effective in postherpetic neuralgia; however, uncontrolled studies suggest that it alleviates acute zoster pain and possibly prevents postherpetic neuralgia. Various neurosurgical procedures have been tried as a last resort for intractable pain (North and Levy, 1994) but are probably of little value, and trigeminal tractotomy for ophthalmic postherpetic pain or dorsal root entry zone lesions are not universally advocated.

### Practical Management

Treatment of acute herpes zoster includes the following:

1. Initiation of nucleoside analog treatment as soon as possible, but within 72 hours, after the appearance

of the rash. Valacyclovir seems to be the drug of choice. In immunocompetent patients 800 mg oral aciclovir five times daily for 7 days is sufficient. In immunocompromised patients 10 mg/kg IV every 8 hours for 7 days should be given. The dose of valacyclovir is 1 g three times daily for 7 days. The dosage of famciclovir is 250 mg three times daily for 7 days, but this should be increased to 500 mg three times daily for 10 days in immunocompromised patients. The dose of all three drugs should be reduced in patients with impaired renal function.

2. Acute pain can be managed with conventional analgesics or sympathetic blockade ideally performed by a trained anesthetist in the context of a pain clinic.

Treatment of postherpetic neuralgia includes the following:

1. Topical application of 5% lidocaine gel can be used as initial therapy, and efficacy can be rapidly determined.
2. Amitriptyline should be started at 10–25 mg with gradual small increments (10–25 mg weekly) until good pain relief or side effects supervene (up to 150 mg/day).
3. Gabapentin can be titrated to a maximum dose of 3600 mg/day (1.2 g three times daily) over a 4-week period.
4. In cases of paroxysmal pain, carbamazepine should be added (slowly increasing from 100 mg to 600–1000 mg/day in divided doses).

## NEUROPATHIES IN HIV INFECTION

### Clinical Aspects

Several forms of peripheral neuropathy are associated with HIV infection. The most common is a distal symmetrical peripheral neuropathy, which accounts for more than 90% of the cases. The typical clinical features are dysesthesia and pain. Motor weakness is minimal or absent, and objective sensory loss frequently mild. Concomitant myelopathy may result in confusing clinical pictures. Neuropathy may become manifest at any time during HIV infection, but its frequency is significantly higher in late stages. The overall prevalence of neuropathy in HIV-seropositive patients was estimated at 10%–20%. Approximately the same proportion have clinically silent abnormalities in electrophysiological investigations (Fuller *et al.*, 1993; Hall *et al.*, 1991; Winer *et al.*, 1992). An autopsy series showed pathological findings in peripheral nerves in almost all individuals dying from AIDS (Mah *et al.*,

1988). Electrophysiological and pathological findings are consistent with axonopathy. Demyelination does not seem to be a major pathology. Guillain-Barré syndrome, chronic inflammatory demyelinating neuropathy, multifocal neuropathy, and lumbosacral polyradiculoneuropathy have been reported to occur in less than 1% of patients with HIV infection. GBS mainly occurs during early HIV infection. HIV seropositivity was significantly more frequent in GBS patients than in blood donors (55% vs 4.3%) in a series from Zimbabwe (Thornton *et al.*, 1991). More direct evidence for an association of HIV and GBS is lacking. Relatively typical is a CSF pleocytosis (up to 50 mononuclear cells/ml) in HIV-seropositive GBS patients; however, the clinical course of GBS is no different. CIDP has mostly been reported in asymptomatic HIV patients and in AIDS-related complex. Multifocal neuropathy is due mostly to vasculitis in AIDS-related complex. The syndrome of lumbosacral polyradiculoneuropathy is associated with cytomegalovirus (CMV) infection usually occurring in the late stages of HIV infection. Asymmetrical weakness with low back pain and disturbance of bladder function are typical. Paraplegia and ascending weakness with cranial nerve involvement occur. The prognosis is often poor. CSF shows pleocytosis and a mildly raised protein content. CMV culture may be positive.

The clinical appearance of herpes zoster infection is indistinguishable from that in HIV-seronegative patients. A high incidence of mild autonomic neuropathy has been reported in HIV-seropositive patients, although its significance is uncertain. A predominantly sensory neuropathy with painful dysesthesia develops after treatment with a number of antiviral drugs used to combat the HIV infecvtion (2,3-dideoxycytidine (ddC) (Berger *et al.*, 1993), 2,3-dideoxyinosine (ddI), and stavudine. Milder symptoms can reverse after stopping the drugs.

Other causes of neuropathy such as drug therapy with dapsone, vincristine, or taxol should be excluded before diagnosing a distal sensory neuropathy caused by the HIV itself. There is little controlled evidence that therapy is particularly effective. Neither mexilitine nor amitriptyline is more effective than placebo, and gabapentin is probably the drug of choice because of its lack of side effects and low risk of interfering with other therapy. There is one report of limited pain relief with recombinant nerve growth factor, but this treatment is not readily available (McArthur, 1998).

### Natural Course and Principles of Therapy

Distal symmetrical polyneuropathy may lead to considerable disability. In some patients pain disappears, but sensory loss remains or progresses. There is no specific treatment for this syndrome. Antiviral treatment for HIV occasionally helps, but usually symptomatic pain therapy with gabapentin is required.

The course of GBS and CIDP is in general no different from that in HIV-seronegative patients. In GBS plasmapheresis has been used without evidence of a higher rate of complications. Little information is available about the efficacy of IVIg in this patient group. In CIDP, corticosteroids, IVIg (Chimowitz *et al.*, 1989), and plasmapherisis seem effective. Long-term steroid treatment should be limited to avoid additional immunosuppression. There is limited experience with multifocal neuropathy. Spontaneous arrest or progression to distal symmetric neuropathy may occur. Immunosuppressive therapy is not justified.

In CMV-associated polyradiculoneuropathy, the outcome is often fatal. Ganciclovir, given within 48 hours after initial presentation, has been described to halt progression in two patients with AIDS (Miller *et al.*, 1990). These patients received 2.5 mg/kg IV every 8 hours for 10 days, and subsequent maintenance treatment with 5–7.5 mg/kg 5 days per week. Dose-related decrease in neutrophils and white blood cell counts are the most common side effects.

## DIPHTHERITIC NEUROPATHY

### Clinical Aspects

Diphtheria is now rare in developed countries but is still common in economically disadvantaged countries. The disease is caused by *Corynebacterium diphtheriae* with an incubation period of 2–6 days. The typical syndrome consists of fever, sore throat, and membranous pharyngitis. The organism elaborates a 62-kDa protein exotoxin that is responsible for the delayed systemic manifestations, including myocarditis and polyneuritis. Diphtheritic neuropathy can be confused with Guillain-Barré syndrome, because the CSF contains a raised protein level. GBS evolves faster and reaches its maximum within 1–3 weeks. Paralysis of accomodation is common in diphtheria but rare in GBS.

### Natural Course

Twenty percent of infected cases have a palatal neuropathy develop 4–30 days after the primary infection. This may be accompanied by impaired pharyngeal sensation and disturbance of pupillary function. Demyelinating sensorimotor polyneuropathy of the trunk and extremities is usually delayed until 1–3 months after the primary infection. Motor symptoms progress from proximal to distal muscles. Sensory abnormalities are of predominantly large fiber type. The incidence of neurological complications is on average 20% but may be up to 70% after severe initial attacks.

### Principles of Therapy

There is no specific treatment for the neurological complications of diphtheria. The administration of antitoxin within 48 hours of the primary infection is the standard treatment and reduces the incidence and severity of complications. Antibiotics have been recommended to eradicate the bacterium and prevent transmission. DL-canithine (100 mg/kg/day) may have a beneficial effect in diphtheritic myocarditis. Glucocorticoids do not reduce the risk of myocarditis or polyneuritis. Prevention against diphtheria requires a primary series of four doses of toxoid, usually given in children together with tetanus and pertussis vaccines. Booster doses should be given at 10-year intervals.

### Practical Management

1. Antitoxin should be given immediately, when diphtheria is clinically suspected. The recommended dose depends on the site of the primary infection and the duration and severity of symptoms. For disease involving the pharynx and larynx and less than 48 hours duration, 250 units/kg should be given IM. If the disease is severe, progressive, and has been present for 3 or more days, up to 4000 units/kg is recommended.

2. Because the antitoxin is produced in horses, conjunctival or intracutaneous testing should help to predict allergic reactions to horse serum before antitoxin administration (0.1 ml of a 1:10 dilution in saline is applied either way with parallel saline control contralaterally and observed for 15 minutes for hypersensitivity reactions).

3. Penicillin G (600,000 units IM twice daily) or erythromycin (500 mg 4 times daily PO or IV) is usually given for 14 days.

## TETANUS

### Clinical Aspects

Tetanus is caused by tetanospasmin, an exotoxin of the anaerobic, spore-forming *Clostridium tetani*. The bacterium is ubiquitous and usually enters the body through a wound. The toxin binds to peripheral motor neuron terminals and reaches the CNS by retrograde axonal transport. It interferes with the secretion of inhibitory transmitters in spinal interneurons, leading to uncontrolled motoneuron activity and also increases sympathetic activity. Because of vaccination programs, the incidence of tetanus has been declining in developed countries (less than 100 cases per year in the United States), where tetanus occurs mainly in elderly persons with inadequate immunization status. By contrast, in developing countries tetanus is still responsible for several hundred thousand deaths per year.

Eighty percent of tetanus cases are of the *generalized type*. Most patients have trismus, a spasm of the masseter muscles. Dysphagia or facial muscle spasms (risus sardonicus) are also common. Severe generalized and extremely painful muscle spasms follow after a few days and may involve axial spinal muscles (opisthotonus) and impair ventilation. Spasms may be spontaneous or triggered by the slightest stimulation. Severe cases show autonomic dysfunction with tachycardia, labile hypertension and hypotension, fever, and profuse sweating.

*Local* and *cephalic tetanus* are rare forms manifesting with stiffness, pain, and muscle spasms in the neighborhood of a wound. Cephalic tetanus frequently becomes generalized. In developing countries *tetanus neonatorum*, caused by contamination of the umbilical cord, is common and often fatal.

The diagnosis of tetanus is made from the typical clinical features and the history of a preceding injury. An EMG may show absence of the silent period after an action potential. Differential diagnosis includes acute meningitis (to be excluded by normal CSF), early focal dyskinesia (responding to IV anticholinergics), strychnine poisoning (begins 5–60 minutes after ingestion, resolving after several hours), rabies (dysphagia, opisthotonus, but no trismus or facial palsy), and stiffman syndrome (longer history of intermittent muscle rigidity).

## Natural Course

The incubation time varies between 1 and more than 30 days. A short incubation time predicts a severe course. In immunized patients tetanus is often benign but may be atypical and difficult to diagnose. The mortality rate of generalized tetanus treated with modern intensive care facilities is now approximately 10% in younger individuals but may be as high as 50% in the elderly. When the most severe symptoms are overcome, the patient recovers completely within a few weeks.

## Principles of Therapy

The importance of prevention by vaccination is evident. Treatment of manifest tetanus is a true challenge to intensive care medicine and nursing and includes (1) administration of antitoxin (human tetanus immunoglobulin) to neutralize the circulating toxin. Toxin bound to neural tissue is unaffected by the antitoxin. Antitoxin should be given before surgical treatment; the value of infiltrating the wound is uncertain. (2) Surgical treatment of the wound is required. Additional antibiotics are recommended, although their value is uncertain. (3) Muscle relaxation may be achieved by many agents and combinations; for example, diazepam, chlorpromazine, barbiturates (not to be combined with diazepam to avoid hypoventilation), or curarization with mechanical ventilation. Recently dantrolene or intrathecal baclofen (by an indwelling catheter or repeated injections) has also been recommended (Saissy *et al.*, 1992). (4) For management of autonomic dysfunction, morphine or magnesium sulfate have mostly been used. Alpha- and beta-blockers are now thought to increase the risk of cardiac arrest and are no longer recommended. (5) Active immunization is necessary, because immunity is not induced by the amount of toxin that causes disease.

## Practical Management

1. An intensive care setting and a dark, quiet room are required to minimize stimulation. Even in mild tetanus the patient should be observed there for at least 1 week. Appropriate management of the airway is the first priority, and early tracheostomy is recommended.

2. Human tetanus immunoglobulin (5000–10,000 units IM) should be given as early as possible. Because of the large volume, this is best given in divided portions. The value of repeated injections on the following days is uncertain. Intravenous administration must be avoided.

3. Surgical wound management after antitoxin administration.

4. Penicillin IV (1 million units 4 times daily) or erythromycin (500 mg 4 times daily) for 10 days.

5. Diazepam 2–10 mg IV every 4–12 hours up to 5 mg/kg/day, a neuroleptic (chlorpromazine 200–300 mg/day), or both. If deep sedation is not sufficient, pancuronium bromide, 0.02–0.08 mg/kg, should be added every 2–4 hours with mechanical ventilation. To control autonomic dysfunction, morphine (2–10 mg/h) can be tried. Also magnesium sulfate can be used: a loading dose of 70 mg/kg IV followed by a continuous infusion of 1–3 g/h to maintain a plasma level of 2.5–4 mmol/L (Mg and Ca serum levels should be measured every 4 hours).

6. Active immunization with three injections. The first dose should be given after the acute disease, the

second dose 4–8 weeks after the first, the third dose 6–12 months after the second. A booster dose every 10 years is required.

## BOTULISM

### Clinical Aspects

*Food-borne botulism* is nowadays uncommon and is associated mainly with home canned food. Botulism is caused by the exotoxin of the anaerobic, spore-forming *Clostridium botulinum*. Seven antigenically different types of toxins are known, types A, B, and E being typically responsible for human food-borne botulism. As little as 0.1 µg of the toxin may be lethal. The toxin is inactivated by heating at 85°C for 5 minutes. Botulinum toxin interferes with the presynaptic release of acetylcholine at the neuromuscular junction and other cholinergic synapses. Thus the defect in botulism is pathophysiologically similar to that in Lambert-Eaton myasthenic syndrome, resulting in neuromuscular weakness and autonomic dysfunction. Botulinum toxin is now used therapeutically to treat focal dystonias.

Symptoms of botulism appear 6–60 hours after ingestion and evolve over 2–4 days and may therefore be mistaken for Guillain-Barré syndrome. Nausea, abdominal cramps, blurred vision, and diplopia are the first symptoms. Ptosis and ophthalmoparesis may be mistaken for acute onset of myasthenia gravis; however, the pupils are large and unreactive in botulism. Weakness is usually symmetrical and descends to other muscles, including the respiratory muscles. Tendon reflexes may be diminished or lost. Dry eyes, dry mouth, and gastrointestinal ileus may occur. In equivocal and oligosymptomatic cases electrophysiological studies may help to show a defect of neuromuscular transmission. Single-fiber EMG reveals abnormally high blocking and increased jitter. Repetitive nerve stimulation with 30 Hz shows low initial amplitudes with increment.

*Wound botulism* is rare, caused by wound infection, and has a longer incubation period. *Infant botulism* is a comparably common form of the disease (after absorption of toxin from intestinal colonies of *C. botulinum*), mainly in children less than 6 months of age, and may be a cause of sudden infant death.

### Natural Course

Mortality is highest in type A and lowest in type B infection but has now fallen to less than 10% because of improved respiratory and intensive care. Complete recovery can be expected but may take many months.

Immunity is not induced, and relapse may occur after re-exposure.

### Principles of Therapy

A high degree of suspicion and an early diagnosis is required. It is most important to recognize clustering of cases and identify the source of infection. Not all people who ingested the toxin will have symptoms. An early onset of symptoms after ingestion predicts severe disease.

The trivalent antiserum should be used, because three types of the toxin may cause the disease. Gastric lavage and enemas may help to eliminate the toxin. Close monitoring and supportive treatment are necessary, including mechanical ventilation, when the vital capacity falls to less than 30% of the normal.

The value of antibiotics is not proven. In wound botulism, surgical exploration and excision of the wound are necessary. Theoretically, a beneficial effect might be obtained from presynaptic acetylcholine–releasing agents such as guanidine or 3,4-diaminopyridine (up to four times 25 mg/day), but there is little evidence of any advantages to these drugs.

### Practical Management

1. Trivalent immunoglobulin (500 ml IV) should be given in severe cases. Intradermal testing for hypersensitivity to horse serum has to be performed as described for diphtheria. In case of further progression, injection with 250 ml can be repeated after 4–6 hours.
2. Gastric lavage, early after ingestion, and high enemas (if there is no ileus) are used to eliminate intestinal toxin.
3. Respiratory monitoring and intensive care are necessary in moderately and severely affected patients.

## REFERENCES

### Guillain-Barré Syndrome

Anonymous. (1997). Randomised trial of plasma exchange, intravenous immunoglobulin, and combined treatments in Guillain-Barre syndrome. Plasma Exchange/Sandoglobulin Guillain-Barré Syndrome Trial Group [see comments]. *Lancet* **349**, 225–230.

Asbury, A. K., and Cornblath, D. R. (1990). Assessment of current diagnostic criteria for Guillain-Barré syndrome. *Ann. Neurol.* **27**, (**Suppl.**), S21–24.

Chiba, A., Kusunoki, S., Obata, H., Machinami, R., and Kanazawa, I. (1993). Serum anti-GQ1b IgG antibody is associated with ophthalmoplegia in Miller Fisher syndrome and Guillain-Barré syndrome: Clinical and immunohistochemical studies. *Neurology* 43, 1911–1971.

Enders, U., Karch, H., Toyka, K. V., Michels, M., Zielaseck, J., Pette, M., Heesemann, J., and Hartung, H. P. (1993). The spectrum of immune responses to Campylobacter jejuni and glycoconjugates in Guillain-Barré syndrome and in other neuroimmunological disorders. *Ann. Neurol.* 34, 136–144.

Feasby, T. E., Gilbert, J. J., Brown, W. F., Bolton, C. F., Hahn, A. F., Koopman, W. F., and Zochodne, D. W. (1986). An acute axonal form of Guillain-Barré polyneuropathy. *Brain* 109, 1115–1126.

French Cooperative Group on Plasma Exchange in Guillain-Barré; Syndrome. (1987). Efficiency of plasma exchange in Guillain-Barré syndrome: Role of replacement fluids. *Ann. Neurol.* 22, 753–761.

Grand'Maison, F., Feasby, T. E., Hahn, A. F., and Koopman, W. J. (1992). Recurrent Guillain-Barré syndrome. Clinical and laboratory features. *Brain* 115, 1093–1106.

Hughes, R. A. C. (1990). "Guillain-Barré syndrome." SpringerVerlag, London.

Hughes, R. A. C. (1991). Ineffectiveness of high-dose intravenous methylprednisolone in Guillain-Barré syndrome. *Lancet* 338, 1142.

Lasky, T., Terracciano, G. J., Magder, L., Koski, C. L., Ballesteros, M., Nash, D., Clark, S., Haber, P., Stolley, P. D., Schonberger, L. B., and Chen, R. T. (1998). The Guillain-Barré syndrome and the 1992–1993 and 1993–1994 influenza vaccines. *N. Engl. J. Med.* 339, 1797–1802.

McKhann, G. M., Cornblath, D. R., Griffin, J. W., Ho, T. W., Li, C. Y., Jiang, Z., Wu, H. S., Zhaori, G., Liu, Y., Jou, L. P., Liu, T. C., Gao, C. Y., Mao, J. Y., Blaser, M. J., Mishu, B., and Asbury, A. K. (1993). Acute motor axonal neuropathy: A frequent cause of acute flaccid paralysis in China. *Ann. Neurol.* 33, 333–342.

Roberts, M., Willison, H., Vincent, A., and Newsom-Davis, J. (1994). Serum factor in Miller-Fisher variant of Guillain-Barré syndrome and neurotransmitter release [see comments]. *Lancet* 343, 454–455.

Ropper, A. H., and Kehne, S. M. (1985). Guillain-Barré syndrome: Management of respiratory failure. *Neurology* 35, 1662–1665.

The Guillain-Barré Syndrome Study Group. (1985). Plasmapheresis and acute Guillain-Barré syndrome. *Neurology* 35, 1096–1104.

Van der Meché, F. G. A., Schmitz, P. I. M., and the Dutch Guillain-Barré Study Group. (1992). A randomized trial comparing intravenous immune globulin and plasma exchange in Guillain-Barré syndrome. *N. Engl. J. Med.* 326, 1123–1129.

Vriesendorp, F. J., Mishu, B., Blaser, M. J., and Koski, C. L. (1993). Serum antibodies to GM1, GB1b, peripheral nerve myelin, and Campylobacter jejuni in patients with Guillain-Barré syndrome and controls: Correlation and prognosis. *Ann. Neurol.* 34, 130–135.

Winer, J. B., and Hughes, R. A. C. (1988). Identification of patients at risk of arrhythmia in the Guillain-Barré syndrome. *Q. J. Med.* 257, 735–739.

Winer, J. B., Hughes, R. A. C., Anderson, M. J., Jones, D. M., Kangro, H., and Watkins, R. P. F. (1988). A prospective study of acute idiopathic neuropathy. II. Antecedent events. *J. Neurol. Neurosurg. Psychiatry* 51, 613–618.

Yu, Z., and Lennon, V. A. (1999). Mechanism of intravenous immune globulin therapy in antibody-mediated autoimmune diseases [see comments]. *N. Engl. J. Med.* 340, 227–228.

## Chronic Inflammatory Demyelinating Polyneuropathy

Barnett, M. H., Pollard, J. D., Davies, L., and McLeod, J. G. (1998). Cyclosporin A in resistant chronic inflammatory demyelinating polyradiculoneuropathy. *Muscle Nerve* 21, 454–460.

Barohn, R. J., Kissel, J. T., Warmolts, J. R., and Mendell, J. R. (1989). Chronic inflammatory demyelinating polyradiculoneuropathy. Clinical characteristics, course, and recommendations for diagnostic criteria. *Arch. Neurol.* 46, 878–884.

Dyck, P. J., Daube, J., O'Brien, P. C., Pineda, A., Low, P. A., Windebank, A. J., and Swanson, C. (1986). Plasma exchange in chronic inflammatory demyelinating polyradiculoneuropathy. *N. Engl. J. Med.* 314, 461–465.

Dyck, P. J., O'Brien, P. C., Oviatt, K. F., Dinapoli, R. P., Daube, J. R., Bartleson, J. D., Mokri, B., Swift, T., Low, P. A., and Windebank, A. J. (1982). Prednisone improves chronic inflammatory demyelinating polyradiculoneuropathy more than no treatment. *Ann. Neurol.* 11, 136–141.

Feasby, T. E., Hahn, A. F., Koopman, R. N., and Lee, D. H. (1990). Central lesions in chronic inflammatory demyelinating polyneuropathy: An MRI study. *Neurology* 40, 476–478.

Good, J. L. C. M., Mayer, R. F., and Koski, C. L. (1998). Pulsed cyclophosphamide therapy in chronic inflammatory demyelinating polyneuropathy. *Neurology* 51, 1735–1738.

Gorson, K. C. R. A., Clark, B. D., Dew, R. B., Simovic, D., and Allam, G. (1998). Treatment of chronic inflammatory demyelinating polyneuropathy with interferon-alpha 2a. *Neurology* 50, 84–87.

Hadden, R. D. M., Sharrack, B., Bensa, S., Soudain, S. E., and Hughes, R. A. C. (1999). Randomized trial of interferon β-1a in chronic inflammatory demyelinating polyradiculoneuropathy. *Neurology* 53, 57–61.

Hahn, A., Bolton, C., Pillay, N., Chalk, C., Benstead, T., Bril, V., Shumak, K., Vandervoort, M., and Feasby, T. (1996a). Plasma-exchange therapy in chronic inflammatory demyelinating polyneuropathy. A double-blind, sham-controlled, cross-over study. *Brain* 119, 1055–1066.

Hahn, A., Bolton, C., Zochodne, D., and Feasby, T. (1996b). Intravenous immunoglobulin treatment in chronic inflammatory demyelinating polyneuropathy. A double-blind, placebo-controlled, cross-over study. *Brain* 119, 1067–1077.

Hughes, R. (1990). "Guillain-Barré Syndrome." Springer-Verlag, London.

McCombe, P. A., Pollard, J. D., and McLeod, J. G. (1987). Chronic inflammatory demyelinating polyradiculoneuropathy. *Brain* 110, 1617–1630.

Mendell, J. R., Kolin, S., Kissel, J. T., Weiss, K. L., Chakeres, D. W., and Rammohan, K. W. (1987). Evidence for central nervous system demyelination in chronic inflammatory demyelinating polyradiculoneuropathy. *Neurology* 37, 1291–1294.

Report from an Ad Hoc Subcomittee of the American Academy of Neurology AIDS Task Force (1991). Research criteria for diagnosis of chronic inflammatory demyelinating polyneuropathy (CIDP). *Neurology* 41, 617–618.

Vermeulen, M., van Doorn, P. A., Brand, A., Strengers, P. F. W., Jennekens, F. G. I., and Busch, H. F. M. (1993). Intravenous immunoglobulin treatment in patients with chronic inflammatory demyelinating polyneuropathy: A double blind, placebo controlled study. *J. Neurol. Neurosurg. Psychiatry* 56, 36–39.

## Multifocal Motor Neuropathy

Azulay, J., Blin, O., Pouget, J., Boucraut, J., Bille-Turc, F., Carles, G., and Serratrice, G. (1994). Intravenous immunoglobulin treatment in patients with motor neuron syndromes associated with anti-GM1 antibodies: a double-blind, placebo-controlled study. *Neurology* 44, 429–432.

Hirota, N., Kaji, R., Bostock, H., Shindo, K., Kawasaki, T., Mizutani, K., Oka, N., Kohara, N., Saida, T., and Kimura, J. (1997). The physiological effect of anti-GM1 antibodies on saltatory conduction and transmembrane currents in single motor axons. *Brain* **120**, 2159–2169.

Leger, J., Chassande, B., Musset, L., Meininger, V., Bouche, P., and Baumann, N. (2001). Intravenous immunoglobulin therapy in multifocal motor neuropathy. A double-blind, placebo-controlled study. *Brain* **124**, 145–153.

Levine, T., and Pestronk, A. (1999). IgM antibody-related polyneuropathies: B-cell depletion chemotherapy using Rituximab. *Neurology* **52**, 1701–1704.

Martina, I., van Doorn, D. P., Schmitz, P., Meulstee, J., and van der Meche, M. F. (1999). Chronic motor neuropathies: response to interferon-beta1a after failure of conventional therapies. *J. Neurol. Neurosurg. Psychiatry* **66**, 197–201.

Meucci, N., Cappellari, A., Barbieri, S., Scarlato, G., and Nobile-Orazio, E. (1997). Long term effect of intravenous immunoglobulins and oral cyclophosphamide in multifocal motor neuropathy. *J. Neurol. Neurosurg. Psychiatry* **63**, 765–769.

Van den Berg-Vos, R., Van den Berg, L., Franssen, H., Van Doorn, P., Merkies, I., and Wokke, J. (2000). Treatment of multifocal motor neuropathy with interferon-beta1A. *Neurology 2000* **54**, 1518–1521.

## Multifocal Acquired Demyelinating Sensory And Motor Neuropathy (MADSAM)— Lewis-Sumner Syndrome

Lewis, R., Sumner, A., Brown, M., and Asbury, A. (1982). Multifocal demyelinating neuropathy with persistent conduction block. *Neurology* **32**, 958–964.

Saperstein, D., Amato, A., Wolfe, G., Katz, J., Nations, S., Jackson, C., Bryan, W., Burns, D., and Barohn, R. (1999). Multifocal acquired demyelinating sensory and motor neuropathy: the Lewis-Sumner syndrome. *Muscle Nerve* **22**(5), 560–566.

Van den Berg-Vos, R., Van den Berg, L., Franssen, H., Vermeulen, M., Witkamp, T., Jansen, G., van Es, H., Kerkhoff, H., and Wokke, J. (2000). Multifocal inflammatory demyelinating neuropathy: a distinct clinical entity? *Neurology* **54**, 26–32.

## Polyneuropathy with Monoclonal Gammopathy

Apartis, E., Leger, J., Musset, L., Gugenheim, M., Cacoub, P., Lyon-Caen, O., Pierrot-Deseilligny, C., Hauw, J., and Bouche, P. (1996). Peripheral neuropathy associated with essential mixed cryoglobulinaemia: a role for hepatitis C virus infection? *J. Neurol. Neurosurg. Psychiatry* **60**, 661–666.

Blume, G., Pestronk, A., and Goodnough, L. (1995). Anti-MAG antibody-associated polyneuropathies: improvement following immunotherapy with monthly plasma exchange and IV cyclophosphamide. *Neurology* **45**, 1577–1580.

Dalakas, M., Flaum, M., Rick, M., Engel, W., and Gralnick, H. (1983). Treatment of polyneuropathy in Waldenstrom's macroglobulinemia: role of paraproteinemia and immunologic studies. *Neurology* **33**, 1406–1410.

Dalakas, M., Quarles, R., Farrer, R., Dambrosia, J., Soueidan, S., Stein, D., Cupler, E., Sekul, E., and Otero, C. (1996). A controlled study of intravenous immunoglobulin in demyelinating neuropathy with IgM gammopathy. *Ann. Neurol.* **40**, 792–795.

Dyck, P., Low, P., Windebank, A., Jaradeh, S., Gosselin, S., Bourque, P., Smith, B., Kratz, K., Karnes, J., Evans, B. *et al.* (1991). Plasma exchange in polyneuropathy associated with monoclonal gammopathy of undetermined significance. *N. Engl. J. Med.* **325**, 1482–1486.

Ellie, E., Vital, A., Steck, A., Boiron, J., Vital, C., and Julien, J. (1996). Neuropathy associated with "benign" anti-myelin-associated glycoprotein IgM gammapathy: clinical, immunological, neurophysiological pathological findings and response to treatment in 33 cases. *J. Neurol.* **243**, 34–43.

Enevoldson, T., and Harding, A. (1992). Improvement in the POEMS syndrome after administration of tamoxifen. *J. Neurol. Neurosurg. Psychiatry* **55**, 71–72.

Kelly Jr, J., Kyle, R., O'Brien, P., and Dyck, P. (1981). Prevalence of monoclonal protein in peripheral neuropathy. *Neurology* **31**, 1480–1483.

Kyle, R. (1992). Monoclonal proteins in neuropathy. *Neurol. Clin.* **10**(3), 713–734.

Kyle, R., and Dyck, P. (1993a). Amyloidosis and neuropathy. *In* "Peripheral Neuropathy," 3rd ed. (P. Dyck, P. Thomas, J. Griffin, P. Low, and J. Poduslo, Eds.), pp. 1294–1309. Saunders, Philadelphia.

Kyle, R., and Dyck, P. (1993b). Osteosclerotic myeloma (POEMS syndrome). *In* "Peripheral Neuropathy" (P. Dyck, P. Thomas, J. Griffin, P. Low, and J. Poduslo, Eds.), pp. 1288–1293. Saunders, Philadelphia.

Kyle, R., Gertz, M., Greipp, P., Witzig, T., Lust, J., Lacy, M., and Therneau, T. (1997). A trial of three regimens for primary amyloidosis: colchicine alone, melphalan and prednisone, and melphalan, prednisone, and colchicine. *N. Engl. J. Med.* **336**, 1202–1207.

Kyle, R., and Lust, J. (1990). The monoclonal gammopathies (paraproteins). *Ad. Clin. Chem.* **28**, 145–218.

Latov, N., and Sherman, W. (1999). Therapy of neuropathy associated with anti-MAG IgM monoclonal gammopathy with Rituxan. *Neurology* **52**(S), A551.

Levine, T., and Pestronk, A. (1999). IgM antibody-related polyneuropathies: B-cell depletion chemotherapy using Rituximab. *Neurology* **52**, 1701–1704.

Misiani, R., Bellavita, P., Fenili, D., Vicari, O., Marchesi, D., Sironi, P., Zilio, P., Vernocchi, A., Massazza, M., Vendramin, G. *et al.* (1994). Interferon alfa-2a therapy in cryoglobulinemia associated with hepatitis C virus. *N. Engl. J. Med.* **330**, 751–756.

Nobile-Orazio, E., Meucci, N., Baldini, L., Di Troia, A., and Scarlato, G. (2000). Long-term prognosis of neuropathy associated with anti-MAG IgM M-proteins and its relationship to immune therapies. *Brain* **123**, 710–717.

Notermans, N., Lokhorst, H., Franssen, H., Van den Graaf, Y., Teunissen, L., Jennekens, F., Van den Berg, B. L., and Wokke, J. (1996). Intermittent cyclophosphamide and prednisone treatment of polyneuropathy associated with monoclonal gammopathy of undetermined significance. *Neurology* **47**, 1227–1233.

Oksenhendler, E., Chevret, S., Leger, J., Louboutin, J., Bussel, A., and Brouet, J. (1995). Plasma exchange and chlorambucil in polyneuropathy associated with monoclonal IgM gammopathy. IgM-associated Polyneuropathy Study Group. *J. Neurol. Neurosurg. Psychiatry* **59**, 43–247.

Quarles, R., and Dalakas, M. (1996). Do anti-ganglioside antibodies cause human peripheral neuropathies? *J. Clin. Invest.* **97**, 1136–1137.

Rajkumar, S. V., Gertz, M. A., and Kyle, R. A. (1998). Prognosis of patients with primary systemic amyloidosis who present with dominant neuropathy. *Am. J. Med.* **104**, 232–237.

Scelsa, S., Herskovitz, S., and Reichler, B. (1998). Treatment of mononeuropathy multiplex in hepatitis C virus and cryoglobulinemia. *Muscle Nerve* **21**, 1526–1529.

Simmons, Z., Albers, J., Bromberg, M., and Feldman, E. (1993a). Presentation and initial clinical course in patients with chronic demyelinating polyradiculopathy: comparison of patients without and with monoclonal gammopathy. *Neurology* 43, 2202–2209.

Simmons, Z., Blaivas, M., Aguilera, A., Feldman, E., Bromberg, M., and Towfighi, J. (1993b). Low diagnostic yield of sural nerve biopsy in patients with peripheral neuropathy and primary amyloidosis. *J. Neurol. Sci.* 120, 60–63.

Smith, I. (1994). The natural history of chronic demyelinating neuropathy associated with benign IgM paraproteinaemia. A clinical and neurophysiological study. *Brain* 117, 949–957.

Watanabe, O., Maruyama, I., Arimura, K., Kitajima, I., Arimura, H., Hanatani, M., Matsuo, K., Arisato, T., and Osame, M. (1998). Overproduction of vascular endothelial growth factor/vascular permeability factor is causative in Crow-Fukase (POEMS) syndrome. *Muscle Nerve* 21, 1390–1397.

Wilson, H. C., Lunn, M. P., Schey, S., and Hughes, R. A. (1999). Successful treatment of IgM paraproteinaemic neuropathy with fludarabine. *J. Neurol. Neurosurg. Psychiatry* 66, 575–580.

## Vasculitic Neuropathy

Hawke, S. H., Davies, L., Pamphlett, R., Guo, Y. P., Pollard, J. D., and McLeod, J. G. (1991). Vasculitic neuropathy. A clinical and pathological study. *Brain* 14, 2175–2190.

## Sarcoid Neuropathy

Gainsborough, N., Hall, S. M., Hughes, R. A. C., and Leibowitz, S. (1991). Sarcoid neuropathy. *J. Neurol.* 238, 177–180.

Matthews, W. B. (1993). Sarcoid neuropathy. *In* "Peripheral Neuropathy" (P. J. Dyck, and P. K. Thomas, Eds.), pp. 1418–1423. Saunders, Philadelphia.

## Neuralgic Amyotrophy

Gil-Neciga, E., Franco, E., Sanchez, A., Donaire, A., Chinchon, I., and Palau, F. (2000). [Recurrent familial brachial plexopathy as the only clinical expression of neuropathy with susceptibility to pressure]. [Spanish]. *Neurologia* 15, 177–181.

van Alfen, N., van Engelen, B., Reinders, J., Kremer, H., and Gabreels, F. (2000). The natural history of hereditary neuralgic amyotrophy in the Dutch population: two distinct types? *Brain* 123, 718–723.

## Leprosy

Noordeen, S. K. (1993). Leprosy today. *Schweiz. Med. Wochenschr.* 123, 1228–1236.

## Herpes Zoster

Beutner, K., Friedman, D., Forszpaniak, C., Andersen, P., and Wood, M. (1995). Valaciclovir compared with acyclovir for improved therapy for herpes zoster in immunocompetent adults. *Antimicrob. Agents Chemother.* 39, 1546–1553.

Kost, R., and Straus, S. (1996). Postherpetic neuralgia—pathogenesis, treatment, and prevention. *N. Engl. J. Med.* 335, 32–42.

Max, M., Schafer, S., Culnane, M., Dubner, R., and Gracely, R. (1988). Association of pain relief with drug side effects in postherpetic neuralgia: a single-dose study of clonidine, codeine, ibuprofen, and placebo. *Clin. Pharmacol. Ther.* 43, 363–371.

McKendrick, M., McGill, J., White, J., and Wood, M. (1986). Oral acyclovir in acute herpes zoster. *BMJ* 293, 1529–1532.

North, R., and Levy, R. (1994). Consensus conference on the neurosurgical management of pain. *Neurosurgery* 34, 756–760.

Ragozzino, M., Melton, L. D., Kurland, L., Chu, C., and Perry, H. (1982). Population-based study of herpes zoster and its sequelae. *Medicine* 61, 310–316.

Rowbotham, M., Davies, P., and Fields, H. (1995). Topical lidocaine gel relieves postherpetic neuralgia. *Ann. Neurol.* 37, 246–253.

Rowbotham, M., Harden, N., Stacey, B., Bernstein, P., and Magnus-Miller, L. (1998). Gabapentin for the treatment of postherpetic neuralgia: a randomized controlled trial [see comments]. *JAMA* 280, 1837–1842.

Rowbotham, M., Reisner-Keller, L., and Fields, H. (1991). Both intravenous lidocaine and morphine reduce the pain of postherpetic neuralgia. *Neurology* 41, 1024–1028.

Tyring, S., Barbarash, R., Nahlik, J., Cunningham, A., Marley, J., Heng, M., Jones, T., Rea, T., Boon, R., and Saltzman, R. (1995). Famciclovir for the treatment of acute herpes zoster: effects on acute disease and postherpetic neuralgia. A randomized, double-blind, placebo-controlled trial. Collaborative Famciclovir Herpes Zoster Study Group. *Ann. Intern. Med.* 123, 89–96.

Watson, C., Evans, R., Reed, K., Merskey, H., Goldsmith, L., and Warsh, J. (1982). Amitriptyline versus placebo in postherpetic neuralgia. *Neurology* 32, 671–673.

Watson, C., Tyler, K., Bickers, D., Millikan, L., Smith, S., and Coleman, E. (1993). A randomized vehicle-controlled trial of topical capsaicin in the treatment of postherpetic neuralgia. *Clin. Ther.* 15, 510–526.

Wood, M., Johnson, R., McKendrick, M., Taylor, J., Mandal, B., and Crooks, J. (1994). A randomized trial of acyclovir for 7 days or 21 days with and without prednisolone for treatment of acute herpes zoster. *N. Engl. J. Med.* 330, 896–900.

## Neuropathy with Human Immunodeficiency Virus Infection

Berger, A. R., Arezzo, J. C., Schaumburg, H. H., Skowron, G., Merigan, T., Bozzette, S., Richman, D., and Soo, W. (1993). 2,3-dideoxycytidine (ddC) toxic neuropathy: A study of 52 patients. *Neurology* 43, 358–362.

Chimowitz, M. I., Audet, A. M. J., Hallet, A., and Kelly, J. J. (1989). HIV-associated CIDP. *Muscle Nerve* 12, 695–696.

Fuller, G. N., Jacobs, J. M., and Guiloff, R. J. (1993). Nature and incidence of peripheral nerve syndromes in HIV infection. *J. Neurol. Neurosurg. Psychiatry* 56, 372–381.

Hall, C. D., Snyder, C. R., Messenheimer, J. A., Wilkins, J. W., Robertson, W. T., Whaley, R. A., and Robertson, K. R. (1991). Peripheral neuropathy in a cohort of human immunodeficiency virus-infected patients. Incidence and relationship to other nervous system dysfunction. *Arch. Neurol.* 48, 1273–1274.

Mah, V., Vartavarian, L. M., Akers, M. A., and Vinters, H. V. (1988). Abnormalities of peripheral nerve in patients with human immunodeficiency virus infection. *Ann. Neurol.* 24, 713–717.

McArthur, J., Yiannoutsos, C., Simpson, D., and the ACTG 291 Study

Team. (1998). Trial of recombinant nerve growth factor for HIV associated sensory neuropathy. *J. Neurovirol.* 4, 359.

Miller, R. G., Storey, J. R., and Greco, C. M. (1990). Gangciclovir in the treatment of progressive AIDS-related polyradiculopathy. *Neurology* 40, 569–574.

Thornton, C. A., Latif, A. S., and Emmanuel, J. C. (1991). Guillain-Barre; syndrome associated with human immunodeficiency virus infection in Zimbabwe. *Neurology* 41, 812–815.

Winer, J. B., Bang, B., Clarke, J. R., Knox, K., Cook, T. J., Gompels, M., Hughes, R. A. C., Hall, S. M., Pinching, A. J., Harris, J. W. R., Kitchen, V., Jeffries, D. J., Leibowitz, S., Smith, S., Cockbain, Z., Ekong, T., and Hughes, C. (1992). A study of neuropathy in HIV infection. *Q. J. Med.* 83, 473–488.

## Tetanus

Saissy, J. M., Demaziere, J., Vitris, M., Seck, M., Marcoux, L., Gaye, M., and Ndiaye, M. (1992). Treatment of severe tetanus by intrathecal injections of baclofen without artificial ventilation. *Intensive Care Med.* 18, 241–244.

*Muscle and Peripheral Nervous System*

CHAPTER 89
# Noninflammatory Polyneuropathy

Michael Donaghy

## CLINICAL ASPECTS AND DIFFERENTIAL DIAGNOSIS

This chapter addresses polyneuropathies resulting from inherited, metabolic, and toxic causes. Most of these disorders are rare, but two are encountered relatively commonly: diabetic polyneuropathy and hereditary motor and sensory neuropathy type 1.

Most metabolic, toxic, and inherited neuropathies involve axonal degeneration or maldevelopment, the noteworthy exception being type 1 hereditary motor and sensory neuropathies that are primarily demyelinating. Axonal polyneuropathy causes distally distributed symptoms and signs. Sensory loss generally predominates and involves modalities served both by myelinated (joint position and vibration senses) and unmyelinated (temperature and pin prick) sensory fibers. Autonomic nerve fiber involvement leads to the warm, dry, autonomically denervated foot, with cutaneous vasodilation and loss of sweating; and hair growth tends to be affected. Motor involvement is often less marked, with weakness and wasting of the foot and distal leg muscles and of the small hand muscles.

Nerve conduction studies are crucial to the differentiation of axonal and demyelinating polyneuropathies. In axonal polyneuropathy electromyography will show denervation of distal muscles. Nerve conduction studies will show diminution or loss of sensory nerve action potentials. Motor conduction velocities may be normal, or if there is substantial dropout of large myelinated motor fibers, conduction velocity may be reduced. However, axonal polyneuropathies do not reduce motor conduction velocity by more than 20% below the lower limit of normal (approximately 40 m/s for the arm), cause conduction block, or significantly prolong distal motor latencies and F waves.

## INHERITED NEUROPATHIES

### Hereditary Motor and Sensory Neuropathy (HMSN)

HMSN is the most frequent cause of the peroneal muscular atrophy syndrome composed of distal leg muscle weakness and wasting, usually accompanied by pes cavus. It is also known as Charcot-Marie-Tooth disease (CMT), peroneal muscular atrophy, the Roussy-Lévy syndrome, and Dejerine-Sottas disease. HMSN consists of a range of demyelinating and axonal loss neuropathies of varied inheritance. Presentation is generally in childhood or adolescence, although symptoms may become evident only much later in life. Asymptomatic affected relatives may be identified and are often crucial to diagnosis.

The usual first symptom is foot deformity or difficulty in walking. Paresthesia, or other positive sensory symptoms, make the diagnosis of HMSN unlikely and are more suggestive of an acquired neuropathy.

HMSN is classified either into types I–V on clinical and electrophysiological grounds or on the basis of identified molecular genetic abnormalities. Increasingly, incongruities are recognized between these two classification systems. The clinical classification has consisted of type I (adult-onset demyelinating), type II (adult-onset axonal), type III (Dejerine-Sottas disease; infantile, or childhood onset), type IV (Refsum disease; phytanic acid accumulation), or type V (HMSN with other nervous system involvement such as spastic paraparesis, optic atrophy, pigmentary retinal degeneration, deafness or mental retardation). This clinical and electrophysiological classification is mainly of practical use in differentiating type I (demyelinating) from type II (axonal). However, extensive subclassifications are developing, often correlating with newly described mutations or linkage loci, and overlap forms are recognized. Three main molecular genetic abnormalities are recognized to underlie most HMSN. Chromosome 17p11.2 reduplications involving the peripheral myelin protein 22 (PMP 22) gene underlie typical autosomal dominantly inherited HMSN type I. Mutations of the Connexin 32 gene underlie sex-linked recessive forms of HMSN I. Myelin protein $P_0$ gene mutations, which affect compaction of the myelin sheath, have been described in association with HMSN I, HMSN II, Dejerine-Sottas syndrome, and congenital hypomyelination.

### HMSN Type I

Generalized areflexia occurs in approximately half of patients. A degree of hand weakness, often minor, is eventually evident in most. Occasionally limb tremor or ataxia develops, the so-called Roussy-Lévy syndrome. Nerve thickening occurs in about a quarter, most reliably palpable on the great auricular nerve. Spinal fluid protein levels are usually normal, which can be helpful if differentiation from chronic inflammatory demyelinating polyneuropathy is proving difficult.

The median nerve motor conduction velocity is 38 m/s or less, and generalized motor slowing has already been evident in infancy and early childhood (Garcia *et al.*, 1998). Motor conduction velocities in HMSN type I rarely lie below 12 m/s, but if so, the diagnosis of HMSN III should be considered (Ouvrier *et al.*, 1987). Sural nerve biopsy shows hypertrophic onion bulb changes and reduced density of myelinated fibers. Clinical and neurophysiological examination of first-degree relatives is often crucial to proving the diagnosis and remains the cornerstone of diagnosis in those patients whose HMSN is not due to recognized and routinely detectable molecular genetic abnormalities. The condition is most usually dominantly inherited, but sex-linked recessive forms are also encountered.

The important differential diagnosis of HMSN type I is from chronic inflammatory demyelinating polyneuropathy, which is treatable. Features favoring HMSN type I are pes cavus, palpable nerve thickening, preservation of power in proximal limb muscles, spinal fluid protein of less than 0.8 g/L, little or no progression of weakness, absence of positive sensory symptoms, and onion bulb changes on nerve biopsy.

### HMSN Type II

HMSN type II, the axonal form, is less common than type I. Motor nerve conduction velocities are normal, or only slightly reduced, and inheritance is usually autosomal dominant (Harding and Thomas, 1980). The clinical features of HMSN type II are heterogeneous, and this probably represents different underlying molecular genetic abnormalities. The most common form involves distal weakness and wasting, with lesser degrees of sensory loss and areflexia, starting in the second or third decade. Other forms can show younger age of onset, foot ulceration, vocal cord or diaphragm paralysis, or relatively more severe arm involvement.

The usual time of presentation is within the second or third decade of life. However, the onset can be delayed until old age, in which case it may prove difficult to differentiate from chronic idiopathic axonal polyneuropathy. The clinical features generally resemble HMSN type I, although palpable nerve thickening does not occur. Median motor nerve conduction velocity usually lies just within the normal range and does not fall into the demyelinating range (less than 39 m/s).

### HMSN Type III

This heterogeneous group of conditions starts in infancy or early childhood and may be inherited either autosomally recessively or dominantly. It is also known as Dejerine-Sottas syndrome, which is seen as delayed onset of walking or congenital hypomyelination neuropathy that presents at birth. Motor conduction velocity is generally less than 12 m/s. The condition should be differentiated from steroid-responsive chronic inflammatory demyelinating polyneuropathy of infancy and childhood; however, elevation of spinal fluid protein is not reliable in this differentiation.

### Hereditary Neuropathy with Liability to Pressure Palsies

Autosomal dominant inheritance of a vulnerability to develop mononeuropathies occurs in some families whose nerves are exposed to pressure or traction. Superficial nerves, such as the radial or lateral popliteal are particularly liable to compression, and brachial plexus lesions may result from sleeping awkwardly or carrying bags with heavy shoulder straps. Nerve conduction studies often show a mild generalized slowing of distal motor latencies or sensory nerve action potentials; this finding should raise suspicion of this disorder when investigating a seemingly uncomplicated peripheral nerve palsy. Nerve biopsy specimens from such patients show sausage-shaped (tomaculous) swellings on teased nerve fibers caused by redundant myelin loops that result from overgrowth of the myelin spiral. Most patients show deletions of the PNP22 gene 17p11.2 (Lenssen *et al.*, 1998); new mutations may account for up to 5% of all patients.

## Hereditary Sensory and Autonomic Neuropathies (HSAN)

These neuropathies are due to failure of development or degeneration of subpopulations of peripheral sensory and autonomic neurones. Poor pain appreciation leads to mutilating acropathy with skin ulceration, long bone fractures, Charcot joints, and digit amputations. These disorders are rare, and at least five clinically distinct forms occur, with different ages of onset, genetics, and affecting different subpopulations of sensory and autonomic neurons (Donaghy et al., 1987). Autosomal dominant or recessive forms may affect all modalities of sensation and also lead to minor distal muscle weakness and wasting; reflexes are absent. More selective forms may affect only unmyelinated sensory neurons, autonomic neurons, or small myelinated sensory fibers. This group of disorders may be seen in infancy or childhood and includes the Riley-Day syndrome of familial dysautonomia. These disorders all produce loss of pain and temperature sensation and autonomic impairments. In some, the preservation of large-diameter sensory fibers means that reflexes are preserved, and sensory nerve action potentials normal.

## Other Inherited Polyneuropathies

A wide range of other rare polyneuropathies arise as a result either of inherited disorders of neuronal development or survival or as a result of inherited metabolic conditions affecting the peripheral nervous system.

*Giant axonal neuropathy* is an autosomal recessively inherited progressive disorder of childhood causing peripheral neuropathy, ataxia, intellectual loss, and pyramidal tract dysfunction. The pathognomonic feature is accumulations of abnormally closely packed neurofilaments within swollen segments of peripheral nerve axons (Donaghy et al., 1988). This disorder affects intermediate filaments in a variety of cell types, including the keratin of hair, which is characteristically tightly curled.

*Multiple symmetrical lipomatosis* involves disfiguring multiple subcutaneous lipomata over the upper trunk and proximal arms. It is also known as a Madelung's disease, and most patients have a mild axonal sensory motor neuropathy generally presenting in middle age (Chalk et al., 1990).

*Refsum disease*, also known as hereditary motor and sensory neuropathy type IV, is due to phytanic acid accumulation. Demyelinating polyneuropathy is associated with retinitis pigmentosa, cerebellar ataxia, and raised spinal fluid protein. The neuropathy improves or stabilizes after dietary restriction of phytate ingestion (dairy products, animal fats, green vegetable chlorophyll), and early plasma exchange may help induce remission before dietary restriction becomes effective (Harari et al., 1991).

*The metachromatic leukodystrophies* are a rare group of autosomal recessive diseases in which sulfatides accumulate in nervous tissue because of various defects in the arylsulfatase A gene (Draghia et al., 1997). The age of onset is variable, and the clinical picture is diverse. Severe demyelinating sensory motor neuropathy is accompanied by psychomotor retardation and seizures, and leukodystrophic changes will be evident on cerebral MRI. The prognosis for survival is poor, particularly in those with infantile onset.

Galactosylceramide-beta-galactosidase deficiency produces the condition known as *Krabbe disease* or globoid cell leukodystrophy. Although demyelinating sensory motor peripheral neuropathy is generally detectable later in life, presentation in infancy or childhood is usually with psychomotor retardation (De Gasperi et al., 1996).

*Anderson-Fabry disease* is on X-linked disorder that is seen in childhood or early adult life with burning pain and paresthesias distally in the limbs, often provoked by exercise or heat. Formal evidence of peripheral neuropathy affecting large sensory motor fibers is not usually evident clinically or electrophysiologically. The clue to diagnosis is a crimson angiokeratoma corporis diffusum rash in the bathing trunks area. Prognosis is poor, because strokes, hypertension, renal failure, and corneal opacification are common. The disorder is due to mutations of the alpha-galactosidase A gene (Blanch et al., 1996).

## AMYLOID NEUROPATHY

Various nonbranching fibrillary proteins, which form beta-pleated sheets, are deposited in the nerves and other body tissues in amyloidotic polyneuropathy. Biopsy of sural nerve, or other tissue such as the rectal mucosa, reveals amyloid material typically staining with Congo-red dye and exhibiting apple-green birefringence in polarizing light. Two main groups of amyloidotic neuropathy patients are recognized: familial amyloidotic neuropathies and the acquired condition of primary amyloidosis.

*Familial amyloidotic polyneuropathies* are inherited autosomal dominantly, and symptom onset is generally from the third to the sixth decade. The original clinical subclassification into types 1–4 is being replaced by classification based on genetic mutations. Most patients with types 1 (Portuguese) and types 2 (Indiana/Swiss) show deposition of abnormal transthyretin (Planté-

Bordeneuve *et al.*, 1998), type 3 (Iowa) is due to deposition of mutant apolipoprotein A-1, and type 4 (Finnish) is due to gelsolin gene mutations and involves the distinctive additional feature of corneal lattice dystrophy.

The polyneuropathy is fundamentally similar in all these different forms of familial amyloidosis, and the early symptoms are of numbness associated with reduced pain and temperature sensation in the hands and feet. Spontaneous limb pains, areflexia, distal weakness, and autonomic failure may develop. Renal and cardiac failure contribute to premature death. Liver transplantation is being introduced to treat transthyretin-related amyloid polyneuropathy, given that 90% of this protein is produced by the liver (Adams *et al.*, 2000).

*Primary amyloidosis* is due to immunoglobulin light chain deposition. It is unusual before middle age. Various acquired underlying lymphoproliferative disorders may be responsible, ranging from malignant myeloma to benign paraproteinemia; serum paraproteinemia or Bence-Jones proteinuria may be detected. The initial symptoms are sensory or autonomic and often involve unpleasant dysesthetic or lancinating pains, postural hypotension, impotence, and constipation. Relentless progression of the neuropathy, coupled with amyloid deposition in other organs, leads to death in 80% of patients within 3 years. Various aggressive chemotherapy regimens are being evaluated to try and prevent further immunoglobulin light chain production in such patients and thus stabilize their neuropathy.

## NEUROPATHY CAUSED BY SYSTEMIC MEDICAL DISORDERS

### Diabetic Polyneuropathy

Two types of diabetic polyneuropathy are recognized, and they may coexist: symmetrical sensorimotor and autonomic. In addition, various focal neuropathies can occur, including mononeuropathies affecting cranial and peripheral nerves, truncal neuropathies, and diabetic proximal neuropathy (Said, 1996; Watkins, 1990). Although neuropathy is detectable in more than half of established insulin-dependent diabetics and non-insulin–dependent diabetics, only about 20% of patients have diabetic symptoms, and less than 5% overall have clinically significant forms of neuropathy (Dyck *et al.*, 1993). Multiple interacting abnormalities of cell biology may be responsible for neuropathy in diabetes, but the pathogenic mechanisms have not been precisely identified.

The initial symptoms are paresthesias, or burning or lancinating pains in the legs. Unmyelinated fiber sensations are generally more severely affected, with stocking distribution loss of pain and temperature sensations. With progression to more severe forms of the neuropathy, myelinated fiber sensations and distal muscles become affected. Autonomic denervation leads to a warm, dry foot, vulnerable to ulceration. Nerve conduction studies show diminished or absent sensory nerve action potentials with relatively milder involvement of motor nerve conduction.

Strict control of glycemia offers the best chance of reducing occurrence and progression of diabetic polyneuropathy (Diabetes Control and Complications Trial Research Group 1993). Pancreas transplantation halts progression of diabetic neuropathy, with minor degrees of recovery appearing within a few years (Kennedy *et al.*, 1990). Acutely painful diabetic polyneuropathy is disabling and difficult to treat, although it generally improves after some months of strict glycemic control. Drugs such as carbamazepine or amitryptiline offer the best chance of reducing pain to tolerable levels.

Autonomic polyneuropathy contributes to ulceration of dry foot skin; is most usually associated with symptoms of abnormal sweating, diarrhea, or sexual impotence; and may produce postural hypotension, vomiting from gastroparesis, micturition difficulties, or retrograde ejaculation. Autonomic neuropathy is most reliably detected at the bedside by measuring postural hypotension, which reflects failure of sympathetic fibers, and measuring sinus arrhythmia during deep breathing, which reflects the parasympathetic innervation of the heart.

*Diabetic proximal neuropathy* generally occurs in elderly non-insulin–dependent diabetics. Usually it consists of acute asymmetrical painful proximal leg muscle weakness developing over days or weeks and is often associated with profound weight loss. Most patients improve neurologically after some months and, although this is generally attributed to improved diabetic control, recovery may reflect the untreated natural history of this condition (Coppack and Watkins, 1991). The pathogenesis of diabetic proximal neuropathy is unclear, but it may be a form of epineurial microvasculitis given recent biopsy evidence from the intermediate cutaneous nerve of the thigh (Said *et al.*, 1994).

### Chronic Renal Failure

Evidence of polyneuropathy is detectable in more than 50% of patients with end-stage renal disease. If symptomatic, restless leg syndrome, burning paresthesias, or muscle cramps are common early symptoms. Subsequently full-blown sensory, motor, and autonomic neuropathy may develop. Nerve conduction studies reflect axonal degeneration. Renal replacement therapy normally prevents further deterioration and may allow some recovery.

### Critical Illness Polyneuropathy

Patients being ventilated for cardiorespiratory disease who have sepsis or multiorgan failure may develop sensorimotor polyneuropathy (Zochodne et al., 1987). It is first noted when patients fail to wean from the ventilator. Nerve conduction studies show reduced amplitude compound muscle action potentials and sensory nerve action potentials. This disorder should be differentiated from Guillain-Barré syndrome, which is associated usually with electrophysiological evidence of demyelination and raised spinal fluid protein and from the critical illness myopathy associated with the use of nondepolarizing neuromuscular blocking agents (Gutmann and Gutmann, 1999).

## TOXIC POLYNEUROPATHY

### Alcohol

Polyneuropathy in alcoholics may reflect either longstanding high consumption of alcohol (Monforte et al., 1995), vitamin $B_1$ (thiamine) deficiency, or treatment with disulfiram (*Antabuse*). Neuropathy is often erroneously attributed to moderately high, but not extreme, alcohol consumption. Typically alcoholic polyneuropathy produces symmetrical burning dysesthesias sensory ataxia, and mild motor involvement. Nerve conduction studies reflect axonal degeneration (Behse and Buchthal, 1977). For improvement to occur, long-term abstinence is necessary, and vitamin $B_1$ replacement therapy should be given.

### Drug-Induced

A wide variety of therapeutic drugs cause neuropathy: almitrine, amiodarone, chloroquine, cisplatin, clioquinol, colchicine, dapsone, didanosine, disulfiram, ethambutol, FK506, gold, isoniazid, lithium, metronidazole, nitrofuranotoin, perhexiline, phenytoin, high-dose pyridoxine, sodium cyanate, suramin, taxol, thalidomide, and vincristine (Donaghy, 2001). Most of these neuropathies are predominantly sensory with varying degrees of additional motor or autonomic involvement. Electrophysiology generally produces features of axonal degeneration, but motor slowing may be evident in amiodarone toxicity and with FK506. Predominantly motor disorders may be seen with high-dose

dapsone treatment of dermatitis herpetiformis. Particular difficulty may involve the differentiation of HIV-associated sensory neuropathy from that induced by the antiretroviral drugs didanosine (ddI), zalcitabine (ddC), and stavudine (d4T). Myokymia is a distinctive feature of gold-induced neuropathy. Isoniazid antagonizes vitamin $B_6$ (pyridoxine), and neuropathy can be prevented by simultaneous administrative of this vitamin.

### Metal Poisoning

Polyneuropathy is recognized to be due to inorganic arsenic; inorganic lead; inorganic, organic, or elemental mercury; and thallium after either chronic industrial exposures or acute homicidal or suicidal ingestion. Arsenic poisoning leads to a predominantly sensory axonal polyneuropathy with anemia and the appearance of lines across the nails (Mee's lines) (Donofrio et al., 1987). Classically, the peripheral neuropathy caused by inorganic lead poisoning has caused a subacute motor neuropathy coupled with abdominal cramping (Cullen et al., 1983). This clinical syndrome closely resembles acute hepatic porphyria, a metabolic disorder that can be induced by lead poisoning. However, in very long-term lead-exposed workers, motor function may be normal, with mild sensory and autonomic features in very long-term lead-exposed workers (Rubens et al., 2001). Organic mercury poisoning produces a marked ataxia and visual field constriction with variable electrophysiological evidence of peripheral nerve involvement (Nierenberg et al., 1998). Thallium neuropathy has been described after poisonings, and the subacute sensory motor neuropathy closely resembles Guillain-Barré syndrome with the additional features of confusional psychosis, involuntary movements, and dark pigmentation of the hair roots (Davis et al., 1981). Treatment of all these polyneuropathies involves removal from the exposure, supportive treatment for acute organ failure, and the use chelating agents.

### Industrial and Agricultural Chemicals

Acrylamide monomer has entered well water during soil stabilization procedures. Mild sensory motor polyneuropathy has been noted some days later (Davenport et al., 1976). Carbon disulfide was used in rubber vulcanization and caused axonal degeneration polyneuropathy, sometimes accompanied by encephalopathy (Vasilescu, 1976). Dimethylaminopropionitrile was used as a catalyst in polyurethane manufacture and led to axonal degeneration sensory motor neuropathy with additional involvement of bladder control and sexual function (Keogh et al., 1980). The herbicide 2,4-D leads to sensory motor neuropathy starting some days after exposure (Goldstein *et al.*, 1959). The hexacarbons *n*-hexane and methyl *n*-butyl ketone are used as glue solvents and in flexographic printing. They are metabolized after ingestion to 2,5-hexanedione, which causes sensory motor polyneuropathy of a severity that may resemble Guillain-Barré syndrome after acute exposures (Altenkirch *et al.*, 1977). Organophosphorous and carbamate pesticide intoxication produces an initial cholinergic poisoning involving acute paralysis, bronchial secretions, bradycardia, and seizures before a predominantly motor neuropathy develops in some patients 1 to 3 weeks later (Senanayake and Johnson, 1982).

### Vitamin Deficiency

Combined vitamin and nutritional deficiency may result in painful polyneuropathy in patients with starvation, chronic gastrointestinal disease, or malnourished alcoholism (Cockerell and Ormerod, 1993). A burning foot syndrome often remains over the long term despite effective reinstitution of a balanced diet. Dry or neuropathic beriberi associated with vitamin $B_1$ (thiamine) deficiency leads to sensorimotor polyneuropathy often sufficient to prevent walking; neuropathic limb pains occur. Supplementation with vitamin $B_1$ leads to steady improvement. Vitamin $B_6$ deficiency causes pellagra, with skin, gastrointestinal, and neuropsychiatric symptoms in addition to polyneuropathy (Bomb *et al.*, 1977). Notably, this neuropathy can be associated with isoniazid therapy (see earlier). Vitamin $B_{12}$ deficiency may cause sensory polyneuropathy, although this is normally overshadowed by the associated spinal cord lesion known as subacute combined degeneration (Hemmer *et al.*, 1998). Some degree of recovery follows vitamin $B_{12}$ replacement injections. Long-standing vitamin E deficiency associated with malabsorption, abetalipoproteinemia, or a familial deficiency of the $\alpha$-tocopherol transfer protein can cause polyneuropathy with prominent large fiber sensory loss and ataxia resembling a spinocerebellar degeneration; vitamin E supplementation prevents further downhill progression (Hentati *et al.*, 1996).

## PRACTICAL MANAGEMENT OF NEUROPATHIC PAIN AND OTHER SYMPTOMATIC THERAPIES

The pathophysiology of neuropathic pain is not fully understood. The symptoms may consist of a variable mixture of paresthesias, protopathic or epicritic pain, hyperalgesia or allodynia, muscle cramps, and fascicu-

lations. Nerve injury can set off causalgia or reflex sympathetic dystrophy, which are the terms traditionally used to describe chronic burning pain in a limb, coupled with various autonomic and trophic changes in the latter case (Schott, 1986). Recent proposals attempt to subsume both entities within the overall term "complex regional pain syndrome" (Stanton-Hicks *et al.*, 1995). These forms of pain can be extremely distressing, and there is no single approach to therapy which is reliably effective.

Antidepressants, anticonvulsants, and neuroleptics are most often used in the pharmacological treatment of neuropathic pain. However, well-controlled double-blind trials showing these drugs to be beneficial are rare. As a rule, anticonvulsants are used for epicritic pain, antidepressants for the more protopathic type of pain. In severe cases, neuroleptics may be added to enhance the effect of the antidepressant. Carbamazepine at approximately 500 mg/day should be used as first choice; even higher doses are often required. Its efficacy in painful diabetic neuropathy has been established in double-blind trials (Rull *et al.*, 1969). Diphenylhydantoin has also been documented to be helpful for neuropathic pain in diabetes and Fabry's disease (Lockman *et al.*, 1973). It may be combined with carbamazepine for severe symptoms.

The effect of tricyclic antidepressants for the treatment of pain is independent of their antidepressant effect and is thought to be mediated mainly by the inhibition of serotonin uptake. Amitriptyline is the best-studied compound for pain therapy. It should be started at 10–25 mg at night, gradually increasing to a maintenance dosage of 75–100 mg/day. The anticholinergic and myocardial side effects are most important, and tricyclics should be avoided in patients with heart disease. An alternative to amitriptyline is nortriptyline, which has less anticholinergic and sedative side effects. The efficacy of selective serotonin uptake inhibitors for the treatment of pain has not yet been established. The use of neuroleptic drugs (e.g., fluphenazine up to 2 mg twice daily) to increase the pain-relieving effect of antidepressants is often recommended but has not been proven in a trial.

For pain in diabetic neuropathy a well-controlled study showed the beneficial effect of mexiletine at 10 mg/kg (Dejard *et al.*, 1988). Quinine, dantrolene, baclofen, procainamide, or tranquilizers may be used for the symptomatic treatment of muscle cramps or painful fasciculations.

Rehabilitation of patients with peripheral neuropathy includes general aspects of nursing, particularly positioning of paretic limbs to prevent contraction, which can be a secondary source of disability. Physical therapy includes active or passive exercises to preserve joint mobility. Re-education of motor performance may help to regain voluntary control over severely, but incompletely, denervated muscles. In milder pareses more vigorous exercise helps to build up strength and stamina (Stillwell and Thorsteinsson, 1993).

Attempts to stimulate nerve regeneration have been disappointing from a clinical standpoint. Thyroid hormones were applied to induce protein synthesis but without any effect on neuropathy. Gangliosides from bovine brain were shown to stimulate regeneration of peripheral nerves and synapses in animal experiments; however, clinical effects are not clearly proven. A number of cases of Guillain-Barré syndrome occurred after parenteral application of gangliosides, and they are no longer used. The clinical efficacy of nucleotides is also not proven. Uncritical use of vitamins for neuropathy is discouraged and should be guided by a clearly defined deficiency state. Attempts to induce axonal regeneration using neuronal growth factors have been generally disappointing.

# REFERENCES

Adams, D., Didier, S., Goulon-Goeau, C. et al. (2000). The course and prognostic factors of familial amyloid polyneuropathy after liver transplantation. Brain 123, 1495–1504.

Altenkirch, H., Mager, J., Stoltenburg, G. et al. (1977). Toxic polyneuropathies after sniffing a glue thinner. J. Neurol. 214, 137–152.

Behse, F., and Buchthal, F. (1977). Alcoholic neuropathy: clinical, electrophysiological, and biopsy findings. Ann. Neurol. 2, 95–110.

Blanch, L. C., Meaney, C., and Morris, C. P. (1996). A sensitive mutation screening strategy for Fabry Disease: detection of nine mutations in the alpha-galactosidase A gene. Hum. Mutat. 8, 38–43.

Bomb, B. S., Bedi, H. K., and Bhatnagar, K. (1977). Postischaemic paraesthesiae in pellagrins. J. Neurol. Neurosurg. Psychiatry 40, 265–267.

Chalk, C. H., Mills, K. R., Jacobs, J. M. et al. (1990). Familial multiple symmetric lipomatosis with peripheral neuropathy. Neurology 40, 1246–1250.

Cockerell, O. C., and Ormerod, I. E. C. (1993). Strachan's syndrome: variation on a theme. J. Neurol. 240, 315–318.

Coppack, S. W., and Watkins, P. J. (1991). The natural history of diabetic femoral neuropathy. Quart. J. Med. 79, 307–313.

Cullen, M. R., Robins, J. M., and Eskenazi, B. (1983). Adult inorganic lead intoxication: presentation of 31 new cases and a review of recent advances in the literature. Medicine 62, 221–247.

Davenport, J. G., Farrell, D. F., and Sumi, S. M. (1976). 'Giant axonal neuropathy' caused by industrial chemicals. Neurology 26, 919–923.

Davis, L. E., Standefer, J. C., Kornfeld, M. et al. (1981). Acute thallium poisoning: toxicological and morphological studies of the nervous system. Ann. Neurol. 10, 38–44.

De Gasperi, R., Sosa, M. A. G., Sartorato, E. L. et al. (1996). Molecular heterogeneity of late-onset forms of globoid-cell leucodystrophy. Am. J. Hum. Genet. 59, 1233–1242.

Dejgard, A., Petersen, P., and Kastrup, J. (1988). Mexiletine for treatment of chronic painful diabetic neuropathy. Lancet 1, 9–11.

Diabetes Control and Complications Trial Research Group. (1993). The effect of intensive treatment of diabetes on the development and progression of long-term complications in insulin-dependent diabetes mellitus. N. Engl. J. Med. 329, 977–986.

Donaghy, M. (2001). Polyneuropathy. In "Brain's Diseases of the Nervous System," 11th ed. (M. Donaghy, Ed.), pp. 337–403. Oxford University Press, Oxford.

Donaghy, M., Brett, E. M., Ormerod, I. E. et al. (1988). Giant axonal neuropathy: observations on a further patient. J. Neurol. Neurosurg. Psychiatry 51, 991–994.

Donaghy, M., Hakin, R. N., Bamford, J. M. et al. (1987). Hereditary sensory neuropathy with neurotrophic keratitis. Description of an autosomal recessive disorder with a selective reduction of small unmyelinated nerve fibres and a discussion of the classification of the hereditary sensory neuropathies. Brain 110, 563–583.

Donofrio, P. D., Wilbourn, A. J., Albers, J. W. et al. (1987). Acute arsenic intoxication presenting as Guillain-Barré-like syndrome. Muscle Nerve 10, 114–120.

Draghia, R., Letourneur, F., Drugan, C. et al. (1997). Metachromatic leucodystrophy: identification of the first deletion in axon 1 and nine novel point mutations in the arylsulfatase gene. Hum. Mutat. 9, 234–242.

Dyck, P. J., Kratz, K. M., Litchy, W. J. et al. (1993). The prevalence of staged severity of various types of diabetic neuropathy, retinopathy, and nephropathy in a population based cohort. Neurology 43, 817–824.

Garcia, A., Combarros, O., Calleja, J. et al. (1998). Charcot-Marie-Tooth disease Type IA with 17p duplication in infancy and early childhood. A longitudinal clinical and electrophysiologic study. Neurology 50, 1061–1067.

Goldstein, N. P., Jones, P. H., and Brown, J. R. (1959). Peripheral neuropathy after exposure to an ester of dichlorophenoxyacetic acid. JAMA 171, 1306–1309.

Gutmann, L., and Gutmann, L. (1999). Critical illness neuropathy and myopathy. Arch. Neurol. 56, 527–528.

Harari, D., Gibberd, F. B., Dick, J. P. R. et al. (1991). Plasma exchange in the treatment of Refsum disease (heredopathia atactica polyneuritiformis). J. Neurol. Neurosurg. Psychiatry 54, 614–617.

Harding, A. E., and Thomas, P. K. (1980). The clinical features of hereditary motor and sensory neuropathy types I and II. Brain 103, 259–280.

Hemmer, B., Glocker, F. X., Schumacher, M. et al. (1998). Subacute combined degeneration: clinical, electrophysiological, and magnetic resonance imaging methods. J. Neurol. Neurosurg. Psychiatry 65, 822–827.

Hentati, A., Deng, H.-X., Hung, W.-Y. et al. (1996). Human α-tocopherol transfer protein: gene structure and mutations in familial vitamin E deficiency. Ann. Neurol. 39, 295–300.

Kennedy, W. R., Navarro, X., Goetz, F. C. et al. (1990). Effects of pancreatic transplantation on diabetic neuropathy. N. Engl. J. Med. 322, 1031–1037.

Keogh, J. P., Pestronk, A., Wertheimer, D. et al. (1980). An epidemic of urinary retention caused by dimethylaminopropionitrile. JAMA 243, 746–749.

Lenssen, P. P. A., Gabreëls-Festen, A. A. W. M., Valentijn, L. J. et al. (1998). Hereditary neuropathy with liability to pressure palsies. Penotypic differences between patients with the common deletion and a PMP22 frame shift mutation. Brain 121, 1451–1458.

Lockman, L. A., Hunninghake, D. B., Krivit, W., and Desnick, R. J. (1973). Relief of pain of Fabry's disease by diphenylhydantoin. Neurology 23, 871–875.

Monforte, R., Estruch, R., Valls-Sole, J. et al. (1995). Autonomic and peripheral neuropathies in patients with chronic alcoholism. Arch. Neurol. 52, 45–51.

Nierenberg, D. W., Nordgren, R. E., Chang, M. B. et al. (1998). Delayed cerebellar disease and death after accidental exposure to dimethylmercury. N. Engl. J. Med. 338, 1672–1676.

Ouvrier, R. A., McLeod, J. G., and Conchin, T. E. (1987). The hypertrophic forms of hereditary motor and sensory neuropathy. A study of hypertrophic Charcot-Marie-Tooth disease (HMSN type I) and Dejerine-Sottas disease (HMSN type III) in childhood. *Brain* **110**, 121–148.

Planté-Bordeneuve, V., Lalu, T., Misrahi, M. *et al.* (1998). Genotypic-phenotypic variations in a series of 65 patients with familial amyloidotic polyneuropathy. *Neurology* **51**, 708–714.

Rubens, O., Logina, I., Kravale, I. *et al.* (2001). Peripheral nerve function in chronic occupational inorganic lead exposure: a clinical and electrophysiological study. *J. Neurol. Neurosurg. Psychiatry* **71**, 200–204.

Rull, J. A., Quibrera, R., Gonzalez-Millan, H., and Lozano Castaneda, O. (1969). Symptomatic treatment of peripheral diabetic neuropathy with carbamazepine (Tegretol®): double blind crossover trial. *Diabetologia* **5**, 215–218.

Said, G. (1996). Diabetic neuropathy: an update. *J. Neurol.* **243**, 431–440.

Said, G., Goulon-Goeau, C., Lacroix, C. *et al.* (1994). Nerve biopsy findings in different patterns of proximal diabetic neuropathy. *Ann. Neurol.* **35**, 559–569.

Schott, G. D. (1986). Mechanisms of causalgia and related clinical conditions. *Brain* **109**, 717–738.

Senanayake, N., and Johnson, M. K. (1982). Acute polyneuropathy after poisoning by a new organophosphate insecticide. *N. Engl. J. Med.* **306**, 155–157.

Stanton-Hicks, M., Janig, W., Hassenusch, S. *et al.* (1995). Reflex sympathetic dystrophy: changing concepts and taxanomy. *Pain* **63**, 127–133.

Stillwell, G. K., and Thorsteinsson, G. (1993). Rehabilitation procedures. *In* "Peripheral Neuropathy" (P. J. Dyck and P. K. Thomas, Eds.), Saunders, Philadelphia.

Vasilescu, C. (1976). Sensory and motor conduction in chronic carbon disulphide poisoning. *Eur. Neurol.* **14**, 447–457.

Watkins, P. J. (1990). Natural history of the diabetic neuropathies. *Q. J. Med.* **77**, 1209–1218.

Zochodne, D. W., Bolton, C. F., Wells, G. A. *et al.* (1987). Critical illness polyneuropathy: a complication of sepsis and multiple organ failure. *Brain* **110**, 819–842.

CHAPTER 90

# Disorders of Nerve Roots Caused by Bony and Disk Diseases

Michael Swash

The anterior motor roots and the posterior sensory roots are components of the peripheral nervous system that are located within the spinal canal. Damage to these nerve roots may result from degenerative, traumatic, neoplastic, infectious, and vertebral disorders (Table I). The topographical relationship of nerve roots with vertebral bodies, intervertebral disks, and spinal joints leads to vulnerability of nerve roots to lesions caused by movement, especially when degenerative changes occurring in the spinal column, such as spondylosis, spondylarthrosis, and intervertebral disk protrusion, reduce the available space for movement of these nerve roots. The lower cervical and lumbar segments, particularly the lumbosacral transitional region, are especially vulnerable to radicular compression syndromes, probably as a result of their comparably high mobility and their vulnerability to static strains. As in lesions caused by pressure on the peripheral nerves, short-term focal pressure on nerve roots leads to segmental demyelination with functional conduction block, from which rapid clinical recovery may occur. Severe longer term compression or entrapment leads to axonal degeneration with denervation in a segmental distribution. Combined lesion patterns with conduction block and partial axonotmesis are frequent. Severe root damage from spinal instability usually causes a lasting functional deficit, because reinnervation is generally relatively ineffective when there is a considerable distance between the nerve root and its target muscles, and the sensory receptors in muscle, tendon, and skin.

The clinical features of radicular syndromes include segmental sensory syndromes, especially pain radiating in a searing or piercing distribution (brachialgia, sciatica). Sensory impairment, pain and temperature sensitivity, weakness of limb muscles in a segmental pattern, and absence of the corresponding tendon reflexes are found corresponding to the segmental innervation of the skin and muscles in the affected limb. Marked weakness of groups of muscles in an affected myotome is less common in single-root lesions. Because the sympathetic sudomotor efferents leave the spinal cord only in the T3 to L2 segments, sweating is unaffected in cervical, sacral, and L3–L5 lumbar nerve root lesions.

## DIAGNOSTIC INVESTIGATIONS

Clinical investigations evaluate the clinical diagnosis and the localization of nerve root lesions. These include radiological, electrodiagnostic, laboratory, and CSF tests.

Imaging techniques assess normal anatomical variations and congenital abnormalities and reveal the morphological changes associated with the disease process (Table II). The conventional initial radiological technique is usually routine radiography of the spinal column in the anteroposterior and lateral planes, often also with oblique radiographs to show the intervertebral foramina. However, modern methods, especially computed tomography (CT) and magnetic resonance imaging (MRI), allow greater resolution of bony and soft tissue structures, respectively. In contemporary practice MRI is often the primary diagnostic imaging method, because it allows direct visualization of the spinal cord, nerve roots, and subarachnoid space in the sagittal and transverse planes. Several segments can be seen in a single image. MRI is noninvasive and, apart from claustrophobia, has no recognized risks. CT offers advantages for elderly patients with degenerative vertebral disease and is preferred as the initial method for investigation of spinal trauma. Enhanced CT, using a

TABLE I   Differential Diagnosis of Radiculopathies

Degenerative
   Intervertebral disk protrusion or prolapse
   Spondylosis
   Spondylolisthesis
   Spinal canal stenosis
Traumatic
   Whiplash injury
   Vertebral fracture
   Root avulsion
   Epidural hemorrhage
Neoplastic
   Metastases from systemic carcinoma (bronchus, breast, prostate,
     kidney, thyroid)
   Multiple myeloma
   Malignant lymphoma
   Neoplastic meningitis (secondary neoplasm of subarachnoid
     space)
   Metastatic seeding from primary CNS tumors, such as
     medullo-blastoma
   Primary tumors of spinal roots
     Neurofibroma
     Neurinoma
     Schwannoma
     Meningioma
   Synovial cysts
Infections
   Herpes zoster
   Meningoradiculitis in Lyme borreliosis
   Epidural abscess
   Purulent osteomyelitis, discitis
   Spondylitis tuberculosa
Rheumatic disorders
   Rheumatoid arthritis
   Ankylosing spondylitis
   Polyarteritis nodosa and Churg-Strauss syndromes
Metabolic
   Diabetic radiculopathy
   Osteitis deformans (Paget)

TABLE II   Imaging Techniques in the Diagnosis of Radiculopathies

| Method | Formulation of the Question |
|---|---|
| Plain X-rays | Spondylosis |
| | Osteoporosis |
| | Congenital abnormalities |
| | Traumatic |
| | Discitis |
| Functional imaging | Spondylolisthesis, spondylosis |
| Computed tomography | Intervertebral disk protrusion or prolapse |
| | PID with free disk fragment |
| | Bony spinal malformations |
| | Differentiation between "soft prolapse" and osteophyte |
| | Ligamentum flavum ossification |
| | Spinal canal stenosis |
| | Bone erosion |
| Magnetic resonance imaging | Intervertebral disk degeneration, protrusion, or prolapse |
| | Spinal canal stenosis |
| | Spondylitis |
| | Discitis |
| | Syringomyelia |
| | Spinal tumor |
| | Spinal angioma |
| | Epidural abscess |
| Myelography | Imaging of nerve root sleeves |
| | Extension of spinal space-occupying lesions |
| | Cerebrospinal fluid abnormalities |
| CT myelography | Imaging of localized narrowing of the spinal cord, nerve roots, and caudal dural sac |
| | Relationship between neural elements and the surrounding anatomy |

water-soluble subarachnoid contrast medium (Scotti *et al.*, 1983), is now little used, except sometimes in planning surgery. Myelography is now an obsolete investigation (Edelmann, 1990).

Clinical neurophysiological studies can evaluate the functional state of the affected nerve root. Evidence from needle EMG of active denervation, such as fibrillation potentials, positive sharp waves, and fasciculations in a myotome, is suggestive of an axonal lesion. Increased latency of an F wave, H reflex, or somatosensory evoked potential is indicative of damage to the myelin sheath of a nerve or nerve root. An intact sensory nerve action potential in the affected segment when there is clinical evidence of a sensory disorder supports the presence of a preganglionic lesion of the sensory neuron, involving the posterior root (Benecke and Conrad, 1980).

Cerebrospinal fluid (CSF) studies with bacteriological, immunological, serological, and cytological studies

may provide diagnostic information, especially in suspected infective or neoplastic nerve root disorders.

## CERVICAL RADICULAR SYNDROMES

### Acute Cervical Prolapsed Intervertebral Disk (PID)

#### Clinical Aspects

Intervertebral disk protrusion with an intact annulus fibrosis and prolapsed intervertebral disk (PID) with prolapse of the nucleus pulposus through the perforated fibrous ring are the result of degenerative changes in the intervertebral disk tissue. The prolapsed intervertebral disk usually protrudes in a posterior direction through the weak posterior longitudinal ligament. In the cervical region, a central disk protrusion can cause spinal cord compression with spastic paraparesis, posterior column syndrome, and bladder dysfunction.

The more common posterolateral PID causes isolated root compression. Because of the horizontal course of the nerve roots through the subarachnoid space, a lateral PID generally compresses only one cervical root. Almost all patients show painful, restricted movement of the cervical vertebral column with paravertebral muscle spasm and intensification of radicular pain with movement, especially when the head is turned to the affected side and the cervical vertical column is laterally extended, thus stretching the damaged nerve root.

CT and MRI can both be used for direct imaging of cervical PID. MRI produces high-resolution images of the disks and can directly demonstrate disk protrusion without the need for further investigation. It can also demonstrate annular tears of the annulus, for example, in whiplash injury. MRI is also suitable for accurate demonstration of extreme lateral intervertebral disk sequestra in the intervertebral canal (Landmann et al., 1984; Wilson et al., 1991).

A C7 root compression syndrome must be distinguished from carpal tunnel syndrome; in the latter, EMG studies reveal focal slowing of motor and sensory nerve conduction in the median nerve across the wrist. A C8 root syndrome may resemble an ulnar nerve lesion, in which a sensory disorder affects the medial surface of the hand, both dorsally and ventrally, with a low-amplitude or absent sensory nerve action potential. A lesion of the lower part of the brachial plexus with an accompanying neurovascular compression syndrome in the thoracic outlet causes clinical and EMG findings more extensive than could be explained by involvement of the C8 root alone. There is a positive Adson's maneuver: that is, reduction of the radial pulse pressure during downward force applied to the arm by pulling downward on the hand, with the arm in the resting state in the erect posture. Pancoast syndrome causes an ipsilateral Horner's syndrome with pain in the arm in the distribution of the lower part of the brachial plexus, in the axilla, and on the inner surface of the upper arm.

**Natural History**

Intervertebral disk prolapse in the cervical spinal column is rare compared with the lumbar spinal column. The peak age of onset is in the fifth decade. There is a 2:1 male preponderance (Lunsford et al., 1980a; Scoville et al., 1976). The most frequent locations are at the C6/7 and C5/6 levels, causing unilateral monoradicular syndromes. Most cases are of acute onset with pain at the nape of the neck. Previous trauma is relatively uncommon. The first symptom often appears on waking. It is followed by typical radicular pain, weakness, and radicular sensory disorder within a matter of hours to days.

It is difficult to compare the results of conservative and surgical treatment, because no controlled, randomized, prospective studies are available. A prospective study of conservative treatment of cervical pain ascribed to cervical disk pathology carried out by the British Association of Physical Medicine (1966) contained many patients (60% of those in the series) without objective radicular symptoms. In a retrospective study on the results of surgery reported by Lunsford (1980a), 41% of patients had radicular motor syndromes. There are no reports of the natural history of MRI-confirmed PID with conservative management.

With conservative treatment, 73% of patients with radicular cervical pain are relieved of symptoms or showed substantial improvement within a period of 4 weeks (British Association of Physical Medicine, 1966). In the series of Lees and Turner (1963), 22 of 51 patients were relieved of symptoms within a few months, another 15 had slight or intermittent residual symptoms, and 10 were described as "moderately disabled." Only one patient had a severe relapse of radicular symptoms. The observation period of this retrospective study ranged from 2–19 years. Surgical procedures are reserved for severe radicular syndromes unresponsive to conservative treatment over several weeks. Surgery is estimated to be necessary in 12% of cases (Chirls, 1978; Hunt, 1980). The results of surgery are beneficial, with marked improvement in 75%–100% of cases. The best results have been reported in patients with a short history (Murphy et al., 1973). However, it is likely that many patients operated on in the early stages would also have improved with conservative treatment (Monro, 1984). The presumption that the postoperative prognosis of acute cervical PID is better than that of chronic, spondylitic compression (Symon and Lavender, 1967) is not confirmed by other studies (Lunsford, 1980a). There is no correlation between the preoperative case history and the severity of the radicular syndrome with the postoperative result, especially regarding the duration of the history and a background of trauma (Gregorius et al., 1976; Lunsford et al., 1980a).

A central cervical PID can cause a painless, progressive, cervical cord compression syndrome that, perhaps because of secondary vascular factors, can resemble intrinsic spinal cord disease: for example, the spinal form of multiple sclerosis. These clinical syndromes can be distinguished reliably only by means of neuroimaging and immunological studies on the CSF.

*Principles of Therapy*

Most patients respond to conservative treatment (Lees and Turner, 1963; Rothman and Marvel, 1975). In the acute phase, treatment focuses on primary mechanical root decompression and on secondary consequences such as muscle tension and disturbed posture. It is especially important to break the vicious circle of pain, muscle tenseness, and abnormal posture, leading to intensification of pain. The basic principles of management are rest, relief of strain and traction of the cervical spinal column, relief of muscle spasm, analgesia (cryotherapy, drugs), and when the acute radicular symptoms subside, muscular stabilization. Later, mobilization of the cervical spinal column is an important goal.

Two techniques are commonly recommended for surgical decompression of nerve roots:

1. Ventral diskectomy with or without spinal fusion (Cloward, 1963; Lunsford *et al.*, 1980a)
2. Dorsolateral approach with a foramenotomy and partial facetectomy (Scoville *et al.*, 1976)

A ventral route of surgical access is used for medial and, sometimes, for lateral intervertebral disk prolapse. A dorsolateral surgical access is particularly suitable for lateral disk prolapse.

### Practical Management

#### In the acute phase:

1. Rest and relief of strain on the cervical vertebral column is important, using a neck collar for an initial period of 2–3 weeks. Establish a correct sitting position. A cervical collar should be worn at night, because muscular relaxation during sleep may carry a danger of uncontrolled cervical movement. Careful manual or mechanical traction may be used to extend the intervertebral space (the principle of Glisson's sling in supine position).
2. Cryotherapy of the paravertebral muscles, prolonged application over 10–20 minutes twice a day.
3. Drugs (see Table III). Muscle relaxants, such as diazepam 2.5–10 mg tds. Effective analgesia, such as paracetamol 500–1000 mg bd is essential. If stronger analgesia is required, prescribe individually in accordance with the course of the illness. Opiates may be used if necessary in the initial phase of the illness. Anti-inflammatory analgesics, such as ibuprofen, 400 mg 2 times daily, are useful.

#### When the acute radicular syndrome subsides:

1. Stop drug therapy.
2. Start isometric exercise treatment to strengthen the paravertebral muscles.
3. Use physiotherapy to mobilize the cervical spinal column.

Other physical treatment measures in common use, such as heat treatment, massage, and electrotherapy,

**TABLE III** Drug Therapy of Acute Compressive Radiculopathies

| Therapy | Substance (generic name) | Dosage |
|---|---|---|
| Muscle relaxants | Diazepam | 10–40 mg/day |
| | Tetrazepam | 50–300 mg/day |
| | Chlormezanone | 400–1000 mg/day |
| | Tizanidine | 4–8 mg/day |
| Analgesics | Paracetamol | 2000–3000 mg/day |
| Opioid analgesics | Tramadol | Single dose: 1 ampule (1 ml = 50 mg) IM/SC; 20 drops (50 mg) PO; 1 capsule (50 mg) PO; 1 suppository (100 mg). Maximum dosage per day 400 mg |
| | Pentazocine | Single dose: 1 ampule (30 mg) IM; 1 capsule (50 mg) PO; 1 suppository (50 mg). Single doses may be repeated every 3–4 h; maximum dosage of 360 mg by parenteral administration |
| Nonsteroidal anti-inflammatory drugs (NSAIDs) | Acetylsalicylic acid | 1000–3000 mg/day PO |
| | Piroxicam | Initial dose 40 mg/day over 2 days; continuing dose 20 mg/day |
| | Diclofenac | Day 1: 150 mg; continuing dosage 50–100 mg/day |
| | Indomethacin | Initiation 2–3 × 1 capsule 25 mg/day or 1 suppository 100 mg in the evening. Recommended maximum dosage 200 mg/day. |
| | Ibuprofen | Daily dose 600–1800 mg |

whose benefits cannot be easily estimated or are doubtful, are described elsewhere (Kramer, 1994).

*Surgical decompression is indicated in the following situations:*

1. Acute central PID with neurological symptoms of spinal cord compression. This is an absolute indication for immediate investigation and surgical decompression.
2. Lateral PID with functionally significant radicular paresis (muscle strength of MRC grade 3 or less) without improvement over a period of 3 weeks and with acute denervation on EMG.
3. Severe radicular syndrome unresponsive to treatment with corresponding PID in the CT or MRI.

### Treatments No Longer Recommended

Manipulation is always contraindicated for acute cervical root syndromes caused by disk prolapse because of the danger of exacerbation of the neurological syndrome or even of acute cord compression.

## Chronic Intervertebral Disk Degeneration

### Clinical Aspects

With increasing age there is attrition of intervertebral disks, so that the disk margins extend beyond the rims of the vertebral bodies. This is associated with narrowing of the intervertebral space, encroachment of the intervertebral foramina, and, sometimes, slippage during movement. Later, osteophytes develop between the normal bulge of the intervertebral disk and the vertebral body (spondylosis) and on the uncinate process. The result is an osseous constriction of the spinal canal and the intervertebral foramina.

Spondylotic bars and exostoses of the uncinate processes can lead to compression of nerve roots, either in the spinal canal or in involved intervertebral foramina. The main symptoms are chronic cervical neck pain with radiation toward the shoulder and arm. This pain is triggered or aggravated by head movements. There is sensitivity to pressure, with focal painful zones in the paravertebral muscles as well and restricted movement of the cervical spinal column. Severe radicular deficits are not a feature of uncomplicated spondylosis, and it is not always possible to distinguish this syndrome from nonradicular cervical pain syndromes.

Neuroimaging demonstrates intervertebral disk degeneration with spondylosis, spondyloarthrosis, exostoses of the uncinate processes, and narrowing of the spinal intervertebral space. Malignant bony destruction and inflammatory processes can be excluded by these studies. MRI or CT examination allows accurate assessment of the extent of osseous constriction of the spinal canal caused by osteophytes.

### Natural History

There is a close relationship between the prevalence of cervical spondylosis and increasing age. Men are affected slightly earlier and more frequently. At the age of 60 years, corresponding x-ray findings are observed among 98% of men and 91% of women. Heavy physical activity and previous cranial and cervical spinal column injuries are correlated with more severe and earlier onset of spondylosis (Irvine et al., 1965). The presenting symptoms correlate poorly with degenerative changes in the spinal column. Neck pain begins insidiously and worsens gradually. In most

patients with radicular pain syndromes, the pain is in a C6 or C7 distribution. The peak age of onset of chronic cervical root compression syndromes is between 50 and 60 years. It is a chronic and remittent syndrome with long symptom-free periods. With more extensive degenerative cervical spinal column changes and intervertebral instability in flexion/extension movements, multiple and bilateral root features, involving sensory and motor functions, may develop.

A prospective study carried out by the British Association of Physical Medicine (1966) on conservative therapy outlined the following number of prognostic factors: old age, severity of symptoms, number of previous in attacks, and bilateral paresthesias on initial examination. Retrospective surgical studies do not support a beneficial effect of early surgery on the long-term course, although surgical treatment is clearly indicated for specific complications, such as root or cord compression. Conservative treatment over a period of weeks or months is justifiable before surgical intervention is recommended (Monro, 1984).

### Principles of Therapy

Compared with acute PID, radicular syndromes caused by osseous degeneration are difficult to treat. Repeated, often prolonged treatments are necessary. Drugs that lead to dependence and sedatives should be avoided, especially when treating outpatients. Rest and traction therapy to the cervical spinal column in the exacerbation phase form the basis of conservative treatment. Intensive postural training may correct pain related to poor posture, Avoidance of maximum movement excursions in flexion and, in particular, extension, and ergonomically designed workplaces are preventive measures that probably help to avoid recurrent symptoms. Evidence-based data to support these therapies are lacking.

Surgical decompression entails the removal of spondylotic osteophytes and of any exostoses of the uncinate processes. New central spondylotic disk protrusion or spondylotic bars should be approached ventrally as described by Cloward. Exostoses encroaching on the intervertebral foramina can be removed by use of a ventrolateral approach through an uncal foramenectomy (Kehr et al., 1981). Both operations are combined with intervertebral fusion by bone grafting.

### Practical Management

*General management includes the following:*

1. Collar immobilization; repeated short-term use for up to 1 week, especially at night. Prolonged use of a

collar tends to prolong symptoms, decreases cervical mobility, and weakens neck muscles.

2. Careful manual or mechanical traction (principle of Glisson's sling in supine position).
3. Cryotherapy of paravertebral muscles for 10–20 minutes, repeated twice daily, for symptomatic relief. This probably reduces muscle spasm.
4. Supported movement: passive, active, and active against resistance. Postural training and strengthening the shoulder and neck muscles.

Surgical intervention should be deferred unless there are clear indications, especially progression and failure to improve spontaneously. Only in the case of a chronic or relapsing, severe monoradicular syndrome that does not respond to treatment over a period of months with concurrent osteophytes, appropriately located, should an operation be considered. The neuroimaging features must be consistent with the clinical syndrome and indicate a role for surgical intervention.

### Treatments No Longer Recommended

Chiropractic manipulations are contraindicated in the case of marked degenerative cervical spinal column changes. Root blocks, using local anesthetic and paravertebral injections, carry a risk of subarachnoid injection. This can lead to a high spinal lesion, with anesthesia and weakness below the level of the lesion, and root blocks should not be used as a treatment.

## Cervical Spondylitic Myelopathy (CSM)

### Clinical Aspects

Progressive degenerative cervical spinal column changes in the middle and lower portion of the cervical spine can lead to narrowing of the spinal canal in the anteroposterior plane. This results mainly from ventrally located osteophytic bar formation together with dorsal thickening and kinking of the ligamenta flava because of the reduction in height of the intervertebral spaces. This kinking occurs especially in full cervical flexion. The intervertebral foramina may also be encroached on by laterally placed osteophytes and by zygoapophysial arthrosis. These changes result in chronic, compression-induced recurrent microtraumatic injuries to the nerve roots and spinal cord. In addition, there may be intermittent reduction of the blood supply through constriction of the radicular arteries and of the anterior spinal artery (Braakman, 1994; Taylor and Aberd, 1964). This results in combined radicular and medullary syndromes with monoradicular or polyradicular patterns of root disturbances in C5–C7, often com-

bined with a spastic paraparesis caused by corticospinal tract compression. Involvement of the dorsal columns is less frequent, so that there is usually no sensory loss below the level of the lesion. There may be radicular pattern sensory disturbance at the level of the lesion, Central cord involvement, causing dysfunction in the ascending spinothalamic tracts that modulate pain and temperature sensation leads to impairment of pain and temperature sensation most evident in the distal lower extremities and gradually extending toward the level of the lesion itself. Cord compression at the C5/C6 level is characterized by a decreased or absent biceps reflex (radicular lesion C6) with an increased triceps reflex from involvement of the pyramidal tract at the same level.

Of the imaging methods available, MRI is the most important. MRI can provide a direct image of the disturbed anatomy in relation to the spinal cord and the spinal roots. Because nerve root and cord injury is at least in part related to movement, there is only a moderate correlation between the image and the clinical features (Modic et al., 1988). Hyperintense areas in the spinal cord in T2-weighted images are interpreted as compression-induced gliosis or edema formation (Matsuda et al., 1991). Tibial somatosensory evoked potentials correlate well with posterior cord involvement in myelopathy (Yu and Jones, 1985) but give no indication of the severity or prognosis of the myelopathy (Aminoff, 1984). Magnetic cortex stimulation and central motor conduction studies allow investigation of corticospinal tract function (Maertens de Noordhout et al., 1991; Snooks and Swash, 1985), but their usefulness in clinical management is unproven.

Differential diagnosis should take into account the spinal form of multiple sclerosis, amyotrophic lateral sclerosis, syringomyelia, and extrinsic cervical tumors, such as meningioma or neurofibroma, as well as inherited degenerative disorders such as Friedreich's ataxia and familial spastic paraparesis. Underlying primary bone disease, such as osteomalacia, should be considered. In many cases there will be a history of previous traumatic injury to the cervical spine, or occupational hazards, such as carrying heavy weights on the shoulder and neck. A narrow spinal canal, as occurs in about 10% of the population, and is especially prominent in achondroplasia and certain other congenital disorders, predisposes to spondylotic cord and root compression.

### Natural History

In patients with a narrow spinal canal and quite modest intervertebral disk degeneration, CSM can develop as early as age 30–40 years, although the common age of onset is between 50 and 60 years. In most patients, chronic compression of the spinal cord and nerve roots particularly involves the middle cervical segments, especially C4/C5, C5/C6, and C6/C7. The clinical course is variable, usually intermittently progressive with long periods of largely unchanged clinical features (Clarke and Robinson, 1956; Lees and Turner, 1963). Some patients show rapid deterioration at first, with a more stable course later on. Unfavorable long-term predictors are a free range of movement of the spondylotic cervical spinal column and onset after the age of 60 years (Barnes and Saunders, 1984; Nurick, 1972b).

The variable natural history makes it difficult to estimate the benefits of the different treatment measures. Rest, as a conservative principle, leads to improvement in 43% of cases; 34% remain unchanged, and 23% become worse (Monro, 1984). The results of surgery after laminectomy are no better than those achieved through conservative treatment (Monro, 1984). The published improvement rates of retrospective studies of surgical treatment, using a ventral route of access, vary between 50% and 82% (Jeffreys, 1986; Lunsford, 1980b). Patients with severe disability benefit if operated on early, and patients with stenosis in one to two segments experience improvement after an anterior surgical approach. The detailed results of surgery and the potential complications can be found in the reviews of Monro (1984), Jeffreys (1986), and Whitecloud (1988).

### Principles of Therapy

The use of a cervical collar to limit flexion and extension of the cervical vertebral column is a reasonable therapy for root pain. The collar relieves the radicular pains that frequently accompany this syndrome. Myelopathy is improved in 43% of the cases (Monro, 1984), and its further progression is halted in another 34%.

Surgery is reserved for those cases showing progressive deterioration in cord function. The most common surgical techniques are:

1. The anterior surgical technique of Cloward or Smith–Robinson, with or without fusion of the vertebral bodies, as well as removal of the compressing structures such as disk, or osteophytes.
2. Cervical laminoplasty of variable extent (Hirabayashi and Saromi, 1988).
3. Decompressive laminectomy, with or without spinal fusion.

### Practical Management

#### General Management
1. Neck collar worn at night, but for no longer than 2 months.

2. Surgery is only recommended in cases of severe disability, such as spastic paraparesis on first examination; or a chronic progressive course, resistant to conservative treatment over a period of several months.
   a. The anterior surgical technique is used when there is:
      i. Spondylosis involving one or two segments only.
      ii. Intervertebral disk protrusion or osseous exostoses projecting more than 4 mm beyond the posterior ring of the spinal body (Jeffreys, 1986).
      iii. Subluxation with instability in at least one spinal segment.
   b. Laminoplasty is used when:
      i. Spinal canal stenosis (sagittal diameter <12 mm at the C4 level).
      ii. Extensive ossification of the posterior longitudinal ligament and spondylosis involving more than four spinal segments.
   c. Laminectomy is used when:
      i. Extensive spondylitic changes over several segments.
      ii. Spinal canal stenosis (constitutional or acquired) with sagittal canal diameter less than 12 mm.

## LUMBAR AND SACRAL RADICULAR SYNDROMES

### Lumbar PID

#### Clinical Aspects

Intervertebral disk protrusion and PID develop as a consequence of chronic intervertebral disk degeneration. The intervertebral spaces L4/L5 and L5/S1 are the most frequently affected. PID occurs less frequently at the L3/L4 levels, and the other lumbar levels are affected only exceptionally. With a posterolateral disk protrusion the nerve root is compressed at the entrance to the intervertebral foramen, where the sensory and motor fibers have come together to form the mixed nerve root. The sensory fibers are more vulnerable to compression at this site, and sensory symptoms, especially neuralgic pain, precede motor features. In general, a lateral PID at L4/L5 compresses the L5 root, and the S1 root and a lateral PID at L5/S1 compress the S1 root. However, PID at only one level in the lumbar spine can damage more than one root, especially when the protrusion is ventrolateral. With severe central disk prolapse, which is uncommon, multiple, bilateral lumbosacral nerve root lesions may occur, sometimes leading to cauda equina dysfunction.

The clinical picture of intervertebral disk protrusion with nerve root compression is characterized by localized pain in the small of the back. The pain is often just off to one side of the midline, and there is usually tenderness to pressure on the tense paravertebral muscles, with restriction of movement in the lumbar spine. Radicular pain, induced by movement of the spine and by leg flexion radiates into one or both legs in the distribution of the sciatic nerve (sciatica). Eventually, sensory loss, weakness, and loss of the ankle reflex (S1 root) may develop. Maneuvers that cause stretching of the nerve root (Lassegue and Bragard signs for L4–S1 and femoral extension pain for L2–L4) or that increase the intraspinal pressure (coughing, sneezing, forced Valsalva maneuver) provoke or intensify this radicular discomfort.

*Cauda equina syndrome* is an important clinical syndrome caused by a central lumbar PID. Acute compression of the lower lumbar and sacral roots causes bilateral sacral pain and bilateral sciatica. The full clinical picture is characterized by weakness of ankle dorsiflexion and plantar flexion, weakness of the knee flexors and abductors of the hip, reduced sensation in the sacral segments (saddle anesthesia), and weakness of the anal and urinary sphincter muscles, with urinary retention and incontinence and also anal incontinence. In men there is erectile failure and in women anorgasmia, with loss of vaginal sensation. Because the disk prolapse is central, the ankle jerks are not always absent in cauda equina syndrome. The syndrome may occur suddenly, but there is usually a background of previous bilateral sciatica. MRI is reliable in imaging this lesion. Investigation and management are urgent, because with cauda equina compression of longer than about 8 hours recovery is only partial.

Extreme lateral intervertebral disk prolapse involving the intervertebral foramen can be accurately demonstrated by MRI (Table IV). Using MRI after an injection of contrast medium (gadolinium DTPA), it is possible to establish a reliable differentiation between postoperative scar tissue and prolapsed intervertebral disk material (Hueftle *et al.*, 1988). Electrodiagnostic techniques (EMG, nerve conduction velocity, SSEP, F wave, H reflex) are helpful for objective assessment of the extent of a disk protrusion and in assessing the prognosis of a lumbosacral radicular lesion. The most useful clinical examination is needle electromyography (Wilbourn and Aminoff, 1988). Because of the frequent involvement of the posterior rami, denervation is found predominantly and early in the paraspinal muscles (Parry, 1993).

In most patients root compression syndromes with PID are characteristic and even clinically obvious, but difficulties can arise with multiple root lesions and with central disk prolapse, causing cauda equina compression. The differential diagnosis of subacute syndromes

**TABLE IV** Comparison of Myelography, CT Scan, and MRI When Diagnosing Lumbar Disk Herniation[a]

| | CT scan | MRI | Myelography[b] |
|---|---|---|---|
| Accuracy: | 90% | 95% | 85% |
| Advantages: | Noninvasive; painless; done on outpatient basis; direct prolapse demonstration; simultaneous demonstration of bones and soft tissues; safe representation of extreme lateral disk herniation | Noninvasive; painless; no radiation exposure; no complications known so far; done on outpatient basis; continuous demonstration of entire lumbar vertebra column with intervertebral and subarachnoid space in the sagittal plane; direct and indirect prolapse demonstration; safe differentiation of recurrent prolapse vs scar tissue (with Gd-DTPA); superior differentiation of soft tissue | Continuous demonstration of subarachnoid space in all sections; search method for clinically unclear height localization; minimum use of equipment |
| Disadvantages: | Discontinuous demonstration; restricted to three segments (radiation exposure); clinical segmental location is a prerequisite; differentiation between relapsed prolapse and scar tissue not reliable | High technical expenditure; cost intensive; capacity not sufficient; insufficient illustration of degenerative bony appositions of spondylosis or spondylarthrosis; *contraindicated* with carriers of pacemakers and carriers of ferromagnetic metal devices, surgical clips, grenade splinters, hearing aid; pregnancy, claustrophobia | Indirect prolapse demonstration; false-negative results possible when prolapse in intervertebral space between L5 and S1 vertebra because of wide anterior epidural space; lack of proof of extreme lateral disk herniation; complications possible: CSF hypertension syndrome; allergic reaction to contrast medium; inpatient treatment required |

[a]From Edelmann (1990).
[b]Myelography is now relatively little used.

includes intraspinal tumors such as neurinoma, meningioma, glioma, ependymoma, chordoma, and metastatic carcinoma, sarcoma and lymphoma. Malignant meningitis should also be considered in the differential diagnosis. Meningopolyradiculitis caused by Lyme borreliosis and herpes zoster radiculitis can be recognized by the additional clinical features, especially the cutaneous manifestations, and by an inflammatory CSF pattern. Diabetic polyradiculopathy (diabetic proximal neuropathy) (Bastron and Thomas, 1981) and idiopathic lumbosacral plexus neuropathy (lumbar plexitis) can also resemble a polyradicular compression syndrome. An acute lumbar disk syndrome is easily overlooked in the differential diagnosis of these illnesses, and MR imaging should always be carried out.

### Natural History

Almost two thirds of all intervertebral disk–induced illnesses affect the lumbar vertebral column. Most of these arise in young and middle-aged adults. The peak age of onset is in the fifth decade. Men are affected more frequently and at an earlier age than women (Kramer, 1994). In 95% of cases the PID is at the L4/L5 or L5/S1 levels. The PID syndrome usually develops without an immediate history of trauma. It is often thought that occupational and sporting activities that involve bending, turning, and lifting heavy loads act as final triggers, but the inducing event is often trivial (e.g., twisting the body while getting out of a car). Congenital abnormalities of the lumbar spinal column are common and predispose toward early intervertebral disk degeneration and thus to PID. The most important of these are spina bifida, assimilation disorders (lumbarization of the S1 vertebral body), spondylosis, and spondylolisthesis.

Most patients with lumbar PID improve with conservative treatment or simply with the passage of time. Consistent rest and use of a multilevel bed over a period of 2–4 weeks led to satisfactory long-term results in 70% of cases (Pearce and Moll, 1967). In modern practice a considerably shorter period of rest (i.e., a few days) is also effective (Deyo et al., 1986). Even when there are signs of root involvement, both the early results (up to 90% improvement and recovery of ability to work) and the later results after initial physical therapy are of the same value as the results of surgery (Saal and Saal, 1989). In the past the proportion of patients managed by surgical intervention reached 10%–20% (Pearce and Moll, 1967), but it is now recognized that most patients will improve equally well without surgery. A key element to this improvement is a progressive aerobic exercise program designed to strengthen paraspinal muscles, improve posture, and restore the full range of movement in the lumbar spine

(Goldie et al., 1970). Only in acute cauda equina syndrome is surgery absolutely indicated.

### Principles of Therapy

Both conservative management and surgical treatment of lumbar intervertebral disk prolapse aim, in the acute stages, to relieve pressure on the affected nerve root.

The principles of conservative treatment in the acute phase should include appropriate positioning to relieve pain, consideration of traction to the lumbar vertebral column, relief of muscle spasm by local measures, and analgesia. When the acute radicular syndrome subsides, emphasis is placed on a program of active physiotherapy and exercise to stabilize the lumbar vertebral column. Postural training is an essential element of this program (Kramer, 1994). Relapses can be prevented by attention to the ergonometrics of the workplace and home; for example, the height of working surfaces, the use of firmly upholstered chairs, appropriately contoured car seats, reduction of body weight toward the predicted ideal weight, and intensive postural training combined with avoidance of dangerous movements, especially lifting loads with arms outstretched, as in unloading the trunk of a car.

The most frequently used surgical techniques are excision of part of the ligamentum flavum and interlaminar foramenotomy. Men have been managed surgically more often than women (gender ratio approximately 2:1; Oppel et al., 1977; Schepelmann et al., 1977). The results of disrectomy performed using microsurgery are better than those obtained using conventional laminectomy, with satisfactory results reported in 88%–98% of cases (Ebeling et al., 1986). However, long-term follow-up of surgically managed disk prolapse has suggested very much less favorable outcomes, both in terms of symptom relief and return to work or previous lifestyle, leading to the recognition of the value of the conservative approach outlined earlier (Harris, 1977; Henderson et al., 1983). The incidence of second operations as a result of a relapse at the operative site is similar for both surgical methods (1.4%–8%). Perioperative complications, such as increased neurological deficit, wound infection, and pulmonary embolism, occur less frequently with microsurgery. The risk of postoperative diskitis is less than 2% for both techniques.

As for any surgical procedure, the results of intervertebral disk surgery are dependent, first, on the careful selection of suitable patients and, second, on surgical technique. Only in typical radicular syndromes, with a close correlation between the clinical syndrome and the imaging findings, are very high levels of success achieved by surgery (Frymoyer, 1988). Surgery for pain without radicular signs is not likely to be successful.

Chemical lytic treatments, through enzymatic proteolysis (chymopapain) or lysis of collagen (collagenasis), called chemonucleolysis, can lead to a reduction in the volume of the nucleus pulposus and therefore relieve of root compression. Although initially promising results were reported, in recent years this technique has been generally abandoned as an alternative to surgical diskectomy, because the long-term results have been poor. The procedure has been used in a diverse range of clinical syndromes related to PID, including patients with pain but without radicular signs. The latter presumably did not have root compression, and the role of disk lysis in this situation is difficult to understand, because active exercise and rehabilitation is probably at least as effective. With an initial success rate of 70%–80% and even without agreed precise indications, it offers no convincing advantages over conventional treatments (Maroon and Abla, 1985). With a failure rate of between 18% and 31%, this procedure seems inferior to microsurgery.

Retrospective assessments of disk lysis therapy, with longer observation periods, indicate particularly unfavorable late results, with a failure rate as high as 60% (Shields, 1986). A randomized clinical multicenter study to compare the results of surgical diskectomy as opposed to chemonucleolysis (78 patients were operated on, 73 treated with chemonucleolysis) in PID at the L4/L5 and L5/S1 levels showed that the results of surgery were markedly superior to those achieved using chemonucleolysis. A quarter of the chemonucleolysis group later required surgery, but only 3% of the diskectomy group were operated on a second time (August et al., 1989). The most frequent, and indeed the most feared, complication of chymopapain chemonucleolysis is anaphylactic reaction, which has an incidence of 0.5% (Sussmann et al., 1981). Other rare but serious neurological complications may occur, including acute transverse myelitis leading to paraplegia, cauda equina root damage, and subarachnoid and subdural bleeding after inadvertent intrathecal injection. In addition, there is a risk of bacterial diskitis and epidural abscess, which may also cause paraplegia or cauda equina syndrome.

Other techniques include percutaneous nucleotomy, in which nerve root pressure is relieved by mechanically shaving the intervertebral disk. This procedure requires only a very small incision with consequent rapid postoperative recovery from the surgical stress. In a study of 109 patients Schreiber et al. (1988) reported improvement after this procedure in 72%, and Kambin and Schaffer (1989) noted improvement in 87%. Onik et al. (1990), in a prospective multicenter study, reported successful results in 75% of cases. This latter study was carefully designed, with clearly defined inclusion and exclusion criteria. Patients who did not fulfil the entry criteria for the study, but were similarly treated, showed a success rate of only 49%. Approximately 5% of patients considered candidates for surgical management of PID are suitable for percutaneous nucleotomy. These are patients with intervertebral disk protrusions in which the posterior longitudinal ligament is intact.

Another percutaneous surgical technique on the threshold of clinical application is percutaneous intradikcal laser nucleotomy. In this procedure the nerve root is relieved of pressure through denaturation of the nucleus pulposus with the help of a laser. Long-term results are lacking as, indeed, is the case for established surgical methods (Quigley and Maroon, 1994).

### Practical Management

*In the acute stage—for severe syndromes:*

In the acute stages of pain, conservative treatment can be administered on an inpatient basis with a view to continuing increasingly active treatment on an outpatient basis as soon as possible. Temporary use of an orthosis (e.g., sacroiliac support bodice) in conjunction with physiotherapy is useful for relief in some cases of chronic radicular syndrome and lumbar postural instability. It is often difficult, however, to wean the patient from the physical and emotional support provided by the orthosis, which comes to assume the status of an icon representing proof of the necessity of illness "behavior" and implying chronicity of the disabled role.

1. Relieve pain by bedrest and comfortable positioning of the spine, if necessary, in a multiadjustable bed. Most cases are relatively mild and can be managed at home.
2. Traction, using a sling table or a horizontal extension table; one or two sessions daily.
3. Movement therapy in a hydrotherapy pool twice daily for 20–30 minutes.
4. Nonpharmacological analgesics such as massage, vibration therapy, interferential therapy, cryotherapy, or local heat, depending on individual tolerance (cryotherapy as prolonged application 10–20 minutes, twice daily; thermotherapy for 30 minutes twice daily).
5. Drug therapy (see Table III) includes spasmolytic drugs, such as diazepam, 10 mg tds; analgesics, such as paracetamol, 500–1000 mg up to four times daily; or other mild analgesics (dose dependent on outcome, no longer than 2 weeks); nonsteroidal anti-inflammatory analgesics, such as indomethacin, 100–200 mg daily, or diclofenac 25–50 mg tds. Corticosteroids are only very rarely indicated; dexamethasone, 2–4 mg tds for 3 days, may be effective in pain relief, but the effect is often only temporary. Antidepressants, such as amitriptyline,

25–75 mg as a single evening dose, help to elevate mood and promote sleep.

*After the acute symptoms have resolved:*

1. Slowly withdraw analgesics.
2. Promote an active exercise program to stabilize and strengthen the lumbar vertebral musculature, using isometric and isotonic techniques. This should be combined with postural training.

**Surgical Management.** Various options are available. Open diskectomy is indicated based on the following findings:

1. Central PID with cauda equina syndrome or polyradicular, sensorimotor signs is an absolute indication for surgery.
2. Lateral PID with more than one radicular lesion and functionally significant weakness (MRC 3 or less).
3. Single root syndrome with pain resistant to conservative therapy and PID confirmed by CT or MRI at the appropriate intervertebral level.
4. In general, a trial of at least 4 weeks of conservative treatment should precede consideration of surgery.

Percutaneous nucleotomy is indicated only in treatment-resistant single root pain with CT- or MRI-proven intervertebral disk protrusion, after unsuccessful inpatient physical therapy. Percutaneous procedures are contraindicated for patients with severe neurological abnormalities, especially when there is a cauda equina syndrome and for multiple root lesions, implying disk protrusions at multiple levels. It is also contraindicated when there is evidence of free disk fragments in the spinal canal, when there is diskitis, and when there is a history of previous operations at the same level.

### Secondary Preventive Measures
1. Individual exercises at home, using the techniques learned for postural training and strengthening of the muscles (Kramer, 1994).
2. Relating learned back-protective behavior to everyday situations.
3. Avoidance of sports that involve high pressure and back strain or excessive flexion, extension, and rotation of the lumbar vertebral column. These include swimming (breaststroke), cricket, alpine skiing, tennis, golf, football, rowing, marathon running, and bodybuilding. Contact sports, such as football and rugby, should also be avoided.
4. Recommended sports are swimming (backstroke), cross-country skiing, volleyball, handball, table tennis, riding, gymnastics, and cross-country running using training shoes with air-cushioned soles.

**Treatments No Longer Recommended.** Spinal manipulation is contraindicated for acute radicular syndromes after PID. Injection treatment, in particular the use of root blockers, is inadvisable because of the danger of subarachnoid injection with spinal anesthesia and the theoretical risk of an epidural infection. Because repeated intragluteal injection with steroids and anti-inflammatory drugs can cause the formation of scar tissue in muscle and atrophy of the subcutaneous fatty tissue with unpleasant cosmetic defects, this treatment is also not recommended.

### Lumbar Spinal Canal Stenosis

#### Clinical Aspects

Lumbar spinal canal stenosis occurs in middle or old age, either with radicular symptoms or in the form of a secondary cauda syndrome, that is, neurogenic intermittent claudication, which may recur with further minor strains. This is evident when symptoms appear as the result of injury of the lumbar vertebral column with forced lordosis; these symptoms and signs quickly subside after rest in the ventral position and probably result from temporary mechanical cauda equina compression. Stress-dependent relative ischemia of the cauda equina is a possible additional factor, particularly in those patients in whom the syndrome cannot be linked to postural abnormalities.

The final diagnosis of lumbar canal stenosis is based on radiological documentation using CT or a water-soluble contrast myelography. Plain radiographs of the lumbar vertebral column show characteristic bony changes in many cases. However, CT or MR imaging is important, because it also visualizes the soft tissue abnormalities that lead to root compression.

A pathological narrowing of the spinal canal sagittal diameter, or its area, occurs in a range of different disorders. As opposed to congenital spinal disorders, developmental canal stenosis is much more frequent. Verbiest, the first physician to systematically describe the clinical picture, introduced the term developmental stenosis and suggested that increased postnatal bone growth, leading to hypertrophy of the spinal laminas and vertebral bodies was the cause (Verbiest, 1976; 1980). However, there is a consistent transition from this increased bone growth to the degenerative spinal changes that develop in many individuals during middle age. Anteroposterior stenosis has the greatest significance for the emergence of neurological symptoms. It is due to shortening of the vertebral pedicles with a reduced median sagittal diameter. A distinction is made between an absolute stenosis (mean saggital diameter, >10 mm) and a relative stenosis (canal diameter, 10–12 mm). Other forms of stenosis, describing the cross-sectional appearance on CT scanning, are concentric and clover leaf–shaped stenosis.

### Treatment

In less severe cases, treatment of canal stenosis should be conservative. The regimen is similar to the physical treatment used for other degenerative spinal column disorders. In patients with severe or progressive neurological symptoms or treatment-resistant pain syndromes, laminectomy to relieve the nerve root pressure (multisegmental hemilaminectomy, complete laminectomy) should be considered. To extend the lateral recess and the intervertebral foramen, the vertebral pedicles and joint process sections must be explored. Relief of symptoms is not usually observed until several months later, occurring in 70%–80% of cases (Tile et al., 1975).

## REFERENCES

Adams, R. D., and Victor, M. (1989). "Principles of Neurology." McGraw-Hill, New York.

van Alphen, H. M., Braakman, R., Bezemer, P. D., Broere, G., and Berfelo, M. W. (1989). Chemonucleolysis versus discectomy: A randomized multicenter trial. J. Neurosurg. 70, 869–875.

Aminoff, M. J. (1984). The clinical role of somatosensory evoked potential studies: a critical appraisal. Muscle Nerve 7, 345–354.

August, H., van Alphen, M., Braakman, R., Bezemer, P. D, Broere, G., and Berfelo, M. W. (1989). Chemonucleolysis versus discectomy: A randomized multicenter trial. J. Neurosurg. 70, 869–875.

Barnes, M. P., and Saunders, M. (1984). The effect of cervical mobility on the natural history of cervical spondylotic myelopathy. J. Neurol. Neurosurg. Psychiat. 47, 17–20.

Bastron, J. A., and Thomas, J. E. (1981). Diabetic polyradiculopathy, clinical and electromyographic findings in 105 patients. Mayo Clinic Proc. 56, 725–732.

Benecke, R., and Conrad, B. (1980). The distal sensory nerve action potential as a diagnostic tool for the differentiation of lesions in dorsal roots and peripheral nerves. J. Neurol. 223, 231–239.

Braakman, R. (1994). Management of cervical spondylotic myelopathy and radiculopathy. J. Neurol. Neurosurg. Psychiat. 57, 257–263.

British Association of Physical Medicine. (1996). Pain in the neck and arm: A multicentre trial of the effects of physiotherapy. BMJ. 1, 253–258.

Chirls, M. (1978). Retrospective study of cervical spondylosis treated by anterior interbody fusion (in 505 patients performed by the Cloward technique). Bull. Hosp. Jt. Dis. 39, 74–82.

Clarke, E., and Robinson, P. K. (1956). Cervical myelopathy: A complication of cervical spondylosis. Brain 79, 483–510.

Cloward, R. B. (1958). The anterior approach for removal of ruptured cervical discs. J. Neurosurg. 15, 602–614.

Cloward, R. B. (1963). Lesions of the intervertebral disks and their treatment by interbody fusion methods. Clin. Orthop. 27, 51–77.

Deyo, R. A., Diehl, A. K., and Rosenthal, M. (1986). How many days of bed rest for acute low back pain? A randomized clinical trial. N. Engl. J. Med. 315, 1064–1070.

Ebeling, U., Reichenberg, W., and Reulen, H. J. (1986). Results of microsurgical lumbar discectomy. Neurochirurgica 81, 45–52.

Edelmann, R. R. (1990). Magnetic resonance imaging of the nervous system. In "Discussions in Neuroscience" (P. J. Magistretti, Ed.). Vol. VII, No. 1, pp. 35–37. Elsevier, Amsterdam.

Frymoyer, J. W. (1988). Back pain and sciatica. N. Engl. J. Med. 318, 291–300.

Goldie I, Landquist A. (1970). Evaluation of the effects of different forms of physiotherapy in cervical pain. Scand. J. Rehabil. Med. 2, 117–121.

Gregorius, F. K., Estrin, T., and Crandall, P. H. (1976). Cervical spondylotic radiculopathy and myelopathy. A long term follow-up study. Arch. Neurol. 33, 618–625.

Harris P. R. (1977). Cervical traction—review of the literature and treatment guidelines. Phys. Ther. 57, 910–914.

Henderson, C. M., Hemmessy, R. G., Shuey, H. M. Jr. et al. (1983). Posterior-lateral foramenotomy as an exclusive operation technique for cervical radiculopathy: a review of 846 consecutively operated cases. Neurosurgery 13, 504–512.

Haughton, V. M., Eldevik, O. P., Magnaes, B., and Amundsen, P. (1982). A prospective comparison of computed tomography and myelography in the diagnosis of herniated lumbar disks. Neuroradiology 142, 103–110.

Hirabayashi, K., and Satomi, K. (1988). Operative procedure and results of expansive open-door laminoplasty. Spine 7, 870–876.

Hueftle, M. G., Modic, M. T., Ross, J. S., Masaryk, T. J., Carter, J. R., Wilber, R. G., Bohlmann, H. H., Steinberg, P. M., and Delamarter, R. B. (1988). Lumbar spine: Postoperative MR imaging with Gd-DTPA. Radiology 167, 817–824.

Hunt, W. E. (1980). Cervical spondylosis: Natural history and rare indications for surgical decompression. Clin. Neurosurg. 27, 446–480.

Irvine, D. H., Foster, J. B., Newell, D. J., and Klukvin, B. N. (1965). Prevalence of cervical spondylosis in a general practice. Lancet 1, 1089–1092.

Jeffreys, R. V. (1986). The surgical treatment of cervical myelopathy due to spondylosis and disc degeneration. J. Neurol. Neurosurg. Psychiat. 49, 353–361.

Kambin, P., and Schaffer, J. L. (1989). Percutaneous lumbar discectomy. Clin. Orthop. 258, 24–34.

Kehr, P., Lang, G., Ternotte, H., and Moncade, N. (1981). Die Unkusektomie und Unkoforaminektomie nach Jung mit oder ohne intersomatische Fusion—Indikation und Resultate. Z. Orthop. 119, 612–619.

Kotilainen, E., Valtonen, S., and Carlson, C.-A. (1993). Microsurgical treatment of lumbar disc herniation: Follow-up of 237 patients. Acta Neurochir. (Wien) 120, 143–149.

Kramer, J. (1994). "Bandscheibenbedingte Erkrankungen: Ursachen, Diagnose, Behandlung, Vorbeugung, Begutachtung." Thieme, Stuttgart.

Landmann, J. A., Hoffmann, J. C., Braun, I. F., and Barrow, D. L. (1984). Value of computed tomographic myelography in the recognition of cervical herniated disk. Am. J. Neuroradiol. 5, 391–394.

Lees, F., and Turner, J. W. A. (1963). Natural history and prognosis of cervical spondylosis. BMJ. 2, 1607–1610.

Lunsford, L. D., Bissonette, D. J., Janetta, P. J., Sheptak, P. E., and Zorub, D. S. (1980a). Anterior surgery for cervical disc disease. part 1: Treatment of lateral cervical disc herniation in 253 cases. J. Neurosurg, 53, 1–11.

Lunsford, L. D., Bissonette, D. J., and Zorub, D. S. (1980b). Anterior surgery for cervical disc disease. part 2: Treatment of spondylotic myelopathy in 32 cases. J. Neurosurg. 53, 12–19.

Maroon, J. C., and Abla, A. (1985). Microdiscectomy versus chemonucleolysis. *Neurosurgery* 16, 644–648.

Maertens de Noordhout, A, Remacle, J. M., Pepin, J. L., Born, J. D., and Delwaide, P. J. (1991). Magnetic stimulation of the motor cortex in cervical spondylosis. *Neurology* 41, 75–80.

Matsuda, Y., Miyazaki, K., Tada, K. *et al.* (1991). Increased MR signal intensity due to cervical myelopathy: Analysis of 29 surgical cases. *J. Neurosurg.* 74, 887–892.

Modic, M. T., Masaryk, T. J., Ross, J. S., and Carter, J. R. (1988). Imaging of degenerative disc disease. *Radiology* 168, 177–186.

Modic, M. T., Rlicek, W., Weinstein, M. A., Boumphrey, F., Ngo, F., Hardy, R., and Duchesneau, P. M. (1984). Magnetic resonance imaging of intervertebral disk disease. *Radiology* 152, 103–111.

Modic, M. T., Weinstein, M. A., Vlicek, W., Boumphrey, F., Starnes, D., and Duchesneau, P. M. (1983). Magnetic resonance imaging of the cervical spine: Technical and clinical observations. *Am. J. Radiol.* 141, 1129–1136.

Monro, P. (1984). What has surgery to offer in cervical spondylosis? *In* "Dilemmas in the Management of the Neurological Patient" (W. Charles and J. Garfield, Eds.), pp. 168–187. Churchill Livingstone, Edinburgh/New York/London/Melbourne.

Mumenthaler, M., and Schliack, H. (1993). "Lasionen peripherer Nerven." Thieme, Stuttgart/New York.

Murphy, F., Simmons, J. C. H., and Brunson, B. (1973). Surgical treatment of laterally ruptured cervical discs: Review of 648 cases 1939–1972. *J. Neurosurg.* 38, 679–683.

North American Spine Society. (1991). Common diagnostic and therapeutic procedures of the lumbosacral spine. *Spine* 16, 161–1167.

Nurick, S. (1972a). The thogenesis of the spinal cord disorder associated with cervical spondylosis. *Brain* 95, 87–100.

Nurick, S. (1972b). The natural history and the results of surgical treatment of the spinal cord disorder associated with cervical spondylosis. *Brain* 95, 101–108.

Onik, G., Mooney, V., Maroon, J. C., Wiltse, L., Helms, C., Schweigel, J., Watkins, R., Kahanovitz, N., Day, A., Morris, J., McCollough, J. A., Reicher, M., Croissant, P., Dunsker, S., Davis, G. W., Brown, C., Hochschuler, S., Saul, T., and Ray, C. (1990). Automated percutaneous discectomy: A prospective multi-institutional study. *Neurosurgery* 26, 228–233.

Onofrio, B. M., and Mih, A. D. (1988). Synovial cysts of the spine. *Neurosurgery* 22, 642–647.

Oppel, F., Schramm, J., Schirmer, M., and Zeitner, M. (1977). Results and complicated course after surgery for lumbar disc herniation. *In* "Lumbar Disc: Adult Hydrocephalus" (R. Wüllenweber, M. Brock, J. Hamer, M. Klinger, and O. Spoerri, Eds.), pp. 36–51. Springer-Verlag, Berlin/Heidelberg/New York.

Parry, G. J. (1993). Diseases of spinal roots. *In* "Peripheral Neuropathy" (P. J. Dyck, P. K. Thomas, J. W. Griffin, P. A. Cow, and J. F. Poduo, Eds.), Vol. 2, pp. 899–910. WB Saunders, Philadelphia/London/Toronto/Mexico City/Rio de Janeiro/Tokyo.

Pearce, J., and Moll, J. M. H. (1967). Conservative treatment and natural history of acute lumbar disc lesions. *J. Neurol. Neurosurg. Psychiat.* 30, 13–17.

Quigley, M. R., and Maroon, J. C. (1994). Laser discectomy: A review. *Spine* 19, 53–56.

Raskin, S. P., and Keating, J. W. (1982). Recognition of lumbar disk disease: comparison of myelography and computed tomography. *Am. J. Radiol.* 139, 349–355.

Roberts, A. H. (1966). Myelopathy due to cervical spondylosis treated by collar immobilization. *Neurology* 16, 951–954.

Rothman, R. H., and Marvel, J. P. (1975). The acute cervical disk. *Clin. Orthop.* 109, 59–68.

Saal, J. A., and Saal, J. S. (1989). Nonoperative treatment of herniated lumbar intervertebral disc with radiculopathy: An outcome study. *Spine* 14, 431–437.

Schepelmann, F., Greiner, L., and Pia, H. W. (1977). Complications following operation of herniated lumbar discs. *In* "Lumbar Disc: Adult Hydrocephalus" (R. Wüllenweber, M. Brock, J. Hamer, M. Klinger, and O. Spoerri, Eds.), pp. 52–54. Springer-Verlag, Berlin/Heidelberg/New York.

Schreiber, A., Suezawa, Y., and Leu, H. (1988). Does percutaneous nucleotomy with discoscopy replace conventional discectomy. *Clin. Orthop.* 238, 35–42.

Scotti, G., Pierallis, E., Boccardi, E., Valsecchi, F., and Tonon, C. (1983). Myelopathy and radiculopathy due to cervical spondylosis: Myelographic-CT correlations. *Am. J. Neuroradiol.* 4, 601–603.

Scoville, W. B., Dohrmann, G. J., and Corkill, G. (1976). Late results of cervical disc surgery. *Neurosurgery* 45, 203–210.

Shields, C. (1986). (Quotation in: Merz, B.) The honeymoon is over: Spinal surgeons begin to divorce themselves from chemonucleolysis. *JAMA.* 256, 317–318.

Snooks, S. J., and Swash, M. (1985). Motor conduction velocity in the human spinal cord: owed conduction in multiple sclerosis. *J. Neurol. Neurosurg. Psychiat.* 48, 1135–1139.

Sussmann, B. J., Bromley, J. W., and Gomez, J. C. (1981). Injection of collagenase in the treatment of herniated lumbar disc. *JAMA.* 245, 730–732.

Swash, M. (1986). Diagnosis of brachial root and plexus lesions. *J. Neurol.* 233, 131–135.

Symon L., and Lavender, P. (1967). The surgical treatment of cervical spondylotic myelopathy. *Neurology* 17, 117–127.

Taylor, A. R., and Aberd, M. B. (1964). Vascular factors in the myelopathy associated with cervical spondylosis. *Neurology* 14, 62–68.

Tile, M., McNeil, S. R., Zarins, R. K., Pennal, G. F., and Garside, S. H. (1975). Spinal stenosis, results of treatment. *Clin. Orthop.* 115, 104–108.

Verbiest, H. (1976). Fallacies of the present definition, nomenclature and classification of the stenoses of the lumbar vertebral canal. *Spine* 1, 217–225.

Verbiest, H. (1980). Stenosis of the lumbar vertebral canal and sciatica. *Neurosurg. Rev.* 3, 75–89.

Whitecloud, T. S. (1988). Anterior surgery for cervical spondylotic myelopathy. *Spine* 13, 861–863.

Wilbourn, A. J., and Aminoff, M. J. (1988). AAEE Minimonograph #32: The electrophysiologic examination in patients with radiculopathies. *Muscle Nerve* 11, 1099–1114.

Wilson, D. W., Pezzuti, R. T., and Place, J. N. (1991). Magnetic resonance imaging in the preoperative evaluation of cervical radiculopathy. *Neurosurgery* 28, 175–179.

Yu, Y. L., and Jones, S. J. (1985). Somatosensory evoked potentials in cervical spondylosis. *Brain* 108, 273–300.

*Muscle and Peripheral Nervous System*

CHAPTER 91

# Compression Neuropathies of Peripheral Nerves and Compartment Syndromes

Manfred Stöhr, David M. Dawson, and Michael Swash

When evaluating nerve compression syndromes and nerve injuries, the neurologist has to decide whether conservative treatment is indicated or whether referral to a neurosurgeon, hand surgeon, or orthopedist is appropriate. This decision depends in part on the severity of the lesion (see Chapter 92). For more detailed information please refer to the following books (monographs) in English: Sunderland (1978), Omer and Spinner (1980), Swash and Schwartz (1997), Mumenthaler and Schliack (1991), and Dawson *et al.* (1999) or these texts in German: Tackmann *et al.* (1989), Stöhr (1996), and Mumenthaler *et al.* (1998).

## COMPRESSION NEUROPATHIES IN THE ARM

### Thoracic Outlet Syndrome

#### Clinical Aspects

In passing from the vertebral column to the arm, the brachial plexus traverses several areas of potential constriction at which nerve entrapments can occur, especially of the inferior trunk—originating from the roots C8 and T1. Depending on the site of the compression, the thoracic outlet syndrome (TOS) may develop as one of several syndromes, such as the scalenus syndrome (compression between the scalenus anterior and medius muscle and the first rib), costoclavicular syndrome (compression between the first rib and the clavicle), and the hyperabduction syndrome (compression of the plexus below the origin of the pectoralis minor muscle at the coracoid). Clinically, differentiation among these syndromes is frequently impossible to make. Therefore,

the term TOS is now widely accepted. TOS also encompasses purely vascular compression syndromes without neurological symptoms. Diagnostic criteria for TOS are controversial (Roos, 1990; Wilbourn, 1990).

TOS is rare; its prevalence is about 1/million (Gilliatt, 1984). The most frequent anomalies leading to TOS are a cervical rib, a prolonged transverse process of the C7 vertebra, or a ligament between C7 and the first rib (Gilliatt, 1984). These structures may compress the inferior trunk of the plexus from below, accompanied by pressure of the plexus against the rim of the scalenus anterior or medius muscle. Susceptibility is conferred by a narrow gap between the muscles as a result of the closely located origins of the muscles. The fact that bony anomalies are not a sufficient cause for a TOS is illustrated by the occurrence of these bony anomalies in 0.5% of the population, although TOS is much rarer (Dawson *et al.*, 1999; Gilliat, 1984; Mumenthaler *et al.*, 1998; Tackmann *et al.*, 1989).

TOS often presents with pain on the ulnar side of the hand and forearm, especially in awkward arm positions such as elevation, at work, sleeping with the arm elevated, or carrying loads with the arm hanging down. Pulling the arm downward may be a useful diagnostic maneuver, provoking pain, but the sensitivity and specificity of this test are unproven. With time, signs of a lower plexus lesion with sensory and motor deficits and predominant atrophy of the abductor pollicis brevis muscle may develop.

Compression of the subclavian artery may lead to disturbances of circulation and even to necrosis of the fingertips. Irritation of sympathetic fibers may provoke peripheral vasoconstriction with signs of Raynaud's disease. Dislocation of small blood clots may lead to painful embolic obliterations of finger arteries. These vascular events are very rare in patients with neurogenic TOS; the vascular syndrome and symptoms of brachial plexus compression seem to occur separately.

Diagnostic hints include palpation of the pulse of the radial artery in the normal relaxed position of the arm hanging down and its disappearance during positioning the hand above the occiput, during hyperabduction, while pulling the shoulder down, and during the so-called Adson maneuver (leading to a diminution of the gap between the scalenus muscles by first tilting the head backward and then rotating it toward the side of the suspected lesion during maximal inspiration). Obliteration of the radial pulse is, however, not specific for TOS, because it frequently occurs in asymptomatic persons.

X-ray examination of the upper thoracic aperture is necessary to demonstrate or exclude pathological processes of the apex of the lung, cervical ribs, or callus after fracture of the clavicle. Doppler sonography of the subclavian artery and the radial artery, and especially digital subtraction angiography in the normal arm position and in provoking positions (described earlier), may sometimes demonstrate vascular compression, but they do not correlate with neural compression. EMG recordings from muscles innervated by the lower plexus, recording of sensory nerve action potentials of the ulnar nerve with an amplitude reduction on the involved side, and the measurement of F waves (Eisen et al., 1977; Gilliatt et al., 1978) may contribute to the diagnosis. Somatosensory evoked potential studies after ulnar nerve stimulation and motor evoked potentials in the abductor pollicis brevis muscle after cervical magnetic or high-voltage electric stimulation in patients with sensorimotor deficits may be helpful in localization (Stöhr, 1998; Yiannikas and Walsh, 1983). However, the neurophysiological diagnosis of TOS is difficult, especially in the early stages.

### Therapy

If symptoms occur only with physical stress, conservative management with prevention of the pain-provoking movements is usually all that is necessary. Physical therapy can strengthen the elevator muscles of the shoulder and lead to an improvement in poor posture of the spine and shoulder (Dale and Lewis, 1975; review of the physiotherapeutic treatments, Mumenthaler et al., 1998). Complaints evoked by habitual sleeping with an elevated arm may be cured by restricting movement of the arm at the trunk during sleep. However, clinical evidence of a plexus lesion or especially any disturbance of blood flow is an indication for surgical management.

In the past, scalenotomy with resection of the cervical rib and any accessory ligaments was the accepted

procedure. More recently, resection of the first rib through a transaxillary approach has become the preferred procedure (Etheredge et al., 1979; Kelly 1979; Roos, 1982). Nonetheless, because there are risks associated with this transaxillary approach, and occasionally symptoms return, a supraclavicular approach, in which there is good visualization of the plexus, is still favored by some surgeons. It is usually carried out through access obtained by an anterior scalenectomy instead of scalenotomy (Sanders et al., 1979). Sometimes this is combined with resection of the first rib. For a discussion of different surgical procedures, see Tackmann et al., 1989 and Dawson et al., 1999.

The differential diagnosis of compression of the brachial plexus includes tumors (e.g., Pancoast tumor, neurinoma, metastasis, lymphogranulomatosis, lymphosarcoma), late sequelae of irradiation, and brachial plexus neuritis. Because compression syndromes of the median and ulnar nerve and compression of cervical roots may also cause brachialgia, they also enter the differential diagnosis. Rare compression syndromes of proximal nerves are listed in Table I.

## Axillary Nerve

### Clinical Aspects

Paresis of the deltoid muscle leads to weakness of abduction of the arm (acromial part of the muscle), elevation (clavicular part), and backward movement of the horizontally elevated arm (spinal part). This lesion may be due to trauma to the shoulder, repositioning after shoulder luxation, or fracture of the proximal humerus. True compression of the nerve in the lateral axillary gap is very rare. It is characterized by pain and paresthesia in the shoulder and upper arm with radiation down to the hand (Aita, 1984; Cahill and Palmer, 1983).

### Therapy

The rare compression of the nerve in the posterior axillary gap is best treated by operative dorsal decompression (Cahill and Palmer, 1983). With other types of lesions, stretching the shoulder joint and painful pericapsular adhesions of this joint should be avoided by carrying the arm in a sling and by passive physiotherapy and movement. If there is no spontaneous reinnervation 5–6 months after a traumatic lesion, an operative revision is advisable, including, if necessary, neurolysis, nerve suture, or nerve transplantation. Surrogate operations frequently proposed for irreversible palsies usually show indifferent results, so that in the presence of isolated axillary nerve lesions, training should be

**TABLE I    Rare Compression Syndromes of Proximal Arm Nerves**

| Nerve | Clinical signs | Etiology | Therapy |
|---|---|---|---|
| *Dorsal scapular nerve* (dorsal branch of the plexus from root C3–C5), innervating the levator scapulae and rhomboid muscles | Slight dislocation of the scapula (medial rim and inferior angle slightly rotated outward, medial rim slightly sticking out, bad fixation of the scapula, pain medial to the scapula) | Gunshot and stab lesions; because of its covered position compressions are rare; rarely seen with hypertrophy of the scalenus medius muscle | Usually not necessary, with stab wounds nerve suture, surgery for substitute of rhomboid function: connection of the lower scapular angle to the latissimus dorsi muscle |
| *Suprascapular nerve* (from roots C4–C6 via the upper primary bundle), innervates supraspinatus and infraspinatus muscles, branches to the scalenus medius muscle | Paresis of lateral rotation and abduction, especially within the first 15 degrees of abduction; pain above the scapula, especially during pull forward and to the opposite side of the body | Chronic compression at the incisura scapulae, especially by frequent pull of the shoulder forward; fractures of the collum scapulae | Incision of the ligamentum transversum scapulae, if necessary, neurolysis; surrogate operation for paresis of lateral rotation: transposition of the teres major muscle to the dorsal side of the humerus |
| *Subscapular nerve* (from roots C5–C8 from the upper primary bundle and posterior fascicle), innervates subscapular and teres major muscle | Atrophy invisible because of deep anatomical position of the muscle; inward rotation of upper arm hardly involved if the function of other inward rotators (pectoralis major muscle, latissimus dorsi, and anterior part of the deltoid muscle) is intact | Isolated lesion very rare because of the deep protected position of the nerve | Usually not necessary in an isolated lesion |
| *Long thoracic nerve* (from roots C5–C7 before the formation of the primary bundle), innervates the serratus anterior muscle | Sticking out of the scapula, especially during arm elevation (scapula alata) displacement of the scapula rostrally and of the lower scapula angle medially, lateral tilt of the acromion | Compression by bandages, carrying of loads (rucksack paresis), iatrogenic tear during thoracotomy, extirpation of lymph nodes in the axilla, paresis as a part of neuralgic shoulder amyotrophy, parainfectious neuritis | Immobilization to avoid pull of the muscle by the weight of the arm; if necessary, fixation of the lower scapula angle at the scapula of the opposite side or at the ninth rib or by the latissimus dorsi and pectoralis major muscle |
| *Thoracodorsal nerve* (from roots C6–C8 and the medium primary bundle), innervates latissimus dorsi muscle and partly also teres major muscle | Weakness of adduction, inward rotation and lowering of the upper arm | Isolated lesions rare | Usually not necessary, loss of function is compensated for by the pectoralis major and teres major muscles |
| *Pectorales mediales and laterales nerves* (from C5–T1), innervate pectoralis major and minor muscle | Weakness of upper arm adduction and anteversion, atrophy visible (differential diagnosis congenital aplasia, muscular dystrophy) | Isolated lesions rare | Usually not necessary |

used to strengthen the supraspinatus muscle for arm abduction, the clavicular part of the pectoralis major muscle, the coracobrachialis muscle, and the long head of the biceps. If the result of physical training is unsatisfactory, arthrodesis of the shoulder joint may be considered, but only if the motion of the scapula is unrestricted.

## Musculocutaneous Nerve

### Clinical Aspects

There are two sites of predisposition to compression of the musculocutaneous nerve. One is a proximal site, where the nerve pierces the coracobrachialis muscle, and the other is more distally located in the upper arm, where the nerve pierces the brachial fascia at the elbow. Proximal lesions lead to a paresis of elbow flexion and supination. Very proximal lesions also lead to a weakness of arm elevation, because the coracobrachialis muscle is weak. Flexion of the elbow can, however, still be accomplished by the brachioradialis and pronator teres muscles and supination by the supinator muscle. Except for very rare traumatic lesions, proximal lesions occur most frequently during anesthesia, when the arm is in abduction. Occasionally, after hard muscle effort, either a proximal or distal type of paresis may be seen, depending on the type of movement. Distal compression is characterized by pain and paresthesias at the lateral elbow and the territory of the lateral cutaneous antebrachial nerve, sometimes with Tinel's sign over the anterior arm. A lesion of the C6 root may be distinguished on the basis of sensory deficits at the thumb and index finger.

### Therapy

With compression syndromes, especially after hard muscle work, immobilization and the use of nonsteroidal anti-inflammatory drugs or, if these are ineffective, oral steroids are advisable (Basset and Nunley, 1982). Acute lesions have a good prognosis. If conservative treatment fails, especially in distal compression syndromes, surgical decompression of the nerve in the cubital fossa with resection of fibrous tissue and incision of the biceps tendon may be successful (Basset and Nunley, 1982). With an isolated traumatic lesion, surgery is the treatment of choice. With irreversible lesions surrogate operations may be considered (Rudigier and Degreif, 1986).

### Lesions of the Radial Nerve at the Upper Arm

#### Clinical Aspects

Proximal lesions of the nerve, for example radial nerve injury in the axilla provoked by using crutches, are relatively uncommon. They can be recognized clinically by the presence of triceps weakness with sensory loss on the posterior surface of the upper arm in addition to the well-known dropped hand and fingers. The relatively frequent site of radial nerve injury at the bony sulcus of the upper arm, caused by upper arm fractures and compression of the nerve against the bone under the influence of alcohol or drugs (Saturday night palsy) or occasionally by pressure of the edge of the operation table during anesthesia (Figure 1), are accompanied by intact triceps function. The same is the case with lesions at the lateral intermuscular septum (midportion of the upper arm) provoked by strong contraction of the triceps or chronic strain of the triceps (Wilhelm and Suden, 1985).

### Therapy

Acute pressure injury (Trojaborg, 1970) and acute and chronic occupational palsy have a good prognosis, if trauma to the nerve is alleviated. If conservative treatment is not successful within 5–6 months, a revision and, if required, neurolysis of the nerve may be advisable after previous precise localization of the lesion by neurophysiological investigation. Lesions of the nerve with upper arm fractures may also initially be treated

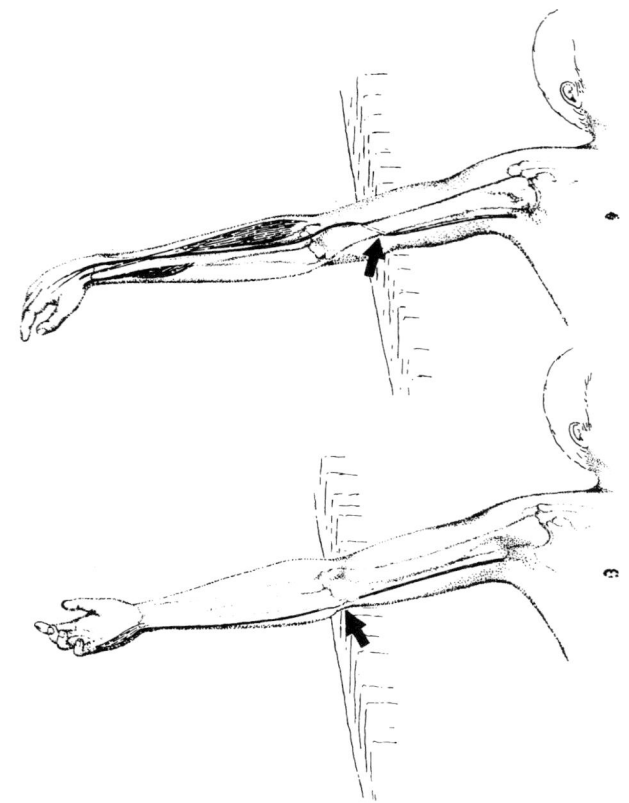

**FIGURE 1** External compression of the radial nerve (above) and the ulnar nerve (below) at the typical sites.

conservatively, because the trauma usually leads to contusion of the nerve but only rarely to a complete nerve section. If signs of reinnervation are missing after 5–6 months, operative revision is indicated. The nerve is usually found compressed by scar tissue and not by callus. If the palsy is irreversible, a transposition of the latissimus dorsi muscle may be considered to compensate for the missing function of the triceps (Du Toit and Levy, 1967).

### Interosseus Posterior Syndrome—Lesion of the Deep Branch of the Radial Nerve

#### Clinical Aspects

Compression syndromes of the deep branch of the radial nerve (posterior interosseus syndrome), where it runs below the brachioradialis muscle from the lateral epicondyle to the supinator muscle, are rare (Lister *et al.*, 1979; Moss and Switzer, 1983). Compression of the posterior interosseus nerve at the site where the nerve runs between the two heads of the supinator muscle below the arcade of Frohse into the supinator muscle is more frequent (supinator muscle syndrome; Benini and Di Martino, 1976; Stille, 1974). The posterior interosseus syndrome may start with pain in the extensor muscles of the forearm or with weakness. Often, extension of the fifth digit is first affected. Because involvement of the extensor carpi ulnaris muscle occurs with more complete lesions and the radial extensor muscles are spared, the hand deviates radially during extension. Sensory deficits are regularly absent. The diagnosis is supported by an increase of pain during the night and during supination against resistance. Pain may be provoked by pressure on the nerve 4–5 cm distal to the lateral epicondyle. This helps to differentiate this nerve lesion from epicondylitis of the humerus, which additionally is never accompanied by motor deficits. Further differential diagnoses are tumors, lead poisoning, and idiopathic mononeuritis, including multifocal neuritis with conduction block. The lesion is associated mostly with an anatomical anomaly (most frequent at the arcade of Frohse) that becomes symptomatic as a result of acute or chronic strain of the supinator or extensor carpi radialis muscle.

Suematsu and Hirayama (1998) differentiate three types: type 2 (drop fingers) with compression of the recurrent branch, type 3 (drop thumb) with compression of the descending branch, and type 1 with compression of both branches. A lesion of the sensory superficial branch of the radial nerve is usually the consequence of a trauma at the forearm, rarely the consequence of repetitive pronation and supination of the forearm or chronic pressure (operating gloves, bracelet) and has a good prognosis for spontaneous recovery.

Persistent tennis elbow has nothing to do with compression of the deep radial branch, and nerve decompression is not indicated in this condition.

### Therapy

Initially, immobilization and the use of anti-inflammatory agents, especially after acute muscular strain, may be an effective therapy for either paresis or pain. As the prognosis, especially of the paresis, becomes poorer with increasing duration of the symptoms, operative decompression should not be delayed more than 6–8 weeks if clinical or electromyographic signs of improvement are absent. If the palsy lasts for more than 16 months, decompression will not be effective. At surgery, the arcade of Frohse is frequently fibrous, and, more rarely, a lipoma, neurinoma, or ganglion may be found. Operation usually leads to improvement within 3–18 months. With permanent paresis, transposition of the flexor carpi ulnaris and palmaris muscles and, if necessary, also of the pronator teres muscle to the tendons of the extensor muscles may be performed.

## The Median Nerve and its Lesions at the Upper Arm

Lesions at the upper arm are rare and may be caused by trauma, sleep paresis, pressure of crutches in the axilla, and pressure of cuffs used for a blood-free field during operation. Of the population, 1%–3% have a supracondylar process 5 cm proximal to the medial epicondyle at the humerus, sometimes accompanied by a fibrous band from the medial epicondyle to the supracondylar process (Struther's ligament). The nerve and the brachial artery, which run behind and below the process and the ligament and cross them, may be irritated here, leading to pain at the elbow, in the volar forearm, and in the sensory innervation territory of the median nerve of the hand and fingers during flexion and pronation of the arm. Tinel's sign may be provoked by pressure 3–5 cm proximal to the medial epicondyle. The degree of sensory loss in the median territory at the hand is variable. Marked paresis caused by compression from a supracondylar process and Struther's ligament is rare. These lesions in the upper arm may cause marked tenderness often associated with a paresis of pronation (pronation is partly carried out by the brachioradialis muscle) and of hand and finger flexion. The flexor carpi ulnaris muscle and the ulnar part of the flexor digitorum profundus muscle are unaffected. During innervation of the hand flexors, the predominance of the flexor carpi ulnaris muscle may lead to an ulnar deviation of the hand.

The prognosis of pressure injury at the upper arm is almost always favorable with conservative management, so that only in rare cases with long-lasting deficit is an operative revision indicated. Compression of the nerve at the distal upper arm by a supracondylar process or Struther's ligament has a good prognosis with operative treatment.

## Pronator Teres Syndrome

The pronator teres syndrome is often caused by repetitive strong pronation of the dominant arm (occupational paresis) and rarely by anatomical variations causing compression of the nerve. Initially, pain occurs on the volar side of the forearm and typically increases during pronation against resistance, radiating proximally. Pain and paresthesias may be present in the sensory territory of the median nerve. Patients may also complain of writer's cramp. Sensitivity to pressure above the pronator muscle is almost invariably present. Tinel's sign is less frequent. Paresis, if present at all, is only slight and spares the pronator teres, the flexor carpi radialis, the superficial flexor digitorum, and the palmaris longus muscle in contrast to compression at the upper arm. Compression of the median nerve at the elbow is easily mistaken for a carpal tunnel syndrome. The neurogenic lesion is easier to document by concentric needle electromyography than by searching for a delay in motor conduction at the proximal forearm. The typical signs are features of chronic partial denervation in the flexor pollicis longus, as well as in the abductor pollicis brevis. In addition, normal motor and sensory conduction in the median nerve above the wrist is important to exclude carpal tunnel syndrome. Apart from a compression at the entrance into the pronator muscle, the nerve may also be compressed at this site by the aponeurosis of the biceps (Laha et al., 1978), below the tendinous arc of the flexor digitorum superficialis muscle, by intramuscular bleeding or by tumors.

A change in occupational activity (avoiding forced pronation), intermittent immobilization, and if necessary, infiltration with corticosteroids is usually helpful. Otherwise, surgical decompression gives good results if a compression of the nerve is found. Intramuscular bleeding requires immediate drainage of the hematoma and incision of the fascia.

## Anterior Interosseus Syndrome (Kiloh–Nevin Syndrome)

The anterior interosseus syndrome (Kiloh and Nevin, 1952) consists of pain in the proximal forearm and weakness of flexion of the most distal joint of the thumb (flexor pollicis longus muscle) and the last joint of the index finger (flexor digitorum profundus muscle), so that during pressure of the index finger against the thumb there is no flexion in the most distal joints (pinch sign or circle sign). Dexterity is reduced, and writing is rendered more difficult. Impairment of flexion of the middle finger and of pronation with a flexed forearm (pronator quadratus muscle) indicates more marked paresis. Sensory deficits are absent. Electromyography of the flexor pollicis longus and measurement of the latency to this muscle after electrical stimulation at the elbow compared with the opposite unaffected side may contribute to the diagnosis. Causes of the lesion include fracture, repositioning after fracture, trauma to the forearm, thrombosis, injection into the cubital rim, anatomical variations of the flexor tendons at the forearm (especially at the tendinous rim of the superficial digital flexor muscle), external pressure from a cast or during sleep, an occupational palsy, weight lifting, or an uncommon manifestation of neuralgic amyotrophy (Staal et al., 1999).

The prognosis depends largely on the etiology and is favorable in closed injury and occupational paresis. In the case of Volkmann's ischemia, urgent decompression is necessary to avoid nerve and muscle necrosis. If there is no improvement after 5–6 months, operative decompression and exterior neurolysis should be carried out. The result of these procedures is usually excellent (Hill et al., 1985). However, Seror (1999) observed spontaneous recovery, with a mean delay of 14.3 months and argued against surgery within the first 12–16 months. Permanent paralysis of the deep finger flexors may be treated by a tendon transfer operation (Omer and Spinner, 1980).

## Carpal Tunnel Syndrome

The carpal tunnel syndrome (CTS) is by far the most frequent peripheral nerve compression syndrome (de Krom et al., 1992). It causes about 20% of all compression syndromes and about 50% of all cases of brachialgia. It occurs twice as frequently in women as in men, usually at ages older than 50. The dominant side is affected more frequently, and if the compression is bilateral (40%–50% of the cases), the dominant side is usually more affected.

### Clinical Aspects

Frequently, carpal tunnel syndrome is seen with brachialgia and paraesthesias occurring at night. The patient is awakened from sleep by paresthesias and a feeling of swelling of the fingers without objective signs of edema. The pain may radiate to the upper arm or even to the shoulder. Massages, changes of arm position, or shaking of the hand may bring relief. Morning stiffness with diminished dexterity and numbness in the finger will decrease in the morning hours (Loong, 1977).

The syndrome is caused by compression of the median nerve in the carpal tunnel, which is enclosed by the carpal bones and the transverse carpal ligament and contains, in addition to the nerve, the tendons of the flexor muscles (Figure 2). Frequently, anatomical variations, especially congenital stenosis of the carpal tunnel, may play a role in eliciting the syndrome. Pressure in the carpal tunnel is least in the mid-position and increases with flexion and even more with extension of the wrist. In patients with CTS, resting pressure is increased even in the mid-position (Gelbermann et al., 1981). The provocation of paresthesia by full flexion or extension of the wrist for 30–40 seconds is of diagnostic significance (McLellan and Swash, 1976).

Among many causes of CTS, we may list these:

1. Frequent flexion and extension of the hand or occupation exposure to vibration
2. Tenosynovitis
3. Infectious diseases like tuberculosis or histoplasmosis
4. Rheumatic inflammation
5. Bleeding, thrombosis, neoplasms

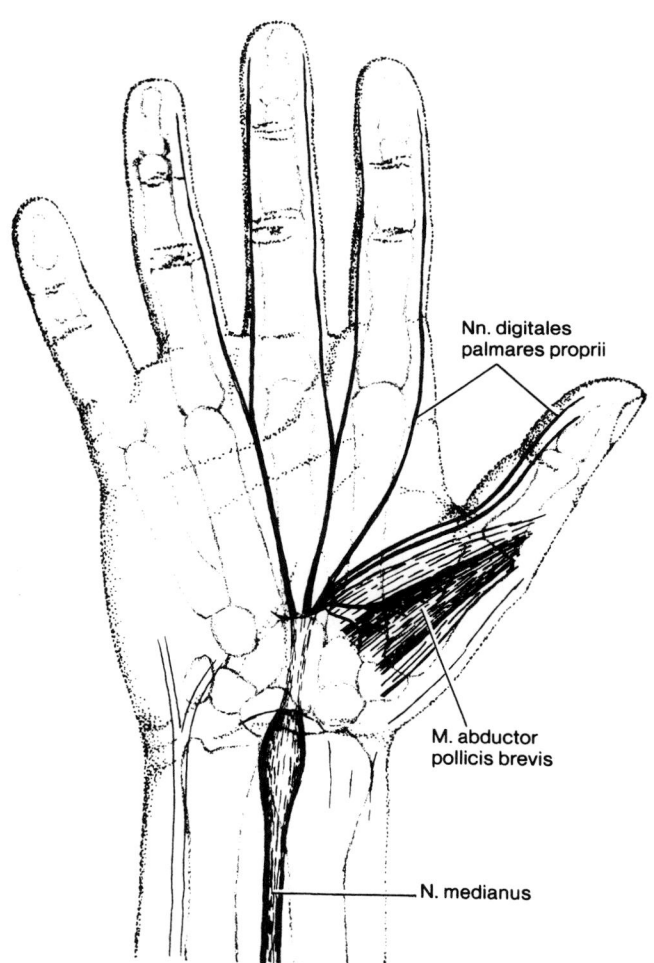

Nn. digitales
palmares proprii

M. abductor
pollicis brevis

N. medianus

**FIGURE 2** Carpal tunnel syndrome. From M. Stöhr [1998] with permission.

6. Disorders of water and lipid contents (obesity, pregnancy, menopause, oral contraceptives, myxedema, acromegaly)
7. Rare hereditary diseases (amyloidosis, mucopolysaccharidosis, mucolipidosis)

All may lead to a relative increase of the contents of the carpal tunnel and thereby to CTS. Frequently, no etiology is found, and the tendons and other canal contents appear normal. In idiopathic CTS, the carpal tunnel may be narrow (measured by computed tomography), or there may be other anatomical variations of muscles, tendons, and blood vessels. Apart from a history, the diagnosis is made on the basis of the typical sensory deficit, paresis, or atrophy of thenar muscles, a positive Phalen test (flexion of the wrist by gravity for 60 seconds while keeping the forearm vertical and the elbow supported on the surface of a table), or by a positive dorsal-extension test of the wrist. Nerve conduction testing and electromyographic examination

strongly support the diagnosis (Stevens, 1987; Stöhr, 1998), which, like all diagnoses, must be clinical. The most sensitive parameter is the sensory conduction from the palm to the wrist. Successive reduction of the distance between stimulating and recording electrode may make it possible to localize the site of the lesion even more precisely. Measuring sensory conduction after stimulation of the fingers is less sensitive, because the distance of reduced conduction in the carpal tunnel is relatively short compared with the overall distance measured with this technique. It is, however, less prone to disturbances of measurement by electrical artifacts of the stimulating impulse. Estimates of the frequency of abnormalities, when measuring the distal motor latency, are variable (Kimura and Ayyar, 1985). It is important to exclude generalized polyneuropathy. A comparison of motor and sensory parameters of the median and ulnar nerve is less reliable because of the relatively frequent simultaneous affection of both nerves at the wrist (Buchthal *et al.*, 1974; Sedal *et al.*, 1973).

The development of high-resolution ultrasound transducers allows detailed imaging of the median nerve within and above the carpal tunnel and provides relevant structural information at lower cost than MRI and may prove useful in the diagnosis (Angerer *et al.*, 2000).

### Clinical Course

The symptoms of a nocturnal paresthesias may occur for years with no neurological deficit, although slowing of sensory conduction may be found at this stage (Thomas *et al.*, 1967). Use of the hand may provoke or worsen clinical symptoms. Increase of symptoms includes loss of dexterity, weakness of the hand, and permanent sensory deficits. Sensory deficits (frequency according to the literature between 9% and 85% of the cases) never include the fifth finger and also usually rarely include the whole sensory territory of the median nerve at the hand, because the palmar branch usually arises before the carpal tunnel, running above the ligamentum carpi transversum. Thus, the index and middle fingers are the most often affected. Autonomic disorders (feeling of coldness, increased sweating, symptoms of Raynaud's disease, Alföldi's nail-bed sign) may also be present. Paresis and muscle atrophy (present in 35%–40% at the first examination), but only slight involvement of the lumbrical muscles usually develop after sensory deficits.

### Therapy

In the presence of painful paresthesias without objective neurological deficits, volar splinting of the wrist in mid-position during the night is the treatment of choice (initial improvement in 50% of all cases, long-lasting relief of complaints, however, only in 1%–14% of patients). Nonsteroidal anti-inflammatory drugs may be used at this stage, although not during pregnancy. If rapid improvement is expected after trauma or heavy use of the wrist, even in the presence of neurological deficits, nonsteroidal anti-inflammatory drugs may be useful. Treatment by vitamin $B_2$ supplements is probably not efficacious when there is no vitamin deficiency. In the presence of incomplete remission of complaints in chronic CTS, especially near the end of pregnancy, the injection of 25 mg of prednisolone into the carpal tunnel after previous injection of 2–3 ml of 1% lidocaine may be tried. This treatment may be repeated after 3–8 days, although the risk of infection increases with each successive injection. This treatment usually leads to initial improvement. However, the long-term results of steroid injections into the carpal tunnel are much less favorable. After 1–2 years only 20%–50% of the patients are free of symptoms. The results of surgical treatment are far superior.

In all cases with definite sensory or motor defects surgery is the treatment of choice, as well as in all patients in whom conservative treatment has failed (Kaplan et al., 1990). Many variations of surgical procedure have been described. Most frequently, the operation is done after plexus or local anesthesia in a blood-free field. Complete section of the retinaculum is carried out to the tendon of the flexor palmaris longus muscle. Visualization of the motor branch is advisable, because it may be subject to isolated compression. In addition to freeing the nerve from the compressing fibrous tissue, some authors, when there is strangulation of the nerve by the epineurium and by interfascicular scar tissue, recommend a microsurgical epineural and endoneural neurolysis (Meese et al., 1980). Resection of the palmaris longus tendon is occasionally done. Apart from the compression of the nerve at the carpal tunnel, tumors of different origin (ganglia, lipoma, hemangioma, chondroma, heterotopic formation of cartilage, ectopic calcification, fibroma) may lead to a distal compression lesion of the median nerve.

After a correct diagnosis, complete incision of the transverse carpal ligament almost invariably leads to instant relief from pain. Sensory deficits show good improvement within about 18 months in most (90%) cases. Improvement of motor function depends mainly on the preoperative functional deficit and the duration of the compression; atrophy rarely improves if symptoms have been present for longer than 12 months. In bilateral involvement, the most affected side should be operated on first, because symptoms on the other side may disappear spontaneously (probably because of decreased strain on this hand, because the operated hand can be used more and better after successful treatment). After the operation, the arm is kept in an elevated position for 24 hours, and the hand is immobilized in the neutral position of the wrist by splinting for 2 weeks. Immediately after the operation, if necessary after analgesics have been given, the patient is asked to move his or her fingers actively to avoid postoperative edema, and reflex sympathetic dystrophy (see Chapter 11). It is also important to avoid impairment of joint motility. A poor response to surgery often indicates that an incomplete ligamentous section was performed, and the nerve may need to be re-explored (O'Malley et al., 1992).

Endoscopic carpal tunnel release decreases postoperative discomfort and allows faster functional recovery. The complication rate of this technique is similar to open surgery (1%) if performed by well-trained surgeons, but the severity of the complications seems to be greater (Dawson et al., 1999). Contraindications to endoscopic techniques are anatomical abnormalities within the carpal tunnel, mass lesions, rheumatoid arthritis, isolated involvement of the motor thenar branch, and previous surgery at the wrist.

### The Ulnar Nerve and its Lesions at the Upper Arm

In considering ulnar nerve lesions, the relatively frequent anomalies of innervation of small hand muscles have to be taken into account (Sunderland, 1978). The frequency of the Martin-Gruber anastomosis, in which motor fibers from the median nerve pass to and travel with the ulnar nerve to the small hand muscles, is estimated to be 10%–44%. In about 4% of the patients, the small hand muscles are completely supplied by the ulnar nerve (Hopf and Hense, 1974). Variations of innervation, with additional or exclusive median innervation of muscles usually innervated by the ulnar nerve, may be present in up to 10% in the first dorsal interosseus muscle and up to 50% in the third lumbrical muscle. Rarely (1%) the radial nerve also takes part in the motor innervation of the hand.

The ulnar nerve is subject to direct trauma at the axilla or upper arm only rarely. Compression at the distal upper arm by a supracondylar process is also rare. More frequent are traumatic lesions at the distal upper arm where the nerve lies quite close to the surface on its way to the cubital tunnel. Apart from direct trauma, the nerve may be damaged by supracondylar fracture, especially in childhood. The prognosis of such lesions (Galbraith and McCullough, 1979) is very favorable, so that operative treatment is required only in very few cases of long-lasting failed reinnervation. The clinical syndrome caused by lesions at the upper arm is the same as in lesions at the cubital tunnel (see later). In cases of a defect in the nerve that needs bridging, a volar transposition of the nerve may give an additional length of 6–7 cm.

## Cubital Tunnel Syndrome (Ulnar Neuropathy at the Elbow, UNE)

### Clinical Aspects

Lesions at the cubital tunnel may develop acutely because of unfavorable arm positions (Figure 1) (Miller, 1991; Mumenthaler *et al.*, 1998; Stöhr, 1996). Compression during anesthesia and coma may be avoided by supination of the forearm. The habit of resting the elbow on the table or mechanical strain on the nerve by frequent flexion and extension of the elbow joint may cause ulnar neuropathy. The nerve may be pushed by the medial head of the triceps in the direction of the medial epicondyle and against the collateral ulnar ligament. Habitual subluxation of the nerve because of a flat cubital sulcus (according to the literature in 2%–36% of operated patients) may also cause ulnar nerve injury. Scar formation in the perineural environment after fractures (in 19% of the patients), deforming arthritis (in 8%–21% of the patients), chondromatosis, ganglia, or anatomical variations with a rudimentary epitrochleoanconeus muscle (in 9% of the patients; as a normal variation in 16%–28% of the general population) may also lead to cubital tunnel syndrome.

Pressure on the nerve frequently elicits paresthesias, numbness of the ulnar fingers, and pressure damage to the nerve. Paresis and muscle atrophy are later signs of nerve compression. Posttraumatic changes in the cubital tunnel, however, may lead to slowly progressive paresis, sometimes years or even decades after the trauma, that involves the small hand muscles more than the flexors in the forearm, so that the clinical signs imitate a distal ulnar lesion. Atrophy is usually first noticed at the first dorsal interosseus muscle. The well-known clinical sign of a clawhand is seen only with severe lesions. When one palpates the nerve in the cubital tunnel, the nerve is frequently found to be thickened, adherent or only slightly movable, and susceptible to pain in the ulnar distribution, as well as locally, during local pressure on the nerve. Electrophysiological measurement of motor (and sensory) conduction velocity in the proximal nerve segment at the cubital tunnel and a distal segment in the forearm is useful in the diagnosis (Benecke and Conrad, 1980; Brown *et al.*, 1980; Eisen and Danon, 1974; Kincaid, 1988; Stöhr, 1998) (Figures 3 and 4). Measurement of the F wave, the ulnar nerve sensory action potential, somatosensory evoked potential studies, and electromyography of paravertebral muscles may help to differentiate a C8 root lesion that may have identical clinical signs.

### Therapy

There are no randomized controlled studies giving clear advice concerning conservative or surgical treat-

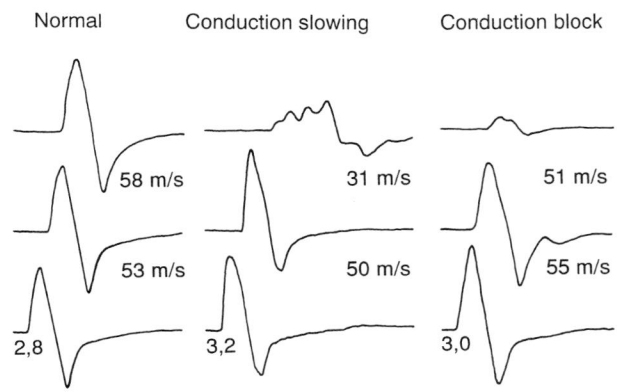

**FIGURE 3** Chronic compression of the ulnar nerve leads to focal slowing of conduction, whereas acute external compression is followed by conduction block.

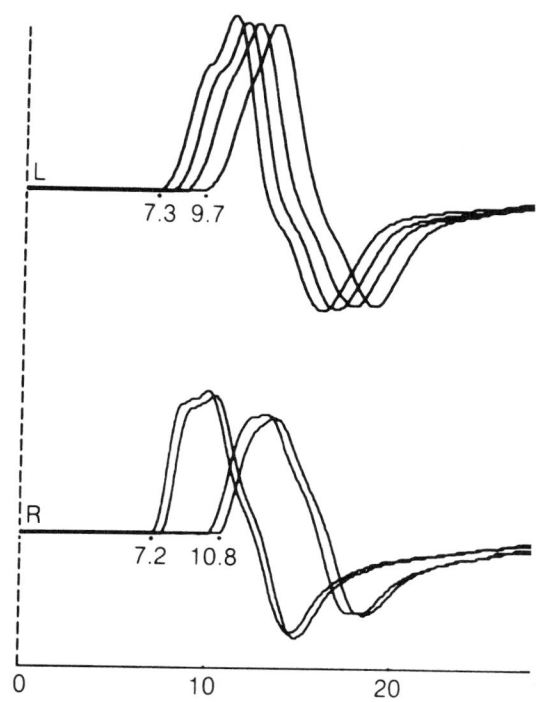

**FIGURE 4** Evidence of focal slowing of conduction within the right ulnar groove by inching technique.

ment (Dellon *et al.*, 1993), so that the following guidelines are based on our personal experience. In acute pressure lesions, spontaneous recovery is frequent, so treatment is always initially conservative. Pressure on the nerve during sleep may be avoided by protecting the elbow with a splint that also restricts extreme flexion of the elbow. Also, irritation of the nerve by frequent flexion and extension of the elbow joint should at first be treated conservatively. If it is possible to avoid the traumatizing movement, the prognosis is favorable (Eisen and Danon, 1974). Repeated pressure injury to the nerve induced by sleeping with an extremely flexed elbow may be managed by splinting the arm during the night in a slightly flexed position (Seror, 1993). If this conservative treatment is not successful or if frequent movement of the elbow joint cannot be avoided, surgery is justified. Traumatic late paresis should be treated surgically, because it is usually progressive. The following surgical methods are recommended:

1. Decompression of the nerve by cutting the tendinous arcade crossing the nerve at the insertion point of the flexor carpi ulnaris muscle (Lugnegard et al., 1982; Miller and Hummel, 1980; Wilson and Krout, 1973).
2. Ventral transposition of the nerve, first suggested by Learmonth (1942), either into the subcutaneous tissue or into the ulnar flexor muscles. The latter procedure has the advantage of better protecting the nerve. The ventral transposition includes cutting the tendinous insertion of the flexor carpi ulnaris muscle at the medial epicondyle.
3. Medial epicondylectomy is carried out, removing most of the bony prominence with an osteotome, which allows the nerve to ride freely into this new area.

The results after submuscular transposition of the nerve show rates of improvement of 80%–90% (Gerl and Schlüter, 1980; Leffert, 1982; Lugnegard et al., 1977); the same is the case for subcutaneous transposition (Dellon, 1989). In two studies (Lugnegard et al., 1977, 1982), the same group of authors (and surgeons) found identical results after submuscular and subcutaneous transposition; however, they favored the submuscular transposition, because preoperative clinical deficits were more pronounced in this group of patients. Submuscular transposition is suggested in patients with especially powerful muscles, because the nerve may be compressed at the entrance under the flexor carpi ulnaris muscle. Ventral transposition, either subcutaneously or submuscularly, is particularly indicated in patients with frequent subluxation of the nerve out of the sulcus. Compression of the nerve under the aponeurosis of the flexor carpi ulnaris muscle may initially be treated by simple decompression.

In summary, simple ulnar decompression showed excellent results in uncomplicated cases (Assmus, 1994; Bimmler and Meyer, 1996; Bradshaw and Shefner, 1999) and has the advantages of simplicity and low morbidity. In addition, no postoperative splinting is required. Anterior transposition seems the method of choice when the evidence indicates pathology in the ulnar groove, as well as in cases with luxation of the

ulnar nerve during elbow flexion. The risks of this procedure are undermobilization of the nerve with secondary kinking or overmobilization with involvement of the segmental blood supply.

After ventral transposition of the nerve, the elbow and wrist should be immobilized by dorsal splinting for about 3 weeks. After that, physiotherapy may be started cautiously. Paresthesias and sensory deficits usually improve rapidly. Prognosis for motor deficits is less favorable, especially if they have been present for a long time. There are, however, rare cases with good recovery of atrophy despite being present for more than 2 years, so that surgery may be the treatment of choice in such cases, as well. The results with respect to motor deficits should not be evaluated earlier than 18 months after surgery.

### Ulnar Lesions at the Wrist

#### Clinical Aspects

In contrast to proximal lesions, ulnar lesions near the wrist usually show no sensory deficits on the ulnar side of the dorsum of the hand. Distal lesions are most frequently the result of ganglia (29%; Shea and McClain, 1969) or of occupational compression (24%). Other causes are compression by crutches or bicycling (Eckmann et al., 1975; Noth et al., 1980), by acute trauma, or late paresis by fracture of the carpal bones or the basis of metacarpalia IV and V, distal fractures of the radius, or compression resulting from arthrosis, lipoma, aneurysms, and anomalies of muscles and tendons. In a few patients, the cause remains unclear, so that a compression of the nerve in the fibrous tunnel (loge de Guyon) is assumed (Shea and McClain, 1969). A complete lesion of the ulnar nerve near the wrist, including the superficial and deep branch, may lead to combined motor and sensory deficits. Compression of the deep branch leads to motor deficits only; and with more distal compression of the deep branch in the palm of the hand, the hypothenar muscles may be spared. Now and then, compression in the loge de Guyon may lead to pure sensory deficits on the volar side of the hand. Pure motor ulnar nerve paresis is usually painless; with sensory deficits pain is usually located at the wrist and may radiate into the ulnar fingers and the forearm. Pain may increase during the night and with movement of the wrist. Electroneurography with measurement of the distal motor latency to the hypothenar and to the first dorsal interosseus muscle and measurement of the sensory nerve conduction velocity in the ulnar nerve to the wrist aid in the localization of the lesion. As a differential diagnosis, an early form of motoneuron disease and cervical amyotrophy caused by focal spinal cord infarction may be hard to distinguish (Stöhr and Scheglmann, 1999).

## Therapy

Therapy depends on the cause. Frequently, it is sufficient to stop the external compression or the motor activity (the repetitive movement) that has been suspected to be the cause. After acute trauma, also after fractures at or near the wrist, enough time should be given for spontaneous recovery. However, if no recovery is apparent within 5–6 months, surgical exploration is advisable. Slowly progressive paresis without acute or chronic trauma is often due to a space-occupying process, which should be subject to surgical exploration. Apart from access from the hypothenar region, surgical access may be from the carpal tunnel, which allows the simultaneous visualization of the ulnar and the median nerves, especially if a combined lesion of both nerves or a lesion distal to the loge de Guyon is suspected.

# COMPRESSION NEUROPATHIES IN THE LEG

## Lumbosacral Plexus

### Clinical Aspects

The lumbosacral plexus derives from the ventral rami of the first lumbar to the fourth sacral spinal nerves. It is made up of the lumbar plexus (L1–L4), which passes downward within the psoas major muscle, and the sacral plexus (L4–S4), which overlies the lateral sacrum and the posterolateral pelvic wall. The main symptom of a lumbar plexus lesion is involvement of the iliopsoas and quadriceps muscles, which may be mistaken for a femoral neuropathy. A lesion of the sacral plexus may initially cause footdrop similar to a sciatic or peroneal nerve lesion.

Recognition that the plexus has been affected usually follows detection of the neurological deficit beyond the distribution of a single nerve or nerve root. Disorders leading to external compression of the lumbosacral plexus include many space-occupying processes (Table II). The initial symptom is usually low back or pelvic pain that often radiates into the leg, followed by progressive sensory and motor deficits. Bilateral damage to the sacral plexus may lead to bladder and bowel dysfunction. The onset of symptoms is subacute in the case of retroperitoneal hemorrhage, whereas the course in the other conditions is usually more insidious.

### Practical Management

The main principle of treatment is decompression of the affected part of the plexus. Invasion or compression by malignant tumors requires surgical excision, radiotherapy, or chemotherapy. Recovery from neuro-

**TABLE II** Disorders Causing Compression of the Lumbosacral Plexus

Tumors
  Primary tumors arising from pelvic, abdominal, or retroperitoneal structures (prostate gland, cervix, intestine, kidney, bladder, retroperitoneal lymph nodes, bone)
  Metastases from extrapelvic tumors (malignant lymphoma, lung and breast cancer, melanoma)
  Primary nerve sheath tumors (neurofibroma, schwannoma)
Retroperitoneal hemorrhage
  As a complication of hemophilia or anticoagulant therapy
  After intrapelvic surgery
  Rupture of an abdominal aortic aneurysm
Retroperitoneal abscesses
  Arising from infection of the vertebrae, perinephritic tissues, or gastrointestinal tract
  Primary abscesses
Aneurysms (abdominal aorta, hypogastric, common and internal iliac arteries)
Pregnancy and childbirth (cephalopelvic disproportion, compression during prolonged or forceps delivery)

**TABLE III** Reversal of Anticoagulation

Heparin
  Discontinuation of IV infusion. Because of the short half-life of heparin, normal coagulation is obtained in hours
  Protamine sulfate IV (1 mg for every 100 units of heparin remaining in the patient)
Oral anticoagulants (warfarin, dicumarol)
  Discontinuation of oral medication
  Vitamin K, 5–10 mg PO (maximal effect takes at least 24 h)
  Fresh frozen plasma, 10–20 ml/kg

logical deficits is often disappointing (Jaeckle et al., 1985).

Sufficient analgesic therapy (see Chapter 11) is very important. Retroperitoneal hemorrhage requires assessment of cardiovascular function, and some patients may need a blood transfusion. Any underlying bleeding disorder should be treated (Table III). The need for surgical evacuation of the hematoma depends on the size of the hematoma and the severity of the neurological deficit (Donaghy, 1993). Early surgical decompression may improve the outcome in patients with severe symptoms (Tysvaer, 1982; Young and Norris, 1976), although a prospective trial has never been carried out. An aneurysm of the abdominal aorta or iliac arteries requires surgical intervention.

Retroperitoneal abscesses should be managed by drainage of pus, either by surgery or percutaneous catheters inserted with ultrasonic or CT guidance (Alexander and Dellinger, 1991). Initial antimicrobial treatment should include a combination of antibiotics with activity against both gram-positive and gram-negative organisms until culture results are available. The possibility of tuberculous abscesses must be taken into account.

## Femoral Nerve

### Clinical Aspects

The femoral nerve descends from the lumbar plexus between the psoas and iliacus muscle and passes under the inguinal ligament, lateral to the femoral artery and vein. In the thigh it divides into branches to the quadriceps muscle and cutaneous branches to the anterior thigh and terminates as the saphenous nerve. Lesions of the femoral nerve produce weakness of the iliopsoas and quadriceps muscles, a diminished or absent knee reflex, and sensory disturbances over the anteromedial side of the thigh and the medial side of the lower leg.

The most common compressive cause of femoral palsy is retroperitoneal hemorrhage caused by hemophilia or anticoagulant therapy and rarely by rupture of an arterial aneurysm or traumatic avulsion of the iliacus muscle (Dawson *et al.*, 1999; Stewart, 1993). The main symptom is severe pain that begins below the inguinal ligament and spreads to the distribution of the saphenous nerve, with local tenderness in the groin. Examination reveals signs of femoral palsy. Extension of the hip (femoral stretch sign) increases pain. Occasionally, ecchymoses are present in the upper thigh.

Other disorders leading to less acute compression of the femoral nerve (Table IV) include iliac abscesses, aneurysms of the internal iliac artery, and tumors arising from the iliopsoas muscle or the ileum. Iatrogenic causes include various surgical procedures (e.g., lithotomy position), leading to kinking and compression of the nerve under the inguinal ligament during hip flexion (Mumenthaler *et al.*, 1998; Stewart, 1993; Stöhr, 1996).

CT scanning will demonstrate acute hemorrhage very well. MRI may give better visualization of nonhemorrhagic lesions such as abscess or tumor.

### Practical Management

The early management of iliacus hematomas depends on the severity of the injury and the patient's condition. Analgesic treatment (Chapter 11) and correction of bleeding disorders (Table III) are essential. Indications for surgery are controversial, because conservative treatment may give satisfactory results, and no controlled trials are available (Silverstein, 1979). In severe cases most authorities would agree that prompt surgical evacuation of the hematoma is useful and improves the outcome (Chiu, 1976). Percutaneous drainage may be a useful alternative (Merrick *et al.*, 1991).

**TABLE IV  Disorders Causing Compression of the Femoral Nerve**

Retroperitoneal hemorrhage
   As a complication of hemophilia or anticoagulant therapy
   After intrapelvic or hip surgery
   Rupture of an arterial aneurysm
Retroperitoneal abscesses
   Secondary infection of a hematoma or primary infection
Tumors
   Tumors of pelvic, abdominal, or retroperitoneal structures
   Primary nerve sheath tumors
Aneurysms (abdominal aorta, external iliac and femoral arteries)
Iatrogenic compression during surgical procedures in lithotomy
   position

**TABLE V  Disorders Causing Compression of the Saphenous Nerve**

Compression in the thigh (fibrous bands, arterial branches, tumors)
*Neuropathia patellae* (entrapment at the exit from Hunter's canal)
*Gonyalgia paresthetica* (entrapment within the sartorius tendon)
External compression of the medial side of the knee

Compression by malignant tumors should be treated by surgical excision, radiotherapy, or chemotherapy. Retroperitoneal abscesses require drainage and antimicrobial treatment (see the section on the lumbosacral plexus), and aneurysms should be removed surgically. In femoral nerve compression caused by positioning, spontaneous recovery is frequent within a few weeks (Stöhr, 1996). Regardless of cause, every femoral nerve lesion with weakness of the quadriceps muscle requires physiotherapy to maintain joint mobility and sometimes orthotic devices, for example, a long leg brace with a spring-loaded knee brace.

## Saphenous Nerve

### Clinical Aspects

The saphenous nerve is the terminal sensory branch of the femoral nerve. It descends through the adductor canal and penetrates the fascia above the level of the knee, supplying the medial calf, the medial malleolus, and the medial portion of the arch of the foot. Compression of the sapheous nerve may occur anywhere in its long course, in the thigh, at the knee, or in the lower leg (Table V). In the thigh, compression results from fibrous bands and from anomalous branches of the femoral artery, and rarely from schwannomas of the nerve (Stewart, 1993). The symptoms are radiating pain and sensory disturbances over the distribution of the nerve. Often Tinel's sign may be elicited along the nerve (Dawson *et al.*, 1999). Entrapment may occur at the exit of the nerve from Hunter's canal, leading to pain near the knee and the medial lower leg that may be increased by walking (Worth *et al.*, 1984). Spontaneous or postoperative entrapment of the nerve where it pierces the sartorius tendon results in pain and paresthesias in the distribution of the infrapatellar branch of the saphenous nerve (*gonyalgia paresthetica*, Wartenberg, 1954; *neuropathia patellae*, Kopell and Thompson, 1960).

External compression of the saphenous nerve may occur as a result of patient positioning, leg holders during gynecological surgery, plaster casts, or leg orthoses (Mumenthaler *et al.*, 1998).

### Practical Management

The treatment of saphenous nerve entrapment syndromes is that of neuropathic pain (see Chapter 11). If a series of injections of local anesthetic agents or corticoids or both is not successful, surgical decompression should be performed. Compression of the infrapatellar ramus may be treated by displacement of the nerve into the subcutaneous fatty tissue (House and Ahmed, 1977) or by neurectomy (Luerssen *et al.*, 1983).

### Sciatic Nerve

### Clinical Aspects

The sciatic nerve consists of a lateral trunk that becomes the peroneal nerve and a medial trunk that becomes the tibial nerve. It leaves the pelvis through the sciatic notch below the piriformis muscle and courses behind the hip joint and deep in the thigh. The two trunks diverge in the upper popliteal fossa, forming the peroneal and tibial nerves. The sciatic nerve innervates the hamstring muscles and all the muscles below the knee. No sensory branches arise from the nerve above the knee. The sensory supply area below the knee is that of the peroneal plus tibial nerves with exception of the medial aspect, which is innervated by the saphenous nerve. Sciatic nerve lesions usually lead to partial damage, more severe in the peroneal division. Complete palsy of the sciatic nerve is extremely rare and not expected from an entrapment disorder.

Entrapment syndromes of the sciatic nerve are rare (Table VI). Some authors believe that compression of the nerve by the piriform muscle may occur because of an anatomical variation of the course of the nerve during its passage through the sciatic notch (*piriformis syndrome*). Usually the peroneal trunk deviates and runs above or between the parts of the piriform muscle. Rarely, nerve entrapment by fibrosis or fibrovascular bands are found.

**TABLE VI**   Disorders Causing Compression of the Sciatic Nerve

Deeply situated lesions
  Tumors (endometriosis, nerve tumors, lipoma)
  Hematoma
  Baker's cyst
  Aneurysm of the iliac artery
  Fibrosis
  Compartment syndrome of the thigh
Childbirth
Positioning (surgical procedures in semilateral or sitting position)
Piriformis syndrome

Masses within the pelvis can compress the sciatic nerve. Endometriosis leads to pain in the hip and buttock, radiating into the leg and foot, sometimes varying with the menstrual circle (cyclical sciatica). Other tumors in the gluteal region include schwannomas, neurofibromas, lipomas, and lymphomas. Hematomas may occur as a complication of anticoagulation or hip surgery (see the section on the lumbosacral plexus). During childbirth, pressure on the nerve may occur from passage of the fetal head.

External compression occurs mainly from prolonged pressure against the buttock or posterior thigh during coma or in surgery with the patient in a sitting position. In severe cases a compartment syndrome with pressure-induced swelling and necrosis of muscles within the posterior compartment of the thigh can occur (Stewart, 1993).

### Practical Management

The piriformis syndrome is extremely rare. Pain relief may be achieved after injections of local anesthetics, corticosteroids, or both into the piriformis and sciatic nerve area (Stewart, 1993). Surgical dissection of a portion of the muscle or tenotomy produces mostly pain relief (Tackmann *et al.*, 1989) and is rarely carried out, because the surgical success rate is low. If there is weakness in hip abduction because of compression of branches of the sciatic nerve by the piriformis muscle, surgical exploraiton is probably indicated. There is no consensus on the best approach, and endoscopic surgery may be an option.

Entrapment of the sciatic nerve by other deeply located structures may require surgical intervention with dissection of fibrous bands or resection of a tumor or hematoma. Therapy of complications of anticoagulation is described in the section on the lumbosacral plexus. Endometriosis should be treated with synthetic hormones or surgical resection of the tissue or both (Salazar-Grueso and Roos, 1986), according to gynecological principles. Postoperative sciatic neuropathies without hematoma, resulting from patient positioning, do not need surgical exploration. Spontaneous recovery is frequent.

## Peroneal Nerve

### Clinical Aspects

After its origin from the sciatic nerve in the popliteal fossa, the common peroneal nerve winds around the neck of the fibula and passes through the attachment of the superficial head of the peroneus longus muscle (fibular tunnel). It then divides into the superficial peroneal nerve, innervating the peroneal muscles and the skin of the lateral lower leg and the deep peroneal nerve that innervates the dorsiflexors of the foot and toes and the skin between the first and second toe. The symptoms of common peroneal nerve palsy include footdrop and partial sensory loss and vary according to the etiology of the lesion. In many peroneal neuropathies the muscles supplied by the deep branch are more severely affected, and sensory loss is variable or lacking.

The most common of the causes for damage of the peroneal nerve (Table VII) is external compression in its course around the head and neck of the fibula during sleep, anesthesia, coma, prolonged bed rest, or by plaster casts (Figure 5). Habitual crossing of the legs or prolonged squatting are frequent causes of common peroneal neuropathy in India. In some patients, there is a history of excessive weight loss caused by cancer or diet, leading to changes in behavior and increased vulnerability to compression (Dawson et al., 1999; Mumenthaler et al., 1998; Stöhr, 1996).

Entrapment of the nerve rarely occurs at the head of the fibula (fibular tunnel), resulting in a slowly progressive and painful peroneal neuropathy (Berry and Richardson, 1976). Other conditions for compression of the nerve are ganglia and cysts arising from the knee joint, lipomas, callus, or tumors of the fibula.

The anterior compartment syndrome (anterior tibial syndrome) is caused by compression of the deep peroneal nerve by acute muscle swelling within the anterior fascial compartment. The syndrome includes severe anterior lower leg pain, swelling, and redness, with motor and sensory dysfunction of the deep peroneal nerve. It results from excessive exercise, soft tissue trauma, fractures, hemorrhage, occlusion of the anterior tibial artery, or restoration of blood flow after acute arterial insufficiency in the leg (Stewart, 1993). A similar disorder, although rare, may occur in the peroneal compartment, where swelling and necrosis are confined to the peroneal muscles and the superficial peroneal nerve (see last section of this chapter).

Compression of the distal part of the deep peroneal nerve can occur where it crosses the anterior ankle (anterior tarsal tunnel syndrome) because of tight shoes or trauma. Symptoms are painful paresthesias. The distal part of the superficial peroneal nerve can be compressed where it pierces the deep fascia of the lower leg. This leads to local tenderness and painful paresthesias in the distribution of the nerve.

### Practical Management

Peroneal neuropathies caused by external compression do not require surgical treatment. Patients with conduction block caused by a demyelinative lesion recover in a few weeks, whereas recovery of axonal lesions will take much longer, at least several months. Patients with mixed axonal and conduction block lesions may show a biphasic recovery (Wilbourn, 1986). Treatment consists of avoidance of further compression, physical therapy, and a lower leg orthosis to compensate for footdrop and to provide stability to the foot.

Progressive and painful peroneal neuropathies caused by entrapment or compression by masses need early surgical exploration. Some smaller masses or soft tissue lesions may not be identifiable on imaging studies (Stewart, 1993). Intraneural ganglia should be merely evacuated, because total resection results in a complete nerve lesion (Tackmann et al., 1989). An acute compartment syndrome of the lower leg is a surgical emergency and requires prompt fasciotomy. Therapy is described in more detail later.

Compression of the distal parts of the deep and superficial peroneal nerve should be followed conservatively, pressure from tight shoes should be avoided. Orthosis to change foot position or local steroid injections may help. If there is no improvement, surgical decompression by incision of constricting fascial bands should be performed (Dawson et al., 1999). Regardless of the cause of the peroneal palsy, patients with footdrop need physical therapy and bracing. In the case of irreversible peroneal palsy, surgical treatment with tendon transfers may be necessary (Mumenthaler et al., 1998).

TABLE VII   Disorders Causing Compression of the Peroneal Nerve

External compression
  Habitual leg crossing
  Plaster casts, braces
  Anesthesia, coma, bed rest
Masses (ganglion, Baker's cyst, callus, fibular tumor, neuroma, lipoma)
Entrapment in the fibular tunnel
Compartment syndrome (anterior tibial syndrome)
Distal compression
  Anterior tarsal tunnel syndrome
  Distal compression of the superficial peroneal nerve

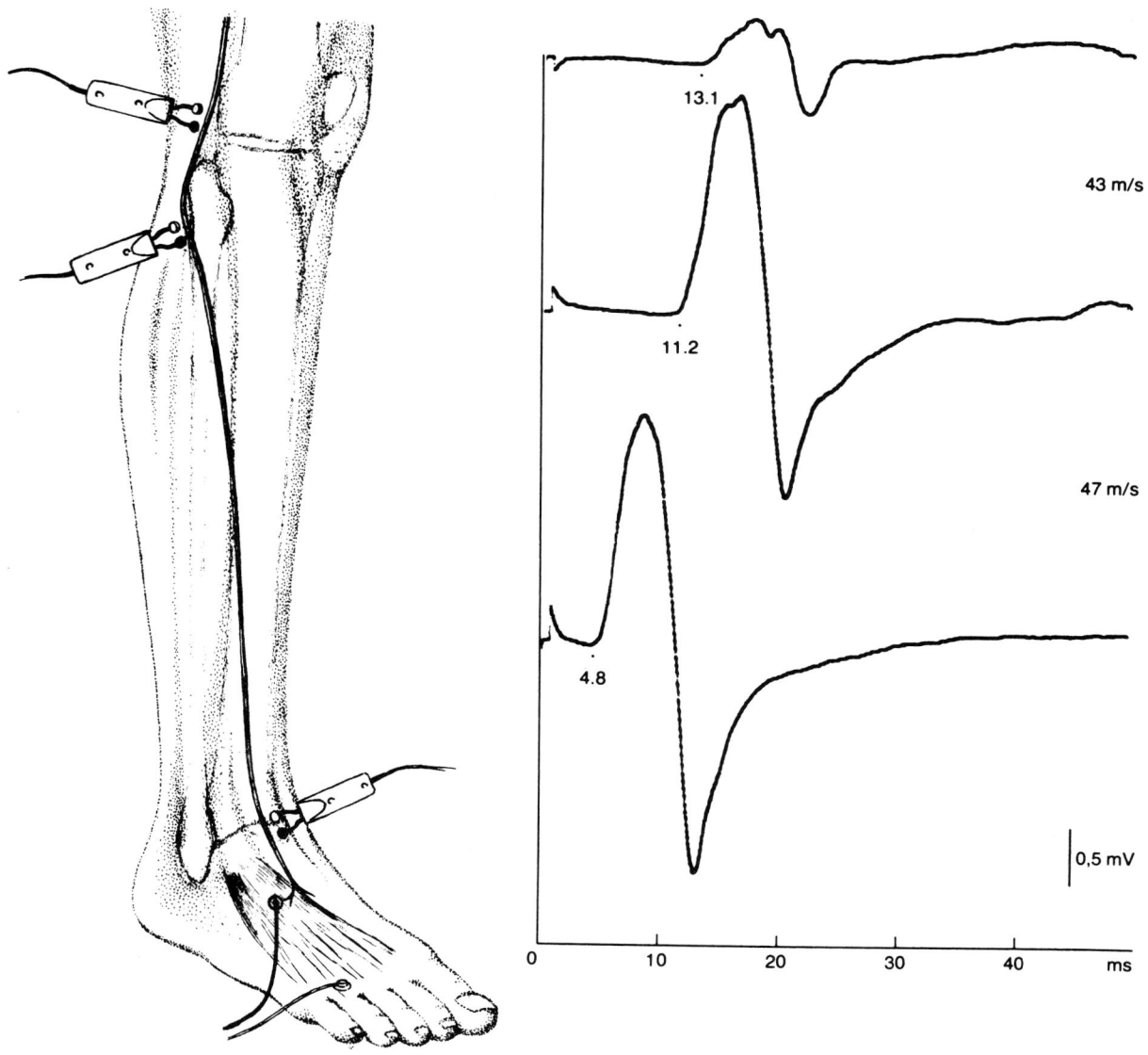

13.1

43 m/s

11.2

47 m/s

4.8

0,5 mV

0    10    20    30    40    ms

**FIGURE 5** External compression of the common peroneal nerve at the head of the fibula with partial conduction block.

## Tibial Nerve

### Clinical Aspects

The nerve passes through the popliteal fossa and then deep to the gastrocnemius muscle. Shortly above the ankle, the nerve becomes superficial and passes under the flexor retinaculum (tarsal tunnel) into the foot. It divides into the two plantar nerves, usually within the tarsal tunnel, but occasionally the division occurs more proximally or distally.

Compression of the proximal tibial nerve is infrequent and may be due to a ganglion, Baker's cyst, or nerve tumor, and sometimes to positioning or plaster casts (Table VIII). Symptoms include weakness of the

**TABLE VIII** Disorders Causing Compression of the Tibial Nerve

External compression
  Plaster casts, braces
  Anesthesia, coma, bed rest
Masses (ganglion, Baker's cyst, tumors)
Entrapment in the tarsal tunnel (posterior tarsal tunnel syndrome)
Distal compression (Morton's metatarsalgia)

plantar flexors, long toe flexors, and the intrinsic foot muscles, and sensory loss in the sole of the foot (Oh and Meyer, 1999).

In *posterior tarsal tunnel syndrome*, entrapment of the tibial nerve at the level of the medial malleolus occurs. Trauma to the ankle, posttraumatic fibrosis, cysts arising from the synovial sheath of the tendons, or rheumatoid arthritis may lead to compression of the nerve. Symptoms include pain in the sole of the foot, Tinel's sign just below and behind the medial malleolus, sensory loss over the plantar surface of the foot, and occasionally weakness of the intrinsic muscles of the foot (Dawson *et al.*, 1999; Mumenthaler *et al.*, 1998).

*Morton's metatarsalgia* results from compression of an interdigital nerve between two adjacent metatarsal heads. Usually the third metatarsal interspace is involved. Symptoms are painful paresthesias in the adjacent toes, induced by movement or weight-bearing. A chronic lesion often leads to swelling of the nerve because of fibrous tissue (Stewart, 1993; Tackmann *et al.*, 1989).

### Practical Management

Lesions caused by external compression usually recover spontaneously. Chronic lesions from masses in the popliteal fossa require surgical exploration and decompression of the nerve. Treatment of the *tarsal tunnel syndrome* consists of immobilization of the ankle with an orthosis and injections of local anesthetics in the tarsal tunnel for 8 weeks. If conservative treatment fails, surgical exploration and neurolysis is indicated, which often leads to permanent relief of symptoms (Oloff *et al.*, 1983).

Conservative therapy of *Morton's metatarsalgia* consists of padding the appropriate metatarsal heads. Local injections of anesthetics and corticosteroids are often successful (Greenfield *et al.*, 1984). If these measures fail, surgical intervention with excision of the nerve (Keh *et al.*, 1992) or incision of the deep intermetatarsal ligament (Dellon, 1992) is required. Both approaches have good results. Nashi *et al.* (1997) recommend excision of the neuroma through a dorsal approach, because of faster remobilization and less frequent occurrence of painful scars.

### Lateral Femoral Cutaneous Nerve

#### Clinical Aspects

The lateral cutaneous nerve of the thigh courses from the lumbar plexus across the iliacus muscle and passes under the lateral end of the inguinal ligament. Usually it crosses the ligament through a small tunnel in its lateral attachment to the anterior superior iliac spine. Anatomical variations are frequent. Because the nerve is entirely sensory, compression leads only to painful paresthesias and sensory loss in the skin of the anterolateral aspect of the thigh, called the syndrome of *meralgia paresthetica*. There are many causes of damage to

**TABLE IX   Disorders Causing Compression of the Lateral Cutaneous Nerve of the Thigh**

Retroperitoneal masses (hematoma, tumors)
Entrapment at the inguinal ligament (meralgia paresthetica)
External compression (tight clothes, belts, surgery in a prone position)

the nerve (Table IX). Proximally the nerve may be compressed by masses in the retroperitoneal space or the iliacus compartment as described earlier for the lumbar plexus. At the inguinal ligament, external compression occurs. This may occur from wearing tight clothes, belts, or trusses and sometimes from surgical procedures in the prone position. In most cases there is no identifiable cause, and the nerve is compressed by the inguinal ligament.

### Practical Management

Symptomatic lesions of the nerve caused by the retroperitoneal masses should be treated as described in the section on the lumbosacral plexus. In *meralgia paresthetica*, symptoms usually disappear spontaneously in a few weeks or months. Therefore, waiting for spontaneous recovery for several months supported by injections of local anesthetics or corticoids or both to relieve the pain is sufficient. Drugs should be injected under the fascia medial and below the anterior superior iliac crest. Constricting belts and binders should be avoided; in adipose patients, weight reduction may help. In a minority of patients, the symptoms persist or become so severe that surgical intervention is required (Williams and Trzil, 1991). The nerve should be decompressed by neurolysis with incision of the iliac fascia and the inguinal ligament (Tackmann *et al.*, 1989), which is highly effective. Sectioning of the nerve is effective in pain relief but may lead to recurrence of symptoms because of neuroma formation.

### Iliohypogastric, Ilioinguinal, and Genitofemoral Nerves

#### Clinical Aspects

The iliohypogastric and ilioinguinal nerves arise from the upper lumbar plexus and follow a course around the abdominal wall similar to the thoracoabdominal nerves. They run above the iliac crest, innervating the inguinal region and part of the genitalia. In their course around the abdomen the nerves may be affected by tumors of the kidneys or abscesses or hematomas in the paranephritic region. Mostly, compression results from scars of surgical incisions (e.g., appendectomy and gynecological operations). Entrapment of the ilioinguinal nerve very rarely occurs where the nerve passes through the muscles of the abdominal wall (Mumenthaler *et al.*, 1965). Symptoms are pain in the iliac fossa and inguinal region radiating into the genitals, sensory abnormalities in the distribution of the nerve, and tenderness below the inguinal ligament (Knockaert *et al.*, 1989).

The genitofemoral nerve passes through the psoas muscle and descends retroperitoneally on its surface to innervate the cremaster muscle and the skin of the scrotum or labium majus. Compression of the nerve results in most cases from scarring and adhesions from appendectomies or herniorrhaphies. Other causes are psoas abscess, direct trauma to the groin, and pressure from tight jeans. Symptoms are painful paresthesias in the distribution of the nerve.

### Practical Management

For symptomatic treatment of painful paresthesias, medications used for neuropathic pain and injections of local anesthetics, corticoids, or both should be tried; these treatments in combination with transcutaneous electrical nerve stimulation may be useful. Local nerve blocks are useful to distinguish between ilioinguinal and genitofemoral neuropathy. If conservative treatment fails, nerve resection is indicated, which produces good pain relief in most patients (Starling and Harms, 1989). Percutaneous rhizotomy may be a useful alternative (Wiegand *et al.*, 1986).

## COMPARTMENT SYNDROMES AND OTHER ISCHEMIC LESIONS OF NERVE AND MUSCLE

### Clinical Characteristics

Peripheral nerve lesions caused by ischemia may be accompanied by ischemic injury affecting muscle and in some instances the overlying skin. Possible causes of nerve ischemia include the following:

1. Interruption of the arterial blood supply by occlusion of the vessel by thrombus or embolus, external compression, or spasm induced by the intra-arterial injection of vasotoxic substances
2. Involvement of microcirculation within the compartment because of increased pressure within compartments encompassed by fascia (compartment syndrome) or by external compression (Table X)

We have excluded mononeuropathies or polyneuropathies associated with diabetes mellitus or the

**TABLE X** Causes of Ischemic Nerve Lesions

Compartment syndromes
  Increase in the intracompartmental volume (hemorrhage, edema, infusion)
  Reduction of the intracompartmental volume (e.g., closure of a fascial gap)
Circulatory impairment caused by external compression
  Constricting splints or bands
  Compression of the extremity caused by positioning during coma (sedative or drug overdose, anesthesia)
Arterial occlusion
  Ligation
  Intra-arterial injection of vasotoxic substances with induction of spasm or thrombosis
  Thrombotic or embolic occlusion

various connective tissue diseases from the following discussion.

### Compartment Syndromes

Regions of the body where muscles and their accompanying peripheral nerves are confined within tough fascial envelopes are particularly vulnerable to ischemic injury. Any increase in the intracompartmental pressure above the critical perfusion pressure of the tissues within the compartment may be followed by initially reversible and later irreversible ischemic injury of those nerves that pass through the compartment, as well as muscles lying within the compartment. Increases in the intracompartmental pressure may follow hemorrhage or edema within the compartment, usually as a consequence of a fracture or, in some instances, operations such as repositioning osteotomies and, much less often, as a complication of coagulopathy. Even vigorous exercise in the presence of a perhaps constitutionally narrow compartment may so critically raise the intracompartmental pressure as to induce ischemic symptoms and signs from tissues lying within the compartment. The initial symptoms are pain, followed by swelling and induration in the region of the compartment. Ischemia of cutaneous nerves that pass through the compartment is signaled by paresthesiae and later numbness in the cutaneous territory of the affected nerve. In the case of motor nerve fibers, there is weakness to which must be added the weakness resulting from the ischemia and perhaps infarction of the muscles themselves within the compartment. Usually, except for rare cases in which the intracompartmental pressure is very high, arterial pulses distal to the compartment remain palpable. Serum muscle enzymes such as the creatine phosphokinase level are elevated to levels that roughly parallel the extent of the muscle necrosis.

Failure to recognize a compartment syndrome may delay treatment long enough to result in infarction and subsequent necrosis of muscles lying within the compartment. The ensuing secondary fibrosis shortens the affected muscles and produces contractures with sometimes severe limitations in the range of movement possible at affected joints (Figure 6).

The two most commonly involved compartments include the deep extensor and flexor compartments of the calf and the flexor compartment of the forearm (Dawson *et al.*, 1999; Matsen, 1980; Mumenthaler *et al.*,1998; Echtermeyer, 1984; Stöhr and Riffel, 1988).

### Circulatory Impairment Caused by External Compression

Bandages, pneumatic splints, antishock trousers wrapped about the limb, or prolonged compression of the arm or leg beneath the body, as commonly happens in comatose patients, may impair the capillary-venous circulation within the compartment. This results in consequences similar to the compartment syndromes, with the added feature of affecting the overlying skin and eventually causing necrosis of the skin as well.

### Failure of the Arterial Blood Supply to the Compartment

Occlusion by thrombosis, embolus, or obstruction of the arterial blood supply by any other means may also cause compartment syndrome. In addition, other signs are usually present such as pallor or blanching of the skin, loss of pulses, and even gangrene (Volkmann's ischemic contracture).

### Prevention and Conservative Treatment

#### The Compartment Syndromes

With accidents and operations known to predispose to the development of the compartment syndrome, measures should be adopted to ensure maintenance of an adequate perfusion pressure to the tissues within the compartment. These include the following:

1. The prevention and treatment of shock
2. Avoidance and removal of occlusive bandages or splints
3. Horizontal positioning of the limb, because elevation induces a lowering of the arterial pressure.

Symptoms caused by the chronic anterior tibial syndrome for the most part are avoidable by ceasing such intensive activities as jogging or extended walks. In some of these cases, some physicians advocate prophylactic fasciotomy (discussed later).

**FIGURE 6**   Volkmann's contracture of the right hand.

## Circulatory Failure from External Compression and Occlusion of the Arterial Blood Supply

The use of pneumatic splints, antishock pants, and circular (or in any way compressive) bandages should be avoided if there is any risk of increasing the intracompartmental pressure. Special care should be taken in cases in which there has been extensive trauma, because the development of posttraumatic edema may be delayed and develop slowly. In such cases monitoring the intracompartmental pressure with a wick catheter may be of help to identify a significant increase in intracompartmental pressure and lead to early surgical prophylactic decompression of the compartment. For cases of acute arterial ischemia, treatment options include anticoagulation and the use of fibrinolytic agents. In those cases caused by the accidental injection of a vasoconstrictive and toxic agent, the following course of management may be effective in helping to restore the circulation and preventing or limiting the damage caused by ischemia (Chase, 1972; Stöhr et al., 1980):

1. Leave the intra-arterial needle or catheter in place
2. Intra-arterial injection of 10 ml procaine (1%) or lidocaine (0.25%), both without adrenaline
3. Intra-arterial injection of 40–80 mg papaverine dissolved in 10–20 ml of isotonic NaCl
4. Intra-arterial injection of 250 mg methylprednisolone
5. Heparin, 10,000 IU IV, followed by 25,000 IU/day by continuous infusion for 1 week, followed by wafarin treatment for several months

## Operative Treatment

### Compartment Syndromes

If, despite conservative management, a compartment syndrome develops, immediate fasciotomy is usually required (Matsen, 1980; Szyszkowitz and Reschauer, 1982). This should be carried out in such a way as to ensure adequate reduction of the intracompartmental pressure in all compartments involved. In the case of the calf, this often involves release of the anterior tibial, peroneal, superficial, and deep flexor compartments (Echtermeyer et al., 1982). Removal of necrotic tissue may help to create additional space and to reduce the risk of infection.

Rehabilitation may be helped by later lengthening tenotomies, tendon transfers, and possibly neurolysis of nerves encased in scar tissue. In the case of the chronic anterior tibial syndrome, a limited fasciotomy of the middle third of the anterolateral calf may be helpful (Echtermeier, 1984).

## Circulatory Impairment Caused by External Compression

Ischemic injury to nerves and muscles caused by external compression of a limb may continue to progress even after the source of the compression (for example, a plaster of paris cast) has been removed because of postischemic edema within the compartment. In such cases, fasciotomy is indicated in most instances.

## Circulatory Impairment Caused by Arterial Occlusion

Depending on the specific cause, thrombectomy, embolectomy, or reconstruction of the arteries may be indicated. However, the development of postischemic edema and the ensuing subfascial increase in pressure within one or more compartments may require fasciotomies in some or all of the involved compartments.

## REFERENCES

Aita, J. (1984). An unusual compressive neuropathy. *Arch. Neurol.* **41**, 341.

Alexander, J. W., and Dellinger, E. P. (1991). Surgical infections and choice of antibiotics. In "Textbook of Surgery: The Biological Basis of Modern Surgical Practice" (D. C. Sabiston, ed.), pp. 221–236. Saunders, Philadelphia.

Angerer, M., Kleudgen, S., Grigo, B., and Bogdahn, U. (2000). Hochauflösende Sonographie des Nervus medianus—neue sonomorphologische Untersuchungstechnik peripherer. *Nerven. Klin. Neurophysiol.* **31**, 53–58.

Assmuss, H. (1994). Simple decompression of the ulnar nerve in cubital tunnel syndrome with and without morphologic changes. *Nervenarzt* **65**, 846–853.

Basset, F. H., and Nunley, J. A. (1982). Compression of the musculocutaneous nerve at the elbow. *J. Bone Joint Surg.* **64**, 1050–1052.

Benecke, R., and Conrad, B. (1980). The value of electrophysiological examination of the flexor carpi ulnaris muscle in the diagnosis of ulnar nerve lesions at the elbow. *J. Neurol.* **223**, 207–217.

Benini, A., and Di Martino, E. (1976). Die Schädigung des Ramus profundus nervi radialis (Supinator-Syndrom). Die dissoziierte Radialislähmung vom proximalen Unterarmtyp. *Schweiz Med. Wschr.* **106**, 639–643.

Berry, H., and Richardson, P. M. (1976). Common peroneal nerve palsy: A clinical and electrophysiological review. *J. Neurol. Neurosurg. Psychiatry* **39**, 1162–1171.

Bimmler, D., and Meyer, V. E. (1996). Surgical treatment of the ulnar nerve entrapment neuropathy: submuscular anterior transposition or simple decompression of the ulnar nerve? Long-term results in 79 cases. *Ann. Chir. Mem. Superieur* **15**, 148–157.

Bradshaw, D. Y., and Shefner, J. M. (1999). Ulnar neuropathy at the elbow. *Neurol. Clin.* **17**, 447–461.

Brown, W. F., Yates, S. K., and Ferguson, G. G. (1980). Cubital tunnel syndrome and ulnar neuropathy. *Ann. Neurol.* **7**, 289–290.

Buchthal, F., Rosenfalck, A., and Trojaborg, W. (1974). Electrophysiological findings in entrapment of the median nerve at wrist and elbow. *J. Neurol. Neurosurg. Psychiatry* **37**, 340–360.

Cahill, B. R., and Palmer, R. E. (1983). Quadrilateral space syndrome. *J. Hand Surg.* **8**, 65–69.

Chase, R. A. (1972). Surgery of the hand. *N. Engl. J. Med.* **287**, 1227–1234.

Chiu, W. S. (1976). The syndrome of retroperitoneal hemorrhage and lumbar plexus neuropathy during anticoagulant therapy. *South. Med. J.* **69**, 595–599.

Dale, W. A., and Lewis, M. R. (1975). Management of thoracic outlet syndrome. *Ann. Surg.* **118**, 575–585.

Dawson, D. M., Hallet, M., and Wilbourn, A. J. (1999). "Entrapment Neuropathies," 3rd ed. Lippincott-Raven, Philadelphia.

de Krom, M. C. T. F. M., Knipschild, G., Kester, A. D. M., Thijs, C. T., Boekkooi, P. F., and Spaans, F. (1992). Carpal tunnel syndrome: Prevalence in the general population. *J. Clin. Epidemiol.* **45**, 373–376.

Dellon, A. L. (1989). Review of treatment results for ulnar nerve compression at the elbow. *J. Hand Surg.* **14**, 688–699.

Dellon, A. L. (1992). Treatment of Morton's neuroma as a nerve compression: The role for neurolysis. *Am. J. Podiatr. Med. Assoc.* **82**, 399–402.

Dellon, A. L., Hament, W., and Gittelshon, A. (1993). Nonoperative management of cubital tunnel syndrome: An 8-year prospective study. *Neurology* **43**, 1673–1677.

Donaghy, M. (1993). Lumbosacral plexus lesions. *In* "Peripheral Neuropathy" (P. J. Dyck, P. K. Thomas, J. W. Griffin, P. A. Low, and J. F. Poduslo, Eds.), pp. 951–960. Saunders, Philadelphia.

Du Toit, G. T., and Levy, S. J. (1967). Transposition of the latissimus dorsi for paralysis of triceps brachii. *J. Bone Joint Surg. B* **49**, 135–137.

Echtermeyer, V. (1984). "The Compartment Syndrome: Diagnosis and Therapy: A Clinical and Experimental Study." Springer-Verlag, Berlin.

Echtermeyer, V., Muhr, G., Oestern, H. J., and Tscherne, H. (1982). Surgical treatment of the compartment syndrome. *Unfallheilkunde* **85**, 144–152.

Eckmann, P. B., Perlstein, G., and Atrocchi, P. H. (1975). Ulnar neuropathy in bicycle riders. *Arch. Neurol.* **32**, 130–131.

Eisen, A., and Danon, J. (1974). The mild cubital tunnel syndrome: Its natural history and indications for surgical intervention. *Neurology (Minneapolis)* **24**, 608–613.

Eisen, A., Schomer, D., and Melmed, C. (1977). The application of F-wave measurements in the differentiation of proximal and distal upper limb entrapments. *Neurology* **27**, 662–668.

Etheredge, S., Wilbur, B., and Stoney, R. J. (1979). Thoracic outlet syndrome. *Am. J. Surg.* **138**, 175–182.

Galbraith, K. A., and McCullough. C. J. (1979). Acute nerve injury as a complication of closed fracture or dislocations of the elbow. *Injury* **11**, 159–164.

Gelberman, R. H., Hergenroeder, P. T., Hargens, A. R., Lundborg, G. N., and Akeson, W. H. (1981). The carpal tunnel syndrome: A study of carpal canal pressures. *J. Bone Joint Surg.* **63A**, 380–383.

Gerl, A., and Schlüter, R. (1980). Postoperative Kompression und Regeneration nach Ulnarisverlagerung. *Zbl. Neurochir. (Leipzig)* **41**, 149–166.

Gilliatt, R. W. (1984). "Thoracic Outlet Syndromes. Peripheral Neuropathy" (P. J. Dyk, P. K. Thomas, E. H. Lambert, and R. Bunge, Eds.), Vol. II, pp. 1409–1424. Saunders, Philadelphia.

Gilliatt, R. W., Willison, R. G., Dietz, V., and Williams, I. R. (1978). Peripheral nerve conduction in patients with a cervical rib and band. *Ann. Neurol.* **4**, 124–129.

Greenfield, J., Rea, J., and Ilfeld, F. (1984). Morton's interdigital neuroma: Indications for treatment by local injections versus surgery. *Clin. Orthop.* **185**, 142–144.

Hill, N. A., Howard, F. M., and Huffer, B. R. (1985). The incomplete anterior interosseus nerve syndrome. *J. Hand Surg.* **10**, 14–16.

Hopf, H. C., and Hense, W. (1974). Anomalien der motorischen Innervation der Hand. *EEG EMG* **5**, 220–224.

House, J. H., and Ahmed, K. (1977). Entrapment neuropathy of the infrapatellar branch of the saphenous nerve: A new peripheral entrapment syndrome? *Am. J. Sports Med.* **5**, 217–224.

Jaeckle, K. A., Young, D. F., and Foley, K. M. (1985). The natural history of lumbosacral plexopathy in cancer. *Neurology* **35**, 8–15.

Kaplan, S. J., Glickel, S. Z., and Eaton, R. G. (1990). Predictive factors in the non-surgical treatment of carpal tunnel syndrome. *J. Hand Surg.* **15**, 106–108.

Keh, R. A., Ballew, K. K., Higgins, K. R., Odom, R., and Harkless, L. B. (1992). Long-term follow-up of Morton's neuroma. *J. Foot Surg.* **31**, 93–95.

Kelly, T. R. (1979). Thoracic outlet syndrome: Current concepts of treatment. *Ann. Surg.* **190**, 657–662.

Kiloh, L., and Nevin, S. (1952). Isolated neuritis of the anterior interosseus nerve. *BMJ* **1**, 850–851.

Kimura, J., and Ayyar, D. R. (1985). The carpal tunnel syndrome: Electrophysiological aspects in 639 symptomatic extremities. *Electromyogr. Clin. Neurophysiol.* **25**, 151–164.

Kincaid, J. C. (1988). AAEE minimonograph #3: The electrodiagnosis of ulnar neuropathy at the elbow. *Muscle Nerve* **11**, 1005–1115.

Knockaert, D. C., D'Heygere, F. G., and Bobbaers, H. J. (1989). Ilioinguinal nerve entrapment: A little-known cause of iliac fossa pain. *Postgrad. Med. J.* **65**, 632–635.

Kopell, H., and Thompson, W. A. L. (1960). Knee pain due to saphenous nerve entrapment. *N. Engl. J. Med.* **263**, 351–353.

Kopell, H. R., and Thompson, W. H. L. (1963). "Peripheral Entrapment Neuropathies." Williams & Wilkins, Baltimore.

Laha, R. K., Lunsford, L. D., and Dujovny, M. (1978). Lacertus fibrosus compression of the median nerve. *J. Neurosurg.* **48**, 838–841.

Lanz, U. (1979). Ischemic necrosis of muscle. *Unfallheilkunde* **139**, 7–57.

Learmonth, J. R. (1942). A technique for transplanting the ulnar nerve. *Surg. Gynecol. Obstet.* **75**, 792–793.

Leffert, R. D. (1982). Anterior submuscular transposition of the ulnar nerves by the Learmonth technique. *J. Hand Surg.* **7**, 147–155.

Lister, G. D., Belsole, R. B., and Kleinert, H. E. (1979). The radial tunnel syndrome. *J. Hand Surg.* **4**, 52–59.

Loong, S. C. (1977). The carpal tunnel syndrome: A clinical and electrophysiological study of 250 patients. *Proc. Aust. Assoc. Neurol.* **14**, 51–65.

Luerssen, T. G., Campbell, R. L., Defalque, R. J., and Worth, R. M. (1983). Spontaneous saphenous neuralgia. *Neurosurgery* **13**, 238–241.

Lugnegard, H., Juhlin, L., and Nilsson, B. Y. (1982). Ulnar neuropathy at the elbow treated with decompression. *Scand. J. Plast. Reconstr. Surg.* **16**, 195–200.

Lugnegard, H., Walheim, G., and Wennberg, A. (1977). Operative treatment of ulnar nerve neuropathy in the elbow region. *Acta Orthop. Scand.* **48**, 168–176.

McLellan, D. L., and Swash, M. (1976). Longitudinal sliding of the median nerve during movements of the upper limb. *J. Neurol. Neurosurg. Psychiatry* **39**, 566–570.

Meese, W., Mauersberger, W., Grumme, T., and Hopfenmüller, W. (1980). Der Wert der mikrochirurgischen Operationstechnik beim Karpaltunnelsyndrom. *Akt. Neurol.* **7**, 9–15.

Merrick, H. W., Zeiss, J., and Woldenberg, L. S. (1991). Percutaneous decompression for femoral neuropathy secondary to heparin-induced retroperitoneal hematoma: Case report and review of the literature. *Am. Surg.* **57**, 706–711.

Miller, R. G. (1991). Ulnar neuropathy at the elbow. *Muscle Nerve* **14**, 97–101.

Miller, R. G., and Hummel, E. E. (1980). The cubital tunnel syndrome: Treatment with simple decompression. *Ann. Neurol.* **7**, 567–569.

Mumenthaler, A., Mumenthaler, M., Luciani, G., and Kramer, J. (1965). Das Ilioinguinalis-Syndrom. *Dtsch. Med. Wochenschr.* **90**, 1073–1078.

Mumenthaler, M., Baasch, E., and Ulrich, J. (1960). The anterior-tibial syndrome. Paresis of foot dorsiflection with an underlying vascular cause. *Schweizer Arch. Neurol. Neurochir. Psychiat.* **86**, 137–181.

Mumenthaler, M., and Schliack, H. (1991). "Peripheral Nerve Lesions: Diagnosis and Therapy." Thieme, Stuttgart and New York.

Mumenthaler, M., Schliack, H., and Stöhr, M. (1998). "Läsionen peripherer Nerven und radikuläre Syndrome." 7. Aufl. Thieme, Stuttgart.

Nashi, M., Venkatachalam, A. K., and Muddu, B. N. (1997). Surgery of Morton's neuroma: dorsal or plantar approach? *J. Roy. Coll. Sur. Edinb.* **42**, 36–37.

Noth, J., Dietz, V., and Mauritiz, K. H. (1980). Cyclists palsy. *J. Neurol. Sci.* **47**, 111–116.

Oh, S. J., and Meyer, R. D. (1999). Entrapment neuropathies of the tibial (posterior tibial) nerve. *Neurol. Clin.* **17**, 593–617.

Oloff, L. M., Jacobs, A. M., and Jaffe, S. (1983). Tarsal tunnel syndrome: A manifestation of systemic disease. *J. Foot Surg.* **22**, 302–307.

O'Malley, M. J., Evanoff, M., Terrono, A. L., and Millender, L. H. (1992). Factors that determine reexploration treatment of carpal tunnel syndrome. *J. Hand Surg.* **17**, 638–641.

Omer, G. E., and Spinner, M. (1980). "Management of Peripheral Nerve Problems." Saunders, Philadelphia.

Reill, P. (1982). Compartment syndrome of the upper extremity and its sequelae: Operative therapy. *Unfallheilkunde* **85**, 153–158.

Roos, D. B. (1982). The place for scalectomy and first rib resection in thoracic outlet syndrome. *Surgery* **92**, 1077–1085.

Roos, D. B. (1990). The thoracic outlet syndrome is underrated. *Arch. Neurol* **47**, 327–328.

Rudigier, J., and Degreif, J. (1986). Ersatzoperationen am Oberarm bei nicht mehr regenerationsfähigen Lähmungen. *Akt. Neurol.* **13**, 109–113.

Salazar-Grueso, E., and Roos, R. (1986). Sciatic endometriosis: A treatable sensorimotor mononeuropathy. *Neurology* **36**, 1360–1363.

Sanders, R. J., Monsour, J. W., Gerber, W. F., Adams, W. R., and Thompson, N. (1979). Scalenectomy versus first rib resection for treatment of the thoracic outlet-syndrome. *Surgery* **85**, 109–121.

Sedal, L., McLeod, J. G., and Walsh, J. C. (1973). Ulnar nerve lesions associated with the carpal tunnel syndrome. *J. Neurol. Neurosurg. Psychiatry* **36**, 118–123.

Seror, P. (1993). Treatment of ulnar nerve palsy at the elbow with a night splint. *J. Bone Joint Surg.* **75**, 322–327.

Seror, P. (1999). Electrodiagnostic examination of the anterior interosseus nerve. Normal and pathologic data (21 cases). *Electromyogr. Clin. Neurophys.* **39**, 183–189.

Shea, D. J., and McClain, E. J. (1969). Ulnar-nerve-compression syndromes at and below the wrist. *J. Bone Joint Surg.* **51**, 1095–1102.

Silverstein, A. (1979). Neurological complications of anticoagulation therapy. A neurologist's review. *Arch. Intern. Med.* **139**, 217–220.

Staal, A., van Gijn, J., and Spaans, F. (1999). "Mononeuropathies." Saunders, London.

Starling, J. R., and Harms, B. A. (1989). Diagnosis and treatment of genitofemoral and ilioinguinal neuralgia. *World J. Surg.* **13**, 586–591.

Stewart, J. D. (1993). "Focal Peripheral Neuropathies." Raven, New York.

Stille, D. (1974). Die distale Radialisparese (Supinator-Syndrom). *Akt. Neurol.* **1**, 5–11.

Stevens, J. C. (1987). AAEE minimonograph #26: The electrodiagnosis of carpal tunnel syndrome. *Muscle Nerve* **10**, 99–113.

Stöhr, M. (1996). "Iatrogene Nervenläsionen," 2nd ed. Thieme, Stuttgart and New York.

Stöhr, M. (1998). "Atlas der klinischen Elektromyographie und Neurographie," 4th ed. Kohlhammer, Stuttgart.

Stöhr, M., Dichgans, J., and Dörstelmann, D. (1980). Ischemic neuropathy of the lumbosacral plexus following intragluteal injection. *J. Neurol. Neurosurg. Psychiatry* **43**, 489–494.

Stöhr, M., and Riffel, B. (1988). "Nerve and Nerve Root Lesions." Edition Medizin VCH, Weinheim, Germany.

Stöhr, M., and Scheglmann, K. (1999). "Zervikogene Amyotrophie" durch Vorderhorninfarkt. *Akt. Neurologie* **26**, 283–285.

Suematsu, N., and Hirayama, T. (1998). Posterior interosseous nerve palsy. *J. Hand Surg.* **23B**, 104–106.

Sunderland, S. (1978). "Nerves and Nerve Injuries." Churchill-Livingstone, Edinburgh, London, and New York.

Swash, M., and Schwartz, M. S. (1997). "Neuromuscular Diseases," 2nd ed. Springer, London.

Szyszkowitz, R., and Reschauer, R. (1982). Etiology, pathophysiology and localisation of the compartment syndrome. *Unfallheilkunde* **85**, 126–132.

Tackmann, W., Richter, H. P., and Stöhr, M. (1989). "Kompressionssyndrome peripherer Nerven." Springer, Berlin and Heidelberg.

Thomas, J. E., Lambert, E. H. and Cseuz, K. A. (1967). Electrodiagnostic aspects of the carpal tunnel syndrome. *Arch. Neurol. (Chicago)* **16**, 635–641.

Trojaborg, W. (1970). Rate of recovery in motor and sensory fibres of the radial nerve: Clinical and electrophysiological aspects. *J. Neurol. Neurosurg. Psychiatry* **33**, 625–638.

Tysvaer, A. T. (1982). Computerized tomography and surgical treatment of femoral compression neuropathy. *J. Neurosurg.* **57**, 137–139.

Wartenberg, R. (1954). Digitalgia paresthetica and gonyalgia paresthetica. *Neurology* **4**, 106–115.

Wiegand, H., Renella, R., and Hussein, S. (1986). Das Ilioinguinaliskompressionssyndrom und seine Therapie durch perkutane Rhizotomie. *Akt. Neurol.* **13**, 58–60.

Wilbourn, A. J. (1986). AAEE Case Report 12: Common peroneal mononeuropathy at the fibular head. *Muscle Nerve* **9**, 825–836.

Wilbourn, A. J. (1990). The thoracic outlet syndrome is overdiagnosed. *Arch. Neurol.* **47**, 328–330.

Wilhelm, A., and Suden, R. (1985). Das proximale Radialiskompressionssyndrom: Behandlung und Ergebnisse. *Handchirurgie* **17**, 219–224.

Williams, P. H., and Trzil, K. P. (1991). Management of meralgia paresthetica. *J. Neurosurg.* **74**, 76–80.

Wilson, D. H., and Krout, R. (1973). Surgery of the ulnar neuropathy at the elbow: 16 cases treated by decompression without transposition. *J. Neurosurg.* **38**, 780–785.

Worth, R. M., Kettelkamp, D. B., Defalque, R. J., and Underwood, D. K. (1984). Saphenous nerve entrapment: A case of medial knee pain. *Am. J. Sports Med.* **12**, 80–81.

Yiannikas, C., and Walsh, J. C. (1983). Somatosensory evoked responses in the diagnosis of thoracic outlet syndrome. *J. Neurol. Neurosurg. Psychiatry* **46**, 234–240.

Young, M. R., and Norris, J. W. (1976). Femoral neuropathy during anticoagulant therapy. *Neurology* **26**, 1173–1175.

# CHAPTER 92
# Nerve Injury

Rolfe Birch and Praveen Anand

## CLINICAL ASPECTS

Nerves can be damaged in a number of ways: physical agents such as traction, severance, injection, cold, heat, ionizing radiation; infection, inflammation, ischemia; drugs and metals; tumors, and the effects of systemic disease. This chapter will focus on nerve trauma. The trauma may be acute or chronic; single, intermittent, or continuing. The lesion may affect the whole or part of the nerve and vary with respect to the motor, sensory, and autonomic fiber subtypes involved. The nerve itself may be healthy, or it may be affected by a pre-existing neuropathy. The damage to the nerve may be "closed" or "open" through a wound of the skin. The term "compound nerve injuries" was proposed (Birch *et al.*, 1998) for those in whom damage to nerves is associated with major damage to other tissues or organs such as skin, muscle, skeleton, viscera, or large blood vessels at the same site. The International Red Cross Wound Classification (Coupland, 1993) is particularly valuable in the analysis of such injuries in the field of war.

### Classification of Nerve Injuries

Seddon (1943) distinguished three types of nerve injury:

1. Neurapraxia—"nerve not working"—myelin damage, conduction block
2. Axonotmesis—"axon cutting"—a degenerative lesion of the axon with basal lamina of Schwann cell intact; no conduction
3. Neurotmesis—"nerve cutting"—a degenerative lesion in which all elements of the nerve are interrupted; no conduction.

None of these classifications fully explain lesions of the brachial plexus or lumbosacral plexus in which there may be intradural rupture of the spinal nerve roots (Figure 1).

Sunderland (1951) introduced a rather more elaborate system of classifying injury. Five degrees of severity were named, ranging from simple conduction block to loss of continuity. We tend to a further simplification, to classification as "degenerative and nondegenerative lesion," a concept elaborated by Thomas and Holdorff (1993) (Table I).

With neurapraxia and axonotmesis, the prognosis is favorable if the cause is removed. Neurapraxic lesions usually improve within days or weeks, rarely as late as 6 months. Both axonotmesis and neurotmesis are degenerative lesions. The clinical features may be identical at presentation, and only time or exposure of the nerve permit distinction. If the lesion leaves the Schwann cell basal lamina intact, axonotmesis, the pathways for axonal regeneration remain (Thomas, 1964; Young, 1949). Guttman and Sanders (1943) showed that "only after crushing as opposed to section was the nerve fully reconstituted." With neurotmesis, recovery can occur only after nerve repair and will usually be imperfect. Furthermore, a favorable lesion may progress to one much less favorable if the cause is not dealt with. If a nerve is accidentally encircled by suture or crushed under a plate, recovery is likely if the cause is urgently removed; if this remains for days or weeks, the outlook is very different. Nerves crushed in the swollen ischemic limb or in a tense compartment progress from conduction block to much less favorable degenerative lesions.

### Anatomical Factors

The structure of peripheral nervous tissue is one of nerve fibers (axons, Schwann cell units) suspended in

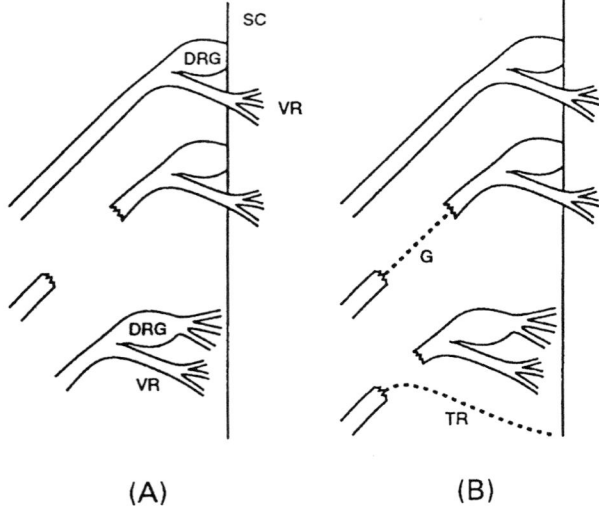

(A)                    (B)

FIGURE 1  Schematic representation of brachial plexus injuries (A, left) and their repair (B, right). DRG, dorsal root ganglion; VR, ventral root; SC, spinal cord; G, nerve graft; TR, nerve transfer. Before repair (A): (i) intact root; (ii) rupture; (iii) avulsion. After repair (B): (i) intact root; (ii) repair of rupture by nerve grafting; (iii) avulsion repaired by nerve transfer. From Berman *et al.*, 1998, with permission.

**TABLE I**  Classification of Focal Mechanical Nerve Injury

1. Focal conduction block
    Transient
        Ischemic
        Other
    More persistent
        Demyelinating
        Axonal constriction
2. Axonal degeneration
    With preservation of basal laminal sheaths of nerve fibers
    With partial section of nerve
    With complete transection of nerve

Derived from Thomas, P. K., Holdorff, B. (1993). Neuropathy due to physical agents. *In* "Peripheral Neuropathy," 3rd ed. (P. J. Dyck, P. K. Thomas, J. W. Griffin, P. A. Low, and J. F. Poduslo, Eds.), p. 991. W. B. Saunders, London. With permission.

collagen-rich extracellular space (Berthold *et al.*, 1993). The spinal nerve roots and rootlets within the spinal canal are fragile and easily damaged; they have lost the supporting envelope of the epineurium, nerve fibers are more densely packed, and extrinsic blood vessels are scanty. Supporting structures outside the foramina make the nerves here more robust: the epineurium, a prolongation of the dural sleeve, is composed of longitudinally directed collagen fibers and fibroblasts (Gamble and Eames, 1964), and the perineurium, which ensheaths the nerve fascicles, of flattened cell processes alternating with layers of collagen.

In brachial plexus injury, there are two salient anatomical features (Figure 1). First, there may be intradural rupture of the spinal nerve roots, which leads to degeneration of somatic efferent and preganglionic autonomic efferent fibers and central processes of sensory fibers; however, the distal axons of somatic afferents remain intact. The latter are electrophysiologically "functional" and continue to mediate cutaneous axon-reflex vasodilatation (Bonney and Gilliatt, 1958). Second, transition from central to peripheral nervous structures takes place in the rootlets, or, less often, in the roots of the spinal nerves in the transitional region. CNS extrudes into the rootlets in a cone-shaped arrangement; thus "each transitional region can be subdivided into an axial central nervous system compartment and a surrounding peripheral nervous system compartment" (Berthold *et al.*, 1993). The axons traverse this region; the blood vessels of the nerves do

not. In traction lesions of the adult brachial plexus the rupture is usually peripheral to the transitional zone (Schenker and Birch, 2001), a finding of significance in the operative repair of these lesions. However, some avulsion injuries cause damage central to the transitional zone.

The topographical arrangement of bundles along the course of the nerve trunk has been extensively mapped (see Sunderland, 1993), showing branching, fusion, and changes in numbers. There is evidence of a degree of topographical segregation of nerves according to function and destination over considerable lengths of the trunk, which is important for diagnosis and repair.

The segmental blood supply is both from without and the axial vessels within the nerve. The intrinsic epineurial, perineurial, and endoneurial plexus and the extrinsic regional vessels in the perineurium form interconnected microvascular systems. McManus *et al.* (1993) observe that "anastomotic vessels confer a resistance to ischaemia in peripheral nerves so that nerves suffer functional or structural changes only when there is widespread vascular or microvascular damage." Clinical examples demonstrate the robust nature of the intrinsic blood supply. In one patient the muscles of the posterior compartment of the thigh were excised for gas gangrene. The sciatic nerve was maintained by the longitudinal supply over the length of 30 cm. There was useful recovery. However, such examples cannot make a general rule. Bell and Weddell (1984) showed that such maneuvers greatly affected blood supply to nerves, and later clinical experience has shown that it is preferable to bridge a gap by interposition of grafts rather than by mobilization. Isolated cranial nerve lesions, such as acute oculomotor nerve palsy in diabetic patients, have been attributed to ischemia and infarction after occlusion of the vasa nervorum.

## Examination and Diagnosis

The early symptoms of nerve injury include abnormal sensations, alteration or loss of sensibility, weakness, motor paralysis, impairment of function, and pain. A patient may note warming and dryness of the affected skin. Pain is a most important symptom; its onset at injury and persistence may suggest that the cause is continuing, such as when nerve is stretched over bone projection, compressed by a hard object, constricted by suture, or involved in an expanding hematoma.

Careful clinical examination and special investigations are critical in the diagnosis of nerve injury. The clinical presentation is of flaccid weakness and wasting, with decreased or absent deep tendon reflexes. Sensory examination may reveal lowered or elevated threshold to one or more modalities; sensory thresholds may be lowered outside the territory of the injured nerve, as a result of changes in spinal cord (secondary allodynia or hyperalgesia). The strongly positive Tinel's sign over a lesion soon after injury indicates rupture of a nerve. Failure of distal progression of Tinel's sign in a closed lesion similarly indicates rupture or at least a lesion that will not recover spontaneously. Autonomic and trophic features may be present.

There are many pitfalls in the diagnosis of injuries to peripheral nerves. Patients may be rendered insensible by drugs or alcohol or injury; examination of the fearful young child in pain is certainly difficult. The site and the nature of the wound is relevant. If there is a wound over the line of a main nerve and if there is any suggestion of loss of sensibility of motor function in the distribution of that nerve, it must be considered as having been transected until proven otherwise. High-energy transfer injuries leading to open fractures, penetrating missile injuries, and wounds, either accidental or surgical, are often complicated by serious lesions of nerves. In closed injuries the presence of swelling and of linear bruising suggests expanding hematoma or rupture of an axial structure. Anhidrosis and loss of vasomotor tone in the skin of the affected part indicates that the lesion must be a degenerative one. In the young child the anesthetic hand or digits may be wholly excluded from function. Testing of sensibility and measurement of motor power is certainly difficult soon after wounding, particularly so when the nerve is injured by fracture or dislocation. In late cases diagnosis is usually straightforward. The changes of disuse appear. There is thinning of the skin and even ulceration from accidental injury. There is atrophy of the tips of the digits with loss of skin marking. There is unmistakable wasting, there may be even stiffness and contractures of joints. In the growing child early disturbance of growth is apparent, especially after birth injuries to the brachial plexus or injuries to major trunk nerves in the lower limb.

## Assessment of Nerve Dysfunction

Clinical neurophysiological studies are important in the detection of nerve injury and any associated generalized neuropathy; they often provide insight into underlying structural and functional mechanisms. Electrophysiological examination is no substitute for clinical observation, rather it confirms clinical diagnosis, detects subclinical changes, and helps in defining the site and degree of the lesion and in the recognition and recording of recovery. Perhaps the most important contribution of these investigations is to distinguish between degenerative and nondegenerative lesions and to reveal conduction block (Smith, 1998). In neurapraxia, there is slowing or conduction block across the affected segment, but conduction is preserved in proximal and distal nerve segments, as is the compound muscle action potentials on stimulation distal to the injury (even after a period of 2 weeks or more). Fibrillations and positive sharp waves on electromyographic examination indicate Wallerian degeneration (present in both axonotmesis and neurotmesis), and by 2 weeks, there is a decline or loss of response to nerve stimulation below the site of injury. Neuromuscular transmission fails earlier than nerve conduction. Quantitative sensory and autonomic testing may provide information about small myelinated and unmyelinated nerve fibers, as well as large fiber function. These tests can be particularly useful in patients with mechanical or thermal hypoalgesia, hyperalgesia, and allodynia, particularly when other investigations are unhelpful, by providing objective assessment of abnormalities within and outside the territory of the affected nerve and by establishing selective involvement of small nerve fibers. Quantitative sensory testing is also useful for epidemiological and therapeutic studies.

## Iatropathic Nerve Injuries

Iatropathic injury of peripheral nerves is a persistent and probably increasing problem. A recent review (Khan and Birch, 2001) recorded a fivefold increase in cases referred to one Unit in the United Kingdom over a period of 10 years. Even more disturbing was the finding that the average time from the incident to establishing diagnosis and taking action had increased from 6 to 10 months.

Nerves are injured by doctors in many ways (reviewed in Birch *et al.*, 1998)

1. Drug-induced neuropathy
2. Drug-induced disturbance of coagulation leading to intraneural or extraneural hemorrhage

3. Injection of a noxious substance into or near a major nerve
4. Pressure from extraneural hemorrhage from arterial puncture
5. By the production of, or failure to recognize, ischemia of a limb
6. By closed pressure or traction during general anesthesia
7. By traction during manipulation under anesthesia or during delivery
8. By direct damage from pressure, traction, heat, clamping, or cutting in the field of operation
9. By ionizing radiation

The most serious damage is usually inflicted during operation by cutting, stretching, or burning. The nerves principally at risk are those of the brachial plexus, the spinal accessory, the radial, the common peroneal, and the ulnar. The femoral, sciatic, and obturator nerves are particularly at risk during arthroplasty of the hip. We may take the spinal accessory nerve as an example. It is usually damaged by a surgeon. Diagnosis is usually delayed. The lesion is crippling. Only exceptionally are urgent efforts made to rectify the situation. Williams *et al.* (1996) made a detailed study of more than 40 cases. Injury to the accessory nerve resulted in a characteristic syndrome: reduced shoulder, abduction, drooped shoulder and pain. Pain in 26 patients was severe at rest and even disturbed sleep. The pain was of two types: appearing as neuralgic pain from the damage to the sensory fibers in the nerve, and a constant dragging pain from the loss of one of the principal muscles supporting the forequarter. When the nerve has been damaged during operation performed with the patient under local anesthesia for lymph node biopsy, patients experience a sharp painful "electric shock" at the time of injury to the nerve. Repair of the nerve reduced symptoms in most cases. One remarkable finding was the speedy resolution of neuropathic pain after repair of the nerve many months before there was functional recovery into the paralyzed trapezius muscle. In this series, delay in diagnosis was usual, ranging from immediate to 32 months. The use of the nerve stimulator ought to prevent this serious complication of surgery in the neck.

A patient's complaint of pain and loss of function after operation must be taken seriously, and facile optimism is to be avoided. A diagnosis of "neurapraxia" should not be made when there is pain, vasomotor or sudomotor paralysis; indeed, it cannot be made unless stimulation of the injured nerve distal to the level of lesion produces a normal motor response at no earlier than 6 days from the incident. If there is a wound over the course of the nerve that has stopped working, then it is most likely that the nerve has been cut. As Bonney (1986) has commented, "it is indeed hard for the operating surgeon to credit that he or she has been responsible for the production of a serious nerve lesion, a feeling which may cloud vision and inhibit action."

## Pain Syndromes after Nerve and Brachial Plexus Injury

Acute pain occurs commonly, and chronic pain in about 5% of patients with peripheral nerve injury (Sunderland, 1993). Spontaneous pain may be accompanied by primary or secondary allodynia to mechanical or thermal stimuli, dysesthesias, and hyperpathia. Mechanical allodynia may be static or dynamic and may display temporal and spatial summation. Most patients with brachial plexus avulsion injuries report a characteristic constant crushing and intermittent shooting "central" pain from the time of injury or within days, which is often intractable (Berman *et al.*, 1998; Birch *et al.*, 1998). Pain is just as likely to follow damage to a "motor" nerve as it is to follow damage to a "mixed" or "sensory" nerve. The different syndromes of neuropathic pain are addressed in Chapter 11.

Pain after nerve or associated tissue injury may be classified as follows:

1. Posttraumatic neuralgia
2. Pain caused by persistent nerve compression/distortion/ischemia, which may be termed "neurostenagia" and would correspond to Seddon's "irritative" lesions (Birch *et al.*, 1998)
3. Causalgia (complex regional pain syndrome, CRPS type 2)—usually presents as burning pain and allodynia, often after a partial nerve injury, with disturbance of skin color, temperature, and sweating
4. Reflex sympathetic dystrophy (RSD, or CRPS type 1)—pain is accompanied after injury by changes that are dependent on the sympathetic nervous system, edema of tissues, trophic changes, and disproportionate loss of function
5. Phantom limb pain
6. "Central" pain (e.g., after brachial plexus spinal root avulsions, see later)
7. "Psychogenic" pain

The mechanisms of neuropathic pain in general are described in Chapter 11; this section will focus on brachial plexus injuries that involve both central and distal axons of sensory neurones at different spinal root levels. At some point in the course of their condition, most patients with spinal cord root avulsion injury report severe constant and shooting pain, referred to a numb (i.e., deafferented) limb region, which originates in the spinal cord and may persist for years (see Berman *et al.*, 1998; Birch *et al.*, 1998).

It is now well established that peripheral nerve injury can lead to central changes in the spinal cord, including windup, disinhibition, sensitization, and altered rostral processing (see Attal and Bouhassira, 1999; Banati *et al.*, 2001; Bennett, 1994). The peripheral mechanisms are better understood in molecular terms. The formation of a neuroma may lead to positive sensory symptoms, following accumulation of sodium channels, which generate spontaneous and evoked discharges. Recently, the distribution of two TTX-resistant "sensory neurone specific" sodium channels—SNS/PN3 and NaN/SNS2, which are preferentially expressed in nociceptors, were identified in injured human neruones; SNS/PN3 was shown to accumulate at the site of nerve injury and also in nerve fibers in skin from patients with mechanical allodynia and hyperalgesia (Coward *et al.*, 2000). Specific potassium channels were found to be decreased in human dorsal root ganglion and nerve after axotomy and may also contribute to paresthesias and hypersensitivity (Boettger *et al.*, 2002). Other peripheral pain mechanisms include ephaptic transmission, nociceptor sensitization, adrenergic chemosensitivity of regenerating axons, and nerve trunk inflammation (see Bennett, 1994; Chen *et al.*, 1996; Devor, 1983).

However, different mechanisms are involved in pain after spinal cord root avulsion. There is evidence that pain experienced after brachial plexus avulsion injury parallels the generation of abnormal activity within the dorsal horn of the spinal cord. After dorsal rhizotomy in cats, increased burst activity is seen in dorsal horn cells as early as 14 days (Loeser and Ward, 1967). Continuous high-frequency activity has also been reported in the cat dorsal horn, and, significantly, this type of firing is characteristic when the denervation is due to avulsion rather than rhizotomy (Ovelmen-Levitt, 1988). A high proportion of root avulsion patients report at least transient pain relief after dorsal root entry zone lesioning (Nashold *et al.*, 1983). In a prospective study, we found a strong correlation and temporal relationship between reduction in pain and successful nerve repair (Berman *et al.*, 1998). It was concluded that nerve repair can reduce pain from spinal root avulsions and that the mechanism may involve successful regeneration and/or restoration of peripheral connections. The findings are different after nerve and brachial plexus injuries in neonates (Anand and Birch, 2002). There was no evidence of chronic pain behavior or neuropathic syndromes, although pain was reported normally to external stimuli in unaffected regions. We proposed that differences in neonates are related to later maturation of injured fibers and that CNS plasticity (e.g., sprouting or maturation of descending inhibitory tracts) may account

for their lack of long-term chronic pain after spinal root avulsion injury.

## NATURAL COURSE

### Local Consequences

Transient ischemia is the mildest form of damage, provoking a transient failure of conduction affecting principally the large myelinated fibers. Lewis *et al.* (1931) described this form of centripetal paralysis after application of a suprasystolic cuff to the arm. Superficial sensibility is lost first; next there is loss of skeletal muscle power. Delayed pain response lingers for about 40 minutes of ischemia; pilomotor and vasomotor function are scarcely affected. Recovery is rapid after release of the cuff. Merrington and Nathan (1949) observed that the paresthesias occurring during recovery arose from the nerve trunks recovering from ischemia and not from the periphery. Lewis *et al.* (1931) noted that the differential response in nerve fibers is reversed in the conduction block of local anesthesia. The first sign is drying and warming of the extremity. This pattern is seen in the nerve paralysis of hematoma or aneurysm: autonomic paralysis is early and deep; loss of power extends over hours or days; deep pressure sense and joint position sense persists. Severe pain is usual. If the cause is not relieved, the lesion of the nerve progresses from one of conduction block to one of degeneration. Commonly, however, mild local nerve compression (e.g., entrapment or sleep palsy of the radial nerve) leads to neurapraxia, is associated with transient localized slowing or conduction block, and focal demyelination.

### Distal Consequences

A lesion deep enough to interrupt the axon leads to Wallerian degeneration (Waller, 1850). This process extends beyond the axon to the cell body, to the Schwann cell envelope and myelin sheath, to the endoneurial cells, and ultimately to the motor and sensory end organs. Landau (1953) found that the interval between injury and the failure of neuromuscular conduction ranged from 66–121 hours. Gilliatt and Taylor (1959) examined motor responses after section of a facial nerve in humans. The visible twitch in response to stimulation disappeared within 3–4 days, although an electrical response persisted for a further 48–70 hours. Gilliatt and Hjorth (1972) showed that failure of transmission of the neuromuscular junction precedes the failure of conduction along the degenerating axons. Landau (1953) said "the distinction between complete Wallerian degeneration and less severe injury can be made on the basis of the disappearance of excitability in the peripheral nerve segment" at the time of disappearance of neuromuscular function. Many misdiagnoses of neurapraxia could be avoided by this simple procedure.

If there is no regeneration of axons into the distal stump, changes occur in the target organs, which over time become irreversible. Most end plates disappear, and the denervated muscle becomes fibrosed. Bowden and Guttman (1994) had difficulty in finding end plates after 3 months of denervation; at 3 years there was fragmentation of structures, and they considered these changes irreversible. The reaction of muscle spindles is rather slower: Sherrington 1894 found them very obvious within muscle denervated for 5 months. Batten 1897 showed that spindle atrophy was much slower than for ordinary muscle fibers in a case of a complete lesion of the brachial plexus. Atrophy of cutaneous sensory organs is slower than it is for muscle fibers (Dellon, 1981).

### Proximal Consequences

Within days of axonotomy there is reduction in the caliber of the proximal axon, which may progress to atrophy (Guttman and Sanders, 1943). The velocity of conduction in the proximal segment drops (Cragg and Thomas, 1961), and there are changes in the cell body. However, section of the central branches of cells within the dorsal root ganglion does not lead to such severe changes as those after interruption of axons peripheral to the dorsal root ganglion, or after axonotomy of the ventral horn neurons in the adult. Despite the degeneration of the proximal process of the cells in the posterior root ganglion, these neurons seem to retain, for a long time, their ability to sustain their distal processes (Bonney and Gilliatt, 1958). It may be different in the neonate. Johnson and Yip (1985) found that section of the dorsal root in the neonatal rat provoked as much death as peripheral axonotomy. Dyck et al. (1984) examined the spinal cord of two patients years after amputation of the lower limb: they found "loss of target tissue by axotomy leads to atrophy and then loss of motor neurones." Bonney found atrophy of ipsilateral segments of the spinal cord after complete avulsion of the brachial plexus (see Birch et al., 1998). Spinal nerve root repair and reimplantation of avulsed ventral roots into spinal cord after brachial plexus injury (Carlstedt et al., 2000) also provides evidence about the harmfulness of delay in reconnection. A major consequence of degeneration of central terminals of afferent fibers after spinal root cord avulsion is severe intractable pain, as described previously.

## PRINCIPLES OF THERAPY

The clinician must always bear in mind that the sooner the distal segment is reconnected to the cell body and proximal segment, the better the result will be. In the extreme case of replantation after traumatic amputation, O'Brien (1975) showed that primary suture of nerves provides the only hope of recovery. Merle et al. (1986), Le Clerq et al. (1985), and Birch and Raji (1991) have all shown the ill effects of delay in the repair of the median and ulnar nerves. However, in three cases reported by Birch and St. Claire Strange (1990), removal of external pressure was rapidly followed by recovery in lesions that had persisted for up to 3 years. Nagano et al. (1996) described "hour glass"–like constriction of bundles within the median nerve from compression with early recovery after operation.

### Prognosis after Decompression or Repair

Review of larger series from both civil and military practice (Birch and Raji, 1991; Kline and Hudson, 1995a; Seddon, 1975; Zachary, 1954) shows that age, level of injury, type of nerve, violence of injury, and delay from injury to repair are all significant. Of these factors violence of injury and delay to repair are particularly so. This is demonstrated in analysis of large recent series of repairs of the radial and the musculocutaneous nerves. Shergill et al. (2001) described 160 repairs of the radial nerve in adults analyzing results according to four patterns of injury. They were: open "tidy" (knife or glass); open "untidy" (open fracture or penetrating missile injury), closed traction, and associated rupture of axillary or brachial artery. Seventy-nine percent of open "tidy" repairs achieved good or fair function, only 36% achieved this grade within the cases of arterial injury. Forty-nine percent of repairs done within 48 days of injury reached a good result: 28% of later repairs did so. Most repairs failed when the defect of the trunk exceeded 10 cm. All repairs performed after 12 months failed. Sixteen of the 18 repairs of the posterior interosseous nerve were successful. The results were no better than those described by Zachary (1954)! Osborne et al. (2000) found similar results in their account of 85 repairs of the musculocutaneous nerve. Twelve of the 13 open "tidy" lesions were successful; this is compared with 30 from the 48 closed traction lesions. Results were better when nerves were repaired within 14 days of injury and when grafts were less than 10 cm long. They were worse in the presence of associated arterial or bony injury. The trend demonstrated by these articles is repeatedly confirmed by published series for repairs of median and ulnar, circumflex, sciatic, and common peroneal nerves.

Recovery after nerve repair in a child is generally better than it is in adults. A properly performed urgent repair of the median and ulnar nerves at the wrist will be followed by function indistinguishable from the normal in infants or young children, a result scarcely seen in adults. However, recovery is perhaps not as good as is often assumed, and some of the most difficult problems in reconstruction follow failure of recovery of either tibial or common peroneal nerves in the growing child because of the rapid onset of most severe deformity. Children are not immune to the deleterious effect of delay or depression of regeneration capacity after violent proximal injury (Birch and Achan, 2000). Results of repair in the obstetrical brachial plexus lesion are unpredictable, and, for motor function, often disappointing. Results of sensory recovery are much better, demonstrated by quantitative sensory testing (Anand and Birch, 2002). Although recovery of function after spinal root avulsion was demonstrably related to surgery, there were remarkable differences from adults, including excellent restoration of sensory function in children and evidence of exquisite CNS plasticity (i.e., perfect localization of restored sensation in avulsed spinal root dermatomes, now presumably routed by way of nerves that had been transferred from a distant spinal region. Sensory recovery exceeded motor or cholinergic sympathetic recovery). The system of recording muscle power proposed by Highet in 1942 for the Nerve Injuries Committee of the Medical Research Council (see Highet, 1954; Medical Research Council, 1954) has not been superseded, but the limitations of the system for recording sensory recovery are severe for this method confined to cutaneous sensation and ignore the function of afferents from muscle spindles, joints, tendons, and ligaments. Most clinical work on recovery of sensibility after nerve injury and repair has been directed to cutaneous sensibility. Yet, stereognosis and proprioception must depend on signals from endings in muscles, tendons, and ligaments. There may be good recovery of sensory function in the hand with only imperfect cutaneous reinnervation. Narakas (1997) and Kline and Hudson (1995b) provided valuable systems for assessment, which may be useful in future studies.

## PRACTICAL MANAGEMENT

### Indications for Surgical Exploration and Repair

The decision whether to operate is easy only in the presence of an open wound or when nerve injury is associated with damage to major bones and blood vessels. Repair of a cut or ruptured nerve is an urgent matter. It is reasonable to defer operation on an uncomplicated open injury of a nerve for 24 hours or so to allow treatment by an experienced surgeon. Such delay is not permissible in the presence of damage to a major vessel or when there is impending ischemia.

The aims of operation include:

1. To confirm or establish diagnosis
2. To restore continuity to a severed or ruptured nerve
3. To remove a noxious agent compressing, distorting, or occupying a nerve
4. To prevent development or produce amelioration of chronic pain

We suggest that indications toward operation include:

1. Deep paralysis after a wound over the course of a main nerve or after injection close to it
2. Deep paralysis after closed injury associated with severe damage to soft tissues or skeleton
3. Deep paralysis after closed traction lesion of the brachial plexus
4. Nerve lesion in the presence of arterial injury
5. Nerve lesion associated with fracture requiring internal fixation
6. Deepening of nerve lesion under observation
7. Failure to show recovery after predicted time for lesion thought to be axonotmesis
8. Failure of recovery of conduction block within 6 weeks of injury
9. Persistent pain at almost any interval after injury

The particular requirements and special techniques of nerve and brachial plexus repair are described in detail elsewhere (Birch et al., 1998). The surgeon must be capable of dealing with associated injuries to major vessels, skin, bones, and joints. The equipment and the experience in using equipment for stimulating and recording is valuable, and in treatment of closed traction brachial plexus lesions or in analysis of incomplete lesions, it is essential. The surgeon should be able to interpret evidence of conduction from exposed nerves to the cortex or spinal cord (SSEP) and across lesions exposed at operation.

### Intractable Pain after Nerve and Brachial Plexus Injury

The general treatment options in patients with neuropathic pain are described in Chapter 11; here we consider the special case of brachial plexus injuries, where pain may be severe and intractable. The general treatment options are summarized in Table II. Understanding the different underlying mechanisms is necessary for interpreting the patient's symptoms and to plan rational and effective treatment. The latter may involve rational

TABLE II    Treatments for Pain after Nerve and Brachial Plexus Injury

| Drug | Starting dose | Maintenance dose | Mechanism |
|---|---|---|---|
| **a. Oral medications** | | | |
| Amitriptyline | 10 mg nocte | 75–150 mg/day | NA/5-HT reuptake inhibition |
| | | | Sodium channel blockade |
| Gabapentin | 300 mg od | 2400–4500 mg/day | Calcium channel blockade? |
| Carbamazepine | 50 mg nocte | 1600–1800 mg/day | Sodium channel blocker |
| Clonazepam | 0.5 mg nocte | 2 mg/day | GABAergic mechanism |
| Mexiletine | 100 mg bd | 600–900 mg/day (max., 1200 mg) | Sodium channel blocker |
| Phenytoin | 50 mg bd | 300 mg/day | Sodium channel blocker |
| Clonidine | 50 µg bd | 75 µg bd | $\alpha_2$-Adrenoreceptor agonist |
| Tramadol | 50–100 mg qid | 200–400 mg/day | Centrally acting opioid, nonopioid analgesic |
| Dextromethorphan | 60 mg | 60 mg bd | NMDA receptor antagonist |

**b. Other therapies**

i. *Topical applications for primary allodynia*
   - Capsaicin cream (0.075%, qid, for 8 wk)
   - Local anesthetic, 5% lignocaine gel or lotion

ii. *Intravenous lignocaine infusion*: 5 mg/kg body weight over 30 minutes with electrocardiogram and blood pressure monitoring. Useful for persistent parasthesias; a response to lignocaine infusion may be a predictor of response to mexiletine and other sodium channel blockers. Epidural infusion with lignocaine may be similarly helpful.

iii. *Sympathetic blocks*: IV guanethidine (10–30 mg in 20–50 ml saline solution), can be used for treatment of continuous burning pain, and to test whether sympathetic system is involved in the generation of pain. Stellate ganglion blocks.

iv. *Neuromodulation*
   - Acupuncture
   - TENS
   - Spinal cord stimulation

v. *Cognitive/behavioral rehabilitation*
   - Pain management programs
   - Relaxation therapy

vi. *Surgery*
   - Decompression/neurolysis/neurotisation
   - Dorsal root entry-zone (DREZ) lesions (for deafferentation pain)

combinations of pharmacological, anesthetic, surgical, psychological, and rehabilitative measures; the surgical procedures are usually a last resort.

There are few systematic studies or randomized clinical trials of pain relief in patients with nerve or brachial plexus injury. Using the "number needed to treat" (NNT) method (numbers of patients needed to treat to obtain one patient with more than 50% pain relief), Sindrup and Jensen (1999) evaluated the efficacy of different pharmacological agents used in the treatment of painful neuropathy. For antidepressants of all types combined, NNT of 3 was observed, with slightly better NNT for tricyclic antidepressants. For the ion channel blockers the NNT values were carbamazepine, 3.3; gabapentin, 3.7; phenytoin, 2.1; and mexiletine, 10.0. The values for other agents (dextromethorphan, tramadol, and capsaicin) were 1.9, 3.4, and 5.9, respectively. Side effects often determine the choice of drug used. Some patients report benefit from recreational drugs, such as alcohol and cannabis.

### Antidepressants

Tricyclic antidepressants are widely used in the treatment of neuropathic pain (McQuay *et al.*, 1996), including brachial plexus injury. It is assumed that antidepressants are effective in pain relief through descending inhibitory aminergic systems and by blocking sodium channels. The effect on pain is believed to be essentially independent of antidepressive, anxiolytic, and sedative effects, which may be helpful. Amitriptyline, imipramine, nortriptyline, and clomipramine are used. Amitriptyline is the most commonly used tricyclic antidepressant. Initial dosages should be low (10 mg/day) and subsequently increased by 10-mg steps. If side effects from amitriptyline are troublesome, desipramine can be used. It has fewer anticholinergic side effects and causes less sedation. If pain relief is not adequate, other drugs like anticonvulsants need to be added.

### Anticonvulsants and Antiarrhythmic Agents

Anticonvulsants reduce the transmembrane transport of $Na^+$ and $K^+$ ions, potentiate presynaptic and postsynaptic inhibition, and reduce posttetanic potentiation and the amplitude of evoked potentials (see Swerdlow, 1984). Carbamazepine, phenytoin, sodium valproate, gabapentin, clonazepam, topiramate, or lamotrigine have been used for pain after nerve (McQuay *et al.*,

1995) and brachial plexus injury. Gabapentin has been shown recently to be effective in the treatment of neuropathic pain associated with diabetic peripheral neuropathy and postherpetic neuralgia in multicenter, double-blind, placebo-controlled, randomized trials (Backonja *et al.*, 1998; Rowbotham *et al.*, 1998). Carbamazepine is often the drug of choice. Phenytoin had been used widely, and there are promising anecdotal reports with topiramate (Potter and Edwards, 1998). Combining an anticonvulsant with antidepressant may prove beneficial when the pain relief with either of them is not adequate. Mexiletine has fewer side effects and has also been reported to be helpful in conditions that are responsive to carbamazepine.

### Other Agents

**Capsaicin.** This is considered in brachial plexus injury patients in the setting of limb hypersensitivity, which may be associated with cutaneous reinnervation (see Berman *et al.*, 1998). Capsaicin is the constituent of chilli peppers responsible for the symptoms of heat, burning, and erythema. Capsaicin has been advocated as a topical agent for the therapy of postherpetic neuralgia and painful diabetic peripheral neuropathy. The underlying mechanism is not yet clear but may rely on the depletion of substance P or an effect on nerve terminals. The initial exacerbation of symptoms may be explained by activation of small sensory fibers and release of substance P. This effect, however, ceases in 3–4 weeks, after which the efficacy of the treatment usually becomes evident. The burning induced by capsaicin can be significantly reduced with pretreatment with the topical anesthetic. Recently, capsaicin may relieve intractable neuropathic pain when used in very high doses (5%–10%) in association with regional anesthesia.

**Tramadol.** Tramadol is a centrally acting analgesic. It is a weak opiate, modulates central serotoninergic and noradrenergic inhibition of pain, and has very low risk of addiction. It was shown to reduce pain in patients with diabetic neuropathy. The side effects commonly seen are nausea, constipation, headache, and somnolence.

**Narcotics.** Severe pain may be helped by narcotics, which are predominantly centrally acting analgesics. The prototype is morphine. Narcotics are generally reserved for severe acute pain states and chronic pain caused by malignancy with poor prognosis, where they usually provide satisfactory pain relief with adequate doses. The application of opioids in diseases of non-malignant origin is restricted by the potential risk of development or dependence. It is difficult to predict whether pain after brachial plexus injury would respond to opioids, and it may be helpful to undertake a short trial in severe refractory pain states to see whether the patient is opioid sensitive. Opioids are less likely to work if pain is in a numb area.

### Other Therapies

**Transcutaneous Electrical Nerve Stimulation (TENS).** TENS is based on the gate-control theory and involves selective activation of nonnociceptor fibers. It may be helpful, if correctly administered, in some patients with neuropathic and postbrachial plexus injury pain (Meyer and Fields, 1972), and needs persistence to find the optimum conditions of stimulation.

**Acupuncture.** Acupuncture, a traditional method to induce analgesia, relies on the placement of needles at specific acupuncture spots, where stimulation of afferents presumably activates endogenous inhibitory systems. Acupuncture is used in a wide range of pain syndromes, including brachial plexus injury. The comparison of acupuncture with sham acupuncture and placebo yields response rates of 60%–75%, 50%, and 30% (Lewith and Machin, 1983).

**Spinal Cord Stimulation.** Spinal cord stimulation was based on the gate-control theory of pain, now linked to many mechanisms. It is thought to activate spinal inhibitory circuits, mainly those concerned with GABAergic mechanisms, and may have a suppressive action on dorsal horn neuronal hyperexcitability. Before permanent implantation of a stimulation device, a trial period with temporary external stimulation is strongly recommended. Spinal cord stimulation has been shown to have long-term benefit in various conditions associated with neuropathic pain caused by cauda equina injury and phantom limb pain. Strict criteria need to be applied for selection of a patient, and all therapeutic modalities should be exhausted before decision to implant a stimulator is made (Kumar *et al.*, 1998).

## REFERENCES

Anand, P., and Birch, R. (2002). Restoration of sensory function and lack of long-term chronic pain syndromes after brachial plexus injury in human neonates. *Brain* 125, 113–122.

Attal, N., and Bouhassira, D. (1999). Mechanisms of pain in peripheral neuropathy. *Acta Neurol. Scand.* 173(**Suppl**), 12–24.

Backonja, M., Beydoun, A., Edwards, K. R., Schwartz, S. L., Fonseca, V., Hes, M., LaMoreaux, L., and Garofalo, E. (1998). Gabapentin for the symptomatic treatment of painful neuropathy in patients with diabetes mellitus: a randomized controlled trial. *JAMA* 280, 1831–1836.

Banati, R. B., Cagnin, A., Brooks, D. J., Gunn, R. N., Myers, R., Jones, T., Birch, R., and Anand, P. (2001). Long-term trans-synaptic glial responses in the human thalamus after peripheral nerve injury. *NeuroReport.* 12, 3439–3442.

Batten, F. B. (1897). The muscle-spindle under pathological conditions. *Brain* **20**, 138–179.

Bell, M. A., and Weddell, A. G. M. (1984). A descriptive study of the blood vessels of the sciatic nerve in the rat, man and other animals. *Brain* **107**, 871–898.

Bennett, G. J. (1994). Neuropathic pain. *In* "Textbook of Pain," (P. D. Wall and R. Melzac, Eds.), pp. 201–224. London, Churchill Livingstone.

Berman, J. S., Birch, R., and Anand, P. (1998). Pain following human brachial plexus injury with spinal cord root avulsion and the effect of surgery. *Pain* **75**, 199–207.

Berthold, C. H., Carlstedt, T., and Corneliuson, O. (1993). Anatomy of the mature transitional zone. *In* "Peripheral Neuropathy," 3rd ed. (P. J. Dyck, P. K. Thomas, J. W. Griffin, P. A. Low, and J. F. Poduslo, Eds.), pp. 75–80. WB Saunders, Philadelphia.

Birch, R., and Achan, P. (2000). Peripheral nerve repairs and their results in children. *Hand Clin.* **16**, 579–595.

Birch, R., Bonney, G., and Wynn Parry, C. B. (1998). "Surgical Disorders of the Peripheral Nerves." pp. 467–490. London, Churchill Livingstone.

Birch, R., and Raji, A. R. M. (1991). Repair of median and ulnar nerves. *J. Bone Joint Surg.* **73B**, 154–157.

Birch, R., and St. Claire Strange, F. G. (1990). A new type of peripheral nerve lesion. *J. Bone Joint Surg.* **72B**, 312–313.

Bonney, G. (1986). Iatrogenic injuries of nerves. *J. Bone Joint Surg.* **68B**, 9–13.

Bonney, G., and Gilliatt, R. W. (1958). Sensory nerve conduction after traction lesion of the brachial plexus. *Proc. Roy. Soc. Med.* **51**, 365–367.

Boettger, M. K., Till, S., Chen, M. X., Anand, U., Otto, W., Plumpton, C., Trezise, D., Tate, S. N., Coward, K., Birch, R., and Anand, P. (2002). Calcium-activated potassium channels are decreased in injured human sensory neurons and regulated selectively by neurotrophic factors. *Brain* **125**, 252–263.

Bowden, R. E. M., and Gutman, E. (1944). Denervation and re-innervation of human voluntary muscle. *Brain* **67**, 273–313.

Carlstedt, T., Anand, P., Hallin, R., Misra, P. V., Noren, G., and Seferlis, T. (2000). Spinal nerve root repair and re-implantation of avulsed ventral roots into spinal cord after brachial plexus injury. *J. Neurosurg. (Spine)* **93**, 237–247.

Chen, Y., Michaelis, M., Janig, W., and Devor, M. (1996). Adreno-receptor subtype mediating sympathetic-sensory coupling in injured sensory neurons. *J. Neurophysiol.* **76**, 3721–3730.

Coupland, R. M. (1993). "War Wounds of Limbs." *Butterworth-Heinemann*, Oxford.

Coward, K., Plumpton, C., Facer, P., Birch, R., Carlstedt, T., Tate, S., Bountra, C., and Anand, P. (2000). Immunolocalisation of SNS/PN3 and NaN/SNS2 sodium channels in human pain states. *Pain* **85**, 41–50.

Cragg, B., and Thomas, P. K. (1961). Changes in conduction velocity and fibre size proximal to peroneal nerve lesions. *J. Physiol.* **157**, 315–327.

Dellon, A. L. (1981). Results of nerve repair in the hand. *In* "Evaluation of Sensibility and Re-education of Sensation in the Hand." (A. L. Dellon, Ed.), pp. 193–202. Williams & Wilkins, Baltimore.

Devor, M. (1983). Nerve pathophysiology and mechanisms of pain in causalgia. *J. Autonom. Nerv. Syst.* **7**, 371–384.

Dyck, P. J., Nukada, H., Lais, C. A. *et al.* (1984). Permanent axotomy: a model of chronic neuronal degeneration produced by axonal atrophy, myelin remodelling and regeneration. *In* "Peripheral Neuropathy," 2nd ed. (P. J. Dyck, P. K. Thomas, E. H. Lambert, and R. Bunge, Eds.), pp. 660–690. WB Saunders, Philadelphia.

Gamble, H. J., and Eames, R. A. (1964). An electron microscope study of the connective tissues of human peripheral nerve. *J. Anatomy* **98**, 655–662.

Gilliatt, R. W., and Hjorth, R. J. (1972). Nerve conduction during Wallerian degeneration in the baboon. *J. Neurol. Neurosurg. Psychiatry* **35**, 335–341.

Gilliatt, R. W., and Taylor, J. C. (1959). Electrical changes following section of the facial nerve. *Proc. Roy. Soc. Med.* **52**, 1080–1083.

Guttman, E., and Sanders, F. K. (1943). Recovery of fibre numbers and diameters in the regeneration of peripheral nerves. *J. Physiol.* **101**, 489–518.

Highet, W. B. (1954). Quoted by Zachary R.B. *In* "Peripheral Nerve injuries." (H. J. Seddon, Ed.), p. 355. *Medical Research Council Special Report* Series No. 282. HMSO, London.

Johnson, E. M. Jr., and Yip, H. K. (1985). Central nervous system and peripheral nerve growth factor provide trophic support critical to mature sensory neuronal survival. *Nature* 1;**314**, 751–752.

Khan, R., and Birch, R. (2001). Iatropathic injuries of peripheral nerves. *J. Bone Joint Surg.* **83**, 1145–1148.

Kline, D. G., and Hudson, A. R. (1995a). "Nerve Injuries," 1st edn. WB Saunders, Philadelphia.

Kline, D. G., and Hudson, A. R. (1995b). Grading results. *In* "Nerve Injuries." pp. 87–99. W. B. Saunders, Philadelphia.

Kumar, K., Toth, C., Nath, R. K., and Laing, P. (1998). Epidural spinal cord stimulation for treatment of chronic pain-some predictors of success. A 15-year experience. *Surg. Neurol.* **50**, 110–120.

Landau, W. M. (1953). The duration of neuromuscular function after nerve section. *Neurosurgery* **10**, 64–68.

Le Clerq, D. C., Carlier, A. J., Khuc, T. *et al.* (1985). Improvement in the results in sixty-four ulnar nerve sections associated with arterial repair. *J. Hand Surg.* **10A** **(Intl Suppl 2)**, 997–999.

Lewis, T., Pickering, G. W., and Rothschild, P. (1931). Centripetal paralysis arising out of arrested blood flow to the limb. *Heart* **16**, 1–32.

Lewith, G. T., Machin, D. (1983). The evaluation of the clinical effects of acupuncture. *Pain* **16**, 111–127.

Loeser, J. D., and Ward, A. A. (1967). Some effects of deafferentation on neurons of the cat spinal cord. *Arch. Neurol.* **17**, 629–636.

McManus, P. G., Low, P. A., and Lagerlund, T. D. (1993). Micro-environment of nerve: blood flow and ischaemia. *In* "Peripheral Neuropathy," 3rd ed. (P. J. Dyck, P. K. Thomas, J. W. Griffin, P. A. Low, and J. F. Poduslo, Eds.), pp. 433–440. WB Saunders, Philadelphia.

McQuay, H., Carroll, D., Jadad, A. R., Wiffen, P., and Moore, A. (1995). Anticonvulsant drugs for management of pain: a systematic review. *BMJ* **311**, 1047–1052.

McQuay, H. J., Tramer, M., Nye, B. A., Carroll, D., Wiffen, P. J., and Moore, R. A. (1996). A systemic review of antidepressants in neuropathic pain. *Pain* **68**, 217–227.

Medical Research Council. (1954). (H. J. Seddon Ed.), "Peripheral Nerve Injuries." *Special Report* Series No 282. HMSO, London.

Merle, M., Amend, P., Cour, C., *et al.* (1986). Microsurgical repair of peripheral nerve lesions: a study of 150 injuries of the median and ulnar nerves. *Peripheral Nerve Repair Regen.* **2**, 17–26.

Merrington, W. R., and Nathan, P. W. (1949). A study of post-ischaemic paraesthesiae. *J. Neurol. Neurosurg. Psychiatry* **12**, 1–18.

Meyer, G. A., and Fields, H. L. (1972). Causalgia treated by selective large fibre stimulation of peripheral nerve. *Brain* **95**, 163–167.

Nagano, A., Shibata, K., Tokimura, H. *et al.* (1996). Spontaneous anterior interosseous nerve palsy with hourglass-like fascicular constriction within the main trunk of the median nerve. *J. Hand Surg.* **21A**, 266–270.

Narakas, A. O. (1977). Indications et resultants du traitement chirurgicale dans les lésions pare élongation du plexus brachial. *Rev. Chir. Orthop. Réparatrice l'Appareil Moteur* **63**, 44–45.

Nashold, B. S., Ostdahl, R. H., Bullitt, E., Friedman, A., and Brophy, B. (1983). Dorsal root entry zone lesions: A new neurosurgical therapy for deafferentation pain. (J. J. Bonica, Ed.), *Adv. Pain Res. Ther.* 5, 739–750.

O'Brien, B. McC. (1975). Microsurgery in the treatment of injuries. *In* "Recent Advances in Orthopaedics." (B. McKibbin, Ed.), pp. 235–279. *Churchill Livingstone*, Edinburgh.

Osborne, A. W., Birch, R. M., Munshi, P., and Bonney, G. (2000). The musculocutaneous nerve. *J. Bone Joint Surg.* 82, 1140–1142.

Ovelmen-Levitt, J. (1988). Abnormal physiology of the dorsal horn as related the deafferentation syndrome. *Appl. Neurophysiol.* 51, 104–116.

Potter, D., and Edwards, K. R. (1998). Potential role of topiramate in relief of neuropathic pain. *Neurology* 50(**Suppl 4**), A255.

Rowbotham, M., Harden, N., Stacey, B., Bernstein, P., and Magnus-Miller, L. (1998). Gabapentin for the treatment of postherpetic neuralgia: a randomized controlled trial. *JAMA* 280, 1837–1842.

Schenker, M., and Birch, R. (2001). Diagnosis of the level of intradural rupture of the rootlets in transaction lesions of the brachial plexus. *J. Bone Joint Surg.* 83, 916–920.

Seddon, H. J. (1943). Three types of nerve injury. *Brain* 66, 237–288.

Seddon, H. J. (1975). "Surgical Disorders of the Peripheral Nerves," 2nd ed. Churchill Livingstone, Edinburgh, London and New York.

Shergill, G., Bonney, G., Munshi, P., and Birch, R. (2001). The radial and posterior interosseous nerves. Results for 260 repairs. *J. Bone Joint Surg.* 83B, 646–649.

Sherrington, C. S. (1984 or 1894). On the anatomical constitution of nerves of skeletal muscles with remarks on recurrent fibres in the ventral spinal nerve-root. *J. Physiol.* 17, 211–258.

Sindrup, S. H., and Jensen, T. S. (1999). Efficacy of pharmacological treatments of neuropathic pain: an update and effect related to mechanism of drug action. *Pain* 83, 389–400.

Smith, S. M. J. (1998). Electrodiagnosis. *In* "Surgical Disorders of the Peripheral Nerves." (R. Birch, G. Bonney, and C. B. Wynn Parry, Eds.), pp. 467–490. Churchill Livingstone, London.

Sunderland, S. (1951). A classification of peripheral nerve injuries producing loss of function. *Brain* 74, 491–516.

Sunderland, S. (1993). "Nerves and Nerve Injuries." Churchill Livingstone, London.

Swerdlow, M. (1984). Anticonvulsant drugs and chronic pain. *Clin. Neuropharmacol.* 7, 55–81.

Thomas, P. K. (1964). Changes in the endoneurial sheaths of peripheral myelinated nerve fibres during Wallerian degeneration. *J. Anatomy* 98, 175–182.

Thomas, P. K., and Holdorff, B. (1993). Neuropathy due to physical agents. *In* "Peripheral Neuropathy," 3rd ed. (P. J. Dyck, P. K. Thomas, J. W. Griffin, P. A. Low, and J. F. Poduslo, Eds.), pp. 990–1013. WB Saunders, Philadelphia.

Waller, A. (1850). Experiments on the section of the glossopharyngeal and hypoglossal nerves of the frog, and observations of the alterations produced thereby in the structure of their primitive fibres. *Philos. Trans. Roy. Soc. London* 140, 423–429.

Williams, W. W., Twyman, R. S., Donell, S. T., and Birch, R. (1996). The posterior triangle and the painful shoulder: spinal accessory nerve injury. *Ann. R. Coll. Surg. Eng.* 78, 521–525.

Young, J. Z. (1949). Factors influencing the regeneration of nerves. *Adv. Surg.* 1, 165–220.

Zachary, R. B. (1954). Results of nerve suture. *In* "Peripheral Nerve Injuries." By the Nerve Injury Committee of the Medical Research Council. (H. J. Seddon, Ed.), pp. 354–386. HMSO, London.

CHAPTER 93

# Delayed Radiotherapy Injury

Manfred Stöhr

## CLINICAL CHARACTERISTICS

Delayed radiotherapy injury to the peripheral nervous system may affect any peripheral or cranial nerve within the field of the radiotherapy. However, by far the most common lesions are those affecting the brachial and lumbosacral plexuses (Dropcho, 1998; Holdorff, 1978; Kori *et al.*, 1981; Mumenthaler, 1964; Stöhr, 1980; Thomas and Holdorff, 1993) (Table I).

Injury to the *brachial plexus* is the most frequent complication of radiotherapy to the peripheral nervous system. The actuarial incidence of a radiation-induced brachial plexus injury in 449 breast cancer patients was 4.9% after 5.5 years of therapy (Powell *et al.*, 1990). The plexopathy occurs after a latent interval that varies from 6 months–26 years; one half of such cases become evident within the first 3 years.

According to McDermont (1971), *the tolerance dose* for the brachial plexus is 1500 ret (nominal standard dose, Ellis). With current radiotherapy techniques, this corresponds to a dose of 50 Gy. However, in one of the author's (Stöhr) own patient groups, definite, delayed radiation damage after a nominal dose between 800 and 900 ret was observed. Possible explanations for this discrepancy include the following:

1. The brachial plexus is located at a more superficial level than the depth for which the doses are calculated.
2. Neighboring fields of radiotherapy overlap.
3. The radiotherapy load of individuals increases as a consequence of stray radiation.
4. The tolerance to radiation falls as the area of the radiation field increases.
5. Large doses of radiation per fraction are less well tolerated than small doses per fraction (Powell *et al.*, 1990).

There is a clear relationship between the latent period, the severity of the resulting disability, and the total dose of radiotherapy administered to the plexus. Plexopathies after doses of more than 50 Gy tend to occur after a shorter latent period, to progress more rapidly, and to result in more severe disability than plexopathies after doses of less than 50 Gy (Stöhr, 1996). In addition, experimental data show a relationship between the dose and the manifestation of nerve lesions. Moreover, there is an approximately inverse linear relationship between the dose and the time to onset of nerve lesions (Kinsella *et al.*, 1991).

The most common *presenting symptoms* of radiation-induced brachial plexopathy include paresthesia, numbness, and weakness in various distributions. Pain is initially uncommon (Dropcho, 1998; Stöhr, 1996; Thomas and Holdorff, 1993). Fasciculations, myokymia, cramps, or even fine myoclonic jerks are frequently present (Stöhr, 1982).

*Lumbosacral plexopathy* and *femoral nerve palsy* are less common than brachial plexopathy; however, their presenting symptoms, latent intervals, and course are similar. The tolerance dose for the lumbosacral plexus is estimated to be 1400 ret (Sadowsky *et al.*, 1976). The risk is increased in patients who receive a combination of external beam treatment plus intracavitary radiation implants (Georgiou *et al.*, 1993) or when large irradiation fields are used.

After irradiation of the para-aortic lymph nodes, a progressive "motor neuron syndrome" with bilateral muscle atrophy and fasciculations may develop ("radiogenic amyotrophy"). For a long time the underlying damage was thought to be in the anterior horn cells, whereas additional sensory and vegetative symptoms (erectile and bladder dysfunction) in some patients, delayed somatosensory-evoked potentials after tibial nerve stimulation, and a few morphological investigations strongly suggest an underlying damage within the cauda equina (Berlit and Schwechheimer, 1987; Bowen *et al.*, 1996; Stöhr, 1996). Therefore, Wohlgemuth *et al.* (1998) suggested the term "post-irradiation cauda equina syndrome."

*Lower cranial nerve palsies*—especially involving hypoglossal, vagus, and spinal accessory nerves—may follow radiation of head and neck tumors. Radiation—induced bulbar palsy (Shapiro *et al.*, 1996) may mimic ALS, if accompanied by a central nervous system involvement (Glenn and Ross, 2000).

## DIAGNOSTIC CRITERIA

Evidence in support of radiation-induced neuropathy includes

1. The affected nerve lies within the field of radiation.
2. The dose of radiation equals or exceeds the level of tissue tolerance.

TABLE I    Radiation-Induced Lesions within the Peripheral Nervous System in 101 Patients

| | |
|---|---|
| Brachial plexus | 59 |
| Lumbosacral plexus | 18 |
| Cauda equina | 9 |
| Femoral nerve | 6 |
| Lower cranial nerves | 5 |
| Cervical plexus | 3 |
| Sciatic nerve | 1 |

3. The latent period and the presenting symptoms are in accordance with the forementioned criteria.

A clinical dilemma that must frequently be faced is to distinguish radiation-induced brachial plexopathy from tumor metastasis to the plexus (Dropcho, 1998; Stöhr, 1996; Thomas and Holdorff, 1993). Horner's syndrome, early and severe pain, and rapid progression speak in favor of tumor infiltration, whereas electromyographically recorded myokymic discharges (recurrent trains of MUP) (Figure 1) are highly suggestive of radiation-induced plexopathy (Albers, 1981; Stöhr, 1982).

## NATURAL COURSE

In approximately two thirds of patients the motor and sensory deficits gradually worsen over several years to reach a level of severe disability. In about 40% the course rapidly progresses, sometimes in a remitting and relapsing manner. In a minority of cases the course is very slow, so that even many years later the sensorimotor deficits remain minor. The rest of these patients (17% of the author's own patients) experience a spontaneous cessation of progression after 1–3 years (Dropcho, 1998; Kori *et al.*, 1981; Stöhr, 1996). A spontaneous recovery of neurological function seems highly unusual, and the author has personally never observed it. Enevoldson *et al.* (1992), however, reported a spontaneous resolution of a postradiation plexopathy.

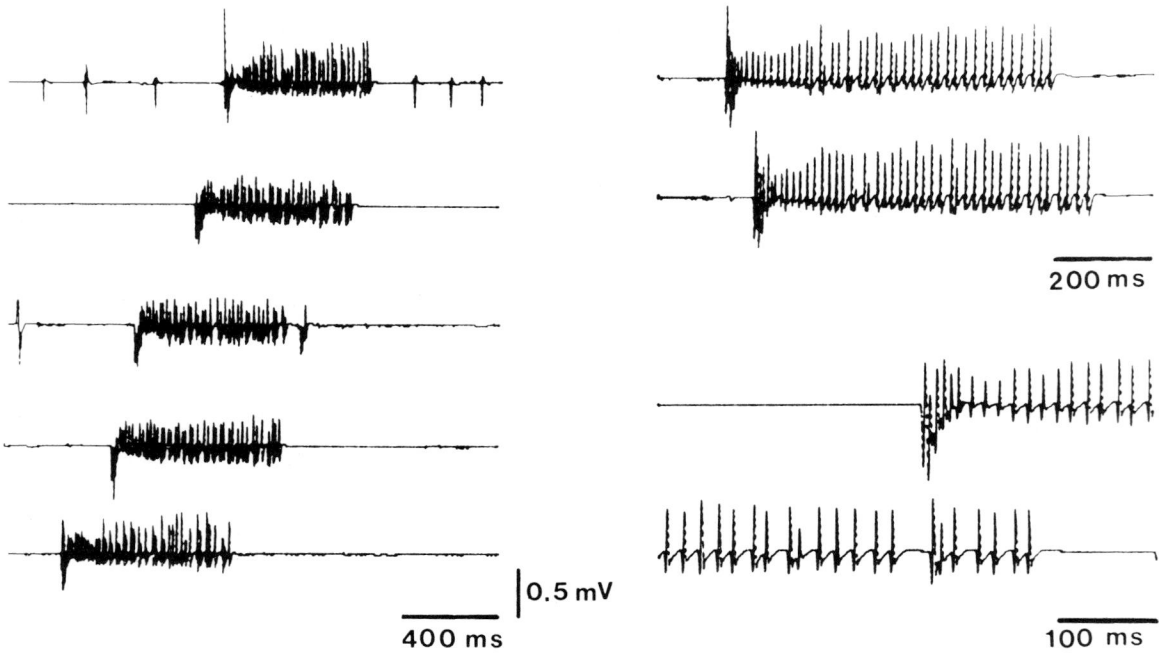

FIGURE 1    "Myokymic discharges" (recurrent trains of motor unit potentials) in radiation-induced neuropathy.

Killer and Hess (1990) observed in 8 of 12 patients with a mean follow-up time of 20 years a slow and steady progression of brachial plexopathy, whose final outcome was severe sensorimotor pareses that rendered the hand useless. Four patients experienced no progression, three had stabilization of the paresis with only slight functional impairment of the affected arm, and one had severe palsy. The course was similar in conservatively and surgically treated patients.

## MANAGEMENT

The pathogenesis of radiogenic injury involves direct injury by radiation to the various cellular elements constituting the peripheral nerves such as the Schwann cells. There is secondary damage to the same cells and axon processes from the late angiomesenchymal changes. To date there are no proven strategies for treating postradiation injury of the peripheral nerves.

Basic treatment includes physiotherapy to prevent and treat joint contractures, strength training, the reduction and prevention of any further tissue swelling, and the prevention of lymphedema. Injections or infusions into the affected limb must be avoided, as well as major injuries or extensive use of the affected limb. Pain may be managed in a variety of ways, sometimes including the use of antidepressants (see Chapter 11).

Improvement of the blood supply of the affected peripheral nerves by using drugs such as pentoxifyllin (Trental) or by inhibiting collagen formation within and around the peripheral nerves with drugs such as L-trijodthyronine or penicillamine has not been proven to effectively reduce the symptoms or to alter the course of postradiation neuritis.

Anticoagulation with intravenous heparin followed by warfarin has been reported to bring about neurologic improvement in a few patients (Glantz et al., 1994). A patient with sacral plexopathy experienced a progressive resolution of the neurological complaints after hyperbaric oxygen therapy 15 years after radiotherapy of a leiomyosarcoma of the bladder (Videtic and Venkatesan, 1999).

There are several reports of surgical attempts to stop progression or even improve neurologic function in brachial plexopathy. For many years different authors have recommended freeing the nerves and fascicles from surrounding tissues as a means of treating the symptoms and possibly favorably altering the course of the disease. Unfortunately, despite claims in support of external and partial internal neurolysis (Cormier and Ferry, 1974; Spiess, 1972), others have been unable to determine any benefit from these surgical procedures (Kunft and Penzholz, 1971; Match, 1975; Stöhr, 1996). More recently, endoneurolysis has been abandoned in favor of ventral epineurotomy (removal of only the ventral epineurium). To improve the blood supply to the liberated nerve and prevent renewal of the constrictive fibrosis, the surgeon envelops the brachial plexus in a free or pedunculated connective tissue flap (Clodius et al., 1984; Le Quang, 1989). The value of this procedure has also not been proven.

Brachial plexopathy was surgically treated either by neurolysis or by neurolysis in combination with omental grafting in 8 of 12 patients of Killer and Hess (1990). The postoperative course of the paresis was progressive in six of the eight patients; thus, their course did not differ from the natural course of the disease. The almost complete pain relief, however, was considered a beneficial effect of the surgical treatment. It is my own experience that pain relief is frequently associated with a significant deterioration of sensorimotor functions.

Currently one reasonable indication for the exploration of the brachial plexus is to help make a diagnosis of suspected malignant invasion of the plexus, when the clinical suspicion is strong but the MRI with gadolinium fails to resolve the differential diagnosis between a postradiation plexitis and malignant invasion of the plexus. Indeed, the two may coexist. In such cases, histological confirmation of the diagnosis may be vital for planning further therapy. Clearly, the decision to explore the brachial plexus is not to be taken lightly. It should be made only when the anticipated benefits of making a more secure diagnosis are in the best interests of the patient.

## REFERENCES

Albers, J. W., Allen, A. A., and Bastron, J. A. (1981). Limb myokymia. *Muscle Nerve* **4**, 494.

Berlit, P., and Schwechheimer, K. (1987). Neuropathological findings in radiation myelopathy of the lumbosacral cord. *Eur. Neurol.* **27**, 29–34.

Bowen, J., Gregory, R., Squier, M., and Donaghy, M. (1996). The post-irradiation lower motor neuron syndrome: neuronopathy or radiculopathy? *Brain* **119**, 1429–1439.

Clodius, L., Uhlschmid, G., and Hess, K. (1984). Irradiation plexitis of the brachial plexus. *Clin. Plast. Surg.* **11**, 161–165.

Cormier, J. M., and Ferry, J. (1974). Brachial plexus paralysis in treated breast neoplasms. *Nouv. Presse Med.* **3**, 1000.

Dropcho, E. J. (1998). Neurologic complications of radiation therapy. *In* "Iatrogenic Neurology" (J. Biller, Ed.), Butterworth-Heinemann, Boston.

Enevoldson, T. P., Scadding, J. W., Rustin, G. J., and Senanayake, L. F. (1992). Spontaneous resolution of a postirradiation lumbosacral plexopathy. *Neurology* **42**, 2224–2225.

Georgiou, A., Grigsby, P. W., and Perez, C. A. (1993). Radiation induced lumbosacral plexopathy in gynecologic tumors: clinical findings and dosimetric analysis. *Int. J. Radiat. Oncol. Biol. Phys.* **26**, 479.

Glantz, M. J., Burger, P. C., and Friedman, A. H. (1994). Treatment of radiation-induced nervous system injury with heparin and warfarin. *Neurology* **44**, 2020.

Glenn, S. A., and Ross, M. A. (2000). Delayed radiation-induced bulbar palsy mimicking ALS. *Muscle Nerve* **23**, 814–817.

Holdorff, B. (1978). Lesions of the cauda equina and the lumbosacral plexus by ionising rays. *Akt. Neurol.* **5**, 23–27.

Killer, H. E., and Hess, K. (1990). Natural history of radiation-induced brachial plexopathy compared with surgically treated patients. *J. Neurol.* **237**, 247–250.

Kinsella, T. J., DeLuca, A. M., Barnes, M., Anderson, W., Terrill, R., and Sindelar, W. F. (1991). Threshold dose for peripheral neuropathy following intraoperative radiotherapy (IORT) in a large animal model. *Int. J. Radiat. Oncol. Biol. Phys.* **20**, 697–701.

Kori, S. H., Foley, K. M., and Posner, J. B. (1981). Brachial plexus lesions in patients with cancer: 100 cases. *Neurology* **31**, 45.

Kunft, H. D., and Penzholz, H. (1971). Limits of surgical treatment of radiation-induced lesions of the brachial plexus. *In* "Present Limits of Neurosurgery" (J. Fusek and Z. Lunc, Eds.), Proceedings of the IVth European Congress of Neurosurgery, Prague.

LeQuang, C. (1989). Postirradiation lesions of the brachial plexus: results of surgical treatment. *Hand. Clin.* **5**, 23.

Match, R. M. (1975). Radiation-induced brachial plexus paralysis. *Arch. Surg.* **110**, 384–386.

McDermont, R. S. (1971). Cobalt 60 beam therapy, postradiation effects in breast cancer patients. *J. Can. Assoc. Radiol.* **22**, 195.

Mumenthaler, M. (1964). Lesions of the brachial plexus following radiotherapy. *Schweiz. Med. Wschr.* **94**, 1069–1075.

Olsen, N. K., Pfeiffer, P., and Johannsen, L. (1993). Radiation-induced brachial plexopathy: neurological follow-up in 161 recurrence-free breast cancer patients. *Int. J. Radiat. Oncol. Biol. Phys.* **26**, 43.

Powell, S., Cooke, J., and Parsons, C. (1990). Radiation-induced brachial plexus injury: follow-up of two different fractionation schedules. *Radiother. Oncol.* **18**, 213–220.

Sadowsky, C. H., Sachs, E., and Ochoa, J. (1976). Postradiation motor neuron syndrome. *Arch. Neurol.* **33**, 786–787.

Shapiro, B. E., Rordorf, G., Schwamm, L., and Preston, D. C. (1996). Delayed radiation-induced bulbar palsy. *Neurology* **46**, 1604–1606.

Spiess, H. (1972). "Peripheral Nerve Damage Due to Ionising Irradiation." Springer-Verlag, Berlin.

Stöhr, M. (1980). "Iatrogenic Nerve Lesions: Injections, Operations, Positioning, Radiotherapy." Thieme-Verlag, Stuttgart.

Stöhr, M. (1982). Special types of spontaneous electrical activity in radiogenic nerve injuries. *Muscle Nerve* **5**, 78–83.

Stöhr, M. (1996). Nerven- und Plexuslähmungen nach Strahlentherapie. *In* "Iatrogene Nervenläsionen" 2nd ed., (M. Stöhr, Ed.), Thieme-Verlag, Stuttgart, New York.

Thomas, P. K., and Holdorff, B. (1993). Neuropathy due to physical agents. *In* "Peripheral Neuropathy," 3rd ed. (P. J. Dyck, P. K. Thomas, and J. W. Griffin, Eds.), p. 1000. Saunders, Philadelphia.

Videtic, G. M., and Venkatesan, V. M. (1999). Hyperbaric oxygen corrects sacral plexopathy due to osteoradionecrosis appearing 15 years after pelvic irradiation. *Clin. Oncol.* **11**, 198–199.

Wohlgemuth, W. A., Rottach, K., Jaenke, G., and Stöhr, M. (1998). Radiogene Amyotrophie. Cauda-equina-Läsion als Strahlenspätfolge. *Nervenarzt* **69**, 1061–1065.

CHAPTER 94

# Therapy of Myasthenia Gravis and Myasthenic Syndromes

Reinhard Hohlfeld, A. Melms, C. Schneider, K. V. Toyka, and D. B. Drachman

## CLINICAL FEATURES AND DIAGNOSTIC TESTS

Myasthenia gravis (MG) and the myasthenic syndromes are diseases that affect the neuromuscular junction (Table I; reviewed in Drachman, 1994; Engel *et al.*, 1999; Hohlfeld and Wekerle, 1999). In autoimmune MG, autoantibodies reduce the number of available acetylcholine receptors (AChRs), and thereby impair neuromuscular transmission. Initially, the most important clues to the diagnosis of MG are obtained from the history and clinical examination. The cardinal features of MG are weakness and fatigability of skeletal muscles, usually in a characteristic distribution. The weakness increases with activity and improves with rest. Typically, the signs and symptoms are more pronounced at the end of the day. Not uncommonly, weakness is exacerbated by infections, certain medications (see later), emotional disturbances, or in the premenstrual period. On physical examination, the findings are limited to the motor system.

The presenting symptoms vary from patient to patient, but in more than half of cases, the initial complaints include diplopia and ptosis, often followed by difficulties in speech, chewing, and swallowing or by weakness of the upper and lower extremities. The distribution of muscle weakness is usually bilateral and is often proximal but may be asymmetrical and may involve different muscle groups over the course of the disease. If weakness of respiration or swallowing becomes so severe as to require mechanical support, the patient is in "myasthenic crisis."

On the basis of the clinical presentation and prognostic factors, different classifications have been proposed (Table II; Task Force of the Medical Scientific Advisory Board of the Myasthenia Gravis Foundation of America *et al.*, 2000). Further evidence for disease heterogeneity is based on differences in age at onset (<45 or >45 years), thymic abnormalities, immunological parameters (HLA-association, AChR-antibody titer), and response to therapy (Compston *et al.*, 1980).

A diagnosis of MG usually commits the patient to long-term medical treatment and/or surgery that entails substantial risks. It is therefore essential to establish the diagnosis unequivocally, to exclude other conditions that mimic MG, and to search for associated conditions that may influence the choice of treatment. The diagnosis may be obscure in patients with coexisting neurological disorders affecting motor performance, such as cerebrovascular disease, Parkinson's disease, multiple sclerosis, and others. Sometimes the diagnosis can be difficult, especially in emergency situations, when the time available for testing is limited. Most importantly, one needs to consider MG in the differential diagnosis of a variety of disorders seen with muscle weakness (Table III).

Acquired autoimmune MG is the most prevalent myasthenic syndrome. Congenital myasthenias are rare. In general, they begin early in childhood and require sophisticated electrophysiological, immunocytological, and molecular investigations for precise diagnosis (Engel *et al.*, 1999). These syndromes should be considered if there is a positive family history or if there remains doubt regarding the autoimmune cause in a patient with the early onset of myasthenic weakness after complete workup for autoimmune MG.

Myasthenic patients have an increased incidence of several associated disorders. Thymic tumors occur in 10%–15% of patients. A thyroid disorder occurs in 3%–8% of myasthenic patients, and either hyperthy-

**TABLE I  Classification of Myasthenic Syndromes**

Acquired
    Autoimmune myasthenia gravis
    Lambert-Eaton myasthenic syndrome (LEMS)
Congenital
    *Presynaptic Defects*
        Paucity of synaptic vesicles
        Defect in ACh synthesis/packaging
        Congenital Lambert-Eaton–like syndrome
    *Synaptic Defect*
        Endplate AChE deficiency
    *Postsynaptic Defects*
        Primary kinetic abnormality with/without AChR deficiency
        Primary AChR deficiency without or with minor kinetic abnormality

**TABLE II  Clinical Classification Scheme for Myasthenia Gravis**

| | |
|---|---|
| Class I | Any ocular muscle weakness |
| | May have weakness of eye closure |
| | All other muscle strength is normal |
| Class II | Mild weakness affecting other than ocular muscles |
| | May also have ocular muscle weakness of any severity |
| IIa | Predominantly affecting limb, axial muscles, or both |
| | May also have lesser involvement of oropharyngeal muscles |
| IIb | Predominantly affecting oropharyngeal, respiratory muscles, or both |
| | May also have lesser or equal involvement of limb, axial muscles, or both |
| Class III | Moderate weakness affecting other than ocular muscles |
| | May also have ocular muscle weakness of any severity |
| IIIa | Predominantly affecting limb, axial muscles, or both |
| | May also have lesser involvement of oropharyngeal muscles |
| IIIb | Predominantly affecting oropharyngeal, respiratory muscles, or both |
| | May also have lesser or equal involvement of limb, axial muscles, or both |
| Class IV | Severe weakness affecting other than ocular muscles |
| | May also have ocular muscle weakness of any severity |
| IVa | Predominantly affecting limb and/or axial muscles |
| | May also have lesser involvement of oropharyngeal muscles |
| IVb | Predominantly affecting oropharyngeal, respiratory muscles, or both |
| | May also have lesser or equal involvement of limb, axial muscles, or both |
| Class V | Defined by intubation, with or without mechanical ventilation, except when employed during routine postoperative management. The use of a feeding tube without intubation places the patient in class IVb. |

From Jaretzki *et al.* (2000), copyright Lippincott Williams & Wilkins.

roidism or hypothyroidism may aggravate myasthenic weakness. Tests for thyroid function and thyroid autoantibodies should be obtained routinely. It is important to screen for other autoimmune diseases and subclinical signs of autoreactivity, because they may add to the diagnostic picture of autoimmune dysregulation, and because they may require additional therapy. Disorders that may interfere with immunosuppressive therapy include unsuspected infections such as tuberculosis (a skin test should always precede immunotherapy), diabetes, peptic ulcer, occult gastrointestinal bleeding, renal disease, hypertension, and occult malignancies.

## Clinical Tests

Myasthenic muscle weakness should be quantified using a clinical score based on the examination of various muscle groups. Two examples of useful quantitative myasthenia scores are shown in Table IV (Besinger *et al.*, 1981; Task Force of the Medical Scientific Advisory Board of the Myasthenia Gravis Foundation of America *et al.*, 2000). For pharmacological testing, the edrophonium (Tensilon) test is most widely used. Edrophonium chloride is a short-acting cholinesterase inhibitor (duration 3–10 minutes). The rapid action after intravenous administration allows repeated interaction between acetylcholine (ACh) and AChR and partially compensates for the functional deficit of receptors. This test should be carried out with objective assessment (scoring) of myasthenic weakness in muscle groups that are unequivocally affected, typically resulting in improvement of ptosis, eye movements, swallowing, speech, vital capacity or endurance of limb muscles. Confidence in an equivocal test is increased if the effect of edrophonium is compared with that of a placebo (saline) or if the results are documented by electrophysiological recording. If the response to repetitive nerve stimulation improves after

edrophonium, this confirms a defect in neuromuscular transmission. Details of the test, which should only be performed for strict diagnostic indication and with great care, are as follows.

For adults, 10 mg edrophonium is diluted in 10 ml saline. Atropine (1–2 mg) should be ready to antagonize possible muscarinic side effects. First, 2 mg (2 ml) of edrophonium is injected IV as a test dose. While comfortably seated, the patient is carefully observed for 60 seconds for objective improvement of function in weak muscles and side effects. Next, 3 mg of edrophonium is injected as a bolus. If the response is still equivocal after 3 minutes, the remaining 5 mg is injected. In case of marked sweating, bradycardia, or bronchospasm, 1–2 mg atropine should be injected. Relative contraindications to the edrophonium test are bradycardia and bronchial asthma.

The edrophonium test is not entirely specific for MG, and equivocal or frankly positive responses, especially of ocular symptoms, have been observed in a variety of

**TABLE III**   Differential Diagnosis of Autoimmune Myasthenia Gravis

| Disorder | Comments |
|---|---|
| Congenital myasthenic syndromes | Positive family history |
| Drug-induced myasthenia | Drug history (e.g., D-penicillamine) |
| Lambert-Eaton myasthenic syndrome | Small cell lung cancer, autonomic nervous system disturbances |
| Polymyositis, dermatomyositis | EMG, muscle biopsy, elevated muscle enzymes |
| Motor neuron disorders | Early atrophy, fasciculations, hyperactive reflexes |
| Ocular symptoms in multiple sclerosis patients | Relapsing course, associated CNS findings |
| Cranial neuritis | Pupillary abnormalities, impaired sensation; CSF examination |
| Guillain–Barré's studies syndrome | Reflexes, sensory symptoms, nerve conduction studies |
| Mitochondrial myopathies (CPEO) | Retinitis, muscle biopsy |
| Endocrine ophthalmopathy | Thyroid abnormalities, CT |
| Botulism | History, gastrointestinal disturbances |
| Intracranial lesion or lesion at the base of the skull | Pattern of cranial nerve involvement; CT or MRI imaging results |
| Vasculitis involving cranial motor nerves | Localized pain |
| Dyskalemic periodic paralyses | Often familial; potassium levels, absence of ocular or oropharyngeal signs |
| Addison's disease | Associated clinical findings, laboratory results |
| Myotonia, Becker type, with weakness and fatigue | Autosomal recessive; myotonic discharges on EMG |
| Nonorganic weakness | Not uncommon in conjunction with MG |
| Chronic fatigue syndrome | Ill-defined disorder, may include cases of MG |

disorders including brainstem glioma or vascular malformations, cranial neuropathies, and orbital tumors. The test is negative in patients with congenital myasthenia caused by endplate acetylcholinesterase (AChE) deficiency and in patients with a defect in ACh resynthesis or packaging. The test is inconsistently positive in the slow-channel syndrome and in the high-conductance fast-channel syndrome. Hence, a negative edrophonium test does not exclude the diagnosis of a congenital myasthenic syndrome, and a positive test does not differentiate congenital and aquired autoimmune myasthenia (Engel *et al.*, 1999).

A useful alternative or addition to the edrophonium test is the oral pyridostigmine (Mestinon) test. The patient receives 30 or 60 mg PO and reports back after 60 and 90 minutes for quantitative testing.

**TABLE IVA**   Clinical Score for Myasthenia Gravis

| Test items | None 0 | Mild 1 | Moderate 2 | Severe 3 | Weakness grades |
|---|---|---|---|---|---|
| 1. Arms outstretched (90°, standing), seconds | >180 | >60–180 | 10–60 | <10 | |
| 2. Leg outstretched (45°, supine), seconds | >45 | >30–45 | 5–30 | <5 | |
| 3. Head lifted (45°, supine), seconds | >90 | >30–90 | 5–30 | <5 | Extremities and trunk muscles |
| 4. Vital capacity | | | | | |
| Male | >4 | >2.5–4 | 1.5–2.5 | <1.5 | |
| Female | >3 | >2–3 | 1.2–2 | <1.2 | |
| Forced vital capacity (%) | >90 | >60–90 | 40–60 | <40 | |
| 5. Chewing/swallowing | Normal | Fatigue on chewing solid foods | Only soft foods possible | Gastric tube | Facial and pharyngeal muscles |
| 6. Facial muscles | Normal | Mild weakness on lid closure | Incomplete lid closure | No mimic expression | |
| 7. Double vision (lateral gaze), seconds | >60 | >10–60 | >0–10 | Spontaneous heterotropia | Ocular symptoms |
| 8. Ptosis (upward gaze), seconds | >60 | | | Spontaneous ptosis | |

*Note:* Ocular symptoms are not scored if there is generalized weakness. Calculate score by adding grades of the two categories with the worst grades. If worst grades are identical for two categories, select category in which deterioration occurred last. If two or more categories are identical with respect to these criteria, select categories in the following order: Vital capacity—swallowing/chewing—arms—legs—head—face.
Modified after Besinger *et al.* (1981).

TABLE IVB  Quantitative MG Score for Disease Severity

| Test item | None | Mild | Moderate | Severe |
|---|---|---|---|---|
| Grade | 0 | 1 | 2 | 3 |
| Double vision on lateral gaze right or left (circle one), seconds | 61 | 11–60 | 1–10 | Spontaneous |
| Ptosis (upward gaze), seconds | 61 | 11–60 | 1–10 | Spontaneous |
| Facial muscles | Normal lid closure | Complete, weak, some resistance | Complete, without resistance | Incomplete |
| Swallowing 4 oz. water (½ cup) | Normal | Minimal coughing or throat clearing | Severe coughing/choking or nasal regurgitation | Cannot swallow (test not attempted) |
| Speech after counting aloud from 1 to 50 (onset of dysarthria) | None at 50 | Dysarthria at 30–49 | Dysarthria at 10–29 | Dysarthria at 9 |
| Right arm outstretched (90 deg sitting), seconds | 240 | 90–239 | 10–89 | 0–9 |
| Left arm outstretched (90 deg sitting), seconds | 240 | 90–239 | 10–89 | 0–9 |
| Vital capacity, % predicted | ≥80 | 65–79 | 50–64 | <50 |
| Rt-hand grip, kgW | | | | |
| Men | ≥45 | 15–44 | 5–14 | 0–4 |
| Women | ≥30 | 10–29 | 5–9 | 0–4 |
| Lt-hand grip, kgW | | | | |
| Men | ≥35 | 15–34 | 5–14 | 0–4 |
| Women | ≥25 | 10–24 | 5–9 | 0–4 |
| Head lifted (45 deg supine), seconds | 120 | 30–119 | 1–29 | 0 |
| Right leg outstretched (45 deg supine), seconds | 100 | 31–99 | 1–30 | 0 |
| Left leg outstretched (45 deg supine), seconds | 100 | 31–99 | 1–30 | 0 |
| | | | Total QMG score (range, 0–39) _____ | |

From Jaretzki et al. (2000), copyright Lippincott Williams & Wilkins.
For each item, the "line score" will be added and the total QMG score is the SUM of all line scores.

## Electrophysiological Tests

The main purpose of electrophysiological investigations in MG is to provide additional support for the diagnosis, demonstrate impaired neuromuscular transmission, aid in differentiating MG from other disorders such as the Lambert-Eaton myasthenic syndrome, and assess the distribution and severity of impairment of neuromuscular transmission (Harper, 2001).

The typical finding in MG is a decremental response of the compound muscle action potential after supramaximal repetitive nerve stimulation at 2–3 Hz. A decrease in the amplitude of more than 10%, usually most pronounced at the fifth evoked potential, is considered abnormal. "Postactivation exhaustion" provides an even more sensitive test. The patient first contracts the muscle maximally for 30–60 seconds. Repetitive nerve stimulation is then carried out as earlier at 1-minute intervals for the next 5 minutes. In generalized MG, a positive decremental response is present in about 50% of patients with mild disease and 80% with moderate to severe weakness. A decremental response

at 2–3 Hz stimulation is not unique to MG. It may be observed in other neuromuscular diseases, such as the congenital myasthenic syndromes and the Lambert-Eaton myasthenic syndrome.

If the results of repetitive nerve stimulation are equivocal, single-fiber electromyography, which is the most sensitive (but not specific) electrophysiological test for detection of disturbances of neuromuscular transmission, may show increased "jitter," that is, variability of the interval between two action potentials from two muscle fibers belonging to the same motor unit (Harper, 2001). Both decrement and jitter measurements are more frequently abnormal in proximal than in distal muscles.

In the Lambert-Eaton myasthenic syndrome, there is also a decremental response after 3 Hz stimulation. However, in contrast to MG, the initial compound muscle action potential is abnormally small. On double stimulation or on high-frequency repetitive nerve stimulation (20–50 Hz) or after the patient strongly contracts the muscle, a characteristic incremental response is seen. An increment of more than 25% suggests and an increment of more than 100% confirms the diagnosis (see section on "Lambert-Eaton Myasthenic Syndrome").

### Laboratory Tests

The detection of anti-AChR autoantibodies is the single most important diagnostic finding in MG. Anti-AChR autoantibodies can be demonstrated by radioimmunoassay in more than 90% of patients with generalized MG but in only about 50% with purely ocular myasthenia (Vincent and Newsom-Davis, 1982). Other tests that use cell lines have been developed (Beeson *et al.*, 1996; Blaes *et al.*, 2000; Kennel *et al.*, 1995; Somnier, 1994; Voltz *et al.*, 1991). Virtually all thymoma patients have anti-AChR autoantibodies.

Positive results (more than ~0.4 nM α-Bungarotoxin binding sites/L) have been found in a few thymoma patients without clinical myasthenia. A low incidence of "false-positive" results has been reported for patients with other autoantibodies and occasionally in motor neuron disease, muscular dystrophy, and tardive dyskinesia. Patients with ocular myasthenia tend to have lower titers than patients with generalized disease. However, absolute AChR-antibody titers do not correlate well with the severity and distribution of symptoms. By contrast, in individual patients, there is a reasonably good correlation between changes in the anti-AChR antibody concentration and the clinical course.

Approximately 10% of all MG patients and 50% of patients with ocular MG are AChR antibody-negative by the standard radioimmunoassay. Most (about 70%) patients with generalized MG who are "seronegative," that is, do not have measurable anti-AChR autoantibody titers, have serum autoantibodies against the muscle specific receptor tyrosine kinase (MuSK) (MuSK mediates the agrin-induced clustering of AchRs during synapse formation and is also expressed at the mature neuromuscular junction [Hoch *et al.*, 2001]).

Autoantibodies to additional skeletal muscle antigens are present in approximately 80% of patients with thymoma, and a positive test is considered suggestive of the presence of thymoma. However, myasthenic patients without thymoma may also have antibodies to skeletal muscle antigens. At least two muscle proteins, titin and the ryanodine receptor of the sarcoplasmic reticulum, are recognized by these "striated muscle autoantibodies" in MG. It is not clear whether these autoantibodies play any pathogenic role in MG (Romi *et al.*, 2000). Regardless of the results of tests for autoantibodies against striated muscle, all MG patients should have a computed tomography (CT) or magnetic resonance imaging (MRI) scan of the thorax to exclude thymoma (see next section).

### Radiological Investigations

Most MG patients have gross or microscopic thymic abnormalities. A CT or MRI scan of the chest and anterior mediastinum should be performed in all patients with MG to search for a *thymoma*, which can be expected in 10%–15% of patients irrespective of their age at onset or severity of symptoms. Small tumors may be difficult to detect. However, given the high resolution of current imaging technology, most thymomas become detectable. MRI is particularly helpful for the detection of tumors in atypical locations, because thymomas may be found throughout the thoracic cavity.

Note that a normal CT or MRI scan does not rule out *thymic hyperplasia*. This is a histological diagnosis characterized by the presence of germinal centers and is found in approximately 60% of myasthenic patients. Patients with purely ocular symptoms and an equivocal response to AchE inhibitors should have a CT or MRI of the head, including the orbital region, to exclude intracranial or orbital neoplastic or vascular disorders as the cause of their symptoms (see "Differential Diagnosis").

## NATURAL COURSE

The annual incidence of MG is 2–4/million, and the prevalence has recently been assessed to be as high as 2–7/10,000 population in Great Britain (MacDonald *et al.*, 2000). The distribution is age and zendic-related, with one peak in the second and third decades affecting mostly women, and a peak in the sixth and seventh decades affecting mostly men. It is rare in children less than 10 years of age. However, in Japanese and Chinese populations there is a relatively high incidence of autoimmune MG in children less than 3 years of age. In about 50%–60% of patients, the presenting symptoms involve extraocular muscles, producing ptosis and diplopia. Ocular muscle weakness shows patterns that usually do not correspond to involvement of individual nerves, and the pupillary reactions are normal. Pronounced fluctuations, and symptoms variably affecting either eye, are characteristic of myasthenic weakness. These locally restricted manifestations are followed by more widespread weakness over the next weeks or months in about 60% of patients who are seen with ocular symptoms (Oosterhuis, 1989). In most cases the disease reaches its maximum severity in the first 3–5 years after onset. If weakness remains localized to the extraocular and eyelid muscles for more than 12 months, generalized MG is unlikely to develop later. About 15% of patients have purely "ocular MG." The symptoms may initially be transient and may reappear later, often affecting additional muscle groups. It should

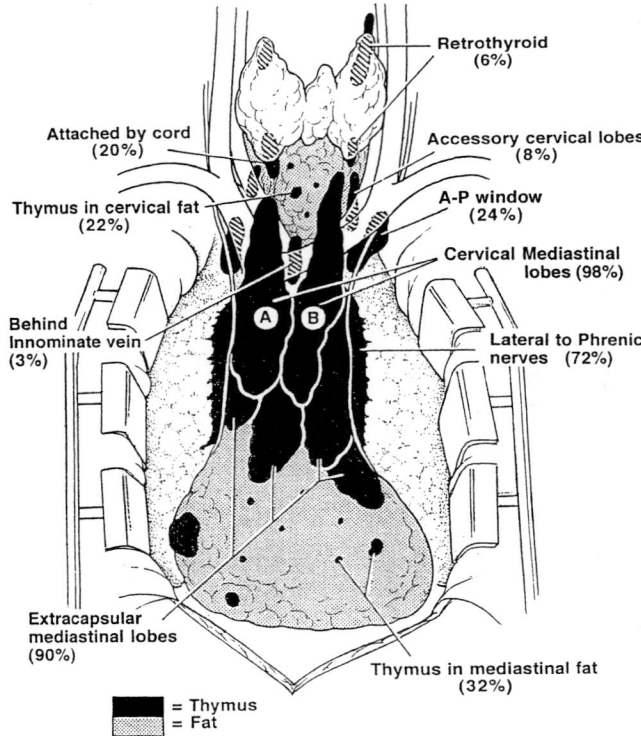

**FIGURE 1** Anatomy of the thymus. This illustration represents what is now generally accepted as the surgical anatomy of the thymus. The frequencies (percent occurrence) of the variations are noted. Black = thymus; gray = fat that may contain islands of thymus and microscopic thymus. A-P window = aortopulmonary window. (From Jaretzki *et al.*, 2000, copyright Lippincott Williams & Wilkins.)

be emphasized that the underlying immune disorder is systemic, regardless of the localization of weakness. Muscle atrophy, as a result of "functional" denervation at the shoulder girdle, upper arms, face, neck, or tongue, although not typical for MG, has been observed in about 10% of very severe generalized cases. This is now very rare in patients treated early with adequate immunosuppression.

Modern treatment has greatly changed the prognosis of MG (Grob, 1999). Before the introduction of immunosuppressive therapy, the mortality rate for generalized MG was 30%, and more than 60% of patients either failed to improve or actually deteriorated. In particular, mortality in patients who real myasthenic crisis at that time was about 70%. With the treatment options available today, most patients can lead essentially normal lives. However, most patients must take immunosuppressive medication for many years or even indefinitely, despite of the risk of adverse effects. Sudden deterioration with respiratory failure (myasthenic crisis) is now rare (less than 2%) in patients treated with long-term immunosuppression. Patients with MG and

thymoma (who are often older than 40 years) are more liable to myasthenic crisis than are patients without thymoma. In thymoma patients, the prognosis is related to the biology of the tumor. The ultimate goal of a cure of the autoimmune process has not yet been achieved, although promising strategies based on new knowledge of the immunobiology of MG are being developed.

## THERAPEUTIC PRINCIPLES

### Pathophysiological Basis

The basic defect in autoimmune MG is a loss of available postsynaptic AChRs at the neuromuscular junction. If the AChR concentration is reduced by more than 50%, the endplate potential may fail to reach the threshold for generation of an action potential on repetitive activation. At the normal motor endplate, the AChR concentration is many fold greater than is required for the generation of an action potential. This excess of available AChRs is called the "safety factor" of neuromuscular transmission. Normally, the safety factor allows high-frequency activation of the endplate up to 40 Hz, which exceeds physiological requirements in skeletal muscles.

MG is one of the few human autoimmune diseases in which the autoantigen is known. Although the cause of the autoimmune response against AChR remains unknown, many steps in the pathogenetic cascade that lead to impairment of neuromuscular transmission are well understood (Drachman, 1994; Hohlfeld and Wekerle, 1999). More than 90% of patients have circulating anti-AChR autoantibodies that impair AChR function, mainly by three different mechanisms. First, the binding of antibodies leads to cross-linking of receptors, which accelerates internalization and degradation of AChR. Second, the binding of antibodies causes local activation of the complement cascade, eventually leading to complement-mediated destruction of the folds of the postsynaptic membrane at the neuromuscular junction. Third, some anti-AChR antibodies block the binding site for acetylcholine. It is less certain whether antibody-mediated alteration of the properties of the acetylcholine-gated ion channel plays a role in the pathophysiology of MG.

The production of anti-AChR autoantibodies is controlled by AChR-specific helper T lymphocytes, which are regulatory T cells that can be isolated from the blood and thymus. Several lines of evidence indicate that the thymus is profoundly involved in the pathogenesis of MG (reviewed in Hohlfeld and Wekerle, 1999). First, more than 70% of patients have lymphofollicular hyperplasia of the thymus, characterized by the presence of germinal centers, and at least 10% have thymic epithelial neoplasia. Second, the thymus contains epithelial cells (myoid, or musclelike cells) that express AChR. Third, the thymus in MG contains AChR-specific B lymphocytes and plasma cells that spontaneously produce anti-AChR antibodies in vitro. Taken together, this suggests that the thymus contains all the elements required to initiate and sustain an autoimme response against the AChR. Finally, there is some evidence that early thymectomy is beneficial for many patients (Gronseth and Barohn, 2000).

Two lines of therapy are usually combined to treat MG. Symptoms can be rapidly ameliorated by anticholinesterase agents, which partially compensate for the reduced safety margin at the neuromuscular junction. Such symptomatic treatment may be sufficient in cases of mild MG. However, this treatment does not affect the underlying autoimmune process. In patients with generalized autoimmune MG, immunosuppression with steroids, azathioprine, or other immunosuppressive drugs is usually required. Immediate removal of autoantibodies by plasma exchange or immunoadsorption is generally very effective as a short-term treatment in crisis, in cases with rapid deterioration, or in instable patients before thymectomy. Following the guidelines below, among various therapeutic options the best available therapy has to be selected for the individual patient.

Detailed knowledge of the interaction of autoimmune lymphocytes with their targets has inspired various approaches to interfere with the activation of lymphocytes as the principal immunological event (Hohlfeld, 1997; Wu et al., 2000). A large number of therapeutic strategies for selective immunointervention have been explored in experimental conditions in vivo and in vitro. Some approaches seem to have much potential and may be further developed to restore immune tolerance to the AChR. But it is still difficult to predict which are the most promising strategies.

## Anticholinesterase Agents

Several classes of AChE inhibitors differ slightly in their mode of action (Millard and Broomfield, 1995). The clinically useful anticholinesterase reagents inhibit AChE reversibly. An example is edrophonium, which has a very brief duration of action because of the reversibility of its binding to AChE and rapid elimination by the kidneys. In contrast, drugs like pyridostigmine and neostigmine are hydrolyzed by AChE but much more slowly than is ACh. In vivo, the duration of inhibition by these carbamylating agents is in the range of hours. Table V compares the anticholinesterases that

are in clinical use. The action of edrophonium chloride is so short-lived that it is used only for diagnostic purpose.

At the neuromuscular junction, a massive overdose of anti-AChE drugs leads to failure of neuromuscular transmission caused by the excess and prolonged action of ACh. Fasciculations or cramps can be observed, and AChRs are desensitized, resulting in increased weakness, fatigability, or "cholinergic crisis" with excessive overdoses. In general, the toxic side effects of anti-AChE drugs are caused by an excess of ACh and actions on muscarinic AChR. These adverse reactions may affect the gastrointestinal system (abdominal cramps, diarrhea, anorexia, nausea, vomiting), the respiratory system (bronchoconstriction and increased bronchial secretion), the eyes (miosis, conjunctival congestion), glandular secretion (lacrimation, salivation, profuse sweating), and the heart (bradycardia, hypotension). Atropine effectively antagonizes the muscarinic side effects but is ineffective at the neuromuscular junction and, therefore, has no influence on the muscle weakness caused by excessive ACh.

Experimental evidence indicates that chronic application of very high doses of neostigmine may lead to changes in the neuromuscular junction that are similar to those caused by MG itself, that is, degeneration of junctional folds, decrease of AChR, and reduced amplitudes of miniature endplate potentials. However, patients do not receive or tolerate the high doses of AChE inhibitors used in the animal experiments, so that the practical importance of this potential side effect is not proven.

## Glucocorticosteroids

The anti-inflammatory and immunosuppressive effects of glucocorticosteroids have several different components (Katz, 2000). Steroids profoundly influence the distribution and trafficking of leukocytes. For example, steroids induce peripheral blood neutrophilia, whereas T cells, monocytes, and eosinophils are depleted from blood. Second, steroids alter the functional properties of T cells and monocytes. Third, steroids act on the synthesis and secretion of cytokines and immune mediators. Fourth, steroids have an effect on microvascular permeability. This may contribute to the effects on leukocyte migration.

A beneficial effect of adrenocorticotrophic hormone (ACTH) in the treatment of MG was first described in the 1930s. In a cooperative study some 300 courses of ACTH were given to 100 patients, and good improvement was achieved (Genkins et al., 1971). A gradual effect of prednisone, which starts after a few days but is usually obvious after 2 weeks or exceptionally after 4–6 weeks, was reported in later studies (Genkins *et al.*, 1971). Initial steroid-induced worsening of myasthenic symptoms may occur despite slowly increasing alternate-day schedules (Genkins *et al.*, 1971). Short courses of large intravenous doses were used with good results to manage exacerbations (Arsura *et al.*, 1985), but high doses carry a considerable risk of steroid-induced worsening.

Two randomized trials of prednisone versus placebo (Howard *et al.*, 1976; Lindberg *et al.*, 1998), two randomized trials of prednisone versus azathioprine (Bromberg *et al.*, 1997; Myasthenia Gravis Clinical Study Group, 1993), and one randomized double-blind trial of prednisolone alone or combined with azathioprine (Palace *et al.*, 1998) can be summarized as follows: Howard and coworkers (Howard *et al.*, 1976) found improvement in three of six patients treated with high-dose oral prednisone but also spontaneous improvement in three of seven patients in the placebo group. Bromberg and coworkers (Bromberg *et al.*, 1997) observed improvement of myasthenic symptoms and a decrease of AChR antibody titers in the prednisone and in the azathioprine group. Methylprednisolone was shown to be effective in moderately severe MG in 80% of the patients receiving 2 g on two consecutive days (Lindberg *et al.*, 1998). A double-blind, randomized study comparing high-dose intravenous immunoglobulins versus medium-dose methylprednisolone indicated that intravenous immunoglobulin seems slightly more effective than methylprednisolone within the first 3–7 days of treatment. After 14 days, both therapies showed equivalent clinical effects (Schuchardt and Toyka, unpublished). For ethical reasons most recent studies lacked a control group treated with placebo alone, but placebo was given in combination with at least one immunosuppressive agent.

Corticosteroids can produce improvement in patients with all degrees of myasthenic weakness, from diplopia to severe respiratory involvement. Most patients improve with steroid treatment; the clinical effect often begins within 2–3 weeks, although it may take longer, and maximal benefit may not be realized until 6–12 months, or more. Although steroids are the most consistently effective immunosuppressive agents for MG, they also have the largest array of potential side effects. Therefore, steroid therapy must be managed with care by clinicians experienced in its use. Patients with moderate to severe generalized MG should be hospitalized for the initiation of steroid therapy because of transient steroid-induced exacerbation of myasthenic weakness, which may occur in up to 48% of patients during the early stages of treatment. This side effect may come from a direct action on neuromuscular transmission (Dudel *et al.*, 1979). This may be avoided by gradually increasing the dose of the steroid medication over the course of weeks. Alternatively, in more severe MG, plasma exchange may reduce the likelihood of early steroid-induced exacerbation and accelerate the attainment of a therapeutic level of medication.

The continued administration of steroids carries the risk of side effects for which patients should be closely monitored. The most important side effects with long-term treatment are osteoporosis, hypertension, exacerbation or precipitation of diabetes, obesity, gastrointestinal ulcers, cataracts, and opportunistic infections. Blood pressure, body weight, serum electrolytes, and glucose level should be closely followed and adequately treated. Regular slit-lamp examinations (every 6 months) help to detect cataracts early. Gastrointestinal discomfort is best dealt with by drinking skim or low-fat milk during the day. If a patient has a history of recurrent ulcers, histamine $H_2$-receptor antagonists such as ranitidine, or $H^+$, $K^+$-ATPase inhibitors such as omeprazole (Antra) may be given to protect the upper gastrointestinal mucosa from ulceration and bleeding.

To minimize steroid-induced osteoporosis, the single most effective measure is reduction of dosage or, if possible, complete discontinuation of the glucocorticoid. All patients, especially the elderly, should be given calcium supplements. The 24-hour urine calcium output should be measured after 1–3 months of glucocorticosteroid therapy, and oral calcium (1 g/day) and vitamin D (50,000 units 1–2x per week) should be given, based on the level of urinary calcium excretion. Bisphosphonate agents are increasingly used for the treatment and prevention of steroid-induced osteoporosis. In postmenopausal women estrogens may be given to reduce the risk of fractures. Potassium replacement is necessary only in patients who are known to have hypokalemia develop.

### Azathioprine (Imurek, Imuran)

Azathioprine acts primarily on proliferating lymphocytes and induces both B- and T-cell lymphopenia (Elion, 1993). Antigen- and mitogen-induced in vitro proliferative responses of T cells are less inhibited in azathioprine-treated patients than in cyclophosphamide-treated patients. Azathioprine also has mild anti-inflammatory properties probably caused by the inhibition of promonocyte cell division.

Azathioprine is used most commonly as an adjunct to reduce the dose of steroids required but may be used alone as maintenance treatment. It is one of the best-tolerated therapeutic agents to use but has two drawbacks: First, patients may show an acute idiosyncratic reaction, with general malaise, fever, skin reactions, and gastrointestinal symptoms of nausea and vomiting, even after the first dose. In this situation, the drug should be discontinued immediately. Second, its beneficial effect in MG begins slowly, requiring many months to 1 year for an adequate trial.

Azathioprine has been administered in MG since the late 60s (Mertens *et al.*, 1969) and is now well established as a standard immunosuppressive agent (Kuks *et al.*, 1991; Myasthenia Gravis Clinical Study Group, 1993). Alternately, cyclosporine A (Tindall *et al.*, 1987), cyclophosphamide (Perez *et al.*, 1981), and methotrexate may be used for long-term immunosuppressive treatment of myasthenia gravis in patients who do not tolerate or respond to azathioprine.

The first extensive experience with cytotoxic immunosuppressive agents in the treatment of MG was reported by Mertens and colleagues (Mertens *et al.*, 1969), who described 38 patients treated with a variety of agents (6-mercaptopurine, azathioprine, methotrexate, and actinomycin D). From the data reported by these authors it was impossible to discriminate whether patients responded to azathioprine or other drugs, but overall improvement was observed in 84% of the patients. Matell and associates found improvement in 78% of 26 patients treated with azathioprine (2 mg/kg/day) who had not responded to ACTH or glucocorticosteroids. The possible role of thymectomy and concurrent use of glucocorticosteroids was not established (Matell *et al.*, 1976). Mertens and coworkers then reported 78 patients treated with azathioprine and other agents, of whom 91% were reported to show definite improvement (Mertens *et al.*, 1981). Witte and coworkers found definite improvement in 15 of 18 patients (83%) treated with azathioprine alone for longer than 6 months (Witte *et al.*, 1984). Hohlfeld *et al.* (1985) evaluated the disease courses of severely affected MG patients who had been treated with long-term azathioprine until they had gone into stable clinical remission or near-remission for longer than 6 months. After discontinuation of azathioprine, 8 of 15 patients had a clinical relapse within 1 year. These 15 patients and another 5 patients who were treated identically were re-evaluated later (Michels *et al.*, 1988). The evaluation of these patients, including some patients from the earlier study, revealed relapses in four additional patients after 27–70 months, leaving about 40% of the patients in remission off treatment. One patient died from pulmonary embolism after plasma exchange. All relapsing patients again achieved remission after combined immunosuppressive treatment including azathioprine (Michels *et al.*, 1988).

Only recently randomized studies addressed the efficacy of azathioprine in the treatment of MG. In two of these trials prednisone was tested versus azathioprine (Bromberg *et al.*, 1997; Myasthenia Gravis Clinical Study Group, 1993). In a randomized, placebo-controlled double-blind study azathioprine in combination with prednisone was tested versus azathioprine and placebo (Palace *et al.*, 1998). None of these studies had a placebo arm, because this was believed to be unethical.

The incidence of serious side effects of azathioprine is surprisingly low. The most frequent adverse reactions encountered during long-term treatment in 104 patients with severe generalized MG in decreasing order of frequency were reversible bone marrow depression with leukopenia, gastrointestinal discomfort, infections, and transient elevation of liver enzymes (Hohlfeld *et al.*, 1988). The most serious observation, which may possibly have been related to 6 years of treatment with azathioprine, was the development of a renal lymphoma in one patient. In another study, 2 of 229 azathioprine-treated patients had primary central nervous system lymphoma develop after 6 and 12 years of azathioprine (Herrlinger *et al.*, 2000). All these observations should alert us that in patients with MG, particularly patients older than 50 years, an attempt should be made to taper and finally discontinue azathioprine therapy after several months or years of stable remission, although the strong possibility of recurrence of symptomatic myasthenia must be closely watched.

Mild intestinal discomfort, reported relatively commonly, can usually be alleviated by splitting the dose into three or more divided doses, taking the drug after meals, starting treatment with a first 50-mg bedtime dose, and reducing the dose temporarily. Elevation of liver enzymes up to three times the baseline is also common and may be tolerated, because it is usually reversible after the dose has been reduced. The risk of malignancies developing, especially lymphoma, is not certain (see earlier) but seems to be lower in patients with MG than in organ transplant recipients. By contrast to what might be expected, serious infections are rarely a problem in myasthenic patients on azathioprine. Azathioprine is potentially teratogenic and mutagenic. Patients should be advised to use contraceptive measures during treatment and for at least several months after its completion whenever this is possible. Data available from mothers treated with azathioprine for kidney transplant have not shown an increased rate of birth defects in their children, but no data on the actual risk are available.

Patients should be monitored carefully for side effects during treatment. Complete blood counts should be obtained at least weekly during the first 2 months and monthly thereafter. If the total white blood count (WBC) is reduced to less than 3000/μL, the medication should be discontinued for a few days and treatment continued at a lower dose after the WBC returns to more than 3500/μL. The long-term dose can be adjusted to maintain the WBC around 4000/μL and lymphocyte counts ranging between 800 and 1000/μL. However, it is not certain whether the immunosuppressive efficacy of azathioprine therapy in autoimmune diseases is directly correlated to the WBC or lymphocyte count.

In patients receiving azathioprine and steroids, the total WBC is usually elevated because of steroid-induced neutrophilia (see earlier). Therefore, the preceding suggestions for monitoring treatment by total WBC do not apply. We use a WBC of 6000–8000/μL as the lower range during combined treatment. The lymphocyte count is less markedly altered by corticosteroids and can also be used for monitoring.

Another measure of a drug effect is the mean corpuscular volume (MCV) of red cells, which is usually but not invariably elevated (up to 15%) during long-term treatment. This may be useful in situations when there is doubt about the compliance of patients in taking their medication.

An important drug interaction occurs with allopurinol. Inhibition of xanthine oxidase by allopurinol impairs the conversion of azathioprine to 6-thiouric acid, which accumulates and eventually leads to potentially deleterious bone marrow suppression. If allopurinol must be administered concurrently, the dose of azathioprine must be reduced to about 25% of the regular dose (approximately 0.5 mg/kg body weight), and the WBC should be closely monitored.

### Mycophenolate Mofetil

Like azathioprine, mycophenolate mofetil is an immunosuppressive agent acting on DNA metabolism. In transplantation medicine, mycophenolate mofetil has proved useful and seems more powerful than AZA (Lipsky, 1996). In patients with neuroimmunological disease (e.g., MG), mycophenolate mofetil has been used as an alternative to AZA. Mycophenolate mofetil inhibits inosine monophosphate dehydrogenase and thereby depletes guanine nucleotides, leading to inhibition of DNA synthesis in lymphocytes, but not in other cells (which have an alternative "salvage pathway" of purine synthesis). Its reported adverse effects include gastrointestinal symptoms, gastrointestinal hemorrhage, leukopenia, and infection. Compared with AZA, its hepatotoxicity is low, but its risk of secondary lymphoma may be slightly higher (Lipsky, 1996). In contrast to AZA, the combination of mycophenolate mofetil and allopurinol is not problematic. Thus far, a number of case reports indicate that mycophenolate mofetil is beneficial in MG (Chaudhry et al., 2001; Ciafaloni et al., 2001; Hauser et al., 1998; Meriggioli

and Rowin, 2000). Preliminary results from several open clinical trials indicated that at least 50% of the patients improved to some degree. In general, clinical benefit from mycophenolate mofetil occurs as late as 3–12 months, or even more (Chaudhry et al., 2001). Because it is not cytotoxic, but only prevents proliferation of lymphocytes, the pre-existing populations of AChR-reactive lymphocytes must gradually die off before a beneficial clinical effect is apparent. Furthermore, mycophenolate mofetil seems to be an effective alternative immunosuppressant in severe refractory MG.

### Cyclosporine A (Sandimmune)

Together with sirolimus (rapamycin) and tacrolimus (FK506), cyclosporine belongs to the trinity of immunophilin-binding drugs. These agents form complexes with abundant intracellular proteins called immunophilins: cyclosporine A with cyclophilin (CP), and tacrolimus and sirolimus with the tacrolimus-binding protein (FKBP). The cyclosporine-CP and tacrolimus-FKBP complexes inhibit the the phosphatase calcineurin and its substrate, the transcription factor NFAT, thus preventing the transcription of messenger RNAs for key cytokines, such as interleukin-2. Interference with NFAT regulation might be expected to yield additional avenues for immunosuppression in the future (Kiani et al., 2000).

The immunophilin-binding agents very efficiently suppress transplant rejection, but presently are only rarely used in neuroimmunological disease. Cyclosporine was demonstrated to be effective in myasthenia gravis (see later) but has remained a second-line agent, mainly because of its adverse effects and higher cost.

Cyclosporine A is a cyclical undecapeptide isolated from two species of soil fungi. It targets the immune system more specifically than azathioprine or cyclophosphamide. Since its immunosuppressive effects were first noted, it has revolutionized transplant surgery, leading to a dramatic increase in graft survival. The effects of cyclosporine focus on T lymphocytes, whereas B cells and macrophages are apparently spared. The only appreciable effect of cyclosporin on accessory cells seems to be the disruption of lymphokine-dependent T lymphocyte/macrophage interactions. Clinical improvement usually begins after 2–4 weeks, not unlike corticosteroids.

The more serious potential side effects of cyclosporine include nephrotoxicity and hepatic disorders. In addition, cyclosporine can affect other organs such as the pancreas, CNS, bone, and skeletal muscle. Further adverse reactions include arterial hypertension, tremor, and hirsutism. Most of the adverse effects correlate with the dose and duration of treatment. Optimal dosage is monitored by measuring "trough" drug levels 12 hours after the last dose (i.e., first thing in the morning). The rate of nephrotoxicity is dose-dependent and low when treatment is initiated with modest doses (about 5 mg/kg body weight). If the creatinine level increases by 50% over baseline levels or by more than 1.5 mg/100 ml during treatment, the dose should be reduced or the drug discontinued. A more sensitive indicator of nephrotoxicity is the measurement of creatinine clearance. Cyclosporin must be discontinued if idiosyncratic or allergic reactions develop. The risk of late malignancies is not established but may be similar to that of azathioprine. With overt malignancy including thymic carcinoma, cyclosporine A is not recommended.

Cyclosporine A was the first effective drug in MG to be studied in a prospective, double-blind and placebo-controlled trial (Tindall *et al.*, 1987). It is about as effective as azathioprine but has a more rapid onset of action (Bonifati and Angelini, 1997; Ciafaloni *et al.*, 2000; Tindall *et al.*, 1987, 1993). However, because of its multiple and potentially serious side effects and its high cost, it is considered to be a second-line drug.

## Cyclophosphamide (Endoxan, Cytoxan, Cyclostin, Cycloblastine)

Cyclophosphamide is an alkylating agent that has occasionally been used in refractory MG (e.g., Flachenecker *et al.*, 1998; Perez *et al.*, 1981). It is much more toxic than azathioprine and should be reserved for patients with severe MG in whom combined treatments with corticosteroids, azathioprine, or cyclosporine A have been ineffective. The immunosuppressive action of cyclophosphamide is secondary to its cytotoxic effect on hematopoietic cells, particularly lymphocytes and monocytes/macrophages.

Cyclophosphamide may have serious side effects. The major and often limiting side effect is bone marrow suppression, which affects leukocytes more than erythrocytes and platelets. The active metabolites are concentrated in the urine, which may cause damage to the epithelium of the bladder (hemorrhagic cystitis, malignant transformation). This kind of toxicity can be reduced by a high fluid intake and, during pulse administration of high doses intravenously, by the simultaneous use of a uroprotective agent (Mesna). Nausea and vomiting is common and can be largely prevented by ondansetron (Zofran) or similar compounds. Patients should be informed that during prolonged therapy serious side effects may develop and should be instructed to report signs and symptoms without delay. Cyclophosphamide is both teratogenic and mutagenic.

Permanent infertility may occur in both male and female patients. Alopecia may develop with short- and long-term therapy but is mild and reversible after treatment is stopped. Additional rarer side effects include myocardial damage and pulmonary fibrosis. After long-term treatment, the potential for the development of leukemia and lymphoma is of particular concern. The risk of malignancy depends on the cumulative dose, which should not exceed a total of 85 g. Like other cytotoxic agents, cyclophosphamide is potentially teratogenic and should be avoided during pregnancy, especially in the first trimester.

Other immunosuppressive or immunomodulatory treatments that have been used in MG include splenic and whole-body irradiation and splenectomy. These treatments have been used in exceptional cases and will not be discussed further.

## Intravenous Immunoglobulins

The potential mechanisms of this therapy are not yet clear but may include interactions with inhibitory Fc receptors on phagocytic and antigen presenting cells (Dwyer, 1992; Samuelsson *et al.*, 2001). There is now convincing evidence that in principle, IV immunoglobulin treatment is effective in MG (Edan and Landgraf, 1994; Gajdos *et al.*, 1997, 1998; Howard, 1998). IgG has a potential role as an acute intervention in rapidly progressive weakness or as a chronic maintenance therapy when all other treatment modalities have failed or are contraindicated. The clinical response to IV IgG is similar to but slower than the response to plasma exchange, but it offers an alternative in myasthenic crisis when therapeutic apheresis is contraindicated or when vascular access is problematic (Howard, 1998).

In the only randomized trial in patients with severe myasthenic exacerbation, IV IgG treatment showed similar efficacy but was better tolerated than plasmapheresis (Gajdos *et al.*, 1997). In a retrospective analysis, plasma exchange more efficiently improved the ventilatory status but was associated with a higher complication rate than IV Ig treatment (Qureshi *et al.*, 1999). The usual total dose is 2 g/kg body weight given in divided doses on 3–5 consecutive days. If patients respond, the onset is usually within 4–5 days. The effect is temporary but may be sustained for weeks or months, allowing intermittent therapy for otherwise refractory patients. Adverse reactions occur in less than 10% of patients and include headache, fluid overload, aseptic meningitis, and rarely, renal failure (Ellis *et al.*, 1994; Meiner *et al.*, 1993; Steg and Lefkowitz, 1994). Patients with IgA deficiency can develop anti-IgA antibodies causing anaphylactic reaction on repeated treatment.

Therefore, IgG, A, and M levels should be measured before treatment. Disadvantages of IV Ig therapy are the inconsistency of the response, high cost, and shortage of supplies.

## Plasmapheresis

There are two basic techniques of therapeutic plasmapheresis: plasma separation by a cell separator (centrifuge) or by membrane filtration. A typical plasmapheresis protocol uses three to five exchanges of 1 or 1.5 plasma volumes over 1 week or longer until the patient shows satisfactory improvement. Usually, plasma exchange therapy is combined with immunosuppressive treatment (e.g., a combination of corticosteroids and azathioprine).

In the treatment of autoimmune diseases, plasmapheresis aims at the removal of circulating autoantibodies, inflammatory mediators, or both. In MG, early clinical effects of plasmapheresis are occasionally observed in less than 24 hours. Such immediate improvement is probably due to the removal of a minor fraction of autoantibodies that have a direct functional effect on the ACh receptor (Bufler et al., 1996). Often the effects of plasmapheresis are more delayed and become apparent only after 2 or more days. This delayed improvement is usually due to the removal of antibodies that act indirectly, for example, by increased receptor turnover or by complement-mediated lysis of the postsynaptic membrane.

Although there is practically no age limit for this treatment if the patient is in good general health, it is the elderly patient with multiorgan disease who carries an increased risk for severe complications developing. The complications of plasmapheresis include cardiovascular systemic reactions, electrolyte disturbances, and sepsis. Thrombophlebitis, thromboses, pulmonary embolism, and subacute bacterial endocarditis have been observed, particularly in patients who have had arteriovenous shunts or grafts placed for vascular access. Bacterial respiratory tract infections are a common problem in myasthenic crisis. Immunoglobulin depletion after plasmapheresis may decrease resistance to or favor these infections. We therefore recommend the early administration of appropriate antibiotics and add IV Igs (IgM/IgG, e.g., Pentaglobin) if overt pneumonia or sepsis is present. Severe infections have been reported after plasma exchange in patients who were immunosuppressed. It is important to note that the removal of clotting factors during plasmapheresis results in impairment of hemostasis for about 24 hours.

To increase the efficiency and selectivity of plasmapheresis, standard plasmapheresis techniques have been combined with (semi-)selective immunoadsorption to tryptophan-linked polyvinylalcohol gels or protein-A columns (Benny et al., 1999; Cornelio et al., 1998; Flachenecker et al., 1998; Grob et al., 1995; Sato et al., 1988; Shibuya et al., 1994; Yeh and Chiu, 2000). These more selective procedures seem to be at least as effective as standard plasmapheresis. Because there is negligible adsorption of albumin, protein substitution is not required.

## Thymectomy

Up to now, thymectomy has not been investigated in a prospective controlled clinical trial in MG. However, this form of treatment has been found useful empirically and is widely applied. For example, a retrospective study (Buckingham et al., 1976) compared computer-matched groups of patients treated with or without thymectomy before the era of immunosuppressive treatment. After thymectomy, clinical remission occurred in about 35% and improvement in another 50% of patients, which was significantly better than the outcome in patients who were not surgically treated.

A detailed review of the existing data by a Quality Standards Committee of the American Academy of Neurology concluded that most of the published controlled but nonrandomized studies showed positive associations between thymectomy and MG remission and improvement (Gronseth and Barohn, 2000). However, in all studies there were confounding differences in baseline characteristics between the thymectomy and nonthymectomy groups. Moreover, different immunosuppressive treatments were given after thymectomy. Because of the serious methodological flaws of the published studies, the Committee concluded that the benefit of thymectomy in nonthymomatous autoimmune MG has not been established conclusively. Therefore, thymectomy is recommended for patients with nonthymomatous autoimmune MG as an option to increase the probability of remission or improvement (Gronseth and Barohn, 2000). A well-designed prospective controlled study of thymectomy would help to resolve this issue but is unlikely to be accomplished in view of the present availability of effective treatment for myasthenia that would undoubtedly confound the results of thymectomy.

Most published reports did not show any correlation between the severity of myasthenic symptoms before surgery and the timing or degree of improvement after surgery. The time of the beginning of clinical improvement after thymectomy was relatively variable, and many patients improved only after several years. Most studies report better responses when thymectomy is performed early in the disease, but the mechanism is still uncertain.

Further to the uncertainties previously mentioned, there is no consensus about the lower and upper age limits for thymectomy, the indication for thymectomy in pure ocular myasthenia, or the benefit of early or late thymectomy compared with the natural course of MG. However, despite of the limited data, a number of clinical views seem reasonable in light of the presently available experience:

1. The best results may be expected in patients between 10 and 40 years of age with relatively recent onset of MG (within 3–5 years).

2. Men have about the same chance to improve as women.

3. Improvement may begin months to years after thymectomy, although it occasionally occurs within days or weeks. Rarely, late relapses occur in patients who experienced remission after thymectomy. It is possible that this is related to immunological activity or regrowth of thymic tissue that may not have been completely removed at surgery.

4. Children between 1 and 5 years are usually not submitted to thymectomy. Between ages 6 and 10, the indication for thymectomy is controversial. Adams et al. (1990) reviewed the long-term outcome of transsternal thymectomy in 24 children with generalized myasthenia (mean age, 11 years; range, 2–16) of mild and moderate severity; all but four of whom underwent surgery within 9 months of onset and the remaining four within 2 years. Two thirds developed complete remission, and no patients deteriorated. Despite surveys such as this, disagreement about the real place of thymectomy in the management of childhood myasthenia will continue unless a prospective study becomes available.

5. Patients with pure ocular myasthenia are usually excluded from thymectomy unless other therapies have proved unsuccessful or generalized symptoms appear. However, one small study reported that thymectomy resulted in significant improvement in 80% of patients with MG confined to the ocular muscles (Schumm et al., 1985).

6. Patients older than 60–65 years are usually not thymectomized, except for thymoma (see section on "Thymoma").

7. Thymoma is generally considered an absolute indication for thymectomy.

8. Thymectomy is usually not recommended in patients with seronegative MG.

9. Thymectomy (and thymomectomy) should be an elective procedure performed after the patient has been treated effectively by other means of treatment (see later) (e.g., by plasma exchange or other immunotherapies).

The most widely performed surgical approach to the thymus is transsternal. A transcervical approach is not recommended, because it does not ensure the removal of ectopic thymic tissue that may be widely distributed in the neck and mediastinum (Jaretzki et al., 2000). Minimally invasive thymectomy is now advocated by some surgeons, but its benefit is not established. When properly performed, thymectomy has a low mortality rate that is essentially that of anesthesia. However, it should be performed in a center with extensive experience and a neuromuscular consultant available. In patients with stable disease and after appropriate preparation, the operation is very safe, and severe perioperative complications are very uncommon (less than 1%). It should be emphasized that thymectomy should never be carried out as an emergency procedure, and the patient's vital functions should be stabilized by treatment with IV IgG or plasmapheresis if necessary before surgery (see later).

There is no need to discontinue azathioprine before or after surgery. During the immediate preoperative and postoperative period, oral Mestinon can be replaced by IV Prostigmine.

## PRAGMATIC THERAPY

### Generalized Myasthenia Gravis

Therapy should aim at complete or near-complete remission. This will allow most patients to return to normal professional and social activities. With the currently available treatment modalities, it is possible to achieve this goal eventually in almost every patient. It is essential that the patient be fully informed about the treatment options and time schedule and that the risks and benefits involved are explained to the patient. In short, the patient should be an active partner in the decision process.

All patients should first be treated with an AChE-inhibiting drug, usually pyridostigmine (Mestinon). Occasional patients tolerate or respond better to neostigmine bromide (Prostigmin) or ambenonium chloride (Mytelase). The optimum dose of any of these drugs and the timing of repeated doses must be determined for each patient. Especially when treatment is initiated, patients should be carefully observed for side effects. In adult patients, treatment begins with 30–60 mg pyridostigmine every 4 hours during the daytime. The action of pyridostigmine begins about 30 minutes after ingestion and lasts for about 4–6 hours, although the half-life is much longer. In infants and children the starting oral dose is 0.5–1.0 mg/kg pyridostigmine. Equivalent doses of AChE-inhibitors are shown in Table V.

**TABLE V**  Comparison of Different Anti-Cholinesterase Agents

| Drug* | Dose | | | Duration | |
|---|---|---|---|---|---|
| | PO | IV | IM | Begins action | Maximum action |
| Pyridostigmine bromide (Mestinon) | 60–90 mg | 1–2 mg | 2 mg | 14–45 min | 3–6 h[†] |
| Slow-release preparation (Mestinon-retard, -timespan) | 90–180 mg | — | — | 60 min | 6–10 h[†] |
| Neostigmine (Prostigmine) | 15 mg | 0.5 mg | 1 mg | 10–30 min | 2–3 h[†] |
| Ambenonium chloride (Mytelase) | 10 mg | — | — | 60 min | 6–8 h[†] |
| Distigmine (Ubretid) | 5–10 mg | — | — | 3 days | 24 h |

*Pyridostigmine is the *drug of choice* for oral long-term use. Daily doses should not exceed 600 mg for prolonged time. Neostigmine is the drug of choice for intravenous application during myasthenic crisis at 0.15–0.3 mg/h by means of an infusion pump.

[†]The total time of action may last up to 12 h, oral/IV ratio may vary among patients.

Adjustments of the drug dose and interval are made on the basis of clinical observations. It is important to quantify the clinical course with a quantitative scoring system (see Table IV). In patients in whom there is evidence of respiratory weakness, the vital capacity should be determined at regular intervals and included in the clinical score.

A fixed dosage schedule will not suit all patients. The needs may vary from day to day and in response to infection, menstruation, or emotional stress. Different muscles may respond differently. The dosage schedule should be tailored to produce the best response in the muscles that are of greatest concern in the patient. It is useful for patients to keep a diary describing the duration of beneficial and adverse effects of the medication.

The intervals at which pyridostigmine is taken can be shortened to every 3 hours. In patients with difficulty chewing or swallowing, the timing should be adjusted so that a dose is given 30–60 minutes before major meals. A sustained-release tablet containing 180 mg (Mestinon Retard/Timespan) may be given at bedtime to patients with marked weakness in the early morning. The sustained-release medication should not be substituted for regular pyridostigmine during the daytime because of its unpredictable pharmacokinetics. The maximum dosage of AChE-inhibitors usually should not exceed 90 mg of pyridostigmine or pyridostigmine-equivalent (Table V) every 3 hours. The need for a steady increase in the dose of AChE-inhibitors indicates deterioration and should alert the physician to the possibility that myasthenic crisis may be imminent and that additional treatment modalities must be considered.

AChE inhibitors may precipitate or exacerbate asthma and peptic ulcer. Other side effects are usually limited and reversible. The most common side effects are gastrointestinal symptoms. Persistent diarrhea can be treated with atropine, 0.125–0.25 mg (rarely 0.5 mg),

probanthine, 7.5–15 mg, or other analogs, occasionally with every dose of Mestinon, if the AChE inhibitor cannot be reduced. Antimuscarinic agents should not be given routinely from the start.

Thymectomy is always an elective procedure. It is indicated in patients with generalized MG. Patients with seronegative MG (no detectable serum anti-AChR antibodies by radioimmunoassay) have only small numbers of germinal centers in their thymus (Willcox *et al.*, 1991). The benefit of thymectomy in these patients is doubtful.

The upper and lower age limits for thymectomy are controversial. Most authorities agree that thymectomy is usually not indicated before age 5–10 and beyond age 60–65. In contrast, thymectomy is indicated in essentially all patients with thymoma. For patients with mild myasthenia, we recommend thymectomy in conjunction with AChE inhibitors. For patients with moderate or severe MG, surgery should be postponed until the symptoms, especially pulmonary function and swallowing, are controlled by medical treatment. It is particularly important to establish good respiratory function, with a forced vital capacity of approximately 2 L or more. There is a difference of opinion as to the relative risks and benefits of immunosuppressive therapy compared with plasma exchange, when it is necessary to improve the myasthenic patient's condition in preparation for surgery. Immunosuppression increases the risk of postoperative infection and slows the rate of wound healing. On the other hand, plasmapheresis may pose difficulties in obtaining venous access and can present potential circulatory problems, especially in older patients. Therefore, treatment with IV Igs is a reasonable alternative in patients at risk to prepare for thymectomy.

In patients whose symptoms persist for more than 3–6 months after thymectomy, in those who deteriorate after thymectomy, or in patients for whom thymectomy is rejected because of high surgical risk or advanced age (>60–65 years), immunosuppressive treatment is advised. If symptoms are not disabling, but sufficiently severe to interfere significantly with daily activities, it seems justified to offer the possibility of treatment with corticosteroids alone for a limited period (e.g., several months) after thymectomy. If symptoms improve during that time, the corticosteroids can be tapered and may eventually be discontinued. If the symptoms do not improve, or increase in severity, azathioprine should be added. If symptoms are mild and if the patient is willing to tolerate them for a limited period of time, it is justified to wait 6–12 months after thymectomy before starting immunosuppressive treatment.

In adults, immunosuppression may be initiated either (1) with corticosteroids alone, followed later with azathioprine as needed, or (2) with a combination of corticosteroids and azathioprine from the beginning. The rationale for (1) is that it permits evaluation of the beneficial and adverse effects of each drug separately. The use of combined immunosuppression (2) takes advantage of the more rapid clinical benefit of corticosteroids, while allowing extra time for azathioprine to take effect. It may be possible to taper the steroids somewhat earlier with this regimen.

1. If steroids are used alone to initiate treatment, the beginning daily dose (in hospital) is usually 15–20 mg of prednisone. It is increased by about 5 mg every 2–3 days, while observing the patient closely. The rate of increase should be guided by the patient's clinical response, and the end point is either a satisfactory clinical response or 50–60 mg/day (whichever occurs first). With this method, azathioprine may be added to the treatment regimen after steroid-induced improvement is established. A test dose of azathioprine (50 mg/day) is given for several days. If tolerated, it is increased to 2–3 mg/kg/day.
2. If the patient's condition is stable, and especially if plasmapheresis has been carried out to enhance the clinical status, the prednisone dosage may be started at a higher level (40–80 mg/day), which is the practice of one of us (KVT). If the patient tolerates a test dose (50 mg/day) of azathioprine, this drug may be added to the regimen at a dosage of 2–3 mg/kg body weight/day with subsequent tapering once remission has been achieved.

In either case, when the improvement in myasthenic weakness is deemed satisfactory (usually after 4–6 months), the steroids may be tapered very slowly, while the patient's clinical status and AChR antibody levels are monitored closely. It should be emphasized that most patients require continued immunosuppression at some level for many years and often for their entire lives. Therefore, the goal of the taper is to find the minimum effective dose required for maintenance of a satisfactory clinical status.

An important issue concerns the duration of immunosuppressive treatment after clinical remission has been achieved. Based on a study by some of us (Hohlfeld et al., 1985) an attempt may be made to taper and eventually discontinue azathioprine over 6–12 months in patients if they have been in stable clinical remission or near remission and have shown stable antibody titers for about 1 year. After withdrawal of azathioprine, patients are monitored monthly for the first 6 months and every 2–3 months thereafter for rising antibody titers and signs of clinical relapse. If antibody titers rise or if there are signs of clinical deterioration, immunosuppressive treatment is reinstituted.

In our experience serial antibody titers have been helpful in predicting an imminent clinical relapse (Hohlfeld et al., 1985; Michels et al., 1988). However, in some patients serial antibody titers paralleled rather than preceded a clinical relapse. Most important, patients must be advised to report immediately if they note any deterioration in their clinical condition.

## Ocular Myasthenia

In patients with pure ocular myasthenia it is advisable to start treatment with a lower dose of AChE inhibitors than is used in generalized myasthenia. Ptosis responds to AChE inhibitors much more favorably than does diplopia. If AChE inhibitors do not correct the symptoms at moderate to high doses, they should be discontinued at that time but may become more effective after successful immunosuppressive treatment. If AChE inhibitors fail, various mechanical devices may be considered. Occlusion of one eye with an adhesive patch applied to a spectacle may be helpful as a means to relieve transient diplopia. In an occasional patient, the diplopia may be sufficiently stable so that prisms may restore single vision. Self-adhesive plastic prisms are relatively inexpensive and may be helpful. Ptosis may be relieved by a custom-made lid "crutch" soldered to the inner side of the spectacle frame or by the use of a strip of transparent "Scotch" tape. If AChE inhibitors fail and mechanical devices are either insufficient or not acceptable to the patient (driving!), corticosteroids should be tried. Often, a relatively low dose results in resolution of ptosis and diplopia (Kupersmith et al., 1996).

The indication for thymectomy and azathioprine in ocular myasthenia is controversial. However, these treatment options may be considered for patients who have disabling ocular symptoms (interfering with their professional activities), need high maintenance doses of steroids, or develop adverse reactions to steroids.

## Thymoma

Tumors of the thymus (thymoma) occur in up to 15% of patients with MG; conversely, about 35% of patients with thymoma have MG. There is a broad consensus that patients with MG and thymoma should have their tumors removed surgically. The prime indication in these cases is the possible growth and spread of the tumor rather than improvement of myasthenia. Older patients (>70 years) with evidence of slowly growing tumors may be exceptions to the rule. In such patients it may be justifiable to follow the tumor size with serial CT or MRI scans of the mediastinum and offer palliative radiotherapy.

Before surgery, the patient's clinical status should be optimized, exactly as for thymectomy without thymoma (see earlier). After removal of the tumor, the principles of treatment for MG, previously described, also apply for patients with thymoma. Further treatment of the thymoma depends on the intraoperative staging of tumor invasion, on the histopathological findings, and on the clinical response after surgery. The most widely used staging of tumor invasion is as follows: stage I, completely encapsulated tumor without microscopic capsular invasion; stage II.A, microscopic invasion into surrounding fatty tissue and/or mediastinal pleura; stage II.B, macroscopic invasion of pleura; stage III, macroscopic invasion into pericardium, great vessels, lung; Stage IVA, pleural or pericardial dissemination; stage IVB, lymphogenous or hematogenous metastasis, which is very uncommon (Masaoka *et al.*, 1981). The new World Health Organization's (WHO) histological classification of thymomas is shown in Table VI (Muller-Hermelink and Marx, 2000). The prognosis of medullary and mixed thymomas (A stages; benign thymoma) seems better than that of the other tumors.

For noninvasive, encapsulated thymomas, radical thymectomy is considered curative. Nonetheless, patients should be followed with regular chest CT or MRI scans. Postoperative radiotherapy is usually not necessary in noninvasive thymomas. Patients with invasive thymomas are commonly treated with surgery, radiotherapy, and chemotherapy in varying combinations and sequences. It should be noted that previous radiation may be a complicating factor if the tumor recurs and a second operation is needed. Several protocols of adjuvant chemotherapy have been evaluated in invasive thymomas. Chemotherapy regimens generally consist of cisplatin in combination with another drug. Recent reviews of the pertinent literature may be found in Cowen *et al.*, 1998; Giaccone, 2000; Graeber and Tamim, 2000; Hejna *et al.*, 1999; Lara, 2000; Loehrer, 1999; and Thomas *et al.*, 1999. Rare reports claim a response to corticosteroids (Tandan *et al.*, 1990).

**TABLE VI** New WHO Classification of Tumors: Comparison with Clinicopathological and Histogenetic Classification and Nomenclature

| WHO type | Clinicopathological classification | Terminology of histogenetic classification for TET histologic subtypes |
|---|---|---|
| A | Benign thymoma | Medullary thymoma |
| AB | | Mixed thymoma |
| B1 | Malignant | Predominantly cortical |
| B2 | thymomas, | Cortical |
| B3 | category I | Well-differentiated thymic carcinoma |
| C | Malignant thymomas, category II | Epidermoid keratinizing (squamous cell) carcinoma |
| | | Epidermoid nonkeratinizing carcinoma |
| | | Lymphoepithelioma-like carcinoma |
| | | Sarcomatoid carcinoma (carcinosarcoma) |
| | | Clear cell carcinoma |
| | | Basaloid carcinoma |
| | | Mucoepidermoid carcinoma |
| | | Undifferentiated carcinoma |

## Neonatal Transient Myasthenia

Myasthenic weakness develops in about 10%–20% of infants born to myasthenic mothers (Papazian, 1992; Seybold, 1999). Neither the mother's anti-AChR antibody titer nor her clinical state is predictive of neonatal MG. Usually, the symptoms last for 2–4 weeks, but elimination of maternal IgG antibodies may take months. Patients should be advised to schedule delivery in a specialized center with experience with this condition. In affected children, myasthenic symptoms usually become apparent 3–72 hours after birth (Papazian, 1992). Apart from acute problems in the first week, the prognosis of neonatal MG is excellent. Recovery is usually complete within 2–4 months after birth.

All children born to a mother with MG should be observed carefully for myasthenic signs during the first 3–6 days of life. If no symptoms have occurred by then, they are unlikely to occur later. When symptoms are mild, small feedings and careful surveillance are sufficient. When symptoms are more severe, with weakness of suckling and swallowing, a feeble cry, general muscle hypotonia, or respiratory difficulties, neostigmine methylsulfate should be given by subcutaneous or intramuscular injection (usually 0.04–0.05 mg/kg), or neostigmine bromide can be given orally through a nasogastric tube (0.5 mg/kg). Alternatively, pyridostigmine bromide intramuscularly (0.05–0.15 mg/kg) or orally (1–2 mg/kg) may be used (Seybold, 1999). In case of severe respiratory problems caused by neonatal myasthenia, exchange transfusion or discontinuous plasma exchange may be considered. IV Igs have been also been used with mixed results (Bassan and Spirer, 1999; Bassan et al., 1998; Tagher et al., 1999).

### Childhood Autoimmune Myasthenia Gravis

One problem in this age group is that some children may suffer from congenital myasthenic syndromes, and in these children it would be inappropriate to use thymectomy or any other form of immunosuppression (Seybold, 1999). Thus, especially in antibody-negative cases, all attempts must be made to exclude a congenital myasthenic syndrome. Occasionally, in severe disease a short course of plasma exchange may help to identify autoimmune MG with low or absent antibodies to AchR. In antibody-positive cases, AChE inhibitors are the drugs of choice before puberty (Andrews, 1998; Seybold, 1999). Immunosuppressive agents should be avoided if possible. However, corticosteroids may be used if necessary. If the symptoms are not adequately controlled, thymectomy may be considered (Seybold, 1998). The lower age limit for thymectomy is as controversial as is the upper limit (see earlier), but we would not usually recommend thymectomy in children younger than 5 years of age and possibly younger than the age of puberty. Azathioprine may be considered in older children in exceptional cases of severe and otherwise refractory MG. In such cases, azathioprine should be combined with corticosteroids as previously described for generalized MG in adults. IV Igs are helpful for short-term treatment for exacerbations, myasthenic crisis, or in preparation for surgery (Lindner et al., 1997; Selcen et al., 2000).

### Myasthenic Crisis

Myasthenic crisis is a neurological emergency requiring prompt treatment. "Crisis" is defined as the inability to maintain adequate respiratory function and a patent airway free of secretions. The management of crisis requires anticipation of problems and straightforward step-by-step decisions to use available therapeutic options. Precipitating factors are often infections, surgery, emotional distress, or too rapid tapering of steroids. Prompt control of infection and careful management of cardiovascular and renal function is essential to overcome the crisis and restore stable respiratory function. Rarely, patients may first be seen in myasthenic crisis. It may be difficult to make the correct diagnosis in the emergency room unless MG had previously been suspected and appropriate diagnostic tests had already been performed. In patients with known MG and long-term immunosuppression, crisis may develop

as a result of opportunistic infections or activation of latent viruses (e.g., cytomegalovirus pneumonitis). "Cholinergic crisis" results from excessive doses of AChE inhibitors inappropriately given to a progressively myasthenic patient (usually more than 600 mg pyridostimine/day), but this is now rare, and even some experienced clinicians have never seen a case.

If progressive deterioration of muscle strength cannot be improved with AChE inhibitors, patients are prone to sudden deterioration, and crisis management should be anticipated. Therefore, any patient who reports difficulty in breathing or swallowing should immediately be examined for vital capacity and impairment of swallowing. Patients in crisis should be hospitalized in an intensive care unit. Ventilatory support should be available during transfer. If the situation cannot be drastically improved by administration of AChE inhibitors (e.g., by IV injection of 1–3 mg pyridostigmine or 0.5–1 mg neostigmine), the need for intubation is imminent. After the respiratory problem is under control, the cause of the crisis should be investigated.

Respiratory assistance should be provided if the forced vital capacity is less than 15 ml/kg body weight, the tidal volume drops to less than 5–6 ml/kg body weight, or if arterial oxygen decreases to less than 75 mmHg and carbon dioxide increases to more than 45 mmHg. It should be noted that arterial gas determinations may give a false assurance and cannot substitute for close and careful clinical observation of the patient. Patients in myasthenic crisis receive the same respiratory support as patients with other neuromuscular breathing disorders.

During assisted ventilation, AChE inhibitors may be maintained by continuous IV infusion (e.g., 0.15–0.5 mg neostigmine/h, i.e., a total dose of 16–20 mg/day). Along with improvement, AChE may be temporarily reduced in order to judge the clinical response to immunological therapy. In patients who are not benefiting greatly from AChE agents, it is sometimes helpful to stop the medication for 48–72 hours ("drug holiday"), resulting in increased responsiveness when it is restarted. In any case, the dose of AChE should be readjusted as necessary during crisis treatment.

It is a cardinal rule that infections in MG should be treated immediately and vigorously with antibiotics. In immunosuppressed patients, prophylactic use of IgG and IgM preparations may be considered as an additional measure if there is an increased risk of infection.

A common error in the management of myasthenic crisis is to wait too long before initiating vigorous antibiotic therapy. The process of selection of antibiotics sometimes leads to unnecessary delay. Appropriate cultures should be obtained as quickly as possible, and empirical antibiotic treatment started immediately, even before the results of the cultures are available. A good principle is to apply the antibiotic regimen in current use in the same hospital for immunocompromised oncology patients with infections. If indicated, third-generation cephalosporins can be given without adverse effect on myasthenic symptoms. Certain antibiotics, such as the aminoglycosides, may have adverse effects on neuromuscular transmission and are generally avoided. However, in an intensive care setting, the primary consideration is successful treatment of the infection, whereas potential neuromuscular effects of antibiotics are clearly less important, especially if the patient is ventilated.

If the crisis cannot be controlled within a few days with the measures described, the patient should be treated with plasmapheresis. This is often combined with corticosteroids, although the use of corticosteroids is controversial in patients with bacterial infection. High-dose IgG (10–15 g/day) and, with ongoing infections IgM (5–10 g/day) may be added to substitute for losses during plasmapheresis. Azathioprine may be added to the regimen if long-term immunosuppressive therapy is indicated. This is usually the case in patients whose symptoms are severe enough to evolve into crisis.

Patients in myasthenic crisis who have severe infection or other contraindications to plasmapheresis should be treated with high-dose IV Igs, which may be similarly effective (Achiron *et al.*, 2000; Gajdos *et al.*, 1997, 1998).

## Anesthetic Management of Patients with Myasthenia Gravis

Myasthenic weakness may worsen after surgery. Depolarizing neuromuscular blocking agents like succinylcholine should not be used. Even low doses may lead to pronounced and long-lasting neuromuscular block, which cannot be antagonized by cholinesterase inhibitors. If administration of a neuromuscular blocking agent is necessary, a nondepolarizing curare-like relaxant, such as alcuronium, vecuronium, or atracurium, is used at one tenth to one half the normal dose. It should be anticipated that patients with MG occasionally require assisted respiration for a longer time than normal after surgery, even if the MG has been treated successfully before the operation. Assisted respiration is continued until the patient is able to cooperate and demonstrates adequate respiratory function (Krucylak and Naunheim, 1999).

Certain local anesthetics like procaine were reported to unmask MG, possibly by interfering with nerve or muscle conduction and/or AChR channel properties. Amide-type local anesthetics like lidocaine are preferable, because anti-AChE therapy may interfere with the degradation of ester-type local anesthetics (procaine).

## Pregnancy and Myasthenia Gravis

The influence of pregnancy on myasthenic symptoms is variable and unpredictable (Batocchi *et al.*, 1999). Frequent adjustments of the anti-AChE medications may be required. During pregnancy, AChE inhibitors should not be given intravenously except in emergencies, because they may cause uterine contractions. There is no firm evidence that glucocorticosteroids have embryotoxic or teratogenic effects in humans. However, most manufacturers recommend that during the first trimester steroids, like almost all medications, should be given only if indicated. Infants born to mothers who have taken steroids during pregnancy should be monitored for adrenal insufficiency during the neonatal period. Also, because glucocorticosteroids are excreted in breast milk, inhibition of endogeneous steroid production and growth suppression can occur in infants who are breastfed by mothers receiving the hormone.

Cytotoxic drugs should definitely be avoided during pregnancy because of their potential teratogenic effects in the first trimester. Myasthenic crisis during pregnancy should be treated as described earlier for nonpregnant patients. Labor and delivery are usually normal but should be scheduled in a specialized center. Sedatives and narcotics may be given in half the usual doses used for nonmyasthenics. Small amounts of anti-AChE may be given orally or intramuscularly as needed. After delivery, the babies should be closely observed for development of neonatal myasthenia. Breast feeding carries a theoretical risk of passage of maternal autoantibodies through the colostrum, but the amount transferred is low compared with placental transfer. Similar to other autoimmune disorders, postpartum exacerbation may occur in up to 30% of patients, and close observation of the patient in the postpartum period is recommended (Plauche, 1991).

## Drugs with Adverse Effects on Neuromuscular Transmission

Many drugs may compromise neuromuscular transmission and exacerbate myasthenic weakness (Howard, 1990). This is clinically relevant in all MG patients with marked systemic weakness. The following agents should be used only if absolutely necessary, and the patient should be closely monitored for any exacerbation of myasthenic symptoms: neuromuscular blocking agents (e.g., curare-like compounds); local anesthetics (prefer amide- over ester-type) and antiarrhythmics (quinine, quinidine, procainamide, verapamil); aminoglycoside and quinolone antibiotics; beta-blockers; calcium channel–blocking agents. Acute deterioration of MG was observed after IV injection of gadolinium DTPA for MRI (Nordenbo and Somnier, 1992). D-penicillamine must not be used in myasthenic patients, because it can itself induce autoimmune MG, which is completely reversible after the drug is discontinued. Many other drugs have been reported to produce worsening or unmasking of myasthenic symptoms (Howard, 1990). It is not always possible to avoid drugs that may adversely affect neuromuscular transmission. In general, all myasthenic patients should be observed for clinical worsening when any new medication is begun.

# LAMBERT-EATON MYASTHENIC SYNDROME

## Clinical Features and Diagnostic Tests

The Lambert-Eaton myasthenic syndrome (LEMS) is a rare disorder of the neuromuscular junction that occurs predominantly in men older than 40 years of age (Newsom-Davis and Lang, 1999). In most (about 60%–70%) patients the disorder occurs as a paraneoplastic syndrome in association with small-cell lung cancer or, much less frequently, non-Hodgkin lymphoma. In some surveys about 1% of patients with a small-cell lung cancer had clinical and electrophysiological features of LEMS. In the occasional patient the tumor may escape diagnosis for 5 years or even up to 15 years, requiring repeated diagnostic workups for many years. Successful treatment of the tumor may lead to concomitant improvement of LEMS. Some LEMS patients without a tumor have other organ-specific autoimmune disorders, such as Hashimoto's thyroiditis, pernicious anemia, vitiligo, and premature ovarian failure.

The weakness results from deficient ACh release from nerve terminals at both nicotinic (neuromuscular junction) and muscarinic (parasympathetic ganglia) sites. Presynaptic transmitter release is compromised by autoantibodies against voltage-gated presynaptic calcium channels. Antibodies directed against P/Q-type voltage-gated calcium channels can be measured by radioimmunoassay, using a radiolabeled omega-conotoxin as the target antigen (Newsom-Davis and Lang, 1999).

LEMS patients commonly have symmetrical proximal limb weakness that begins in the lower extremities, particularly affecting the gait. Oropharyngeal or ocular symptoms are not prominent, and only in occasional cases does weakness significantly affect respiratory function. Patients may note improvement in strength with repeated effort ("warming-up"). As indicators of the cholinergic autonomic disorder, dryness of the mouth, sexual impotence, and occasional sphincter disorders are reported. Tendon reflexes are hypoactive or absent but may come out on repeated tapping, a useful clue to the diagnosis. Some patients complain of muscle pain and paresthesias in the hands and feet.

Electrodiagnostic studies in LEMS show a characteristic pattern that is different from that of MG. The first evoked potential has an abnormally low amplitude, which declines even further at low rates of repetitive stimulation (2–5 Hz). The best sites for stimulation are the ulnar, median, peroneal, and accessory nerves. Characteristically, there is a marked (2–20 fold) increase of the amplitude at stimulation rates above 10 Hz. An increment of 25% is highly suggestive, and an increment of >100% confirms the diagnosis. A similar increase in amplitude can be observed after brief maximum voluntary contraction or simply by two successive supramaximal impulses (double stimulus test) or by tetanic nerve stimulation. This incremental response results from facilitation of transmitter release at high-frequency stimulation. The response to AChE inhibitors is much less pronounced than in MG. Very rarely, both diseases may coexist. Patients are also abnormally sensitive to neuromuscular blocking agents, and some cases are recognized because of prolonged apnea after exposure to such drugs during surgery. "Myasthenic crisis" is less common in LEMS patients and usually only occurs in those with a long history of weakness and fatigue.

The prognosis for survival in LEMS depends largely on whether the patient has an underlying carcinoma, which is fatal in a high proportion of cases. In primary autoimmune LEMS without an associated neoplasm, the prognosis is obviously much more favorable.

## Therapeutic Principles and Management

The preferred preparation for producing rapid symptomatic relief in LEMS with and without cancer is 3,4-diaminopyridine. It prolongs the duration of the presynaptic action potential by blocking the outward potassium current and thereby improving ACh release from the nerve terminal. This agent is effective in relieving both the motor and the autonomic symptoms of LEMS (McEvoy et al., 1989). It should be started in graded doses, beginning with 10–20 mg/day and can usually be safely increased up to 60–80/day (Newsom-Davis and Lang, 1999). Side effects are dose-dependent and reversible and include paresthesias, fatigue, gastrointestinal discomfort with diarrhea, and abdominal cramps, especially when 3,4-diaminopyridine is used in combination with pyridostigmine. High doses may provoke seizures and precipitate asthma attacks. Pyridostigmine should be tried but usually has only a mild effect in LEMS.

Immunosuppression with steroids and azathioprine may be used in patients with nonneoplastic LEMS. This therapy should be closely monitored, because in some LEMS patients who were treated with immunosuppressive drugs a lung tumor was later discovered. In paraneoplastic LEMS, corticosteroid treatment may be combined with antitumor therapy, which will often improve the neurological disorder. In severe refractory cases treatment with plasma exchange may be useful, although its effect is more delayed and less striking than in MG. IV Ig treatment is also effective in LEMS, as was established in a double-blind placebo-controlled crossover trial (Bain *et al.*, 1996).

## CONGENITAL MYASTHENIC SYNDROMES

The treatment options for this heterogenous group of rare disorders include AChE inhibitors, calcium channel blockers, 3,4 diaminopyridine, and corticosteroids. An excellent review of the different congenital syndromes and their treatment may be found in Engel *et al.* (1999).

## REFERENCES

Achiron, A., Barak, Y., Miron, S., and Sarova-Pinhas, I. (2000). Immunoglobulin treatment in refractory myasthenia gravis. *Muscle Nerve* 23, 551–555.

Adams, C., Theodorescu, D., Murphy, E. G., and Shandling, B. (1990). Thymectomy in juvenile myasthenia gravis. *J. Child. Neurol.* 5, 215–218.

Andrews, P. I. (1998). A treatment algorithm for autoimmune myasthenia gravis in childhood. *Ann. N Y Acad. Sci.* 841, 789–802.

Arsura, E., Brunner, N. G., Namba, T., and Grob, D. (1985). High-dose intravenous methylprednisolone in myasthenia gravis. *Arch. Neurol.* 42, 1149–1153.

Bain, P. G., Motomura, M., Newsom-Davis, J., Misbah, S. A., Chapel, H. M., Lee, M. L., Vincent, A., and Lang, B. (1996). Effects of intravenous immunoglobulin on muscle weakness and calcium-channel autoantibodies in the Lambert-Eaton myasthenic syndrome. *Neurology* 47, 678–683.

Bassan, H., Muhlbaur, B., Tomer, A., and Spirer, Z. (1998). High-dose intravenous immunoglobulin in transient neonatal myasthenia gravis. *Pediatr. Neurol.* 18, 181–183.

Bassan, H., and Spirer, Z. (1999). Intravenous immunoglobulin in neonatal myasthenia gravis. *J. Pediatr.* 135, 790.

Batocchi, A. P., Majolini, L., Evoli, A., Lino, M. M., Minisci, C., and Tonali, P. (1999). Course and treatment of myasthenia gravis during pregnancy. *Neurology* 52, 447–452.

Beeson, D., Jacobson, L., Newsom-Davis, J., and Vincent, A. (1996). A transfected human muscle cell line expressing the adult subtype of the human muscle acetylcholine receptor for diagnostic assays in myasthenia gravis. *Neurology* 47, 1552–1555.

Benny, W. B., Sutton, D. M., Oger, J., Bril, V., McAteer, M. J., and Rock, G. (1999). Clinical evaluation of a staphylococcal protein A

immunoadsorption system in the treatment of myasthenia gravis patients. *Transfusion* 39, 682–687.

Besinger, U. A., Toyka, K. V., Heininger, K., Fateh-Moghadam, A., Schumm, F., Sandel, P. C., and Birnberger, K. L. (1981). Long term correlation of clinical course and acetylcholine receptor antibody in patients with myasthenia gravis. *Ann. N Y Acad. Sci.* 377, 812–813.

Blaes, F., Beeson, D., Plested, P., Lang, B., and Vincent, A. (2000). IgG from "seronegative" myasthenia gravis patients binds to a muscle cell line, TE671, but not human acetylcholine receptor. *Ann. Neurol.* 47, 504–510.

Bonifati, D. M., and Angelini, C. (1997). Long-term cyclosporine treatment in a group of severe myasthenia gravis patients. *J. Neurol.* 244, 542–547.

Bromberg, M. B., Wald, J. J., Feldman, E. L., and Albers, J. W. (1997). Randomized trial of azathioprine or predisone for initial immunosuppressive treatment of myasthenia gravis. *J. Neurol. Sci.* 150, 59–62.

Buckingham, J. M., Howard, F. M., Jr., Bernatz, P. E., Payne, W. S., Harrison, E. G., Jr., O'Brien, P. C., and Weiland, L. H. (1976). The value of thymectomy in myasthenia gravis: a computer-assisted matched study. *Ann. Surg.* 184, 453–458.

Bufler, J., Kahlert, S., Tzartos, S., Toyka, K. V., Maelicke, A., and Franke, C. (1996). Activation and blockade of mouse muscle nicotinic channels by antibodies directed against the binding site of the acetylcholine receptor. *J. Physiol.* 492(Pt 1), 107–114.

Chaudhry, V. V., Cornblath, D. R., Griffin, J. W., O'Brien, R., and Drachman, D. B. (2001). Mycophenolate mofetil: A safe and promising immunosuppressant in neuromuscular diseases. *Neurology* 56, 94–96.

Ciafaloni, E., Massey, J. M., Tucker-Lipscomb, B., and Sanders, D. B. (2001). Mycophenolate mofetil for myasthenia gravis: An open-label pilot study. *Neurology* 56, 97–99.

Ciafaloni, E., Nikhar, N. K., Massey, J. M., and Sanders, D. B. (2000). Retrospective analysis of the use of cyclosporine in myasthenia gravis. *Neurology* 55, 448–450.

Compston, D. A. S., Vincent, A., Newsom-Davis, J., and Batchelor, J. R. (1980). Clinical, pathological, HLA antigen and immunological evidence for disease heterogeneity in myasthenia gravis. *Brain* 103, 579–601.

Cornelio, F., Antozzi, C., Confalonieri, P., Baggi, F., and Mantegazza, R. (1998). Plasma treatment in diseases of the neuromuscular junction. *Ann. N Y Acad. Sci.* 841, 803–810.

Cowen, D., Hannoun-Levi, J. M., Resbeut, M., and Alzieu, C. (1998). Natural history and treatment of malignant thymoma. *Oncology (Huntingt.)* 12, 1001–1005.

Drachman, D. B. (1994). Myasthenia gravis. *N. Engl. J. Med.* 330, 1797–1810.

Dudel, J., Birnberger, K. L., Toyka, K. V., Schlegel, C., and Besinger, U. (1979). Effects of myasthenic immunoglobulins and of prednisolone on spontaneous miniature end-plate potentials in mouse diaphragms. *Exp. Neurol.* 66, 365–380.

Dwyer, J. M. (1992). Manipulating the immune system with immune globulin. *N. Engl. J. Med.* 326, 107–116.

Edan, G., and Landgraf, F. (1994). Experience with intravenous immunoglobulin in myasthenia gravis: a review. *J. Neurol. Neurosurg. Psychiatry* 57 Suppl, 55–56.

Elion, G. B. (1993). The George Hitchings and Gertrude Elion Lecture. The pharmacology of azathioprine. *Ann. N Y Acad. Sci.* 685, 400–407.

Ellis, R. J., Swendson, M. R., and Bajorek, J. (1994). Aseptic meningitis as a complication of intravenous immunoglobulin therapy for myasthenia gravis. *Muscle Nerve* 17, 683–684.

Engel, A. G., Ohno, K., and Sine, S. M. (1999). Congenital myasthenic syndromes. *In* "Myasthenia Gravis and Myasthenic Disorders" (A. G. Engel, Ed.), pp. 251–297. Oxford University Press, Oxford, UK.

Flachenecker, P., Taleghani, B. M., Gold, R., Grossmann, R., Wiebecke, D., and Toyka, K. V. (1998). Treatment of severe myasthenia gravis with protein A immunoadsorption and cyclophosphamide. *Transfus. Sci.* **19 Suppl**, 43–46.

Gajdos, P., Chevret, S., Clair, B., Tranchant, C., and Chastang, C. (1997). Clinical trial of plasma exchange and high-dose intravenous immunoglobulin in myasthenia gravis. Myasthenia Gravis Clinical Study Group. *Ann. Neurol.* **41**, 789–796.

Gajdos, P., Chevret, S., Clair, B., Tranchant, C., and Chastang, C. (1998). Plasma exchange and intravenous immunoglobulin in autoimmune myasthenia gravis. *Ann. N Y Acad. Sci.* **841**, 720–726.

Genkins, G., Kornfeld, P., Osserman, K. E., Namba, T., Grob, D., and Brunner, N. G. (1971). The use of ACTH and corticosteroids in myasthenia gravis. *Ann. N Y Acad. Sci.* **183**, 369–374.

Giaccone, G. (2000). Treatment of thymoma and thymic carcinoma. *Ann. Oncol.* **11 Suppl 3**, 245–246.

Graeber, G. M., and Tamim, W. (2000). Current status of the diagnosis and treatment of thymoma. *Semin. Thorac. Cardiovasc. Surg.* **12**, 268–277.

Grob, D. (1999). Natural history of myasthenia gravis. *In* "Myasthenia Gravis and Myasthenic Disorders" (A. G. Engel, Ed.), pp. 131–145. Oxford University Press, Oxford, UK.

Grob, D., Simpson, D., Mitsumoto, H., Hoch, B., Mokhtarian, F., Bender, A., Greenberg, M., Koo, A., and Nakayama, S. (1995). Treatment of myasthenia gravis by immunoadsorption of plasma. *Neurology* **45**, 338–344.

Gronseth, G. S., and Barohn, R. J. (2000). Practice parameter: Thymectomy for autoimmune myasthenia gravis (an evidence-based review). Report of the Quality Standards Subcommittee of the American Academy of Neurology. *Neurology* **55**, 7–15.

Harper, C. M. (2001). Electrodiagnosis of end plat disease. *In* "Myasthenia Gravis and Myasthenic Disorders" (A. G. Engel, Ed.), pp. 65–84. Oxford University Press, Oxford, UK.

Hauser, R. A., Malek, A. R., and Rosen, R. (1998). Successful treatment of a patient with severe refractory myasthenia gravis using mycophenolate mofetil. *Neurology* **51**, 912–913.

Hejna, M., Haberl, I., and Raderer, M. (1999). Nonsurgical management of malignant thymoma. *Cancer* **85**, 1871–1884.

Herrlinger, U., Weller, M., Dichgans, J., and Melms, A. (2000). Association of primary central nervous system lymphoma with long-term azathioprine therapy for myasthenia gravis. *Ann. Neurol.* **47**, 682–683.

Hoch, W., McConville, J., Helms, S., Newsom-Davis, J., Melms, A., and Vincent, A. (2001). Auto-antibodies to the receptor tyrosine kinase MuSK in patients with myasthenia gravis without acetylcholine receptor antibodies. *Nature Med.* **7**, 365–368.

Hohlfeld, R. (1997). Biotechnological agents for the immunotherapy of multiple sclerosis. Principles, problems and perspectives. *Brain* **120**, 865–916.

Hohlfeld, R., Michels, M., Heininger, K., Besinger, U., and Toyka, K. V. (1988). Azathioprine toxicity during long-term immunosuppression of generalized myasthenia gravis. *Neurology* **38**, 258–261.

Hohlfeld, R., Toyka, K. V., Besinger, U. A., Gerhold, B., and Heininger, K. (1985). Myasthenia gravis: Reactivation of clinical disease and of autoimmune factors after discontinuation of long-term azathioprine. *Ann. Neurol.* **17**, 238–242.

Hohlfeld, R., and Wekerle, H. (1999). The immunopathogenesis of myasthenia gravis. *In* "Myasthenia Gravis and Myasthenic Syndromes" (A. G. Engel, Ed.), pp. 87–110. Oxford University Press, Oxford, UK.

Howard, F. M., Duane, D. D., Lambert, E. H., and Daube, J. R. (1976). Alternate-day prednisone: preliminary report of a double-blind controlled study. *Ann. N Y Acad. Sci.* **274**, 596–607.

Howard, J. F., Jr. (1990). Adverse drug effects on neuromuscular transmission. *Semin. Neurol.* **10**, 89–102.

Howard, J. F., Jr. (1998). Intravenous immunoglobulin for the treatment of acquired myasthenia gravis. *Neurology* **51**, S30–S36.

Jaretzki, A., III, Barohn, R. J., Ernstoff, R. M., Kaminski, H. J., Keesey, J. C., Penn, A. S., and Sanders, D. B. Task Force of the Medical Scientific Advisory Board of the Myasthenia Gravis Foundation of America. (2000). Myasthenia gravis: Recommendations for clinical research standards. *Neurology* **55**, 16–23.

Katz, P. (2000). Glucocorticoids in relation to inflammatory disease. *In* "Cecil Textbook of Medicine" (L. Goldman and J. C. Bennett, Eds.), pp. 111–114. W.B. Saunders, Philadelphia.

Kennel, P. F., Vilquin, J.-T., Braun, S., Fonteneau, P., Warter, J.-M., and Poindron, P. (1995). Myasthenia gravis: Comparative autoantibody assays using human muscle, TE671, and glucocortiicoid-treated TE671 cells as sources of antigen. *Clin. Immunol. Immunopathol.* **74**, 293–296.

Kiani, A., Rao, A., and Aramburu, J. (2000). Manipulating immune responses with immunosuppressive agents that target NFAT. *Immunity* **12**, 359–372.

Krucylak, P. E., and Naunheim, K. S. (1999). Preoperative preparation and anesthetic management of patients with myasthenia gravis. *Semin. Thorac. Cardiovasc. Surg.* **11**, 47–53.

Kuks, J. B., Djojoatmodjo, S., and Oosterhuis, H. J. (1991). Azathioprine in myasthenia gravis: observations in 41 patients and a review of literature. *Neuromuscul. Disord.* **1**, 423–431.

Kupersmith, M. J., Moster, M., Bhuiyan, S., Warren, F., and Weinberg, H. (1996). Beneficial effects of corticosteroids on ocular myasthenia gravis. *Arch. Neurol.* **53**, 802–804.

Lara, P. N., Jr. (2000). Malignant thymoma: current status and future directions. *Cancer Treat. Rev.* **26**, 127–131.

Lindberg, C., Andersen, O., and Lefvert, A. K. (1998). Treatment of myasthenia gravis with methylprednisolone pulse: a double blind study. *Acta Neurol. Scand.* **97**, 370–373.

Lindner, A., Schalke, B., and Toyka, K. V. (1997). Outcome in juvenile-onset myasthenia gravis: a retrospective study with long-term follow-up of 79 patients. *J. Neurol.* **244**, 515–520.

Lipsky, J. J. (1996). Mycophenolate mofetil. *Lancet* **348**, 1357–1359.

Loehrer, P. J., Sr. (1999). Current approaches to the treatment of thymoma. *Ann. Med.* **31 Suppl 2**, 73–79.

MacDonald, B. K., Cockerell, O. C., Sander, J. W., and Shorvon, S. D. (2000). The incidence and lifetime prevalence of neurological disorders in a prospective community-based study in the UK. *Brain* **123**(Pt 4), 665–676.

Masaoka, A., Monden, Y., Nakahara, K., and Tanioka, T. (1981). Follow-up study of thymomas with special reference to their clinical stages. *Cancer* **48**, 2485–2492.

Matell, G., Bergstrom, K., Franksson, C., Hammarstrom, L., Lefvert, A. K., Moller, E., von Reis, G., and Smith, E. (1976). Effects of some immunosuppressive procedures on myasthenia gravis. *Ann. N Y Acad. Sci.* **274**, 659–676.

McEvoy, K. M., Windebank, A. J., Daube, J. R., and Low, P. A. (1989). 3,4-Diaminopyridine in the treatment of Lambert-Eaton myasthenic syndrome. *N. Engl. J. Med.* **321**, 1567–1571.

Meiner, Z., Ben Hur, T., River, Y., and Reches, A. (1993). Aseptic meningitis as complication of intravenous immunoglobulin therapy for myasthenia gravis. *J. Neurol. Neurosurg. Psychiatry* **56**, 830–831.

Meriggioli, M. N., and Rowin, J. (2000). Treatment of myasthenia gravis with mycophenolate mofetil: a case report. *Muscle Nerve* **23**, 1287–1289.

Mertens, H. G., Balzereit, F., and Leipert, M. (1969). The treatment of severe myasthenia gravis with immunosuppressive agents. *Eur. Neurol.* **2**, 321–339.

Mertens, H. G., Hertel, G., Reuther, P., and Ricker, K. (1981). Effect of immunosuppressive drugs (azathioprine). *Ann. N Y Acad. Sci.* **377**, 691–699.

Michels, M., Hohlfeld, R., Hartung, H. P., Heininger, K., Besinger, U. A., and Toyka, K. V. (1988). Myasthenia gravis: discontinuation of long-term azathioprine. *Ann. Neurol.* 24, 798.

Millard, C. B., and Broomfield, C. A. (1995). Anticholinesterases: medical applications of neurochemical principles. *J. Neurochem.* 64, 1909–1918.

Muller-Hermelink, H. K., and Marx, A. (2000) Thymoma. *Curr. Opin. Oncol.* 12, 426–433.

Myasthenia Gravis Clinical Study Group. (1993). A randomized clincial trial comparing prednisone and azathioprine in myasthenia gravis. Results of the second interim analysis. *J. Neurol. Neurosurg. Psychiatry* 56, 1157–1163.

Newsom-Davis, J., and Lang, B. (1999). The Lambert-Eaton syndrome. *In* "Myasthenia Gracis and Myasthenic Syndromes" (A. G. Engel, Ed.), pp. 205–228. Oxford University Press, Oxford, UK.

Nordenbo, A. M., and Somnier, F. E. (1992). Acute deterioration of myasthenia gravis after intravenous administration of gadolinium-DTPA. *Lancet* 340, 1168.

Oosterhuis, H. J. G. H. (1989). The natural course of myasthenia gravis. *J. Neurol. Neurosurg. Psychiatry* 52, 1121–1127.

Palace, J., Newsom-Davis, J., Lecky, B., and The Myasthenia Gravis Study Group. (1998). A randomized double-blind trial of prednisolone alone or with azathioprine in myasthenia gravis. *Neurology* 50, 1778–1783.

Papazian, O. (1992). Transient neonatal myasthenia gravis. *J. Child. Neurol.* 7, 135–141.

Perez, M. C., Buot, W. L., Mercado-Danguilan, C., Bagabaldo, Z. G., and Renales, L. D. (1981). Stable remissions in myasthenia gravis. *Neurology* 31, 32–37.

Plauche, W. C. (1991). Myasthenia gravis in mothers and their newborns. *Clin. Obstet. Gynecol.* 34, 82–99.

Qureshi, A. I., Choudhry, M. A., Akbar, M. S., Mohammad, Y., Chua, H. C., Yahia, A. M., Ulatowski, J. A., Krendel, D. A., and Leshner, R. T. (1999). Plasma exchange versus immunoglobulin treatment in myasthenic crisis. *Neurology* 52, 629–632.

Romi, F., Skeje, G. O., Aarli, J. A., and Gilhus, N. E. (2000). Muscle autoantibodies in subgroups of myasthenia gravis patients. *J. Neurol.* 247, 369–375.

Samuelsson, A., Towers, T. L., and Ravetch, J. V. (2001). Anti-inflammatory activity of IVIG mediated through the inhibitory Fc receptor. *Science* 291, 484–486.

Sato, T., Ishigaki, Y., Komiya, T., and Tsuda, H. (1988). Therapeutic immunoadsorption of acetylcholine receptor antibodies in myasthenia gravis. *Ann. N Y Acad. Sci.* 540, 554–556.

Schumm, F., Wietholter, H., Fateh-Moghadam, A., and Dichgans, J. (1985). Thymectomy in myasthenia with pure ocular symptoms. *J. Neurol. Neurosurg. Psychiatry* 48, 332–337.

Selcen, D., Dabrowski, E. R., Michon, A. M., and Nigro, M. A. (2000). High-dose intravenous immunoglobulin therapy in juvenile myasthenia gravis. *Pediatr. Neurol.* 22, 40–43.

Seybold, M. E. (1998). Thymectomy in childhood myasthenia gravis. *Ann. N Y Acad. Sci.* 841, 731–741.

Seybold, M. E. (1999). Treatment of myasthenia gravis. *In* "Myasthenia Gravis and Myasthenic Disorders" (A. G. Engel, Ed.), pp. 167–201. Oxford University Press, Oxford, UK.

Shibuya, N., Sato, T., Osame, M., Takegami, T., Doi, S., and Kawanami, S. (1994). Immunoadsorption therapy for myasthenia gravis. *J. Neurol. Neurosurg. Psychiatry* 57, 578–581.

Somnier, F. E. (1994). Anti-acetylcholine receptor (AChR) antibodies measurement in myasthenia gravis: The use of cell line TE671 as a source of AChR antigen. *J. Neuroimmunol.* 51, 63–68.

Steg, R. E., and Lefkowitz, D. M. (1994). Cerebral infarction following intravenous immunoglobulin therapy for myasthenia gravis. *Neurology* 44, 1180–1181.

Tagher, R. J., Baumann, R., and Desai, N. (1999). Failure of intravenously administered immunoglobulin in the treatment of neonatal myasthenia gravis. *J. Pediatr.* 134, 233–235.

Tandan, R., Taylor, R., DiCostanzo, D. P., Sharma, K., Fries, T., and Roberts, J. (1990). Metastasizing thymoma and myasthenia gravis. Favorable response to glucocorticoids after failed chemotherapy and radiation therapy. *Cancer* 65, 1286–1290.

Task Force of the Medical Scientific Advisory Board of the Myasthenia Gravis Foundation of America, Jaretzki, A., Barohn, R. J., Ernstorff, R. M., Kaminski, H. J., Keesey, J. C., Penn, A. S., and Sanders, D. B. (2000). Myasthenia gravis. Recommendations for clinical research standards. *Neurology* 55, 16–23.

Thomas, C. R., Wright, C. D., and Loehrer, P. J. (1999). Thymoma: state of the art. *J. Clin. Oncol.* 17, 2280–2289.

Tindall, R. S., Phillips, J. T., Rollins, J. A., Wells, L., and Hall, K. (1993). A clinical therapeutic trial of cyclosporine in myasthenia gravis. *Ann. N Y Acad. Sci.* 681, 539–551.

Tindall, R. S., Rollins, J. A., Phillips, J. T., Greenlee, R. G., Wells, L., and Belendiuk, G. (1987). Preliminary results of a double-blind, randomized, placebo-controlled trial of cyclosporine in myasthenia gravis. *N. Engl. J. Med.* 316, 719–724.

Vincent, A., and Newsom-Davis, J. (1982). Acetylcholine receptor antibody characteristics in myasthenia gravis. I. Patients with generalized myasthenia or disease restricted to ocular muscles. *Clin. Exp. Immunol.* 49, 257–265.

Voltz, R., Hohlfeld, R., Fateh-Moghadam, A., Witt, T. N., Wick, M., Reimers, C., Siegele, B., and Wekerle, H. (1991). Myasthenia gravis: Measurement of anti-AChR autoantibodies using cell line TE671. *Neurology* 41, 1836–1838.

Willcox, N., Schluep, M., Ritter, M. A., and Newsom-Davis, J. (1991). The thymus in seronegative myasthenia gravis patients. *J. Neurol.* 238, 256–261.

Witte, A. S., Cornblath, D. R., Parry, G. J., Lisak, R. P., and Schatz, N. J. (1984). Azathioprine in the treatment of myasthenia gravis. *Ann. Neurol.* 15, 602–605.

Wu, J. M., Wu, B., Guarnieri, F., August, J. T., and Drachman, D. B. (2000). Targeting antigen-specific T cells by genetically engineered antigen presenting cells. A strategy for specific immunotherapy of autoimmune disease. *J. Neuroimmunol.* 106, 145–153.

Yeh, J. H., and Chiu, H. C. (2000). Comparison between double-filtration plasmapheresis and immunoadsorption plasmapheresis in the treatment of patients with myasthenia gravis. *J. Neurol.* 247, 510–513.

## CHAPTER 95
# Inflammatory Myopathies

Marinos C. Dalakas and Dieter Pongratz

Inflammatory myopathies are a heterogeneous group of acquired muscle diseases, in which muscle weakness and inflammatory infiltrates within the skeletal muscle are the principal clinical and myopathological findings. There are three different forms:

- Polymyositis (PM)
- Dermatomyositis (DM)
- Inclusion body myositis (IBM)

## CLINICAL ASPECTS

All three forms have in common muscle weakness and atrophy, which develop more acutely (DM), subacutely (PM) or slowly (IBM).

As a rule, there is a predominant involvement of the proximal muscles of the limbs and arms. In IBM, however, the presence of distal muscle weakness, especially of the foot extensors and finger flexors, often from the beginning of the illness, is a diagnostic clue.

In all forms the pharyngeal and neck flexor muscles are often affected causing dysphagia and, at times, difficulties in holding up the head.

Muscular wasting develops during the course of the illness and is pronounced in chronic polymyositis and especially in IBM. Myalgia and muscle tenderness are most frequent in acute cases of DM. In PM it is a fluctuating symptom, whereas in IBM muscle pain is rarely seen. As a rule, fasciculations do not occur. Tendon reflexes are preserved, but sometimes reduced in DM and PM. In IBM, however, they can be absent because of atrophy of the respective muscles. Sensation is always normal. The most important clinical signs are summarized in Table I.

### Special Signs of DM

DM is clinically characterized by skin lesions accompanying or, more often, preceeding muscle symptoms.

There is the typical *heliotrope erythema* (lilac disease), involving especially the eyelids, face, and upper trunk. It can extend to other body surfaces, including the elbows, knees, neck, and upper chest. More chronic skin lesions show *depigmentation* and *hyperpigmentations*. In the skin near the knuckles violaceous scaly eruptions (*sign of Gottron*) occur. Dilated painful capillary loops at the base of the fingernails (*sign of Keinig*) are also typical. Rough and cracked areas with irregular, dirty, horizontal lines (mechanic's hands) are also frequently found on the palmar areas of the fingers. As the disease progresses, *subcutaneous calcifications* occur.

### Laboratory Diagnosis and Differential Diagnosis

Laboratory diagnosis (Table II) is based on:

1. Measurement of muscle enzymes
2. Electromyography
3. Muscle biopsy
4. In some cases, clinical or laboratory signs of an associated connective tissue disease may be an additional aid.

The level of *muscle enzymes*, especially creatine kinase (CK), usually parallels disease activity and can be elevated in acute stages as much as 50 times above normal. In rare cases of active PM and DM, however, CK can also be within normal range. In IBM the level of CR is not usually elevated more than 10-fold, and in some cases it may be normal.

*Needle electromyography* shows myopathic potentials characterized by short duration, low amplitude, and polyphasic configuration on voluntary activation in combination with increased spontaneous activity (fibrillations, positive sharp waves, complex repetitive discharges). This pattern occurs also in a variety of active myopathic processes of other origins and should there-

*Neurological Disorders: Course and Treatment, Second Edition*

1363

**TABLE I    Clinical Aspects of Dermatomyositis, Polymyositis, and Inclusion Body Myositis**

| | DM | PM | IBM |
|---|---|---|---|
| Age of onset | Childhood and adulthood | Above 18 years | Above 50 years |
| Development of muscle symptoms | Acute or subacute | Subacute | Slowly |
| Predominant involvement of muscle weakness | Proximal muscles | Proximal muscles | Proximal and distal muscles |
| Muscle wasting | Not prominent | Present in chronic forms | Nearly always pronounced in selected muscles (triceps, finger flexors, quadriceps) |
| Myalgia | Often (especially in acute cases) | Sometimes | Never |
| Rash or calcinosis | Present | Absent | Absent |

fore not be considered diagnostic for the inflammatory myopathies. Mixed myopathic and neurogenic potentials may be present in some cases as a consequence of regeneration and chronicity of the disease. Some patients with IBM may have electromyographic signs of sensory axonal neuropathy.

*Muscle biopsy* is the definitive test, not only for establishing the diagnosis of DM, PM, or IBM, but also for excluding other neuromuscular disorders.

In DM (Table III), inflammatory infiltrates are found predominantly in perivascular and perifascicular regions, producing the characteristic picture of a myositis of the *perifascicular type.* There are striking lesions of the *small intramuscular blood vessels* with endothelial proliferation and so-called tubulovesicular inclusions seen by electron microscopy. In more severe cases, especially in childhood, signs of active vasculitis and microinfarcts within the muscle can be seen. *Perifascicular atrophy and fiber damage* is diagnostic of DM even in the absence of infiltration. It is found in more than 50% of adults and lymphocytic in nearly all cases of childhood DM.

Immunhistological methods show that the cellular infiltrates in this disease consist of B lymphocytes, CD4+

**TABLE II    Diagnostic Criteria for Inflammatory Myopathies***

| | Polymyositis | | Dermatomyositis | | Inclusion body myositis (definite) |
|---|---|---|---|---|---|
| Criterion | Definite | Probable[†] | Definite | Mild or early | |
| Muscle strength | Myopathic muscle weakness[‡] | Myopathic muscle weakness[‡] | Myopathic muscle weakness[‡] | Seemingly normal strength[§] | Myopathic muscle weakness with early involvement of distal muscles[‡] |
| Electromyographic findings | Myopathic | Myopathic | Myopathic | Myopathic or nonspecific | Myopathic with mixed potentials |
| Muscle enzymes | Elevated (up to 50-fold) | Elevated (up to 50-fold) | Elevated (up to 50-fold) | Elevated (up to 10-fold) or normal | Elevated (up to 10-fold) or normal |
| Muscle biopsy findings[‖] | Diagnostic for this type of inflammatory myopathy | Nonspecific or without signs of primary inflammation | Diagnostic | Nonspecific or diagnostic | Diagnostic |

*From Dalakas (1991). *N. Engl. J. Med.* **325**, 1487–1498, with permission. Copyright © 1991, Massachusetts Medical Society. All rights reserved.

[†]An adequate trial of prednisone or other immunosuppressive drugs is warranted in probable cases. If, in retrospect, the disease is unresponsive to therapy, another muscle biopsy should be considered to exclude other diseases or possible evolution to inclusion body myositis.

[‡]Myopathic muscle weakness, affecting proximal muscles more than distal ones and sparing eye and facial muscles, is characterized by subacute onset (weeks to months) and rapid progression in patients who have no family history of neuromuscular disease, no endocrinopathy, no exposure to myotonic drugs or toxins, and no biochemical muscle disease (excluded on the basis of muscle biopsy findings).

[§]Although strength is seemingly normal, patients often have new onset of easy fatigue, myalgia, and reduced endurance. Careful muscle testing may reveal mild muscle weakness.

[‖]See the text for details.

TABLE III  Immunopathologic Mechanisms in Dermatomyositis and Polymyositis*

|  | Dermatomyositis | Polymyositis |
|---|---|---|
| Necrosis of capillaries | + | − |
| Muscle infarcts | + | − |
| Tubulovesicular inclusions in endothelial cells | + | − |
| C5b9-complement depositions in small vessels | + | − |
| Perimysial infiltrates | ++ | + |
| Endomysial infiltrates | + | ++ |
| B lymphocytes | ++ | − |
| CD4+ lymphocytes | ++ | + |
| CD8+ lymphocytes | + | ++ |
| Invasion of CD8+ lymphocytes in nonnecrotic muscle fibers | − | ++ |

*From Pongratz (1992) with permission.

more than CD8$^+$ cells, and macrophages. *C5b9 complement deposits* within small blood vessels and capillaries of muscle are very characteristic, causing destruction of the capillaries and muscle ischemia.

In PM (Table III), infiltrates are predominantly endomysial, producing the picture of a diffuse myositis. A perifascicular atrophy does not develop. There is also no evidence of microangiopathy. Immunhistologically, cytotoxic CD8$^+$ lymphocytes are the predominant cells that invade nonnecrotic muscle fibers. The muscle fibers themselves aberrantly express the *major histocompatibility complex class I* (MHC I) *antigen*, which is absent in normal muscle.

IBM is characterized by *endomysial inflammation* with CD8$^+$ lymphocytes in particular, similar to those seen in PM. In addition, *rimmed vacuoles with eosinophilic cytoplasmatic inclusions* can be found, as a rule, within many muscle fibers. Some of the nuclei can be prominent. On electronmicroscopy, the vacuoles correspond to the so-called *autophagic vacuoles*. In addition, *filamentous inclusions in the cytoplasm and nuclei* are prominent but not pathognomonic.

High levels of *myositis-associated autoantibodies* are found, especially in cases of overlap syndromes. PM-Scl antibodies are characteristic of cases with scleroderma-DM overlap syndromes. Mi-2 autoantibodies may be a rare serologic finding in DM. Jo-1 antibodies are seen in 80% of patients with PM and DM, who have an associated interstitial lung disease.

### Differential Diagnosis

Differential diagnostic considerations are especially important in cases classified as probable PM (see Table II). They include the following:

- Sporadic cases of muscular dystrophy
- Toxic myopathies
- Infectious myopathies
- Some metabolic myopathies (especially acid maltase and phosphorylase deficiency)
- Endocrinopathies
- Possible evolution to IBM

## NATURAL COURSE

The incidence of PM, DM, and IBM is approximately 1/100,000. DM affects both children and adults, but females more often than males. PM is mostly seen in the second decade of life and very rarely in childhood. IBM is three times more frequent in men than in women and most likely to affect persons older than 50 years. The natural course of PM and DM is relatively unknown, because patients are almost always treated with steroids. The mortality rates reported 30 years ago are obviously outdated. IBM is generally resistant to all therapies and, as a rule, is slowly but steadily progressive.

## PRINCIPLES OF THERAPY

Because corticosteroids and immunosuppressive drugs seem to be of benefit, PM and DM belong to the treatable group of myopathies. In contrast, IBM is resistant to most therapies, although treatment with intravenous immunoglobulins may be of some transient benefit in some patients (Dalakas, 1993; Dalakas *et al.*, 1997, 2001; Walter *et al.*, 2000). The treatment regimens of PM and DM are empirical (Dalakas, 1991; Engel, 1992). Only recently, separate, large-scale, prospective, controlled clinical studies of both diseases in adults and children have been carried out. In most series reported, the patients have been adequately examined to confirm the diagnosis of DM or PM by modern criteria and, in particular, to exclude IBM.

Although the immunopathogenesis of DM and PM is different, specific therapeutic approaches such as monoclonal antibodies or T-cell vaccinations, especially in PM, have not been attempted.

## PRACTICAL MANAGEMENT OF PM AND DM

### Corticosteroids

Prednisone is the *first-line drug* for the empirical treatment of PM and DM (Dalakas, 1988, 1989, 1990; Henriksson and Sandstedt, 1982; Oddis and Medsger, 1988; Uchino *et al.*, 1985). Because in many cases the drug's potential efficacy determines the future need for potent immunosuppressive drugs, high-dose prednisone therapy is preferable in the early stages of the disease. The standard initial dose is at least 1 mg/kg of body weight/day for 3–4 weeks. This dose is then tapered slowly over a period of 10 weeks to 1 mg/kg every other day. Then, if there is evidence of efficacy and no serious side effects are seen, the dose is further reduced by 5 or 10 mg every 3 (or 4) weeks until the lowest possible dose that controls the disease is reached. This should be less than 15 mg/day, if possible. The efficacy of prednisone is determined by an objective increase in muscle strength, which almost always occurs by the third month of therapy. A reduction of CK activity without a concomitant increase in muscle strength is not a realiable sign of improvement.

If prednisone provides no objective benefit after 3 months of high-dose therapy, the disease is probably unresponsive to the drug, and tapering should be accelerated while the next immunosuppressive drug is started.

### Azathioprine or Methotrexate

Azathioprine or Methotrexate are the *second-line drug* treatments of PM/DM in adults (Bunch, 1981; Bunch *et al.*, 1980; Dalakas, 1991; Mertens and Rohkamm, 1990; Pongratz, 1992). This additional treatment is recommended:

1. From the beginning of the illness in severe, rapidly progressive forms of the disease, especially when there is general weakness, respiratory failure, or severe dysphagia.
2. For a "steroid-sparing" effect, if after 3 months of high-dose steroid therapy pronounced side effects have developed, the high-dose steroid cannot be tapered without a flareup, or very high doses of steroids are required to control the disease.

The use of azathioprine (up to 3 mg/kg daily) is based on long-term experience of many neurologists in Europe and the United States. It is a relatively safe treatment compared with other immunosuppressive drugs.

Methotrexate (10–20 mg/wk po) may be considered instead of azathioprine or if azathioprine is ineffective (Dalakas, 1988; Metzger *et al.*, 1974; Niakan *et al.*, 1980; Wallace *et al.*, 1885).

In childhood DM, many pediatricians prefer methotrexate as a drug of second choice instead of azathioprine. Our preference, however, for children is intravenous immunoglobulin.

### Other Second-Line Immunosuppressive Drugs

The effect of cyclophosphamide and cyclosporin is generally disappointing, despite occasional reports of beneficial effects (Bombardieri *et al.*, 1989; Cronin *et al.*, 1989; Heckmatt *et al.*, 1989; Jones *et al.*, 1987).

Only some patients with severe concomitant extramuscular involvement, such as interstitial lung disease, require aggressive treatment with cyclophosphamide (al-Janidi *et al.*, 1989; Hochberg *et al.*, 1986; Salmeron *et al.*, 1981; Tazelaar *et al.*, 1990; Tymms and Webb, 1985). The effect of Mycophenolate remains unclear but the first indications are promising.

### Plasmapheresis and Total Body or Lymph Node Irradiation

Plasmapheresis in a double-blind study was ineffective (Miller *et al.*, 1992). Total-body or lymph node irradiation is not generally recommended but sometimes has been effective as a last resort (Dalakas and Engel, 1888, Engel *et al.*, 1981; Hubbard *et al.*, 1982; Kelly *et al.*, 1986).

### High-Dose Intravenous Immunoglobulins

High-dose intravenous immunoglobulin is a promising, although expensive, new therapy for PM, DM, or IBM (Cherin *et al.*, 1990; Jan *et al.*, 1990; Roifman *et al.*, 1987; Dalakas *et al.*, 1993). Efficacy should be based on controlled studies. In a controlled study on DM, IVIg was effective and reversed the immunopathological findings on repeated muscle biopsies (Dalakas *et al.*, 1993). In IBM three controlled studies have been performed and showed minimal or no benefit (Dalakas *et al.*, 1997, 2001; Walter *et al.*, 2000).

### Long-Term Treatment

In contrast to the acute stages of PM and DM, it is difficult to give recommendations for long-term treatment to prevent relapses of the disease. If the acute phase is under control, we prefer a low-dose steroid maintenance treatment, often combined with azathioprine, for a period of time that may vary from 1–3 years.

During the long-term use of prednisone, an increase of weakness associated with a normal or unchanged creatine kinase level may be seen. This effect, referred to as steroid myopathy, is one possible symptom of hypercorticism.

In a patient who previously responded to high doses of prednisone, the development of increased weakness may be related to steroid myopathy or to disease activity that either will respond to a higher dose of steroids or has become resistant to steroids. It may be difficult to distinguish one cause from the other, because the two can coexist (Dalakas, 1988; Engel and Dalakas, 1982) or be complicated by other factors, such as decreased mobility or associated systemic illnesses.

In these circumstances the decision to raise or lower the prednisone dosage may be influenced by reviewing the patient's history of muscle strength, creatine kinase levels, and especially changes of medication for the preceding 2 months. It should be noted that, as a rule, steroid myopathy is not the only symptom of hypercorticism but rather one of them. If the increase of a patient's weakness is thought to be a symptom of hypercorticism, the dose of prednisone should be lowered.

Another difficult clinical condition is the treatment of primary chronic PM, whereby the patient has considerable weakness but no evidence of active disease in a muscle biopsy (no inflammation) or the blood (normal CK). There are very few options here. First, the clinician should make every effort possible (with repeat biopsy or detailed history) to exclude another disease (e.g., toxic or metabolic myopathy or IBM), which has been erroneously considered to be PM. If the diagnosis of probable PM is still considered, prednisone (perhaps combined with azathioprine) or IVIg could be tried for a 3-month period. Some physicians continue these drugs, even if they are ineffective, hoping that they may slow down the progress of the disease. If nothing is effective, our preference in these cases is to discontinue these drugs, because their long-term use is more harmful.

## REFERENCES

al-Janadi, M., Smith, C. D., and Karsh, J. (1989). Cyclophosphamide treatment of interstitial pulmonary fibrosis in polymyositis/dermatomyositis. *J. Rheumatol.* **16**, 1592–1596.

Bombardieri, S., Hughes, G. R. V., Neri, R., Del Bavo, P., and Del Bavo, L. (1989). Cychlophosphamide in severe polymyositis. *Lancet* **1**, 1138–1139.

Bunch, T. W. (1981). Prednisone and azathioprine for polymyositis: Long-term follow-up. *Arthritis Rheum.* **24**, 5–8.

Bunch, T. W., Worthington, J. W., Combs, J. J., Duane, M., Istrup, M. S., and Engel, A. G. (1980). Azathioprine with prednisone for polymyositis: A controlled, clinical trial. *Ann. Intern. Med.* **92**, 365–369.

Cherin, P., Herson, S., Wechsler, B. *et al.* (1990). Intravenous immunoglobulin for polymyositis and dermatomyositis. *Lancet* **336**, 116.

Cronin, M. E., Miller, P. W., Hicks, J. E., Dalakas, M. C., and Plotz, P. H. (1989). The failure of intravenous cyclophosphamide therapy in refractory idiopathic inflammatory myopathy. *J. Rheumatol.* **16**, 1225–1228.

Dalakas, M. C. (1988). Treatment of polymyositis and dermatomyositis with corticosteroids: A first therapeutic approach. *In* "Polymyositis and Dermatomyositis" (M. Dalakas, Ed.), pp. 235–253. Butterworth, Boston.

Dalakas, M. C. (1989). Treatment of polymyositis and dermatomyositis. *Curr. Opin. Rheumatol.* **1**, 443–449.

Dalakas, M. C. (1990). Pharmacologic concerns of corticosteroids in the treatment of patients with immune-related neuromuscular diseases. *Neurol. Clin.* **8**, 93–118.

Dalakas, M. C. (1991). Polymyositis, dermatomyositis, and inclusion body myositis. *N. Engl. J. Med.* **325**, 1487–1498.

Dalakas, M. C., Illa, I., Dambrosia, J. M., Soueidan, S. A., Stein, D. P., Otero, C., Dinsmore, S. T., McCrosky, S. (1993). A controlled trial of high-dose intravenous immune globulin infusions as treatment for dermatomyositis. *N. Engl. J. Med.* **329**, 1993–2000.

Dalakas, M. C., and Engel, W. K. (1988). Total body irradiation in the treatment of intractable polymyositis/dermatomyositis. *In* "Polymyositis and Dermatomyositis" (M. C. Dalakas, Ed.), pp. 281–291. Butterworth, Boston.

Dalakas, M. C., Koffmann, B., Fujii, M., Spector, S., Sivakumar, K., and Cupler, E. (2001). A controlled study of intravenous immunoglobulin combined with prednisone in the treatment of IBM. *Neurology* **56**, 323–327.

Dalakas, M. C., Sonies, B., Dambrosia, J., Sekul, E., Cupler, E., and Sivakumar, K. (1997). Treatment of inclusion-body myositis with IVIg: a double-blind, placebo-controlled study. *Neurology* **48**, 712–716.

Engel, A. G. (1992). Immunoeffektormechanismen bei entzündlichen Myopathien: Klinische Bedeutung und therapeutische Auswirkungen. *In* "Aktuelle Myologie" (D. Pongratz, C. D. Reimers, and M. Schmidt-Achert, Eds.), pp. 42–55, Urban & Schwarzenberg, Munich, Vienna, and Baltimore, Maryland.

Engel, W. K., and Dalakas, M. C. (1982). Treatment of neuromuscular diseases. *In* "Therapy of Neurologic Diseases" (W. C. Wiederhold, Ed.) pp. 51–101, Wiley, New York.

Engel, W. K., Lichter, A. S., and Galdi, A. P. (1981). Polymyositis: Remarkable response to total body irradiation. *Lancet*, **1**, 658.

Heckmatt, J., Hasson, N., Saunders, C. *et al.* (1989). Cyclosporine in juvenile dermatomyositis. *Lancet* **1**, 1063–1066.

Henriksson, K. H., and Sandstedt, P. (1982). Polymyositis—treatment and prognosis: A study of 107 patients. *Acta Neurol. Scand.* **65**, 280–300.

Hochberg, M. C., Feldman, D., and Stevens, M. B. (1986). Adult onset polymyositis/dermatomyositis: An analysis of clinical and laboratory features and survival in 76 patients with a review of the literature. *Semin. Arthritis Rheum.* **15**, 168–178.

Hubbard, W. N., Walport, M. J., Halman, K. E., Beaney, R. P., and Hughes, G. R. V. (1982). Remission from polymyositis after total body irradiation. *BMJ.* **284**, 1915–1916.

Jan, S., Beretta, S., Maggio, M., Alobbati, L., and Pellegrini, G. (1990). High-dose intravenous human immunoglobulin in treatment-resistant polymyositis. *Neurology* **40**(Suppl. 1), 120 (Abstract).

Jones, D. W., Snaith, M. L., and Isenberg, D. A. (1987). Cyclosporine treatment for intractable polymyositis. *Arthritis Rheum.* **30**, 959–960.

Kelly, J. J., Jr., Madoc-Jones, H., Adelman, L. S., Andrs, P. L., and Munsat, T. L. (1986). Total body irradiation is not effective in inclusion body myositis. *Neurology* **36**, 1264–1266.

Mertens, H. G., and Rohkamm, R. (1990). "Therapie neurologischer Krankheiten und Syndrome." Georg Thieme Verlag, Stuttgart and New York.

Metzger, A. L., Bohan, A., Goldberg, L. S., Bluestone, R., and Pearson, C. M. (1974). Polymyositis and dermatomyositis: Combined methotrexate and corticosteroid therapy. *Ann. Intern. Med.* **81**, 182–189.

Miller, F. W., Leitman, S. F., Cronin, M. E., Hicks, J. F., Leff, R., Wesley, R., Fraser, D. D., Dalakas, M. C., Plotz, P. H. (1992). Controlled trial of plasma exchange and leukopheresis in polymyositis and dermatomyositis. *N. Engl. J. Med.* **326**, 1380–1384.

Niakan, E., Pitner, S. E., Whitaker, J. N., and Bertorini, T. E. (1980). Immunusuppressive agents in corticosteroid-refractory childhood dermatomyositis. *Neurology* **30**, 286–291.

Oddis, C. V., and Medsger, T. A. (1988). Relationship between serum creatine kinase level and corticosteroid therapy in polymyositis/dermatomyositis. *J. Rheumatol.* **15**, 807–811.

Pongratz, D. E. (1992). Myositiden. *In* "Therapiehandbuch." Urban & Schwarzenberg, Munich, Vienna, and Baltimore, Maryland.

Roifman, C. M., Schaffer, F. M., Wachsmuth, S. E., Murphy, G., and Gland, E. W. (1987). Reversal of chronic polymyositis following intravenous immune serum globulin therapy. *JAMA.* **258**, 513–515.

Salmeron, G., Greensberg, S. D., and Lidsky, M. D. (1981). Polymyositis and diffuse interstitial lung disease: A review of the pulmonary histopathologic findings. *Arch. Intern. Med.* **141**, 1005–1010.

Soueidan, S. A., and Dalakas, M. C. (1993). Treatment of inclusion body myositis with high-dose intravenous immunoglobulins. *Neurology* **43**, 876–879.

Tazelaar, H. D., Viggiano, R. W., Pickersgill, J., and Colby, T. V. (1990). Interstitial lung disease in polymyositis and dermatomyositis: Clinical features and prognosis as correlated with histologic findings. *Am. Rev. Respir. Dis.* **141**, 727–733.

Tymms, K. E., and Webb, J. (1985). Dermatopolymyositis and other connective tissue diseases: A review of 105 cases. *J. Rheumatol.* **12**, 1140–1148.

Uchino, M., Araki, S., Yoshida, O., Uekawa, K., and Nagata, J. (1985). High single-dose alternate-day corticosteroid regimens in treatment of polymyositis. *J. Neurol.* **232**, 175–178.

Wallace, D. J., Metzger, A. L., and White, K. K. (1985). Combination immunosuppresive treatment of steroid-resistant dermatomyositis/polymyositis. *Arthritis Rheum.* **28**, 590–592.

Walter, M. C., Lochmuller, H., Toepfer, M., Schlotter, B., Reilich, P., Schroder, M., Muller-Felber, W., and Pongratz, D. (2000). High-dose immunoglobulin therapy in sporadic inclusion body myositis: a double-blind, placebo-controlled study. *J. Neurol.* **247**(1), 22–28.

*Muscle and Peripheral Nervous System*

CHAPTER 96

# Myopathies

Richard W. Orrell and Robert C. Griggs

## INTRODUCTION

There is a wide range of disorders of skeletal muscle, or myopathies. The most readily treatable are the inflammatory myopathies (Chapter 95), but there are many other myopathies with successful therapy—metabolic myopathies (Chapter 97), myotonias (Chapter 98), channelopathies (Chapter 99)—all dealt with elsewhere in this text. Myopathies may be secondary to other systemic disorders, medication, and infection. Treatment of these is that of the systemic disorder or infection or avoidance of the medication or toxin. In this chapter we present the muscular dystrophies, or progressive myopathies with a genetic basis.

A broad spectrum of muscular dystrophies is recognized (Table I) (Kaplan and Fontaine, 2000; Orrell and Griggs, 1999). Historically, these have been categorized by their most severely involved muscles at clinical presentation, for example, limb girdle muscular dystrophy, distal myopathy, facioscapulohumeral dystrophy, and scapulohumeral dystrophy. Increasingly, many are identified by their genetic basis and their underlying abnormal protein product. Although there remains no curative treatment for any of the muscular dystrophies, several have treatment that slows their course, and all have valuable supportive care strategies. Many of the muscular dystrophies are individually rare, so that it will require multicenter studies to conduct randomized placebo-controlled clinical trials of sufficient power in each individual disorder. It may be, however, that in the future, treatments with benefit for one muscular dystrophy may be applicable to other muscular dystrophies.

Two relatively common muscular dystrophies have been subjected to the most detailed clinical investigation and therapeutic trials. These are Duchenne dystrophy (DD) and facioscapulohumeral dystrophy (FSHD). We will discuss these separately. The general principles of

treatment and management—pharmacological, physical therapy, aids and appliances, respiratory support, surgery, and genetic counseling, may be extended to other muscular dystrophies and myopathies.

In thinking about developing new treatments, it is helpful to focus on the goals of treatment (Figure 1). Although cure, a total reversal of weakness and a return to normal strength, is the strategy that is easiest to conceptualize, other goals are both more realistic and equally important to pursue. Because all muscular dystrophies are genetic and because the genetic cause for most can now be deducted at or before birth, arrest of disease would be tantamount to cure (Figure 1). A slowing of course or an improvement with a plateau (such as is seen in prednisone treatment of Duchenne's muscular dystrophy) can have a major effect on morbidity.

The difficulties in reconciling clinical and molecular classifications may be illustrated by the muscular dystrophies. More than 20 genes for limb girdle muscular dystrophy (LGMD) have been mapped, and at least 10 of these identified. In a recent review of more than 216 patients from 85 families with LGMD, similar clinical courses among patients with different molecular forms of LGMD, but also dissimilar course among patients with the same molecular form, including the same mutation within a family, were demonstrated (Zatz *et al.*, 2000). LGMD2F patients ($\delta$ sarcoglycanopathy) typically had a severe "Duchenne dystrophy" type presentation. Some patients with LGMD2G (telethoninopathy) had distal weakness resembling Miyoshi myopathy, but others had calf hypertrophy with typical clinical features of LGMD.

Phenotypic classifications are now limited by the understanding of phenotypic and genotypic heterogeneity. For a clinical differential diagnosis the classifications may remain helpful (Table I). These include Duchenne/Becker Dystrophy (see later), the limb girdle muscular

**TABLE I**  Differential Diagnosis of Muscular Dystrophies—Genetic Correlations of Clinical Phenotypes

| Presentation | Abnormal gene product | Gene product (linkage) | Inheritance | Distinctive features |
|---|---|---|---|---|
| Duchenne/Becker muscular dystrophy | | | | |
| DMD | Dystrophin | DYS | XR | See text |
| Limb girdle muscular dystrophies | | | | |
| LGMD1A | | (5q22–34) | AD | |
| LGMD1B | Lamin A/C | LMNA | AD | Cardiac involvement |
| LGMD1C | Caveolin-3 | CAV3 | AD | |
| LGMD1D | | (6q23) | AD | Dilated cardiomyopathy with conduction defect |
| LGMD1E | | (7q) | AD | |
| LGMD2A | Calpain 3 | CAPN3 | AR | |
| LGMD2B | Dysferlin | | AR | Allelic to Myoshi myopathy |
| LGMD2C | γ-Sarcoglycan | SGCG | AR | |
| LGMD2D | α-Sarcoglycan (adhalin) | SGCA | AR | |
| LGMD2E | β-Sarcoglycan | SGCG | AR | |
| LGMD2F | δ-Sarcoglycan | SGCD | AR | |
| LGMD2G | Telethonin | | AR | |
| LGMD2H | | (9q31–34.1) | AR | |
| LGMD2I | | (19q13.3) | AR | |
| Distal myopathies | | | | |
| Miyoshi myopathy | | | | |
| MM | Dysferlin | | AR | Onset in posterior compartment of lower leg. Allelic to LGMD2B |
| Distal myopathy with rimmmed vacuoles (Nonaka) | | | | |
| DMRV | | (9p1-q1) | AR | Onset in anterior compartment of lower leg |
| Hereditary inclusion body myopathy | | | | |
| HIBM | | (9p1-q1) | AR | |
| Autosomal dominant distal myopathy | | | | |
| MPD1 | | (14) | AD | |
| Tibial muscular dystrophy (Markesbery-Griggs-Udd) | | | | |
| TMD | | (2q31) | AD | Onset in legs |
| Welander distal myopathy | | (2p13) | AD | Onset in hands |
| Congenital myopathies | | | | |
| Myotubular myopathy | | | | |
| MTMX | Myotubularin | | XR | |
| Central core disease | | | | |
| CCD | Ryanodine receptor | RYR1 | AD | |
| Nemaline myopathy | | | | |
| NEM1 | α tropomyosin | TPM3 | AD | |
| NEM2 | Nebulin | | AR | |
| ACTA1 | Actin alpha, skeletal muscle | | AD | |
| Fukuyama congenital muscular dystrophy | | | | |
| FCMD | Fukutin | | AR | Severe mental retardation and other central nervous system involvement |
| Congenital muscular dystrophy with merosin deficiency | | | | |
| LAMA2 | Laminin alpha2 chain of merosin | | AR | Generalized muscle weakness |
| Congenital muscular dystrophy with secondary merosin deficiency | | | | |
| CMD1B | 1q42 | | AR | |

*continues*

**TABLE I** *Continued*

| Presentation | Abnormal gene product | Gene product (linkage) | Inheritance | Distinctive features |
|---|---|---|---|---|
| Congenital muscular dystrophy with integrin deficiency | | | | |
| ITGA7 | Integrin α7 | | ARa | |
| Congenital muscular dystrophy with rigid spine | | | | |
| RSMD-1 | 1p35–36 | | AR | |
| Muscular dystrophies with early prominent conractures | | | | |
| Emery-Dreifuss muscular dystrophy | | | | |
| EMD | Emerin | | XR | Cardiomyopathy with conduction defects |
| EMD-AD | Lamin A/C | | AD | |
| Bethlem myopathy | | | | |
| COL6A1, COL6A2 | Collagen type VI subunit a1 or a2 | | AD | No cardiac involvement |
| COL6A3 | Collagen type VI subunit a3 | | AD | |
| Facioscapulohumeral muscular dystrophy | | | | |
| FSHD | (4q35) | | AD | See text |
| Oculopharyngeal muscular dystrophy | | | | |
| OPMD | Poly (A) binding protein 2 | | AD | |
| Other genetically identified muscular dystrophies | | | | |
| Vocal cord and pharyngeal weakness with autosomal dominant distal myopathy | | | | |
| VPDMD | (5q31) | | AD | |
| Autosomal dominant myopathy with proximal muscle weakness and early respiratory muscle involvement (Edstrom) | | | | |
| MPRM2 | (2q21) | | AD | |
| Epidermolysis bullosa simplex associated with late onset muscular dystrophy | | | | |
| MD-EBS | Plectin | | AR | |
| Desmin-related myopathy | | | | |
| DRM | α B crystallin | | AD | |
| Myopathy with excessive autophagy | | | | |
| MEAX | Xq28 | | XR | |

dystrophies, the distal myopathies, the congenital myopathies, myopathies with early prominent contractures, oculopharyngeal muscular dystrophy, and facioscapulohumeral dystrophy (see later).

## DUCHENNE MUSCULAR DYSTROPHY

### Clinical Aspects (Differential Diagnosis)

Duchenne dystrophy (DD) is an X-linked muscular dystrophy caused by a mutation in the dystrophin gene on chromosome Xp21. Mutations leading to total absence of dystrophin (out-of-frame mutations) lead to classical features of DD. As a rule, in-frame deletions lead to diminished or truncated forms of dystrophin, with milder and variable phenotypes, usually termed Becker dystrophy (BD). A variety of clinical phenotypes are recognized to be associated with mutations in the dystrophin gene (Table II).

The DD gene, which encodes the dystrophin protein, is the largest known human gene, with around 2,400,000 base pairs (Molnar and Karpati, 1999). It is located at Xp21.1. The coding sequence of the gene is 13,900 base pairs, with 79 exons. The dystrophin gene produces several isoforms of dystrophin, of which the full-length 427-kD dystrophins are relevant to DD. These are muscle, brain, and Purkinje types.

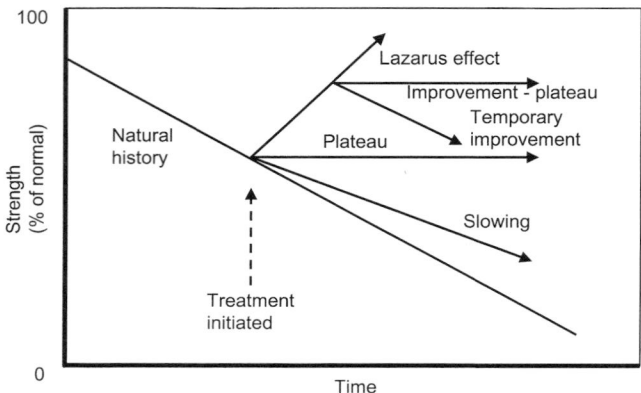

**FIGURE 1**  Potential effects of treatment of a muscle disease. The underlying disorder is slowly progressive. A dramatic improvement ("Lazarus effect") is the ideal, but a slowing in the course, or plateau, may have a major effect on long-term morbidity. (Based on Griggs 1994, with permission from Lippincott Williams & Wilkins.).

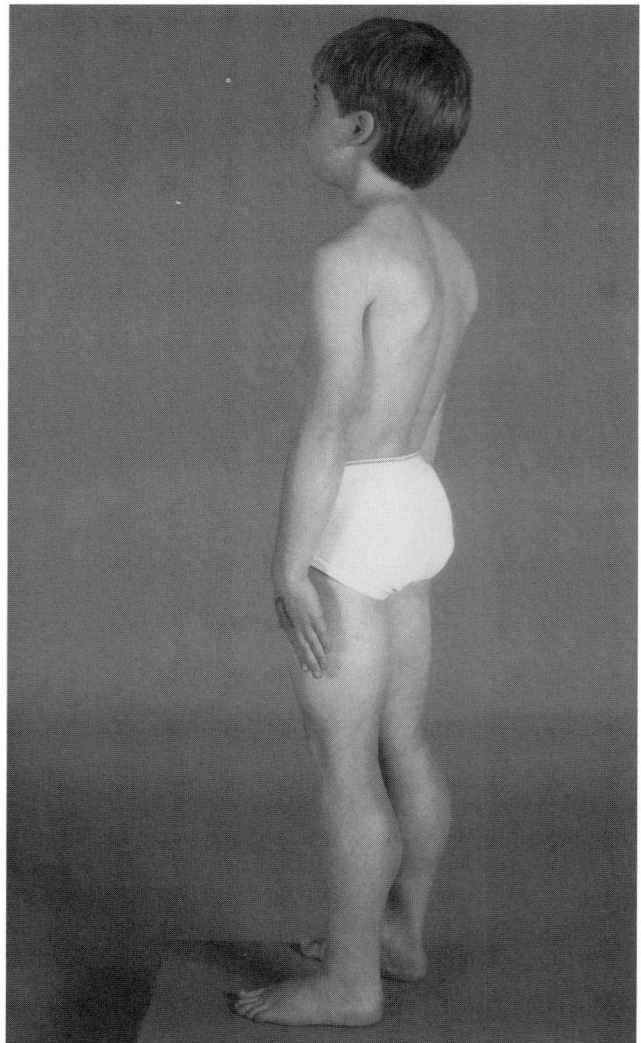

**FIGURE 2**  A boy with Duchenne muscular dystrophy, demonstrating calf hypertrophy.

The dystrophin protein is rod shaped and constitutes approximately 5% of membrane-associated cytoskeletal proteins. (The other clinically relevant sarcolemma-related cytoskeletal proteins include α and β dystroglycans, α, β, γ, and δ sarcoglycans, laminin $α_2$ [merosin], utrophin, and syntrophins). The complex of these proteins connects the intracellular cytoskeletal actin to the extracellular matrix. Dystrophinopathies, sarcoglycanopathies, and merosin deficiency are the most common causes of muscular dystrophy.

### Natural Course

DD has an incidence of 1/3500 live births. Inheritance is X-linked, being manifested by boys inheriting the gene from heterozygote nonmanifesting mothers. In approximately two thirds of patients the dystrophinopathy is transmitted by a carrier female, but in one third it is a new mutation.

DD typically presents at around 3–5 years of age with difficulty running, rising from a lying position, climbing stairs, and a waddling gait. Muscle weakness is pre-

**TABLE II    Clinical Presentations of Dystrophinopathy**

1. Typical Duchenne muscular dystrophy
2. Milder Duchenne muscular dystrophy
3. Manifesting Duchenne muscular dystrophy heterozygotes in females
4. Becker muscular dystrophy
5. Quadriceps myopathy
6. Cramps with myoglobinuria
7. Phenotypically normal individuals

dominantly proximal in the pelvic and shoulder girdles but also occurs more generally, including the trunk. Proximal weakness accounts for Gowers' sign, when the boy uses his hands to push against his legs when rising. There may be exaggerated lumber lordosis and thoracolumbar scoliosis. The cranial muscles are unaffected until late in the disease. In addition there is hypertrophy of the calf muscles which, together with wasting of other muscles, gives a typical appearance of the legs (Figure 2).

Muscle weakness progresses, typically with loss of independent mobility around age 9–11 years. In the mid-to-late teens, respiratory support may be necessary, because of respiratory muscle weakness, and in the early 20s cardiomyopathy may develop. Generally cardiorespiratory failure leads to death in the 20s.

Significant additional complications include contractures, especially in the legs and thoracic scoliosis, which may contribute to respiratory insufficiency. Primary cardiomyopathy may occur in a subset of patients with DD or BD (Brooke *et al.*, 1989). It is important to consider the possibility of cardiac failure being secondary to respiratory failure (cor pulmonale).

In Becker dystrophy (BD), a reduced amount of truncated dystrophin is produced. The clinical progress is much slower than in DD. The incidence of BD is around 10% that of DD, but the prevalence is higher, because individuals live longer. Age of onset is usually later (3–20 years), loss of independent mobility occurs between 12 and 40 years, and death from respiratory insufficiency or cardiomyopathy can occur between 30 and 60 years or even later.

Milder phenotypes include quadriceps myopathy and muscle cramps without weakness. There is also an isolated increase in creatine kinase termed "hyperCKemia." Because DD is X-linked, carrier females generally do not manifest the disease but, depending on X chromosome inactivation (Lyon hypothesis), may manifest milder forms of the disease.

Diagnosis is initially clinical, supported by molecular testing, and an increased serum creatine kinase, at least 20 and sometimes up to 200 times normal. Muscle biopsy shows features of a myopathy. Immunohistochemistry demonstrates lack of dystrophin on the surface of muscle fibers in DD. The dystrophin protein is large, and in BD immunostaining may be patchy, or even normal, depending on the the point of protein truncation and the location of epitopes recognized by the antibody. Antibodies are available for the NH$_2$, COOH, and mid rod domains. There is associated reduction in immunostaining of α and β dystroglycans and sarcoglycans, but merosin is preserved. Western blot analysis of the dystrophin protein may be important in the diagnosis of BD, demonstrating the truncated protein.

In approximately 65% of DD patients an out-of-frame deletion is detectable by multiplex polymerase chain reaction, including hot spots for mutation. New techniques are becoming available that detect more than 90% of mutations (Mendell *et al.*, 2001). These usually produce an unstable dystrophin protein that is not detectable. In 5% there is reduplication of one or more exons and in the remainder point mutations or other abnormalities.

## Principles of Therapy

Dystrophin is involved in providing the mechanical strength of the sarcolemma, and deficient fibers are vulnerable to stretch, causing damage to the muscle. Dystrophin also acts as a stabilizer of other proteins of the complex, which include sarcoglycans. Aditional possible roles of dystrophin include signal transduction.

Goals of therapy include replacement with normal dystrophin protein, gene replacement or correction, and stabilization of the muscle fiber with prevention of necrosis. Other considerations include the effects of abnormalities in other isoforms of dystrophin, for example, in the brain.

In practice, gene replacement has not yet proven clinically effective (Fletcher *et al.*, 2000; Kakulas, 1997). Trials of myoblast transfer, with normal dystrophin, were clinically ineffective. An initial double-blind study of eight boys with DD demonstrated that transplanted myoblasts from a normal first-degree relative donor persisted and produced dystrophin in muscle fibers of DD patients (Gussoni, 1992). The myoblasts were injected into 80–100 sites in the tibialis anterior of one leg, with placebo injected into the other leg. Muscle biopsies were taken for analysis after 1 month. A further blinded study injected 55 sites of one biceps muscle of 8 DD boys with a total of 55,000,000 normal myoblasts cultured from their father. The patients received immunosuppression with cyclophosphamide for 6 or 12 months. The procedure was considered to have been proven safe. Unfortunately, there was no evidence of persistent dystrophin-positive fibers at 12 months and no significant clinical benefit (Karpati *et al.*, 1993).

Dystrophin gene transfer is being studied in animals. Full-length cDNA can be introduced into muscle fibers with adenovirus vector. This procedure has not yet progressed to human clinical trials. Utrophin gene upregulation, or protein overexpression, is an alternative strategy. Utrophin is a protein structurally related to dystrophin, expressed in muscle, and attached to the same proteins in the complex but encoded by a different gene (on chromosome 6q24).

Another strategy has been based on the ability of several aminoglycosides to allow incorporation of an amino acid at stop codons during protein synthesis. In the *mdx* mouse, the dystrophin gene contains a premature stop codon, leading to early termination of translation and absence of dystrophin. Injection of gentamicin led to full-length dystrophin being present at 10%–20% of normal levels. There was also a reduction in eccentric contraction injury to muscle (Barton-Davis *et al.*, 1999). In human trials, four patients with DD caused by stop codon sequences were given intravenous gentamicin, 7.5 mg/kg/day, for 2 weeks. Full-length dystrophin remained absent from muscle biopsies after treatment. There was a reduction in serum creatine kinase but no significant change in muscle strength (Wagner *et al.*, 2001). There was concern of toxicity with higher doses or more prolonged treatment. A preliminary report of a similar study in 12 patients with dystrophin and sarcoglycan deficiencies caused by stop codons demonstrated a similar lowering of serum creatine kinase (Serrano et al., 2001). It remains to be proven whether similar strategies will have clinical benefit.

A number of studies have demonstrated benefit from prednisone in DD. The pathophysiological basis of this effect is uncertain. Prednisone was originally used because of its anti-inflammatory effect. Other possible mechanisms include reduced muscle catabolism, delayed muscle apoptosis, or growth suppression.

Sustained treatment, over periods up to 3 years, shows maintained improvement in strength but with side effects in at least 30% of patients, including weight gain, cushingoid appearance, behavioral disturbances, and asymptomatic cataracts. Osteopenia is frequent in DMD and is worsened by corticosteroid treatment.

Prednisone was first demonstrated to have benefit in an uncontrolled unblinded study of 14 boys with DD over a period of up to 28 months (Drachman, 1974). A further study of 16 boys, unblinded, nonrandomized, demonstrated similar benefit (DeSilva et al., 1987). Prednisone, 1.5/mg/kg/day, was studied in 33 boys with DD, using natural history cohorts. After 6 months, there was improvement in muscle strength and function (Brooke et al., 1987). In a randomized, double-blind, controlled trial of prednisone, 0.75 mg/kg/day or 1.5 mg/kg/day, similar improvements in muscle strength and function were found after 6 months treatment (Mendell et al., 1989). A further randomized, placebo-controlled trial was performed in 99 boys with DD. Prednisone was given at daily doses of 0.3 mg/kg and 0.75 mg/kg and also placebo. Patients were studied using manual muscle testing and myometry, functional testing, pulmonary function, and laboratory measurements. There was significant increase in strength in those taking prednisone, more marked in the higher dose group. However, at 6 months there were side effects of weight gain, cushingoid appearance, and excessive hair growth in the high-dose group, with only weight gain in the lower dose group (Griggs et al., 1991). Subsequent study with an alternate-day regimen of prednisone demonstrated decline of strength with no significant improvement in side effects (Fenichel et al., 1991a). The double-blind study using prednisone, 0.75 mg/day, was extended for 18 months, demonstrating that prednisone arrests the disease for at least 3 years (Fenichel et al., 1991b).

A randomized controlled trial with addition of azathioprine (2–2.5 mg/kg/day) to prednisone (0.3 and 0.75 mg/kg/day) did not demonstrate additional benefit of azathioprine (Griggs et al., 1993). This also raises doubts over the proposed immunosuppressive mechanism of prednisone in DD. Studies of muscle protein synthesis and breakdown have shown that prednisone slows breakdown, resulting in a 20% increase in muscle mass (Rifai et al., 1995).

To explore possible reduction of steroid side effects, a multicenter, double-blind, randomized trial of deflazacort (0.9 mg/kg/day) against prednisone (0.75/mg/kg/day) was performed in 18 boys with DD. At 12 months the two steroids had equal benefit in improving motor function and functional performance but with reduced weight gain in the patients taking deflazacort (Bonifati et al., 2000).

Ten boys with DD were treated for 3 months with oxandrolone, an anabolic steroid. Functional benefit similar to prednisone was suggested (Fenichel et al., 1997). However, a recent, larger 6-month, randomized, double-blind study failed to show clinical efficacy (Fenichel, 2001). There was no significant change in average muscle strength score of oxandrolone-treated patients compared with placebo. There was, however, a significant improvement in the average of four quantitative muscle tests. No adverse effects were attributable to oxandrolone, and this suggested a possibility of a beneficial effect in slowing progression of weakness before starting corticosteroid therapy, with the inevitable associated side effects.

Studies to determine the optimum dose regimen for prednisone or alternative medications continue (Dubowitz, 1997). At this point it is clear that prednisone in a dose of 0.75 mg/kg/day delays progression of Duchenne muscular dystrophy and enables patients to maintain motor function for 30%–35% longer than untreated patients. Prednisone delays respiratory failure and prolongs life, making gainful employment possible. As many as 10%–15% of patients have weight gain or other side effects that require reduction or discontinuation of treatment. Strategies for preventing corticosteroid complications are indicated in all patients.

Creatine monohydrate has been studied in neuromuscular disorders, including muscular dystrophy. Creatine supplementation is used by athletes and increases muscle mass and strength. Possible mechanisms include increased intramuscular phosphocreatine, improved respiratory chain function, and protein synthesis. An initial study (Tarnopolsky and Martin, 1999) used a single-blind design in 21 patients with neuromuscular disease, with short-term increased muscle power. A further double-blind placebo-controlled study of 36 patients (12 with FSHD, 10 with BD, 8 with DD, and 6 with sarcoglycan deficient limb girdle dystrophy) gave creatine or placebo for 8 weeks, with a washout period of 3 weeks (Walter et al., 2000). There was a mild, but significant, increase in muscle strength (3%) and daily life activities (10%). The creatine (CREAPURE; SKN, Trotsberg, Germany) was well tolerated. Longer term studies are needed to assess whether benefit is maintained.

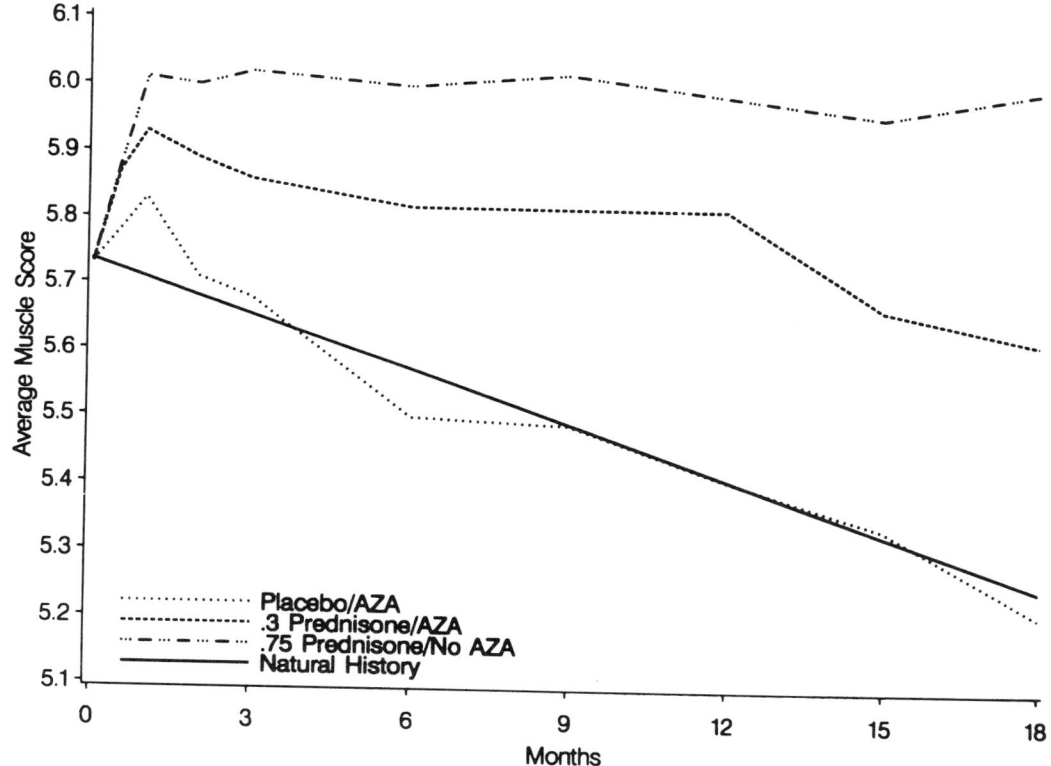

**FIGURE 3** Comparison of average muscle score on manual muscle testing in patients with Duchenne muscular dystrophy. Significant improvement was seen in patients treated with prednisone, 0.75 mg/kg/day. Patients treated with prednisone, 0.3 mg/kg/day, and addition of azathioprine, showed no significant difference from natural history mean at 12 months. (Griggs *et al.*, 1993, copyright Lippincott Williams & Wilkins.)

Studies of surgical and other therapeutic approaches are often uncontrolled and anecdotal, but the findings in large series may be helpful (Scheuerbrandt, 1998). A review of 144 boys with DD suggested that operative treatment for contractures and bracing of the legs provided improved control of contractures and prolonged ability to walk. This was a retrospective, nonrandomized, unblinded study. Contractures of the lower extremeties were best managed by a combination of daily passive stretching exercises, prescribed periods of standing and walking, tenotomy of the Achilles tendon, posterior tibial-tendon transer, and application of knee-ankle-foot orthoses (Vignos *et al.*, 1996).

A review of spine surgery in 30 patients with DD concluded that spine surgery corrects the scoliosis, providing a comfortable sitting position and preventing more severe progressive deformity, which may cause further functional disadvantage and poor body image. Spine surgery did not improve or stabilize respiratory function or prolong life. Quality of life was improved. The surgical result is best if performed before the scoliosis exceeds 40 degrees. There is occasional, but low, mortality associated with the procedure (Granata *et al.*, 1996).

**Practical Management**

In ambulatory patients with DD, older than 5 years of age, we recommend consideration of prednisone, 0.75 mg/kg/day\*\*\*, to be continued if side effects are not severe. The most significant side effect is weight gain, which may be minimized by dietary advice. Other side effects include hypertension, behavioral changes, growth retardation, osteoporosis, and cataract. Addtional prophylactic medication for prevention of osteoporosis is important. Deflazacort\*\*\* may be a useful alternative, with less side effects (but is not available in the United States).

The patient with DD should attend a specialist neuromuscular clinic and requires assessment and management by a team that includes a physiotherapist, occupational therapist, and physicians skilled as to consideration of social and psychological issues. It may be helpful to have contact with a patient support group (Table III). Orthopedic consultation and appropriate orthoses and mobility aids including a wheelchair may be necessary.

Surgical management includes careful consideration of tendon surgery for contractures and surgery for scoliosis.

**TABLE III**  Patient Support Groups and Charities

| Country | Organization | Address | Website |
|---|---|---|---|
| Australia | Muscular Dystrophy Australia | GPO Box 9932, Melbourne 3001, Australia | http://www.mda.org.au |
| Canada | Muscular Dystrophy Association of Canada | 2345 Yonge Street, Suite 900, Toronto, Ontario, M4P 2E5, Canada | http://www.mdac.ca |
| Europe | EAMDA | 7–11 Prescott Place, Clapham, London SW4 6BS, England | http://www.sonnet.co.uk/eamda |
| Ireland | Muscular Dystrophy Ireland | Carmichael Centre, Coleraine House, Coleraine Street, Dublin 7, Ireland. | http://www.mdi.ie |
| The Netherlands | Vereniging Spierziekten Nederland | Lt Gen van Hetszlaan 6, 3743 JN Baarn, The Netherlands | http://www.vsn.nl |
| United Kingdom | Muscular Dystrophy Campaign | Nattrass House, 7–11 Prescott Place, London SW4 6BS, England | http://www.muscular-dystrophy.org |
|  | FSH MD Support Group | 8 Caldecote Gardens, Bushey Heath, WE2 3RA, England | http://www.fsh-group.org |
| USA | Muscular Dystrophy Association | National Headquarters, 3300 E Sunrise Drive, Tucson, AZ 85718, USA | http://mdausa.org |
|  | Muscular Dystrophy Family Foundation | 2330 North Meridian Street, Indianapolis, Indiana 46208-5730, USA | http://www.mdff.org |
|  | FSH Society | 3 Westwood Road, Lexington, MA 02420, USA | http://www.fshsociety.org |

Graded respiratory support may be required. Published studies include the effects of noninvasive intermittent positive pressure ventilation and assisted coughing compared with tracheostomy intermittent positive pressure ventilation (Bach *et al.*, 1997). Pulmonary morbidity and hospitalization were reduced with noninvasive ventilation. Sleep-disordered breathing may be a predictor of mortality (Phillips *et al.*, 1999).

Cardiac failure should be treated using standard medical methods. Patients with signs of heart failure should be studied carefully for evidence of hypoxia and cor pulmonale. Anticoagulation with warfarin should be considered in patiens with an ejection fraction of less than 50% to prevent emboli. Symptomatic cardiomyopathy usually occurs late in the course of DD, when skeletal muscle weakness and general disability is marked. In these circumstances cardiac transplantation is usually inappropriate. In some milder dystrophinopathies, cardiomyopathy may be a prominent early feature, without skeletal muscle weakness. Cardiac transplantation may satisfactorily restore cardiac function in these patients. Ankle edema may due to cardiac failure but may also be related to immobility, dependent limbs, and loss of muscle pump function returning venous blood in the legs. Treatment with limb elevation and, occasionally, elastic stockings or diuretics may be helpful.

Gastrointestinal smooth muscle may be involved in DD (Barohn, 1988). This may cause gastric dilation and intestinal pseudo-obstruction, requiring nasogastric tube insertion and aspiration. Metaclopramide (10–20 mg/day) has been used in these circumstances. Constipation may also require treatment. Dietary advice may be important to minimize obesity resulting from an imbalance between dietary intake and physical activity.

Malignant hyperthermia may occur during anesthesia (see later).

Genetic counseling should be offered to the patient and family. Prenatal diagnosis is possible for DD and other muscular dystrophies in which the gene has been identified.

## Treatments No Longer Recommended

Treatments that have previously been considered in DD, for which there is no demonstrated significant benefit and that are no longer recommended, include allopurinol, adenine nucleotides, superoxide dismutase, vasoactive and antiserotonin drugs, leucine, penicillamine and vitamin E, mazindol, verapamil, flunarizine, nifedipine, diltiazem, and dantrolene (which may, however, be indicative if there is malignant hyperthermia).

# FACIOSCAPULOHUMERAL DYSTROPHY

## Clinical Aspects (Differential Diagnosis)

Facioscapulohumeral dystrophy (FSHD) is an autosomal-dominant muscular dystrophy, presenting with weakness predominantly in the muscles of the face, shoulder girdle, and upper arms (excluding deltoid) (Tawil *et al.*, 1998). Weakness may also affect the peroneal muscles in the leg. The differential diagnosis is now aided by molecular genetic diagnostics (see later). In a typical patient, the clinical diagnosis is usually apparent on the basis of the distribution of muscle weakness and possibly additional affected family members. Approximately 30% of patients have no other affected family members, and in the milder forms a range of other muscular dystrophies may be considered. Differential diagnosis includes limb girdle dystrophies and scapuloperoneal myopathies, together with desmin myopathy, polymyositis, inclusion body myositis, mitochondrial myopathy, and congenital myopathies.

## Natural Course

FSHD is present worldwide, with a prevalence of approximately 1/20,000. It is the third most common muscular dystrophy in adults. The disease is caused by an unusual mechanism resulting from a genetic abnormality in the telomeric region of chromosome 4q (van Deutekom *et al.*, 1993). There is a reduction in the number of large (3300 base) repeats in this region. Normal individuals have more than 10, and affected individuals 10 or fewer, repeats. The mechanism by which this reduction in repeat number causes the disease remains uncertain at present. The number of repeats may be measured by DNA analysis. This allows diagnostic and predictive molecular testing (Orrell *et al.*, 1999). As with other genetic conditions, the presence of a diagnostic test has extended the phenotype of FSHD. Milder forms, in which the clinical presentation may be incomplete, are associated with borderline or minor truncation of repeat number (Ricci *et al.*, 1999).

FSHD affects both males and females. The disease may be first seen at any age from infancy to adulthood, but typically some features of the disease are present by age 20s. This may require careful clinical assessment of mild weakness, for example, difficulty completely burying the eyelids on eye closure. Progression of disease is relatively slow and may at times seem to be stable, although the underlying muscle condition persists. Lifespan is generally not significantly affected, but around 20% of patients may require a wheelchair for mobility at some stage.

Shoulder weakness is the most common presenting symptom, because facial weakness is often overlooked as a result of the more prominent disability from loss of shoulder girdle function. The scapular fixator muscles—latissimus dorsi, trapezius, rhomboids, and serratus anterior—give a typical appearance. Scapular winging is present (Figure 4). The deltoid muscle is relatively unaffected (Figure 5). Pectoral muscle weakness and wasting gives a flattened appearance to the chest wall. Facial weakness involves predominantly the oribicularis oculi and oris muscles, with weakness of eyelid closure and mouth function. Masseter, temporalis, extraocular, and pharyngeal muscles are usually unaffected. In the

legs, tibialis anterior weakness may cause footdrop. Other muscles may also be involved to a lesser extent. The weakness is generally qualitatively symmetrical in distribution, although quantitative studies have documented major side-to-side asymmetry as a typical feature of the disease (FSH-DY Group, 1997).

Upper limb girdle weakness causes disability, with reduced arm function, especially lifting the arms over the head. Peroneal weakness causes difficulty walking with tripping. Facial weakness may include inability to close the eyes and difficulty speaking and eating. Systemic complications are uncommon but include a retinopathy termed Coats' disease. There is a range of retinal vasculopathies, which may be demonstrated with fluoroscein angiography (Fitzsimons *et al.*, 1987). Macular exudates may threaten vision. Sensorineural hearing loss is present in a small proportion of patients and is especially important to detect in children, because this may impair learning and be misinterpreted as mental retardation. Cardiac disease is generally not associated with FSHD. Children who are severely affected may be developmentally delayed and have epilepsy (Miura *et al.*, 1998).

## Principles of Therapy

Although the genetic basis, a 3.3-kb repeat truncation on chromosome 4q, is identified, the molecular pathology of FSHD remains uncertain. Historically, it was noticed that a proportion of patients with FSHD had inflammatory features on the muscle biopsy. In view of this, and the success of steriods in Duchenne muscular dystrophy, a pilot study of prednisone was performed (Tawil *et al.*, 1997). Prednisone (1.5 mg/kg/day; maximum 80 mg/day) was administered for 12 weeks. There was no significant change in strength or muscle mass measured by manual muscle testing, maximum voluntary isometric contraction testing, dual energy x-ray absorptiometry, and muscle mass. The conclusion of the study was that there was no improvement in strength or muscle mass. The study did not have sufficient power or length of follow-up to assess the possibility of prednisone arresting or slowing disease progression. However, FSHD is a relatively slowly progressive disease, and the goal of any treatment for FSHD should be to reverse the condition. Lifelong treatment with a medication such as prednisone, if only slowing progression, was considered to be unjustified in view of the long-term adverse effects. It was concluded that unless a subgroup of FSHD patients with poor outcome could be identified, it was difficult to justify a long-term study of prednisone in FSHD.

Beta-agonists, albuterol (salbutamol), and clenbuterol have been demonstrated to increase muscle strength in

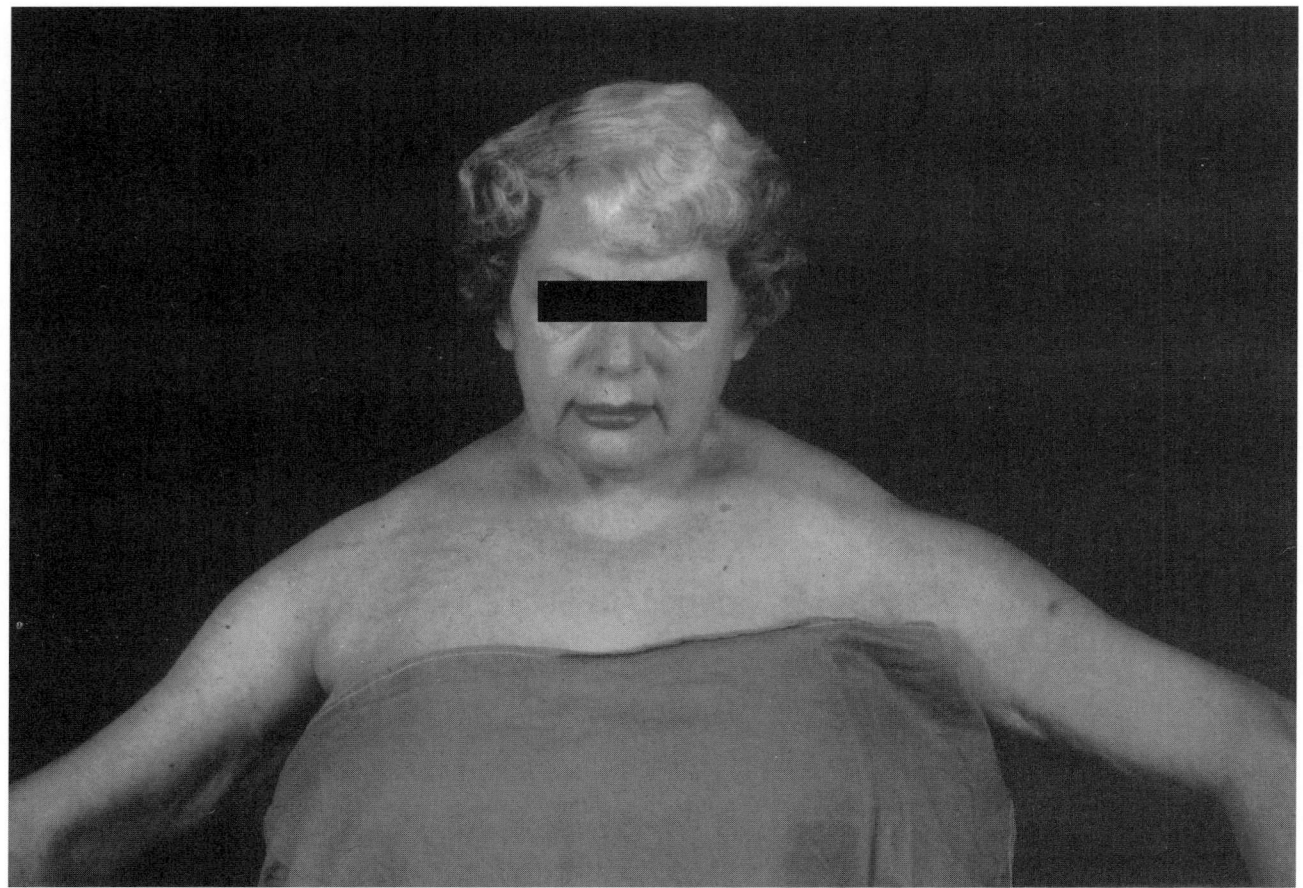

**FIGURE 4** A woman with facioscapulohumeral dystrophy, demonstrating wasting of the shoulder girdle and humeral muscles, with preservation of deltoid.

healthy individuals and in animals. Clenbuterol has been used in animals to increase meat production and abused by athletes to improve performance. Clenbuterol was demonstrated to increase relative muscle strength in orthopedic patients (Maltin, 1993) and albuterol (salbutamol) to increase skeletal muscle strength in young men (Martineau, 1992). This suggested that beta-agonists may have a therapeutic benefit in neuromuscular disorders. In particular, a relatively slowly progressive condition such as FSHD may be most responsive. Clenbuterol is not readily available in many countries. An open-label pilot study of albuterol (salbutamol) was performed in 15 patients with FSHD (Kissel, 1998). The patients were given sustained-release albuterol (salbutamol) (16 mg/day) for 3 months. The primary outcome was lean body mass assessed by dual-energy x-ray absorptiometry (DEXA). Strength was assessed by maximal voluntary isometric contraction testing (MVICT) and manual muscle testing. Salbutamol significantly increased DEXA lean body mass (the skeletal muscle compartment) by 1.3 ± 1.2 kg (mean +

SD, $p = 0.001$). Strength assessed through composite MVICT scores increased by an average of 0.3 ± 0.6 ($p = 0.05$), an overall increase of 12% in strength.

These encouraging results prompted a randomized, controlled trial of 90 patients with facioscapulohumeral dystrophy. Comparing two doses of albuterol, 8 mg bd and 4 mg bd, and placebo, Kissel *et al.* (2001) again noted an increase in muscle mass and an increase in grip strength (Figure 6) but no improvement in the primary outcome variable—a composite muscle strength score. Further studies are in progress. An additional study is being performed in Europe (Padberg *et al.*, unpublished).

Important background research has indicated that to provide 80% power to detect complete arrest of disease progression, a two-armed clinical trial would require 160 patients per group and 1 year of follow-up. Smaller studies are unlikely to detect improvements in strength, unless follow-up duration is prolonged (The FSH-DY Group, 1997). This has important implications for the design and interpretation of clinical trials in FSHD. This

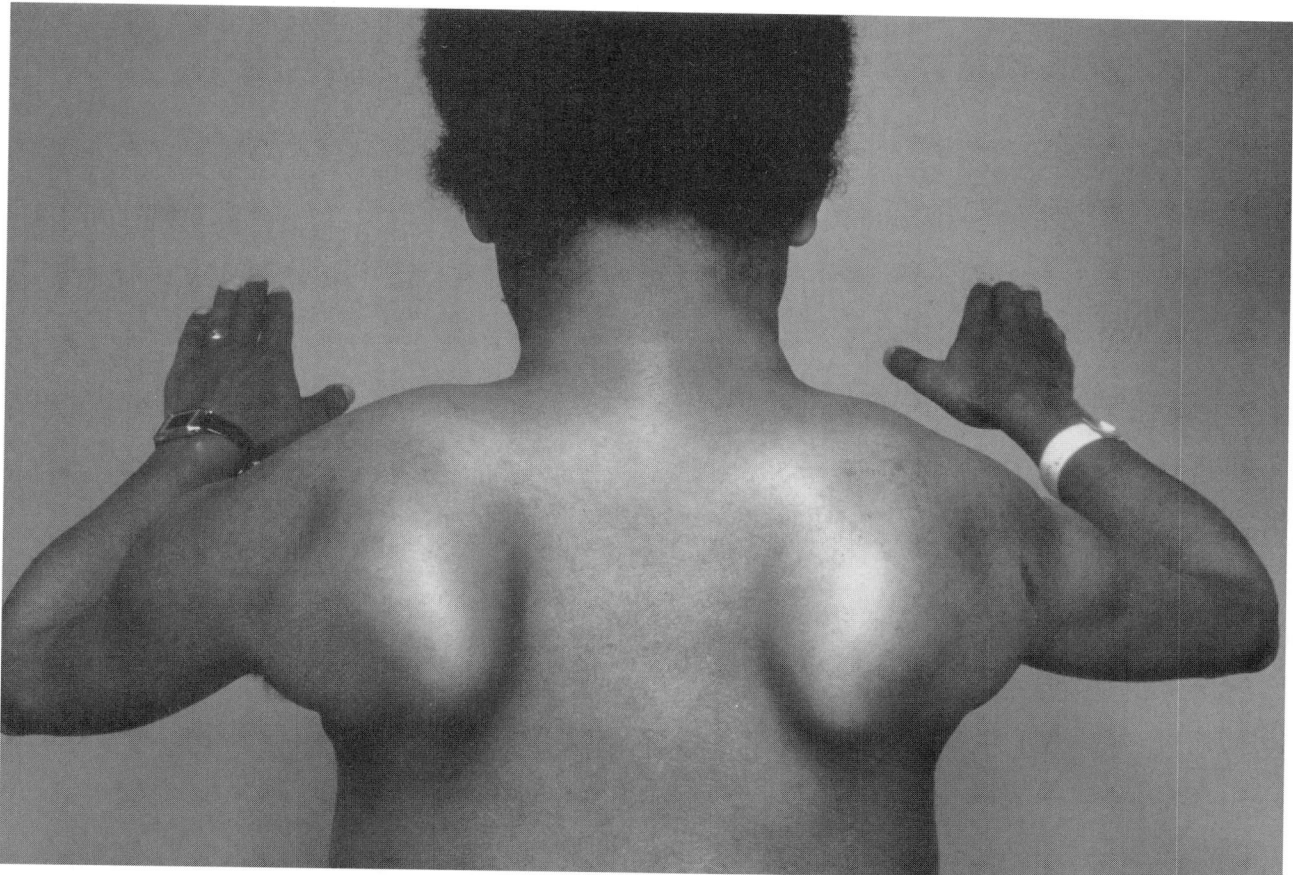

FIGURE 5  A man with facioscapulohumeral dystrophy, demonstrating scapular winging.

also emphasizes the difficulties in designing clinical trials in rarer muscular dystrophies.

## Practical Management

There is no specific medication recommended at present that alters the disease course, in particular, no medication that halts or reverses the condition. In view of the relatively slow progression of the disease, with normal longevity, any medication with potential toxic effects is to be avoided, unless benefit is substantial and proven.

Physiotherapy, occupational therapy, rehabilitation, and orthotic services may all be helpful in managing specific difficulties related to limb weakness, mobility and function. For example, an ankle support orthosis may aid mobility related to footdrop. Patients benefit from being seen in a specialist neuromuscular clinic for diagnosis and continuing management and advice. Many patients benefit from contact with a voluntary organization, for example the FSH Society (USA) or FSHD group of the Muscular Dystrophy Association (UK) (Table III).

Surgical interventions include scapular fixation (Birch, 1998; Bunch and Siegel, 1993; Jakab and Gledhill, 1993; Letournel et al., 1990). The results of this are largely anecdotal, and the surgical procedure may differ between surgeons. This usually involves fixation of the scapula to the ribs, possibly with screws, wires, or bone graft. Complications include the screws or wires working loose. Some patients gain benefit from this procedure, and it may be helpful for them to discuss their expectations with others who have had the procedure performed.

Incomplete eyelid closure (lagophthalmos) caused by muscle weakness may lead to prolonged exposure of the sclera and exposure keratitis. This has been corrected using small 24-carat gold weight implants in the upper lid to facilitate closure (Sansone et al., 1997) or tarsorrhaphy. Sling procedures may help correct ptosis.

## Treatments No Longer Recommended

On the basis of present evidence, we do not recommend treatment with prednisone or albuterol (salbuta-

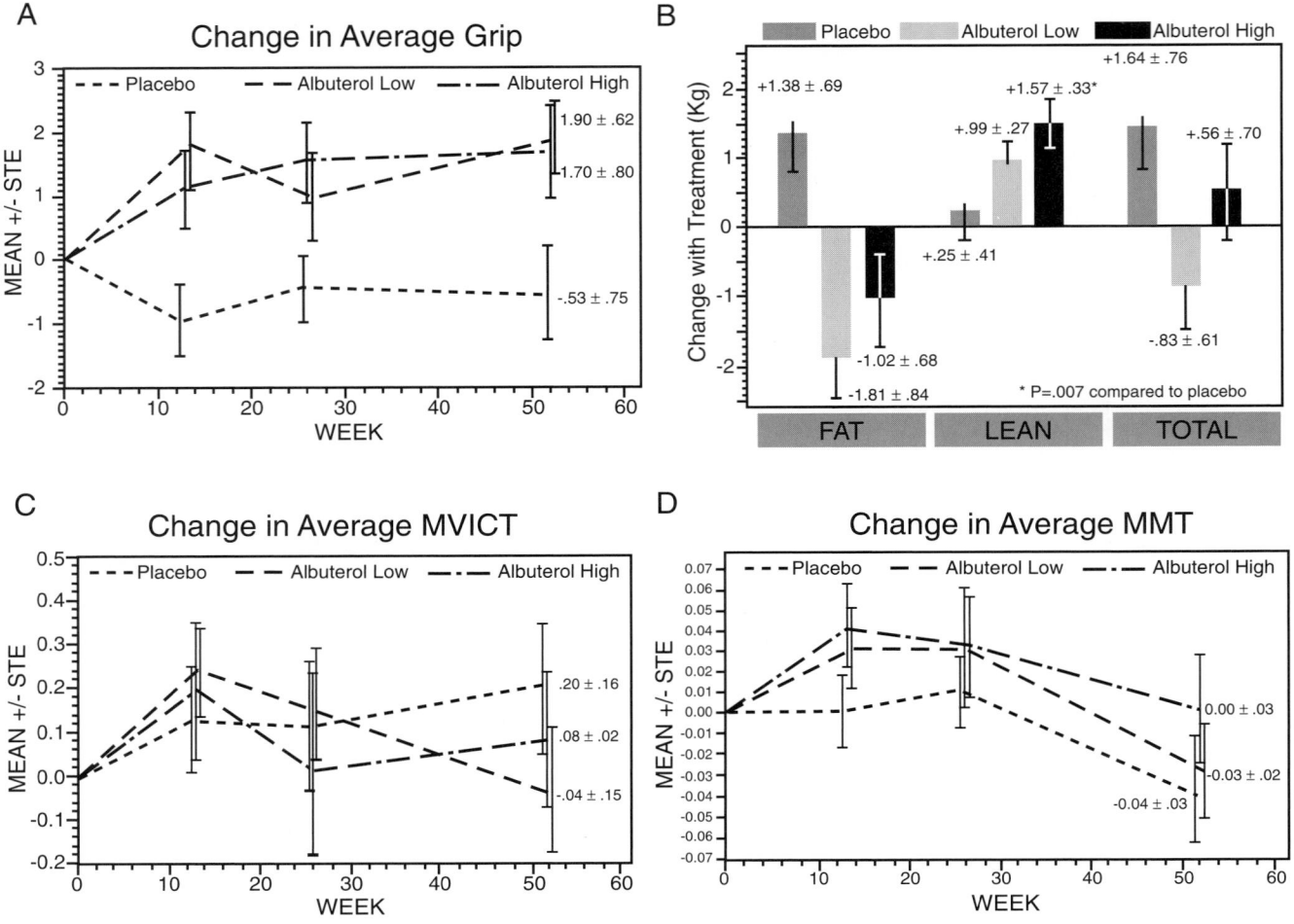

**FIGURE 6**  (A) Change in average grip strength in patients with facioscapulohumeral dystrophy treated with albuterol. There is a significant difference between albuterol and placebo groups at 1 year. (B) There is a significant increase in lean body mass, representing muscle, in FSH patients treated with albuterol. (C) There was no change in primary outcomes of maximum voluntary isometric contraction (MVICT) or (D) average manual muscle testing score (MMT). (Kissel *et al.*, 2001, with permission from Lippincott Williams & Wilkins.).

mol) except in the possible circumstances of a well-designed clinical trial.

## OTHER MUSCULAR DYSTROPHIES

The general principles of diagnosis and management described for Duchenne muscular dystrophy and facioscapulohumeral dystrophy apply to the less common muscular dystrophies. There is, however, no specific therapeutic medication indicated for other muscular dystrophies at present. The importance of making as accurate a clinical diagnosis as possible is that this is most likely to lead to performing the appropriate molecular tests. There are now an increasing number of recognized genetic causes of muscular dystrophy, and the individual tests may be complex, costly, and time consuming, especially in large genes, where the muta-

tion may be unique to individual families. Having made the molecular diagnosis, the most accurate advice may then be given to the patient. The patient and family should be aware of the genetic implications and possible implications for other family members. Patients with identified molecular disorders may be able to participate in therapeutic trials or take advantage of any treatment that may become available for their molecular condition.

Two important additional therapeutic considerations in muscular dystrophy are anesthesia and physical training.

### Anesthesia

There is an inceased risk of malignant hyperthermia in certain muscular dystrophies. Malignant hyperthermia is an autosomal-dominant trait. A hypercatabolic reaction occurs induced by anesthetic drugs or physical or emotional stress. Around half of malignant hyperthermia reactions have been preceded by uneventful anesthetics (Britt, 1996).

The peak incidence is around 30 years of age, and it is more common in men than women. It occurs worldwide. In some families, mutations are found in the ryanodine gene (RYR1) on chromosome 19q. This is related to the calcium release channel in the muscle sarcoplasmic reticulum. Mutations are also recognized in the calcium channel dihydropyridine receptor (CACNL1A3) on chromosome 1q. Additional channelopathies may be associated. Duchenne muscular dystrophy occurs in approximately 4% of patients with malignant hyperthermia. Myotonic dystrophy and central core disease have also been associated.

Succinylcholine may precipitate an early, transient, masseter muscle rigidity, which is usually not lethal. Other agents may have more intense effects, including halothane, enflurane, isoflurane, desflurane, sevoflurane, and methoxyflurane. Physical or emotional stress may trigger or exacerbate the reaction. The muscle rigidity may make it impossible to open the jaw, making intubation difficult or impossible. Fever is due to increased heat production from the muscles in spasm and reduced heat loss due from peripheral vasospasm. Occasionally, a late-onset reaction may occur in the recovery room or on the ward.

Diagnosis is often by the Caffeine Halothane Contraction Test, performed on muscle biopsy. The contractile response of the muscle when exposed to halothane is studied.

Dantrolene sodium is the medication of choice for treatment and prevention of acute malignant hyperthermia. It inhibits the tension of the muscle fascicles and prevents release of calcium from the sarcoplasmic reticulum. Prophylactically, this may be given orally or intravenously. Patients with known susceptibility should be carefully monitored during and after anesthesia, including temperature measurement.

In the acute reaction, the triggering drug should be discontinued. Active body cooling seems to have no effect on outcome. Oxygen, carbon dioxide, and base deficit should be corrected. Dantrolene is administered at 1 mg/kg/min, intravenously, until the temperature and muscle stiffness start to resolve. A maximum of 10 mg/kg may be given over a 15-minute period. The dose should be repeated if the reaction continues or recurs. This should later be replaced by oral dantrolene. In addition electrolytes should be corrected, and tachycardia should be controlled. Mannitol infusion may reverse cerebral and muscle edema. Mannitol is also of benefit in preventing renal damage from myoglobinuria. Mortality is around 3%.

There is increased risk of cardiac abnormality, mainly disorders of conduction, but also cardiomyopathy in some, but not all, muscular dystrophies. Sudden cardiac arrest is a significant cause of death in some patients with muscular dystrophy. These abnormalities should be anticipated during anesthesia. There may also be an increased risk of respiratory complications during and following anesthesia.

### Physical Training

Physical training will not reverse the muscle weakness and wasting of muscular dystrophies but may have benefit in improving endurance and cardiovascular fitness, in addition to some improvement in strength. The overall effect on fitness and respiratory function may improve the performance and abilities of the patient. There has often been some concern as to whether exercise may further damage muscle and accelerate the muscular dystrophy. As with other treatments of muscular dystrophy, many of the individual conditions are rare, and clinical trials in an individual muscular dystrophy may not have been performed. The studies have generally been of small numbers of patients, with mixed muscular diseases. An increase in muscle strength is reported in exercise programs performed for at least 1–3 months, but there may be some positive reporting bias. In general, resisted strength training and aerobic training programs seem to be of benefit to patients with myopathy, but convincing evidence from good controlled trials is lacking (Phillips and Mastaglia, 2000)

## REFERENCES

Bach, J. R., Ishikawa, Y., and Kim, H. (1997). Prevention of pulmonary morbidity for patients with Duchenne Muscular Dystrophy. *Chest* 112, 1024–1028.

Barohn, R. J., Levin, E. J., Olson, J. O., and Mendell, J. R. (1988). Gastric motility in Duchenne's muscular dystrophy. *N. Engl. J. Med.* 319, 15–18.

Barton-Davis, E. R., Corier, L., Shoturma, D. I. *et al.* (1999). Aminoglycoside antibiotics restore dystrophin function to skeletal muscles of mdx mice. *J. Clin. Invest.* 104, 375–381.

Birch, J. G. (1998). Orthopedic management of neuromuscular disorders of children. *Semin. Pediatr. Neurol.* 5, 78–91.

Bonifati, M. D., Ruzza, G., Bonometto, P. *et al.* (2000). A multicenter, double-blind, randomized trial of deflazacort versus prednisone in Duchenne muscular dystrophy. *Muscle Nerve* 23, 1344–1347.

Britt, B. A. (1996). Malignant hyperthermia. *In* "Handbook of Muscle Disease." (R. J. M. Lane, Ed.), pp. 451–471. Marcel Dekker, New York.

Brooke, M. H., Fenichel, G. M., Griggs, R. C. *et al.* (1987). Clinical investigation of Duchenne muscular dystrophy. *Arch. Neurol.* 44, 812–817.

Brooke, M. H., Fenichel, G. M., Griggs, R. C. et al. (1989). Duchenne muscular dystrophy: patterns of clinical progression and effects of supportive therapy. Neurology 39, 475–481.

Bunch, W. H., and Siegel, I. M. (1993). Scapulothoracic arthrodesis in facioscapulohumeral muscular dysrophy. Review of seventeen procedures with three to twenty-one-year followup. J. Bone. Joint. Surg. Am. 73, 372–376.

DeSilva, S., Drachman, D. B., Mellits, D., and Kuncl, R. W. (1987). Prednisone treatment in Duchenne muscular dystrophy. Arch. Neurol. 44, 818–822.

Drachman, D. B., Toyka, K. V., and Myer, E. (1974). Prednisone in Duchenne muscular dystrophy. Lancet 2, 1409.

Dubowitz, V. (1997). 47th ENMC International Workshop: treatment of muscular dystropy. Neuromusc. Disord. 7, 261–267.

Fenichel, G., Pestronk, A., Florence, J., Robison, V., and Hemelt, V. (1997). A beneficial effect of oxandrolone in the treatment of Duchenne muscular dystrophy: a pilot study. Neurology 48, 1225–1226.

Fenichel, G. M., Florence, J. M., Pestronk, A. et al. (1991b). Long-term benefit from prednisone therapy in Duchenne muscular dystrophy. Neurology 41, 1874–1877.

Fenichel, G. M., Griggs, R. C, Kissel, J. et al. (2001). A randomized efficacy and safety trial of oxandrolone in the treatment of Duchenne dystrophy. Neurology 56, 1075–1079.

Fenichel, G. M., Mendell, J. R., Moxley III, R. T. et al. (1991a). A comparison of daily and alternate-day prednisone therapy in the treatment of Duchenne Muscular Dystrophy. Arch. Neurol. 48, 575–579.

Fitzsimons, R. B., Gurwin, E. B., and Bird, A. C. (1987). Retinal vascular abnormalities in facioscapulohumeral muscular dystrophy. A general association with genetic and therapeutic implications. Brain 110, 631–648.

Fletcher, S., Wilson, S. D., and Howell, J. M. (2000). Gene therapy and molecular approaches to the treatment of hereditary muscular disorders. Curr. Opin. Neuro. 13, 553–560.

Granata, C., Merlini, L., Cervellati, S. et al. (1996). Long-term results of spine surgery in Duchenne muscular dystrophy. Neuromusc. Disord. 6, 61–68.

Griggs, R. C., Moxley III, R. T., Mendell, J. R. et al. (1991). Prednisone in Duchenne dystrophy: a randomized, controlled trial defining the time course and dose response. Arch. Neurol. 48, 383–388.

Griggs, R. C., Moxley III, R. T., Mendell, J. R. et al. (1993). Duchenne dystrophy: randomized, controlled trial of prednisone (18 months) and azathioprine (12 months). Neurology 43, 520–527.

Gussoni, E., Pavlath, G. K., Lanctot, A. D. et al. (1992). Normal dystrophin transcripts detected in Duchenne muscular dystrophy patients after myoblast transplantation. Nature 356, 435–438.

Jakab, E., and Gledhill, R. B. (1993). Simplified techique for scapulocostal fusion in facioscapulohumeral dystrophy. J. Pediatr. Orthop. 13, 749–751.

Kakulas, B. A. (1997). Problems and potential for gene therapy in Duchenne muscular dystrophy. Neuromusc. Disord. 7, 319–324.

Kaplan, J.-C., and Fontaine, B. (2000). Neuromuscular disorders: gene location. Neuromusc. Disord. 10, I–IX.

Karpati, G. K., Ajdukovic, D., Arnold, D. et al. (1993). Myoblast transfer in Duchenne muscular dystrophy. Ann. Neurol. 34, 8–17.

Kissel, J. T., McDermott, M. P., Mendell, J. R. et al. (2001). Randomized, double-blind, placebo-controlled trial of albuterol in facioscapulohumeral dystrophy. Neurology 57, 1434–1440.

Kissel, J. T., McDermott, M. P., Natarajan, R. et al. (1998). Pilot trial of albuterol in facioscapulohumeral muscular dystrophy. Neurology 50, 1402–1406.

Letournel, E., Fardeau, M., Lytle, J. O., Serrault, M., and Gosselin, R. A. (1990). Scapulothoracic arthrodesis for patients who have facioscapulohumeral muscular dystrophy. J. Bone. Joint. Surg. Am. 72, 78–84.

Maltin, C. A., Delday, M. I., Watson, J. S. et al. (1993). Clenbuterol, a β-adrenoceptor agonist, increases relative muscle strength in orthopaedic patients. Clin. Sci. 84, 651–654.

Martineau, L, Horan, M. A., Rothwell, N. J., and Little, R. A. (1992). Salbutamol, a β2-adrenoceptor agonist, increases skeletal muscle strength in young men. Clin. Sci. 83, 615–621.

Mendell, J. R., Buzin, C. H., Feng, J., Yan, J., Serrano, C., Sangani, D. S., Wall, C., Prior, T. W., and Sommer, S. S. (2001). Diagnosis of Duchenne dystrophy by enhanced detection of small mutations. Neurology 57, 645–650.

Mendell, J. R., Moxley, R. T., Griggs, R. C. et al. (1989). Randomized, double-blind six-month trial or prednisone in Duchenne muscular dystrophy. N. Engl. J. Med. 320, 1592–1597.

Miura, K., Kumagai, T., Matsumoto, A. et al. (1998). Two cases of chromosome 4q35-linked early onset facioscapulohumeral muscular dystrophy with mental retardation and epilepsy. Neuropediatrics 29, 239–241.

Molnar, M. J., and Karpati, G. (1999). Muscular dystrophies related to defiency of sarcolemmal proteins. In "Muscle Diseases," (A. H. V. Schaira, and R. C. Griggs, Eds.), pp. 83–114. Butterworth-Heinemann, Boston.

Orrell, R. W., and Griggs, R. C. (1999). Muscular dystrophies: overview of clinical and molecular approaches. In "Muscle Diseases." (A. H. V. Schapira, and R. C. Griggs, Eds.), pp. 59–82. Butterworth Heinemann, Boston.

Orrell, R. W., Tawil, R., Forrester, J., Kissel, J. T., Mendell, J. R., and Figlewicz, D. A. (1999). Definitive molecular diagnosis of facioscapulohumeral muscular dystrophy. Neurology 52, 1822–1826.

Phillips, B. A., and Mastaglia, F. L. (2000). Exercise therapy in patients with myopathy. Curr. Opin. Neurol. 13, 547–552.

Phillips, M. F., Smith, P. E. M., Carroll, N., Edwards, R. H. T., and Calverley, P. M. A. (1999). Nocturnal oxygenation and prognosis in Duchenne muscular dystrophy. Am. J. Respir. Crit. Care. Med. 160, 198–202.

Ricci, E., Galluzzi, G., Deidda, G. et al. (1999). Progress in the molecular diagnosis of facioscapulohumeral muscular dystrophy and correlation between the numberof Kpn1 repeats at the 4q35 locus and clinical phenotype. Ann. Neurol. 45, 751–757.

Rifai, Z., Welle, S., Moxley, R. T., Lorenson, M., and Griggs, R. C. (1995). Effect of prednisone on protein metabolism in Duchenne dystrophy. Am. J. Physiol. 268, E67–E74.

Sansone, V., Boynton, J., and Palenski, C. (1997). Use of gold weights to correct lagophthalmos in neuromuscular disease. Neurology 48, 1500–1503.

Scheuerbrandt, G. (1998). DMD Meeting. First meeting of the Duchenne Parent Project in Europe: Treatment of Duchenne Muscular Dystrophy. Neuromusc. Disord. 213–219.

Serrano, C., Wall, C., Moore, S. A. et al. (2001). Gentamicin treatment for muscular dystrophy patients with stop codon mutations. Neurology 56, Supp 3, A79.

Tarnopolsky, M., and Martin, J. (1999). Creatine monohydrate increases strength in patients with neuromuscular disease. Neurology 52, 854–857.

Tawil, R., Figlewicz, D. A., Griggs, R. C., Weiffenbach, B., and the FSH Consortium. (1998). Facioscapulohumeral dystrophy: a distinct regional myopathy with a novel pathogenesis. Ann. Neurol. 43, 279–282.

Tawil, R., McDermott, M. P., Pandya, S. et al. (1997). A pilot trial of prednisone in facioscapulohumeral muscular dystrophy. Neurology 48, 46–49.

The FSH-DY Group. (1997). A prospective, quantitative study of the natural history of facioscapulohumeral muscular dystrophy (FSHD): implications for therapeutic trials. Neurology 48, 38–46.

van Deutekom, J. C. T., Wijmenga, C., van Tienhoven, E. A. E. *et al.* (1993). FSHD associated DNA arrangements are due to deletions of integral copies of a 3.2 kb tandemly repeated unit. *Hum. Mol. Genet.* **2**, 2037–2042.

Vignos, P. J., Wagner, M. B., Karlinchak, B., and Katirji, B. (1996). Evaluation of a program for long-term treatment of Duchenne muscular dystrophy. *J. Bone Joint Surg.* **78A**, 1844–1852.

Wagner, K. R., Hamed, S., Hadley, D. W. *et al.* (2001). Gentamicin treatment of Duchenne and Becker muscular dystrophy due to nonsense mutations. *Ann. Neurol.* **49**, 706–711.

Walter, M. C., Lochmuller, H., Reilich, P. *et al.* (2000). Creatine monohydrate in muscular dystrophies: a double-blind placebo-controlled study. *Neurology* **54**, 1848–1850.

Zatz, M., Vainzof, M., and Passos-Bueno, M. R. (2000). Limb-girdle muscular dystrophy: one gene with different phenotypes, one phenotype with different genes. *Curr. Opin. Neurol.* **13**, 511–517.

*Muscle and Peripheral Nervous System*

CHAPTER 97

# Metabolic Myopathies

Thomas Klopstock and Salvatore DiMauro

Metabolic myopathies "sensu stricto" are those with defined blocks of metabolic pathways caused by deficient or nonfunctional enzymes. Mitochondrial encephalomyopathies are multisystemic disorders that usually include a metabolic myopathy. In addition, myopathies caused by drugs, toxins, and nutritional deficiency will also be discussed in this chapter.

## DISORDERS OF CARBOHYDRATE METABOLISM (GLYCOGENOSES)

Muscle contraction depends on the provision of energy in the form of adenosine triphosphate (ATP). Carbohydrates are the most important source of energy for brief, intense exercise in anaerobic conditions. The **glycogenoses** are hereditary disorders caused by enzyme deficiencies in glycogen metabolism or glycolysis. They have been assigned Roman numerals in the order of their discovery but are now also specified by the underlying enzyme defect (Table I). Because glycogenoses types I and VI do not affect muscle, they will not be discussed here. Type II is the only lysosomal glycogenosis: as such, it causes glycogen storage but no impairment of energy metabolism.

### Clinical Features and Course

Except for type IX, which is X-linked recessive, all other glycogenoses have autosomal recessive inheritance. Symptoms begin most often in childhood or adolescence, but adult onset may also occur. Cardiac failure causes early death in the infantile form of type II, and hepatic failure is the typical cause of death in type IV. Hepatopathy in types III and VIII is more benign. The other glycogenoses (types V, VII, X, and XI) are confined to muscle.

Muscle damage results from glycogen storage on the one hand and from energy failure on the other. The clinical manifestation of glycogen storage is fixed, progressive, mostly proximal muscle weakness. This phenotype predominates in glycogenoses II, III, and IV and is indistinguishable from limb-girdle muscular dystrophy. In the other glycogenoses, dynamic features caused by energy failure prevail: **exercise-related muscle cramps** and pain and eventually **rhabdomyolysis** and myoglobinuria. Approximately 50% of patients with glycogenosis type V manifest with myoglobinuria after intense exercise, and half of those progress to renal failure (DiMauro and Tsujino, 1994). The "second wind" phenomenon is mainly found in glycogenoses V and VII and denotes an improvement of exercise intolerance after a short break. This is due to the muscle switching from glycogen use to the oxidation of free fatty acids.

Between attacks of myoglobinuria, serum creatine kinase (CK) levels are mildly to moderately elevated, but after rhabdomyolysis, there is a massive increase of serum CK. The forearm ischemic exercise test shows absent or decreased rise of venous lactate in glycogenoses caused by blocks in glycogen breakdown or glycolysis, except for type VIII, in which the ischemic exercise is often normal. Ischemic exercise is normal in the lysosomal glycogenosis (type II) and in blocks of glycogen synthesis (type IV). Electromyography is usually myopathic and often accompanied by pathological spontaneous activity. The typical finding in muscle biopsy is vacuolar myopathy with glycogen-containing vacuoles. Histochemical stains are available for glycogenoses type V and VII. Biochemical documentation of specific enzyme defects requires muscle in glycogenoses V, VIII, X, and XI. More easily accessible nonmuscle tissues can be used for biochemical assays in the other glycogenoses.

**TABLE I** Glycogenoses

| Type gene locus | Deficient enzyme | Clinical symptoms and course |
|---|---|---|
| Type II Pompe Chr 17q21–23 | Acid maltase (= acid α-glucosidase) | Infantile form: floppy infant, cardiohepatomegaly, death before age 2 y Juvenile form: progressive muscle weakness, death in the 2nd decade; Adult form: myopathy, 30% respiratory insufficiency |
| Type III Forbes Chr 1p21 | Debranching enzyme | Infantile form: benign course with hepatomegaly, hypoglycemia; prominent liver involvement Adult form: progressive muscle weakness, cardiomyopathy |
| Type IV Andersen Chr 3 | Branching enzyme | Congenital, prominent liver involvement, death mostly before age 4 y caused by hepatic coma; a clinical variant is "adult polyglucosan body disease," a late-onset neurodegenerative disease with dementia and motor neuron loss |
| Type V McArdle Chr 11q12–13 | Myophosphorylase | Exercise-related muscle weakness, pain, and cramps; in 50% episodic rhabdomyolysis with myoglobinuria, in 25% renal failure; permanent muscle weakness in 30%; second wind-phenomenon |
| Type VII Tarui Chr 1cen–q32 | Phosphofructokinase | Similar to type V, but rhabdomyolysis, myoglobinuria, and renal failure are less common |
| Type VIII Chr 7, 16 and X | Phosphorylase-b-kinase | Variable forms: X-chromosomal hepatopathy, autosomal recessive hepatomyopathy, or pure myopathy with exercise-related symptoms |
| Type IX Chr Xq13 | Phosphoglycerate kinase | Most frequent presentation with epileptic seizures, mental retardation, and hemolytic anemia; few patients with exercise-related muscle symptoms |
| Type X Chr 7p13 | Phosphoglycerate mutase | Exercise-related muscle weakness, cramps and pain; myoglobinuria |
| Type XI Chr 11p13 | Lactate dehydrogenase | Exercise-related muscle weakness, cramps and pain; myoglobinuria |
| Type XII Chr 16q22–24 | Aldolase A | Only two patients described so far with proximal myopathy and episodic exercise intolerance |

Modified from DiMauro and Tsujino (1994).

## Principles of Therapy

### Enzyme Replacement Therapy

A series of in vitro and in vivo studies has demonstrated the feasibility of enzyme replacement therapy in glycogenosis II (Bijvoet *et al.*, 1999). Recombinant human acid α-glucosidase has been produced on an industrial scale in the milk of transgenic rabbits. Intravenous administration of this enzyme in a knockout mouse model of glycogenosis II led to correction of acid α-glucosidase deficiency, degradation of excessive glycogen in heart and muscle, and improved tissue morphology. In a recent open-label study, this same recombinant human acid α-glucosidase led to normalization of α-glucosidase and to improved motor and cardiac function in four human babies with infantile acid maltase deficiency (Pompe's disease)* (Van den Hout *et al.*, 2001).

### Diet

The results of several forms of diet have been contradictory. Because the enzymatic block in glycogenoses III, IV, V, and VIII is in the use of glycogen, but not in glycolysis, it may be circumvented by supplementation of glucose or fructose. Oral supplementation of these monosaccharides, however, was ineffective in glycogenosis type V, and indirect elevation of blood glucose after glucagon was similarly ineffective in types IV and V (DiMauro and Tsujino, 1994). Those glycogenoses that are due to blocks of glycolysis (types VII, IX, X, and XI) may even worsen with elevated glucose, because glucose inhibits lipolysis and therefore impairs the availability of free fatty acids as an alternative substrate (Haller and Lewis, 1991). Attempts to raise the concentration of free fatty acids by means of a fat-rich diet were also unsuccessful. A protein-rich diet (providing amino acids as an alternative substrate) had some beneficial effect on clinical symptoms (Slonim and Goans, 1985) and spectroscopic features (Jensen *et al.*, 1990) in glycogenosis type V, but a controlled trial is still lacking*. A high-protein diet led to inconclusive results in types II (Padberg *et al.*, 1989; Slonim *et al.*, 1983), III, and IV. Frequent small meals and nocturnal gastric infusion of glucose and uncooked cornstarches are recommended in type III to avoid hypoglycemia.

### Other Metabolic Therapies

A double-blind, placebo-controlled crossover study of oral creatine supplementation showed alleviation of symptoms in nine patients with glycogenosis V** (Vorgerd *et al.*, 2000). In a small open study, oral alanine supplementation reduced resting energy expenditure and whole-body protein breakdown in five subjects with the late-onset form of glycogenosis II* (Bodamer *et al.*, 2000).

### Physical Therapy

In physical activity, brief, intense muscle exercise in anaerobic conditions should be avoided, whereas aerobic training is of therapeutic benefit. The metabolic block may be circumvented by burning of free fatty acids during moderate exercise (imitating what nature does with the second wind).

### Liver Transplantation

Glycogenoses types I, III, and IV can be associated with severe liver disease. The possible development of hepatic failure or hepatocellular carcinoma makes patients with these disorders potential candidates for liver transplantation. Three patients with glycogenosis type III have received liver transplants because of end-stage cirrhosis or carcinoma. The long-term effects of this procedure in terms of other organ involvement is not yet known (Matern *et al.*, 1999). Thirteen patients with glycogenosis type IV underwent liver transplantation at a mean age of 3.2 years because of progressive cirrhosis. Although 3 died in the first postoperative year, 10 patients have had no neuromuscular or cardiac manifestations in follow-up periods as long as 13.5 years (Matern *et al.*, 1999).

### Gene Therapy

Intravenous administration of a modified adenovirus vector encoding human acid α-glucosidase resulted in efficient hepatic secretion of the precursor enzyme into the plasma of a knockout mouse model of glycogenosis type II. Subsequently, the proenzyme was taken up by skeletal and cardiac muscles of the animal and glycogen accumulation decreased in several muscles (Amalfitano *et al.*, 1999). In McArdle's disease, myoblast cultures from phosphorylase-deficient human and sheep muscle were efficiently transduced with an adenoviral recombinant containing the human phosphorylase cDNA, resulting in restoration of the enzyme activity. A similar correction of the genetic defect in muscles of McArdle's patients in vivo seems feasible, preferably with the use of an adeno-associated viral vector (Pari *et al.*, 1999). These results indicate that gene therapy is promising for the glycogenoses.

### Symptomatic Measures

Therapy of cardiomyopathy or respiratory failure follows the principles of internal medicine. Rhabdomyolysis should be treated under intensive care conditions. Most important is the maintenance of respiration, the prevention or treatment of renal failure, and the correction of hyperkalemia.

### Treatments No Longer Recommended

There was no benefit with steroid administration in glycogenoses type IV and VIII, with glucagon in types IV and V, and with ribose in type V. Ineffective dietary regimens have been mentioned previously.

## DISORDERS OF LIPID METABOLISM

After approximately 40 minutes of sustained exercise of mild to modest intensity, aerobic combustion of long-chain fatty acids becomes the main energy source for muscle contraction. To be transported into mitochondria, long-chain fatty acyl-CoAs are bound to carnitine by carnitine palmitoyl transferase 1 (CPT1), transported through the inner mitochondrial membrane as acyl carnitine ester, and reconverted in the mitochondrial matrix to carnitine and fatty acids acyl-CoA by CPT2. Fatty acyl-CoAs are used for energy production in the β-oxidation spiral. Accordingly, diseases of lipid metabolism are compound of carnitine deficiency, defects of CPT I or CPT II, and defects of β-oxidation. All these disorders are inherited as autosomal recessive traits.

### Carnitine Deficiency

#### Clinical Features and Course

Muscle carnitine deficiency manifests as progressive, painless, proximal muscle weakness, sometimes associated with cardiomyopathy, with onset between age 2 and 50. Exercise intolerance is rare.

Primary systemic carnitine deficiency causes myopathy, cardiomyopathy, hepatopathy, and metabolic crises with vomiting, somnolence, hypoglycemia, and hyperammonemia in early childhood. Cardiomyopathy is the predominant feature and the main cause of death in untreated patients.

Secondary carnitine deficiency can be the consequence of other metabolic disorders, such as defects of β-oxidation or mitochondrial diseases and of excessive renal leakage, as in the renal Fanconi syndrome. It can also accompany chronic hemodialysis, parenteral nutrition, and therapy with valproate or pivampicilline (Di Donato, 1994).

Serum CK is normal or mildly elevated, EMG usually shows a myopathic pattern and pathological spontaneous activity. In patients with systemic carnitine deficiency, both total and free carnitine levels are markedly decreased in blood. In all forms of carnitine deficiency, muscle biopsy shows lipid storage, especially in type I fibers. Definitive diagnosis requires biochemical demonstration of carnitine deficiency in muscle.

## Principles of Therapy

Oral or parenteral administration of L-carnitine leads to rapid and marked improvement of the life-threatening cardiomyopathy, increased strength, and resolution of the metabolic crises in systemic carnitine deficiency (Penn, 1994). Carnitine replacement enables patients to live a normal life.

In the muscle form of carnitine deficiency, L-carnitine supplementation usually improves muscle strength (Shapira et al., 1993). Some patients responded to the administration of prednisone (Engel and Siekert, 1972), propranolol, or riboflavin (Griggs et al., 1995). Some patients with secondary carnitine deficiency benefit from L-carnitine supplementation (Di Donato, 1994). Hemodialysis-induced loss of L-carnitine should be corrected in all cases.

All forms of carnitine deficiency respond to a fat-reduced diet and small, frequent, carbohydrate-rich meals. Prolonged fasting should be avoided, because it inhibits the release of glycogen and glucose while stimulating the release of fatty acids.

## Practical Management

Apart from the dietary measures discussed earlier, carnitine deficiency is treated by the supplementation of L-carnitine in the dose of 2–4 g/day in adults and 100 mg/kg body weight per day in children (Griggs et al., 1995). Carnitine loss caused by chronic hemodialysis is treated by oral or parenteral administration of 1–2 g L-carnitine at the end of each dialysis. There are no side effects except mild gastrointestinal discomfort (Campos et al., 1993b).

## Carnitine Palmitoyl Transferase II Deficiency

### Clinical Features and Course

CPT II deficiency (DiMauro and DiMauro, 1973) is a common inherited disorder of fatty acid oxidation. CPT I is located on the outer mitochondrial membrane and has a liver (L-CPT I) and a muscle (M-CPT I) isoform. Although L-CPT I deficiency causes recurrent attacks of fasting hypoketotic hypoglycemia, M-CPT I deficiency has not yet been described (Bonnefont et al., 1999). CPT II is a ubiquitous protein of the inner mitochondrial membrane. CPT II deficiency may present in infancy with severe recurrent hypoketotic hypoglycemia, but the more frequent clinical phenotype is the "adult myopathic" form, characterized by episodes of rhabdomyolysis. This disorder often begins in childhood with episodes of myalgia and muscle stiffness (Zierz, 1994). Although CPT II deficiency is inherited as an autosomal recessive trait, 80% of

patients are men. Severe attacks cause rhabdomyolysis, and the attending myoglobinuria may cause renal failure. Although some patients experience only one or two attacks in their lifetime, others have frequent attacks. Most episodes begin a few hours after prolonged exercise, but other precipitating factors include fasting, cold exposure, fatty meals, lack of sleep, emotional stress, infections, fever, general anesthesia, and drugs such as ibuprofen and diazepam. Neurological examination, serum CK, forearm ischemic exercise test, and EMG are typically unremarkable between attacks. The diagnosis is difficult to establish, because muscle biopsy is usually normal; only biopsies taken shortly after an attack of rhabdomyolysis may show scattered necrotic fibers. CPT deficiency is proven by biochemical analysis of muscle or by molecular documentation of common mutations in blood (Kaufmann et al., 1997).

### Principles of Therapy

The main goal in the CPT deficiencies is to provide sufficient glucose supply to prevent lipolysis (Saudubray et al., 1999). In the neonatal period and during acute metabolic attacks (L-CPT I or severe CPT II deficiencies), intravenous infusion of glucose or continuous nasogastric feeding may be needed. L-carnitine supplementation may be useful in the severe form of CPT II deficiency. In the muscular form of CPT II deficiency, prevention of myoglobinuria attacks includes avoidance of the precipitating factors discussed previously and implementation of a low-fat, high-carbohydrate diet with frequent small meals. Prolonged exercise is better tolerated on a high-carbohydrate diet. During general anesthesia, glucose should be infused, because fasting may induce attacks. The management of myoglobinuria is discussed later.

### Defects of β-Oxidation Enzymes

Defects of the acyl-CoA dehydrogenases specific for long-, medium-, and short-chain fatty acids cause life-threatening pediatric disorders with metabolic crises similar to those caused by systemic carnitine deficiency. Patients who survive these metabolic storms may later have myopathy or exercise-induced muscle symptoms develop. All these disorders are associated with secondary carnitine deficiency and benefit from carnitine administration.

In glutaraciduria type II, deficiency of the electron transfer flavoprotein (ETF) may be associated with severe infantile syndromes or, less frequently, with lipid storage myopathy in adults. In patients with riboflavin-responsive defects, life-long oral supplementation of riboflavin (vitamin B$_2$), 100 mg/day, abolishes all symptoms (Di Donato and Deal, 1989).

# DISORDERS OF PURINE METABOLISM: MYOADENYLATE DEAMINASE DEFICIENCY

## Clinical Features and Course

Myoadenylate deaminase (MADA) is an enzyme of the purine nucleotide cycle, which deaminates adenosine monophosphate (AMP) to inosine monophosphate (IMP) and ammonia. MADA deficiency (Fishbein *et al.*, 1978) is found in 1%–2% of all muscle biopsies and is therefore the most frequent metabolic disease of muscle. The clinical importance of this defect, however, is controversial. Most homozygous individuals are asymptomatic or have other neuromuscular disorders; only a few have exertional myalgia and, rarely, myoglobinuria. Serum CK is mildly elevated in approximately half of the cases. The diagnosis is suggested by the forearm ischemic exercise test, which causes no rise in ammonia, contrasting with a normal rise in lactate.

MADA deficiency is an autosomal recessive trait. The gene is located on chromosome 1p13–21 (Sabina *et al.*, 1990). Most patients are homozygous for a nonsense mutation at nucleotide 34 in the second exon (Morisaki *et al.*, 1992).

## Principles of Therapy

There is no effective therapy for MADA deficiency. In the rare patients with severe clinical symptoms, administration of D-ribose (up to 50 g/day) may be tried* (Reimers *et al.*, 1987). Xylitol was beneficial in one patient* (Bruyland and Ebinger, 1994). The best remedy seems to be avoidance of strenuous physical activity.

# RHABDOMYOLYSIS AND MYOGLOBINURIA

## Clinical Features and Course

Many metabolic myopathies discussed in this chapter and several other conditions (Table II), especially exposure to therapeutic or recreational drugs, may lead to rhabdomyolysis and myoglobinuria. Even otherwise healthy individuals may have myoglobinuria when they subject themselves to unusually strenuous physical exertion for which they have not adequately trained (military boot camp, bodybuilding, marathon running). Other causes of myoglobinuria include toxic, ischemic, mechanical, or thermal damage to muscle.

Patients typically have weakness, swelling, and pain of affected muscles, and they produce discolored urine (variously described as coca cola or marsala color, depending on the cultural environment). Serum CK

---

TABLE II    Causes of Rhabdomyolysis and Myoglobinuria

| | |
|---|---|
| **Metabolic** | **Ischemic** |
| **Carnitine palmitoyl transferase deficiency** | **Arterial thrombosis or embolism** |
| Glycogenoses types V, VII to XI | Sickle cell anemia |
| Enzyme deficiencies of β-oxidation | |
| Myoadenylate deaminase deficiency | **Toxic** |
| Mitochondrial diseases | **Alcohol** |
| Electrolyte imbalance (e.g., hypokalemia, acidosis, | Cocaine, heroine, amphetamine, "ecstasy," "angel dust" |
| hypernatremia, hypophosphatemia) | Snake, spider, bee venoms |
| Unaccustomed physical exertion | Tetanus, typhus, staphylococcus toxins |
| | Carbomonoxide, organophosphate |
| **Mechanical** | Strychnine, toluole, cyanide, arsenic |
| **Status epilepticus, delirium** | Cicuta virosa (water hemlock toxin) |
| Direct muscle trauma by "crush injury" | Quail-eater's disease |
| Compartment syndrome | Haff disease (after ingestion of fish) |
| | |
| **Thermic** | **Drugs** |
| Malignant hyperthermia | **Lipid-lowering agents (e.g., clofibrate)** |
| Malignant neuroleptic syndrome | **HMG-CoA reductase inhibitors (e.g., lovastatin)** |
| Fever, heat stroke | Neuroleptics |
| | Barbiturates, doxylamine |
| **Inflammatory** | Potassium-lowering drugs (e.g., diuretics, theophylline, |
| **Polymyositis, dermatomyositis** | carbenoxolone, amphotericin B) |
| Viral infections (e.g., influenza, Coxsackie, HIV) | Suxamethoniume |
| Bacterial infections (e.g., staphylococcal) | Azathioprin, 5′-azacytidine |
| Rickettsial infections | ε-Aminocapronsäure |

Bold type = frequent causes.
Modified from Penn (1994).

levels often surpass 50,000 U/L. When pigmenturia is visible to the naked eye, it means that 200 g of muscle have been damaged, and myoglobin concentration in serum is greater than 1 mg/ml (Griggs *et al.*, 1995).

Despite the massive muscle necrosis occurring during the acute attack, fixed weakness after recovery is unusual, unless there is an underlying myopathic disorder (Penn, 1994). Release of potassium from and sequestration of calcium into necrotic muscle result in **hyperkalemia** and **hypocalcemia**. This combination may lead to life-threatening **cardiac arrhythmia**. The most serious complication of rhabdomyolysis, however, is acute oliguric **renal failure** caused by precipitation of myoglobin in the renal tubuli. If a patient survives the acute stage, permanent renal dysfunction is uncommon (David, 2000).

### Principles of Therapy

Initial treatment includes stabilization of the patient and removal of the triggering factors, whenever possible. Because precipitation of myoglobin in the renal tubuli is favored by acidosis and dehydration, prevention of acute tubular necrosis includes volume substitution, alkalinization of the urine, and forced diuresis under intensive care conditions. Hemodialysis may be required if renal failure ensues (David, 2000). Treatment of hyperkalemia, hypocalcemia, cardiac arrhythmia, respiratory problems, and disseminated intravascular coagulopathy follows standard principles of internal medicine and intensive care (Penn, 1994).

## MITOCHONDRIAL DISEASES

### Biological Basis, Clinical Features, and Course

The mitochondrial matrix contains the final pathways of lipid metabolism (the β-oxidation spiral) and the final pathway of carbohydrate and protein metabolism (the Krebs cycle). These pathways generate reducing equivalents in the form of NADH and $FADH_2$, which, in turn, feed electrons to the respiratory chain located in the inner mitochondrial membrane. The electrons are carried along a series of complexes (I–IV) and small carriers (coenzyme Q10 and cytochrome-c) all the way to oxygen, and the energy so generated is used to pump protons outside of the inner membrane. The flow of protons back into the matrix through complex V (ATP synthetase) is coupled to the production of ATP. This process, called oxidative phosphorylation (OXPHOS), is the main source of energy in eucaryotes.

Most subunits of respiratory chain enzymes are encoded by the nuclear genome and imported into mito-

chondria. However, mitochondria also have their own DNA (mtDNA), which encodes 13 polypeptides of the respiratory chain and the ribosomal RNAs (rRNA) and transfer RNAs (tRNA) necessary for a semiautonomic transcription and translation.

A group of mitochondrial diseases are caused by mutations of the mtDNA and are, as a rule, maternally inherited. The stochastic distribution of normal and mutant DNA (mitotic segregation) results in the coexistence of normal and mutant mtDNAs in cells and tissues, a phenomenon known as heteroplasmy. Various degrees of heteroplasmy in different tissues and in different individuals explain the variable clinical manifestations seen in members of the same family or in different organs of the same individual. On the other hand, the high rate of spontaneous mutation of mtDNA explains the frequent occurrence of sporadic disorders.

Mitochondrial diseases as a whole also comprise nuclear-encoded defects of mitochondrial proteins, including enzymes of β-oxidation, of the Krebs cycle, and of the respiratory chain. In addition, because mitochondria are the slaves of the nucleus, a group of disorders (defects of intergenomic signaling) are due to mutations in nuclear genes encoding factors needed for mtDNA replication, transcription, and translation (Table III).

### *Chronic Progressive External Ophthalmoplegia and Kearns-Sayre Syndrome*

These disorders are most often due to large-scale single deletions of mtDNA (Holt *et al.*, 1989; Moraes *et al.*, 1989), rarely to other mutations (Ciafaloni *et al.*, 1992; Seibel *et al.*, 1994). Patients with chronic progressive external ophthalmoplegia (CPEO) have ptosis and slowly progressive limitation of eye movements, usually without diplopia. Weakness of limb muscles, dysphagia, and dysarthria may also occur. Kearns-Sayre syndrome (KSS) is defined as external ophthalmoplegia and pigmentary retinal degeneration, with onset before age 20, plus at least one of the following signs: ataxia, cardiac conduction block, or CSF protein >1 g/L (Rowland *et al.*, 1983). Frequent additional symptoms include neurosensory hearing loss, dementia, short stature, neuropathy, diabetes mellitus, and other endocrine problems. Complications may derive from the reduced respiratory drive that may exist without muscle weakness (Barohn *et al.*, 1990b). The frequent occurence of some of these symptoms in patients with CPEO (labeled CPEO-plus) points to the fact that CPEO and KSS are the extreme manifestations of a clinical continuum. The course is variable and depends on the degree of multisystem involvement.

EMG shows a myopathic pattern, except in patients with peripheral neuropathy. Serum CK and lactate are

**TABLE III**  Mitochondrial Diseases

| Disease | Inheritance | Gene defect/ protein deficiency | Clinical symptoms and course |
|---|---|---|---|
| I. Defects of mitochondrial DNA | | | |
| A. Mutations in genes affecting mitochondrial protein synthesis | | | |
| 1. Large-scale rearrangements: deletions, duplications | | | |
| a. CPEO | Sporadic | Del, Dupl | External ophthalmoplegia and ptosis, may include proximal myopathy, dysphagia and dysarthria; RRF |
| b. CPEO plus | Sporadic | as CPEO | CPEO plus additional symptoms of a multisystem disease, not fulfilling the criteria of KSS; RRF |
| c. Kearns-Sayre syndrome (KSS) | Sporadic | Del, Dupl | Obligatory: CPEO, retinitis pigmentosa, onset before age 20, plus at least one of the following: ataxia, heart block, CSF protein >1g/L; nonobligatory but frequent: myopathy, deafness, dementia, etc.; RRF |
| d. Pearson syndrome | Sporadic | Del | Bone marrow (pancytopenia) and pancreas affected; when the children survive childhood, they may develop KSS (McShane et al., 1991) |
| 2. Transfer-RNA point mutations | | | |
| a. MELAS Mitochondrial Encephalomyopathy, Lactic acidosis and strokelike episodes | Maternal | mt3243-Leu, mt3271-Leu and other mutations | Onset in childhood; retardation, myopathy, migraine-like headache with vomiting, focal and generalized seizures, hearing loss, dementia, strokelike episodes with hemiparesis and hemianopia, short stature etc.; focal lesions and basal ganglia calcifications in MRI; RRF; death often as young adults, but very variable course (Ciafaloni et al., 1992) |
| b. MERRF Myoclonic epilepsy with ragged-red fibers | Maternal | mt8344-Lys, mt8356-Lys and other mutations | Onset mostly in childhood; myoclonus, generalized epileptic seizures, myopathy, ataxia, deafness, dementia, optic atrophy, neuropathy, and others; MERRF-MELAS-overlap; lactic acidosis, RRF; death often as young adults, but very variable course (Silvestri et al., 1993) |
| c. Mitochondrial myopathy | Maternal > sporadic | Several mutations | Exercise-dependent muscle weakness of the limbs with muscle pain, without PEO; often cardiomyopathy and respiratory weakness (Zeviani et al., 1991) |
| d. Multiple symmetric lipomatosis | Sporadic > maternal | mt8344-Lys | Multiple lipomas, particularly in neck and shoulder areas; often with peripheral neuropathy, partly CNS involvement (Klopstock et al., 1994) |
| e. CPEO | Maternal | e.g., mt3243-Leu | Clinically indistinguishable from sporadic CPEO (cf. I.A.1.a.) |
| f. CPEO plus | Maternal | e.g., mt3243-Leu | Clinically indistinguishable from sporadic CPEO plus (cf. I.A.1.b.) |
| 3. Ribosomal RNA point mutations | | | |
| a. Maternal deafness | Maternal | mt1555-rRNA | Maternally inherited deafness, with or without exposure to aminoglycosides |
| B. Mutations in protein-coding genes | | | |
| a. Leber's hereditary optic neuropathy | Maternal | mt11778-ND4, mt3460-ND1, mt14484-ND6 and other mutations | Subacute, often bilateral, extensive central scotomas, most often in young men (80%); mostly severely impaired vision, rarely improvement, infrequent additional symptoms like tremor, dystonia, athetosis, spasticity, and others, in females partly MS-like course; microangiopathy; no RRF (Riordan-Eva et al., 1995) |
| b. NARP/MILS | Maternal | mt8993-ATPase | Neuropathy, ataxia, retinitis pigmentosa, no RRF (Holt et al., 1990) or maternally inherited Leigh syndrome |
| c. Exercise intolerance | Maternal | Mitochondrial gene for cytochrome b | Severe and progressive exercise intolerance, mild proximal limb weakness, myoglobinuria, and lactic acidosis (Andreu et al., 1999) |

*continues*

TABLE III  *Continued*

| Disease | Inheritance | Gene defect/protein deficiency | Clinical symptoms and course |
|---|---|---|---|
| II. Defects of nuclear DNA | | | |
| A. Mutations in genes encoding respiratory chain subunits | | | |
| 1. Complex I | | | |
|   a. Leigh syndrome | ar | Complex I subunits | Subacute necrotizing encephalomyelopathy with motor and mental retardation, hypotonic myopathy, anorexia, brainstem signs, epileptic seizures, and others; death most often before age 2 from respiratory insufficiency (Van Coster et al., 1991) |
| 2. Complex II | | | |
|   a. Leigh syndrome | ar | Complex II subunits | Leigh syndrome as above (cf. II.A.1.a) (Bourgeron et al., 1995) |
|   b. Optic atrophy | | | Late-onset optic atrophy, ataxia, and myopathy (Birch-Marchin et al., 2000) |
| B. Mutations in genes encoding ancillary proteins of Complex IV | | | |
|   a. Leigh syndrome | ar | SURF-1 | Leigh syndrome as above (cf. II.A.1.a) (Zhu et al., 1998) |
|   b. Infantile cardioencephalomyopathy | ar | SCO2 | Encephalopathy and hypertrophic cardiomyopathy; infantile onset, very rapid progression to death (Papadopoulou et al., 1999) |
| C. Defects of intergenomic signaling | | | |
| 1. leading to mtDNA depletion and multiple mtDNA deletions | | | |
|   a. MNGIE | ar | Thymidine phosphorylase deficiency | Myopathy, neuropathy, gastrointestinal pseudoobstruction, encephalopathy; ophthalmoplegia, lactic acidosis, and others (Hirano et al., 1994) |
| 2. leading to multiple mtDNA deletions | | | |
|   a. ad-PEO | ad | | Clinically indistinguishable from sporadic CPEO (cf. A.1.a.) |
|   b. ar-PEO | ar | | Clinically indistinguishable from sporadic CPEO (cf. A.1.a.) |
|   c. ARCO | ar | | Autosomal recessive cardiomyopathy and ophthalmoplegia (Bohlega et al., 1996) |
| D. Other nuclear mitochondrial diseases | | | |
|   a. Luft disease | ar | Uncoupling of oxidative phosphorylation | Euthyroid hypermetabolism with sweating, polyphagia and polydipsia, mild muscle weakness (Luft et al., 1962) |
|   b. CoQ deficiency | ar | CoQ | Onset at age 3, proximal muscle weakness and fatiguability, later tremor, ataxia, seizures, myoglobinuria; therapy with CoQ (Ogasahara et al., 1989) |
|   c. Alpers syndrome | Sporadic | COX | Like Leigh syndrome plus cortical and hepatic involvement (Egger et al., 1987) |
|   d. Fatal infantile myopathy | ar | COX | Hypotonia, generalized muscle weakness from birth, renal tubular defect; death in the first year from respiratory insufficiency (Bresolin et al., 1985) |
|   e. Benign infantile myopathy | Sporadic | COX | In the beginning as the fatal form, without renal tubular defect, improvement from age 2–3 (Zeviani et al., 1987) |

Del, deletion in mtDNA; Dupl, duplication in mtDNA; mt3243-Leu, point mutation at position 3243 of the mtDNA in the tRNA gene for leucine; Lys, lysine; ND, NADH dehydrogenase; CoQ, coenzyme Q; COX, cytochrome-c oxidase.

elevated in only half of the cases. CT scan and MRI of the brain often show generalized atrophy, hyperintensity of the white matter, and, sometimes, basal ganglia calcification. Muscle biopsy almost invariably reveals ragged red fibers (RRF), which are histochemically cytochrome-c oxidase (COX)–negative. In each patient, a single species of deleted mtDNA is found in muscle. Blood is not a reliable tissue for molecular analysis, because mtDNA deletions are not found in patients with CPEO and may be undetectable even in patients with KSS.

## MELAS

Mitochondrial encephalomyopathy with lactic acidosis and strokelike episodes (MELAS) (Pavlakis *et al.*, 1984) is caused by a variety of mutations in mtDNA, the most common being the A3243G mutation in the tRNA$^{Leu(UUR)}$ (Goto *et al.*, 1990). Typically, there is normal early development, followed by strokelike episodes in childhood or young adult life (Ciafaloni *et al.*, 1992). In the full-blown syndrome, patients may have short stature, exercise intolerance, seizures, hearing loss, dementia, pigmentary retinal degeneration, migraine-like headache with vomiting, and strokelike episodes often leading to hemiparesis and hemianopia. Death often occurs in early adulthood. There are also mono- or oligosymptomatic forms, particularly in maternal relatives of a typical patient, often manifesting only as deafness, diabetes mellitus, or migraine. EMG is myopathic (except in cases with coexistent neuropathy), and serum CK may be modestly increased. Resting lactate in blood and CSF is often elevated. CT scan and MRI of the brain often show brain atrophy, basal ganglia calcifications, and, in typical cases, strokelike lesions most often in the posterior regions and not corresponding to the distribution of major vessels. As a rule, muscle biopsy shows RRF, which are usually COX-positive, and mitochondrial proliferation in the walls of intramuscular blood vessels best revealed with the SDH stain. The A3243G mutation is almost always detectable in blood of typical cases.

## MERRF

Myoclonic epilepsy with ragged red fibers (MERRF) (Fukuhara *et al.*, 1980) is most commonly due to a A8344G mutation in the mtDNA tRNA$^{Lys}$ gene (Shoffner *et al.*, 1990). As in MELAS, the onset and course of the disease are variable. Usually MERRF begins in the second or third decade (Silvestri *et al.*, 1993), and the course is more benign than in MELAS. The main symptoms are myoclonus, ataxia, seizures, and myopathy with RRF. In addition, deafness, dementia, neuropathy, optic atrophy, multiple lipomas, cataract, and strokelike

episodes may occur. Maternal relatives are frequently oligosymptomatic. Differential diagnosis includes other causes of progressive myoclonus epilepsy, such as Unverricht-Lundborg disease, Lafora disease, or neuronal ceroid lipofuscinoses.

EMG is usually myopathic (banning the coexistence of neuropathy), and serum CK may be increased. Resting lactate is frequently elevated in serum and CSF. CT scan and MRI of the brain may show brain atrophy. Muscle biopsy shows RRF that are COX negative. The point mutation is usually detectable in blood and in muscle.

## NARP/MILS

Neuropathy, ataxia, and retinitis pigmentosa (NARP) (Holt *et al.*, 1990) is associated with a T8993G point mutation in the gene for subunit 6 of mitochondrial ATPase. Onset is usually in adolescence. Apart from neuropathy, ataxia, and retinitis pigmentosa, patients may have pyramidal signs, mental deterioration, and proximal limb weakness. Lactic acidosis and RRF in muscle are not present. If the percentage of the 8993 mutation is very high, it leads to maternally inherited Leigh syndrome (MILS). In contrast to the different forms of Leigh syndrome with Mendelian inheritance, retinitis pigmentosa is a major sign in MILS.

### Other Mitochondrial Disorders

Leber's hereditary optic neuropathy (LHON) does not cause myopathy and is not discussed here. In keeping with the notion that most proteins of the respiratory chain are encoded by nuclear DNA (nDNA), many mitochondrial disorders are inherited as Mendelian, usually autosomal recessive, traits. These disorders are usually multisystemic and are frequent causes of Leigh syndrome. This is a devastating encephalopathy of infancy or early childhood characterized by symmetrical lesions in the basal ganglia and in the brainstem. Biochemical and molecular studies are defining increasing numbers of defects in respiratory chain complexes I, II, and IV. These entities are summarized in Table III.

## Principles of Therapy

### Metabolic Therapy

Because of the poorly investigated, but strikingly variable, natural course of mitochondrial disorders, statistically meaningful therapeutical trials are difficult to perform.

**Coenzyme Q10.** CoQ is a component of the respiratory chain that transfers electrons from complexes I and II to complex III. In a number of anecdotal reports (summarized in Griggs *et al.*, 1995), there was improvement of symptoms, MR spectroscopic features, or lactate levels with 30–300 mg CoQ/day in individual patients. An open trial in 44 patients with CPEO or KSS demonstrated lower postexercise lactate after 6 months of CoQ (2 mg/kg/day) administration in 16 patients, but a subsequent double-blind study of the 16 responders did not show significant benefit of CoQ over placebo* (Bresolin *et al.*, 1990). In another open trial, treatment with 300 mg CoQ/day plus multivitamins raised CoQ concentration in serum but did not affect symptoms, lactate concentration, or MR spectroscopy in 16 patients with different mitochondrial diseases* (Matthews *et al.*, 1993). Finally, yet another open study showed normalization of postexercise lactate/pyruvate ratios in three of nine patients after 6 months of CoQ therapy* (150 mg/day) (Chan *et al.*, 1998). Thus, there is no rigorously controlled study showing a significant effect of CoQ in mitochondrial disorders. However, it is noteworthy that both the anecdotal reports showing beneficial effects and the negative trials had methodical shortcomings. Therefore, a final statement on the therapeutic value of CoQ in mitochondrial disorders is not possible at present.

From a theoretical point of view, one argument against CoQ therapy could be that CoQ concentration in muscle is normal or elevated in most mitochondrial diseases (Matthews *et al.*, 1993). However, exceptions include the few described cases of unambiguous (presumably primary) CoQ deficiency (Table III). In these cases, CoQ led to marked improvement* (Boitier *et al.*, 1998; Ogasahara *et al.*, 1989; Rotig *et al.*, 2000; Sobreira *et al.*, 1997).

**Creatine (Cr).** Cr is a naturally occuring compound that plays a pivotal role in the regulation of energy metabolism. The highest Cr concentrations are found in muscle and brain, where the enzyme creatine kinase catalyzes the reversible phosphorylation of Cr and phosphocreatine (PCr). This reaction acts as a high-energy phosphate buffering system allowing the resynthesis of ATP. Supplemental Cr ingestion in doses of 10–20 g/day for 4–6 days is well tolerated by healthy subjects, causes an approximately 20% increase of muscle Cr and PCr (Harris *et al.*, 1992; Hultman *et al.*, 1996), and may increase maximal power output in anaerobic activities up to 20% (Greenhaff *et al.*, 1993). There are controlled trials on Cr supplementation in mitochondrial diseases, but their results are still contradictory. A controlled trial of Cr monohydrate (10 g daily for 14 days to 4 g daily for 7 days) in seven patients, six with MELAS and one with isolated myopathy, showed increased strength of

both anaerobic and aerobic activities** (Tarnopolsky *et al.*, 1997). An open trial by the same authors of 81 patients with various neuromuscular disorders (including 17 with mitochondrial diseases) showed significant improvement of ischemic isometric handgrip strength and nonischemic isometric dorsiflexion torque* (Tarnopolsky and Martin, 1999). Another placebo-controlled, double-blind, randomized crossover trial in 16 patients with CPEO or mitochondrial myopathy, however, did not find significant effects on exercise performance, eye movements, or activities of daily life** (Klopstock *et al.*, 2000). Taken together, these data suggest that Cr may be effective in some, but not all, mitochondrial diseases. Because Cr is virtually free of adverse effects, its administration may be warranted in patients with muscle weakness even before a large controlled trial resolves the issue of its efficacy.

**Vitamins.** Thiamine (vitamin $B_1$) and α-lipoic acid are cofactors of pyruvate dehydrogenase. They may improve lactic acidosis in cases of pyruvate dehydrogenase deficiency. Succinate is able to transfer electrons directly to complex II. Riboflavin (vitamin $B_2$), vitamin C, and vitamin K are electron acceptors and may improve electron transfer in the respiratory chain. In single case reports and in different mitochondrial diseases, improvement was reported with the administration of 100–300 mg/day thiamine (Lou, 1981), 600 mg/day α-lipoic acid (Barbiroli *et al.*, 1995), 6 g/day succinate (Shoffner *et al.*, 1989), 25–300 mg/day riboflavin (Bernsen *et al.*, 1991), 2–3 g/day vitamin C alone (Przyrembel, 1987) or in combination with 60–150 mg/day vitamin K (Eleff *et al.*, 1984). However, the larger trial by Matthews *et al.* mentioned earlier found no significant improvement of clinical or laboratory features with a multivitamin therapy that included CoQ, thiamine, riboflavin, niacin, vitamin C, and vitamin K (Matthews *et al.*, 1993). Again, the lack of controlled studies makes it difficult to assess the value of multivitamin supplementation.

**Dichloroacetate (DCA).** DCA is a powerful and specific inhibitor of the phosphorylation of pyruvate dehydrogenase (PDH), thereby keeping the enzyme in the dephosphorylated, active state. DCA improves lactic acidosis not only in PDH deficiency but also in other disorders of energy metabolism (Taivassalo *et al.*, 1996). The only double-blind trial so far reported significant decreases of blood lactate, pyruvate, and alanine and significant improvement of MR-spectroscopic parameters in 11 patients with various mitochondrial disorders** (De Stefano *et al.*, 1995). Single case reports described clinical and laboratory improvements in patients with a variety of disorders, including Leigh syndrome caused by PDH deficiency* (Kimura *et al.*, 1995), complex I deficiency* (Kimura *et al.*, 1995), the NARP/MILS mutation* (Takanashi *et al.*, 1997), MELAS* (Saitoh *et al.*, 1998), and congenital acidosis of other etiology* (Kuroda *et al.*, 1986). Because DCA can cause peripheral neuropathy, it has to be administered together with thiamine (8.6 mg/kg body weight).

**Carnitine.** Campos *et al.* found carnitine deficiency in 21 of 48 patients with various mitochondrial disorders and described improvement after carnitine supplementation* (Campos *et al.*, 1993).

### Other Pharmacological Measures

**Antiepileptic Drugs (AED).** There are no studies addressing the issue of which AED are best suited for the treatment of seizures in mitochondrial diseases. However, valproate seems to be inappropriate, because it causes reduction in serum carnitine (Campistol *et al.*, 2000; Tein *et al.*, 1993), inhibition of beta-oxidation (Kibayashi *et al.*, 1999), oxidative phosphorylation (Silva *et al.*, 1997), and ultrastructural abnormalities of mitochondria with lipid deposition (Melegh and Trombitas, 1997). Seizures in mitochondrial diseases may worsen with valproate (Ponchaut and Veitch, 1993). Moreover, mitochondrial diseases may be considered a risk factor for valproate-induced liver failure (Krahenbuhl *et al.*, 2000). Other AED (e.g., carbamazepine) may be appropriate for the treatment of seizures in mitochondrial disorders.* Because status epilepticus is a major threat in mitochondrial disorders, patients should be carefully monitored for this development and promptly referred to a neurological intensive care unit.

**Corticosteroids.** Some improvement was reported in single cases of MERRF, MELAS, and mitochondrial myopathy with corticosteroids* (Griggs *et al.*, 1995). The mechanism is unknown, and there are no controlled trials.

**Acetylcholine Esterase Inhibitors.** In some patients, low doses of pyridostigmine may cause a mild improvement of muscle symptoms* (Reichmann, 1993).

### General Measures

**Physical Therapy.** Exercise intolerance is a major complaint of patients with mitochondrial disorders. Limited exercise capacity is probably due in large measure to the impairment in energy metabolism, but it may also be exacerbated by reduced physical activity and sedentary living patterns (chronic deconditioning). There have been concerns that endurance training may be unfavorable to mitochondrial patients, because

mutant mtDNA may have a replicative advantage compared with wild-type mtDNA (Grossman and Shoubridge, 1996). However, an open trial in 10 patients with mitochondrial exercise intolerance revealed a marked improvement in aerobic capacity, resting and exercise heart rate, blood lactate, and ADP recovery after 8 weeks of aerobic training on a treadmill* (Taivassalo *et al.*, 1998). Physical training in these patients should, however, be closely supervised and individualized to ensure that the lower limit of intensity sufficient to produce an effect is reached, while safe upper limits are not surpassed (Taivassalo *et al.*, 1998).

**Pacemaker.** The early implantation of a cardial pacemaker may be lifesaving in patients with CPEO-plus or KSS*.

**Respiratory Function.** In all mitochondrial diseases, the possibility of a reduced respiratory drive should be kept in mind, particularly in the context of anesthesia, respiratory infections, or application of sedating drugs* (Barohn *et al.*, 1990).

**Surgical Therapy.** Ptosis in CPEO and KSS may compromise vision besides creating esthetic problems. Surgical shortening of the levator muscle is indicated, when the eyelid covers the pupil*. The procedure, however, leaves patients at risk of exposure keratopathy (Daut *et al.*, 2000). Strabismus surgery is rarely indicated, because most patients do not have double vision despite ophthalmoparesis*. Severe muscular dysphagia may be treated by surgical cricopharyngotomy or injections of botulinum toxin into the cricopharyngeal muscle*.

**Other Therapies.** The treatment of myoclonus, strokes, and other symptoms is discussed in other chapters of this book. Treatment of endocrine abnormalities should be performed in consultation with an endocrinologist.

### Gene Therapy

MtDNA-related disorders are candidates for future gene therapy, because they often have fatal outcomes and are not adequately treated by other means. Compared with gene therapy of nuclear diseases, there is the additional problem of transporting the construct into mitochondria. A number of experimental strategies are currently being pursued. These include introducing modified genes into mitochondria by means of the physiological protein import mechanism (Seibel *et al.*, 1995) and the specific inhibition of the replication of mutant mtDNA by sequence-specific antigenomic peptide nucleic acids (Taylor *et al.*, 2000). These trials, however, are not yet clinically relevant.

Novel approaches aim to correct the genetic defect in situ. These procedures take advantage of the fact that muscle satellite cells may contain remarkably low levels of a mitochondrial mutation that is abundant in mature muscle. Therefore, induction of muscle damage and subsequent regeneration by proliferation of satellite cells may lead to a functionally restored muscle (Fu *et al.*, 1996). The feasibility of this approach was indeed shown in two patients with mitochondrial myopathy (Clark *et al.*, 1997) and KSS (Shoubridge *et al.*, 1997) after induction of muscle necrosis by bupivacaine*. The hope to improve ptosis in five patients using this approach, however, was frustrated* (Andrews *et al.*, 1999). Most interestingly, Taivassalo *et al.* could induce a remarkable increase in the ratio of wild-type to mutant mtDNAs and in the proportion of muscle fibers with normal respiratory chain activity by stimulating muscle regeneration through concentric exercise training* (Taivassalo *et al.*, 1999). This form of "gene shifting" may be a promising therapy for the future.

# MYOPATHIES CAUSED BY DRUGS, TOXINS, OR MALNUTRITION

Skeletal muscle is remarkably resistant to exogenous or toxic influences. Although therapeutic doses of drugs rarely cause muscle damage, different types and degrees of muscle symptoms have been reported in small groups of patients or in single cases. Drugs or toxic substances can induce muscle disease, precipitate as yet undiagnosed neuromuscular conditions, or aggravate existing disorders. The clinical pictures caused by myotoxic effects are diverse. An individual drug may cause isolated myalgia, myalgia with marked increase of serum creatine kinase levels, slowly progressive painless weakness, or necrotizing myopathy with myoglobinuria. The underlying pathophysiological mechanisms are diverse, and often there is no convincing, or generally accepted, explanation.

For diagnosis, a detailed history of medication (or substance abuse) is mandatory in each patient. The single most important diagnostic feature of a drug- or toxin-induced neuromuscular disorder is the chronological relationship between exposure and development of symptoms. Lack of personal or family history of previous neuromuscular disease, predisposing features such as renal impairment (which might increase plasma drug levels), and, particularly, resolution of symptoms and signs after removal of the suspect agent must be looked for.

Although relatively rare, drug-induced myopathies are therapeutically important. Some are iatrogenic, but abuse of alcohol and opiates is the most common cause.

Most are easily relieved through identification and removal of the offending agent.

This chapter focuses on the more common types of drug- and toxin-induced myopathies. More extended information can be found in published reviews (Blain and Lane, 1991; Lane and Mastaglia, 1978; Mastaglia, 1991; Victor and Sieb, 1994). Table IV provides a comprehensive list of drugs and associated clinical and pathological syndromes.

## Drug-Induced Muscle Disorders

### Focal Myopathy Caused by Local Toxicity

The local toxicity of a drug depends on the physicochemical properties of both the pharmacological agent and its galenic preparation, such as carrier solvents. An acute local reaction, including complications such as hematoma or infection, is observed in about 0.4% of hospitalized patients receiving at least one intramuscular injection. Cephalothin, tetracycline, and paraldehyde are the most common offenders.

Chronic intermittent injection in the same site is also an important cause of focal myopathy. Local induration and replacement of muscle and subcutaneous fat may cause fibrosis and contractures, even fixed-limb deformities, which may require surgery. This type of reaction has been associated most commonly with narcotic abuse and with intramuscular administration of antibiotics in children.

However, localized muscle damage is not invariably associated with intramuscular injection and has rarely been observed after topical external application of some ointments.

### General Myopathy

Weakness (with or without overt atrophy), diffuse myalgia, stiffness, or cramping are the most common complaints. Electromyographic, laboratory (serum CK), or histological signs of a myopathic lesion may be evident or not. The clinical continuum runs from benign impairment to severe and possibly life-threatening muscle necrosis. In this chapter, drug and toxic myopathies are discussed in order of their frequency in the population.

**Myalgia.** Myalgia, with or without cramping, is a common drug-related symptom, which, however, is rarely investigated, unless it is associated with weakness or significant increase of serum CK levels. Diverse and chemically unrelated agents have been implicated (Table IV). Suxamethonium, a depolarizing muscle relaxant, is commonly involved. It causes myalgia in about half of the patients the day after surgery, often associated with an increase of serum CK activity (Blain and Lane, 1991). The problem may be caused by calcium-dependent phospholipid hydrolysis, generation of prostanoids, and free radicals. Treatment with 600 mg acetylsalicylic acid preoperatively has been shown to reduce the incidence (McLoughlin *et al.*, 1988). Interferon-β, now widely used in immunomodulation of multiple sclerosis, frequently causes flulike symptoms, including myalgia, 3 to 6 hours after injection. These symptoms improve with dose reduction and administration of ibuprofen, low-dose steroids, or pentoxifylline (Walther and Hohlfeld, 1999).

**Chronic Painless Proximal Myopathy.** This is the most common form of drug-induced myopathy. Slowly progressive, symmetrical proximal weakness, and variable degree of wasting are typical. Serum muscle enzymes are normal in most cases. The muscle biopsy usually shows either type 2 fiber atrophy or vacuolar myopathy.

• *Type 2 Fiber Atrophy.* Type 2 fiber atrophy is the typical pathological manifestation of chronic alcoholic myopathy (discussed later in this chapter) and steroid myopathy, but it is also seen with drug-induced hypokalemia. Myopathy may occur with any type of steroid, but fluorinated steroids (triamcinolone, betamethasone, dexamethasone) are the most common culprits. There is no relation with dose or duration of therapy, but myopathy is rare with daily doses equivalent to 10 mg prednisone or less. Steroid myopathy results principally from disturbances of protein and carbohydrate metabolism and may be related to impaired RNA synthesis, reduced ribosomal function, and decreased peptide chain synthesis, particularly in type 2 fibers. An additional mechanism of steroid myotoxicity may involve induction of lysosomal proteases, again largely in type 2 fibers. Steroids also inhibit myophosphorylase activity, which further increases the vulnerability of type 2 fibers. Some weakness and wasting is likely to occur in every patient taking steroids, but isokinetic exercise will combat these side effects to some extent (Horber *et al.*, 1985).

• *Vacuolar Myopathy.* A number of drugs induce autophagic vacuoles containing membranous (myeloid) bodies. This drug-induced lysosomal storage myopathy presents as a chronic painless proximal myopathy. In such cases, an associated drug-induced peripheral neuropathy is almost always present. The most frequently involved agents include amphophilic drugs, such as chloroquine and its derivatives, used as antimalarials and immunosuppressants. Their positively charged groups interact with anionic residues of phospholipid membranes, disturbing their geometry and ultimately forming myeloid bodies that are resistant to lysosomal degradation. The typical cytoplasmic vacuolation is associated with various degrees of fiber necrosis. Acid phosphatase activity is increased, and, in contrast to steroid and hypokalemic myopathy, type 1 fibers are preferentially affected. Myopathy is seen much less commonly, since lower maintenance doses of these drugs are used, and hydroxychloroquine, which seems to be less myotoxic than chloroquine, is used.

Similar muscle biopsy changes have been reported with amiodarone and perhexilene, although the associated neuropathy tends to dominate the picture and may occur in isolation. Similarly, colchicine and vincristine may produce a vacuolar myopathy with membranous inclusions, in addition to the usually more prominent neuropathy.

• *Myalgia with Weakness and Increased CK.* Some drugs induce an acute or subacute syndrome with myalgia, tenderness, and weakness, particularly of the proximal limb girdle musculature, and marked rise of serum CK activity. The syndrome may develop within days to weeks of starting the drug and resolves rapidly on withdrawal. If the responsible drug is not stopped, rhabdomyolysis may ensue with potentially life-threatening consequences (discussed later).

• *Vacuolar Myopathy Associated with Hypokalemia.* Severe hypokalemia (serum $K^+$ < 2 mM) induced by, among others, diuretics, laxatives, amphotericin B, or licorice and derivatives of glycyrrhinic acid such as carbenoxolone, can cause remitting or persistent painful flaccid weakness with areflexia, reminiscent of hypokalemic periodic paralysis but often associated with high serum CK activity. Vacuolar myopathy with relatively little necrosis may be the pathological counterpart of this syndrome. In rare cases, rapid progression to fiber necrosis and rhabdomyolysis can occur. Vacuoles, if present, are not membrane bound, unlike those seen in periodic paralysis (Jensen *et al.*, 1977).

• *Mitochondrial Myopathy.* Patients on zidovudine (azidothymidine, AZT) treatment for HIV infection may develop a syndrome akin to a subacute painful proximal myopathy, with degenerative changes in muscle fibers but little inflammatory reaction. CK levels may increase progressively, with rapid resolution after withdrawing the drug. Such symptoms usually develop only in patients who have been taking about 1 g AZT daily for around a year. Because maintenance doses have been halved, the syndrome is much less common. Muscle biopsies in cases of AZT myopathy may show RRF, with abnormal accumulations of subsarcolemmal mitochondria on ultrastructural examination. Mitochondrial damage to muscle fibers, associated with mtDNA depletion (Arnaudo *et al.*, 1991) and decreased respiratory chain enzyme activities, resolved when the drug was discontinued. The drug probably inhibits mtDNA

# TABLE IV  More extended information can be found in published reviews: Lane & Mastaglia 1978, Victor & Sieb 1994, Blain & Lane 1991, Mastaglia 1991

| Substance | A | B | C | D | E | F | G | H |
|---|---|---|---|---|---|---|---|---|
| Acebutole | | | ■ | ■ | | | | |
| Acetazolamide | | | | | | | | ■ |
| ACTH | | | ■ | | ■ | | ■ | |
| Adriamycin | | | | | ■ | | | |
| Albuterol | | | ■ | ■ | | | | |
| Alcohol | | | ■ | ■ | ■ | ■ | | |
| Allopurinol | | | | ■ | | | | |
| Alprenolol | | | ■ | ■ | | | | |
| Aluminium hydroxide | | | | | | | | |
| Aminocaproic acid | | ■ | | ■ | | | | |
| Amiodarone | | | | | | | | |
| Ammonium chloride | | | ■ | | | ■ | | |
| Amoxapine | | | | ■ | | | | |
| Amphetamine | | | ■ | | | | | |
| Amphotericin B | | | ■ | | | | | |
| Anthranilic acid derivates | | | ■ | | | ■ | | |
| Aprindine | | | ■ | | | | | |
| Aprotinin | | ■ | | ■ | | | | |
| Aurokeratinate | | | | ■ | | | | |
| Aurothioglucose | | | | ■ | | | | |
| Aurothiomalate | | | | ■ | | | | |
| Aurothiopolypeptide | | | | ■ | | | | |
| 5-Azacytidine | | ■ | | | | | | |
| Azathioprine | | | ■ | | ■ | | | |
| Barbiturates | | | ■ | ■ | | | | |
| Barium intoxication | | | ■ | | | ■ | | |
| Bencyclan | | | ■ | | | | | |
| Benzodiazepines | | | ■ | | | | | |
| Benzothiadiazines | | | | | ■ | | | |
| Beta-sympathicomimetics | | | ■ | | | | | |
| Bezafibrate | | | | ■ | | | | |
| Buformine | | | | | | | | |
| Bumetanide | | | ■ | ■ | | ■ | | |
| Bunitrolol | | | ■ | | | | | |
| Bupivacaine | | | ■ | | | | | |
| Bupranolol | | | | ■ | | | | |
| Calcium salts | | ■ | | | | | | |
| Calvazine | | | | | | | | |
| Captopril | | | ■ | | | | | |
| Carazolol | | | ■ | | | | | |
| Carbenoxolone | | | | | ■ | ■ | | |
| Carbimazole | | | ■ | | ■ | | | |
| Carbon monoxide poisoning | | | | | | | | |
| Carbutamide | | | | | | | ■ | |
| Carteolol | | | ■ | | | | | |
| Cephalothin | ■ | | | | | | | |
| Chinidine | | | ■ | | | | | |
| Chinine | | | | ■ | | | | |
| Chloramphenicol | ■ | | | | | | | |
| Chloromycetine | ■ | | | | | | | |
| Chloroquine | | | | ■ | ■ | | | |
| Chlorpromazine | | ■ | ■ | | | | | |
| Chlorthalidone | | | | | | | ■ | |
| Cimetidine | | | ■ | ■ | | ■ | | |
| Clofibrate | | | ■ | ■ | | | | |
| Clofibride | | | | ■ | | | | |
| Clomifen | | | | ■ | | | | |
| Clonidine | | ■ | | | | | | |
| Codeine | | | | ■ | | | | |
| Colchicine | | ■ | | ■ | | | | |
| Corticosteroids | ■ | | | ■ | | | | |
| Cromoglycate | | | | | | | | |
| Cycloserine | | | | ■ | ■ | | | |
| Cyclosporin A | | | | ■ | | | | |
| D-Penicillamine | | | | ■ | | | | |
| Danazol | | ■ | | | | | | |
| Decamethonium | | | | ■ | | | | ■ |
| Diazepam | | | | | | | | |
| Diazoxide | | ■ | | | | | | |
| Digoxin | | | ■ | | | | | |
| Dihydroergotamine | | | | ■ | | | | |
| Dimenhydrinate | | | | | | | | |
| Diphenhydramine | | | | | | | | |
| Dipyridamole | | | | | | | | |
| Disopyramide | | | ■ | | | | | |
| Diuretics | | | ■ | | | | | ■ |
| Emetine (Ipecac) | | | ■ | | | | | |
| Encainide | | | ■ | | | | | |
| Ergotamine | ■ | | | ■ | | | | |
| Etacrynic acid | | | ■ | | | | ■ | |
| Ethanol | | | | | | | | |
| Etofibrate | | | | ■ | | | | |
| Etofyllinclofibrate | | | | ■ | | | | |
| Etretinate | | | | | | | | |
| Fat emulsions iv | | | ■ | | | | | |
| Fenofibrate | | | | ■ | | | | |
| Fenoterol | | | | | | | | ■ |
| Ferrum | | | ■ | | | | | |
| Furosemide | | | ■ | | | | ■ | |
| Gemfibrozil | | | | ■ | | | | |
| Germanium | | | | ■ | | | | |
| Glutethimide | | | ■ | | | | | |
| Guanethidine | | | ■ | ■ | | | | |
| Heparin | | | | | | | | |
| Heroin | ■ | | ■ | | | | | |
| Hydralazine | | | | | | | | |
| Hydroxychloroquine | | | | ■ | | | | |
| Imipramine | | | | | | | | |
| Insulin | | ■ | | | | ■ | | |
| Interferon ß | | | ■ | | | | | |
| Isoetharine | | | ■ | ■ | | | | |
| Isoniazid | | | | ■ | | | | |
| Isotretinoin | | | | ■ | | | | |
| Ketoconazole | | | | ■ | | | | |
| L-Tryptophan | | | | | | | ■ | |
| Labetalol | | | ■ | ■ | | | | |
| Laxatives | | | | | | | | |
| Levodopa | | | ■ | | | | | |
| Lidocaine | | | ■ | | | | | |
| Liquorice | | | | | ■ | | | |
| Lithium | | | ■ | | | | | |
| Lofepramine | | | | | | | | |
| Lonidamine | | ■ | | | | | | |
| Lovastatin | | | | ■ | | | | |
| Mannitol | | ■ | | | | | | |
| Meperidine | ■ | | | | | | | |

*continues*

**TABLE IV** *(Continued)*

| Substance | A | B | C | D | E | F | G | H |
|---|---|---|---|---|---|---|---|---|
| Mepindolol |  |  | X | X |  |  |  |  |
| Meprobamate |  |  | X | X |  |  |  |  |
| Metamizole |  |  | X |  |  |  |  |  |
| Metformine |  |  | X | X |  |  |  |  |
| Methyldopa |  | X |  |  |  |  |  |  |
| Methylphenidate |  | X | X |  |  |  |  |  |
| Methysergide |  | X |  |  |  |  |  |  |
| Metipranolol |  |  | X | X |  |  |  |  |
| Metolazone |  |  | X | X |  |  |  |  |
| Metoprolol |  |  | X | X |  |  |  |  |
| Metronidazole |  |  | X |  |  |  |  |  |
| Mexiletine |  |  | X |  |  |  |  |  |
| Miconazole |  |  | X |  |  |  |  |  |
| Mithramycin |  |  | X |  |  |  |  |  |
| Na-pentosan-polysulphate |  |  |  |  |  | X |  |  |
| Nalidixic acid |  |  |  |  | X |  |  |  |
| Neostigmine |  |  | X | X |  |  |  |  |
| Nifedipine |  |  | X | X |  |  |  |  |
| Nimorazole |  |  | X |  |  |  |  |  |
| Nitroxoline |  |  | X |  |  |  |  |  |
| Organophosphates |  |  |  |  | X |  |  |  |
| Ornidazole |  |  | X |  |  |  |  |  |
| Oxolinic acid |  |  | X |  |  |  |  |  |
| Oxoprenolol |  |  | X | X |  |  |  |  |
| Para-aminosalicylic acid |  |  |  |  |  |  | X |  |
| Paraldehyde | X |  | X |  |  |  |  |  |
| Penbutolol |  |  | X | X |  |  |  |  |
| Penicillamine |  | X | X |  | X |  |  |  |
| Penicillins |  | X |  |  |  |  |  |  |
| Pentazocine | X |  | X | X |  |  |  |  |
| Perhexilene |  |  |  |  | X |  |  |  |
| Perphenazine |  |  | X |  |  |  |  |  |
| Pethidine | X |  |  |  |  |  |  |  |
| Phencyclidine |  |  |  |  | X |  |  |  |
| Phenformin |  |  | X |  |  |  |  |  |
| Phenprobamate |  |  | X |  |  |  |  |  |
| Phenytoin |  |  |  |  |  | X |  |  |
| Physostigmine |  |  |  |  |  |  |  | X |
| Pindolol |  |  | X | X |  |  |  |  |
| Pipemidic acid |  |  | X | X |  |  |  |  |
| Piperazine |  |  | X | X |  |  |  |  |
| Piretanide |  |  | X | X |  |  | X |  |
| Piromidic acid |  |  | X |  |  |  |  |  |
| Pizotifen |  | X |  | X |  |  |  |  |
| Potassium |  |  |  |  |  |  |  | X |
| Pralidoxime | X |  |  |  |  |  |  |  |
| Pravastatine |  |  |  |  | X |  |  |  |
| Praziquantel |  | X |  |  |  |  |  |  |
| Prazosin |  |  | X |  |  |  |  |  |
| Pridinol |  |  | X |  |  |  |  |  |
| Procainamide |  | X | X | X |  |  |  |  |
| Propranolole |  |  | X | X |  |  |  |  |
| Propiothiouracil |  |  |  |  |  | X |  |  |
| Propranolol |  |  | X | X |  |  |  |  |
| Pyrimethamine |  |  | X |  |  |  |  |  |
| Quinacrine |  |  |  |  | X |  |  |  |
| Ritodrine |  |  |  |  |  |  |  | X |
| Rosoxacine |  |  | X |  |  |  |  |  |
| Salbutamol |  |  |  |  | X |  |  |  |
| Salicylic acid | X |  |  |  |  |  |  |  |

| Substance | A | B | C | D | E | F | G | H |
|---|---|---|---|---|---|---|---|---|
| Secobarbitol | X |  |  |  |  |  |  |  |
| Simvastatine |  |  |  |  | X |  |  |  |
| Sodium |  |  |  |  |  |  | X |  |
| Spironolactone |  |  |  |  |  |  |  |  |
| Streptokinase |  |  | X |  |  |  |  |  |
| Streptomycin | X |  |  |  |  |  |  |  |
| Succinylcholine |  |  | X |  |  |  |  | X |
| Suxamethonium chloride |  |  | X | X |  |  |  |  |
| Tamoxifen |  |  |  |  |  |  |  |  |
| Terbutaline |  |  |  |  |  |  |  |  |
| Tetracosactid |  | X |  |  | X |  |  |  |
| Tetracyclines |  | X |  |  |  |  |  |  |
| Thiethylperazine |  |  |  | X |  |  |  |  |
| Thyroid hormones |  |  |  |  |  |  |  |  |
| Timolol |  |  | X | X |  |  |  |  |
| Tinidazol |  |  | X |  |  |  |  |  |
| Tocainide |  |  | X |  |  |  |  |  |
| Tolbutamide |  |  |  |  |  |  | X |  |
| Toliprolol |  |  | X | X |  |  | X |  |
| Toluene abuse |  |  |  |  |  |  | X |  |
| Triamterene |  |  | X |  | X |  |  |  |
| Trifluoroperazine |  |  |  | X |  |  |  |  |
| Urokinase |  | X |  |  |  |  |  |  |
| Vincristine |  |  |  |  | X |  |  |  |
| Vinylchloride |  |  |  |  | X |  |  |  |
| Vitamin E |  |  |  |  | X |  |  |  |
| Zidovudine |  | X |  |  | X |  |  |  |
| Zimelidine |  |  |  |  | X |  |  |  |

| Substance | A | B | C | D | E | F | G | H |
|---|---|---|---|---|---|---|---|---|
| A=Local Muscle Damage |  |  |  |  |  |  |  |  |
| B=Myalgia |  |  |  |  |  |  |  |  |
| C=Painful Cramps |  |  |  |  |  |  |  |  |
| D=Overt Pareses |  |  |  |  |  |  |  |  |
| E=Myopathy |  |  |  |  |  |  |  |  |
| F=Myositis |  |  |  |  |  |  |  |  |
| G=Dyskalemic Paralyses |  |  |  |  |  |  |  |  |
| H=Myotonia |  |  |  |  |  |  |  |  |

polymerase γ. Mitochondrial abnormalities may also be seen in HIV-infected patients not taking AZT (Lane *et al.*, 1993).

• *Inflammatory Myopathy.* Inflammation is unusual in drug-induced myopathies. It may be seen as a non-specific feature in general allergic reactions to drugs such as penicillin and phenytoin. Isolated examples have been reported with drugs such as levodopa, cromogly-cate, propiothiouracil, cimetidine, or after vaccination. Myositis or dermatomyositis have both been observed after D-penicillamine treatment of systemic sclerosis, cystinuria, Wilson's disease, or rheumatoid disease. Procainamide has been linked to the development of an inflammatory myopathy as part of a drug-induced lupus syndrome.

• *Necrotizing Myopathy.* Necrotizing myopathy occurs most frequently in association with either alcohol or opiate abuse; however, it may also be encountered with therapeutic agents. This syndrome has been reported rather commonly with the new cholesterol- and lipid-lowering drugs: the 3-hydroxy-3-methyl-glutaryl-coenzyme A (HMG-CoA) inhibitors, and fibric acid derivatives. The first drug to be implicated was the fibric acid derivative clofibrate. Of the reported cases, 80% had myalgia develop within 3 weeks, and weakness, often severe, occurred in more than half (Rimon *et al.*, 1984). Because renal insufficiency prolongs the half-life of the drug, most cases have been reported in association with uremia. Low plasma protein levels, such as occur in liver cirrhosis or nephrotic syndrome, also increase the risk of myotoxicity, because the drug is bound to albumin.

However, myopathy is rare with fibric acid derivatives alone. The usual patient developing myopathy has been on a combination therapy with an HMG-CoA reductase inhibitor, such as lovastatin. Myalgia with mildly increased CK activities occurs in about 0.5% of patients taking lovastatin alone. The incidence is 10 times higher when lovastatin is combined with gemfibrozil and still higher with cyclosporin. In this situation, necrotizing myopathy may occur in up to 30% of cases. Cyclosporin seems to inhibit the biliary excretion of lovastatin, causing much higher plasma and tissue levels of the drug.

Some β-receptor blockers (labetolol, sotalol), the vitamin A analogs etretinate and isotretinoin, and the fibrinolytic inhibitor epsilon-aminocaproic acid may also be involved in necrotizing myopathy.

Steroids usually produce a chronic painless proximal myopathy. However, an acute, painful necrotizing myopathy has been observed after parenteral application of gram doses of prednisone in asthma patients. Another peculiar myotoxic complication of high-dose steroid therapy has been observed in the presence of functional denervation by neuromuscular blocking agents, such as pancuronium and vecuronium, or in association with myasthenia gravis. Such patients are usually treated for status asthmaticus in an ICU setting, and the clinical syndrome with flaccid tetraparesis may be misdiagnosed as Guillain-Barré syndrome or critical illness neuropathy. The selective loss of myosin on muscle biopsies is a unique histological finding and may result from up-regulation of steroid receptors, increased catabolic metabolism, and a predisposition to myosin degradation under denervating conditions. Withdrawal of steroids usually results in gradual improvement.

• *Drug-Induced Rhabdomyolysis (see "Rhabdomyolysis and Myoglobinuria").* If the offending drug is not withdrawn, the pathological processes leading to acute and subacute painful proximal myopathies may culminate with rhabdomyolysis, the most serious manifestation of drug-induced myopathy. The widely spread muscle necrosis is potentially life-threatening because of the serious metabolic disturbances it brings about. In clinical practice, rhabdomyolysis is encountered most frequently with drug abuse, particularly with alcohol or opiates (discussed later).

• *Drug-Induced Myotonia.* Drug-induced clinical myotonia occurs most commonly when a drug unmasks or exacerbates this symptom in a patient with hitherto undiagnosed myotonic syndromes, such as myotonic dystrophy or congenital myotonia. Beta-blockers, barbiturates, diuretics, and acetazolamide are most commonly implicated (see Chapter 98). Electromyographic myotonia and pseudomyotonia reflect electrical destabilization of the sarcolemma. Both findings are common in patients with acute necrotizing myopathies or myopathies caused by lipid-soluble drugs that affect membrane structure, such as chloroquine and clofibrate.

• *Periodic Paralysis.* Several drugs will precipitate periodic paralysis in patients with latent, pre-existing dyskalemic paresis or induce such symptoms by chronic changes of the potassium homeostasis. These effects must be differentiated from the possible induction of a myopathy through chronic hypokalemia (Comi *et al.*, 1985), see Chapter 99.

• *Myasthenic Syndromes.* Myasthenia gravis may be severely worsened by certain drugs, especially antibiotics. This can be easily avoided. Some drugs will induce myasthenia gravis in the occasional patient without prior signs of the disease. These and related questions are discussed in Chapter 94.

## Toxic Myopathies

### Alcoholic Myopathies

The considerable variability in the myotoxic manifestations of individual drugs is particularly well illustrated by alcohol, whose deleterious effects range from asymptomatic elevation of CK levels to fatal rhabdomyolysis after binges.

**Acute Alcoholic Myopathies.** Acute painful necrotizing myopathy, often preceded by cramps, with myoglobinuria and markedly increased serum CK activity, is the most dramatic presentation of alcoholic myopathy. It is usually observed after a drinking binge. In many cases, there will be a history of previous similar, milder episodes. Muscle swelling may be prominent and even require emergency fasciotomy. The swelling is often local (e.g., in a leg or a calf) and may be misdiagnosed as deep venous thrombosis.

In most patients, the condition is relatively mild, and recovery will occur in 2–3 weeks. However, renal failure and other complications of rhabdomyolysis may follow more severe attacks.

The spectrum of muscle biopsy findings reflects the clinical diversity: it ranges from generalized fiber swelling, to patchy segmental necrosis and hyalinization of individual fibers, to massive necrosis.

Nutritional abnormalities and deficits, recurrent vomiting, and diarrhea may lead to severe hypokalemia in the alcoholic patient. The clinical and pathological manifestations are similar to those described in other forms of *acute hypokalemic myopathy*: severe proximal weakness, marked increase in CK levels, and vacuolar fiber necrosis. Muscle pain, cramping, and swelling are usually inconspicuous.

Such *acute symptomatic muscle disorders* probably appear in 1% of alcoholics but in about 10% of those admitted to the hospital because of acute alcohol intoxication. Such patients are particularly sensitive to rechallenges with alcohol. Significant increases of serum CK levels are seen 2–6 hours after ingestion in this group but not in normal controls. If such an elevation of CK activity after alcohol exposure is taken as indicating alcoholic myopathy, some 80% of alcoholics, unselected for additional muscle symptoms, may be affected.

**Chronic Alcoholic Myopathy.** Chronic alcoholic myopathy is probably the most common neuromuscular disorder. It has been estimated that some 10% of European and North American adults abuse alcohol. Some two thirds of such chronic alcohol misusers develop the slowly progressive, painless, proximal myopathy with insidious wasting and weakness that is typical of alcoholic myopathy.

The condition's histological hallmark is selective type 2 fiber atrophy. This finding, however, as well as other pathological changes, such as scattered fiber necrosis or increased central nuclei and tubular aggregates, is generally regarded as nonspecific. These pathological features may also occur, for example, in patients with peripheral neuropathy, which is also common in alcoholics. Thus, the very existence of chronic alcoholic myopathy has been questioned. Arguments in favor of a primarily myopathic lesion include data from quantitative electromyography or from animal models and similarities to alcoholic cardiomyopathy, when denervation atrophy is not a factor.

The myotoxic effects of chronic alcohol exposure are poorly understood. Several factors may be operative. Alcohol is intrinsically myotoxic, causing ultrastructural damage to mitochondria and the sarcoplasmic reticulum and impairing ATPase activities. Such immediate, toxic effects may be potentiated by the additional burden of malnutrition. A drinker's diet has many deficits, including lack of antioxidants. Thus, alcohol-generated free radicals may not be scavenged normally, and they will damage the integrity of cell membranes, disturb ion homeostasis, and impair glycolytic function and energy metabolism in general. Fiber necrosis during stress will affect type 2 fibers in particular, because their antioxidant capacities are lower compared with type 1 fibers. Furthermore, protein metabolism is abnormal in alcoholics, with a negative nitrogen balance and loss of amino acids that would normally be destined for incorporation into muscle.

Therapy is complete abstinence, of course, which is always effective.

*Other Toxic Myopathies*

**Acute Necrotizing Myopathies.** Severe myopathy with rhabdomyolysis may occur after use of heroin and cocaine. Initial reports could not clarify whether the myopathy was a direct effect of the substance abused, of an adulterant, or whether it was due to indirect sequelae, such as trauma, coma, or infection. Reports of myopathy after intravenous use of relatively pure heroin, prescribed dihydrocodeine and morphine, after the administration of opiates during anesthesia, and after ingestion of codeine-linctus seem to have settled the question.

Phencyclidine and the related drug ketamine were widely abused a decade ago, and rhabdomyolysis was reported in a number of cases as a result of extreme motor excitation. Hyperpyrexia, profuse sweating, and rhabdomyolysis have been observed after consumption of the more modern designer drug methylenedioxymethylamphetamine (MDMA, "Ecstasy"), but these symptoms may be related to the circumstances under which the drug is taken rather than to direct myotoxic effects.

Overdoses of drugs such as methylxanthine inhibitors (aminophylline, theophylline) and pentamidine have also occasionally induced rhabdomyolysis. But, again, direct toxic effects cannot necessarily be inferred, because mechanical pressure on muscle in comatose patients may also induce myoglobinuria.

Acute poisoning with organophosphorus insecticides results in irreversible inhibition of cholinesterases at neuromuscular junctions and severe weakness, not only through neuromuscular blockade but also as a result of necrotizing myopathy.

Finally, some snake venoms are potent myotoxins. Toxicity results from different components, including crotamine, myotoxin A, and the A2-phospholipases, which have strong predilection for skeletal muscle. Severe myonecrosis with myoglobinuria may result.

**Acute and Chronic Hypokalemic Myopathy.** Both acute and chronic myopathies have been reported in toluene-sniffing abusers. The myotoxic action seems to be mediated through the hypokalemia that results from renal tubular damage. The effects are usually reversible within days of withdrawal.

**Inflammatory Myopathy.** Outbreaks of eosinophilia-myalgia syndrome induced by contaminated L-tryptophan (1974 and 1989) and the toxic oil syndrome in Spain (1981) caused severe and persistent disabilities, including myopathy, and shared a number of characteristics.

Clinical findings in eosinophilia-myalgia syndrome include very severe myalgia and perivascular and perimysial inflammatory infiltration without myonecrosis, tending to spare the endomysium, with normal CK activity. Intramuscular nerves and muscle spindles tend to be particularly affected, and the neuromuscular problems are probably related more to nerve than to muscle involvement.

The condition seems to be due to contamination of L-tryptophan by aniline and anthranilic acid, which were used as reactants for L-tryptophan synthesis by a single company. Similar components may have been adulterants of rapeseed oil in the toxic oil syndrome, which was also associated with marked myalgia and eosinophilia.

Finally, polymyositis has been linked to poisoning by ciguatera, a low-molecular-weight saponin derived from a coral reef dinoflagellate, *Gambierdicus toxicus*, which enters the human food chain through seafood.

**Myopathies Caused by Nutritional Deficits**

Chronically deficient nutrition alone will not usually lead to myopathy, and isolated vitamin deficiencies have become a rarity. Functional deficits usually arise through other mechanisms; for example, the osteomalacic myopathy of vitamin D deficiency (Victor and Sieb, 1994). Whether vitamin E deficiency causes a myopathy in humans is not yet established. A myopathy induced by nicotinic acid has been reported (Litin and Anderson, 1989).

The myopathy seen in some patients with anorexia nervosa/bulimia indicates electrolyte imbalances caused by abuse of laxatives or induced vomiting. The chronic ingestion of *radix ipecacuanae* extracts may induce a myopathy *sui generis* in addition to the concomitant hypokalemia.

## REFERENCES

Amalfitano, A., McVie-Wylie, A. J., Hu, H., Dawson, T. L., Raben, N., Plotz, P., and Chen, Y. T. (1999). Systemic correction of the muscle disorder glycogen storage disease type II after hepatic targeting of a modified adenovirus vector encoding human acid-alpha-glucosidase. *Proc. Natl. Acad. Sci. USA* **96**, 8861–8866.

Andreu, A. L., Hanna, M. G., Reichmann, H., Bruno, C., Penn, A. S., Tanji, K., Pallotti, F., Iwata, S., Bonilla, E., Lach, B., Morgan-Hughes, J., and DiMauro, S. (1999). Exercise intolerance due to mutations in the cytochrome b gene of mitochondrial DNA. *N. Engl. J. Med.* **341**, 1037–1044.

Andrews, R. M., Griffiths, P. G., Chinnery, P. F., and Turnbull, D. M. (1999). Evaluation of bupivacaine-induced muscle regeneration in the treatment of ptosis in patients with chronic progressive external ophthalmoplegia and Kearns-Sayre syndrome. *Eye* **13(Pt 6)**, 769–772.

Arnaudo, E., Dalakas, M., Shanske, S., Moraes, C. T., DiMauro, S., and Schon, E. A. (1991). Depletion of muscle mitochondrial DNA in AIDS patients with zidovudine-induced myopathy. *Lancet* **337**, 508–510.

Barbiroli, B., Medori, R., Tritschler, H. J., Klopstock, T., Seibel, P., Reichmann, H., Iotti, S., Lodi, R., and Zaniol, P. (1995). Lipoic (thioctic) acid increases brain energy availability and skeletal muscle performance as shown by in vivo 31P-MRS in a patient with mitochondrial cytopathy. *J. Neurol.* **242**, 472–477.

Barohn, R. J., Clanton, T., Sahenk, Z., and Mendell, J. R. (1990). Recurrent respiratory insufficiency and depressed ventilatory drive complicating mitochondrial myopathies. *Neurology* **40**, 103–106.

Bernsen, P. L., Gabreels, F. J., Ruitenbeek, W., Sengers, R. C., Stadhouders, A. M., and Renier, W. O. (1991). Successful treatment of pure myopathy, associated with complex I deficiency, with riboflavin and carnitine. *Arch. Neurol.* **48**, 334–338.

Bijvoet, A. G., Van Hirtum, H., Kroos, M. A., Van de Kamp, E. H., Schoneveld, O., Visser, P., Brakenhoff, J. P., Weggeman, M., van Corven, E. J., Van der Ploeg, A. T., and Reuser, A. J. (1999). Human acid alpha-glucosidase from rabbit milk has therapeutic effect in mice with glycogen storage disease type II. *Hum. Mol. Genet.* **8**, 2145–2153.

Birch-Machin, M. A., Taylor, R. W., Cochran, B., Ackrell, B. A., and Turnbull, D. M. (2000). Late-onset optic atrophy, ataxia, and myopathy associated with a mutation of a complex II gene. *Ann. Neurol.* **48**, 330–335.

Blain, P. G., and Lane, R. J. M. (1991). Neurological disorders. *In* "Textbook of Adverse Drug Reactions." (D. M. Davies, Ed.), pp. 535–566. Oxford Medical Publications, Oxford.

Bodamer, O. A., Halliday, D., and Leonard, J. V. (2000). The effects of l-alanine supplementation in late-onset glycogen storage disease type II. *Neurology* **55**, 710–712.

Bohlega, S., Tanji, K., Santorelli, F. M., Hirano, M., al-Jishi, A., and DiMauro, S. (1996). Multiple mitochondrial DNA deletions associated with autosomal recessive ophthalmoplegia and severe cardiomyopathy. *Neurology* 46, 1329–1334.

Boitier, E., Degoul, F., Desguerre, I., Charpentier, C., Francois, D., Ponsot, G., Diry, M., Rustin, P., and Marsac, C. (1998). A case of mitochondrial encephalomyopathy associated with a muscle coenzyme Q10 deficiency. *J. Neurol. Sci.* 156, 41–46.

Bonnefont, J. P., Demaugre, F., Prip-Buus, C., Saudubray, J. M., Brivet, M., Abadi, N., and Thuillier, L. (1999). Carnitine palmitoyltransferase deficiencies. *Mol. Genet. Metab.* 68, 424–440.

Bourgeron, T., Rustin, P., Chretien, D., Birch-Machin, M., Bourgeois, M., Viegas-Pequignot, E., Munnich, A., and Rotig, A. (1995). Mutation of a nuclear succinate dehydrogenase gene results in mitochondrial respiratory chain deficiency. *Nat. Genet.* 11, 144–149.

Bresolin, N., Doriguzzi, C., Ponzetto, C. *et al.* (1990). Ubidecarenone in the treatment of mitochondrial myopathies: a multi-center double-blind trial. *J. Neurol. Sci.* 100, 70–78.

Bresolin, N., Zeviani, M., Bonilla, E., Miller, R. H., Leech, R. W., Shanske, S., Nakagawa, M., and DiMauro, S. (1985). Fatal infantile cytochrome c oxidase deficiency: decrease of immunologically detectable enzyme in muscle. *Neurology* 35, 802–812.

Bruyland, M., and Ebinger, G. (1994). Beneficial effect of a treatment with xylitol in a patient with myoadenylate deaminase deficiency. *Clin. Neuropharmacol.* 17, 492–493.

Campistol, J., Chavez, B., Vilaseca, M. A., and Artuch, R. (2000). [Antiepileptic drugs and carnitine]. *Rev. Neurol.* 30 Suppl 1, S105–109.

Campos, Y., Huertas, R., Lorenzo, G., Bautista, J., Gutierrez, E., Aparicio, M., Alesso, L., and Arenas, J. (1993a). Plasma carnitine insufficiency and effectiveness of L-carnitine therapy in patients with mitochondrial myopathy. *Muscle Nerve* 16, 150–153.

Chan, A., Reichmann, H., Kogel, A., Beck, A., and Gold, R. (1998). Metabolic changes in patients with mitochondrial myopathies and effects of coenzyme Q10 therapy. *J. Neurol.* 245, 681–685.

Ciafaloni, E., Ricci, E., Shanske, S., Moraes, C. T., Silvestri, G., Hirano, M., Simonetti, S., Angelini, C., Donati, M. A., Garcia, C., *et al.* (1992). MELAS: clinical features, biochemistry, and molecular genetics. *Ann. Neurol.* 31, 391–398.

Clark, K. M., Bindoff, L. A., Lightowlers, R. N., Andrews, R. M., Griffiths, P. G., Johnson, M. A., Brierley, E. J., and Turnbull, D. M. (1997). Reversal of a mitochondrial DNA defect in human skeletal muscle. *Nat. Genet.* 16, 222–224.

Comi, G., Testa, D., Cornelio, F., Comola, M., and Canal, N. (1985). Potassium depletion myopathy: a clinical and morphological study of six cases. *Muscle Nerve* 8, 17–21.

Daut, P. M., Steinemann, T. L., and Westfall, C. T. (2000). Chronic exposure keratopathy complicating surgical correction of ptosis in patients with chronic progressive external ophthalmoplegia. *Am. J. Ophthalmol.* 130, 519–521.

David, W. S. (2000). Myoglobinuria. *Neurol. Clin.* 18, 215–243.

De Stefano, N., Matthews, P. M., Ford, B., Genge, A., Karpati, G., and Arnold, D. L. (1995). Short-term dichloroacetate treatment improves indices of cerebral metabolism in patients with mitochondrial disorders. *Neurology* 45, 1193–1198.

DiDonato, S. (1994). Disorders of lipid metabolism affecting skeletal muscle: Carnitine deficiency syndromes, defects in the catabolic pathway, and chanarin disease. *In* "Myology" (A. G. Engel and C. Franzini-Armstrong, Eds.), pp. 1587–1609, McGraw-Hill, New York/St. Louis/San Francisco.

DiDonato, S., Gellera, C., Peluchetti, D., Uziel, G., Antonelli, A., Lus, G., and Rimoldi, M. (1989). Normalization of short-chain acylcoenzyme A dehydrogenase after riboflavin treatment in a girl with multiple acylcoenzyme A dehydrogenase-deficient myopathy. *Ann. Neurol.* 25, 479–484.

DiMauro, S., and DiMauro, P. M. (1973). Muscle carnitine palmityltransferase deficiency and myoglobinuria. *Science* 182, 929–931.

DiMauro, S., and Tsujino, S. (1994). Nonlysosomal glycogenoses. *In* "Myology" (A. G. Engel and C. Franzini-Armstrong, Eds.), pp. 1554–1576. McGraw-Hill, New York/St. Louis/San Francisco.

Egger, J., Harding, B. N., Boyd, S. G., Wilson, J., and Erdohazi, M. (1987). Progressive neuronal degeneration of childhood (PNDC) with liver disease. *Clin. Pediatr. (Phila).* 26, 167–173.

Eleff, S., Kennaway, N. G., Buist, N. R., Darley-Usmar, V. M., Capaldi, R. A., Bank, W. J., and Chance, B. (1984). 31P NMR study of improvement in oxidative phosphorylation by vitamins K3 and C in a patient with a defect in electron transport at complex III in skeletal muscle. *Proc. Natl. Acad. Sci. USA* 81, 3529–3533.

Engel, A. G., and Siekert, R. G. (1972). Lipid storage myopathy responsive to prednisone. *Arch. Neurol.* 27, 174–181.

Fishbein, W. N., Armbrustmacher, V. W., and Griffin, J. L. (1978). Myoadenylate deaminase deficiency: a new disease of muscle. *Science* 200:545–548.

Fu, K., Hartlen, R., Johns, T., Genge, A., Karpati, G., and Shoubridge, E. A. (1996). A novel heteroplasmic tRNAleu(CUN) mtDNA point mutation in a sporadic patient with mitochondrial encephalomyopathy segregates rapidly in skeletal muscle and suggests an approach to therapy. *Hum. Mol. Genet.* 5, 1835–1840.

Fukuhara, N., Tokiguchi, S., Shirakawa, K., and Tsubaki, T. (1980). Myoclonus epilepsy associated with ragged-red fibres (mitochondrial abnormalities): disease entity or a syndrome? Light-and electron-microscopic studies of two cases and review of literature. *J. Neurol. Sci.* 47, 117–133.

Goto, Y., Nonaka, I., and Horai, S. (1990). A mutation in the tRNA(Leu)(UUR) gene associated with the MELAS subgroup of mitochondrial encephalomyopathies. *Nature* 348, 651–653.

Greenhaff, P. L., Casey, A., Short, A. H., Harris, R., Soderlund, K., and Hultman, E. (1993). Influence of oral creatine supplementation of muscle torque during repeated bouts of maximal voluntary exercise in man. *Clin. Sci. (Colch).* 84, 565–571.

Griggs, R. C., Mendell, J. R., and Miller, R. G. (1995). Evaluation and Treatment of Myopathies. FA Davis, Philadelphia.

Grossman, L. I., and Shoubridge, E. A. (1996). Mitochondrial genetics and human disease. *Bioessays* 18, 983–991.

Haller, R. G., and Lewis, S. F. (1991). Glucose-induced exertional fatigue in muscle phosphofructokinase deficiency. *N. Engl. J. Med.* 324, 364–369.

Harris, R. C., Soderlund, K., and Hultman, E. (1992). Elevation of creatine in resting and exercised muscle of normal subjects by creatine supplementation. *Clin. Sci. (Colch).* 83, 367–374.

Hirano, M., Silvestri, G., Blake, D. M., Lombes, A., Minetti, C., Bonilla, E., Hays, A. P., Lovelace, R. E., Butler, I., Bertorini, T. E., *et al.* (1994). Mitochondrial neurogastrointestinal encephalomyopathy (MNGIE): clinical, biochemical, and genetic features of an autosomal recessive mitochondrial disorder. *Neurology* 44, 721–727.

Holt, I. J., Harding, A. E., Cooper, J. M., Schapira, A. H., Toscano, A., Clark, J. B., and Morgan, H. J. (1989). Mitochondrial myopathies: clinical and biochemical features of 30 patients with major deletions of muscle mitochondrial DNA. *Ann. Neurol.* 26, 699–708.

Holt, I. J., Harding, A. E., Petty, R. K., and Morgan-Hughes, J. A. (1990). A new mitochondrial disease associated with mitochondrial DNA heteroplasmy. *Am. J. Hum. Genet.* 46, 428–433.

Horber, F. F., Scheidegger, J. R., Grunig, B. E., and Frey, F. J. (1985). Evidence that prednisone-induced myopathy is reversed by physical training. *J. Clin. Endocrinol. Metab.* 61, 83–88.

Hultman, E., Soderlund, K., Timmons, J. A., Cederblad, G., and Greenhaff, P. L. (1996). Muscle creatine loading in men. *J. Appl. Physiol.* 81, 232–237.

Jensen, K. E., Jakobsen, J., Thomsen, C., and Henriksen O. (1990). Improved energy kinetics following high protein diet in McArdle's syndrome. A 31P magnetic resonance spectroscopy study. *Acta. Neurol. Scand.* 81, 499–503.

Jensen, O. B., Mosdal, C., and Reske-Nielsen, E. (1977). Hypokalemic myopathy during treatment with diuretics. *Acta. Neurol. Scand.* 55, 465–482.

Kaufmann, P., el-Schahawi, M., and DiMauro, S. (1997). Carnitine palmitoyltransferase II deficiency: diagnosis by molecular analysis of blood. *Mol. Cell. Biochem.* 174, 237–239.

Kibayashi, M., Nagao, M., and Chiba, S. (1999). Influence of valproic acid on the expression of various acyl-CoA dehydrogenases in rats. *Pediatr. Int.* 41, 52–60.

Kimura, S., Osaka, H., Saitou, K., Ohtuki, N., Kobayashi, T., and Nezu, A. (1995). Improvement of lesions shown on MRI and CT scan by administration of dichloroacetate in patients with Leigh syndrome. *J. Neurol. Sci.* 134, 103–107.

Klopstock, T., Naumann, M., Schalke, B., Bischof, F., Seibel, P., Kottlors, M., Eckert, P., Reiners, K., Toyka, K. V., and Reichmann, H. (1994). Multiple symmetric lipomatosis: abnormalities in complex IV and multiple deletions in mitochondrial DNA. *Neurology* 44, 862–866.

Klopstock, T., Querner, V., Schmidt, F., Gekeler, F., Walter, M., Hartard, M., Henning, M., Gasser, T., Pongratz, D., Straube, A., Dieterich, M., and Muller-Felber, W. (2000). A placebo-controlled crossover trial of creatine in mitochondrial diseases. *Neurology* 55, 1748–1751.

Krahenbuhl, S., Brandner, S., Kleinle, S., Liechti, S., and Straumann, D. (2000). Mitochondrial diseases represent a risk factor for valproate-induced fulminant liver failure. *Liver* 20, 346–348.

Kuroda, Y., Ito, M., Toshima, K., Takeda, E., Naito, E., Hwang, T. J., Hashimoto, T., Miyao, M., Masuda, M., Yamashita, K., *et al.* (1986). Treatment of chronic congenital lactic acidosis by oral administration of dichloroacetate. *J. Inherit. Metab. Dis.* 9, 244–252.

Lane, R. J., and Mastaglia, F. L. (1978). Drug-induced myopathies in man. *Lancet* 2, 562–566.

Lane, R. J., McLean, K. A., Moss, J., and Woodrow, D. F. (1993). Myopathy in HIV infection: the role of zidovudine and the significance of tubuloreticular inclusions. *Neuropathol. Appl. Neurobiol.* 19, 406–413.

Litin, S. C., and Anderson, C. F. (1989). Nicotinic acid-associated myopathy: a report of three cases. *Am. J. Med.* 86, 481–483.

Lou, H. C. (1981). Correction of increased plasma pyruvate and plasma lactate levels using large doses of thiamine in patients with Kearns-Sayre syndrome. *Arch Neurol.* 38, 469.

Luft, R., Ikkos, D., Palmieri, G., Ernster, L., and Afzelius, B. (1962). A case of severe hypermetabolism of nonthyroid origin with a defect in the maintenance of mitochondrial respiratory control: a correlated clinical, biochemical, and morphological study. *J. Clin. Invest.* 41, 1776–1804.

Mastaglia, F. L. (1991). Toxic myopathies. *In* "Handbook of Clinical Neurology." Vol. 18. (L. P. Rowland and S. DiMauro, Eds.), pp. 595–622. Elsevier, Amsterdam.

Matern, D., Starzl, T. E., Arnaout, W., Barnard, J., Bynon, J. S., Dhawan A., Emond, J., Haagsma, E. B., Hug, G., Lachaux, A., Smit, G. P., and Chen, Y. T. (1999). Liver transplantation for glycogen storage disease types I, III, and IV. *Eur. J. Pediatr.* 158 Suppl 2, S43–48.

Matthews, P. M., Ford, B., Dandurand, R. J., Eidelman, D. H., O'Connor, D., Sherwin, A., Karpati, G., Andermann, F., and Arnold, D. L. (1993). Coenzyme Q10 with multiple vitamins is generally ineffective in treatment of mitochondrial disease. *Neurology* 43, 884–890.

McLoughlin, C., Nesbitt, G. A., and Howe, J. P. (1988). Suxamethonium induced myalgia and the effect of pre-operative administra-

tion of oral aspirin. A comparison with a standard treatment and an untreated group. *Anaesthesia* 43, 565–567.

McShane, M. A., Hammans, S. R., Sweeney, M., Holt, I. J., Beattie, T. J., Brett, E. M., and Harding, A. E. (1991). Pearson syndrome and mitochondrial encephalomyopathy in a patient with a deletion of mtDNA. *Am. J. Hum. Genet.* 48, 39–42.

Melegh, B., and Trombitas, K. (1997). Valproate treatment induces lipid globule accumulation with ultrastructural abnormalities of mitochondria in skeletal muscle. *Neuropediatrics* 28, 257–261.

Moraes, C. T., DiMauro, S., Zeviani, M., Lombes, A., Shanske, S., Miranda, A. F., Nakase, H., Bonilla, E., Werneck, L. C., Servidei, S., *et al.* (1989). Mitochondrial DNA deletions in progressive external ophthalmoplegia and Kearns-Sayre syndrome. *N. Engl. J. Med.* 320, 1293–1299.

Morisaki, T., Gross, M., Morisaki, H., Pongratz, D., Zollner, N., and Holmes, E. W. (1992). Molecular basis of AMP deaminase deficiency in skeletal muscle. *Proc. Natl. Acad. Sci. USA* 89, 6457–6461.

Ogasahara, S., Engel, A. G., Frens, D., and Mack, D. (1989). Muscle coenzyme Q deficiency in familial mitochondrial encephalomyopathy. *Proc. Natl. Acad. Sci. USA* 86, 2379–2382.

Padberg, G. W., Wintzen, A. R., Giesberts, M. A., Sterk, P. J., Molenaar, A. J., and Hermans, J. (1989). Effects of a high-protein diet in acid maltase deficiency. *J. Neurol. Sci.* 90, 111–117.

Papadopoulou, L. C., Sue, C. M., Davidson, M. M., Tanji, K., Nishino, I., Sadlock, J. E., Krishna, S., Walker, W., Selby, J., Glerum, D. M., Coster, R. V., Lyon, G., Scalais, E., Lebel, R., Kaplan, P., Shanske, S., De Vivo, D. C., Bonilla, E., Hirano, M., DiMauro, S., and Schon, E. A. (1999). Fatal infantile cardioencephalomyopathy with COX deficiency and mutations in SCO2, a COX assembly gene. *Nat. Genet.* 23, 333–337.

Pari, G., Crerar, M. M., Nalbantoglu, J., Shoubridge, E., Jani, A., Tsujino, S., Shanske, S., DiMauro, S., Howell, J. M., and Karpati, G. (1999). Myophosphorylase gene transfer in McArdle's disease myoblasts in vitro. *Neurology* 53, 1352–1354.

Pavlakis, S. G., Phillips, P. C., DiMauro, S., De Vivo, D. C., and Rowland, L. P. (1984). Mitochondrial myopathy, encephalopathy, lactic acidosis, and strokelike episodes: a distinctive clinical syndrome. *Ann. Neurol.* 16, 481–488.

Penn, A. S. (1994). Myoglobinuria. *In* "Myology" (A. G. Engel and C. Franzini-Armstrong, Eds.), pp. 1679–1696. McGraw-Hill, New York/St. Louis/San Francisco.

Ponchaut, S., and Veitch, K. (1993). Valproate and mitochondria. *Biochem. Pharmacol.* 46, 199–204.

Przyrembel, H. (1987). Therapy of mitochondrial disorders. *J. Inherit. Metab. Dis.* 10 Suppl 1, 129–146.

Reichmann, H. (1993). [Therapy of metabolic myopathies]. *Nervenarzt* 64, 627–632.

Reimers, C. D., Pongratz, D., Paetzke, I., and Zöllner, N. (1987). Therapeutische Beeinflußbarkeit des Myoadenylatdeaminase-Mangels durch D-Ribose. Bericht über 7 Fälle. *Klin Wochenschrift* 65, 75–76.

Rimon, D., Ludatscher, R., and Cohen, L. (1984). Clofibrate-induced muscular syndrome. Case report with ultrastructural findings and review of the literature. *Isr. J. Med. Sci.* 20, 1082–1086.

Riordan-Eva, P., Sanders, M. D., Govan, G. G., Sweeney, M. G., Da Costa, J., and Harding, A, E. (1995). The clinical features of Leber's hereditary optic neuropathy defined by the presence of a pathogenic mitochondrial DNA mutation. *Brain* 118(Pt 2), 319–337.

Rotig, A., Appelkvist, E. L., Geromel, V., Chretien, D., Kadhom, N., Edery, P., Lebideau, M., Dallner, G., Munnich, A., Ernster, L., and Rustin, P. (2000). Quinone-responsive multiple respiratory-chain dysfunction due to widespread coenzyme Q10 deficiency. *Lancet* 356, 391–395.

Rowland, L. P., Hays, A. P., DiMauro, S., DeVivo, D. C., and Behrens, M. (1983). Diverse clinical disorders associated with morphological abnormalities of mitochondria. *In* "Mitochondrial Pathology in Muscle Diseases." (G. Scarlato, Ed.), pp. 142–158. Piccin, Padua.

Sabina, R. L., Morisaki, T., Clarke, P., Eddy, R., Shows, T. B., Morton, C. C., and Holmes, E. W. (1990). Characterization of the human and rat myoadenylate deaminase genes. *J. Biol. Chem.* **265**, 9423–9433.

Saitoh, S., Momoi, M. Y., Yamagata, T., Mori, Y., and Imai, M. (1998). Effects of dichloroacetate in three patients with MELAS. *Neurology* **50**, 531–534.

Saudubray, J. M., Martin, D., de Lonlay, P., Touati, G., Poggi-Travert, F., Bonnet, D., Jouvet, P., Boutron, M., Slama, A., Vianey-Saban, C., Bonnefont, J. P., Rabier, D., Kamoun, P., and Brivet, M. (1999). Recognition and management of fatty acid oxidation defects: a series of 107 patients. *J. Inherit. Metab. Dis.* **22**, 488–502.

Seibel, P., Lauber, J., Klopstock, T., Marsac, C., Kadenbach, B., and Reichmann, H. (1994). Chronic progressive external ophthalmoplegia is associated with a novel mutation in the mitochondrial tRNA(Asn) gene. *Biochem. Biophys. Res. Commun.* **204**, 482–489.

Seibel, P., Trappe, J., Villani, G., Klopstock, T., Papa, S., and Reichmann, H. (1995). Transfection of mitochondria: strategy towards a gene therapy of mitochondrial DNA diseases. *Nucleic Acids Res.* **23**, 10–17.

Shapira, Y., Glick, B., Harel, S., Vattin, J. J., and Gutman, A. (1993). Infantile idiopathic myopathic carnitine deficiency: treatment with L-carnitine. *Pediatr. Neurol.* **9**, 35–38.

Shoffner, J. M., Lott, M. T., Lezza, A. M., Seibel, P., Ballinger, S. W., and Wallace, D. C. (1990). Myoclonic epilepsy and ragged-red fiber disease (MERRF) is associated with a mitochondrial DNA tRNA(Lys) mutation. *Cell* **61**, 931–937.

Shoffner, J. M., Lott, M. T., Voljavec, A. S., Soueidan, S. A., Costigan, D. A., and Wallace, D. C. (1989). Spontaneous Kearns-Sayre/chronic external ophthalmoplegia plus syndrome associated with a mitochondrial DNA deletion: a slip-replication model and metabolic therapy. *Proc. Natl. Acad. Sci. USA* **86**, 7952–7956.

Shoubridge, E. A., Johns, T., and Karpati, G. (1997). Complete restoration of a wild-type mtDNA genotype in regenerating muscle fibres in a patient with a tRNA point mutation and mitochondrial encephalomyopathy. *Hum. Mol. Genet.* **6**, 2239–2242.

Silva, M. F., Ruiter, J. P., Illst, L., Jakobs, C., Duran, M., de Almeida, I. T., and Wanders, R. J. (1997). Valproate inhibits the mitochondrial pyruvate-driven oxidative phosphorylation in vitro. *J. Inherit. Metab. Dis.* **20**, 397–400.

Silvestri, G., Ciafaloni, E., Santorelli, F. M., Shanske, S., Servidei, S., Graf, W. D., Sumi, M., and Di, M. S. (1993). Clinical features associated with the A–G transition at nucleotide 8344 of mtDNA ("MERRF mutation"). *Neurology* **43**, 1200–1206.

Slonim, A. E., Coleman, R. A., McElligot, M. A., Najjar, J., Hirschhorn, K., Labadie, G. U., Mrak, R., Evans, O. B., Shipp, E., and Presson R. (1983). Improvement of muscle function in acid maltase deficiency by high-protein therapy. *Neurology* **33**, 34–38.

Slonim, A. E., and Goans, P. J. (1985). Myopathy in McArdle's syndrome. Improvement with a high-protein diet. *N. Engl. J. Med.* **312**, 355–359.

Sobreira, C., Hirano, M., Shanske, S., Keller, R. K., Haller, R. G., Davidson, E., Santorelli, F. M., Miranda, A. F., Bonilla, E., Mojon, D. S., Barreira, A. A., King, M. P., and DiMauro, S. (1997). Mitochondrial encephalomyopathy with coenzyme Q10 deficiency. *Neurology* **48**, 1238–1243.

Taivassalo, T., De Stefano, N., Argov, Z., Matthews, P. M., Chen, J., Genge, A., Karpati, G., and Arnold, D. L. (1998). Effects of aerobic training in patients with mitochondrial myopathies. *Neurology* **50**, 1055–1060.

Taivassalo, T., Fu, K., Johns, T., Arnold, D., Karpati, G., and Shoubridge, E. A. (1999). Gene shifting: a novel therapy for mitochondrial myopathy. *Hum. Mol. Genet.* **8**, 1047–1052.

Taivassalo, T., Matthews, P. M., De Stefano, N., Sripathi, N., Genge, A., Karpati, G., and Arnold, D. L. (1996). Combined aerobic training and dichloroacetate improve exercise capacity and indices of aerobic metabolism in muscle cytochrome oxidase deficiency. *Neurology* **47**, 529–534.

Takanashi, J., Sugita, K., Tanabe, Y., Maemoto, T., and Niimi, H. (1997). Dichloroacetate treatment in Leigh syndrome caused by mitochondrial DNA mutation. *J. Neurol. Sci.* **145**, 83–86.

Tarnopolsky, M., and Martin, J. (1999). Creatine monohydrate increases strength in patients with neuromuscular disease. *Neurology* **52**, 854–857.

Tarnopolsky, M. A., Roy, B. D., and MacDonald, J. R. (1997). A randomized, controlled trial of creatine monohydrate in patients with mitochondrial cytopathies. *Muscle Nerve* **20**, 1502–1509.

Taylor, R. W., Chinnery, P. F., Turnbull, D. M., and Lightowlers, R. N. (2000). In-vitro genetic modification of mitochondrial function. *Hum. Reprod.* **15 Suppl 2,** 79–85.

Tein, I., DiMauro, S., Xie, Z. W., and De Vivo, D. C. (1993). Valproic acid impairs carnitine uptake in cultured human skin fibroblasts. An in vitro model for the pathogenesis of valproic acid-associated carnitine deficiency. *Pediatr. Res.* **34**, 281–287.

Van Coster, R., Lombres, A., De Vivo, D. C., Chi, T. L., Dodson, W. E., Rothman, S., Orrechio, E. J., Grover, W., Berry, G. T., Schwartz, J. F. *et al.* (1991). Cytochrome c oxidase-associated Leigh syndrome: phenotypic features and pathogenetic speculations. *J. Neurol. Sci.* **104**, 97–111.

Van den Hout, J. M., Reuser, A. J., de Klerk, J. B., Smeitink, W. F., Smeitink, J. A., and Van der Ploeg, A. T. (2001). Enzyme therapy for Pompe disease with recombinant human alpha-glucosidase from rabbit milk. *J. Inherit. Metab. Dis.* **24**, 266–274.

Victor, M., Sieb J. P. (1994). Myopathies due to drugs, toxins, and nutritional deficiency. *In* "Myology" (A. G. Engel and C. Franzini-Armstrong, Eds.), pp. 1697–1725. McGraw-Hill, New York/St. Louis/San Francisco.

Vorgerd, M., Grehl, T., Jager, M., Muller, K., Freitag, G., Patzold, T., Bruns, N., Fabian, K., Tegenthoff, M., Mortier, W., Luttmann, A., Zange, J., and Malin, J. P. (2000). Creatine therapy in myophosphorylase deficiency (McArdle disease): a placebo-controlled crossover trial. *Arch. Neurol.* **57**, 956–963.

Walther, E. U., and Hohlfeld, R. (1999). Multiple sclerosis: side effects of interferon beta therapy and their management. *Neurology* **53**, 1622–1627.

Zeviani, M., Gellera, C., Antozzi, C., Rimoldi, M., Morandi, L., Villani, F., Tiranti, V., and DiDonato, S. (1991). Maternally inherited myopathy and cardiomyopathy: association with mutation in mitochondrial DNA tRNA(Leu)(UUR). *Lancet* **338**, 143–147.

Zeviani, M., Peterson, P., Servidei, S., Bonilla, E., and DiMauro, S. (1987). Benign reversible muscle cytochrome c oxidase deficiency: a second case. *Neurology* **37**, 64–67.

Zhu, Z., Yao, J., Johns, T., Fu, K., De Bie, I., Macmillan, C., Cuthbert, A. P., Newbold, R. F., Wang, J., Chevrette, M., Brown, G. K., Brown, R. M., and Shoubridge, E. A. (1998). SURF1, encoding a factor involved in the biogenesis of cytochrome c oxidase, is mutated in Leigh syndrome. *Nat. Genet.* **20**, 337–343.

Zierz, S. (1994). Carnitine palmitoyltransferase deficiency. *In* "Myology" (A. G. Engel and C. Franzini-Armstrong, Eds.), pp. 1577–1586, McGraw-Hill, New York/St. Louis/San Francisco.

*Muscle and Peripheral Nervous System*

CHAPTER 98

# Myotonias

Kenneth Ricker

## INTRODUCTION

The term "myotonia" is used for a distinct clinical symptom in several genetically caused disorders of intrinsic muscle function. It means the impaired ability of skeletal muscle fibers to relax instantly after a muscle contraction. The resulting problem for the patient is myotonic muscle stiffness restricting the proper control of voluntary movements. On the electromyogram (EMG) myotonia is revealed by a special type of electrical muscle fiber activity, myotonic runs (or myotonia-like runs).

Historically, disorders with the symptom of myotonia have been grouped together as "myotonias" or "myotonic disorders." However, these disorders turned out to have a very different background regarding the genetics and the mechanisms causing myotonia. The same holds true for the clinical phenotype presentation. In some such disorders myotonia may be the leading clinical symptom, and it may be sometimes quite disabling for the patient in rare cases. In others the myotonia is only felt from time to time or hardly at all, or it is obvious only on the electromyogram; other clinical symptoms such as muscle weakness may sometimes be much more troublesome for the patient with "myotonia." None of the myotonias can be cured. Only in a few instances it may be necessary to lessen the occurrence of myotonic muscle stiffness by medication. However, it is helpful for the patient to receive proper advice on how to best get along and to avoid medical (and social) complications.

The myotonias are comprised of two groups of diseases: muscle ion channel disorders and multisystemic myotonic myopathies also named "myotonic dystrophies" (DM, from dystrophia myotonica). The two groups of myotonic diseases are caused by two different types of genetic mutation. Ion channel myotonias are caused by point mutation or deletion within the protein coding exons of a channel gene. At present, no routine molecular genetic diagnosis is available, because the number of possible point mutations to look for exceeds the capacity of the laboratory. This may change in the future with the arrival of DNA microarray technology. On the other hand, both known types of myotonic dystrophy, DM1 and DM2, are caused by dynamic mutations, an expansion of a nucleotide repeat located outside the protein coding region of a particular gene. The presence of such a repeat expansion makes a reliable diagnostic confirmation in the routine molecular genetics laboratory.

## MUSCLE ION CHANNEL MYOTONIAS

### Chloride Channel Myotonia

Historically, chloride channel myotonia happened to be the first myotonic disorder on record. It was described in 1876 by the German physician A. J. Thomsen. He and his family had the disease, which was later named Thomsen's disease or myotonia congenita (MC). The mode of inheritance is autosomal dominant or recessive. In MC myotonic stiffness in the legs is usually discovered at the age of 2 to 6 years; these children stumble and fall quite often and seem to be "lazy." Growing up, some patients may have a rather "athletic" appearance. Myotonia usually exists to a similar degree in a given patient's lifetime. Pregnancy or untreated hypothyroidism may sometimes worsen the myotonia temporarily. Recessive MC is rather similar in appearance, although some degree of progression during adolescence can be observed (Becker, 1977). In severe recessive cases myotonic stiffness is intermingled with transient muscle weakness ("transient" meaning that initial muscle contractions after rest are weak; with further contractions strength improves). Any skeletal muscle may be involved, but no other organ system is affected. Diagnosis is made by (family-) history, physical examination, and EMG.

Skeletal muscle fiber membrane has a rather high chloride conductance, which is needed to restore the resting membrane potential (RMP) immediately after an action potential. The chloride conductance is reduced in MC because of dysfunctional chloride channels. Because of that, the restoration of the RMP is delayed. The voltage-gated sodium channels respond with abnormal functioning to the abnormal restoration of the RMP (repetitive discharges = myotonia, or unresponsiveness = transient weakness). Substances with a "normalizing" effect on voltage-gated sodium channels (some used as cardiac antiarrhythmic drugs) have been demonstrated to reduce or abolish the repetitive myotonic discharges in vitro. Medication that directly influences the primary dysfunction of muscle chloride channels in MC is not available (De Luca et al., 1997; Lehmann-Horn and Jurkat-Rott, 1999; Zhang et al., 2000).

### Practical Management

There is no doubt that antimyotonic medication can reduce myotonic stiffness to some extent in a given patient. Quinine (200–1000 mg/day), procainamide (125–500 mg/day), phenytoin (300–400 mg/day), or mexiletine (150–600 mg/day) have been recommended. Mexiletine is more effective but should not be given to children or to older patients at risk of cardiac conduction block. Tocainide is similarly effective as mexiletine but has been abandoned because of side effects (Griggs et al., 1995).

Unfortunately, MC is a lifelong condition. There is no observation on record that MC patients really have had any substantial benefit over their lifetime by constantly taking antimyotonic medication. Most MC patients do not want medication anyway, or they try and stop it after some time. In many cases the myotonic stiffness is only mild to moderate, and patients learn how to handle it by slightly moving their muscles and keeping them in a state of "warm-up." Sometimes teen-aged MC patients involved in athletics, dancing lessons, and courtship may ask for antimyotonic medication, being ambitious or anxious as to the social consequences of their muscle stiffness. Such a request should be carefully considered. Phenytoin is less effective than mexiletine; it may not really be of help but has fewer side effects. Mexilitine has a dangerously small safety margin. The young patient should be expected to show reliable compliance. The author lost a 16-year-old girl with recessive MC because of unmanageable lethal intoxication. The girl, being very ambitious in athletics, had constantly increased the mexiletine dosage on her own. In severe recessive MC transient weakness is only improved to some extent by a high dosage of medication. In my experience, almost all MC patients give up medication sooner or later, because they feel that it does not really pay off. After 10 years, only two of my patients were still taking antimyotonic medication (mexiletine). With constant medication of mexiletine, serum concentration of the drug should be checked from time to time. After months or years, withdrawal of mexiletine should be done gradually, reducing the dose over 10–14 days. Otherwise, an unpleasant and painful rebound of myotonic stiffness will occur.

In school it is important that teachers and friends know of the nature of myotonic stiffness in a child. MC patients have some occupational restriction. In Germany they do not get a license to drive a bus or a truck. They will also not get accepted for military service. In case of a necessary surgical procedure with general anesthesia, depolarizing muscle relaxants (succinylcholine type) should be avoided to prevent a sudden increase of myotonic stiffness (not to be mistaken for malignant hyperthermia). Some pregnant women experience deterioration of myotonia, others do not. In case of premature labor, the infusion of fenoterol should be avoided, because this drug triggers severe generalized myotonic stiffness in women with MC. Delivery has never been complicated because of myotonia in MC.

### Sodium Channel Myotonia

Sodium channel myotonia is autosomal dominantly inherited. For reasons not yet entirely understood, three different clinical phenotypes exist, which may or may not show some symptomatic overlap: paramyotonia congenita with mainly cold (and exercise)-induced myotonia and weakness; myotonia fluctuans (potassium-aggravated myotonia, PAM) with mainly exercise-induced myotonia; hyperkalemic period paralysis. For historical and practical reasons the latter phenotype is being dealt with in Chapter 99, "Dyskalemic Periodic Paralysis." In all three of these phenotypes, patients are very sensitive to ("diagnostic") potassium loading, which may result in severe generalized myotonic stiffness and/or generalized weakness.

In the past paramyotonia congenita often was detected in the newborn by the knowledgeable parent, when the baby's face became masklike and stiff after being washed with cold water. In paramyotonia any skeletal muscle may develop cold-induced myotonic stiffness, but patients most often mention problems with their fingers and face. Prolonged exercise in the cold may lead to local muscle weakness, which may take hours to restore. In addition, some patients may have attacks of hyperkalemic periodic paralysis. Paramyotonia is a condition that does not change much over a lifetime. No other organ system is affected. The diagnosis depends on the (family-) history, physical examination, and the EMG. After local cooling and exercise of a hand muscle, the EMG reveals abundant "fibrillation-like" spontaneous activity. Creatine kinase may be elevated in serum. In myotonia fluctuans myotonic stiffness is induced by prolonged and heavy muscle work; the feature of cold-induced stiffness and weakness is missing; there are no attacks of hyperkalemic periodic paralysis in these families. These patients sometimes report attacks of myotonically impaired eye movement with short-lasting double vision.

In these disorders sodium conductance of the muscle fiber membrane is increased because of dysfunction of the voltage-gated muscle sodium channel. The delayed closing of the channel after an action potential increases the influx of sodium ions, which in turn decreases the resting membrane potential (depolarization), resulting at first in repetitive discharges (myotonia) and then in unresponsiveness (weakness, paralysis). The (poisonous) sodium channel–blocking substance tetrodotoxin completely prevents the membrane depolarization of a patient's muscle fibers in vitro. The cardiac antiarrhythmic drugs tocainide and mexiletine have been shown in vitro to directly improve the paramyotonic dysfunction of the sodium channel, and these substances also work quite effectively in the patient (Cannon, 2000; Lehmann-Horn and Jurkat-Rott, 1999; Weckbecker *et al.*, 2000).

### Practical Management

Patients with paramyotonia usually know quite well that they have to avoid cold environments and/or excessive muscle effort to avoid becoming stiff (and weak). There is an obvious job restriction for some patients (in Germany, for instance, one cannot obtain a licence for bus or truck driving nor be accepted for military service). Still, lifelong constant medication with mexiletine is not recommended (cost, possible long-term side effects) and is usually not asked for. However, medication may be helpful to bridge short-lasting exposure to cold in special situations; for instance, playing a musical instrument in winter time in the open air or in church; outdoor recreational activity during winter time like skiing or sailing. Starting daily medication of mexiletine (360 mg in one daily dose after a meal; this slow-release type only rarely produces mild side effects at the beginning like nausea or dizziness) about 2–3 days before the day of the event or the beginning of vacation and continuing this medication for days or a few weeks has been proven (in my experience) to be helpful in a number of younger or middle-aged patients.

In case of general anesthesia and surgery, several precautions may be recommended. Depolarizing muscle relaxants (succinylcholine type) should not be used; otherwise resulting myotonic stiffness of jaw muscles may prevent intubation. The patient's body should be carefully kept warm, and infusion fluids containing substantial amounts of potassium should be avoided to prevent stiffness and/or weakness (which of course will resolve again usually within some hours; but still causing discomfort to the patient and possible irritation in routine caretaking). Contrary to widespread belief, there is no case of sodium channel disorder on record with an incidence of true malignant hyperthermia. Local or spinal anesthesia usually cause no special risk in these patients. The same holds true for pregnancy and delivery.

In rare instances (I have only three such patients on record) myotonia fluctuans may be unusually severe, starting in the newborn baby and showing almost constantly fluctuating generalized stiffness ("myotonia permanens"). Attacks of stiffness may even temporarily impair breathing mechanics and may therefore become life-threatening. These are the only cases of sodium channel myotonia constantly needing antimyotonic medication, probably lifelong. A combined medication of acetazolamide, carbamazepine, and phenytoin has been used by pediatricians, with the addition of mexiletine later on. With constant medication of mexiletine, serum concentration of the drug should be checked from time to time. One of the three patients with "myotonia permanens" under constant mexiletine medication went successfully through pregnancy and delivery of a healthy baby. Because of dangerously increasing myotonic stiffness, it had not been possible to withdraw the medication during this time.

## MULTISYSTEMIC MYOTONIC MYOPATHIES (MYOTONIC DYSTROPHIES)

### Myotonic Dystrophy Type I, Steinert Disease (DM1)

These disorders have been known for 100 years. The type number has been added only recently (IDMC, 2000). It is dominantly inherited, displaying in addition marked anticipation (usually more severe and occurring at a younger age in the younger generation). It is said to be one of the most frequent "muscle disorders" in adults, with an approximate prevalence of 5/100,000. Core features are myotonia, progressive muscle weakness and wasting, and cataract. In the individual patient there may (or may not) be a surprising list of other features like hypogonadism, cardiac involvement, and mental disability. Symptoms of the disease often appear at the age of 20–40 years, although virtually any age is possible. In the adult, muscle weakness usually manifests itself distally in the hands and/or feet and in the facial muscles. Because of myotonic stiffness and weakness, manual skills are impaired. Because of ptosis, facial weakness, and "nasal" speech, the appearance of the patient may change early on. The disease has been extensively described (Harper, 2001; Mathieu *et al.*, 1999; Moxley, 1992; Thornton, 1999).

DM1 is caused by the expansion of a trinucleotide repeat (CTG) on chromosome 19q. This unstable expansion is located outside the protein coding region of what has been named the DMPK gene. However, it seems likely that dysfunction of this gene and two other neighboring genes plays only a minor role in the pathogenesis of DM1. Apparently, the RNA transcript of the repeat expansion causes a widespread, and as yet not well understood, impairment in the expression of very many genes (Mankodi *et al.*, 2000). The size of the expansion correlates roughly with the severity of the phenotype (Marchini *et al.*, 2000). Normally 5–30 CTG repeats are found. In DM1 the number of repeats may be increased from 50 to more than 1000. Many laboratories routinely perform this sensitive diagnostic test using a patient's blood sample (IDMC, 2000).

### Practical Management

Patients with DM1 cannot be cured, and there are only a few measures to relieve some of the symptoms. In principle, myotonic muscle stiffness is treatable by medication. However, these patients quite often do not even realize their grip myotonia, although it may be obvious to the examiner. I do not recommend giving "antimyotonic" drugs (see earlier) to patients with DM1 to avoid any additional risk of possibly dangerous cardiac complications. In any case, cardiac arrhythmia occurring suddenly and afflicting the otherwise only mildly affected patient is a major threat in DM1 (Antonini *et al.*, 2000). An electrocardiogram should be used to test for partial conduction block. Cardiac pacemaker insertion may be needed in an otherwise only mildly affected patient. Cataracts may be removed surgically. In a few male patients, primary or secondary hypogonadism may be detected and partially corrected by hormonal treatment, which may relieve sexual impotence. About one third of patients with DM1 complain about excessive daytime sleepiness. The wake-promoting drug modafinil has been shown to be useful for some of these patients (Damian *et al.*, 2001).

In general anesthesia, depolarizing muscle-relaxing medication (succinylcholine type) should be avoided to prevent sudden myotonic stiffness of jaw muscles. Cardiac monitoring is mandatory because of the increased risk of sudden arrhythmia (Mathieu *et al.*, 1997).

Patients with advanced muscle atrophy sometimes need prolonged respiratory artificial ventilation after surgery because of retarded recovery of their respiratory muscle function. Short-term corticosteroid medication may be tried in such an unfortunate situation.

Pregnancy and delivery may be rather uneventful if the fetus does not carry the mutation. The risk is, of course, 50%. On the other hand, there may be many miscarriages and complications during pregnancy like hydramnios. A newborn carrying the mutation may be normal as a baby but will develop the diseases later on in life. However, a smaller number will show the dreaded congenital type of DM2, appearing floppy with facial weakness, and sometimes with cardiac and brain malformation. This particularly occurs when the mother is the carrier of the mutation, but sometimes it also occurs with the father as the carrier. Genetic counseling therefore is highly recommendable. Unfortunately, many couples turn up with a pregnancy already under way are seen. Intrauterine checking for the DM1 mutation, and, if present, abortion of the fetus, has been practiced. In vitro fertilization and selection of a product without the mutation for reimplantation is technically difficult but has been done. (In Germany the latter procedure is unlawful.)

### Myotonic Dystrophy Type 2 (DM2, PROMM Syndrome)

Until 1992 myotonic dystrophy was believed to be one single clinical and genetic entity (Harper, 2001). With the arrival of the diagnostic CTG repeat test, it became clear that a second type of myotonic dystrophy–like disorder exists, at first named atypical myotonic dystrophy and/or PROMM (proximal myotonic myopathy). This disease was mapped genetically to a locus named "DM2" located on chromosome 3q (Ranum *et al.*, 1998; Ricker *et al.*, 1999). The disease-causing mutation (Liquori *et al.*, 2001) is a huge tetranucleotide repeat expansion, CCTG. The disorder in patients carrying this DM2 mutation is now called myotonic dystrophy type 2. As of March 2002, this mutation had been found in Germany in 132 families. It is still open to question whether a third type of myotonic dystrophy exists.

Symptoms of DM2 may first appear roughly between 20 and 50 years of age (Ricker, 2000). In younger patients occasional mild myotonic stiffness of fingers after firm grip or of a leg muscle while climbing stairs may occur. Episodes of peculiar muscle and joint pain lasting days or weeks are reported by many patients. Middle-aged or older patients may experience weakness of their proximal leg muscle while trying to get up from a kneeling position or climbing stairs. Cataract may (or may not) develop at almost any age in the adult patient. Usually, the course of the disease seems to be more benign compared with DM1. Anterior neck muscles may be weak. However, facial looks, speech, manual skills, and mental abilities usually are not impaired to a recognizable degree. Congenitally affected newborns with DM2 have so far not been detected. On the other hand, cardiac involvement with unpredictable sudden death in middle-aged patients and slowly developing cardiomyopathy and progression of weakness in older patients (older than 75 years) causing inability to walk within a few years have been observed.

### Practical Management

In most patients clinical myotonia is minimal or absent. In a few patients even electrical myotonia in the EMG is difficult to detect or apparently not present. Therefore the question regarding antimyotonic treatment does not arise. The problem is different with episodic or even constant pain. A broad spectrum of medication has been tried in one or the other patient. Unfortunately, no recommended scheme for pain treatment in DM2 could be outlined so far. Some measures like surgery for cataract or monitoring cardiac function to rule out conduction block are the same as described for DM1 (see earlier). Pitfalls and measurements to be avoided include considering vertebral disk or hip surgery because of episodes of pain, which in reality are caused by the DM2 disease; considering muscle biopsy or cortisone medication, because muscle pain, weakness, and elevated creatine kinase in serum are suggestive of polymyositis; accusing a patient of alcohol abuse or considering invasive liver diagnostics because of elevated liver enzymes in serum (in particular γ-GT) caused by DM2 (also by DM1) quite often even in young patients.

In regard to information and genetic counseling, the situation has much improved with the discovery of the DM2 mutation. Details regarding the disease mechanisms and the genotype-phenotype correlation will be described in the future.

## REFERENCES

Antonini, G., Giubilei, F., Mammarella, A., Amicucci, P., Fiorelli, M., Gragnani, F., Morino, S., Geschin, P. V., and Gennarelli, M. (2000). Natural history of cardiac involvement in myotonic dystrophy: correlation with CTG repeats. *Neurology* 24, 1207–1209.

Becker, P. E. (1977). "Myotonia and Syndromes Associated with Myotonia." G. Thieme, Stuttgart.

Cannon, S. C. (2000). Spectrum of sodium channel disturbances in the nondystrophic myotonias and periodic paralyses. *Kidney Int.* 57, 772–779.

Damian, M. S., Gerlach, A., Schmidt, F., Lehmann, E., and Reichmann, H. (2001). Modafinil for eyxcessive datime sleepiness in myotonic dystrophy. *Neurology* 56, 794–796.

De Luca, A., Pierno, S., Natuzzi, F., Franchini, C., Duranti, A., Lentini, G., Tortorella, V., Jockusch, H., and Camerino, D. C. (1997). Evaluation of the antimyotonic activity of mexiletine and some new analogs on sodium currents of single muscle fibers and on the abnormal excitability of the myotonic ADR mouse. *J. Pharmacol. Exp. Ther.* 282, 93–100.

Griggs, R. C., Mendell, J. R., and Miller, R. G. (1995). "Evaluation and Treatment of Myopathies." F. A. Davis, Philadelphia.

Harper, P. (2001). "Myotonic Dystrophy." W. B. Saunders, London.

IDMC, The International Myotonic Dystrophy Consortium. (2000). New nomenclature and DNA testing guidelines for myotonic dystrophy type 1 (DM1). *Neurology* 54, 1218–1221.

Lehmann-Horn, F., and Jurkat-Rot, K. (1999). Voltage-gated ion channels and hereditary disease. *Physiol. Rev.* 79, 1317–1372.

Liquori, C. L., Ricker, K., Moseley, M. L., Jacobsen, J. F., Kress, W., Naylor, S. L., Day, J. W., and Ranum, L. P. W. (2001). Myotonic dystrophy type 2 caused by a CCTG expansion in intron 1 of ZNF9. *Science* 293, 864–867.

Mankodi, A., Logigian, E., Callahan, L., McClain, C., White, R., Henderson, D., Krym, M., and Thornton, C. A. (2000). Myotonic dystrophy in transgenic mice expressing an expanded CUG repeat. *Science* 289, 1769–1772.

Marchini, C., Lonigro, R., Verriello, L., Pellizzari, L., Bergonzi, P., and Damante, G. (2000). Correlations between individual clinical manifestations and CTG repeat amplification in myotonic dystrophy. *Clin. Genet.* 57, 74–82.

Mathieu, J., Allard, P., Gobeil, G., Girard, MK., De Braekeleer, M., and Bégin, P. (1997). Anesthetic and surgical complications in 219 cases of myotonic dystrophy. *Neurology* 49, 1646–1659.

Mathieu, J., Allard, P., Potvin, L., Prevost, C., and Bégin, P. (1999). A 10-year study of mortality in a cohort of patients with myotonic dystrophy. *Neurology* 52, 1658–1662.

Moxley III, R. T. (1992). Myotonic muscular dystrophy. *In* "Handbook of Clinical Neurology" (L. P. Rowland and S. DiMauro, Eds.), Vol.18, pp. 209–259. Elsevier, Amsterdam.

Ranum, L. P. W., Rasmussen, P. F., Benzow, K. A., Koop, M. D., and Day, J. W. (1998). Genetic mapping of a second myotonic dystrophy locus (DM2). *Nature Genet.* 19, 196–198.

Ricker, K. (2000). The expanding clinical and genetic spectrum of the myotonic dystrophies. *Acta Neurol. Belg.* 100, 151–155.

Ricker, K., Grimm, T., Koch, M. C., Schneider, C., Kress, W., Reimers, C. D., Schulte-Mattler, W., Mueller-Myhsok, B., Toyka, K. V., and Mueller, C. R. (1999). Linkage of proximal myotonic myopathy to chromosome 3q. *Neurology* 52, 170–171.

Thornton, C. (1999). The myotonic dystrophies. *Semin. Neurol.* 19, 25–33.

Weckbecker, K., Würz, A., Mohammadi, B., Mansuroglu, T., George Jr., A. L., Lerche, H., Dengler, R., Lehmann-Horn, F., and Mitrovic, N. (2000). Different effects of mexiletine on two mutant sodium channels causing paramyotonia congenita and hyperkalemic periodic paralysis. *Neuromusc. Disord.* 10, 31–39.

Zhang, J., Bendahhou, S., Sanguinetti, M. C., and Ptácek, L. J. (2000). Functional consequences of chloride channel gene (CLCN1) mutations causing myotonia congenita. *Neurology* 54, 937–942.

*Muscle and Peripheral Nervous System*

CHAPTER 99

# Dyskalemic Periodic Paralyses

Frank Lehmann-Horn and Karin Jurkat-Rott

## CLINICAL ASPECTS

The clinical presentation of dyskalemic periodic paralyses (PP) is characterized by episodes of flaccid paralysis of variable duration, severity, and frequency that develop over a period ranging from a few minutes to several hours and occur without disturbing the patient's consciousness. During attacks the stretch reflexes of the paralyzed muscles are diminished or unobtainable. Paralysis usually begins in the proximal leg muscles and spreads distally and partly asymmetrically to the arms, with those muscles closest to the trunk affected most severely. Swallowing, facial, and respiratory muscles are rarely affected, however. The external ocular muscles remain unaffected, and the sphincter functions are preserved.

During an attack, an increasing number of muscle fibers become electrically (directly and indirectly) unexcitable. The EMG findings range from a reduction of amplitude and abnormally small motor unit potentials through progressive rarefaction in the activity pattern, to the complete disappearance of electromyographic activity.

## PRIMARY PERIODIC PARALYSES

Primary, familial periodic paralyses (FPP) are myogenic and to be distinguished from secondary muscle impairment (i.e., forms of different origin). Electrophysiological in vitro and genetic linkage studies made the identification of the responsible genes possible (Fontaine *et al.*, 1990, 1994; Jurkat-Rott *et al.*, 2000; Lehmann-Horn *et al.*, 1987). In familial hypokalemic periodic paralysis (HypoPP), sensory symptoms do not occur, and there is no myotonia. Although loss-of-function mutations in four different ion channels (calcium, sodium, and two potassium channels) can

cause the disease, the clinical picture is the same. In a proportion of patients with hyperkalemic familial periodic paralysis (HyperPP), the attacks often begin with paresthesia, myalgia, or mild myotonic symptoms. Nevertheless, frank muscle stiffness is atypical. In the rare normokalemic periodic paralysis (NormoPP), serum potassium levels remain unchanged during the attack. However, paralyses can normally be induced by potassium, which is why NormoPP is generally regarded as a variation of the hyperkalemic form. HyperPP is clinically overlapping with two further diseases, paramyotonia congenita (PC) and potassium-aggravated myotonia (PAM). PC patients often have spontaneous episodes of weakness, which may go along with an elevated serum potassium level, and PAM patients are also sensitive to potassium. As a clear distinction, HyperPP patients never show substantial stiffness when cooled, and muscle weakness never occurs in PAM. Although HyperPP, PC, and PAM are caused by gain-of-function mutations in the same channel, the muscle sodium channel alpha subunit, and clinically intermediate forms are frequent, it seems reasonable to maintain the classification of separate nosological entities, because, in the pure forms, not only the symptoms but also the recommended treatments differ (for review see Lehmann-Horn and Jurkat-Rott, 1999). Another primary form is the rare X-chromosomal episodic hyperkalemic paralysis (see Table I). In addition to the attacks, a persistent muscle weakness can appear in all forms of FPP (Bradley *et al.*, 1990). This is characterized histologically by central vacuolation of the muscle fibers and ultrastructurally by a proliferation and dilation of the sarcoplasmic reticulum (Links *et al.*, 1990).

In addition to the family history and the exclusion of other disorders of potassium metabolism, the diagnosis of FPP requires confirmatory investigations during an attack (including clinical examination, serial serum potassium, EMG, ECG) and tests to provoke an attack

TABLE I   Clinical Features and Natural Course of Familial Periodic Paralysis

| | Hypokalemic FPP | Hyperkalemic FPP |
|---|---|---|
| Genetics and gene products | Autosomal dominant<br>Type 1: L-type calcium channel alpha1S (Jurkat-Rott et al., 1994; Ptacek et al., 1994)<br>Type 2: sodium channel alpha (Jurkat-Rott et al., 2000)<br>Type 3: potassium channel beta (Abbott et al., 2001)<br>Type 4 (Andersen's syndrome): potassium inward rectifier channel (Plaster et al., 2001) | Autosomal dominant (or rarely X-chromosomal)<br>Type 1: Locus on chromosome 17q23<br>Gene product: Alpha subunit of the sodium channel of sarcolemma (Rojas et al., 1991)<br>Type 2: Locus on X.p22.3 (Ryan et al., 1999)<br>Gene product: unknown |
| Penetrance | Male: complete<br>Female: reduced | Complete in both sexes |
| Age at onset | First or second decade | First decade, occasionally infancy |
| Paralytic attacks | | |
|   Frequency | Low<br>Once in lifetime to daily attacks<br>Marked inter- and intrafamilial variability<br>Tendency toward fewer attacks after age 30 | High<br>Repeated daily to once a year<br>Marked inter- and intrafamilial variability<br>Tendency toward fewer attacks after age 30 |
|   Duration | Hours to 3 days | Minutes to hours |
|   Severity | Usually severe generalized paresis or plegia, more pronounced in males | Usually mild paresis, frequently isolated muscle groups after preceding physical exercise |
|   Time of day | Early morning | Morning, daytime |
| Ictal serum potassium | Declines, not always below the normal range | Usually significantly elevated, occasionally up to 8 mmol/L |
| Provoking factors | Carbohydrate and sodium intake; rest after strenuous physical exercise; exposure to cold; emotional stress; beta-sympathicomimetics | Potassium load; fasting state; rest after strenuous physical exercise; exposure to cold; emotional stress; glucocorticoids; pregnancy |
| Provocative testing | Oral glucose 1.5–2 g/kg body weight plus 10 to 20 IU of crystalline insulin subcutaneously; attack probably occurs within 1–3 h (for detail, see Lehmann-Horn et al., 1994) | Oral potassium chloride 2–10 g (40–120 mmol) |
| Life expectancy | Not impaired today | Not impaired (for exceptions see text) |

(Table I). Between the paralytic attacks, the serum potassium level is normal, in contrast to that in the acquired forms of PP.

## SECONDARY PERIODIC PARALYSES

### Thyrotoxic Periodic Paralysis

Of the patients with thyrotoxicosis, 2%–6% have episodic paralyses. Thyrotoxic PP corresponds clinically to hypokalemic FPP, except that (1) 95% of the cases are sporadic, (2) 75% of the cases occur in Orientals, (3) the first attacks appear later in life (20th to 39th year). Other symptoms of hyperthyroidism are generally present, and thyrotoxic PP may be associated with hyperthyroid myopathy and other autoimmune disorders including myasthenia gravis.

### Acquired Hypokalemic Periodic Paralysis

Renal, gastrointestinal, endocrine, drug or toxicity-related potassium deficiency can lead to acute (with rhabdomyolysis) and chronic myopathies, as well as to secondary periodic paralysis (serum potassium <3 mmol/L). The underlying diseases or conditions are chronic diarrhea, prolonged vomiting, villous rectum adenoma, primary hyperaldosteronism (Conn's syndrome), potassium-wasting kidney, recovery phase of diabetic coma, recovery phase of acute tubular necrosis, bilateral ureterocolostomy, alcoholism, substances causing a mineralocorticoid effect (liquorice, glycyrrhizic acid), secondary RTA (amphotericin B, toluene abuse), barium and ammonium chloride intoxication. The attacks of the acquired forms can be distinguished from those of familial HypoPP by the accompanying symptoms of the respective primary disease and by the serum and urine electrolyte disturbances that are present between attacks.

## Familial Potassium Deficiency Syndromes

Various dominant and recessive syndromes lead to acute and chronic potassium deficiency myopathies characterized by adynamia and muscle fatiguability; typical hypokalemic PP is rare. To this group belong the various Bartter's syndromes, Fanconi's syndrome, Liddle syndrome, familial hypokalemia Gullner, and distal, renal tubular acidosis (RTA). The responsible genes and gene products have been already identified. Treatment of RTA consists of the administration of potassium and bicarbonate. *Bicarbonate is contraindicated* in hypokalemic FPP, because it promotes the influx of potassium into the cells. AAA (acetazolamide), on the other hand, is contraindicated in RTA, because the illness is accompanied by acidosis. Hypokalemic FPP and PP in distal RTA cannot be differentiated with certainty based on the patient's history and the clinical aspects. The demonstration of hypokalemia in combination with hyperchloremic acidosis and a urine pH value of 6.0 or more, however, ensures the diagnosis of distal RTA (Christensen, 1985).

## Secondary Hyperkalemic "Periodic Paralysis"

This type of adynamia can occur at serum potassium levels of 7 mmol/L and upward and also is accompanied by paresthesia. The differential diagnosis includes all acquired and familial diseases in which potassium retention is associated: chronic renal insufficiency and chronic heparin therapy, Gordon's syndrome (Pasman *et al.*, 1989), Addison's disease, hyporeninemic hypoaldosteronism, corticosterone methyl oxidase (CMO) deficiency, and pseudohypoaldosteronisms I and II. Myopathies associated with paroxysmal myoglobinuria (e.g., McArdle's syndrome, carnitine-palmityltransferase deficiency) also can damage the kidney and lead to potassium retention.

### Natural Course

Overall, periodic dyskalemic paralyses are rare. FPP show a prevalence of about 1/80,000 and are transmitted as autosomal-dominant traits. Patients experience generalized attacks of flaccid weakness that usually begin in the first or second decade of life. The attacks usually occur after short (HyperPP) or long rest (HypoPP), after strenuous physical work, and/or after a carbohydrate-rich meal (HypoPP). Sustained mild exercise may postpone or prevent the attack of weakness (working off the attack). Cold environment, emotional stress, and pregnancy also provoke or worsen the attacks. Symptoms commonly last 1 (HyperPP) to several hours (HypoPP) and are usually accompanied by significant drop (HypoPP) or increase (HyperPP) in serum potassium. As soon as potassium level normalizes, muscle strength recovers to preictal values. A progressive muscle weakness may develop, independently of the number of attacks, starting in most cases in the 40s, an age at which attacks usually become rarer or disappear. A third of HypoPP patients are stricken with this persistent weakness, whereas only HyperPP patients carrying a specific sodium channel mutation (T704M) are disposed to.

The weakness usually develops over years, particularly in the region of the pelvic girdle. Although the weakness progresses very slowly, afflicted patients may become wheelchair-bound. Cardiac complications occur with the very rare *Andersen's syndrome*, which is caused by mutations in a potassium channel expressed in both skeletal and cardiac muscle or with the other FPP, as far as the electrolyte shifts are large enough to impair cardiac excitability and therefore the ECG.

The course of thyrotoxic PP is not correlated with the duration or the intensity of hyperthyroidism. The attacks stop, however, when normal thyroid function is restored. Men are 20-fold more frequently affected by this form than women (Riggs, 1989). The risk of cardiac arrhythmia is higher than that in FPP.

The course and the prognosis of secondary PP is determined by the type or treatability of the primary disorder.

### Principles of Therapy

Although the paralysis of secondary hyperkalemic PP is probably nerve related (inadequately investigated), paralytic attacks in the other forms of PP result from a prolonged depolarization block of the muscle fibers (Lehmann-Horn *et al.*, 1987; Rüdel *et al.*, 1984). In contrast, normal muscle hyperpolarizes and does not become paretic in acute hypokalemia. The fact that episodic paralyses are associated with a depolarization block in chronic hypokalemia and in hypokalemic FPP clearly indicates increased membrane permeability to sodium compared with potassium (Layzer, 1984). This, however, does not explain why potassium flows against the concentration gradient in the muscle at the beginning of hypokalemic paralysis. The cause of this may be related to the relative overactivity of the membrane $Na^+/K^+$ pump. Insulin or beta-sympathomimetics, which can trigger hypokalemic PP, reduce potassium conductance and stimulate the $Na^+/K^+$ pump.

**TABLE II**   Principles of Drug Therapy in Dyskalemic Periodic Paralysis

| | Prophylaxis | | Acute attack | |
|---|---|---|---|---|
| | Effective agents | Mechanism | Effective agents | Mechanism |
| Hypokalemic familial periodic paralysis | Carbonic anhydrase inhibitors (CAI) Acetazolamide (AAA), Dichlorophenamide<br><br>The preventive effect is proven by controlled studies (Griggs *et al.*, 1970; Tawil *et al.*, 2000)<br><br>Few patients are refractory to AAA<br><br>In two families AAA induced muscle weakness (Torres *et al.*, 1981; Vern *et al.*, 1987)<br><br>AAA also improves persistent weakness<br><br>AAA-refractory persistent weakness may respond sufficiently to dichlorophenamide (Dalakas and Engel, 1983). | Most likely by causing metabolic acidosis that leads to a decrease of an intracellular potassium shift; decrease of insulin secretion by CAI (Riggs *et al.*, 1984). | Potassium chloride (KCl). Before starting potassium substitution, hypokalemia should be confirmed.<br>Oral administration is preferable.<br>Caution with patients using potassium-saving diuretics, diabetics, and patients with restricted renal function.<br>Glucose-containing preparations should be avoided (Riggs, 1989).<br>Intravenous potassium is restricted to cases in which oral administration is not possible.<br>Saline or glucose diluents may cause an initial worsening of hypokalemia; therefore KCl in 5% mannitol is preferable for intravenous potassium administration (Griggs *et al.*, 1983). | High ingestion of potassium leads to displacement of sodium from the intracellular compartment. |
| | Diazoxide | Insulin secretion inhibitor; opens ATP-sensitive potassium channels (Henquin *et al.*, 1982) | | |
| | Potassium-saving diuretics Triamterene, Spironolactone<br>Spironolactone is effective in preventing attacks, on long-term medication, serious side effects are to be expected.<br>No supplemental potassium chloride taken during therapy with potassium-saving diuretics. | Renal potassium retention; inducing mild acidosis | | |
| | Lithium Worth trying in resistant to CAI and potassium-saving diuretics (Confavreux *et al.*, 1991) | Enhancement of membrane Na–K pump activity continues. | | |

*continues*

TABLE II    *Continued*

| | Prophylaxis | | Acute attack | |
| | Effective agents | Mechanism | Effective agents | Mechanism |
|---|---|---|---|---|
| Thyrotoxic periodic paralysis | Best proven are propranolol and spironolactone. Acetazolamide is not effective and may induce paralytic attacks. | The intracellular potassium uptake is decreased by beta-blocking drugs. | Potassium chloride (KCl). | See hypokalemic familial periodic paralysis. |
| Hyperkalemic familial periodic paralysis | Carbonic anhydrase inhibitors (CAI) Acetazolamide (AAA) Just a few nonresponders. | Kaliuretic effect | Prompt intake of carbohydrate-rich meals or beverages will often abort or attenuate attacks. | Quick lowering of serum potassium level. |
| | Thiazide diuretics Comparable effective to AAA Compared with AAA thiazide diuretics are preferable in respect to possible complications. | Kaliuretic effect | Beta-sympathicomimetics Orciprenalin, salbutamol, are also effective for prevention (Bendheim *et al.*, 1985). | Stimulation of the Na–K pump leading to an increase of cellular potassium uptake. |
| Diet | Frequent carbohydrate-rich meals; low-potassium diet. | | Glucose Intravenously for treating severe attacks. | |

Differentiation between the treatment of acute attacks and the prophylaxis of attacks must be made.

**Causal Treatment.** For secondary forms of PP, treatment consists of therapy for the primary disorder. The removal of provoking substances may be effective as well. For this, the reader is referred to textbooks on internal medicine or the section on drug-induced myopathy (Chapter 96). Thus far, there is no primary treatment for the genetically determined membrane defect of FPP.

**Symptomatic Treatment.** Because persistent weakness develops in most patients with FPP who are left untreated, the goal of therapy is the prevention of attacks. For prophylaxis, provoking factors must be avoided (Table I). If general measures (see earlier) are not sufficient, drug therapy is indicated; principles of such therapy are presented in Table II. In thyrotoxic PP, the attacks stop on normalization of the thyroid function, so that long-term prophylaxis is not necessary.

*Practical Management*

**Prophylaxis.** For hypokalemic forms, general measures include a low-carbohydrate (60–80 g/day) and low-salt (2–3 g/day) diet and the avoidance of strenuous physical exertion and cold. For potassium-inducible

PP, patients should allow for brief intervals between carbohydrate-rich meals. Complete rest after physical exertion is to be avoided, as are cold temperatures and potassium-rich medications and foods (fruit juices, bananas). Because the attacks in hyperkalemic FPP usually subside on ingestion of carbohydrates, many patients learn to eat sweets at the first symptoms; prophylactic drugs are thus rarely needed for this form of dyskalemic PP. Hyperthyroidism can be effectively treated in most cases. Therefore, further measures for thyrotoxic PP are often superfluous. Like treatment for hypokalemic FPP, a low-carbohydrate diet and the avoidance of potassium-uretic substances are helpful. If the attacks cannot be prevented by the aforementioned measures or if persistent weakness is present, prophylactic medication is indicated (Table III).

In addition to therapy with acetazolamide or dichlorophenamide, some patients with hypokalemic FPP require a regular potassium substitute. Therefore, frequent measurement of serum potassium levels, particularly at the beginning of treatment, is recommended. For hypokalemic pareses that are present at awakening, the additional ingestion of potassium effervescent tablets in the middle of the night can be helpful.

**Treatment of Attacks.** The duration of attacks of paralysis in hyperkalemic FPP is often so brief that therapy at the outset may be superfluous. The patient should be made aware that continuous, mild exercise

**TABLE III    Drug Prophylaxis of Dyskalemic Periodic Paralyses**

|  | Drug | Side effects, complications |
|---|---|---|
| Familial hypokalemic periodic paralysis | 1. Acetazolamide: 125–1500 mg/day; dosage should be kept as low as possible. | Paresthesias; dysgeusia for carbonated beverages; nausea; anorexia; mild weight loss. Long-term medication has been complicated by nephrolithiasis: high fluid intake, urine status twice a year, ultrasound imaging once a year, avoiding sulfonamide administration is recommended. Worsening of metabolic state in diabetics. |
|  | 2. Dichlorophenamide: 2–3 times; 25–50 mg/day. | Lack of concentration; impaired reaction ability; otherwise see acetazolamide. |
|  | If 1 and 2 fail, |  |
|  | 3. Triamterene: 50–150 mg/day | Gastrointestinal disorders; acidosis. |
|  | 4. Spironolactone: 100–200 mg/day | Gynecomastia; hirsutism; impotence; fever; exanthema. |
| Thyrotoxic periodic paralysis | 1. Propranolol: 2 to 4 times; 40 mg/day | Exanthema; muscle cramps; fatigue; alopecia; thrombopenia. |
|  | 2. Spironolactone: 100–200 mg/day |  |
| Familial hyperkalemic periodic paralysis | 1. Hydrochlorothiazide: 25–75 mg/day; single dose in the very early morning. | Gastrointestinal disorders; allergic reactions; arterial hypotension; loss of electrolytes; hypercalcemia; hyperglycemia; uremia; pancreatitis. |
|  | 2. Acetazolamide: see 1 |  |
|  | 3. Salbutamol | Long-term experience is lacking. |

improves the paralysis. The drug therapy for attacks in FPP is presented in Table IV.

With the consistent use of available measures tailored to the individual patient and the cooperation of the patient in certain aspects of lifestyle, the prognosis of FPP is favorable, with the ability to work usually remaining preserved.

### Ineffective or Obsolete Therapies

Monotherapy with potassium, even in high doses, does not provide sustained prevention of attacks in hypokalemic FPP. With this treatment, muscle weakness progresses.

Acetazolamide administered during an attack is ineffective; the preventive effect does not begin until 24–48 hours after administration. The administration of AAA only rarely leads to aggravation of FPP, but in thyrotoxic PP it typically results in exacerbation of symptoms. The antimyotonic agent tocainide (Xylotocan) had no effect on PP in the hypo- and hyperkalemic forms (Ricker *et al.*, 1986; Rüdel *et al.*, 1984). Cation exchangers failed to bring results in hyperkalemic FPP.

The bidirectional, ventricular tachyarrhythmia of a child with hyperkalemic PP could not be improved by AAA or the usual antiarrhythmic agents but responded well to imipramine (Gould *et al.*, 1985).

### Related Clinical Presentations and Differential Diagnosis

A case of hypermagnesemic PP that lasted several days was described by Emser (1982). This patient also had hyperkalemic FPP with mild, brief attacks. Digoxin or lithium proved effective in treating the hypermagnesemic PP, but AAA and diuretics were not found to be of use (Durlach, 1984; Emser, 1982).

**TABLE IV    Practical Management of Acute Attacks in Dyskalemic Periodic Paralyses**

| Familial hypokalemic periodic paralysis | 2–10 g potassium chloride (60–120 mEq) orally in an unsweetened aqueous solution. May be repeated after 2 h if serum potassium is not elevated and ECG within normal ranges. If oral administration is impossible (swallowing disorder, vomiting), intravenous potassium by a central catheter at a maximum rate of 20 mEq KCl/h in a 5% mannitol diluent. |
|---|---|
| Thyrotoxic periodic paralysis | Potassium chloride orally (see preceding). |
| Familial hyperkalemic periodic analysis | 1. Carbohydrate intake: sweets, sugar-containing juices.<br>2. Glucose, 2 g/kg of body weight orally and insulin 15–20 U subcutaneously.<br>3. Salbutamol, 4 times; 2 puffs (salbutamol inhaler) within 1 h (1 puff 0.1 mg salbutamol) (Wang and Clausen, 1976).<br>4. Calcium gluconate, 0.5–2 g intravenously. |

Various etiologically different diseases may cause episodic weakness or paralysis representing a differential diagnosis: guanidine intoxication, sleep paralyses within the scope of narcolepsy, myasthenia gravis, myasthenic syndrome, Guillain-Barré syndrome, acute intermittent porphyria, multiple sclerosis, transient ischemic attacks, hypocalcemic tetany.

# REFERENCES

Abbott, G. W., Butler, M. H., Bendahhou, S., Dalakas, M. C., Ptacek, L. J., and Goldstein, S. A. (2001). MiRP2 forms potassium channels in skeletal muscle with Kv3.4 and is associated with periodic paralysis. *Cell* 104, 217–231.

Bendheim, P. E., Obstarczyk, R., and Berg, B. O. (1985). Beta-adrenergic treatment of hyperkalemic periodic paralysis. *Neurology* 35, 746–749.

Bradley, W. G., Taylor, R., Rice, E. R., Hausmanowa-Petruzewics, I., Adelman, L. S., Jenkinsons, M., Jedrzejowska, H., Drac, H., and Pendlebury, W. W. (1990). Progressive myopathy in hyperkalemic periodic paralysis. *Arch. Neurol.* 47, 1013–1017.

Christensen, K. S. (1985). Hypokalemic periodic paralysis secondary to renal tubular acidosis. *Eur. Neurol.* 24, 303–305.

Confavreux, C., Garassus, P., Vighetto, A., and Aimard, G. (1991). Familial hypokalaemic periodic paralysis: Prevention of paralytic attacks with lithium gluconate. *J. Neurol. Neurosurg. Psychiatry* 54, 87–88.

Dalakas, M. C., and Engel, W. K. (1983). Treatment of permanent muscle weakness in familial hypokalemic periodic paralysis. *Muscle Nerve* 6, 182–186.

Durlach, J. (1984). Hypermagnesemic paralysis, digitalis, and acetylcholine release. *Arch. Neurol.* 41, 134–135.

Emser, W. E. (1982). Hypermagnesemic periodic paralysis. *Arch. Neurol.* 39, 727–730.

Fontaine, B., Khurana, T. S., Hoffmann, E. P., Bruns, G. A. P., Haines, J. L., Trofatter, J. A., Hanson, M. P., Rich, J., McFarlene, H., McKenna-Yasek, D., Romano, D., Gusella, J. F., and Brown, R. H., Jr. (1990). Hyperkalemic periodic paralysis and the adult muscle sodium channel-subunit gene. *Science* 250, 1000–1002.

Fontaine, B., Vale-Santos, J., Jurkat-Rott, K., Reboul, J., Plassart, E., Rime, C.-S., Elbaz, A., Heine, R., Guimaraes, J., Weissenbach, J., Baumann, N., Fardeau, M., and Lehmann-Horn, F. (1994). Mapping of the hypokalaemic periodic paralysis (HypoPP) locus to chromosome 1q31-32 in three European families. *Nature Genet.* 6, 267–272.

Gould, R. J., Steeg, C. N., Eastwood, A. B., Penn, A. S., Rowland, L. P., and DeVivo, D. C. (1985). Potentially fatal cardiac dysrhythmia and hyperkalemic periodic paralysis. *Neurology* 35, 1208–1212.

Griggs, R. C., Engel, W. K., and Resnick, J. S. (1970). Acetazolamide treatment of hypokalemic periodic paralysis. *Ann. Intern. Med.* 73, 39–48.

Griggs, R. C., Resnick, J., and Engel, W. K. (1983). Intravenous treatment of hypokalemic periodic paralysis. *Arch. Neurol.* 40, 539–540.

Henquin, J. C., Charles, S., Nenquin, M., Mathot, F., and Tamagawa, T. (1982). Diazoxide and D600 inhibition of insulin release. *Diabetes* 31, 776–783.

Johnson, T. (1981). Familial periodic paralysis with hypokalaemia. *Dan. Med. Bull.* 28, 1–27.

Jurkat-Rott, K., Lehmann-Horn, F., Elbaz, A., Heine, R., Gregg, R. G., Hogan, K., Powers, P., Lapie, P., Vale-Santos, J. E.,

Weissenbach, J., and Fontaine, B. (1994). A calcium channel mutation causing hypokalemic periodic paralysis. *Hum. Mol. Genet.* 3, 1415–1419.

Jurkat-Rott, K., Mitrovic, N., Hang, C., Kouzmenkin, A., Iaizzo, P., Herzog, J., Lerche, H., Nicole, N., Vale-Santos, J., Chauveau, D., Fontaine, B., and Lehmann-Horn, F. (2000). Voltage sensor sodium channel mutations cause hypokalemic periodic paralysis type 2 by enhanced inactivation and reduced current. *Proc. Natl. Acad. Sci. USA* 97, 9549–9554.

Layzer, R. B. (1984). Pathophysiology of the periodic paralysis: Overview and theoretical aspects. *In* "Neuromuscular Diseases" (G. Serratice, Ed.), pp. 173–177. Raven, New York.

Lehmann-Horn, F., Engel, A. G., Ricker, K., and Rüdel, R. (1994). The periodic paralyses and paramyotonia congenita. *In* "Myology" (A. G. Engel and C. Franzini-Armstrong, Eds.), 2nd ed., Vol. 2., pp. 1303–1334. McGraw-Hill, New York.

Lehmann-Horn, F., and Jurkat-Rott, K. (1999). Voltage-gated ion channels and hereditary disease. *Physiol. Rev.* 79, 1317–1371.

Lehmann-Horn, F., Küther, G., Ricker, K., Grafe, P., Ballanyi, K., and Rüdel, R. (1987). Adynamia episodica hereditaria with myotonia: A non-activating sodium current and the effect of extracellular pH. *Muscle Nerve* 10, 363–374.

Links, T. P., Zwarts, M. J., Wilnink, J. T., Molenaar, W. M., and Oosterhuis, H. J. G. H. (1990). Permanent muscle weakness in familial hypokalaemic periodic paralysis. *Brain* 113, 1873–1889.

Pasman, J. W., Gabreels, F. J. M., Semmekrot, B., Renier, W. O., and Monnens, L. A. H. (1989). Hyperkalemic periodic paralysis in Gordon's syndrome: A possible defect in atrial natriuretic peptide function. *Ann. Neurol.* 26, 392–395.

Plaster, N. M., Tawil, R., Tristani-Firouzi, M., Canun, S., Bendahhou, S., Tsunoda, A., Donaldson, M. R., Iannaccone, S. T., Brunt, E., Barohn, R., Clark, J., Deymeer, F., George, A. L., Jr., Fish, F. A., Hahn, A., Nitu, A., Özdemir, C., Serdaroglu, P., Subramony, S. H., Wolfe, G., Fu, Y. H., and Ptacek, L. J. (2001). Mutations in Kir2.1 cause the developmental and episodic electrical phenotypes of Andersen's syndrome. *Cell* 105, 511–519.

Ptacek, L. J., Tawil, R., Griggs, R. C., Engel, A. G., Layzer, R. B., Kwiecinski, H., McManis, P. G., Santiago, L., Moore, M., Fouad, G., Bradley, P., and Leppert, M. F. (1994). Dihydropyridine receptor mutations cause hypokalemic periodic paralysis. *Cell* 77, 863–868.

Ricker, K., Böhlen, R., and Rohkamm, R. (1983). Different effectiveness of tocainide and hydrochlorothiazide in paramyotonia congenita with hyperkalemic episodic paralysis. *Neurology* 33, 1615–1618.

Ricker, K., Camacho, L. M., Grafe, P., Lehmann-Horn, F., and Rüdel, R. (1989). Adynamia episodica hereditaria: What causes the weakness? *Muscle Nerve* 12, 883–891.

Ricker, K., Rohkamm, R., and Böhlen, R. (1986). Adynamia episodica and paralysis periodica paramytonica. *Neurology* 36, 682–686.

Riggs, J. E. (1989). Periodic paralysis. *Clin. Neuropharmacol.* 4, 249–257.

Riggs, J. E., Griggs, R., and Moxley, R. T. (1984). Dissociation of glucose and potassium arterial-venous differences across the forearm by acetazolamide. *Arch. Neurol.* 41, 35–38.

Rojas, C. V., Wang, J. Z., Schwartz, L. S., Hoffman, E. P., Powell, B. R., and Brown, R. H., Jr. (1991). A Met-to-Val mutation in the skeletal muscle Na+ channel alpha-subunit in hyperkalaemic periodic paralysis. *Nature* 354, 387–389.

Rosa, R. M., Silva, P., Young, J. B., Landsberg, L., Brown, R. S., Rowe, J. W., and Epstein, F. H. (1980). Adrenergic modulation of extrarenal potassium disposal. *N. Engl. J. Med.* 302, 431–434.

Rüdel, R., and Ricker, K. (1985). The primary periodic paralysis. *Trends Neurosci.* 8, 467–470.

Rüdel, R., Lehmann-Horn, F., Ricker, K., and Küther, G. (1984). Hypokalemic periodic paralysis: In vitro investigation of muscle fiber membrane parameters. *Muscle Nerve* 7, 110–120.

Ryan, M. M., Taylor, P., Donald, J. A., Ouvrier, R. A., Morgan, G., Danta, G., Buckley, M. F., and North, K. N. (1999). A novel syndrome of episodic muscle weakness maps to xp22.3. *Am. J. Hum. Genet.* 65, 1104–1113.

Tawil, R., McDermott, M. P., Brown, R., Jr., Shapiro, B. C., Ptacek, L. J., McManis, P. G., Dalakas, M. C., Spector, S. A., Mendell, J. R., Hahn, A. F., and Griggs, R. C. (2000). Randomized trials of dichlorophenamide in the periodic paralyses. Working Group on Periodic Paralysis. *Ann. Neurol.* 47, 46–53.

Torres, C. F., Griggs, R. C., Moxley, R. T., and Bender, A. N. (1981). Hypokalemic periodic paralysis exacerbated by acetazolamide. *Neurology* 31, 1423–1428.

Vern, B. A., Danon, M. J., and Hanlon, K. (1987). Hypokalemic periodic paralysis with unusual responses to acetazolamide and sympathomimetics. *J. Neurol. Sci.* 81, 159–172.

Wang, P., and Clausen, T. (1976). Treatment of attacks in hyperkalaemic familial periodic paralysis by inhalation of salbutamol. *Lancet* 1, 221–223.

CHAPTER 100

# Cramps

Helge Topka and E. Logigian

## CLINICAL ASPECTS

Muscle cramps are characterized by painful, involuntary spasms involving part or the whole of a skeletal muscle lasting from seconds to minutes. They may be triggered by exercise or mild voluntary muscle contraction but more frequently occur at rest, in particular at night (nocturnal cramps). Muscle cramp is distinct from other related conditions such as "contracture," "spasm," "spasticity," or "myalgia." Cramps most frequently involve foot or calf muscles and, more rarely, affect the biceps muscle, finger extensor muscles, or the myohyoideus muscle. Alcohol and drugs among other metabolic or physical factors may predispose to cramping. Lengthening of the affected muscle by passive stretch or activation of antagonistic muscles tends to relieve symptoms.

Electromyographically, cramps are associated with high-frequency motor unit discharges (>25 Hz) that may be preceded and followed by muscle fasciculations. In contrast, painful muscle contractures that occur in McArdle's disease (myophosphorylase deficiency) and sometimes clinically mimic cramps are due to intrinsic muscle dysfunction and, therefore, are electrically silent. Severe episodes of cramping may be associated with elevated serum levels of creatine kinase (CK). In those cases, hardening of the affected muscles and muscle pain may persist for several days. Exercise-induced muscle cramps during sporting activities, for example, are associated with drastic increases in serum nitric oxide; however, the causal relationship between increased nitric oxide levels and muscle cramps is yet to be established (Maddalli *et al.*, 1998).

The exact pathophysiological basis of cramping is unknown. Even though some myopathies may be associated with cramping, true muscle cramps are thought to be of neurogenic origin. Recent electromyographical studies have demonstrated that cramps are initiated by contraction of a slowly moving fraction of muscle fibers, indicating that the spatial arrangement of muscle fibers and motor neurons are closely correlated or that cramps may actually originate very close to the level of the muscle (Roeleveld *et al.*, 2000). Nevertheless, the exact site of lower motor neuron hyperexcitability has yet to be determined. Pathoanatomical studies reveal type 2 fiber predominance and tubular aggregates in patients with muscle cramps and exertional myalgia (Telerman-Toppet *et al.*, 1985), but the pathophysiologic significance of this finding is uncertain.

The etiology of muscle cramp is remarkably heterogeneous. In most cases, cramps are infrequent and harmless in nature. However, in some patients, cramps may not be a benign complaint but represent a symptom of an unsuspected underlying neuromuscular condition. Hence, unusually frequent or severe cramping requires further investigation. A variety of neurological or metabolic disorders have been associated with muscle cramps (Table I). In general, metabolic disorders are more frequently associated with muscle cramps than disorders of the peripheral or, rarely, the central nervous system. Overall, cramps are more frequent in the elderly, in patients on dialysis, during pregnancy, and in patients with significant medical comorbidity. For example, in 80% of cancer patients experiencing first-time cramps, a metabolic, neurogenic (neuropathy, radiculopathy), or even myogenic cause (e.g., polymyositis) was determined (Steiner and Siegal, 1989).

Rarely, muscle cramping may point to an underlying hereditary disorder. Familial cramps are characterized by autosomal-dominant inheritance, teenage onset, and distal muscle involvement. In some families, what appeared to be autosomal-dominant muscle cramp syndrome turned out to be related to a hereditary form of polyneuropathy (Chiba *et al.*, 1999).

TABLE I   Cramps and Associated Disorders

| | |
|---|---|
| Idiopathic | Central nervous system disorders |
|    Cramps with no known etiology |    Tetanus / strychnine intoxication |
|    Cramps associated with prolonged or |    Satoyoshi syndrome |
|       excessive muscle use | |
| | |
| Myopathies | Medical disorders |
|    Glycogen storage myopathies |    β-Thalassaemia |
|    Carnitine palmityltransferase deficiency |    Liver cirrhosis |
|    Carnitine deficiency |    Viral enteritis |
|    Xanthine oxidase deficiency |    Hyperparathyroidism |
|    Myoadenylate deaminase deficiency |    Hexosaminidase A or B deficiency |
|    Myophosphorylase deficiency (McArdle |    Ferritine deficiency |
|       disease) |    Vascular claudication (arterial insufficiency) |
|    Phosphofructokinase deficiency |    Uremia |
|    Brody syndrome (disordered calcium uptake) |    Peripheral edema |
|    Hypothyroidism | |
|    Malignant hyperthermia | Peripheral nervous system disorders |
|    Progressive muscular dystrophy |    Polyneuropathies |
|    Familial cramps |    Radiculopathies |
|    Central-core disease |    Motoneuron disease (ALS) |
|    Polymyositis |    Spinal muscular atrophy |
|    Paraneoplastic myopathy |    Poliomyelitis |
| |    Cramp-Fasciculation-Myalgia syndrome |

## NATURAL COURSE

Occasional nocturnal cramps occur in 15% of healthy college students (Norris *et al.*, 1957). At least 75% of adults have experienced at least one muscle cramp (Jerusalem, 1984). In a survey of outpatient veterans, 56% reported repeated nocturnal leg cramps (Oboler *et al.*, 1991). Other surveys report repeated nocturnal muscle cramps in up to 42% of otherwise healthy adults (Jansen *et al.*, 1991; Naylor and Young, 1994). One report suggests slightly increased frequencies of nocturnal muscle cramps in women (Jansen *et al.*, 1991). The incidence of muscle cramps seems to increase with age.

In symptomatic forms, the course of the condition is determined by the underlying disorder. In amyotrophic lateral sclerosis (ALS), as well as in other conditions, cramps may precede other symptoms (Fleet and Watson, 1986).

## PRINCIPLES OF THERAPY

Because the exact pathophysiology of muscle cramps has yet to be determined, specific physiological treatment is not available. Nonspecific therapy includes physical, behavioral, and drug therapy. In most instances, physical therapy is very effective and includes manual compression of the affected muscle (Helin, 1985), passive lengthening of the affected muscle (Bertolasi *et al.*, 1993), or voluntary activation of the antagonist muscles (Norris *et al.*, 1957). Cramps most frequently occur during or after unusual and prolonged muscle use; therefore, moderate and regular muscle training may be beneficial. Underlying orthopaedic problems such as pes planovalgus or other forefoot abnormalities may eventually lead to nonphysiological muscle use and subsequent cramps and may therefore require correction. For a summary of metabolic and physical factors that predispose to cramping see Table II.

Table III outlines the drug therapy of cramps in order of efficacy. The notion that hyperexcitable terminal motor axons cause muscle cramps has prompted numerous therapeutic trials of membrane-stabilizing drugs (Layzer, 1979). Most studies have tested quinine, which is thought to increase the refractory period of muscle membrane and decrease the excitability of the motor endplate to repetitive nerve stimulation and to acetylcholine (Rollo, 1980). Worldwide, quinine is the most frequently used substance in the treatment of muscle cramps. However, the results of several double-blind, placebo-controlled trials of quinine sulfate for the treatment of nocturnal leg cramps are conflicting. Several studies of 8–25 patients and daily quinine doses of 200 and 300g were not effective in preventing cramps (Lim, 1986; Sidorov, 1993; Warburton *et al.*, 1987). Other investigators reported a reduction of number, duration, or severity of nocturnal cramps at slightly higher doses of 200–500mg quinine sulfate

**TABLE II**  Predisposing Factors

| Physical factors | Drugs |
|---|---|
| During or after prolonged use of muscles | β-Blockers (Labetolol, Pindolol) |
| Nonphysiological use of muscles (e.g., as in joint deformity) | β-Adrenergic agents (Albuterol, Terbutaline) |
| Warmth, cold | Suxamethonium |
| Alcohol, lack of sleep, nicotine | Creatine |
| | Quinidine |
| Imbalance of water or | Metozalone |
| electrolyte metabolism | Danazole |
| (lowered calcium, | Nifedipine |
| potassium, sodium, or | Carbimazole |
| magnesium levels; often | Diuretics |
| associated with | Laxatives |
| dehydration, excessive | Clofibrate |
| sweating, vomiting, | Cyclosporine |
| diarrhea, diuresis) | Chemotherapeutic agents |
| | Corticosteroids |
| | |
| | Other |
| | Plasmapheresis |
| | Dialysis |
| | Pregnancy |

(Connolly *et al.*, 1992; Fung and Holbrook, 1989; Jones and Castleden, 1983) or hydroquinine hydrobromide dihydrate (Jansen *et al.*, 1997). One large clinical study by Görlich and colleagues (1991) compared quinine sulfate combined with theophylline ethylene diamine (QTED) versus QTED alone and quinine sulfate alone vs placebo in a multicenter trial involving 164 patients. Although they found both QTED and quinine sulfate to be superior to placebo, QTED alone was more effective than quinine sulfate alone. Quinine sulfate has also been found to be effective in patients with muscle cramps and liver cirrhosis, a relative contraindication for this medication (Lee *et al.*, 1991). Finally, two large meta-analyses of short-term efficacy of quinine for muscle cramps (Man-Son-Hing *et al.*, 1995, 1998) both showed only a modest benefit in reducing the number of muscle cramps by four to eight attacks per month.

In recent years it has been appreciated that the therapeutic usefulness of quinine sulfate is limited by serious side effects. These include life-threatening hypersensitivity reactions such as rash, pruritius, anaphylaxis, thrombocytopenia (Freiman, 1990), hemolytic uremic syndrome (Crum and Gable, 2000), and hepatitis (Punukollu *et al.*, 1990) that may occur even at relatively low doses from over-the-counter preparations (Beyens *et al.*, 1999). The classical syndrome of cinchonism tends to occur after large single doses or after chronic use (Goldenberg and Wexler, 1988) and includes tinnitus, headache, nausea, and disturbed vision. After chronic quinine therapy, optic atrophy, renal insuffi-

ciency, gastrointestinal symptoms, and cardiac arrhythmias have been observed. In light of these potentially severe side effects and the absence of compelling evidence of therapeutic efficacy, the FDA in 1994 rescinded its approval for quinine therapy of nocturnal leg cramps in the United States. Therefore, its use as an off-label medication for this indication should be considered very carefully before prescription.

The alternative drug therapies for muscle cramps have not been extensively investigated. Other oral membrane-stabilizing agents include procainamide (Joekes, 1979) and tocainide (Puniani and Bertorini, 1991). Calcium antagonist drugs have been hypothesized to compensate for a $Ca^{2+}$ ATPase deficiency in individuals with familial exertional muscle pain syndrome (Taylor *et al.*, 1988). In an open trial, the calcium antagonistic drug verapamil was found to be beneficial in eight elderly patients who did not show improvement with quinine sulfate (Baltodano *et al.*, 1988). Finally, long-term ergoloid mesylate (Hydergin) was found in a placebo-controlled double-blind study to provide an alternative to quinine sulfate therapy (Huber *et al.*, 1986).

Recently, local injection of Botulinum toxin, widely used for symptomatic treatment of hyperkinetic movement disorders, has been proposed as an alternative to oral medication for the treatment of muscle cramps. Bertolasi and colleagues (1997) reported significant benefit (without notable side effects) lasting some 3 months in patients with inherited benign cramp–fasciculation syndrome after injections of botulinum toxin into calf muscles and small flexor muscles of the foot.

Hemodialysis-associated muscle cramps are ascribed to a reduction in plasma volume during dialysis and an insufficient sympathetic nervous system response to volume stress (Kaplan *et al.*, 1992). In accord with this view, Canzanello and colleagues (1991) found that hypertonic solutions consisting of dextrose, mannitol, or saline were effective therapy for hemodialysis-associated cramps. Quinine sulfate and vitamin E have also been found effective in treating hemodialysis-asociated muscle cramps (Roca *et al.*, 1992).

Cramps associated with occlusive arterial diseases are probably not caused by ischemia but rather by electrolyte imbalance and accompanying edema.

Cramps occurring during venography may be prevented by adding xylocaine to the contrast material.

## PRACTICAL MANAGEMENT

The most important initial step in the management of muscle cramps is to determine whether there is an underlying cause. In particular, medications known to

TABLE III   Drug Treatment of Muscle Cramps

|  | Drug | Daily dosage | Author |
|---|---|---|---|
| Benign or nocturnal cramps | Hydroquinine (see text) | 200–500 mg | Connolly *et al.*, 1992; Fung and Holbrook, 1989; Jones and Castleden, 1983; Jansen *et al.*, 1997 |
|  | Verapamil | 120 mg | Walton, 1981 |
|  | Hydergin | 1.5 mg tid | Huber *et al.*, 1986 |
|  | Procainamide | 250 mg | Joekes, 1979 |
|  | Vitamin E | 400 IU | Ayres and Mihan, 1974 |
|  | Vitamin $B_2$ | 20 mg | Morgan, 1983 |
|  | Diazepam | 2–10 mg | Warne, 1984 |
|  | Phenytoin | 100–300 mg | Layzer, 1979 |
|  | Dantrolene | 25 mg | Myers, 1977 |
| Cramp/fasciculation | Carbamazepine | max. 1600 mg | Tahmoush *et al.*, 1991 |
| ALS/myotonia | Tocainide | 200–400 mg bid | Puniani and Bertorini, 1991 |
| Pregnancy cramps | Calcium | 1000–2000 mg | Hammar *et al.*, 1981 |
|  | Magnesium | 5 mmol tid | Riss *et al.*, 1983; Dahle *et al.*, 1995 |
| Hemodialysis-associated cramps | 50% dextrose water, 25% mannitol, or 23.5% saline IV infusion |  | Canzanello *et al.*, 1991 |
|  | Vitamin E | 400 IU | Roca *et al.*, 1992 |

precipitate muscle cramps should be discontinued if possible. Underlying metabolic disorders should be treated. For benign idiopathic muscle cramps, patients should be advised as to the harmless nature of the condition and instructed to use symptomatic physical therapy.

## Physical Treatment

In the absence of predisposing factors (see Table II), physical therapy should be initiated. For benign muscle cramps, passive stretching of the affected muscles relieves symptoms (Eaton, 1989). Voluntary activation of antagonist muscles is presumed to stop muscle cramps by reciprocal inhibition (Fowler, 1973). Weiner and Weiner (1980) therefore, recommend ankle dorsiflexion when stretching or while swimming. Raising the foot of the bed to prevent venous edema is also sometimes recommended. In general, benefits of physical treatment can be judged after a few days. If physical therapy is not effective, drug treatment should be considered.

## Drug Therapy

As indicated previously, quinine sulfate alone or in combination with theophylline ethylene diamine is still frequently prescribed for nocturnal muscle cramps***. However, given its minimal efficacy (Man-Son-Hing *et al.*, 1995, 1998) and its potential serious side effects, quinine can no longer be recommended as first-line therapy. However, the evidence for the efficacy of alternative drugs is also largely lacking. A number of different medications have been investigated in open or small

controlled trials (Table III) that may be helpful for some patients.

During *pregnancy*, quinine sulfate is contraindicated because of an increase in stillbirths. Muscle cramps during pregnancy have been linked to latent magnesium deficiency (Riss *et al.*, 1983). Treatment with magnesium is often required in pre-eclampsia or premature labor to reduce complications (spontaneous abortions, premature delivery), as well as the number of muscle cramps (Classen and Helbig, 1984). A recent *Cochrane Review* summarized the results of three trials investigating the therapeutic efficacy of calcium and sodium chloride on leg cramps in 217 pregnant women (Young and Jewell, 2000). According to this review, both calcium and sodium chloride seemed to be superior to placebo**. Trials investigating the use of magnesium for treatment of pregnancy cramps, however, were not included in this analysis.

Muscle cramps occurring during *hemodialysis* are frequently accompanied by muscle pain and depend both on the dialysis procedure (hemodialysis, peritoneal dialysis, hemofiltration) and the solution used. Infusion of hypertonic solutions such as 50% dextrose water, 25% mannitol, or 23.5% saline are beneficial in preventing hemodialysis-associated muscle cramps without the risk of significant side effects* (Canzanello *et al.*, 1991). Considering the potential toxicity of quinine sulfate, Roca *et al.* (1992) believe that if drug therapy of hemodialysis-associated cramps is necessary, initial therapy should be started with vitamin E. In their study, vitamin E proved to be as effective as quinine sulfate in preventing cramps.

Patients with *liver cirrhosis* frequently experience muscle cramps, usually during sleep. Systematic studies are lacking, however, but the results of an open study involving 12 patients, suggest that oral zinc sulfate replacement may have beneficial effects (Kugelmas, 2000). Others emphasize the reduction in plasma volume caused by ascites and, therefore, recommend intravenous infusion of human albumin (Angeli *et al.*, 1996). As an oral medication, a trial using vitamin E (3 × 200 mg/day) (Konikoff *et al.*, 1991) or taurine (Yamamoto *et al.*, 1994) may be helpful, although controlled studies are lacking.

*Muscle cramps associated with exertional myalgia* may improve after administration of the calcium blocking agent nifedipine (Sufit and Peters, 1984).

Cramps associated with *intermittent vascular claudication or other peripheral vascular disorders* may respond to oral administration of pentoxifylline (Ward and Clissold, 1987).

In a subset of young and otherwise healthy adults, *diurnal cramps* occur, mostly during exercise, and can affect a variety of different muscles. In these patients, quinine sulfate seems to be less effective. Although not supported by large-scale clinical studies, in some patients, exercise-induced muscle cramps may be related to excessive sweating and, as a consequence, excessive sodium loss. This type of exercise-induced muscle cramping, therefore, may respond favorably to an increase in the daily dietary intake of sodium (Bergeron, 1996).

Alternative drug therapy with phenytoin, carbamazepine, or amitriptyline may be tried in patients not responding to other medications (Layzer, 1979).

## TREATMENTS NO LONGER RECOMMENDED

Because of an increasing number of reports on potentially severe side effects and limited evidence of therapeutic efficacy, the use of quinine as a nonprescription drug is no longer recommended for treatment of nocturnal leg cramps.

In an open trial, TRH analogs have been found to improve muscle cramping, bulbar symptoms, and spasticity in amyotrophic lateral sclerosis (Modarres-Sadeghi *et al.*, 1988). However, because the therapeutic efficacy of TRH is considered very limited, its therapeutic use is no longer recommended.

## RELATED SYNDROMES

Severe myalgia associated with cramplike sensations are typical symptoms in *compartment syndrome*, a malignant form of ischemic contracture (Martens and Moeyersoons, 1990) (see Table IV). The condition usually is caused by intracompartmental hemorrhage

TABLE IV   Related Syndromes

| |
|---|
| Compartment syndrome (see Chapters 91 and 92) |
| Hypocalcemic tetany |
| Stiff-person syndrome (see Chapter 101) |
| Myotonia (see Chapter 98) |
| Isaacs' syndrome |
| Focal dystonia (see Chapter 78) |
| Abdominal cramping |
| Dysmenorrhea |

and should be treated by immediate decompression to prevent permanent nerve damage or Volkmann's ischemic contracture (see Chapter 91).

*Tetany*, resulting from severe hypocalcemia or a reduction in the serum ionized Ca fraction without marked hypocalcemia, is characterized by carpopedal spasm, spasm of facial musculature, and sensory symptoms such as paresthesia of the lips, tongue, and fingers. The condition is presumably caused by hyperexcitability not only of terminal axons as in cramps, but of the motoneuron as a whole and parts of the central nervous system.

The *stiff-person syndrome* is a rare neurological disorder characterized by fluctuating and progressive rigidity of axial and limb muscles. Additional painful spasms are precipitated by voluntary movement, by startle, or by emotional stimuli (see Chapter 101). Many patients have associated autoimmune endocrinopathies, particularly type I diabetes mellitus. Immunological studies have demonstrated antibodies directed against the GABA synthesizing enzyme glutamic acid decarboxylase (GAD) in most patients. Electromyography reveals continuous motor unit activity of agonist and antagonist muscle groups, including paraspinal muscles and abnormal exteroceptive reflexes that seem to be due to a disinhibition of spinal GABAergic interneurons. Dramatic improvement with administration of benzodiazepines has been suggested as a diagnostic criterion. Accordingly, symptomatic treatment consists of oral administration of benzodiazepines, such as diazepam (20–100 mg/day) or clonazepam (3–10 mg/day). Clinical trials involving plasma exchange or immunosuppression have yielded controversial results.

*Focal dystonias* such as writer's cramp may present with cramp-like symptoms (see Chapter 78). As opposed to *true muscle cramp*, the involuntary muscle contractions in focal dystonia are not mediated by high-frequency motor unit discharges, affect agonist and antagonist muscle simultaneously, and are usually less painful. Symptomatic treatment includes anticholinergic drugs or local injection of botulinum toxin.

Patients with *spasticity* can have painful muscle spasms, but upper motor neuron findings such as velocity-dependent increase in muscle tone are present on neurological examination. Therapy with antispastic (see Chapter 68) or analgesic medications (see Chapter 11) is typically required.

*Pain associated with dysmennorrhea* and *crampy abdominal pain* are caused by a different pathomechanism and, therefore, require a different pharmacological approach. Pain associated with dysmenorrhea is ascribed to uterine contractions and ischemia, probably mediated by prostaglandins.

## REFERENCES

Angeli, P., Albino, G., Carraro, P., Dalla Pria, M., Merkel, C., Caregaro, L. *et al.* (1996). Cirrhosis and muscle cramps: evidence of a causal relationship. *Hepatology* 23, 264–273.

Ayres, S. J., and Mihan, R. (1974). Nocturnal leg cramps (systremma): a progress report on response to vitamin E. *South Med. J.* 67, 1308–1312.

Baltodano, N., Gallo, B. V., and Weidler, D. J. (1988). Verapamil vs quinine in recumbent nocturnal leg cramps in the elderly. *Arch. Intern. Med.* 148, 1969–1970.

Beard, T. C. (1989). Effect of salt restriction on hypertension. *Lancet* 2, 801.

Beaver, W. T., and McMillan, D. (1980). Methodological considerations in the evaluation of analgesic combinations: acetaminophen (paracetamol) and hydrocodone in postpartum pain. *Br. J. Clin. Pharmacol.* 10 Suppl 2, 215S–223S.

Bergeron, M. F. (1996). Heat cramps during tennis: a case report. *Int. J. Sport. Nutr.* 6, 62–68.

Bertolasi, L., De Grandis, D., Bongiovanni, L. G., Zanette, G. P., and Gasperini, M. (1993). The influence of muscular lengthening on cramps. *Ann. Neurol.* 33, 176–180.

Bertolasi, L., Priori, A., Tomelleri, G., Bongiovanni, L. G., Fincati, E., Simonati, A. *et al.* (1997). Botulinum toxin treatment of muscle cramps: a clinical and neurophysiological study. *Ann. Neurol.* 41, 181–186.

Beyens, M. N., Guy, C., Ollagnier, M. (1999). Adverse effects of quinine in the treatment of leg cramps. *Therapie.* 54, 59–62.

Canzanello, V. J., Hylander-Rossner, B., Sands, R. E., Morgan, T. M., Jordan, J., and Burkart, J. M. (1991). Comparison of 50% dextrose water, 25% mannitol, and 23.5% saline for the treatment of hemodialysis-associated muscle cramps. *ASAIO Trans.* 37, 649–652.

Chiba, S., Saitoh, M., Hatanaka, Y., Kashiwagi, M., Imai, T., Matsumoto, H. *et al*. (1999). Autosomal dominant muscle cramp syndrome in a Japanese family. *J. Neurol. Neurosurg. Psychiatry* 67, 116–119.

Classen, H. G., and Helbig, J. (1984). Magnesium therapy in pregnancy. Pharmacologic and toxicologic aspects of magnesium supplementation and use in pre-eclampsia and threatened premature labor. *Fortschr. Med.* 102, 841–844.

Connolly, P. S., Shirley, E. A., Wasson, J. H., and Nierenberg, D. W. (1992). Treatment of nocturnal leg cramps. A crossover trial of quinine vs vitamin E. *Arch. Intern. Med.* 152, 1877–1880.

Crum, N. F., and Gable, P. (2000). Quinine-induced hemolytic-uremic syndrome. *South. Med. J.* 93, 726–728.

Dahle, L. O., Berg, G., Hammar, M., Hurtig, M., and Larsson, L. (1995). The effect of oral magnesium substitution on pregnancy-induced leg cramps. *Am. J. Obstet. Gynecol.* 173, 175–180.

Eaton, J. M. (1989). Is this really a muscle cramp? *Postgrad. Med.* 86, 227–232.

Fleet, W. S., and Watson, R. T. (1986). From benign fasciculations and cramps to motor neuron disease. *Neurology* 36, 997–998.

Fossel, M., and Rosen, P. (1984). Naloxone treatment for codeine-induced gastrointestinal symptoms. *J. Emerg. Med.* 2, 107–110.

Fowler, A. W. (1973). Relief of cramp. *Lancet* 1, 99.

Freiman, J. P. (1990). Fatal quinine-induced thrombocytopenia. *Ann. Intern. Med.* 112, 308–309.

Fung, M. C., and Holbrook, J. H. (1989). Placebo-controlled trial of quinine therapy for nocturnal leg cramps. *West. J. Med.* 151, 42–44.

Goldenberg, A. M., and Wexler, L. F. (1988). Quinine overdose: review of toxicity and treatment. *Clin. Cardiol.* 11, 716–718.

Görlich, H. D., von Gablenz, E., and Steinberg, H. W. (1991). Treatment of nocturnal leg cramps. A multicenter, double blind, placebo controlled comparison between the combination of quinine and theophylline ethylene diamine with quinine. *Arzneimittelforschung* 41, 167–175.

Hammar, M., Larsson, L., and Tegler, L. (1981). Calcium treatments of leg cramps in pregnancy. *Acta. Obstet. Gynaecol. Scand.* 60, 345–347.

Helin, P. (1985). Physiotherapy and electromyography in muscle cramp. *Br. J. Sports. Med.* 19, 230–231.

Huber, F., Koberle, S., Prestele, H., and Spiegel, R. (1986). Effects of long-term ergoloid mesylates ("Hydergine") administration in healthy pensioners: 5-year results. *Curr. Med. Res. Opin.* 10, 256–279.

Jansen, P. H., Joosten, E. M., Van Dijck, J., Verbeek, A. L., and Durian, F. W. (1991). The incidence of muscle cramp. *J. Neurol. Neurosurg. Psychiatry* 54, 1124–1125.

Jansen, P. H., Veenhuizen, K. C., Wesseling, A. I., de Boo, T., and Verbeek, A. L. (1997). Randomised controlled trial of hydroquinine in muscle cramps. *Lancet* 349, 528–532.

Jerusalem, F. (1984). Schmerzhafte nächtliche Wadenkrämpfe. *Dtsch. Med. Wochenschr.* 109, 34–35.

Joekes, A. M. (1979). Cramp. *Clin. Rheum. Dis.* 5, 873–881.

Jones, K., and Castleden, C. M. (1983). A double-blind comparison of quinine sulphate and placebo in muscle cramps. *Age Ageing* 12, 155–158.

Kaplan, B., Wang, T., Rammohan, M., del Greco, F., Molteni, A., and Atkinson, A. J. (1992). Response to head-up tilt in cramping and noncramping hemodialysis patients. *Int. J. Clin. Pharmacol. Ther. Toxicol.* 30, 173–180.

Konikoff, F., Ben-Amitay, G., Halpern, Z., Weisman, Y., Fishel, B., Theodor, E. *et al.* (1991). Vitamin E and cirrhotic muscle cramps. *Isr. J. Med. Sci.* 27, 221–223.

Kugelmas, M. (2000). Preliminary observation: oral zinc sulfate replacement is effective in treating muscle cramps in cirrhotic patients. *J. Am. Coll. Nutr.* 19, 13–15.

Layzer, R. B. (1979). Motor unit hyperactivity states. *In* "Handbook of Clinical Neurology," Volume 41. (P. J. Vinken and G. W. Bruyn, Eds.), pp. 295–316. Amsterdam, New York, and Oxford.

Lee, F. Y., Lee, S. D., Tsai, Y. T., Lai, K. H., Chao, Y., Lin, H. C. *et al.* (1991). A randomized controlled trial of quinidine in the treatment of cirrhotic patients with muscle cramps. *J. Hepatol.* 12, 236–240.

Lim, S. H. (1986). Randomised double-blind trial of quinine sulphate for nocturnal leg cramp. *Br. J. Clin. Pract.* 40, 462.

Maddalli, S., Rodeo, S. A., Barnes, R., Warren, R. F., and Murell, G. A. C. (1998). Postexercise increase in nitric oxide in football players with muscle cramps. *Am. J. Sports Med.* 26, 820–824.

Man-Son-Hing, M., and Wells, G. (1995). Meta-analysis of efficacy of quinine for treatment of nocturnal leg cramps in elderly people. *BMJ* 310, 13–17.

Man-Son-Hing, M., Wells, G., and Lau, A. (1998). Quinine for nocturnal leg cramps: a meta-analysis including unpublished data. *J. Gen. Intern. Med.* **13**, 600–606.

Martens, M. A., and Moeyersoons, J. P. (1990). Acute and recurrent effort-related compartment syndrome in sports. *Sports Med.* **9**, 62–68.

Modarres-Sadeghi, H., Rogers, H., Emami, J., and Guiloff, R. J. (1988). Subacute administration of a TRH analogue (RX77368) in motorneuron disease: an open study. *J. Neurol. Neurosurg. Psychiatry* **51**, 1146–1157.

Morgan, A. A. (1983). Treatment of cramp. *J. R. Soc. Med.* **76**, 712.

Naylor, J. R., and Young, J. B. (1994). A general population survey of rest cramps. *Age Ageing* **23**, 418–420.

Norris, F. H., Gasteiger, E. L., and Chatfield, P. O. (1957). An electromyographic study of induced and spontaneous muscle cramps. *Electroencephalogr. Clin. Neurophysiol.* **9**, 139–147.

Oboler, S. K., Prochazka, A. V., and Meyer, T. J. (1991). Leg symptoms in outpatient veterans. *West. J. Med.* **155**, 256–259.

Puniani, T. S., and Bertorini, T. E. (1991). Tocainide therapy in muscle cramps and spasms due to neuromuscular disease. *Muscle Nerve* **14**, 280–285.

Punukollu, R. C., Kumar, S., and Mullen, K. D. (1990). Quinine hepatotoxicity: an underrecognized or rare phenomenon? *Arch. Intern. Med.* **150**, 1112–1113.

Riss, P., Barth, W., and Jelinic, D. (1983). Zur Klinik und Therapie von Wadenkrämpfen in der Schwangerschaft. *Geburtshilfe Frauenheilkd* **43**, 329.

Roca, A. O., Jarjoura, D., Blend, D., Cugino, A., Rutecki, G. W., Nuchikat, P. S. *et al.* (1992). Dialysis leg cramps. Efficacy of quinine versus vitamin E. *Asaio J.* **38**, M481–M485.

Roeleveld, K., Van Engelen, B. G. M., and Stegeman, D. F. (2000). Possible mechanisms of muscle cramp from temporal and spatial surface EMG characteristics. *J. Appl. Physiol.* **88**, 1698–1706.

Rollo, I. (1980). Drugs used in chemotherapy of malaria. *In* "The Pharmacological Basis of Therapeutics." (A. Gillman, L. Goodman, and A. Gilman, Eds.), p. 1056. New York: MacMillan.

Sidorov, J. (1993). Quinine sulfate for leg cramps: does it work? *J. Am. Geriatr. Soc.* **41**, 498–500.

Steiner, I., and Siegal, T. (1989). Muscle cramps in cancer patients. *Cancer* **63**, 574–577.

Sufit, R. L., and Peters, H. A. (1984). Nifedipine relieves exercise-exacerbated myalgias. *Muscle Nerve* **7**, 647–649.

Tahmoush, A. J., Alonso, R. J., Tahmoush, G. P., and Heiman, P. T. (1991). Cramp-fasciculation syndrome: a treatable hyperexcitable peripheral nerve disorder. *Neurology* **41**, 1021–1024.

Taylor, D. J., Brosnan, M. J., Arnold, D. L., Bore, P. J., Styles, P., Walton, J. *et al.* (1988). Ca2+-ATPase deficiency in a patient with an exertional muscle pain syndrome. *J. Neurol. Neurosurg. Psychiatry* **51**, 1425–1433.

Telerman-Toppet, N., Bacq, M., Khoubesserian, P., and Coers, C. (1985). Type 2 fiber predominance in muscle cramp and exertional myalgia. *Muscle Nerve* **8**, 563–567.

Warburton, A., Royston, J. P., Nicholson, P. W., Jee, R. D., Denham, M. J. *et al.* (1987). A quinine a day keeps the leg cramps away? *Br. J. Clin. Pharmacol.* **23**, 459–465.

Ward, A., and Clissold, S. P. (1987). Pentoxifylline: A review of its pharmacodynamic and pharmacokinetic properties, and its therapeutic efficacy. *Drugs* **34**, 50–97.

Walton, J. (1981). Diffuse exercise-induced pain of undetermined cause relieved by Verapamil. *Lancet* **I**, 933.

Warne, R. W. (1984). Cramps, stiffness and restless legs. *Curr. Therap.* **25**, 35–39.

Weiner, I. H., and Weiner, H. L. (1980). Nocturnal leg muscle cramps. *JAMA* **244**, 2332–2333.

Yamamoto, S., Ohmoto, K., Ideguchi, S., Yamamoto, R., Mitsui, Y., Shimabara, M. *et al.* (1994). Painful muscle cramps in liver cirrhosis and effects of oral taurine administration. *Nippon Shokakibyo Gakkai Zasshi* **91**, 1205–1209.

Young, G. L., and Jewell, D. (2000). Interventions for leg cramps in pregnancy. *Cochrane Database Syst. Rev.* **48**, CD000121.

*Muscle and Peripheral Nervous System*

## CHAPTER 101
# Muscle Stiffness

Philip D. Thompson and Hans-Michael Meinck

## CLINICAL ASPECTS AND DIFFERENTIAL DIAGNOSIS

Muscle stiffness and cramps are common complaints that may arise from the central or peripheral nervous system (Table I). Within this large differential diagnosis, many conditions have additional features that provide valuable diagnostic clues. Accordingly, the assessment of these symptoms requires a detailed history and careful physical examination.

Central nervous system diseases such as those affecting the basal ganglia produce rigidity and muscle spasm as one part of a syndrome of dystonia and parkinsonism. The rigidity of parkinsonism is accompanied by hypokinesia and bradykinesia of voluntary hand and finger movements, walking, and axial movement. In addition there is a reduction in automatic movements such as facial expression when talking and arm swing during walking. In dystonia, rigidity and spasm diminish at rest and disappear during sleep. The increase in muscle tone and abnormal postures of dystonia typically occurs on action, varies during movement, and subsides with rest.

Peripheral neuropathy may be complicated by acquired neuromyotonia with muscle stiffness, cramps, continuous twitching or rippling of muscles, and delayed relaxation after muscle contraction. Occasionally, these features are the sole manifestation of a peripheral nerve disorder as in Isaacs' syndrome. Electromyography provides valuable diagnostic information and is discussed later.

Delayed relaxation after muscle contraction is also a feature of the myotonic myopathies and gives rise to symptoms of stiffness and limitation of movement. Myotonia may be detected clinically as a delay in muscle relaxation, as persistent contraction after percussion of the muscle, or on electromyography as waxing and waning myotonic discharges. Cold weather may exacerbate myotonia.

Exercise-induced cramps are characteristic of metabolic myopathies such as McArdle's disease and phosphofructokinase deficiency. These cramps are painful, develop after or during exercise, and are accompanied by fatigue and myalgia. The painful cramps of McArdle's disease are due to muscle contracture resulting from a failure of muscle relaxation after contraction and are electrically silent on electromyography. Similar physiological mechanisms are responsible for exertion-related cramps and contracture in Brody's disease (Brody, 1969).

Several other myopathies are associated with muscle fibrosis and contracture. Contracture, or an electrically silent shortening of muscle, results in a limitation in the range of limb movement and may give rise to symptoms of muscle tightness or stiffness. Examples include polymyositis (elbow contracture), the rigid-spine syndrome (limb and paraspinal muscle contracture), Bethlem muscular dystrophy (elbow, fingers, and ankle contracture), and Emery-Dreifuss muscular dystrophy (spine, elbows, and finger contracture). The fixed muscle shortening of contracture persists during sleep or anesthesia.

Symptoms of muscle aches, stiffness, and cramps are common in endocrine myopathies, particularly hypothyroidism. Delayed muscle relaxation after contraction, muscle stiffness during and after voluntary contraction, localized swelling of muscle after percussion (myoedema), and muscle enlargement are signs of hypothyroid myopathy. Muscle contractures, which may be painful, also occur in Addison's disease.

Benign physiological cramps are a common cause of muscle cramp. Physiological muscle cramps typically occur at night after vigorous exercise, dehydration, and electrolyte disturbance. The cramps are painful and can be "broken" by stretching the cramping muscle but

TABLE I    The Differential Diagnosis of Muscle Stiffness, Rigidity, Cramps and Spasms

*Muscle stiffness, rigidity and spasms*
  Stiff-man syndrome
  Progressive encephalomyelitis with rigidity
  Rigidity associated with spinal cord lesions
  Axial torsion dystonia
  Akinetic rigid basal ganglia syndromes

*Muscle cramps, stiffness, and delayed muscle relaxation*
  Isaacs' syndrome (neuromyotonia)
  Schwartz-Jampel syndrome
  Myotonic syndromes
  Metabolic myopathies
    McArdle's disease
    Phosphofructokinase deficiency
    Brody's disease (deficiency of sarcoplasmic reticulum $Ca^{++}$ transport ATPase)
  Endocrine myopathies
    Hypothyroidism
    Addison's disease
  Benign (physiological) cramps

*Myopathies with contractures*
  Inflammatory myopathies (e.g., polymyositis-elbow contractures)
  Inherited myopathies
    The rigid-spine syndrome (cervical and thoracic spine)
    Bethlem muscular dystrophy (elbow, fingers, and ankles)
    Emery-Dreifuss muscular dystrophy (spine, elbows, and fingers)
  Muscle ischemia with contractures (Volkmann's ischemic contracture)

TABLE II    The Main Clinical and Electromyographic Features of the Stiff-Man Syndrome

*Clinical features*
  Gradual onset of stiffness in axial muscles
  Stiffness spreads to proximal limbs, legs more affected than arms
  Exaggerated lumbar lordosis caused by contraction of thoracolumbar paraspinal muscles
  Abdominal wall rigidity caused by contraction of abdominal muscles
  Rigidity abolished by sleep
  Stimulus sensitive painful muscle spasms

*Electromyographic features*
  Continuous motor unit activity
  Normal motor unit morphology
  Normal peripheral nerve conduction
  EMG activity abolished by sleep, peripheral nerve block, spinal or general anesthesia
  Enhanced exteroceptive (cutaneomuscular) reflexes

may recur if the muscle shortens again. Motor unit morphology and discharge patterns during cramps are normal.

Symptoms of muscle spasm, stiffness, and cramps are commonly reported in muscular or musculoskeletal pain syndromes (fibromyalgia). These complaints are typically less prominent than back pain, either cervical or lumbar, and are accompanied by tender and painful muscles but no other neurological signs. Evidence of rigidity or spasm on physical examination is unusual. In patients with considerable back pain and an abnormal trunkal or limb posture, the distinction between an antalgic or voluntary posture and neurological disease such as the stiff-man syndrome may be difficult. In this situation, electromyographical examination of paraspinal muscles is useful, because patients with fibromyalgia do not exhibit continuous motor unit activity when relaxed.

There remains a group of conditions in which muscle stiffness and rigidity are the sole complaints and result from continuous muscle activity. The stiff-man syndrome and related conditions that originate in the central nervous system and the peripheral syndromes of acquired neuromyotonia or Isaacs' syndrome are discussed in more detail later.

## STIFF-MAN SYNDROME

The earliest symptoms of the stiff-man syndrome (SMS) are the insidious onset of trunkal stiffness and rigidity most often in the fourth and fifth decades. Trunkal stiffness progresses slowly and is followed by involvement of proximal limb muscles, especially the legs. A characteristic finding is rigidity of lumbar and abdominal muscles, which produces lumbar hyperlordosis and "boardlike" rigidity of the abdominal wall (Moersch and Woltman, 1956). The range of trunkal movement may be severely restricted, limiting trunkal flexion and the normal fluidity of movement during walking. The axial rigidity persists when supine, but disappears during sleep. Spontaneous and reflex spasms to auditory or unexpected stimuli occur frequently and may lead to falls "like a wooden man" (Moersch and Woltman, 1956). Cutaneous stimuli such as light touch or stroking the sole of the foot while testing the plantar response may elicit reflex leg and trunk spasms. Brisk tendon reflexes and loss of abdominal skin reflexes may occur. The remainder of the neurological examination is normal.

The uniform increase in muscle tone (rigidity) is caused by continuous discharge of normal motor units that disappear during sleep and anesthesia indicating a central origin. Peripheral nerve conduction is normal. Enhanced excitability within spinal interneuronal networks that mediate cutaneomuscular or exteroceptive reflexes seem responsible for the continuous motor activity (Meinck *et al.*, 1984). These reflexes are exaggerated, fail to habituate, and contribute to the reflex jerks and spasms (Figure 1). Brainstem reflexes such as the blink reflexes and auditory or tactile startle response also may be exaggerated in the SMS (Figure 1).

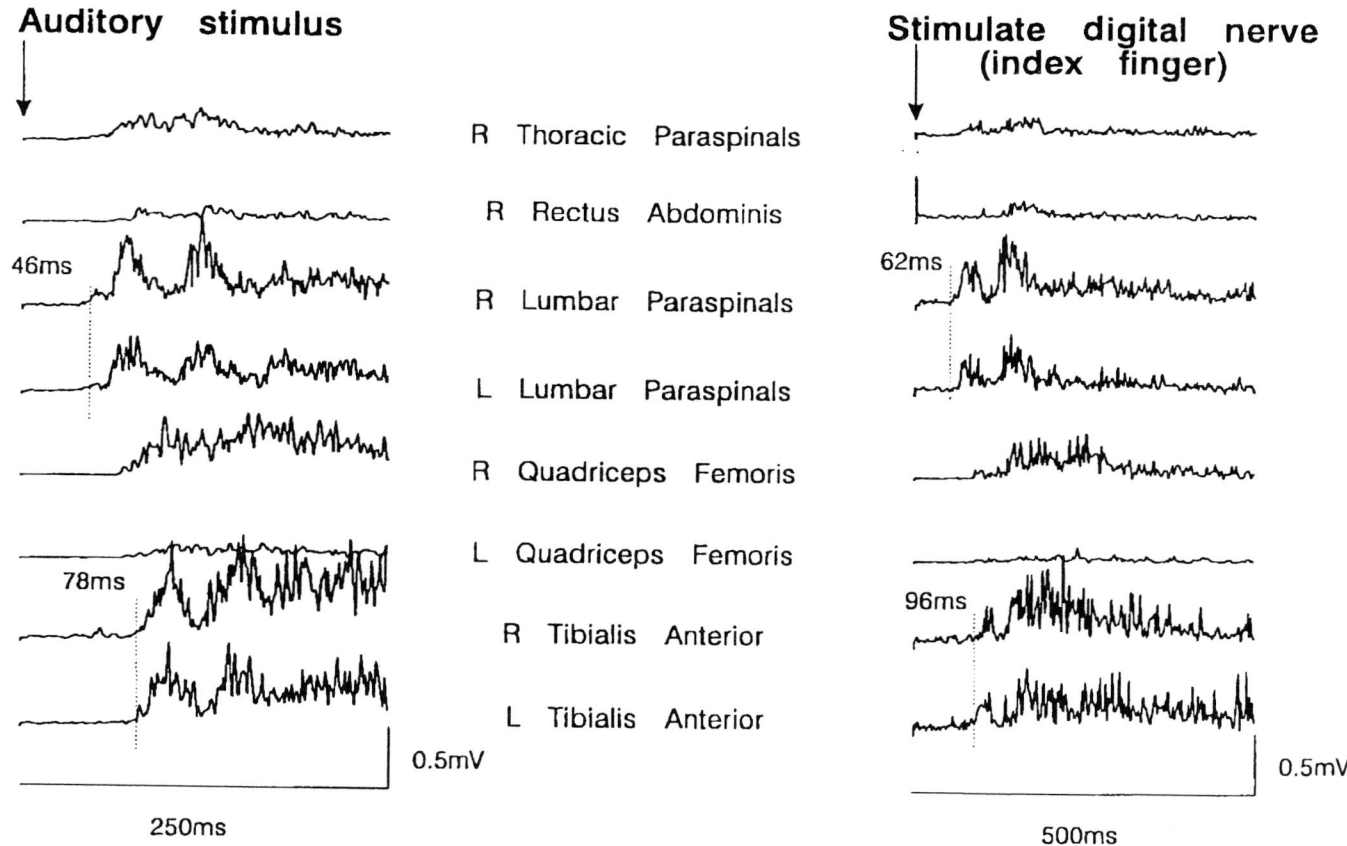

**FIGURE 1** Surface electromyographic (EMG) recordings from trunkal and leg muscles in a patient with SMS after an auditory stimulus (left panel) and digital nerve stimulation to the index finger (right panel). For both forms of stimulation, two or three phasic bursts of activity are followed by prolonged tonic muscle contraction corresponding to the typical muscle spasms of the SMS. Reproduced from Thompson (2001) with permission of the publisher.

Pharmacological evidence suggests reduced GABA-ergic and relatively enhanced noradrenergic activity within the central nervous system. This imbalance forms the rationale for treatment with drugs that enhance GABA activity (diazepam, baclofen) or reduce catecholamine effects (clonidine, tizanidine). Both reduce the severity of spasms (Meinck and Conrad, 1986).

AntiGAD antibodies in serum and cerebrospinal fluid are found in more than 60% of patients, and autoimmune endocrinopathies are present in 10%–20% of cases (Solimena et al., 1990). Antipancreatic islet cell (50%–60%), gastric parietal cell (50%), thyroid microsomal (30%–40%), and thyroglobulin antibodies (15%) also are found in those with antiGAD antibodies (Solimena et al., 1990). Oligoclonal IgG in cerebrospinal fluid is present in more than 50%.

The neuropathological findings in early case reports were inconsistent (Asher, 1958; Goetz and Klawans, 1983; Moersch and Woltman, 1956; Trethowan et al., 1960). Several reports have described spinal neuronal and interneuronal loss (Ishizawa et al., 1999; Martinelli

et al., 1978; Mitsumoto et al., 1991; Saiz et al., 1999; Sikes, 1959; Warich-Kirches et al., 1997), and brainstem or spinal perivascular lymphocytic cuffing (Armon et al., 1996; Meinck et al., 1994; Mitsumoto et al., 1991; Nakamura et al., 1986). These findings do not necessarily relate to the presence of antiGAD antibodies (Ishizawa et al., 1999; Mitsumoto et al., 1991; Saiz et al., 1999; Warich-Kirches et al., 1997).

Symptomatic treatment with benzodiazepines, baclofen, clonidine, tizanidine, or valproate alleviate spontaneous and stimulus-sensitive spasms, but axial rigidity and continuous motor unit activity respond less well. Immunological-based therapies with plasmapheresis, intravenous immunoglobulin, and high-dose methylprednisolone have been helpful (Meinck et al., 1994).

## VARIANTS OF STIFF-MAN SYNDROME

In the *jerking stiff-man* syndrome myoclonus and stimulus-sensitive spasms are prominent (Alberca et al.,

1982; Leigh *et al.*, 1980). Some reports of progressive encephalomyelitis with rigidity and myoclonus (PERM) have also used this term (Rothwell *et al.*, 1986). The concept of a *focal stiff-man syndrome* was introduced to describe cases in which lower limb rigidity and spasm remained confined to one limb for many years. These cases exhibited antiGAD antibodies and enhanced exteroceptive reflexes (Saiz *et al.*, 1998). The SMS also may begin focally then spread to the trunk and other limbs (Thompson, 2001). The *"stiff-leg syndrome"* refers to chronic nonprogressive stiffness of one limb attributed to "chronic spinal interneuronitis" without additional clinical or autoimmune features of the SMS (Brown *et al.*, 1997). A *paraneoplastic stiff-man syndrome* of muscle stiffness, particularly beginning in the upper limb, occurs in women with breast cancer and antibodies to amphiphysin I, a 128-kD neuronal protein concentrated in nerve terminals (Folli *et al.*, 1993). Neoplasia of the colon, Hodgkin's disease, thymoma (Grimaldi *et al.*, 1993), and lung (Bateman *et al.*, 1990) have all been described in association with SMS. Segmental rigidity and spasm affecting the trunk and leg may also occur as a paraneoplastic phenomenon (Roobol *et al.*, 1987).

## Progressive Encephalomyelitis with Rigidity and Myoclonus (PERM)

Axial rigidity and muscle jerks, similar to the SMS, are also the presenting features of this rare condition. However, the onset is subacute, evolving over weeks to months, and is often followed by a relentlessly progressive course. Sensory symptoms, severe rigidity, lower motor neuron signs of wasting, weakness and segmental denervation, brainstem myoclonus, areflexia, extensor plantar responses, cranial nerve, and brain stem involvement (nystagmus, opsoclonus, ophthalmoplegia, deafness, dysarthria, dysphagia) all suggest PERM (Kasperek and Zebrowski, 1971; Whiteley *et al.*, 1976). Investigations may reveal cerebrospinal fluid lymphocytic pleocytosis, elevated protein and oligoclonal IgG bands, and abnormal signal intensity in the brainstem and cervical spinal cord on MRI (McCombe *et al.*, 1989).

Pathological findngs include encephalomyelitis, perivascular lymphocyte cuffing and infiltration (Barker *et al.*, 1998; Campbell and Garland, 1956; Howell *et al.*, 1979), and neuronal and interneuronal loss in spinal central gray matter. In some cases of PERM, there is evidence of an autoimmune process with antiGAD antibodies, gastric parietal, thyroid microsomal, thyroglobulin, or the acetylcholine receptor (Burn *et al.*, 1991). The precise relationship of PERM to the SMS remains the subject of debate.

Diazepam and baclofen may provide symptomatic treatment for the jerks and spasms. Methylpredisolone may also be beneficial (McCombe *et al.*, 1989; Meinck *et al.*, 1994).

### Spinal Cord Lesions and Rigidity

Structural lesions of the central spinal cord, extensive demyelination, and watershed infarction of central spinal gray matter isolate spinal alpha motor neurons from inhibitory interneuronal control, leading to continuous motor unit activity and spinal or "alpha" rigidity (Gelfan and Tarlov, 1959). There may be additional lower motor neuron signs of wasting and weakness, depending on the extent of segmental disease. Accordingly, it is important to exclude the possibility of a structural lesion with appropriate imaging in all cases of segmental rigidity and spasm.

## ISAACS' SYNDROME AND ACQUIRED NEUROMYOTONIA

Inherited or acquired neuromyotonia, also referred to as Isaacs' syndrome, is characterised by continuous spontaneous muscle activity with myokymia (rippling of muscles) and fasciculations. Delay in muscle relaxation after voluntary contraction produces muscle stiffness and cramps (Isaacs, 1964; Newsom-Davis and Mills, 1993). All muscle groups are affected with some preponderance in distal and facial muscles. The muscle contraction may result in abnormal postures of the extremities and trunk. Continuous muscle contraction may be accompanied by muscle aches and sweating. Tendon reflexes usually are absent. Muscle activity persists during narcosis but lessens with peripheral nerve or spinal nerve root anaesthesia and is abolished by neuromuscular blockade, indicating a peripheral nerve origin. Neuromyotonia may be inherited without other evidence of peripheral nerve disease (Ashizawa *et al.*, 1983; Auger *et al.*, 1984) and occurs in association with hereditary (Lance *et al.*, 1979; Vasilescu *et al.*, 1984) and inflammatory demyelinating (Valenstein *et al.*, 1978) neuropathies. Neuromyotonia has also been described in association with small-cell lung cancer (Walsh, 1976), thymoma (Halbach *et al.*, 1987), and antibodies to voltage-gated potassium channels (Newsom-Davis and Mills, 1993).

The electrophysiological findings of fibrillations, fasciculations, myokymic, and high-frequency doublet or triplet discharges are diagnostic of neuromyotonia (Figure 2). After discharges follow percussion of peripheral nerves, gentle voluntary muscle activation, and the compound muscle action potential (M-wave) to periph-

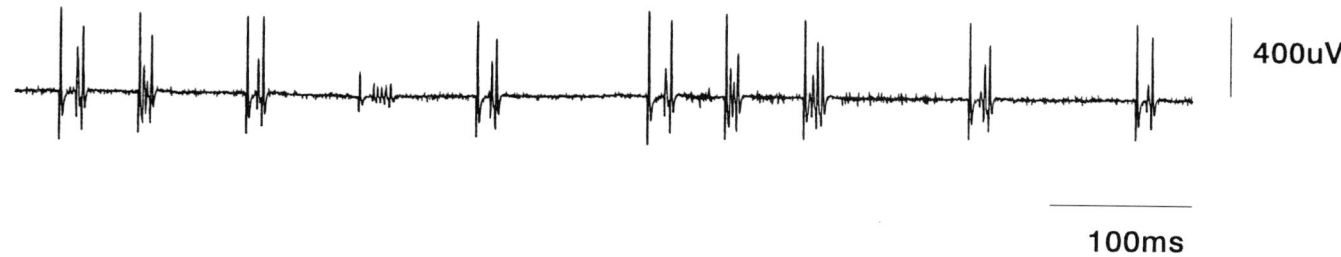

**FIGURE 2** Needle EMG recording of spontaneous muscle fiber (small repetitive discharge) and motor unit discharges from the medial gastrocnemius in a patient with acquired neuromyotonia. Reproduced from Thompson (2001) with permission of the publisher.

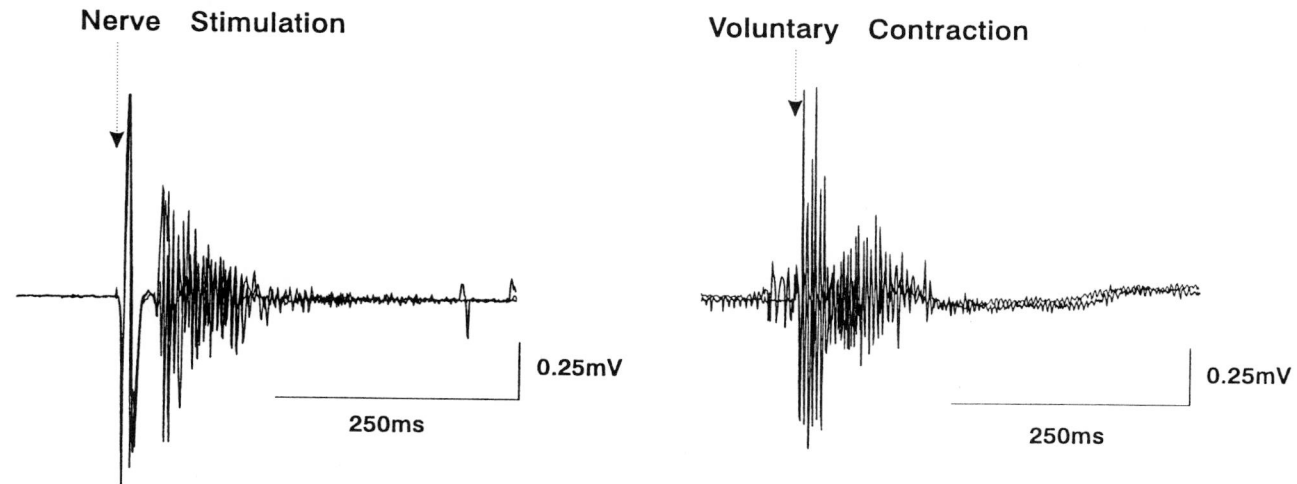

**FIGURE 3** After-discharges in tibialis anterior after peroneal nerve stimulation (left panel) and voluntary dorsiflexion of the ankle (right panel) in a patient with acquired neuromyotonia. Reproduced from Thompson (2001) with permission of the publisher.

eral nerve stimulation (Figure 3). Abnormal peripheral nerve conduction suggests an underlying neuropathy. The peripheral neuromuscular hyperexcitability and multiple muscle fiber and motor unit discharges may arise from ephaptic excitation or local circuits of re-excitation within peripheral nerves and terminal motor axons or enhanced membrane excitability caused by potassium channel blockade (Newsom-Davis and Mills, 1993).

Carbamazepine and phenytoin abolish the continuous motor activity, and muscle stiffness resolves. Immunological-based therapies including plasmapheresis have been used when an autoimmune basis has been suspected (Newsom-Davis and Mills, 1993).

## SCHWARTZ-JAMPEL SYNDROME

The characteristic skeletal abnormalities of this syndrome, short stature and spondyloepiphyseal dysplasia,

are caused by high-frequency muscle discharges and muscle stiffness. Spasm of facial muscles narrows the palpebral fissures (blepharophimosis), closes the mouth, and produces a dimpled chin. The muscle activity is thought to be of peripheral muscular origin and related to abnormal $Na^+$ channel opening in muscle fibers (Lehmann-Horn *et al.*, 1990).

## REFERENCES

Alberca, R., Romero, M., and Chaparro, J. (1982). Jerking stiff-man syndrome. *J. Neurol. Neurosurg. Psychiatry* **45**, 1159–1160.

Armon, C., Swanson, J. W., McLean, J. M. *et al.* (1996). Subacute encephalomyelitis presenting as stiff-person syndrome: clinical, polygraphic and pathologic correlations. *Mov. Disord.* **11**, 701–709.

Asher, R. (1958). A woman with the stiff-man syndrome. *BMJ* **1**, 265–266.

Ashizawa, T., Butler, I. J., Harati, Y., and Roongta, S. M. (1983). A dominantly inherited syndrome with continuous motor neuron discharges. *Ann. Neurol.* **13**, 285–290.

Auger, R. G., Daube, J. R., Gomez, M. R., and Lambert, E. H. (1984). Hereditary form of sustained muscle activity of peripheral nerve origin causing generalized myokymia and muscle stiffness. *Ann. Neurol.* **15**, 13–21.

Barker, R. A., Revesz, T., Thom, M., Marsden, C. D., and Brown, P. (1998). Review of 23 patients affected by the stiff man syndrome: clinical subdivision into stiff trunk (man) syndrome, stiff limb syndrome, and progressive encephalomyelitis with rigidity. *J. Neurol. Neurosurg. Psychiatry* **65**, 633–640.

Bateman, D. E., Weller, R. O., and Kennedy, P. (1990). Stiff-man syndrome: a rare paraneoplastic disorder. *J. Neurol. Neurosurg. Psychiatry* **53**, 695–696.

Brody, I. (1969). Muscle contracture induced by exercise. A syndrome attributable to decreased relaxing factor. *N. Engl. J. Med.* **281**, 187–192.

Brown, P., Rothwell, J. C., and Marsden, C. D. (1997). The stiff-leg syndrome. *J. Neurol. Neurosurg. Psychiatry* **62**, 31–37.

Burn, D. J., Ball, J., Lees, A. J., Behan, P. O., and Morgan-Hughes, J. A. (1991). A case of progressive encephalomyelitis with rigidity and positive antiglutamic acid dehydrogenase antibodies. *J. Neurol. Neurosurg. Psychiatry* **54**, 449–451.

Butler, M. H., Hayashi, A., Ohkoshi, N. *et al.* (2000). Autoimmunity to gephyrin in stiff-man syndrome. *Neuron* **26**, 307–312.

Campbell, A. M. G., and Garland, H. (1956). Subacute myoclonic spinal neuronitis. *J. Neurol. Neurosurg. Psychiatry* **19**, 268–274.

Folli, F., Solimena, M., Cofiell, R. *et al.* (1993). Autoantibodies to a 128-kd synaptic protein in three women with the stiff-man syndrome and breast cancer. *N. Engl. J. Med.* **328**, 546–551.

Gelfan, S., and Tarlov, I. M. (1959). Interneurones and rigidity of spinal origin. *J. Physiol.* **146**, 594–617.

Goetz, C. G., and Klawans, H. L. (1983). On the mechanism of sudden death in Moersch-Woltman syndrome. *Neurology* **33**, 930–932.

Grimaldi, L. M. E., Martino, G., Braghi, S. *et al.* (1993). Heterogeneity of autoantibodies in stiff-man syndrome. *Ann. Neurol.* **34**, 57–64.

Halbach, M., Homberg, V., and Freund, H.-J. (1987). Neuromuscular, automomic and central cholinergic hyperactivity associated with thymoma and acetylcholine receptor-binding antibody. *J. Neurol.* **234**, 433–436.

Howell, D. A., Lees, A. J., and Toghill, P. J. (1979). Spinal internuncial neurones in progressive encephalomyelitis with rigidity. *J. Neurol. Neurosurg. Psychiatry* **42**, 773–785.

Isaacs, H. (1964). A syndrome of continuous muscle-fibre activity. *J. Neurol. Neurosurg. Psychiatry* **24**, 319–325.

Ishizawa, K., Komori, T., Okayama, K. *et al.* (1999). Large motor neuron involvement in stiff man syndrome: a qualitative and quantitative study. *Acta Neuropathologica* **97**, 63–70.

Kasperek, S., and Zebrowski, S. (1971). Stiff-man syndrome and encephalomyelitis. *Arch. Neurol.* **24**, 22–31.

Lance, J. W., Burke, D., and Pollard, J. (1979). Hyperexcitability of motor and sensory axons in neuromyotonia. *Ann. Neurol.* **5**, 523–532.

Lehmann-Horn, F., Iaizzo, P. A., Franke, C., Hatt, H., and Spaans, F. (1990). Schwartz-Jampel syndrome 2. Na+ channel defect causes myotonia. *Muscle Nerve* **13**, 528–535.

Leigh, P. N., Rothwell, J. C., Traub, M., and Marsden, C. D. (1980). A patient with reflex myoclonus and muscle rigidity: "Jerking stiff-man syndrome." *J. Neurol. Neurosurg. Psychiatry* **43**, 1125–1131.

Martinelli, P., Pazzaglia, P., Montagna, P. *et al.* (1978). Stiff-man syndrome associated with nocturnal myoclonus and epilepsy. *J. Neurol. Neurosurg. Psychiatry* **41**, 458–462.

McCombe, P. A., Chalk, J. B., Searle, J. W., Tannenberg, A. E. G., Smith, J. J., and Pender, M. P. (1989). Progressive encephalomyelitis with rigidity: a case report with magnetic resonance imaging findings. *J. Neurol. Neurosurg. Psychiatry* **52**, 1429–1431.

Meinck, H. M., and Conrad, B. (1986). Neuropharmacological investigations in the stiff man syndrome. *J. Neurol.* **233**, 114–122.

Meinck, H. M., Ricker, K., and Conrad, B. (1984). The stiff-man syndrome: new pathophysiological aspects from abnormal exteroceptive reflexes and the response to clomipramine, clonidine and tizanidine. *J. Neurol. Neurosurg. Psychiatry* **47**, 280–287.

Meinck, H. M., Ricker, K., Hülser, P. J., Schmid, E., Peiffer, J., and Solimena, M. (1994). Stiff man syndrome: clinical and laboratory findings in eight patients. *J. Neurol.* **241**, 157–166.

Mitsumoto, H., Schwartzman, M. J., Estes, M. L. *et al.* (1991). Sudden death and paroxysmal autonomic dysfunction in stiff-man syndrome. *J. Neurol.* **238**, 91–96.

Moersch, F. P., and Woltman, H. W. (1956). Progressive fluctuating muscular rigidity and spasm ("stiff man syndrome"): report of a case and some observations in 13 other cases. *Mayo Clin. Proc.* **31**, 421–427.

Nakamura, N., Fujiya, S., Yahara, O., Fujioka, Y., and Kawakami, Y. (1986). Stiff-man syndrome with spinal cord lesion. *Clin. Neuropathol.* **5**, 40–46.

Newsom-Davis, J., and Mills, K. R. (1993). Immunological associations of acquired neuromyotonia. *Brain* **116**, 453–469.

Roobol, T. H., Kazzaz, B. A., and Vecht, C. J. (1987). Segmental rigidity and spinal myoclonus as a paraneoplastic syndrome. *J. Neurol. Neurosurg. Psychiatry* **50**, 628–631.

Rothwell, J. C., Obeso, J. A., and Marsden, C. D. (1986). Electrophysiology of somatosensory reflex myoclonus. *Adv. Neurol.* **43**, 385–398.

Saiz, A., Graus, F., Valldeoriola, F., Valls-Sole, J., and Tolosa, E. (1998). Stiff leg syndrome: A focal form of stiff man syndrome. *Ann. Neurol.* **43**, 400–403.

Saiz, A., Minguez, A., Graus, F., Marin, C., Tolosa, E., and Cruz-Sanchez, F. (1999). Stiff man syndrome with vacuolar degeneration of anterior horn motor neurons. *J. Neurol.* **246**, 858–860.

Sikes, S. Z. (1959). Stiff man syndrome (analysis and case report with spinal cord autopsy). *Dis. Nerv. Syst.* **20**, 254–258.

Solimena, M., Folli, F., Morello, F., Bottazzo, G. P., Toso, V., and DeCamilli, P. (1990). Autoantibodies to GABAergic neurons and pancreatic beta cells in stiff-man syndrome. *N. Engl. J. Med.* **322**, 1555–1560.

Thompson, P. D. (2001). Stiff man syndrome and related disorders. *Parkinsonism Rel. Disord.* **8**, 147–153.

Trethowan, W. H., Allsop, J. L., and Turner, B. (1960). The stiff-man syndrome. *Arch. Neurol.* **3**, 114–122.

Valenstein, E., Watson, R. T., and Parker, J. L. (1978). Myokymia, muscle hypertrophy and percussion "myotonia" in chronic recurrent polyneuropathy. *Neurology* **28**, 1130–1134.

Vasilescu, C., Alexianu, M., and Dan, A. (1984). Neuronal type of Charcot-Marie-Tooth disease with a syndrome of continuous motor unit activity. *J. Neurol. Sci.* **63**, 11–25.

Walsh, J. C. (1976). Neuromyotonia: an unusual presentation of intrathoracic malignancy. *J. Neurol. Neurosurg. Psychiatry* **39**, 1086–1091.

Warich-Kirches, M., Von Bossanyi, P., Treuheit, T. *et al.* (1997). Stiff man syndrome: possible autoimmune aetiology targeted against GABA-ergic cells. *Clin. Neuropathol.* **16**, 214–219.

Whiteley, A. M., Swash, M., and Urich, H. (1976). Progressive encephalomyelitis with rigidity: its relation to subacute myoclonic spinal neuronitis and to the stiff man syndrome. *Brain* **99**, 27–42.

*Endocrine and Autonomic Function*

## CHAPTER 102
# Neuroendocrine Disorders

Oliver Kastrup and A. Bernard Pleet[†]

This chapter deals with disorders of the pituitary and hypothalamus. Other endocrine disorders are treated insofar as they affect the nervous system. The section on acromegaly provides detailed information about therapeutic approaches to pituitary adenoma that will be referred to in other sections. Diagnostic aspects of hypothalamic or pituitary disease are compiled in a section on hypopituitarism and hypothalamic dysfunction. Diabetes mellitus, pediatric, and psychiatric endocrine syndromes will not be discussed. The treatment of neuroendocrine disorders requires cooperation between neurologist and endocrinologist.

## ACROMEGALY

### Clinical Aspects

Acromegaly (AM) is caused by excessive secretion of growth hormone (GH), resulting in bony and soft tissue overgrowth. The clinical findings are insidious in onset and protean in nature. Tumor growth may cause headache (50%), visual field deficits (14%–49%), hypopituitarism (9%–20%), hyperprolactinemia (20%–40%) with loss of libido, impotence, and amenorrhea. GH excess may cause growth of hands and feet and coarsening of facial features (in children gigantism), osteoporosis, arthropathy, hyperhidrosis (60%–80%), abnormal glucose tolerance (50%), hypertension (30%), hypertrophic cardiomyopathy (70%), and goiter (50%). Thyroid blockage should be performed before the application of iodine-containing x-ray contrast media. Neurological symptoms of AM include median and ulnar nerve compression syndrome (40%), neuropathy, proximal myopathy (50%, with hypertrophy followed by atrophy, myopathic EMG changes, and elevated creatine kinase [CK]), hypersomnolence, and depression.

[†]Deceased.

The diagnosis is suspected when basal GH levels are >2 ng/ml and is likely at >5 ng/ml. The standard screening test is the determination of GH 60 minutes and 120 minutes after oral glucose loading (100 g); GH levels <5 ng/ml are considered normal. The tests may be affected by the pulsatile releasing pattern and the short half-life time of GH (20–30 minutes). GH levels are usually very low and peak at irregular times of day after exercise, meals, or sleep. Somatomedin C (SmC) is a protein synthesized by the liver in response to stimulation by GH. It is regularly elevated when GH is overproduced, even when the nadir of the pulse of GH is normal. Simultaneous measurement of SmC is confirmatory for abnormally high production of GH and obviates false-negative results when pulsatile GH is measured during a trough. GH levels may be elevated in anorectics.

Differential diagnosis includes abnormal hypothalamic growth hormone releasing hormone (GRH) secretion (<1%) and GRH-producing tumors of, for example, lung or thyroid.

### Natural Course

The incidence of AM is 5.3/100,000/year; it occurs most frequently in the fourth and fifth decades. The most frequent cause is an eosinophilic adenoma (10%–20%) with resulting partial hypopituitarism. Growth hormone–releasing hormone (GHRH) secreting tumors are rare: 15% of all pituitary adenomas secrete GH, 15%–20% are microadenomas (diameter <10 mm); 40%–50% are macroadenomas (>10 mm); 20%–40% are invasive (although not necessarily malignant!) macroadenomas. Twenty-five percent to 40% of GH-secreting adenomas also produce prolactine. Ectopic GH secretion is rare; ectopic GHRH secetetion is found in cases of carcinoid or islet cell tumors. Familial AM occurs in the syndrome of multiple endocrine neoplasia (MEN).

The median time elapsing before diagnosis is 6.6 years. As a consequence of hypertrophic cardiomyopathy, arrhythmia, and vascular disease, only 26%–50% of AM patients reach 50 years of age. Treatment of AM is therefore essential. Successful therapy (GH levels <5 ng/ml) may restore life expectancy to normal (Barkan, 1989; Bates et al., 1993).

## Principles of Therapy

The goal is to achieve normal GH levels. GH acts through a series of intermediary hormones (somatomedins [Sm]). SmC or insulin-like growth factor1 (IGF-1) is the most important Sm and is synthesized in the liver, with a half-life of 5–18 hours. Normal ranges for IGF-1 are not established and may vary between laboratories. IGF-1 levels are useful for both diagnosis and for monitoring therapy.

## Surgery

Surgical removal of the pituitary adenoma is clearly required in the presence of a mass effect and when pharmacological treatment is unavailable or proves ineffective. Acute loss of visual acuity caused by pituitary apoplexy or hemorrhage necessitates emergency treatment. Surgery is most often done through a transsphenoidal route. Faster access, better visualization of the sellar content and, a lower risk of lesioning cranial nerves I and II argue in favor of this procedure, as does the absence of risk for epilepsy or damage to the frontal lobes***.

Complications of transsphenoidal surgery include rhinorrhea (5%), meningitis (2.5%), rarely ischemic infarction, and diabetes insipidus. Perioperative mortality is 0.9% (Molitch, 1988).

A transcranial operative approach is preferable when the growth occurs parasellar, strictly suprasellar, invades the subfrontal or retrochiasmal region, or when a nonadenomatous lesion is suspected (Adams and Burke, 1993; Stolke, 1989)***.

For AM cases, the rates of success cited vary widely, because different outcome measures are applied. If postoperative GH levels <5 ng/ml are accepted, 63% of surgically treated patients will be found to be in remission (David et al., 1993). Anticipated rates of remission after 6 years are microadenoma (78%), macroadenoma (33%), intrasellar tumor (27%–90%), intrasellar and suprasellar growth (34%–68%) (Molitch, 1988; Ross and Wilson, 1988; Zervas, 1984).

Recurrence of AM is unlikely when glucose loading and thyroid-releasing hormone tests are normal and GH levels are <5 ng/ml; otherwise recurrence occurs in 50% of the cases (Arafah et al., 1980). Visual field deficits existing before surgery will recover completely in 30% and partially in 50%.

## Radiotherapy

Radiotherapy in addition to surgery is recommended when, 3 months after surgery, GH levels are still >5 mg/ml, when removal is incomplete, or when the adenoma is invasive or malignant. Conventional high-voltage radiotherapy uses 45 Gy over 5 weeks in single doses of 1.8 Gy. Therapeutic effects are detected within 6 months and are maximum at 4 years. Large adenomas must be surgically reduced before radiotherapy. When primary radiotherapy is used, basal GH levels after 2 years are found to be less than 5 ng/ml in 17%, after 5 years in 44%, and after 10 years in 65%.

Grigsby et al. (1989) and Clarke et al. (1993) reviewed the efficacy of combined surgical and radiological treatment. For all pituitary adenomas they found an average remission rate of 85%–89% after 10 years, without increased mortality. There was a strong correlation between radiation dosage and outcome: remission rates were 94% for 50–54 Gy, 85% for 40–49 Gy, and 75% for 30–39 Gy. The overall progression-free survival was 94% in 10 years and 88% in 20 years, whereas the relative risk for death was 1.76, compared with a normal control population (Brada et al., 1993)***.

Radiation damage to the optic chiasm and optic nerve is avoidable when single doses do not exceed 2 Gy. Hypopituitarism is not avoidable and may appear many years after successful treatment (Bamberg et al., 1989); see Table I.

The risk of radiation-associated glioma is estimated at 1.9%–2.7% (within 15–20 years, 45–50 Gy) according to two large retrospective studies of patients undergoing surgery and radiotherapy for pituitary adenoma (Brada et al., 1993; Tsang et al., 1993)***.

Transsphenoidal radionuclide implantation (TRI) is advisable only for localized intrasellar adenomas. Local

TABLE I    Risk of Endocrine Failure after Combined Surgery and Radiotherapy

|  | Preoperative | Years after surgery | |
| --- | --- | --- | --- |
|  |  | 5 | 10 |
| Hypothyroidism | 9% | 12% | 19% |
| Adrenal failure | 6% | 30% | 38% |
| Male ypogoadism | 13% | 41% | 58% |
| Female hypogonadism | 19% | 44% | 50% |

Note: 100%, all irradiated patients.

doses of 140–160 Gy may be used (Mundinger, 1985). The adverse effects are damage to the chiasm and the cavernous sinus. The remission rate (GH <5 ng/ml) for TRI after surgery is 50% within 6 months. When TRI is used as the sole therapy, the remission rate is 18% after 1 year, 53% after 10 years, and 76% after 20 years (Jadresic et al., 1987; Quabbe, 1982). Hypopituitarism occurs in 30% of cases. Heavy particle radiation therapy entails the application of 120–150 Gy through alpha particle or proton radiation. It is effective only with strictly intrasellar adenomas that are well defined and have not extended beyond the confines of the sella turcica. It takes some years to develop its effects (75% achieve GH levels <5 ng/ml within 10 years), but the treatment is accomplished in one single radiation session (Kliman et al., 1984). The facilities for applying proton or alpha particle beams are sparse (in Europe, Schweizer Institut für Nuklearforschung, Villingen; Karolinska-Institut, Stockholm, in the United States, Harvard Cyclotron). One study comparing both heavy-particle and conventional high-voltage radiation found that both methods are equally effective, but fewer complications arose from conventional radiotherapy (Ludecke et al., 1989)**.

### Pharmacotherapy

Bromocriptine acts on dopamine receptors present on some adenomas. It is most effective when the adenomas secrete both GH and prolactin. In 70% of patients GH levels are lowered, but only 10%–20% achieve GH levels <5 ng/ml, and the tumor mass is rarely significantly reduced (Molitch, 1988; Molitch et al., 1985). Therapy must be started with doses of 1.25 mg/day and may be increased by 1.25 mg/day every second week to a total dose of 40 mg/day. Side effects include hypotension, psychosislike states, and nausea. Other dopaminergic agents such as lisuride, pergolide, and cabergoline fare similarly. Cabergoline has the advantage of a once- or twice-weekly dosage (range from 0.5–4.5 mg)***.

Octreotide is a long-acting analog of somatostatin. At doses of 100 mg (range, 50–250 mg) tid, a reduction of tumor mass occurs in 30%–40%, and GH levels <5 ng/ml are achieved in 80%–90% of microadenomas and in 50% of macroadenomas (Melmed, 1993). The drug must be administered subcutaneously, and continuous treatment seems to be superior to intermittent (Riedel et al., 1992). Octreotide treatment before surgery produces better remission rates because of adenoma mass reduction (Stevenaert and Beckers, 1993). When treatment has become effective, it should not be discontinued; even after 3–4 years in remission, relapse will occur within 4 weeks (Plockinger et al., 1993). The treatment is begun at doses of 2 × 50 μg/day and gradually increased to 3 × 100 μg. The maximal dose is 3 × 500 μg. Side effects include gastritis, cholelithiasis, abnormal glucose tolerance, steatorrhea, and headache.

### Practical Management

If feasible, GH-producing adenomas should be removed by transsphenoidal surgery. Surgery for non-invasive adenomas should be preceded by 3–4 months of treatment with octreotide.

When surgery does not lead to GH levels <5 ng/ml, long-term treatment with octreotide should be instituted and if ineffective or not tolerated with bromocriptine or cabergoline.

If after 3–4 months elevated GH levels persist, radiotherapy should be used. Pharmacotherapy should be continued during and after radiotherapy. To evaluate whether radiotherapy has been effective, pharmacotherapy is discontinued for 4 weeks each year, and GH levels are measured (Young, 1990). If the patient is not a surgical candidate, primary pharmacotherapy with octreotide, or alternatively with bromocriptine or cabergoline, should be begun. Pregnancy and lactation do not preclude pharmacotherapy but require critical evaluation. If not effective, radiotherapy should follow after 3–4 months. Primary radiotherapy is advised only when surgery is deemed too dangerous and pharmacotherapy is ineffective.

Routine follow-up in all cases includes a neurological and endocrinological examination twice a year, including evaluation of visual acuity and the visual fields.

### Treatments No Longer Recommended

Therapy with estrogens has been abandoned. Estrogens act as peripheral GH antagonists but stimulate adenoma growth.

## CUSHING'S DISEASE AND CUSHING'S SYNDROME

### Clinical Aspects

Cushing's syndrome is due to excess secretion of cortisol. In 85% of cases the cause is a pituitary adenoma secreting adrenocorticotropic hormone (ACTH). In the case of an ACTH-secreting pituitary adenoma, the constellation of symptoms and signs is termed Cushing's disease (CD). Clinical findings are increase of facial and central adipose tissue (97%), skin atrophy, acne, hirsutism, striae, pigmentation (80%), hypertension (82%), edema (62%), thromboembolism, amenorrhea, impotence (77%), polyuria (23%), nephrolithiasis, osteoporosis, abnormal glucose tolerance (90%); in children, frequent viral and bacterial infections and growth retardation. Neurologic and psychiatric signs may manifest before telltale physical signs evolve: steroid myopathy (87%, proximal muscle weakness and type II fiber atrophy with normal CK), personality disorders (66%), and depressive or manic bouts, sleeplessness.

Differential diagnosis includes iatrogenic Cushing's syndrome, ectopic ACTH or corticotropin-releasing hormone (CRH)–secreting tumors (carcinoma of the bronchus or pancreas, thymoma, hypothalamic growths), as well as cortisol-releasing adrenocortical tumors (50% are adenomas, 50% carcinomas). In cases of suspected pituitary adenomas, petrosal sinus catheterization may confirm the source of the excess ACTH secretion. Occasionally, hypothalamic lesions (granulomatous inflammation, metastases, craniopharyngiomas) may cause CD.

Patients with chronic alcoholism may have the clinical signs of CD, including elevated urine 17-hydroxysterolds, absent diurnal rhythmicity, and impaired suppression of cortisol production after dexamethasone administration. The laboratory findings typical of CD may be encountered in depressed patients who also exhibit elevated CSF levels of corticotropin-releasing factor (CRF).

## Natural Course

CD is four times more common in women than in men. The course depends strongly on the cause, ranging from benign in the case of pituitary adenomas to fatal with pancreatic carcinoma. Patients with ACTH-secreting bronchial carcinomas fare worst, with a median life expectancy of <3 years. In untreated CD, life expectancy is reduced because of hypertension, abnormal glucose tolerance, and thromboembolism.

## Laboratory Findings

The diagnosis is established by excess urinary free cortisol secretion, loss of circadian secretion pattern, or failure of 17-hydroxycorticosteroid suppression after dexamethasone administration. Pathological laboratory findings may include hypokalemia and leukopenia. Single exploratory determinations of plasma ACTH or cortisol are useless and the following routine is suggested: (1) Determination of 24-hour urinary cortisol. CD is unlikely with urinary cortisol excretion <75 μg/24h and likely above 150 g/24h. (2) Diurnal profile for cortisol and ACTH secretion. ACTH >400 pg/ml indi-

cates a pituitary adenoma or ectopic ACTH secretion. (3) Dexamethasone suppression test. Because ACTH-secreting adenomas are often small (about 3 mm) and may evade conventional CT, high-resolution gadolinium MRI is recommended, which shows microadenoma in about 50% of cases. In cases that still remain unclear, catheterization of the petrosal sinus can be performed to localize the ACTH production.

Both with pituitary adenomas and with ectopic ACTH or CRH production, plasma ACTH is elevated; however, plasma ACTH is low in the case of adrenal tumors. Pituitary adenomas may be distinguished from ectopic ACTH production by greatly diminished response to dexamethasone, metyrapone, and CRH of ectopic tumors.

## Principles of Therapy

Therapy is directed at normalizing cortisol secretion. General aspects of surgery and radiotherapy for pituitary adenornas are as outlined for acromegaly.

## Surgery

CD (caused by intrasellar adenoma) can be cured by transsphenoidal adenomectomy in 90% of cases, success being apparent immediately after surgery. There may be an immediate postoperative glucocorticoid deficiency, because nontumorous ACTH-producing cells are suppressed by the adenoma's hormonal activity and may take months to recover. Long-term monitoring of successfully operated patients is advised, because 8%–14% have a recurrence within 2–3 years.

Bilateral adrenalectomy is resorted to when the pituitary tumor cannot be removed or the source of ACTH secretion cannot be established. In this case, postoperative replacement of glucocorticoids and mineralocorticoids is required. A further complication of bilateral adrenalectomy is the occurrence of Nelson's syndrome (hyperpigmentation and chromophobic ACTH/MSH-secreting pituitary adenoma), which occurs in 10% of adrenalectornized patients. Nelson's tumors must be treated by surgery or radiation, because they are prone to malignant degeneration (Martin and Reichlin, 1987).

## Radiotherapy

Primary radiotherapy is indicated only in patients with contraindications to surgery or in those refusing surgery and is effective in 50%–83%. In contrast to radiation for acromegaly, it is rapidly effective (within 1–8 months), and a recurrence of symptoms within 10 years is rare (Grigsby *et al.*, 1989). Patients with CD seem to be more susceptible to radiation-associated necrosis of brain tissue (8% within 6 years, 50 Gy; Grattan-Smith *et al.*, 1992)***.

### Pharmacotherapy

Pharmacotherapy is limited to preoperative treatment, or when the source of ACTH is either unknown or cannot be surgically approached. Bromocriptine is effective in <10% of the cases. Cyproheptadine is a centrally active 5HT-antagonist but works in less than 50% of the patients (Al-Damluji and Rees, 1988). In some patients, sodium valproate may reduce ACTH secretion. Metapyrone and aminogluthedimide are fast-acting inhibitors of adrenal steroid synthesis. Side effects include hirsutism, depression, and drowsiness (Jeffcoate *et al.*, 1977). Irreversible chemical adrenalectomy can be performed using mitotane or ketoconazole. Adverse effects are hepatotoxicity and neurotoxicity; lifelong steroid substitution is required.

### Practical Management

When the source of excess ACTH production is known, it should be removed. After successful surgery, intermediate glucocorticoid deficiency may occur. It is therefore advised to routinely supply 100 mg hydrocortisone every 8 hours intravenously on the day of surgery. Within 2–3 days, this may be reduced to 37.5 mg and in most cases discontinued after 1–2 weeks (Carpenter, 1990). When surgery cannot be performed, radiotherapy (conventional, alpha beam, or radionuclide implantation) is recommended. When the source cannot be located, pharmacotherapy acting either on ACTH or cortisol production may be attempted. If this fails, the remaining alternatives must be evaluated together with the patient (i.e., surgical or chemical adrenalectomy, meaning lifelong substitution of mineralocorticoids and glucocorticoids).

## DIABETES INSIPIDUS

### Clinical Aspects

Diabetes insipidus (DI) manifests with excessive thirst, polydipsia, and polyuria. The cause is defective regulation or secretion of antidiuretic hormone (ADH). Complaints often arise only when daily urine excretion surpasses 3–4 L daily, which correlates less with the extent of ADH deficiency than with the thirst response elicited. The cardinal diagnostic finding is hypo-osmolar (50–150 mosm/L) polyuria. However, adrenocortical insufficiency, such as secondary to pituitary surgery, may mask hypoosmolality (Verbalis *et al.*, 1984). Thalamic lesions may blunt the thirst response with ensuing severe hypernatremia. During pregnancy deterioration of DI may erroneously be assumed, because blood osmolality is physiologically reduced and thus the remaining stimulus for ADH secretion is removed. Differential diagnosis includes diabetes mellitus, psychogenic polydipsia, some psychotrophic drugs (lithium, phenothiazines, anticholinergics), some narcotics (penthrane, methoxyflurane), chronic renal disease, familial nephrogenic diabetes insipidus (refractoriness of renal tubules to ADH), primary aldosteronism, and hypercalcemia (hyperparathyroidism).

### Laboratory Findings

Spontaneous daily urine volume and osmolality should be ascertained. Five International Units of ADH are then administered (to a well-hydrated patient); urine osmolality should double. If not, nephrogenic DI is suspected. The dehydration test (Moses, 1984) with administration of 5 IU ADH after the plateau phase of urine osmolality and simultaneous determination of plasma (pOsm) and urine osmolality (uOsm) may distinguish between psychogenic and central DI. In the case of psychogenic DI the ratio of uOsm/pOsm remains normal or increases during dehydration, and there is no response to ADH. In the case of central DI the ratio decreases, and urine osmolality rises after ADH injection.

### Natural Course

The most frequent causes of central DI are brain tumor (30%), brain injury (20%), postoperatively (after lesioning of the posterior hypophysis or the pituitary stalk, often transient; 11%), histiocytosis or sarcoidosis (5%), other inflammatory disease (with swelling of the pituitary stalk and pituitary noted on MRI), post-encephaliticly and idiopathic (30%); other causes; see section on hypopituitarism.

Traumatic or postoperative DI becomes apparent within a few days. The course is typically triphasic. After the initial symptoms and signs caused by a lack of ADH, spurious clinical improvement may be noted when ADH is released from decaying neuron endings, followed by a severe relapse of DI after 3–4 days. Because autoregulation is impeded, the patients are endangered by uncontrolled volume substitution (Verbalis *et al.*, 1984). Untreated severe DI ends with hypertonic dehydration, hyperthermia, and death.

## Principles of Therapy

The dehydration test will differentiate between complete and partial deficiency of ADH secretion. Complete deficiency must be treated by ADH replacement. Therapy for incomplete deficiency depends on (1) whether the thirst response is preserved and (2) the subjective distress suffered from polydipsia and polyuria. Basically, there are two arms of therapy: replacement of ADH and administration of nonhormonal agents. ADH may be given as an aqueous solution (5–10 IU) subcutaneously, which will act for 3–6 hours. This route is preferable for unconscious or acutely ill patients, whose demand for ADH may change rapidly. For chronic replacement therapy a nasal application of desmopressin (10–20 μg) is used, which lasts for 12–24 hours. The efficacy decreases when infections and edema of the nasal mucosa occur. When medication is changed from intranasal to parenteral application, the dose must be reduced to 1/5–1/20. Vasopressin tannate is an oily solution for intramuscular injection (2.5–5 IU) with a duration of action of 24–72 hours. It is essential that the preparation be thoroughly warmed and shaken. Promising oral applications of ADH are being developed (Fjellestad-Paulsen et al., 1993).

Nonhormonal agents eliciting ADH secretion are clofibrate (4 × 500 mg/day), carbamazepine (400–600 mg/day), and chlorpropramide (200–500 mg/day). Chlorpropramide also improves the thirst response but may cause hypoglycemia. Hydrochlorothiazide (50–100 mg/day) and chlorthalidone (50 mg) act on the renal tubules and thus are effective in nephrogenic DI.

## Practical Management

### Patients with Residual Releasable ADH

This group responds to nonhormonal agents stimulating ADH release. Chlorpropramide is well tolerated and effective as a single daily dose. The agents named previously may be combined, but if therapy is inefficient, ADH has to be substituted.

### Patients without Residual Releasable ADH

Treatment is begun with an initial dose of 2.5 μg desmopressin in the evening and is increased by steps of 2.5 μg until nocturia is prevented. This dose is given twice a day (Jorkasky, 1990). Pregnancy and coronary heart disease are relative contraindications because of the vasoconstrictor action of ADH.

Patients with nephrogenic DI are treated with hydrochlorothiazide or chlorthalidone. ADH is without effect.

Therapy is controlled by balancing fluid volumes and monitoring central venous pressure. This is especially important in unconscious or severely ill patients, who should be treated by subcutaneous administration of aqueous ADH. All patients should carry a document or wear a medical alert bracelet or necklace stating the diagnosis of DI and their usual ADH requirement.

## SYNDROME OF INAPPROPRIATE ADH SECRETION

### Clinical Aspects

The Schwartz-Bartter's syndrome or syndrome of inappropriate ADH secretion (SIADH) is caused by excessive ADH release. The clinical findings include rapid weight gain, weakness, nausea (plasma sodium <120–125 mmol/L), lethargy (plasma sodium <110 mmol/L), confusion, cerebral edema, convulsions, and coma.

Notably, peripheral edema is not part of the syndrome, and the symptoms encountered may vary considerably, depending on the rate with which the syndrome evolves.

### Laboratory Findings

Typical are hypervolemia, hyponatremia (<130 mmol/L), hypo-osmolality (<275 mOsm/kg), and at the onset of the disease, increased urinary sodium (>30 mmoU/L). Plasma ADH may not be increased, but a disturbed regulation of ADH secretion is found in the water excretion test. The test should be used only when serum sodium is >125 mmol/L. Compared with normal regulation, the ratio of plasma ADH to osmolality is increased during dehydration, and plasma ADH does not vary with osmolality. The most frequent cause of SIADH is ectopic ADH production by bronchial carcinomas (80%). Other ectopic sources are carcinomas of the pancreas, thymoma, and malignant lymphomas, including Hodgkin's disease.

### Differential Diagnosis

The following conditions may cause hyponatremia and secondary ADH hypersecretion: renal loss of sodium in Addison's disease, chronic renal disease, diuretics, and misuse of diuretics; loss of sodium by vomiting, diarrhea, excessive sweating; central nervous system disorders, head trauma, subarachnoid hemorrhage, encephalitis, tuberculous or granulomatous meningitis; medication with cytostatic drugs, chlorpropramide, tricyclic antidepressants, and carbamazepine, oxcarbezepine, barbiturates; and chronic (nonmalignant) lung disease, such as tuberculosis, empyema.

## Natural Course

The course and prognosis are determined by the underlying cause, therefore no general outlook is afforded. SIADH occurring with cerebral or pulmonary disorders often improves gradually and necessitates a continuous adjustment of therapy.

## Principles of Therapy

Therapy is directed toward removing the source of ADH hypersecretion. The first approach to the patient, however, and in cases in which aberrant ADH secretion cannot be corrected is normalization of fluid balance and osmolality. Pharmacological options include demeclocycline and lithium, which antagonize ADH effects at the renal tubule.

## Practical Management

In most cases restriction of fluid intake to <800–1000 ml/day will result in gradual recovery of plasma sodium and osmolality. Therapy is sufficient when subjective well-being is restored. In addition, demeclocyclin (600–1200 mg/day) may be administered. The treatment reaches maximum efficacy after 3–4 days. Liver disease is a contraindication and renal failure a major potential adverse event.

In patients requiring rapid intervention (somnolence, convulsions, plasma sodium <110 mmol/L) plasma sodium is carefully restored to 120–125 mmol/L by intravenous administration of 5% sodium solution (3 ml/kg/h). Plasma sodium should not rise more than 10 mmol in 24/h. This is in most cases sufficient to ward off immediate danger.

Care must be taken to ensure that normalization of serum osmolality occurs slowly, because equilibration of intracellular and extracellular compartment must be awaited. Otherwise, pontine myelinolysis may ensue (Verbalis and Robinson, 1985). To avoid fluid overload, solutions of dextrans are contraindicated. Monitoring of central venous pressure and fluid balance is recommended.

## Treatments No Longer Recommended

Lithium salts have only moderate effects and may cause severe adverse reactions in hyponatremic patients.

# DISORDERS OF GONADOTROPIN SECRETION

## Clinical Aspects

The clinical findings will vary with the extent of gonadotropin deficiency (GD), ranging from shortening of the luteal phase or irregular menstrual cycles with loss of LH/FSH periodicity to estrogen deficiency and amenorrhea with atrophy of breasts and uterus, osteoporosis, and baldness. In men loss of libido, impotence, and muscle wasting may arise. Laboratory findings include lowered LH and FSH, and in women loss of LH/FSH periodicity. However, the release of LH and FSH is pulsatile, with considerable individual variation. Hence laboratory results may be equivocal. LHRH administration may be used to test the hypothalamic-pituitary system.

### Differential Diagnosis

Secondary GD occurs because of hyperprolactinemia caused by inhibition of LHRH release, physical or psychological stress, anorexia nervosa, polyglandular endocrine failure, and hemochromatosis.

In primary hypogonadism (refractoriness of the gonads to gonadotropins) in men and in postmenopausal women FSH and LH are increased.

## Natural Course

Gonadotropin secretion is a rather sensitive pituitary function and may be impaired in many of the disorders listed in the section on hypopituitarism. GD may then not be accompanied by any other neuroendocrine deficiency and remit spontaneously. Hypogonadism secondary to GD deficiency is called central or hypogonadotropic hypogonadism, which may exist as a congenital disorder. Kallmann's syndrome is characterized by GD and anosmia caused by a genetic defect leading to olfactory bulb and hypothalamic dysplasia with LHRH deficiency.

## Principles of Therapy

If the cause of GD is known, specific treatment should be so directed. In most cases only symptomatic therapy can be offered (i.e., hormonal replacement). When treatment of GD is discussed with the patient, psychological factors such as alterations of the self-concept with loss of secondary sexual characteristics and impaired libido in men and women should be addressed. The induction of puberty is not discussed in this book.

## Practical Management

### Women

**When Fertility Is Not Desired.** Secondary hypogonadism in women should be treated by cyclic estrogen administration to prevent osteoporosis and premature involution. Usually estrogens, ethinylestradiol, or its precursors are used. Addition of medroxyprogesterone acetate (5–10 mg/day) during the last 3–5 days of monthly estrogen therapy avoids endometrial hyperplasia. The adverse effects are the same as with oral contraceptives. The libido is not restored by this regimen and may require low-dose androgen administration.

**When Fertility Is Desired.** When ovulation is to be achieved, gonadotropins (human menopausal gonadotropins [hMG] or human chorionic gonadotropins [hCG] or LHRH [gonadorelin]) are administered in a pulsatile pattern by a portable infusion pump. The treatments are equipotent. A risk of 20% for induction of multiple pregnancies is reported.

### Men

Libido, potency, and sexual characteristics (beard, body hair, muscle build) are restored by testosterone enanthate (100–200 mg) every 1–3 weeks; however, fertility is not. As in women, this requires gonadotropin administration (see preceding). Restoration of spermatogenesis may take 1 year to become effective.

## EXCESSIVE GONADOTROPIN SECRETION

### Clinical Aspects and Natural Course

The clinical findings are precocious puberty (of iso, or heterosexual type) in children, gynecomastia in men, and frequently, the absence of symptoms in women. In adults the cause is either a pituitary adenoma or ectopic gonadotropin secretion (see later). Sexual precocity in 10% of cases is caused by affections of the hypothalamic region (glioma, germinoma, hamartoma, encephalitis, meningitis, neurofibromatosis); in the remainder it is caused by ectopic hormone secretion in a variety of ovarian or adrenal tumors and congenital errors of metabolism. The differential diagnosis and symptomatic treatment of sexual precocity will not be discussed here.

Laboratory findings in men are decreased serum testosterone and normal prolactin levels when an FSH-secreting adenoma is present. Of these adenomas, 50% will respond on stimulation by TRH adminis-tration with increased FSH secretion. In the case of LH-secreting adenomas, testosterone and LH are increased.

### Differential Diagnosis

Ectopic gonadotropin secretion may occur with germinomas, carcinomas of the lung, and other tumors. Gonadotropin secretion is physiologically increased in postmenopausal women and in primary hypogonadism.

### Practical Management

Because pituitary adenomas secreting gonadotropins are usually large, they should be excised. When this cannot be done, radiotherapy is indicated. Specific or symptomatic pharmacotherapy is not known.

## HYPERPROLACTINEMIA

### Clinical Aspects

Hyperprolactinemia (HPRL) is the most common cause (13%–40%) of secondary amenorrhea (Aisaka et al., 1992; Asukai et al., 1993). Amenorrhea may occur when prolactin (PRL) levels exceed 40 ng/ml and will always be present above 100 ng/ml. The supposed mechanism is suppression of LHRH by PRL. Other findings are galactorrhea (73%), headache, visual field deficits, osteoporosis, and hirsutism.

Malignancies of the breast are not more frequent in HPRL. In men HPRL manifests with impotence (8% of impotence cases), gynecomastia, and galactorrhea (13%).

### Laboratory Findings

The hallmark finding is elevated PRL. Normal ranges are 1–25 ng/ml for females of reproductive age and 1–20 ng/ml for men and postmenopausal women. There is no influence of age or phase of the menstrual cycle. Pregnant and lactating females may have PRL levels of 200 ng/ml that normalize within weeks after the lactation period. A prolactinoma is to be suspected only with PRL levels >150 ng/ml. Pseudoprolactinoma denotes a pituitary adenoma that causes a rise in PRL secretion because of pituitary stalk compression. PRL release is pulsatile with peaks during sleep (independent of the daytime), after meals, and under physical or psychological stress. Venipuncture may cause a rise in PRL levels, hence sampling from an indwelling catheter is recommended. Diagnosis may be confounded by transient HPRL peaks or abnormal secretion patterns. Although abnormal PRL secretion on TRH administration is found in only 15% of infertile women, transient HPRL at night occurs in 42% of normal responders to TRH with the presenting symptoms of infertility and luteal insufficiency (Aisaka *et al.*, 1992). Hence, without appropriate testing, disorders of PRL secretion will be overlooked.

## Differential Diagnosis

Secondary HPRL is encountered in hypothyroidism (normalizes when hypothyroidism is corrected), Addison's disease, renal disease (HPRL occurs in 80% of dialyzed patients), liver disease and alcohol abuse, polycystic ovaries, after breast amputation as a transient disorder, and with the following drugs: dopamine antagonists (under chronic medication the rise in PRL tapers off after a few weeks and normalizes 24–96 hours after the drug is discontinued), MAO inhibitors, methyldopa, reserpine, verapamil, metoclopramide, domperidon, $H_2$-receptor antagonists, and abuse of opiates or cocaine.

Transient HPRL is noted after brain surgery and epileptic seizures.

## Natural Course

PRL secretion is regulated by dopaminergic inhibition and is released when dopamine access (by way of the portal circulation) is impeded by compression of the pituitary stalk or of the pituitary or by lesions of the hypothalamus. HPRL is the most common endocrine symptom associated with pituitary tumors (60%–70%), although only 28% of pituitary adenoma are true prolactinomas.

PRL cosecretion is found in 25%–40% of patients with acromegaly and in 25% of cases with Cushing's disease. Prolactinomas may arise in the syndrome of multiple endocrine neoplasms (MEN). Ectopic PRL secretion is rare (Molitch, 1992).

The prognosis of untreated HPRL is good. In one series of 43 patients with intrasellar prolactinoma, 3 regressed spontaneously, 2 enlarged, and 38 did not change within 5 years (March *et al.*, 1981). Koppelman *et al.* (1984) report one case of adenoma enlargement in 18 macroadenoma and 7 microadenoma patients (median follow-up, 11.3 years). Within the microadenoma group, PRL levels halved and remained unchanged in the macroadenoma patients. Martin *et al.* (1985) followed "idiopathic HPRL" patients (HPRL of unknown cause) over a median 5.5 years: PRL levels fell in most patients, normalized in 14, and 1 patient had a prolactinoma develop.

## Principles of Therapy

### Pharmacotherapy

Dopamine agonists like bromocriptine, lisuride, and pergolide lower PRL levels in more than 80% of patients within 48 hours and lead to tumor shrinkage (by 76%) in 35%–100% of cases treated with 7.5 mg/day bromocriptine (Frantz, 1988). Shrinkage does not occur in pseudoprolactinomas. PRL levels are lowered promptly, and ovulatory menses may follow within 2–10 months. Adverse effects are nausea, hypotension, and osteoporosis. After tumor shrinkage and normalization of PRL levels (usually after 2 years), the dose can be reduced but not discontinued, because this will provoke recurrence (Martin and Reichlin, 1987)***.

Bromocriptine is begun with doses of 1.25 mg in the evening and increased every 4 weeks by 1.25 mg until PRL levels fall to less than 20 ng/ml (usually 2.5 mg bromocriptine twice daily). Up to 30 mg/day of bromocriptine may be required.

Lisuride is started with 0.2 mg in the evening, and the dose is increased every 2–3 days by 0.2 mg up to 1.6 mg/day. Usually, a maintenance dose of 0.8 mg/day is sufficient. Cabergoline is started with 0.5 mg twice weekly, maintenance dose is up to 4 mg. Monthly injectable and slow-release forms of bromocriptine, quinagolide, and carbegoline are not yet licensed but promise good results (Besser, 1993; Sarapura and Schlaff, 1993). An available alternative for oral bromocriptine, which avoids gastrointestinal irritation, is a vaginally applicable form**.

### Surgery

Transsphenoidal surgery achieves normalization of PRL secretion in 50%–88% of microadenomas and 11%–44% of macroadenomas. When presurgery PRL levels are <250 ng/ml, cure is effected in 85% and in 50% when beyond. Perioperative mortality is 0.9% for macroadenomas and 0.27% for microadenomas. The risk of recurrence within 4 years is 24%–40% for microadenoma and 27%–80% for macroadenoma (Frantz, 1988).

### Radiotherapy

Conventional radiation with 45 Gy leads to a gradual reduction of PRL levels with a normalization in 30% of cases. Radiotherapy is indicated only when pharmacotherapy and surgery have failed, as in some macro-prolactinomas with suprasellar extension (Bamberg *et al.*, 1989).

### Practical Management

Because of the benign prognosis, therapy is formulated according to the goal envisaged.

### Microprolactinoma and Fertility Desired

Pharmacotherapy with bromocriptine is effective and sufficient: ovulatory menses are restored in 90% of padents. When infertility is the only complaint, the medication may be discontinued after conception. Complications during pregnancy arise in less than 1%–5% of unmedicated HPRL patients (Gemzell and Wang, 1974). An increase of pituitary size during pregnancy is physiological. In 1410 pregnancies during which bromocriptine was continued, no increased risk for congenital anomalies was found (Turkali *et al.*, 1982).

When a prolactinoma is suspected, the approach is identical. Surgery is advised only when pharmacotherapy fails to suppress HPRL.

### Microprolactinoma and Fertility Not Desired

There is no need for therapy, but PRL levels should be assayed every 6 months, and every 2–4 years an MRI scan of the pituitary should be done. When the patient's complaints require therapy, bromocriptine is prescribed.

### Macroprolactinoma and Fertility Desired

First, an attempt to reduce the tumor size by 6–18 weeks on bromocriptine should be made. Then adenomectomy is performed. When PRL levels are normalized, no further therapy is required, but patients should be monitored regularly. When PRL levels are not normalized, pharmacotherapy is reinstituted.

### Macroprolactinoma and Fertility Not Desired

When there are symptoms caused by a mass effect of the adenoma (visual field deficits, hypopituitarism), tumor shrinkage by bromocriptine during 2 weeks before adenomectomy is advised. During this period the patient must be monitored closely, and immediate adenomectomy is mandatory if symptoms progress.

In the absence of mass effects, bromocriptine may be given and further development awaited. Again, regular clinical and laboratory controls are mandatory. When complaints persist, transsphenoidal surgery is performed. Radiotherapy is used in otherwise refractory cases.

### Pseudoprolactinoma

When HPRL occurs in association with a pituitary or peripituitary tumor (PRL usually <200 ng/ml) PRL levels may be lowered by bromocriptine. However, the tumor will not shrink and should be excised (Besser, 1993; Brabant, 1989; Christy, 1990; Koppelman *et al.*, 1984; Molitch, 1992).

## HYPOPITUITARISM AND HYPOTHALAMIC DYSFUNCTION

### Clinical Aspects

Hypothalamic function will be impaired only when the damage is bilateral. Hypothalamic dysfunction is assumed when, in addition to endocrine symptoms, altered food and fluid intake (ventromedial nucleus), impaired temperature regulation (anterior hypothalamus), disturbed sleep-wake cycle, and impaired autonomic nervous system function are present; and when fits of rage, apathy, and working memory deficits appear, depending on the hypothalamic areas involved.

The reserve functional capacity of the pituitary gland is ample, and clinical symptoms become manifest only after significant loss of parenchyma. In adults hypogonadism and hypothyroidism often herald hypopituitarism (HP). The clinical entity is usually defined by hypogonadism, hypothyroidism, and secondary adrenocortical deficiency: loss of interest and vitality, decay of physical ability and weight loss, bradycardia, decreased skin pigmentation, and loss of axillary and pubic hair. In addition there may be antidiuretic hormone deficiency with moderate diabetes insipidus.

Growth hormone deficiency: in children growth deficits are evident. In adults GH has important anabolic effects, such as regulation of amino acid metabolism and lipolysis in muscle tissue, cartilage, red blood cells, liver, kidney, and spleen. GH secretion is controlled by a feedback circuit involving hepatically synthesized insulin-like growth factor 1 (IGF-1). Although untreated GH deficiency leads to increased mortality, GH replacement has been shown to be efficient in reverse catabolic alterations in malnourished elderly patients, during prednisone treatment, as well as in young GH-deficient patients (Amato *et al.*, 1993; Bengtsson, 1993; review by Moxley, 1994).

Prolactin deficiency manifests as inability to lactate. This may be the first sign of HP after postpartum pituitary infarction (*Sheehan's syndrome*).

## Etiology

There are numerous causes for impaired hypothalamic-pituitary function. A discussion of some of these follows.

### Tumors

Intrasellar hormone–secreting and nonsecreting adenomas, carcinoma metastasis, chondromas, and craniopharyngiomas may be found. Of pituitary adenomas, 15%–25% are nonsecreting. Pituitary enlargement occurs physiologically during pregnancy and adolescence. Enlargement is found in 10% of adrenalectornized patients (Nelson's syndrome) and in chronic primary hypothyroidism caused by hypertrophy of thyreotropic cells. Perisellar craniopharyngioma is the second most common tumor of the region after pituitary adenomas. The incidence shows a bimodal age distribution, with one peak between 5 and 10 years and a second one in the sixth decade. They occur both intrasellar and suprasellar. Other perisellar tumors are meningiomas, chordomas, cysts of Rathke's pouch, arachnoid cysts, epidermoids, gliomas of the hypothalamus, and hamartomas. Epiphyseal tumors (0.2%–1% of primary brain tumors) typically cause either precocious or delayed puberty because of secretion of LHRH or choriongonadotropins by tumor cells or by inducing dysfunction of hypothalamic cells through a mass effect. Hormonally active tumors commonly are dysgenetic (germinomas, teratomas, choriocarcinomas). In advanced stages, diabetes insipidus, visual field deficits, and Parinaud's syndrome will ensue.

Tumors in other brain regions may cause endocrine abnormalities by compression of midbrain areas.

### Inflammation

Histiocytosis may be seen with diabetes insipidus, exophthalmos, lytic bone lesions, and the Hand-SchüllerChristian syndrome (25% of cases). Viral meningoencephalitis may affect the hypothalamus and give rise to endocrine dysfunction. Cerebral toxoplasmosis in immunocompromised patients may involve the hypothalamus and spread to the anterior pituitary gland. Granulomatous inflammatory disease (syphilis, tuberculosis) and autoimmune hypophysitis (mostly affecting women) may all lead to neuroendocrine dysfunction. Apart from CSF alterations, a swelling of the pituitary and pituitary stalk on MRI imaging suggests the diagnosis (Chong and Newton, 1993). These structures may also be afflicted by cerebral manifestation of Whipple's disease.

### Trauma

In postmortem examinations of severe brain trauma casualties, trauma-induced hypothalamic lesions were found in 43% and lesions of the pituitary in 62%. Conflicting data are offered for the incidence of hypothalamic-pituitary damage in surviving brain trauma patients, ranging from 20%–100%. In the absence of direct brain trauma, hemorrhagic shock may cause acute pituitary ischemic infarction.

### Pituitary Infarction

Diabetes mellitus may cause partial ischemic necrosis of pituitary tissue and provoke abrupt hypopituitarism. In the absence of impressive local symptoms and as a result of impaired glucocorticoid synthesis, a sudden drop in insulin requirement may be the only apparent manifestation. Postpartum partial pituitary necrosis (Sheehan's syndrome) is assumed to result from increased vulnerability of the enlarged pituitary gland.

### Other Causes

Aneurysms of the intracavernous carotid artery (<1.9% of aneurysms) and subarachnoid hemorrhage (in 65%), pseudotumor cerebri, amyloidosis, and skull irradiation (1.3%) may lead to endocrine dysfunction. Hemorrhagic infarction of the pituitary (a complication of anticoagulant therapy) manifests with sudden visual loss and may be confused with optic neuritis. Pituitary hemochromatosis preferentially impairs gonadotropin secretion and is seen as a dark pituitary gland on the T2-weighted MRI (Chong and Newton, 1993).

A number of hereditary syndromes include hypothalamic dysfunction, the Laurence-Moon-Biedl syndrome (retinitis pigmentosa, polydactyly, renal defects, obesity, hypogonadotropic hypogonadism, and mental retardation), the similar Biemond syndrome, and the Prader-Willi syndrome (hypogonadism, obesity, mental retardation).

## Diagnosis

The patient's record should include a detailed medical history, family history, and a complete list of drugs prescribed. The clinical evaluation includes an internal and ophthalmologic (visual field) examination. Additional tests are anthropometry and determination of bone age and density.

## Radiological Examination

MRI with gadolinium is superior to enhanced high-resolution CT with regard to recognition of tumors and blood vessels (Peck et al., 1989). Sellar tomography is obsolete. The normal sellar volume is 23.5–110.0 ml and is usually increased only from macroadenomas, not from microadenomas. Pituitary hyperplasia (normal height is 3–7 mm, maximum 9 mm in adolescents) as found in pregnancy and after long-standing hypothyroidism appears in the CT as a homogeneous enlarged sella without enhancement. Enhancement is homogeneous in 60%, heterogeneous in 40% of normals; 20% may show small low-density areas. Adenomas appear hypodense on enhanced CT. On MRI they demonstrate low intensity on T1-weighted images and high intensity on T2-weighted scans. Pituitary infarctions are hypodense in CT scans, hemorrhage is hyperdense in CT and MRI. Angiography will allow recognition of aneurysms or cavernous sinus invasion by tumors.

## Laboratory Tests

The following laboratory tests should be done: plasma GH before, 30, 60, and 120 minutes after stimulation by either 0.1 IU/kg insulin IV or by 10 mg/kg levodopa orally; serum prolactin levels basal and 20 minutes after application of 250 µg TRH IV; $T_3$, $T_4$, free $T_4$, and TSH; FSH, LH, testosterone, FSH response after TRH administration; and free urine cortisol excretion over 24 hours, plasma cortisol at 8 AM after 30 mg/kg metyrapone at midnight; serum sodium and osmolality, urine sodium and osmolality (see also the section on ADH deficiency).

## Natural Course

The average life expectancy is shortened by acute steroid or ADH deficiency, and over the long term by myocardial ischemic dysfunction and peripheral arterial disease, associated with impaired cholesterol metabolism (Johnston et al., 1992).

## Principles of Therapy

Because of the diversity of causes, no general approach can be offered. When hormone replacement is required, this must not be started until exact laboratory evaluation has been accomplished, otherwise a balanced hormone deficiency may decompensate. For gonadotropin deficiency, see the section on "Disorders in Gonadotropin Secretion." For antidiuretic hormone deficiency, see the section on "Diabetes Insipidus."

Glucocorticoid deficiency is less severe than in Addison's disease. Hypothyroidism is in most cases more prominent than adrenocortical insufficiency. Inadvertent replacement of thyroid hormones in compensated HP will lead to accelerated steroid metabolism and eventually to coma caused by adrenal insufficiency. Other drugs increasing steroid metabolism include barbiturates, diphenylhydantoin, and rifampin. Smoking also accelerates steroid metabolism.

Steroids with a long half-life like dexamethasone should be avoided, because no diurnal rhythm in glucocorticoid levels can be achieved. Steroid requirements are lower in the elderly and when liver function is impaired. Mineralocorticoid secretion is regulated predominantly by the Renin-angiotensin system and by the potassium level and is thus largely independent of pituitary ACTH.

## Practical Management

### Chronic Hypopituitarism

For female hypogonadism and male hypogonadism, see those subsections in the section on 'Gonadotropin Disorders." For hypothyroidism, see that section. For ADH deficiency, see the section on "Diabetes Insipidus." Thyroid hormone deficiency must be gradually corrected according to the clinical findings and $T_3$ and $T_4$ levels (with L-thyroxine begun with 25 µg/day and gradually increased to 100–150 µg/day, while $T_4$ is monitored). See also the section on "Hypothyroidism."

In secondary adrenocortical deficiency, the glucocorticoid deficiency is corrected with oral hydrocortisone 15–10–5 mg (morning–noon–evening). The dosage should be doubled when minor surgery is planned or when other acute illnesses supervene. As a rule, infectious disease requires an increase in corticoid medication when body temperature reaches 39°C. For extended surgery and for severe illness, 200–300 mg is given on the first and second days, with a gradual reduction to 30 mg/day. Vomiting necessitates intravenous administration of glucocorticoids. Blood glucose and electrolytes should be monitored. Patients should carry a card or wear a medical alert bracelet or necklace that defines the diagnosis and current treatment.

### Acute Hypopituitarism

The patients are endangered by acute adrenocortical failure. Therefore, on day 1 after an initial bolus of 100 mg hydrocortisone IV, 5 mg/h is infused continuously, on day 2 hydrocortisone is reduced to 2.5 mg/h intravenously, on day 3 treatment is continued as for chronic HP (see earlier). Thyroid hormone is replaced as indicated in the next section. To detect ADH deficiency, central venous pressure should be monitored. ADH is replaced as indicated in the section on diabetes insipidus.

## HYPOTHYROIDISM

### Clinical Aspects

Primary hypothyroidism (HT) is defined as deficiency of thyroid hormone with the intact secretion of thyroid-stimulating hormone (TSH). It is the most common (95%) form of HT and the only one associated with goiter because of excess TSH production. Secondary HT is rare and caused by TSH deficiency. Tertiary HT designates a hypothalamic disorder with lack of thyrotropin-releasing hormone (TRH).

The clinical presentation of the disorders is similar: fatigue, constipation; exercise and cold intolerance; loss of hair, dry skin, and myxedema; and anginal complaints, enlarged heart. Neurological findings include apathy, depression, demyelinating and axonal neuropathy, nerve compression syndromes, cerebellar ataxia, and proximal myopathy with weakness, cramps, delayed, and prolonged contraction with "hung-up" tendon reflex, myoedema, initially increased muscle bulk followed by atrophy, and elevated CK (Abend and Tyler, 1990). Untreated acute HT may deteriorate into hypothyroid coma with hypothermia, arrhythmia, hypotonia, respiratory failure, epileptic seizures, electrolyte dysregulation, and brain edema (Perlmutter and Cohn, 1964).

### Laboratory Findings and Differential Diagnosis

In primary HT, $T_4$ is <57 nmoUl, TSH is elevated, and the TRH response on stimulation is normal. In secondary HT, both $T_4$ and TSH are low, and the TRH response is excessive. In tertiary HT, both $T_4$ and TSH are low, the TRH response is normal. The latter two disorders are usually accompanied by a deficiency of other pituitary hormones.

### Natural Course

The discussion of primary HT is beyond the scope of this book. Secondary and tertiary HT are rare, and readers are referred to the previous section on "Hypopituitarism and Hypothalamic Dysfunction."

### Practical Management

In secondary and tertiary HT, therapy must be directed against the underlying lesion (e.g., pituitary adenoma, aneurysm). If this cannot be accomplished, patients are treated as with primary HT by thyroid hormone replacement (100–150 µg/day L-thyroxine). The therapy is started with 25 µg/day and increased by 25 µg every 3 weeks, depending on $T_4$ levels and clinical findings. Elderly patients need less and children more L-thyroxine. Caution must be exerted when there is cardiovascular disease or when severe, untreated HT has persisted for a long time. Therapy should in these cases be started with 12.5 µg/day. In primary HT replacement therapy is sufficient when TSH levels normalize.

When secondary or tertiary HT is suspected, glucocorticold hormone deficiency must be corrected before administration of thyroid hormones. Otherwise, adrenocortical coma may be provoked.

During pregnancy, HT must be corrected preferably by administration of L-thyroxine to prevent HT-related fetal defects and obstetrical complications.

## HYPERTHYROIDISM

### Clinical Aspects

In primary hyperthyroidism (Graves' disease) $T_4$ levels are elevated, and TSH is lowered. Secondary hypersecretion of thyroid hormone caused by TSH-secreting pituitary adenoma is quite rare.

The clinical findings are alike: ophthalmopathia with optic nerve neuropathy; increased sweating, palmar erythema; loss of appetite and weight; dyspnea; and tachycardia, atrial arrhythmia, and heart failure. Neurological signs include fine tremor, headache, hyperkinesia, myopathy (80%) with pronounced proximal muscle weakness, slight atrophy, normal or hyperactive tendon reflexes, normal CK, decrease of duration of motor unit potentials, and increase of polyphasic potentials. Weakness of the bulbar musculature and flaccid paralysis of the legs (Basedow's paraplegia) occur rarely; and nervousness, depression, disturbance of sleep, emotional lability occur.

### Laboratory Findings and Differential Diagnosis

There are many causes for primary hyperthyroidism, ranging from thyroid adenoma, Graves' disease, thyroiditis, to trophoblastic tumors, which are not discussed here. Factitious hyperthyroidism is caused by misuse of thyroid hormone preparations (TSH is low and radioactive iodine uptake reduced). In secondary hyperthyroidism (a TSH-secreting pituitary adenoma) TSH and T$_4$ levels are elevated, TSH response to TRH is blunted and free $\alpha$-subunits of TSH may be found. Elevated TSH and T$_4$ levels are also found in familial hypothalamic resistance to thyroid hormones (a condition that responds to bromocriptine; Martin and Reichlin, 1987). Thyroid radioactive iodine uptake is increased in primary and secondary hyperthyroidism. TSH levels are reduced in primary hyperthyroidism and ectopic thyroid hormone production.

### Natural Course

TSH-secreting pituitary adenomas are rare. Hyperthyroidism of the autoimmune type is present in 6% of myasthenia gravis patients. Exophthalmic orbitopathia is associated with HT in 50% of cases. The orbit is filled with an inflammatory exudate, which may lead to atrophy of the optic nerve and ocular muscles.

Thyrotoxic coma may evolve when other acute illness intervenes: the patient will show fever, cardiac arrhythmia, vomiting, agitation then stupor, finally bulbar signs and seizures (Abend and Tyler, 1990).

### Practical Management

Hyperthyroidism is prone to spontaneous remissions and exacerbations, thus close monitoring is advised. TSH-secreting pituitary adenomas should be surgically removed or irradiated (see the first section of this chapter). They do not respond to pharmacotherapy. If this cannot be achieved, treatment is the same as for primary hyperthyroidism, either by subtotal thyroidectomy (for patients younger than 40 years) or by radioactive ablation (in older patients). Pharmacotherapy is accomplished with 3 × 100–150 mg/day propylthiouracil or the slower acting methimazole. Leukopenia is the predominant adverse effect. The effects of iodide administration occur rapidly but are often transient.

Hyperkinesia and tremor improve on administration of $\beta$-adrenergic antagonists or reserpine, but this is rarely necessary once thyroid hormone levels are normalized. During pregnancy, antithyroid drugs are continued. There is no evidence for fetal toxicity, but there may be neonatal hypothyroidism (Lazarus, 1993).

Hypothalamic resistance to thyroid hormones is treated with bromocriptine, begun with doses of 1.25 mg in the evening and increased every 4 weeks by 1.25 mg.

Thyrotoxic coma must be treated with $\beta$-adrenergic blocking agents, dexamethasone, thyrostatic medication, and cooling the patients (sprinkle with warm [!] water; Newcomer *et al.*, 1983). Optic nerve neuropathy may be treated with steroids, surgical decompression, and radiotherapy. The modalities are equivalent in efficacy***.

## HYPOPARATHYROIDISM

### Clinical Aspects

Hypoparathyroidism (HPT) is caused by deficiency of parathyroid hormone (parathormone or PTH). Peripheral resistance to PTH is named Albright's osteodystrophy. Clinical signs are tetanus, bronchial and urinary bladder spasms, Raynaud's syndrome, and cataracts. Neurological findings are Chvostek and Trousseau signs, seizures (40%), myalgia, increased CK with normal histopathology, paresthesias, dementia, and occasional psychotic states.

Also, there may be increased intracranial pressure with papilledema (pseudotumor cerebri) (Sugar, 1983). Basal ganglia calcification may be found and is rarely associated with choreoarthetosis, parkinsonism, or tremor (Muenter and Whisnant, 1968). In most cases the disorder arises after thyroidectomy (4%).

Laboratory investigations include serum calcium (low), phosphates (increased), and an abnormal PTH stimulation test (Ellsworth-Howard test).

### Practical Management

Muscle cramps, paresthesias, and a serum calcium less than 8 mg/dl necessitate treatment. Epileptic seizures will cease only when HPT is controlled, whereas antiepileptic drugs are futile. Mild forms may be treated by calcium-rich food or oral calcium (1–2 g/day). Severe forms require replacement of vitamin D (calcitriole 2 × 0.25–0.50 mg/day) during the first days. Long-term therapy is accomplished with dihydrotachysterol, 400 mg/day. Treatment is monitored by measurement of urinary calcium (which should be <200–20 mg/24 h) and serum calcium (8–8.5 mg/dl). Normalization of serum calcium should not be attempted to avoid nephrocalcinosis (Nanes and Caterwood, 1990).

Tetanic crises are managed by a slow infusion of 20–50 ml of 10% calcium gluconate.

## HYPERPARATHYROIDISM

### Clinical Aspects

Although half of hyperparathyroid patients are asymptomatic, the disorder may manifest with nephrolithiasis and renal colic (60%), osteoporosis, soft tissue calcification, peptic ulcers, and polydipsia.

Neurological symptoms include proximal muscle weakness and atrophy, with hyperactive tendon reflexes and absence of fasciculation and fibrillation in EMG; depression; and in severe cases hallucinations and hypercalcemic coma. Uremic hyperparathyroidism causes metastatic calcification of blood vessels with gangrenous skin lesions and necrotizing myopathy.

### Laboratory Findings and Differential Diagnosis

In primary hyperparathyroidism PTH, serum calcium, and alkaline phosphatase are increased as is urinary calcium; serum phosphate is reduced. Secondary hyperparathyroidism denotes a reactive increase of PTH encountered in hypocalcemia, vitamin D deficiency, or resistance and renal failure. In 10% of lithium-treated patients hypercalcemia with elevated PTH levels occurs and remits after the discontinuation of lithium.

### Natural Course

Hyperparathyroidism has an incidence of 1/10,000/year in men older than 60 years of age and of 2/1000/year in women and evolves most frequently in the fifth decade. This course is generally benign and may run over years with subtle symptoms. The patients are endangered by the sudden evolution of parathyroid coma. Hyperparathyroidism in 80% of cases is provoked by an adenoma of the parathyroid and in the remaining cases by diffuse hyperplasia. Ectopic PTH production is rare.

### Practical Management

Patients with a serum calcium <12 mg/dl do not require treatment, unless the symptoms necessitate. They are monitored by assessment of bone density and renal function and advised to avoid calcium-rich food, dehydration, immobilization, and thiazide medication. Often, and especially in elderly patients, hydration alone will suffice.

Acute hypercalcemia is treated by infusion of isotonic saline with furosemide. Biphosphonates are effective in management of acute hypercalcemia and in preventing osteoporosis but do not maintain normal calcium levels with long-term therapy. When long-term treatment is necessary, adenomectomy is preferred for young symptomatic patients and for patients older than 50 years with bone loss attributed to HPT (Abrams and Schipper, 1990). The success rate of surgically exploring all four glands and removing one enlarged gland is 90%.

In asymptomatic hypercalcemia and HPT caused by lithium treatment, no therapy is required. When symptoms supervene, lithium must be discontinued.

## EMPTY SELLA SYNDROME

### Clinical Aspects

Empty sella (ES) syndrome is defined as enlargement of the sella turcica with a radiological volume of >11–12 ml. The subarachnoid space extends into the sellar lumen and forces the pituitary against the sellar wall. Typical findings are mild endocrine dysfunction (51%), hypertension (29%), adiposity (93%), visual disturbance (34%), pseudotumor cerebri (10%), headache (70%), rhinorrhea (11%).

The laboratory findings are similar to those of hypopituitarism (see that section). A CT or MRI should be performed, because a standard skull x-ray film will, in most cases, be inadequate for diagnosis.

Differential diagnosis consists in excluding an intrasellar tumor.

### Natural Course

Of the patients, 87% are women and most are multiparous. Hence, one possible cause for ES is postpartum pituitary necrosis (Sheehan's syndrome). Dysgenesis of the diaphragma sellae with progressive extension of the subarachnoid space mediated by CSF pulsations may be etiologic. In some cases, ES develops after radiation therapy.

### Practical Management

Hormone deficiencies are managed as discussed in the section on "Hypopituitarism." Transsphenoidal "packing" (i.e., interpolation of muscle tissue) led in one series to improvement of headache in 71%, of rhinorrhea in 50%, and of visual impairment in 46%. However, spontaneous improvement of headache was noted in 64% (Gallardo *et al.*, 1992). Surgical intervention is definitely advised when there are progressive visual deficits, often caused by arachnoidal adhesions pulling the optic nerve into the sellar lumen (Martin and Reichlin, 1987)**.

## NONSECRETING PITUITARY ADENOMA

### Clinical Aspects

Of pituitary adenomas, 15%–25% are nonsecreting and detected only when they have become large and cause either hypopituitarism, hyperprolactinernia, or visual field deficits.

## Practical Management

Microadenoma or macroadenoma detected by serendipity does not require treatment. However, the patient should be monitored every 6–12 months by MRI and endocrine evaluation. Endocrine and ophthalmological symptoms are indications for adenomectomy. Emergency surgery is required when apoplexy or hemorrhage of the adenoma is suspected. Visual impairment improves in 80% of cases after surgery (Sassolas *et al.*, 1993). In patients with contraindications to surgery or for those who refuse surgery, or when adenoma resection is incomplete, radiation therapy is advised. With radiation treatment, recurrence occurs in 20% of cases. Pharmacotherapy with bromocriptine will decrease tumor size in less than 15% of patients.

## REFERENCES

Abend, W. K., and Tyler, H. R. (1990). Thyroid disease and the nervous system. *In* "Neurology and General Medicine: The Neurological Aspects of Medical Disorders" (M. J. Aminoff, Ed.), pp. 257–271. Churchill Livingstone, New York, Edinburgh, London, and Melbourne.

Abrams, G. M., and Schipper, H. M. (1990). Other endocrinopathies and the nervous system. *In* "Neurology and General Medicine: The Neurological Aspects of Medical Disorders" (M. J. Aminoff, Ed.), pp. 305–321. Churchill Livingstone, New York, Edinburgh, London, and Melbourne.

Adams, C. B. T., and Burke, C. W. (1993). Current modes of treatment of pituitary tumors. *Br. J. Neurosurg.* 7, 123–128.

Aisaka, K., Yoshida, K., and Mori, H. (1992). Analysis of clinical backgrounds and pathogenesis of luteal-phase defect. *Horm. Res.* 37(Suppl. 1), 41–47.

Al-Damluji, S., and Rees, L. H. (1988). The neuroendorine control of corticotropin secretion in normal humans and in Cushing's disease. *In* "Clinical Neuroendocrinology" (R. Collu, G. M. Brown, and G. R. van Loon, Eds.) pp. 251–285. Blackwell, Boston, Oxford, and London.

Amato, G., Carella, C., Fazio, S., La Montagna, G., Cirradini, A., Sabatini, D., Marciano-Mone, S., Sacca, L., and Bellastella, A. (1993). Body composition, bone metabolism and heart structure and function in GH-deficient adults before and after GH replacement therapy at low doses. *J. Clin. Endocrinol. Metab.* 1993, 1671–1676.

Arafah, B. M., Brodkey, J. S., Kaufmann, B., Velasco, M., Manni, A., and Pearson, O. H. (1980). Transsphenoidal microsurgery in the treatment of acromegaly and gigantism. *J. Clin. Endocrinol. Metab.* 48, 578–585.

Asukai, K., Uemura, T., and Minaguchi, H. (1993). Occult hyperprolactinernia in infertile women. *Fertil. Steril.* 60, 423–427.

Bamberg, M., Rauhut, F., Budach, V., and Stuschke, M. (1989). Die Radiotherapie von Hypophysentumoren. *Aktuel. Neurol.* 16, 61–64.

Barkan, A. L. (1989). Acromegaly: Diagnosis and therapy. *Endocrinol. Metab. Clin. North Am.* 18, 277–310.

Bates, A. S., Van't Hoff, W., Jones, J. m., and Clayton, R. N. (1993). An audit of outcome of treatment in acromegaly. *Q. J. MEd.* 86, 293–299.

Bengtsson, B. A. (1993). The consequences of growth hormone deficiency in adults. *Acta Endocrinol.* 128(Suppl. 2), 2–5.

Besser, M. (1993). Criteria for medical as opposed to surgical treatment of prolactinomas. *Acta Endocrinol.* 129(Suppl. 1), 27–30.

Brabant, G. (1989). Medikamentöse Therapie von Hypophysentumoren. *Aktuel. Neurol.* 16, 65–67.

Brada, M., Rajan, B., Traish, D., Ashley, S., Holmes-Sellors, P. J., Nussey, S., and Urtley, D. (1993). The long-term efficacy of conservative surgery and ratiotherapy in the control of pituitary adenomas. *Clin. Endocrinol.* 38, 571–578.

Carpenter, P. C. (1990). Cushing's syndrome. *In* "Conn's Current Therapy" (R. E. Rakel, Ed.), pp. 563–567. Saunders, Philadelphia, Pennsylvania.

Chong, B. W., and Newton, T. H. (1993). Hypothalamic and pituitary pathology. *Endocr. Radiol.* 31, 1147–1183.

Christy, N. P. (1990). Hyperprolactinernia. *In* "Conn's Current Therapy" (R. E. Rakel, Ed.), pp. 578–584. Saunders, Philadelphia, Pennsylvania.

Clarke, S. D., Woo, S. Y., Butler, E. B., Dennis, W. S., Lu, H., Carpenter, L. S., Chiu, J. K., Thomby, J. L., and Baskin, D. S. (1993). Treatment of secretory pituitary adenoma with radiation therapy. *Radiology* 188, 759–763.

David, D. H., Laws, E. R. Jr., Ilstrup, D. M., Speed, J. K., Caruso, M., Shaw, E. G., Abboud, C. F., Scheithauer, B. W., Roor, L. M., and Schleck, C. (1993). Results of surgical treatment for growth hormone-secreting pituitary adenomas. *J. Neurorusg.* 79, 70–75.

Fiellestad-Paulsen, A., Paulsen, O., d'Agay-Abensour, L., Lundin, S., and Czernichow, P. (1993). Central diabetes insipidus: Oral treatment with dDAVP. *Regul. Pept.* 45, 303–307.

Frantz, A. G. (1988). Hyperprolactinernia. *In* "Clinical Neuroendocrinology" (R. Collu, G. M. Brown, and G. R. van Loon, Eds.), pp. 311–332. Blackwell, Boston, Oxford, and London.

Gallardo, E., Schachter, D., Caceres, E., Becker, P., Colin, E., Martinez, C., and Henriquez, C. (1992). The empty sella: Results of treatment in 76 successive cases and high frequency of endocrine and neurological disturbances. *Clin. Endocrinol.* 37, 529–533.

Gemzell, C., and Wang, C. F. (1974). Outcome of pregnancy in women with pituitary adenoma. *Fertil. Steril.* 31, 363–372.

Grattan-Smith, P. J., Morris, J. G., and Langlands, A. O. (1992). Delayed radiation necrosis of the central nervous system in patients irradiated for pituitary turnouts. *J. Neurol. Neurosurg. Psychiatry* 55, 949–955.

Grigsby, P. W., Simpson, J. R., Enami, B. N., Fineberg, B. B., and Schwartz, H. G. (1989). Prognostic factors and results of surgery and postoperative irradiation in the treatment of pituitary adenomas. *Int. J. Radiat. Oncol. Biol. Phys.* 16, 1411–1417.

Jadresic, A., Jimenez, L. E., and Joplin, G. F. (1987). Long term effect of 90Y pituitary implantation in acromegaly. *Acta Endocrinol.* 115, 301–306.

Jeffcoare, W. J., Rees, L. H., Tomlin, S., Jones, A. E., Edwards, C. R. W., and Besser, G. M. (1977). Metyrapone in long-term management of Cushing's disease. *Br. MEd. J.* 2, 215–217.

Johnston, D. G., Beshyah, S. A., Markussis, V., Shahi, M., Sharp, P. S., Foale, R. A., and Skinner, E. M. (1992). Metabolic changes and vascular risk factors in hypopituitarism. *Horm. Res.* 38(Suppl. 1), 68–72.

Jorkasky, D. K. (1990). Diabetes insipiclus. *In* "Conn's Current Therapy" (R. E. Rakel, Ed.), pp. 567–568. Saunders, Philadelphia, Pennsylvania. Kliman, B., Kjellberg, R. N., Swisher, B., and Butler, W. (1984). Proton therapy of acromegaly: A 20-year experience.

*In* "Secretory Tumors of the Pituitary Gland" (N. T. Zervas, E. C. Ridgeway, and J. B. Martin, Eds.), pp. 191–211. Raven, New York.

Koppelman, M. C. S., Jaffe, M. J., Rieth, K. G., Caruso, R. C., and Loriaux, D. L. (1984). Hyperprolactinemia, amenorrhea and galactorrhea. *Ann. Intern. MEd.* **100,** 115–121.

Lazarus, J. H. (1993). Treatment of hyper- and hypothyroidism in pregnancy. *J. Endocrinol. Invest.* **16,** 391–396.

Ludecke, D. K., Lutz, B. S., and Niedworok, G. (1989). The choice of treatment after incomplete adenomectomy in acromegaly: Proton versus high voltage radiation. *Acta Neurochir.* **96,** 32–38.

March, C. M., Klerzky, O. A., Davajan, V., Teal, J., Weiss, M., Apuzzo, M. L., Marrs, R. P., and Mishell, D. R. Jr. (1981). Longitudinal evaluation of patients with untreated prolactin-secreting pituitary adenomas. *Am. J. Obstet. Gynecol.* **139,** 835–844.

Martin, J. B., and Reichlin, S. (1987). "Clinical Neuroendocrinology," 2nd Ed. Davis, Philadelphia.

Martin, T. L., Kim, M., and Malarkey, W. B. (1985). The natural history of idiopathic hyperprolactinernia. *J. Clin. Endocrinol. Metab.* **60,** 855–858.

Melmed, S. (1993). Medical management of acromegaly-what and when? *Acta Endocrinol.* **129**(Suppl. 1), 13–17.

Molitch, M. E. (1988). Acromegaly. *In* "Clinical Neuroendocrinology" (R. Collu, G. M. Brown, and G. R. van Loon, Eds.), pp. 189–227. Blackwell, Boston, Oxford, and London.

Molitch, M. E. (1992). Pathologic hyperprolactinemia. *Endocrinol. Metab. Clin. North Am.* **21,** 877–901.

Molitch, M. E., Elton, R. L., Blackwell, R. E., Caldwell, B., Chang, R. J., Jaffe, R., Joplin, G., Robbins, R. J., Tyson, J., and Thorner, M. O. (1985). Bromocriptine as primary therapy for prolactinsecreting macroadenomas: Results of a prospective multicenter study. *J. Clin. Endocrinol. Metab.* **60,** 698–705.

Moses, A. M. (1984). Clinical and laboratory observations in the adult with diabetes insipidus and related syndromes. *In* "Diabetes Insipidus in Man" (P. Czernichov and A. G. Robinson, Eds.), pp. 156–175. Karger, Basel.

Moxley, R. T. (1994). Potential for growth factor treatment of muscle disease. *Curr. Opin. Neurol.* **7,** 427–434.

Muenter, M. D., and Whisnant, J. P. (1968). Basal ganglia calcification, hypoparathyroidism and extrapyramidal motor manifestation. *Neurology* **18,** 1075.

Mundinger, F. (1985). Technik und Ergebnisse der interstitiellen Hirnrumorbestrahlung. *In* "Handbuch der Medizinischen Radiologie" (H. P. Heilmann, Ed.), Vol. 19, Part 4, Spezielle Strahlentherapie maligner Tumoren, pp. 179–214. Springer-Verlag, Heidelberg.

Nanes, M. S., and Catherwood, B. D. (1990). Hyperparathyroidism and hypoparathyroidism. *In* "Conn's Current Therapy" (R. E. Rakel, Ed.), pp. S70–574. Saunders, Philadelphia.

Newcomer, J., Haire, W., and Hartmann, C. R. (1983). Coma and thyreotoxicosis. *Ann. Neurol.* **14,** 689.

Peck, W. W., Dillon, W. P., Norman, D., Newton, T. H., and Wilson, C. B. (1989). High-resolution MR imaging of pituitary microadenomas at 1.5 T: Experience with Cushing disease. *AJR, Am. J. Roentgenol.* **152,** 145–151.

Perlmutter, M., and Cohn, H. (1964). Myxedema crisis of pituitary or thyroid origin. *Am. J. Med.* **36,** 883.

Plockinger, U., Liehr, R. M., and Quabbe, H. J. (1993). Octreotide long term treatment of acromegaly: Effect of drug withdrawal on serum growth hormone/insulin-like growth factor-I concentrations and on serum gastrin/24-hour intragastric pH values. *J. Clin. Endocrinol. Metab.* **77,** 157–162.

Quabbe, H. J. (1982). Treatment of acromegaly by transsphenoidal operation, 90-Yttrium implantation and bromocriptine: Results in 230 patients. *Clin. Endocrinol.* **16,** 107–119.

Riedel, M., Gunther, T., von zur Muhlen, A., and Brabant, G. (1992). The pulsatile GH secretion in acromegaly: Hypothalamic or pituitary origin? *Clin. Endocrinol.* **37,** 233–239.

Ross, D. A., and Wilson, C. B. (1988). Results of transsphenoidal microsurgery for growth-hormone secreting pituitary adenoma in a series of 214 patients. *J. Neurosurg.* **68,** 854–867.

Sarapura, V., and Schlaff, W. D. (1993). Recent advances in the understanding of the pathophysiology and treatment of hyperprolactinemia. *Curr. Opin. Obstet. Gynecol.* **5,** 360–367.

Sassolas, G., Trouillas, J., Treluyer, C., and Perrin, G. (1993). Management of nonfunctioning pituitary adenomas. *Acta Endocrinol.* **129**(Suppl. 1), 21–26.

Stevenaert, A., and Beckers, A. (1993). Presurgical octreotide treatment in acromegaly. *Acta Endocrinol.* **129**(Suppl. 1), 18–20.

Stolke, D. (1989). Chirurgische Therapie von Hypophysentumoren. *Aktuell. Neurol.* **16,** 58–60.

Sugar, O. (1983). Central neurological complications of hypoparathyroidism. *Arch. Neurol. Psychiatry* **70,** 86.

Tsang, R. W., Laperriere, N. J., Simpson, W. J., Brierley, J., Panzarella, T., and Smyth, H. S. (1993). Glioma arising after radiation therapy for pituitary adenoma. A report of four patients and estimation of risk. *Cancer* **72,** 2227–2233.

Turkali, L, Braun, P., and Krupp, P. (1982). Surveillance of bromocriptine in pregnancy. *JAMA* **247,** 1589–1591.

Verbalis, J. G., and Robinson, A. G. (1985). Neurophysin and Vasopressin: Newer concepts of secretion and regulation. *In* "The Pituitary Gland" (H. Imure, Ed.), pp. 307–339. Raven, New York.

Verbalis, J. G., Robinson, A. G., and Moses, A. M. (1984). Postoperative and posttraumatic diabetes insipidus. *In* "Diabtes Insipidus in Man" (P. Czernichov and A. G. Robinson, Eds.), pp. 247–265. Karger, Basel.

Young, W. F. (1990). Acromegaly. *In* "Conn's Current Therapy" (R. E. Rakel, Ed.), pp. 557–560. Saunders, Philadelphia.

Zervas, N. T. (1984). Surgical results for pituitary adenomas: Results of an international survey. *In* "Secretory Tumors of the Pituitary Gland" (N. T. Zervas, E. C. Ridgeway, and J. B. Martin, Eds.), pp. 377–385. Raven, New York.

## CHAPTER 103
# Autonomic Dysfunction

C. J. Mathias and R. Freeman

## INTRODUCTION

The autonomic nervous system innervates all organs of the body. Autonomic dysfunction thus can affect every organ system: the cardiovascular system, the respiratory system, the gastrointestinal system, the urogenital system, the skin (temperature regulation, sweating), and the endocrine system. Patients with disorders resulting in autonomic failure may present to various specialists, although the main disciplines dealing with them are in neurology, cardiology, and internal medicine. It is not possible in this chapter to cover all disorders of the autonomic system. We will focus mainly on autonomic failure, especially the chronic primary autonomic failure syndromes, including multiple system atrophy (MSA), synonymous with the Shy–Drager syndrome. Treatment of the following will be discussed: orthostatic hypotension; nausea and vomiting; esophageal spasm and achalasia; esophagus, bowel, and gut distension; constipation; neurogenic diarrhea; and disorders of sweating. The relationship between arterial hypertension, coronary spasm, respiratory disorders, and autonomic dysfunction will not be discussed. Pupillary disorders are of greater interest in diagnosis than in treatment. Certain disturbances associated with, overlapping, or related to specific aspects of autonomic system function are dealt with in separate chapters of this book: vertigo (Chapter 13), hiccup (Chapter 15), sleep disorders (Chapter 16), dysphagia (Chapter 20), malignant hyperthermia (Chapter 49), neuroendocrine syndromes (Chapter 90), disturbances of sphincter activity and sexual function (Chapter 92), and syncope (Chapter 93). Involvement of the autonomic nervous system in cases of polyneuropathy are referred to in Chapters 61, 69, 70, 75, and 76 in this volume, as well as in Strian and Haslbeck (1986), Thomas (1988), and Mathias and Bannister (1999a). Autonomic failure associated with Parkinson's disease (Chapter 64) or with syndromes due to brainstem degeneration (e.g., OPCA, Chapter 63) are covered elsewhere. Detailed information on clinical aspects of autonomic failure is found in books by Johnson and Spalding (1974), Sturm and Birkmayer (1977), Schiffter (1985), Appenzeller and Oribe (1997), Low (1997), and Mathias and Bannister (1999b). Neuroanatomical, physiological, and pathophysiological principles are discussed in Monnier (1963), Brooks *et al.* (1979), Thews and Vaupel (1990), and Mathias and Bannister (1999c).

## CLASSIFICATION

The autonomic system emerges from the central nervous system at different sites: the sympathetic system at the spinal thoracolumbar level and the parasympathetic system at cranial and spinal sacral levels. Involvement of the autonomic system results in a wide variety of disorders. In some, only the autonomic nervous system may be affected. The dysfunction may be highly specific, as in the isolated deficiency of dopamine betahydroxylase (DBH), with failure to synthesize noradrenaline and adrenaline (Mathias and Bannister, 1999a); it may involve only one efferent pathway, affecting only one neurotransmitter, as in pure cholinergic dysautonomia (Thomashefsky *et al.*, 1972); or it may encompass the entire autonomic nervous system (without other neurological involvement), as in pure autonomic failure (PAF) (Mathias, 2000a). An autonomic neuropathy may complicate neurological disorders, as in the Guillain–Barre syndrome, metabolic and endocrine diseases such as diabetes mellitus, and various medical conditions, ranging from renal failure to human immunodeficiency virus (HIV) infection.

Autonomic dysfunction can be divided into *primary*, where the etiology is not known, and secondary, where there is a clear association with disease or where the site of the lesion is known (Table I). Drugs are a common cause of autonomic dysfunction (Table II).

Intermittent disorders of autonomic function include neurally mediated syncope and the postural tachycardia syndrome (PoTS).

## CLINICAL MANIFESTATIONS

The localized autonomic disorders (Table III) may affect one organ or a single system. These result in specific presenting features, in contrast to the generalized forms of autonomic failure, where there may be a variety of features. Certain disorders, although localized, such as the Holmes–Adie pupil, may be associated with more widespread autonomic defects that include baroreceptor dysfunction, sudomotor deficits (Ross' syndrome), and chronic cough (Kimber *et al.*, 1998).

Chronic forms of autonomic failure may begin with symptoms of tiredness and weakness, orthostatic dizziness, or syncope while upright (see Table IV). These symptoms may be concealed for years by compensatory mechanisms of the autonomic and endocrine system. Orthostatic (postural) hypotension is a cardinal manifestation of autonomic failure (AF) (Fig. 1); sometimes, however, AF may present with urinary bladder symptoms or impotence. Another group of patients, as in MSA, may present initially with features suggestive of Parkinson's disease (PD) (Mathias and Williams, 1994; Wenning *et al.*, 1994).

## INVESTIGATION

Autonomic screening tests are directed mainly toward blood pressure and heart rate regulation as part of cardiovascular autonomic testing. There are a variety of screening tests that include orthostatic change using a tilt table, deep breathing, hyperventilation, Valsalva maneuver, and stress to cause pressor responses as induced by isometric exercise, cutaneous cold, and cortical arousal elicited by sudden noise or mental arithmetic (Table V). These screening tests form a useful first step. They are abnormal in antonomic failure; however, they are usually normal between attacks in neurally mediated syncope (Mathias *et al.*, 2001). This emphasizes the need for a detailed history and clinical examination to plan specific testing, in addition to the screening tests. In autonomic failure affecting systems other than the circulation, such as the urinary bladder or bowel, additional testing will be needed (Table V).

If the tests indicate that autonomic function is abnormal, it is then important to determine the degree of dysfunction, the site of the lesion, and whether the

**TABLE I    Classification of Disorders Resulting in Autonomic Dysfunction**

Primary (Etiology unknown)
  Acute/subacute dysautonomias
    Pure cholinergic dysautonomia
    Pure pandysautonomia
    Pandysautonomia with neurological features
  Chronic autonomic failure syndromes
    Multiple system atrophy (Shy–Drager syndrome)
    Autonomic failure with Parkinson's disease
    Pure autonomic failure
Secondary
  Congenital
    Nerve growth factor deficiency
  Hereditary
    Autosomal dominant trait
      Familial amyloid neuropathy
      Porphyria
    Autosomal recessive trait
      Familial dysautonomia—Riley–Day syndrome
      Dopamine β-hydroxylase deficiency
      Aromatic L-amino acid decarboxylase deficiency
    X-linked recessive
      Fabry's disease
  Metabolic diseases
    Diabetes mellitus
    Chronic renal failure
    Chronic liver disease
    Vitamin B$_{12}$ deficiency
    Alcohol-induced
  Inflammatory
    Guillain–Barre syndrome
    Transverse myelitis
  Infections
    Bacterial—tetanus
    Viral—human immunodeficiency virus
    Parasitic—*Trypanosomiasis cruzi*; Chagas' disease
    Prion—fatal familial insomnia
  Neoplasia
    Brain tumors—especially of third ventricle or posterior fossa
    Paraneoplastic, to include adenocarcinomas—lung, pancreas, and Lambert–Eaton syndrome
  Connective tissue disorders
    Rheumatoid arthritis
    Systemic lupus erythematosus
    Mixed connective tissue disease
  Surgery
    Regional sympathectomy—upper limb and splanchnic denervation
    Vagotomy and drainage procedures—"dumping syndrome"
    Organ transplantation—heart, kidney
  Trauma
    Spinal cord transection
  Miscellaneous
    Subarachnoid hemorage
    Syringobulbia and syringomyelia
Drugs, chemicals, toxins (see Table II)
Neurally mediated syncope
  Vasovagal syncope
  Carotid sinus hypersensitivity
  Situational syncope
Postural tachycardia syndrome

*Note.* Reprinted with permission from Mathias (2000a).

**TABLE II**  Drugs/Chemicals/Poisons/Toxins Causing Autonomic Dysfunction

Decreasing sympathetic activity
  Centrally acting
    Clonidine
    Methyldopa
    Reserpine
    Barbiturates
    Anaesthetics
  Peripherally acting
    Sympathetic nerve ending (guanethidine, bethanadine)
    α-Adrenoceptor blockade (phenoxybenzamine)
    β-Adrenoceptor blockade (propranolol)
Increasing sympathetic activity
  Amphetamines
  Releasing noradrenaline (tyramine)
  Uptake blockers (imipramine)
  Monoamine oxidase inhibitors (tranylcypromine)
  β-Adrenoceptor stimulants (isoprenaline)
Decreasing parasympathetic activity
  Antidepressants (imipramine)
  Tranquillisers (phenothiazines)
  Antidysrhythmics (disopyramide)
  Anticholinergics (atropine, probanthine, benzotropine)
  Toxins (botulinum)
Increasing parasympathetic activity
  Cholinomimetics (carbachol, bethanechol, pilocarpine, mushroom poisoning)
  Anticholinesterases
    Reversible carbonate inhibitors (pyridostigmine, neostigmine)
    Organophosphorus inhibitors (parathion)
Miscellaneous
  Alcohol, thiamine (vitamin $B_1$ deficiency)
  Vincristine, perhexiline maleate
  Thallium, arsenic, mercury
  Mercury poisoning ("Pink" disease)
  Ciguatera toxicity
  Jellyfish and marine animal venoms
  First dose of certain drugs (prazosin, captopril)
  Withdrawal of chronically used drugs (clonidine, opiates, alcohol)

*Note.* Reprinted with permission from Mathias (2000a).

disorder is of the primary or secondary variety. This may entail a range of additional tests (from neuroimaging and electrophysiological studies to sural nerve biopsy and genetic typing), designed to exclude or confirm the underlying disorder. Such tests are helpful not only for diagnosis and treatment but also for predicting outcome and prognosis, which vary in the different disorders. They are necessary in evolving appropriate management strategies, which should encompass complications resulting from the autonomic deficits and also the other systems affected by the primary disorder.

The following sections deal with a brief description of the autonomic disorders followed by treatment of autonomic failure affecting different organs.

**TABLE III**  Examples of Localized Autonomic Disorders

Horner's syndrome
Holme–Adie pupil
Crocodile tears (Bogorad's syndrome)
Gustatory sweating (Frey's syndrome)
Reflex sympathetic dystropy
Idiopathic palmar or axillary hyperhidrosis
Chagas' disease (*Trypanosomiasis cruzi*)[a]
Surgical procedures[b]
  Sympathectomy (regional)
  Vagotomy and gastric drainage procedures in "dumping" syndrome
  Organ transplantation (heart, lungs)

*Note.* Reprinted with permission from Mathias (2000a).
[a]Listed here because it specifically targets intrinsic cholinergic plexuses in the heart and gut.
[b]Surgery also may cause other localized disorders, such as Frey's syndrome after parotid surgery.

# PRIMARY AUTONOMIC FAILURE

These disorders can be broadly divided into the rare acute/subacute autonomic neuropathies and the more common chronic disorders that consist mainly of MSA, allied parkinsonian disorders, and PAF.

## Acute and Subacute Autonomic Neuropathies

Autonomic manifestations may be the sole or predominant feature of an acute or subacute neuropathy. They present with varying combinations of orthostatic hypotension, constipation, bladder atony, impotence, secretomotor paralysis, and blurring of vision associated with tonic pupils. The disorder may involve both the sympathetic and parasympathetic divisions of the autonomic nervous system (pure pandysautonomia) (Young *et al.*, 1975) or the parasympathetic and sympathetic cholinergic systems (pure cholinergic dysautonomia) (Hart and Kanter, 1990; Thomashetsky *et al.*, 1972).

**TABLE IV**  Some of the Clinical Manifestations in Primary Chronic Autonomic Failure

Cardiovascular: postural (orthostatic) hypotension.
Sudomotor: anhidrosis, heat intolerance.
Gastrointestinal: constipation, occasionally diarrhea, oropharyngeal dysphagia.
Renal and urinary bladder: nocturia, frequency, urgency, retention, incontinence.
Sexual: erectile and ejaculatory failure in the male.
Ocular: anisocoria, Horner's syndrome.
Respiratory: stridor, involuntary inspiratory gasps, apneic episodes.
Other neurological deficits: parkinsonian, cerebellar or pyramidal features.

*Note.* Adapted from Mathias (1997).

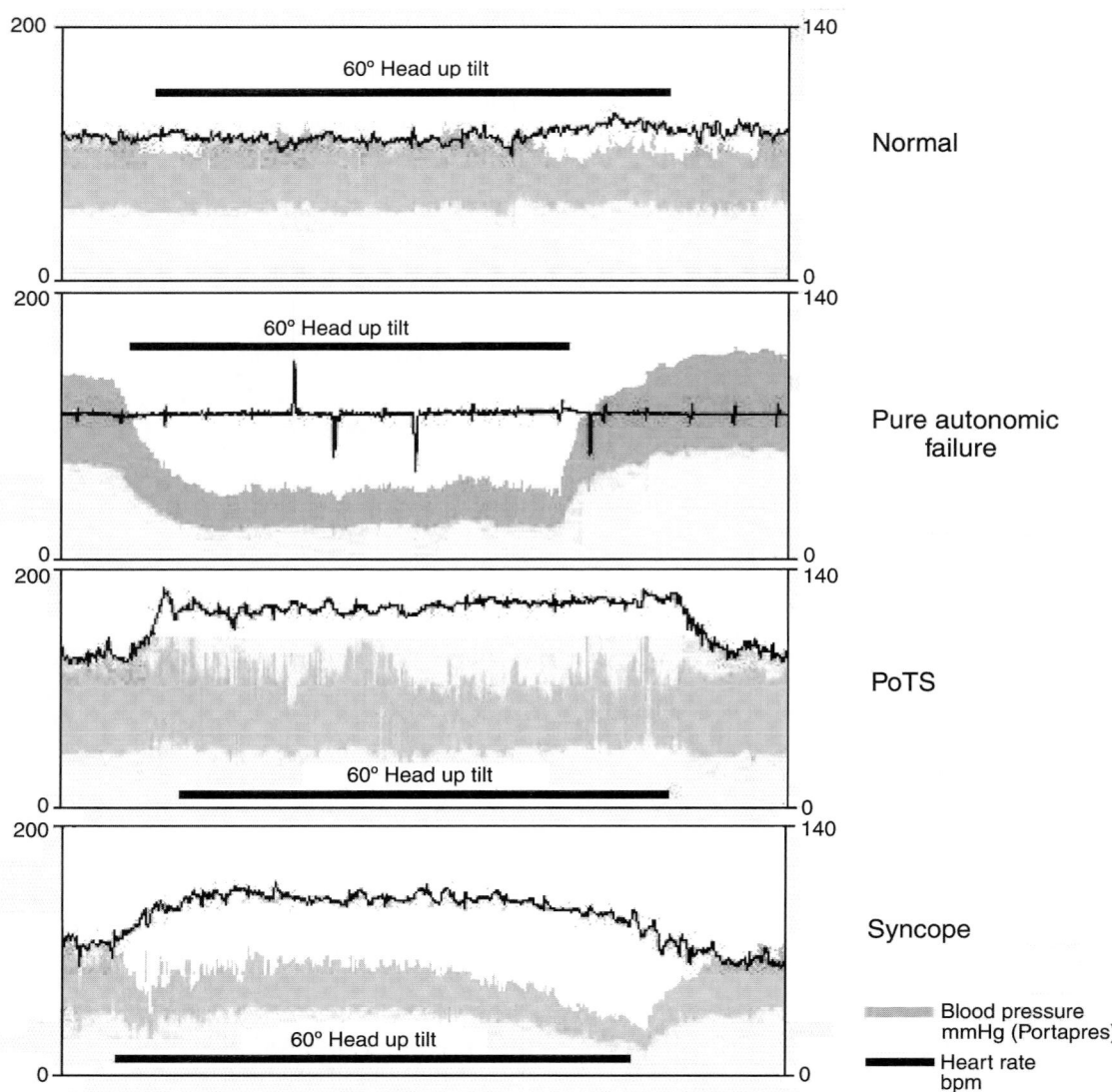

**FIGURE 1**    Blood pressure and heart rate measured (noninvasively) continuously before, during, and after 60° head-up tilt (by Portapres II) in a normal subject and in subjects with three different autonomic disorders; with pure autonomic failure (PAF), postural tachycardia syndrome (PoTS), and vasovagal syncope.

Sensorimotor manifestations may accompany the autonomic manifestations but usually are not a major feature (Colan *et al.*, 1980; Serratrice, 1988).

Autonomic manifestations may accompany the Guillain–Barre syndrome although they are usually overshadowed by motor features (Asahina *et al.*, 2001; Ropper *et al.*, 1991; Tuck and McLeod, 1981). The most frequently noted autonomic features include sinus tachycardia and other cardiac arrhythmias, sustained hypertension, blood pressure lability, bowel and bladder dysfunction, pupillomotor disturbances, sudomotor dysfunction, and vasomotor abnormalities (Appenzeller and Marshall, 1963; Birchfield and Shaw, 1964; Cortelli *et al.*, 1990; Davies and Dingle, 1972; Edmonds and

Sturrock, 1979; Flachenecker *et al.*, 1997; Keane, 1977; Lichtenfeld, 1971; Persson and Solders, 1983; Ropper *et al.*, 1991; Tuck and McLeod, 1981; Zochodne, 1994). Bradyarrhythmias and heart block may necessitate emergency intervention. Sixty-five percent of patients exhibit dysautonomic features with tachycardia and 50% of cases have loss of heart rate variability (Ropper *et al.*, 1991). Recent evidence suggests cardiovascular autonomic abnormalities in most Guillain–Barre syndrome patients (Flachenecker *et al.*, 1997). Some authors (Feasby *et al.*, 1986; Ropper *et al.*, 1991; Winer and Hughes, 1988), unlike others (Tuck and McLeod, 1981), have suggested that dysautonomia is more frequent in patients with respiratory failure, severe

**TABLE V** Outline of Investigations in Autonomic Failure

| | |
|---|---|
| **Cardiovascular** | |
| Physiological | Head-up tilt (45°); standing; Valsalva maneuver |
| | Pressor stimuli—isometric exercise, cold pressor, mental arithmetic |
| | Heart rate responses—deep breathing, hyperventilation, standing, head-up tilt, 30:15 ratio |
| | Liquid meal challenge |
| | Exercise testing |
| | Carotid sinus massage |
| Biochemical | Plasma noradrenaline—supine and head-up tilt or standing; urinary catecholamines; plasma renin activity and aldosterone |
| Pharmacological | Noradrenaline—α-adrenoceptors—vascular |
| | Isoprenaline—β-adrenoceptors—vascular and cardiac |
| | Tyramine—pressor and noradrenaline response |
| | Edrophonium—noradrenaline response |
| | Atropine—parasympathetic cardiac blockade |
| Sudomotor | Central regulation—thermoregulatory sweat test |
| | Sweat gland response—intradermal acetylcholine—quantitative sudomotor axon reflex test (Q-SART)—localized sweat test |
| | Sympathetic skin response |
| Gastrointestinal | Barium studies, video-cine-fluoroscopy, endoscopy, gastric emptying studies |
| Renal function and urinary tract | Day and night urine volumes and sodium/potassium excretion |
| | Urodynamic studies, intravenous urography, ultrasound examination, sphincter electromyography |
| Sexual function | Penile plethysmography |
| | Intracavernosal papaverine |
| Respiratory | Laryngoscopy |
| | Sleep studies to assess apnoea/oxygen desaturation |
| Eye | Schirmer's test |
| | Pupil function—pharmacological and physiological |

*Note.* Reprinted from Mathias and Bannister (1999b), copyright Oxford University Press, Oxford.

motor deficits, and the axonal variant of Guillain–Barre syndrome. Autonomic abnormalities include reduced heart rate variability with deep respiration (Flachenecker *et al.*, 1997; Persson and Solders, 1983; Singh *et al.*, 1987) and an abnormal Valsalva ratio (Flachenecker *et al.*, 1997; Singh *et al.*, 1987; Tuck and McLeod, 1978). The autonomic abnormalities improve in parallel with regaining motor function (Flachenecker *et al.*, 1997; Persson and Solders, 1983).

## Multiple System Atrophy and Allied Parkinsonian Disorders

Multiple system atrophy (MSA) with autonomic failure (synonymous with the Shy–Drager syndrome; Shy and Drager, 1960), and Parkinson's disease (PD) are the most prevalent neurodegenerative diseases that result in autonomic disturbances. These disorders often are phenotypically similar and despite recent advances to find clinical and laboratory markers to differentiate these two disorders, diagnosis at an early stage is often difficult (Fig. 2); definitive diagnosis may only be made at autopsy.

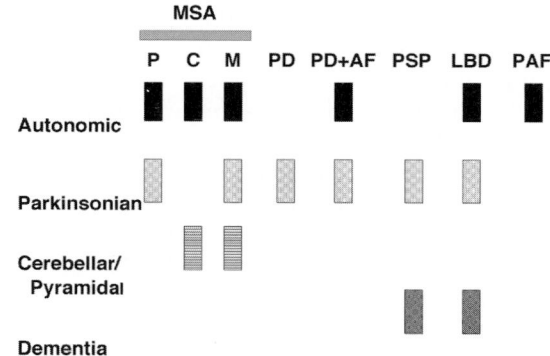

**FIGURE 2** The major clinical features in parkinsonian syndromes and allied disorders with autonomic failure. These include the three major neurological forms of multiple system atrophy: the parkinsonian form (MSA-P, also called striatonigral degeneration), the cerebellar form (MSA-C, also called olivopontocerebellar atrophy), and the multiple or mixed form (MSA-M, which has features of both other forms); idiopathic Parkinson's disease (PD), Parkinson's disease plus autonomic failure (PD + AF), progressive supranuclear palsy (PSP), diffuse Lewy body disease (LBD), and pure autonomic failure (PAF). Adapted from Mathias (1997).

MSA is the most common degenerative disorder of the central nervous system that results in generalized autonomic failure. It affects extrapyramidal (striatonigral degeneration), cerebellar (olivopontocerebellar atrophy), and autonomic neurons. Patients with MSA typically present with autonomic symptoms such as orthostatic hypotension, bowel and bladder dysfunction, anhidrosis, and impotence, together with motor dysfunction that can involve the extrapyramidal, cerebellar, and pyramidal systems (Table IV) (Bannister and Oppenheimer, 1972; Cohen et al., 1987, 1996; Gilman et al., 1998). Impotence in males, urinary incontinence, and orthostatic hypotension are the most frequent symptoms of MSA (Wenning et al., 1994). Symptoms of orthostatic hypotension usually lead patients to seek medical attention.

There are few community-based estimates of the prevalence of this disorder. A recent study recruited eligible subjects from the computerized records of 15 general practices in London, United Kingdom, and confirmed the diagnosis by review of records, interview, physical examination, and video recordings of neurological signs. In this report, the age-adjusted prevalence for MSA was 4.4 per 100,000 (Schrag et al., 1999).

When extrapyramidal features dominate, this disorder is frequently misdiagnosed as idiopathic PD. In contrast, MSA is more likely to have a symmetrical onset. The pill-rolling, resting tremor that is characteristic of PD is infrequent in MSA.

Antecollis and bulbar symptoms are common features of MSA. MSA is less responsive to levodopa and dopamine agonists than PD although a number, particularly those with onset at a young age, experience a temporary improvement in motor symptoms (Wenning et al., 1994). In progressive supranuclear palsy (PSP) there may also be a poor response to L-lopa and recurrent falls; cardiovascular autonomic failure, however, does not occur and is considered an exclusionary feature (Kimber et al., 2000). Cold blue hands and Raynaud's phenomenon are common in MSA (Mallipedi and Mathias, 1998).

Inspiratory stridor, vocal cord abductor paresis and central hypoventilation are important respiratory manifestations of MSA that commonly occur. The presence of nocturnal stridor is associated with a poor prognosis. Tracheostomy should be considered for patients with stridor or vocal cord weakness (Harcourt et al., 1996; Silber and Levine, 2000).

Rapid eye movement (REM) sleep behavior disorder, consisting of excessive motor activity during dreaming in association with loss of skeletal muscle atonia of REM sleep is often associated with MSA. Self and/or spousal injury may occur (Plazzi et al., 1997).

The external anal sphincter muscle is innervated by fibers that originate in Onuf's nucleus in segments S2–S4 of the spinal cord. This region is among central nervous system sites affected by neuronal cell loss in MSA but not in PD. Thus features of denervation on anal sphincter electromyography have been used to differentiate MSA from Parkinson's disease. Features thought to be due to an increased duration of the motor unit potentials and an increased polyphasia may reflect neuronal loss of lower motor neurons characteristic for MSA (Palace et al., 1997).

The median survival of MSA is usually less than 10 years (Bannister et al., 1988; Testa et al., 1996; Wenning et al., 1994). These reports may be biased toward the more severely affected; some patients survive for considerably longer.

Pathological studies have demonstrated cell loss and gliosis that involves the striatonigral, olivopontocerebellar and autonomic systems (Daniel, 1999). Neuroanatomical sites of pathological change include the putamen, caudate nucleus, external pallidum, substantia nigra, locus ceruleus, inferior olives, pontine nuclei, cerebellar Purkinje cells, and intermediolateral cell columns of the spinal cord. The thalamus, vestibular nucleus, dorsal vagal nucleus, corticospinal tracts, and anterior horn cells are less frequently involved. Autonomic dysfunction in MSA is due to loss of preganglionic autonomic neurons in the brainstem and spinal cord (Daniel, 1999; Matthews, 1999; Wenning et al., 1995, 1997).

Argyrophilic, intracytoplasmic inclusion glial and neuronal inclusions appear to be specific pathological hallmarks of MSA (Papp and Lantos, 1994). These inclusions are composed of 10- to 15-nm-diameter coated filaments that are immunoreactive for ubiquitin and α-synuclein. α-Synuclein is a structural component of the filaments in Lewy bodies of PD, dementia with Lewy bodies (DLB), and the Lewy body variant of Alzheimer's disease. Two mutations in the α-synuclein gene have been shown to be pathogenic for familial PD. The role played by α-synuclein in the pathogenesis of MSA is unknown (Dickson et al., 1999; Giasson et al., 2000; Tu et al., 1998).

In contrast, the autonomic features of PD are usually not as severe as those seen in MSA. They characteristically occur late in the course of the illness and are often associated with levodopa and dopamine agonist therapy (Mathias, 1998). Nevertheless, autonomic dysfunction is frequently the source of significant morbidity in the parkinsonian patient. Cardiovascular, gastrointestinal, urogenital, and sudomotor dysfunction may occur (Goetz et al., 1986; Mathias, 1998; Senard et al., 1997). In patients with PD, Lewy bodies are found not only in central but also in peripheral sympathetic ganglia and neurons suggesting that autonomic dysfunction in this disorder may be caused by both pre- and postganglionic neuronal dysfunction (Rajput and Rozdilsky, 1976; Rajput *et al.*, 1990). Autonomic failure may precede dementia in Lewy body disease (Larner *et al.*, 2000).

Decreased myocardial [$^{123}$I]meta-iodobenzylguanidine-derived radioactivity in PD has been demonstrated in several studies (Braune *et al.*, 1999; Druschky *et al.*, 2000; Iwasa *et al.*, 1998). This does not occur in MSA (where the lesion is preganglionic), suggesting that the cardiac denervation in PD is postganglionic. Thoracic positron imaging tomography studies confirm decreased myocardial sympathetic innervation in PD, with and without cardiovascular autonomic dyfunction, based on the uptake of the positron emitter 6-[$^{18}$F]fluorodopamine in the myocardial septum (Goldstein *et al.*, 1997, 2000). Brain MRI scans have features that distinguish atypical parkinsonian syndromes (Schrag *et al.*, 1998 and 2000) as do positron and emmission tomography scans using different ligands (Rinne *et al.*, 1995; Brooks, 1999), but this may not be useful in individual cases. With the clonidine-growth hormone stimulation test, levels of growth hormone do not rise in MSA, unlike PD (Kimber *et al.*, 1997; Thomaides *et al.*, 1992) (Figs. 3a and 3b). Whether this will be of value in differentiating these disorders at an early stage is not known.

## Pure Autonomic Failure

Pure autonomic failure is an idiopathic peripheral autonomic nervous system degeneration that, in contrast to MSA, has no motor manifestations (Bradbury and Eggleston, 1925; Freeman, 1995; Mathias, 2000a). There are also no other neurological signs. This disorder is not progressive and has a significantly better prognosis than MSA (Bannister *et al.*, 1988). The autonomic features of MSA, however, may precede the other neurological signs by several years, preventing an early definitive diagnosis (Bannister and Mathias, 1999). Thus it is important to differentiate between these two disorders especially in the early stages. In PAF there usually is a low resting plasma noradrenaline level due to degeneration of the postganglionic sympathetic neuron. These levels increase in normal subjects, by 100–200% when moving from the supine to the upright position, and do not change in PAF. There is, however, a wide scatter of results within PAF that often makes it difficult to classify individual patients (Cohen *et al.*, 1987; Goldstein *et al.*, 1989). Cardiovascular autonomic tests show impairment with a severity that is equivalent to, or greater than, in MSA. In a controlled study of autonomic function, 77% of PAF had abnormal measures of heart rate variability in response to

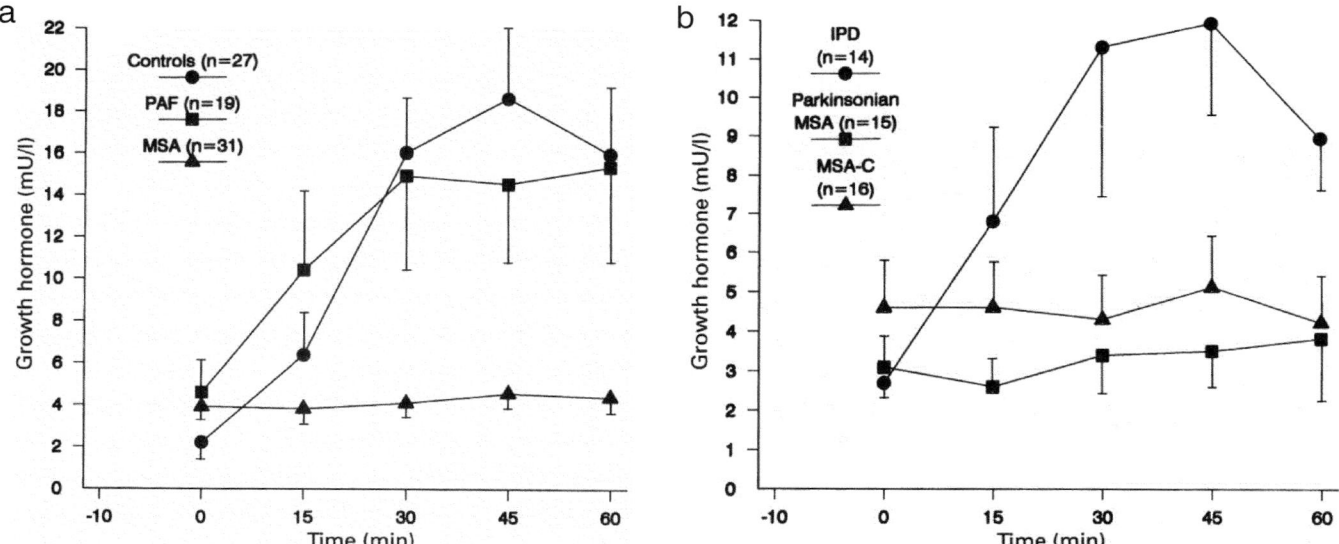

**FIGURE 3** (a) Serum growth hormone (GH) concentrations before (0) and at 15 min intervals for 60 min after clonidine (2 µg/kg/min) in normal subjects (controls) and in patients with pure autonomic failure (PAF) and multiple system atrophy (MSA). GH concentrations rise in controls and in patients with PAF with a peripheral lesion; there is no rise in patients with MSA with a central lesion. Reprinted with permission from Kimber *et al.* (1997). (b) Indicates lack of serum GH response to clonidine in MSA (the cerebellar form (MSA-C) and the parkinsonian form) in contrast to patients with idiopathic Parkinson's disease with no autonomic deficit (IPD), in whom there is a significant rise in GH levels. Reprinted with permission from Kimber *et al.* (1997).

deep respiration (less than the first percentile) and 91% had an abnormal Valsalva ratio (Cohen *et al.*, 1987). The clonidine-growth-hormone stimulation test differentiates central from peripheral autonomic failure; in PAF there is a rise in growth hormone levels, similar to normal subjects (Fig. 3a) (Kimber *et al.*, 1997; Thomaides *et al.*, 1992).

## SECONDARY AUTONOMIC FAILURE

This chapter will cover those peripheral neuropathies and disorders in which autonomic dysfunction is of clinical significance. Autonomic dysfunction occurs with many peripheral neuropathies—particularly those that predominantly involve the small or unmyelinated fibers (McLeod, 1999; Miyawaki and Freeman, 1994).

### Diabetes Mellitus

Diabetic autonomic neuropathy is the most common autonomic neuropathy in the developed world. A constellation of autonomic features occur that affect cardiovascular, gastrointestinal, urogenital, thermoregulatory, sudomotor, and pupillomotor function (Hilsted and Low, 1997; Tarsy and Freeman, 1994; Watkins, 1998; Watkins and Edmonds, 1999). The cardiovascular autonomic manifestations include an increased resting heart rate followed by a fixed heart rate that responds only minimally to physiological stimuli. The initial tachycardia is most likely due to a vagal cardiac neuropathy. The ensuing decrease in heart rate and ultimately fixed heart rate is predominanltly due to progressive cardiac sympathetic nervous system dysfunction. Orthostatic hypotension occurs in diabetes as a consequence of efferent sympathetic vasomotor denervation, causing reduced vasoconstriction of the splanchnic and other peripheral vascular beds. There is an association between increased mortality and cardiovascular autonomic dysfunction in diabetics. The 5-year cumulative mortality rate of patients with cardiovascular autonomic test abnormalities ranges form 27 to 56% (Ewing *et al.*, 1980; Navarro *et al.*, 1990; O'Brien *et al.*, 1991; Rathmann *et al.*, 1993; Sampson *et al.*, 1990).

### Amyloid Neuropathy

Autonomic dysfunction frequently accompanies the polyneuropathy of both primary and familial amyloidosis. Autonomic dysfunction appears less common in myeloma-associated amyloidosis. The variant transthyretin, in which methionine substitutes for valine at position 30, is the point mutation that is a common cause of familial amyloid polyneuropathy (Reilly and Thomas, 1999; Saraiva *et al.*, 1985). Other transthyretin mutations as well as mutations in apolipoprotein A1 and gelsolin also give rise to familial amyloid polyneuropathy. Amyloid is thought to cause autonomic and somatosensory dysfunction by deposition of amyloid, causing pressure in peripheral nerves, dorsal nerve root ganglia, or autonomic ganglia and by ischemic damage due to amyloid infiltration of epineural and intraneural blood vessel walls (Kyle and Dyck, 1993). In familial amyloid polyneuropathy the relentless progress of the disease can be halted by transplantation of the liver, where the abnormal protein mainly is produced. This reduces levels of variant transthyretin and its deposition in nerves.

Patients with amyloid neuropathy typically present with distal sensory symptoms such as numbness, paresthesiae, and dysesthesiae. Autonomic manifestations are less common presenting features of amyloid neuropathy. On examination there are signs of a sensorimotor polyneuropathy that predominantly involve the small nerve fibers that mediate pain and temperature sensation. Characteristic autonomic signs and symptoms include postural hypotension, diarrhea, constipation, fecal incontinence, disturbances in bladder function, pupillary abnormalities, and erectile failure. These autonomic manifestations have similarities to those described with diabetic autonomic neuropathy (Kyle and Dyck, 1993). Sick sinus syndrome and A-V conduction deficits may occur. Tests assessing cardiac vagal function are often abnormal (Niklasson *et al.*, 1989).

### Infectious Diseases

The peripheral neuropathies associated with a number of infectious diseases have accompanying autonomic manifestations. Autonomic dysfunction may occur in human immunodeficiency virus infection. The symptoms include orthostatic hypotension, syncope, presyncope, sweating disturbances, bladder and bowel dysfunction, and impotence (Freeman *et al.*, 1990).

Autonomic testing of seropositive and AIDS patients has been the subject of several reports and case studies (Cohen *et al.*, 1991; Cohen and Laudenslager, 1989; Craddock *et al.*, 1987; Evenhouse *et al.*, 1987; Lin-Greenberger and Taneja-Uppal, 1987; Mulhall and Jennens, 1987; Villa *et al.*, 1987, 1990). In a controlled study, there were significant differences in autonomic function between controls and HIV-infected patients, using multiple tests of the autonomic nervous system. A steady decline in autonomic function was noted across diagnostic groups (controls, ARC, AIDS) with the most severe autonomic dysfunction found in the patients with AIDS (Freeman *et al.*, 1990). These test results have been replicated in studies of similar design (Rüttimann *et al.*, 1991; Villa *et al.*, 1990, 1992; Welby *et al.*, 1991). These observations may underlie the predisposition of HIV-infected patients to cardiac arrhythmias such as ventricular tachycardia and torsades de pointes (often in association with medication) and the observed incidence of unexpected cardiorespiratory arrest (Cohen *et al.*, 1990; Craddock *et al.*, 1987; Eisenhauer *et al.*, 1994; Luginbuhl *et al.*, 1993; Stein *et al.*, 1990; Wharton *et al.*, 1987).

Chagas' disease, due to a parasitic infection by *Trypanosoma cruzi*, is associated in the late stages of illness with severe cardiovascular and gastrointestinal dysautonomia. The pathogenesis of the autonomic dysfunction is unresolved and may be due to direct neural injury during the acute illness or a persisting immune mediated response. Reduced bowel motility, sialorrhea, megaesophagus, and megacolon are the most frequent gastrointestinal manifestations of this disease. Cardiovascular manifestations include resting bradycardia, anhidrosis, conduction abnormalities, arrhythmias, cardiac failure, cardiomegaly, and impairment in the blood pressure response to standing (Iosa *et al.*, 1989, 1990).

Some of the prion diseases have also been associated with autonomic disturbances. In fatal familial insomnia (Cortelli *et al.*, 1991), the manifestations usually result from autonomic overactivity, rather than autonomic failure. Autonomic overactivity may also occur in tetanus, with marked hypertension and tachycardia especially in patients who are severely ill and need respirator support.

## Hereditary Disorders

Hereditary disorders include the Riley–Day syndrome (familial dysautonomia), which occurs in children of Ashkenazi-Jewish extraction (Axelrod, 1999). At birth the diagnosis is made on the basis of absent fungiform papillae, lack of corneal reflexes, decreased deep tendon reflexes, and a diminished response to pain. Pupillary hypersensitivity to cholinomimetics and an abnormal intradermal histamine skin test (with an absent flare response) confirm the diagnosis. The defective gene has been mapped to the long arm of chromosome nine (*q*31). These patients have features resulting from both autonomic underactivity and overactivity. There are associated neurological problems, skeletal problems (such as scoliosis), and renal failure that previously contributed to the poor prognosis.

## Dopamine β-Hydroxylase Deficiency

This disorder was first recognized in the 1980s and has been described in seven patients, two of whom are siblings (Man in't Veld *et al.*, 1987; Mathias *et al.*, 1990; Robertson *et al.*, 1986; Thompson *et al.*, 1995). Symptoms begin in childhood, although in the reported cases an autonomic disorder was not considered until their teenage years, when orthostatic hypotension was recognised. The clinical features indicate sympathetic adrenergic failure with sparing of sympathetic cholinergic and parasympathetic function. Sweating along with urinary bladder and bowel function is preserved. In one of the males erection was possible, but ejaculation was difficult to achieve. Plasma noradrenaline and adrenaline levels are usually undetectable, while plasma dopamine levels are elevated. The enzymatic defect is highly specific, with the sympathetic nerve pathways and terminals otherwise intact as has been demonstrated by electron microscopy, immunohistochemical staining for various neuropeptides (Mathias *et al.*, 1990), and preservation of muscle sympathetic nerve activity using microneurography (Rea *et al.*, 1990; Thompson *et al.*, 1995). Treatment is with the drug L-dihydroxyphenylserine (Fig. 4a), which has a structure similar to noradrenaline except for a carboxyl group that is acted upon by the enzyme dopadecarboxylase. This enzyme is abundantly present in tissue such as liver and kidney and transforms the prodrug into noradrenaline. This reduces orthostatic hypotension and has been remarkably beneficial in helping these patients lead reasonably active lives (Fig. 4b) (Mathias and Bannister, 1999a).

## Spinal Cord Lesions

From the spinal cord emerge the entire sympathetic and also the sacral parasympathetic pathways. In spinal cord injuries, the degree of autonomic dysfunction depends on the level and completeness of the lesion (Mathias and Frankel, 1999). In cervical and high thoracic spinal-cord lesions, autonomic dysfunction can result from underactivity and overactivity. In the acute stage of "spinal shock" in high lesions, lack of sympathetic activity in the presence of excessive cardiac vagal activity can result in bradycardia or even cardiac arrest (Frankel *et al.*, 1975; Mathias, 1976). As patients recover from spinal shock the isolated spinal cord can function independently of the brain and, when activated, causes the syndrome "autonomic dysreflexia." This can be induced by a variety of stimuli from cutaneous, skeletal muscle or visceral sources below the level of the lesion. It results in marked hypertension, usually with bradycardia as the afferent and vagal efferent

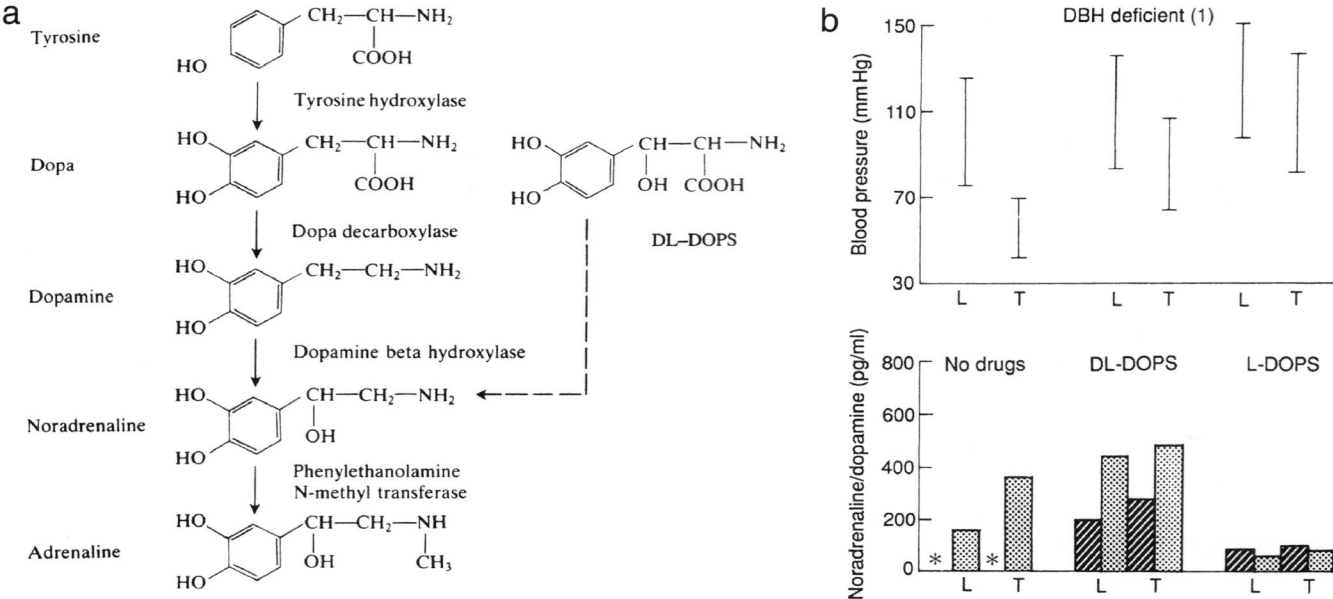

FIGURE 4   (a) Biosynthetic pathway in the formation of noradrenaline and adrenaline. The structure of DL-DOPS is indicated on the right. It is converted directly to noradrenaline by dopa decarboxylase, thus bypassing dopamine β-hydroxylase. (b) Blood pressure (systolic and diastolic) while lying (L) and during head-up tilt (T) in a patient with DBH deficiency, before and during treatment with DL-DOPS and L-DOPS. Plasma noradrenaline (hatched histogram) and dopamine (stippled histogram) levels are indicated before and during tilt. Plasma noradrenaline was undetectable (*, <5 pg/mL) in both while off drugs. Reprinted from Mathias et al. (1990), with permission from Oxford University Press, Oxford.

components of the baroreceptor reflex are preserved (Fig. 5). Paroxysmal hypertension may result in intracerebral hemorrhage. The reverse, orthostatic hypotension, may occur because of the inability of the brain to activate sympathetic pathways in response to gravitational stress.

## DRUGS, CHEMICALS, AND TOXINS

These can cause autonomic dysfunction by their pharmacological effects or by causing of autonomic neuropathy (Table II).

## NEURALLY MEDIATED SYNCOPE

These disorders are characterized by a transient cardiovascular autonomic abnomality that results in syncope (fainting, blackouts, loss of consciousness) (Hainsworth, 1999; Kapoor, 2000; Mathias et al., 2001). Autonomic nervous system function usually is perserved between attacks. The autonomic abnormalities consist of an increase in cardiac parasympathetic activity resulting in bradycardia or cardiac arrest (the

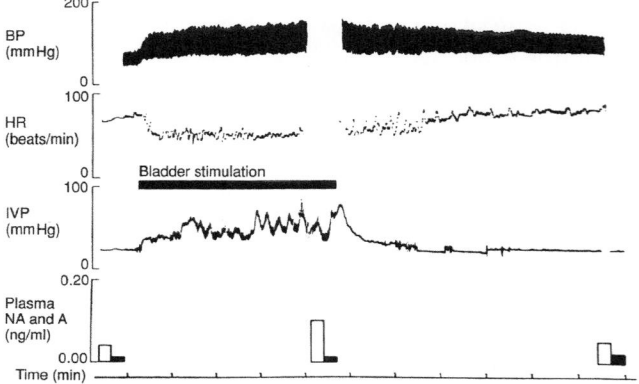

FIGURE 5   Blood pressure (BP), heart rate (HR), intravesical pressure (IVP), and plasma noradrenaline (NA) and adrenaline (A) levels in a tetraplegic patient before, during, and after bladder stimulation induced by suprapubic percussion of the anterior abdominal wall. The rise in BP is accompanied by a fall in heart rate as a result of increased vagal activity in response to the rise in blood pressure. Level of plasma NA (open histograms), but not A (filled histograms) rise, suggesting an increase in sympathetic neural activity independently of adrenomedullary activation. Reprinted from Mathias and Frankel (1986), copyright Oxford University Press, Oxford.

cardioinhibitory form) and withdrawal of sympathetic neural activity causing hypotension (the vasodepressor form). The two may occur together (the mixed form) (see Fig. 1).

Neurally mediated syncope made be broadly considered under the following groups. The most common is vasovagal syncope, also known as common faints. Subjects may present in their teenage years, often with a family history of syncope (Mathias *et al.*, 1998, 2000). The history may provide clues to provoking factors that range from standing still at school assembly to venepuncture, the sight of needles and blood, and sometimes even discussions of these stimuli; hence the alternative term "emotional" syncope. The prognosis is usually excellent, although the condition can be extremely disruptive to both the subject and the family, especially if there are recurrent episodes that may occur without warning or result in convulsions. There also may be diagnostic difficulties in separating the conditon from epilepsy. Fainting may cause concerns about driving and operating machinery and result in loss of confidence, especially in younger subjects. The newer techniques enabling mapping of cerebral autonomic centers in humans (Critchley *et al.*, 2000, 2001, 2002) may cast light on whether these subjects show a differential central autonomic response to conditioned responses.

In the elderly, the increasingly recognized form of neurally mediated syncope is carotid sinus hypersensitivity, especially in those with falls of unknown etiology (McIntosh *et al.*, 1993). Some may provide a classical history of activation of afferents in the neck, with buttoning of the collar, shaving, or cervical movements. Testing with carotid sinus massage should also be performed in the head-up position, ideally on a tilt table, as the vasodepressor form can be missed with stimulation while supine (Mathias *et al.*, 1991a). The cardioinhibitory and mixed forms often respond favorably to a cardiac pacemaker. The diagnosis actively should be sought as falls, especially in the frail and elderly, can result in trauma that contributes to morbidity and mortality.

There are many other factors that cause neurally mediated syncope. These are sometimes grouped under the term "situational" syncope. They may occur in response to physiological stimuli, such as induced by the "mess trick," during trumpet blowing (causing an exaggerated Valsalva maneuver) or during micturition usually while upright and in the presence of alcohol (micturition syncope). There may be a pathological enhancement of stimuli as in swallow syncope associated with glossopharyngeal neuralgia, neoplastic infiltration, or drugs (Deguchi and Mathias, 1999). Syncope due to severe bradycardia may occur during tracheal stimulation in high cervical lesions needing respirator support (Frankel *et al.*, 1975; Mathias, 1976).

## POSTURAL TACHYCARDIA SYNDROME (PoTS)

This is a disorder predominantly affecting women between the ages of 20 and 50 years. They have orthostatic intolerance with symptoms of light-headedness and palpitations (Low *et al.*, 1995). The symptoms often disappear on sitting or lying down. Investigations exclude orthostatic hypotension and autonomic failure. The heart rate rises by 30 beats per minute or more during head-up postural change (Fig. 1). There are similarities with the syndromes initially described by da Costa and Lewis (also known as the soldier's heart syndrome or neurocirculatory esthenia) and associations with mitral valve prolapse, hyperventilation and the chronic fatigue syndrome. The condition is heterogenous (Khurana, 1995). In some it may follow a viral infection; there have been descriptions of a partial autonomic neuropathy mainly involving the lower limbs (Jacob *et al.*, 2000; Schondorf and Low, 1993). A gene mutation encoding the noradrenaline transporter has been described and is thought to be responsible for a hyperadrenergic state (Shannon *et al.*, 2000).

The prognosis is variable, with spontaneous recovery in some. A variety of drugs that include beta blockers have been used along with other drugs used to treat orthostatic hypotension. An increase in intravascular volume appears also to be beneficial.

## ORTHOSTATIC (POSTURAL) HYPOTENSION

### Clinical Aspects

Orthostatic hypotension is arbitrarily defined as a fall of more than 20 mm Hg in systolic orthostatic blood pressure or 10 mm Hg diastolic on standing or on head-up tilt to at least 60° for 3 min (Schatz *et al.*, 1996). The diagnosis initially can be made by measuring the blood pressure lying and standing, or sitting if the latter is not possible. Heart rate and blood pressure responses to head-up posture are divided into five categories by Wieling and Karamaker (1999):

1. orthostatic dizziness in healthy subjects (an increase in heart rate with variable short-lived changes in systolic and diastolic blood pressure);
2. hyperadrenergic orthostatic response (a large heart rate increase, no change or rise in systolic blood pressure, and a rise in diastolic blood pressure);

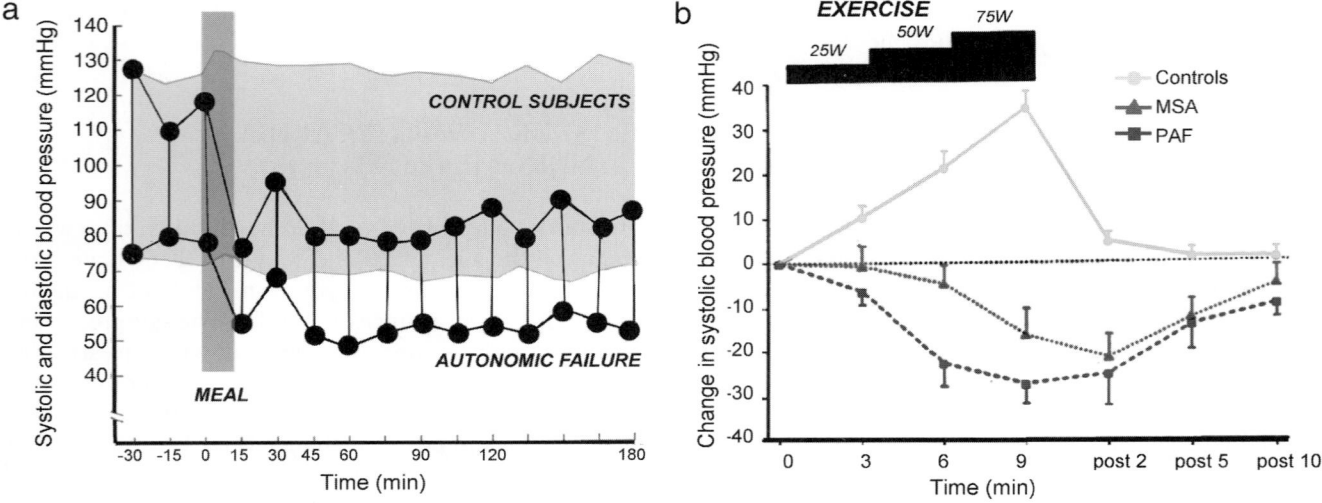

**FIGURE 6** (a) Supine systolic and diastolic blood pressure before and after a standard meal while horizontal in normal subjects (stippled area) and in a patient with autonomic failure. Blood pressure does not change in normal subjects after a meal. In the patient it rapidly falls to around 80/50 mm Hg and remains low over 3 h Reprinted with permission from Mathias and Bannister (1999c). (b) Changes in systolic pressure during horizontal bicycle exercise at three incremental levels in normal subjects (controls) and in patients with multiple system atrophy (MSA) and pure autonomic failure (PAF). In both MSA and PAF, unlike controls, there is a fall in blood pressure. Adapted from Smith *et al.* (1995).

3. vasovagal orthostatic response (heart rate decrease, fall of both systolic and diastolic blood pressure);
4. hypoadrenergic orthostatic response with intact heart rate control (large heart rate rise, decline in systolic and diastolic blood pressure);
5. hypoadrenergic orthostatic response with impairment of autonomic heart rate control (no change in heart rate, fall in systolic and diastolic blood pressure).

Orthostatic hypotension reflects a disturbance of the many regulatory mechanisms that maintain arterial blood pressure. Normally regulation is mediated through carotid baroreceptors with afferent sensory fibers in the glossopharyngeal and vagus nerves, the medullary centers for circulatory control, and sympathetic efferent pathways that increase the performance of the heart (rate and contraction force), and peripheral resistance by controlling vasoconstriction.

Thus orthostatic hypotension appears distinct from a "normally" low supine (horizontal) blood pressure (constitutional hypotension), which appears to be more common in young subjects (Pemberton, 1989). Although not precisely evaluated, subjects with constitutional hypotension (but without orthostatic hypotension) have no underlying disorder; however, they may have overlapping symptoms including fatigue. Orthostatic hypotension may cause a variety of symptoms, some of which are due to underperfusion of organs,

especially those above the level of the heart (Mathias *et al.*, 1999) (Table VI). The mechanisms for more recently recognized symptoms, such as "coathanger" neck pain, are not clear (Bleasdale-Barr and Mathias, 1998; Cariga *et al.*, 2002). Many factors in daily life, that range from food ingestion and mild exercise to prolonged recumbency can influence orthostatic hypotension (Table VII). The pathophysiological mechanisms continue to be elu-

**TABLE VI** Some of the Symptoms Resulting from Orthostatic Hypotension and Impaired Perfusion of Various Organs

| | |
|---|---|
| Cerebral hypoperfusion | |
|   Dizziness | |
|   Visual disturbances | |
|     Blurred | Color defects |
|     Scotoma | Tunnel vision |
|     Graying out | Blacking out |
|   Loss of consciousness | |
|   Impaired cognition | |
| Muscle hypoperfusion | |
|   Paracervical and suboccipital ("coathanger") ache | |
|   Lower back/buttock ache | |
|   Calf claudication | |
| Cardiac hypoperfusion | |
|   Angina pectoris | |
| Spinal cord hypoperfusion | |
| Renal hypoperfusion | |
|   Oliguria | |
| Nonspecific | |
|   Weakness, lethargy, fatigue | |

## TABLE VII  Factors Influencing Postural (Orthostatic) Hypotension

Speed of positional change
Time of day (worse in the morning)
Prolonged recumbency
Warm environment (hot weather, central heating, hot bath)
Raising intrathoracic pressure—micturition, defecation, or coughing
Food and alcohol ingestion
Water ingestion[a]
Physical exertion
Maneuvers and positions (bending forward, abdominal
    compression, leg crossing, squatting, activating calf muscle
    pump)[b]
Drugs with vasoactive properties (including dopaminergic agents)

*Note.* Adapted from Mathias (2000a).
[a]Can raise supine blood pressure in chronic autonomic failure (Mathias, 2000b).
[b]These maneuvres usually reduce the postural fall in blood pressure, unlike the others.

cidated, and these have a bearing on treatment (Mathias and Bannister, 1999c; Puvi-Rajasingham *et al.*, 1997). Orthostatic hypotension in the elderly often coexists with disease (cerebrovascular disease, heart disease, impaired baroreceptor function, brain-stem ischemia, autonomic failure). There are both neurogenic and non-neurogenic causes of orthostatic hypotension (see Table VIII).

Plasma noradrenaline levels are useful in confirming a neurogenic cause for orthostatic hypotension. Either minimal or no change in plasma noradrenaline levels with head-up orthostatic change provides biochemical evidence of sympathetic failure. Nonneurogenic causes of orthostatic hypotension may need to be considered.

Plasma catecholamine measurements may differentiate the central disorder, MSA, from peripheral disorders such as PAF; they also may diagnose DBH deficiency

(Fig. 7). Basal levels in MSA (with a preganglionic lesion) are often similar to levels in normal subjects, but in PAF (with a postganglionic lesion) levels are considerably lower than normal. In DBH deficiency, plasma noradrenaline and adrenaline are undetectable, and plasma dopamine is abnormally elevated (Man in't Veld *et al.*, 1987; Mathias *et al.*, 1990; Robertson *et al.*, 1983). In nerve growth factor (NGF) deficiency syndrome, which includes absence of DBH, biochemical changes are similar to isolated DBH deficiency, except that plasma dopamine levels are low, suggesting tyrosine hydroxylase deficiency (Anand *et al.*, 1991). The clonidine-growth hormone stimulation test separates central from peripheral autonomic failure syndromes. (Kimber *et al.*, 1997; Thomaides *et al.*, 1992). Growth hormone levels rise in normal subjects and PAF with peripheral involvement, but not in those with the central impairment caused by MSA.

### Principles of Therapy

Orthostatic hypotension should be treated, especially with drugs, only if the patient is symptomatic. The management includes a combination of nonpharmacological and pharmacological measures (Mathias and Kimber, 1998, 1999), which ideally should be tailored individually (Table IX). Some of the former are directly related to the various factors, listed in Table VII, that influence orthostatic hypotension. Advice about physical countermeasures that increase muscle activity (such as crossing the legs or tensing calf muscles) often reduce symptoms (Wieling *et al.*, 1993). Head-up tilt at night, if tolerated, can improve orthostatic hypotension and reduce supine hypertension. Smaller and more frequent

## TABLE VIII  Nonneurogenic Causes of Postural Hypotension

| | |
|---|---|
| Low intravascular volume | |
|   Blood/plasma loss | Hemorrhage, burns, hemodialysis |
|   Fluid/electrolyte | Inadequate intake—anorexia nervosa |
| | Fluid loss—vomiting, diarrhoea, losses from ileostomy |
| | Renal endocrine—salt losing nepropathy, adrenal insufficiency |
| | (Addison's disease), diabetes insipidus, diuretics |
| Vasodilatation | |
|   Drugs—glyceryl trinitrate | |
|   Alcohol | |
|   Heat, pyrexia | |
|   Hyperbradykininism | |
|   Systemic mastocytosis | |
|   Extensive varicose veins | |
| Cardiac impairment | |
|   Myocardial | Myocarditis |
|   Impaired filling | Atrial myxoma, constrictive pericarditis |
|   Impaired output | Aortic stenosis |

*Note.* Reprinted with permission from Mathias (2000a).

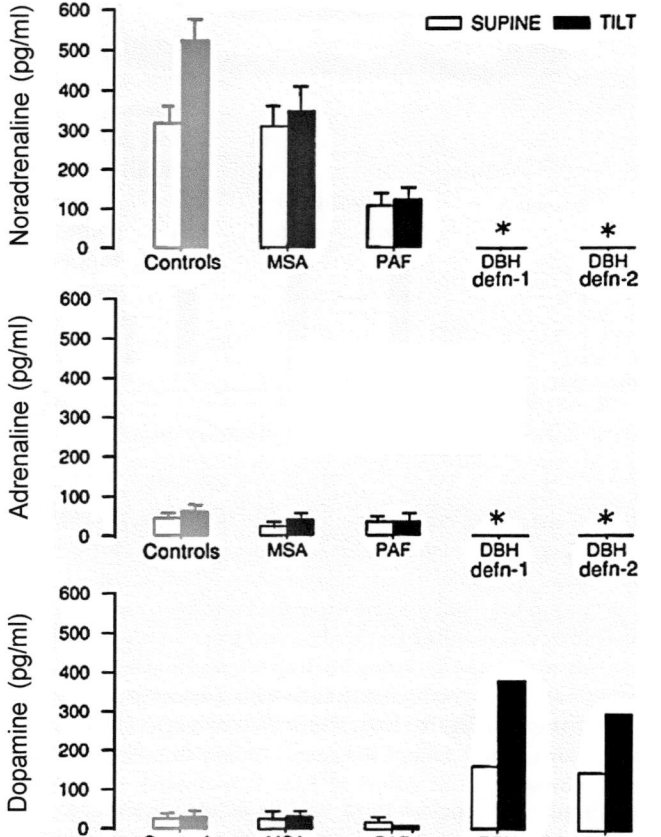

FIGURE 7    Plasma noradrealine, adrenaline, and dopamine levels (measured by high pressure liquid chromatography) in normal subjects (controls), patients with multiple system atrophy (MSA), pure autonomic failure (PAF), and two individual patients with dopamine β-hydroxlase deficiency (DBH def) while supine and after head-up tilt to 45° for 10 min. The asterisk indicates levels below the detection limits for the assay, which are less than 5 pg/mL for noradrenaline and adrenaline and less than 20 pg/mL for dopamine. Bars indicate ± SEM Reprinted from Mathias *et al.* (1990), with permission from Oxford University Press, Oxford).

TABLE IX    Summary Outline of Nonpharmacological and Pharmacological Measures in the Management of Orthostatic Hypotension Due to Neurogenic Failure

Nonpharmalogical
    To be avoided
        Sudden head-up postural change (especially on waking)
        Prolonged recumbency
        Straining during micturition and defaecation
        High environmental temperature (including hot baths)
        "Severe" exertion
        Large meals (especially with refined carbohydrate)
        Alcohol
        Drugs with vasodepressor properties
    To be introduced
        Head-up tilt during sleep
        Small frequent meals
        High salt intake
        Judicious exercise (including swimming)
        Body positions and maneuvers
    To be considered
        Water ingestion
        Elastic stockings
        Abdominal binders
Pharmacological
    Starter drug                    — fludrocortisone
    Sympathomimetics         — ephedrine or midodrine
    Specific targeting             — octreotide, desmopressin, or erythropoietin

*Note.* It should be emphasized that nonneurogenic factors (such as fluid loss due to vomiting or diarrhoea) may substantially worsen orthostatic hypotension. Adapted from Mathias and Kimber (1998, 1999).

meals should be taken. Salt intake should be increased. Water ingestion (500 mL) may help (Mathias, 2000b); physical methods either may not work adequately or, as is the case with antigravity suits, be effective only when in operation. They usually make the subject more susceptible to symptoms when not in use. A variety of drugs used to treat the associated disorder (such as L-dopa for parkinsonism) or autonomic impairment (such as sildenafil for male erectile failure (Hussain *et al.*, 2001)) may lower pressure further.

Supplementation of nonpharmacological approaches with anti-hypotensive drugs often is needed (Table X). A valuable starter drug is fludrocortisone, in a dose of between 100 and 200 μg at night. The drug probably acts by combining its salt-retaining effects with increas-

ing sensitivity of blood vessels to pressor substances. Side effects, usually with larger doses, include ankle edema and hypokalemia. Fludrocortisone alone may be useful in mild orthostatic hypotension. In moderate to severe cases the addition of sympathomimetics, to take the place of the deficient neurotransmitter noradrenaline, should be considered. These drugs include ephedrine that acts both directly and indirectly and often is beneficial in central and incomplete autonomic lesions, such as MSA. It is used in a dose of 15 mg three times daily and can be increased to 30 or 45 mg thrice daily. Higher doses may induce insomnia, reduce appetite, and cause tremor or tachycardia. In more distal lesions such as in PAF, a directly acting drug, such as midodrine, is of value. This is converted to the active metabolite, desglymidodrine (Low *et al.*, 1997), which is an alpha-adrenoceptor agonist. Other sympathomimitics include the pro-drug, L-dihydroxyphenylserine, which successfully reduces orthostatic hypotension in dopamine β-hydroxylase deficiency; it also has been successfully used in familial amyloid polyneuropathy (Suzuki *et al.*, 1989) and chronic autonomic failure (Freeman *et al.*, 1996; Mathias *et al.*, 2001). The alkalide dihydroergotamine acts predominantly on venous capacitance vessels

**TABLE X** Outline of the Major Actions by Which a Variety of Drugs May Reduce Orthostatic Hypotension

Reducing salt loss/plasma volume expansion
    Mineralocorticoids (fludrocortisone)
Reducing nocturnal polyuria
    V$_2$-receptor agonists (desmopressin)
Vasoconstriction—sympathetic
    Directly
        on resistance vessels (midodrine, phenylephrine, noradrenaline, clonidine)
        on capacitance vessels (dihydroergotamine)
    Indirectly (ephedrine, tyramine with monoamine oxidase inhibitors, yohimbine)
    Prodrug (L-dihydroxyphenylserine)
Vasoconstrictor—nonsympathomimetic
    V$_1$ receptor agents—terlipressin
Preventing vasodilatation
    Prostaglandin synthetase inhibitors (indomethacin, flurbiprofen)
    Dopamine receptor blockade (metoclopramide, domperidone)
    β$_2$-Adrenoceptor blockade (propranolol)
Preventing postprandial hypotension
    Adenosine receptor blockade (caffeine)
    Peptide release inhibitors (somatostatin analogue: octreotide)
Increasing cardiac output
    Beta blockers with intrinsic sympathomimetic activity (pindolol, xamoterol)
    Dopamine agonists (ibopamine)
Increasing red cell mass
    Erythropoietin

but its effects are limited by poor absorption. Other agents include the V1 receptor agonists, which cause vasoconstriction.

Drugs acting on pre- and post α-2 adrenoceptors have also been used but have limited application in practice. The α-2 adrenoceptor against clonidine predominantly lowers blood through its central effects in reducing sympathetic outflow (Kooner *et al.*, 1991; Reid *et al.*, 1976); it also has actions on postsynaptic α-2 adrenoceptors and in the presence of supersensitivity may raise blood pressure. This may account for its modest success in severe distal sympathetic lesions. Drugs such as yohimbine, which block presynaptic α-2 adrenoceptors which normally suppress release of noradrenaline, should theoretically be of benefit in incomplete sympathetic lesions, as has been observed in single dose studies.

The risk–benefit ratio of these agents needs consideration, especially in the presence of coronary artery and peripheral vascular disease. Supine hypertension is another factor. With the combination of drugs such as tyramine and monoamine oxidase inhibitors, severe hypertension may occur with serious consequences, resulting in cerebral hemorrhage, aortic dissection, myocardial infarction, and cardiac failure, especially if the pressor effects of these drugs are prolonged. To reduce severe supine hypertension that may be a particular problem at night, head-up tilt, a pre-bedtime snack to induce postprandial hypotension and even the

nocturnal use of short-acting vasodilators may need to be considered.

Attempts have been made to administer sympathomimetics using a subcutaneous infusion pump. There was initial success (Polinsky *et al.*, 1982); a more recent report also favors its potential especially in severely disabled patients (Oldenburg *et al.*, 2001).

In addition to fludrocortisone and sympathomimetics, specific targeting using a variety of drugs to counteract the mechanisms contributing to orthostatic hypotension should be employed. These include reducing nocturnal polyuria with the vasopressin-2 agonist, desmopressin (Mathias *et al.*, 1986), 5–40 mg intranasally or 100–400 mg orally at night, preferably in combination with fludrocortisone so that nocturnal natriuresis is reduced. PAF patients require smaller doses and are more sensitive than MSA. Sodium should be monitored to exclude hyponatremia and water intoxication.

In subjects with postprandial hypotension, a variety of approaches need to be used. This includes caffeine, although its effects appear to be limited by tolerance, especially in more severe lesions. Octreotide is of value, but has to be given subcutaneously about half an hour before each meal. It has side-effects that include abdominal colic and diarrhea. It is of value in severe postprandial hypotension. It acts by inhibiting release of vasodilatatory gastrointestinal peptides and preventing splanchnic vasodilation in response to food ingestion (Kooner *et al.*, 1989a,b). It also may partly reduce postural and exercise-induced hypotension (Smith and Mathias, 1995). It does not appear to enhance nocturnal (supine) hypertension (Alam *et al.*, 1995). Erythropoietin has been used with success in patients who are anemic (especially in diabetes and amyloidosis with renal failure); it increases red cell mass (Hoeldtke and Streeten, 1993; Perera *et al.*, 1995; Watkins, 1998).

The above relate predominantly to primary autonomic failure. In secondary autonomic failure, additional strategies should include the effects of the primary disorder and its treatment. In spinal-cord injuries, paroxysmal hypertension may complicate autonomic dysreflexia. Caution therefore is needed with the benefits provided by treatment weighed against the drawbacks of excessive, albeit short-lived, hypertension. In diabetes mellitus the same applies as an elevated supine blood pressure may accelerate renal damage.

The management of orthostatic hypertension should include education not only of the patient, but also the patient's relatives and carers. In rare disorders, briefing of medical and paramedical support staff may be needed. It is essential that the patient is given a realistic indication of the goals of treatment and limitations of current management. The symptoms of orthostatic hypertension, not the level of blood pressure, are of crucial importance, especially as the two may be dissociated. It should be emphasized that a major aim is to improve the quality of life and help subjects lead a full and independent life. Preventing falls and trauma is of prime importance, especially in the elderly. Treatment should ideally be integrated seamlessly with the help of autonomic liaison nurses and autonomic nurse specialists.

## NAUSEA AND VOMITING

### Clinical Aspects

Nausea and vomiting is associated with dizziness (see Chapter 10). It may be caused by irritation of afferent sensory fibers (transmission mostly via the vagus nerve) from the pharynx or the gut (with gastric ulceration, pyloric obstruction, or stenosis); by impaired motor activity of the digestive tract (associated with diabetes mellitus or after vagotomy); by intoxication by bacterial, viral, or chemically contaminated food (in most cases by endotoxin from staphylococcal infection); by irritation of afferent pathways from the urogenital tract or the cardiovascular system (cardiac infarction); and with metabolic disorders (uremia, diabetic ketoacidosis). Vestibular stimulation with nonphysiological inputs from other sensory systems or with functional deficits in the semicircular canals or the vestibular nuclei lead to nausea and vomiting (see Chapter 13).

Stimulation of taste or smell may have adverse effects through emotional associations. Vomiting may be caused iatrogenically, by radiation treatment or side effects of drugs such as those used in chemotherapy. Vomiting can also occur especially in the first trimester of pregnancy or following drug intoxication (e.g., digitalis). Chemotherapeutic agents for neoplasms may cause vomiting by releasing 5-hydroxytryptamine (serotonin) from the enterochromaffin cells of the small intestine. The emetic effects of the morphine derivatives are often an unwanted side effect of the drug in pain-reducing treatment (as in myocardial infarction). A variety of drugs used in neurological practice, from L-dopa to carbamazepine can cause nausea and vomiting as a side effect.

Vomiting is triggered by the vomiting center in the lateral reticular formation (near the dorsal vagal nucleus), which gets its input from the vagal afferent fibers from higher centers and from the area postrema (in the floor of the fourth ventricle), where there is a chemoreceptive trigger zone sensitive to emetics in blood and cerebrospinal fluid (CSF; Borison, 1974; Borison and McCarthy, 1983). Vomiting as a result of raised intracranial pressure is presumed to result from a rise in pressure in the fourth ventricle (Kuntz, 1953). Posture-dependent vomiting in patients with lesions of the posterior fossa is unlikely to be induced by direct irritation of the vomiting area (Drachman et al., 1977), as the concomitant nystagmus and dizziness (see Chapter 13) suggest that vestibular irritation is caused by lesions close to the vestibular nuclei. Direct cerebral causes of vomiting include intracerebral and subarachnoid bleeding, encephalitis, and migraine.

### Principles of Therapy

If the cause of vomiting is known, therapy should be directed to the cause, as in cases of intoxication and elevated intracranial pressure. Symptomatic therapy (review (Andrews, 1999; Peroutka and Snyder, 1982)) is directed toward different receptors at centers influencing vomiting; $H_2$ and muscarinic receptors (in the vomiting center and vestibular nuclei) and dopamine $D_2$ and muscarinic receptors (in the area postrema). Antiemetic agents have different sites of action. Phenothiazines (e.g., chlorpromazine) act simultaneously on the $H_1$ receptors (vomiting center) and on the dopamine receptors (area postrema). The latter are also the site of action for metoclopramide and domperidone; drugs that provoke vomiting such as morphine, apomorphine, and levodopa also act at these sites. In high doses, metoclopramide seems to affect the 5-hydroxytryptamine-5-HT$_3$-receptors rather than dopamine receptors (Gralla et al., 1981). These 5-HT receptors are also the sites of action for selectively acting serotonin antagonists such as ondansetrone and granisetron. These drugs are superior to metoclopramide in vomiting induced by cytostatic or chemotherapeutic agents and do not cause extrapyramidal side effects (Marty et al., 1990). A combination of granisetron and dexamethasone was more effective in preventing emesis induced by moderately emetogenic chemotherapy than was granisetron alone (Roila et al., 1995). Metoclopramide (Harrington et al., 1983) and domperidone (Brodgen et al., 1982) prevent distension of the bowel preceding vomiting by increasing gastrointestinal motility. Muscarinic cholinergic receptors are the sites of action for scopolamine, meclocine, and dimenhydrinate. In the treatment of motion sickness, scopolamine and dimenhydrinate inhibit the frequency modulation of the vestibular nuclei in response to vestibular stimulation (see Chapter 13).

## AUTONOMIC DYSFUNCTION AND THE GUT

Autonomic dysfunction of the gastrointestinal tract results in features that may be specific to the organ affected but not necessarily to the disease (Albanese et al., 2000). The clinical picture of diseases, not usually dealt with by neurologists (such as Hirschsprung's disease, Chagas' disease, irritable bowel syndrome (IBS)), will not be covered. Disordered neural control of the gut may result in uncoordinated motility, with spasm or dilatation. The investigation is dependent on possible diagnosis and includes endoscopy, imaging by ultrasound examination, X-ray contrast examination, radionuclide scintigraphy (of gastric emptying and of the ejection fraction of the gallbladder), and manometric recordings (Wingate, 1999).

## Oesophageal Spasm and Achalasia

Esophageal spasm is characterized by repetitive, segmental uncoordinated contractions of the esophagus (without peristaltic waves), independent of the act of swallowing. The spasm can be provoked by swallowing bigger boluses or cold liquids. In most cases the autonomically innervated distal third of the esophagus (unlike the upper third without, and the middle third with mixed autonomic innervation) is affected. Often there is only insufficient relaxation of the lower esophageal sphincter (cardiospasm achalasia). Dysphagia, retrosternal feeling of pressure, and angina–pectoris-like pain often occur. Regurgitation and aspiration may result. With longer duration, weight loss and malnutrition may occur. Loss of postganglionic parasympathetic neurons of the myenteric plexus, occasionally also of the extrinsic vagal nerve fibers and lesions at the dorsal motor vagal nucleus (rarely seen with amyotrophic lateral sclerosis with bulbar signs), may be the cause. Compression, e.g., by mediastinal tumors, and esophagitis need to be excluded. The diagnosis can be achieved by X-ray contrast examination studies and a provocation test with bethanechol, which is based on cholinomimetics provoking painful segmental contractions because of receptor hypersensitivity as a result of the cholinergic deficit.

Mechanical aids include head-up tilt at night, liquid food, and adequate mastication. The calcium antagonist nifedipine, 10 mg before each meal, reduces pressure in the lower esophageal sphincter from 50 to 30 mm Hg after 10–40 min, the effect continuing for 1 h (Bortolotti and Labo, 1981). Anticholinergic drugs may be used. Relaxation of the lower esophageal sphincter can also be achieved by glyceryl trinitrate (Schiffter, 1985). Surgical procedures (like Heller operation) are of unproven value. Endoscopic balloon dilatation may help.

## Esophageal Distension

With esophageal distension there is delayed emptying. Reduced tone of the lower esophageal sphincter may lead to reflux esophagitis. Neurogenic causes include diabetic autonomic neuropathy (Clarke et al., 1979), Riley–Day syndrome, acute pandysautonomia, and acute cholinergic dysautonomia. Myogenic causes include scleroderma, dermatomyositis, and familial visceral myopathy (Faulk et al., 1978).

Simple mechanical procedures, which include head-up tilting are initial measures. To enhance the tone of the lower esophageal sphincter, bethanechol (after a test dose of 2.5 mg) in single doses of 5–10 mg 3–4 times a day), metoclopramide (5–20 mg before meals; Harrington et al., 1983; Malagelada et al., 1980), and cisapride (before meals; Brogna et al., 1989) have been tried. Cisapride is not available in the UK because of an interaction with drugs that causes cardiac dysrhythmia.

## Gastric Stasis

Symptoms of gastric stasis and distension include lack of appetite, epigastric fullness, nausea, and vomiting (Malagaleda, 1982). Abnormal gastric emptying can be detected by X-ray contrast examination (the contrast medium is normally visible in the duodenum after 30 min, but in gastric stasis the contrast medium may remain in the stomach for several days) or by using radionuclide techniques with scintigraphy. Gastric stasis may occur in diabetis mellitus (diabetic gastroparesis) (Clarke et al., 1979; Watkins and Edmonds, 1999), acute pandysautonomia, familiar visceral myopathy (Faulk et al., 1978), acute intermittent porphyria, and when the vagus is damaged.

In addition to drug treatment, advice should include small meals more frequently. Effective drugs are metoclopramide, domperidone (Brogden et al., 1982), cisapride (Brogna et al., 1989), and cholinomimetic agents. Metoclopramide (5–20 mg 3 times a day, 30 min before meals) appears to stimulate release of acetylcholine from cholinergic intramural neurons, sensitize the muscarinic receptors (Albibi and McCallum, 1983), and provoke, independent of the plasma level of motilin, interdigestive motor complexes (Achem-Karam et al., 1985). Improvement has been reported with domperidone (10 mg every 4 h; Heer, 1980), which is advantageous compared to metoclopramide, as it has no central effects and is unlikely to cause or interfere with treatment of extrapyramidal movement disorders. Cisapride stimulates the production and release of acetylcholine in postganglionic nerve terminals of the myenteric plexus but, in contrast to metoclopramide, is not a dopamine antagonist (McCullum et al., 1988). Cholinomimetics such as bethanechol in doses of 10 mg before meals (after a test dose of 2.5 mg) may be tried (Malagelada et al., 1980). In insulin-dependent diabetes mellitus (IDDM), the macrolide, erythromycin that stimulates motilin receptors (200 mg intravenously or 250 mg 3 times daily orally) may improve gastric emptying (Janssens *et al.*, 1990).

## INTESTINAL DISTENSION AND INTESTINAL PSEUDOOBSTRUCTION

Intestinal distension may present with abdominal discomfort, meteorism, or constipation. Some of the causes are similar to those of gastric distension; thus intestinal pseudoobstruction may be caused by familial visceral myopathy (Faulk *et al.*, 1978).

The drugs used in intestinal distension are based on the same principles as in gastric distension; it is important to determine the cause and exclude mechanic obstruction and paralytic ileus; the latter may complicate high spinal cord lesion in the early phases aftr injury. Metoclopramide, domperidone, cisapride, and parasympathomimetics have been used. Bethanechol in idiopathic intestinal pseudoobstruction (Maldonado *et al.*, 1970) and neostigmine in familiar visceral myopathy appear to be of no value. However, blocking the effect of the endorphins on the opiate receptors with naloxone (0.4 mg intravenously) is of benefit (Schang and Devroede, 1985). Chlorpromazine (25–50 mg orally or intravenously), and trifluperidol (0.5–2.5 mg intravenously or orally) is of benefit in intestinal distension. If associated with bacterial overgrowth in the small intestines, broad spectrum antibiotic treatment is recommended (e.g., tetracycline 500–1000 mg daily for 10 days; Clarke *et al.*, 1979).

## NEUROGENIC DIARRHEA

Bloating, abdominal pain, diarrhea without symptoms of malabsorption, and pseudoincontinence because of insuppressible urgency of defecation are symptoms in neurogenic diarrhea. The course of neurogenic diarrhea is usually chronic and intermittent with spontaneous remissions of variable duration. Diabetes mellitus and alcoholic enteropathy are some of the causes, along with anxiety neurosis.

### Principles of Therapy

Loperamide, codeine phosphate, and anticholinergics (e.g., scopolamine) reduce intestinal motility. Clonidine (Fedorak *et al.*, 1985) may also help and may improve reabsorption of fluids. Bacterial overgrowth in

the small intestines may be prevented by giving tetracycline or metronidazole. In addition to treatment with antibiotics, cholestyramine binds bile acids, the release of which is increased bacterial overgrowth, which may contribute to diarrhea.

## DISTURBANCES OF SWEATING

### Physiology and Pathophysiology

The regulation of body temperature is dependent on a hierarchical neuronal network (Collins, 1999). Afferent pathways from peripheral warm and cold receptors via C-fibers reach the posterior root ganglion cells, the spinothalamic tract and the preoptic area/anterior hypothalamus and septum which are important sites in thermoregulation. Limbic afferent pathways that are responsible for emotional sweating project to the hypothalamus. Eccrine sweat glands play an important role in thermoregulation; their innervation is sympathetic cholinergic.

Disorders of sweating can result in either anhidrosis or hyperhidrosis; both have neurological and nonneurological causes. Anhidrosis may result from central or peripheral neurological diseases. The pattern is usually global in PAF with sparing in MSA. Brain tumors and strokes may cause segmental anhidrosis. In spinal cord injury, anhidrosis usually occurs below the level of the lesion. In peripheral neuropathies with postganglionic denervation, anhidrosis is usually distal; this can vary as in diabetic neuropathy. Localized abnormalities occur in leprosy. Radiation injury causes anhidrosis by sweat gland damage. Anticholinergics often cause global reduction in sweating, while after sympathectomy regional anhidrosis occurs. Isolated anhidrosis without other signs of autonomic failure may be congenital or acquired (Houlden *et al.*, 2001; Low *et al.*, 1985).

Hyperhidrosis may be either primary or secondary. The latter has numerous causes that include pyrexia, shock, hypoglycemia, the dumping syndrome, severe pain, malignant tumors that include Hodgkins disease and lymphomas, hyperthyroidism, withdrawal from alcohol and other agents including opiates, intoxication of mercury, and pheochromocytoma. It may be iatrogenic and caused by cholinergic substances like pilocarpine and diaphoretics such as paracetamol and acetylsalicylic acid. It may be compensatory, following upper thoracic sympathectomy and in the variant of the Holmes–Adie syndrome (Ross's syndrome with patchy anhidrosis). Other pathological forms include gustatory sweating (Frey's syndrome) because of abberant innervation when parasympathetic fibers to the salivary glands connect with sympathetic cholinergic fibers. This may occur after surgery for parotid tumors, in diabetic autonomic neuropathy, and with neuroregeneration during recovery from acute autonomic neuropathy. Gustatory sweating is provoked by oral stimulation and sometimes even by the smell or thought of food. A form of compensatory hyperhidrosis may also affect tetraplegics above the lesion, in the face and neck during autonomic dysreflexia.

In primary or essential hyperhidrosis there is no specific cause although in some stress is a provoking factor. Situations causing sweating physiologically (such as in response to heat, exercise, stress and emotion) cause excessive sweating, and this may occur over the palmar and plantar regions, the axilla, the head and neck, and sometimes over the whole body. In some there may be a family history.

## Investigation

Sudomotor function may be assessed in various ways (Low and Fealey, 1999). These include the thermoregulatory sweat test, during which body temperature is raised, ideally by 1°C. The extent of sweating can be observed visually and by the use of dyes that change color on exposure to moisture. Various modifications of the original quinizarin dye (Guttmann, 1947), currently not available in its present form but with allied substances, result in change from pale to a bright red or purple color. The quantitative sudomotor axon reflex test (Q-SART) (Low et al., 1983) and various other evaporimeters, determine the volume of sweating. The number and size of sweat droplets can be determined using a silastic inprint material (Kennedy et al., 1984a, b) and can test responses to drugs such as pilocarpine. The sympathetic skin response is dependent upon changes in conductance following activation of sweat glands, producing a change in skin potential. It is thus thought to be a measure of sympathetic cholinergic function, and is abnormal in distal sympathetic cholinergic failure (Magnifico et al., 1998) and in spinal injuries (Cariga et al., 2002).

Depending upon the possible cause of sweating, tests may need to be modified to determine further the abnormality. Thus in spinal cord lesions assessments may need to be made before and after induction of autonomic dysreflexia, and in gustatory sweating before and after food challenge. In suspected Ross's syndrome, investigation into the other manisfestations of the Holme–Adie syndrome need to be pursued.

## Treatment of Hyperhidrosis

In mild to moderate hyperhidrosis, intermittent application of antihidrotics (e.g., hexamethylenetetramine) in the form of ointments and powders have been used. Water iontophoresis (Hölzle, 1988; Raulin et al., 1988) has been used for palmar and plantar sweating, utilizing devices available for home use. It is claimed that euhidrosis may be reached and maintained after 10–15 sessions lasting for 20–30 min, with continuing one or two maintenance sessions per week. Alcoholic or aqueous solutions of aluminum chloride have been used for axillary hyperhidrosis (Hölzle, 1988; Scholes, 1978). The application should be every second day initially, and then once per week prior to going to bed as there is less sweat gland activity and the solution may enter the secretory ducts of the sweat glands by diffusion. Aluminum chloride should be only be applied to intact skin as it may cause skin irritation.

Drugs given systemically include the anticholinergics (propantheline bromide, biperidine, bornaprine, trihexyphenidyl). These act on postsynaptic muscarinic receptors on sweat glands. Side effects include a dry mouth due to hypostomia. They have the potential to raise intraoccular pressure. The sympatholytic agent, clonidine, which acts predominantly centrally, is of value, even in low doses of 25–50 μg thrice daily (Kuritzky et al., 1984). Beta blockers, such as propranolol, may be of value. In emotional or stress-induced hyperhidrosis, cognitive behavioral therapy is of benefit. In some antidepressant drugs, including the SSRIs, may be needed.

When the above measures fail surgical therapy has been utilized. With localized problems, such as axillary hyperhidrosis (Blank and Eichmann, 1984) excision of glands may be successful; complications include necrosis of the wound margin and dehiscence of the suture.

Sympathectomy, particularly for palmar hyperhidrosis, with destruction of T2–T4 segments, often leads to permanent relief (Kux, 1960). The development of thoracoscopic surgical procedures for localized and preganglionic sympathectomy (Wittmoser, 1978) makes it possible to treat palmar, axillary, and even facial hyperhidrosis, with a reduced risk compared to conventional surgery. The complications, however, include compensatory hyperhidrosis in the spared areas, involving the middle and lower part of the trunk and the lower limbs which can be as or more disconcerting (Adar, 1997; Masters and Rennie, 1992). Sympathectomy also may be used for hyperhidrosis of the lower limb. The combination of thoracic and lumbar sympathectomy should not be carried out as this can cause orthostatic hypotension. Surgical procedures for gustatory sweating include auriculotemporal denervation (Gardner and McCubbin, 1956). The recent introduction of botulinum toxin, has been of value in the management of localised hyperhidrosis. It has been used successfully to treat axilliary, palmar and gustatory hyperhidrosis

(Naumann *et al.*, 1997; Schulze-Bonhage *et al.*, 1996). The disadvantage is multiple injections that may have to be repeated.

### Anhidrosis

Anhidrosis may be caused by an interruption of sympathetic fibers at various levels (cerebral or spinal cord lesions, sympathetic trunk lesions, or peripheral lesions associated with polyneuropathies). Anhidrosis may also occur congenitally, without other signs of autonomic failure (Houlden *et al.*, 2001; Low *et al.*, 1985). In regional anhidrosis with involvment of small areas no specific therapy is required. With larger areas control of body temperature may be affected and with advice on appropriate reduction of physical exercise and exposure to heat to prevent overheating. Subjects need to avoid of hot drinks and food, use relevant clothing, and ingest cold drinks. With hyperthermia heat stroke may occur; symptomatic treatment may be needed immediately to include a cold bath, ingestion of cold drinks, preventing dehydration and checking electrolyte imbalance, and considering drugs such as chlorpromazine, that reduce temperature, probably by central mechanisms.

## REFERENCES

Achem-Karam, S. R., Funakoski, A., Vinik, A. I., and Owyang, C. (1985). Plasma motilin concentrations and interdigestive migrating motor complex in diabetic gastroparesis. *Gastroenterology* **88**, 492–499.

Adar, R. (1997). Compensatory hyperhidrosis after thoracic sympathectomy. *Lancet* **351**, 231–232.

Alam, M., Smith, G. D. P., Bleasdale-Barr, K., Pavitt, D. V., and Mathias, C. J. (1995). Effects of the peptide release inhibitor, Octreotide, on daytime hypotension and on nocturnal hypertension in primary autonomic failure. *J. Hypertens.* **13**, 1664–1669.

Albanese, A., Brisinda, G., and Mathias, C. J. (2000). The autonomic nervous system and gastrointestinal disorders. *In* "Handbook of Clinical Neurology" (P. J. Vinken and G. W. Bruyn, Eds.), Vol. 75, pp. 613–663. Elsevier NV, Amsterdam, The Netherlands.

Albibi, R., and McCallum, R. W. (1983). Metoclopramide: Pharmacology and clinical application. *Ann. Intern. Med.* **98**, 86–95.

Anand, P., Rudge, P., Mathias, C. J., Springall, D. R., Ghatei, M. A., Natter-Noe, M., Sharief, M., Misra, V. P., Polak, J. M., Bloom, S. R., and Thomas, P. K. (1991). New autonomic and sensory neuropathy with loss of adrenergic sympathetic functions and sensory neuropeptides. *Lancet* **337**, 1253–1254.

Andrews, P. L. R. (1999). Nausea, vomiting, and the autonomic nervous system. *In* "Autonomic Failure: A Textbook of Clinical Disorders of the Autonomic Nervous System" (C. J. Mathias and R. Bannister, Eds.), 4th Ed., pp. 126–135. Oxford University Press, Oxford.

Appenzeller, O., and Marshall, J. (1963). Vasomotor disturbance in Landry–Guillain–Barre syndrome. *Arch. Neurol.* **9**, 368–372.

Appenzeller, O., and Oribe, E. (1997). "The Autonomic Nervous System," 5th Ed. Elsevier Biomedical, Amsterdam/New York/Oxford.

Asahina, M., Kuwabara, S., Suzuki, A., and Hattori, T. (2001). Autonomic function in demyelinating and axonal subtypes of Guillain–Barre syndrome. *Acta Neurol. Scand.* **105**, 1–7.

Axelrod, F. B. (1999). Familial dysautonomia. *In* "Autonomic Failure: A Textbook of Clinical Disorders of the Autonomic Nervous System" (C. J. Mathias and R. Bannister, eds.), 4th Ed., pp. 402–409. Oxford University Press, Oxford.

Bannister, R., and Oppenheimer, D. R. (1972). Degenerative diseases of the nervous system associated with autonomic failure. *Brain* **95**, 457–474.

Birchfield, R. I., and Shaw, C. M. (1964). Postural hypotension in the Guillain–Barre syndrome. *Arch. Neurol.* **40**, 149–157.

Blank, A. A., and Eichmann, F. (1984). Operative Therapie der Hyperhidrosis axillaris: Indikationsabwägung und Komplikationen. *In* "Komplikationen in der operativen Dermatologie" (B. Konz and O. Braun-Falco, Hrsg.). Springer, Berlin/Heidelberg/New York.

Bleasdale-Barr, K., and Mathias, C. J. (1998). Neck and other muscle pains in autonomic failure: Their association with orthostatic hypotension. *J. R. Soc. Med.* **91**, 355–359.

Borison, H. (1974). Area postrema: Chemoreceptor trigger zone for vomiting—Is that all? *Life Sci.* **14**, 1807–1817.

Borison, H., and McCarthy, L. E. (1983). Neuropharmacology of chemotherapy-induced emesis. *Drugs* **25**(Suppl. 1), 8–17.

Bortolotti, M., and Labo, G. (1981). Clinical and manometric effects of nifedipine in patients with esophageal achalasia. *Gastroenterology* **80**, 39–44.

Bradbury, S., and Eggleston, C. (1925). Orthostatic hypotension: A report of three cases. *Am. Heart J.* **1**, 73–86.

Braune, S., Reinhardt, M., Schnitzer, R., Riedel, A., and Lucking, C. H. (1999). Cardiac uptake of (123I)MIBG separates Parkinson's disease from multiple system atrophy. *Neurology* **53**, 1020–1025. [see comments].

Brogden, R. N., Carmine, A. A., Heel, R. C., Speight, T. M., and Avery, G. S. (1982). Domperidone. *Drugs* **24**, 360–400.

Brogna, A., Ferrara, R., Scornavacca, G., Lombardo, A., Bucceri, A., Catalano, F., Paradisi, V., Onorato, S. (1989). Cisapride and gastric emptying of a solid meal in dyspeptic diabetics without autonomic neuropathy and in healthy volunteers. *Eur. J. Clin. Pharmacol.* **37**, 411–413.

Brooks, C. M., Koizumi, K., and Sato, A. (1979). "Integrative Functions of the Autonomic Nervous System. " Univ. of Tokyo Press and Elsevier, Amsterdam/New York.

Brooks, D. J. (1999). Neuroimaging and allied studies in autonomic failure syndromes. *In* "Autonomic Failure: A Textbook of Clinical Disorders of the Autonomic Nervous System. " (C. J. Mathias and R. Bannister, eds.), 4th Ed., pp. 317–320. Oxford University Press, Oxford.

Cariga, P., Ahmed, S., Mathias, C. J., and Gardner, B. P. (2002). The prevalence and association of neck (coat-hanger) pain and orthostatic (postural) hypotension in human spinal cord injury. *Spinal Cord* **40**, 77–82.

Cariga, P., Catley, M., Savic, G., Frankel, H. L., Mathias, C. J., and Ellaway, P. H. (2002). Organisation of the sympathetic skin response in spinal cord injury. *J. Neurol. Neurosurg. Psychiatry* **72**, 356–360.

Clarke, B. F., Ewing, D. J., and Campbell, I. W. (1979). Diabetic autonomic neuropathy. *Diabetologia* **17**, 195–212.

Cohen, A. J., Weiser, B., Afzal, Q., and Fuhrer, J. (1990). Ventricular tachycardia in two patients with AIDS receiving ganciclovir (DHPG). *AIDS* **4**, 807–809.

Cohen, J., Low, P., Fealey, R., Sheps, S., and Jiang, N. S. (1987). Somatic and autonomic function in progressive autonomic failure and multiple system atrophy. *Ann. Neurol.* **22**, 692–699.

Cohen, J. A., and Laudenslager, M. (1989). Autonomic nervous system involvement in patients with human immunodeficiency virus infection. *Neurology* **39**, 1111–1112.

Cohen, J. A., Miller, L., and Polish, L. (1991). Orthostatic hypotension in human immunodeficiency virus infection may be the result of generalized autonomic nervous system dysfunction. *J. Acquir. Immune. Defic. Syndr.* **4,** 31–33.

Colan, R. V., Snead, O. C., Oh, S. J., and Kashlan, M. B. (1980). Acute autonomic and sensory neuropathy. *Ann. Neurol.* **8,** 441–444.

Collins, K. J. (1999). Temperature regulation and the autonomic nervous system. *In* "Autonomic Failure: A Textbook of Clinical Disorders of the Autonomic Nervous System" (C. J. Mathias and R. Bannister, Eds.), 4th Ed., pp. 92–98. Oxford University Press, Oxford.

Cortelli, P., Contin, M., Lugaresi, A., Baruzzi, A., and Montagna, P. (1990). Severe dysautonomic onset of Guillain–Barre syndrome with good recovery. A clinical and autonomic follow-up study. *It. J. Neurol. Sci.* **11,** 159–162.

Cortelli, P., Parchi, P., and Contin, M., *et al.* (1991). Cardiovascular dysautonomia in fatal familial insomnia. *Clin. Auton. Res.* **1,** 15–22.

Cortelli, P., Perani, D., Parchi, P., Grassi, F., Montagna, P., De Martin, M., Castellani, R., Tinuper, P., Gambetti, P., Lugaresi, E., and Fazio, F. (1997). Cerebral metabolism in fatal familial insomnia: Relation to duration, neuropathology, and distribution of protease-resistant prion protein. *Neurology* **49,** 126–133.

Craddock, C., Pasvol, G., Bull, R., Protheroe, A., and Hopkin, J. (1987). Cardiorespiratory arrest and autonomic neuropathy in AIDS. *Lancet* **2,** 16–18.

Critchley, H. D., Corfield, D. R., Chandler, M. P., Mathias, C. J., and Dolan, R. J. (2000). Cerebral correlates of autonomic cardiovascular arousal: A functional neuroimaging investigation in humans. *J. Physiol.* **523,** 259–270.

Critchley, H. D., Mathias, C. J., and Dolan, R. J. (2001). Neural correlates of first and second-order representation of bodily states. *Nature Neurosci.* **2,** 207–212.

Critchley, H. D., Mathias, C. J., and Dolan, R. J. (2002). Fear conditioning in humans: The influence of awareness and autonomic arousal on functional neuroanatomy. *Neuron* **33,** 653–663.

Daniel, S. E. (1999). The neuropathology and neurochemistry of multiple system atrophy. *In* "Autonomic Failure: A Textbook of Clinical Disorders of the Autonomic Nervous System" (C. J. Mathias and R. Bannister, Eds.), 4th Ed., pp. 321–328. Oxford University Press, Oxford.

Davies, A. G., and Dingle, H. R. (1972). Observations on cardiovascular and neuroendocrine disturbance in the Guillain–Barre syndrome. *J. Neurol. Neurosurg. Psychiatry* **35,** 176–179.

Deguchi, K., and Mathias, C. J. (1999). Continuous haemodynamic monitoring in an unusual case of swallow-induced syncope. *J. Neurol. Neurosurg. Psychiatry* **67,** 220–222.

Dickson, D. W., Lin, W., Liu, W. K., and Yen, S. H. (1999). Multiple system atrophy: A sporadic synucleinopathy. *Brain Pathol.* **9,** 721–732.

Drachman, D. A., Diamond, E. R., and Hart, C. W. (1977). Orthostaticly evoked vomiting: Association with posterior fossa lesions. *Ann. Otol. Rhin. Laryngol.* **86,** 97–101.

Druschky, A., Hilz, M. J., Platsch, G. *et al.* (2000). Differentiation of Parkinson's disease and multiple system atrophy in early disease stages by means of I-123-MIBG-SPECT. *J. Neurol. Sci.* **175,** 3–12. [see comments]

Edmonds, M. E., and Sturrock, R. D. (1979). Autonomic neuropathy in the Guillain–Barre syndrome. *Br. Med. J.* **2,** 668.

Eisenhauer, M. D., Eliasson, A. H., Taylor, A. J., Coyne, P. E., Jr., and Wortham, D. C. (1994). Incidence of cardiac arrhythmias during intravenous pentamidine therapy in HIV-infected patients. *Chest* **105,** 389–395.

Evenhouse, M., Haas, E., Snell, E., Visser, J., Paul, L., and Gonzalez, R. (1987). Hypotension infection with the human immunodeficiency virus. *Ann. Intern. Med.* **107,** 598–599. (letter)

Ewing, D. J., Campbell, I. W., and Clarke, B. F. (1980). The natural history of diabetic autonomic neuropathy. *Q. J. Med.* **49,** 95–108.

Faulk, D. L., Anuras, S., Gardner, D., Mitros, F. A., Summers, R. W., and Christensen, J. (1978). A familiar visceral myopathy. *Ann. Intern. Med.* **89,** 600–606.

Feasby, T. E., Gilbert, J. J., Brown, W. F. *et al.* (1986). An acute axonal form of Guillain–Barre polyneuropathy. *Brain* **109,** 1115–1126.

Fedorak, R. N., Field, M., and Chang, E. B. (1985). Treatment of diabetic diarrhea with Clonidine. *Ann. Intern. Med.* **102,** 197–199.

Flachenecker, P., Wermuth, P., Hartung, H. P., and Reiners, K. (1997). Quantitative assessment of cardiovascular autonomic function in Guillain–Barre syndrome. *Ann. Neurol.* **42,** 171–179.

Frankel, H. L., Mathias, C. J., and Spalding, J. M. K. (1975). Mechanisms of reflex cardiac arrest in tetraplegic patients. *Lancet* **ii,** 1183–1185.

Freeman, R. (1995). Pure autonomic failure. *In* (D. Robertson and I. Biaggioni, Eds.), pp. 83–106. "Disorders of the Autonomic Nervous System" Harwood Academic, Luxenbourg.

Freeman, R., Roberts, M. S., Friedman, L. S., and Broadbridge, C. (1990). Autonomic function and human immunodeficiency virus infection. *Neurology* **40,** 575–580.

Freeman, R., Young, J., Landsbert, L., and Lipsitz, L. (1996). The treatment of postprandial hypotension in autonomic failure with 3,4-DL-threo-dihydroxphenylserine. *Neurology* **47,** 1414–1420.

Gardner, J. W., and McCubbin, J. W. (1956). Auriculo-temporal syndrome. *J. Am. Med. Assoc.* **160,** 272.

Giasson, B. I., Duda, J. E., Murray, I. V. *et al.* (2000). Oxidative damage linked to neurodegeneration by selective alpha-synuclein nitration in synucleinopathy lesions. *Science* **290,** 985–989.

Gilman, S., Low, P. A., Quinn, N. *et al.* (1998). Consensus statement on the diagnosis of multiple system atrophy. *Clin. Auton. Res.* **8,** 359–362.

Goetz, C. G., Lutge, W., and Tanner, C. M. (1986). Autonomic dysfunction in Parkinson's disease. *Neurology* **36,** 73–75.

Goldstein, D. S., Holmes, C., Cannon, R. O., Eisenhofer, G., and Kopin, I. J. (1997). Sympathetic cardioneuropathy in dysautonomias. *N. Engl. J. Med.* **336,** 696–702.

Goldstein, D. S., Holmes, C., Li, S. T., Bruce, S., Metman, L. V., and Cannon, R. O., III (2000). Cardiac sympathetic denervation in Parkinson disease. *Ann. Intern. Med.* **133,** 338–347.

Goldstein, D. S., Polinsky, R. J., Garty, M. *et al.* (1989). Patterns of plasma levels of catechols in neurogenic orthostatic hypotension. *Ann. Neurol.* **26,** 558–563.

Gralla, R. J., Itri, L. M., Pisko, S. E., Squilante, A. E., Kelsen, D. P., Braun, D. W., Bordin, L. A., Braun, T. J., and Young, C. W. (1981). Antiemetic efficacy of high-dose metoclopramide: Randomized trials with placebo and prochlorperazine in patients with chemotherapy-induced nausea and vomiting. *N. Engl. J. Med.* **305,** 905–909.

Guttmann, L. (1947). The management of the quinizarin sweat test (QST). *Postgrad. Med. J.* **23,** 353–366.

Hainsworth, R. (1999). Syncope and fainting. *In* "Autonomic Failure: A Textbook of Clinical Disorders of the Autonomic Nervous System" (C. J. Mathias and R. Bannister, Eds.), 4th Ed. pp. 428–436. Oxford Univ. Press, Oxford.

Harcourt, J., Spraggs, P., Mathias, C.J., and Brookes, G. (1996). Sleep-related breathing disorders in the Shy-Drager syndrome. Observations on investigation and management. *Eur. J. Neurol.* **3,** 186–190.

Harrington, R. A., Hamilton, C. W., Brogden, R. N., Luikewich, J. A., Romanciewicz, J. A., and Heel, R. C. (1983). Metoclopramide. *Drugs* **25,** 451–494.

Hart, R. G., Kanter, M. C. (1990). Acute autonomic neuropathy. Two cases and a clinical review. *Arch. Intern. Med.* **150,** 2373–2376.

Heer, M., Pirovino, M., Japp, H., Bühler, H., and Schmid, M. (1980). Diabetic gastroparesis and colonic dilatation treated with domperidone. *Lancet* II, 1145–1146.

Hilsted, J., and Low, P. A. (1997). Diabetic autonomic neuropathy. *In* "Clinical Autonomic Disorders" (P. A. Low, Ed.), pp. 487–508. Lipincott-Raven, Philadelphia.

Hoeldtke, R. D., and Streeten, D. H. P. (1993). Treatment of orthostatic hypotension with erythropoietin. *N. Engl. J. Med.* **329**, 611–615.

Hölzle, E. (1988). Axilla̅re und palmoplantare Hyperhidrosis. *Dt. A̅rztebl.* **85**, 2135–2139.

Houlden, H., King, R. H. M., Hashemi-Nejad, A., Wood, N. W., Mathias, C. J., Reilly, M., and Thomas, P. K. (2001). A novel TRK A (*NTRK1*) mutation associated with hereditary sensory and autonomic neuropathy Type V. *Ann. Neurol.* **49**, 521–525.

Hughes, A. J., Daniel, S. E., Kilford, L., and Lees, A. J. (1992). The accuracy of clinical diagnosis of idiopathic Parkinson's disease: A clinico-pathological study of 100 cases. *J. Neurol. Neurosurg. Psychiatry.* **55**, 181–182.

Hussain, I. F., Brady, C., Swinn, M. J., Mathias, C. J., and Fowler, C. (2001). Treatment of erectile dysfunction with sildenafil citrate (Viagra) in parkinsonism due to Parkinson's disease or multiple system atrophy with observations on orthostatic hypotension. *J. Neurol. Neurosurg. Psychiatry* **71**, 371–374.

Iosa, D., Dequattro, V., De-Ping Lee, D., Elkayam, U., Caeiro, T., and Palmero, H. (1990). Pathogenesis of cardiac neuro-myopathy in Chagas' disease and the role of the autonomic nervous system. *J. Auton. Nerv. Syst.* **30**, S83–S88.

Iosa, D., Dequattro, V., De-Ping Lee, D., Elkayam, U., and Palmero, H. (1989). Plasma norepinephrine in Chagas' cardioneuromyopathy: A marker of progressive dysautonomia. *Am. Heart. J.* **117**, 882–887.

Iwasa, K., Nakajima, K., Yoshikawa, H., Tada, A., Taki, J., and Takamori, M. (1998). Decreased myocardial 123I-MIBG uptake in Parkinson's disease. *Acta Neurol. Scand.* **97**, 303–306.

Jacob, G., Costa, F., Shannon, J. R. *et al.* (2000). The neuropathic postural tachycardia syndrome. *N. Engl. J. Med.* **343**, 1008–1014.

Janssens, J., Peeters, T. I., Vantrappen, G., Tack, J., Urban, J. L., de Roo, M., Muls, E., and Boutillon, R. (1990). Improvement of gastric emptying in diabetic gastroparesis by erythromycin: Preliminary studies. *N. Engl. J. Med.* **322**, 1028–1031.

Johnson, R. H., and Spalding, J. M. K. (1974). "Disorders of the Autonomic Nervous System." Blackwell Scientific, Oxford/London.

Kapoor, W. N. (2000). Syncope. *N. Engl. J. Med.* **343**, 1856–1862.

Keane, J. R. (1977). Tonic pupils with acute ophthalmoplegic polyneuritis. *Ann. Neurol.* **2**, 393–396.

Kennedy, W. R., Sakuta, M., Sutherland, D., and Goetz, F. C. (1984a). Quantitation of the sweating deficit in diabetes mellitus. *Ann. Neurol.* **15**, 482–488.

Kennedy, W. R., Sakuta, M., Sutherland, D., and Goetz, F. C. (1984b). The sweating deficiency in diabetes mellitus: Methods of quantitation and clinical correlation. *Neurology* **34**, 758–763.

Khurana, R. K. (1995). Orthostatic intolerance and orthostatic tachycardia: a heterogeneous disorder. *Clin. Auton. Res.* **5**, 2–18.

Kimber, J., Mitchell, D., and Mathias, C. J. (1998). Chronic cough in the Holmes-Adie syndrome: Association in five cases with autonomic dysfunction. *J. Neurol. Neurosurg. Psychiatry* **65**, 583–586.

Kimber, J., Mathias, C. J., Lees, A. J., Bleasdale-Barr, K., Chang, H. S., Churchyard, A., and Watson, L. (2000). Physiological, pharmacological and neurohormonal assessment of autonomic function in progressive supranuclear palsy. *Brain* **123**, 1422–1430.

Kimber, J., Watson, L., and Mathias, C. J. (in press). (2001). Cardiovascular and neurohormonal responses to i.v. l-arginine I two groups with primary autonomic failure. *J. Neurol.* **248**.

Kimber, J. R., Watson, L., and Mathias, C. J. (1997). Distinction of idiopathic Parkinson's disease from multiple system atrophy by stimulation of growth hormone release with clonidine. *Lancet*, **349**, 1877–1881.

Kooner, J. S., Birch, R., Frankel, H. L., Peart, W. S., and Mathias, C. J. (1991). Haemodynamic and neurohormonal effects of clonidine in patients with preganglionic and postganglionic sympathetic lesions. Evidence for a central sympatholytic action. *Circulation* **34**, 75–83.

Kooner, J. S., Frankel, H. L., Mirando, N., Peart, W. S., and Mathias, C. J. (1988). Haemodynamic, hormonal and urinary responses to postural change in tetraplegic and paraplegic man. *Paraplegia* **26**, 233–237.

Kooner, J. S., Peart, W. S., and Mathias, C. J. (1989a). The peptide release inhibitor, octreotide (SMS201-995) prevents the haemodynamic changes following food ingestion in normal human subjects. *Q. J. Exp. Physiol.* **74**, 569–572.

Kooner, J. S., Raimbach, S. J., Watson, L., Bannister, R., Peart, W. S., and Mathias, C. J. (1989b). Relationship between splanchnic vasodilatation and post-prandial hypotension in patients with primary autonomic failure. *J. Hypertens.* **7**(Suppl. 6), s40–s41.

Kuntz, A. (1953). "The Autonomic Nervous System." Lea & Febiger, Philadelphia.

Kuritzky, A., Hering, R., Goldhammer, G., and Bechar, M. (1984). Clonidine treatment in paroxysmal localised hyperhidrosis. *Arch. Neurol.* **41**, 1210–1211.

Kux, E. (1960). Über die thorakoskopische vegetative Denervation. *Münch. Med. Wschr.* **102**, 637–639.

Kyle, R. A., and Dyck, P. J. (1993). Amyloidosis and neuropathy. *In* (P. J. Dyck, P. K. Thomas, J. W. Griffin, P. A. Low, and J. F. Podulso, Eds.), pp. 1294–1309. "Peripheral Neuropathy" Saunders Philadelphia.

Larner, A. J., Mathias, C. J., and Rossor, M. N. (2000). Autonomic failure preceding dementia with Lewy bodies. *J. Neurol.* **247**, 229–231.

Lichtenfeld, P. (1971). Autonomic dysfunction in the Guillian–Barre syndrome. *Am. J. Med.* **50**, 772–780.

Lin-Greenberger, A., and Taneja-Uppal, N. (1987). Dysautonomia and infection with the human immunodeficiency virus. *Ann. Intern. Med.* **106**, 167. [letter]

Low, P. A. (ed.) (1997). "Clinical Autonomic Disorders," 2nd Ed. Little, Brown, Boston.

Low, P. A., Caskey, P. E., Tuck, R. R., Fealey, R. D., and Dyck, P. J. (1983). Quantitative sudomotor axon reflex test in normal and neuropathic subjects. *Ann. Neurol.* **14**, 573–580.

Low, P. A., and Fealey, R.D. (1999). Evaluation of sudomotor function. *In* "Autonomic Failure: A Textbook of Clinical Disorders of the Autonomic Nervous System" (C. J. Mathias and R. Bannister, Eds.), 4th Ed., pp. 263–270. Oxford University Press, Oxford.

Low, P. A., Fealey, R. D., Sheps, S. G., Su, W. P. D., Trautmann, J. C., and Kuntz, N. C. (1985). Chronic idiopathic anhidrosis. *Ann. Neurol.* **18**, 344–348.

Low, P. A., Opfer-Gehrking, T. L., Textor, S. C. *et al.* (1995). Postural tachycardia syndrome (POTS). *Neurology* **45**, S19–S25.

Luginbuhl, L. M., Orav, E. J., McIntosh, K., and Lipshultz, S. E. (1993). Cardiac morbidity and related mortality in children with HIV infection. *JAMA* **269**, 2869–2875.

Magnifico, F., Misra, V. P., Murray, N. M. F., and Mathias, C. J. (1998). The sympathetic skin response in peripheral autonomic failure—Evaluation in pure autonomic failure, pure cholinergic dysautonomia and dopamine-beta-hydroxylase deficiency. *Clin. Auton. Res.* **8**, 133–138.

Malagelada, J. R. (1982). Gastric emptying disorders: Clinical significance and treatment. *Drugs* **24**, 353–359.

Malagelada, J. R., Rees, W. D. W., Mazzotta, L. J., and Go, V. L. W. (1980). Gastric motor abnormalities in diabetic and postvagotomy gastroparesis: Efficacy of metoclopramide and betanechol. *Gastroenterology* **78**, 286–293.

Maldonado, J. E., Gregg, J. A., Green, P. A., and Brown, A. L. (1970). Chronic idiopathic intestinal pseudo-obstruction. *Am. J. Med.* **49**, 203–212.

Mallipedi, R., and Mathias, C. J. (1998). Raynaud's phenomenon after sympathetic denervation in patients with primary autonomic failure: Questionnaire survey. *Bri. Med. J.* **316**, 438–439.

Man in't Veld, A. J., Boomsma, F., Moleman, P., and Schalekamp, M. A. D. H. (1987). Congenital dopamine beta-hydroxylase deficiency: A novel orthostatic syndrome. *Lancet* **i**, 183–187.

Man in't Veld, A. J., Van den Meirackers, A. H., Boomsma, F., and Schalekamp, M. A. D. H. (1987). Effect of unnatural noradrenaline precursor on sympathetic control and orthostatic hypotension in dopamine beta-hydroxylase deficiency. *Lancet* **ii**, 1172–1175.

Markman, M., Sheidler, V., Ettinger, D. S., Quaskey, S. A., and Mellits, E. D. (1984). Antiemetic efficacy of dexamethasone. *N. Engl. J. Med.* **311**, 549–552.

Marty, M., Pouillart, P., Scholl, S., Droz, J. P., Azab, M., Brion, N., Pujade-Lauraine, E., Paule, B., Paes, D., and Bons, J. (1990). Comparison of the 5HT3 (Serotonin)-antagonist ondansetron (GR 38032 F) with high-dose metoclopramide in the control of cisplatin-induced emesis. *N. Engl. J. Med.* **322**, 816–821.

Masters, A., and Rennie, J. A. (1992). Endoscopic transthoracic sympathectomy for idiopathic upper limb hyperhidrosis. *Clin. Auton. Res.* **2**, 349–352.

Mathias, C. J. (1976). Bradycardia and cardiac arrest during tracheal suction—mechanisms in tetraplegic patients. *Eur. J. Intens. Care Med.* **2**, 147–156.

Mathias, C. J. (1997). Autonomic disorders and their recognition. *N. Engl. J. Med.* **10**, 721–724.

Mathias, C. J. (1998). Cardiovascular autonomic dysfunction in parkinsonian patients. *Clin Neurosci.* **5**, 153–166.

Mathias, C. J. (2000a). Disorders of the autonomic nervous system. *In* "Neurology in Clinical Practice" (W. G. Bradley, R. B. Daroff, G. M. Fenichel, and C. D. Marsden, Eds.), 3rd Ed., pp. 2131–2165. Butterworth-Heinemann, Boston.

Mathias, C. J. (2000b). A 21st century water cure. *Lancet* **356**, 1046–1048.

Mathias, C. J., Armstrong, E., Browse, N., Chaudhuri, K. R., Enevoldson, P., and Ross Russell, R. (1991a). Value of non-invasive continuous blood pressure monitoring in the detection of carotid sinus hypersensitivity. *Clin. Autonom. Res.* **2**, 157–159.

Mathias, C. J, and Bannister (1999). "Autonomic Failure: A Textbook of Clinical Disorders of the Autonomic Nervous System," 4th Ed. Oxford Univ. Press, Oxford.

Mathias, C. J., and Bannister, R. (1999a). Dopamine-beta-hydroxylase deficiency and other genetically determined autonomic disorders. *In* "Autonomic Failure: A Textbook of Clinical Disorders of the Autonomic Nervous System, 4th Ed., pp. 387–401. Oxford University Press, Oxford.

Mathias, C. J., and Bannister, R. (1999b). Investigation of autonomic disorders. *In* "Autonomic Failure: A Textbook of Clinical Disorders of the Autonomic Nervous System" (R. Bannister and C. J. Mathias, Eds. ), 4th Ed., pp. 169–195. Oxford Univ. Press, Oxford.

Mathias, C. J., and Bannister, R. (1999c). Postprandial hypotension in autonomic disorders. *In* "Autonomic Failure: A Textbook of Clinical Disorders of the Autonomic Nervous System" (C. J. Mathias and R. Bannister), 4th Ed. pp. 283–295. Oxford University Press, Oxford.

Mathias, C. J., Bannister, R., Cortelli, P., Heslop, K., Polak, J., Raimbach, S. J., Springall, D. B., and Watson, L. (1990). Clinical autonomic and therapeutic observations in two siblings with orthostatic hypotension and sympathetic failure due to an inability to synthesize noradrenaline from dopamine because of a deficiency of dopamine beta hydroxylase. *Q. J. Med.* **278**, 617–633.

Mathias, C. J., da Costa, D. F., Fosbraey, P., Bannister, R., Wood, S. M., Bloom, S. R., and Christensen, N. J. (1989). Cardiovascular, biochemical and hormonal changes during food induced hypotension in chronic autonomic failure. *J. Neurol. Sci.* **94**, 255–269.

Mathias, C. J, Deguchi, K., Bleasdale-Barr, K., and Kimber, J. R. (1998). Frequency of family history in vasovagal syncope. *Lancet* **352**, 33–34.

Mathias, C. J., Deguchi, K., Bleasdale-Barr, K., and Smith, S. (2000). Familial vasovagal syncope and pseudosyncope: Observations in a case with both natural and adopted siblings. *Clin. Auton. Res.* **10**, 43–45.

Mathias, C. J., Deguchi, K., and Schatz, I. (2001). Observations on recurrent syncope and presyncope in 641 patients. *Lancet* **357**, 348–353.

Mathias, C. J., Fosbraey, P., da Costa, D. F., Thornley, A., and Bannister, R. (1986). Desmopressin reduces nocturnal polyuria, reverses overnight weight loss and improves morning orthostatic hypotension in autonomic failure. *Br. Med. J.* **293**, 353–354.

Mathias, C. J., and Frankel, H. (1999). Autonomic disturbances in spinal cord lesions. *In* "Autonomic Failure: A Textbook of Clinical Disorders of the Autonomic Nervous System" (C. J. Mathias and R. Bannister, Eds.), 4th Ed., pp. 494–513. Oxford University Press, Oxford.

Mathias, C. J., Holly, E., Armstrong, E., Shareef, M., and Bannister, R. (1991b). The influence of food on postural hypotension in three groups with chronic autonomic failure: Clinical and therapeutic implications. *J. Neurol. Neurosurg. Psychiatry* **54**, 726–730.

Mathias, C. J., and Kimber, J. R. (1998). Treatment of postural hypotension. *J. Neurol. Neurosurg. Psychiatry* **65**, 285–289.

Mathias, C. J., and Kimber, J. R. (1999). Postural hypotension—Causes, clinical features, investigation and management. *Ann. Rev. Med.* **50**, 317–336.

Mathias, C. J., Mallipeddi, R., and Bleasdale-Barr, K. (1999). Symptoms associated with orthostatic hypotension in pure autonomic failure and multiple system atrophy. *J. Neurol.* **246**, 893–898.

Mathias, C. J., Senard, J., Braune, S., Watson, L., Aragishi, A., Keeling, J., and Taylor, M. (2001). L-Theo-dihydroxphenylserine (L-threo-DOPS; droxidopa) in the management of neurogenic orthostatic hypotension: A multi-national, multi-centre, dose-ranging study in multiple system atrophy and pure autonomic failure. *Clin. Auton. Res* **11**, 235–242.

Mathias, C. J., and Williams, A. C. (1994). The Shy–Drager syndrome (and multiple system atrophy). *In* "Neurodegenerative Disorders" (D. Calne, Ed.), pp. 743–767. Saunders, Philadelphia.

Matthews, M. R. (1999). Autonomic ganglia and preganglionic neurones in autonomic failure. *In* "Autonomic Failure: A Textbook of Clinical Disorders of the Autonomic Nervous System" (C. J. Mathias, and R. Bannister, Eds.), 4th Ed. pp. 329–339. Oxford Univ. Press, Oxford.

McCullum, R. W., Prakash, C., Campoli-Richards, D. M., and Goa, K. L. (1988). Cisapride: A preliminary review of its pharmacodynamics and pharmacokinetic properties and therapeutic use as a prokinetic agent in gastrointestinal motility disorders. *Drugs* **36**, 652–681.

McIntosh, S. J., Lawson, J., and Kenny, R. A. (1993). Clinical characteristics of vasodepressor, cardioinhibitory and mixed carotid sinus syndrome in the elderly. *Am. J. Med.* **95**, 203–208.

McLeod, J. G. (1999). Autonomic dysfunction in peripheral nerve disease. *In* "Autonomic Failure: A Textbook of Clinical Disorders of the Autonomic Nervous System," 4th Ed. Oxford Univ. Press, Oxford.

Miyawaki, E., and Freeman, R. (1994). Peripheral autonomic neuropathy. *In* "Handbook of Autonomic Nervous Sytem Dysfunction," (A. Korczyn, Ed.), pp. 253–282. Dekker, New York.

Monnier, M. (1963). "Physiologie und Pathophysiologie des vegetativen Nervensystems," 1. Band Physiologie. 2. Band Pathophysiologie. Hippokrates, Stuttgart.

Mulhall, B., and Jennens, I. (1987). Testing for neurological involvement in HIV infection. *Lancet* II, 1531. [letter]

Naumann, M., Flachenecker, P., Brocker, E.-B., Toyka, K. V., and Reines, K. (1997). Botulinum toxin for palmar hyperhidrosis. *Lancet*, 349, 252.

Navarro, X., Kennedy, W. R., Loewenson, R. B., and Sutherland, D. E. (1990). Influence of pancreas transplantation on cardiorespiratory reflexes, nerve conduction, and mortality in diabetes mellitus. *Diabetes* 39, 802–806.

Niklasson, U., Olofsson, B. O., and Bjerle, P. (1989). Autonomic neuropathy in familial amyloidotic polyneuropathy. A clinical study based on heart rate variability. *Acta Neurol. Scand.* 79, 182–187.

O'Brien, I. A., McFadden, J. P., and Corrall, R. J. (1991). The influence of autonomic neuropathy on mortality in insulin-dependent diabetes. *Q. J. Med.* 79, 495–502.

Oldenburg, O., Mitchell, A. N., Nurnberger, J. *et al.* (2001). Ambulatory norepinephrine treatment of severe autonomic orthostatic hypotension. *J. Am. Coll. Cardiol.* 37, 219–223.

Palace, J., Chandiramani, V. A., and Fowler, C. J. (1997). Value of sphincter electromyography in the diagnosis of multiple system atrophy. *Muscle Nerve* 20, 1396–1403.

Papp, M. I., and Lantos, P. L. (1994). The distribution of oligodendroglial inclusions in multiple system atrophy and its relevance to clinical symptomatology. *Brain* 117, 235–243.

Pemberton, J. (1989). Does constitutional hypotension exist? *Br. Med. J.* 298, 660–662.

Perera, R., Isola, L., and Kaufmann, H. (1995). Effect of recombinant erythropoietin on anemia and orthostatic hypotension in primary autonomic failure. *Clin. Auton. Res.* 5, 211–214.

Peroutka, S. J., and Snyder, S. H. (1982). Antiemetics: Neurotransmitter binding predicts therapeutic actions. *Lancet* II, 658–659.

Persson, A., and Solders, G. (1983). R-R variations in Guillain–Barre syndrome: A test of autonomic dysfunction. *Acta Neurol. Scand.* 67, 294–300.

Plazzi, G., Corsini, R., Provini, F. *et al.* (1997). REM sleep behavior disorders in multiple system atrophy. *Neurology* 48, 1094–1097.

Polinsky, R. J., Samaras, G. M., and Kopin, I. J. (1983). Sympathetic neural prosthesis for managing orthostatic hypotension. *Lancet* i, 901–904.

Puvi-Rajasingham, S., Smith, G. D. P., Akinola, A., and Mathias, C. J. (1997). Abnormal regional blood flow responses during and after exercise in human sympathetic denervation. *J. Physiol.* 505, 481–489.

Rajput, A. H., and Rozdilsky, B. (1976). Dysautonomia in Parkinsonism: A clinicopathological study. *J. Neurol. Neurosurg. Psychiatry* 39, 1092–1100.

Rajput, A. H., Rozdilsky, B., Rajput, A., and Ang, L. (1990). Levodopa efficacy and pathological basis of Parkinson's syndrome. *Clin. Neuropharmacol.* 13, 553–558.

Rathmann, W., Ziegler, D., Jahnke, M., Haastert, B., and Gries, F. A. (1993). Mortality in diabetic patients with cardiovascular autonomic neuropathy. *Diabet. Med.* 10, 820–824.

Raulin, C., Rösing, S., and Petzoldt, D. (1988). Heimbehandlung der Hyperhidrosis manuum er pedum durch Leitungswasser-Iontophorese. *Hautarzt* 39, 504–508.

Rea, R., Biaggioni, I., Robertson, R. M., Haile, V., and Robertson, D. (1990). Reflex control of sympathetic nerve activity in dopamine-beta-hydroxylase deficiency. *Hypertension* 1, 107–112.

Reid, J. L., Wing, L. M. H., Mathias, C. J., Frankel, H. L., and Neill, E. (1977). The central hypotensive effect of clonidine: studies in tetraplegic subjects. *Clin. Pharmacol. Ther.* 21, 375–381.

Reilly, M. M., and Thomas, P. K. (1999). Amyloid neuropathy. *In* "Autonomic Failure: A Textbook of Clinical Disorders of the Autonomic Nervous System" (C. J. Mathias and R. Bannister, Eds.), 4th Ed., pp. 402–409. Oxford University Press, Oxford.

Rinne, J. O., Burn, D. J., Mathias, C. J., Quinn, N. P., Marsden, C. D., and Brooks, N. J. (1995). Positron emission tomography studies on the dopaminergic system and striatal opioid binding in the olivopontocerebellar atrophy variant of multiple system atrophy. *Ann. Neurol.* 37, 568–573.

Robertson, D., Goldberg, M. R., Onrot, J. *et al.* (1986). Isolated failure of autonomic noradrenergic neurotransmission. Evidence for impaired beta-hydroxylation of dopamine. *N. Engl. J. Med.* 314, 1494–1497.

Roila and the Italian Group for Antiemetic Research (1995). Dexamethasone, granisetron, or both for the prevention of nausea and vomiting during chemotherapy for cancer. *N. Engl. J. Med.* 332, 1–5.

Ropper, A. H., Wijdicks, E. F. M., and Truax, B. T. (1991). "Guillain Barre Syndrome." Davis, Philadelphia.

Rüttimann, S., Hilti, P., Spinas, G. A., and Dubach, U. C. (1991). High frequency of human immunodeficiency virus-associated autonomic neuropathy and more severe involvement in advanced stages of human immunodeficiency virus disease. *Arch. Intern. Med.* 151, 2441–2443. [see comments].

Sampson, M. J., Wilson, S., Karagiannis, P., Edmonds, M., and Watkins, P. J. (1990). Progression of diabetic autonomic neuropathy over a decade in insulin-dependent diabetics. *Q. J. Med.* 75, 635–646.

Saraiva, M. J. M., Costa, P. P., and Goodman, D. S. (1985). Biochemical marker in familial amyloidotic polyneuropathy, Portuguese type: Family studies of transthyretin (prealbumin)—methionine-30 variant. *J. Clin. Invest.* 76, 2171–2177.

Schondorf, R., and Low, P. A. (1993). Idiopathic postural tachycardia syndrome: An attenuated form of acute pandysautonomia? *Neurology* 43, 132–137.

Schang, J. C., and Devroede, G. (1985). Beneficial effects of naloxone in a patient with intestinal pseudoobstruction. *Am. J. Gastroenterol.* 80, 407–411.

Schatz, I. J., Bannister, R., Freeman, R. L. *et al.* (1996). Consensus statement on the definition of orthostatic hypotension, pure autonomic failure and multiple system atrophy. *Clin. Auton. Res.* 6, 125–126.

Schiffter, R. (1985). "Neurologie des vegetativen Nervensystems." Springer, Berlin/Heidelberg/New York.

Scholes, K. T., Crow, R. D., Ellis, J. P., Harman, D. R., and Saikan, E. M. (1978). Axillary hyperhidrosis treated with alcoholic solution aluminium chloride hexahydrate. *Br. Med. J.* II, 84–85.

Schrag, A., Ben Shlomo, Y., and Quinn, N. P. (1999). Prevalence of progressive supranuclear palsy and multiple system atrophy: A cross-sectional study. *Lancet* 354, 1771–1775. [see comments]

Schrag, A., Good, C. D., Miszkiel, K., Morris, H. R., Mathias, C. J., Lees, A. J., and Quinn, N. P. (2000). Differentiation of atypical parkinsonian syndromes with routine MRI. *Neurology* 54, 697–702.

Schrag, A., Kingsley, D., Phatouros, C., Mathias, C. J., Lees, A. J., Daniel, S. E., and Quinn, N. P. (1998). Clinical usefulness of magnetic resonance imaging in multiple system atrophy. *J. Neurol. Neurosurg. Psychiatry* 65, 65–71.

Schulze-Bonhage, A., Schroder, M., and Ferbert, A. (1996). Botulinum toxin in the therapy of gustatory sweating. *J. Neurol.* 243, 143–146.

Seigel, L. J., and Longo, D. L. (1981). The control of chemotherapy-induced emesis. *Am. Intern. Med.* 95, 352–359.

Senard, J.-M., Rai, S., Lapeyre-Mestre, M. *et al.* (1997). Prevalence of orthostatic hypotension in Parkinson's disease. *J. Neurol. Neurosurg. Psychiatry.* 63, 578–559.

Serratrice, G. (1988). Acute pandysautonomia. *N. Issues Neurosci.* **1**, 311–315.

Shannon, J. R., Flatten, N. L., Jordan, J. *et al.* (2000) Orthostatic intolerance and tachycardia associated with norepinephrine-transporter deficiency. *N. Engl. J. Med.* **342**, 541–549.

Shy, G. M., and Drager, G. M. (1960). A neurological syndrome associated with orthostatic hypotension. *Arch. Neurol. Psychiat.* **2**, 511–527.

Silber, M. H., and Levine, S. (2000). Stridor and death in multiple system atrophy. *Mov. Disord.* **15**, 699–704.

Singh, N. K., Jaiswal, A. K., Misra, S., and Srivastava, P. K. (1987). Assessment of autonomic dysfunction in Guillain–Barre syndrome and its prognostic implications. *Acta Neurol. Scand.* **75**, 101–105.

Smith, G. D. P., and Mathias, C. J., (1995). Postural hypotension enhanced by exercise in patients with chronic autonomic failure. *Q. J. Med.* **88**, 251–256.

Smith, G. D. P., Watson, L. P., Pavitt, D. V., and Mathias, C. J. (1995). Abnormal cardiovascular and catecholamine responses to supine exercise in human subjects with sympathetic dysfunction. *J. Physiol. (London)* **485**, 255–265.

Stein, K. M., Haronian, H., Mensah, G. A., Acosta, A., Jacobs, J., and Kligfield, P. (1990). Ventricular tachycardia and torsades de pointes complicating pentamidine therapy of Pneumocystis carinii pneumonia in the acquired immunodeficiency syndrome. *Am. J. Cardiol.* **66**, 888–889.

Strian, F., and Haslbeck, M. (1986). "Autonome Neuropathie bei Diabetes mellitus." Springer, Berlin/Heidelberg/New York/Tokyo.

Sturm, A., and Birkmayer, W. (1977). "Klinische Pathologie des vegetativen Nervensystems." Gustav Fischer, Stuttgart/New York.

Suzuki, S., Higa, S., Tsuga, I., Sakoda, S., Hayashi, A., Yamamura, Y., Takaba, Y., and Nakajima, A. (1980). Effects of infused L-threo-3, 4-dihydroxyphenylserine in patients with familial amyloid polyneuropathy. *Eur. J. Clin. Pharmacol.* **17**, 429–435.

Tarsy, D., and Freeman, R. (1994). The nervous system and diabetes. *In* "Joslin's Diabetes Mellitus" (G. Weir and R. Kahn, Eds.), pp. 794–816. Lea and Febiger, Philadelphia.

Testa, D., Filippini, G., Farinotti, M., Palazzini, E., and Caraceni, T. (1996). Survival in multiple system atrophy: A study of prognostic factors in 59 cases. *J. Neurol.* **243**, 401–404.

Thews, G., and Vaupel, R. (1990). "Vegetative Physiologie." Springer, Berlin.

Thomaides, T., Chaudhuri, K. R., Maule, S., Watson, L., Marsden, C. D., and Mathias, C. J. (1992). The growth hormone response to clonidine in central and peripheral primary autonomic failure. *Lancet* **340**, 263–266.

Thomas, P. K. (1988). Autonomic neuropathies. *N. Issues Neurosci.* **1**, 267–461.

Thomashefsky, A. J., Horwitz, S. J., and Feingold, M. H. (1972). Acute autonomic neuropathy. *Neurology* **22**, 251–255.

Thompson, J. M., O'Callaghan, C. J., Kingwell, B. A., Lambert, G. W., Jennings, G. L. and Esler, M. D. (1995). Total norepinephrine spillover, muscle sympathetic nerve activity and heart rate spectral analysis in a patient with dopamine β-hydroxylase deficiency. *J. Auton. Nerv. Syst.* **55**, 198–206.

Tu, P. H., Galvin, J. E., and Baba, M. *et al.* (1998). Glial cytoplasmic inclusions in white matter oligodendrocytes of multiple system atrophy brains contain insoluble alpha-synuclein. *Ann. Neurol.* **44**, 415–422.

Tuck, R. R., and McLeod, J. G. (1978). Autonomic dysfunction in the Landry–Guillain–Barre syndrome. *Clin. Exp. Neurol.* **15**, 197–203.

Tuck, R. R., and McLeod, J. G. (1981). Autonomic dysfunction in Guillain–Barre syndrome. *J. Neurol. Neurosurg. Psychiatry* **44**, 983–990.

Villa, A., Cruccu, V., Foresti, V., Guareschi, G., Tronchi, M., and Confalonieri, F. (1990). HIV related functional involvement of autonomic nervous system. *Acta Neurol.* **12**, 14–18.

Villa, A, Foresti, V., and Confalonieri, F. (1987). Autonomic neuropathy and HIV infection. *Lancet* **1**, 915. [letter]

Villa, A., Foresti, V., and Confalonieri, F. (1992). Autonomic nervous system dysfunction associated with HIV infection in intravenous heroin users. *AIDS* **6**, 85–89.

Watkins, P. J. K. (1998). The enigma of autonomic failure in diabetes. *J. R. Coll Phys.* **32**, 360–365.

Watkins, P. J., and Edmonds, M. E. (1999). Diabetic autonomic failure. *In* "Autonomic Failure: A Textbook of Clinical Disorders of the Autonomic Nervous System" (C. J. Mathias and R. Bannister, Eds.), 4th Ed., pp. 378–386. Oxford Univ. Press, Oxford.

Welby, S. B., Rogerson, S. J., and Beeching, N. J. (1991). Autonomic neuropathy is common in human immunodeficiency virus infection. *J. Infect.* **23**, 123–128.

Wenning, G. K., Ben Shlomo, Y., Magalhaes, M., Daniel, S. E., and Quinn, N. P. (1994). Clinical features and natural history of multiple system atrophy. An analysis of 100 cases. *Brain* **117** (Pt. 4), 835–845.

Wenning, G. K., Ben Shlomo, Y., Magalhaes, M., Daniel, S. E., and Quinn, N. P. (1995). Clinicopathological study of 35 cases of multiple system atrophy. *J. Neurol. Neurosurg. Psychiatry* **58**, 160–166.

Wenning, G. K., Tison, F., Ben Shlomo, Y., Daniel, S. E., and Quinn, N. P. (1997). Multiple system atrophy: a review of 203 pathologically proven cases. *Mov. Disord.* **12**, 133–147.

Wharton, J. M., Demopulos, P. A., and Goldschlager, N. (1987). Torsade de pointes during administration of pentamidine isethionate. *Am. J. Med.* **83**, 571–576.

Wieling, W, and Karamaker, J. (1999). Non-invasive continuous recording of heart rate and blood pressure in the evaluation of neurocardiovascular control. *In* "Autonomic Failure: A Textbook of Clinical Disorders of the Autonomic Nervous System" 4th Ed., pp. 196–210. Oxford University Press, Oxford.

Wieling, W., Van Lieshout, J. J., and Van Leuwen, A. M. (1993). Physical manoeuvres that reduce orthostatic hypotension in autonomic failure. *Clin. Autonom. Res.* **3**, 57–66.

Wingate, D. L. (1999). Autonomic dysfunction and the gut. *In* "Autonomic Failure: A Textbook of Clinical Disorders of the Autonomic Nervous System" 4th Ed., pp. 271–282. Oxford University Press, Oxford.

Winer, J. B., and Hughes, R. A. C. (1988). Identification of patients at risk of arrhythmia in the Guillain–Barre syndrome. *Q. J. Med.* **257**, 735–739.

Wittmoser, R. (1978). Operative Methoden zur Behebung des krankhaften Schwitzens (Hyperhidrosis). *Ärztl Kosmetologie* **8**, 343–362.

Young, R. R., Asbury, A. K., Corbett, J. L., and Adams, R. D. (1975). Pure pandysautonomia with recovery: Description and diagnosis criteria. *Brain* **98**, 613–636.

Zochodne, D. W. (1994). Autonomic involvement in Guillain–Barre syndrome: A review. *Muscle Nerve* **17**, 1145–1155. [review]

CHAPTER 104

# Neurogenic Disorders of Micturition, Defecation, and Sexual Function

M. Harper, D. E. Andrich, and C. J. Fowler

## NEUROGENIC DISORDERS OF MICTURITION

### Clinical Aspects

The variety of bladder complaints that can occur as a consequence of neurological disease are somewhat limited and consist mainly of disorders of the two phases of bladder activity (i.e., a failure to store urine and a failure to empty the bladder). Patients may experience storage failure as urgency and frequency, and if the problem is more severe, urge incontinence. Problems with emptying may be experienced as hesitancy, interrupted stream or frank urinary retention. However, the consequences of incomplete emptying are likely to be frequency and urgency, because the bladder reaches capacity sooner. Although this range of possible symptoms is not extensive, the neurological disorders that can cause bladder dysfunction are numerous. To understand the possible pathogenesis of a bladder complaint it is necessary to have a clear concept of the neural mechanism of bladder control.

Urine storage and bladder emptying are dependent upon coordinated interaction between the bladder and urethra. Initiation of voiding requires relaxation of the external and internal urethral sphincters before a sustained detrusor contraction can empty the bladder quickly and completely. During the storage phase, on the other hand, the detrusor muscle must be kept relaxed and the bladder outlet contracted above intravesical pressure to maintain continence. Mechanisms also have to be in place to prevent leakage during sudden increases in abdominal pressure (coughing or sneezing) and to maintain continence until it is socially convenient to void.

The innervation of the lower urinary tract represents a complex interaction between efferent and afferent pathways containing sympathetic, parasympathetic and somatic pathways under supraspinal control, as well as local spinal reflex arcs. The importance of the frontal regions for social bladder control has been known for some time (Andrew and Nathan, 1964), and studies of patients with brain injury secondary to trauma or cerebrovascular accident have added to our understanding of the role of these areas (Khan *et al.*, 1990; Sakakibara *et al.*, 1996). Recently published functional imaging studies of brain activation on voiding (Blok *et al.*, 1997; Nour *et al.*, 2000) and urine storage (Athwal *et al.*, 2001) have shown that diffuse neural networks are also involved in these processes.

In 1925, Barrington first described the pontine micturition center (PMC) (Barrington, 1925), an area in the medial part of the dorsolateral pons that projects to the intermediolateral and intermediomedial cell columns of the sacral spinal cord. Recent positron emission tomography (PET) studies show that the PMC in humans is analogous to that described in the cat (Blok and Holstege, 1998; Blok *et al.*, 1997) (Figure 1). A separate area in the pons, known as the "L-region" is active during the storage phase of urine and lies ventrolateral to the PMC. Neurons from this area project to motoneurons of the urethral sphincter in the ventral horn at the S1–S2 level of the spinal cord (Onuf's nucleus) and travel by way of the pudendal nerve to the external urethral sphincter (Holstege *et al.*, 1986). Stimulation of the L-region results in pelvic

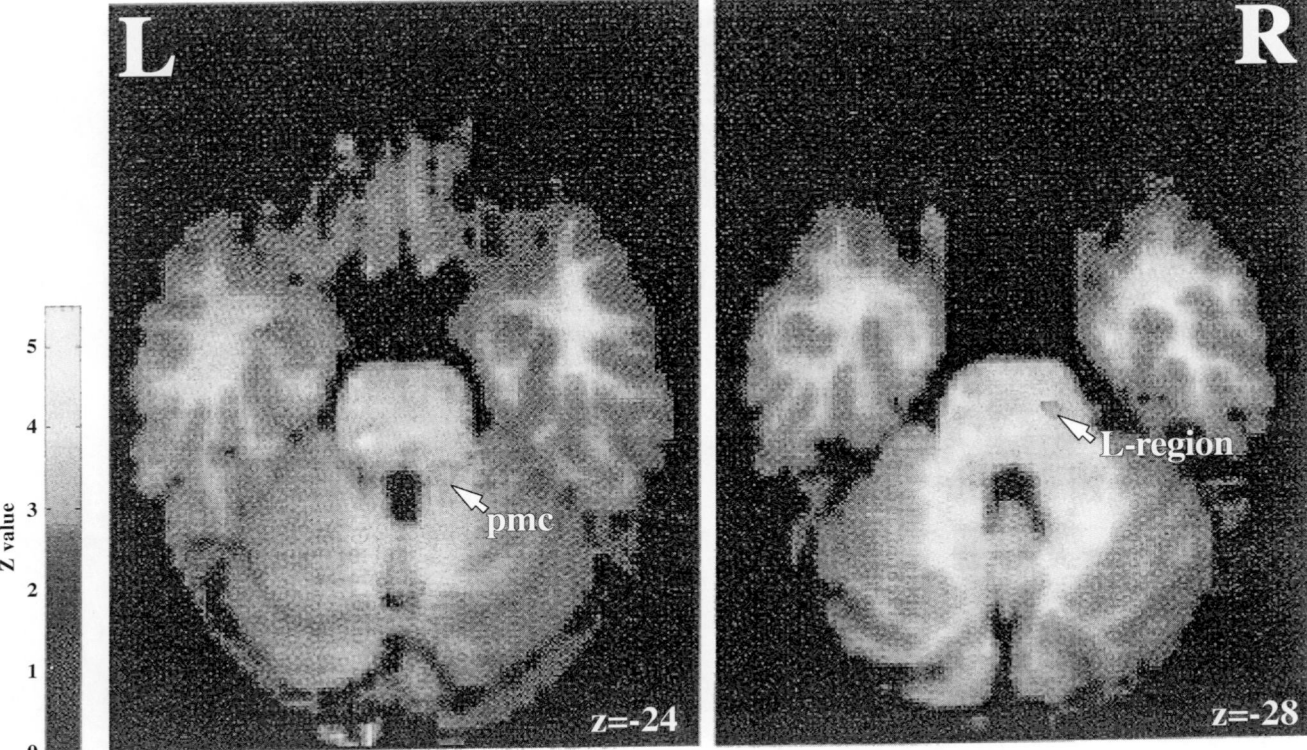

**FIGURE 1**  PET studies in man demonstrating increased regional blood flow in the PMC during micturition (left) and the L-region in those unable to void (right) (Blok *et al.*, 1997, by permission of Oxford University Press).

floor contraction and leads to an increase in urethral pressure.

The micturition reflex pathway is, in turn, modulated by higher centers in the cerebral cortex, which are, presumably, involved in the voluntary control of micturition. Several neurotransmitters, including GABA, opioid peptides, and glutamic acid, seem to have a role in these central pathways (de Groat, 1990).

The bladder detrusor muscle is innervated by parasympathetic postganglionic fibers. The preganglionic motor neurons are located in the sacral intermediolateral column of S2–S4 and their axons reach the bladder by means of the pelvic nerves. The postganglionic fibers are located within the bladder wall. Sympathetic nerve fibers (T9–L2) reach the bladder neck and male sexual organs by way of the hypogastric nerves and have an important role in the prevention of retrograde ejaculation into the bladder.

Interruption of the descending neurons from the pons results in dyssynergic micturition, whereby simultaneous contraction of the sphincter and detrusor, known as detrusor sphincter dyssynergia (DSD), prevents effective and complete bladder emptying. Brain lesions rostral to the pons rarely show detrusor sphincter dyssynergia but

tend to exhibit urgency as their main neuropathic bladder symptom.

### Natural Course

Detrusor hyperreflexia (DH) is the most common disorder of bladder function seen in patients with neurological disease and is defined as detrusor overactivity secondary to a neurological cause. Table I contains the latest definitions as agreed by the International Continence Society (ICS).

Involuntary contractions of the detrusor are reflexly triggered by bladder filling, often by only low volumes. Their clinical presentation is with urinary frequency and nocturia, urgency, and often urge incontinence. DH is characteristic of spinal cord lesions, when there has been an interruption of the connections between the pontine regions and the sacral cord, but it may also occur with suprapontine lesions. When resulting from a spinal cord lesion, DH is frequently accompanied by incomplete emptying caused by DSD (see earlier). The combination of hyperreflexia and DSD may lead to high intravesical pressures with resultant vesicoureteric reflux and

TABLE I    ICS Definitions Used in Detrusor Hyperreflexia

| Symptom | Latest ICS definitions |
|---|---|
| Urinary frequency | The complaint by the patient who considers that he/she voids too often by day. |
| Nocturia | The complaint that the individual has to wake at night one or more times to void. |
| Urinary urgency | The complaint of a sudden compelling desire to pass urine, which is difficult to defer. |
| Urge incontinence | The complaint of involuntary leakage accompanied by or immediately preceded by urgency. |

TABLE II    Comparison of Common Symptoms of Neuropathic Bladder Dysfunction and Bladder Outflow Obstruction

| | Neurogenic bladder dysfunction | Bladder outflow obstruction |
|---|---|---|
| Poor stream | + | +++++ |
| Hesitancy | + | +++++ |
| Frequency (daytime, night-time) | ++ | ++ |
| Urgency | +++++ | ++ |
| Incomplete emptying/urinary residue | ++ | +++ |
| Urinary incontinence | +++++ | + |

eventual renal impairment. This is common in patients after traumatic spinal cord injury or with spina bifida and spinal dysraphism but rare in those with progressive neurological disease. The reason for this difference in natural history is not known.

Detrusor overactivity in the absence of a neurological lesion is termed detrusor instability (DI). This is the most common cause of the overactive bladder (OAB) seen in the general population (Abrams and Wein, 2000), and whether it has a single or multiple causes remains to be determined. DH and DI produce similar symptoms and are urodynamically indistinguishable. It is the job of the neurologist to make the distinction, on the basis of the neurological findings.

Incomplete emptying caused by DSD frequently accompanies DH when the latter is due to spinal cord disease but may also result from impaired detrusor contractility (caused, for example, by diabetic neuropathy) or other forms of outlet obstruction. The condition may be relatively asymptomatic, with an abnormally raised postmicturition residual volume being an incidental finding. In the presence of DH, however, a persistent residual volume will tend to further exacerbate frequency and urgency by constantly triggering reflex detrusor contractions.

Urinary retention may be acute and painful or chronic and painless. It occurs in the initial stages of spinal shock after traumatic injury and may also be a feature of an acute cauda equina lesion. However, there is also a condition of isolated acute urinary retention in young women that neurologists may encounter. Retention in these women may be either of spontaneous onset or secondary to some intervention, commonly a general anaesthetic (Swinn et al., 2002). The cause is thought to be an abnormality of the striated urethral sphincter, and the EMG recorded from the muscle using a needle electrode consists characteristically of complex repetitive discharges. Many of these young women have polycystic ovaries, and a syndrome encompassing these features was described by Fowler et al. (1988).

It is hypothesised that there is a hormonally sensitive channelopathy affecting the striated muscle of the urethral sphincter that results in impaired relaxation and thus urinary retention. The same EMG abnormality is found in some young women with idiopathic obstructed voiding.

## Principles of Therapy

Patients with an established neurological disease who have voiding dysfunction will need some form of urological investigation, even if only a measurement of the postmicturition residual volume. Although a working diagnosis of the bladder dysfunction can often be made on the basis of the underlying neurology and empirical treatment started, further investigation may be indicated in more complex cases. Urodynamic investigations should be performed in the older male in whom the clinical differentiation between bladder outflow obstruction caused by benign prostatic hyperplasia and neuropathic bladder overactivity is often difficult, due to similarities in symptoms (see Table II).

Occasionally, anal sphincter EMG may be used to recognize changes of chronic reinnervation in multiple system atrophy (Vodusek, 2001) or to investigate the cause of urinary retention in young women (Fowler and Kirby, 1986).

### Urodynamics

The term "urodynamics" encompasses many investigations of the lower urinary tract, although it is often used as a synonym for pressure-flow cystometry. In the following section a brief overview of different investigations of the lower urinary tract is given (see also Table III). Flow rate and postvoid residual volume measurements are the simplest urodynamic investigations.

For a flow rate trace to be meaningful, the voided volume should, ideally, be more than 150 ml. Figure 2

**TABLE III** Urodynamic Investigations of the Lower Urinary Tract in the Assessment of Neuropathic Bladder Dysfunction

| Investigation | Indication | Advantage |
|---|---|---|
| Flow rate/residual volume measurement | — Initial voiding assessment<br>— Follow-up urinary residues | — Easy<br>— Quick<br>— Cheap<br>— Noninvasive |
| Pressure flow cystometry (CMG) | — Second line investigation<br>— Any lower urinary tract dysfunction | — Assessment of bladder function<br>— Cheaper, quicker, and less invasive than videourodynamics |
| Videourodynamics | — Obstructed outflow<br>— Stress incontinence<br>— Previous surgery<br>— Neurogenic disorders with suspected reflux<br>— Children | — Assessment of bladder detrusor and urinary sphincter function |

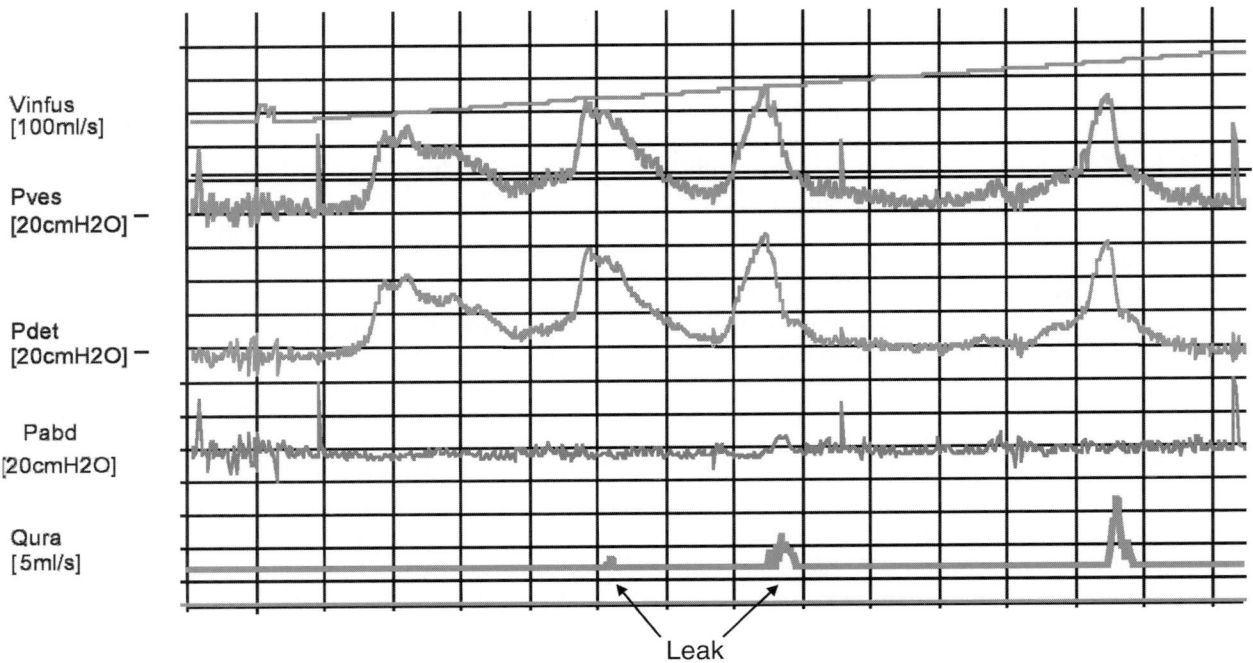

Vinfus [100ml/s]

Pves [20cmH2O]

Pdet [20cmH2O]

Pabd [20cmH2O]

Qura [5ml/s]

Leak

**FIGURE 2** Filling cystometrogram demonstrating detrusor hyperreflexia. During the filling phase the patient develops involuntary detrusor contractions (Pdet), which are strong enough to lead to involuntary loss of urine (Qura) at the 2nd, 3rd and 4th involuntary detrusor contraction.

shows a normal urinary flow curve. The postvoid residual volume is measured using ultrasonography or by "in-out" catheterization. Ultrasonic measurement of postvoid residual volume is helpful in the follow-up of impaired bladder emptying, with residuals greater than 100 ml generally accepted to be significant. In association with incontinence or recurrent urinary tract infection, a raised residual volume is an indication for intermittent self-catheterization.

Simple pressure-flow cystometry measures detrusor pressure during filling and voiding. Figure 3 shows a tyical cystometric trace of a patient with marked DH associated with incontinence. Videocystometrography (VCMG) combines cystometry with fluoroscopy of the lower urinary tract to allow observation of bladder anatomy, vesicoureteric reflux, bladder neck, and sphincter function, as well as other relevant factors. VCMG may indicate the cause of voiding dysfunction

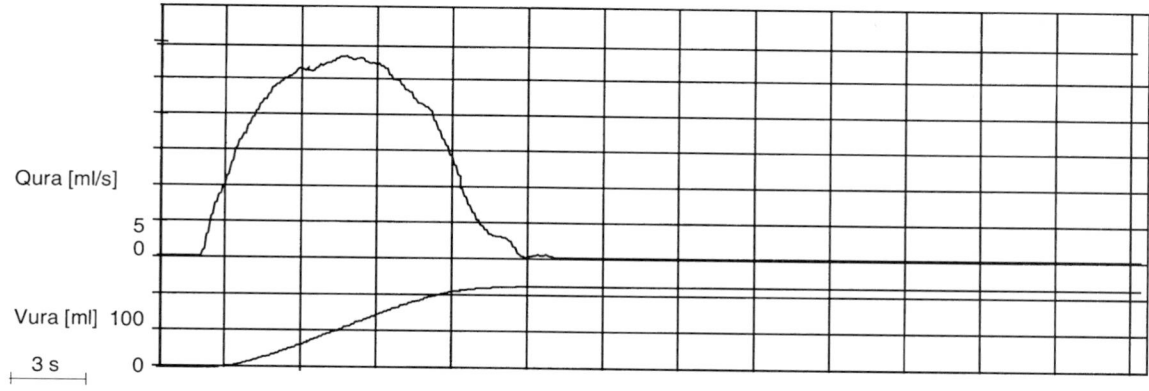

**FIGURE 3** Normal bell-shaped flow curve with no residual volume of urine.

and needs to be interpreted by a radiologist or urologist with specialist skills.

### Imaging

Ultrasonography is the investigation of choice in assessing the upper and lower urinary tract. Postvoid residual volume estimation is critical in determining the management of neurogenic bladder dysfunction and can be measured using a small and relatively inexpensive bladder scanner. Assessment of the kidneys and ureters is indicated in the presence of renal impairment or recurrent urinary tract infection.

Neurological imaging may be important if the patient was referred from a urologist to exclude a neurogenic basis for their lower urinary tract symptoms. Imaging is particularly indicated to exclude a suprapontine abnormality or subsacral lesion. MRI is now the investigation of choice, but neurological imaging has no role in determining the management of neurogenic voiding dysfunction in a patient with a known neurological diagnosis.

### Treatment of Detrusor Hyperreflexia

Because acetylcholine is the major excitatory neurotransmitter involved in detrusor contraction, anticholinergic agents are the primary compounds used to decrease the involuntary detrusor contractions of DH. They act to decrease the frequency of contractions and to increase bladder capacity. It is advisable to start at a low dose and to increase the dosage as necessary to minimize side effects. Many patients will experience a dry mouth. Blurred vision, constipation, and other side effects are much less common but may be seen at higher dosages. Unwanted anticholinergic effects may be reduced by the use of selective preparations, such as

oxybutynin or tolterodine, which should now be considered as the first line of treatment for control of hyperreflexia. Slow-release preparations offer the convenience of having to medicate only once a day with comparable efficacy and fewer side effects. The addition of a tricyclic antidepressant with antimuscarinic properties, such as imipramine, is sometimes helpful in patients who fail to respond to anticholinergics alone.

Those who do not tolerate anticholinergic side effects may benefit from propiverine hydrochloride or propantheline, whereas troublesome nocturia may respond well to the ADH analog, desmopressin, which acts on the collecting tubules of the kidney to increases water reabsorption and reduce nocturnal urine production. This, however, is contraindicated in the elderly and those with incipient congestive heart failure.

The sensory arm of the micturition reflex in patients with detrusor hyperreflexia is a potential therapeutic target. Agents to desensitize afferent nerve endings within the bladder suburothelium include the C-fiber neurotoxins (Fowler, 2000). Capsaicin is no longer in clinical use, having been replaced by the more potent and less pungent analog, resiniferatoxin (RTX). This has been shown to decrease urinary frequency and increase bladder capacity without causing bladder pain during instillation (Silva *et al.*, 2000). However, a number of large trials have been hampered by the propensity of RTX to bind to the plastics used in its packaging and administration, although work is in progress to overcome this. Despite these problems, the principle of deafferenting the bladder in DH seems promising. Most recently, the injection of botulinum toxin into the detrusor to block acetylcholine release has been reported to be effective in the treatment of refractory DH (Schurch *et al.*, 2000).

Urinary collecting devices like condom sheath catheters are both simple and safe and have their place in the management of urge incontinence in patients who do not respond to other treatments.

### Treatment of Incomplete Emptying

Unfortunately, many patients with neurogenic urinary incontinence are managed initially by an indwelling urethral catheter. The preferred option for incomplete emptying is intermittent self-catheterization (ISC). It is a safe and effective method but relies on reasonable manual dexterity or else a willing partner to perform it. Specialist nurse continence advisors offer valuable expert advice and support for the patient while they master the technique. Wherever possible, ISC should be considered preferable to an indwelling catheter, because it preserves bladder volume, reduces infection rates, and gives a better chance of recovery of normal bladder function in patients with improving neurology. Only if manual dexterity is insufficient should an indwelling catheter be contemplated. In such circumstances a suprapubic catheter, formally inserted in the operating theater provides the best long-term management option for patient and care giver alike.

Whichever is used, regular change and meticulous catheter care are essential to minimize encrustation, blockage, and recurrent infections. Long-term urethral catheters need changing approximately every 6–8 weeks, whereas a suprapubic catheter may last several months. Leakage of urine around the catheter (bypassing) is most commonly due to blockage of the catheter drainage holes or insufficiently controlled detrusor hyperreflexia. Recurrent catheter problems need to be managed actively with urological input.

### Surgical Treatment

Surgical management should usually be reserved for patients who do not have progressive neurological disease and in whom urgency and urge incontinence are not controlled with pharmacotherapy alone.

Enterocystoplasty using a patch of bowel to enlarge the bladder and reduce its pressures, combined with insertion of an artificial urinary sphincter (AUS), provides excellent results for spina bifida patients, resulting in a marked improvement of their quality of life. Patients with progressive neurological conditions, however, are unsuitable for these procedures, because awareness and manual dexterity are required to perform regular ISC and to manipulate AUS control buttons.

For those severely disabled by their disease, surgery does have a place in the form of urinary diversion (ileal conduit), which can bring relief from other-wise intractable incontinence and recurrent catheter problems.

### Practical Management

A simple algorithm for the first-time management of neurogenic bladder dysfunction is shown in Figure 4 and makes the point that once an anticholinergic drug is begun, bladder emptying may be adversely affected and, therefore, follow-up assessment of postmicturition volumes is important.

### Treatments No Longer Recommended

- Flavoxate. No proven anticholinergic effect and probably no better than placebo (Chapple *et al.*, 1990).
- Nonspecific anticholinergics, side effects limit use.
- Enterocystoplasty in progressive neurological disease.
- Early placement of long-term indwelling catheters. ISC should be implemented wherever possible.

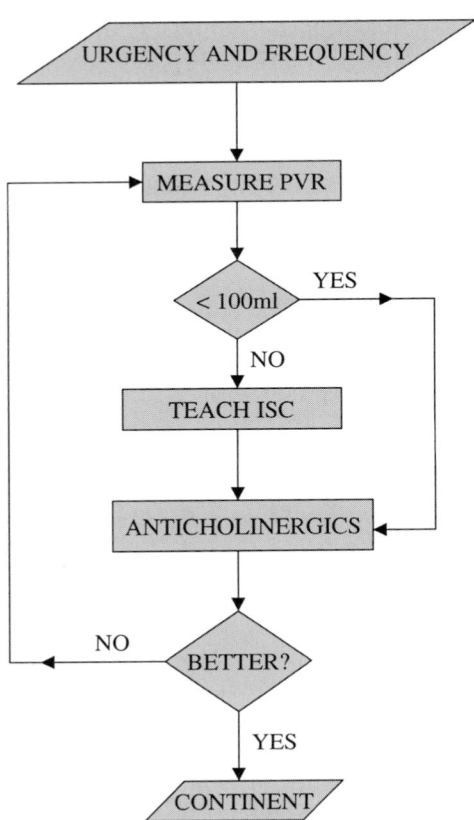

FIGURE 4   Algorithm for management of neurogenic bladder disorders.

# BOWEL DYSFUNCTION IN NEUROLOGICAL DISEASE

## Clinical Aspects

The distal colon, rectum and anal canal serve to store and eliminate fecal matter. The control of both of these processes may be affected by neurological disease, leading to constipation or fecal incontinence, or a combination of both. Bowel dysfunction is much more common in those with neurological disease than in the general population, and furthermore, it is likely to become more troublesome with increasing neurological disability. As with bladder disorders, an understanding of the neural control of the bowel is critical to understanding bowel complaints in patients with neurological disease.

Colorectal contractions are mainly controlled by the enteric nervous system (ENS) within the myenteric and submucosal plexus of the bowel. This is under the control of prokinetic parasympathetic vagal efferents and efferents from spinal cord segments S2–S4 and inhibitory sympathetic fibers from T9–L2. Local reflex arcs within the bowel wall have been referred to as a "brain in the gut" (Wood, 1999). Afferent parasympathetic nerves transmit nonconscious sensory information from the colon with painful stimuli conveyed by way of sympathetic afferents. The immune and endocrine systems, as well as psychological mental state, also influence gut function, particularly in relation to projections of discomfort and pain in the digestive tract (Wood, 1999).

Anal continence is achieved by a combination of anal resting pressure, anorectal sensibility, and anal squeeze pressure. The anal resting pressure is maintained by the internal anal sphincter, the external sphincter, and the venous pressure in the hemorrhoid plexus. Parasympathetic neurons (S2–S4) and sympathetic fibers from the hypogastric plexus (T9–L2) innervate the internal sphincter, whereby parasympathetic activity relaxes the smooth muscle. The external anal rhabdosphincter is under voluntary control and innervated by the pudendal nerve (S2–S4). Anal sensation is also conveyed by the pudendal nerve. Therefore, the integrity of the parasympathetic, sympathetic, and somatic nervous system is crucial for the maintenance of fecal continence.

## Natural Course

Constipation can be defined as a decreased frequency of bowel action (less than one bowel action per week), necessitating straining and resulting in incomplete evacuation. It may present with abdominal pain, distention, nausea, or even vomiting. Patients may also have fecal incontinence, which is the involuntary or inappropriate passage of stool. The term "anal incontinence" refers to the inability to control flatus. By far, the most common bowel complaint in patients with neurological disease is constipation, the contributing factors of which are shown in Table IV.

Constipation in patients with central neurological disease occurs by two main mechanisms. The first is increased bowel transit time, the origin of which is unclear, and the second is failure of relaxation of the pelvic floor and anal sphincters. Poor diet, inadequate hydration, immobility, and drugs often compound the problem.

Fecal incontinence tends to occur in the context of advanced neurological disability and in association with the inability to evacuate at appropriate times. It is, however, also seen in those with fecal impaction and overflow diarrhea. Constipation affects about 70% of patients with multiple sclerosis (MS) and is mostly due to slow transit in the early stages of the disease. Difficulty with evacuation of stool is a feature of advanced disability in MS, usually associated with a severe paraplegia (Wiesel et al., 2001).

In patients with Parkinson's disease (PD) and MSA, constipation and difficulty with defecation are not uncommon (Pfeiffer, 1998). Immunostaining of colonic musculature has been found to show Lewy bodies in the myenteric plexus (Singaram et al., 1995) and a reduc-

TABLE IV  Common Contributing Factors in Constipation

| Colorectal | Anal fissure |
| | Colon cancer |
| | Colonic/anal strictures |
| | Extraluminal tumors |
| | Hemorrhoids |
| Drugs | Anticholinergics |
| | Antidepressants |
| | Bismuth |
| | Barium sulfate |
| | Iron |
| | Opiates |
| Neurogenic | Autonomic neuropathy |
| | Diabetic polyneuropathy |
| | Spina bifida |
| | Multiple sclerosis |
| | Parkinson's disease |
| | CNS lesions |
| Psychogenic | Depression |
| | Anorexia nervosa |
| Metabolic | Hypokalemia |
| | Hypercalcemia |
| | Hypothyroidism |
| Idiopathic | Colonic hypoactivity |

tion in dopamine-containing neurons. A paradoxical increase in EMG activity in puborectalis was shown in some patients with PD and defecatory complaints (Mathers *et al.*, 1988). Fecal incontinence with loose stool may occur in MSA, presumably because of denervation of the anal sphincter, but it is uncommon (Wenning *et al.*, 1994).

Most patients with complete cauda equina lesions and some with spina bifida have anal incontinence caused by sphincter weakness, particularly if they have loose stool. Paradoxically, they may also have difficulty in initiating defecation. In diabetic neuropathy, bowel dysfunction may comprise constipation, diarrhea, and anorectal incontinence but is usually only seen in those with advanced autonomic neuropathy (Feldman and Schiller, 1983).

## Principles of Therapy

Patients with chronic constipation and fecal incontinence need plenty of support to deal with these distressing problems. They should be given a simple explanation of the normal mechanisms of bowel function, together with an understanding of what has gone wrong in their own case. Emphasis should be on preventing fecal impaction, with the importance of diet emphasized and the patient's toilet facilities properly adapted to enable comfort and ease of use.

Management tends to be empirical rather than evidence-based, because there is little research in this area. For simple constipation, an increase in dietary fiber with adequate fluid intake is often all that is required. Oral laxatives are classified according to their mode of action, and Table V gives an overview. Bulking preparations are particularly useful in patients with hard, small-volume stools but should be avoided in those with slow transit constipation, because marked and painful distension may result. Unprocessed wheat bran is one of the more effective bulk-forming preparations, but ispaghula, polysaccharides, and methylcellulose are alternatives in patients who do not tolerate it well. Osmotic laxatives loosen the stool by retaining fluid in the bowel. Magnesium salts are useful when the aim is to empty the bowels rapidly. Lactulose is a nonabsorbable disaccharide, which is suitable for regular use in moderate constipation and which can be combined with phosphate enemas for severe constipation. Stimulant laxatives such as bisacodyl and senna should be prescribed with caution, because prolonged use can precipitate the onset of an atonic nonfunctioning colon and hypokalemia. They increase intestinal motility and may cause abdominal cramps or discomfort and are contraindicated in intestinal obstruction. There is some evidence that chronic use of senna may increase the risk of bowel tumors by acting as a promoter (Mereto *et al.*, 1996). Glycerol suppositories stimulate the rectum locally and also have lubricant action for very hard stools. Enemas containing arachis oil soften and lubricate impacted feces and promote bowel movement. Sodium docusate probably acts both as a stimulant and as a softening agent.

Conservative treatment of fecal incontinence is based on dietary measures, constipating agents like codeine phosphate, and regular bowel evacuations with enemas. Other than pads, very few devices are available for fecal leakage. Disposable anal plugs are sometimes used but tend to be uncomfortable and are poorly tolerated. Biofeedback techniques may have a role.

### Surgery

Surgery should be reserved for patients in whom all other therapies have failed. A discrete stoma may be formed using a reverse reimplanted appendix or a segment of tubularized ileum, and this can be used to administer antegrade colonic enemas with saline solutions every 2–3 days (ACE procedure). Patients with spina bifida may be suitable for this procedure, however, it is only recommended for highly motivated patients with reasonable dexterity (Wedderburn *et al.*, 2001). In selected and very disabled patients, a colostomy may be indicated and can make a big difference to their quality of life.

Surgical inplantation of a sacral nerve stimulator has been reported (Krogh *et al.*, 2001), but the clinical usefulness of this intervention has yet to be established in randomized controlled studies.

### Practical Management

For constipation, a progressive, stepwise plan of management should be implemented (see Figure 5) and its

**TABLE V   Laxative Agents for Treatment of Constipation**

| | |
|---|---|
| Bulk-forming laxatives | Bran |
| | Ispaghula |
| | Mucilaginous polysaccharides |
| | Methylcellulose |
| Fecal softeners | Arachis oil enema |
| Stimulant laxatives | Bisacodyl |
| | Senna |
| | Docusate sodium |
| | Dantron |
| | Glycerol suppositories |
| Osmotic agents | Lactulose |
| | Magnesium salts |
| | Rectal phosphates |
| | Rectal sodium citrate |

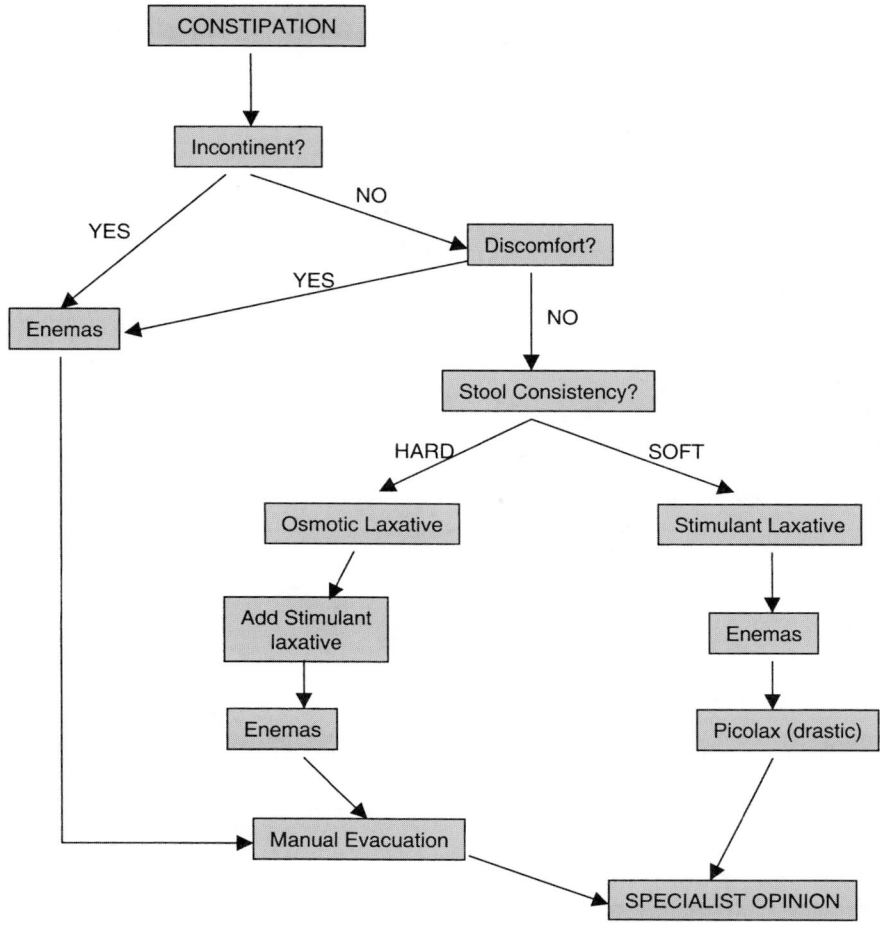

**FIGURE 5**    Algorithm for management of constipation (modified from Edwards, C. An overview of adult constipation and faecal incontinence. *In* "Promoting Continence." (K. Getliffe, and M. Dolman, Eds.) Bailliere Tindall, 1997).

effectiveness monitored. Reasonable hydration should be ensured, and unnecessary constipating drugs should be discontinued. Mobilization and activity should be encouraged as much as is possible. The distribution of fecal loading on a plain abdominal x-ray films may help distinguish between generalized slow transit and a rectal evacuation problem.

For overflow incontinence secondary to impaction, the management is shown in Figure 5. For incontinence associated with loose stool, low-dose constipating agents (loperamide or codeine phosphate) may help gain some degree of control. Other treatments will require a specialist referral.

### Treatments No Longer Recommended

- Excessive fiber in those with slow transit caused by neurological disease; may cause marked colic and distension.

- Chronic senna use; suspected but unproven long-term risk of bowel malignancy (Mereto *et al.*, 1996; van Gorkom *et al.*, 2000).
- Cisapride for constipation; little effect, and has been withdrawn because of potential cardiac side effects.

## SEXUAL DYSFUNCTION IN NEUROLOGICAL DISEASE

### Clinical Aspects

In 1993, the National Institute of Health Consensus Panel defined erectile dysfunction (ED) as "the inability to achieve or maintain an erection satisfactory for sexual function."

Until the late 1960s, sexual problems were not freely discussed, and ED was largely an unquantified problem, but with changing attitudes and, more recently, the advent of effective oral treatments (sildenafil and apo-

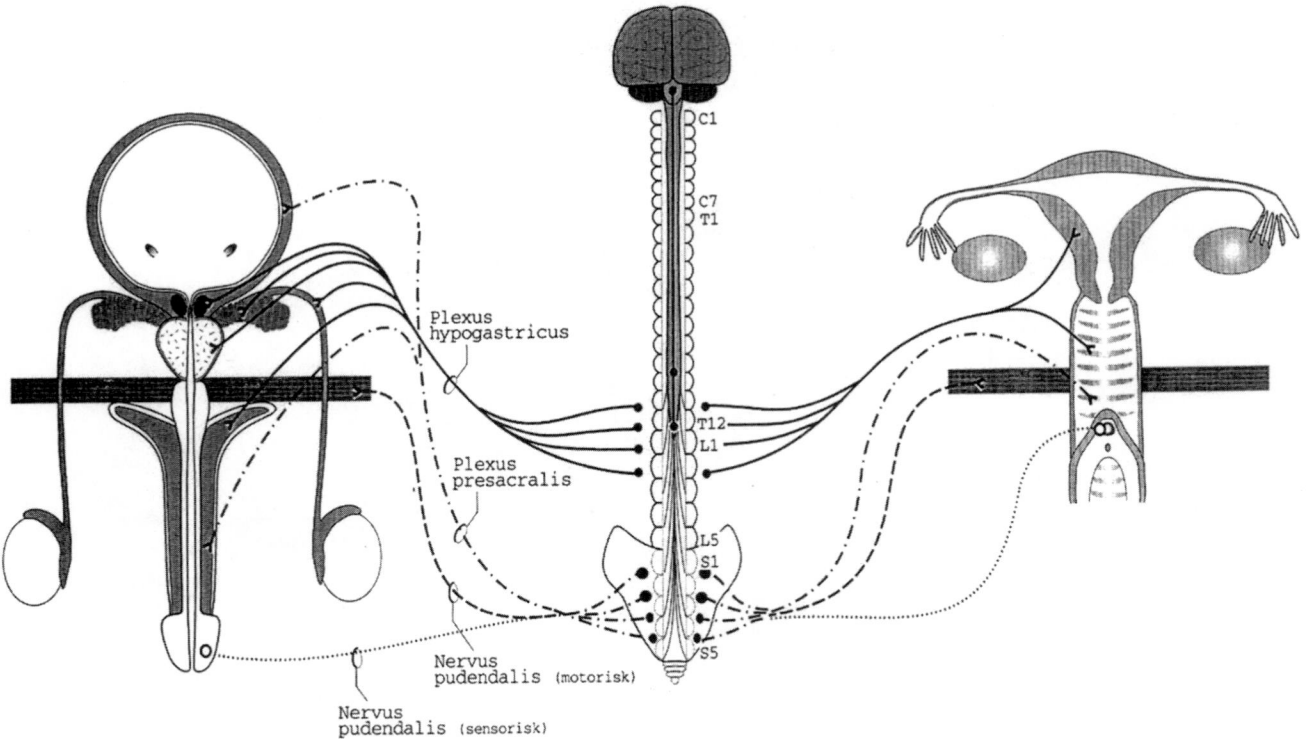

**FIGURE 6**  Peripheral innervation of male and female genitalia (reproduced with permission of Petter Hedlund and Axel Brattburg.

morphine) its true extent is becoming apparent. Studies have shown 8% of healthy men to have erectile dysfunction by the age of 55 years (Morley and Kaiser, 1992), and an overall prevalence of 55% in those between 40 and 70 years of age (Feldman *et al.*, 1994). Although common in the general population, ED particularly affects men with neurological disease, for example, 70% of men with MS (Zorzon *et al.*, 1999), and although not a life-threatening problem, it causes considerable reduction in quality of life (Krane *et al.*, 1989).

Research into female sexual dysfunction (FSD) lags years behind that of their male counterparts, and there is little in the way of firm data on which to base treatment recommendations.

The way in which neurological disease interferes with normal sexual function can be appreciated once the normal neurophysiology of erection has been outlined.

### Neural Control of Erection and Ejaculation

Penile erection is a reflex-mediated neurovascular event associated with tumescence of the cavernous bodies. It is the end result of a complex series of integrated physiological processes, which involves spinal and supraspinal pathways. Some of the anatomical areas of the brain that relate to sexual function have been defined, including the medial amygdala, medial preoptic area, paraventricular nucleus, the periaqueductal gray, and ventral tegmentum (for a review see Andersson, [2001]). Visual, olfactory, and imaginative stimuli influence the central regulation of penile erection. Erection of the penis requires parasympathetic neural activity, whereas ejaculation is under sympathetic nervous control. The cell bodies of the autonomic neurons are located in the spinal cord at the thoracolumbar level for the sympathetic nuclei (T9–L2) and at the sacral level (S2–4) for the parasympathetic and somatic nuclei (Figure 6). The efferent sympathetic nerves emerge from the spinal column at T9–L2 and descend along the aorta on each side before coming together in the midline to form the hypogastric plexus just below the bifurcation of the aorta. They pass through the pelvis as hypogastric nerves to innervate the bladder neck, prostate, vasa deferentia, and seminal vesicles. Damage to the sympathetic nerves, as in spinal cord injury, may affect central reflex pathways and may lead to loss of ejaculation.

Parasympathetic nerve fibers emerge from the sacral segments S2–S4 and travel in the pelvic nerves (nervi erigentes) to innervate the corpora cavernosa, prostate, base of the bladder, and bladder neck. Injury to the parasympathetic nerve fibers may lead to loss of erection. The pudendal nerve (S2–S4) carries somatic nerve fibers that innervate the periurethral striated muscles, and penile sensation is also conveyed by way of the pudendal nerve. Rhythmic contraction of the bulbospongiosus expels the contents of the urethra in the final act of ejaculation, and contraction of the bladder neck prevents retrograde ejaculation into the bladder. Orgasm is a psychic and sensory phenomenon in which the rhythmic contractions of the pelvic floor and periurethral muscles are perceived as pleasurable.

Local vascular events leading to penile erection involve relaxation of corporal smooth muscle and arteries, which leads to expansion of the corporal lacunar spaces with blood. The increased blood volume raises penile pressure above systemic systolic blood pressure, resulting in venous outflow obstruction with tapering arterial inflow until a steady state is reached. Rigid erection is achieved by contraction of the bulbocavernosus and ischiocavernosus muscles.

Numerous central and peripheral neurotransmitters are involved in the regulation of penile erection (Andersson, 2001). Dopamine, nitric oxide, oxytocin, and ACTH/α-MSH seem to facilitate erection, whereas enkephalins are inhibitory. Peripherally, noradrenaline, endothelins, and angiotensins act on the smooth muscle of the corpora, resulting in contraction of trabecular smooth muscle. Nitric oxide plays a major role in smooth muscle relaxation of penile vessels and the corpora.

### Neural Control of Female Sexual Function

The innervation and erectile mechanisms of the female genitalia are comparable to those in the male (Lundberg, 1999). Arousal and hyperalgesia of the genital tissues can be provoked by imagery and fantasy or by stimulation of vaginal or cervical tissues or other erogenous zones. During this process, blood flow is increased to the vagina that results in lubrication and erection of the female cavernous tissues and clitoris. Lubrication is derived mainly from a transudate from the vaginal tissues, although secretion from Bartholin's glands, the cervix, and uterus also contribute (Lundberg, 1999). Diminished lubrication can impair sexual function and can be caused by estrogen deficiency states, pituitary insufficiency, antiestrogen therapy, and anticholinergic agents.

The clitoris is equivalent to the glans penis and is the most densely innervated area of skin, with an innervation density approximately twice that of the male glans and equivalent to the fingertips. Clitoral sensory thresholds are lower than that of the glans penis. The clitoral

nerves join the pudendal nerves, which ultimately transmit information to the sacral spinal cord.

With appropriate stimulation and rising sexual excitement, orgasm may be reached. The female orgasm is characterized by the experience of a spreading sensation of warmth followed by up to 20 synchronous contractions of the vaginal musculature and sphincters lasting for 10–50 seconds. Some women can experience multiple sequential orgasms in response to repeated or persistent stimulation. Recent studies of women with spinal cord injury suggest that female orgasmic function is mediated by the sympathetic nervous system (Sipski et al., 2001).

### Natural Course

Sexual dysfunction affects both men and women with neurological disease. Men may develop difficulty with initiating and maintaining erection and with ejaculation, whereas women may experience failure of vaginal lubrication, reduced sensation, and loss of orgasmic function.

Because penile erection is a neurovascular event, any dysfunction or disease affecting the brain, spinal cord, cavernous and pudendal nerves, or their terminal branches can cause erectile dysfunction. Table VI lists neurological conditions associated with erectile dysfunction.

The medial preoptic area, the paraventricular nucleus, and the dorsal hypothalamic areas of the brain all contain neurons that project to the corpus cavernosum (Giuliano et al., 1995), and the limbic system seems to be a supraspinal regulatory site of penile erection (Chen et al., 1992). We know that dopamine is one of the most important regulatory neurotransmitters (Andersson, 2001; Heaton, 2000) and that patients with parkinsonism experience erectile dysfunction, presumably through derangement of the dopaminergic transmitter system (Wermuth and Stenager, 1992). Other brain lesions such as tumors, cerebrovascular accidents, head trauma, and Alzheimer's disease are also associated with erectile dysfunction (Zeiss et al., 1990), probably through an imbalance of the hypothalamic center or overinhibition of the spinal center (Carrier et al., 1993).

The spinal cord is at the origin of efferent parasympathetic, sympathetic, and somatic pathways to the penis and related striated muscles. It receives information from the glans and corpora and represents a site of integration for information of peripheral and central origin. Lesions of the cord are the most common cause of neurogenic erectile dysfunction, with the degree of dysfunction dependent on the nature, location, and extent of the lesion (Lue, 1995). Reflexogenic erection is preserved in 95% of patients with complete upper

**TABLE VI  Neurogenic Causes of Erectile Dysfunction**

| | |
|---|---|
| Suprapontine lesions | Parkinson's disease |
| | Multiple system atrophy |
| | After head injury |
| | Cerebrovascular disease |
| | Alzheimer's disease |
| | Temporal lobe epilepsy |
| Suprasacral spinal cord disease | Multiple sclerosis |
| | Traumatic spinal cord injury |
| | Tumors |
| Conus and cauda equina | Central disk prolapse |
| | Trauma |
| | Spinal dysraphism (tethered cord, meningomyelocele) |
| | Vascular malformations |
| | Iatrogenic |
| Peripheral nerve damage | Trauma |
| | Transection injury (pelvic surgery) |
| Small fiber autonomic neuropathies | Diabetes mellitus |
| | Amyloidosis |
| | Alcoholism |
| | Uremia |
| | Leprosy |
| | Fabry's disease |
| | Tangier disease |
| | Inherited autonomic neuropathies (type I–IV) |

cord lesions but only in 25% of those with complete lower cord lesions (Bors and Comarr, 1960). It seems, however, that the erections in those with upper cord lesions are short lasting and insufficient for sexual intercourse (Benet and Melman, 1995), and these patients should be considered as having ED.

ED is the most common complaint in men with MS, with an incidence of approximately 70% (Zorzon *et al.*, 1999), although it is rarely what they complain of at presention. Typically men with ED caused by MS find they can get erections on waking or in response to genital stimulation but are unable to sustain an adequate erection for intercourse. As in men after spinal cord trauma, this is thought to be due to preservation of segmental mechanisms but interruption of spinally mediated psychogenic erections.

Surgical procedures are a major cause of iatrogenic neurogenic erectile dysfunction. Surgery close to the spine, such as retroperitoneal lymph node dissection and aortoiliac repair, is associated with ED caused by sympathetic nerve damage. Pelvic surgery, such as abdominoperineal resection, is associated with damage to the cavernosal nerves and has been reported to cause dysfunction in up to 30% of patients (Cunsolo *et al.*, 1990).

Peripheral neuropathies that involve small-caliber nerve fibers may affect the cavernosal nerves. ED is likely in men with alcoholic neuropathy, vitamin deficiencies (folate, vitamin $B_6$, and $B_{12}$), amyloidosis, AIDS,

Guillan-Barré syndrome, and diabetic neuropathy. Diabetes is the most common cause of ED because of its vascular effects on the erectile process in addition to the neuropathy.

### Female Neurogenic Sexual Dysfunction

FSD in women with neurological disease has been not been extensively studied, but similar neurogenic etiologies to those that cause male sexual dysfunction may be implicated in women. Spinal injury and peripheral neuropathy may lead to failure of lubrication, dyspareunia, loss of sensation, and loss of orgasmic function. Women with incomplete spinal injury may retain some capacity for psychogenic arousal and vaginal lubrication (Sipski *et al.*, 1995).

### Principles of Therapy

The introduction of effective oral pharmacotherapy has revolutionized the treatment of erectile dysfunction. The convenience and relative safety of today's proerectile medications means that a trial of these should be offered as a first line of treatment before investigation and regardless of etiology (WHO, 1999). There is, however, a significant failure rate, and referral to a specialist center for second-line treatments then becomes appropriate.

If the problem is mainly related to a decreased libido, an endocrinological cause such as hyperprolactinemia or low serum testosterone should be excluded. There will be a psychological component in most cases of sexual dysfunction, whether male or female, and this must be addressed with sensitivity to maximize benefit from medical treatment.

### Oral Treatments

Selective phosphodiesterase (PDE) inhibitors were introduced in 1998 for the treatment of erectile dysfunction. The first commercially available PDE inhibitor, sildenafil (Viagra), has proved to be safe and well tolerated. Side effects tend to be mild and transient, consisting mainly of headache, flushing, and dyspepsia (Morales *et al.*, 1998). Its use is contraindicated in patients taking nitrates and those with severe hypertension, hepatic impairment, degenerative retinal disorders, recent stroke, or myocardial infarction.

Sildenafil should be taken 1 hour before anticipated intercourse starting at a low dose, usually 50 mg, and increasing to 75 mg or 100 mg according to response.

The mode of action is to increase the levels of nitrous oxide available in the corpora cavernosa in response to appropriate stimuli and so to bring about relaxation of the smooth muscle of the penile arterial inflow. Thus, in neurologic causes of ED, the best response can be expected in those conditions in which some pathways for an erectile response are intact, that is, spinal cord disease in which segmental reflex erections are preserved.

In a placebo-controlled crossover study of men with spinal cord injury, 76% of 168 men reported improved erections with sildenafil (Giuliano *et al.*, 1999). A multicenter, placebo-controlled trial of the effect of the medication in men with MS showed that it was highly efficacious, producing a response in 90% of 104 men on active treatment compared with 24% of 113 patients on placebo. An improvement in quality of life was also demonstrated (Fowler *et al.*, 1999). Sildenafil has also been shown to be effective in men with PD, but a small number of men who had parkinsonism caused by MSA and who also had autonomic failure had a hazardous exacerbation of hypotension (Hussain *et al.*, 2001).

Recently, the centrally acting selective dopamine agonist apomorphine (Uprima) has been introduced as an alternative first-line agent. Apomorphine activates specific neural events in the hypothalamus leading to an erectile response. The development of a sublingual formulation has improved the side effect profile, which is mainly nausea and headache (Heaton, 2001). Unfortunately, patients with neurological disease were excluded from the clinical trials, and because its mode of action is through spinally mediated pathways, it has been thought to be of limited use in conditions in which spinal pathways are not intact. Against this, however, a recent study has shown that in spinal-injured rats apomorphine elicits erections through spinal and supraspinal targets (Giuliano *et al.*, 2002). A trial of apomorphine may, therefore, be warranted in men with neurogenic ED.

### Injection Therapy

Intracavernosal prostaglandin injections are indicated if oral agents are not effective and provided the patient has good enough eyesight and manual dexterity and is sufficiently motivated. The local action of prostaglandins ensures onset of erection within 10 minutes, and the response is not dependent on intact neural pathways. Patients are advised to alternate the side of injection to prevent penile fibrosis at the injection site and to use it only once per week. Pain as a result of injection may also be a problem. Patients need to be given clear instructions when to seek medical advice in the rare case of treatment-induced priapism. The definition of priapism is imprecise, but it is generally accepted that painful erections beyond 1 hour and any erection beyond 4 hours should be treated as an emergency.

### Vacuum Devices

These consist of an external cylinder that fits over the penis from which air is pumped out, resulting in increased blood flow into the corpora. After this, a constriction ring is fitted around the base of the penis to maintain the erection. The erection, caused by venous congestion rather than arterial inflow, lacks the warmth and rigidity of a physiological erection. The most common complaints are of premature loss of rigidity and difficulty in placing and removing the constriction bands. Common complications are bruising and skin edema, but serious complications are rare. Vacuum devices may suit patients refractory to oral or cavernosal therapy.

### Surgical Options

Penile prostheses are semirigid, inflatable, or malleable implants that are surgically inserted into the corporal cavities. They require significant manual dexterity to operate and are unsuitable for patients with progressive neurological disease.

### Treatment of Female Sexual Dysfunction

The treatments that have been found to be effective for treating ED in men are now being trialed in women. Sildenafil has been shown to be effective in women with FSD secondary to SSRI use (Nurnberg *et al.*, 1999), and also in postmenopausal women (Kaplan *et al.*, 1999). A recent study by an Italian group suggests that it may also prove helpful in younger women with FSD (Caruso *et al.*, 2001). Studies are currently in progress looking at the effect of sildenafil in women with FSD caused by MS.

Before the advent of the erectogenic agents, some men with ED were given hormonal treatments, and the treatment of FSD is at that equivalent stage now. Lack of estrogen is associated with vaginal dryness, burning, and dyspareunia, and topical estrogen creams can improve clitoral sensitivity and reduce pain during intercourse. This is a treatment that is indicated in postmenopausal women in whom estrogen replacement (systemically) has other benefits such as preventing osteoporosis and reducing hot flushes. Methyltestosterone can be combined with estrogen for symptoms of reduced libido and lack of vaginal lubrication. However, the benefits of this have to be considered against the potential side effects of testosterone, which include clitoral enlargement, increased weight, and facial hair.

It is hoped that in the near future agents that have specific effects on female sexual dysfunction will be discovered and found to be of benefit in both healthy women and those with neurogenic FSD.

## Practical Management

- Provide a suitably quiet and private environment to discuss such sensitive matters.
- Explain to the patient how their disease has affected their sexual function.
- Offer a trial of oral therapy.
- Exclude hypotension in men with parkinsonism and therefore possible MSA.
- Offer a referral for specialist advice and secondary treatment if oral agents fail.

## Treatments No Longer Recommended

- Specialist investigation of erectile dysfunction before a trial of oral therapy in men with neurological disease.
- Penile prostheses in patients with progressive neurological disease.
- Sildenafil treatment in men with autonomic failure (may cause hypotensive crisis).

## REFERENCES

Abrams, P., and Wein, A. (2000). Introduction: Overactive bladder and its treatment. *Urology* 55, 1–2.

Andersson, K. E. (2001). Neurophysiology/pharmacology of erection. *Int. J. Impot. Res.* 13, S8–S17.

Andrew, J., and Nathan, P. W. (1964). Lesions of the anterior frontal lobes and disturbances of micturition and defaecation. *Brain* 87, 233–262.

Athwal, B. S., Berkley, K. J. *et al.* (2001). Brain responses to changes in bladder volume and urge to void in healthy men. *Brain* 124, 369–377.

Barrington, R. J. F. (1925). The effect of lesions of the hind and midbrain on micturition in the cat. *Q. J. Physiol.* 15, 81–102.

Benet, A. E., and Melman, A. (1995). The epidemiology of erectile dysfunction. *Urol. Clin. North Am.* 22, 699–709.

Blok, B. F., and Holstege, G. (1998). The central nervous system control of micturition in cats and humans. *Behav. Brain Res.* 92, 119–125.

Blok, B. F., Willemsen, A. T. *et al.* (1997). A PET study on brain control of micturition in humans. *Brain* 120(Pt 1), 111–121.

Bors, E., and Comarr, A. (1960). Neurological disturbances of sexual function with special references to 529 patients with spinal cord injury. *Urol. Survey* 10, 191–222.

Carrier, S., Brock, G. *et al.* (1993). Pathophysiology of erectile dysfunction. *Urology* 42, 468–481.

Caruso, S., Intelisano, G. *et al.* (2001). Premenopausal women affected by sexual arousal disorder treated with sildenafil: a double-blind, cross-over, placebo-controlled study. *Br. J. Obstet. Gynaecol.* 108, 623–628.

Chapple, C. R., Parkhouse, H. *et al.* (1990). Double-blind, placebo-controlled, cross-over study of flavoxate in the treatment of idiopathic detrusor instability. *Br. J. Urol.* 66, 491–494.

Chen, K. K., Chan, J. Y. *et al.* (1992). Elicitation of penile erection following activation of the hippocampal formation in the rat. *Neurosci. Lett.* 141, 218–222.

Cunsolo, A., Bragaglia, R. B. *et al.* (1990). Urogenital dysfunction after abdominoperineal resection for carcinoma of the rectum. *Dis. Colon Rectum* 33, 918–922.

de Groat, W. C. (1990). Central neural control of the lower urinary tract. *Ciba Found. Symp.* 151, 27–44; discussion 44–56.

Feldman, H., Goldstein, I. *et al.* (1994). Impotence and its medical and psychosocial correlates: results of the Massachusetts Male Aging Study. *J. Urology* 151, 54–61.

Feldman, M., and Schiller, L. (1983). Disorders of gastrointestinal motility associated with diabetes mellitus. *Ann. Intern. Med.* 98, 378–384.

Fowler, C., Miller, J. *et al.* (1999). Viagra (sildenafil citrate) for the treatment of erectile dysfunction in men with multiple sclerosis. *Ann. Neurol.* 46(3), 497.

Fowler, C. J. (2000). Intravesical treatment of overactive bladder. *Urology* 55, 60–64; discussion 66.

Fowler, C. J., Christmas, T. J. *et al.* (1988). Abnormal electromyographic activity of the urethral sphincter, voiding dysfunction, and polycystic ovaries: a new syndrome? *Br. J. Med.* 297, 1436–1438.

Fowler, C. J., and Kirby, R. S. (1986). Electromyography of the urethral sphincter in women with urinary retention. *Lancet* 1(8496), 1455–1456.

Giuliano, F., Allard, J. *et al.* (2002). Pro-erectile effect of systemic apomorphine: existence of a spinal site of action. *J. Urol.* 167, 402–406.

Giuliano, F., Hultling, C. *et al.* (1999). Randomized trial of sildenafil for the treatment of erectile dysfunction in spinal cord injury. *Ann. Neurol.* 46, 15–21.

Giuliano, F. A., Rampin, O. *et al.* (1995). Neural control of penile erection. *Urol. Clin. North Am.* 22, 747–766.

Heaton, J. P. (2000). Central neuropharmacological agents and mechanisms in erectile dysfunction: the role of dopamine. *Neurosci. Biobehav. Rev.* 24, 561–569.

Heaton, J. P. (2001). Characterising the benefit of apomorphine SL (Uprima(R)) as an optimised treatment for representative populations with erectile dysfunction. *Int. J. Impot. Res.* 13, S35–39.

Holstege, G., Griffiths, D. *et al.* (1986). Anatomical and physiological observations on supraspinal control of bladder and urethral sphincter muscles in the cat. *J. Comparative Neurol.* 250, 449–461.

Hussain, I., Brady, C. *et al.* (2001). Exacerbation of orthostatic hypotension with sildenafil citrate (Viagra) in patients with autonomic failure due to multiple system atrophy (MSA). *J. Neurol. Neurosurg. Psychiatry* 71, 371–374.

Kaplan, S. A., Reis, R. B. *et al.* (1999). Safety and efficacy of sildenafil in postmenopausal women with sexual dysfunction. *Urology* 53, 481–486.

Khan, Z., Starer, P. *et al.* (1990). Analysis of voiding disorders in patients with cerebrovascular accidents. *Urology* 35, 263–270.

Krane, R., Goldstein, I. *et al.* (1989). Impotence. *N. Engl. J. Med.* 321, 1648–1659.

Krogh, K., Christensen, P. *et al.* (2001). Colorectal symptoms in patients with neurological diseases. *Acta Neurol. Scand.* 103, 335–343.

Lue, T. (1995). Physiology of penile erection and pathophysiology of erectile dysfunction and priapsm. *Campbell's Urology* 7th Edition, 1157–1179.

Lundberg, P. (1999). Physiology of female sexual function and how it is affected in neurological disease. *In* Neurology of Bladder,

Bowel and Sexual Dysfunction. (C. J. Fowler, ed.). Butterworth Heinemann, Newton, MA.

Mathers, S., Kempster, P. *et al.* (1988). Constipation and paradoxical puborectalis contractions in anismus and Parkinson's disease: a dystonic phenomenon? *J. Neurol. Neurosurg. Psychiatry* **51**, 1503–1507.

Mereto, E., Ghia, M. *et al.* (1996). Evaluation of the potential carcinogenic activity of Senna and Cascara glycosides for the rat colon. *Cancer Lett.* **101**, 79–83.

Morales, A., Gingell, C. *et al.* (1998). Clinical safety of oral sildenafil citrate (VIAGRA) in the treatment of erectile dysfunction. *Int. J. Impot. Res.* **10**, 69–73; discussion 73–64.

Morley, J. E., and Kaiser, F. E. (1992). Impotence in elderly men. *Drugs Aging* **2**, 330–344.

Nour, S., Svarer, C. *et al.* (2000). Cerebral activation during micturition in normal men. *Brain* **123**(Pt 4), 781–789.

Nurnberg, H. G., Lauriello, J. *et al.* (1999). Sildenafil for sexual dysfunction in women taking antidepressants. *Am. J. Psychiatry* **156**, 1664.

Pfeiffer, R. F. (1998). Gastrointestinal dysfunction in Parkinson's disease. *Clin. Neurosci.* **5**, 136–146.

Sakakibara, R., Hattori, T. *et al.* (1996). Micturitional disturbance after acute hemispheric stroke: analysis of the lesion site by CT and MRI. *J. Neurol. Sci.* **137**, 47–56.

Schurch, B., Schmid, D. M. *et al.* (2000). Treatment of neurogenic incontinence with botulinum toxin A. *N. Engl. J. Med.* **342**, 665.

Silva, C., Rio, M. E. *et al.* (2000). Desensitization of bladder sensory fibers by intravesical resiniferatoxin, a capsaicin analog: long-term results for the treatment of detrusor hyperreflexia. *Eur. Urol.* **38**, 444–452.

Singaram, C., Ashraf, W. *et al.* (1995). Dopaminergic defect of enteric nervous system in Parkinson's disease patients with chronic constipation. *Lancet* **346**, 861–864.

Sipski, M., Alexander, C. *et al.* (2001). Sexual arousal and orgasm in women: effects of spinal cord injury. *Ann. Neurol.* **49**, 35–44.

Sipski, M. L., Alexander, C. J. *et al.* (1995). Orgasm in women with spinal cord injuries: a laboratory-based assessment. *Arch. Phys. Med. Rehabil.* **76**, 1097–1102.

Swinn, M. J., Wiseman, O. *et al.* (2002). The cause and natural history of isolated urinary retention in young women. *J. Urol.* **167**(3), 1348–1351.

van Gorkom, B. A., Karrenbeld, A. *et al.* (2000). Influence of a highly purified senna extract on colonic epithelium. *Digestion* **61**, 113–120.

Vodusek, D. B. (2001). Sphincter EMG and differential diagnosis of multiple system atrophy. *Mov. Disord.* **16**, 600–607.

Wedderburn, A., Lee, R. S. *et al.* (2001). Synchronous bladder reconstruction and antegrade continence enema. *J. Urol.* **165**, 2392–2393.

Wenning, G., Shlomo, Y. *et al.* (1994). Clinical features and natural history of multiple system atrophy. *Brain* **117**, 835–845.

Wermuth, L., and Stenager, E. (1992). Sexual aspects of Parkinson's disease. *Semin. Neurol.* **12**, 125–127.

WHO. (1999). First International Consultation on Erectile Dysfunction. Sponsored by World Health Organisation (WHO), International Consutation on Urological Diseases (ICUD) and Societe Internationale d'Urology (SIU). Paris, 1–3 July.

Wiesel, P. H., Norton, C. *et al.* (2001). Pathophysiology and management of bowel dysfunction in multiple sclerosis. *Eur. J. Gastroenterol. Hepatol.* **13**, 441–448.

Wood, J. D. (1999). Enteric nervous control of motility in the upper gastrointestinal tract in defensive states. *Dig. Dis. Sci.* **44**, 44S–52S.

Zeiss, A. M., Davies, H. D. *et al.* (1990). The incidence and correlates of erectile problems in patients with Alzheimer's disease. *Arch. Sexual Behav.* **19**, 325–331.

Zorzon, M., Zivadinov, R. *et al.* (1999). Sexual dysfunction in multiple sclerosis: a case-control study. I. Frequency and comparison of groups. *Mult. Scler.* **5**, 418–427.

## CHAPTER 105

# Syncope

N. Colman, W. Wieling, and R. Freeman

## INTRODUCTION

Syncope is a transient, self-limited loss of consciousness caused by acute impairment of cerebral blood flow. There are a number of neurological and cardiac causes of syncope. The clinical presentation of syncope depends in large part on the underlying cause. The first important issue is to differentiate syncope from other conditions like seizures and hypoglycemia associated with real or apparent transient loss of consciousness. The initial evaluation of a patient with transient loss of consciousness includes a detailed history, a thorough physical examination, and an electrocardiogram. This will lead to a diagnosis in approximately 50% of the patients. If no diagnosis has been established during the initial evaluation, additional steps to be considered are assessment of underlying structural heart disease and testing for arrhythmias. The principles of management of patients with syncope are based on the pathophysiological knowledge of the diverse causes of transient loss of consciousness. The physiological understanding of the basic mechanisms underlying syncope, in combination with clinical experience and common sense, provide sufficient evidence for firm nonpharmacological management recommendations. In a subgroup of patients pharmacological therapy may be helpful.

## CLINICAL ASPECTS

Syncope is a transient, self-limited loss of consciousness caused by acute impairment of cerebral blood flow. Recovery occurs spontaneously without the need for cardioversion or other medical interventions. There are a number of neurological and cardiac causes of syncope (Table I). The pathophysiology of neurological disorders that cause syncope is disease specific. The most common cause is the group of conditions in which cerebral per-

fusion is reduced by reflex disorders. This group, which is due to an episodic reflex-mediated vasodepressor-bradycardic response, is called neurally mediated syncope, neurocardiogenic syncope, or vasovagal syncope (Freeman and Rutkove, 2000). Cerebral blood flow also may be reduced because of chronic failure of the autonomic reflexes that maintain systemic vascular resistance (e.g., orthostatic hypotension and postprandial hypotension) or by an increase in intracranial pressure caused by a space-occupying mass or maneuvers that raise the intracranial pressure. Impaired brainstem perfusion as a result of intrinsic or functional disease of the vertebrobasilar vascular system (e.g., vertebrobasilar ischemia, and basilar migraine) may also lead to syncope. Syncope in most cardiac disorders is a consequence of a reduction in cardiac output. Common cardiac causes of syncope include cardiac arrhythmias, intrinsic heart disease, and pulmonary emboli.

The clinical presentation of syncope depends in large part on the underlying cause. Close attention to the clinical features preceding, during, and after a syncopal episode provides essential diagnostic information. Vasovagal syncope is typically preceded by a trigger such as a strong emotion, pain, or prolonged standing. Premonitory symptoms and signs such as blurred vision, muffled hearing, light-headedness, weakness, sweating, pallor, and nausea are usually present. Some of these symptoms and signs may still be prominent shortly after regaining consciousness. Patients appear pale, motionless with decreased muscle tone, and a slow, weak pulse. However, syncope can also occur abruptly without warning. This is a feature of cardiac syncope. Such patients may complain of palpitations, angina, or dyspnea. Sudden loss of consciousness without premonitory symptoms can also be seen in elderly patients with carotid sinus hypersensitivity. Syncope typically occurs while the patient is upright or seated, but Adams-Stokes syndrome caused by an intermittent atrioven-

TABLE I Pathophysiological Classification of Syncope

Reflex Syncopal Syndromes*
  Vasovagal faint
  Carotid sinus syncope
  Glossopharyngeal and trigeminal neuralgia
  Situational faint
    Cough, sneeze
    Gastrointestinal (swallow, defecation, visceral pain)
    Micturition (postmicturition)
    Postexercise, postprandial
    Others (e.g., brass instrument playing, weightlifting, arising
      from squatting)
Autonomic failure*
  Primary autonomic failure syndromes (e.g., pure autonomic
    failure, multiple system atrophy, Parkinson's disease with
    autonomic failure)
  Secondary autonomic failure syndromes (e.g., diabetic
    neuropathy, amyloid neuropathy)
  Drugs and alcohol
Insufficient circulatory volume
  Hemorrhage, diarrhea, Addison's disease, pheochromocytoma
Cardiac arrhythmias as primary cause
  Sinus node dysfunction (including bradycardia/tachycardia
    syndrome)
  Paroxysmal supraventricular and ventricular tachycardias
  AV conduction system disease
  Inherited syndromes (e.g., long QT syndrome, Brugada syndrome)
  Implanted device (pacemaker, ICD) malfunction
Impaired ability of the heart to increase output
  Valvular disease, obstructive cardiomyopathy, atrial myxoma
  Acute myocardial infarction/ischemia
  Acute aortic dissection
  Pericardial disease/tamponade
  Pulmonary embolus/pulmonary hypertension
Disorders in the cerebral circulation
  Vascular "steal" syndrome
  Transient ischemic attack basilaris region

  *May be exacerbated by drugs or heat, which causes volume
depletion or vasodilation.

TABLE II Nonsyncopal Causes of Transient Loss of Consciousness

Disorders with impaired or lost consciousness
  Epilepsy
  Metabolic disorders, including hypoglycemia and hypoxia
  Intoxication
Disorders resembling syncope with intact consciousness
  Cataplexy
  Vertigo
  Falls
  Psychiatric disorders
    Somatization disorders, conversion
    Anxiety
    Major depression

tricular block may occur while the patient is in the supine position. Resumption of the supine position often is sufficient to reverse the symptoms, especially in the reflex-mediated types of syncope.

## DIFFERENTIAL DIAGNOSIS

Syncope is easily diagnosed when the characteristic features are present. Atypical presentations, however, lead to diagnostic confusion and costly and often unnecessary evaluations (Brignole et al., 2001a; Kapoor, 2000; Linzer et al., 1997a). Syncope is associated with significant morbidity and even mortality, and, particularly when episodes occur frequently, may adversely affect the quality of life (Linzer et al., 1991a). The diagnostic approach should be thorough but individualized to accurately define the cause of syncope while avoiding large numbers of expensive diagnostic tests of low diagnostic yield.

The first important issue is to differentiate syncope from other conditions associated with real or apparent transient loss of consciousness (Brignole et al., 2001a; Kapoor, 2000; Linzer et al., 1997a) (Table II).

Seizures are frequently confused with syncope. Tonic-clonic movements are the hallmark of a generalized seizure. An attentive eyewitness may confirm their presence. However, myoclonus also may accompany a syncopal episode. These myoclonic jerks may be multifocal or generalized. They are typically arrhythmic, of short duration (<15 seconds), and start after the loss of consciousness. The tonic-clonic movements in patients with seizures in contrast are usually prolonged, and their onset coincides with the loss of consciousness (Brignole et al., 2001a; Hoefnagels et al., 1991b; Lempert et al., 1994). Partial or partial complex seizures may be preceded by an aura. Common auras include an unpleasant smell, fear, anxiety, abdominal discomfort, and other visceral sensations. These auras should be differentiated from the premonitory features of syncope. Loss of consciousness associated with a seizure usually lasts longer than 5 minutes and is associated with postictal drowsiness and disorientation. Muscle aches lasting for hours or days also indicate a prior seizure (Hoefnagels et al., 1991b). Urinary incontinence, however, is not useful in the distinction between seizures and syncope. Its rare to have triggering events like emotions or pain involved in seizures, and sweating or nausea before the loss of consciousness is exceptionally rare with seizures (Hoefnagels et al., 1991b).

Hypoglycemia may cause transient loss of consciousness as well, but its onset is slow after a period of confusion, disturbance of concentration, restlessness, and sensations of hunger. Normally, loss of consciousness is of longer duration than in syncope. The environmental setting in which loss of consciousness caused by hypoglycemia occurs usually leaves little difficulty in differentiating it from other forms of syncope (Linzer et al., 1997a).

Patients with cataplexy experience complete loss of muscular tone triggered by strong emotions, usually laughter. Consciousness, however, remains unaffected

(Brignole *et al.*, 2001a). Cataplexy is associated with narcolepsy in 75% of cases. Acute vestibular disorders causing falls because of severe vertigo also are characterized by the absence of loss of consciousness. In addition, nausea, vomiting, and nystagmus occur (Kapoor, 2000). Falls in elderly subjects may be difficult to distinguish from syncope. In fact, recent data indicate that up to one third of elderly patients with unexplained falls do have syncope. Amnesia for loss of consciousness is common in this age group and may explain the diagnostic difficulty (Dey and Kenny, 1997; Shaw and Kenny, 1997).

Apparent loss of consciousness can be a manifestation of psychiatric disorders such as generalized anxiety, panic disorders, major depression, and somatization. These causes of syncope should be considered in young people who faint frequently with prodromal symptoms and multiple unexplained physical complaints (Kapoor *et al.*, 1995). The patients involved are rarely hurt despite numerous falls and have no heart disease (Kapoor *et al.*, 1998; Linzer *et al.*, 1990a, 1997a). This phenomenon should be considered a manifestation of a conversion disorder, a somatoform-related illness in which neurological deficit is psychologically mediated (Kapoor *et al.*, 1998; Linzer *et al.*, 1992). These patients should be differentiated from patients in whom transient loss of consciousness is due to vasovagal reactions precipitated by fear, stress, anxiety, and emotional distress (Kapoor *et al.*, 1995; Linzer *et al.*, 1992). Tilt-table testing is particularly helpful in the differentiation of these two diagnostic possibilities. Patients with psychogenic syncope will seem to lose consciousness without concurrent changes in blood pressure or heart rate.

## DIAGNOSTIC FEATURES

The initial evaluation of a patient with transient loss of consciousness includes a detailed history, a thorough physical examination, and an electrocardiogram (Brignole *et al.*, 2001a; Kapoor, 2000; Linzer *et al.*, 1997a) (Figure 1). The initial assessment alone can lead to the identification of a cause of syncope in almost 50% of patients (Linzer *et al.*, 1997a; Oh *et al.*, 1999). In the remaining 50% of patients, history, physical examination, and ECG may suffice as risk stratifiers for acute cardiac death (Kapoor *et al.*, 1996; Martin *et al.*, 1997).

### History Taking

Detailed information about the episode is by far the most important aspect of the initial evaluation of a patient with transient loss of consciousness. The history

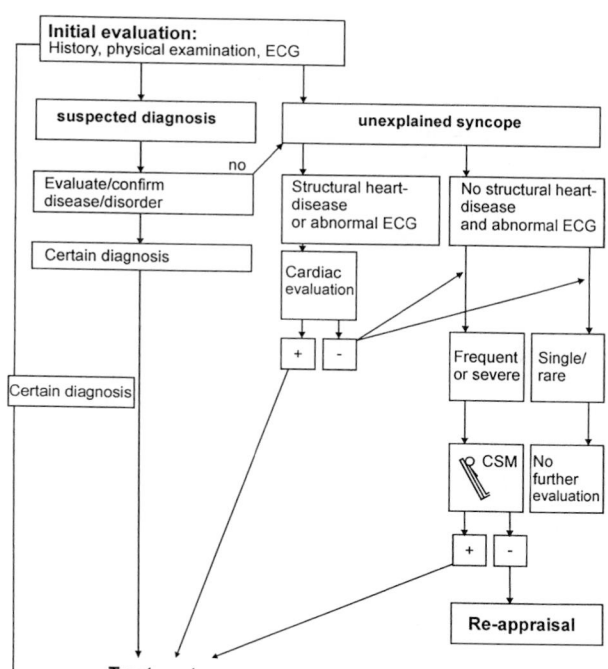

**Syncope**

**FIGURE 1** Diagnostic evaluation of syncope.

enables the physician to define an event as syncope and differentiate it from other forms of transient loss of consciousness (Tables I and II). If available, an accurate history obtained from an eyewitness is very important. The history provides a description of the clinical features of the event, which will help establish specific causes of syncope (Brignole *et al.*, 2001a; Kapoor, 2000). The use of a standard questionnaire may help examiners ask the key questions in a limited time (Petch, 1994).

### Physical Examination

The routine physical examination should include a blood pressure measurement in the supine and standing position. A postural drop in systolic blood pressure after 3 minutes standing of at least 20 mmHg or a systolic blood pressure less than 80 mmHg is considered to be diagnostic for orthostatic hypotension (Anonymous, 1996). Auscultation of the heart may reveal a murmur susceptible for aortic stenosis (Brignole *et al.*, 2001a).

### Electrocardiogram

Although a cause of syncope is only rarely assigned on the basis of an electrocardiogram (in less than 5% of the patients), it should not be omitted. An electrocardiogram is easy to perform, not expensive, and in case of abnormalities may lead to important underlying heart disease (Linzer *et al.*, 1997a). Examples of important findings are bradyarrhythmias or tachyarrhythmias, atrioventricular block, ischemia, old infarction, long QT syndrome, and bundle branch block. Such abnormalities imply the need for further cardiac evaluation.

## Laboratory Tests

Baseline laboratory blood tests are rarely helpful in finding the cause of syncope. They should only be done when specific disorders like anemia or myocardial infarction are suspected (Brignole *et al.*, 2001a; Kapoor, 2000; Linzer *et al.*, 1997a). Measuring creative phasphakinase (CPK) can be helpful when epilepsy is suspected (Libman *et al.*, 1991).

## Additional Testing

If no diagnosis has been established during the initial evaluation, additional steps to be considered are assessment of underlying structural heart disease and testing for arrhythmias (Brignole *et al.*, 2001a; Kapoor, 2000; Linzer *et al.*, 1997a) (Figure 1).

### *Evaluation for Underlying Heart Disease*

Echocardiography of the heart should be performed in patients with a history of cardiac disease or if abnormalities are found on physical examination or on the electrocardiogram (Recchia and Barzilai, 1995). Diagnoses that may be responsible for syncope include valvular heart disease or hypertrophic cardiomyopathy.

Treadmill exercise testing should be performed only when exercise-induced syncope is suspected. The presence of exertional syncope increases the likelihood that organic heart disease (e.g., aortic stenosis, hypertrophic cardiomyopathy) is the cause of syncope (Linzer *et al.*, 1997b).

### *Testing for Arrhythmias*

In patients with a high pretest likelihood of arrhythmias (e.g., brief loss of consciousness with short or absent prodromal symptoms, abnormal ECG, or organic heart disease), further evaluation for arrhythmias should be pursued.

In studies that evaluated syncope or presyncope during 24-hour electrocardiographic monitoring, 4% of the patients had symptomatic correlation with arrhythmias (leading to a diagnosis of arrhythmic syncope). In 15% of the patients symptoms were not associated with arrhythmias (potentially ruling out arrhythmic syncope) (Linzer *et al.*, 1997b).

Continuous loop recorders are used for long-term ECG monitoring. Monitors can be worn for weeks. They continually record and erase cardiac rhythm. Patients depress a record button after a syncopal spell and record in memory the previous 4 to 20 minutes of cardiac rhythm. Studies with loop electrocardiographic recorders in patients with unexplained syncope and relatively high recurrence of syncope have reported arrhythmias in association with symptoms in 8%–20% and symptoms during normal rhythm in 12%–27% (Kinlay *et al.*, 1996; Linzer *et al.*, 1990b).

A novel approach is an implantable continuous-loop recorder that is inserted subcutaneously. It has the capability of performing cardiac monitoring for up to 18 months. It is most beneficial when a patient has recurrent but infrequent unexplained syncope (Brignole *et al.*, 2001b; Kenny and Krahn, 1999).

Electrophysiological studies are indicated in patients who are at high risk for ventricular tachycardias or acute cardiac death, such as patients with a structural heart disease or abnormalities on ECG. The primary reason to perform this test is to exclude or diagnose ventricular tachycardia as the cause of syncope (Bachinsky *et al.*, 1992; Linzer *et al.*, 1991b, 1997b). The diagnostic value of this method is not always clear, because the clinical significance of many subtle abnormal electrophysiological findings (unless clearly associated with symptoms) remains controversial (Doherty *et al.*, 1985; Fujimura *et al.*, 1989).

### *Cardiovascular Reflex Testing*

Evaluation for reflex syncopal syndromes must also be considered in patients with recurrent or severe syncope (Brignole *et al.*, 2001a; Linzer *et al.*, 1997b).

Head-up tilt table testing, either alone or with pharmacological provocation such as intravenous isoproterenol or sublingual nitroglycerine, has been proposed as a useful investigation in the diagnosis of patients with unexplained syncope and suspected vasovagal syncope. Positive responses are reported in 30%–85% of patients, depending on the protocol used (Fenton *et al.*, 2000). Studies using chemical stimulation protocols reported approximately 66% positive responses on isoproterenol among patients with unexplained syncope (Kapoor, 2000). The specificity of most currently used tilt tests involving chemical stimulation approaches ranges from 80%–90% (Fenton *et al.*, 2000; Kapoor, 2000).

Carotid sinus massage should be considered in patients with spontaneous symptoms suggestive of carotid sinus syncope and in patients older than 50 years of age with recurrent syncope of unknown origin. The method of carotid sinus massage and the criteria for abnormal results are described elsewhere (Brignole et al., 2001a; Parry et al., 2000). This test should be avoided in patients with carotid bruits, plaques, or stenosis. Furthermore, carotid sinus massage should only be carried out under continuous ECG and blood pressure monitoring. Overall, carotid sinus massage is a safe procedure (Davies and Kenny, 1998).

In patients suspected of having autonomic failure, specific testing is indicated. Details are given elsewhere (Mathias and Bannister, 1999).

### Neurological Testing

Syncope during transient ischemic attacks is extremely uncommon and almost exclusively occurs in disorders of the posterior circulation. In these cases syncope is usually accompanied by neurological signs and symptoms such as vertigo, diplopia, ataxia, nystagmus, or dysarthria. The diagnosis is frequently made clinically but may be supported by imaging studies of the brainstem and posterior circulation (Brignole et al., 2001a). There are no studies demonstrating that transcranial or carotid Doppler and duplex ultrasonography are useful in evaluating syncope (Brignole et al., 2001a; Kapoor, 2000; Linzer et al., 1997a).

In cases of suspected epileptic seizures, an EEG, may be helpful. A normal EEG, however, does not exclude underlying epilepsy. Hoefnagels et al. (1991a) reported a 40% sensitivity and a 95% specificity for establishing the diagnosis of a seizure through EEG.

### Psychiatric Evaluation

Screening for psychiatric disorders has been recommended in patients with unexplained syncope with recurrent episodes and multiple physical symptoms (Kapoor et al., 1995, 1998b).

The Diagnostic Interview Schedule (DIS) is based on the DSM-III-R criteria for specific psychiatric diagnoses and can be used by trained physicians to evaluate patients with unexplained syncope. Tilt table testing might be useful in reproducing syncope in patients with suspected psychogenic syncope (Brignole et al., 2001; Grubb et al., 1992).

## NATURAL HISTORY

Few studies have evaluated the epidemiology of syncope in the general population. The Framingham study found a 3% prevalence of "isolated" syncope in men and 3.5% in women aged 30–62 years during a 26-year observation period (Savage et al., 1985).

Studies among teenagers, adolescents, and institutionalized elderly have reported a much higher prevalence. The prevalence in young male adolescents is reported as high as 20%–30% (Lamb et al., 1960), and in humans between 71 and 107 years a 10-year prevalence of 23% was found (Bonema and Maddens, 1991; Lipsitz et al., 1985).

Vasovagal syncope is the most common cause of syncope in most settings and occurs mostly in young individuals. The mean prevalence of vasovagal syncope for all age groups in clinical settings is 18% (range, 8%–37%) (Linzer et al., 1997a). The prevalence not only depends on age but also on the specific clinical setting. Among patients seen in an emergency department, Martin et al. (1984) found a 37% rate of vasovagal syncope. Whereas vasovagal syncope mainly occurs in adolescents, in elderly patients situational syncope and orthostatic hypotension are common causes of syncope. Syncope related to micturation, defecation, postural changes, or meals was found in 20% of institutionalized elderly patients (Linzer et al., 1997b).

Cardiac causes are found in approximately 8%–15% of cases (Linzer et al., 1997a; Oh et al., 1999; Sarasin et al., 2001). Psychiatric disorders seem to have a much higher correlation with syncope than previously known. Kapoor et al. (1995) reported that 20% of a group of patients with unexplained syncope met the criteria for at least one psychiatric disorder or alcohol/drug abuse. Sarasin et al. (2001) recently found 11% of psychiatric causes in a group of 650 patients with syncope seen in the emergency department. Until recently, in a considerable number of patients the cause of syncope (26%–41%) remained unclear despite elaborate investigations (Kapoor, 1990; Linzer et al., 1997a). With the use of new techniques such as tilt table tests, implantable loop recorders, and strict diagnostic protocols, this group will probably be much smaller (Kapoor, 1998a).

The prognosis for all age groups after a single syncope episode is generally benign (Savage et al., 1985). Syncope of noncardiac and unexplained origin in younger individuals especially has an excellent prognosis; there is no reduction in long-term survival. A cardiac cause of syncope, however, is associated with an increased risk of sudden death. Kapoor et al. (1996) compared the mortality rate of a group of patients with syncope and without syncope. The overall and cardiac mortality rate was not significantly different between the two groups. Recently Oh et al. (1999) found that cardiac diseases and an abnormal electrocardiogram were the only predictors of mortality in syncopal patients. These results seem to imply that the underlying heart disease is the most important risk factor for cardiac death, independent of whether the patient experiences syncope or not.

Recurrence rates depend mainly on the cause of syncope and the number of previous episodes but also on age and gender. Recurrence rates during 5 years of follow-up range between 33% (cardiac) and 51% (unknown) (Kapoor, 1990). Grimm *et al.* (1997) reported a recurrence rate of 54% after 23 ± 8 months in a group of patients with suspected neurally mediated syncope with a history of two or more episodes of syncope and 10% recurrence in a group with one episode of syncope. This confirms the findings of Sheldon *et al.* (1996), who reported earlier that the logarithm of the number of preceding syncopal spells was the most powerful predictor of a recurrence. Furthermore, they found that patients with a recurrence were younger and more likely to be women. The 18-month recurrence rate in a group of 611 patients studied by Sarasin *et al.* (2001) varied from 9% in cardiac syncope to 17% in vasovagal syncope. One-year recurrence rate for syncope in patients with any psychiatric diagnosis is 35% according to Kapoor *et al.* (1995) opposed to 15% in patients without a psychiatric disorder. Recurrent syncope can have a profound impact on physical, social, and mental function (Linzer *et al.*, 1991a).

## PRINCIPLES OF THERAPY

The principles of management of patients with syncope are based on the pathophysiological knowledge of the diverse causes of transient loss of consciousness (Tables I and II). The physiological understanding of the basic mechanisms underlying syncope, in combination with clinical experience and common sense, provide sufficient evidence for firm nonpharmacological management recommendations.

For advice about medications and application of devices in patients with syncope, formal randomized controlled trials are needed. However, most studies dealing with these aspects of management of patients with syncope have used a nonrandomized design, which makes the interpretation almost impossible. General principles of management are based on experience with patients with autonomic failure, and the management in these patients will therefore be discussed first.

### Autonomic Failure

The mainstay of management of a patient with orthostatic hypotension consists of advice and education on the various factors that influence systemic blood pressure and of the chronic expansion of the intravascular volume or reducing the vascular volume into which pooling occurs (Table III). Management advice should be specific to the individual patient. The goal of man-

**TABLE III    General Principles and Recommendations**

To be avoided*
- Sudden head-up postural change (especially on waking)
- Standing still
- Prolonged recumbence during daytime
- Straining during micturition and defecation, hyperventilation
- High environmental temperature (including hot baths and showers)
- More than severe exertion
- Large meals (especially with refined carbohydrate)
- Alcohol
- Drugs with vasodepressor properties
- Diet and cold preparations containing sympathomimetic amines

To be introduced
- Salt intake of at least 8 g (150 mmol) a day*,**
- 2–2.5 L fluid/day*
- Small frequent meals with a reduced carbohydrate content*
- Head-up tilt during sleep*
- Judicious exercise (including swimming)*
- Physical counter maneuvers*
- Air-conditioning or fan in summer*

To be considered
- Abdominal binders*
- Elastic stockings*
- Support garment*
- Portable chairs*

From Wieling *et al.* (2000), with permission.
*Retrospective studies, nonrandomized studies, or empirical, recommendation without scientific proof.
**Randomized, prospective study.

agement is to reduce the symptoms of insufficient cerebral perfusion. The actual upright blood pressure level is less relevant (Mathias *et al.*, 1998).

Patients should increase salt and fluid intake to expand blood volume. Regular exercise is important to avoid deconditioning. If these measures fail to reverse symptoms, raising the head of the bed 4 inches and administration of fludrocortisone (from 0.05 mg daily to 0.1 mg twice a day) is usually effective (ten Harkel *et al.*, 1992; van Lieshout *et al.*, 2000). Physical maneuvers like leg crossing or squatting may be helpful (Table IV) (van Lieshout *et al.*, 1992; Wieling *et al.*, 1993).

Elastic support garments may be helpful in the treatment of a patient with incapacitating orthostatic hypotension, but they have several disadvantages. The motivation of the patient must be strong. Donning of elastic body garments is difficult, and they are uncomfortable to wear, especially in warm weather (Mathias *et al.*, 1998). Recent work has shown that when using compression in patients with orthostatic hypotension, the abdomen is the most important single site for compression (Tanaka *et al.*, 1997). We have found abdominal binders in particular useful as a temporary expedient to achieve mobility in our most severely affected patients. The beneficial effect of compression is

TABLE IV  Physical Countermaneuvers

| Maneuver | Instruction |
|---|---|
| Leg crossing | Cross one leg over the other at thigh level while standing and sitting. |
| Limb and/or abdominal contraction | Contract leg musculature and/or abdomen while standing. This can also be done while leg crossing. |
| Bending forward | Lowering the head to heart level while standing or sitting. |
| Squatting | Sit on the heels. A variant is to kneel down, pretending to lace a shoe or standing on one foot while placing the other on an elevated surface with bending of the trunk (foot on chair maneuver). |
| Toe raising | Rise onto the front of the feet, maintain calf contraction and return to flat-stance position. |
| Knee flexion | March on the spot. |
| Muscle tension | Tension of the muscles of the abdomen, buttocks, and legs. |

reported to be more pronounced in combination with volume expansion, but no objective data are available as yet. A small portable folding chair is a handy mechanical aid for severely affected patients who are still ambulatory (Smit *et al.*, 1997).

In patients with severe orthostatic hypotension who do not react on the management advice listed in Table IV, drug treatment is indicated. However, such treatment may aggravate supine hypertension or have other undesirable effects. The compound most widely used presently is the alpha-agonist midodrine, although other alpha-adrenoreceptor agonists may be equally effective. It is administered in doses of 2.5–10 mg thrice daily and is best taken on awaking when upright blood pressure is lowest, with further doses before lunch and a last dose

in the afternoon. A 10-mg midodrine dose is reported to increase standing systolic blood pressure by 20–30 mmHg (Table V) (Low *et al.*, 1997; Wright *et al.*, 1998). Supine hypertension is a common side effect of midodrine, and this can be severe in some patients. Other alpha-adrenoceptor agonist side effects include goose bumps (cutis anserina); tingling of the skin; pruritus, especially of the scalp; and in the male impaired urine flow, hesitancy, and urinary retention. If the combination of fludrocortisone and midodrine does not produce the desired effect, supplementary agents should be used. These agents may target specific pathophysiological abnormalities. Thus, demopressin may be used in nocturnal polyuria, octreotide in postprandial hypotension, and erythropoietin in anemia (Table V) (Mathias *et al.*, 1998).

### Vasovagal Syncope

For acute management, resumption of the supine position is usually sufficient. Elevation of the legs may be performed to increase venous return. There are few satisfactory clinical trials on the treatment of vasovagal syncope. Long-term management strategies of patients with vasovagal syncope are based on clinical experience. Patients require principally reassurance and explanation regarding the nature of the condition. Modification of drug treatment for concomitant conditions (diuretics and vasodilators) should be considered. Initial specific advice should include avoidance of triggering events, volume depletion, and prolonged exposure to upright posture and/or hot confining environments. Patients should be taught to recognize warning symptoms (Table III). The tilt table test often allows patients to appreciate the premonitory symptoms in a controlled environment.

TABLE V  Pharmacological Treatment

| Medication | Indication | Dose | (Relative) contraindication |
|---|---|---|---|
| Fludrocortisone*,*** | Autonomic failure Vasovagal syncope | From 0.05 mg daily to 0.1 mg twice a day | Long-term treatment in elderly, hypertension, congestive heart failure |
| Midodrine*,*** | Autonomic failure Vasovagal syncope | 2.5–10 mg thrice daily, during daytime only | Hypertension |
| Desmopressin* | Autonomic failure —Nocturnal polyuria | 10–40 μg nasal or 100–400 μg orally | |
| Erythropoietin* | Autonomic failure —Anemia | 50 μg/kg 3 times a week for 6–8 weeks | |
| Octreotide* | Autonomic failure —Postprandial hypotension | 25–50 μg SC ½ h before a meal | |
| *Paroxetine*** | Vasovagal syncope | | |

Adapted from Mathias *et al.* (1998).
*Retrospective studies, nonrandomized studies, or empirical recommendation without scientific proof.
**Randomized, prospective study.
***More than one prospective, randomized, placebo-controlled study or meta-analysis.

In most patients with few episodes, additional steps are hardly needed. When additional measures are necessary, chronic expansion of total body water by encouraging a higher than normal salt (about 10 g NaCl) and fluid intake (El Sayed and Hainsworth, 1996; Mtinangi and Hainsworth, 1998a,b), exercise training (Hainsworth, 1999), and instructions in maneuvers that improve orthostatic tolerance (van Dijk et al., 2000) seem to be among the safest approaches. In highly motivated patients with recurrent vasovagal syncope, raising the head of the bed on blocks to permit gravitational exposure during sleep (Hainsworth, 2000) or the prescription of progressively prolonged periods of enforced upright posture (so-called tilt-training) have been advocated (Di Girolamo et al., 1999a; Ector et al., 1998).

If the abovementioned measures are ineffective, drugs may be used (Table V). Although there are limited data on the use of fludrocortisone to treat vasovagal syncope, it is often used early in the management of this disorder in younger subjects because of its low side effect profile (Calkins, 1999; Parry and Kenny, 1999). Elderly patients may experience many side effects, especially when using fludrocortisone for a long period (Hussain et al., 1996). Beta-blockers are widely used, and atenolol was effective in one randomized trial (Mahanonda et al., 1995). However, the follow-up in that study was very short (1 month), and several controlled studies have been negative. Overall the evidence seems to fail to support beta-blocker efficacy (Brignole et al., 2001a). Etilephrine, a new alpha-stimulating agent, also proved to be ineffective in a randomized placebo-controlled trial (Raviele et al., 1999). The general picture seems to be that while the results have been satisfactory in uncontrolled trials, several placebo-controlled prospective trials have been unable to show benefit of the active drug over placebo (Brignole et al., 2001a). An exception is the serotonin inhibitor paroxetine, which has been shown to be effective in a randomized trial with long-term follow-up (Di Girolamo et al., 1999b). However, long-term prophylactic therapy with this compound should not be advised, because vasovagal syncope is a self-limited, non-life-threatening condition. Recently, a study in a subgroup of patients with at least one syncope a month showed a significant benefit of midodrine. At a 6-month follow-up, 81% of the midodrine-treated patients had remained syncope free (Perez-Lugones et al., 2001).

Permanent pacing has been advocated for refractory patients. Three recent studies have reported a beneficial treatment effect (Ammirati et al., 2001; Connolly et al., 1999; Sutton et al., 2000). Unfortunately, these studies were open-label, and the medical regiments were not standardized or controlled. A placebo effect could not be excluded. A controlled study, which is now in progress, should provide guidance as to which, if any,

patients will benefit from this intervention (Brignole et al., 2001a; Raviele et al., 2001).

## Carotid Sinus Hypersensitivity

Treatment of patients with carotid sinus hypersensitivity must be guided by the results of the carotid sinus massage. Cardiac pacing seems to be beneficial and is acknowledged to be the treatment of choice when bradycardia has been documented (Brignole et al., 1991, 1992; Kenny et al., 2001). For the most part, dual-chamber cardiac pacing is preferred. There are as yet no randomized studies that examine treatment of carotid sinus syncope in which hypotension is predominantly of vasodepressor origin. Certain therapies used for vasovagal syncope may be expected to be of some benefit; volume expansion and vasoconstrictors are the most likely in this regard, but supine hypertension is a concern (Brignole et al., 2001a).

## Situational Syncope

Treatment of situational syncope relies heavily on avoiding or ameliorating the triggering event. In conditions in which trigger avoidance is not feasible, certain general treatment strategies may be advocated. These strategies include maintenance of central blood volume, protected posture (e.g., sitting rather than standing during micturation), or slower change of posture (e.g., waiting after bowel movement before arising).

Postprandial syncope is a disease in elderly individuals, probably caused by age-related abnormal vasomotor compensatory mechanisms for pooling blood in the splanchnic circulation (Lipsitz et al., 1993). Patients with this disorder should ingest frequent small meals, ideally with a low carbohydrate content. Standing after meals should be avoided.

## Cardiac Syncope

In brief, arrhythmias are usually treated with medications, pacemaker implantation, or an implantable cardiac defibrillator. Recently, some rapid supraventricular tachycardias have been treated with radiofrequency ablation of bypass tracts (Sorbera et al., 1999). Valvular disease resulting in syncope, in particular aortic stenosis, usually necessitates surgery.

## Psychiatric Causes

Syncope caused by psychiatric disorders (such as anxiety and depressive disorders) often responds to treatment with antidepressants (Linzer *et al.*, 1990a). Hyperventilation syndrome might be considered to be part of an anxiety disorder or a broader complex of psychological complaints that are accompanied by somatic symptoms (Hornsveld and Garssen, 1997). In these cases reassurance or behavioral therapy is probably the best treatment. In patients with specific phobia's psychological deconditioning should be considered (van Dijk *et al.*, 2001). In patients with somatization and conversions psychiatric consultation is indicated.

## Syncope and Driving

Patients with unexplained syncope without warning signals should be cautioned not to drive until a cause is found or the probability of recurrence is minimal.

## CONCLUSION

It is important to distinguish true syncope from other causes of loss of consciousness. History is the most diagnostic tool and will lead to a diagnosis in almost 50% of the patients. In the remaining patients history will help in stratifying patients at high and low risk for cardiac death and should be used to guide the use of diagnostic tests. Knowledge of the pathophysiology of the different types of syncope will not only contribute to an efficient evaluation of patients with syncope but also to the appropriate treatment.

## REFERENCES

Ammirati, F., Colivicchi, F., and Santini, M. (2001). Permanent Cardiac Pacing Versus Medical Treatment for the Prevention of Recurrent Vasovagal Syncope; a Multicenter, Randomized, Controlled Trial. *Circulation* 104, 52–57.

Anonymous. (1996). Consensus statement on the definition of orthostatic hypotension, pure autonomic failure, and multiple system atrophy. The Consensus Committee of the American Autonomic Society and the American Academy of Neurology. *Neurology* 46, 1470.

Bachinsky, W. B., Linzer, M., Weld, L., and Estes, N. A. (1992). Usefulness of clinical characteristics in predicting the outcome of electrophysiologic studies in unexplained syncope. *Am. J. Cardiol.* 69, 1044–1049.

Bonema, J. D., and Maddens, M. E. (1991). Syncope in elderly patients. Why their risk is higher. *Postgrad. Med.* 91, 129–132.

Brignole, M., Alboni, P., Benditt, D. G., Bergfeldt, l., Blanc, J. J., Bloch Thomsen, P. E., Dijk, J. G. van, Fitzpatrick, A., Hohnloser, S., Janousek, J., Kapoor, W. N., Kenny, R. A., Kulakowski, P., Moya, A., Raviele, A., Sutton, R., Theodorakis, G., and Wieling, W. (2001a). Guidelines on management (diagnosis and treatment) of syncope. *Eur. Heart J.* 22, 1256–1306.

Brignole, M., Menozzi, C., Lolli, G., Bottoni, N., and Gaggioli, G. (1992). Long-term outcome of paced and nonpaced patients with severe carotid sinus syndrome. *Am. J. Cardiol.* 69, 1039–1043.

Brignole, M., Menozzi, C., Lolli, G., Oddone, D., Gianfranchi, L., and Bertulla, A. (1991). Validation of a method for choice of pacing mode in carotid sinus syndrome with or without sinus bradycardia. *Pacing Clin. Electrophysiol.* 14, 196–203.

Brignole, M., Menozzi, C., Moya, A., and Garcia-Civera, R. (2001b). Implantable loop recorder: towards a gold standard for the diagnosis of syncope? *Heart* 85, 610–612.

Calkins, H. (1999). Pharmacologic approaches to therapy for vasovagal syncope. *Am. J. Cardiol.* 84, 20Q–25Q.

Connolly, S. J., Sheldon, R., Roberts, R. S., and Gent, M. (1999). The North American Vasovagal Pacemaker Study (VPS). A randomized trial of permanent cardiac pacing for the prevention of vasovagal syncope. *J. Am. Coll. Cardiol.* 33, 16–20.

Davies, A. J., and Kenny, R. A. (1998). Frequency of neurologic complications following carotid sinus massage. *Am. J. Cardiol.* 81, 1256–1257.

Dey, A. B., and Kenny, R. A. (1997). Drop attacks in the elderly revisited. *Q. J. Med.* 90, 1–3.

Di Girolamo, E., Di Iorio, C., Leonzio, L., Sabatini, P., and Barsotti, A. (1999a). Usefulness of a tilt training program for the prevention of refractory neurocardiogenic syncope in adolescents: A controlled study. *Circulation* 100, 1798–1801.

Di Girolamo, E., Di Iorio, C., Sabatini, P., Leonzio, L., Barbone, C., and Barsotti, A. (1999b). Effects of paroxetine hydrochloride, a selective serotonin reuptake inhibitor, on refractory vasovagal syncope: A randomized, double-blind, placebo-controlled study. *J. Am. Coll. Cardiol.* 33, 1227–1230.

Dijk, N. van, Harms, M. P., Linzer, M., and Wieling, W. (2000). Treatment of vasovagal syncope: pacemaker or crossing legs. *Clin. Aut. Res.* 10, 347–349.

Dijk, N. van, Velzeboer, S. C. J. M., Destree-Vonk, A., Linzer, M., and Wieling, W. (2001). Psychological treatment of malignant vasovagal syncope due to blood phobia. *PACE* 24, 122–124.

Doherty, J. U., Pembrook-Rogers, D., Grogan, E. W., Falcone, R. A., Buxton, A. E., Marchlinski, F. E., Cassidy, D. M., Kienzle, M. G., Almendral, J. M., and Josephson, M. E. (1985). Electrophysiologic evaluation and follow-up characteristics of patients with recurrent unexplained syncope and presyncope. *Am. J. Cardiol.* 55, 703–708.

Ector, H., Reybrouck, T., Heidbuchel, H., Gewillig, M., Van de Werf, F. (1998). Tilt training: a new treatment for recurrent neurocardiogenic syncope and severe orthostatic intolerance. *Pacing Clin. Electrophysiol.* 21, 193–196.

El-Sayed, H., and Hainsworth, R. (1996). Salt supplement increases plasma volume and orthostatic tolerance in patients with unexplained syncope. *Heart* 75, 134–140.

Fenton, A. M., Hammill, S. C., Rea, R. F., Low, P. A., and Shen, W. K. (2000). Vasovagal syncope. *Ann. Intern. Med.* 133, 714–725.

Freeman, R., and Rutkove, S. (2000). Syncope. *In* "Handbook of Clinical Neurology. The Autonomic Nervous System. Part II. Dysfunctions." (P. J. Vincken and G. W. Bruyn, Eds.), pp. 203–229. Elsevier Press, New York.

Fujimura, O., Yee, R., Klein, G. J., Sharma, A. D., and Boahene, K. A. (1989). The diagnostic sensitivity of electrophysiologic testing in patients with syncope caused by transient bradycardia. *N. Engl. J. Med.* 321, 1703–1707.

Grimm, W., Degenhardt, M., Hoffman, J., Menz, V., Wirths, A., and Maisch, B. (1997). Syncope recurrence can better be predicted by history than by head-up tilt testing in untreated patients with suspected neurally mediated syncope. *Eur. Heart J.* 18, 1465–1469.

Grubb, B. P., Gerard, G., Wolfe, D. A., Samoil, D., Davenport, C. W., Homan, R. W., and Temesy-Armos, P. (1992). Syncope and seizures of psychogenic origin: Identification with head- upright tilt table testing. *Clin. Cardiol.* 15, 839–842.

Hainsworth, R. (1999). Syncope and fainting; classification and pathophysiological basis. *In* "A Textbook of Clinical Disorders of the Autonomic Nervous System" (C. J. Mathias and R. Bannister, Eds.), pp. 428–436. Oxford University Press, Oxford.

Hainsworth, R. (2000). Head-up sleeping for posturally related syncope. *Clin. Auton. Res.* **10**, 225–226.

Hoefnagels, W. A., Padberg, G. W., Overweg, J., Roos, R. A., van Dijk, J. G., and Kamphuisen, H. A. (1991a). Syncope or seizure? The diagnostic value of the EEG and hyperventilation test in transient loss of consciousness. *J. Neurol. Neurosurg. Psychiatry* **54**, 953–956.

Hoefnagels, W. A., Padberg, G. W., Overweg, J., van der Velde, E. A., and Roos, R. A. (1991b). Transient loss of consciousness: the value of the history for distinguishing seizure from syncope. *J. Neurol.* **238**, 39–43.

Hornsveld, H., and Garssen, B. (1997). Hyperventilation syndrome: an elegant but scientifically untenable concept. *Neth. J. Med.* **50**, 13–20.

Hussain, R. M., McIntosh, S. J., Lawson, J., and Kenny, R. A. (1996). Fludrocortisone in the treatment of hypotensive disorders in the elderly. *Heart* **76**, 507–509.

Kapoor, W. N. (1998). An overview of the evaluation and management of syncope. *In* "Syncope: Mechanisms and Management." (B. P. Grubb and B. Olshansky, Eds.), pp. 1–13. Futura Publishing Co., Armonk, NY.

Kapoor, W. N. (1990). Evaluation and outcome of patients with syncope. *Medicine (Baltimore)* **69**, 160–175.

Kapoor, W. N. (2000). Syncope. *N. Engl. J. Med.* **343**, 1856–1862.

Kapoor, W. N., Fortunato, M., Hanusa, B. H., and Schulberg, H. C. (1995). Psychiatric illnesses in patients with syncope. *Am. J. Med.* **99**, 505–512.

Kapoor, W. N., and Hanusa, B. H. (1996). Is syncope a risk factor for poor outcomes? Comparison of patients with and without syncope. *Am. J. Med.* **100**, 646–655.

Kapoor, W. N., Karpf, M., Wieand, S., Peterson, J. R., and Levey, G. S. (1983). A prospective evaluation and follow-up of patients with syncope. *N. Engl. J. Med.* **309**, 197–204.

Kapoor, W. N., and Schulberg, H. C. (1998b). Psychiatric disorders in patients with syncope. *In* "Syncope; Mechanisms an Management." (B. P. Grubb and B. Olshansky, Eds.), pp. 253–263. Futura Publishing Co., Armonk, NY.

Kenny, R. A., and Krahn, A. D. (1999). Implantable loop recorder: evaluation of unexplained syncope. *Heart* **81**, 431–433.

Kenny, R. A., Richardson, D. A., Steen, N., Bexton, R. S., Shaw, F. E., and Bond, J. (2001). Carotid sinus syndrome: A modifiable risk factor for nonaccidental falls in older adults (SAFE PACE). *J. Am. Coll. Cardiol.* **38**, 1491–1496.

Kinlay, S., Leitch, J. W., Neil, A., Chapman, B. L., Hardy, D. B., and Fletcher, P. J. (1996). Cardiac event recorders yield more diagnoses and are more cost-effective than 48-hour Holter monitoring in patients with palpitations. A controlled clinical trial. *Ann. Intern. Med.* **124**, 16–20.

Lamb, L. E., Green, H. C., Combs, J. J., Cheeseman, S. A., and Hammond, J. (1960). Incidence of Loss of Consciousness in 1980 Air Force Personnel. *Aerospace Med.* **31**, 973–988.

Lempert, T., Bauer, M., and Schmidt, D. (1994). Syncope: a videometric analysis of 56 episodes of transient cerebral hypoxia. *Ann. Neurol.* **36**, 233–237.

Libman, M. D., Potvin, L., Coupal, L., and Grover, S. A. (1991). Seizure vs. syncope: measuring serum creatine kinase in the emergency department. *J. Gen. Intern. Med.* **6**, 408–412.

Linzer, M., Felder, A., Hackel, A., Perry, A. J., Varia, I., Melville, M. L., and Krishnan, K. R. (1990a). Psychiatric syncope: a new look at an old disease. *Psychosomatics* **31**, 181–188.

Linzer, M., Pontinen, M., Gold, D. T., Divine, G. W., Felder, A., and Brooks, W. B. (1991a). Impairment of physical and psychosocial function in recurrent syncope. *J. Clin. Epidemiol.* **44**, 1037–1043.

Linzer, M., Pritchett, E. L., Pontinen, M., McCarthy, E., and Divine, G. W. (1990b). Incremental diagnostic yield of loop electrocardiographic recorders in unexplained syncope. *Am. J. Cardiol.* **66**, 214–219.

Linzer, M., Prystowsky, E. N., Divine, G. W., Matchar, D. B., Samsa, G., Harrell, F. Jr, Pressley, J. C., and Pryor, D. B. (1991b). Predicting the outcomes of electrophysiologic studies of patients with unexplained syncope: preliminary validation of a derived model. *J. Gen. Intern. Med.* **6**, 113–120.

Linzer, M., Varia, I., Pontinen, M., Divine, G. W., Grubb, B. P., and Estes, M., III. (1992). Medically unexplained syncope: Relationship to psychiatric illness. *Am. J. Med.* **92**, 18S–25S.

Linzer, M., Yang, E. H., Estes, M., III, Wang, P., Vorperian, V. R., and Kapoor, W. N. (1997a). Diagnosing syncope part 1: Value of history, physical examination, and electrocardiography. *Ann. Intern. Med.* **126**, 989–996.

Linzer, M., Yang, E. H., Estes, M., III, Wang, P., Vorperian, V. R., and Kapoor, W. N. (1997b). Diagnosing syncope part 2: Unexplained syncope. *Ann. Intern. Med.* **127**, 76–86.

Lipsitz, L. A., Ryan, S. M., Parker, J. A., Freeman, R., Wei, J. Y., and Goldberger, A. L. (1993). Hemodynamic and autonomic nervous system responses to mixed meal ingestion in healthy young and old subjects and dysautonomic patients with postprandial hypotension. *Circulation* **87**, 391–400.

Lipsitz, L. A., Wei, J. Y., and Rowe, J. W. (1985). Syncope in an elderly, institutionalised population: prevalence, incidence, and associated risk. *Q. J. Med.* **55**, 45–54.

Low, P. A., Gilden, J. L., Freeman, R., Sheng, K. N., and McElligott, M. A. (1997). Efficacy of midodrine for neurogenic orthostatic hypotension. Reply. *JAMA* **278**, 388.

Mahanonda, N., Bhuripanyo, K., Kangkagate, C., Wansanit, K., Kulchot, B., Nademanee, K., and Chaithiraphan, S. (1995). Randomized double-blind, placebo-controlled trial of oral atenolol in patients with unexplained syncope and positive upright tilt table test results. *Am. Heart J.* **130**, 1250–1253.

Martin, G. J., Adams, S. L., Martin, H. G., Mathews, J., Zull, D., and Scanlon, P. J. (1984). Prospective evaluation of syncope. *Ann. Emerg. Med.* **13**, 499–504.

Martin, T. P., Hanusa, B. H., and Kapoor, W. N. (1997). Risk stratification of patients with syncope. *Ann. Emerg. Med.* **29**, 459–466.

Mathias, C. J., and Bannister, R. (1999). Investigation of autonomic disorders. *In* "Autonomic Failure" (C. J. Mathias and R. Bannister, Eds.), pp. 169–195. Oxford University Press, Oxford.

Mathias, C. J., and Kimber, J. R. (1998). Treatment of postural hypotension. *J. Neurol. Neurosurg. Psychiatry* **65**, 285–289.

Mtinangi, B. L., and Hainsworth, R. (1998a). Early effects of oral salt on plasma volume, orthostatic tolerance, and baroreceptor sensitivity in patients with syncope. *Clin. Auton. Res.* **8**, 231–235.

Mtinangi, B. L., and Hainsworth, R. (1998b). Increased orthostatic tolerance following moderate exercise training in patients with unexplained syncope. *Heart* **80**, 596–600.

Oh, J. H., Hanusa, B. H., and Kapoor, W. N. (1999). Do symptoms predict cardiac arrhythmias and mortality in patients with syncope? *Arch. Intern. Med.* **159**, 375–380.

Parry, S. W., and Kenny, R. A. (1999). The management of vasovagal syncope. *Q. J. Med.* **92**, 697–705.

Parry, S. W., Richardson, D. A., O'Shea, D., Sen, B., and Kenny, R. A. (2000). Diagnosis of carotid sinus hypersensitivity in older adults: carotid sinus massage in the upright position is essential. *Heart* **83**, 22–23.

Perez-Lugones, A., Schweikert, R., Pavia, S., Sra, J., Akhtar, M., Jaeger, F., Tomassoni, G. F., Saliba, W., Leonelli, F. M., Bash, D.,

Beheirry, S., Shewchik, J., Tchou, P. J., and Natale, A. (2001). Usefulness of midodrine in patients with severly symptomatic neurocardiogenic syncope: A randomized control study. *J. Cardiovasc. Electrophysiol.* **12**, 935–938.

Petch, M. C. (1994). Syncope. *BMJ* **308**, 1251–1252.

Raviele, A., Brignole, M., Sutton, R., Alboni, P., Giani, P., Menozzi, C., and Moya, A. (1999). Effect of etilefrine in preventing syncopal recurrence in patients with vasovagal syncope: A double-blind, randomized, placebo-controlled trial. *Circulation* **99**, 1452–1457.

Raviele, A., Giada, F., Sutton, R., Alboni, P., Brignole, M., Del Rosso, A., Di Girolamo, E., Luise, R., and Menozzi, C. (2001). The vasovagal syncope and pacing (Synpace) trial: rationale and study design. *Europace* **3**, 336–341.

Recchia, D., and Barzilai, B. (1995). Echocardiography in the evaluation of patients with syncope [see comments]. *J. Gen. Intern. Med.* **10**, 649–655.

Sarasin, F. P., Louis-Simonet, M., Carballo, D., Slama, S., Rajeswaran, A., Metzger, J. T., Lovis, C., Unger, P. F., and Junod, A. F. (2001). Prospective evaluation of patients with syncope: A population-based study. *Am. J. Med.* **111**, 177–184.

Savage, D. D., Corwin, L., McGee, D. L., Kannel, W. B., and Wolf, P. A. (1985). Epidemiologic features of isolated syncope: the Framingham Study. *Stroke* **16**, 626–629.

Shaw, F. E., and Kenny, R. A. (1997). The overlap between syncope and falls in the elderly. [Review] [30 refs]. *Postgrad. Med. J.* **73**, 635–639.

Sheldon, R., Rose, S., Flanagan, P., Koshman, M. L., and Killam, S. (1996). Risk factors for syncope recurrence after a positive tilt-table test in patients with syncope. *Circulation* **93**, 973–981.

Smit, A. A., Hardjowijono, M. A., and Wieling, W. (1997). Are portable folding chairs useful to combat orthostatic hypotension? *Ann. Neurol.* **42**, 975–978.

Sorbera, C., Cohen, M., and Christiana, J. (1999). Radiofrequency catheter ablation: the first decade. *Heart Dis.* **1**, 210–220.

Sutton, R., Brignole, M., Menozzi, C., Raviele, A., Alboni, P., Giani, P., and Moya, A. (2000). Dual-chamber pacing in the treatment of neurally mediated tilt-positive cardioinhibitory syncope: Pacemaker versus no therapy: A multicenter randomized study. *Circulation* **102**, 294–299.

Tanaka, H., Yamaguchi, H., and Tamai, H. (1997). Treatment of orthostatic intolerance with inflatable abdominal band [letter]. *Lancet* **349**, 175.

Ten Harkel, A. D., van Lieshout, J. J., and Wieling, W. (1992). Treatment of orthostatic hypotension with sleeping in the head-up tilt position, alone and in combination with fludrocortisone. *J. Intern. Med.* **232**, 139–145.

Van Lieshout, J. J., ten Harkel, A. D., and Wieling, W. (1992). Physical manoeuvres for combating orthostatic dizziness in autonomic failure. *Lancet* **339**, 897–898.

Van Lieshout, J. J., ten Harkel, A. D., and Wieling, W. (2000). Fludrocortisone and sleeping in the head-up position limit the postural decrease in cardiac output in autonomic failure. *Clin. Auton. Res.* **10**, 35–42.

Wieling, W., Cortelli, P., and Mathias, C. J. (2000). Treating neurogenic orthostatic hypotension. *In* "Handbook of Clinical Neurology: The Autonomic Nervous System. Part II." (O. Appenzeller, Ed.), pp. 1–17. Elsevier Press, New York.

Wieling, W., van Lieshout, J. J., and van Leeuwen, A. M. (1993). Physical manoeuvres that reduce postural hypotension in autonomic failure. *Clin. Auton. Res.* **3**, 57–65.

Wright, R. A., Kaufmann, H. C., Perera, R. *et al.* (1998). A double-blind, dose-response study of midodrine in neurogenic orthostatic hypotension. *Neurology* **51**, 120–124.

CHAPTER 106

# Neurological and General Side Effects of Drug Therapy

H. Christoph Diener and Oliver Kastrup

In this chapter, the most important side effects of drugs used in neurology are presented. Their knowledge is important, because serious side effects are the cause of 2%–4% of all admissions to hospitals (Dikkey and Morrow, 1990). Frequent and familiar side effects are listed (**f**). Most emphasis is put on the peripheral and central nervous system, as well as muscular side effects (**n**). Absolute contraindications (**c**) and the most important interactions with other drugs (**i**) are listed, as well as reversibility (**r**). Side effects of drugs that are listed in other chapters of this book (e.g., antiepileptics) will not be presented here.

## NEUROLEPTICS

Neuroleptics have sedative and antipsychotic properties. They block dopamine receptors in the mesolimbic and extrapyramidal motor systems (except clozapine and fluperlapine) and the hypothalamus. Side effects are related to antidopaminergic potency. Further side effects are caused by antimuscarinergic, antinicotinergic, antiserotoninergic, antihistaminergic, and alpha-blocking effects of neuroleptics.

### Frequent and Familiar Side Effects

Because of the blockade of alpha receptors, neuroleptics cause orthostatic arterial hypotension. Cardiotoxic effects are due to disturbances of repolarization. In patients with pre-existing cardiac diseases, phenothiazines may cause atrioventricular (AV) block and tachycardias, including ventricular fibrillation. Additional frequent side effects include dry mouth, constipation, weight gain, urinary retention in patients with prostate hypertrophy, disturbances of accomodation, and narrow-angle glaucoma. The following skin reactions may occur: rash, photodermatitis, erythema. Rare side effects are respiratory failure (in pre-existing respiratory insufficiency), disturbances of temperature regulation with hyperthermia, hyperprolactinemia with galactorrhea, and impotence or amenorrhea. Agranulocytosis and liver damage are extremely rare. An isolated mild leukopenia may occur in up to 10% of patients treated. The risk of venous embolism is increased with typical neuroleptics, not with atypical new substances.

### Peripheral, Central Nervous System, and Muscular Side Effects

All typical neuroleptics may cause movement disorders. The frequency of extrapyramidal side effects increases with antidopaminergic potency (see Table I). The following side effects may occur:

1. Parkinsonism (4%–30%) with hypokinesia, rigidity, axial apraxia, and tremor.
2. Acute dystonic reactions (2%–10%) that appear within the first 24 hours of therapy with tonic cramps of tongue and throat muscles, laryngeal dystonia, rarely facial dyskinesias, oculogyric crisis, retrocollis, torsiondystonia, and choreiform movements. Acute dystonic reactions respond to treatment with anticholinergic drugs such as biperiden IV; younger male patients are most prone to acute dystonia.
3. Akathisia (7%–10%): restlessness of the legs with inability to stand or sit still. Anticholinergics are little helpful, dose reduction or switching to an atypical neuroleptic is often necessary.

The preceding side effects cease (>95%) when the patient stops taking neuroleptics.

4. Tardive dyskinesia may occur after use of neuroleptics over several years. Hyperkinesias predominate in orofacial muscles with movements of the tongue, chewing, and smacking. Furthermore, choreiform, ballistic, and athetoid movements occur, as well as akathisia and tardive dystonia, which can also involve the respiratory muscles. Tardive dyskinesia is more common and develops earlier in patients with pre-existing cerebral dysfunction (prenatal and traumatic brain damage, age) who are taking highly potent neuroleptics. The symptoms increase when the patient abruptly stops taking the drugs. Therapy consists of reduction and cessation of neuroleptics. Switching to clozapine or possibly other atypical neuroleptics is often necessary. Adjunct therapy with clonazepam can be helpful. Trials with high-dose vitamin E (1600 IU daily for 8 weeks) and sulpiride, up to 600 mg daily, have been repoprted with variable effects.

Clozapine has no extrapyramidal motor side effects and can be used to treat patients with pre-existing Parkinsons disease. The risk of agranulocytosis with this agent has to be considered, weekly blood counts are mandatory. Clozapine lowers the epileptic threshold and can induce seizures in predisposed patients. Severe EEG changes with profuse or rhythmic delta activity are seen also without clinical seizures. Other atypical neuroleptics are apparently free of these effects.

Centrally acting neuroleptics reduce seizure threshold and may cause grand mal seizures (<1%), especially at onset of therapy or when the dose is increased. The effect of diphenylhydantoin is antagonized. Less potent neuroleptics may cause depression. This also applies to depot-neuroleptics that are given intramuscularly as tranquilizers.

The neuroleptic malignant syndrome, a very rare but severe side effect, is characterized by hyperthermia, muscular rigidity, tachycardia, incontinence, and altered consciousness, including coma. Creatine phosphokinase (CPK) is elevated. Mortality is >50% in the untreated patient. Dantrolene (4–10 mg/kg/day), bromocriptine or amantadine (IV) are recommended for therapy; in refractory cases electroconvulsive therapy (ECT) is often used. Neuroleptics have to be discontinued.

With use of highly potent neuroleptics like haloperidol cases of sudden and unexplained death have been reported (cardiac arrest? apnea?).

## Contraindications

Contraindications are parkinsonism, epilepsy, closed-angle glaucoma, prostatic hypertrophy, bone marrow suppression, myasthenia gravis. Relative contraindications are arterial hypotension, depression, severely impaired liver, kidney, cardiovascular, or cerebrovascular function.

## Interactions with Other Drugs

Neuroleptics may potentiate effects of antihypertensive drugs, antihistamines, and sedatives. Absorption of neuroleptics is reduced with simultaneous intake of coffee, tea, cola, milk, antacids, and anticholinergics. The pharmacological effect may be potentiated by simultaneous use of anticonvulsants, phenylbutazone, lithium, doxycycline, and nicotine.

## Reversibility

Side effects mostly disappear rapidly after cessation of the offending drug. Tardive dyskinesia can persist for longer periods or permanently.

## ANTIDEPRESSANT DRUGS

Tricyclic and tetracyclic antidepressants have similar side effects. They are mixed inhibitors of predominantly serotonin reuptake (e.g., imipramine) or norepinephrine reuptake (e.g., desipramine).

The tricyclic and tetracyclic antidepressants mentioned here are frequently used in neurology: amitriptyline, doxepin, imipramine, clomipramine, trimipramine, and maprotiline. Newer antidepressants with different structures can cause different symptoms. At initiation of therapy, postural hypotension, sinus tachycardia, arrhythmias in patients with pre-existing heart disease and AV block can occur with tricyclic antidepressants. Frequent side effects of sedating antidepressants are sedation and reduced level of consciousness, including coma in cases of intoxication. Activating antidepressants (desipramine and others) can produce insomnia and agitation. Constipation, reduced gastrointestinal motility, weight gain, dry mouth, sweating, blurred vision, urinary hesitancy (especially in patients with prostate hypertrophy), impotence, and increased intraocular pressure are common complaints. Rare side effects are thrombocytopenia, agranulocytosis and development of cholestatic jaundice.

Special side effects of other antidepressants are as follows: viloxazine may trigger migraine attacks. Mianserine may cause agranulocytosis in extremely rare cases. Standard inhibitors of monamine oxidase inhibitors (phenelzine, tranylcypromine) should be restricted in use because of their marked side effects (hypertensive crisis, hepatocellular damage, postural hypotension, dry mouth, constipation, weight gain). More recently developed selective MAO-A inhibitors (e.g., moclobemide) have fewer side effects and do not require a special diet. The main problems are insomnia, agitation, and paresthesias. Selective serotonin reuptake inhibitors (SSRIs) (e.g., fluoxetine, fluvoxamine, citalopram, sertraline, paroxetine) can cause nausea, diarrhea, nervousness, insomnia, anxiety, headache, tremor, drowsiness, and sweating in lessening degrees from fluoxetine to paroxetine.

## Peripheral, Central Nervous System, and Muscular Side Effects

Antidepressants may alter cerebral activity leading to sedation (amitriptyline) or result in increased activity with restlessness, agitation, and disturbances of sleep (imipramine). In the elderly, disorientation and delirium may occur, in most cases preceded by nightmares and hallucinations (amitriptyline, clomipramine). In predisposed individuals (age and/or brain damage) the anticholinergic properties of the drug can lead to the central anticholinergic intoxication syndrome (CAS) with delirium and vegetative changes. The combination of several anticholinergic drugs should be avoided, because this increases the risk of CAS. Antidepressants reduce seizure theshold. At initiation of therapy, paresthesias and headache may occur (5-HT reuptake inhibitors).

With long-term use and high doses, a postural tremor is not infrequent; this necessitates dose reduction. Should this not be possible, β-blockers (propranolol) can be helpful in small doses. Tricyclics and notably tetracyclics (maprotiline) lower the epileptic threshold, can lead to seizures, and produce EEG changes. Extrapyramidal side effects have rarely been reported with fluoxetine and with overdosage of tricyclics. Newer antidepressants like the noradrenalin serotonin reuptake inhibitor (NSRI) venlafaxin can lead to agitation, dizziness, and visual disturbances; noradrenergic and specifically serotonergic antidepressants (NaSSA) like mirtazapin have similar side effects but can lead to marked sedation, like the new substance nefazodone. Reports of hepatotoxicity may restrict its use. The first new selective noradrenalin reuptake inhibitor (selective NARI) reboxetine seems to be devoid of CNS toxicity except for a mild tremor.

## Contraindications

Contraindications are epilepsy, acute schizophrenia, delirium, congestive heart failure, cardiac arrhythmias, severe liver and kidney disease.

## Interactions with Other Drugs

Combination with MAO inhibitors, SSRIs, sedatives, corticosteroids, and potent neuroleptics should be avoided with tricyclic antidepressants. The effects of antidepressants are potentiated by the simultaneous use of alcohol, neuroleptics, hormonal contraceptives, and cimetidine. Their effects are inhibited by phenobarbital and smoking. A combination of antidepressants with antiparkinson drugs amplifies anticholinergic effects in the bladder, eye, and CNS. Thyroid hormones potentiate the effects of antidepressants. With MAO inhibitors: amphetamines, ephedrine, levodopa, reserpine and 5-HT reuptake inhibitors may lead to hypertensive crisis. Amine-containing food (e.g., broad beans, cheese, chicken liver, chocolate, packaged soups) should be avoided.

**L-tryptophan** was used as a sedative and mild antidepressant and for the treatment of degenerative cerebellar diseases. This substance may cause eosinophilia myalgia syndrome, characterized by muscle pain, muscular atrophy, paresis, shortness of breath, skin rash, abdominal pain, and eosinophilia. Its license was suspended. Similar side effects have been reported for 5-hydroxytryptophan.

## Reversibility

Complete remission of symptoms is the rule after discontinuation or dose reduction.

### Lithium Salts

Lithium can lead to marked cardiotoxicity and CNS toxicity; combination with NSAIDs can lead to renal damage. Findings can include drowsiness with high blood levels and coma with lethal course with toxic levels. Seizures occur; marked EEG-changes can be apparent even with normal blood levels.

### Benzodiazepines

Long-term use leads to sedation and impaired cognition and concentration, mostly in elderly patients who also have an increase in muscle hypotonia and falls.

## SIMPLE ANALGESICS

Combination drugs should not be used, because side effects can be potentiated without a significant increase in efficacy. Derivatives of pyrazolone (metamizole, propyphenazone, and phenazone) should be avoided, because they may induce headache and agranulocytosis. Aminophenazone has been prohibited in most countries, because it is metabolized to carcinogenic nitrosamines. The combination of several analgesics, or analgesics with caffeine, codeine, barbiturates, or tranquilizers enhances the risk of addiction and may lead to drug-induced headache (see Chapter 6).

## Acetylsalicylic Acid (ASA)

### Frequent and Familiar Side Effects

In 0.2% of all patients and 20% of asthmatic patients, acetylsalicylic acid (ASA) may cause bronchospasm (asthma), urticaria, angioedema, and vasomotor rhinitis. In diabetic patients, acetylsalicylic acid reduces blood glucose levels. ASA inhibits platelet aggregation and thus increases bleeding time (a positive effect for the prophylaxis of stroke and myocardial infarction). In 10% of patients with chronic high-dose intake (>1000 mg/day) occult blood in the feces is found. Other frequent side effects are dyspepsia, nausea, and chronic gastric ulceration. Toxic doses of acetylsalicylic acid lead to lethargy, confusion, liver damage, nausea, and vomiting.

### Peripheral, Central Nervous System, and Muscular Side Effects

Tinnitus, reduced hearing and dizziness are fully reversible after the patient stops taking acetylsalicylic acid. The frequency of these side effects for low doses (50–100 mg) is not known. Seizures and confusion result from overdosage. ASA may lead to headache in up to 3% of patients with pre-existing headache (e.g., migraine).

### Contraindications

Absolute contraindications are coagulopathies, treatment with anticoagulants, gastrointestinal ulcers, bronchial asthma, and the first two trimesters of pregnancy. Viral infections with pyrexia and dehydration in children should not be treated with acetylsalicylic acid or its derivatives, because Reye's syndrome may result (vomiting, hypoglycemia, hepatitis, altered consciousness, and coma). Relative contraindications are liver and renal insufficiency and third trimester of pregnancy.

### Interactions with Other Drugs

The risk of gastric bleeding in chronic use is 2.5% and is increased in chronic alcoholics taking acetylsalicylic acid. ASA leads to increased activity of methotrexate, sulfonamides, co-trimoxazole, and sodium valproate. ASA diminishes the antihypertensive effects of captopril.

## Paracetamol (Acetaminophen)

Paracetamol is the least toxic analgesic with the fewest side effects.

### Frequent and Familiar Side Effects

Skin rash and bronchospasm are rare. Liver damage, including hepatic failure may result from treatment with very high doses, more than 6 g/day.

### Peripheral, Central Nervous System, and Muscular Side Effects

Central side effects are not known. In very few cases, headache results from long-term treatment with paracetamol in patients with primary headache.

### Contraindications

Liver disease and bronchial asthma are contraindications.

### Interactions with Other Drugs

Combination with interferon (alpha-2a) or vinblastine may lead to liver damage. Concomitant administration with zidovudine may lead to increased toxicity.

### Reversibility

All side effects except acute hepatic failure are reversible.

## DRUGS THAT ARE SAID TO IMPROVE CEREBRAL CIRCULATION OR METABOLISM

*Bencyclane*: nausea, dizziness, tremor, sedation, memory disturbances, reduced blood pressure, angina pectoris.

*Cinnarizine*: numbness, fatigue, apathy, weakness, extrapyramidal side effects.

*Cylandelate*: nausea, gastrointestinal distress, flushing, tachycardia, sweating, dizziness, headache.

*Dihydroergotoxinmesylate*: postural hypotension, nausea, vomiting, headache, weight loss, dizziness, bradycardia, blurred vision, stuffy nose.

*Ginkgo-biloba*: headache, dizziness, reduced blood pressure, skin rash, palpitations, nausea.

*Meclofenoxat*: nausea, vomiting, hypomania.

*Naftidrofuryl oxalate*: nausea, diarrhea, headache, insomnia, liver dysfunction, epileptic seizures, anaphylactic reactions.

*Nimodipine*: reduction of blood pressure, phlebitis, headache, muscle cramps, facial erythema.

*Pentoxyfylline*: headache, numbness, insomnia, postural hypotension, nausea, vomiting, dizziness.

*Piracetam*: tremor, dizziness, insomnia, fear, irrational behavior, hyperirritability, increase of libido.

*Vincamine*: loss of appetite, nausea, diarrhea. Ventricular arrhythmias may occur after parenteral application.

*Xanithol nicotinate*: orthostatic hypotension, flushing, facial erythema, urticaria, loss of appetite, nausea, vomiting, abdominal pain, disturbed liver function, pruritus, increased demand of insulin.

## Contraindications

Contraindications for any of these substances are intracerebral bleeding, recent stroke, diabetes mellitus, renal failure, and hypotension.

## Reversibility

Most side effects are reversible.

## ANTISPASTIC MEDICATION AND MUSCLE RELAXANTS

### Baclofen

#### Frequent and Familiar Side Effects

General side effects are hypotension, nausea, constipation, and incontinence.

#### Peripheral, Central Nervous System, and Muscular Side Effects

Fatigue, depression, hypotonia, headache, dizziness, and increase of paresis are side effects. Rare side effects are disorientation and hypomania and provocation of seizures in epileptics.

#### Contraindications

Contraindications are gastrointestinal ulcer, liver damage, epilepsy, psychiatric disorder.

#### Interactions with Other Drugs

Baclofen potentiates the effects of sedatives and alcohol. Tricyclic antidepressants enhance the effect of baclofen. The combination of tricyclics with baclofen can result in short-term memory impairment.

### Dantrolene Sodium

#### Frequent and Familiar Side Effects

Hepatitis may occur with the use of doses >300 mg/day. Further side effects are loss of appetite, nausea, vomiting, diarrhea, or constipation. Very rare side effects are respiratory depression, seizures, pleuropericardial reaction.

#### Peripheral, Central Nervous System, and Muscular Side Effects

Drowsiness, dizziness, fatigue, weakness, and dysarthria are rarely observed.

#### Contraindications

Acute liver disease is a contraindications.

#### Interactions with Other Drugs

Dantrolene amplifies the effects of sedatives and alcohol.

### Tizanidine

#### Frequent and Familiar Side Effects

Rare side effects are dry mouth, nausea, reduced blood pressure, bradycardia, and increase of liver enzymes.

#### Peripheral, Central Nervous System, and Muscular Side Effects

Fatigue, insomnia, ataxia, disorientation, hallucinations, headache, and muscle pain can occur.

#### Contraindications

Liver or renal insufficiency is a contraindications.

#### Interactions with Other Drugs

CNS effects are potentiated by alcohol.

## Muscle Relaxants

Nondepolarizing muscle relaxants (pancuronium, vecuronium, pipecuronium, atracurium) can induce an acute myopathy with rhabdomyolysis.

# ANTI-PARKINSON DRUGS

## Levodopa Plus Benserazide or Plus Carbidopa

### Frequent and Familiar Side Effects

Up to 50% of patients complain of nausea, vomiting, anorexia, and tachycardia at the beginning of therapy. The gastrointestinal side effects can be minimized by slowly increasing the dose and administration of the peripheral antiemetic domperidone. The following side effects are less frequent: postural hypotension, increase of blood pressure, ventricular arrhythmias in patients with existing cardiac disorders, increase of uric acid, hypokalemia, and glucose intolerance.

### Peripheral, Central Nervous System, and Muscular Side Effects

Depression, confusion, visual or auditory hallucinations, paranoid psychosis and disturbed sleep are mostly seen in elderly patients. After long-term treatment, fluctuations of efficacy with "on-off" effect, wearing off, early morning dyskinesias and hyperkinesia, dyskinesia of the lips and tongue, as well as choreatic and athetoid movements of the extremities. All side effects are less severe and less frequent when the dose is slowly increased. Abrupt cessation of drug intake after years may cause hyperthermia, akinesia, muscle rigidity, tachycardia, tachypnea, and altered consciousness (resembling the neuroleptic malignant syndrome).

### Contraindications

L-Dopa is contraindicated in florid psychosis, recent myocardial infarction, cardiac arrhythmias, severe hyperthyroidism, renal insufficiency, liver cirrhosis. Relative contraindications are arterial hypotension, tachycardia, asthma, and diabetes mellitus.

## Dopamine Agonists (Bromocriptine, Lisuride, Pergolide, Pramipexole, Cabergoline, Ropinirole)

### Frequent and Familiar Side Effects/Peripheral, Central Nervous System, and Muscular Side Effects

Side effects are similar to L-dopa. Further side effects are Raynaud syndrome (bromocriptine), leg cramps,

bradycardia, congestive heart failure, angina pectoris, nasal congestion, moderate increase of liver transaminases, and headache. Notably pramipexole may induce daytime sedation.

### Reversibility

All side effects are reversible.

## Selegiline

### Frequent and Familiar Side Effects

Gastrointestinal side effects include anorexia, nausea, diarrhea, constipation, and postural hypotension. Rare side effects are dyspnea, dizziness, disturbed sleep, and headache.

### Peripheral, Central Nervous System, and Muscular Side Effects

In patients with long-standing Parkinsons disease, treatment with L-dopa or dopamine agonists in combination with selegeline may lead to psychosis, hypomania, depression, restlessness, anxiety, and aggressive behavior. Dyskinesia may worsen (see Chapter 71).

### Interactions with Other Drugs

Selegeline enhances the effects of alcohol, amantadine, and anticholinergics. Combination with the antidepressant fluoxetine should be avoided.

## Amantadine Hydrochloride

### Frequent and Familiar Side Effects

Frequent side effects are nausea, dizziness, headache, ankle edema, dry mouth, and skin rash. Rare side effects are livedo reticularis, urinary retention, ataxia, lethargy, and neutropenia.

### Peripheral, Central Nervous System, and Muscular Side Effects

Psychotic reactions, hallucinations, and confusion are side effects.

## Anticholinergics: Benztropine, Biperiden, Bornaprine, Methixene, Trihexyphenidyl

### Frequent and Familiar Side Effects

Typical side effects are tachycardia; urinary retention; especially with prostate hypertrophy; dryness of the mouth; constipation; nausea; vomiting; disturbed accomodation; mydriasis; photophobia; and provocation of an acute attack of glaucoma.

### Peripheral, Central Nervous System, and Muscular Side Effects

Rare side effects are disturbed short-term memory and concentration, restlessness, disorientation, mostly visual hallucinations, paranoid reactions and, after long-term treatment, memory dysfunction. In the elderly, in patients with pre-existing cerebral damage or during rapid increase of dose, delirium and CAS may develop accompanied by ataxia, dysarthria, and tremor.

### Contraindications

Contraindications are glaucoma and urinary retention.

### Reversibility

All symptoms are fully reversible are discontinuation or lowering the dose.

## ACTH AND CORTICOSTEROIDS

Side effects of ACTH and corticosteroids are similar and therefore are presented together. Long-term treatment with steroids leads to iatrogenic Cushing's syndrome with bilateral atrophy of the adrenal glands. Therefore, the dose should be as low as possible and for long-term treatment should not exceed the so called "cushing-dose" (daily dose, 1–2 mg dexamethasone, 30–40 mg hydrocortisone, 7.5–10 mg prednisone or prednisolone). Dosing should be adjusted to the circadian rhythm, with the highest dose given in the morning hours, a reduced dose at noon, and an intermediate dose at night. This regimen can be changed to a single administration in the morning for minimal suppression of the adrenal glands. During long-term treatment, cortisone should be given in an alternating fashion every other day. Cortisone must not be abruptly stopped but slowly reduced. Very-high doses of prednisone given for 3–5 days (1 g/day) are well tolerated, and treatment can be stopped without tapering the dose. The mineralocorticoid effect is lowest for dexamethasone and betamethasone, markedly higher with prednisone, prednisolone, hydrocortisone, and ACTH. Before long-term treatment with ACTH or cortisone, active tuberculosis has to be excluded. At the start of treatment, routine tests every 4 days include measurement of body weight, blood pressure, temperature, blood count, electrolytes, and blood glucose. Prophylaxis of gastrointestinal ulcer is nec-

essary during long-term treatment, especially when NSAIDs are given additionally or in patients with pre-existing gastric or duodenal ulcers. An osteoporotic fracture of vertebral bodies must be considered when a patient complains of new intense back pain after prolonged steroid treatment. Long-term treatment requires regular ophthalmological evaluation for glaucoma, and cataract. The following side effects can occur.

### Frequent and Familiar Side Effects

Retention of salt and water causes ankle edema and arterial hypertension. Hypokalemia may lead to cardiac arrhythmia. Blood count is characterized by leukocytosis with a relative lymphocytopenia. The increased risk of opportunistic infections is due to the reduced ability of monocytes to phagocytose. The reduction in fibrinolytic activity causes an increased risk of deep vein thrombosis and embolism. So-called steroid-diabetes is due to enhanced gluconeogenesis. Triglyceride and cholesterol levels increase. Pre-existing but still latent gastrointestinal ulcers can be activated. Furthermore, an increase of body weight and appetite occurs. Side effects of the skin include acne, hirsutism, and striae. The catabolic effect of glucocorticoids causes osteoporosis, and thus the risk of spontaneous fractures and aseptic necrosis of the femur and the femoral head (prophylaxis with sodium fluoride, calcium and vitamin D). Pancreatits with increase of serum amylase may rarely occur.

### Peripheral, Central Nervous System, and Muscular Side Effects

Large doses of cortisone can lead to behavioral or personality changes with nervousness, insomnia, euphoria, and mania or depression (up to 10% of patients treated for an acute attack of multiple sclerosis). Rarely, paranoid states are observed. More frequently, restlessness, headache, dizziness, sweating, and insomnia occur. Long-term treatment with high doses can cause confusion, disorientation, apathy, confabulation and slow mentation. Seizure threshold is reduced. Children treated with high doses may have papilledema and increased intracranial pressure develop. Treatment may lead to steroid myopathy, which presents in two distinct forms. The acute steroid myopathy appears relatively quickly about 1 week after administration of high doses of steroids. The effect seems to be dose-dependent; pretreatment with vancuronium and sepsis are predisposing factors. The rapidly ascending proximal and distal weakness can also involve the respiratory muscles. CK is markedly elevated, histological findings show muscle fiber necrosis. Chronic myopathy appears months after daily treatment as progressive proximal weakness and wasting, the EMG is sometimes normal or mildly myopathic, CK is normal. Histological findings show type-2-fiber atrophy. Long-term treatment may induce cataract, glaucoma, papilledema, and (rarely) exophthalmus. A pre-existing physiological or essential tremor may be worsened. A severe withdrawal syndrome, caused by an abrupt cessation of drug intake, is characterized by muscle ache, headache, dizziness, nausea, lethargia, fever, and disturbances of electrolyte balance.

### Contraindications

Absolute contraindications are gastric or duodenal ulcer, severe osteoporosis, acute infections, psychosis, active tuberculosis. With quiescent tuberculosis and vital indications for steroid, treatment may be carried out with simultaneous tuberculostatic therapy. Relative contraindications are diabetes mellitus, increased risk of thrombosis, pregnancy, congestive heart failure, and chronic renal insufficiency. Long-term treatment requires coadministration of antacids or $H_2$-receptor blockers (cimetidine, ranitidine) in patients with a history of ulcer. The intake of sodium should be limited; potassium and vitamin D should be substituted.

### Interactions with Other Drugs

Diphenylhydantoin and phenobarbital reduce the half-life of corticosterids by 50%. In combination with acetylsalicylic acid or NSAIDs the risk of gastrointestinal ulcer or hemorrhage is increased. The interaction of corticosterids and dicumarol requires frequent testing of coagulation parameters.

## IMMUNOSUPPRESSIVE DRUGS

### Azathioprine

#### Frequent and Familiar Side Effects

Rarely, alterations of hemato-, leuco-, or thrombocytogenesis and liver damage (liver cell necrosis, intrahepatic cholestasis, or fatty liver) may occur. On initiation of therapy, regular blood counts are necessary, once per week during the first month, decreasing to twice per month in the second month, and once per month from then on. Liver enzymes should be checked once per month. When an altered blood count is found, the dose should be reduced. If the leukocyte count drops

further, the drug should be temporarily withdrawn. Even minimal doses may lead to gastrointestinal side effects, especially vomiting. An increased rate of lymphomas has been observed in patients after kidney transplantation but not in neurological patients.

#### Peripheral, Central Nervous System, and Muscular Side Effects

Neurotoxicity has not yet been observed.

#### Contraindications

Contraindications are pregnancy, defective hematopoesis, disturbed liver and kidney functions and increased risk of infections.

#### Interactions with Other Drugs

Interaction with allopurinol may cause myelosuppression, hemolysis, and pancreatitis. In case of comedication with allopurinol, the dose of azathioprine has to be reduced to 25%.

### Cyclosporin

#### Frequent and Familiar Side Effects

Side effects are nephrotoxicity, disturbances of liver function, loss of appetite, gum hyperplasia, hypertrichosis, edema, arterial hypertension. Disturbances of renal function are reversible after drug cessation.

#### Peripheral, Central Nervous System, and Muscular Side Effects

Cyclosporin may induce seizures. Further side effects are tremor and burning paresthesias in the extremities. Some patients develop reversible regions of brain edema.

#### Contraindications

Other immunosuppressives, nephrotoxic drugs are contraindications.

#### Interactions with Other Drugs

Combination with certain antibiotics (aminoglycosides, doxycycline) may disturb renal function. Plasma levels of ketoconazole are enhanced, levels of phenytoin, phenobarbital, rifampicin, and isoniazid are reduced. The concentration of cyclosporin is increased by coadministration of erythromycin, amphotericin, ketoconazole, diltiazem, verapamil, and oral contraceptives. Myopathy and rhabdomyolysis can occur a month after beginning of treatment, mostly with normal CK levels. The risk is increased when lovastatin, an antihyperlipemic agent, is coadministered.

### Tacrolimus (FK 506)

This newer immunosuppressant can also show marked neurotoxicity. The most common effects are tremor, headache, insomnia, irritability, confusion, depression, and ataxia. Polyneuropathy, dysarthria and extrapyramidal side effects have been reported.

Both cyclosporin and tacrolimus can lower the epileptic threshold and cause EEG changes. A rarer and severe complication notably of cyclosporin and tacrolimus therapy is the so called toxic posterior leukoencephalopathy. Patients have subacute or acute confusion, seizures, and cortical blindness; MRI shows occipital lesions in the white matter extending into the cortex. Discontinuation of the drug often leads to reversal of the symptoms, yet the mortality and morbidity are high. The pathogenesis of this syndrome, which resembles hypertensive encephalopathy, remains poorly understood. Direct toxic or vascular effects have been discussed.

### Immunoglobulins

High-dose intravenous immunoglobulins can cause headache and aseptic (Mollaret's) meningitis and benign intracranial hypertension.

## ANTIBIOTICS

### Aminoglycoside Antibiotics: Amikacin, Gentamicin, Kanamycin, Sisomicin, Streptomycin, Tobramycin

#### Frequent and Familiar Side Effects

Side effects are contact dermatitis, malabsorption, and kidney damage.

#### Peripheral, Central Nervous System, and Muscular Side Effects

Ototoxicity hearing loss or vestibular damage occurs in 0.4%–16% of treated patients. It is characterized by tinnitus, reduced hearing of high-frequency tones, even deafness. Damage of the vestibular part of the nerve is less frequent and results in dizziness and vestibular ataxia. Electronystagmography demonstrates a reduced

vestibulo-ocular reflex gain and reduced caloric responses. Gentamicin, tobramycin, and streptomycin predominantly damage the vestibular part of the VIIIth nerve, whereas neomycin, amikacin, and kanamycin predominantly affect the acoustic part. An audiogram should be performed before and during treatment with aminoglycosides in patients with renal insufficiency, long-term therapy with high doses, pretreatment with aminoglycosides, simultaneous treatment with diuretics, or age >50 years.

#### Contraindications

Contraindications are pre-existing damage of the VIIIth nerve, myasthenia gravis (neuromuscular blockade), Guillain-Barré syndrome, respiratory insufficiency, and hypokalemia.

### Quinolones

#### Peripheral, Central Nervous System, and Muscular Side Effects

Cinoxacine, norfloxacine, ofloxacine and rosoxacine can induce restlessness, disturbed sleep, hallucinations, convulsions, and headache. Frequent side effects are dizziness, numbness, tinnitus, and nonspecific disturbances of vision.

### Chloramphenicol

#### Frequent and Familiar Side Effects

Aplastic anemia (1/30,000), disturbances of hematopoesis, and local inflammation after oral intake are side effects frequently encounterial.

#### Peripheral, Central Nervous System, and Muscular Side Effects

Visual symptoms can be caused by long-term intake for weeks and months at doses >2 mg/kg. These symptoms include optic neuritis with reduced visual acuity, central scotoma, disturbed red-green vision, and papillitis. Polyneuropathy can rarely occur and is reversible after cessation of drug intake. Local application can be ototoxic.

#### Contraindications

Contraindications are pregnancy, disturbances of hematopoesis, leukopenia, neutropenia, or thrombocytopenia.

## Interactions with Other Drugs

Chloramphenicol increases concentrations of phenytoin and phenobarbital.

## Fosfomycin

### Peripheral, Central Nervous System, and Muscular Side Effects

Headache and disturbed vision occur.

## Macrolides (Erythromycin)

### Peripheral, Central Nervous System, and Muscular Side Effects

Treatment with erythromycin can cause reversible loss of hearing, disturbed vision, and neuromuscular blockade. Rare side effects are confusion, paranoia, fear, and nightmares.

## Nitrofuranes (Nitrofurantoin)

### Frequent and Familiar Side Effects

Side effects are exanthemas, urticaria, fever, cholestasis, pleuritis, and gastrointestinal complaints.

### Peripheral, Central Nervous System, and Muscular Side Effects

Treatment may cause headache, dizziness, and (rarely) predominantly sensory polyneuropathy with paresthesias and, occasionally, distal weakness.

## Beta-Lactam Antibiotics/Carbapenems

The neurotoxic side effects of the different substance groups are significantly different. Some carbapenemes are highly neurotoxic (imipeneme induces seizures, psychosis, and encephalopathy), meropeneme is less toxic. Penicillins are generally less neurotoxic (benzylpenicillin, penicillin G, amoxicillin, ampicillin, phenoxypenicillins, isoxazolyl penicillins, meticillin, sodium nafcillin, ureidopenicillins), and cephalosporins (cefalexine, cefodizim, cefotaxime, cefpodoximproxetile, ceftacidime, ceftriaxone) show only minor neurotoxicity.

### Frequent and Familiar Side Effects

Anaphylaxis, allergic reactions with asthma, urticaria, and anaphylactic shock can occur. Cytotoxic reactions may induce hemolysis, fever, vasculitis, or arthritis. Contact dermatitis or eczema may develop. Sensitization mostly takes place after local application. Accidental intra-arterial application may cause gangrene. After intravascular injection of a depot penicillin, the so-called Hoigné syndrome develops, which is characterized by epileptic seizures, disorientation, and hallucinations.

### Peripheral, Central Nervous System, and Muscular Side Effects

After application of high doses (30–40 Mega units IV or 5–10,000 units intrathecal) or in patients with renal insufficiency, generalized myoclonus, and focal or generalized epileptic seizures (refractory to anticonvulsants) have been observed.

### Contraindications

Allergy to penicillin or procaine. A relative contraindication is epilepsy.

## Polymyxins

### Peripheral, Central Nervous System, and Muscular Side Effects

After parenteral administration, paresthesias, disturbed vision, ataxia, and headache may develop. Sudden respiratory failure caused by neuromuscular blockade is possible.

## Sulfonamides (Sulfamethoxazole, Salazosulfapyridine)

### Frequent and Familiar Side Effects

Rarely, allergic reactions of delayed type with serum sickness, angiitis, skin reactions, hepatitis, and agranulocytosis can occur. Renal damage may be observed. The frequency of exanthems is up to 3%.

### Peripheral, Central Nervous System, and Muscular Side Effects

Neurological side effects are rare (0.1%): desorientation, dizziness, ataxia, tremor, epileptic seizures, and toxic psychosis. Sometimes, polyneuropathy and headache are observed. Single cases of aseptic meningitis have been reported.

### Contraindications

Allergy, renal insufficiency, and last trimester of pregnancy are contraindications.

*Interactions with Other Drugs*

Additive effects of sulfonylurea and sulfonamides may cause hypoglycemia.

## Co-Trimoxazole and Trimethoprim

*Frequent and Familiar Side Effects*

Increased serum creatinine level, mild leukopenia, skin reactions can occur.

*Peripheral, Central Nervous System, and Muscular Side Effects*

Aseptic meningitis very rarely occurs.

## Tetracyclines (Doxycycline, Minocycline, Tetracycline)

*Frequent and Familiar Side Effects*

Predominant side effects are gastrointestinal with nausea, vomiting, and photodermatosis. In children and during pregnancy, deposition of tetracyclines in bones and teeth may occur.

*Peripheral, Central Nervous System, and Muscular Side Effects*

Pseudotumor cerebri (benign intracranial hypertension) may develop, characterized by vomiting, headache, visual disturbances, and papilledema. Minocycline may cause vestibular toxicity (dizziness and ataxia of gait) in up to 5% of patients. Furthermore, nausea and headache occur.

*Contraindications*

Allergy to tetracyclines, drugs containing magnesium in patients with myasthenia gravis, last trimester of pregnancy, and children are contraindications.

## DRUGS USED IN TREATMENT OF TUBERCULOSIS (RESTRICTED TO NEUROTOXIC EFFECTS)

### Ethambutol

At doses >25 mg/kg/day, damage of the optic nerve or retina with reduced vision, central scotoma, and disturbed red-green vision may develop. Therefore, neurological and ophthalmological examinations should be performed before and during therapy. Rare side effects are polyneuropathy, paresthesias, headache, and dizziness.

### Isoniazid

One percent to 2% of affected patients have polyneuropathy with distal paresthesias, burning feet, and rarely motor deficits of small hand muscles. Disturbances of accommodation result from involvement of autonomic nerve fibers. Prophylaxis with pyridoxine (vitamin $B_6$) is possible. More than 20 mg/kg/day of isoniazid may cause central side effects like headache, dizziness, ataxia, seizures, and paranoid psychosis. Isoniazid amplifies the effects of diphenylhydantoin and acetylsalicylic acid. Isoniazid should not be given with MAO inhibitors.

### Rifampicin

Rare side effects are headache, dizziness, drowsiness, weakness, disorientation, and reduced hearing. An increase in the frequency of leg vein thrombosis has been observed.

### Pyrazinamide

Rare side effects are polyneuropathies, headache, and photophobia.

### Thioamides

Side effects are polyneuropathies (substitution of vitamin $B_6$), depression, paranoia, and dizziness. Very rare side effects include diplopia, headache, and tremor.

## ANTIMYCOTICS (RESTRICTED TO NEUROTOXIC EFFECTS) AND ANTIVIRAL MEDICATION

### Amphotericin B

Initially chills, vomiting, headache, and muscle and joint pain can occur. After intrathecal or intraventricular administration, paresthesias, rarely arachnoiditis with consequent myelopathy and paraplegia have been observed. Myalgia and muscle weakness can occur as a result of hypocalcemia.

## Flucytosine

Rare side effects are polyneuropathy, disorientation, and headache.

## Griseofulvin

Treatment with griseofulvin may cause headache (50%), depression (15%), fatigue, dizziness, insomnia, and disorientation. Rarely ataxia and disturbed vision occur. Side effects are potentiated by alcohol.

## Miconazole

Increased ADH secretion may lead to hyponatremia with disorientation and coma. Fear, euphoria, hyperesthesias, and disturbed vision are less frequent.

## Clotrimazole

Treatment with clotrimazole frequently causes fatigue, numbness, and depression. Rare side effects are desorientation and hallucinations.

## Ketoconazole

Most important side effects are headache, numbness, and, very rarely, somnolence, and restlessness.

## Antiviral Medication

Acyclovir, famciclovir, foscarnet-sodium, and ganciclovir can all induce psychosis and encephalopathy ranging from irritability and mania to depression, seizures, and coma. Antiretroviral medication (didanosin, stavudin, zalcitabin) can lead to polyneuropathy; zidovudin can induce myopathy and myalgia. CK is normal, histological findings show signs of mitochondrial dysfunction. Discontinuation leads to regression of myopathy. Headache is common with all drugs.

# CYTOSTATICS

Neurotoxic side effects of systemic cytostatics are rare. Most of these substances do not penetrate the blood–brain barrier. The following drugs are suitable for intrathecal administration into the CSF: methotrex- ate, cytosine-arabinoside, and thiotepa. Substances used for the treatment of brain tumors and those leading to specific neurotoxicity are discussed.

## Alkylating Agents

Neurotoxicity has not been observed for melphalane or busulfane. Predominant side effects of cyclophosphamide are immunodepression, bone marrow depression, and azoospermia. Headache is rarely observed. Chlorambucil may trigger epileptic seizures. Trofosfamide and Ifosfamide may cause disorientation and psychosis. After intrathecal administration of thiotepa, myelopathy has been observed.

## Antimetabolites

Methotrexate is not neurotoxic when applied in standard doses. After high-dose IV administration, stroke with hemiparesis, aphasia, and partial epileptic seizures has been observed. Aseptic meningitis is a rare side effect after intrathecal administration.

Paraplegia associated with transverse lesions of the spinal cord and epileptic seizures are very rare side effects. In association with radiation, subacute leukencephalopathy may develop over a few months because of toxic vascular damage. It is characterized by disorientation, dementia, bulbar paralysis, tetraparesis, tremor, and ataxia. The risk increases with the cumulative dose; 160 mg should not be exceeded. Further side effects are nausea, vomiting, stomatitis, diarrhea, leukopenia, and alopecia. The combination of methotrexate and ketoprofene may be lethal.

5-Fluorouracil sometimes causes cerebellar deficits. After high-dose therapy, encephalopathy and coma, parkinsonism, and disturbed vision have been observed.

Cytarabine rarely causes polyneuropathy. After intrathecal administration, transverse myelopathy and optic nerve atrophy have been observed. Patients older than 50 are at increased risk of polyneuropathy and cerebellar dysfunction.

5-Azacytidine frequently causes weakness and gait disturbances. Up to 50% of patients have lethargy and desorientation develop.

No severe neurotoxic effects have been described for 6-mercaptopurine and thioguanine.

## Alkaloids

Vincristine may cause nausea, vomiting, loss of hair, leukopenia, and increased uric acid levels. Axonal neuropathy with sensory deficits is common after a cumulative dose of 4 mg. Autonomic polyneuropathy may result in ileus, constipation, impotence, urinary retention, and orthostatic hypotension. Cranial nerves are rarely affected, producing ptosis and diplopia. A proximal myopathy may develop. Epileptic seizures are very rare and may be due to hyponatremia.

Vinblastine causes depression or polyneuropathy.

Vindesine has side effects similar to those of vincristine but predominantly causes proximal weakness.

### Podophyllotoxins

Treatment with VM26 and VP16 occasionally causes polyneuropathy. Neurotoxic effects are rare.

### Cytostatic Antibiotics

Doxorubicin and dactinomycin are not neurotoxic. Bleomycin may cause polyneuropathy and depression. Taxoles can cause severe neuropathy.

### Nitrosourea

BCNU, CCNU, ACNU, and meCCNU may lead to nausea, vomiting, bone marrow suppression with marked thrombocytopenia, progessive pulmonary fibrosis, requiring monitoring of chest x-ray and liver function tests for BCNU. These substances are hepatotoxic, nephrotoxic, and teratogenic. Furthermore, they may induce secondary tumors. Severe complications were observed after injection into the carotid artery: stroke, coma, and death. Multifocal demyelination in the CNS was observed after treatment with very high doses in bone marrow transplant patients.

### Other Cytostatics

Cisplatin is ototoxic, producing tinnitus and hearing loss. Seizures, diffuse encephalopathy with ataxia, and polyneuropathy may develop. The tractus gracilis and cuneatus are very rarely affected; optic neuritis is very rare. In up to 15% of patients treated with L-asparaginase, encephalopathy with disturbed short- and long-term memory, disorientation, or delirium may develop. Seizures rarely occur. Procarbazine causes nausea, vomiting, bone marrow suppression, and leukopenia. More frequently, polyneuropathy occurs. In 10% disorientation, mania, or depression develop.

### Adjunct Cancer Therapy

Granulocyte-macrophage–stimulating factors like filgrastim and molgramostim have potential neurotoxic side effects and can cause seizures and encephalopathy; this is reversible with discontinuation.

## CARDIOVASCULAR DRUGS

### Glycosides

Digitoxin and digoxin have a small therapeutic window and a high risk of toxicity.

#### Frequent and Familiar Side Effects

AV block, extrasystole, loss of appetite, and nausea may occur.

#### Peripheral, Central Nervous System, and Muscular Side Effects

Side effects during treatment with standard doses are photophobia, visual disturbances with flickering and flashing lights, disturbed red-green vision, and scotomas. Intoxication causes drowsiness, disorientation, headache, hallucinations, psychosis, seizures and, ultimately, stupor and coma. A delirium may be the first sign of digitalis overdose.

#### Contraindications

AV block is an absolute contraindication. Relative contraindications are renal insufficiency, hypokalemia, and hypocalcemia.

#### Interactions with Other Drugs

Phenytoin and phenobarbital inhibit absorption of glycosides and reduce their effect. Intravenous injection of calcium increases toxicity of glycosides.

### Coronary Vasodilators

#### Nitrates

Glyceryl nitrate and isosorbide dinitrate cause diffuse headache and may worsen pre-existing migraine, especially during the first 2 weeks of treatment. Numbness, dizziness, and orthostatic hypotension may occur.

## Calcium Antagonists

### Frequent and Familiar Side Effects

Calcium antagonists like nifedipine, nimodipine and nimodipine can cause arterial hypotension, constipation, tremor, restlessness, ankle edema, muscle pain, and rarely tachycardia and thoracic pain. Verapamil and diltiazem can cause tremor and bradycardia.

### Peripheral, Central Nervous System, and Muscular Side Effects

Neurological side effects are fatigue, disturbed sleep, headache (especially nifedipine), dizziness, and depression. Treatment with flunarizine may cause reversible parkinsonism; nifedipine may cause dystonia. Cases of myoclonic dystonia has been reported after verapamil therapy; this substance can also deteriorate muscle weakness in myasthenia gravis and Lambert-Eaton syndrome.

### Contraindications

Contraindications are AV block and hypotension. During treatment with beta-blockers, calcium antagonists should not be administrated IV, because they may result in cardiac arrest.

### Interactions with Other Drugs

Verapamil and diltiazem reduce the catabolism of carbamazepine. Cimetidine enhances the effects of nifedipine and diltiazem.

## Beta-Blockers

### Propranolol (Nonselective); Metoprolol (Beta₁-Selective)

**Frequent and Familiar Side Effects.** Bradycardia, bronchospasm, constipation, diarrhea, and impotence are rare side effects. In diabetics, hypoglycemia may occur. Abrupt cessation of drug intake after long-term treatment causes tachycardia and occasionally angina pectoris.

**Peripheral, Central Nervous System, and Muscular Side Effects.** Diplopia, fatigue, disturbed sleep, and increased appetite may occur. Disorientation in the elderly, deterioration of pre-existing migraine, and depression are very rare side effects. Single cases of reversible carpal tunnel syndrome and myopathy with elevated CK have been reported under propranolol and metoprolol.

**Contraindications.** Congestive Heart Disease, AV block and bronchial asthma are contraindications.

**Interactions with Other Drugs.** The effect of insulin and oral hypoglycemics is enhanced. Smoking may cause a paradoxical increase of blood pressure.

## Diuretics

Disturbances of electrolyte balance may be caused by any diuretic. Hyponatremia (<2.4 mmol/L) causes muscular weakness and lethargy. With further decreasing sodium concentration, asterixis, myoclonus, disorientation, and decreasing level of consciousness develop. Hypokalemia causes progressive weakness of proximal muscles, paralytic ileus, polyuria, vomiting, loss of consciousness and coma, seizures, and arrhythmias.

### Loop Diuretics, Furosemide

**Frequent and Familiar Side Effects.** Diuresis may cause marked disturbances of postural blood pressure regulation. Frequent side effects are nausea and vomiting. Thrombocytopenia is very rare.

**Peripheral, Central Nervous System, and Muscular Side Effects.** Disturbed vision and disorientation are sometimes observed. The substances may cause irreversible ototoxicity.

**Interactions with Other Drugs.** Combination of furosemide and aminoglycoside antibiotics increases the risk of ototoxic damage.

. . .

Treatment with thiazide diuretics rarely causes neurotoxic effects such as disturbed vision. Potassium-sparing diuretics such as spironolactone, an aldosterone antagonist, may cause hyperkalemia with fatigue and disorientation. Further side effects are nausea and vomiting, gynecomastia, and impotence.

## ACE Inhibitors

### Frequent and Familiar Side Effects

Renal failure, potassium retention, cough, and angioedema may occur.

### Peripheral, Central Nervous System, and Muscular Side Effects

Depression, polyneuropathy, taste disturbances, and polymyalgia (rare) are sometimes seem.

## Clonidine

### Peripheral, Central Nervous System, and Muscular Side Effects

Sedation, lethargy, fatigue, insomnia, vivid dreams, hallucinations may occur.

## Alpha-Blockers (Doxazosin, Prazosin, Terazosin)

### Peripheral, Central Nervous System, and Muscular Side Effects

Drowsiness, fatigue, depression, mood changes, headache. Prazosine can excacerbate pre-existing narcolepsy.

## Antiarrhythmics

Disopyramide and flecainide can cause neuropathy, dizziness, and visual disturbances. Mexiletine and lidocaine can cause tremor, ataxia, diplopia, and encephalopathy with visual hallucinations and seizures. Propafenone can induce myoclonus and muscle fatigue. Chinidine causes delirium, headache, and visual disturbances in case of overdosage. Amiodarone can cause a vacuolar myopathy, which is reversible after discontinuation. Tremor, ataxia, and neuropathy are also common.

## Hypolipidemic Drugs

Fibrates (clofibrate, to a lesser extent bezafibrate, fenofibrate, and gemfibrozil) cause rhabdomyolysis from painful myopathy. Patients with renal insufficiency are most at risk, CK is elevated and often returns to normal after discontinuation before clinical improvement is seen.

Also newer lipid-lowering drugs like statins (HMG-CoA reductase inhibitors) cause the same type of myopathy, mostly after lovastatin and simvastatin, less frequently with pravastatin. Reversibility seems to be better than after fibrates.

Because hypothyroidism seems to be a contributing factor, screening thyroid function before institution of therapy seems advisable.

## NONSTEROIDAL ANTI-INFLAMMATORY DRUGS (NSAIDS)

### Frequent and Familiar Side Effects

All NSAIDs potentially reduce platelet aggregation and coagulation. The following gastrointestinal symptoms may occur: heartburn, nausea, and gastric ulcer with frequent lack of pain. Both prescription and non-prescription NSAIDs cause ulcers. In patients with pre-existing renal disease, renal insufficiency, hyperkalemia, and edema may develop. Asthma exacerbations may occur. Modern COX-2 inhibitors have fewer gastrointestinal side effects but are not completely without risk of ulcers.

### Peripheral, Central Nervous System, and Muscular Side Effects

All NSAIDs can cause or worsen pre-existing headache. In the elderly, hallucinations and psychotic reactions can occur. Indomethacin may cause paresthesias. Irreversible retinopathy and renal damage may result from long-term treatment with indomethacin. Dizziness and nausea are rare side effects.

### Contraindications

Gastrointestinal ulcers, gastritis, asthma, reduced renal function and hypertension are contraindications.

### Interactions with Other Drugs

NSAIDs enhance serum levels of phenytoin and methotrexate. Simultaneous intake of lithium may rapidly lead to lithium intoxication. NSAIDs may antagonize beta-blockers, thiazides, ACE inhibitors, and vasodilators.

## ANTIHISTAMINICS (RESTRICTED TO NEUROLOGICAL SIDE EFFECTS)

### Peripheral, Central Nervous System, and Muscular Side Effects

All antihistaminics initially act as sedatives, usually only for several days. Loratadine is said to be less sedating. High doses in the elderly cause paradoxical reactions with disturbed sleep, hallucinations, and delirium. Seizures are rare. Extrapyramidal side effects are very rare: orofacial dyskinesias, blepharospasm, and throat spasms. Antihistaminics may exert an anticholinergic effect causing tachycardia, dry mouth, constipation, urinary retention, disturbed accomodation, and impotence.

## Interactions with Other Drugs

Antihistaminics potentiate the effect of anticholinergics, antidepressants, neuroleptics, barbiturates, and alcohol.

**TABLE I  Drugs That Reduce Seizure Threshold (in Alphabetical Order, Groups of Drugs Are Underlined)**

| | |
|---|---|
| Amantadine | Lithium |
| Amphotericin B | Local anesthetics |
| Anticholinergics | Lysergide |
| Antihistaminics | Methylphenidate |
| Baclofen | Metronidazole |
| Beta-lactams | Nalidixic acid |
| Caffeine | Naloxone |
| | Narcotics (initially) |
| Chloramphenicol | Naftidrofuryl |
| Clozapine | Neuroleptics |
| Cocaine | Oxytocin |
| Cyclosporine A | Penicillamine |
| Dantrolene | Penicillin |
| Didanosine | Pentazocine |
| Disopyramide | Pethidine |
| Fenfluramine | Praziquantel |
| Fluconazol | Prostaglandines |
| Glucocorticoids | Sympathomimetics |
| Indomethacin | Theophylline |
| Isoniazid | Tricyclic antidepressants |
| | X-ray contrast agents |
| | Zidovudine |

**TABLE II  Drugs with Extrapyramidal Side Effects**

| | | |
|---|---|---|
| Antihistaminics | | Metoclopramide |
| Bromocriptine | Doxorubicin | Ondansetron |
| Carbamazepine | Fenfluramine | Papaverine |
| Chloroquine | Flunarizine | Perhexiline |
| Cimetidine | Fluoxetine | Pethidine |
| Cinnarizine | Fluvoxamin | Phenytoin |
| | Levo-Dopa | Reserpine |
| | Lithium | Selegeline |
| | Methylphenidate | Valproic acid |

Neuroleptics with extrapyramidal side effects.

| Rarely | Frequently | Very frequently |
|---|---|---|
| Chlorprothixene | Chlorpromazine | Benperidol |
| Levomepromazine | | Flupentixole |
| Pipamperone | Triflupromazine | Fluphenazine |
| Perazine | Pimozide | Fluspirilene |
| Promethazine | | Haloperidol |
| | Perphenazine | |
| Sulpiride | | |
| Thioridazine | | |

# ANTICOAGULANTS AND ANTIPLATELET DRUGS

## Coumarin Derivatives (Warfarin, Phenprocumon)

### Frequent and Familiar Side Effects

Bleeding is the most frequent side effect, involving most frequently the gastrointestinal tract and the kidneys (macrohematuria or microhematuria). Skin reactins are rare.

### Peripheral, Central Nervous System, and Muscular Side Effects

Intracerebral hemorrhages (0.5%–5% per year), subdural hematomas, hematomyelia, and epidural spinal hematomas have been observed. Damage of the lum-

**TABLE III  Drugs that May Cause Polyneuropathy (5% of All Polyneuropathies)**

| Drug | Clinical symptoms |
|---|---|
| Allopurinol | S, A |
| Amiodarone | SM, D |
| Chlorambucil | S, M |
| Chloramphenicol | S, D |
| Chloroquine | A, M, P |
| Chlorprothixene | D, M |
| Cisplatin | A, S, D |
| Coumarin | |
| Dapsone | A, M, D, P |
| Didanosine | |
| Disulfiram | A, S, M, D, P (pain) |
| Doxorubicin | S |
| Ergotamine | DM, SM, D |
| Ethambutol | S, M |
| Gentamicin | A, M, D |
| Gold | M, D (pain) |
| Heparin | |
| Hydralazine | S, M |
| Isoniazid | A, S, D, V |
| Laxative abuse | |
| Lithium | A, M, P |
| Metronidazole | A, S (pain) |
| Nalidixinic acid | |
| Nitrofurantoin | A, S, M, D (pain) |
| Penicillamine | |
| Perhexiline | DM, S, M, D, P |
| Phenytoin | D, S |
| Podophyllotoxines | |
| Streptomycin | A, M, D |
| Stavudin | |
| Tetanus vaccine | |
| Vincristine | A, M, D, P, V |
| Vindesine | A, M, D, P, V |
| Zidovudine | |

S, sensory; M, motor; A, axonal; DM, demyelinating; D, distal; P, proximal; V, vegetative/autonomic.

**TABLE IV   Drugs that May Cause Disorientation or Delirium**

| | |
|---|---|
| Amantadine | Indomethacin |
| Amphetamines | Isoniazid |
|   Amfepramone | Levo-Dopa |
|   Fenfluramine | Local anesthetics |
|   Mefenorex | Meprobamate |
|   Phentermine | Methyldopa |
| Antihistamines | Methysergide |
|   Antiarrhythmics | Miconazole |
|   Anticholinergics | Morphine |
|   Atropine | Neuroleptics |
|   Baclofen |   (notably tricyclic |
|   Barbiturates |   phenothiazines and |
|   Beta-blockers |   thioxanthines) |
|   Benzodiazepine withdrawal | Penicillin |
|   Bromine | Pentazocine |
|   Bromocriptine | Pergolide |
|   Captopril | Reserpine |
|   Chinidine | Scopolamine |
|   Chloral hydrate | Spironolactone |
|   Chloroquine | Sympathomimetics |
|   Clonidine |   Etilefrine/norfenefrine |
|   Digitalis |   Synephrine |
|   Dihydralazine |   Theophylline |
|   Disulfiram | Tricyclic antidepressants |
|   Ephedrine | Valproic acid |
|   Glucocorticoids | Vigabatrin |

**TABLE V   Drugs that May Induce Depression**

| | |
|---|---|
| Analgesics | Neurological drugs |
|   Ibuprofen |   Baclofen |
|   Indomethacin |   Bromocriptine |
|   Ketoprofen |   Carbamazepine |
|   Opioids |   Ethosuximide |
|   Pentazocine |   Levo-dopa |
|   Phenylbutazone |   Methysergide |
| Antibiotics |   Phenytoin |
|   Clotrimazole |   Valproic acid |
|   Griseofulvin | Psychiatric drugs |
|   Metronidazole |   Withdrawal of amphetamines |
|   Nitrofurantoin |   Neuroleptics |
|   Nalidixic acid | Sedatives and hypnotics |
|   Sulfonamides |   Barbiturates |
|   Streptomycin |   Benzodiazepines |
| Cytostatics |   Clomethiazole |
|   Azathioprine | Steroids and hormones |
|   Bleomycin |   ACTH |
|   L-asparaginase |   Corticosteroids |
|   Mithramycin |   Oral contraceptives |
|   Vincristine | Diverse drugs |
| Cardiovascular drugs |   Chloroquine |
|   ACE-inhibitors |   Cimetidine |
|   Beta-blockers |   Cyproheptadine |
|   Clonidine |   Diphenoxylate |
|   Guanethidine |   Withdrawal of fenfluramine |
|   Hydralazine |   Mebeverine |
|   Lidocaine |   Meclozine |
|   Methyl-dopa |   Metoclopramide |
|   Procainamide |   Salbutamol |
|   Reserpine | |

bosacral plexus, femoral nerve, or sciatic nerve occur spontaneously or are caused by iatrogenic hematoma of the psoas and gluteus muscles by means of IM injections.

## Contraindications

Contraindications are hemorrhagic diathesis, gastrointestinal ulcer, nephrolithiasis, liver cirrhosis, pulmonary tuberculosis, diabetic retinopathy, acute pancreatitis, hypertension, and chronic alcoholism.

## Interactions with Other Drugs

The effects of derivatives of coumarin are enhanced by baclofen, cimetidine, danazol (treatment of endometriosis), disulfiram, immunosuppressives, indom-

**TABLE VI   Drugs that May Induce or Worsen Pre-existing Headache (F Signifies Frequent Complaints of Headache)**

| | | | |
|---|---|---|---|
| Acetazolamide | F | Interferons | |
| Acetylsalicylic acid | | Isoniazid | |
| Ajmaline | | Lamotrigine | |
| Amantadine | | Meprobamate | |
| Antihistaminics | | Methaqualone | |
| Drugs that reduce | | Methysergide | |
|   appetite | | Metronidazole | |
| Atenolol | F | Morphine and | |
| Barbiturates | F |   derivatives | |
| Benzodiazepines | | Muscle relaxants | |
| Bromocriptine | | Nalidixic acid | F |
| Caffeine | | Nifedipine | F |
| Calcium antagonists | | Nimodipine | |
| Carbamazepine | | Nitrofurantoin | |
| Carbimazol | | Nitrates | F |
| Captopril | F | Nonsteroidal | |
| Chinidine | |   antiinflammatory | |
| Chloroquine | |   drugs | |
| Cimetidine | | Octreotide | F |
| Clofibrate | F | Omeprazole | |
| Codeine | | Ondansetron | |
| Didanosine | | Paroxetin | |
| Dihydralazine | | Pentoxifylline | |
| Dihydroergotamine | | Perhexiline | |
| Dipyridamol | F | Phenazone | |
| Disopyramide | | Primidone | F |
| Disulfiram | | Prostacyclins | |
| Diuretics | | Ranitidine | |
| Ergotamine | | Rifampicin | |
| Estrogens | | Sildenafil | |
| Etofibrate | | Sumatriptan | |
| Ferrum parenteral | | Theophylline and | |
| Gestagens | |   derivatives | |
| Glucocorticoids | | Thiamazole | |
| Glycosides | | Trimethoprim + | |
| Griseofulvin | |   sulfamethoxazole | |
| Guanethidine | | Vitamin A | F |
| Immunoglobulins | F | | |

**TABLE VII    Drugs with Otoxic Effects**

| | |
|---|---|
| Acetylsalicylic acid | Chinine |
| Aminoglycoside antibiotics | Cloroquine |
|   Amikacin | Diuretics |
|   Framicetine |   Ethacrynic acid |
|   Gentamicin |   Furosemide |
|   Kanamycin | Erythromycin |
|   Neomycin | Indomethacin |
|   Paromomycin | Minocycline |
|   Streptomycin | Polypeptide antibiotics |
|   Tobramycin |   Capreomycin |
| Carboplatin/cisplatin |   Vancomycin |

**TABLE VIII    Drugs with Oculotoxicity or That May Lead to Retinopathy**

| | |
|---|---|
| Amiodarone | Indomethacin |
| Anticonvulsants (vigabatrine) | Interferons |
| Chloramphenicol | Nicotinic acid |
| Chloroquine | Neuroleptics |
| Corticosteroids | Penicillamine |
| Deferoxamine | Prostaglandins |
| Digitalis | Retinoids |
| Ethambutol | |

ethacin, macrolide antibiotics, paracetamol, phenothiazines, phenytoin, and salicylates.

The effects are reduced by antacids, barbiturates, carbamazepine, glucocorticoids, dextran, and haloperidol.

**Ticlopidine** can cause neutropenia and rarely thrombocytopenic purpura.

### Frequent and Familiar Side Effects

Diarrhea and rash may occur. Up to 1% of patients have reversible neutropenia develop within the first 3 months of treatment (blood count required every 2 weeks for 12 weeks). Clopidogrel is free of relevant side effects.

## FURTHER SIDE EFFECTS

Drugs that may reduce seizure threshold are listed in Table I. Drugs with extrapyramidal motor side effects can be seen in Table II. Table III shows drugs that may cause polyneuropathy, Table IV shows drugs that may cause disorientation or delirium, Table V shows drugs that may induce depression, and Table VI shows drugs that may induce or worsen pre-existing headache. Tables VII and VIII list drugs with ototoxic effects and drugs with oculotoxicity or that may lead to retinopathy.

Side effects of hypnotics and tranquilizers are mentioned in Chapters 19 and 27, side effects of antiepileptics in Chapter 20, and side effects of antiemetics in Chapter 63.

## REFERENCES

Biller, J. (1998). "Iatrogenic Neurology." Butterworth Heinemann, Woburn, MA.

Brust, J. C. M. (1996). "Neurotoxic Effects of Prescription Drugs." Butterworth Heinemann, Boston, MA.

Dickey, W., and Morrow, J. I. (1990). Drug induced neurological disorders. *Progr. Neurobiol.* 4, 331–342.

Dukes, M. N. G., ed. (2000). "Meyler's Side Effects of Durgs," 14th edn. Elsevier, Amsterdam, Oxford, Princeton.

Rowland, L. (1998). "Current Neurologic Drugs." Lippincott, New York.

Young, L. L., and Koda Kimble, M. A. (1996). "Applied Therapeutics: The Clinical Use of Drugs," 6th ed. Applied Therapeutics Inc., Vancouver.

CHAPTER 107

# Molecular Genetic Diagnosis of Neurological Diseases

Thomas Gasser and N.W. Wood

Recent progress in molecular genetics has greatly improved our understanding of the molecular basis of many inherited neurological diseases. With the Human Genome Project nearing completion, the genomic sequence of a large number of genes which, when mutated, can cause neurological disorders, is now known. This increasing wealth of knowledge has allowed the reclassification of a number of formerly heterogeneous clinical syndromes, opens up novel diagnostic possibilities, and allows characterization of the pathological gene products, thereby providing further insight into the molecular pathogenesis of these disorders. This will eventually lead toward new approaches to therapy and prevention.

The following section briefly outlines the molecular genetic basis of inherited diseases, describes some fundamental methods of genetic analysis, and gives an overview of the present role of molecular diagnosis in neurological diseases (see Table I).

## METHODS OF GENE MAPPING

### Genome

The human genome consists of 23 pairs of chromosomes (22 autosomes plus the sex chromosomes X and Y). One of each pair is inherited from the mother, the other from the father. The backbone of the chromosome consists of a continuous DNA double-strand, approximately 50–200 million base pairs (bp) in length, depending on the size of the chromosome. The entire genome spans about 3 billion bp. The genetic information is contained within roughly 30–40,000 genes (i.e., DNA-sequences coding for a specific protein plus regulatory sequences). The genes themselves have an average length

of several thousand bp each and account for only a small proportion of the entire genomic DNA. They are separated by larger "noncoding" segments, some of which have regulatory functions. There are, however, large regions of the genome for which, at present, no function is known (Figure 1). In a major international collaborative effort, the human genome project (HUGO), the sequence of more than 90% of the human genome, has now been determined. However, it will probably take many more years to unravel its function.

Up-to-date information can be obtained through the internet (*www.ncbi.gov*).

### Mutations

The genetic information contained in the DNA sequence of a gene is translated into the amino-acid sequence of the corresponding protein (gene product). Mutations (i.e., changes of the DNA sequence of a gene) can result in an altered or absent gene product. These alterations may in turn cause disease. If a mutation occurs within the germ line, it will be passed on from generation to generation. DNA sequence variations outside the coding or regulatory regions of a gene usually have no direct influence on cellular function. They are much more common than functionally relevant mutations, occuring on average once in every 500 bp. They are also transmitted from one generation to the next and can serve as valuable "genetic markers."

### Gene Mapping

For most inherited neurological diseases, the primary metabolic or structural defect (i.e., the pathological gene

**TABLE I** Molecular Diagnosis in Neurogenetic Disorders

| Disease | Symbol | Inheritance | Position | Gene product | Mutation | Molecular diagnosis | Reference | Remarks | MIM Number |
|---|---|---|---|---|---|---|---|---|---|
| **Ataxias** | | | | | | | | | |
| Friedreich's ataxia | FRDA | AR | 9q13–21.1 | Frataxin | Trinuc/Pm | A | (Campuzano et al., 1996) | Most common form of recessive ataxia | 229 300 |
| Spinocerebellar ataxia | SCA1 | AD | 6p21.3 | Ataxin 1 | Trinuc | A | (Orr et al., 1993) | SCA1, 2, and 3 comprises approx. 60% of the dominant hereditary spinocerebellar atrophies | 164 400 |
| | SCA2 | AD | 12q23–24.1 | Ataxin 2 | Trinuc | A | (Gispert et al., 1993) | | 183 090 |
| | SCA3/MJD | AD | 14q24 | Ataxin 3 | Trinuc | A | (Kawaguchi et al., 1994) | | 109 150 |
| | SCA6 | AD | 19p13 | Calcium-channel | Trinuc | A | (Zhuchenko et al., 1997) | Allelic to FHM and EA2 | 183 086 |
| | SCA7 | AD | 3p12–21.1 | Ataxin 7 | Trinuc | A | (Benomar et al., 1995) | | 164 500 |
| | SCA12 | AD | 5q31–q33 | Protein phosphatase 2 | Trinuc | A | (Holmes et al., 1999) | | 604 326 |
| Episodic ataxia with myokymia | EA1 | AD | 12p13 | Potassium-channel | Pm | B | (Browne et al., 1994) | | 160 120 |
| Episodic ataxia without myokymia | EA2 | AD | 19p13 | Calcium-channel | Pm | B | (Ophoff et al., 1996) | Allelic to FHM and SCA6 | 108 500 |
| Ataxia with vitamin deficiency | AVED | AR | 8q13.1–13.3 | α-Tocopherol transfer protein | Pm | C | (Ouahchi et al., 1995) | | 277 460 |
| **Movement disorders** | | | | | | | | | |
| Huntington's chorea | HD | AD | 4p16.3 | Huntingtin | Trinuc | A | (Huntington's Disease Collaborative Research Group, 1993) | | 143 100 |
| Wilson's disease | WND | AR | 13q14.1 | Copper transport protein | Pm/Del | B | (Tanzi et al., 1993) | | 277 900 |
| Primary torsion dystonia | DYT1 | AD | 9q34 | Torsin A | GAG-Deletion | A | (Ozelius et al., 1997) | Early onset, generalized, rarely isolated writer's cramp | 128 100 |
| Dopa-responsive dystonia | DYT5, DRD | AD | 14q22 | GTP-cyclohydrolase I | Pm | B/C | (Ichinose et al., 1994) | No mutations found in some cases | 600 225 |
| Dopa-responsive dystonia | DYT5, DRD | AR | 11p15.5 | Tyrosine hydroxylase | Pm | C | (Knappskog et al., 1995) | Individual case reports | 191 290 |
| Myoclonus-dystonia syndrome | DYT11 MDS | AD | 7q21 | Epsilon-sarcoglycan | Pm, Del | C | (Zimprich et al., 2001) | Probably major locus for MDS | 605 408 |
| Dentatorubropallidoluysian atrophy | DRPLA | AD | 12p13.31 | DRPLA protein | Pm | A | (Yazawa et al., 1995) | Rare in Europe | 125 370 |
| Familial Parkinson's disease | PARK1 | AD | 4q21 | Alpha-synuclein | Pm | C | (Polymeropoulos et al., 1997) | Very rare, Mediterranean founder effect | 601 508 |
| Autosomal-recessive juvenile parkinsonism | PARK2, AR-JP | AR | 6q25–27 | Parkin | Del | B | (Kitada et al., 1998) | No Lewy-body pathology, relatively common in juvenile PD | 602 544 |

## Neuromuscular diseases and myopathies

### Spinal muscular atrophy

| Disease | Symbol | Inher. | Locus | Gene/Protein | Mutation | Class | Reference | Comments | MIM |
|---|---|---|---|---|---|---|---|---|---|
| Infantile (Werdnig-Hoffmann) | SMA I | AR | 5q11.2–13 | Survival motoneuron | Del | A | (Lefebvre et al., 1995) | | 253 300 |
| | SMA II | AR | 5q11.2–13 | SMN | Del | | (Roy et al., 1995) | | 253 400 |
| Juvenile (Kugelberg-Welander) | SMA III | AR | 5q11.2–13 | SMN | Del | | (Kausch et al., 1991) | | |
| Adult | SMA IV | (?) | ? | | ? | | | | |
| Bulbospinal muscular atrophy | XBSN | X | Xq13–22 | Androgen receptor | Trinuc | A | (La Spada et al., 1991) | | 313 200 |
| Duchenne | DMD | XL | Xp21.2 | Dystrophin | Del/Dupl/Pm | A | (Koenig et al., 1987) | | 310 200 |
| Becker | BMD | XL | Xp21.2 | Dystrophin | Del/Dupl/Pm | | | | |
| Emery Dreyfuss myopathy | EDMD | XL | Xq28 | Emerin | Del/Ins/Pm | C | (Bione et al., 1994) | | 310 300 |
| | EDMD-AD | AD | 1q11–q23 | Lamin A/C | PM | C | (Bonne et al., 1999) | | 181 350 |
| Myotonic dystrophy (Curschmann's disease) | DM | AD | 19q13.3 | Myotonin | Trinuc | A | (Brook et al., 1992) | Most common inherited myopathy | 160 900 |
| Facioscapulohumeral dystrophy | FSHD | AD | 4qter | Unknown | unknown | A | (Wijmenga et al., 1992) | | 158 900 |

## Muscle diseases caused by defects in ion channels

| Disease | Symbol | Inher. | Locus | Gene/Protein | Mutation | Class | Reference | Comments | MIM |
|---|---|---|---|---|---|---|---|---|---|
| Potassium sensitive myotonia | SCN4A | AD | 17q23 | Sodium channel α-subunit | Pm | B | (Lerche et al., 1993) | | 603 967 |
| Hyperkalemic periodic paralysis | SCN4A | AD | 17q23 | Sodium channel α-subunit | Pm | B | (Fontaine et al., 1990) | | 168 300 |
| Paramyotonia congenita | SCN4A | AD | 17q23 | Sodium channel α-subunit | Pm | B | (McClatchey et al., 1992) | | 168 300 |
| Potassium-aggravated myotonia | SCN4A | AD | 17q23 | Sodium channel α-subunit | Pm | B | (Lerche et al., 1993) | | 168 300 |
| Hypokalemic paresis 1 | CACN1AS | AD | 1q31–32 | Calcium channel | Pm | B | (Jurkat-Rott et al., 1994) | | 600 304 |
| Hypokalemic paresis 2 | SNCA4 | AD | 17q23 | Sodium channel α-subunit | Pm | B | (Jurkat-Rott et al., 2000) | One family | 600 304 |
| Myotonia congenita Thomsen | CLCN1 | AD | 7q35 | Chloride channel | Pm | B | (Koch et al., 1992) | | 160 800 |
| Myotonia congenita Becker | CLCN1 | AR | 7q35 | Chloride channel | Pm, Del, Ins | B | (Koch et al., 1992) | | 255 700 |

## Neuropathies

| Disease | Symbol | Inher. | Locus | Gene/Protein | Mutation | Class | Reference | Comments | MIM |
|---|---|---|---|---|---|---|---|---|---|
| Charcot-Marie-Tooth type Ia | CMT1a | AD | 17p11.2 | PMP-22 | Dupl/Pm | A | (Lupski et al., 1991) | In 70% duplication of a 1.5-Mb fragment | 118 220 |
| Charcot-Marie-Tooth type Ib | CMT1b | AD | 1q22–23 | P₀ | Pm | B | (Hayasaka et al., 1993) | | 118 200 |
| Charcot-Marie-Tooth, X-chromosomal | CMTX | XL | Xq13.1 | Connexin-32 | Pm | B | (Bergoffen et al., 1993) | | 302 800 |
| Tomaculous neuropathy (liability to pressure palsies) | HNPP | AD | 17p11.2 | PMP-22 | Del/Pm | A | (Chance et al., 1993) | Mostly deletion of a 1.5-Mb fragment (complementary to CMT1) | 162 500 |

*continues*

**TABLE I** *continued*

| Disease | Symbol | Inheritance | Position | Gene product | Mutation | Molecular diagnosis | Reference | Remarks | MIM Number |
|---|---|---|---|---|---|---|---|---|---|
| **Inherited tumor syndromes** | | | | | | | | | |
| Neurofibromatosis 1 (v. Recklinghausen) | NF1 | AD | 17q11.2 | Neurofibromin | Del/Pm | B | (Wallace et al., 1991) | | 162 200 |
| Neurofibromatosis 2 | NF2 | AD | 22q12.2 | Merlin | Del/Pm | B | (Trofatter et al., 1993) | | 101 100 |
| von Hippel-Lindau disease | VHL | AD | 3p25 | | Pm/Del | B | (Richards et al., 1993) | | 193 300 |
| Tuberous sclerosis | TSC1 | AD | 9q34 | Hamartin | Pm/Del/Ins | B | (van Slegtenhorst et al., 1997) | | 191 100 |
| | TSC2 | AD | 16p13 | Tuberin | Del | B | (European Chromosome 16 Tuberous Sclerosis Consortium, 1993) | | 191 092 |
| **Dementias** | | | | | | | | | |
| Familial Alzheimer's disease | AD1 | AD | 21q21 | Amyloid precursor protein | Pm | C | (Goate et al., 1991) | Very rare | 104 760 |
| | AD3 | AD | 14q24.3 | Presenilin 1 | Pm | C | (Sherrington et al., 1995) | Most common early-onset AD form | 104 311 |
| | AD4 | AD | 1q31-q42 | Presenilin 2 | Pm | C | (Rogaev et al., 1995) | Very rare | 600 759 |
| Frontotemporal dementia with parkinsonism | FTPD-17 | AD | 17q21 | MAPTAU | Pm | C | (Hutton et al., 1998) | 5%–10% of FTD-cases | 601 630 |
| Familial Creutzfeld-Jakob disease | PRNP | AD | 20pter-p12 | Prion-Protein | Pm/ins | B | (Owen et al., 1989) | 5%–10% of CJD cases | 123 400 |
| Gerstmann-Sträussler-syndrome | PRNP | AD | 20pter-p12 | Prion-Protein | Ins/Pm | B | (Hsiao et al., 1989) | Part of the CJD spectrum | 137 440 |
| Fatal familial insomnia | PRNP | AD | 20pter-p12 | Prion-Protein | Pm | B | (Medori et al., 1992) | Part of the CJD spectrum | 600 072 |
| **Other neurodegenerative disorders** | | | | | | | | | |
| Familial amyotrophic lateral sclerosis | SOD1 | AD | 21q22 | Superoxide dismutase 1 | Pm | C | (Rosen et al., 1993) | 20% of hereditary ALS | 105 400 |
| **Epilepsies** | | | | | | | | | |
| Progressive myoclonic epilepsy of Unverricht-Lundborg type | EPM1 | AR | 21q22.3 | Cystatin B, CSTB | 12-bp repeat expansion, pm, del | A | (Pennacchio et al., 1996) (Lalioti et al., 1997) | The dodecamer repeat expansion accounts for approx. 90% of disease alleles worldwide | 254 800 |
| Lafora's disease | MELF | AR | 6q23-q25 | Laforin | Microdel, pm, ins | B | (Minassian et al., 1998) (Serratosa et al., 1999) | Not all families are linked to chromosome 6q | 254 780 |

| Disease | Inheritance | Locus | Gene/Protein | Mutation | Reference | Avail. | OMIM | Comments |
|---|---|---|---|---|---|---|---|---|
| **Neurovascular disorders** | | | | | | | | |
| CADASIL | AD | 19p13.1 | Notch3 | Pm | (Joutel *et al.*, 1996) | A | 125 310 | 70% of mutations in exons 3 and 4 |
| HCHWA-D — Hereditary cerebral hemorrhage with amyloidosis | AD | 21q21 | Amyloid precursor protein | Pm | (Levy *et al.*, 1990) | C | 104 760 | |
| **Migraine** | | | | | | | | |
| FHM1 — Familial hemiplegic migraine | AD | 19p13 | Calcium channel | Pm | (Ophoff *et al.*, 1996) | C | 141 500 | Allelic to SCA6 and EA2, see Channelopathies |
| **Mitochondrial diseases** | | | | | | | | |
| CPEO — Chronic progressive external ophthalmoplegia | spor | | mtDNA | Del, Dupl, Pm | (Holt *et al.*, 1988) | A | 530 000 | Molecular genetic analysis should be performed in muscle |
| KSS — Kearns-Sayre syndrome | spor | | mtDNA | Del, Dupl, Pm | (Zeviani *et al.*, 1988) | A | 530 000 | Molecular genetic analysis should be performed in muscle |
| MELAS — Mitochondrial encephalomyopathy with lactacidosis and "strokelike episodes" | mat | np 3243 (np 3271) | $^{mt}$-tRNA$^{Leu}$ | Pm | (Goto *et al.*, 1990) | A | 540 000 | Other point mutations have been described in rare cases |
| MERRF — Myoclonus epilepsy with ragged red fibers | mat | np 8344 | $^{mt}$-tRNA$^{Lys}$ | Pm | (Shoffner *et al.*, 1990) | A | 545 000 | Other point mutations have been described in rare cases |
| LHON — Leber's hereditary optic neuropathy | mat | np 11778 np 3460 np 14484 | Complex I of the respiratory chain | Pm | (Wallace *et al.*, 1988) | A | 535 000 | Other point mutations have been described in rare cases |
| NARP — Neurogenic weakness, ataxia, and retinitis pigmentosa | mat | np 8993 | ATPase 6 of the respiratory chain | Pm | (Holt *et al.*, 1990) | A | 551 500 | High percentage of this mutation may lead to MILS |
| MILS — Maternally inherited Leigh syndrome | mat | np 8993 | ATPase 6 of the respiratory chain | Pm | (Tatuch *et al.*, 1992) | A | 516 060 | Lower percentage of this mutation may lead to NARP |

*Note.* This compilation of inherited neurologic disorders is not complete. The selection of the listed diseases was chosen according to practical relevance of molecular genetic diagnosis. The speed in which progress is made in molecular genetic research, however, causes such tables to become outdated very quickly. In case of doubt, it is recommended that current publications or special centers be consulted.

For many, if not for most, of the diseases listed here, mutations in other currently unknown genes may also be responsible.

AD, autosomal dominant; AR, autosomal recessive; X, X-chromosomal; mat, maternal (mitochondrial) transmission; Pm, Point mutation; Del, Deletion; Ins, Insertion; Trinuc, Trinucleotid-repeat expansion. Availability of molecular diagnosis: (A) Routine procedure, commercially available, results usually within 4 weeks; (B) routine procedure, but may be time-consuming and expensive, usually because of occurrence of multiple mutations; results may take several months; (C) usually available only within research setting.

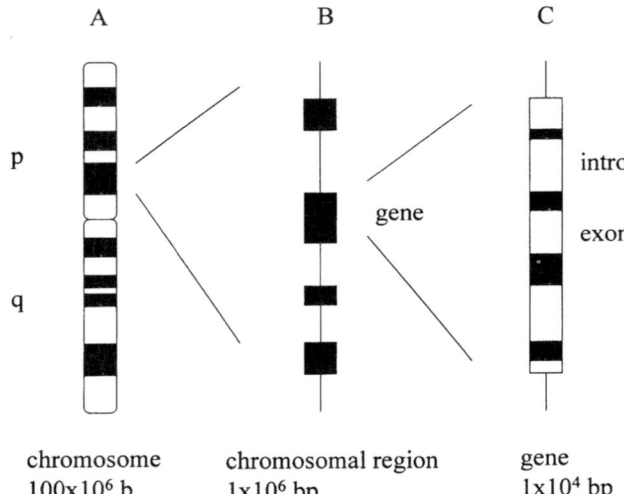

chromosome
100x10$^6$ b

chromosomal region
1x10$^6$ bp

gene
1x10$^4$ bp

**FIGURE 1** Organization of the genetic information. Each chromosome contains a continuous DNA double-strand, approximately 50–200 million basepairs (bp) in length. Genetic linkage analysis allows localization of a gene to a chromosomal region of 1–10 million bp. This region may contain dozens of genes, which themselves contain information-bearing (exons) and intervening (introns) sequences.

product) has escaped detection by biochemical or pathological methods. Most of the disease genes known today have been identified by genetic methods, which allow determination of the chromosomal position of a gene causing an inherited disease without prior knowledge of its gene product. This procedure is called *genetic linkage mapping* and *positional cloning*. Because genetic linkage mapping has been by far the most successful strategy in the analysis of hereditary diseases over the past 20 years and because it is of direct importance for molecular genetic diagnosis, the basic concepts of this method will be briefly outlined in the following sections.

## Linkage and Recombination

During meiosis, the homologous maternal and paternal chromosomes of the stem cell are separated to form the haploid set of chromosomes of the gamete. During this process, corresponding fragments of homologous chromosomes are exchanged (recombination or crossing-over). An average of one to three recombinations occur on each chromosome during meiosis. The recombination breakpoints are, to a large extent, randomly placed along the length of the chromosome. Therefore, genes that are located in proximity to each other on the same chromosome are passed on together to the following generation (i.e., they are genetically linked). On the other hand, if genes are located far apart on a chromosome, they are likely to be separated by recombination events during meiosis

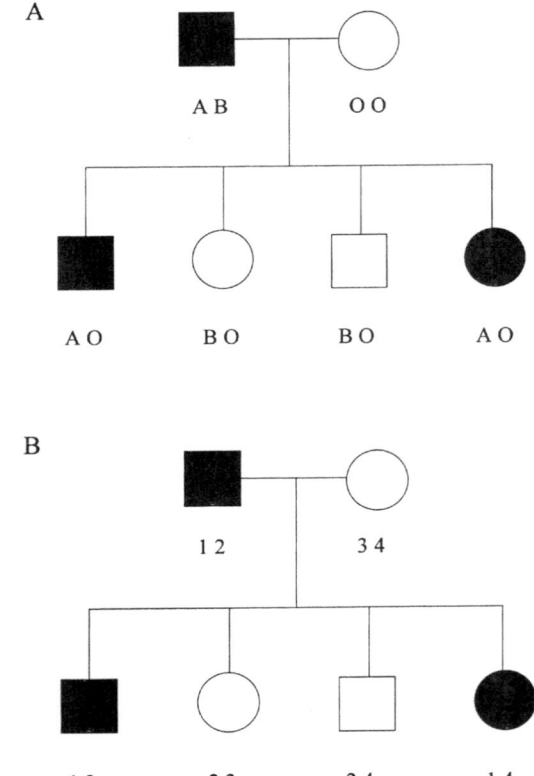

**FIGURE 2** (A) Cosegregation of two inherited phenotypic traits, the ABO-blood group and an inherited disease: all affecteds in this family carry the A-allele. (B) Cosegregation of a DNA marker and an inherited disease: all affecteds in this family carry the marker allele "1."

and are therefore inherited independently of each other. This means that the frequency of observed recombination events between two gene loci is a measure of their distance on a chromosome. On average, a recombination frequency of 1% (1 recombination in 100 observed meioses) corresponds roughly to a physical distance of 1 million bp. If a large number of meioses is studied for different gene loci, the order and distance of these loci can be determined and genetic "maps" can be established.

Figure 2 shows an example for genetic linkage of two inherited traits: tuberous sclerosis and the ABO blood-group system. The genes for both traits are located on the long arm of chromosome 9. All affected family members, indicated by black symbols, have a blood group allele in common (allele A in this example). This cosegregation of two traits in a small family could well be observed by chance alone. However, if it is found in large families or in a large number of small families, it provides statistical evidence that the two genes are located within a relatively small region of the same chromosome. If the chromosomal position of one of these genes is already known, the position of the linked gene locus can be inferred.

## DNA Polymorphisms

Only polymorphic traits (i.e., those that are present in two or more distinguishable forms [alleles]), can be used for linkage mapping. Polymorphic phenotypic traits, which can be detected by biochemical or immunological means and are inherited in a mendelian fashion, such as blood groups or HLA types, are rare. The major contribution of molecular genetics to linkage analysis has been the discovery of polymorphisms that can be distinguished on the DNA level (DNA markers). The most useful polymorphisms are short repetitive DNA motifs, which can be highly variable in length (variable number of tandem repeats = VNTRs or dinucleotide, trinucleotide or tetranucleotide repeats = microsatellites). A large number of these polymorphic DNA sequences has been identified and placed on genetic maps, covering the entire genome. Most of these polymorphisms are situated in noncoding DNA regions (introns) and seem to have no functional significance. The chromosomal segments bearing these DNA markers can be distiguished, and their segregation can be followed through the generations of a family (Figure 2B). In principle, any disease showing mendelian inheritance can be mapped to a chromosomal region by this method, provided that a sufficient number of meioses can be observed.

## Linkage Analysis

Linkage analysis is the statistical method used to determine the recombination frequency, and consequently the genetic distance, between two (or more) genetic loci (genes or DNA markers).

The statistical measure for the reliability of these observations is the lod score (lod = log of the odds). The lod score is calculated as the log of the ratio of the two probabilities that linkage does or does not exist between two gene loci. A lod score of greater than three is generally considered to be positive evidence for linkage, provided that the genetic parameters (mode of inheritance, penetrance, frequency of phenocopies) are known. If this is not the case (as in diseases with presumed polygenic inheritance or multifactorial etiology), lod scores have to be interpreted with caution.

## Positional Cloning

The approach of linkage analysis allows the localization of a gene locus to a segment of about 1–2 million bp. Dozens of genes may be contained within such a region. To identify the gene under consideration, all the genes of this segment may have to be sequenced and searched, base by base, for mutations. After completion, at least in draft form, of the human genome, gene identification is much simpler than it was. There are now a number of search engines available free to anyone who wishes to investiagate a genetic region (e.g., the Human Genome Browser at http://genome.ucsc.edu/goldenPath/septTracks.html). One can now rapidly assimilate a "virtual" contig across the region of interest. This can then be searched for possible candidates, initially on the computer but eventually sequencing in the laboratory is required to identify the pathogenic mutation.

## MOLECULAR GENETIC DIAGNOSIS

In clinical practice, the availability and the limitations of molecular diagnosis depend on our knowledge of the molecular genetic basis of the respective disease but also on the degree of genetic complexity of the disorder under investigation. Some diseases, such as Huntington's disease, are caused by a specific mutation in a single gene (Huntington's Disease Collaborative Research Group, 1993), and routine molecular diagnosis can be provided by a simple polymerase chain reaction (PCR)-based assay. In other cases, however, many different mutations in a given gene may underlie a disorder (allelic heterogeneity). Depending on the size of the gene(s), this may render molecular diagnosis very costly and time-consuming. Molecular diagnosis may be further complicated by the fact that mutations in a number of different genes may cause similar or indistinguishable phenotypes (genetic heterogeneity). For example, tuberose sclerosis may be caused by mutations in a gene either on chrosome 9 or chrosome 16. Clinicaly these disorders are indistinguishable.

### Molecular Genetic Diagnosis in Diseases with Known Genetic Defect

If the gene causing a neurological disorder is known, direct molecular diagnosis can be performed by mutational analysis. Only DNA from the affected individual is required. Usually, exons that are known to harbor mutations in the particular disorder will be amplified from DNA, which has been extracted from peripheral blood leukocytes by the PCR. Depending on its type, the mutation will then be detected either directly by gel electrophoresis (e.g., in the case of trinucleotide repeat expansions or large deletions) or after digestion by restriction enzymes or by direct sequencing (point mutations, small deletions, or insertions).

## Diseases Caused by Expansions of Triplet Repeat Sequences

An increasing number of inherited neurological disorders are recognized to be caused by expansions of unstable triplet repeat sequences within or adjacent to the coding region of genes, among them Huntington's disease (Huntington's Disease Collaborative Research Group, 1993), the spinocerebellar ataxias (Brice, 1998), X-linked bulbospinal neuronopathy (Kennedy's disease) (La Spada *et al.*, 1991), as well as myotonic dystrophy (Brook *et al.*, 1992) and Friedreich's ataxia (Campuzano *et al.*, 1996).

Most triplet repeat disorders are caused by the expansion of a CAG-repeat sequence within the coding region of the gene, leading to an elongated polyglutamine domain in the encoded protein. These expansions are likely to be "gain of function" mutations, with the mutation causing a novel toxic function that dominates the normal gene/protein function. Friedreich's ataxia is an exception. The expansion of a GAA-repeat in the Frataxin gene leads to a loss of functional protein and consequently to a recessive mode of inheritance. All repeat expansions can be detected by a simple assay based on PCR or Southern blotting. Routine molecular diagnosis is possible for confirmation of a clinical diagnosis or for the determination of genetic status presymptomatically or prenatally.

Age at onset of disorders caused by triplet repeat expansions is usually inversely correlated with the length of the triplet repeat. The number of triplets may increase from one generation to the next (dynamic mutation); this observation explains the phenomenon of anticipation (i.e., progressively earlier age of onset of an inherited disease in successive generations), which had been described in Huntington's disease and myotonic dystrophy long before the discovery of triplet repeats.

## Diseases Caused by Point Mutations, Deletions, and Insertions

Most inherited neurological disorders can be caused by a number of different mutations (base exchanges, small deletions, insertions or other alterations of the DNA sequence). This is called "allelic heterogeneity." In principle, these sequence variations can be detected by direct sequencing of the exons of a gene.

However, if a gene is very large (genes with more than 30 exons are not uncommon) and mutations are scattered throughout the entire gene, direct mutational analysis can be very costly and time-consuming. In these cases, routine sequence analysis is sometimes offered only for portions of a gene, where mutations may be clustered (e.g., in CADASIL, where 70% of mutations are found in exons 3 and 4 of the Notch3 gene (Joutel *et al.*, 1997).

In other cases, indirect molecular diagnosis by family studies with linked DNA markers (see later) will remain an important tool for the presymptomatic detection of gene carriers.

One example is Duchenne and Becker muscular dystrophy, which are both caused by a variety of mutations in the dystrophin gene on chromosome Xp21.1, only about 60% of all mutations (usually structural alterations such as large deletions or duplications) are detected by routine mutational analysis (PCR amplification of exons or sets of exons). In the remaining cases, which may be caused by more subtle gene defects, such as point mutations, monoclonal antibodies against dystrophin can be used to demonstrate a lack or abnormal size of this structural protein in muscle biopsy specimens by immunohistochemistry or Western blot analysis, thus allowing a diagnosis at the protein level. In such families, studies with linked DNA markers may be used to determine carrier status at the DNA level.

## Genetic Diagnosis in Diseases with Known Gene Location ("Indirect" Molecular Diagnosis)

Knowledge of the chromosomal position of a disease gene allows molecular support for the diagnosis, even if the disease gene itself is unknown or if analysis is unfeasible. "Indirect" molecular diagnosis is limited to risk determination for an individual in whose family an inherited neurological disease has already been diagnosed clinically. The method is based on the analysis of DNA markers known to be closely linked to the disease under investigation. Determination of these marker alleles in healthy and affected family members allows the identification of the disease-gene–bearing chromosome in this particular family. It must be emphasized that accurate clinical diagnosis in at least one affected family member is an absolute prerequisite for this type of molecular diagnosis, and there is a small but definite error rate with this sort of linkage analysis. Until the gene for Huntington's disease was identified in 1993, predictive testing for Huntington's disease was the most widely used application for indirect molecular diagnosis. As more and more disease genes are identified, direct mutational analysis will become increasingly important.

## Genetic Heterogeneity

Many hereditary neurological disorders are genetically heterogeneous (i.e., the disease can be caused by mutations in a number of different genes). This may complicate molecular genetic diagnosis in individuals who do not carry one of the frequent mutations.

The most common form of inherited neuropathy, Charcot-Marie-Tooth disease (CMT, formerly called hereditary motor and sensory neuropathies, HMSN), is an example (Pareyson, 1999). Most cases of the demyelinating form of the disease, CMT I, are caused by the duplication of a large segment (approximately 1.5 megabases) of chromosome 17 (CMT type Ia). The duplicated DNA segment contains the gene for a component of the myelin sheath of peripheral nerves, peripheral myelin protein-22 (PMP-22). This duplication can easily be detected by routine molecular methods. In some families, however, a point mutation, rather than a duplication, of PMP-22 is responsible (i.e., allelic heterogeneity, see earlier). In still other families (about 5%–10% of cases), mutations in another component of myelin (P0-Protein) on chromosome 1 cause a clinically indistinguishable disorder (CMT Ib). Another form of CMT is due to mutations of the gene for connexin 32, a gap junction protein on the X-chromosome, and other forms, collectively called CMT-1c, exist for which the genetic defects are still unknown (Table I).

Another example of genetic heterogeneity are the spinocerebellar ataxias. Although there are group differences between the SCAs with respect to their clinical features, the overlap between the different entities makes prediction of the molecular origin impossible and emphasizes the usefulness of molecular analysis (Stevanin et al., 2000).

### Mitochondrial Diseases

In recent years, several syndromes have been described in association with mutations of the mitochondrial genome (mtDNA). Deletions, and occasionally duplications, of mitochondrial DNA are found in most patients with mitochondrial encephalomyopthies, a spectrum of disorders ranging from mild chronic progressive external ophthalmoplegia (CPEO), sometimes with associated neurological symptoms (CPEO-plus), to severe cases of the Kearns-Sayre syndrome (Simon and Johns, 1999). Deletions are sometimes not detectable in DNA prepared from lymphocytes, so that muscle DNA has to be examined. Recurrence risk for relatives, including offspring, is very low in patients with deletions (Klopstock and Gasser, 1999).

Point mutations of mitochondrial DNA, the two most common (at positions 3243 and 8344) of which are associated with two well-described maternally inherited syndromes, mitochondrial encephalopathy with lactic acidosis and strokelike episodes (MELAS) and myoclonic epilepsy with ragged red fibers (MERRF), can be detected in a blood sample in most cases. Other mutations have been found to cause Leber's hereditary optic neuropathy (LHON), multiple symmetric lipomatosis, and an increasing number of other phenotypes (Scharfe et al., 2000).

### Neurological Diseases with Complex (Polygenic or Multifactorial) Inheritance

Neurological disorders exhibiting monogenic mendelian inheritance are relatively rare. In a number of much more common diseases, such as Alzheimer's disease, Parkinson's disease, the epilepsies, or multiple sclerosis, genetic factors seem to contribute to the etiology, but only a small minority of cases (usually 5% or less) are inherited in a clear mendelian fashion. Several genes for rare mendelian subgroups of these disorders have been identified, and their analysis has provided important insight into their molecular pathogenesis. For example, the crucial role of amyloid-β protein and β-secretase in the formation of amyloid plaques and neurodegeneration in Alzheimer's disease (Hardy et al., 1998), or of alpha-synuclein in Parkinson's disease (Farrer et al., 1999), has been defined an the basis of the analysis of monogenic variants of these common diseases.

Nevertheless, the etiology in most sporadic cases is still unknown. This may be because, in addition to genetic susceptibilty, other nongenetic factors (e.g., toxic, infectious, or immunological factors) may play a role in the etiology of the disorder (multifactorial etiology). Alternatively, several inherited traits may be required to act together to determine a particular phenotype (polygenic inheritance). A number of difficulties limit the applicability of linkage analysis to these disorders.

A major prerequisite for "classical" linkage analysis is that the genetic parameters of the disease under investigation, such as the mode of inheritance, penetrance, and gene frequency, are known. For diseases with complex inheritance, these parameters usually cannot be determined with any degree of confidence.

Furthermore, a clear distinction between affected and unaffected individuals, which is the basis for all forms of linkage analysis, may not always be made. For example, it is still unclear whether asymptomatic individuals with some degree of neuropathological changes suggestive of Alzheimer's disease or Parkinson's disease ("incidental Lewy-body brains") should be considered as presymptomatically affected. Similarly, it is not known whether certain EEG changes in unaffected relatives of patients with epilepsy should be considered as a sign of the disease.

Despite these difficulties, recent advances in our understanding of disorders with complex inheritance show that they are in principle amenable to molecular genetic analysis. The E4-allele of the gene for apolipoprotein E has been identified as the first major risk factor for sporadic late-onset Alzheimer's disease (Strittmatter et al., 1993) and more are likely to follow (Du et al., 2000). Out of the heterogeneous clinical syndromes, more and more genetically homogeneous subgroups will be isolated, and their study will provide an increasing understanding of the cascade of molecular and cellular events leading to neuronal dysfunction and disease.

## ETHICAL CONSIDERATIONS

The primary goal of molecular diagnosis is to provide help for the individual patient, client, and/or their families. Reducing the prevalence of inherited disorders in a population or in subsequent generations may be a secondary effect but must never be allowed to guide the process of genetic counseling (Pulst, 2000).

### Genetic Counseling

It must always be kept in mind that the molecular genetic diagnosis of an inherited disorder affects not only the patient but also the entire family. Therefore, genetic counseling is an essential component of diagnosis of inherited disorders. Sensitive and informed counseling provides patients and families a foundation for decisions about testing. Patients should be counseled as to the clinical features and course of the respective disease, as well as to potential consequences for the family, taking into consideration the most important genetic parameters such as mode of inheritance and penetrance. If the treating neurologist does not have thorough experience with inherited disorders, counseling by an experienced counselor from a department of human genetics or another genetic counseling unit is strongly advised. In most situations, genetic testing should not be performed until adequate counseling has been provided (The Practice Committee Genetics Testing Task Force of the American Academy of Neurology, 1996; The American College of Medical Genetics/American Society of Human Genetics Huntington Disease Genetic Testing Working Group, 1998).

In the case of predictive testing (see later), psychological counseling by appropriately trained persons is essential and mandatory before testing and after results have been disclosed.

### Informed Consent

As is true for all diagnostic procedures, the essential prerequisite for molecular diagnosis is the informed and voluntary consent of the patient. Therefore, the neurologist should establish that a patient or lawful surrogate is capable of comprehending relevant information and capable of exercising informed choices. Molecular genetic diagnostic tests should not be performed at the request of members of the patients' families or other third parties (e.g., insurers, employers) without the express written consent of the patient.

### Confidentiality

Test results suggesting that patients or family members carry mutations that indicate or predict a major neurological disorder or a susceptibility to a neurological disorder are highly sensitive. Therefore, rigorous measures to ensure confidentiality should be taken. Test results should never be disclosed to a third party without explicit written consent from the patient or their lawful surrogates.

### Presymptomatic Diagnosis

The identification of disease genes allows presymptomatic (predictive) diagnosis in many cases. Guidelines for presymptomatic diagnosis have been issued for Huntington's disease by The World Federation of Neurology Research Group on Huntington's Disease (1993). These guidelines, which include extensive pretest and posttest counseling and thorough psychological support during the extensive process, should be followed in all cases of presymptomatic diagnosis. Generally, presymptomatic diagnosis should be provided within the setting of a department of Human Genetics. If no clear therapeutic consequences can be envisioned, presymptomatic diagnosis should not be performed on minors.

Molecular genetic diagnosis allows accurate presymptomatic and prenatal determination of an individual's disease-carrying status. It does not indicate whether a person will necessarily have the disease develop. This is governed by the penetrance of the mutation. For example, Huntington's disease is virtually 100% penetrant; that is, all gene mutation carriers will have the disease develop should they live long enough. Whereas in the case of primary generalized dystonia caused by a GAG deletion in the DYT1 gene, the penetrance is reduced to 30%. Therefore even carrying the mutation in this case means that only a minority will have the disease develop. There is an age dependency to this penetrance such that older than the age of 28 years the chance of being affected with dystoina is very small. The underlying mechanisms behind reduced penetrance are not clear. Predictive diagnosis is especially important if it allows early therapeutic intervention. For example, early initiation of therapy in Wilson's disease can prevent serious neurological complications. Most inherited neurological diseases, however, are still not amenable to specific therapy. In these instances, the decision to determine the risk of an individual has to be considered carefully in each case.

In diseases with onset during adult life, such as Huntington's disease, the reason for presymptomatic testing is usually the planning of partnership, family, and career. However, a presymptomatic diagnosis may mean that the individual has to spend many years of productive life with the knowledge of an incurable disease developing later in life.

The experience of genetic counselling centers shows that a significant proportion of at-risk individuals seeking presymptomatic testing will change their mind during the counseling process. The initiative for presymptomatic testing should always arise from the proband, not from the counselor, and informed consent must be given. Extensive genetic counseling is required before the test and support should be available afterwards.

Molecular genetic techniques have opened up a new approach to many scientific questions and have improved our diagnostic repertoire, but they also pose new problems and dilemmas. The diagnostic approach in an individual patient should always be guided by a careful deliberation of the possible consequences for the patient and his or her relatives. Individual discussion with a physician who is knowledgable about the possibilities and limitations of genetic diagnosis is of paramount importance.

In summary, it can be seen that progress in our understanding of molecular pathogenesis is already having a major impact on the clinical neurologist with improved diagnosis and counseling of families. However, this progress will be dwarfed by the changes over the next few years. One can foresee not only improved understanding of the genetic factors involved in relatively rare mendelian disorders but also a much deeper knowledge of those factors involved in common disorders, such as stroke, multiple sclerosis and Parkinson's disease. Pharmacogenomics is an emerging field that will produce major benefits in terms of drug discovery, side effect management, and drug responsiveness. In the coming 5–10 years there will be no neurological field that will not be affected by these developments.

# REFERENCES

Benomar, A., Krols, L., Stevanin, G., Cancel, G., LeGuern, E., David, G., Ouhabi, H., Martin, J. J., Durr, A., and Zaim, A. (1995). The gene for autosomal dominant cerebellar ataxia with pigmentary macular dystrophy maps to chromosome 3p12–p21.1. *Nat. Genet.* 10, 84–88.

Bergoffen, J., Scherer, S. S., Wang, S., Scott, M. O., Bone, L. J., Paul, D. L., Chen, K., Lensch, M. W., Chance, P. F., and Fischbeck, K. H. (1993). Connexin mutations in X-linked Charcot-Marie-Tooth disease. *Science* 262, 2039–2042.

Bione, S., Maestrini, E., Rivella, S., Mancini, M., Regis, S., Romeo, G., and Toniolo, D. (1994). Identification of a novel X-linked gene responsible for Emery-Dreifuss muscular dystrophy. *Nat. Genet.* 8, 323–327.

Bonne, G., Di Barletta, M. R., Varnous, S., Becane, H. M., Hammouda, E. H., Merlini, L., Muntoni, F., Greenberg, C. R., Gary, F., Urtizberea, J. A., Duboc, D., Fardeau, M., Toniolo, D., and Schwartz, K. (1999). Mutations in the gene encoding lamin A/C cause autosomal dominant Emery-Dreifuss muscular dystrophy. *Nat. Genet.* 21, 285–288.

Brice, A. (1998). Unstable mutations and neurodegenerative disorders. *J. Neurol.* 245, 505–510.

Brook, J. D., McCurrach, M. E., Harley, H. G., Buckler, A. J., Church, D., Aburatani, H., Hunter, K., Stanton, V. P., Thirion, J. P., and Hudson, T. (1992). Molecular basis of myotonic dystrophy: expansion of a trinucleotide (CTG) repeat at the 3′ end of a transcript encoding a protein kinase family member. *Cell* 69, 385.

Browne, D. L., Gancher, S. T., Nutt, J. G., Brunt, E. R., Smith, E. A., Kramer, P., and Litt, M. (1994). Episodic ataxia/myokymia syndrome is associated with point mutations in the human potassium channel gene, KCNA1. *Nat. Genet.* 8, 136–140.

Campuzano, V., Montermini, L., Molto, M. D., Pianese, L., Cossee, M., Cavalcanti, F., Monros, E., Rodius, F., Duclos, F., and Monticelli, A. (1996). Friedreich's ataxia: autosomal recessive disease caused by an intronic GAA triplet repeat expansion. *Science* 271, 1423–1427.

Chance, P. F., Alderson, M. K., Leppig, K. A., Lensch, M. W., Matsunami, N., Smith, B., Swanson, P. D., Odelberg, S. J., Disteche, C. M., and Bird, T. D. (1993). DNA deletion associated with hereditary neuropathy with liability to pressure palsies. *Cell* 72, 143–151.

Du, Y., Dodel, R. C., Eastwood, B. J., Bales, K. R., Gao, F., Lohmuller, F., Muller, U., Kurz, A., Zimmer, R., Evans, R. M., Hake, A., Gasser, T., Oertel, W. H., Griffin, W. S., Paul, S. M., and Farlow, M. R. (2000). Association of an interleukin 1 alpha polymorphism with Alzheimer's disease. *Neurology* 55, 480–483.

European Chromosome 16 Tuberous Sclerosis Consortium. (1993). Identification and characterization of the tuberous sclerosis gene on chromosome 16. *Cell* 75, 1305–1315.

Farrer, M., Gwinn-Hardy, K., Hutton, M., and Hardy, J. (1999). The genetics of disorders with synuclein pathology and parkinsonism. *Hum. Mol. Genet.* 8, 1901–1905.

Fontaine, B., Khurana, T. S., Hoffman, E. P., Bruns, G. A., Haines, J. L., Trofatter, J. A., Hanson, M. P., Rich, J., McFarlane, H., and Yasek, D. M. (1990). Hyperkalemic periodic paralysis and the adult muscle sodium channel alpha-subunit gene. *Science* 250, 1000–1002.

Gispert, S., Twells, R., Orozco, G., Brice, A., Weber, J., Heredero, L., Scheufler, K., Riley, B., Allotey, R., and Nothers, C. (1993). Chromosomal assignment of the second locus for autosomal dominant cerebellar ataxia (SCA2) to chromosome 12q23–24.1. *Nat. Genet.* 4, 295–299.

Goate, A., Chartier-Harlin, M. C., Mullan, M., Brown, J., Crawford, F., Fidani, L., Giuffra, L., Haynes, A., Irving, N., and James, L. (1991). Segregation of a missense mutation in the amyloid precursor protein gene with familial Alzheimer's disease. *Nature* 349, 704–706.

Goto, Y., Nonaka, I., and Horai, S. (1990). A mutation in the tRNA(Leu)(UUR) gene associated with the MELAS subgroup of mitochondrial encephalomyopathies. *Nature* 348, 651–653.

Hardy, J., Duff, K., Hardy, K. G., Perez-Tur, J., and Hutton, M. (1998). Genetic dissection of Alzheimer's disease and related dementias: amyloid and its relationship to tau. *Nat. Neurosci.* **1**, 355–358.

Hayasaka, K., Himoro, M., Sato, W., Takada, G., Uyemura, K., Shimizu, N., Bird, T. D., Conneally, P. M., and Chance, P. F. (1993). Charcot-Marie-Tooth neuropathy type 1B is associated with mutations of the myelin P0 gene. *Nat. Genet.* **5**, 31–34.

Holmes, S. E., O'Hearn, E. E., McInnis, M. G., Gorelick-Feldman, D. A., Kleiderlein, J. J., Callahan, C., Kwak, N. G., Ingersoll-Ashworth, R. G., Sherr, M., Sumner, A. J., Sharp, A. H., Ananth, U., Seltzer, W. K., Boss, M. A., Vieria-Saecker, A. M., Epplen, J. T., Riess, O., Ross, C. A., and Margolis, R. L. (1999). Expansion of a novel CAG trinucleotide repeat in the 5′ region of PPP2R2B is associated with SCA12. *Nat. Genet.* **23**, 391–392.

Holt, I. J., Harding, A. E., and Morgan-Hughes, J. A. (1988). Deletions of muscle mitochondrial DNA in patients with mitochondrial myopathies. *Nature* **331**, 717–719.

Holt, I. J., Harding, A. E., Petty, R. K., and Morgan-Hughes, J. A. (1990). A new mitochondrial disease associated with mitochondrial DNA heteroplasmy. *Am. J. Hum. Genet.* **46**, 428–433.

Hsiao, K., Baker, H. F., Crow, T. J., Poulter, M., Owen, F., Terwilliger, J. D., Westaway, D., Ott, J., and Prusiner, S. B. (1989). Linkage of a prion protein missense variant to Gerstmann- Straussler syndrome. *Nature* **338**, 342–345.

Huntington's Disease Collaborative Research Group. (1993). A novel gene containing a trinucleotide repeat that is expanded and unstable on Huntington's disease chromosomes. *Cell* **72**, 971–983.

Hutton, M., Lendon, C. L., Rizzu, P., Baker, M., Froelich, S., Houlden, H., Pickering-Brown, S., Chakraverty, S., Isaacs, A., Grover, A., Hackett, J., Adamson, J., Lincoln, S., Dickson, D., Davies, P., Petersen, R. C., Stevens, M., de Graaff, E., Wauters, E., van Baren, J., Hillebrand, M., Joosse, M., Kwon, J. M., Nowotny, P., and Heutink, P. (1998). Association of missense and 5′-splice-site mutations in tau with the inherited dementia FTDP-17. *Nature* **393**, 702–705.

Ichinose, H., Ohye, T., Takahashi, E., Seki, N., Hori, T., Segawa, M., Nomura, Y., Endo, K., Tanaka, H., Tsuji, S. *et al.* (1994). Hereditary progressive dystonia with marked diurnal fluctuation caused by mutations in the GTP cyclohydrolase I gene. *Nat. Genet.* **8**, 236–242.

Joutel, A., Corpechot, C., Ducros, A., Vahedi, K., Chabriat, H., Mouton, P., Alamowitch, S., Domenga, V., Cecillion, M., Marechal, E., Maciazek, J., Vayssiere, C., Cruaud, C., Cabanis, E. A., Ruchoux, M. M., Weissenbach, J., Bach, J. F., Bousser, M. G., and Tournier-Lasserve, E. (1996). Notch3 mutations in CADASIL, a hereditary adult-onset condition causing stroke and dementia. *Nature* **383**, 707–710.

Joutel, A., Vahedi, K., Corpechot, C., Troesch, A., Chabriat, H., Vayssiere, C., Cruaud, C., Maciazek, J., Weissenbach, J., Bousser, M. G., Bach, J. F., and Tournier-Lasserve, E. (1997). Strong clustering and stereotyped nature of Notch3 mutations in CADASIL patients. *Lancet* **350**, 1511–1515.

Jurkat-Rott, K., Lehmann-Horn, F., Elbaz, A., Heine, R., Gregg, R. G., Hogan, K., Powers, P. A., Lapie, P., Vale-Santos, J. E., and Weissenbach, J. (1994). A calcium channel mutation causing hypokalemic periodic paralysis. *Hum. Mol. Genet.* **3**, 1415–1419.

Jurkat-Rott, K., Mitrovic, N., Hang, C., Kouzmekine, A., Iaizzo, P., Herzog, J., Lerche, H., Nicole, S., Vale-Santos, J., Chauveau, D., Fontaine, B., and Lehmann-Horn, F. (2000). Voltage-sensor sodium channel mutations cause hypokalemic periodic paralysis type 2 by enhanced inactivation and reduced current. *Proc. Natl. Acad. Sci. U S A* **97**, 9549–9554.

Kausch, K., Muller, C. R., Grimm, T., Ricker, K., Rietschel, M., Rudnik-Schoneborn, S., and Zerres, K. (1991). No evidence for linkage of autosomal dominant proximal spinal muscular atrophies to chromosome 5q markers. *Hum. Genet.* **86**, 317–318.

Kawaguchi, Y., Okamoto, T., Taniwaki, M., Aizawa, M., Inoue, M., Katayama, S., Kawakami, H., Nakamura, S., Nishimura, M., and Akiguchi, I. (1994). CAG expansions in a novel gene for Machado-Joseph disease at chromosome 14q32.1. *Nat. Genet.* **8**, 221–228.

Kitada, T., Asakawa, S., Hattori, N., Matsumine, H., Yamamura, Y., Minoshima, S., Yokochi, M., Mizuno, Y., and Shimizu, Y. (1998). Mutations in the parkin gene cause autosomal recessive juvenile parkinsonism. *Nature* **392**, 605–608.

Klopstock, T., and Gasser, T. (1999). Genetic counseling and prenatal diagnosis in mitochondrial diseases. *Nervenarzt* **70**, 504–508.

Knappskog, P. M., Flatmark, T., Mallet, J., Ludecke, B., and Bartholome, K. (1995). Recessively inherited L-DOPA-responsive dystonia caused by a point mutation (Q381K) in the tyrosine hydroxylase gene. *Hum. Mol. Genet.* **4**, 1209–1212.

Koch, M. C., Steinmeyer, K., Lorenz, C., Ricker, K., Wolf, F., Otto, M., Zoll, B., Lehmann Horn, F., Grzeschik, K. H., and Jentsch, T. J. (1992). The skeletal muscle chloride channel in dominant and recessive human myotonia. *Science* **257**, 797–800.

Koenig, M., Hoffman, E. P., Bertelson, C. J., Monaco, A. P., Feener, C., and Kunkel, L. M. (1987). Complete cloning of the Duchenne muscular dystrophy (DMD) cDNA and preliminary genomic organization of the DMD gene in normal and affected individuals. *Cell* **50**, 509–517.

La Spada, A. R., Wilson, E. M., Lubahn, D. B., Harding, A. E., and Fischbeck, K. H. (1991). Androgen receptor gene mutations in X-linked spinal and bulbar muscular atrophy. *Nature* **352**, 77–79.

Lalioti, M. D., Scott, H. S., Buresi, C., Rossier, C., Bottani, A., Morris, M. A., Malafosse, A., and Antonarakis, S. E. (1997). Dodecamer repeat expansion in cystatin B gene in progressive myoclonus epilepsy. *Nature* **386**, 847–851.

Lefebvre, S., Burglen, L., Reboullet, S., Clermont, O., Burlet, P., Viollet, L., Benichou, B., Cruaud, C., Millasseau, P., and Zeviani, M. (1995). Identification and characterization of a spinal muscular atrophy-determining gene. *Cell* **80**, 155–165.

Lerche, H., Heine, R., Pika, U., George, A. L. J., Mitrovic, N., Browatzki, M., Weiss, T., Rivet-Bastide, M., Franke, C., and Lomonaco, M. (1993). Human sodium channel myotonia: slowed channel inactivation due to substitutions for a glycine within the III-IV linker. *J. Physiol. (Lond.)* **470**, 13–22.

Levy, E., Carman, M. D., Fernandez-Madrid, I. J., Power, M. D., Lieberburg, I., van Duinen, S. G., Bots, G. T., Luyendijk, W., and Frangione, B. (1990). Mutation of the Alzheimer's disease amyloid gene in hereditary cerebral hemorrhage, Dutch type. *Science* **248**, 1124–1126.

Lupski, J. R., de Oca-Luna, R. M., Slaugenhaupt, S., Pentao, L., Guzzetta, V., Trask, B. J., Saucedo-Cardenas, O., Barker, D. F., Killian, J. M., and Garcia, C. A. (1991). DNA duplication associated with Charcot-Marie-Tooth disease type 1A. *Cell* **66**, 219–232.

McClatchey, A. I., Van den Bergh, P., Pericak-Vance, M. A., Raskind, W., Verellen, C., McKenna-Yasek, D., Rao, K., Haines, J. L., Bird, T., and Brown, R. H. J. (1992). Temperature-sensitive mutations in the III-IV cytoplasmic loop region of the skeletal muscle sodium channel gene in paramyotonia congenita. *Cell* **68**, 769–774.

Medori, R., Tritschler, H. J., LeBlanc, A., Villare, F., Manetto, V., Chen, H. Y., Xue, R., Leal, S., Montagna, P., and Cortelli, P. (1992). Fatal familial insomnia, a prion disease with a mutation at codon 178 of the prion protein gene. *N. Engl. J. Med.* **326**, 444–449.

Minassian, B. A., Lee, J. R., Herbrick, J. A., Huizenga, J., Soder, S., Mungall, A. J., Dunham, I., Gardner, R., Fong, C. Y., Carpenter, S., Jardim, L., Satishchandra, P., Andermann, E., Snead, O. C., Lopes-Cendes, I., Tsui, L. C., Delgado-Escueta, A. V., Rouleau, G. A., and Scherer, S. W. (1998). Mutations in a gene encoding a novel protein tyrosine phosphatase cause progressive myoclonus epilepsy. *Nat. Genet.* **20**, 171–174.

Ophoff, R. A., Terwindt, G. M., Vergouwe, M. N., van Eijk, R., Oefner, P. J., Hoffman, S. M., Lamerdin, J. E., Mohrenweiser, H. W., Bulman, D. E., Ferrari, M., Haan, J., Lindhout, D., van Ommen, G. J., Hofker, M. H., Ferrari, M. D., and Frants, R. R. (1996). Familial hemiplegic migraine and episodic ataxia type-2 are caused by mutations in the Ca2+ channel gene CACNL1A4. *Cell* **87**, 543–552.

Orr, H. T., Chung, M. Y., Banfi, S., Kwiatkowski, T. J., Jr., Servadio, A., Beaudet, A. L., McCall, A. E., Duvick, L. A., Ranum, L. P., and Zoghbi, H. Y. (1993). Expansion of an unstable trinucleotide CAG repeat in spinocerebellar ataxia type 1. *Nat. Genet.* **4**, 221–226.

Ouahchi, K., Arita, M., Kayden, H., Hentati, F., Ben Hamida, M., Sokol, R., Arai, H., Inoue, K., Mandel, J. L., and Koenig, M. (1995). Ataxia with isolated vitamin E deficiency is caused by mutations in the alpha-tocopherol transfer protein. *Nat. Genet.* **9**, 141–145.

Owen, F., Poulter, M., Lofthouse, R., Collinge, J., Crow, T. J., Risby, D., Baker, H. F., Ridley, R. M., Hsiao, K., and Prusiner, S. B. (1989). Insertion in prion protein gene in familial Creutzfeldt-Jakob disease. *Lancet* **1**, 51–52.

Ozelius, L., Hewett, J. W., Page, C. E., Bressman, S. B., Kramer, P. L., Shalish, C., deLeon, D., Brin, M. F., Raymond, D., Corey, D., Fahn, S., Risch, N. J., Buckler, A. J., Gusella, J. F., and Breakefield, X. O. (1997). The early-onset torsion dystonia gene (Dyt1) encodes an ATP-binding protein. *Nat. Genet.* **17**, 40–48.

Pareyson, D. (1999). Charcot-marie-tooth disease and related neuropathies: molecular basis for distinction and diagnosis. *Muscle Nerve* **22**, 1498–1509.

Pennacchio, L. A., Lehesjoki, A. E., Stone, N. E., Willour, V. L., Virtaneva, K., Miao, J., D'Amato, E., Ramirez, L., Faham, M., Koskiniemi, M., Warrington, J. A., Norio, R., de la Chapelle, A., Cox, D. R., and Myers, R. M. (1996). Mutations in the gene encoding cystatin B in progressive myoclonus epilepsy (EPM1). *Science* **271**, 1731–1734.

Polymeropoulos, M. H., Lavedan, C., Leroy, E., Ide, S. E., Dehejia, A., Dutra, A., Pike, B., Root, H., Rubenstein, J., Boyer, R., Stenroos, E. S., Chandrasekharappa, S., Athanassiadou, A., Papapetropoulos, T., Johnson, W. G., Lazzarini, A. M., Duvoisin, R. C., Di Iorio, G., and Nussbaum, R. L. (1997). Mutation in the α-synuclein gene identified in families with Parkinson's disease. *Science* **276**, 2045–2047.

Pulst, S. M. (2000). Ethical issues in DNA testing. *Muscle Nerve* **23**, 1503–1507.

Richards, F. M., Phipps, M. E., Latif, F., Yao, M., Crossey, P. A., Foster, K., Linehan, W. M., Affara, N. A., Lerman, M. I., and Zbar, B. (1993). Mapping the Von Hippel-Lindau disease tumour suppressor gene: identification of germline deletions by pulsed field gel electrophoresis. *Hum. Mol. Genet.* **2**, 879–882.

Rogaev, E. I., Sherrington, R., Rogaeva, E. A., Levesque, G., Ikeda, M., Liang, Y., Chi, H., Lin, C., Holman, K., and Tsuda, T. (1995). Familial Alzheimer's disease in kindreds with missense mutations in a gene on chromosome 1 related to the Alzheimer's disease type 3 gene. *Nature* **376**, 775–778.

Rosen, D. R., Siddique, T., Patterson, D., Figlewicz, D. A., Sapp, P., Hentati, A., Donaldson, D., Goto, J., O'Regan, J. P., and Deng, H. X. (1993). Mutations in Cu/Zn superoxide dismutase gene are associated with familial amyotrophic lateral sclerosis. *Nature* **362**, 59–62.

Roy, N., Mahadevan, M. S., McLean, M., Shutler, G., Yaraghi, Z., Farahani, R., Baird, S., Besner-Johnston, A., Lefebvre, C., and Kang, X. (1995). The gene for neuronal apoptosis inhibitory protein is partially deleted in individuals with spinal muscular atrophy. *Cell* **80**, 167–178.

Scharfe, C., Zaccaria, P., Hoertnagel, K., Jaksch, M., Klopstock, T., Dembowski, M., Lill, R., Prokisch, H., Gerbitz, K. D., Neupert, W., Mewes, H. W., and Meitinger, T. (2000). MITOP, the mitochondrial proteome database: 2000 update. *Nucleic Acids Res.* **28**, 155–158.

Serratosa, J. M., Gomez-Garre, P., Gallardo, M. E., Anta, B., de Bernabe, D. B., Lindhout, D., Augustijn, P. B., Tassinari, C. A., Malafosse, R. M., Topcu, M., Grid, D., Dravet, C., Berkovic, S. F., and de Cordoba, S. R. (1999). A novel protein tyrosine phosphatase gene is mutated in progressive myoclonus epilepsy of the Lafora type (EPM2). *Hum. Mol. Genet.* **8**, 345–352.

Sherrington, R., Rogaev, E. I., Liang, Y., Rogaeva, E. A., Levesque, G., Ikeda, M., Chi, H., Lin, C., Li, G., and Holman, K. (1995). Cloning of a gene bearing missense mutations in early-onset familial Alzheimer's disease. *Nature* **375**, 754–760.

Shoffner, J. M., Lott, M. T., Lezza, A. M., Seibel, P., Ballinger, S. W., and Wallace, D. C. (1990). Myoclonic epilepsy and ragged-red fiber disease (MERRF) is associated with a mitochondrial DNA tRNA(Lys) mutation. *Cell* **61**, 931–937.

Simon, D. K., and Johns, D. R. (1999). Mitochondrial disorders: clinical and genetic features. *Annu. Rev. Med.* **50**, 111–127.

Stevanin, G., Durr, A., and Brice, A. (2000). Clinical and molecular advances in autosomal dominant cerebellar ataxias: from genotype to phenotype and physiopathology. *Eur. J. Hum. Genet.* **8**, 4–18.

Strittmatter, W. J., Saunders, A. M., Schmechel, D., Pericak Vance, M., Enghild, J., Salvesen, G. S., and Roses, A. D. (1993). Apolipoprotein E: high-avidity binding to beta-amyloid and increased frequency of type 4 allele in late-onset familial Alzheimer disease. *Proc. Natl. Acad. Sci. U S A* **90**, 1977–1981.

Tanzi, R. E., Petrukhin, K., Chernov, I., Pellequer, J. L., Wasco, W., Ross, B., Romano, D. M., Parano, E., Pavone, L., Brzustowicz, L. M. *et al.* (1993). The Wilson disease gene is a copper transporting ATPase with homology to the Menkes disease gene. *Nat. Genet.* **5**, 344–350.

Tatuch, Y., Christodoulou, J., Feigenbaum, A., Clarke, J. T., Wherret, J., Smith, C., Rudd, N., Petrova-Benedict, R., and Robinson, B. H. (1992). Heteroplasmic mtDNA mutation (T—G) at 8993 can cause Leigh disease when the percentage of abnormal mtDNA is high. *Am. J. Hum. Genet.* **50**, 852–858.

The American College of Medical Genetics/American Society of Human Genetics Huntington Disease Genetic Testing Working Group (1998). ACMG/ASHG statement. Laboratory guidelines for Huntington disease genetic testing. *Am. J. Hum. Genet.* **62**, 1243–1247.

The Practice Committee Genetics Testing Task Force of the American Academy of Neurology (1996). Practice parameter: genetic testing alert. Statement of (1996). *Neurology* **47**, 1343–1344.

The World Federation of Neurology Research Group on Huntington's Disease. (1993). Presymptomatic testing for Huntington's disease: a world wide survey. *J. Med. Genet.* **30**, 1020–1022.

Trofatter, J. A., MacCollin, M. M., Rutter, J. L., Murrell, J. R., Duyao, M. P., Parry, D. M., Eldridge, R., Kley, N., Menon, A. G., and Pulaski, K. (1993). A novel moesin-, ezrin-, radixin-like gene is a candidate for the neurofibromatosis 2 tumor suppressor. *Cell* **72**, 791–800.

van Slegtenhorst, M., de Hoogt, R., Hermans, C., Nellist, M., Janssen, B., Verhoef, S., Lindhout, D., van den Ouweland, A., Halley, D., Young, J., Burley, M., Jeremiah, S., Woodward, K., Nahmias, J., Fox, M., Ekong, R., Osborne, J., Wolfe, J., Povey, S., Snell, R. G., Cheadle, J. P., Jones, A. C., Tachataki, M., Ravine, D., and Kwiatkowski, D. J. (1997). Identification of the tuberous sclerosis gene TSC1 on chromosome 9q34. *Science* **277**, 805–808.

Wallace, D. C., Singh, G., Lott, M. T., Hodge, J. A., Schurr, T. G., Lezza, A. M., Elsas, L. J. 2., and Nikoskelainen, E. K. (1988). Mitochondrial DNA mutation associated with Leber's hereditary optic neuropathy. *Science* **242**, 1427–1430.

Wallace, M. R., Andersen, L. B., Saulino, A. M., Gregory, P. E., Glover, T. W., and Collins, F. S. (1991). A de novo Alu insertion results in neurofibromatosis type 1. *Nature* **353**, 864–866.

Wijmenga, C., Hewitt, J. E., Sandkuijl, L. A., Clark, L. N., Wright, T. J., Dauwerse, H. G., Gruter, A. M., Hofker, M. H., Moerer, P., and Williamson, R. (1992). Chromosome 4q DNA rearrangements associated with facioscapulohumeral muscular dystrophy. *Nat. Genet.* **2**, 26–30.

Yazawa, I., Nukina, N., Hashida, H., Goto, J., Yamada, M., and Kanazawa, I. (1995). Abnormal gene product identified in hereditary dentatorubral- pallidoluysian atrophy (DRPLA) brain. *Nat. Genet.* **10**, 99–103.

Zeviani, M., Moraes, C. T., DiMauro, S., Nakase, H., Bonilla, E., Schon, E. A., and Rowland, L. P. (1988). Deletions of mitochondrial DNA in Kearns-Sayre syndrome. *Neurology* **38**, 1339–1346.

Zhuchenko, O., Bailey, J., Bonnen, P., Ashizawa, T., Stockton, D. W., Amos, C., Dobyns, W. B., Subramony, S. H., Zoghbi, H. Y., and Lee, C. C. (1997). Autosomal dominant cerebellar ataxia (SCA6) associated with small polyglutamine expansions in the alpha 1A-voltage-dependent calcium channel. *Nat. Genet.* **15**, 62–69.

Zimprich, A., Grabowski, M., Asmus, F., Naumann, M., Berg, D., Bertram, L., Scheidtmann, K., Winkelmann, J., Müller-Myhsok, B., Bauer, M., Castro, M., Meitinger, T., Strom, T., and Gasser, T. (2001). Mutations in the gene for epsilon-sarcoglycan (SGCE) cause myoclonus-dystonia syndrome. *Nat. Genet.* **29**, 66–69.

# INDEX

## A

Abetalipoteinemia, ataxia, 101
Abscess
 brain, *see* Brain abscess
 spinal, *see* Spinal epidural abscess; Spinal
  intramedullary abscess
Acamprosate, alcohol dependency
  management, 986
ACE inhibitors, *see* Angiotensin-converting
  enzyme inhibitors
Acetaminophen
 contraindications, 1510
 drug–drug interactions, 1510
 migraine treatment, 7
 poisoning and management, 795–796
 radiculopathy management, 1292
 side effects, 1510
Acetazolamide, episodic ataxia management,
  1016
*N*-Acetylcysteine (NAC), acetaminophen
  poisoning management, 795
Acquired immunodeficiency syndrome
  (AIDS), *see* Human
  immunodeficiency virus
Acridine, prion disease management, 718
Acromegaly
 clinical presentation, 1435
 course, 1435–1436
 diagnosis, 1435
 differential diagnosis, 1435
 treatment
  bromocriptine, 1437
  cabergoline, 1437
  obsolete treatments, 1437
  octreotide, 1437
  principles, 1436
  radiotherapy, 1436–1437
  surgery, 1436–1437
Acrophobia, *see* Height vertigo
Acrylamide polyneuropathy, features, 1285
ACTH, *see* Adrenocorticotropic hormone
Actinomycosis, central nervous system
  infection and management, 540
Acupuncture
 migraine prevention, 12
 pain management, 107, 1333
 tension-type headache treatment, 26
Acute intoxication, *see also specific poisons*
 differential diagnosis, 791–792
 incidence, 791
 monitoring, 795
 screening and toxicological analysis, 795

severity, 791
 syndromes
  anticholinergic, 792
  cholinergic, 792
  opiate, 792
  sedative-hypnotic, 793
  sympathomimetic, 791–792
 treatment
  emesis, 794
  gastric lavage, 794
  hemodialysis and hemoperfusion,
   794–795
  nonspecific therapy, 793
  specific antidotes, 793–794
  vital function stabilization, 793
  whole bowel irrigation, 794
Acute ischemic stroke
 course, 328–329
 critical care
  course of space-occupying middle
   cerebral artery infarction, 735–
   736
  management
   decompressive surgery, 737
   edema, 737
   general management, 736–737
   hypothermia, 737–738
  prognosis, 736
 diagnosis
  evaluation questions, 340
  imaging, 328, 340–341
  laboratory tests, 328, 340
  patient history, 328, 340
  physical examination, 340
 excitotoxicity hypothesis, 338
 mechanisms, 327–328
 prevention, *see* Stroke
 thrombosis versus embolism, 327–328
 treatment
  ancrod, 337–338
  angioplasty and stenting, 334
  aspirin, 336, 342–343
  blood pressure management, 341
  calcium channel blockers, 337
  edema, 338–339
  eicosopentanoic acid, 338
  fever and infection management, 341
  fluids and electrolytes, 341
  glycoprotein IIb/IIIa inhibitors, 336–
   337
  heparin, 334–336, 342–343
  hypothermia, 338–339

insulin, 338, 341
  low-molecular weight heparin, 335,
   342–343
  neuroprotection, 338–339
  obsolete treatments, 342–343
  plasmapheresis, 338
  principles, 329
  pulmonary function and airway
   protection, 341
  stroke units and services, 339–340
  surgery, 333–334
  thrombolysis
   efficacy, 329–330
   imaging, 342
   indications, 341–342
   intra-arterial thrombolysis, 332–333
   intravenous thrombolysis, 330–332
  vasodilating agents, 337
  volume expansion, 337
  warfarin, 334, 342–343
Acute pain, *see* Pain
Acute transverse myelitis
 clinical features, 934
 treatment, 934
Acyclovir
 Bell's palsy management, 137
 herpes simplex encephalitis management,
  610–611
 mechanism of action, 608
 side effects, 1518
 spectrum of activity, 608
 zoster management, 1270–1271
AD, *see* Alzheimer's disease
Adenovirus, clinical features and course of
  infection, 605
ADH, *see* Antidiuretic hormone
Adrenocorticotropic hormone (ACTH)
 contraindications, 1514
 dosing, 1513
 drug–drug interactions, 1514
 opsoclonus management, 130
 side effects, 1513–1514
Advanced sleep phase syndrome (ASPS)
 clinical features, 202
 treatment, 202
Agenesis of the corpus callosum, features,
  954–955
Ageusia
 classification, 171
 course, 176–179
 definition, 171
 etiology, 177–179

Ageusia (continued)
  hypergeusia, 177
  management
    antifungals and antibiotics, 181
    clinical evaluation, 179
    history, 179
    imaging, 180
    laboratory tests, 180
    physical examination, 179–180
  tests, 171–173
Aggramatism, definition, 266
Agyria, see Lissencephaly
AIDS, see Human immunodeficiency virus
Albendazole
  adverse reactions, 655
  diphyllobothriasis management, 597
  neurocysticercosis management, 652–653, 655
  neurofilariasis management, 669
  toxocariasis management, 597, 599
Albuterol, facioscapulohumeral dystrophy management, 1377–1378
Alcohol dependency
  drug therapy
    acamprosate, 986
    disulfiram, 986
    gamma-hydroxybutyric acid, 986–987
    naltrexone, 986
    serotonin modulators, 987
  motivation of patients for treatment, 985–986
  rationale for treatment, 985
Alcoholic cerebellar atrophy
  clinical presentation, 979–980
  course, 980
  treatment, 980
Alcoholic cerebellar degeneration, ataxia features and management, 1017
Alcoholic hallucinosis
  clinical presentation, 972
  course, 972–973
  treatment
    anxiolytics, 973
    neuroleptics, 973
Alcoholic myelopathy
  clinical presentation, 982
  course, 982
  treatment, 982
Alcoholic myopathies
  acute alcoholic rhabdomyolysis, 984
  acute myopathies, 1401
  chronic alcoholic myopathy, 985, 1401
  hypokalemic alcoholic myopathy, 984–985
Alcoholic pellagra encephalopathy, features, 982
Alcoholic polyneuropathy
  clinical presentation, 982–983, 1285
  course, 983
  treatment, 983
Alcohol intoxication
  classification and features, 971–972
  treatment, 972
Alcohol poisoning, see Ethanol; Methanol
Alcohol-related cerebral atrophy, features, 980

Alcohol withdrawal
  clinical features, 973
  course, 973–974
  delirium
    critical care, 745–746
    pathogenesis, 745
  treatment
    benzodiazepines, 974–975
    clomethiazole, 975–976
    clonidine, 976
    gamma-hydroxybutyric acid, 976
    haloperidol with clomethiazole, 976–977
    ineffective treatment, 977
    principles, 974
    supportive measures, 977
ALS, see Amyotrophic lateral sclerosis
Aluminum
  dialysis dementia
    clinical features, 997–998
    course, 998
    pathophysiology, 998
    treatment, 998–999
  encephalopathy, 1003–1004
Alzheimer's disease (AD)
  aphasia, 268–269
  course, 312, 314
  diagnostic criteria, 312, 314
  epidemiology, 312
  neuropathology, 314
  treatment
    antioxidants, 318
    behavioral pharmacotherapy, 318–319
    cholinesterase inhibitors, 317–318
    complementary alternative medicine, 318
    obsolete treatments
      estrogen, 321
      hydergine, 321
      tacrine, 321
    prospects, 319
Amantadine
  akinetic crisis management, 788–789
  mechanism of action, 609
  multiple system atrophy management, 1016
  neuroleptic malignant syndrome management, 787
  Parkinson's disease management, 1046–1047
  progressive supranuclear palsy management, 1092
  side effects, 1512
  spectrum of activity, 609
Amebiasis, see Cerebral amebiasis
4-Aminopyridine, multiple sclerosis management, 702
Amitriptyline
  migraine prevention, 10
  neuropathic pain management, 1286
  pain management after nerve injury, 1332
  poisoning and management, 797
  tension-type headache prevention, 27
  zoster management, 1271–1272

Amphotericin B
  adverse effects and contraindications, 618, 1517
  cerebral mycosis treatment, 619
  dose escalation, 620
  formulations, 618–619
  intrathecal application, 620
  laboratory testing during therapy, 620
  leishmaniasis management, 672
  pharmacological properties, 618
  primary amebic meningoencephalitis management, 663
  solution preparation for intravenous infusion, 620
Amyloid neuropathy
  autonomic dysfunction, 1460
  familial amyloidotic polyneuropathies, 1283–1284
  primary amyloidosis, 1284
Amyotrophic lateral sclerosis (ALS)
  clinical presentation, 1165
  course, 1167
  diagnosis, 1165–1166
  familial disease, 1166
  lay resources, 1173
  primary lateral sclerosis, 1166
  progressive bulbar palsy, 1166
  treatment
    alternative medicine, 1168
    counseling, 1168
    depression, 1169
    dysarthria, 1169
    dysphagia, 1168–1169
    dyspnea, 1169
    mucous secretions, 1169
    muscle weakness, 1168
    obsolete treatments, 1170–1171
    prevention, 1168
    principles, 1167–1168
    psychosocial care, 1169–1170
    riluzole, 1167–1168
    sleep disorders, 1169
    spiritual care and bereavement, 1170
    terminal phase, 1170
    xaliproden, 1168
Anaplastic astrocytoma
  clinical presentation, 838
  course, 839
  treatment
    chemotherapy, 840–841
    radiochemotherapy, 842–843
    radiotherapy, 840, 844
    surgery, 839, 844
Ancrod, acute ischemic stroke management, 337–338
Anderson-Fabry disease, features, 1283
Aneurysm, see Subarachnoid hemorrhage
Angioplasty and stenting, see also Percutaneous transluminal balloon angioplasty
  acute ischemic stroke management, 334
  brain embolism management, 377
  carotid artery
    clinical trials, 498–500
    complications, 499, 501–502

indications, 502
  patient evaluation, 501
  safety, 499
  stent designs, 501
principles, 498
stroke prevention, 358
subclavian artery, 503–504
thrombolytic therapy combination, 508
vertebral artery stenosis
  course, 502
  frequency, 502
  practical management, 503
  principles of therapy, 502–503
Angiostrongyliasis, features and management, 598
Angiotensin-converting enzyme (ACE) inhibitors, side effects, 1520
Anhidrosis
  investigation, 1471
  pathophysiology, 1470–1471
  treatment, 1471
Anosmia
  classification, 171
  course, 173, 176
  definition, 171
  etiology, 173–176
  hyperosmia rarity, 173
  management
    clinical evaluation, 179
    history, 179
    imaging, 180
    laboratory tests, 180
    physical examination, 179–180
    sensorineural damage, 181
    systemic steroids, 180–181
  tests, 171–173
Anterior interosseus syndrome, features and management, 1308
Antibiotic therapy
  ageusia, 181
  bacterial meningitis
    dosing, 537
    duration, 537
    initiation, 534
    selection of drug, 535–537
  brain abscess
    prophylaxis, 550
    treatment, 549–550
  brain embolism prevention, 383
  diphtheria, 1273
  intracranial pressure elevation, 757
  leptospirosis, 590
  Lyme neuroborreliosis, 587–589
  myasthenic crisis, 1357–1358
  relapsing fever, 591
  side effects and contraindications
    aminoglycosides, 1515
    beta-lactams, 1516
    chloramphenicol, 1515–1516
    fosfomycin, 1516
    macrolides, 1516
    nitrofuranes, 1516
    polymyxins, 1516
    quinolones, 1515
    sulfonamides, 1516–1517

tetracyclines, 1517
    trimethoprim, 1517
  spinal epidural abscess, 554
  Sydenham's chorea, 1154
  syphilis, 580–581
  tetanus, 1274
Antidepressants
  classification, 1508
  contraindications, 1508–1509
  drug–drug interactions, 1509
  side effects, 1508–1509
Antidiuretic hormone (ADH), diabetes insipidus management, 1440
Antihistamines
  drug–drug interactions, 1522
  side effects, 1521
Aphasia
  assessment, 269–270
  Broca's aphasia, 267
  cognitive neurolinguistics, 265–266
  completion phenomenon, 267
  conduction aphasia, 267
  cortical lesions, 266–267
  definition, 265
  etiology, 268–269
  global aphasia, 267
  logorrhea, 268
  phonetic disintegration, 266
  subcortical aphasic syndromes, 267–268
  transcortical mixed aphasia, 267
  treatment
    pharmacotherapy, 272–273
    practical approach, 272–273
    pragmatic therapy, 271–272
    prospects, 273–274
    reactivation of linguistic functions, 270–271
  Wernicke's aphasia, 267
Apomorphine
  Parkinson's disease management, 1042–1043
  sexual dysfunction management, 1491
  tremor management, 1237
Ara-C, leptomeningeal metastasis intrathecal chemotherapy, 902, 906
Arachnoid cyst
  clinical presentation, 958
  course, 958
  magnetic resonance imaging, 958–959
  treatment, 958
Arbovirus, clinical features and course of infection, 605
Arenavirus, clinical features and course of infection, 605
Arnold-Chiari formation, headache association, 42, 44
Arrhythmias
  control in brain embolism prevention, 383, 386
  critical care
    atrial flutter and fibrillation, 735
    sinus tachycardia, 735
    tachycardia, 735
  subarachnoid hemorrhage complication, 436–437

ARSACS, see Autosomal recessive ataxia of Charlevoix-Saguenay
Artemether, cerebral malaria management, 661–662
Arteriovenous malformation (AVM)
  cerebral, see Cerebral arteriovenous malformation
  spinal, see Spinal arteriovenous malformation
Artesunate, cerebral malaria management, 661–662
Ascariasis, features and management, 598
Aspergillosis
  cerebral mycosis treatment, 619
  clinical features, 616–617
Aspirin
  acute ischemic stroke management, 336, 342–343
  brain embolism
    management, 379
    prevention, 379–383
  cerebral ischemia prevention following subarachnoid hemorrhage, 432–433
  contraindications, 1510
  drug–drug interactions, 1510
  medication overuse headache, see Medication overuse headache
  migraine
    prevention, 10
    treatment, 3–4, 6–7
  poisoning and management, 796
  radiculopathy management, 1292
  side effects, 1510
  stroke prevention
    primary prevention, 351
    secondary prevention, 354–356, 358
  Susac syndrome management, 367–368
  tension-type headache treatment, 26, 28
  vasculitides management, 464
ASPS, see Advanced sleep phase syndrome
Astroblastoma, features and management, 848
AT, see Ataxia telangiectasia
Ataxia, see also specific ataxias
  classification, 1011–1012
  hypothyroidism, 1018
  immune-mediated ataxias, 1018
  sporadic adult-onset ataxia of unknown origin, 1016–1017
  toxic etiology, 1017
  vitamin deficiencies
    thiamine, 1017
    vitamin $B_{12}$, 1017
    vitamin E, 1013, 1017–1018
Ataxia telangiectasia (AT)
  clinical features, 1012
  course, 1012
  gene mutations, 1012
  treatment, 1012–1013
Atropine, palliative care, 818
Atypical facial pain
  clinical presentation, 59–60
  course, 59
  diagnosis, 59–60
  pathogenesis, 60–61
  related syndromes, 63

Atypical facial pain (continued)
    treatment
        amitriptyline, 62
        carbamazepine, 63
        clomipramine, 62
        obsolete treatments, 63
        principles, 60–62
        selective serotonin reuptake inhibitors,
            62
Atypical parkinsonism, see Corticobasal
        degeneration; Dementia with Lewy
        bodies; Multiple system atrophy;
        Progressive supranuclear palsy
Autonomic failure
    amyloid neuropathy, 1460
    anhidrosis
        investigation, 1471
        pathophysiology, 1470–1471
        treatment, 1471
    classification of disorders, 1453–1454
    clinical manifestations, 1454–1455
    diabetes mellitus, 1460
    diarrhea, 1470
    dopamine β-hydroxylase deficiency, 1461,
        1466
    drug induction, 1454–1455, 1462
    esophageal dysfunction
        distension, 1469
        spasm, 1469
    gastric stasis and distension, 1469–1470
    Guillain-Barré syndrome, 1456–1457
    human immunodeficiency virus,
        1460–1461
    hyperhidrosis
        investigation, 1471
        pathophysiology, 1470–1471
        treatment, 1471–1472
    intestine
        distension, 1470
        pseudoobstruction, 1470
    local versus global dysfunction, 1453,
        1455
    multiple system atrophy, 1457–1458
    nausea and vomiting, mechanisms and
        treatment, 1468
    orthostatic hypotension, see Orthostatic
        hypotension
    Parkinson's disease, 1457–1459
    postural tachycardia syndrome, 1463
    pure autonomic failure, 1459–1460
    Riley–Day syndrome, 1461
    sleeping sickness, 1461
    spinal cord injury, 1461–1462, 1467
    syncope, see Syncope
    tests, 1454–1455, 1457
Autosomal dominant optic atrophy, clinical
        features, 122
Autosomal recessive ataxia of Charlevoix-
        Saguenay (ARSACS), features,
        1013
AVM, see Arteriovenous malformation
Axillary nerve compression, features and
        management, 1304–1305
Azathioprine
    chronic inflammatory demyelinating
        polyneuropathy management, 1264

contraindications, 1514
dermatomyositis/polymyositis management,
        1366
drug–drug interactions, 1514
Duchenne's muscular dystrophy
        management, 1374
multiple sclerosis management, 688–689,
        696
myasthenia gravis management,
        1348–1350, 1355
sarcoid neuropathy management, 1267
sarcoidosis management, 571–572
side effects, 1514
vasculitic neuropathy management,
        1268

B

Baclofen
    contraindications, 1511
    drug–drug interactions, 1511
    generalized dystonia management, 1127
    hiccup management, 187
    nystagmus management, 127–128
    pain management, 104
    side effects, 1511
    spasticity management
        intrathecal, 1252–1253
        oral, 1251
    trigeminal neuralgia management, 69,
        71–72
    vertigo management, 141–142
Bacterial meningitis, see also Tuberculous
        meningitis
    chemoprophylaxis
        Haemophilus influenzae meningitis, 538
        meningococcal meningitis, 538
    clinical presentation, 529–530, 532
    complications
        central nervous system, 532–533
        systemic, 533
        temporal profile of major cerebral
            complications, 533
    diagnosis, 529–530
    differential diagnosis, 530
    epidemiology, 530–531
    immunoprophylaxis, 538, 542
    morbidity and mortality, 532
    pathogenesis, 533
    pathogens, 529–530
    risk factors, 531
    treatment
        adjunctive therapy, 534–535, 537–538
        antibiotic therapy
            dosing, 537
            duration, 537
            initiation, 534
            selection of drug, 535–537
        corticosteroids, 534–535, 537–538
        general management, 533–534
        patient isolation, 537
        principles, 533
        prospects, 535
BAL, see 2,3-Dimercaptopropanol
Balint's syndrome
    clinical presentation, 279

course, 279–280
    treatment, 280
Ballism
    clinical features, 1137
    course, 1138
    etiology, 1137–1138
    treatment, 1138
Barbiturates
    intracranial pressure control, 409, 730,
        759
    palliative care, 823
    poisoning and management, 799
Behavioral therapy, migraine prevention,
        11–12
Behçet's disease, see Vasculitides
Bell's palsy
    clinical presentation, 135
    course, 135–136
    differential diagnosis, 135
    electrophysiological testing, 135–136
    pathogenesis, 136–137
    treatment
        acyclovir, 137
        botulinum toxin, 137
        corticosteroids, 137
        eye care, 137
        eyelid surgery, 137
        nerve decompression, 137
        obsolete treatments, 137
Bencyclane, side effects, 1510
Benign paroxysmal positioning vertigo
        (BPPV)
    frequent posterior semicircular canal type
        course, 143–144
        diagnostic criteria, 143
        pathogenesis, 144
        treatment
            obsolete treatments, 147
            physical therapy, 145–147
            principles, 144–145
            surgery, 147
    rare horizontal semicircular canal type
        clinical presentation, 147
        course, 147
        treatment, 147–148
Benzodiazepines, see also specific drugs
    advantages and limitations, 218
    adverse reactions, 217, 1509
    alcohol withdrawal management, 974–975
    dosing, 216
    epilepsy management, 215, 217
    generalized dystonia management, 1127
    pain management, 101
    palliative care, 818–820, 823
    pharmacokinetics, 216
    poisoning and management, 799
    spasticity management, 1252
    tremor management, 1238
Benzotropine, side effects and
        contraindications, 1512–1513
Bereavement, palliative care, 816
Beta blockers
    contraindications, 1520
    drug–drug interactions, 1520
    migraine prevention, 8–9
    side effects, 1520

Betahistine, Meniére's disease management, 151

Biperiden, side effects and contraindications, 1512–1513

Bladder dysfunction
  algorithm for management, 1484–1485
  catheter drainage, 1484
  detrusor hyperreflexia
    course, 1480–1481
    definition, 1480–1481
    treatment, 1483–1484
  detrusor instability, 1481
  investigations
    electromyography, 1481
    imaging, 1483
    urodynamics, 1481, 1483
  micturition mechanism, 1479–1480
  multiple sclerosis, 701
  neural prosthesis
    contraindications, 1212
    indications, 1212
    side effects, 113
    systems, 1212–1213
    urinary incontinence, 1213
  neuroanatomy, 1479–1480
  obsolete treatments, 1485
  Parkinson's disease, 1062
  spinal cord injury and drainage, 930, 932
  surgical management, 1484

Blastomycosis
  cerebral mycosis treatment, 619
  clinical features, 617

Blepharospasm
  botulinum toxin therapy, 1131
  clinical presentation, 1130
  course, 1130

Bornaprine, side effects and contraindications, 1512–1513

Borreliosis, see Lyme neuroborreliosis

Botulinum toxin
  antibody resistance, 1133
  Bell's palsy management, 137
  corticobasal degeneration management, 1095
  cramp management, 1423
  focal dystonia management
    blepharospasm, 1131
    cervical dystonia, 1133–1134
    cranial dystonia, 1131–1132
    hand cramp, 1135
    overview, 1128–1130
    spasmodic dystonia, 1134
  formulations and dosing, 1130
  hemifacial spasm management, 1142
  mechanism of action, 1128–1129
  multiple system atrophy management, 1085
  serotypes, 1128
  spasticity management
    lower limbs, 1191–1192
    overview, 1253
    upper limbs, 1189
  stiff-man syndrome management, 1254
  tension-type headache prevention, 27

Botulism
  clinical features, 1275

course, 1275
toxins, 1275
treatment, 1275
types, 1275

Bowel dysfunction
  course, 1485–1486
  defecation mechanism, 1485
  multiple sclerosis, 701–702
  neural prosthesis
    contraindications, 1212
    indications, 1212
    side effects, 113
    systems, 1212–1213
  neuroanatomy, 1485
  spinal cord injury, 930, 932
  treatment
    algorithm, 1487–1488
    laxatives, 1486
    obsolete treatments, 1487
    principles, 1486
    surgery, 1486

BPPV, see Benign paroxysmal positioning vertigo

Brachial plexus injury
  course
    distal consequences, 1329–1330
    local consequences, 1329
    proximal consequences, 1330
  delayed radiotherapy injury, 1337–1339
  pain syndromes, 1328–1329
  prognosis after decompression or repair, 1330–1331
  treatment
    indications for surgical exploration and repair, 1331
    intractable pain
      acupuncture, 1333
      anticonvulsants, 1332–1333
      antidepressants, 1332
      capsaicin, 1333
      dosing of drugs, 1331–1332
      opioids, 1333
      spinal cord stimulation, 1333
      tramadol, 1333
      transcutaneous electrical nerve stimulation, 1333

Brachial plexus neuropathy, see Neuralgic amyotrophy

Brain abscess
  amebic brain abscess
    clinical presentation, 664
    course, 664
    definition, 664
    diagnosis, 664
    treatment
      metronidazole, 666
      obsolete treatments, 666
      principles, 664, 666
      tinidazole, 666
  classification, 545
  course, 548
  definition, 545
  diagnosis, 546–547
  differential diagnosis, 547–548
  etiology, 545–546
  fungal abscesses, 621

location, 546
prophylaxis, 550
treatment
  antibiotics, 549–550
  anticonvulsants, 550
  conservative treatment, 550
  corticosteroids, 550
  surgery, 548–549

Brain edema
  acute ischemic stroke management, 338–339
  brain embolism management, 379, 386
  high-altitude cerebral edema, 760

Brain embolism
  arterial distribution and patterns, 374–375
  course, 375–376
  definition, 373
  diagnosis, 374–375
  distribution of infarcts, 375
  donor sources, 375
  hemorrhagic infarction, 374
  imaging, 374, 384
  pathophysiology, 373
  prevention
    antiarrhythmic agents, 383, 386
    antibiotics, 383
    anticoagulants, 379–383
    cardiac surgery, 383–384
    carotid endarterectomy, 384
    myocardial ischemia and cardiomyopathies, 387–388
    prosthetic heart valve patients, 383, 386–387
    rheumatic heart disease patients, 386
  size of infarcts, 375
  symptoms and signs, 373
  treatment
    angioplasty and stenting, 377
    aortic source embolism, 388
    arterial source embolism, 388
    aspirin, 379
    blood flow maximization, 385
    brain edema, 379, 386
    Danaparoid, 378
    heparin, 378, 385
    mechanical clot lysis or removal, 377
    neuroprotection, 379
    obsolete treatments, 388
    surgery, 377–378
    thrombolysis, 376–377, 384–385
    warfarin, 379

Brain tumor, see also specific tumors
  epidemiology, 827, 831
  genetic predisposition, 827, 830
  imaging, 827, 829, 832
  metastasis from systemic solid tumors
    clinical presentation, 881–882
    course, 882–884
    imaging, 881
    markers, 881–882
    primary tumor epidemiology, 881–882
    treatment
      anticonvulsants, 883, 885, 889–890
      chemotherapy, 887–889, 890
      concomitant leptomeningeal metastasis, 889

Brain tumor *(continued)*
    corticosteroids, 883, 889
    obsolete treatments, 891
    radiotherapy, 885–887, 890
    recurrent metastasis, 889
    side effects, 890–891
    surgery, 885, 890
    treatment principles
        anticonvulsants, 837–838
        chemotherapy, 833–836
        corticosteroids, 837
        deep vein thrombosis, 838
        gene therapy, 836
        immunotherapy, 836
        pulmonary embolism, 838
        radiotherapy, 832–833
        side effects, 836–837
        surgery, 832
        symptomatic treatment, 832
    World Health Organization classification, 827–829
Brandt–Daroff exercises, benign paroxysmal positioning vertigo management, 145–148
Breast cancer metastasis, *see* Brain tumor; Leptomeningeal metastasis
Broca's aphasia, *see* Aphasia
Bromocriptine
    acromegaly management, 1437
    corticobasal degeneration management, 1094
    hyperprolactinemia management, 1443–1444
    hyperthyroidism management, 1448
    menstrual migraine prevention, 49
    Parkinson's disease management, 1037–1041
    progressive supranuclear palsy management, 1091
    side effects, 1512
    tremor management, 1238
Bromopride, hiccup management, 187
Bulbar palsy, delayed radiotherapy injury, 1338
Bupidine, Parkinson's disease management, 1047–104
Buspirone, spinocerebellar ataxia management, 1015

# C

CAA, *see* Cerebral amyloid angiopathy
Cabergoline
    acromegaly management, 1437
    Parkinson's disease management, 1037–1041
CADASIL, *see* Cerebral autosomal dominant arteriopathy with subcortical infarcts and leukoencephalopathy
Calcitonin, hypercalcemic encephalopathy management, 1002
Calcium channel blockers
    acute ischemic stroke management, 337
    cerebral ischemia prevention following subarachnoid hemorrhage, 431–432

contraindications, 1520
drug–drug interactions, 1520
migraine prevention, 9
side effects, 1520
Candidiasis
    cerebral mycosis treatment, 619
    clinical features, 616
Cannabinoids
    spasticity management, 1252
    toxicity and management, 800
Capillary teleangiectases, spinal, 527
Capsaicin cream, pain management, 104, 1333
Carbamazepine
    advantages and limitations, 218
    adverse reactions, 217, 654
    brain tumor management, 837–838
    dosing, 216
    epilepsy management, 215, 218–219
    neurocysticercosis management, 654
    neuropathic pain management, 1286
    pain management, 101–102, 1332
    pharmacokinetics, 216
    poisoning and management, 796
    schistosomiasis management, 659
    Sydenham's chorea management, 1154
    trigeminal neuralgia management, 68–69
    vestibular paroxysmia management, 153
    zoster management, 1272
Carbohydrate metabolism disorders, *see* Glycogenoses
Carbon disulfide polyneuropathy, features, 1285
Carbon monoxide, parkinsonism induction, 1068
Cardiac arrhythmias, *see* Arrhythmias
Cardiac pacemaker
    mitochondrial disease management, 1395
    syncope management, 1502
Carnitine
    deficiency
        clinical presentation, 1387
        diagnosis, 1387
        treatment, 1388
    mitochondrial disease management, 1395
Carnitine palmitoyl transferase II deficiency, features and management, 1388
Carotid artery angioplasty and stenting
    clinical trials, 498–500
    complications, 499, 501–502
    indications, 502
    patient evaluation, 501
    safety, 499
    stent designs, 501
Carotid artery jugular fistula, management, 497
Carotid cavernous sinus fistula (CCF)
    classification, 494–495
    diagnosis
        computed tomography, 493
        digital subtraction angiography, 493–494
        magnetic resonance imaging, 493
    epidemiology, 494
    pathogenesis, 494

signs and symptoms, 493
treatment, 494–495, 497
Carotid endarterectomy
    indications, 359
    morbidity, 500
    stroke prevention
        primary prevention, 351–352
        secondary prevention, 357–358
Carpal tunnel syndrome (CTS)
    causes, 1308–1309
    clinical presentation, 1308
    course, 1309
    diagnosis, 1309
    epidemiology, 1308
    treatment, 1309–1310
Catechol-O-methyltransferase (COMT), inhibitors
    adverse effects, 1044–1045
    entacapone, 1043–1044
    Parkinson's disease management rationale, 1043
    tolcapone, 1043–1044
Cavernoma, spinal, 523, 527
CBD, *see* Corticobasal degeneration
CBGD, *see* Cortical basal ganglia degeneration
CBT, *see* Cognitive behavioral therapy
CCF, *see* Benign paroxysmal positioning vertigo
CEE, *see* Central European encephalitis
Celiac disease, myoclonus, 1226
Central European encephalitis (CEE), treatment, 611
Central nervous system malformations, *see also specific malformations*
    cerebellar malformations, 955–959
    classification, 947–948
    differentiation disturbances, 954–955
    encephaloclastic lesions, 955
    etiology, 947
    gestational timing, 947–948
    midline malformations, 951
    neurocutaneous disorders, 947–948, 960–966
    neurulation disorders, 950–951
    proliferation or migration disorders, 952–954
    treatment principles, 948, 950
Central neurocytoma, features and management, 849
Central pain, management, 101, 110
Central pontine myelinolysis
    clinical presentation, 980–981
    course, 981
    treatment, 981
Central pontine myelinosis (CPM)
    clinical features, 995
    course, 995
Central venous catherization
    critical care, 726
    spinal cord injury, 929
Central vestibular vertigo
    migraine, basilar and vestibular
        clinical presentation, 155
        course, 155

related clinical syndromes, 155–156
treatment, 155
syndromes, overview, 154
traumatic vertigo, 156
vestibular epilepsy
clinical presentation, 154
pathogenesis, 155
related clinical syndromes, 155
treatment, 155
vestibulo-ocular reflex planes and
disorders, 153
Cerebellar neurolipocytoma, features and
management, 849
Cerebellar tremor, deep brain stimulation,
1106–1107
Cerebral amebiasis
amebic brain abscess
clinical presentation, 664
course, 664
definition, 664
diagnosis, 664
treatment
metronidazole, 666
obsolete treatments, 666
principles, 664, 666
tinidazole, 666
granulomatous amebic encephalitis
clinical presentation, 663
course, 664
definition, 663
diagnosis, 663–664
treatment, 664
pathogens, 662
primary amebic meningoencephalitis
clinical presentation, 662
course, 662–663
definition, 662
diagnosis, 662
prevention, 663
treatment
amphotericin B, 663
miconazole, 663
Cerebral amyloid angiopathy (CAA)
clinical manifestations, 366
course, 366
treatment, 367
Cerebral arteriovenous malformation
course, 490
diagnosis
computed tomography, 490
digital subtraction angiography, 489
magnetic resonance imaging, 489
signs and symptoms, 489–490
treatment
embolization
anesthesia, 492
anticoagulation, 492
complications, 492–493
efficacy, 491
principles, 490–491
Cerebral autosomal dominant arteriopathy
with subcortical infarcts and
leukoencephalopathy (CADASIL)
course, 365–366
diagnosis, 365

gene mutations, 365–366
treatment, 366
Cerebral blindness
clinical presentation, 255–256
course, 256–257
etiology, 255–256
treatment
blindsight, 258
head shifts, 258
oculomotor compensation
overview, 257
reading, 257–258
visual exploration, 257
optical aids, 258
principles, 257
Cerebral dural arteriovenous fistula
classification, 494
diagnosis
computed tomography, 493
digital subtraction angiography, 493–494
magnetic resonance imaging, 493
epidemiology, 494
pathogenesis, 494
signs and symptoms, 493
treatment, 494, 497
Cerebral fibrosarcoma, features and
management, 854
Cerebral malaria
clinical presentation, 659–660
course, 660
definition, 659
diagnosis, 660
treatment
ancillary treatment, 661
Artemether, 661–662
Artesunate, 661–662
combination therapy, 661
mefloquine, 661
obsolete treatments, 662
principles, 660
quinine, 660–662
Cerebral perfusion pressure (CPP),
maintenance in critical care,
728–729, 757, 761
Cerebral small vessel disease
clinical presentation, 363
course, 364
imaging, 363–364
treatment
obsolete treatments, 365
practical management, 365
principles, 364–365
Cerebral venous and sinus thrombosis
(CVST)
course, 449, 451–453, 508–509
diagnosis
cerebrospinal fluid findings, 449
computed tomography, 448–449
magnetic resonance angiography,
447–448, 451, 509
magnetic resonance imaging, 447–448,
451, 509
etiologies and risk factors, 449, 452, 457
frequency, 447
symptoms and signs, 447–448

treatment
analgesia, 456–457
endovascular thrombolysis, 454, 509
heparin, 453, 455–456
intracranial pressure elevation, 457
low-molecular weight heparin, 453–454
obsolete treatments, 459
pregnancy, 457
seizures, 456
septic disease, 457–459
thrombectomy, 455
Cerebroretinal vasculopathy, features, 70
Cerebrospinal fluid (CSF)
absorption, 806–807
bacterial meningitis findings, 529
cerebral venous and sinus thrombosis
findings, 449
drainage
hydrocephalus management, 435,
809–810
intracerebral hemorrhage, 409
intracranial pressure control, 730–732,
757 findings
spinal cord ischemia management, 399
formation, 806
fungal infection of nervous system, 615
hydrocephalus, see Hydrocephalus
imaging of dynamics, 807
intracranial pressure, see Intracranial
pressure
leptomeningeal metastasis findings,
898–899
Lyme neuroborreliosis findings, 585–586,
589
multiple sclerosis findings, 679–680
normal pressure hydrocephalus
diagnostics
drainage, 1115
dynamics, 1115
markers, 1116
shunting
outcomes, 1119–1120
technique, 1118–1119
pressure regulation, 807–808
radiculopathy, 1290
sarcoidosis findings, 567–568
syphilis findings, 576–578
toxoplasmosis findings, 595–596
tuberculous meningitis findings, 559–560
viral infection diagnostics, 603
Cerebrotendinous xanthomatosis, features,
1014
Cervical dystonia
clinical presentation, 1132
course, 1132
treatment
botulinum toxin, 1132–1134
surgery, 1133
trihexyphenidyl, 1132
Cervical spondylitic myelopathy (CSM)
clinical features, 1294
course, 1294–1295
diagnosis, 1294
differential diagnosis, 1294
treatment, 1295

Charcot-Marie-Tooth disease, *see* Hereditary
        motor and sensory neuropathy
Chemotherapy
    administration routes, 835
    agent indications, dosing, and toxicity, 834
    astrocytic tumors, 840–841
    blood–brain barrier disruption, 835
    choroid plexus tumors, 848
    combination protocols, 834–835
    drug transport, 833
    efficacy evaluation, 836
    embryonal tumors, 850
    germ cell tumors, 855–856
    high-dose chemotherapy, 835
    Hodgkin's disease, 875–876
    leptomeningeal metastasis
        intrathecal chemotherapy
            Ara-C, 902, 906
            combination chemotherapy, 902
            duration and termination, 906–907
            methotrexate, 901–902, 905–906
            rationale, 899, 901
            technique, 905
            thiotepa, 902
        outcomes, 900–901
        systemic chemotherapy, 901, 907
    metastatic brain tumor management,
        887–889, 890
    non-Hodgkin's lymphoma, 872–874
    primary central nervous system lymphoma,
        870–871
    radiochemotherapy, 835
    resistance, 833
    side effects
        adjunct therapy, 1519
        alkaloids, 1518–1519
        alkyating agents, 1518
        antibiotics, 1519
        antimetabolites, 1518
        nitrosourea, 1519
        overview, 837, 890–891
        podophyllotoxins, 1519
Chest X-ray, critical care monitoring, 733
Chiari malformation
    type I
        clinical presentation, 956
        course, 956
        treatment, 956–957
    type II
        clinical presentation, 957
        course, 957
        treatment, 957
    type III, 957
    type IV, 957
Chlordiazepoxide, alcoholic hallucinosis
        management, 973
Chloride channel myotonia, *see* Myotonia
        congenita
Chlormezanone, radiculopathy management,
        1292
Chlorpromazine, poisoning and management,
        798
Chorambucil, vasculitides management,
        467
Chordoma, features and management,
        856–857

Choroid plexus tumors
    carcinomas, 848
    clinical features, 848
    papillomas, 848
    treatment, 848
Chronic cervical myelopathy
    clinical features, 934
    treatment, 934–935
Chronic inflammatory demyelinating
        polyneuropathy (CIDP)
    clinical presentation, 1262–1263
    course, 1263
    diagnosis, 1263
    incidence, 1262
    treatment, 1263–1264
Chronic pain, *see* Pain
Chronic paroxysmal hemicrania, hormone-
        sensitive headache in women,
        54–55
Chronic progressive external
        ophthalmoplegia, *see* Mitochondrial
        disease
Chronic renal failure, polyneuropathy
        features, 1284–1285
Churg–Strauss, *see* Vasculitides
Cidofovir
    mechanism of action, 609
    spectrum of activity, 609
CIDP, *see* Chronic inflammatory
        demyelinating polyneuropathy
Cinnarizine, side effects, 1510
CJD, *see* Creutzfeldt-Jakob disease
Cladribine, multiple sclerosis management,
        690
Clenbuterol, facioscapulohumeral dystrophy
        management, 1377–1378
Clobazam
    advantages and limitations, 218
    adverse reactions, 217
    dosing, 216
    epilepsy management, 215
    pharmacokinetics, 216
Clofazimine, leprosy management,
        1269–1270
Clofibrate, side effects, 1521
Clomethiazole, alcohol withdrawal
        management, 975–977
Clonazepam
    acquired pendular nystagmus management,
        129
    advantages and limitations, 218
    adverse reactions, 217
    dementia with Lewy bodies management,
        1089
    dosing, 216
    epilepsy management, 215
    pain management, 101–102, 1332
    pharmacokinetics, 216
    restless legs syndrome management,
        1178–1179, 1181
    trigeminal neuralgia management, 72
Clonidine
    alcohol withdrawal management, 976
    orthostatic hypotension management, 1467
    pain management, 104
    pain management after nerve injury, 1332

restless legs syndrome management, 1179,
        1181
    side effects, 1521
    tic disorder management, 1140
Clopidogrel, stroke prevention, 357–359
*Clostridium botulinum*, *see* Botulism
*Clostridium tetani*, *see* Tetanus
Clotrimazole, side effects, 1518
Clozapine
    dementia with Lewy bodies management,
        1087–1089
    dopaminergic psychosis management,
        1066
    Parkinson's disease management, 1048
    poisoning and management, 798–799
Cluster headache
    course, 17–18
    diagnostic criteria, 17–18
    differential diagnosis, 17–18
    hormone-sensitive headache in women, 54
    patient education, 19
    prevention
        corticosteroids, 19–20
        ergotamine, 20
        lithium, 20
        long-term, 20
        methysergide, 20
        short-term, 19–20
        surgery, 20
        verapamil, 20
    treatment
        lignocaine, 19
        obsolete treatments, 19
        oxygen therapy, 19
        principles, 18–19
        sumatriptan, 19
CMV, *see* Cytomegalovirus
Cocaine, poisoning and management, 800
Coccidioidomycosis
    cerebral mycosis treatment, 619
    clinical features, 617
Cochlear implant
    auditory brainstem implant comparison,
        1209
    complications, 1209
    electrode design, 1206–1207
    outcomes, 1209
    patient selection and indications,
        1207–1208
    popularity, 1206
    preoperative procedures, 1208–1209
    signal transmission, 1207
    stimulation strategies, 1207
    stimulator design, 1207
Codeine, metabolism, 106
Coenzyme Q10, mitochondrial disease
        management, 1394
Cogan's syndrome, *see* Vasculitides
Cognitive behavioral therapy (CBT), pain
        management, 107–108
Collet-Sicard syndrome, *see*
        Glossopharyngeal neuralgia
Colorectal cancer metastasis, *see* Brain tumor
Color vision impairment
    clinical features and course, 258–259
    treatment, 259

radiculopathy, 1290
  stiff-man syndrome, 1430–1431
Eletriptan, migraine treatment, 4–5
Embolism, *see* Brain embolism
Embryonal tumors
  clinical presentation, 850
  course, 850
  treatment, 850
  types, 849–850
EMG, *see* Electromyography
Empty sella syndrome, features and
      management, 1449
Empyema, *see* Spinal subdural empyema;
      Subdural empyema
Encephalitis, *see* Paraneoplastic syndromes;
      Viral encephalitis
Encephaloceles, features, 951
Encephalopathy, *see specific encephalopathies*
Entacapone, Parkinson's disease
      management, 1043–1044
EOCA, *see* Early-onset cerebellar ataxia
Eosinophilia-myalgia syndrome, features,
      1402
EPA, *see* Eicosopentanoic acid
Ependymomas
  clinical features, 845, 847
  treatment
      radiotherapy, 847–848
      surgery, 847
  types, 845
Ephedrine, orthostatic hypotension
      management, 1466
Epidermoid tumors, features and
      management, 857
Epidural abscess
  definition, 552
  diagnosis, 552
  etiology, 552
  treatment, 552
Epidural blood patch, postlumbar puncture
      headache treatment, 79–80
Epilepsy
  aphasia, 268
  classification
      epilepsy syndromes, 209–211
      seizure type, 208–209
  course and prognosis, 213–214, 235
  definition, 207
  differential diagnosis, 210–211
  etiology, 210
  febrile seizures and risks, 226
  frequency, 210
  gene mutations, 207–208
  investigations
      electroencephalography, 211–212, 236
      magnetic resonance imaging, 212, 236
      neuropsychological testing, 236–237
      serum drug monitoring, 212–213
  palliative care of epileptic seizures, 820
  pathophysiology, 207–208
  posttraumatic epilepsy, 776–777
  status epilepticus
      causes of tonic–clonus status, 229–230
      classification, 229
      emergency management, 229–230
      failure of emergency drug therapy, 231

medical complications, 230
  status epilepticus
      clinical features, 744
      critical care, 744–745
      focal status management, 745
      grand mal seizure management, 744–745
  treatment
      adverse reactions, 217, 221
      benzodiazepines, 217
      carbamazepine, 218–219
      comparison of drugs, 218
      dosing and pharmacokinetics, 216, 221
      efficacy, 214
      ethosuximide, 219
      felbamate, 220
      fosphenytoin, 220
      gabapentin, 222
      initial treatment and drug selection, 214
      lamotrigine, 222
      levetiracetam, 223
      oxcarbazepine, 223
      phenobarbitol, 219–220
      phenytoin, 220
      primidone, 220
      refractory epilepsy, 224–225
      stopping of drugs, 215
      surgical therapy
          deep brain stimulation, 242
          extramesial temporal epilepsies,
              240–241
          gamma knife radiosurgery, 242
          generalized epilepsies, 241
          indications, 235–237
          mesial temporal epilepsies, 238, 240
          morbidity and mortality, 240
          outcome classification, 239–240
          overview of procedures, 225–226,
              238
          presurgical evaluation, 236–238
          principles, 236–237
          vagus nerve stimulation, 226,
              241–242
      tiagabine, 223
      topiramate, 223
      valproic acid, 219
      vigabatrin, 223–224
      zonisamide, 224
  women
      fertility reduction, 226–227
      oral contraceptives, considerations, 227
      pregnancy management
          breastfeeding, 229
          fetal effects of seizures, 228–229
          obstetric complications, 229
          overview, 227
          prepregnancy counseling, 227
          teratogenicity of drugs, 227–228
Episodic ataxia (EA)
  course, 1015
  pathogenesis, 1015
  treatment, 1016
  types and gene mutations, 1015
Epistaxis
  causes, 492
  embolization, 482–483

Erectile dysfunction, *see* Sexual dysfunction
Ergotamine
  cluster headache prevention, 20
  dependence, 32
  medication overuse headache, *see*
      Medication overuse headache
  migraine treatment, 4, 7
Esophageal autonomic dysfunction
  distension, 1469
  spasm, 1469
Essential tremor
  deep brain stimulation, 1105–1106
  diagnostic criteria, 1234–1235
  epidemiology, 1235–1236
  gene loci, 1236
  pathophysiology, 1236
  possible cases, types, 1234–1235
  treatment, 1236
Etanercept, vasculitides management, 468
Ethambutol
  adverse effects, 564
  side effects, 1517
  tuberculous meningitis management,
      562–563
Ethanol, poisoning and management, 802
Ethosuximide
  advantages and limitations, 218
  adverse reactions, 217
  dosing, 216
  epilepsy management, 215, 219
  pharmacokinetics, 216
Ethylene glycol, poisoning and management,
      802–803
Etomidate, intracranial pressure control, 730
Evoked potential, critical care monitoring,
      732
Exertional headache, clinical features, 43
Experimental autoimmune encephalitis
      (EAE), pathogenesis, 683–684

**F**

Fabry's disease
  clinical manifestations, 369
  course, 369
  treatment, 369–370
Facioscapulohumeral dystrophy (FSHD)
  clinical presentation, 1376–1377
  course, 1377
  gene mutations, 1377
  treatment
      albuterol, 1377–1378
      clenbuterol, 1377–1378
      clinical trial design, 1378–1379
      corticosteroids, 1377
      obsolete treatments, 1379–1380
      surgery, 1379
Famcilovir
  mechanism of action, 608
  spectrum of activity, 608
Fatigue, palliative care, 822–823
Felbamate
  adverse reactions, 222
  dosing, 221
  epilepsy management, 220
  pharmacokinetics, 221

Femoral nerve compression, features and management, 1314
Femoral nerve palsy, delayed radiotherapy injury, 1337–1338
Fentanyl
    administration routes, 105
    pain management, 103
FES, see Functional electrical stimulation
Fever, see also specific diseases
    definitions, 783
    hyperthermia comparison, 783
    management
        antibiotics and relapsing fever management, 591
        intracranial pressure elevation, 756
        stroke, 341
    pathophysiology, 783
Filariasis, see Neurofilariasis
FK506, see Tacrolimus
Fluconazol
    adverse effects and contraindications, 618
    cerebral mycosis treatment, 619
    dosing, 621
    indications, 619, 621
    pharmacological properties, 618–619
Flucytosine
    adverse effects and contraindications, 618
    cerebral mycosis treatment, 619
    dosing, 621
    indications, 619
    pharmacological properties, 618–619
    side effects, 1518
Fludrocortisone
    orthostatic hypotension management, 1466–1467
    syncope management, 1501–1502
Fluphenazine, tic disorder management, 1140
Focal dystonia
    blepharospasm, 1130–1131
    botulinum toxin therapy, 1128–1130
    cervical dystonia, 1132–1134
    clinical features, 1128
    course, 1128
    cranial dystonia, 1131–1132
    hand cramp, 114–1135
    spasmodic dystonia, 1134
Foscarnet
    cytomegalovirus treatment, 611
    mechanism of action, 608
    spectrum of activity, 608–609
Fosphenytoin
    adverse reactions, 222
    dosing, 221
    epilepsy management, 220, 222
    pharmacokinetics, 221
FRDA, see Friedreich's ataxia
Friedreich's ataxia (FRDA)
    clinical presentation, 1011
    course, 1011
    gene mutations, 1011
    treatment, 1011–1012
Frontotemporal degeneration (FTD)
    classification, 315
    diagnostic criteria, 315–316
    neuropathology, 316
    treatment, 320

FSHD, see Facioscapulohumeral dystrophy
FTD, see Frontotemporal degeneration
Fulminant hepatic encephalopathy, see Hepatic encephalopathy
Functional electrical stimulation (FES), see also Neural prosthesis
    advantages, 1200–1201
    ambulation, 1203
    ambulation with stance hybrid systems, 1203–1204
    contraindications, 1201
    cough induction, 1211–1212
    fecal continence, 1213
    gait in hemiparesis, 1204
    grasp restoration in tetraplegia, 1205–1206
    lower limbs, 1192
    muscle fatigue and training, 1200
    prerequisites, 1201
    stance and gait in paraplegia, 1202–1203
    stimulators and electrodes, 1201–1202
    stroke and upper extremity function, 1204–1205
    upper limbs, 1189–1190, 1204–1205
Fungal meningitis
    clinical presentation, 615
    diagnosis, 615
    differential diagnosis, 615

G
Gabapentin
    acquired pendular nystagmus management, 129
    advantages and limitations, 218
    adverse reactions, 222
    dosing, 221
    epilepsy management, 222
    pain management after nerve injury, 1332
    pharmacokinetics, 221
    trigeminal neuralgia management, 69, 72
    vertigo management, 142
Galantamine, Alzheimer's disease dementia management, 317–318
Gamma-hydroxybutyrate (GHB)
    alcohol dependency management, 986–987
    alcohol withdrawal management, 976
    poisoning and management, 801
Gamma knife radiosurgery
    epilepsy management, 242
    thalamotomy, 1108
Ganciclovir
    cytomegalovirus treatment, 611
    mechanism of action, 608
    spectrum of activity, 608
Ganglioblioma, features and management, 848–849
Gangliocytoma, features and management, 848–849
Ganglioneuroblastoma, features and management, 849
Ganglioneuroma, features and management, 849
Gastric autonomic dysfunction, stasis and distension, 1469–1470

Gaucher's disease, progressive myoclonic epilepsy, 1225
GCA, see Giant cell arteritis
GCS, see Glasgow Coma Scale
Gene mapping
    DNA polymorphisms, 1531
    genome sequencing, 1525
    linkage analysis, 1530–1531
    mutation, 1526
    neurogenic disorders and genes, table, 1526–1529
    positional cloning, 1530–1531
Gene therapy
    brain tumor, 836
    Duchenne's muscular dystrophy management, 1373
    glycogenoses, 1387
    leptomeningeal metastasis, 903
Generalized dystonia
    clinical features, 1124–1126
    course, 1126
    paroxysmal kinesiogenic choreoathetosis, 1127
    paroxysmal nonkinesiogenic dystonia, 1127
    secondary dystonias, 1125–1126
    treatment
        anticholinergic agents, 1127
        baclofen, 1127
        benzodiazepines, 1127
        dopaminergic therapy, 1126–1127
Genitofemoral nerve compression, features and management, 1318–1319
Gentamicin, intratympanic therapy for Meniére's disease, 150–151
Germ cell tumors
    clinical presentation, 855
    course, 855
    treatment, 855–856
    types, 855
GHB, see Gamma-hydroxybutyrate
Giant axonal neuropathy, features, 1283
Giant cell arteritis (GCA)
    biopsy, 476
    clinical presentation, 475
    course, 477
    diagnosis, 475–476
    treatment, 477–478
Gingko biloba, side effects, 1511
Glasgow Coma Scale (GCS), traumatic brain injury assessment, 739–740
Glatiramer acetate, multiple sclerosis management, 687, 695–696
Glaucomatous optic neuropathy, clinical features, 123
Glioblastoma
    clinical presentation, 838–839
    course, 839
    treatment
        chemotherapy, 840–841, 844
        radiochemotherapy, 842–843
        radiotherapy, 840, 844
        surgery, 839, 844
Gliomatosis cerebri, features and management, 848

Glossopharyngeal neuralgia
  clinical presentation, 73
  course, 73
  differential diagnosis, 73
  pathogenesis, 73
  treateoment
    computed tomography-guided
      tracteotomy, 74
    drugs, 74
    microvascular decompression, 73
    nerve section, 74
    percutaneous thermocoagulation, 74
Glucocorticoids, see Corticosteroids
Glutaraciduria type II, features and
  management, 1388–1389
Glycine, spasticity management, 1252
Glycogenoses
  course, 1385
  treatment
    creatine, 1386
    diet, 1386
    enzyme replacement therapy, 1386
    gene therapy, 1387
    liver transplantation, 1387
    obsolete treatments, 1387
    physical therapy, 1387
    symptomatic therapy, 1387
  types and presentation, 1385–1386
Glycoprotein IIb/IIIa inhibitors
  acute ischemic stroke management,
    336–337
  stroke prevention, 357
  thrombolytic therapy utilization, 508
Gnathostomiasis, features and management,
  598
Gonadotropin deficiency
  clinical presentation, 1441
  course, 1441
  diagnosis, 1441
  differential diagnosis, 1441
  treatment
    men, 1442
    principles, 1441
    women, 1442
Gonadotropin excess, features and
  management, 1442
Granular cell tumors of the neurohypophysis,
  features and management, 856
Granulomatous amebic encephalitis, see
  Cerebral amebiasis
Griseofulvin, side effects, 1518
Growth hormone disorders, see Acromegaly;
  Hypopituitarism
Guillain-Barré syndrome
  autonomic dysfunction, 1456–1457
  clinical presentation, 1259
  course, 1261
  diagnosis, 1260
  differential diagnosis, 1260
  epidemiology, 1259
  human immunodeficiency virus
    neuropathy, 1272–1273
  pathogenesis, 1259–1260
  treatment
    general medical and nursing care,
      1261

    intravenous immunoglobulin,
      1261–1262
    plasmapheresis, 1261–1262
    rehabilitation, 1262
    variants, 1260

H

HAART, see Highly-active antiretroviral
  therapy
Hallucination
  course, 261
  definition, 260
  etiology, 260–261
Haloperidol
  alcoholic hallucinosis management, 973
  alcohol intoxication management, 972
  alcohol withdrawal management, 976–977
  delirium management, 308
  Huntington's disease management, 1152
  palliative care, 819
  poisoning and management, 798
  Sydenham's chorea management, 1154
  tic disorder management, 1140
Hand cramp
  clinical presentation, 1134
  course, 1134
  treatment, 1134–1135
HD, see Hodgkin's disease; Huntington's
  disease
Headache, see also Atypical facial pain;
    Cluster headache; Cough headache;
    Exertional headache; Hemicrania
    continua; Hormone-sensitive
    headache in women; Idiopathic
    stabbing headache; Medication
    overuse headache; Migraine;
    Paroxysmal hemicrania; Postlumbar
    puncture headache; Sexual
    headache; Short-lasting unilateral
    neuralgiform headache with
    conjuctival injection and tearing;
    Subarachnoid hemorrhage; Tension-
    type headache
  drug induction, 1523
  exertion and exacerbation, 44
  neurogenic plasma protein extravasation,
    2–3
  neuropeptide release, 3
  trigeminocervical complex, 3
Heavy metal encephalopathies
  aluminum, 1003–1004
  antidotes, 1004–1005
  clinical features
    acute encephalopathy, 1002
    chronic encephalopathy, 1002–1003
  lead
    overview, 1003
    treatment, 1005–1006
  manganese
    overview, 1003
    treatment, 1006
  mercury
    overview, 1003
    treatment, 1006
  selenium, 1003

  thallium
    overview, 1003
    treatment, 1006
Heavy metal polyneuropathy, features, 1285
Height vertigo
  clinical presentation, 158–159
  course, 159
  treatment, 160–161
Hemangioblastoma, features and
  management, 854–855
Hemangiopericytoma, features and
  management, 854
Hematoma, critical care
  acute subdural hematoma, 743
  chronic subdural hematoma, 743
  epidural hematoma, 742–743
  intracerebral hematoma, 743
  monitoring, 741
Hematopoietic stem cell transplantation
  multiple sclerosis management, 691
  primary central nervous system lymphoma
    management, 871
Hemicrania continua
  clinical features, 37–38
  diagnostic criteria, 38
  differential diagnosis, 39
  hormone-sensitive headache in women, 55
  treatment, 38–39
Hemifacial spasm (HFS)
  clinical features, 1140–1141
  course, 1141
  treatment
    botulinum toxin, 1142
    microvascular decompression,
      1141–1142
Heparin
  acute ischemic stroke management,
    334–336, 342–343
  anticoagulation reversal, 405–406,
    414–415
  brain embolism management, 378, 385
  cerebral venous and sinus thrombosis
    management, 453, 455–456
  reversal of anticoagulation, 1313
  thrombolytic therapy utilization, 508
Hepatic encephalopathy
  clinical presentation, 991–992
  course
    fulminant hepatic encephalopathy, 992
    portal systemic encephalopathy,
      992–993
  pathogenesis, 993
  stages, 991–992
  treatment
    fulminant hepatic encephalopathy, 993
    portal systemic encephalopathy,
      993–994
Hereditary endotheliopathy with retinopathy,
  nephropathy, and stroke (HERNS),
  features, 370
Hereditary motor and sensory neuropathy
  (HMSN)
  classification, 1281
  gene mutations, 1282
  genetic heterogeneity, 1533
  type I, 1282

Hereditary motor and sensory neuropathy
        (HMSN) *(continued)*
    type II, 1282
    type III, 1282
Hereditary sensory and autonomic
        neuropathies (HSAN), features,
        1283
Hereditary spastic paraparesis, features, 1172
Hereditary neuropathy with liability to
        pressure palsies (HNPP), features,
        1282
HERNS, *see* Hereditary endotheliopathy with
        retinopathy, nephropathy, and
        stroke
Herpes simplex encephalitis
    diagnosis, 610
    myoclonus, 1226
    treatment, 610–611
Herpes simplex meningitis, management,
        611
Herpes simplex virus (HSV)
    Bell's palsy, role, 136
    human immunodeficiency virus association
        and management, 631, 635
Herpesvirus, clinical features and course of
        infection, 605
Herpes zoster virus
    human immunodeficiency virus association
        and management, 631, 635
    zoster, *see* Zoster
Heterotopic ossification, spinal cord injury
        management, 933
Hexane, poisoning and management, 802
HFS, *see* Hemifacial spasm
Hiccup
    course, 186
    definition, 185
    etiology, 185–186
    myoclonus, 185
    primitive motor pattern, 186
    treatment
        folk remedies, 187
        invasive therapy, 188
        noninvasive therapy, 187
        pharmacological therapy, 187
        principles, 186–187
    vagal reflex, 185
High-altitude cerebral edema, management,
        760
Highly-active antiretroviral therapy (HAART),
        *see* Human immunodeficiency virus
Histoplasmosis
    cerebral mycosis treatment, 619
    clinical features, 617
HIV, *see* Human immunodeficiency virus
HMSN, *see* Hereditary motor and sensory
        neuropathy
HNPP, *see* Hereditary neuropathy with
        liability to pressure palsies
Hodgkin's disease (HD)
    central nervous system metastases, 874
    course, 874–875
    imaging, 874
    markers, 874
    treatment
        chemotherapy, 875–876

corticosteroids, 875–876
    radiotherapy, 875–876
Holoprosencephaly (HPE), features, 951–952
Hormone replacement therapy
    Alzheimer's disease trials, 321
    benefits and risks, 50
    migraine treatment, 50–51
Hormone-sensitive headache in women
    chronic paroxysmal hemicrania, 54–55
    cluster headache, 54
    hemicrania continua, 55
    menopausal migration
        hormonal replacement therapy and
            migraine treatment, 49–51
        hormone levels, 49
    menstrual migraine
        clinical presentation, 47–48
        pathogenesis, 48
        prevention, 48–49
        treatment, 48
    migraine types, 47
    oral contraceptives induction of migraine,
        51–52
    postpartum headache, 54
    pregnancy
        drug safety, 53–54
        migraine frequency effects, 52–53
        migraine treatment and prevention, 53
        postnatal headache, 52–53
        secondary headaches and causes, 55
    tension-type headache, 54
HPE, *see* Holoprosencephaly
HSAN, *see* Hereditary sensory and
        autonomic neuropathies
HSV, *see* Herpes simplex virus
Human African trypanosomiasis, *see* Sleeping
        sickness
Human immunodeficiency virus (HIV)
    autonomic dysfunction, 1460–1461
    Centers for Disease Control definition of
        AIDS, 630
    cerebrovascular disease, 635
    course, 636
    diagnosis, 623–624
    highly-active antiretroviral therapy
        development, 637–638
        drug classes, 638
        regimens, 638
        toxicity of drugs, 631, 638
    laboratory classification and clinical
        category of disease, 629
    lentivirus features, 623
    malignancies and neurological involvement
        Kaposi's sarcoma, 633, 635
        non-Hodgkin's lymphoma, 633,
            641–642
        primary central nervous system
            lymphoma, 633–634, 641, 868–870
        treatment, 641–642
    neurological complications, features and
        treatment
        dementia, 625, 632, 637, 639
        differential diagnosis, 624–625
        frequency of types, 628
        mononeuritis multiplex, 632, 640
        myopathy, 627, 632

peripheral nerve syndromes, 626–627,
            632, 636–637, 639–640
        seroconversion illness, 624
        timing of onset, 624, 636
        vacuolar myelopathy, 625–626, 632,
            640
    neuropathies
        course, 1272–1273
        treatment, 1272–1273
        types and presentation, 1272
    opportunistic infections and neurological
        complications
        cryptococcal meningitis, 628–629, 634,
            637, 640–641
        cytomegalovirus, 629–630, 634–635
        herpes simplex virus, 631, 635
        herpes zoster virus, 631, 635
        JC virus, 630
        miscellaneous infections, 633
        syphilis, 578–579, 633, 635
        toxoplasmosis, 628, 634, 637, 640
        tuberculous meningitis, 631
    progressive multifocal leukencephalopathy
        features and management, 634, 637,
            641
    receptors, 623
    serotypes, 623
Huntington's disease (HD)
    clinical features, 1149
    course, 1151
    diagnosis, 1149
    differential diagnosis, 1150
    etiology, 1150
    lay organizations, 1153
    pathogenesis, 1150–1151
    treatment
        akinetic-rigid symptoms, 1152
        experimental treatments, 1152–1153
        general measures, 1152
        hyperkinesias, 1152
        ineffective treatments, 1153
        principles, 1151–1152
        psychiatric symptoms, 1152
Hydatid disease, *see* Echinococcosis
Hydergine, Alzheimer's disease trials, 321
Hydranencephaly, features, 955
Hydrocephalic dementia, features and
        management, 321
Hydrocephalus
    clinical presentation, 808–809
    definition, 805
    historical perspective, 805–806
    imaging, 809
    mechanisms, 808
    normal pressure hydrocephalus, *see*
        Normal pressure hydrocephalus
    secondary parkinsonism, 1068
    treatment
        cerebrospinal fluid drainage, 435,
            809–810
        fungal infection complication
            management, 621
        pharmacotherapy, 809
        sarcoidosis complication management,
            570, 572
        shunts, 810

subarachnoid hemorrhage association and acute management, 428, 434–435
surgical correction, 565
Hydromorphone, pain management, 104
γ-Hydroxybutyrate, *see* Gamma-hydroxybutyrate
Hypercalcemic encephalopathy
clinical features, 1000
treatment, 1000–1002
Hyperhidrosis
investigation, 1471
pathophysiology, 1470–1471
treatment, 1471–1472
Hyperkalemic periodic paralysis
clinical presentation of familial disease, 1413–1414
diagnosis, 1413–1414
differential diagnosis, 1418–1419
secondary disease
course, 1415
pathophysiology, 1415
treatment
attacks, 1417–1418
obsolete treatments, 1418
principles, 1415, 1417
prophylaxis, 1417–1418
sodium channel myotonia, 1408
Hypernatremia, critical care management, 725
Hyperparathyroidism
clinical presentation, 1448–1449
course, 1449
diagnosis, 1449
differential diagnosis, 1449
treatment, 1449
Hyperprolactinemia
clinical presentation, 1442
course, 1443
diagnosis, 1442–1443
differential diagnosis, 1443
treatment
bromocriptine, 1443–1444
lisuride, 1443
macroprolactinoma, 1444
microprolactinoma, 1444
pergolide, 1443
pseudoprolactinoma, 1444
radiotherapy, 1443–1444
surgery, 1443
Hypersensitivity vasculitis, *see* Vasculitides
Hypertension
cerebral ischemia prevention following subarachnoid hemorrhage, 431
critical care management, 727–728
intracerebral hemorrhage risks and control, 406–407, 415
stroke risks and control, 349, 352–353
Hyperthyroidism
clinical presentation, 1447
course, 1448
diagnosis, 1447–1448
differential diagnosis, 1448
treatment
bromocriptine, 1448
methimazole, 1448

propylthiouracil, 1448
thyrotoxic coma, 1448
Hyperventilation, intracranial pressure control, 730, 758
Hypokalemic myopathy, features, 1402
Hypokalemic periodic paralysis
acquired periodic paralysis, 1414–1415
clinical presentation, 1413–1414
diagnosis, 1413–1414
differential diagnosis, 1418–1419
familial potassium deficiency syndromes, 1415
thyrotoxic periodic paralysis, 1414
treatment
attacks, 1417–1418
obsolete treatments, 1418
principles, 1416
prophylaxis, 1417–1418
Hyponatremia
central pontine myelinolysis correction, 981
critical care management, 724–725
management following subarachnoid hemorrhage, 435–436
Hypoparathyroidism
features and management, 1448
parkinsonism induction, 1068–1069
Hypopituitarism
causes
inflammation, 1445
pituitary infarction, 1445
rare causes, 1445
trauma, 1445
tumors, 1445
clinical presentation, 1444
course, 1446
diagnosis, 1445–1446
growth hormone deficiency, 1444–1445
treatment
acute disease, 1446–1447
chronic disease, 1446
principles, 1446
Hypotension, critical care management, 728
Hypothalamic dysfunction, *see* Hypopituitarism
Hypothermia
acute ischemic stroke management, 338–339, 737–738
spinal cord ischemia management, 399
Hypothyroidism
ataxia, 1018
clinical presentation, 1447
course, 1447
diagnosis, 1447
differential diagnosis, 1447
treatment
hypopituitarism, 1446–1447
thyroid hormone, 1447
Hypoxia, myoclonus, 1226–1227

**I**

IBM, *see* Inclusion body myositis
ICH, *see* Intracerebral hemorrhage
ICS, *see* Intracerebral stimulation

Idazoxan, progressive supranuclear palsy management, 1092
Idebenone, Friedreich's ataxia management, 1012
Idiopathic facial palsy, *see* Bell's palsy
Idiopathic stabbing headache
clinical features and diagnosis, 40–42
treatment, 40–41
Iliohypogastric nerve compression, features and management, 1318–1319
Ilioinguinal nerve compression, features and management, 1318–1319
Imipramine, pain management, 101
Inclusion body myositis (IBM)
clinical presentation, 1363
course, 1365
diagnosis
creatine kinase levels in muscle, 1363
criteria, 1364
electromyography, 1363–1364
muscle biopsy, 1364–1365
differential diagnosis, 1365
treatment, 1365
Indomethacin
hemicrania continua treatment, 38
intracranial pressure elevation management, 762
paroxysmal hemicrania treatment, 21
radiculopathy management, 1292
Infliximab, vasculitides management, 468
Insomnia
psychiatric insomnia
clinical features, 194
treatment, 194
psychophysiological insomnia
clinical features, 195
treatment, 195–196
Insulin
acute ischemic stroke management, 338, 341
intracranial pressure elevation management, 757
Intensive care, *see* Critical care neurology
Interferon-β
mechanism of action, 609
multifocal motor neuropathy management, 1264
multiple sclerosis management, 685–686, 693–695
optic neuritis management, 119
spectrum of activity, 609
Interferon-γ, vasculitides management, 467–468
Intestinal autonomic dysfunction
distension, 1470
pseudoobstruction, 1470
Intoxication, *see* Acute intoxication
Intracerebral hemorrhage (ICH)
causes, 403
clinical features
hemorrhagic location, 403–404
increased intracranial pressure, 403
course, 404–405
intracranial pressure response
hypertension, 407
hyperthermia, 408
hypoxia, 407

Intracerebral hemorrhage (ICH) (continued)
    intrathoracic pressure elevation, 408
    seizures, 407–408
  risk factors, 403–404
  treatment
    anticoagulation reversal, 405–406,
      414–415
    hypertension control, 406–407, 415
    intracranial pressure control
      cerebrospinal fluid drainage, 409
      corticosteroids, 409
      hyperventilation, 408
      intravenous barbiturates, 409
      monitoring, 410
      osmotic diuretics, 408–409, 416
    obsolete treatments, 417
    principles, 405
    seizures, 415–416
    surgery
      cerebellar hemorrhage, 412–413
      clinical trials of surgery versus
        conservative treatment, 410–412
      course monitoring, 416–417
      endoscopic drainage of hematomas,
        413–414
      indications, 416
      lobar hemorrhage, 412
      mass effect and drainge patterns, 416
      stereotactic drainage of hematomas,
        413
      thalamic hemorrhage, 412
      timing, 416
      ultra-early, minimally invasive,
        thrombolytic-assisted evacuation of
        hematomas, 414
  vascular malformation evaluation, 406
Intracerebral stimulation (ICS), pain
    management, 109
Intracranial pressure
  anatomy, 749–750
  causes of elevation, 750
  cerebral perfusion pressure relationship,
    728–729, 757, 761
  compensatory mechanisms, 752
  contributing factors, 753–754
  critical care
    cerebral perfusion pressure maintenance,
      728–729
    control
      barbiturates, 730
      body position, 729
      cerebrospinal fluid drainage, 730–732
      corticosteroids, 730
      etomidate, 730
      hyperventilation, 730
      lidocaine, 731
      osmotherapy, 729–730
      propofol, 731
      traumatic brain injury, 740–741,
        776–777
      tromethamine, 730
    monitoring, 728, 731
  diagnosis of elevation
    clinical presentation, 750–751
    differential diagnosis, 751
    imaging, 751

herniation syndromes, 752–753
idiopathic intracranial hypertension, see
    Pseudotumor cerebri
intracerebral hemorrhage
  control of pressure
    cerebrospinal fluid drainage, 409
    corticosteroids, 409
    hyperventilation, 408
    intravenous barbiturates, 409
    monitoring, 410
    osmotic diuretics, 408–409, 416
  response
    hypertension, 407
    hyperthermia, 408
    hypoxia, 407
    intrathoracic pressure elevation, 408
    seizures, 407–408
management of elevation
  airway management, 757–758
  anesthetics, 759
  anoxia, 760–761
  antibiotics, 757
  anticonvulsants, 756–757
  barbiturates, 759
  blood pressure, 757
  body position, 756
  cerebrospinal fluid drainage, 757
  corticosteroids, 759
  decompressive craniotomy, 761
  deep venous thrombosis prophylaxis,
    757
  fever, 756
  fluids, 756
  fulminant hepatic failure, 759–760
  high-altitude cerebral edema, 760
  hypertonic saline, 761–762
  hyperventilation, 758
  indomethacin, 762
  insulin, 757
  Lund protocol, 761
  nutrition, 757
  obsolete treatments, 761
  osmotic diuretics, 758–759
  pseudotumor cerebri, 760, 922–923
  resection of swollen tissue, 757
  sedation and analgesia, 756
  tromethamine, 759
  ulcer prophylaxis, 757
monitoring, 754–756
normal values, 749
outcomes of elevation, 753
viral infection effects, 610
Intravenous immunoglobulin
  chronic inflammatory demyelinating
    polyneuropathy management, 1263
  contraindications and adverse reactions,
    697
  dermatomyositis/polymyositis management,
    1366
  Guillain-Barré syndrome management,
    1261–1262
  multifocal motor neuropathy management,
    1264
  multiple sclerosis management, 687
  myasthenia gravis management,
    1351–1352, 1358

vasculitides management, 468
virus infection management, 605, 607
Invertebral disk degeneration
  clinical features, 1293
  course, 1293
  treatment, 1293–1294
Isaacs' syndrome
  clinical presentation, 1432
  diagnosis, 1432–1433
  treatment, 1433
Ischemic nerve lesion
  causes, 1319–1320
  clinical presentation, 1319–1320
  operative treatment, 1321
  prevention, 1320–1321
Ischemic stroke, see Acute ischemic stroke
Isocarboxazid, poisoning and management,
    798
Isoniazid
  acquired pendular nystagmus management,
    129
  adverse effects, 564
  side effects, 1517
  tuberculous meningitis management,
    562–563
Itraconazol
  adverse effects and contraindications, 618
  cerebral mycosis treatment, 619
  dosing, 621
  indications, 619
  pharmacological properties, 618
Ivermectin, neurofilariasis management, 669

## J

JC virus, human immunodeficiency virus
    association and management, 630
Juvenile nasopharyngeal angiofibroma,
    preoperative embolization, 481–482

## K

Kaposi's sarcoma, human immunodeficiency
    virus association and management,
    633, 635
Kawasaki disease, see Vasculitides
Kearns-Sayre syndrome, see Mitochondrial
    disease
Ketamine, poisoning and management,
    800–801
Ketoconazole, side effects, 1518
Kidney cancer metastasis, see Brain tumor
Kiloh–Nevin syndrome, see Anterior
    interosseus syndrome
Korsakoff's psychosis
  clinical features, 978
  treatment, 979
Krabbe's disease, features, 1283

## L

Lafora body disease, progressive myoclonic
    epilepsy, 1225
Lambert-Eaton myasthenic syndrome (LEMS)
  clinical presentation, 1359
  diagnosis, 1359

paraneoplastic syndrome, *see*
Paraneoplastic syndromes
treatment, 1359–1360
Lamotrigine
advantages and limitations, 218
adverse reactions, 222
dosing, 221
epilepsy management, 220, 222
pharmacokinetics, 221
trigeminal neuralgia management, 69, 72
Lateral femoral cutaneous nerve
compression, features and
management, 1318
Lead, encephalopathy
overview, 1003
treatment, 1005–1006
Leber's hereditary optic neuropathy (LHON)
clinical presentation, 121
course, 122
heredity, 121–122
Leflunomide, vasculitides management, 467
Leishmaniasis
clinical presentation, 669–670
course, 670
definition, 669
diagnosis, 672
differential diagnosis, 670
prognosis, 673
transmission, 670
treatment
amphotericin B, 672
cutaneous disease, 672–673
goals, 670–671
miltefosine, 671–672
paromoxine, 672
pentamidine, 672
sodium stibogluconate, 672–673
visceral disease, 672
LEMS, *see* Lambert-Eaton myasthenic
syndrome
Leprosy
course, 1269
diagnosis, 1269
epidemiology, 1269
pathogen, 1269
treatment, 1269–1270
Leptomeningeal metastasis, *see also* Non-
Hodgkin's lymphoma
chemotherapy
intrathecal chemotherapy
Ara-C, 902, 906
combination chemotherapy, 902
duration and termination, 906–907
methotrexate, 901–902, 905–906
rationale, 899, 901
technique, 905
thiotepa, 902
outcomes, 900–901
systemic chemotherapy, 901, 907
clinical presentation, 897–878
concomitant brain metastasis management,
889
course, 899
diagnosis, 898–899
gene therapy, 903
immunotherapy, 903

individualization of treatment plan,
904–905
markers, 899
mechanisms, 897
primary tumor epidemiology, 897–898
radiotherapy
outcomes, 900
rationale, 902
techniques, 907
toxicity, 903
supportive care, 907–908
tumor-specific tumor considerations in
treatment, 903–904
Leptospirosis
antibiotic therapy, 590
clinical presentation, 590
diagnosis, 590
epidemiology, 589–590
Levetiracetam
advantages and limitations, 218
adverse reactions, 222
dosing, 221
epilepsy management, 220, 223
pharmacokinetics, 221
Levodopa, *see* L-Dopa
LHON, *see* Leber's hereditary optic
neuropathy
Lidocaine, intracranial pressure control,
731
Lignocaine, cluster headache treatment,
19
Lissencephaly
gene mutations, 953–954
imaging, 953–954
types, 953–954
Listeriosis, central nervous system infection
and management, 539
Lisuride
hyperprolactinemia management, 1443
migraine prevention, 10
Parkinson's disease management,
1037–1041
tremor management, 1238
Lithium
cluster headache prevention, 20
side effects, 1509
Long thoracic nerve compression, features
and management, 1305
Low-molecular weight heparin
acute ischemic stroke management, 335,
342–343
cerebral venous and sinus thrombosis
management, 453–454
Lumbar puncture
acute hydrocephalus management, 435
findings, *see* Cerebrospinal fluid
headache, *see* Postlumbar puncture
headache
hydrocephalus management, 809–810
subarachnoid hemorrhage, 424, 483
Lumbosacral plexus
compression neuropathy
clinical features, 1313
treatment, 1313–1314
delayed radiotherapy injury, 1337,
1339

Lung cancer metastasis, *see* Brain tumor;
Leptomeningeal metastasis
Lyme neuroborreliosis
cerebrospinal fluid findings, 585–586, 589
course, 586–587
diagnostic criteria, 583–584
differential diagnosis, 585
pathogen strains, 583
pathogenesis, 587
serology, 558–586
stages
I, 583–584
II, 584
III, 584–585
transmission, 583, 586–587
treatment
antibiotics, 587–589
monitoring, 588

## M

Magnetic resonance imaging (MRI)
acute ischemic stroke, 328, 340–342
arachnoid cysts, 958–959
arteriovenous malformation, 489
brain abscess, 546–547
brain metastasis from systemic solid
tumors, 881
brain tumor diagnosis and monitoring,
827, 829, 832
capillary teleangiectases, 527
carotid cavernous sinus fistula, 493
cerebral venous and sinus thrombosis,
447–448, 451
cerebrospinal fluid dynamics, 807
cervical spondylitic myelopathy, 1295
Chiari malformation, 956
Dandy-Walker malformation, 956–957
delirium, 306
dementia, 311
dural arteriovenous fistula, 493
epilepsy, 212, 236
fungal infection of nervous system, 615
hydrocephalus, 809
hypopituitarism, 1446
idiopathic demyelinating optic neuritis,
118
intracranial pressure elevation, 751
leptomeningeal metastasis, 898
lumbar disk herniation, 1296
multiple sclerosis
diagnostics, 679–680
prognostic imaging, 678–681
treatment monitoring, 682
neurocysticercosis, 653–654
normal pressure hydrocephalus, 1115
radiculopathy, 1289–1290
sarcoidosis, 567
schistosomiasis, 658
spinal arteriovenous malformation,
523–526
spinal cavernomas, 527
spinal cord injury, 925
spinal cord ischemia, 397–398
spinal epidural abscess, 553
subarachnoid hemorrhage, 424–425, 483

Magnetic resonance imaging (MRI)
    (continued)
    subdural empyema, 551
    syringomyelia, 939, 942
    thymoma, 1345
    toxoplasmosis, 595
    traumatic brain injury, 767
    tuberculous meningitis, 560–561
    viral infection diagnostics, 60
    whiplash injury, 85–87, 92
    Wilson's disease, 1158
Magnetic resonance angiography (MRA)
    cerebral venous and sinus thrombosis,
        447–448, 451
    subarachnoid hemorrhage, 426
Maintenance of wakefulness test (MWT),
        principles, 193–194
Malaria, see Cerebral malaria
Malignant dopamine deficiency
    clinical features, 788
    course, 788
    treatment, 788–789
Malignant glioma, see Anaplastic
        astrocytoma; Glioblastoma
Malignant hyperthermia
    clinical presentation, 744, 783–784
    course, 784
    critical care, 744
    epidemiology, 743
    gene mutations, 784
    levodopa withdrawal syndrome,
        1060
    muscular dystrophy and anesthesia,
        1380–1381
    screening, 784
    treatment, 785
Malignant peripheral nerve sheath tumor
        (MPNST), features and
        management, 852
Manganese
    encephalopathy
        overview, 1003
        treatment, 1006
    parkinsonism induction, 1068
Marchiafava-Bignami disease
    clinical presentation, 981–982
    course, 982
    treatment, 982
Marijuana, toxicity and management, 800
MBP, see Myelin basic protein
MC, see Myotonia congenita
MDMA, see Methylenedioxyamphetamine
Measles, clinical features and course of
        infection, 606
Mebendazole
    diphyllobothriasis management, 597
    echinococcosis management, 597
    toxocariasis management, 599
    trichinosis management, 599
Meclofenoxate, side effects, 1511
Median nerve compression, features and
        management, 1307
Medication overuse headache
    clinical features, 33
    course, 33

definition, 31
epidemiology, 31–32
pathophysiology, 32–33
treatment
    inpatient treatment, 34–35
    long-term treatment, 35
    outpatient treatment, 35
    principles, 34
Mediterranean Spotted Fever, central nervous
        system infection and management,
        540
Mefloquine, cerebral malaria management,
        661
Melanocytic lesions, features and
        management, 854
Melanoma metastasis, see Brain tumor;
        Leptomeningeal metastasis
Melarsoprol, sleeping sickness management,
        667
Memantine
    acquired pendular nystagmus management,
        129
    spasticity management, 1252
Memory
    assessment, 294
    disorders
        etiology, 288, 294
        hypoxic brain damage case study,
            292–293
        overview, 287–288
        personality dimensions, 291–293
        trauma induction, 289, 291
        treatment
            aims, 295
            medical support, 299
            overview, 294
            prospects, 299
            training, 296–298
    HERA model, 289
    long-term memory systems, 288–289
    training techniques
        external memory, 296, 298
        internal memory, 296
        overview, 296–297
Meniére's disease
    clinical presentation, 149
    course, 149–150
    pathogenesis, 150
    treatment
        betahistine, 151
        diuretics, 151
        intratympanic gentamicin therapy,
            150–151
        obsolete treatments, 151–152
        principles, 150
        surgery, 151
Meningioma
    clinical presentation, 852
    course, 852–853
    grades, 852
    preoperative embolization, 481
    treatment, 852–854
Meningitis, see Bacterial meningitis; Fungal
        meningitis; Tuberculous meningitis;
        Viral meningitis

Meningitis, see Cryptococcal meningitis
Meningomyeloceles, features, 951
Menopausal migraine, see Hormone-sensitive
        headache in women
Menstrual migraine, see Hormone-sensitive
        headache in women; Migraine
Mental State Examination (MSE)
    delirium assessment, 305
    parts
        appearance and behavior, 305
        delusions and hallucinations, 305
        mood, 305
        talk, 305
Mercury, encephalopathy
    overview, 1003
    treatment, 1006
Metabolic encephalopathy, mechanisms,
        991–992
Metachromatic leukodystrophy, features,
        1283
Methadone
    metabolism, 106
    pain management, 103–104
Methanol, poisoning and management, 803
Methimazole, hyperthyroidism management,
        1448
Methixene, side effects and contraindications,
        1512–1513
Methotrexate
    dermatomyositis/polymyositis management,
        1366
    giant cell arteritis management, 478
    leptomeningeal metastasis, intrathecal
        chemotherapy, 901–902, 905–906
    multiple sclerosis management, 690–691
    polymyalgia rheumatica management,
        478
    vasculitides management, 467
Methyl-N-butyl ketone, poisoning and
        management, 802
Methylendioxymetanphetamine (MDMA),
        poisoning and management, 801
N-Methyl-3-phenyl-1,2,3,6-
        tetrahydropyridine (MPTP),
        parkinsonism induction, 1068
Methysergide
    cluster headache prevention, 20
    progressive supranuclear palsy
        management, 1092
Metoclopramide
    hiccup management, 187
    Parkinson's disease management,
        1048
Metoprolol
    contraindications, 1520
    drug–drug interactions, 1520
    side effects, 1520
    tension-type headache prevention, 27
Metronidazole, amebic brain abscess
        management, 666
Mexiletine
    acquired pendular nystagmus management,
        129
    pain management after nerve injury,
        1332

MG, see Myasthenia gravis
Miconazole
    primary amebic meningoencephalitis
        management, 663
    side effects, 1518
Microangiopathies, see also Cerebral amyloid
        angiopathy; Cerebral autosomal
        dominant arteriopathy with
        subcortical infarcts and
        leukoencephalopathy; Cerebral
        small vessel disease; Fabry's disease;
        Susac's syndrome; Thrombotic
        thrombocytopenic purpura
    causes, 363
    cerebroretinal vasculopathy, 70
    classification, 363–364
    hereditary endotheliopathy with
        retinopathy, nephropathy, and
        stroke, 370
Microscopic polyarteritis, see Vasculitides
Microvascular decompression
    glossopharyngeal neuralgia management, 73
    hemifacial spasm management, 1141–1142
    hiccup management, 188
    trigeminal neuralgia management, 70
    vestibular paroxysmia management, 153
Micturition, see Bladder dysfunction
Midodrine
    orthostatic hypotension management, 1466
    syncope management, 1501–1502
Migraine, see also Hormone-sensitive
        headache in women
    anatomy, 2
    children, 1
    complications, 1
    course, 2, 8
    diagnosis, 2
    migraine with aura, 1
    migraine without aura, 1
    neurogenic plasma protein extravasation,
        2–3
    neuropeptide release, 3
    patient education, 5
    prevention
        acupuncture, 12
        amitryptyline, 10
        aspirin, 10
        behavioral therapy, 11–12
        beta blockers, 8–9
        calcium channel blockers, 9
        children, 11
        clinical aspects, 8
        dihydroergotamine, 9
        drug priorities, 10–11
        elderly, 11
        lisuride, 10
        menstrual migraine, 11
        nonsteroidal anti-inflammatory drugs, 10
        obsolete treatments, 12
        pitfalls, 11
        serotonin receptor antagonists, 9–10
        valproic acid, 10
    treatment
        antiemetics, 5
        aspirin, 3–4, 6–7

dihydroergotamine, 4, 7
eletriptan, 4–5
ergotamine, 4, 7
naratriptan, 4–5
nonpharmacological management, 6
nonspecific treatments, 7
nonsteroidal anti-inflammatory drugs,
    6–7
obsolete treatments, 8
paracetamol, 7
principles of therapy, 2
rizatriptan, 4–5
specific treatments, 7–8
stepped versus stratified care, 6
sumatriptan, 4–5
triptan selection, 7–8
zolmitriptan, 4–5
trigeminocervical complex, 3
Miltefosine, leishmaniasis management,
        671–672
Mirtazapine, poisoning and management,
        797–798
Mitochondrial disease
    chronic progressive external
        ophthalmoplegia, 1390, 1393
    drug-induced myopathy, 1397, 1400
    Kearns-Sayre syndrome, 1390, 1393
    mitochondrial encephalopathy with lactic
        acidosis and strokelike episodes,
        1393
    molecular genetic diagnosis, 1533
    myoclonic epilepsy with ragged red fibers,
        1393
    neuropathy, ataxia, and retinitis
        pigmentosa, 1393
    overview of types, gene defects, and
        features, 1391–1392
    pathophysiology, 1390
    progressive myoclonic epilepsy, 1225
    treatment
        acetylcholinesterase inhibitors, 1395
        anticonvulsants, 1395
        cardiac pacemaker, 1395
        carnitine, 1395
        coenzyme Q10, 1394
        corticosteroids, 1395
        creatine, 1394
        dichloroacetate, 1394–1395
        gene therapy, 1395–1396
        physical therapy, 1395
        respiratory function, 1395
        surgery, 1395
        vitamins, 1394
Mitoxantrone
    contraindications and adverse reactions, 697
    multiple sclerosis management, 688, 697
MLST, see Multiple sleep latency test
MMN, see Multifocal motor neuropathy
Moclobemide, poisoning and management,
        798
Molecular genetic diagnosis, see also Gene
        mapping
    ethics
        confidentiality, 1534
        genetic counseling, 1534

informed consent, 1534
    presymptomatic diagnosis, 1534–1535
genetic heterogeneity, 1532–1533
indirect diagnosis, 1532
mitochondrial diseases, 1533
point mutations, deletions, and insertions,
    1532
polygenic disorders, 1533–1534
polymerase chain reaction, 1531
trinucleotide repeat expansions, 1532
Monoclonal gammopathy
    clinical features, 1265–1266
    course, 1266
    treatment, 1266–1267
    types and incidence, 1265
Mononeuritis multiplex, human
        immunodeficiency virus association
        and management, 632, 640
Morphine
    administration routes, 105
    pain management, 103
Morphology, definition, 266
Motion sickness
    clinical presentation, 157–158
    course, 158
    treatment
        dimenhydrinate, 158
        obsolete treatments, 158
        physical prevention, 158, 160
        principles, 158
        scopolamine, 158
Motor rehabilitation, see Rehabilitation
MPNST, see Malignant peripheral nerve
        sheath tumor
MPTP, see N-Methyl-3-phenyl-1,2,3,6-
        tetrahydropyridine
MRA, see Magnetic resonance angiography
MRI, see Magnetic resonance imaging
MS, see Multiple sclerosis
MSA, see Multiple system atrophy
MSE, see Mental State Examination
Multifocal motor neuropathy (MMN)
    clinical presentation, 1264
    course, 1264
    diagnosis, 1264
    treatment, 1264
Multiple sclerosis (MS)
    course, 677–678, 680–681
    diagnosis, 679–680
    differential diagnosis, 681
    epidemiology, 681
    exacerbation
        heat intolerance, 699
        professional activity, 700
        surgery and anesthesia, 699
        vaccination considerations, 699
    experimental autoimmune encephalitis
        model, 683–684
    genetics, 681
    histopathology, 677
    magnetic resonance imaging
        diagnostics, 679–680
        prognostic imaging, 678–681
        treatment monitoring, 682
    pathogenesis, 684–685

Multiple sclerosis (MS) *(continued)*
  prognosis, 678–679
  rehabilitation, 1193–1194
  survival, 679
  treatment
    4-aminopyridine, 702
    azathioprine, 688–689, 696
    bladder dysfunction, 701
    bowel dysfunction, 701–702
    cladribine, 690
    complications of disease, 702
    corticosteroids, 684–685, 692–693
    cyclophosphamide, 689, 697–698
    cyclosporine A, 689–690, 696–697
    diaminopyridine, 702
    efficacy assessment, 682
    experimental approaches, 691
    fatigue, 700
    glatiramer acetate, 687, 695–696
    hematopoietic stem cell autologous
      transplantation, 691
    interferon-β, 685–686, 693–695
    intravenous immunoglobulin, 687
    methotrexate, 690–691
    mitoxantrone, 688, 697
    nutrition and daily activity, 700
    nystagmus, 701
    obsolete treatments, 702
    pain, 701
    physiotherapy, 700
    plasmapheresis, 690
    pregnancy and contraception,
      698–699
    principles, 682–684
    psychotherapy, 700
    seizures, 699
    sexual dysfunction, 702
    spasticity, 700–701
    special forms of disease, 703
    total lymphoid irradiation, 690
    tremor, 701
  tremor, 1240–1241
Multiple sleep latency test (MLST),
    principles, 193
Multiple symmetrical lipomatosis, features,
    1283
Multiple system atrophy (MSA)
  autonomic dysfunction, 1457–1458
  clinical presentation, 1016, 1081
  course, 1016, 1082
  diagnosis, 1081–1082
  treatment
    autonomic failure, 1082–1085
    cerebellar ataxia, 1084
    overview, 1016
    parkinsonism, 1083–1085
Muscle cramps, *see* Cramps
Muscular dystrophy, *see also* Duchenne's
    muscular dystrophy;
    Facioscapulohumeral dystrophy
  anesthesia and malignant hyperthermia,
    1380–1381
  dystrophin mutations and phenotypes,
    1371–1372
  limb girdle muscular dystrophy gene
    mutations, 1369

physical training, 1381
support groups, 1376
treatment goals, 1369, 1372
types, 1369–1371
Musculocutaneous nerve compression, fea-
    tures and management, 1305–1306
MWT, *see* Maintenance of wakefulness test
Myasthenia gravis (MG)
  anesthetic management, 1358
  associated disorders, 1341–1342
  classification, 1341–1342
  course, 1345–1346
  diagnosis
    autoantibody assays, 1345
    edrophonium test, 1342–1343
    electromyography, 1344–1345
    imaging, 1345
    pyridostigmine test, 1343
  differential diagnosis, 1341, 1343
  drug exacerbation, 1358–1359, 1400
  grading
    clinical score, 1343
    quantitative score, 1344
  ocular myasthenia gravis
    epidemiology, 1345
    management, 1355
  paraneoplastic syndrome, *see*
    Paraneoplastic syndromes
  pathophysiology, 1346–1347
  treatment
    acetylcholinesterase inhibitors, 1347,
      1353–1355, 1357
    azathioprine, 1348–1350, 1355
    childhood autoimmune disease, 1357
    corticosteroids, 1347–1348, 1355
    cyclophosphamide, 1351
    cyclosporine A, 1350–1351
    generalized disease, 1353–1355
    intravenous immunoglobulin,
      1351–1352, 1358
    myasthenic crisis, 1357–1358
    mycophenolate mofetil, 1350
    neonatal transient myasthenia,
      1356–1357
    plasmapheresis, 1352, 1358
    pregnancy, 1358
    thymectomy, 1352–1354, 1356
*Mycobacterium leprae, see* Leprosy
Mycophenolate mofetil
  myasthenia gravis management, 1350
  vasculitides management, 467
Mycoplasmosis, central nervous system
    infection and management, 541
Myelin basic protein (MBP), autoimmunity,
    683–684
Myelography, lumbar disk herniation, 1296
Myoadenylate deaminase deficiency, features
    and management, 1389
Myocardial infarction
  critical care, 733–734
  stroke risks, 350
Myoclonic epilepsy with ragged red fibers,
    *see* Mitochondrial disease
Myoclonus
  asterixis, 1229
  clinical features, 130, 1221–1222

corticobasal degeneration management,
    1095
course, 131
critical care, 739
diagnosis, 1223–1224
differential diagnosis, 1222–1223
dystonic myoclonus, 1227
electroencephalography, 1223–1224
electromyography, 1223–1224
epileptic myoclonus
  causes
    basal ganglia disorders, 1226
    dementias, 1225–1226
    focal central nervous system damage,
      1227
    hypoxia, 1226–1227
    metabolic encephalopathies, 1226
    progressive myoclonic epilepsy,
      1224–1225
    storage diseases, 1225
    toxic encephalopathies, 1226
    viral encephalopathies, 1226
  cortical reflex myoclonus, 1223
  diagnosis, 1222
  primary generalized myoclonus, 1223
  reticular reflex myoclonus, 1223
  treatment, 1224
essential myoclonus, 1227
etiology of syndromes, 1222
exaggerated startle, 1227–1228
nonepileptic myoclonus, 1227
nocturnal myoclonus, 1228
palliative care, 820
pathogenesis, 130–131
physiological myoclonus, 1224
prion disease management, 718
psychogenic myoclonus, 1229
segmental myoclonus, 1228–1229
treatment, 131
Myoglobinuria
  causes, 1389
  clinical features, 1389–1390
  management, 1390
Myotonia, *see also specific diseases*
  definition, 1407
  drug induction, 1400
Myotonia congenita (MC)
  clinical presentation, 1407–1408
  diagnosis, 1408
  treatment, 1408
Myotonia fluctuans, features and
    management, 1408–1409
Myotonic dystrophy type I
  clinical features, 1409–1410
  gene mutations, 1410
  treatment, 1410
Myotonic dystrophy type II
  clinical features, 1410–1411
  management, 1411
Myxovirus, clinical features and course of
    infection, 605–606

**N**

NAC, *see* N-Acetylcysteine
Naftidrofuryl, side effects, 1511

Naltrexone, alcohol dependency
  management, 986
Naratriptan, migraine treatment, 4–5
Narcolepsy
  clinical presentation, 199–200
  diagnosis, 200
  genetics, 200
  prevalence, 199
  treatment, 200
Nausea
  mechanisms and treatment, 1468
  palliative care, 821
NCC, see Neurocysticercosis
Necrotizing myopathy
  acute myopathies, 1401–1402
  drug induction, 1400
Nerve entrapment syndromes, see specific
    nerves
Nerve injury, see Brachial plexus injury;
    Ischemic nerve lesion; Peripheral
    nerve injury; Spinal cord injury
Neuralgic amyotrophy
  clinical presentation, 1268
  course, 1268–1269
  differential diagnosis, 1268
  treatment, 1269
Neural prosthesis, see also Cochlear implant;
    Phrenic pacemaker
  auditory brain stem implant, 1209
  bladder and bowel function
    contraindications, 1212
    fecal continence, 1213
    indications, 1212
    side effects, 113
    systems, 1212–1213
    urinary incontinence, 1213
  dynamic cardiomyoplasty, 1210
  functional electrical stimulation
    advantages, 1200–1201
    ambulation, 1203
    ambulation with stance hybrid systems,
        1203–1204
    contraindications, 1201
    cough induction, 1211–1212
    fecal continence, 1213
    gait in hemiparesis, 1204
    grasp restoration in tetraplegia,
        1205–1206
    muscle fatigue and training, 1200
    prerequisites, 1201
    stance and gait in paraplegia,
        1202–1203
    stimulators and electrodes, 1201–1202
    stroke and upper extremity function,
        1204–1205
  government regulation, 1199
  indications, 1199
  sexual function, 1213–1214
  visual prosthesis, 1206
Neuroblastoma, features and management,
    849
Neuroborreliosis, see Lyme neuroborreliosis
Neurobrucellosis, central nervous system
    infection and management, 539
Neurocysticercosis (NCC)
  clinical presentation, 647–648

course, 650–651
diagnosis, 653–654
differential diagnosis, 648, 650
parasite features, 647
transmission, 647
treatment
  albendazole, 652–653, 655
  carbamazepine, 654
  corticosteroids, 654–655
  obsolete treatments, 655
  phenytoin, 654
  praziquantel, 655
  principles, 652–653
  supportive therapy, 653
Neurofibromas, features and management,
    851–852
Neurofibromatosis
  clinical presentation
    craniocerebral manifestations, 949,
        960–961
    neurological symptoms, 961
    skeletal abnormalities, 961
    tumor sites, 961
    vascular malformations, 961
  course, 961
  diagnostic criteria of types, 949
  gene mutations, 960
  treatment, 949, 961–962
Neurofilariasis
  clinical presentation, 668
  course, 668
  definition, 668
  diagnosis, 668
  pathogens, 668
  treatment
    albendazole, 669
    diethylcarbamazine, 668–669
    ivermectin, 669
    suramin, 668
Neuroleptic malignant syndrome
  clinical presentation, 786
  course, 786–787
  diagnosis, 786
  nomenclature, 785
  pathogenesis, 786
  treatment, 787–788
Neuroleptics, see also specific drugs
  contraindications, 1508
  drug–drug interactions, 1508
  side effects
    frequent side effects, 1507
    peripheral and central nervous system
        side effects, 1507–1508
    reversibility
Neuromyotonia
  acquired disease, see Isaacs' syndrome
  clinical presentation, 1254
  course, 1254–1255
  treatment, 1255
Neuropathic pain
  brachial plexus injury, 1328–1329
  management, 109, 1286–1287
  treatment after nerve injury, 1332–1333
Neuropathy, ataxia, and retinitis
    pigmentosa, see Mitochondrial
    disease

Neurosarcoidosis, see Sarcoidosis, central
    nervous system
Neurosyphilis, see Syphilis
Nevoid basal cell carcinoma syndrome
  clinical features, 965–966
  course, 965
  treatment, 966
NHL, see Non-Hodgkin's lymphoma
Nicaraven, cerebral ischemia prevention
    following subarachnoid
    hemorrhage, 432
Nimodipine, side effects, 1511
Nitric oxide (NO), tension-type headache
    role, 25
Nitrous oxide, poisoning and management,
    802
NO, see Nitric oxide
Nocardiosis, central nervous system infection
    and management, 540
Non-Hodgkin's lymphoma (NHL)
  central nervous system manifestations, 871
  course, 871–872
  diagnosis, 874
  human immunodeficiency virus association
      and management, 633, 641–642
  treatment in central nervous system
    central nervous system prophylaxis, 873
    chemotherapy, 872–874
    corticosteroids, 874
    epidural spinal metastasis, 872
    leptomeningeal tumor, 872
    radiotherapy, 872–874
    surgery, 873
Nonsteroidal anti-inflammatory drugs
    (NSAIDs)
  acute pain management, 100
  administration routes, 100
  adverse effects, 99–100, 1521
  contraindications, 1521
  dosing, 100
  drug–drug interactions, 1521
  migraine
    prevention, 10
    treatment, 6–7
  opioid combination therapy, 100
  palliative care, 821
  radiculopathy management, 1292
  tension-type headache treatment, 26–27
  vasculitides management, 464–465
Normal pressure hydrocephalus (NPH)
  cerebrospinal fluid shunting
    outcomes, 1119–1120
    technique, 1118–1119
  clinical features, 810–811, 1113–1114
  diagnosis
    cerebrospinal fluid
      drainage, 1115
      dynamics, 1115
      markers, 1116
    computed tomography, 1115
    magnetic resonance imaging, 1115
    positron emission tomography,
        1115–1116
  differential diagnosis, 1114–1115
  epidemiology, 1116
  pathogenesis, 1116–1118

Nortriptyline, neuropathic pain management, 1286
NPH, *see* Normal pressure hydrocephalus
NSAIDs, *see* Nonsteroidal anti-inflammatory drugs
Nutrition
  critical care
    enteral nutrition, 726
    parenteral nutrition, 726
    requirement determinants, 725
    supplements, 725–726
  delirium, 307–308
  dysphagia complications, 249
  intracranial pressure elevation management, 757
  multiple sclerosis, 700
Nutritional optic neuropathy, *see* Optic neuropathy
Nystagmus
  acquired pendular nystagmus
    course, 128
    pathogenesis, 128
    treatment, 128–129
  congenital nystagmus
    clinical presentation, 129
    course, 129
    treatment, 129–130
  downbeat nystagmus
    course, 127
    pathogenesis, 126
    treatment, 127
  latent nystagmus
    clinical presentation, 129
    course, 129
    treatment, 130
  periodic alternating nystagmus
    course, 128
    pathogenesis, 128
    treatment, 128
  seesaw nystagmus
    pathogenesis, 127
    treatment, 127–128
  upbeat nystagmus
    course, 127
    pathogenesis, 126–127
    treatment, 127
  vestibulo-ocular reflex, central disorders
    clinical presentation, 125–126
    course, 126
    treatment, 126

# O

Obstructive sleep apnea (OSA)
  clinical features, 198–199
  treatment
    continuous positive airway pressure, 199
    surgery, 199
    theophylline, 199
Octreotide
  acromegaly management, 1437
  orthostatic hypotension management, 1467
Ocular motor disorders, *see also* Myoclonus; Nystagmus; Opsoclonus; Superior oblique myokymia

isolated paresis of ocular muscles
  course, 132
  nerve palsies, 132
  treatment, 132
paroxysmal disorders
  pathogenesis, 131–132
  treatment
    basilar migraine, 132
    hereditary paroxysmal ataxia and nystagmus, 132
    neurovasular compression, 132
    ocular neuromyotonia, 132
  types, 131–132
steady gaze mechanisms, 125
Oculopalatal tremor, *see* Myoclonus
Oculotoxicity, drugs, 1524
Olanzapine
  dementia with Lewy bodies management, 1087
  dopaminergic psychosis management, 1066
  poisoning and management, 799
  tic disorder management, 1140
Olfactory dysfunction, *see* Anosmia
Olfactory neuroblastoma, features and management, 849
Oligoastrocytoma
  clinical presentation, 844
  course, 845
  treatment, 845
Oligodendroglioma
  clinical presentation, 844
  course, 845
  treatment, 845–846
Ondansetron, dementia with Lewy bodies management, 1087
Opioids, *see also specific drugs*
  acute pain management, 102
  administration routes, 105–106
  chronic pain management, 102–104
  metabolism, 106
  nonsteroidal anti-inflammatory drug combination therapy, 100
  pain management after nerve injury, 1333
  palliative care, 817–818, 823
  poisoning and management, 800
  restless legs syndrome management, 1178–1181
Opsoclonus
  clinical presentation, 130
  course, 130
  pathogenesis, 130
  treatment, 130
Optic neuropathy
  anterior ischemic optic neuropathy
    arteritic anterior ischemic optic neuropathy
      course, 119
      treatment, 119–120
    clinical features, 119
    diabetic papillopathy, 119
    nonarteritic anterior ischemic optic neuropathy
      course, 120
      treatment, 120
  autosomal dominant optic atrophy, 122
  compressive optic neuropathy, 121

differential diagnosis, 115–116
  glaucomatous optic neuropathy, 123
  infiltrative optic neuropathy, 121
  Leber's hereditary optic neuropathy, 121–122
  nutritional optic neuropathy
    clinical presentation, 120–121, 983–984
    course, 984
    treatment, 984
  optic neuritis
    clinical presentation, 115–116
    idiopathic demyelinating optic neuritis, 116, 118
    inflammatory disease causes, 116, 118
    treatment, 118–119
  overview of types, 115, 117
  papilledema, 123
  toxic optic neuropathy, 120–121
  traumatic optic neuropathy
    clinical presentation, 122
    course, 122
    treatment, 122–123
  visual defects, 115
  Wolfram's syndrome, 122
Oral contraceptives
  adverse events, 51
  cerebral venous and sinus thrombosis risks, 457
  epilepsy considerations, 227
  formulations, 51
  menstrual migraine prevention, 49
  migraine induction, 51–52
  multiple sclerosis patients, 698–699
Orthostatic hypotension
  classification, 1463
  clinical presentation, 1463
  factors affecting, 1464–1465
  nonneurogenic causes, 1465
  treatment
    clonidine, 1467
    dihydroergotamine, 1466–1467
    ephedrine, 1466
    fludrocortisone, 1466–1467
    goals, 1467–1468
    midodrine, 1466
    octreotide, 1467
    physical measures, 1465–1466
    yohimbine, 1467
  work-up, 1465
Orthostatic hypotension
  multiple system atrophy, 1083–1084
  Parkinson's disease, 1061–1062
OSA, *see* Obstructive sleep apnea
Osteoporosis, screening in corticosteroid therapy patients, 572
Ototoxicity, drugs, 1524
Oxcarbazepine
  advantages and limitations, 218
  adverse reactions, 222
  dosing, 221
  epilepsy management, 220, 223
  pharmacokinetics, 221
  trigeminal neuralgia management, 68–69, 71
Oxycodone, pain management, 102

Oxygen therapy
  cluster headache treatment, 19
  Susac's syndrome management, 367–368

**P**

Pachygyria, features, 953
PAF, *see* Pure autonomic failure
Pain, *see also* Neuropathic pain
  acute pain
    barriers to treatment, 98–99
    definition, 95
  chronic pain definition, 95
  critical care analgesia, 727
  definition and terminology, 95–96
  neurobiology, 96–97
  palliative care, 820–821
  prion disease management, 718
  treatment
    analgesic efficacy assessment, 98
    anticonvulsants, 101–102
    antidepressants, 100–101
    baclofen, 104
    benzodiazepines, 101
    capsaicin cream, 104
    central pain, 110
    clonidine, 104
    complex regional pain syndrome, 110
    corticosteroids, 104
    invasive therapies, 108–109
    local anesthetics, 104
    management plan, 96
    neuropathic pain, 109, 1286–1287
    nonsteroidal anti-inflammatory drugs,
      99–100
    opioids
      acute pain, 102
      administration routes, 105–106
      chronic pain, 102–104
      metabolism, 106
      nonsteroidal anti-inflammatory drug
        combination, 100
    phantom pain, 109–110
    physical therapy, 106–107
    placebo response, 98
    principles, 97–98
    psychiatry, 107–108
    WHO analgesic ladder, 99, 102
Palliative care
  advance care planning, 815
  bereavement, 816
  cancer versus neurology patients, 813
  communication issues, 814–815
  deficiency and physician-assisted suicide,
    813
  definition, 813–814
  psychosocial care, 815–816
  spiritual care, 818
  symptom management
    death rattle, 818
    delirium, 819–820
    depression, 822
    drowsiness, 820
    drug administration routes, 817
    dyspnea, 817–818
    epileptic seizures, 820

    fatigue, 822–823
    myoclonus, 820
    nausea and vomiting, 821
    pain, 820–821
    prevalence of symptom types, 816
    restlessness, 818–819
    sedation, 823
    thirst, 822
Pallidotomy, thermocoagulation, 1107–1108
Pamidronate, hypercalcemic encephalopathy
    management, 1001
PAN, *see* Polyarteritis nodosa
Papilledema, clinical features, 123
Paracetamol, *see* Acetaminophen
Paracoccidioidomycosis, clinical features, 617
Paragangliomas, features and management,
    849
Paraglioma, preoperative embolization, 482
Paragonimiasis, features and management,
    598
Paragrammatism, definition, 266
Paramyotonia congenita, features and
    management, 1408–1409
Paraneoplastic syndromes
  antibodies, 913–914
  autoimmune pathogenesis, 913–914
  cerebellar degeneration and ataxia, 1018
  definition, 911
  diagnosis, 911, 914–915
  incidence, 912–913
  prognosis, 917
  treatment
    immunosuppression, 916–917
    tumor removal, 916
  types, 911–912
Parasomnias
  NREM/REM sleep parasomnias
    clinical features, 198
    treatment, 198
  NREM sleep parasomnias
    clinical features, 197
    treatment, 197–198
  REM sleep parasomnia
    clinical features, 198
    treatment, 198
Parkinson's disease (PD)
  akinetic crisis
    clinical features, 788
    course, 788
    treatment, 788–789, 1059–1060
  atypical parkinsonism, *see* Corticobasal
      degeneration; Dementia with Lewy
      bodies; Multiple system atrophy;
      Progressive supranuclear palsy
  autonomic dysfunction, 1457–1459
  classification, 1022–1023, 1025–1026
  clinical presentation
    akinesia, 1012
    autonomic symptoms, 1022
    complex disorders of movement,
      1021–1022
    neuropsychiatric symptoms, 1022
    rest tremor, 1021
    rigidity, 1021
  course, 1026–1027
  definition, 1021

  dementia, 317, 1064–1065
  diagnosis, 1022, 1024–1026
  differential diagnosis, 1022, 1024–1035
  drug-induced dopaminergic psychosis
    etiology, 1066
    frequency, 1065–1066
    treatment
      clozapine, 1066
      olanzapine, 1066
      quetiapine, 1048, 1066–1067
      risperidone, 1066
  drug-induced parkinsonism, 1069
  epidemiology, 1026
  etiology and pathogenesis
    hereditary disease, 1027
    sporadic disease, 1027–1028
  mortality causes, 1027
  neuropathology, 1028
  neurotransmitter changes
    cholinergic neurotransmission,
      1028–1029
    dopamine, 1028
    excitatory amino acids and glutamate
      receptors, 1029–1031
  paraneoplastic syndrome, *see*
      Paraneoplastic syndromes
  rehabilitation, 1194
  secondary parkinsonism
    dihydrobiopterin reductase deficiency,
      1069
    hydrocephalus-associated parkinsonism,
      1068
    hypoparathyroidism, 1068–1069
    postencephalitic parkinsonism, 1067
    toxins
      carbon monoxide, 1068
      manganese, 1068
      *N*-methyl-3-phenyl-1,2,3,6-
        tetrahydropyridine, 1068
    tumor-associated parkinsonism, 1068
    vascular pseudoparkinsonism,
      1067–1068
    Wilson's disease, 1069
  stages
    Hoehn and Yahr Scale, 1048, 1050
    rating scales, 1050
  surgery considerations in patients
    anesthesia, 1061
    oral medication, 1060–1061
    parenteral medication, 1061
  treatment
    amantadine, 1046–1047
    anticholinergics, 1048
    bladder dysfunction, 1062
    bupidine, 1047–1048
    catechol-O-methyltransferase inhibitors
      adverse effects, 1044–1045
      entacapone, 1043–1044
      rationale, 1043
      tolcapone, 1043–1044
    clozapine, 1048
    deep brain stimulation
      indications, 1103
      initial parameter selection, 1104
      medication adjustment, 1105
      outcomes, 1102–1103

Parkinson's disease (PD) (continued)
    stimulation parameter setting,
        1103–1105
    tremor management, 1106
depression, 1063–1064
domperidone, 1048–1049
L-Dopa
    combination with dopamine agonists,
        1053–1054
    controlled-release preparations and
        indications, 1034–1036
    decarboxylase inhibitor combination,
        1031–1032
    drug holiday, 1037
    formulations, 1031–1034
    initiation of treatment, 1052–1053
    intravenous administration,
        1036–1037
    malignant levodopa withdrawal
        syndrome, 1060
    rapid-acting preparation, 1036
    response, 1032–1033
    side effects, 1033–1034, 1063
dopamine agonists
    apomorphine, 1042–1043
    bromocriptine, 1037–1041
    cabergoline, 1037–1041
    combination with L-Dopa, 1053–1054
    α-dihydroergocriptine, 1037–1041
    dosing, 1040
    lisuride, 1037–1041
    monotherapy, de novo, 1053
    pergolide, 1037–1041
    pharmacokinetics, 1038
    piribedil, 1041–1042
    pramipexole, 1037–1040, 1042
    receptor specificity, 1038
    ropinirole, 1037–1040, 1042
    side effects, 1040
    switching drugs, 1039
L-threo DOPS, 1055
duration of effect, drug comparison,
    1037
dyskinesias
    fixed dyskinesia, 1058–1059
    mobile dyskinesia, 1057–1058
dysphagia, 1061
end-of-dose akinesia, 1056–1057
freezing, 1057
gastrointestinal dysfunction, 1062–1063
initiation guidelines for symptomatic
    drug therapy, 1050–1052
metoclopramide, 1048
neural transplantation, 1108
orthostatic hypotension, 1061–1062
pallidotomy, 1107–1108
paroxysmal on–off phenomenon, 1057
physical therapy, 1049
pregnant patients, 1067
protein-restricted diet, 1049
psychosocial support, 1049–1050
response fluctuations, 1055–1056
rest tremor, 1054–1055
selegiline, 1045–1046, 1054
sexual dysfunction, 1062
sleep disorders, 1065

stereotactic neurosurgery, 1049
subthalamotomy, 1108
thalamotomy, 1108
weight loss, 1063
tremor
    clinical features, 1236–1237
    treatment, 1237–1238, 1242
    types, 1236
Paromoxine, leishmaniasis management, 672
Paroxetine, syncope management, 1502
Paroxysmal hemicrania
    course, 21
    diagnostic criteria, 20–21
    treatment
        indomethacin, 21
        principles, 21
        prospects, 21–22
Paroxysmal kinesiogenic choreoathetosis,
    features, 1127
Paroxysmal nonkinesiogenic dystonia,
    features, 1127
Patent foramen ovale (PFO)
    anticoagulant prevention of recurrent
        stroke, 382
    repair, 384
    stroke risk, 351
PCNSL, see Primary central nervous system
    lymphoma
PCP, see Phencyclidine
PCR, see Polymerase chain reaction
PD, see Parkinson's disease
Pectorales nerve compression, features and
    management, 1305
Penicillamine
    chelation therapy, 1006
    Wilson's disease management, 1160
Pentamidine
    leishmaniasis management, 672
    sleeping sickness management, 667
Pentazocine, radiculopathy management,
    1292
Pentosan, prion disease management,
    716–717
Pentoxyfylline, side effects, 1511
Percutaneous transluminal balloon
    angioplasty (PTA), see also
    Angioplasty and stenting
    principles, 498
    vasospasm management following
        subarachnoid hemorrhage, 488–489
Pergolide
    corticobasal degeneration management,
        1094
    hyperprolactinemia management, 1443
    Parkinson's disease management,
        1037–1041
    tremor management, 1238
Perilymph fistula
    anterior semicircular canal, 152
    clinical presentation, 152
    course, 152
    treatment, 152
Periodic limb movements during sleep
    (PLMS)
    clinical features, 196–197, 1177, 1179,
        1181–1182

treatment, 197
Peripheral nerve block, trigeminal neuralgia
    management
    phenol in glycerol, 70
    tetracaine, 70–71
Peripheral nerve injury, see also specific
    nerves
    anatomical factors, 1325–1326
    assessment of nerve dysfunction, 1327
    classification, 1325–1326
    course
        distal consequences, 1329–1330
        local consequences, 1329
        proximal consequences, 1330
    diagnosis, 1327
    iatropathic injuries, 1327–1328
    pain syndromes, 1328–1329
    prognosis after decompression or repair,
        1330–1331
    treatment
        indications for surgical exploration and
            repair, 1331
        intractable pain
            acupuncture, 1333
            anticonvulsants, 1332–1333
            antidepressants, 1332
            capsaicin, 1333
            dosing of drugs, 1331–1332
            opioids, 1333
            spinal cord stimulation, 1333
            tramadol, 1333
            transcutaneous electrical nerve
                stimulation, 1333
PERM, see Progressive encephalitis with
    rigidity and myoclonus
Peroneal nerve compression, features and
    management, 1316
Perphenazine, Huntington's disease
    management, 1152
Pesticide polyneuropathy, features, 1286
PET, see Positron emission tomography
Pethidine
    metabolism, 106
    pain management, 104
PFO, see Patent foramen ovale
Phantom pain, management, 109–110
Phencyclidine (PCP), poisoning and
    management, 800–801
Phenelzine, poisoning and management, 798
Phenobarbitol
    advantages and limitations, 218
    adverse reactions, 217
    dosing, 216
    epilepsy management, 215, 219–220
    pharmacokinetics, 216
Phenytoin
    advantages and limitations, 218
    adverse reactions, 217, 654
    dosing, 216
    epilepsy management, 215, 220
    neurocysticercosis management, 654
    pain management, 101–102, 1332
    pharmacokinetics, 216
    poisoning and management, 796
    schistosomiasis management, 659
    trigeminal neuralgia management, 71

Phobic postural vertigo, *see* Psychogenic vertigo
Phonology, definition, 266
Phrenic pacemaker
    contraindications, 1210
    indications, 1210
    surgery and rehabilitation, 1210–1211
    systems, 1211
Pick's disease, *see* Frontotemporal degeneration
Picornavirus, clinical features and course of infection, 606
PID, *see* Prolapsed invertebral disk
Pilocytic astrocytoma
    clinical presentation, 838
    course, 839
    treatment
        radiotherapy, 840–841
        surgery, 839
Pimozide
    Huntington's disease management, 1152
    Sydenham's chorea management, 1154
    tic disorder management, 1140
Pineal parenchymal tumors, features and management, 849
Piracetam
    aphasia management, 273
    side effects, 1511
Piribedil, Parkinson's disease management, 1041–1042
Piroxicam, radiculopathy management, 1292
Pituitary adenoma, nonsecreting tumor features and management, 1449–1450
Plasmapheresis
    acute ischemic stroke patients, 338
    chronic inflammatory demyelinating polyneuropathy management, 1263
    dermatomyositis/polymyositis management, 1366
    Guillain-Barré syndrome management, 1261–1262
    multiple sclerosis management, 690
    myasthenia gravis management, 1352, 1358
    vasculitides management, 468
Pleomorphic xanthroastrocytoma, features and management, 844
PLMS, *see* Periodic limb movements during sleep
PLP, *see* Proteolipid protein
PM, *see* Polymyositis
PML, *see* Progressive multifocal leukencephalopathy
PMR, *see* Polymyalgia rheumatica
Poisoning, *see* Acute intoxication
Poliomyelitis, clinical presentation, 602
Polyarteritis nodosa (PAN), vasculitic neuropathy features and treatment, 469, 1267–1268
Polymerase chain reaction (PCR)
    molecular genetic diagnosis, 1531
    viral infection diagnostics, 603–604
Polymicrogyria, features, 954

Polymyalgia rheumatica (PMR)
    clinical presentation, 475–476
    course, 477
    diagnosis, 475–477
    treatment, 477–478
Polymyositis (PM)
    clinical presentation, 1363
    course, 1365
    diagnosis
        creatine kinase levels in muscle, 1363
        criteria, 1364
        electromyography, 1363–1364
        muscle biopsy, 1364–1365
    differential diagnosis, 1365
    immunopathology, 1365
    treatment
        azathioprine, 1366
        corticosteroids, 1365–1366
        intravenous immunoglobulin, 1366
        long-term treatment, 1366–1367
        methotrexate, 1366
        plasmapheresis, 1366
        principles, 1365
Polyradiculitis, critical care, 738
Polysomnography
    electrophysiology recordings, 192
    portable equipment, 192–193
    scoring, 193
    variables, 192
Porencephaly, features, 955
Portal systemic encephalopathy, *see* Hepatic encephalopathy
Positron emission tomography (PET)
    brain abscess, 547
    cluster headache, 18
    dementia, 311–312
    normal pressure hydrocephalus, 1115–1116
    paraneoplastic syndrome diagnosis, 915
    sarcoidosis, 567
Postherpetic neuralgia
    clinical presentation, 1271
    treatment, 1271–1272
Postlumbar puncture headache
    aggravation by exertion, 44
    clinical presentation, 77
    course, 77–78
    differential diagnosis, 77
    pathophysiology, 78
    prevention
        needle factors
            bevel direction, 79
            design, 79
            size, 78
        postural maneuvers, 79
    treatment
        epidural blood patch, 79–80
        medication, 79
        obsolete treatments, 80
        posture, 79
Postpolio syndrome, features, 1172–1173
Postural tachycardia syndrome, features, 1463
Potassium sulfate, Wilson's disease management, 116
Pragmatics, definition, 266

Pramipexole
    Parkinson's disease management, 1037–1040, 1042
    tremor management, 1237–1238
Praziquantel
    adverse reactions, 655
    neurocysticercosis management, 655
    schistosomiasis management, 658–659
Pregnancy
    cerebral venous and sinus thrombosis management, 457
    cramp management, 1425
    epilepsy management
        breastfeeding, 229
        fetal effects of seizures, 228–229
        obstetric complications, 229
        overview, 227
        prepregnancy counseling, 227
        teratogenicity of drugs, 227–228
    headache
        drug safety, 53–54
        migraine frequency, 52–53
        migraine treatment and prevention, 53
        postnatal headache, 52–53
        postpartum headache, 54
        secondary headaches and causes, 55
    multiple sclerosis management, 698–699
    myasthenia gravis management, 1358
    Parkinson's disease management, 1067
    sarcoidosis management, 572–573
    toxoplasmosis management, 596
    Wilson's disease management, 1162
Primary amebic meningoencephalitis, *see* Cerebral amebiasis
Primary central nervous system lymphoma (PCNSL)
    clinical presentation, 866–867
    course, 867–869
    cytogenetics and molecular biology, 865–866
    diagnosis, 870
    human immunodeficiency virus association and management, 633–634, 641, 868–870
    pathology, 865–866
    treatment
        chemotherapy, 870–871
        hematopoietic stem cell transplantation, 871
        immunocompetent patients, 869
        immunocompromised patients, 869–870
        radiochemotherapy, 870
        radiotherapy, 870–871
Primary writing tremor (PWT)
    clinical presentation, 1238–1239
    pathophysiology, 1239
    treatment, 1239
    types, 1239
Primidone
    advantages and limitations, 218
    adverse reactions, 217
    dosing, 216
    epilepsy management, 215, 220
    pharmacokinetics, 216
    tremor management, 1236, 1241

Prion disease, *see also* Creutzfeldt-Jakob disease
  dementia, 317
  pathogenesis, 713–714
  treatment
    acridine, 718
    antiamyloidogenic agents, 717
    antibodies, 717
    dysphagia, 718
    mental and behavioral symptoms, 718
    myoclonus, 718
    nucleic acids, 717
    obsolete treatments, 718
    pain, 718
    pentosan, 716–717
    strategies
      neurotoxicity reduction, 716–717
      overview, 707, 715
      PrP$^C$ posttranslational alteration and accumulation reduction, 715
      PrP$^C$ production reduction, 715–717
    tetrapyrroles, 717
    types, 707–708
Progressive encephalitis with rigidity and myoclonus (PERM), features and management, 1432
Progressive multifocal leukencephalopathy (PML), human immunodeficiency virus association and management, 634, 637, 641
Progressive supranuclear palsy (PSP)
  clinical features, 1090
  course, 1092
  diagnosis, 1090
  differential diagnosis, 1091–1092
  treatment
    aids, 1092–1093
    cognitive disturbances, 1092
    dysphagia, 1092–1093
    oculomotor disturbances, 1092
    parkinsonism, 1091–1093
Prolapsed invertebral disk (PID)
  cervical disk
    clinical features, 1290–1291
    course, 1291
    diagnosis, 1291
    treatment, 1291–1293
  lumbar disk
    cauda equina syndrome, 1296
    clinical features, 1295–1296
    course, 1297
    diagnosis, 1296–1297
    treatment
      acute stage, 1298–1299
      chemical lytic treatments, 1298
      surgery, 1297–1299
PROMM syndrome, *see* Myotonic dystrophy type II
Pronator teres syndrome, features and management, 1307–1308
Propantheline, detrusor hyperreflexia management, 1483
Propiverine, detrusor hyperreflexia management, 1483

Propofol, intracranial pressure control, 731
Propranolol
  contraindications, 1520
  drug–drug interactions, 1520
  side effects, 1520
  tremor management, 1236–1237, 1241–1242
Propylthiouracil, hyperthyroidism management, 1448
Prostaglandin injections, sexual dysfunction management, 1491
Protamine sulfate, anticoagulation reversal, 414–415
Proteolipid protein (PLP), autoimmunity, 683–684
Pseudotumor cerebri
  clinical presentation, 921
  course, 921–922
  diagnostic criteria, 931–922
  differential diagnosis, 921–922
  treatment, 760, 922–923
PSP, *see* Progressive supranuclear palsy
Psychogenic vertigo
  course, 156–157
  phobic postural vertigo clinical features, 156
  treatment, 157
PTA, *see* Percutaneous transluminal balloon angioplasty
Pulmonary edema
  management following subarachnoid hemorrhage, 437
  neurogenic pulmonary edema, 723
Pulmonary embolism, brain tumor patients, 838
Pure autonomic failure (PAF), features, 1459–1460
Pusher syndrome
  assessment, 283
  clinical presentation, 283
  course, 283
  treatment, 283–284
PWT, *see* Primary writing tremor
Pyrazinamide
  adverse effects, 564
  side effects, 1517
  tuberculous meningitis management, 562–563
Pyridostigmine, myasthenia gravis management, 1347, 1353–1355, 1357
  testing, 1343
Pyridoxine, tuberculous meningitis management, 562–563
Pyrimethamine, toxoplasmosis management, 596

## Q

Q fever, central nervous system infection and management, 541
Quetiapine
  dementia with Lewy bodies management, 1088–1089
  dopaminergic psychosis management, 1048, 1066–1067

Quinine
  cerebral malaria management, 660–662
  cramp management, 1422–1423, 1425

## R

Rabies
  clinical features, 612
  postexposure treatment, 605, 607
  transmission, 612
Radial nerve compression
  clinical features, 1306
  interosseus posterior syndrome, 1306–1307
  management, 1306
Radiculopathy
  cervical radicular syndromes, *see* Cervical spondylitic myelopathy; Invertebral disk degeneration; Prolapsed invertebral disk
  diagnosis
    cerebrospinal fluid findings, 1290
    electromyography, 1290
    imaging, 1289–1290
  differential diagnosis, 1289–1209
  lumbar and sacral radicular syndromes, *see* Prolapsed invertebral disk; Spinal canal stenosis
Radiotherapy
  acromegaly management, 1436–1437
  astrocytic tumors, 840–841
  brachytherapy, 832
  chordoma, 857
  choroid plexus tumors, 848
  Cushing's syndrome, 1438–1439
  delayed radiotherapy injury
    course, 1338–1339
    diagnostic criteria, 1338
    management, 1339
    sites and presentation, 1337–1338
  embryonal tumors, 850
  ependymomas, 847–848
  fractionated radiotherapy, 832
  germ cell tumors, 855
  Hodgkin's disease, 875–876
  hyperprolactinemia, 1443–1444
  leptomeningeal metastasis
    outcomes, 900
    rationale, 902
    techniques, 907
    toxicity, 903
  mechanisms of action, 833
  meningiomas, 852–854
  metastatic brain tumor management
    overview, 885–886
    prophylactic cranial radiation, 886
    radiosensitizing agents, 886
    radiosurgery, 886–887
    side effects, 890–891
  non-Hodgkin's lymphoma, 872–874
  primary central nervous system lymphoma, 870–871
  radiochemotherapy, 835
  radiosurgery, 832
  schwannomas, 851

side effects, 836–837
Recombination, chromosomes, 1530
Refsum's disease
  clinical features, 1013, 1283
  treatment, 101–1014
Rehabilitation
  delirium, 308
  Guillain-Barré syndrome, 1262
  motor rehabilitation
    brain plasticity promotion, 1185–1186
    detrimental drugs, 1187–1188
    indications, 1185
    initiation after stroke, 1186–1187
    intensity of therapy, 1187
    learned nonuse, 1186
    lower limbs
      botulinum toxin treatment of
        spasticity, 1191–1192
      electrical stimulation, 1192
      locomotor therapy, 1190–1191
      technical aids, 1192–1193
    task-specific repetition, 1186
    upper limbs
      bilateral practice, 1188–1189
      botulinum toxin treatment of
        spasticity, 1189
      constrained induced movement
        therapy, 1188
      electrical stimulation, 1189–1190
      repetitive training of isolated
        movements, 1188
      robot-aided simulation, 1190
  peripheral neuropathy, 1286–1287
  spinal cord injury, 931–933
  traumatic brain injury, 777–778
Relapsing fever
  antibiotic therapy, 591
  clinical presentation, 590
  differential diagnosis, 590
  epidemiology, 590
Renal transplantation, encephalopathies
  posterior leukencephalopathy syndrome,
    1000
  rejection encephalopathy, 1000
Resiniferatoxin, detrusor hyperreflexia
  management, 1484
Restless legs syndrome (RLS)
  clinical presentation, 1177
  course, 1177–1178
  diagnostic criteria, 1177–1178
  pathogenesis, 1178
  periodic limb movements of sleep features,
    1177, 1179, 1181–1182
  treatment
    anticonvulsants, 1178–1179, 1181
    clonazepam, 1178–1179, 1181
    clonidine, 1179, 1181
    L-Dopa, 1178–1180
    obsolete treatment, 1181
    opioids, 1178–1181
Restlessness, palliative care, 818–819
Retinocochleocerebral vasculopathy, see
  Susac's syndrome
Rhabdomyolysis
  acute alcoholic rhabdomyolysis, 984

causes, 1389
clinical features, 1389–1390
drug induction, 1400
management, 1390
Rhabdovirus, clinical features and course of
  infection, 606
Rhinocerebral syndrome, clinical features,
  616
Rhizotomy, trigeminal neuralgia management
  glycerol rhizotomy, 69–70
  partial sensory trigeminal rhizotomy, 70
  radiofrequency thermal rhizotomy, 69
Ribavirin
  mechanism of action, 609
  spectrum of activity, 609
Rifampicin
  leprosy management, 1269–1270
  side effects, 1517
Rifampin
  adverse effects, 564
  tuberculous meningitis management,
    562–563
Riley–Day syndrome, autonomic dysfunction,
  1461
Riluzole, amyotrophic lateral sclerosis
  management, 1167–1168
Risperidone
  dementia with Lewy bodies management,
    1087
  dopaminergic psychosis management, 1066
  poisoning and management, 798
  tic disorder management, 1140
Rituximab, multifocal motor neuropathy
  management, 1264
Rivastigmine
  Alzheimer's disease dementia management,
    317–318
  dementia with Lewy bodies management,
    1086–1087, 1089
Rizatriptan, migraine treatment, 4–5
RLS, see Restless legs syndrome
Robinul, drooling management, 253
Rocky Mountain Spotted Fever, central
  nervous system infection and
  management, 540
Ropinirole, Parkinson's disease management,
  1037–1040, 1042
RS-86, progressive supranuclear palsy
  management, 1092
Rubella, clinical features and course of
  infection, 606

S

SAH, see Subarachnoid hemorrhage
Saphenous nerve compression, features and
  management, 1314–1315
Sarcoid neuropathy, features and treatment,
  1267
Sarcoidosis, central nervous system
  clinical presentation, 567–568
  complications, 573
  course, 569
  diagnosis
    cerebrospinal fluid findings, 567–568

classification based on certainty,
  568–569
  magnetic resonance imaging, 567
  positron emission tomography, 567
differential diagnosis, 567–568
epidemiology, 569
pathophysiology, 569
systemic disease, 567–568
treatment
  aseptic meningitis, 570
  azathioprine, 571–572
  corticosteroids, 569–571
  hydrocephalus, 570, 572
  immunosuppressants, 571–572
  irradiation, 572
  parenchymal disease, 570
  peripheral facial nerve palsy, 569–570
  peripheral neuropathy and myopathy,
    570
  pregnant patients, 572–573
  supportive care, 572
  surgery, 572
SCA, see Spinocerebellar ataxia
Schistosomiasis
  clinical manifestations, 656
  course, 657–658
  definition, 655
  diagnosis, 658
  differential diagnosis, 657
  neuroschistosomiasis features, 656
  treatment
    carbamazepine, 659
    obsolete treatments, 659
    phenytoin, 659
    praziquantel, 658–659
Schizencephaly, features, 952–953
Schwannoma
  clinical features, 850–851
  course, 851
  treatment, 851
Schwartz-Bartter's syndrome, see Syndrome
  of inappropriate antidiuretic
  hormone secretion
Schwartz-Jampel syndrome, features and
  management, 1433
SCI, see Spinal cord injury
Sciatic nerve compression, features and
  management, 1315–1316
Scopolamine
  acquired pendular nystagmus management,
    129
  motion sickness management, 158
  palliative care, 818
SCS, see Spinal cord stimulation
Sedation, critical care, 726–727
Seizure, see Epilepsy
Selective serotonin reuptake inhibitors
  (SSRIs)
  atypical facial pain management, 62
  pain management, 101
  palliative care, 822
  poisoning and management, 797
Selegeline
  Alzheimer's disease dementia management,
    318

Selegeline *(continued)*
  corticobasal degeneration management,
    1094
  drug–drug interactions, 1512
  Parkinson's disease management,
    1045–1046, 1054
  poisoning and management, 798
  side effects, 1512
Selenium, encephalopathy, 1003
Semantics, definition, 265
Senile dyskinesias, *see* Spontaneous oral
    dyskinesias
Septic encephalopathy
  clinical presentation, 1002
  pathophysiology, 1002
Septo-optic dysplasia (SOD), features, 952
Sexual dysfunction
  chronic renal failure patients, 999–1000
  course, 1489–1490
  erectile dysfunction definition, 1487
  multiple sclerosis, 702
  multiple system atrophy, 1083
  neural control
    ejection and ejaculation, 1487–1488
    female sexual function, 1488–1489
  neural prosthesis, 1213–1214
  neurogenic causes, 1490
  Parkinson's disease, 1062
  spinal cord injury, 933
  treatment
    apomorphine, 1491
    female sexual dysfunction, 1491–1492
    obsolete treatments, 1492
    principles, 1490
    prostaglandin injections, 1491
    sildenafil, 1491
    surgery, 1491
    vacuum devices, 1491
Sexual headache, clinical features, 43
Short-lasting unilateral neuralgiform
    headache with conjuctival injection
    and tearing (SUNCT)
  clinical presentation, 39–40
  diagnostic criteria, 39
  treatment, 40
Shy–Drager syndrome, *see* Multiple system
    atrophy
SIADH, *see* Syndrome of inappropriate
    antidiuretic hormone secretion
Sildenafil, sexual dysfunction management,
    1491
Single-photon emission computed
    tomography (SPECT), Parkinson's
    disease, 1025
Sjögren's disease, *see* Vasculitides
Sleep
  age effects on patterns, 191
  cycles, 191
  disorders, *see* Sleep disorders
  hygiene instructions, 196
  hypnotic classification and influences, 196
  testing
    maintenance of wakefulness test,
      193–194
    multiple sleep latency test, 193
    polysomnography, 192–193

Sleep disorders
  amyotrophic lateral sclerosis management,
    1169
  circadian disorders
    advanced sleep phase syndrome, 202
    delayed sleep phase syndrome, 201–202
    shift workers, 202
  disorders of initiating and maintaining
    sleep
    drug and alcohol use, 196
    periodic limb movements during sleep,
      196–197
    psychiatric insomnia, 194
    psychophysiological insomnia, 195–196
  overview of types, 194–195
  parasomnias
    NREM/REM sleep parasomnias, 198
    NREM sleep parasomnias, 197–198
    REM sleep parasomnia, 198
  Parkinson's disease management, 1065
  sleepiness syndromes
    idiopathic central nervous system
      hypersomnia, 200–201
    insufficient sleep, 198
    narcolepsy, 199–200
    nervous system disorders, 201
    psychiatric disorders, 201
    sleep-related breathing disorders,
      198–199
Sleeping sickness
  autonomic dysfunction, 1461
  clinical presentation, 666
  course, 667
  definition, 666
  diagnosis, 666
  treatment
    eflornithine, 667
    melarsoprol, 667
    pentamidine, 667
    suramin, 667
SMA, *see* Spinal muscular atrophy
Smallpox, clinical features and course of
    infection, 606
Small vessel disease, *see* Cerebral small vessel
    disease
Smoking, stroke risks and control, 349, 352
SMS, *see* Stiff-man syndrome
SOD, *see* Septo-optic dysplasia
Sodium channel myotonia, *see* Paramyotonia
    congenita; Myotonia fluctuans;
    Hyperkalemic period paralysis
Sodium stibogluconate, leishmaniasis
    management, 672–673
Sodium valproate, *see* Valproic acid
Sparganosis, features and management, 598
Spasmodic dystonia
  clinical presentation, 1134
  course, 1134
  treatment, 1134
Spasticity
  definition, 1247
  hereditary spastic paraparesis, 1172
  pathophysiology, 1247–1248
  signs and symptoms, evolution,
    1247–1248
  spastic paresis, 1247

  treatment
    baclofen
      intrathecal, 1252–1253
      oral, 1251
    benzodiazepines, 1252
    botulinum toxin
      lower limbs, 1191–1192
      overview, 1253
      upper limbs, 1189
    cannabinoids, 1252
    dantrolene, 1252
    dorsal rhizotomy, 1253
    drug mechanisms, 1248–1249
    glycine, 1252
    locomotor training, 1250–1251
    memantine, 1252
    multiple sclerosis, 700–701
    nonspecific procedures, 1249–1250
    physiotherapy, 1250
    principles, 1248
    spinal cord injury, 932–933
    surgery, 1253–1254
    tizanidine, 1251–1252
    transcutaneous electrical stimulation,
      1253
Spatial contrast sensitivity impairment
  clinical features and course, 258–259
  treatment, 259
Spatial neglect
  clinical presentation, 280
  course, 280–281
  treatment
    active orienting to contralesional side,
      281
    eye patching, 282
    limb activation and contralesional
      cueing, 281–282
    neck muscle vibration, 282
    prism adaptation, 282–283
SPECT, *see* Single-photon emission computed
    tomography
Spinal arteriovenous malformation
  evaluation, 527
  glomerular type and treatment, 521–523
  overview, 517–518
  perimedullary fistula types, 520–521
Spinal canal stenosis
  clinical features, 1299
  diagnosis, 1299–1300
  treatment, 1300
Spinal cord injury (SCI)
  acute stage features, 925–926
  autonomic dysfunction, 1461–1462,
    1467
  autonomic nervous system dysfunction,
    927–928, 931
  course, 926–927
  evolution of signs and symptoms, 928
  magnetic resonance imaging, 925
  motor rehabilitation
    brain plasticity promotion,
      1185–1186
    detrimental drugs, 1187–1188
    indications, 1185
    intensity of therapy, 1187
    learned nonuse, 1186

lower limbs
 botulinum toxin treatment of
  spasticity, 1191–1192
 electrical stimulation, 1192
 locomotor therapy, 1190–1191
 technical aids, 1192–1193
 task-specific repetition, 1186
upper limbs
 bilateral practice, 1188–1189
 botulinum toxin treatment of
  spasticity, 1189
 constrained induced movement
  therapy, 1188
 electrical stimulation, 1189–1190
 repetitive training of isolated
  movements, 1188
 robot-aided simulation, 1190
rehabilitation, 931–933
sexual function, 933
treatment
 acute stage, 925–926, 929–931
 bladder drainage, 930, 932
 body positioning, 930
 bowel motility, 930, 932
 bradycardia, 930
 corticosteroids, 928–929
 deep vein thrombosis, 930–931
 functional electrical stimulation, see
  Functional electrical stimulation
 heterotopic ossification, 933
 ineffective treatments, 929
 respiration, 930–932
 skin problems, 932
 spasticity, 932–933
 surgery, 930–931
 syringomyelia, 933
Spinal cord ischemia
 causes
  aortic diseases and procedures, 396
  cartilaginous disc embolism, 396–397
  hypoxic–ischemic injury, 397
  overview, 394–395
  rare causes, 397
  spinal dural arteriovenous fistulas,
   395–396
  venous spinal cord infarction, 397
 course, 398
 diagnosis, 397–398
 imaging, 397–398
 treatment, 398–399
 vascular anatomy, 393–394
Spinal cord stimulation (SCS), pain
 management, 109, 1333
Spinal cord syndromes
 acute transverse myelitis
  clinical features, 934
  treatment, 934
 chronic cervical myelopathy
  clinical features, 934
  treatment, 934–935
 etiology, 925
 rehabilitation, 925, 931–933
 tumors
  clinical features, 935
  treatment, 935
 types, 925–926

vascular lesions
 clinical features, 933
 treatment, 933–934
Spinal dural arteriovenous fistula
 epidemiology, 517
 evaluation, 527
 segmental distribution, 517–518
 symptoms, 519
 treatment, 519
Spinal epidural abscess
 course, 554
 definition, 552
 diagnosis, 553
 differential diagnosis, 553–554
 etiology, 552–553
 location, 553
 treatment
  antibiotics, 554
  conservative treatment, 554–555
  corticosteroids, 554
  surgery, 554
Spinal intramedullary abscess, features and
 management, 555
Spinal muscular atrophy (SMA)
 distal forms, 1171–1172
 monomelic amyotrophy, 1172
 proximal forms
  bulbospinal muscular atrophy, 1171
  types I–IV, 1171
 scapulo-peroneal form, 1172
 treatment, 1172
Spinal subdural empyema, features and
 management, 555
Spinocerebellar ataxia (SCA)
 course, 1014
 genetic heterogeneity, 1533
 treatment, 1014–1015
 types and gene mutations, 1014
Spontaneous oral dyskinesias
 clinical features, 1135
 course, 1135
 treatment
  tetrabenazine, 1135
  tiapride, 1135
SSRIs, see Selective serotonin reuptake
 inhibitors
Status epilepticus, see Epilepsy
Steele-Richardson-Olszewski disease, see
 Progressive supranuclear palsy
Steinert's disease, see Myotonic dystrophy
 type I
Stenting, see Angioplasty and stenting
Stiff-man syndrome (SMS)
 clinical presentation, 1254, 1430
 course, 1254
 diagnosis, 1254, 1430–1431
 electromyography, 1430–1431
 pathophysiology, 1426
 treatment, 1254, 1426, 1431
 variants
  focal stiff-man syndrome, 1432
  jerking stiff-man syndrome, 1431–1432
  paraneoplastic stiff-man syndrome, 1432
  stiff-leg syndrome, 1432
Streptomycin
 adverse effects, 564

tuberculous meningitis management,
 562–563
Stroke, see also Acute ischemic stroke; Brain
 embolism; Intracerebral hemorrhage;
 Subarachnoid hemorrhage
 course, 349, 352
 delirium induction, 304–305
 hemodialysis patients, 999
 motor rehabilitation
  brain plasticity promotion, 1185–1186
  detrimental drugs, 1187–1188
  indications, 1185
  initiation after stroke, 1186–1187
  intensity of therapy, 1187
  learned nonuse, 1186
  lower limbs
   botulinum toxin treatment of
    spasticity, 1191–1192
   locomotor therapy, 1190–1191
   technical aids, 1192–1193
   task-specific repetition, 1186
  upper limbs
   bilateral practice, 1188–1189
   botulinum toxin treatment of
    spasticity, 1189
   constrained induced movement
    therapy, 1188
   functional electrical stimulation,
    1204–1205
   repetitive training of isolated
    movements, 1188
   robot-aided simulation, 1190
 multiple sclerosis, 1193–1194
 obsolete treatments, 1194
 Parkinson's disease, 1194
 prevention
  aspirin, 351
  carotid endarterectomy, 351–352
  hypertension control, 352
  obsolete treatments, 359
  secondary prevention
   angioplasty and stenting, 358
   anticoagulants, 357
   antiplatelet drugs, 353–357
   carotid endarterectomy, 357–358
   glycoprotein IIb/IIIa inhibitors, 357
   practical management, 358–359
   principles, 352–353
   risk factor modification, 353
   smoking cessation, 352
 risk factors
  atrial fibrillation, 350
  blood lipids, 349–350
  hypertension, 349
  myocardial infarction, 350
  patent foramen ovale, 351
  prosthetic heart valves, 350
  smoking, 349
 spinal cord, see Spinal cord ischemia
Strongyloidiasis, features and management,
 598
Sturge-Weber syndrome (SWS)
 clinical presentation, 964
 course, 964
 differential diagnosis, 964
 treatment, 964–965

Subarachnoid hemorrhage (SAH)
  aneurysm morphology, 486–487
  causes
    nonaneurysmal perimesencephalic
      hemorrhage, 423–424
    overview, 422–423
    rare causes, 424
    saccular aneurysm, 422–423
  clinical presentation, 421, 483
  epidemiology, 421–422, 483–484
  investigations
    angiography, 425–426, 483
    computed tomography, 424–425, 483
    lumbar puncture, 424, 483
    magnetic resonance imaging, 424–425,
      483
  outcomes, 422, 426
  pathogenesis, 484
  prevention of rebleeding
    aneurysm clipping, 429, 488
    aneurysm treatment guidelines, 489
    coiling, 429–430, 485, 487–488
    drug treatment, 429
    incidence, 428–429
  prognostic factors, 426–427
  secondary cerebral ischemia
    prevention
      antiplatelet agents, 432–433
      calcium channel blockers, 431–432
      fluid balance and electrolytes, 431
      hypertension management, 431
      incidence, 430
      nicaraven, 432
      pathophysiology, 430
      tirilazad, 432
      volume expansion, 433
    treatment, 433–434
  treatment
    acute hydrocephalus, 428, 434–435
    acute subdural hematoma, 428
    complications
      arrhythmias, 436–437
      hyponatremia, 435–436
      pulmonary edema, 437
    early rebleeding, 427–428
    global cerebral ischemia, 428
    intracerebral hematoma, 428
    nursing and general management,
      426–427
    ruptured aneurysms, 485–487
    unruptured aneurysms, 484–485
  vasospasm management, 488–489
Subclavian artery, angioplasty, 503–504
Subdural empyema
  definition, 551
  diagnosis, 551
  differential diagnosis, 551–552
  etiology, 551
  management, 552
Subependymal giant cell astrocytoma,
    features and management, 844
Subscapular nerve compression, features and
    management, 1305
Subthalamotomy, thermocoagulation, 1108
Sulfadiazine, toxoplasmosis management,
    596

Sulpiride, Huntington's disease management,
    1152
Sumatriptan
  cluster headache treatment, 19
  migraine treatment, 4–5
SUNCT, see Short-lasting unilateral
    neuralgiform headache with
    conjuctival injection and tearing
Superior oblique myokymia
  clinical features, 131
  course, 131
  pathogenesis, 131
  treatment, 131
Suprascapular nerve compression, features
    and management, 1305
Suramin
  neurofilariasis management, 668
  sleeping sickness management, 667
Susac's syndrome
  clinical presentation, 367
  course, 367
  treatment, 367–368
Swallowing, see Dysphagia
Sweating disorders, see Anhidrosis;
    Hyperhidrosis
SWS, see Sturge-Weber syndrome
Sydenham's chorea
  clinical features, 1153
  course, 1154
  treatment, 1154
Syncope
  clinical features, 1462–1463, 1495–1496
  course, 1499–1500
  diagnosis
    cardiovascular reflex testing, 1498–1499
    echocardiography, 1498
    electrocardiogram, 1497–1498
    electroencephalography, 1499
    laboratory tests, 1498
    patient history, 1497
    physical examination, 1497
    psychiatric evaluation, 1499
  differential diagnosis, 1496–1497
  etiology, 1499
  treatment
    autonomic failure, 1500–1501
    cardiac syncope, 1502
    carotid sinus hypersensitivity, 1502
    driving avoidance, 1503
    principles, 1500
    psychiatric causes, 1502–1503
    situational syncope, 1502
    vasovagal syncope, 1501–1502
  types, 1463, 1495–1496
Syndrome of inappropriate antidiuretic
    hormone secretion (SIADH)
  clinical presentation, 1440
  course, 1441
  diagnosis, 1440
  differential diagnosis, 1440–1441
  treatment, 1441
Syntax, definition, 266
Syphilis
  course, 579
  diagnosis of neurosyphilis
    cerebrospinal fluid findings, 576–578

imaging, 578
serology, 577
  epidemiology, 575
  human immunodeficiency virus association
      and management, 578–579, 633,
      635
  latency, 575
  manifestations in central nervous system,
      576
  pathogenesis, 579–580
  stages, 575
  treatment
    antibiotics, 580–581
    complication management, 581
    monitoring, 581
    pharmacology and pharmacokinetics,
      580
Syringomyelia
  clinical findings, 941–942
  course, 943
  definition, 939
  differential diagnosis, 943
  etiology and classification
    idiopathic, 941
    inflammatory conditions with
      arachnoiditis, 940–941
    neoplastic, 940
    posterior fossa lesion association, 940
    posttraumatic cavities, 941
  frequency, 939–940
  hydromyelia comparison, 939
  imaging
    computed tomography, 942
    magnetic resonance imaging, 939, 942
  pathology, 939
  treatment
    obsolete treatments, 945
    principles, 943
    spinal cord injury management, 933
    surgery, 943–944
    symptomatic treatment, 945
Syrup of ipecac, dosing, 794
Systemic lupus erythematosus, see
    Vasculitides

T

Tacrine
  Alzheimer's disease trials, 321
  dementia with Lewy bodies management,
      1087, 1089
Tacrolimus
  paraneoplastic syndrome management,
      916–917
  side effects, 1515
Taenia solium, see Neurocysticercosis
Takayasu's disease, see Vasculitides
Taste disorder, see Ageusia
TBI, see Traumatic brain injury
TCD, see Transcranial Doppler
    ultrasonography
Temporomandibular joint (TMJ) syndrome,
    atypical facial pain relationship, 59,
    63
TENS, see Transcutaneous electrical nerve
    stimulation

Tension-type headache
course, 24–25
diagnostic criteria, 23–24
differential diagnosis, 23–24
hormone-sensitive headache in women, 54
pain mechanisms, 25–26
prevention
amitriptyline, 27
botulinum toxin, 27
metoprolol, 27
treatment
acupuncture, 26
aspirin, 26, 28
muscle relaxation, 26
nonsteroidal anti-inflammatory drugs, 26–27
obsolete treatments, 28
practical management, 27–28
principles, 25–26
stress management, 26
tizanidine, 27
Tetanus
clinical features, 1274
course, 1274
immunization, 1274–1275
pathogenesis, 1273–1274
treatment, 1274–1275
Tetany, features, 1425–1426
Tetrabenazine
Huntington's disease management, 1152
spontaneous oral dyskinesia management, 1135
Sydenham's chorea management, 1154
tardive dyskinesia management, 1137
Tetrathiomolybdate, Wilson's disease management, 1161
Tetrazepam, radiculopathy management, 1292
Thalamotomy
gamma knife, 1108
thermocoagulation, 1108
Thalidomide, vasculitides management, 467
Thallium, encephalopathy
overview, 1003
treatment, 1006
Theophylline
cramp management, 1423, 1425
obstructive sleep apnea management, 199
Thiamine deficiency, see Alcoholic cerebellar atrophy; Wernicke's encephalopathy
Thiotepa, leptomeningeal metastasis intrathecal chemotherapy, 902
Thirst, palliative care, 822
Thomsen's disease, see Myotonia congenita
Thoracic outlet syndrome (TOS)
clinical presentation, 1303
diagnosis, 1304
prevalence, 1303
treatment, 1304
Thoracodorsal nerve compression, features and management, 1305
Thrombolysis
acute ischemic stroke management
efficacy, 329–330
imaging, 342

indications, 341–342
intra-arterial thrombolysis, 332–333
intravenous thrombolysis, 330–332
brain embolism management, 376–377, 384–385
cerebral venous and sinus thrombosis management, 454, 509
intra-arterial thrombolytic therapy
angioplasty and stenting, 508
indications, 505
lysis, 507–508
principles, 506–507
Thrombotic thrombocytopenic purpura (TTP)
clinical presentation, 368
course, 368
treatment, 368–369
Thymoma
histological classification, 1356
imaging, 1345
thymectomy, 1352–1354, 1356
TIA, see Transient ischemic attack
Tiagabine
advantages and limitations, 218
adverse reactions, 222
dosing, 221
epilepsy management, 220, 223
pharmacokinetics, 221
Tiapride
spontaneous oral dyskinesia management, 1135
tic disorder management, 1140
Tibial nerve compression, features and management, 1317–1318
Ticlopidine
side effects, 1524
stroke prevention, 356, 358
Tics
classification, 1138–1139
clinical features, 1138–1139
course, 1139
treatment, 1139–1140
Tinidazole, amebic brain abscess management, 666
Tinnitus
course, 165
objective tinnitus etiology and treatment
continual swishing noise, 166
sharp clicks, 165–166
Souffle during inspiration/expiration, 165
spontaneous otoacoustic emissions, 166
vascular bruit, 166
retraining therapy, 168
subjective tinnitus
rotational vertigo, see Meniére's disease
with hearing defects
acoustic neuroma, 166–167
head trauma, 166
noise-induced trauma, 166
otosclerosis, 166
ototoxic medication, 167
sudden hearing loss, 166
without hearing defects
clinical presentation, 167
course, 167–168
etiology, 167

obsolete treatments, 168
treatment, 168
Tirilazad, cerebral ischemia prevention following subarachnoid hemorrhage, 432
Tizanidine
contraindications, 1511
drug–drug interactions, 1511
radiculopathy management, 1292
side effects, 1511
spasticity management, 1251–1252
tension-type headache treatment, 27
TLI, see Total lymphoid irradiation
TMJ syndrome, see Temporomandibular joint syndrome
TNF-α, see Tumor necrosis factor-α
Tolcapone, Parkinson's disease management, 1043–1044
Toluene, poisoning and management, 801–802
Topiramate
advantages and limitations, 218
adverse reactions, 222
dosing, 221
epilepsy management, 220, 223
pharmacokinetics, 221
TOS, see Thoracic outlet syndrome
Total lymphoid irradiation (TLI)
dermatomyositis/polymyositis management, 1366
multiple sclerosis management, 690
Tourette's syndrome
clinical features, 1138–1139
course, 1139
treatment, 1139–1140
Toxocariasis
clinical features, 597
course, 597
diagnosis, 597
treatment, 597, 599
Toxoplasmosis
clinical presentation, 595
congenital infection, 596
diagnosis, 595–596
differential diagnosis, 596
epidemiology, 595
human immunodeficiency virus encephalopathy association and management, 628, 634, 637, 640
prevention, 596
treatment, 596
Tracheotomy, critical care, 722
Tramadol
pain management, 103, 1333
radiculopathy management, 1292
Transcranial Doppler ultrasonography (TCD), critical care monitoring, 732
Transcutaneous electrical nerve stimulation (TENS)
pain management, 106–107, 1333
spasticity management, 1253
zoster management, 1271
Transient ischemic attack (TIA)
embolism, see Brain embolism
prevention, see Stroke

Transmissible spongiform encephalopathy, *see* Prion disease

Tranylcypromine, poisoning and management, 798

Traumatic brain injury (TBI)
  classification, 739, 765
  critical care
    cerebral oxygenation monitoring, 741
    electroencephalography, 741
    emergency measures, 740, 773–774
    hematoma management
      acute subdural hematoma, 743
      chronic subdural hematoma, 743
      epidural hematoma, 742–743
      intracerebral hematoma, 743
      monitoring, 741
    intracranial pressure control, 740–741, 776–777
    mild brain trauma, 740
  economic impact, 768
  evaluation, 765–766
  Glasgow Coma Scale, 739–740, 765–766
  imaging, 767, 772
  mild traumatic brain injury
    features, 767–768
    management, 772–773
  pathology
    diffuse axonal injury, 770–771
    diffuse gray matter dysfunction, 771
    diffuse microvascular injury, 770
    focal injury, 770
    hypoxia-ischemia, 770
    neuronal excitotoxicity, 771
  prognosis in severe injury
    acute, 768
    indicators, 769
    long-term outcome, 768–769
  treatment of moderate and severe disease
    delayed secondary injury and neuroprotection, 771–772
    epilepsy, 776–777
    operative management, 775–776
    principles, 769–770
    rehabilitation, 777–778
    resuscitation, 773–774

Tremor, *see also* Cerebellar tremor; Essential tremor; Parkinson's disease; Primary writing tremor
  assessment, 1233
  cerebellar disease, 1240
  classification
    etiology classification, 1234
    state of activation classification, 1233–1234
  definition, 1233
  drug-induced tremor, 1241
  dystonic tremor syndromes, 1240
  Holmes' tremor syndrome, 1240, 1242
  multiple sclerosis, 1240–1241
  neuropathic tremor, 1239–1240
  physiological tremor, 1234
  primary orthostatic tremor, 1238
  psychogenic tremor, 1241
  treatment, 1241–1242

*Treponema pallidum*, *see* Syphilis

Trichinosis

clinical features, 599
course, 599
diagnosis, 599
treatment, 599

Trientine hydrochloride, Wilson's disease management, 1160–1161

Trigeminal neuralgia
  causes, 67–68
  clinical presentation, 67
  course, 68
  diagnosis, 67
  differential diagnosis, 67–68
  treatment
    drugs, 68–72
    gamma knife radiosurgery, 71
    gasserian ganglion, percutaneous microcompression, 69
    glycerol rhizotomy, 69–70
    microvascular decompression, 70
    nonmedical treatments, 68–69, 72
    obsolete treatments, 72
    partial sensory trigeminal rhizotomy, 70
    peripheral nerve block
      phenol in glycerol, 70
      tetracaine, 70–71
    peripheral neurectomy, 70
    radiofrequency thermal rhizotomy, 69

Trihexyphenidyl
  acquired pendular nystagmus management, 128
  cervical dystonia management, 1132
  side effects and contraindications, 1512–1513

Tromethamine, intracranial pressure elevation management, 730, 759

Trypanosomiasis, *see* Sleeping sickness

TS, *see* Tuberous sclerosis

TTP, *see* Thrombotic thrombocytopenic purpura

Tuberculous meningitis
  cerebrospinal fluid findings, 559–560
  clinical presentation, 559–560
  course, 562
  diagnosis, 559–560
  human immunodeficiency virus association and management, 631
  magnetic resonance imaging, 560–561
  stages, 560, 562
  treatment
    adverse effects, 564
    children, 562–563
    corticosteroids, 563–564
    ethambutol, 562–563
    isoniazid, 562–563
    pyrazinamide, 562–563
    pyridoxine, 562–563
    rifampin, 562–563
    streptomycin, 562–563
    surgery for hydrocephalus, 565

Tuberous sclerosis (TS)
  clinical presentation, 962–963
  course, 963
  diagnosis and screening, 950
  diagnostic criteria, 949
  treatment, 963–964

Tumor necrosis factor-α (TNF-α), inhibition in vasculitides treatment, 468

U

Ulnar nerve compression
  upper arm, 1310
  wrist, 1312–1313

University of Pennsylvania Smell Identification Test (UPSIT)
  administration, 172
  cognitive impairment patient testing, 176
  interpretation, 172–173

UPSIT, *see* University of Pennsylvania Smell Identification Test

Uremic encephalopathy
  clinical features, 995–996
  course, 996
  treatment, 996

Urokinase, intra-arterial thrombolytic therapy, 507–508

V

Vaccination
  Alzheimer's disease dementia management, 319
  multiple sclerosis studies, 699

Vacuolar myelopathy, human immunodeficiency virus association and management, 625–626, 632, 640

Vacuolar myopathy, drug-induced myopathy, 1397

Vagus nerve, stimulation for epilepsy management, 226, 241–242

Valacyclovir
  mechanism of action, 608
  spectrum of activity, 608
  zoster management, 1272

Valproic acid
  advantages and limitations, 218
  adverse reactions, 217
  brain tumor management, 837–838
  dosing, 216
  epilepsy management, 215, 219
  migraine prevention, 10
  pain management, 101–102
  pharmacokinetics, 216
  poisoning and management, 797
  Sydenham's chorea management, 1154
  trigeminal neuralgia management, 69, 72

Varicella zoster virus, *see* Zoster

Vascular dementia
  clinical findings, 316
  course, 316–317
  diagnostic criteria, 316
  pathology, 317
  treatment, 320–321

Vasculitic neuropathy
  associated diseases, 1267
  clinical presentation, 1267–1268
  course, 1268
  treatment, 1268

Vasculitides
  Behçet's disease features, 471

Churg–Strauss syndrome features, 469
classification, 461–462
clinical presentation, 461
Cogan's syndrome features, 471
definition, 461
diagnosis, 462
Eales disease features, 471
hypersensitivity vasculitis features, 469
isolated angiitis of the central nervous system features, 470
Kawasaki disease features, 471
microscopic polyarteritis features, 469
pathophysiology
    antibody-mediated vasculitis, 463
    immune complex-mediated vasculitis, 462–463
    nonspecific vascular injury, 463
    overview, 462
    T cell-mediated vasculitis, 463
polyarteritis nodosa features, 469
Sjögren's disease features, 472
systemic lupus erythematosus features, 471–472
Takayasu's disease features, 470
treatment
    aspirin, 464
    chorambucil, 467
    complications, 464
    corticosteroids, 465–467
    cyclophosphamide, 466–467
    etanercept, 468
    infliximab, 468
    interferon-γ, 467–468
    intravenous immunoglobulin, 468
    leflunomide, 467
    medications and regimens by specific disease, 464–465
    methotrexate, 467
    mycophenolate mofetil, 467
    nonsteroidal anti-inflammatory drugs, 464–465
    plasmapheresis, 468
    principles, 463–464
    surgery, 469
    thalidomide, 467
Venlafaxine, poisoning and management, 797–798
Ventilation
critical care
    modes, 722–723
    settings, 722
hyperventilation and intracranial pressure control, 730, 758
viral infection indications, 609–610
Verapamil, cluster headache prevention, 20
Vertebral artery jugular fistula, management, 497
Vertebral artery stenosis
angioplasty
    practical management, 503
    principles of therapy, 502–503
course, 502
frequency, 502
Vertigo, see also Benign paroxysmal positioning vertigo; Central

vestibular vertigo; Height vertigo; Meniére's disease; Motion sickness; Perilymph fistula; Psychogenic vertigo; Vestibular neuritis; Vestibular paroxysmia
clinical manifestations, 139
prognosis, 139
treatment
    antiemetic drugs, 139–141
    baclofen, 141–142
    corticosteroids, 141
    gabapentine, 142
    physical therapy, 140, 142–143
    surgical interventions, 140, 142
    vestibular compensation, 142–143
    vestibular suppressants, 139–141
Vestibular neuritis
    clinical presentation, 148
    course, 148
    pathogenesis, 148–149
    related clinical syndromes, 149
    treatment, 148–149
Vestibular paroxysmia
    clinical presentation, 152
    course, 153
    diagnostic criteria, 152–153
    treatment, 153
Vestibulo-ocular reflex disorders, see Central vestibular vertigo; Nystagmus
VHL, see Von Hippel-Lindau disease
Vidarabine
    herpes simplex encephalitis management, 611
    mechanism of action, 608
    spectrum of activity, 608
Vigabatrin
    advantages and limitations, 218
    adverse reactions, 222
    dosing, 221
    epilepsy management, 220, 223–224
    pharmacokinetics, 221
Vincamine, side effects, 1511
Viral encephalitis
    central European encephalitis, 611
    clinical presentation, 602
    differential diagnosis, 604
    herpes simplex encephalitis, 610–611
    herpes zoster encephalitis, 611
    seizure management, 610
Viral meningitis
    clinical presentation, 601–602
    herpes simplex meningitis, 611
Visual adaptation impairment
    clinical features and course, 258–259
    treatment, 259
Visual agnosia
    course, 259–260
    etiology, 259
    treatment, 260
    types, 259
Visual illusion
    course, 261
    definition, 260
    etiology, 260–261
Visual space perception disorders

clinical features, 277–278
    course, 278
    etiology, 277–278
    localization, 277
    orientation, 278
    stereopsis, 277–278
    treatment, 278–279
Visuoconstructive ability disorders
    clinical aspects, 279
    course, 279
    treatment, 279
Vitamin B$_{12}$, deficiency and ataxia, 1017
Vitamin D deficiency, myopathy, 1402
Vitamin deficiency polyneuropathy, features, 1286
Vitamin E
    Alzheimer's disease dementia management, 318
    deficiency and ataxia, 1013, 1017–1018
    tardive dyskinesia management, 1137
Vomiting
    mechanisms and treatment, 1468
    palliative care, 821
Von Hippel-Lindau disease (VHL)
    clinical features, 965
    course, 965
    gene mutations, 965
    treatment, 965

## W

Warfarin
    acute ischemic stroke management, 334, 342–343
    brain embolism management, 379
    contraindications, 1523
    drug–drug interactions, 1523–1524
    reversal of anticoagulation, 1313
    side effects, 1522–1523
    stroke prevention, 357
Wernicke's aphasia, see Aphasia
Wernicke's encephalopathy
    ataxia, 1017
    clinical presentation, 977–978
    course, 978
    treatment, 978–979
Whiplash injury
    course, 87–88
    definition, 83
    grading, 83–84
    imaging of complications, 85–87
    pathogenesis, 83–84
    prognostic factors, 87–88
    sequelae, 84–85
    symptoms and signs, 84–86
    treatment
        acute pain, 88–89, 91
        nonpharmacological management, 90
        obsolete treatments, 90–91
        principles, 88
        prolonged pain, 90–91
        secondary illnesses, 91
Whipple's disease
    central nervous system infection and management, 539
    X-ray findings, 85

Wilson's disease
  course, 1159
  diagnosis, 1158
  differential diagnosis, 1159
  frequency, 1157
  gene mutations, 1157
  parkinsonism induction, 1069
  symptoms and signs, 1157–1158
  treatment
    approaches
      copper resorption reduction, 1160
      copper storage reduction, 1159
    asymptomatic patients, 1162
    dietary restriction of copper, 1161
    2,3-dimercaptopropanol, 1161
    liver transplantation, 1162
    penicillamine, 1160
    potassium sulfate, 1161

    pregnancy, 1162
    symptomatic treatment, 1162
    tetrathiomolybdate, 1161
    trientine hydrochloride, 1160–1161
    zinc therapy, 1161–1162
Wolfram's syndrome, clinical features, 122
Writer's cramp, features, 1426

X
Xaliproden, amyotrophic lateral sclerosis
    management, 1168
Xanithol nicotinate, side effects, 1511

Y
Yohimbine, orthostatic hypotension
    management, 1467

Z
Zinc, Wilson's disease management,
    1161–1162
Zolmitriptan, migraine treatment, 4–
    5
Zolpidem, progressive supranuclear palsy
    management, 1092
Zonisamide
  adverse reactions, 222
  dosing, 221
  epilepsy management, 224
  pharmacokinetics, 221
Zoster
  clinical presentation, 1270
  course, 1270–1271
  incidence, 1270
  treatment, 1271–1272
Zycomycosis, clinical features, 617

ISBN 0-12-125831-9

9 780121 258313

90038